Marks'

Standard Handbook for Mechanical Engineers

Marks'

Standard Handbook for Mechanical Engineers

ALI M. SADEGH *Editor*

Consulting Engineer; Professor of Mechanical Engineering and
Director, Center for Advanced Engineering Design and Development
The City College of the City University of New York

WILLIAM M. WOREK *Editor*

Executive Director, Eagle Ford Center for Research, Education and Outreach
Texas A&M University—Kingsville

Twelfth Edition

Mc Graw Hill Education

New York Chicago San Francisco Athens London
Madrid Mexico City Milan New Delhi
Singapore Sydney Toronto

The Library of Congress cataloged the first issue of this title as follows:

Standard handbook for mechanical engineers. 1st-ed.;
1916–

New York, McGraw-Hill.

v. illus. 18–24 cm.

Title varies: 1916–58; Mechanical engineers' handbook.
Editors: 1916–51, L. S. Marks.—1958– T. Baumeister.
Includes bibliographies.
1. Mechanical engineering—Handbooks, manuals, etc. I. Marks, Lionel Simeon, 1871– ed. II. Baumeister, Theodore, 1897– ed. III. Title; Mechanical engineers' handbook.
TJ151.S82 502'.4'621 16–12915

Library of Congress Catalog Card Number: 87-641192

Library of Congress Control Number: 2017954426

Marks' Standard Handbook for Mechanical Engineers, Twelfth Edition

1 2 3 4 5 6 7 8 9 LWI 23 22 21 20 19 18 17

ISBN 978-1-259-58850-1
MHID 1-259-58850-5

This book is printed on acid-free paper.

Sponsoring Editor
Robert Argentieri

Editing Supervisor
Stephen M. Smith

Production Supervisor
Lynn M. Messina

Acquisitions Coordinator
Lauren Rogers

Project Manager
Kritika Kaushik, Cenveo® Publisher Services

Copy Editor
Syed Haider, Cenveo Publisher Services

Proofreader
Cenveo Publisher Services

Indexer
Jack Lewis

Art Director, Cover
Jeff Weeks

Composition
Cenveo Publisher Services

Contents

For the detailed contents of any section, consult the title page of that section.

Contributors

Mehdi Ahmadian *Professor and Director, Center for Vehicle Systems & Safety and Railway Technologies Laboratory, Virginia Tech* (Secs. 1, 8)

Michael D. Allega *The Timken Corporation* (Sec. 6)

Farid M. Amirouche *Professor of Mechanical and Industrial Engineering, University of Illinois at Chicago* (Sec. 12)

Yiannis Andreopoulos *Professor, Department of Mechanical Engineering, The City College of the City University of New York* (Sec. 12)

Glenn E. Asauskas *Lubrication Engineer, Chevron Corp.* (Sec. 4)

Dennis N. Assanis *President, University of Delaware* (Sec. 7)

E. A. Avallone *Ardsley, NY* (Sec. 7)

Frederick G. Baily *Retired Engineer/Manager, General Electric, Schenectady, NY* (Sec. 7)

Nael Barakat *Professor, Mechanical Engineering, Texas A&M University—Kingsville* (Sec. 13)

Robert D. Bartholomew *Associate, Sheppard T. Powell Associates, LLC* (Sec. 4)

Charles E. Baukal, Jr. *Director, John Zink Institute* (Sec. 5)

Asfaw Beyene *Professor of Mechanical Engineering, San Diego State University* (Sec. 7)

Donald B. Bivens *Refrigerant Applications Consultant, Du Pont Engineering, E. I. du Pont de Nemours and Company, retired* (Sec. 11)

Malcolm Blair *Technical and Research Director, Steel Founders' Society of America* (Sec. 4)

Heinz P. Block *Process Machinery Consulting, Houston, TX and Westminster, CO* (Sec. 7)

Donald E. Bolt *Late Engineering Manager, Heat Transfer Products Dept., Foster Wheeler Energy Corp.* (Sec. 7)

William J. Bow *Late Director, Heat Transfer Products Dept., Foster Wheeler Energy Corp.* (Sec. 7)

Charles F. Bowman *President, Chuck Bowman Associates, Inc.* (Sec. 7)

Steven V. Boyd *The Timken Corporation* (Sec. 6)

Walter H. Boyes, Jr. *Editor and Publisher, The Industrial Automation and Process Control INSIDER* (Sec. 10)

Richard L. Brazill *Technology Specialist, ALCOA* (Sec. 4)

Richard J. Burke *ABS Professor of Naval Architecture & Marine Engineering, State University of New York Maritime College* (Sec. 8)

Thomas Butcher *Brookhaven National Laboratory* (Secs. 5, 7)

Charles P. Butterfield *Chief Engineer, National Wind Technology Center, National Renewable Energy Laboratory* (Sec. 7)

Zhiyong Cai *Project Leader, USDA Forest Service* (Sec. 4)

C. L. Carlson *Late Fellow Engineer, Research Labs, Westinghouse Electric Corp.* (Sec. 4)

Scott W. Case *Professor of Engineering Science and Mechanics, Virginia Polytechnic Institute and State University* (Sec. 3)

Vittorio (Rino) Castelli *Senior Research Fellow, Retired, Xerox Corp.; Engineering Consultant* (Sec. 6)

Paul V. Cavallaro *Senior Mechanical Research Engineer, Naval Undersea Warfare Center* (Sec. 12)

Yunus A. Çengel *Professor Emeritus, University of Nevada, Reno* (Sec. 1)

Robin O. Cleveland *Professor of Engineering Science, Tutorial Fellow of Magdalen College, University of Oxford* (Sec. 10)

Gary L. Cloud *Professor of Mechanical Engineering, Michigan State University* (Sec. 3)

David E. Cole *Chairman Emeritus, Center for Automotive Research (CAR)* (Sec. 7)

Piotr Cyklis *Professor, Mechanical Engineering, Cracow University of Technology* (Sec. 9)

Benjamin C. Davenny *Acoustical Consultant, Acentech Inc., Cambridge, MA* (Sec. 10)

William H. Day *President, Longview Energy Associates LLC* (Sec. 7)

James Drago *Manager, Engineering, Garlock Sealing Technologies, an EnPro Industries Company* (Sec. 6)

F. J. Edeskuty *Los Alamos National Laboratory, retired* (Sec. 11)

Alex Elvin *Professor, School of Civil and Environmental Engineering, University of the Witwatersrand, South Africa* (Sec. 10)

Niell Elvin *Professor of Mechanical Engineering, The City College of the City University of New York* (Sec. 10)

Robert E. Eppich *Vice President, Technology, American Foundry Society* (Sec. 4)

Bryan A. Erler *President, Erler Engineering Ltd.* (Sec. 7)

Hakan Ertürk *Associate Professor of Mechanical Engineering, Boğaziç University, Istanbul, Turkey* (Sec. 2)

Heimir Fannar *Chief Design Engineer, Ariel Corp.* (Sec. 9)

Erich A. Farber *Distinguished Professor Emeritus, Mechanical Engineering, University of Florida* (Sec. 7)

Sanford Fleeter *Late Professor Emeritus, Purdue University* (Sec. 8)

Michael D. Florczykowski *Lincoln Electric Co.* (Sec. 4)

Luc G. Fréchette *Professor of Mechanical Engineering, Université de Sherbrooke, Canada* (Sec. 10)

Thomas A. Frederick *Retired Senior Scientist, Lyondell Basell; Principal Scientist, TF Solutions LLC* (Sec. 4)

Francis H. (Sam) Froes *Retired and Consultant to the Light Metals Industry* (Sec. 4)

Ivo A. Garza *Manager/Consultant, Sargent and Lundy* (Sec. 7)

Afshin J. Ghajar *Regents Professor and John Brammer Professor of Mechanical Engineering, Oklahoma State University* (Sec. 2)

Andrew A. Goldenberg *Professor of Mechanical and Industrial Engineering, University of Toronto; President and CEO of Engineering Service Inc. (ESI)* (Sec. 10)

Frank E. Goodwin *Executive Vice President, ILZRO, Inc.* (Sec. 4)

Don Graham *Manager, Turning Products, Carboloy, Inc.* (Sec. 4)

Leonard M. Grillo *Grillo Engineering Company* (Sec. 5)

Julio C. Guerrero *Principal, Exponent Inc.* (Sec. 7)

Marsbed Hablanian *Retired Manager of Engineering and R&D, Varian Vacuum Technologies* (Sec. 9)

Laura Hasburgh *General Engineer, USDA Forest Service* (Sec. 4)

Keith L. Hawthorne *Late Vice President—Technology, Transportation Technology Center, Inc.* (Sec. 8)

V. Terrey Hawthorne *Late Senior Engineer, LTK Engineering Services* (Sec. 8)

Terry L. Henshaw *Consulting Engineer, Magnolia, TX* (Sec. 9)

Peyman Honarmandi *Professor of Mechanical Engineering, Manhattan College, Bronx, New York* (Secs. 1, 3)

Hoyt C. Hottel *Late Professor Emeritus, Massachusetts Institute of Technology* (Sec. 2)

Michael W. Hyer *Professor of Engineering Science and Mechanics, Virginia Polytechnic Institute and State University* (Sec. 3)

Timothy J. Jacobs *Associate Professor, Department of Mechanical Engineering, Texas A&M University* (Sec. 7)

Ankur Jain *Associate Professor of Mechanical and Aerospace Engineering, University of Texas at Arlington* (Sec. 7)

M. W. M. Jenkins *Professor, Aerospace Design, Georgia Institute of Technology* (Sec. 8)

Peter K. Johnson *Consultant* (Sec. 4)

Robert Jorgensen *Engineering Consultant* (Sec. 9)

Serope Kalpakjian *Professor of Mechanical, Materials, and Aerospace Engineering, Illinois Institute of Technology* (Sec. 4)

Michael B. Kane *Professor, Department of Civil and Environmental Engineering, Northeastern University* (Sec. 10)

Jonathan D. Kemp *Vibration Consultant, Acentech, Inc., Cambridge, MA* (Sec. 10)

J. Randolph Kissell *President, The TGB Partnership* (Sec. 4)

John B. Kitto, Jr. *Consulting Engineer, Retired, The Babcock and Wilcox Company* (Sec. 7)

Andrew C. Klein *Professor, Nuclear Science and Engineering, Oregon State University* (Sec. 7)

Doyle Knight *Professor, Department of Mechanical and Aerospace Engineering, Rutgers University* (Sec. 12)

Glen Kowach *Professor of Chemistry, The City College of the City University of New York* (Sec. 4)

Ezra S. Krendel *University of Pennsylvania* (Sec. 7)

David Kretschmann *President, American Lumber Standard Committee, Incorporated* (Sec. 4)

Francis A. Kulacki *Professor of Mechanical Engineering, University of Minnesota* (Sec. 2)

Srirangam Kumaresan *Biomechanics Institute, Santa Barbara, California* (Sec. 12)

L. D. Kunsman *Late Fellow Engineer, Research Labs, Westinghouse Electric Corp.* (Sec. 4)

Francesco Lanza di Scalea *Professor of Structural Engineering, University of California San Diego* (Sec. 3)

Stan Lebow *Research Forest Products Technologist, USDA Forest Service* (Sec. 4)

John H. Lewis *Late Engineering Consultant; Formerly, Engineering Staff, Pratt & Whitney Division, United Technologies Corp.* (Sec. 7)

Jacqueline J. Li *Professor of Mechanical Engineering, The City College of the City University of New York* (Sec. 4)

Gene Liao *Professor, Engineering Technology, Electric-Drive Vehicle Engineering, and Director, Alternative Energy Technology Program, Wayne State University, Detroit, Michigan* (Sec. 8)

Sitirios Mamalis *Assistant Professor, Mechanical Engineering, Stony Brook University* (Sec. 7)

Joaquim R. R. A. Martins *Professor of Aerospace Engineering, University of Michigan* (Sec. 8)

Lawrence M. Matta *Stress Engineering Services, Inc.* (Sec. 3)

Roy McAlister *Founder and President, American Hydrogen Association* (Sec. 7)

Mark O. McLinden *National Institute of Standards and Technology, Applied Materials Division, Boulder, CO* (Sec. 2)

Leonard Meirovitch *Late Professor of Engineering Science, Virginia Tech* (Sec. 1)

Sarah A. Meyer *The Timken Corporation* (Sec. 6)

James H. Michel *Copper Development Association, Inc.* (Sec. 4)

Duane K. Miller *Engineering Services and Welding Design Consultant, Lincoln Electric Co.* (Sec. 4)

Glenn J. Miller *National Aeronautics and Space Administration* (Sec. 8)

Patrick C. Mock *Principal Electron Optical Scientist, Science and Engineering Associates, Inc.* (Sec. 11)

Eduard Muljadi *Senior Engineer, National Wind Technology Center, National Renewable Energy Laboratory* (Sec. 7)

J. W. Murdock *Late Consulting Engineer* (Sec. 1)

Gregory V. Murphy *Head and Professor, Department of Electrical Engineering, College of Engineering, Tuskegee University* (Sec. 10)

Trung Nguyen *Professor of Chemical and Petroleum Engineering, University of Kansas* (Sec. 7)

James J. Noble *Former Research Associate Professor of Chemical and Biological Engineering, Tufts University* (Sec. 2)

Donald J. Patterson *Emeritus Professor, Mechanical Engineering, University of Michigan* (Sec. 7)

Harold W. Paxton *United States Steel Professor Emeritus, Carnegie Mellon University* (Sec. 4)

Courtney A. Peckens *Professor, Department of Engineering, Hope College, Holland, MI* (Sec. 10)

Kenneth A. Phair *Global Power Solutions LLC, Golden, Colorado* (Sec. 7)

Patrick E. Phelan *Professor of Mechanical and Aerospace Engineering, Arizona State University* (Sec. 7)

Rama Ramakumar *PSO/Albrecht Naeter Professor and Director, Engineering Energy Laboratory, Oklahoma State University* (Sec. 7)

James D. Redmond *Principal, Technical Marketing Resources, Inc.* (Sec. 4)

Robert J. Ross *Project Leader, USDA Forest Service, and Research Professor, Michigan Technological University* (Sec. 4)

A. J. Rydzewski *Consultant, Du Pont Engineering, E. I. du Pont de Nemours and Company* (Sec. 11)

Parisa Saboori *Professor of Mechanical Engineering, Manhattan College* (Sec. 4)

Ali M. Sadegh *Professor of Mechanical Engineering, The City College of the City University of New York* (Secs. 3, 4, 6, 12)

Jung-Yang San *Professor, Mechanical Engineering Department, National Chung Hsing University, Taiwan, Republic of China* (Sec. 7)

Anthony Sances, Jr. *Late of Biomechanics Institute, Santa Barbara, California* (Sec. 12)

Adel F. Sarofim *Late Presidential Professor of Chemical Engineering, University of Utah* (Sec. 2)

John R. Schley *Consultant* (Sec. 4)

Steven R. Schmid *Professor of Aerospace & Mechanical Engineering, University of Notre Dame* (Sec. 4)

C. Adam Senalik *Research General Engineer, USDA Forest Service* (Sec. 4)

David A. Shifler *Director of Cellular Materials Program, Office of Naval Research* (Sec. 4)

Peter J. Shirron *National Aeronautics and Space Administration/Goddard Space Flight Center* (Sec. 11)

Rajiv Shivpuri *Professor of Integrated Systems Engineering, Ohio State University* (Sec. 4)

Simon Simms *Professor of Chemistry, The City College of the City University of New York* (Sec. 4)

James G. Speight *Consultant* (Sec. 5)

Robert D. Steele *Project Manager, Voith Siemens Hydro Power Generation, Inc.* (Sec. 7)

Stephen R. Swanson *Professor of Mechanical Engineering, University of Utah* (Sec. 4)

Arvind C. Thekdi *President, Energy and Environmental Efficiency Management (E3M), Inc.* (Sec. 5)

Erik T. Thostenson *Professor of Mechanical Engineering, University of Delaware* (Sec. 4)

Matthew Tones *Sr. Applications Engineer, Garlock Sealing Technologies, an EnPro Industries Company* (Sec. 6)

John H. Tundermann *Former Vice President, Research and Technology, INCO Intl.* (Sec. 4)

John M. (Jack) Tuohy, Jr. *Energy Industry Consultant* (Sec. 7)

Israel Urieli *Professor Emeritus, Mechanical Engineering, Ohio University* (Sec. 7)

Charles O. Velzy *Past President, American Society of Mechanical Engineers* (Sec. 5)

Xiping Wang *Research Forest Products Technologist, USDA Forest Service* (Sec. 4)

Larry F. Wieserman *Retired Senior Technical Supervisor, ALCOA* (Sec. 4)

William M. Worek *Professor of Mechanical Engineering, Texas A&M University—Kingsville* (Sec. 7)

Mike Yoon *Fellow, American Society of Mechanical Engineers* (Sec. 8)

M. Zebarjadi *Professor of Electrical & Computer Engineering and Materials Science Engineering, University of Virginia* (Sec. 7)

M. Zenouzi *Professor of Mechanical Engineering and Technology, Wentworth Institute of Technology* (Sec. 7)

The Editors

ALI M. SADEGH, PH.D., is Professor of Mechanical Engineering, Founder and Director of the Center for Advanced Engineering Design and Development, and former Chairman of the Department of Mechanical Engineering at The City College of the City University of New York. He is a Fellow of both the American Society of Mechanical Engineers and the Society of Manufacturing Engineers. He is also the co-editor of the eleventh edition of *Marks'*.

WILLIAM M. WOREK, PH.D., is Executive Director of the Eagle Ford Center for Research, Education and Outreach (EFCREO) at Texas A&M University—Kingsville. He is a Fellow of both the American Society of Mechanical Engineers and the American Society of Heating, Refrigerating and Air-Conditioning Engineers.

Preface to the Twelfth Edition

The rapid advancement of information technology and the evolutionary trends underlying mechanical engineering practice are grounded not only on the tried and true principles and techniques of the past, but also on more recent and current advances. At the same time, we are faced with a vast shift in the way we access our information both professionally and personally. Because of this, the preparation of this 100th anniversary, twelfth edition of *Marks' Standard Handbook for Mechanical Engineers* involved a great deal of effort to move the content away from the incredibly broad array of subjects that has been covered in past editions and focus on the subjects that represent the core of the discipline. Thus, the Editors have considered the broad enterprise falling under the rubric of "mechanical engineering" and have added to and amended the Handbook to emphasize subject areas that will be of maximum utility to today's practicing engineer.

For many years, "Marks'" and "Handbook for Mechanical Engineers" have become synonymous to a wide readership which includes mechanical engineers, engineers in the associated disciplines, and others. Our readership derives from a wide spectrum of interests, and it appears that many find *Marks'* useful as they pursue their professional endeavors. The Editors recognize that the current engineering practitioner learns quickly that the revolutionary developments of the recent past soon become standard practice. By the same token, it is prudent to realize that as a consequence of rapid developments, some cutting-edge technologies prove to have a short shelf life and soon are regarded as obsolescent, if not obsolete. To this end, for this twelfth edition, the Editors have made a special effort to eliminate topics that are available through the Internet and other information sources, but include such topics as micro- and nano-engineering, robotic vision, alternative energy production, biological materials, biomechanics, composite materials, engineering ethics, and much more.

Finally, the utmost care has been exercised to avoid errors, but if any inadvertently have been included, the Publisher and Editors will appreciate being so informed. Corrections will be incorporated into subsequent printings and errata will be posted on the Internet.

Ali M. Sadegh
William M. Worek

Preface to the First Edition*

This Handbook is intended to supply both the practicing engineer and the student with a reference work which is authoritative in character and which covers the field of mechanical engineering in a comprehensive manner. It is no longer possible for a single individual or a small group of individuals to have so intimate an acquaintance with any major division of engineering as is necessary if criticial judgment is to be exercised in the statement of current practice and the selection of engineering data. Only by the cooperation of a considerable number of specialists is it possible to obtain the desirable degree of reliability. This Handbook represents the work of fifty specialists.

Each contributor is to be regarded as responsible for the accuracy of his section. The number of contributors required to ensure sufficiently specialized knowledge for all the topics treated is necessarily large. It was found desirable to enlist the services of thirteen specialists for an adequate handling of the "Properties of Engineering Materials." Such topics as "Automobiles," "Aeronautics," "Illumination," "Patent Law," "Cost Accounting," "Industrial Buildings," "Corrosion," "Air Conditioning," "Fire Protection," "Prevention of Accidents," etc., though occupying relatively small spaces in the book, demanded each a separate writer.

A number of the contributions which deal with engineering practice, after examination by the Editor-in-Chief, were submitted by him to one or more specialists for criticism and suggestions. Their cooperation has proved of great value in securing greater accuracy and in ensuring that the subject matter does not embody solely the practice of one individual but is truly representative.

An accuracy of four significant figures has been assumed as the desirable limit; figures in excess of this number have been deleted, except in special cases. In the mathematical tables only four significant figures have been kept.

The Editor-in-Chief desires to express here his appreciation of the spirit of cooperation shown by the Contributors and of their patience in submitting to modifications of their sections. He wishes also to thank the Publishers for giving him complete freedom and hearty assistance in all matters relating to the book from the choice of contributors to the details of typography.

LIONEL S. MARKS
Cambridge, Mass.
April 23, 1916

*Excerpt.

Acknowledgments

The Editors would like to recognize the contributions of the Editors of the eleventh and previous editions, the late Eugene A. Avallone and Theodore Baumeister III. It is only through their past diligence and stewardship that this twelfth edition has been possible.

We also recognize, with pleasure, the input from our many contributors—past, continuing, and those newly engaged. Their contributions have been prepared with care, and are authoritative, informative, and concise. And we thank Michael Worek for his input on content during the initial stages of this revision.

Finally, the Editors would like to recognize the team of Robert Argentieri, Editorial Director; Stephen M. Smith, Editing Manager; and Lynn M. Messina, Senior Production Supervisor, at McGraw-Hill Education and Kritika Kaushik at Cenveo Publisher Services for their tireless work to make this 100th anniversary edition a reality.

Symbols and Abbreviations

Pairs of parentheses, brackets, etc., are frequently used in this work to indicate corresponding values. For example, the statement "The cost per kW of a 30,000-kW plant is $86; of a 15,000-kW plant, $98; and of an 8,000-kW plant, $112" is condensed as follows: The cost per kW of a 30,000 (15,000) [8,000]-kW plant is $86 (98) [112].

When references are cited, readers should always attempt to consult the latest edition of the referenced publications.

A or Å	Angstrom unit = 10^{-10} m; 3.937×10^{-11} in
A	mass number = N + Z; ampere
AA	arithmetical average
AAA	Am. Automobile Assoc.
AAMA	American Automobile Manufacturers' Assoc.
AAR	Assoc. of Am. Railroads
AAS	Am. Astronautical Soc.
ABAI	Am. Boiler & Affiliated Industries
abs	absolute
a.c.	aerodynamic center
a-c, ac	alternating current
ACI	Am. Concrete Inst.
ACM	Assoc. for Computing Machinery
ACRMA	Air Conditioning and Refrigerating Manufacturers Assoc.
ACS	Am. Chemical Soc.
ACSR	aluminum cable steel-reinforced
ACV	air cushion vehicle
A.D.	anno Domini (in the year of our Lord)
AEC	Atomic Energy Commission (U.S.)
a-f, af	audio frequency
AFBMA	Anti-friction Bearings Manufacturers' Assoc.
AFS	Am. Foundrymen's Soc.
AGA	Am. Gas Assoc.
AGMA	Am. Gear Manufacturers' Assoc.
ahp	air horsepower
AIChE	Am. Inst. of Chemical Engineers
AIEE	Am. Inst. of Electrical Engineers (see IEEE)
AIME	Am. Inst. of Mining Engineers
AIP	Am. Inst. of Physics
AISC	American Institute of Steel Construction, Inc.
AISE	Am. Iron & Steel Engineers
AISI	Am. Iron and Steel Inst.
Al. Assn.	Aluminum Association
a.m.	ante meridiem (before noon)
a-m, am	amplitude modulation
Am. Mach.	Am. Machinist (New York)
AMA	Acoustical Materials Assoc.
AMCA	Air Moving & Conditioning Assoc., Inc.
amu	atomic mass unit
AN	ammonium nitrate (explosive); Army-Navy Specification
AN-FO	ammonium nitrate-fuel oil (explosive)
ANC	Army-Navy Civil Aeronautics Committee
ANS	Am. Nuclear Soc.
ANSI	American National Standards Institute
antilog	antilogarithm of
API	Am. Petroleum Inst.
approx	approximately
APWA	Am. Public Works Assoc.
AREA	Am. Railroad Eng. Assoc.
ARI	Air Conditioning and Refrigeration Inst.
ARS	Am. Rocket Soc.
ASCE	Am. Soc. of Civil Engineers
ASHRAE	Am. Soc. of Heating, Refrigerating, and Air Conditioning Engineers
ASLE	Am. Soc. of Lubricating Engineers
ASM	Am. Soc. of Metals
ASME	Am. Soc. of Mechanical Engineers
ASST	Am. Soc. of Steel Treating
ASTM	Am. Soc. for Testing and Materials
ASTME	Am. Soc. of Tool & Manufacturing Engineers
atm	atmosphere
Auto. Ind.	Automotive Industries (New York)
avdp	avoirdupois
avg, ave	average
AWG	Am. Wire Gage
AWPA	Am. Wood Preservation Assoc.
AWS	American Welding Soc.
AWWA	American Water Works Assoc.
b	barns
bar	barometer
B&S	Brown & Sharp (gage); Beams and Stringers
bbl	barrels
B.C.	before Christ
B.C.C.	body centered cubic
Bé	Baumé (degrees)
B.G.	Birmingham gage (hoop and sheet)
bgd	billions of gallons per day
BHN	Brinell Hardness Number
bhp	brake horsepower
BLC	boundary layer control
B.M.	board measure; bench mark
bmep	brake mean effective pressure
B of M, BuMines	Bureau of Mines

BOD	biochemical oxygen demand
bp	boiling point
Bq	becquerel
bsfc	brake specific fuel consumption
BSI	British Standards Inst.
Btu	British thermal units
Btub, Btu/h	Btu per hr
bu	bushels
Bull.	Bulletin
Buweaps	Bureau of Weapons, U.S. Navy
BWG	Birmingham wire gage
c	velocity of light
°C	degrees Celsius (centigrade)
C	coulomb
CAB	Civil Aeronautics Board
CAGI	Compressed Air & Gas Inst.
cal	calories
C-B-R	chemical, biological & radiological (filters)
CBS	Columbia Broadcasting System
cc, cm^3	cubic centimetres
CCR	critical compression ratio
c to c	center to center
cd	candela
c.f.	centrifugal force
cf.	confer (compare)
cfh, ft^3/h	cubic feet per hour
cfm, ft^3/min	cubic feet per minute
C.F.R.	Cooperative Fuel Research
cfs, ft^3/s	cubic feet per second
cg	center of gravity
cgs	centimetre-gram-second
Chm. Eng.	Chemical Eng'g (New York)
chu	centigrade heat unit
C.I.	cast iron
cir	circular
cir mil	circular mils
cm	centimetres
CME	Chartered Mech. Engr. (IMechE)
C.N.	cetane number
coef	coefficient
COESA	U.S. Committee on Extension to the Standard Atmosphere
col	column
colog	cologarithm of
const	constant
cos	cosine of
cos^{-1}	angle whose cosine is, inverse cosine of
cosh	hyperbolic cosine of
$cosh^{-1}$	inverse hyperbolic cosine of
cot	cotangent of
cot^{-1}	angle whose cotangent is (see cos^{-1})
coth	hyperbolic cotangent of
$coth^{-1}$	inverse hyperbolic cotangent of
covers	coversed sine of
c.p.	circular pitch; center of pressure
cp	candle power
cp	coef of performance
CP	chemically pure
CPH	close packed hexagonal
cpm	cycles per minute
cycles/min	
cps, cycles/s	cycles per second
CSA	Canadian Standards Assoc.
csc	cosecant of
csc^{-1}	angle whose cosecant is (see cos^{-1})
csch	hyperbolic cosecant of
$csch^{-1}$	inverse hyperbolic cosecant of
cu	cubic
cyl	cylinder
db, dB	decibel
d-c, dc	direct current
def	definition
deg	degrees
diam. (dia)	diameter
DO	dissolved oxygen
D_2O	deuterium (heavy water)
d.p.	double pole
DP	Diametral pitch
DPH	diamond pyramid hardness
DST	daylight saving time
d^2 tons	breaking strength, d = chain wire diam. in.
DX	direct expansion
e	base of Napierian logarithmic system (= 2.7182+)
EAP	equivalent air pressure
EDR	equivalent direct radiation
EEI	Edison Electric Inst.
eff	efficiency
e.g.	exempli gratia (for example)
ehp	effective horsepower
EHV	extra high voltage
El. Wld.	Electrical World (New York)
elec	electric
elong	elongation
emf	electromotive force
Engg.	Engineering (London)
Engr.	The Engineer (London)
ENT	emergency negative thrust
EP	extreme pressure (lubricant)
ERDA	Energy Research & Development Administration (successor to AEC; see also NRC)
Eq.	equation
est	estimated
etc.	et cetera (and so forth)
et seq.	et sequens (and the following)
eV	electron volts
evap	evaporation
exp	exponential function of
exsec	exterior secant of
ext	external
°F	degrees Fahrenheit
F	farad
FAA	Federal Aviation Administration
F.C.	fixed carbon, %
FCC	Federal Communications Commission; Federal Constructive Council
F.C.C.	face-centered-cubic (alloys)
ff.	following (pages)
fhp	friction horsepower
Fig.	figure
F.I.T.	Federal income tax
f-m, fm	frequency modulation
F.O.B.	free on board (cars)
FP	fore perpendicular
FPC	Federal Power Commission
fpm, ft/min	feet per minute
fps	foot-pound-second system
ft/s	feet per second
F.S.	Federal Specifications
FSB	Federal Specifications Board
fsp	fiber saturation point
ft	feet
fc	foot candles
fL	foot lamberts
ft·lb	foot-pounds
g	acceleration due to gravity
g	grams
gal	gallons

gc	gigacycles per second
GCA	ground-controlled approach
g · cal	gram-calories
gd	Gudermannian of
G.E.	General Electric Co.
GEM	ground effect machine
GFI	gullet feed index
G.M.	General Motors Co.
GMT	Greenwich Mean Time
GNP	gross national product
gpcd	gallons per capita day
gpd	gallons per day, grams per denier
gpm, gal/min	gallons per minute
gps, gal/s	gallons per second
gpt	grams per tex
H	henry
h	Planck's constant = 6.624×10^{-27} org-sec
$\hbar$	Planck's constant, $\hbar = h/2\pi$
HEPA	high-efficiency particulate matter
h-f, hf	high frequency
hhv	high heat value
horiz	horizontal
hp	horsepower
h-p	high pressure
HPAC	Heating, Piping, & Air Conditioning (Chicago)
hp · hr	horsepower-hour
hr, h	hours
HSS	high speed steel
H.T.	heat-treated
HTHW	high temperature hot water
Hz	hertz = 1 cycle/s (cps)
IACS	International Annealed Copper Standard
IAeS	Institute of Aerospace Sciences
ibid.	ibidem (in the same place)
ICAO	International Civil Aviation Organization
ICC	Interstate Commerce Commission
ICE	Inst. of Civil Engineers
ICI	International Commission on Illumination
I.C.T.	International Critical Tables
I.D., ID	inside diameter
i.e.	id est (that is)
IEC	International Electrotechnical Commission
IEEE	Inst. of Electrical & Electronics Engineers (successor to AIEE, *q.v.*)
IES	Illuminating Engineering Soc.
i-f, if	intermediate frequency
IGT	Inst. of Gas Technology
ihp	indicated horsepower
IMechE	Inst. of Mechanical Engineers
imep	indicated mean effective pressure
Imp	Imperial
in., in	inches
in.· lb, in · lb	inch-pounds
INA	Inst. of Naval Architects
Ind. & Eng. Chem.	Industrial & Eng'g Chemistry (Easton, PA)
int	internal
i-p, ip	intermediate pressure
ipm, in/min	inches per minute
ipr	inches per revolution
IPS	iron pipe size
IRE	Inst. of Radio Engineers (see IEEE)
IRS	Internal Revenue Service
ISO	International Organization for Standardization
isoth	isothermal
ISTM	International Soc. for Testing Materials
IUPAC	International Union of Pure & Applied Chemistry
J	joule
J&P	joists and planks
Jour.	Journal
JP	jet propulsion fuel
k	isentropic exponent; conductivity
K	degrees Kelvin (Celsius abs)
K	Knudsen number
kB	kilo Btu (1,000 Btu)
kc	kilocycles
kcps	kilocycles per second
kg	kilograms
kg · cal	kilogram-calories
kg · m	kilogram-metres
kip	1,000 lb or 1 kilo-pound
kips	thousands of pounds
km	kilometres
kmc	kilomegacycles
kmcps	kilomegacycles per second
kpsi	thousands of pounds per sq in
ksi	one kip per sq in, 1,000 psi (lb/in^2)
kts	knots
kVA	kilovolt-amperes
kW	kilowatts
kWh	kilowatt-hours
L	lamberts
l, L	litres
£	Laplace operational symbol
lb	pounds
L.B.P.	length between perpendiculars
lhv	low heat value
lim	limit
lin	linear
ln	Napierian logarithm of
loc. cit.	loco citato (place already cited)
log	common logarithm of
LOX	liquid oxygen explosive
l-p, lp	low pressure
LPG	liquefied petroleum gas
lpw, lm/W	lumens per watt
lx	lux
L.W.L.	load water line
lm	lumen
m	metres
M	thousand; Mach number; moisture, %
mA	milliamperes
Machy.	Machinery (New York)
max	maximum
MBh	thousands of Btu per hr
mc	megacycles
m.c.	moisture content
Mcf	thousand cubic feet
mcps	megacycles per second
Mech. Eng.	Mechanical Eng'g (ASME)
mep	mean effective pressure
METO	maximum, except during take-off
me V	million electron volts
MF	maintenance factor
mhc	mean horizontal candles
mi	mile
MIL-STD	U.S. Military Standard
min	minutes; minimum
mip	mean indicated pressure
MKS	metre-kilogram-second system
MKSA	metre-kilogram-second-ampere system
mL	millilamberts
ml, mL	millilitre = 1.000027 cm^3
mlhc	mean lower hemispherical candles
mm	millimetres

mm-free	mineral matter free
mmf	magnetomotive force
mol	mole
mp	melting point
MPC	maximum permissible concentration
mph, mi/h	miles per hour
MRT	mean radiant temperature
ms	manuscript; milliseconds
msc	mean spherical candles
MSS	Manufacturers Standardization Soc. of the Valve & Fittings Industry
mu	micron, micro
MW	megawatts
MW day	megawatt day
MWT	mean water temperature
n	polytropic exponent
N	number (in mathematical tables)
N	number of neutrons; newton
N_s	specific speed
NA	not available
NAA	National Assoc. of Accountants
NACA	National Advisory Committee on Aeronautics (see NASA)
NACM	National Assoc. of Chain Manufacturers
NASA	National Aeronautics and Space Administration
nat.	natural
NBC	National Broadcasting Company
NBFU	National Board of Fire Underwriters
NBS	National Bureau of Standards (see NIST)
NCN	nitrocarbonitrate (explosive)
NDHA	National District Hearing Assoc.
NEC®	National Electric Code® (National Electrical Code® and NEC® are registered trademarks of the National Fire Protection Association, Inc., Quincy, MA)
NEMA	National Electrical Manufacturers Assoc.
NFPA	National Fire Protection Assoc.
NIST	National Institute of Standards and Technology
NLGI	National Lubricating Grease Institute
nm	nautical miles
No. (Nos.)	number(s)
NPSH	net positive suction head
NRC	Nuclear Regulatory Commission (successor to AEC; see also ERDA)
NTP	normal temperature and pressure
O.D., OD	outside diameter (pipes)
O.H.	open-hearth (steel)
O.N.	octane number
op. cit.	opere citato (work already cited)
OSHA	Occupational Safety & Health Administration
OSW	Office of Saline Water
OTS	Office of Technical Services, U.S. Dept. of Commerce
oz	ounces
p. (pp.)	page (pages)
Pa	pascal
P.C.	propulsive coefficient
PE	polyethylene
PEG	polyethylene glycol
P.E.L.	proportional elastic limit
PETN	an explosive
pf	power factor
PFI	Pipe Fabrication Inst.
PIV	peak inverse voltage
p.m.	post meridiem (after noon)
PM	preventive maintenance
P.N.	performance number
ppb	parts per billion
PPI	plan position indicator
ppm	parts per million
press	pressure

Proc.	Proceedings
PSD	power spectral density, g^2/cps
psi, lb/in^2	lb per sq in
psia	lb per sq in. abs
psig	lb per sq in. gage
pt	point; pint
PVC	polyvinyl chloride
Q	10^{18} Btu
qt	quarts
q.v.	quod vide (which see)
r	roentgens
R	gas constant
R	deg Rankine (Fahrenheit abs); Reynolds number
rad	radius; radiation absorbed dose; radian
RBE	see rem
R-C	resistor-capacitor
RCA	Radio Corporation of America
R&D	research and development
RDX	cyclonite, a military explosive
rem	Roentgen equivalent man (formerly RBE)
rev	revolutions
r-f, rf	radio frequency
RMA	Rubber Manufacturers Assoc.
rms	square root of mean square
rpm, r/min	revolutions per minute
rps, r/s	revolutions per second
RSHF	room sensible heat factor
ry.	railway
s	entropy
s	seconds
S	sulfur, %; siemens
SAE	Soc. of Automotive Engineers
sat	saturated
SBI	Steel Boiler Inst.
scfm	standard cu ft per min
SCR	silicon controlled rectifier
sec	secant of
$\sec^{-1}$	angle whose secant is (see $\cos^{-1}$)
Sec.	Section
sech	hyperbolic secant of
sech^{-1}	inverse hyperbolic secant of
segm	segment
SE No.	steam emulsion number
SEI	Structural Engineering Institute
sfc	specific fuel consumption, lb per hphr
sfm, sfpm	surface feet per minute
shp	shaft horsepower
SI	International System of Units (Le Systéme International d'Unites)
sin	sine of
$\sin^{-1}$	angle whose sine is (see $\cos^{-1}$)
sinh	hyperbolic sine of
$\sinh^{-1}$	inverse hyperbolic sine of
SME	Society of Manufacturing Engineers (successor to ASTME)
SNAME	Soc. of Naval Architects and Marine Engineers
SP	static pressure
sp	specific
specif	specification
sp gr	specific gravity
sp ht	specific heat
spp	species unspecified (botanical)
SPS	standard pipe size
sq	square
sr	steradian
SSF	sec Saybolt Furol
SSU	seconds Saybolt Universal (same as SUS)
std	standard

SUS	Saybolt Universal seconds (same as SSU)
SWG	Standard (British) wire gage
T	tesla
TAC	Technical Advisory Committee on Weather Design Conditions (ASHRAE)
tan	tangent of
$\tan^{-1}$	angle whose tangent is (see $\cos^{-1}$)
tanh	hyperbolic tangent of
$\tanh^{-1}$	inverse hyperbolic tangent of
TDH	total dynamic head
TEL	tetraethyl lead
temp	temperature
THI	temperature-humidity (discomfort) index
thp	thrust horsepower
TNT	trinitrotoluene (explosive)
torr	= 1 mm Hg = 1.332 millibars (1/760) atm = (1.013250/760) dynes per cm^2
TP	total pressure
tph	tons per hour
tpi	turns per in
TR	transmitter-receiver
Trans.	Transactions
T.S.	tensile strength; tensile stress
tsi	tons per sq in
ttd	terminal temperature difference
UHF	ultra high frequency
UKAEA	United Kingdom Atomic Energy Authority
UL	Underwriters' Laboratory
ult	ultimate
UMS	universal maintenance standards
USAF	U.S. Air Force
USCG	U.S. Coast Guard
USCS	U.S. Commercial Standard; U.S. Customary System
USDA	U.S. Dept. of Agriculture
USFPL	U.S. Forest Products Laboratory
USGS	U.S. Geologic Survey
USHEW	U.S. Dept. of Health, Education & Welfare
USN	U.S. Navy
USP	U.S. Pharmacopoeia
USPHS	U.S. Public Health Service
USS	United States Standard
USSG	U.S. Standard Gage
UTC	Coordinated Universal Time
V	volt
VCF	visual comfort factor
VCI	visual comfort index
VDI	Verein Deutscher Ingenieure
vel	velocity
vers	versed sine of
vert	vertical
VHF	very high frequency
VI	viscosity index
viz.	videlicet (namely)
V.M.	volatile matter, %
vol	volume
VP	velocity pressure
vs.	versus
W	watt
Wb	weber
W&M	Washburn & Moen wire gage
w.g.	water gage
WHO	World Health Organization
W.I.	wrought iron
W.P.A.	Western Pine Assoc.
wt	weight
yd	yards
Y.P.	yield point
yr	year(s)
Y.S.	yield strength; yield stress
z	atomic number, figure of merit
Zeit.	Zeitschrift
Δ	mass defect
μc	microcurie
σ, s	Boltzmann constant
μ	micro (= 10^{-6}), as in μs
μm	micrometre (micron) = 10^{-6} m (10^{-3} mm)
Ω	ohm

MATHEMATICAL SIGNS AND SYMBOLS

$+$	plus (sign of addition)
$+$	positive
$-$	minus (sign of subtraction)
$-$	negative
$\pm(\mp)$	plus or minus (minus or plus)
$\times$	times, by (multiplication sign)
$\cdot$	multiplied by
$\div$	sign of division
$/$	divided by
$:$	ratio sign, divided by, is to
$::$	equals, as (proportion)
$<$	less than
$>$	greater than
$\ll$	much less than
$\gg$	much greater than
$=$	equals
$\equiv$	identical with
$\sim$	similar to
$\approx$	approximately equals
$\cong$	approximately equals, congruent
$\leq$	equal to or less than
$\geq$	equal to or greater than
$\neq$ $\neq$	not equal to
$\rightarrow$ $\doteq$	approaches
$\propto$	varies as
∞	infinity
$\sqrt{}$	square root of
$\sqrt[3]{}$	cube root of
$\therefore$	therefore
$\parallel$	parallel to
( ) [ ] { }	parentheses, brackets, and braces; quantities enclosed by them to be taken together in multiplying, dividing, etc.
$\overline{AB}$	length of line from A to B
π	pi (= 3.14159^{+})
$^{\circ}$	degrees
$'$	minutes
$''$	seconds
$\angle$	angle
dx	differential of x
Δ	(delta) difference
Δx	increment of x
$\partial u/\partial x$	partial derivative of u with respect to x
$\int$	integral of
$\int_b^a$	integral of, between limits a and b
$\oint$	line integral around a closed path

Σ	(sigma) summation of
$f(x)$, $F(x)$	functions of x
$\exp x = e^x$	[e = 2.71828 (base of natural, or Napierian, logarithms)]
∇	del or nabla, vector differential operator
∇^2	Laplacian operator
£	Laplace operational symbol
4!	factorial $4 = 4 \times 3 \times 2 \times 1$
$\lvert x \rvert$	absolute value of x
$\dot{x}$	first derivative of x with respect to time
$\ddot{x}$	second derivative of x with respect to time
A × B	vector product; magnitude of A times magnitude of B times sine of the angle from A to B; $AB \sin \overline{AB}$
A · B	scalar product; magnitude of A times magnitude of B times cosine of the angle from A to B; $AB \cos \overline{AB}$

Section 1

Mechanics of Solids and Fluids

BY

PEYMAN HONARMANDI *Professor of Mechanical Engineering, Manhattan College, Bronx, New York*

J. W. MURDOCK *Late Consulting Engineer*

YUNUS A. ÇENGEL *Professor Emeritus, University of Nevada, Reno*

LEONARD MEIROVITCH *Late Professor of Engineering Science, Virginia Tech*

MEHDI AHMADIAN *Professor and Director, Center for Vehicle Systems & Safety and Railway Technologies Laboratory, Virginia Tech*

1.1 MECHANICS OF SOLIDS
Revised by Peyman Honarmandi

1.2 FRICTION
Revised by Peyman Honarmandi

1.3 MECHANICS OF FLUIDS
by J. W. Murdock, Reviewed by Yunus A. Çengel

1.4 VIBRATION
by Leonard Meirovitch, Revised by Mehdi Ahmadian

1.1 MECHANICS OF SOLIDS

Revised by Peyman Honarmandi

REFERENCES: Beer and Johnston, "Mechanics for Engineers," McGraw-Hill. Ginsberg and Genin, "Statics and Dynamics," Wiley. Higdon and Stiles, "Engineering Mechanics," Prentice-Hall. Holowenko, "Dynamics of Machinery," Wiley. Housner and Hudson, "Applied Mechanics," Van Nostrand. Meriam, "Statics and Dynamics," Wiley. Hamilton, Ocvirk, and Mabie, "Mechanisms and Dynamics of Machinery," Wiley. Synge and Griffith, "Principles of Mechanics," McGraw-Hill. Timoshenko and Young, "Advanced Dynamics," McGraw-Hill. Timoshenko and Young, "Engineering Mechanics," McGraw-Hill.

PHYSICAL MECHANICS

Definitions

Force is the action of one body on another which will cause acceleration of the second body unless acted on by an equal and opposite action counteracting the effect of the first body. It is a vector quantity.

Time is a measure of the sequence of events. In newtonian mechanics it is an absolute quantity. In relativistic mechanics it is relative to the *frames of reference* in which the sequence of events is observed. The common unit of time is the second.

Inertia is a property of matter which causes a resistance to any change in the motion of a body.

Mass is a quantitative measure of *inertia.*

Acceleration of Gravity Every object which falls in a vacuum at a given position on the earth's surface will have the same acceleration g. Accurate values of the acceleration of gravity as measured *relative* to the earth's surface include the effect of the earth's rotation and flattening at the poles. The international gravity formula for the acceleration of gravity at the earth's surface is $g = 32.0881(1 + 0.005288 \sin^2 \phi - 0.0000059 \sin^2 2\phi)$ ft/s², where ϕ is latitude in degrees which ranges from 0° at the Equator to 90° at the North or South poles. For extreme accuracy, the local acceleration of gravity must also be corrected for the presence of large water or land masses and for height above sea level. The absolute acceleration of gravity for a nonrotating earth discounts the effect of the earth's rotation and is rarely used, except outside the earth's atmosphere. If g_0 represents the absolute acceleration at sea level, the absolute value at an altitude h is $g = g_0R^2/(R + h)^2$, where R is the radius of the earth, approximately 3,960 mi (6,373 km).

Weight is the resultant force of attraction on the mass of a body due to a gravitational field. On earth, unit of weight is pound-force lbf (newton N) based upon an acceleration of gravity of 32.1740 ft/s² (9.80665 m/s²).

Linear momentum is the product of mass and the linear velocity of a particle and is a vector. The moment of the linear-momentum vector about a fixed axis is the **angular momentum** of the particle about that fixed axis. For a rigid body rotating about a fixed axis, angular momentum is defined as the product of the moment of inertia and angular velocity, each measured about the fixed axis.

An increment of **work** is defined as the product of an incremental displacement and the component of the force vector in the direction of the displacement, or the component of the displacement vector in the direction of the force. The increment of work done by a force acting on a body during a displacement of dr in the direction of the force is $dU = F\,dr$. The increment of work done by a couple acting on a body during a rotation of $d\theta$ in the plane of the couple is $dU = M\,d\theta$.

Energy is defined as the capacity of a body to do work by reason of its motion or configuration (see **Work and Energy**).

A **vector** is a directed line segment that has both magnitude and direction, unlike scalar that has only magnitude. In script or text, a vector is distinguished from a scalar V by a boldface-type $\mathbf{V}$. The magnitude of the scalar is the magnitude of the vector, $V = |\mathbf{V}|$.

A frame of reference is a specified set of geometric conditions to which other locations, motion, and time are referred. In newtonian mechanics, the fixed stars are referred to as the **primary (inertial) frame of reference** that describes time and space homogeneously and in a time-independent manner. Relativistic mechanics denies the existence of a primary reference frame and holds that all reference frames must be described relative to each other.

SYSTEMS AND UNITS OF MEASUREMENTS

In *absolute systems*, the units of **length, mass,** and **time** are considered fundamental quantities, and all other units including that of **force** are derived.

In *gravitational systems*, the units of **length, force,** and **time** are considered fundamental qualities, and all other units including that of **mass** are derived.

In the SI system of units, the unit of mass is in kilogram (kg) and the unit of length is in meter (m). A force of one newton (N) is derived as the force that will give a mass of 1 kg an acceleration of 1 m/s².

In the English engineering system of units, the unit of mass is in pound mass (lbm) and the unit of length is in foot (ft). A force of one pound (1 lbf) is the force that gives a pound mass (1 lbm) an acceleration equal to the standard acceleration of gravity on the earth, 32.1740 ft/s² (9.80665 m/s²). A slug is the mass that will be accelerated 1 ft/s² by a force of 1 lbf. Therefore, 1 slug = 32.1740 lbm. When described in the gravitational system, mass is a derived unit, being the constant of proportionality between force and acceleration, as determined by Newton's second law.

General Laws

NEWTON'S LAWS

i. If a balanced force system acts on a particle at rest, it will remain at rest. If a balanced force system acts on a particle in motion, it will remain in motion in a straight line without acceleration.

ii. If an unbalanced force system acts on a particle, it will accelerate in proportion to the magnitude and in the direction of the resultant force.

iii. When two particles exert forces on each other, these forces are equal in magnitude, opposite in direction, and collinear.

Fundamental Equation The basic relation between mass, acceleration, and force is contained in Newton's second law of motion. The force applied to a particle can be described as, $\mathbf{F} = m\mathbf{a}$, force = mass × acceleration. This equation is a vector equation, since the direction of $\mathbf{F}$ must be the direction of $\mathbf{a}$, as well as having $\mathbf{F}$ equal in magnitude to $m\mathbf{a}$. An alternative form of Newton's second law states that the resultant force is equal to the time rate of change of momentum, $\mathbf{F} = d(m\mathbf{v})/dt$.

Law of the Conservation of Mass The mass of a body remains unchanged by any ordinary physical or chemical change to which it may be subjected. (True in classic mechanics.)

Law of the Conservation of Energy The law of conservation of energy states that the total energy of an isolated system remains constant. Energy can be neither created nor destroyed, but it transforms from one form to another.

Law of the Conservation of Mechanical Energy The principle of conservation of mechanical energy states that the total mechanical energy of a system remain unchanged if it is subjected only to conservative forces or forces which depend on position or configuration.

Law of the Conservation of Momentum The linear momentum of a system of bodies is unchanged if there is no resultant external force on the system. The angular momentum of a system of bodies about a fixed axis is unchanged if there is no resultant external moment about this axis.

Law of Mutual Attraction (Universal Gravitation) Two particles attract each other with a force F proportional to their masses m_1 and m_2 and inversely proportional to the square of the distance r between them, or $F = Gm_1m_2/r^2$, in which G is the gravitational constant. The value of the gravitational constant is $G = 6.67408 \times 10^{-11}$ m³/kg·s² in SI or absolute units, or $G = 3.4397 \times 10^{-8}$ lbf·ft²/slug² in English system of units. It should be pointed out that the unit of force F in the SI system is in **newton** and is derived, while the unit force in the English system is in **pound-force** and is a fundamental quantity.

EXAMPLE. Each of two solid steel spheres 6 in in diameter will weigh 32.0 lbf on the earth's surface. This is the force of attraction between the earth and the steel sphere. The force of mutual attraction between the spheres if they are just touching is 0.000000136 lbf. (For spheres, center-to-center distance may be used.)

STATICS OF RIGID BODIES

General Considerations

If the forces acting on a rigid body do not produce any acceleration, they must neutralize each other, i.e., form a **system of forces in equilibrium.** Equilibrium is said to be **stable** when the body with the forces acting upon it returns to its original position after being displaced a very small amount from that position; **unstable** when the body tends to move still farther from its original position than the very small displacement; and **neutral** when the forces retain their equilibrium when the body is in its new position.

External and Internal Forces The forces by which the individual particles of a body act on each other are known as internal forces. All other forces are called external forces. If a body is supported by other bodies while subject to the action of forces, deformations and forces will be produced at the points of support or contact and these internal forces will be distributed throughout the body until equilibrium exists and the body is said to be in a state of tension, compression, or shear. The forces exerted by the body on the supports are known as **reactions.** They are equal in magnitude and opposite in direction to the forces with which the supports act on the body, known as **supporting forces.** The supporting forces are external forces applied to the body.

In considering a body at a definite section, it will be found that all the internal forces act in pairs, i.e., the two forces being equal and opposite. The external forces act singly.

General Law When a body is at rest, the forces acting externally to it must form an equilibrium system. This law will hold for any part of the body, in which case the forces acting at any section of the body become external forces when the part on either side of the section is considered alone. In the case of a **rigid body,** any two forces of the same magnitude, but acting in opposite directions in any straight line, may be added or removed without change in the action of the forces acting on the body, provided the strength of the body is not affected.

Principle of Transmissibility Conditions of equilibrium or motion are not affected by transmitting a force along its line of action.

Composition, Resolution, and Equilibrium of Forces

The **resultant** of several forces acting at a point is a force which will produce the same effect as all the individual forces acting together.

Forces Acting on a Body at the Same Point The resultant R of two forces F_1 and F_2 applied to a rigid body at the same point is represented in magnitude and direction by the diagonal of a parallelogram formed by F_1 and F_2 (see Figs. 1.1.1 and 1.1.2).

$$R = \sqrt{F_1^2 + F_2^2 + 2F_1F_2 \cos a}$$

$$\sin a_1 = (F_2 \sin a)/R \qquad \sin a_2 = (F_1 \sin a)/R$$

When $a = 90°$, $R = \sqrt{F_1^2 + F_2^2}$, $\sin a_1 = F_2/R$, and $\sin a_2 = F_1/R$.

$$\left.\begin{aligned} &\text{When } a = 0°, \ R = F_1 + F_2 \\ &\text{When } a = 180°, \ R = F_1 - F_2 \end{aligned}\right\} \text{Forces act in same straight line.}$$

A **force R may be resolved into two component forces** intersecting anywhere on R and acting in the same plane as R, by the reverse of the operation shown by Figs. 1.1.1 and 1.1.2; and by repeating the operation with the components, R may be resolved into any number of component forces intersecting R at the same point and in the same plane.

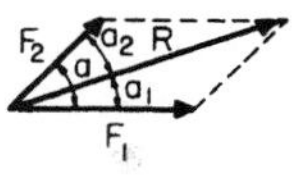

Figure 1.1.1 Resultant of forces using parallelogram.

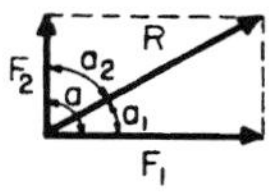

Figure 1.1.2 Resolving vector R into two components.

Resultant of Any Number of Forces Applied to a Rigid Body at the Same Point Resolve each of the given forces F into components along three rectangular coordinate axes. If A, B, and C are the angles made with XX, YY, and ZZ, respectively, by any force F, the components will be $F \cos A$ along XX, $F \cos B$ along YY, $F \cos C$ along ZZ; add the components of all the forces along each axis algebraically and obtain $\Sigma F \cos A = \Sigma X$ along XX, $\Sigma F \cos B = \Sigma Y$ along YY, and $\Sigma F \cos C = \Sigma Z$ along ZZ.

The resultant $R = \sqrt{(\Sigma X)^2 + (\Sigma Y)^2 + (\Sigma Z)^2}$. The angles made by the resultant with the three axes are A_r with XX, B_r with YY, C_r with ZZ, where

$$\cos A_r = \Sigma X/R \qquad \cos B_r = \Sigma Y/R \qquad \cos C_r = \Sigma Z/R$$

These cosines are called **direction cosines** by which the **direction** of the **resultant** can be determined by plotting the algebraic sums of the components.

If the forces are all in the same plane, the components of each of the forces along one of the three axes (say ZZ) will be 0; i.e., angle $C_r = 90°$ and $R = \sqrt{(\Sigma X)^2 + (\Sigma Y)^2}$, $\cos A_r = \Sigma X/R$, and $\cos B_r = \Sigma Y/R$.

For equilibrium, it is necessary that $R = 0$; i.e., ΣX, ΣY, and ΣZ must each be equal to zero.

General Law In order that a number of forces acting at the same point shall be in equilibrium, the algebraic sum of their components along any *three* coordinate axes must each be equal to zero. When the forces all act in the same plane, the algebraic sum of their components along any *two* coordinate axes must each equal zero.

When the Forces Form a System in Equilibrium Three unknown forces can be determined if the lines of action of the forces are *all* known and are in different planes. If the forces are all in the same plane, the lines of action being known, only *two* unknown forces can be determined. If the lines of action of the unknown forces are *not* known, only *one* unknown force can be determined in either case.

Couples and Moments

Couple Two parallel forces of equal magnitude (Fig. 1.1.3) which act in opposite directions and are not collinear form a couple. **A couple cannot be reduced to a single force.**

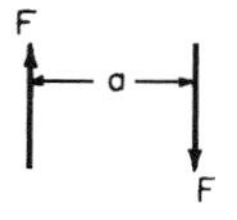

Figure 1.1.3 A couple force.

Displacement and Change of a Couple The forces forming a couple may be moved and their magnitude and direction changed, provided they always remain parallel to each other and remain in either the original plane or one parallel to it, and provided the product of one of the forces and the perpendicular distance between the two is constant and the direction of rotation remains the same. Therefore, the couples have the same effect on a body and are called **equivalent couples**.

Moment of a Couple The moment of a couple is the product of the magnitude of one of the forces and the perpendicular distance between the lines of action of the forces. Fa = moment of couple; a = arm of couple. If the forces are measured in pounds and the distance a in feet, the **unit of rotation moment** is the pound-foot. If the force is measured in newtons and the distance in meters, the unit is the newton-meter. In the cgs system the unit of rotation moment is 1 dyne-cm.

Rotation moments of couples acting in the same plane are conventionally considered to be positive for counterclockwise moments and negative for clockwise moments, although it is only necessary to be consistent within a given problem. The magnitude, direction, and sense of rotation of a couple are completely determined by its moment axis, or moment vector, which is a line drawn perpendicular to the plane in which the couple acts, with an arrow indicating the direction from which the couple will appear to have right-handed rotation; the length of the line represents the magnitude of the moment of the couple. Figure 1.1.4*a* shows a counterclockwise couple, designated +. Figure 1.1.4*b* shows a clockwise couple, designated –.

Composition of Couples A couple may be represented in space by a single vector (having no specific line of action) perpendicular to the plane of the couple. If you let the relaxed fingers of your right hand represent the "swirl" of the forces in the couple, your stiff right thumb will indicate the direction of the vector. Vector couples may be added vectorially in the same manner by which concurrent forces are added.

Couples lying in the same or parallel planes are added algebraically. Let +28 lbf · ft (+38 N · m), –42 lbf · ft (–57 N · m), and +70 lbf · ft (95 N · m) be the moments of three couples in the same or parallel planes; their resultant is a single couple lying in the same or in a parallel plane, whose moment is $\Sigma M = +28 - 42 + 70 = +56$ lbf · ft ($\Sigma M = +38 - 57 + 95 = 76$ N × m).

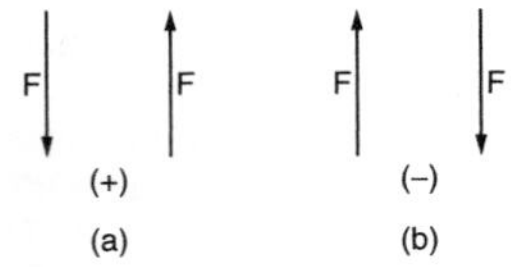

Figure 1.1.4 (a) Positive couple (b) Negative couple.

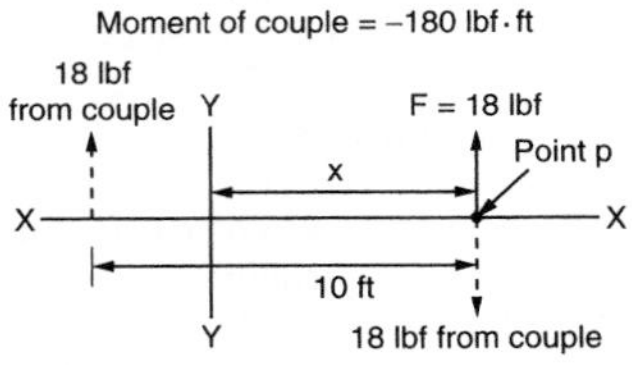

Figure 1.1.5 Combination of a couple and a single force.

If the polygon formed by the moment vectors of several couples closes itself, the couples form an equilibrium system. Two couples will balance each other when they lie in the same or parallel planes and have the same moment in magnitude, but opposite in sign.

Combination of a Couple and a Single Force in the Same Plane As shown in Fig. 1.1.5, given a force $F = 18$ lbf (80 N) acting at distance x from YY, and a couple whose moment is –180 lbf · ft (244 N × m) in the same or parallel plane, to find the resultant. A couple may be changed to any other couple in the same or a parallel plane having the same moment and same sign. Let the couple consist of two forces of 18 lbf (80 N) each and let the arm be 10 ft (3.05 m). Place the couple in such a manner that one of its forces is opposed to the given force at p. This force of the couple and the given force being of the same magnitude and opposite in direction will neutralize each other, leaving the other force of the couple acting at a distance of 10 ft (3.05 m) from p and parallel and equal to the given force 18 lbf (80 N).

General Rule The resultant of a couple and a single force lying in the same or parallel planes is a single force, equal in magnitude, in the same direction and parallel to the single force, and acting at a distance from the line of action of the single force equal to the moment of the couple divided by the single force. The moment of the resultant force about any point on the line of action of the given single force must be of the same sense as that of the couple, positive if the moment of the couple is positive, and negative if the moment of the couple is negative. If the moment of the couple in Fig. 1.1.5 had been + instead of –, the resultant would have been a force of 18 lbf (80 N) acting in the same direction and parallel to F, but at a distance of 10 ft (3.05 m) to the right of it, making the moment of the resultant about any point on F positive.

To effect a parallel displacement of a single force F over a distance a, a couple whose moment is Fa must be added to the system. The sense of the couple will depend upon which way it is desired to displace force F.

The moment of a force with respect to a point is the product of the force F and the perpendicular distance from the point to the line of action of the force.

The Moment of a Force with Respect to a Straight Line If the force is resolved into components parallel and perpendicular to the given line, the moment of the force with respect to the line is the product of the magnitude of the perpendicular component and the distance from its line of action to the given line.

Varigon's Theorem The moment about a give point O of the resultant of several concurrent forces is equal to the sum of the moments of the various moments about the same point O. Varigon's Theorem makes it possible to replace the direct determination of the moment of a force F by the moments of two or more component forces of F.

Forces with Different Points of Application

Composition of Forces If each force **F** is resolved into components parallel to three rectangular coordinate axes XX, YY, ZZ, the magnitude of the resultant is $R = \sqrt{(\Sigma X)^2 + (\Sigma Y)^2 + (\Sigma Z)^2}$, and its line of action makes angles A_r, B_r, and C_r with axes XX, YY, and ZZ, where $\cos A_r = \Sigma X/R$, $\cos B_r = \Sigma Y/R$, and $\cos C_r = \Sigma Z/R$; and there are three couples which may be combined by their moment vectors into a single resultant couple having the moment $M_r = \sqrt{(M_x)^2 + (M_y)^2 + (M_z)^2}$, whose moment vector makes angles of A_m, B_m, and C_m with axes XX, YY, and ZZ, such that $\cos A_m = M_x/M_r$, $\cos B_m = M_y/M_r$, $\cos C_m = M_z/M_r$. If this single resulting couple is in the same plane as the single resulting force at the origin or a plane parallel to it, the system may be reduced to a single force **R** acting at a distance from **R** equal to M_r/R. If the couple and force are not in the same or parallel planes, it is impossible to reduce the system to a single force. If $\mathbf{R} = 0$, i.e., if ΣX, ΣY, and ΣZ all equal zero, the system will reduce to a single couple whose moment is $\mathbf{M}_r$. If $\mathbf{M}_r = 0$, i.e., if M_x, M_y, and M_z all equal zero, the resultant will be a single force **R**.

When the forces are all in the same plane, the cosine of one of the angles A_r, B_r, or $C_r = 0$, say, $C_r = 90°$. Then $R = \sqrt{(\Sigma X)^2 + (\Sigma Y)^2}$, $M_r = \sqrt{M_x^2 + M_y^2}$, and the final resultant is a force equal and parallel to R, acting at a distance from R equal to M_r/R.

A system of forces in the same plane can always be replaced by either a couple or a single force. If $\mathbf{R} = 0$ and $\mathbf{M}_r > 0$, the resultant is a couple. If $\mathbf{M}_r = 0$ and $\mathbf{R} > 0$, the resultant is a single force.

A rigid body is in equilibrium when acted upon by a system of forces whenever $R = 0$ and $M_r = 0$, i.e., when the following six conditions hold true: $\Sigma X = 0$, $\Sigma Y = 0$, $\Sigma Z = 0$, $M_x = 0$, $M_y = 0$, and $M_z = 0$. When the system of forces is in the same plane, equilibrium prevails when the following three conditions hold true: $\Sigma X = 0$, $\Sigma Y = 0$, $\Sigma M = 0$.

Forces Applied to Support Rigid Bodies

The external forces in equilibrium acting upon a body may be statically determinate or indeterminate according to the number of unknown forces existing. When the forces are all in the same plane and act at a common point, two unknown forces may be determined if their lines of action are known, or one unknown force if its line of action is unknown.

When the forces are all in the same plane and are parallel, two unknown forces may be determined if the lines of action are known, or one unknown force if its line of action is unknown.

When the forces are anywhere in the same plane, three unknown forces may be determined if their lines of action are known, if they are not parallel or do not pass through a common point; if the lines of action are unknown, only one unknown force can be determined.

If the forces all act at a common point but are in different planes, three unknown forces can be determined if the lines of action are known, one if unknown.

If the forces act in different planes but are parallel, three unknown forces can be determined if their lines of action are known, one if unknown.

The first step in the solution of problems in statics is the determination of the supporting forces. The following data are required for the complete knowledge of supporting forces: magnitude, direction, and point of application. According to the nature of the problem, none, one, or two of these quantities are known.

One Fixed Support The point of application, direction, and magnitude of the load are known. See Fig. 1.1.6. As the body on which the forces act is in equilibrium, the supporting force P must be equal in magnitude and opposite in direction to the resultant of the loads L.

In the case of a **rolling surface,** the point of application of the support is obtained from the center of the connecting bolt A (Fig. 1.1.7), both the direction and magnitude being unknown. The point of application and line of action of the support at B are known, being determined by the rollers.

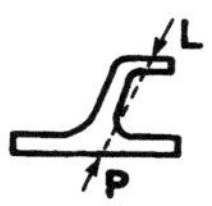

Figure 1.1.6 Fixed support at P.

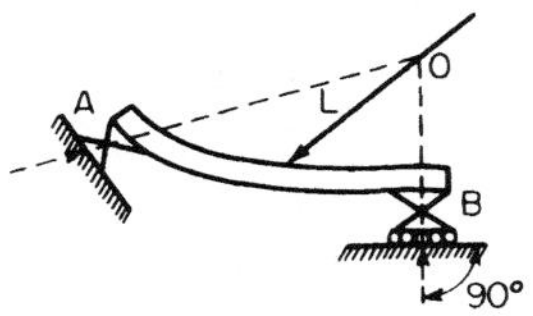

Figure 1.1.7 Simply supported at A and roller support at B.

When three forces acting in the same plane on the same rigid body are in equilibrium, their lines of action must pass through the same point O. The load L is known in magnitude and direction. The line of action of the support at B is known on account of the rollers. The point of application of the support at A is known. The three forces are in equilibrium and are in the same plane; therefore, the lines of action must meet at the point O.

In the case of the rolling surfaces shown in Fig. 1.1.8, the direction of the support at A is known, the magnitude unknown. The line of action and point of application of the supporting force at B are known, its magnitude unknown. The lines of action of the three forces must meet in a point, and the supporting force at A must be perpendicular to the plane XX.

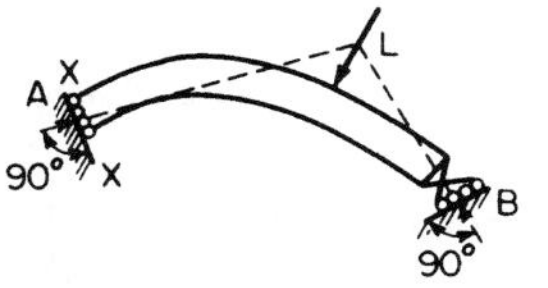

Figure 1.1.8 Member supported at two points by roller supports.

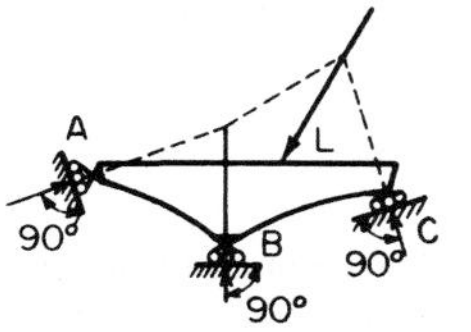

Figure 1.1.9 Member supported at three points by roller supports.

If a member of a truss or frame in equilibrium is pinned at two points and loaded at these two points only, the line of action of the forces exerted on the member or by the member at these two points must be along a line connecting the pins.

If the external forces acting upon a rigid body in equilibrium are all in the same plane, the equations $\Sigma X = 0$, $\Sigma Y = 0$, and $\Sigma M = 0$ must be satisfied. When trusses, frames, and other structures are under discussion, these equations are usually used as $\Sigma V = 0$, $\Sigma H = 0$, $\Sigma M = 0$, where V and H represent vertical and horizontal components, respectively.

In Fig. 1.1.9, the directions and points of application of the supporting forces are known, but all three are of unknown magnitudes. Including load L, there are four forces, and the three unknown magnitudes may be determined by $\Sigma X = 0, \Sigma Y = 0,$ and $\Sigma M = 0$.

The **supports** are said to be **statically determinate** when the laws of equilibrium are sufficient for their determination. When the conditions are not sufficient for the determination of the supports or other forces, the structure is said to be **statically indeterminate;** the unknown forces can then be determined from considerations involving the deformation of the material.

When several bodies are so connected to one another as to make up a rigid structure, the forces at the points of connection must be considered as internal forces and are not taken into consideration in the determination of the supporting forces for the entire structure.

The distortion of any practically rigid structure under its working loads is so small as to be negligible when determining supporting forces. When the forces acting at the different joints in a built-up structure cannot be determined by dividing the structure up into parts, the structure is said to be **statically indeterminate internally.** A structure may be statically indeterminate internally and still be statically determinate externally.

Fundamental Problems in Graphical Statics

A force may be represented by a straight line in a determined position, and its magnitude by the length of the straight line. The direction in which it acts may be indicated by an arrow.

Polygon of Forces The parallelogram of two forces intersecting each other (see Figs. 1.1.4 and 1.1.5) leads directly to the graphic composition by means of the triangle of forces. In Fig. 1.1.10, R is called the **closing side,** and represents the resultant of the forces $\mathbf{F}_1$ and $\mathbf{F}_2$ in magnitude and direction. Its position is given by the point of application O. By means of repeated use of the triangle of forces and by omitting the closing sides of the individual triangles, the magnitude and direction of the resultant $\mathbf{R}$ of any number of forces in the same plane and intersecting at a single point can be found. In Fig. 1.1.11 the lines representing the forces start from point O, and in the force polygon (Fig. 1.1.12) they are joined in any order from the head of one to the toe of the next vector, the arrows showing their directions following around the polygon in the same direction. The magnitude of the resultant at the point of application of the forces is represented by the closing side $\mathbf{R}$ of the force polygon; its direction, as shown by the arrow, is counter to that in the other sides of the polygon.

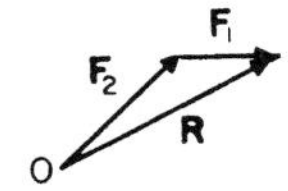

Figure 1.1.10 Resultant of two forces using triangle method.

If the forces are in equilibrium, then R must equal zero, i.e., **the force polygon will become a closed polygon. If one of the forces is reversed in direction in the closed polygon, this force becomes the resultant of all the others.**

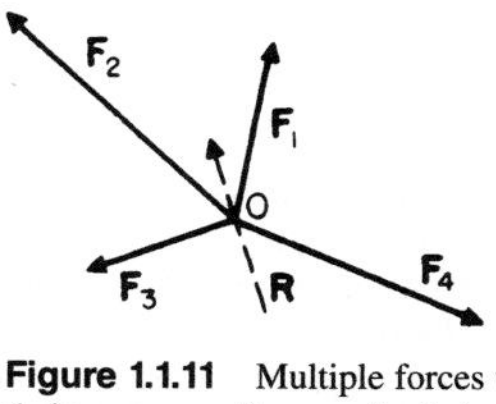

Figure 1.1.11 Multiple forces with their corresponding resultant at a single point.

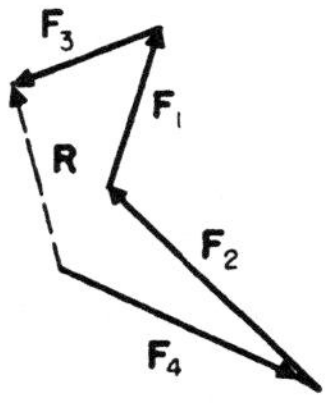

Figure 1.1.12 Resultant of multiple forces using polygon method.

Determination of Stresses in Members of Truss Structure in Equilibrium under External Loads

Truss structures are statically determinate plane structures whose members are connected through joints. It will be assumed that the loads are applied at the joints of the structure, i.e., at the points where the different members are connected and that the connections are pins with no friction. If a member is connected to its adjacent members with only two joints and carries no loads in-between, it is called a **two-force member.** The load or stress of the two-force member must then be along a center line drawn between two joints, unless any member is loaded at more than two joints by pin connections. A **three-force**

member is generally a member under application of three forces, or a member with three joints in the case of truss structures. The three forces should be coplanar either concurrent or parallel.

Equilibrium In order the whole structure be in equilibrium, it is necessary that the external forces (external loads and support reactions) shall form a balanced system. Graphical and analytical methods are both of service.

Supporting Forces When the supporting forces or reactions are to be determined, it is not necessary to pay any attention to the makeup of the structure under consideration so long as it is practically rigid; the loads may be taken as they occur, or the resultant of the loads may be used instead. When the stresses in the members of the structure are being determined, the loads *must* be distributed at the joints where they belong.

Method of Joints When all the external forces have been determined, any joint at which there are not more than two unknown forces may be taken and these unknown forces determined by the methods of the stress polygon, resolution, or moments. In Fig. 1.1.13, let O be the joint of a structure and **F** be the only known force; but let O_1 and O_2 be two members of the structure joined at O. Then the lines of action of the unknown forces are known and their magnitude may be determined (1) by a **stress polygon** which, for equilibrium, must close; (2) by resolution into H and V components, using the condition of equilibrium $\Sigma H = 0$, $\Sigma V = 0$; or (3) by moments, using any convenient point on the line of action of O_1 and O_2 and the condition of equilibrium $\Sigma M = 0$. No more than *two* unknown forces can be determined from each joint. In this manner, proceeding from joint to joint, the stresses in all the members of the truss can usually be determined if the structure is statically determinate internally.

Method of Sections The structure may be divided into parts by passing a cutting section through some of its members; one part may then be treated as a rigid body with the external forces acting upon it. Some of these forces will be the stresses in the members themselves. For example, let xx (Fig. 1.1.14) be a section taken through a truss loaded at P_1, P_2, and P_3, and supported on rollers at S with a vertical reaction force. As the whole truss is in equilibrium, any part of it must be also in equilibrium, and consequently the part shown to the left of xx must be in equilibrium under the action of the forces acting externally

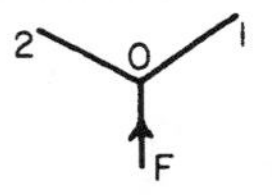

Figure 1.1.13 Method of Joints.

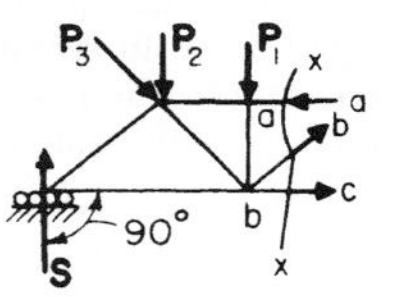

Figure 1.1.14 Method of Sections.

to it. Three of these forces are the stresses in the members aa, bb, and bc, which are the unknown forces to be determined. They can be determined by applying the conditions of equilibrium of forces acting in the same plane but not at the same point, i.e., $\Sigma H = 0$, $\Sigma V = 0$, $\Sigma M = 0$. The three unknown forces can be determined only if they are not parallel or do not pass through the same point; if, however, the forces are parallel or meet in a point, two unknown forces only can be determined. Section cuts may be passed through the structure members in any convenient manner, as a rule, however, cutting not more than three members, unless members are unloaded.

For the determination of stresses in framed structures, see Sec. 2.12.

CENTER OF GRAVITY

Consider a three-dimensional body of any size, shape, and weight, but preferably thin and flat. If it is suspended as in Fig. 1.1.15 by a cord from any point A, it will be in equilibrium under the action of the tension in the cord and the weight W. If the experiment is repeated by suspending the body from point B, it will again be in equilibrium. Use a plumb bob through point A and draw the "first line" on the body. If the lines of action of the weight were marked in each case, they would be concurrent at a point G known as the **center of gravity** or **center of mass.** Whenever the density of the body is uniform, it will be a constant factor and like geometric shapes of different densities will have the same center of gravity. The term **centroid** is used in this case since the location of the center of gravity is of geometric concern only. If densities are nonuniform, like geometric shapes will have the same centroid but different centers of gravity.

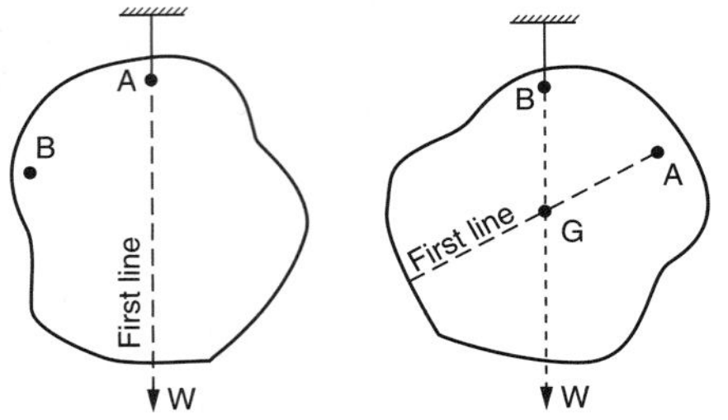

Figure 1.1.15 Experimental technique to locate center of gravity.

Centroids of Technically Important Lines, Areas, and Solids

CENTROIDS OF LINES

Straight Lines The centroid is at its middle point.

Circular Arc *AB* (Fig. 1.1.16) $x_0 = r \sin c/\text{rad } c$; $y_0 = 2r \sin^2 \tfrac{1}{2}c/\text{rad } c$. (rad c = angle c measured in radians.)

Circular Arc *AC* (Fig. 1.1.17) $x_0 = r \sin c/\text{rad } c$; $y_0 = 0$.

Quadrant, *AB* (Fig. 1.1.18*a*) $x_0 = y_0 = 2r/\pi = 0.6366r$.

Semicircumference, *AC* (Fig. 1.1.18*b*) $y_0 = 2r/\pi = 0.6366r$; $x_0 = 0$.

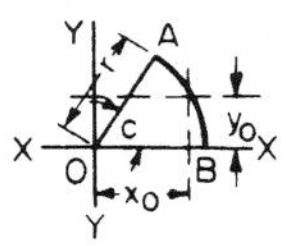

Figure 1.1.16

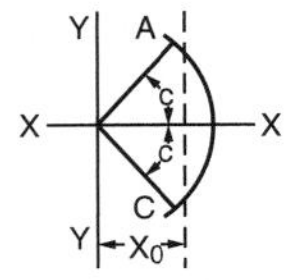

Figure 1.1.17

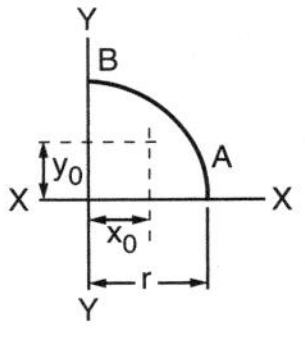

Figure 1.1.18*a*

Combination of Arcs and Straight Line (Fig. 1.1.19) AD and BC are two quadrants of radius r. $y_0 = \{(AB)r + 2[0.5\pi r(r - 0.6366r)]\} \div [AB + 2(0.5\pi r)]$ (Symmetrical about YY).

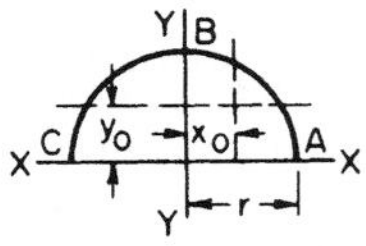

Figure 1.1.18*b*

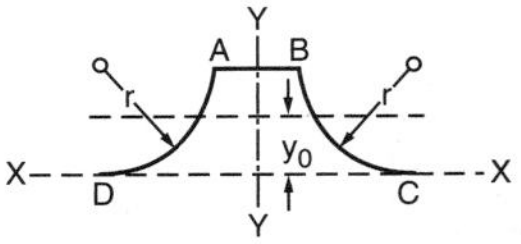

Figure 1.1.19

CENTROIDS OF PLANE AREAS

Triangle Centroid lies at the intersection of the lines joining the vertices with the midpoints of the sides, and at a distance from any side equal to one-third of the corresponding altitude.

Parallelogram Centroid lies at the point of intersection of the diagonals.

Trapezoid (Fig. 1.1.20) Centroid lies on the line joining the middle points m and n of the parallel sides. The distances h_a and h_b are

$$h_a = h(a + 2b)/3(a + b) \qquad h_b = h(2a + b)/3(a + b)$$

Draw $BE = a$ and $CF = b$; EF will then intersect mn at centroid.

Any Quadrilateral The centroid of any quadrilateral may be determined by the general rule for areas, or graphically by dividing it into two sets of triangles by means of the diagonals. Find the centroid of each of the four triangles and connect the centroids of the triangles belonging to the same set. The intersection of these lines will be centroid of area. For example, in Fig. 1.1.21, O, O_1, O_2, and O_3 are the centroids of the triangles ABD, ABC, BDC, and ACD, respectively. The intersection of O_1O_3 with OO_2 gives the centroid of the quadrilateral area.

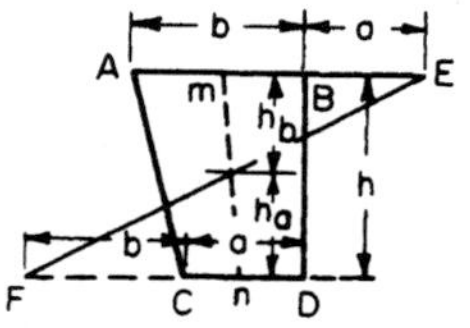

Figure 1.1.20

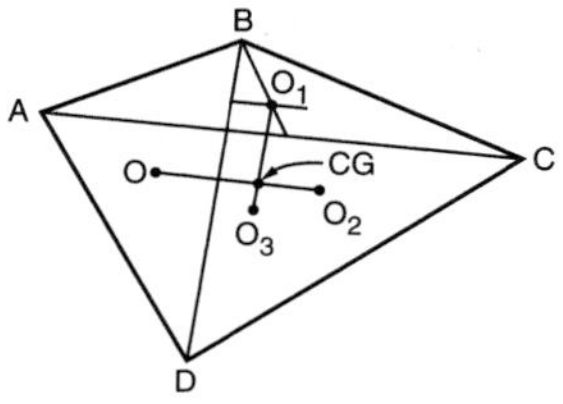

Figure 1.1.21

Segment of a Circle (Fig. 1.1.22) $x_0 = \frac{2}{3} r \sin^3 c/(\text{rad } c - \cos c \sin c)$. A segment may be considered to be a sector from which a triangle is subtracted, and the general rule applied.

Sector of a Circle (Fig. 1.1.23) $x_0 = \frac{2}{3} r \sin c/\text{rad } c$; $y_0 = \frac{4}{3} r \sin^2 \frac{1}{2} c/\text{rad } c$.

Semicircle $x_0 = \frac{4}{3} r/\pi = 0.4244r$; $y_0 = 0$.

Quadrant (90° sector) $x_0 = y_0 = \frac{4}{3} r/\pi = 0.4244r$.

Parabolic Half Segment (Fig. 1.1.24) Area ABO: $x_0 = \frac{3}{5} x_1$; $y_0 = \frac{3}{8} y_1$.

Parabolic Spandrel (Fig. 1.1.24) Area AOC: $x_0' = \frac{3}{10} x_1$; $y_0 = \frac{3}{4} y_1$.

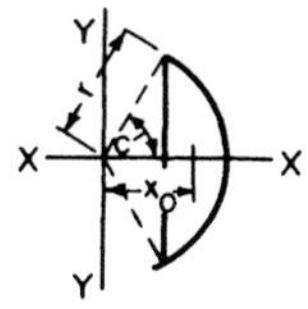

Figure 1.1.22

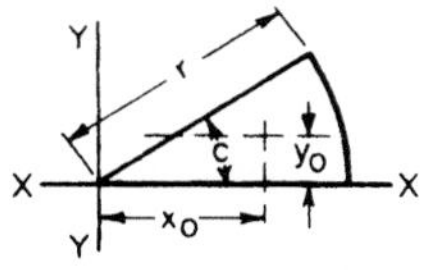

Figure 1.1.23

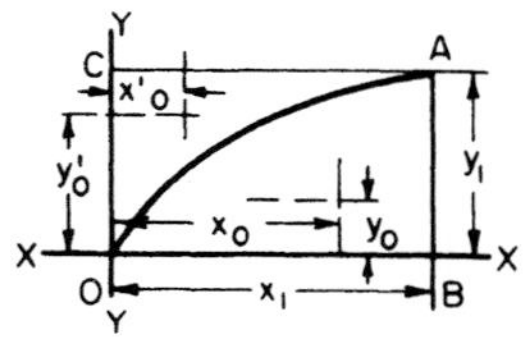

Figure 1.1.24

Quadrant of an Ellipse (Fig. 1.1.25) Area OAB: $x_0 = \frac{4}{3}(a/\pi)$; $y_0 = \frac{4}{3}(b/\pi)$.

The centroid of a figure such as that shown in Fig. 1.1.26 may be determined as follows: Divide the area $OABC$ into a number of parts

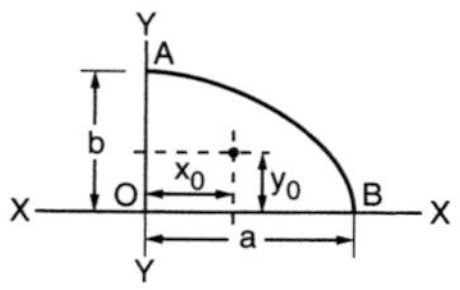

Figure 1.1.25

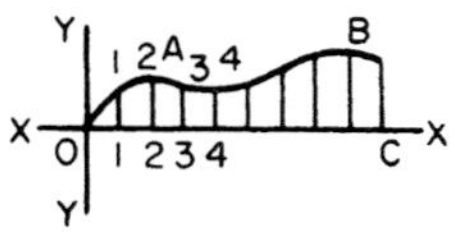

Figure 1.1.26

by lines drawn perpendicular to the axis XX, e.g., 11, 22, 33, etc. These parts will be approximately either triangles, rectangles, or trapezoids. The area of each division may be obtained by taking the product of its mean height and its base. The centroid of each area may be obtained as previously shown. The sum of the moments of all the areas about XX and YY, respectively, divided by the sum of the areas will give approximately the distances from the center of gravity of the whole area to the axes XX and YY. The greater the number of areas taken, the more nearly exact the result.

Centroids of Solids

Prism or Cylinder with Parallel Bases The centroid lies in the center of the line connecting the centers of gravity of the bases.

Oblique Frustum of a Right Circular Cylinder (Fig. 1.1.27) Let 1, 2, 3, 4 be the plane of symmetry. The distance from the base to the centroid is $\frac{1}{2} h + (r^2 \tan^2 c)/8h$, where c is the angle of inclination of the oblique section to the base. The distance of the centroid from the axis of the cylinder is $r^2 \tan c/4h$.

Pyramid or Cone The centroid lies in the line connecting the centroid of the base with the vertex and at a distance of one-fourth of the altitude above the base.

Truncated Pyramid If h is the height of the truncated pyramid and A and B the areas of its bases, the distance of its centroid from the surface of A is

$$h(A + 2\sqrt{AB} + 3B)/4(A + \sqrt{AB} + B)$$

Truncated Circular Cone If h is the height of the frustum and R and r the radii of the bases, the distance from the surface of the base whose radius is R to the centroid is $h(R^2 + 2Rr + 3r^2)/4(R^2 + Rr + r^2)$.

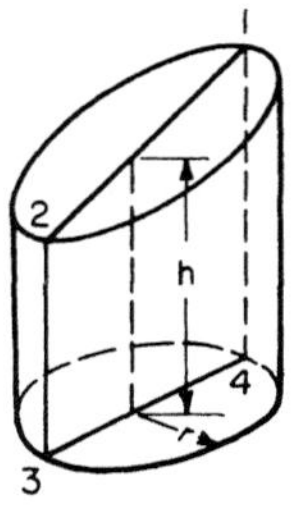

Figure 1.1.27

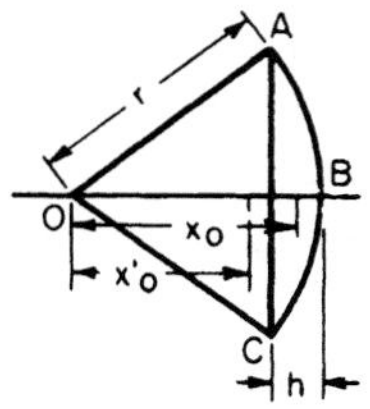

Figure 1.1.28

Segment of a Sphere (Fig. 1.1.28) Volume ABC: $x_0 = 3(2r - h)^2/4(3r - h)$.

Hemisphere $x_0 = 3r/8$.

Hollow Hemisphere $x_0 = 3(R^4 - r^4)/8(R^3 - r^3)$, where R and r are, respectively, the outer and inner radii.

Sector of a Sphere (Fig. 1.1.28) Volume $OABCO$: $x_0' = \frac{3}{8}(2r - h)$.

Ellipsoid, with Semiaxes *a*, *b*, and *c* For each octant, distance from center of gravity to each of the bounding planes = $\frac{3}{8}$ × length of semiaxis perpendicular to the plane considered.

The formulas given for the determination of the centroid of lines and areas can be used to determine the areas and volumes of surfaces and solids of revolution, respectively, by employing the theorems of Pappus Guldinus.

Determination of Center of Gravity of a Body by Experiment The center of gravity may be determined by hanging the body up from different points and plumbing down; the point of intersection of the plumb lines will give the center of gravity. It may also be determined as shown in Fig. 1.1.29. The body is placed on knife-edges which

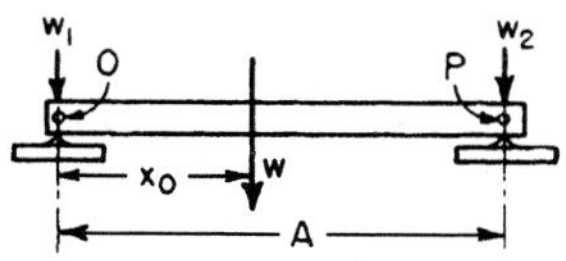

Figure 1.1.29 Determination of Center of Gravity of a Body.

rest on platform scales. The sum of the weights registered on the two scales ($w_1 + w_2$) must equal the weight (w) of the body. Taking a moment axis at either end (say, O), $w_2A/w = x_0$ = distance from O to plane containing the center of gravity. (See also Fig. 1.1.15 and accompanying text.)

Graphical Determination of the Centroids of Plane Areas See Fig. 1.1.40.

MOMENT OF INERTIA

The **moment of inertia of a solid body** with respect to a given axis is the limit of the sum of the products of the masses of each of the elementary particles into which the body may be conceived to be divided and the square of their distance from the given axis.

If $dm = dw/g$ represents the mass of an elementary particle and y its distance from an axis, the moment of inertia I of the body about this axis will be $I = \int y^2\,dm = \int y^2\,dw/g$.

The moment of inertia may be expressed in weight units ($I_w = \int y^2\,dw$), in which case the moment of inertia in weight units, I_w, is equal to the moment of inertia in mass units, I, multiplied by g.

If $I = k^2m$, the quantity k is called the **radius of gyration** or the **radius of inertia.**

If a body is considered to be composed of a number of parts, its moment of inertia about an axis is equal to the sum of the moments of inertia of the several parts about the same axis, or $I = I_1 + I_2 + I_3 + \cdots + I_n$.

The **moment of inertia of an area** with respect to a given axis that lies in the plane of the area is the limit of the sum of the products of the elementary areas into which the area may be conceived to be divided and the square of their distance (y) from the axis in question. $I = \int y^2\,dA = k^2A$, where k = **radius of gyration.**

The quantity $\int y^2\,dA$ is more properly referred to as the *second moment of area* since it is not a measure of *inertia* in a true sense.

Formulas for moments of inertia and radii of gyration of various areas follow later in this section.

Relation between the Moments of Inertia of an Area and a Solid The moment of inertia of a solid of elementary thickness about an axis is equal to the moment of inertia of the area of one face of the solid about the same axis multiplied by the mass per unit volume of the solid times the elementary thickness of the solid.

Moments of Inertia about Parallel Axes The moment of inertia of an area or solid about any given axis is equal to the moment of inertia about a parallel axis through the center of gravity plus the square of the distance between the two axes times the area or mass.

In Fig. 1.1.30*a*, the moment of inertia of the area $ABCD$ about axis YY is equal to I_0 (or the moment of inertia about Y_0Y_0 through the center of gravity of the area and parallel to YY) plus $x_0^2\,A$, where A = area of $ABCD$. In Fig. 1.1.30*b*, the moment of inertia of the mass m about axis $YY = I_0 + x_0^2\,m$. Y_0Y_0 passes through the centroid of the mass and is parallel to YY.

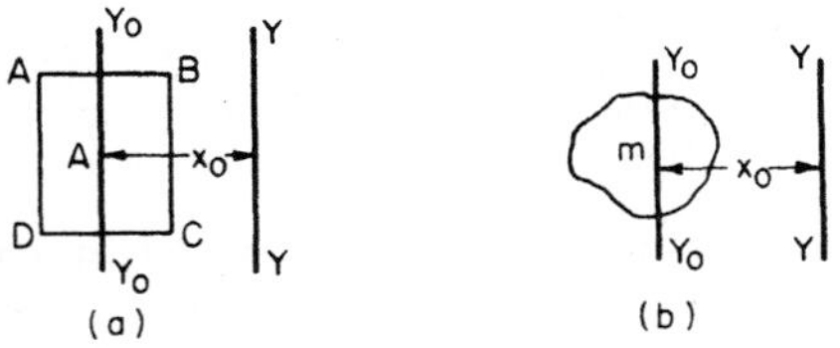

Figure 1.1.30 Parallel Axes Theorem.

Polar Moment of Inertia The polar moment of inertia is taken about an axis perpendicular to the plane of the area. Referring to Fig. 1.1.31, if I_y and I_x are the moments of inertia of the area A about YY and XX, respectively, then the polar moment of inertia $I_p = I_x + I_y$, or the polar moment of inertia is equal to the sum of the moments of inertia about any two axes perpendicular to each other in the plane of the area and intersecting at the pole.

Product of Inertia This quantity will be represented by I_{xy}, and is $\iint xy\,dy\,dx$, where x and y are the coordinates of any elementary part into which the area may be conceived to be divided. I_{xy} may be positive or negative, depending upon the position of the area with respect to the coordinate axes XX and XY.

Relation between Moments of Inertia about Axes Inclined to Each Other Referring to Fig. 1.1.32, let I_y and I_x be the moments of inertia of the area A about YY and XX, respectively, I'_y and I'_x the moments about $Y'Y'$ and $X'X'$, and I_{xy} and I'_{xy} the products of inertia for XX and YY, and $X'X'$ and $Y'Y'$, respectively. Also, let c be the angle between the respective pairs of axes, as shown. Then,

$$I'_y = I_y \cos^2 c + I_x \sin^2 c + I_{xy} \sin 2c$$

$$I'_x = I_x \cos^2 c + I_y \sin^2 c + I_{xy} \sin 2c$$

$$I'_{xy} = \frac{I_x - I_y}{2} \sin 2c + I_{xy} \cos 2c$$

Principal Moments of Inertia In every plane area, a given point being taken as the origin, there is at least one pair of rectangular axes

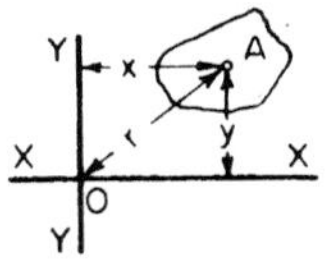

Figure 1.1.31 Polar Moment of Inertia.

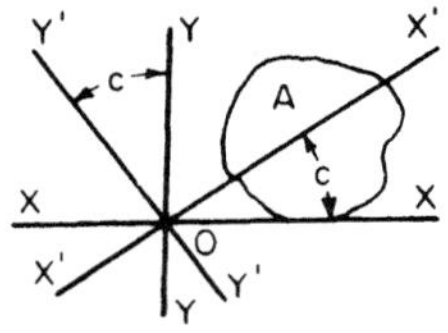

Figure 1.1.32 Relation between Moments of Inertia about two inclined axes.

in the plane of the area about one of which the moment of inertia is a maximum, and a minimum about the other axis. These moments of inertia are called the **principal moments of inertia,** and the axes about which they are taken are the **principal axes of inertia.** One of the conditions for principal moments of inertia is that the product of inertia I_{xy} shall equal zero. **Axes of symmetry** of an area are always principal axes of inertia.

Relation between Products of Inertia and Parallel Axes In Fig. 1.1.33, X_0X_0 and Y_0Y_0 pass through the center of gravity of the area parallel to the given axes XX and YY. If I_{xy} is the product of inertia for XX and YY, and $I_{x_0y_0}$ that for X_0X_0 and Y_0Y_0, then, $I_{xy} = I_{x_0y_0} + abA$.

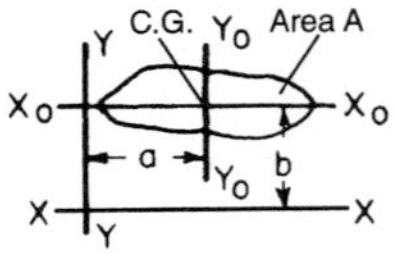

Figure 1.1.33 Parallel Axes Theorem for product of inertia.

Mohr's Circle The **principal moments of inertia** and the location of the **principal axes of inertia** for any point of a plane area may be established graphically as follows.

Given at any point A of a plane area (Fig. 1.1.34), the moments of inertia I_x and I_y about axes X and Y, and the product of inertia I_{xy} relative to X and Y. The graph shown in Fig. 1.1.34*b* is plotted on rectangular

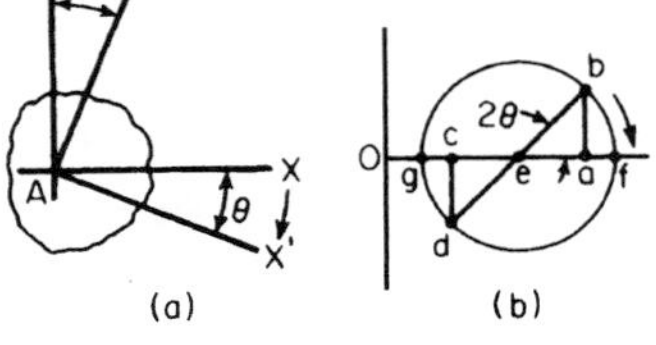

Figure 1.1.34 Mohr's Circle and principle moments of inertia.

coordinates with moments of inertia as abscissas and products of inertia as ordinates. Lay out $Oa = I_x$ and $ab = I_{xy}$ (upward for positive products of inertia, downward for negative). Lay out $Oc = I_y$ and $cd =$ *negative of* I_{xy}. Draw a circle with *bd* as diameter. This is **Mohr's circle.** The *maximum* moment of inertia $I'_x = Of$; the *minimum* moment of inertia is $I'_y = Og$. The principal axes of inertia are located as follows. From axis *AX* (Fig. 1.1.34*a*) lay out angular distance $\theta = ½ < bef$. This locates axis *AX*′, one principal axis ($I'_x = Of$). The other principal axis of inertia is *AY*′, perpendicular to *AX*′ ($I'_x = Og$).

The **moment of inertia of any area** may be considered to be made up of the sum or difference of the known moments of inertia of simple figures. For example, the dimensioned figure shown in Fig. 1.1.35 represents the section of a rolled shape with hole *oprs* and may be divided into the semicircle *abc*, rectangle *edkg*, and triangles *mfg* and *hkl*, from which the rectangle *oprs* is to be subtracted. Referring to axis *XX*,

$$I_{xx} = \pi 4^4/8 \text{ for semicircle } abc + (2\times 11^3)/3 \text{ for rectangle } edkg$$
$$+ 2[(5\times 3^3)/36 + 10^2(5\times 3)/2] \text{ for the two triangles } mfg \text{ and } hkl$$

From the sum of these there is to be subtracted $I_{xx} = [(2 \times 3^2)/12 + 4^2(2 \times 3)]$ for the rectangle *oprs*.

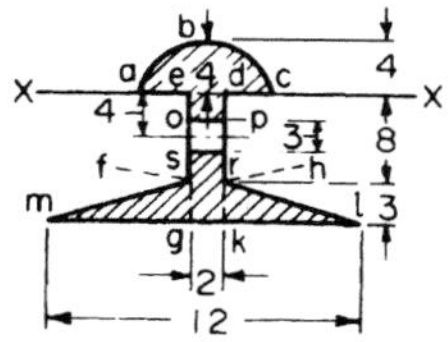

Figure 1.1.35 Example of rolled shape with inner hole.

If the moment of inertia of the whole area is required about an axis parallel to *XX*, but passing through the center of gravity of the whole area, $I_0 = I_{xx} - x_0^2 A$, where x_0 = distance from *XX* to center of gravity. The **moments of inertia of built-up sections** used in structural work may be found in the same manner, the moments of inertia of the different **rolled sections** being given in Sec. 2.12.

Moments of Inertia of Solids For moments of inertia of solids about parallel axes, $I_x = I_0 + x_0^2 m$.

Moment of Inertia with Reference to Any Axis Let a mass particle *dm* of a body have *x*, *y*, and *z* as coordinates, *XX*, *YY*, and *ZZ* being the coordinate axes and *O* the origin. Let *X*′*X*′ be any axis passing through the origin and making angles of *A*, *B*, and *C* with *XX*, *YY*, and *ZZ*, respectively. The moment of inertia with respect to this axis then becomes equal to

$$I'_x = \cos^2 A \int (y^2 + z^2)\, dm + \cos^2 B \int (z^2 \times x^2)\, dm$$
$$+ \cos^2 C \int (x^2 + y^2)\, dm - 2 \cos B \cos C \int yz\, dm$$
$$- 2 \cos C \cos A \int zx\, dm - 2 \cos A \cos B \int xy\, dm$$

Let the moment of inertia about $XX = I_x = \int (y^2 + z^2)\, dm$, about $YY = I_y = \int (z^2 + x^2)\, dm$, and about $ZZ = I_z = \int (x^2 + y^2)\, dm$. Let the products of inertia about the three coordinate axes be

$$I_{yz} = \int yz\, dm \qquad I_{zx} = \int zx\, dm \qquad I_{xy} = \int xy\, dm$$

Then the moment of inertia I'_x becomes equal to

$$I'_x = I_x \cos^2 A + I_y \cos^2 B + I_z \cos^2 C - 2I_{yz} \cos B \cos C - 2I_{zx}$$
$$\cos C \cos A - 2I_{xy} \cos A \cos B$$

The moment of inertia of any solid may be considered to be made up of the sum or difference of the moments of inertia of simple solids of which the moments of inertia are known.

Moments of Inertia of Important (Homogeneous) Solids

w = weight per unit of volume of the body
$m = w/g$ = mass per unit of volume of the body
$M = W/g$ = total mass of body
r = radius
I = moment of inertia (mass units)
$I_w = I \times g$ = moment of inertia (weight units)

Solid circular cylinder about its axis: $I = \pi r^4 ma/2 = Mr^2/2$. ($a$ = length of axis of cylinder.)

Solid circular cylinder about an axis through the center of gravity and perpendicular to axis of cylinder: $I = M[r^2 + (a^2/3)]/4$.

Hollow circular cylinder about its axis: $I = M(r_1^2 + r_2^2)/2$ (r_1 and r_2 = outer and inner radii.)

Thin-walled hollow circular cylinder about its axis: $I = Mr^2$.

Solid sphere about a diameter: $I = 8m\pi r^5/15 = 2Mr^2/5$.

Thin hollow sphere about a diameter: $I = 2Mr^2/3$.

Thick hollow sphere about a diameter: $I = 8m\pi(r_1^5 - r_2^5)/15$. ($r_1$ and r_2 are outer and inner radii.)

Rectangular prism about an axis through center of gravity and perpendicular to a face whose dimensions are a and b: $I = M(a^2 + b^2)/12$.

Solid right circular cone about an axis through its apex and perpendicular to its axis: $I = 3M[(r^2/4) + h^2]/5$. (h = altitude of cone, r = radius of base.)

Solid right circular cone about its axis of revolution: $I = 3Mr^2/10$.

Ellipsoid with semiaxes a, b, and c: I about diameter $2c$ (z axis) $= 4m\pi abc\,(a^2 + b^2)/15$. [Equation of ellipsoid: $(x^2/a^2) + (y^2/b^2) + (z^2/c^2) = 1$.]

Ring with Circular Section Referring to Fig. 1.1.36, $I_{yy} = ½\, m\pi^2 Ra^2(4R^2 + 3a^2)$; $I_{xx} = m\pi^2 Ra^2[R^2 + (5a^2/4)]$.

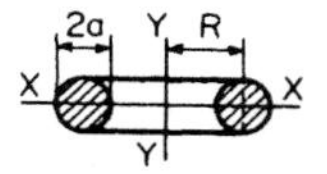

Figure 1.1.36 Ring with circular section.

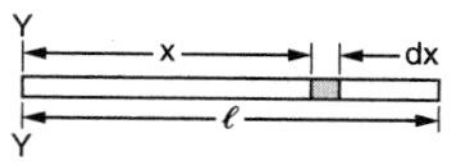

Figure 1.1.37 Rod.

Approximate Moments of Inertia of Solids In order to determine the moment of inertia of a solid, it is necessary to know all its dimensions. In the case of a rod of mass M (Fig. 1.1.37) and length l, with shape and size of the cross section unknown, making the approximation that the weight is all concentrated along the axis of the rod, the moment of inertia about *YY* will be $\int_0^l (M/l)x^2\, dx = Ml^2/3$.

A **thin plate** may be treated in the same way (Fig. 1.1.38): $I_{yy} = \int_0^l (M/l)x^2\, dx = \frac{1}{3}Ml^2$.

Thin Ring, or Cylinder Assume the mass M of the ring or cylinder to be concentrated at a distance r from O, see Fig. 1.1.39. The moment of inertia about an axis through O perpendicular to plane of ring or along the axis of the cylinder will be $I = Mr^2$; this will be greater than the exact moment of inertia, and r is sometimes taken as the distance from O to the center of gravity of the cross section of the rim.

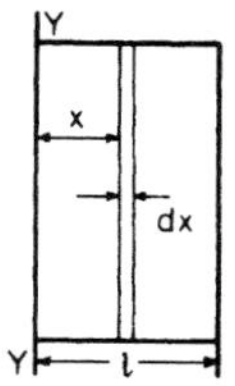

Figure 1.1.38 Thin plate.

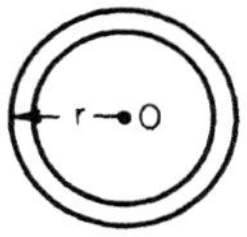

Figure 1.1.39 Thin ring or cylinder.

Flywheel Effect The moment of inertia of a solid is often called flywheel effect in the solution of problems dealing with rotating bodies.

Graphical Determination of the Centroids and Moments of Inertia of Plane Areas Required to find the center of gravity of the area *MNP* (Fig. 1.1.40) and its moment of inertia about any axis *XX*.

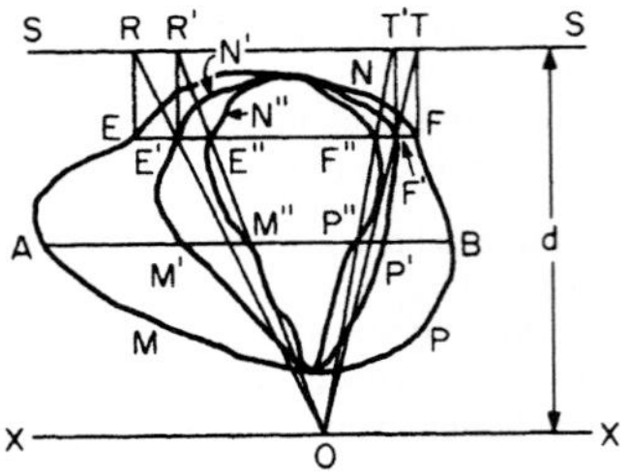

Figure 1.1.40 Graphical Determination of the Centroid.

Draw any line *SS* parallel to *XX* and at a distance *d* from it. Draw a number of lines such as *AB* and *EF* across the figure parallel to *XX*. From *E* and *F* draw *ER* and *FT* perpendicular to *SS*. Select as a pole any point on *XX*, preferably the point nearest the area, and draw *OR* and *OT*, cutting *EF* at *E′* and *F′*. If the same construction is repeated, using other lines parallel to *XX*, a number of points will be obtained, which, if connected by a smooth curve, will give the area *M′N′P′*. Project *E′* and *F″* onto *SS* by lines *E′R′* and *F′T′*. Join *F′* and *T′* with *O*, obtaining *E″* and *F′*; connect the points obtained using other lines parallel to *XX* and obtain an area *M″N″P″*. The area $M'N'P' \times d$ = moment of area *MNP* about the line *XX*, and the distance from *XX* to the centroid *MNP* = area $M'N'P' \times d$/area *MNP*. Also, area $M''N''P'' \times d^2$ = moment of inertia of *MNP* about *XX*. The areas *M′N′P′* and *M″N″P″* can best be obtained by use of a planimeter.

Dynamics of Rigid Bodies

Dynamics is the study of ***kinematics*** and ***kinetics*** of particles and rigid bodies. Kinematics is the study of the geometry of motion in that it is used to relate displacement, velocity, acceleration, and time without reference to the cause of motion. Kinetics is the study of the relations existing between the forces acting on a body, the mass of the body, and the motion of the body. Kinetics is used to predict the motion caused by given forces or to determine the forces required to produce a given motion.

KINEMATICS OF PARTICLES

Kinematics is the study of the motion of bodies without reference to the forces causing that motion or the mass of the bodies.

The **displacement** of a point is the directed distance that a point has moved on a geometric path from a convenient origin. It is a **vector,** having both magnitude and direction, and is subject to all the laws and characteristics attributed to vectors. In Fig. 1.1.41, the displacement of the point *A* from the origin *O* is the directed distance *O* to *A*, symbolized by the vector **s**.

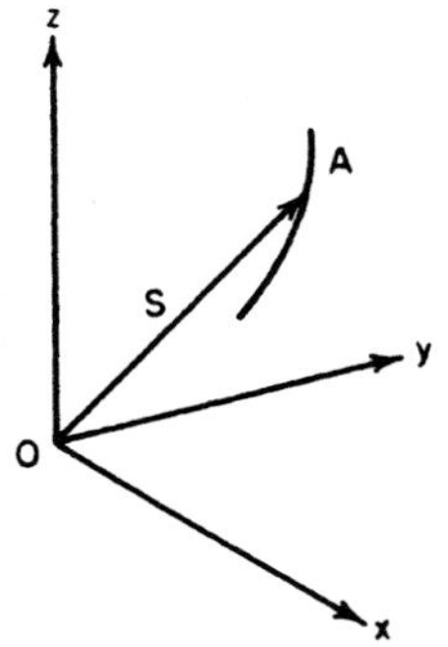

Figure 1.1.41 Trajectory path of a point and its corresponding position vector.

The **velocity** of a point is the time rate of change of displacement, or $\mathbf{v} = d\mathbf{s}/dt$.

The **acceleration** of a point is the time rate of change of velocity, or $\mathbf{a} = d\mathbf{v}/dt$.

The kinematic definitions of velocity and acceleration involve the four variables, *displacement, velocity, acceleration,* and *time.* If we eliminate the variable of time, a third equation of motion is obtained, $ds/v = dt = dv/a$. This differential equation, together with the definitions of velocity and acceleration, make up the *three kinematic equations* of motion, $v = ds/dt$, $a = dv/dt$, and $a\ ds = v\ dv$. These differential equations are usually limited to the scalar form when expressed together, since the last can only be properly expressed in terms of the scalar *dt.* The first two, since they are definitions for velocity and acceleration, are **vectorial equations.**

A **space-time curve** offers a convenient means for the study of the motion of a point that moves in a straight line. The **slope of the curve** at any point will represent the **velocity** at that time. In Fig. 1.1.42*a* the slope is constant, as the graph is a straight line; the velocity is therefore uniform. In Fig. 1.1.42*b* the slope of the curve varies from point to point, and the velocity must also vary. At *p* and *q* the slope is zero; therefore, the velocity of the point at the corresponding times must also be zero.

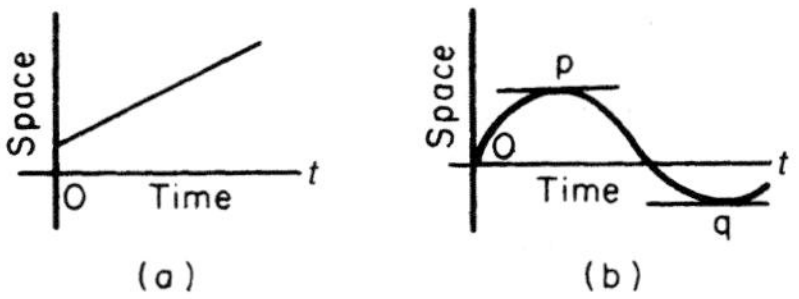

Figure 1.1.42 Space-time curves.

A **velocity-time curve** offers a convenient means for the study of acceleration. The **slope of the curve** at any point will represent the **acceleration** at that time. In Fig. 1.1.43*a* the slope is constant; so the acceleration must be constant. In the case represented by the full line, the acceleration is positive; so the velocity is increasing. The dotted line shows a negative acceleration and therefore a decreasing velocity. In Fig. 1.1.43*b* the slope of the curve varies from point to point; so the acceleration must also vary. At *p* and *q* the slope is zero; therefore, the acceleration of the point at the corresponding times must also be zero. The area under the velocity-time curve between any two ordinates such as *NL* and *HT* will represent the distance moved in time interval *LT.* In the case of the uniformly accelerated motion shown by the full line in Fig. 1.1.43*a*, the area *LNHT* is $\frac{1}{2}(NL + HT) \times (OT - OL)$ = mean velocity multiplied by the time interval = space passed over during this time interval. In Fig. 1.1.43*b* the mean velocity can be obtained from the equation of the curve by means of the calculus, or graphically by approximation of the area.

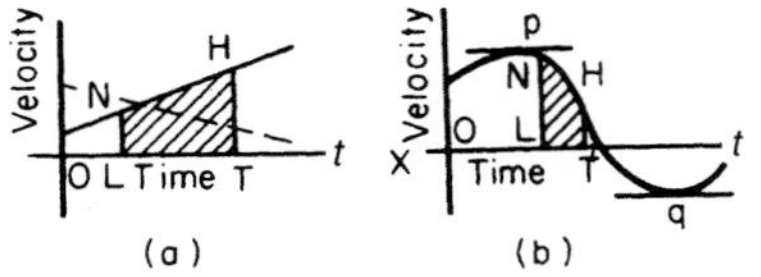

Figure 1.1.43 Velocity-time curves.

An **acceleration-time curve** may be constructed by plotting accelerations as ordinates, and times as abscissas. The area under this curve between any two ordinates will represent the total change in velocity during the time interval. The area *ABCD* (Fig. 1.1.44) represents the total increase in velocity between time t_1 and time t_2.

The conditions of motion are typically specified by the type of acceleration experienced by the particle. Three classes of motion may be defined for acceleration given as a function of time, $a = f(t)$, as a function of position, $a = f(x)$, or as a function of velocity, $a = f(v)$.

Then, the determination of velocity and position requires two successive integrations.

General Expressions Showing the Relations between Space, Time, Velocity, and Acceleration for Rectilinear Motion

SPECIAL MOTIONS

Uniform Motion If the **velocity** is **constant,** the **acceleration** must be **zero,** and the point has uniform motion. The space-time curve becomes a straight line inclined toward the time axis (Fig. 1.1.42*a*). The velocity-time curve becomes a straight line parallel to the time axis. For this motion $a = 0$, $v = \text{constant}$, and $s = s_0 + vt$.

Uniformly Accelerated or Retarded Motion If the **velocity** is **not uniform** but the **acceleration** is **constant,** the point has uniformly accelerated motion; the acceleration may be either positive or negative. The space-time curve becomes a parabola and the velocity-time curve becomes a straight line inclined toward the time axis (Fig. 1.1.43*a*). The acceleration-time curve becomes a straight line parallel to the time axis. For this motion $a = \text{constant}$, $v = v_0 + at$, $s = s_0 + v_0t + \frac{1}{2}at^2$.

If the point starts from rest, $v_0 = 0$. Care should be taken concerning the sign + or − for acceleration.

Composition and Resolution of Velocities and Acceleration

Resultant Velocity A velocity is said to be the resultant of two other velocities when it is represented by a vector that is the geometric sum of the vectors representing the other two velocities. This is the **parallelogram of motion.** In Fig. 1.1.45, **v** is the resultant of $\mathbf{v}_1$ and $\mathbf{v}_2$ and is represented by the diagonal of a parallelogram of which $\mathbf{v}_1$ and $\mathbf{v}_2$ are the sides; or it is the third side of a triangle of which $\mathbf{v}_1$ and $\mathbf{v}_2$ are the other two sides.

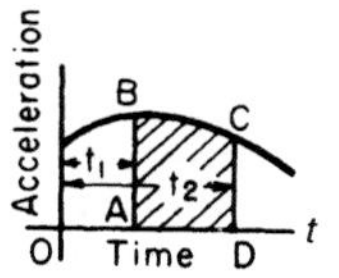

Figure 1.1.44 Acceleration-time curve.

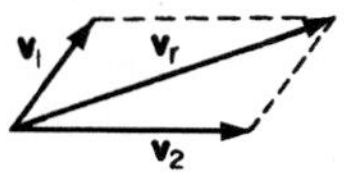

Figure 1.1.45 Resultant velocity using parallelogram.

Polygon of Motion The parallelogram of motion may be extended to the polygon of motion. Let $\mathbf{v}_1$, $\mathbf{v}_2$, $\mathbf{v}_3$, $\mathbf{v}_4$ (Fig. 1.1.46*a*) show the directions of four velocities imparted in the same plane to point *O*. If the lines $\mathbf{v}_1$, $\mathbf{v}_2$, $\mathbf{v}_3$, $\mathbf{v}_4$ (Fig. 1.1.46*b*) are drawn parallel to and proportional to the velocities imparted to point *O*, **v** will represent the resultant velocity imparted to *O*. It will make no difference in what order the velocities are taken in constructing the motion polygon. As long as the arrows showing the direction of the motion follow each other in order about the polygon, the resultant velocity of the point will be represented in magnitude by the closing side of the polygon, but opposite in direction.

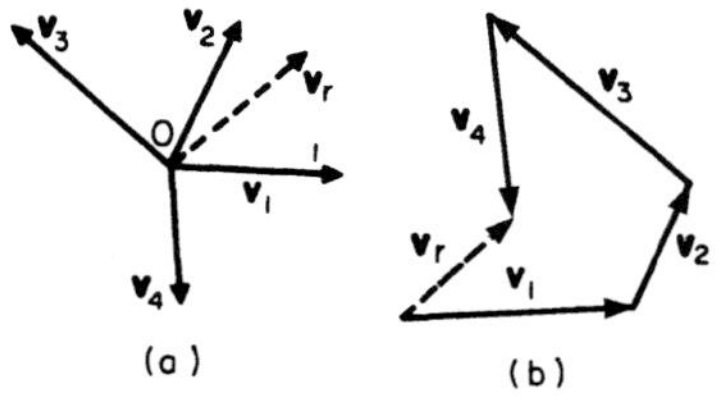

Figure 1.1.46 Resultant velocity using polygon method.

Resolution of Velocities Velocities may be resolved into **component velocities** in the same plane, as shown by Fig. 1.1.47. Let the velocity of point *O* be v_r. In Fig. 1.1.47*a* this velocity is resolved into two components in the same plane as v_r and at right angles to each other.

$$v_r = \sqrt{(v_1)^2 + (v_2)^2}$$

In Fig. 1.1.47*b* the components are in the same plane as v_r, but are not at right angles to each other. In this case,

$$v_r = \sqrt{(v_1)^2 + (v_2)^2 + 2v_1v_2 \cos B}$$

If the components v_1 and v_2 and angle *B* are known, the direction of v_r can be determined. $\sin bOc = (v_1/v_r) \sin B$. $\sin cOa = (v_2/v_r) \sin B$. Where v_1 and v_2 are at right angles to each other, $\sin B = 1$.

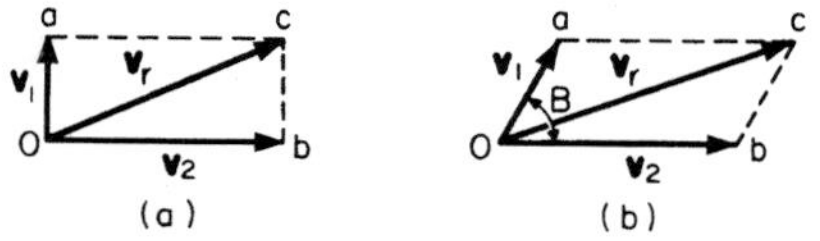

Figure 1.1.47 Resolving velocity into components.

Resultant Acceleration Accelerations may be combined and resolved in the same manner as velocities, but in this case the lines or vectors represent accelerations instead of velocities. If the acceleration had components of magnitude a_1 and a_2, the magnitude of the resultant acceleration would be $a = \sqrt{(a_1)^2 + (a_2)^2 + 2a_1a_2 \cos B}$, where *B* is the angle between the vectors $\mathbf{a}_1$ and $\mathbf{a}_2$.

Curvilinear Motion in a Plane

The linear velocity $v = ds/dt$ of a point in curvilinear motion is the same as for rectilinear motion. Its direction is tangent to the path of the point. In Fig. 1.1.48*a*, let P_1, P_2, P_3 be the path of a moving point and $\mathbf{V}_1$, $\mathbf{V}_2$, $\mathbf{V}_3$ represent its velocity at points P_1, P_2, P_3, respectively. If *O* is taken as a pole (Fig. 1.1.48*b*) and vectors $\mathbf{V}_1$, $\mathbf{V}_2$, $\mathbf{V}_3$ representing the velocities of the point at P_1, P_2, and P_3 are drawn, the curve connecting the terminal points of these vectors is known as the **hodograph** of the motion. This velocity diagram is applicable only to motions all in the same plane.

Acceleration Tangents to the curve (Fig. 1.1.48*b*) indicate the directions of the **instantaneous velocities.** The direction of the tangents does not, as a rule, coincide with the direction of the accelerations as represented by tangents to the path. If the acceleration *a* at some point in the path is resolved by means of a parallelogram into components tangent and normal to the path, the normal acceleration $a_n = v^2/\rho$, where ρ = radius of curvature of the path at the point in question, and the tangential acceleration $a_t = dv/dt$, where v = velocity tangent to the path at the same point. $a = \sqrt{a_n^2 + a_t^2}$. The normal acceleration is constantly directed toward the center of curvature on the concave side of the path.

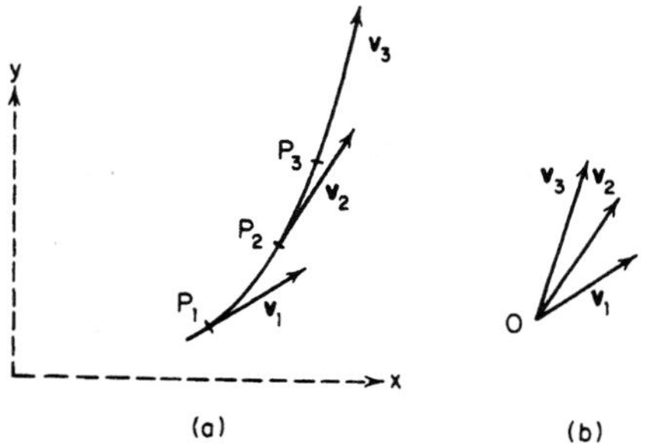

Figure 1.1.48 Curvilinear motion and velocity at different trajectory points.

EXAMPLE. Figure 1.1.49 shows a point moving in a curvilinear path. At p_1 the velocity is $\mathbf{v}_1$; at p_2 the velocity is $\mathbf{v}_2$. If these velocities are drawn from pole *O* (Fig. 1.1.49*b*), $\Delta\mathbf{v}$ will be the difference between $\mathbf{v}_2$ and $\mathbf{v}_1$. The acceleration during travel p_1p_2 will be $\Delta\mathbf{v}/\Delta t$, where Δt is the time interval. The approximation becomes closer to instantaneous acceleration as shorter intervals Δt are employed. The acceleration $\Delta\mathbf{v}/\Delta t$ can be resolved into normal and tangential components leading to $\mathbf{a}_n = \Delta\mathbf{v}_n/\Delta t$, normal to the path, and $\mathbf{a}_r = \Delta\mathbf{v}_p/\Delta t$, tangential to the path.

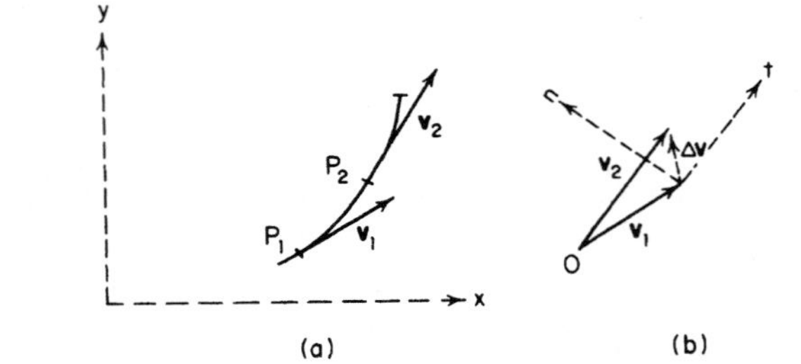

Figure 1.1.49 Change of velocity in curvilinear motion between two points of path.

Velocity and *acceleration* may be expressed in **radial and transverse coordinates (polar coordinates)** such that $v = \sqrt{v_r^2 + v_\theta^2}$ and $a = \sqrt{a_r^2 + a_\theta^2}$. Figure 1.1.50 may be used to explain the r and θ coordinates.

EXAMPLE. At P_1 the velocity is $\mathbf{v}_1$, with components $\mathbf{v}_{1r}$ in the r direction and $\mathbf{v}_{1\theta}$ in the θ direction. At P_2 the velocity is $\mathbf{v}_2$, with components $\mathbf{v}_{2r}$ in the radial or r direction and $\mathbf{v}_{2\theta}$ in the transverse or θ direction. It is evident that the difference in velocities $\mathbf{v}_2 - \mathbf{v}_1 = \Delta\mathbf{v}$ will have components $\Delta\mathbf{v}_r$ and $\Delta\mathbf{v}_\theta$, giving rise to accelerations $\mathbf{a}_r$ and $\mathbf{a}_\theta$ in a time interval Δt.

In polar coordinates, $v_r = dr/dt$, $a_r = d^2r/dt^2 - r(d\theta/dt)^2$, $v_\theta = r(d\theta/dt)$, and $a_\theta = r(d^2\theta/dt^2) + 2(dr/dt)(d\theta/dt)$.

If a point P moves on a circular path of radius r with an angular velocity of ω and an angular acceleration of α, the linear velocity of the point P is $v = \omega r$ and the two components of the linear acceleration are $a_n = v^2/r = \omega^2 r = v\omega$ and $a_t = \alpha r$.

If the angular velocity is constant, the point P travels equal circular paths in equal intervals of time. The projected displacement, velocity, and acceleration of the point P on the x and y axes are sinusoidal functions of time, and the motion is said to be **harmonic motion.** Angular velocity is usually expressed in radians per second, and when the number (N) of revolutions traversed per minute (N/min) by the point P is known, the angular velocity of the radius r is $\omega = 2\pi N/60 = 0.10472N$. In Fig. 1.1.51, let the angular velocity of the line OP be a constant ω. Let the point P start at X' and move to P in time t. Then the angle $\theta = \omega t$. If $OP = r$, $OA = s = r \cos \omega t$. The velocity v of the point A on the x axis will equal $ds/dt = -\omega r \sin \omega t$, and the acceleration $a = dv/dt = -\omega^2 r \cos \omega t$. The period τ is the time necessary for the point P to complete one cycle of motion $\tau = 2\pi/\omega$, and it is also equal to the time necessary for A to complete a full cycle on the x axis from X' to X and return.

Curvilinear Motion in Space

If **three dimensions** are used, velocities and accelerations may be resolved into components not in the same plane by what is known as the **parallelepiped of motion.** Three coordinate systems are widely used, cartesian, cylindrical, and spherical. In **cartesian coordinates,** $v = \sqrt{v_x^2 + v_y^2 + v_z^2}$ and $a = \sqrt{a_x^2 + a_y^2 + a_z^2}$. In **cylindrical coordinates,** the radius vector **R** of displacement lies in the rz plane, which is at an angle with the xz plane. Referring to (a) of Fig. 1.1.52, the θ coordinate is perpendicular to the rz plane. In this system $v = \sqrt{v_r^2 + v_\theta^2 + v_z^2}$ and $a = \sqrt{a_r^2 + a_\theta^2 + a_z^2}$ where $v_r = dr/dt$, $a_r = d^2r/dt^2 - r(d\theta/dt)^2$, $v_\theta = r(d\theta/dt)$, and $a_\theta = r(d^2\theta/dt^2) + 2(dr/dt)(d\theta/dt)$. In **spherical coordinates,** the three coordinates are the R coordinate, the θ coordinate, and the ϕ coordinate as in (b) of Fig. 1.1.52. The velocity and acceleration are $v = \sqrt{v_R^2 + v_\theta^2 + v_\phi^2}$ and $a = \sqrt{a_R^2 + a_\theta^2 + a_\phi^2}$, where $v_R = dR/dt$, $v_\phi = R(d\phi/dt)$, $v_\theta = R\cos\phi(d\theta/dt)$, $a_R = d^2R/dt^2 - R(d\phi/dt)^2 - R\cos^2\phi(d\theta/dt)^2$, $a_\phi = R(d^2\phi/dt^2) + R\cos\phi\sin\phi(d\theta/dt)^2 + 2(dR/dt)(d\phi/dt)$, and $a_\theta = R\cos\phi\,(d^2\theta/dt^2) + 2[(dR/dt)\cos\phi - R\sin\phi(d\phi/dt)]\,d\theta/dt$.

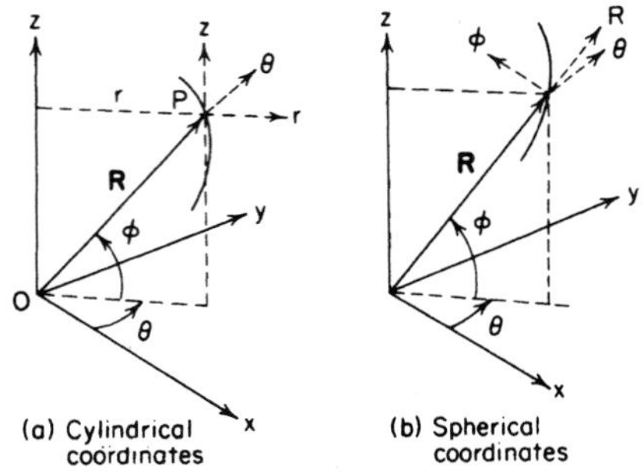

Figure 1.1.52 Curvilinear motion in space.

Motion of Rigid Bodies

A body is said to be **rigid** when the distances between all its particles are invariable. Theoretically, rigid bodies do not exist, but materials used in engineering are rigid under most practical working conditions. The motion of a rigid body can be completely described by knowing the **angular motion** of a line on the rigid body and the **linear motion** of a point on this line and relating the motion of all other parts of the rigid body to these motions. If a rigid body moves so that a straight line connecting any two of its particles remains parallel to its original position at all times, it is said to have **translation.** In **rectilinear translation,** all points move in straight lines. In **curvilinear translation,** all points move on congruent curves but without rotation. **Rotation** is defined as angular motion about an axis, which may or may not be fixed. Rigid body motion in which the paths of all particles lie on parallel planes is called **plane motion.**

Angular Motion

Angular displacement is the change in angular position of a given line as measured from a convenient reference line. In Fig. 1.1.53, consider the motion of the line AB as it moves from its original position $A'B'$. The angle between lines AB and $A'B'$ is the angular displacement of line **AB,** symbolized as θ. It is a directed quantity and is a vector. The usual notation used to designate angular displacement is a vector normal to the plane in which the angular displacement occurs. The length of the

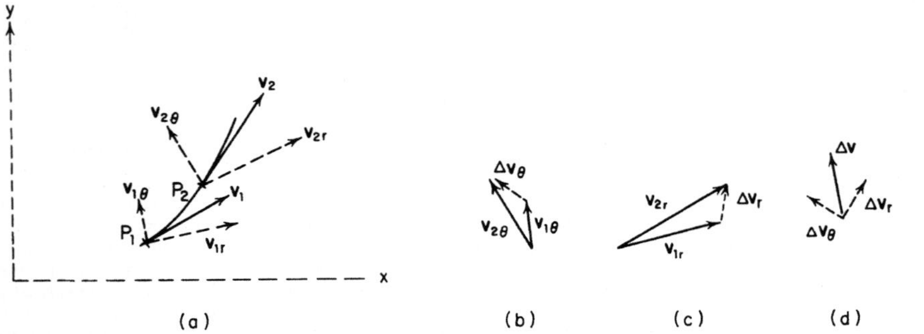

Figure 1.1.50 Change of radial and transverse velocities between two points on curvilinear path.

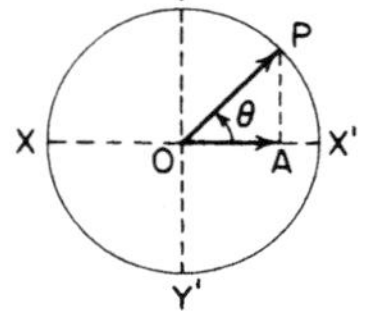

Figure 1.1.51 Harmonic motions on XX′ or YY′ axis.

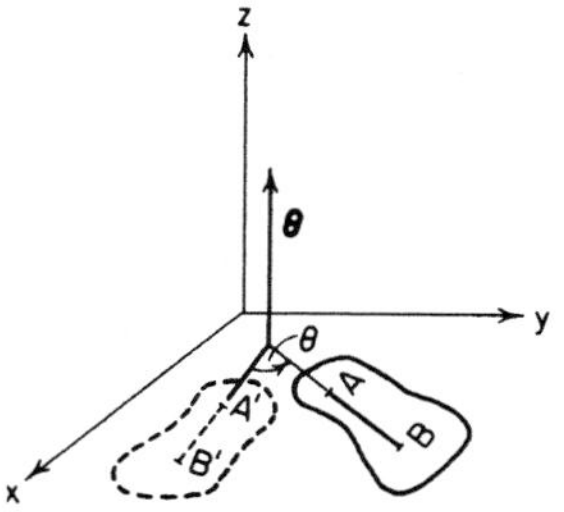

Figure 1.1.53 Angular displacement of line AB of a body.

vector is proportional to the magnitude of the angular displacement. For a rigid body moving in three dimensions, the line *AB* may have angular motion about any three orthogonal axes. For example, the angular displacement can be described in cartesian coordinates as a vectorial equation $\theta = \theta_x + \theta_y + \theta_z$, where the angular magnitude is $\theta = \sqrt{\theta_x^2 + \theta_y^2 + \theta_z^2}$.

Angular velocity is defined as the time rate of change of angular displacement, $\omega = d\theta/dt$. Angular velocity may also have components about any three orthogonal axes.

Angular acceleration is defined as the time rate of change of angular velocity, $\alpha = d\omega/dt = d^2\theta dt^2$. Angular acceleration may also have components about any three orthogonal axes.

The *kinematic equations* of angular motion of a line are analogous to those for the motion of a point. In referring to Table 1.1.1, $\omega = d\theta/dt$ $\alpha = d\omega/dt$, and $\alpha\, d\theta = \omega\, d\omega$. Substitute θ for *s*, ω for *v*, and α for *a*.

Motion of a Rigid Body in a Plane

Plane motion is the motion of a rigid body such that the paths of all particles of that rigid body lie on parallel planes.

Instantaneous Axis An axis about which all points of a body rotate at that moment is called instantaneous axis of rotation. In the next time step, the position of the instantaneous axis may change for the body; wherein the curve passing through all positions of the instantaneous axis is called the **centrode.**

When the velocity of two points in the same plane of a rigid body having plane motion is known, the instantaneous axis for the body will be at the intersection of the lines drawn from each point and perpendicular to its velocity. See Fig. 1.1.54, in which *A* and *B* are two points on the rod *AB*, $\mathbf{v}_1$ and $\mathbf{v}_2$ representing their velocities. *O* is the instantaneous axis for *AB*; therefore point *C* will have velocity shown in a line perpendicular to *OC*.

Linear velocities of points in a body rotating about an instantaneous axis are proportional to their distances from this axis. In Fig. 1.1.54, $\mathbf{v}_1$: $\mathbf{v}_2$: $\mathbf{v}_3$ = *AO*:*OB*:*OC*. If the velocities of *A* and *B* were parallel, the lines *OA* and *OB* would also be parallel and there would be no instantaneous axis. The motion of the rod would be translation, and all points would be moving with the same velocity in parallel straight lines.

If a rigid body has plane motion, the components of the velocities of any two points in the body along the straight line joining them must be equal. *Ax* must be equal to *By* and *Cz* in Fig. 1.1.54 since their distance won't change during rigid body motion.

EXAMPLE. In Fig. 1.1.55*a*, the velocities of points *A* and *B* are known—they are $\mathbf{v}_1$ and $\mathbf{v}_2$, respectively. To find the instantaneous axis of the body, perpendiculars *AO* and *BO* are drawn. *O*, at the intersection of the perpendiculars, is the **instantaneous axis** of the body. To find the velocity of any other point, like *C*, line *OC* is drawn and v_3 erected perpendicular to *OC* with magnitude equal to v_1 (*CO*/*AO*). The **angular velocity** of the body will be $\omega = v_1/AO$ or v_2/BO or v_3/CO.

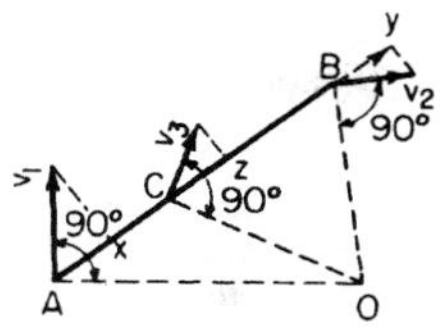

Figure 1.1.54 General planar motion of rigid body.

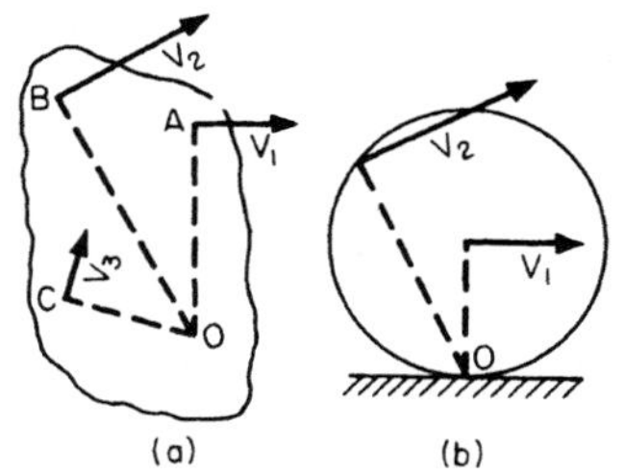

Figure 1.1.55 Concept of instantaneous point of rotation.

The instantaneous axis of a wheel rolling on a rack without slipping (Fig. 1.1.55*b*) lies at the point of contact *O*, which has zero linear velocity. All points of the wheel will have velocities perpendicular to radii to *O* and proportional in magnitudes to their respective distances from *O*.

Another way to describe the plane motion of a rigid body is with the use of **relative motion.** In Fig. 1.1.56, the velocity of point *A* is $\mathbf{v}_1$. The angular velocity of the line *AB* is ω_{AB}. The velocity of *B* relative to *A* is $\omega_{AB} \times r_{AB}$. Point *B* is considered to be moving on a circular path around *A* as a center. The direction of relative velocity of *B* to *A* would be tangent to the circular path in the direction that ω_{AB} would make *B* move. The velocity of *B* is the vector sum of the velocity *A* added to the velocity of *B* relative to *A*, $\mathbf{v}_B = \mathbf{v}_A + \mathbf{v}_{B/A}$.

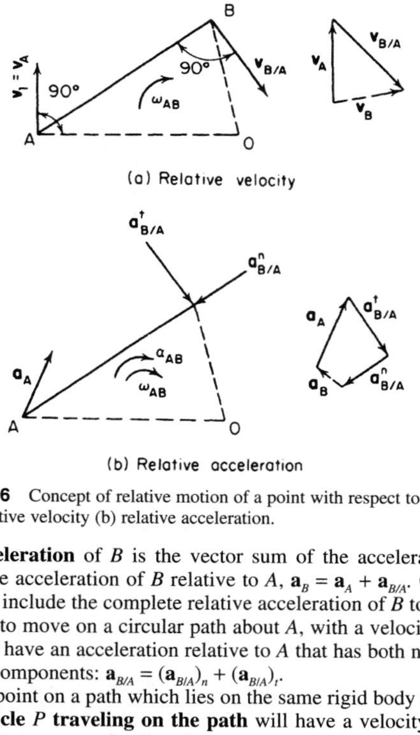

Figure 1.1.56 Concept of relative motion of a point with respect to a reference point (a) relative velocity (b) relative acceleration.

The **acceleration** of *B* is the vector sum of the acceleration of *A* added to the acceleration of *B* relative to *A*, $\mathbf{a}_B = \mathbf{a}_A + \mathbf{a}_{B/A}$. Care must be taken to include the complete relative acceleration of *B* to *A*. If *B* is considered to move on a circular path about *A*, with a velocity relative to *A*, it will have an acceleration relative to *A* that has both normal and tangential components: $\mathbf{a}_{B/A} = (\mathbf{a}_{B/A})_n + (\mathbf{a}_{B/A})_t$.

If *B* is a point on a path which lies on the same rigid body as the line *AB*, a **particle *P* traveling on the path** will have a velocity $\mathbf{v}_p$ at the instant *P* passes over point *B* such that $\mathbf{v}_P = \mathbf{v}_A + \mathbf{v}_{B/A} + \mathbf{v}_{P/B}$, where the velocity $\mathbf{v}_{P/B}$ is the velocity of *P* relative to point *B*.

Table 1.1.1 Kinematic Equations

Variables	$s = f(t)$	$v = f(t)$	$a_2 = f(t)$	$a = f(s, v)$
Displacement		$s = s_0 + \int_{t_0}^{t} v\,dt$	$s = s_0 + \int_{t_0}^{t}\int_{t_0}^{t} a\,dt\,dt$	$s = s_0 + \int_{v_0}^{v} (v/a)\,dv$
Velocity	$v = ds/dt$		$v = v_0 + \int_{t_0}^{t} a\,dt$	$\int_{v_0}^{v} v\,dv \int_{s}^{s_0} a\,ds$
Acceleration	$a = d^2s/dt^2$	$a = dv/dt$		$a = v\,dv/ds$

The particle P will have an acceleration $\mathbf{a}_P$ at the instant P passes over the point B such that $\mathbf{a}_P = \mathbf{a}_A + \mathbf{a}_{B/A} + \mathbf{a}_{P/B} + 2\omega_{AB} \times \mathbf{v}_{P/B}$. The term $\mathbf{a}_{P/B}$ is the acceleration of P relative to the path at point B. The last term $2\omega_{AB}\ \mathbf{v}_{P/B}$ is frequently referred to as the **coriolis acceleration.** The direction is always normal to the path in a sense which would rotate the head of the vector $\mathbf{v}_{P/B}$ about its tail in the direction of the angular velocity of the rigid body ω_{AB}.

EXAMPLE. In Fig. 1.1.57, arm AB is rotating counterclockwise about A with a constant angular velocity of 38 r/min or 4 rad/s, and the slider moves outward with a velocity of 10 ft/s (3.05 m/s). At an instant when the slider P is 30 in (0.76 m) from the center A, the acceleration of the slider will have two components. One component is the normal acceleration directed toward the center A. Its magnitude is $\omega^2 r = 4^2\ (30/12) = 40$ ft/s² [$\omega^2 r = 4^2\ (0.76) = 12.2$ m/s²]. The second is the coriolis acceleration directed normal to the arm AB, upward and to the left. Its magnitude is $2\omega v = 2(4)(10) = 80$ ft/s² [$2\omega v = 2(4)(3.05) = 24.4$ m/s²].

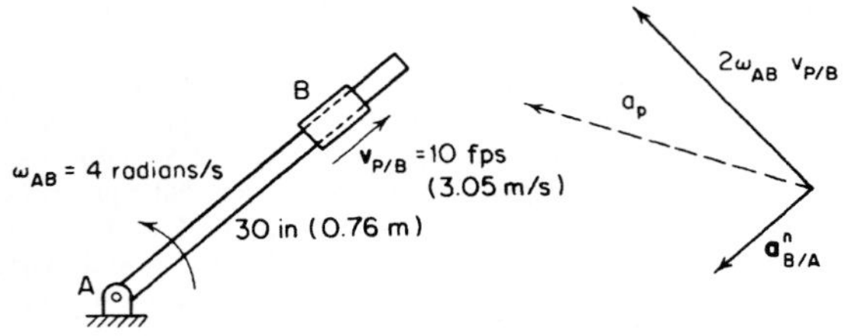

Figure 1.1.57 Example of coriolis acceleration.

General Motion of a Rigid Body

The general motion of a point moving in a coordinate system which is itself in motion is complicated and can best be summarized by using vector notation. Referring to Fig. 1.1.58, let the point P be displaced a vector distance $\mathbf{R}$ from the origin O of a moving reference frame x, y, z which has a velocity $\mathbf{v}_o$ and an acceleration $\mathbf{a}_o$. If point P has a velocity and an acceleration relative to the moving reference plane, let these be $\mathbf{v}_r$ and $\mathbf{a}_r$. The angular velocity of the moving reference fame is ω, and the origin of the moving reference frame is displaced a vector distance $\mathbf{R}_1$ from the origin of a primary (fixed) reference frame X, Y, Z. The velocity and acceleration of P are $\mathbf{v}_P = \mathbf{v}_o + \omega \times \mathbf{R} + \mathbf{v}_r$ and $\mathbf{a}_P = \mathbf{a}_o + (d\omega/dt) \times \mathbf{R} + \omega \times (\omega \times \mathbf{R}) + 2\omega \times \mathbf{v}_r + \mathbf{a}_r$.

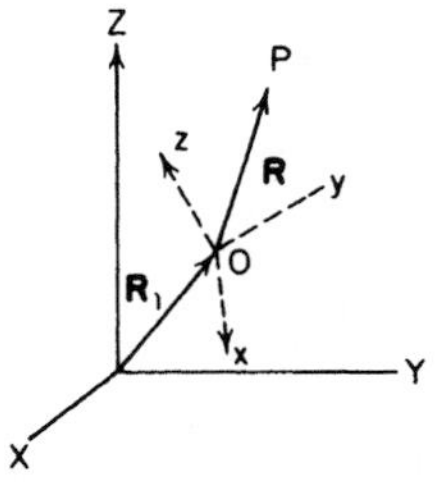

Figure 1.1.58 General motion of a point P in moving reference frame.

KINETICS OF PARTICLES

Consider a particle of mass m subjected to the action of forces $\mathbf{F}_1$, $\mathbf{F}_2$, $\mathbf{F}_3$, . . . , whose vector resultant is $\mathbf{R} = \Sigma\mathbf{F}$. According to Newton's first law of motion, if $\mathbf{R} = 0$, the body is acted on by a balanced force system, and it will either remain at rest or move uniformly in a straight line. If $\mathbf{R} \neq 0$, Newton's second law of motion states that the body will accelerate in the direction of and proportional to the magnitude of the resultant R. This may be expressed as $\Sigma\mathbf{F} = m\mathbf{a}$. If the resultant of the force system has components in the x, y, and z directions, the resultant acceleration will have proportional components in the x, y, and z direction so that $F_x = ma_x$, $F_y = ma_y$, and $F_z = ma_z$. If the resultant of the force system varies with time, the acceleration will also vary with time.

In **rectilinear motion,** the acceleration and the direction of the unbalanced force must be colinear with the direction of motion. In any direction other than the direction of motion, **forces must be in balance and the acceleration equal to zero.**

EXAMPLE 1. The body in Fig. 1.1.59 has a mass of 90 lbm (40.8 kg) and is subjected to an external horizontal force of 36 lbf (160 N) applied in the direction shown. The coefficient of friction between the body and the inclined plane is 0.1. Required, the velocity of the body at the end of 5 s, if it starts from rest.

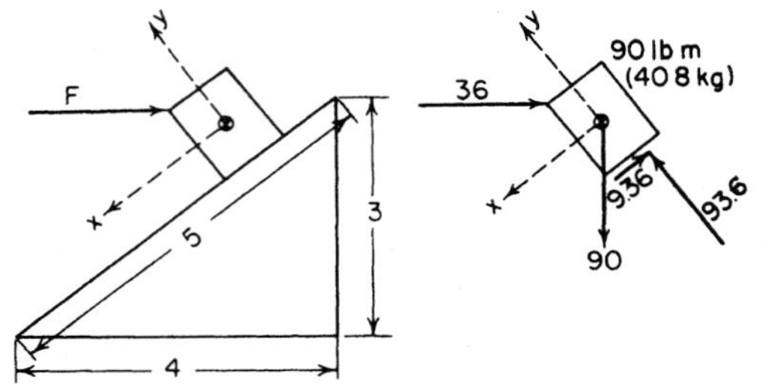

Figure 1.1.59 Example of motion of a body on an inclined surface.

First determine all the forces acting externally on the body. These are the applied force $F = 36$ lbf (106 N), the weight $W = 90$ lbf (400 N), and the force with which the plane reacts on the body. The latter force can be resolved into component forces, one normal and one parallel to the surface of the plane. Motion will be downward along the plane since a static analysis will show that the body will slide downward unless the static coefficient of friction is greater than 0.269. In the direction normal to the surface of the plane, the forces must be balanced. The normal force is $(3/5)(36) + (4/5)(90) = 93.6$ lbf (416 N). The frictional force is $93.6 \times 0.1 = 9.36$ lbf (41.6 N). The unbalanced force acting on the body along the plane is $(3/5)(90) - (4/5)(36) - 9.36 = 15.84$ lbf (70.46 N) downward. $F = (W/9)\ a = (90/g)\ a$; therefore, $a = 0.176\ g = 5.66$ ft/s² (1.725 m/s²). In SI units, $F = ma = 70.46 = 40.8a$; and $a = 1.725$ m/s². The body is acted upon by constant forces and starts from rest; therefore, $v = \int_0^5 a\,dt$ and at the end of 5 s, the velocity would be 28.35 ft/s (8.91 m/s).

EXAMPLE 2. The force with which a rope acts on a body is equal and opposite to the force with which the body acts on the rope, and each is equal to the tension in the rope. In Fig. 1.1.60a, neglecting the weight of the pulley and the rope, the tension in the cord must be the force of 27 lbf. For the 18-lb mass, the unbalanced force is $27 - 18 = 9$ lbf in the upward direction, i.e., $27 - 18 = (18/g)a$, and $a = 16.1$ ft/s² upward. In Fig. 1.1.60b the 27-lb force is replaced by a 27-lb mass. The unbalanced force is still $27 - 18 = 9$ lbf, but it now acts on two masses so that $27 - 18 = (45a/g)$ and $a = 6.44$ ft/s². The 18-lb mass is accelerated upward, and the 27-lb mass is accelerated downward. The tension in the rope is equal to 18 lbf plus the unbalanced force necessary to give it an upward acceleration of $g/5$ or $T = 18 + (18/g)(g/5) = 21.6$ lbf. The tension is also equal to 27 lbf less the unbalanced force necessary to give it a downward acceleration of $g/5$ or $T = 27 - (27/g) \times (g/5) = 21.6$ lbf.

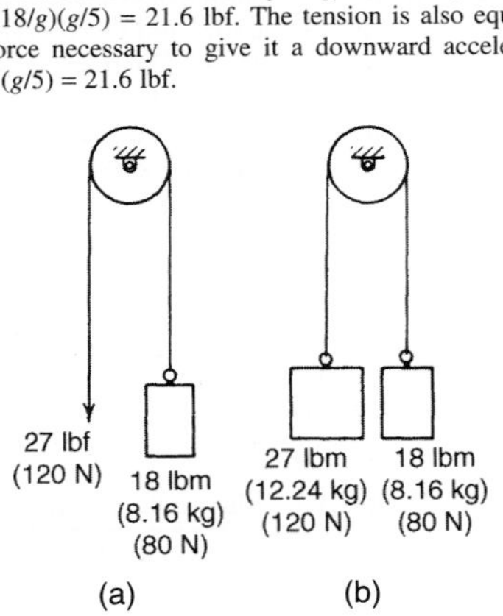

Figure 1.1.60 Example of motion of connected masses over a pulley.

In SI units, in Fig. 1.1.60a, the unbalanced force is $120 - 80 = 40$ N, in the upward direction, i.e., $120 - 80 = 8.16a$, and $a = 4.9$ m/s² (16.1 ft/s²).

In Fig. 1.1.60*b* the unbalanced force is still 40 N, but it now acts on the two masses so that $120 - 80 = 20.4a$ and $a = 1.96$ m/s^2 (6.44 ft/s^2). The tension in the rope is the weight of the 8.16-kg mass in newtons plus the unbalanced force necessary to give it an upward acceleration of 1.96 m/s^2, $T = 9.807(8.16) + (8.16)(1.96) =$ 96 N (21.6 lbf).

General Formulas for the Motion of a Body under the Action of a Constant Unbalanced Force

If the resultant force is constant, the acceleration of center of gravity is also remain constant. Let s = distance, ft; a = acceleration, ft/s^2; v = velocity, ft/s; v_0 = initial velocity, ft/s; h = height, ft; F = force; m = mass; w = weight; g = acceleration due to gravity.

Initial velocity = 0

$$F = ma = (w/g)a$$
$$v = at$$
$$s = \tfrac{1}{2} at^2 = \tfrac{1}{2} vt$$
$$v = \sqrt{2as}$$
$$= \sqrt{2gh} \text{ (falling freely from rest)}$$

Initial velocity = v_0

$$F = ma = (w/g)a$$
$$v = v_0 + at$$
$$s = v_0 t + \tfrac{1}{2} at^2 = \tfrac{1}{2} v_0 t + \tfrac{1}{2} vt$$

If a body is to be moved in a straight line by the resultant force, the line of action of this force must pass through its center of gravity.

General Rule for the Solution of Problems When the Forces Are Constant in Magnitude and Direction

Resolve all the forces acting on the body into two components, one in the direction of the body's motion and one perpendicular to it. Add the components in the direction of the body's motion algebraically and find the **unbalanced force,** if any exists.

In **curvilinear motion,** a particle moves along a curved path, and the resultant of the unbalanced force system may have components in directions other than the direction of motion. **The acceleration in any given direction is proportional to the component of the resultant in that direction.** It is common to utilize orthogonal coordinate systems such as **cartesian coordinates, polar coordinates,** and **normal and tangential coordinates** in analyzing forces and accelerations.

EXAMPLE. A conical pendulum consists of a weight suspended from a cord or light rod and made to rotate in a horizontal circle about a vertical axis with a constant angular velocity of N r/min. For any given constant speed of rotation, the angle θ, the radius r, and the height h will have fixed values. Looking at Fig. 1.1.61, we see that the forces in the vertical direction must be balanced, $T \cos \theta = w$. The forces in the direction normal to the circular path of rotation are unbalanced such that $T \sin \theta = (w/g)a_n = (w/g)\omega^2 r$. Substituting $r = l \sin \theta$ in this last equation gives the value of the tension in the cord $T = (w/g)l\omega^2$. Dividing the second equation by the first and substituting $\tan \theta = r/h$ yields the additional relation that $h = g/\omega^2$.

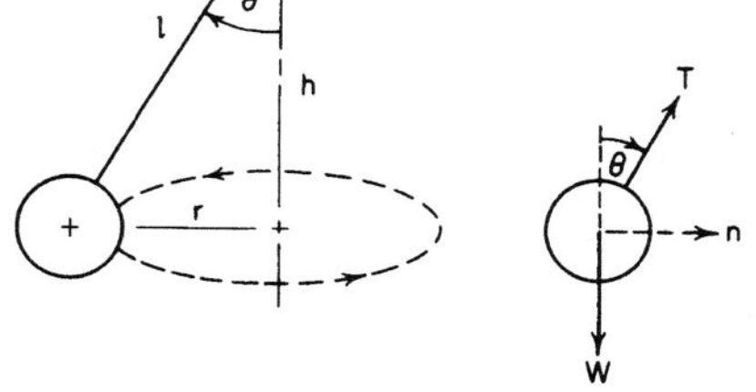

Figure 1.1.61 Example of circular motion of conical pendulum.

An unresisted **projectile** has a motion compounded of the vertical motion of a falling body, and of the horizontal motion due to the horizontal component of the velocity of projection. In Fig. 1.1.62 the only force acting after the projectile starts is gravity, which causes an accelerating downward. The horizontal component of the original velocity v_0 is not changed by gravity. The projectile will rise until the velocity

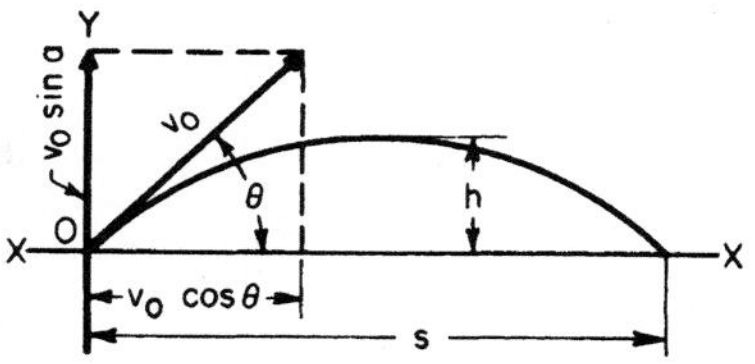

Figure 1.1.62 A projectile motion with an initial velocity V_0.

given to it by gravity is equal to the vertical component of the starting velocity v_0, and the equation $v_0 \sin \theta = gt$ gives the time t required to reach the highest point in the curve. The same time will be taken in falling if the surface XX is level, and the projectile will therefore be in flight $2t$ s. The distance $s = v_0 \cos \theta \times 2t$, and the maximum height of ascent $h = (v_0 \sin \theta)^2/2g$. The expressions for the coordinates of any point on the path of the projectile are: $x = (v_0 \cos \theta)t$, and $y = (v_0 \sin \theta)t - \tfrac{1}{2} gt^2$, giving $y = x \tan\theta - \left(gx^2/2v_0^2 \cos^2\theta\right)$ as the equation for the curve of the path. The radius of curvature of the highest point may be found by using the general expression $v^2 = gr$ and solving for r, v being taken equal to $v_0 \cos \theta$.

Simple Pendulum The period of oscillation is $\tau = 2\pi\sqrt{l/g}$, where l is the length of the pendulum. Note that the arc-length of the swing is not large compared to l.

Centrifugal and Centripetal Forces When a body revolves about an axis, some connection must exist capable of applying force enough to the body to constantly deviate it toward the axis. This deviating force is known as **centripetal force.** The equal and opposite resistance offered by the body to the connection is called the **centrifugal force.** The acceleration toward the axis necessary to keep a particle moving in a circle about that axis is v^2/r; therefore, the force necessary is $ma = mv^2/r = wv^2/gr = w\pi^2N^2r/900g$, where N = r/min. This force, the centripetal force, is constantly directed toward the axis.

The centrifugal force of a solid body revolving about an axis is the same as if the whole mass of the body were concentrated at its center of gravity. Centrifugal force = $wv^2/gr = mv^2/r = w\omega^2 r/g$, where w and m are the weight and mass of the whole body, r is the distance from the axis about which the body is rotating to the center of gravity of the body, ω the angular velocity of the body about the axis in radians, and v the linear velocity of the center of gravity of the body.

Balancing

A rotating body is said to be in **standing balance** when its center of gravity coincides with the axis upon which it revolves. Standing balance may be obtained by resting the axle carrying the body upon two horizontal plane surfaces, as in Fig. 1.1.63. If the center of gravity of the wheel A lies on the axis of the shaft B, there will be no movement,

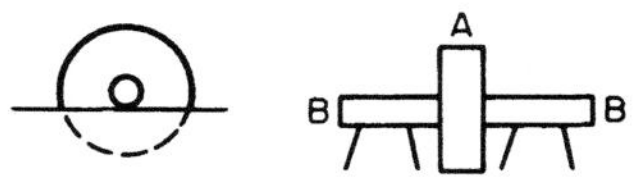

Figure 1.1.63 Balancing test of a rotating body on two bearing supports B.

but if the center of gravity does not lie on the axis of the shaft, the shaft will roll until the center of gravity of the wheel comes directly under the axis of the shaft. The center of gravity may be brought to the axis of the shaft by adding or taking away weight at proper points on the diameter passing through the center of gravity and the center of the shaft. Weights may be added to or subtracted from any part of the wheel so long as its center of gravity is brought to the center of the shaft.

A rotating body may be in standing balance and not in **dynamic balance.** In Fig. 1.1.64, *AA* and *BB* are two disks whose centers of gravity are at *o* and *p*, respectively. The shaft and the disks are in standing balance if the disks are of the same weight and the distances of *o* and *p* from the center of the shaft are equal, and *o* and *p* lie in the same axial plane but on opposite sides of the shaft. Let the weight of each disk be

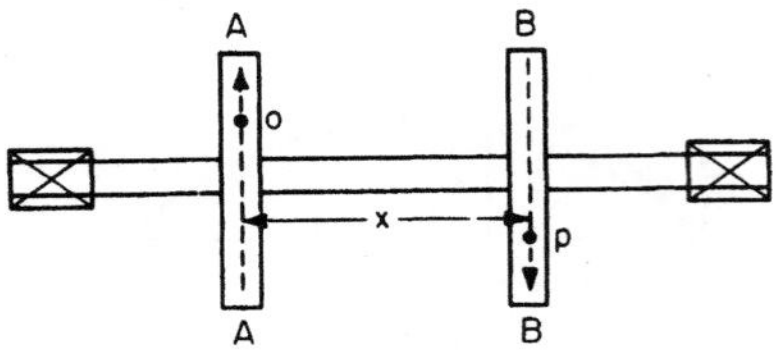

Figure 1.1.64 Schematic difference between static and dynamic balance.

w and the distances of *o* and *p* from the center of the shaft each be equal to *r*. The force exerted on the shaft by *AA* is equal to $w\omega^2 r/g$, where ω is the angular velocity of the shaft. Also, the force is exerted on the shaft by $BB = w\omega^2 r/g$. These two equal and opposite parallel forces act at a distance *x* apart and constitute a couple with a moment tending to rotate the shaft, as shown by the arrows, of $(w\omega^2 r/g)x$. A couple cannot be balanced by a single force; so there must be at least two forces to add to or subtract from the system to achieve dynamic balance.

Systems of Particles The principles of motion for a single particle can be extended to cover a **system of particles.** In this case, **the vector resultant of all external forces acting on the system of particles must equal the total mass of the system times the acceleration of the mass center, and the direction of the resultant must be the direction of the acceleration of the mass center.** This is the **principle of motion of the mass center.**

Rotation of Solid Bodies in a Plane about Fixed Axes

For a rigid body revolving **in a plane** about a fixed axis, **the resultant moment about that axis must be equal to the product of the moment of inertia (about that axis) and the angular acceleration,** $\Sigma M_0 = I_0\alpha$. This is a general statement which includes the particular case of rotation about an axis that passes through the center of gravity.

Rotation about an Axis Passing through the Center of Gravity The rotation of a body about its center of gravity can only be caused or changed by a **couple.** See Fig. 1.1.65. If a single force *F* is applied to the wheel, the axis immediately acts on the wheel with an equal force to prevent translation, and the result is a couple (moment *Fr*) acting on the body and causing rotation about its center of gravity.

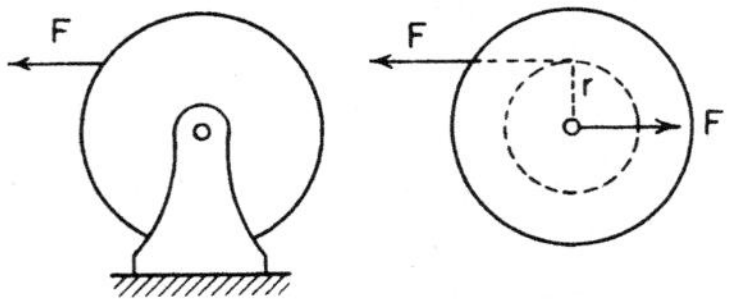

Figure 1.1.65 Pure rotation of a body about an axis passing through its center of gravity.

General formulas for rotation of a body about a fixed axis through the center of gravity, if a constant unbalanced moment is applied (Fig. 1.1.65).

Let θ = angular displacement, rad; ω = angular velocity, rad/s; α = angular acceleration, rad/s²; *M* = unbalanced moment, ft·lb; *I* = moment of inertia (mass); *g* = acceleration due to gravity; *t* = time of application of *M*.

Initial angular velocity = 0	Initial angular velocity = ω_0
$M = I\alpha$	$M = I\alpha$
$\theta = \frac{1}{2}\alpha t^2$	$\theta = \omega_0 t + \frac{1}{2}\alpha t^2$
$\omega = \sqrt{2\alpha\theta}$	$\omega = \sqrt{\omega_0^2 + 2\alpha\theta}$

General Rule for Rotating Bodies Determine all the external forces acting and their moments about the axis of rotation. If these moments are balanced, there will be no change of motion. If the moments are unbalanced, this unbalanced moment, or **torque,** will cause an angular acceleration about the axis.

Rotation about an Axis Not Passing through the Center of Gravity **The resultant force acting on the body must be proportional to the acceleration of the center of gravity and directed along its line of action.** If the axis of rotation does not pass through the center of gravity, the center of gravity will have a resultant acceleration with a component $a_n = \omega^2 r$ directed toward the axis of rotation and a component $a_t = \alpha r$ tangential to its circular path. The resultant force acting on the body must also have two components, one directed normal and one directed tangential to the path of the center of gravity. The line of action of this resultant does not pass through the center of gravity because of the unbalanced moment $M_0 = I_0\alpha$ but at a point *Q*, as in Fig. 1.1.66. The point of application of this resultant is known as the **center of percussion** and may be **defined** as the point of application of the resultant of all the forces tending to cause a body to rotate about a certain axis. It is the point at which a suspended body may be struck without causing any force on the axis passing through the point of suspension.

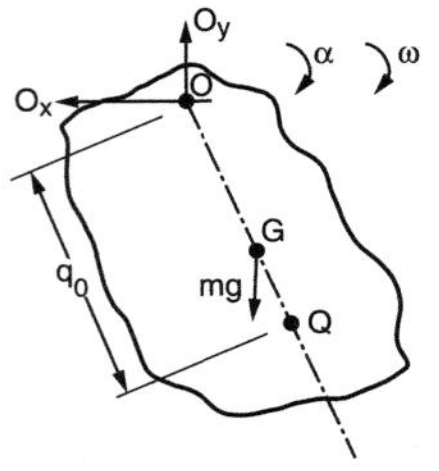

Figure 1.1.66 Pure rotation of a body about an axis not passing through its center of gravity.

Center of Percussion The distance from the axis of suspension to the center of percussion is $q_0 = I/mr_G$, where *I* = moment of inertia of the body about its axis of suspension and r_G is the distance from the axis of suspension to the center of gravity of the body.

EXAMPLES. 1. Find the center of percussion of the homogeneous rod (Fig. 1.1.67) of length *L* and mass *m*, suspended at *XX*.

$$q_0 = \frac{I}{mr_G}$$

$$I = \frac{m}{L}\int_0^L r^2\,dr = mL^2/3 \qquad r_G = L/2 \qquad \therefore\; q_0 = \frac{mL^2/3}{mL/2} = 2L/3$$

2. Find the center of percussion of a solid cylinder, of mass *m*, resting on a horizontal plane. In Fig. 1.1.68, the instantaneous center of the cylinder is at *O*, and serves temporarily as a fixed axis of rotation. The center of percussion will therefore be a height above the plane equal to $q_0 = I/mx_0$. Since $I = (mr^2/2) + mr^2$ and $x_0 = r$, $q_0 = 3r/2$.

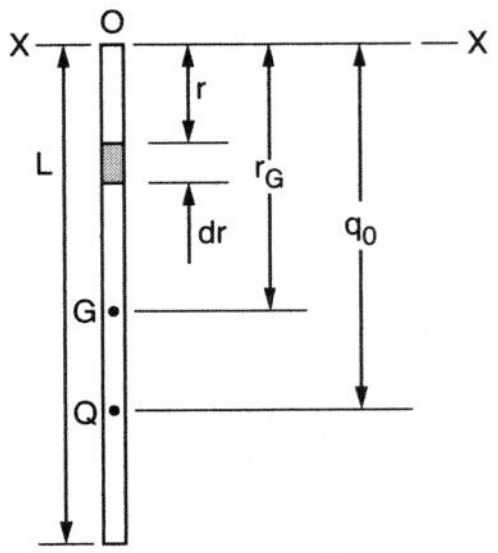

Figure 1.1.67 Center of percussion of homogeneous rod.

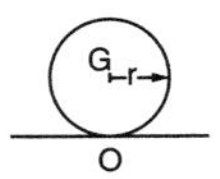

Figure 1.1.68 Center of percussion of solid cylinder.

Wheel or Cylinder Rolling down a Plane In this case the component of the weight along the plane tends to make it roll down and is

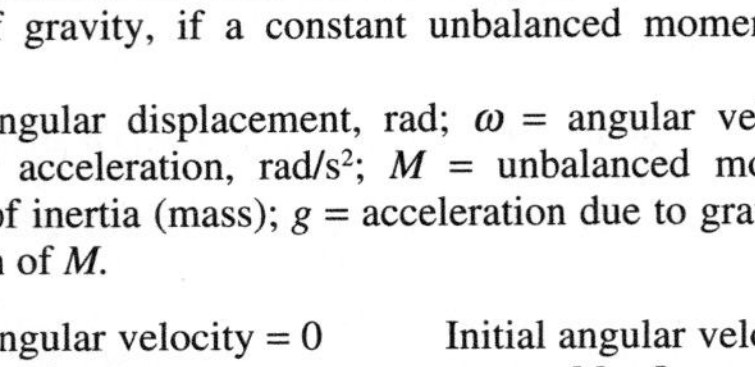

treated as a force causing rotation. The forces acting on the body should be resolved into components along the line of motion and perpendicular to it. If the forces are all known, their resultant is at the center of percussion. If one force is to be determined (the exact conditions as regards slipping or not slipping must be known), the center of percussion can be determined and the unknown force found.

Relation between the Center of Percussion and Radius of Gyration $q_0 = I/mr_G = k^2/r_G \therefore k^2 = r_G q_0$ where k = radius of gyration. Therefore, the radius of gyration is a mean proportional between the distance from the axis of oscillation to the center of percussion and the distance from the same axis to the center of gravity.

Interchangeability of Center of Percussion and Axis of Oscillation If a body is suspended from an axis, the center of percussion for that axis can be found. If the body is suspended from this center of percussion as an axis, the original axis of suspension will then become the center of percussion. The center of percussion is sometimes known as the **center of oscillation.**

Period of Oscillation of a Compound Pendulum The length of an equivalent simple pendulum is the distance from the axis of suspension to the center of percussion of the body in question. To find the **period of oscillation** of a body about a given axis, find the distance $q_0 = I/mr_G$ from that axis to the center of percussion of the swinging body. The length of the simple pendulum that will oscillate in the same time is this distance q_0. The period of oscillation for the equivalent single pendulum is $\tau = 2\pi\sqrt{q_0/g}$.

Determination of Moment of Inertia by Experiment To find the moment of inertia of a body, suspend it from some axis not passing through the center of gravity and, by swinging it, determine the period of one complete oscillation in seconds. The known values will then be τ = time of one complete oscillation, r_G = distance from axis to center of gravity, and m = mass of body. The length of the equivalent simple pendulum is $q_0 = I/mr_G$. Substituting this value of q_0 in $\tau = 2\pi\sqrt{q_0/g}$ gives $\tau = 2\pi\sqrt{I/mr_G g}$, from which $\tau^2 = 4\pi^2 I/mr_G g$, or $I = mr_G g\tau^2/4\pi^2$.

Plane Motion of a Rigid Body

Plane motion may be considered to be a combination of translation and rotation (see "Kinematics"). For translation, Newton's second law of motion must always be satisfied, and the resultant of the external force system must be equal to the product of the mass times the acceleration of the center of gravity in any system of coordinates. In rotation, the body moving in plane motion will not have a fixed axis. When the methods of relative motion are being used, any point on the body may be used as a reference axis to which the motion of all other points is referred.

The sum of the moments of all external forces about the reference axis must be equal to the vector sum of the centroidal moment of inertia times the angular acceleration and the amount of the resultant force about the reference axis.

EXAMPLE. Determine the forces acting on the piston pin A and the crankpin B of the connecting rod of a reciprocating engine shown in Fig. 1.1.69 for a position of 30° from TDC. The crankshaft speed is constant at 2,000 r/min. Assume that the pressure of expanding gases on the 4-lbm (1.81-kg) piston at this point is 145 lb/in² (10^6 N/m²). The connecting rod has a mass of 5 lbm (2.27 kg) and has a centroidal radius of gyration of 3 in (0.076 m).

The kinematics of the problem are such that the angular velocity of the crank is ω_{OB} = 209.4 rad/s clockwise, the angular velocity of the connecting rod is ω_{AB} = 45.7 rad/s counterclockwise, and the angular acceleration is α_{AB} = 5,263 rad/s² clockwise. The linear acceleration of the piston is 7,274 ft/s² in the direction of the crank. From the free-body diagram of the piston, the horizontal component of the piston-pin force is $145 \times (\pi/4)(5^2) - P = (4/32.2)(7{,}274)$, P = 1,943 lbf. The acceleration of the center of gravity G is the vector sum of the component accelerations $a_G = a_B + a^n_{G/B} + a^1_{G/B}$ where $a^n_{G/B} = \omega^2_{GB} \cdot r_{GB} = 3/12(45.7)^2$ = 522 ft/s² and $a^1_{G/B} = \alpha_{GB} \cdot r_{GB}$ = 3/12(5.263) = 1,316 ft/s². The resultant acceleration of the center of gravity is 6,685 ft/s² in the x direction and 2,284 ft/s² in the negative y direction. The resultant of the external force system will have corresponding components such that ma_{Gx} = (5/32.2)(6,685) = 1,039 lbf and ma_{Gy} = (5/32.2)(2,284) = 355 lbf. The three remaining unknown forces can be found from the three equations of motion for the connecting rod.

Taking the sum of the forces in the x direction, $\in F = ma_{Gx}$; $P - R_x = ma_{Gx}$, and R_x = 905.4 lbf. In the y direction, $\Sigma F = ma_{Gy}$; $R_y - N = ma_{gy}$; this has two unknowns, R_y and N. Taking the sum of the moments of the external forces about the center of mass g, $\Sigma M_G = I_G \alpha AB$; $(N)(5)\cos(7.18°) - (P)(5)\sin(7.18°) + (R_y)(3)\cos(7.18°) - R_x(3)\sin(7.18°) - (5/386.4)(3)^2(5{,}263)$. Solving for R_y and N simultaneously, R_y = 494.7 lbf and N = 140 lbf. We could have avoided the solution of two simultaneous algebraic equations by taking the moment summation about end A, which would determine R_y independently, or about end B, which would determine N independently.

In SI units, the kinematics would be identical, the linear acceleration of the piston being 2,217 m/s² (7,274 ft/s²). From the free-body diagram of the piston, the horizontal component of the piston-pin force is $(10^6) \times (\pi/4)(0.127)^2 - P$ = (1.81)(2,217), and P = 8,640 N. The components of the acceleration of the center of gravity G are $a^n_{G/B} = 522$ ft/s² and $a^T_{G/B} = 1{,}315$ ft/s². The resultant acceleration of the center of gravity is 2,037.5 m/s² (6,685 ft/s²) in the x direction and 696.3 m/s² (2,284 ft/s²) in the negative y direction. The resultant of the external force system will have the corresponding components; ma_{Gx} = (2.27) (2,037.5) = 4,620 N; ma_{Gy} = (2.27)(696.3) = 1,579 N. R_x = 4,027 N, R_y = 2,201 N, force N = 623 newtons.

WORK AND ENERGY

Work When a body accelerates or is displaced against resistance, work must be done upon it. An increment of work is defined as the product of an incremental displacement times the component of the force vector in the direction of the displacement, or the product of a force times the component of the incremental displacement vector in the direction of the force. In other words, an increment of the work is the inner or dot product of the force and the incremental displacement: $dU = F \cdot ds \cos \alpha$, where α is the angle between the displacement and force vectors. The increment of work done by a couple M acting on a body during an increment of angular rotation $d\theta$ in the plane of the couple is $dU = M\, d\theta$. Force-displacement or moment-angle diagram is called a **work diagram,** as shown in Fig.1.1.70. In such diagram, force or moment is plotted as a function of linear or angular displacement, respectively. The area under the curve represents the work done, which is equal to $\int_{S_1}^{S_2} F\, ds \cos\alpha$ or $\int_{\theta_1}^{\theta_2} M\, d\theta$.

Units of Work When the force of 1 lb acts through the distance of 1 ft, 1 lb·ft of work is done. In SI units, a force of 1 newton acting through 1 meter is 1 joule of work. 1.356 N·m = 1 lb·ft.

Energy A body is said to possess energy when it can do work. A body may possess this capacity through its **position** or **condition.** When a body is so held that it can do work, if released, it is said to possess energy of position or **potential energy.** When a body is moving

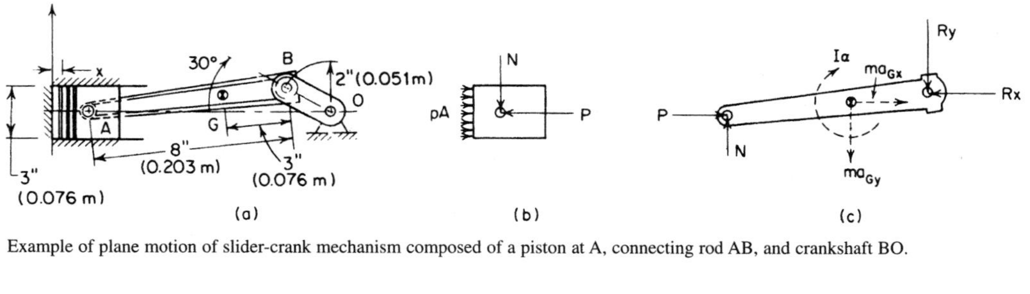

Figure 1.1.69 Example of plane motion of slider-crank mechanism composed of a piston at A, connecting rod AB, and crankshaft BO.

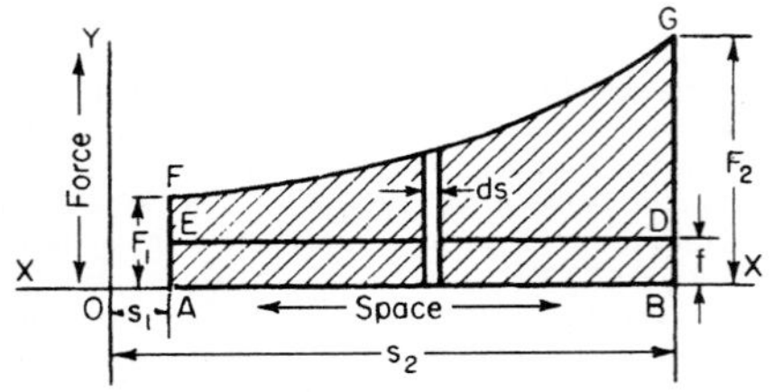

Figure 1.1.70

with some velocity, it is said to possess energy of motion or **kinetic energy.** An example of potential energy is a body held suspended by a rope; the position of the body is such that if the rope is removed work can be done by the body.

Energy is expressed in the same units as work. The kinetic energy of a particle is expressed by the formula $E = \frac{1}{2}mv^2 = \frac{1}{2}(w/g)v^2$. The kinetic energy of a rigid body in translation is also expressed as $E = \frac{1}{2}mv^2$. Since all particles of the rigid body have the same identical velocity v, the velocity v is the velocity of the center of gravity. The kinetic energy of a rigid body, rotating about a fixed axis is $E = \frac{1}{2}I_0\omega^2$, where I_0 is the mass moment of inertia about the axis of rotation. In plane motion, a rigid body has both translation and rotation. The kinetic energy is the algebraic sum of the translating kinetic energy of the center of gravity and the mass rotating kinetic energy about the center of gravity, $E = \frac{1}{2}mv^2 + \frac{1}{2}I\omega^2$. Here the velocity v is the velocity of the center of gravity, and the mass moment of inertia I is the centroidal moment of inertia.

If a force which varies acts through a space on a body of mass m, the work done is $\int_{S_1}^{S} F\,ds$ and if the work is all used in giving kinetic energy to the body it is equal to $\frac{1}{2}m(v_2^2 - v_1^2)$ = **change in kinetic energy,** where v_2 and v_1 are the velocities at distances s_2 and s_1, respectively. This is a specific statement of the law of conservation of energy. **The principle of conservation of energy requires that the mechanical energy of a system remain unchanged if it is subjected only to forces which depend on position or configuration.**

Certain problems in which the velocity of a body at any point in its straight-line path when acted upon by varying forces is required can be easily solved by the use of a **work diagram.**

In Fig. 1.1.70, let a body start from rest at A and be acted upon by a force that varies in accordance with the diagram $AFGBA$. Let the resistance to motion be a constant force $= f$. Find the velocity of the body at point B. The area $AFGBA$ represents the work done upon the body and the area $AEDBA$ (= force $f \times$ distance AB) represents the work that must be done to overcome resistance. The difference of these areas, or $EFGDE$, will represent work done in excess of that required to overcome resistance, and consequently is equal to the increase in kinetic energy. Equating the work represented by the area $EFGDE$ to $\frac{1}{2}wv^2/g$ and solving for v will give the required velocity at B. If the body did not start from rest, this area would represent the change in kinetic energy, and the velocity could be obtained by the formula: Work $= \frac{1}{2}(w/g)(v_1^2 - v_0^2)$, v_1 being the required velocity.

General Rule for Rectilinear Motion Resolve each force acting on the body into components, one of which acts along the line of motion of the body and the other perpendicular to the line of motion. Take the sum of all the components acting in the direction of the motion and multiply this sum by the distance moved through for constant forces. (Take the average force times distance for forces that vary.) This product will be the total work done upon the body. If there is no unbalanced component, there will be no change in kinetic energy and consequently no change in velocity. If there is an unbalanced component, the change in kinetic energy will be this unbalanced component multiplied by the distance moved through.

The **work done by a system of forces acting on a body** is equal to the algebraic sum of the work done by each force taken separately.

Power is the rate at which work is performed, or the number of units of work performed in unit time. In the English engineering system, the units of power are the horsepower, or 33,000 lb·ft/min = 550 lb·ft/s, and the kilowatt = 1.341 hp = 737.55 lb·ft/s. In SI units, the unit of power is the watt, which is 1 newton-meter per second or 1 joule per second.

Friction Brake In Fig. 1.1.71 a pulley revolves under the band and in the direction of the arrow, exerting a pull of T on the spring. The friction of the band on the rim of the pulley is $(T - w)$, where w is the weight attached to one end of the band. Let the pulley make N r/min; then the work done per minute against friction by the rim of the pulley is $2\pi RN(T - w)$, and the horsepower absorbed by brake = $2\pi RN(T - w)/33{,}000$.

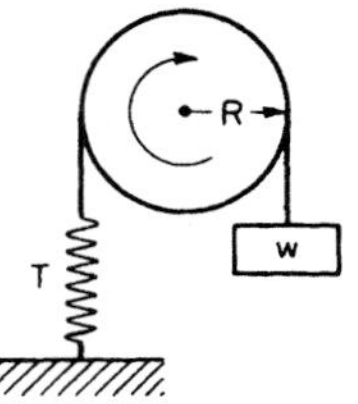

Figure 1.1.71 Schematic of friction brake system.

IMPULSE AND MOMENTUM

The product of force and time is defined as linear impulse. The impulse of a constant force over a time interval $t_2 - t_1$ is $F(t_2 - t_1)$. If the force is not constant in magnitude but is constant in direction, the impulse is $\int_{t_1}^{t_2} F\,dt$. The dimensions of linear impulse are (force) × (time) in pound-seconds, or newton-seconds.

Impulse is a **vector** quantity which has the direction of the resultant force. Impulses may be added vectorially by means of a vector polygon, or they may be resolved into components by means of a parallelogram. The **moment** of a **linear impulse** may be found in the same manner as the moment of a force. The linear impulse is represented by a directed line segment, and the moment of the impulse is the product of the magnitude of the impulse and the perpendicular distance from the line segment to the point about which the moment is taken. **Angular impulse** over a time interval $t_2 - t_1$ is a product of the sum of applied moments on a rigid body about a reference axis and time. The dimensions for angular impulse are (force) × (time) × (displacement) in foot-pound-seconds or newton-meter-seconds. **Angular impulse and linear impulse cannot be added.**

Momentum is also a vector quantity and can be added and resolved in the same manner as force and impulse. The dimensions of linear momentum are (force) × (time) in pound-seconds or newton-seconds, and are identical to linear impulse. An alternate statement of Newton's second law of motion is that the resultant of an unbalanced force system must be equal to the time rate of change of linear momentum, $\Sigma \mathbf{F} = d(mv)/dt$.

If a variable force acts for a certain time on a body of mass m, the quantity $\int_{t_1}^{t_2} F\,dt = m(v_2 - v_1)$ = the change of momentum of the body.

The **moment of momentum** can be determined by the same methods as those used for the moment of a force or moment of an impulse. The dimensions of the moment of momentum are (force) × (time) × (displacement) in foot-pound-seconds, or newton-meter-seconds.

In **plane motion** the angular momentum of a rigid body about a reference axis perpendicular to the plane of motion is the sum of the moments of linear momenta of all particles in the body about the reference axes. Specifically, **the angular momentum of a rigid body in plane motion is the vector sum of the angular momentum about the reference axis and the moment of the linear momentum of the center of gravity about the reference axis,** $\mathbf{H}_0 = I_0\omega + \mathbf{d} \times mv$.

In three-dimensional rotation about a fixed axis, the angular momentum of a rigid body has components along three coordinate axes, which involve both the moments of inertia about the x, y, and z axes, $I_{0_{xx}}$, $I_{0_{yy}}$, and $I_{0_{zz}}$, and the products of inertia, $I_{0_{xy}}$, $I_{0_{zz}}$, and $I_{0_{yz}}$; $H_{0_x} - I_{0_{xx}} \times \omega_x - I_{0_{xy}} \cdot \omega_y - I_{0_{xy}} \cdot \omega_z$, $H_{0_y} = -I_{0_{xy}} \cdot \omega_x + I_{0_{yy}} \cdot \omega_y - I_{0_{yz}} \cdot \omega_z$, and $H_{0_z} = I_{0_{xz}} \cdot \omega_x - I_{0_{zy}} \cdot \omega_y + I_{0_{zz}} \cdot \omega_z$ where $\mathbf{H}_0 = \mathbf{H}_{0_x} + \mathbf{H}_{0_y} + \mathbf{H}_{0_z}$.

Impact

The collision between two bodies, where relatively large forces result over a comparatively short interval of time, is called **impact.** A straight line perpendicular to the plane of contact of two colliding bodies is called the **line of impact.** If the centers of gravity of the two bodies lie on the line of contact, the impact is called **central impact,** in any other case, **eccentric impact.** If the linear momenta of the centers of gravity are also directed along the line of impact, the impact is **collinear or direct central impact.** In any other case impact is said to be **oblique.**

Collinear Impact When two masses m_1 and m_2, having respective velocities u_1 and u_2, move in the same line, they will collide if $u_2 > u_1$ (Fig. 1.1.72*a*). During collision (Fig. 1.1.72*b*), kinetic energy is absorbed in the deformation of the bodies. There follows a period of restoration

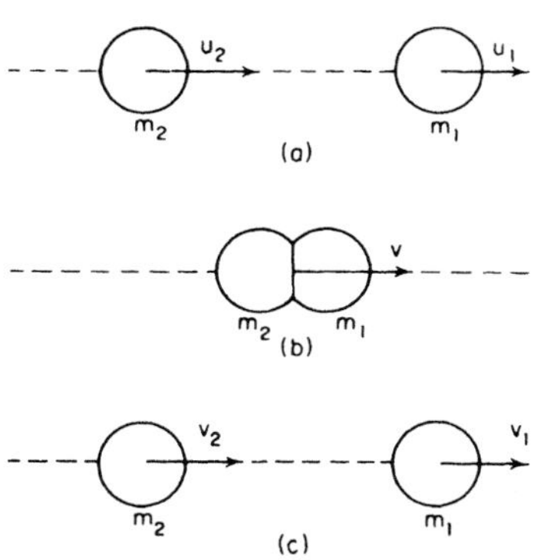

Figure 1.1.72 Stages of impact of two objects (a) before impact or approaching stage, (b) during impact or restitution stage and (c) after impact or separation stage.

which may or may not be complete. If complete restoration of the energy of deformation occurs, the impact is **elastic.** If the restoration of energy is incomplete, the impact is referred to as **inelastic.** After collision (Fig. 1.1.72*c*), the bodies continue to move with changed velocities of v_1 and v_2. Since the contact forces on one body are equal to and opposite the contact forces on the other, the sum of the linear momenta of the two bodies is conserved; $m_1u_1 + m_2u_2 = m_1v_1 + m_2v_2$.

The law of conservation of momentum states that the linear momentum of a system of bodies is unchanged if there is no resultant external force on the system.

Coefficient of Restitution The ratio of the velocity of separation $v_1 - v_2$ to the velocity of approach $u_2 - u_1$ is called the **coefficient of restitution** *e* that express by $e = (v_1 - v_2)/(u_2 - u_1)$.

The value of *e* will depend on the shape and material properties of the colliding bodies. In elastic impact, the coefficient of restitution is unity and there is no energy loss. A coefficient of restitution of zero indicates perfectly inelastic or plastic impact, where there is no separation of the bodies after collision and the energy loss is a maximum. In **oblique impact,** the coefficient of restitution applies only to those components of velocity along the line of impact or normal to the plane of impact. The coefficient of restitution between two materials can be measured by making one body greatly larger than the other so that m_2 is infinitely large in comparison to m_1. The velocity of m_2 is unchanged for all practical purposes during impact and $e = v_1/u_1$. For a small ball dropped from a height *H* upon an extensive horizontal surface and rebounding to a height *h*, $e = \sqrt{h/H}$.

Impact of Jet of Water on Flat Plate When a jet of water strikes a flat plate perpendicularly to its surface, the force exerted by the water on the plate is wv/g, where *w* is the weight of water striking the plate in a unit of time and *v* is the velocity. When the jet is inclined to the surface by an angle, *A*, the force is $(wv/g) \cos A$.

Variable Mass

If the mass of a body is variable such that mass is being either added or ejected, an alternate form of Newton's second law of motion must be used which accounts for changes in mass:

$$\mathbf{F} = m\frac{d\mathbf{v}}{dt} + \frac{dm}{dt}\,\mathbf{u}$$

The mass *m* is the instantaneous mass of the body, and dv/dt is the time rate of change of the absolute of velocity of mass *m*. The velocity *u* is the velocity of the mass *m* relative to the added or ejected mass, and dm/dt is the time rate of change of mass. In this case, care must be exercised in the choice of coordinates and expressions of sign. If mass is being added, dm/dt is plus, and if mass is ejected, dm/dt is minus.

Fields of Force—Attraction

The space within which the action of a physical force comes into play on bodies lying within its boundaries is called the **field of the force.**

The **strength** or **intensity of the field** at any given point is the relation between a force *F* acting on a mass *m* at that point and the mass. Intensity of field = $i = F/m$; $F = mi$.

The **unit of field intensity** is the same as the unit of acceleration, i.e., 1 ft/s^2 or 1 m/s^2. The intensity of a field of force may be represented by a line (or **vector**).

A field of force is said to be **homogeneous** when the intensity of all points is uniform and in the same direction.

A field of force is called a **central field of force** with a center *O*, if the direction of the force acting on the mass particle *m* in every point of the field passes through *O* and its magnitude is a function only of the distance *r* from *O* to *m*. A line drawn so that its direction coincides at every point with that of the force prevailing at that point is called a **line of force.**

Rotation of Solid Bodies about Any Axis

The general moment equations for three-dimensional motion are usually expressed in terms of the angular momentum. For a reference axis *O*, which is either a fixed axis of the center of gravity, $M_{0_x} = (dH_{0_x}/dt) - H_{0_y} \cdot \omega_z + H_{0_z} \cdot \omega_y$, $M_{0_y} = (dH_{0_y}/dt) - H_{0_z} \cdot \omega_x + H_{0_x} \cdot \omega_z$, and $M_{0_z} = (dH_{0_z}/dt) - H_{0_x}\omega_y + H_{0_y}\omega_x$. If the coordinate axes are oriented to coincide with the principal axes of inertia, $I_{0_{xx}}, I_{0_{yy}}$, and $I_{0_{zz}}$, a similar set of three differential equations results, involving moments, angular velocity, and angular acceleration; $M_{0_x} = I_{0_{xx}}(d\omega_x/dt) + (I_{0_{zz}} - I_{0_{yy}} 2\omega_y \cdot \omega_z$, $M_{0_y} = I_{0_{yy}}(d\omega_y/dt) + (I_{0_{xx}} - I_{0_{zz}})\omega_z \cdot \omega_x$, and $I_{0_{zz}}(d\omega_z/M_{0_z} = dt) + (I_{0_{yy}} - I_{0_{xx}})\omega_x\omega_y$. These equations are known as **Euler's equations of motion** and may apply to any rigid body.

GYROSCOPIC MOTION AND THE GYROSCOPE

Gyroscopic motion can be explained in terms of Euler's equations. Let I_1, I_2, and I_3 represent the principal moments of inertia of a gyroscope spinning with a constant angular velocity ω, about axis 1, the subscripts 1, 2, and 3 representing a right-hand set of reference axes (Figs. 1.1.73 and 1.1.74). If the gyroscope is precessed about the third axis, a vector moment results along the second axis such that

$$M_2 = I_2(d\omega_2/dt) + (I_1 - I_3)\omega_3\omega_1$$

Where the precession and spin axes are at right angles, the term $(d\omega_2/dt)$ equals the component of $\omega_3 \times \omega_1$ along axis 2. Because of this, in the simple case of a body of symmetry, where $I_2 = I_3$, the gyroscopic moment can be reduced to the common expression $M = I\omega\Omega$, where Ω is the rate of precession, ω the rate of spin, and *I* the moment of inertia about the spin axis. It is important to realize that these are equations of motion and relate the applied or resulting gyroscopic moment due to forces which act *on* the rotor, as disclosed by a free-body diagram, to the resulting motion of the rotor.

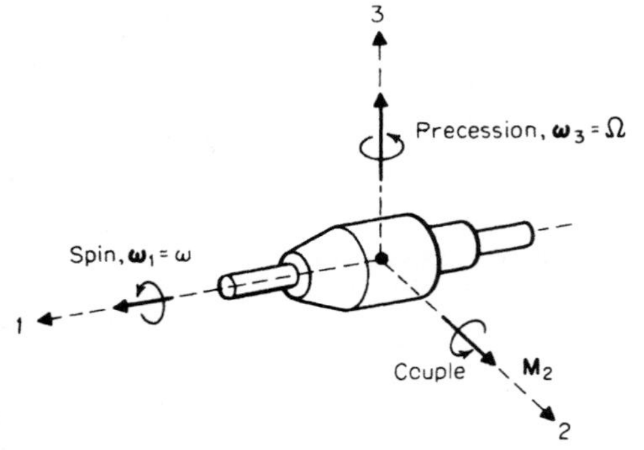

Figure 1.1.73 Terminology of gyroscopic motion.

Physical insight into the behavior of a steady precessing gyro with mutually perpendicular moment, spin, and precession axes is gained by recognizing from Fig. 1.1.74 that the change dH in angular momentum H is equal to the angular impulse $M\,dt$. In time dt, the angular-momentum vector swings from H *to* H', because of the velocity of precession ω_3. The vector change dH in angular momentum is in the direction of the applied moment M. This fact is inherent in the basic moment-momentum equation and can always be used to establish the correct spatial relationships between the moment, precessional, and spin vectors. It is seen, therefore, from Fig. 1.1.74 that the spin axis always turns toward the moment axis. Just as the change in direction of the mass-center velocity is in the same direction as the resultant force, so does the change in angular momentum follow the direction of the applied moment.

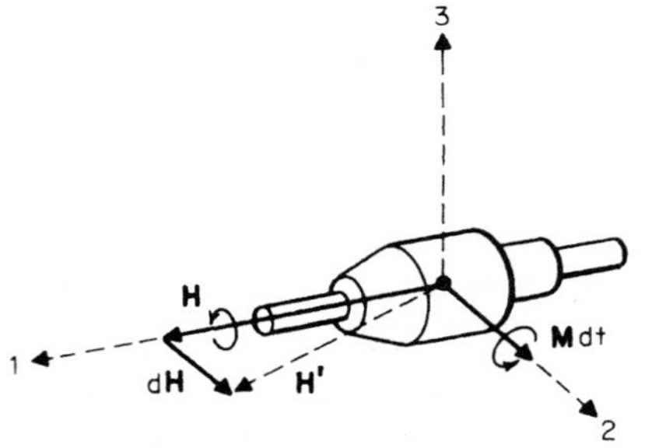

Figure 1.1.74 Demonstration of gyroscopic motion, i.e. change in angular momentum cause impulse angular moment.

For example, suppose an airplane is driven by a right-handed propeller (turning like a right-handed screw when moving forward). If a gust of wind or other force turns the airplane to the *left,* the gyroscopic action of the propeller will make the forward end of the shaft strive to *rise;* if the wing surface is large, this motion will be practically prevented by the resistance of the air, and the gyroscopic forces become effective merely as internal stresses, whose maximum value can be computed by the formula above. Similarly, if the airplane is dipped *downward,* the gyroscopic action will make the forward end of the shaft strive to turn to the *left.*

Modern **applications** of the gyroscope are based on one of the following properties: (1) a gyroscope mounted in three gimbal rings so as to be entirely free angularly in all directions will retain its direction *in space* in the absence of outside couples; (2) if the axis of rotation of a gyroscope turns or precesses in space, a couple or torque acts on the gyroscope (and conversely on its frame).

Devices operating on the first principle are satisfactory only for short durations, say less than half an hour, because no gyroscope is entirely without outside the couple. The friction couples at the various gimbal bearings, although small, will precess the axis of rotation so that after a while the axis of rotation will have changed its direction in space. The chief device based on the first principle is the **airplane compass,** which is a freely mounted gyro, keeping its direction in space during fast maneuvers of a fighting airplane. No magnetic compass will indicate correctly during such maneuvers. After the plane is back on an even keel in steady flight, the magnetic compass once more reads the true magnetic north, and the gyro compass has to be reset to point north again.

An example of a device operating on the second principle is the **automatic pilot** for keeping a vehicle on a given course. This device has been installed on torpedoes, ships, airplanes. When the ship or plane turns from the chosen course, a couple is exerted on the gyro axis, which makes it precess and this operates electric contacts or hydraulic or pneumatic valves. These again operate on the rudders, through relays, and bring the ship back to its course.

Another application is the ship **antirolling** gyroscope. This very large gyroscope spins about a vertical axis and is mounted in a ship so that the axis can be tipped fore and aft by means of an electric motor, the precession motor. The gyro can exert a large torque on the ship about the fore-and-aft axis, which is along the "rolling" axis. The sign of the torque is determined by the direction of rotation of the precession motor, which in turn is controlled by electric contacts operated by a small pilot gyroscope on the ship, which feels which way the ship rolls and gives the signals to apply a countertorque.

The **turn indicator** for airplanes is a gyro, the frame of which is held by springs. When the airplane turns, it makes the gyro axis turn with it, and the resultant couple is delivered by the springs. Thus the elongation of the springs is a measure of the rate of turn, which is suitably indicated by a pointer.

The most complicated and ingenious application of the gyroscope is the **marine compass.** This is a pendulously suspended gyroscope which is affected by gravity and also by the earth's rotation so that the gyro axis is in equilibrium only when it points north, i.e., when it lies in the plane formed by the local vertical and by the earth's north-south axis. If the compass is disturbed so that it points away from north, the action of the earth's rotation will restore it to the correct north position in a few hours.

1.2 FRICTION

Revised by Peyman Honarmandi

REFERENCES: Bowden and Tabor, "The Friction and Lubrication of Solids," Oxford. Fuller, "Theory and Practice of Lubrication for Engineers," 2nd ed., Wiley. Shigley, "Mechanical Design," McGraw-Hill. Rabinowicz, "Friction and Wear of Materials," Wiley. Ling, Klaus, and Fein, "Boundary Lubrication—An Appraisal of World Literature," ASME, 1969. Dowson, "History of Tribology," Longman, 1979. Petersen and Winer, "Wear Control Handbook," ASME, 1980. M. M. Khonsari and E. R. Booser, "Applied Tribology," Wiley, 2001.

Friction is the resistance that is encountered when two solid surfaces slide or tend to slide relative to each other. The relative motion could be the motion of solid surfaces, fluid layers, and material elements sliding against each other. The surfaces may be either dry or lubricated. In the first case, when the surfaces are free from contaminating fluids or films, the resistance is called **dry friction**, also known as **coulomb friction.** The friction of brake shoes on the rim of a railroad wheel is an example of dry friction.

When the rubbing surfaces are separated from each other by a very thin film of lubricant, the friction is that of **boundary** (or **greasy**)

lubrication. This kind of fluid friction is also called **lubricated friction.** In this case, the lubrication depends on the strong adhesion of the lubricant to the material of the rubbing surfaces; the layers of lubricant slip over each other instead of the dry surfaces. A journal when starting, reversing, or turning at very low speed under a heavy load is an example of the condition that will cause boundary lubrication. Other examples are gear teeth (especially hypoid gears), cutting tools, wire-drawing dies, power screws, bridge trunnions, and the running-in process of most lubricated surfaces.

When the lubrication is arranged so that the rubbing surfaces are separated by a fluid film, and the load on the surfaces is carried entirely by the hydrostatic or hydrodynamic pressure in the film, the friction is that of **complete** (or **viscous**) **lubrication.** In this case, the frictional losses are solely due to the internal fluid friction in the film. Oil ring bearings, bearings with forced feed of oil, pivoted shoe-type thrust and journal bearings, bearings operating in an oil bath, hydrostatic oil pads, oil lifts, and step bearings are instances of complete lubrication.

Incomplete lubrication or mixed lubrication takes place when the load on the rubbing surfaces is carried partly by a fluid viscous film and partly by areas of boundary lubrication. The friction is intermediate between that of fluid and boundary lubrication. Incomplete lubrication exists in bearings with drop-feed, waste-packed, or wick-fed lubrication, or on parallel-surface bearings.

Other than dry and fluid frictions, which are between two surfaces or layers, there are two more friction types so-called Skin and Internal frictions. **Skin friction** is a component of drag force that resists the motion of a fluid across the surface of a body. **Internal friction** is the resisting force against the motion between the molecular structure of a solid material while undergoes elastic or plastic deformation.

STATIC AND KINETIC COEFFICIENTS OF FRICTION

When the surfaces of two bodies are in contact the resultant of the forces between the surfaces is normal to the surface of contact. This force that pressing outward upon each body is called **normal force.** However, with the presence of friction force, the resultant deviates from the normal.

If one body is pressed against another by a force P, as in Fig. 1.2.1, the first body will not move, provided the angle a_0 included between the line of action of the force and a normal to the surfaces in contact does not exceed a certain value which depends upon the nature of the surfaces. In case of equilibrium, the reaction force R has the same magnitude and line of action as the force P. In Fig. 1.2.1, R is resolved into two components: a force N normal to the surfaces in contact and a force F_r parallel to the surfaces in contact. Since motion does not occur it follows,

$$F_r = N \tan a_0 = Nf_0$$

where $f_0 = \tan a_0$ is called the **coefficient of friction at rest** (or of **static friction**) and a_0 is the **angle of friction at rest.**

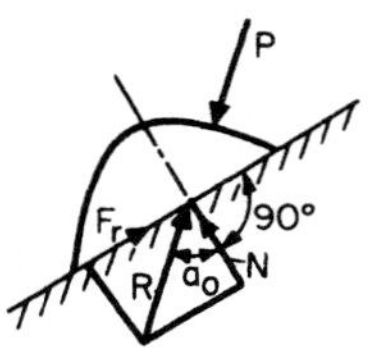

Figure 1.2.1 Free body diagram of a body on inclined surface with an external force P.

If the normal force N between the surfaces is kept constant, and the tangential force F_r is gradually increased, there will be no motion while $F_r < Nf_0$. A state of *impending motion* is reached when F_r nears the value of Nf_0. If sliding motion occurs, a frictional force F resisting the motion must be overcome. The force F is commonly expressed as $F = fN$, where f is the **coefficient of sliding friction,** or **kinetic friction.** In general, the coefficients of sliding friction are smaller than the coefficients of static friction. With small velocities of sliding and very clean surfaces, the two coefficients do not differ appreciably.

Table 1.2.4 demonstrates the typical reduction of sliding coefficients of friction below corresponding static values. Figure 1.2.2 indicates results of tests on lubricated machine tool ways showing a reduction of friction coefficient with increasing sliding velocity.

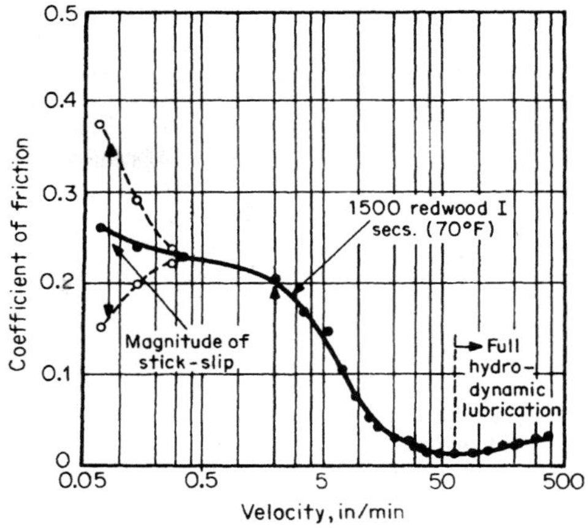

Figure 1.2.2 Typical relationship between kinetic friction and sliding velocity for lubricated cast iron on cast iron slideways (load, 20 lb/in^2; upper slider, scraped; lower slideway, scraped). *(From Birchall, Kearny, and Moss, Intl. J. Machine Tool Design Research, 1962.)*

This behavior is normal with dry friction, some conditions of boundary friction, and with the break-away friction in ball and roller bearings. This condition is depicted in Fig. 1.2.3, where the friction force decreases with relative velocity. This negative slope leads to locally

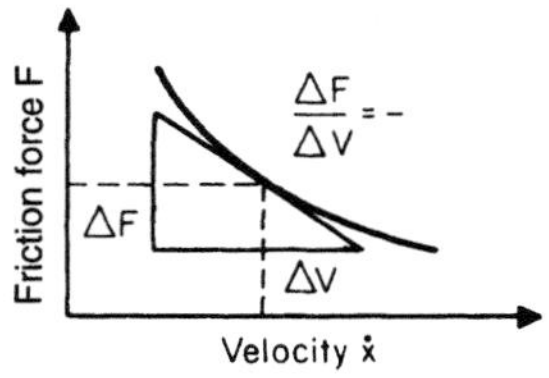

Figure 1.2.3 Friction force decreases as velocity increases.

unstable equilibrium and **self-excited vibrations** in systems such as the one of Fig. 1.2.4. This phenomenon takes place because, for small amplitudes, the oscillatory system displays damping in which the damping factor is equal to the slope of the friction curve and thus is termed **negative damping.** When the slope of the friction force versus sliding velocity is positive (**positive damping**) this type of instability is not possible. This is typical of fluid damping, squeeze films, dash pots, and fluid film bearings in general.

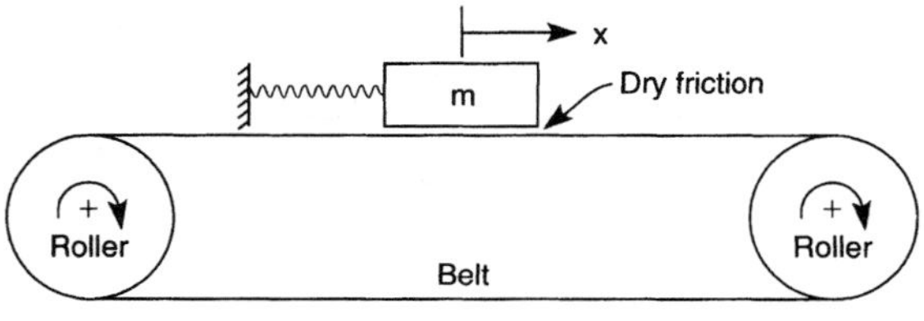

Figure 1.2.4 Belt friction apparatus with possible self-excited vibrations.

It is interesting to note that these self-excited systems vibrate at close to their natural frequency over a large range of frictional levels and speeds. This symptom is a helpful means of identification. Another characteristic is that the moving body comes periodically to momentary relative rest, that is, zero sliding velocity. For this reason, this phenomenon is also called **stick-slip vibration.** Common examples are violin strings, chalk on blackboard, water-lubricated rubber stern tube ship bearings at low speed, squeaky hinges, and oscillating rolling element bearings, especially if they are supporting large flexible structures such as radar antennas. Control requires the introduction of fluid film bearings, viscous seals, or viscous dampers into the system with sufficient positive damping to override the effects of negative damping.

Under moderate pressures, the frictional force is proportional to the normal load on the rubbing surfaces. It is independent of the pressure per unit area of the surfaces. The direction of the frictional force opposing the sliding motion is exactly opposite to the local relative velocity. Therefore, it takes very little effort to displace transversally two bodies which have a major direction of relative sliding. This behavior, **compound sliding,** is exploited when easing the extraction of a nail by simultaneously rotating it about its axis, and accounts for the ease with which an automobile may skid on the road or with which a plug gage can be inserted into a hole if it is rotated while being pushed in.

The coefficients of friction for dry surfaces (dry friction) depend on the materials sliding over each other and on the finished condition of the surfaces. With greasy (boundary) lubrication, the coefficients depend not only on both the materials and conditions of the surfaces but also on the lubricants employed itself.

Coefficients of friction are sensitive to atmospheric dust and humidity, oxide films, surface finish, velocity of sliding, temperature, vibration, and the extent of contamination. In many instances the degree of contamination is perhaps the most important single variable. For example, in Table 1.2.1, values for the static coefficient of friction of steel on steel are listed, and, depending upon the degree of contamination of the specimens, the coefficient of friction varies effectively from ∞ (infinity) to 0.013.

Table 1.2.1 Coefficients of Static Friction for Steel on Steel

Test condition	f_0	Ref.
Degassed at elevated temp in high vacuum	∞ (weld on contact)	1
Grease-free in vacuum	0.78	2
Grease-free in air	0.39	3
Clean and coated with oleic acid	0.11	2
Clean and coated with solution of stearic acid	0.013	4

SOURCES: (1) Bowden and Young, *Proc. Roy. Soc.*, 1951. (2) Campbell, *Trans. ASME,* 1939. (3) Tomlinson, *Phil. Mag.*, 1929. (4) Hardy and Doubleday, *Proc. Roy. Soc.*, 1923.

The most effective boundary lubricants are generally those which react chemically with the solid surface and form an adhering film that is attached to the surface with a chemical bond. This action depends upon the nature of the lubricant and upon the reactivity of the solid surface. Table 1.2.2 indicates that a fatty acid, such as found in animal, vegetable, and marine oils, reduces the coefficient of friction markedly only if it can react effectively with the solid surface. Paraffin oil is almost completely nonreactive.

Values in Table 1.2.4 of sliding and static coefficients have been selected largely from investigations where these variables have been very carefully controlled. They are representative values for smooth surfaces. It has been generally observed that sliding friction between hard materials is smaller than that between softer surfaces.

Table 1.2.2 Coefficients of Static Friction at Room Temperature

Surfaces	Clean	Paraffin oil	Paraffin oil plus 1% lauric acid	Degree of reactivity of solid
Nickel	0.7	0.3	0.28	Low
Chromium	0.4	0.3	0.3	Low
Platinum	1.2	0.28	0.25	Low
Silver	1.4	0.8	0.7	Low
Glass	0.9		0.4	Low
Copper	1.4	0.3	0.08	High
Cadmium	0.5	0.45	0.05	High
Zinc	0.6	0.2	0.04	High
Magnesium	0.6	0.5	0.08	High
Iron	1.0	0.3	0.2	Mild
Aluminum	1.4	0.7	0.3	Mild

SOURCE: From Bowden and Tabor, "The Friction and Lubrication of Solids," Oxford.

Effect of Surface Films Campbell observed a lowering of the coefficient of friction when oxide or sulfide films were present on metal surfaces (*Trans. ASME,* 1939; footnotes to Table 1.2.4). The reductions listed in Table 1.2.3 were obtained with oxide films formed by heating in air at temperatures from 100 to 500°C, and sulfide films produced by immersion in a 0.02 percent sodium sulfide solution.

Table 1.2.3 Static Coefficient of Friction f_0

	Clean and dry	Oxide film	Sulfide film
Steel-steel	0.78	0.27	0.39
Brass-brass	0.88		0.57
Copper-copper	1.21	0.76	0.74

Effect of Sliding Velocity It has generally been observed that coefficients of friction reduce on dry surfaces as sliding velocity increases. (See results of railway brake-shoe tests below.) Dokos measured this reduction in friction for mild steel on medium steel. Values are for the average of four tests with high contact pressures (*Trans. ASME,* 1946; see footnotes to Table 1.2.4).

Sliding velocity, in/s	0.0001	0.001	0.01	0.1	1	10	100
f	0.53	0.48	0.39	0.31	0.23	0.19	0.18

Effect of Surface Finish The degree of surface roughness has been found to influence the coefficient of friction. Burwell evaluated this effect for conditions of boundary or greasy friction (*Jour. SAE,* 1942; see footnotes to Table 1.2.4). The values listed in Table 1.2.5 are for sliding or kinetic coefficients of friction, hard steel on hard steel. The friction coefficient and wear rates of **polymers against metals** are often lowered by decreasing the surface roughness. This is particularly true of composites such as those with polytetrafluoroethylene (PTFE) which function through transfer to the counterface.

Solid Lubricants In certain applications solid lubricants are used successfully. Boyd and Robertson with pressures ranging from 50,000 to 400,000 lb/in^2 (344,700 to 2,757,000 kN/m^2) found sliding coefficients of friction f for hard steel on hard steel as follows: powdered mica, 0.305; powdered soapstone, 0.306; lead iodide, 0.071; silver sulfate, 0.054; graphite, 0.058; molybdenum disulfide, 0.033; tungsten disulfide, 0.037; stearic acid, 0.029 (*Trans. ASME,* 1945; see footnotes to Table 1.2.4).

Coefficients of Static Friction for Special Cases

Masonry and Earth Dry masonry on brickwork, 0.6 to 0.7; timber on polished stone, 0.40; iron on stone, 0.3 to 0.7; masonry on dry clay, 0.51; masonry on moist clay, 0.33.

Earth on Earth Dry sand, clay, mixed earth, 0.4 to 0.7; damp clay, 1.0; wet clay, 0.31; shingle and gravel, 0.8 to 1.1.

Natural Cork On cork, 0.59; on pine with grain, 0.49; on glass, 0.52; on dry steel, 0.45; on wet steel, 0.69; on hot steel, 0.64; on oiled steel, 0.45; water-soaked cork on steel, 0.56; oil-soaked cork on steel, 0.42.

Coefficients of Sliding Friction for Special Cases

Soapy Wood Lesley gives for wood on wood, copiously lubricated with tallow, stearine, and soft soap (as used in launching practice), a starting coefficient of friction equal to 0.036, diminishing to an average value of 0.019 for the first 50 ft of motion of the ship. Rennie gives 0.0385 for wood on wood, lubricated with soft soap, under a load of 56 lb/in^2.

Table 1.2.4 Coefficients of Static and Sliding (Kinetic) Friction

(Reference letters indicate the lubricant used; numbers in parentheses give the sources. See footnote.)

Materials	Static		Sliding	
	Dry	Greasy	Dry	Greasy
Hard steel on hard steel	0.78 (1)	0.11 (1, *a*)	0.42 (2)	0.029 (5, *h*)
		0.23 (1, *b*)		0.081 (5, *c*)
		0.15 (1, *c*)		0.080 (5, *i*)
		0.11 (1, *d*)		0.058 (5, *j*)
		0.0075 (18, *p*)		0.084 (5, *d*)
		0.0052 (18, *h*)		0.105 (5, *k*)
				0.096 (5, *l*)
				0.108 (5, *m*)
				0.12 (5, *a*)
Mild steel on mild steel	0.74 (19)		0.57 (3)	0.09 (3, *a*)
				0.19 (3, *u*)
Hard steel on graphite	0.21 (1)	0.09 (1, *a*)		
Hard steel on babbitt (ASTM No. 1)	0.70 (11)	0.23 (1, *b*)	0.33 (6)	0.16 (1, *b*)
		0.15 (1, *c*)		0.06 (1, *c*)
		0.08 (1, *d*)		0.11 (1, *d*)
		0.085 (1, *e*)		
Hard steel on babbitt (ASTM No. 8)	0.42 (11)	0.17 (1, *b*)	0.35 (11)	0.14 (1, *b*)
		0.11 (1, *c*)		0.065 (1, *c*)
		0.09 (1, *d*)		0.07 (1, *d*)
		0.08 (1, *e*)		0.08 (11, *h*)
Hard steel on babbitt (ASTM No. 10)		0.25 (1, *b*)		0.13 (1, *b*)
		0.12 (1, *c*)		0.06 (1, *c*)
		0.10 (1, *d*)		0.055 (1, *d*)
		0.11 (1, *e*)		
Mild steel on cadmium silver				0.097 (2, *f*)
Mild steel on phosphor bronze			0.34 (3)	0.173 (2, *f*)
Mild steel on copper lead				0.145 (2, *f*)
Mild steel on cast iron		0.183 (15, *c*)	0.23 (6)	0.133 (2, *f*)
Mild steel on lead	0.95 (11)	0.5 (1, *f*)	0.95 (11)	0.3 (11, *f*)
Nickel on mild steel			0.64 (3)	0.178 (3, *x*)
Aluminum on mild steel	0.61 (8)		0.47 (93)	
Magnesium on mild steel			0.42 (3)	
Magnesium on magnesium	0.6 (22)	0.08 (22, *y*)		
Teflon on Teflon	0.04 (22)			0.04 (22, *f*)
Teflon on steel	0.04 (22)			0.04 (22, *f*)
Tungsten carbide on tungsten carbide	0.2 (22)	0.12 (22, *a*)		
Tungsten carbide on steel	0.5 (22)	0.08 (22, *a*)		
Tungsten carbide on copper	0.35 (23)			
Tungsten carbide on iron	0.8 (23)			
Bonded carbide on copper	0.35 (23)			
Bonded carbide on iron	0.8 (23)			
Cadmium on mild steel			0.46 (3)	
Copper on mild steel	0.53 (8)		0.36 (3)	0.18 (17, *a*)
Nickel on nickel	1.10 (16)		0.53 (3)	0.12 (3, *w*)
Brass on mild steel	0.51 (8)		0.44 (6)	
Brass on cast iron			0.30 (6)	
Zinc on cast iron	0.85 (16)		0.21 (7)	
Magnesium on cast iron			0.25 (7)	
Copper on cast iron	1.05 (16)		0.29 (7)	
Tin on cast iron			0.32 (7)	
Lead on cast iron			0.43 (7)	
Aluminum on aluminum	1.05 (16)		1.4 (3)	
Glass on glass	0.94 (8)	0.01 (10, *p*)	0.40 (3)	0.09 (3, *a*)
		0.005 (10, *q*)		0.116 (3, *v*)
Carbon on glass			0.18 (3)	
Garnet on mild steel			0.39 (3)	
Glass on nickel	0.78 (8)		0.56 (3)	

(*a*) Oleic acid; (*b*) Atlantic spindle oil (light mineral); (*c*) castor oil; (*d*) lard oil; (*e*) Atlantic spindle oil plus 2 percent oleic acid; (*f*) medium mineral oil; (*g*) medium mineral oil plus ½ percent oleic acid; (*h*) stearic acid; (*i*) grease (zinc oxide base); (*j*) graphite; (*k*) turbine oil plus 1 percent graphite; (*l*) turbine oil plus 1 percent stearic acid; (*m*) turbine oil (medium mineral); (*n*) olive oil; (*p*) palmitic acid; (*q*) ricinoleic acid; (*r*) dry soap; (*s*) lard; (*t*) water; (*u*) rape oil; (*v*) 3-in-1 oil; (*w*) octyl alcohol; (*x*) triolein; (*y*) 1 percent lauric acid in paraffin oil.

SOURCES: (1) Campbell, *Trans. ASME,* 1939; (2) Clarke, Lincoln, and Sterrett, *Proc. API,* 1935; (3) Beare and Bowden, *Phil. Trans. Roy. Soc.,* 1935; (4) Dokos, *Trans. ASME,* 1946; (5) Boyd and Robertson, *Trans. ASME,* 1945; (6) Sachs, *Zeit f. angew. Math. und Mech.,* 1924; (7) Honda and Yamaha, *Jour. I of M,* 1925; (8) Tomlinson, *Phil. Mag.,* 1929; (9) Morin, *Acad. Roy. des Sciences,* 1838; (10) Claypoole, *Trans. ASME,* 1943; (11) Tabor, *Jour. Applied Phys.,* 1945; (12) Eyssen, General Discussion on Lubrication, *ASME,* 1937; (13) Brazier and Holland-Bowyer, General Discussion on Lubrication, *ASME,* 1937; (14) Burwell, *Jour. SAE.,* 1942; (15) Stanton, "Friction," Longmans; (16) Ernst and Merchant, Conference on Friction and Surface Finish, M.I.T., 1940; (17) Gongwer, Conference on Friction and Surface Finish, M.I.T., 1940; (18) Hardy and Bircumshaw, *Proc. Roy. Soc.,* 1925; (19) Hardy and Hardy, *Phil. Mag.,* 1919; (20) Bowden and Young, *Proc. Roy. Soc.,* 1951; (21) Hardy and Doubleday, *Proc. Roy. Soc.,* 1923; (22) Bowden and Tabor, "The Friction and Lubrication of Solids," Oxford; (23) Shooter, *Research,* 4, 1951.

(Continued)

Table 1.2.4 Coefficients of Static and Sliding (Kinetic) Friction (*Continued*)
(Reference letters indicate the lubricant used; numbers in parentheses give the sources. See footnote.)

Materials	Static		Sliding	
	Dry	Greasy	Dry	Greasy
Copper on glass	0.68 (8)		0.53 (3)	
Cast iron on cast iron	1.10 (16)		0.15 (9)	0.070 (9, *d*) 0.064 (9, *n*)
Bronze on cast iron			0.22 (9)	0.077 (9, *n*)
Oak on oak (parallel to grain)	0.62 (9)		0.48 (9)	0.164 (9, *r*) 0.067 (9, *s*)
Oak on oak (perpendicular)	0.54 (9)		0.32 (9)	0.072 (9, *s*)
Leather on oak (parallel)	0.61 (9)		0.52 (9)	
Cast iron on oak			0.49 (9)	0.075 (9, *n*)
Leather on cast iron			0.56 (9)	0.36 (9, *t*) 0.13 (9, *n*)
Laminated plastic on steel			0.35 (12)	0.05 (12, *t*)
Fluted rubber bearing on steel				0.05 (13, *t*)

(*a*) Oleic acid; (*b*) Atlantic spindle oil (light mineral); (*c*) castor oil; (*d*) lard oil; (*e*) Atlantic spindle oil plus 2 percent oleic acid; (*f*) medium mineral oil; (*g*) medium mineral oil plus ½ percent oleic acid; (*h*) stearic acid; (*i*) grease (zinc oxide base); (*j*) graphite; (*k*) turbine oil plus 1 percent graphite; (*l*) turbine oil plus 1 percent stearic acid; (*m*) turbine oil (medium mineral); (*n*) olive oil; (*p*) palmitic acid; (*q*) ricinoleic acid; (*r*) dry soap; (*s*) lard; (*t*) water; (*u*) rape oil; (*v*) 3-in-1 oil; (*w*) octyl alcohol; (*x*) triolein; (*y*) 1 percent lauric acid in paraffin oil.

SOURCES: (1) Campbell, *Trans. ASME,* 1939; (2) Clarke, Lincoln, and Sterrett, *Proc. API,* 1935; (3) Beare and Bowden, *Phil. Trans. Roy. Soc.,* 1935; (4) Dokos, *Trans. ASME,* 1946; (5) Boyd and Robertson, *Trans. ASME,* 1945; (6) Sachs, *Zeit f. angew. Math. und Mech.,* 1924; (7) Honda and Yamaha, *Jour. I of M,* 1925; (8) Tomlinson, *Phil. Mag.,* 1929; (9) Morin, *Acad. Roy. des Sciences,* 1838; (10) Claypoole, *Trans. ASME,* 1943; (11) Tabor, *Jour. Applied Phys.,* 1945; (12) Eyssen, General Discussion on Lubrication, *ASME,* 1937; (13) Brazier and Holland-Bowyer, General Discussion on Lubrication, *ASME,* 1937; (14) Burwell, *Jour. SAE.,* 1942; (15) Stanton, "Friction," Longmans; (16) Ernst and Merchant, Conference on Friction and Surface Finish, M.I.T., 1940; (17) Gongwer, Conference on Friction and Surface Finish, M.I.T., 1940; (18) Hardy and Bircumshaw, *Proc. Roy. Soc.,* 1925; (19) Hardy and Hardy, *Phil. Mag.,* 1919; (20) Bowden and Young, *Proc. Roy. Soc.,* 1951; (21) Hardy and Doubleday, *Proc. Roy. Soc.,* 1923; (22) Bowden and Tabor, "The Friction and Lubrication of Solids," Oxford; (23) Shooter, *Research,* 4, 1951.

Table 1.2.5 Coefficient of Kinetic Friction of Hard Steel on Hard Steel

	Surface					
	Superfinished	Ground	Ground	Ground	Ground	Grit-blasted
Roughness, microinches	2	7	20	50	65	55
Mineral oil	0.128	0.189	0.360	0.372	0.378	0.212
Mineral oil + 2% oleic acid	0.116	0.170	0.249	0.261	0.230	0.164
Oleic acid	0.099	0.163	0.195	0.222	0.238	0.195
Mineral oil + 2% sulfonated sperm oil	0.095	0.137	0.175	0.251	0.197	0.165

Asbestos-Fabric Brake Material The coefficient of sliding friction f of asbestos fabric against a cast-iron brake drum, according to Taylor and Holt (*NBS,* 1940) is 0.35 to 0.40 when at normal temperature. It drops somewhat with rise in brake temperature up to 300°F (149°C). With a further increase in brake temperature from 300 to 500°F (149 to 260°C) the value of f may show an increase caused by disruption of the brake surface.

Steel Tires on Steel Rails (Galton)

Speed, mi/h	Start	6.8	13.5	27.3	40.9	54.4	60
Values of f	0.242	0.088	0.072	0.07	0.057	0.038	0.027

Railway Brake Shoes on Steel Tires Galton and Westinghouse give, for cast-iron brakes, the following values for f, which decrease rapidly with the speed of the rim; the coefficient f decreases also with time, as the temperature of the shoe increases.

Speed, mi/h	10	20	30	40	50	60
f, when brakes were applied	0.32	0.21	0.18	0.13	0.10	0.06
f, after 5 s	0.21	0.17	0.11	0.10	0.07	0.05
f, after 12 s		0.13	0.10	0.08	0.06	0.05

Schmidt and Schrader confirm the marked decrease in the coefficient of friction with the increase of rim speed. They also show an irregular slight decrease in the value of f with higher shoe pressure on the wheel, but they did not find the drop in friction after a prolonged application of the brakes. Their observations are as follows:

Speed, mi/h	20	30	40	50	60
Coefficient of friction	0.25	0.23	0.19	0.17	0.16

Friction of Steel on Polymers A useful list of friction coefficients between steel and various polymers is given in Table 1.2.6.

Grindstones The coefficient of kinetic friction between coarse-grained sandstone and cast iron is $f = 0.21$ to 0.24; for steel, 0.29; for wrought iron, 0.41 to 0.46, according as the stone is freshly trued or dull; for fine-grained sandstone (wet grinding) $f = 0.72$ for cast iron, 0.94 for steel, and 1.0 for wrought iron.

Honda and Yamada give $f = 0.28$ to 0.50 for carbon steel on emery, depending on the roughness of the wheel.

Rubber Tires on Pavement Arnoux gives $f = 0.67$ for dry macadam, 0.71 for dry asphalt, and 0.17 to 0.06 for soft, slippery roads. For a cord tire on a sand-filled brick surface in fair condition. Agg (*Bull.* 88, *Iowa State College Engineering Experiment Station,* 1928) gives the following values of f depending on the inflation of the tire:

Inflation pressure, lb/in^2	Dry pavement		Wet pavement	
	Static f_0	Sliding f	Static f_0	Sliding f
40	0.90	0.85	0.74	0.69
50	0.88	0.84	0.64	0.58
60	0.80	0.76	0.63	0.56

Table 1.2.6 Coefficient of Kinetic Friction of Steel on Polymers
Room temperature, low speeds.

Material	Surface Condition	f
Nylon	Dry	0.4
Nylon	Wet with water	0.15
Plexiglas	Dry	0.5
Polyvinyl chloride (PVC)	Dry	0.5
Polystyrene	Dry	0.5
Low-density (LD) polyethylene, no plasticizer	Dry	0.4
LD polyethylene, no plasticizer	Wet	0.1
High-density (HD) polyethylene, no plasticizer	Dry or wet	0.15
Soft wood	Natural	0.25
Lignum vitae	Natural	0.1
PTFE, low speed	Dry or wet	0.06
PTFE, high speed	Dry or wet	0.3
Filled PTFE (15% glass fiber)	Dry	0.12
Filled PTFE (15% graphite)	Dry	0.09
Filled PTFE (60% bronze)	Dry	0.09
Polyurethane rubber	Dry	1.6
Isoprene rubber	Dry	3–10
Isoprene rubber	Wet (water and alcohol)	2–4

Tests of the Goodrich Company on wet brick pavement with tires of different treads gave the following values of f:

	Coefficients of friction			
	Static (before slipping)		Sliding (after slipping)	
Speed, mi/h	5	30	5	30
Smooth tire	0.49	0.28	0.43	0.26
Circumferential grooves	0.58	0.42	0.52	0.36
Angular grooves at 60°	0.75	0.55	0.70	0.39
Angular grooves at 45°	0.77	0.55	0.68	0.44

Development continues using various manufacturing techniques (bias ply, belted, radial, studs), tread patterns, and rubber compounds, so that it is not possible to present average values applicable to present conditions.

Sleds For **unshod wooden runners** on smooth wood or stone surfaces, $f = 0.07$ (0.15) when tallow (dry soap) is used as a lubricant (= 0.38 when not lubricated); on snow and ice, $f = 0.035$. For **runners with metal shoes** on snow and ice, $f = 0.02$. Rennie found for steel on ice, $f = 0.014$. However, as the temperature falls, the coefficient of friction will get larger. Bowden cites the following data for brass on ice:

Temperature, °C	f
0	0.025
−20	0.085
−40	0.115
−60	0.14

ROLLING FRICTION

Rolling is substituted frequently for sliding friction, as in the case of wheels under vehicles, balls or rollers in bearings, rollers under skids when moving loads; frictional resistance to the rolling motion is substantially smaller than to sliding motion. The fact that a resistance arises to rolling motion is due to several factors: (1) the contacting surfaces are elastically deflected, so that, on the finite size of the contact region, relative sliding occurs, (2) the deflected surfaces dissipate energy due to internal friction (hysteresis), (3) the surfaces are imperfect so that contact takes place on asperities ahead of the line of centers, and (4) surface adhesion phenomena. The **coefficient of rolling friction** $f_r = P/L$ where L is the vertical load and P is the frictional resistance.

The frictional resistance P to the rolling of a cylinder under a load L applied at the center of the roller (Fig. 1.2.5) is inversely proportional to the radius r of the roller; $P = (k/r)L$. Note that k has the dimension of length and one may call it the **rolling resistance coefficient.** Quite often k increases with load, particularly for cases involving plastic deformations. Values of k, in inches, are as follows: hardwood on hardwood, 0.02; iron on iron, steel on steel, 0.002; hard polished steel on hard polished steel, 0.0002 to 0.0004.

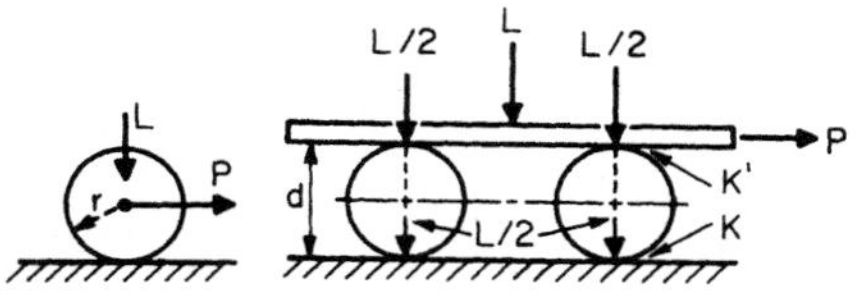

Figure 1.2.5 Rolling resistance of cylinder under vertical load.

Data on rolling friction are scarce. Noonan and Strange give, for steel rollers on steel plates and for loads varying from light to those causing a permanent set of the material, the following values of k in inches: surfaces well finished and clean, 0.0005 to 0.001; surfaces well oiled, 0.001 to 0.002; surfaces covered with silt, 0.003 to 0.005; surfaces rusty, 0.005 to 0.01.

If a load L is moved on rollers (Fig. 1.2.5) and if k and k' are the respective coefficients of friction for the lower and upper surfaces, the frictional force $P = (k + k')L/d$.

McKibben and Davidson (*Agri. Eng.*, 1939) give the data in Table 1.2.7 on the rolling resistance of various types of wheels for typical road and field conditions. Note that the coefficient f_r is the ratio of resistance force to load.

Moyer found the following average values of f_r for pneumatic rubber tires properly inflated and loaded: hard road, 0.008; dry, firm, and wellpacked gravel, 0.012; wet loose gravel, 0.06.

FRICTION OF MACHINE ELEMENTS

Work of Friction—Efficiency In a simple machine or assemblage of two elements, the work done by an applied force P acting through the distance s is measured by the product Ps. The useful work is less and is measured by the product of Ll where L is the resistance and l is the distance through which it acts. The **efficiency** e of the machine is the quotient of the useful work performed to the total work received, or $e = Ll/Ps$. The **work expended in friction** W_f is the difference between the total work and the useful work, or $W_f = Ps - Ll$. The lost-work ratio = $V = W_f/Ll$, and $e = 1/(1 + V)$.

If a machine consists of a train of mechanisms having the respective efficiencies $e_1, e_2, e_3 \ldots e_n$, the combined efficiency of the machine is equal to the product of these efficiencies.

Efficiencies of Machines and Machine Elements The values for machine elements in Table 1.2.8 are from "Elements of Machine Design," by Kimball and Barr. Those for machines are from Goodman's "Mechanics Applied to Engineering." The given quantities are in percentage efficiencies.

Wedges

Sliding in V Guides If a wedge-shaped slide having an angle $2b$ is pressed into a V guide by a force P (Fig. 1.2.6), the total force normal to the wedge faces will be $N = P/\sin b$. A friction force F, opposing motion along the longitudinal axis of the wedge, arises by virtue of the coefficient of friction f between the contacting surface of the wedge and guides longitudinally: $F = fN = fP/\sin b$. This friction force is along the V guide or perpendicular to Fig. 1.2.6 and cannot be shown in this

Table 1.2.7 Coefficients of Rolling Friction f_r for Wheels with Steel and Pneumatic Tires

Wheel	Inflation press, lb/in²	Load, lb	Concrete	Bluegrass sod	Tilled loam	Loose sand	Loose snow 10–14 in deep
2.5 × 36 steel		1,000	0.010	0.087	0.384	0.431	0.106
4 × 24 steel		500	0.034	0.082	0.468	0.504	0.282
4.00–18 4-ply	20	500	0.034	0.058	0.366	0.392	0.210
4 × 36 steel		1,000	0.019	0.074	0.367	0.413	
4.00–30 4-ply	36	1,000	0.018	0.057	0.322	0.319	
4.00–36 4-ply	36	1,000	0.017	0.050	0.294	0.277	
5.00–16 4-ply	32	1,000	0.031	0.062	0.388	0.460	
6 × 28 steel		1,000	0.023	0.094	0.368	0.477	0.156
6.00–16 4-ply	20	1,000	0.027	0.060	0.319	0.338	0.146
6.00–164-ply*	30	1,000	0.031	0.070	0.401	0.387	
7.50–10 4-ply†	20	1,000	0.029	0.061	0.379	0.429	
7.50–16 4-ply	20	1,500	0.023	0.055	0.280	0.322	
7.50–28 4-ply	16	1,500	0.026	0.052	0.197	0.205	
8 × 48 steel		1,500	0.013	0.065	0.236	0.264	0.118
7.50–36 4-ply	16	1,500	0.018	0.046	0.185	0.177	0.0753
9.00–10 4-ply†	20	1,000	0.031	0.060	0.331	0.388	
9.00–16 6-ply	16	1,500	0.042	0.054	0.249	0.272	0.099

* Skid-ring tractor tire.
† Ribbed tread tractor tire.
All other pneumatic tires with implement-type tread.

Table 1.2.8 Efficiencies of Machines and Machine Elements

Common bearing (singly)	96–98
Common bearing, long lines of shafting	95
Roller bearings	98
Ball bearings	99
Spur gear, including bearings	
Cast teeth	93
Cut teeth	96
Bevel gear, including bearings	
Cast teeth	92
Cut teeth	95
Worm gear	
Thread angle, 30°	85–95
Thread angle, 15°	75–90
Belting	96–98
Pin-connected chains (bicycle)	95–97
High-grade transmission chains	97–99
Weston pulley block (½ ton)	30–47
Epicycloidal pulley block	40–45
1-ton steam hoist or windlass	50–70
Hydraulic windlass	60–80
Hydraulic jack	80–90
Cranes (steam)	60–70
Overhead traveling cranes	30–50
Locomotives (drawbar hp/ihp)	65–75
Hydraulic couplings, max	98

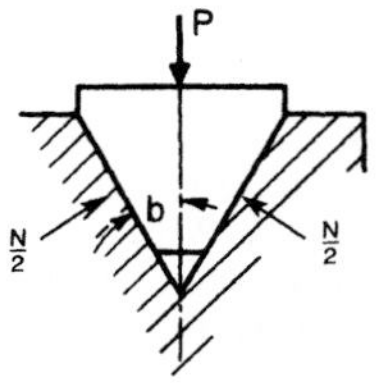

Figure 1.2.6 Wedge-shaped slide.

figure. In these formulas, the fact that the elasticity of the materials permits an advance of the wedge into the guide under the load P has been neglected. If one considers the upward friction force along the side of the V grove, the normal force can be obtained by $N = P/(\sin b + f \cos b)$. The common efficiency for V guides is $e = 0.88$ to 0.90.

Taper Keys In Fig. 1.2.7 if the key is moved in the direction of the force P, the force H must be overcome. The supporting reactions K_1, K_2, and K_3 together with the required force P may be obtained by drawing the force polygon (Fig. 1.2.8). The friction angles of these faces are a_1, a_2, and a_3, respectively. In Fig. 1.2.8, draw AB parallel to H in Fig. 1.2.7, and lay it off to scale to represent H. From the point A, draw

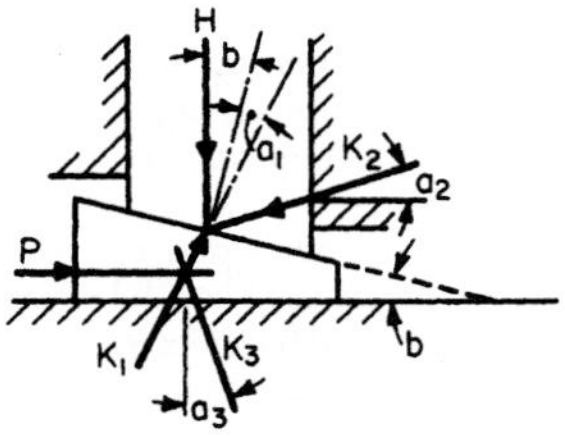

Figure 1.2.7 Free body diagram of Taper Keys.

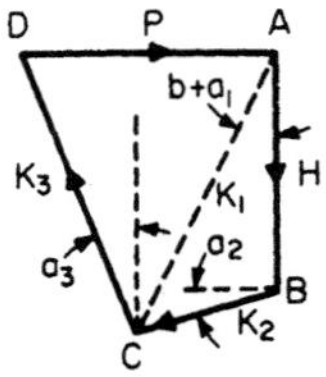

Figure 1.2.8 Force polygon.

AC parallel to K_1, i.e., making the angle $b + a_1$ with AB; from the other extremity of AB, draw BC parallel to K_2 in Fig. 1.2.7. AC and BC then give the magnitudes of K_1 and K_2, respectively. Now through C draw CD parallel to K_3 to its intersection with AD which has been drawn through A parallel to P. The magnitudes of K_3 and P are then given by the lengths of CD and DA.

By calculation,

$$K_1/H = \cos a_2/\cos (b + a_1 + a_2)$$
$$P/K_1 = \sin (b + a_1 + a_3)/\cos a_3$$
$$\text{Thus, } P/H = \cos a_2 \sin (b + a_1 + a_3)/\cos a_3 \cos (b + a_1 + a_2)$$

If $a_1 = a_2 = a_3 = a$, then $P = H \tan (b + 2a)$, and efficiency $e = \tan b/\tan (b + 2a)$. Force required to loosen the key $= P_1 = H \tan (2a - b)$. In order for the key not to slide out when force P is removed, it is necessary that $b < (a_1 + a_3)$, or $b < 2a$.

The forces acting upon the taper key of Fig. 1.2.9 may be found in a similar way (see Fig. 1.2.10).

$$P = 2H \cos a \sin (b + a)/\cos (b + 2a)$$
$$= 2H \tan (b + a)/[1 - \tan a \tan (b + a)]$$
$$= 2H \tan (b + a) \text{ approx}$$

The force to loosen the key is $P_1 = 2H \tan (a - b)$ approx, and the efficiency $e = \tan b/\tan (b + a)$. The key will be self-locking when $b < a$, or, more generally, when $2b < (a_1 + a_3)$.

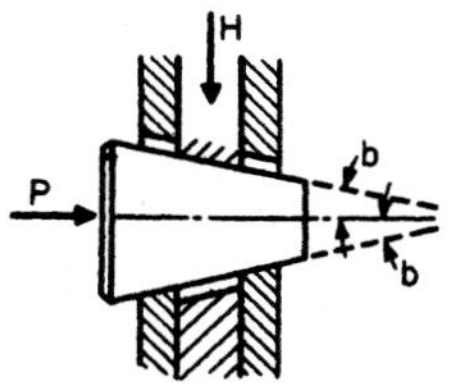

Figure 1.2.9 Free body diagram of Taper Keys.

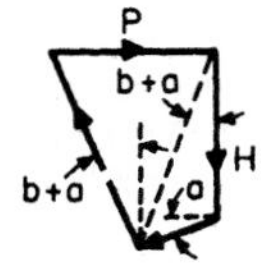

Figure 1.2.10 Force polygon.

Screws

Screws with Square Threads Let r = mean radius of the thread = ½ (radius at root + outside radius), and l = pitch (or lead of a single-threaded screw), both in inches; b = angle of inclination of thread to a plane at right angles to the axis of screw ($\tan b = l/2\pi r$); and f = coefficient of sliding friction = $\tan a$. Note that the friction of the root and outside surfaces are neglected. Then for a screw in uniform motion, there is required a force P acting at right angles to the axis at the distance r. $P = L \tan (b \pm a) = L(l \pm 2\pi rf)/(2\pi r \pm fl)$, where the upper signs are for motion in a direction opposed to that of L and the lower signs for motion in the same direction as that of L. When $b \leq a$, the screw will not "overhaul" or move under the action of the load L.

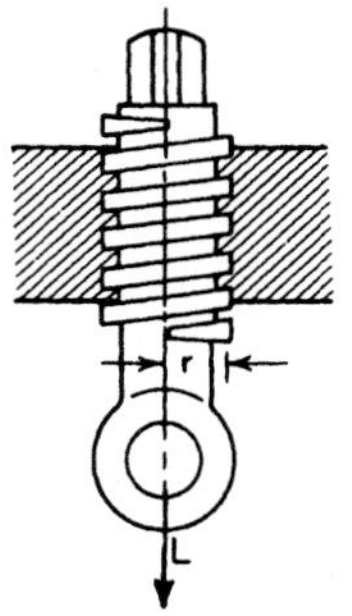

Figure 1.2.11 Screws with Square Threads.

The **efficiency** for motion opposed to direction in which L acts = $e = \tan b/\tan(b + a)$; for motion in the same direction in which L acts, $e = \tan (b - a)/\tan b$.

The value of e is a maximum when $b = 45° - ½a$; e.g., $e_{max} = 0.81$ for $b = 42°$ and $f = 0.1$. Since e increases rapidly for values of b up to 20°, this angle is generally not exceeded; for $b = 20°$, and $f_1 = 0.10$, $e = 0.74$. In presses, where the mechanical advantage is required to be great, b is taken down to 3°, for which value $e = 0.34$ with $f = 0.10$.

Kingsbury found for square-threaded screws running in loose-fitting nuts, the following coefficients of friction: lard oil, 0.09 to 0.25; heavy mineral oil, 0.11 to 0.19; heavy oil with graphite, 0.03 to 0.15.

Ham and Ryan give for screws the following values of coefficients of friction, with medium mineral oil: high-grade materials and workmanship, 0.10; average quality materials and workmanship, 0.12; poor workmanship, 0.15. The use of castor oil as a lubricant lowered f from 0.10 to 0.066. The coefficients of static friction (at starting) were 30 percent higher. Table 1.2.8 gives representative values of efficiency.

Screws with V-Threads (Fig. 1.2.12) Let c = half the angle between the faces of a thread. Then, using the same notation as for square-threaded screws, for a V-thread screw in motion (neglecting friction of root and outside surfaces),

$$P = L(l \pm 2\pi rf \sec d)/(2\pi r \pm lf \sec d)$$

d is the angle between a plane normal to the axis of the screw through the point of the resultant thread friction, and a plane which is tangent to the surface of the thread at the same point (see Groat, *Proc. Engs. Soc. West. Penn,* **34**). $\sec d = \sec c\sqrt{1-(\sin b \sin c)^2}$. For small values of b this reduces practically to $\sec d = \sec c$, and, for all cases the approximation, $P = L(l \pm 2\pi rf \sec c)/(2\pi r \pm lf \sec c)$ is within the limits of probable error in estimating values to be used for f.

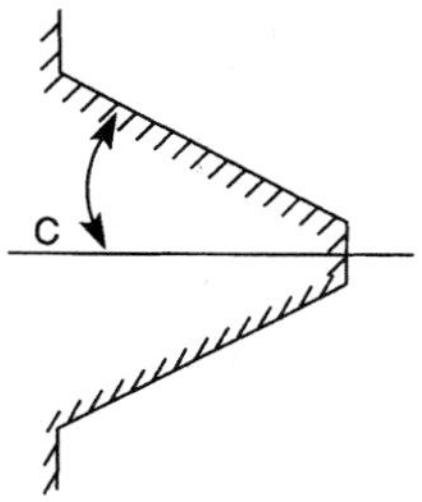

Figure 1.2.12 Screws with V-Threads.

The **efficiencies** are: $e = \tan b(1 - f \tan b \sec d)/(\tan b + f \sec d)$ for motion opposed to L, and $e = (\tan b - f \sec d)/\tan b(1 + f \tan b \sec d)$ for motion with L. If we let $\tan d' = f \sec d$, these equations reduce to $e = \tan b/\tan (b + d')$ and $e = \tan (b - d')/\tan b$, respectively. Negative values in the latter case merely mean that the thread will not overhaul. Subtract the values from unity for actual efficiency, considering the external moment and not the load L as being the driver. The efficiency of a V-thread is lower than that of a square thread of the same helix angle, since $d' > a$.

For a **V-threaded screw and nut,** let r_1 = outside radius of thread, r_2 = radius at root of thread, $r = (r_1 + r_2)/2$, $\tan d' = f \sec d$, r_0 = mean radius of nut seat = $1.5r$ (approx) and f' = coefficient of friction between nut and seat.

To tighten up the nut, the turning moment or required torque is $M = Pr + Lr_0 f = Lr[\tan (d' + b) + 1.5f']$. **To loosen** $M = Lr[\tan (d' - b) + 1.5f']$.

The **total tension in a bolt** due to tightening up with a moment M is $T = 2\pi M/(l + fl \sec b \sec d \operatorname{cosec} b + f'3\pi r)$. $T \div$ area at root gives unit pure tensile stress induced, S_t. There is also a unit torsional stress: $S_s = 2(M - 1.5rf'T)/\pi r_2^3$. The equivalent combined stress is $S = 0.35S_t + 0.65\sqrt{S_t^2 + 4S_s^2}$.

Kingsbury, from tests on U.S. standard bolts, finds efficiencies for tightening up nuts from 0.06 to 0.12, depending upon the roughness of the contact surfaces and the character of the lubrication.

Toothed and Worm Gearing

The efficiency of **spur and bevel gearing** depends on the material and the workmanship of the gears and on the lubricant employed. For highspeed gears of good quality the efficiency of the gear transmission is 99 percent; with slow-speed gears of average workmanship the efficiency of 96 percent is common. On average, the efficiencies of 97 to 98 percent can be considered normal.

In **helical gears,** where considerable transverse sliding of the meshing teeth on each other takes place, the friction is much greater. If b and c are, respectively, the spiral angles of the teeth of the driving and driven helical gears (i.e., the angle between the teeth and the axis of rotation), $b + c$ is the shaft angle of the two gears, and $f = \tan a$ is the coefficient of sliding friction of the teeth, the efficiency of the gear transmission is $e = [\cos b \cos (c + a)]/[\cos c \cos (b - a)]$.

In the case of **worm gearing** when the shafts are normal to each other ($b + c = 90$), the efficiency is $e = \tan c/\tan (c + a) = (1 - pf/2\pi r)/(1 + 2\pi rf/p)$, where c is the spiral angle of the worm wheel, or the lead angle of the worm; p the lead, or pitch of the worm thread; and r the mean radius of the worm. Typical values of f are shown in Table 1.2.9.

Journals and Bearings

Friction of Journal Bearings If P = total load on journal, l = journal length, and $2r$ = journal diameter, then $p = P/2rl$ = mean normal pressure on the projected area of the journal. P is often called journal bearing pressure. If f_1 is the **coefficient of journal friction,** the resistant

Table 1.2.9 Coefficients of Friction for Worm Gears

Rubbing speed of worm, ft/min	100	200	300	500	800	1200
(m/min)	(30.5)	(61)	(91.5)	(152)	(244)	(366)
Phosphor-bronze wheel, polished-steel worm	0.054	0.045	0.039	0.030	0.024	0.020
Single-threaded cast-iron worm and gear	0.060	0.051	0.047	0.034	0.025	

moment of journal friction for a cylindrical journal is $M = f_1Pr$. The **work expended in friction** at angular velocity ω is

$$W_f = \omega M = f_1 Pr\omega$$

For the **conical bearing**, the mean radius $r_m = (r + R)/2$ is to be used as shown in Fig. 1.2.13.

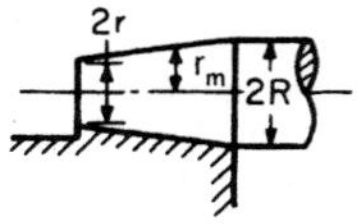

Figure 1.2.13 Conical bearing in contact with surface.

Values of Coefficient of Friction For very low velocities of rotation (e.g., below 10 r/min), high loads, and with good lubrication, the coefficient of friction approaches the value of greasy friction, 0.07 to 0.15 (see Table 1.2.4). This is also the "pullout" coefficient of friction *on starting* the journal. With higher velocities, a fluid film is established between the journal and bearing, and the values of the coefficient of friction depend on the speed of rotation, the pressure on the bearing, and the viscosity of the oil. For journals running in complete bearing bushings, with a small clearance, i.e., with the diameter of the bushing slightly larger than the diameter of the journal, the experimental data of McKee give approximate values of the coefficient of friction as in Fig. 1.2.14.

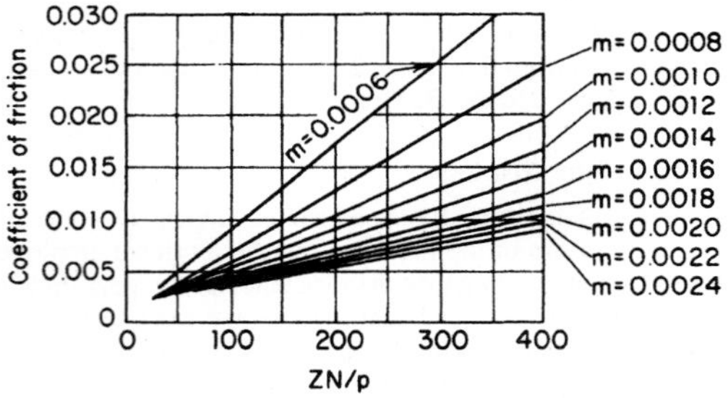

Figure 1.2.14 Coefficient of friction of journal. *Z* is the viscosity of the oil, *N* the speed of the journal, *m* the clearance ratio, and *p* the bearing pressure.

If d_1 is the diameter of the bushing in inches, d the diameter of the journal in inches, then $(d_1 - d)$ is the diametral clearance and $m = (d_1 - d)/d$ is the clearance ratio. The diagram of McKee (Fig. 1.2.14) gives the coefficient of friction as a function of the characteristic number ZN/p, where N is the speed of rotation in revolutions per minute, $p = P/(dl)$ is the average pressure in lb/in^2 on the projected area of the bearing, P is the total load acting on the bearing, l is the axial length of the bearing, and Z is the absolute viscosity of the oil in centipoises. (1 centipoise = 10^{-3} Pa·s) Approximate values of Z at 100°F (130)are as follows: light machine oil, 30 centiposes (16); medium machine oil, 60 (25); medium-heavy machine oil, 120 (40); heavy machine oil, 160 (60).

For purposes of design of ordinary machinery with bearing pressures from 50 to 300 lb/in^2 (344.7 to 2,068 kN/m^2) and speeds of 100 to 3,000 rpm, values for the coefficient of journal friction can be taken from 0.008 to 0.020.

Thrust Bearings

Frictional Resistance for Flat Ring Bearing Step bearings or pivots may be used to resist the end thrust of shafts. Let L = total load in the direction of the shaft axis and f = coefficient of sliding friction.

For a **ring-shaped flat step bearing** such as that shown in Fig. 1.2.15 (or a **collar bearing**), the **moment of thrust friction** $M = \frac{1}{3} fL(D^3 - d^3)/(D^2 - d^2)$. For a **flat circular step bearing,** $d = 0$, and $M = \frac{1}{3} fLD$.

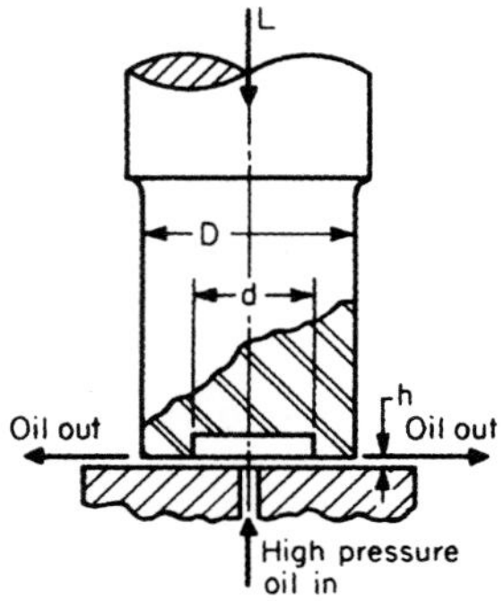

Figure 1.2.15 Ring-shaped flat step bearing.

The value of the coefficient of sliding friction is 0.08 to 0.15 when the speed of rotation is very slow. At higher velocities when a collar or step bearing is used, f = 0.04 to 0.06. If the design provides for the formation of a load carrying oil film, as in the case of the Kingsbury thrust bearing, the coefficient of friction has values f = 0.001 to 0.0025.

Where oil is supplied from an external pump with such pressure as to separate the surfaces and provide an oil film of thickness h (Fig. 1.2.15), the frictional moment is

$$M = \frac{Zn(D^4 - d^4)}{67 \times 10^7\, h} = \frac{\pi\mu\omega(D^4 - d^4)}{32\, h}$$

where D and d are in inches, μ is the absolute viscosity, ω is the angular velocity, h is the film thickness, in, Z is viscosity of lubricant in centipoises, and n is rotation speed, r/min. With this kind of lubrication the frictional moment depends upon the speed of rotation of the shaft and actually approaches zero for zero shaft speeds. The thrust load will be carried on a film of oil regardless of shaft rotation for as long as the pump continues to supply the required volume and pressure (see also Secs. 6 and 10).

EXAMPLE. A hydrostatic thrust bearing carries 101,000 lb, D is 16 in, d is 10 in, oil-film thickness h is 0.006 in, oil viscosity Z, 30 centipoises at operating temperature, and n is 750 r/min. Substituting these values, the frictional torque M is 310 in·lb (358 cm·kg). The oil supply pressure was 82.5 lb/in^2 (569 kN/m^2); the oil flow, 12.2 gal/min (46.2 l/min).

Frictional Forces in Pin Joints of Mechanisms

In the absence of friction, or when the effect of friction is negligible, the force transmitted by the link b from the driver a to the driven link c (Figs. 1.2.16 and 1.2.17) acts through the centerline OO of the pins connecting the link b with links a and c. With friction, this line of action shifts to the line AA, tangent to small circles of diameter d. For

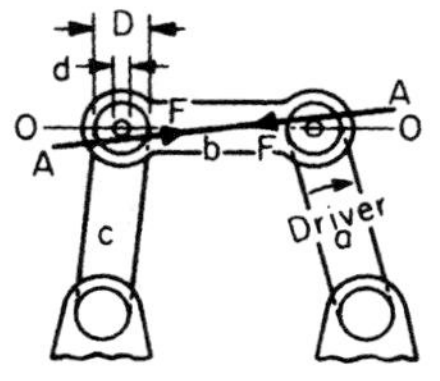

Figure 1.2.16 Frictional force in linkage, driver and driven on same side.

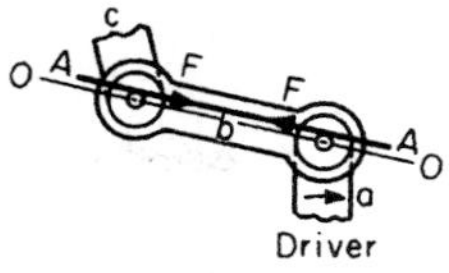

Figure 1.2.17 Driver and driven on opposite side.

each individual joint, the circle with diameter d is called the **friction circle** that is equal to fD, where D is the diameter of the pin and f is the coefficient of friction between the pin and the link. The choice of the proper disposition of the tangent AA with respect to the two friction circles is dictated by the consideration that friction always opposes the action of the linkage. The force f opposes the motion of a; therefore, with friction it acts on a longer lever than without friction (Figs. 1.2.16 and 1.2.17). On the other hand, the force F drives the link c; friction hinders its action, and the equivalent lever is shorter with friction than without friction; the friction throws the line of action toward the center of rotation of link c.

EXAMPLE. An engine eccentric (Fig. 1.2.18) is a joint where the friction loss may be large. For the dimensions shown and with a torque of 250 in · lb applied to

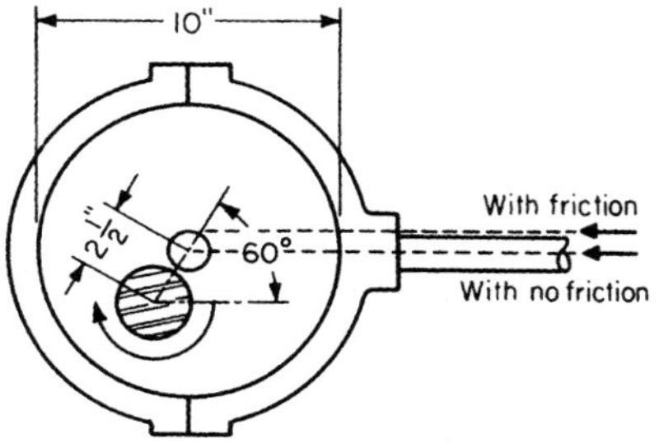

Figure 1.2.18 Example of an engine eccentric.

the rotating shaft, the resultant horizontal force, with no friction, will act through the center of the eccentric and be 250/(2.5 sin 60) or 115.5 lb. With friction coefficient 0.1, the resultant force (which for a long rod remains approximately horizontal) will be tangent to the friction circle of radius 0.1 × 5, or 0.5 in, and have a magnitude of 250/(2.5 sin 60 + 0.5), or 93.8 lb (42.6 kg).

Tension Elements

Frictional Resistance In Fig. 1.2.19, let T_1 and T_2 be the tensions with which a rope, belt, chain, or brake band is strained over a drum, pulley, or sheave, and let the rope or belt be on the point of slipping from T_2 toward T_1 by reason of the difference of tension $T_1 - T_2$. Then $T_1 - T_2$ = circumferential force P transferred by friction must be equal to the frictional resistance W of the belt, rope, or band on the drum or

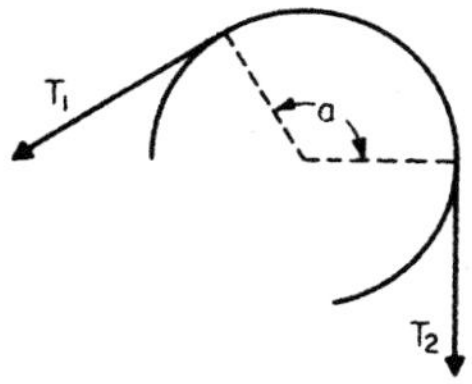

Figure 1.2.19 Tension forces on a drum or pulley.

pulley. Also, let a = angle subtending the arc of contact between the drum and tension element. Then, disregarding centrifugal forces,

$$T_1 = T_2 e^{fa} \text{ and } P = (e^{fa} - 1)T_1/e^{fa} = (e^{fa} - 1)T_2 = W$$

where e = base of the napierian system of logarithms = 2.178 +.

f is the static coefficient of friction (f_0) when there is no slip of the belt or band on the drum and the coefficient of kinetic friction (f) when slip takes place. For ease of computation, the values of the quantity e^{fa} are tabulated on Table 1.2.10.

Average values of f_0 for belts, ropes, and brake bands are as follows: for leather belt on cast-iron pulley, very greasy, 0.12; slightly greasy,

Table 1.2.10 Values of e^{fa}

$\frac{a°}{360°}$	f = 0.1	0.15	0.2	0.25	0.3	0.35	0.4	0.45	0.5
0.1	1.06	1.1	1.13	1.17	1.21	1.25	1.29	1.33	1.37
0.2	1.13	1.21	1.29	1.37	1.46	1.55	1.65	1.76	1.87
0.3	1.21	1.32	1.45	1.60	1.76	1.93	2.13	2.34	2.57
0.4	1.29	1.46	1.65	1.87	2.12	2.41	2.73	3.10	3.51
0.425	1.31	1.49	1.70	1.95	2.23	2.55	2.91	3.33	3.80
0.45	1.33	1.53	1.76	2.03	2.34	2.69	3.10	3.57	4.11
0.475	1.35	1.56	1.82	2.11	2.45	2.84	3.30	3.83	4.45
0.5	1.37	1.60	1.87	2.19	2.57	3.00	3.51	4.11	4.81
0.525	1.39	1.64	1.93	2.28	2.69	3.17	3.74	4.41	5.20
0.55	1.41	1.68	2.00	2.37	2.82	3.35	3.98	4.74	5.63
0.6	1.46	1.76	2.13	2.57	3.10	3.74	4.52	5.45	6.59
0.7	1.55	1.93	2.41	3.00	3.74	4.66	5.81	7.24	9.02
0.8	1.65	2.13	2.73	3.51	4.52	5.81	7.47	9.60	12.35
0.9	1.76	2.34	3.10	4.11	5.45	7.24	9.60	12.74	16.90
1.0	1.87	2.57	3.51	4.81	6.59	9.02	12.35	16.90	23.14
1.5	2.57	4.11	6.59	10.55	16.90	27.08	43.38	69.49	111.32
2.0	3.51	6.59	12.35	23.14	43.38	81.31	152.40	285.68	535.49
2.5	4.81	10.55	23.14	50.75	111.32	244.15	535.49	1,174.5	2,575.9
3.0	6.59	16.90	43.38	111.32	285.68	733.14	1,881.5	4,828.5	12,391
3.5	9.02	27.08	81.31	244.15	733.14	2,199.90	6,610.7	19,851	59,608
4.0	12.35	43.38	152.40	535.49	1,881.5	6,610.7	23,227	81,610	286,744

NOTE: e^{π} = 23.1407, log e^{π} = 1.3643764.

0.28; moist, 0.38. For hemp rope on cast-iron drum, 0.25; on wooden drum, 0.40; on rough wood, 0.50; on polished wood, 0.33. For iron brake bands on cast-iron pulleys, 0.18. For wire ropes, Tichvinsky reports coefficients of static friction, f_0, for a ⅝ rope (8 × 19) on a worn-in cast-iron groove: 0.113 (dry); for mylar on aluminum, 0.4 to 0.7.

Belt Transmissions; Effects of Belt Compliance

In the configuration of Fig. 1.2.20, pulley A drives a belt at angular velocity ω_A. Pulley B, here assumed to be of the same radius R as A, is driven at angular velocity ω_B. If the belt is extensible and the resistive torque $M = (T_1 - T_2)R$ is applied at B, ω_B will be smaller than ω_A and power will be dissipated at a rate $W = M(\omega_A - \omega_B)$. Likewise, the surface velocity V_1 of the more stretched belt will be larger than V_2. No slip will take place over the wraps A_T-A_S and B_T-B_S where they are called no-slip or stick zones. The slip zones or angles a_A and a_B can be calculated from

$$a_A = [\ln (T_1/T_2)]/f_A \qquad a_B = [\ln (T_1/T_2)]/f_B$$

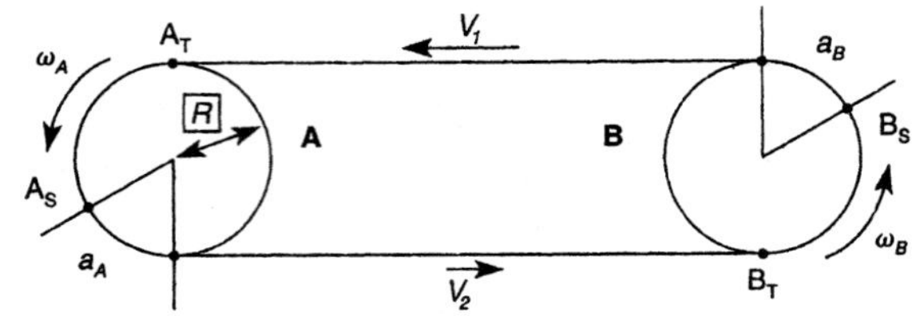

Figure 1.2.20 Pulley transmission with extensible belt. Pulley A is driver, a_A and a_B are slip zones, A_T-A_S and B_T-B_S are stick zones.

where f_A and f_B are the coefficients of friction on pulleys A and B, respectively. To calculate the above values, it is necessary to know the mean tension of the belt, $T = (T_1 + T_2)/2$. Then, $T_1/T_2 = [T + M/(2R)]/[T - M/(2R)]$. In this configuration, when the slip angles become equal to π (180°), complete slip occurs.

It is interesting to note that torque is transmitted only over the slip arcs a_A and a_B since there is no tension variation in the arcs A_T-A_S and B_T-B_S where the belt is in a uniform state of stretch.

1.3 MECHANICS OF FLUIDS

by J. W. Murdock, Reviewed by Yunus A. Çengel

REFERENCES: *Specific.* "Handbook of Chemistry and Physics," Chemical Rubber Company. "Smithsonian Physical Tables," Smithsonian Institution. "Petroleum Measurement Tables," ASTM. "Steam Tables," ASME. "American Institute of Physics Handbook," McGraw-Hill. "International Critical Tables," McGraw-Hill. "Tables of Thermal Properties of Gases," *NBS Circular 564.* Murdock, "Fluid Mechanics and Its Applications," Houghton Mifflin, 1976. "Pipe Friction Manual," Hydraulic Institute. "Flow of Fluids," ASME, 1971. "Fluid Meters," 6th ed., ASME, 1971. "Measurement of Fluid Flow in Pipes Using Orifice, Nozzle, and Venturi," ASME Standard MFC-3M-1984. Murdock, ASME 64-WA/FM-6. Horton, *Engineering News,* 75, 373, 1916. Belvins, ASME 72/WA/FE-39. Staley and Graven, ASME 72PET/30. "Temperature Measurement," PTC 19.3, ASME. Moody, *Trans. ASME,* 1944, pp. 671–684. *General.* Binder, "Fluid Mechanics," Prentice-Hall. Langhaar, "Dimensional Analysis and Theory of Models," Wiley. Murdock, "Fluid Mechanics," Drexel University Press. Rouse, "Elementary Mechanics of Fluids," Wiley. Shames, "Mechanics of Fluids," McGraw-Hill. Streeter, "Fluid Mechanics," McGraw-Hill.

Notation

a = acceleration, area, exponent
A = area
c = velocity of sound
C = coefficient
C = Cauchy number
$\mathbf{C}_p$ = pressure coefficient
d = diameter, distance
E = bulk modulus of elasticity, modulus of elasticity (Young's modulus), velocity of approach factor, specific energy
E = Euler number
f = frequency, friction factor
F = dimension of force, force
F = Froude number
g = acceleration due to gravity
g_c = proportionality constant = 32.1740 lbm/(lbf) (ft/s²)
G = mass velocity
h = head, vertical distance below a liquid surface
H = geopotential altitude
i = ideal
I = moment of inertia
J = mechanical equivalent of heat, 778.169 ft · lbf
k = isentropic exponent, ratio of specific heats
K = constant, resistant coefficient, weir coefficient
K = flow coefficient
L = dimension of length, length
m = mass, lbm
$\dot{m}$ = mass rate of flow, lbm/s
M = dimension of mass, mass (slugs)
$\dot{M}$ = mass rate of flow, slugs/s
M = Mach number
n = exponent for a polytropic process, roughness factor
N = dimensionless number
p = pressure
P = perimeter, power
q = heat added
q = flow rate per unit width
Q = volumetric flow rate
r = pressure ratio, radius
R = gas constant, reactive force
R = Reynolds number
R_h = hydraulic radius
s = distance, second
sp. gr. = specific gravity
S = scale reading, slope of a channel
S = Strouhal number
t = time
T = dimension of time, absolute temperature
u = internal energy
U = stream-tube velocity
v = specific volume
V = one-dimensional velocity, volume
V = velocity ratio
W = work done by fluid
W = Weber number
x = abscissa
y = ordinate
Y = expansion factor
z = height above a datum
Z = compressibility factor, crest height
α = angle, kinetic energy correction factor
β = ratio of primary element diameter to pipe diameter
γ = specific weight

δ = boundary-layer thickness
ε = absolute surface roughness
θ = angle
μ = dynamic viscosity
ν = kinematic viscosity
π = 3.14159 . . . , dimensionless ratio
ρ = density
σ = surface tension
τ = unit shear stress
ω = rotational speed

FLUIDS AND OTHER SUBSTANCES

Substances may be classified by their response when at rest to the imposition of a shear force. Consider the two very large plates, one moving, the other stationary, separated by a small distance y as shown in Fig. 1.3.1. The space between these plates is filled with a substance whose surfaces adhere to these plates in such a manner that its upper surface moves at the same velocity as the upper plate and the lower surface is stationary. The upper surface of the substance attains a velocity of U as the result of the application of shear force F_s. As y approaches dy, U approaches dU, and the rate of deformation of the substance becomes dU/dy. The **unit shear stress** is defined by $\tau = F_s/A_s$, where A_s is the shear or surface area. The deformation characteristics of various substances are shown in Fig. 1.3.2.

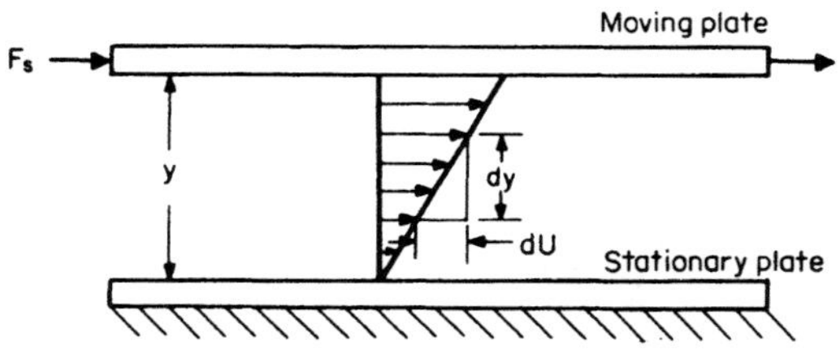

Figure 1.3.1 Flow of a substance between parallel plates.

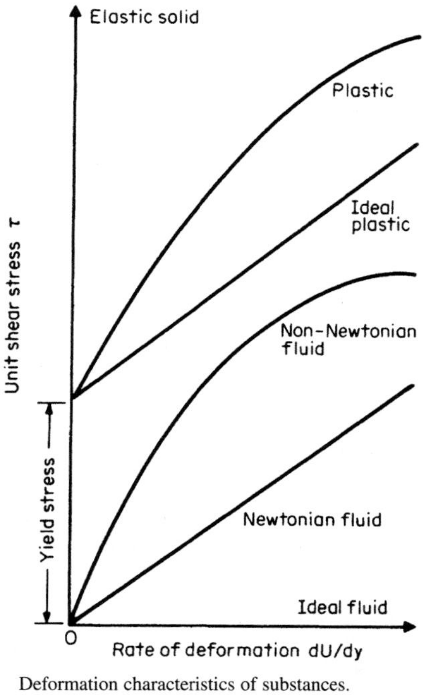

Figure 1.3.2 Deformation characteristics of substances.

An ideal or **elastic solid** will resist the shear force, and its rate of deformation will be zero regardless of loading and hence is coincident with the ordinate of Fig. 1.3.2. A **plastic** will resist the shear until its yield stress is attained, and the application of additional loading will cause it to deform continuously, or flow. If the deformation rate is directly proportional to the flow, it is called an **ideal plastic.**

If the **substance** is unable to resist even the slightest amount of shear without flowing, it is a **fluid.** An **ideal fluid** has no internal friction, and hence its deformation rate coincides with the abscissa of Fig. 1.3.2. All real fluids have internal friction so that their rate of deformation is proportional to the applied shear stress. If it is directly proportional, it is called a **Newtonian fluid;** if not, a **non-Newtonian fluid.**

Two kinds of **fluids** are considered in this section, incompressible and compressible. A **liquid** except at very high pressures and/or temperatures may be considered incompressible. **Gases** and **vapors** are compressible fluids, but only **ideal gases** (those that follow the ideal-gas laws) are considered in this section. All others are covered in Secs. 2.1 and 2.2.

FLUID PROPERTIES

The **density** ρ of a fluid is its mass per unit volume. Its dimensions are M/L^3. In **fluid mechanics,** the units are slugs/ft^3 and lbf·s^2/ft^4 (515.3788 kg/m^3), but in **thermodynamics** (Sec. 2.1), the units are lbm/ft^3 (16.01846 kg/m^3). Numerical values of densities for selected liquids are shown in Table 1.3.1. The temperature change at 68°F (20°C) required to produce a 1 percent change in density varies from 12°F (6.7°C) for kerosene to 99°F (55°C) for mercury.

The **specific volume** v of a fluid is its volume per unit mass. Its dimensions are L^3/M. The units are ft^3/lbm. Specific volume is related to density by $v = 1/\rho g_c$, where g_c is the proportionality constant [32.1740 (lbm/lbf)(ft/s^2)]. **Specific volumes** of ideal gases may be computed from the equation of state: $v = RT/p$, where R is the gas constant in ft·lbf/(lbm)(°R) (see Sec. 2.1), T is the temperature in degrees Rankine (°F + 459.67), and p is the pressure in lbf/ft^2 abs.

The **specific weight** γ of a fluid is its weight per unit volume and has dimensions of F/L^3 or $M/(L^2)(T^2)$. The units are lbf/ft^3 or slugs/(ft^2)(s^2) (157.087 N/m^3). Specific weight is related to density by $\gamma = \rho g$, where g is the acceleration of gravity.

The **specific gravity** (sp. gr.) of a substance is a dimensionless ratio of the density of a fluid to that of a reference fluid. Water is used as the reference fluid for solids and liquids, and air is used for gases. Since the density of liquids changes with temperature for a precise definition of **specific gravity,** the temperature of the fluid and the reference fluid should be stated, for example, 60/60°F, where the upper temperature pertains to the liquid and the lower to water. If no temperatures are stated, reference is made to water at its maximum density, which occurs at 3.98°C and atmospheric pressure. The maximum density of water is 1.9403 slugs/ft^3 (999.973 kg/m^3). For gases, it is common practice to use the ratio of the molecular weight of the gas to that of air (28.9644), thus eliminating the necessity of stating the pressure and temperature for ideal gases.

The **bulk modulus of elasticity** E of a fluid is the ratio of the pressure stress to the volumetric strain. Its dimensions are F/L^2. The units are lbf/in^2 or lbf/ft^2. E depends upon the thermodynamic process causing the change of state so that $E_x = -v(\partial p/\partial v)_x$, where x is the process. For ideal gases, $E_T = p$ for an isothermal process and $E_s = kp$ for an isentropic process where k is the ratio of specific heats. Values of E_T and E_s for liquids are given in Table 1.3.2. For liquids, a mean value is used by integrating the equation over a finite interval, or $E_{xm} = -v_1(\Delta p/\Delta v)_x = v_1(p_2 - p_1)/(v_1 - v_2)_x$.

EXAMPLE. What pressure must be applied to ethyl alcohol at 68°F (20°C) to produce a 1 percent decrease in volume at constant temperature?

$$\Delta p = -E_T(\Delta v/v) = -(130{,}000)(-0.01)$$
$$= 1{,}300 \text{ lbf/in}^2 \ (9\times 10^6 \text{ N/m}^2)$$

In a like manner, the pressure required to produce a 1 percent decrease in the volume of mercury is found to be 35,900 lbf/in^2 (248 × 10^6 N/m^2). For most engineering purposes, liquids may be considered as incompressible fluids.

Table 1.3.1 Density of Liquids at Atmospheric Pressure

Temp: °C / °F	0 / 32	20 / 68	40 / 104	60 / 140	80 / 176	100 / 212
Liquid	ρ, slugs/ft³ (515.4 kg/m³)					
Alcohol, ethyl[f]	1.564	1.532	1.498	1.463		
Benzene[a,b]	1.746	1.705	1.663	1.621	1.579	
Carbon tetrachloride[a,b]	3.168	3.093	3.017	2.940	2.857	
Gasoline,[c] sp. gr. 0.68	1.345	1.310	1.275	1.239		
Glycerin[a,b]	2.472	2.447	2.423	2.398	2.372	2.346
Kerosene,[c] sp. gr. 0.81	1.630	1.564	1.536	1.508	1.480	
Mercury[b]	26.379	26.283	26.188	26.094	26.000	25.906
Oil, machine,[c] sp. gr. 0.907	1.778	1.752	1.727	1.702	1.677	1.651
Water, fresh[d]	1.940	1.937	1.925	1.908	1.885	1.859
Water, salt[e]	1.995	1.988	1.975			

SOURCES: Computed from data given in:
[a] "Handbook of Chemistry and Physics," 52nd ed., Chemical Rubber Company, 1971–1972.
[b] "Smithsonian Physical Tables," 9th rev. ed., 1954.
[c] ASTM-IP, "Petroleum Measurement Tables."
[d] "Steam Tables," ASME, 1967.
[e] "American Institute of Physics Handbook," 3rd ed., McGraw-Hill, 1972.
[f] "International Critical Tables," McGraw-Hill.

Table 1.3.2 Bulk Modulus of Elasticity, Ratio of Specific Heats of Liquids and Velocity of Sound at One Atmosphere and 68°F (20°C)

	E in lbf/in² (6,895 N/m²)			
Liquid	Isothermal E_T	Isentropic E_s	$k = c_p/c_v$	c in ft/s (0.3048 m/s)
Alcohol, ethyl[a,e]	130,000	155,000	1.19	3,810
Benzene[a,c,d,f]	154,000	223,000	1.45	4,340
Carbon tetrachloride[a,b]	139,000	204,000	1.47	3,080
Glycerin[f]	654,000	719,000	1.10	6,510
Kerosene,[a,c,d,e] sp. gr. 0.81	188,000	209,000	1.11	4,390
Mercury[e]	3,590,000	4,150,000	1.16	4,770
Oil, machine,[f] sp. gr. 0.907	189,000	219,000	1.13	4,240
Water, fresh[a]	316,000	319,000	1.01	4,860
Water, salt[a,e]	339,000	344,000	1.01	4,990

SOURCES: Computed from data given in:
[a] "Handbook of Chemistry and Physics," 52nd ed., Chemical Rubber Company, 1971–1972.
[b] "Smithsonian Physical Tables," 9th rev. ed., 1954.
[c] ASTM-D1250, "Petroleum Measurement Tables."
[d] "Steam Tables," ASME, 1967.
[e] "American Institute of Physics Handbook," 3rd ed., McGraw-Hill, 1972.
[f] "International Critical Tables," McGraw-Hill.

The **acoustic velocity,** or velocity of sound in a fluid, is given by $c = \sqrt{E_s/\rho}$. For an ideal gas $c = \sqrt{kp/\rho} = \sqrt{kg_c pv} = \sqrt{kg_c RT}$. Values of the speed of sound in liquids are given in Table 1.3.2.

EXAMPLE. Check the value of the velocity of sound in benzene at 68°F (20°C) given in Table 1.3.2 using the isentropic bulk modulus. $c = \sqrt{E_s/\rho} = \sqrt{144 \times 223{,}000/1.705} = 4{,}340$ ft/s (1,320 m/s). Additional information on the velocity of sound is given in Secs. 2 and 8.

Application of shear stress to a fluid results in the continual and permanent distortion known as flow. **Viscosity** is the resistance of a fluid to shear motion—its internal friction. This resistance is due to two phenomena: (1) cohesion of the molecules and (2) molecular transfer from one layer to another, setting up a tangential or shear stress. In liquids, cohesion predominates, and since cohesion decreases with increasing temperature, the viscosity of liquids does likewise. Cohesion is relatively weak in gases; hence increased molecular activity with increasing temperature causes an increase in molecular transfer with corresponding increase in viscosity.

The **dynamic viscosity** μ of a fluid is the ratio of the shearing stress to the rate of deformation. From Fig. 1.3.1, $\mu = \tau/(dU/dy)$. Its dimensions are $(F)(T)/L^2$ or $M/(L)(T)$. The units are lbf · s/ft² or slugs/(ft)(s) [47.88026(N · s)/m²].

In the cgs system, the unit of dynamic viscosity is the **poise,** $2{,}089 \cdot 10^{-6}$ (lbf · s)/ft² [0.1 (N · s)/m²], but for convenience the **centipoise** (1/100 poise) is widely used. The dynamic viscosity of water at 68°F (20°C) is approximately 1 centipoise.

Table 1.3.3 gives values of dynamic viscosity for selected liquids at atmospheric pressure. Values of viscosity for fuels and lubricants are given in Sec. 4. The effect of pressure on liquid viscosity is generally unimportant in fluid mechanics except in lubricants (Sec. 4). The viscosity of water changes little at pressures up to 15,000 lbf/in², but for animal and vegetable oils it increases about 350 percent and for mineral oils about 1,600 percent at 15,000 lbf/in² pressure.

The dynamic viscosity of gases is primarily a temperature function and essentially independent of pressure. Table 1.3.4 gives values of dynamic viscosity of selected gases.

The **kinematic viscosity** ν of a fluid is its dynamic viscosity divided by its density, or $\nu = \mu/\rho$. Its dimensions are L^2/T. The units are ft²/s (9.290304×10^{-2} m²/s).

In the cgs system, the unit of kinematic viscosity is the **stoke** (1×10^{-4} m²/s), but for convenience, the **centistoke** (1/100 stoke) is widely

Table 1.3.3 Dynamic Viscosity of Liquids at Atmospheric Pressure

Temp: °C	0	20	40	60	80	100
°F	32	68	104	140	176	212
Liquid	μ, (lbf · s)/(ft²) [47.88 (N · s)/(m²)] × 10⁶*					
Alcohol, ethyl[a,e]	37.02	25.06	17.42	12.36	9.028	
Benzene[a]	19.05	13.62	10.51	8.187	6.871	
Carbon tetrachloride[e]	28.12	20.28	15.41	12.17	9.884	
Gasoline,[b] sp. gr. 0.68	7.28	5.98	4.93	4.28		
Glycerin[d]	252,000	29,500	5,931	1,695	666.2	309.1
Kerosene,[b] sp. gr. 0.81	61.8	38.1	26.8	20.3	16.3	
Mercury[a]	35.19	32.46	30.28	28.55	27.11	25.90
Oil, machine,[a] sp. gr. 0.907						
"Light"	7,380	1,810	647	299	164	102
"Heavy"	66,100	9,470	2,320	812	371	200
Water, fresh[c]	36.61	20.92	13.61	9.672	7.331	5.827
Water, salt[d]	39.40	22.61	18.20			

* $\mu \times 10^6$ = tabulated value
μ = tabulated value $\times 10^{-6}$
SOURCES: Computed from data given in:
[a] "Handbook of Chemistry and Physics," 52nd ed., Chemical Rubber Company, 1971–1972.
[b] "Smithsonian Physical Tables," 9th rev. ed., 1954.
[c] "Steam Tables," ASME, 1967.
[d] "American Institute of Physics Handbook," 3rd ed., McGraw-Hill, 1972.
[e] "International Critical Tables," McGraw-Hill.

Table 1.3.4 Dynamic Viscosity of Gases at One Atmosphere

Temp: °C	0	20	60	100	200	400	600	800	1000
°F	32	68	140	212	392	752	1112	1472	1832
Gas	μ, (lbf · s)/(ft²) [47.88(N · s)/(m²)] × 10⁸*								
Air[a]	35.67	39.16	41.79	45.95	53.15	70.42	80.72	91.75	100.8
Carbon dioxide[a]	29.03	30.91	35.00	38.99	47.77	62.92	74.96	87.56	97.71
Carbon monoxide[b]	34.60	36.97	41.57	45.96	52.39	66.92	79.68	91.49	102.2
Helium[a]	38.85	40.54	44.23	47.64	55.80	71.27	84.97	97.43	
Hydrogen[a,b]	17.43	18.27	20.95	21.57	25.29	32.02	38.17	43.92	49.20
Methane[a]	21.42	22.70	26.50	27.80	33.49	43.21			
Nitrogen[a,b]	34.67	36.51	40.14	43.55	51.47	65.02	76.47	86.38	95.40
Oxygen[b]	40.08	42.33	46.66	50.74	60.16	76.60	90.87	104.3	116.7
Steam[c]		18.49	21.89	25.29	33.79	50.79	67.79	84.79	

* $\mu \times 10^8$ = tabulated value
μ = tabulated value $\times 10^{-8}$
SOURCES: Computed from data given in:
[a] "Handbook of Chemistry and Physics," 52nd ed., Chemical Rubber Company, 1971–1972.
[b] "Tables of Thermal Properties of Gases," *NBS Circular* 564, 1955.
[c] "Steam Tables," ASME, 1967.

used. The kinematic viscosity of water at 68°F (20°C) is approximately 1 centistoke.

The standard device for **experimental determination of kinematic viscosity** in the United States is the **Saybolt Universal** viscometer. It consists essentially of a metal tube and an orifice built to rigid specifications and calibrated. The time required for a gravity flow of 60 cubic centimeters is called the Saybolt Seconds Universal (SSU). Approximate conversions of SSU to stokes may be made as follows:

$$32 < \text{SSU} < 100 \text{ seconds, stokes} = 0.00226\,(\text{SSU}) - 1.95/(\text{SSU})$$
$$\text{SSU} > 100 \text{ seconds, stokes} = 0.00220\,(\text{SSU}) - 1.35/(\text{SSU})$$

For viscous oils, the **Saybolt Furol** viscometer is used. Approximate conversions of Saybolt Seconds Furol (SSF) may be made as follows:

$$25 < \text{SSF} < 40 \text{ seconds, stokes} = 0.0224\,(\text{SSF}) - 1.84/(\text{SSF})$$
$$\text{SSF} > 40 \text{ seconds, stokes} = 0.0216\,(\text{SSF}) - 0.60/(\text{SSF})$$

For exact conversions of Saybolt viscosities, see ASTM D445–71 and Sec. 4.8.

The **surface tension** σ of a fluid is the work done in extending the surface of a liquid one unit of area or work per unit area. Its dimensions are F/L. The units are lbf/ft (14.5930 N/m).

Values of σ for various interfaces are given in Table 1.3.5. Surface tension decreases with increasing temperature. Surface tension is of importance in the formation of bubbles and in problems involving atomization.

Capillary action is due to surface tension, **cohesion** of the liquid molecules, and the **adhesion** of the molecules on the surface of a solid. This action is of importance in fluid mechanics because of the formation of a meniscus (curved section) in a tube. When the adhesion is greater than the cohesion, a liquid "wets" the solid surface, and the liquid will rise in the tube and conversely will fall if the reverse. Figure 1.3.3 illustrates this effect on manometer tubes. In the reading of a manometer, all data should be taken at the center of the meniscus.

Table 1.3.5 Surface Tension of Liquids at One Atmosphere and 68°F (20°C)

Liquid	δ, lbf/ft (14.59 N/m) × 10^3 In vapor	In air	In water
Alcohol, ethyl*	1.56	1.53	
Benzene*†	2.00	1.98	2.40
Carbon tetrachloride*	1.85	1.83	3.08
Gasoline,*† sp. gr. 0.68	1.3–1.6		2.7–3.6
Glycerin*	4.30	4.35	
Kerosene,*† sp. gr. 0.81	1.6–2.2		
Mercury*	32.6§	32.8	25.7
Oil, machine,‡ sp. gr. 0.907	2.5	2.6	2.3–3.7
Water, fresh‡		4.99	
Water, salt‡		5.04	

SOURCES: Computed from data given in:
* "International Critical Tables," McGraw-Hill.
† ASTM-IP, "Petroleum Measurement Tables."
‡ "American Institute of Physics Handbook," 3rd ed., McGraw-Hill, 1972.
§ In vacuum.

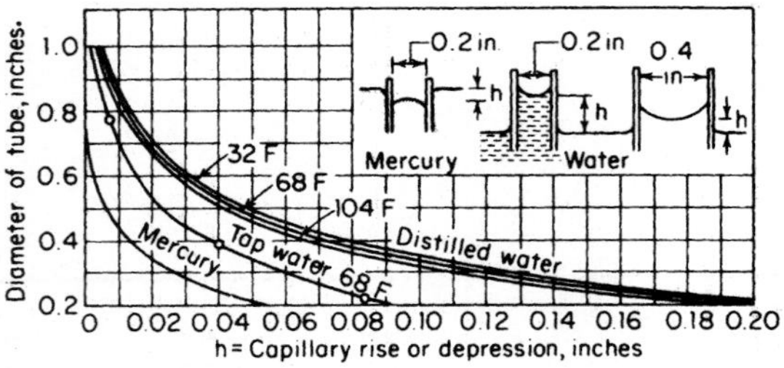

Figure 1.3.3 Capillarity in circular glass tube.

The **vapor pressure** *pv* of a fluid is the pressure at which its liquid and vapor are in equilibrium at a given temperature. See Secs. 2.1 and 2.2 for further definitions and values.

FLUID STATICS

Pressure *p* is the force per unit area exerted on or by a fluid and has dimensions of F/L^2. In fluid mechanics and in thermodynamic equations, the units are lbf/ft² (47.88026 N/m²), but engineering practice is to use units of lbf/in² (6,894.757 N/m²).

The relationship between **absolute pressure, gage pressure,** and **vacuum** is shown in Fig. 1.3.4. Most fluid-mechanics equations and **all** thermodynamic equations require the use of **absolute pressure,** and unless otherwise designated, a pressure should be understood to be **absolute pressure.** Common practice is to denote absolute pressure as lbf/ft² abs, or psfa, lbf/in² abs or psia; and in a like manner for **gage pressure** lbf/ft² g, lbf/in² g, and psig.

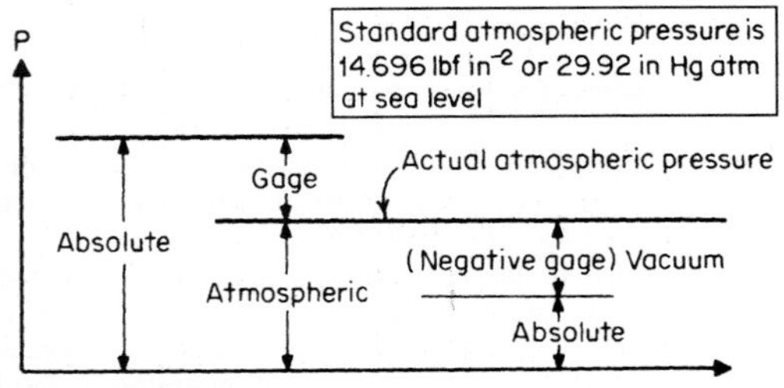

Figure 1.3.4 Pressure relations.

According to **Pascal's** principle, the pressure in a static fluid is the same in all directions.

The **basic equation of fluid statics** is obtained by consideration of a fluid particle at rest with respect to other fluid particles, all being subjected to body-force accelerations of a_x, a_y, and a_z opposite the directions of *x*, *y*, and *z*, respectively, and the acceleration of gravity in the *z* direction, resulting in the following:

$$dp = -\rho[a_x\,dx + a_y\,dy + (a_z + g)dz]$$

Pressure-Height Relations For a fluid at rest and subject only to the gravitational force, a_x, a_y, and a_z are zero and the **basic equation for fluid statics reduces to** $dp = -\rho g\,dz = \gamma\,dz$.

Liquids (Incompressible Fluids) The pressure-height equation integrates to $(p_1 - p_2) = \rho g(z_2 - z_1) = \gamma(z_2 - z_1) = \Delta p = \gamma h$, where *h* is measured from the liquid surface (Fig. 1.3.5).

EXAMPLE. A large closed tank is partly filled with 68°F (20°C) benzene. If the pressure on the surface is 10 lb/in², what is the pressure in the benzene at a depth of 11 ft below the liquid surface?

$$p_1 = \rho g h + p_2 = \frac{1.705 \times 32.17 \times 11}{144} + 10$$
$$= 14.19\ \text{lbf/in}^2 (9.784 \times 10^4\ \text{N/m}^2)$$

Ideal Gases (Compressible Fluids) For problems involving the upper atmosphere, it is necessary to take into account the variation of gravity with altitude. For this purpose, the **geopotential altitude** *H* is used, defined by $H = Z/(1 + z/r)$, where *r* is the radius of the earth (≈21 × 10⁶ ft ≈ 6.4 × 10⁶ m) and *z* is the height above sea level. The integration of the pressure-height equation depends upon the thermodynamic process. For an **isothermal process** $p_2/p_1 = e^{-(H_2-H_1)/RT}$ and for a **polytropic process** ($n \neq 1$)

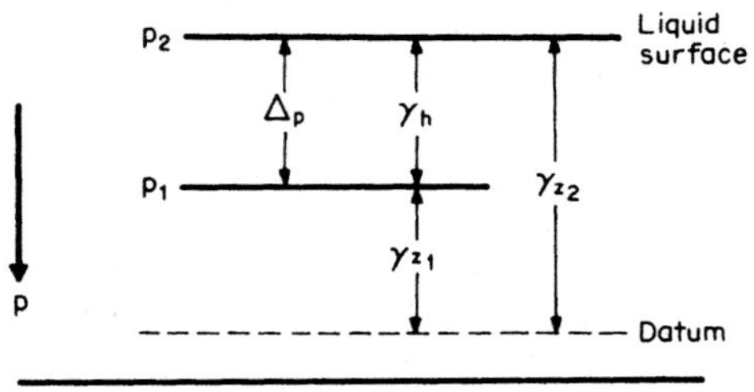

Figure 1.3.5 Pressure equivalence.

$$\frac{p_2}{p_1} = \left[1 - \frac{(n-1)(H_2 - H_1)}{nRT_1}\right]^{n/(n-1)}$$

Temperature-height relations for a polytropic process ($n \neq 1$) are given by

$$\frac{n}{1-n} = \frac{H_2 - H_1}{R(T_2 - T_1)}$$

Substituting in the **pressure-altitude** equation,

$$p_2/p_1 = (T_2/T_1)^{(H_2-H_1) \div R(T_1-T_2)}$$

EXAMPLE. The U.S. Standard Atmosphere 1962 (Sec. 8) is defined as having a sea-level temperature of 59°F (15°C) and a pressure of 2,116.22 lbf/ft². From sea level to a geopotential altitude of 36,089 ft (11,000 m) the temperature decreases linearly with altitude to −69.70°F (−56.5°C). Check the value of pressure ratio at this altitude given in the standard table.

Noting that $T_1 = 59 + 459.67 = 518.67$, $T_2 = -69.70 + 459.67 = 389.97$, and $R = 53.34$ ft·lbf/(lbm)(°R),

$$p_2/p_1 = (T_2/T_1)^{(H_2-H_1)/R(T_1-T_2)}$$
$$= (389.97/518.67)^{(36{,}089-0)/53.34(518.67-389.97)}$$
$$= 0.2233 \text{ vs. tabulated value of } 0.2234$$

Pressure-Sensing Devices The two principal devices using liquids are the **barometer** and the **manometer.** The barometer senses absolute pressure and the manometer senses pressure differential. For discussion of the barometer and other pressure-sensing devices, refer to Sec. 12.

Manometers are a direct application of the basic equation of fluid statics and serve as a pressure standard in the range of $\frac{1}{10}$ in of water to 100 lbf/in². The most familiar type of manometer is the **U tube** shown in Fig. 1.3.6*a*. Because of the necessity of observing both legs simultaneously, the **well** or **cistern** type (Fig. 1.3.6*b*) is sometimes used. The **inclined manometer** (Fig. 1.3.6*c*) is a special form of the well-type manometer designed to enhance the readability of small pressure differentials. Application of the basic equation of fluid statics to each of the types results in the following equations. For the **U tube,** $p_1 - p_2 = (\gamma_m - \gamma_f)h$, where γ_m and γ_f are the specific weights of the manometer and sensed fluids, respectively, and h is the vertical distance between the liquid interfaces. For the **well type,** $p_1 - p_2 = (\gamma_m - \gamma_f)(z_2) \times (1 + A_2/A_1)$, where A_1 and A_2 are as shown in Fig. 1.3.6*b* and z_2 is the vertical distance from the fill line to the upper interface. Commercial manufacturers of well-type manometers correct for the area ratios so that $p_1 - p_2 = (\gamma_m - \gamma_f)S$, where S is the scale reading and is equal to $z_1(1 + A_2/A_1)$. For this reason, scales should not be interchanged between U type or well type or between well types without consulting the manufacturer. For inclined manometers,

$$p_1 - p_2 = (\gamma_m - \gamma_f)(A_2/A_1 + \sin\theta)R$$

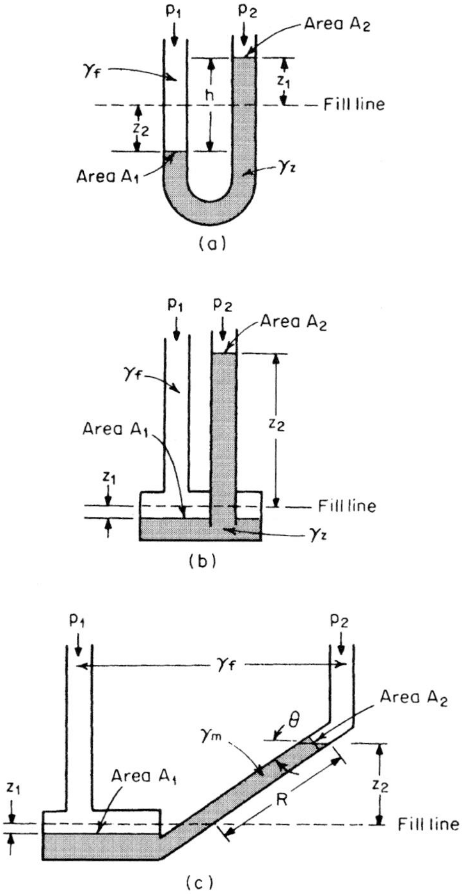

Figure 1.3.6 (*a*) U-tube manometer; (*b*) well or cistern-type manometer; (*c*) inclined manometer.

where R is the distance along the inclined tube. Commercial inclined manometers also have special scales so that $p_1 - p_2 = (\gamma_m - \gamma_f)S$, where $S = (A_2/A_1 + \sin\theta)R$.

EXAMPLE. A U-tube manometer containing mercury is used to sense the difference in water pressure. If the height between the interfaces is 10 in and the temperature is 68°F (20°C), what is the pressure differential?

$$\begin{aligned} p_1 - p_2 &= (\gamma_m - \gamma_f)h = g(\rho_m - \rho_f)h \\ &= 32.17(26.283 - 1.937)(10/12) \\ &= 652.7 \text{ lbf/ft}^2 (3.152 \times 10^4 \text{ N/m}^2) \end{aligned}$$

Liquid Forces The force exerted by a liquid on a **plane submerged surface** (Fig. 1.3.7) is given by $F = \int p\, dA = \gamma \int h \cdot dA = \gamma h_c A$, where h_c is the distance from the liquid surface to the center of gravity of the surface, and A is the area of the surface. The location of the center of this force is given by

$$s_F = s_c + I_G / s_c A$$

where s_F is the inclined distance from the liquid surface to the center of force, s_c the inclined distance to the center of gravity of the surface, and I_G the moment of inertia around its center of gravity. Values of I_G are given in Sec. 3.2. See also Sec. 3.1. From Fig. 1.3.7, $h = R \sin\theta$, so that the vertical center of force becomes

$$h_F = h_c + I_G (\sin\theta)^2 / h_c A$$

EXAMPLE. Determine the force and its location acting on a rectangular gate 3 ft wide and 5 ft high at the bottom of a tank containing 68°F (20°C) water, 12 ft deep, (1) if the gate is vertical, and (2) if it is inclined 30° from horizontal.

1. *Vertical gate*

$F = \gamma g h_c A = \rho g h_c A$
$= 1.937 \times 32.17(12 - 5/2)(5 \times 3)$
$= 8{,}800 \text{ lb}_f (3.914 \times 10^4 \text{ N})$
$h_F = h_c + I_G(\sin\theta)^2/h_c A$, from Sec. 3.2, I_G for a rectangle = (width)(height)³/12
$h_F = (12 - 5/2) + (3 \times 5^3/12)(\sin 90°)^2/(12 - 5/2)(3 \times 5)$
$h_F = 9.719$ ft (2.962 m)

2. *Inclined gate*

$F = \gamma h_c A = \rho g h_c A$
$= 1.937 \times 32.17(12 - 5/2 \sin 30°)(5 \times 3)$
$= 10{,}048$ lbf (4.470×10^4 N)
$h_F = h_c + I_G(\sin\theta)^2/h_c A$
$= (12 - 5/2 \sin 30°) + (3 \times 5^3/12)(\sin 30°)^2/(12 - 5/2 \sin 30°)(3 \times 5)$
$= 10.80$ ft (3.291 m)

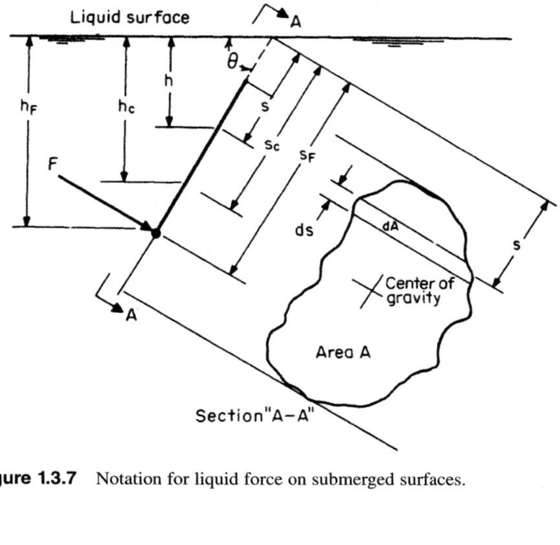

Figure 1.3.7 Notation for liquid force on submerged surfaces.

Forces on irregular surfaces may be obtained by considering their horizontal and vertical components. The vertical component F_z equals the weight of liquid above the surface and acts through the centroid of the volume of the liquid above the surface. The horizontal component F_x equals the force on a vertical projection of the irregular surface. This force may be calculated by $F_x = \gamma h_{cx} A_x$, where h_{cx} is the distance from the surface center of gravity of the horizontal projection, and A_x is the projected area. The forces may be combined by $F = \sqrt{F_x^2 + F_z^2}$.

When fluid masses are **accelerated** without relative motion between fluid particles, the basic equation of fluid statics may be applied. For **translation** of a **liquid** mass due to uniform acceleration, the basic equation integrates to

$$p_2 - p_1 = -\rho[(x_2 - x_1)a_x + (y_2 - y_1)a_y + (z_2 - z_1)(a_z + g)]$$

Example. An open tank partly filled with a liquid is being accelerated up an inclined plane as shown in Fig. 1.3.8. The uniform acceleration is 20 ft/s² and the angle of the incline is 30°. What is the angle of the free surface of the liquid? Noting that on the free surface $p_2 = p_1$ and that the acceleration in the y direction is zero, the basic equation reduces to

$$(x_2 - x_1)a_x + (z_2 - z_1)(a_z + g) = 0$$

Solving for tan θ,

$$\tan\theta = \frac{z_1 - z_2}{x_2 - x_1} = \frac{a_x}{a_z + g} = \frac{a\cos\alpha}{a\sin\alpha + g}$$
$$= (20\cos 30°)/(20\sin 30° + 32.17) = 0.4107$$
$$\theta = 22°20'$$

For **rotation** of liquid masses with uniform rotational acceleration, the basic equation integrates to

$$p_2 - p_1 = \rho\left[\frac{\omega^2}{2}(x_2^2 - x_1^2) - g(z_2 - z_1)\right]$$

where ω is the rotational speed in rad/s and x is the radial distance from the axis of rotation.

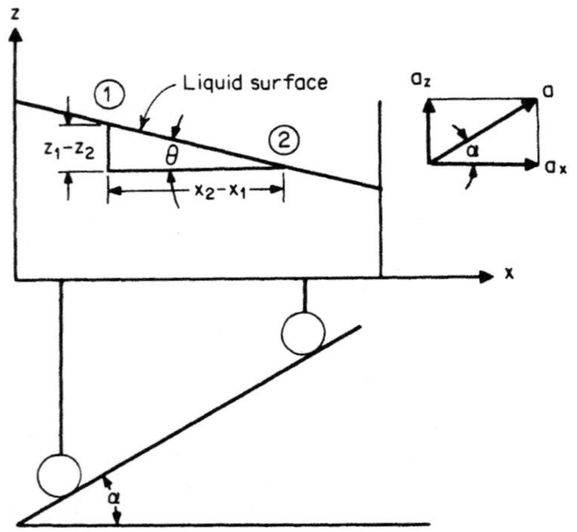

Figure 1.3.8 Notation for translation example.

Example. The closed cylindrical tank shown in Fig. 1.3.9 is 4 ft in diameter and 10 ft high and is filled with 104°F (40°C) benzene. The tank is rotated at 250 r/min about an axis 3 ft from its centerline. Compute the maximum pressure differential in the tank. Analysis of the rotation equation indicates that the maximum pressure will occur at the maximum rotational radius and the minimum elevation and, conversely, the minimum at the minimum rotational radius and maximum elevation. From Fig. 1.3.9, $x_1 = 3 - 4/2 = 1$ ft, $x_2 = 1 + 4 = 5$ ft, $z_2 - z_1 = -10$ ft, and the rotational speed $\omega = 2\pi N/60 = 2\pi(250)/60 = 26.18$ rad/s. Substituting into the rotational equation,

$$p_2 - p_1 = \rho\left[\frac{\omega^2}{2}(x_2^2 - x_1^2) - g(z_2 - z_1)\right]$$
$$= \frac{1.663}{144}\left[\frac{(26.18)^2}{2}(5^2 - 1^2) - 32.17(-10)\right]$$
$$= 98.70\ \text{lbf/in}^2 (6.805 \times 10^5\ \text{N/m}^2)$$

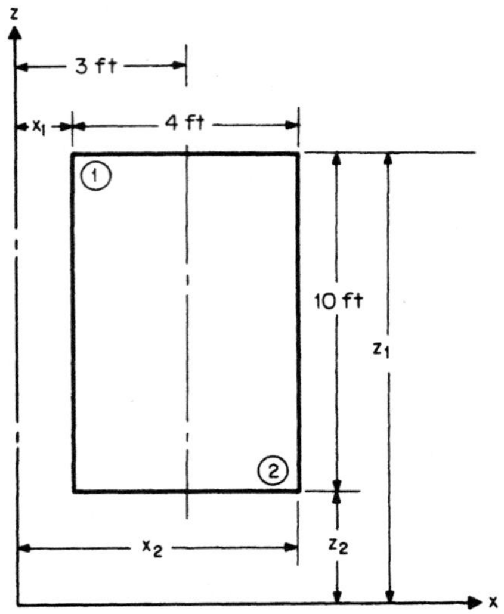

Figure 1.3.9 Notation for rotation example.

Buoyancy Archimedes' principle states that a body immersed in a fluid is buoyed up by a force equal to the weight of the fluid displaced. If an object immersed in a fluid is heavier than the fluid displaced, it will sink to the bottom, and if lighter, it will rise. From the free-body diagram of Fig. 1.3.10, it is seen that for vertical equilibrium,

$$\Sigma F_z = 0 = F_B - F_g - F_D$$

where F_B is the buoyant force, F_g the gravity force (weight of body), and F_D the force required to prevent the body from rising. The buoyant force being the weight of the displaced liquid, the equilibrium equation may be written as

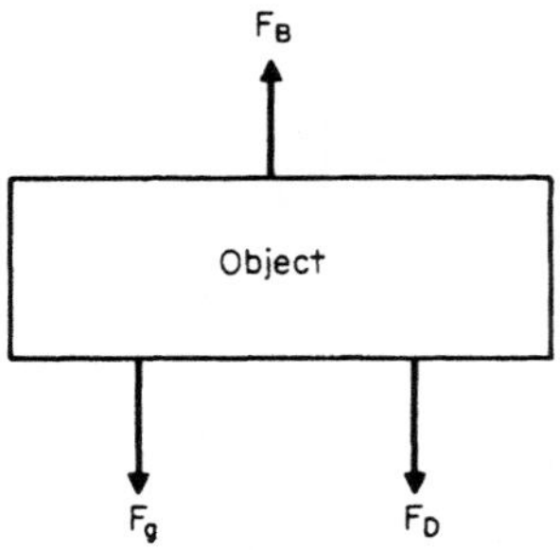

Figure 1.3.10 Free body diagram of an immersed object.

$$F_D = F_B - F_g = \gamma_f V - \gamma_0 V = (\gamma_f - \gamma_0)V$$

where γ_f is the specific weight of the fluid, γ_0 is the specific weight of the object, and V is the volume of the object.

EXAMPLE. An airship has a volume of 3,700,000 ft^3 and is filled with hydrogen. What is its gross lift in air at 59°F (15°C) and 14.696 psia? Noting that $\gamma = p/RT$,

$$F_D = (\gamma_f - \gamma_0)V = \left(\frac{p}{R_a T} - \frac{p}{R_{H_2} T}\right)V$$

$$= \frac{pV}{T}\left(\frac{1}{R_a} - \frac{1}{R_{H_2}}\right)$$

$$= \frac{144 \times 14.696 \times 3,700,000}{59 + 459.7}\left(\frac{1}{53.34} - \frac{1}{766.8}\right)$$

$$= 263,300 \text{ lbf } (1.71 \times 10^6 \text{ N})$$

Flotation is a special case of buoyancy where $F_D = 0$, and hence $F_B = F_g$.

EXAMPLE. A crude **hydrometer** consists of a cylinder of ½ in diameter and 2 in length surmounted by a cylinder of ⅛ in diameter and 10 in long. Lead shot is added to the hydrometer until its total weight is 0.32 oz. To what depth would this hydrometer float in 104°F (40°C) glycerin? For flotation, $F_B = F_g = \gamma_f V = \rho_f g V$ or $V = F_B/\rho_f g = (0.32/16)/(2.423 \times 32.17) = 2.566 \times 10^{-4}$ ft^3. Volume of cylindrical portion of hydrometer = $V_c = \pi D^2 L/4 = \pi(0.5/12)^2(2/12)/4 = 2.273 \times 10^{-4}$ ft^3. Volume of stem immersed = $V_S = V - V_C = 2.566 \times 10^{-4} - 2.273 \times 10^{-4} = 2.930 = 10^{-5}$ ft^3. Length of immersed stem = $L_S = 4\ V_S/\pi D^2 = (4 \times 2.930 \times 10^{-5})/\pi(0.125/12)^2 = 0.3438$ ft $= 0.3438 \times 12 = 4.126$ in. Total immersion = $L + L_S = 2 + 4.126 = 6.126$ in (0.156 m).

Static Stability A body is in static equilibrium when the imposition of a small displacement brings into action forces that tend to restore the body to its original position. For **completely submerged bodies,** the center of buoyancy and the center of gravity must lie on the same vertical line and the center of buoyancy must be located above the center of gravity. Figure 1.3.11*a* shows a balloon and its basket in its normal position with the center of buoyancy *B* above and on the same vertical line as the center of gravity *G*. Figure 1.3.11*b* shows the balloon displaced from its normal position. In this position, there is a couple $F_g x$ which tends

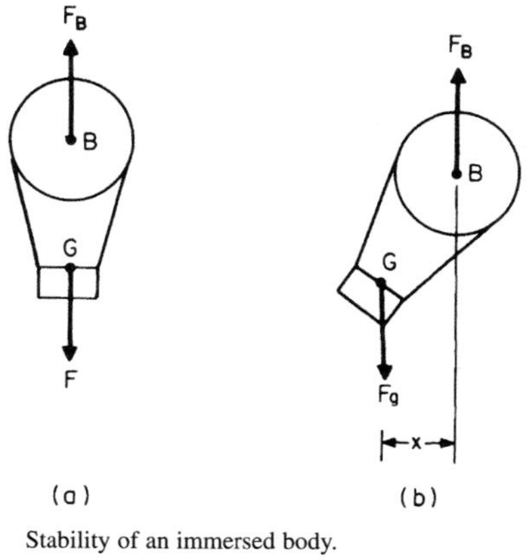

Figure 1.3.11 Stability of an immersed body.

to restore the balloon and its basket to its original position. For **floating bodies,** the center of gravity and the center of buoyancy must lie on the same vertical line, but the center of buoyancy may be below the center of gravity, as is common practice in surface-ship design. It is required that when displaced, the line of action of the buoyant force intersect the centerline above the center of gravity. Figure 1.3.12*a* shows a floating body in its normal position with its center of gravity *G* on the same vertical line and above the center of buoyancy *B*. Figure 1.3.12*b* shows the object displaced. The intersection of the line of action of the buoyant force with the centerline of the body at *M* is called the metacenter. As shown, this is above the center of buoyancy and sets up a restoring couple. When the metacenter is below the center of gravity, the object will capsize (see Sec. 8.3).

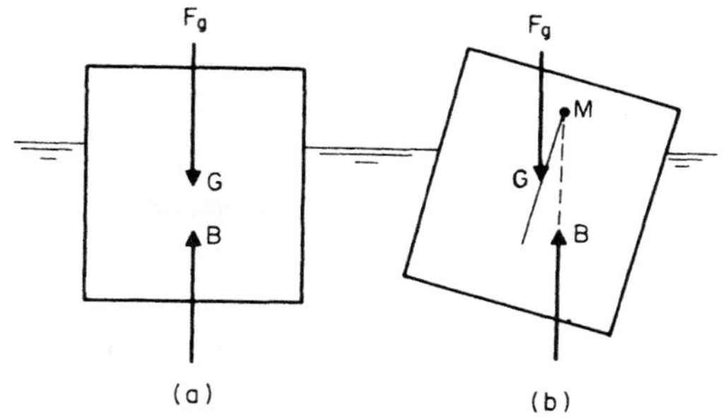

Figure 1.3.12 Stability of a floating body.

FLUID KINEMATICS

Steady and Unsteady Flow If at every point in the fluid stream none of the local fluid properties changes with time, the flow is said to be steady. The mathematical conditions for steady flow are met when ∂(fluid properties)$/\partial t = 0$. While flow is generally unsteady by nature, many real cases of unsteady flow may be treated as steady flow by using average properties or by changing the space reference. The amount of error produced by the averaging technique depends upon the nature of the unsteady flow, but the latter technique is error-free when it can be applied.

Streamlines and Stream Tubes Velocity has both magnitude and direction and hence is a vector. A **streamline** is a line which gives the *direction* of the velocity of a fluid particle at each point in the flow stream. When streamlines are connected by a closed curve in steady flow, they will form a boundary through which the fluid particles cannot pass. The space between the streamlines becomes a **stream tube.** The stream-tube concept broadens the application of fluid-flow principles; for example, it allows treating the flow inside a pipe and the flow around an object with the same laws. A stream tube of decreasing size approaches its own axis, a central streamline; thus equations developed for a stream tube may also be applied to a streamline.

Velocity and Acceleration In the most general case of fluid motion, the resultant **velocity** *U* along a streamline is a function of both distance *s* and time *t*, or $U = f(s, t)$. In differential form,

$$dU = \frac{\partial U}{\partial s} ds + \frac{\partial U}{\partial t} dt$$

An expression for **acceleration** may be obtained by dividing the velocity equation by *dt*, resulting in

$$\frac{dU}{dt} = \frac{\partial U}{\partial s}\frac{ds}{dt} + \frac{\partial U}{\partial t}$$

for steady flow $\partial U/\partial t = 0$.

Velocity Profile In the flow of real fluids, the individual streamlines will have different velocities past a section. Figure 1.3.13 shows the steady flow of a fluid past a section (*A*-*A*) of a circular pipe. The **velocity profile** is obtained by plotting the velocity *U* of each streamline as it passes *A*-*A*. The stream tube that is formed by the space

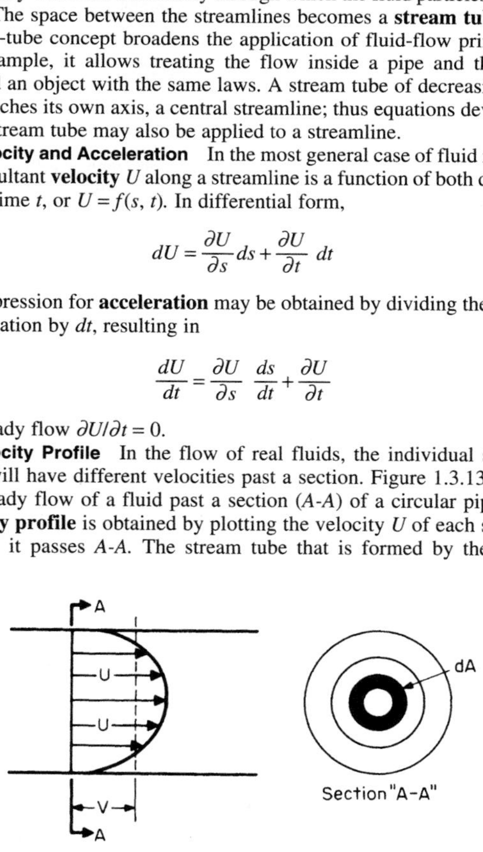

Figure 1.3.13 Velocity profile.

between the streamlines is the annulus whose area is dA, as shown in Fig. 1.3.13 for the stream tube whose velocity is U. The volumetric rate of flow Q for the flow past section A-A is $Q = \int U\, dA$. All flows take place between boundaries that are **three-dimensional.** The terms **one-dimensional, two-dimensional,** and **three-dimensional** flow refer to the number of dimensions required to describe the velocity profile. For **three-dimensional** flow, a **volume** (L^3) is required; for example, the flow of a fluid in a circular pipe. For **two-dimensional** flow, an **area** (L^2) is necessary; for example, the flow between two parallel plates. For **one-dimensional** flow, a **line** (L) describes the profile. In cases of two- or three-dimensional flow, $\int U dA$ can be integrated either mathematically if the equations are known or graphically if velocity-measurement data are available. In many engineering applications, the **average velocity** V may be used where $V = Q/A = (1/A)\int U dA$.

The **continuity equation** is a special case of the general physical law of the conservation of mass. It may be simply stated for a control volume:

Mass rate entering = mass rate of storage + mass rate leaving

This may be expressed mathematically as

$$\rho U\ dA = \left[\frac{\partial}{\partial t}(\rho\ dA\ ds)\right] + \left[\rho U\ dA + \frac{\partial}{\partial s}(\rho U\ dA) sd\right]$$

where ds is an incremental distance along the control volume. For steady flow, $\partial/\partial t(\rho\ dA\ ds) = 0$, the general equation reduces to $d(\rho U\ dA) = 0$. Integrating the steady-flow **continuity equation** for the average velocity along a flow passage:

$$\rho VA = \text{a constant} = \rho_1 V_1 A_1 = \rho_2 V_2 A_2 = \cdots = \rho_n V_n A_n = \dot{M}$$

where $\dot{M}$ is the mass flow rate in slugs/s (14.5939 kg/s). In many engineering applications, the flow rate in pounds mass per second is desired, so that

$$\dot{m} = \frac{V_1 A_1}{v_1} = \frac{V_2 A_2}{v_2} = \cdots = \frac{V_n A_n}{v_n}$$

where $\dot{m}$ is the flow rate in lbm/s (0.4535924 kg/s).

EXAMPLE. Air discharges from a 12-in-diameter duct through a 4-in-diameter nozzle into the atmosphere. The pressure in the duct is 20 lbf/in^2, and atmospheric pressure is 14.7 lbf/in^2. The temperature of the air in the duct just upstream of the nozzle is 150°F, and the temperature in the jet is 147°F. If the velocity in the duct is 18 ft/s, compute (1) the mass flow rate in lbm/s and (2) the velocity in the nozzle jet. From the equation of state

$$v = RT/p$$
$$v_D = RT_D/p_D = 53.34\ (150 + 459.7)/(144 \times 20)$$
$$= 11.29\ \text{ft}^3/\text{lbm}$$
$$v_J = RT_J/p_J = 53.34\ (147 + 459.7)/(144 \times 14.7)$$
$$= 15.29\ \text{ft}^3/\text{lbm}$$
$$(1)\ \dot{m} = V_D A_D / v_D = 18[(\pi/4)(12/12)^2]/11.29$$
$$\dot{m}_J = 1.252\ \text{lbm/s}\ (0.5680\ \text{kg/s})$$
$$(2)\ VJ = \dot{m}v_J/A_J = (1.252)(15.29)/[(\pi/4)(4/12)^2]$$
$$v_J = 219.2\ \text{ft/s}(66.82\ \text{m/s})$$

FLUID DYNAMICS

Equation of Motion For steady one-dimensional flow, consideration of forces acting on a fluid element of length dL, flow area dA, boundary perimeter in fluid contact dP, and change in elevation dz with a unit shear stress τ moving at a velocity of V results in

$$v\ dp + \frac{V\ dV}{g_c} + \frac{g}{g_c} dz + v\tau\left(\frac{dP}{dA}\right) dL = 0$$

Substituting $v = g/g_c\gamma$ and simplifying,

$$\frac{dp}{\gamma} + \frac{V\ dV}{g} + dz + dh_f = 0$$

where $dh_f = (\tau/\gamma)(dP/dA)\ dL = \tau\ dL/\gamma R_h$.

The expression $1/(dP/dA)$ is the **hydraulic radius** R_h and equals the flow area divided by the perimeter of the solid boundary in contact with the fluid. This perimeter is usually called the "wetted" perimeter. The hydraulic radius of a pipe flowing full is $(\pi D^2/4)/\pi D = D/4$. Values for other configurations are given in Table 1.3.6. Integration of the equation of motion for an incompressible fluid results in

$$\frac{p_1}{\gamma} + \frac{V_1^2}{2g} + z_1 = \frac{p_2}{\gamma} + \frac{V_2^2}{2g} + z_2 + h_{1f2}$$

Each term of the equation is in feet and is equivalent to the height the fluid would rise in a tube if its energy were converted into potential energy. For this reason, in hydraulic practice, each type of energy is

Table 1.3.6 Values of Flow Area A and Hydraulic Radius R_h for Various Cross Sections

Cross section	Condition		Equations
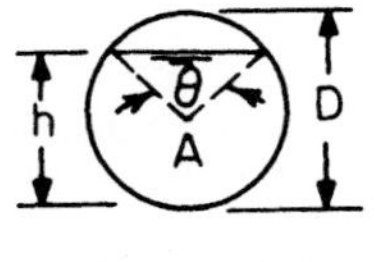	Flowing full	$h/D = 1$	$A = \pi D^2/4 \quad R_h = D/4$
	Upper half partly full	$0.5 < h/D < 1$	$\cos(\theta/2) = (2h/D - 1)$ $A = [\pi(360 - \theta) + 180 \ \sin\theta](D^2/1{,}440)$ $R_h = [1 + (180 \sin\theta)/(\pi\theta)](D/4)$
		$h/D = 0.8128$	$A = 0.6839\ D^2 \quad R_h \ \max = 0.3043D$
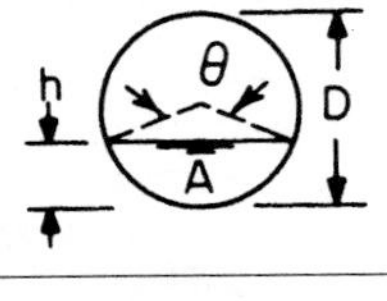	Lower half partly full	$h/D = 0.5$	$A = \pi D^2/8 \quad R_h \max = h/2$
		$0 < h/D < 0.5$	$\cos(\theta/2) = (1 - 2h/D)$ $A = [\pi\theta - 180 \sin\theta](D^2/1{,}440)$ $R_h = [1 - (180 \sin\theta)/(\pi\theta)](D/4)$
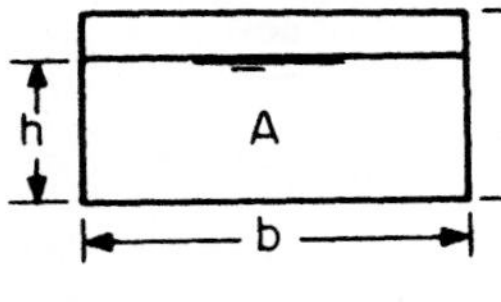	Flowing full	$h/D = 1$	$A = bD \quad R_h = bD/2(b + D)$
		Square $b = D$	$A = D^2 \quad R_h = D/4$
	Partly full	$h/D < 1$ $h/b = 0.5$ $b \to \infty,\ h \to 0$	$A = bh \quad R_h = bh/(2h + b)$ $A = b^2/2 \quad R_h \ \max = h/2$ $R_h \to h$ (wide shallow stream)

Table 1.3.6 Values of Flow Area *A* and Hydraulic Radius R_h for Various Cross Sections (*Continued*)

Cross section	Condition			Equations
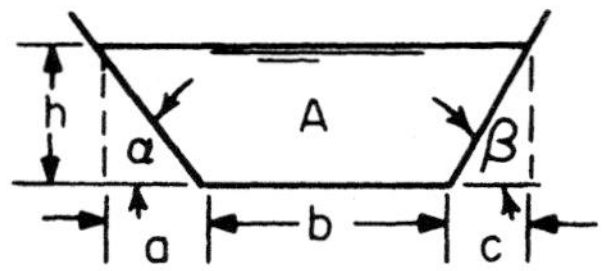	$\alpha \neq \beta$			R_h max $= h/2$ $A = [b + 1/2h(\cot\alpha + \cot\beta)]h$ $R_h = A/[b + h(\csc\alpha + \csc\beta)]$
	$\alpha = \beta$	$\frac{h}{a} = \frac{1}{2}$	$\alpha = 26°34'$	$A = (b + 2h)h$ $R_h = (b + 2h)h/(b + 4.472h)$
		$\frac{h}{a} = \frac{\sqrt{3}}{3}$	$\alpha = 30°$	$A = (b + 1.732h)h$ $R_h = (b + 1.732h)h/(b + 4h)$
		$\frac{h}{a} = \frac{2}{3}$	$\alpha = 33°41'$	$A = (b + 1.5h)h$ $R_h = (b + 1.5h)h/(b + 3.606h)$
		$\frac{h}{a} = 1$	$\alpha = 45°$	$A = (b + h)h$ $R_h = (b + h)h/(b + 2.828h)$
		$\frac{h}{a} = \frac{3}{2}$	$\alpha = 56°19'$	$A = (b + 0.6667h)h$ $R_h = (b + 0.6667h)h/(b + 2.404h)$
		$\frac{h}{a} = \sqrt{3}$	$\alpha = 60°$	$A = (b + 0.5774h)h$ $R_h = (b + 0.5774h)h/(b + 2.309h)$
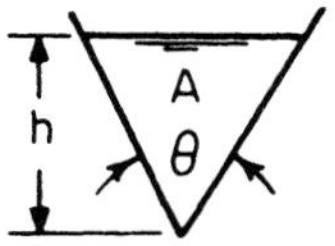	θ = any angle			$A = \tan(\theta/2)h^2$ $R_h = \sin(\theta/2)h/2$
	θ = 30			$A = 0.2679h^2$ $R_h = 0.1294h$
	θ = 45			$A = 0.4142h^2$ $R_h = 0.1913h$
	θ = 60			$A = 0.5774h^2$ $R_h = 0.2500h$
	θ = 90			$A = h^2$ $R_h = 0.3536h$

referred to as a **head.** The **static pressure head** is p/γ. The **velocity head** is $V^2/2g$, and the **potential head** is z. The energy loss between sections h_{1f2} is called the **lost head** or **friction head.** The **energy grade line** at any point $\Sigma(p/\gamma + V^2/2g + z)$, and the **hydraulic grade line** is $\Sigma(p/\gamma + z)$ as shown in Fig. 1.3.14.

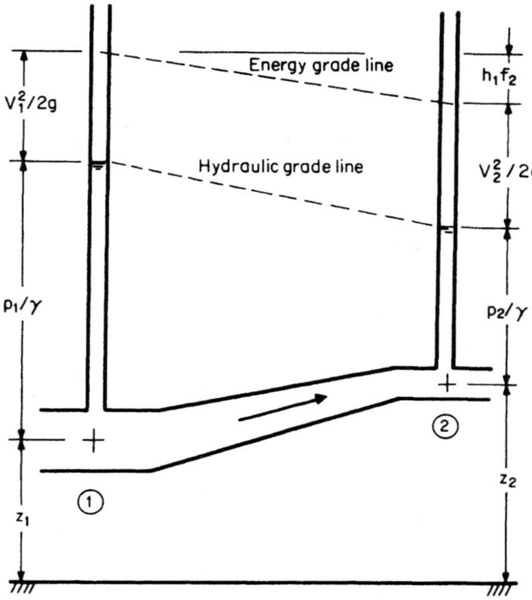

Figure 1.3.14 Energy relations.

EXAMPLE. A 12-in pipe (11.938 in inside diameter) reduces to a 6-in pipe (6.065 in inside diameter). Benzene at 68°F (20°C) flows steadily through this system. At section 1, the 12-in pipe centerline is 10 ft above the datum, and at section 2, the 6-in pipe centerline is 15 ft above the datum. The pressure at section 1 is 20 lbf/in² and the velocity is 4 ft/s. If the head loss due to friction is 0.05 $V_2^2/2g$, compute the pressure at section 2. Assume $g = g_c$, $\gamma = \rho g = 1.705 \times 32.17 = 54.85$ lbf/ft³. From the continuity equation,

$$M = \rho_1 A_1 V_1 = \rho_2 A_2 V_2 \qquad (\rho_1 = \rho_2)$$
$$V_2 = V_1(A_1/A_2) = V_1(\pi D_1^2)/4)/(\pi D_2^2/4) = V_1(D_1/D_2)^2$$
$$V_2 = 4(11.938/6.065)^2 = 15.50 \text{ ft/s}$$

From the equation of motion,

$$\frac{p_2}{\gamma} = \frac{p_1}{\gamma} + \frac{V_1^2}{2g} + z_1 - \left(\frac{V_2^2}{2g} + z_2 + h_{1f2}\right)$$
$$\frac{p_2}{\gamma} = \frac{p_1}{\gamma} + \frac{V_1^2 - V_2^2 - 0.05V_2^2}{2g} + z_1 - z_2$$
$$\frac{p_2}{\gamma} = \frac{144 \times 20}{54.85} + \frac{4^2 - 1.05(15.50)^2}{2 \times 32.17} + 10 - 15$$
$$\frac{p^2}{\gamma} = 43.83 \text{ ft}$$
$$p_2 = \frac{54.88 \times 43.83}{144} = 16.70 \text{ lbf/in}^2 \ (1.151 \times 10^5 \text{ N/m}^2)$$

Energy Equation Application of the principles of conservation of energy to a control volume for one-dimensional flow results in the following for steady flow:

$$J\,dq = dW + \frac{V\,dV}{g_c} + \frac{g}{g_c}dz + J\,du + d(pv)$$

where J is the mechanical equivalent of heat, 778.169 ft·lbf/Btu; q is the heat added, Btu/lbm (2,326 J/kg); W is the steady-flow shaft work done *by* the fluid; and u is the internal energy. Btu/lbm (2,326 J/kg). If the energy equation is integrated for an incompressible fluid,

$$J_1q_2 = {}_1W_2 + \frac{V_2^2 - V_1^2}{2g_c} + \frac{g}{g_c}(z_2 - z_1) + J(u_2 - u_1) + v(p_2 - p_1)$$

The equation of motion does not consider thermal energy or steady-flow work; the energy equation has no terms for friction. Subtracting the differential equation of motion from the energy equation and solving for friction results in

$$dh_f = (dW + J\ du + p\ dv - J\ dq)(g_c/g)$$

Integrating for an incompressible fluid ($dv = 0$),

$$h_{1f2} = [{}_1W_2 + J(u_2 - u_1) - J_1q_2](g_c/g)$$

In the absence of steady-flow work in the system, the effect of friction is to increase the internal energy and/or to transfer heat from the system.

For steady frictionless, incompressible flow, both the equation of motion and the energy equation reduce to

$$\frac{p_1}{\gamma} + \frac{V_1^2}{2g} + z_1 = \frac{p_2}{\gamma} + \frac{V_2^2}{2g} + z_2$$

which is known as the **Bernoulli equation.**

Area-Velocity Relations The continuity equation may be written as $\log_e \dot{M} = \log_e V + \log_e A + \log_e \rho$, which when differentiated becomes

$$\frac{dA}{A} = -\frac{dV}{V} - \frac{d\rho}{\rho}$$

For incompressible fluids, $d\rho = 0$, so

$$\frac{dA}{A} = -\frac{dV}{V}$$

Examination of this equation indicates:

1. If the area increases, the velocity decreases.
2. If the area is constant, the velocity is constant.
3. There are no critical values.

For the frictionless flow of compressible fluids, it can be demonstrated that

$$\frac{dA}{A} = -\frac{dV}{V}\left[1 - \left(\frac{V}{c}\right)^2\right]$$

Analysis of the above equation indicates:

1. *Subsonic velocity* $V < c$. If the area increases, the velocity decreases. Same as for incompressible flow.
2. *Sonic velocity* $V = c$. Sonic velocity can exist only where the change in area is zero, i.e., at the *end* of a convergent passage or at the *exit* of a constant-area duct.
3. *Supersonic velocity* $V > c$. If area increases, the velocity increases, the reverse of incompressible flow. Also, supersonic velocity can exist only in the **expanding** portion of a passage **after** a constriction where **sonic** velocity existed.

Frictionless adiabatic compressible flow of an ideal gas in a horizontal passage must satisfy the following requirements:

1. *Conservation of mass.* As expressed by the continuity equation $M = \rho_1 A_1 V_1 = \rho_2 A_2 V_2$.
2. *Conservation of energy.* As expressed by the energy equation

$$\frac{V^2}{2g_c} + Ju + pv = \frac{V_1^2}{2g_c} + Ju_1 + p_1v_1 = \frac{V_2^2}{2g_c} + Ju_2 + p_2v_2$$

3. *Process relationship.* For an ideal gas undergoing a frictionless adiabatic (isentropic) process,

$$pv^k = p_1v_1^k = p_2v_2^k$$

4. *Ideal-gas law.* The equation of state for an ideal gas

$$pv = RT$$

In an expanding supersonic flow, a **compression shock wave** will be formed if the requirements for the conservation of mass and energy are not satisfied. This type of wave is associated with large and sudden rises in pressure, density, temperature, and entropy. The shock wave is so thin that for computation purposes it may be considered as a single line. For compressible flow of gases and vapors in passages, refer to Sec. 2.1; for steam-turbine passages, Sec. 7.4; for compressible flow around immersed objects, see Sec. 8.4.

The **impulse-momentum equation** is an application of the principle of conservation of momentum and is derived from Newton's second law. It is used to calculate the forces exerted on a solid boundary by a moving stream. Because velocity and force have both magnitude and direction, they are vectors. The impulse-momentum equation may be written for all three directions:

$$\Sigma F_x = \dot{M}(V_{x2} - V_{x1})$$
$$\Sigma F_y = \dot{M}(V_{y2} - V_{y1})$$
$$\Sigma F_z = \dot{M}(V_{z2} - V_{z1})$$

Figure 1.3.15 shows a free-body diagram of a control volume. The pressure forces shown are those imposed by the boundaries on the fluid and on the atmosphere. The reactive force R is that imposed by the downstream boundary on the fluid for equilibrium. Application of the impulse-momentum equation yields

$$\Sigma F = (F_{p1} + F_{a2}) - (F_{a1} + F_{p2} + R) = \dot{M}(V_2 - V_1)$$

Solving for R,

$$R = (p_1 - p_a)A_1 - (p_2 - p_a)A_2 = \dot{M}(V_2 = V_1)$$

The impulse-momentum equation is often used in conjunction with the continuity and energy equations to solve engineering problems. Because of the wide variety of possible applications, some examples are given to illustrate the methods of attack.

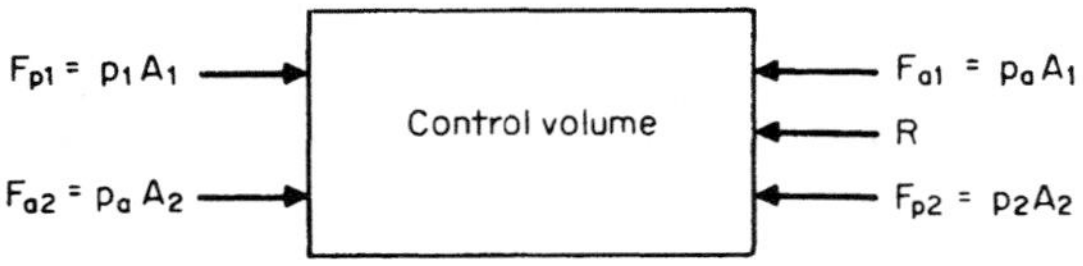

Figure 1.3.15 Notation for impulse momentum.

EXAMPLE. **Compressible Fluid in a Duct.** Nitrogen flows steadily through a 6-in (5.761 in inside diameter) straight, horizontal pipe at a mass rate of 25 lbm/s. At section 1, the pressure is 120 lbf/in^2 and the temperature is 100°F. At section 2, the pressure is 80 lbf/in^2 and the temperature is 110°F. Find the friction force opposing the motion. From the equation of state,

$$v = RT/p$$
$$v_1 = 55.16(459.7 + 100)/(144 \times 120) = 1.787 \text{ ft}^3/\text{lbm}$$
$$v_2 = 55.16(459.7 + 110)/(144 \times 80) = 2.728 \text{ ft}^3/\text{lbm}$$

$$\text{Flow area of pipe} \times \pi D^2/4 = \pi(5.761/12)^2/4 = 0.1810 \text{ ft}^2$$

From the continuity equation,

$$v = \dot{m}V/A$$
$$V_1 = (25 \times 1.787)/0.1810 = 246.8 \text{ ft/s}$$
$$V_2 = (25 \times 2.728)/0.1810 = 376.8 \text{ ft/s}$$

Applying the free-body equation for impulse momentum ($A = A_1 = A_2$),

$$R = (p_1 - p_a)A_1 - (p_2 - p_a)A_2 - \dot{M}(V_2 - V_1)$$
$$= (p_1 - p_2)A - M(V_2 - V_1) = 144(120 - 80)\ 0.1810$$
$$-(25/32.17)(376.8 - 246.8) = 941.5 \text{ lbf } (4.188 \times 10^3 \text{ N})$$

EXAMPLE. **Water Flow through a Nozzle.** Water at 68°F (20°C) flows through a horizontal 12- by 6-in-diameter nozzle discharging into the atmosphere.

The pressure at the nozzle inlet is 65 lbf/in² and barometric pressure is 14.7 lbf/in². Determine the force exerted by the water on the nozzle.

$$A = \pi D^2/4$$
$$A_1 = \pi(12/12)^2/4 = 0.7854 \text{ ft}^2$$
$$A_2 = \pi(6/12)^2/4 = 0.1963 \text{ ft}^2$$
$$\gamma = \rho g = 1.937 \times 32.17 = 62.31 \text{ lb/ft}^3$$

From the continuity equation $\rho_1 A_1 V_1 = \rho_2 A_2 V_2$ for $\rho_1 = \rho_2$, $V_2 = V_1 A_1/A_2 = (0.7854/0.1963)V_1 = 4V_1$. From Bernoulli's equation ($z_1 = z_2$).

$$p_1/\gamma + V_1^2/2g = p_2/\gamma + V_2^2/2g = p_2/\gamma + (4V_1)^2/2g$$

or
$$V_1 = \sqrt{2g(p_1 - p_2)/15\gamma}$$
$$= \sqrt{2 \times 32.17 \times 144(65 - 14.7)/15 \times 62.31} = 22.33 \text{ ft/s}$$
$$V_2 = 4 \times 22.33 = 89.32 \text{ ft/s}$$

Again from the equation of continuity

$$\dot{M} = \rho_1 A_1 V_1 = 1.937 \times 0.7854 \times 22.33 = 33.97 \text{ slugs/s}$$

Applying the free-body equation for impulse momentum,

$$R = (p_1 - p_a)A_1 - (p_2 - p_a)A_2 - \dot{M}(V_2 - V_1)$$
$$= 144(65 - 14.7)0.7854 - 144(14.7 - 14.7)0.1963$$
$$= (33.97)(89.32 - 22.33)$$
$$= 3{,}413 \text{ lbf } (1.518 \times 10^4 \text{ N})$$

Example. **Incompressible Flow through a Reducing Bend.** Carbon tetrachloride flows steadily without friction at 68°F (20°C) through a 90° horizontal reducing bend. The mass flow rate is 4 slugs/s, the inlet diameter is 6 in, and the outlet is 3 in. The inlet pressure is 50 lbf/in² and the barometric pressure is 14.7 lbf/in². Compute the magnitude and direction of the force required to "anchor" this bend.

$$A = \pi D^2/4$$
$$A_1 = (\pi/4)(6/12)^2 = 0.1963 \text{ ft}^2$$
$$A_2 = (\pi/4)(3/12)^2 = 0.04909 \text{ ft}^2$$

From continuity,

$$V = M/\rho A$$
$$V_1 = 4/(3.093)(0.1963) = 6.588 \text{ ft/s}$$
$$V_2 = 4/(3.093)(0.04909) = 26.35 \text{ ft/s}$$

From the Bernoulli equation ($z_1 = z_2$),

$$\frac{p_2}{\gamma} = \frac{p_1}{\gamma} + \frac{V_1^2}{2g} - \frac{V_2^2}{2g} = \frac{144 \times 50}{3.093 \times 32.17} + \frac{(6.588)^2 - (26.35)^2}{2 \times 32.17} = 62.24 \text{ ft}$$
$$p_2 = \frac{(3.093 \times 32.17)(62.24)}{144} = 43.01 \text{ lbf/in}^2$$

From Fig. 1.3.16,

$$\Sigma F_x = (p_1 - p_a)A_1 - (p_2 - p_a)A_2 \cos\alpha - R_x$$
$$= M(V_2 \cos\alpha - V_1)$$

or
$$R_x = (p_1 - p_a)A_1 - (p_2 - p_a)A_2 \cos\alpha - \dot{M}(V_2 \cos\alpha - V_1)$$
$$\Sigma F_y = 0 - (p_2 - p_a)A_2 \sin\alpha + R_y = \dot{M}(V_2 \sin\alpha - 0)$$

or
$$R_y = (p_2 - p_a)A_2 \sin\alpha + MV_2 \sin\alpha$$
$$R_x = 144(50 - 14.7)0.1963 - 144(43.01 - 14.7)(\cos 90°)$$
$$- 4(26.35 \cos 90° - 6.588)$$
$$= 1{,}024 \text{ lbf}$$
$$R_y = 144(43.01 - 14.7)(0.04909) \sin 90° + 4(26.35)(\sin 90°)$$
$$= 305.5 \text{ lbf}$$
$$R = \sqrt{R_x^2 + R_y^2} = \sqrt{(1{,}024)^2 + (305.5)^2}$$
$$= 1{,}068 \text{ lbf}_f\,(4.753 \times 10^3 \text{ N})$$
$$\theta = \tan^{-1}(F_y/F_x) = \tan^{-1}(305.5/1{,}024)$$
$$= 16°37'$$

Forces on Blades and Deflectors The forces imposed on a fluid jet whose velocity is V_J by a blade moving at a speed of V_b away from the jet are shown in Fig. 1.3.17. The following equations were developed from the application of the impulse-momentum equation for an open jet ($p_2 = p_1$) and for **frictionless** flow:

$$F_x = \rho A_J (V_J - V_b)^2 (1 - \cos\alpha)$$
$$F_y = \rho A_J (V_J - V_b)^2 \sin\alpha$$
$$F = 2\rho A_J (V_J - V_b)^2 \sin(\alpha/2)$$

Example. In the nozzle-blade system of Fig. 1.3.17, water at 68°F (20°C) enters a 3- by 1½-in-diameter horizontal nozzle with a pressure of 23 lbf/in² and discharges at 14.7 lbf/in² (atmospheric pressure). The blade moves away from the nozzle at a velocity of 10 ft/s and deflects the stream through an angle of 80°. For frictionless flow, calculate the total force exerted by the jet on the blade. Assume $g = g_c$; then $\gamma = \rho g$. From the continuity equation ($\rho_I = \rho_J$), $\rho_I A_I V_I = \rho_J A_J V_J$, $V_I = (A_J/A_I)V_J$,

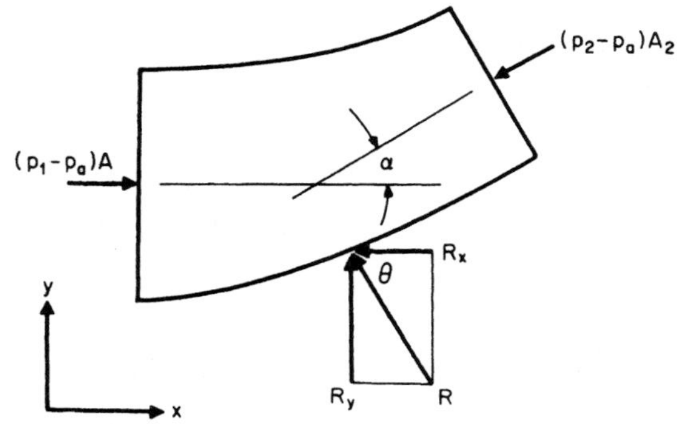

Figure 1.3.16 Forces on a bend.

$$V_I = \frac{\pi D_J^2/4}{\pi D_I^2/4} V_J = \left(\frac{D_J}{D_I}\right)^2 V_J$$
$$V_I = (1.5/3)^2 V_J = V_J/4$$

From the Bernoulli equation ($z_2 = z_1$),

$$\frac{V_J^2}{2g} = \frac{V_I^2}{2g} + \frac{p_I - p_J}{\rho g}$$
$$\frac{V_J^2 - (V_J/4)^2}{2g} = \frac{(p_I - p_J)}{\rho g}$$
$$V_J = \sqrt{\frac{2(16/15)(p_I - p_J)}{\rho}}$$
$$= \sqrt{\frac{2 \times (16/15)144(23 - 14.7)}{1.937}} = 36.28 \text{ ft/s}$$

The total force $F = 2\rho A_J (V_J - V_b)^2 \sin(\alpha/2)$

$$F = 2 \times 1.937(\pi/4)(1.5/12)^2(36.28 - 10)^2 \sin(80/2)$$
$$= 21.11 \text{ lbf } (93.90 \text{ N})$$

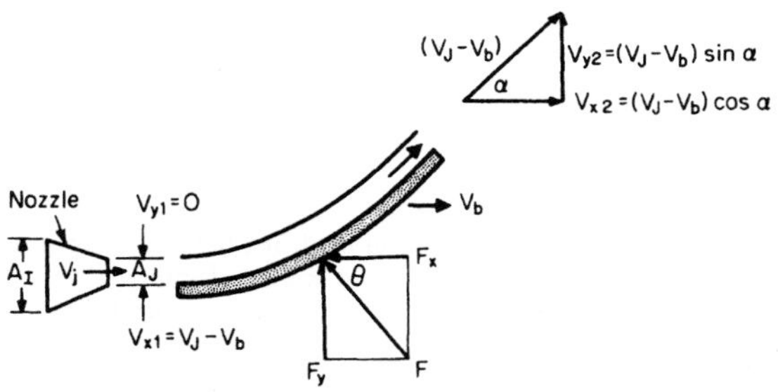

Figure 1.3.17 Notation for blade study.

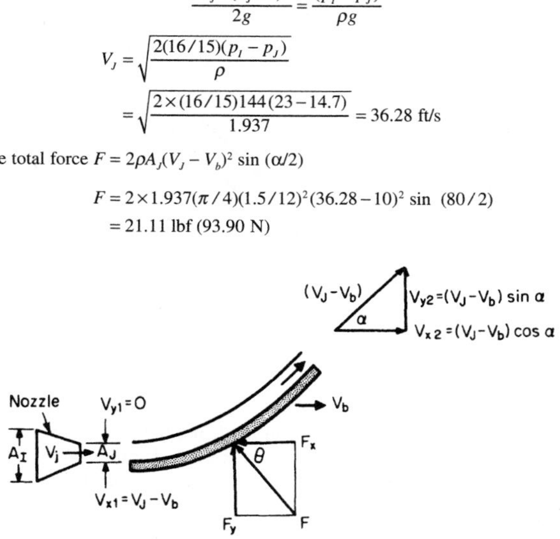

Impulse Turbine In a turbine, the total of the separate forces acting simultaneously on each blade equals that caused by the combined mass flow rate $\dot{M}$ discharged by the nozzle or

$$\Sigma P = \Sigma F_x V_b = \dot{M}(V_J - V_b)(1 - \cos\alpha)V_b$$

The maximum value of power P is found by differentiating P with respect to V_b and setting the result equal to zero. Solving for V_b yields $V_b = V_J/2$, so that maximum power occurs when the velocity of the jet is equal to twice the velocity of the blade. Examination of the power equation also indicates that the angle α for a maximum power results when $\cos\theta = -1$ or $\alpha = 180°$. For theoretical maximum power of a blade, $2V_b = V_J$ and $\alpha = 180°$. It should be noted that in any practical impulse-turbine application, α cannot be 180° because the discharge interferes with the next set of blades. Substituting $V_b = V_J/2$, $\alpha = 180°$ in the power equation,

$$\Sigma P_{max} = \dot{M}(V_J - V_J/2)[1-(-1)]V_J/2 = MV_J^2/2 = \dot{m}V_J^2/2g_c$$

or the maximum power per unit mass is equal to the total power of the jet. Application of the Bernoulli equation between the surface of a reservoir and the discharge of the turbine shows that $\Sigma P_{max} = \dot{M}\sqrt{2g(z_2 - z_1)}$. For design details, see Sec. 7.9.

Flow in a Curved Path When a fluid flows through a bend, it is also rotated around an axis and the energy required to produce rotation must be supplied from the energy already in the fluid mass. This fluid rotation is called a **free vortex** because it is free of outside energy. Consider the fluid mass $\rho(r_o - r_i)dA$ of Fig. 1.3.18 being rotated as it flows through a bend of outer radius r_o, inner radius r_i, with a velocity of V. Application of Newton's second law to this mass results in

$$dF = p_o\, dA - p_i\, dA = [\rho(r_o - r_i)\, dA][V^2/(r_o + r_i)/2]$$

which reduces to

$$p_o - p_i = 2(r_o - r_i)\rho V^2/(r_o + r_i)$$

Because of the difference in fluid pressure between the inner and outer walls of the bend, secondary flows are set up, and this is the primary cause of friction loss of bends. These secondary flows set up turbulence that require 50 or more straight pipe diameters downstream to dissipate.

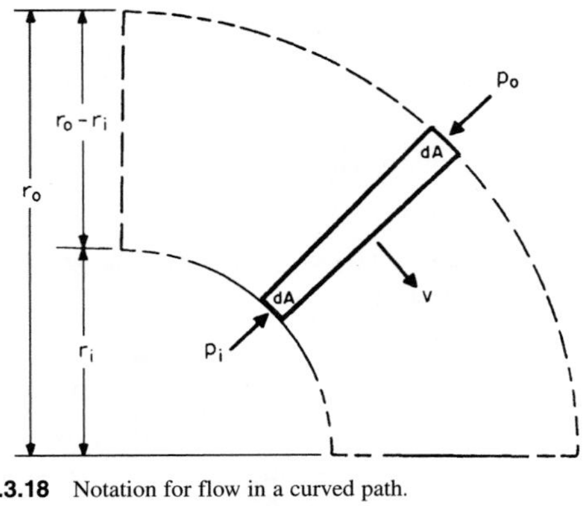

Figure 1.3.18 Notation for flow in a curved path.

Thus this loss does not take place in the bend, but in the downstream system. These losses may be reduced by the use of splitter plates which help minimize the secondary flows by reducing $r_o - r_i$ and hence $p_o - p_i$.

EXAMPLE. 104°F (40°C) benzene flows at a rate of 8 ft³/s in a square horizontal duct. This duct makes a 90° turn with an inner radius of 1 ft and an outer radius of 2 ft. Calculate the difference between the walls of this bend. The area of this duct is $(r_o - r_i)^2 = (2-1)^2 = 1$ ft². From the continuity equation $V = Q/A = 8/1 = 8$ ft/s. The pressure difference

$$\begin{aligned} p_o - p_i &= 2(r_o - r_i)\rho V^2/(r_o + r_i) \\ &= 2(2-1)1.663\,(8)^2/(2+1) = 70.95\ \text{lbf/ft}^2 \\ &= 70.95/144 = 0.4927\ \text{lbf/in}^2\ (3.397\times10^3\ \text{N/m}^2) \end{aligned}$$

DIMENSIONLESS PARAMETERS

Modern engineering practice is based on a combination of theoretical analysis and experimental data. Often the engineer is faced with the necessity of obtaining practical results in situations where for various reasons, physical phenomena cannot be described mathematically and experimental data must be considered. The generation and use of **dimensionless parameters** provides a powerful and useful tool in (1) reducing the number of variables required for an experimental program, (2) establishing the principles of model design and testing, (3) developing equations, and (4) converting data from one system of units to another. Dimensionless parameters may be generated from (1) **physical equations,** (2) the principles of **similarity,** and (3) **dimensional analysis.** All **physical equations** must be dimensionally correct so that a dimensionless parameter may be generated by simply dividing one side of the equation by the other. A minimum of two dimensionless parameters will be formed, one being the inverse of the other.

EXAMPLE. It is desired to generate a series of dimensionless parameters to describe the ratios of static pressure head, velocity head, and potential head to total head for frictionless incompressible flow. From the Bernoulli equation,

$$\frac{p}{\gamma} + \frac{V^2}{2g} + z = \Sigma h = \text{total head}$$

$$N_1 = \frac{p/\gamma + V^2/2g + z}{\Sigma h} = \frac{p/\gamma}{\Sigma h} + \frac{V^2/2g}{\Sigma h} + \frac{z}{\Sigma h} = N_p + N_V + N_z$$

or

$$N_2 = \frac{\Sigma h}{p_2/\gamma + V^2/2g + z} = N_1^{-1}$$

N_1 and N_2 are total energy ratios and N_p, N_V, and N_z are the ratios of the static pressure head, velocity head, and potential head, respectively, to the total head.

Models versus Prototypes There are times when for economic or other reasons it is desirable to determine the performance of a structure or machine by testing another structure or machine. This type of testing is called model testing. The equipment being tested is called a **model,** and the equipment whose performance is to be predicted is called a **prototype.** A **model** may be smaller than, the same size as, or larger than the **prototype.** Model experiments on aircraft, rockets, missiles, pipes, ships, canals, turbines, pumps, and other structures and machines have resulted in savings that more than justified the expenditure of funds for the design, construction, and testing of the model. In some situations, the model and the prototype may be the same piece of equipment, for example, the laboratory calibration of a flowmeter with water to predict its performance with other fluids. Many manufacturers of fluid machinery have test facilities that are limited to one or two fluids and are forced to test with what they have available in order to predict performance with nonavailable fluids. For towing-tank testing of ship models and for wind-tunnel testing of aircraft and aircraft-component models, see Secs. 8.4 and 8.5.

Similarity Requirements For complete similarity between a model and its prototype, it is necessary to have **geometric, kinematic,** and **dynamic** similarity. **Geometric similarity** exists between model and prototype when the **ratios** of all corresponding dimensions of the model and prototype are equal. These ratios may be written as follows:

Length: $L_{model}/L_{prototype} = L_{ratio} = L_m/L_p = L_r$

Area: $L^2_{model}/L^2_{prototype} = L^2_{ratio} = L^2_m/L^2_p = L^2_r$

Volume: $L^3_{model}/L^3_{prototype} = L^3_{ratio} = L^3_m/L^3_p = L^3_r$

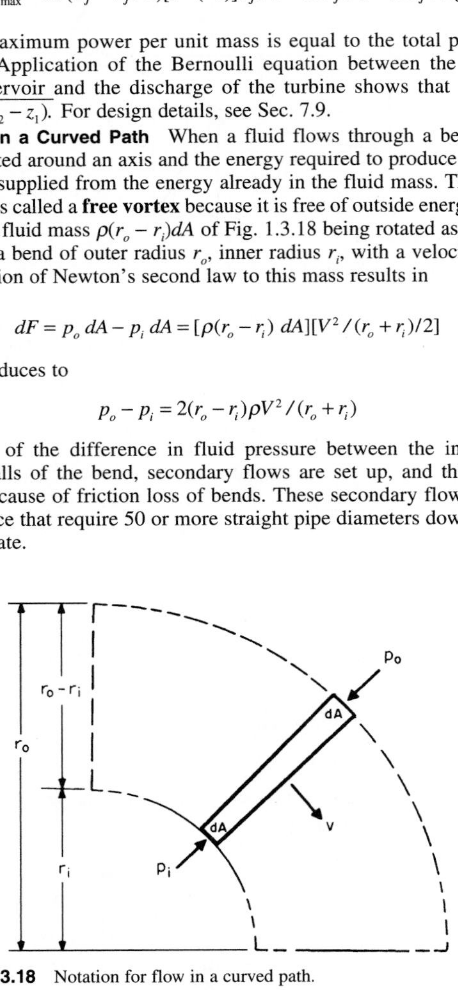

Kinematic similarity exists between model and prototype when their streamlines are geometrically similar. The kinematic ratios resulting from this condition are

Acceleration: $a_r = a_m / a_p = L_m T_m^{-2} / L_p T_p^{-2} = L_r / T_r^{-2}$

Velocity: $V_r = V_m / V_p = L_m T_m^{-1} / L_p T_p^{-1} = L_r / T_r^{-1}$

Volume flow rate: $Q_r = Q_m / Q_p = L_m^3 T_m^{-1} / L_p^3 T_p^{-1} = L_r^3 / T_r^{-1}$

Dynamic similarity exists between model and prototype having geometric and kinematic similarity when the ratios of all forces are the same. Consider the model/prototype relations for the flow around the object shown in Fig. 1.3.19. For **geometric similarity** $D_m/D_p = L_m/L_p = L_r$ and for **kinematic similarity** $U_{Am}/U_{Ap} = U_{Bm}/U_{Bp} = V_r = L_r T_r^{-1}$. Next consider the three forces acting on point C of Fig. 1.3.19 without specifying their nature. From the geometric similarity of their vector polygons and Newton's law, for **dynamic similarity** $F_{1m}/F_{1p} = F_{2m}/F_{2p} = F_{3m}/F_{3p} = M_m a_{Cm}/M_p a_{Cp} = F_r$. For dynamic similarity, these force ratios must be maintained on all corresponding fluid particles throughout the flow pattern. From the force polygon of Fig. 1.3.19, it is

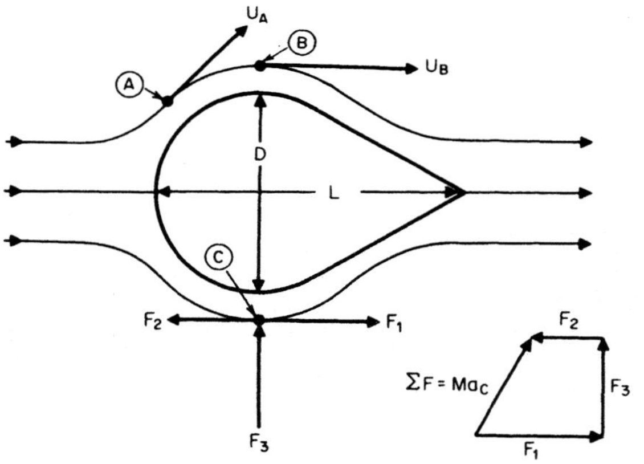

Figure 1.3.19 Notation for dynamic similarity.

evident that $F_1 +\!\!\rightarrow F_2 +\!\!\rightarrow F_3 = Ma_C$. For total model/prototype force ratio, comparisons of force polygons yield

$$F_r = \frac{F_{1m} +\!\!\rightarrow F_{2m} +\!\!\rightarrow F_{3m}}{F_{1p} +\!\!\rightarrow F_{2p} +\!\!\rightarrow F_{3p}} = \frac{M_m a_{Cm}}{M_p a_{Cp}}$$

Fluid Forces The fluid forces that are considered here are those acting on a fluid element whose mass = ρL^3, area = L^2, length = L, and velocity = L/T.

Inertia force F_i = (mass)(acceleration) $= (\rho L^3)(L/T^2) = \rho L(L^2/T^2) = \rho L^2 V^2$

Viscous force F_μ = (viscous shear stress)(shear area) $= \tau L^2 = \mu(dU/dy)L^2 = \mu(V/L)L^2 = \mu LV$

Gravity force F_g = (mass)(acceleration due to gravity) $= (\rho L^3)(g) = \rho L^3 g$

Pressure force F_p = (pressure)(area) $= pL^2$

Centrifugal force F_ω = (mass)(acceleration) $= (\rho L^3)(L/T^2) = (\rho L^3)(L\omega^2) = \rho L^4 \omega^2$

Elastic force F_E = (modulus of elasticity)(area) $= EL^2$

Surface-tension force F_σ = (surface tension)(length) $= \sigma L$

Vibratory force F_f = (mass)(acceleration) $= (\rho L^3)(L/T^2) = (\rho L^4)(T^{-2}) = \rho L^4 f^2$

If all fluid forces were acting on a fluid element,

$$F_r = \frac{F_{\mu m} +\!\!\rightarrow F_{gm} +\!\!\rightarrow F_{pm} +\!\!\rightarrow F_{\omega m} +\!\!\rightarrow F_{Em} +\!\!\rightarrow F_{\sigma m} +\!\!\rightarrow F_{fm}}{F_{\mu p} +\!\!\rightarrow F_{gp} +\!\!\rightarrow F_{pp} +\!\!\rightarrow F_{\omega p} +\!\!\rightarrow F_{Ep} +\!\!\rightarrow F_{\sigma p} +\!\!\rightarrow F_{fp}} = \frac{F_{im}}{F_{ip}}$$

Examination of the above equation and the force polygon of Fig. 1.3.19 lead to the conclusion that dynamic similarity can be characterized by an equality of force ratios one less than the total number involved. Any force ratio may be eliminated, depending upon the quantities which are desired. Fortunately, in most practical engineering problems, not all of the eight forces are involved because some may not be acting, may be of negligible magnitude, or may be in opposition to each other in such a way as to compensate. In each application of similarity, a good understanding of the fluid phenomena involved is necessary to eliminate the irrelevant, trivial, or compensating forces. When the flow phenomenon is too complex to be readily analyzed, or is not known, only experimental verification with the prototype of results from a model test will determine what forces should be considered in future model testing.

Standard Numbers With eight fluid forces that can act in flow situations, the number of dimensionless parameters that can be formed from their ratios is 56. However, conventional practice is to ratio the inertia force to the other fluid forces, usually by division because the inertia force is the vector sum of all the other forces involved in a given flow situation. Results obtained by dividing the inertia force by each of the other forces are shown in Table 1.3.7 compared with the standard numbers that are used in conventional practice.

DYNAMIC SIMILARITY

Vibration In the flow of fluids around objects and in the motion of bodies immersed in fluids, **vibration** may occur because of the formation of a wake caused by alternate shedding of eddies in a periodic fashion or by the vibration of the object or the body. The **Strouhal number S** is the ratio of the velocity of vibration Lf to the velocity of the fluid V. Since the vibration may be fluid-induced or structure-induced, two frequencies must be considered, the wake frequency $f\omega$ and the natural frequency of the structure f_n. Fluid-induced forces are usually of small magnitude, but as the wake frequency approaches the natural frequency of the structure, the vibratory forces increase very rapidly. When $f\omega = f_n$, the structure will go into resonance and fail. This imposes on the model designer the requirement of matching to scale the natural-frequency characteristics of the prototype. This subject is treated later under Wake Frequency. All further discussions of model/prototype relations are made under the assumption that either vibratory forces are absent or they are taken care of in the design of the model or in the test program.

Incompressible Flow Considered in this category are the flow of fluids around an object, motion of bodies immersed in incompressible fluids, and the flow of incompressible fluids in conduits. It includes, for example, a submarine traveling under water but not partly submerged, and liquids flowing in pipes and passages when the liquid completely fills them, but not when partly full as in open-channel flow. It also includes aircraft moving in atmospheres that may be considered incompressible. Incompressible flow in rotating machinery is considered separately.

In these situations the gravity force, although acting on all fluid particles, does not affect the flow pattern. Excluding rotating machinery, centrifugal forces are absent. By definition of an incompressible fluid, elastic forces are zero, and since there is no liquid-gas interface, surface-tension forces are absent.

The only forces now remaining for consideration are the inertia, viscous, and pressure. Using standard numbers, the parameters are **Reynolds**

Table 1.3.7 Standard Numbers

Force ratio	Equations	Result	Conventional practice: Form	Symbol	Name
$\frac{\text{Inertia}}{\text{Viscous}}$	$\frac{F_i}{F_\mu}=\frac{\rho L^2V^2}{\mu LV}$	$\frac{\rho LV}{\mu}$	$\frac{\rho LV}{\mu}$	**R**	Reynolds
$\frac{\text{Inertia}}{\text{Gravity}}$	$\frac{F_i}{F_g}=\frac{\rho L^2V^2}{\rho L^3 g}$	$\frac{V^2}{Lg}$	$\frac{V}{\sqrt{Lg}}$	**F**	Froude
$\frac{\text{Inertia}}{\text{Centrifugal}}$	$\frac{F_i}{F_p}=\frac{\rho L^2V^2}{\rho L^3}$	$\frac{\rho V^2}{p}$	$\frac{\rho V^2}{p}$	**E**	Euler
			$\frac{2\Delta p}{\rho V^2}$	**Cp**	Pressure coefficient
$\frac{\text{Inertia}}{\text{Centrifugal}}$	$\frac{F_i}{F_\omega}=\frac{\rho L^2V^2}{\rho L^4\omega^2}$	$\frac{V^2}{L^2\omega^2}$	$\frac{V}{DN}$	**V**	Velocity ratio
$\frac{\text{Inertia}}{\text{Elastic}}$	$\frac{F_i}{F_E}=\frac{\rho L^2V^2}{EL^2}$	$\frac{\rho V^2}{E}$	$\frac{\rho V^2}{E}$	**C**	Cauchy
			$\frac{V}{\sqrt{E/\rho}}$	**M**	Mach
$\frac{\text{Inertia}}{\text{Surface tension}}$	$\frac{F_i}{F_\sigma}=\frac{\rho L^2V^2}{\sigma^L}$	$\frac{\rho LV^2}{\sigma}$	$\frac{\rho LV^2}{\sigma}$	**W**	Weber
$\frac{\text{Inertia}}{\text{Vibration}}$	$\frac{F_i}{F_f}=\frac{\rho L^2V^2}{\rho L^4 f^2}$	$\frac{V^2}{L^2f^2}$	$\frac{Lf}{V}$	**S**	Strouhal

SOURCE: Computed from data given in Murdock, "Fluid Mechanics and Its Applications," Houghton Mifflin, 1976.

number and **pressure coefficient.** The Reynolds number may be converted into a kinematic ratio by noting that by definition $\nu = \mu/\rho$ and substituting in $\mathbf{R} = \rho LV/\mu = LV/\nu$. In this form, Reynolds number is the ratio of the fluid velocity V and the "shear velocity" ν/L. For this reason, Reynolds number is used to characterize the velocity profile. Forces and pressure losses are then determined by the pressure coefficient.

EXAMPLE. A submarine is to move submerged through 32°F (0°C) seawater at a speed of 10 knots. (1) At what speed should a 1:20 model be towed in 68°F (20°C) fresh water? (2) If the thrust on the model is found to be 42,500 lbf, what horsepower will be required to propel the submarine?

1. Speed of model for Reynolds-number similarity

$$R_m = R_p = \left(\frac{\rho VL}{\mu}\right)_m = \left(\frac{\rho VL}{\mu}\right)_p$$

$$V_m = V_p(\rho_p/\rho_m)(L_p/L_m)(\mu_m/\mu_p)$$

$$V_m = (10)(1.995/1.937)(20/1)(20.92\times10^{-6}/39.40\times10^{-6})$$

$$= 109.4 \text{ knots } (56.27 \text{ m/s})$$

2. Prototype horsepower

$$\mathbf{C}_{pp} = \mathbf{C}_{pm} = \left(\frac{2\Delta p}{\rho V^2}\right)_p = \left(\frac{2\Delta p}{\rho V^2}\right)_m$$

$$F = \Delta pL^2,\ \Delta p = \frac{F}{L^2},\ \text{so that}\ \left(\frac{2F}{\rho V^2L^2}\right)_p = \left(\frac{2F}{\rho V^2L^2}\right)_m$$

$$F_p = F_m(\rho_p/\rho_m)(V_p/V_m)^2(L_p/L_m)^2$$

$$= 42{,}500(1.995/1.937)(10/109.4)^2(20/1)^2 = 146{,}300 \text{ lbf}$$

$$P = FV = \left(\frac{146{,}300}{550}\right)\left(\frac{10\times6{,}076}{3{,}600}\right)$$

$$= 4{,}490 \text{ hp } (3.35\times10^6 \text{ W})$$

Compressible Flow Considered in this category are the flow of compressible fluids under the conditions specified for incompressible flow in the preceding paragraphs. In addition to the forces involved in incompressible flow, the elastic force must be added. Conventional practice is to use the square root of the inertia/elastic force ratio or **Mach number.**

Mach number is the ratio of the fluid velocity to its speed of sound and may be written $\mathbf{M} = V/c = V\sqrt{E_s/\rho}$. For an ideal gas, $\mathbf{M} = V/\sqrt{kg_cRT}$. In compressible-flow problems, practice is to use the Mach number to characterize the velocity or kinematic similarity, the Reynolds number for dynamic similarity, and the pressure coefficient for force or pressure-loss determination.

EXAMPLE. An airplane is to fly at 500 mi/h in an atmosphere whose temperature is 32°F (0°C) and pressure is 12 lbf/in². A 1:20 model is tested in a wind tunnel where a supply of air at 392°F (200°C) and variable pressure is available. At (1) what speed and (2) what pressure should the model be tested for dynamic similarity?

1. *Speed for Mach-number similarity*

$$\mathbf{M}_m = \mathbf{M}_p = \left(\frac{V}{\sqrt{E/\rho}}\right)_m = \left(\frac{V}{\sqrt{E/\rho}}\right)_p = \left(\frac{V}{\sqrt{kg_cRT}}\right)_m = \left(\frac{V}{\sqrt{kg_cRT}}\right)_p$$

$$V_m = V_p(k_m/k_p)^{1/2}(R_m/R_p)^{1/2}(T_m/T_p)^{1/2}$$

For the same gas $k_m = k_p$, $R_m = R_p$, and

$$V_m = V_p\sqrt{T_m/T_p} = 500\sqrt{(851.7/491.7)} = 658.1 \text{ mi/h}$$

2. *Pressure for Reynolds-number similarity*

$$\mathbf{R}_m = \mathbf{R}_p = \left(\frac{\rho VL}{\mu}\right)_m = \left(\frac{\rho VL}{\mu}\right)_p$$

$$\rho_m = \rho_p(V_p/V_m)(L_p/L_m)(\mu_m/\mu_p)$$

Since $\rho = p/g_cRT$

$$\left(\frac{p}{g_cRT}\right)_m = \left(\frac{p}{g_cRT}\right)_p\left(\frac{V_p}{V_m}\right)\left(\frac{L_p}{L_m}\right)\left(\frac{\mu_m}{\mu_p}\right)$$

$$p_m = p_p(T_m/T_p)(V_p/V_m)(L_p/L_m)(\mu_m/\mu_p)$$

$$p_m = 12(851.7/491.7)(500/658.1)(20/1)(53.15\times10^{-6}/35.67\times10^{-6})$$

$$p_m = 470.6 \text{ lbf/in}^2\ (3.245\times10^6 \text{ N/m}^2)$$

For information about wind-tunnel testing and its limitations, refer to Sec. 8.4.

Centrifugal Machinery   This category includes the flow of fluids in such centrifugal machinery as compressors, fans, and pumps. In addition to the inertia, pressure, viscous, and elastic forces, centrifugal forces must now be considered. Since centrifugal force is really a special case of the inertia force, their ratio as shown in Table 1.3.7 is **velocity ratio** and is the ratio of the fluid velocity to the machine tangential velocity. In model/prototype relations for centrifugal machinery, DN (D = diameter, ft, N = rotational speed) is substituted for the velocity V, and D for L, which results in the following:

$$\mathbf{M} = DN/\sqrt{kg_cRT} \qquad \mathbf{R} = \rho D^2N/\mu \qquad \mathbf{C}_p = 2\Delta p/\rho D^2N^2$$

EXAMPLE. A centrifugal compressor operating at 100 r/min is to compress methane delivered to it at 50 lbf/in² and 68°F (20°C). It is proposed to test this compressor with air from a source at 140°F (60°C) and 100 lbf/in². Determine compressor speed and inlet-air pressure required for dynamic similarity. Find speed for Mach-number similarity:

$$\mathbf{M}_m = \mathbf{M}_p = (DN/\sqrt{kg_cRT})_m = (DN/\sqrt{kg_cRT})_p$$
$$N_m = N_p(D_p/D_m)\sqrt{(k_m/k_p)(R_m/R_p)(T_m/T_p)}$$
$$= 100(1)\sqrt{(1.40/1.32)(53.34/96.33)(599.7/527)}$$
$$= 81.70 \text{ r/min}$$

Find pressure for Reynolds-number similarity:

$$\mathbf{R}_m = \mathbf{R}_p = \left(\frac{\rho D^2N}{\mu}\right)_m = \left(\frac{\rho D^2N}{\mu}\right)_p$$

For an ideal gas $\rho = p/g_cRT$, so that

$$\left(pD^2N/g_cRT\mu\right)_m = \left(pD^2N/g_cRT\mu\right)_p$$
$$p_m = p_p(D_p/D_m)^2(N_p/N_m)(R_m/R_p)(T_m/T_p)(\mu_m/\mu_p)$$
$$= 50(1)^2(100/81.70)(53.34/96.33)(599.7/527.7)\times(41.79$$
$$\times 10^{-8}/22.70 \times 10^{-8}) = 70.90 \text{ lbf/in}^2(4.888\times10^5 \text{ N/m}^2)$$

See Sec. 10 for specific information on pump and compressor similarity.

Liquid Surfaces   Considered in this category are ships, seaplanes during takeoff, submarines partly submerged, piers, dams, rivers, openchannel flow, spillways, harbors, etc. Resistance at liquid surfaces is due to surface tension and wave action. Since wave action is due to gravity, the gravity force and surface-tension force are now added to the forces that were considered in the last paragraph. These are expressed as the square root of the inertia/gravity force ratio or **Froude number** $F = V/\sqrt{Lg}$ and as the inertia/surface tension force ratio or **Weber number** $W = \rho LV^2/\sigma$. On the other hand, elastic and pressure forces are now absent. Surface tension is a minor property in fluid mechanics and it normally exerts a negligible effect on wave formation *except* when the waves are small, say less than 1 in. Thus the effects of surface tension on the model might be considerable, but negligible on the prototype. This type of "scale effect" must be avoided. For accurate results, the inertia/surface tension force ratio or Weber number should be considered. It is never possible to have complete dynamic similarity of liquid surfaces unless the model and prototype are the same size, as shown in the following example.

EXAMPLE. An ocean vessel 500 ft long is to travel at a speed of 15 knots. A 1:25 model of this ship is to be tested in a towing tank using seawater at design temperature. Determine the model speed required for (1) wave-resistance similarity, (2) viscous or skin-friction similarity, (3) surface-tension similarity, and (4) the model size required for complete dynamic similarity.

1. *Speed for **Froude-number** similarity*

$$\mathbf{F}_m = \mathbf{F}_p = (V/\sqrt{Lg})_m = (V/\sqrt{Lg})_p$$

or

$$V_m = V_p\sqrt{L_m/L_p} = 15\sqrt{1/25} = 3 \text{ knots}$$

2. *Speed for **Reynolds-number** similarity*

$$\mathbf{R}_m = \mathbf{R}_p = (\rho LV/\mu)_m = (\rho LV/\mu)_p$$
$$V_m = V_p(\rho_p/\rho_m)(L_p/L_m)(\mu_m/\mu_p)$$
$$V_m = 15(1)(25/1)(1) = 375 \text{ knots}$$

3. *Speed for **Weber-number** similarity*

$$\mathbf{W}_m = \mathbf{W}_p = (\rho LV^2/\sigma)_m = (\rho LV^2/\sigma)_p$$
$$V_m = V_p\sqrt{(\rho_p/\rho_m)(L_p/L_m)(\sigma_m/\sigma_p)}$$
$$V_m = 15\sqrt{(1)(25)(1)} = 75 \text{ knots}$$

4. *Model size for complete similarity.* First try Reynolds and Froude similarity; let

$$V_m = V_p(\rho_p/\rho_m)(L_p/L_m)(\mu_m/\mu_p) = V_p\sqrt{L_m/L_p}$$

which reduces to

$$L_m/L_p = (\rho_p/\rho_m)^{2/3}(\mu_m/\mu_p)^{2/3}$$

Next try Weber and Froude similarity; let

$$V_m = V_p\sqrt{(\rho_p/\rho_m)(L_p/L_m)(\sigma_m/\sigma_p)} = V_p\sqrt{L_m/L_p}$$

which reduces to

$$L_m/L_p = (\rho_p/\rho_m)^{1/2}(\sigma_m/\sigma_p)^{1/2}$$

For the same fluid at the same temperature, either of the above solves for $L_m = L_p$, or the model must be the same size as the prototype. For use of different fluids and/or the same fluid at different temperatures.

$$L_m/L_p = (\rho_p/\rho_m)^{2/3}(\mu_m/\mu_p)^{2/3} = (\rho_p/\rho_m)^{1/2}(\sigma_m/\sigma_p)^{1/2}$$

which reduces to

$$(\mu^4/\rho\sigma^3)_m = (\mu^4/\rho\sigma^3)_p$$

No practical way has been found to model for complete similarity. Marine engineering practice is to model for wave resistance and correct for skin-friction resistance. See Sec. 8.3.

DIMENSIONAL ANALYSIS

Dimensional analysis is the mathematics of dimensions and quantities and provides procedural techniques whereby the variables that are assumed to be significant in a problem can be formed into dimensionless parameters, the number of parameters being less than the number of variables. This is a great advantage, because fewer experimental runs are then required to establish a relationship between the parameters than between the variables. While the user is not presumed to have any knowledge of the fundamental physical equations, the more knowledgeable the user, the better the results. If any significant variable or variables are omitted, the relationship obtained from dimensional analysis will not apply to the physical problem. On the other hand, inclusion of all possible variables will result in losing the principal advantage of dimensional analysis, i.e., the reduction of the amount of experimental data required to establish relationships. Two formal methods of dimensional analysis are used, the **method of Lord Rayleigh** and **Buckingham's Π theorem.**

Dimensions used in mechanics are mass M, length L, time T, and force F. Corresponding **units** for these dimensions are the slug (kilogram), the foot (metre), the second (second), and the pound force (newton). Any system in mechanics can be defined by three fundamental dimensions. Two systems are used, the force (*FLT*) and the mass (*MLT*). In the force system, mass is a derived quantity and in the mass system, force is a derived quantity. Force and mass are related by Newton's law: $F = MLT^{-2}$ and $M = FL^{-1}T^2$. Table 1.3.8 shows common variables and their dimensions and units.

Lord Rayleigh's method uses algebra to determine interrelationships among variables. While this method may be used for any number of variables, it becomes relatively complex and is not generally used for more than four. This method is most easily described by example.

EXAMPLE. In laminar flow, the unit shear stress τ is some function of the fluid dynamic viscosity μ, the velocity difference dU between adjacent laminae separated by the distance dy. Develop a relationship.

1. Write a functional relationship of the variables:

$$\tau = f(\mu, \, dU, \, \mathbf{dy})$$

Assume $\tau = K(\mu^a \, dU^b \, dy^c)$.

Table 1.3.8 Dimensions and Units of Common Variables

		Dimensions		Units	
Symbol	Variable	MLT	FLT	USCS*	SI
		Geometric			
L	Length	L		ft	m
A	Area	L^2		ft^2	m^2
V	Volume	L^3		ft^3	m^3
		Kinematic			
t	Time	T		s	s
ω	Angular velocity	T^{-1}		s^{-1}	s^{-1}
f	Frequency				
V	Velocity	LT^{-1}		ft/s	m/s
ν	Kinematic viscosity	L^2T^{-1}		ft^2/s	m^2/s
Q	Volume flow rate	L^3T^{-1}		ft^3/s	m^3/s
α	Angular acceleration	T^{-2}		s^{-2}	s^2
a	Acceleration	LT^{-2}		ft/s^{-2}	m/s^2
		Dynamic			
ρ	Density	ML^{-3}	$FL^{-4}T^2$	$slug/ft^3$	kg/m^3
M	Mass	M	$FL^{-1}T^2$	slug	kg
I	Moment of inertia	ML^2	FLT^2	$slug \cdot ft^3$	$kg \cdot m^2$
μ	Dynamic viscosity	$ML^{-1}T^{-1}$	$FL^{-2}T$	$slug/ft \cdot s$	$kg/m \cdot s$
M	Mass flow rate	MT^{-1}	$FL^{-1}T^{-1}$	slug/s	kg/s
MV	Momentum	MLT^{-1}	FT	$lbf \cdot s$	$N \cdot s$
Ft	Impulse				
$M\omega$	Angular momentum	ML^2T^{-1}	FLT	$slug \cdot ft^2/s$	$kg \cdot m^2/s$
γ	Specific weight	$ML^{-2}T^{-2}$	FL^{-3}	lbf/ft^3	N/m^3
p	Pressure				
τ	Unit shear stress	$ML^{-1}T^{-2}$	FL^{-2}	lbf/ft^2	N/m^2
E	Modulus of elasticity				
σ	Surface tension	MT^{-2}	FL^{-1}	lbf/ft	N/m
F	Force	MLT^{-2}	F	lbf	N
E	Energy				
W	Work	ML^2T^{-2}	FL	$lbf \cdot ft$	J
FL	Torque				
P	Power	ML^2T^{-3}	FLT^{-1}	$lbf \cdot ft/s$	W
v	Specific volume	$M^{-1}L^3$	$F^{-1}L^4T^{-2}$	ft^3/lbm	m^3/kg

*United States Customary System.

2. Write a dimensional equation in either *FLT* or *MLT* system:

$$(FL^{-2}) = K(FL^{-2}T)^a(LT^{-1})^b(L)^c$$

3. Solve the dimensional equation for exponents:

		τ	$\mu\ dU\ dy$
Force	F	$1 =$	$a + 0 + 0$
Length	L	$-2 =$	$-2a + b + c$
Time	T	$0 =$	$a - b + 0$

Solution: $a = 1$, $b = 1$, $c = -1$

4. Insert exponents in the functional equation: $\tau = K(\mu^a\ dU^b\ dy^c) = K(\mu^1\ du^1\ dy^{-1})$, or $K = (\mu\, dU/\tau\, dy)$. This was based on the assumption of $\tau = K(\mu^a\ dU^b\ dy^c)$. The general relationship is $K = f(\mu\, dU/\tau\, dy)$. The functional relationship cannot be obtained from dimensional analysis. Only physical analysis and/or experiments can determine this. From both physical analysis and experimental data,

$$\tau = \mu\ dU/dy$$

The Buckingham Π theorem serves the same purpose as the method of Lord Rayleigh for deriving equations expressing one variable in terms of its dependent variables. The Π theorem is preferred when the number of variables exceeds four. Application of the Π theorem results in the formation of dimensionless parameters called π ratios. These π ratios have no relation to 3.14159. . . . The Π theorem will be illustrated in the following example.

EXAMPLE. Experiments are to be conducted with gas bubbles rising in a still liquid. Consider a gas bubble of diameter D rising in a liquid whose density is ρ, surface tension σ, viscosity μ, rising with a velocity of V in a gravitational field of g. Find a set of parameters for organizing experimental results.

1. List all the physical variables considered according to type: geometric, kinematic, or dynamic.
2. Choose either the *FLT* or *MLT* system of dimensions.
3. Select a "basic group" of variables characteristic of the flow as follows:

   a. B_G, a geometric variable
   b. B_K, a kinematic variable
   c. B_D, a dynamic variable (if three dimensions are used)

4. Assign A numbers to the remaining variables starting with A_1.

Type	*Symbol*	*Description*	*Dimenstions*	*Number*
Geometric	D	Bubble diameter	L	B_G
Kinematic	V	Bubble velocity	LT^{-1}	B_K
	g	Acceleration of gravity	LT^{-2}	A_1
Dynamic	ρ	Liquid density	ML^{-3}	B_D
	σ	Surface tension	MT^{-2}	A_2
	μ	Liquid viscosity	$ML^{-1}T^{-1}$	A_3

5. Write the basic equation for each π ratio as follows:

$$\pi_1 = (B_G)^{x1}(B_K)^{y1}(B_D)^{z1}(A_1)$$
$$\pi_2 = (B_G)^{x2}(B_K)^{y2}(B_D)^{z2}(A_2)\ \ldots \pi_n = (B_G)^{xn}(B_K)^{yn}(B_D)^{zn}(A_n)$$

Note that the number of π ratios is equal to the number of A numbers and thus equal to the number of variables less the number of fundamental dimensions in a problem.

6. Write the dimensional equations and use the algebraic method to determine the value of exponents x, y, and z for each π ratio. Note that for all π ratios, the sum of the exponents of a given dimension is zero.

$$\pi_1 = (B_G)^{x1}(B_K)^{y1}(B_D)^{z1}(A_1) = (D)^{x1}(V)^{y1}(\rho)^{z1}(g)$$
$$(M^0L^0T^0) = (L^{x1})(L^{y1}T^{-y1})(M^{z1}L^{-3z1})(LT^{-2})$$

Solution: $x_1 = 1,\ y_1 = -2,\ z_1 = 0$

$$\pi_1 = D^1V^{-2}\rho^0 g = Dg/V^2$$
$$\pi_2 = (B_G)^{x2}(B_K)^{y2}(B_D)^{z2}(A_2) = (D)^{x2}(V)^{y2}(\rho)^{z2}(\sigma)$$
$$(M^0L^0T^0) = (L^{x2})(L^{y2}T^{-y2})(M^{z2}L^{-3z2})(MT^{-2})$$

Solution: $x_2 = -1,\ y_2 = -2,\ z_2 = -1$

$$\pi_2 = D^{-1}V^{-2}\rho^{-1}\sigma = \sigma/DV^2\rho$$
$$\pi_3 = (B_G)^{x3}(B_K)^{y3}(B_D)^{z3}(A_3) = (D)^{x3}(V)^{y3}(\rho)^{z3}(\mu)$$
$$(M^0L^0T^0) = (L^{x3})(L^{y3}T^{2y3})(M^{z3}L^{-3z3})(ML^{-1}T^{-1})$$

Solution: $x_3 = -1,\ y_3 = -1,\ z_3 = -1$

$$\pi_3 = D^{-1}V^{-1}\rho^{-1}\mu = \mu/DV\rho$$

7. Convert π ratios to conventional practice. One statement of the Buckingham II theorem is that any π ratio may be taken as a function of all the others, or $f(\pi_1, \pi_2, \pi_3, \ldots, \pi_n) = 0$. This equation is mathematical shorthand for a functional statement. It could be written, for example, as $\pi_2 = f(\pi_1, \pi_3, \ldots, \pi_n)$. This equation states that π_2 is some function of π_1 and π_3 through π_n but is not a statement of *what* function π_2 is of the other π ratios. This can be determined only by physical and/or experimental analysis. Thus we are free to substitute *any* function in the equation; for example, π_1 may be replaced with $2\pi_1^{-1}$ or π_n with $a\pi_n^b$.

The procedures set forth in this example are designed to produce π ratios containing the same terms as those resulting from the application of the principles of similarity so that the physical significance may be understood. However, any other combinations might have been used. The only real requirement for a "basic group" is that it contain the same number of terms as there are dimensions in a problem and that each of these dimensions be represented in it.

The π ratios derived for this example may be converted into conventional practice as follows:

$$\pi_1 = Dg/V^2$$

is recognized as the inverse of the square root of the Froude number **F**

$$\pi_2 = \sigma/DV^2\rho$$

is the inverse of the Weber number **W**

$$\pi = \mu/DV\rho$$

is the inverse of the Reynolds number **R**

Let $\pi_1 = f(\pi_2, \pi_3)$
Then $V = K(Dg)^{1+2}$
where $K = f(\mathbf{W}, \mathbf{R})$

This agrees with the results of the dynamic-similarity analysis of liquid surfaces. This also permits a reduction in the experimental program from variations of six variables to three dimensionless parameters.

FORCES OF IMMERSED OBJECTS

Drag and Lift When a fluid impinges on an object as shown in Fig. 1.3.20, the undisturbed fluid pressure p and the velocity V change. Writing Bernoulli's equation for two points on the surface of the object, the point S being the most forward point and point A being any other point, we have, for horizontal flow,

$$p + \rho V^2/2 = p_S + \rho V_S^2/2 = p_A + \rho V_A^2/2$$

At point S, $V_S = 0$, so that $p_S = p + \rho V^2/2$. This is called the **stagnation** point, and p_S is the **stagnation pressure.** Since point A is any other

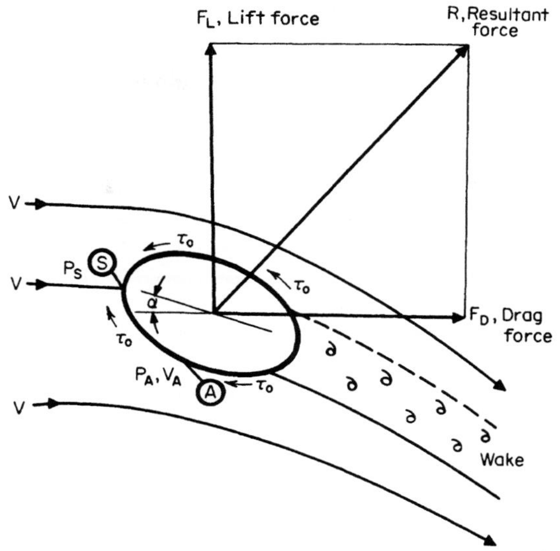

Figure 1.3.20 Notation for drag and lift.

point, the result of the fluid impingement is to create a pressure $p_A = p + \rho(V^2 - V_A^2)/2$ acting normal to every point on the surface of the object. In addition, a frictional force $F_f = \tau_0 A_s$ tangential to the surface area A_s opposes the motion. The sum of these forces gives the resultant force R acting on the body. The resultant force R is resolved into the **drag** component F_D parallel to the flow and **lift** component F_L perpendicular to the fluid motion. Depending upon the shape of the object, a wake may be formed which sheds eddies with a frequency of f. The angle α is called the angle of attack. (See Secs. 8.4 and 8.5.)

From dimensional analysis or dynamic similarity,

$$f(\mathbf{C}_p, \mathbf{R}, \mathbf{M}, \mathbf{S}) = 0$$

The formation of a wake depends upon the Reynolds number, or $\mathbf{S} = f(\mathbf{R})$. This reduces the functional relation to $f(\mathbf{C}_p, \mathbf{R}, \mathbf{M}) = 0$. Since the drag and lift forces may be considered independently,

$$F_D = C_D\rho V^2(A)/2$$

where $C_D = f(\mathbf{R}, \mathbf{M})$, and A = characteristic area.

$$F_L = C_L\rho V^2(A)/2$$

where $C_L = f(\mathbf{R}, \mathbf{M})$.

It is evident from Fig. 1.3.20 that C_D and C_L are also functions of the angle of attack. Since the drag force arises from two sources, the pressure or shape drag F_p and the skin-friction drag F_f due to wall shear stress τ_0, the drag coefficient is made up of two parts:

$$F_D = F_p + F_f = C_D\rho AV^2/2 = C_p\rho AV^2/2 + C_f\rho A_s V^2/2$$

or $$C_D = C_p + C_f A_s/A$$

where C_p is the coefficient of pressure, C_f the skin-friction coefficient, and A_s the characteristic area for shear.

Skin-Friction Drag Figure 1.3.21 shows a fluid approaching a smooth flat plate with a uniform velocity profile of V. As the fluid passes over the plate, the velocity at the plate surface is zero and increases to V at some distance δ from the surface. The region in which the velocity varies from 0 to V is called the **boundary layer.** For some distance along the plate, the flow within the boundary layer is laminar, with viscous forces predominating, but in the transition zone as the inertia forces become larger, a turbulent layer begins to form and increases as the laminar layer decreases.

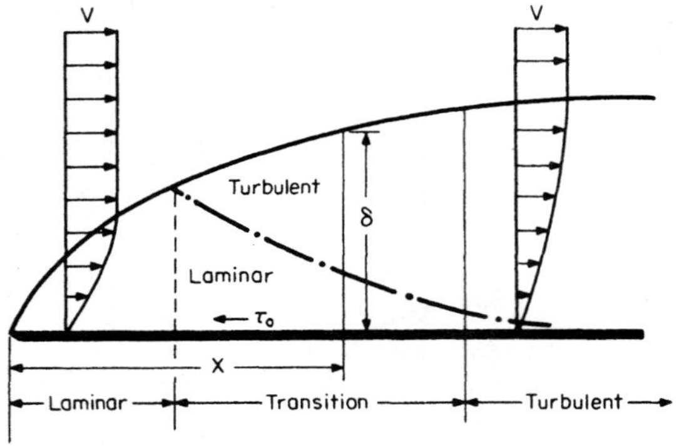

Figure 1.3.21 Boundary layer along a smooth flat plate.

Boundary-layer thickness and skin-friction drag for incompressible flow over smooth flat plates may be calculated from the following equations, where $\mathbf{R}_X = \rho VX/\mu$:

Laminar

$$\delta/X = 5.20\,\mathbf{R}_X^{-1/2} \qquad 0 < \mathbf{R}_X < 5\times10^5$$

$$C_f = 1.328\,\mathbf{R}_X^{-1/2} \qquad 0 < \mathbf{R}_X < 5\times10^5$$

Turbulent

$$\delta/X = 0.377\,\mathbf{R}_X^{-1/5} \qquad 5\times10^4 < \mathbf{R}_X < 10^6$$

$$\delta/X = 0.220\,\mathbf{R}_X^{-1/6} \qquad 10^6 < \mathbf{R}_X < 5\times10^8$$

$$C_f = 0.0735\,\mathbf{R}_X^{-1/5} \qquad 2\times10^5 < \mathbf{R}_X < 10^7$$

$$C_f = 0.455\,(\log_{10}\mathrm{R}_X)^{-2.58} \qquad 10^7 < \mathbf{R}_X < 10^8$$

$$C_f = 0.05863\,(\log_{10}C_f R_X)^{-2} \qquad 10^8 < \mathbf{R}_x < 10^9$$

Transition The Reynolds number at which the boundary layer changes depends upon the roughness of the plate and degree of turbulence. The generally accepted number is 500,000, but the transition can take place at Reynolds numbers higher or lower. (Refer to Secs. 8.4 and 8.5.) For transition at any Reynolds number $\mathbf{R}_X$,

$$C_f = 0.455\,(\log_{10}\mathbf{R}_X)^{-2.58} - \left(0.0735\,\mathbf{R}_t^{4/5} - 1.328\,8\,\mathbf{R}_t^{1/2}\right)\mathbf{R}_X^{-1}$$

For $\mathbf{R}_t = 5\times10^5, C_f = 0.455\,(\log_{10}\mathbf{R}_X)^{-2.58} - 1{,}725\,\mathbf{R}_X^{-1}$.

Pressure Drag Experiments with sharp-edged objects placed perpendicular to the flow stream indicate that their drag coefficients are essentially constant at Reynolds numbers over 1,000. This means that the drag for $\boldsymbol{R}_X > 10^3$ is pressure drag. Values of C_D for various shapes are given in Sec. 8 along with the effects of Mach number.

Wake Frequency An object in a fluid stream may be subject to the downstream periodic shedding of vortices from first one side and then the other. The frequency of the resulting transverse (lift) force is a function of the stream Strouhal number. As the wake frequency approaches the natural frequency of the structure, the periodic lift force increases asymptotically in magnitude, and when resonance occurs, the structure fails. Neglecting to take this phenomenon into account in design has been responsible for failures of electric transmission lines, submarine periscopes, smokestacks, bridges, and thermometer wells. The wake-frequency characteristics of **cylinders** are shown in Fig. 1.3.22. At a Reynolds number of about 20, vortices begin to shed alternately. Behind the cylinder is a staggered stable arrangement of vortices known as the "Kármán vortex trail." At a Reynolds number of about 10^5, the flow changes from laminar to turbulent. At the end of the transition zone ($\mathbf{R} \approx 3.5\times10^5$), the flow becomes turbulent, the alternate shedding stops, and the wake is aperiodic. At the end of the supercritical zone ($\mathbf{R} \approx 3.5\times10^6$), the wake continues to be turbulent, but the shedding again becomes alternate and periodic.

The alternating **lift force** is given by

$$F_L(t) = C_L\rho V^2 A\,\sin\,(2\pi ft)/2$$

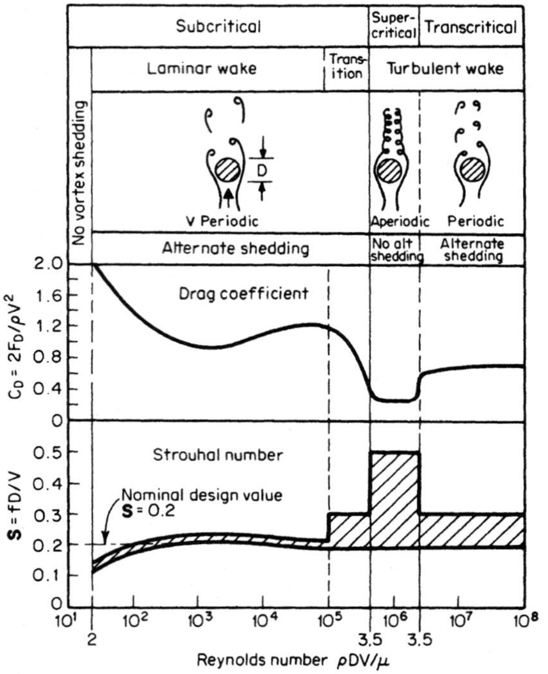

Figure 1.3.22 Flow around a cylinder. (*From Murdock, "Fluid Mechanics and Its Applications," Houghton Mifflin, 1976.*)

where t is the time. For an analysis of this force in the subcritical zone, see Belvins (Murdock, "Fluid Mechanics and Its Applications," Houghton Mifflin, 1976). For design of steel stacks, Staley and Graven (ASME 72PET/30) recommend $C_L = 0.8$ for $10^4 < \mathbf{R} < 10^5$, $C_L = 2.8 - 0.4\log_{10}\mathbf{R}$ for $\mathbf{R} = 10^5$ to 10^6, and $C_L = 0.4$ for $10^6 < \mathbf{R} < 10^7$.

The **Strouhal number** is nearly constant to $\mathbf{R} = 10^5$, and a nominal design value of 0.2 is generally used. Above $\mathbf{R} = 10^5$, data from different experimenters vary widely, as indicated by the crosshatched zone of Fig. 1.3.22. This wide zone is due to experimental and/or measurement difficulties and the dependence on surface roughness to "trigger" the boundary layer. Examination of Fig. 1.3.22 indicates an inverse relation of Strouhal number to drag coefficient.

Observation of actual structures shows that they vibrate at their natural frequency and with a mode shape associated with their fundamental (first) mode during vortex excitation. Based on observations of actual stacks and wind-tunnel tests, Staley and Graven recommend a constant Strouhal number of 0.2 for all ranges of Reynolds number. The ASME recommends $\mathbf{S} = 0.22$ for thermowell design ("Temperature Measurement," PTC 19.3). Until such time as the value of the Strouhal number above $\mathbf{R} = 10^5$ has been firmly established, designers of structures in this area should proceed with caution.

FLOW IN PIPES

Parameters for Pipe Flow The forces acting on a fluid flowing through and completely filling a horizontal pipe are inertia, viscous, pressure, and elastic. If the surface roughness of the pipe is ε, either similarity or dimensional analysis leads to $\mathbf{C}_p = f(\mathbf{R}, \mathbf{M}, L/D, \varepsilon/D)$, which may be written for incompressible fluids as $\Delta p = \mathbf{C}_p V^2/2 = K\rho V^2/2$, where K is the **resistance coefficient** and ε/D the **relative roughness** of the pipe surface, and the resistance coefficient $K = f(\mathbf{R}, L/D, \varepsilon/D)$. The pressure loss may be converted to the terms of **lost head:** $h_f = \Delta p/\gamma = KV^2/2g$. Conventional practice is to use the **friction factor** f, defined as $f = KD/L$ or $h_f = KV^2/2g = (fL/D)V^2/2g$, where $f = f(\mathbf{R}, \varepsilon/D)$. When a fluid flows into a pipe, the boundary layer starts at the entrance, as shown in Fig. 1.3.23, and grows continuously until it fills the pipe.

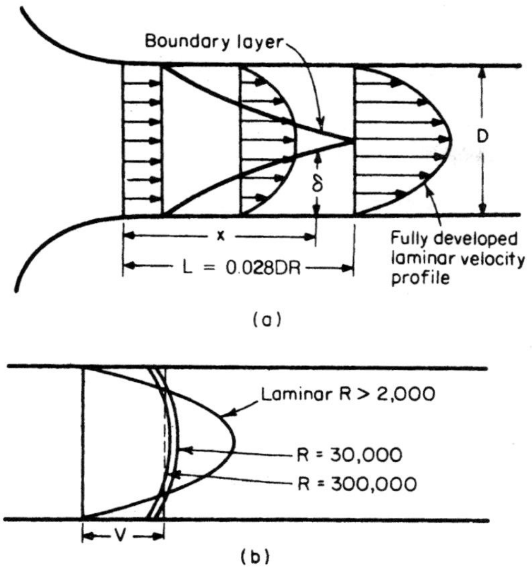

Figure 1.3.23 Velocity profiles in pipes.

From the equation of motion $dh_f = \tau dL/\gamma R_h$ and for circular ducts $R_h = D/4$. Comparing wall shear stress τ_0 with friction factor results in the following: $\tau_0 = f\rho V^2/8$.

Laminar Flow In this type of flow, the resistance is due to viscous forces only so that it is independent of the pipe surface roughness, or $\tau_0 = \mu\, dU/dy$. Application of this equation to the equation of motion and the friction factor yields $f = 64/\mathbf{R}$. Experiments show that it is possible to maintain laminar flow to very high Reynolds numbers if care is taken to increase the flow gradually, but normally the slightest disturbance will destroy the laminar boundary layer if the value of Reynolds number is greater than 4,000. In a like manner, flow initially turbulent can be maintained with care to very low Reynolds numbers, but the slightest upset will result in laminar flow if the Reynolds number is less than 2,000. The Reynolds-number range between 2,000 and 4,000 is called the **critical zone** (Fig. 1.3.24). Flow in the zone is **unstable,** and designers of piping systems must take this into account.

EXAMPLE. Glycerin at 68°F (20°C) flows through a horizontal pipe 1 in in diameter and 20 ft long at a rate of 0.090 lbm/s. What is the pressure loss? From the continuity equation $V = Q/A = (m/\rho g)/(\pi D^2/4) = [0.090/(2.447 \times 32.17)]/[(\pi/4)(1/12)^2] = 0.2096$ ft/s. The Reynolds number $\mathbf{R} = \rho VD/\mu = (2.447)(0.2096)(1/12)/(29{,}500 \times 10^{-6}) = 1.449$. $\mathbf{R} < 2{,}000$; therefore, flow is laminar and $f = 64/\mathbf{R} = 64/1.449 = 44.17$. $K = fL/D = 44.17 \times 20(1/12) = 10{,}600$. $\Delta p = K\rho V^2/2 = 10{,}600 \times 2.447\,(0.2096)^2/2 = 569.8$ lbf/ft^2 = 569.8/144 = 3.957 lbf/in^2 (2.728×10^4 N/m^2).

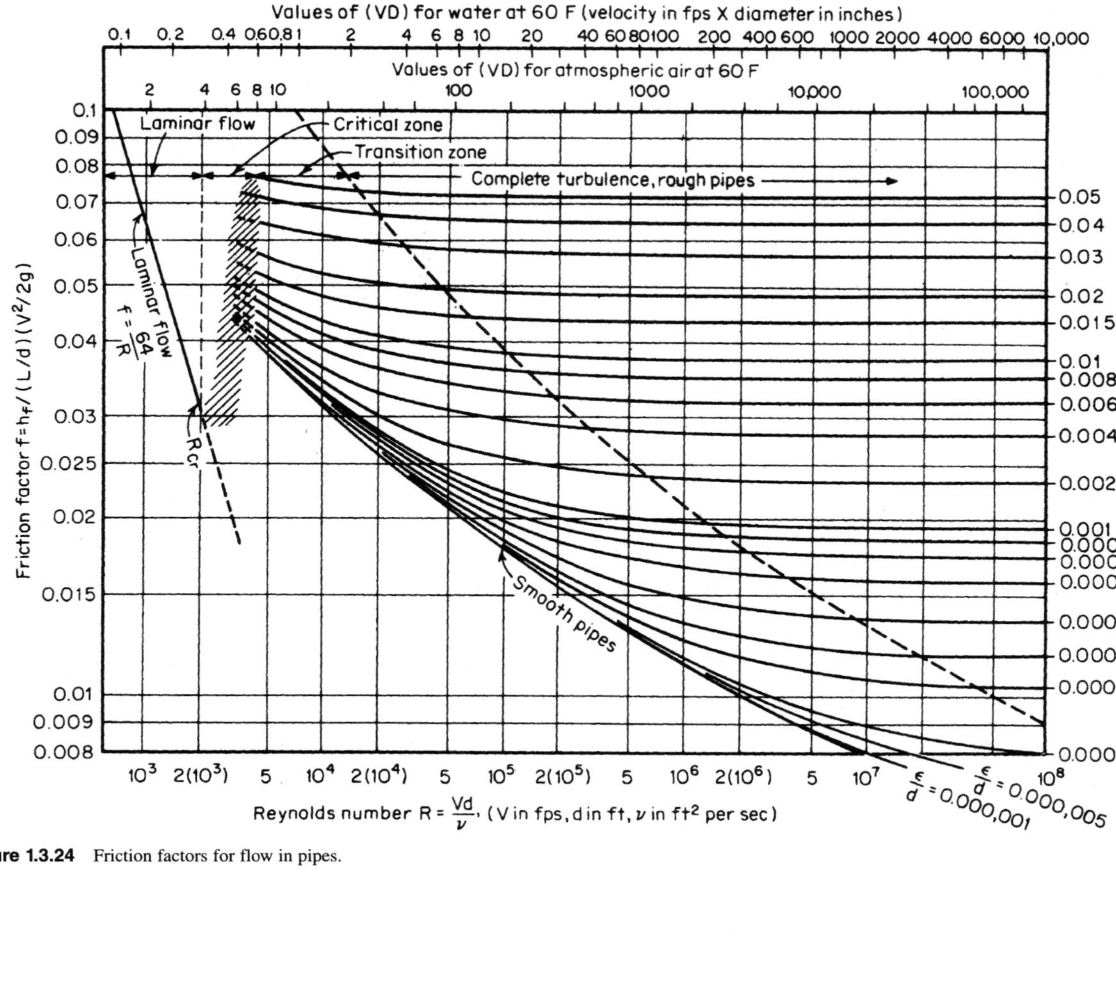

Figure 1.3.24 Friction factors for flow in pipes.

Turbulent Flow The friction factor for Reynolds number over 4,000 is computed using the Colebrook equation:

$$\frac{1}{\sqrt{f}} = -2\log_{10}\left(\frac{\varepsilon/D}{3.7} + \frac{2.51}{\mathbf{R}\sqrt{f}}\right)$$

Figure 1.3.24 is a graphical presentation of this equation (Moody, *Trans. ASME,* 1944, pp. 671–684). Examination of the Colebrook equation indicates that if the value of surface roughness ε is small compared with the pipe diameter ($\varepsilon/D \to 0$), the friction factor is a function of Reynolds number only. A **smooth pipe** is one in which the ratio $(\varepsilon/D)/3.7$ is small compared with $2.51/\mathbf{R}\sqrt{f}$. On the other hand, as the Reynolds number increases so that $2.51/\mathbf{R}\sqrt{f} \to 0$, the friction factor becomes a function of relative roughness only and the pipe is called a **rough pipe.** Thus the same pipe may be smooth under one flow condition, and rough under another. The reason for this is that as the Reynolds number increases, the thickness of the laminar sublayer decreases as shown in Fig. 1.3.21, exposing the surface roughness to flow. Values of absolute roughness ε are given in Table 1.3.9. The variation of friction factor shown in Fig. 1.3.9 is for new, clean pipes. The change of friction factor with **age** depends upon the chemical properties of the fluid and the piping material. Published data for flow of water through wrought-iron or cast-iron pipes show as much as 20 percent increase after a few months to 500 percent after 20 years. When necessary to allow for service life, a study of specific conditions is recommended. The calculation of friction factor to four significant figures in the examples to follow is only for numerical comparison and should not be construed to mean accuracy.

Table 1.3.9 Values of Absolute Roughness, New Clean Commercial Pipes

Type of pipe or tubing	ε ft (0.3048 m) × 10^{-6} Range		ε ft (0.3048 m) × 10^{-6} Design	Probable max variation of f from design, %
Asphalted cast iron		400	400	−5 to + 5
Brass and copper		5	5	−5 to + 5
Concrete	1,000	10,000	4,000	−35 to 50
Cast iron		850	850	−10 to + 15
Galvanized iron		500	500	0 to + 10
Wrought iron		150	150	−5 to 10
Steel		150	150	−5 to 10
Riveted steel	3,000	30,000	6,000	−25 to 75
Wood stave	600	3,000	2,000	−35 to 20

SOURCE: Compiled from data given in "Pipe Friction Manual," Hydraulic Institute, 3rd ed., 1961.

Engineering Calculations Engineering pipe computations usually fall into one of the following classes:

1. Determine pressure loss Δp when Q, L, and D are known.
2. Determine flow rate Q when L, D, and Δp are known.
3. Determine pipe diameter D when Q, L, and Δp are known.

Pressure-loss computations may be made to engineering accuracy using an expanded version of Fig. 1.3.24. Greater precision may be obtained by using a combination of Table 1.3.9 and the Colebrook equation, as will be shown in the example to follow. Flow rate may be determined by direct solution of the Colebrook equation. Computation of pipe diameter necessitates the trial-and-error method of solution.

EXAMPLE. *Case 1:* 2,000 gal/min of 68°F (20°C) water flow through 500 ft of cast-iron pipe having an internal diameter of 10 in. At point 1 the pressure is 10 lbf/in² and the elevation 150 ft, and at point 2 the elevation is 100 ft. Find p_2.

From continuity $V = Q/A = [2{,}000 \times (231/1{,}728)/60]/[(\pi/4)(10/12)^2] = 8.170$ ft/s. Reynolds number $\mathbf{R} = \rho VD/\mu = (1.937)(8.170)(10/12)/(20.92 \times 10^{-6}) = 6.304 \times 10^5$. $\mathbf{R} > 4{,}000$ ∴ flow is turbulent. $\varepsilon/D = (850 \times 10^{-6})/(10/12) = 1.020 \times 10^{-3}$.

Determine f: from Fig. 1.3.24 by interpolation $f = 0.02$. Substituting this value on the right-hand side of the Colebrook equation,

$$\frac{1}{\sqrt{f}} = -2\log_{10}\left(\frac{\varepsilon/D}{3.7} + \frac{2.51}{R\sqrt{f}}\right)$$

$$= -2\log_{10}\left[\frac{1.020\times10^{-3}}{3.7} + \frac{2.51}{(6.305\times10^5)\sqrt{0.02}}\right]$$

$$\frac{1}{\sqrt{f}} = 7.035 \qquad f = 0.02021$$

Resistance coefficient $K = \dfrac{fL}{D} = \dfrac{0.02021\times500}{10/12}$

$$K = 12.13$$

$$h_{1f2} = KV^2/2g = 12.13\times(8.170)^2/2\times32.17$$

$$h_{1f2} = 12.58 \text{ ft}$$

Equation of motion: $p_1/\gamma + V_1^2/2g + z_1 = p_2/\gamma + V_2^2/2g + z_2 + h_{1f2}$. Noting that $V_1 = V_2 = V$ and solving for p_2,

$$p_2 = p_1 + \gamma(z_1 - z_2 - h_{1f2})$$

$$= 144\times10 + (1.937\times32.17)(150-100-12.58)$$

$$p_2 = 3{,}772 \text{ lbf/ft}^2 = 3{,}772/144 = 26.20 \text{ lbf/in}^2\ (1.806\times10^5 \text{ N/m}^2)$$

EXAMPLE. *Case 2:* Gasoline (sp. gr. 0.68) at 68°F (20°C) flows through a 6-in schedule 40 (ID = 0.5054 ft) welded steel pipe with a head loss of 10 ft in 500 ft. Determine the flow. This problem may be solved directly by deriving equations that do not contain the flow rate Q.

$$\text{From } h_f = \left(\frac{fL}{D}\right)\frac{V^2}{2g}, \qquad V = (2gh_fD)^{1/2}(fL)^{1/2}$$

$$\text{From } \mathbf{R} = \rho VD/\mu, \qquad V = \mathbf{R}\mu/\rho D$$

Equating the above and solving,

$$\mathbf{R}\sqrt{f} = (\rho D/\mu)(2gh_fD/L)^{1/2}$$

$$= (1.310\times0.5054/5.98\times10^{-6})$$

$$\times(2\times32.17\times10\times0.5054/500)^{1/2} = 89{,}285$$

which is now in a form that may be used directly in the Colebrook equation:

$$\varepsilon/D = 150\times10^{-6}/0.5054 = 2.968\times10^{-4}$$

From the Colebrook equation,

$$\frac{1}{\sqrt{f}} = -2\log_{10}\left(\frac{\varepsilon/D}{3.7} + \frac{2.51}{\mathbf{R}\sqrt{f}}\right)$$

$$= -2\log_{10}\left(\frac{2.968\times10^{-4}}{3.7} + \frac{2.51}{89{,}285}\right)$$

$$\frac{1}{\sqrt{f}} = 7.931 \qquad f = 0.01590$$

$$\mathbf{R} = 89{,}285/\sqrt{f} = 89{,}285\times7.93 = 7.08\times10^5$$

$$\mathbf{R} > 4{,}000 \ \therefore \text{ flow is turbulent}$$

$$V = \mathbf{R}\mu/\rho D = (7.08\times10^5\times5.98$$

$$\times10^{-6})/(1.310\times0.5054) = 6.396 \text{ ft/s}$$

$$Q = AV = (\pi/4)(0.5054)^2(6.396)$$

$$Q = 1.283 \text{ ft}^3\text{/s } (3.633\times10^{-2} \text{ m}^3/\text{s}^1)$$

EXAMPLE. *Case 3;* Water at 68°F (20°C) is to flow at a rate of 500 ft³/s through a concrete pipe 5,000 ft long with a head loss not to exceed 50 ft. Determine the diameter of the pipe. This problem may be solved by trial and error using methods of the preceding example. First trial: Assume any diameter (say 1 ft).

$$\mathbf{R}\sqrt{f} = (\rho D/\mu)(2gh_fD/L)^{1/2}$$

$$= (1.937D/20.92\times10^{-6})\times(2\times32.17\times50D/5{,}000)^{1/2}$$

$$= 74{,}269D^{3/2} = 74{,}269(1)^{3/2} = 74{,}269$$

$$\varepsilon/D_1 = 4{,}000\times10^{-6}/D = 4{,}000\times10^{-6}/(1) = 4{,}000\times10^{-6}$$

$$\frac{1}{\sqrt{f_1}} = -2\log_{10}\left(\frac{\varepsilon/D_1}{3.7} + \frac{2.51}{\mathbf{R}\sqrt{f_1}}\right)$$

$$= -2 \log_{10}\left(\frac{4{,}000\times10^{-6}}{3.7}+\frac{2.51}{74{,}269}\right)$$

$$\frac{1}{\sqrt{f_1}} = 5.906 \qquad f_1 = 0.02867$$

$$\mathbf{R}_1 = 74{,}269/\sqrt{f_1} = 74{,}269\times5.906 = 438{,}600$$

$$V_1 = \mathbf{R}\mu/\rho D_1 = (438{,}600 = 20.92\times10^{-6})/(1.937\times1)$$

$$V_1 = 4.737 \text{ ft/s}$$

$$Q_1 = A_1V_1 = [\pi(1)^2/4]4.737 = 3.720 \text{ ft}^3/\text{s}$$

For the same loss and friction factor,

$$D_2 = D_1(Q/Q_1)^{2/5} = (1)(500/3.720)^{2/5} = 7.102 \text{ ft}$$

For the second trial use D_2 = 7.102, which results in Q = 502.2 ft³/s. Since the nearest standard size would be used, additional trials are unnecessary.

Velocity Profile Figure 1.3.23*a* shows the formation of a laminar velocity profile. As the fluid enters the pipe, the boundary layer starts at the entrance and grows continuously until it fills the pipe. The flow while the boundary is growing is called **generating flow.** When the boundary layer completely fills the pipe, the flow is called **established flow.** The distance required for establishing **laminar flow** is $L/D \approx 0.028$ **R.** For **turbulent flow,** the distance is much shorter because of the turbulence and not dependent upon Reynolds number, L/D being from 25 to 50.

Examination of Fig. 1.3.23*b* indicates that as the Reynolds number increases, the velocity distribution becomes "flatter" and the flow approaches **one-dimensional.** The velocity profile for laminar flow is parabolic, $U/V = 2[1 - (r/r_o)^2]$ and for **turbulent flow, logarithmic** (except for the very thin laminar boundary layer), $U/V = 1+1.43\sqrt{f} + 2.15\sqrt{f}\,\log_{10}(1 - r/r_o)$. The use of the average velocity produces an error in the computation of kinetic energy. If α is the **kinetic-energy correction factor,** the true kinetic-energy change per unit mass between two points on a flow system $\Delta KE = \alpha_1V_1^2/2g_c - \alpha_2V_2^2/2g_c$, where $\alpha = (1/AV^3)\int U^3dA$. For **laminar flow,** α = 2 and for **turbulent flow,** $\alpha \approx 1 + 2.7f$. Of interest is the **pipe factor** V/U_{max}; for **laminar flow,** V/U_{max} = 1/2 and for **turbulent flow,** $V/U_{max} = 1+1.43\sqrt{f}$. The location at which the local velocity equals the average velocity for **laminar flow** is $U = V$ at r/r_o = 0.7071 and for **turbulent flow** is $U = V$ at r/r_o = 0.7838.

Compressible Flow At the present time, there are no true analytical solutions for the computation of actual characteristics of compressible fluids flowing in pipes. In the real flow of a compressible fluid in a pipe, the amount of heat transferred and its direction are dependent upon the amount of insulation, the temperature gradient between the fluid and ambient temperatures, and the heat-transfer coefficient. Each condition requires an individual application of the principles of thermodynamics and heat transfer for its solution.

Conventional engineering practice is to use one of the following methods for flow computation.

1. Assume **adiabatic** flow. This approximates the flow of compressible fluids in short, insulated pipelines.
2. Assume **isothermal** flow. This approximates the flow of gases in long, uninsulated pipelines where the fluid and ambient temperatures are nearly equal.

Adiabatic Flow If the Mach number is less than ¼, results within normal engineering-accuracy requirements may be obtained by considering the fluid to be incompressible. A detailed discussion of and methods for the solution of compressible adiabatic flow are beyond the scope of this section, and any standard gas-dynamics text should be consulted.

Isothermal Flow The equation of motion for a horizontal piping system may be written as follows:

$$dp + \rho V\,dV + \gamma\,dh_f = 0$$

noting, from the continuity equation, that $\rho V = M/A = G$, where G is the **mass velocity** in slugs/(ft²)(s), and that $\gamma\,dh_f = [(f/D)\rho V^2/2]dL = [(f/D)GV/2]dL$. Substituting in the above equation of motion and dividing by $GV/2$ results in

$$\frac{2\rho\,dp}{G^2}+\frac{2dV}{V}+\left(\frac{F}{D}\right)dL = 0$$

Integrating for an isothermal process ($p/\rho = C$) and assuming f is a constant,

$$\frac{\rho_1p_1}{G^2}\left[\left(\frac{p_2}{p_1}\right)^2-1\right]+2\log_e\left(\frac{V_2}{V_1}\right)+\frac{fL}{D} = 0$$

Noting that $A_1 = A_2$, $V_2/V_1 = \rho_1/\rho_2 = p_1/p_2$, and solving for G,

$$G = \left\{\frac{\rho_1p_1[1-(p_2/p_1)^2]}{2\log_e(p_1/p_2)+fL/D}\right\}^{1/2}$$

The Reynolds number may be written as

$$\mathbf{R} = \frac{\rho VD}{\mu} = \frac{GD}{\mu} \qquad \text{and} \qquad G = \mathbf{R}\frac{\mu}{D}$$

The value of $\mathbf{R}\sqrt{f}$ may be obtained from the simultaneous solution of the two equations for G, assuming that 2 $\log_e p_1/p_2$ is small compared with fL/D.

$$\mathbf{R}\sqrt{f} \approx \left\{\frac{D^3\rho_1p_1}{\mu^2L}\left[1-\left(\frac{p_2}{p_1}\right)\right]\right\}^{1/2}$$

EXAMPLE. Air at 68°F (20°C) is flowing isothermally through a horizontal straight standard 1-in steel pipe (inside diameter = 1.049 in). The pipe is 200 ft long, the pressure at the pipe inlet is 74.7 lbf/in², and the pressure drop through the pipe is 5 lbf/in². Find the flow rate in lbm/s. From the equation of state $\rho_1 = p/g_cRT = (144\times74.7)/(32.17\times53.34\times527.7) = 0.01188$ slugs/ft³.

$$\mathbf{R}\sqrt{f} = \{[(D^3\rho_1p_1/\mu^2L)][1-(p_2/p_1)^2]\}^{1/2} = \{[(1.049/12)^3(0.01188)$$
$$\times(144\times74.7)/(39.16\times10^{-8})^2(200)[1-(69.7/74.7)^2]\}^{1/2}$$
$$= 18{,}977$$

For steel pipe $\varepsilon = 150\times10^{-6}$ ft, $\varepsilon/D = (150\times10^{-6})/(1.049/12) = 1.716\times10^{-3}$. From the Colebrook equation,

$$\frac{1}{\sqrt{f}} = -2\log_{10}\left(\frac{\varepsilon/D}{3.7}+\frac{2.51}{\mathbf{R}\sqrt{f}}\right)$$

$$= 2\log_{10}[(1.716\times10^{-3}/3.7)+(2.51)/(18{,}977)] = 6.449$$

$$f = 0.02404$$

$$\mathbf{R} = (\mathbf{R}\sqrt{f})(1/\sqrt{f}) = (18{,}953)(6.449) = 122{,}200$$

$$\mathbf{R} > 4{,}000 \;\therefore\; \text{flow is turbulent}$$

$$G = \left\{\frac{\rho_1p_1[1-(p_2/p_1)^2]}{2\log_e(p_1/p_2)+fL/D}\right\}^{1/2}$$

$$= \left\{\frac{(0.01188)(144\times74.7)[1-(69.7/74.7)^2]}{2\log_e(74.7/69.7)+(0.02404)(200)/(1.049/12)}\right\}^{1/2}$$

$$= 0.5476 \text{ slug/(ft}^2\text{)(s)}$$

$$\dot m = g_cAG = (32.17)(\pi/4)(1.049/12)^2(0.5476)$$

$$\dot m = 0.1057 \text{ lbm/s } (47.94\times10^{-3}\text{ kg/s})$$

Noncircular Pipes For the flow of fluids in noncircular pipes, the **hydraulic diameter** D_h is used. From the definition of hydraulic radius, the diameter of a circular pipe was shown to be four times its **hydraulic radius;** thus $D_h = 4R_h$. The Reynolds number thus may be written as $\mathbf{R} = \rho VD_h/\mu = GD_h/\mu$, the relative roughness as ε/D_h, and the resistance coefficient $K = fL/D_h$. With the above modifications, flows through noncircular pipes may be computed in the same manner as for circular pipes.

EXAMPLE. Air at 68°F (20°C) and 100 lbf/in² enters a rectangular duct 1 by 3 ft at a rate of 720 lbm/s. The duct is horizontal, 100 ft long, and made of galvanized iron. Assuming isothermal flow, estimate the pressure loss due to friction in this line. From the equation of state, $\rho_1 = \rho_1/g_cRT_1 = (144\times100)/(32.17)(53.34)(527.7) = 0.01590$ slug/ft³. From Table 1.3.6, $R_h = bD/2(b+D) = 3\times1/2(3+1) = 0.375$ ft, and $D_h = 4R_h = 4\times0.375 = 1.5$ ft. For galvanized iron, $\varepsilon/D_h = 500\times10^{-6}/1.5 = 3.333\times10^{-4}$

$$G = (\dot m/g_c)/A = (720/32.17)/(1\times3) = 7.460 \text{ slugs/(ft}^2\text{)(s)}$$

$$\mathbf{R} = GD_h/\mu = (7.460)(1.5)/(39.16\times10^{-8})$$

$$= 28{,}580{,}000 > 4{,}000 \;\therefore\; \text{flow is turbulent}$$

From Fig. 1.3.24, $f \approx 0.015$

$$\frac{1}{\sqrt{f}} = -2\log_{10}\left(\frac{\varepsilon/D_h}{3.7} + \frac{2.51}{\mathbf{R}\sqrt{f}}\right)$$

$$= -2\log_{10}\left(\frac{3.333\times10^{-4}}{3.7} + \frac{2.51}{28{,}580{,}000\sqrt{0.015}}\right)$$

$$f = 0.01530$$

Solving the isothermal equation for p_2/p_1,

$$\frac{p_2}{p_1} = \left\{1 - \left(\frac{G^2}{\rho_1 p_1}\right)\left[2\log_e\left(\frac{p_1}{p_2}\right) + \frac{fL}{D_h}\right]\right\}^{1/2}$$

For first trial, assume $2\log_e(p_1/p_2)$ is small compared with fL/D:

$$p_2/p_1 = \{1 - [(7.460)^2/(0.01590)(144\times100)][0 + (0.01530)(100)/1.5]\}^{1/2}$$
$$= 0.8672$$

Second trial using first-trial values results in 0.8263. Subsequent trials result in a balance at p_2/p_1 = 0.8036, p_2 = 100 × 0.8036 = 80.36 lbf/in^2 (5.541 × 10^5 N/m^2).

PIPING SYSTEMS

Resistance Parameters The resistance to flow of a piping system is similar to the resistance of an object immersed in a flow stream and is made up of pressure (inertia) or shape drag and skin-friction (viscous) drag. For long, straight pipes the pressure drag is characterized by the **relative roughness** ε/D and the skin friction by the **Reynolds number R.** For other piping components, two parameters are used to describe the resistance to flow, the **resistance coefficient** $K = fL/D$ and the **equivalent length** $L/D = K/f$. The resistance-coefficient method assumes that the component loss is all due to pressure drag and that the flow through the component is completely turbulent and independent of Reynold's number. The equivalent-length method assumes that resistance of the component varies in the same manner as does a straight pipe. The basic assumption then is that its pressure drag is the same as that for the relative roughness ε/D of the pipe and that the friction drag varies with the Reynolds number **R** in the same manner as the straight pipe. Both methods have the inherent advantage of simplicity in application, but neither is correct except in the fully developed turbulent region. Two excellent sources of information on the resistance of piping-system components are the Hydraulic Institute "Pipe Friction Manual," which uses the resistance-coefficient method, and the Crane Company Technical Paper 410 ("Fluid Meters," 6th ed. ASME, 1971), which uses the equivalent-length concept.

For valves, branch flow through tees, and the type of components listed in Table 1.3.10, the pressure drag is predominant, is "rougher" than the pipe to which it is attached, and will extend the completely turbulent region to lower values of Reynolds number. For bends and elbows, the loss is made up of pressure drag due to the change of direction and the consequent secondary flows which are dissipated in 50 diameters or more downstream piping. For this reason, loss through adjacent bends will not be twice that of a single bend.

In long pipelines, the effect of bends, valves, and fittings is usually negligible, but in systems where there is little straight pipe, they are the controlling factor. Under-design will result in the failure of the system to deliver the required capacity. Over-design will result in inefficient operation because it will be necessary to "throttle" one or more of the valves. For estimating purposes, Tables 1.3.10 and 1.3.11 may be used as shown in the examples. When available, the manufacturers' data should be used, particularly for valves, because of the wide variety of designs for the same type.

Series Systems In a single piping system made of various sizes, the practice is to group all of one size together and apply the continuity equation, as shown in the following example.

Table 1.3.10 Representative Values of Resistance Coefficient *K*

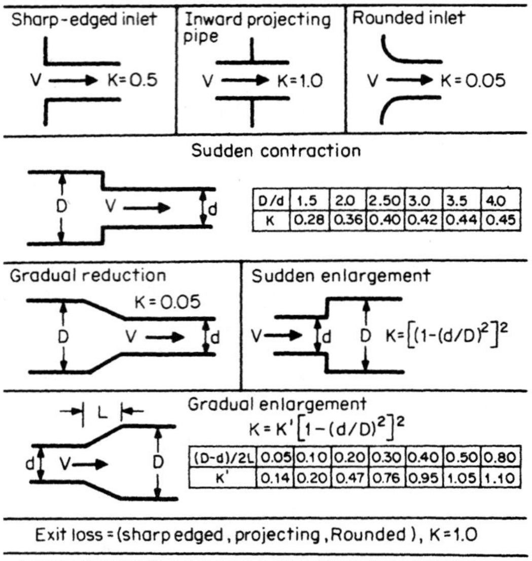

SOURCE: Compiled from data given in "Pipe Friction Manual," 3rd ed., Hydraulic Institute, 1961.

Table 1.3.11 Representative Equivalent Length in Pipe Diameters (L/D) of Various Valves and Fittings

Globe valves, fully open	450
Angle valves, fully open	200
Gate valves, fully open	13
¾ open	35
½ open	160
¼ open	900
Swing check valves, fully open	135
In line, ball check valves, fully open	150
Butterfly valves, 6 in and larger, fully open	20
90° standard elbow	30
45° standard elbow	16
90° long-radius elbow	20
90° street elbow	50
45° street elbow	26
Standard tee:	
Flow through run	20
Flow through branch	60

SOURCE: Compiled from data given in "Flow of Fluids," Crane Company Technical Paper 410, ASME, 1971.

EXAMPLE. Water at 68°F (20°C) leaves an open tank whose surface elevation is 180 ft and enters a 2-in schedule 40 steel pipe via a sharp-edged entrance. After 50 ft of straight 2-in pipe that contains a 2-in globe valve, the line enlarges suddenly to an 8-in schedule 40 steel pipe which consists of 100 ft of straight 8-in pipe, two standard 90° elbows and one 8-in angle valve. The 8-in line discharges below the surface of another open tank whose surface elevation is 100 ft. Determine the volumetric flow rate.

$$D_1 = 2.067/12 = 0.1723 \text{ ft} \quad \text{and} \quad D_2 = 7.981/12 = 0.6651 \text{ ft}$$
$$\varepsilon/D_1 = 150\times10^{-6}/0.1723 = 8.706\times10^{-4}$$
$$\varepsilon/D_2 = 150\times10^{-6}/0.6651 = 2.255\times10^{-4}$$

For turbulent flow,

$$\frac{1}{\sqrt{f_1}} = -2\log_{10}\left(\frac{\varepsilon/D}{3.7}\right)$$

$$\frac{1}{\sqrt{f_1}} = -2\log_{10}\left(\frac{8.706\times10^{-4}}{3.7}\right) \qquad f_1 = 0.01899$$

$$\frac{1}{\sqrt{f_1}} = -2\log_{10}\left(\frac{2.255\times10^{-4}}{3.7}\right) \qquad f_2 = 0.01407$$

1. *2-in components* — K

	K
Entrance loss, sharp-edged	= 0.5
50 ft straight pipe = f_1 (50/0.1723)	= $290.2f_1$
Globe valve = f_1 (L/D)	= $450.0f_1$
Sudden enlargement $k = [-(D/D_2)^2]^2$ = $[1 - (2.067/7.981)^2]^2$	= 0.87
	$\Sigma K_1 = 1.37 + 740.2f_1$

2. *8-in components*

	K
100 ft of straight pipe f_2 (100/0.6651)	= $150.4f_2$
2 standard 90° elbows $2 \times 30 f_2$	= $60\ f_2$
1 angle valve $200 f_2$	= $200\ f_2$
Exit loss	= 1
	$\Sigma K_2 = 1 + 410.4f_2$

3. *Apply equation of motion*

$$h_{1f2} = z_1 - z_2 = (\Sigma K_1)\frac{V_1^2}{2g} + (\Sigma K_2)\frac{V_2^2}{2g}$$

From continuity, $\rho_1 A_1 V_1 = \rho_2 A_2 V_2$ for $\rho_1 = \rho_2$

$$V_2 = V_1(A_1/A_2) = V_1(D_1/D_2)^2$$

$$h_{1f2} = z_1 - z_2 = \left[\Sigma K_1 + \Sigma K_2(D_1/D_2)^4\right]V_1^2/2g$$

$$V_1 = \left\{[2g(z_1 - z_2)]/[\Sigma K_1 + \Sigma K_2(D_1/D_2)^4]\right\}^{1/2}$$

$$V_1 = \left[\frac{2\times32.17\times(180-100)}{(1.37+740.2f_1)+(1+410.4f_2)(2.067/7.981)^4}\right]^{1/2}$$

$$V_1 = \frac{71.74}{(1.374+740.2f_1+1.846f_2)^{1/2}}$$

4. For first trial assume f_1 and f_2 for complete turbulence

$$V_1 = \frac{71.74}{(1.374+740.2\times0.01899+1.846\times0.01407)^{1/2}}$$

$$V_1 = 18.25 \text{ ft/s}$$

$$V_2 = 18.25(2.067/7.981)^2 = 1.224 \text{ ft/s}$$

$$\mathbf{R}_1 = \rho_1 V_1 D_1/\mu = (1.937)(18.25)(0.1723)/(20.92\times10^{-6})$$

$$\mathbf{R}_1 = 291{,}100 > 4{,}000 \therefore \text{ flow is turbulent}$$

$$\mathbf{R}_2 = \rho_2 V_2 D_2/\mu_2 = (1.937)(1.224)(0.6651)/(20.92\times10^{-6})$$

$$\mathbf{R}_2 = 75{,}420 > 4{,}000 \therefore \text{ flow is turbulent}$$

5. For second trial use first trial V_1 and V_2. From Fig. 1.3.24 and the Colebrook equation,

$$\frac{1}{\sqrt{f_1}} = -2\log_{10}\left(\frac{8.706\times10^{-4}}{3.7} + \frac{2.51}{291{,}100\sqrt{0.020}}\right)$$

$$f_1 = 0.02008$$

$$\frac{1}{\sqrt{f_2}} = -2\log_{10}\left(\frac{2.255\times10^{-4}}{3.7} + \frac{2.51}{75{,}420\sqrt{0.020}}\right)$$

$$f_2 = 0.02008$$

$$V_1 = \frac{71.74}{(1.374+740.2\times0.02008+1.864\times0.02008)^{1/2}}$$

$$V_1 = 17.78$$

A third trial results in $V = 17.77$ ft/s or $Q = A_1V_1 = (\pi/4)(0.1723)^2(17.77) = 0.4143$ ft³/s (1.173×10^{-2} m³/s).

Parallel Systems In solution of problems involving two or more parallel pipes, the head loss for all of the pipes is the same as shown in the following example.

Example. Benzene at 68°F (20°C) flows at a rate of 0.5 ft³/s through two parallel straight, horizontal pipes connecting two pressurized tanks. The pipes are both schedule 40 steel, one being 1 in, the other 2 in. They both are 100 ft long and have connections that project inwardly in the supply tank. If the pressure in the supply tank is maintained at 100 lbf/in², what pressure should be maintained on the receiving tank?

$$D_1 = 1.049/12 = 0.08742 \text{ ft} \quad \text{and} \quad D_2 = 2.067/12 = 0.1723 \text{ ft}$$

$$\varepsilon/D_1 = 150\times10^{-6}/0.08742 = 1.716\times10^{-3}$$

$$\varepsilon/D_2 = 150\times10^{-6}/0.1723 = 8.706\times10^{-4}$$

For turbulent flow,

$$\frac{1}{\sqrt{f}} = -2\log_{10}\left(\frac{\varepsilon/D}{3.7}\right)$$

$$\frac{1}{\sqrt{f_1}} = -2\log_{10}\left(\frac{1.716\times10^{-3}}{3.7}\right) \qquad f_1 = 0.02249$$

$$\frac{1}{\sqrt{f_2}} = -2\log_{10}\left(\frac{8.706\times10^{-4}}{3.7}\right) \qquad f_1 = 0.01899$$

1. *1-in. components*

	K
Entrance loss, inward projection	= 1.0
100 ft straight pipe f_1 (100/0.08742)	= $1{,}144 f_1$
Exit loss	= 1.0
	$\Sigma K_1 = 2.0 + 1{,}144 f_1$

2. *2-in components*

	K
Entrance loss, inward projection	= 1.0
100 ft straight nine f_2 (100/0.1723)	= $580.4 f_2$
Exit loss	= 1.0
	$\Sigma K_2 = 2.0 + 580.4 f_2$

$$h_f = \Sigma K_1\, V_1^2/2g = \Sigma K_2\, V_2^2/2g$$

From the continuity equation, $Q = AV$

$$\Sigma K_1 \frac{Q_1^2}{2gA_1^2} = \Sigma K_2 \frac{Q_2^2}{2gA_2^2}$$

Solving for Q_1/Q_2,

$$\frac{Q_1}{Q_2} = \frac{A_1}{A_2}\sqrt{\frac{\Sigma K_2}{\Sigma K_1}} = \left(\frac{D_1}{D_2}\right)^2\sqrt{\frac{\Sigma K_2}{\Sigma K_1}} = \left(\frac{D_1}{D_2}\right)^2\sqrt{\frac{2.0+580.4\,f_2}{2.0+1{,}144\,f_1}}$$

For first trial assume flow is completely turbulent,

$$\frac{Q_1}{Q_2} = \left(\frac{0.08742}{0.1723}\right)^2\sqrt{\frac{2.0+580.4\times0.01899}{2.0+1{,}144\times0.02249}}$$

$$\frac{Q_1}{Q_2} = 0.1764 \qquad Q = Q_1 + Q_2 = 0.1764\,Q_2 + Q_2$$

$$0.5000 = 1.1764 Q_2 \qquad Q_2 = 0.4250$$

$$Q_1 = 0.5000 - 0.4250 = 0.0750$$

for the second trial use first-trial values,

$$V_1 = Q_1/A_1 = 0.0750/(\pi/4)(0.08742)^2 = 12.50$$

$$V_2 = Q_2/A_2 = 0.4250/(\pi/4)(0.1723)^2 = 18.23$$

$$\mathbf{R}_1 = \rho_1 V_1 D_1/\mu_1 = (1.705)(12.50)(0.08742)/(13.62\times10^{-6})$$

$$\mathbf{R}_1 = 136{,}800 > 4{,}000 \therefore \text{ flow is turbulent}$$

$$\mathbf{R}_2 = \rho_2 V_2 D_2/\mu_2 = (1.705)(18.23)(0.1723)/(13.62\times10^{-6})$$

$$\mathbf{R}_2 = 393{,}200 > 4{,}000 \therefore \text{ flow is turbulent}$$

Using the Colebrook equation and Fig. 1.3.24,

$$\frac{1}{\sqrt{f_1}} = -2\log_{10}\left(\frac{1.716\times10^{-3}}{3.7} + \frac{2.51}{136{,}800\sqrt{0.024}}\right)$$

$$f_1 = 0.02389$$

$$\frac{1}{\sqrt{f_2}} = -2\log_{10}\left(\frac{8.706\times10^{-4}}{3.7} + \frac{2.51}{393{,}200\sqrt{0.020}}\right)$$

$$f_2 = 0.01981$$

$$h_f = \Sigma K_1\frac{V_1^2}{2g} = \Sigma K_2\frac{V_2^2}{2g}$$

$$\Sigma K_1 V_1^2/2g = (2.0+1{,}144\times0.02389)(12.50)^2/(2\times32.17) = 71.23$$

$$\Sigma K_2 V_2^2/2g = (2.0+580.4\times0.01981)(18.23)^2/(2\times32.17) = 69.80$$

71.23 = 69.80; further trials not justifiable because of accuracy of f, K, L/D. Use average or 70.52, so that $\Delta p = \rho g h_f = (1.705 \times 32.17 \times 70.52)/144 = 26.86$ lbf/in² $= p_1 - p_2 = 100 - p_2$, $p_2 = 100 - 26.86 = 73.40$ lbf/in² (5.061×10^5 N/m²).

Branch Flow Problems of a single line feeding several points may be solved as shown in the following example.

EXAMPLE. Ethyl alcohol at 68°F (20°C) flows from tank A, which is maintained at a constant pressure of 100 lb/in^2 through 200 ft of 2-in cast-iron schedule 40 pipe to a Y branch connection ($K = 0.5$) where 100 ft of 2-in pipe goes to tank B, which is maintained at 80 lbf/in^2 and 50 ft of 2-in pipe to tank C, which is also maintained at 80 lbf/in^2. All tank connections are flush and sharp-edged and are at the same elevation. Estimate the flow rate to each tank.

$$D = 2.067/12 = 0.1723 \text{ ft}$$
$$\varepsilon/D = 850\times10^{-6}/0.1723 = 4.933\times10^{-3}$$

For turbulent flow, $\dfrac{1}{\sqrt{f}} = -2\log_{10}\left(\dfrac{\varepsilon/D}{3.7}\right)$

$$\frac{1}{\sqrt{f}} = -2\log_{10}\left(\frac{4.933\times10^{-3}}{3.7}\right) \qquad f = 0.03025$$
$$h_{AfB} = (p_A - p_B)/\rho g = 144(100-80)/(1.532\times32.17) = 58.44$$
$$h_{Afc} = (p_A - p_C)/\rho g = h_{AfB} = 58.44$$

Let point X be just before the Y; then

1. *From tank A to Y*

	K
Entrance loss, sharp-edged	$= 0.5$
200 ft straight pipe $= f_{AX}(200/0.1723)$	$= 1{,}161\, f_{AX}$
	$\Sigma K_{AX} = 0.5 + 1{,}161\, f_{AX}$

2. *From Y to tank B*

Y branch	$= 0.5$
100 ft straight pipe $= f_{XB}(100/0.1723)$	$= 580.4\, f_{XB}$
Exit loss	$= 1.0$
	$\Sigma K_{XB} = 1.5 + 580.4\, f_{XB}$

3. *From Y to tank C*

Y branch	$= 0.5$
50 ft straight pipe $= f_{XC}(50/0.1723)$	$= 290.2\, f_{XC}$
Exit loss	$= 1.0$
	$\Sigma K_{XC} = 1.5 + 290.2\, f_{XC}$

Balance of flows:

$$Q_{AX} = Q_{XB} + Q_{XC}$$

and from continuity, $(A_{AX} = A_{XB} = A_{XC})$, $V_{AX} = V_{XB} + V_{XC}$; then

$$h_{AfB} = \Sigma K_{AX}\frac{V_{AX}^2}{2g} + \Sigma K_{XB}\frac{V_{XB}^2}{2g}$$
$$h_{AfC} = \Sigma K_{AX}\frac{V_{AX}^2}{2g} + \Sigma K_{XC}\frac{V_{XC}^2}{2g}$$

For first trial assume completely turbulent flow

$$h_{AfB} = \frac{(0.5+1{,}161\ f_{AX})V_{AX}^2}{2g} + \frac{(1.5+580.4\ f_{XB})V_{XB}^2}{2g}$$
$$58.44 = \frac{(0.5+1{,}161\times0.03025)V_{AX}^2}{2\times32.17} + \frac{(1.5+580.4\times0.03025)V_{XB}^2}{2\times32.17}$$
$$58.44 = 0.5536\ V_{AX}^2 + 0.2962\ V_{XB}^2$$

and in a like manner

$$h_{AfC} = 58.44 = 05536V_{XC}^2 + 0.1598\ V_{XC}^2$$

Equating $h_{AfB} = h_{AfC}$,

$$0.5536V_{AX}^2 + 0.2962\ V_{XB}^2 = 0.5536\ V_{AX}^2 + 0.1598\ V_{XC}^2$$

or $\quad V_{XC} = 1.3615\ V_{XB} \quad$ and since $\quad V_{AX} = V_{XB} + V_{XC}$

$$V_{AX} = V_{XB} + 1.3615\ V_{XB} = 2.3615\ V_{XB}$$

so that

$$h_{AfB} = 58.44 = 0.5536\left(2.3615\ V_{XB}^2\right) + 0.2962\ V_{XB}^2$$
$$V_{XB} = 4.156$$
$$V_{XC} = 1.3615(4.156) = 5.658$$
$$V_{AX} = 4.156 + 5.658 = 9.814$$

Second trial,

$$\mathbf{R}_{AX} = \frac{\rho V_{AX} D}{\mu} = \frac{1.532\times9.814\times0.1723}{25.06\times10^{-6}}$$
$$\mathbf{R}_{AX} = 103{,}400 > 4{,}000 \ \therefore \text{ flow is turbulent}$$

In a like manner,

$$\mathbf{R}_{XB} = 43{,}780 \qquad \mathbf{R}_{XC} = 59{,}600$$

Using the Colebrook equation and Fig. 1.3.24,

$$\frac{1}{\sqrt{f_{AX}}} = -2\log_{10}\left(\frac{4.933\times10^{-3}}{3.7} + \frac{2.51}{103{,}400\sqrt{0.031}}\right)$$
$$f_{AX} = 0.03116$$

In a like manner,

$$f_{XB} = 0.03231 \qquad f_{XC} = 0.03179$$
$$h_{AfB} = \frac{(0.5+1{,}161\times0.03116)V_{AX}^2}{2\times32.17} + \frac{(1.5+580.4\times0.03231)V_{XB}^2}{2\times32.17}$$
$$h_{AfB} = 0.5700\ V_{AX}^2 + 0.3148\ V_{XB}^2$$
$$h_{AfC} = 0.5700\ V_{AX}^2 + \frac{(1.5+290.2\times0.03179)V_{XC}^2}{2\times32.17}$$
$$h_{AfC} + 0.5700\ V_{AX}^2 + 0.1667\ V_{XC}^2$$
$$0.3148\ V_{XB}^2 = 0.1667\ V_{XC}^2$$
$$V_{XC} = 1.374\ V_{XB}$$
$$V_{AX} = V_{XB} + 1.374\ V_{XB} = 2.374\ V_{XB}$$

so that

$$h_{AfB} = 58.44 = 0.5700(2.374\ V_{XB})^2 + 0.3148\ V_{XB}^2$$
$$V_{XB} = 4.070 \qquad V_{XC} = 5.592 \qquad V_{AX} = 9.663$$

Further trials are not justified.

$$A = \pi D^2/4 = (\pi/4)(0.1723)^2 = 0.02332 \text{ ft}^2$$
$$Q_{AB} = V_{AB}A = 4.070\times0.02332 = 0.09491 \text{ ft}^1/\text{s } (2.686\times10^{-1}\text{ m}^1/\text{s})$$
$$Q_{XC} = V_{XC}A = 5.592\times0.02332 = 0.1304 \text{ ft}^3/\text{s } (3.693\times10^{-3}\text{ m}^3/\text{s})$$

Siphons are arrangements of hose or pipe which cause liquids to flow from one level A in Fig. 1.3.25 to a lower level C over an intermediate summit B. Performance of siphons may be evaluated from the equation of motion between points A and B:

$$\frac{p_A}{\gamma} + \frac{V_A^2}{2g} + z_A = \frac{p_B}{\gamma} + \frac{V_B^2}{2g} + z_B + h_{AfB}$$

Noting that on the surface $V_A = 0$ and the minimum pressure that can exist at point B is the vapor pressure pv, the maximum elevation of point B is

$$z_B - z_A = \frac{p_A}{\gamma} - \left(\frac{p_v}{\gamma} + \frac{V_B^2}{2g} + h_{AfB}\right)$$

The friction loss $h_f = \Sigma K_{AB}V_B^2/2g$, and let $V_B = V$; then

$$z_B - z_A = \frac{p_A - p_V}{\rho g} - (1 - \Sigma K_{AB})\frac{V^2}{2g}$$

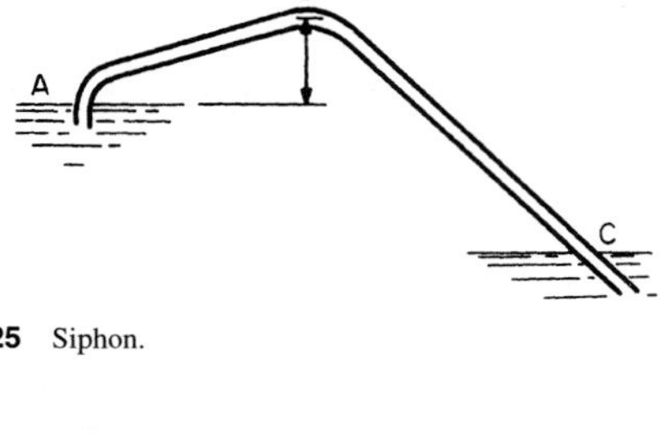

Figure 1.3.25 Siphon.

Flow under this maximum condition will be uncertain. The air pump or ejector used for priming the pipe (flow will not take place unless the siphon is full of water) might have to be operated occasionally to remove accumulated air and vapor. Values of $z_B - z_A$ less than those calculated by the above equation should be used.

EXAMPLE. The siphon shown in Fig. 1.3.25 is composed of 2,000 ft of 6-in schedule 40 cast-iron pipe. Reservoir A is at elevation 800 ft and C at 600 ft. Estimate the maximum height for $z_B - z_A$ if the water temperature may reach 104°F (40°C), and the amount of straight pipe from A to B is 100 ft. For the first bend $L/D = 25$ and the second (at B) $L/D = 50$. Atmospheric pressure is 14.70 lbf/in^2. For 6-in schedule 40 pipe $D = 6.065/12 = 0.5054$ ft, $\varepsilon/D = 850 \times 10^{-6}/0.5054 = 1.682 \times 10^{-3}$. Turbulent friction factor

$$1/\sqrt{f} = -2 \log_{10}\left(\frac{\varepsilon D}{3.7}\right) = -2 \log_{10}(1.682 \times 10^{-3}/3.7) = 0.02238$$

1. *Components from* A *to* B. (Note loss in second bend takes place in downstream piping.)

	K
Entrance (inward projection)	$= 1.0$
100 ft straight pipe $f\,(100/0.5054)$	$= 197.9f$
First bend	$= 25f$
ΣK_{AB}	$= 1.0 + 227.9f$

2. *Components from* A *to* C

ΣK_{AB}	$= 1.0 + 2{,}229f$
1,900 ft of straight pipe $f\,(1{,}900/0.5054)$	$= 3{,}759.4f$
Second bend	$= 50f$
Exit loss	$= 1$
ΣK_{AC}	$= 2.0 + 4{,}032f$

First trial assume complete turbulence. Writing the equation of motion between A and C.

$$\frac{p_A}{\gamma} + \frac{V_A^2}{2g} + z_A = \frac{p_C}{\gamma} + \frac{V_C^2}{2g} + z_C + \Sigma K_{AC}\frac{V^2}{2g}$$

Noting $V_A = V_C = 0$, and $p_A = p_C = 14.7$ lbf/in^2,

$$V = \sqrt{\frac{2g(z_A - z_C)}{\Sigma K_{AC}}} = \sqrt{\frac{2g(z_A - z_C)}{2.0 + 4{,}032f}}$$
$$= \sqrt{\frac{2 \times 32.17(800 - 600)}{2.0 + 4.032f}}$$
$$= \frac{113.44}{\sqrt{2.0 + 4{,}032 \times 0.02238}}$$
$$= 11.81$$

Second trial, use first-trial values,

$$\mathbf{R} = \frac{\rho VD}{\mu} = (1.925)(11.81)(0.5054)/(13.61 \times 10^{-6})$$
$$\mathbf{R} = 846{,}200 > 4{,}000 \quad \therefore \text{ flow is turbulent}$$

From Fig. 1.3.24 and the Colebrook equation,

$$\frac{1}{\sqrt{f}} = -2 \log_{10}\left(\frac{1.682 \times 10^{-3}}{3.7} + \frac{2.51}{844{,}200\sqrt{0.023}}\right)$$
$$f = 0.02263$$
$$V = \frac{113.44}{\sqrt{2.0 + 4{,}032 \times 0.02263}} = 11.75 \qquad \text{(close check)}$$

From Sec. 2.2 steam tables at 104°F, $p_v = 1.070$ lbf/in^2, the maximum height

$$z_B - z_A = \frac{p_A - p_v}{\rho g} - (1 + \Sigma K_{AB})\frac{V^2}{2g}$$
$$= \frac{144(14.70 - 1.070)}{1.925 \times 32.17} - (1 + 1 + 227.9$$
$$\times\, 0.02262)\frac{(11.75)^2}{2 \times 32.17} = 16.58 \text{ ft}(5.053\text{ m})$$

Note that if a ±10 percent error exists in calculation of pressure loss, maximum height should be limited to ~ 15 ft (5 m).

ASME PIPELINE FLOWMETERS

Parameters Dimensional analysis of the flow of an incompressible fluid flowing in a pipe of diameter D, surface roughness ε, through a primary element (venturi, nozzle or orifice) whose diameter is d with a velocity of V, producing a pressure drop of Δp; sensed by pressure taps located a distance L apart results in $f(\mathbf{C}_p, \mathbf{R}_d, \varepsilon/D, d/D) = 0$, which may be written as $\Delta p = \mathbf{C}_p \rho V^2/2$. Conventional practice is to express the relations as $V = \mathbf{K}\sqrt{2\Delta p/\rho}$, where $\mathbf{K}$ is the **flow coefficient,** $\mathbf{K} = 1\sqrt{\mathbf{C}_p}$, and $\mathbf{K} = f(\mathbf{R}_d, L/D, \varepsilon/d, d/D)$. The ratio of the diameter of the primary element to meter tube (pipe) diameter D is known as the **beta ratio,** where $\beta = d/D$. Application of the continuity equation leads to $Q = \mathbf{K}A_2\sqrt{2\Delta p/\rho}$, where A_2 is the area of the primary element.

Conventional practice is to base flowmeter computations on the assumption of one-dimensional frictionless flow of an incompressible fluid in a horizontal meter tube and to correct for actual conditions by the use of a coefficient for viscous effects and a factor for elastic effects. Application of the Bernoulli equation for horizontal flow from section 1 (inlet tap) to section 2 (outlet tap) results in $p_1/\rho g + V_1^2/2g = p_2/\rho g + V_2^2/2g$ or $(p_1 - p_2)/\rho = V_2^2 - V_1^2 = \Delta p/\rho$. From the equation of continuity, $Q_i = A_1V_1 = A_2V_2$, where Q_i is the **ideal flow rate.** Substituting, $2\Delta p/\rho = Q_i^2/A_i^2 - Q_i^2/A_2^2$, and solving for Q_i, $Q_i = A_2\sqrt{2\Delta p/\rho}\,/\sqrt{1-(A_2/A_1)^2}$, noting that $A_2/A_1 = (d/D)^2 = \beta^2$, $Q_i = A_2\sqrt{2\Delta p/\rho}\,/\sqrt{1-\beta^4}$. The **discharge coefficient** C is defined as the ratio of the actual flow Q to the ideal flow Q_i, or $C = Q/Q_i$, so that $Q = CQ_i = CA_2\sqrt{2\Delta p/\rho}\,/\sqrt{1-\beta^4}$. It is customary to write the volumetric-flow equation as $Q = CEA_2\sqrt{2\Delta p/\rho}$, where $E = 1/\sqrt{1-\beta^4}$. E is called the **velocity-of-approach factor** because it accounts for the one-dimensional kinetic energy at the upstream tap. Comparing the equation from dimensional analysis with the modified Bernoulli equation, $Q = \mathbf{K}A_2\sqrt{2\Delta p/\rho} = CEA_2\sqrt{2\Delta p/\rho}$, or $\mathbf{K} = CE$ and $C = f(\mathbf{R}_d, L/D, \beta)$.

For compressible fluids, the incompressible equation is modified by the **expansion factor** Y, where Y is defined as the ratio of the flow of a compressible fluid to that of an incompressible fluid at the same value of Reynolds number. Calculations are then based on inlet-tap-fluid properties, and the compressible equation becomes

$$Q_1 = \mathbf{K}YA_2\sqrt{2\Delta p/\rho_1} = CEYA_2\sqrt{2\Delta p/\rho_1}$$

where $Y = f(L/D, \varepsilon/D, \beta, \mathbf{M})$. Reynolds number $\mathbf{R}_d$ is also based on inlet-fluid properties, but on the primary-element diameter or

$$\mathbf{R}_d = \rho_1 V_2 d/\mu_1 = \rho_1(Q_1/A_2)d/\mu_1 = 4\rho_1 Q_1/\pi\, d\mu_1$$

Caution The numerical values of coefficients for flowmeters given in the paragraphs to follow are based on experimental data obtained with long, straight pipes where the velocity profile approaching the primary element was fully developed. The presence of valves, bends, and fittings upstream of the primary element can cause serious errors. For approach and discharge, straight-pipe requirements, "Fluid Meters," (6th ed., ASME, 1971) should be consulted.

Venturi Tubes Figure 1.3.26 shows a typical venturi tube consisting of a cylindrical inlet, convergent cone, throat, and divergent cone.

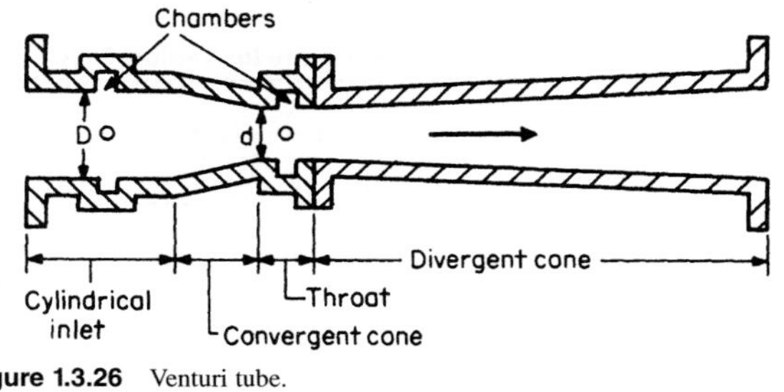

Figure 1.3.26 Venturi tube.

The convergent entrance has an included angle of about 21° and the divergent cone 7 to 8°. The purpose of the divergent cone is to reduce the overall pressure loss of the meter; its removal will have no effect on the coefficient of discharge. Pressure is sensed through a series of holes in the inlet and throat. These holes lead to an annular chamber, and the two chambers are connected to a pressure-differential sensor. Discharge coefficients for venturi tubes as established by the American Society of Mechanical Engineers are given in Table 1.3.12. Coefficients of discharge outside the tabulated limits must be determined by individual calibrations.

EXAMPLE. Benzene at 68°F (20°C) flows through a machined-inlet venturi tube whose inlet diameter is 8 in and whose throat diameter is 3.5 in. The differential pressure is sensed by a U-tube manometer. The manometer contains mercury under the benzene, and the level of the mercury in the throat leg is 4 in. Compute the volumetric flow rate. Noting that $D = 8$ in (0.6667 ft) and $\beta = 3.5/8 = 0.4375$ are within the limits of Table 1.3.12, assume $C = 0.995$, and then check $\mathbf{R}_d$ to verify if it is within limits. For a U-tube manometer (Fig. 1.3.6*a*), $p_2 - p_1 = (\gamma_m - \gamma_f)h = \Delta p$ and $\Delta p/\rho_1 = (\rho_m g - \rho_f g)h/\rho_f = g(\rho_m/\rho_f - 1)h = 32.17(26.283/1.705 - 1)(4/12) = 154.6$. For a liquid, $Y = 1$ (incompressible fluid), $E = 1/\sqrt{1-\beta^4} = 1/\sqrt{1-(0.4375)^4} = 1.019$.

$$Q_1 = CEY\,A_d\sqrt{2\Delta p/\rho_1}$$
$$= (0.995)(1.019)(\pi/4)(3.5/12)^2\sqrt{2\times 154.6}$$
$$= 1.192\ \text{ft}^3/\text{s}\,(3.373\times 10^{-3}\ \text{m}^3/\text{s})$$

$$\mathbf{R}_d = 4\rho_1 Q_1/\pi\, d\mu_1 = 4(1.705)(1.192)/\pi(3.5/12)(13.62\times 10^{-6})$$

$\mathbf{R}_d = 651{,}400$, which lies between 200,000 and 1,000,000 of Table 1.3.12 ∴ solution is valid

Flow Nozzles Figure 1.3.27 shows an ASME flow nozzle. This nozzle is built to rigid specifications, and pressure differential may be sensed by either throat taps or pipe-wall taps. Taps are located one pipe diameter upstream and one-half diameter downstream from the nozzle inlet. Discharge coefficients for ASME flow nozzles may be computed from $C = 0.9975 - 0.00653\,(10^6/\mathbf{R}_d)^a$, where $a = 1/2$ for $\mathbf{R}_d < 10^6$ and $a = 1/5$ for $\mathbf{R}_d > 10^6$. Most of the data were obtained for D between 2 and 15.75 in, $\mathbf{R}_d$ between 10^4 and 10^6, and beta between 0.15 and 0.75. For values of C within these ranges, a tolerance of 2 percent may be anticipated, and outside these limits, the tolerance may be greater than 2 percent. Because slight variations in form or dimension of either pipe or nozzle may affect the observed pressures, and thus cause the exponent a and the slope term (−0.00653) to vary considerably, nozzles should be individually calibrated.

EXAMPLE. An ASME flow nozzle is to be designed to measure the flow of 400 gal/min of 68°F (20°C) water in a 6-in schedule 40 (inside diameter = 6.065 in) steel pipe. The pressure differential across the nozzle is not to exceed 75 in of water. What should be the throat diameter of the nozzle? $\Delta p = h\rho_1 g$, $\Delta p/\rho_1 = hg = (75/12)(32.17) = 201.1$, $Q = (400/60)(231/1{,}728) = 0.8912$ ft³/s. A trial-and-error solution is necessary to establish the values of C and E because they are dependent upon β and $\mathbf{R}_d$, both of which require that d be known. Since $\mathbf{K} = CE \approx 1$, assume for first trial that $CE = 1$. Since a liquid is involved, $Y = 1$, $A_2 = Q_1/(CE)(Y)\sqrt{2\Delta p/\rho_1} = (0.8912)/(1)(1)\sqrt{2\times 201.1} = 0.04444$ ft², $d = \sqrt{4A_2/\pi} = \sqrt{4(0.04444)/\pi} = 0.2379$ ft or $d = 0.2379 \times 12 = 2.854$ in, $\beta = d/D = 2.854/6.065 = 0.4706$.

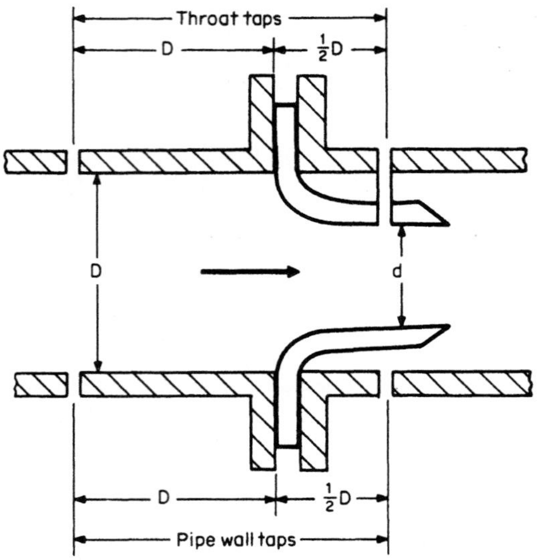

Figure 1.3.27 ASME flow nozzle.

For second trial use first-trial value:

$$E = 1/\sqrt{1-\beta^4} = 1/\sqrt{1-(0.4706)^4} = 1.025$$

$\mathbf{R}_d = 4\rho_1 Q_1/\pi d\mu_1 = 4(1.937)(0.8912)/\pi(0.2379)(20.92 \times 10^{-6}) = 442{,}600 < 10^6$ ∴ $a = 1/2$ and $C = 0.9975 - 0.00653(10^6/\mathbf{R}_d)^{1/2}$. $C = 0.9975 - 0.00653(10^6/442{,}600)^{1/2} = 0.9877$. $A_2 = (0.8912)/(0.9877 \times 1.025)\sqrt{2\times 201.1} = 0.04389$, $d_2 = \sqrt{4\times(0.04389/\pi)} = 0.2364$, $d_2 = 0.2364 \times 12 = 2.837$ in (7.205×10^{-2} m). Further trials are not necessary in view of the ± 2 percent tolerance of C.

Compressible Flow—Venturi Tubes and Flow Nozzles The expansion factor Y is computed based on the assumption of a frictionless adiabatic (isentropic) expansion of an ideal gas from the inlet to the throat of the primary element, resulting in (see Sec. 2.1)

$$Y = \left[\frac{kr^{2/k}(1-r^{(k-1)/k})(1-\beta^4)}{(1-r)(k-1)(1-\beta^4 r^{2/k})}\right]^{1/2}$$

where $r = p_2/p_1$.

Maximum flow is obtained when the critical pressure ratio is reached. The critical pressure ratio r_c may be calculated from

$$r^{(1-k)/k} + \frac{k-1}{2}\beta^4 r^{2/k} = \frac{k+1}{2}$$

Table 1.3.13 gives selected values of Y_c and r_c.

EXAMPLE. A piping system consists of a compressor, a horizontal straight length of 2-in-inside-diameter pipe, and a 1-in-throat-diameter ASME flow nozzle attached to the end of the pipe, discharging into the atmosphere. The compressor is operated to maintain a flow of air with 115 lbf/in² and 140°F (60°C)

Table 1.3.12 ASME Coefficients for Venturi Tubes

<table>
<tr><th rowspan="2">Type of inlet cone</th><th colspan="2">Reynolds number $\mathbf{R}_d$</th><th colspan="2">Inlet diam D in (2.54×10^{-2} m)</th><th colspan="2">β</th><th rowspan="2">C</th><th rowspan="2">Tolerance, %</th></tr>
<tr><th>Min</th><th>Max</th><th>Min</th><th>Max</th><th>Min</th><th>Max</th></tr>
<tr><td>Machined</td><td rowspan="3">5×10^5</td><td>1×10^6</td><td>2</td><td>10</td><td rowspan="2">0.4</td><td>0.75</td><td>0.995</td><td>±1.0</td></tr>
<tr><td>Rough welded sheet metal</td><td>2×10^6</td><td>8</td><td>48</td><td>0.70</td><td>0.985</td><td>±1.5</td></tr>
<tr><td>Rough cast</td><td></td><td>4</td><td>32</td><td>0.3</td><td>0.75</td><td>0.984</td><td>±0.7</td></tr>
</table>

SOURCE: Compiled from data given in "Fluid Meters," 6th ed., ASME, 1971.

Table 1.3.13 Expansion Factors and Critical Pressure Ratios for Venturi Tubes and Flow Nozzles

		Critical values		Expansion factor Y			
β	k	r_c	Y_c	$r = 0.60$	$r = 0.70$	$r = 0.80$	$r = 0.90$
0	1.10	0.5846	0.6894	0.7021	0.7820	0.8579	0.9304
	1.20	0.5644	0.6948	0.7228	0.7981	0.8689	0.9360
	1.30	0.5457	0.7000	0.7409	0.8119	0.8783	0.9408
	1.40	0.5282	0.7049	0.7568	0.8240	0.8864	0.9449
0.20	1.10	0.5848	0.6892	0.7017	0.7817	0.8577	0.9303
	1.20	0.5546	0.6946	0.7225	0.7978	0.8687	0.9359
	1.30	0.5459	0.6998	0.7406	0.8117	0.8781	0.9407
	1.40	0.5284	0.7047	0.7576	0.8237	0.8862	0.9448
0.50	1.10	0.5921	0.6817	0.6883	0.7699	0.8485	0.9250
	1.20	0.5721	0.6872	0.7094	0.7864	0.8600	0.9310
	1.30	0.5535	0.6923	0.7248	0.8007	0.8699	0.9361
	1.40	0.5362	0.6973	0.7440	0.8133	0.8785	0.9405
0.60	1.10	0.6006	0.6729		0.7556	0.8374	0.9186
	1.20	0.5808	0.6784	0.6939	0.7727	0.8495	0.9250
	1.30	0.5625	0.6836	0.7126	0.7875	0.8599	0.9305
	1.40	0.5454	0.6885	0.7292	0.8006	0.8689	0.9352
0.70	1.10	0.6160	0.6570		0.7290	0.8160	0.9058
	1.20	0.5967	0.6624	0.6651	0.7469	0.8292	0.9131
	1.30	0.5788	0.6676	0.6844	0.7626	0.8405	0.9193
	1.40	0.5621	0.6726	0.7015	0.7765	0.8505	0.9247
0.80	1.10	0.6441	0.6277		0.6778	0.7731	0.8788
	1.20	0.6238	0.6331		0.6970	0.7881	0.8877
	1.30	0.6087	0.6383		0.7140	0.8012	0.8954
	1.40	0.5926	0.6433	0.6491	0.7292	0.8182	0.9021

SOURCE: Murdock, "Fluid Mechanics and Its Applications," Houghton Mifflin, 1976.

conditions in the pipe just one pipe diameter before the nozzle inlet. Barometric pressure is 14.7 lbf/in^2. Estimate the flow rate of the air in lbm/s.

From the equation of state, $\rho_1 = p_1/g_cRT_1 = (144 \times 115)/(32.17)(53.34)(140 + 459.7) = 0.01609$ slug/ft^3, $\beta = d/D = 1/2 = 0.5$, $E = 1/\sqrt{1-\beta^4} = 1/\sqrt{1-(0.5)^4} = 1.033$, $r = p_2/p_1 = 14.7/115 = 0.1278$, but from Table 1.3.13 at $\beta = 0.5$, $k = 1.4$, $r_c = 0.5362$, and $Y_c = 0.6973$, so that because of critical flow the throat pressure $p_c = 115 \times 0.5362 = 61.66$ lbf/in^2. $\Delta p_c/\rho_1 = 144(115 - 61.66)/0.01609 = 477{,}375$. A trial-and-error solution is necessary to obtain C. For the first trial assume $10^6/\mathbf{R}_d = 0$ or $C = 0.9975$. Then $Q_1 = CEY_cA_2\sqrt{2\Delta p_c/\rho_1} = (0.9975)(1.033)(0.6973)(\pi/4)(1.12)^2\sqrt{2\times 477{,}375} = 3.829$ ft^3/s, $\mathbf{R}_d = 4\rho_1 Q/\pi d\mu_1 = (4)(0.01609)(3.828)/\pi(1/12)(41.79 \times 10^{-8}) = 2{,}252{,}000$.

Second trial, use first-trial values:

$$\mathbf{R} > 10^6,\ a = 115,\ C = 0.9975 - (0.00653)(16^6/2{,}252{,}000)^{1/5}$$

$$C = 0.9919,\ Q_1 = 3.828(0.9919/0.9975) = 3.806\ \text{ft}^3/\text{s}$$

Further trials are not necessary in view of ±2 percent tolerance on C.

$$\dot{m} = Q_1\rho_1 g = 3.806 \times 0.01609 \times 32.17 = 1.970\ \text{lbm/s}\ (0.8935\ \text{kg/s})$$

Orifice Meters When a fluid flows through a square-edged thin-plate orifice, the minimum-flow area is found to occur downstream from the orifice plate. This minimum area is called the vena contracta, and its location is a function of beta ratio. Figure 1.3.28 shows the relative pressure difference due to the presence of the orifice plate. Because the location of the pressure taps is vital, it is necessary to specify the exact position of the downstream pressure tap. The jet contraction amounts to about 60 percent of the orifice area; so orifice coefficients are in the order of 0.6 compared with the nearly unity obtained with venturi tubes and flow nozzles.

Three pressure-differential-measuring tap locations are specified by the ASME. These are the flange, vena contracta, and the 1 D and 1/2 D. In the flange tap, the location is always 1 in from either face of the orifice plate regardless of the size of the pipe. In the vena contracta tap, the upstream tap is located one pipe diameter from the inlet face of the orifice plate and the downstream tap at the location of the vena contracta. In the 1 D and 1/2 D tap, the upstream tap is located one pipe diameter from the inlet face of the orifice plate and downstream one-half pipe diameter from the inlet face of the orifice plate.

Flange taps are used because they can be prefabricated, and flanges with holes drilled at the correct locations may be purchased as off-the-shelf items, thus saving the cost of field fabrication. The disadvantage of flange taps is that they are not symmetrical with respect to pipe size. Because of this, coefficients of discharge for flange taps vary greatly with pipe size.

Vena contracta taps are used because they give the maximum differential for any given flow. The disadvantage of the vena contracta tap is that if the orifice size is changed, a new downstream tap must be drilled. The 1 D and 1/2 D taps incorporate the best features of the vena contracta taps and are symmetrical with respect to pipe size.

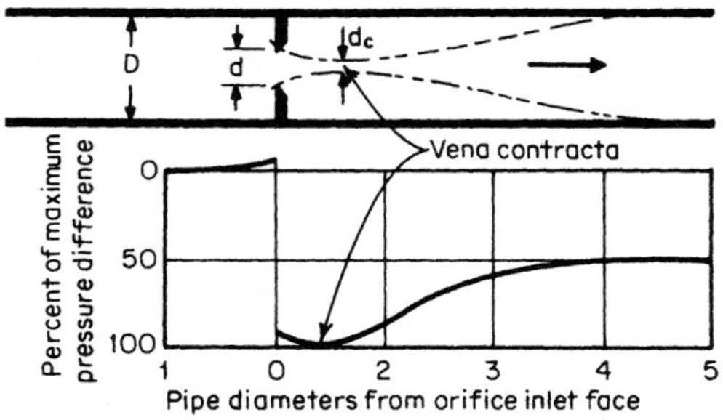

Figure 1.3.28 Relative-pressure changes due to flow through an orifice.

Discharge coefficients for orifices may be calculated from

$$C = C_o + \Delta C \mathbf{R}_d^{-0.75} \qquad (\mathbf{R}_d > 10^4)$$

where C_o and ΔC are obtained from Table 1.3.14.

Tolerances for uncalibrated orifice meters are in the order of ±1 to ±2 percent depending upon β, D, and $\mathbf{R}_d$.

Compressible Flow through ASME Orifices As shown in Fig. 1.3.28, the minimum flow area for an orifice is at the vena contracta located downstream of the orifice. The stream of compressible fluid is not restrained as it leaves the orifice throat and is free to expand transversely and longitudinally to the point of minimum-flow area. Thus the contraction of the jet will be less for a compressible fluid than for a liquid. Because of this, the theoretical-expansion-factor equation may not be used with orifices. Neither may the critical-pressure-ratio equation be used, as the phenomenon of critical flow has not been observed during testing of orifice meters.

For orifice meters, the following equation, which is based on experimental data, is used:

$$Y = 1 - (0.41 + 0.35\beta^4)(\Delta p / p_1)/k$$

EXAMPLE. Air at 68°F (20°C) and 150 lbf/in² flows in a 2-in schedule 40 pipe (inside diameter = 2.067 in) at a volumetric rate of 15 ft³/min. A 0.5500-in ASME orifice equipped with flange taps is used to meter this flow. What deflection in inches could be expected on a U-tube manometer filled with 60°F water? From the equation of state, $\rho_1 = p_1/g_c RT_1 = (144 \times 150)/(32.17)(53.34)(68 + 459.7) = 0.02385$ slug/ft³, $\beta = 0.5500/2.067 = 0.2661$. $Q_1 = 15/60 = 0.25$ ft³/s, $A_2 = (\pi/4)(0.5500/12)^2 = 1.650 \times 10^{-3}$ ft². $E = 1/\sqrt{1-\beta^4} = 1/\sqrt{1-(0.2661)^4} = 1.003$. $\mathbf{R}_d = 4\rho_1 Q_1/\pi\ d\mu_1 = 4(0.02385)(0.25)/\pi(0.5500/12)(39.16 \times 10^{-8})$. $\mathbf{R}_d = 423{,}000$.

From Table 1.3.14 at $\beta = 0.2661$, $D = 2.067$-in flange taps, by interpolation, $C_o = 0.5977$, $\Delta C = 9.087$, from orifice-coefficient equation $C = C_o + \Delta C \mathbf{R}_d^{-0.75}$. $C = 0.5977 + (9.087)(423{,}000)^{-0.75} = 0.5982$. A trial-and-error solution is required because the pressure loss is needed in order to compute Y. For the first trial, assume $Y = 1$, $\Delta p = (Q_1/CEYA_2)^2(\rho_1/2) = [(0.25)/(0.5982)(1.003)(Y)(1.650 \times 10^{-3})]^2(0.02385/2) = 760.5/Y^2 = 760.5/(1)^2 = 760.5$ lbf/ft².

For the second trial we use first-trial values.

$$Y = 1 - (0.41 + 0.35\ \beta^4)\frac{\Delta p/p_1}{k}$$

$$= 1 - [0.41 + 0.35(0.2661)^4]\ \frac{760.5/144 \times 150}{1.4} = 0.9896$$

$$\Delta p = \frac{760.5}{Y^2} = \frac{760.5}{(0.9896)^2} = 776.1 \text{ lbf/ft}^2$$

For the third trial we use second-trial values.

$$Y = 1 - [0.41 + 0.35(0.2661)^4]\ \frac{776.1/144 \times 150}{1.4} = 0.9894$$

$$\Delta p = \frac{776.1}{(0.9894)^2} = 793.3 \text{ lbf/ft}^2$$

Resubstitution does not produce any further change in Y. From the U-tube-manometer equation:

$$h = \frac{\Delta p}{\gamma_m = \gamma_f} = \frac{\Delta p}{g_c(\rho_m - \rho_f)}$$

$$= \frac{793.3}{32.17}(1.937 - 0.02385) = 12.89 \text{ ft}$$

$$= 12.89 \times 12 = 154.7 \text{ in } (3.929 \text{ m})$$

PITOT TUBES

Definition A Pitot tube is a device that is shaped in such a manner that it senses stagnation pressure. The name "Pitot tube" has been applied to two general classifications of instruments, the first being a tube that measures the impact or stagnation pressures only, and the second a combined tube that measures both impact and static pressures with a single primary instrument. The combined sensor is called a Pitot-static tube.

Table 1.3.14 Values of C_o and ΔC for Use in Orifice Coefficient Equation

	β										
Pipe ID, in	0.20	0.25	0.30	0.35	0.40	0.45	0.50	0.55	0.60	0.65	0.70
					ΔC, all taps						
All	5.486	8.106	11.153	14.606	18.451	22.675	27.266	32.215	37.513	45.153	49.129
					C_o, vena contracta and 1D and ½D taps						
All	0.5969	0.5975	0.5983	0.5992	0.6003	0.6016	0.6031	0.6045	0.6059	0.6068	0.6069
					C_o, flange taps						
2.0	0.5969	0.5975	0.5982	0.5992	0.6003	0.6016	0.6030	0.6044	0.6056	0.6065	0.6066
2.5	0.5969	0.5975	0.5983	0.5993	0.6004	0.6017	0.6032	0.6046	0.6059	0.6068	0.6068
3.0	0.5969	0.5975	0.5983	0.5993	0.6004	0.6017	0.6031	0.6044	0.6055	0.6061	0.6057
3.5	0.5969	0.5975	0.5983	0.5993	0.6004	0.6016	0.6030	0.6042	0.6052	0.6056	0.6049
4.0	0.5969	0.5976	0.5983	0.5993	0.6004	0.6016	0.6029	0.6041	0.6050	0.6052	0.6043
5.0	0.5969	0.5976	0.5983	0.5993	0.6004	0.6016	0.6028	0.6039	0.6047	0.6047	0.6034
6.0	0.5969	0.5976	0.5983	0.5993	0.6004	0.6016	0.6028	0.6038	0.6045	0.6044	0.6029
8.0	0.5969	0.5976	0.5984	0.5993	0.6004	0.6015	0.6027	0.6037	0.6042	0.6040	0.6022
10.0	0.5969	0.5976	0.5984	0.5993	0.6004	0.6015	0.6026	0.6036	0.6041	0.6037	0.6017
12.0	0.5970	0.5976	0.5984	0.5993	0.6004	0.6015	0.6026	0.6035	0.6040	0.6035	0.6015
16.0	0.5970	0.5976	0.5984	0.5993	0.6003	0.6015	0.6026	0.6035	0.6039	0.6033	0.6011
24.0	0.5970	0.5976	0.5984	0.5993	0.6003	0.6015	0.6025	0.6034	0.6037	0.6031	0.6007
48.0	0.5970	0.5976	0.5984	0.5993	0.6003	0.6014	0.6025	0.6033	0.6036	0.6029	0.6004
∞	0.5970	0.5976	0.5984	0.5993	0.6003	0.6014	0.6025	0.6032	0.6035	0.6027	0.6000

SOURCE: Compiled from data given in ASME Standard MFC-3M-1984 "Measurement of Fluid Flow in Pipes. Using Orifice, Nozzle and Venturi."

Tube Coefficient From Fig. 1.3.29, it is evident that the Pitot tube can sense only the stagnation pressure resulting from the local stream-tube velocity U. The local ideal velocity U_i for an incompressible fluid is obtained by the application of the Bernoulli equation ($z_S = z$), $U_i^2/2g + p/\rho g = U_S^2/2g + p_S/\rho g$. Solving for U_i and noting that by definition $U_S = 0$, $U_i = \sqrt{2(p_S - p)/\rho}$. Conventional practice is to define the **tube coefficient** C_T as the ratio of the actual stream-tube velocity to the ideal stream-tube velocity, or $C_T = U/U_i$ and $U = C_T U_i = C_T\sqrt{2\Delta p/\rho}$. The numerical value of C_T depends primarily upon its geometry. The value of C_T may be established (1) by calibration with a uniform velocity, (2) from published data for similar geometry, or (3) in the absence of other information, may be assumed to be unity.

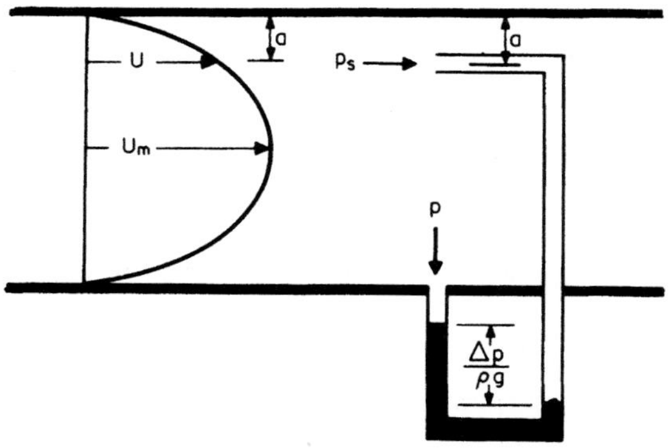

Figure 1.3.29 Notation for Pitot tube study.

Pipe Coefficient For the calculation of volumetric flow rate, it is necessary to integrate the continuity equation, $Q = \int U\, da = AV$. The **pipe coefficient** C_P is defined as the ratio of the average velocity to the stream-tube velocity, or $C_P = V/U$, and $Q = C_P A_1 V = C_P C_T A_1\sqrt{2\Delta p/\rho}$. The numerical value of C_P is dependent upon the location of the tube and the **velocity profile.** The values of C_P may be established by (1) making a "traverse" by taking data at various points in the flow stream and determining the velocity profile experimentally (see "Fluid Meters," 6th ed., ASME, 1971, for locations of traverse points), (2) using standard velocity profiles, (3) locating the Pitot tube at a point where $U = V$, and (4) assuming one-dimensional flow of $C_P = 1$ **only** in the absence of other data.

Compressible Flow For compressible flow, the **compression factor** Z is based on the assumption of a frictionless adiabatic (isentropic) compression of an ideal gas from the moving stream tube to the stagnation point (see Sec. 2.1), which results in

$$Z = \left[\frac{k}{k-1}\,\frac{(p_S/p)^{(k-1)/k} - 1}{(p_S/p) - 1}\right]^{1/2}$$

and the volumetric flow rate becomes

$$Q = C_P C_T Z A_1\sqrt{2\Delta p/\rho}$$

EXAMPLE. Carbon dioxide flows at 68°F (20°C) and 20 lbf/in² in an 8-in schedule 40 galvanized-iron pipe. A Pitot tube located on the pipe centerline indicates a pressure differential of 6.986 lbf/in². Estimate the mass flow rate. For 8-in schedule 40 pipe D = 7.981/12 = 0.6651, $\varepsilon/D = 500 \times 10^{-6}/0.6651 = 7.518 \times 10^{-4}$, $A_1 = \pi D^2/4 = (\pi/4)(0.6651)^2 = 0.3474$ ft², $p_S = p + \Delta p = 20 + 6.986 = 26.986$ lbf/in². From the equation of state, $\rho = p/g_c R T_o = (20 \times 144)/(32.17)(35.11)(68 + 459.7) = 0.004832$,

$$Z = \left[\frac{k}{k-1}\,\frac{(p_S/p)^{(k-1)/k} - 1}{(p_S/p) - 1}\right]^{1/2}$$

$$= \left\{[1.3/(1.3-1)] \times \frac{(26.986/20)^{(1.3-1)/1.3} - 1}{(26.986/20) - 1}\right\}^{1/2} = 0.9423$$

In the absence of other data, C_T may be assumed to be unity. A trial-and-error solution is necessary to determine C_p, since f requires flow rate. For the first trial assume complete turbulence.

$$1/\sqrt{f} = -2\log_{10}(7.518 \times 10^{-4}/3.7) \qquad \sqrt{f} = 0.1354$$

$$C_P = V/U = V/U_{max} = 1/(1 + 1.43\sqrt{f}) = 1/(1 + 1.43 \times 0.1354) = 0.8378$$

$$V = C_P C_T Z\sqrt{2\Delta P/\rho} = (0.8378)(1)(0.9423)\sqrt{2 \times 144(6.987)/(0.004832)}$$

$$V = 509.4 \text{ ft/s}$$

$$\mathbf{R} = \rho VD/\mu = (0.004832)(509.4)(0.6651)/(30.91 \times 10^{-18})$$

$$\mathbf{R} = 5{,}296{,}000 > 4{,}000 \ \therefore \text{ flow is turbulent}$$

From the Colebrook equation and Fig. 1.3.24,

$$\frac{1}{\sqrt{f}} = -2\log_{10}\left(\frac{7.518 \times 10^{-4}}{3.7} + \frac{2.51}{5{,}296{,}000\sqrt{0.018}}\right)$$

$$\sqrt{f} = 0.1357$$

$$C_P = 1/(1 + 1.43 \times 0.1357) = 0.8375 \qquad \text{(close check)}$$

$$V = 509.4(0.8375/8378) = 509.2 \text{ ft/s}$$

From the continuity equation, $m = \rho A_1 V g_c = (0.004832)(0.3474)(509.2)(32.17) = 27.50$ lbm/s (12.47 kg/s).

ASME WEIRS

Definitions A weir is a dam over which liquids are forced to flow. Weirs are used to measure the flow of liquids in open channels or in conduits which do not flow full; i.e., there is a free liquid surface. Weirs are almost exclusively used for measuring water flow, although small ones have been used for metering other liquids. Weirs are classified according to their notch or opening as follows: (1) **rectangular notch** (original form); (2) V or **triangular notch;** (3) **trapezoidal notch,** which when designed with end slopes one horizontal to four vertical is called the **Cipolletti weir;** (4) the **hyperbolic weir** designed to give a constant coefficient of discharge; and (5) the **parabolic weir** designed to give a linear relationship of head to flow. As shown in Fig. 1.3.30, the top of the weir is the crest and the distance from the liquid surface to the crest h is called the **head.**

The sheet of liquid flowing over the weir crest is called the **nappe.** When the nappe falls downstream of the weir plate, it is said to be free, or **aerated.** When the width of the approach channel L_c is greater than the crest length L_w, the nappe will contract so that it will have a minimum width less than the crest length. For this reason, the weir is known as a **contracted weir.** For the special case where $L_w = L_c$, the contractions do not take place, and such weirs are known as **suppressed weirs.**

Parameters The forces acting on a liquid flowing over a weir are inertia, viscous, surface tension. and gravity. If the weir head produced by the flow is h, the characteristic length of the weir is L_w, and the channel width is L_c, either similarity or dimensional analysis leads to $f(\mathbf{F}, \mathbf{W}, \mathbf{R}, L_w/L_c) = 0$, which may be written as $V = K\sqrt{2gh}$, where K is the **weir coefficient** and $K = f(\mathbf{W}, \mathbf{R}, L_w/L_c)$. Since the weir has been almost exclusively used for metering water flow over limited temperature ranges, the effects of surface tension and viscosity have not been adequately established by experiment.

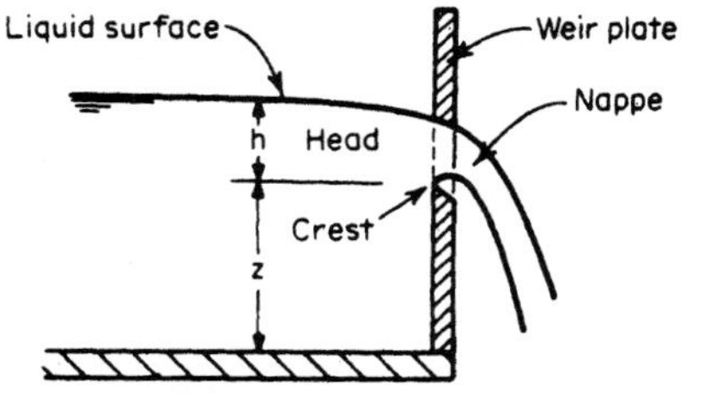

Figure 1.3.30 Notation for weir study.

Caution The numerical values of coefficients for weirs are based on experimental data obtained from calibration of weirs with long approaches of straight channels. Head measurement should be made at a distance at least three or four times the expected maximum head h. Screens and baffles should be used as necessary to ensure steady uniform flow without waves or local eddy currents. The approach channel should be relatively wide and deep.

Rectangular Weirs Figure 1.3.31 shows a rectangular weir whose crest width is L_w. The volumetric flow rate may be computed from the continuity equation: $Q = AV = (L_w h)(K\sqrt{2gh}) = KL_w\sqrt{2g}\,h^{3/2}$. The ASME "Fluid Meters" report recommends the following equation for rectangular weirs: $Q = (2/3)CL_a\sqrt{2g}\,h_a^{3/2}$, where C is the **coefficient of discharge** $C = f(L_w/L_c, h/Z)$, L_a is the adjusted crest length $L_a = L_w + \Delta L$, and h_a is the adjusted weir head $h_a = h + 0.003$ ft. Values of C and ΔL may be obtained from Table 1.3.15. To avoid the possibility that the liquid drag along the sides of the channel will affect side contractions, $L_c - L_w$ should be at least $4h$. The minimum crest length should be 0.5 ft to prevent mutual interference of the end contractions The minimum head for free flow of the nappe should be 0.1 ft.

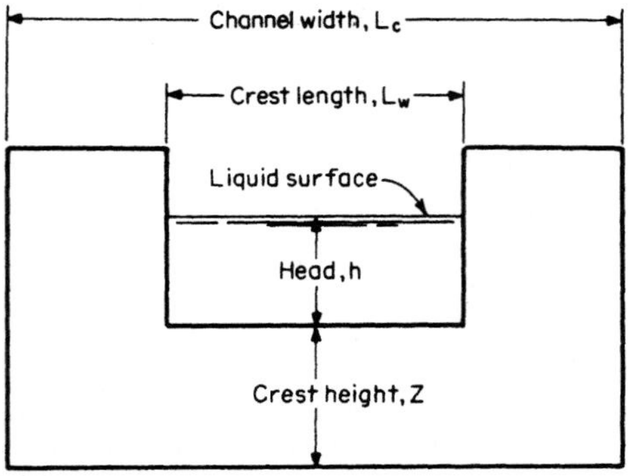

Figure 1.3.31 Rectangular weir.

EXAMPLE. Water flows in a channel whose width is 40 ft. At the end of the channel is a rectangular weir whose crest width is 10 ft and whose crest height is 4 ft. The water flows over the weir at a height of 3 ft above the crest of the weir. Estimate the volumetric flow rate. $L_w/L_c = 10/40 = 0.25$, $h/Z = 3/4 = 0.75$, from Table 1.3.15 (interpolated), $C = 0.589$, $\Delta L = 0.008$, $L_a = L_w + \Delta L = 10 + 0.008 = 10.008$ ft, $h_a = h + 0.003 = 3 + 0.003 = 3.003$ ft, $Q = (2/3)\ CL_a\sqrt{2g}\ h^{3/2}$, $Q = (2/3)(0.589)(10.008)(2 \times 32.17)^{1/2}(3.003)^{3/2}$ $Q = 164.0$ ft³/s (4.644 m³/s).

Triangular Weirs Figure 1.3.32 shows a triangular weir whose notch angle is θ. The volumetric flow rate may be computed from the continuity equation $Q = AV = (h^2 \tan\theta/2)(K\sqrt{2gh}) = K\tan(\theta/2)\sqrt{2g}\,h^{5/2}$. The ASME "Fluid Meters" report recommends the following for triangular weirs: $Q = (8/15)\ C\tan(\theta/2)\sqrt{2g}(h + \Delta h)^{5/2}$, where C is the coefficient of discharge $C = f(\theta)$ and Δh is the correction for head/crest ratio $\Delta h = f(\theta)$. Values of C and Δh may be obtained from Table 1.3.16.

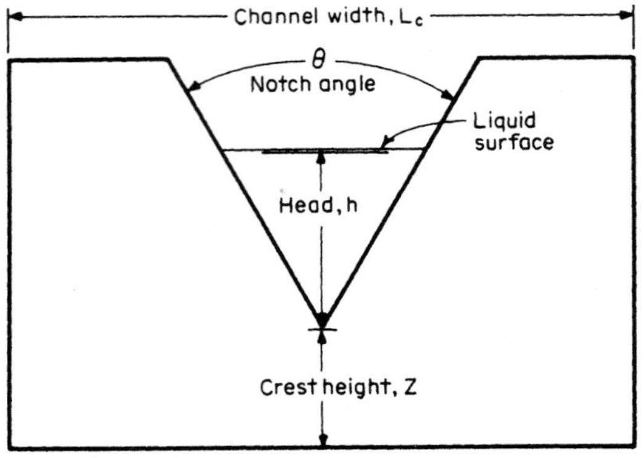

Figure 1.3.32 Triangular weir.

Table 1.3.16 Values of C and Δh for Use in Triangular-Weir Equation

	Weir notch angle θ, deg						
Item	20	30	45	60	75	90	100
C	0.592	0.586	0.580	0.576	0.576	0.579	0.581
Δh, ft	0.010	0.007	0.005	0.004	0.003	0.003	0.003

SOURCE: Compiled from data given in "Fluid Meters." ASME, 1971.

EXAMPLE. It is desired to maintain a flow of 167 ft³/s in an open channel whose width is 20 ft at a height of 7 ft by locating a triangular weir at the end of the channel. The weir has a crest height of 2 ft. What notch angle is required to maintain these conditions? A trial-and-error solution is required. For the first trial assume $\theta = 60°$ (mean value 20 to 100°): then $C = 0.576$ and $\Delta h = 0.004$.

$$h + Z = 7 = h + 2 \ \therefore\ h = 5$$

$$Q = (8/15)\ C\tan(\theta/2)\sqrt{2g}(h + \Delta h)^{5/2}$$

$167 = (8/15)(0.576)\ \tan(\theta/2)\sqrt{2 \times 32.17}\ (5 + 0.004)^{5/2}$, $\tan^{-1}(\theta/2) = 1.20993$, $\theta = 100°51'$.

Second trial, using $\theta = 100$, $C = 0.581$, $\Delta h = 0.003$, $167 = (8/15)(0.581)\tan(\theta/2)\sqrt{2 \times 32.17}\ (5 + 0.003)^{5/2}$, $\tan^{-1}(\theta/2) = 1.20012$, $\theta = 100°39'$ (close check).

Table 1.3.15 Values of C and ΔL for Use in Rectangular-Weir Equation

	Crest length/channel width = L_w/L_c							
h/Z	0	0.2	0.4	0.6	0.7	0.8	0.9	1.0
	Coefficient of discharge C							
0	0.587	0.589	0.591	0.593	0.595	0.597	0.599	0.603
0.5	0.586	0.588	0.594	0.602	0.610	0.620	0.631	0.640
1.0	0.586	0.587	0.597	0.611	0.625	0.642	0.663	0.676
1.5	0.584	0.586	0.600	0.620	0.640	0.664	0.695	0.715
2.0	0.583	0.586	0.603	0.629	0.655	0.687	0.726	0.753
2.5	0.582	0.585	0.608	0.637	0.671	0.710	0.760	0.790
3.0	0.580	0.584	0.610	0.647	0.687	0.733	0.793	0.827
	Adjustment for crest length ΔL, ft							
Any	0.007	0.008	0.009	0.012	0.013	0.014	0.013	–0.005

SOURCE: Compiled from data given in "Fluid Meters," ASME, 1971.

OPEN-CHANNEL FLOW

Definitions An **open channel** is a conduit in which a liquid flows with a free surface subjected to a constant pressure. Flows of water in natural streams, artificial canals, irrigation ditches, sewers, and flumes are examples where the water surface is subjected to atmospheric pressure. The flow of any liquid in a pipe where there is a free liquid surface is an example of open-channel flow where the liquid surface will be subjected to the pressure existing in the pipe. The **slope** S of a channel is the change in elevation per unit of horizontal distance. For small slopes, this is equivalent to dividing the change in elevation by the distance L measured along the channel bottom between two sections. For steady uniform flow, the velocity distribution is the same at all sections of the channel, so that the energy grade line has the same angle as the bottom of the channel, thus:

$$S = h_f/L$$

The distance between the liquid surface and the bottom of the channel is sometimes called the **stage** and is denoted by the symbol y in Fig. 1.3.33. When the stages between the sections are not uniform, that is, $y_1 \neq y_2$ or the cross section of the channel changes, or both, the flow is said to be **varied.** When a liquid flows in a channel of uniform cross section and the slope of the surface is the same as the slope of the bottom of the channel ($y_1 = y = y_2$), the flow is said to be **uniform.**

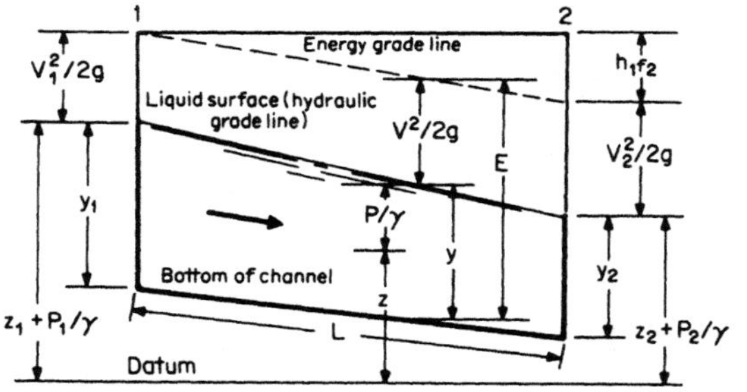

Figure 1.3.33 Notation for open channel flow.

Parameters The forces acting on a liquid flowing in an open channel are inertia, viscous, surface tension, and gravity. If the channel has a surface roughness of ε, a hydraulic radius of R_h, and a slope of S, either similarity or dimensional analysis leads to $f(\mathbf{F}, \mathbf{W}, \mathbf{R}, \varepsilon/4R_h) = 0$, which may be written as $V = C\sqrt{R_h S}$, where $C = f(\mathbf{W}, \mathbf{R}, \varepsilon/4R_h)$ and is known as the **Chézy coefficient.** The relationship between the Chézy coefficient C and the friction factor may be determined by equating

$$V = \sqrt{8R_h h_f g/fL} = C\sqrt{R_h S} = C\sqrt{(R_h h_f)/L}$$

or $C = (8g/f)^{1/2}$. Although this establishes a relationship between the Chézy coefficient and the friction factor, it should be noted that $f = f(\mathbf{R}, \varepsilon/4R_h)$ and $C = f(\mathbf{W}, \mathbf{R}, \varepsilon/4R_h)$, because in open-channel flow, pressure forces are absent and in pipe flow, surface-tension and gravity forces are absent. For these reasons, data obtained in pipe flow should not be applied to open-channel flow.

Roughness Factors For open-channel flow, the Chézy coefficient is calculated by the Manning equation, which was developed from examination of experimental results of water tests. The **Manning relation** is stated as

$$C = \frac{1.486}{n} R_h^{1/6}$$

where n is a roughness factor and should be a function of Reynolds number, Weber number, and relative roughness. Since only water-test data obtained at ordinary temperatures support these values, it must be assumed that n is the value for turbulent flow only. Since surface tension is a weak property, the effects of Weber-number variation are negligible, leaving n to be some function of surface roughness. Design values of n are given in Table 1.3.17. Maximum flow for a given slope will take place when R_h is a maximum, and values of R_h max are given in Table 1.3.6.

Table 1.3.17 Values of Roughness Factor n for Use in Manning Equation

Surface	n	Surface	n
Brick	0.015	Earth, with stones and weeds	0.035
Cast iron	0.015	Gravel	0.029
Concrete, finished	0.012	Riveted steel	0.017
Concrete, unfinished	0.015	Rubble	0.025
Brass pipe	0.010	Wood, planed	0.012
Earth	0.025	Wood, unplaned	0.013

SOURCE: Compiled from data given in R. Horton, *Engineering News,* 75, 373, 1916.

EXAMPLE. It is necessary to carry 150 ft³/s of water in a rectangular unplaned timber flume whose width is to be twice the depth of water. What are the required dimensions for various slopes of the flume? From Table 1.3.6, $A = b^2/2$ and $R_h = h/2 = b/4$. From Table 1.3.17, $n = 0.013$ for unplaned wood. From Manning's equation, $C = 1.486/n$, $R_h^{1/6} = (1.486/0.013)(b^{1/6}/(4)^{1/6} = 90.73\, b^{1/6}$. From the continuity equation, $V = Q/A = 150/(b^2/2)$, $V = 300/b^2$. From the Chézy equation, $V = C\sqrt{R_h S} = 300/b^2 = 90.73 b^{1/6}\sqrt{(b/4)S}$; solving for b, $b = 2.0308/S^{3/16}$.

Assumed S:	1×10^{-1}	1×10^{-2}	1×10^{-3}	1×10^{-4}	1×10^{-5}	1×10^{-6} ft/ft
Required b:	3.127	4.816	7.416	11.42	17.59	27.08 ft

EXAMPLE. A rubble-lined trapezoidal canal with 45° sides is to carry 360 ft³/s of water at a depth of 4 ft. If the slope is 9×10^{-4} ft/ft, what should be the dimensions of the canal? From Table 1.3.17, $n = 0.025$ for rubble. From Table 1.3.6 for $\alpha = 45°$, $A = (b + h)h = 4(b + 4)$, and $R_h = (b + h)h/(b + 2.828h) = 4(b + 4)/(b + 11.312)$. From the Manning relation, $C = (1.486/n)\left(R_h^{1/6}\right) = (1.486/0.025)R_h^{1/6} = 59.44\, R_h^{1/6}$. For the first trial, assume $R_h = R_h$ max $= h/2 = 4/2 = 2$; then $C = 59.44(2)^{1/6} = 66.72$ and $V = C\sqrt{R_h S} = 66.72\sqrt{2 \times 9 \times 10^{-4}} = 2.831$. From the continuity equation, $A = Q/V = 360/2.831 = 127.2 = 4(b + 4)$; $b = 27.79$ ft. Second trial, use the first trial, $R_h = 4(27.79 + 4)/(27.79 + 11.312)$, $R_h = 3.252$, $V = 59.44(3.252)^{1/6}\sqrt{3.252 \times 9 \times 10^{-4}} = 3.914$. From the equation of continuity, $Q/V = 360/3.914 = 91.97 = 4(b + 4)$, $b = 18.99$. Subsequent trial-and-error solutions result in a balance at $b = 19.93$ ft (6.075 m).

Specific Energy Specific energy is defined as the energy of the fluid referred to the bottom of the channel as the datum. Thus the specific energy E at any section is given by $E = y + V^2/2g$; from the continuity equation $V = Q/A$ or $E = y + (Q/A)^2/2g$. For a rectangular channel whose width is b, $A = by$; and if $\mathbf{q}$ is defined as the flow rate per unit width, $\mathbf{q} = Q/b$ and $E = y + (\mathbf{q}b/by)^2/2g = y + (\mathbf{q}/y)^2/2g$.

Critical Values For rectangular channels, if the specific-energy equation is differentiated and set equal to zero, critical values are obtained; thus $dE/dy = d/dy[y + (\mathbf{q}/y)^2/2g] = 0 = 1 - \mathbf{q}^2/y^3g$ or $\mathbf{q}_c^2 = y_c^3 g$. Substituting in the specific-energy equation, $E = y_c + y_c^3 g/2gy_c^2 = 3/2y_c$. Figure 1.3.34 shows the relation between depth and specific energy for a constant flow rate. If the depth is greater than critical, the flow is *subcritical;* at critical depth it is *critical* and at depths below critical the flow is *supercritical.* For a given specific energy, there is a maximum unit flow rate that can exist.

The **Froude number** $\mathbf{F} = V/\sqrt{gy}$, when substituted in the specific-energy equation, yields $E = y + (\mathbf{F}^2 gy)/2g = y(1 + \mathbf{F}^2/2)$ or $E/y = 1 + \mathbf{F}^2/2$. For critical flow, $E_c/y_c = 3/2$. Substituting $E_c/y_c = 3/2 = 1 + \mathbf{F}_c^2/2$, or $\mathbf{F} = 1$,

$\mathbf{F} < 1$	Flow is subcritical
$\mathbf{F} = 1$	Flow is critical
$\mathbf{F} > 1$	Flow is supercritical

It is seen that for open-channel flow the Froude number determines the type of flow in the same manner as Mach number for compressible flow.

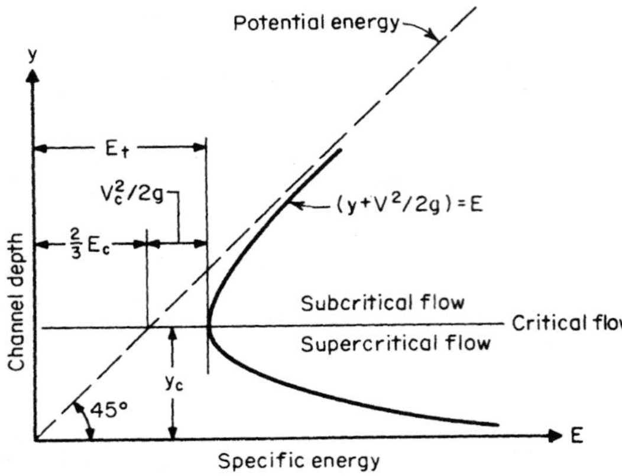

Figure 1.3.34 Specific energy diagram, constant flow rate.

EXAMPLE. Water flows at a rate of 600 ft³/s in a rectangular channel 10 ft wide at a depth of 4 ft. Determine (1) specific energy and (2) type of flow.

1. from the continuity equation,

$$V = Q/A = 600/(10\times 4) = 15 \text{ ft/s}$$

$$E = y + V^2/2g = 4 + (15)^2/2(2\times 32.17) = 7.497 \text{ ft}$$

2. $\mathbf{F} = V/\sqrt{gy} = 15/\sqrt{32.17\times 4} = 1.322$; $\mathbf{F} > 1$ ∴ flow is supercritical.

FLOW OF LIQUIDS FROM TANK OPENINGS

Steady State Consider the jet whose velocity is V discharging from an open tank through an opening whose area is a, as shown in Fig. 1.3.35. The liquid height above the centerline is h, and the cross-sectional area of the tank at h is A. The ideal velocity of the jet is $V_i = \sqrt{2gh}$. The ratio of the actual velocity V to the ideal velocity V_i is the **coefficient of velocity** Cv, or $V = C_vV_i = C_v\sqrt{2gh}$. The ratio of the actual opening a to the minimum area of the jet a_c is the **coefficient of contraction** C_c, or $a = C_ca_c$. The ratio of the actual discharge Q to the ideal discharge Q_i is the **coefficient of discharge** C, or $Q = CQ_i = C_aV_i = C_cC_va\sqrt{2gh}$, and $C = C_cC_v$. Nominal values of coefficients for various openings are given in Fig. 1.3.36.

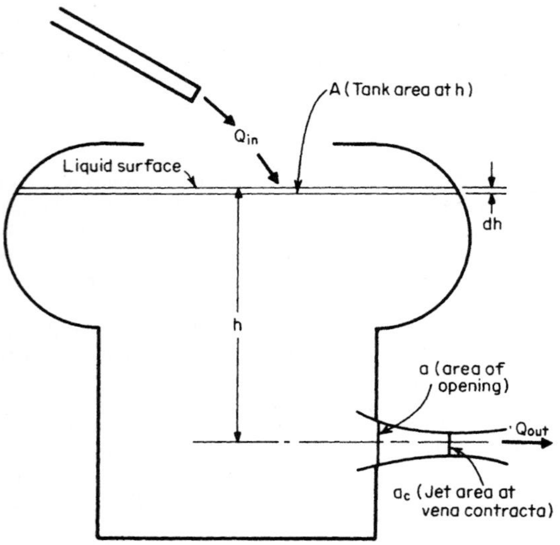

Figure 1.3.35 Notation for tank flow.

Unsteady State If the rate of liquid entering the tank Q_{in} is different from that leaving, the level h in the tank will change because of the change in storage. For liquids, the conservation-of-mass equation may be written as $Q_{in} - Q_{out} = Q_{stored}$; for a time interval dt, $(Q_{in} - Q_{out})dt = A\,dh$, neglecting fluid acceleration,

Type		Coefficient		
		C	Cc	Cv
Sharp-edged orifice	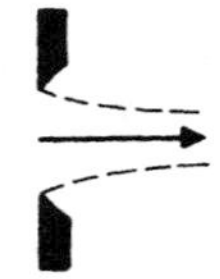	0.61	0.62	0.98
Rounded-edged orifice	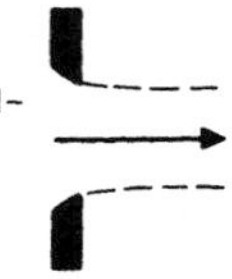	0.98	1.00	0.98
Short tube	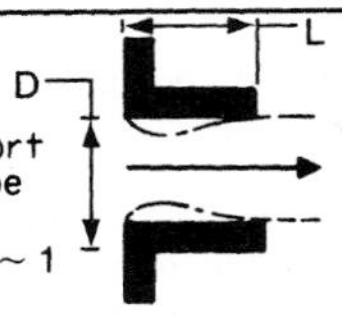	0.80	1.00	0.80
Borda	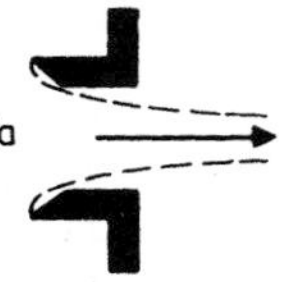	0.51	0.52	0.98

Figure 1.3.36 Nominal coefficients of orifices.

$$Q_{out}dt = Ca\sqrt{2gh}dt, \text{ or } (Q_{in} - Ca\sqrt{2gh})dt$$

$$= A\,dh, \text{ or } \int_{t_1}^{t_2} dt = \int_{h_1}^{h_2} \frac{A\,dh}{Q_{in} - Q_{out}}$$

$$= \int_{h_1}^{h_2} \frac{A\,dh}{Q_{in} - Ca\sqrt{2gh}}$$

EXAMPLE. An open cylindrical tank is 6 ft in diameter and is filled with water to a depth of 10 ft. A 4-in-diameter sharp-edged orifice is installed on the bottom of the tank. A pipe on the top of the tank supplies water at the rate of 1 ft³/s. Estimate (1) the steady-state level of this tank, (2) the time required to reduce the tank level by 2 ft.

1. *Steady-state level.* From Fig. 1.3.36, $C = 0.61$ for a sharp-edged orifice, $a = (\pi/4)d^2 = (\pi/4)(4/12)^2 = 0.08727$ ft². For steady state, $Q_{in} = Q_{out} = Ca\sqrt{2gh} = 1 = (0.61)(0.08727)(2\times 32.17h)^{1/2}$; $h = 5.484$ ft.

2. Time required to lower level 2 ft, $A = (\pi/4)D^2 = (\pi/4)(6)^2 = 28.27$ ft²

$$t_2 - t_1 = \int_{h_1}^{h_2} \frac{A\,dh}{Q_{in} - Ca\sqrt{2gh}}$$

This equation may be integrated by letting $Q = Ca\sqrt{2g}\,h^{1/2}$; then $dh = 2Q\,dQ/(Ca\sqrt{2g})^2$; then

$$t_2 - t_1 = \frac{2A}{(Ca\sqrt{2g})^2}\left[Q_{in}\log_e\left(\frac{Q_{in} - Q_1}{Q_{in} - Q_2}\right) + Q_1 - Q_2\right]$$

$$\text{At } t_1:\ Q_1 = 0.61\times 0.08727\sqrt{2\times 32.17\times 10} = 1.350 \text{ ft}^3/\text{s}$$

$$\text{At } t_2:\ Q_2 = 0.61\times 0.08727\sqrt{2\times 32.17\times 8} = 1.208 \text{ ft}^3/\text{s}$$

$$t_2 - t_1 = \frac{2\times 28.27}{(0.61\times 0.08727\sqrt{2\times 32.17}^{\,2})} \times\left[(1)\log_e\left(\frac{1-1.350}{1-1.208}\right) + 1.350 - 1.208\right]$$

$$t_2 - t_1 = 205.4 \text{ s}$$

WATER HAMMER

Equations Water hammer is the series of shocks, sounding like hammer blows, produced **by suddenly reducing the flow** of a fluid in a pipe. Consider a fluid flowing frictionlessly in a rigid pipe of uniform area A with a velocity V. The pipe has a length L, and inlet pressure p_1 and a pressure p_2 at L. At length L, there is a valve which can suddenly reduce the velocity at L to $V - \Delta V$. The equivalent mass rate of flow of a pressure wave traveling at sonic velocity c, $M = \rho Ac$. From the impulse-momentum equation, $M(V_2 - V_1) = p_2A_2 - p_1A_1$; for this application, $(\rho Ac)(V - \Delta V - V) = p_2A - p_1A$, or the increase in pressure $\Delta p = -\rho c\Delta V$. When the liquid is flowing in an elastic pipe, the equation for pressure rise must be modified to account for the expansion of the pipe; thus

$$c = \sqrt{\frac{E_s}{\rho[1 + (E_s/E_p)(D_o + D_i)/(D_o - D_i)]}}$$

where ρ = mass density of the fluid, E_s = bulk modulus of elasticity of the fluid, E_p = modulus of elasticity of the pipe material, D_o = outside diameter of pipe, and D_i = inside diameter of pipe.

Time of Closure The time for a pressure wave to travel the length of pipe L and return is $t = 2L/c$. If the time of closure $t_c \le t$, the approximate pressure rise $\Delta p \approx 22\ \rho V(L/t_c)$. When it is not feasible to close the valve slowly, **air chambers** or **surge tanks** may be used to absorb all or most of the pressure rise. Water hammer can be very dangerous. See Sec. 7.9.

EXAMPLE. Water flows at 68°F (20°C) in a 3-in steel schedule 40 pipe at a velocity of 10 ft/s. A valve located 200 ft downstream is suddenly closed. Determine (1) the increase in pressure considering pipe to be rigid, (2) the increase considering pipe to be elastic, and (3) the maximum time of valve closure to be considered "sudden."

For water, $\rho = -1.937$ slugs/ft³ = 1.937 lb · sec²/ft⁴; E_s = 319,000 lb/in²; $E_p = 28.5 \times 10^6$ lb/in² (Secs. 3.1 and 4); c = 4,860 ft/s; from Sec. 6.7, D_o = 3.5 in, D_i = 3.068 in.

1. *Inelastic pipe*

$$\Delta p = -\rho c\Delta V = 2(1.937)(4{,}860)(-10) = 94{,}138 \text{ lbf/ft}^2$$
$$= 94{,}138/144 = 653.8 \text{ lbf/in}^2\ (4.507 \times 10^6 \text{ N/m}^2)$$

2. *Elastic pipe*

$$c = \sqrt{\frac{E_s}{\rho[1 + (E_s/E_p)(D_o + D_i)/(D_o - D_i)]}}$$
$$= \sqrt{\frac{319{,}000 \times 144}{1.937\left[1 + \frac{(319{,}000/28.5 \times 10^6)(3.500 + 3.067)}{(3.500 - 3.067)}\right]}}$$
$$= 4{,}504$$
$$\Delta p = -(1.937)(4{,}504)(-10)$$
$$= 87{,}242 \text{ lbf/ft}^2 = 605.9 \text{ lbf/in}^2\ (4.177 \times 10^6 \text{ N/m}^2)$$

3. *Maximum time for closure*

$$t = 2L/c = 2 \times 200/4{,}860 = 0.08230 \text{ s or less than } 1/10 \text{ s}$$

COMPUTATIONAL FLUID DYNAMICS (CFD)

The partial differential equations of fluid motion are very difficult to solve. CFD methods are utilized to provide discrete approximations for the solution of those equations. A brief introduction to CFD is included in Section 12.3. The references therein will guide the reader further.

1.4 VIBRATION

by Leonard Meirovitch, Revised by Mehdi Ahmadian

REFERENCES: Harris, "Shock and Vibration Handbook," 3rd ed., McGraw-Hill. Thomson, "Theory of Vibration with Applications," Prentice Hall. Meirovitch, "Fundamentals of Vibration," McGraw-Hill. Meirovitch, "Principles and Techniques of Vibrations," Prentice-Hall.

SINGLE-DEGREE-OF-FREEDOM SYSTEMS

Discrete System Components A system is defined as an aggregation of components acting together as one entity. The components of a vibratory mechanical system are of three different types, and they relate forces to displacements, velocities, and accelerations. The component relating forces to displacements is known as a spring (Fig. 1.4.1*a*). For a **linear spring** the force F_s is proportional to the elongation $\delta = x_2 - x_1$, or

$$F_s = k\delta = k(x_2 - x_1) \tag{1.4.1}$$

where k represents the **spring constant,** or the **spring stiffness,** and x_1 and x_2 are the displacements of the end points. The component relating forces to velocities is called a **viscous damper** or a **dashpot** (Fig. 1.4.1*b*). It consists of a piston fitting loosely in a cylinder filled with liquid so that the liquid can flow around the piston when it moves relative to the cylinder. The relation between the damper force and the velocity of the piston relative to the cylinder is

$$F_d = c(\dot{x}_2 - \dot{x}_1) \tag{1.4.2}$$

in which c is the **coefficient of viscous damping;** note that dots denote derivatives with respect to time. Finally, the relation between forces and accelerations is given by Newton's second law of motion:

$$F_m = m\ddot{x} \tag{1.4.3}$$

where m is the **mass** (Fig. 1.4.1*c*).

The spring constant k, coefficient of viscous damping c, and mass m represent physical properties of the components and are the **system parameters.** By implication, these properties are concentrated at points, thus they are **lumped,** or **discrete, parameters.** Note that springs and dampers are assumed to be massless and masses are assumed to be rigid.

Springs can be arranged in parallel and in series. Then, the proportionality constant between the forces and the end points is known as an **equivalent spring constant** and is denoted by k_{eq}, as shown in Table 1.4.1. Certain elastic components, although distributed over a given line segment, can be regarded as lumped with an equivalent spring

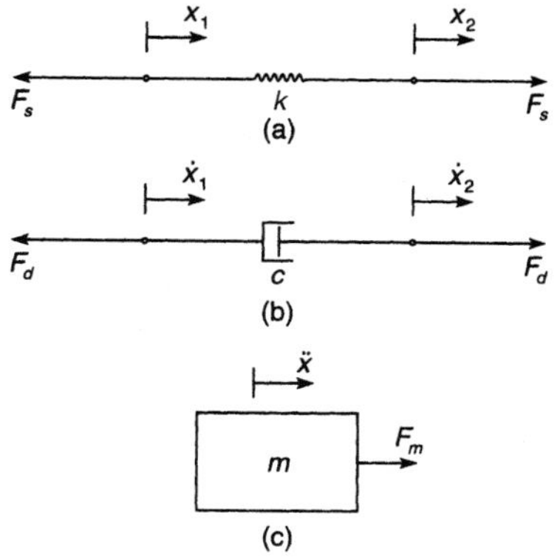

Figure 1.4.1 Mechanical Vibration Elements.

Table 1.4.1 Equivalent Spring Constants

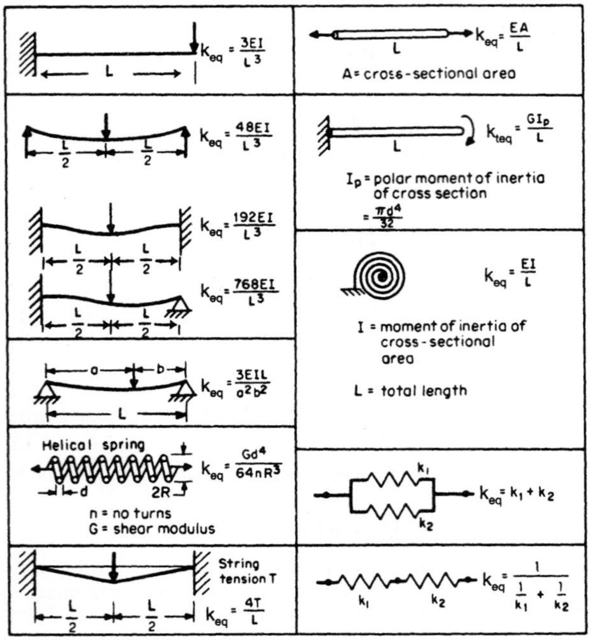

constant given by $k_{eq} = F/\delta$, where δ is the deflection at the point of application of the force F. A similar relation can be given for springs in torsion. Table 1.4.1 lists the equivalent spring constants for a variety of components.

Equation of Motion The dynamic behavior of many engineering systems can be approximated with good accuracy by the **mass-damper-spring model** shown in Fig. 1.4.2. Using Newton's second law in conjunction with Eqs. (1.4.1)–(1.4.3) and measuring the displacement $x(t)$ from the static equilibrium position, we obtain the differential equation of motion

$$m\ddot{x}(t) + c\dot{x}(t) + kx(t) = F(t) \quad (1.4.4)$$

which is subject to the **initial conditions** $x(0) = x_0$, $\dot{x}(0) = v_0$, where x_0 and v_0 are the **initial displacement** and **initial velocity,** respectively. Equation (1.4.4) is in terms of a single coordinate. namely $x(t)$; the system of Fig. 1.4.2 is therefore said to be a **single-degree-of-freedom system.**

Free Vibration of Undamped Systems Assuming zero damping and external forces and dividing Eq. (1.4.4) through by m, we obtain

$$\ddot{x} + \omega_n^2 x = 0 \qquad \omega_n = \sqrt{k/m} \quad (1.4.5)$$

In this case, the vibration is caused by the initial excitations alone. The solution of Eq. (1.4.5) is

$$x(t) = A\cos(\omega_n t - \phi) \quad (1.4.6)$$

which represents **simple sinusoidal,** or **simple harmonic oscillation** with **amplitude** A, **phase angle** ϕ, and **frequency**

$$\omega_n = \sqrt{k/m} \qquad \text{rad/s} \quad (1.4.7)$$

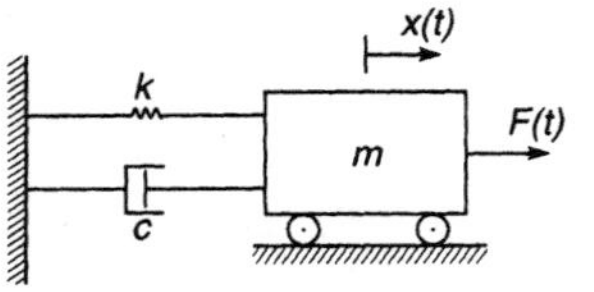

Figure 1.4.2 Mass-Spring-Damper Mechanical System Model.

Systems described by equations of the type (1.4.5) are called **harmonic oscillators.** Because the frequency of oscillation represents an inherent property of the system, independent of the initial excitation, ω_n is called the **natural frequency.** On the other hand, the amplitude and phase angle do depend on the initial displacement and velocity, as follows:

$$A = \sqrt{x_0^2 + (v_0/\omega_n)^2} \qquad \phi = \tan^{-1} v_0/x_0\omega_n \quad (1.4.8)$$

The time necessary to complete one cycle of motion defines the **period**

$$T = 2\pi/\omega_n \qquad \text{seconds} \quad (1.4.9)$$

The reciprocal of the period provides another definition of the **natural frequency,** namely,

$$f_n = \frac{1}{T} = \frac{\omega_n}{2\pi} \qquad \text{Hz} \quad (1.4.10)$$

where Hz denotes *hertz* [1 Hz = 1 cycle per second (cps)].

A large variety of vibratory systems behave like harmonic oscillators, many of them when restricted to small amplitudes. Table 1.4.2 shows a variety of harmonic oscillators together with their respective natural frequency.

Free Vibration of Damped Systems Let $F(t) = 0$ and divide through by m. Then, Eq. (1.4.4) reduces to

$$\ddot{x}(t) + 2\zeta\omega_n\dot{x}(t) + \omega_n^2 x(t) = 0 \quad (1.4.11)$$

where

$$\zeta = c/2m\omega_n \quad (1.4.12)$$

is the **damping factor,** a nondimensional quantity. The nature of the motion depends on ζ. The most important case is that in which $0 < \zeta < 1$.

In this case, the system is said to be **underdamped** and the solution of Eq. (1.4.11) is

$$x(t) = Ae^{-\zeta\omega_n t}\cos(\omega_d t - \phi) \quad (1.4.13)$$

where

$$\omega_d = (1 - \zeta^2)^{1/2}\omega_n \quad (1.4.14)$$

is the **frequency of damped free vibration** and

$$T = 2\pi/\omega_d \quad (1.4.15)$$

is the **period of damped oscillation.** The amplitude and phase angle depend on the initial displacement and velocity, as follows:

$$A = \sqrt{x_0^2 + (\xi\omega_n x_0 + v_0)^2/\omega_d^2}$$

$$\phi = \tan^{-1}(\xi\omega_n x_0 + v_0)/x_0\omega_d \quad (1.4.16)$$

The motion described by Eq. (1.4.13) represents **decaying oscillation,** where the term $Ae^{-\zeta\omega_n t}$ can be regarded as a time-dependent amplitude, providing an envelope bounding the harmonic oscillation.

When $\zeta \geq 1$, the solution represents **aperiodic decay.** The case $\zeta = 1$ represents **critical damping,** and

$$c_c = 2m\omega_n \quad (1.4.17)$$

is the **critical damping coefficient,** although there is nothing critical about it. It merely represents the borderline between oscillatory decay and aperiodic decay. In fact, c_c is the smallest damping coefficient for which the motion is aperiodic. When $\zeta > 1$, the system is said to be **overdamped.**

Logarithmic Decrement Quite often the damping factor is not known and must be determined experimentally. In the case in which the system is underdamped, this can be done conveniently by plotting $x(t)$ versus t (Fig. 1.4.3) and measuring the response at two different times separated by a complete period. Let the times be t_1 and $t_1 + T$, introduce the notation $x(t_1) = x_1$, $x(t_1 + T) = x_2$, and use Eq. (1.4.13) to obtain

$$\frac{x_1}{x_2} = \frac{Ae^{-\zeta\omega_n t_1}\cos(\omega_d t_1 - \phi)}{Ae^{-\zeta\omega_n(t_1+T)}\cos[\omega_d(t_1+T) - \phi]} = e^{\zeta\omega_n T} \quad (1.4.18)$$

Table 1.4.2 Harmonic Oscillators and Natural Frequencies

$$\omega_n = \sqrt{\frac{k}{m + \frac{1}{3}m_s}} \qquad \omega_n = \sqrt{\frac{K_s}{J + \frac{1}{3}J_s}} \qquad \omega_n = \sqrt{\frac{K_s(J_1 + J_2)}{J_1 J_2}}$$

If $m_s = 0$

$$\omega_n = \sqrt{\frac{k}{m}}$$

If $J_s = 0$

$$\omega_n = \sqrt{\frac{K_s}{J}}$$

$$\omega_n = \sqrt{\frac{k(m_1 + m_2)}{m_1 m_2}}$$

Speed ratio of gears $= n$

$$\omega_n = \sqrt{\frac{K_1 K_2(J_1 + n^2 J_2)}{J_1 J_2(K_1 + n^2 K_2)}}$$

$$\omega_n = \sqrt{\frac{3EI}{(m + 0.23 m_b)\ell^3}} \qquad \omega_n = \sqrt{\frac{48EI}{(m + 0.5 m_b)\ell^3}}$$

String tension T

$$\omega_n = \sqrt{\frac{4T}{m\ell}} \qquad \omega_n = \sqrt{\frac{192 EI}{(m + 0.37 m_b)\ell^3}}$$

$$\omega_n = \sqrt{\frac{g}{\ell}} \qquad \omega_n = \Omega$$

$$\omega_n^4 - \left[\frac{k_1}{m_2} + \frac{k_2}{m_2}\left(1 + \frac{m_2}{m_1}\right)\right]\omega_n^2 + \frac{k_1}{m_1}\frac{k_2}{m_2} = 0$$

$$\omega_n^4 - \left[\frac{K_1}{J_1} + \frac{K_2}{J_2}\left(1 + \frac{J_2}{J_1}\right)\right]\omega_n^2 + \frac{K_1}{J_1}\frac{K_2}{J_2} = 0$$

$$\omega^4 - \left[\frac{k_1(J_1 + J_2)}{J_1 J_2} + \frac{k_2(J_2 + J_3)}{J_2 J_3}\right]\omega^2$$
$$+ \frac{k_1 k_2(J_1 + J_2 + J_3)}{J_1 J_2 J_3} = 0 \qquad \text{where } k = \frac{GI_p}{\ell}$$

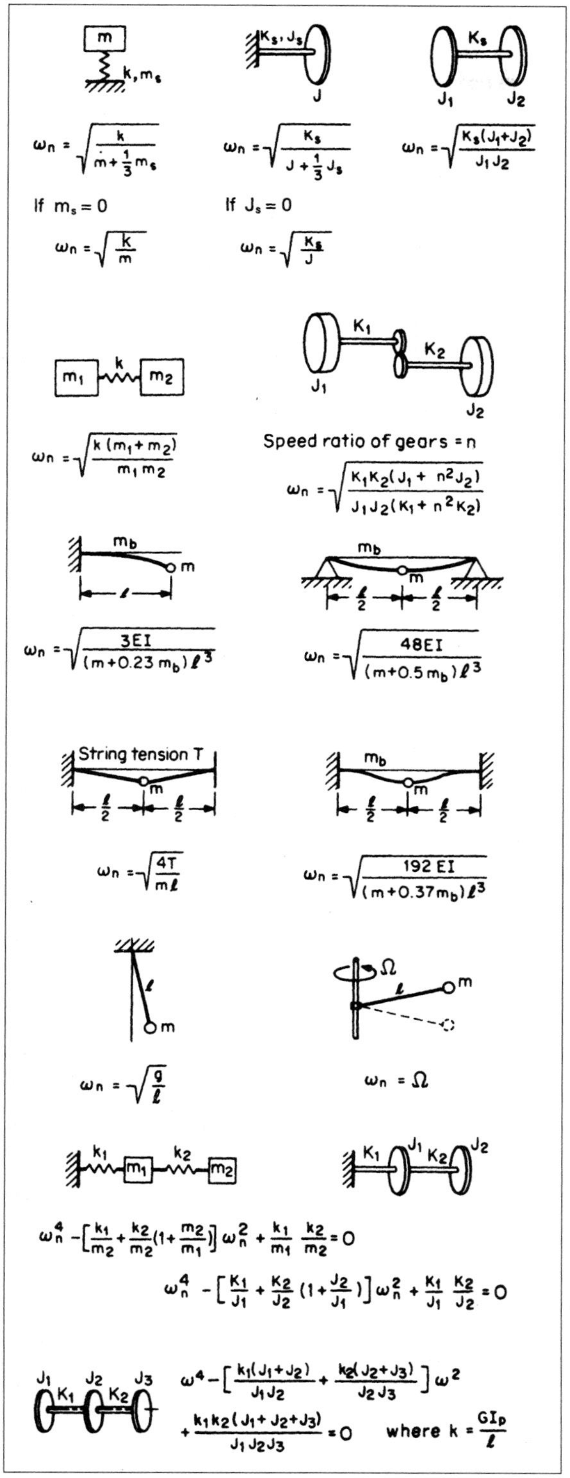

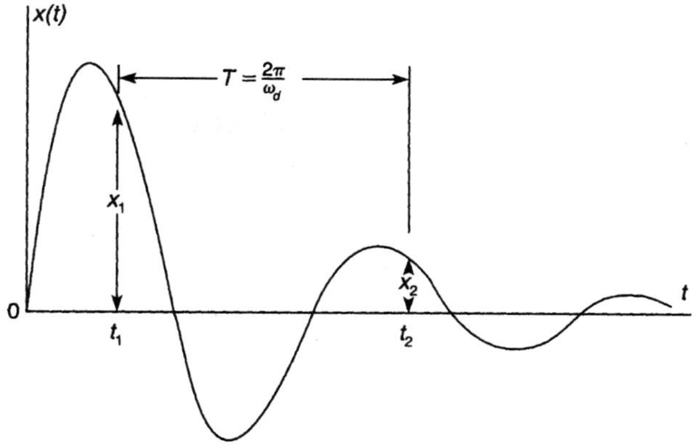

Figure 1.4.3 Oscillatory displacement response of an under-damped system versus time.

where $\cos[\omega_d(t_1 + T) - \phi] = \cos(\omega_d t_1 - \phi + 2\pi) = \cos(\omega_d t_1 - \phi)$. Equation (1.4.18) yields the **logarithmic decrement**

$$\delta = \ln\frac{x_1}{x_2} = \zeta\omega_n T = \frac{2\pi\zeta}{\sqrt{1 - \zeta^2}} \qquad (1.4.19)$$

which can be used to obtain the damping factor

$$\zeta = \frac{\delta}{\sqrt{(2\pi)^2 + \delta^2}} \qquad (1.4.20)$$

For small damping, the logarithmic decrement is also small, and the damping factor can be approximated by

$$\zeta \approx \frac{\delta}{2\pi} \qquad (1.4.21)$$

Response to Harmonic Excitations Consider the case in which the excitation force $F(t)$ in Eq. (1.4.4) is harmonic. For convenience, express $F(t)$ in the form $kA\cos\omega t$, where k is the spring constant, A is an amplitude with units of displacement and ω is the **excitation frequency.** When divided through by m, Eq. (1.4.4) has the form

$$\ddot{x} + 2\xi\omega_n\dot{x} + \omega_n^2 x = \omega_n^2 A\cos\omega t \qquad (1.4.22)$$

The solution of Eq. (1.4.22) can be expressed as

$$x(t) = A|G(\omega)|\cos(\omega t - \phi) \qquad (1.4.23)$$

where

$$|G(\omega)| = \frac{1}{\sqrt{[1 - (\omega/\omega_n)^2]^2 + (2\zeta\omega/\omega_n)^2}} \qquad (1.4.24)$$

is a **nondimensional magnitude factor[1]** and

$$\phi(\omega) = \tan^{-1}\frac{2\zeta\omega/\omega_n}{1 - (\omega/\omega_n)^2} \qquad (1.4.25)$$

is the phase angle; note that both the magnitude factor and phase angle depend on the excitation frequency ω.

Equation (1.4.23) shows that the response to harmonic excitation is also harmonic and has the same frequency as the excitation, but different amplitude $A|G(\omega)|$ and phase angle $\phi(\omega)$. Not much can be learned by plotting the response as a function of time, but a great deal of information can be gained by plotting $|G|$ versus ω/ω_n and ϕ versus ω/ω_n. They are shown in Fig. 1.4.4 for various values of the damping factor ζ.

---

* The term $|G(\omega)|$ is often referred to as *magnification factor,* but this is a misnomer, as we shall see shortly.

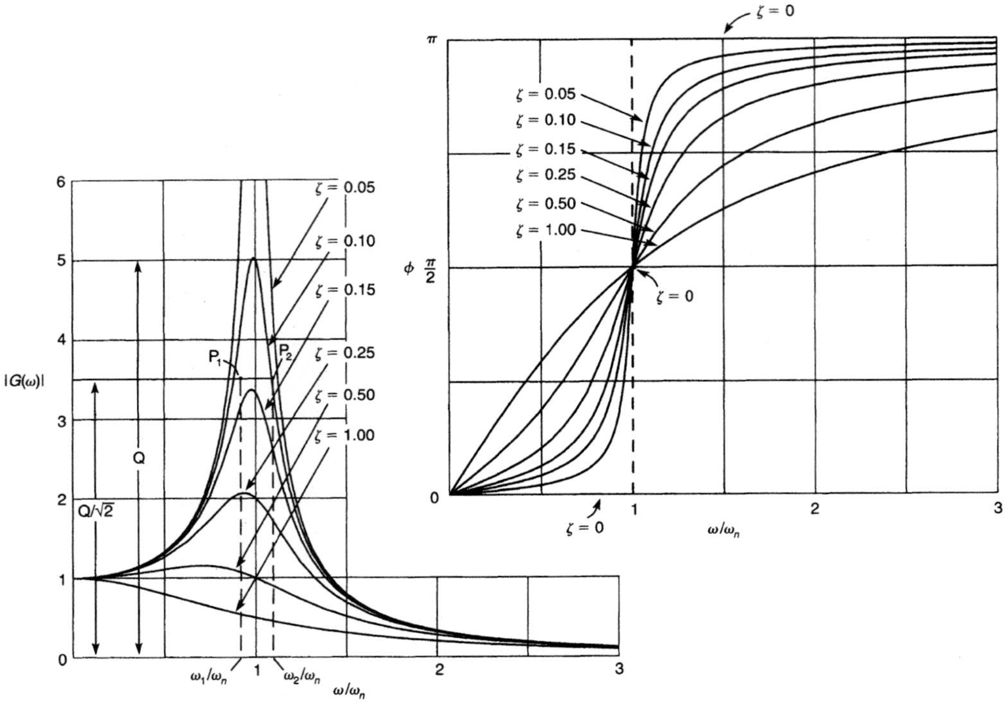

Figure 1.4.4 Magnification Factor and Phase Angle versus Frequency Ratio for System in Figure 1.4.2.

In Fig. 1.4.4, for low values of ω/ω_n, the nondimensional magnitude factor $|G(\omega)|$ approaches unity and the phase angle $\phi(\omega)$ approaches zero. For large values of ω/ω_n, the magnitude approaches zero (see accompanying footnote about magnification factor) and the phase angle approaches 180°. The magnitude experiences peaks for $\omega/\omega_n = \sqrt{1-2\zeta^2}$, provided $\zeta < 1/\sqrt{2}$. The peak values are $|G(\omega)|_{max} = 1/2\zeta\sqrt{1-\zeta^2}$. For small ζ, the peaks occur approximately at $\omega/\omega_n = 1$ and have the approximate values $|G(\omega)|_{max} = Q \approx 1/2\zeta$, where Q is known as the **quality factor.** In such cases, the phase angle tends to 90°. Clearly, for small ζ the system experiences large-amplitude vibration, a condition known as **resonance.** The points P_1 and P_2, where $|G|$ falls to $Q/\sqrt{2}$, are called **half-power points.** The increment of frequency associated with the half-power points P_1 and P_2 represents the **bandwidth** $\Delta\omega$ of the system. For small damping, it has the value

$$\Delta\omega = \omega_2 - \omega_1 \approx 2\zeta\omega_n \tag{1.4.26}$$

The case $\zeta = 0$ deserves special attention. In this case, referring to Eq. (1.4.22), the response is simply

$$x(t) = \frac{A}{1-(\omega/\omega_n)^2} \cos \omega t \tag{1.4.27}$$

For $\omega/\omega_n < 1$, the displacement is in the same direction as the force, so that the phase angle is zero; the **response is in phase** with the excitation. For $\omega/\omega_n > 1$, the displacement is in the direction opposite to the force, so that the phase angle is 180° **out of phase** with the excitation. Finally, when $\omega = \omega_n$ the response is

$$x(t) = \frac{A}{2}\omega_n t \sin \omega_n t \tag{1.4.28}$$

This is typical of the **resonance** condition, when the response increases without bounds as time increases. Of course, at a certain time the displacement becomes so large that the spring ceases to be linear, thus violating the original assumption and invalidating the solution. In practical terms, unless the excitation frequency varies, passing quickly through $\omega = \omega_n$, the system can break down.

When the excitation is $F(t) = kA \sin \omega t$, the response is

$$x(t) = A|G(\omega)| \sin(\omega t - \phi) \tag{1.4.29}$$

One concludes that in **harmonic response,** time plays a secondary role to the frequency. In fact, the only significant information is extracted from the magnitude and phase angle plots of Fig. 1.4.4. They are referred to as **frequency-response plots.**

Since time plays no particular role, the harmonic response is called **steady-state response.** In general, for linear systems with constant parameters, such as the mass-damper-spring system under consideration, the response to the initial excitations is added to the response to the excitation forces. The response to initial excitations, however, represents **transient response.** This is due to the fact that every system possesses some amount of damping, so that the response to initial excitations disappears with time. In contrast, steady-state response persists with time. Hence, in the case of harmonic excitations, it is meaningless to add the response to initial excitations to the harmonic response.

Vibration Isolation A problem of great interest is the magnitude of the force transmitted to the base by a system of the type shown in Fig. 1.4.2 subjected to harmonic excitation. This force is a combination of the spring force kx and the dashpot force $c\dot{x}$. Recalling Eq. (1.4.23), write

$$\begin{aligned} kx &= kA|G|\cos(\omega t - \phi) \\ c\dot{x} &= -c\omega A|G|\sin(\omega t - \phi) \\ &= c\omega A|G|\cos\left(\omega t - \phi + \frac{\pi}{2}\right) \end{aligned} \tag{1.4.30}$$

so that the dashpot force is 90° out of phase with the spring force. Hence, the magnitude of the force is

$$\begin{aligned} F_{tr} &= \sqrt{(kA|G|)^2 + (c\omega A|G|)^2} = kA|G|\sqrt{1+(c\omega/k)^2} \\ &= kA|G|\sqrt{1+(2\zeta\omega/\omega_n)^2} \end{aligned} \tag{1.4.31}$$

Let the magnitude of the harmonic excitation be $F_0 = kA$; the measure of the force transmitted to the base is then

$$T = \frac{F_{tr}}{F_0} = |G| \sqrt{1 + (2\zeta\omega/\omega_n)^2}$$
$$= \sqrt{\frac{1 + (2\zeta\omega/\omega n)^2}{[1 - (\omega/\omega_n)^2]^2 + (2\zeta\omega/\omega_n)^2}} \qquad (1.4.32)$$

which represents a nondimensional ratio called **transmissibility.** Figure 1.4.5 plots F_{tr}/F_0 versus ω/ω_n for various values of ζ.

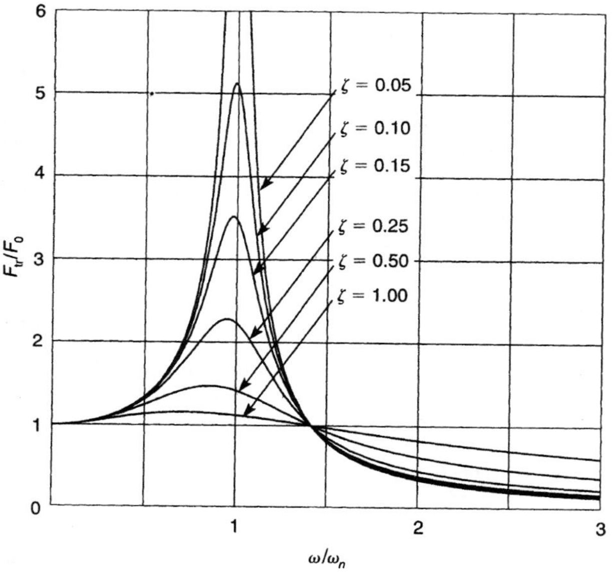

Figure 1.4.5 Transmissibility versus Frequency Ratio for System in Figure 1.4.2.

The transmissibility is less than 1 for $\omega/\omega_n > \sqrt{2}$, and decreases as ω/ω_n increases. Hence, for **an isolator to perform well,** its natural frequency must be much smaller than the excitation frequency. However, for very low natural frequencies, difficulties can be encountered in isolator design. Indeed, the natural frequency is related to the static deflection δ_{st} by $\omega_n = \sqrt{k/m} = \sqrt{g/\delta_{st}}$, where g is the gravitational constant. For the natural frequency to be sufficiently small, the static deflection may have to be impractically large. The relation between the excitation frequency f measured in rotations per minute and the static deflection δ_{st} measured in inches is

$$f = 187.7\sqrt{\frac{2 - R}{\delta_{st}(1 - R)}} \quad \text{rpm} \qquad (1.4.33)$$

where $R = 1 - T$ **represents the percent reduction** in vibration. Figure 1.4.6 shows a logarithmic plot of f versus δ_{st} with R as a parameter.

Figure 1.4.7 depicts two types of isolators. In Fig. 1.4.7*a*, isolation is accomplished by means of springs and in Fig. 1.4.7*b* by rubber rings supporting the bearings. Isolators of all shapes and sizes are available commercially.

Rotating Unbalanced Masses Many appliances, machines, etc., involve components spinning relative to a main body. A typical example is the clothes dryer. Under certain circumstances, the mass of the spinning component is not symmetric relative to the center of rotation, as when the clothes are not spread uniformly in the spinning drum, giving rise to harmonic excitation. The behavior of such systems can be simulated adequately by the single-degree-of-freedom model shown in Fig. 1.4.8, which consists of a main mass $M - m$, supported by two springs of combined stiffness k and a dashpot with coefficient of viscous

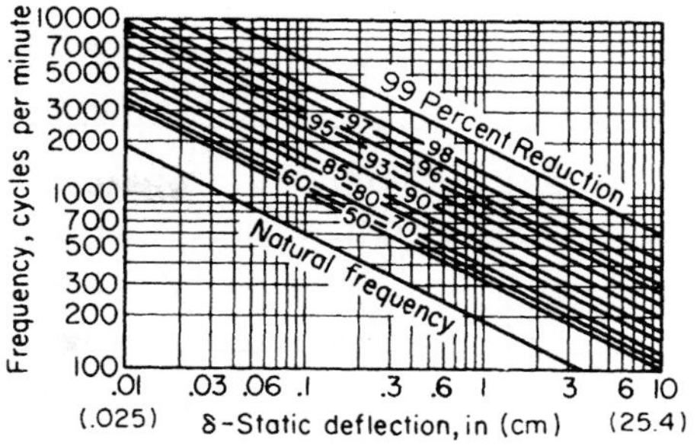

Figure 1.4.6 Logarithmic Plot of Excitation frequency, f, (Eq. (1.4.33)) versus Static Deflection for various Reduction Ratios, R.

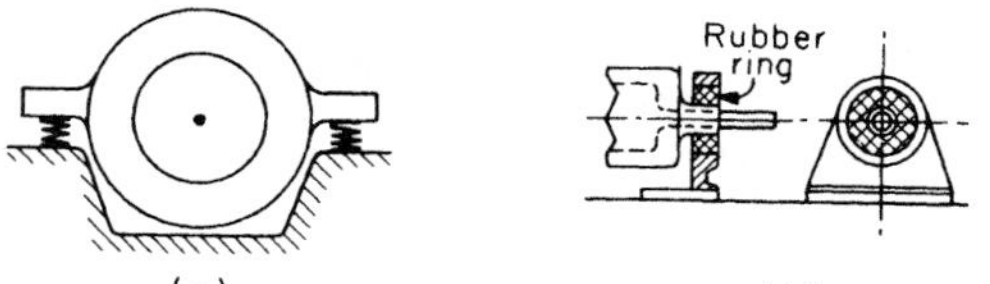

(a) (b)

Figure 1.4.7 Common Methods of Vibration Isolation; (a) spring suspension; (b) rubber ring.

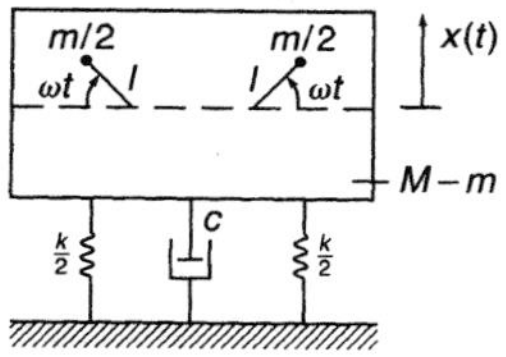

Figure 1.4.8 System with Unbalanced Rotating Mass.

damping c, and two eccentric masses $m/2$ rotating in opposite sense with the constant angular velocity ω. Although there are three masses, the motion of the eccentric masses relative to the main mass is prescribed, so that there is only one degree of freedom. The equation of motion for the system is

$$M\ddot{x} + c\dot{x} + kx = m/\omega^2 \sin \omega t \qquad (1.4.34)$$

Using the analogy with Eq. (1.4.29), the solution of Eq. (1.4.34) is

$$x(t) = \frac{m}{M}\, l \left(\frac{\omega}{\omega_n}\right)^2 |G(\omega)| \sin(\omega t - \phi) \qquad \omega_n^2 = \frac{k}{M} \qquad (1.4.35)$$

The magnitude factor in this case is $(\omega/\omega_n)^2|G(\omega)|$, where $|G(\omega)|$ is given by Eq. (1.4.24); it is plotted in Fig. 1.4.9. On the other hand, the phase angle remains as in Fig. 1.4.4.

Whirling of Rotating Shafts Many mechanical systems involve rotating shafts carrying disks. If the disk has some eccentricity, then the centrifugal forces cause the shaft to bend, as shown in Fig. 1.4.10*a*. The rotation of the plane containing the bent shaft about the bearing axis is called **whirling.** Figure 1.4.10*b* shows a disk with the body axis x, y rotating about the origin O with the angular velocity ω. The geometrical center of the disk is denoted by S and the mass center by C. The distance between the two points is the eccentricity e. The shaft is massless and of stiffness k_{eq} and the disk is rigid and of mass m. The x and y components of the displacement of S relative to O are independent from one another and, for no damping, satisfy the equations of motion

$$\ddot{x} + \omega_n^2 x = e\omega^2 \cos \omega t \qquad \ddot{y} + \omega_n^2 y = e\omega^2 \sin \omega t \qquad \omega_n^2 = k_{eq}/m \qquad (1.4.36)$$

where, assuming that the shaft is simply supported (see Table 1.4.1), $k_{eq} = 48EI/L^3$, in which E is the modulus of elasticity, I the cross-sectional

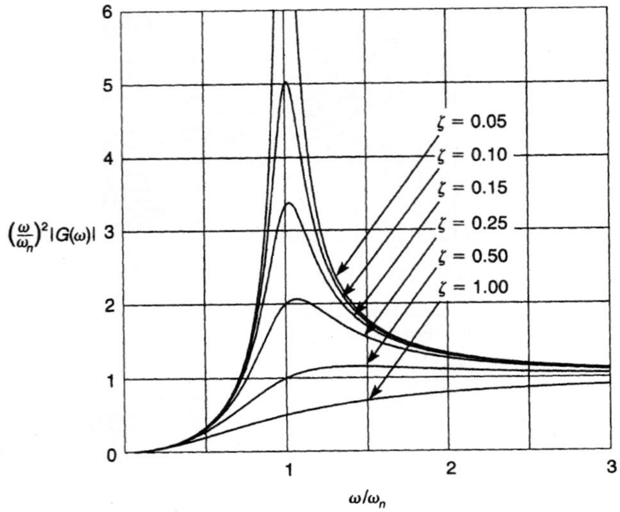

Figure 1.4.9 Magnitude Factor for System with Unbalanced Masses in Figure 1.4.8.

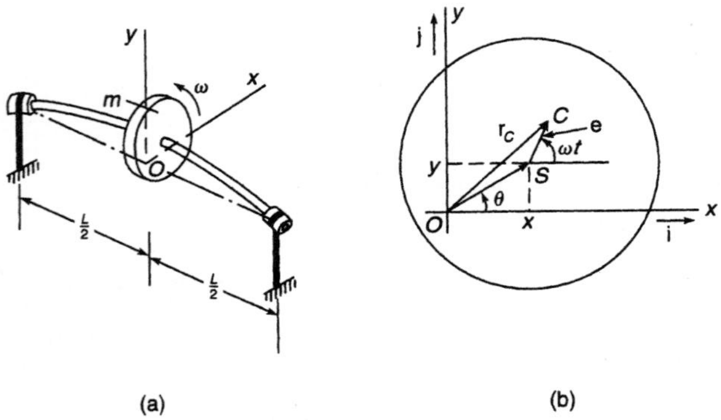

Figure 1.4.10 Rotating System with Whirling Dynamics.

area moment of inertia, and L the length of the shaft. By analogy with Eq. (1.4.27), Eq. (1.4.36) have the solution

$$x(t) = \frac{e(\omega/\omega_n)^2}{1-(\omega/\omega_n)^2} \cos \omega t \qquad y(t) = \frac{e(\omega/\omega_n)^2}{1-(\omega/\omega_n)^2} \sin \omega t \quad (1.4.37)$$

Clearly, resonance occurs when the whirling angular velocity coincides with the natural frequency. In terms of rotations per minute, it has the value

$$f_c = \frac{60}{2\pi}\omega_n = \frac{60}{2\pi}\sqrt{\frac{48EI}{mL^3}} \quad \text{rpm} \qquad (1.4.38)$$

where f_c is called the **critical speed.**

Structural Damping Experience shows that energy is dissipated in all real systems, including those assumed to be undamped. For example, because of internal friction, energy is dissipated in real springs undergoing cyclic stress. This type of damping is called **structural damping** or **hysteretic damping** because the energy dissipated in one cycle of stress is equal to the area inside the hysteresis loop. Systems possessing structural damping and subjected to harmonic excitation with the frequency ω can be treated as if they possess viscous damping with the equivalent coefficient

$$c_{eq} = \alpha/\pi\omega \qquad (1.4.39)$$

where α is a material constant. In this case, the equation of motion is

$$m\ddot{x} + \frac{\alpha}{\pi\omega}\dot{x} + kx = kA\cos\omega t \qquad (1.4.40)$$

The solution of Eq. (1.4.40) is

$$x(t) = A\,|G|\cos(\omega t - \phi) \qquad (1.4.41)$$

where this time the magnitude factor and phase angle have the values

$$G = \frac{1}{\sqrt{[1-(\omega/\omega_n)^2]^2 + \gamma^2}} \qquad \phi = \tan^{-1}\frac{\gamma\omega_n^2}{\omega[1-(\omega/\omega_n)^2]} \qquad (1.4.42)$$

in which

$$\gamma = \frac{\alpha}{\pi k} \qquad (1.4.43)$$

is known as the **structural damping factor.** One word of caution is in order: **the analogy between structural and viscous damping is valid only for harmonic excitation.**

Balancing of Rotating Machines Machines such as electric motors and generators, turbines, compressors, etc., contain rotors with journals supported by bearings. In many cases, the rotors rotate relative to the bearings at very high rates, reaching into tens of thousands of revolutions per minute. Ideally the rotor is rigid and the axis of rotation coincides with one of its principal axes; by implication, the rotor center of mass lies on the axis of rotation. Such a rotor does not wobble and the only forces exerted on the bearings are due to the weight of the rotor. Such a rotor is said to be **perfectly balanced.** These ideal conditions are seldom realized, and in practice the mass center lies at a distance e (eccentricity) from the axis of rotation, so that there is a net centrifugal force $F = me\omega^2$ acting on the rotor, where m is the mass of the rotor and ω is the rotational speed. This centrifugal force is balanced by reaction forces in the bearings, which tend to wear out the bearings with time.

The rotor unbalance can be divided into two types, static and dynamic. Static unbalance can be detected by placing the rotor on a pair of parallel rails. Then, the mass center will settle in the lowest position in a vertical plane through the rotation axis and below this axis. To balance the rotor statically, it is necessary to add a mass m' in the same plane at a distance r from the rotation axis and above this axis, where m' and r must be such that $m'r = me$. In this manner, the net centrifugal force on the rotor is zero. The net result of **static balancing** is to cause the mass center to coincide with the rotation axis, so that the rotor will remain in any position placed on the rails. However, unless the mass m' is placed on a line containing m and at right angles with the bearings axis, the centrifugal forces on m and m' will form a couple (Fig. 1.4.11). Static balancing is suitable when the rotor is in the form of a thin disk, in which case the couple tends to be small. Automobile tires are at times balanced statically, although strictly speaking they are neither thin nor rigid.

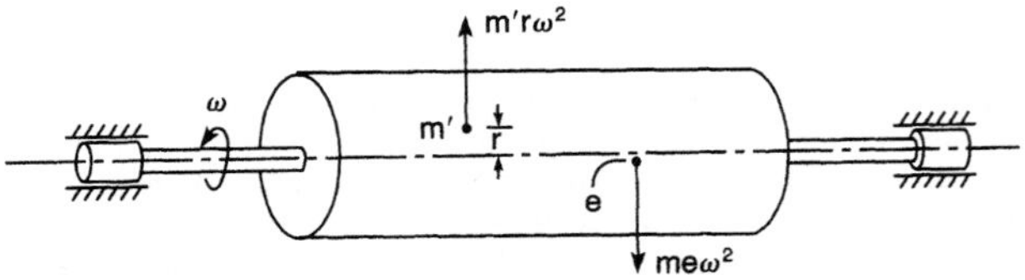

Figure 1.4.11 Couple caused by Static Balancing of a Rotating System.

In general, for practical reasons, the mass m' cannot be placed on an axis containing m and perpendicular to the bearing axis. Hence, although in static balancing the mass center lies on the rotation axis, the rotor principal axis does not coincide with the bearing axis, as shown in Fig. 1.4.12, causing the rotor to wobble during rotation. In this case, the rotor is said to be **dynamically unbalanced.** Clearly, it is highly desirable to place the mass m' so that the rotor

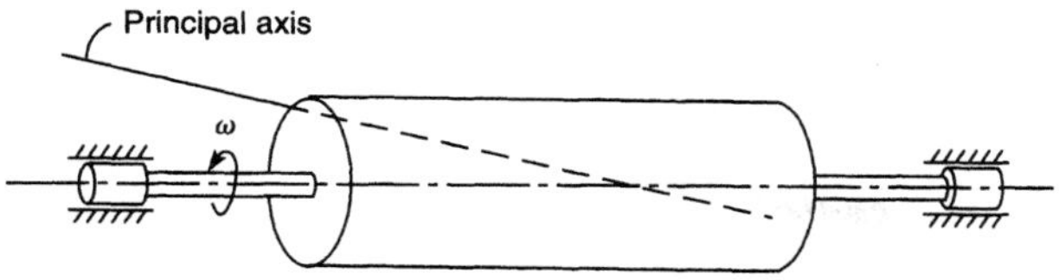

Figure 1.4.12 Wobble caused by Principal Axis and Rotational Axis not Coinciding.

is both statically and dynamically balanced. In this regard, note that the end planes of the rotor are convenient locations to place correcting masses. In Fig. 1.4.13, if the mass center is at a distance a from the right end, then **dynamic balance** can be achieved by placing masses $m'a/L$ and $m'(L - a)/L$ on the intersection of the plane of unbalance and the rotor left end plane and right end plane, respectively. In this manner, the resultant centrifugal force is zero and the two couples thus created are equal in value to $m'a(L - a)\omega^2/L$ and opposite in sense, so that they cancel each other. This results in a rotor completely balanced, that is, balanced statically and dynamically.

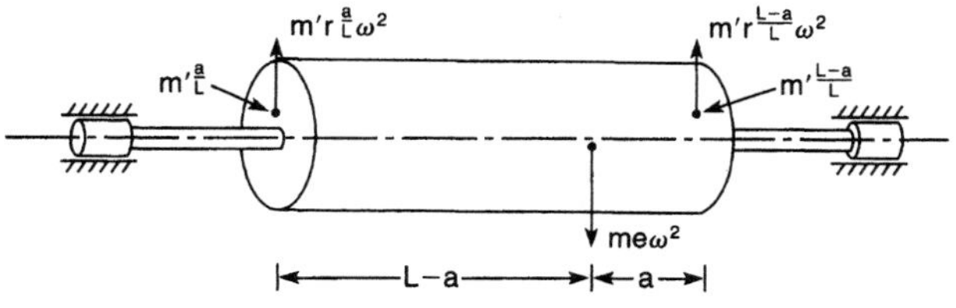

Figure 1.4.13 Dynamic Balancing through placing masses at the end planes of the rotor.

The task of determining the magnitude and position of the unbalance is carried out by means of a balancing machine provided with elastically supported bearings permitting the rotor to spin (Fig. 1.4.14). The unbalance causes the bearings to oscillate laterally so that electrical pickups and stroboflash light can measure the amplitude and phase of the rotor with respect to an arbitrary rotor.

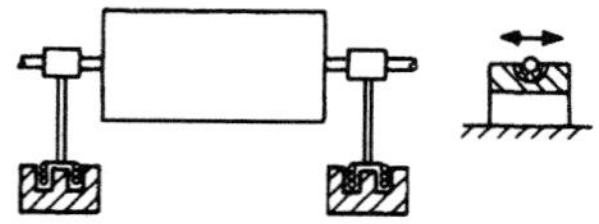

Figure 1.4.14 Balancing Machine.

In cases in which the rotor is very long and flexible. The position of the unbalance depends on the elastic configuration of the rotor, which in turn depends on the speed of rotation, temperature, etc. In such cases, it is necessary to balance the rotor under normal operating conditions by means of a portable balancing instrument.

Inertial Unbalance of Reciprocating Engines The crank-piston mechanism of a reciprocating engine produces dynamic forces capable of causing undesirable vibrations. Rotating parts, such as the crankshaft, can be balanced. However, translating parts, such as the piston, cannot be easily balanced, and the same can be said about the connecting rod, which executes a more complex motion of combined rotation and translation.

In the calculation of the unbalanced forces in a single-cylinder engine, the mass of the moving parts is divided into a reciprocating mass and a rotating mass. This is done by apportioning some of the mass of the connecting rod to the piston and some to the crank end. In general, this division of the connecting rod into two lumped masses tends to cause errors in the moment of inertia, and hence in the torque equation. On the other hand, the force equation can be regarded as being accurate.

Assuming that the rotating mass is counterbalanced, only the reciprocating mass is of concern, and the inertia force for a single-cylinder engine is

$$F = m_{\text{rec}} r\omega^2 \cos\omega t + m_{\text{rec}} \frac{r^2}{L}\omega^2 \cos 2\omega t \tag{1.4.44}$$

where m_{rec} is the reciprocating mass, r the radius of the crank, ω the angular velocity of the crank, and L the length of the connecting rod. The first component on the right side, which alternates once per revolution, is denoted by X_1 and referred to as the **primary force,** and the second component, which is smaller and alternates twice per revolution, is denoted by X_2 and is called the **secondary force.**

In addition to the inertia force, there is an unbalanced torque about the crankshaft axis due to the reciprocating mass. However, this torque is considered together with the torque created by the power stroke, and the torsional oscillations resulting from these excitations can be mitigated by means of a pendulum-type absorber (see "Centrifugal Pendulum Vibration Absorbers" below) or a torsional damper.

The analysis for the single-cylinder engine can be extended to multicylinder in-line and V-block engines by superposition. For the in-line engine or one block of the V engine, the inertia force becomes

$$F = m_{\text{rec}} r\omega^2 \sum_{j=1}^{n} \cos(\omega t + \phi_j) + m_{\text{rec}} \frac{r^2}{L}\omega^2 \sum_{j=1}^{n} \cos 2(\omega t + \phi_j) \tag{1.4.45}$$

where ϕ_j is a phase angle corresponding to the crank position associated with cylinder j and n is the number of cylinders. The vibration force can be eliminated by proper spacing of the angular positions $\phi_j (j = 1, 2, \ldots, n)$.

Even if $F = 0$, there can be pitching and yawing moments due to the spacing of the cylinders. Table 1.4.3 gives the inertial unbalance and pitching of the primary and secondary forces for various crank-angle arrangements of n-cylinder engines.

Centrifugal Pendulum Vibration Absorbers For a rotating system, such as the crank mechanism just discussed, the exciting torques are proportional to the rotational speed ω, which varies over a wide range. Hence, for a vibration absorber to be effective, its natural frequency must be proportional to ω. The centrifugal pendulum shown in Fig. 1.4.15 is ideally suited to this task. Strictly speaking, the system of Fig. 1.4.15 represents a two-degree-of-freedom nonlinear system. However, assuming that the motion of the wheel consists of a steady rotation ω and a small harmonic oscillation at the frequency Ω, or

$$\theta(t) = \omega t + \theta_0 \sin \Omega t \tag{1.4.46}$$

and that the pendulum angle ϕ is relatively small, then the equation of motion of the pendulum reduces to the linear single-degree-of-freedom system

$$\ddot{\phi} + \omega_n^2 \phi = \frac{R + r}{r} \Omega^2 \theta_0 \sin \Omega t \tag{1.4.47}$$

where

$$\omega_n = \omega\sqrt{R/r} \tag{1.4.48}$$

is the natural frequency of the pendulum. The torque exerted by the pendulum on the wheel is

$$T = \frac{m(R + r^2)}{1 - r\Omega^2/R\omega^2}\ddot{\theta} \tag{1.4.49}$$

so that the system behaves like a wheel with the effective mass moment of inertia

$$J_{\text{eff}} = -\frac{m(R + r)^2}{1 - r\Omega^2/R\omega^2} \tag{1.4.50}$$

which becomes infinite when Ω is equal to the natural frequency ω_n. To suppress disturbing torques of frequency Ω several times

Table 1.4.3 Inertial Unbalance of Four-Stroke-per-Cycle Engines

No. n of cylinders	Crank phase angle ϕ_j	Unbalanced forces		Unbalanced pitching moments about 1st cylinder	
		Primary	Secondary	Primary	Secondary
1		X_1	X_2	—	—
2	0–180°	0	$2X_2$	ℓX_1	$2\ell X_2$
4	0–180°–180°–0	0	$4X_2$	0	$6\ell X_2$
4	0–90°–270°–180°	0	0	$\ell X_1\sqrt{1+3^2}$	0
6	0–120°–240°–240°–120°–0	0	0	0	0
8	0–180°–90°–270°–270°–90°–180°–0	0	0	0	0
90° V-8	0–90°–270°–180°	0	0	Rotating primary couple of constant magnitude $\sqrt{10,X_1}$ which may be completely counterbalanced	

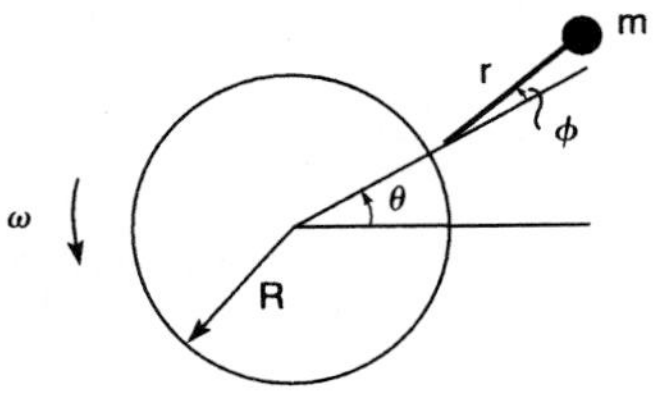

Figure 1.4.15 Centrifugal Pendulum.

larger than the rotational speed ω, the ratio r/R must be very small, which requires a short pendulum. The **bifilar pendulum** depicted in Fig. 1.4.16, which consists of a U-shaped counterweight that fits loosely and rolls on two pins of radius r_2 within two larger holes of equal radius r_1, represents a suitable design whereby the effective pendulum length is $r = r_1 - r_2$.

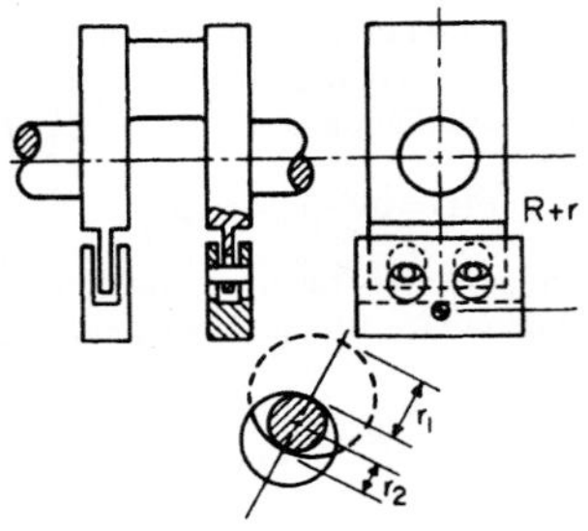

Figure 1.4.16 Bifilar Pendulum.

Response to Periodic Excitations A problem of interest in mechanical vibrations concerns the response $x(t)$ of the cam and follower system shown in Fig. 1.4.17. As the cam rotates at a constant angular rate, the follower undergoes the periodic displacement $y(t)$, where $y(t)$ has the period T. The equation of motion is

$$m\ddot{x} + (k_1 + k_2)x = k_2 y \tag{1.4.51}$$

Any periodic function can be expanded in a series of harmonic components in the form of the Fourier series

$$y(t) = \frac{1}{2}a_0 + \sum_{p=1}^{\infty}(a_p \cos \omega_0 t + b_p \sin p\omega_0 t) \qquad \omega_0 = 2\pi/T \tag{1.4.52}$$

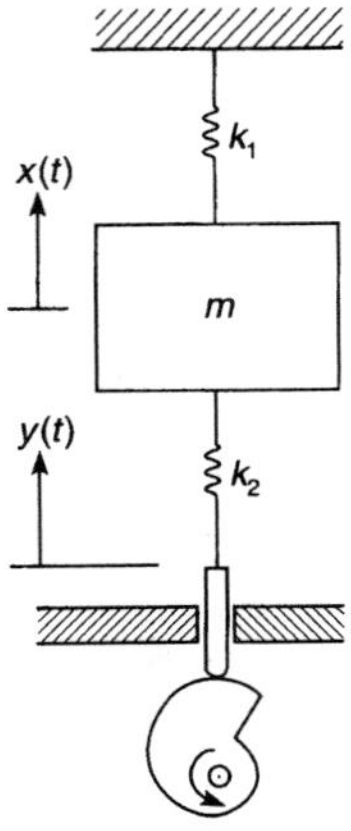

Figure 1.4.17 Cam and Follower System.

where ω_0 is called the **fundamental harmonic** and $p\omega_0$ ($p = 1, 2, \ldots$) are called **higher harmonics,** in which p is an integer. The coefficients have the expressions

$$a_p = \frac{2}{T}\int_0^T y(t)\cos p\omega_0 t \quad p = 0, 1, 2\ldots$$
$$b_p = \frac{2}{T}\int_0^T y(t)\sin p\omega_0 t \quad p = 1, 2\ldots \tag{1.4.53}$$

Note that the limits of integration can be changed, as long as the integration covers one complete period. From Eq. (1.4.27), and a companion equation for the sine counterpart, the response is

$$x(t) = \frac{k_2}{k_1 + k_2}\left[\frac{1}{2}a_0 + \sum_{p=1}^{\infty}\frac{1}{1-(p\omega_0/\omega_n)^2} \times (a_p \cos p\omega_0 t + b_p \sin p\omega_0 t)\right] \tag{1.4.54}$$

where

$$\omega_n = \sqrt{(k_1 + k_2)/m} \tag{1.4.55}$$

is the natural frequency of the system. Equation (1.4.54) describes a steady-state response, so that a description in terms of time is not very informative. More significant information can be extracted by plotting the amplitudes of the harmonic components versus the harmonic number. Such plots are called **frequency spectra,** and there is one for the excitation and one for the response. Equation (1.4.54) leads to the conclusion that **resonance occurs for** $p\omega_0 = \omega_n$.

As an example, consider the periodic excitation shown in Fig. 1.4.18 and use Eq. (1.4.53) to obtain the coefficients

$$a_0 = 2A,\ a_p = 0,\ b_p = \begin{cases} 4B/p\pi & p \text{ odd} \\ 0 & p \text{ even} \end{cases} \tag{1.4.56}$$

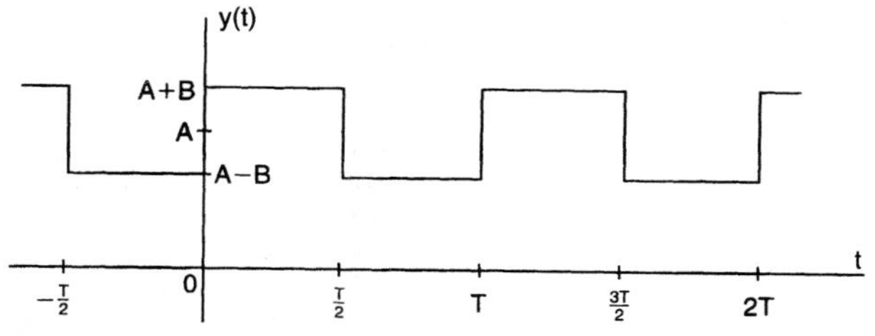

Figure 1.4.18 Example of periodic excitation.

The excitation and response frequency spectra are displayed in Figs. 1.4.19*a* and *b*, the latter for the case in which $\omega_n = 4\omega_0$.

Unit Impulse and Impulse Response Harmonic and periodic forces represent **steady-state excitations;** they **persist indefinitely.** The response to such forces is also steady state. An entirely different class of forces consists of **arbitrary,** or **transient, forces.** The term *transient* is not entirely appropriate, as some of these forces can also persist indefinitely. Concepts pivotal to the response to arbitrary forces are the

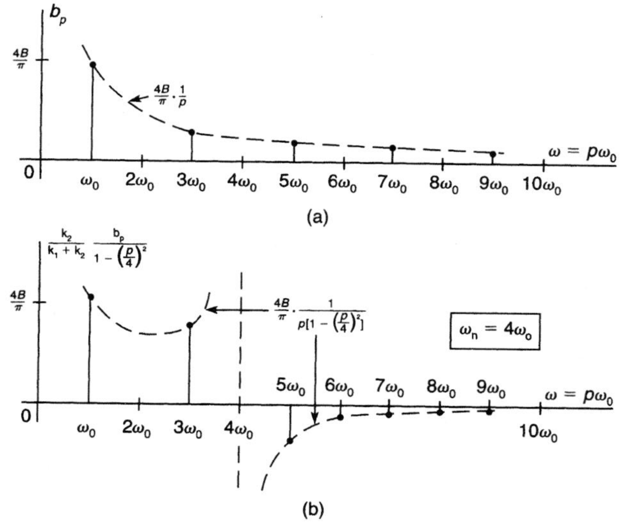

Figure 1.4.19 (*a*) Excitation frequency spectrum; (*b*) response frequency spectrum for the periodic excitation of Fig. 1.4.18.

unit impulse and the impulse response. The **unit impulse,** denoted by $\delta(t-a)$, represents a function of very high amplitude and defined over a very small time interval at $t = a$ such that the area enclosed is equal to 1 (Fig. 1.4.20). The **impulse response,** denoted by $g(t)$, is defined as the response of a system to a unit impulse applied at $t = 0$, with the initial conditions being equal to zero. For the mass-damper-spring system of Fig. 1.4.2, the impulse response is

$$g(t) = \frac{1}{m\omega_d} e^{-\zeta\omega_n t} \sin \omega_d t \qquad t > 0 \tag{1.4.57}$$

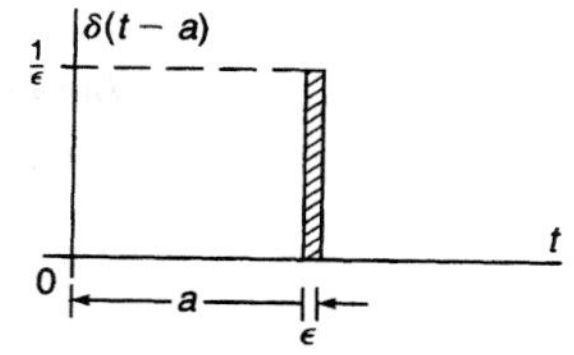

Figure 1.4.20 Unit Impulse.

Convolution Integral An arbitrary force $F(t)$ as shown in Fig. 1.4.21 can be regarded as a superposition of impulses of magnitude $F(\tau)d\tau$ and applied at $t = \tau$. Hence, the response to an arbitrary force can be regarded as a superposition of impulse responses $g(t-\tau)$ of magnitude $F(\tau)d\tau$, or

$$x(t) = \int_0^t F(\tau)g(t-\tau)d\tau$$

$$= \frac{1}{m\omega_d}\int_0^t F(\tau)e^{-\xi\omega_n(t-\tau)} \sin \omega_d(1-\tau)d\tau \tag{1.4.58a}$$

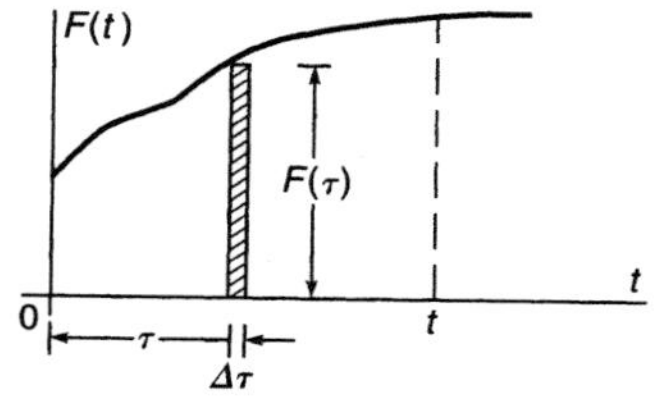

Figure 1.4.21 Convolution of an Arbitrary Function $F(t)$ as superposition of impulses $F(\tau)$.

which is called the **convolution integral** or the **superposition integral;** it can also be written in the form

$$x(t) = \int_0^t F(t-\tau)g(\tau)d\tau$$

$$= \frac{1}{m\omega_d}\int_0^t F(t-\tau)e^{-\xi\omega_n\tau} \sin \omega_d\tau\, d\tau \tag{1.4.58b}$$

Shock Spectrum Many systems are subjected on occasions to large forces applied suddenly and over periods of time that are short compared to the natural period. Such forces are capable of inflicting serious damage on a system and are referred to as **shocks.** The severity of a shock is commonly measured in terms of the maximum value of the response of a mass-spring system. The plot of the peak response versus the natural frequency is called the **shock spectrum** or **response spectrum.**

A shock $F(t)$ is characterized by its maximum value F_0, its duration T, and its shape. It is common to approximate the force by the half-sine pulse

$$F(t) = \begin{cases} F_0 \sin \omega t & \text{for } 0 < t < T = \pi/\omega \\ 0 & \text{for } t < 0 \text{ and } t > T \end{cases} \tag{1.4.59}$$

Using the convolution integral, Eq. (1.4.58*b*) with $\zeta = 0$, the response of a mass-spring system *during the duration of the pulse* is

$$x(t) = \frac{F_0}{k[1-(\omega/\omega_n)^2]}\left(\sin \omega t - \frac{\omega}{\omega_n}\sin \omega_n t\right) \qquad 0 < t < \pi/\omega \tag{1.4.60}$$

The maximum response is obtained when $\dot{x} = 0$ and has the value

$$x_{\max} = \frac{F_0}{k(1-\omega/\omega_n)} \sin \frac{2i\pi}{1+\omega_n/\omega} \qquad i = 1, 2, \ldots;\ i < \frac{1}{2}\left(1+\frac{\omega_n}{\omega}\right) \tag{1.4.61}$$

On the other hand, the response *after the termination of the pulse* is

$$x(t) = \frac{F_0\omega_n^2/\omega}{k[1-(\omega_n/\omega)^2]}\,[\cos\omega_n t + \cos\omega_n(t-T)] \qquad (1.4.62)$$

which has the maximum value

$$x_{\max} = \frac{2F_0\omega_n/\omega}{k[1-(\omega_n/\omega)^2]}\cos\frac{\pi\omega_n}{2\omega} \qquad (1.4.63)$$

The shock spectrum is the plot $x_{\max}$ versus ω_n/ω. For $\omega_n < \omega$, the maximum response is given by Eq. (1.4.63) and for $\omega_n > \omega$ by Eq. (1.4.61). The shock spectrum is shown in Fig. 1.4.22 in the form of the nondimensional plot $x_{\max}k/F_0$ versus ω_n/ω.

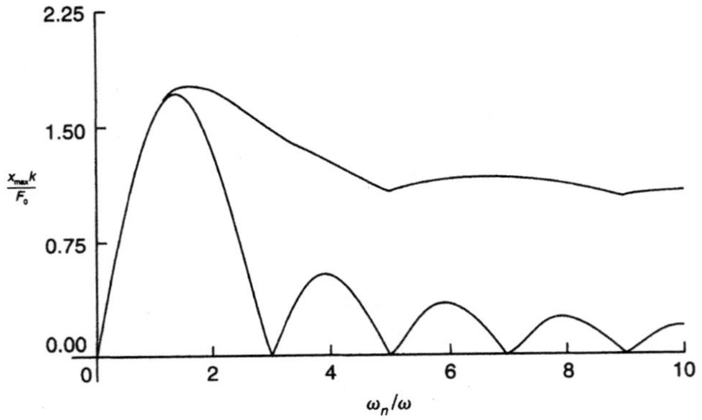

Figure 1.4.22 Nondimensional plot of shock spectrum.

MULTI-DEGREE-OF-FREEDOM SYSTEMS

Equations of Motion Many vibrating systems require more elaborate models than a single-degree-of-freedom system, such as the **multi-degree-of-freedom system** shown in Fig. 1.4.23. By using Newton's second law for each of the n masses m_i $(i = 1, 2, \ldots, n)$, the equations of motion can be written in the form

$$m_i\ddot{x}_i(t) + \sum_{j=1}^{n} c_{ij}\dot{x}_j(t) + \sum_{j=1}^{n} k_{ij}x_j(t) = F_i(t) \qquad i = 1, 2, \ldots, n \qquad (1.4.64)$$

where $x_i(t)$ is the displacement of mass m_i, $F_i(t)$ is the force acting on m_i, and c_{ij} and k_{ij} are **damping** and **stiffness coefficients,** respectively. The matrix form of Eq. (1.4.64) is

$$M\ddot{\mathbf{x}}(t) + C\dot{\mathbf{x}}(t) + K\mathbf{x}(t) = \mathbf{F}(t) \qquad (1.4.65)$$

in which $\mathbf{x}(t)$ is the n-dimensional displacement vector, $\mathbf{F}(t)$ the corresponding force vector, M the **mass matrix,** C the **damping matrix,** and K the **stiffness matrix,** all three symmetric matrices. (In the present case the mass matrix is diagonal, but in general it is not, although it is symmetric.)

Response of Undamped Systems to Harmonic Excitations Let the harmonic excitation have the form

$$\mathbf{F}(t) = \mathbf{F}_0 \sin\omega t \qquad (1.4.66)$$

where $\mathbf{F}_0$ is a constant vector and ω is the excitation, or driving frequency. The response to the harmonic excitation is a **steady-state** response and can be expressed as

$$\mathbf{x}(t) = Z^{-1}(\omega)\mathbf{F}_0 \sin\,\omega t \qquad (1.4.67)$$

where $Z^{-1}(\omega)$ is the inverse of the **impedance matrix** $Z(\omega)$. In the absence of damping, the impedance matrix is

$$Z(\omega) = K - \omega^2 M \qquad (1.4.68)$$

Undamped Vibration Absorbers When a mass-spring system m_1, k_1 is subjected to a harmonic force with the frequency equal to the natural frequency, resonance occurs. In this case, it is possible to add a second mass-spring system m_2, k_2 so designed as to produce a two-degree-of-freedom system with the response of m_1 equal to zero. We refer to m_1, k_1 as the **main system** and to m_2, k_2 as the **vibration absorber.** The resulting two-degree-of-freedom system is shown in Fig. 1.4.24 and has the impedance matrix

$$Z(\omega)\begin{bmatrix} k_1 + k_2 - \omega^2 m_1 & -k_2 \\ -k_2 & k_2 - \omega^2 m_2 \end{bmatrix} \qquad (1.4.69)$$

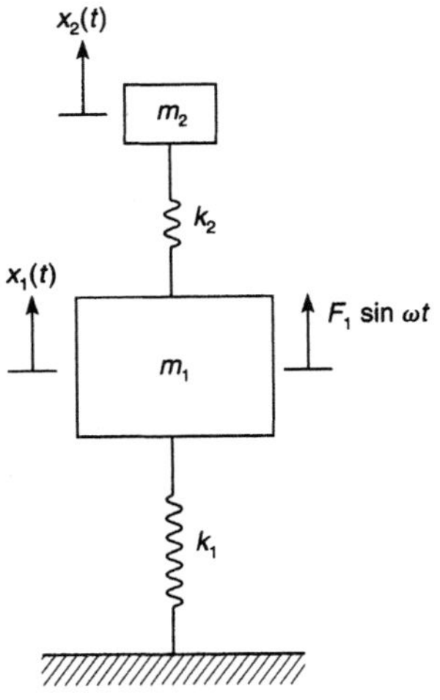

Figure 1.4.24 Two-Degree-of-Freedom System.

Inserting Eq. (1.4.69) into Eq. (1.4.67), together with $F_1(t) = F_1 \sin\omega t$, $F_2(t) = 0$, write the steady-state response in the form

$$x_1(t) = X_1(\omega)\sin\omega t \qquad (1.4.70a)$$

$$x_2(t) = X_2(\omega)\sin\omega t \qquad (1.4.70b)$$

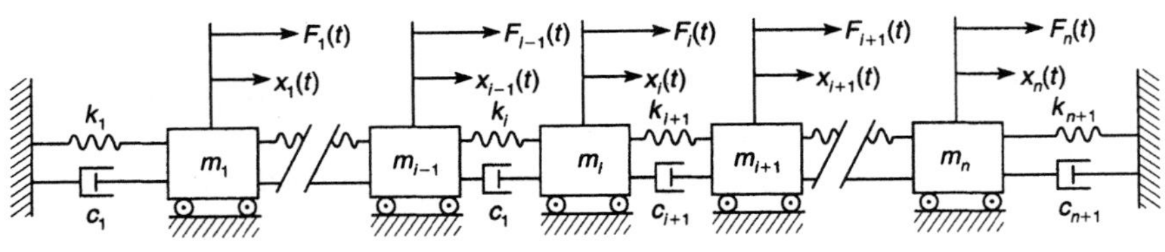

Figure 1.4.23 Multi-Degree-of-Freedom System.

where the amplitudes are given by

$$X_1(\omega) = \frac{[1-(\omega/\omega_a)^2]x_{st}}{[1+\mu(\omega_a/\omega_n)^2-(\omega/\omega_n)^2][1-(\omega/\omega_a)^2]-\mu(\omega_a/\omega_n)^2} \quad (1.4.71a)$$

$$X_2(\omega) = \frac{x_{st}}{[1+\mu(\omega_a/\omega_n)^2-(\omega/\omega_n)^2][1-(\omega/\omega_a)^2]-\mu(\omega_a/\omega_n)^2} \quad (1.4.71b)$$

in which

$\omega_n = \sqrt{k_1/m_1}$ = the natural frequency of the main system alone

$\omega_a = \sqrt{k_2/m_2}$ = the natural frequency of the absorber alone

$x_{st} = F_1/k_1$ = the static deflection of the main system

$\mu = m_2/m_1$ = the ratio of the absorber mass to the main mass

From Eqs. (1.4.70a) and (1.4.71a), we conclude that if we choose m_2 and k_2 such that $\omega_a = \omega$, the response $x_1(t)$ of the main mass is zero. Moreover, from Eqs. (1.4.70b) and (1.4.71b),

$$x_2(t) = -\left(\frac{\omega_n}{\omega_a}\right)^2 \frac{x_{st}}{\mu} \sin \omega t = -\frac{F_1}{k_2} \sin \omega t \quad (1.4.72)$$

so that the force in the absorber spring is

$$k_2 x_2(t) = -F_1 \sin \omega t \quad (1.4.73)$$

Hence, *the absorber exerts a force on the main mass balancing exactly the applied force $F_1 \sin \omega t$.*

A vibration absorber designed for a given operating frequency ω can perform satisfactorily for operating frequencies that vary slightly from ω. In this case, the motion of m_1 is not zero, but its amplitude tends to be very small, as can be verified from a frequency response plot $X_1(\omega)/x_{st}$ versus ω/ω_n; Fig. 1.4.25 shows such a plot for $\mu = 0.2$ and $\omega_n = \omega_a$. The shaded area indicates the range in which the performance can be regarded as satisfactory. Note that the thin line in Fig. 1.4.25 represents the frequency response of the main system alone. Also note that the system resulting from the combination of the main system and the absorber has two resonance frequencies, but they are removed from the operating frequency $\omega = \omega_n = \omega_a$.

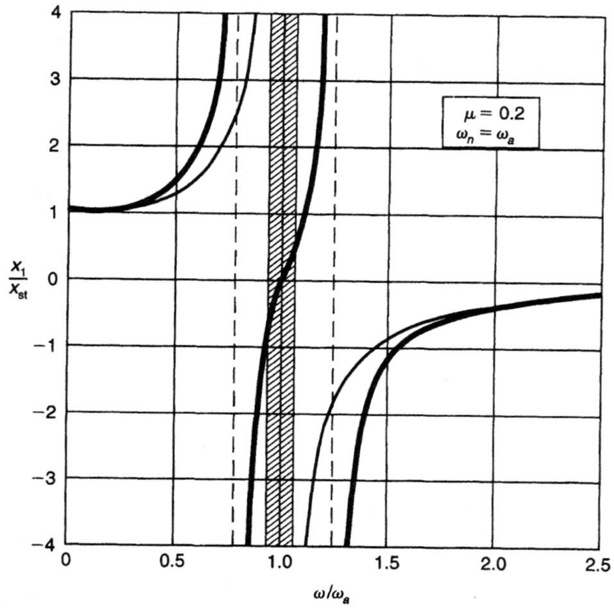

Figure 1.4.25 Frequency Response versus Frequency Ratio.

Natural Modes of Vibration In the absence of damping and external forces, Eq. (1.4.65) reduces to the free-vibration equation

$$M\ddot{\mathbf{x}}(t) + K\mathbf{x}(t) = \mathbf{0} \quad (1.4.74)$$

which has the harmonic solution

$$\mathbf{x}(t) = \mathbf{u} \cos (\omega t - \phi) \quad (1.4.75)$$

where $\mathbf{u}$ is a constant vector, ω a frequency of oscillation, and ϕ a phase angle. Introduction of Eq. (1.4.75) into Eq. (1.4.74) and division through by $\cos (\omega t - \phi)$ results in

$$K\mathbf{u} = \omega^2 M\mathbf{u} \quad (1.4.76)$$

which represents a set of n simultaneous algebraic equations known as the **eigenvalue problem.** It has n solutions consisting of the **eigenvalues** ω_r^2; the square roots represent the **natural frequencies** ω_r ($r = 1, 2, \ldots, n$). Moreover, to each natural frequency ω_r there corresponds a vector $\mathbf{u}_r$ ($r = 1, 2, \ldots, n$) called **eigenvector,** or **modal vector,** or **natural mode.** The modal vectors possess the **orthogonality property,** or

$$\mathbf{u}_s^T M \mathbf{u}_r = 0 \quad (1.4.77a)$$

$$\mathbf{u}_s^T K \mathbf{u}_r = 0 \quad (1.4.77b)$$

(for $r, s = 1, 2, \ldots, n$; $r \neq s$), in which $\mathbf{u}_s^T$ is the **transpose** of $\mathbf{u}_s$, a row vector. It is convenient to adjust the magnitude of the modal vectors so as to satisfy

$$\mathbf{u}_r^T M \mathbf{u}_r = 1 \quad (1.4.78a)$$

$$\mathbf{u}_r^T K \mathbf{u}_r = \omega_r^2 \quad (1.4.78b)$$

(for $r = 1, 2, \ldots, n$), a process known as normalization, in which case $\mathbf{u}_r$ are called **normal modes.** Note that the normalization process involves Eq. (1.4.78a) alone, as Eq. (1.4.78b) follows automatically. The solution of the eigenvalue problem can be obtained by a large variety of computational algorithms (Meirovitch, "Principles and Techniques of Vibrations," Prentice-Hall). Commercially, they are available in software packages for numerical computations, such as MATLAB.

The actual solution of Eq. (1.4.74) is obtained below in the context of the transient response.

Transient Response of Undamped Systems From Eq. (1.4.65), the vibration of undamped systems satisfies the equation

$$M\ddot{\mathbf{x}}(t) + K\mathbf{x}(t) = \mathbf{F}(t) \quad (1.4.79)$$

where $\mathbf{F}(t)$ is an arbitrary force vector. In addition, the displacement and velocity vectors must satisfy the initial conditions $\mathbf{x}(0) = \mathbf{x}_0$, $\dot{\mathbf{x}}(0) = \mathbf{v}_0$. The solution of Eq. (1.4.79) has the form

$$\mathbf{x}(t) = \sum_{r=1}^{n} \mathbf{u}_r \, q_r(t) \quad (1.4.80)$$

in which $\mathbf{u}_r$ are the modal vectors and $\mathbf{q}_r(t)$ are associated **modal coordinates.** Inserting Eq. (1.4.80) into Eq. (1.4.79), premultiplying the result by $\mathbf{u}_s^T$, and using Eqs. (1.4.77) and (1.4.78) we obtain the modal equations

$$\ddot{q}_r(t) + \omega_r^2 q_r(t) = Q_r(t) \qquad r = 1, 2, \ldots, n \quad (1.4.81)$$

where

$$Q_r(t) = \mathbf{u}_r^T \mathbf{F}(t) \qquad r = 1, 2, \ldots, n \quad (1.4.82)$$

are **modal forces.** Equations (1.4.81) resemble the equation of a single-degree-of-freedom system and have the solution

$$q_r(t) = \frac{1}{\omega_r}\int_0^t Q_r(1-\tau)\sin\omega_r\tau\, d\tau + q_r(0)\cos\omega_r t + \frac{\dot{q}_r(0)}{\omega_r}\sin\omega_r t \qquad r = 1, 2, \ldots, n \quad (1.4.83)$$

where

$$q_r(0) = \mathbf{u}_r^T M\mathbf{x}_0 \qquad (1.4.84a)$$

$$\dot{q}_r(0) = \mathbf{u}_r^T M\mathbf{v}_0 \qquad (1.4.84b)$$

(for $r = 1, 2, \ldots, n$) are **initial modal displacements and velocities,** respectively. The solution to both external forces and initial excitations is obtained by inserting Eq. (1.4.83) into Eq. (1.4.80).

Systems with Proportional Damping When the system is damped, the response does not in general have the form of Eq. (1.4.80), and a more involved approach is necessary (Meirovitch, "Fundamentals of Vibration," McGraw-Hill). In the special case in which the damping matrix C is proportional to the mass matrix M or the stiffness matrix K, or is a linear combination of M and K, the preceding approach yields the modal equations

$$\ddot{q}_r(t) + 2\zeta_r\omega_r\dot{q}_r(t) + \omega_r^2 q_r(t) = Q_r(t) \qquad r = 1, 2, \ldots, n \quad (1.4.85)$$

where ζ_r are **modal damping factors.** Equation (1.4.85) have the solution

$$q_r(t) = \frac{1}{\omega_{dr}}\int_0^t Q_r(t-\tau)e^{-\zeta_r\omega_r\tau}\sin\omega_{dr}\tau d\tau + \frac{q_r(0)}{\sqrt{1-\zeta_r^2}}e^{-\zeta_r\omega_r t}\cos(\omega_{dr}t - \psi_r) + \frac{\dot{q}_r(0)}{\omega_{dr}}\sin\omega_{dr}t$$

$$r = 1, 2, \ldots, n \qquad (1.4.86)$$

in which

$$\omega_{dr} = \omega_r\sqrt{1-\zeta_r^2} \qquad r = 1, 2, \ldots, n \qquad (1.4.87)$$

is the damped frequency in the rth mode and

$$\psi_r = \tan^{-1}\frac{\zeta_r}{\sqrt{1-\zeta_r^2}} \qquad r = 1, 2, \ldots, n \qquad (1.4.88)$$

is a phase angle associated with the rth mode. The quantities $Q_r(t)$, $q_r(0)$, and $\dot{q}_r(0)$ remain as defined by Eqs. (1.4.82), (1.4.84*a*), and (1.4.84*b*), respectively.

DISTRIBUTED-PARAMETER SYSTEMS

Vibration of Rods, Shafts, and Strings The axial vibration of rods is described by the equation

$$-\frac{\partial}{\partial x}\left[EA(x)\frac{\partial u(x,t)}{\partial x}\right] + m(x)\frac{\partial^2 u(x,t)}{\partial t^2} = f(x,t)$$

$$0 < x < L \qquad (1.4.89)$$

where $u(x, t)$ is the axial displacement, $f(x, t)$ the axial force per unit length, E the modulus of elasticity, $A(x)$ the cross-sectional area, and $m(x)$ the mass per unit length. The solution $u(x, t)$ is subject to one boundary condition at each end.

Before attempting to solve Eq. (1.4.89), consider the free vibration problem, $f(x, t) = 0$. The solution of the latter problem is harmonic and can be expressed as

$$u(x,t) = U(x)\cos(\omega t - \phi) \qquad (1.4.90)$$

where $U(x)$ is the amplitude, ω the frequency, and ϕ an inconsequential phase angle. Inserting Eq. (1.4.90) into Eq. (1.4.89) with $f(x, t) = 0$ and dividing through by $\cos(\omega t - \phi)$, we conclude that $U(x)$ and ω must satisfy the *eigenvalue problem*

$$-\frac{d}{dx}\left[EA(x)\frac{dU(x)}{dx}\right] = \omega^2 m(x)U(x) \qquad 0 < x < L \qquad (1.4.91)$$

where $U(x)$ must satisfy one boundary condition at each end. At a fixed end the displacement U must be zero and at a free end the axial force $EA\, dU/dx$ is zero.

Exact solutions of the eigenvalue problem are possible in only a few cases, mostly for uniform rods, in which case Eq. (1.4.91) reduces to

$$\frac{d^2U(x)}{dx^2} + \beta^2 U(x) = 0 \qquad \beta^2 = \frac{\omega^2 m}{EA} \qquad 0 < x < L \qquad (1.4.92)$$

whose solution is

$$U(x) = A\sin\beta x + B\cos\beta x \qquad (1.4.93)$$

where A and B are constants of integration, determined from specified boundary conditions. In the case of a **fixed-free** rod, the boundary conditions are

$$U(0) = 0 \qquad (1.4.94a)$$

$$EA\frac{dU}{dx}\bigg|_{x=L} = 0 \qquad (1.4.94b)$$

Condition (1.4.94*a*) gives $B = 0$ and condition (1.4.94*b*) yields the **characteristic equation**

$$\cos\beta L = 0 \qquad (1.4.95)$$

which has the infinity of solutions

$$\beta_r L = \frac{(2r-1)\pi}{2} \qquad r = 1, 2, \ldots \qquad (1.4.96)$$

where β_r represent the *eigenvalues;* they are related to the *natural frequencies* ω_r by

$$\omega_r = \beta_r\sqrt{\frac{EA}{m}} = \frac{(2r-1)\pi}{2}\sqrt{\frac{EA}{mL^2}} \qquad r = 1, 2, \ldots \qquad (1.4.97)$$

From Eq. (1.4.93), the **normal modes** are

$$U_r(x) = \sqrt{\frac{2}{mL}}\sin\frac{(2r-1)\pi x}{2L} \qquad r = 1, 2, \ldots \qquad (1.4.98)$$

For a **fixed-fixed** rod, the natural frequencies and normal modes are

$$\omega_r = r\pi\sqrt{\frac{EA}{mL^2}} \qquad U_r(x) = \sqrt{\frac{2}{mL}}\sin\frac{r\pi x}{L}$$

$$r = 1, 2, \ldots \qquad (1.4.99)$$

and for a **free-free** rod they are

$$\omega_0 = 0 \qquad U_0 = \sqrt{\frac{1}{mL}} \qquad (1.4.100a)$$

$$\omega_r = r\pi\sqrt{\frac{EA}{mL^2}} \qquad U_r(x) = \sqrt{\frac{2}{mL}}\cos\frac{r\pi x}{L}$$

$$r = 1, 2, \ldots \qquad (1.4.100b)$$

Note that U_0 represents a **rigid-body mode,** with zero natural frequency.

In every case the modes are orthogonal, satisfying the conditions

$$\int_0^L mU_s(x)U_r(x)dx = 0 \qquad (1.4.101a)$$

$$-\int_0^L U_s(x)\frac{d}{dx}\left[EA\frac{dU_r(x)}{dx}\right]dx = 0 \qquad (1.4.101b)$$

(for $r, s = 0, 1, 2, \ldots, r \neq s$) and have been normalized to satisfy the relations

$$\int_0^L mU_r^2(x)dx = 1 \qquad (1.4.102a)$$

$$-\int_0^L U_r(x)\frac{d}{dx}\left[EA\frac{dU_r(x)}{dx}\right]dx = \omega_r^2 \qquad (1.4.102b)$$

(for $r = 0, 1, 2, \ldots$). Note that the orthogonality of the normal modes extends to the rigid-body mode.

The response of the rod has the form

$$u(x,t)=\sum_{r=1}^{\infty}U_r(x)q_r(t) \tag{1.4.103}$$

Introducing Eq. (1.4.103) into Eq. (1.4.89), multiplying through by $U_s(x)$, integrating over the length of the rod, and using Eqs. (1.4.101) and (1.4.102) we obtain the **modal equations**

$$\ddot{q}_r(t)+\omega_r^2 q_r(t)=Q_r(t) \qquad r=1,2,\ldots \tag{1.4.104}$$

where

$$Q_r(t)=\int_0^L U_r(x)f(x,t)dx \qquad r=1,2,\ldots \tag{1.4.105}$$

are the **modal forces.** Equation (1.4.104) resemble Eq. (1.4.81) entirely; their solution is given by Eq. (1.4.83). The displacement of the rod is obtained by inserting Eq. (1.4.83) into Eq. (1.4.103).

As an example, consider the response of a uniform fixed-free rod to the uniformly distributed impulsive force

$$f(x,t)=\hat{f}_0\delta(t) \tag{1.4.106}$$

Inserting Eqs. (1.4.98) and (1.4.106) into Eq. (1.4.105), we obtain the modal forces

$$Q_r(t)=\sqrt{\frac{2}{mL}}\int_0^L \sin\frac{(2r-1)\pi x}{2L}\hat{f}_0\delta(t)dx$$

$$=\frac{2}{(2r-1)\pi}\sqrt{\frac{2L}{m}}\hat{f}_0\delta(t) \qquad r=1,2,\ldots \tag{1.4.107}$$

so that, from Eq. (1.4.83), the modal displacements are

$$q_r(t)=\frac{1}{\omega_r}\frac{2}{(2r-1)\pi}\sqrt{\frac{2L}{m}}\hat{f}_0\int_0^t\delta(1-\tau)\sin\omega_r\tau\, d\tau$$

$$=\left[\frac{2}{(2r-1)\pi}\right]^2\sqrt{\frac{2L^3}{EA}}\hat{f}_0\sin\frac{(2r-1)\pi}{2}\sqrt{\frac{EA}{mL^2}}\,t$$

$$r=1,2,\ldots \tag{1.4.108}$$

Finally, from Eq. (1.4.103), the response is

$$u(x,t)=\frac{8\hat{f}_0L}{\pi^2}\sqrt{\frac{1}{mEA}}\sum_{r=1}^{\infty}\frac{1}{(2r-1)^2}\sin\frac{(2r-1)\pi x}{2L}$$

$$\times\sin\frac{(2r-1)\pi}{2}\sqrt{\frac{EA}{mL^2}}\,t \tag{1.4.109}$$

The torsional vibration of shafts and the transverse vibration of strings are described by the same differential equation and boundary conditions as the axial vibration of rods, except that the nature of the displacement, inertia and stiffness parameters, and external excitations differs, as indicated in Table 1.4.4.

Bending Vibration of Beams The procedure for evaluating the response of beams in transverse vibration is similar to that for rods, the main difference arising in the stiffness term. The differential equation for beams in bending is

$$\frac{\partial^2}{\partial x^2}\left[EI(x)\frac{\partial^2 w(x,t)}{\partial x^2}\right]+m(x)\frac{\partial^2 w(x,t)}{\partial t^2}$$

$$=f(x,t) \qquad 0<x<L \tag{1.4.110}$$

in which $w(x, t)$ is the transverse displacement, $f(x, t)$ the force per unit length, $I(x)$ the cross-sectional area moment of inertia, and $m(x)$ the mass per unit length. The solution $w(x, t)$ must satisfy two boundary conditions at each end.

The eigenvalue problem is described by the differential equation

$$\frac{d^2}{dx^2}\left[EI(x)\frac{d^2W(x)}{dx^2}\right]=\omega^2 m(x)W(x) \qquad 0<x<L \tag{1.4.111}$$

and two boundary conditions at each end, depending on the type of support. Some possible boundary conditions are given in Table 1.4.5. The solution of the eigenvalue problem consists of the natural frequencies

Table 1.4.5 Quantities Equal to Zero at Boundary

Boundary type	Displacement W	Slope dW/dx	Bending moment EId^2W/dx^2	Shearing force $d(EId^2W/dx^2)/dx$
Hinged	✓		✓	
Clamped	✓	✓		
Free			✓	✓

ω_r and natural modes $W_r(x)$ ($r = 1, 2, \ldots$). The first five normalized natural frequencies of uniform beams with six different boundary conditions are listed in Table 1.4.6. The normal modes for the hinged-hinged beam are

$$W_r(x)=\sqrt{\frac{2}{mL}}\sin\frac{r\pi x}{L} \qquad r=1,2,\ldots \tag{1.4.112}$$

The normal modes for the remaining beam types are more involved and they involve both trigonometric and hyperbolic functions (Meirovitch, "Fundamentals of Vibrations"). The modes for every beam type are orthogonal and can be used to obtain the response $w(x, t)$ in the form of a series similar to Eq. (1.4.103).

Vibration of Membranes **A membrane** is a very thin sheet of material stretched over a two-dimensional domain enclosed by one or two nonintersecting boundaries. It can be regarded as the two-dimensional counterpart of the string. Like a string, it derives the ability to resist transverse displacements from tension, which acts in all directions in the plane of the membrane and at all its points. It is commonly assumed that the tension is uniform and does not change as the membrane deflects. The general procedure for calculating the response of membranes remains the same as for rods and beams, but there is one significant new factor, namely, the shape of the boundary, which dictates the type of coordinates to be used. For rectangular membranes cartesian coordinates must be used, and for circular membranes polar coordinates are indicated.

Table 1.4.4 Analogous Quantities for Rods, Shafts, and Strings

	Rods	Shafts	Strings
Displacement	Axial—$u(x, t)$	Torsional—$\theta(x, t)$	Transverse—$w(x, t)$
Inertia (per unit length)	Mass—$m(x)$	Mass polar moment of inertia—$I(x)$	Mass—$\rho(x)$
Stiffness	Axial—$EA(x)$ E = Young's modulus $A(x)$ = cross-sectional area	Torsional—$GJ(x)$ G = shear modulus $J(x)$ = area polar moment of inertia	Tension—$T(x)$
Load (per unit length)	Force—$f(x, t)$	Moment—$m(x, t)$	Force—$f(x, t)$

Table 1.4.6 Normalized Natural Frequencies for Various Beams

Beam type	$\omega_1\sqrt{mL^4/EI}$	$\omega_2\sqrt{mL^4/EI}$	$\omega_3\sqrt{mL^4/EI}$	$\omega_4\sqrt{mL^4/EI}$	$\omega_5\sqrt{mL^2/EI}$
Hinged-hinged	π^2	$4\pi^2$	$9\pi^2$	$16\pi^2$	$25\pi^2$
Clamped-free	1.875^2	4.694^2	7.855^2	10.996^2	14.137^2
Free-free	0	0	$(1.506\pi)^2$	$(2.500\pi)^2$	$(3.500\pi)^2$
Clamped—clamped	$(1.506\pi)^2$	$(2.500\pi)^2$	$(3.500\pi)^2$	$(4.500\pi)^2$	$(5.500\pi)^2$
Clamped-hinged	3.927^2	7.069^2	10.210^2	13.352^2	16.493^2
Hinged-free	0	3.927^2	7.069^2	10.210^2	13.352^2

The differential equation for the transverse vibration of membranes is

$$-T\nabla^2 w + \rho\frac{\partial^2 w}{\partial t^2} = f \tag{1.4.113}$$

which must be satisfied at every interior point of the membrane, where w is the transverse displacement, f the transverse force per unit area, T the tension, and ρ the mass per unit area. Moreover, ∇^2 is the Laplacian operator, whose expression depends on the coordinates used. The solution w must satisfy one boundary condition at every boundary point. Using the established procedure, the eigenvalue problem is described by the differential equation

$$-T\nabla^2 W = \omega^2 \rho W \tag{1.4.114}$$

where W is the displacement amplitude; it must satisfy one boundary condition at every point of the boundary.

Consider a **rectangular membrane fixed** at $x = 0$, a and $y = 0$, b, in which case the Laplacian operator in terms of the cartesian coordinates x and y has the form

$$\nabla^2 = \frac{\partial^2}{\partial x^2} + \frac{\partial^2}{\partial y^2} \tag{1.4.115}$$

The boundary conditions are $W(0, y) = W(a, y) = W(x, 0) = W(x, b) = 0$. The natural frequencies are

$$\omega_{mn} = \pi\sqrt{\left[\left(\frac{m}{a}\right)^2 + \left(\frac{n}{b}\right)^2\right]\frac{T}{\rho}} \qquad m, n = 1, 2, \ldots \tag{1.4.116}$$

and the normal modes are

$$W_{mn}(x, y) = \frac{2}{\sqrt{\rho ab}}\sin\frac{m\pi x}{a}\sin\frac{n\pi y}{b} \qquad m, n = 1, 2, \ldots \tag{1.4.117}$$

The modes satisfy the orthogonality conditions

$$\int_0^a\int_0^b \rho W_{mn}(x,y)W_{rs}(x,y)dx\,dy = 0, \qquad m \neq r \text{ and/or } n \neq s \tag{1.4.118a}$$

$$-\int_0^a\int_0^b W_{mn}(x,y)T\nabla^2 W_{rs}(x,y)dx\,dy = 0, \qquad m \neq r \text{ and/or } n \neq s \tag{1.4.118b}$$

and have been normalized so that $\int_0^a\int_0^b \rho W_{mn}^2(x, y)dxdy = 1 (m, n = 1,2, \ldots)$. Note that, because the problem is two-dimensional, it is necessary to identify the natural frequencies and modes by two subscripts. With this exception, the procedure for obtaining the response is the same as for rods and beams.

Next, consider a **uniform circular membrane fixed at** $r = a$. In this case, the Laplacian operator in terms of the polar coordinates r and θ is

$$\nabla^2 = \frac{\partial^2}{\partial r^2} + \frac{1}{r}\frac{\partial}{\partial r} + \frac{1}{r^2}\frac{\partial^2}{\partial \theta^2} \tag{1.4.119}$$

The natural modes for circular membranes are appreciably more involved than for rectangular membranes. They are products of Bessel functions of $\omega_{mn}r$ and trigonometric functions of $m\theta$, where $m = 0, 1, 2, \ldots$ and $n = 1, 2, \ldots$. The modes are given in Meirovitch, "Principles and Techniques of Vibrations," Prentice-Hall. Table 1.4.7 gives the normalized natural frequencies $\omega_{mn}^* = (\omega_{mn}/2\pi)\sqrt{\rho a^2/T}$ corresponding to $m = 0, 1, 2$, and $n = 1, 2, 3$. The modes satisfy the orthogonality relations

$$\int_0^a\int_0^{2\pi} \rho W_{mn}(r,\theta)W_{rs}(r,\theta)r\,dr\,d\theta = 0 \qquad m \neq r \text{ and/or } n \neq s \tag{1.4.120a}$$

$$-\int_0^a\int_0^{2\pi} W_{mn}(r,\theta)T\nabla^2 W_{rs}(r,\theta)r\,dr\,d\theta = 0 \qquad m \neq r \text{ and/or } n \neq s \tag{1.4.120b}$$

The response of circular membranes is obtained in the usual manner.

Table 1.4.7 Circular Membrane Normalized Natural Frequencies $\omega_{mn}^* = (\omega_{mn}/2\pi)\sqrt{\rho a^2/T}$

	n		
m	1	2	3
0	0.3827	0.8786	1.3773
1	0.6099	1.1165	1.6192
2	0.8174	1.3397	1.8494

Bending Vibration of Plates Consider **plates** whose behavior is governed by the elementary plate theory, which is based on the following assumptions: (1) deflections are small compared to the plate thickness; (2) the normal stresses in the direction transverse to the plate are negligible; (3) there is no force resultant on the cross-sectional area of a plate differential element: the middle plane of the plate does not undergo deformations and represents a neutral plane; and (4) any straight line normal to the middle plane remains so during bending. Under these assumptions, the differential equation for the bending vibration of plates is

$$D\nabla^4 w + m\frac{\partial^2 w}{\partial t^2} = f \tag{1.4.121}$$

and is to be satisfied at every interior point of the plate, where w is the transverse displacement, f the transverse force per unit area, m the mass per unit area, $D = Eh^3/12(1 - \nu^2)$ the plate flexural rigidity, E Young's modulus, h the plate thickness, and ν Poisson's ratio. Moreover, ∇^4 is the biharmonic operator. The solution w must satisfy two boundary conditions at every point of the boundary. The eigenvalue problem is defined by the differential equation

$$D\nabla^4 W = \omega^2 m W \tag{1.4.122}$$

and corresponding boundary conditions.

Consider a **rectangular plate simply supported** at $x = 0$, a and $y = 0$, b. Because of the shape of the plate, we must use cartesian coordinates, in which case the biharmonic operator has the expression

$$\nabla^4 = \nabla^2\nabla^2 = \left(\frac{\partial^2}{\partial x^2} + \frac{\partial^2}{\partial y^2}\right)\left(\frac{\partial^2}{\partial x^2} + \frac{\partial^2}{\partial y^2}\right)$$

$$= \frac{\partial^4}{\partial x^4} + 2\frac{\partial^4}{\partial x^2\,\partial y^2} + \frac{\partial^4}{\partial y^4} \qquad (1.4.123)$$

Moreover, the boundary conditions are $W = 0$ and $\partial^2 W/\partial x^2 = 0$ for $x = 0$, a and $W = 0$ and $\partial^2 W/\partial y^2 = 0$ for $y = 0$, b. The natural frequencies are

$$\omega_{mn} = \pi^2\left[\left(\frac{m}{a}\right)^2 + \left(\frac{n}{b}\right)^2\right]\sqrt{\frac{D}{m}} \qquad m, n = 1, 2, \ldots . \quad (1.4.124)$$

and no confusion should arise because the same symbol is used for one of the subscripts and for the mass per unit area. The corresponding normal modes are

$$W_{mn}(x, y) = \frac{2}{\sqrt{mab}} \sin\frac{m\pi x}{n} \sin\frac{n\pi y}{b} \qquad m, n = 1, 2, \ldots \quad (1.4.125)$$

and they are recognized as being the same as for rectangular membranes fixed at all boundaries.

A **circular plate** requires use of polar coordinates, so that the biharmonic operator has the form

$$\nabla^4 = \nabla^2\nabla^2 = \left(\frac{\partial^2}{\partial r^2} + \frac{1}{r}\frac{\partial}{\partial r} + \frac{1}{r^2}\frac{\partial^2}{\partial \theta^2}\right)\left(\frac{\partial^2}{\partial r^2} + \frac{1}{r}\frac{\partial}{\partial r} + \frac{1}{r^2}\frac{\partial^2}{\partial \theta^2}\right) \quad (1.4.126)$$

Consider a **plate clamped** at $r = a$, in which case the boundary conditions are $W(r, \theta) = 0$ and $\partial W(r, \theta)/\partial r = 0$ at $r = a$. In addition, the solution must be finite at every interior point in the plate, and in particular at $r = 0$. The natural modes have involved expressions; they are given in Meirovitch, "Principles and Techniques of Vibrations," Prentice-Hall. Table 1.4.8 lists the normalized natural frequencies $\omega^*_{mn} = \omega_{mn}(a/\pi)^2\sqrt{m/D}$ corresponding to $m = 0, 1, 2$, and $n = 1, 2, 3$.

Table 1.4.8 Circular Plate Normalized Natural Frequencies $\omega^*_{mn} = \omega_{mn}(a/\pi)^2\sqrt{m/D}$

	n		
m	1	2	3
0	1.015^2	2.007^2	3.000^2
1	1.468^2	2.483^2	3.490^2
2	1.879^2	2.992^2	4.000^2

The natural modes of the plates are orthogonal and can be used to obtain the response to both initial and external excitations.

APPROXIMATE METHODS FOR DISTRIBUTED SYSTEMS

Rayleigh's Energy Method The eigenvalue problem contains vital information concerning vibrating systems, namely, the natural frequencies and modes. In the majority of practical cases, exact solutions to the eigenvalue problem for distributed systems are not possible, so that the interest lies in **approximate solutions.** This is often the case when the mass and stiffness are distributed nonuniformly and/or the boundary conditions cannot be satisfied, the latter in particular for two-dimensional systems with irregularly shaped boundaries.

When the objective is to estimate the lowest natural frequency, Rayleigh's energy method has few equals. As discussed earlier, free vibration of undamped systems is harmonic and can be expressed as

$$w(x, t) = W(x)\cos(\omega t - \phi) \qquad (1.4.127)$$

where $W(x)$ is the displacement amplitude, ω the free vibration frequency, and ϕ an inconsequential phase angle. The kinetic energy represents an integral involving the velocity squared. Hence, using Eq. (1.4.127), the kinetic energy can be written in the form

$$T(t) = \frac{1}{2}\int_0^L m(x)\left[\frac{\partial w(x, t)}{\partial t}\right]^2 dx = \omega^2 T_{ref}\sin^2(\omega t - \phi) \quad (1.4.128)$$

where

$$T_{ref} = \frac{1}{2}\int_0^L m(x)W^2(x)dx \qquad (1.4.129)$$

is called the **reference kinetic energy.** The form of the potential energy is system-dependent, but in general is an integral involving the square of the displacement and of its derivatives with respect to the spatial coordinates (see Table 1.4.9). It can be expressed as

$$V(t) = V_{max}\cos^2(\omega t - \phi) \qquad (1.4.130)$$

where V_{max} is the maximum potential energy, which can be obtained by simply replacing $w(x, t)$ by $W(x)$ in $V(t)$. Using the principle of conservation of energy in conjunction with Eqs. (1.4.128) and (1.4.130), we can write

$$E = T + V = T_{max} + 0 = 0 + V_{max} \qquad (1.4.131)$$

in which

$$T_{max} = \omega^2 T_{ref} \qquad (1.4.132)$$

It follows that

$$\omega^2 = \frac{V_{max}}{T_{ref}} \qquad (1.4.133)$$

Equation (1.4.133) represents **Rayleigh's quotient,** which has the remarkable property that it has a minimum value for $W(x) = W_1(x)$, the minimum value being ω_1^2. Rayleigh's energy method amounts to selecting a trial function $W(x)$ reasonably close to the lowest natural mode $W_1(x)$, inserting this function into Rayleigh's quotient, and carrying out the indicated integrations. Then, ω^2 will be one order of magnitude closer to the lowest eigenvalue ω_1^2 than $W(x)$ is to $W_1(x)$, thus providing a good estimate ω of the lowest natural frequency ω_1. Quite often, the static deformation of the system acted on by loads proportional to the mass distribution is a good choice. In some cases, the lowest mode of a related simpler system can yield good results.

Table 1.4.9 Potential Energy for Various Systems

System	Potential energy* $V(t)$
Rod (also shafts and strings)	$\frac{1}{2}\int_0^L EA(x)[\partial u(x,t)/\partial x]^2 dx$
Beams	$\frac{1}{2}\int_0^L EI(x)[\partial^2 w(x,t)/\partial x^2]^2 dx$
Beams with axial force	$\frac{1}{2}\int_0^L \{EI(x)[\partial^2 w(x,t)/\partial x^2]^2 + P(x)[\partial \omega(x,t)/\partial x]^2\}dx$
Membranes	$\frac{1}{2}\int_{Area} T\{[\partial w(x,y,t)/\partial x]^2 + [\partial w(x,y,t)/\partial y]^2\}dx\,dy$
Plates	$\frac{1}{2}\int_{Area} D\{\nabla^2 w(x,y,t)^2 + 2(1-\nu)[\partial^2 w(x,y,t)/\partial x\,\partial y]^2 - [\partial^2 w(x,y,t)/\partial x^2][\partial^2 w(x,y,t)/\partial y^2]\}dx\,dy$

*If the distributed system has a spring at the boundary point a, then add a term $kw^2(a, t)/2$.

As an example, estimate the lowest natural frequency of a uniform bar in axial vibration with a mass M attached at $x = L$ (Fig. 1.4.26) for the three trial functions (1) $U(x) = x/L$; (2) $U(x) = (1 + M/mL)(x/L) - (x/L)^2/2$, representing the static deformation; and (3) $U(x) = \sin \pi x/2L$,

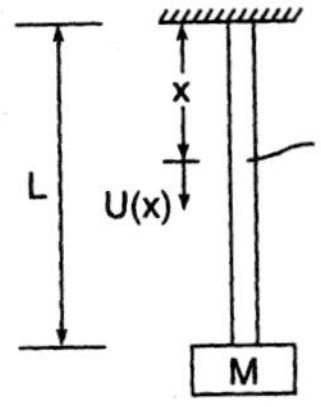

Figure 1.4.26 Uniform bar with lumped mass attached at the end.

representing the lowest mode of the bar without the mass M. The Rayleigh quotient for this bar is

$$\omega^2 = \frac{\int_0^L EA(x)[dU(x)/dx]^2\,dx}{\int_0^L m(x)U^2(x)\,dx + MU^2(L)} \tag{1.4.134}$$

The results are:

$$1.\quad \omega^2 = \frac{\int_0^L EA(1/L)^2 dx}{\int_0^L m(x/L)^2 dx + M} = \frac{EA}{(M + mL/3)L}$$

$$2.\quad \frac{\int_0^L EA(1 + M/mL - x/L)^2(1/L)^2 dx}{\int_0^L m[(1 + M/mL)(x/L) - (x/L)^2/2]^2 dx + M(1 + 2M/mL)^2/4}$$

$$= \frac{\left(\dfrac{M}{mL}\right)^2 + \dfrac{M}{mL} + \dfrac{1}{3}}{\dfrac{1}{3}\left(\dfrac{M}{mL}\right)^2 + \dfrac{5}{12}\dfrac{M}{mL} + \dfrac{2}{15} + \dfrac{M}{mL}\left(\dfrac{1}{2} + \dfrac{M}{mL}\right)^2}\,\frac{EA}{mL^2} \tag{1.4.135}$$

$$3.\quad \omega^2 = \frac{\int_0^L EA\left(\dfrac{\pi}{2L}\right)^2 \cos^2 \dfrac{\pi x}{2L} dx}{\int_0^L m \sin^2 \dfrac{\pi x}{2L} dx + M} = \frac{\pi^2}{8\left(\dfrac{1}{2} + \dfrac{M}{mL}\right)}\frac{EA}{mL^2}$$

For comparison purposes, let $M = mL$, which yields the following estimates for the lowest natural frequency:

$$1.\quad \omega = 0.8660\sqrt{\frac{EA}{mL^2}}$$

$$2.\quad \omega = 0.8629\sqrt{\frac{EA}{mL^2}} \tag{1.4.136}$$

$$3.\quad \omega = 0.9069\sqrt{\frac{EA}{mL^2}}$$

The best estimate is the lowest one, which corresponds to case 2, with the trial function in the form of the static displacement. Note that the estimate obtained in case 1 is also quite good. It corresponds to the first case in Table 1.4.2, representing a mass-spring system in which the mass of the spring is included.

Rayleigh-Ritz Method **Rayleigh's quotient,** Eq. (1.4.133), corresponding to any trial function $W(x)$ is always larger than the lowest eigenvalue ω_1^2, and it takes the minimum value of ω_1^2 when $W(x)$ coincides with the lowest natural mode $W_1(x)$. However, this possibility must be ruled out by virtue of the assumption that W_1 is not available. The **Rayleigh-Ritz method** is a procedure for minimizing Rayleigh's quotient by means of a sequence of approximate solutions converging to the actual solution of the eigenvalue problem. The minimizing sequence has the form

$$\begin{aligned} W(x) &= a_1\phi_1(x) \\ W(x) &= a_1\phi_1(x) + a_2\phi_2(x) = \sum_{j=1}^{2} a_j\phi_j(x) \\ &\cdots \\ W(x) &= a_1\phi_1(x) + a_2\phi_2(x) + \cdots + a_n\phi_n(x) = \sum_{j=1}^{n} a_j\phi_j(x) \end{aligned} \tag{1.4.137}$$

where a_j are undetermined coefficients and $\phi_j(x)$ are suitable trial functions satisfying all, or at least the geometric boundary conditions. The coefficients $a_j (j = 1, 2, \ldots, n)$ are determined so that Rayleigh's quotient has a minimum. With Eq. (1.4.137) inserted into Eq. (1.4.133), Rayleigh's quotient becomes

$$\Omega^2 = \frac{\sum_{i=1}^{n}\sum_{j=1}^{n} k_{ij}a_ia_j}{\sum_{i=1}^{n}\sum_{j=1}^{n} m_{ij}a_ia_j} \qquad n = 1, 2, \ldots \tag{1.4.138}$$

where $k_{ij} = k_{ji}$ and $m_{ij} = m_{ji}$ $(i, j = 1, 2, \ldots, n)$ are symmetric stiffness and mass coefficients whose nature depends on the potential energy and kinetic energy, respectively. The special case in which $n = 1$ represents Rayleigh's energy method. For $n \geq 2$, minimization of Rayleigh's quotient leads to the solution of the eigenvalue problem

$$\sum_{j=1}^{n} k_{ij}a_j = \Omega^2 \sum_{j=1}^{n} m_{ij}a_j$$

$$i = 1, 2, \ldots, n;\ n = 2, 3, \ldots \tag{1.4.139}$$

Equation (1.4.139) can be written in the matrix form

$$\boldsymbol{Ka} = \Omega^2 \boldsymbol{Ma} \tag{1.4.140}$$

in which $K = [k_{ij}]$ is the symmetric stiffness matrix and $M = [m_{ij}]$ is the symmetric mass matrix. Equation (1.4.140) resembles the eigenvalue problem for multi-degree-of-freedom systems, Eq. (1.4.76), and its solutions possess the same properties. The eigenvalues Ω_r^2 provide approximations to the actual eigenvalues ω_r^2, and approach them from above as n increases. Moreover, the eigenvectors $\mathbf{a}_r = [a_{r1}\ a_{r2}\ \ldots\ a_{rn}]^T$ can be used to obtain the approximate natural modes by writing

$$W_r(x) = a_{r1}\phi_1(x) + a_{r2}\phi_2(x) + \cdots + a_{rn}\phi_n(x) = \sum_{j=1}^{n} a_{rj}\phi_j(x)$$

$$r = 1, 2, \ldots, n;\ n = 2, 3, \ldots \tag{1.4.141}$$

As an illustration, consider the same rod in axial vibration used to demonstrate Rayleigh's energy method. Insert Eq. (1.4.137) with $W(x)$ replaced by $U(x)$ into the numerator and denominator of Eq. (1.4.134) to obtain

$$\int_0^L EA(x)\left[\frac{dU(x)}{dx}\right]^2 dx = \sum_{i=1}^{n}\sum_{j=1}^{n}\left[\int_0^L EA(x)\frac{d\phi_i(x)}{dx}\frac{d\phi_j(x)}{dx}dx\right]a_ia_j \tag{1.4.142a}$$

$$\int_0^L m(x)U^2(x)dx + MU^2(L) = \sum_{i=1}^{n}\sum_{j=1}^{n}\left[\int_0^L m(x)\phi_i(x)\phi_j(x)dx + M\phi_i(L)\phi_j(L)\right]a_ia_j \tag{1.4.142b}$$

so that the stiffness and mass coefficients are

$$k_{ij} = \int_0^L EA(x)\frac{d\phi_i(x)}{dx}\frac{d\phi_j(x)}{dx}dx \qquad i, j = 1, 2, \ldots, n \tag{1.4.143a}$$

$$m_{ij} = \int_0^L m(x)\phi_i(x)\phi_j(x)dx + M\phi_i(L)\phi_j(L) \qquad i, j = 1, 2, \ldots, n \tag{1.4.143b}$$

respectively. As trial functions, use

$$\phi_j(x) = (x/L)^j \qquad j = 1, 2, \ldots, n \tag{1.4.144}$$

which are zero at $x = 0$, thus satisfying the geometric boundary condition. Hence, the stiffness and mass coefficients are

$$k_{ij} = \frac{EAij}{L^{i+j}} \int_0^L x^{i-1} x^{j-1}\, dx = \frac{ij}{i+j-1} \frac{EA}{L} \qquad i, j = 1, 2, \ldots, n \tag{1.4.145a}$$

$$m_{ij} = \frac{m}{L^{i+j}} \int_0^L x^i x^j dx + M = \frac{mL}{i+j+1} + M \qquad i, j = 1, 2, \ldots, n \tag{1.4.145b}$$

so that the stiffness and mass matrices are

$$K = \frac{EA}{L} \begin{bmatrix} 1 & 1 & 1 & \cdots & 1 \\ 1 & 4/3 & 3/2 & \cdots & 2n/(n+1) \\ 1 & 3/2 & 9/5 & \cdots & 3n/(n+2) \\ \cdots & \cdots & \cdots & \cdots & \cdots \\ 1 & 2n/(n+1) & 3n/(n+2) & \cdots & n^2/(2n-1) \end{bmatrix} \tag{1.4.146a}$$

$$M = mL \begin{bmatrix} 1/3 & 1/4 & 1/5 & \cdots & 1/(n+2) \\ 1/4 & 1/5 & 1/6 & \cdots & 1/(n+3) \\ 1/5 & 1/6 & 1/7 & \cdots & 1/(n+4) \\ \cdots & \cdots & \cdots & \cdots & \cdots \\ 1/(n+2) & 1/(n+3) & 1/(n+4) & \cdots & 1/(2n+1) \end{bmatrix} + M \begin{bmatrix} 1 & 1 & 1 & \cdots & 1 \\ 1 & 1 & 1 & \cdots & 1 \\ 1 & 1 & 1 & \cdots & 1 \\ \cdots & \cdots & \cdots & \cdots & \cdots \\ 1 & 1 & 1 & \cdots & 1 \end{bmatrix} \tag{1.4.146b}$$

For comparison purposes, consider the case in which $M = mL$. Then, for $n = 2$, the eigenvalue problem is

$$\begin{bmatrix} 1 & 1 \\ 1 & 4/3 \end{bmatrix} \begin{bmatrix} a_1 \\ a_2 \end{bmatrix} = \lambda \begin{bmatrix} 4/3 & 5/4 \\ 5/4 & 6/5 \end{bmatrix} \begin{bmatrix} a_1 \\ a_2 \end{bmatrix}$$

$$\lambda = \Omega^2 \frac{mL^2}{EA} \tag{1.4.147}$$

which has the solutions

$$\lambda_1 = 0.7407 \qquad \mathbf{a}_1 = [1 \quad -0.1667]^T$$
$$\lambda_1 = 12.0000 \qquad \mathbf{a}_1 = [1 \quad -1.0976]^T \tag{1.4.148}$$

Hence, the computed natural frequencies and modes are

$$\Omega_1 = 0.8607\sqrt{\frac{EA}{mL^2}} \qquad U_1(x) = \frac{x}{L} - 0.1667\left(\frac{x}{L}\right)^2$$
$$\Omega_2 = 3.4641\sqrt{\frac{EA}{mL^2}} \qquad U_2(x) = \frac{x}{L} - 1.0976\left(\frac{x}{L}\right)^2 \tag{1.4.149}$$

Comparing Eq. (1.4.149) with the estimates obtained by Rayleigh's energy method, Eq. (1.4.136), note that the Rayleigh-Ritz method has produced a more accurate approximation for the lowest natural frequency. In addition, it has produced a first approximation for the second lowest natural frequency, as well as approximations for the two lowest modes, which Rayleigh's energy method is unable to produce. The approximate solutions can be improved by letting $n = 3, 4, \ldots$.

Finite Element Method In the Rayleigh-Ritz method, the trial functions extend over the entire domain of the system and tend to be complicated and difficult to work with. More importantly, they often cannot be produced, particularly for two-dimensional problems. Another version of the Rayleigh-Ritz method, the **finite element method,** does not suffer from these drawbacks. Indeed, the trial functions extending only over small subdomains, referred to as **finite elements,** are known low-degree polynomials and permit easy computer coding. As in the Rayleigh-Ritz method, a solution is assumed in the form of a linear combination of trial functions, known as **interpolation functions,** multiplied by undetermined coefficients. In the finite element method the coefficients have physical meaning, as they represent "nodal" displacements, where "nodes" are boundary points between finite elements. The computation of the stiffness and mass matrices is carried out for each of the elements separately and then the element stiffness and mass matrices are assembled into global stiffness and mass matrices. One disadvantage of the finite element method is that it requires a large number of degrees of freedom.

To illustrate the method, and for easy visualization, consider the transverse vibration of a string fixed at $x = 0$ and with a spring of stiffness K attached at $x = L$ (Fig. 1.4.27) and divide the length L into n elements of width h, so that $nh = L$. Denote the displacements of the nodal points x_e by a_e and assume that the string displacement is linear between any two nodal points. Figure 1.4.28 shows a typical element e. The process can be simplified greatly by introducing the nondimensional local coordinate $\xi = j - x/h$. Then, considering the two linear interpolation functions

$$\phi_1(\xi) = \xi \qquad \phi_2(\xi) = 1 - \xi \tag{1.4.150}$$

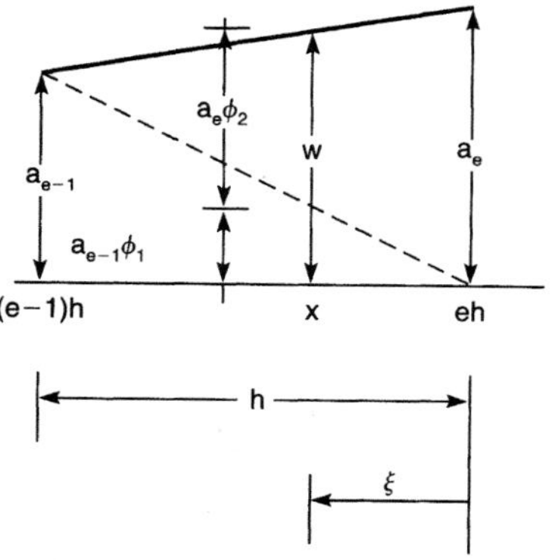

Figure 1.4.28 A two-node element with linear displacement between the nodes.

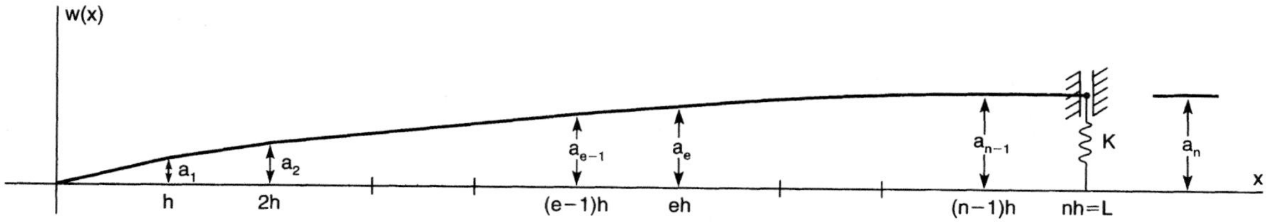

Figure 1.4.27 Transverse vibration of string fixed on one end and attached to a spring K at the opposite end.

the displacement at point ξ can be expressed as

$$\omega(\xi) = a_{e-1}\phi_1(\xi) + a_e\phi_2(\xi) \tag{1.4.151}$$

where a_{e-1} and a_e are the nodal displacements for element e. Using Eq. (1.4.143) and changing variables from x to ξ, we can write the element stiffness and mass coefficients

$$k_{eij} = \frac{1}{h}\int_0^1 EA\frac{d\phi_i}{d\xi}\frac{d\phi_j}{d\xi}d\xi \quad m_{eij} = h\int_0^1 m\phi_i\phi_j\,d\xi, \qquad i, j = 1, 2 \tag{1.4.152}$$

Introducing Eq. (1.4.151) into Eq. (1.4.152) and considering the boundary conditions, we obtain the element stiffness and mass matrices

$$K_1 = \frac{EA}{h} \qquad K_e = \frac{EA}{h}\begin{bmatrix} 1 & -1 \\ -1 & 1 \end{bmatrix} \quad e = 2, 3, \ldots, n-1$$

$$K_n = \frac{EA}{h}\begin{bmatrix} 1 & -1 \\ -1 & Kh/EA \end{bmatrix} \qquad M_1 = \frac{hm}{3}$$

$$M_e = \frac{hm}{6}\begin{bmatrix} 2 & 1 \\ 1 & 2 \end{bmatrix} \qquad e = 2, 3, \ldots, n \tag{1.4.153}$$

where K_1 and M_1 are really scalars, because the left end of the first element is fixed, so that the displacement is zero. Then, since the nodal displacement a_e is shared by elements e and $e+1$ ($e = 1, 2, \ldots, n-2$), the element stiffness and mass matrices can be assembled into the global stiffness and mass matrices

$$K = \frac{EA}{h}\begin{bmatrix} 2 & -1 & 0 & \cdots & 0 & 0 \\ -1 & 2 & -1 & \cdots & 0 & 0 \\ 0 & -1 & 2 & \cdots & 0 & 0 \\ \cdots & \cdots & \cdots & \cdots & \cdots & \cdots \\ 0 & 0 & 0 & \cdots & 2 & -1 \\ 0 & 0 & 0 & \cdots & -1 & Kh/EA \end{bmatrix}$$

$$M = \frac{hm}{6}\begin{bmatrix} 4 & 1 & 0 & \cdots & 0 & 0 \\ 1 & 4 & 1 & \cdots & 0 & 0 \\ 0 & 1 & 4 & \cdots & 0 & 0 \\ \cdots & \cdots & \cdots & \cdots & \cdots & \cdots \\ 0 & 0 & 0 & \cdots & 4 & 1 \\ 0 & 0 & 0 & \cdots & 1 & 2 \end{bmatrix} \tag{1.4.154}$$

For beams in bending, the displacements consist of one translation and one rotation per node; the interpolation functions are the Hermite cubics

$$\phi_1(\xi) = 3\xi^2 - 2\xi^3, \ \phi_2(\xi) = \xi^2 - \xi^3$$
$$\phi_3(\xi) = 1 - 3\xi^2 + 2\xi^3, \ \phi_4(\xi) = -\xi + 2\xi^2 - \xi^3 \tag{1.4.155}$$

and the element stiffness and mass coefficients are

$$k_{eij} = \frac{1}{h^3}\int_0^1 EI\frac{d^2\phi_i}{d\xi^2}\frac{d^2\phi_j}{d\xi^2}d\xi \quad m_{eij} = h\int_0^1 m\phi_i\phi_j d\xi \qquad i, j = 1, 2, 3, 4 \tag{1.4.156}$$

yielding typical element stiffness and mass matrices

$$K_e = \frac{EI}{h^3}\begin{bmatrix} 12 & 6 & -12 & 6 \\ 6 & 4 & -6 & 2 \\ -12 & -6 & 12 & -6 \\ 6 & 2 & -6 & 4 \end{bmatrix}$$

$$M_e = \frac{hm}{420}\begin{bmatrix} 156 & 22 & 54 & -13 \\ 22 & 4 & 13 & -3 \\ 54 & 13 & 156 & -22 \\ -13 & -3 & -22 & 4 \end{bmatrix} \tag{1.4.157}$$

The treatment of two-dimensional problems, such as for membranes and plates, is considerably more complex (see Meirovitch, "Principles and Techniques of Vibration," Prentice-Hall) than for one-dimensional problems.

The various steps involved in the finite element method lend themselves to ready computer programming. There are many computer codes available commercially; one widely used is NASTRAN.

SEMIACTIVE VIBRATION ISOLATION SYSTEMS

The idea of semiactive dampers (or shock absorbers) has been in existence since early 1970s. Introduced by Crosby and Karnopp, semiactive isolation systems have most often been studied and used for vehicle suspension systems. Semiactive dampers draw a small amount of energy to adjust the suspension force, most often through adjusting the damper's resistance to motion. Therefore, semiactive dampers do not add any energy to the system—they only dissipate energy in a manner exactly the same as passive dampers. In contract, active systems use a relatively larger amount of energy to provide the force needed to control the bodies connected across the suspension, commonly according to an optimal control strategy that is independent of the suspension motion.

Semiactive vibration isolation systems are often used to control various rigid and flexible modes of a multibody system. The most common use of such systems is in vehicular primary suspension applications, where they are used to control wheelhop (axle vertical motion) and vehicle-body acceleration. Other applications of such systems are in vehicle secondary suspensions, such as engine, cab, or seat suspensions. Secondary suspensions increase ride comfort by reducing vibration transmission to the operator compartment.

As per Fig. 1.4.5, passive suspensions compromise between controlling the dynamic peak at the resonance frequency and isolating at the higher frequencies (i.e., frequencies higher than the resonance frequency $\times\sqrt{2}$). Larger damping ratios reduce the resonance peak, but they increase transmissibility at the higher frequencies. Conversely, smaller damping ratios decrease transmissibility at the higher frequencies, at the expense of a larger resonance peak. With semiactive and active suspensions, it is possible to improve this compromise. An interesting attribute of semiactive isolation systems is that they are able to isolate at frequencies well below passive suspensions, even though similar to passive suspension they do not add any energy to the system (i.e., they only dissipate energy).

Skyhook Control The original control policy for semiactive vibration isolation systems is known as the "skyhook" policy. It assumes that the system damping can be adjusted between a minimum and maximum level. The two common types of skyhook dampers are on-off skyhook and continuous skyhook (shown in Fig. 1.4.29).

As per Fig. 1.4.30, for semiactive isolators with on-off skyhook control, damping force is switched between a minimum and maximum level by means of a solenoid or a similar type of driver. The switching is performed based on:

$$\begin{aligned} v_1 \times v_{12} \geq 0 &\Rightarrow \quad c_s = c_{\text{maximum}} \\ v_1 \times v_{12} \leq 0 &\Rightarrow \quad c_s = c_{\text{minimum}} \end{aligned} \tag{1.4.158}$$

where v_1 represents the absolute velocity of the suspended body, and v_{12} indicates the relative velocity across the suspension. Equation (1.4.158) implies that if the relative velocity across the damper and the body velocity have the same sign, then a high damping is desirable, as illustrated in Fig. 1.4.30. Otherwise, a minimal amount of damping (theoretically zero) is needed in order to provide a better compromise between the conflicting requirements on the suspension—for instance, ride comfort and vehicle stability (or handling), for vehicle primary suspensions.

For semiactive isolators with continuous skyhook control, the damping force is adjusted in the range between the minimum and maximum levels, as shown in Fig. 1.4.29, according to:

$$\begin{aligned} v_1 \times v_{12} \geq 0 &\qquad c_s = \min(Gv_1, c_{\text{maximum}}) \\ v_1 \times v_{12} < 0 &\qquad c_s = c_{\text{minimum}} \end{aligned} \qquad (1.4.159$$

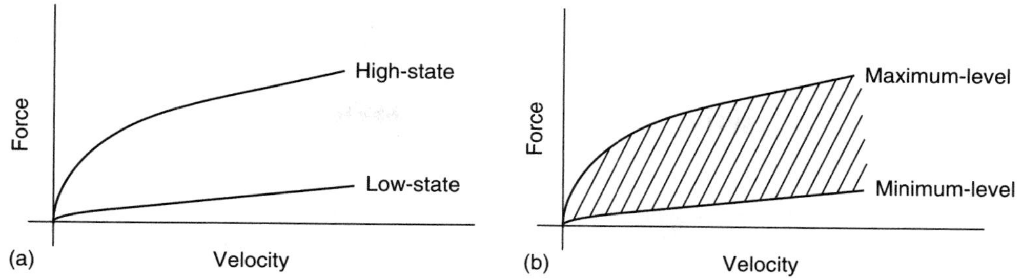

Figure 1.4.29 Suspension damping force according to skyhook control. (*a*) On-off skyhook control; (*b*) continuous skyhook control.

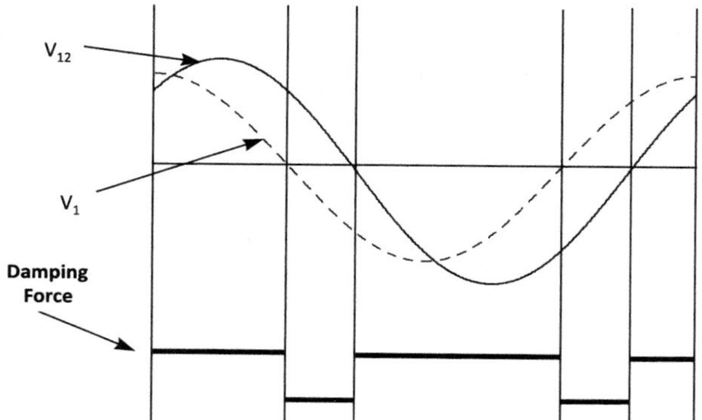

V_1	+	−	−	+	+
V_{12}	+	+	−	−	+
$V_1 \times V_{12}$	"+"	"−"	"+"	"−"	"+"
Damping	*high*	*low*	*high*	*low*	*high*

Figure 1.4.30 Damping force variation for on-off semiactive vibration isolator.

The gain G relates the damping force to the absolute velocity of the suspended body. Equation (1.4.159) attempts to more closely simulate an ideal passive damper that connects the suspended body to an "imaginary sky," as illustrated in Fig. 1.4.31.

Comparing passive and semiactive vibration isolation system, through using the single-degree-of-freedom model in Fig. 1.4.31,

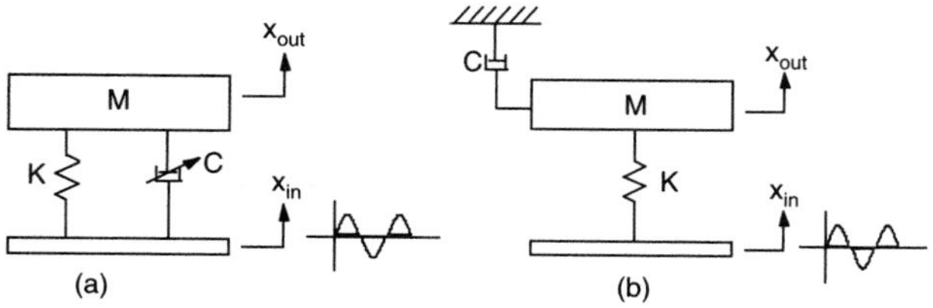

Figure 1.4.31 Base-excited single-degree-of-freedom system: (*a*) with semiactive vibration isolation system; (*b*) with equivalent passive suspension.

indicates base-excited semiactive system is able to provide better isolation across a broad frequency range, as compared with passive systems. More specifically, semiactive systems are able to isolate at frequencies far below the fixed frequency of passive dampers (i.e., $\sqrt{2}\omega_n$), the natural frequency of the suspension.

As indicated in Fig. 1.4.32, isolators can produce isolation only at frequencies larger than $\sqrt{2}\omega_n$, irrespective of the damping ratio. Semiactive suspensions are also passive, in the sense that they can only dissipate energy; in contrast to fully active suspensions that both add and dissipate energy from the system. Yet semiactive suspensions do not follow the same isolation principal as the passive suspensions.

Semiactive Isolation System Transmissibility Let us consider a base-excited system representing a single suspension, as shown in Fig. 1.4.5. The response of the system is shown in Fig. 1.4.33. The response of this system to a harmonic excitation, such as

$$x_{\text{in}} = A\sin(\omega t) \tag{1.4.160}$$

is given by

$$x_{\text{out}} = A\,|H(\omega)|\,\sin(\omega t - \phi)$$

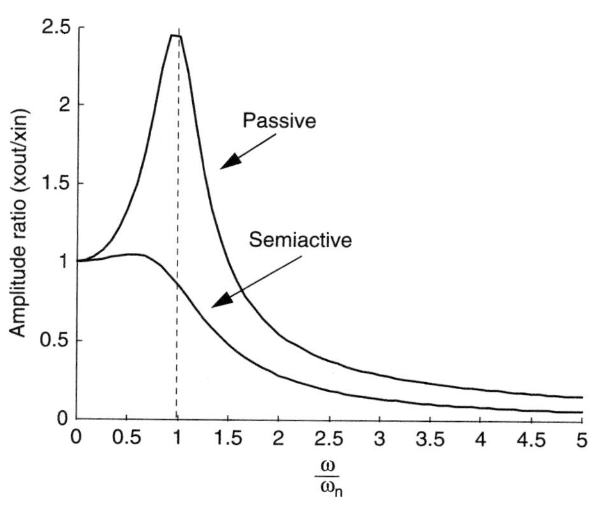

Figure 1.4.32 Simulation transmissibility data for a single suspension.

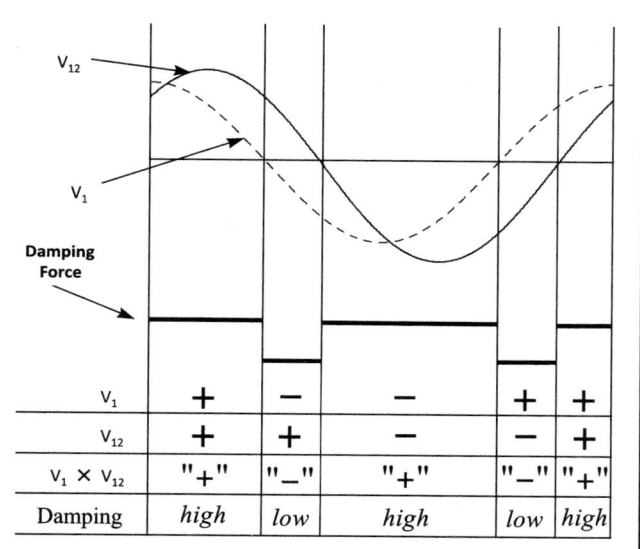

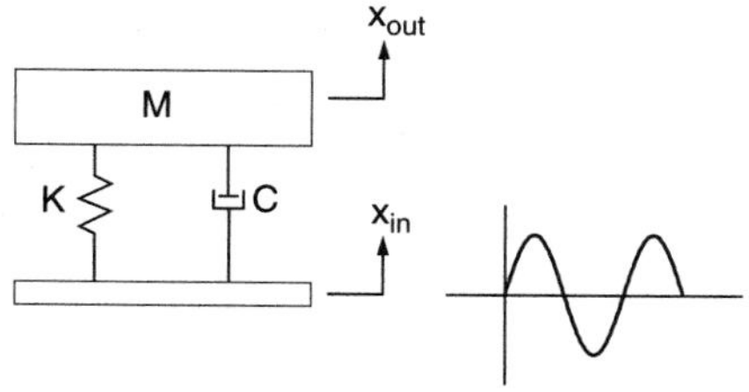

Figure 1.4.33 Single-degree-of-freedom suspension model.

where $|H(\omega)|$ is the same as T in Eq. (1.4.32) and

$$\phi = \tan^{-1}\left(\frac{2\xi\left(\frac{\omega}{\omega_n}\right)^3}{1-\left(\frac{\omega}{\omega_n}\right)^2+\left(2\xi\frac{\omega}{\omega_n}\right)^2}\right) \qquad (1.4.161)$$

The variables $|H(\omega)|$ and ϕ represent the transmissibility amplitude and phase shift between the input and output. Further, the variables ξ and ω_n indicate the damping ratio and natural frequency of the system, respectively.

As shown in Fig. 1.4.5, for all damping ratios the transmissibility plots pass through a fixed point at $\frac{\omega}{\omega_n} > \sqrt{2}$ commonly known as the "fixed" point of the suspension. Increasing damping reduces the resonance amplitude, but it increases transmissibility in the isolation range (i.e., $\frac{\omega}{\omega_n} > \sqrt{2}$). Additionally, as the phase diagram in Fig. 1.4.34 shows, increasing damping contributes to a lower phase difference (i.e., stronger coupling) across the suspension, as expected.

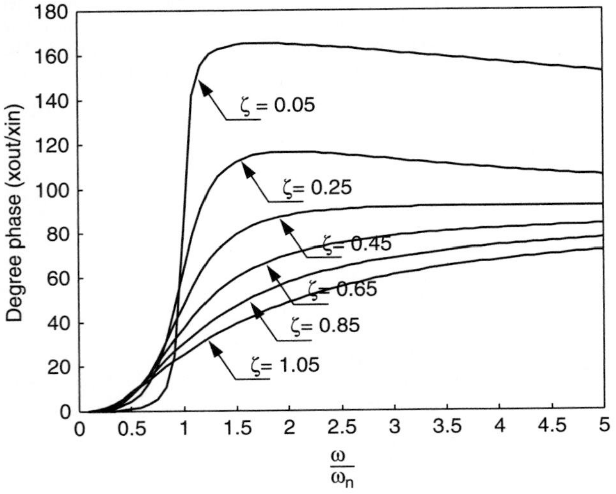

Figure 1.4.34 Transmissibility phase for a passive suspension.

Let us replace the passive damper in Fig. 1.4.33 with a semiactive damper, as shown in Fig. 1.4.35. The transmissibility amplitude and phase for the semiactive suspension in Figs. 1.4.36 and 1.4.37 behave completely different from the passive system in Fig. 1.4.33.

Some of the dissimilarities are:

1. The semiactive system always provides a better isolation than a passive suspension. As the damping ratio increases, the difference between the two systems becomes more obvious. While the passive suspension cannot provide isolation below the fixed frequency of $\sqrt{2}\omega_n$, the semiactive suspension can isolate at frequencies well below that. Isolation refers to the transmissibility ratios of less than one.

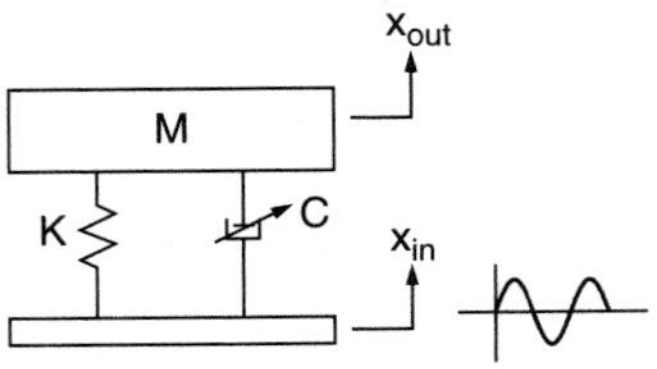

Figure 1.4.35 Base-excited system with semiactive damper.

2. The compromise between resonance control and isolation that is inherent in passive suspensions does not exist for the semiactive suspension. As Fig. 1.4.36 shows, the reduction in the resonance peak does not occur at the cost of reduced isolation, unlike Fig. 1.4.5. In fact, with a sufficiently large damping ratio, one can completely eliminate the resonance peak and actually achieve isolation across the whole frequency spectrum. This feature is useful for many applications, particularly for sensitive machinery that cannot tolerate any overshoot in power-up or power-down, and yet must have good isolation during normal operation. Passive suspensions offer only one of the two aspects, whereas semiactive suspensions offer both.

3. The phase plots 1.4.34 and 1.4.37 further confirm the differences between semiactive and passive isolation. Increasing damping causes far less coupling across the semiactive suspension than for passive suspension. The coupling is judged by the phase angle ϕ. A large (close to π) phase angle indicates that the bodies across the suspension are out

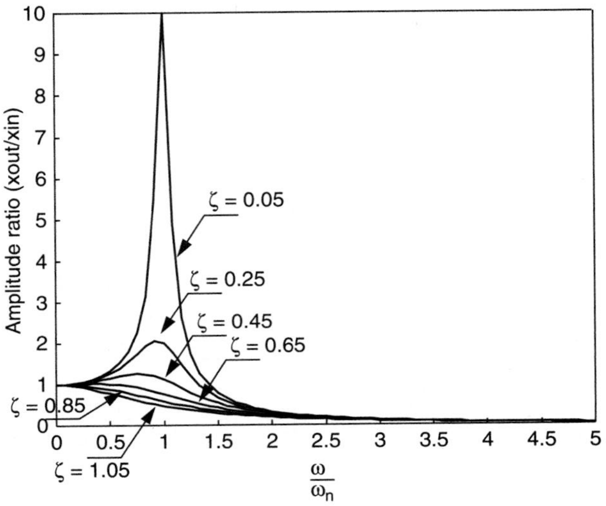

Figure 1.4.36 Transmissibility amplitude for the semiactive suspension.

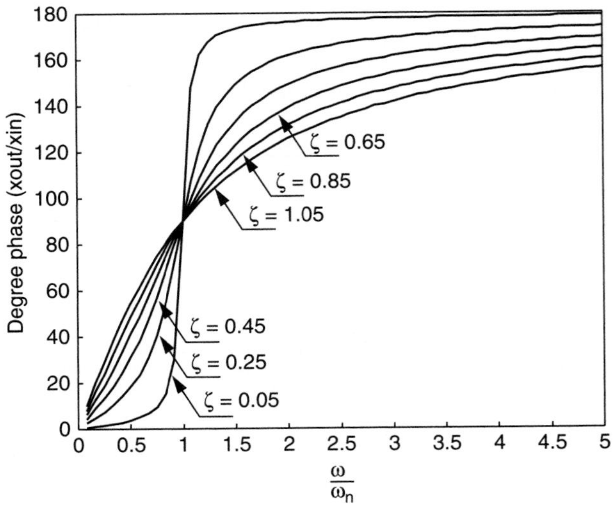

Figure 1.4.37 Transmissibility phase for the semiactive suspension.

of phase, and they are relatively uncoupled. Conversely, a small (close to zero) phase angle means that the bodies connected by the suspension are strongly coupled and move in unison.

Transmissibility Derivation Unlike the passive system in Fig. 1.4.33, it is not possible to derive the transmissibility equation for the semiactive system in Fig. 1.4.35, because the damping coefficient is time-dependent. The damping coefficient changes between a minimum and a maximum level according to a semiactive control policy, such as Eqs. (1.4.158) and (1.4.159).

We consider the passive suspension that is equivalent to the semiactive suspension. Recall that the semiactive control in Eqs. (1.4.158) and (1.4.159) attempt to emulate the optimal configuration of a passive damper, where the damper is connected between the mass and an "imaginary sky," shown in Fig. 1.4.31*b*. Therefore, for a harmonic base excitation, the transmissibility equations for a semiactive suspension are nearly the same as the transmissibility equation for the passive system of Fig. 1.4.31*b*. For such a system, the transmissibility amplitude and phase are given by Eqs. (1.4.24) and (1.4.25), respectively.

Comparing the transmissibility amplitude and phase equations for the passive and semiactive isolation in Fig. 1.4.31 reveals that the transmissibility amplitudes that is, Eqs. (1.4.24) and (1.4.37)—have the same denominators, but different numerators. For passive suspensions, the numerator is a function of the damping ratio ξ, whereas for the semiactive suspension in Eq. (1.4.24) it is a constant. Table 1.4.10 shows the effect of ξ on transmissibility in different frequency ranges for a passive suspension, effectively duplicating the results in transmissibility plot shown in Fig. 1.4.6. The table is derived from evaluating the numerator and denominator of Eq. (1.4.167). This evaluation is relatively straightforward, and is not shown here for brevity. Table 1.4.10 shows that the frequency range in which amplification, direct transmission, and attenuation occur is independent of the damping ratio ξ.

Table 1.4.10 Effect ξ on Transmissibility in Different Frequency Ranges for a Passive Suspension

$\frac{\omega}{\omega_n}$	$\left[1-\left(\frac{\omega}{\omega_n}\right)^2\right]^2$	$\lvert H(\omega)\rvert$	Transmissibility
<1	<1	>1	Amplification
=1	=0	>1	Amplification
$1<\cdot<\sqrt{2}$	<1	>1	Amplification
$=\sqrt{2}$	=1	=1	Direct transmission
$>\sqrt{2}$	>1	<1	Attenuation (or isolation)

Examining Eq. (1.4.24) yields completely different results. The numerator is a constant, and therefore the transmissibility magnitude relative to unity (i.e., direct transmission) depends on whether the denominator is less than 1, greater than 1, or equal to 1. Evaluating the denominator for one of these three possibilities will allow us to make conclusions on the effect of ξ on the transmissibility magnitude. Therefore, let us consider

$$\left[1-\left(\frac{\omega}{\omega_n}\right)^2\right]^2+\left[2\xi\left(\frac{\omega}{\omega_n}\right)\right]^2>1 \qquad (1.4.162)$$

which can be reduced to

$$4\xi^2>2-\left(\frac{\omega}{\omega_n}\right)^2 \qquad (1.4.163)$$

or

$$\left(\frac{\omega}{\omega_n}\right)>\sqrt{2-4\xi^2}\geq 0 \qquad (1.4.164$$

The inequality "≥ 0" is added to Eq. (1.4.164), because it is not possible for $\left(\frac{\omega}{\omega_n}\right)$ to be negative. Equation (1.4.164) proves that, for semiactive suspension, attenuation occurs at lower frequencies as ξ increases, unlike passive suspensions in which attenuation is independent of ξ.

Two special cases of Eq. (1.4.164) are worthwhile examining further. First, the case where $\xi = 0$, which reduces Eq. (1.4.164) to

$$\left(\frac{\omega}{\omega_n}\right)>\sqrt{2} \qquad (1.4.165)$$

This indicates that when no damping is present, then attenuation starts at $\omega=\sqrt{2}\omega_n$, the fixed frequency of the passive suspension. Of course, for $\xi > 0$, attenuation starts at frequencies smaller than $\sqrt{2}\omega_n$. Therefore, Eq. (1.4.165) indicates that for a given ξ, a semiactive suspension always isolates better than a passive suspension.

Next, let us consider the case where

$$2-4\xi^2=0$$

or

$$\xi=\frac{\sqrt{2}}{2}=0.707 \qquad (1.4.166)$$

This is the damping ratio at which the semiactive suspension provides attenuation at *all* frequencies. Therefore, unlike passive suspensions, it is possible to tune semiactive suspensions such that they isolate across the whole frequency spectrum.

The passive representation of the semiactive system in Fig. 1.4.31 assumes that the off-state damping—(C_{off}) in Eqs. (1.4.158) and (1.4.159)—is zero. In practice, however, this is not possible, and for some applications it may not even be desirable. In most cases, C_{off} is a small portion of the on-state damping (C_{on}). Therefore, in reality the passive representation of the semiactive dampers appears as shown in Fig. 1.4.38.

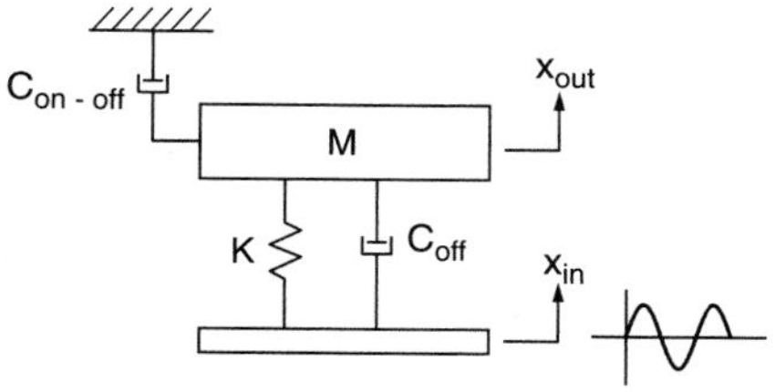

Figure 1.4.38 Actual passive representation of a semiactive suspension.

This modifies the transmissibility equations to

$$|H(\omega)|=\left\{\frac{1+\left[2\xi_{off}\frac{\omega}{\omega_n}\right]^2}{\left[1-\left(\frac{\omega}{\omega_n}\right)^2\right]^2+\left[2\xi_{on}\left(\frac{\omega}{\omega_n}\right)\right]^2}\right\}^{1/2} \qquad (1.4.167)$$

$$\phi=\tan^{-1}\left[\frac{2\left[\xi_{on}\frac{\omega}{\omega_n}-\xi_{off}\frac{\omega}{\omega_n}+\xi_{off}\left(\frac{\omega}{\omega_n}\right)^3\right]}{1-\left(\frac{\omega}{\omega_n}\right)^2+4\xi_{on}\xi_{off}\left(\frac{\omega}{\omega_n}\right)^2}\right] \qquad (1.4.168)$$

where

$$\xi_{on}=\frac{C_{on}}{2m\omega_n} \qquad (1.4.169)$$

$$\xi_{off}=\frac{C_{off}}{2m\omega_n} \qquad (1.4.170)$$

Using Eqs. (1.4.167) and (1.4.168), one can derive a relationship between damping ratios ξ_{on} and ξ_{off}, and the frequencies at which attenuation occurs, similar to Eqs. (1.4.162) to (1.4.164). This yields

$$1+\left[2\xi_{off}\left(\frac{\omega}{\omega_n}\right)\right]^2<\left[1-\left(\frac{\omega}{\omega_n}\right)^2\right]^2+\left[2\xi_{on}\left(\frac{\omega}{\omega_n}\right)\right]^2 \qquad (1.4.171)$$

or

$$4\left(\xi_{on}^2-\xi_{off}^2\right)>2-\left(\frac{\omega}{\omega_n}\right)^2 \qquad (1.4.172)$$

or

$$\left(\frac{\omega}{\omega_n}\right)>\sqrt{2-4\left(\xi_{on}^2-\xi_{off}^2\right)} \qquad (1.4.173)$$

Equation (1.4.173) implies that the frequencies at which attenuation occurs are a function of the difference between ξ_{on}^2 and ξ_{off}^2. As ξ_{on} is increased or ξ_{off} is lowered, attenuation starts at lower frequencies. Two special cases that are worthwhile mentioning are:

$$\xi_{on}=\xi_{off} \qquad (1.4.174)$$

and

$$2-4\left(\xi_{on}^2-\xi_{off}^2\right)=0 \qquad (1.4.175)$$

The first case indicates a passive suspension. The second case implies that for

$$\xi_{on}^2-\xi_{off}^2\geq\frac{1}{2} \qquad (1.4.176)$$

the semiactive suspension provides attenuation at *all* frequencies. Equation (1.4.176) is the same as Eq. (1.4.166) for $\xi_{off}=0$.
Figs. 1.4.39 and 1.4.40 show that increasing C_{off} relative to C_{on} will diminish the isolation effectiveness of semiactive isolation systems. This agrees with intuition, in the sense that as the separation between C_{on} and C_{off} becomes smaller, the semiactive suspension behaves more like a passive suspension.

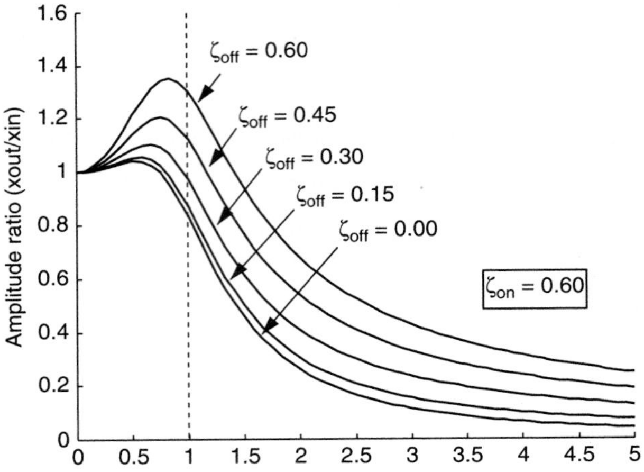

Figure 1.4.39 Transmissibility amplitude for a realistic semiactive suspension.

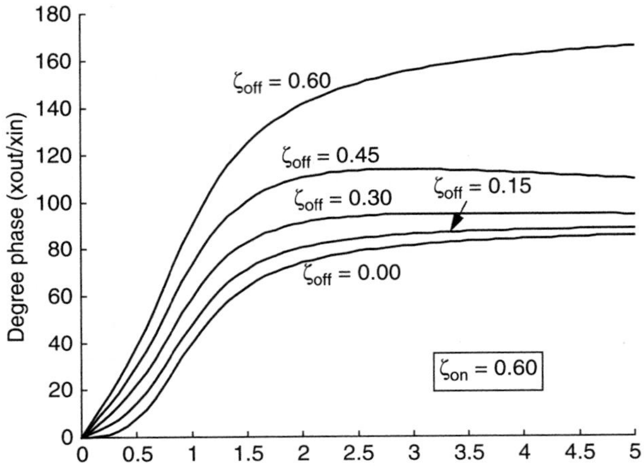

Figure 1.4.40 Transmissibility phase for a realistic semiactive suspension.

Section 2
Heat

BY

FRANCIS A. KULACKI *Professor of Mechanical Engineering, University of Minnesota*
MARK O. MCLINDEN *National Institute of Standards and Technology*
HOYT C. HOTTEL *Late Professor Emeritus, Massachusetts Institute of Technology*
ADEL F. SAROFIM *Late Presidential Professor of Chemical Engineering, University of Utah*
JAMES J. NOBLE *Former Research Associate Professor of Chemical and Biological Engineering, Tufts University*
HAKAN ERTÜRK *Associate Professor of Mechanical Engineering, Boğaziçi University, Turkey*
AFSHIN J. GHAJAR *Regents Professor and John Brammer Professor of Mechanical Engineering, Oklahoma State University*

2.1 THERMODYNAMICS
by Francis A. Kulacki

2.2 THERMODYNAMIC PROPERTIES OF SUBSTANCES
by Mark O. McLinden

2.3 RADIANT HEAT TRANSFER
by Hoyt C. Hottel, Adel F. Sarofim, James J. Noble, and Hakan Ertürk

2.4 HEAT TRANSFER BY CONDUCTION AND CONVECTION
by Afshin J. Ghajar

2.1 THERMODYNAMICS

by Francis A. Kulacki

DEFINITIONS

Dimensions and Units

In thermodynamics, the primary dimensions most frequently used are mass, m, length, L, time, t, and temperature, T. The respective units in the Systѐme International d'Unités (SI) are the kilogram, kg, meter, m, second, s, and Kelvin, K. Corresponding English units are respectively the pound mass, lbm, foot, ft, second, s, and Rankine, R.

Force, mass, length, and time are related by Newton's Second Law of motion, $F \propto ma$, where a is acceleration. An equality is obtained with the introduction of the constant $1/g_c$ (mL/t^2F), $F = ma/g_c$. For example, 1 lbm in the earth's gravitational field, where the acceleration is 32.174 ft/s^2, experiences a force exerted on it defined as the pound force, lbf. This system of units thus gives $g_c = 32.174$ lbm·ft/lbf·s^2 = 1 kg·m/N·s^2. Alternative systems of units give,

$$g_c = 32.174\frac{\text{lbm}\cdot\text{ft}}{\text{lbf}\cdot\text{s}^2} = 1\frac{\text{slug}\cdot\text{ft}}{\text{lbf}\cdot\text{s}^2} = 1\frac{\text{g}\cdot\text{cm}}{\text{dyn}\cdot\text{s}^2} = 1\frac{\text{kg}\cdot\text{m}}{\text{N}\cdot\text{s}^2}$$

The relation which involves weight, w, a gravitational force, and a fixed mass acted upon by a gravitational force, g, and no other forces is obtained from Newton's Second Law. The acceleration is the acceleration of gravity, g. Thus the relation between weight and mass is $w = mg/g_c$. For constant gravitational acceleration, a proportionality is therefore obtained between weight and mass.

Temperature

Temperature, T, is an intensive thermodynamic property. Intensive properties are not dependent on either mass or volume of a system. Thermometric devices (thermometers) are used to establish a temperature scale in accordance with the Zeroth Law of Thermodynamics. The Zeroth Law states that two systems each in thermal equilibrium with a third system must be in thermal equilibrium with each other. Thermometry and the temperature scale are founded on the Zeroth Law. The Celsius (centigrade) and Fahrenheit temperature scales are related by,

$$^\circ\text{C} = \frac{5}{9}(^\circ\text{F} - 32) \qquad ^\circ\text{F} = \frac{9}{5}\,^\circ\text{C} + 32$$

The boiling point of water at 1 atm is assigned the value of 100°C, and this establishes congruity for the two scales.

If the pressure of a constant volume hydrogen thermometer is extrapolated to zero pressure, the corresponding temperatures are –273.15°C and –459.67°F. An absolute temperature scale was formerly used on which zero temperature corresponded to zero pressure on the hydrogen thermometer. The current basis is to define the numerical value T = 0.018°C at the triple point of water where the solid, liquid and vapor states are in thermodynamic equilibrium. The modern absolute temperature scales are,

$$\text{Kelvins (K)} = \text{Degrees Celsius } (^\circ\text{C}) + 273.15$$

$$\text{Degrees Rankine } (^\circ\text{R}) - \text{Degrees Fahrenheit } (^\circ\text{F}) + 459.67$$

Pressure

Pressure is an intensive thermodynamic property. The SI dimensions of pressure are N/m^2. Normal atmospheric pressure is 1.01325 × 10^5 N/m^2. While some use has been made of pressure expressed in kN/m^2, or kPa (1 Pa = 1 N/m^2), and in MN/m^2, or MPa, general technical usage now favors the bar, 1 bar = 10^5 N/m^2 = 10^5 Pa so that 1 atm = 1.01325 bar. For many approximate calculations the atmosphere and the bar can be equated.

Energy

Energy is an extensive thermodynamic property and is generally defined as the ability to do work.

The units of energy most used in engineering practice are the calorie (cal), British thermal unit (Btu), and watt-hour (Wh). The International Steam Table Conference (London, 1929) defines the Steam Table (IT) calorie as the watt-hour. One British thermal unit is defined as 1 Btu 251.996 IT cal = 778.26 ft·lbf, and this establishes the quantitative equivalence of energy and work. Traditionally the Btu has been defined as the energy necessary to raise one pound mass of water one degree Fahrenheit at some arbitrarily chosen temperature. Similarly, the calorie has been defined as the energy required to raise one gram of water one degree Celsius at 15°C (or at 17.5°C). In the SI system, the joule (J) is the unit of energy, and the Newton-meter (N·m) is the work equivalent. The two are equal, so that 1 J = 1 N·m, and the mechanical equivalent of work and energy is unity.

Heat Capacity and Specific Heat

The heat capacity of a material is the energy needed to raise a unit mass one degree in temperature. The ratio of the energy transferred to raise a unit mass one degree to that required to raise a unit mass of water one degree at some specified temperature is denoted the specific heat of the material. For most engineering purposes, heat capacities may be assumed numerically equal to specific heats. Heat capacities for constant pressure, c_P, and constant volume, c_v, processes are defined in terms of the specific enthalpy, $h(T, p)$, and specific internal energy, $u(T,v)$, where v is specific volume, m^3/kg. Specific heats are defined,

$$c_P = \left(\frac{\partial h}{\partial T}\right)_p \qquad c_V = \left(\frac{\partial u}{\partial T}\right)_v$$

Generally, specific heats vary with temperature, and over a range of temperature, the average (mean) values are,

$$c_{p,m} = \frac{1}{T - T_0}\int_{T_0}^{T} c_P\,dT \qquad c_{v,m} = \frac{1}{T - T_0}\int_{T_0}^{T} c_v\,dT$$

where T_0 is a suitably chosen reference temperature. For a moderate temperature range, a constant mean value of specific heat can be used. In that case the energy required to raise a given mass, m, to the temperature T is $mc(T - T_0)$, where no distinction here is made for a constant pressure or constant volume process. If the specific heat is either a strong function of temperature or the temperature range is large, the energy required is $m\int_{T_0}^{T} c(T)\,dT$, where $c(T)$ is a given function obtained either empirically or by theory. Data for the specific heats of some solids, liquids, and gases are found in Tables 2.2.19, 2.2.20, 2.2.24, and 2.2.26.

For elements near room temperature, specific heat may be approximated by the Dulong-Petit law, which states that the specific heat at constant volume approaches $3R$, where R is the universal gas constant (3R ~25 J/mol·K). At low temperatures, Debye's theory for crystalline solids employing the concept that lattice vibrations correspond to phonons in a box leads to the equation,

$$\frac{c_v}{R} = 3\left(\frac{T}{\theta}\right)\int_0^{\Theta/T} \frac{\xi^4 e^\xi}{(e^\xi - 1)^2}\,d\xi$$

where θ is the Debye temperature and the integral on the l.h.s. is the Debye function. (Debye also derived this result using classical continuum mechanics and the theory of the harmonic oscillator.) The Debye temperatures for several crystalline solids are listed in Table 2.1.1, and Fig. 2.1.1 shows the variation of the specific heat at constant volume versus T/θ.

Table 2.1.1 Debye Temperatures in Degrees Kelvin for Selected Crystalline Solids

Aluminum	428	Nickel	450
Carbon	2230	Platinum	240
Copper	343.5	Silver	215
Iron	470	Titanium	420
Lead	105	Tungsten	400

SOURCE: *Kittel C* (2004). *Introduction to Solid State Physics, 8e. John Wiley & Sons, New York.*

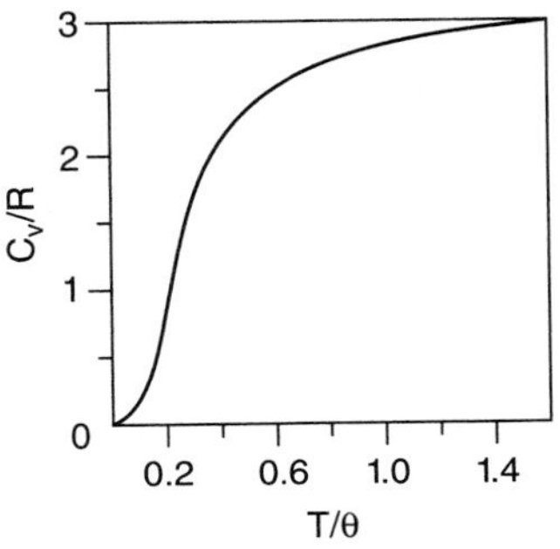

Figure 2.1.1 Specific heat at constant volume according to Debye's theory.

Specific heats of solids used in engineering practice are generally determined by measurement. Values of specific heat for a wide range of solids are given in Table 2.2.24.

Owing to the wide range of liquids employed in engineering practice, measured data sets are the best resource for values of specific heat. Generally, the specific heat of a liquid is temperature dependent. Table 2.2.19 provides an additional resource.

Table 2.1.2 lists specific heats obtained from the kinetic theory of gases for perfect gases, i.e., gases comprising point masses for molecules and exhibiting no intermolecular forces. Specific heats for such gases are determined by the degrees of freedom allowed at the molecular level. Determination of the effective number of degrees of freedom limits extending the theory to more complex gases.

In Table 2.1.2, R is the universal gas constant, $R = 8.3145$ (J/mol K), and the specific heats are expressed as molar specific heats, J/mol K. Specific heats may be conveniently expressed in terms of the ratio $k = c_P/c_V$,

$$c_V = \frac{R}{k-1}, \quad c_P = Rk(k-1)$$

The gas constant for specific gases is $R_{\text{specific}} = MR$, where M is the molecular weight. For dry air (rounded values) $M_{\text{air}} \approx 29$ (kg/kg–mol), and $R_{\text{specific}} = 287.06$ (J/kg·K) = 53.353 (ft-lbf/lbm·R).

Properties of gases are generally available on a molar (mol), the kg-mol or lbm-mol basis. On a molar basis, 1 lbm-mol of oxygen is 32 lbm. At the same pressure and temperature, the volume of 1 mol is the same for all perfect gases, i.e., following the gas laws. Avogadro's principle states that 1 mol of any perfect gas contains the same number of molecules, 6.02214×10^{26} molecules/kg-mol, which is denoted as Avogadro's number.

Table 2.1.2 Molar Specific Heats of Gases via Kinetic Theory

Gas	c_P/R	c_V/R	c_P/c_V
Monatomic	5/2	3/2	5/3
Diatomic	7/2	5/2	7/5
N degrees of freedom	(n + 2)/2	n/2	1 + 2n

The specific heat of a mixture of several masses of total mass m with mass fractions $x_i = m_i/m$, is given by $c_{\text{mixture}} = \Sigma x_i c_i$, where c_i is the specific heat of the constituent. The temperature of the mixture in the absence of chemical reaction is, $T_{\text{mixture}} = \Sigma x_i c_i T_i$.

For aqueous solutions of salts, the specific heat may be estimated by assuming the specific heat of the solution equal to that of the water. Thus, for a 20 percent by mass solution of sodium chloride in water, the specific heat would be ~0.8. Although approximate calculations of mixture properties often consist simply of multiplying the molar fraction of each constituent by the property of each constituent, more accurate calculations are in some cases necessary.

For a pure substance, the energy associated with a change in state of a unit mass at constant pressure is denoted the latent heat because no temperature change occurs. The latent heats of fusion, vaporization, sublimation, and change in crystalline form are familiar examples and of practical value. Values of latent heat for the heat of fusion and the latent heat of vaporization are presented in Table 2.2.18.

References

1. Benedict RP. "Fundamentals of Temperature, Pressure and Flow Measurements," John Wiley & Sons, New York, 2009. doi:10.1002/9780470172698

2. Cezairliyan A, Maglic KD, Peletsky VE, eds. "Compendium of Thermophysical Property Measurement Methods," vol. 1: Survey of Measurement Techniques, Springer, New York, 2009.

3. Green DW, Perry RH, eds. "Perry's Chemical Engineers' Handbook," 8e, McGraw-Hill, New York, 2007.

4. Kittel C. "Introduction to Solid State Physics," 8e, John Wiley & Sons, New York, 2004.

5. "ASHRAE handbook." American Society of Heat Refrigeration and Air Conditioning Engineers, Atlanta, 2016, http://www.ashrae.org/

6. "Metrologia," 2016, http://iopscience.iop.org/journal/0026-1394

7. "NIST Chemistry Webbook." National Institute of Standards and Technology, 2016, http://webbook.nist.gov, http://webbook.nist.gov/chemistry/fluid

8. Thermophysical Properties Laboratory, Inc. Purdue University, 2016, http://www.tprl.com/; http://cindasdata.com

9. Thermodynamics Research Center of the National Institute of Standards and Technology, 2016, http://trc.nist.gov/

PRINCIPLES OF MACROSCOPIC THERMODYNAMICS

Fundamental Concepts

Macroscopic thermodynamics deals with systems in equilibrium states and with processes that produce changes of state. A process that returns a system to its original equilibrium state is denoted as a cycle. Equilibrium states are described by time-independent values of the relevant thermodynamic properties, or state variables. An equilibrium state implies constancy of the state variables, and in the broadest sense thermal, chemical, and mechanical equilibrium. No assumptions or descriptions of the structure of matter at the molecular, or atomic, level are necessary to describe an equilibrium state in macroscopic thermodynamics.

System, Surroundings, and Universe A system is defined by either a given region of space or a quantity (mass) of matter contained within a boundary that separates it from the surroundings. The system plus the surroundings constitute the universe that is subject to thermodynamic analysis. Systems can be either closed such that no mass crosses its boundary or open such that mass and energy can cross the boundary at either specified points or areas. An isolated system restricts both energy and mass flows across its boundary and thus has no energetic communication with the surroundings. The molecular nature of matter is not considered in describing a system and its properties, and therefore macroscopic thermodynamics is the appropriate descriptive term. The use of the term thermodynamics in the present context of engineering principles and applications implies the macroscopic assumption.

Thermodynamic properties are used to describe the state of a system, and conversely the state of a system is determined by its properties. Extensive properties are dependent on the system mass or volume. Intensive properties are independent of system mass or volume. Unless otherwise noted, the change in any thermodynamic property, Ψ, is denoted as the final state value minus the initial state value, e.g., $\Delta\Psi_{12} = \Psi_2 - \Psi_1$. Subscripts are used to denote a given thermodynamic state, e.g., Ψ_i, for any property. Specific quantities are denoted by lower case symbols, e.g., for internal energy per unit mass u (J/kg), and upper case symbols denote total quantities, e.g., internal energy U (J) for a given mass, m, of a substance.

Processes and Cycles Thermodynamic processes produce changes in equilibrium states via energetic interactions between the system and the surroundings. Energetic interactions can take the form of work transfer and heat across the system boundary. For open systems, work transfer and heat can be accompanied by energy transfer with mass that crosses the system boundary, e.g., fluid flow with a given energy content. Once an energetic interaction occurs, the equilibrium state of the system is altered and a new equilibrium state is produced. Work transfers can include compression of the system boundary, electrical energy transfer, and transfer of shaft work produced by rotating machinery.

Thermodynamic processes can be either reversible or irreversible. A reversible process is one in which both the system and the surroundings may be returned to their original states. After an irreversible process takes place, this is not possible, and there is a permanent change in either the system or surroundings. For example, processes involving friction, an unbalanced potential, or the sudden expansion of a gas are irreversible. No loss in the ability of a system to do work is suffered because of a reversible process, but there is always a loss in ability to do work because of an irreversible process. All actual processes are irreversible.

A series of processes that starts and finishes with the system in the same state is called a cycle. Cyclic processes can be either reversible or irreversible.

Heat The general term applied to energy that is transferred due only to a temperature difference is denoted as heat. Temperature, T, is an intensive thermodynamic property, but heat (or heat transfer as is commonly used) implies thermal disequilibrium wherein a difference in temperature provides the motive force for the energy transfer. The energy transferred to or from a system in the form of heat results in a change in the total energy of the system.

The State Principle Equilibrium states are described by thermodynamic properties uniquely related to the number of relevant interactions between the system and its surroundings. This is termed the State Principle, and it is empirically based. The State Principle asserts: (1) there is one independent thermodynamic property for each way that the total energy of the system can be varied independently, and (2) the number of independent properties that describes the equilibrium state is the number of relevant energetic interactions plus one. For example, a compressible substance in a closed system can have its equilibrium state altered by either work or heat transfer at the boundary, and thus the number of state variables needed is three.

First and Second Laws

First Law The First Law of thermodynamics states that all adiabatic processes produce the same change in the total energy of a closed system. The total energy includes the internal energy, U, the potential energy, PE, and the kinetic energy, KE, and thus $E \equiv U + PE + KE$. An adiabatic process involving a work transfer across the system boundary is one for which there is no heat interaction between the system and surroundings. Therefore adiabatic work, W_{adb}, defines the energy of the system as a thermodynamic property describing the change of the equilibrium state. Thus $W_{adb} = \Delta E$.

The sign convention for work transfer is that it is positive for work done by the system and negative for work done on the system, i.e., $W_{adb,by} = -\Delta E$. The First Law is empirically based and has never been refuted within the realm of macroscopic thermodynamics.

The First Law has several important consequences: (1) the energy of the system, E, is a thermodynamic property defined by a process that results in a unique change in the equilibrium state of the system; (2) internal energy, U, potential energy, PE, and kinetic energy, KE, are thermodynamic properties and thus describe the equilibrium state of the system; (3) the energy of the system is conserved for adiabatic processes when internal, potential, and kinetic energy changes are accounted for; (4) non-adiabatic work transfers between two equilibrium states define a process-dependent quantity denoted as heat, q; and (5) energy changes due to nuclear reactions are excluded, and thus mass conservation is maintained during state changes.

The difference between adiabatic work and non-adiabatic work done by a system defines the transfer of energy across the system boundary as heat, q (J). Non-adiabatic work is dependent on the process by which energy as heat crosses the system boundary, and thus heat is not a thermodynamic property. This principle is expressed,

$$-W_{by,12} + q_{12} = E_2 - E_1$$

where $W_{by,12}$ and q_{12} are the work and heat transfers between the initial and final equilibrium states. This relation defines heat in terms of energy quantities, and the appearance of the process-dependent work transfer implies that the heat transfer is also process dependent. For a differential process between two closely associated equilibrium states, this equation is rewritten in differential form,

$$-\delta W_{by} + \delta q = dE$$

where δW_{by} and δq denote process-dependent quantities. Both of these forms of the First Law imply finite quantities for a given process. They can also be written as rate equations for particular applications.

If a closed system undergoes a cyclic process in which there is non-adiabatic work transfer across the system boundary via a succession of equilibrium states, i.e., a succession of near equilibrium states constituting a quasi-static processes forming the cycle, the cyclic integral applied to the above equation produces the equivalence between cyclic work and cyclic heat transfer,

$$\oint \delta q = \oint \delta W_{by}$$

The quasi-static process is necessarily slow and not of practical value. It serves as conceptual framework and provides a mathematical construct.

Second Law The Second Law of thermodynamics is motivated by the observation that there is a natural tendency for an isolated (closed) system to seek an equilibrium state and neither naturally nor spontaneously return to its initial state without the action of either an external agent or auxiliary system, e.g., an external work or heat transfer to the system. This observation was expressed by W. F. Sears in the following way: *Everyone realizes that the reverse process cannot happen. But why not? The total energy of [the isolated] system would remain constant in the reversed process as it did in the original, and there would be no violation of the first law. There must therefore be some other natural principle, in addition to the first law and not derivable from it, which determines the direction in which a process can take place in an isolated system.* This observation of the impossibility of the reversibility of natural processes for an isolated system leads to two equivalent statements of the Second Law.

Clausius Statement It is impossible for any system to operate in such a way that the sole result would be an energy transfer by heat from a cooler to a hotter body. The aforementioned Zeroth Law is developed from this principle, and heat transfer from a high to low temperature is the everyday experience.

Kelvin-Planck Statement It is impossible for any system to operate in a thermodynamic cycle (a cyclic heat engine) and deliver a net amount of work to its surroundings while receiving energy as heat transferred from a single thermal reservoir. A thermal reservoir is defined as an energy source for which the temperature remains constant while energy in the form of heat is transferred to another system. The Kelvin-Plank statement of the First Law implies that two thermal reservoirs are necessary for the operation of all work-producing heat engines.

The two statements of the Second Law can be shown to be equivalent, but neither can be proved independently. A violation of one produces a violation of the other. The Second Law has never been refuted experimentally within the realm of macroscopic thermodynamics. It applies equally well to work neutral (e.g., an expander) and work producing and work consuming cyclic systems. Moreover, the Second Law denies the possibility of perpetual motion machines.

The important consequences of the Second Law are: (1) two thermal reservoirs are necessary for the operation of all engines; (2) no cycle can be more efficient than a reversible cycle operating between given temperature limits (the. efficiency of such a cycle denoted the Carnot efficiency); and (3) the thermal efficiency of all reversible cycles absorbing heat at a single high temperature, T_H, and rejecting heat at a single low temperature, T_C, must be the same.

For a reversible work producing cycle where the heat transfer at a high operating temperature, T_H, is q_H and a portion of it is converted to work, $W = q_H - q_C$, where q_C is the heat rejected at the low operating temperature, T_C. The Carnot efficiency is given by,

$$\eta = \frac{W}{q_H} = \frac{T_H - T_C}{T_H}$$

The Carnot efficiency represents the highest possible efficiency of work producing engines because all practical power cycles are irreversible, and it depends only on the temperatures of the high and low temperature reservoirs. From the expressions for work and efficiency,

$$\frac{q_H}{T_H} = \frac{q_C}{T_C}$$

where the absolute values of q are implied in this equation. This last expression is the basis of The Thermodynamic Temperature Scale.

Entropy

Entropy is a thermodynamic property defined within the context of the Second Law. For reversible cyclical processes in which the temperature varies during heat absorption and rejection, i.e., for any reversible cycle,

$$\int_{\substack{\text{reversible}\\ \text{cycle}}} \frac{dq_{\text{rev}}}{T} = 0$$

where dq_{rev} is the reversible heat transfer to the system from thermal reservoirs at temperatures T along the closed cycle. This expression is denoted the Clausius Equality, and implies that all reversible paths can comprise the cycle. For all cycles,

$$\oint \frac{\delta q}{T} \le 0$$

which is denoted the Clausius Theorem and applies for all cycles, reversible and irreversible.

For any two reversible processes forming a cycle, the integral of dq_{rev}/T has the same value, and thus defines the change of a state variable denoted as entropy, S,

$$\int_1^2 \frac{dq_{\text{rev}}}{T} = S_2 - S_1$$

where T denotes the reservoir temperatures along the process path. Entropy is therefore a thermodynamic state variable, or property, and can be included in the description of an equilibrium state under the mentioned State Principle. The quantity dq_{rev}/T is thus mathematically a perfect differential, dS, which is denoted as entropy (J/K).

All actual processes are irreversible and therefore occur with an increase of entropy and a decrease in the amount of energy available for doing work, i.e., with an increase in unavailable energy. The net (or total) entropy change for a process comprises that of the system and surroundings,

$$\Delta S_{\text{net}} = \Delta S_{\text{system}} + \Delta S_{\text{surroundings}}$$

This is the increase of entropy principle and applies to all processes and cycles. As a result of the entropy increase, the increase in unavailable energy for the production of useful work is the product of the lowest available temperature, T_0, for heat rejection to the surroundings and ΔS_{net}, i.e., the increase in unavailable energy, $T_0 \Delta S_{\text{net}}$. Any spontaneous process will proceed in such a direction so as to produce a net increase in entropy.

The Clausius Theorem motivates the concept of entropy production. When a closed system undergoes an irreversible process, the increase of entropy principle in concert with the Clausius Theorem provides the basis for the introduction of entropy production, σ_s,

$$S_2 - S_1 = \oiint_{\text{boundary}} \frac{\delta q}{T} + \sigma_s$$

T-dS Equations

When a closed system undergoes a reversible process without a change in both kinetic and potential energies,

$$\Delta U = \delta q_{\text{rev}} + \delta W_{\text{rev}}$$

For systems comprising a single compressible substance (simple chemical systems), the equilibrium state can be described in terms of temperature, T, pressure, p, and volume, V, and the First Law can be written,

$$TdS = dU + pdV$$

The state property enthalpy, $H \equiv U + pV$, is next introduced, and $dH = dU + pdV + Vdp$. With substitution and rearrangement the second T-dS equation is written,

$$TdS = dH - Vdp$$

The T-dS equations now are in a form that contains only state variables, and thus they apply to both reversible and irreversible processes. Moreover, the calculation of entropy is facilitated by either equation,

$$S = S_0 + \int_{U_0} \frac{dU}{T} + \int_{V_0} \frac{p}{T} dV$$

$$S = S_0 + \int_{H_0} \frac{dH}{T} - \int_{p_0} \frac{V}{T} dp$$

where the reference state is denoted by the subscript 0. Evaluation of the integrals in either equation is obtained from the state equations, $U = U(T,V)$ and $H = H(T,p)$.

Maxwell Relations

Three thermodynamic potentials for a system comprising a single component can be defined in terms of the specific state properties u, T, p, v, and s. Specific enthalpy is defined $h = u - Ts$. The specific Helmholtz function, or free energy, is defined $f = u - Ts$ (also written $a = u - Ts$). The specific Gibbs function, or either free enthalpy or Gibbs free energy, is defined $g = h - Ts$. Similar expressions are written for these potentials in terms of the respective extensive properties. Each of these potentials can be considered a state variable, or property, as it comprises only properties in its definition.

The Maxwell relations arise owing to the above definitions, the First Law, the T-dS equations, and the rules of calculus. For a point function $z = z(x,y)$, the total derivative is $dz = Mdx + Ndy$, here $M = (\partial z/\partial x)_y$ and $N = (\partial z/\partial y)_x$. It follows that $\partial N/\partial x = \partial M/\partial y$, which is the Euler criterion for integrability. Several thermodynamic relations can be thus derived from the functions and differentials for u, h, f, and g and are listed in Table 2.1.3.

From the third column of the bottom half of the table, by equating various terms which are equal,

$$\left(\frac{\partial u}{\partial s}\right)_v = \left(\frac{\partial h}{\partial s}\right)_p \qquad \left(\frac{\partial u}{\partial v}\right)_s = \left(\frac{\partial f}{\partial v}\right)_T$$

$$\left(\frac{\partial h}{\partial p}\right)_s = \left(\frac{\partial g}{\partial p}\right)_T \qquad \left(\frac{\partial g}{\partial T}\right)_p = \left(\frac{\partial f}{\partial T}\right)_v$$

Table 2.1.3 The Maxell Relations and Subsidiary Relations

Function	Differential	Maxwell relation
$\Delta u = q + W$	$du = T\,ds - p\,dv$	$\left(\frac{\partial T}{\partial v}\right)_s = -\left(\frac{\partial p}{\partial s}\right)_v$
$h = u + pv$	$dh = T\,ds + v\,dp$	$\left(\frac{\partial T}{\partial p}\right)_s = \left(\frac{\partial v}{\partial s}\right)_p$
$f = u - Ts$	$df = -s\,dT - p\,dv$	$\left(\frac{\partial s}{\partial v}\right)_T = \left(\frac{\partial p}{\partial T}\right)_v$
$g = h - Ts$	$dg = -s\,dT + v\,dp$	$\left(\frac{\partial s}{\partial p}\right)_T = -\left(\frac{\partial v}{\partial T}\right)_p$

By holding certain variables constant, a second set of relations is obtained:

Differential	Independent variable held constant	Relation
$du = T\,ds - p\,dv$	s	$\left(\frac{\partial u}{\partial v}\right)_s = -p$
	v	$\left(\frac{\partial u}{\partial s}\right)_v = T$
$dh = T\,ds + v\,dp$	s	$\left(\frac{\partial h}{\partial p}\right)_s = v$
	p	$\left(\frac{\partial h}{\partial s}\right)_p = T$
$df = -s\,dT - p\,dv$	T	$\left(\frac{\partial f}{\partial v}\right)_T = -p$
	v	$\left(\frac{\partial f}{\partial T}\right)_v = -s$
$dg = -s\,dT + v\,dp$	T	$\left(\frac{\partial g}{\partial p}\right)_T = v$
	p	$\left(\frac{\partial g}{\partial T}\right)_p = -s$

With further mathematical manipulation, the following important relations are obtained,

$$c_v = \left(\frac{\partial q}{\partial T}\right)_v = T\left(\frac{\partial s}{\partial T}\right)_v = \left(\frac{\partial u}{\partial T}\right)_v$$

$$c_p = \left(\frac{\partial q}{\partial T}\right)_p = T\left(\frac{\partial s}{\partial T}\right)_p = \left(\frac{\partial u}{\partial T}\right)_p$$

$$c_p - c_v = T\left(\frac{\partial v}{\partial T}\right)_p\left(\frac{\partial p}{\partial T}\right)_v$$

$$\left(\frac{\partial c_v}{\partial v}\right)_T = \left(\frac{\partial^2 p}{\partial T^2}\right)_v \qquad \left(\frac{\partial c_p}{\partial T^2}\right)_p = -T\left(\frac{\partial^2 v}{\partial T^2}\right)_p$$

Additional relations involving q, u, h, and s are,

$$dq = c_v dT + T\left(\frac{\partial p}{\partial T}\right)_v dv = c_p dT - T\left(\frac{\partial v}{\partial T}\right)_p dp$$

$$du = c_v dT + \left[T\left(\frac{\partial p}{\partial T}\right)_v - p\right]dv$$

$$dh = c_p dT - \left[T\left(\frac{\partial v}{\partial T}\right)_p - v\right]dp$$

$$ds = c_v\frac{dT}{T} + \left(\frac{\partial p}{\partial T}\right)_v dv = c_p\frac{dT}{T} - \left(\frac{v}{\partial T}\right)_p dp$$

Since $\delta q + \delta w = du$ and $h = u + pv$, for reversible processes,

$$du = Tds - pdv \quad \text{and} \quad dh = du + pdv + vdp$$

It follows that,

$$v = T\left(\frac{\partial s}{\partial p}\right)_T + \left(\frac{\partial h}{\partial p}\right)_T$$

Substituting the Maxwell relation $(\partial v/\partial T)_p = -(\partial s/\partial p)_T$,

$$\left(\frac{\partial h}{\partial p}\right)_T = v - T\left(\frac{\partial v}{\partial T}\right)_p$$

Similarly,

$$\left(\frac{\partial u}{\partial v}\right)_T = -\left[p - T\left(\frac{\partial p}{\partial T}\right)_v\right]$$

The last two equations for specific enthalpy and specific internal energy in terms of p, v, and T must hold for any system. An equation in p, v, and T for the properties of a substance is called an equation of state. These two equations are the thermodynamic equations of state and are applicable to either any substance or single component system. A system comprising a compressible, single component is termed as a simple chemical system.

Phase Change

First Order Phase Change The familiar processes of evaporation, sublimation, and fusion take place at constant pressure and temperature. From Table 2.1.3 it is seen that for these processes the Gibbs function remains constant, i.e., $dg = 0$. Consequently the molar Gibbs functions (denoted by over bars) for the saturated liquid, $\bar{g}_{liq}$, vapor, $\bar{g}_{vap}$, and solid, $\bar{g}_s$, define the vaporization, sublimation and fusion *curves,* e.g., $\bar{g}_{liq} = \bar{g}_{vap}$ etc. At the triple point where the liquid, vapor, and solid are in equilibrium, $\bar{g}_{liq} = \bar{g}_{vap} = \bar{g}_s$. The molar Gibbs function can be regarded as functions of T and p, and equilibrium fixes the temperature and pressure at the triple point.

A first order phase change gives rise to the concept of latent heat described equivalently by either a change of entropy and volume or discontinuous changes in the derivatives of the Gibbs function. The first *T-dS* equation for a substance undergoing a reversible, constant pressure phase transition is,

$$Tds = c_V dT + T\left(\frac{\partial p}{\partial T}\right)_v dv$$

where $(\partial p/\partial T)_v$ is independent of volume. Upon integration over the phase change process,

$$T(s_2 - s_1) = T\frac{dp}{dT}(v_2 - v_1)$$

and upon rearrangement,

$$\frac{dp}{dT} = \frac{\lambda}{T(v_2 - v_1)}$$

The latent heat is defined as $\lambda = T(s_2 - s_1)$ (J/kg) when specific volume, v, is expressed in m^3/kg. This equation is the Clapeyron equation and applies to any first order phase change. The slope of the phase transition curve is determined by the change in the molar volumes of the phases.

The Clapeyron equation is used to determine the volume change of a substance during a phase change. For evaporation from the fluid to gas phase, $\lambda \equiv h_{fg}$, $v_2 - v_1 \equiv v_{fg}$, and upon rearrangement,

$$v_{fg} = \frac{h_{fg}}{T\left(dp/dT\right)}$$

Second Order Phase Change A second order phase change takes place at constant temperature and pressure with no entropy and no change in volume. Second order phase change is characterized by discontinuities in specific heat, c_P, thermal conductivity, k (W/m·K), and coefficient of thermal expansion, β (K^{-1}). Discontinuities occur in the second derivatives of the Gibbs function. Second order phase changes are seen in ferromagnetic materials, superconductors, and super fluids.

References

1. Çengel Y, Boles MA. "Thermodynamics: An Engineering Approach," 5e, McGraw-Hill, New York, 2006.
2. Hall NA, Ibele WE. "Engineering Thermodynamics," Prentice-Hall, Englewood Cliffs, 1960.
3. Kestin J. "A Course in Thermodynamics," Blaisdell Publishing Company, Waltham, 1966.
4. Kestin J. "A Course in Thermodynamics," vol. 2, Blaisdell Publishing Company, Waltham, 1968.
5. Moran MJ, Shaprio HN. "Fundamentals of Engineering Thermodynamics," 5e, John Wiley & Sons, New York, 2006.
6. Sears WF. "Thermodynamics," Addison Wesley, Reading, 1953, p. 110.

OPEN SYSTEM ANALYSIS

First Law Analysis

Steady Flow Systems Open systems are widely encountered in engineering practice. Figure 2.1.2 illustrates an open system, *CV*, with work transfer and heat transfer across its boundary. The equilibrium state for the fluid is assumed to be described by the properties p, v, and T (a simple chemical substance). In the case of one steady inlet and one steady outlet flow, the First Law and mass continuity are respectively,

$$\dot{q} - \dot{W} = \dot{m}\left[(u_2 - u_1) + \frac{\bar{V}_2^2 - \bar{V}_1^2}{2g_c} + \frac{g}{g_c}(z_2 - z_1)\right]$$

$$\dot{m}_1 - \dot{m}_2 = 0$$

where $\dot{m} = \rho A\bar{V}$, ρ is the fluid density, $\bar{V}$ is the mass average velocity, A is the area through which mass flows, and z is the elevation with respect to a reference elevation. Magnetic, electrical, and surface tension effects are not included. The above equation indicates that the internal energy, U, is constant as all terms on the r.h.s. are evaluated the boundary. All terms in the energy equation are in units of Watts (J/s).

The work term includes the shaft work, or net work, and the work required to move the fluid into and out of the system. The latter is denoted as flow work, and for the above equation is $p_1A_1\bar{V}_1 - p_2A_2\bar{V}_2$. Upon substitution and noting that $pA\bar{V} = \dot{m}pv$, the energy balance becomes

$$\dot{q} - \dot{W}_{shaft} = \dot{m}\left[(h_2 - h_1) + \frac{V_2^2 - V_1^2}{2g_c} + \frac{g}{g_c}(z_2 - z_1)\right]$$

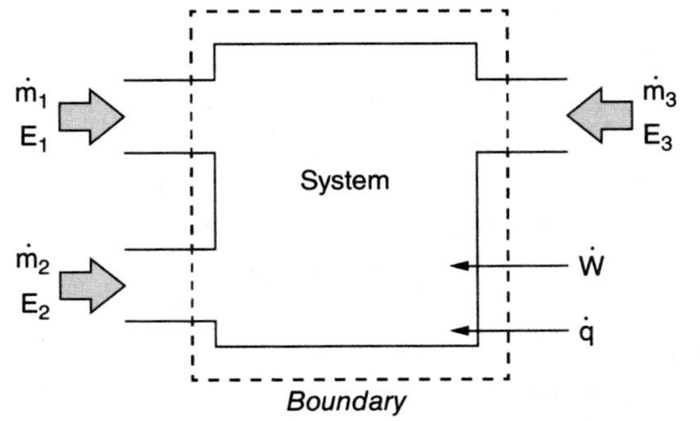

Figure 2.1.2 An open system with mass, heat, and work transfer across the system boundary. The control volume is shown with the dashed line.

The flow work is thus included in the enthalpy terms which are evaluated at the equilibrium states specified at the boundary. The work and heat transfers are evaluated over the boundary where they take place.

Joule-Thompson Flow When a fluid passes adiabatically through a conduit without doing any work and with negligible kinetic and potential energy effects, $h_2 = h_1$. This process is denoted Joule-Thomson flow and in real gases can produce either a cooling or heating effect on the fluid. The Joule-Thompson coefficient is denoted as $\mu_{JT} = (\partial T/\partial p)_h$. For real gases, the Joule-Thompson coefficient is not a constant. For certain ranges of pressure ratio it is negative, i.e., the expanding gas undergoes a temperature reduction, or a cooling effect is achieved. The states at which $\mu_{JT} = 0$ is denoted as the inversion line.

In practice, the cooling effect is achieved by throttling the fluid, usually a gas, through either a constriction or a valve inserted into the conduit. Throttling valves can be as simple as a porous plug, or an orifice, in the conduit. Typically throttling valves are designed for specific fluids and applications, such as the expansion of refrigerants.

For a large number of gases, throttling produces a decrease in temperature. For some gases however, notably hydrogen, the temperature rises after throttling over ordinary ranges of temperature and pressure. Whether there is a rise or fall in temperature depends on the particular range of pressure and temperature over which the change occurs. Throttling is an irreversible process, and thus entropy increases as a result of flow from the high pressure to low pressure state irrespective of the rise or fall in temperature.

Figure 2.1.3 and Table 2.1.4 show the variation of inversion temperature with reduced pressure and temperature. The reduced pressure and temperature are obtained by dividing given values by the critical temperature and pressure, T_c and p_c, respectively. The critical values define a state at which the specific volume of the liquid and that of the gas are the same. (See discussion below on equation of state.)

Unsteady Flow Systems In many engineering processes involving open systems, fluid flow is time dependent. For example, the process of a gas discharging from a storage tank represents a transient situation. The pressure within the bottle changes as the amount of oxygen in the tank decreases. The analysis of some transient processes is very complex; however, in order to show the general approach, a simple case will be considered.

The quantity of material flowing into and out of the engineering system in Fig. 2.1.2 varies with time. According to the law of conservation of mass, the rate of change of mass within the system is equal to the rate of mass flow into and out of it, and

$$\frac{dm_{\text{system}}}{dt} = \frac{dm_1}{dt} - \frac{dm_2}{dt}$$

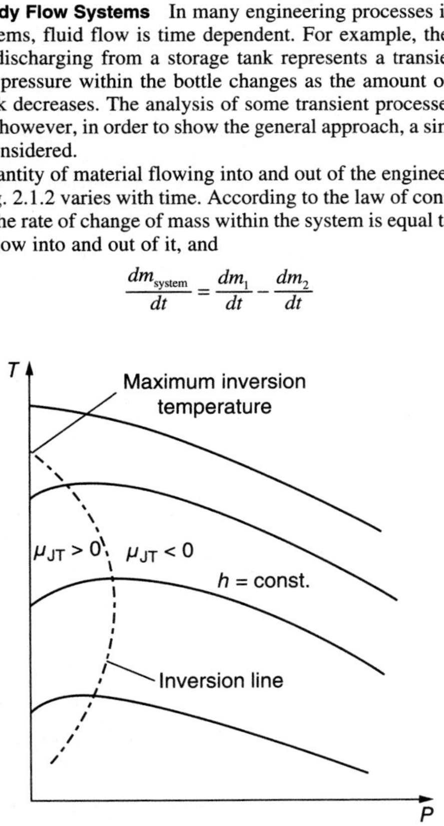

Figure 2.1.3 Inversion line in reduced temperature and pressure. (*Çengel YA, Boles MA. Thermodynamics: An Engineering Approach, 5e, McGraw-Hill.*)

Table 2.1.4 (a) Approximate Inversion Curve Locus in Reduced Pressure and Temperature. $T_r = T/T_C$ and $p_r = p/p_C$. (b) Approximate Inversion Curve Locus for Air

P_r	0	0.5	1	1.5	2	2.5	3	4
T_{RL}	0.782	0.800	0.818	0.838	0.859	0.880	0.903	0.953
T_{RU}	4.984	4.916	4.847	4.777	4.706	4.633	4.550	4.401
P_r	5	6	7	8	9	10	11	11.79
T_{RL}	1.01	1.08	1.16	1.25	1.35	1.50	1.73	2.24
T_{RU}	4.23	4.06	3.88	3.68	3.45	3.18	2.86	2.24

Calculated from the best three-constant equation recommended by Miller, *Ind. Eng. Chem. Fundam.* **9,** 1970, p. 585. T_{RL} refers to the lower curve and T_{RU} to the upper curve.

(a)

P_r, bar	0	25	50	75	100	125	150	175	200	225
T_L, K	(112)*	114	117	120	124	128	132	137	143	149
T_U, K	653	641	629	617	606	594	582	568	555	541
P, bar	250	275	300	325	350	375	400	425	432	
T_L, K	156	164	173	184	197	212	230	265	300	
T_U, K	526	509	491	470	445	417	386	345	300	

* Hypothetical low-pressure limit.

(b)

here for simplicity there being one inlet and one outlet flow. For a finite period of time integration gives,

$$\Delta m_{\text{system}} = \Delta m_1 - \Delta m_2$$

The First Law is then written,

$$\frac{dU}{dt} = \frac{dq}{dt} + \frac{dW}{dt} + \dot{m}_1\left(h_1 + \frac{\bar{V}_1^2}{2g_c} + z_1\frac{g}{g_c}\right) - \dot{m}_2\left(h_2 + \frac{\bar{V}_2^2}{2g_c} + z_2\frac{g}{g_c}\right)$$

Under non-steady flow conditions the enthalpy, h, and elevation, z, may also change with time, in which case the solution of the energy balance and mass continuity equations may have to be obtained numerically.

Second Law Analysis

Steady Flow Systems Entropy is carried with the fluid as it crosses the boundary of the system. Consider an open system for which fluid flow and heat cross the system boundary. The increase of entropy principle and the Clausius Theorem are extended to an entropy balance for a control volume,

$$\frac{dS}{dt} = \sum_j \frac{q_j}{T_j} + \sum_i \dot{m}_i s_i - \sum_e \dot{m}_e s_e + \dot{\sigma}$$

where finite rates of heat and mass transfer take place at specific locations (j, i, and e) on the boundary. The introduction of the entropy generation rate, $\dot{\sigma}$ (J/K·s), produces the equality under the Clausius Theorem. Entropy generation is always positive under the increase of entropy principle.

AVAILABILITY AND EXERGY

Availability

Availability of either a system or quantity of energy is defined $G = H - T_0S$, or for specific quantities, $g = h - T_0s$, where all quantities except T_0 refer to the system, and T_0 is the lowest temperature available for heat rejection in the surroundings. This definition assumes the absence of kinetic energy, potential energy, and other effects. When these are not negligible, proper allowance must be made, e.g., $g = h - T_0s + \bar{V}^2/2g_c + zg/g_c$.

By substitution of $q = T_0\Delta S$ in the appropriate First Law expressions for a process, a decrease in availability for either a steady flow process or a constant-pressure non-flow (closed system) process is equal to the maximum possible (reversible) net work effect with a heat sink T_0. A heat sink is a thermal reservoir in which temperature does not change upon a transfer of energy in the form of heat.

The Gibbs function, G, or g, is of particular value in the thermodynamic analysis of changes occurring in constant-pressure steady flow processes and where chemical changes occur, and it is of general utility in determining thermodynamic efficiencies, i.e., the ratio of actual work performed during a process to that which theoretically should have been performed. For reversible isothermal steady flow processes, or for reversible constant-pressure isothermal non-flow processes, the change in free energy is equal to the net work. The specific Helmholtz free energy, $f = u - Ts$ (or $a = u - Ts$) is equal to the net work during a constant-volume isothermal reversible non-flow process.

Loss due to Throttling The throttling process in a cycle of operations always introduces a loss of efficiency due to the entropy increase, i.e., throttling is an irreversible process. If T_0 is the temperature corresponding to the reference state, or the back pressure, the loss of available energy is $T_0\Delta s$.

EXAMPLE. Liquid ammonia at 70°F passes through a throttling value into a heat exchanger in which the temperature is 20°F, and the pressure is 48.21 psia. The initial enthalpy is $h_{f1} = 120.5$ Btu/lbm, and therefore the quality after throttling is $x_2 = 0.101$ Btu/°F. The initial entropy is $s_{f1} = 0.254$ Btu/lbm·°F. The final entropy is $s_2 = s_{f2} + (x_2h_{fg2}/T_2) = 0.144 + 0.101 \times 1.153 = 0.260$ Btu/lbm·°F.

The loss of cooling effect is $T_0\Delta s_{12} = 460 \times (0.260 - 0.254) = 2.9$ Btu/lbm.

Exergy

Exergy is an alternative expression for availability and is the preferred usage in the thermodynamics literature. It is defined as the maximum useful work obtainable from a system and its surroundings.

For a system exchanging heat q at its boundary with a thermal reservoir of variable temperature T, the exergy is,

$$\epsilon = q - T_0\left(\iint_{\substack{\text{system}\\\text{boundary}}} \frac{dq}{T}\right)$$

where the subscript T_0 refers to the surroundings and is the lowest temperature for heat rejection. For a system with specific enthalpy h and entropy s, the exergy is,

$$\epsilon = (h + ke + pe) - [h_0 + T_0(s - s_0)]$$

where the subscript 0 refers to the surroundings.

The loss of exergy in a system is denoted the irreversibility, I,

$$I = \epsilon_{in} - \epsilon_{out} - (useful\ work) - [H_0 + T_0(S - S_0)]$$

where the useful work is the total work minus the work done against the surroundings at p_0 and T_0. The exergy function is defined $\phi_e \equiv U + p_0V - T_0S$, and thus $\Delta\varepsilon = \Delta\phi$. Exergy transported across the boundary of an open system is expressed via the exergy function. Neglecting kinetic and potential energy effects, the specific free stream exergy is $(h - T_0s) - (h_0 - T_0s_0)$.

The irreversibility of a process is $I = T_0\Sigma\Delta S$, where ΔS is the entropy change for all systems including the surroundings and is also called the loss of availability. If $T = T_0$ and $V = V_0$, $\varepsilon = A - A_0$. If $T = T_0$ and $p = p_0$, $\varepsilon = G - G_0$. For either reversible isothermal steady-flow processes or reversible constant-pressure isothermal non-flow processes, the change in Gibbs free energy is equal to the net work. The change in the Helmhotz free energy is equal to the work during a constant-volume isothermal reversible non-flow process.

EXAMPLE. As an example of exergy change, consider the heat transfer between two identical solids each of mass m and constant specific heat c. Let block A be initially at temperature T_a and block B at T_b. The surrounding atmosphere is at p_0 and T_0. When the two blocks are brought into contact with no other energy transfer to the surroundings, the exergy change when thermal equilibrium is reached is,

$$\Delta \in = -mcT_0 \ln\left[\frac{(T_a+T_b)^2}{4T_aT_b}\right]$$

and the irreversibility $I = -\Delta\varepsilon > 0$. The exchange of heat between systems at different temperatures is irreversible, and hence there is a destruction of exergy. In a reversible process the loss of exergy is exactly equal to the useful work done. If the process is irreversible, the loss in exergy exceeds the useful work done. The exergy of a vacuum of volume V in an atmosphere (p_0,T_0) is p_0V, while that of an ideal gas at (p,V,T_0) in an atmosphere (p_0,T_0) is,

$$p_0V\left\{1-\frac{p}{p_0}\left[1-\ln\left(\frac{p}{p_0}\right)\right]\right\}$$

If air from an atmosphere at p_0 and T_0 slowly flows into an initially exhausted insulted tank of volume V until the pressure becomes p_0, the irreversibility of the process is,

$$I = \left(\frac{p_0V}{k-1}\right)\ln(k)$$

where $k = c_p/c_v$.

References

1. Bejan A. "Advanced Engineering Thermodynamics," John Wiley & Sons, New York, 1988.
2. Bejan A. "Entropy Generation Minimization: The Method of Thermodynamic Optimization of Finite-size Systems and Finite-time Processes," CRC Press, Boca Raton, 1995.
3. Bejan A. "Entropy generation minimization: the new thermodynamics of finites-size devices and finite time processes," *J. Appl. Phys.*, **79**, 1996, p. 1191. http://dx.doi.org/10.1063/1.362674
4. Callen HB. "Thermodynamics and an Introduction to Thermostatics," 2e, John Wiley & Sons, New York, 1985.
5. Çengel YA, Boles MA. Thermodynamics," 5e, McGraw-Hill, New York, 2006.
6. Gyftopoulos E, Beretta GP. "Thermodynamics: Fundamentals and Applications," Dover, Mineola, 2005.
7. Haywood J. "Equilibrium Thermodynamics for Engineers and Scientists," Wiley, New York, 1980.
8. Lior N. "The Exergy Fields in Processes," Departmental Papers (MEAM), University of Pennsylvania, Paper 265, 2008. http://repository.upenn.edu/meam_papers/265
9. Moran MJ. "Availability Analysis," Prentice-Hall, Englewood Cliffs, 1982; "Availability Analysis: A Guide to Efficient Use of Energy," ASME Press, New York, 1990.
10. Moran MJ, Shapiro HN. "Fundamentals of Engineering Thermodynamics," 5e, John Wiley & Sons, New York, 2006.
11. Rant Z. "Exergie ein neues Wort für technische Arbeitsfähigkeit," *Forsch. Ing-Wes.*, **22,** no. 1, 1956, pp. 36-37.
12. Sato N. "Chemical Energy and Exergy," Elsevier, Amsterdam, 2004.
13. Stodola A. "The Cycle Processes of the Gas Engine (in German)," 1898.
14. Szargut J, Morris DR, Steward FF. "Exergy Analysis of Thermal, Chemical and Metallurgical Processes," Hemisphere, New York, 1988.

EQUATION OF STATE

A simple chemical system comprises a single compressible substance, and equilibrium states are described in terms of pressure, volume, and temperature under the State Principle. The various properties of such a system at equilibrium may be expressed by an equation of state, e.g., $f(p,v,T) = 0$, or $p = f(v,T)$, which defines the surface comprising all equilibrium states. Two-dimensional graphs, such as p-versus-v, p-versus-h, T-versus-s, etc., show phase relations and are useful in cycle analysis. Equations of state are typically obtained via measurement.

The Ideal Gas

At low pressure and high enough temperature (well above absolute zero but not so high as to cause molecular dissociation) and in the absence of chemical reaction, all gases approach a condition such that their p-v-T surface can expressed by the relation.

$$pv = R_{\text{specific}}T$$

where T is the absolute temperature (K), R_{specific} is the gas constant specific to the given gas, and v is the specific volume (m^3/kg). If the volume is expressed as the specific molar volume, V/N (m^3/mol),

$$pV = NMR_{\text{specific}}T = NRT = \frac{m}{M}RT$$

where R is the universal gas constant and is the same for all gases, and N is the number moles of the gas. This last equation can be cast as $pv = RT/M$ which is convenient in practice. For all ideal gases, $R = MR_{\text{specific}}$ (8.3145 J/mol·K, 1545.4 ft·lbf/mol·R). One pound-mol of any perfect gas occupies a volume of 359 ft^3 at ~32°F (0°C, or 273.15 K) and 1 atm.

For many engineering purposes, use of the ideal gas equation of state is acceptable up to pressures of 100-200 lbf/in^2 (~6.7-13.5 atm) provided the absolute temperature is at least twice the critical temperature, T_c. The critical point is the state where the specific volume of the liquid and vapor are equal and above which the gas cannot be liquefied regardless of the applied pressure. The critical point of the gas is denoted by both critical pressure, p_c, and temperature, T_c, and $v_g = v_l$. For $T < T_c$, errors introduced by use of the ideal gas equation of state may be neglected for gauge pressure up to 15 lbf/in^2 (~1 atm). For a saturated vapor, errors of 5 percent are often encountered.

Specific heats of ideal gases are functions of temperature. Specific heats for perfect gases are not functions of temperature, and perfect gases are described by the same equation of state as for the ideal gas.

Real Gasses

Reduced Equation of State For real gases, the equation of state is based on the observation that most gases approximate a common reduced equation of state in terms of the reduced properties. For a single component gas, reduced temperature, pressure, and volume are $T_r = T/T_c$, $p_r = p/p_c$ and $v_r = v/v_c$. Table 2.1.5 lists critical point properties for selected gases.

The ideal gas equation of state can be made to apply to a real gas by the use of a the compressibility factor, Z,

$$pV = ZNRT$$

For an ideal gas, $Z = 1$. On a plot of Z versus reduced pressure, p_r, lines of constant reduced temperature, T_r, fall in narrow bands for different gases. Thus single curves for a given reduced temperature may be drawn to represent approximately the various bands. Figure 2.1.4 is the generalized compressibility graph for Z in terms of p_r for selected gasses. To use the graph, T_c and p_c are needed for a particular gas.

EXAMPLE. Find the volume of 1 lbm of steam at 5,500 psia and 1,200.
For water, T_c = 705.4°F, and p_c = 3,206.4 psia. Thus T_r =1,660/1,165 = 1.43, and $Z \approx 0.83$.
v = (0.83 × 1,546 × 1,660)/(18 × 5,500 × 144) = 0.149 ft^3/lbm.
From the steam tables, v = 0.1516 ft^3/lbm, and thus the error is 1.7 percent.
Based on the ideal gas law had been used, the error is ~17 percent.

Principle of Corresponding States Figure 2.1.4 suggests that the critical state quantities (p_c, T_c, and ρ_c) and molecular weight, M, can be used to generally characterize the differences between various gases. The generalization is denoted as the Principle of Corresponding States and means that the equation of state for all gases is a universal function of these quantities and the universal gas constant, or $F(p, T, \rho, R, M, p_c, T_c, \rho_c, Z_c) = 0$. Five dimensionless groups can be formed: p_r, T_r, ρ_r, Z, and Z_c $(= p_cM/\rho_cMT_c)$. Only three of these groups are independent because $Z = Z_c(p_r/\rho_rT_r)$. Consequently, gases with the same reduced pressure, temperature, and density can be represented by $F(p_r, T_r, \rho_r) = 0$.

Table 2.1.5 Critical Point Properties for Selected Gases

Substance	Formula	Molar mass, M kg/kmol	Gas constant, R kJ/kg · K*	Critical-point properties Temperature, K	Pressure, MPa	Volume, m³/kmol
Air	—	28.97	0.2870	132.5	3.77	0.0883
Ammonia	NH_3	17.03	0.4882	405.5	11.28	0.0724
Argon	Ar	39.948	0.2081	151	4.86	0.0749
Benzene	C_6H_6	78.115	0.1064	562	4.92	0.2603
Bromine	Br_2	159.808	0.0520	584	10.34	0.1355
n-Butane	C_4H_{10}	58.124	0.1430	425.2	3.80	0.2547
Carbon dioxide	CO_2	44.01	0.1889	304.2	7.39	0.0943
Carbon monoxide	CO	28.011	0.2968	133	3.50	0.0930
Carbon tetrachloride	CCl_4	153.82	0.05405	556.4	4.56	0.2759
Chlorine	Cl_2	70.906	0.1173	417	7.71	0.1242
Chloroform	$CHCl_3$	119.38	0.06964	536.6	5.47	0.2403
Dichlorodifluoromethane (R-12)	CCl_2F_2	120.91	0.06876	384.7	4.01	0.2179
Dichlorofluoromethane (R-21)	$CHCl_2F$	102.92	0.08078	451.7	5.17	0.1973
Ethane	C_2H_6	30.070	0.2765	305.5	4.48	0.1480
Ethyl alcohol	C_2H_5OH	46.07	0.1805	516	6.38	0.1673
Ethylene	C_2H_4	28.054	0.2964	282.4	5.12	0.1242
Helium	He	4.003	2.0769	5.3	0.23	0.0578
n-Hexane	C_6H_{14}	86.179	0.09647	507.9	3.03	0.3677
Hydrogen (normal)	H_2	2.016	4.1240	33.3	1.30	0.0649
Krypton	Kr	83.80	0.09921	209.4	5.50	0.0924
Methane	CH_4	16.043	0.5182	191.1	4.64	0.0993
Methyl alcohol	CH_3OH	32.042	0.2595	513.2	7.95	0.1180
Methyl chloride	CH_3Cl	50.488	0.1647	416.3	6.68	0.1430
Neon	Ne	20.183	0.4119	44.5	2.73	0.0417
Nitrogen	N_2	28.013	0.2968	126.2	3.39	0.0899
Nitrous oxide	N_2O	44.013	0.1889	309.7	7.27	0.0961
Oxygen	O_2	31.999	0.2598	154.8	5.08	0.0780
Propane	C_3H_8	44.097	0.1885	370	4.26	0.1998
Propylene	C_3H_6	42.081	0.1976	365	4.62	0.1810
Sulfur dioxide	SO_2	64.063	0.1298	430.7	7.88	0.1217
Tetrafluoroethane (R-134a)	CF_3CH_2F	102.03	0.08149	374.2	4.059	0.1993
Trichlorofluoromethane (R-11)	CCl_3F	137.37	0.06052	471.2	4.38	0.2478
Water	H_2O	18.015	0.4615	647.1	22.06	0.0560
Xenon	Xe	131.30	0.06332	289.8	5.88	0.1186

*The unit kJ/kg · K is equivalent to kPa · m³/kg · K. The gas constant is calculated from $R = R_u/M$, where R_u = 8.31447 kJ/kmol · K and M is the molar mass.

SOURCE: Çengel YA, Boles MA. Thermodynamics: An Engineering Approach, 5th ed., McGraw-Hill.

The compressibility factor can thus be written in the convenient form

$$\frac{pM}{\rho RT} = Z(p_r, T_r, Z_c)$$

Typically, the critical compressibility is of secondary importance in the above equation and is often omitted, implying that Z_c is invariant (or approximately so), which can be deduced from Table 2.1.5.

Van der Waals Equation of State Alternative equations of state have been developed both theoretically and empirically for real gases and gas mixtures, such as encountered with hydrocarbon gases. The van der Walls equation of state is a modification of the ideal gas equation of state and attempts to correct for intermolecular forces and finite molecular size,

$$\left[p + a\left(\frac{N}{V}\right)^2\right](V - Nb) = NRT$$

where V is the volume (L), the constant a (bar·L²/mol²) is a correction for intermolecular forces, b (L/mol) is a correction for the molecular size, and N is the number of moles. The constants a and b can be determined from critical point data: $a = (27R^2T_c^2)/(64p_c)$ and $b = (RT_c)/(8p_c)$. The equation approaches the ideal gas equation of state when a and b approach zero. Representative values are listed in Table. 2.1.6.

Beattie-Bridgeman Equation of State The Beattie-Bridgeman equation of state extends the van der Waals equation of state by the introduction of five empirical constants.

$$p = \frac{RT}{\bar{v}^2}\left(1 - \frac{c}{\bar{v}^2 T^3}\right)(\bar{v} + B) - \frac{A}{\bar{v}^2}$$

where $a = A_0\left(1 - a/\bar{v}\right)$, and $b = B_0\left(1 - b/\bar{v}\right)$, and the molar specific volume is represented by the over bar. See Table 2.1.7a for the values of the constants.

Benedict-Webb-Rubin Equation of State The Benedict-Webb-Rubin equation of state is empirically based and contains eight constants that are determined by curve fitting of measured p-v-T data.

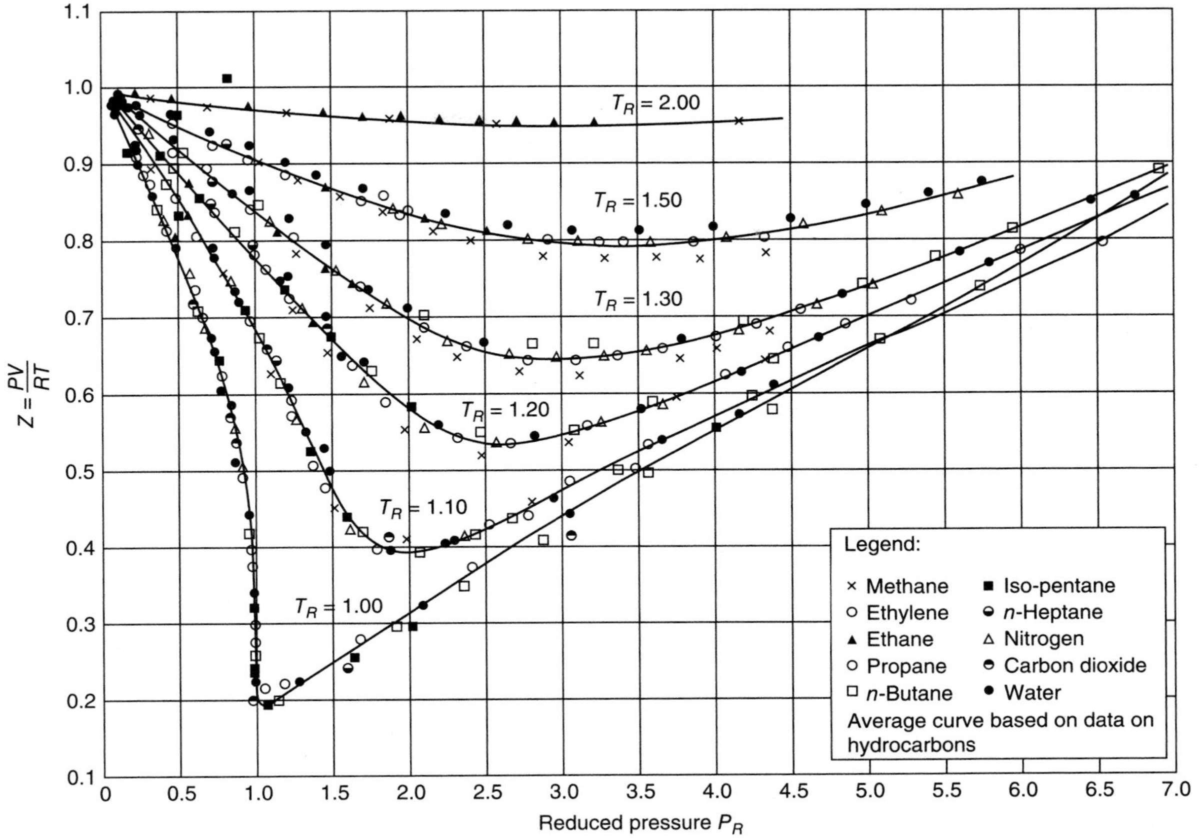

Figure 2.1.4 Compressibility factors for selected gases and vapors. (*Çengel YA, Boles MA. Thermodynamics: An Engineering Approach, 5e, McGraw-Hill.*)

Table 2.1.6 Selected van der Waals Constants*

Gas	a (bar·L²·mol⁻²)	b (L/mol)
Helium	0.0346	0.0238
Hydrogen	0.245	0.0265
Carbon Dioxide	4.658	0.0429
Water Vapor	5.537	0.0305

*CRC Handbook of Chemistry and Physics, 75e, CRC Press, Ann Arbor (1994).

$$p = \frac{RT}{\bar{v}} + \left(B_0RT - A_0 - \frac{C_0}{T^2}\right)\frac{1}{\bar{v}^2} + \frac{BRT - a}{\bar{v}^3} + \frac{a\alpha}{\bar{v}^6} + \frac{c}{\bar{v}^3T^2}\left(1 + \frac{\gamma}{\bar{v}^2}\right)e^{-\gamma/\bar{v}^2}$$

The constants are given in Table 2.1.7b for selected gases.

Virial Equation of State The compressibility factor, Z, for all gases and low density can be conveniently expressed

$$Z = 1 + B_r\rho_r + C_r\rho_r^2 + D_r\rho_r^3 + \cdots$$

This expansion is termed the viral equation of state, and the coefficients are denoted as the viral coefficients. They are formally functions of reduced temperature and the critical compressibility factor, Z_c, and are conceptually grounded on intermolecular forces. For the ideal gas, $Z \rightarrow 1$. Relations between the empirically based constants in alternative equations of state and the virial coefficients can be algebraically derived. The methods of statistical mechanics must be used to drive the virial coefficients from first principles.

Equation of State for Dry Air and Mixtures When no water vapor is present, air is a mixture gases comprising 20 percent oxygen and 79 percent nitrogen and 1 percent argon on a mole fraction basis. The gas constant for the mixture is R_{specific} = 287.058 J/kg·K. The p-v-T equation of state for each component can be represented as an ideal gas over a wide range of temperature and pressure, and thus the mixture is usually treated as single gas in design and analysis. Compressibility factors for air for $75 \leq T \leq 1{,}000$ K and $1 \leq p \leq 400$ bar (1 bar = 1,000 Pa = 0.987 atm) are listed Perry's Chemical Engineers' Handbook, 6th ed., McGraw-Hill, New York.

The critical temperature and pressure for real gas mixtures can be estimated by weighting the critical state variables by the mole fractions, y, of each component of the mixture. The results for the mixture are termed pseudocritical temperature, $T_{c,\text{mixture}}$, and pressure, $p_{c,\text{mixture}}$, and are given by,

$$T_{c,\text{mixture}} = \sum_i T_{c,i}y_i$$

$$p_{c,\text{mixture}} = \sum_i p_{c,i}y_i$$

where the summation is taken over all constituents, i = 1, 2,…n. The mixture is then treated for computation as if it were a single gas.

Equation of State for Multiphase Substances

A substance that exhibits solid, liquid, and gas phases generally does not have a simple closed form for its equation of state. Instead equilibrium states are represented by complex three-dimensional surfaces

Table 2.1.7 Constants of the Beattie-Bridgeman and Benedict-Webb-Rubin Equations of State Where (a) p is in kPa, $\bar{v}$ is in m³/kmol, T is in·K and (b) p is in kPa, $\bar{v}$ is in m³/kmol, R = 8.314 kPa·³/kmol

(a) **Beattie-Bridgeman and Benedict-Rubin equations of state**

Gas	A_0	a	B_0	b	c
Air	131.8441	0.01931	0.04611	−0.001101	4.34×10^4
Argon, Ar	130.7802	0.02328	0.03931	0.0	5.99×10^4
Carbon dioxide, CO_2	507.2836	0.07132	0.10476	0.07235	6.60×10^5
Helium, He	2.1886	0.05984	0.01400	0.0	40
Hydrogen, H_2	20.0117	−0.00506	0.02096	−0.04359	504
Nitrogen, N_2	136.2315	0.02617	0.05046	−0.00691	4.20×10^4
Oxygen, O_2	151.0857	0.02562	0.04624	0.004208	4.80×10^4

(b) **Benedict-Rubin equation of state**

Gas	a	A_0	b	B_0	c	C_0	a	γ
n-Butane, C_4H_{10}	190.68	1021.6	0.039998	0.12436	3.205×10^7	1.006×10^8	1.101×10^{-3}	0.0340
Carbon dioxide, CO_2	13.86	277.30	0.007210	0.04991	1.511×10^6	1.404×10^7	8.470×10^{-5}	0.00539
Carbon monoxide, CO	3.71	135.87	0.002632	0.05454	1.054×10^5	8.673×10^5	1.350×10^{-4}	0.0060
Methane, CH_4	5.00	187.91	0.003380	0.04260	2.578×10^5	2.286×10^6	1.244×10^{-4}	0.0060
Nitrogen, N_2	2.54	106.73	0.002328	0.04074	7.379×10^4	8.164×10^5	1.272×10^{-4}	0.0053

SOURCE: Çengel YA, Boles MA. Thermodynamics: An Engineering Approach, 5th ed., McGraw-Hill.

on which equilibria of vapor-liquid, vapor-solid, liquid, solid and vapor states are distinct surface elements. The state principle holds for a single substance (a simple chemical system), and therefore equilibrium state surfaces can be constructed in terms of p-v-T, h-s-T, etc., surfaces. Water, refrigerants, and hydrocarbon compounds exhibit complex equilibrium state surfaces. Tables of thermodynamic properties have been developed for a wide variety of substances. See Sec. 2.2 for various tabulations: T-versus-s, h-versus-s (the Mollier chart), and h-versus-log(p). Equilibrium state surfaces for substances that contract and expand on freezing are shown in Fig. 2.1.5 where the various liquid, liquid-vapor mixtures, vapor, etc., regions are identified.

Water Water expands on freezing, but removed from the vapor-solid and solid regions. Water is used as the working fluid in both subcritical and supercritical vapor power cycles. Figure 2.1.6 shows the vapor liquid region and the critical isotherm for water. For p-v-T data of other engineering fluids see Sec. 2.2.

Within the vapor-liquid region between the states of saturated liquid and saturated vapor (steam), the quality of the two-phase mixture is,

$$x = \frac{v - v_f}{v_g - v_f}$$

where v_f is the specific volume of the saturated liquid, v_g is the specific volume of the saturated vapor (or gas). The denominator is generally donated $v_{fg} = v_g - v_f$ along the constant pressure and temperature line in the vapor-liquid region. Generally for any property in the vapor-liquid region at a given pressure and temperature, $\Psi = (1 - x)\Psi_f - x\Psi_g = \Psi_f + x\Psi_{fg}$, where $\Psi_{fg} = \Psi_g - \Psi_f$. The saturated state has $x = 1$, and the saturated liquid has $x = 0$. The pressure in the vapor-liquid region is a function of temperature alone, and the specific volume of the mixture is a function of T and x.

Critical State If the temperature of the gas phase lies above the critical temperature, T_c, the gas cannot be liquefied by compression. At the critical state, $v_f = v_g$, and for $T > T_c$ it is impossible to have a mixture of vapor and liquid. Table 2.1.5 lists critical state data for selected gases. There is renewed interest in processes at or near the critical point in connection with heat transfer behavior in refrigeration systems and heat exchangers.

Estimating Vapor Pressure At a specified temperature, a liquid can exist in equilibrium with its vapor at but only at the vapor pressure. A plot of vapor pressure versus the corresponding temperature is denoted by the vapor pressure curve. The shape of this curve on ln(p)-versus-1/T coordinates, where temperature is in Kelvins, is an S-shape. However, if the curvature is slight and is neglected, the equation of the curve becomes $\ln(p) = A + B/T$, where A and B are constants. In terms of any two pairs of values (p_1,T_1) and (p_2,T_2), $A = (T_2 \ln p_2 - T_1 \ln p_1)/(T_2 - T_1)$, and $B = T_1T_2/(T_2 - T_1)\ln(p_1/p_2)$. Note that B is always negative. Once the values of A and B are determined, the equation can be used to determine either p_3 at T_3, or T_3 at p_3. Algebraically, A and B can be eliminated to yield $\ln(p_3/p_1) = [T_2(T_3 - T_1)/T_3(T_2 - T_1)] \ln(p_2/p_1)$, and at any temperature the slope of the vapor pressure curve is $dp/dT = (1/T_2)[T_1T_2/(T_1 - T_2)] \ln(p_2/p_1)$.

Changes of State

Table 2.1.8 contains a summary of the essential equations and relations for various processes involving superheated vapors and vapor-liquid mixtures.

For water vapor, the relation between p and V during an isentropic change can be represented approximately by $pV^n = Constant$. The exponent is not constant and varies with initial quality and initial pressure (Table 2.1.9). The isentropic expansion of superheated steam is adequately represented with $n = 1.315$. The volume at the end of the expansion (or compression) is $V_2 = V_1(p_1/p_2)^{1/n}$. The external work is,

$$W_{21} = \frac{p_2V_2 - p_1V_1}{n-1} = -p_1V_1\left[\frac{1 - \left(p_2/p_1\right)^{(n-1)/n}}{n-1}\right]$$

If the initial state is in the region of superheat and the final state is in the mixture region, two values of n must be used: $n = 1.315$ for expansion to the saturated state, and the appropriate value taken from the first row of Table 2.1.9 for the expansion of the mixture.

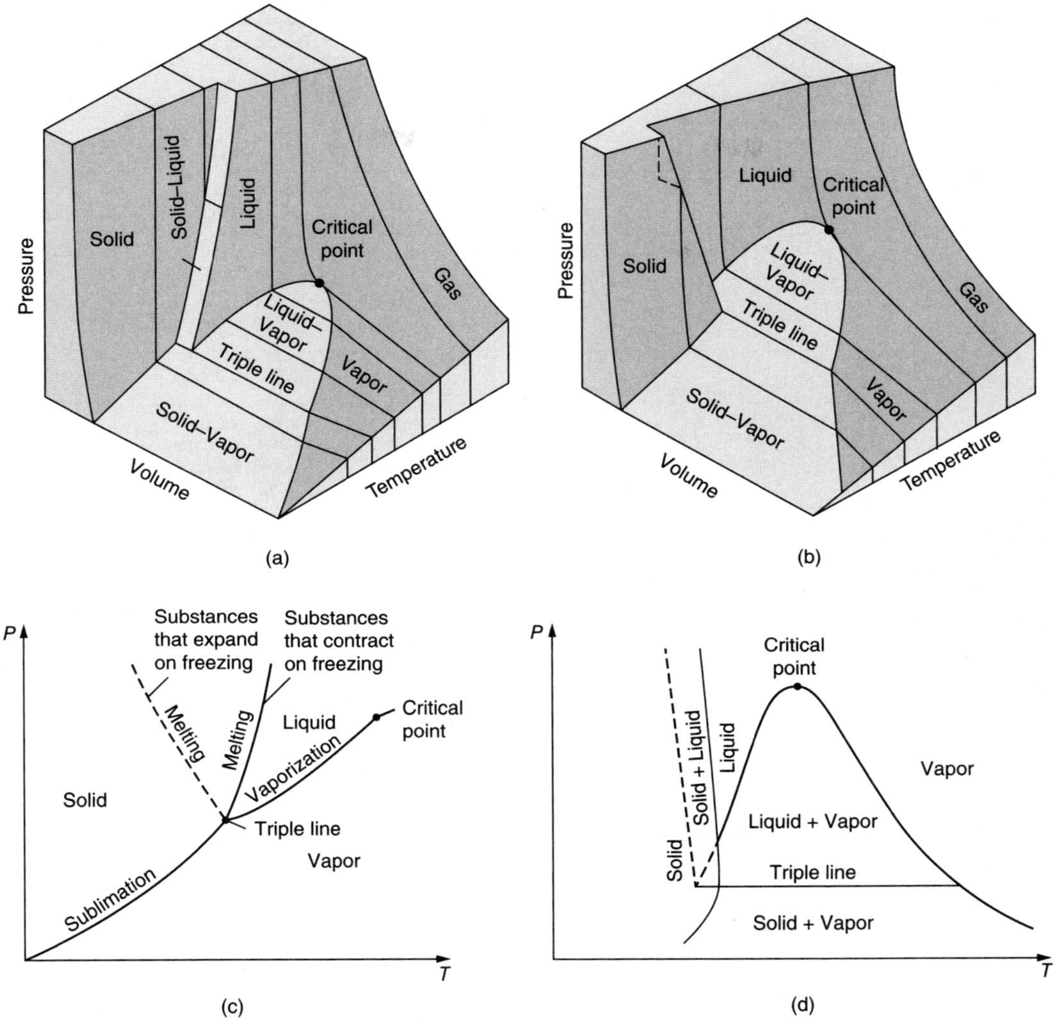

Figure 2.1.5 (*a*) p-v-T surface of a substance that contracts on freezing. (*b*) p-v-T surface that expands on freezing. (*c*) p-T diagram for surfaces depicted in (*a*) and (*b*). (*d*) p-v diagram for surface depicted in (*b*). (*Çengel YA, Boles MA. Thermodynamics: An Engineering Approach, 5e, McGraw-Hill.*)

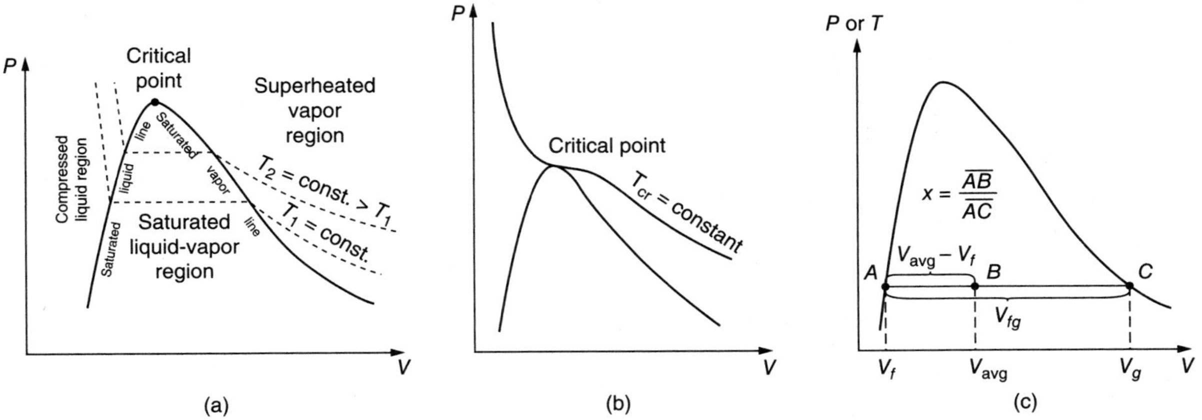

Figure 2.1.6 (*a*) p-v diagram showing regions of compressed liquid, liquid-vapor equilibrium and superheated vapor; (*b*) p-v diagram with critical isotherm; (*c*) p-v or T-v diagram showing determination of quality in the liquid-vapor region. (*Diagrams from Çengel YA, Boles MA. Thermodynamics: An Engineering Approach, 5e, McGraw-Hill.*)

Table 2.1.8 Process Equations for Superheated Vapors and Vapor-liquid Mixtures

Process (1 → 2)	Pressure, Temperature, Mass	Process Equations
Isothermal	T = Constant p = Constant m = system mass	$q_{12} = mh_{fg}(x_2 - x_1)$ $W_{12} = mpv_{fg}(x_2 - x_1)$ $\Delta U_{12} = mu_{fg}(x_2 - x_1)$ $\Delta S_{12} = q_{12}/T$
Constant Pressure	p = Constant m = System mass Initial state is a vapor-liquid mixture and final state is superheated vapor	$q_{12} = m\Delta h_{12}$ $\Delta U_{12} = m\Delta u_{12}$ $\Delta S_{12} = m\Delta s_{12}$ $W_{12} = -mp\Delta v_{12}$ $h_1 = h_{f1} + x_1 h_{fg1}$ $u_1 = u_{f1} + x_1 u_{fg1}$ $s_1 = s_{f1} + x_1 s_{fg1}$ $v_1 = v_{f1} + x_1 v_{fg1}$
Isentropic	s = Constant Initial and final states are a vapor-liquid mixture	$s_{f1} + x_1 s_{fg1} = s_{f2} + x_2 s_{fg2}$
Isentropic	s = Constant Initial state is superheated vapor	$s_1 = s_{f2} + x_2 s_{fg2}$ x_2 is determined from $u_1 = u_{f2} + x_2 u_{fg2}$ $q_{12} = 0, \quad W_{12} = U_2$

Table. 2.1.9 Values of the Polytropic Constant *n* for Water Vapor

Initial	Initial pressure, psia											
quality	20	40	60	80	100	120	140	160	180	200	220	240
1.00	1.131	1.132	1.133	1.134	1.136	1.137	1.138	1.139	1.141	1.142	1.143	1.145
0.95	1.127	1.128	1.129	1.130	1.131	1.131	1.132	1.133	1.134	1.135	1.136	1.137
0.90	1.123	1.123	1.124	1.124	1.125	1.125	1.126	1.126	1.127	1.127	1.128	1.129
0.85	1.119	1.119	1.119	1.119	1.120	1.120	1.120	1.120	1.120	1.120	1.120	1.121
0.80	1.115	1.115	1.114	1.114	1.114	1.114	1.113	1.113	1.113	1.113	1.112	1.112
0.75	1.111	1.110	1.110	1.109	1.109	1.108	1.107	1.106	1.106	1.105	1.104	1.104

References

1. Çengel YA, Boles MA. "Thermodynamics: An Engineering Approach," 5e, McGraw-Hill, New York, 2006.
2. Hall NA, Ibele WE. "Engineering Thermodynamics," Prentice-Hall, Englewood Cliffs, 1960.
3. Irvine TF, Liley PE. "Steam and Gas Tables with Computer Programs," Academic Press, New York, 1984.
4. Kondaputi D. "Introduction to Modern Thermodynamics," John Wiley & Sons, West Sussex, 2008.
5. Libre Texts. "van der Waals constant for real gases," 2016. http://chem.libretexts.org/Reference/Reference_Tables/Atomic_and_Molecular_Properties/A8%3A_van_der_Waal's_Constants_for_Real_Gases
6. __________ "NIST Data Gateway, National Institute of Standards and Technology," 2016. http://srdata.nist.gov/gateway/gateway?keyword=equation+of+state

IDEAL GAS MIXTURES

Mixture Rules

Many fluids involved in engineering systems are either gas mixtures or a gas mixed with a either superheated or saturated vapor. Normal atmospheric air is a mixture of oxygen and nitrogen with traces of other gases, plus superheated or saturated water vapor, or at times saturated vapor and liquid. A mixture of gases can be regarded as an equivalent gas with properties dependent upon the kind and proportion of each of the constituents.

Dalton's Law For a mixture of chemically inert ideal gases occupying a total volume V at a total pressure p, each constituent occupies the entire volume by virtue of the properties of the ideal gas, i.e., no intermolecular forces and no effects due to molecular size. The pressure exerted by a given constituent is given by its equation of state and the pressure is denoted as the partial pressure. $p_i = m_i R_{i, \text{specific}} T/V$, where the subscript denotes the constituent gas. Dalton's Law states that the total pressure of the mixture is the sum of the partial pressures, $p = \Sigma p_i$. This is also denoted as the additive pressure rule.

The total number of moles of the mixture is similarly $n = \Sigma n_i$ and the total mass of the mixture is $m = \Sigma m_i$. The mole fraction of each constituent is $y_i = n_i/n$, and it follows that $1 = \Sigma y_i$. The mole fraction is also given by $y_i = p_i/p$. The mass fraction of each constituent is $x_i = m_i/m$, and $1 = \Sigma x_i$. A mixture molecular weight is defined $M = m/n = (\Sigma n_i m_i)/n = \Sigma y_i M_i$. Similarly the gas constant for the mixture is $R_{\text{mixture}} = (\Sigma R_{i,\text{specific}} m_i)/m = \Sigma x_i R_{i,\text{specific}}$. If the mixture deviates from ideal gas behavior, a compressibility factor for the mixture, Z_{mixture}, can be used. In this case, $p_i = n_i Z_i RT/V$ and $p = nZRT/V$. Under the additive pressure rule, $Z_{\text{mixture}} = \Sigma y_i Z_i$.

Amagat's Law The Amagat model considers an ideal gas mixture initially comprising constituents all at total pressure p and temperature T but occupying volumes V_i proportionate to their mole numbers and masses. Under this condition, the total volume of the mixture is $V_{\text{mixture}} = \Sigma V_i$. This is the additive volume rule. With each constituent gas described by the ideal gas equation state, $y_i = V_i/V$.

Gibbs-Dalton Law The Gibbs-Dalton law states that the extensive properties of a mixture are determined under the assumption that the gases do not interact as they occupy the total volume. Thus the extensive representation of internal energy, enthalpy, entropy, Gibbs function, and Helmholtz function are respectively,

$$U = \sum_{i=1}^{n} U_i,\ H = \sum_{i=1}^{n} H_i,\ S = \sum_{i=1}^{n} S_i,\ G = \sum_{i=1}^{n} G_i,\ F = \sum_{i=1}^{n} F_i$$

Mixture properties and specific heats can be expressed on either a mass or a mole fraction basis. For example, specific heats on a mass basis are,

$$c_p = \sum_i \frac{m_i c_{pi}}{m}, \quad c_v = \sum_i \frac{m_i c_{vi}}{m}$$

The Gibbs-Dalton law cannot be derived from the First Law and the Second Law. It holds for ideal gases as a limiting case but does not hold for mixtures of all gases.

Entropy of Mixing

The mixing of gases is an irreversible process. The separation of the mixture into its separate constituents does not occur without the action of an external agent and the expenditure of work with the consequent increase of entropy of the surroundings under the increase of entropy principle. If an ideal gas with constant specific heats changes from an initial state (p_1,V_1,T_1) to a final state (p_2,V_2,T_2), the changes in specific internal energy and enthalpy are,

$$\Delta u_{12} = c_v(T_2 - T_1) = (p_2v_2 - p_1v_1)(k-1)$$

$$\Delta h_{12} = c_p(T_2 - T_1) = (p_2v_2 - p_1v_1)\left(\frac{k}{k-1}\right)$$

where $k = c_p/c_v$. On a mass basis, the integration of the *T*-dS equations with constant specific heats gives,

$$\Delta s_{12} = c_v ln\frac{T_2}{T_1} + Rln\frac{v_2}{v_1}$$

$$\Delta s_{12} = c_p ln\frac{T_2}{T_1} + Rln\frac{p_2}{p_1}$$

On a molar basis, the integration of the second *T*-dS equation and application of the Gibbs-Dalton law gives,

$$\Delta S = \sum_i n_i R\left[ln\frac{p_1}{p_0} - ln\frac{p}{p_0}\right] = \sum_i n_i Rln\frac{p_i}{p}$$

$$\Delta S = \sum_i n_i Rlny_i > 0$$

where p is the total pressure, p_i is the partial pressure of each constituent, and the mole fraction is $y_i = p_i/p$. Thus, for mixing of ideal gases in a closed isolated system at temperature, T, the increase of entropy of the universe is always greater than zero, $\Delta S_{\text{universe}} > 0$. Entropy production due to mixing is thus $\sigma = \Delta S_{\text{universe}}$.

Gibbs Paradox Entropy increase due to the mixing to two dissimilar gases is intuitively and mathematically realized under the Second Law and the *T*-dS equations. However, if the two gases are not dissimilar and mixing is at constant temperature and pressure, the question arises as to whether there is an entropy increase, or if mixing has any physical meaning. This virtual mixing process can be realized conceptually when a container with a membrane separates two similar gases which then mix when the membrane is removed. This is known as the Gibbs Paradox and is the anomalous result when the two gases become indistinguishable.

IDEAL GAS PROCESSES

Polytropic Processes

When a simple chemical system of constant mass comprising a compressible substance undergoes a process, the change of state is generally described by $pV^n = Constant$. This process is described by as a polytropic process, and n is termed the polytropic exponent. The underlying assumption is that the process is a succession of infinitesimal state changes such that integration of the state variables can be carried out along the (reversible) process path. The special cases for polytropic processes are: constant volume (isochoric), constant pressure (isobaric), constant temperature (isothermal), constant entropy (isentropic). The values of the exponent n and the relation between pressure, volume, and temperature for each process with constant specific heats are listed in Table 2.1.10 with processes quantities determined from the First Law and the *T*-dS equations.

For a polytropic change of an ideal gas with a constant c_V, the specific heat at constant pressure is $c_P = c_V(n - k) \times (n - 1)$; hence for $1 < n < k$, c_P is negative. This is approximately the case in air compression up to a pressure of a few hundred pounds force per square inch. The principal relations for determining work and heat transfer are given in Table 2.1.10. If two representative points (p_1,V_1) and (p_2,V_2) on the process path be chosen, then

$$n = (log\ p_1 - log\ p_2)/(\log V_2 - \log V_1)$$

Several pairs of points should be used to test the constancy of n.

Changes of State with Variable Specific Heat

The assumption of constant specific heat is generally not permissible when the temperature range is large, and the equations referring to changes of state must be suitably modified with the exception of inert and monatomic gases. Experiments show that the specific heats may be taken as linear functions of the temperature, $c_V = a + bT$, $c_P = a' + b'T$. In that case,

$$\Delta U_{12} = m\left[a(T - T) + 0.5b\left(T_2^2 - T_1^2\right)\right]$$

$$\Delta S_{12} = m\left[aln\left(\frac{T_2}{T_1}\right) + b\left(T_2 - T_1\right) + Rln\left(\frac{V_2}{V_1}\right)\right]$$

and for an isentropic process,

$$W_{12} = \Delta U_{12}$$

$$Rln\left(\frac{V_2}{V_1}\right) = aln\left(\frac{T_2}{T_1}\right) + b(T_2 - T_1)$$

IDEAL GAS CYCLES

Gases are used in several important types of power producing cycles. In air compressors, air engines and air refrigerating machines, atmospheric air is the fluid. In internal-combustion engines, the fluid medium is a mixture of products of combustion and excess air. In such cycles the conversion of fuel energy by heat transfer to the product gases and the production of work is limited by temperature, and thus they are denoted as temperature-limited cycles. Analysis of such cycles using a perfect gas (constant specific heats) provides approximately valid results to assess the thermal performance of the actual cycle. If the ideal gas is assumed, specific heats can vary with temperature and thus polytropic processes are used for expansion and compression.

Carnot Cycle

The Carnot cycle is an internally reversible cycle comprising an isothermal heat addition from a high temperature thermal reservoir at T_H, heat rejection to a low temperature (cold) thermal reservoir at T_C, and isentropic expansion and compression processes. Figure 2.1.7 shows the cycle on T-s and p-V coordinates. The area within the diagram is the work produced by the cycle. The heat absorbed at T_H $(1 \rightarrow 2)$ is $q_{12} = mRT \ln(V_2/V_1)$, and the work production per cycle is $W = -mR(T_H - T_C) \ln(V_2/V_1) = q_{12}(1 - T_C/T_H)$. The heat rejects at T_C $(3 \rightarrow 4)$ is $q_{34} = mRT \ln(V_4/V_3)$, The thermal efficiency of the cycle, $\eta = (1 - T_C/T_H)$, is the ratio of work produced to the heat transfer at T_H and is the highest possible efficiency of all heat engines operating between these same temperatures.

Table 2.1.10 Polytropic Processes for the Ideal Gas with Constant Specific Heats

Process	Exponent n	Equation	Process Quantities
Isothermal	1	$p_1V_1 = p_2V_2$	$\Delta U_{12} = 0,\ q_{12} = -W_{12}$ $W_{12} = -mRT\ln(V_2/V_1) = -p_1V_1\ln(V_2/V_1)$ $\Delta S_{12} = q_{12}/T = mR\ln(V_2/V_1)$
Isentropic	$k = c_p/c_v$	$p_1V_1^k = p_2V_2^k$	$T_2/T_1 = (V_2/V_1)^{k-1} = (p_2/p_1)^{(k-1)/k}$ $W_{12} = U_1 - U_1 = mc_v(T_2 - T_1)$ $q_{12} = 0,\ \Delta S_{12} = 0$ $W_{12} = (p_2V_2 - p_1V_1)/(k-1)$ $= -p_1V_1\dfrac{(p_2/p_1)^{(k-1)/k} - 1}{k-1}$
Isobaric	0	$p_1 = p_2$ $V_2/V_1 = T_2/T_1$	$W = -p(V_2 - V_1) = -mR(T_2 - T_1)$ $q_{12} = mc_p(T_2 - T_1) = kW_{12}/(k-1)$ $\Delta S_{12} = mc_p ln(T_2/T_1)$
Isochoric	$n = \infty$	$V_1 = V_2$ $p_2/p_1 = T_2/T_1$	$q = U_2 - U_1 = mc_v(T_2 - T_1) = V(p_2 - p_1)/(k-1)$ $W_{12} = 0,\ \Delta S_{12} = mc_v ln(T_2/T_1)$
Polytropic	n	$p_1V_1^n = p_2V_2^n$	$T_2/T_1 = (V_2/V_1)^{n-1} = (p_2/p_2)^{(n-1)/n}$ $W_{12} = (p_2V_2 - p_1V_1)/(n-1)$ $= -p_1V_1\dfrac{(p_2/p_1)^{(n-1)/n} - 1}{n-1}$ $q_{12} = mc_n(T_2 - T_1)$

A reversed Carnot cycle represents the ideal refrigerator operating on the cycle of Fig. 2.1.7. The coefficient of performance (COP) of the cycle is defined as the ratio of heat transferred at T_C to the work in put W. Thus $COP = q_{43}/W = T_C/(T_H - T_C)$, where $q_{43} = mRT_0\ln(V_3/V_4)$. This is the largest COP that can be obtained by all refrigeration cycles operating between these two temperatures.

The thermal efficiency of a heat engine producing the maximum possible work per cycle consistent with its operating temperatures results in efficiencies equal to or approximated by $\eta = 1 - \sqrt{T_C/T_H}$. If the work output is fixed, the thermal efficiency can be increased by operating the engine at less than maximum work output per cycle, the limit being an engine of infinite size having a Carnot efficiency. Figure 2.1.8 illustrates the difference between maximum power and maximum efficiency.

Otto and Diesel Cycles

Otto Cycle The Otto cycle is a power cycle in which a fuel-air mixture is ignited by a spark that initiates an expansion process due to the energy release of combustion. The internally reversible cycle comprises an isentropic compression, a constant volume pressure rise (ignition and rapid heating), an isentropic expansion, and a constant volume

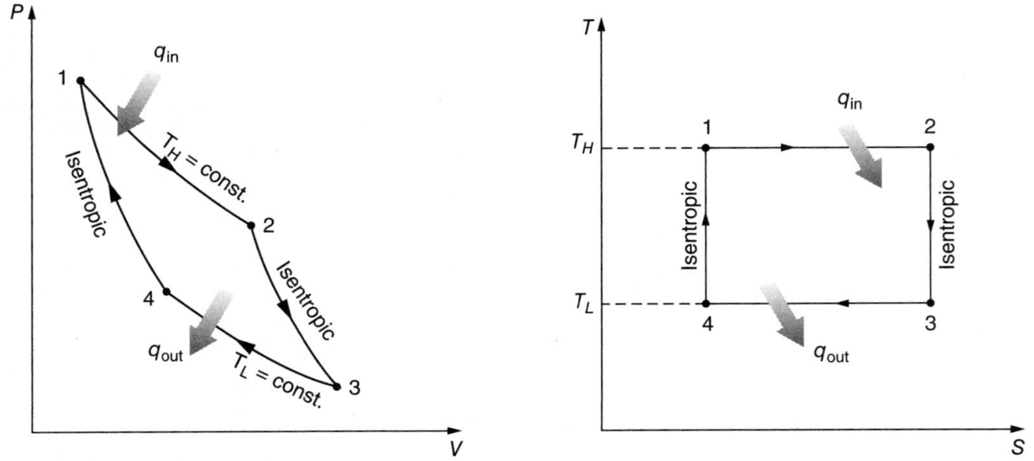

Figure 2.1.7 The Carnot cycle on p-V and T-s coordinates. (*Diagrams from Çengel YA, Boles MA. Thermodynamics: An Engineering Approach, 5e, McGraw-Hill.*)

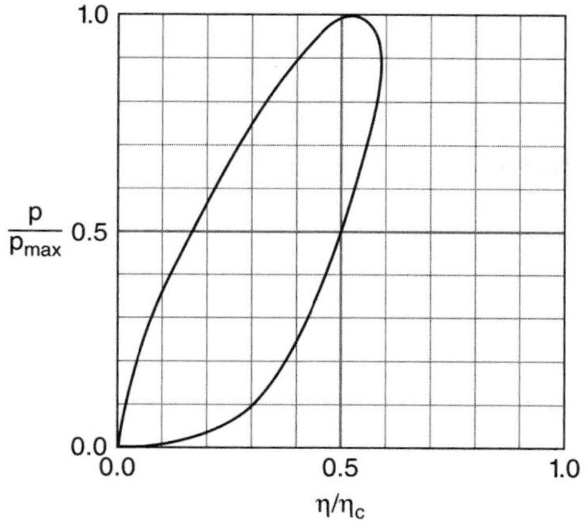

Figure 2.1.8 Ratio of actual power to maximum power as a function of the ratio of actual thermal efficiency to Carnot efficiency.

pressure reduction. The cycle is realized in practice in a reciprocating cylinder which transfers power to a rotating shaft to deliver the work produced externally, e.g., in automotive applications. An internally reversible cycle is depicted in Fig. 2.1.9 on p-V and T-s coordinates.

Approximate analysis of the Otto cycle is accomplished by assuming the fuel-air mixture behaves as air which is treated as an ideal gas, hence the frequently used term *air-standard* Otto cycle. Non-ideal behavior is approximated by a variation of the ratio $k = c_P/c_V$ in the isentropic expansion and compression processes. Assuming constant specific heats the following relations hold where the states are as denoted in Fig. 2.1.9,

$$\frac{T_2}{T_1} = \frac{T_3}{T_4} = \left(\frac{p_2}{p_1}\right)^{(k-1)/k} = \left(\frac{p_3}{p_4}\right)^{(k-1)/k} = \left(\frac{V_1}{V_2}\right)^{(k-1)/k}$$

$$q_{12} = mc_v(T_1 - T_2)$$

$$W_{\text{cycle}} = q_{23}[1-(T_1/T_2)] = mc(T_3 - T_4 - T_2 + T_1)$$

The thermal efficiency of the cycle is,

$$\eta = 1 - \frac{T_1}{T_2} = 1 - \left(\frac{V_2}{V_1}\right)^{k-1} = 1 - \left(\frac{p_1}{p_2}\right)^{(k-1)/k}$$

where V_1/V_2 is termed the compression ratio. Thus the efficiency of the air-standard Otto cycle is independent of the working fluid and independent of the temperatures in the cycle.

If the compression and expansion processes are polytropic with n replacing k, the work production of the cycle is,

$$W = \frac{(p_3V_3 - p_4V_4) - (p_2V_2 - p_1V_1)}{n-1} = \frac{mR(T_3 - T_4 - T_2 + T_1)}{n-1}$$

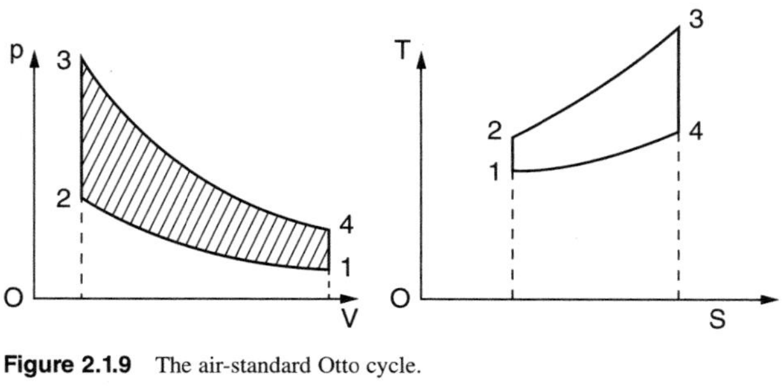

Figure 2.1.9 The air-standard Otto cycle.

The mean effective pressure (MEP) of the cycle is given by,

$$p = ap_1\left(\frac{p_3}{p_2} - 1\right)$$

where a has the values given in Table 2.1.11. The work produced by the cycle is given by W = MEP × A × Cylinder Displacement, where A is the cylinder cross sectional area, and cylinder displacement is obtained from a scaled measurement, e.g., it would be obtained from an indicator diagram.

Table 2.1.11 Mean Effective Pressures for the Otto Cycle with Polytropic Expansion and Compression

	$p_2/p_1 = 3$	4	5	6	8	10	12	14	16
(n = 1.4)	a = 1.70	1.94	2.13	2.31	2.62	2.88	3.10	3.31	3.50
(n = 1.3)	a = 1.69	1.92	2.11	2.28	2.57	2.81	3.03	3.22	3.39
(n = 1.2)	a = 1.68	1.90	2.08	2.25	2.51	2.74	2.94	3.12	3.27

Diesel Cycle In the air-standard diesel cycle, air in the cylinder is compressed to a high pressure and temperature. Fuel in the form of atomized liquid droplets is then injected into the air mass which is at a temperature above the ignition point, and it burns at nearly constant pressure (2 → 3, Fig. 2.1.10). The Diesel cycle is termed a compression ignition cycle owing to the explosive combustion of the isentropic expansion of the products of combustion is followed by exhaust and suction of fresh air, as in the Otto cycle.

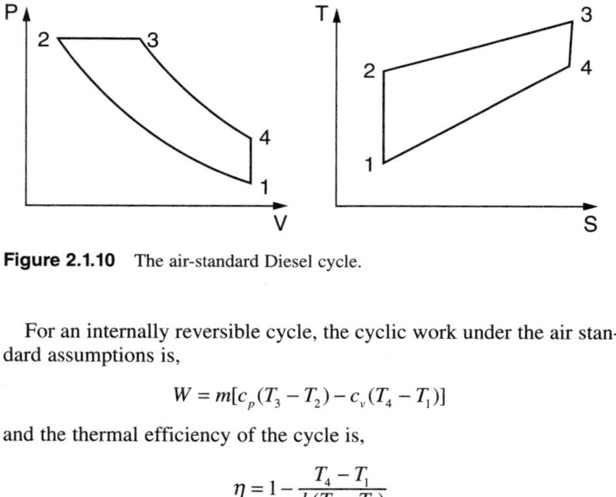

Figure 2.1.10 The air-standard Diesel cycle.

For an internally reversible cycle, the cyclic work under the air standard assumptions is,

$$W = m[c_p(T_3 - T_2) - c_v(T_4 - T_1)]$$

and the thermal efficiency of the cycle is,

$$\eta = 1 - \frac{T_4 - T_1}{k(T_3 - T_2)}$$

Deviations from ideality are accommodated via polytropic expansion and compression processes. Alternatively, the thermal efficiency can be based on the internal energies and enthalpies obtained from tables of thermodynamic properties which take into account variable specific heats.

The key parameters of the Diesel cycle are the compression ratio, $r_v = V_1/V_2$, the expansion ratio, $r_e = V_4/V_3$, and the cut off ratio, $r_c = V_3/V_2$. For the air-standard cycle, the thermal efficiency expressed in these ratio is,

$$\eta = 1 - \left(\frac{1}{r_v^{k-1}}\right)\left(\frac{r_c^k - 1}{k(r_c - 1)}\right)$$

When $r_c = 1$, the Otto and Diesel cycles have the same thermal efficiency. The physical implication for the Diesel cycle is that there is no change of volume when heat is applied. When $r_c > 1$, the Diesel cycle is less efficient that the Otto cycle.

Dual Cycle The dual cycle, or combined cycle, incorporates features of the Otto and Diesel cycles to take advantage of both. The air-standard combined cycle is a better approximation to the operation of practical compression ignition engines than the air-standard Otto cycle. Figure 2.1.11 shows the states of an internally reversible combined cycle n p-V coordinates. Energy addition is accomplished at both constant volume and constant pressure, a → b and b → c respectively. Energy addition is followed by an isentropic expansion (c → d) and a constant volume heat transfer (d → e) followed by an isentropic compression (e → a).

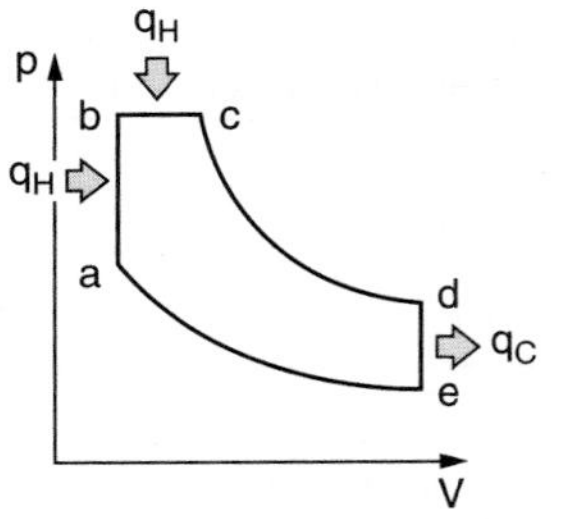

Figure 2.1.11 The duel cycle.

The thermal efficiency of the combined cycle is,

$$\eta = \frac{c_V(T_d - T_e)}{c_V(T_b - T_a) + c_b(T_c - T_b)} = 1 - \frac{T_d - T_e}{(T_b - T_a) + k(T_c - T_b)}$$

where $k = c_p/c_V$. The thermal efficiency can be represented in terms of the volumetric ratios pertinent to the heat transfer and compression/expansion processes,

$$\eta = 1 - \frac{1}{r_V^{k-1}}\left[\frac{r_p r_c^k - 1}{r_p - 1 + k r_p (r_c - 1)}\right] r$$

where $r_v = V_c/V_a$, $r_c = V_c/V_b$, and $r_p = p_b/p_a$

Brayton Cycle

The Brayton cycle, also called the Joule cycle, is an internally reversible cycle comprising two isentropic and two constant-pressure processes (Fig. 2.1.12). It is the basis for analysis of the modern gas turbine which finds extensive use in aircraft, marine, and automotive applications. The Brayton cycle can be conceptualized as either an open cycle, such as is realized in aircraft applications, or a closed cycle. Air enters the compressor wherein the pressure is increased (1 → 2). Fuel burning in the combustion chamber raises the temperature of the compressed air under constant-pressure conditions (2 → 3). The key assumption in the combustion process is that the addition of fuel and products of combustion do not appreciably change the mixture properties. However, more

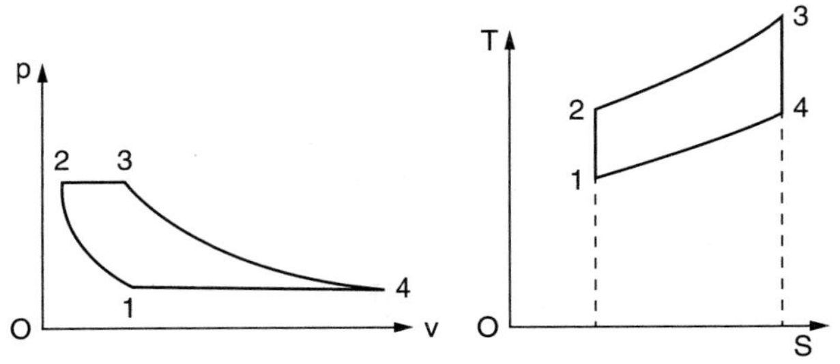

Figure 2.1.12 The air-standard Brayton cycle.

detailed analysis is often employed in the design of the combustor in practice. The resulting high temperature gases are then introduced to the turbine where they expand and perform work (3 → 4). The excess work of the engine over that required to compress the air is available for operating other devices, such as a generator. The simple gas turbine cycle is the same as the Brayton cycle, except that the compressor and engine are replaced by an axial flow compressor and gas turbine. A portion of the work produced during expansion is used to power the compressor, and ratio of the compressor work to net turbine work is termed the work ratio. The thermal objective is to minimize the work ratio consistent with the desired net work production and cycle thermal efficiency.

Under the assumptions of air-standard cycles, the relations of the four states, work and thermal efficiency are,

$$\frac{V_3}{V_2} = \frac{V_4}{V_1} = \frac{T_3}{T_2} = \frac{T_4}{T_1}$$

$$\frac{T_2}{T_1} = \frac{T_3}{T_4} = \left(\frac{V_1}{V_2}\right)^{k-1} = \left(\frac{V_4}{V_3}\right)^{k-1} = \left(\frac{p_2}{p_1}\right)^{(k-1)/k}$$

$$W_{\text{cycle}} = mc_p(T_3 - T_2 - T_4 + T_1)$$

$$\eta = \frac{W_{\text{cycle}}}{q_{23}} = 1 - \frac{T_1}{T_2} = 1 - \left(\frac{p_2}{p_1}\right)^{(1-k)/k}$$

The thermal performance of the Brayton cycle is limited in practice by irreversibility in each process of the cycle. These include heat losses, compressor inefficiency, and expansion inefficiency. The most important factors in overall cycle efficiency are the efficiencies of the compressor and the turbine. Figure 2.1.13 illustrates the trend of cycle efficiency based on equal compression and expansion efficiencies.

EXAMPLE. Consider the Brayton cycle with irreversibility in the compression and expansion processes as shown in the diagram. The working fluid is air. The compressor and turbine thermal efficiencies are 80 percent. The inlet to the compressor is 15 psia at 530 R and the compression ratio is 6:1. Heat addition raises the working fluid to 1,700 R. The turbine exhaust is a 15 psia.

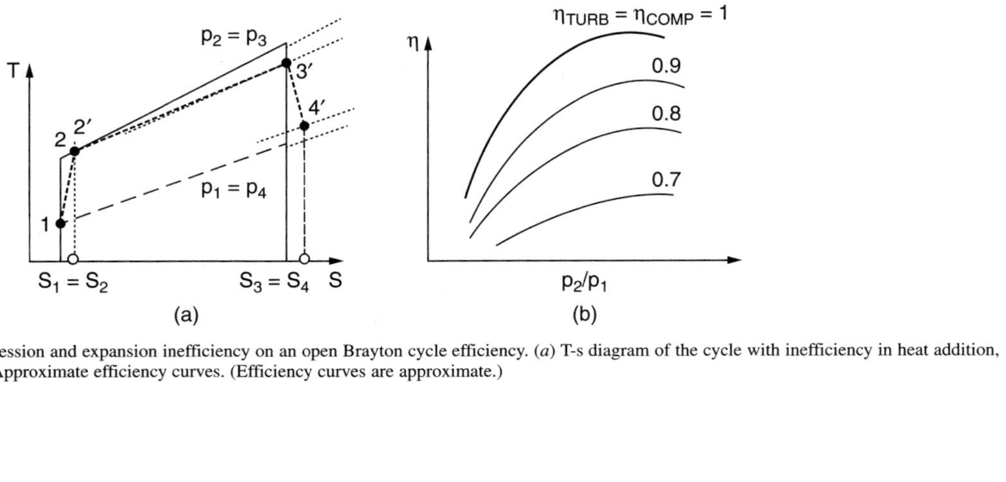

Figure 2.1.13 Effects of compression and expansion inefficiency on an open Brayton cycle efficiency. (*a*) T-s diagram of the cycle with inefficiency in heat addition, compression, and expansion. (*b*) Approximate efficiency curves. (Efficiency curves are approximate.)

Determine the cycle efficiency, the work ratio, and entropy production.

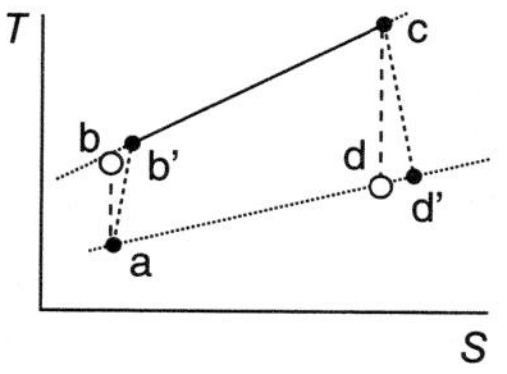

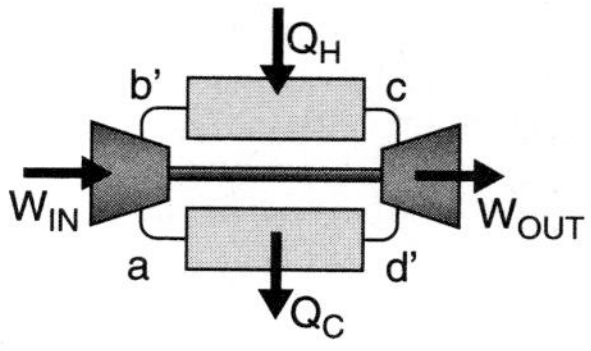

Taking data from the air tables, the following states apply.

State	T (R)	p (psia)	h (Btu/lbm)	s^0 (Btu/lbmR)
a	530	15	126.7	0.5963
b	881	90	211.6	0.7191
b'	967	90	232.7	0.7421
c	1700	90	422.6	0.8876
d	1059	15	255.7	0.7647
d'	1192	15	289.3	0.7946

Process calculations yield the following for enthalpy, work, heat transfer, and entropy.

Process	Δh (Btu/lbm)	w (Btu/lbm)	Q (Btu/lbm)	Δs (Btu/lbmR)
a-b	84.9	−84.9	0	0
a-b'	106	−106	0	0.0230
b'-c	189.9	0	189.9	0.1455
c-d	−166.9	166.9	0	0
c-d'	−133.3	133.3	0	0.0299
d'-a	−162.6	0	−162.6	−0.1983
NET	0	27.3	0	0

Cycle efficiency and the work ratio are,

$$\eta = \frac{W_{net} - W_{comp}}{q_{b'c}} = \frac{133.3 - 106}{189} = 0.144$$

$$\frac{W_{comp}}{W_{turb}} = \frac{106}{133.3} = 0.795$$

By way of comparison, the Carnot efficiency for the cycle is,

$$\eta_{Carnot} = 1 - \frac{530}{1700} = 0.68$$

Entropy production is determined by application of the Second Law entropy balance for open systems for each cycle process. Heat transfers take place at the high and low temperatures respectively. Thus these processes product entropy from the external heat transfers which are not included in the following table.

Process	Q (Btu/lbm)	Δs (Btu/ lbm-R)	Q/T (Btu/ lbm-R)	σ (Btu/ lbm-R)
a-b	0	0	0	0
a-b'	0	0.0230	0	0.0230
b'-c	189.9	0.1455	0.1455	0
c-d	0	0	0	0
c-d'	0	0.0299	0	0.0299
d'-a	−162.6	−0.1983	−0.1983	0
NET	27.3	0	0.0528	0.0529

Improvements to the Brayton Cycle

Various methods have been developed to improve the thermal efficiency of the Brayton cycle. The goals are to minimize the work of the compression process and increase the average temperature of heat addition. The theoretical limit of cycle efficiency is the Carnot efficiency based on the highest temperature and lowest temperatures between which the cycle operates. The key variable is the turbine inlet temperature which is limited by the material properties and turbine blade cooling capability.

Regeneration A Brayton cycle with regeneration is illustrated below (Fig. 2.1.14).

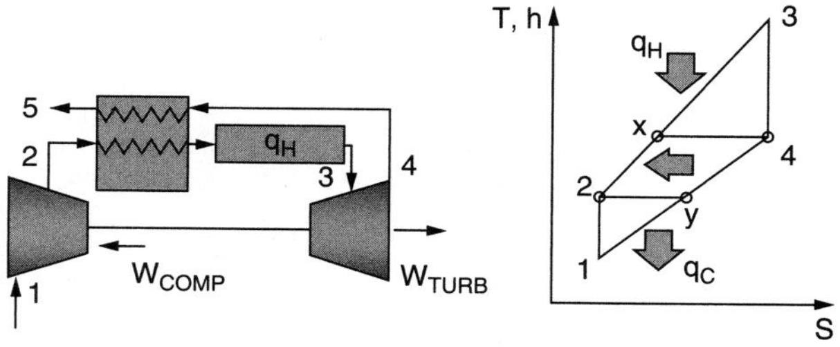

Figure 2.1.14 Air-standard Brayton cycle with regeneration.

A heat exchanger captures energy from the turbine exhaust (4 → y) and under ideal conditions raises the inlet temperature in the combustor (2 → x). Cycle efficiency is increased as energy addition is reduced without sacrificing work output. Thus cycle efficiency is given by $\eta = 1 - (h_2 - h_1)/(h_3 - h_4)$. Less efficient heat exchange reduces T_x and raises T_y.

Intercooling Compression work can be reduced by employing multiple stages of compression with heat rejection between each stage. The number of stages of compression is usually limited by economics and technical factors (size, weight, and power). Figure 2.1.15 illustrates an open Brayton cycle with one state of intercooling between a two-stage compression process (1 → 5, 6 → 7).

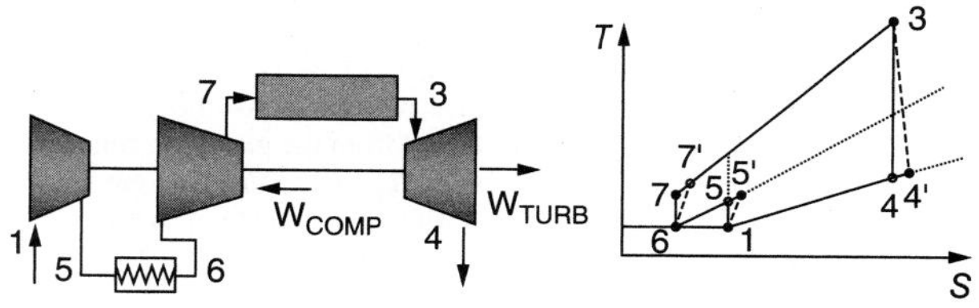

Figure 2.1.15 Air-standard Brayton cycle with intercooling in the compression process. Dashed lines show effect of irreversibility.

The total compression work, $W_{comp} = (h_7 - h_6) + (h_5 - h_6)$ is less than that which would have been required if a single state of compression would have been used starting at state 1. Compressor inefficiency increases compression work above the isentropic values as shown.

Reheat Turbine work can be increased by multiple stages of expansion with heat addition between them. The reheating of the gas at an intermediate pressure raises the enthalpy input to the cycle without raising the temperature of the external thermal reservoir, i.e., the maximum temperature of the combustion process. Figure 2.1.16 shows the T-s diagram for a Brayton cycle with one stage of reheating (4 → 5). Work output is increased by the additional expansion process (5 → 6). The number of stages of reheat is usually limited by economics and technical restrictions on the plant (size, weight, and power).

Stirling Cycle

The Stirling engine may be visualized as a cylinder with a piston at each end. Between the pistons is a regenerator. The cylinder is assumed to be insulated except for a contact with a hot reservoir at one end and a contact with a cold reservoir at the other end. A reversible Stirling cycle on p-V coordinates is shown in Fig. 2.1.17.

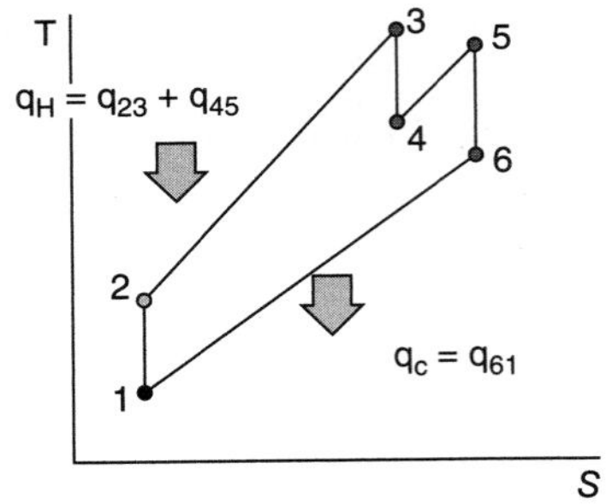

Figure 2.1.16 Air-standard Brayton cycle with one stage of reheat in the expansion process.

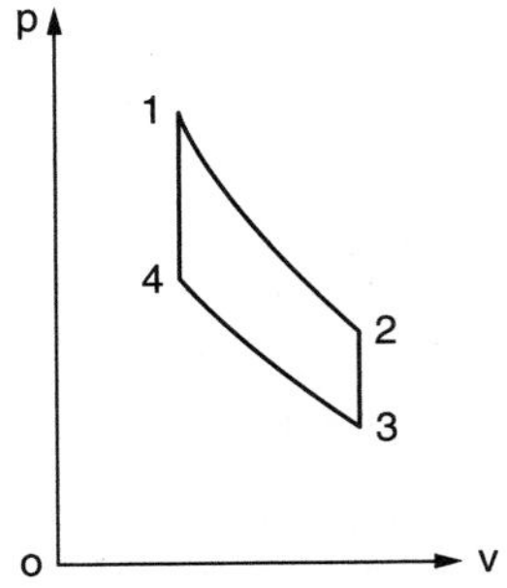

Figure 2.1.17 The air-standard Stirling cycle.

Starting with state 1, heat from the hot reservoir is added to the gas at T_H (1 → 2). During the reversible isothermal process, the left piston moves outward, doing work as the system volume increases and the pressure falls. Both pistons are then moved to the right at the same rate to keep the system volume constant process (2 → 3).

No heat transfer occurs with either reservoir. As the gas passes through the regenerator, heat is transferred from the gas to the regenerator, causing the gas temperature to fall to T_C by the time the gas leaves the right end of the regenerator. For this heat-transfer process to be reversible, the temperature of the regenerator at each point must equal the gas temperature at that point. Hence there is a temperature gradient through the regenerator from T_H at the left end to T_L at the right end. No work is accomplished during this process. In process 3 → 4, heat is removed from the gas at T_C to the reservoir at T_C. To hold the gas temperature constant, the right piston is moved inward doing work on the gas with a resulting increase in pressure. During process 4 → 1, both pistons are moved to the left at the same rate to keep the system volume constant. The pistons are closer together during this process than they were during process 2 → 3, since $V_4 = V_1 < V_2 < V_3$. No heat is transferred to either reservoir. As the gas passes back through the regenerator, the energy stored in the regenerator during 2 → 3 is returned to the gas. The gas emerges from the left end of the regenerator at the temperature T_H. No work is performed during this process since the volume remains constant. Thus the cycle is completed and is externally reversible. The system exchanges a net amount of heat with only the two energy reservoirs T_H and T_C.

Two types of Stirling engines are shown in Fig. 2.1.18. Extensive research-and development effort has been devoted to the Stirling engines for future use as prime movers in space power systems operating on solar energy.

Gas Compression

In ideal gas compression, the process is taken to be reversible (i.e., no friction losses), and the compressor works without a clearance volume. When the ideal gas laws cannot be used, analysis can be acceptably done in using the ideal gas equation of state with the compressibility factor. If the compression process $p_1 \rightarrow p_2$ is polytropic, $pV^n = Constant$, and work of compression is,

$$W_{\text{compression}} = np_1V_1 \frac{\left(\frac{p_2}{p_1}\right)^{(n-1)/n} - 1}{n-1}$$

The temperature at the end of compression is,

$$\frac{T_2}{T_1} = \left(\frac{p_2}{p_1}\right)^{\frac{n-1}{n}}$$

The compression work is smaller the value on n, and the cooling the compression chamber, e.g., with water jackets, reduces n from the isentropic value 1.4. Under usual working conditions, $n > 1.3$.

Intercooling When the pressure p_2 is high, it is advantageous to divide the process into two or more stages and cool the gas between the stages. Such cooling is termed intercooling, and in theory any number

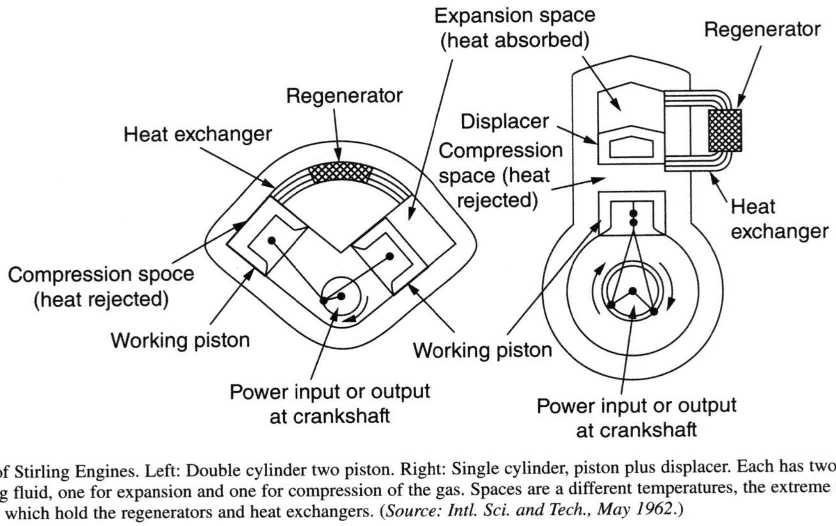

Figure 2.1.18 Two type of Stirling Engines. Left: Double cylinder two piston. Right: Single cylinder, piston plus displacer. Each has two variable volume working space filled with the working fluid, one for expansion and one for compression of the gas. Spaces are a different temperatures, the extreme temperatures of the cycle, and are connected by a duct which hold the regenerators and heat exchangers. (*Source: Intl. Sci. and Tech., May 1962.*)

of stages of intercooling can be used. Figure 2.1.19 illustrates the effect of two stages of intercooling on the overall compression process.

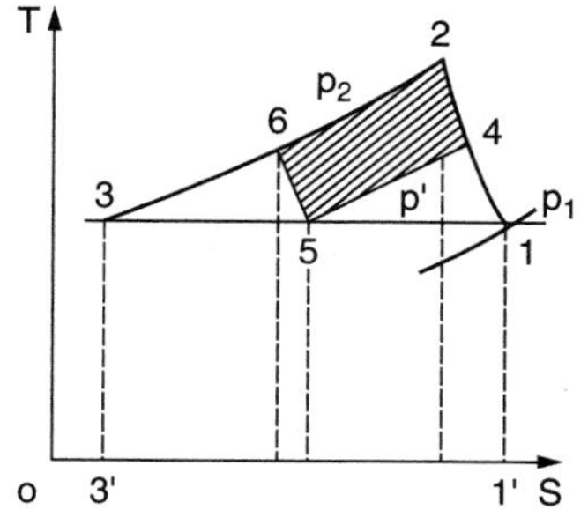

Figure 2.1.19 Compression with intercooling.

With single-stage compression, process 1 → 2 represents the compression from p_1 to p_2, and if the constant-pressure line 2 → 3 is drawn cutting the isothermal through point 1 in point 3, the area 1-2-3-1 represents the work *W*. When two stages are used, process 1 → 4 represents the compression from p_1 to an intermediate pressure $p_4 < p_2$; 4 → 5, cooling at constant pressure in the intercooler between the cylinders; and 5 → 6, the compression in the second stage. The area under 1-4-5-6-3 represents the work of the two stages and the area 4-5-6-2 the savings effected by intercooling. This saving is a maximum when $T_4 = T_6$, and this is the case when the intermediate pressure, p_4, is given by $p_4 = \sqrt{p_1 p_2}$.

The total work in two-stage compression is,

$$W_{\text{comp}} = v p_1 V_1 \left[\frac{\left(\frac{p_4}{p_1}\right)^{(n-1)/n} + \left(\frac{p_2}{p_4}\right)^{(n-1)/n} - 2}{n-1} \right]$$

References

1. Boyce MP. "Gas Turbine Engineering Handbook," 4e, Elsevier, Waltham, 2011. doi: http://dx.doi.org/10.1016/B978-0-12-383842-1.00025-1

2. Curzon FL, Ahlborn B. "Efficiency of a carnot engine at maximum power output," *Amer. J. Phys.*, **43**, no. 1, 1975, pp. 22-24

3. Decher R. "Energy Conversion," Oxford University Press, New York, 1994.

4. Leff HS. Thermal efficiency at maximum work output: new results for old heat engines," *Amer. J. Phys.*, **55**, no. 7, 1987, pp. 602-609. doi: http://dx.doi.org/10.1119/1.15071

5. McDonald CF. The increasing role of heat exchangers in gas turbine plants. Paper 89-GT-103, ASME 1989. *Int Gas Turbine Aero Congr Expo,* 1989. doi: 10.1115/89-GT-103

6. McDonald CF. Gas turbine recuperator renaissance. *Heat Recovery Sys & CHP*, **10**, no. 1, 1990, pp. 1-30. doi: http://dx.doi.org/10.1016/0890-4332(90)90246-G

VAPOR POWER CYCLES

Rankine Cycle

Ideal Rankine Cycle The ideal Rankine cycle is a standard reference for comparing the performance of actual vapor power cycles. Figure 2.1.20 shows two internally reversible cycles on T-s coordinates in which the fluid is heated from the compressed liquid state to a superheated state and then expanded isentropic ally to produce work. Vapor power cycles are generally classified as property-limited cycles owing to the limits on the work produced because of the thermodynamic properties of the working fluid. Cycles in which energy addition begins below the critical point are denoted as subcritical cycles and thermal performance is determined by the peak pressure and temperature of heat transfer to the working fluid. Cycles for which heat transfer to the working fluid is at or above the critical point are denoted as supercritical cycles.

Stationary power plants for production of grid-delivered electricity mainly use the Rankine cycle with water as the working fluid, although other compressible fluids and fluid mixtures may be used in certain applications. An internally reversible subcritical cycle is shown in Fig. 2.1.20*a*. Process 0 → 1 → 2 represents the isentropic pumping of compressed liquid from the saturated state to the pressure where heat is transferred to it in a boiler. The pump work reduces the work output of the cycle, and pumping of the compressed liquid minimizes pump work. The ratio of pump work to turbine work per unit mass of the working fluid, $W_r = W_{\text{pump}}/W_{\text{turbine}}$, is denoted the work ratio. Process 2 → 4 represents energy in addition to the fluid from the compressed liquid state at constant pressure to a superheated state. Process 4 → 5 represents isentropic expansion through a turbine in which work is produced. State 5 is shown to lie within the vapor-liquid region at a high quality. Process 5 → 0 represents heat rejection at constant pressure and temperature to return the fluid to the saturated liquid at state 0. The condensing temperature, $T_5 = T_0$, and the temperature of the fluid entering

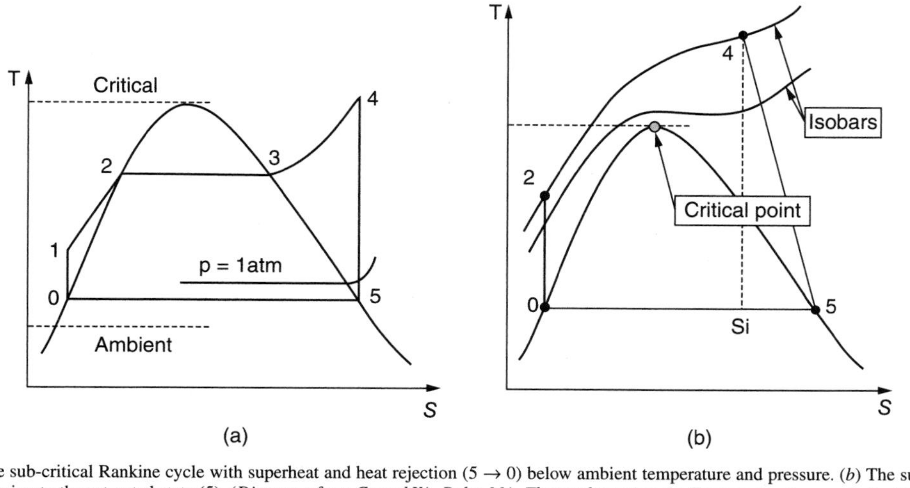

Figure 2.1.20 (*a*) The sub-critical Rankine cycle with superheat and heat rejection (5 → 0) below ambient temperature and pressure. (*b*) The super-critical Rankine cycle with turbine expansion to the saturated state (5). (*Diagrams from Çengel YA, Boles MA. Thermodynamics: An Engineering Approach, 5e, McGraw-Hill.*)

the turbine, T_4, represent the limits that define the Carnot efficiency for the cycle ($\eta_{Carnot} = 1 - T_5/T_4$). Modern Rankine power systems employ a steam turbine for the expansion process, and superheating is common as condensed water is to be avoided because it causes damage to turbine blades. The cycle is closed, and each process of it is described with a steady open system analysis.

Thermal Efficiency The isentropic work produced per unit mass of working fluid via the expansion process 4 → 5 is $h_4 - h_4$. The isentropic work input for pumping is $h_2 - h_1$. The energy input is $h_3 - h_2$. The cycle efficiency is therefore,

$$\eta = \frac{(h_3 - h_4) - (h_2 - h_1)}{h_3 - h_2}$$

The work ratio for the cycle is $W_r = (h_2 - h_1)/(h_3 - h_4)$. Because of the pumping of the compressed fluid, work ratios are generally small, i.e., $\approx v_1(p_2 - p_1)$ per unit mass of fluid.

Figure 2.1.21 illustrates an internally reversible subcritical cycle without superheat and shows the variation of thermal efficiency with the pressure of heat addition when heat rejection takes place at 1 atm (~14 psia). The turbine expansion process in this case takes place within the liquid vapor region, which is to be avoided in practice. The thermal efficiencies illustrate the effect of superheating on raising cycle efficiency.

Steam Consumption Steam consumption via the heating process in pounds mass per horsepower-hour is $2{,}544/(h_3 - h_4)$ (lbm/hph). Expressed in pounds mass per kilowatt-hour, $3{,}412.7/(h_3 - h_4)$ (lbm/Kwh). The thermal performance of the cycle is frequently stated in terms of the energy use per horsepower-hour,

$$\eta' = 2544\eta = 2544\frac{[(h_3 - h_4) - (h_2 - h_1)]}{h_3 - h_2}$$

Reheat Cycle Reheating the working fluid after an initial expansion and adding additional stages of expansion increases the work output per unit mass and raises the cycle thermal efficiency. Reheating the working fluid essentially raises the average temperature of heat addition and thus raises the cycle efficiency.

Figure 2.1.22 is a representative schematic and the T-s diagram for an internally reversible Rankine cycle with one stage of reheat between the high and low pressure turbines. Reheat is achieved by passing the exhaust of the high pressure turbine (state 4) through the boiler where its temperature is raised a constant pressure to state 5 (4 → 5). The fluid is then expanded through a low pressure turbine (5 → 6) to extract additional work per unit mass. The energy absorbed per unit mass is $(h_3 - h_4) + (h_5 - h_6)$. The cyclic work is $(h_3 - h_4) + (h_5 - h_6) - (h_2 - h_1)$. The cycle efficiency is increased with the work gained from additional heat transfer at the intermediate pressure p_4. The number of stages of reheat that can be added in practice is generally limited by economic considerations involving initial plant cost, operating costs, and annual maintenance. The following example illustrates how reheat increases cycle thermal efficiency.

EXAMPLE. An internally reversible Rankine cycle with one stage of reheat has a maximum operating temperature of 800°F, a quality at the high and low pressure turbine discharge of 0.9, and a minimum condensing temperature of 70°F. Reheat brings the fluid up to the maximum temperature at an intermediate pressure p_c. The T-s diagram below shows additional operating parameters of the cycle. Determine the work produced and the cycle efficiency.

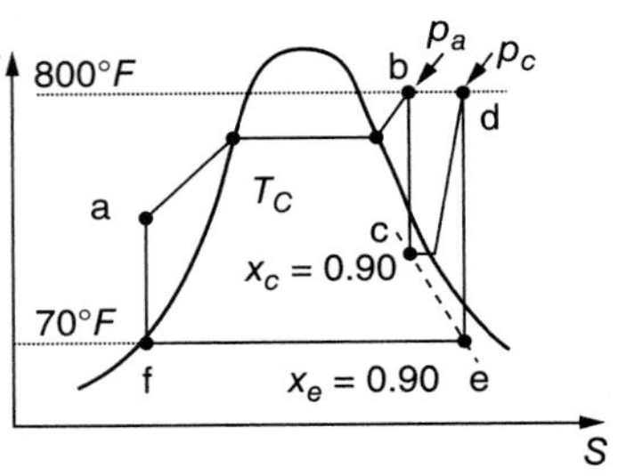

The state properties and process quantities are summarized in the following tables.

State	T R	P psi	h BTU/Ibm	s BTU/Ibm,-R	X
a	530	1600	42.8	0.0745	...
b	1260	1600	1358	1.499	...
c	775	84	1094	1.499	0.9
d	1260	84	1430	1.864	...
e	530	0.3631	988	1.864	0.9
f	530	0.3631	38	0.0745	0

Process	ΔH BTU/lbm	δW BTU/lbm	δQ BTU/lbm
a-b	1315	0	1315
b-c	−264	264	0
c-d	336	0	336
d-e	−442	442	0
e-f	−950	0	−950
f-a	5	−5	0
Net	0	701	701

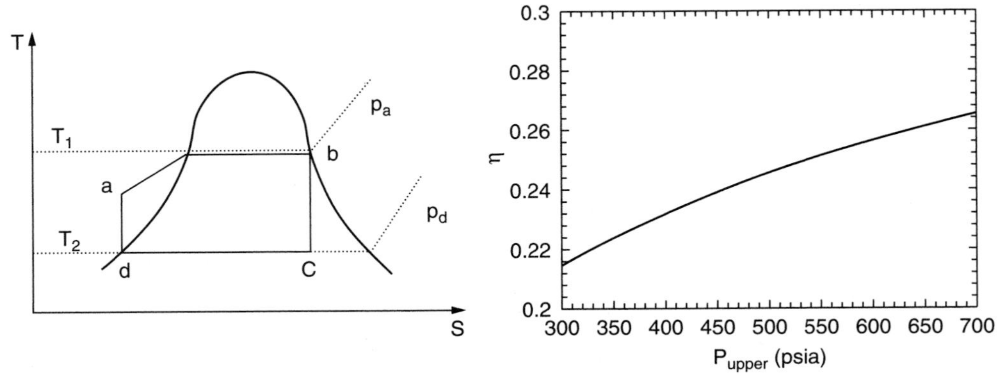

(a) (b)

Figure 2.1.21 (*a*) A sub-critical internally reversible Rankine cycle without superheat. (*b*) Variation of efficiency of an internally reversible Rankine cycle with pressure of heat addition, p_a, and condensing pressure of 14 psia. The working fluid is water.

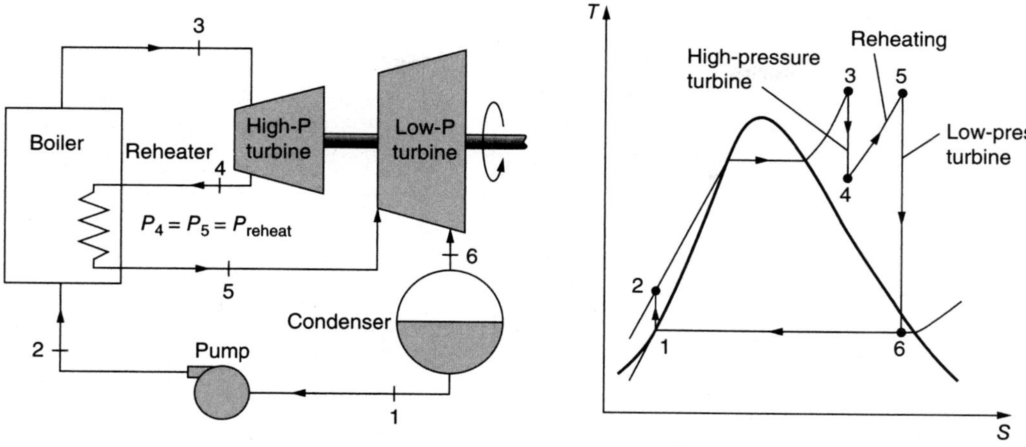

Figure 2.1.22 Rankine cycle with one stage of reheat and T-s diagram. (*Diagrams from Çengel YA, Boles MA. Thermodynamics: An Engineering Approach, 5e, McGraw-Hill.*)

The thermal efficiency is η = 0.424, the Carnot efficiency is η = 0.579, and the work ratio is W_r = 0.007. In this example, the additional stage of reheat has significantly increased work output with the addition of 336 Btu/lbm at the intermediate pressure.

The pressure at which reheat is initiated effects thermal efficiency. In Fig. 2.1.23, an internally reversible Rankine cycle with one stage of reheat is illustrated and water as the working fluid. Reheat begins at an intermediate pressure p_{middle} where $x = 1$. The exhaust of the low pressure turbine (state 4) begins when $x_4 = 1$. The variation of cycle efficiency with p_{middle} is shown for a fixed maximum pressure, $p_{\text{up}} = p_2$. The nonlinear behavior of efficiency with the reheat pressure is a result of the complex nature of the state surface for water.

Effect of Irreversibility reduces cycle efficiency. Irreversibility arises in pumping, turbine expansion (heat losses), and pressure drop in the boiler and condenser. Figure 2.1.24 shows how a cycle with superheat might deviate from the ideal when these irreversible effects are present. The following example illustrates the effect of irreversibility in pumping and turbine expansion on thermal efficiency of a basic Rankine cycle without reheat.

EXAMPLE. An internally reversible Rankine cycle with water as the working fluid operates with $p_a = p_b$ = 500 psi and a condensing temperature of 70°F at 0.361 psi. The figure below shows the T-s diagram, and tables list the state properties and process quantities. Determine the work put, work ratio, and thermal efficiency.

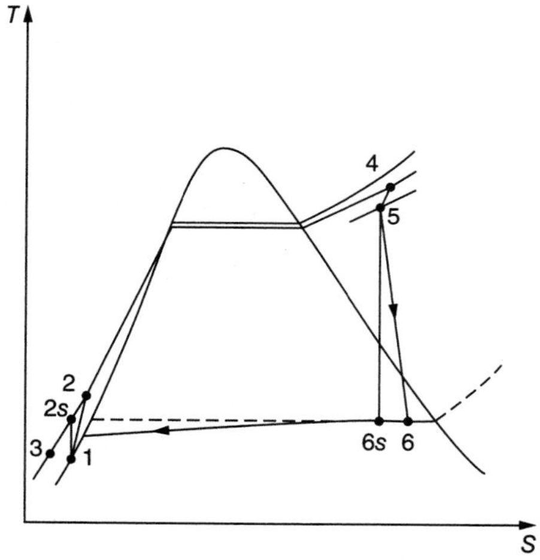

Figure 2.1.24 Deviation of the actual Rankine cycle with reheat from the ideal and internally reversible cycle. Subscript *s* denotes states for isentropic processes. (*Diagram from Çengel YA, Boles MA. Thermodynamics: An Engineering Approach, 5e, McGraw-Hill.*)

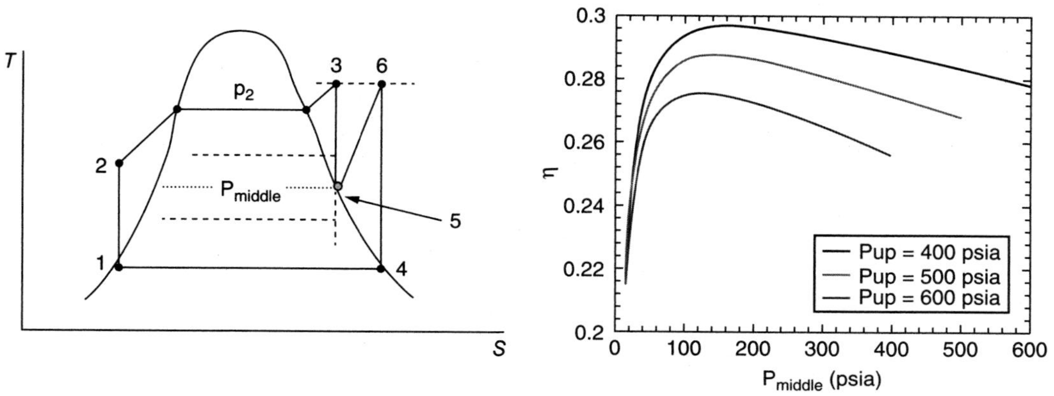

Figure 2.1.23 Effect of reheat pressure on cycle efficiency of an internally reversible Rankine cycle with water as the working fluid and reheat at p_{middle} beginning at saturation (state 5).

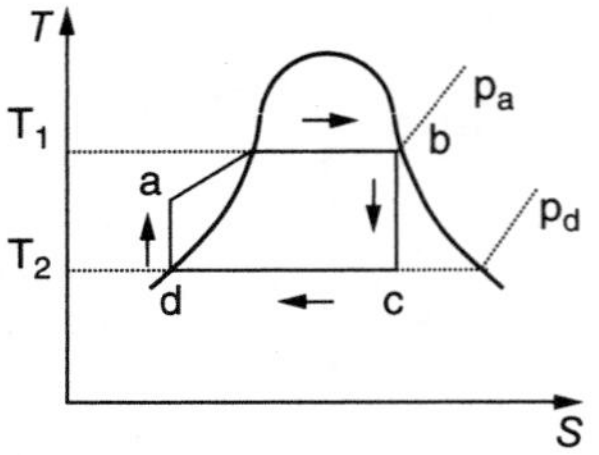

State	T R	P psi	h BTU/lbm	S BTU/lbm-R	X
a	927	500	39.5	0.0745	0
b	927	500	1204	1.4634	1
c	530	0.361	774	1.4634	0.698
d	530	0.361	38.0	0.0745	0

Process	ΔH BTU/lbm	δW BTU/lbm	δQ BTU/lbm	ΔS BTU/ lbm-R	δQ/T BTU/ lbm-R	σ. BTU/ lbm-R
a-b	1164.5	0	1164.5	1.3889	1.3889	0
b-c	−430	430	0	0	0	0
c-d	−736	0	−736	−1.3889	−1.3889	0
d-a	1.5	−1.5	0	0	0	0
Net	0	428.5	428.5	0	0	0

The thermal efficiency is 36.8 percent, and $W_r = 0.0035$. The Carnot efficiency is 42.8 percent.

If the turbine and pump isentropic efficiencies are 80 percent, the state data and process quantities are as follows.

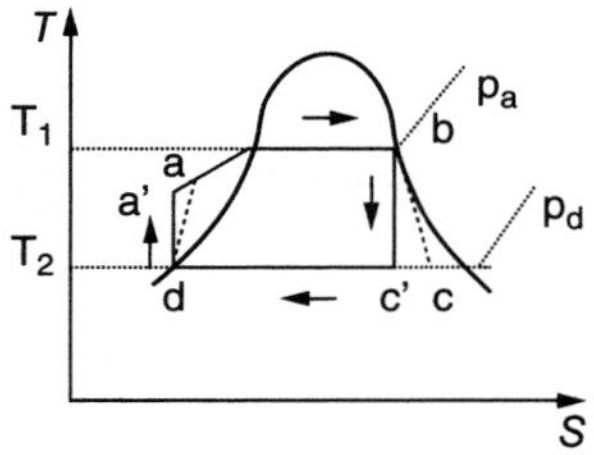

State	T R	P psi	h BTU/lbm	s BTU/lbm-R	X
a	530	500	39.5	0.0745	...
a'	530	500	39.9	0.0753	
b	927	500	1204	1.4364	1
c	530	0.3631	774	1.4364	0.698
c'	530	0.3631	860	1.6262	0.780
d	530	0.3631	38.0	0.0745	0

Process	ΔH BTU/lbm	δW BTU/lbm	δQ BTU/lbm	ΔS BTU/ lbm-R	δQ/T BTU/ lbm-R	σ. BTU/ lbm-R
(a'-a)	(0.4)	(0)	(0.4)	(0.0008)	(0.0008)	(0)
a-b	**1164**	**0**	**1164**	**1.3881**	**1.3881**	0
(b-c')	(−430)	(0)	(0)	(0)	(0)	(0)
b-c	**−344**	**344**	**0**	**0.1628**	**0**	**0.1628**
c-d	**−822**	**0**	**−822**	**−1.5517**	**−1.5517**	**0**
(d-a')	(1.5)	(−1.5)	(0)	(0)	(0)	(0)
d-a	**1.9**	**−1.9**	**0**	**0.008**	**0**	**0.0008**
Net	0	342	342	0	−0.1636	+0.1636

The work output is 342 Btu/lbm and the work ratio is 0.0054. Thermal efficiency is reduced to 29.4 percent, and entropy production is 0.1636 Btu/lbm. Irreversible pumping and expansion has significantly reduced thermal efficiency.

Regenerative Cycle In the regenerative cycle, the working fluid is drawn from the turbine at one or more stages of the expansion process and used to heat the feed water prior to pumping. In Fig. 2.1.25 an open feed water heater (FWH) implies the mixing of the two feed water streams and simultaneous condensation of the vapor withdrawn from the turbine. Closed FWHs are also employed. In a closed FWH, direct mixing of the regenerative flow from the turbine and the condensing fluid does not occur. For either open or closed feed water heating, the number of stages of regeneration is limited by capital and maintenance costs.

Figure 2.1.25*b* shows the T-s diagram for one stage of regeneration in a cycle without reheat. Entering the turbine is $1 + y$ lbm at (p_5, T_5). Regeneration begins for y lbm of fluid at (p_6, T_6), and h_2 enters the FWH. The remaining mass passes through the turbine and condenser and enters the FWH as water at temperature (p_3, T_3). The work done by the regenerating fluid is $y(h_5 - h_6)$ in the initial expansion in the turbine and that by the remaining fluid, $h_5 - h_7$. Total work output is therefore $y(h_5 - h_6) + (h_5 - h_7)$

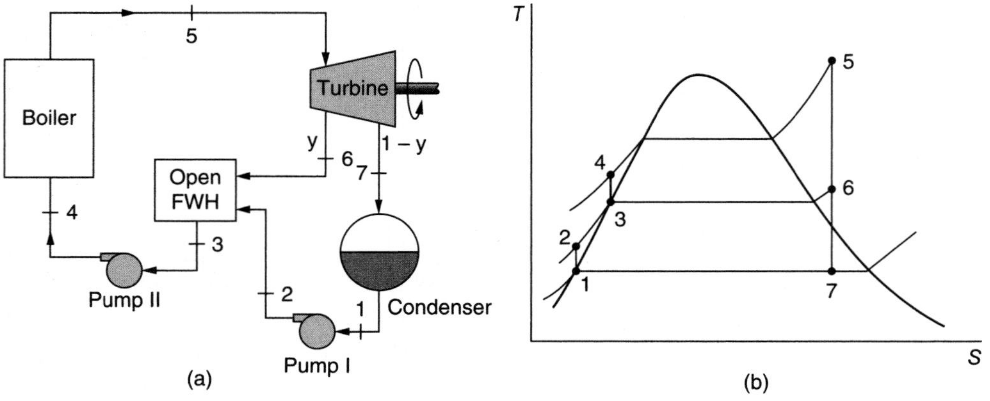

Figure 2.1.25 Regenerative feed water heating. (*Diagrams from Çengel YA, Boles MA. Thermodynamics: An Engineering Approach, 5e, McGraw-Hill.*)

if pump work is neglected. The energy supplied between FWH and turbine is $(1 + y)(h_5 - h_4)$. Hence the ideal cyclic efficiency is,

$$\eta = \frac{y(h_5 - h_6) + (h_5 - h_7)}{(1 + y)(h_5 - h_4)}$$

References

1. Cengel YA, Boles MA. "Thermodynamics: An Engineering Approach," 8e, McGraw-Hill, New York, 2015.
2. Decher R. "Energy Conversion," Oxford University Press, New York, 1994.
3. Hall N, Ibele WE. "Engineering Thermodynamics," Prentice-Hall, Englewood Cliffs, 1960.
4. Moran MJ, Shapiro HN. "Fundamentals of Engineering Thermodynamics," 5e, John Wiley & Sons, Hoboken, 2004.
5. Radermacher R. "Thermodynamic and heat transfer implications of working fluid mixtures in Rankine cycles," *Int. J. Heat Fluid Flow,* **10**, no. 2, 1989, pp. 90-102.

PSYCHROMETRICS

Psychrometrics broadly applies to the description of air-water vapor mixtures. The control of water content and temperature for human comfort is the objective of air conditioning systems design and operation, and analysis of processes involving air-water vapor mixtures is a necessity to enable the design of heating, ventilating, and air conditioning equipment and systems.

Air-water vapor mixtures are treated as non-reacting ideal gas mixtures in which the water is near the saturated vapor line on the p-v-T surface. Environmental applications are generally considered to take place at one atmosphere total pressure, or slightly removed from it. Dry air is a mixture of approximately 21 percent oxygen, 79 percent nitrogen, and 1 percent argon. The average molecular weight is $M = 28.97$, and the specific gas constant is $R_{specific} = 287$ (J/kg·K) = 53.34 (ft-lbf/lbm·R). Water vapor has a molecular weight $M = 18.02$ and a specific gas constant $R_{specific} = 461.5$ (J/kg·K).

Definitions

Descriptions of air-water vapor mixtures are generally given in terms of specific humidity, degree of saturation, and relative humidity. The mixture temperature recorded by an exposed thermometer is denoted as the dry bulb temperature, T_{db}.

Specific Humidity The specific humidity, or humidity, is defined as the mass of water vapor present per unit mass of dry air, $\omega = m_v/m_a$. An equivalent relation is $\omega = p_v/1.61(p - p_v)$ (lbm/lbm-dry air).

Degree of Saturation The degree of saturation is defined as the ratio of the humidity at a given total pressure and temperature to that at saturation, $\mu = (\omega/\omega_{sat})_{T,p}$.

Relative Humidity The relative humidity is the ratio of the partial pressure to the vapor to the saturation pressure of the vapor at the same temperature and same total pressure, $\phi = (p_v/p_{sat})_{T,p} = (\rho_v/\rho_{v,sat})_{T,p}$. For saturated air $\phi = 1$. In terms of mole fractions, $\phi = y_v/y_{v,sat}$.

Dew Point The dew point is the temperature at which the vapor condenses (or solidifies) when it is cooled at constant pressure. Thus, at a given vapor state (T, p_v), the dew point is $T_{sat}(p_v)$.

Figure 2.1.26 illustrates the relation of the several state quantities in the definitions these definitions.

Many methods are used to determine the dew point: (1) The dew point method measures the temperature at which condensation begins; water-vapor pressure can then be found from steam tables. A dew point apparatus can either cool a surface or compress and expand moist air. (2) Hygrometers measure relative humidity, often by using the change in dimensions of a hygroscopic material; these instruments are simple and inexpensive but require frequent calibration. The electrical resistance of an electrolytic film can also be used as an indication of relative humidity. (3) The wet and dry bulb psychrometer is widely used. Humidity measurements of air flowing in ducts can be made with psychrometers that use mercury-in-glass thermometers, thermocouples, or resistance thermometers. Humidity measurements in still air can be made with sling psychrometers as aspiration psychrometers. Psychrometric wet-bulb temperatures must be corrected to obtain thermodynamic wet-bulb temperatures, or there must be adequate air motion pass the wet-bulb thermometer, 800 to 900 ft/min (with duct walls at air temperature), to ensure a proper balance between radiation and convection. (4) Chemical analysis by the use of desiccants, such as sulfuric acid, phosphorus pentoxide, lithium chloride, or silica gel, can be used as primary standards of humidity measurement.

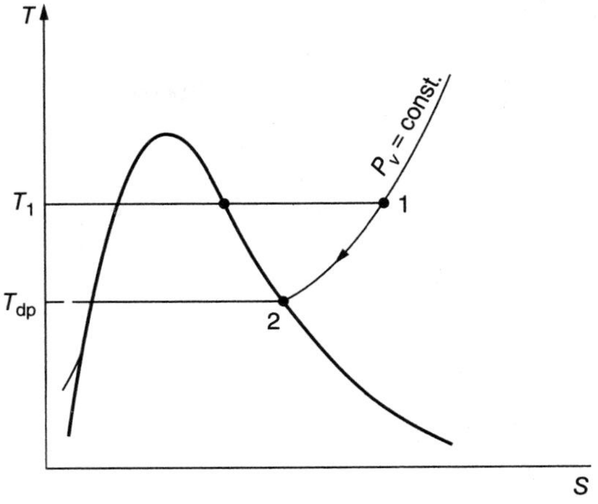

Figure 2.1.26 The T-s diagram for water. The dew point, T_{dp}, is at State 2, (s_2, T_2) where $p_v = p_g$. The relative humidity is determined by the pressure at State 1 and the saturation pressure at T_1. At State 2, f = 1. (*Diagram from Çengel YA, Boles MA. Thermodynamics: An Engineering Approach, 5e, McGraw-Hill.*)

Useful Relations

The following relations are useful in describing air-water vapor mixtures and in design and analysis of air conditioning systems.

The volume of mixture in ft³/lbm-dry air, v_a, is,

$$v_a = \frac{1}{\rho_a} = 0.754\frac{T + 460}{p - \phi p_d}$$

where T is the dry bulb temperature and ρ_d is the density of dry air.

The volume of mixture in ft³/lbm, v_m, of mixture is,

$$v_m = \frac{v_a}{1 + \omega}$$

The enthalpy of a mixture of dry air and water vapor is the sum of the enthalpy of a unit mass of dry air and the enthalpy of ω mass units of water vapor mixed with the dry air. The specific enthalpy of dry air (above 0°F) is $h_a = 0.240T$ (Btu/lbm) for $T < 130$°F. The specific enthalpy of low-pressure steam (saturated or superheated) is nearly independent of the vapor pressure and depends only on temperature. An empirical equation for the specific enthalpy of low-pressure steam and hence water vapor for –40 to 250°F in Btu/lbm is,

$$h_v = 1.062 + 0.44T$$

The specific enthalpy of a mixture of air and water vapor in Btu/lbm-dry air is,

$$h_m = 0.240T + \omega(1.062 + 0.44T)$$

The specific heat of a mixture of dry air and water vapor per pound mass of dry air may be called the humid specific heat and is 0.240 + 0.44ω (Btu/lbm-dry air). For a steady-flow process without change of specific humidity, heat transfer per pound mass of dry air may be computed as the product of humid specific heat and change in dry-Bulb Temperature, $\dot{q}/\dot{m} = (0.240 + 0.44\omega)\Delta T_{db}$.

Wet Bulb Temperature

The thermodynamic wet bulb temperature, T^*, is an important property of state of air-water vapor mixtures. It is the temperature at which water (or ice), by evaporating into a mixture of air and water vapor, will bring the mixture to saturation ($\phi = 1$) at the same temperature in a steady-flow process in the absence of external heat transfer. The process on T-s coordinates is illustrated in Fig. 2.1.27.

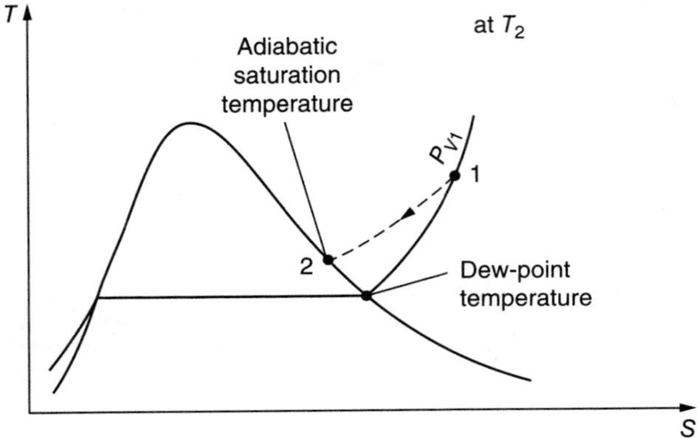

Figure 2.1.27 The T-s diagram for the adiabatic saturation process 1 → 2. (*Diagram from Çengel YA, Boles MA. Thermodynamics: An Engineering Approach, 5e, McGraw-Hill.*)

For a mixture of dry air and saturated vapor only, $T^* = T_{db}$, and when $\phi < 1$, $T^* < T_{db}$. By writing energy and mass balances for the adiabatic saturation process with water supplied at T^*, the specific humidity in terms of the dry and wet bulb temperatures is,

$$\omega = \omega^* - \frac{(0.240 + 0.44\omega^*)(T - T^*)}{1.094 + 0.44T - T^*}$$

where ω^* is the specific humidity for saturation at the total pressure, p.

The enthalpy of a mixture of dry air and saturated water vapor at total pressure p and T^* exceeds the enthalpy of a mixture of dry air and superheated steam at the same p and T^* and is,

$$h_m^* = h_m + (\omega^* - \omega)h_f^*$$

A property of the mixture that remains constant for constant p and T^* is denoted the S-function,

$$\Sigma^* = h_m^* - \omega^* h_f^* = \Sigma = h_m - \omega h_f^*$$

EXAMPLE. Given a mixture of dry air and saturated water vapor at p = 24 in.-Hg, T = 76°F. Determine the volume of mixture per lbm of dry air, volume of mixture per lbm of mixture, and enthalpy of mixture.

The partial pressure of water vapor from tables is $p_v = p_d = 0.905$ in.-Hg

The partial pressure of dry air is $p_a = p - p_v = 23.095$ in,-Hg.

Specific humidity is $\omega = 0.905/(1.61 \times 23.095) = 0.0243$ lbm/lbm-dry air.

Volume of mixture per lbm of dry air is $v_a = 0.754 \times 536/23.095 = 17.5$ ft^3

Volume of mixture per lbm of mixture is $v_m = 17.5/1.0243 = 17.1$ ft^3

Enthalpy of mixture is $h_m = 0.240 \times 76 + 0.0243 \times 1.095 = 44.85$ Btu/lbm-dry air.

EXAMPLE. Given a mixture of dry air and superheated water vapor at 24 in.-Hg total pressure, T = 76°F, and T^* = 62°F. Determine the specific humidity, relative humidity, volume of mixture per lbm-dry air, volume of mixture per lbm-mixture, and specific enthalpy.

The pressure of saturated water vapor at T^* is 0.560 in.-Hg. The specific humidity at T^* is $\omega^* = 0.560/(1.61 \times 23.44) = 0.01484$ lbm/lbm-dry air.

The mixture humidity is $\omega = 0.01484 - (0.2465 \times 14/1065.4) = 0.0116$ lbm/lbm-dry air.

The partial pressure of water vapor is determined from $0.0116 = p_v/[1.61(24 - p_v)]$, and $p_v = 0.44$ in.-Hg.

Relative humidity is $\phi = 0.44/0.905 = 0.486$.

The volume of mixture per lbm of dry air is $v_a = 0.754 \times 536/23.56 = 17.2$ ft^3/lbm-dry air.

The volume of mixture per lbm of mixture is v_m =17.2/1.0116 = 17.0 ft^3/lbm-mixture. The enthalpy of mixture is $h_m = 0.240 \times 76 + 0.0116 \times 1095 = 30.95$ Btu/lbm-dry air.

Pyschrometric Chart

For occasional use, algebraic equations are broadly applicable and highly reliable. For frequent use, the psychometric chart at a given total pressure may be preferable. Figure 2.1.28 shows the several variables plotted on the chart. For commonly encountered applications, a total pressure of 760 mm Hg, 30 in-Hg, or 1 atm.

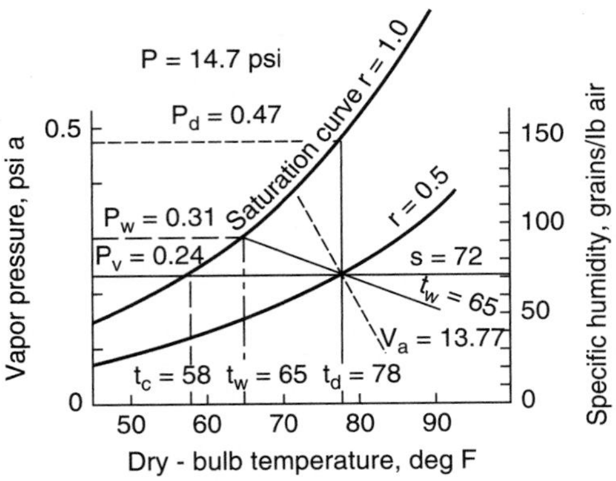

Figure 2.1.28 Essential features of the psychrometric chart.

Psychrometric charts are usually plotted with the dry-bulb temperature as the abscissa and specific humidity as the ordinate. The specific humidity is determined by the vapor pressure and the barometric pressure, which is constant for a given chart and is nearly proportional to the vapor pressure, a second ordinate scale, departing slightly from uniform graduations, gives the vapor pressure. The saturation curve ($\phi = 1.0$) gives the specific humidity and vapor pressure for a mixture of air and saturated vapor. Similar curves below it give results for various values of relative humidity. Inclined lines of one set carry fixed values of the wet-bulb temperature, T^*, and those of another set carry fixed values of v_a, the specific volume per pond mass of dry air. Many charts carry additional scales of enthalpy, or the S function. Any two values will locate the point representing the state of the atmosphere, and the desired values can be read directly.

Psychrometric charts are generally available for sea level applications and temperatures normally encountered in air conditioning applications. Charts for different altitudes and temperature ranges are available from the American Society of Heating Refrigeration and Air Conditioning Engineers (http://www.ashrae.org/resources—publications/bookstore/psychrometrics#other).

AIR CONDITIONING

Air conditioning processes alter the temperature and specific humidity of air-water vapor mixtures. In these processes, the mass of dry air remains constant, and consequently computations are best based upon a unit mass of dry air. Thus the specific humidity under goes a change, and either an increase (humidification) or decrease (dehumidification) can take place depending on the application. The specific enthalpy of the liquid, h_f (Btu/lbm-liquid) at the temperature of supply or removal, T_f, is $h_f = T_f - 32$.

Most air conditioning processes are steady flow processes in which there is no work transfer at the boundary. The analysis of them is based on an open system and is written in terms of a unit mass of dry air, e.g.,

Btu/lbm-dry air. Using subscript 1 for entering mixture and 2 for the departing mixture, the energy balance (First Law) is,

$$h_{m1} + m_{f1}h_{f1} + q_1 = h_{m2} + m_{f2}h_{f2} + q_2$$

where the subscript m denotes the mixture. Either or both values of m_f or q may be zero.

Processes involved in air conditioning include heating and cooling an air-water vapor mixture above its dew point, cooling below the dew point, adiabatic saturation, and mixing of two mixture flows. These in various sequences make it possible to start with a given mixture and produce one with any required characteristics.

Heating and Cooling above the Dew Point

Heating and cooling above the dew point (Fig. 2.1.29) do not produce condensation of water vapor. The barometric pressure and composition are unaltered, and partial pressure remains constant. Cooling below the dew point with dehumidification (Fig. 2.1.30) results in condensation of water vapor, and the final state will be saturated liquid.

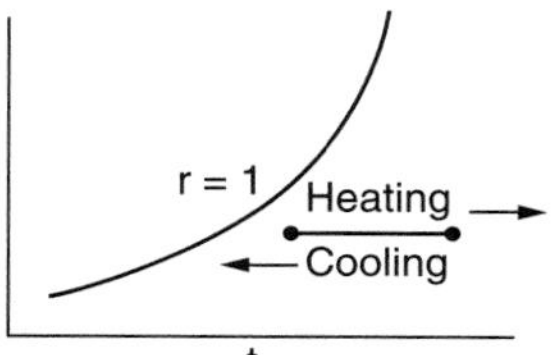

Figure 2.1.29 Heating and cooling above the dew point without condensation.

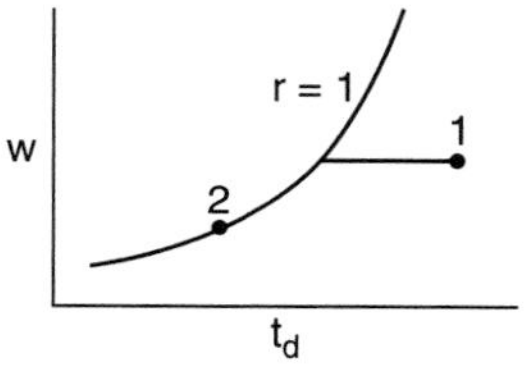

Figure 2.1.30 Cooling above the dew point with condensation.

EXAMPLE. Given an air-water vapor mixture at p = 28 in.-Hg, T = 60°F, T^* = 50°F, p_v = 0.26 in.-Hg and V = 1,200 ft³. The mixture under goes a process such that the final dry bulb temperature T = 82°F. Determine the final values of relative humidity, mixture density, mixture volume, mixture enthalpy, and heat transfer to the mixture.

Initial computed values: $\phi = 0.50$; $\omega = 0.0058$ lbm-vapor/lbm-dry air; $\rho_{ac} = 0.0707$ lbm-dry air/ft³; $m_a = V \times \rho_a = 1{,}200 \times 0.0707 = 84.9$ lbm-dry air; $h_m = 20.7$ Btu/lbm-mixture.

Final computed values: p_v, ω, and m_a are unaltered; $\phi = 0.24$; $\rho_a = 0.0679$ lbm-dry air/ft³; $V = m_a/\rho_a = 84.9/0.0679 = 1{,}250$ ft³; $h_m = 26.1$ Btu/lbm-mixture.

Heat transfer: $q_{12} = h_{m2} - h_{m1} = 26.1 - 20.7 = 5.4$ Btu/lbm-mixture. Heat transfer to the mixture, $Q = q_{12} \times m_a = 5.4 \times 84.9 = 458$ Btu.

EXAMPLE. Given an air-water vapor mixture at p = 29 in.-Hg; T = 75°F; T^* = 65°F, and V = 1,500 ft³. The mixture undergoes a process so that the dry bulb temperature at the state is T = 45°F. Determine the mass of liquid formed and the energy transfer.

Initial computed values: $\omega = 0.0113$ lbm-water/lbm-dry air; $\rho_a = 0.0706$ lbm-dry air/ft³.

$m_a = 1{,}500 \times 0.0706 = 106.0$ lbm-dry air; $h_m = 30.4$ Btu/lbm-dry air; the condensing temperature T_c = 60°F.

Final computed values: T = 45°F; p_v = 0.30 in.-Hg; $\phi = 1.0$; $\omega = 0.0065$ lbm-water/lbm-dry air; $\rho_a = 0.0754$ lbm-dry air/ft³; $V = 106.0/0.0754 = 1{,}406$ ft³; $h_m = 17.8$ Btu/lbm-dry air.

Liquid formed: $m_f = \omega_1 - \omega_2 = 0.0113 - 0.0065 = 0.0048$ lbm-liquid/lbm-dry air; $h_f = 50 - 32 = 18$ Btu/lbm-liquid (assuming that the liquid is drained out at an average temperature T_f = 50°F).

Heat transfer $q_{12} = h_{m1} - h_{m2} - m_1h_f = 30.4 - 17.8 - 0.0048 \times 18 = 12.5$ Btu/lbm-dry air; $Q = q_{12} \times m_a = 12.5 \times 106.0 = 1{,}325$ Btu.

Dehumidification

Dehumidification may be accomplished in a surface cooler, in which the air passes over tubes cooled by either brine or refrigerant flowing through them. The solution of this type of problem is most easily solved using the psychrometric chart (Fig. 2.1.31). The point representing the state of the entering atmosphere is first located. A straight line is then drawn to a point on the saturation curve (ϕ = 1) at the temperature of the cooling surface. The final state is approximated by a point on this line determined by the heat transferred by the cooling medium. This depends upon the extent of surface and the overall heat transfer coefficient.

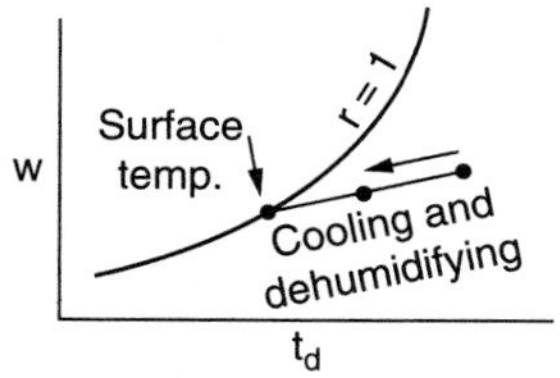

Figure 2.1.31 Cooling with dehumidification.

Adiabatic Saturation

Adiabatic saturation (humidification) may be conducted in a spray chamber through which an air-water vapor mixture flows. A large excess of water is supplied via spraying, and evaporation is made up by a suitable water supply integral with the throughflow. After the process has been operating for some time, the water in the spray chamber will have been cooled to the temperature of adiabatic saturation, which differs from the wet bulb temperature owing to the effects of radiation and velocity (convective heat and mass transfer). No heat transfer takes place and thus process is adiabatic. The heat of vaporization for the evaporated water is supplied by the cooling of the air passing through the chamber. The wet-bulb temperature of the atmosphere is constant throughout the chamber (Fig. 2.1.32). If the chamber is sufficiently large, the resulting mixture will be saturated at the wet bulb temperature of the entering mixture, i.e., as it passes through the chamber, T^* remains constant, and T_{db} is reduced from its initial value to T^*. In a chamber of commercial size, the process can terminate somewhat short of this state, and the endpoint is determined by the duration and effectiveness of contact between mixture and spray water. The evaporated water equals the increase in the specific humidity of the mixture.

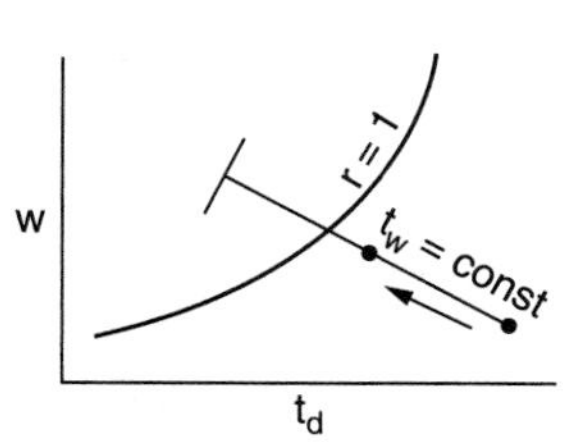

Figure 2.1.32 The adiabatic saturation process on the psychrometric chart.

EXAMPLE. Given an air-water vapor mixture at p = 30 in.-Hg; T = 78°F; T^* = 55°F; $\phi = 0.20$; ω = 28 grains-water/lbm-dry air. The mixture undergoes a process to produce the final conditions: $T = T^*$ = 55°F; $\phi = 1.0$; ω = 64 grains-water/lbm-dry air. Determine the water evaporated.

Water evaporated: $\omega_2 - \omega_1 = 64 - 28 = 36$ grains-water/lbm-dry air.

The spray chamber is preceded and followed by heat exchangers in line with the flow, the first to warm the entering atmosphere to the desired value of T^* determined by the prescribed final specific humidity, the second to warm the issuing mixture to the desired dry bulb temperature, T, and simultaneously reduce its relative humidity to the desired value.

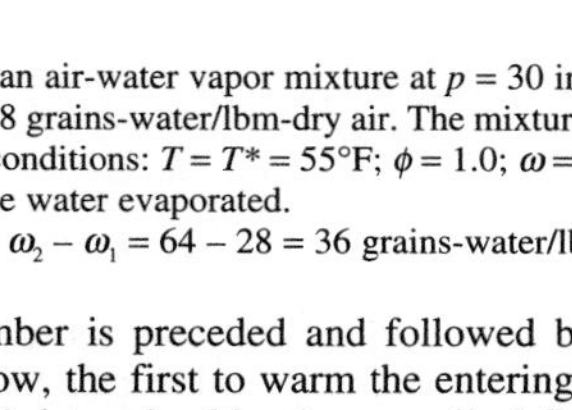

The spray chamber that is used for adiabatic saturation (humidification) in winter may be used for dehumidification in summer by supplying the spray nozzles with refrigerated water instead of recirculated water. In this case, the exiting mixture will be saturated at the temperature of the spray water, which will be held at the desired dew point. Subsequent heating of the atmosphere to an acceptable temperature will simultaneously reduce the relative humidity to the desired value.

Mixing

In recirculating ventilation systems, two air-water vapor mixture are mixed to form a third. The state of the final mixture is determined graphically on the psychrometric chart (Fig. 2.1.33). Locate the points 1 and 2 representing the states of the initial atmospheres. Connect these points by a straight line. Locate a point that divides this line into segments inversely proportional to the air masses in the respective mixtures. The division point represents the state of the final mixture, so long as it falls below the saturation curve ($\phi = 1$). If the final point falls above the saturation curve (Fig. 2.1.34) condensation will ensue.

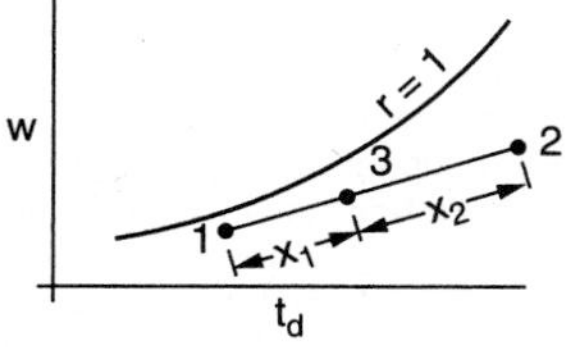

Figure 2.1.33 Mixing process of two air-water vapor streams ($1 + 2 \rightarrow 3$) diagramed on the psychrometric chart. The final state (3) is below saturation.

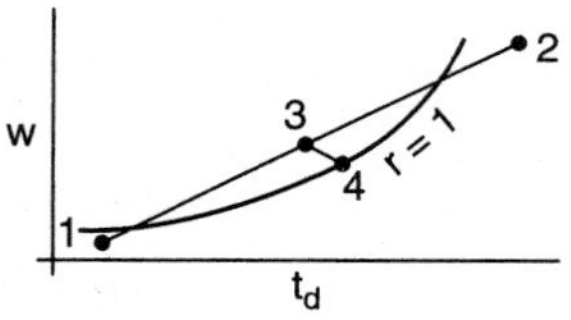

Figure 2.1.34 Mixing process of two air-water vapor streams ($1 + 2 \rightarrow 3$) diagramed on the psychrometric chart. The predicted final state (3) is above saturation, and the actual final state (4) is located on a line of constant wet bulb temperature.

The true final state is determined on a line of constant wet bulb temperature at state 3 at its intersection with the saturation curve. From all the points involved, readings of specific humidity may be taken, including state 3 when it falls above the saturation curve, and in this case the difference between ω_3 and ω_4 will be the specific condensate. For adiabatic mixing in a steady-flow process of two masses of moist air, each at the total pressure p,

$$\dot{m}_{a3} = \dot{m}_{a1} + \dot{m}_{a3}$$

In the absence of condensation,

$$\dot{m}_{a3}\omega_3 = \dot{m}_{a1}\omega_1 + \dot{m}_{a2}\omega_2$$

$$\dot{m}_{a3}h_{m3} = \dot{m}_{a1}h_{m1} + \dot{m}_{a2}h_{m2}$$

When condensation occurs, it is assumed that the condensate is removed at the final temperature, T_4, and that the final mixture consists of dry air and saturated water vapor at this same temperature. The mass of condensate is,

$$\dot{m}_{\text{condensate}} = \dot{m}_{a1}\omega_1 + \dot{m}_{a2}\omega_2 - \dot{m}_{a3}\omega_4$$

where ω_4 is the specific humidity for saturation at temperature T_4 and total pressure p. Also,

$$\dot{m}_{a1}h_{m1} + \dot{m}_{a2}h_{m2} = \dot{m}_{a2}h_{m4} + \dot{m}_{\text{condensate}}h_{f4}$$

In the case of condensation, a trial solution is necessary to find the temperature T_4 that will satisfy these relations.

EXAMPLE. Two thousand cubic feet of air per min at $T_1 = 80°F$ and $T_1^* = 65°F$ are mixed in an adiabatic, steady-flow process with 1,000 ft³ of air per min at $T_2 = 95°F$ and $T_2^* = 75°F$. The total pressure of each mixture is 29 in.-Hg. Determine the specific humidity, mixture enthalpy, and dry bulb temperature of the resulting mixture.

By computation, $\dot{m}_{a1} = 140$ lbm-dry air/min; $\omega_1 = 0.010$ lbm-water/lbm-dry air; $\dot{m}_{a2} = 67.6$ lbm-dry air/min; $\omega_2 = 0.0146$ lbm-water/lbm-dry air.

For the resulting mixture: $\dot{m}_{a3} = 207.6$ lbm-dry air/min; $\omega_3 = 0.0116$ lbm-water/lbm-dry air; $h_{m1} = 30.3$ Btu/lbm-mixture and $h_{m2} = 38.9$ Btu/lbm-mixture; $h_{m3} = 33.1$ Btu/lbm-mixture; $T_3 = 84.9°F$.

EXAMPLE. Fifteen hundred cubic feet of air per min at $T_1 = 0°F$ and $\phi_1 = 0.8$ are mixed in an adiabatic, steady-flow process with 1,000 ft³ of air per min at $T_2 = 100°F$ and $\phi_2 = 0.9$. The total pressure of each mixture is 30 in.-Hg. Determine the final mixture dry bulb temperature.

By computation, $\dot{m}_{a1} = 129.6$ lbm-dry air/min; $\omega_1 = 0.000626$ lbm-water/lbm-dry air; $\dot{m}_{a2} = 66.9$ lbm-dry air/min; $\omega_2 = 0.03824$ lbm-water/lbm-dry air; $h_{m1} = 0.09$ Btu/lbm-dry air; $h_{m2} = 66.29$ Btu/lbm-dry air.

Three equations must be satisfied by a choice of the terminal temperature,

$$m_{\text{condensate}} = 2.63 - 196.54\omega_4$$

$$4551 = 196.5h_{m4} + m_{\text{condnsate}}h_{f4}$$

$$\omega = \frac{p_{v4}}{1.61(30 - p_{v4})}, \quad \text{for } \phi_4 = 1$$

The value of T_4 that satisfies these equations is 55°F; condensation amounts to 0.84 lbm-water/min.

Cooling Towers

The cooling tower is a chamber for cooling liquid water. Outdoor air flows through a through a spray of entering hot water, and the temperature of the water is reduced in part by the warming of the air and by the evaporation of a portion of the water. The external air at ambient conditions, emerges at a higher temperature and is usually saturated ($\phi = 1$). It is commonly possible to cool the water below the temperature of the entering air to approximately half way between T_{db} and T_{wb}. The volume of inlet air per unit mass of supplied water and the mass of water evaporated are to be computed.

Evaporative Condensers In an evaporative condenser, vapor is condensed within tubes that are cooled by the evaporation of water flowing over the outside of the tubes; the water evaporates into the atmosphere. The computation of results is similar to that for the cooling tower.

EXAMPLE. A cooling tower is to receive water at 120°F and an atmosphere at $T = 90°F$, and $T^* = 80°F$, Thus $p_v = 0.92$ in.-Hg, $\omega = 0.0196$ lbm-water/lbm-dry air, $\rho_a = 0.0702$ lbm-dry air/ft³, and $h_m = 43.2$. The water is to be cooled to 85°F. What volume of atmosphere must be passed through the tower, and what mass of water will be lost by evaporation?

The issuing air-water vapor mixture will be assumed to be saturated at 115°F. Then $T = 115°F$, $p_v = 3.0$ in.-Hg, $\omega = 0.0690$ lbm-water/lbm-dry air, $\rho_a = 0.0623$ lbm-dry air/ft³, and $h_m = 104.4$ Btu/lbm-dry air.

The two unknowns are the mass flow through the tower and the mass of water to be evaporated. The two equations are the water-mass balance and the enthalpy balance (the steady-flow equation assuming zero heat transfer to or from the outside. Assume that 1 lbm of water enters, of which x lbm are evaporated. The water-mass balance $1 + m_a w_1 = 1 - x + m_a w_2$ becomes $x = m_a(w_2 - w_1) = m_a(0.0690 - 0.0196) = 0.0494m_a$. The enthalpy balance $1 \times (120 - 32) + m_a h_{m1} = (1 - x)(85 - 32) + m_a h_{m2}$ becomes $88 + 43.2m_a = 53(1 - x) + 104.4m_a$; whence $53x = 53 - 88 + m_a(104.4 - 43.2) = -35 + 61.2m_a$. Solving these simultaneous equations, $x = 0.0295$ lbm-water water evaporated/lbm-water entering, and $m_a = 0.597$ lbm-dry air/lbm-water entering.

REFRIGERATION

Vapor Compression Systems

Vapor compression refrigeration systems utilize the cooling effect achieved by throttling a saturated liquid refrigerant to a low temperature in the liquid-vapor region where heat transfer from the refrigerated medium takes place. An ideal cycle with one stage of compression is best shown on T-s coordinates (Fig. 2.1.35). State 1 is saturated vapor from which the compression process $1 \rightarrow 2$ raises the vapor to a superheated state (state 2). Heat transfer, Q_H, takes place by condensation in process $2 \rightarrow 3$ to produce saturated liquid at state 3. The throttling process ($3 \rightarrow 4$) lowers the temperature to T_4 with an increase of entropy Δs_{34}. The refrigeration effect is realized by heat transfer, Q_L at T_4.

Under the assumption of adiabatic throttling, the energy transferred, or absorbed, at T_1 per unit mass of refrigerant is $Q_L = h_1 - h_4 = h_1 - h_3$. The heat transferred per unit mass of refrigerant in the condensation process is $Q_H = h_2 - h_3$, and the work of compression per unit mass of refrigerant is, $W_C = h_2 - h_1$. The COP for the cycle is the ratio of the energy transferred from the low temperature to the work of compression,

$$COP = \frac{h_1 - h_4}{h_2 - h_1}$$

A reversed Carnot cycle represents the most efficient refrigeration cycle operating between a given high temperature, T_H, and low temperature, T_C,

$$COP_{Carnot} = \frac{T_C}{T_H - T_C}$$

Letting $\dot{q}$ to denote the rate of energy absorption, the fluid circulated is $\dot{m} = \dot{q}/(h_1 - h_4)$, which also termed the refrigeration capacity. The power is $\dot{W} = \dot{q}(h_2 - h_1)/(h_1 - h_4)$, and in English units the horsepower is $HP = \dot{W}/2554$. The U.S. ton of refrigeration represents a cooling rate of 200 Btu/min, approximately the rate of cooling one U.S. ton of ice per day (~50 kcal/min, 210 kJ/min, or 3.5 kW). If v_1 is the specific volume of the saturated vapor at T_1, and n is the number of working strokes per minute in a reciprocating compressor, the volumetric displacement of the compressor cylinder is $\dot{m}v_1/60n$.

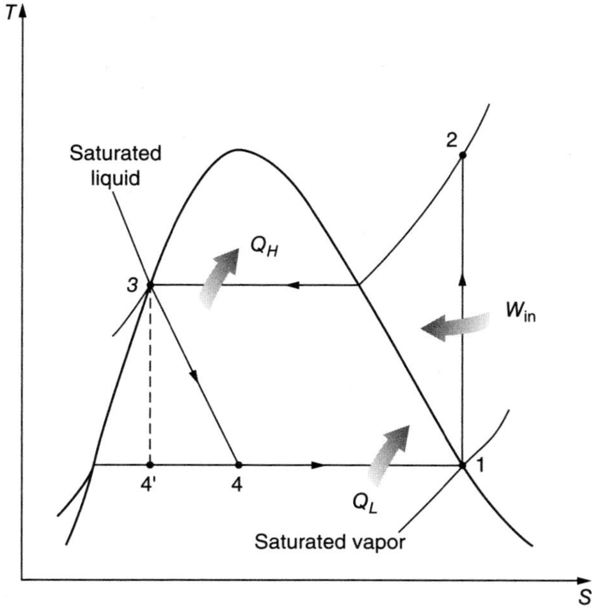

Figure 2.1.35 Vapor compression refrigeration cycle. (*Diagram from Çengel YA, Boles MA. Thermodynamics: An Engineering Approach, 5e, McGraw-Hill.*)

The work necessary for operating a refrigerator, although usually supplied through the compressor, may be supplied in other ways. An absorbent, usually water, absorbs the refrigerant, typically ammonia. Owing to its affinity for the ammonia, the water has, in a thermodynamic sense, ability to do work. Having absorbed the ammonia and thereby lost its ability to do work, the water may have its work capacity restored by passing the ammonia-water solution through a rectifying column from which water and ammonia emerge. With operation under a suitable pressure, the ammonia is condensed to a liquid and in turn can be evaporated to yield refrigeration. Thermodynamically the analysis for these absorption cycles is similar to that for compression cycles.

EXAMPLE. Given the refrigeration cycle as shown below with ammonia as the working fluid and the state properties shown in the table. The compression process is reversible. Determine the COP and entropy production in each of the processes.

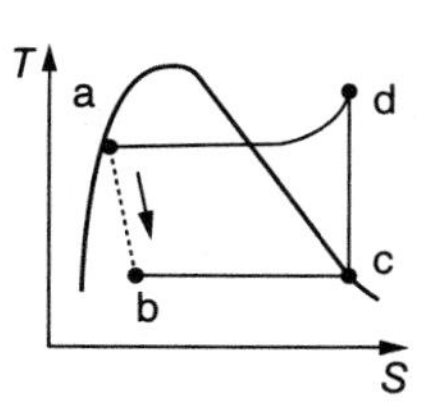

State	T (°F)	p (psia)	h (Btu/lbm)	s (Btu/lbmR)
a	86	169.2	138.9	0.285
b	5	34.3	138.9	0.298
c	5	34.3	613	1.32
d	210	169.2	713	1.32

$$COP = \frac{Q_L}{W} = \frac{h_c - h_b}{h_d - h_c} = \frac{613 - 138.9}{713 - 613} = \frac{474.1}{100} = 4.74$$

Coefficient of Performance,

$$COP_{\text{Carnot}} = \beta_r = \frac{T_L}{T_H - T_L} = \frac{T_b}{T_a - T_b} = \frac{465}{86 - 5} = 5.74$$

Entropy at State b,

$$x_b = \frac{h_a - h_{fb}}{h_c - h_{fb}} = 0.16$$

$$s_b = s_{fb} + x_b s_{fgb} = 0.303 \qquad Btu/lbm - R$$

Entropy production:

Process	Q (Btu/min)	Δs (Btu/min/lbm-R)	Q/T (Btu/min/lbm-R)	σ (Btu/min/lbm-R)
a-b	0	0.018	0	0.018
b-c	474.1	1.022	1.019	~0
c-d	0	0	0	0
d-a	–574.1	–1.040	–1.083	0.043

The internal entropy production is in the throttling process $a \rightarrow b$ is $\Delta\dot{s}_{ab} = 0.018$ (Btu/min/lbm-R). The external entropy production in the condensing process $d \rightarrow a$ is $\Delta\dot{s}_{da} = 0.041$ (Btu/min/lbm-R).

Deviations from ideal thermal and mechanical performance reduce the COP, in increase irreversibility and entropy production. Sources of irreversibility include frictional (piping) losses, heat transfer (heat leaks), and mechanical inefficiency. Figure 2.1.36 shows the effects of these losses on vapor compression system with one stage of refrigeration.

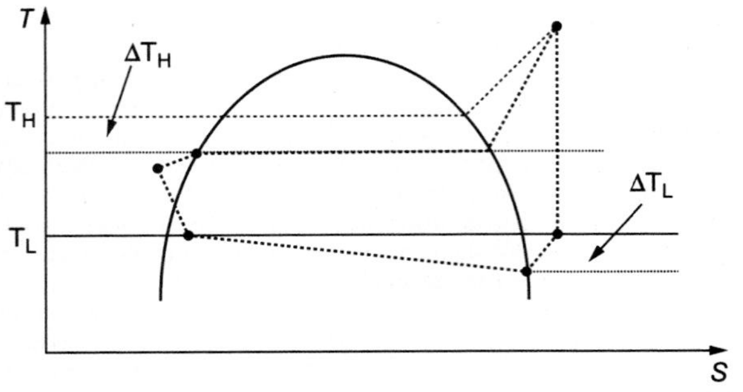

Figure 2.1.36 A vapor compression refrigeration system with effects of piping, thermal and mechanical losses.

Cascade Systems Cascade refrigeration systems comprise multiple stages of compression and one or more cycles with heat exchangers. Figure 2.1.37 illustrates the concept with two stages of compression coupled with a closed heat exchanger and the same working fluid in each stage. Typically, cascade systems reduce compressor work and increase refrigeration capacity.

Intercooling Systems Intercooling refrigeration systems typically comprise one or more cycles in which the refrigerant is separated in a flash chamber after throttling in the high pressure loop. The liquid fraction is then throttled to the low pressure cycle where heat transfer Q_L at T_L takes place. The two refrigerant streams are combined in a direct contact heat exchanger prior to the compression stage in the high pressure loop. Figure 2.1.38 illustrates the concept with one stage of intercooling.

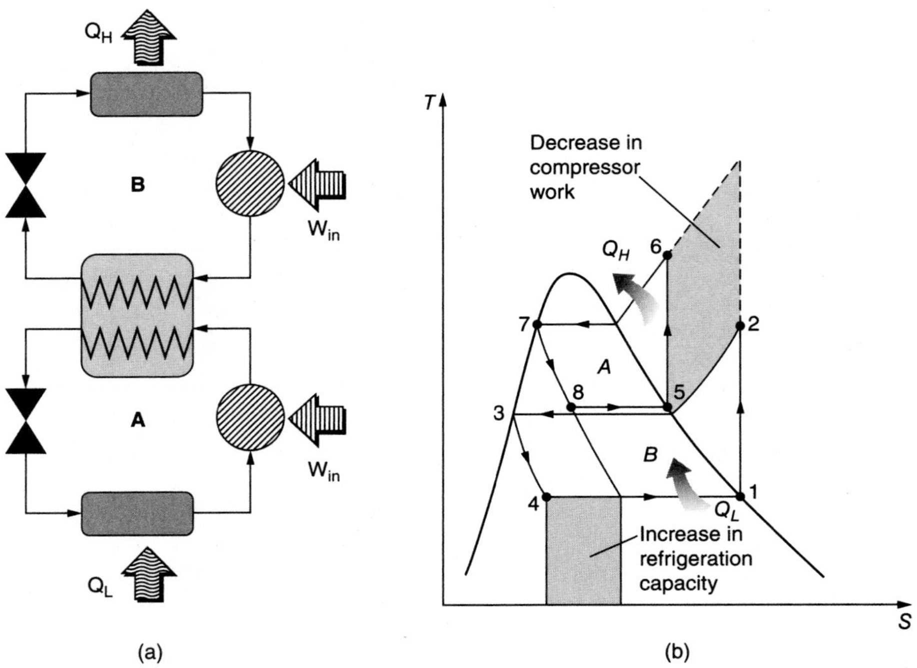

Figure 2.1.37 (*a*) A two-stage cascade vapor compression refrigeration system with a closed heat exchanger. In this case, the working fluids in each cycle may be different. (*b*) The T-s diagram when the working fluids are the same in each cycle. The shaded areas indicate the decrease in compressor work and the increase of refrigeration capacity. (*Diagrams from Çengel YA, Boles MA. Thermodynamics: An Engineering Approach, 5e, McGraw-Hill.*)

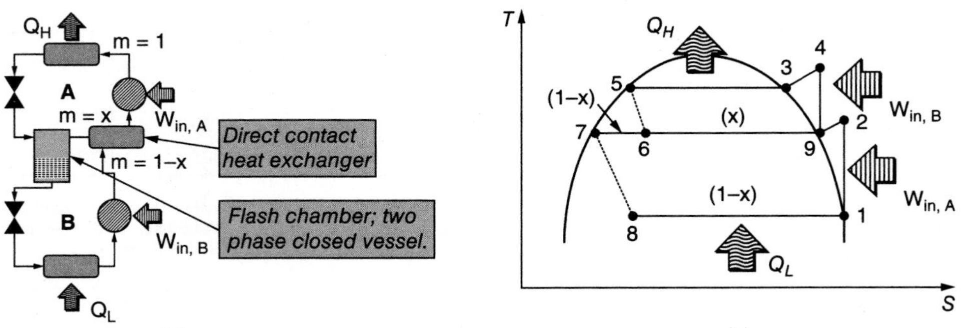

Figure 2.1.38 (*a*) A vapor compression refrigeration system with intercooling. (*b*) the T-s diagram corresponding to (*a*). (*Diagrams from Çengel YA, Boles MA. Thermodynamics: An Engineering Approach, 5e, McGraw-Hill.*)

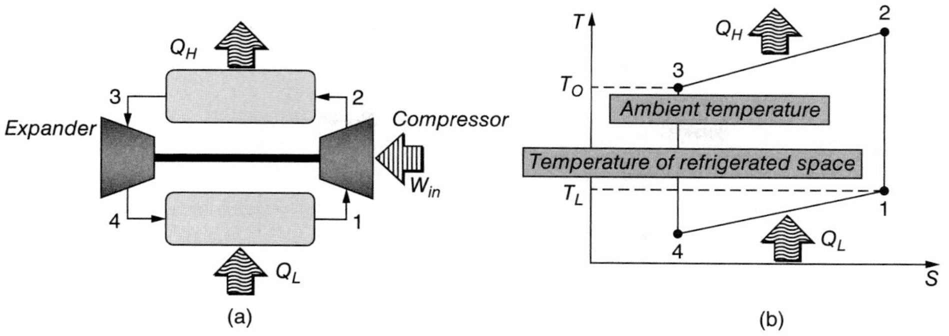

Figure 2.1.39 (*a*) Basic gas refrigeration cycle. (*b*) T-s diagram for the basic cycle showing the ambient temperature and temperature of the refrigerated space.

Gas Refrigeration

Reversed Brayton cycles are employed in gas refrigeration systems. Applications include aircraft air conditioning and liquefaction of gases. Advantages include the use of lightweight centrifugal compressors and environmentally benign refrigerants. The primary disadvantage is high power consumption and hence low COP. A basic cycle is shown in Fig. 2.1.39.

The COP of the cycle is,

$$\beta = \frac{Q_L}{W_{in}} = \frac{h_1 - h_4}{(h_2 - h_3) - (h_1 - h_4)}$$

For adiabatic compression of a constant property gas,

$$COP = \frac{1}{\left(\frac{P_2}{P_1}\right)^{\frac{k-1}{k}} - 1}$$

where $k = c_P/c_V$. The limiting performance of the Carnot refrigerator is determined by the low temperature, T_1 and ambient temperature, T_3, i.e., $COP_{\text{Carnot}} = 1/(1 - T_1/T_3)$. The lowest possible pressure ratio is obtained with the Carnot refrigerator for the two temperature limits. When $T_2 \rightarrow T_3$ and $T_1 \rightarrow T_4$, the Carnot limit for COP is obtained.

Space Cooling Open and closed gas refrigeration cycles have proved practical for space cooling. Figure 2.1.40 illustrates an open cycle and Fig. 2.1.41 illustrates a closed cycle with regeneration.

EXAMPLE. An internally reversible gas refrigeration system operates on the air standard Brayton cycle with a compression ratio of 4:1. The space being cooled is at 300 R. The ambient temperature is at 400 R and the low temperature is 300 R. Assume $k = c_p/c_v = 3.5$.

Determine the compression ratio needed for operation as a Carnot refrigerator.

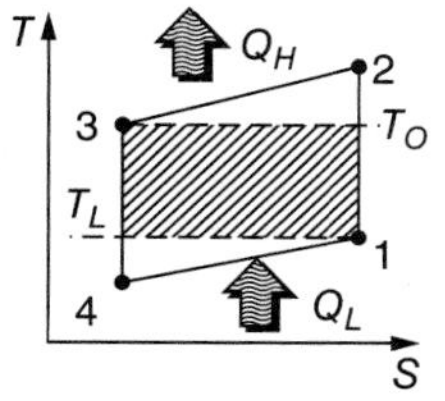

$$\frac{T_2}{T_1} = \left(\frac{P_2}{P_1}\right)^{1/3.5} = 1.486$$

$$\beta = \frac{1}{\left(\frac{P_2}{P_1}\right)^{1/3.5} - 1} = \frac{1}{1.482 - 1} = 2.06$$

$$T_2 = 446 \quad R$$

$$\beta_C = \frac{T_1}{T_3 - T_1} = \frac{300}{400 - 300} = 3$$

$$\beta_C = 3 = \frac{1}{\left(\frac{P_2}{P_1}\right)^{1/k} - 1} \quad \Rightarrow \frac{P_2}{P_1} = 2.74$$

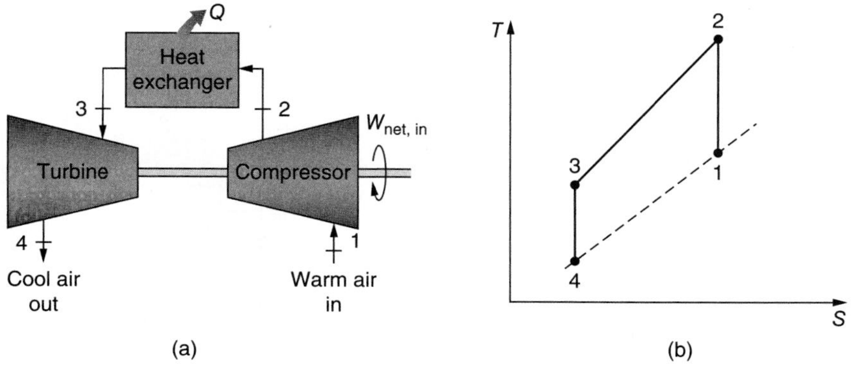

Figure 2.1.40 (*a*) Basic gas refrigeration system. (*b*) T-s diagram for the cycle. (*Diagrams from Çengel YA, Boles MA. Thermodynamics: An Engineering Approach, 5e, McGraw-Hill.*)

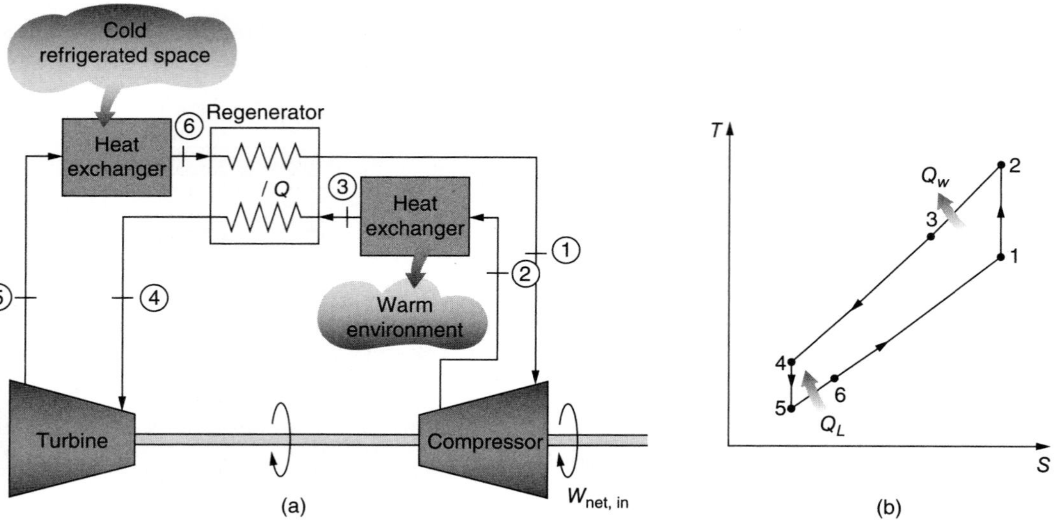

Figure 2.1.41 (*a*) Gas refrigeration system with regeneration. (*b*) T-s diagram for the cycle. (*Diagrams from Çengel YA, Boles MA. Thermodynamics: An Engineering Approach, 5e, McGraw-Hill.*)

Gas Liquefaction

Industrially important processes at cryogenic temperatures generally require use of liquefied gases. Gas separation processes, materials processing, preparation, and storage of liquid propellant are typical examples. Liquefaction generally involves a throttling process to produce cooling below the critical temperature. Throttling gives rise to cooling only when the temperature of the gas is below the point where the inversion curve intersects the temperature axis on a temperature-versus-pressure graph. Thus in some cases, precooling of the gas is necessary to start the liquefaction process. The maximum inversion temperature and the magnitude of the Joule-Thompson coefficient, μ_{JT}, are thus important parameters in the production of a liquefied gas. Table 2.1.12 lists the maximum inversion temperature for several gases, and Table 2.1.13 lists Joule-Thompson coefficients and specific heats for Helium and Nitrogen at 1 atm.

At temperatures above the critical-point, a substance exists in the gas phase only. Critical temperatures for helium, hydrogen, nitrogen, and air (commonly encountered liquefied gases) are 5.3, 33.3, and 126.2 K respectively (Table 2.1.5). Thus, liquefaction of these gases requires precooling them. Figure 2.1.42 illustrates the basic elements of the gas liquefaction cycle. Makeup gas is mixed with the uncondensed portion of the gas from the previous cycle, and the mixture is compressed (1 → 2).

The compression process approaches an isothermal process due to intercooling. The high-pressure gas is cooled in an aftercooler by a cooling medium or by a separate external refrigeration system (2 → 3). It is then throttle to state 3 where the liquefied fraction is collected. The vapor fraction is then mixed with incoming gas (4 → 1) to maintain steady operation of the cycle.

Table 2.1.12 Maximum Inversion Temperatures for Selected Gases

Gas	Maximum Inversion Temperature (K)
Carbon dioxide	~1,500
Argon	723
Nitrogen	621
Air	603
Hydrogen	202
Helium	~26 – 60

SOURCE: Zemansky MW (1957). Heat and Thermodynamics. McGraw-Hill, New York.

Table 2.1.13 The Joule-Thompson Coefficient and Heat Capacity at 1 atm

T (°C)	Helium		Nitrogen	
	c_p (cal/g·°C)	μ (°C/atm)	c_p (cal/g·°C)	μ (°C/atm)
300	1.271	–0.0597	0.2501	+0.0139
200	1.264	–0.0641	0.2490	+0.0558
100	1.257	–0.0638	0.2476	+0.1291
75	1.255	–0.0635	0.2472	+0.1555
50	1.254	–0.0631	0.2469	+0.1854
25	1.252	–0.0624	0.2467	+0.2216
0	1.250	–0.0616	0.2466	+0.2655
100	1.243	–0.0584	0.2473	+0.6487
155	1.239	–0.0503	0.2478	+1.449
180	1.237	–0.0412	0.2480	+2.391

SOURCE: Zemansky MW (1957). Heat and Thermodynamics. McGraw-Hill, New York.

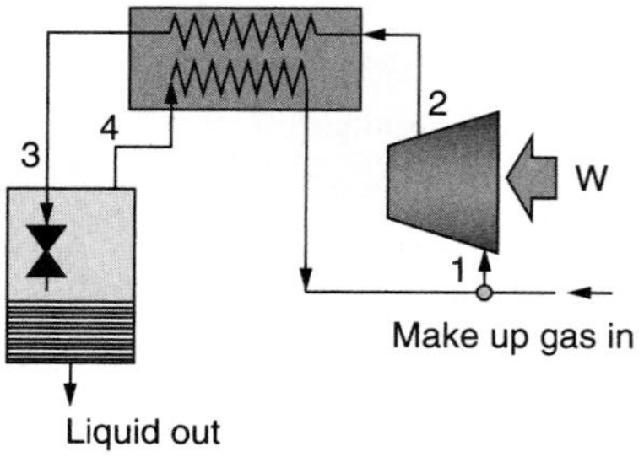

Figure 2.1.42 Basic gas liquefaction cycle.

References

1. Cengel YA, Boles MA. "Thermodynamics: An Engineering Approach," 5e, McGraw-Hill, New York, 2006.
2. De'Rossi F, Mastrullo R, Mazzei R. "Working fluids thermodynamic behavior for vapor compression cycles," *Appl. Energy*, **38**, 1991, no. 3, pp. 163-180.
3. Jordan RC, Priester GB. "Refrigeration and Air Conditioning," Prentice-Hall, New York, 1948.
4. Kuehn TH, Ramsey JW, Threlkeld JL. "Thermal Environmental Engineering," 3e, Prentice-Hall, Englewood Cliffs, 1998.
5. Moran MJ, Shaprio HN. "Fundamentals of Engineering Thermodynamics," John-Wiley and Sons, New York, 2000.
6. Steocker WF, Jones JW. "Refrigeration and Air Conditioning," McGraw-Hill, New York, 1982.
7. Wang SK. "Handbook of Air Conditioning and Refrigeration," 2e, McGraw-Hill, New York, 2000.
8. Wood D. "Applications of Thermodynamics," Addison-Wesley, Boston, 1969.
9. __________. "Refrigeration, ASHRAE Handbook, American Society of Heating, Refrigeration and Air-conditioning Engineers," Atlanta, 2002.
10. __________. "HVAC Systems and Equipment. ASHRAE Handbook, American Society of Heating, Refrigeration and Air-conditioning Engineers," Atlanta, 2016.

THERMODYNAMICS OF FLUID FLOW

Compressible Fluids

Important examples of the flow of compressible fluids include: (1) the flow of air and steam through orifices and short tubes or nozzles, (2) the flow of compressed air, steam, and natural gas in long mains, (3) the flow of low-pressure gases, as furnace gases in ducts and chimneys or air in ventilating ducts, and (4) the flow of gases in moving channels, as in the centrifugal fan. Conduits for these flows can be either uniform or variable. The cross sections of a variable area conduit are denoted by A_1, A_2, etc., (Fig. 2.1.43), and the various magnitudes of flow and thermodynamic properties pertaining to them are denoted by the corresponding subscripts.

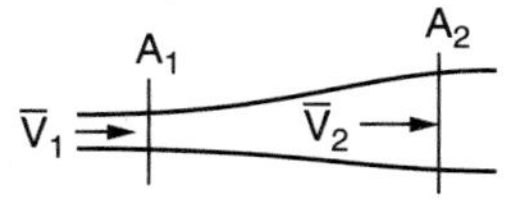

Figure 2.1.43 A variable are conduit.

Fundamental Equations For steady flow in the absence of external and shaft work, three fundamental equations are of use in the analysis of compressible fluid flow: mass continuity,

$$A_1\bar{V}_1\rho_1 = A_2\bar{V}_2\rho_2 \quad \text{or} \quad \frac{dv}{v} = \frac{dA}{A} + \frac{d\bar{V}}{\bar{V}}$$

where $\rho - 1/v$ is the fluid density, and $\bar{V}$ is the average velocity; the energy balance (the First Law),

$$q = (h_2 - h_1) + \frac{\bar{V}_1^2 - \bar{V}_2^2}{2g_c} + \frac{g}{g_c}(z_2 - z_1)$$

where z is the elevation; and the available mechanical energy balance,

$$vdp + \frac{\bar{V}d\bar{V}}{g} + dF + dz = 0$$

where dF_f is the energy expended in overcoming friction. In most practical processes, the third equation cannot be integrated because the actual thermodynamic process is not known. Usually adiabatic flow is assumed, but occasionally the assumption of isothermal flow may be appropriate. In the case of adiabatic, frictionless flow of an incompressible fluid, the equation reduces to the familiar form,

$$\frac{p_2 - p_1}{\rho} + \frac{1}{2}\left(\bar{V}_1^2 - \bar{V}_2^2\right) + \frac{g}{g_c}(z_2 - z_1) = 0$$

Flow through Orifices and Nozzles As a compressible fluid passes through a nozzle, a decrease in pressure and simultaneous increase in velocity result. By assuming the type of flow, e.g., adiabatic, it is possible to calculate from the properties of the fluid at a given cross section the area required to maintain mass continuity. It is found that the nozzle form must first be converging, but if the pressure drops sufficiently, the nozzle must become diverging to accommodate the increased volume due to fluid expansion. The smallest cross section of the nozzle is called the throat, and the pressure at the throat is the *critical flow pressure* denoted by p_m. (The critical flow pressure is not to be confused with the critical state pressure.) If the nozzle is cut off at the throat with no diverging section and the pressure at the discharge end is progressively decreased, with fixed inlet pressure, the flow increases until the discharge pressure equals the critical. However, a further decrease in the discharge pressure does not result in increased flow. (This is not true for thin plate orifices.) For any gas, the ratio of critical to inlet pressure is approximately constant. For gases $p_m/p_1 \cong 0.53$. For saturated steam $p_m/p_1 \cong 0.575$, and for moderately superheated steam $p_m/p_1 \cong 0.55$.

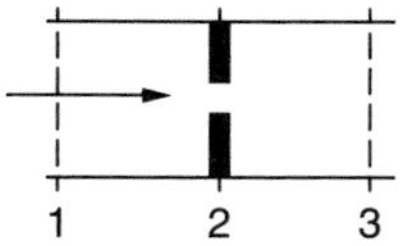

Figure 2.1.44 A typical plate orifice.

A typical plate orifice is shown in Fig. 2.1.44. The energy balance for process 1 → 2 in the absence of heat transfer (adiabatic flow) is,

$$\frac{\bar{V}_1^2 - \bar{V}_2^2}{2g_c} + h_1 - h_2 = 0$$

Substituting the continuity equation and rearranging,

$$\bar{V}_2 = \frac{C\sqrt{2g_c(h_1 - h_2)}}{\sqrt{1 - \left(\frac{A_2}{A_1}\right)^2\left(\frac{\rho_2}{\rho_1}\right)^2}}$$

where C is the coefficient of discharge specific the orifice. For ideal gases and assuming reversible adiabatic expansion through the orifice,

$$\bar{V}_2 = C\frac{\sqrt{2g_c\, p_1\bar{V}_1 \frac{k}{k-1}\left[1 - \left(\frac{p_2}{p_1}\right)^{(k-1)/k}\right]_1}}{\sqrt{1 - \left(\frac{A_2}{A_1}\right)^2\left(\frac{p_2}{p_1}\right)^{2/k}}}$$

$$\dot{m} = CA_2p_2\frac{\sqrt{\frac{2g_c}{RT_1}\frac{k}{k-1}\left(\frac{p_2}{p_1}\right)^{(k-1)/k}\left[\left(\frac{p_1}{p_2}\right)^{(k-1)/k} - 1\right]}}{\sqrt{1 - \left(\frac{A_2}{A_1}\right)^2\left(\frac{p_2}{p_1}\right)^{2/k}}}$$

where $\dot{m}$ is the mass flow (kg/s), T_1 is the inlet temperature, and $k = c_P/c_V$.

Often $\bar{V}_1 \ll \bar{V}_2$, and under this condition the denominators in the preceding equations become approximately unity. For air, assuming

$R_{\text{specific}} = 53.3$ (ft-lbf/lbm·R), $k = 1.3937$, and $\bar{V}_1 \approx 0$,

$$\dot{m} = 2.05CA_2p_2\sqrt{\left(\frac{1}{T_1}\right)\left(\frac{p_1}{p_2}\right)^{0.283}\left[\left(\frac{p_1}{p_2}\right)^{0.283} - 1\right]}$$

The preceding relations are generally applicable under the assumed conditions. However, p_2 cannot become less than p_m whatever the value of the exit pressure, p_3. When $p_3 < p_m$, the flow rate becomes independent of the downstream pressure. For the ideal gas,

$$\dot{m} = CA_2p_1\sqrt{\left(\frac{g_c}{RT_1}\right)k\left(\frac{2}{k+1}\right)^{(k+1)/(k-1)}}$$

and for air,

$$\dot{m} = 0.53Cp_1\frac{A_2}{\sqrt{T_1}}$$

The following formula is useful for calculating the volumetric flow rate, Q (ft³/min), of any gas (provided no condensation occurs) through a nozzle for pressure drops less than the critical range,

$$Q_1 = \frac{31.5CY'd_n^2}{\rho_1}\sqrt{\rho_1\Delta p}$$

where

$$Y' = \left(\frac{k}{k-1}\right)^{1/2}\left(\frac{p_2}{p_1}\right)^{1/k}\sqrt{\frac{1-\left(\dfrac{p_2}{p_1}\right)^{(k-1)/k}}{\left(1-\dfrac{p_2}{p_1}\right)^{(k-1)/k}\left[1-\left(\dfrac{d_n}{d_1}\right)^4\left(\dfrac{p_2}{p_1}\right)^{2/k}\right]}}$$

and p_1 is the static pressure on upstream side of nozzle; p_2 is the static pressure on downstream side of nozzle, d_1 is the diameter upstream of nozzle, d_n is the nozzle throat diameter, and ρ_1 is the density of the gas at the upstream side of nozzle. Values of Y' are listed in Table 2.1.14. When the pressure drop through the orifice is small, the hydraulic relations applicable to incompressible fluids may be employed for gases and other compressible fluids.

The above relations are applicable to nozzles with the relevant discharge coefficient. For steam nozzles, this value for C may be as large as 0.94 to 0.96, although for many orifice installations it is as low as 0.50 to 0.60. Steam nozzles constitute a most important type, and calculations for them are best carried out with the aid of a Mollier or similar chart.

Saturated Steam Discharge When the back pressure $p_3 < p_m$, the discharge depends upon the area of orifice A_2 and p_1. Three widely used relations express approximately the discharge $\dot{m}$ of saturated steam in terms of A_2 and p_1:

Napier's equation, $m = A_2p_1/70$

Grashof's formula, $\dot{m} = 0.0165A_2p_1^{0.97}$

Rateau's formula, $\dot{m} = A_2p_1\,(16.367 - 0.96 log p_1)/1000$

where A_2 is in in.², and p_1 is in psi. Napier's formula is convenient as a rough check. The Grashof and Rateau formulas are applicable to well-rounded convergent orifices in which case $C \approx 1$. When the back pressure $p_2 > p_m$, the velocity and discharge are found most conveniently from the general formulas of flow. From either the steam tables or the Mollier chart, find the initial enthalpy h_1 and the enthalpy h_2 after isentropic expansion and the specific volume $v_2 = 1/\rho_2$ (see Fig. 2.1.31). Then,

$$\bar{V}_2 = 223.7\sqrt{h_1 - h_2} \quad \text{and} \quad \dot{m} = A_2\bar{V}_2\rho_2$$

Table 2.1.14 Values for Y′

	$k = 1.40$					$k = 1.35$					$k = 1.30$				
	d_n/d_1					d_n/d_1					d_n/d_1				
P_2/P_1	0	0.2	0.3	0.4	0.5	0	0.2	0.3	0.4	0.5	0	0.2	0.3	0.4	0.5
1.00	1.000	1.001	1.004	1.013	1.033	1.00	1.001	1.004	1.013	1.033	1.000	1.001	1.004	1.013	1.033
0.99	0.995	0.995	0.999	1.007	1.027	0.994	0.995	0.999	1.007	1.027	0.994	0.995	0.998	1.007	1.026
0.98	0.989	0.990	0.993	1.002	1.021	0.989	0.990	0.993	1.001	1.020	0.988	0.989	0.992	1.001	1.020
0.97	0.984	0.985	0.988	0.996	1.015	0.983	0.984	0.987	0.995	1.014	0.983	0.983	0.986	0.995	1.013
0.96	0.978	0.979	0.982	0.990	1.009	0.978	0.978	0.981	0.990	1.008	0.977	0.977	0.980	0.989	1.007
0.95	0.973	0.974	0.977	0.985	1.002	0.972	0.973	0.976	0.984	1.001	0.971	0.972	0.974	0.982	1.000
0.94	0.967	0.968	0.971	0.979	0.996	0.966	0.967	0.970	0.978	0.995	0.965	0.966	0.968	0.976	0.993
0.93	0.962	0.963	0.965	0.973	0.990	0.961	0.961	0.964	0.972	0.989	0.959	0.960	0.962	0.970	0.987
0.92	0.956	0.957	0.960	0.967	0.984	0.955	0.955	0.958	0.966	0.982	0.953	0.954	0.956	0.964	0.980
0.91	0.951	0.951	0.954	0.961	0.978	0.949	0.950	0.952	0.960	0.976	0.947	0.948	0.950	0.957	0.973
0.90	0.945	0.946	0.948	0.956	0.971	0.943	0.944	0.946	0.953	0.969	0.941	0.942	0.944	0.951	0.966
0.89	0.939	0.940	0.943	0.950	0.965	0.937	0.938	0.940	0.947	0.963	0.935	0.935	0.938	0.945	0.959
0.88	0.934	0.934	0.937	0.944	0.959	0.931	0.932	0.934	0.941	0.956	0.929	0.929	0.932	0.938	0.953
0.87	0.928	0.928	0.931	0.938	0.953	0.925	0.926	0.926	0.935	0.950	0.922	0.923	0.926.	0.932	0.946
0.86	0.922	0.923	0.925	0.932	0.946	0.919	0.920	0.922	0.929	0.943	0.916	0.917	0.919	0.926	0.939
0.85	0.916	0.917	0.919	0.926	0.940	0.913	0.914	0.916	0.923	0.936	0.910	0.911	0.913	0.923	0.932
0.84	0.910	0.911	0.913	0.920	0.933	0.907	0.908	0.910	0.916	0.930	0.904	0.904	0.907	0.919	0.925
0.83	0.904	0.905	0.907	0.913	0.927	0.901	0.902	0.904	0.910	0.923	0.897	0.898	0.900	0.916	0.918
0.82	0.898	0.899	0.901	0.907	0.920	0.895	0.895	0.898	0.904	0.917	0.891	0.891	0.894	0.900	0.911
0.81	0.892	0.893	0.895	0.901	0.914	0.889	0.889	0.891	0.897	0.910	0.885	0.885	0.887	0.893	0.904
0.80	0.886	0.887	0.889	0.895	0.907	0.883	0.883	0.885	0.891	0.903	0.878	0.879	0.880	0.886	0.897
0.79	0.880	0.881	0.883	0.889	0.901	0.876	0.877	0.879	0.834	0.896	0.872	0.872	0.874	0.880	0.890
0.78	0.874	0.875	0.877	0.882	0.894	0.870	0.870	0.872	0.878	0.889	0.865	0.865	0.868	0.873	0.883
0.77	0.868	0.869	0.871	0.876	0.887	0.864	0.864	0.866	0.871	0.882	0.859	0.859	0.861	0.866	0.876
0.76	0.862	0.862	0.864	0.869	0.881	0.857	0.858	0.859	0.865	0.876	0.852	0.852	0.854	0.859	0.869
0.75	0.856	0.856	0.858	0.863	0.874	0.851	0.851	0.853	0.858	0.869	0.845	0.846	0.848	0.852	0.862
0.74	0.849	0.850	0.852	0.857	0.867	0.844	0.845	0.846	0.851	0.862	0.839	0.839	0.841	0.845	0.855
0.73	0.843	0.844	0.845	0.850	0.860	0.838	0.838	0.840	0.845	0.855	0.832	0.832	0.834	0.838	0.848
0.72	0.837	0.837	0.839	0.844	0.854	0.831	0.831	0.833	0.838	0.848	0.825	0.825	0.827	0.831	0.841
0.71	0.820	0.831	0.832	0.837	0.847	0.825	0.825	0.827	0.831	0.840	0.818	0.819	0.820	0.824	0.834
0.70	0.824	0.824	0.826	0.830	0.840	0.818	0.818	0.820	0.824	0.833	0.811	0.812	0.813	0.817	0.826

If the velocity of approach is zero (as with a nozzle taking in air from the outside), d_1 is infinite and d_n/d_1 is zero.

The same method is used in the case of steam initially superheated.

EXAMPLE. Determine the steam discharge through an orifice of 0.5 in. diameter at 140 psi superheated to 110°F, back, and pressure of 90 psia.

From the Mollier chart, or steam tables: h_1 = 1,255.7 Btu/lbm, h_2 = 1,214 Btu/lbm, and $v_2 = 5.30$ ft³.

$$V_2 = 223.7\sqrt{1255.7 - 1214} = 1{,}444.6 \text{ (ft/s)}$$

$$A_2 = 0.1962 \text{ (in.}^2\text{)}$$

$$m = A_2V_2\rho_2 = (0.1964/144)(1444.6/5.30) = 0.372 \ (lbm/s)$$

This calculation assumes ideal conditions, and the results must be multiplied by the correct coefficient of discharge to get actual results.

Converging-Diverging Nozzles At the throat, or smallest cross section of the nozzle (Fig. 2.1.45), the pressure of saturated steam takes the value $p_m = 0.57p_1$. The discharge is fixed by the area A_2 and p_1. For saturated steam, either the Grashof or Rateau formula may be used. The diverging part of the nozzle permits further expansion to the break pressure p_3, the velocity of the jet meanwhile increasing from $\bar{V}_m = \bar{V}_2$, the critical velocity at the throat, to $\bar{V}_3$ given by the fundamental equation $\bar{V}_2 = 223.7\sqrt{h_1 - h_3}$.

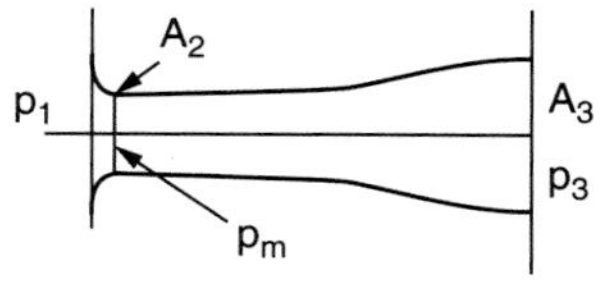

Figure 2.1.45 A converging-diverging nozzle.

Frictional resistance in the nozzle has the effect of decreasing the jet energy, $\bar{V}_3^2/2g_c$ and correspondingly increasing the enthalpy of the flowing fluid. Thus, if h_3 is the enthalpy in the final state with frictionless expansion, h_3' (> h_3) is the enthalpy when friction is taken into account; hence $\bar{V}_3' < V_3$. The loss of kinetic energy is $h_3' - h_3$, and the ratio of this loss to the available kinetic energy is denoted by $y = (h_3' - h_3)/(h_1 - h_3)$.

The design of a nozzle for a given discharge with pressures p_1 and p_3 specified is most conveniently affected with the aid of the Mollier chart. Determine p_m, h_1, h_m, and h_3 assuming frictionless flow. Then $\bar{V}_m = 223.7\sqrt{h_1 - h_m}$, and $\bar{V}_3' = 223.7\sqrt{(1-y)(h_1 - h_3)}$. Next find vm and v_3'. From the continuity equation, $A_m = mv_m/\bar{V}_m$, and $A_3' = mv_3'/\bar{V}_3'$. The following example illustrates the method.

EXAMPLE. Determine the throat and end section area of a nozzle to deliver steam at 0.7 lbm/s. The initial pressure is 160 psia, the back pressure 15 psia, and the steam is initially superheated at 100°F; $y = 0.15$.

The critical pressure is 160 × 0.55 = 88 lbm. On the Mollier (see below) the point *A* representing the initial state is located, and line of constant entropy (frictionless adiabatic flow) is drawn from *A*.

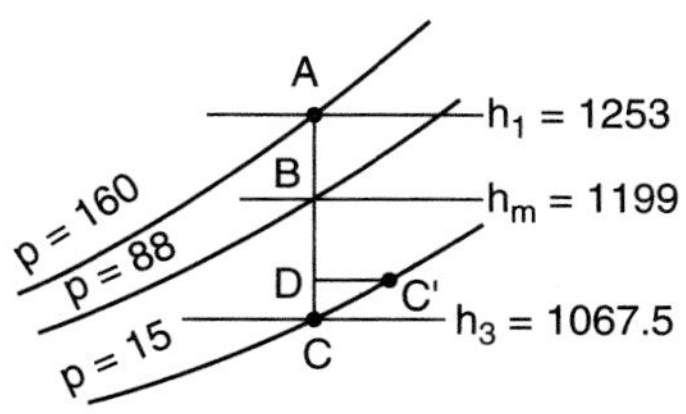

This cuts the curves $p = 88$ psia and $p = 15$ psia at *B* and *C* respectively. The three values of specific enthalpy are: h_1 = 1,253 Btu/lbm, h_m - 1,199 Btu/lbm, and h_3 = 1,067 Btu/lbm. Of the available drop in enthalpy, $h_1 - h_3$ = 185.5 Btu/lbm, ~15 percent or 27.9 Btu/lbm are lost through friction. Hence, $CD = 27.9$ is laid off and *D* is projected horizontally to point C' on the curve $p = 15$ psia. Then C' represents the final state, and the quality is found to be $x = 0.943$. The specific volume at C' is $v_3 = 26.29 \times 0.943 = 24.8$ ft³/lbm. Likewise, the specific volume for the state *B* is found to be 5.29 ft³/lbm. For the velocities at throat and end sections,

$$\bar{V}_m = 223.7\sqrt{1253 - 1199} = 1{,}643 \text{ (ft/s)}$$

$$V_m = 223.7\sqrt{1855 - 27.9} = 2{,}813 \text{ (ft/s)}$$

$$A_m = (0.7 \times 5.29)/1643 = 0.00225 \ \ (\text{ft}^2) = 0.324 \text{ (in.}^2\text{)}$$

$$A_3 = (0.7 \times 24.8)/2813 = 0.00617 \ (\text{ft}^2)0.89 \text{ (in.}^2\text{)}$$

The diameters are d_m = 0.643 in. and d_3 = 1.064 in.

Divergence of Nozzles for Various Expansion Ratios Figure 2.1.46 shows the required divergence of the nozzle, i.e., the ratio of the area of any section to the throat area. Thus in the case of saturated steam, if the final pressure is 1/15th of the initial pressure, the ratio of the areas is 3.25. The curves apply to frictionless flow; the effect of friction is to increase the divergence.

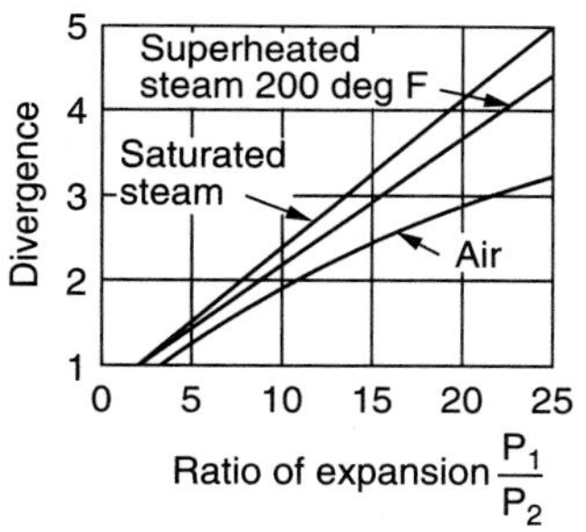

Figure 2.1.46 Divergence ratio for superheated steam at 200°F.

Theory of Super Saturation Certain discrepancies between the discharge of saturated steam through an orifice as calculated from the preceding theory and the discharge actually observed are explained by a hypothesis that steam when expanded rapidly, as in turbine nozzles, becomes supersaturated, i.e., the condensation required by the ordinary theory of adiabatic expansion does not occur on account of the rapidity of the expansion. The effect of super saturation in turbines is a loss of energy, the amount of which may be 1.5 to 3 percent of the available energy of the steam.

Flow of Wet Steam When the steam entering a nozzle is wet ($x < 1$), the speed of the water particles at exit is not the same as the speed of the steam. Denoting the speed of the steam by $\bar{V}$, the speed of the water drop is $\beta\bar{V}$, where $\beta \sim 0.20 - 0.05$ or less depending on the pressure. The actual velocity of the steam is greater than the velocity calculated on the usual assumption that both the steam and water droplets have the same velocity. The ratio of these velocities is

$$\frac{\bar{V}}{\bar{V}_0} = \frac{1}{\sqrt{x + \beta^2(1-x)}}$$

Thus with $x = 0.92$ and $\beta = 0.15$, $\bar{V}/\bar{V}_0 = 1.036$. Since the discharge is practically proportional to the steam velocity, the actual discharge in this case is 3.6 percent greater than the discharge computed on the usual assumptions.

On account of frictional losses, the actual velocity attained at the outlet is less than the velocity calculated under ideal conditions. That is, $\bar{V} = \tilde{x}\bar{V}_0$, where $\tilde{x}$ (< 1) is denoted as the velocity coefficient. The coefficient $\tilde{x}$ is connected to the coefficient y, giving the loss of energy by $y = 1 - \tilde{x}^2$.

The elaborate and accurate experiments of the General Electric Co. on turbine nozzles give convergent nozzles values of $\tilde{x} > 0.98$, with a corresponding loss of energy $y = 0.025$ to 0.04. For similar nozzles, the experiments of the Steam Nozzles Research Committee (of England) by a different method give values of $\tilde{x} \sim 0.96$, or $y \sim 0.08$. In the case of divergent nozzles, the velocity coefficient may be somewhat lower.

Flow with Friction and Heating Friction and heat transfer are present in the flow of all real fluids. Industrially important flows pertain to straight (constant cross section) ducts in which the flow may be either adiabatic or diabatic, and both cases imply limits on the applicability of the following analysis. However, many flows to a good approximation can be handled within these limits, e.g., isothermal flow in long conduits is generally encountered in gas flow in long, uninsulated pipelines.

Fanno Line Flow For steady one-dimensional adiabatic flow of a perfect gas with no work transfer, the energy equation is written,

$$h^o = h + \frac{\bar{V}^2}{2g_c}$$

where $\bar{V}$ is the mass average velocity, and $h^o(p^o, T^o)$ is the total enthalpy. As the cross section in constant, mass conservation requires that $\varrho\bar{V} = Constant$. Denote the reference state as (p_1, T_1), and with the T-dS equation for a perfect gas, the temperature-versus-entropy relation is,

$$\frac{s_1 - s}{C_v} = \ln\left(\frac{T}{T_1}\right) + \frac{\gamma - 1}{2}\ln\left(\frac{T^o - T}{T^o - T_1}\right)$$

or

$$\frac{s - s_1}{C_V} = \ln T + \frac{k-1}{2}\ln(T^o - T) + \text{Constant}$$

where $k = C_P/C_V$.

On a T-s diagram, the locus of states is termed the Fanno line and the flow as Fanno flow. For a perfect gas, the T-s and h-s diagrams are similar, and Fig. 2.1.47 is an h-s diagram for Fanno flow at a given mass flow. At the point where $ds/dT = 0$, the velocity is precisely the speed of sound, $\bar{V} = \sqrt{kg_cRT}$, or the Mach number $M = \bar{V}/\sqrt{kg_cRT} = 1$. From the energy equation, higher velocities correspond to lower enthalpy, and conversely for lower velocities. In subsonic flow, $M < 1$, the flow is irreversible, entropy increases in the flow direction ($ds > 0$), and $M \to 1$ at which point $ds = 0$. In supersonic flow ($M > 1$), $ds > 1$ along the flow direction, and the limit is reached at $ds = 0$ and $M = 1$. The sonic limit fixes the flow because entropy would have to increase otherwise, which would be violation of the Second Law. For a given mass flow, the limit of $M = 1$ implies that the flow is limited, or choked, due to friction. Fanno lines exist for a given mass flow, and increasing mass flow moves the limiting points and the curves to the right on the h-s diagram, i.e., to larger entropies.

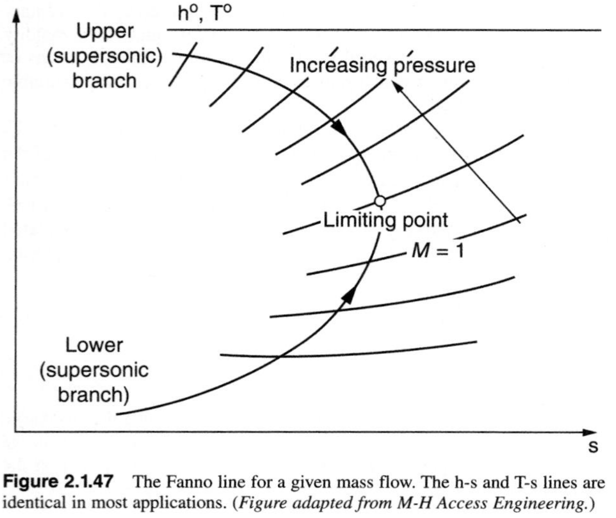

Figure 2.1.47 The Fanno line for a given mass flow. The h-s and T-s lines are identical in most applications. (*Figure adapted from M-H Access Engineering.*)

The change of properties of the fluid along the flow length is determined by the momentum balance with frictional effects taken into account. With the friction factor given by $f = 4\tau_f/(\rho\bar{V}^2/2g_c)$, where f is dependent on the Reynolds number and wall roughness, and τ_f is the wall shear stress. The pressure gradient along the flow length, x, is given by,

$$-dP - \frac{\rho\bar{V}^2}{2g_c} f \frac{dx}{D} = \frac{\varrho\bar{V}d\bar{V}}{g_c}$$

where D is the hydraulic diameter. In terms of Mach number this equation is recast as,

$$\frac{dP}{P} + \frac{kM^2 f}{2}\frac{dx}{D} + kM^2\frac{d\bar{V}}{\bar{V}} = 0$$

Assuming that the perfect gas law holds and with the definition of the Mach number, this equation is integrated to give,

$$\frac{fdx}{D} = \frac{2dM}{M}\left[\frac{1 - M^2}{kM^2\left(1 - \frac{k-1}{2}M^2\right)}\right]$$

where T^o is constant as the flow is adiabatic.

The Mach number as a function of length is determined by integration,

$$\frac{fL_{\max}}{D} = \int_M^1 \frac{2(1 - M^2)}{\gamma k\left(1 + \frac{k-1}{2}M^2\right)}\frac{dM}{M}$$

wherein $M = 1$ and $L = L_{\max}$ are selected as reference points. Similarly, the pressure versus Mach number or fdx/D is,

$$\int_P^{P^*}\frac{dP}{P} = -\int_M^1\left[\frac{1 + (k-1)M^2}{1 + \frac{k-1}{2}M^2}\right]\frac{dM}{M}$$

where P^* denotes the pressure at $M = 1$. These equations have been evaluated numerically, and tabulated values are available (e.g., see John, JEA, Gas Dynamics, Appendix E, Allyn and Bacon, 1969).

Isothermal Flow In long, uninsulated pipes, isothermal flow is generally encountered. For such a flow, friction and heat transfer are involved, and the continuity, momentum, and energy equations must be combined to determine the local Mach number and heat transfer. In differential form,

$$\frac{d\rho}{\rho} + \frac{d\bar{V}}{\bar{V}} = 0$$

$$dP + \frac{\rho\bar{V}d\bar{V}}{g_c} + \frac{fL}{D}\left(\frac{\rho V^2}{2g_c}\right) = 0$$

$$\frac{dP}{P} + \frac{kM^2}{2}\left(\frac{fdx}{D}\right) + kM^2\frac{d\bar{V}}{\bar{V}} = 0$$

With the perfect gas law and a constant speed of sound (T = Constant), $dP/P = dM/M$, this equation becomes,

$$\frac{dM}{M} = \frac{fdx}{D}\left[\frac{kM^2/2}{1-kM^2}\right]$$

Thus in isothermal flow, $M = \sqrt{\gamma}$ is the critical condition. For $M < \sqrt{\gamma}$, Mach number increases with flow length, x, and conversely.

Heat transfer per unit mass of flow is $dq = C_p dT^o$ and since $T^o = T\left[1 + (k - 1/2)M^2\right]$,

$$dq = \left[\frac{C_p T(k-1)M^4(k/2)}{\left(1+\frac{k-1}{2}M^2\right)(1-kM^2)}\right]\frac{fdx}{D}$$

This result shows that an infinite rate of heat transfer is needed as $M \to \sqrt{\gamma}$ and isothermal flow is maintained with an infinite heat transfer rate. Thus for low Mach number, isothermal flow can be maintained. For such flows,

$$\frac{fL}{D} = \frac{1-kM_1^2}{\gamma M_1^2} - \frac{1-kM_2^2}{kM_1^2} + \ln\frac{M_1^2}{M_2^2}$$

where the subscripts refer to the states over the length of the flow, L. With $P_2/P_1 = M_1/M_2$ in isothermal flow, the pressure drop (or pressure ratio) can be determined.

Adiabatic Flow with Friction and Area Change In pipe flows with friction and a variable cross section, analysis similar to that for Fanno flow is carried out. With the assumption of a perfect gas,

$$\frac{kM^2}{2}\frac{f}{D} = \frac{1}{A}\frac{dA}{dx} + \frac{dM}{dx}\frac{\frac{1-M^2}{M}}{1+\frac{k-1}{2}M^2}$$

This result must be evaluated numerically as direct integration is not possible. For the case of $dA/dx = 0$, Fanno flow is obtained. If $dM/dx = 0$, $(kM^2/4)f = dD/dx$, and constant Mach number flow is thus obtained.

For nozzle flow with friction approximation is needed owing to the impossibility of direct integration of the above result. For isentropic flow, $f = 0$, in a converging-diverging nozzle, and $M = 1$ when $dA/dx = 0$, or at the throat. From the above equation, $M = 1$ when $dA/dx > 0$, or in the diverging section of the nozzle. Thus at the throat, $dA/dx = 0$, $M < 1$, and $dM/dx > 0$. Thus to solve the problem for nozzle flow with friction, coefficients $\hat{C}_V$ and $\hat{C}_h$ are introduced. The velocity coefficient, $\hat{C}_V = \left(\overline{V}_{\text{exit,actual}}/\overline{V}_{\text{exit,isentropic}}\right)$ and $\hat{C}_h = \Delta h_{\text{actual}}/\Delta h_{\text{isentropic}}$. When the nozzle inlet velocity is negligible, $\Delta h = \overline{V}_{\text{exit}}^2/2g_c$ for adiabatic flow, and thus $\hat{C}_h = \hat{C}_V^2$. When Reynolds numbers based on nozzle diameter are greater than 10^6, losses are mainly a result of wall friction, and values of $\hat{C}_V \sim 0.95$ or greater can be obtained with proper design.

Flow with Heat Transfer in Constant Area Ducts Steady duct flow with friction, heat transfer, and constant cross section area is analyzed in a similar manner as for adiabatic flow. Under the assumptions of a perfect gas and one-dimensional flow, the basic equations are,

$$\rho\overline{V} = \text{Constant}$$

$$P + \frac{\rho\overline{V}^2}{2g_c} = \text{Constant}$$

$$dq = C_p dT^o$$

The momentum equation can be expressed in terms of Mach number as,

$$P(1+kM^2) = \text{Constant}$$

which for the perfect gas becomes,

$$T = \frac{M^2}{(1+kM^2)^2} \times \text{Constant}$$

For flow between two points in a constant diameter duct,

$$P_1(1+kM_1^2) = P_2(1+kM_2^2)$$

$$\frac{T_1(1+kM_1^2)^2}{M_1^2} = \frac{T_2(1+kM_2^2)^2}{M_2^2}$$

$$q = C_p(T_2^o - T_1^o)$$

With state 2 chosen as the reference with $M = 1$, the relations are recast as,

$$\frac{P}{P^*} = \frac{1+k}{1+kM^2}$$

$$\frac{T}{T^*} = \frac{(1+k)^2M^2}{(1+kM^2)^2} \quad \text{and} \quad \frac{T^o}{T^{o*}} = \frac{(1+k)^2M^2}{(1+kM^2)^2}\left(\frac{1+\frac{k-1}{2}M^2}{1+\frac{k-1}{2}}\right)$$

Evaluation of these equations for a range of Mach numbers is presented elsewhere (e.g., see J. E. A. John, Gas Dynamics, Allyn and Bacon, 1969).

Rayleigh Flow A T-s diagram reveals the effect of heat transfer on Mach number in diabatic flow without friction. For the perfect gas and integration of the T-dS equation with the reference state at $M = 1$,

$$s - s^* = C_p\ln\frac{T}{T^*} - R\ln\frac{P}{P^*}$$

which is recast in terms of Mach number as,

$$\frac{s-s^*}{C_p} = \ln\frac{T}{T^*} - \frac{k-1}{k}\ln\left[\frac{(k+1)\pm\sqrt{(k+1)^2 - 4k(T/T^*)}}{2}\right]$$

This relation produces the locus of states associated with diabatic flow and no friction in constant area ducts and is termed the Rayleigh line (Fig. 2.1.48*a*). At the point where $ds/dT = 0$, $M = 1$, and a state of maximum entropy is reached. The point at which T/T^* is a maximum, $M = 1/\sqrt{k}$ and $(T/T^*)_{\max} = (1+k)^2/4k$. For subsonic flow, heat addition increases the temperature, but as sonic flow is approached temperature decreases on approach to the point of maximum entropy. In supersonic flow, heating increases entropy until the point of maximum entropy is reached. Figure 2.1.48*b* shows corresponding states for the Raleigh line in terms of pressure-versus-specific volume. The states on the Rayleigh line are available in tabulated form (e.g., see J. E. A. John, Gas Dynamics, Allyn and Bacon, 1969).

Flow with Heat Addition and Area Change With area change and no friction, combination of the continuity, momentum, and energy equations gives,

$$-\frac{dA}{A} + (kM^2 - 1)\frac{dM}{M} + \frac{kM^2+1}{2}\frac{dT}{T} = 0$$

Since $dq = C_p dT^o$ and $T^o = T[1 + (k-1)(M^2/2)]$,

$$\frac{dT}{T} = \frac{\frac{dq}{C_pT}}{1+\frac{k-1}{2}M^2} - \frac{M^2(k-1)\frac{dM}{m}}{1+\frac{k-1}{2}M^2}$$

Combination of these equations produces the relation between area, heat transfer, and Mach number, which cannot be integrated directly. For the special case $dq = 0$, the isentropic flow equation is obtained, and $M = 1$ when A is a minimum, or at the throat of a converging-diverging nozzle. With heat addition, sonic velocity occurs where $dA/A > 0$, or in the diverging part of a converging-diverging nozzle, and for heat removal ($dq < 0$), sonic velocity is obtained in the converging part of the nozzle.

Flow with Friction and Heat Addition For flow in a constant area duct with friction and heat transfer, the governing equations are

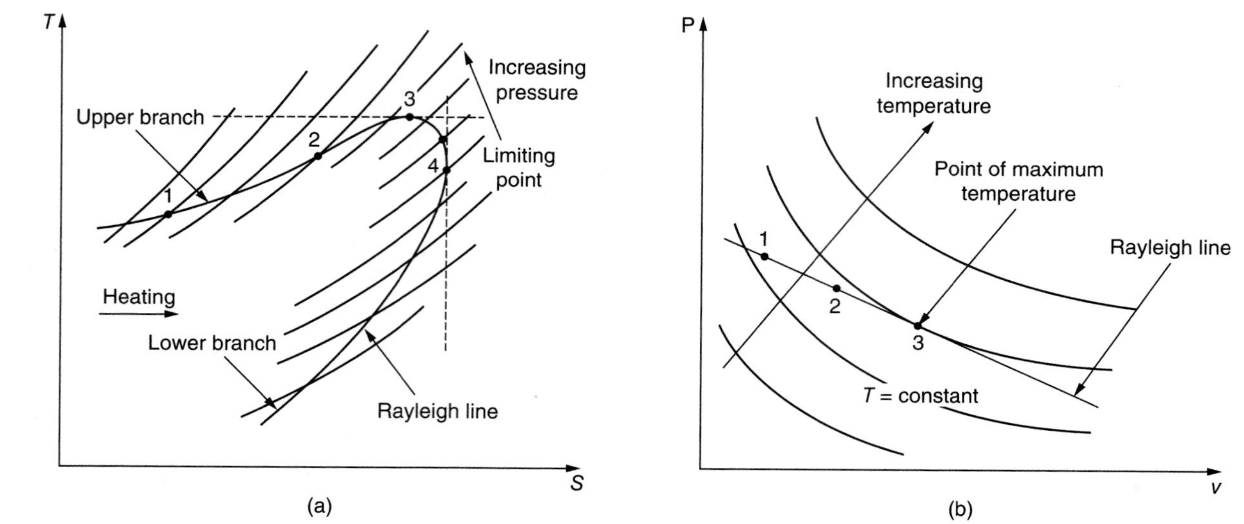

Figure 2.1.48 Rayleigh flow with numbered corresponding states for a given mass flow. State 3 is the maximum temperature attained by heating, and $M = 1/\sqrt{k}$. State 4 is the sonic limit $M = 1$ where entropy is a maximum. (*Figures adapted from M-H Access Engineering.*)

modified accordingly. The continuity, momentum, and energy equations in differential form are respectively,

$$\frac{d\rho}{\rho}+\frac{d\bar{V}}{\bar{V}}=0$$

$$dP+\frac{\rho\bar{V}^2}{2g_c}\frac{fdx}{D}+\rho\frac{\bar{V}d\bar{V}}{g_c}=0,\ \text{or}\ \frac{dP}{P}+\frac{kM^2}{2}\frac{fdx}{D}+kM^2\frac{dM}{M}+\frac{kM^2}{2}\frac{dT}{T}=0$$

$$\frac{dq}{C_pT}=\left(1+\frac{k-1}{2}M^2\right)\frac{dT}{T}+(k-1)MdM$$

With the differential forms of the ideal gas law, the definition of the speed of sound, and the definition of the Mach number, an equation relating to heat transfer, Mach number, and pressure drop (or friction factor) is obtained,

$$\frac{dq}{C_pT^o}\frac{1+kM^2}{2}+\frac{kM^2}{2}\frac{fdx}{D}=\left[\frac{1-kM^2}{M}+\frac{(k-1)(1+kM^2)M}{2\left(1+\frac{k-1}{2}M^2\right)}\right]dM$$

This equation shows the relative effects of friction and heat transfer, and it must be numerically integrated to obtain a solution.

Typically the heat transfer is expressed in terms of convective heat transfer coefficient $dq=\hat{h}(T_w-T_{aw})dA_p$, where $\hat{h}$ is the convective heat transfer coefficient (W/m²K), A_p is the perimeter of the duct, and T_{aw} is the adiabatic wall temperature (i.e., the temperature of a perfectly insulated wall in high speed flow). For gases with $Pr \sim 1$, $T_{aw} > T^o$. With a given mass flow, $\dot{m}$, $dq = d\dot{q}/\dot{m} = d\dot{q}/\rho A\bar{V}$. The convective heat transfer coefficient is related to the friction factor via the Reynolds analogy $\hat{h}=\rho\bar{V}C_pf/8$. Thus the energy equation above can be written,

$$\frac{T_w-T^o}{T^o}\frac{1+kM^2}{2}+\frac{kM^2}{2}\frac{fdx}{D}=\left[\frac{1-kM^2}{M}+\frac{(k-1)(1+kM^2)M}{2\left(1+\frac{k-1}{2}M^2\right)}\right]dM$$

where the roles of wall temperature, friction, and Mach number are emphasized. Heat transfer dominates friction when the wall temperature is larger than the total temperature, but friction is most important when the wall temperature is little removed from the total temperature.

When wall temperature and its variation with distance along the flow are given (or known via measurement), the total temperature must be obtained via integration of the equation for the convective heat transfer coefficient. The result for wall temperature is then used in a numerical integration of the energy equation.

Incompressible Fluids

Section 1.3 presents a thorough exposition of the fundamentals of the flow of incompressible fluids, and the reader should consult it in connection with the following.

The mechanical energy equation given above can be written in the case of a horizontal conduct,

$$vdp+\frac{\bar{V}d\bar{V}}{g}dF=0$$

The friction term, dF, includes not only losses due to frictional flow along the pipe but also those due to fittings, valves, etc., as well as losses occasioned by any enlargement or contraction of the pipe as, e.g., the loss occurring when a fluid passes from a pipe into a tank. For long straight pipes of uniform diameter, D, $dF/dL \approx 2f'/gD$, where the friction factor $f = 4f'$ and L is the pipe length (see Section 1 for definition of the friction factors). It is usual to express friction due to fittings, etc., in terms of additional length of pipe, adding this to the actual pipe length to get the equivalent pipe length. Integration of the fundamental equation leads to two sets of formulas.

1. For pressure drops that are small relative to the initial pressure, the specific volume v and the velocity may be assumed constant, and approximately

$$p_1-p_2=\frac{2f'\bar{V}^2L}{vgD}$$

Expressing pressure in pounds force per square inch (psi), the pipe diameter in inches, and as a function of wv/d^2,

$$p_1-p_2=\frac{174.2f'w^2vL}{D^5}$$

where w is the weight of fluid flow (lbf/s).

2. For considerable pressure drops, when dealing with approximately isothermal flow of gases and vapors to which the gas laws are applicable, the fundamental equation on a weight basis may be integrated to give,

$$p_1^2-p_2^2=\frac{2w^2RT}{gA^2}\ln\left(\frac{v_2}{v_1}\right)+\frac{4f'RTw^2L}{gA^2D}$$

12
11
10
9
8
7
6
5
4
3
2
$p\bar{y}/x$
Smooth pipe
Clean iron and steel pipe
Streomline flow
10^2
10^3
10^4
10^5
10^6
10^7
$\frac{D_x}{\mu}$ Where x = $\sqrt{2g\ Dp^2\ dF'/dL}$

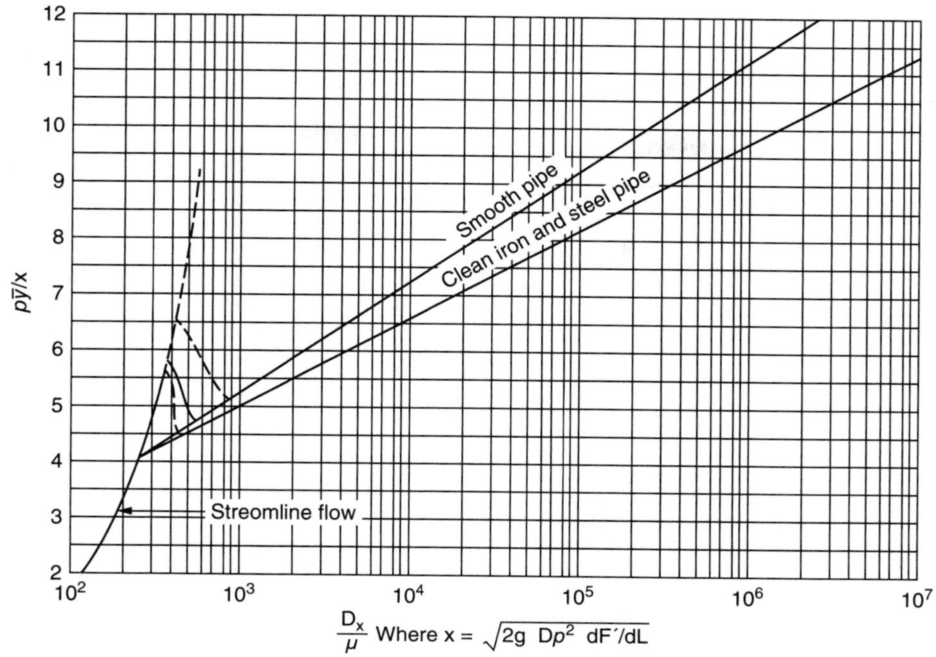

Figure 2.1.49 Chart for estimating the rate of flow from the pressure gradient.

Friction Factor The friction factor, f, is not a constant but is a function of the dimensionless expression, or the reciprocal of the Reynolds number. For water and other fluids,

$$f' = 0.0054 + 0.375\left(\frac{\mu v}{\bar{V}d}\right)$$

where μ is the dynamic viscosity and d is the pipe diameter in inches. For high-pressure steam, the second term in the expression is small, and $f' \approx 0.0054$. An alternative approximation for steam is $f' = 0.0027(1 + 3.6/d)$.

Values of $f = 4f'$ as a function of the pipe surface are given in Section 1. For predicting the capacity of a given pipe operating on a chosen fluid with fixed pressure drop, the use of Fig. 2.1.49 eliminates the trial and error methods usually involved. Commonly used resistances due to fittings, expressed in terms of L/D, are shown in Table 2.1.15.

Table 2.1.15 Resistances for Circular Pipes Expressed in Equivalent Pipe Length for Common Fittings

Fitting		Equivalent Length (L/D)
90° Elbow	D = 1 – 2.5 in.	30
	D = 3 – 6 in.	40
	D = 7 – 10 in.	50
90° Curves	Center line radius of curve – 2 – 8 D	10
Globe Valve	D = 1 – 2.5 in.	45
	D = 3 – 6 in.	60
	D = 7 – 10 in.	75
Tee	D = 1 – 4 in.	60

The resistance in energy units, due to sudden enlargement in a pipe, is ~ $\left(\bar{V}_1^2 - \bar{V}_2^2\right)/2g$. For a sudden contraction it is $1.5(1-r)\bar{V}_2^2/[2g(3-r)]$, where $r = A_2/A_1$.

References

1. Çengel YA, Cimbala JM. "Fluid Mechanics," McGraw-Hill, New York, 2006.
2. Cengel YA, Boles MA. "Thermodynamics: An Engineering Approach," 5e, McGraw-Hill, New York, 2006.
3. Heald CC. "Camerons Hydraulic Data," 19e, Flowserve Corporation, Irving, 2002.
4. Houghton EL, Brock AE. "Tables for the Compressible Flow of Dry Air," Butterworth-Heinemann, Waltham, 1975.
5. John JEA. "Gas Dynamics," Allyn and Bacon, Boston, 1969.
6. John JAE, Keith TG. "Gas Dynamics," 3e, Pearson Prentice Hall, Upper Saddle River, 2006.
7. Shapiro AH. "Dynamics and Thermodynamics of Compressible Fluid," vol. 1, The Ronald Press, New York, 1958.

COMBUSTION

Hydrocarbon Fuels

Hydrocarbon fuels are fossil based and derived from decayed plant life that has been rendered in gaseous, liquid, or solid form within the earth. The combustible elements that characterize hydrocarbon fuels are carbon, hydrogen, and, in some cases, sulfur. Solid, liquid, and gaseous hydrocarbon fuels are generally analyzed for their energy value in the same manner, i.e., employing the conservation of chemical species, energy, and mass. The general chemical representation of a hydrocarbon fuel is C_nH_m.

The complete combustion of carbon produces carbon dioxide, CO_2, and the complete combustion of hydrogen gives water, H_2O. The complete combustion of a hydrocarbon fuel occurs when all carbon atoms have been consumed, or oxidized. Special properties of various fuels are given in Sec. 2.2. All chemical reactions conserve chemical species, mass, and energy. Exothermic reactions release energy, which is the important property of a fuel. Endothermic reactions absorb energy.

Stoichiometry

The approximate molecular weights of the several important elements and compounds relevant to combustion calculations are listed in Table 2.1.16.

Table 2.1.16 Approximate Molecular Weights of Gaseous, Liquid and Solid Fuels

Molecule	C	H_2	O_2	N_2	CO	CO_2	H_2O
M	12	2	32	28	28	44	18
Molecule	CH_4	C_3H_8	C_2H_6O	S	NO	NO_2	SO_2
M	12	28	46	32	30	46	64

The complete combustion of a hydrocarbon fuel is generally represented by,

$$C_nH_m + \left(n + \frac{m}{4}\right)O_2 \rightarrow nCO_2 + \left(\frac{m}{2}\right)H_2O$$

The stoichiometric ratio is denoted as the ratio of the number of fuel molecules to oxygen molecules is $(n + m/4)$. The stoichiometric mass ratio is obtained by multiplying each mole number by its respective molecular weight: *oxygen mass/fuel mass* $= (32n + 8)/(12n + 1.008m)$. The stoichiometric mass ratio ranges from 2.667 for carbon to 7.937 for hydrogen.

The stoichiometric equation considers only oxygen, but in practice combustion takes place in air, a combination of ~21 percent molecular oxygen and ~79 percent molecular nitrogen. The stoichiometric air mass/fuel mass ratio, AF_{st}, is obtained by multiply in the stoichiometric mass ratio by the ratio of the mass of air to the mass of oxygen,

$$AF_{st} = 4.14\left(\frac{32n + 8m}{12n + 1.008m}\right)$$

The coefficients in the combustion equation also indicate the combining volumes of the gaseous reactants. For example, one volume of CH_4 requires two volumes of O_2 for complete combustion, and the resulting gaseous products of combustion are one volume of CO_2 and two volumes of H_2O. The coefficients multiplied by the corresponding molecular weights give the combining masses. Thus for CH_4, 1 kg requires $64/16 = 4$ kg of O_2 for complete combustion, and the products are $44/16 = 2.75$ kg of CO_2 and $36/16 = 2.25$ kg of H_2O.

The composition of air is approximately 0.232 O_2 and 0.768 N_2 on a mass basis, or ~0.21 O_2 and ~0.79 N_2 by volume. On a molar basis, $N_2{:}O_2 = 3.76$. For exact analyses, it may be necessary to account for water vapor mixed with the air, but ordinarily this may be neglected when the humidity is very low. The minimum amount of air required for the combustion of a unit mass of fuel is the quantity of oxygen, as found from the combustion equation, divided by 0.232. Likewise, the minimum volume of air required for the combustion of 1 m^3 of a fuel gas is the volume of oxygen divided by 0.21. For example, in the combustion of CH_4 the air required per pound mass is $4/0.232 = 17.24$ lbm and the volume of air per cubic foot of CH_4 is $2/0.21 = 9.52$ ft^3. In practice, more air is provided, i.e., excess air, than is required for complete combustion. Letting a denote the minimum amount required and xa the quantity of air admitted; then $x - 1$ is denoted the excess air coefficient.

Products of Combustion

The products arising from the complete combustion of a hydrocarbon fuel are CO_2, H_2O, and if sulfur is present, SO_2. Accompanying these are the nitrogen and the oxygen in the excess of air. Hence the products of complete combustion are principally CO_2, H_2O, N_2, and O_2. The presence of CO indicates incomplete combustion. In simple calculations, the reaction of nitrogen with oxygen to form nitrogen oxides, termed NOX, such as nitric oxide, NO, nitrogen peroxide, NO_2, etc., is neglected. In practice, an internal combustion engine is run at a lower compression ratio to reduce NOX. The reduced pollution is bought at the expense of reduced operating efficiency. The composition of the products of combustion is readily calculated from the combustion equations. (See also Table 2.1.18.)

EXAMPLE. A producer gas having the volume composition given is burned with 20 percent excess of air; required the volume composition of the exhaust gases.

	V
H_2	0.08
CO	0.22
CH_4	0.024
CO_2	0.066
N_2	0.61
	1.0

Coefficients in chemical balance			Coefficients × V		
O_2	CO_2	H_2O	O_2	CO_2	H_2O
0.5	0	1	0.04	0	0.08
0.5	1	0	0.11	0.22	0
2	1	2	0.048	0.024	0.048
0	1	0	0	0.066	0
0	0	0	0	0	0
			0.198	0.31	0.128

For 1 ft^3 of the producer gas, 0.198 ft^3 of O_2 is required for complete combustion. For 1 ft^3 of the producer gas, 0.198 ft^3 of O^2 is required for complete combustion. The minimum volume of air required is $0.198/0.21 = 0.943$ ft^3 and with 20 percent excess the air supplied is $0.943 \times 1.2 = 1.132$ ft^3. Of this, 0.238 ft^3 is O_2 and 0.894 ft^3 is N_2. Consequently, for 1 ft^3 of the fuel gas, the exhaust gas contains,

CO_2		0.31 ft^3	15.7%
H_2O		0.128 ft^3	6.5%
N_2	0.61 + 0.894 =	1.504 ft^3	75.8%
O_2 (excess)	0.238 − 0.198 =	0.040 ft^3	2.0%
		1.982 ft^3	100%

Volume Contraction and Expansion As a result of chemical action, there is often a change of volume. For example, in the reaction $2H_2 + O_2 = 2H_2O$, three volumes (two of H_2 and one of O_2) contract to two volumes of water vapor. In the example above, the volume of producer gas and air supplied is 1 ft^3 gas + 1.132 ft^3 air = 2.132 ft^3, and the corresponding volume of the exhaust gas is 1.982 ft^3, showing a contraction of ~7 percent. For a hydrocarbon having the molecular composition C_mH_n, the relative volume contraction is $1 - n/4$. Thus for CH_4 and C_2H_4 there is no change of volume, for C_2H_2 the contraction is half the volume, and for C_2H_6 there is an increase of one-half in volume. The change of volume accompanying a chemical reaction, such as in combustion, causes a corresponding change in the gas constant R. Let R' denote the constant for the mixture of gas and air, e.g., 1 mass unit of gas and xa mass unites of air, before combustion, and R'' the constant of the mixture of the products of combustion. If y is the resulting contraction of volume, $R''/R' = (1 + xa - y)/(1 + xa)$.

Fuel Heating Value

A chemical reaction is accompanied by either release (exothermic reaction) or absorption (endothermic reaction) of energy in the form of heat. The combustion of a useful fuel with oxygen is exothermic, and the released energy per complete combustion of a unit mass of fuel is denoted the heat of combustion, or the heating value. The higher heating value (HHV) is determined via calorimetry in which the products of combustion are cooled to the initial temperature, the product water is fully condensed to the liquid state, and the heat transfer to the cooling medium is measured. The lower heating value (LHV) of the fuel is determined when the product water remains in the vapor state. At atmospheric pressure, the partial pressure of the water vapor in the resulting products of combustion will usually be sufficiently high to cause the water to condense if the temperature is below ~120 to 140°F. This causes liberation of the heat of vaporization of the condensed water.

To facilitate calculations of the temperature attained by combustion, it is desirable to make use of the LHV. The necessity of taking into account the heat of vaporization of the water and the difference between the specific heats of liquid water and water vapor is thus avoided. A constant volume calorimeter gives practically correct values of the HHV, and a gas constant pressure calorimeter gives values which may be incorrect by a fraction of 1 percent. The quantity to be subtracted from the HHV to obtain the LHV will vary with the composition of the fuel. An approximate value is 1,050*m*, where *m* is the number of mass units of H_2O formed per mass unit mass fuel consumed.

Heating Value at Constant Volume When a fuel is burned at constant volume, the energy released is,

$$Q_V = U_P - U_R$$

where the internal energies of the reactant product mixtures can be expressed on a mass or molar basis, and the subscripts P and R denote products and reactants respectively. The heat of combustion at constant volume may also be described as the energy in the form of heat transferred from a calorimeter to the external surroundings when the temperature and volume of the combustion products are brought to the temperature and volume of the reactants.

Heating Value at Constant Pressure For a constant pressure reaction, the energy released is,

$$Q_P = H_P - H_R$$

where the enthalpies of the reactant and product mixtures can be expressed on either a mass or molar basis. Assuming that the internal energy change for a constant volume reaction is approximately equal to that for a constant pressure reaction,

$$Q_P = Q_V + p(V_P - V_R)$$

Since Q_V is equal to the change in internal energy, ΔU_{RP}, this relation may be written in terms of enthalpy, $Q_P = \Delta H_{RP}$. The heat of combustion at constant pressure may also be described as the heat transferred from a calorimeter when the pressure and temperature of the products are brought back to the pressure and temperature of the reactants.

If the reactants and products are assumed to be ideal gases,

$$Q_P = Q_V + RT\Delta n$$

where Δn is the change in number of moles, and R is the universal gas constant. From this relation the heat of combustion at constant pressure is obtained from the heat of reaction at constant volume, or vice versa, if the temperature and molar-volume change are known.

If there is no change of volume due to the combustion, $Q_P = Q_V$. When there is a contraction of volume, Q_P exceeds Q_V by the energy equivalent of the work done on the gas during the contraction. For example, in the stoichiometric combustion of CO, $CO + 0.5\ O_2 = CO_2$, there is a contraction of one-half the volume. At 62°F, the volume of 1 lbm of CO at 1 atm is 13.6 ft^3, and the energy equivalent of the work done at atmospheric pressure is $0.5 \times 13.6 \times 2{,}116/778 = 18.5$ Btu, which is ~0.4 percent of the heating value of CO. Because the difference between Q_P and Q_V is small in most fuels, it is usually neglected. Additionally, heating values vary with the initial temperature which is also the final temperature, but the variation is usually negligible.

Heating Value per Unit Volume of Fuel Gas The consumption of a fuel gas is more easily measured by volume than by mass, and it is convenient to express heating values in terms of volumes. For this purpose, a standard temperature and pressure must be assumed. It is customary to take 1 atm (14.696 psia = 101.325 kPa) as a standard pressure, but there are a diversity of definition and practice for the standard temperature and pressure. Because designation of a standard temperature has changed over the years, caution is recommended in using combustion data in the literature, and source materials used by various professions and industry associations must be checked prior to use in design and analysis.

The standard state (standard temperature and pressure, STP) established by the International Union of Pure and Applied Chemistry (IUPAC) is 273.5 K (0°C) and 100 kPa (1 bar). The standard ambient temperature and pressure (SATP) is 298.15 K (25°C, 77°F) and 100 kPa (1 bar). The U.S. National Institute of Standards and Technology (NIST) standard is 20°C (293.15 K, 66°F) and 14.696 psi, or 101.325 kPa, which is also referred to as normal temperature and pressure (NTP). The American Gas Association (AGA) currently stipulates 60°F (15.56°C) and 101.32 kPa as the standard reference. The International Standard Metric Conditions for natural gas and similar fluids are 288.15 K (15.00°C, 59.0°F) and 101.325 kPa.

Conversion of density and heating values from 68 to 60°F of dry gas is obtained by multiplying by the factor 1.0154 and for saturated gas, 1.0212. Conversion of specific volumes of dry gas is obtained by multiplying by the factor 0.9848 and for saturated gas, 0.9792.

If the gas is at some other pressure and temperature, say p_1 (psia) and T_1 (R), the heating value per cubic foot is found by multiplying the heating value per cubic foot under standard conditions by $35.9p_1/T_1$. The heating values of a few of the more common fuels at 14.7 psia and 68°F are given in Table 2.1.17 for complete (stoichiometric) combustion.

Table 2.1.17 Heating Values of Common Hydrocarbon Fuels

Fuel	Chemical symbol	High heat value		Low heat value	
		Btu/lb	*Btu/ft^3	Btu/lb	*Btu/ft^3
Carbon to CO_2	C	14,096			
Carbon to CO	C	3,960			
CO to CO_2	CO	4,346	316.0		
Sulfur to SO_2	S	3,984			
Hydrogen	H_2	61,031	319.4	51,593	270.0
Methane	CH_4	23,890	994.7	21,518	896.0
Ethane	C_2H_6	22,329	1,742.6	20,431	1,594.5
Propane	C_3H_8	21,670	2,480.1	19,944	2,282.6
Butane	C_4H_{10}	21,316	3,215.6	19,679	2,968.7
Pentane	C_5H_{12}	21,095	3,950.2	19,513	3,654.0
Hexane (liquid)	C_6H_{14}	20,675	....	19,130	
Octane (liquid)	C_8H_{18}	20,529	....	19,029	
n-Decane (liquid)	$C_{10}H_{22}$	20,371	....	19,175	
Ethylene	C_2H_4	21,646	1,576.1	20,276	1,477.4
Propene (propylene)	C_3H_6	21,053	2,299.4	19,683	2,151.3
Acetylene (ethyne)	C_2H_2	21,477	1,451.4	20,734	1,402.0
Benzene	C_6H_6	18,188	3,687.5	17,446	3,539.3
Toluene (methyl benzene)	C_7H_8	18,441	4,410.1	17,601	4,212.6
Methanol (methyl alcohol. liquid)	CH_4O	9,758	....	8,570	
Ethanol (ethyl alcohol, liquid)	C_2H_6O	12,770	....	11,531	
Naphthalene (solid)	$C_{10}H_8$	17,310	....	13,110	

* Measured as a gas at 68°F and 14.70 psia. Multiply by 1.0154 for 60°F and 14.70 psia.

Heating Value per Unit Volume of Mixture Let a denote the volume of air required for the combustion of unit volume of fuel gas and xa the value of air actually admitted, $(x - 1)a$ being therefore the excess air. Then the volume of the mixture of fuel gas and air is $1 + xa$, and the quotient $Q_{RP}/(1 + xa)$ may be called the heating value per unit volume of the mixture. This quantity is useful in comparing the relative volumes of mixture required with different fuel gases. A lean gas, e.g., blast-furnace gas or producer gas, has a low heating value, but the value of a is correspondingly low. On the other hand a rich gas, e.g., natural gas, has a high heating value but requires a large volume of air for combustion (Table 2.1.18).

Heat of Formation The change in enthalpy resulting when a compound is formed from its elements isothermally and at constant pressure is numerically equal to the heat of formation, $\Delta H_f^0 = -Q_f$. It is equal to the difference between the heats of combustion of the constituents forming the compound and the heat of combustion of the compound itself. Table 2.1.19 lists values of the heat of formation for selected fuels and inorganic compounds. A plus sign indicates that energy (heat) is evolved on formation the compound, and a minus sign indicates that energy (heat) is absorbed on formation. Heats of formation in chemical literature are usually listed on a molar basis.

Internal Energy and Enthalpy of Gases

Table 2.1.20 lists the internal energy of common encountered gases in Btu/(lbm-mol) measured above 520 R (60°F). The corresponding values of the enthalpy are obtained by adding the value of Apv from the last column.

Temperature Attained By Combustion

Excluding the effect of dissociation of the reactants and products, the temperature attained at the end of combustion may be calculated by an energy balance. The heat of combustion less the heat lost by conduction and radiation during the process is equal to the increase in internal energy of the products mixture if the combustion is at constant volume; or, if the combustion is at constant pressure, the increase in enthalpy of the products mixture.

When there is no heat transfer and for complete combustion, adiabatic combustion results, and the temperature of reactants is denoted the adiabatic flame temperature. The adiabatic flame temperature is generally calculated by iteration on the energy balance. In adiabatic combustion, specific heats are not constant. For extremely high adiabatic flame temperature, dissociation of the reactants and products can occur. Dissociation lowers the temperature of the products owing to the endothermic nature of the dissociation reactions.

EXAMPLE. To calculate the temperature of combustion of a fuel gas having the composition $H_2 = 0.50$, $CO = 0.46$, and $CO_2 = 0.04$. The gas is burned with 15 percent excess air at constant volume, and the initial temperature is 62°F (522 R).

The volume compositions of the initial mixture of fuel gas and air and of the mixture of products are,

Table 2.1.18 Products of Combustion for Common Fuels

Fuel	Chemical formula	Molecular weight $O_2 = 32$	Specific weight, lb/ft³ at 68°F and 14.70 lb/in²	Volume of air necessary for combustion of unit volume of fuel at same temperature and pressure	Products of combustion of 1 ft³ of fuel in theoretical amount of air, ft³			Weight of air necessary for combustion of unit weight of fuel	Products of combustion of 1 lb of fuel in theoretical amount of air, lb		
					CO_2	H_2O	N_2		CO_2	H_2O	N_2
Oxygen	O_2	32	0.0831								
Nitrogen	N_2	28.08	0.0727								
Air			0.0753								
Hydrogen	H_2	2.016	0.0052	2.39	0	1	1.89	34.2	0.0	8.94	26.28
Steam	H_2O	18.016									
Carbon monoxide	CO	28.00	0.0727	2.39							
Carbon dioxide	CO_2	44.00	0.1142								
Methane	CH_4	16.03	0.0416	9.55	1	2	7.55	17.21	2.75	2.248	13.22
Ethane	C_2H_6	30.05	0.0779	16.71	2	3	13.21	16.07	2.93	1.799	12.34
Propane	C_3H_8	44.06	0.1142	23.87	3	4	18.87	15.65	3.00	1.635	12.02
Butane	C_4H_{10}	58.1	0.1506	30.94	4	5	24.53	15.44	3.03	1.551	11.86
Pentane	C_5H_{12}	72.1	0.1869	38.08	5	6	30.2	15.31	3.05	1.499	11.76
Hexane	C_6H_{14}	86.1	0.2232	45.3	6	7	35.8	15.22	3.07	1.465	11.69
Heptane	C_7H_{16}	100.1	0.2596	52.5	7	8	41.5	15.15	3.08	1.439	11.64
Octane	C_8H_{18}	114.1	0.2959	59.7	8	9	47.2	15.11	3.08	1.421	11.60
Nonane	C_9H_{20}	128.2	0.3323	66.8	9	10	52.8	15.07	3.09	1.406	11.57
Benzene	C_6H_6	78.0	0.2025	35.8	6	3	28.3	13.26	3.38	0.693	10.18
Toluene	C_7H_8	92.1	0.2388	42.9	7	4	34.0	13.50	3.35	0.783	10.36
Xylene	C_8H_{10}	106.2	0.2752	50.1	8	5	39.6	13.57	3.31	0.845	10.42
Cyclohexane	C_6H_{12}	84.0	0.2180	43.0	6	6	34.0	14.76	3.14	1.285	11.34
Ethylene	C_2H_4	28.03	0.0728	14.32	2	2	11.32	14.76	3.14	1.285	11.34
Propylene	C_3H_6	42.0	0.1090	21.48	3	3	16.98	14.76	3.14	1.285	11.34
Butylene	C_4H_8	64.1	0.1454	28.64	4	4	22.64	14.76	3.14	1.285	11.34
Acetylene	C_2H_2	26.02	0.0675	11.93	2	1	9.43	13.26	3.38	0.693	10.18
Allylene	C_3H_4	40.0	0.1038	19.09	3	2	15.09	13.78	3.30	0.900	10.59
Naphthalene	$C_{10}H_8$	128.1	0.3322	57.3	10	4	45.28	12.93	3.44	0.563	9.93
Methyl alcohol	CH_4O	32.0	0.0830	7.16	1	2	5.66	6.46	1.37	1.125	4.96
Ethyl alcohol	C_2H_6O	46.0	0.1194	14.32	2	3	11.32	8.99	1.91	1.174	6.90

SOURCE: Marks, "The Airplane Engine."

Table 2.1.19 Heats of Formation in Btu/lbm at 65 °F and 1 atm

Fuel (g = gas, l = liquid, v = vapor)	ΔH_f^0 (Btu/lbm)	Inorganic Compound (g = gas, l = liquid, v = vapor)	ΔH_f^0 (Btu/lbm)
Methane, CH_4 (g)	2,001.4	Aluminum Oxide, Al_2O_3 (s)	6,710
Ethane, C_2H_6 (g)	1,206.1	Calcium Oxide, CaO (s)	4,869
Propane, C_3H_8 (v)	1,008.5	Calcium Carbonate, $CaCO_3$ (s)	5,206
Acetylene, C_2H_2 (g)	−3,747	Iron (II) Oxide, FeO (s)	1,611
Ethylene, C_2H_4 (g)	−805.3	Iron (III)Oxide, Fe_2O_3 (s)	2,238
Benzene, C_6H_6 (v)	−459	Iron (II,III) Oxide , Fe_3O_4 (s)	2,075
Toluene, C_7H_8 (v)	−234.9	Iron Sulfide, FeS_2 (s)	532.7
Methyl Alcohol, CH_3OH (l)	3,227.3	Hydrochloric Acid, HCl (g)	1,089
Ethyl Alcohol, C_2H_5OH (l)	2,623.3	Nitric Acid, HNO_3 (l)	3,555.8
		Potassium Oxide, K_2O (s)	164.7
		Magnesium Oxide, MgO (s)	6,522
		Manganese Oxide, MnO (s)	2,449
		Nitric Oxide, NO (g)	−1,296
		Nitrous Oxide, N_2O (g)	−803.5
		Sodium Oxide, Na_2O (s)	2,888
		Ammonia, NH_3 (g)	1,163
		Ammonium Chloride, NH_4Cl (s)	1,480
		Nickel Oxide, NiO (s)	1,407
		Lead Oxide, Pb_2O_5 (s)	5,394
		Lead (II) Oxide, PbO (red) (s)	423
		Lead Dioxide, PbO_2 (s)	489.1
		Sulfur Dioxide, SO_2 (g)	1.933
		Sulfur Trioxide, SO_3 (g)	2,112
		Tin Oxide, SnO (s)	904.7
		Zinc Oxide, ZnO (s)	1,847

Initial: $H_2 = 0.50$, $CO = 0.46$, $CO = 0.04$, $O_2 = 0.552$, $N_2 = 2.098$
Products: $H_2O = 0.50$, $CO_2 = 0.50$, $O_2 = 0.072$, $N_2 = 2.098$

Since a volume composition is also a molar composition, the product mixture may be regarded as made up of 0.5 mol each of H_2O and CO_2, 0.072 mol of O_2, and 2.098 mols of N_2. If values are taken from Tables 2.1.6 and 2.1.7, the heat generated by combustion of the fuel mixture is $0.50 \times 2 \times 51{,}593 + 0.46 \times 28 \times 4{,}346 = 107{,}569$ Btu. The internal energy u of the products mixture at 522 R is now calculated (Table 2.1.8). For 0.5 mol H_2O + 0.5 mol CO_2 + 0.072 mol O_2 + 2.098 mol N_2 this is $6.1 + 6.95 + 0.72 + 20.34 = 34.11$ Btu.

The energy u of the mixture is next calculated for various assumed temperatures, the proper values being taken from Table 2.1.20. If the heat of combustion, 107,569 Btu, is entirely used in the increase of energy, the temperature attained lies somewhere between 5,000 R and 5,100 R; by interpolation, the value 5,073 R is obtained.

Loss of heat during combustion may readily be taken into account; thus, if 10 percent of the heat of combustion is lost, the amount available for increasing the energy of the products is $107{,}569 \times 0.90 = 96{,}812$ Btu, and this increase gives $T_2 = 4{,}671°$. If the fuel is burned at constant pressure, Q_p is used instead of Q_v and values of h are determined from Table 2.1.17 instead of values of u.

T_2 assumed (R)	4,700	4,800	4,900	5,000	5,100
u, 0.5 mol H_2O	18,308	18,851	19,396	19,943	20,429
u, 0.5 mol CO_2	23,742	24,388	25,035	25,683	26,351
u, 0.072 mol O_2	1,973	2,026	2,079	2,132	2,185
u, 2.098 mil N_2	54,596	55,018	56,447	57,882	59,321
u_2 =	97,619	100,283	102,957	105,640	108,286
u_1 =	34	34	34	34	34
	97,585	100,249	102,923	105,606	108,252

Effect of Dissociation

The maximum temperature that can be obtained by the combustion of any fuel is limited by the dissociation of the products formed. Dissociation and the chemical equilibrium associated with in high-temperature combustion are exceedingly complex, involving such chemical species as CO_2, CO, H_2O, H_2, H, OH, N_2, NO, N, O_2, and O. The equilibrium reached is a direct consequence of the Second Law. However, the calculation of the equilibrium constant even for simple reactions is tedious. For all possible reactions $aA + bB \rightarrow cC + dD$ ($a, b, c, d \leq 2$), values of the equilibrium constant K_p are contained in the American Institute of Physics Handbook, 3e, McGraw-Hill, New York.

Calculated flame temperatures, allowing for dissociation, for gaseous fuels with stated amounts of air present are given in Table 2.1.21. The combustion is assumed to be adiabatic and at 14.7 psia. In the case of explosion in the internal-combustion engine, the figures in Table 2.1.21 will be altered somewhat. The effect of compression is to increase both the initial temperature and the initial pressure. The resulting increase in the explosion temperature tends to increase the degree of dissociation, and the increase of pressure tends to reduce it. The net effect will be a small reduction.

Combustion of Liquid Fuels

For properties of fuel oils, heating values, etc., see Sec. 5.1. Calculations for the burning of liquid fuels are fundamentally the same as for gaseous fuels. Liquid fuels are almost always gasified before or during actual combustion.

Combustion of Solid Fuels

For properties of solid fuels, heat values, etc., see Sec. 5.1.

Combustion Air Required Let c, h, and o, denote, respectively, the parts of carbon, hydrogen, and oxygen in a unit mass of the fuel. The minimum

Table 2.1.20 Internal Energy of Gases in Btu/lbm-mol above 520 R

Temp. °R	O_2	N_2	Air	CO_2	H_2O	H_2	CO	*Apv*
520	0	0	0	0	0	0	0	1,033
540	100	97	97	139	122	96	97	1,072
560	200	196	196	280	244	193	196	1,112
580	301	295	295	424	357	291	295	1,152
600	402	395	395	570	490	390	396	1,192
700	920	896	897	1,320	1,110	887	896	1,390
800	1,449	1,399	1,403	2,120	1,734	1,386	1,402	1,589
900	1,989	1,905	1,915	2,965	2,366	1,886	1,913	1,787
1,000	2,539	2,416	2,431	3,852	3,009	2,387	2,430	1,986
1,100	3,101	2,934	2,957	4,778	3,666	2,889	2,954	2,185
1,200	3,675	3,461	3,492	5,736	4,399	3,393	3,485	2,383
1,300	4,262	3,996	4,036	6,721	5,030	3,899	4,026	2,582
1,400	4,861	4,539	4,587	7,731	5,740	4,406	4,580	2,780
1,500	5,472	5,091	5,149	8,764	6,468	4,916	5,145	2,979
1,600	6,092	5,652	5,720	9,819	7,212	5,429	5,720	3,178
1,700	6,718	6,224	6,301	10,896	7,970	5,945	6,305	3,376
1,800	7,349	6,805	6,889	11,993	8,741	6,464	6,899	3,575
1,900	7,985	7,393	7,485	13,105	9,526	6,988	7,501	3,773
2,000	8,629	7,989	8,087	14,230	10,327	7,517	8,109	3,972
2,100	9,279	8,592	8,698	15,368	11,146	8,053	8,722	4,171
2,200	9,934	9,203	9,314	16,518	11,983	8,597	9,339	4,369
2,300	10,592	9,817	9,934	17,680	12,835	9,147	9,961	4,568
2,400	11,252	10,435	10,558	18,852	13,700	9,703	10,588	4,766
2,500	11,916	11,056	11,185	20,033	14,578	10,263	11,220	4,965
2,600	12,584	11,682	11,817	21,222	15,469	10,827	11,857	5,164
2,700	13,257	12,313	12,453	22,419	16,372	11,396	12,499	5,362
2,800	13,937	12,949	13,095	23,624	17,288	11,970	13,144	5,561
2,900	14,622	13,590	13,742	24,836	18,217	12,549	13,792	5,759
3,000	15,309	14,236	14,394	26,055	19,160	13,133	14,443	5,958
3,100	16,001	14,888	15,051	27,281	20,117	13,723	15,097	6,157
3,200	16,693	15,543	15,710	28,513	21,086	14,319	15,754	6,355
3,300	17,386	16,199	16,369	29,750	22,066	14,921	16,414	6,554
3,400	18,080	16,855	17,030	30,991	23,057	15,529	17,078	6,752
3,500	18,776	17,512	17,692	32,237	24,057	16,143	17,744	6,951
3,600	19,475	18,171	18,356	33,487	25,067	16,762	18,412	7,150
3,700	20,179	18,833	19,022	34,741	26,085	17,385	19,082	7,348
3,800	20,887	19,496	19,691	35,998	27,110	18,011	19,755	7,547
3,900	21,598	20,162	20,363	37,258	28,141	18,641	20,430	7,745
4,000	22,314	20,830	21,037	38,522	29,178	19,274	21,107	7,944
4,100	23,034	21,500	21,714	39,791	30,221	19,911	21,784	8,143
4,200	23,757	22,172	22,393	41,064	31,270	20,552	22,462	8,341
4,300	24,482	22,845	23,073	42,341	32,326	21,197	23,140	8,540
4,400	25,209	23,519	23,755	43,622	33,389	21,845	23,819	8,738
4,500	25,938	24,194	24,437	44,906	34,459	22,497	24,499	8,937
4,600	26,668	24,869	25,120	46,193	35,535	23,154	25,179	9,136
4,700	27,401	25,546	25,805	47,483	36,616	23,816	25,860	9,334
4,800	28,136	26,224	26,491	48,775	37,701	24,480	26,542	9,533
4,900	28,874	26,905	27,180	50,069	38,791	25,418	27,226	9,731
5,000	29,616	27,589	27,872	51,365	39,885	25,819	27,912	9,930
5,100	30,361	28,275	28,566	52,663	40,983	26,492	28,600	10,129
5,200	31,108	28,961	29,262	53,963	42,084	27,166	29,289	10,327
5,300	31,857	29,648	29,958	55,265	43,187	27,842	29,980	10,526
5,400	32,607	30,337	30,655	56,569	44,293	28,519	30,674	10,724

SOURCE: L. C. Lichty, "Internal Combustion Engines," p. 582, derived from data given by Hershey, Eberhardt, and Hottel, *Trans. SAE*, **31**, 1936, p. 409.

Table 2.1.21 Flame Temperatures at 14.7 psia Allowing for Dissociation.
(a) Flame Temperatures (R) at 14.7 psia Allowing for Dissociatin
(b) Volumetric Composition of the Fuels in (a)

(a)

Fuel	Percent of theoretical air				
	80	90	100	120	140
Hydrogen	4,210	4,330	4,390	4,000	3,670
Carbon monoxide	4,280	4,370	4,320	4,140	3,850
Methane	4,050	.....	4,010	3,660	3,330
Carbureted water gas	3,940	.....	4,150	3,820	3,510
Coal gas	3,920	.....	4,050	3,780	3,440
Natural gas	4,010	.....	4,180	3,840	3,520
Producer gas	3,040	.....	3,330	3,130	2,970
Blast furnace gas	2,810	.....	3,060	2,920	2,750

SOURCE: Satterfield, "Generalized Thermodynamics of High-Temperature Combustion," Sc.D. thesis, M.I.T., 1946.

(b)

Fuels	CO	H_2	CH_4	C_2H_4	Illuminants (assumed C_2H_4)	CO_2	O_2	N_2
H_2	....	100.0						
CO	100.0							
CH_4	....	....	100.0					
Carbureted water gas	24.1	32.5	9.0	2.2	10.3	4.6	0.6	16.7
Coal gas	5.9	53.2	29.6	....	2.7	1.4	0.7	6.5
Natural gas	....	....	78.8	14.0	......	0.4	.....	6.8
Producer gas	26.0	3.0	0.5	....	......	2.50	.....	56.0
Blast furnace gas	26.5	3.5	0.2	....	......	12.8	0.1	56.9

mass of oxygen required for complete combustion is $2.67c + 8h - o$ lbm, and the minimum mass air required is $a = (2.67c + 8h - o)/0.23 = 11.6[c + 3(h - o/8)]$ lbm.

With air at 62°F and 1 atm, the minimum volume of air required is $147[c + 3(h - o/8)]$ ft^3. In practice, an excess of air over that required for combustion is admitted to the furnace. The actual quantity admitted per pound mass of fuel may be denoted by xa. Then x = amount admitted/minimum amount.

Combustion Products If V_{min} is the minimum volume of air required for complete combustion and xV_{min} the actual volume supplied, then the products will contain per unit mass of fuel $O_2 = 0.21V_{min}(x - 1)$, and $N_2 = 0.79xV_{min}$.

From the reaction equation $C + O_2 = CO_2$, the volume of CO_2 formed is equal to the volume of oxygen required for the carbon constituent alone; hence the volume of CO_2 is $0.21V_{min}c/[c + 3(h - 0.125o)]$.

Of the *dry* gaseous products, i.e., without water, the CO_2 content by volume is therefore given by the expression,

$$CO_2 = \frac{0.21c}{xc + 3(x - 0.21)(h - 0.125o)}$$

The combined CO_2 and O_2 content is,

$$CO_2 + O_2 = 0.21\left[1 - \frac{0.79}{\left(\dfrac{x + cx}{3(h - 0.125o) - 0.21}\right)}\right]$$

If the fuel is all carbon, the combined CO_2 and O_2 content is by volume 21 percent of the gaseous products. The more hydrogen contained in the fuel, the smaller is the $CO_2 + O_2$ content. The CO_2 content depends in the first instance on the excess of air. Thus for pure carbon, it is $CO_2 = 0.21/x$.

The excess of air may be calculated from the composition of the gases and that of the fuel,

$$x = 0.21\left[\frac{\dfrac{c}{[CO_2]} + 3(h - 0.125o)}{c + 3(h - 0.125o)}\right]$$

where $[CO_2]$ denotes the percent by volume of carbon dioxide in the dry gas. The temperature of combustion is calculated by the same method as for gaseous fuels.

Losses Due to Incomplete Combustion The loss due to incomplete combustion of the carbon in the fuel (Btu/lbm-fuel) is,

$$L = 10{,}136C \times \frac{CO}{CO + CO_2}$$

where 10,136 is the difference in heat evolved in burning 1 lbm of carbon to CO_2 to CO, and CO and CO_2 are percentages by volume of carbon monoxide and carbon dioxide as found by analysis, and C is the fraction of quantity of carbon in the fuel actually burned and passes up the stack either as CO or CO_2. The presence of 1 percent CO in the flue gases will represent a decrease in boiler efficiency of 4.5 percent. An additional loss is caused by either unburned or partly burned fuel.

It is generally assumed that high CO_2 readings are indicative of good combustion and, hence, of high efficiencies. Such readings are not satisfactory when considered apart from the CO determination. The best percentage of CO_2 to maintain varies with different fuels and is lower for those with a high hydrogen content than for fuel mainly composed of carbon.

Hydrogen in a fuel increases the nitrogen content of the flue gases because the water vapor formed by the combustion of hydrogen will condense at the temperature at which the analysis is made, while the nitrogen which accompanied the oxygen maintains its gaseous form and passes in that form into the sampling apparatus. When highly volatile coals containing considerable hydrogen are burned, the flue gas contains an apparently increased amount of nitrogen. The effect is even more pronounced when burning gaseous or liquid hydrocarbon fuels.

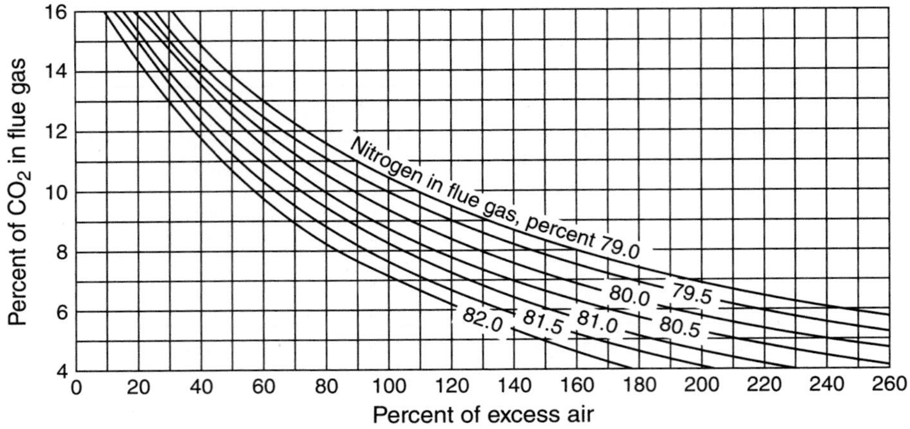

Figure 2.1.50 Ratio of air supplied per unit mass of combustible to that theoretically required.

The amount of flue gas per unit mass of fuel, including moisture formed by the hydrogen component, is approximately $W_1 = 3.02[N/(CO_2 + CO)]C + (1 - A)$, where A is the percent of ash found in test.

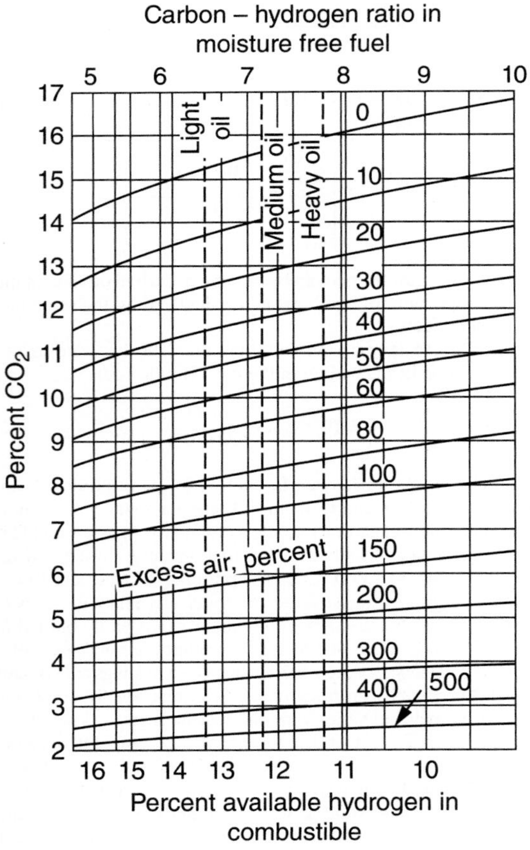

Figure 2.1.51 Relation of CO_2 to excess air.

The quantity of dry flue gas per unit mass of fuel is approximately $W_2 = C[11CO_2 + 8O + 7(CO + N)]/3(CO_2 + CO)$. In these formulas, the amount of gas is per unit mass of dry or moist fuel as the percentage of C is referred to a dry or moist basis.

The ratio of air supplied per pound of fuel to the air theoretically required is,

$$\frac{W_1}{W} = \frac{3.02C[N/(CO_2 + CO)]}{34.56(C/3 + H - 0.8)}$$

The ratio of air supplied per unit mass of combustible to that theoretically required is $N/[N - 3.782\,(O - 0.5CO)]$ on the assumption that all the nitrogen in the flue gas comes from the air supplied. Figure 2.1.50 gives the value of this ratio for varying flue-gas analyses where there is no CO present. For petroleum fuels with hydrogen content from 9 to 16 percent, the excess air can be determined from the CO_2 content of the flue gases (with no CO present) by the use of Fig. 2.1.51. The curves are based on the assumption of 0.4 percent sulfur in the oil.

References

1. Glassman I, Yetter RA. "Combustion," 4e, Academic Press, Burlington, 2008.

2. Moran MJ, Shapiro RN. "Fundamentals of Engineering Thermodynamics," 5e, John Wiley & Sons, Hoboken, 2006.

3. Çengel Y, Boles MS. "Thermodynamics: An Engineering Approach," 5e, McGraw-Hill, New York, 2006.

4. Turns S. "An Introduction to Combustion: Concepts and Applications," 3e, McGraw-Hill, New York, 2012.

5. Ferguson CR, Kirkpatrick AT. "Internal Combustion Engines," 3e, John Wiley & Sons, New York, 2015.

6. Denbigh K. "The Principles of Chemical Equilibrium," Cambridge University Press, London, 1964.

7. Hall NA, Ibele WE. "Engineering Thermodynamics," Prentice-Hall, Englewood Cliffs, 1960.

8. Strehlow RA. "Combustion Fundamentals," McGraw-Hill, New York, 1984.

8. Gordon. NASA Technical Paper, 1906.

9. __________. "NIST-JANAF Thermochemical Tables," 4e, National Institute of Standards and Technology, 2016. http://www.nist.gov/

2.2 THERMODYNAMIC PROPERTIES OF SUBSTANCES

by Mark O. McLinden

NOTE: Thermodynamic properties of a variety of other specific materials are listed also in Secs. 2.1, 4.1, and 7.8.

Table 2.2.1 International Standard Atmosphere*

Z, m	T, K	P, bar	ρ, kg/m³	g, m/s²	M	a, m/s	l, m	H, m
0	288.15	1.01325	1.2250	9.80665	28.964	340.29	6.63×10^{-8}	0
1,000	281.65	0.89876	1.1117	9.8036	28.964	336.43	7.31×10^{-8}	1,000
2,000	275.15	0.79501	1.0066	9.8005	28.964	332.53	8.07×10^{-8}	1,999
3,000	268.66	0.70121	0.90925	9.7974	28.964	328.58	8.94×10^{-8}	2,999
4,000	262.17	0.61660	0.81935	9.7943	28.964	324.59	9.92×10^{-8}	3,997
5,000	255.68	0.54048	0.73643	9.7912	28.964	320.55	1.10×10^{-7}	4,996
6,000	249.19	0.47217	0.66011	9.7882	28.964	316.45	1.23×10^{-7}	5,994
7,000	242.70	0.41105	0.59002	9.7851	28.964	312.31	1.38×10^{-7}	6,992
8,000	236.22	0.35651	0.52579	9.7820	28.964	308.11	1.55×10^{-7}	7,990
9,000	229.73	0.30800	0.46706	9.7789	28.964	303.85	1.74×10^{-7}	8,987
10,000	223.25	0.26499	0.41351	9.7759	28.964	299.53	1.97×10^{-7}	9,984
15,000	216.65	0.12111	0.19476	9.7605	28.864	295.07	4.17×10^{-7}	14,965
20,000	216.65	0.05529	0.08891	9.7452	28.964	295.07	9.14×10^{-7}	19,937
25,000	221.55	0.02549	0.04008	9.7300	28.964	298.39	2.03×10^{-6}	24,902
30,000	226.51	0.01197	0.01841	9.7147	28.964	301.71	4.42×10^{-6}	29,859
40,000	250.35	2.87×10^{-3}	4.00×10^{-3}	9.6844	28.964	317.19	2.03×10^{-5}	39,750
50,000	270.65	8.00×10^{-4}	1.03×10^{-3}	9.6542	28.964	329.80	7.90×10^{-5}	49,610
60,000	247.02	2.20×10^{-4}	3.10×10^{-4}	9.6241	28.964	315.07	2.62×10^{-4}	59,439
70,000	219.59	5.22×10^{-4}	8.28×10^{-5}	9.5942	28.964	297.06	9.81×10^{-4}	69,238
80,000	198.64	1.05×10^{-5}	1.85×10^{-5}	9.5644	28.964	282.54	4.40×10^{-3}	79,006
90,000	186.87	1.84×10^{-6}	3.43×10^{-6}	9.5348	28.95		2.37×10^{-2}	88,744
100,000	195.08	3.20×10^{-7}	5.60×10^{-7}	9.5052	28.40		0.142	98,451
150,000	634.39	4.54×10^{-9}	2.08×10^{-9}	9.3597	24.10		33	146,542
200,000	854.56	8.47×10^{-10}	2.54×10^{-10}	9.2175	21.30		240	193,899
250,000	941.33	2.48×10^{-10}	6.07×10^{-11}	9.0785	19.19		890	240,540
300,000	976.01	8.77×10^{-11}	1.92×10^{-11}	8.9427	17.73		2,600	286,480
400,000	995.83	1.45×10^{-11}	2.80×10^{-12}	8.6799	15.98		1.6×10^{4}	376,320
500,000	999.24	3.02×10^{-12}	5.22×10^{-13}	8.4286	14.33		7.7×10^{4}	463,540
600,000	999.85	8.21×10^{-13}	1.14×10^{-13}	8.1880	11.51		2.8×10^{5}	548,252
800,000	999.99	1.70×10^{-13}	1.14×10^{-14}	7.7368	5.54		1.4×10^{6}	710,574
1,000,000	1,000.00	7.51×10^{-14}	3.56×10^{-15}	7.3218	3.94		3.1×10^{6}	864,071

* Extracted from *U.S. Standard Atmosphere*, 1976, *National Oceanic and Atmospheric Administration*, National Aeronautics and Space Administration, and the U.S. Air Force, Washington, 1976. Z = geometric altitude, T = temperature, ρ = density, P = pressure, g = acceleration of gravity, M = molecular weight, a = velocity of sound, l = mean free path, and H = geopotential altitude.

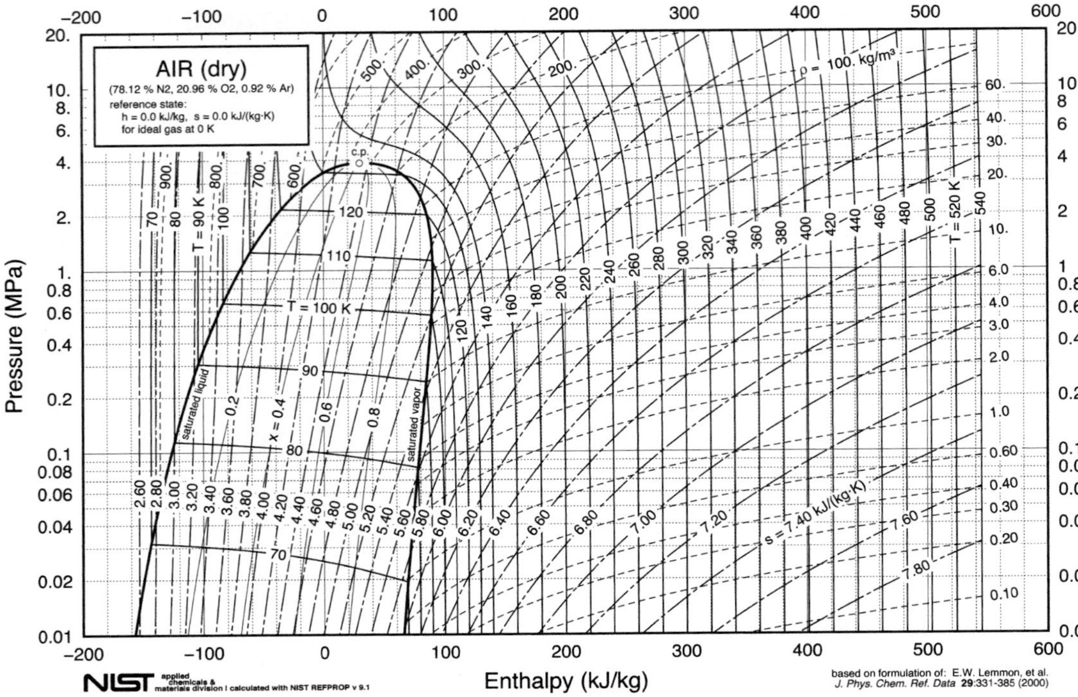

Figure 2.2.1 Enthalpy—log pressure diagram for dry air.

Table 2.2.2 Dry Air

P	T_{bubble}	T_{dew}	ρ_f	v_g	h_f	h_g	s_f	s_g	$C_{p,f}$	$C_{p,g}$	α_f	α_g	μ_f	μ_g	k_f	k_g	σ
MPa	°C	°C	kg/m³	m³/kg	kJ/kg		kJ/(kg·K)		kJ/(kg·K)		m/s		μPa·s		mW/(m·K)		mN/m
0.008	−211.3	−208.0	951.5	2.3258	−159.2	64.7	2.508	6.062	1.91	1.02	1037.	161.	380.	4.63	182.	5.7	14.6
0.010	−210.1	−206.9	946.6	1.8913	−157.0	65.8	2.544	6.015	1.91	1.02	1026.	163.	357.	4.71	179.	5.9	14.3
0.015	−207.9	−204.6	937.1	1.2996	−152.6	67.9	2.611	5.930	1.91	1.02	1003.	165.	317.	4.88	175.	6.1	13.7
0.020	−206.1	−203.0	929.8	0.9962	−149.3	69.4	2.661	5.870	1.92	1.03	987.	167.	291.	5.00	172.	6.3	13.3
0.030	−203.5	−200.4	918.7	0.6852	−144.3	71.7	2.734	5.787	1.92	1.03	962.	170.	258.	5.19	167.	6.5	12.6
0.040	−201.6	−198.5	910.1	0.5255	−140.5	73.3	2.788	5.729	1.92	1.04	943.	171.	236.	5.33	163.	6.7	12.1
0.050	−199.9	−196.9	903.0	0.4278	−137.4	74.7	2.831	5.684	1.93	1.04	927.	173.	220.	5.45	160.	6.9	11.7
0.060	−198.5	−195.6	896.8	0.3616	−134.7	75.8	2.867	5.647	1.93	1.05	914.	174.	208.	5.56	158.	7.1	11.4
0.080	−196.2	−193.4	886.4	0.2773	−130.2	77.6	2.926	5.590	1.94	1.06	892.	176.	190.	5.73	153.	7.3	10.8
0.100	−194.3	−191.5	877.8	0.2256	−126.5	79.0	2.973	5.545	1.95	1.07	873.	177.	177.	5.87	150.	7.5	10.4
0.150	−190.6	−187.9	860.5	0.1549	−119.2	81.6	3.063	5.465	1.96	1.09	837.	180.	154.	6.15	143.	8.0	9.5
0.200	−187.7	−185.1	846.7	0.1185	−113.5	83.4	3.130	5.408	1.98	1.12	809.	182.	140.	6.37	137.	8.3	8.8
0.300	−183.3	−180.8	824.8	0.0810	−104.7	85.9	3.230	5.327	2.01	1.16	765.	184.	122.	6.72	129.	8.9	7.8
0.400	−179.9	−177.5	807.0	0.0617	−97.6	87.4	3.305	5.269	2.04	1.21	730.	185.	109.	7.00	123.	9.4	7.0
0.500	−177.0	−174.8	791.7	0.0498	−91.7	88.5	3.367	5.223	2.08	1.25	700.	186.	101.	7.25	117.	9.8	6.4
0.600	−174.6	−172.4	778.1	0.0417	−86.5	89.3	3.419	5.185	2.11	1.30	674.	186.	94.	7.47	113.	10.2	5.9
0.800	−170.4	−168.4	753.9	0.0314	−77.6	90.0	3.505	5.122	2.18	1.39	628.	186.	83.	7.85	105.	11.0	5.0
1.000	−166.9	−165.0	732.4	0.0250	−69.8	90.7	3.577	5.071	2.26	1.50	589.	186.	76.	8.20	98.	11.8	4.3
1.500	−160.0	−158.4	684.8	0.0162	−53.5	88.7	3.719	4.968	2.52	1.83	505.	184.	62.	8.99	86.	13.9	3.0
2.000	−154.6	−153.2	640.6	0.0116	−39.5	85.4	3.833	4.881	2.90	2.34	432.	182.	53.	9.78	76.	16.3	2.0
2.500	−150.1	−148.9	595.7	0.0087	−26.3	80.3	3.936	4.798	3.54	3.22	364.	178.	45.	10.66	67.	19.5	1.2
3.000	−146.2	−145.2	545.6	0.0066	−12.9	72.8	4.036	4.708	4.90	5.13	297.	174.	38.	11.79	60.	24.3	0.7
3.80c	−140.6	−140.6	342.6	0.0029	29.8	29.8	4.350	4.350									0.0

c = critical point.
Properties computed with NIST REFPROP version 9.1, based on a composition of 0.7812 (mol frac) N_2 + 0.2096 O_2 + 0.0092 Ar.

Table 2.2.3 Properties of Dry Air as a Function of Temperature and Pressure

P, MPa, (T_{sat}, °C)		T (°C)									
		−100.0	0.0	100.0	200.0	300.0	400.0	500.0	600.0	700.0	800.0
0.04 (−198.51)	v	1.24094	1.96020	2.67857	3.39666	4.11464	4.83254	5.55041	6.26826	6.98609	7.70392
	h	173.0	273.5	374.2	475.9	579.4	685.0	793.1	903.5	1,016.1	1,130.6
	s	6.582	7.040	7.354	7.595	7.794	7.963	8.113	8.247	8.369	8.481
0.06 (−195.59)	v	0.82665	1.30665	1.78576	2.26459	2.74330	3.22194	3.70055	4.17914	4.65770	5.13626
	h	172.9	273.4	374.2	475.9	579.4	685.0	793.1	903.5	1,016.1	1,130.7
	s	6.465	6.923	7.237	7.479	7.677	7.847	7.997	8.131	8.253	8.365
0.08 (−193.35)	v	0.61951	0.97988	1.33935	1.69855	2.05763	2.41664	2.77562	3.13457	3.49351	3.85244
	h	172.8	273.4	374.1	475.9	579.4	685.0	793.1	903.5	1,016.1	1,130.7
	s	6.382	6.840	7.155	7.396	7.594	7.764	7.914	8.048	8.170	8.282
0.10 (−191.51)	v	0.49522	0.78381	1.07151	1.35893	1.64623	1.93346	2.22066	2.50783	2.79499	3.08214
	h	172.6	273.3	374.1	475.9	579.4	685.0	793.1	903.5	1,016.1	1,130.7
	s	6.317	6.776	7.090	7.332	7.530	7.700	7.850	7.984	8.106	8.218
0.20 (−185.13)	v	0.24665	0.39169	0.53583	0.67969	0.82343	0.96710	1.11074	1.25436	1.39796	1.54155
	h	172.0	273.0	374.0	475.8	579.3	685.1	793.1	903.6	1,016.2	1,130.7
	s	6.116	6.576	6.891	7.133	7.331	7.501	7.651	7.785	7.907	8.019
0.30 (−180.84)	v	0.16379	0.26098	0.35727	0.45327	0.54916	0.64498	0.74077	0.83653	0.93228	1.02802
	h	171.4	272.7	373.8	475.7	579.3	685.1	793.2	903.6	1,016.2	1,130.8
	s	5.997	6.459	6.774	7.016	7.215	7.385	7.534	7.669	7.791	7.903
0.40 (−177.52)	v	0.12236	0.19562	0.26799	0.34007	0.41203	0.48392	0.55578	0.62762	0.69944	0.77125
	h	170.7	272.5	373.7	475.7	579.3	685.1	793.2	903.7	1,016.3	1,130.9
	s	5.911	6.375	6.691	6.933	7.132	7.302	7.452	7.586	7.708	7.820
0.60 (−172.38)	v	0.08093	0.13027	0.17871	0.22686	0.27489	0.32286	0.37080	0.41871	0.46660	0.51449
	h	169.4	271.9	373.4	475.5	579.3	685.1	793.3	903.8	1,016.4	1,131.0
	s	5.790	6.257	6.574	6.816	7.015	7.185	7.335	7.470	7.592	7.704
0.80 (−168.36)	v	0.06021	0.09760	0.13407	0.17026	0.20633	0.24233	0.27830	0.31425	0.35018	0.38611
	h	168.1	271.4	373.2	475.4	579.2	685.1	793.3	903.9	1,016.5	1,131.1
	s	5.702	6.173	6.491	6.733	6.932	7.103	7.252	7.387	7.509	7.621
1.00 (−165.02)	v	0.04778	0.07800	0.10729	0.13630	0.16519	0.19402	0.22281	0.25158	0.28033	0.30908
	h	166.8	270.8	372.9	475.3	579.2	685.1	793.4	903.9	1,016.6	1,131.3
	s	5.632	6.107	6.426	6.669	6.868	7.038	7.188	7.323	7.445	7.557
2.00 (−153.20)	v	0.02290	0.03880	0.05373	0.06838	0.08291	0.09738	0.11182	0.12623	0.14063	0.15502
	h	160.0	268.1	371.6	474.7	579.0	685.2	793.7	904.4	1,017.2	1,131.9
	s	5.405	5.900	6.223	6.468	6.668	6.838	6.989	7.123	7.245	7.358
3.00 (−145.23)	v	0.01460	0.02575	0.03589	0.04575	0.05549	0.06518	0.07482	0.08445	0.09406	0.10367
	h	152.8	265.5	370.3	474.1	578.8	685.4	794.0	904.9	1,017.8	1,132.6
	s	5.258	5.775	6.102	6.349	6.550	6.721	6.871	7.006	7.129	7.241
4.00 (*)	v	0.01044	0.01924	0.02697	0.03444	0.04178	0.04907	0.05633	0.06356	0.07078	0.07799
	h	145.1	262.8	369.1	473.6	578.7	685.5	794.3	905.3	1,018.4	1,133.3
	s	5.143	5.684	6.016	6.264	6.466	6.637	6.788	6.923	7.046	7.158
6.00 (*)	v	0.00631	0.01275	0.01807	0.02313	0.02808	0.03297	0.03783	0.04267	0.04750	0.05232
	h	128.8	257.7	366.7	472.5	578.5	685.8	795.0	906.3	1,019.5	1,134.6
	s	4.955	5.552	5.892	6.144	6.347	6.519	6.671	6.806	6.929	7.041
8.00 (*)	v	0.00433	0.00953	0.01363	0.01749	0.02123	0.02493	0.02859	0.03223	0.03586	0.03948
	h	111.9	252.9	364.5	471.6	578.3	686.1	795.7	907.3	1,020.7	1,136.0
	s	4.798	5.454	5.802	6.057	6.262	6.435	6.587	6.722	6.845	6.958
10.00 (*)	v	0.00328	0.00761	0.01098	0.01410	0.01713	0.02010	0.02304	0.02597	0.02888	0.03178
	h	96.9	248.3	362.4	470.8	578.2	686.5	796.5	908.3	1,022.0	1,137.3
	s	4.667	5.375	5.732	5.989	6.195	6.369	6.521	6.657	6.781	6.893
20.00 (*)	v	0.00194	0.00394	0.00574	0.00738	0.00894	0.01047	0.01196	0.01345	0.01492	0.01638
	h	66.5	230.6	354.5	468.0	578.7	689.2	800.7	913.7	1,028.3	1,144.4
	s	4.355	5.115	5.503	5.772	5.985	6.162	6.317	6.454	6.578	6.692
30.00 (*)	v	0.00166	0.00285	0.00406	0.00517	0.00624	0.00727	0.00828	0.00928	0.01027	0.01125
	h	62.5	221.4	350.4	467.4	580.5	692.7	805.6	919.6	1,035.0	1,151.8
	s	4.229	4.960	5.363	5.642	5.859	6.039	6.195	6.334	6.459	6.573
40.00 (*)	v	0.00153	0.00236	0.00325	0.00409	0.00490	0.00568	0.00645	0.00720	0.00795	0.00869
	h	64.5	218.3	349.2	468.6	583.5	697.1	811.1	925.9	1,042.0	1,159.4
	s	4.148	4.854	5.264	5.547	5.768	5.951	6.108	6.248	6.374	6.489

P = pressure; v = specific volume, $m^3 \cdot kg^{-1}$; h = enthalpy, $kJ \cdot kg^{-1}$; s = entropy, $kJ \cdot kg^{-1} \cdot K^{-1}$; T_{sat} is saturation temperature at indicated pressure; * = supercritical state.
Properties computed with NIST REFPROP version 9.1, based on a composition of 0.7812 (mol frac) N_2 + 0.2096 O_2 + 0.0092 Ar.

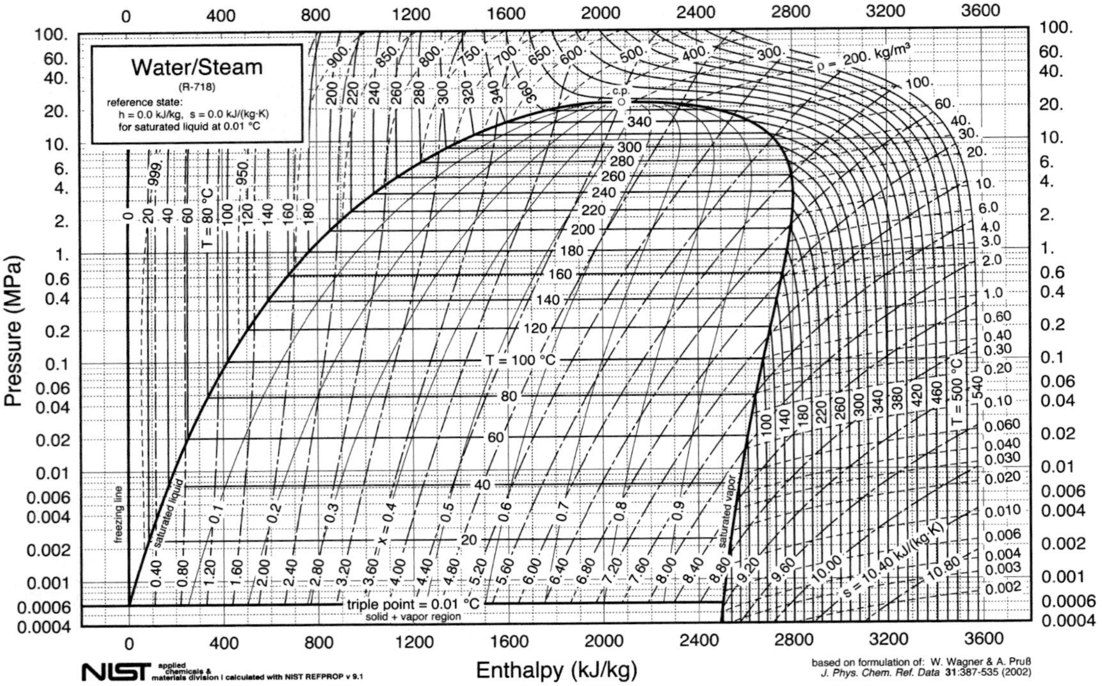

Figure 2.2.2 Enthalpy—log pressure diagram for water/steam.

Table 2.2.4 Saturated Water/Steam

T	P	ρ_f	v_g	h_f	h_g	s_f	s_g	$C_{p,f}$	$C_{p,g}$	α_f	α_g	μ_f	μ_g	k_f	k_g	σ
°C	MPa	kg/m³	m³/kg	kJ/kg		kJ/(kg·K)		kJ/(kg·K)		m/s		μPa·s		mW/(m·K)		mN/m
0.1a	0.007	999.8	205.99	0.0	2500.9	0.000	9.156	4.22	1.88	1402.	409.	1791.	9.0	556.	16.8	75.7
10.0	0.001	999.7	106.30	42.0	2519.2	0.151	8.900	4.20	1.90	1447.	416.	1306.	9.2	579.	17.4	74.2
20.0	0.002	998.2	57.757	83.9	2537.4	0.297	8.666	4.18	1.91	1482.	423.	1002.	9.5	598.	18.1	72.7
30.0	0.004	995.6	32.878	125.7	2555.6	0.437	8.452	4.18	1.92	1509.	430.	797.	9.9	614.	18.8	71.2
40.0	0.007	992.2	19.515	167.5	2573.5	0.572	8.256	4.18	1.93	1529.	437.	653.	10.2	628.	19.5	69.6
50.0	0.012	988.0	12.027	209.3	2591.3	0.704	8.075	4.18	1.95	1542.	443.	547.	10.5	641.	20.3	67.9
60.0	0.020	983.2	7.6672	251.2	2608.8	0.831	7.908	4.19	1.97	1551.	450.	466.	10.9	651.	21.0	66.2
70.0	0.031	977.7	5.0395	293.1	2626.1	0.955	7.754	4.20	1.99	1555.	456.	404.	11.2	660.	21.9	64.5
80.0	0.047	971.8	3.4052	335.0	2643.0	1.076	7.611	4.20	2.01	1554.	461.	354.	11.5	667.	22.7	62.7
90.0	0.070	965.3	2.3591	377.0	2659.5	1.193	7.478	4.21	2.04	1550.	467.	314.	11.9	673.	23.6	60.8
100.0b	0.101	958.4	1.6732	419.1	2675.5	1.307	7.354	4.22	2.08	1543.	472.	282.	12.2	677.	24.6	58.9
100.0	0.101	958.3	1.6718	419.2	2675.6	1.307	7.354	4.22	2.08	1543.	472.	282.	12.2	677.	24.6	58.9
110.0	0.143	950.9	1.2093	461.4	2691.1	1.419	7.238	4.23	2.12	1533.	477.	255.	12.6	680.	25.6	57.0
120.0	0.199	943.1	0.8912	503.8	2705.9	1.528	7.129	4.24	2.18	1520.	482.	232.	12.9	682.	26.7	55.0
130.0	0.270	934.8	0.6680	546.4	2720.1	1.635	7.026	4.26	2.24	1504.	486.	213.	13.3	683.	27.8	52.9
140.0	0.362	926.1	0.5085	589.2	2733.4	1.739	6.929	4.28	2.31	1486.	490.	197.	13.6	683.	29.0	50.9
150.0	0.476	917.0	0.3925	632.2	2745.9	1.842	6.837	4.31	2.39	1466.	493.	187.	14.0	681.	30.3	48.7
160.0	0.618	907.4	0.3068	675.5	2757.4	1.943	6.749	4.34	2.49	1443.	496.	170.	14.3	679.	31.7	46.6
170.0	0.792	897.5	0.2426	719.1	2767.9	2.042	6.665	4.37	2.59	1419.	499.	160.	14.6	676.	33.2	44.4
180.0	1.003	887.0	0.1938	763.1	2777.2	2.139	6.584	4.41	2.71	1392.	501.	150.	15.0	671.	34.8	42.2
190.0	1.255	876.1	0.1564	807.4	2785.3	2.236	6.506	4.45	2.84	1363.	503.	142.	15.3	666.	36.6	40.0
200.0	1.555	864.7	0.1272	852.3	2792.0	2.331	6.430	4.50	2.99	1332.	504.	135.	15.7	660.	38.4	37.7
220.0	2.320	840.2	0.0861	943.6	2801.0	2.518	6.284	4.62	3.33	1264.	505.	122.	16.4	645.	42.6	33.1
240.0	3.347	813.4	0.0597	1037.6	2803.0	2.702	6.142	4.77	3.75	1189.	503.	111.	17.1	627.	47.5	28.4
260.0	4.692	783.6	0.0422	1135.0	2796.6	2.885	6.002	4.99	4.31	1105.	499.	102.	17.8	606.	53.5	23.7
280.0	6.417	750.3	0.0302	1236.9	2779.9	3.069	5.858	5.29	5.07	1013.	492.	94.	18.6	581.	60.9	19.0
300.0	8.588	712.1	0.0217	1345.0	2749.6	3.255	5.706	5.75	6.22	909.	481.	86.	19.6	553.	70.9	14.4
320.0	11.284	667.1	0.0155	1462.2	2700.6	3.449	5.537	6.54	8.16	793.	464.	78.	20.8	520.	86.2	9.9
340.0	14.601	610.7	0.0108	1594.5	2621.9	3.660	5.336	8.21	12.24	658.	441.	70.	22.5	482.	114.5	5.6
360.0	18.666	527.6	0.0070	1761.7	2481.5	3.917	5.054	15.00	27.46	480.	402.	60.	25.6	439.	191.4	1.9
374.0c	22.064	322.0	0.0031	2084.3	2084.3	4.407	4.407	∞	∞	0.	0.	-	-	∞	∞	0.0

a = triple point; b = normal boiling point; c = critical point.
Properties computed with NIST REFPROP version 9.1, based on the formulation of Wagner and Pruß, *J. Phys. Chem. Ref. Data* **31**:387-535 (2002).

Table 2.2.5 Properties of Water/Steam as a Function of Temperature and Pressure

P, MPa, (T_{sat}, °C)		T (°C)									
		100.0	150.0	200.0	250.0	300.0	350.0	400.0	500.0	600.0	700.0
0.10	v	1.69587	1.93665	2.17243	2.40615	2.63885	2.87095	3.10269	3.56553	4.02788	4.48997
(99.61)	h	2,675.8	2,776.6	2,875.5	2,974.5	3,074.5	3,175.8	3,278.6	3,488.7	3,705.6	3,929.4
	s	7.361	7.615	7.836	8.035	8.217	8.387	8.545	8.836	9.100	9.342
0.20	v	0.00104	0.95986	1.08048	1.19889	1.31623	1.43297	1.54933	1.78142	2.01301	2.24433
(120.21)	h	419.2	2,769.1	2,870.7	2,971.2	3,072.1	3,173.9	3,277.0	3,487.7	3,704.8	3,928.8
	s	1.307	7.281	7.508	7.710	7.894	8.064	8.224	8.515	8.779	9.022
0.30	v	0.00104	0.63401	0.71642	0.79644	0.87534	0.95363	1.03154	1.18671	1.34139	1.49579
(133.52)	h	419.3	2,761.2	2,865.9	2,967.9	3,069.6	3,172.0	3,275.5	3,486.6	3,704.0	3,928.2
	s	1.307	7.079	7.313	7.518	7.704	7.875	8.035	8.327	8.591	8.834
0.40	v	0.00104	0.47088	0.53433	0.59520	0.65489	0.71396	0.77264	0.88936	1.00557	1.12152
(143.61)	h	419.4	2,752.8	2,860.9	2,964.5	3,067.1	3,170.0	3,273.9	3,485.5	3,703.2	3,927.6
	s	1.307	6.931	7.172	7.380	7.568	7.740	7.900	8.193	8.458	8.701
0.60	v	0.00104	0.00109	0.35212	0.39390	0.43442	0.47427	0.51374	0.59200	0.66976	0.74725
(158.83)	h	419.5	632.3	2,850.6	2,957.6	3,062.0	3,166.1	3,270.8	3,483.4	3,701.7	3,926.4
	s	1.307	1.842	6.968	7.183	7.374	7.548	7.710	8.004	8.270	8.513
0.80	v	0.00104	0.00109	0.26088	0.29320	0.32416	0.35442	0.38428	0.44332	0.50185	0.56011
(170.41)	h	419.7	632.4	2,839.7	2,950.4	3,056.9	3,162.2	3,267.6	3,481.3	3,700.1	3,925.3
	s	1.307	1.841	6.818	7.040	7.234	7.411	7.573	7.869	8.135	8.379
1.00	v	0.00104	0.00109	0.20602	0.23275	0.25799	0.28250	0.30661	0.35411	0.40111	0.44783
(179.88)	h	419.8	632.5	2,828.3	2,943.1	3,051.6	3,158.2	3,264.5	3,479.1	3,698.6	3,924.1
	s	1.307	1.841	6.696	6.926	7.125	7.303	7.467	7.764	8.031	8.275
2.00	v	0.00104	0.00109	0.00116	0.11150	0.12551	0.13860	0.15121	0.17568	0.19961	0.22326
(212.38)	h	420.6	633.1	852.5	2,903.2	3,024.2	3,137.7	3,248.3	3,468.2	3,690.7	3,918.2
	s	1.306	1.840	2.330	6.547	6.768	6.958	7.129	7.434	7.704	7.951
3.00	v	0.00104	0.00109	0.00116	0.07063	0.08118	0.09056	0.09938	0.11620	0.13245	0.14841
(233.85)	h	421.3	633.7	852.9	2,856.5	2,994.3	3,116.1	3,231.7	3,457.2	3,682.8	3,912.2
	s	1.305	1.839	2.328	6.289	6.541	6.745	6.923	7.236	7.510	7.759
4.00	v	0.00104	0.00109	0.00115	0.00125	0.05887	0.06647	0.07343	0.08644	0.09886	0.11098
(250.35)	h	422.1	634.4	853.3	1,085.8	2,961.7	3,093.3	3,214.5	3,446.0	3,674.9	3,906.3
	s	1.304	1.838	2.327	2.793	6.364	6.584	6.771	7.092	7.371	7.621
6.00	v	0.00104	0.00109	0.00115	0.00125	0.03619	0.04225	0.04742	0.05667	0.06527	0.07355
(275.58)	h	423.6	635.6	854.1	1,085.7	2,885.5	3,043.9	3,178.2	3,423.1	3,658.7	3,894.3
	s	1.303	1.836	2.324	2.789	6.070	6.336	6.543	6.883	7.169	7.425
8.00	v	0.00104	0.00109	0.00115	0.00124	0.02428	0.02997	0.03434	0.04177	0.04846	0.05483
(295.01)	h	425.1	636.9	854.9	1,085.7	2,786.5	2,988.1	3,139.4	3,399.5	3,642.4	3,882.2
	s	1.301	1.834	2.320	2.784	5.794	6.132	6.366	6.727	7.022	7.282
10.00	v	0.00104	0.00108	0.00115	0.00124	0.00140	0.02244	0.02644	0.03281	0.03838	0.04360
(311.00)	h	426.6	638.1	855.8	1,085.8	1,343.3	2,924.0	3,097.4	3,375.1	3,625.8	3,870.0
	s	1.300	1.831	2.317	2.779	3.249	5.946	6.214	6.599	6.904	7.169
20.00	v	0.00103	0.00108	0.00114	0.00123	0.00136	0.00166	0.00995	0.01479	0.01818	0.02113
(365.75)	h	434.2	644.4	860.3	1,086.7	1,334.4	1,646.0	2,816.9	3,241.2	3,539.0	3,807.8
	s	1.292	1.821	2.303	2.757	3.209	3.729	5.553	6.145	6.507	6.799
30.00	v	0.00103	0.00107	0.00113	0.00121	0.00133	0.00155	0.00280	0.00869	0.01144	0.01365
(*)	h	441.7	650.9	865.0	1,088.4	1,328.9	1,608.8	2,152.8	3,084.7	3,446.7	3,743.9
	s	1.285	1.811	2.289	2.737	3.176	3.644	4.476	5.796	6.237	6.560
40.00	v	0.00102	0.00107	0.00112	0.00120	0.00131	0.00149	0.00191	0.00562	0.00809	0.00993
(*)	h	449.3	657.4	870.0	1,090.7	1,325.6	1,588.8	1,931.4	2,906.5	3,350.4	3,679.1
	s	1.278	1.801	2.275	2.719	3.147	3.587	4.114	5.474	6.017	6.374
60.00	v	0.00102	0.00106	0.00111	0.00118	0.00127	0.00141	0.00163	0.00295	0.00483	0.00626
(*)	h	464.6	670.7	880.6	1,096.8	1,323.5	1,567.5	1,843.2	2,570.3	3,156.8	3,551.3
	s	1.264	1.782	2.251	2.685	3.099	3.507	3.932	4.936	5.653	6.081
80.00	v	0.00101	0.00105	0.00109	0.00116	0.00124	0.00135	0.00152	0.00219	0.00338	0.00452
(*)	h	479.8	684.2	891.7	1,104.4	1,325.1	1,557.7	1,808.8	2,397.4	2,988.1	3,432.7
	s	1.250	1.764	2.228	2.655	3.058	3.447	3.834	4.647	5.367	5.851
100.00	v	0.00100	0.00104	0.00108	0.00114	0.00121	0.00131	0.00144	0.00189	0.00267	0.00355
(*)	h	495.1	697.9	903.4	1,113.1	1,329.1	1,554.0	1,791.1	2,316.2	2,865.1	3,330.7
	s	1.237	1.748	2.206	2.628	3.022	3.398	3.764	4.490	5.158	5.664

P = pressure; v = specific volume, $m^3 \cdot kg^{-1}$; h = enthalpy, $kJ \cdot kg^{-1}$; s = entropy, $kJ \cdot kg^{-1} \cdot K^{-1}$; T_{sat} is saturation temperature at indicated pressure; * = supercritical state.
Properties computed with NIST REFPROP version 9.1, based on the formulation of Wagner and Pruß, *J. Phys. Chem. Ref. Data* **31**:387-535 (2002).

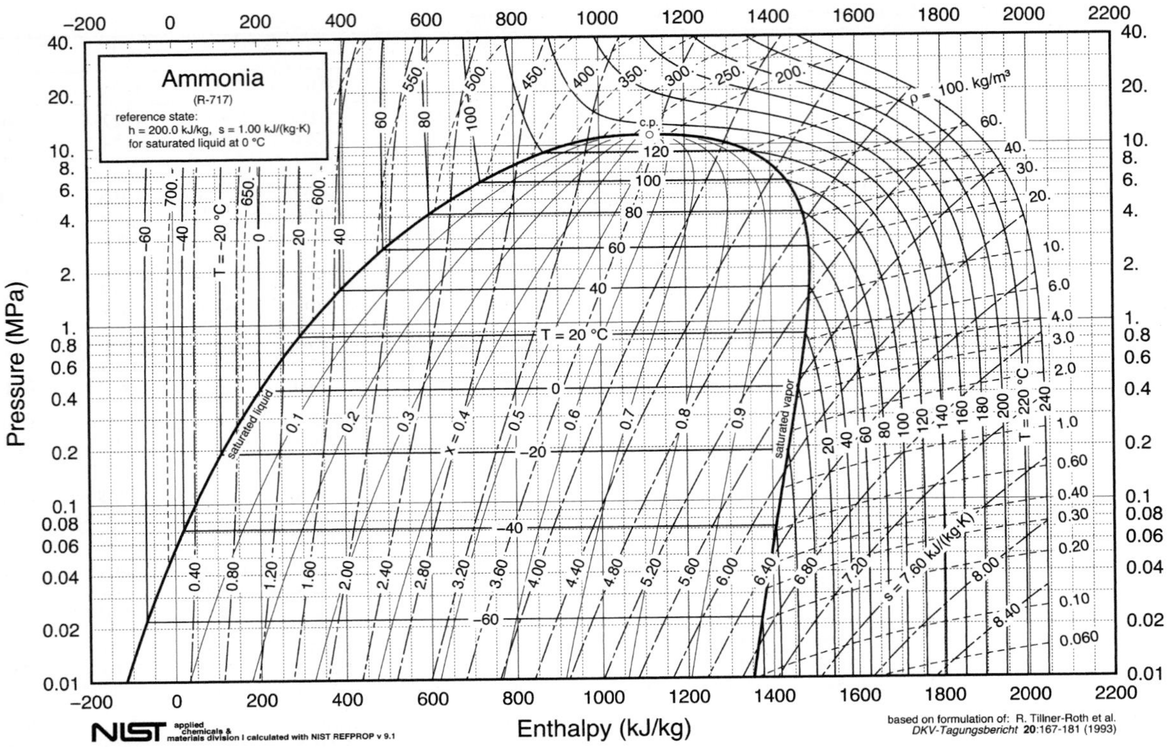

Figure 2.2.3 Enthalpy—log pressure diagram for ammonia.

Table 2.2.6 Saturated Ammonia (R-717)

T	P	ρ_f	v_g	h_f	h_g	s_f	s_g	$C_{p,f}$	$C_{p,g}$	α_f	α_g	μ_f	μ_g	k_f	k_g	σ
°C	MPa	kg/m³	m³/kg	kJ/kg		kJ/(kg·K)		kJ/(kg·K)		m/s		μPa·s		mW/(m·K)		mN/m
−77.7a	0.006	732.9	15.602	−143.2	1341.2	−0.472	7.121	4.20	2.06	2124.	354.	560.	6.84	819.	19.6	43.93
−70.0	0.011	724.7	9.0079	−110.8	1355.6	−0.309	6.909	4.25	2.09	2051.	361.	475.	7.03	792.	19.7	42.34
−60.0	0.022	713.6	4.7057	−68.1	1373.7	−0.104	6.660	4.30	2.13	1967.	368.	391.	7.30	757.	19.9	40.18
−50.0	0.041	702.1	2.6278	−24.7	1391.2	0.095	6.440	4.36	2.18	1890.	376.	329.	7.57	722.	20.2	37.95
−40.0	0.072	690.2	1.5533	19.2	1407.8	0.287	6.243	4.41	2.24	1816.	382.	281.	7.86	688.	20.6	35.67
−33.3b	0.101	682.0	1.1242	48.8	1418.3	0.412	6.122	4.45	2.30	1768.	386.	256.	8.05	666.	21.0	34.12
−30.0	0.119	677.8	0.9640	63.6	1423.3	0.473	6.065	4.47	2.33	1744.	388.	244.	8.15	655.	21.2	33.35
−20.0	0.190	665.1	0.6237	108.6	1437.7	0.654	5.904	4.51	2.43	1673.	393.	214.	8.45	622.	21.8	31.00
−10.0	0.291	652.1	0.4183	154.0	1450.7	0.829	5.757	4.56	2.54	1602.	398.	190.	8.75	590.	22.5	28.65
0.0	0.429	638.6	0.2893	200.0	1462.2	1.000	5.621	4.62	2.68	1531.	401.	170.	9.06	559.	23.4	26.29
10.0	0.615	624.6	0.2054	246.6	1472.1	1.166	5.495	4.68	2.84	1458.	403.	153.	9.36	529.	24.4	23.95
20.0	0.857	610.2	0.1492	293.8	1480.2	1.329	5.376	4.75	3.03	1384.	405.	138.	9.68	500.	25.5	21.64
30.0	1.167	595.2	0.1105	341.8	1486.2	1.488	5.263	4.83	3.25	1309.	405.	126.	10.00	471.	26.9	19.35
40.0	1.555	579.4	0.0831	390.6	1489.9	1.645	5.155	4.93	3.51	1232.	404.	114.	10.33	444.	28.4	17.09
50.0	2.034	562.9	0.0634	440.6	1491.1	1.799	5.050	5.06	3.82	1153.	402.	104.	10.67	416.	30.2	14.88
60.0	2.616	545.2	0.0488	492.0	1489.3	1.952	4.946	5.24	4.21	1070.	398.	95.	11.05	390.	32.3	12.73
70.0	3.314	526.3	0.0379	545.0	1483.9	2.105	4.842	5.47	4.70	984.	393.	86.	11.47	363.	34.8	10.63
80.0	4.142	505.7	0.0295	600.3	1474.3	2.260	4.734	5.78	5.36	895.	387.	78.	11.95	337.	38.0	8.60
90.0	5.117	482.8	0.0230	658.6	1459.2	2.417	4.621	6.25	6.29	800.	378.	71.	12.55	311.	42.2	6.65
100.0	6.255	456.6	0.0178	721.0	1436.6	2.580	4.498	6.99	7.76	701.	367.	63.	13.32	285.	48.4	4.79
110.0	7.578	425.6	0.0136	789.7	1403.1	2.753	4.354	8.36	10.46	594.	353.	56.	14.42	258.	58.3	3.06
120.0	9.113	385.5	0.0100	869.9	1350.2	2.950	4.172	11.94	17.21	477.	335.	48.	16.21	231.	78.4	1.48
130.0	10.898	312.3	0.0064	992.0	1239.3	3.244	3.857	–	–	334.	307.	37.	20.63	222.	160.4	0.19
132.3c	11.333	225.0	0.0044	1119.2	1119.2	3.554	3.554	∞	∞	0.	0.	–	-	∞	∞	0.00

a = triple point; b = normal boiling point; c = critical point.
Properties computed with NIST REFPROP version 9.1, based on the formulation of Tillner-Roth *et al.*, *DKV-Tagungsbericht* **20**:167-181 (1993).

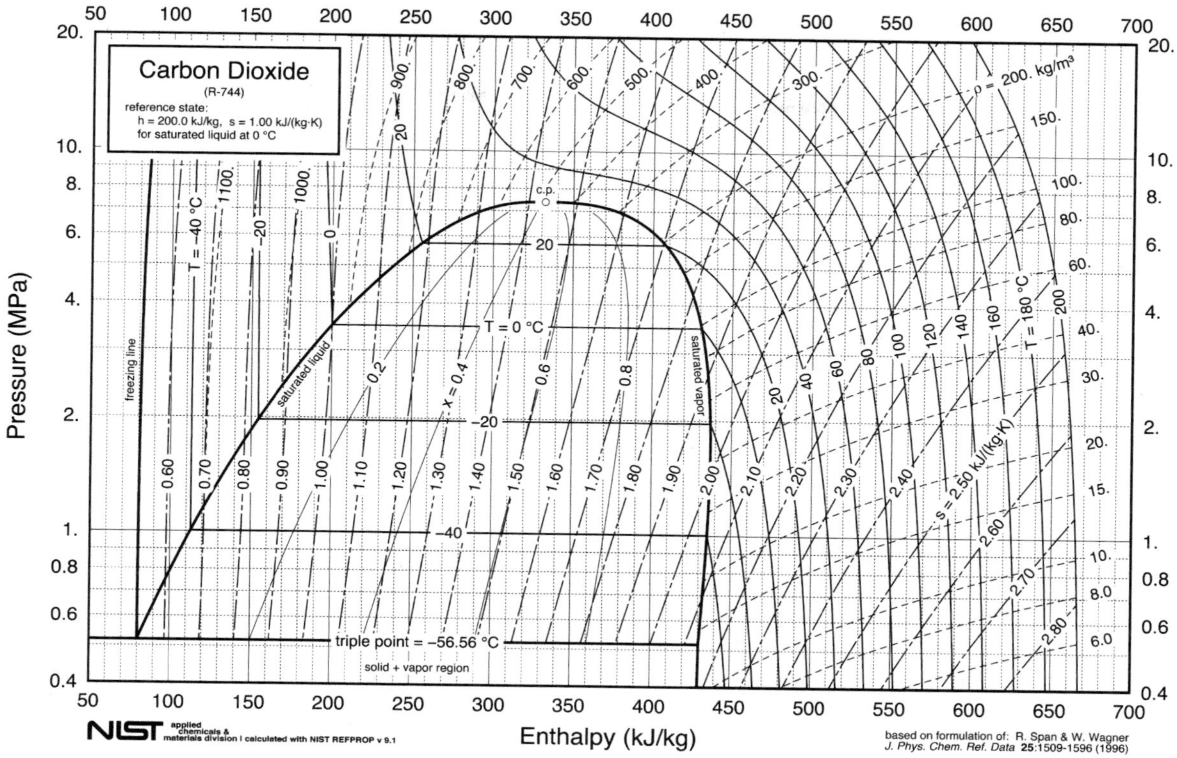

Figure 2.2.4 Enthalpy—log pressure diagram for carbon dioxide (R-744).

Table 2.2.7 Carbon Dioxide (R-744)

T	P	ρ_f	v_g	h_f	h_g	s_f	s_g	$C_{p,f}$	$C_{p,g}$	α_f	α_g	μ_f	μ_g	k_f	k_g	σ
°C	MPa	kg/m³	m³/kg	kJ/kg		kJ/(kg·K)		kJ/(kg·K)		m/s		μPa·s		mW/(m·K)		mN/m
−56.6a	0.518	1178.5	0.0727	80.0	430.4	0.521	2.139	1.95	0.91	976.	223.	257.	11.0	180.6	11.0	16.5
−55.0	0.554	1172.9	0.0682	83.1	431.0	0.535	2.130	1.96	0.92	965.	223.	250.	11.0	178.6	11.1	16.1
−50.0	0.682	1154.6	0.0558	92.9	432.7	0.579	2.102	1.97	0.95	928.	223.	229.	11.3	172.1	11.6	15.0
−45.0	0.832	1135.8	0.0461	102.9	434.1	0.623	2.075	1.99	0.99	892.	224.	211.	11.6	165.6	12.0	13.8
−40.0	1.005	1116.4	0.0383	112.9	435.3	0.666	2.049	2.01	1.03	856.	224.	194.	11.9	159.3	12.5	12.7
−35.0	1.202	1096.4	0.0320	123.1	436.2	0.708	2.023	2.04	1.08	820.	223.	178.	12.2	153.0	13.1	11.6
−30.0	1.428	1075.7	0.0270	133.3	436.8	0.750	1.998	2.07	1.14	783.	223.	164.	12.5	146.9	13.7	10.5
−25.0	1.683	1054.2	0.0228	143.8	437.1	0.791	1.973	2.11	1.21	746.	222.	151.	12.8	140.7	14.3	9.4
−20.0	1.970	1031.7	0.0193	154.5	436.9	0.833	1.949	2.17	1.29	708.	220.	139.	13.1	134.6	15.1	8.4
−15.0	2.291	1008.0	0.0165	165.3	436.3	0.874	1.924	2.23	1.39	668.	219.	128.	13.5	128.6	16.0	7.4
−10.0	2.649	982.9	0.0141	176.5	435.1	0.916	1.899	2.31	1.51	626.	217.	118.	13.9	122.5	17.0	6.4
−5.0	3.046	956.2	0.0120	188.1	433.4	0.958	1.873	2.41	1.66	582.	215.	108.	14.3	116.5	18.2	5.4
0.0	3.485	927.4	0.0102	200.0	430.9	1.000	1.845	2.54	1.87	536.	212.	99.	14.8	110.4	19.7	4.5
5.0	3.970	896.0	0.0087	212.5	427.5	1.043	1.816	2.73	2.14	489.	209.	91.	15.4	104.3	21.6	3.6
10.0	4.502	861.1	0.0074	225.7	422.9	1.088	1.785	3.00	2.56	441.	205.	83.	16.1	98.1	24.2	2.8
15.0	5.087	821.2	0.0062	240.0	416.6	1.136	1.749	3.44	3.24	391.	201.	74.	17.0	91.9	28.0	2.0
20.0	5.729	773.4	0.0052	255.9	407.9	1.188	1.706	4.26	4.56	338.	196.	66.	18.2	85.7	34.0	1.2
25.0	6.434	710.5	0.0041	274.8	394.4	1.249	1.650	6.47	8.21	274.	189.	57.	20.2	80.8	45.5	0.6
30.0	7.214	593.3	0.0029	304.6	365.1	1.344	1.543	–	–	177.	171.	44.	25.2	95.4	98.0	0.1
31.0c	7.377	467.6	0.0021	332.3	332.3	1.434	1.434	∞	∞	0.	0.	–	–	∞	∞	0.0

a = triple point; c = critical point.
Properties computed with NIST REFPROP version 9.1, based on the formulation of Span and Wagner, *J. Phys. Chem. Ref. Data* **25**:1509-1596 (1996).

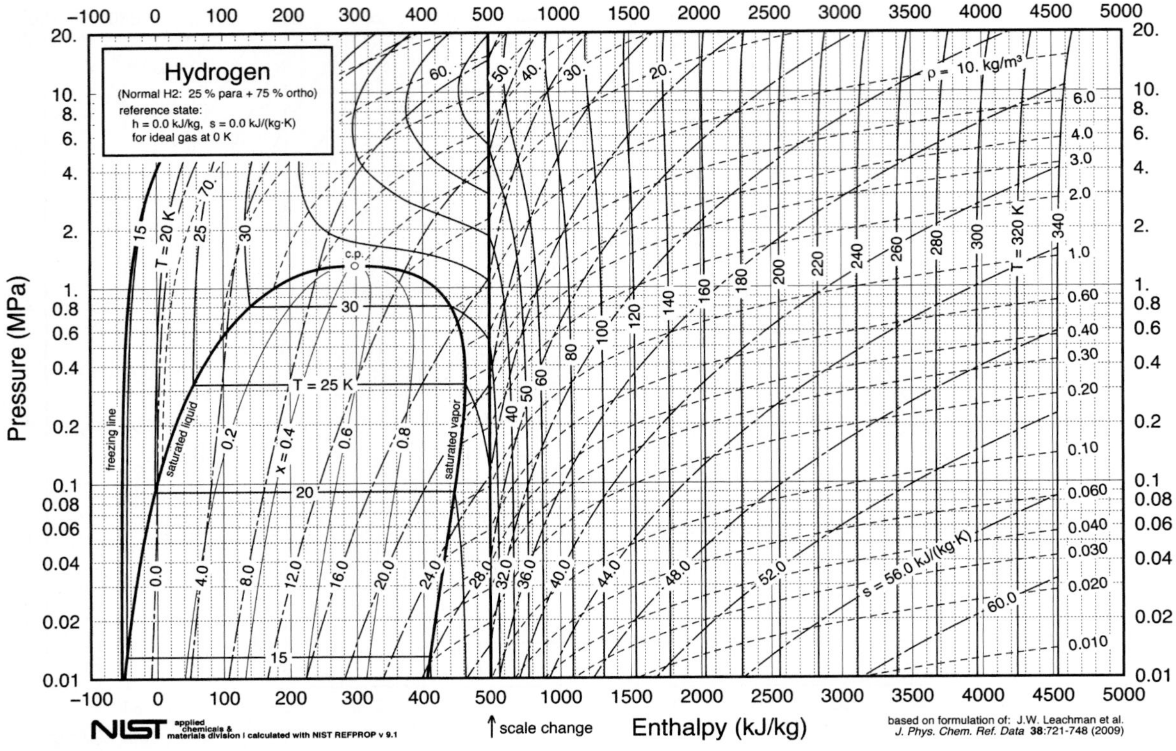

Figure 2.2.5 Enthalpy—log pressure diagram for hydrogen.

Table 2.2.8 Saturated Normal Hydrogen

T	P	ρ_f	v_g	h_f	h_g	s_f	s_g	$C_{p,f}$	$C_{p,g}$	α_f	α_g	μ_f	μ_g	k_f	k_g	σ
K	MPa	kg/m³	m³/kg	kJ/kg		kJ/(kg·K)		kJ/(kg·K)		m/s		μPa·s		mW/(m·K)		mN/m
14.0a	0.007	77.0	7.701	−53.9	399.8	−3.072	29.438	7.02	10.56	1269.	307.	25.6	0.65	98.	10.6	3.03
14.0	0.008	77.0	7.535	−53.6	400.2	−3.051	29.367	7.03	10.57	1268.	308.	25.4	0.65	98.	10.6	3.02
15.0	0.013	76.1	4.685	−46.4	409.3	−2.557	27.821	7.34	10.70	1247.	317.	22.6	0.70	99.	11.6	2.85
16.0	0.021	75.3	3.076	−38.8	417.9	−2.072	26.471	7.72	10.86	1225.	326.	20.2	0.76	101.	12.6	2.67
17.0	0.032	74.3	2.112	−30.7	426.1	−1.594	25.276	8.14	11.05	1203.	334.	18.2	0.81	102.	13.6	2.50
18.0	0.047	73.4	1.503	−22.2	433.6	−1.119	24.204	8.58	11.27	1180.	341.	16.6	0.87	103.	14.7	2.32
19.0	0.066	72.3	1.103	−13.2	440.5	−0.646	23.233	9.05	11.55	1156.	348.	15.2	0.92	103.	15.8	2.15
20.0	0.091	71.3	0.830	−3.7	446.6	−0.174	22.341	9.57	11.89	1129.	354.	13.9	0.98	104.	17.0	1.98
20.4b	0.101	70.8	0.751	−0.0	448.7	−0.000	22.029	9.77	12.04	1119.	356.	13.5	1.00	104.	17.5	1.91
21.0	0.122	70.1	0.637	6.5	452.0	0.299	21.514	10.14	12.31	1101.	360.	12.8	1.03	104.	18.3	1.80
22.0	0.159	68.9	0.498	17.2	456.4	0.775	20.738	10.77	12.83	1070.	365.	11.8	1.09	104.	19.6	1.63
23.0	0.204	67.6	0.395	28.7	459.9	1.256	20.002	11.49	13.47	1037.	369.	11.0	1.15	103.	21.1	1.46
24.0	0.258	66.2	0.317	41.0	462.2	1.744	19.296	12.32	14.27	1002.	372.	10.2	1.21	103.	22.6	1.29
25.0	0.321	64.7	0.257	54.2	463.4	2.242	18.610	13.30	15.29	964.	375.	9.5	1.27	102.	24.3	1.13
26.0	0.394	63.1	0.210	68.4	463.1	2.753	17.936	14.49	16.60	923.	377.	8.8	1.34	101.	26.2	0.96
27.0	0.478	61.3	0.172	83.7	461.2	3.283	17.262	15.99	18.36	878.	379.	8.2	1.42	100.	28.4	0.80
28.0	0.574	59.3	0.142	100.6	457.3	3.837	16.578	17.98	20.80	829.	380.	7.6	1.50	99.	30.9	0.64
29.0	0.682	57.1	0.117	119.2	450.9	4.425	15.864	20.81	24.45	775.	380.	7.0	1.60	97.	33.9	0.49
30.0	0.804	54.5	0.096	140.3	441.2	5.066	15.096	25.28	30.43	713.	380.	6.4	1.72	95.	37.8	0.35
31.0	0.942	51.4	0.077	165.1	426.4	5.793	14.224	33.76	42.07	641.	378.	5.8	1.86	92.	43.5	0.21
32.0	1.096	47.1	0.061	196.7	402.3	6.698	13.122	–	–	553.	376.	5.0	2.08	90.	53.6	0.10
33.2c	1.297	31.3	0.032	298.2	298.2	9.644	9.644	∞	∞	0.	0.	–	–	∞	∞	0.00

* Hydrogen exists in two different spin isomers; this table tabulates properties for "normal" hydrogen comprising 25% para and 75% ortho. This is the equilibrium distribution at normal temperatures and for hydrogen that has been quickly liquefied. Hydrogen at very low temperatures is almost entirely para at equilibrium, but the ortho/para conversion is slow unless catalyzed.
a = triple point; b = normal boiling point; c = critical point.
Properties computed with NIST REFPROP version 9.1, based on the formulation of Leachman *et al.*, *J. Phys. Chem. Ref. Data* **38**:721-748 (2009).

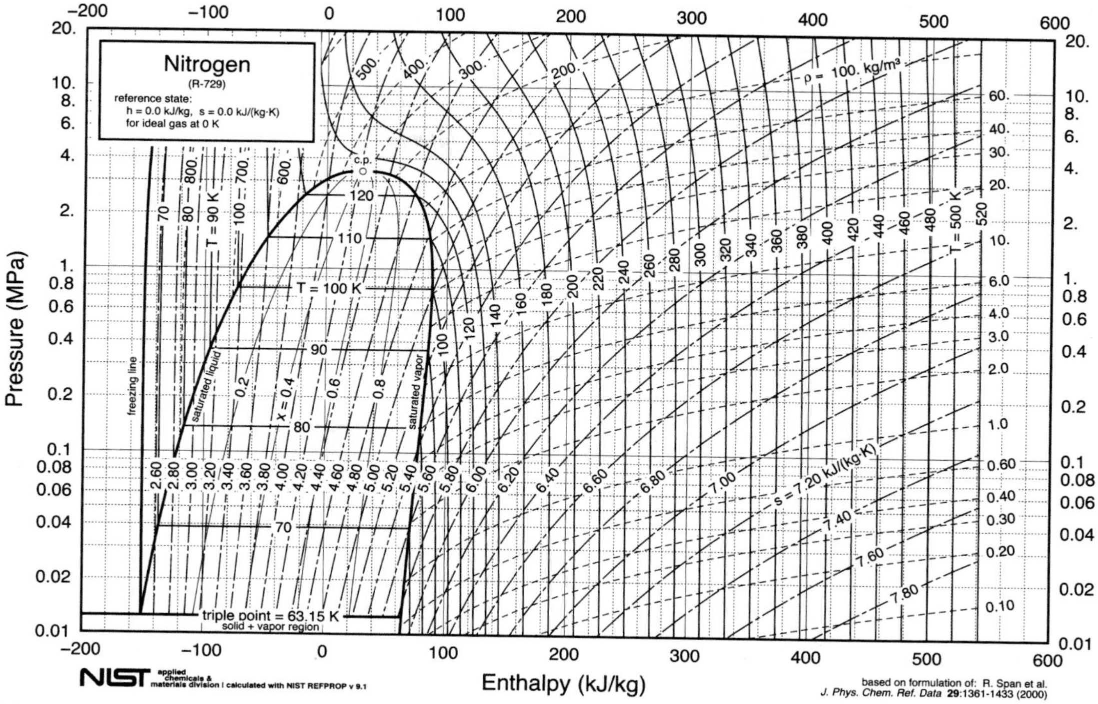

Figure 2.2.6 Enthalpy—log pressure diagram for nitrogen.

Table 2.2.9 Saturated Nitrogen

T	P	ρ_f	v_g	h_f	h_g	s_f	s_g	$C_{p,f}$	$C_{p,g}$	α_f	α_g	μ_f	μ_g	k_f	k_g	σ
K	MPa	kg/m³	m³/kg	kJ/kg		kJ/(kg·K)		kJ/(kg·K)		m/s		μPa·s		mW/(m·K)		mN/m
63.2a	0.013	867.2	1.483	−150.7	64.8	2.426	5.838	2.00	1.06	995.	161.	311.6	4.38	173.	5.6	12.21
65.0	0.017	859.6	1.095	−147.0	66.5	2.483	5.769	2.00	1.06	976.	163.	282.1	4.51	170.	5.8	11.76
70.0	0.039	838.5	0.527	−137.0	71.1	2.632	5.605	2.01	1.08	926.	168.	220.2	4.88	160.	6.4	10.58
72.0	0.051	829.9	0.406	−133.0	72.8	2.689	5.547	2.02	1.09	906.	170.	201.1	5.03	156.	6.6	10.11
74.0	0.067	821.1	0.318	−129.0	74.5	2.744	5.493	2.03	1.10	885.	172.	184.3	5.19	152.	6.8	9.65
76.0	0.086	812.2	0.252	−125.0	76.1	2.798	5.442	2.04	1.11	865.	174.	169.6	5.34	148.	7.0	9.19
77.4b	0.101	806.1	0.217	−122.0	77.2	2.834	5.409	2.04	1.12	851.	175.	160.7	5.44	145.	7.2	8.88
78.0	0.109	803.1	0.202	−120.7	77.6	2.851	5.394	2.05	1.13	845.	175.	156.6	5.49	144.	7.3	8.73
80.0	0.137	793.9	0.164	−116.6	79.1	2.903	5.349	2.06	1.15	824.	177.	145.1	5.65	140.	7.5	8.28
82.0	0.169	784.6	0.135	−112.4	80.5	2.953	5.306	2.07	1.16	804.	178.	134.8	5.81	136.	7.8	7.84
84.0	0.208	775.0	0.112	−108.3	81.8	3.003	5.265	2.08	1.18	783.	179.	125.6	5.97	132.	8.0	7.40
86.0	0.252	765.2	0.0931	−104.0	82.9	3.052	5.226	2.10	1.21	762.	180.	117.2	6.14	128.	8.3	6.97
88.0	0.303	755.2	0.0783	−99.8	84.0	3.100	5.189	2.12	1.24	741.	181.	109.7	6.31	124.	8.6	6.54
90.0	0.360	745.0	0.0663	−95.5	85.0	3.147	5.153	2.14	1.27	719.	182.	102.8	6.48	120.	8.9	6.11
92.0	0.426	734.5	0.0565	−91.2	86.0	3.194	5.118	2.17	1.30	697.	182.	96.5	6.66	116.	9.2	5.69
94.0	0.500	723.8	0.0484	−86.8	87.0	3.240	5.084	2.20	1.34	675.	183.	90.7	6.84	112.	9.5	5.28
96.0	0.583	712.7	0.0417	−82.3	87.1	3.286	5.051	2.23	1.39	652.	183.	85.4	7.03	108.	9.9	4.88
98.0	0.676	701.2	0.0361	−77.8	87.5	3.331	5.018	2.27	1.44	629.	183.	80.4	7.23	104.	10.3	4.48
100.0	0.778	689.4	0.0313	−73.2	87.8	3.376	4.986	2.32	1.50	605.	183.	75.8	7.43	100.	10.7	4.09
105.0	1.083	657.5	0.0222	−61.3	87.6	3.488	4.906	2.48	1.71	543.	183.	65.3	7.98	90.	12.0	3.14
110.0	1.466	621.5	0.0160	−48.5	85.8	3.602	4.823	2.74	2.06	476.	181.	56.0	8.63	80.	13.8	2.24
115.0	1.937	578.7	0.0115	−34.4	81.9	3.720	4.731	3.24	2.75	403.	178.	47.3	9.44	71.	16.6	1.42
120.0	2.511	523.4	0.0080	−17.9	74.2	3.851	4.619	4.51	4.63	317.	173.	38.4	10.62	61.	21.7	0.68
125.0	3.207	426.1	0.0049	6.4	55.0	4.037	4.426	–	–	–	–	–	–	–	–	0.09
126.2c	3.396	313.3	0.0032	29.2	29.2	4.215	4.215	∞	∞	0.	0.	–	–	∞	∞	0.00

a = triple point; b = normal boiling point; c = critical point.
Properties computed with NIST REFPROP version 9.1, based on the formulation of Span *et al.*, *J. Phys. Chem. Ref. Data* **29**:1361-1433 (2000).

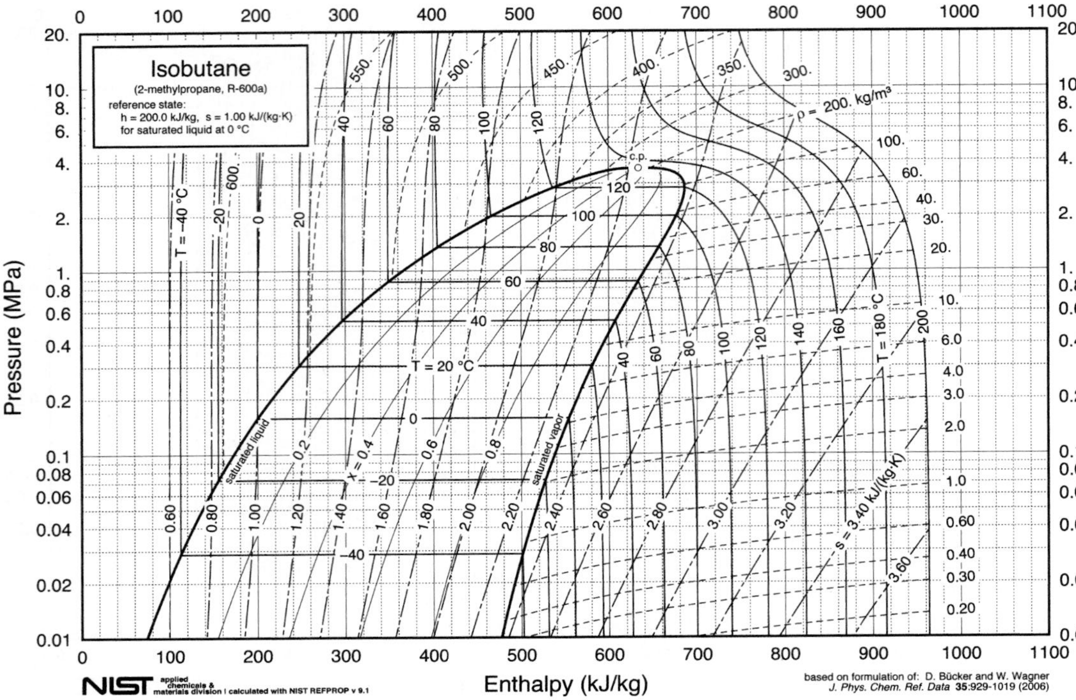

Figure 2.2.7 Enthalpy—log pressure diagram for isobutane (R-600a).

Table 2.2.10 Saturated Isobutane (R600a)

T	P	ρ_f	v_g	h_f	h_g	s_f	s_g	$C_{p,f}$	$C_{p,g}$	α_f	α_g	μ_f	μ_g	k_f	k_g	σ
°C	MPa	kg/m³	m³/kg	kJ/kg		kJ/(kg·K)		kJ/(kg·K)		m/s		μPa·s		mW/(m·K)		mN/m
−100.0	0.000	683.9	65.234	−6.4	428.2	0.067	2.577	1.88	1.13	1558.	168.	937.	4.40	140.	6.0	24.64
−90.0	0.001	674.2	26.648	12.6	439.6	0.173	2.505	1.91	1.17	1494.	173.	753.	4.65	136.	6.8	23.45
−80.0	0.002	664.5	12.140	31.8	451.4	0.276	2.448	1.94	1.21	1431.	177.	619.	4.91	132.	7.5	22.26
−70.0	0.005	654.6	6.059	51.4	463.5	0.375	2.403	1.98	1.25	1370.	181.	518.	5.16	128.	8.3	21.06
−60.0	0.009	644.6	3.266	71.4	475.9	0.471	2.368	2.02	1.30	1309.	184.	440.	5.41	124.	9.1	19.87
−50.0	0.017	634.4	1.879	91.8	488.5	0.564	2.342	2.05	1.34	1250.	188.	378.	5.65	119.	9.9	18.68
−40.0	0.029	624.1	1.143	112.5	501.4	0.655	2.323	2.09	1.39	1191.	190.	327.	5.90	115.	10.7	17.50
−30.0	0.047	613.6	0.728	133.7	514.4	0.744	2.310	2.14	1.44	1133.	193.	286.	6.14	111.	11.6	16.32
−20.0	0.072	602.9	0.483	155.3	527.6	0.831	2.301	2.18	1.50	1075.	195.	252.	6.38	107.	12.5	15.15
−11.8b	0.101	593.8	0.354	173.5	538.6	0.901	2.298	2.22	1.55	1028.	197.	228.	6.58	103.	13.2	14.19
−10.0	0.108	591.9	0.332	177.4	540.9	0.916	2.298	2.23	1.56	1018.	197.	223.	6.62	103.	13.4	13.99
0.0	0.157	580.6	0.235	200.0	554.3	1.000	2.297	2.28	1.62	961.	198.	199.	6.86	99.	14.3	12.84
10.0	0.221	568.9	0.170	223.2	567.8	1.083	2.300	2.34	1.69	905.	198.	178.	7.11	95.	15.3	11.69
20.0	0.302	556.9	0.126	246.9	581.2	1.165	2.305	2.40	1.76	849.	198.	159.	7.37	91.	16.3	10.56
30.0	0.405	544.3	0.0954	271.2	594.6	1.246	2.312	2.46	1.84	793.	197.	143.	7.63	88.	17.4	9.45
40.0	0.531	531.2	0.0732	296.3	607.8	1.326	2.321	2.54	1.92	736.	196.	129.	7.91	84.	18.5	8.35
50.0	0.685	517.4	0.0568	322.1	620.8	1.406	2.331	2.62	2.02	680.	193.	117.	8.22	81.	19.8	7.28
60.0	0.869	502.7	0.0446	348.7	633.5	1.486	2.341	2.71	2.13	622.	190.	106.	8.56	78.	21.2	6.23
70.0	1.088	487.0	0.0353	376.2	645.8	1.566	2.352	2.81	2.25	564.	185.	95.	8.94	75.	22.7	5.20
80.0	1.344	469.9	0.0280	404.7	657.3	1.647	2.362	2.94	2.41	505.	180.	86.	9.39	72.	24.5	4.21
90.0	1.642	451.1	0.0223	434.5	667.9	1.728	2.371	3.11	2.63	444.	173.	77.	9.95	69.	26.7	3.27
100.0	1.987	429.6	0.0176	465.9	676.9	1.811	2.377	3.35	2.95	381.	164.	68.	10.65	67.	29.4	2.37
110.0	2.383	404.3	0.0138	499.4	683.7	1.897	2.379	3.74	3.52	313.	153.	59.	11.62	64.	33.2	1.53
120.0	2.837	372.0	0.0106	536.3	686.5	1.989	2.371	4.59	4.81	240.	140.	50.	13.09	62.	39.1	0.79
130.0	3.358	321.0	0.0074	581.3	678.4	2.099	2.340	–	–	–	–	–	–	–	–	0.18
134.7c	3.629	225.5	0.0044	633.9	633.9	2.226	2.226	∞	∞	0.	0.	–	–	∞	∞	0.00

b = normal boiling point; c = critical point.
Properties computed with NIST REFPROP version 9.1, based on the formulation of Bücker and Wagner, *J. Phys. Chem. Ref. Data* **35**:929-1019 (2006).

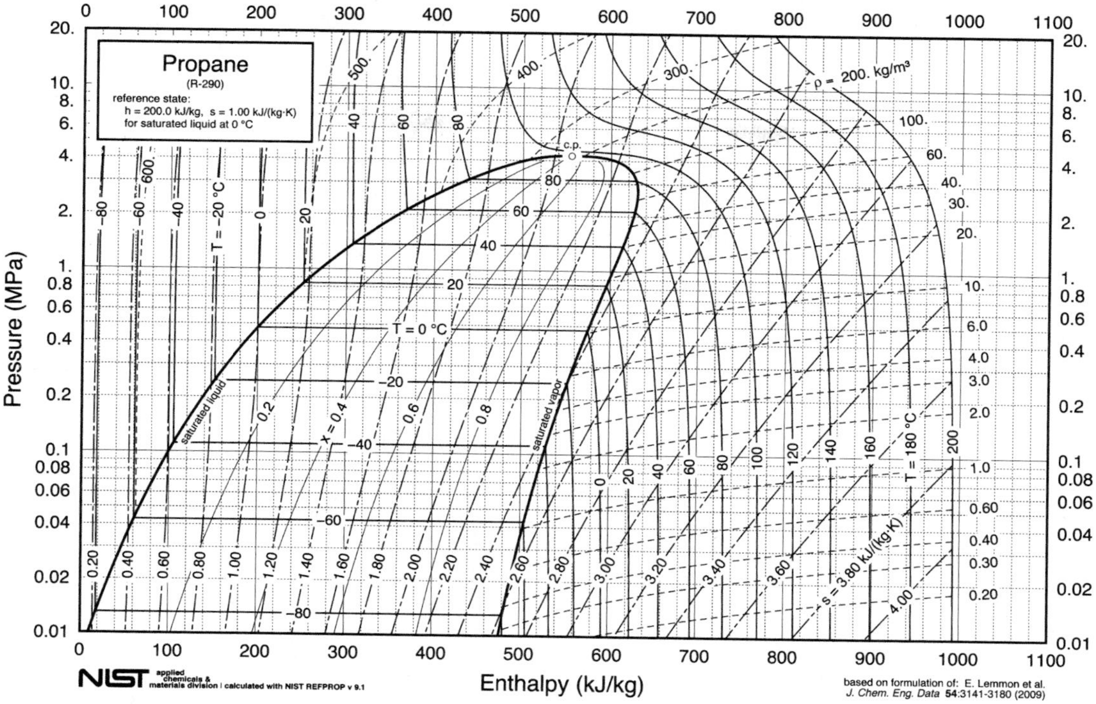

Figure 2.2.8 Enthalpy—log pressure diagram for propane (R-290).

Table 2.2.11 Saturated Propane (R-290)

T	P	ρ_f	v_g	h_f	h_g	s_f	s_g	$C_{p,f}$	$C_{p,g}$	α_f	α_g	μ_f	μ_g	k_f	k_g	σ
°C	MPa	kg/m³	m³/kg	kJ/kg		kJ/(kg·K)		kJ/(kg·K)		m/s		μPa·s		mW/(m·K)		mN/m
−150.0	0.00001	694.6	4316.4	−123.8	402.1	−0.690	3.580	1.96	1.02	1880.	169.	1343.	3.55	193.	3.68	29.41
−140.0	0.00003	684.5	864.49	−104.0	412.4	−0.537	3.343	1.98	1.05	1813.	175.	985.	3.80	188.	4.28	28.29
−130.0	0.0001	674.4	223.53	−84.2	423.1	−0.393	3.151	1.99	1.09	1745.	181.	762.	4.05	182.	4.90	27.12
−120.0	0.0004	664.3	70.785	−64.2	434.1	−0.258	2.996	2.01	1.12	1679.	186.	612.	4.31	176.	5.55	25.91
−110.0	0.0012	654.0	26.386	−44.0	445.4	−0.130	2.870	2.03	1.15	1612.	192.	505.	4.56	170.	6.23	24.65
−100.0	0.003	643.7	11.231	−23.6	456.9	−0.008	2.766	2.05	1.18	1545.	197.	426.	4.82	164.	6.94	23.38
−90.0	0.006	633.3	5.330	−2.9	468.6	0.108	2.682	2.08	1.22	1478.	202.	365.	5.08	158.	7.67	22.07
−80.0	0.013	622.8	2.768	18.0	480.4	0.219	2.613	2.11	1.26	1411.	206.	316.	5.34	152.	8.43	20.75
−70.0	0.024	612.0	1.549	39.3	492.4	0.326	2.557	2.14	1.30	1345.	210.	276.	5.60	146.	9.22	19.42
−60.0	0.043	601.1	0.923	60.8	504.4	0.429	2.511	2.17	1.35	1278.	214.	244.	5.85	140.	10.04	18.07
−50.0	0.071	589.9	0.579	82.8	516.5	0.530	2.473	2.21	1.40	1213.	217.	216.	6.11	134.	10.88	16.73
−40.0	0.111	578.4	0.380	105.1	528.5	0.628	2.443	2.26	1.45	1147.	219.	193.	6.36	128.	11.76	15.39
−30.0	0.168	566.6	0.259	128.0	540.4	0.723	2.419	2.31	1.51	1082.	221.	172.	6.62	122.	12.67	14.05
−20.0	0.245	554.5	0.182	151.4	552.1	0.817	2.400	2.36	1.58	1016.	222.	155.	6.89	117.	13.63	12.73
−42.1b	0.101	580.9	0.414	100.4	526.0	0.607	2.449	2.25	1.44	1161.	218.	197.	6.31	129.	11.57	15.67
−10.0	0.345	541.8	0.131	175.4	563.7	0.909	2.385	2.42	1.66	951.	222.	139.	7.16	111.	14.65	11.42
0.0	0.474	528.6	0.0966	200.0	574.9	1.000	2.372	2.49	1.74	885.	221.	126.	7.45	106.	15.74	10.13
10.0	0.637	514.7	0.0726	225.4	585.7	1.090	2.363	2.57	1.84	819.	220.	113.	7.75	101.	16.93	8.87
20.0	0.836	500.1	0.0553	251.6	596.0	1.180	2.354	2.67	1.95	753.	217.	102.	8.09	96.	18.24	7.63
30.0	1.079	484.4	0.0426	278.8	605.5	1.270	2.347	2.78	2.09	685.	214.	92.	8.46	91.	19.72	6.43
40.0	1.369	467.5	0.0332	307.2	614.2	1.359	2.340	2.91	2.26	617.	209.	83.	8.89	87.	21.43	5.26
50.0	1.713	448.9	0.0259	336.8	621.7	1.450	2.332	3.09	2.50	547.	202.	74.	9.40	83.	23.47	4.14
60.0	2.117	428.0	0.0202	368.1	627.4	1.543	2.321	3.34	2.84	474.	194.	66.	10.03	78.	26.00	3.08
70.0	2.587	403.6	0.0157	401.8	630.4	1.639	2.305	3.74	3.42	398.	184.	57.	10.86	74.	29.41	2.08
80.0	3.132	373.3	0.0119	438.9	628.7	1.742	2.279	4.55	4.71	315.	172.	49.	12.07	70.	34.75	1.17
90.0	3.764	328.8	0.0084	483.7	616.5	1.862	2.227	–	–	–	–	–	–	–	–	0.38
96.7c	4.251	220.5	0.0045	555.2	555.2	2.052	2.052	∞	∞	0.	0.	–	–	∞	∞	0.00

b = normal boiling point; c = critical point.
Properties computed with NIST REFPROP version 9.1, based on the formulation of Lemmon *et al.*, *J. Chem. Eng. Data* **54**:3141-3180 (2009).

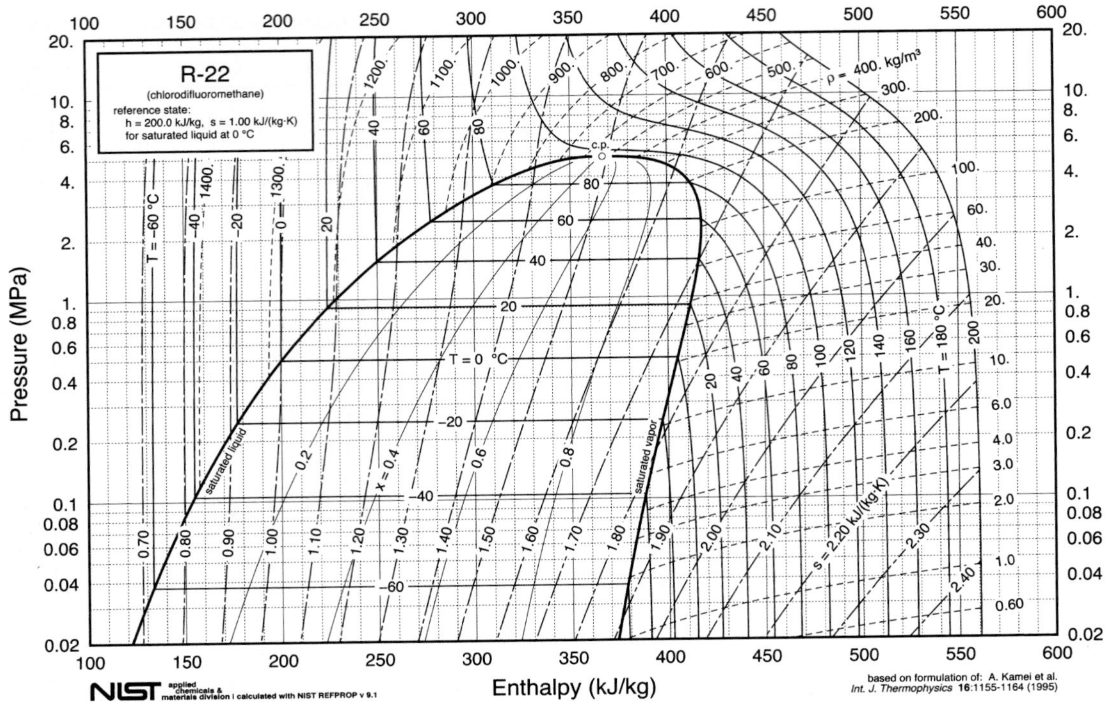

Figure 2.2.9 Enthalpy—log pressure diagram for refrigerant-22 (chlorodifluoromethane).

Table 2.2.12 Saturated Refrigerant 22 (chlorodifluoromethane)

T	P	ρ_f	v_g	h_f	h_g	s_f	s_g	$C_{p,f}$	$C_{p,g}$	α_f	α_g	μ_f	μ_g	k_f	k_g	σ
°C	MPa	kg/m³	m³/kg	kJ/kg		kJ/(kg·K)		kJ/(kg·K)		m/s		μPa·s		mW/(m·K)		mN/m
–100.0	0.002	1571.3	8.266	90.7	359.0	0.505	2.054	1.06	0.50	1127.	144.	896.	7.24	143.	4.46	28.24
–90.0	0.005	1544.9	3.645	101.3	363.9	0.565	1.998	1.06	0.51	1080.	147.	731.	7.66	138.	4.84	26.51
–80.0	0.010	1518.2	1.778	111.9	368.8	0.621	1.951	1.06	0.53	1033.	150.	609.	8.08	133.	5.25	24.79
–70.0	0.020	1491.2	0.943	122.6	373.7	0.675	1.911	1.07	0.55	986.	153.	518.	8.50	128.	5.68	23.09
–60.0	0.038	1463.7	0.537	133.3	378.6	0.726	1.877	1.07	0.56	940.	156.	446.	8.91	123.	6.12	21.41
–50.0	0.065	1435.6	0.324	144.0	383.4	0.775	1.848	1.08	0.59	893.	158.	389.	9.32	118.	6.59	19.74
–40.8b	0.101	1409.2	0.213	154.0	387.8	0.819	1.825	1.09	0.61	851.	160.	345.	9.70	113.	7.05	18.24
–40.0	0.105	1406.8	0.205	154.9	388.1	0.823	1.823	1.09	0.61	847.	160.	342.	9.73	113.	7.09	18.10
–30.0	0.164	1377.2	0.136	165.9	392.7	0.869	1.802	1.11	0.64	800.	162.	303.	10.13	108.	7.61	16.49
–20.0	0.245	1346.5	0.0927	177.0	397.1	0.914	1.783	1.12	0.67	754.	163.	270.	10.54	104.	8.17	14.89
–15.0	0.296	1330.8	0.0775	182.7	399.2	0.935	1.774	1.13	0.68	730.	163.	255.	10.74	102.	8.46	14.11
–10.0	0.355	1314.7	0.0653	188.4	401.2	0.957	1.766	1.14	0.70	707.	163.	241.	10.95	99.	8.76	13.33
–5.0	0.422	1298.3	0.0553	194.2	403.2	0.979	1.758	1.16	0.72	683.	163.	228.	11.15	97.	9.08	12.56
0.0	0.498	1281.5	0.0471	200.0	405.1	1.000	1.751	1.17	0.74	660.	163.	216.	11.36	95.	9.41	11.80
5.0	0.584	1264.3	0.0403	205.9	406.9	1.021	1.744	1.18	0.76	636.	163.	205.	11.58	93.	9.75	11.04
10.0	0.681	1246.7	0.0347	211.9	408.6	1.042	1.737	1.20	0.79	613.	163.	194.	11.80	90.	10.11	10.29
15.0	0.789	1228.6	0.0300	217.9	410.2	1.063	1.730	1.22	0.81	589.	162.	184.	12.03	88.	10.50	9.56
20.0	0.910	1209.9	0.0260	224.1	411.7	1.084	1.724	1.24	0.84	565.	161.	174.	12.26	86.	10.90	8.83
25.0	1.044	1190.7	0.0226	230.3	413.0	1.105	1.717	1.26	0.87	541.	160.	164.	12.51	84.	11.34	8.11
30.0	1.192	1170.7	0.0197	236.6	414.3	1.125	1.711	1.28	0.91	517.	159.	156.	12.77	81.	11.81	7.41
35.0	1.355	1150.1	0.0173	243.1	415.3	1.146	1.705	1.31	0.95	492.	158.	147.	13.05	79.	12.32	6.72
40.0	1.534	1128.5	0.0151	249.7	416.3	1.167	1.699	1.34	1.00	468.	156.	139.	13.36	77.	12.89	6.03
45.0	1.729	1106.0	0.0133	256.4	417.0	1.187	1.692	1.38	1.05	443.	155.	131.	13.69	74.	13.52	5.37
50.0	1.943	1082.3	0.0116	263.3	417.4	1.208	1.685	1.42	1.11	417.	153.	123.	14.06	72.	14.23	4.72
55.0	2.175	1057.2	0.0102	270.3	417.7	1.229	1.678	1.47	1.19	391.	150.	115.	14.47	70.	15.06	4.08
60.0	2.428	1030.4	0.0089	277.6	417.6	1.250	1.671	1.54	1.29	364.	148.	108.	14.95	67.	16.03	3.46
65.0	2.701	1001.4	0.0079	285.2	417.1	1.272	1.662	1.63	1.41	337.	145.	101.	15.50	65.	17.19	2.87
70.0	2.997	969.7	0.0069	293.1	416.1	1.295	1.653	1.74	1.58	309.	142.	93.	16.16	62.	18.63	2.30
75.0	3.318	934.4	0.0060	301.5	414.5	1.318	1.642	1.91	1.83	280.	138.	86.	16.98	59.	20.47	1.75
80.0	3.664	893.7	0.0051	310.4	412.0	1.342	1.630	2.18	2.23	249.	134.	78.	18.02	57.	22.95	1.24
90.0	4.442	780.1	0.0036	332.1	401.9	1.400	1.592	3.98	4.98	177.	125.	61.	21.69	52.	32.82	0.36
96.2c	4.990	523.8	0.0019	366.9	366.9	1.493	1.493	∞	∞	0.	0.	–	–	∞	∞	0.00

b = normal boiling point; c = critical point.
Properties computed with NIST REFPROP version 9.1, based on the formulation of Kamei *et al.*, *Int. J. Thermophysics* **16**:1155-1164 (1995).

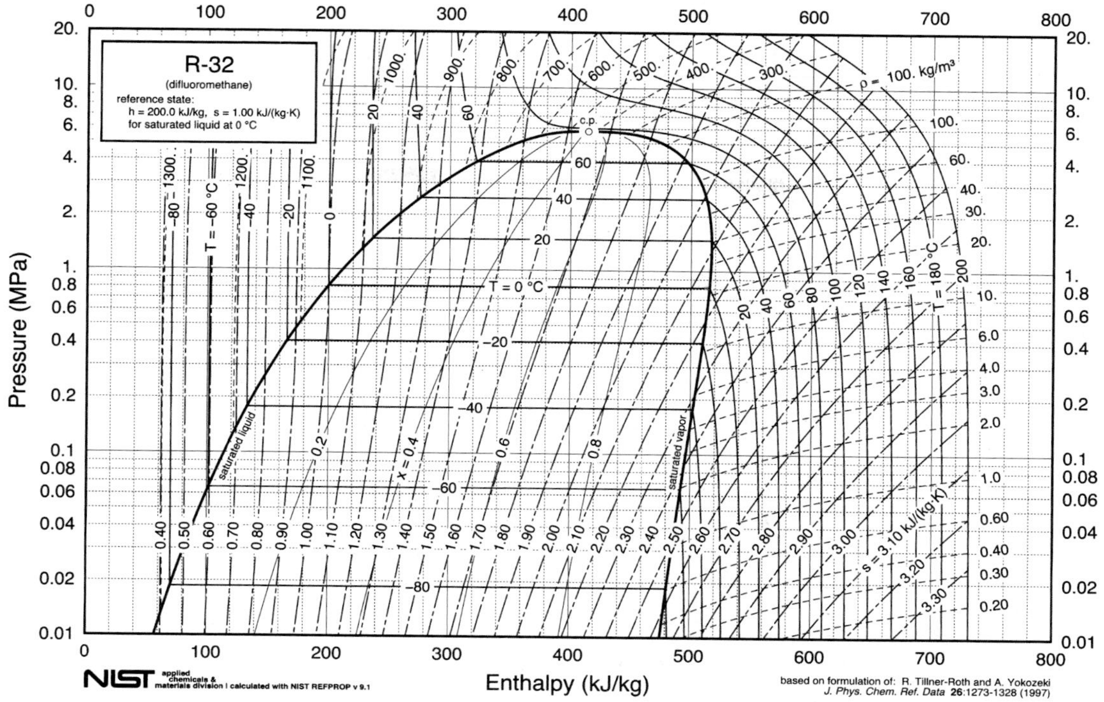

Figure 2.2.10 Enthalpy—log pressure diagram for refrigerant-32 (difluoromethane).

Table 2.2.13 Saturated Refrigerant 32 (difluoromethane)

T	P	ρ_f	v_g	h_f	h_g	s_f	s_g	$C_{p,f}$	$C_{p,g}$	α_f	α_g	μ_f	μ_g	k_f	k_g	σ
°C	MPa	kg/m³	m³/kg	kJ/kg		kJ/(kg·K)		kJ/(kg·K)		m/s		μPa·s		mW/(m·K)		mN/m
−136.8a	0.000	1429.3	453.85	−19.1	444.3	−0.105	3.294	1.59	0.66	1414.	170.	1221.	5.70	243.	6.95	38.75
−130.0	0.000	1412.7	174.36	−8.3	448.8	−0.028	3.165	1.58	0.67	1378.	174.	1019.	5.97	241.	6.97	37.23
−120.0	0.000	1388.4	51.184	7.5	455.3	0.079	3.003	1.57	0.67	1326.	179.	809.7	6.39	236.	7.02	35.01
−110.0	0.001	1363.8	17.907	23.2	461.9	0.178	2.867	1.57	0.69	1273.	185.	662.9	6.80	231.	7.12	32.82
−100.0	0.004	1339.0	7.222	38.8	468.3	0.271	2.752	1.56	0.70	1221.	190.	554.6	7.22	224.	7.27	30.66
−90.0	0.009	1313.9	3.272	54.4	474.6	0.359	2.653	1.56	0.73	1169.	194.	471.2	7.64	217.	7.45	28.53
−80.0	0.019	1288.4	1.632	70.0	480.7	0.442	2.568	1.56	0.75	1118.	198.	405.1	8.06	210.	7.68	26.43
−70.0	0.036	1262.4	0.881	85.7	486.6	0.520	2.494	1.57	0.79	1066.	202.	351.5	8.48	202.	7.96	24.37
−60.0	0.065	1235.7	0.508	101.4	492.1	0.596	2.429	1.58	0.83	1014.	205.	307.1	8.91	194.	8.28	22.33
−51.7b	0.101	1212.9	0.335	114.6	496.5	0.657	2.381	1.59	0.88	971.	207.	275.7	9.26	187.	8.60	20.67
−50.0	0.110	1208.4	0.309	117.2	497.3	0.668	2.371	1.59	0.88	962.	208.	270.0	9.33	186.	8.66	20.34
−40.0	0.177	1180.2	0.197	133.2	502.0	0.738	2.320	1.61	0.94	910.	210.	238.6	9.75	178.	9.10	18.38
−30.0	0.273	1151.0	0.131	149.5	506.3	0.806	2.274	1.63	1.00	858.	211.	211.7	10.18	170.	9.60	16.46
−20.0	0.406	1120.6	0.0846	165.9	510.0	0.872	2.231	1.66	1.08	805.	212.	188.5	10.61	161.	10.19	14.59
−15.0	0.488	1104.9	0.0749	174.3	511.6	0.904	2.211	1.68	1.11	778.	212.	178.1	10.83	157.	10.53	13.67
−10.0	0.583	1088.8	0.0630	182.8	513.0	0.937	2.192	1.70	1.16	751.	212.	168.3	11.05	153.	10.89	12.76
−5.0	0.691	1072.2	0.0533	191.3	514.3	0.968	2.173	1.72	1.20	724.	211.	159.1	11.28	149.	11.29	11.86
0.0	0.813	1055.3	0.0453	200.0	515.3	1.000	2.154	1.75	1.25	697.	211.	150.5	11.51	145.	11.73	10.98
5.0	0.951	1037.7	0.0386	208.8	516.1	1.031	2.136	1.77	1.31	669.	210.	142.3	11.75	141.	12.23	10.11
10.0	1.107	1019.7	0.0331	217.7	516.7	1.063	2.119	1.81	1.37	641.	209.	134.6	12.00	137.	12.79	9.26
15.0	1.281	1000.9	0.0284	226.8	516.9	1.094	2.101	1.84	1.44	613.	207.	127.3	12.26	134.	13.43	8.42
20.0	1.475	981.4	0.0245	236.1	516.9	1.125	2.083	1.89	1.51	584.	206.	120.3	12.53	130.	14.17	7.59
25.0	1.690	961.0	0.0211	245.6	516.5	1.157	2.065	1.94	1.60	554.	204.	113.7	12.82	126.	15.02	6.79
30.0	1.928	939.6	0.0183	255.3	515.7	1.188	2.047	2.00	1.71	524.	202.	107.2	13.13	122.	16.03	6.00
35.0	2.190	917.0	0.0158	265.3	514.5	1.220	2.029	2.07	1.84	493.	199.	101.0	13.46	118.	17.24	5.23
40.0	2.478	893.0	0.0137	275.6	512.7	1.252	2.009	2.16	2.00	461.	196.	95.0	13.83	115.	18.71	4.49
45.0	2.795	867.3	0.0118	286.3	510.3	1.285	1.989	2.28	2.21	428.	193.	89.1	14.24	111.	20.52	3.77
50.0	3.141	839.3	0.0102	297.5	507.1	1.318	1.967	2.44	2.48	394.	189.	83.2	14.72	107.	22.82	3.07
55.0	3.520	808.3	0.0087	309.3	502.9	1.353	1.943	2.66	2.86	358.	185.	77.4	15.28	103.	25.82	2.41
60.0	3.933	773.3	0.0074	321.9	497.4	1.390	1.917	3.00	3.44	320.	180.	71.4	15.97	99.	29.88	1.78
70.0	4.877	680.9	0.0051	351.7	479.5	1.474	1.846	4.87	6.64	236.	168.	58.2	18.14	92.	45.05	0.65
78.1c	5.782	424.0	0.0024	414.2	414.2	1.649	1.649	∞	∞	0.	0.	–	–	∞	∞	0.00

a = triple point; b = normal boiling point; c = critical point.
Properties computed with NIST REFPROP version 9.1, based on the formulation of Tillner-Roth and Yokozeki, *J. Phys. Chem. Ref. Data* **26**:1273-1328 (1997).

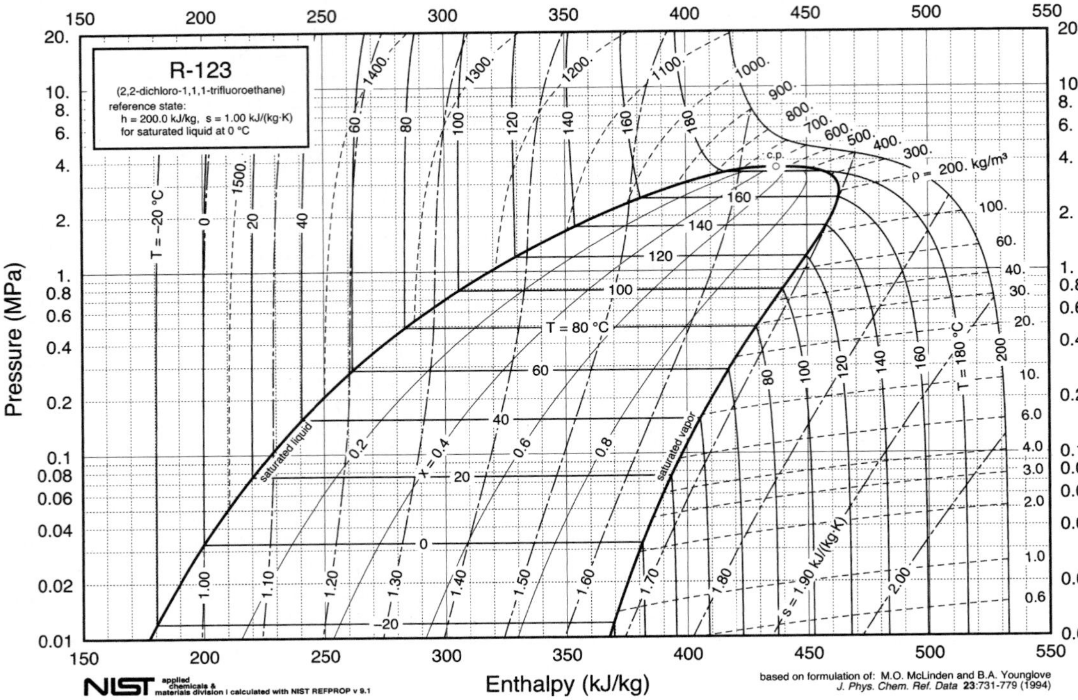

Figure 2.2.11 Enthalpy—log pressure diagram for refrigerant-123 (2,2-dichloro-1,1,1-trifluoroethane).

Table 2.2.14 Saturated Refrigerant 123 (2,2-dichloro-1,1,1-trifluoroethane)

T	P	ρ_f	v_g	h_f	h_g	s_f	s_g	$C_{p,f}$	$C_{p,g}$	α_f	α_g	μ_f	μ_g	k_f	k_g	σ
°C	MPa	kg/m³	m³/kg	kJ/kg		kJ/(kg·K)		kJ/(kg·K)		m/s		μPa·s		mW/(m·K)		mN/m
−20.0	0.012	1573.8	1.136	180.4	369.5	0.926	1.673	0.97	0.62	881.	122.	735.	9.09	89.8	6.61	20.68
−10.0	0.020	1550.1	0.697	190.2	375.5	0.963	1.668	0.98	0.63	841.	124.	642.	9.47	86.7	7.18	19.43
0.0	0.033	1526.1	0.446	200.0	381.4	1.000	1.664	0.99	0.65	801.	125.	565.	9.84	83.7	7.74	18.20
5.0	0.041	1513.9	0.362	205.0	384.4	1.018	1.663	1.00	0.66	781.	126.	530.	10.02	82.2	8.03	17.59
10.0	0.051	1501.6	0.296	210.0	387.5	1.036	1.663	1.00	0.67	761.	127.	499.	10.20	80.7	8.31	16.98
15.0	0.062	1489.2	0.245	215.0	390.5	1.053	1.662	1.01	0.68	742.	127.	470.	10.38	79.3	8.60	16.38
20.0	0.076	1476.6	0.203	220.1	393.5	1.071	1.662	1.01	0.69	723.	128.	443.	10.56	77.8	8.89	15.78
25.0	0.091	1463.9	0.170	225.1	396.5	1.088	1.663	1.02	0.70	703.	129.	418.	10.74	76.4	9.18	15.19
27.8b	0.101	1456.6	0.155	228.0	398.2	1.098	1.663	1.02	0.70	693.	129.	404.	10.84	75.6	9.35	14.85
30.0	0.110	1451.0	0.144	230.3	399.5	1.105	1.663	1.03	0.71	684.	129.	394.	10.91	75.0	9.48	14.60
35.0	0.131	1438.0	0.122	235.4	402.5	1.122	1.664	1.03	0.71	666.	129.	373.	11.09	73.7	9.78	14.01
40.0	0.154	1424.8	0.104	240.6	405.5	1.138	1.665	1.04	0.72	647.	130.	352.	11.26	72.4	10.08	13.43
45.0	0.182	1411.4	0.0891	245.8	408.5	1.155	1.666	1.05	0.73	628.	130.	334.	11.43	71.1	10.39	12.85
50.0	0.212	1397.8	0.0767	251.1	411.5	1.171	1.668	1.05	0.75	610.	130.	316.	11.60	69.8	10.70	12.28
55.0	0.247	1384.0	0.0664	256.3	414.5	1.187	1.669	1.06	0.76	591.	130.	299.	11.77	68.5	11.02	11.72
60.0	0.286	1370.0	0.0578	261.7	417.4	1.203	1.671	1.07	0.77	573.	130.	284.	11.94	67.3	11.34	11.16
65.0	0.329	1355.7	0.0504	267.0	420.3	1.219	1.673	1.07	0.78	555.	130.	269.	12.11	66.1	11.67	10.60
70.0	0.377	1341.2	0.0442	272.4	423.2	1.235	1.674	1.08	0.79	536.	129.	256.	12.28	64.9	12.01	10.05
75.0	0.430	1326.4	0.0389	277.9	426.1	1.251	1.676	1.09	0.80	518.	129.	243.	12.45	63.8	12.36	9.51
80.0	0.489	1311.2	0.0343	283.4	428.9	1.266	1.678	1.10	0.82	500.	128.	231.	12.63	62.6	12.73	8.97
90.0	0.624	1279.9	0.0269	294.5	434.4	1.297	1.682	1.12	0.85	464.	127.	208.	12.98	60.4	13.48	7.91
100.0	0.786	1246.9	0.0213	305.8	439.8	1.327	1.686	1.14	0.88	427.	125.	188.	13.37	58.2	14.29	6.88
110.0	0.976	1211.9	0.0170	317.3	444.9	1.357	1.690	1.17	0.92	391.	123.	170.	13.80	56.0	15.17	5.88
120.0	1.199	1174.4	0.0136	329.2	449.7	1.387	1.694	1.21	0.96	354.	120.	153.	14.29	53.9	16.14	4.91
130.0	1.458	1133.6	0.0109	341.3	454.1	1.417	1.697	1.25	1.03	317.	116.	138.	14.89	51.7	17.22	3.97
140.0	1.756	1088.3	0.0088	353.9	457.9	1.448	1.699	1.32	1.11	279.	112.	124.	15.65	49.5	18.44	3.08
150.0	2.099	1036.8	0.0070	367.1	461.1	1.478	1.700	1.42	1.24	239.	106.	110.	16.68	47.2	19.87	2.23
160.0	2.490	975.7	0.0056	381.1	463.0	1.510	1.699	1.58	1.47	198.	99.	97.	18.19	44.8	21.63	1.44
170.0	2.937	896.9	0.0043	396.6	462.9	1.544	1.694	1.98	2.03	154.	91.	83.	20.71	42.3	24.05	0.73
180.0	3.451	765.9	0.0029	416.2	456.8	1.587	1.676	4.55	5.66	102.	81.	64.	26.59	39.7	28.82	0.14
183.7c	3.662	550.0	0.0018	437.4	437.4	1.633	1.633	∞	∞	0.	0.	–	–	∞	∞	0.00

b = normal boiling point; c = critical point.
Properties computed with NIST REFPROP version 9.1, based on the formulation of Younglove and McLinden, *J. Phys. Chem. Ref. Data* **23**:731-779 (1994).

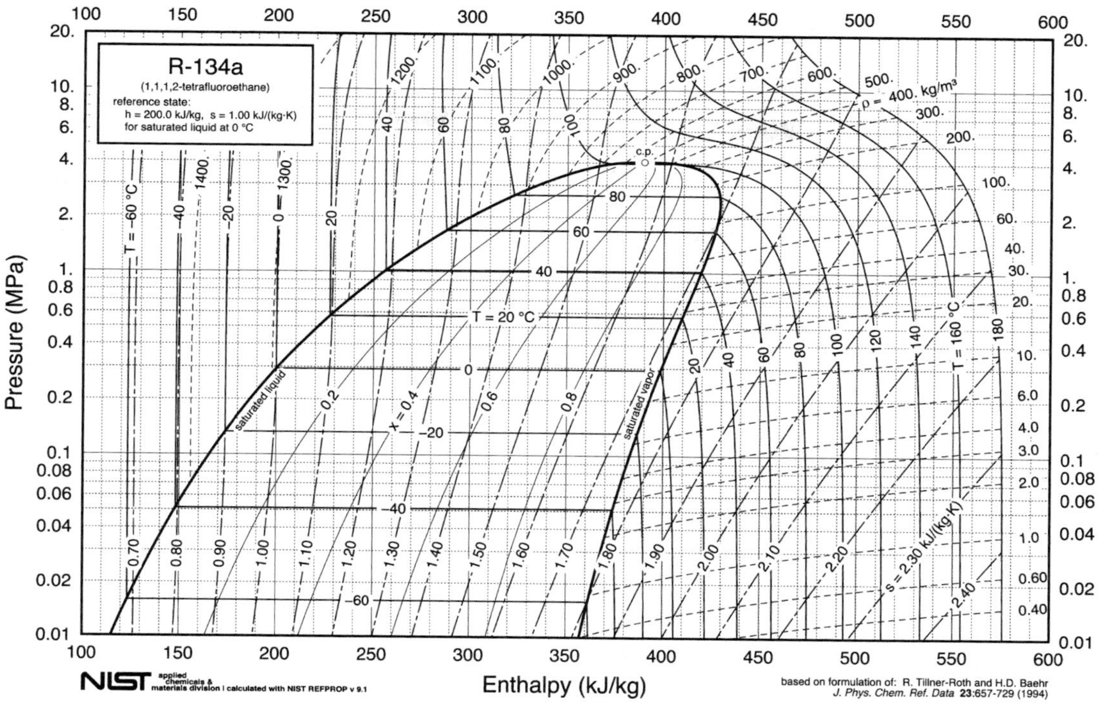

Figure 2.2.12 Enthalpy—log pressure diagram for refrigerant-134a (1,1,1,2-tetrafluoroethane).

Table 2.2.15 Saturated Refrigerant 134a (1,1,1,2-tetrafluoroethane)

T	P	ρ_f	v_g	h_f	h_g	s_f	s_g	$C_{p,f}$	$C_{p,g}$	α_f	α_g	μ_f	μ_g	k_f	k_g	σ
°C	MPa	kg/m³	m³/kg	kJ/kg		kJ/(kg·K)		kJ/(kg·K)		m/s		μPa·s		mW/(m·K)		mN/m
−103.3a	0.000	1591.1	35.496	71.5	334.9	0.413	1.964	1.18	0.59	1120.	127.	2154.	6.83	145.	3.08	27.38
−80.0	0.004	1529.0	4.268	99.2	348.8	0.565	1.858	1.20	0.64	1002.	134.	1020.	7.75	132.	4.95	23.56
−60.0	0.016	1474.3	1.079	123.4	361.3	0.685	1.801	1.22	0.69	903.	139.	660.5	8.52	121.	6.56	20.38
−50.0	0.029	1446.3	0.606	135.7	367.7	0.741	1.781	1.24	0.72	855.	142.	550.9	8.90	116.	7.36	18.82
−40.0	0.051	1417.7	0.361	148.1	374.0	0.796	1.764	1.26	0.75	807.	144.	467.0	9.27	111.	8.17	17.29
−30.0	0.084	1388.4	0.226	160.8	380.3	0.849	1.752	1.27	0.78	760.	145.	401.0	9.64	106.	8.99	15.78
−20.0	0.133	1358.3	0.147	173.6	386.6	0.900	1.741	1.29	0.82	714.	146.	347.6	10.00	101.	9.82	14.30
−26.1b	0.101	1376.7	0.190	165.8	382.8	0.869	1.747	1.28	0.79	742.	146.	378.7	9.78	104.	9.31	15.19
−15.0	0.164	1342.8	0.121	180.1	389.6	0.926	1.737	1.30	0.84	691.	147.	324.6	10.18	99.	10.23	13.57
−10.0	0.201	1327.1	0.100	186.7	392.7	0.951	1.733	1.32	0.85	668.	147.	303.6	10.36	97.	10.66	12.85
−5.0	0.243	1311.1	0.0828	193.3	395.7	0.975	1.730	1.33	0.88	645.	147.	284.3	10.54	94.	11.08	12.13
0.0	0.293	1294.8	0.0693	200.0	398.6	1.000	1.727	1.34	0.90	622.	147.	266.5	10.73	92.	11.51	11.43
5.0	0.350	1278.1	0.0584	206.8	401.5	1.024	1.725	1.36	0.92	599.	147.	250.1	10.91	90.	11.95	10.73
10.0	0.415	1261.0	0.0494	213.6	404.3	1.049	1.722	1.37	0.95	576.	146.	234.9	11.10	88.	12.40	10.04
15.0	0.488	1243.4	0.0421	220.5	407.1	1.072	1.720	1.39	0.97	553.	146.	220.7	11.29	85.	12.86	9.36
20.0	0.572	1225.3	0.0360	227.5	409.8	1.096	1.718	1.41	1.00	530.	145.	207.4	11.49	83.	13.33	8.69
25.0	0.665	1206.7	0.0309	234.6	412.3	1.120	1.716	1.43	1.03	506.	144.	194.9	11.69	81.	13.82	8.03
30.0	0.770	1187.5	0.0266	241.7	414.8	1.144	1.715	1.45	1.07	483.	143.	183.1	11.91	79.	14.34	7.38
35.0	0.887	1167.5	0.0230	249.0	417.2	1.167	1.713	1.47	1.10	460.	142.	172.0	12.13	77.	14.87	6.74
40.0	1.017	1146.7	0.0200	256.4	419.4	1.191	1.711	1.50	1.15	436.	140.	161.4	12.37	75.	15.45	6.11
45.0	1.160	1125.1	0.0173	263.9	421.5	1.214	1.709	1.53	1.19	413.	139.	151.4	12.63	73.	16.06	5.50
50.0	1.318	1102.3	0.0151	271.6	423.4	1.238	1.707	1.57	1.25	389.	137.	141.8	12.92	70.	16.73	4.90
55.0	1.492	1078.3	0.0131	279.5	425.2	1.261	1.705	1.61	1.31	365.	134.	132.5	13.23	68.	17.48	4.31
60.0	1.682	1052.9	0.0114	287.5	426.6	1.285	1.702	1.66	1.39	340.	132.	123.6	13.59	66.	18.33	3.74
70.0	2.117	996.2	0.0087	304.3	428.7	1.333	1.696	1.80	1.61	290.	126.	106.5	14.48	62.	20.47	2.64
80.0	2.633	928.2	0.0065	322.4	428.8	1.384	1.685	2.07	2.01	237.	118.	89.8	15.77	57.	23.74	1.63
90.0	3.244	837.8	0.0040	342.9	425.4	1.439	1.666	2.76	3.12	176.	108.	72.5	18.02	53.	29.82	0.73
100.0	3.972	651.2	0.0027	373.3	407.7	1.519	1.611	17.59	25.35	101.	94.	47.4	25.42	59.	58.98	0.04
101.1c	4.059	511.9	0.0020	389.6	389.6	1.562	1.562	∞	∞	0.	0.	–	–	∞	∞	0.00

a = triple point; b = normal boiling point; c = critical point.
Properties computed with NIST REFPROP version 9.1, based on the formulation of Tillner-Roth and Baehr, *J. Phys. Chem. Ref. Data* **23**:657-729 (1994).

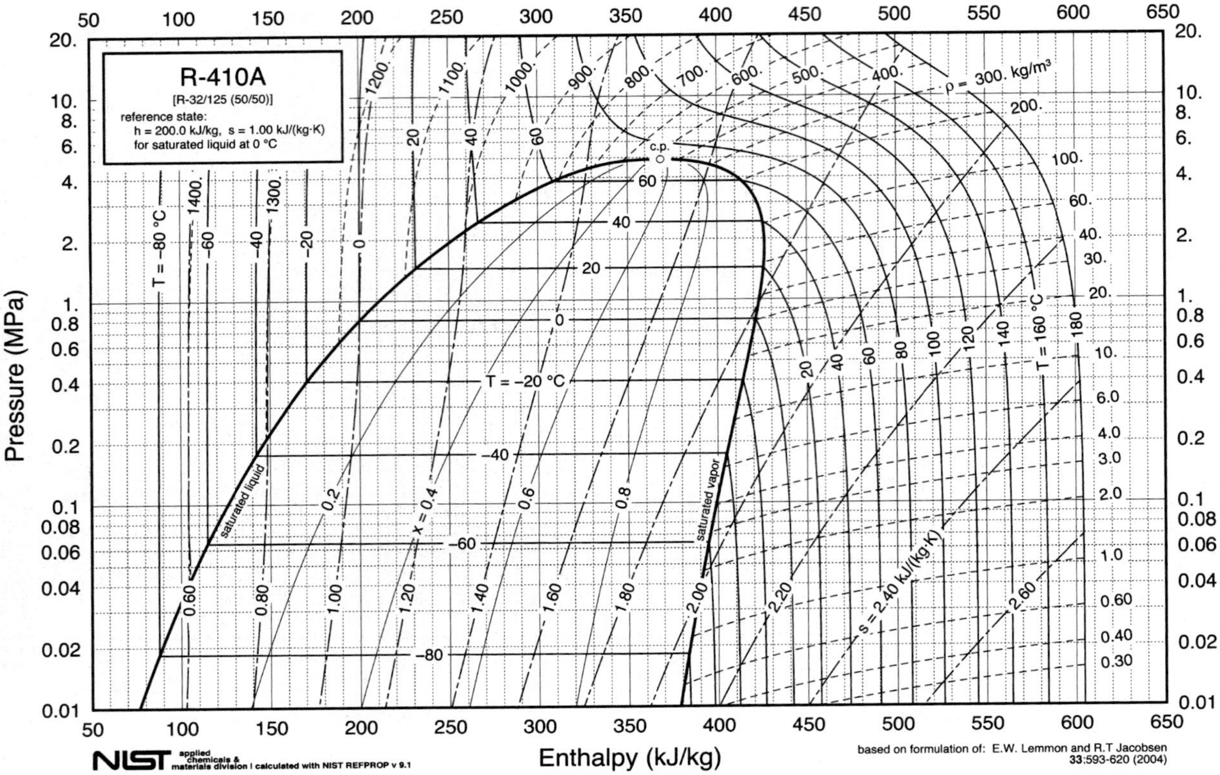

Figure 2.2.13 Enthalpy—log pressure diagram for refrigerant-410A [R-32/125 (50/50)].

Table 2.2.16 Saturated Refrigerant 410A [R32/125 (50/50)]

T	P	ρ_f	v_g	h_f	h_g	s_f	s_g	$C_{p,f}$	$C_{p,g}$	α_f	α_g	μ_f	μ_g	k_f	k_g	σ
°C	MPa	kg/m³	m³/kg	kJ/kg		kJ/(kg·K)		kJ/(kg·K)		m/s		μPa·s		mW/(m·K)		mN/m
−60.0	0.064	1376.3	0.369	114.7	394.7	0.650	1.965	1.36	0.77	866.	168.	353.	9.49	140.	8.28	19.37
−51.4b	0.101	1349.7	0.240	126.3	399.3	0.704	1.935	1.37	0.81	824.	170.	311.	9.90	134.	8.70	17.76
−50.0	0.109	1345.1	0.225	128.3	400.0	0.713	1.931	1.37	0.81	817.	170.	306.	9.95	133.	8.75	17.56
−40.0	0.175	1313.0	0.143	142.2	405.0	0.773	1.901	1.39	0.86	768.	171.	267.	10.41	127.	9.27	15.78
−35.0	0.219	1296.6	0.116	149.1	407.4	0.803	1.888	1.40	0.89	743.	172.	250.	10.64	124.	9.56	14.90
−30.0	0.270	1279.8	0.0948	156.2	409.8	0.832	1.875	1.41	0.92	718.	172.	235.	10.87	121.	9.85	14.04
−25.0	0.330	1262.7	0.0781	163.3	412.0	0.861	1.863	1.42	0.95	693.	172.	220.	11.10	118.	10.17	13.19
−20.0	0.400	1245.1	0.0649	170.4	414.1	0.889	1.852	1.44	0.98	668.	172.	207.	11.32	115.	10.51	12.34
−15.0	0.481	1227.1	0.0542	177.7	416.1	0.917	1.841	1.46	1.02	642.	172.	194.	11.55	112.	10.86	11.51
−10.0	0.574	1208.7	0.0456	185.0	418.0	0.945	1.830	1.47	1.05	616.	171.	183.	11.78	109.	11.25	10.69
−5.0	0.679	1189.6	0.0385	192.5	419.7	0.973	1.820	1.50	1.09	590.	171.	172.	12.00	106.	11.67	9.88
0.0	0.799	1170.0	0.0327	200.0	421.3	1.000	1.810	1.52	1.14	564.	170.	162.	12.23	103.	12.13	9.08
5.0	0.935	1149.6	0.0279	207.7	422.7	1.027	1.801	1.55	1.18	537.	169.	152.	12.48	100.	12.64	8.30
10.0	1.087	1128.4	0.0239	215.5	424.0	1.055	1.791	1.58	1.23	510.	167.	143.	12.76	97.	13.21	7.53
15.0	1.256	1106.3	0.0205	223.4	425.0	1.082	1.782	1.62	1.29	483.	166.	134.	13.05	95.	13.86	6.78
20.0	1.445	1083.1	0.0176	231.5	425.7	1.109	1.772	1.66	1.36	455.	164.	126.	13.35	92.	14.60	6.04
25.0	1.655	1058.6	0.0152	239.9	426.2	1.137	1.762	1.71	1.45	427.	162.	118.	13.68	89.	15.5	5.32
30.0	1.886	1032.7	0.0131	248.4	426.3	1.165	1.752	1.77	1.55	399.	159.	110.	14.05	86.	16.5	4.61
35.0	2.142	1005.1	0.0113	257.2	426.0	1.193	1.741	1.84	1.68	370.	157.	103.	14.45	84.	17.7	3.93
40.0	2.422	975.3	0.0097	266.3	425.3	1.221	1.729	1.94	1.84	340.	154.	96.	14.91	81.	19.2	3.27
45.0	2.730	942.9	0.0083	275.8	423.9	1.250	1.716	2.07	2.07	309.	150.	89.	15.45	78.	21.1	2.64
50.0	3.067	906.8	0.0071	285.9	421.7	1.280	1.701	2.26	2.40	277.	146.	82.	16.10	76.	23.6	2.03
55.0	3.435	865.5	0.0060	296.6	418.5	1.312	1.683	2.56	2.93	243.	141.	75.	16.92	73.	27.1	1.46
60.0	3.838	815.5	0.0050	308.4	413.5	1.346	1.662	3.15	3.93	206.	136.	67.	18.03	71.	32.2	0.93
71.3c	4.901	459.1	0.0022	368.6	368.6	1.518	1.518	∞	∞	0.	0.	–	–	∞	∞	0.00

* R-410A is a nearly azeotropic mixture; tabulated vapor pressures are the average of the bubble and dew point pressures.
b = normal boiling point; c = critical point.
Properties computed with NIST REFPROP version 9.1, based on the formulation of Lemmon and Jacobsen, *J. Phys. Chem. Ref. Data* **33**:593-620 (2004).

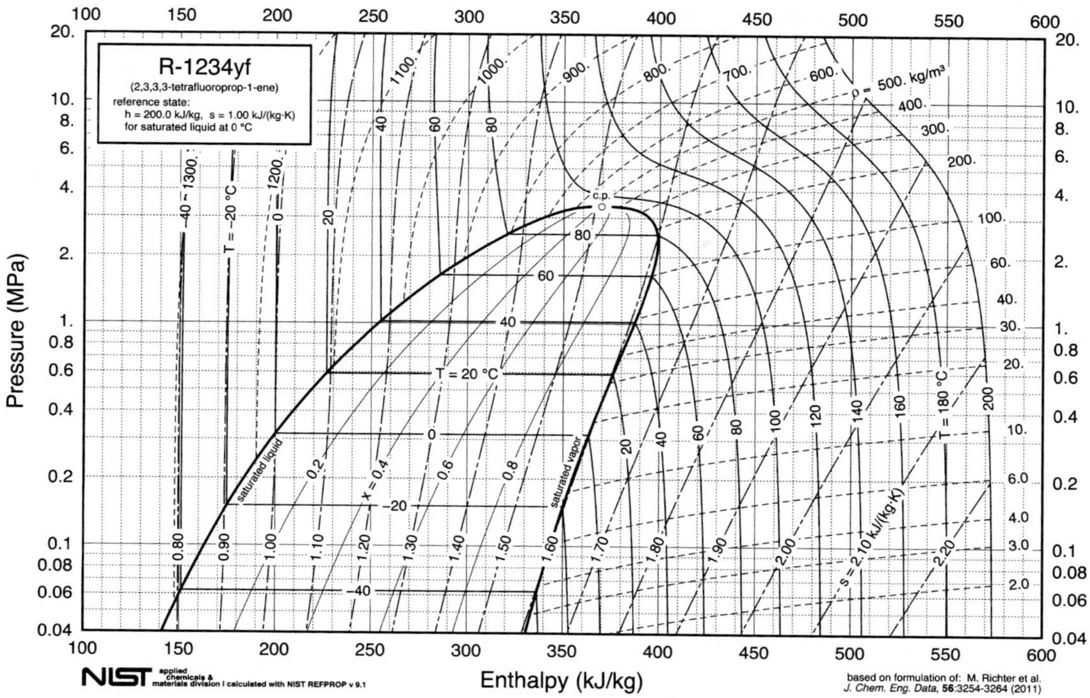

Figure 2.2.14 Enthalpy—log pressure diagram for refrigerant-1234yf (2,3,3,3-tetrafluoroprop-1-ene).

Table 2.2.17 Saturated Refrigerant 1234yf (2,3,3,3-tetrafluoroprop-1-ene)

T	P	ρ_f	v_g	h_f	h_g	s_f	s_g	$C_{p,f}$	$C_{p,g}$	α_f	α_g	μ_f	μ_g	k_f	k_g	σ
°C	MPa	kg/m³	m³/kg	kJ/kg		kJ/(kg·K)		kJ/(kg·K)		m/s		μPa·s		mW/(m·K)		mN/m
−50.0	0.037	1318.4	0.425	139.6	329.9	0.757	1.610	1.13	0.75	783.	132.	370.	7.68	89.0	7.64	17.09
−40.0	0.062	1291.9	0.264	151.1	336.6	0.807	1.603	1.16	0.78	736.	134.	327.	8.23	85.3	8.43	15.46
−30.0	0.099	1264.5	0.171	162.8	343.3	0.857	1.599	1.19	0.81	690.	135.	290.	8.75	81.7	9.22	13.89
−29.5b	0.101	1263.1	0.167	163.4	343.7	0.859	1.599	1.19	0.81	688.	135.	288.	8.78	81.5	9.26	13.81
−25.0	0.123	1250.5	0.139	168.8	346.7	0.881	1.598	1.20	0.83	668.	135.	274.	9.01	80.0	9.62	13.12
−20.0	0.151	1236.3	0.115	174.9	350.1	0.905	1.597	1.22	0.85	646.	136.	258.	9.25	78.2	10.02	12.36
−15.0	0.184	1221.8	0.0953	181.0	353.4	0.929	1.597	1.24	0.87	624.	136.	243.	9.50	76.5	10.41	11.62
−10.0	0.222	1207.0	0.0796	187.3	356.7	0.953	1.597	1.25	0.89	602.	136.	230.	9.74	74.8	10.82	10.88
−5.0	0.266	1191.8	0.0670	193.6	360.0	0.977	1.597	1.27	0.91	581.	136.	217.	9.98	73.1	11.22	10.17
0.0	0.316	1176.3	0.0567	200.0	363.3	1.000	1.598	1.29	0.93	559.	135.	205.	10.21	71.5	11.63	9.46
5.0	0.373	1160.4	0.0482	206.5	366.5	1.023	1.599	1.31	0.95	538.	135.	193.	10.45	69.8	12.05	8.77
10.0	0.438	1144.0	0.0412	213.1	369.7	1.047	1.600	1.33	0.97	516.	134.	182.	10.69	68.2	12.48	8.10
15.0	0.510	1127.2	0.0354	219.8	372.8	1.070	1.601	1.35	1.00	495.	134.	172.	10.93	66.6	12.92	7.44
20.0	0.592	1109.9	0.0305	226.6	375.9	1.093	1.602	1.37	1.02	473.	133.	162.	11.17	65.1	13.38	6.80
25.0	0.683	1091.9	0.0264	233.5	378.9	1.116	1.604	1.39	1.05	451.	131.	153.	11.43	63.5	13.86	6.17
30.0	0.784	1073.3	0.0229	240.5	381.8	1.139	1.605	1.42	1.09	429.	130.	144.	11.69	62.0	14.37	5.56
35.0	0.895	1054.0	0.0200	247.6	384.5	1.162	1.606	1.44	1.12	407.	128.	135.	11.96	60.5	14.92	4.97
40.0	1.018	1033.8	0.0173	254.9	387.2	1.185	1.608	1.47	1.17	385.	127.	127.	12.25	59.0	15.50	4.40
45.0	1.154	1012.6	0.0151	262.3	389.7	1.208	1.609	1.51	1.22	363.	124.	119.	12.56	57.6	16.15	3.85
50.0	1.302	990.4	0.0132	269.9	392.0	1.231	1.609	1.55	1.28	340.	122.	112.	12.89	56.1	16.87	3.32
55.0	1.465	966.7	0.0115	277.6	394.1	1.255	1.610	1.60	1.35	317.	119.	104.	13.26	54.7	17.69	2.82
60.0	1.642	941.3	0.0100	285.5	395.9	1.278	1.609	1.66	1.44	293.	116.	97.	13.67	53.3	18.65	2.33
65.0	1.835	913.7	0.0087	293.7	397.5	1.302	1.609	1.73	1.56	267.	113.	90.	14.15	51.8	19.79	1.88
70.0	2.045	883.2	0.0076	302.2	398.6	1.326	1.607	1.84	1.72	240.	109.	83.	14.70	50.4	21.19	1.45
75.0	2.272	848.8	0.0065	311.1	399.1	1.351	1.604	1.99	1.96	211.	105.	76.	15.38	48.9	23.00	1.06
80.0	2.519	809.0	0.0056	320.5	398.9	1.377	1.599	2.23	2.36	182.	100.	68.	16.26	47.6	25.48	0.71
85.0	2.788	760.4	0.0046	330.8	397.4	1.405	1.591	2.70	3.17	150.	95.	60.	17.47	46.4	29.23	0.39
90.0	3.080	694.1	0.0037	342.8	393.3	1.437	1.576	4.19	5.69	116.	88.	51.	19.49	46.6	36.40	0.14
94.7c	3.382	475.6	0.0021	369.6	369.6	1.509	1.509	∞	∞	0.	0.	–	–	∞	∞	0.00

b = normal boiling point; c = critical point.
Properties computed with NIST REFPROP version 9.1, based on the formulation of Richter *et al.*, *J. Chem. Eng. Data* **56**:3254-3264 (2011).

Table 2.2.18 Phase Transition and Other Data for 100 Fluids*

Name	Formula	M	T_m, K	Δh_{fus}, kJ/kg	T_b, K	Δh_{vap}, kJ/kg	P_c, bar	v_c, m³/kg	T_c, K	Z_c
Acetaldehyde	C_2H_4O	44.053	149.7	73.2	293.7	584.0	55.4	0.00382	461	0.243
Acetic acid	$C_2H_4O_2$	60.053		195.2	391.2	404.7	57.9	0.00285	594.5	0.200
Acetone	C_3H_6O	58.080	178.5	98.5	329.3	500.9	47.2	0.00360	508.2	0.232
Acetylene	C_2H_2	26.038	179.0	96.5	189.2	687.0	62.4	0.00435	308.3	0.276
Air	Mixed	28.966	60.0		79, 82	206.5	37.7	0.00320	132.6	0.263
Ammonia	NH_3	17.031	195.4	331.9	239.7	1,368.0	112.9	0.00427	405.7	0.244
Aniline	C_6H_7N	93.129	266.8	113.2	457.5	484.9	53.0	0.00340	698.8	0.289
Argon	A	39.948	83.8	29.4	87.5	163.2	48.7	0.00187	150.8	0.290
Benzene	C_6H_6	78.114	278.7	125.9	353.3	394.0	49.2	0.00332	562.1	0.273
Bromine	Br_2	159.81	264.9	66.2	331.6	187.7	103.4	0.00079	584.2	0.270
Butane, *n*	C_4H_{10}	58.124	113.7	78.2	261.5	366.4	36.5	0.00452	408.1	0.283
Butane, iso	C_4H_{10}	58.124	137.0	80.2	272.7	385.5	38.0	0.00439	425.2	0.274
Butanol	$C_4H_{10}O$	74.123	183.9	121.2	390.8	593.2	44.1	0.00370	562.9	0.259
Batylene	C_4H_8	56.108	87.8	68.6	266.9	391.0	40.2	0.00428	419.6	0.277
Carbon dioxide	CO_2	44.010	216.6	18.4	194.7	573.2	73.8	0.00214	304.1	0.274
Carbon disulfide	CS_2	76.131	161.1	57.7	319.4	351.6	79.0	0.00223	552.0	0.293
Carbon monoxide	CO	28.010	68.1	29.8	81.6	215.1	35.0	0.00332	132.9	0.295
Carbon tetrachloride	CCl_4	153.82	250.3	16.3	349.8	195.0	45.6	0.00170	556.4	0.258
Carbon tetrafluoride	CF_4	88.005	89.5	8.0	145.2	138.0	37.4	0.00156	228.0	0.272
Cesium	Cs	132.91	301.6	16.4	942.4	494.3	153.7	0.00230	2,043.0	0.240
Chlorine	Cl_2	70.906	172.2	90.4	238.6	287.5	77.0	0.00175	417.2	0.278
Chloroform	$CHCl_3$	119.38	209.7	77.1	334.5	248.5	54.5	0.00202	536.0	0.294
o-Cresol	C_7H_8O	108.14	303.8		464.1		50.0	0.00291	697.6	0.271
Cyclohexane	C_6H_{12}	84.162	279.8	31.7	353.9	357.4	40.7	0.00366	553.4	0.273
Cyclopropane	C_3H_6	42.081	145.5	129.4	240.3	477.0	55.0	0.00387	397.8	0.271
Decane	$C_{10}H_{22}$	142.29	243.4	201.8	447.3	276.2	21.0	0.00424	617.5	0.247
Deuterium	D_2	4.028	18.71	48.9	23.7	304.4	16.7	0.00143	38.34	0.301
Diphenyl	$C_{12}H_{10}$	154.21	342.4	120.6	527.6	317.3	38.5	0.00326	789.0	0.295
Ethane	C_2H_6	30.070	89.9	45.0	184.6	488.4	48.8	0.00486	305.4	0.281
Ethanol	C_2H_6O	46.069	159.0	109.0	351.5	840.9	63.8	0.00362	516.2	0.248
Ethyl acetate	$C_4H_8O_2$	88.107	189.4	119.0	350.3	366.2	38.3	0.00325	523.2	0.252
Ethyl bromide	C_2H_5Br	108.97	153.5	54.0	311.5	218.0	62.3	0.00197	503.9	0.320
Ethyl chloride	C_2H_5Cl	64.515	134.9	68.9	285.4	382.5	52.0	0.00309	460.4	0.270
Ethyl ether	$C_4H_{10}O$	74.123	150.0	93.1	307.8	358.9	36.8	0.00377	466.8	0.262
Ethyl formate	$C_3H_6O_2$	74.080	193.8	124.3	327.4	405.8	47.4	0.00310	508.5	0.257
Ethylene	C_2H_4	28.054	104.0	119.5	169.5	480.0	50.5	0.00455	283.1	0.274
Ethylene oxide	C_2H_4O	44.054	160.6	117.5	283.6	580.0	71.9	0.00318	468.9	0.258
Fluorine	F_2	37.997	53.5	13.4	85.1	172.1	52.2	0.00174	144.3	0.288
Helium 4	He	4.003			4.3	20.6	2.3	0.0144	5.189	0.303
Heptane	C_7H_{16}	100.20	182.5	139.9	371.6	316.3	27.4	0.0043	540.2	0.263
Hexane	C_6H_{14}	86.178	177.8	151.2	341.9	334.8	30.3	0.0043	507.6	0.265
Hydrazine	N_2H_4	32.045	274.7	395.0	386.7	1,207.0	147.0		653.2	0.284
Hydrogen	H_2	2.016	14.0	58.0	20.4	454.0	13.0	0.0323	33.3	0.305
Hydrogen bromide	HBr	80.912	186.3	37.4	206.4	217.5	85.5	0.00124	363.2	0.284
Hydrogen chloride	HCl	36.461	160.0	54.7	188.1	443.0	83.1	0.00022	324.7	0.249
Hydrogen fluoride	HF	20.006	181.8	196.3	272.7	374.3	64.9	0.00345	461.2	0.120
Hydrogen iodide	HI	127.91	222.4	22.4	237.8	154.0	83.1	0.00106	423.9	0.318
Hydrogen sulfide	H_2S	34.076	187.5	69.8	213.0	248.0	89.4	0.00289	373.1	0.283
Iodine	I_2	253.81	387.0	62.1	457.5	164.3	117.5	0.00054	785.0	0.248
Krypton	Kr	83.80	116.0	19.5	121.4	107.9	55.0	0.00109	209.4	0.288
Lithium	Li	6.940	453.8		1,615.0	1,945.0			3,750.0	
Mercury	Hg	200.59	234.3	11.4	630.1	295.6	1,510.0	0.00018	1,763.0	
Methane	CH_4	16.043	90.7	58.4	111.5	511.8	46.0	0.00617	190.5	0.287
Methanol	CH_4O	32.042	175.5	98.9	337.7	1,104.0	79.5	0.00368	512.6	0.220
Methyl acetate	$C_3H_6O_2$	74.080	175.0		330.3	410.0	46.9	0.00308	506.9	0.254
Methyl bromide	CH_3Br	94.939	179.5	62.8	276.7	252.0		0.00173	467.2	

* M = molecular weight, T_m = normal melting temperature, Δh_{fus} = enthalpy (or latent heat) of fusion, T_b = normal boiling point temperature, Δh_{vap} = enthalpy (or latent heat) of vaporization, P_c = critical pressure, v_c = critical volume, T_c = critical temperature, Z_c = critical compressibility factor.

SOURCE: Prepared by P.E. Liley, Purdue University and abstracted from Rohsenow et al., "Handbook of Heat Transfer Fundamentals," McGraw-Hill.

Table 2.2.18 Phase Transition and Other Data for 100 Fluids* *(Continued)*

Name	Formula	M	T_m, K	Δh_{fus}, kJ/kg	T_b, K	Δh_{vap}, kJ/kg	P_c, bar	v_c, m³/kg	T_c, K	Z_c
Methyl chloride	CH_3Cl	50.49	175.4	127.4	249.4	428.5	66.8	0.00270	416.3	0.277
Methyl formate	$C_2H_4O_2$	60.053	173.4	125.5	304.7	481.2	60.0	0.00287	487.2	0.255
Methylene chloride	CH_2Cl_2	84.922	176.5	54.4	312.9	328.0	61.3	0.00197	510.	0.255
Naphthalene	$C_{10}H_8$	128.17	353.4	148.1	491.1	341	40.5	0.00321	748.4	0.270
Neon	Ne	20.179	24.5	16.6	27.3	91.3	27.6	0.00207	44.4	0.311
Nitric oxide	NO	30.006	111.0	76.6	121.4	460	64.9	0.00192	180.0	0.249
Nitrogen	N_2	28.013	63.1	25.7	77.3	197.6	34.0	0.00318	126.2	0.287
Nitrogen peroxide	NO_2	46.006	263.0	159.5	294.5	414.4	101.3	0.00180	431.4	0.233
Nitrous oxide	N_2O	44.013	176.0	148.6	184.7	376.0	72.4	0.00221	309.6	0.274
Octane	C_8H_{18}	114.23	216.4	161.6	398.9	302.7	25.0	0.00426	508.9	0.258
Oxygen	O_2	31.999	54.4	13.9	90.0	212.5	50.4	0.00229	154.6	0.288
Pentane, iso	C_5H_{12}	72.151	113.7	71.1	301.0	341.0	33.5	0.00427	460.4	0.270
Pentane, *n*	C_5H_{12}	72.151	143.7	116.6	309.2	357.2	33.7	0.00431	469.6	0.268
Potassium	K	39.098	336.4	59.8	1,030.0		167.0		2,265.0	
Propane	C_3H_8	44.097	86.0	80.0	231.1	425.7	42.6	0.00453	369.8	0.277
Propanol	C_3H_8O	60.096	147.0	86.5	370.4	695.8	51.7	0.00364	536.7	0.253
Propylene	C_3H_6	42.081	87.9	71.4	225.5	437.5	46.0	0.00429	365.1	0.275
Radon	Rn	224	201.0	12.3	211.0	82.8	65.5	0.00063	377.0	0.293
Refrigerant 11	$CFCl_3$	137.37	162.2	50.2	296.9	180.2	44.1	0.00181	471.2	0.280
Refrigerant 12	CF_2Cl_2	120.91	115.4	34.3	243.4	165.1	41.2	0.00179	385.2	0.278
Refrigerant 13	CF_3Cl	104.46	92.1		191.7	148.4	38.7	0.00173	302.0	0.279
Refrigerant 13B1	CF_3Br	148.91	105.4		215.4	118.7	39.6	0.00134	340.2	0.280
Refrigerant 21	$CHFCl_2$	102.91	138.2		282.1	242.1	51.7	0.00192	451.4	0.271
Refrigerant 22	CHF_2Cl	86.469	113.2	47.6	232.4	233.6	49.8	0.00191	369.2	0.267
Refrigerant 23	CHF_3	70.014	118.0	58.0	191.2	239.0	48.4	0.00190	299.1	0.259
Refrigerant 113	$C_2F_3Cl_3$	187.38	238.2		320.8	146.8	34.1	0.00174	487.3	0.274
Refrigerant 114	$C_2F_4Cl_2$	170.92	179.2		276.7	136.0	32.6	0.00172	418.9	0.275
Refrigerant 115	C_2F_5Cl	154.47	171.0	12.2	234.0	124.1	31.6	0.00163	353.1	0.271
Refrigerant 142b	$C_2F_2H_3Cl$	100.50	142.4	26.7	263.9	223	41.5	0.00232	410.0	0.279
Refrigerant 152a	$C_2F_2H_4$	66.051	156		248	326	45.0	0.00274	386.7	0.253
Refrigerant 216	$C_3F_6Cl_2$	220.93			308	117.3	27.5	0.00174	453.2	0.281
Refrigerant C318	C_4F_8	200.03	233.0		267	116	27.8	0.00161	388.5	0.272
Refrigerant 500	Mix	99.303	114.3		239.7	201.1	44.3	0.00201	378.7	0.281
Refrigerant 502	Mix	111.63			237	172.2	40.7	0.00178	355.3	0.275
Refrigerant 503	Mix	87.267			184	172.9			293	
Refrigerant 504	Mix	79.240			216	242.9			339	
Refrigerant 505	Mix	103.43			243.6		47.3		391	
Refrigerant 506	Mix	93.69			260.7		51.6		416	
Rubidium	Rb	85.468	312.6		959.4	811.3		0.00288	2,083	
Sodium	Na	22.990	371.0		1,155	3,880			2,730	
Sulfur dioxide	SO_2	64.059	197.8		268.4	368.3	78.8	0.00190	430.7	0.269
Sulfur hexafluoride	SF_6	146.051					37.8	0.00137	318.7	0.285
Toluene	C_7H_8	92.141	178.2		383.8	339.0	40.5	0.00345	594.0	0.260
Water	H_2O	18.015	273.2		373.2	2,256	221.2	0.00315	647.3	0.234
Xenon	Xe	131.36	161.5		165	99.2	58.7	0.00091	290	0.290

* M = molecular weight, T_m = normal melting temperature, Δh_{fus} = enthalpy (or latent heat) of fusion, T_b = normal boiling point temperature, Δh_{vap} = enthalpy (or latent heat) of vaporization, P_c = critical pressure, v_c = critical volume, T_c = critical temperature, Z_c = critical compressibility factor.

SOURCE: Prepared by P.E. Liley, Purdue University and abstracted from Rohsenow et al., "Handbook of Heat Transfer Fundamentals," McGraw-Hill.

Table 2.2.19 Specific Heat at Constant Pressure [kJ/(kg K)] of Liquids and Gases

Substance	Temperature, K												
	200	225	250	275	300	325	350	375	400	425	450	475	500
Acetylene	1.457	1.517	1.575	1.635	1.695	1.747							
Air	1.048	1.036	1.029	1.025	1.021	1.021	1.020	1.021	1.022	1.024	1.027	1.030	1.035
Ammonia	4.605	4.360	2.210	2.176	2.169	2.183	2.210	2.247	2.289	2.331	2.381	2.429	2.477
Argon	0.524	0.523	0.522	0.522	0.522	0.521	0.521	0.521	0.521	0.521	0.521	0.521	0.521
Butane, *i*	1.997	2.099	2.207	1.590	1.694	1.810	1.921	2.035	2.149	2.258	2.367	2.436	2.571
Butane, *n*	2.066	2.134	2.214	1.642	1.731	1.835	1.941	2.047	2.154	2.253	2.361	2.461	2.558
Carbon dioxide		0.763	0.791	0.818	0.845	0.870	0.894	0.917	0.938	0.958	0.978	0.995	1.014
Ethane	1.419	1.492	1.574	1.658	1.762	1.871	1.968	2.081	2.184	2.287	2.393	2.492	2.590
Ethylene	1.296	1.340	1.397	1.475	1.560	1.656	1.748	1.839	1.928	2.014	2.097	2.179	2.258
Fluorine	0.785	0.793	0.803	0.815	0.827								
Helium	5.193	5.193	5.193	5.193	5.193	5.193	5.193	5.193	5.193	5.193	5.193	5.193	5.193
Hydrogen, *n*	13.53	13.83	14.05	14.20	14.31	14.38	14.43	14.46	14.48	14.49	14.50	14.50	14.51
Krypton	0.252	0.251	0.250	0.249	0.249	0.249	0.249	0.249	0.249	0.249	0.249	0.249	0.248
Methane	2.105	2.122	2.145	2.184	2.235	2.297	2.375	2.454	2.534	2.617	2.709	2.797	2.892
Neon	1.030	1.030	1.030	1.030	1.030	1.030	1.030	1.030	1.030	1.030	1.030	1.030	1.030
Nitrogen	1.043	1.042	1.042	1.041	1.041	1.041	1.042	1.043	1.045	1.047	1.050	1.053	1.056
Oxygen	0.915	0.914	0.915	0.917	0.920	0.924	0.929	0.935	0.942	0.949	0.956	0.964	0.972
Propane	2.124	2.220	1.500	1.596	1.695	1.805	1.910	2.030	2.140	2.249	2.355	2.458	2.558
Propylene	2.094	2.132	1.436	1.481	1.536	1.625	1.709	1.793	1.890	1.981	2.070	2.153	2.245
Refrigerant 12			0.549	0.578	0.602	0.625	0.647	0.666	0.685	0.701	0.716	0.728	0.741
Refrigerant 21	0.973	0.982	0.991	1.015	0.594	0.617	0.641	0.661	0.683	0.701	0.720	0.735	0.752
Refrigerant 22	1.065	1.085	0.588	0.617	0.647	0.676	0.704	0.729	0.757	0.782	0.806	0.826	0.848
Water substance	1.545	1.738	1.935	4.211	4.179	4.182	4.195	2.035	1.996	1.985	1.981	1.980	1.983
Xenon	0.165	0.163	0.162	0.161	0.160	0.160	0.160	0.159	0.159	0.159	0.159	0.159	0.159

Values for the saturated liquid are tabulated up to the normal boiling point. Higher temperature values are for the dilute gas. Values for water substance below 275 K are for ice.

Table 2.2.20 Specific Heat Ratio c_p/c_v for Gases at Atmospheric Pressure

Substance	Temperature, K												
	200	225	250	275	300	325	350	375	400	425	450	475	500
Acetylene	1.313	1.289	1.269	1.250	1.234	1.219	1.205						
Air	1.399	1.399	1.399	1.399	1.399	1.399	1.398	1.397	1.395	1.393	1.391	1.388	1.386
Ammonia					1.327	1.302	1.295	1.285	1.278	1.269	1.262	1.256	1.249
Argon	1.663	1.665	1.666	1.666	1.666	1.666	1.666	1.666	1.666	1.666	1.666	1.666	1.666
Butane, *i*	1.357	1.356	1.359	1.116	1.103	1.094	1.086	1.080	1.075	1.070	1.066	1.063	1.060
Butane, *n*	1.418	1.412	1.407	1.114	1.103	1.094	1.086	1.080	1.075	1.071	1.067	1.063	1.061
Carbon dioxide		1.344	1.323	1.302	1.290	1.279	1.269	1.260	1.252	1.245	1.239	1.233	1.229
Carbon monoxide	1.405	1.404	1.403	1.402	1.401	1.401	1.400	1.398	1.396	1.394	1.392	1.390	1.387
Ethane		1.246	1.226	1.210	1.193	1.180	1.167	1.157	1.148	1.139	1.132	1.126	1.120
Ethylene			1.275	1.254	1.236	1.220	1.206	1.194	1.183	1.173	1.165	1.158	1.151
Fluorine	1.393	1.386	1.377	1.370	1.362								
Helium	1.667	1.667	1.667	1.667	1.667	1.667	1.667	1.667	1.667	1.667	1.667	1.667	1.667
Hydrogen, *n*	1.439	1.426	1.415	1.410	1.406	1.403	1.401	1.400	1.399	1.398	1.398	1.397	1.397
Krypton	1.649	1.655	1.658	1.660	1.662	1.662	1.662	1.662	1.662	1.663	1.664	1.665	1.667
Methane	1.337	1.332	1.325	1.316	1.306	1.295	1.282	1.271	1.258	1.247	1.237	1.228	1.219
Neon	1.667	1.667	1.667	1.667	1.667	1.667	1.667	1.667	1.667	1.667	1.667	1.667	1.667
Nitrogen	1.399	1.399	1.399	1.399	1.399	1.399	1.399	1.398	1.397	1.396	1.394	1.393	1.391
Oxygen	1.398	1.397	1.396	1.395	1.394	1.391	1.388	1.385	1.381	1.377	1.373	1.369	1.365
Propane	1.513	1.504	1.164	1.148	1.135	1.124	1.114	1.107	1.100	1.094	1.089	1.085	1.081
Propylene			1.171	1.160	1.150	1.140	1.131	1.123	1.116	1.110	1.105	1.100	1.096
Refrigerant 12			1.165	1.114	1.101	1.088	1.077	1.065	1.055	1.044	1.034	1.025	1.071
Refrigerant 21					1.179	1.164	1.152	1.144	1.137	1.132	1.127	1.124	1.120
Refrigerant 22				1.207	1.190	1.172	1.164	1.155	1.148	1.143	1.138	1.133	1.129
Steam								1.322	1.319	1.317	1.314	1.312	1.309
Xenon	1.623	1.634	1.642	1.650	1.655	1.659	1.662	1.662	1.662	1.662	1.662	1.662	1.662

Table 2.2.21 Saturation Temperature, in Kelvins, of Selected Substances

Pressure, bar	Acetone, C_3H_6O	Acetylene, C_2H_2	Air, sat. liquid	Air, sat. vapor	Ammonia, NH_3	Aniline, C_6H_7N	Argon, Ar	Benzene, C_6H_6	Butanol, $C_4H_{10}O$	*n*-Butane, C_4H_{10}	Carbon dioxide, CO_2	Carbon tetrachloride, CCl_4	Carbon tetrafluoride, CF_4
0.010	237.8					337.2			299.1				
0.015	243.8					344.2			305.4	196.5		255.1	
0.020	247.1					350.0			310.0	200.3		259.7	106.4
0.025	251.6					353.6			313.8	203.1		263.2	107.9
0.030	254.3					357.1			316.9	205.7		266.0	109.3
0.04	259.0					363.1		276.5	321.8	209.8		270.9	111.5
0.05	262.7					368.1		280.0	325.6	212.9		275.0	113.2
0.06	266.8					372.3		283.2	328.8	215.8		278.5	114.7
0.08	270.9				198.9	379.0		288.6	333.9	220.3		284.6	117.2
0.10	275.1		62.3	66.3	201.9	385.2		293.0	337.8	224.1		289.1	119.1
0.15	283.2		64.6	68.5	207.6	397.1		301.7	345.8	231.0		298.1	122.9
0.20	289.6		66.4	70.2	211.8	404.5		308.3	351.8	236.3		304.8	125.8
0.25	294.1		67.9	71.5	215.2	410.6		313.7	356.8	240.7		310.2	128.1
0.30	298.0		69.1	72.7	218.1	415.8		318.2	360.8	244.4		314.9	130.1
0.40	304.7		71.2	74.6	222.8	424.8		325.8	367.4	250.5		322.2	133.3
0.5	310.1		72.8	76.2	226.6	432.1		331.8	372.8	255.4		327.8	136.0
0.6	315.0		74.3	77.5	229.9	438.2		337.1	377.2	259.6		333.1	138.2
0.8	322.8		76.6	79.7	235.2	448.8	85.1	345.7	384.3	266.6		342.0	142.0
1.0	329.0		78.6	81.6	239.6	456.5	87.2	352.7	390.0	272.2		349.2	145.0
1.5	341.0	195.1	82.3	85.3	247.9	472.1	91.2	366.5	402.6	283.4		363.8	151.0
2.0	350.1	200.1	85.2	88.1	254.3	484.2	94.3	377.0	412.0	291.9		374.7	155.6
2.5	358.9	204.8	87.6	90.4	259.5	494.0	96.8	385.7	419.8	299.0		383.0	159.3
3.0	365.9	208.8	89.7	92.4	263.9	502.0	99.0	393.1	426.1	305.0		390.2	162.1
4.0	377.1	215.3	93.1	95.7	271.3	516.0	102.6	405.6	436.8	315.1		402.0	168.0
5.0	386.1	220.9	96.0	98.4	277.3	527.0	105.8	415.8	445.1	323.4		412.8	172.5
6	393.9	225.8	98.4	100.8	282.4	536.2	108.4	424.7	452.9	330.6	220.0	422.0	175.5
8	406.4	233.9	102.7	104.8	291.0	551.5	110.8	439.9	465.4	342.6	227.1	433.0	182.9
10	417.3	240.3	106.1	108.1	298.1	564.7	116.6	451.6	475.8	352.6	233.0	449.5	187.7
15	435.9	252.5	113.1	114.7	311.9	593.0	124.0	475.8	496.0	372.2	244.6	474.3	199.0
20	452.0	262.3	118.5	119.9	322.5	613.6	129.8	494.6	510.1	387.6	253.6	495.0	207.4
25	465.8	270.3	123.0	124.1	331.3	631.2	134.6	510.1		400.2	260.6	511.2	214.3
30	476.9	277.6	127.0	227.8	338.9	645.8	138.7	523.5		410.9	269.6	524.0	220.2
40	496.0	289.5			351.6	672.0	145.7	545.8			278.5	547.6	
50		299.0			362.1	693.0					287.4		
60		306.9			371.1						295.1		
80					386.1								
100					398.4								

Table 2.2.21 Saturation Temperature, in Kelvins, of Selected Substances *(Continued)*

Pressure, bar	Chlorine, Cl_2	Cyclohexane, C_6H_{12}	Cyclopropane, C_3H_6	Diphenyl, $C_{12}H_{10}$	Ethane, C_2H_6	Ethanol, C_2H_6O	Ethyl chloride, C_2H_5Cl	Ethylene, C_2H_4	Fluorine, F_2	Helium, He	Heptane, C_7H_{16}	Hexane, C_6H_{14}	Hydrazine, N_2H_4
0.010			172.9	384.3	127.6	266.7		117.4	58.2		266.9	244.2	285.9
0.015	173.4		176.9	393.5	131.1	272.7	207.9	120.4	59.8		273.2	250.0	293.2
0.020	176.6		179.9	400.1	133.6	277.1	212.0	122.8	61.0		277.9	254.2	298.3
0.025	179.1		182.3	405.9	135.7	280.6	215.1	124.7	62.0		281.8	257.7	302.4
0.030	181.2		184.4	410.2	137.3	283.1	217.6	126.1	62.7		285.0	260.7	306.0
0.04	184.7		187.7	418.0	140.4	287.7	221.9	128.8	64.1		290.1	265.6	311.7
0.05	186.4		190.1	423.8	142.5	291.2	225.0	130.8	65.1		294.4	269.7	315.9
0.06	189.8	282.2	192.5	428.9	144.6	294.3	227.8	132.6	66.1	2.25	298.1	273.0	319.7
0.08	193.8	288.0	196.2	436.8	147.6	299.2	232.4	136.9	67.6	2.38	304.2	278.4	325.5
0.10	197.0	292.6	199.2	443.6	150.4	303.1	236.1	138.0	68.8	2.49	309.0	282.6	330.1
0.15	203.0	301.4	205.0	456.3	155.3	310.4	243.4	143.0	71.1	2.71	318.3	291.5	339.1
0.20	207.7	308.0	209.5	466.1	158.9	315.7	248.8	146.6	72.9	2.82	325.1	298.3	345.9
0.25	211.4	313.6	213.2	473.9	161.9	320.1	253.1	149.3	74.3	3.03	330.5	304.0	350.8
0.30	214.8	318.1	216.1	480.4	164.6	323.9	256.9	151.7	75.5	3.15	335.0	308.6	355.2
0.40	219.9	325.7	221.2	491.5	168.8	330.1	263.1	155.3	77.6	3.37	343.0	316.6	362.3
0.5	224.0	331.8	225.1	499.6	172.4	335.2	267.9	158.3	79.2	3.55	349.2	321.4	368.1
0.6	227.7	337.3	228.7	506.6	175.3	339.2	272.1	161.0	80.6	3.71	354.8	326.4	372.8
0.8	233.8	346.9	234.6	518.2	180.2	345.9	279.5	165.6	82.9	3.98	363.9	334.6	380.2
1.0	238.8	353.7	239.3	528.7	184.3	351.4	285.1	169.2	84.8	4.21	371.0	341.3	386.9
1.5	248.3	368.3	249.0	548.0	192.2	362.3	297.6	176.5	88.6	4.67	385.3	355.2	399.1
2.0	255.9	378.9	260.0	562.2	198.2	370.6	306.9	181.9	91.4	5.03	396.7	361.0	408.1
2.5	262.1	387.7	269.0	573.8	203.1	377.5	314.6	186.3	93.7		406.0	373.7	415.6
3.0	267.2	395.0	275.0	583.8	207.4	382.9	319.6	190.6	95.8		413.5	381.7	422.0
4.0	275.9	407.9	282.0	600.3	214.5	392.0	329.4	197.7	99.2		426.0	393.3	432.4
5.0	283.0	418.6	287.0	614.3	220.4	399.1	337.0	202.9	102.0		431.9	404.0	441.2
6	289.2	427.8	292.0	626.6	225.5	405.0	343.5	207.2	104.4		441.5	412.2	449.0
8	290.6	443.0	301.0	646.6	234.0	415.6	354.7	215.0	108.4		462.1	427.0	462.5
10	307.8	455.2	309.0	664.2	241.1	424.1	364.1	221.2	111.8		474.9	438.9	473.7
15	325.0	481.1	327.0	697.8	255.1	442.4	385.9	234.1	118.5		498.9	462.0	494.3
20	338.0	499.9	342.0	722.5	266.0	456.0	400.4	243.9	123.7		518.9	479.1	510.2
25	349.4	516.3	355.0	744.2	275.1	467.0	412.8	252.3	128.1		535.0	495.0	522.2
30	359.2	528.8	365.0	763.4	282.9	476.0	422.2	259.9	131.8			506.3	531.7
40	376.0	551.1	379.0	794.0	296.0	490.4	442.4	272.7	138.1				547.8
50	389.0			818.2		502.8	456.5	282.7	143.3				561.8
60	400.7					513.4							574.0
80													598.0
100													616.2

Table 2.2.21 Saturation Temperature, in Kelvins, of Selected Substances (*Continued*)

Pressure, bar	*n*-Hydrogen, H_2	Hydrochloric acid, HCl	Hydrogen sulfide, H_2S	Mercury, Hg	Methane, CH_4	Methanol	Methyl chloride, CH_3Cl	Naphthalene	Nitrogen, N_2	Octane, C_8H_{18}	Oxygen, O_2	Pentane, C_5H_{12}	Potassium, K
0.010				448.6		252.6	173.2	353.0		286.4	62.2	219.1	699
0.015				460.5		258.2	178.4	361.2		293.8	63.0	225.0	719
0.020				468.9		262.3	182.6	367.6		299.4	64.3	228.8	732
0.025				475.3		265.6	185.9	372.8		303.7	65.2	232.0	743
0.030				481.6		268.4	188.6	376.9		306.8	66.2	234.7	752
0.04				491.2		272.9	192.7	384.3		312.4	67.6	239.2	764
0.05				498.2		276.5	196.0	389.9		316.9	68.8	242.8	776
0.06				504.7		278.8	198.6	394.8		320.8	69.8	245.9	788
0.08	14.0			515.2		284.4	202.8	402.7		327.1	71.4	250.8	811
0.10	14.4			523.6		288.4	206.0	408.9		332.0	72.7	254.9	829
0.15	15.2	161.0		538.9	92.6	295.9	211.9	420.9	64.1	341.6	75.2	263.0	861
0.20	15.8	164.5		551.0	95.0	301.4	216.5	429.5	65.8	348.7	77.1	269.1	883
0.25	16.3	167.2	189.8	561.0	97.0	305.9	220.5	437.1	67.2	354.6	78.7	274.0	899
0.30	16.7	169.3	192.5	569.0	98.6	309.7	223.8	443.3	68.3	359.7	80.0	277.9	916
0.40	17.5	172.7	197.0	582.2	101.4	315.9	229.1	454.0	70.2	368.3	82.2	284.9	940
0.5	18.1	175.2	200.4	592.7	103.7	320.8	233.4	462.1	71.8	374.9	83.9	290.0	960
0.6	18.6	177.5	203.1	601.8	105.6	325.0	237.0	469.4	73.2	380.6	85.5	294.7	977
0.8	19.5	181.0	208.0	616.8	108.8	331.9	243.1	481.1	75.4	390.2	88.0	302.1	1,006
1.0	20.2	187.8	212.5	628.9	111.5	337.5	248.0	490.2	77.2	398.0	90.1	308.6	1,029
1.5	21.7	195.6	222.0	652.1	116.6	348.1	257.9	509.0	80.8	413.3	94.1	321.6	1,074
2.0	22.8	201.5	227.9	670.0	120.6	356.1	265.4	523.3	83.6	424.6	97.2	331.1	1,108
2.5	23.8	206.1	234.3	684.5	123.9	362.6	271.5	535.0	85.9	434.4	99.8	339.2	1,137
3.0	24.6	210.0	237.9	696.7	126.7	368.1	276.6	544.8	87.9	442.1	102.0	345.7	1,176
4.0	26.0	216.7	245.1	717.1	131.4	377.2	285.5	560.3	91.2	456.3	105.7	356.7	1,203
5.0	27.1	222.0	250.9	733.5	135.3	384.6	292.6	573.2	94.0	467.4	108.8	365.4	1,238
6	28.1	226.2	256.1	747.8	138.7	390.9	298.9	584.7	96.4	477.0	111.4	373.6	1,268
8	29.8	233.6	264.9	770.8	144.4	401.4	309.4	605.0	100.4	494.0	115.9	387.6	1,318
10	31.2	241.6	272.1	790.2	149.1	410.0	318.1	620.7	103.8	507.0	119.6	398.2	1,359
15		254.2	283.3	827.5	158.8	426.5	332.7	652.6	110.4	532.4	127.0	420.7	1,440
20		263.7	299.8	857.2	165.8	439.1	348.6	677.9	115.6	553.6	132.8	436.8	1,505
25		272.3	309.4	879.2	172.0	449.4	359.4	699.0	119.9		137.6	450.4	1,558
30		279.9	316.7	901.0	177.2	458.2	368.6	718.4	123.6		141.7	462.1	1,606
40		291.9	330.0	934.4	186.1	473.0	384.3				148.6		1,689
50		301.1	340.4	962.8		484.9	397.2				154.4		1,759
60		309.5	350.3	987.6		495.3	408.5						1,819
80		322.9	366.9	1,028.0		511.9							1,920
100				1,065.0									2,000

Table 2.2.21 Saturation Temperature, in Kelvins, of Selected Substances (*Continued*)

Pressure, bar	Propane, C_3H_8	Propanol, C_3H_8O	Propylene, C_3H_6	Refrigerant 11, $CFCl_3$	Refrigerant 12, CF_2Cl_2	Refrigerant 13, CF_3Cl	Refrigerant 21, $CHFCl_2$	Refrigerant 22, CHF_2Cl	Sodium, Na	Sulfur dioxide, SO_2	Toluene, C_7H_8	Water, H_2O
0.010	162.0	284.0	157.8	209.9	171.5	134.1	201.8	165.7	804	190.6	275.1	280.1
0.015	166.2	289.8	161.9	215.0	175.8	137.6	206.6	170.0	825	195.9	282.0	286.1
0.020	169.4	294.0	165.1	219.1	179.1	140.3	210.5	173.1	841	199.6	286.3	290.6
0.025	171.9	297.3	167.4	222.2	181.8	142.4	212.9	175.6	854	202.6	290.0	294.2
0.030	174.1	299.9	169.6	224.7	184.1	144.2	215.5	177.8	865	204.8	293.4	297.2
0.04	177.4	304.2	172.8	229.0	187.8	147.1	219.6	180.9	883	208.7	298.9	302.1
0.05	180.1	307.8	175.7	232.6	190.3	149.4	222.9	183.5	898	211.8	303.3	306.0
0.06	182.4	310.7	177.9	235.5	193.0	151.3	225.7	185.8	910	214.3	307.1	309.3
0.08	186.2	315.7	181.7	240.7	197.0	154.6	230.2	189.6	930	218.6	313.7	314.7
0.10	189.4	319.7	184.9	244.7	200.1	157.1	233.9	192.6	946	222.0	318.6	319.0
0.15	195.6	327.4	190.7	252.1	206.3	162.1	241.0	198.5	977	228.2	328.5	327.3
0.20	200.1	333.2	195.2	257.8	211.1	165.9	246.3	202.9	1,000	232.9	335.1	333.2
0.25	203.9	337.8	198.7	262.4	214.9	169.0	250.6	206.6	1,019	236.2	341.6	338.1
0.30	207.0	341.7	201.7	266.4	218.2	171.7	254.2	209.6	1,034	239.8	346.3	342.3
0.40	212.1	348.1	206.8	273.1	223.5	176.0	260.2	214.7	1,061	244.8	354.1	349.0
0.5	216.3	352.9	211.0	278.2	227.9	179.5	265.1	218.6	1,082	248.7	360.3	354.5
0.6	219.9	357.3	214.5	282.8	231.7	182.5	269.2	220.2	1,099	252.2	366.1	359.1
0.8	225.9	364.6	221.4	290.3	237.9	187.1	276.1	227.7	1,129	258.0	375.6	366.7
1.0	230.7	370.1	225.2	296.6	243.0	191.4	281.7	232.2	1,153	262.8	383.2	372.8
1.5	240.3	381.4	234.5	308.6	253.0	199.3	292.6	241.2	1,199	272.6	398.2	384.5
2.0	247.7	389.9	241.6	317.9	260.6	205.4	299.4	248.1	1,235	279.6	409.3	393.4
2.5	253.2	396.9	247.3	325.5	266.9	210.4	307.8	253.8	1,264	285.4	419.0	400.6
3.0	258.9	402.3	252.6	331.8	272.3	214.6	313.6	258.6	1,289	290.0	427.0	406.7
4.0	267.6	411.9	262.6	342.7	281.3	221.8	323.5	266.5	1,330	298.6	441.7	416.8
5.0	274.8	419.8	267.2	351.1	288.8	227.1	331.6	273.0	1,364	305.3	451.4	425.0
6	281.0	426.8	273.9	358.8	295.2	232.8	338.5	279.0	1,393	311.1	460.9	432.0
8	291.4	438.5	284.0	372.2	306.0	241.4	350.2	289.1	1,430	320.7	476.8	445.6
10	300.0	447.8	292.3	382.7	314.9	248.5	361.0	297.1	1,480	328.6	490.0	453.0
15	317.0	466.6	308.9	403.7	332.6	262.7	379.1	312.3	1,556	345.2	515.8	472.0
20	330.3	481.4	321.8	419.2	346.3	273.7	394.0	324.9	1,623	357.8	535.5	485.5
25		493.6	332.5	432.1	357.5	282.8	406.4	335.1	1,676	368.2	552.3	497.1
30		503.8	341.7	444.8	367.2	290.7	417.1	343.4	1,720	377.0	566.4	507.0
40		511.3	357.0	463.9	383.3		434.7	358.3	1,795	391.6	589.6	523.5
50		536.1					449.2		1,859	403.8		537.1
60									1,913	414.5		548.7
80									2,010			568.1
100									2,085			584.1

Table 2.2.22 Color Scale of Temperature for Iron or Steel

Color	Temperature °F	Temperature K
Dark blood red, black red	1,000	810
Dark red, blood red, low red	1,050	840
Dark cherry red	1,175	910
Medium cherry red	1,250	950
Cherry, full red	1,375	1,020
Light cherry, light red	1,550	1,120
Orange, free scaling heat	1,650	1,170
Light orange	1,725	1,210
Yellow	1,825	1,270
Light yellow	1,975	1,350
White	2,200	1,475

Table 2.2.23 Melting Points of Refractory Materials

Material	Temperature °F	Temperature K
Aluminum nitride, AlN	4,060	2,500
Aluminum oxide, Al_2O_3	3,720	2,320
Aluminum oxide–beryllium oxide, Al_2O_3-BeO	3,400	2,140
Beryllium carbide, Be_2C	3,810	2,370
Beryllium nitride, Be_3N_4	4,000	2,480
Beryllium oxide, BeO	4,570	2,790
Beryllium silicide, 2BeO · SiO_2	3,630	2,270
Borazon, BN	5,430	3,270
Calcia (lime), CaO	4,660	2,840
Graphite, C	6,700	3,980
Hafnia, HfO_2	5,090	3,090
Magnesia, MgO	5,070	3,070
Niobium carbide, NbC	6,330	3,770
Silica, SiO_2	3,110	1,980
Silicon carbide, SiC	3,990	2,470
Thoria, ThO_2	5,830	3,490
Titanium carbide, TiC	5,680	3,410
Zirconium aluminide, $ZrAl_2$	3,000	1,920
Zirconium beryllide, $ZrBe_{13}$	3,180	2,020
Zirconium carbide, ZrC	6,400	3,810
Zirconium disilicide, $ZrSi_2$	3,090	1,970
Zirconium nitride, ZrN	5,400	3,260
Zirconium oxide, ZrO_2	4,900	2,980
Zirconium silicides, Zr_3Si_2, Zr_4Si_3, Zr_6Si_5	≃4,050	≃2,500

Table 2.2.24 Mean Specific Heats of Various Solids (32-212°F, 273-373 K)

Solid	c, Btu/(lb · °F)	c, kJ/(kg · K)
Alumina	0.183	0.77
Asbestos	0.20	0.84
Ashes	0.20	0.84
Bakelite	≃0.35	1.50
Basalt (lava)	0.20	0.84
Bell metal	0.086	0.36
Bismuth-tin	0.043	0.18
Borax	0.229	0.96
Brass, yellow	0.088	0.37
Brass, red	0.090	0.38
Brick	0.22	0.92
Bronze	0.104	0.44
Carbon-coke	0.203	0.85
Chalk	0.215	0.90
Charcoal	0.20	0.84
Cinders	0.18	0.75
Coal	≃0.30	≃1.25
Concrete	0.156	0.65
Constantan	0.098	0.41
Cork	0.485	2.03
Corundum	0.198	0.83
D'Arcet's metal	0.050	0.21
Dolomite	0.222	0.93
Ebonite	0.33	1.38
German silver	0.095	0.40
Glass, crown	0.16	0.70
Glass, flint	0.12	0.50
Glass, normal	0.20	0.84
Gneiss	0.18	0.75
Granite	0.20	0.84
Graphite	0.20	0.84
Gypsum	0.26	1.10
Hornblende	0.20	0.84
Humus (soil)	0.44	1.80
India rubber (para)	≃0.37	1.50
Kaolin	0.224	0.94
Lead oxide (PbO)	0.055	0.23
Limestone	0.217	0.91
Lipowitz's metal	0.040	0.17
Magnesia	0.222	0.93
Magnesite (Fe_3O_4)	0.168	0.70
Marble	0.210	0.88
Nickel steel	0.109	0.46
Paraffin wax	0.69	2.90
Porcelain	0.22	0.92
Quartz	≃0.23	0.96
Quicklime	0.217	0.91
Rose's metal	0.050	0.21
Salt, rock	0.21	0.88
Sand	0.195	0.82
Sandstone	0.22	0.92
Serpentine	0.25	1.05
Silica	0.191	0.80
Soda	0.231	0.97
Solders (Pb + Sn)	0.043	0.18
Sulfur	0.180	0.75
Talc	0.209	0.87
Tufa	0.33	1.40
Type metal	0.039	0.16
Vulcanite	0.331	1.38
Wood, fir	0.65	2.70
Wood, oak	0.57	2.40
Wood, pine	0.67	2.80
Wood's metal	0.040	0.17

Table 2.2.25 Phase Transition and Other Data for the Elements

Name	Symbol	Formula weight	T_m, K	Δh_{fus}, kJ/kg	T_b, K	Δh_{vap}, kJ/kg	T_c, K	P_c, bar	T_{tr}, K
Actinium	Ac	227.028	1,323	63	3,475	1,750			
Aluminum	Al	26.9815	933.5	398	2,750	10,875	7,850	4,800	
Antimony	Sb	121.75	903.9	163	1,905		5,700	3,200	368; 686
Argon	Ar	39.948	83	30	87.2	163	151	50	
Arsenic	As	74.9216				1,703	2,100		
Barium	Ba	137.33	1,002	55.8		1,099	4,450	720	643
Beryllium	Be	9.01218	1,560	1,355	2,750	32,450	6,200	4,600	1,530
Bismuth	Bi	208.980	544.6	54.0	1,838	725	4,450	1,400	
Boron	B	10.81	2,320	1,933	4,000		3,300		1,473
Bromine	Br	159.808	266	66.0	332	188	584		
Cadmium	Cd	112.41	594	55.1	1,040	886	2,690	1,680	
Calcium	Ca	40.08	1,112	213.1	1,763	3.833	4,300	1,000	720
Carbon	C	12.011	3,810		4,275		7,200	11,500	
Cerium	Ce	140.12	1,072	390		2,955	9,750	3,350	103; 263; 1003
Cesium	Cs	132.905	301.8	16.4	951	496	2,015	125	
Chlorine	Cl_2	70.906	172	180.7	239	576	417		
Chromium	Cr	51.996	2,133	325.6	2,950	6.622	5,500		2,113
Cobalt	Co	58.9332	1,766	274.7	3,185	6,390	6,300		700; 1,400
Copper	Cu	63.546	1,357	206.8	2,845	4,726	8,280	7,400	
Dysprosium	Dy	162.50	1,670	68.1	2,855	1,416	6,925	2,500	1,659
Erbium	Er	167.26	1,795	119.1	3,135	1,563	7,250		1,643
Europium	Eu	151.96	1,092	60.6	1,850	944	4,350	690	
Fluorine	F_2	37.997	53.5	13.4	85.0	172	144		46
Gadolinium	Gd	157.25	1,585	63.8	3,540	2,285	8,670		1,537
Gallium	Ga	69.72	303	80.1	2,500	3,688	7,125	4,150	276
Germanium	Ge	72.59	1,211	508.9	3,110	4,558	8,900	5,300	
Gold	Au	196.967	1,337	62.8	3,130	1,701	7,250	5,450	
Hafnium	Hf	178.49	2,485	134.8	4,885	3,211	10,400		2,000
Helium	He	4.00260	3.5	2.1	4.22	21	5.2	2.3	2.2
Holmium	Ho	164.930	1,744	73.8	2,968	1,461	7,575		1,703
Hydrogen	H_2	2.0159	14.0		20.4				
Indium	In	114.82	430	28.5	2,346	2,019	6,150	2,550	
Iodine	I_2	253.809	387	125.0	457		785		
Iridium	Ir	192.22	2,718	13.7	4,740	3,185	7,800		
Iron	Fe	55.847	1,811	247.3	3,136	6,259	8,500	10,000	1,183; 1,671
Krypton	Kr	83.80	115.8	19.6	119.8	108	209.4	55	
Lanthanum	La	138.906	1,194	44.6	3,715	2,978	10,500		550; 1,134
Lead	Pb	207.2	601	23.2	2,025	858	5,500	1,650	
Lithium	Li	6.941	454		1,607	21,340	3,700	1,000	80
Lutetium	Lu	174.967	1,937	106.6	3.668	2,034			
Magnesium	Mg	24.305	922	368.4	1,364	5,252	3,850	1,750	
Manganese	Mn	54.9380	1,518	219.3	2,334	4,112	4,325	560	1,374; 1,447
Mercury	Hg	200.59	234.6	11.4	630	293	1,720	1,500	194
Molybdenum	Mo	95.94	2,892	290.0	4,900		1,450	12,000	
Neodymium	Nd	144.24	1,290	49.6	3,341	1,891	7,900		1,132; 1,297
Neon	Ne	20.179	24.5	16.4	27.1	89	44.5	26.6	
Neptunium	Np	237.048	910		4,160		12,000		551; 847
Nickel	Ni	58.70	1,728	297.6	3,190	6,308	8,000	11,100	631
Niobium	Nb	92.9064	2,740	283.7	5,020	7,341	12,500		
Nitrogen	N_2	28.013	63.2	25.7	77.3	198	126.2	34.0	35.6
Osmium	Os	190.2	3,310	150.0	5,300	3,310	12,700		
Oxygen	O_2	31.9988	54.4	13.8	90.2	213	154.8		23.8, 43.8

Formula weight = molecular weight, T_m = normal melting temperature, Δh_{fus} = enthalpy (or latent heat) of fusion, T_b = normal boiling-point temperature, Δh_{vap} = enthalpy (or latent heat) of vaporization, T_c = critical temperature, P_c = critical pressure, T_{tr} = transition temperature.

SOURCE: Prepared by P.E. Liley, Purdue University, and abstracted from Rohsenow et al., "Handbook of Heat Transfer Fundamentals," McGraw-Hill.

Table 2.2.25 Phase Transition and Other Data for the Elements ***(Continued)***

Name	Symbol	Formula weight	T_m, K	Δh_{fus}, kJ/kg	T_b, K	Δh_{vap}, kJ/kg	T_c, K	P_c, bar	T_{tr}, K
Palladium	Pd	106.4	1,826	165.0	3,240	3,358	7,700	7,100	
Phosphorus	P	30.9738	317		553		995	81	196, 298
Platinum	Pt	195.09	2,045	101	4,100	2,612	10,700	11,000	
Plutonium	Pu	244	913	11.7	3,505	1,409	10,500	3,250	395; 480; 588; 730
Potassium	K	39.0983	336.4	60.1	1,032	2,052	2,210	170	
Praseodymium	Pr	140.908	1,205	49	3,785	2,05	8,900		1,066
Promethium	Pm	145	1,353		2,730				
Protactinium	Pa	231	1,500	64.8	4,300	2,036			
Radium	Ra	226.025	973		1,900				
Radon	Rn	222	202	12.3	211	83	377	66	
Rhenium	Re	186.207	3,453	177.8	5,920	3,842	18,900	14,900	
Rhodium	Rh	102.906	2,236	209.4	3,980	4,798	7,000		
Rubidium	Rb	85.4678	312.6	26.4	964	810	2,070	168	
Ruthenium	Ru	101.07	2,525	256.3	4,430	5,837	9,600		1,300; 1,475; 1,775
Samarium	Sm	150.4	1,345	57.3	2,064	1,107	5,050	1,780	1,190
Scandium	Sc	44.9559	1,813	313.6	3,550	6,989	6,410	3,750	1,608
Selenium	Se	78.96	494	66.2	958	1,210	1,810	320	398; 425
Silicon	Si	28.0855	1,684	1,802	3,540	14,050	5,160	540	
Silver	Ag	107.868	1,234	104.8	2,435	2,323	6,400	4,450	
Sodium	Na	22.9898	371	113.1	1,155	4,263	2,500	370	
Strontium	Sr	87.62	1,043	1,042	1,650	1,585	4,275	375	505; 893
Sulfur	S	32.06	388	53.4	718		1,210	130	369; 374
Tantalum	Ta	180.948	3,252	173.5	5,640	4,211	16,500	12,000	
Technetium	Tc	98	2,447	232	4,550	5,830	11,500		
Tellurium	Te	127.60	723	137.1	1,261	895	2,330		
Terbium	Tb	158.925	1,631	67.9	3,500	2,083	8,470		228; 1,575
Thallium	Tl	204.37	577	20.1	1,745	806	4,550	1,700	507
Thorium	Th	232.038	2,028	69.4	5,067	2,218	14,400	6,165	1,670
Thulium	Tm	168.934	1,819	99.6	2,220	1,129	6,450		
Tin	Sn	118.69	505	58.9	2,890	2,496	7,700	2,250	286; 476
Titanium	Ti	47.90	1,943	323.6	3,565	8,787	5,850		1,162; 1,353
Tungsten	W	183.85	3,660	192.5	5,890	4,483	15,500	15,000	
Uranium	U	238.029	1,406	35.8	4,422	1,949	12,500	5,000	938; 1,046
Vanadium	V	50.9415	2,191	410.7	3,680	8,870	11,300	10,300	
Xenon	Xe	131.30	161.3	17.5	164.9	96	290	58	
Ytterbium	Yb	173.04	1,098	44.2	1,467	745	4,080	1,150	1,050
Yttrium	Y	88.9059	1,775	128.2	3,610	4,485	8,950		1,758
Zinc	Zn	65.38	692.7	113.0	1,182	1,768			
Zirconium	Zr	91.22	2,125	185.3	4,681	6,376	10,500		1,135

Formula weight = molecular weight, T_m = normal melting temperature, Δh_{fus} = enthalpy (or latent heat) of fusion, T_b = normal boiling-point temperature, Δh_{vap} = enthalpy (or latent heat) of vaporization, T_c = critical temperature, P_c = critical pressure, T_{tr} = transition temperature.

SOURCE: Prepared by P.E. Liley, Purdue University, and abstracted from Rohsenow et al., "Handbook of Heat Transfer Fundamentals," McGraw-Hill.

Table 2.2.26 Thermophysical Properties of Selected Solid Elements

Element	Property	Temperature, K							
		100	200	300	400	500	600	800	1,000
Al	P, bar			2.1×10^{-43}	4.9×10^{-31}	1.1×10^{-23}	1.0×10^{-18}	1.4×10^{-12}	6.6×10^{-9}
	ρ, kg/m^3	2,732	2,719	2,701	2,681	2,661	2,639	2,591	2,365
	h, kJ/kg			4,623	4,716	4,812	4,913	5,131	5,768
	s, kJ/(kg × K)			1.056	1.323	1.539	1.723	2.035	2.728
	c_p, kJ/(kg × K)	0.481	0.797	0.902	0.949	0.997	1.042	1.134	0.921
	λ, W/(m × K)	300	237	237	240	236	231	218	
	α, m^2/s	2.3×10^{-4}	1.1×10^{-4}	9.7×10^{-5}	9.4×10^{-5}	8.9×10^{-5}	8.4×10^{-5}	7.4×10^{-5}	6.6×10^{-5}
Cr	P, bar			4.6×10^{-62}	8.9×10^{-45}	2.1×10^{-34}	1.6×10^{-27}	6.3×10^{-19}	9.1×10^{-14}
	ρ, kg/m^3	7,155	7,145	7,135	7,120	7,110	7,080	7,040	7,000
	h, kJ/kg			4,113	4,160	4,210	4,263	4,375	4,495
	s, kJ/(kg × K)			0.457	0.591	0.703	0.800	0.962	1.094
	c_p, kJ/(kg × K)	0.190	0.382	0.450	0.501	0.537	0.565	0.611	0.653
	λ, W/(m × K)	160	110	94	91	86	81	71	65
	α, m^2/s	1.2×10^{-4}	4.1×10^{-5}	2.9×10^{-5}	2.6×10^{-5}	2.3×10^{-5}	2.0×10^{-5}	1.7×10^{-5}	1.4×10^{-5}
Cu	P, bar			1.1×10^{-52}	9.0×10^{-38}	5.5×10^{-29}	3.8×10^{-23}	7.6×10^{-16}	1.7×10^{-11}
	ρ, kg/m^3	9,009	8,973	8,930	8,884	8,837	8,787	8,686	8,568
	h, kJ/kg			5,067	5,106	5,146	5,188	5,273	5,361
	s, kJ/(kg × K)			0.524	0.637	0.726	0.802	0.924	1.022
	c_p, kJ/(kg × K)	0.254	0.357	0.386	0.396	0.406	0.431	0.448	0.466
	λ, W/(m × K)	480	413	401	393	386	379	366	352
	α, m^2/s	2.2×10^{-4}	1.3×10^{-4}	1.2×10^{-4}	1.1×10^{-4}	1.1×10^{-4}	1.0×10^{-4}	9.0×10^{-5}	8.0×10^{-5}
Au	P, bar			6.7×10^{-58}	6.3×10^{-42}	2.8×10^{-32}	3.6×10^{-25}	6.5×10^{-18}	3.7×10^{-13}
	ρ, kg/m^3	19,460	19,380	19,300	19,210	19,130	19,040	18,860	18,660
	h, kJ/kg			6,046	6,059	6,072	6,086	6,113	6,142
	s, kJ/(kg × K)			0.241	0.279	0.309	0.333	0.373	0.404
	c_p, kJ/(kg × K)	0.109	0.124	0.129	0.131	0.133	0.136	0.141	0.147
	λ, W/(m × K)	327	323	317	311	304	298	284	270
	α, m^2/s	1.5×10^{-4}	1.34×10^{-4}	1.27×10^{-4}	1.23×10^{-4}	1.19×10^{-4}	1.15×10^{-4}	1.07×10^{-4}	9.8×10^{-5}
Fe	P, bar			3.1×10^{-65}	6.3×10^{-54}	3.9×10^{-47}	1.5×10^{-36}	6.6×10^{-20}	1.5×10^{-14}
	ρ, kg/m^3	7,900	7,880	7,860	7,830	7,800	7,760	7,690	7,650
	h, kJ/kg			4,523	4,570	4,621	4,676	4,801	4,958
	s, kJ/(kg × K)			0.491	0.626	0.740	0.840	1,018	1.193
	c_p, kJ/(kg × K)	0.216	0.384	0.450	0.491	0.524	0.555	0.692	1.034
	λ, W/(m × K)	134	94	80	70	61	55	43	32
	α, m^2/s	8.2×10^{-5}	3.1×10^{-5}	2.2×10^{-5}	1.8×10^{-5}	1.5×10^{-5}	1.3×10^{-5}	1.1×10^{-5}	1.0×10^{-5}
Pb	P, bar			6.3×10^{-29}	1.8×10^{-20}	2.2×10^{-15}	5.4×10^{-12}	6.2×10^{-8}	1.6×10^{-5}
	ρ, kg/m^3	11,520	11,430	11,330	11,230	11,130	11,010	10,430	10,190
	h, kJ/kg			6,929	6,942	6,955	6,969	7,022	7,050
	s, kJ/(kg × K)			0.314	0.351	0.381	0.406	0.487	0.519
	c_p, kJ/(kg × K)	0.118	0.125	0.129	0.132	0.137	0.142	0.145	0.142
	λ, W/(m × K)	39.7	36.7	35.3	34.0	32.8	31.4		
	α, m^2/s	2.9×10^{-5}	2.6×10^{-5}	2.4×10^{-5}	2.3×10^{-5}	2.2×10^{-5}	2.0×10^{-5}	1.3×10^{-5}	1.5×10^{-5}
Li	P, bar								
	ρ, kg/m^3	546	541	533	526	492	482	462	442
	h, kJ/kg			4,615	4,919	5,844	6,274	7,113	7,945
	s, kJ/(kg × K)			4.214	5.289	7.182	7.967	9.173	10.102
	c_p, kJ/(kg × K)	1.923	3.105	3.54	3.76	4.34	4.26	4.17	4.15
	λ, W/(m × K)	105	90	85	80				
	α, m^2/s	1.0×10^{-4}	5.4×10^{-5}	4.5×10^{-5}	3.2×10^{-5}	2.1×10^{-5}	2.3×10^{-5}	2.8×10^{-5}	3.3×10^{-5}
Ni	P, bar			1.1×10^{-67}	5.8×10^{-49}	9.8×10^{-38}	2.7×10^{-30}	5.5×10^{-21}	2.1×10^{-15}
	ρ, kg/m^3	8,960	8,930	8,900	8,860	8,820	8,780	8,690	8,610
	h, kJ/kg			4,837	4,883	4,934	4,990	5,099	5,207
	s, kJ/(kg × K)			0.512	0.645	0.758	0.859	1.017	1.137
	c_p, kJ/(kg × K)	0.232	0.383	0.444	0.490	0.540	0.590	0.530	0.556
	λ, W/(m × K)	165	105	91	80	72	66	68	72
	α, m^2/s	8.0×10^{-5}	3.1×10^{-5}	2.3×10^{-5}	1.9×10^{-5}	1.5×10^{-5}	1.3×10^{-5}	1.4×10^{-5}	1.5×10^{-5}
Pt	P, bar			3.2×10^{-91}	1.3×10^{-66}	7.3×10^{-52}	5.0×10^{-42}	9.7×10^{-30}	2.3×10^{-22}
	ρ, kg/m^3	21,550	21,500	21,450	21,380	21,330	21,270	21,140	21,010
	h, kJ/kg			5,837	5,850	5,864	5,878	5,907	5,937
	s, kJ/(kg × K)			0.214	0.253	0.283	0.309	0.350	0.383
	c_p, kJ/(kg × K)	0.101	0.127	0.134	0.136	0.138	0.140	0.146	0.152
	λ, W/(m × K)	78	73	72	72	72	73	76	79
	α, m^2/s	3.6×10^{-5}	2.7×10^{-5}	2.5×10^{-5}	2.5×10^{-5}	2.5×10^{-5}	2.5×10^{-5}	2.5×10^{-5}	2.5×10^{-5}

P = saturation vapor pressure; ρ = density; h = enthalpy; s = entropy; c_p = specific heat at constant pressure; λ = thermal conductivity; α = thermal diffusivity.

Table 2.2.26 Thermophysical Properties of Selected Solid Elements *(Continued)*

Element	Property	Temperature, K 100	200	300	400	500	600	800	1,000
Rh	P, bar			5.5×10^{-89}	6.5×10^{-65}	1.8×10^{-50}	7.1×10^{-41}	7.0×10^{-29}	1.1×10^{-21}
	ρ, kg/m³	12,480	12,460	12,430	12,400	12,360	12,330	12,250	12,170
	h, kJ/kg			4,921	4,946	4,972	4,999	5,055	5,115
	s, kJ/(kg × K)			0.308	0.380	0.437	0.487	0.568	0.635
	c_p, kJ/(kg × K)	0.147	0.220	0.246	0.257	0.265	0.274	0.290	0.307
	λ, W/(m × K)	190	154	150	146	141	136	127	121
	α, m²/s	1.0×10^{-4}	5.6×10^{-5}	4.9×10^{-5}	4.6×10^{-5}	4.3×10^{-5}	4.0×10^{-5}	3.6×10^{-5}	3.2×10^{-5}
Ag	P, bar			2.1×10^{-43}	4.9×10^{-31}	1.1×10^{-23}	1.0×10^{-18}	1.4×10^{-12}	6.6×10^{-9}
	ρ, kg/m³	10,600	10,550	10,490	10,430	10,360	10,300	10,160	10,010
	h, kJ/kg			5,791	5,815	5,839	5,864	5,915	5,968
	s, kJ/(kg × K)			0.3959	0.4641	0.5180	0.5630	0.6365	0.6964
	c_p, kJ/(kg × K)	0.187	0.225	0.236	0.240	0.245	0.251	0.264	0.276
	λ, W/(m × K)	450	430	429	425	419	412	396	379
	α, m²/s	2.3×10^{-4}	1.8×10^{-4}	1.7×10^{-4}	1.7×10^{-4}	1.7×10^{-4}	1.6×10^{-4}	1.5×10^{-4}	1.3×10^{-4}
Ti	P, bar			1.0×10^{-74}	4.6×10^{-54}	5.2×10^{-42}	7.2×10^{-35}	1.2×10^{-23}	1.4×10^{-17}
	ρ, kg/m³	4,530	4,520	4,510	4,490	4,480	4,470	4,440	4,410
	h, kJ/kg			4,857	4,911	4,967	5,025	5,147	5,278
	s, kJ/(kg × K)			0.643	0.797	0.922	1.028	1.205	1.350
	c_p, kJ/(kg × K)	0.295	0.464	0.525	0.555	0.578	0.597	0.627	0.670
	λ, W/(m × K)	31	25	21	20	20	19	19	21
	α, m²/s								
W	P, bar			3.2×10^{-141}	2.9×10^{-104}	4.3×10^{-82}	2.7×10^{-67}	8.7×10^{-49}	1.1×10^{-37}
	ρ, kg/m³	19,310	19,290	19,270	19,240	19,220	19,190	19,130	19,080
	h, kJ/kg			6,255	6,268	6,282	6,296	6,325	6,354
	s, kJ/(kg × K)			0.178	0.217	0.248	0.273	0.315	0.347
	c_p, kJ/(kg × K)	0.089	0.125	0.135	0.137	0.139	0.140	0.144	0.148
	λ, W/(m × K)	208	185	174	159	146	137	125	118
	α, m²/s								
V	P, bar			3.0×10^{-82}	7.1×10^{-60}	1.9×10^{-46}	1.6×10^{-37}	2.4×10^{-26}	1.2×10^{-19}
	ρ, kg/m³	6,074	6,062	6,050	6,030	6,010	6,000	5,960	5,920
	h, kJ/kg			4,740	4,790	4,843	4,896	5,006	5,121
	s, kJ/(kg × K)			0.571	0.716	0.832	0.930	1.088	1.216
	c_p, kJ/(kg × K)	0.257	0.434	0.483	0.512	0.528	0.540	0.563	0.598
	λ, W/(m × K)	36	31	31	31	32	33	36	38
	α, m²/s	2.3×10^{-5}	1.2×10^{-5}	1.1×10^{-5}	1.0×10^{-5}	1.0×10^{-5}	1.0×10^{-5}	1.1×10^{-5}	1.1×10^{-5}
Zn	P, bar			3.7×10^{-17}	1.6×10^{-11}	3.7×10^{-8}	6.7×10^{-6}	3.4×10^{-3}	1.2×10^{-1}
	ρ, kg/m³	7,260	7,200	7,135	7,070	7,000	6,935	6,430	6,260
	h, kJ/kg			5,690	5,730	5,771	5,813	5,970	6,114
	s, kJ/(kg × K)			0.639	0.753	0.844	0.922	1.216	1.323
	c_p, kJ/(kg × K)	0.295	0.366	0.389	0.404	0.419	0.435	0.479	0.479
	λ, W/(m × K)	117	118	116	111	107	103		
	α, m²/s	5.5×10^{-5}	4.7×10^{-5}	4.1×10^{-5}	3.9×10^{-5}	3.7×10^{-5}	3.4×10^{-5}	1.8×10^{-5}	2.2×10^{-5}
Zr	P, bar			2.8×10^{-99}	8.6×10^{-73}	6.6×10^{-57}	2.7×10^{-46}	4.6×10^{-33}	4.1×10^{-25}
	ρ, kg/m³	6,535	6,525	6,515	6,510	6,490	6,480	6,450	6,420
	h, kJ/kg			5,540	5,569	5,600	5,632	5,698	5,768
	s, kJ/(kg × K)			0.429	0.513	0.590	0.640	0.735	0.813
	c_p, kJ/(kg × K)	0.120	0.126	0.130	0.136	0.143	0.153	0.153	0.153
	λ, W/(m × K)	33	25	23	22	21	21	21	23
	α, m²/s								

P = saturation vapor pressure; ρ = density; h = enthalpy; s = entropy; c_p = specific heat at constant pressure; λ = thermal conductivity; α = thermal diffusivity.

2.3 RADIANT HEAT TRANSFER

by Hoyt C. Hottel, Adel F. Sarofim, James J. Noble, and Hakan Ertürk

REFERENCE: Baukal, "The John Zink Combustion Handbook," CRC Press. Blokh, "Heat Transfer in Steam Boiler Furnaces," Taylor & Francis. Brown, Revzan, and Frenklach, Twenty-seventh Symposium (International) on Combustion, 1998, pp. 1573-1580. Ertürk and Howell "Monte Carlo Methods for Radiative Transfer," in "Handbook of Thermal Science and Engineering," Springer. Hottel and Sarofim, "Radiative Transfer," McGraw-Hill. Howell, Siegel and Mengüç, "Thermal Radiation Heat Transfer," 6th ed., CRC Press. Modest "Radiative Heat Transfer," 2d ed., Academic Press. Mulholland and Croarkin, Fire and Materials 24, 2000, pp. 227-230. Stultz and Kitto, "Steam," 41st ed., Babcock and Wilcox. Noble, *Int. J. Heat Mass Transfer*, **18**, 1975, pp. 261-269. Zucca, Marchisio, Barresi, and Fox, "Chemical Engineering Science," 2005.

A heated body loses energy continuously by radiation, which is referred as emission, at a rate dependent on the shape, the size, its properties, and, particularly, the temperature of the body. In contrast to conductive energy transport, such emitted radiation is capable of passing to a distant body without the need of a medium, where it may be absorbed, reflected, scattered, or transmitted.

Consider a pencil of radiation, defined as all the rays passing through each of two infinitesimally small, widely separated areas dA_1 and dA_2. The rays at dA_1 will have a solid angle of divergence $d\Omega_1$; equal to the apparent area of dA_2 viewed from dA_1, divided by the square of the separating distance. Let the normal to dA_1 make the angle θ_1 with the pencil. The flux density q [energy/(time)(area normal to beam)] per unit solid angle of divergence is called the **intensity *I*,** and the flux dQ_1 (energy/time) through area dA_1 (of apparent area dA_1 cos θ_1 normal to the beam) is therefore given by

$$dQ_1 = dA_1 \cos\theta_1 q_1 = I \; dA_1 \cos\theta_1 d\Omega_1 \tag{2.3.1}$$

The intensity I along a pencil, in the absence of absorption or scattering is constant, unless the beam passes into a medium of different refractive index n; $I_1/n_1^2 = I_2/n_2^2$. The **emissive power** of a surface is the flux density [energy/(time)(surface area)] due to emission from it throughout a hemisphere. If the intensity I of emission from a surface is independent of the angle of emission, Eq. (2.3.1) may be used to show that the surface emissive power is πI, though the emission is throughout a solid angle of 2π steradians.

BLACKBODY RADIATION

Engineering calculations of thermal radiation from surfaces are best keyed to the radiation characteristics of the **blackbody,** or **ideal emitter.** The characteristic properties of a blackbody are that it absorbs all the radiation incident on its surface and the quality and intensity of the radiation it emits are completely determined by its temperature. The total radiative flux throughout a hemisphere from a black surface of area A and absolute temperature T is given by the **Stefan-Boltzmann law:** $Q_e = A\ \sigma T^4$ or $q = \sigma T^4$. The Stefan-Boltzmann constant σ has the value 5.67 × 10^{-8} W/m²-K⁴. From the above definition of emissive power, σT^4 is the total emissive power of a blackbody, called E_b; and the intensity I_b of emission from a blackbody is E_b/π, or $\sigma T^4/\pi$.

The spectral distribution of energy flux from a blackbody is expressed by Planck's law

$$E_{b,\lambda} d\lambda = \frac{2\pi hc^2 n^2 \lambda^{-5}}{e^{hc/(k\lambda T)} - 1} d\lambda = \frac{n^2 c_1 \lambda^{-5}}{e^{c_2/\lambda T} - 1} d\lambda \tag{2.3.2}$$

where $E_{b,\lambda} d\lambda$ is the spectral blackbody emissive power in W/m² lying in the wavelength range λ to $\lambda + d\lambda$; h is Planck's constant, 6.6262 × 10^{-34} J·s; c is the velocity of light in vacuum, 2.9979 × 10^8 m/s; k is the Boltzmann constant, 1.3807 × 10^{-23} J/K; λ is the wavelength measured in vacuum, n is the refractive index of the emitter; c_1 and c_2, the first and second Planck's law constants, are 3.7418 × 10^{-16} W·m² and 1.4388 × 10^{-2} m·K. To show how $E_{b,\lambda}$ varies with wavelength or temperature, Planck's law may be cast in the form

$$\frac{E_{b,\lambda}}{n^2 T^5} = \frac{c_1(\lambda T)^{-5}}{e^{c_2/\lambda T} - 1} \tag{2.3.3}$$

Therefore, when $n > 1$, as in the case of a gas, $E_{b,\lambda}/T^5$ is a unique function of λT. According to Wien's displacement law, $E_{b,\lambda}$ is a maximum at λT = 2,898 μm·K. Based on other displacement laws; half of blackbody radiation lies on either side of λT = 4,107 μm·K and the maximum intensity per unit fractional change in wavelength or frequency is at λT = 3,670 μm·K. Integration of $E_{b,\lambda}$ over λ shows that the fraction f of blackbody radiation lying at wavelengths below λ depends only on λT. Values of f versus λT appear in Table 2.3.1.

Table 2.3.1 Fraction *f* of blackbody radiation below λT (λT in μm-K)

λT	1,200	1,600	1,800	2,000	2,200	2,400	2,600	2,800
f	0.002	0.02	0.039	0.067	0.101	0.14	0.183	0.228
λT	3,000	3,200	3,400	3,600	3,800	4,000	4,200	4,500
f	0.273	0.318	0.362	0.404	0.443	0.48	0.516	0.564
λT	4,800	5,100	5,500	6,000	6,500	7,000	7,600	8,400
f	0.608	0.646	0.691	0.738	0.776	0.808	0.839	0.871
λT	10,000	12,000	14,000	20,000	50,000			
f	0.914	0.945	0.963	0.986	0.999			

A limiting form of the Planck equation as $\lambda T \to 0$ is known as the Wien equation, $E_{b,\lambda} = n^2 c_1 \lambda^{-5} e^{c_2/\lambda T}$, which approximates Planck's distribution with less than 1% error when λT is less than 3,000 μm·K. This relation can be useful for optical pyrometry (red screen λ = 0.65 mm) when T < 4,800 K.

RADIATIVE EXCHANGE BETWEEN SURFACES OF SOLIDS

The ratio of the total radiating power of a real surface to that of a black surface at the same temperature is called the emittance of the surface (for a perfectly plane surface, the emissivity), designated by ε. Subscripts λ, θ, and n may be assigned to differentiate monochromatic, directional, and surface-normal values, respectively, from the total hemispherical value. If radiation is incident on a surface, the fraction absorbed is called the absorptance (absorptivity), a term in which two subscripts may be appended, the first to identify the temperature of the surface and the second to identify the quality of the incident radiation. According to Kirchhoff's law, the emissivity and absorptivity of a surface in surroundings at its own temperature are the same, for both monochromatic and total radiation. When the temperatures of the surface and its surroundings differ, the total emissivity and absorptivity of the surface are found often to be different, but because absorptivity is substantially independent of irradiation density, the monochromatic emissivity and absorptivity of surfaces are for all practical purposes the same. The difference between total emissivity and absorptivity depends on the variation, with wavelength, of ε_λ and on the difference between the emitter temperature and the effective source temperature.

Consider radiative exchange between a body of area A_1 and temperature T_1 and its black surroundings at T_2. The net interchange is given by

$$\dot{Q}_{1-2} = A_1\int_0^{\infty}[\varepsilon_\lambda E_{b,\lambda}(T_1)-\alpha_\lambda E_{b,\lambda}(T_2)]d\lambda$$
$$= A_1(\varepsilon_1\sigma T_1^4 - \alpha_{12}\sigma T_2^4) \quad (2.3.4)$$

where

$$\varepsilon_1 = \frac{1}{\sigma T_1^4}\int_0^{\infty}\varepsilon_\lambda E_{b,\lambda}(T_1)d\lambda = \int_0^1 \varepsilon_\lambda\, df_{\lambda T_1}$$
and
$$\alpha_{12} = \frac{1}{\sigma T_2^4}\int_0^{\infty}\varepsilon_\lambda E_{b,\lambda}(T_2)d\lambda = \int_0^1 \varepsilon_\lambda\, df_{\lambda T_2} \quad (2.3.5)$$

i.e., ε_1 (or α_{12}) is the area under a curve of ε_λ versus f, read as a function of λT at T_1 (or T_2) from Table 2.3.1. If ε_λ does not change with wavelength, the surface is called gray, and $\varepsilon_1 = \alpha_{12} = \varepsilon_\lambda$. A selective surface is one whose ε_1 changes dramatically with wavelength. If this change is monotonic, ε_1, and α_{12} are, according to Eqs. (2.3.4)-(2.3.6), markedly different when the absolute temperature ratio is far from 1; e.g., when T_1 = 293 K (ambient temperature) and T_2 = 5,800 K (effective solar temperature), $\varepsilon_1 = 0.9$ and α_{12}= 0.1-0.2 for a white paint, but ε_1 can be as low as 0.12 and α_{12} above 0.9 for a thin layer of copper oxide on bright aluminum, or of chromic oxide on bright nickel.

Although values of emittances and absorptances depend in very complex ways on the real and imaginary components of the refractive index and on the geometric structure of the surface layer, some generalizations are possible:

Polished Metals (1) ε_λ is quite low in the infrared and, for $\lambda > 8$ μm, can be adequately approximated by $0.00365\sqrt{r/\lambda}$, where r is the resistivity in ohm·cm and λ is in μm; at shorter wavelengths, ε_λ increases and, for many metals, has values of 0.4 to 0.8 in the visible (0.4-0.7 μm). ε_λ is approximately proportional to the square root of the absolute temperature ($\varepsilon_\lambda \propto \sqrt{r}$ and $r \propto T$) in the far infrared ($\lambda > 8$ μm), is temperature insensitive in the near infrared (0.7-1.5 μm) and, in the visible, decreases slightly as temperature increases. (2) Total emittance is substantially proportional to absolute temperature; at moderate temperature, $\varepsilon_n = 0.58T\sqrt{r_0/T_0}$, where T is in K. (3) Total absorptance of a metal at T_1 for radiation from a black or gray source at T_2 is equal to the emissivity evaluated at the geometric mean of T_1 and T_2. (4) The ratio of hemispherical to normal emittance (absorptance) varies from 1.33 at very lower ε's (α's) to about 1.03 at an ε (α) of 0.4.

Metal surfaces are subject to oxidation unless extraordinary care are taken to prevent oxidation. An oxidized metallic surface may exhibit several times the emittance or absorptance of a polished specimen. The emittance of iron and steel, for example, varies widely with degree of oxidation and roughness—clean metallic surfaces have an emittance of from 0.05-0.45 at ambient temperatures to 0.4-0.7 at high temperatures; oxidized and/or rough surfaces range from 0.6-0.95 at low temperatures to 0.9-0.95 at high temperatures.

Refractory Materials Grain size and concentration of trace impurities are important. (1) Most refractory materials have an ε_λ of 0.8 to 1.0 at wavelengths beyond 2 to 4 μm; ε_λ decreases rapidly toward shorter wavelengths for materials that are white in the visible but retains its high value for black materials such as FeO and Cr_2O_3. Small concentrations of FeO and Cr_2O_3 or other colored oxides can cause marked increases in the emittance of materials that are normally white. ε_λ for refractory materials varies little with temperature. (2) Refractory materials generally have a total emittance which is high (0.7 to 1.0) at ambient temperatures and decreases with increase in temperature; a change from 1,000 to 1,600°C may cause a decrease in ε of one-fourth to one-third. (3) The emittance and absorptance increase with increase in grain size over a grain size range of 1-200 μm. (4) The ratio $\varepsilon/\varepsilon_n$ of hemispherical to normal emissivity of polished surfaces varies with refractive index from 1 at n = 1 to 0.95 at n = 1.5 (common glass) and back to 0.98 at n = 3.5. (5) The ratio $\varepsilon/\varepsilon_n$ for a surface composed of particulate matter which scatters isotropically varies with ε from 1 when ε = 1 to 0.8 when ε = 0.07. (6) The total absorptance shows a decrease with increase in temperature of the radiation source similar to the decrease in emittance with increase in the specimen temperature. Figure 2.3.1 shows the effect of the temperature of the radiation source on the absorptance of surfaces of various materials at room temperature. It should be noted that polished aluminum (line 15) and anodized aluminum (line 13), representative of metals and nonmetals, respectively, respond oppositely to a change in the temperature of the radiation source. The absorptance of surfaces for sunlight may be read from the right of Fig. 2.3.1, assuming sunlight to consist of blackbody radiation from a source at 5,800 K.

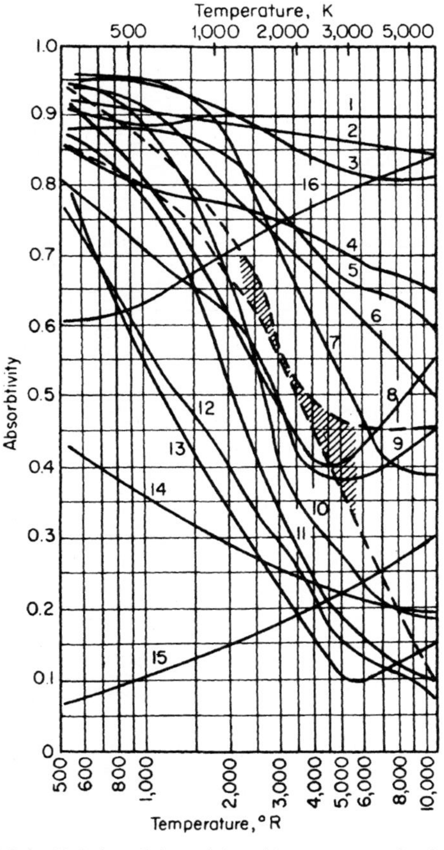

Figure 2.3.1 Variation of absorptivity with temperature of radiation source. (1) Slate composition roofing; (2) linoleum, red-brown; (3) asbestos slate (asbestos use is obsolete, but may be encountered in existing construction); (4) soft rubber, gray; (5) concrete; (6) porcelain; (7) vitreous enamel, white; (8) red brick; (9) cork; (10) white Dutch tile; (11) white chamotte; (12) MgO, evaporated; (13) anodized aluminum; (14) aluminum paint; (15) polished aluminum; (16) graphite. The two dashed lines bound the limits of data for gray paving brick, asbestos paper (asbestos use is obsolete, but may be encountered in existing construction), wood, various cloths, plaster of paris, lithopone, and paper.

When T_2 is not too different from T_1, α_{12} may be expressed as $\varepsilon_1(T_2/T_1)^n$, with n determined from Fig. 2.3.1. For this case, Eq. (2.3.4) becomes

$$\dot{Q}_{1,net} = \sigma A_1 \varepsilon_{AV}(1+n/4)(T_1^4 - T_2^4) \qquad (2.3.6)$$

where ε_{AV} is evaluated at the arithmetic mean of T_1 and T_2.

Table 2.3.2 gives the emittance of various surfaces and emphasizes the variation possible in a single material. The values in the table apply, with a few exceptions, to normal radiation from the surface.

For opaque materials, the reflectance ρ is the complement of the absorptance. The directional distribution of the reflected radiation depends on the material, its degree of roughness or grain size, and if a metal, its state of oxidation. Polished surfaces of homogeneous materials reflect specularly. In contrast, the intensity of the radiation reflected from a perfectly diffuse, or Lambert, surface is independent of direction. The directional distribution of reflectance of many oxidized

Table 2.3.2 Emissivity of Surfaces

Surface	Temp.,* °C	Emissivity*
Metals and their oxides		
Aluminum:		
Highly polished	230-580	0.039-0.057
Polished	23	0.040
Rough plate	26	0.055-0.07
Oxidized at 600°C	200-600	0.11-0.19
Oxide	280-830	0.63-0.26
Alloy 75ST	24	0.10
75ST, repeated heating	230-480	0.22-0.16
Brass:		
Highly polished	260-380	0.03-0.04
Rolled plate, natural	22	0.06
Chromium	40-540	0.08-0.26
Copper:		
Electrolytic, polished	80	0.02
Comm'l plate, polished	20	0.030
Heated at 600°C	200-600	0.57-0.57
Thick oxide coating	25	0.78
Cuprous oxide	800-1,100	0.66-0.54
Molten copper	1,080-1,280	0.16-0.13
Dow metal, cleaned, heated	230-400	0.24-0.20
Gold, highly polished	230-630	0.02-0.04
Iron and steel:		
Pure Fe, polished	180-980	0.05-0.37
Wrought iron, polished	40-250	0.28
Smooth sheet iron	700-1,040	0.55-0.60
Rusted plate	20	0.69
Smooth oxidized iron	130-530	0.78-0.82
Strongly oxidized	40-250	0.95
Molten iron and steel	1,500-1,770	0.40-0.45
Lead:		
99.96%, unoxidized	130-230	0.06-0.08
Gray, oxidized	24	0.28
Oxidized at 190°C	190	0.63
Mercury, pure clean	0-100	0.09-0.12
Molybdenum filament	730-2,590	0.10-0.29
Monel metal, K5700		
Washed, abrasive soap	24	0.17
Repeated heating	230-875	0.46-0.65

Table 2.3.2 Emissivity of Surfaces *(Continued)*

Surface	Temp.,* °C	Emissivity*
Nickel and alloys:		
Electrolytic, polished	23	0.05
Electroplated, not polished	20	0.11
Wire	90-1010	0.10-0.19
Plate, oxid. at 600°C	200-600	0.37-0.48
Nickel oxide	650-1,250	0.59-0.86
Copper-nickel, polished	100	0.06
Nickel-silver, polished	100	0.14
Nickelin, gray oxide	21	0.26
Nichrome wire, bright	50-1,000	0.65-0.79
Nichrome wire, oxide	50-500	0.95-0.98
ACI-HW (60Ni, 12Cr); firm black ox, coat	270-560	0.89-0.82
Platinum, polished plate	230-1,630	0.05-0.17
Silver, pure polished	230-630	0.02-0.03
Stainless steels:		
Type 316, cleaned	24	0.28
316, repeated heating	230-870	0.57-0.66
304, 42 h at 520°C	220-530	0.62-0.73
310, furnace service	220-530	0.90-0.97
Allegheny #4, polished	100	0.13
Tantalum filament	1,330-3,000	0.194-0.33
Thorium oxide	280-830	0.58-0.21
Tin, bright	24	0.04-0.06
Tungsten, aged filament	25-3,320	0.03-0.35
Zinc, 99.1%, comm'l, polished	230-330	0.05
Galv. iron, bright	28	0.23
Galv. iron, gray oxide	24	0.28
Refractories, building materials, paints, misc.		
Alumina	260-680	0.6-0.33
Alumina, 50-mm grain size	1,010-1,570	0.39-0.28
Alumina-silica, cont'g	1,010-1,570	
0.4% Fe_2O_3		0.61-0.43
1.7% Fe_2O_3		0.73-0.62
2.9% Fe_2O_3		0.78-0.68
Al paints (vary with amount of lacquer body, age)	100	0.27-0.67
Asbestos	40-370	0.93-0.95
Calcium oxide	750-1,100	0.29-0.28
Candle soot; lampblack-waterglass	20-370	0.95 ± 0.01
Carbon plate, heated	130-630	0.81-0.79
Ferric oxide (Fe_2O_3)	500-900	0.8-0.43
Magnesium oxide, 1 mm	260-760	0.67-0.41
Oil layers		
Lube oil, 0.01 in on pol. Ni	20	0.82
Linseed, 1-2 coats on A1	20	0.56-0.57
Rubber, soft gray reclaimed	24	0.86
Silica, 3 μm	260-740	0.7-0.5
Misc. I: Shiny black lacquer, planed oak, white enamel, serpentine, gypsum, white enamel paint, roofing paper, lime plaster, black matte shellac	21	0.87-0.91
Misc. II: Glazed porcelain, white paper, fused quartz, polished marble, rough red brick, smooth glass, hard glossy rubber, flat black lacquer, water, electrographite	21	0.92-0.96

* When two temperatures and two emissivities are given they correspond, first to first and second to second, and linear interpolation is suggested.

metals, refractory materials, and natural products approximates that of a perfectly diffuse reflector. A better model, adequate for many calculational purposes, is achieved by assuming that the total reflectance ρ is the sum of diffuse and specular components ρ_D and ρ_S (Hottel and Sarofim, p. 180).

Black Surface Enclosures When several surfaces are present, the need arises for evaluating a geometric factor F, called the direct-view factor (view factor or configuration factor). Restriction is temporarily to black surfaces, the intensity from which is independent of angle of emission. Define F_{12} as the fraction of the radiation leaving surface A_1 in all directions which is intercepted by surface A_2. Since the net interchange between A_1 and A_2 must be zero when their temperatures are alike, it follows that $A_1F_{12} = A_2F_{21}$. From the definition of F and Eq. (2.3.1),

$$A_1F_{12} = \int_{A_1}\int_{\Omega}\frac{dA_1\cos\theta_1 d\Omega_1}{\pi} = \int_{A_1}\int_{A_2}\frac{dA_1\cos\theta_1 dA_2\cos\theta_2}{\pi r^2} \qquad (2.3.7)$$

where $dA\cos\theta$ is the projection of dA normal to θ, with respect to the line connecting dA_1 and dA_2. The product A_1F_{12}, having the dimensions of area, will be called the direct-interchange (or exchange) area and be designated by $\overline{s_1s_2}$, sometimes for brevity by $\overline{12}(=\overline{21})$. Clearly, $F_{11} + F_{12} + F_{13} + \ldots = 1$ (or $\overline{11}+\overline{12}+\overline{13}+\ldots = A_1$); and when A_1 cannot "see" itself, $F_{11} = 0$ ($\overline{11} = 0$). Expressions for F or $\overline{ss}$ have been derived for various surface arrangements. A comprehensive list of these expressions can be found in a web catalog supplementing Howell et al. (2016). Some of these expressions are presented in the following section.

Direct-View Factors and Direct Interchange Areas

CASE 1. Directly opposed parallel rectangles of equal dimensions, and with lengths of sides X and Y divided by separating distance z:

$$\overline{s_1s_2} = A_1F_{12} = A_2F_{21} = \frac{z^2}{\pi}\left[\ln\left\{\frac{(1+X^2)(1+Y^2)}{1+X^2+Y^2}\right\}^{1/2} + X\sqrt{1+Y^2}\tan^{-1}\frac{X}{\sqrt{1+Y^2}} + Y\sqrt{1+X^2}\tan^{-1}\frac{Y}{\sqrt{1+X^2}} - X\tan^{-1}X - Y\tan^{-1}Y\right]$$

See also Fig. 2.3.2.

CASE 2. Parallel circular disks with centers on a common normal and with radii R_1 and R_2 divided by separating distance z:

$$\overline{s_1s_2} = A_1F_{12} = A_2F_{21} = \frac{\pi z^2}{2}\left[1 + R_1^2 + R_2^2 - \sqrt{(1+R_1^2+R_2^2)^2 - 4R_1^2R_2^2}\right]$$

CASE 3. Rectangles in perpendicular planes of area A_1 and A_2, with a common edge l and with other dimension divided by $l = W_1$ and W_2:

$$\overline{s_1s_2} = A_1F_{12} = A_2F_{21} = \frac{l^2}{\pi}\left(\frac{1}{4}\ln\left\{\frac{(1+W_1^2)(1+W_2^2)}{1+W_1^2+W_2^2}\left[\frac{W_1^2(1+W_1^2+W_2^2)}{(1+W_1^2)(W_1^2+W_2^2)}\right]^{W_1^2}\left[\frac{W_2^2(1+W_1^2+W_2^2)}{(1+W_2^2)(W_1^2+W_2^2)}\right]^{W_2^2}\right\} + W_1\tan^{-1}\frac{1}{W_1} + W_2\tan^{-1}\frac{1}{W_2} - \sqrt{W_1^2+W_2^2}\tan^{-1}\frac{1}{\sqrt{W_1^2+W_2^2}}\right)$$

CASE 4. Circular cylinder of radius r_1 surrounded by cylinder of radius r_2, both of equal length l and on a common axis: Here; $A_1 = 2\pi r_1 l$, $A_2 = 2\pi r_2 l$. Let $R_1 = r_1/l$; $R_2 = r_2/l$; $B = [1/R_2{}^2 - (R_2/R_1)^2 + 1]$; $D = [1/R_1{}^2 + (R_2/R_1)^2 + 1]$; and $R = r_2/r_1$.

$$\overline{s_1s_2} = A_1F_{12} = A_2F_{21} = l^2\left\{R_1^2\left[\sqrt{(D+2)^2-4R^2}\cos^{-1}\frac{B}{DR} + B\sin^{-1}\frac{1}{R} - \frac{\pi}{2}D\right] + 2R_1\left(\pi - \cos^{-1}\frac{B}{D}\right)\right\}$$

$$\overline{s_2s_2} = A_2F_{22} = l^2R_1\left\{2\pi(R-1) + 4\tan^{-1}\left(2R_1\sqrt{R^2-1}\right) - \sqrt{4R^2 - \frac{1}{R_1^2}}\left(\frac{\pi}{2} + \sin^{-1}\frac{4(R^2-1) + 1/[R_1^2(1-2/R^2)]}{4(R^2-1+1/R_1^2)}\right) - \frac{1}{R_1}\left[\sin^{-1}\left(1-\frac{2}{R^2}\right) + \frac{\pi}{2}\right]\right\}$$

CASE 5. Two closed surfaces, one enclosing the other and neither having any negative curvature; A_1 is inside. Since $F_{12} = 1$,

$$\overline{s_1s_2} = A_1F_{12} = A_2F_{21} = A_1$$

$$F_{21} = \frac{A_1}{A_2} \qquad F_{22} = 1 - \frac{A_1}{A_2}$$

CASE 6. Sphere of total inside area A_T; radiative exchange between sphere segments of areas A_1 and A_2. Application of Eq. (2.3.7) shows that, independent of relative position,

$$\overline{s_1s_2} = A_1F_{12} = A_2F_{21} = \frac{A_1A_2}{A_T} \qquad F_{12} = \frac{A_2}{A_T}$$

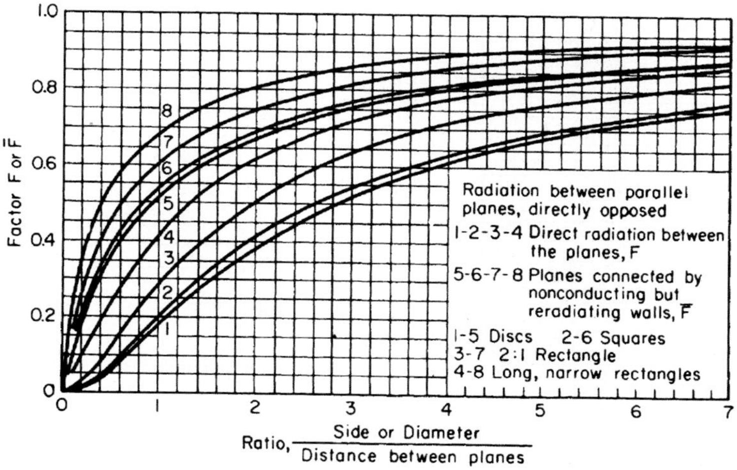

Figure 2.3.2 Variation of the factor F or F for parallel planes directly.

CASE 7. Two dimensional surfaces A_1 and A_2 per unit length normal to cross section, with each area defined by the length of string, stretched on inside face, between ends (i.e., elimination of negative curvature). Graphical exact solution: $\overline{s_1s_2}$ (or A_1F_{12} or A_2F_{21} per unit normal length) = sum of lengths of crossed stretched strings between ends of A_1 and A_2 minus sum of uncrossed strings, all divided by 2. If an obstruction lies between A_1 and A_2, there may be two sets of strings to represent views on both sides of the obstruction, with results added. The relations for cases 8, 9, and 10 are the results of three among many applications of this principle.

CASE 8. Exchange among inside surfaces of hollow triangular shape of infinite length and areas A_1, A_2, and A_3:

$$\overline{s_1s_2} = A_1F_{12} = A_2F_{21} = \frac{A_1 + A_2 - A_3}{2} \qquad F_{12} = \frac{A_1 + A_2 - A_3}{2A_1}$$

CASE 9. Exchange between two long parallel circular tubes of diameter D and center-to-center distance C, having areas A_{1a} and A_{1b} per unit length:

$$\overline{s_{1a}s_{2b}}(\text{per unit length}) = D\left[\sin^{-1}\frac{D}{C} + \sqrt{\left(\frac{C}{D}\right)^2 - 1} - \frac{C}{D}\right]$$

$$F_{1a-1b} = \frac{1}{\pi}\left[\sin^{-1}\frac{D}{C} + \sqrt{\left(\frac{C}{D}\right)^2 - 1} - \frac{C}{D}\right]$$

CASE 10. Exchange between a row of tubes and a plane parallel to it. Consider a unit length along tube axes, with single tube area $A_{1a} = \pi D$ and associated plane area $A_p = C$. A tube sees two tubes and two plane areas:

$$A_{1a} = 2\overline{s_{1a}s_{2b}} + 2\overline{s_{1a}s_p}$$

$$\overline{s_{1a}s_p} = \overline{s_ps_{1a}} = A_pF_{p1} = \frac{A_{1a}}{2} - \overline{s_{1a}s_{1b}}$$

Substituting from previous example (case 9) yields

$$F_{p1} = 1 - \frac{D}{C}\left[\sin^{-1}\frac{D}{C} + \sqrt{\left(\frac{C}{D}\right)^2 - 1} - \frac{\pi}{2}\right]$$

The value from case 10 appears as line 1 of Fig. 2.3.3. The same figure gives the fraction going to the second row. Additional curves in Fig. 2.3.3 can be obtained by considering the refractory backing as radiatively adiabatic, i.e., by assuming that the radiation that is not absorbed directly is reflected or reradiated, undergoing the same fractional absorption as the incoming beam. In a furnace chamber, one or two rows of tubes backed by a refractory can be considered as a zone that may be visualized as a continuous plane of area A_p at a temperature T_T, the tube surface-temperature, and having an effective absorptivity or emissivity ε (= $\overline{F}_{pT}$) that is equal to the value read from Fig. 2.3.3, line 5 or 6; in total exchange area nomenclature (see below), it is $\left(\overline{S_pS_T}\right)_R/A_p$. Its complement is headed back toward the emitter, which is whatever faces the replaced tube zone—radiating gas or surfaces or a mixture of them.

When the tubes are gray,

$$\frac{A_p}{\left(\overline{S_pS_T}\right)_R} = \left.\frac{A_p}{\left(\overline{S_pS_T}\right)_R}\right)_{\text{black tubes}} + \frac{\rho_T}{\varepsilon_T} \tag{2.3.8}$$

When C/D = 2, the treatment of a single tube row system with the tubes divided into two zones, front and rear half, reduces $\left(\overline{S_pS_T}\right)_R$ or $\overline{F}_{pT}$ below the value given by Eq. (2.3.9) by only 1.7 percent (3 percent) when ε_T is 0.8 (0.6). For other cases, see References.

The view factor F may often be evaluated from that for simpler configurations by the application of three principles: that of reciprocity, $A_i F_{ij} = A_jF_{ji}$; that of conservation, $\Sigma F_{ij} = 1$; and that due to Yamauti, showing that the exchange areas AF between two pairs of surfaces are equal when there is a one-to-one correspondence for all sets of symmetrically placed pairs of elements in the two surface combinations (Hottel and Sarofim, p. 60).

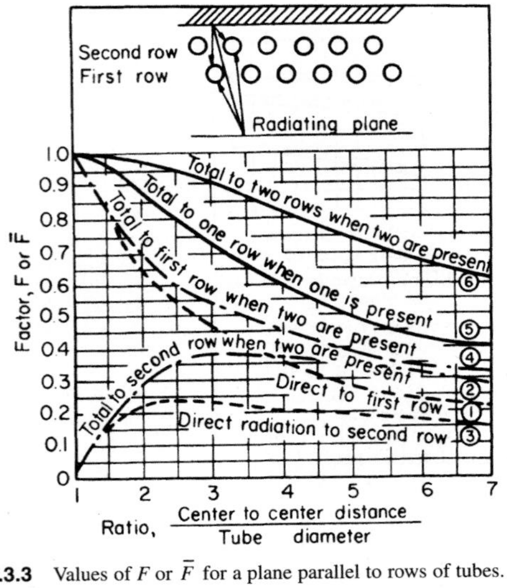

Figure 2.3.3 Values of F or $\overline{F}$ for a plane parallel to rows of tubes.

EXAMPLE. The exchange area between the two squares 1 and 4 of Fig. 2.3.4 is to be evaluated. The following exchange areas may be obtained from the values of F for common-side rectangles (case 3, direct-view factors): $A_1F_{13} = \overline{13} = 0.24$, $A_2F_{24} = \overline{24} = 2 \times 0.29 = 0.58$, $F_{(1+2)-(3+4)}A_{(1+2)} = \overline{(1+2)(3+4)} = 3 \times 0.32 = 0.96$. Expression of $A_{(1+2)}F_{(1+2)-(3+4)}\left(= \overline{(1+2)(3+4)}\right)$ in terms of its components yields $A_{(1+2)}F_{(1+2)-(3+4)} = A_1(F_{13} + F_{14}) + A_2(F_{23} + F_{24})\left(\overline{(1+2)(3+4)} = \overline{13} + \overline{14} + \overline{23} + \overline{24}\right)$. And by the Yamauti principle $F_{14} = F_{13}\left(\overline{14} = \overline{13}\right)$, since for every pair of elements in 1 and 4, there is a corresponding pair in 2 and 3. Therefore,

$$F_{14}A_1 = [F_{(1+2)-(3+4)}A_{(1+2)} - F_{13}A_1 - F_{24}A_2]/2 = \overline{14}$$
$$= \left[\overline{(1+2)(3+4)} - \overline{13} - \overline{24}\right]/2 = 0.07$$

Case 1 may be modified in the same way. Another example is the evaluation of AF for exchange between the outside of the smaller of two coaxial cylinders and the inside of the larger when they are not coextensive, given the view factor for coextensive cylinders (case 4).

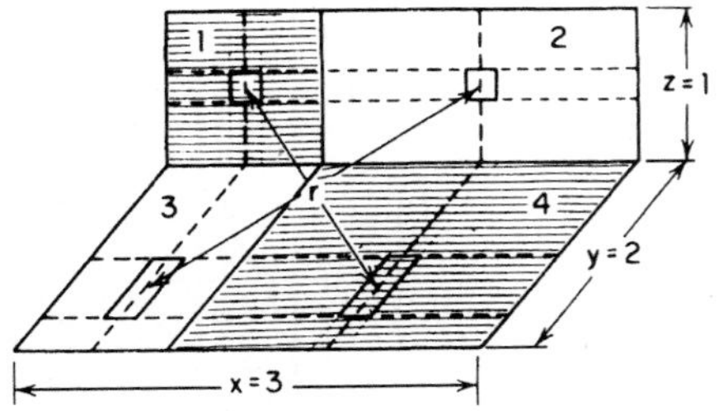

Figure 2.3.4 Illustration of the Yamauti principle.

Enclosures Containing Gray Surfaces, and No Absorbing Gas

The radiation exchange between two surfaces of an enclosure containing gray surfaces is more complicated as it involves multiple reflections from difference surfaces. The net radiation heat transfer to a surface can be described as

$$\dot{Q}_i = q_i A_i = (J_i - G_i)A_i \tag{2.3.9}$$

where J_i is referred as radiosity and G_i is referred as irradiation. J_i comprises of all outgoing radiation from surface i, including emission and reflection and is defined as

$$J_i = \varepsilon_i E_{b,i} + \rho_i G_i \tag{2.3.10}$$

and if all the surfaces are diffuse, the irradiation G_i is comprised as a sum of all radiosities from all surfaces weighted by the respective configuration factors in the enclosure as

$$G_i A_i = \sum_j J_j F_{ji} A_j = A_i \sum_j J_j F_{ij} \tag{2.3.11}$$

The irradiation term, G_i, in Eqs. (2.3.9) and (2.3.11) can be eliminated to yield

$$J_i = q_i + \sum_j J_j F_{ij} \tag{2.3.12}$$

that is often referred as the radiosity equation. In enclosure problems, some surfaces are radiative sources or sinks, with known or specified radiant heat fluxes and others are re-radiating with zero net radiant-heat flux (fulfilled by the average refractory wall where difference between internal convection and external loss is negligible compared with incident radiation). For such an enclosure with known radiative heat fluxes, Eq. (2.3.12) can be written for all the surfaces. However, a modified version of the equation must be used for surfaces where temperature is specified rather than heat flux. For a diffuse, gray, opaque surface; $\rho_i = 1 - \alpha_i = 1 - \varepsilon_i$, and from Eqs. (2.3.9) and (2.3.10):

$$J_i = E_{b,i} + \frac{1-\varepsilon_i}{\varepsilon_i} q_i \tag{2.3.13}$$

Replacing the radiant flux term in Eq. (2.3.12) using Eq. (2.3.13), an alternate version relating radiosity with blackbody emissive power can be defined as:

$$J_i + \frac{1-\varepsilon_i}{\varepsilon_i} \sum_j (J_i - J_j) F_{ij} = E_{b,i} \tag{2.3.14}$$

Once Eqs. (2.3.12) and (2.3.14) are written for all surfaces of a system comprised of N surfaces, $i = 1..N$; the N unknown radiosities can be estimated by solving the resulting linear system.

However, in most of the cases the surface temperature or blackbody emissive power is sought after rather than radiosity or specified rather than the radiant heat flux. Using an alternative formulation eliminating radiosity terms is more practical in such cases. Substituting Eq. (2.3.13) into Eq. (2.3.12) to replace all radiosity terms, yields

$$\sum_j \left[\frac{\delta_{ij}}{\varepsilon_j} - \left(\frac{1-\varepsilon_j}{\varepsilon_j} \right) F_{ij} \right] q_j = \sum_i (\delta_{ij} - F_{ij}) E_{b,j} \tag{2.3.15}$$

where δ_{ij} is the Knoecker delta, with $\delta_{ij} = 1$ for $i = j$ and $\delta_{ij} = 0$ for $i \neq j$. For an enclosure problem either surface heat fluxes (q_j) or surface temperatures/blackbody emissive powers ($E_{b,j} = \sigma T_j^4$) are specified, whereas the other is sought after. For an enclosure of N surfaces with known radiation configuration factors (F_{ij}); N-equations relying on Eq. (2.3.15) can be written either with radiant heat flux term (left hand side of equation), or blackbody emissive power term (right hand side of equation) being unknown, whereas the other is known. This yields a linear system with N-equations and N-unknowns (either q_i or $E_{b,i}$) that can be solved simultaneously using any linear system solver.

The formulation presented so far relies on configuration factors, F_{ij}, (or direct exchange areas, $\overline{s_i s_j}$ or $\overline{ij}$), and is limited to diffuse, gray surfaces. A more generalized formulation can be defined based on exchange factors, Γ_{ij} (or total exchange areas $\overline{S_i S_j}$, or transfer factors, $\tilde{F}_{ij}$). The exchange factor Γ_{ij} represents the fraction of all the radiation emitted by surface i that is absorbed by j by all means including direct and indirect through multiple reflections. The net radiant heat transferred from surface i to j can be described in terms of exchange factors as

$$\dot{Q}_{i-j} = \varepsilon_i E_{b,i} A_i \Gamma_{ij} - \varepsilon_j E_{b,j} A_j \Gamma_{ji} \tag{2.3.16}$$

The transfer factor, $\tilde{F}_{ij}$, and the total exchange area $\overline{S_i S_j}$, are related to exchange factor, Γ_{ij}, as: $\overline{S_i S_j} = \tilde{F}_{ij} A_i = \varepsilon_i \Gamma_{ij} A_i$. The equivalent expression can be defined in terms of transfer factor or total exchange areas as

$$\dot{Q}_{i-j} = E_{b,i} A_i \tilde{F}_{ij} - E_{b,j} A_j \tilde{F}_{ji} = E_{b,i} \overline{S_i S_j} - E_{b,j} \overline{S_j S_i} \tag{2.3.17}$$

The net radiant heat transfer for surface i is then defined in terms of exchange factors as

$$\dot{Q}_i = \sum_j \dot{Q}_{i-j} = \varepsilon_i E_{b,i} A_i - \sum_j \varepsilon_j E_{b,j} A_j \tilde{F}_{ji} \tag{2.3.18}$$

whereas the equivalent relation in terms of transfer factor and total exchange areas can be defined as

$$\dot{Q}_i = \varepsilon_i E_{b,i} A_i - \sum_j E_{b,j} A_j \tilde{F}_{ji} = \varepsilon_i E_{b,i} A_i - \sum_j E_{b,j} \overline{S_j S_i} \tag{2.3.19}$$

From the definitions of F, Γ, and, $\tilde{F}$, or of $\overline{ss}$, and $\overline{SS}$ it is to be noted that

$$\tilde{F}_{ij} = \varepsilon_i \Gamma_{ij}$$
$$\overline{S_i S_j} = \varepsilon_i A_i \Gamma_{ij} = A_i \tilde{F}_{ij}$$
$$\varepsilon_i A_i \Gamma_{ij} = \varepsilon_j A_j \Gamma_{ij}$$
$$A_i \tilde{F}_{ij} = A_j \tilde{F}_{ji}$$
$$\overline{S_i S_j} = \overline{S_j S_i}$$

$$F_{11} + F_{12} + \cdots + F_{1N} = 1$$
$$\Gamma_{11} + \Gamma_{12} + \cdots + \Gamma_{1N} = 1$$
$$\tilde{F}_{11} + \tilde{F}_{12} + \cdots + \tilde{F}_{1N} = \varepsilon_1$$

$$\overline{s_1 s_1} + \overline{s_1 s_2} + \cdots + \overline{s_1 s_N} = A_1$$
$$\text{or} \quad \overline{11} + \overline{12} + \cdots + \overline{1N} = A_1$$
$$\overline{S_1 S_1} + \overline{S_1 S_2} + \cdots + \overline{S_1 S_N} = A_1 \varepsilon_1$$

The exchange factors can be defined based on configuration factors for diffuse, gray surfaces. Considering the radiant heat transferred from surface i to j:

$$\dot{Q}_{i \to j} = \varepsilon_i E_{b,i} A_i \Gamma_{ij} = \varepsilon_i E_{b,i} A_i F_{ij} \alpha_j + \sum_k \varepsilon_i E_{b,i} A_i F_{ik} \rho_k \Gamma_{kj} \tag{2.3.20}$$

which is comprised of the direct exchange represented by the first term in the right hand side of Eq. (2.3.20) and the sum of all indirect terms represented by the second term. Considering the Kirchoff's law for opaque, diffuse, gray surfaces an equation relating exchange factors with configuration factors can be defined as

$$\Gamma_{ij} - \sum_k F_{ik}(1 - \varepsilon_k)\Gamma_{kj} = F_{ij}\varepsilon_j \tag{2.3.21}$$

For a system with N surfaces, $N \times N$ equations can be written for $N \times N$ unknown exchange factors (Γ_{ij} with $i = 1..N$, and $j = 1..N$). The exchange factors can be estimated from solution of the resulting linear system.

For systems with directional properties, the equations presented based on exchange or transfer factors are still valid, whereas those relying on configuration factors need alteration. For such systems exchange or transfer factors can be estimated by more sophisticated methods such as the Monte Carlo method (Ertürk and Howell, 2016) that relies on statistical sampling. While Monte Carlo methods require more computational time than utilizing the configuration factor formulations presented, they are capable of predicting the exchange or transfer factors of systems with directional properties or very complex geometries.

For simpler systems, direct derivations for exchange or transfer factors are possible considering the definitions and the properties of transfer and configuration factors listed above. As an example, consider radiation between two surfaces A_1 and A_2 which together form a complete enclosure. Equation (2.3.21) takes the form

$$A_1\tilde{F}_{12} = \frac{A_1}{1/F_{12} + 1/\varepsilon_1 - 1 + (A_1/A_2)(1/\varepsilon_2 - 1)} \tag{2.3.22}$$

Only one direct-view factor F_{12} is needed because F_{11} equals 1-F_{12} and F_{22} equals 1-F_{21} or 1-$F_{12}\,A_1/A_2$.

Special cases include:

1. Parallel plates, large compared to clearance. Substitution of $F_{12} = 1$ and $A_1 = A_2$ gives

$$A_1\tilde{F}_{12} = \frac{A_1}{1/\varepsilon_1 + 1/\varepsilon_2 - 1} \tag{2.3.23}$$

2. Sphere of area A_1 concentric with surrounding sphere of area A_2. $F_{12} = 1$. Then

$$A_1\tilde{F}_{12} = \frac{A_1}{1/\varepsilon_1 + (A_1/A_2)(1/\varepsilon_2 - 1)} \tag{2.3.24}$$

3. Body of surface A_1 having no negative curvature, surrounded by very much larger surface A_2. $F_{12} = 1$ and $A_1/A_2 \rightarrow 0$. Then

$$\tilde{F}_{12} = \varepsilon_1 \tag{2.3.25}$$

Many furnace problems are adequately handled by dividing the enclosure into but two source-sink zones A_1 and A_2 and any number of no-flux zones, A_r, A_s, ... For this case Eq. (2.3.15) yields

$$\frac{1}{\tilde{F}_{12}A_1} = \frac{1}{\tilde{F}_{21}A_2} = \frac{1}{A_1}\left(\frac{1}{\varepsilon_1} - 1\right) + \frac{1}{A_2}\left(\frac{1}{\varepsilon_2} - 1\right) + \frac{1}{\bar{F}_{12}A_1} \tag{2.3.26}$$

Here the expression $\bar{F}_{12} = \bar{F}_{21}$ represents the blackbody transfer factor for the limiting case of a black source and black sink (the refractory emissivity is of no moment). The blackbody transfer factor is known exactly for a few geometrically simple cases and may be approximated for others. If A_1 and A_2 are equal parallel disks, squares, or rectangles, connected by nonconducting but re-radiating refractory walls, then $\bar{F}$ is given by lines 5 to 8 of Fig. 2.3.2. If A_1 represents an infinite plane and A_2 is one or two rows of infinite parallel tubes in a parallel plane, and if the only other surface is a refractory surface behind the tubes, $\bar{F}_{12}$ is given by line 5 or 6 of Fig. 2.3.3. If an enclosure may be divided into several radiant-heat sources or sinks A_1, A_2, etc., and the rest of the enclosure (re-radiating refractory surface) may be lumped together as A_r at a uniform temperature T_r, then the total interchange area for zone pairs in the black system is given by

$$\bar{F}_{12} = F_{12} + \frac{F_{1r}F_{r2}}{1 - F_{rr}} \tag{2.3.27}$$

For the two-source-sink-zone system to which Eq. (2.3.27) applies, Eq. (2.3.27) simplifies to $\left(\overline{S_1S_2}\right)_B = \overline{12} + 1/[1/\overline{1r} + 1/\overline{2r}]$; and if A_1 and A_2 each can see none of itself, there is further simplification to

$$\bar{F}_{12} = F_{12} + \frac{1}{1/(1 - F_{12}) + A_1/(A_2 - A_1F_{12})}$$
$$= \frac{A_2 - A_1F_{12}^2}{1 + A_2/A_1 - 2F_{12}} \tag{2.3.28}$$

which necessitates the evaluation of one direct-view or configuration factor F.

Equation (2.3.26) covers many problems of radiant heat interchange between source and sink in furnace enclosures involving no radiating gas. The error due to single zoning of source and sink is small even if the "views" of the enclosure from different parts of each zone are quite different, provided the emissivity is fairly high; the error in F is zero if it is obtainable from Fig. 2.3.2 or 2.3.3, small if Eq. (2.3.27) is used and the variation in temperature over the refractory is small.

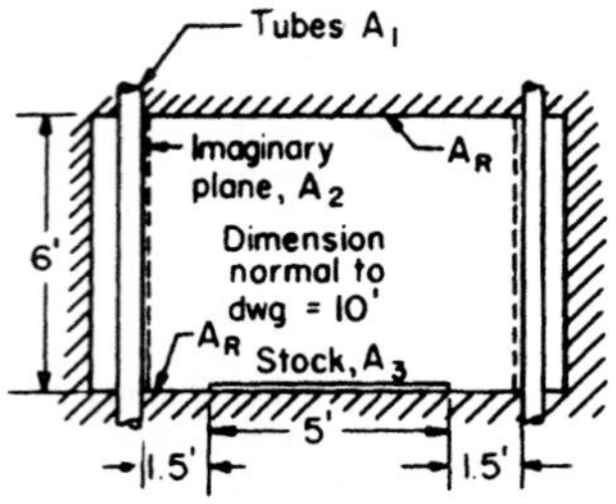

Figure 2.3.5 Dimensions of a furnace chamber.

EXAMPLE. A furnace chamber of rectangular parallelepiped form is heated by the combustion of gas inside vertical radiant tubes lining the side walls. The tubes are on centers 2.4 diameters apart. The stock forms a continuous plane on the hearth. Roof and end walls are refractory. Dimensions are shown in Fig. 2.3.5. The radiant tubes and stock are gray bodies having emissivities 0.8 and 0.9, respectively. What is the net rate of heat transmission to the stock by radiation when the mean temperature of the tube surface is 1,089 K and that of the stock is 922 K?

This problem must be broken up into two parts, first considering the walls with their refractory-backed tubes. To imaginary planes A_2 of area 6 × 10 ft and located parallel to and inside the rows of radiant tubes, the tubes emit radiation $\sigma T_1^4 A_1\tilde{F}_{12}$, which equals $\sigma T_1^4 A_2\tilde{F}_{21}$. To find $\bar{F}_{21}$ use Fig. 2.3.3, line 5, from which $\bar{F}_{21} = 0.81$. Then from Eq. (2.3.21).

$$\tilde{F}_{12} = 1/[(1/0.81) + (1/1 - 1) + (2.4/\pi)(1/0.8 - 1)] = 0.702$$

This amounts to saying that the system of refractory-backed tubes is equal in radiating power to a continuous plane A_2 replacing the tubes and refractory back of them, having a temperature equal to that of the tubes and an equivalent or effective emissivity of 0.702.

The new simplified furnace now consists of an enclosure formed by two 1.83 × 3.05 m² radiating side walls (area A_2, of emissivity 0.702), a 1.52 × 3.05 m² receiving plane on the floor (A_3), and refractory surfaces (A_R) to complete the enclosure (ends, roof, and floor side strips); the desired heat transfer is

$$q_{2\text{-}3} = \sigma(T_2^4 - T_3^4)A_2\tilde{F}_{23}$$

To evaluate $\tilde{F}_{23}$, start with the direct interchange factor F_{23}. $F_{23} = F$ from A_2 to (A_3 + a strip of A_R alongside A_3 which has a common edge with A_2) minus F from A_2 to the strip only. These two Fs may be evaluated from case 3 for direct-view factors. For the first F, $Y/X = 1.83/3.05$, $Z/X = 1.98/3.05$, $F = 0.239$; for the second F, $Y/X = 1.83/3.05$, $Z/X = 0.46/3.05$, $F = 0.100$. Then $F_{23} = 0.239\text{-}0.10 = 0.139$. Now F may be evaluated. From Eq. (2.3.22) et seq.,

$$A_2\bar{F}_{23} = \overline{23} + \frac{1}{1/\overline{2r} + 1/\overline{3r}}$$

$$\bar{F}_{23} = \bar{F}_{23} + \frac{1}{1/F_{2r} + (A_2/A_3)(1/F_{3r})}$$

Since A_2 "sees" A_r, A_3, and some of itself (the plane opposite), $F_{2r} = 1 - F_{22} - F_{23}$. F_{22}, the direct interchange factor between parallel 1.83 × 3.05 m rectangles separated by 2.44 m, may be taken as the geometric mean of the factors for 1.83-m squares separated by 2.44 m, and 3.05-m squares separated by 2.44 m. These come from Fig. 2.3.2, line 2, according to which $F_{22} = \sqrt{0.13 \times 0.255} = 0.182$. Alternatively, the first of the 10 cases listed above under "Direct-View Factors" may be used. Then F_{2r} = 1 − 0.182 − 0.139 = 0.679. The other required direct factor is F_{3r} = 1 − F_{32} = 1 − $F_{23}\,A_2/A_3$ = 1 − 0.139 × (120/50) = 0.666. Then $\bar{F}_{23} = 0.139 + \{1/[(1/0.679) + (120/50)(1/0.666)]\} = 0.336$. Having $\bar{F}_{23}$, we may now evaluate the factor $\tilde{F}_{23}$ using Eq. (2.3.22) with $A_1 \rightarrow A_2$, $A_2 \rightarrow A_3$, and $\left[\overline{S_1S_2}\right]_{1B} \rightarrow A_2\bar{F}_{23}$.

$$\tilde{F}_{23} = \frac{1}{1/0.336 + 1/0.702 - 1 + (120/50)(1/0.9 - 1)} = 0.273$$

$$\dot{Q}_{net} = \sigma(T_1^4 - T_3^4)A_2\tilde{F}_{23} = 5.67(10.89^4 - 9.22^4)(11.15)(0.273) = 118000\ W$$

A result of interest is obtained by dividing the term $A_2\tilde{F}_{23}(11.15 \times 0.273)$, or 3.04 m² by the actual area A_1 of the radiating tubes [(π/2.4) × 60 × 2 = 14.6 m²]. This is 3.04/14.6 = 0.208; i.e., the net radiation from a tube to the stock is 20.8 percent as much as if the tube were black and completely surrounded by black stock.

Enclosures of Surfaces That Are Not Diffuse Reflectors The total interchange-area concept has been generalized to include surfaces the reflectance ρ of which can be divided into a diffuse, or Lambert-reflecting, component ρ_D and a specular component ρ_S independent of angle of incidence, with $\varepsilon + \rho_S + \rho_D = 1$. In application to concentric spheres or infinite cylinders, with A_1 the inner surface, the method yields (Hottel and Sarofim, p. 181)

$$A_1\tilde{F}_{12} = \frac{1}{\dfrac{1}{A_1\varepsilon_1} + \dfrac{1}{A_2}\left(\dfrac{1}{\varepsilon_2} - 1\right) + \dfrac{\rho_{s2}}{1-\rho_{s2}}\left(\dfrac{1}{A_1} - \dfrac{1}{A_2}\right)} \quad (2.3.29)$$

When there is no specular reflectance, the third term in the denominator drops out, in agreement with Eq. (2.3.24). When the reflectance is exclusively specular, the denominator becomes $1/(A_1\varepsilon_1)+\rho_{S2}/[A_1(1-\rho_{S2})]$, easily derivable from first principles.

RADIATION FROM FLAMES, COMBUSTION PRODUCTS, AND PARTICLE CLOUDS

The radiation from a flame consists of (1) radiation throughout the spectrum from burning soot particles of microscopic and submicroscopic dimensions, from suspended larger particles of coal, coke, or ash, all contributing to what is spoken of as flame luminosity, (2) infrared radiation, mostly from the water vapor and carbon dioxide in the hot gaseous combustion products, and (3) nonequilibrium radiation associated with the combustion process itself, called chemiluminescence and not a significant contributor to the total radiation. A major problem is the effect of the shape of the emitting volume on the radiative flux; this will be considered first.

Mean Beam Lengths Evaluation of radiation from a nonisothermal volume is beyond the scope of this section (see Howell et al., Chaps. 11 and 12). If a volume emitter is isothermal and at a temperature T, the ratio of the emission from an element of its volume subtending the solid angle $d\Omega$ at a receiver element dA, and making the angle θ with the normal thereto, to blackbody radiation arriving from within the same solid angle is called the gas emissivity. Clearly, ε depends on the path length L through the volume to dA. A hemispherical volume radiating to a spot on the center of its base represents the case in which L is independent of direction. Flux at that spot relative to hemispherical blackbody flux is thus an alternative way to visualize emissivity. The flux density to an area of interest on the envelope of an emitter volume of any shape can be matched by that at the base of a hemispherical volume of some radius L, which will be called the mean beam length. It is found that, although the ratio of L to a characteristic dimension D of the shape varies with opacity, the variation is small enough for most engineering purposes to permit use of a constant ratio, L_M/D, where L_M is the average mean beam length. L_M can be defined to apply either to a spot on the envelope or to any finite portion of its area. An important limiting case is that of opacity approaching zero ($pD \rightarrow 0$, where p = partial pressure of the emitter constituent). For this case, L (called L_0) equals 4 × ratio of gas volume to bounding area when interest is in radiation to the entire envelope. For the range of pD encountered in practice, L (now L_M) is always less. For various shapes, 0.8 to 0.95 times L_0 has been found optimum (see Table 2.3.3); for shapes not reported in Table 2.3.3, a factor of 0.88 (or $L_M = 0.88$, $L_0 = 3.5V/A$) is recommended.

Luminous Flames Luminosity conventionally refers to soot radiation. Soot is formed in locally fuel-rich portions of flames in amounts that usually correspond, at atmospheric pressure, to amounts less than 1 percent of the carbon in the fuel. Because soot particles are small relative to the wavelength of the radiation of interest in flames (soot primary particle diameters on the order of 20 nm compared to wavelengths of interest of 500 to 8,000 nm), the incident radiation permeates the particles and the absorption is proportional to the volume of the particles. In the limit of $r/\lambda << 1$, the so-called Rayleigh limit, the monochromatic emissivity ε_l is given by

$$\varepsilon_\lambda = 1 - e^{-Kf_vL/\lambda} \quad (2.3.30)$$

where f_v is the volume of soot per unit volume of space, L is the path length in the same units as λ, and K is dimensionless. The value K will vary with fuel type, experimental conditions, and the temperature history of the soot. The values from a wide range of systems are within a factor of about 2 of one another. The single most important factor governing the value of K is the hydrogen/carbon ratio of the soot with the value of K increasing as the H/C ratio decreases. A value of 9.9 is recommended on the basis of seven studies involving 29 fuels (Mulholland and Croarkin, 2000).

The total emissivity of soot ε_s can be shown to be

$$\varepsilon_s = \int_0^1 \varepsilon_\lambda\, df(\lambda T) = 1 - \frac{15}{4}\left[\Psi^{(3)}\left(1 + \frac{Kf_vLT}{c_2}\right)\right] \quad (2.3.31)$$

where $\Psi^{(3)}(x)$ is the pentagamma function of x. An excellent approximation to Eq. (2.3.31) is

$$\varepsilon_s \approx 1 - \left(1 + \frac{Kf_vLT}{c_2}\right)^{-4} \quad (2.3.32)$$

Expression of ε_s in an e power for combination with the gas emissivity correlation is feasible. In that form

$$\varepsilon_s = 1 - e^{-3.83Kf_vLT/c_2} \quad (2.3.33)$$

The approximate relation is good to better than 5 percent for arguments giving values of ε_s less than 0.5.

The biggest uncertainty at present in estimating soot emissivities lies in the estimation of the soot volume fraction f_v. Soot forms in the fuel-rich zones of flames. Soot formation rates are a function of fuel type, mixing rate, local equivalence ratio, temperature, and pressure. Soot formation increases with the aromaticity or C/H ratio of fuels with benzene, a-methyl naphthalene, and acetylene having a high propensity to form soot and methane having a low soot formation propensity; oxygenated fuels, such as alcohols, emit little soot. In practical turbulent diffusion flames, soot forms on the fuel side of the flame front; in premixed flames, at a given temperature, the rate of soot formation increases rapidly as the equivalence ratio exceeds 2.0. Soot burns out rapidly, in less than 0.1 s, under fuel-lean conditions (equivalence ratios less than 1) at temperature above 1,500 K. Because of this rapid soot burnout, soot is usually localized in a relatively small fraction of a furnace or combustor volume. Long, poorly mixed diffusion flames promote soot formation while highly backmixed combustors can burn soot-free. The peak volumetric soot concentration f_v in a typical atmospheric pressure flame is of the range of 1×10^{-7} to 1×10^{-6}, corresponding to a conversion to soot of about 1.5 to 15 percent of the carbon in the fuel. (In calculating f_v at high pressures, allowance needs to be made for the significant increase with pressure of soot formation and for the inverse proportionality of f_v and pressure). Great progress has been made in the ability to calculate soot in premixed flames [see, for examples, the comparisons of predictions of soot concentration in a well-stirred reactor operated over a wide range of equivalent ratios and temperatures in Brown et al. (1998) and in a turbulent diffusion flame (Zucca et al., 2005)].

The importance of soot radiation varies widely between combustors: In large boilers the soot is confined to a small volume and is of only local importance; in gas turbines minimizing the heat flux to the combustor lining is of paramount importance so that only small incremental soot radiation is of concern; in high-temperature glass tanks, the added 0.1 to 0.2 due to soot emissivity is valued in oil-fired flames, and efforts, mostly unsuccessful, had been made to augment the emissivity of natural gas–fired flames; soot contributions to the radiation of pool fires are often dominant and impact the safe separation distances from dikes around oil tanks and the distance from a rig where a flare is located.

Clouds of Large Black Particles The emissivity of a cloud of particles depends on their area projected along the line of sight. The projected area per unit volume of space is the projected area A of a particle times the particle number concentration c, or the volume fraction f_v of space

Table 2.3.3 Mean Beam Lengths for Volume Radiation

Shape	Characteristic dimension D	L_0/D	L_M/D
Sphere	Diameter	0.67	0.63
Infinite cylinder	Diameter	1	0.94
Semi-infinite cylinder, radiating to:			
Center of base	Diameter	1	0.90
Entire base	Diameter	0.81	0.65
Right-circle cylinder, ht = diam, radiating to:			
Center of base	Diameter	0.76	0.71
Whole surface	Diameter	0.67	0.60
Right-circle cylinder, ht = 0.5 diam, radiating to:			
End	Diameter	0.47	0.43
Side	Diameter	0.52	0.46
Total surface	Diameter	0.50	0.45
Right-circle cylinder, ht = 2 × diam, radiating to:			
End	Diameter	0.73	0.60
Side	Diameter	0.82	0.76
Total surface	Diameter	0.80	0.73
Infinite cylinder, half-circle cross section, radiating to spot on middle of flat side	Radius		1.26
Rectangular parallelepipeds			
1:1:1 (cube)	Edge	0.67	0.60
1:1:4, radiating to:			
1 × 4 face	Shortest edge	0.90	0.82
1 × 1 face	Shortest edge	0.86	0.71
Whole surface	Shortest edge	0.89	0.81
1:2:6, radiating to:			
2 × 6 face	Shortest edge	1.18	
1 × 6 face	Shortest edge	1.24	
1 × 2 face	Shortest edge	1.18	
Whole surface	Shortest edge	1.2	
Infinite parallel planes	Clearance	2.00	1.76
Space outside infinite bank of tubes, centers on equilateral triangles; tube diam = clearance	Clearance	3.4	2.8
Same, except tube diam = 0.5 clearance	Clearance	4.45	3.8
Same, except tube centers on squares, diam = clearance	Clearance	4.1	3.5

occupied by particles times b/d, the projected-surface/volume ratio, where d is the characteristic dimension. [For any randomly oriented particles without dimples, A/(total area) is 1/4; for spheres, b = 3/2.] The emissivity of a particle cloud is then given by the alternative formulations

$$\varepsilon = 1 - e^{-bf_v L/d} = 1 - e^{-cA} \tag{2.3.34}$$

As an example, consider heavy fuel oil ($CH_{1.5}$, s.g. 0.95) atomized to a surface mean particle diameter of d μm, burned with 20 percent excess air to produce coke residue particles having the original drop diameter, and suspended in combustion products at 1,500 K. From stoichiometry, $f_v = 1.27 \times 10^{-5}$. For spherical particles $b = 3/2$, and the flame emmisivity due to the particles along a path L will be $1 - e^{-1.9\times10^{-5} L/d}$. With 200-$\mu$m particles and an L of 3 m, the particle contribution to emissivity will be 0.25. Soot luminosity will increase this; particle burnout will decrease it. The combined emissivity due to several kinds of emitters will be treated later. The correction for nonblackness of the particles is complicated by multiple scatter of the radiation reflected by each particle. The emissivity ε_M of a cloud of gray particles of individual surface emissivity ε_1 can be estimated by the use of Eq. (2.3.34) with its exponent multiplied by ε_1 if the optical thickness cAL does not exceed about 2.

Gaseous Combustion Products Radiation from water vapor and carbon dioxide occurs in spectral bands in the infrared. Its magnitude is 3 to 10 times that of convection at furnace temperatures. It depends on gas temperature T_G, on the partial pressure-beam length products p_wL and p_cL (subscripts w and c refer to water vapor and carbon dioxide), and to a much lesser extent on total and partial pressure. The gas emissivity ε_G is the sum of the separate contributions due to H_2O and CO_2, corrected for pressure broadening of the spectral bands and for band overlap (Hottel and Sarofim, Chap. 6, Howel et al., Chap. 9). The elaborate calculations can be combined for a restricted set of conditions, here taken to be the practically important cases of 1-atm total pressure and partial pressures representative of fossil fuel combustion in air. In the range of furnace operating conditions the product $\overline{\varepsilon_G T_G}$ varies much less than ε_G with T_G, and $\overline{\varepsilon_G T_G}$ depends primarily on $(p_w + p_c)L$, much less on $p_w/(p_w + p_c)$, and so little on T_G as to permit linear interpolation between widely separated T_G's. An equation of the form

$$\log \overline{\varepsilon_G T_G} = a_0 + a_1 \log pL + a_3 \log^3 pL \tag{2.3.35}$$

where p is the sum of partial pressures $p_w + p_c$ atm and L is the mean beam length, has been found capable of fitting emissivity data over a 1,000-fold range of pL, from 0.01 to 10 m · atm. Table 2.3.4, section 2, gives values of the constants representing the results of an averaging of all the available total and integrated spectral data on CO_2 and H_2O, together with corrections for spectral band broadening and overlap. Equation (2.3.35) represents the original data with a precision greater than their accuracy. The constants are given for computation in meters and Kelvins for mixtures, in non-radiating gases, of water vapor alone, CO_2 alone, and four p_w/p_c mixtures. Four suffice, since a change halfway from one mixture ratio to the adjacent one changes the emissivity

Table 2.3.4 Emissivity, ε_G, of H_2O-CO_2 Mixtures

Section 1: Limited range for furnaces, valid over 25-fold range of $p_{w+c}L$, 0.046-1.15 m · atm

p_w/p_c	0		0.5		1		2		3		oo	
$\frac{p_w}{p_w+p_c}$	0(0-0.2)		1/3(0.2-0.42)		1/2(0.42-0.6)		2/3(0.6-0.7)		3/4(0.7-0.8)		1(0.8-1.0)	
	CO_2 only		Corresponding to $(CH)_x$, covering coal, heavy oils, pitch		Corresponding to $(CH_2)_x$, covering distillate oils, paraffins, olefins		Corresponding to CH_4, covering natural gas and refinery gas		Corresponding to $(CH_6)_x$, covering future high-H_2 fuels		H_2O only	
	Constant b and n of equation $\overline{\varepsilon_G T_G} = b(pL - 0.015)^n$, pL in m · atm, T in K											
T, K	b	n	b	n	b	n	b	n	b	n	b	n
1,000	188	0.209	384	0.33	416	0.34	444	0.34	455	0.35	416	0.400
1,500	252	0.256	448	0.38	495	0.40	540	0.42	548	0.42	548	0.523
2,000	267	0.316	451	0.45	509	0.48	572	0.51	594	0.52	632	0.640

Section 2: Full range, valid over 2,000-fold range of $p_{w+c}L$, 0.005-10.0 m · atm Constants of equation, $\log \overline{\varepsilon_G T_G} = a_0 + a_1 \log pL + a_2 \log^2 pL + a_3 \log^3 pL$

$\frac{p_w}{p_c}$	$\frac{p_w}{p_w+p_c}$	T, K	a_0	a_1	a_2	a_3
				pL in m · atm, T in K		
0	0	1,000	2.2661	0.1742	−0.0390	0.0040
		1,500	2.3954	0.2203	−0.0433	0.00562
		2,000	2.4104	0.2602	−0.0651	−0.00155
$\frac{1}{2}$	$\frac{1}{3}$	1,000	2.5754	0.2792	−0.0648	0.0017
		1,500	2.6461	0.3418	−0.0685	−0.0043
		2,000	2.6504	0.4279	−0.0674	−0.0120
1	$\frac{1}{2}$	1,000	2.6090	0.2799	−0.0745	−0.0006
		1,500	2.6862	0.3450	−0.0816	−0.0039
		2,000	2.7029	0.4440	−0.0859	−0.0135
2	$\frac{2}{3}$	1,000	2.6367	0.2723	−0.0804	0.0030
		1,500	2.7178	0.3386	−0.0990	−0.0030
		2,000	2.7482	0.4464	−0.1086	−0.0139
3	$\frac{3}{4}$	1,000	2.6432	0.2715	−0.0816	0.0052
		1,500	2.7257	0.3355	−0.0981	0.0045
		2,000	2.7592	0.4372	−0.1122	−0.0065
∞	1	1,000	2.5995	0.3015	−0.0961	0.0119
		1,500	2.7083	0.3969	−0.1309	0.00123
		2,000	2.7709	0.5099	−0.1646	−0.0165

NOTE: Values of $p_w/(p_w + p_c)$ of 1/3, 1/2, 2/3, 3/4 may be used to cover the ranges 0.2-0.42, 0.42-0.6, 0.6-0.7, and 0.7-0.8, respectively, with a maximum error in ε_G of 5 percent at $pL = 6.5$ m · atm, less at lower pL's. Linear interpolation reduces the error generally to less than 1 percent. Linear interpolation or extrapolation on T introduces an error generally below 2 percent, less than the accuracy of the original data.

by a maximum of only 5 percent; linear interpolation may be used if necessary. The constants are given for only three temperatures, which is adequate for linear interpolation since $\overline{\varepsilon_G T}$ changes a maximum of only one-sixth due to a change from one temperature base halfway to the adjacent one. Based on meter atmospheres and Kelvins, the interpolation relation, with T_H and T_L representing the higher and lower base temperatures bracketing T, and with the brackets in the term $[A(x)]$ indicating that the parentheses refer not to a multiplier but to an argument, is

$$\overline{\varepsilon_G T_G} = \frac{\left[\overline{\varepsilon_G T_H}(pL)\right](T_G - T_L) + \left[\overline{\varepsilon_G T_L}(pL)\right](T_H - T_G)}{500} \tag{2.3.36}$$

Extrapolation to a temperature which is above the highest or below the lowest of the three base temperatures in Table 2.3.4 uses the same formulation, but one of the terms becomes negative. Linearization on the constants a_0 to a_3 rather than on $\overline{\varepsilon_G T}$ may be preferable if fuel quality is unchanging.

When pL lies in the 25-fold range of 0.046 to 1.15 m · atm, adequate for furnaces, a much simpler two-constant relation is adequate.

$$\overline{\varepsilon_G T} = b(pL - 0.015)^n \text{ with } T = K,\ pL = \text{m} \cdot \text{atm}$$

The constants are given in Table 2.3.4, section 1.

Combined Radiation from Gases and Suspended Solids The total emissivity of gases and suspended solids is less than the sum of the separate contributions because of interference between overlapping spectral emissions. The spectral overlap of H_2O and CO_2 radiation has been taken into account by the constants of Table 2.3.4 used for obtaining ε_G. Additional overlap occurs when soot emissivity ε_s is added. If the emission bands of water vapor and CO_2 were randomly placed in the spectrum and soot radiation were gray, the combined emissivity would be $\varepsilon_G + \varepsilon_s$ minus an overlap correction $\varepsilon_G\varepsilon_s$. Mono-chromatic soot emissivity is higher as the wavelength gets shorter, and in a highly sooted flame at 1,500 K half the soot emission lies below 2.5 μm where H_2O and CO_2 emission is negligible. Then the correction $\varepsilon_G\varepsilon_s$ must be reduced, and the following is recommended:

$$\varepsilon_{G+s} = \varepsilon_G + \varepsilon_s - M\varepsilon_G\varepsilon_s \tag{2.3.37}$$

where M depends mostly on T_G and to a much less extent on the optical density $Kf_v L$. Values that have been calculated from this simple model can be represented with acceptable error by

$$M = 1.07 + 2.7 \times 10^{-5} f_v L - 0.27(T/1{,}000) \qquad (0 < M < 1)$$

If, in addition to gas and soot, massive particles such as fly ash, coal char, or carbonaceous cenospheres from heavy fuel oil of emissivity ε_M are present, it is recommended that the total emissivity be approximated by

$$\varepsilon_{\text{total}} = \varepsilon_{G+s} + \varepsilon_M - \varepsilon_{G+s}\varepsilon_M \tag{2.3.38}$$

Radiant interchange between a gas and a completely bounding black surface at T_1 produces a surface flux density q given by

$$q = \sigma(\varepsilon_G T_G^4 - \alpha_{G1} T_1^4) \tag{2.3.39}$$

where α_{G1} is the **absorptivity** of the gas at T_G for radiation from a surface at T_1. The absorptivity of water vapor-CO_2 mixtures may also be obtained from the constants for emissivities. The product $\overline{\alpha_{G1}T_1}$—the absorptivity of gas at T_G for black radiation at T_1 times the surface temperature—is the product $\overline{\varepsilon_G T_1}$ with ε_G evaluated at surface temperature T_1 instead of T_G and at pLT_1/T_G instead of pL, then multiplied by $(T_G/T_1)^{0.5}$, or

$$\overline{\alpha_{G1}T_1} = \left[\overline{\varepsilon_G T_1}(pLT_1/T_G)\right](T_G/T_1)^{0.5} \tag{2.3.40}$$

The exponent 0.5 is an adequate average of the exponents for the pure components. The interpolation relation for absorptivity is

$$\overline{\alpha_{G1}T_1} = \left[\overline{\varepsilon_G T_1}\left(\frac{pLT_H}{T_G}\right)\right]\left(\frac{T_G}{T_H}\right)^{0.5}\frac{T_1 - T_L}{500} + \left[\overline{\varepsilon_G T_L}\left(\frac{pLT_L}{T_G}\right)\right]\left(\frac{T_G}{T_L}\right)^{0.5}\frac{T_H - T_1}{500} \tag{2.3.41}$$

The base temperature pair T_H and T_L can be different for the evaluation of ε_G and α_{G1} if T_G and T_1 are far enough apart. Extrapolation from the lowest T_G in Eq. (2.3.41) to a much lower T_1 to obtain α_{G1} may yield too high a value for it. That occurs, however, only when $T_1 << T_G$, and the fourth-power temperature relation makes the error in q negligible.

If the surface is not black, the right-hand side of Eq. (2.3.39) must be modified. If the surface is gray, multiplication by $\alpha_1(=\varepsilon_1)$ allows for reduction in the primary beam from gas to surface and surface to gas, but some of the gas radiation initially reflected from the surface has further opportunity for absorption at the surface because the gas is but incompletely opaque to the reflected beam. Consequently, the factor to allow for surface grayness lies between absorptance α_1 and unity, nearer the latter the more transparent the gas (low pL) and the more convoluted the surface. In the absorptance range of most industrial surfaces, 0.7 to 1.0, an adequate approximation consists in use of an effective absorptance α_1' halfway between the actual value and unity. If the surface is not gray, q depends much more on surface absorptance, which modifies $\varepsilon_G T_G$, than on emittance, which modifies $\alpha_{G1}T_1$. Absorption is treated more rigorously later in the section.

Example. Flue gas containing 9.5 percent CO_2 and 7.1 percent H_2O, wet basis, flows through a bank of tubes of 3.8 cm OD on equilateral triangular centers 11.43 cm apart. In a section in which the gas and tube surface temperatures are 1,200 and 811 K, what is the heat transfer rate per tube area, due to gas radiation only? Tube surface absorptance = 0.8.

$T_G = 1{,}200$ K; $T_S = 811$ K

$p_w/(p_w + p_c) = 7.1/16.6 = 0.428$; use 0.5

$pL = 0.166\,[3.8(11.43 - 3.8) \times 10^{-2}] = 0.0480$ m · atm

$pL(T_s/T_G) = 0.0480(811/1{,}200) = 0.0325$ m · atm

From Table 2.3.4, for $T_G = 1{,}500$ K and $pL = 0.0480$ and 0.0325, $\varepsilon T = 125$ and 101 K, and for $T_G = 1{,}000$ K and $pL = 0.0480$ and 0.0325, $\varepsilon T = 129$ and 107 K. Then

$$\varepsilon_G = \frac{1}{1{,}200}\frac{125(1{,}200 - 1{,}000) + 129(1{,}500 - 1{,}200)}{1{,}500 - 1{,}000} = 0.106$$

$$\alpha_{G1} = \frac{1}{811}\frac{101(811 - 1{,}000) + 107(1{,}500 - 811)}{1{,}500 - 1{,}000} = 0.135$$

The effective surface absorptance factor $\alpha_1 = (0.8 + 1)/2 = 0.9$. From Eq. (2.3.34), modified, components. The interpolation relation for absorptivity is

$$q = 0.9 \times 5.67(0.106 \times 12^4 - 0.135 \times 8.111^4) = 8{,}235\ \text{W/m}^2$$

This is equivalent to a convection coefficient of 21.2 W/(m²·K). The emissivity of an equivalent gray flame is $(0.106 \times 21.6^4 - 0.135 \times 14.6^4)/(21.6^4 - 14.6^4) = 0.098$.

RADIATIVE EXCHANGE IN ENCLOSURES OF RADIATING GAS

The so-called radiant section of a furnace presents a heat-transfer problem in which there enters the combined action of direct radiation from the flame to the stock or heat sink and radiation from the flame to refractory surfaces and thence back through the flame (with partial absorption) to the sink, convection, and external losses. Solutions of the problem based on varying degrees of simplification are available, including allowance for temperature variation in both gas and refractory walls (Hottel and Sarofim, Chap. 14). A less rigorous treatment suffices, however, for handling many problems. There are two limiting cases: the long chamber with gas temperature varying only in the direction of gas flow and the compact chamber containing a gas or flame at a uniform temperature. The latter, with variations, will be considered first.

Total Exchange Areas $\overline{SS}$ and $\overline{GS}$ The arguments leading to the development of the exchange or transfer factor $\varepsilon_i A_i \Gamma_{ij} \left(= A_i \bar{F}_{ij} = \overline{S_i S_j}\right)$ between surface zones [Eq. (2.3.15) et seq.] apply to the case of absorption within the gas volume if, in the evaluation of the exchange factor (or transfer factor or direct exchange area), allowance is made for attenuation of the radiant beam through the gas. This necessitates nothing more than the redefinition, in Eqs. (2.3.7) to (2.3.28), of every term $A_i F_{ij} \left(= \overline{ij} = \overline{s_i s_j}\right)$ to represent, per unit of blackbody emissive power, flux from A_i through an absorbing gas to A_j; that is, the prior F_{ij} must be multiplied by a mean transmittance τ_{ij} of the gas (= 1-α_{ij} = 1-ε_G for a gray gas). In a system containing an isothermal gas and source-sink boundaries of areas $A_1, A_2, \ldots, A_n$, the total emission from A_1 is $\varepsilon_1 E_{b,1} A_1$, of which $\Gamma_{11} + \Gamma_{12} + \ldots + \Gamma_{1n}$ $\left[= \left(\overline{S_1S_1} + \overline{S_1S_2} + \ldots + \overline{S_1S_n}\right)/(\varepsilon_1 A_1)\right]$ is absorbed in the surfaces by all mechanisms, direct and indirect. The difference has been absorbed in the gas; it is called the gas surface total exchange factor $\Gamma_{1G} \left(= \overline{S_1G}/\varepsilon_1 A_1\right)$:

$$\begin{aligned} \Gamma_{1G} &= 1 - \sum\nolimits_i \Gamma_{1i} \\ \overline{S_1G} &= \varepsilon_1 A_1 - \sum\nolimits_i \overline{S_1S_i} \end{aligned} \tag{2.3.42}$$

Here using exchange areas introduce an additional practicality as the letters identifying total exchange areas are commutative; $\overline{S_1G} = \overline{GS_1}$. Note that although Γ_{11} (or $\overline{S_1S_1}$) is never used in calculating radiative interchange, its value is needed for use of Eq. (2.3.42) in calculating Γ_{1G} or $\overline{GS_1}$. $\overline{GS_1}$ embraces the full effect of radiation complexities on radiative exchange between gas and A_1, including multiple reflection at all surfaces, and it is capable of including the effects of gas nongrayness and of assistance given by refractory surfaces to gas-A_1 interchange. It is but mildly temperature-sensitive and is independent of any changes in conduction, convection, mass flow, and energy balances except for their effect on the temperature used in evaluating it.

If the gas volume is not isothermal, the principles used here can be extended to setting up balances on a zoned gas volume (see, e.g., Hottel and Sarofim, Chap. 11, Howell et al., Chap.12).

Systems with a Single Gas Zone and Two Surface Zones

An enclosure consisting of but one isothermal gas zone and two gray surface zones, when properly specified, can model so many industrially important radiation problems as to merit detailed presentation. One can evaluate the total radiation flux between any two of the three zones, including multiple reflection at all surfaces.

$$\begin{aligned} \dot{Q}_{G-1} &= \overline{GS_1}\sigma(T_G^4 - T_1^4) \\ \dot{Q}_{1-2} &= \overline{S_1S_2}\sigma(T_1^4 - T_2^4) \end{aligned} \tag{2.3.43}$$

The total exchange area takes a relatively simple closed form, even when important allowance is made for gas radiation not being gray and when a reduction of the number of system parameters is introduced by assuming that one of the surface zones, if refractory, is radiatively adiabatic (see later). Before allowance is made for these factors, the case of a gray gas enclosed by two source-sink surface zones will be presented. Modification of Eqs. (2.3.7) to (2.3.28), discussed previously, combined with the assumption that a single mean beam length applies to all transfers, i.e., that there is but one gas transmittance $\tau\,(=1-\varepsilon_G)$, gives

$$\overline{S_1S_2} = \frac{A_1\varepsilon_1\varepsilon_2F_{12}}{1/\tau+\tau\rho_1\rho_2(1-F_{12}/C_2)-\rho_1(1-F_{12})-\rho_2(1-F_{21})} \tag{2.3.44}$$

$$\overline{S_1S_1} = \frac{A_1\varepsilon_1^2[F_{11}+\rho_2\tau(F_{12}/C_2-1)]}{1/\tau+\tau\rho_1\rho_2(1-F_{12}/C_2)-\rho_1(1-F_{12})-\rho_2(1-F_{21})} \tag{2.3.45}$$

$$\overline{GS_1} = \frac{A_1\varepsilon_1\varepsilon_G[1/\tau+\rho_2(F_{12}/C_2-1)]}{1/\tau+\tau\rho_1\rho_2(1-F_{12}/C_2)-\rho_1(1-F_{12})-\rho_2(1-F_{21})} \tag{2.3.46}$$

Here C is the area expressed as a ratio to the total enclosure area A_T; $C_1 = A_1/A_T$, $C_2 = A_2/A_T$; $C_1 + C_2 = 1$. The three equations above suffice to formulate total exchange areas for gas-enclosing arrangements which include, e.g., the four geometric cases illustrated in Table 2.3.5, to be discussed later.

An additional surface arrangement of importance is a single zone surface fully enclosing gas. With the gas assumed gray, the simplest derivation of $\overline{GS_1}$ is to note that the emission from surface A_1 per unit of its blackbody emissive power is $A_1\varepsilon_1$, of which the fractions ε_G and $(1-\varepsilon_G)\varepsilon_1$ are absorbed by the gas and the surface, respectively, and the surface reflected residue always repeats this distribution. Therefore,

$$\overline{GS}_{\substack{\text{single surface}\\ \text{zone surrounding}\\ \text{gray gas}}} = \overline{GS_1} = A_1\varepsilon_1\frac{\varepsilon_G}{\varepsilon_G+(1-\varepsilon_G)\varepsilon_1} = \frac{A_1}{1/\varepsilon_G+1/\varepsilon_1-1} \tag{2.3.47}$$

Alternatively, $\overline{GS_1}$ could be obtained from case 1 of Table 2.3.5 by letting plane area A_1 approach 0, leaving A_2 as the sole surface zone. Although departure of gas from grayness has a marked effect on radiative transfer, the subject is complex and will be presented in stages, as the cases shown in Table 2.3.5 are discussed.

Partial Allowance for Spectral Effects of Gas on Total Exchange Areas

A radiating gas departs from grayness in two ways: (1) Gas emissivity ε_G and absorptivity α_{G1} are not the same unless T_1 equals T_G. (2) The fractional transmittance τ of radiation through successive path lengths L_m due to surface reflection, instead of being constant, keeps increasing because at the wavelengths of high absorption the incremental absorption decreases with increasing path length. The first of these effects is sufficiently straightforward to be introduced at this point, coupled with allowance for refractory surfaces being substantially radiatively adiabatic. The second, much more complicated effect will be introduced later; it sometimes changes the computed flux significantly.

In the simplest case of gas-surface radiative exchange—a gas at T_G completely enclosed by a black surface at T_1—the net flux $\dot{Q}_{G-1}$ is given by

$$\dot{Q}_{G-1} = \sigma(\varepsilon_G T_G^4 - \alpha_{G1}T_1^4) = \sigma\varepsilon_{G,e}(T_G^4 - T_1^4)$$

The evaluation of the absorptivity α_G was covered in Eqs. (2.3.40) and (2.3.41). The second form of the above equation defines $\varepsilon_{G,e}$, the equivalent gray-gas emissivity

Table 2.3.5 Total Exchange Areas for Four Arrangements of Two-Zone-Surface Enclosures of a Gray Gas*

Case 1	Case 2	Case 3	Case 4
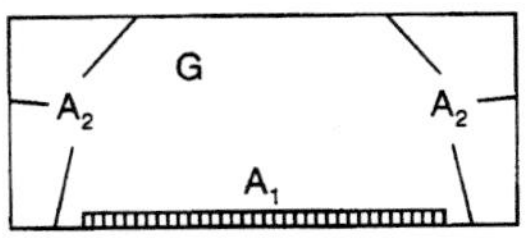	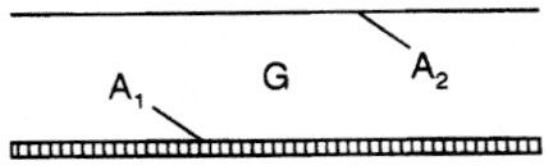	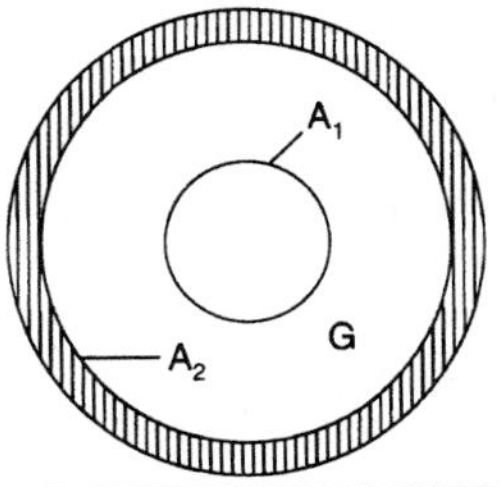	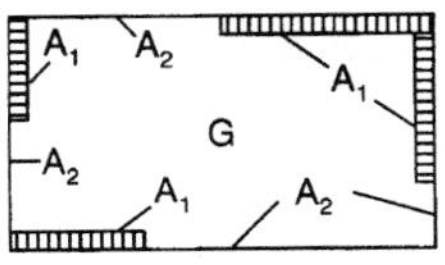
A plane surface A_1 and a surface A_2 complete the enclosure	Infinite parallel planes	Concentric spherical or infinite cylindrical surface zones, A_1 inside	Two-surface-zone enclosure, each zone in one or more parts, any shape Assume that enclosing surface is a speckled enclosure, or spherical
$F_{12}=1$ $\frac{\overline{S_1S_2}}{A_1}=\frac{\varepsilon_1\varepsilon_2}{D_1}$ $\frac{\overline{GS_1}}{A_1}=\frac{\varepsilon_1\varepsilon_G(1/\tau+\rho_2C_1/C_2)}{D_1}$ $\frac{\overline{GS_1}}{A_1}=\frac{\varepsilon_1\varepsilon_G(1/\tau+\rho_2C_1/C_2)}{D_1}$ $\frac{\overline{S_1S_1}}{A_1}=\frac{\varepsilon_1^2\tau\rho_2C_1/C_2}{D_1}$ $D_1=\frac{1}{\tau}-\rho_2\left[1-\frac{C_1}{C_2}(1-\tau\rho_1)\right]$ $\tau=1-\varepsilon_G$	$F_{12}=F_{21}=1$ $\frac{\overline{S_1S_2}}{A_1}=\frac{\varepsilon_1\varepsilon_2}{D_2}$ $\frac{\overline{GS_1}}{A_1}=\frac{\varepsilon_1\varepsilon_G(1/\tau+\rho_2)}{D_2}$ $\frac{\overline{S_1S_1}}{A_1}=\frac{\varepsilon_1^2\tau\rho_2}{D_2}$ $D_2=\frac{1}{\tau}-\tau\rho_1\rho_2$ $\tau=1-\varepsilon_G$	$F_{12}=1 \quad F_{21}=\frac{A_1}{A_2}=\frac{C_1}{C_2}$ $\frac{\overline{S_1S_2}}{A_1}=\frac{\varepsilon_1\varepsilon_2}{D_3}$ $\frac{\overline{GS_1}}{A_1}=\frac{\varepsilon_1\varepsilon_G(1/\tau+\rho_2C_1/C_2)}{D_3}$ $\frac{\overline{GS_2}}{A_2}=\frac{\varepsilon_2\varepsilon_G(1/\tau+\rho_1C_1/C_2)}{D_3}$ $\frac{\overline{S_1S_1}}{A_1}=\frac{\varepsilon_1^2\tau\rho_2C_1/C_2}{D_3}$ $D_3=\frac{1}{\tau}-\rho_2\left[1-\frac{C_1}{C_2}(1-\tau\rho_1)\right]$ $\tau=1-\varepsilon_G$	$F_{12}=F_{22}=C_2$ $F_{21}=F_{11}=C_1$ $\frac{\overline{S_1S_2}}{A_1}=\frac{\varepsilon_1\varepsilon_2C_2}{D_4}$ $\frac{\overline{GS_1}}{A_1}=\frac{\varepsilon_1\varepsilon_G/\tau}{D_4}$ $\frac{\overline{S_1S_1}}{A_1}=\frac{\varepsilon_1^2C_1}{D_4}$ $D_1=\frac{1}{\tau}-\rho_1C_1-\rho_2C_2$ $\tau=1-\varepsilon_G$

* All equations above come from Eqs. (2.3.39) to (2.3.41), with substitutions for view factor F given before equations.

$$\varepsilon_{G,e} = \frac{\varepsilon_G - \alpha_{G1}(T_1^4/T_G^4)}{1-(T_1^4/T_G^4)} \tag{2.3.48}$$

Although this introduction of $\varepsilon_{G,e}$ has added no information, the evaluation of $\dot{Q}_{G-1}$ in terms of $\varepsilon_{G,e}$ rather than ε_G and gives a better structure for trial-and-error solutions of problems in which either T_G or T_1 is not known and a second energy relation is available. With partial allowance for gas nongrayness having been made, the evaluation of radiative flux [or Eq. (2.3.48)] for cases falling in one of the categories of Table 2.3.5 is straightforward if both A_1 and A_2 are source-sink surfaces. Wherever ε_G or τ appears in the table, or in Eqs. (2.3.44) to (2.3.47), use $\varepsilon_{G,e}$ or $1 - \varepsilon_{G,e}$ instead.

EXAMPLE (FIRST APPROXIMATION to NONGRAYNESS). Methane is burned to completion with 20 percent excess air (air half saturated with water vapor at 289 K, 0.0088 mol H_2O/mol dry air) in a furnace chamber with floor dimensions of 3 × 10 m and 5 m high. The whole surface is a gray energy sink of emissivity 0.8 at 1,000 K, surrounding gas at 1,500 K, well stirred. Find the effective gas emissivity $\varepsilon_{G,e}$ and the surface radiative flux density, assuming that the only correction necessary for gas nongrayness is use of $\varepsilon_{G,e}$ rather than ε_G.

Solution: Combustion is 1 CH_4 + 2 × 1.2 O_2 + 1.2 × (79/21)N_2 + 2 × 1.2 × 100/21 × 0.0088 H_2O going to 1 CO_2 + [2 + 2 × 1.2 × (100/21) × 0.0088] H_2O + 0.4 O_2 + 9.03 N_2 = 12.53 mol/mol of CH_4. And $P_C + P_W = (1 + 2.1)/12.53 = 0.2474$ atm. The mean beam length $L_m = 0.88 \times 4V/A_T = 0.88 \times 4(10 \times 3 \times 5)/\{2[2 \times (10 \times 3 + 10 \times 5 + 3 \times 5)]\} = 2.779$ m. And $pL_m = 0.2474 \times 2.779 = 0.6875$ m · atm. From emissivity Table 2.3.4, $b(1{,}500) = 540$; $n(1{,}500) = 0.42$; $b(1{,}000) = 444$; $n(1{,}000) = 0.34$. Also $\varepsilon_G(pL) = 540(0.6875 - 0.015)^{0.42}/1{,}500 = 0.3047$, and $\alpha_{G1}(pL) = 444\,(0.6875 \times 1{,}000/1{,}500 - 0.015)^{0.34}(1{,}500/1{,}000)\,0.5/1{,}000 = 0.4124$. Then $\varepsilon_{G,e}(pL) = [0.3047 - 0.4124(1{,}000/1{,}500)^4]/[1 - (1{,}000/1{,}500)^4] = 0.2782$. From Eq. (2.3.42), with ε_G replaced by $\varepsilon_{G,e}$, $\left(\overline{GS_1}/A_1\right) = 1/(1/0.2782 + 1/0.8 - 1) = 0.2601$. Then $\dot{Q}_{G-1}/A_1 = 56.7 \times 0.260\,[(1{,}500/1{,}000)^4 - (1{,}000/1{,}000)^4] = 59.91$ kW/m².

Refractory Surfaces If one of the surfaces A_r of an enclosure of gas is refractory, an extra temperature T_{refr} and an extra heat transfer equation are needed to determine the fluxes unless A_r can be assumed to be radiatively adiabatic. Consider the facts that irradiation of A_r plus convection from gas to it must equal backbody radiation plus conduction through it if steady state exists, and irradiation is enormous compared to convection. It then follows that the difference between convection and conduction is so minute compared to irradiation or back radiation as to make A_r substantially radiatively adiabatic; assume that A_1 is a source-sink zone and A_2 a radiatively adiabatic zone, and call it A_r. The condition for adiabaticity of A_r is

$$\overline{GS_r}(T_G^4 - T_r^4) = \overline{S_rS_1}(T_r^4 - T_1^4)$$

or, to eliminate T_r,

$$\frac{T_G^4 - T_r^4}{1/\overline{GS_r}} = \frac{T_r^4 - T_1^4}{1/\overline{S_rS_1}} = \frac{T_G^4 - T_1^4}{1/\overline{GS_r} + 1/\overline{S_rS_1}} \tag{2.3.49}$$

The net flux from gas G is $\overline{GS_1}(T_G^4 - T_1^4) + \overline{GS_r}(T_G^4 - T_r^4)$, which can be represented by using a single total exchange area term multiplying $(T_G^4 - T_1^4)$, by replacing the second term using Eq. (2.3.49) as:

$$\dot{Q}_{G-1} = \sigma(T_G^4 - T_1^4)\left[\overline{GS_1} + \frac{1}{1/\overline{GS_r} + 1/\overline{S_rS_1}}\right] = \left(\overline{GS_1}\right)_R \sigma(T_G^4 - T_1^4) \tag{2.3.50}$$

The bracketed term, $\left(\overline{GS_1}\right)_R$, is the total exchange area from G to A_1 with assistance from a refractory surface. Table 2.3.5 supplies the forms for the three total exchange area terms needed to formulate with A_r substituted for A_2 and with ε_G or τ replaced by $\varepsilon_{G,e}$ or $1\text{-}\varepsilon_{G,e}$.

A general expression for a gray gas enclosure of two surfaces, one of which is radiatively adiabatic, comes from Eq. (2.3.50), in which $\left(\overline{GS_1}\right)_R$ becomes $\overline{GS_1}$ because $\overline{GS_1}$ and $\overline{GS_r}$ are zero, and then from Eq. (2.3.42), which becomes $\overline{GS_1}\left[=\left(\overline{GS_1}\right)_R\right] = A_1\varepsilon_1 - \overline{S_1S_1}$. With $\overline{S_1S_1}$ coming from Eq. (2.3.45), one finally obtains

$$\frac{\left(\overline{GS_1}\right)_R}{A_1} = \frac{1}{\rho_1/\varepsilon_1 + 1/\{\varepsilon_G[1 + 1/(C_1/C_2 + \varepsilon_G/\tau F_{1r})]\}} \tag{2.3.51}$$

Note that since the first denominator term is zero when A_1 is black, the denominator of the second term is $\left(\overline{GS_1}\right)_R/A_1$ for a black surface.

The above equation is perfectly general when the gas is gray and the two enclosing surfaces are a sink and a radiatively adiabatic surface. When A_1 is a plane (simulation of slab or billet heating furnaces and glass tanks), $\left(\overline{GS_1}\right)_R/A_1$ is represented by Eq. (2.3.51) with $F_{1r} = 1$. A different but completely equivalent form is

$$\left[\frac{\left(\overline{GS_1}\right)_R}{A_1}\right]_{A_1 \text{ is plane}} = \frac{1}{1/\varepsilon_1 + (1/\varepsilon_G - 1)^2/[1/(C_1\varepsilon_G) - 1]} \tag{2.3.52}$$

For most refinery processing furnaces, with sink and refractory assumed to form a speckled enclosure, $\left(\overline{GS_1}\right)_R/A_1$ is represented by Eq. (2.3.51) with $F_{1r} = C_r$. A better but completely equivalent form is

$$\left[\frac{\left(\overline{GS_1}\right)_R}{A_1}\right]_{\text{surface is speckled}} = \frac{1}{1/\varepsilon_1 + C_1(1/\varepsilon_G - 1)} \tag{2.3.53}$$

As previously stated, partial allowance for the gas not being gray is made by evaluating $\overline{GS}$ or $\left(\overline{GS}\right)_R$ with use of $\varepsilon_{G,e}$ rather than ε_G, and $1\text{-}\varepsilon_{G,e}$ rather than τ, in Eqs. (2.3.51) to (2.3.53). A slightly better but more tedious allowance for partial nongrayness in evaluating $\dot{Q}_{rad}$ is to replace $\overline{GS_1}\sigma(T_G^4 - T_1^4)$ with $\overline{GS_1}$ evaluated by using $\varepsilon_{G,e}$, by $\overline{GS_1}T_G^4 - \overline{S_1G}T_1^4$ where $\overline{GS}$ and $\overline{SG}$ are evaluated by using ε_G and $\alpha_{G,1}$, respectively.

This completes the presentation of procedures for evaluating the total exchange area between gas and sink surface when the gas is gray or by making approximate allowance for the nongrayness by using ε_G and $\alpha_{G,1}$. These exchange areas can be used in the formulations below on furnace chamber performance. Methods will be presented first for a more rigorous treatment of gas nongrayness.

Full Allowance for Gas Spectral Effects The above paragraphs failed to allow for the previously discussed change in gas transmittance on successive passages of reflected radiation through the gas. In many radiative transfer problems, interchange between many different pairs of radiators creates a system of simultaneous equations to be solved for the energy fluxes, and a shift from use of the Stefan-Boltzmann law to the Planck equation would enormously increase the difficulty of solution. Total emissivity of a real gas, whose spectral emissivity e_l and absorptivity λ can be expressed as the a-weighted mean of a gray gas emissivity or absorptivity terms ($\varepsilon_{G,i}$ or $\alpha_{G,i}$) represents the corresponding fractions a_i of the blackbody spectrum. This approach is known as weighted sum of gray gas model (Howell et al., Chap. 9). Then

$$\varepsilon_G = \sum_0^n a_i\varepsilon_{G,i} = \sum_0^n a_i(1 - e^{-k_i pL}) \tag{2.3.54}$$

The linearity between flux $\dot{Q}$ and blackbody emissive power E_b allows the above relations to be used for converting $\overline{GS}$ or $\overline{SS}$ for a gray gas to a form allowing for nongrayness. For a real gas $\overline{GS}$ or $\overline{SS}$ is the a_i-weighted sum of its values based on each of the $\varepsilon_{G,i}$ values of Eq. (2.3.54). $\overline{GS}_{\text{real gas}}$ can be obtained by replacing ε_G by $\varepsilon_{G,i}$, τ by $1\text{-}\varepsilon_{G,i}$ in an expression for $\overline{GS}_{\text{gray}}$, multiplying the result by a_i and summing the resultant values.

Obviously, the number of terms n should be as small as possible while consistent with small error. Consider an n of 2, with the gas modeled as the sum of one gray gas plus a clear gas, with the gray gas of absorption coefficient k occupying the energy fraction a of the blackbody spectrum and the clear gas ($k = 0$) the fraction $1 - a$. Then, at path lengths L and $2L$,

$$[\varepsilon_G(pL)] = a(1-e^{-kpL}) + (1-a)(0)$$
$$[\varepsilon_G(2pL)] = a(1-e^{-2kpL}) + (1-a)(0) \qquad (2.3.55)$$

Solution of these gives

$$a = \frac{\varepsilon_G(PL)}{2-\varepsilon_G(2pL)/\varepsilon_G(pL)}$$
$$kpL = -ln\left[1-\frac{\varepsilon_G(pL)}{a}\right] \qquad (2.3.56)$$

The equivalent gray gas emissivity in the spectral range a is $1 - e^{-kpL}$, from Eq. (2.3.54), and from Eq. (2.3.55) that is $\varepsilon_G(pL)/a$; in the spectral range $1 - a$, the equivalent gray gas emissivity is zero. This simple model will be correct for the contribution of the direct gas emission from path length L and for that of the once-reflected emission (path length $2L$); and the added contributions due to increasing numbers of reflections will be attenuated sufficiently by surface reflections to make errors in them unimportant. Note that a and k are not general constants; they are specific to the subject mean beam length L_m and come from basic data, such as Table 2.3.4. Note also that when full allowance for spectral effects is to be made by replacing ε_G by $\varepsilon_{G,e}$, then a is also changed to a_e, which comes from Eq. (2.3.56), with $\varepsilon_{G,e}$ replacing ε_G.

Conversion of gray gas total exchange areas $\overline{GS}$ and $\overline{SS}$ to their nongray forms is carried out as follows when the nongray model is gray-plus-clear gas: From Eq. (2.3.55) the equivalent gray gas emissivity in the spectral energy fraction a_e is $\varepsilon_{G,e}(pL)/a_e$, which replaces ε_G wherever it or its complement τ occurs in $\overline{GS}$; the result is then multiplied by a_e. There is no contribution from the clear gas energy fraction. Conversion of $\overline{SS}$ from gray to gray plus clear involves making the same substitution, but for $\overline{SS}$ another term must be added. For the clear gas contribution, 0 and 1 are substituted for ε_G and τ, and the result is multiplied by the weighting factor $1-a_e$; this $\overline{SS}$ is added to the preceding one to give $\overline{SS}_{g+c}$.

The simplest application of this gray-plus-clear model of gas radiation is the case of a single gas zone surrounded by a single surface zone, the case covered for a gray gas by Eq. (2.3.57) and illustrated in the last numerical example, where radiation from methane combustion products is surrounded by a single-zone sink surface. That example will be repeated using the gray-plus-clear gas model, for which the total exchange area is

$$\frac{\overline{GS_1}}{A_1} = \frac{a_e}{a_e/\varepsilon_{G,e} + 1/\varepsilon_1 - 1} \qquad (2.3.57)$$

EXAMPLE (GRAY-PLUS-CLEAR GAS RADIATION FROM METHANE COMBUSTION PRODUCTS). The computations of the previous example, radiative flux from methane combustion products, will be repeated with the more rigorous treatment of spectral effects, and the results will be compared with the more approximate calculations for the case of a wall emissivity ε_1 of 0.4 and 0.8.

Solution: Repeat the calculations given, in the earlier example, of $\varepsilon_G(pL)$, $\alpha_{G1}(pL)$, and $\varepsilon_{G,e}(pL)$ for $pL = 2 \times 0.6875$, to give $\varepsilon_G(2pL) = 0.4096$, $\alpha_{G1}(2pL) = 0.5250$, and $\varepsilon_{G,e}(2pL) = 0.3812$. Then $a_e = 0.2782/(2 - 0.3812/0.2782) = 0.4418$, and the emissivity substitute for the gray gas portion of the gray-plus-clear gas model is $0.2782/0.4418 = 0.6297$. For a single enveloping surface zone, the total exchange area comes from Eq. (2.3.52): $\overline{GS_1}/A_1 = a_e/(a_e/\varepsilon_{G,e} + 1/\varepsilon_1 - 1) = 0.4418/(0.4418/0.2782 + 1/0.8 - 1) = 0.2404$. The flux density is $\dot{Q}/A = q = (\overline{GS_1}/A)\sigma(T_G^4 - T_1^4) = 0.2404 \times 56.7 \times [(1{,}500/1{,}000)^4 - (1{,}000/1{,}000)^4] = 55.37$ kW/m^2. This is 7.6 percent lower than it is when only the difference between ε_G and $\alpha_{G,1}$ is allowed for in finding the spectral effects of gas. In some problems the difference is as high as 20 percent. (Note that allowing for average humidity in air adds 5 percent to H_2O and about 2 percent to the gas emissivity.) Changing ε_1 from 0.8 to 0.4 changes the approximate solution for $\overline{GS_1}/A_1$ from 0.2501 to 0.1963 and the gray-plus-clear treatment from 0.2404 to 0.1431.

The procedures for introducing the nongray gas model can be used to convert the total exchange areas for the basic one-gas two-surface model, Eqs. (2.3.44) to (2.3.46), as used to evaluate the cases in Table 2.3.5, to the following gray-plus-clear-gas model forms:

$$\frac{\overline{S_1S_2}}{A_1} = F_{12}\varepsilon_1\varepsilon_2\left(\frac{a_e}{D_a} + \frac{1-a_e}{D_b}\right) \qquad (2.3.58)$$

$$\frac{\overline{S_1S_2}}{A_1} = F_{12}\varepsilon_1^2\left\{\frac{a_e[1-F_{12}+\rho_2(1-\varepsilon_{G,e}/a_e)(F_{12}/C_2-1)]}{D_a} + \frac{(1-a_e)[1-F_{12}+\rho_2(F_{12}/C_2-1)]}{D_b}\right\} \qquad (2.3.59)$$

$$\frac{\overline{GS_2}}{A_1} = \frac{\varepsilon_1\varepsilon_{G,e}[1/(1-\varepsilon_{G,e}/a_e)] + \rho_2(F_{12}/C_2-1)}{D_a} \qquad (2.3.60)$$

$$D_a = \frac{1}{1-\varepsilon_{G,e}/a_e} + \left(1-\frac{\varepsilon_{G,e}}{a_e}\right)\rho_1\rho_2(1-F_{12}C_2) - \rho_1(1-F_{12}) - \rho_2(1-F_{21})$$

$$D_b = 1 + \rho_1\rho_2\left(1-\frac{F_{12}}{C_2}\right) - \rho_1(1-F_{12}) - \rho_2(1-F_{21})$$
$$= \varepsilon_1\varepsilon_2 + F_{12}\left[\varepsilon_2 + \frac{\varepsilon_1(C_1-\varepsilon_2)}{C_2}\right]$$

The above relations, with the view factor F_{12} specified, may be used to convert the geometric cases of Table 2.3.5 to their more nearly correct forms with gray gas replaced by the gray-plus-clear gas model. That has been done in Table 2.3.6, which covers a moderate idealization of many practical industrial systems.

Effect of Gas Nongrayness on Refractory Zones Full allowance for the effect of gas non grayness on enclosures in which part of the enclosing surface is radiatively adiabatic is straightforward but sometimes tedious. The term of Eq. (2.3.50) must be evaluated. It is tempting to use Eq. (2.3.51), but that is invalid because, although total radiative interchange at zone A_r is 0, the gas nongrayness makes A_r a net absorber in the spectral energy fraction a (or a_e) and a net emitter in the clear gas fraction $1-a$. It is necessary, then, to use the basic equation

$$(\overline{GS_1})_R = \left(\overline{GS_1} + \frac{1}{1/\overline{GS_r} + 1/\overline{S_rS_1}}\right) \qquad (2.3.61)$$

evaluating each of the right-hand members of a geometric system of interest, such as found in Table 2.3.5 (where A_r is A_2). As previously discussed, $\varepsilon_{G,e}/a_e$ is substituted for ε_G (or its complement for τ), and the result is weighted by the factor a_e; and for $\overline{S_rS_1}$ an additional term based on ε_G being replaced by 0 or τ by 1, with weighing $1 - a_e$, is added.

Of the cases covered in Table 2.3.5, only two will be evaluated to make A_2 represent the radiatively adiabatic zone A_r. The first is for the case of heat sink A_1 in a plane—the simulation of a slab-heating furnace. Insertion into Eq. (2.3.50) of the gray-plus-clear terms and from Table 2.3.5 (with subscript r replacing subscript 2) and rearrangement gives:

$$\left[\frac{(\overline{GS_1})_R}{A_1}\right]_{A_1\text{ in a plane}} = \frac{\varepsilon_G}{D_1}\left[\varepsilon_1\left(\rho_r\frac{C_1}{C_r} + \frac{1}{1-\varepsilon_{G/a}}\right) + \frac{\varepsilon_r}{\dfrac{1}{\rho_1 + \dfrac{C_r}{C_1}\left(1-\dfrac{\varepsilon_G}{a}\right)} + \dfrac{\varepsilon_G/\varepsilon_1}{a + \dfrac{(1-a)D_1}{\varepsilon_r + \rho_r\varepsilon_1C_1/C_r}}}\right] \qquad (2.3.62)$$

where $D_1 = 1/(1-\varepsilon_G/a) - \rho_r[1-(C_1/C_r)(\varepsilon_1 + \rho_1\varepsilon_G/a)]$. Although ε_G and a are used here, $\varepsilon_{G,e}$ and a_e should be used if allowance is to be made for the difference between gas emissivity and absorptivity. Comparison of Eq. (2.3.62) with it, gray gas equivalent, Eq. (2.3.52), shows the

Table 2.3.6 Conversion of Total Exchange Areas for Cases of Table 2.3.5 to Their Gray-plus-Clear Values

Case 1: Plane slab A_1 and surface A_2 completing an enclosure of gas; $F_{12} = 1$

$$\frac{\overline{S_1S_2}}{A_1} = \frac{a\varepsilon_1\varepsilon_2}{D_1} + \frac{(1-a)\varepsilon_1\varepsilon_2}{1-\rho_2(1-\varepsilon_1C_1/C_2)}$$

where
$$D_1 = \frac{1}{1-\varepsilon_G/a} - \rho_2\left[1-\frac{C_1}{C_2}\left(\varepsilon_1+\frac{\rho_1\varepsilon_G}{a}\right)\right]$$

$$\frac{\overline{GS_1}}{A_1} = \varepsilon_1\varepsilon_G\left[\frac{1/(1-\varepsilon_G/a)+\rho_2C_1/C_2}{D_1}\right]$$

$$\frac{\overline{GS_2}}{A_2} = \varepsilon_2\varepsilon_G\left[\frac{1/(1-\varepsilon_G/a)+\rho_1C_1/C_2}{D_1}\right]$$

Case 2: Infinite parallel planes, gas between; $F_{12} = F_{21} = 1$

$$\frac{\overline{S_1S_2}}{A_1} = \frac{a\varepsilon_1\varepsilon_2}{D_2} + \frac{(1-a)\varepsilon_1\varepsilon_2}{1-\rho_1\rho_2}$$

where
$$D_2 = \frac{1}{1-\varepsilon_G/a} - \left(1-\frac{\varepsilon_G}{a}\right)\rho_1\rho_2$$

$$\frac{\overline{GS_1}}{A_1} = \varepsilon_1\varepsilon_G\left[\frac{1/(1-\varepsilon_G/a)+\rho_2}{D_2}\right]$$

Case 3: Concentric spherical or infinite cylindrical surface zones, A_1 inside; $F_{12} = 1$; $F_{21} = A_1/A_2 = C_1/C_2$

$$\frac{\overline{S_1S_2}}{A_1} = \frac{a\varepsilon_1\varepsilon_2}{D_3} + \frac{(1-a)\varepsilon_1\varepsilon_2}{1-\rho_2(1-\varepsilon_1C_1/C_2)}$$

where
$$D_3 = \frac{1}{1-\varepsilon_G/a} - \rho_2\left[1-\left(\frac{C_1}{C_2}\right)\left(\varepsilon_1+\frac{\rho_1\varepsilon_G}{a}\right)\right]$$

$$\frac{\overline{GS_1}}{A_1} = \varepsilon_1\varepsilon_G\left[\frac{1/(1-\varepsilon_G/a)+\rho_2C_1/C_2}{D_3}\right]$$

$$\frac{\overline{GS_2}}{A_2} = \varepsilon_2\varepsilon_G\left[\frac{1/(1-\varepsilon_G/a)+\rho_1C_1/C_2}{D_3}\right]$$

Case 4: Spherical enclosure of two surface zones or speckled A_1:A_2 enclosure; $F_{12} = F_{22} = C_2$; $F_{21} = F_{11} = C_1$

$$\frac{\overline{S_1S_2}}{A_1} = \frac{a\varepsilon_1\varepsilon_2C_2}{D_4} + \frac{(1-a)\varepsilon_1\varepsilon_2C_2}{1-\rho_1C_1-\rho_2C_2}$$

where
$$D_4 = \frac{1}{1-\varepsilon_G/a} - \rho_2\left[1-\left(\frac{C_1}{C_2}\right)\left(\varepsilon_1+\frac{\rho_1\varepsilon_G}{a}\right)\right]$$

$$C_1 = \frac{A_1}{A_1+A_2}$$

$$\frac{\overline{GS_1}}{A_1} = \frac{\varepsilon_2\varepsilon_G/(1-\varepsilon_G/a)}{D_4}$$

complexity introduced by allowance for gas spectral effects. [$(\overline{GS_1})_R$ for the gray-plus-clear gas model is about 15 percent higher than for gray gas when $\varepsilon_1 = 0.8$, $\varepsilon_G = 0.3$, $C_1 = 1/3$, $a = 0.4$, and $\varepsilon_r = 0.6$, but only 1 percent higher when $\varepsilon_r = 1$.]

The second conversion of $\overline{GS_1}$ to $(\overline{GS_1})_R$ will be case 4 of Table 2.3.5, the two-surface-zone enclosure with the computation simplified by assuming that the direct-view factor from any spot to a surface equals the fraction of the whole enclosure which the surface occupies (the speckled furnace model). This case can be considered as an idealization of many processing furnaces such as distilling and cracking coil furnaces, with parts of the enclosure tube-covered and part left refractory. (The refractory under the tubes is not to be classified as part of the refractory zone.) Again, one starts with substitution, into Eq. (2.3.50), of the terms $\overline{GS_1}$, $\overline{GS_r}$, and $\overline{S_rS_1}$, from Table 2.3.5, case 4, with all terms first converted to their gray-plus-clear form. To indicate the procedure, one of the components, $\overline{S_rS_1}$ will be formulated.

$$\frac{\overline{S_rS_1}}{A_1} = a\frac{C_r\varepsilon_1\varepsilon_r}{D_4'} + (1-a)\frac{C_r\varepsilon_1\varepsilon_r}{1-\rho_1C_1-\rho_rC_r}$$
$$= \frac{C_r\varepsilon_1\varepsilon_r}{D_4'}\left[1+\frac{\varepsilon_G(1-a)(a-\varepsilon_G)}{1-\rho_1C_1-\rho_rC_r}\right]$$

With $D_4' = 1/(1-\varepsilon_G/a)-\rho_1C_1-\rho_rC_r$, the result of the full substitution simplifies to

$$\frac{\left(\overline{GS_1}\right)_R}{A_1} = \frac{1}{C_1\left(\frac{1}{\varepsilon_G} - \frac{1}{a}\right) + \frac{1}{\varepsilon_1} + \frac{1/a - 1}{\varepsilon_1 + \varepsilon_r (C_r / C_1)}} \tag{2.3.63}$$

For a gray gas ($a = 1$) the above becomes

$$\frac{\left(\overline{GS_1}\right)_R}{A_1} = \frac{1}{C_1(1/\varepsilon_G - 1) + 1/\varepsilon_1} \tag{2.3.64}$$

Equation (2.3.63) has wide applicability.

The beginning of this subsection mentions compact chambers (just treated) and long chambers as limiting cases. The latter will now be treated.

The Long Combustion Chamber If a chamber is long enough in the x direction compared to its mean hydraulic radius, the local flux from gas to wall sink comes substantially from gas at its local temperature, with $\overline{GS_1}$ [or $(\overline{GS_1})_R$] calculated by methods just described but based on a two-dimensional structure; i.e., the opposed upstream and downstream fluxes through the flow cross section will substantially cancel. That limiting case will be considered, with $\left(\overline{GS_1}\right)_R / A_1$ evaluated by using local mean values of T_G and T_1. The local $\left(\overline{GS_1}\right)_R$ applicable to a surface element of length dx and perimeter P is then $\left[\left(\overline{GS_1}\right)_R / A_1\right] P dx$. Let $T_{G,in}$, $T_{G,out}$, $T_{1,in}$, and $T_{1,out}$ be specified; furnace length L is to be determined. Assume a constant sink temperature T_1. The equation of heat transfer in the furnace length element $P\,dx$ is then

$$-\dot{m}C_p dT_G = P dx \left[\frac{\left(\overline{GS_1}\right)_R}{A_1}\right] [\sigma(T_G^4 - T_1^4) + h(T_G - T_1)] \tag{2.3.65}$$

The second of the heat-transfer terms is an order of magnitude smaller than the first, and to permit ready integration, $h(T_G - T_1)$ will be set equal to $b\sigma(T_G^4 - T_1^4)$ from which

$$b\sigma = \frac{h}{T_G^4 + T_G^2 T_1 + T_G T_1^2 + T_1^3} \approx \frac{h}{4T_{G1}^3}$$

T_{G1} is the mean value of T_G and T_1, and a 10 percent error in T_1 will lead to a 1 percent error in the calculated heat transfer. Then Eq. (2.3.65) becomes

$$-\dot{m}C_p dT_G = P dx \left[\frac{\left(\overline{GS_1}\right)_R}{A_1} + \frac{h}{4\sigma T_{G1}^3}\right] \sigma(T_G^4 - T_1^4) \tag{2.3.66}$$

Integration of T_G from $T_{G,in}$ to $T_{G,out}$ and of x from 0 to L, and solution for L give

$$L = \frac{\dot{m}C_p\left(\tan^{-1}\frac{T_{G,out}}{T_1} - \tan^{-1}\frac{T_{G,in}}{T_1} - \frac{1}{2}\ln\frac{T_{G,out} - T_1 T_{G,in} + T_1}{T_{G,out} + T_1 T_{G,in} - T_1}\right)}{2PT_1^3\sigma\left[\frac{\left(\overline{GS_1}\right)_R}{A_1} + \frac{h}{4\sigma T_{G1}^3}\right]} \tag{2.3.67}$$

A trial and error based iterative solution is necessary if L is specified and $T_{G,out}$ is to be found. If L is not long, axial radiative flux becomes important and a much more complex treatment is necessary. Use of a multi-gas zone system is one possibility.

Partially Stirred Model of Furnace Chamber Performance An equation representing an energy balance on a combustion chamber of two surface zones (a heat sink A_1 at temperature T_1 and a refractory surface A_r assumed radiatively adiabatic at T_r) is most simply solved if the total enthalpy input H is expressed as $\dot{m}C_p(T_F - T_0)$; $\dot{m}$ is the mass rate of fuel plus air, and T_F is a pseudo adiabatic flame temperature based on a mean specific heat from base temperature T_0 up to the gas exit temperature T_E rather than up to T_F. Assume that enough stirring occurs in the chamber to produce two temperatures (the heat-transfer temperature T_G and the leaving gas enthalpy temperature T_E), the two differing by an empirical amount, zero if the stirring were perfect.

A practical way of introducing this empiricism is to assume that $T_G - T_E$, expressed as a ratio to T_F, is a constant Δ. Although Δ will vary with burner type, the effects of excess air and firing rate are small. At the limits of very high and low firing densities, Δ tends toward zero. These conditions excepted and in the absence of performance data on the subject furnace type, assume $\Delta = 0.08$, or

$$\frac{T_G - T_E}{T_F} = \Delta = 0.08$$

This assumption bypasses complex allowance for temperature variations in the chamber gas and for the effects of fluid mechanics and combustion kinetics, but at the cost of not permitting evaluation of flux distribution over the surface. The heat-transfer rate $\dot{Q}$ out of the gas is then $\dot{m}C_p(T_F - T_0)$ or $\dot{m}C_p(T_F - T_E)$. A combination of energy balance and heat transfer, with the ambient temperature taken as the enthalpy base temperature T_0, gives

$$\begin{aligned}(\dot{Q} =) \dot{H} - \dot{m}C_p(T_E - T_0) &= \dot{m}C_p(T_F - T_E) \\ &= \left(\overline{GS_1}\right)_R \sigma(T_G^4 - T_1^4) + h_1 A_1 (T_G - T_1)\end{aligned} \tag{2.3.68}$$

Dimensionless relations can be derived by dividing all terms by $\left(\overline{GS_1}\right)_R \sigma T_F^4$, and expressing all temperatures as ratios to T_F that are represented as T^*. For simplification, the following definitions of D and N_c are introduced:

$$\dot{m}C_p / \left(\overline{GS_1}\right)_R \sigma T_F^3 \left[= \dot{H} / \left(\overline{GS_1}\right)_R \sigma_F^4 (1 - T_0^*)\right] = D \text{ (dimensionless firing density)}$$

$$\frac{h_1 A_1}{\left(\overline{GS_1}\right)_R \sigma T_F^3} = N_c, \text{ convection number (dimensionless)}$$

After the division, the left-hand side term of Eq. (2.3.68) $= D(1 - T_E^*)$ and the first right-hand side term $= T_G^{*4} - T_1^{*4}$. The equation then becomes

$$D(1 - T_E^*) = T_E^{*4} - T_E^{*4} + N_c(T_G^* - T_1^*) \tag{2.3.69}$$

The two unknowns T_G^* and T_E^* are reduced to one by expressing T_E^* in terms of T_G^* and Δ. Equation (2.3.69), with coefficients of T_G^{*4} and T_G^* collected, then becomes

$$T_G^{*4} + (D + N_c)T_G^* - [T_1^{*4} + N_c T_1^* + D(1 + \Delta)] = 0 \tag{2.3.70}$$

Although Eq. (2.3.70) is a quartic equation, an explicit solution is possible. However, an iterative solution is required as $\left(\overline{GS_1}\right)_R$ and C_p are mild functions of T_G and related T_E, respectively, and a preliminary guess of T_G is necessary. Besides, ambiguity can exist in the interpretation of terms. If part of the enclosure surface consists of screen tubes over the chamber gas exit to a convection section, radiative transfer to those tubes is included in the chamber energy balance, whereas convection is not, as it has no effect on the chamber gas temperature. Although the results must be considered as approximate, as they depend on empirical Δ, the equation may be used to find the effect of firing rate, excess air, and air preheat on efficiency. Moreover, the effect of various factors on Δ may be found with available performance data.

For the commonly encountered case of the sink consisting of a row of tubes mounted on a refractory wall, A_1 is the area of the whole plane in which the tubes lie, T_1 is tube surface temperature, and ε_1 is the effective emissivity of the tube-row-refractory-wall combination, as in the earlier numerical example associated with Fig. 2.3.6, where $\varepsilon_1 = 0.702$. A further simplification is to replace A_1/A_T by C, the "cold" fraction of the wall ($A_T = A_1 + A_r$).

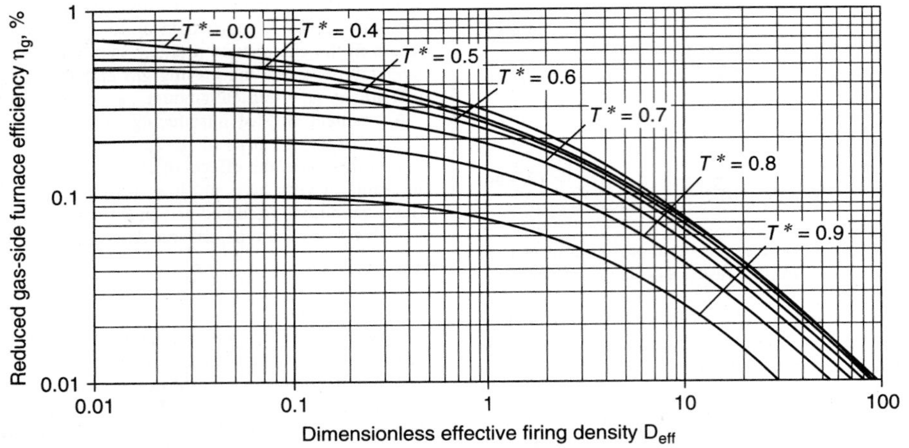

Figure 2.3.6 Reduced gas-side furnace efficiency versus effective firing density for well-stirred combustion chamber model. $\Delta = 0$; $T^* = 0.0, 0.4, 0.5, 0.6, 0.7, 0.8, 0.9$.

With T_G^* known, the chamber efficiency η_G based on heat transfer from the gas is defined as

$$\eta_G = \frac{(1 - T_G^* + \Delta)}{1 - T_O^*} \tag{2.3.71}$$

In the absence of heat losses, this is equal to the efficiency based on energy to the sink, which is expressed as

$$\eta_G = \frac{\left(\overline{GS_1}\right)_R \sigma(T_G^4 - T_1^4) + h_1 A_1 (T_G - T_1)}{\dot{H}} \tag{2.3.72}$$

Solution of Eq. (2.3.69) relates T_G^* and D, and substituting the value of T_G^* into either Eq. (2.3.71) or Eq. (2.3.72) provides the dependence of η_G on D.

Furnace Chamber Performance—General Although the chamber efficiency η depends on D, N_c, and T_1^*, the reduced firing density D is the dominant factor; it makes allowance for such operating variables as fuel type, excess air or air preheat—which affect flame temperature or gas emissivity, for fractional occupancy of the walls by sink surfaces, and for sink emissivity. Variation in the normalized sink temperature T_1^* has little effect until it exceeds 0.3; although the effect of N_c, is significant, it is secondary.

Solution of Eq. (2.3.69) gives the relation between D and T_G^*, and Eqs. (2.3.71) and (2.3.72) give the relation between η_G and D. As an example, Fig. 2.3.6 gives D versus η_G. Approximate operating regimes of various classes of furnaces are shown in Table 2.3.7. Note the significant properties of the functions presented: (1) As firing rate D goes down, η_G rises; T_G^* approaches T_1^* in the limit as D decreases to zero. (2) Changing T_1^* has a large effect only when it exceeds about 0.4. (3) As the furnace walls approach complete coverage by a black sink ($C = \varepsilon_1 = 1$), $\overline{GS_1}$ becomes $\varepsilon_G A_T$ and $\propto 1/\varepsilon_G$. Thus, at very high firing rates where η_G approaches inverse proportionality to D, the efficiency of heat transfer varies directly as ε_G (gas-turbine chambers), but at low firing rates ε_G has relatively little effect. (4) When $C\varepsilon_1 << 1$ because of a non-black sink or much refractory surface, the effect of changing flame emissivity is to produce a much less than proportional effect on heat flux. The regimes of operation of different practical combustion systems are given in Table 2.3.7.

Equations (2.3.69) to (2.3.72) predict the effects of excess air, air preheat, and fuel quality on performance, through the effect on T_F; through $\overline{GS_1}$ they show the effect of gas and sink emissivity and the fraction of the chamber walls occupied by heat sink. They serve as a framework for correlating the performance of furnaces with flow patterns (plug flow, parabolic profile, and recirculatory flow) differing from the well-stirred model (Hottel and Sarofim, Chap. 14). As expected, plug-flow furnaces show somewhat higher efficiency, mild recirculation types somewhat lower efficiency, and strong recirculation furnaces an efficiency closely similar to that of the well-stirred model.

Refractory Temperature T_r Though the average value of T_r in a combustion chamber is not involved in the evaluation of T_G or h, it is sometimes of interest. It assumes a mean value between T_G and T_1, given by

$$\left(\frac{T_r}{T_G}\right)^4 = \frac{1 + E(T_1/T_G)^4}{1 + E}$$

For the speckled furnace gray gas model

$$E = C\varepsilon_1\left(\frac{1}{\varepsilon_G} - 1\right)$$

Allowance is made for the difference between emissivity and absorptivity by changing ε_G to $\varepsilon_{G,e}$ in the above equation. For the gray-plus-clear-gas model,

$$E = C\varepsilon_1\left[\frac{1}{\varepsilon_G} - \frac{1}{a} + \frac{1/a - 1}{C\varepsilon_1 + (1 + C)\varepsilon_R}\right]$$

Table 2.3.7 Operational Domains for Representative Process Furnaces and Combustors

Domain	Furnace or combustor type	Dimensionless sink temperature	Dimensionless firing density
A	Oil processing furnaces; radiant section of oil tube stills and cracking coils	$T^* < 0.4$	$0.1 < D_{eff} < 1.0$
B	Domestic boiler combustion chambers	$T^* < 0.2$	$0.5 < D_{eff} < 1.1$
C	Glass furnaces	$0.7 < T^* < 0.8$	$0.035 < D_{eff} < 0.8$
D	Soaking pits	$T^* < 0.6$	$0.7 < D_{eff} < 1.1$
E	Gas-turbine combustors	$0.4 < T^* < 0.7$	$4.0 < D_{eff} < 25.0$

2.4 HEAT TRANSFER BY CONDUCTION AND CONVECTION

by Afshin J. Ghajar

HEAT TRANSFER MECHANISMS

Heat is defined as the form of energy that can be transferred from one system to another as a result of temperature difference. A thermodynamic analysis is concerned with the *amount* of heat transfer as a system undergoes a process from one equilibrium state to another. The science that deals with the determination of the *rates* of such energy transfers is the *heat transfer*. The transfer of energy as heat is always from the higher-temperature medium to the lower-temperature one, and heat transfer stops when the two mediums reach the same temperature.

Heat can be transferred in three different modes: *conduction, convection,* and *radiation.* All modes of heat transfer require the existence of a temperature difference, and all modes are from the high-temperature medium to a lower-temperature one. In this section heat transfer by conduction, convection (forced and natural), combined conduction-convection, combined convection-radiation, boiling and condensation, and application of nanofluids in heat transfer will be briefly discussed.

CONDUCTION

Conduction is the heat transfer mechanism that takes place when the media is stationary. It may be thought of as the transfer of energy from the more energetic particles of a substance to the adjacent less energetic ones as a result of interactions between the particles. Conduction can take place in solids, liquids, or gases. In solids, it is due to the combination of vibrations of the molecules in a lattice and the energy transport by free electrons. In liquids and gases, conduction is due to the collisions and diffusion of the molecules during their random motion.

The *rate* of heat conduction through a medium depends on the *geometry* of the medium, its *thickness,* and the *material* of the medium, as well as the *temperature difference* across the medium.

Consider steady heat conduction through a large plane wall of thickness $\Delta x = L$ and area A, as shown in Fig. 2.4.1. The temperature difference across the wall is $\Delta T = T_2 - T_1$. Experiments have shown that the rate of heat transfer $\dot{Q}$ through the wall is *doubled* when the temperature difference ΔT across the wall or the area A normal to the direction of heat transfer is doubled, but is *halved* when the wall thickness L is doubled. Thus we conclude that *the rate of heat conduction through a plane layer is proportional to the temperature difference across the layer and the heat transfer area, but is inversely proportional to the thickness of the layer.* That is,

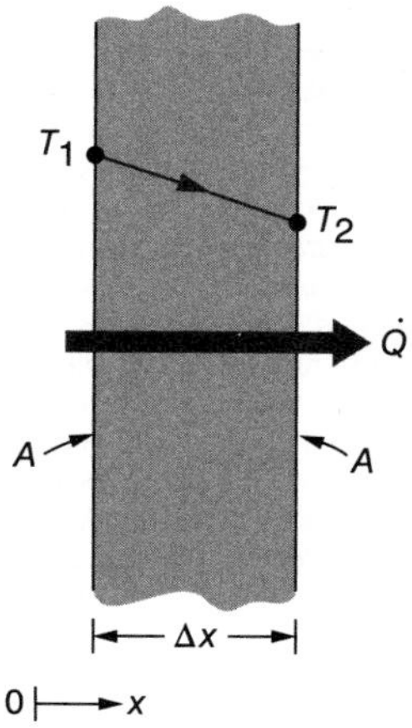

Figure 2.4.1 Heat conduction through a large plane wall of thickness Δx and area A. (*From Çengel and Ghajar, 2015.*)

$$\dot{Q}_{cond} = kA\,\frac{T_1 - T_2}{\Delta x} = -kA\,\frac{\Delta T}{\Delta x} \tag{2.4.1}$$

where the constant of proportionality k is the **thermal conductivity** of the material, which is a *measure of the ability of a material to conduct heat.* Gold and iron are two substances with large values of thermal conductivity, and they are called good conductors. Others with small conductivities like air and wood are good thermal insulators. In the limiting case of $\Delta x \to 0$, the Eq. (2.4.1) reduces to the differential form

$$\dot{Q}_{cond} = -kA\,\frac{dT}{dx} \tag{2.4.2}$$

which is called **Fourier's law of heat conduction** after J. Fourier. Here dT/dx is the **temperature gradient**, which is the slope of the temperature curve on a T-x diagram (the rate of change of T with x), at location x. The relation above indicates that the rate of heat conduction in a given direction is proportional to the temperature gradient in that direction. Heat is conducted in the direction of decreasing temperature, and the temperature gradient becomes negative when temperature decreases with increasing x. The *negative sign* in Eq. (2.4.2) ensures that heat transfer in the positive x direction is a positive quantity. The heat transfer area A is always *normal* to the direction of heat transfer.

To obtain a general relation for Fourier's law of heat conduction, consider a medium in which the temperature distribution is three-dimensional. Figure 2.4.2 shows an isothermal surface in that medium. The heat transfer vector at a point P on this surface must be perpendicular to the surface, and it must point in the direction of decreasing temperature. If n is the normal of the isothermal surface at point P, the rate of heat conduction at that point can be expressed by Fourier's law as

$$\dot{Q}_n = -kA\,\frac{\partial T}{\partial n} \tag{2.4.3}$$

In rectangular coordinates, the heat conduction vector can be expressed in terms of its components as

$$\vec{\dot{Q}}_n = \dot{Q}_x\,\vec{i} + \dot{Q}_y\,\vec{j} + \dot{Q}_z\,\vec{k} \tag{2.4.4}$$

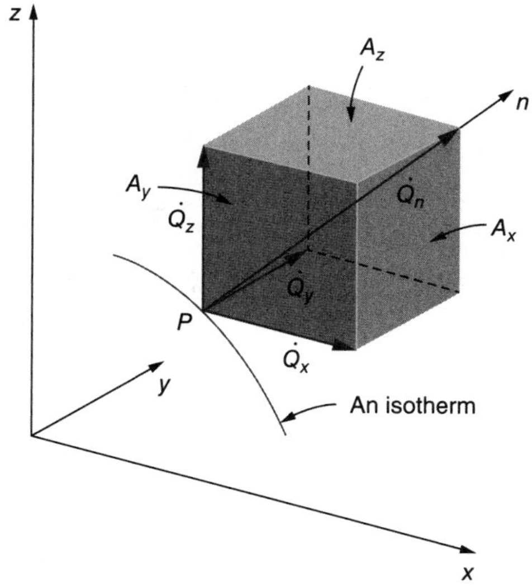

Figure 2.4.2 The heat transfer vector is always perpendicular to an isothermal surface and can always be resolved into its components. (*From Çengel and Ghajar, 2015.*)

where i, j, and k are the unit vectors, and $\dot{Q}_x$, $\dot{Q}_y$, and $\dot{Q}_z$ are the magnitudes of the heat transfer rates in the x-, y-, and z-directions, which again can be determined from Fourier's law as

$$\dot{Q}_x = -kA_x \frac{\partial T}{\partial x}, \quad \dot{Q}_y = -kA_y \frac{\partial T}{\partial y}, \quad \dot{Q}_z = -kA_z \frac{\partial T}{\partial z} \qquad (2.4.5)$$

Here A_x, A_y and A_z are heat conduction areas normal to the x-, y-, and z-directions, respectively (Fig. 2.4.2).

Most engineering materials are *isotropic* in nature, and thus they have the same properties in all directions. For such materials we do not need to be concerned about the variation of properties with direction. But in *anisotropic* materials such as the fibrous or composite materials, the properties may change with direction. For example, some of the properties of wood along the grain are different than those in the direction normal to the grain. In such cases the thermal conductivity may need to be expressed as a tensor quantity to account for the variation with direction.

Thermal Conductivity The thermal conductivity k is a measure of a material's ability to conduct heat. For example, k = 0.607 W/m·K for water and k = 80.2 W/m·K for iron at room temperature, which indicates that iron conducts heat more than 100 times faster than water can. Thus we say that water is a poor heat conductor relative to iron, although water is an excellent medium to store thermal energy.

Equation (2.4.1) for the rate of conduction heat transfer under steady conditions can also be viewed as the defining equation for thermal conductivity. Thus the **thermal conductivity** of a material can be defined as *the rate of heat transfer through a unit thickness of the material per unit area per unit temperature difference.* The thermal conductivity of a material is a measure of the ability of the material to conduct heat. A high value for thermal conductivity indicates that the material is a good heat conductor, and a low value indicates that the material is a poor heat conductor or *insulator.* The thermal conductivities of some common materials at room temperature are given in Table 2.4.1. The thermal conductivity of pure copper at room temperature is k = 401 W/m·K, which indicates that a 1-m-thick copper wall will conduct heat at a rate of 401 W/m^2 area per K temperature difference across the wall. Note that materials such as copper and silver that are good electric conductors are also good heat conductors, and have high values of thermal conductivity. Materials such as rubber, wood, and Styrofoam are poor conductors of heat and have low conductivity values.

A layer of material of known thickness and area can be heated from one side by an electric resistance heater of known output. If the outer surfaces of the heater are well insulated, all the heat generated by the resistance heater will be transferred through the material whose conductivity is to be determined. Then measuring the two surface temperatures of the material when steady heat transfer is reached and substituting them into Eq. (2.4.1) together with other known quantities give the thermal conductivity (Fig. 2.4.3).

Table 2.4.1 The Thermal Conductivities of Some Materials at Room Temperature

Material	k, W/m·K*
Diamond	2,300
Silver	429
Copper	401
Gold	317
Aluminum	237
Iron	80.2
Mercury (l)	8.54
Glass	0.78
Brick	0.72
Water (l)	0.607
Human skin	0.37
Wood (oak)	0.17
Helium (g)	0.152
Soft rubber	0.13
Glass fiber	0.043
Air (g)	0.026
Urethane, rigid foam	0.026

*Multiply by 0.5778 to convert to Btu/h·ft·°F.

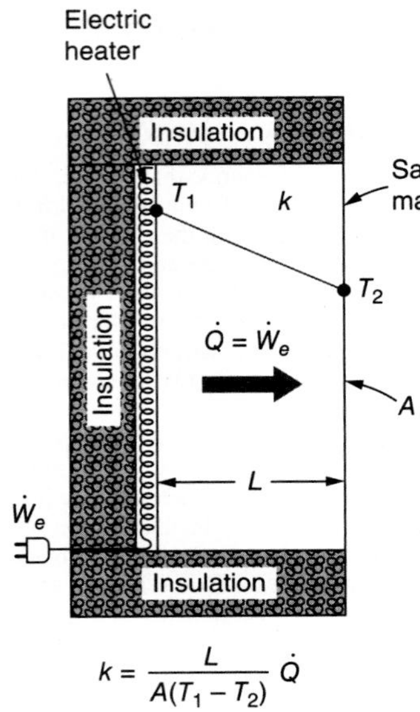

Figure 2.4.3 A simple experimental setup to determine the thermal conductivity of a material. (*From Çengel and Ghajar, 2014.*)

The thermal conductivities of materials vary over a wide range, as shown in Fig. 2.4.4. The thermal conductivities of gases such as air vary by a factor of 10^4 from those of pure metals such as copper. Note that pure crystals and metals have the highest thermal conductivities, and gases and insulating materials the lowest.

The *kinetic theory* of gases predicts and the experiments confirm that the thermal conductivity of gases is proportional to the *square root of the thermodynamic temperature T,* and inversely proportional to the *square root of the molar mass M.* Therefore, for a particular gas (fixed M), the thermal conductivity increases with increasing temperature and at a fixed temperature the thermal conductivity decreases with increasing M. For example, at a fixed temperature of 1,000 K, the thermal conductivity of helium (M = 4) is 0.343 W/m·K and that of air (M = 29) is 0.0667 W/m·K, which is much lower than that of helium.

The thermal conductivities of *gases* at 1 atm pressure are listed in Table 2.4.2. However, they can also be used at pressures other than 1 atm, since the thermal conductivity of gases is *independent of pressure* in a wide range of pressures encountered in practice.

The mechanism of heat conduction in a *liquid* is complicated by the fact that the molecules are more closely spaced, and they exert a stronger intermolecular force field. The thermal conductivities of liquids usually lie between those of solids and gases. The thermal conductivity of a substance is normally highest in the solid phase and lowest in the gas phase. The thermal conductivity of liquids is generally insensitive to pressure except near the thermodynamic critical point. Unlike gases, the thermal conductivities of most liquids decrease with increasing temperature, with water being a notable exception. Like gases, the conductivity of liquids decreases with increasing molar mass. Liquid metals such as mercury and sodium have high thermal conductivities and are very suitable for use in applications where a high heat transfer rate to a liquid is desired, as in nuclear power plants.

The thermal conductivities of materials vary with temperature. The variation of thermal conductivity over certain temperature ranges is negligible for some materials, but significant for others, as shown in

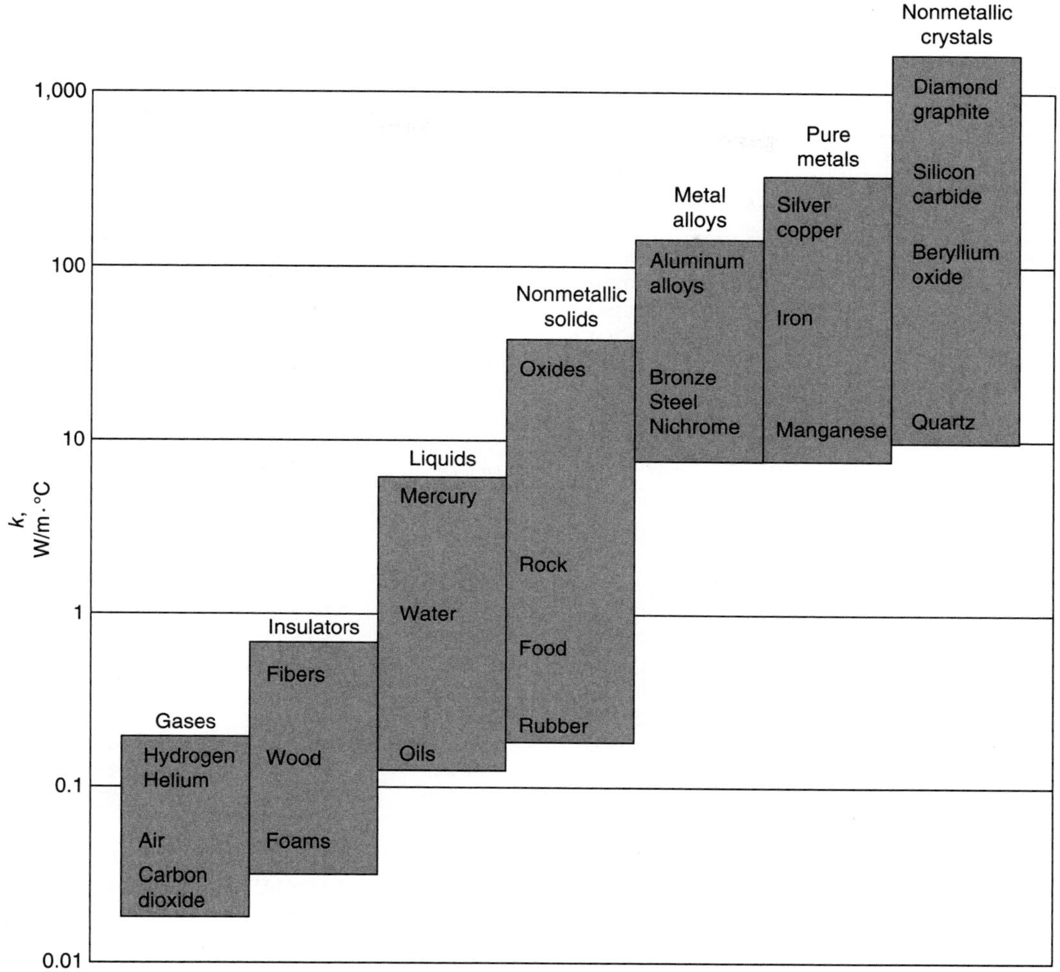

Figure 2.4.4 The range of thermal conductivity of various materials at room temperature. (*From Çengel and Ghajar, 2015.*)

Fig. 2.4.5. The thermal conductivities of certain solids exhibit dramatic increases at temperatures near absolute zero, when these solids become *superconductors.* For example, the conductivity of copper reaches a maximum value of about 20,000 W/m·K at 20 K, which is about 50 times the conductivity at room temperature.

The temperature dependence of thermal conductivity causes considerable complexity in conduction analysis. Therefore, it is common practice to evaluate the thermal conductivity k at the *average temperature* and treat it as a *constant* in calculations.

In heat transfer analysis, a material is normally assumed to be *isotropic;* that is, to have uniform properties in all directions. This assumption is realistic for most materials, except those that exhibit different structural characteristics in different directions, such as laminated composite materials and wood. The thermal conductivity of wood across the grain, for example, is different than that parallel to the grain.

Variable Thermal Conductivity When the variation of thermal conductivity with temperature in a specified temperature interval for some materials is large (Fig. 2.4.6), use of an average value for thermal conductivity and treating it is as a constant is not recommended. In this case, it is necessary to account for this variation to minimize the error. Accounting for the variation of the thermal conductivity with temperature, in general, complicates the analysis. But in the case of simple one-dimensional cases, we can obtain heat transfer relations in a straightforward manner.

When the variation of thermal conductivity with temperature $k(T)$ is known, the average value of the thermal conductivity in the temperature range between T_1 and T_2 can be determined from

$$k_{avg} = \frac{\int_{T_1}^{T_2} k(T)dT}{T_2 - T_1} \tag{2.4.6}$$

Note that in the case of constant thermal conductivity $k(T) = k$, Eq. (2.4.6) reduces to $k_{avg} = k$, as expected.

Then the rate of steady heat transfer through a plane wall of Fig. 2.4.1 for the case of variable thermal conductivity can be determined by replacing the constant thermal conductivity k in Eq. (2.4.1) by the k_{avg} expression (or value) from Eq. (2.4.6):

$$\dot{Q}_{cond} = k_{avg} A \frac{T_1 - T_2}{\Delta x} = \frac{A}{\Delta x}\int_{T_2}^{T_1} k(T)dT \tag{2.4.7}$$

General Heat Conduction Equation Equation (2.4.5) provided the expressions for $\dot{Q}_x$, $\dot{Q}_y$, and $\dot{Q}_z$ which are the magnitudes of the heat

Table 2.4.2 Properties of Gases at 1 atm Pressure (*From Çengel and Ghajar, 2015*)

Temp. T, °C	Density ρ, kg/m^3	Specific Heat c_p, J/kg·K	Thermal Conductivity k, W/m·K	Thermal Diffusivity α, m^2/s	Dynamic Viscosity μ, kg/m·s	Kinematic Viscosity ν, m^2/s	Prandtl Number Pr
			Carbon Dioxide, CO_2				
–50	2.4035	746	0.01051	5.860×10^{-6}	1.129×10^{-5}	4.699×10^{-6}	0.8019
0	1.9635	811	0.01456	9.141×10^{-6}	1.375×10^{-5}	7.003×10^{-6}	0.7661
50	1.6597	866.6	0.01858	1.291×10^{-5}	1.612×10^{-5}	9.714×10^{-6}	0.7520
100	1.4373	914.8	0.02257	1.716×10^{-5}	1.841×10^{-5}	1.281×10^{-5}	0.7464
150	1.2675	957.4	0.02652	2.186×10^{-5}	2.063×10^{-5}	1.627×10^{-5}	0.7445
200	1.1336	995.2	0.03044	2.698×10^{-5}	2.276×10^{-5}	2.008×10^{-5}	0.7442
300	0.9358	1,060	0.03814	3.847×10^{-5}	2.682×10^{-5}	2.866×10^{-5}	0.7450
400	0.7968	1,112	0.04565	5.151×10^{-5}	3.061×10^{-5}	3.842×10^{-5}	0.7458
500	0.6937	1,156	0.05293	6.600×10^{-5}	3.416×10^{-5}	4.924×10^{-5}	0.7460
1,000	0.4213	1,292	0.08491	1.560×10^{-4}	4.898×10^{-5}	1.162×10^{-4}	0.7455
1,500	0.3025	1,356	0.10688	2.606×10^{-4}	6.106×10^{-5}	2.019×10^{-4}	0.7745
2,000	0.2359	1,387	0.11522	3.521×10^{-4}	7.322×10^{-5}	3.103×10^{-4}	0.8815
			Carbon Monoxide, CO				
–50	1.5297	1,081	0.01901	1.149×10^{-5}	1.378×10^{-5}	9.012×10^{-6}	0.7840
0	1.2497	1,048	0.02278	1.739×10^{-5}	1.629×10^{-5}	1.303×10^{-5}	0.7499
50	1.0563	1,039	0.02641	2.407×10^{-5}	1.863×10^{-5}	1.764×10^{-5}	0.7328
100	0.9148	1,041	0.02992	3.142×10^{-5}	2.080×10^{-5}	2.274×10^{-5}	0.7239
150	0.8067	1,049	0.03330	3.936×10^{-5}	2.283×10^{-5}	2.830×10^{-5}	0.7191
200	0.7214	1,060	0.03656	4.782×10^{-5}	2.472×10^{-5}	3.426×10^{-5}	0.7164
300	0.5956	1,085	0.04277	6.619×10^{-5}	2.812×10^{-5}	4.722×10^{-5}	0.7134
400	0.5071	1,111	0.04860	8.628×10^{-5}	3.111×10^{-5}	6.136×10^{-5}	0.7111
500	0.4415	1,135	0.05412	1.079×10^{-4}	3.379×10^{-5}	7.653×10^{-5}	0.7087
1,000	0.2681	1,226	0.07894	2.401×10^{-4}	4.557×10^{-5}	1.700×10^{-4}	0.7080
1,500	0.1925	1,279	0.10458	4.246×10^{-4}	6.321×10^{-5}	3.284×10^{-4}	0.7733
2,000	0.1502	1,309	0.13833	7.034×10^{-4}	9.826×10^{-5}	6.543×10^{-4}	0.9302
			Methane, CH_4				
–50	0.8761	2,243	0.02367	1.204×10^{-5}	8.564×10^{-6}	9.774×10^{-6}	0.8116
0	0.7158	2,217	0.03042	1.917×10^{-5}	1.028×10^{-5}	1.436×10^{-5}	0.7494
50	0.6050	2,302	0.03766	2.704×10^{-5}	1.191×10^{-5}	1.969×10^{-5}	0.7282
100	0.5240	2,443	0.04534	3.543×10^{-5}	1.345×10^{-5}	2.567×10^{-5}	0.7247
150	0.4620	2,611	0.05344	4.431×10^{-5}	1.491×10^{-5}	3.227×10^{-5}	0.7284
200	0.4132	2,791	0.06194	5.370×10^{-5}	1.630×10^{-5}	3.944×10^{-5}	0.7344
300	0.3411	3,158	0.07996	7.422×10^{-5}	1.886×10^{-5}	5.529×10^{-5}	0.7450
400	0.2904	3,510	0.09918	9.727×10^{-5}	2.119×10^{-5}	7.297×10^{-5}	0.7501
500	0.2529	3,836	0.11933	1.230×10^{-4}	2.334×10^{-5}	9.228×10^{-5}	0.7502
1,000	0.1536	5,042	0.22562	2.914×10^{-4}	3.281×10^{-5}	2.136×10^{-4}	0.7331
1,500	0.1103	5,701	0.31857	5.068×10^{-4}	4.434×10^{-5}	4.022×10^{-4}	0.7936
2,000	0.0860	6,001	0.36750	7.120×10^{-4}	6.360×10^{-5}	7.395×10^{-4}	1.0386
			Hydrogen, H_2				
–50	0.11010	12,635	0.1404	1.009×10^{-4}	7.293×10^{-6}	6.624×10^{-5}	0.6562
0	0.08995	13,920	0.1652	1.319×10^{-4}	8.391×10^{-6}	9.329×10^{-5}	0.7071
50	0.07603	14,349	0.1881	1.724×10^{-4}	9.427×10^{-6}	1.240×10^{-4}	0.7191
100	0.06584	14,473	0.2095	2.199×10^{-4}	1.041×10^{-5}	1.582×10^{-4}	0.7196
150	0.05806	14,492	0.2296	2.729×10^{-4}	1.136×10^{-5}	1.957×10^{-4}	0.7174
200	0.05193	14,482	0.2486	3.306×10^{-4}	1.228×10^{-5}	2.365×10^{-4}	0.7155

Table 2.4.2 Properties of Gases at 1 atm Pressure *(Continued)*

Temp. T, °C	Density ρ, kg/m³	Specific Heat c_p, J/kg·K	Thermal Conductivity k, W/m·K	Thermal Diffusivity α, m²/s	Dynamic Viscosity μ, kg/m·s	Kinematic Viscosity ν, m²/s	Prandtl Number Pr
300	0.04287	14,481	0.2843	4.580×10^{-4}	1.403×10^{-5}	3.274×10^{-4}	0.7149
400	0.03650	14,540	0.3180	5.992×10^{-4}	1.570×10^{-5}	4.302×10^{-4}	0.7179
500	0.03178	14,653	0.3509	7.535×10^{-4}	1.730×10^{-5}	5.443×10^{-4}	0.7224
1,000	0.01930	15,577	0.5206	1.732×10^{-3}	2.455×10^{-5}	1.272×10^{-3}	0.7345
1,500	0.01386	16,553	0.6581	2.869×10^{-3}	3.099×10^{-5}	2.237×10^{-3}	0.7795
2,000	0.01081	17,400	0.5480	2.914×10^{-3}	3.690×10^{-5}	3.414×10^{-3}	1.1717
				Nitrogen, N_2			
−50	1.5299	957.3	0.02001	1.366×10^{-5}	1.390×10^{-5}	9.091×10^{-6}	0.6655
0	1.2498	1,035	0.02384	1.843×10^{-5}	1.640×10^{-5}	1.312×10^{-5}	0.7121
50	1.0564	1,042	0.02746	2.494×10^{-5}	1.874×10^{-5}	1.774×10^{-5}	0.7114
100	0.9149	1,041	0.03090	3.244×10^{-5}	2.094×10^{-5}	2.289×10^{-5}	0.7056
150	0.8068	1,043	0.03416	4.058×10^{-5}	2.300×10^{-5}	2.851×10^{-5}	0.7025
200	0.7215	1,050	0.03727	4.921×10^{-5}	2.494×10^{-5}	3.457×10^{-5}	0.7025
300	0.5956	1,070	0.04309	6.758×10^{-5}	2.849×10^{-5}	4.783×10^{-5}	0.7078
400	0.5072	1,095	0.04848	8.727×10^{-5}	3.166×10^{-5}	6.242×10^{-5}	0.7153
500	0.4416	1,120	0.05358	1.083×10^{-5}	3.451×10^{-5}	7.816×10^{-4}	0.7215
1,000	0.2681	1,213	0.07938	2.440×10^{-4}	4.594×10^{-5}	1.713×10^{-4}	0.7022
1,500	0.1925	1,266	0.11793	4.839×10^{-4}	5.562×10^{-5}	2.889×10^{-4}	0.5969
2,000	0.1502	1,297	0.18590	9.543×10^{-4}	6.426×10^{-5}	4.278×10^{-4}	0.4483
				Oxygen, O_2			
−50	1.7475	984.4	0.02067	1.201×10^{-5}	1.616×10^{-5}	9.246×10^{-6}	0.7694
0	1.4277	928.7	0.02472	1.865×10^{-5}	1.916×10^{-5}	1.342×10^{-5}	0.7198
50	1.2068	921.7	0.02867	2.577×10^{-5}	2.194×10^{-5}	1.818×10^{-5}	0.7053
100	1.0451	931.8	0.03254	3.342×10^{-5}	2.451×10^{-5}	2.346×10^{-5}	0.7019
150	0.9216	947.6	0.03637	4.164×10^{-5}	2.694×10^{-5}	2.923×10^{-5}	0.7019
200	0.8242	964.7	0.04014	5.048×10^{-5}	2.923×10^{-5}	3.546×10^{-5}	0.7025
300	0.6804	997.1	0.04751	7.003×10^{-5}	3.350×10^{-5}	4.923×10^{-5}	0.7030
400	0.5793	1,025	0.05463	9.204×10^{-5}	3.744×10^{-5}	6.463×10^{-5}	0.7023
500	0.5044	1,048	0.06148	1.163×10^{-4}	4.114×10^{-5}	8.156×10^{-5}	0.7010
1,000	0.3063	1,121	0.09198	2.678×10^{-4}	5.732×10^{-5}	1.871×10^{-4}	0.6986
1,500	0.2199	1,165	0.11901	4.643×10^{-4}	7.133×10^{-5}	3.243×10^{-4}	0.6985
2,000	0.1716	1,201	0.14705	7.139×10^{-4}	8.417×10^{-5}	4.907×10^{-4}	0.6873
				Water Vapor, H_2O			
−50	0.9839	1,892	0.01353	7.271×10^{-6}	7.187×10^{-6}	7.305×10^{-6}	1.0047
0	0.8038	1,874	0.01673	1.110×10^{-5}	8.956×10^{-6}	1.114×10^{-5}	1.0033
50	0.6794	1,874	0.02032	1.596×10^{-5}	1.078×10^{-5}	1.587×10^{-5}	0.9944
100	0.5884	1,887	0.02429	2.187×10^{-5}	1.265×10^{-5}	2.150×10^{-5}	0.9830
150	0.5189	1,908	0.02861	2.890×10^{-5}	1.456×10^{-5}	2.806×10^{-5}	0.9712
200	0.4640	1,935	0.03326	3.705×10^{-5}	1.650×10^{-5}	3.556×10^{-5}	0.9599
300	0.3831	1,997	0.04345	5.680×10^{-5}	2.045×10^{-5}	5.340×10^{-5}	0.9401
400	0.3262	2,066	0.05467	8.114×10^{-5}	2.446×10^{-5}	7.498×10^{-5}	0.9240
500	0.2840	2,137	0.06677	1.100×10^{-4}	2.847×10^{-5}	1.002×10^{-4}	0.9108
1,000	0.1725	2,471	0.13623	3.196×10^{-4}	4.762×10^{-5}	2.761×10^{-4}	0.8639
1,500	0.1238	2,736	0.21301	6.288×10^{-4}	6.411×10^{-5}	5.177×10^{-4}	0.8233
2,000	0.0966	2,928	0.29183	1.032×10^{-3}	7.808×10^{-5}	8.084×10^{-4}	0.7833

Note: For ideal gases, the properties c_p, k, μ, and Pr are independent of pressure. The properties ρ, ν, and α at a pressure P (in atm) other than 1 atm are determined by multiplying the values of P at the given temperature by ρ, and by dividing ν and α by P.

SOURCE: Data generated from the EES software developed by S. A. Klein and F. L. Alvarado. Originally based on various sources.

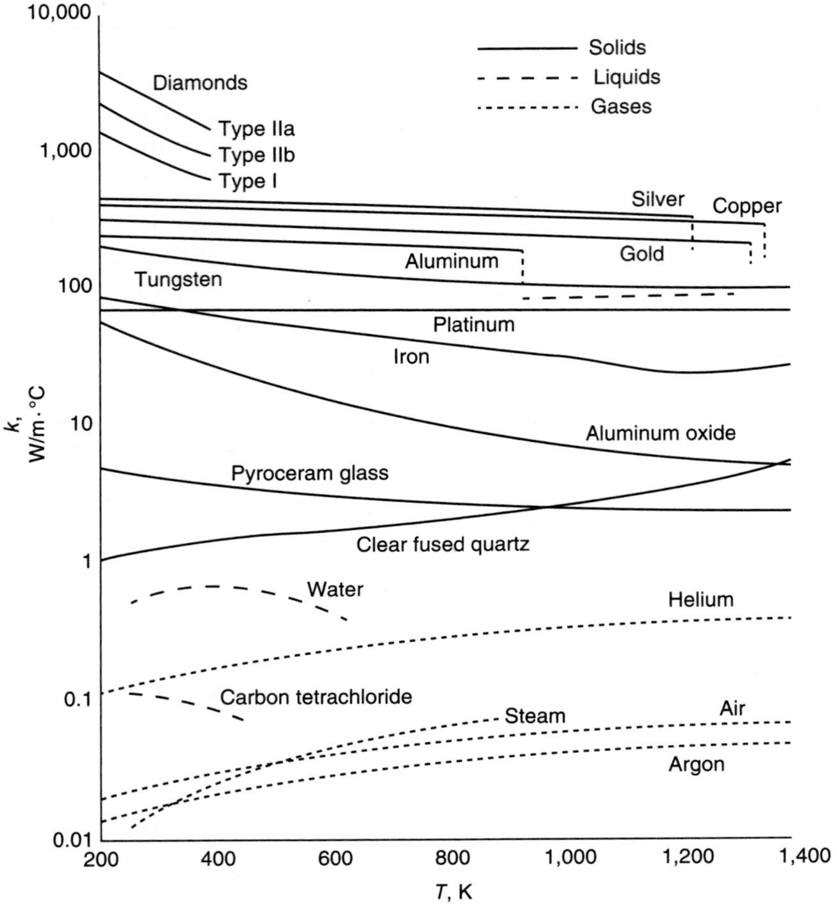

Figure 2.4.5 The variation of the thermal conductivity of various solids, liquids, and gases with temperature. (*From Çengel and Ghajar, 2015.*)

transfer rates in the x-, y-, and z-directions, and were determined based on Fourier's law. In development of these equations we considered one-dimensional heat conduction and assumed heat conduction in other directions to be negligible. Most heat transfer problems encountered in practice can be approximated as being one-dimensional. However, this is not always the case, and sometimes we need to consider heat transfer in other directions as well. In such cases heat conduction is said to be *multidimensional,* and in this section we develop the governing differential equation in such systems in rectangular, cylindrical, and spherical coordinate systems.

Consider a small rectangular element of length Δx, width Δy, and height Δz, as shown in Fig. 2.4.7. Assume the density of the body is ρ and the specific heat is c. An *energy balance* (energy in + energy generated − energy out = energy stored) on this element during a small time interval Δt can be expressed as

$$\dot{Q}_x + \dot{Q}_y + \dot{Q}_z - \dot{Q}_{x+\Delta x} - \dot{Q}_{y+\Delta y} - \dot{Q}_{z+\Delta z} + \dot{E}_{gen} = \dot{E}_{st} \tag{2.4.8}$$

where the conduction heat rates at the opposite surfaces can then be expressed as a Taylor series expression where, neglecting the higher order terms,

$$\dot{Q}_{x+\Delta x} = \dot{Q}_x + \frac{\partial \dot{Q}_x}{\partial x}x, \quad \dot{Q}_{y+\Delta y} = \dot{Q}_y + \frac{\partial \dot{Q}_y}{\partial y}y, \quad \dot{Q}_{z+\Delta z} = \dot{Q}_z + \frac{\partial \dot{Q}_z}{\partial z}\Delta z \tag{2.4.9}$$

Noting that the volume of the element is $V_{\text{element}} = \Delta x \Delta y \Delta z$, the rate of thermal energy generation ($\dot{E}_{gen}$) and the energy storage term ($\dot{E}_{st}$) can be expressed as

$$\dot{E}_{gen} = \dot{e}_{gen}\, V_{element} = \dot{e}_{gen} A\ \Delta x\ ; \ \dot{E}_{st} = \rho c \frac{\partial T}{\partial t} \Delta x \Delta y \Delta z \tag{2.4.10}$$

where ($\dot{e}_{gen}$) is the rate at which energy is generated per unit volume of the element (W/m^3). $\dot{E}_{gen}$ and $\dot{E}_{st}$ are total heat generation and storage rates within the element (W), respectively.

Substituting Eqs. (2.4.9) and (2.4.10) in the energy balance Eq. (2.4.8) and dividing by Δx, Δy, Δz and recognizing that the conduction heat rates can be evaluated from the Fourier's law, we obtain the *general heat conduction equation* or *heat equation*

$$\frac{\partial}{\partial x}\left(k\frac{\partial T}{\partial x}\right) + \frac{\partial}{\partial y}\left(k\frac{\partial T}{\partial y}\right) + \frac{\partial}{\partial z}\left(k\frac{\partial T}{\partial z}\right) + \dot{e}_{gen} = \rho c \frac{\partial T}{\partial t} \tag{2.4.11}$$

In case of constant thermal conductivity, Eq. (2.4.11) reduces to

$$\frac{\partial^2 T}{\partial x^2} + \frac{\partial^2 T}{\partial y^2} + \frac{\partial^2 T}{\partial z^2} + \frac{\dot{e}_{gen}}{k} = \frac{1}{\alpha}\frac{\partial T}{\partial t} \tag{2.4.12}$$

where the property α = *heat conduction/heat storage* = $k/\rho c$ is the *thermal diffusivity* of the material and represents how fast heat diffuses through the material. The larger the α, the faster the propagation of heat into the medium. See Table 2.4.2 for values of thermal diffusivity of

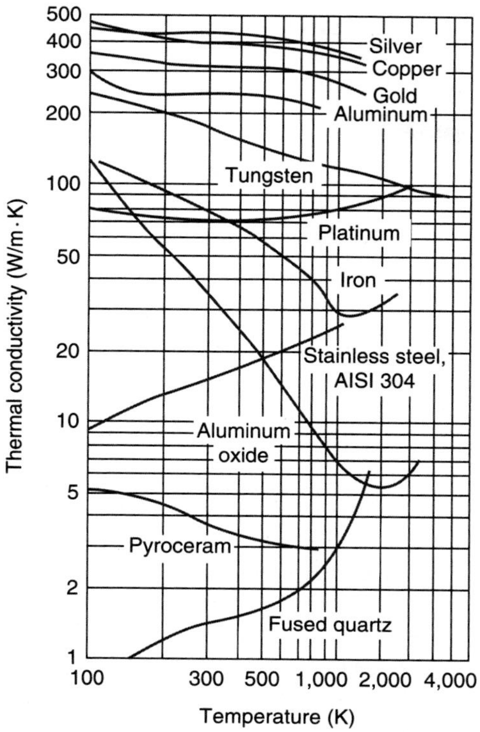

Figure 2.4.6 Variation of the thermal conductivity of some solids with temperature. *(From Çengel and Ghajar, 2015.)*

gases at 1 atm pressure and Table 2.4.3 for the thermal diffusivities of some common materials at 20°C (room temperature).

Equation (2.4.12) is known as the **Fourier-Biot equation** and it reduces to the following forms under specified conditions:

1. Steady-state (called the **Poisson equation**):

$$\frac{\partial^2 T}{\partial x^2}+\frac{\partial^2 T}{\partial y^2}+\frac{\partial^2 T}{\partial z^2}+\frac{\dot{e}_{gen}}{k}=0 \qquad (2.4.13a)$$

2. Transient, no heat generation (called the **diffusion equation**):

$$\frac{\partial^2 T}{\partial x^2}+\frac{\partial^2 T}{\partial y^2}+\frac{\partial^2 T}{\partial z^2}=\frac{1}{\alpha}\frac{\partial T}{\partial t} \qquad (2.13b)$$

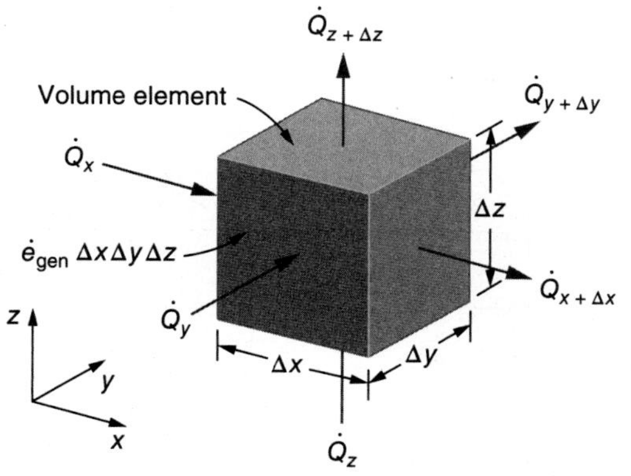

Figure 2.4.7 Three dimensional heat conduction through a rectangular element. *(From Çengel and Ghajar, 2015.)*

Table 2.4.3 The Thermal Diffusivities of Some Materials at Room Temperature

Material	α, m²/s*
Silver	149×10^{-6}
Gold	127×10^{-6}
Copper	113×10^{-6}
Aluminum	97.5×10^{-6}
Iron	22.8×10^{-6}
Mercury (l)	4.7×10^{-6}
Marble	1.2×10^{-6}
Ice	1.2×10^{-6}
Concrete	0.75×10^{-6}
Brick	0.52×10^{-6}
Heavy soil (dry)	0.52×10^{-6}
Glass	0.34×10^{-6}
Glass wool	0.23×10^{-6}
Water (l)	0.14×10^{-6}
Beef	0.14×10^{-6}
Wood (oak)	0.13×10^{-6}

*Multiply by 10.76 to convert to ft²/s.

3. Steady-state, no heat generation (called the **Laplace equation**):

$$\frac{\partial^2 T}{\partial x^2}+\frac{\partial^2 T}{\partial y^2}+\frac{\partial^2 T}{\partial z^2}=0 \qquad (2.4.13c)$$

The general heat conduction equation in cylindrical and spherical coordinates can be obtained from an energy balance on a volume element in cylindrical or spherical coordinates, as shown in Figs. 2.4.8*a* and 2.4.8*b*, respectively. It can also be obtained directly from Eq. (2.4.11) by coordinate transformation. After lengthy manipulations, we obtain Eqs. (2.4.14) and (2.4.15) for cylindrical and spherical coordinates, respectively.

$$\frac{1}{r}\frac{\partial}{\partial r}\left(kr\frac{\partial T}{\partial r}\right)+\frac{1}{r^2}\frac{\partial}{\partial \phi}\left(k\frac{\partial T}{\partial \phi}\right)+\frac{\partial}{\partial z}\left(k\frac{\partial T}{\partial z}\right)+\dot{e}_{gen}=\rho c\frac{\partial T}{\partial t} \qquad (2.4.14)$$

$$\frac{1}{r^2}\frac{\partial}{\partial r}\left(kr^2\frac{\partial T}{\partial r}\right)+\frac{1}{r^2\sin^2\theta}\frac{\partial}{\partial \phi}\left(k\frac{\partial T}{\partial \phi}\right)+\frac{1}{r^2\sin\theta}\frac{\partial}{\partial \theta}\left(k\sin\theta\frac{\partial T}{\partial \theta}\right)+\dot{e}_{gen}=\rho c\frac{\partial T}{\partial t} \qquad (2.4.15)$$

Obtaining analytical solutions to these differential equations (2.4.11)-(2.4.15) requires knowledge of the solution techniques of partial differential equations. For details refer to Hahn and Ozisik (2012). However, these equations can be considerably simplified if we limit our consideration to one-dimensional steady-state cases, since they result in *ordinary differential equations*. The solution procedure for solving heat conduction problems can be summarized as (1) *formulate* the problem by obtaining the applicable differential equation in its simplest form and specifying the boundary conditions, (2) obtain the *general solution* of the differential equation, and (3) apply the *boundary conditions* and determine the arbitrary constants in the general solution. Çengel and Ghajar (2015) in their textbook demonstrate the solution procedure through a wide range of heat conduction problems in rectangular, cylindrical, and spherical geometries.

The heat conduction equations above were developed using an energy balance on a differential element inside the medium, and they remain the same regardless of the *thermal conditions* on the *surfaces* of the medium. That is, the differential equations do not incorporate any information related to the conditions on the surfaces such as the surface temperature or

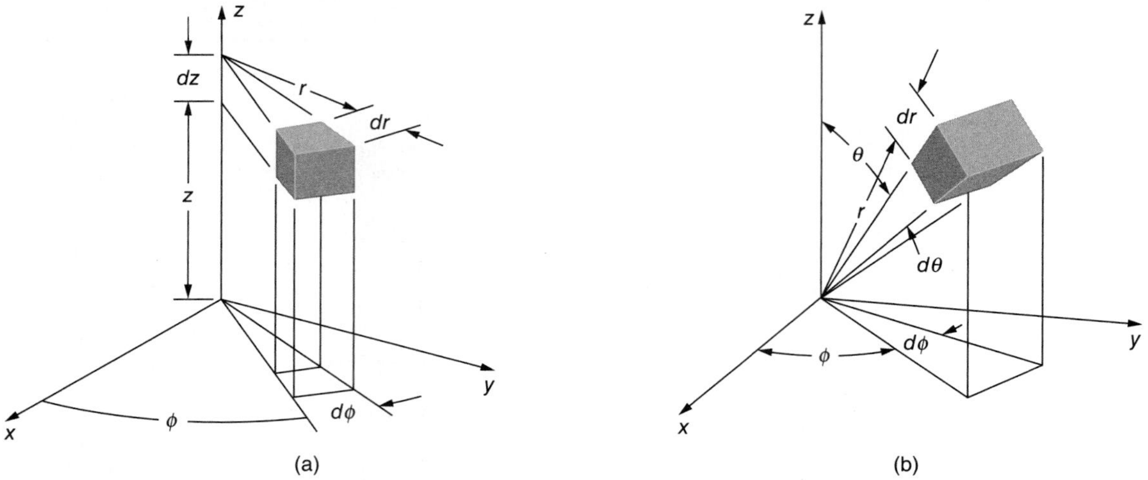

Figure 2.4.8 A differential volume element in (*a*) cylindrical coordinates and (*b*) in spherical coordinates. (*From Çengel and Ghajar, 2015.*)

a specified heat flux. Yet we know that the heat flux and the temperature distribution in a medium depend on the conditions at the surfaces, and the description of a heat transfer problem in a medium is not complete without a full description of the thermal conditions at the bounding surfaces of the medium. The *mathematical expressions* of the thermal conditions at the boundaries are called the **boundary conditions**. To describe a heat transfer problem completely, we need to specify *two boundary conditions* for one dimensional problem, *four boundary conditions* for two-dimensional problems, and *six boundary conditions* for three-dimensional problems. Boundary conditions most commonly encountered in practice are the *specified temperature, specified heat flux, convection,* and *radiation* boundary conditions. For details on different boundary conditions refer to Çengel and Ghajar (2015). In addition to boundary conditions if the temperature varies with time (transient heat conduction), the temperature at any point on the wall at a specified time also depends on the condition of the wall at the beginning of the heat conduction process. Such a condition, which is usually specified at time $t = 0$, is called the **initial condition**, which is a mathematical expression for the temperature distribution of the medium initially. Note that we need only one initial condition for a heat conduction problem regardless of the dimension since the conduction equation is first order in time (it involves the first derivative of temperature with respect to time).

CONVECTION

Convection is the mode of energy transfer between a solid surface and the adjacent liquid or gas that is in motion, and it involves the combined effects of *conduction* and *fluid motion.* The faster the fluid motion, the greater the convection heat transfer. In the absence of any bulk fluid motion, heat transfer between a solid surface and the adjacent fluid is by pure conduction. The presence of bulk motion of the fluid enhances the heat transfer between the solid surface and the fluid, but it also complicates the determination of heat transfer rates.

Consider the cooling of a hot block by blowing cool air over its top surface (Fig. 2.4.9). Heat is first transferred to the air layer adjacent to the block by conduction. This heat is then carried away from the surface by convection, that is, by the combined effects of conduction within the air that is due to random motion of air molecules and the bulk or macroscopic motion of the air that removes the heated air near the surface and replaces it by the cooler air.

Convection is called **forced convection** if the fluid is forced to flow over the surface by external means such as a fan, pump, or the wind. In contrast, convection is called **natural** (or **free**) **convection** if the fluid motion is caused by buoyancy forces that are induced by density differences due to the variation of temperature in the fluid. For example, in the absence of a fan, heat transfer from the surface of the hot block in Fig. 2.4.9 is by natural convection since any motion in the air in this case is due to the rise of the warmer (and thus lighter) air near the surface and the fall of the cooler (and thus heavier) air to fill its place. Heat transfer between the block and the surrounding air is by conduction if the temperature difference between the air and the block is not large enough to overcome the resistance of air to movement and thus to initiate natural convection currents.

Heat transfer processes that involve *change of phase* of a fluid are also considered to be convection because of the fluid motion induced during the process, such as the rise of the vapor bubbles during boiling or the fall of the liquid droplets during condensation.

Despite the complexity of convection, the rate of *convection heat transfer* is observed to be proportional to the temperature difference, and is conveniently expressed by **Newton's law of cooling** as

$$\dot{Q}_{conv} = hA_s\,(T_s - T_\infty) \tag{2.4.16}$$

where h is the *convection heat transfer coefficient* in W/m²·K or Btu/h·ft²·°F, A_s is the surface area through which convection heat transfer takes place, T_s is the surface temperature, and T_∞ is the temperature of the fluid sufficiently far from the surface. Note that at the surface, the fluid temperature equals the surface temperature of the solid.

The convection heat transfer coefficient h is not a property of the fluid. It is an experimentally determined parameter whose value

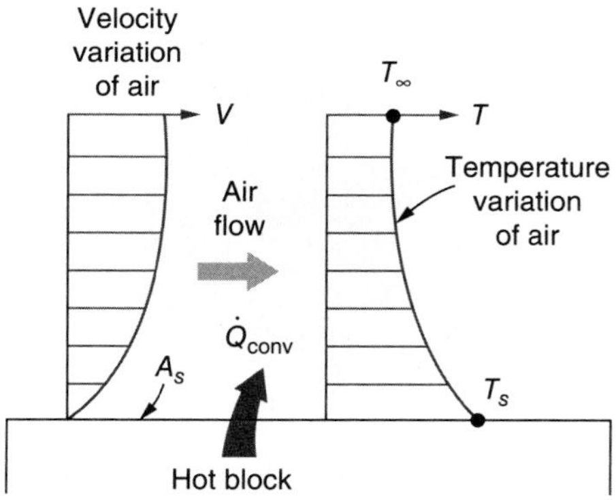

Figure 2.4.9 Heat transfer from a hot surface to air by convection. (*From Çengel and Ghajar, 2015.*)

depends on all the variables influencing convection such as the surface geometry, the nature of fluid motion, the properties of the fluid, and the bulk fluid velocity. Typical values of h are given in Table 2.4.4.

Table 2.4.4 Typical Values of Convection Heat Transfer Coefficient

Type of convection	h, W/m²·K*
Free convection of gases	2-25
Free convection of liquids	10-1,000
Forced convection of gases	25-250
Forced convection of liquids	50-20,000
Boiling and condensation	2,500-100,000

*Multiply by 0.176 to convert to Btu/ft²·°F.

Heat Transfer from Finned Surfaces According to the Newton's law of cooling (Eq. 2.4.16), there are *two ways* to increase the rate of heat transfer by convection from a surface with a fixed temperature T_s to the surrounding medium at a fixed temperature of T_∞: to increase the *convection heat transfer coefficient h* or to increase the *surface area* A_s. Increasing h may require the installation of a pump or fan, or replacing the existing one with a larger one, but this approach may or may not be practical. Besides, it may not be adequate. The alternative is to increase the surface area by attaching to the surface *extended surfaces* called *fins* made of highly conductive materials such as aluminum. Finned surfaces are manufactured by extruding, welding, or wrapping a thin metal sheet on a surface. Fins enhance heat transfer from a surface by exposing a larger surface area to convection and radiation. Finned surfaces are commonly used in practice to enhance heat transfer, and they often increase the rate of heat transfer from a surface several fold. The car radiator is an example of a finned surface. The closely packed thin metal sheets attached to the hot-water tubes increase the surface area for convection and thus the rate of convection heat transfer from the tubes to the air many times. There are a variety of innovative fin designs available in the market, and they seem to be limited only by imagination.

Fin Equation The analysis in principal is very similar to the development of the *general heat conduction equation* (Eq. 2.4.11). Consider a volume element of a fin at location x having a length of Δx, cross-sectional area of A_c, and a perimeter of p, as shown in Fig. 2.4.10. Under steady conditions, the energy balance on this volume element can be expressed as

$$\dot{Q}_{cond,\,x} = \dot{Q}_{cond,\,x+\Delta x} + \dot{Q}_{conv} \tag{2.4.17}$$

where $\dot{Q}_{conv} = h(p\ \Delta x)\,(T - T_\infty)$
Substituting and dividing by Δx and taking the limit as $\Delta x \to 0$ gives

$$\frac{d\dot{Q}_{cond}}{dx} + hp\ (T - T_\infty) = 0 \tag{2.4.18}$$

From Fourier's law of heat conduction (Eq. 2.4.1), we have

$$\dot{Q}_{cond} = -kA_c\,\frac{dT}{dx} \tag{2.4.19}$$

where A_c is the cross-sectional area of the fin at location x. Substitution of Eq. (2.4.19) into Eq. (2.4.18) gives the differential equation governing heat transfer in fins,

$$\frac{d}{dx}\left(kA_c\,\frac{dT}{dx}\right) - hp\ (T - T_\infty) = 0 \tag{2.4.20}$$

In general, the cross-sectional area A_c and the perimeter p of a fin vary with x, which makes this differential equation difficult to solve. In the special case of *constant cross section* and *constant thermal conductivity,* the differential equation (Eq. 2.4.20) reduces to

$$\frac{d^2T}{dx^2} - \frac{hp}{kA_c}\,(T - T_\infty) = 0 \tag{2.4.21}$$

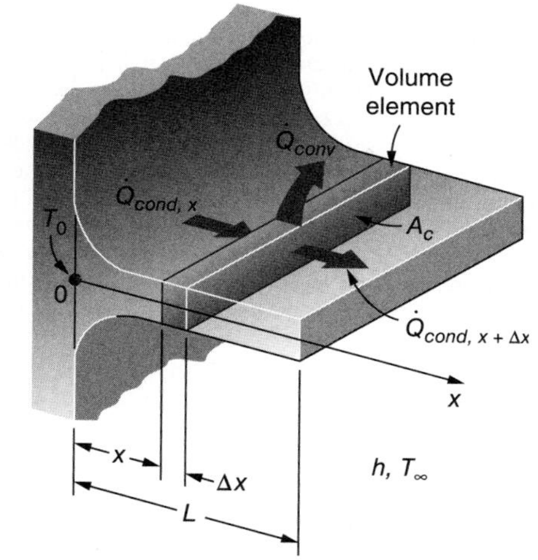

Figure 2.4.10 Volume element of a fin at location x having a length of Δx, cross-sectional area A_c, and perimeter of p. (*From Çengel and Ghajar, 2015.*)

Equation (2.4.21) is a linear, homogeneous, second-order differential equation with constant coefficients. A fundamental theory of differential equations states that such an equation has two linearly independent solution functions, and its general solution is the linear combination of those two solution functions. The general solution of Eq. (2.4.21) is

$$\theta(x) = C_1 e^{mx} + C_2 e^{-mx} \tag{2.4.22}$$

where $\theta = T - T_\infty$ is called *temperature excess* (at the fin base we have $\theta_b = T_b - T_\infty$) and $m = \sqrt{hp/kA_c}$. The constants C_1 and C_2 are arbitrary constants whose values are to be determined from the boundary conditions at the base and at the tip of the fin. Note that we need only two conditions to determine C_1 and C_2. The temperature of the plate to which the fins are attached is normally known in advance. Therefore, at the fin base we have a *specified temperature* boundary condition, expressed as $\theta(0) = \theta_b = T_b - T_\infty$. At the fin tip we have several possibilities, including infinitely long fins, negligible heat loss (idealized as an adiabatic tip), specified temperature, and convection. The solutions to these four special cases are available in Çengel and Ghajar (2015).

Fin Efficiency To maximize the heat transfer from a fin the temperature of the fin should be *uniform (maximized) at the base value of* T_b. However, since the fin itself represents a conduction resistance to heat transfer from the original surface, a temperature gradient must exist along the fin and the above condition is an idealization. To account for the effect of this decrease in temperature on heat transfer, the **fin efficiency** is defined as a measure of the fin thermal performance.

$$\eta_{fin} = \frac{\dot{Q}_{fin}}{\dot{Q}_{fin,\,max}} = \frac{\dot{Q}_{fin}}{hA_{fin}\,(T_b - T_\infty)} \tag{2.4.23a}$$

or

$$\dot{Q}_{fin} = \eta_{fin}\,\dot{Q}_{fin,\,max} = \eta_{fin}\,hA_{fin}\,(T_b - T_\infty) \tag{2.4.23b}$$

where A_{fin} is total surface area of the fin. Equation (2.4.23*b*) enables us to determine the heat transfer from the fin when its efficiency is known. Fin efficiency relations and graphs of common fin configurations with uniform and non-uniform cross section are provided in Çengel and Ghajar (2015) and Bergman et al. (2011).

The Overall Heat Transfer Coefficient (Combined Conduction and Convection) In many practical cases of heat transfer, for example, heat exchangers, heat is transmitted from one fluid to another through a wall separating the two. A heat exchanger typically involves two flowing fluids separated by a solid wall. Heat is first transferred from the hot fluid to the wall by *convection,* through the wall by *conduction,* and from the wall to the cold fluid again by *convection*. Any radiation effects are usually included in the convection heat transfer coefficients.

These types of problems under steady state conditions can be solved easily without involving any differential equations by the introduction of the *thermal resistance concept* in an analogous manner to electrical circuit problems. In this case, the thermal resistance corresponds to electrical resistance, temperature difference corresponds to voltage, and the heat transfer rate corresponds to electric current. The thermal resistance network associated with this heat transfer process involves two convection and one conduction resistances, as shown in Fig. 2.4.11(*a*). Here the subscripts *i* and *o* represent the inner and outer surfaces of the inner tube. For a double-pipe heat exchanger, the *total thermal resistance* becomes

$$R = R_{total} = R_i + R_{wall} + R_o = \frac{1}{h_i A_i} + \frac{\ln\left(\frac{D_o}{D_i}\right)}{2\pi kL} + \frac{1}{h_o A_o} \quad (2.4.24)$$

where k is the thermal conductivity of the wall material and L is the length of the tube. As shown in Fig. 2.4.11(*b*), for the double pipe heat exchanger with one fluid flowing inside a circular tube and the other outside of it, the $A_i = \pi D_i L$ is the area of the inner surface of the wall that separates the two fluids, and $A_o = \pi D_o L$ is the area of the outer surface of the wall. In other words, A_i and A_o are surface areas of the separating wall wetted by the inner and the outer fluids, respectively. For thin tubes $D_i \approx D_o$ and thus $A_i \approx A_o$.

In the analysis of heat exchangers, it is convenient to combine all the thermal resistances in the path of heat flow from the hot fluid to the cold one into a single resistance R, and to express the rate of heat transfer between the two fluids as

$$\dot{Q} = \frac{\Delta T}{R} = UA_s\Delta T = U_iA_i\Delta T = U_oA_o\Delta T \quad (2.4.25)$$

where A_s is the surface area and U is the **overall heat transfer coefficient**, whose unit is W/m²·K (Btu/h·ft²·°F), which is identical to the unit of the ordinary convection coefficient h. Canceling ΔT, Eq. (2.4.25) reduces to

$$\frac{1}{UA_s} = \frac{1}{U_iA_i} = \frac{1}{U_oA_o} = R = \frac{1}{h_iA_i} + \frac{\ln\left(\frac{D_o}{D_i}\right)}{2\pi kL} + \frac{1}{h_oA_o} \quad (2.4.26)$$

Note that $U_i A_i = U_o A_o$, but $U_i \neq U_o$ unless $A_i = A_o$. Therefore, the overall heat transfer coefficient U of a heat exchanger is meaningless unless the area on which it is based is specified. This is especially the case when one side of the tube wall is finned and the other side is not, since the surface area of the finned side is several times that of the unfinned side.

When the wall thickness of the tube is small and the thermal conductivity of the tube material is high, as is usually the case, the thermal resistance of the tube is negligible ($R_{wall} \approx 0$) and the inner and outer surfaces of the tube are almost identical ($A_i \approx A_o \approx A_s$). Then Eq. (2.4.26) for the overall heat transfer coefficient simplifies to

$$\frac{1}{U} \approx \frac{1}{h_i} + \frac{1}{h_o} \quad (2.4.27)$$

where $U \approx U_i \approx U_o$. The individual convection heat transfer coefficients inside and outside the tube, h_i and h_o, are determined using the convection relations for internal and external forced convection flows, respectively. These will be presented in the upcoming sub-sections.

Representative values of the overall heat transfer coefficient U are given in Table 2.4.5. Note that the overall heat transfer coefficient ranges from about 10 W/m²·K for gas-to-gas heat exchangers to about

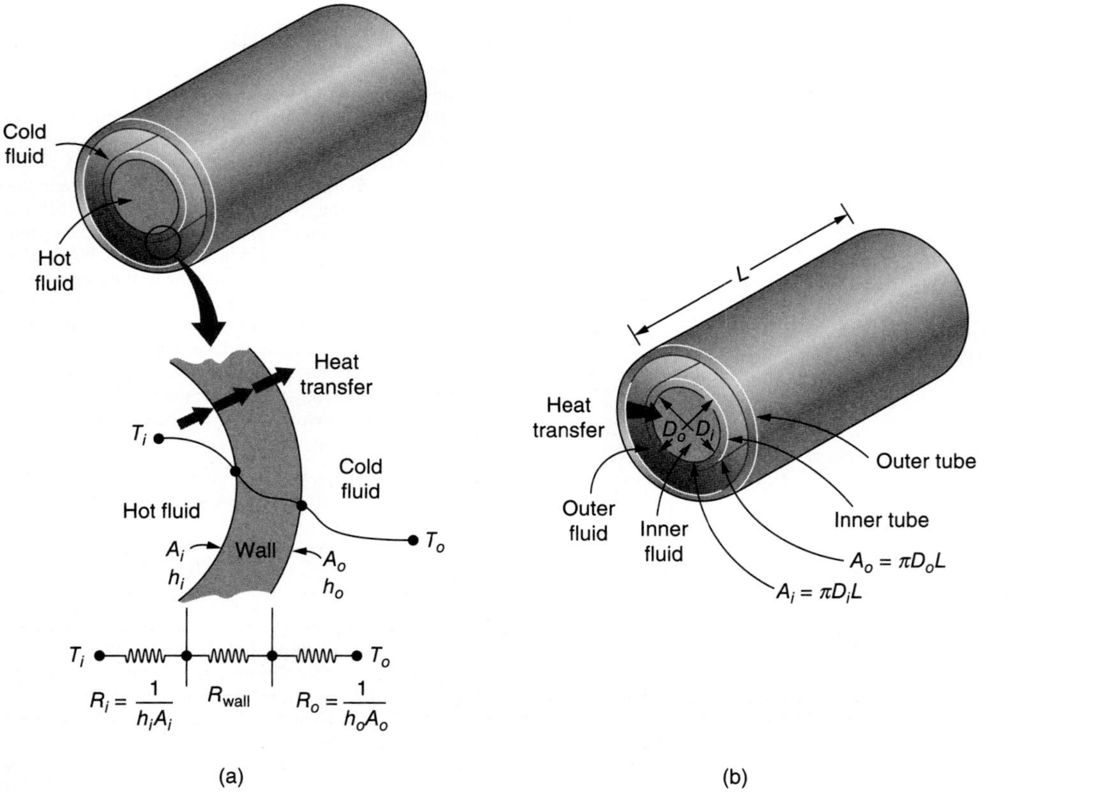

Figure 2.4.11 A double-pipe heat exchanger (*a*) Thermal resistance network associated with heat transfer and (*b*) The two heat transfer surface areas. (*From Çengel and Ghajar, 2015.*)

10,000 W/m²·K for heat exchangers that involve phase changes. This is not surprising, since gases have very low thermal conductivities, and phase-change processes involve very high heat transfer coefficients.

Table 2.4.5 Representative Values of the Overall Heat Transfer Coefficients in Heat Exchanger

Type of heat exchanger	U, W/m²·K*
Water-to-water	85-1,700
Water-to-oil	100-350
Water-to-gasoline or kerosene	300-1,000
Feedwater heaters	1,000-8,500
Steam-to-light fuel oil	200-400
Steam-to-heavy fuel oil	50-200
Steam condenser	1,000-6,000
Freon condenser (water cooled)	300-1,000
Ammonia condenser (water cooled)	800-1,400
Alcohol condensers (water cooled)	250-700
Gas-to-gas	10-40
Water-to-air in finned tubes (water in tubes)	30-60†
	400-850†
Steam-to-air in finned tubes (steam in tubes)	30-300†
	400-4,000‡

*Multiply the listed values by 0.176 to convert them to Btu/h·ft²·°F.
†Based on air-side surface area.
‡Based on water- or steam-side surface area.

Fouling Factor The performance of heat exchangers usually deteriorates with time as a result of accumulation of *deposits* on heat transfer surfaces. The layer of deposits represents *additional resistance* to heat transfer and causes the rate of heat transfer in a heat exchanger to decrease. The net effect of these accumulations on heat transfer is represented by a **fouling factor** R_f, which is a measure of the *thermal resistance* introduced by fouling.

The most common type of fouling, which is common in power and process plants, is the *precipitation* of solid deposits in a fluid on the heat transfer surfaces. Another form of fouling, which is common in the chemical process industry, is *corrosion* and other *chemical fouling*. Heat exchangers may also be fouled by the growth of algae in warm fluids. This type of fouling is called *biological fouling*.

In applications where it is likely to occur, fouling should be considered in the design and selection of heat exchangers. In such applications, it may be necessary to select a larger and thus more expensive heat exchanger to ensure that it meets the design heat transfer requirements even after fouling occurs. The periodic cleaning of heat exchangers and the resulting down time are additional penalties associated with fouling.

The fouling factor is obviously zero for a new heat exchanger and increases with time as the solid deposits build up on the heat exchanger surface. The fouling factor depends on the *operating temperature* and the *velocity* of the fluids, as well as the length of service. Fouling increases with *increasing temperature* and *decreasing velocity*.

The overall heat transfer coefficient relation Eq. (2.4.26) or (2.4.27), for an unfinned double-pipe heat exchanger is valid for clean surfaces and needs to be modified to account for the effects of fouling on both the inner and the outer surfaces of the tube. For an unfinned double-pipe heat exchanger, it can be expressed as

$$\frac{1}{UA_s} = \frac{1}{U_iA_i} = \frac{1}{U_oA_o} = R = \frac{1}{h_iA_i} + \frac{R_{f,i}}{A_i} + \frac{\ln\left(\frac{D_o}{D_i}\right)}{2\pi kL} + \frac{R_{f,o}}{A_o} + \frac{1}{h_oA_o} \tag{2.4.28}$$

where $R_{f,i}$ and $R_{f,o}$ are the fouling factors at those surfaces.

Representative values of fouling factors are given in Table 2.4.6. More comprehensive tables of fouling factors are available in handbooks (see list of references). As you would expect, considerable uncertainty exists in these values, and they should be used as a guide in the selection and evaluation of heat exchangers to account for the effects of anticipated fouling on heat transfer. Note that most fouling factors in the table are of the order of 10^{-4} m²·K/W, which is equivalent to the thermal resistance of a 0.2-mm-thick limestone layer ($k = 2.9$ W/m·K) per unit surface area. Therefore, in the absence of specific data, we can assume the surfaces to be coated with 0.2 mm of limestone as a starting point to account for the effects of fouling.

Table 2.4.6 Representative Fouling Factors, Thermal Resistance Due to Fouling for a Unit Surface Area

Fluid	R_f, m²·K/W
Distilled water, sea-water, river water, boiler feedwater:	
Below 50°C	0.0001
Above 50°C	0.0002
Fuel oil	0.0009
Transformer, lubricating or hydraulic oil	0.0002
Quenching oil	0.0007
Vegetable oil	0.0005
Steam (oil-free)	0.0001
Steam (with oil traces)	0.0002
Organic solvent vapors, natural gas	0.0002
Engine exhaust and fuel gases	0.0018
Refrigerants (liquid)	0.0002
Refrigerants (vapor)	0.0004
Ethylene and methylene glycol (antifreeze) and amine solutions	0.00035
Alcohol vapors	0.0001
Air	0.0004

SOURCE: Tubular Exchange Manufacturers Association.

Analysis of Heat Exchangers Heat exchangers are commonly used in practice, and an engineer often finds himself or herself in a position to *select a heat exchanger* that will achieve a *specified temperature change* in a fluid stream of known mass flow rate, or to *predict the outlet temperatures* of the hot and cold fluid streams in a *specified heat exchanger*. Two methods are used in the analysis of heat exchangers. Of these, the *log mean temperature difference* (or LMTD) method is best suited for the first task and the *effectiveness*–NTU (the number of transfer units) method for the second task.

The rate of heat transfer in a heat exchanger can also be expressed in an analogous manner to Eq. (2.4.25) as

$$\dot{Q} = UA_s\Delta T_m \tag{2.4.29}$$

where U is the overall heat transfer coefficient, A_s is the heat transfer surface area, and ΔT_m is an appropriate mean temperature difference between the two fluids. Here the surface area A_s can be determined precisely using the dimensions of the heat exchanger. However, the overall heat transfer coefficient U and the temperature difference ΔT between the hot and cold fluids, in general, may vary along the heat exchanger.

The average value of the overall heat transfer coefficient can be determined as described in the preceding section by using the average convection coefficients for each fluid. It turns out that the appropriate form of the average temperature difference between the two fluids is *logarithmic* in nature, and its determination is presented in the next section. It should be noted that the mean temperature difference ΔT_m is dependent on the heat exchanger flow arrangement and its type of construction.

The Log Mean Temperature Difference Method (LMTD) As mentioned earlier, the temperature difference between the hot and cold fluids varies along the heat exchanger, and it is convenient to have a *mean temperature difference* ΔT_m for use in the Eq. (2.4.29).

From energy balance analysis on a *parallel-flow double-pipe* heat exchanger, see Fig. 2.4.12(*a*), or a *counter-flow double-pipe* heat

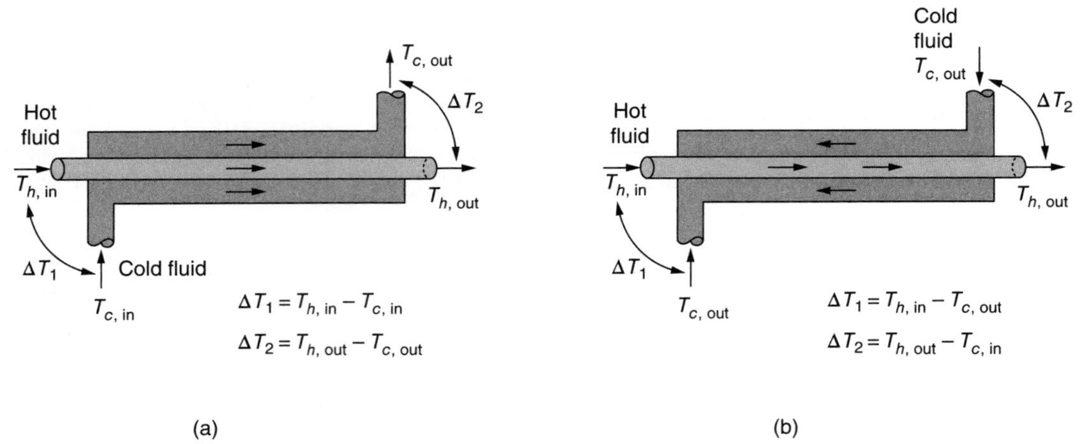

Figure 2.4.12 The ΔT_1 and ΔT_2 expressions in (*a*) parallel-flow double-pipe heat exchanger and (*b*) counter-flow double-pipe heat exchanger. (*From Çengel and Ghajar, 2015.*)

exchanger, see Fig. 2.4.12(*b*), the following equation for ΔT_{lm} **(LMTD)** which is the suitable form of the mean temperature difference (ΔT_m) for use in the analysis of heat exchangers can be obtained

$$\Delta T_{lm} = \frac{\Delta T_2 - \Delta T_1}{\ln\left(\dfrac{\Delta T_2}{\Delta T_1}\right)} = \frac{\Delta T_1 - \Delta T_2}{\ln\left(\dfrac{\Delta T_1}{\Delta T_2}\right)} \tag{2.4.30}$$

Here ΔT_1 and ΔT_2 represent the temperature difference between the two fluids at the two ends (inlet and outlet) of the heat exchanger. It makes no difference which end of the heat exchanger is designated as the inlet or the outlet (Fig. 2.4.12). It should be noted that Eqs. (2.4.29) and (2.4.30) are good for any heat exchanger provided that the end point temperature differences are defined properly. For multipass and cross-flow heat exchangers, as will be shown next, the LMTD should be corrected through a correction factor.

Multipass and Cross-Flow Heat Exchangers—Use of a Correction Factor The LMTD ΔT_{lm} relation, Eq. (2.4.30), is limited to parallel-flow and counter-flow heat exchangers only. Similar relations are also developed for *cross-flow* and *multipass shell-and-tube* heat exchangers, but the resulting expressions are too complicated because of the complex flow conditions.

In such cases, it is convenient to relate the equivalent temperature difference to the LMTD relation for the counter-flow case as

$$\Delta T_{lm} = F\ \Delta T_{lm,\,CF} \tag{2.4.31}$$

where F is the **correction factor**, which depends on the *geometry* of the heat exchanger and the inlet and outlet temperatures of the hot and cold fluid streams. The $\Delta T_{\text{lm, CF}}$ is the LMTD for the case of a *counter-flow* heat exchanger with the same inlet and outlet temperatures and is determined from Eq. (2.4.30) by taking $\Delta T_1 = T_{h,\text{in}} - T_{c,\text{out}}$ and $\Delta T_2 = T_{h,\text{out}} - T_{c,\text{in}}$, see Fig. 2.4.12(*b*).

The correction factor is less than unity for a cross-flow and multipass shell-and-tube heat exchanger. That is, $F \le 1$. The limiting value of $F = 1$ corresponds to the counter-flow heat exchanger. Thus, the correction factor F for a heat exchanger is *a measure of deviation of the* ΔT_{lm} *from the corresponding values for the counter-flow case.*

The correction factor F for common cross-flow and shell-and-tube heat exchanger configurations is given in Fig. 2.4.13 versus two temperature ratios P and R defined in the figure.

The Effectiveness–NTU Method The LMTD method discussed in the previous section is easy to use in heat exchanger analysis when the inlet and the outlet temperatures of the hot and cold fluids are known or can be determined from an energy balance. Once ΔT_{lm}, the mass flow rates, and the overall heat transfer coefficient are available, the heat transfer surface area of the heat exchanger can be determined from Eq. (2.4.29). Therefore, the LMTD method is very suitable for determining the *size* of a heat exchanger to realize prescribed outlet temperatures when the mass flow rates and the inlet and outlet temperatures of the hot and cold fluids are specified.

A second kind of problem encountered in heat exchanger analysis is the determination of the *heat transfer rate* and the *outlet temperatures* of the hot and cold fluids for prescribed fluid mass flow rates and inlet temperatures when the *type* and *size* of the heat exchanger are specified. The heat transfer surface area of the heat exchanger in this case is known, but the *outlet temperatures* are not. Here the task is to determine the heat transfer performance of a specified heat exchanger or to determine if a heat exchanger available in storage will do the job.

The LMTD method could still be used for this alternative problem, but the procedure would require tedious iterations, and thus it is not practical. In an attempt to eliminate the iterations from the solution of such problems, Kays and London (1984) came up with a method called the **effectiveness–NTU method**, which greatly simplified heat exchanger analysis. For further details on this method refer to Çengel and Ghajar (2015) and Bergman et al. (2011).

INTERNAL FORCED CONVECTION

Liquid or gas flow through pipes or ducts is commonly used in heating and cooling applications. The fluid in such applications is forced to flow by a fan or pump through a flow section that is sufficiently long to accomplish the desired heat transfer. As established earlier for determination of the rate of heat transfer by convection, see Eq. (2.4.16), we need to know the *convection heat transfer coefficient.*

There is a fundamental difference between external and internal flows. In *external flow,* to be presented in an upcoming section, the fluid has a free surface, and thus the boundary layer over the surface is free to grow indefinitely. In *internal flow,* however, the fluid is completely confined by the inner surfaces of the tube, and thus there is a limit on how much the boundary layer can grow.

Heat Transfer Correlations for Internal Flow The important physical properties which affect the convection heat transfer coefficients are thermal conductivity, viscosity, density, and specific heat. Factors within the control of the designer include fluid velocity and shape and arrangement of the heating surface. With forced flow of gases or water, under the conditions usually met in practice, the flow is turbulent and under these conditions the convection heat transfer coefficient can be greatly increased by increasing the velocity of the fluid at the expense of a greater power requirement. For a given velocity and fluid, the convection heat transfer coefficient depends upon the direction of flow of fluid relative to the heating surface. With free or natural convection,

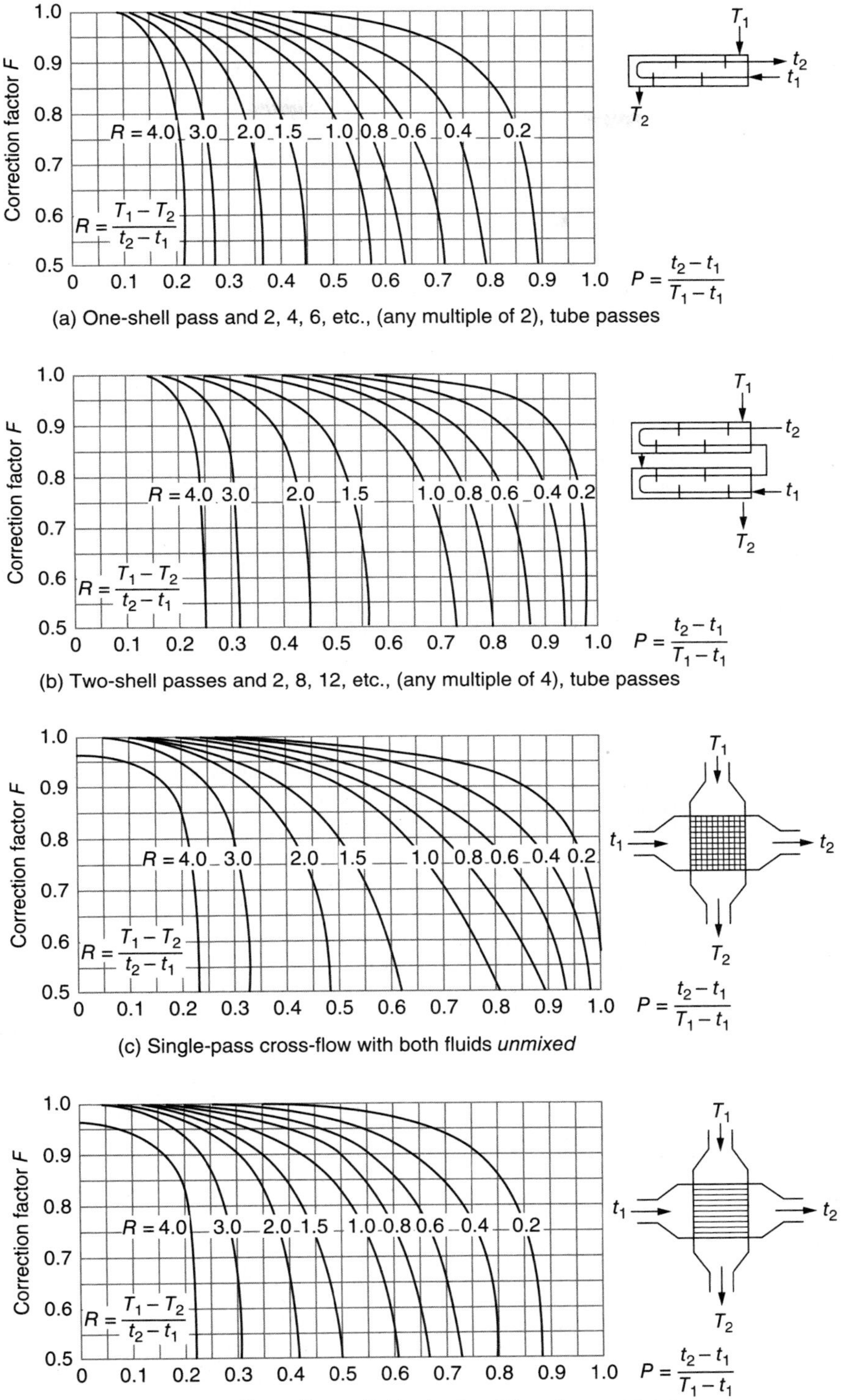

(a) One-shell pass and 2, 4, 6, etc., (any multiple of 2), tube passes

(b) Two-shell passes and 2, 8, 12, etc., (any multiple of 4), tube passes

(c) Single-pass cross-flow with both fluids *unmixed*

(d) Single-pass cross-flow with one fluid *mixed* and the other *unmixed*

Figure 2.4.13 Correction factor F charts for common shell-and-tube and cross-flow heat exchangers. (*From Çengel and Ghajar, 2015.*)

for a given arrangement of surface, the heat transfer coefficient depends on an additional fluid property, the coefficient of thermal expansion, on the temperature difference between surface and fluid, and on the local gravitational acceleration. With forced convection at low rates of flow, particularly with viscous fluids such as oils, laminar motion may prevail and the convection heat transfer coefficient depends on thermal conductivity, specific heat, mass rate of flow per tube, and length and diameter of the tube. In any event, the forced convection heat transfer coefficients h are correlated in terms of dimensionless groups of the controlling factors such as Nusselt number (Nu = hD/k), where k is the thermal conductivity of the fluid and D is the pipe inside diameter, Prandtl number (Pr = ν/α), where ν and α are the kinematic viscosity and thermal diffusivity of the fluid, respectively, and Reynolds number (Re = VD/ν), where V is the average flow velocity.

The experimental data for forced convection heat transfer is often represented by a simple power-law relation of the form

$$\text{Nu} = C\ \text{Re}^m\ \text{Pr}^n \tag{2.4.32}$$

where m and n are constant exponents (usually between 0 and 1), and the value of the constant C depends on the geometry. Sometimes more complex relations are used for better accuracy.

Entry Lengths The hydrodynamic entry length is usually taken to be the distance from the tube entrance where the wall shear stress (and thus the friction factor) reaches within about 2 percent of the fully developed value. In *laminar flow,* the hydrodynamic and thermal entry lengths are given approximately as, see Kays and Crawford (2005) and Shah and Bhatti (1987).

$$L_{\text{h, laminar}} \approx 0.05\ \text{Re}\ D \tag{2.4.33}$$

$$L_{\text{t, laminar}} \approx 0.05\ \text{Re Pr}\ D = \text{Pr}\ L_{\text{h, laminar}} \tag{2.4.34}$$

In *turbulent flow,* the intense mixing during random fluctuations usually overshadows the effects of molecular diffusion, and therefore the hydrodynamic and thermal entry lengths are of about the same size and independent of the Prandtl number. The entry length is much shorter in turbulent flow, as expected, and its dependence on the Reynolds number is weaker. In many tube flows of practical interest, the entrance effects become insignificant beyond a tube length of 10 diameters, and the hydrodynamic and thermal entry lengths are approximately taken to be

$$L_{\text{h, turbulent}} \approx L_{\text{t, turbulent}} \approx 10D \tag{2.4.35}$$

Hydrodynamically and Thermally Fully Developed Laminar Flow There are two analytical solutions for this case depending on the wall boundary condition. If the wall temperature is uniform, for example, if steam is condensing on the outside of the tube wall, the Nusslet number has a constant value given by

$$\textit{Uniform wall temperature } (T_s = \text{constant}):\quad \text{Nu} = \frac{hD}{k} = 3.66 \tag{2.4.36}$$

If, on the other hand, the heat flux through the tube wall is uniform, for example, if the tube is wound with an electrical resistance wire, then

$$\textit{Uniform wall heat flux } (\dot{q}_s = \text{constant}):\quad \text{Nu} = \frac{hD}{k} = 4.36 \tag{2.4.37}$$

Developing Laminar Flow in the Entrance Region There are a limited number of empirical correlations available in the literature for the average Nusselt number under constant surface temperature boundary condition. For example for a circular tube of length L subjected to uniform wall temperature, the average Nusselt number for hydrodynamically developed and thermally developing flow can be determined from Edwards et al. (1979):

$$\textit{Entry region } (T_s = \text{constant}):\quad \text{Nu} = \frac{hD}{k} = 3.66 + \frac{0.065\ (D/L)\text{Re Pr}}{1 + 0.04\ [(D/L)\text{Re Pr}]^{2/3}} \tag{2.4.38}$$

Note that the average Nusselt number is larger at the entrance region, as expected, and it approaches asymptotically to the fully developed value of 3.66 as $L \to \infty$, see Eq. (2.4.36). Equation (2.4.38) assumes that the flow is hydrodynamically developed when the fluid enters the heating section, but it can also be used for flow developing hydrodynamically when Pr ≥ 5.

The average Nusselt number for hydrodynamically and thermally developing laminar flow with uniform wall temperature in a circular tube can be determined from Sieder and Tate (1936):

$$\textit{Entry region } (T_s = \text{constant}):\quad \text{Nu} = \frac{hD}{k} = 1.86\left(\frac{Re\,\text{Pr}\,D}{L}\right)^{1/3}\left(\frac{\mu_b}{\mu_s}\right)^{0.14} \tag{2.4.39}$$

The above equation is recommended for 0.60 ≤ Pr ≤ 5 and 0.0044 ≤ (μ_b/μ_s) ≤ 9.75. Note that the term (D/L)RePr in both Eqs. (2.4.38) and (2.4.39) is the Graetz number (Gz) for a circular tube of length L. All properties appearing in Eqs. (2.4.38) and (2.4.39) should be evaluated at the *bulk mean fluid temperature* $T_b = (T_i + T_e)/2$ except μ_s, which is evaluated at T_s.

Developing Turbulent Flow in the Entrance Region The entry lengths for turbulent flow are typically short, often just 10 tube diameters long, refer to Eq. (2.4.35), and thus the Nusselt number determined for fully developed turbulent flow can be used approximately for the entire tube. This simple approach gives reasonable results for heat transfer for long tubes and conservative results for short ones. Correlations for heat transfer coefficients for the entrance regions are available in the literature for better accuracy.

Hydrodynamically and Thermally Fully Developed Turbulent Flow Transition to turbulence takes place around Re ≈ 2,300, although the turbulence becomes fully established only for Re > 10,000. In contrast to laminar flow, the effect of wall boundary condition (T_s = constant or $\dot{q}_s$ = constant) is unimportant for turbulent flow for all fluids except low Prandtl number liquid metals. For hydrodynamically and thermally fully developed turbulent flow, a simple power law correlation, similar to Eq. (2.4.32), which is due to Sieder and Tate (1936) is

$$\text{Nu} = \frac{hD}{k} = 0.027\ \text{Re}^{0.8}\ \text{Pr}^{1/3}\left(\frac{\mu_b}{\mu_s}\right)^{0.14}\quad \begin{pmatrix} 0.7 \le Pr \le 16{,}700 \\ Re \ge 10{,}000 \end{pmatrix} \tag{2.4.40}$$

Here all the properties are evaluated at the T_b except μ_s, which is evaluated at T_s.

The Nusselt number relation, Eq. (2.4.40), is fairly simple, but it may give errors as large as 25 percent. This error can be reduced considerably to less than 10 percent by using more complex but accurate relations such as the Gnielinski (1976) correlation:

$$\text{Nu} = \frac{hD}{k} = \frac{(f/8)(Re - 1{,}000)\text{Pr}}{1 + 12.7\ (f/8)^{0.5}\ (\text{Pr}^{2/3} - 1)}\quad \begin{pmatrix} 0.5 \le Pr \le 2{,}000 \\ 3\times 10^3 < Re < 5\times 10^6 \end{pmatrix} \tag{2.4.41}$$

where the friction factor f is evaluated from

$$f = (0.79\ln\text{Re} - 1.64)^{-2}\qquad 3\times 10^3 < \text{Re} < 5\times 10^6 \tag{2.4.42}$$

Again properties should be evaluated at the bulk mean fluid temperature (T_b).

The relations given so far do not apply to liquid metals because of their very low Prandtl numbers. For liquid metals (0.004 < Pr < 0.01), the following relations are recommended by Sleicher and Rouse (1975) for 10^4 < Re < 10^6:

$$\textit{Liquid metal } (T_s = \text{constant}):\quad \text{Nu} = 4.8 + 0.0156\ \text{Re}^{0.85}\ \text{Pr}_s^{0.93} \tag{2.4.43}$$

$$\textit{Liquid metal } (\dot{q}_s = \text{constant}):\quad \text{Nu} = 6.3 + 0.0167\ \text{Re}^{0.85}\ \text{Pr}_s^{0.93} \tag{2.4.44}$$

where the subscript s indicates that the Prandtl number is to be evaluated at the surface temperature.

Hydrodynamically and Thermally Developing or Fully Developed Flow in the Transition Region There is a regime of fluid flow which is neither fully laminar nor fully turbulent. Generally, internal flows with Reynolds numbers less than 2,300 are considered fully laminar. As the Reynolds number rises, it contains increasingly more turbulent motion until Re ~ 4,000; at this point, it is mostly turbulent. By the time the Reynolds number is increased to ~10,000, it is normally fully turbulent. The methods to handle fully laminar and fully turbulent heat transfer have already been discussed, however in some cases; the flow is in this transitional zone. Fortunately, the methods for handling turbulent flow can easily be adopted to deal with in this region. A simple approach is discussed in detail in Abraham et al. (2011). The recommendation is to continue to use Gnielinski's (1976) correlation (2.4.41) along with f values determined from the following expression for the smooth round tube, Eq. (2.4.45). This correlation can be applied over the Reynolds number ranges listed and is reasonably accurate for any thermal boundary condition, including the uniform wall temperature and uniform wall heat flux cases. A slightly more accurate approach is available for transitional flows in round tubes [Abraham et al. (2009)] but for most engineering applications, the following expression which is valid for 2,300 < Re < 4,500 is suitable.

$$f = 3.03 \times 10^{-12}\ \text{Re}^3 - 3.67 \times 10^{-8}\ \text{Re}^2 + 1.46 \times 10^{-4}\ \text{Re} - 0.151 \tag{2.4.45}$$

Ghajar and coworkers in a series of publications, for example see Ghajar and Tam (1994), Tam and Ghajar (2006), have provided heat transfer correlations for transitional flows with particular emphasis on the entrance geometry and on the development region. Their proposed correlations also handle flows which may experience buoyant motion and property variations.

EXTERNAL FORCED CONVECTION

In this section, we consider the practical aspects of forced convection to or from flat or curved surfaces subjected to *external flow,* characterized by the freely growing boundary layers surrounded by a free-flow region that involves no velocity and temperature gradients.

We first present average heat transfer correlations for laminar and turbulent *parallel flow over the entire flat plate.* Next, we present average heat transfer correlations for *cross flow over cylinders and spheres.* Finally, we consider *cross flow over tube banks* in aligned and staggered configurations, and present correlations for the average Nusselt number for both configurations.

Heat Transfer Correlations for External Flow The discussion for internal forced convection that leads to the Eq. (2.4.32) also holds for external forced convection. The major difference is the characteristic length used in the definitions of the Nusselt number ($\text{Nu} = hL/k$) and Reynolds number ($\text{Re} = VL/\nu$). In this case L the characteristic length is either plate length (for flow over a flat plate) or the outside diameter (for cross flow over cylinders and spheres).

Laminar and Turbulent Parallel Flow over a Flat Plate The *average* Nusselt number over the entire plate for *uniform wall temperature* is determined from the following relations:

$$\textit{Laminar:}\quad \text{Nu} = \frac{hL}{k} = 0.664\ \text{Re}^{0.5}\text{Pr}^{1/3} \quad \text{Re} < 5 \times 10^5,\ \text{Pr} > 0.6 \tag{2.4.46}$$

$$\textit{Turbulent:}\quad \text{Nu} = \frac{hL}{k} = 0.037\ \text{Re}^{0.8}\text{Pr}^{1/3} \quad 5 \times 10^5 \le \text{Re} \le 10^7,\ 0.6 \le \text{Pr} \le 60 \tag{2.4.47}$$

For the case of *uniform wall heat flux, the average* Nusselt number over the entire plate for the laminar flow is obtained from:

$$\textit{Laminar:}\quad \text{Nu} = \frac{hL}{k} = 0.680\ \text{Re}^{0.5}\text{Pr}^{1/3} \quad \text{Re} < 5 \times 10^5,\ \text{Pr} > 0.6 \tag{2.4.48}$$

However, for turbulent flow with uniform wall heat flux, the average Nusselt number equation (2.4.47) for the constant surface temperature condition is sufficiently accurate to determine the heat transfer coefficient since as discussed earlier, turbulent flow is not sensitive to the wall boundary condition.

Equation (2.4.46) for laminar flow on an isothermal flat plate is not applicable to liquid metals (Pr << 1). Churchill and Ozoe (1973) proposed the following relation which applies to any fluid in laminar flow over an isotheral flat plate and is claimed to be accurate to ±1 percent,

$$\text{Nu} = \frac{hL}{k} = \frac{0.677\ \text{Re}^{0.5}\ \text{Pr}^{1/3}}{\left[1 + \left(\dfrac{0.0468}{\text{Pr}}\right)^{2/3}\right]^{1/4}} \tag{2.4.49}$$

In some cases, a flat plate is sufficiently long for the flow to become turbulent, but not long enough to disregard the laminar flow region. In such cases, the *average* heat transfer coefficient over the entire plate is determined from the following relation,

$$\text{Nu} = \frac{hL}{k} = (0.037\ \text{Re}^{4/5} - A)\ \text{Pr}^{1/3} \quad \text{Re}_{cr} \le \text{Re} \le 10^7,\ 0.6 \le \text{Pr} \le 60 \tag{2.4.50}$$

where the constant A is determined by the value of the *critical Reynolds number* (Re_{cr}),

$$A = 0.037\ \text{Re}_{cr}^{4/5} - 0.664\ \text{Re}_{cr}^{1/2} \tag{2.4.51}$$

Critical Reynolds number is the Reynolds number at which the flow transitions from laminar to turbulent and for flat plate this value is typically taken as 5×10^5. For $\text{Re}_{cr} = 5 \times 10^5$, from Eq. (2.4.51), A = 871. For a completely turbulent flow ($\text{Re}_{cr} = 0$), Eq. (2.4.50) simplifies to Eq. (2.4.47).

Flow across Cylinders and Spheres Flows across cylinders and spheres, in general, involve *flow separation,* which is difficult to handle analytically. Therefore, such flows must be studied experimentally or numerically. Indeed, flow across cylinders and spheres have been studied experimentally by numerous investigators, and several empirical correlations have been developed for the heat transfer coefficient. Of the several such relations available in the literature for the average Nusselt number for cross flow over a cylinder, we present the one proposed by Churchill and Bernstein (1977):

$$\text{Nu}_{cyl} = \frac{hD}{k} = 0.3 + \frac{0.62\ \text{Re}^{0.5}\ \text{Pr}^{1/3}}{[1 + (0.4/\text{Pr})^{2/3}]^{1/4}}\left[1 + \left(\frac{\text{Re}}{282{,}000}\right)^{5/8}\right]^{4/5} \tag{2.4.52}$$

This relation is quite comprehensive in that it correlates available data well for RePr > 0.2. The fluid properties are evaluated at the *film temperature* $T_f = (T_\infty + T_s)/2$, which is the average of the free-stream and surface temperatures.

For flow over a *sphere,* Whitaker (1972) recommends the following comprehensive correlation:

$$\text{Nu}_{sph} = \frac{hD}{k} = 2 + [0.4\ \text{Re}^{0.5} + 0.06\ \text{Re}^{2/3}]\ \text{Pr}^{0.4}\left(\frac{\mu_\infty}{\mu_s}\right)^{1/4} \tag{2.4.53}$$

which is valid for $3.5 \le \text{Re} \le 8 \times 10^4$, $0.7 \le \text{Pr} \le 380$ and $1.0 \le (\mu_\infty/\mu_s) \le 3.2$. The fluid properties in this case are evaluated at the free-stream temperature T_∞, except for μ_s, which is evaluated at the surface temperature T_s. Although the two relations above are considered to be quite accurate, the results obtained from them can be off by as much as 30 percent.

Flow across Tube Banks Cross-flow over tube banks is commonly encountered in practice in heat transfer equipment such as the condensers and evaporators of power plants, refrigerators, and air conditioners. In such equipment, one fluid moves through the tubes while the other moves over the tubes in a perpendicular direction.

In a heat exchanger that involves a tube bank, the tubes are usually placed in a *shell* (and thus the *name shell-and-tube heat exchanger*), especially when the fluid is a liquid, and the fluid flows through the space between the tubes and the shell. There are numerous types of shell-and-tube heat exchangers, some of which were briefly presented in an earlier section on heat exchanger analysis. In this section we consider the general aspects of flow over a tube bank, and try to develop a better and more intuitive understanding of the performance of heat exchangers involving a tube bank.

Flow *through* the tubes can be analyzed by considering flow through a single tube, and multiplying the results by the number of tubes. This is not the case for flow *over* the tubes, however, since the tubes affect the flow pattern and turbulence level downstream, and thus heat transfer to or from them. Therefore, when analyzing heat transfer from a tube bank in cross flow, we must consider all the tubes in the bundle at once.

The tubes in a tube bank are usually arranged either *in-line* or *staggered* in the direction of flow, as shown in Fig. 2.4.14. The outer tube diameter D is taken as the characteristic length. The arrangement of the tubes in the tube bank is characterized by the *transverse pitch* S_T, *longitudinal pitch* S_L, and the *diagonal pitch* S_D between tube centers. The diagonal pitch is determined from $S_D = \sqrt{S_L^2 + (S_T/2)^2}$.

As the fluid enters the tube bank, the flow area decreases between the tubes (see Fig. 2.4.14), and thus flow velocity increases. In staggered arrangement, the velocity may increase further in the diagonal region if the tube rows are very close to each other. In tube banks, the flow characteristics are dominated by the maximum velocity V_{max} that occurs within the tube bank rather than the approach velocity V. Therefore, the Reynolds number is $\mathrm{Re} = V_{max} D/\nu$. The following relations should be used for the calculation of V_{max} in different tube arrangements:

In-line arrangement:
$$V_{max} = \frac{S_T}{S_T - D} V \tag{2.4.54}$$

Staggered arrangement: If $2A_D > A_T$ use Eq. (2.4.54). But if $2A_D < A_T$ or $S_D < (S_T + D)/2$ use the following equation

$$V_{max} = \frac{S_T}{2\,(S_D - D)} V \tag{2.4.55}$$

The nature of flow around a tube in the first row resembles flow over a single tube discussed earlier, especially when the tubes are not too close to each other. Therefore, each tube in a tube bank that consists of a single transverse row can be treated as a single tube in cross-flow. The nature of flow around a tube in the second and subsequent rows is very different, however, because of wakes formed and the turbulence caused by the tubes upstream. The level of turbulence, and thus the heat transfer coefficient, increases with row number because of the combined effects of upstream rows. But there is no significant change in turbulence level after the first few rows, and thus the heat transfer coefficient remains constant.

Flow through tube banks is studied experimentally since it is too complex to be treated analytically. We are primarily interested in the average heat transfer coefficient for the entire tube bank, which depends on the number of tube rows along the flow as well as the arrangement and the size of the tubes.

Several correlations, all based on experimental data, have been proposed for the average Nusselt number for cross flow over tube banks. More recently, Zukauskas (1987) has proposed correlations similar to Eq. (2.4.32) whose general form is

$$\mathrm{Nu} = \frac{hD}{k} = C\,\mathrm{Re}^m\,\mathrm{Pr}^n \left(\frac{\mathrm{Pr}}{\mathrm{Pr}_s}\right)^{0.25} \tag{2.4.56}$$

where the values of the constants C, m, and n depend on Reynolds number. Such correlations are given in Table 2.4.7 for tube banks with more than 16 rows ($N_L > 16$), $0.7 < \mathrm{Pr} < 500$ and $0 < \mathrm{Re} < 2 \times 10^6$.

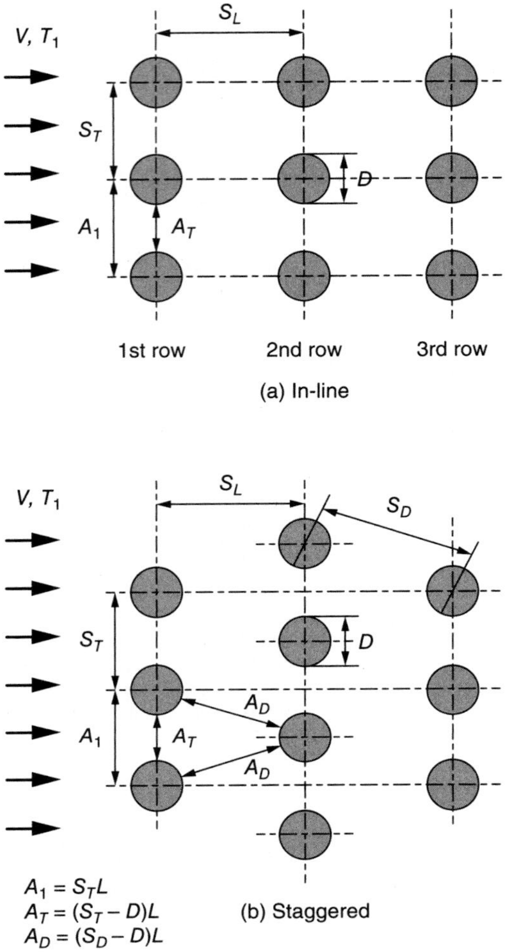

Figure 2.4.14 Arrangement of the tubes in in-line and staggered tube banks (A_1, A_T, and A_D are flow areas at indicated locations, and L is the length of the tubes). (*From Çengel and Ghajar, 2015.*)

The uncertainty in the values of Nusselt number obtained from these relations is ±15 percent. Note that all properties except Pr_s are to be evaluated at the arithmetic mean temperature of the fluid $T_m = (T_i + T_e)$, where T_i and T_e are the fluid temperatures at the inlet and the exit of the tube bank, respectively.

Table 2.4.7 Nusselt Number Correlations for Cross Flow over Tube Banks for $N_L > 16$, $0.7 < \mathrm{Pr} < 500$, and $0 < \mathrm{Re} < 2 \times 10^6$*

Arrangement	Range of Re	Correlation
In-line	0-100	$\mathrm{Nu} = 0.9\,\mathrm{Re}^{0.4}\mathrm{Pr}^{0.36}(\mathrm{Pr}/\mathrm{Pr}_s)^{0.25}$
	100-1,000	$\mathrm{Nu} = 0.52\,\mathrm{Re}^{0.5}\mathrm{Pr}^{0.36}(\mathrm{Pr}/\mathrm{Pr}_s)^{0.25}$
	1,000-2 × 10^5	$\mathrm{Nu} = 0.27\,\mathrm{Re}^{0.63}\mathrm{Pr}^{0.36}(\mathrm{Pr}/\mathrm{Pr}_s)^{0.25}$
	2 × 10^5-2 × 10^6	$\mathrm{Nu} = 0.033\,\mathrm{Re}^{0.8}\mathrm{Pr}^{0.4}(\mathrm{Pr}/\mathrm{Pr}_s)^{0.25}$
Staggered	0-500	$\mathrm{Nu} = 1.04\,\mathrm{Re}^{0.4}\mathrm{Pr}^{0.36}(\mathrm{Pr}/\mathrm{Pr}_s)^{0.25}$
	500-1,000	$\mathrm{Nu} = 0.71\,\mathrm{Re}^{0.5}\mathrm{Pr}^{0.36}(\mathrm{Pr}/\mathrm{Pr}_s)^{0.25}$
	1,000-2 × 10^5	$\mathrm{Nu} = 0.35(S_T/S_L)^{0.2}\mathrm{Re}^{0.6}\mathrm{Pr}^{0.36}(\mathrm{Pr}/\mathrm{Pr}_s)^{0.25}$
	2 × 10^5-2 × 10^6	$\mathrm{Nu} = 0.031(S_T/S_L)^{0.2}\mathrm{Re}^{0.8}\mathrm{Pr}^{0.36}(\mathrm{Pr}/\mathrm{Pr}_s)^{0.25}$

*All properties except Pr_s are to be evaluated at the arithmetic mean of the inlet and outlet temperatures of the fluid (Pr_s is to be evaluated at T_s).

The average Nusselt number relations in Table 2.4.7 are for tube banks with more than 16 rows. Those relations can also be used for tube banks with $N_L < 16$ provided that they are modified as

$$\text{Nu}_{N_L<16} = F \text{ Nu} \tag{2.4.57}$$

where F is a *correction factor* whose values are given in Table 2.4.8. For Re > 1,000, the correction factor is independent of Reynolds number.

Table 2.4.8 Correction Factor F to Be Used in $Nu_{N_L<16} = F\ Nu$ for $N_L < 16$ and Re > 1,000

N_L	1	2	3	4	5	7	10	13
In-line	0.70	0.80	0.86	0.90	0.93	0.96	0.98	0.99
Staggered	0.64	0.76	0.84	0.89	0.93	0.96	0.98	0.99

Once the Nusselt number and thus the average heat transfer coefficient for the entire tube bank is known, the heat transfer rate can be determined from Newton's law of cooling similar to Eq. (2.4.16), using a suitable temperature difference ΔT. The proper temperature difference for internal flow (flow over tube banks is still internal flow through the shell) is the *LMTD* ΔT_{lm} defined earlier as Eq. (2.4.30). To summarize, the equations to be used for heat transfer calculations are:

$$\dot{Q} = hA_s\ \Delta T_{lm} \tag{2.4.58}$$

and

$$\Delta T_{lm} = \frac{(T_s - T_e) - (T_s - T_i)}{\ln\left[\dfrac{(T_s - T_e)}{(T_s - T_i)}\right]} = \frac{\Delta T_e - \Delta T_i}{\ln\left(\dfrac{\Delta T_e}{\Delta T_i}\right)} \tag{2.4.59}$$

where $A_s = N\pi DL$ is the heat transfer surface area. Here N is the total number of tubes in the tube bank which is the product of N_T (number of tubes in the transverse plane) and N_L (number of rows in the flow direction), and L is the length of the tubes.

NATURAL (FREE) CONVECTION

In the previous sections, we considered heat transfer by *forced convection,* where a fluid was *forced* to move in a tube or over a surface by external means such as a pump or a fan. In this section, we consider *natural convection,* alternatively called free convection, where any fluid motion occurs by natural means such as buoyancy. The fluid motion in forced convection is quite *noticeable,* since a fan or a pump can transfer enough momentum to the fluid to move it in a certain direction. The fluid motion in natural convection, however, is often not noticeable because of the low velocities involved.

The convection heat transfer coefficient is a strong function of *velocity:* the higher the velocity, the higher the convection heat transfer coefficient. The fluid velocities associated with natural convection are low, typically less than 1 m/s. Therefore, the heat transfer coefficients encountered in natural convection are usually much lower than those encountered in forced convection (see Table 2.4.4). Yet several types of heat transfer equipment are designed to operate under natural convection conditions instead of forced convection, because natural convection does not require the use of a fluid mover.

As was the case with forced convection, natural convection heat transfer coefficients h are correlated in terms of dimensionless groups of the controlling factors such as Nusselt number ($\text{Nu} = hL/k$), where L is the characteristic length, Prandtl number ($\text{Pr} = \nu/\alpha$), and Grashof number ($\text{Gr} = g\ \beta\ \Delta T\ L^3/\ \nu^2$), where g is the gravitational acceleration, β coefficient of thermal expansion, ΔT is typically the difference between the surface temperature (T_s) and the temperature of the fluid sufficiently far from the surface (T_∞).

The experimental data for natural convection heat transfer is often represented by a simple power-law relation similar to Eq. (2.4.32) of the form

$$\text{Nu} = C\ (\text{Gr Pr})^n = C\ \text{Ra}^n \tag{2.4.60}$$

where Ra is the **Rayleigh number**, which is the product of the Grashof number (Gr), which describes the relationship between buoyancy and viscosity within the fluid, and the Prandtl number (Pr), which describes the relationship between momentum diffusivity and thermal diffusivity. Hence the Rayleigh number itself may also be viewed as the ratio of buoyancy forces and (the products of) thermal and momentum diffusivities.

The values of the constants C and n depend on the *geometry* of the surface and the *flow regime,* which is characterized by the range of the Rayleigh number. The value of n is usually ¼ for laminar flow and 1/3 for turbulent flow. The value of the constant C is normally less than 1.

Simple relations for the average Nusselt number for various geometries are given in Table 2.4.9, together with sketches of the geometries. Also given in this table are the characteristic lengths of the geometries and the ranges of Rayleigh number in which the relation is applicable. All fluid properties are to be evaluated at the film temperature $T_f = (T_s + T_\infty)/2$. For additional relations for the average Nusselt number for more complicated geometries refer to Çengel and Ghajar (2015).

When the average Nusselt number and thus the average convection heat transfer coefficient is known, the rate of heat transfer by natural convection from a solid surface at a uniform temperature T_s to the surrounding fluid T_∞ is expressed by Newton's law of cooling, refer to Eq. (2.4.16).

COMBINED NATURAL AND FORCED CONVECTION

The presence of a temperature gradient in a fluid in a gravity field always gives rise to natural convection currents, and thus heat transfer by natural convection. Therefore, forced convection is always accompanied by natural convection. As mentioned earlier that the convection heat transfer coefficient, natural or forced, is a strong function of the fluid velocity. Heat transfer coefficients encountered in forced convection are typically much higher than those encountered in natural convection because of the higher fluid velocities associated with forced convection. As a result, we tend to ignore natural convection in heat transfer analyses that involve forced convection, although we recognize that natural convection always accompanies forced convection. The error involved in ignoring natural convection is negligible at high velocities but may be considerable at low velocities. Therefore, it is desirable to have a criterion to assess the relative magnitude of natural convection in the presence of forced convection.

For a given fluid, it is observed that the parameter Gr/Re^2 represents the importance of natural convection relative to forced convection. This is not surprising since the convection heat transfer coefficient is a strong function of the Reynolds number Re in forced convection and the Grashof number Gr in natural convection.

From the above discussion and the Nusselt number correlations presented in earlier sections for internal and external forced convection, and for natural convection, it can be concluded that for forced convection $\text{Nu} = f\ (\text{Re, Pr})$, for natural convection $\text{Nu} = f\ (\text{Gr, Pr})$, and for combined natural and forced (mixed) convection $\text{Nu} = f\ (\text{Re, Gr, Pr})$.

Natural convection may *help* or *hurt* forced convection heat transfer, depending on the relative directions of *buoyancy-induced* and the *forced convection* motions. In *assisting flow,* the buoyant motion is in the *same* direction as the forced motion. Therefore, natural convection assists forced convection and *enhances* heat transfer. An example is upward forced flow over a hot surface. In *opposing flow,* the buoyant motion is in the *opposite* direction to the forced motion. Therefore, natural convection resists forced convection and *decreases* heat transfer. An example is upward forced flow over a cold surface. In *transverse flow,* the buoyant motion is *perpendicular* to the forced motion. Transverse flow enhances fluid mixing and thus *enhances* heat transfer. An example is horizontal forced flow over a hot or cold cylinder or sphere. When determining heat transfer under combined natural and forced (mixed) convection conditions, it is tempting to add the contributions of natural and forced convection in assisting flows and to subtract them in opposing flows. However, the evidence indicates differently. A review of experimental data suggests a correlation of the form

$$\text{Nu}_{combined} = \text{Nu}^n_{forced} \pm \text{Nu}^n_{natural} \tag{2.4.61}$$

Table 2.4.9 Empirical Correlations for the Average Nusselt Number for Natural Convection over Surfaces

Geometry	Characteristic length L	Range of Ra	Nu
Vertical plate 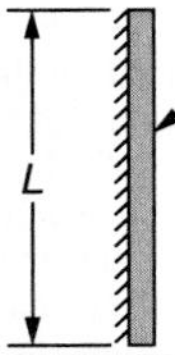	L	10^4-10^9 10^9-10^{13} Entire range	$Nu = 0.59Ra^{1/4}$ $Nu = 0.1Ra^{1/3}$ $Nu = \left\{0.825 + \frac{0.387Ra^{1/6}}{[1+(0.492/Pr)^{9/16}]^{8/27}}\right\}^2$ (complex but more accurate)
Inclined plate 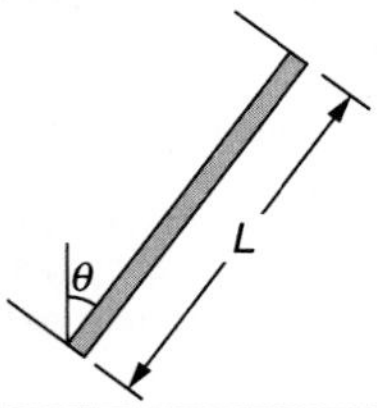	L		Use vertical plate equations for the upper surface of a cold plate and the lower surface of a hot plate Replace g by $g\cos\theta$ for $0 < \theta < 60°$
Horizontal plate (Surface area A and perimeter p) (*a*) Upper surface of a hot plate (or lower surface of a cold plate) 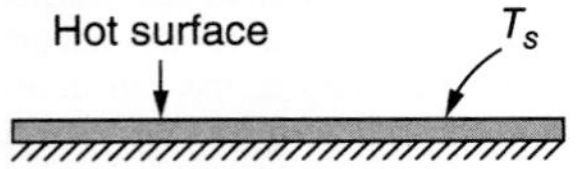 (*b*) Lower surface of a hot plate (or upper surface of a cold plate) 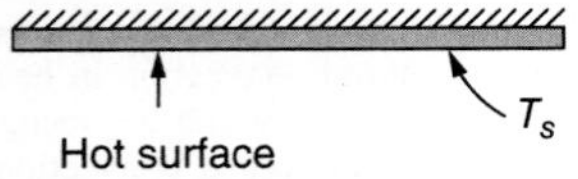	A_s/p	10^4-10^7 10^7-10^{11} 10^5-10^{11}	$Nu = 0.54Ra^{1/4}$ $Nu = 0.15Ra^{1/3}$ $Nu = 0.27Ra^{1/4}$
Vertical cylinder 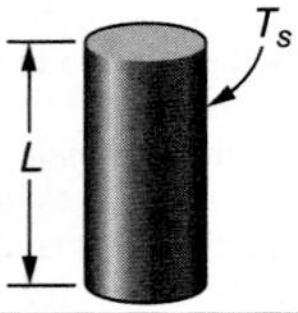	L		A vertical cylinder can be treated as a vertical plate when $D \geq \frac{35L}{Gr^{1/4}}$
Horizontal cylinder 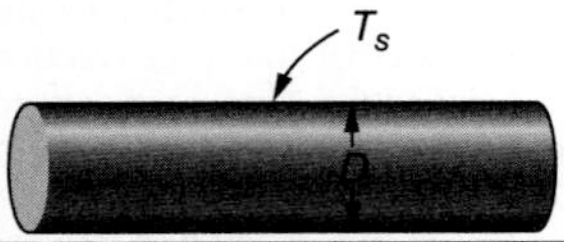	D	$Ra \leq 10^{12}$	$Nu = \left\{0.6 + \frac{0.387Ra^{1/6}}{[1+(0.559/Pr)^{9/16}]^{8/27}}\right\}^2$
Sphere 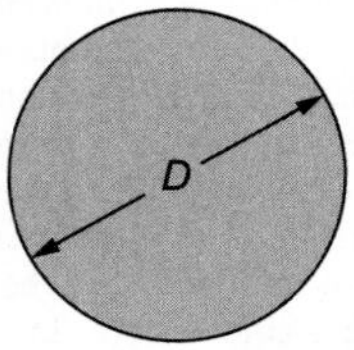	D	$Ra \leq 10^{11}$ $(Pr \geq 0.7)$	$Nu = 2 + \frac{0.589Ra^{1/4}}{[1+(0.469/Pr)^{9/16}]^{4/9}}$

where Nu_{forced} and $Nu_{natural}$ are determined from the correlations for *pure forced* and *pure natural convection,* respectively. The plus sign is for *assisting* and *transverse* flows and the minus sign is for *opposing* flows. The value of the exponent n varies between 3 and 4, depending on the geometry involved. It is observed that n = 3 correlates experimental data for vertical surfaces well. Larger values of n are better suited for horizontal surfaces.

COMBINED CONVECTION AND RADIATION

In some case of heat loss (or gain), such as that from bare and insulated pipes, the surface exposed to the surrounding gas such as air involves convection and radiation simultaneously, and the total heat transfer at the surface is determined by *adding* (or subtracting, if in the opposite direction) the radiation and convection components. For simplicity and convenience, this is often done by defining a **combined heat transfer coefficient** $h_{combined}$ that includes the effects of both convection and radiation. Then the *total* heat transfer rate to or from a surface by convection and radiation is expressed as

$$\dot{Q}_{total} = \dot{Q}_{conv} + \dot{Q}_{rad} = h_{conv}\, A_s\, (T_s - T_{surr}) + \varepsilon\sigma A_s\left(T_s^4 - T_{surr}^4\right)$$

$$\dot{Q}_{total} = h_{combined}\, A_s\, (T_s - T_{surr}) \qquad (2.4.62)$$

$$h_{combined} = h_{conv} + h_{rad} = h_{conv} + \varepsilon\sigma(T_s + T_{surr})\left(T_s^2 + T_{surr}^2\right)$$

where ε is the **emissivity** of the radiating surface ($0 \le \varepsilon \le 1$), $\sigma = 5.670 \times 10^{-8}$ W/m²·K⁴ is the *Stefan-Boltzmann constant*, T_{surr} is the temperature of the large surrounding area in which the heat transfer surface is enclosed in (in most cases $T_{surr} \approx T_\infty$), and h_{conv} is obtained from the earlier correlations presented for the external forced or natural convection. The temperatures in this case must be expressed in K, not °C. Note that the combined heat transfer coefficient is essentially a convection heat transfer coefficient modified to include the effects of radiation. Radiation is usually significant relative to conduction or natural convection, but negligible relative to forced convection. Thus radiation in forced convection applications is usually disregarded, especially when the surfaces involved have low emissivities and low to moderate temperatures.

BOILING AND CONDENSATION

We know from thermodynamics that when the temperature of a liquid at a specified pressure is raised to the saturation temperature T_{sat} at that pressure, *boiling* occurs. Likewise, when the temperature of a vapor is lowered to T_{sat}, *condensation* occurs. In this section we study the rates of heat transfer during such liquid-to-vapor and vapor-to-liquid phase transformations.

Although boiling and condensation exhibit some unique features, they are considered to be forms of *convection* heat transfer since they involve fluid motion (such as the rise of the bubbles to the top and the flow of condensate to the bottom). Boiling and condensation differ from other forms of convection in that they depend on the *latent heat of vaporization* h_{fg} of the fluid and the *surface tension* σ at the liquid–vapor interface, in addition to the properties of the fluid in each phase. Noting that under equilibrium conditions the temperature remains constant during a phase-change process at a fixed pressure, large amounts of heat (due to the large latent heat of vaporization released or absorbed) can be transferred during boiling and condensation essentially at constant temperature. In practice, however, it is necessary to maintain some difference between the surface temperature T_s and T_{sat} for effective heat transfer. This temperature difference is called the *excess temperature* ΔT_{excess}, which represents the temperature excess of the surface above the saturation temperature of the fluid. Heat transfer coefficients h associated with boiling and condensation are typically much higher than those encountered in other forms of convection processes that involve a single phase (see Table 2.4.4).

Boiling Curve The **boiling curve** is a plot of boiling heat flux versus the excess temperature, see Fig. 2.4.15. The plot shows the four different boiling regimes which are observed: *natural convection boiling, nucleate boiling, transition boiling,* and *film boiling. Boiling takes different forms, depending on the value of excess temperature* ΔT_{excess}. Although the boiling curve given in Fig. 2.4.15 is for water, the general shape of the boiling curve remains the same for different fluids. The specific shape of the curve depends on the fluid–heating surface material combination and the fluid pressure, but it is practically independent of the geometry of the heating surface.

Heat Transfer Correlations in Pool Boiling Boiling regimes mentioned above differ considerably in their character, and thus different heat transfer relations need to be used for different boiling regimes (Fig. 2.4.15).

Natural Convection Boiling In this regime ($\Delta T_{excess} \le 5$°C), boiling is governed by natural convection currents, and heat transfer rates in

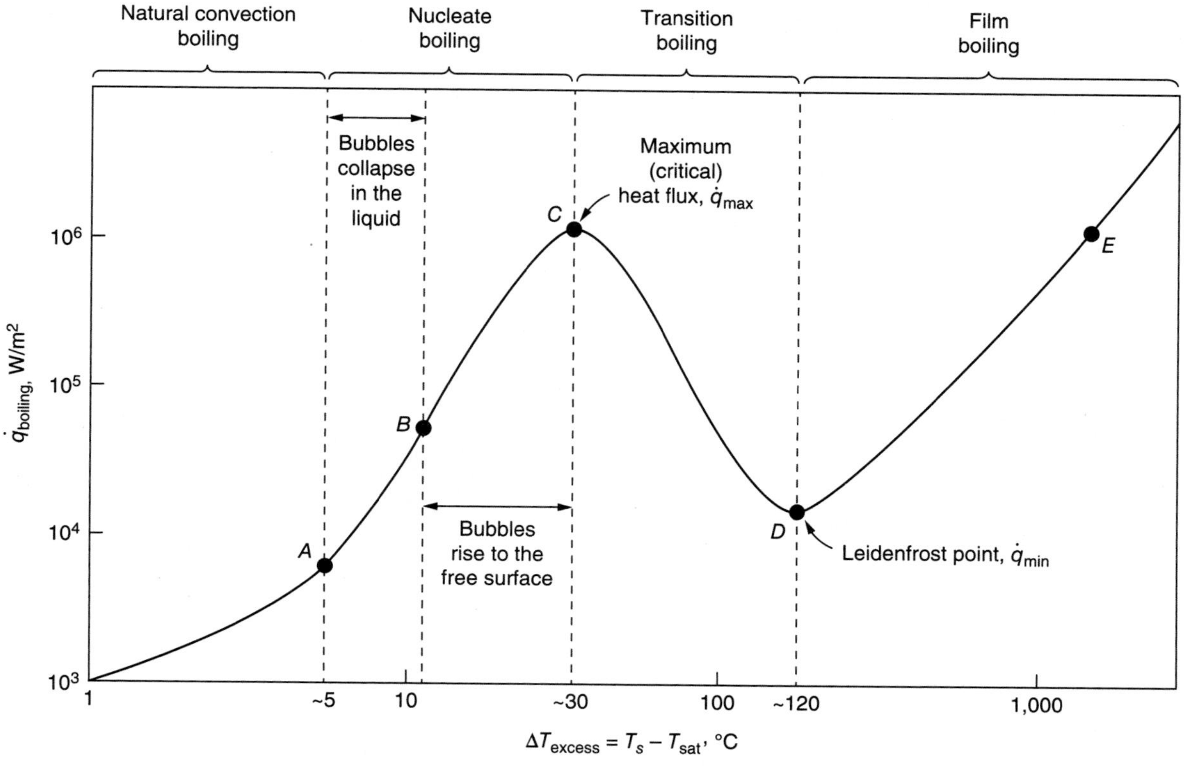

Figure 2.4.15 Typical boiling curve for water at 1 atm. (*From Çengel and Ghajar, 2015.*)

this case can be determined accurately using natural convection relations presented earlier.

Nucleate Boiling In this regime, region A-C of Fig. 2.4.15 (5°C ≤ ΔT_{excess} ≤ 30°C), the rate of heat transfer strongly depends on the nature of nucleation (the number of active nucleation sites on the surface, the rate of bubble formation at each site, etc.), which is difficult to predict. The type and the condition of the heated surface also affect the heat transfer. These complications made it difficult to develop theoretical relations for heat transfer in the nucleate boiling regime, and we had to rely on relations based on experimental data. The most widely used correlation for the rate of heat transfer in the nucleate boiling regime was proposed by Rohsenow (1952), and expressed as

$$\dot{q}_{nucleate} = \mu_l \, h_{fg} \left[\frac{g\,(\rho_l - \rho_v)}{\sigma}\right]^{1/2} \left[\frac{c_{pl}\,(T_s - T_{sat})}{C_{sf} h_{fg}\, \mathrm{Pr}_l^n}\right]^3 \tag{2.4.63}$$

where subscripts l and v refer to liquid and vapor, σ is the surface tension of liquid-vapor interface (see Table 2.4.10), n is an experimental constant that depends on the fluid and C_{sf} is the experimental constant that depends on surface-fluid combination (see Table 2.4.11). The fluid properties in Eq. (2.4.63) are to be evaluated at the saturation temperature T_{sat}.

Peak Heat Flux Referring to Fig. 2.4.15, the heat flux increases with an increase in ΔT_{excess} and peaks at point C. The heat flux at this point is called the **critical** (or **maximum**) **heat flux** (**CHF**). In the design of boiling heat transfer equipment, it is extremely important for the designer to have a knowledge of the maximum heat flux in order to avoid the danger of burnout. The *maximum* (or *critical*) *heat flux* in nucleate pool boiling is expressed as

$$\dot{q}_{max} = C_{cr} \, h_{fg} \left[\sigma \, g \, \rho_v^2 \, (\rho_l - \rho_v)\right]^{1/4} \tag{2.4.64}$$

where C_{cr} is a constant whose value depends on the heater geometry. Specific values of C_{cr} for different heater geometries are listed in Table 2.4.12. Note that the heaters are classified as being large or small based on the value of the parameter $L^* = L\,[g\,(\rho_l - \rho_v)/\sigma]^{1/2}$.

Transition Boiling In this regime, region C-D of Fig. 2.4.15 (30°C ≤ ΔT_{excess} ≤ 120°C), as the ΔT_{excess} is increased past point C, the heat flux decreases. This is because a large fraction of the heater surface is covered by a vapor film, which acts as an insulation due to the low thermal conductivity of the vapor relative to that of the liquid. In this regime nucleate boiling at point C is completely replaced by film boiling at point D. Operation in the transition boiling regime, which is also called the *unstable film boiling regime*, is avoided in practice, and thus no major attempt has been made to develop general correlations for boiling heat transfer in this regime. However, the upper (*peak heat flux*, $\dot{q}_{max}$) and the lower (*minimum heat flux*, $\dot{q}_{min}$) limits of this region are of interest to heat transfer equipment designers.

Minimum Heat Flux Minimum heat flux, which occurs at point D of Fig. 2.4.15 is referred to as the **Leidenfrost point**, is of practical interest since it represents the lower limit for the heat flux in the film boiling regime. Zuber (1958) derived the following expression for the minimum heat flux for a *large horizontal plate,*

$$\dot{q}_{min} = 0.09 \; \rho_v \; h_{fg} \left[\frac{\sigma \; g \; (\rho_l - \rho_v)}{(\rho_l + \rho_v)^2}\right]^{1/4} \tag{2.4.65}$$

Film Boiling In region beyond point D of Fig. 2.4.15 (ΔT_{excess} ≥ 120°C), the heater surface is completely covered by a continuous stable vapor film. The heat transfer rate increases with increasing excess temperature as a result of heat transfer from the heated surface to the liquid through the vapor film by radiation, which becomes significant at high temperatures. Bromley (1950) developed the following equation for the heat flux for film boiling on a *horizontal cylinder* or *sphere* of diameter D

$$\dot{q}_{film} = C_{film} \left[\frac{g k_v^3 \rho_v (\rho_l - \rho_v)\,[h_{fg} + 0.4 c_{pv}\,(T_s - T_{sat})]}{\mu_v \, D\,(T_s - T_{sat})}\right]^{1/4} (T_s - T_{sat}) \tag{2.4.66}$$

where C_{film} = 0.62 for horizontal cylinders and C_{film} = 0.67 for spheres. In Eq. (2.4.66) the *vapor* properties are to be evaluated at the *film temperature* $T_f = (T_s + T_{sat})/2$, which is the *average temperature* of

Table 2.4.10 Surface Tension of Liquid-water Vapor Interface for Water

T, °C	σ, N/m*
0	0.0757
20	0.0727
40	0.0696
60	0.0662
80	0.0627
100	0.0589
120	0.0550
140	0.0509
160	0.0466
180	0.0422
200	0.0377
220	0.0331
240	0.0284
260	0.0237
280	0.0190
300	0.0144
320	0.0099
340	0.0056
360	0.0019
374	0.0

*Multiply by 0.06852 to convert to lbf/ft or by 2.2046 to convert to lbm/s².

Table 2.4.11 Values of the Coefficient C_{sf} and n for Various Fluid-surface Combinations

Fluid-heating surface combination	C_{sf}	n
Water-copper (polished)	0.0130	1.0
Water-copper (scored)	0.0068	1.0
Water-stainless steel (mechanically polished)	0.0130	1.0
Water-stainless steel (ground and polished)	0.0060	1.0
Water-stainless steel (teflon pitted)	0.0058	1.0
Water-stainless steel (chemically etched)	0.0130	1.0
Water-brass	0.0060	1.0
Water-nickel	0.0060	1.0
Water-platinum	0.0130	1.0
n-Pentane-copper (polished)	0.0154	1.7
n-Pentane-chromium	0.0150	1.7
Benzene-chromium	0.1010	1.7
Ethyl alcohol-chromium	0.0027	1.7
Carbon tetrachloride-copper	0.0130	1.7
Isopropanol-copper	0.0025	1.7

Table 2.4.12 Values of the Coefficient C_{cr} for Use in Eq. (2.4.64) for Maximum Heat Flux (Dimensionless Parameter $L = L[g(\rho_l - \rho_v)/\sigma]^{1/2}$)

Heater geometry	C_{cr}	Characteristic dimension of heater, L	Range of L^*
Large horizontal flat heater	0.149	Width or diameter	$L^* > 27$
Small horizontal flat heater[1]	$18.9K_1$	Width or diameter	$9 < L^* < 20$
Large horizontal cylinder	0.12	Radius	$L^* > 1.2$
Small horizontal cylinder	$0.12L^{*-0.25}$	Radius	$0.15 < L^* < 1.2$
Large sphere	0.11	Radius	$L^* > 4.26$
Small sphere	$0.227L^{*-0.5}$	Radius	$0.15 < L^* < 4.26$

[1] $K_1 = \sigma/[g(\rho_l - \rho_v)A_{heater}]$

the vapor film. The liquid properties and h_{fg} are to be evaluated at the saturation temperature at the specified pressure.

At high surface temperatures (typically above 300°C), heat transfer across the vapor film by *radiation* becomes significant and needs to be considered. Treating the vapor film as a transparent medium sandwiched between two large parallel plates and approximating the liquid as a blackbody, *radiation heat transfer* can be determined from

$$\dot{q}_{rad} = \varepsilon\sigma\left(T_s^4 - T_{sat}^4\right) \tag{2.4.67}$$

Equation (2.4.67) is similar to the radiation part of Eq. (2.4.62), where ε is the emissivity of the heating surface and σ is the Stefan-Boltzmann constant. Note that the surface tension and the Stefan-Boltzmann constant share the same symbol.

You may be tempted to simply add the convection and radiation heat transfers to determine the total heat transfer during film boiling. However, these two mechanisms of heat transfer adversely affect each other, causing the total heat transfer to be less than their sum. For example, the radiation heat transfer from the surface to the liquid enhances the rate of evaporation, and thus the thickness of the vapor film, which impedes convection heat transfer. For $\dot{q}_{rad} < \dot{q}_{film}$, Bromley (1950) determined that the experimental data can be correlated well with the relation $\dot{q}_{rad} = \dot{q}_{film} + 0.75\ \dot{q}_{rad}$.

Condensation Heat Transfer Condensation occurs when the temperature of a vapor is reduced *below* its saturation temperature T_{sat}. This is usually done by bringing the vapor into contact with a solid surface whose temperature T_s is below the saturation temperature T_{sat} of the vapor. But condensation can also occur on the free surface of a liquid or even in a gas when the temperature of the liquid or the gas to which the vapor is exposed is below T_{sat}. In the latter case, the liquid droplets suspended in the gas form a fog. In this section, we consider condensation on solid surfaces only.

Two distinct forms of condensation are observed: *film condensation* and *dropwise condensation*. In **film condensation**, the condensate wets the surface and forms a liquid film on the surface that slides down under the influence of gravity. The thickness of the liquid film increases in the flow direction as more vapor condenses on the film. This is how condensation normally occurs in practice. In **dropwise condensation**, the condensed vapor forms droplets on the surface instead of a continuous film, and the surface is covered by countless droplets of varying diameters.

In film condensation, the surface is blanketed by a liquid film of increasing thickness, and this "liquid wall" between solid surface and the vapor serves as a *resistance* to heat transfer. The heat of vaporization h_{fg} released as the vapor condenses must pass through this resistance before it can reach the solid surface and be transferred to the medium on the other side. In dropwise condensation, however, the droplets slide down when they reach a certain size, clearing the surface and exposing it to vapor. There is no liquid film in this case to resist heat transfer. As a result, heat transfer rates that are more than 10 times larger than those associated with film condensation can be achieved with dropwise condensation. Therefore, dropwise condensation is the preferred mode of condensation in heat transfer applications, and people have long tried to achieve sustained dropwise condensation by using various vapor additives and surface coatings. These attempts have not been very successful, however, since the dropwise condensation achieved did not last long and converted to film condensation after some time. Therefore, it is common practice to be conservative and assume film condensation in the design of heat transfer equipment. For better heat transfer, it is desirable to use short surfaces because of the lower thermal resistance.

Flow Regimes in Film Condensation As was the case in forced convection involving a single phase, heat transfer in condensation also depends on whether the condensate flow is *laminar* or *turbulent*. Again the criterion for the flow regime is provided by the Reynolds number, which is defined as

$$Re = \frac{D_h\,\rho_l\,V_l}{\mu_l} = \frac{4\,A_c\,\rho_l\,V_l}{p\,\mu_l} = \frac{4\rho_l\,V_l\,\delta}{\mu_l} = \frac{4\,\dot{m}}{p\,\mu_l} \tag{2.4.68}$$

where $D_h = 4A_c/p = 4\delta$ = hydraulic diameter of the condensate flow, p = wetted perimeter of the condensate, $A_c = p\delta$ = wetted perimeter × film thickness = cross-sectional area of the condensate flow at the lowest part of the flow, ρ_l = density of the liquid, μ_l = viscosity of the liquid, V_l = average velocity of the condensate at the lowest part of the flow, and $\dot{m} = \rho_l\,V_l\,A_c$ = mass flow rate of the condensate at the lowest part. The evaluation of the hydraulic diameter D_h for some common geometries is illustrated in Fig. 2.4.16. Note that the hydraulic diameter is defined such that it reduces to the ordinary diameter for flow in a circular tube, and it is equivalent to 4 times the thickness of the condensate film at the location where the hydraulic diameter is evaluated. That is, $D_h = 4\delta$.

The Reynolds number for condensation on the outer surfaces of vertical tubes or plates increases in the flow direction due to the increase of the liquid film thickness δ. The flow of liquid film exhibits *different regimes*, depending on the value of the Reynolds number. It is observed that the outer surface of the liquid film remains *smooth* and *wave-free* for about Re ≤ 30, and thus the flow is clearly *laminar*. Ripples or waves appear on the free surface of the condensate flow as the Reynolds number increases, and the condensate flow becomes fully *turbulent* at about Re ≈ 1,800. The condensate flow is called *wavy-laminar* in the range of 30 < Re < 1,800 and *turbulent* for Re > 1,800. However, some disagreement exists about the value of Re at which the flow becomes wavy-laminar or turbulent.

Heat Transfer in Film Condensation The rate of heat transfer in film condensation can be expressed as

$$\dot{Q}_{conden} = h\ A_s\ (T_{sat} - T_s) = \dot{m}\ [h_{fg} + 0.68\ c_{pl}\ (T_{sat} - T_s)] \tag{2.4.69}$$

where A_s is the heat transfer area (the surface area on which condensation occurs). Solving for $\dot{m}$ from the equation above and substituting it into Eq. (2.4.68) gives yet another relation for the Reynolds number,

$$Re = \frac{4\ \dot{Q}_{conden}}{p\ \mu_l\ [h_{fg} + 0.68\ c_{pl}\ (T_{sat} - T_s)]} = \frac{4\ h\ A_s\ (T_{sat} - T_s)}{p\ \mu_l\ [h_{fg} + 0.68\ c_{pl}\ (T_{sat} - T_s)]} \tag{2.4.70}$$

This relation is convenient to use to determine the Reynolds number when the condensation heat transfer coefficient or the rate of heat transfer is known. The temperature of the liquid film varies from T_{sat} on the liquid–vapor interface to T_s at the wall surface. Therefore, the properties of the liquid should be evaluated at the *film temperature* $T_f = (T_{sat} + T_s)/2$, which is approximately the *average* temperature of the liquid. The h_{fg}, however, should be evaluated at T_{sat} since it is not affected by the subcooling of the liquid.

Heat Transfer Correlations in Film Condensation Below we present relations for the average heat transfer coefficient h for the case of *laminar* and *turbulent* film condensation for various geometries.

Laminar Flow on Vertical Plates For Re ≤ 30, the *average heat transfer coefficient* for laminar film condensation over a vertical flat plate of height L from theoretical analysis is determined to be

$$h_{vert,\,laminar} = 0.943\left[\frac{g\rho_l\ (\rho_l - \rho_v)[h_{fg} + 0.68\ c_{pl}\ (T_{sat} - T_s)]\ k_l^3}{\mu_l\ (T_{sat} - T_s)\ L}\right]^{1/4} \tag{2.4.71}$$

For algebraic simplicity, we assume $\rho_v << \rho_l$, and thus $\rho_l - \rho_v \approx \rho_l$, which is valid except near the critical point of the substance. Using this approximation, through a series of manipulations of the governing equations for the film condensation over a vertical flat plate, the following alternative correlation for the *average heat transfer coefficient* for laminar film condensation over a vertical flat plate of height L is recommended, for the details of the derivation refer to Çengel and Ghajar (2015)

$$h_{vert,\,laminar} \cong 1.47\ k_l\ Re^{-1/3}\left(\frac{g}{\nu_l^2}\right)^{1/3} \tag{2.4.72}$$

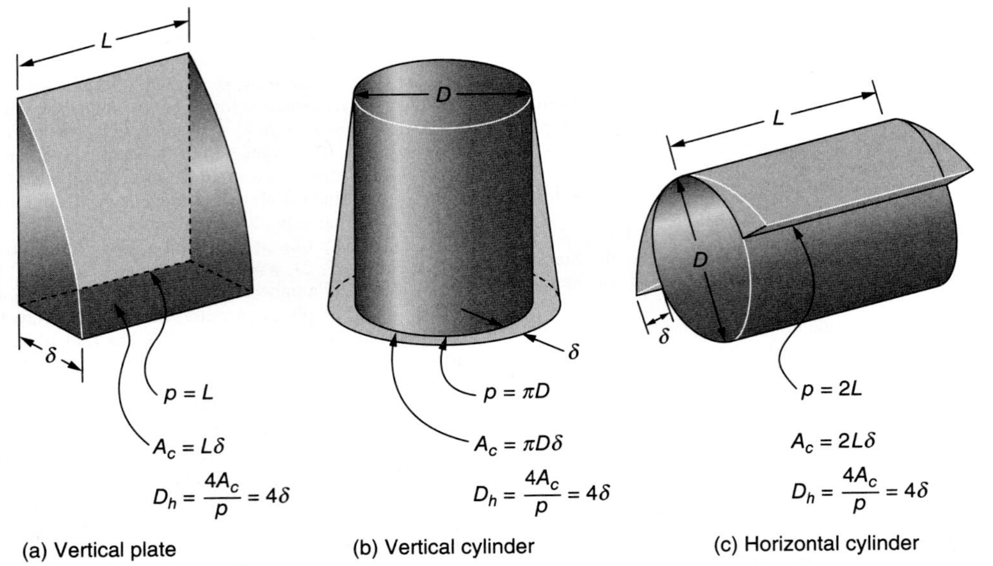

Figure 2.4.16 The wetted perimeter p, the condensate cross-sectional area A_c, and the hydraulic diameter D_h for some common geometries. (*From Çengel and Ghajar, 2015.*)

A relationship for the Reynolds number in the laminar region can be determined by substituting the h relation in Eq. (2.4.72) into the Re relation in Eq. (2.4.70) and simplifying. It yields

$$Re_{vert,\ laminar} = 3.78 \left[\frac{k_l\ L\ (T_{sat} - T_s) \left(\dfrac{g}{\nu_l^2} \right)^{1/3}}{\mu_l\ [h_{fg} + 0.68\ c_{pl}\ (T_{sat} - T_s)]} \right]^{3/4} \tag{2.4.73}$$

Wavy Laminar Flow on Vertical Plates At Reynolds numbers greater than about 30, it is observed that waves form at the liquid–vapor interface although the flow in liquid film remains laminar. The flow in this case is said to be *wavy laminar*. The waves at the liquid–vapor interface tend to increase heat transfer. But the waves also complicate the analysis and make it very difficult to obtain analytical solutions. Therefore, we have to rely on experimental studies. The increase in heat transfer due to the wave effect is, on average, about 20 percent, but it can exceed 50 percent. The exact amount of enhancement depends on the Reynolds number. Kutateladze (1963) based on his experimental studies, recommended the following relation for the average heat transfer coefficient in wavy laminar condensate flow for $\rho_v << \rho_l$ and $30 \leq \text{Re} \leq 1{,}800$,

$$h_{vert,\ wavy\ laminar} = \frac{Re\ k_l}{1.08\ Re^{1.22} - 5.2} \left(\frac{g}{\nu_l^2} \right)^{1/3} \tag{2.4.74}$$

A relation for the Reynolds number in the wavy laminar region can be determined by substituting the h relation in Eq. (2.4.74) into the Re relation in Eq. (2.4.70) and simplifying. It yields

$$Re_{vert,wavy\ laminar} = \left[4.81 + \frac{3.70\ k_l\ L\ (T_{sat} - T_s) \left(\dfrac{g}{\nu_l^2} \right)^{1/3}}{\mu_l\ [h_{fg} + 0.68\ c_{pl}\ (T_{sat} - T_s)]} \right]^{0.82} \tag{2.4.75}$$

Turbulent Flow on Vertical Plates At a Reynolds number of about 1,800, the condensate flow becomes turbulent. Several empirical relations of varying degrees of complexity are proposed for the heat transfer coefficient for turbulent flow. Again assuming $\rho_v << \rho_l$ for simplicity, Labuntsov (1957) proposed the following relation for the turbulent flow ($\text{Re} \geq 1{,}800$) of condensate on *vertical plates*:

$$h_{vert,\ turbulent} = \frac{Re\ k_l}{8{,}750 + 58\ Pr_l^{-0.5}\ (Re^{0.75} - 253)} \left(\frac{g}{\nu_l^2} \right)^{1/3} \tag{2.4.76}$$

The Re relation in this case is obtained by substituting the h relation in Eq. (2.4.76) into the Re relation in Eq. (2.4.70) which gives

$$Re_{vert,turbulent} = \left[\frac{0.069\ k_l\ L\ Pr_l^{0.5}\ (T_{sat} - T_s) \left(\dfrac{g}{\nu_l^2} \right)^{1/3}}{\mu_l\ [h_{fg} + 0.68\ c_{pl}\ (T_{sat} - T_s)]} - 151\ Pr_l^{0.5} + 253 \right]^{4/3} \tag{2.4.77}$$

Inclined Plates Equation (2.4.71) was developed for vertical plates, but it can also be used for laminar film condensation on the upper surfaces of plates that are *inclined* by an angle θ from the *vertical*, by replacing g in that equation by $g \cos \theta$. This approximation gives satisfactory results especially for $\theta \leq 60°$.

Vertical Tubes Equation (2.4.71) for vertical plates can also be used to calculate the average heat transfer coefficient for laminar film condensation on the outer surfaces of vertical tubes provided that the tube diameter is large relative to the thickness of the liquid film ($D >> \delta$).

Horizontal Tubes and Spheres The film condensation analysis on vertical plates can also be extended to horizontal tubes and spheres. The average heat transfer coefficient for film condensation on the outer surfaces of a *horizontal tube* is determined to be

$$h_{horiz} = 0.729 \left[\frac{g \rho_l (\rho_l - \rho_v) [h_{fg} + 0.68\ c_{pl}\ (T_{sat} - T_s)]\ k_l^3}{\mu_l\ (T_{sat} - T_s)\ D} \right]^{1/4} \tag{2.4.78}$$

where D is the diameter of the horizontal tube. Equation (2.4.78) can easily be modified for a *sphere* by replacing the constant 0.729 by 0.815.

Horizontal Tube Banks Horizontal tubes stacked on top of each other are commonly used in condenser design. The average thickness of the liquid film at the lower tubes is much larger as a result of condensate falling on top of them from the tubes directly above. Therefore, the average heat transfer coefficient at the lower tubes in such arrangements is smaller. Assuming the condensate from the tubes above to the ones below drain smoothly, the average film condensation heat transfer coefficient for all tubes in a vertical tier can be expressed as

$$h_{horiz,\,N\ tubes} = 0.729\left[\frac{g\rho_l\,(\rho_l - \rho_v)[h_{fg} + 0.68\ c_{pl}\ (T_{sat} - T_s)]\ k_l^3}{\mu_l\ (T_{sat} - T_s)\ ND}\right]^{1/4}$$

$$= \frac{1}{N^{1/4}}\ h_{horz,\,1\ tube} \qquad (2.4.79)$$

Note that Eq. (2.4.79) can be obtained from the heat transfer coefficient relation for a horizontal tube (Eq. 2.4.78) by replacing D in that relation by ND, where N is the number of horizontal tubes in the tier. This relation does not account for the increase in heat transfer due to the ripple formation and turbulence caused during drainage, and thus generally yields conservative results.

SPECIAL TOPIC

Application of Nanofluids in Heat Transfer The significant growth of different technologies such as electronics, communications, computing, and energy in the last few decades has caused serious problems in the thermal management of the devices associated with these technologies. The increasing power along with decreasing size of these devices has lead into innovative cooling technologies. The demand for better cooling technology is not limited to miniaturized devices. Large devices, such as transportation trucks, also require more efficient cooling.

Several methods have been used to enhance heat transfer. Some of the methods are usage of micro channels and utilization of extended surfaces (fins). One of the more recent methods is to enhance the heat transfer capability of the working fluid itself by increasing its thermal conductivity. A survey of the thermal properties of commonly used heat transfer fluids (such as water, engine coolants, and organic coolants) reveals that they all have extremely low thermal conductivities when compared to the thermal conductivity of solids. The thermal conductivity of a working fluid can be increased by adding small solid particles to the fluid. The name *nanofluid* refers to suspension of nanometer-sized particles (with typical length scales of 1 to 100 nm) in conventional fluids. Nanofluids have been shown to enhance the thermal conductivity and convective heat transfer performance of the working (base) liquids.

Nanofluids have been the subject of numerous studies worldwide since the discovery of the anomalous thermal behavior of these fluids. An excellent review article by Das et al. (2006) presents an exhaustive review of these studies. The review article by Yu et al. (2008) provides a detailed review of experimental results regarding the enhancement of the thermal conductivity and convective heat transfer of nanofluids relative to conventional heat transfer fluids. Their assessment of the experimental results showed that heat transfer enhancement for available nanofluids was in the 15-40 percent range, with a few situations resulting in orders of magnitude enhancement. A review on the mechanisms of heat transfer enhancement in nanofluids was conducted by Chandrasekar and Suresh (2009). Their review focused on the inconsistencies in the reported experimental values and controversies in the mechanisms proposed for heat transport in nanofluids. In their review article, they provided an overview of the various possible mechanisms reported in the literature that contribute to enhanced thermal conductivity of nanofluids. The other factors affecting the nanofluids thermal conductivity and the mathematical models for estimating nanofluids thermal conductivity with their limitations were also discussed. Numerous other review articles available in the literature focus on the use of nanofluids in different heat transfer applications. For example a review paper by Huminic and Huminic (2012) focuses on the application of nanofluids in heat exchangers. A review article by Salman et al. (2013) looks into the application of nanofluids in microtubes and microchannels. A review of conductive, convective, and radiative experimental heat transfer results is presented by Lomascolo et al. (2015). A recent review article by Sarkar et al. (2015) focuses on the possible applications of hybrid nanofluids for further heat transfer improvements.

NOMENCLATURE

A	Area, m^2
A_s	Surface area, m^2
A_c	Cross-sectional area, m^2
c	Specific heat, kJ/kg·K
c_p	Constant pressure specific heat, kJ/kg·K
D	Diameter, m
D_h	Hydraulic diameter, m
$\dot{e}_{gen}$	Heat generation rate, W/m^3
$\dot{E}_{gen}$	Total heat generation rate, W
$\dot{E}_{st}$	Total heat storage rate, W
f	Friction factor
F	Correction factor
g	Gravitational acceleration, m/s^2
Gz	Graetz number
Gr	Grashof number
h	Convection heat transfer coefficient, W/m^2·K
h_{fg}	Latent heat of vaporization, kJ/kg
k	Thermal conductivity, W/m·K
L	Length or characteristic length, m
L_h	Hydrodynamic entry length, m
L_t	Thermal entry length, m
LMTD	Log mean temperature difference
$\dot{m}$	Mass flow rate, kg/s
M	Molar mass, kg/kmol
N	Total number of tubes in the tube bank
N_L	Number of tube rows in the flow direction in the tube bank
N_T	Number of tubes in the transverse plane in the tube bank
NTU	Number of transfer units
Nu	Nusselt number
p	Perimeter, m
Pr	Prandtl number
$\dot{q}$	Heat flux, W/m^2
$\dot{Q}$	Heat transfer rate, kW
R	Thermal resistance, K/W
R_f	Fouling factor, m^2·K/W
Ra	Rayleigh number
Re	Reynolds number
S_D	Diagonal pitch
S_L	Longitudinal pitch
S_T	Transverse pitch
t	Time, s
T	Temperature, °C
T_b	Bulk fluid temperature, °C
T_c	Temperature of cold stream, °C
T_f	Film temperature, °C
T_h	Temperature of hot stream, °C
T_s	Surface (wall) temperature, °C
T_{sat}	Saturation temperature, °C
T_∞	Free stream temperature, °C
U	Overall heat transfer coefficient, W/m^2·K
$V_{element}$	Volume of element, m^3
V	Velocity, m/s

GREEK LETTERS

α	Thermal diffusivity, m^2/s
β	Coefficient of thermal expansion, 1/K
δ	Film thickness, m
ΔT_{lm}	Log mean temperature difference, °C or K
ΔT_m	Mean temperature difference, °C or K
ε	Emissivity
μ	Dynamic viscosity, kg/m·s or N·s/m^2

ν	Kinematic viscosity, m^2/s
ρ	Density, kg/m^3
σ	Stefan-Boltzmann constant
σ	Surface tension, N/m
θ	Excess temperature, °C
θ	Inclination angle

SUBSCRIPTS

avg	Average
b	Boundary, bulk fluid
cond	Conduction
conden	Condensation
conv	Convection
cr	Critical
cyl	Cylinder
e	Exit conditions
horiz	Horizontal
i	Inner or Inlet conditions
in	Inlet
l	Liquid
m	Mean (average)
max	Maximum
min	Minimum
o	Outer
out	Outlet
rad	Radiation
s	Surface
surr	Surrounding surfaces
sat	Saturated
sph	Sphere
v	Water vapor
vert	Vertical
1	Inlet state
2	Exit state
∞	Far from a surface; free-flow conditions

SUPERSCRIPTS

· (over dot)	Quantity per unit time

REFERENCES

1. Abraham JP, Sparrow EM, and Minkowycz WJ. Internal-Flow Nusselt Numbers for the Low-Reynolds-Number End of the Laminar-to-Turbulent Transition Regime, *International Journal of Heat and Mass Transfer*, 54, pp. 584-588, 2011.

2. Abraham JP, Sparrow EM, and Tong JCK. Heat Transfer in All Pipe Flow Regimes—Laminar, Transitional/Intermittent, and Turbulent, *International Journal of Heat and Mass Transfer*, 52, pp. 557-563, 2009.

3. Bergman TL, Lavine AS, Incropera FP, and Dewitt DP. Introduction to Heat Transfer, 6th ed., John Wiley & Sons, New York, 2011.

4. Bowman RA, Mueller AC, and Nagle WM. Mean Temperature Difference in Design, *Trans. ASME*, 62, pp. 283-294, 1940.

5. Bromley LA. Heat Transfer in Stable Film Boiling, *Chemical Engineering Prog*, 46, pp. 221-227, 1950.

6. Çengel YA and Ghajar AJ. Heat and Mass Transfer: Fundamentals & Applications, 5th ed., McGraw-Hill, New York, 2015.

7. Chandrasekar M and Suresh S. A Review on the Mechanisms of Heat Transport in Nanofluids, *Heat Transfer Engineering*, 30, no. 14, pp. 1136-1150, 2009.

8. Churchill SW and Bernstein M. A Correlating Equation for Forced Convection from Gases and Liquids to a Circular Cylinder in Cross Flow, *Journal of Heat Transfer*, 99, pp. 300-306, 1977.

9. Churchill SW and Ozoe H. Correlations for Laminar Forced Convection in Flow over an Isothermal Flat Plate and in Developing and Fully Developed Flow in an Isothermal Tube, *Journal of Heat Transfer*, 95, pp. 78-84, 1973.

10. Das SK, Choi SUS, and Patel HE. Heat Transfer in Nanofluids—A Review, *Heat Transfer Engineering*, 27, no. 10, pp. 3-19, 2006.

11. Edwards DK, Denny VE, and Mills AF. Transfer, Processes, 2nd ed., Hemisphere, Washington, DC, 1979.

12. Fraas AP. Heat Exchanger Design, 2nd ed., John Wiley & Sons, New York, 1989.

13. Ghajar AJ and Tam LM. Heat Transfer Measurements and Correlations in the Transition Region for a Circular Tube with Three Different Inlet Configurations, *Experimental Thermal and Fluid Science*, 8, pp. 79-90, 1994.

14. Gnielinski V. New Equations for Heat and Mass Transfer, in Turbulent Pipe and Channel Flow, *International Chemical Engineering*, 16, pp. 359-368, 1976.

15. Hahn DW and Ozisik MN. Conduction Heat Transfer, 3rd ed., John Wiley & Sons, New York, 2012.

16. Huminic G and Huminic A. Application of Nanofluids in Heat Exchangers: A Review, *Renewable and Sustainable Energy Reviews*, 16, pp. 5625-5638, 2012.

17. Kays WM, Crawford ME, and Weigand B. Convective Heat and Mass Transfer. 4th ed., McGraw-Hill, New York, 2005.

18. Kays WM and London AL. Compact Heat Exchangers, 3rd ed., McGraw-Hill, New York, 1984.

19. Kutateladze SS. Fundamentals of Heat Transfer, Academic Press, New York, 1963.

20. Labuntsov DA. Heat Transfer in Film Condensation of Pure Steam on Vertical Surfaces and Horizontal Tubes, *Teploenergetika*, 4, pp. 72-80, 1957.

21. Lomascolo M, Colangelo G, Milanese M, and de Risi A. Review of Heat Transfer in Nanofluids: Conductive, Convective and Radiative Experimental Results, *Renewable and Sustainable Energy Reviews*, 43, pp. 1182-1198, 2015.

22. Rohsenow WM. A Method of Correlating Heat Transfer Data for Surface Boiling of Liquids, *ASME Transactions*, 74, pp. 969-975, 1952.

23. Salman BH, Mohammed HA, Munisamy KM, and Kherbeet A Sh. Characteristics of Heat Transfer and Fluid Flow in Microtube and Microchannel using Conventional Fluids and Nanofluids: A Review, *Renewable and Sustainable Energy Reviews*, 28, pp. 848-880, 2013.

24. Sarkar J, Ghosh P, and Adil A. A Review on Hybrid Nanofluids: Recent Research, Development and Applications, *Renewable and Sustainable Energy Reviews*, 43, pp. 164-177, 2015.

25. Shah RK and Bhatti MS. Laminar Convective Heat Transfer in Ducts, In: Kakaç S, Shah RK, and Aung W (Eds). Handbook of Single-Phase Convective Heat Transfer, Wiley Interscience, New York, 1987.

26. Sieder EN and Tate GE. Heat Transfer and Pressure Drop of Liquids in Tubes, *Industrial Engineering Chemistry*, 28, pp. 1429-1435, 1936.

27. Sleicher CA and Rouse MW. A Convenient Correlation for Heat Transfer to Constant and Variable Property Fluids in Turbulent Pipe Flow, *International Journal of Heat Mass Transfer*, 18, pp. 1429-1435, 1975.

28. *Standards of Tubular Exchanger Manufacturers Association*, Tubular Exchanger Manufacturers Association, 9th ed., New York, 2007.

29. Taborek J, Hewitt GF, and Afgan N. Heat Exchangers: Theory and Practice, Hemisphere, New York, 1983.

30. Tam LM and Ghajar AJ. Transitional Heat Transfer in Plain Horizontal Tubes, *Heat Transfer Engineering*, 27, no. 5, pp. 23-38, 2006.

31. Walker G. Industrial Heat Exchangers, Hemisphere, Washington, DC, 1982.

32. Whitaker S. Forced Convection Heat Transfer Correlations for Flow in Pipe, Past Flat Plates, Single Cylinders, Single Spheres, and for Flow in Packed Beds and Tube Bundles, *AICHE Journal*, 18, pp. 361-371, 1972.

33. Yu W, France DM, Routbort JL, and Choi SUS. Review and Comparison of Nanofluid Thermal Conductivity and Heat Transfer Enhancements, *Heat Transfer Engineering*, 29, no. 5, pp. 432-460, 2008.

34. Zuber N. On the Stability of Boiling Heat Transfer, *ASME Transactions*, 80, pp. 711-720, 1958.

35. Zukauskas A. Heat Transfer from Tubes in Cross Flow, In: Kakac S, Shah RK, and Aung W (Eds.), Handbook of Single Phase Convective Heat Transfer, Wiley Interscience, New York, 1987.

Section **3**

Strength of Materials

BY

PEYMAN HONARMANDI *Professor of Mechanical Engineering, Manhattan College, Bronx, New York*
ALI M. SADEGH *Professor of Mechanical Engineering, The City College of the City University of New York*
LAWRENCE M. MATTA *Stress Engineering Services, Inc.*
FRANCESCO LANZA DI SCALEA *Professor of Structural Engineering, University of California San Diego*
GARY L. CLOUD *Professor of Mechanical Engineering, Michigan State University*
MICHAEL W. HYER *Professor of Engineering Science and Mechanics, Virginia Polytechnic Institute and State University*
SCOTT W. CASE *Professor of Engineering Science and Mechanics, Virginia Polytechnic Institute and State University*

3.1 MECHANICAL PROPERTIES OF MATERIALS

Revised by Peyman Honarmandi

REFERENCES: Davis, H.E. et al., "Testing and Inspection of Engineering Materials," McGraw-Hill. Timoshenko, S. "Strength of Materials," Part II, Van Nostrand. Richards, C.W. "Engineering Materials Science," Wadsworth. Nadai, A. "Plasticity," McGraw-Hill. Tetelman, A.S. and McEvily, A.J. "Fracture of Structural Materials," Wiley. Sanford, R.J. "Fracture Mechanics," ASTM STP-833. McClintock, F.A. and Argon, A.S. (eds.), "Mechanical Behavior of Materials," Addison-Wesley. Dieter, G.E. "Mechanical Metallurgy," McGraw-Hill. "Creep Data," ASME. ASTM Standards, ASTM International. Blaznynski, T.Z. (ed.), "Plasticity and Modern Metal Forming Technology," Springer.

STRESS-STRAIN DIAGRAMS

The Stress-Strain Curve There are two types of stress-strain curves: engineering stress-strain curve and true stress-strain curve. The engineering tensile stress-strain curve is obtained by **static loading** of a standard specimen, that is, by applying the load slowly enough that all parts of the specimen are nearly in equilibrium at any instant. The curve is usually obtained by controlling a loading rate in a tensile machine. ASTM Standards require the loading rate not exceeding 100,000 lb/in^2 (70 kgf/mm^2)/min. An alternative method of obtaining the curve is to specify the strain rate as the independent variable, in which case the loading rate is continuously adjusted to maintain the required strain rate. A strain rate of 0.05 in/in/(min) is commonly used. It is measured usually by an extensometer attached to the gage length of the specimen. The engineering stress is calculated by having the load divided by initial cross section of specimen. Figure 3.1.1 shows several engineering stress-strain curves.

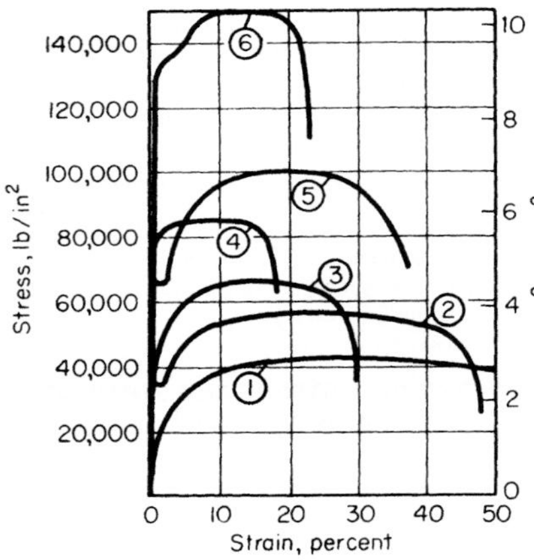

Figure 3.1.1 Comparative stress-strain diagrams. (1) Soft brass; (2) low carbon steel; (3) hard bronze; (4) cold rolled steel; (5) medium carbon steel, annealed; (6) medium carbon steel, heat treated.

For most engineering materials, the curve will have an initial linear elastic region (Fig. 3.1.2) in which deformation is reversible and time-independent. The slope in this region is called **Young's modulus** or elastic modulus, E. The **proportional elastic limit** (PEL) is the point where the curve starts to deviate from a straight line. The **elastic limit** (frequently indistinguishable from PEL) is the point on the curve beyond which plastic deformation is present after release of the load. If the stress is increased further, the stress-strain curve departs more and more from the straight line and material goes to plastic region. Unloading the specimen at point X (Fig. 3.1.2), the portion XX' is linear and is essentially parallel to the original line OX''. The horizontal distance OX' is called the **permanent set** corresponding to the stress at X. To determine **yield strength**, a straight line XX' is drawn parallel to the initial elastic line OX'' but displaced from it by an arbitrary value of permanent or plastic strain. For most metals, the permanent strain commonly used is 0.20 percent of the original gage length. The intersection of this line with the curve determines the stress value called the **yield strength**. In reporting the yield strength, the amount of permanent set should be specified. The arbitrary yield strength is used especially for those materials not exhibiting a natural yield point such as nonferrous metals; but it is not limited to these. Plastic behavior is somewhat time-dependent, particularly at high temperatures. Also at high temperatures, a small amount of time-dependent reversible strain may be detectable, indicative of **anelastic** behavior.

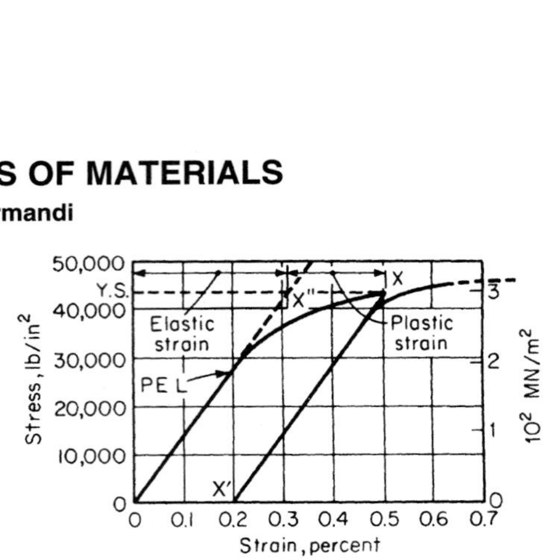

Figure 3.1.2 General stress-strain diagram.

The **ultimate tensile strength** (UTS) is the maximum load sustained by the specimen divided by the original specimen cross-sectional area. The percent **elongation at failure** is the plastic extension of the specimen at failure expressed as (the change in original gage length × 100) divided by the original gage length. This extension is the sum of the **uniform** and **nonuniform** elongations. The uniform elongation occurs prior to the UTS and has an unequivocal significance. The uniform elongation is associated with uniaxial stress, whereas the nonuniform elongation which occurs during localized extension (necking) is associated with triaxial stress. The nonuniform elongation will depend on geometry, particularly the ratio of specimen gage length L_0 to diameter D or square root of crosssectional area A. ASTM Standards specify test-specimen geometry for a number of specimen sizes. The ratio $L_0/\sqrt{A}$ is maintained at 4.5 for flat- and round-cross-section specimens. The original gage length should always be stated in reporting elongation values.

The specimen percent **reduction in area** (RA) is the percentage of contraction in cross-sectional area at the fracture divided by the original area. See Table 3.1.1. It is obtained by measurement of the initial cross section and the cross section of the broken specimen at the fracture location. The RA along with the load at fracture can be used to obtain the **fracture stress,** which fracture load divided by cross-sectional area at the fracture.

The type of fracture in tension gives some indications of the quality of the material, but this is considerably affected by the testing temperature, speed of testing, the shape and size of the test piece, and other conditions. Contraction is greatest in ductile materials and least in brittle materials. In general, fractures are either of the **shear** or of the **separation** (loss of cohesion) type. Flat tensile specimens of ductile metals often show shear failures if the ratio of width to thickness is greater than 6:1. A completely shear-type failure may terminate in a

Table 3.1.1 Typical Mechanical Properties at Room Temperature
(Based on ordinary stress-strain values)

Metal	Tensile strength, 1,000 lb/in²	Yield strength, 1,000 lb/in²	Ultimate elongation, %	Reduction of area, %	Brinell no.
Cast iron	18-60	8-40	0	0	100-300
Wrought iron	45-55	25-35	35-25	55-30	100
Commercially pure iron, annealed	42	19	48	85	70
Hot-rolled	48	30	30	75	90
Cold-rolled	100	95			200
Structural steel, ordinary	50-65	30-40	40-30		120
Low-alloy, high-strength	65-90	40-80	30-15	70-40	150
Steel, SAE 1,300, annealed	70	40	26	70	150
Quenched, drawn 1,300°F	100	80	24	65	200
Drawn 1,000°F	130	110	20	60	260
Drawn 700°F	200	180	14	45	400
Drawn 400°F	240	210	10	30	480
Steel, SAE 4,340, annealed	80	45	25	70	170
Quenched, drawn 1,300°F	130	110	20	60	270
Drawn 1,000°F	190	170	14	50	395
Drawn 700°F	240	215	12	48	480
Drawn 400°F	290	260	10	44	580
Cold-rolled steel, SAE 1,112	84	76	18	45	160
Stainless steel, 18-S	85-95	30-35	60-55	75-65	145-160
Steel castings, heat-treated	60-125	30-90	33-14	65-20	120-250
Aluminum, pure, rolled	13-24	5-21	35-5		23-44
Aluminum-copper alloys, cast	19-23	12-16	4-0		50-80
Wrought, heat-treated	30-60	10-50	33-15		50-120
Aluminum die castings	30		2		
Aluminum alloy 17ST	56	34	26	39	100
Aluminum alloy 51ST	48	40	20	35	105
Copper, annealed	32	5	58	73	45
Copper, hard-drawn	68	60	4	55	100
Brasses, various	40-120	8-80	60-3		50-170
Phosphor bronze	40-130		55-5		50-200
Tobin bronze, rolled	63	41	40	52	120
Magnesium alloys, various	21-45	11-30	17-0.5		47-78
Monel 400, Ni-Cu alloy	79	30	48	75	125
Molybdenum, rolled	100	75	30		250
Silver, cast, annealed	18	8	54		27
Titanium 6-4 alloy, annealed	130	120	10	25	352
Ductile iron, grade 80-55-06	80	55	6		225-255

NOTE: Compressive strength of cast iron, 80,000 to 150,000 lb/in².
Compressive yield strength of all metals, except those cold-worked = tensile yield strength.
Stress 1,000 lb/in² × 6.894 = stress, MN/m².

chisel edge, for a flat specimen, or a point rupture, for a round specimen. Separation failures occur in brittle materials, such as certain cast irons. Combinations of both shear and separation failures are common on round specimens of ductile metal. In the ductile metals, failure often starts at the axis in a necked region and produces a relatively flat area which grows until the material shears along a cone-shaped surface at the outside of the specimen, resulting in what is known as the cup-and-cone fracture. Double cup-and-cone and rosette fractures sometimes occur. Several types of tensile fractures are shown in Fig. 3.1.3.

Annealed or hot-rolled mild steels generally exhibit a **yield point** (see Fig. 3.1.4). Here, in a constant strain-rate test, a large increment

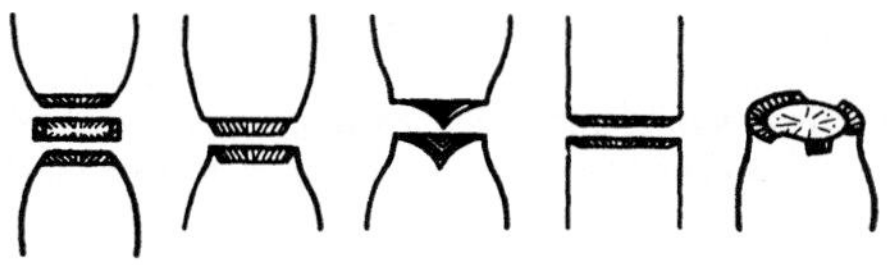

Figure 3.1.3 Typical metal fractures in tension.

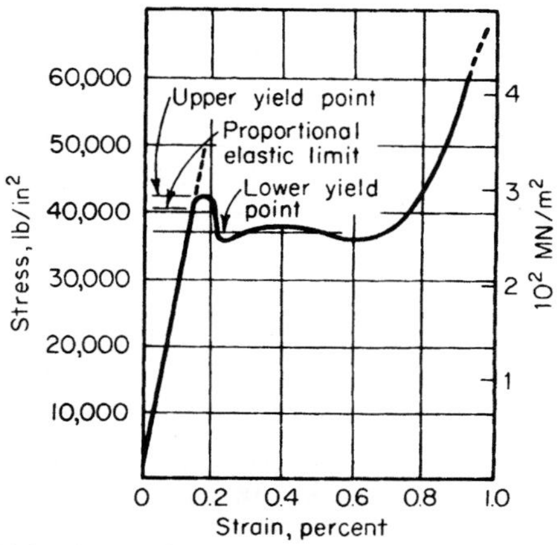

Figure 3.1.4 Yielding of annealed steel.

of extension occurs under constant load at the elastic limit or at a stress just below the elastic limit. In the latter event the stress drops suddenly from the **upper yield point** to the **lower yield point.** Subsequent to the drop, the yield-point extension occurs at constant stress, followed by a rise to the UTS. Plastic flow during the yield-point extension is discontinuous; successive zones of plastic deformation, known as **Luder's bands** or **stretcher strains,** appear until the entire specimen gage length has been uniformly deformed at the end of the yield-point extension. This behavior causes a banded or stepped appearance on the metal surface. The exact form of the stress-strain curve for this class of material is sensitive to test temperature, test strain rate, and the characteristics of the tensile machine employed.

The plastic behavior in a uniaxial tensile test can be represented as the **true stress-strain curve.** The **true stress** σ is based on the instantaneous cross section A, so that σ = load/A. The instantaneous **true strain** increment is $-dA/A$, or dL/L prior to necking. Total true strain ε is

$$\int_{A_0}^{A} -\frac{dA}{A} = \ln\left(\frac{A_0}{A}\right)$$

or $\ln (L/L_0)$ prior to necking. The true stress-strain curve or flow curve obtained has the typical form shown in Fig. 3.1.5 which is usually a monotonic behavior, unlike engineering stress-strain curve. In the part of the test subsequent to the maximum load point (UTS), when necking occurs, the true strain of interest is that which occurs in an infinitesimal length at the region of minimum cross section. True strain for this element can still be expressed as $\ln (A_0/A)$, where A refers to the minimum cross section. Methods of constructing the true stress-strain curve are described in the technical literature. In the range between initial yielding and the neighborhood of the maximum load point the relationship between plastic strain ε_p and true stress often approximates

$$\sigma = k\varepsilon_p^n$$

where k is the **strength coefficient** and n is the **work-hardening exponent.** For a material which shows a yield point, the relationship applies only to the rising part of the curve beyond the lower yield. It can be shown that at the maximum load point the slope of the true stress-strain curve equals the true stress, from which it can be deduced that for a material obeying the above exponential relationship between ε_p and n, $\varepsilon_p = \varepsilon_{ult} = n$

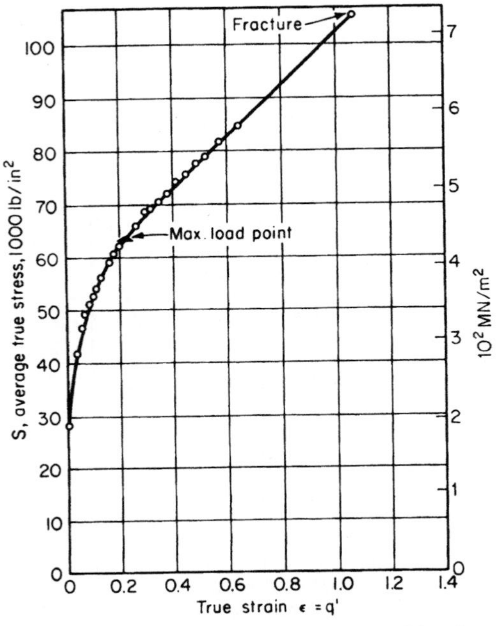

Figure 3.1.5 True stress-strain curve for 20°C annealed mild steel.

Table 3.1.2 Room-Temperature Plastic-Flow Constants for a Number of Metals

Material	Condition	k, 1,000 in² (MN/m²)	n
0.40% C steel	Quenched and tempered at 400°F (478K)	416 (2,860)	0.088
0.05% C steel	Annealed and temper-rolled	72 (49.6)	0.235
2024 aluminum	Precipitation-hardened	100 (689)	0.16
2024 aluminum	Annealed	49 (338)	0.21
Copper	Annealed	46.4 (319)	0.54
70-30 brass	Annealed	130 (895)	0.49

SOURCE: Reproduced by permission from "Properties of Metals in Materials Engineering," ASM, 1949.

at the maximum load point. The exponent strongly influences the spread between YS and UTS on the engineering stress-strain curve. Values of n and k for some materials are shown in Table 3.1.2. A point on the flow curve indentifies the **flow stress** corresponding to a certain strain, that is, the stress required to bring about this amount of plastic deformation. The concept of true strain is useful for accurately describing large amounts of plastic deformation. The linear strain definition $(L - L_0)/L_0$ fails to correct for the continuously changing gage length, which leads to an increasing error as deformation proceeds.

During extension of a specimen under tension, the change in the specimen cross-sectional area is related to the elongation by **Poisson's ratio** μ, which is the ratio of strain in a transverse direction to that in the longitudinal direction. Values of μ for the elastic region are shown in Table 3.1.3. For plastic strain it is approximately 0.5.

The general effect of increased strain rate is to increase the resistance to plastic deformation and thus to raise the flow curve. Decreasing test temperature also raises the flow curve. The effect of strain rate is expressed as **strain-rate sensitivity** m. Its value can be measured in the tension test if the strain rate is suddenly increased by a small increment during the plastic extension. The flow stress will then jump to a higher value. The strain-rate sensitivity is the ratio of incremental changes of $\log \sigma$ and $\log \dot{\varepsilon}$

$$m = \left(\frac{\delta \log \sigma}{\delta \log \dot{\varepsilon}}\right)_{\varepsilon}$$

The strain-rate sensitivity varies with the temperature and the grain size of the specimen. For most engineering materials at room temperature the strain rate sensitivity is of the order of 0.01. The effect becomes more significant at elevated temperatures, with values ranging to 0.2 and sometimes higher.

Compression Testing The compressive stress-strain curve is similar to the tensile stress-strain curve up to the yield strength. Thereafter, the progressively increasing specimen cross section causes the compressive stress-strain curve to diverge from the tensile curve. Some ductile metals will not fail in the compression test. Complex behavior occurs when the direction of stressing is changed, because of the **Bauschinger effect**, which can be described as follows: If a specimen is first plastically strained in tension, its yield stress in compression is reduced and vice versa.

Combined Stresses This refers to the situation in which stresses are present on each of the faces of a cubic element of the material. For a given cube orientation the applied stresses may include shear stresses over the cube faces as well as stresses normal to them. By a suitable rotation of axes the problem can be simplified: applied stresses on the new cubic element are equivalent to three mutually orthogonal **principal stresses** σ_1, σ_2, σ_3 alone, each acting normal to a cube face. Combined stress behavior in the elastic range is described in Sec. 3.2, Mechanics of Materials.

Prediction of the conditions under which plastic yielding will occur under combined stresses can be made with the help of several empirical

Table 3.1.3 Elastic Constants of Metals
(Mostly from tests of R. W. Vose)

Metal	*E* Modulus of elasticity (Young's modulus). 1,000,000 lb/in²	*G* Modulus of rigidity (shearing modulus). 1,000,000 lb/in²	*K* Bulk modulus. 1,000,000 lb/in²	μ Poisson's ratio
Cast steel	28.5	11.3	20.2	0.265
Cold-rolled steel	29.5	11.5	23.1	0.287
Stainless steel 18-8	27.6	10.6	23.6	0.305
All other steels, including high-carbon, heat-treated	28.6-30.0	11.0-11.9	22.6-24.0	0.283-0.292
Cast iron	13.5-21.0	5.2-8.2	8.4-15.5	0.211-0.299
Malleable iron	23.6	9.3	17.2	0.271
Copper	15.6	5.8	17.9	0.355
Brass, 70-30	15.9	6.0	15.7	0.331
Cast brass	14.5	5.3	16.8	0.357
Tobin bronze	13.8	5.1	16.3	0.359
Phosphor bronze	15.9	5.9	17.8	0.350
Aluminum alloys, various	9.9-10.3	3.7-3.9	9.9-10.2	0.330-0.334
Monel metal	25.0	9.5	22.5	0.315
Inconel	31	11		0.27-0.38
Z-nickel	30	11		0.36
Beryllium copper	17	7		0.21-0.285
Elektron (magnesium alloy)	6.3	2.5	4.8	0.281
Titanium (99.0 Ti), annealed bar	15-16	6.5	16	0.34
Zirconium, crystal bar	11-14	4.8	13.2	0.34
Molybdenum, arc-cast	48-52	17.4	33.3	0.31

theories. In the **maximum-shear-stress theory** the criterion for yielding is that yielding will occur when

$$\sigma_1 - \sigma_3 = \sigma_{ys}$$

in which σ_1 and σ_3 are the largest and smallest principal stresses, respectively, and σ_{ys} is the uniaxial tensile yield strength that is twice the value of maximum shear stress in failure. This is the simplest theory for predicting yielding under combined stresses. A more accurate prediction can be made by the **distortion-energy theory,** according to which the criterion is

$$(\sigma_1 - \sigma_2)^2 + (\sigma_1 - \sigma_3)^2 + (\sigma_2 - \sigma_3)^2 = 2(\sigma_{ys})^2$$

Stress-strain curves in the plastic region for combined stress loading can be constructed. However, a particular stress state does not determine a unique strain value. The latter will depend on the stress-state path which is followed.

Plane strain is a condition where strain is confined to two dimensions. There is generally stress in the third direction, but because of mechanical constraints, strain in this dimension is prevented. Plane strain occurs in certain metalworking operations. It can also occur in the neighborhood of a crack tip in a tensile loaded member if the member is sufficiently thick. The material at the crack tip is then in triaxial tension, which condition promotes brittle fracture. On the other hand, ductility is enhanced and fracture is suppressed by triaxial compression.

Stress Concentration In a structure or machine part having a notch or any abrupt change in cross section, the maximum stress will occur at this location and will be greater than the stress calculated by elementary formulas based upon simplified assumptions as to the stress distribution. The ratio of this maximum stress to the nominal stress (calculated by the elementary formulas) is the **stress-concentration factor K_t.** This is a constant for the particular geometry and is independent of the material, provided it is isotropic. The stress-concentration factor may be determined experimentally or, in some cases, theoretically from the mathematical theory of elasticity. The factors shown in Figs. 3.1.6 to 3.1.13 were determined from both photoelastic tests and the theory

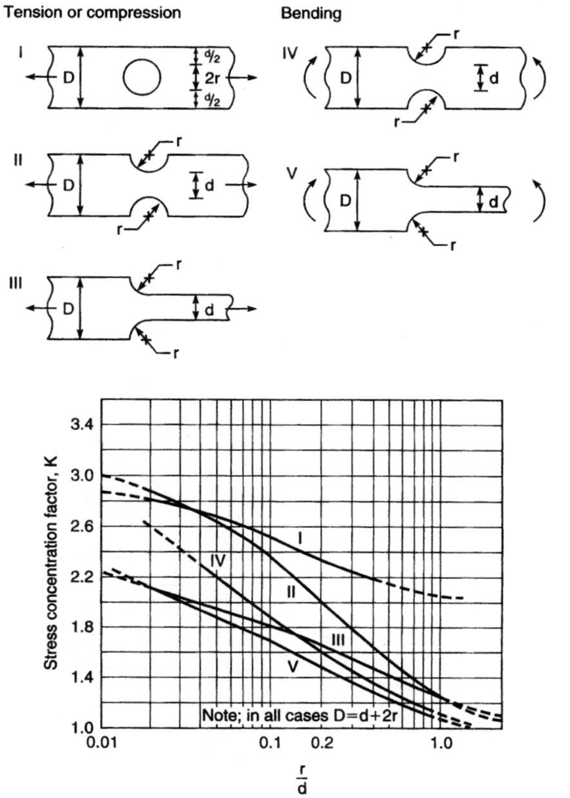

Figure 3.1.6 Flat plate with semicircular fillets and grooves or with holes. I, II, and III are in tension or compression; IV and V are in bending.

of elasticity. Stress concentration will cause failure of brittle materials if the concentrated stress is larger than the ultimate strength of the material. In ductile materials, concentrated stresses higher than the yield strength will generally cause local plastic deformation and redistribution of stresses (rendering them more uniform). On the other hand, even with ductile materials areas of stress concentration are possible sites for fatigue if the component is cyclically loaded.

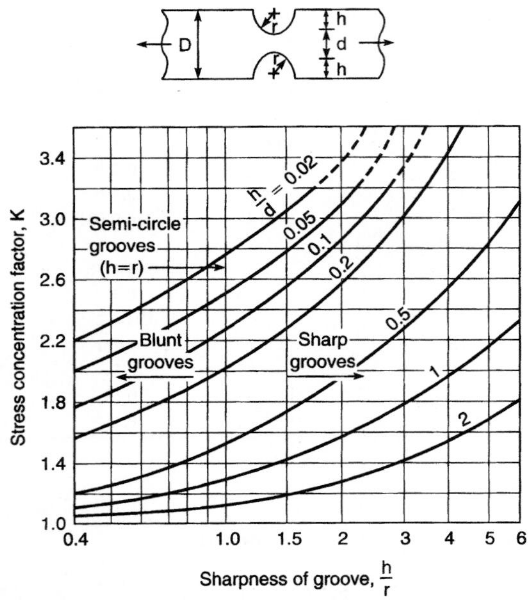

Figure 3.1.7 Flat plate with grooves, in tension.

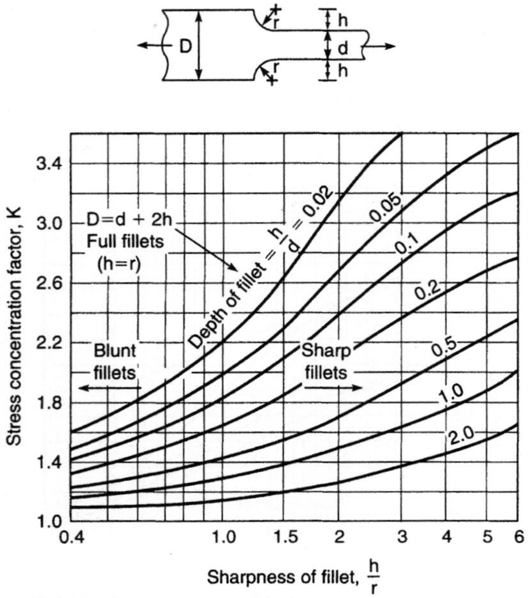

Figure 3.1.8 Flat plate with fillets, in tension.

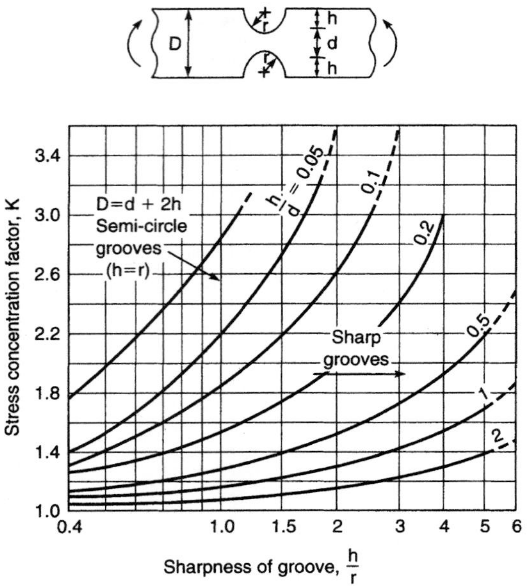

Figure 3.1.9 Flat plate with grooves, in bending.

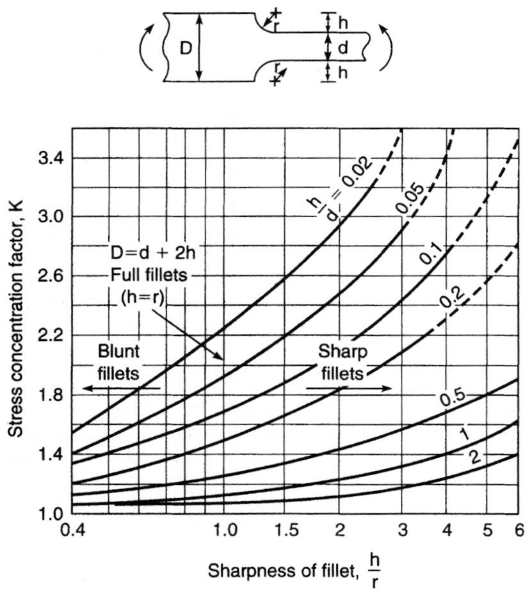

Figure 3.1.10 Flat plate with fillets, in bending.

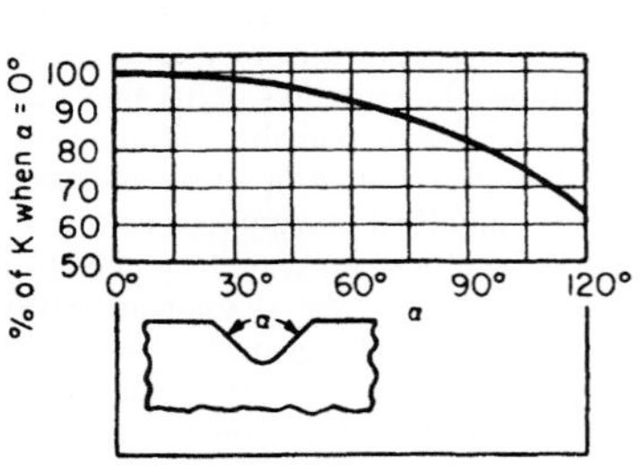

Figure 3.1.11 Flat plate with angular notch, in tension or bending.

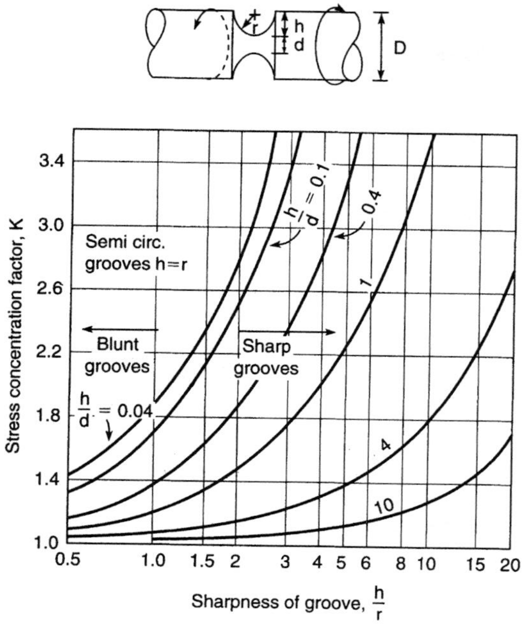

Figure 3.1.12 Grooved shaft in torsion.

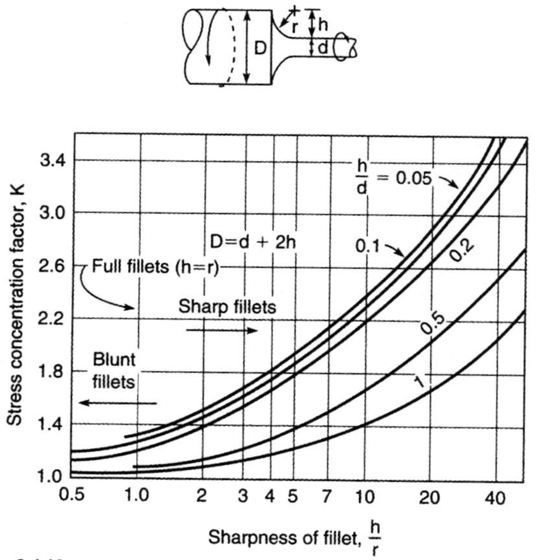

Figure 3.1.13 Filleted shaft in torsion.

FRACTURE AT LOW STRESSES

Materials under tension sometimes fail by rapid fracture at stresses much below their strength level as determined in tests on carefully prepared specimens. These **brittle, unstable,** or **catastrophic failures** originate at pre-existing stress-concentrating flaws which may be inherent in a material.

The transition-temperature approach is often used to ensure fracture-safe design in structural-grade steels. These materials exhibit a characteristic temperature, known as the **ductile brittle transition** (DBT) temperature, below which they are susceptible to brittle fracture. The transition-temperature approach to fracture-safe design ensures that the transition temperature of a material selected for a particular application is suitably matched to its intended use temperature. The DBT can be detected by plotting certain measurements from tensile or impact tests against temperature. Usually the transition to brittle behavior is complex, being neither fully ductile nor fully brittle. The range may extend over 200°F (110 K) interval. The **nil-ductility temperature** (NDT), determined by the **drop weight test** (see ASTM Standards), is an important reference point in the transition range. When NDT for a particular steel is known, temperature-stress combinations can be specified which define the limiting conditions under which catastrophic fracture can occur.

In the **Charpy V-notch** (CVN) impact test, a notched-bar specimen (Fig. 3.1.26) is used which is loaded in bending (see ASTM Standards). The energy absorbed from a swinging pendulum in fracturing the specimen is measured. The pendulum strikes the specimen at 16 to 19 ft (4.88 to 5.80 m)/s so that the specimen deformation associated with fracture occurs at a rapid strain rate. This ensures a conservative measure of toughness, since in some materials, toughness is reduced by high strain rates. A CVN impact energy vs. temperature curve is shown in Fig. 3.1.14, which also shows the transitions as given by percent brittle fracture and by percent lateral expansion. The CVN energy has no analytical significance. The test is useful mainly as a guide to the fracture behavior of a material for which an empirical correlation has been established between impact energy and some rigorous fracture criterion. For a particular grade of steel the CVN curve can be correlated with NDT. (See ASME Boiler and Pressure Vessel Code.)

Fracture Mechanics This analytical method is used for ultra-high-strength alloys, transition-temperature materials below the DBT temperature, and some low-strength materials in heavy section thickness.

Fracture mechanics theory deals with crack extension where plastic effects are negligible or confined to a small region around the crack tip. The present discussion is concerned with a through-thickness crack in a tension-loaded plate (Fig. 3.1.15) which is large enough so that the crack-tip stress field is not affected by the plate edges. Fracture mechanics theory states that unstable crack extension occurs when the work required for an increment of crack extension, namely, surface energy and energy consumed in local plastic deformation, is exceeded by the elastic-strain energy released at the crack tip. The elastic-stress field surrounding one of the crack tips in Fig. 3.1.15 is characterized by the **stress intensity** K_I, which has units of $(\text{lb}\sqrt{\text{in}})/\text{in}^2$ or $(\text{N}\sqrt{\text{m}})/\text{m}^2$. It is a function of applied nominal stress σ, crack half-length a, and a geometry factor Q:

$$K_I^2 = Q\sigma^2\pi a \tag{3.1.1}$$

For a particular material it is found that as K_I is increased, a value K_c is reached at which unstable crack propagation occurs. K_c depends on plate thickness B, as shown in Fig. 3.1.16. It attains a constant value when B is great enough to provide plane-strain conditions at the crack tip. The low plateau value of K_c is an important material property known as the **plane-strain critical stress intensity** or **fracture toughness** K_{Ic}. Values for a number of materials are shown in Table 3.1.4. They are influenced strongly by processing and small changes in composition, so that the values shown are not necessarily typical. K_{Ic} can be used in the critical form of Eq. (3.1.1):

$$(K_{Ic})^2 = Q\sigma^2\pi a_{cr} \tag{3.1.2}$$

to predict failure stress when a maximum flaw size in the material is known or to determine maximum allowable flaw size when the stress is set. The predictions will be accurate so long as plate thickness B satisfies the **plane-strain criterion:** $B \geq 2.5(K_{Ic}/\sigma_{ys})^2$. They will be conservative if a plane-strain condition does not exist. A big advantage of the fracture mechanics approach is that stress intensity can be calculated by equations analogous to (3.1.1) for a wide variety of geometries, types of crack, and loadings (Paris and Sih, "Stress Analysis of Cracks," STP-381, ASTM, 1965). Failure occurs in all cases when K_t reaches K_{Ic}. Fracture mechanics also provides a framework for predicting the occurrence of **stress-corrosion cracking** by using Eq. (3.1.2) with K_{Ic} replaced by K_{Iscc}, which is the material parameter denoting resistance to stress-corrosion-crack propagation in a particular medium.

Two standard test specimens for K_{Ic} determination are specified in ASTM standards, which also detail specimen preparation and test procedure. Recent developments in fracture mechanics permit treatment of crack propagation in the ductile regime. (See "Elastic-Plastic Fracture," ASTM.)

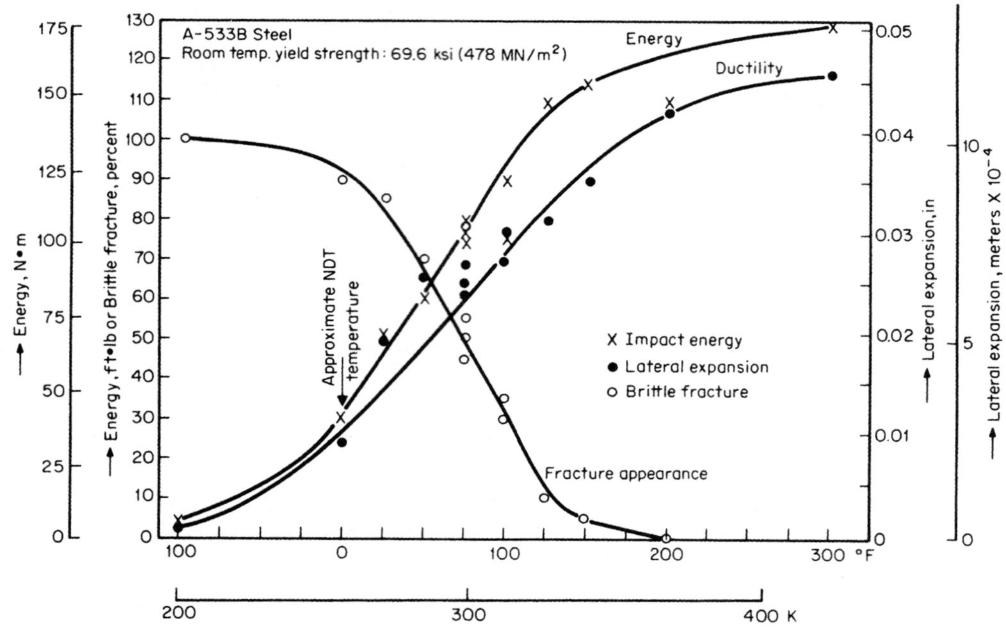

Figure 3.1.14 CVN transition curves. (*Data from Westinghouse R & D Lab.*)

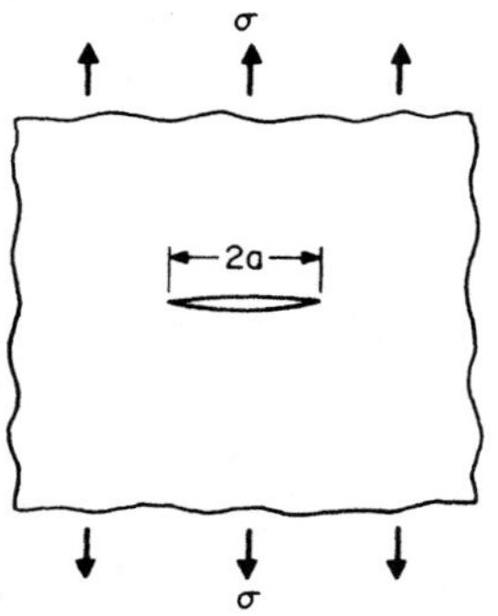

Figure 3.1.15 Through-thickness crack geometry.

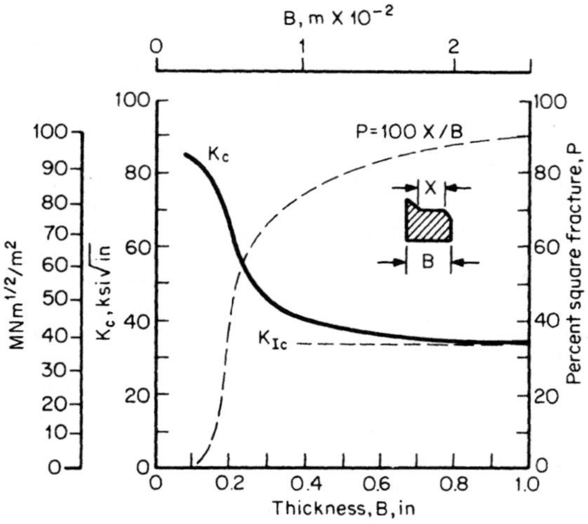

Figure 3.1.16 Dependence of K_c and fracture appearance (in terms of percentage of square fracture) on thickness of plate specimens. Based on data for aluminum 7075-T6. (*From Scrawly and Brown, STP-381, ASTM.*)

Table 3.1.4 Room-Temperature K_{Ic} Values on High-Strength Materials*

Material	0.2% YS, 1,000 lb/in² (MN/m²)	K_{Ic}, 1,000 lb$\sqrt{\text{in/in}^2}$ (MN m$^{1/2}$/m²)
18% Ni maraging steel	300 (2,060)	46 (50.7)
18% Ni maraging steel	270 (1,850)	71 (78)
18% Ni maraging steel	198 (1,360)	87 (96)
Titanium 6-4 alloy	152 (1,022)	39 (43)
Titanium 6-4 alloy	140 (960)	75 (82.5)
Aluminum alloy 7075-T6	75 (516)	26 (28.6)
Aluminum alloy 7075-T6	64 (440)	30 (33)

* Determined at Westinghouse Research Laboratories.

FATIGUE

Fatigue is generally understood as the gradual deterioration of a material which is subjected to repeated loads. In fatigue testing, a specimen is subjected to periodically varying constant-amplitude stresses by means of mechanical or magnetic devices. The applied stresses may alternate between equal positive and negative values, from zero to maximum positive or negative values, or between unequal positive and negative values. The most common loading is alternate tension and compression of equal numerical values obtained by rotating a smooth cylindrical specimen while under a bending load. A series of fatigue tests are made on a number of specimens of the material at different stress levels. The stress endured is then plotted against the number of cycles sustained. By choosing lower and lower stresses, a value may be found which will not produce failure, regardless of the number of applied cycles. This stress value is called the **fatigue limit or endurance limit.** The diagram is called the stress-cycle diagram or ***S-N* diagram.** Instead of recording the data on cartesian coordinates, either stress is plotted vs. the logarithm of the number of cycles (Fig. 3.1.17) or both stress and cycles are plotted to logarithmic scales. Both diagrams show a relatively sharp bend in the curve near the fatigue limit for ferrous metals. The fatigue limit may be established for most steels between 2 and 10 million cycles. Nonferrous metals usually show no clearly defined fatigue limit. The *S-N* curves

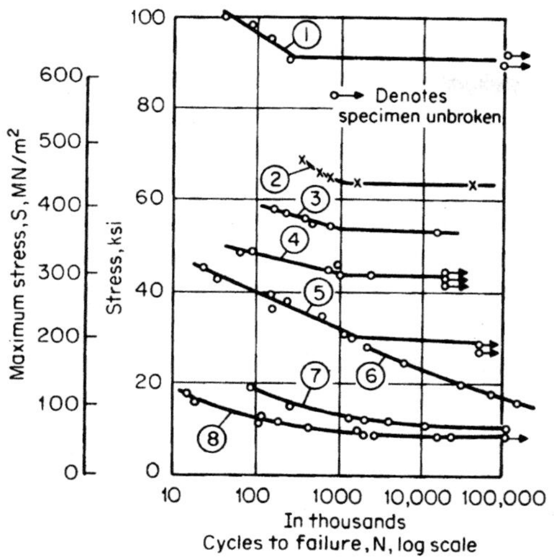

Figure 3.1.17 The *S-N* diagrams from fatigue tests. (1) 1.20% C steel, quenched and drawn at 860°F (460°C); (2) alloy structural steel; (3) SAE 1050, quenched and drawn at 1,200°F (649°C); (4) SAE 4130, normalized and annealed; (5) ordinary structural steel; (6) Duralumin; (7) copper, annealed; (8) cast iron (reversed bending).

in these cases indicate a continuous decrease in stress values to several hundred million cycles, and both the stress value and the number of cycles sustained should be reported. See Table 3.1.5.

The mean stress (the average of the maximum and minimum stress values for a cycle) has a profound influence on the stress range (the algebraic difference between the maximum and minimum stress values). Several empirical formulas and graphical methods such as the "modified Goodman diagram" have been developed to show the influence of the mean stress on the stress range for failure. A simple but conservative approach (see Soderberg, Working Stresses, *Jour. Appl. Mech.,* **2,** Sept. 1935) is to plot the variable stress S_v (one-half the stress range) as ordinate vs. the mean stress S_m as abscissa (Fig. 3.1.18). At zero mean stress, the ordinate is the fatigue limit under completely reversed stress. Yielding will occur if the mean stress exceeds the yield stress S_o, and this establishes the extreme right-hand point of the diagram. A straight line is drawn between these two points. The coordinates of any other point along this line are values of S_m and S_v which may produce failure.

Surface defects, such as roughness or scratches, and notches or shoulders all reduce the fatigue strength of a part. With a notch of prescribed geometric form and known concentration factor, the reduction in strength is appreciably less than would be called for by the concentration factor itself, but the various metals differ widely in their susceptibility to the effect of roughness and concentrations, or **notch sensitivity.**

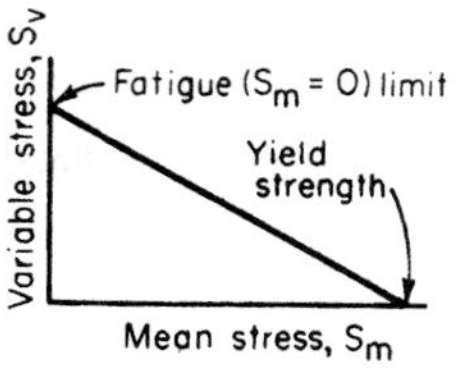

Figure 3.1.18 Effect of mean stress on the variable stress for failure.

For a given material subjected to a prescribed state of stress and type of loading, notch sensitivity can be viewed as the ability of that material to resist the concentration of stress incidental to the presence of a notch. Alternately, notch sensitivity can be taken as a measure of the degree to which the geometric stress concentration factor is reduced. An attempt is made to rationalize notch sensitivity through the equation $q = (K_f - 1)/(K - 1)$, where q is the notch sensitivity, K is the geometric stress concentration factor (from data similar to those in Figs. 3.1.5 to 3.1.13 and the like), and K_f is defined as the ratio of the strength of unnotched material to the strength of notched material. Ratio K_f is obtained from laboratory tests, and K is deduced either theoretically or from laboratory tests, but both must reflect the same state of stress and type of loading. The value of q lies between 0 and 1, so that (1) if $q = 0$, $K_f = 1$ and the material is not notch-sensitive (soft metals such as copper, aluminum, and annealed low-strength steel); (2) if $q = 1$, $K_f = K$, the material is fully notch-sensitive and the full value of the geometric stress concentration factor is not diminished (hard, high-strength steel). In practice, q will lie somewhere between 0 and 1, but it may be hard to quantify. Accordingly, the pragmatic approach to arrive at a solution to a design problem often takes a conservative route and sets $q = 1$. The exact material properties at play which are responsible for notch sensitivity are not clear.

Further, notch sensitivity seems to be higher, and ordinary fatigue strength lower in large specimens, necessitating full-scale tests in many cases (see Peterson, Stress Concentration Phenomena in Fatigue of Metals, *Trans. ASME,* **55,** 1933, p. 157, and Buckwalter and Horger, Investigation of Fatigue Strength of Axles, Press Fits, Surface Rolling and Effect of Size, *Trans. ASM,* **25,** Mar. 1937, p. 229). **Corrosion** and **galling** (due to rubbing of mating surfaces) cause great reduction of fatigue strengths, sometimes amounting to as much as 90 percent of the original endurance limit. Although any corroding agent will promote severe corrosion fatigue, there is so much difference between the effects of "sea water" or "tap water" from different localities that numerical values are not quoted here.

Overstressing specimens above the fatigue limit for periods shorter than necessary to produce failure at that stress reduces the fatigue limit in a subsequent test. Similarly, **understressing** below the fatigue limit may increase it. Shot peening, nitriding, and cold work usually improve fatigue properties.

Table 3.1.5 Typical Approximate Fatigue Limits for Reversed Bending

Metal	Tensile strength, 1,000 lb/in²	Fatigue limit, 1,000 lb/in²	Metal	Tensile strength, 1,000 lb/in²	Fatigue limit, 1,000 lb/in²
Cast iron	20-50	6-18	Copper	32-50	12-17
Malleable iron	50	24	Monel	70-120	20-50
Cast steel	60-80	24-32	Phosphor bronze	55	12
Armco iron	44	24	Tobin bronze, hard	65	21
Plain carbon steels	60-150	25-75	Cast aluminum alloys	18-40	6-11
SAE 6150, heat-treated	200	80	Wrought aluminum alloys	25-70	8-18
Nitralloy	125	80	Magnesium alloys	20-45	7-17
Brasses, various	25-75	7-20	Molybdenum, as cast	98	45
Zirconium crystal bar	52	16-18	Titanium (Ti-75A)	91	45

NOTE: Stress, 1,000 lb/in² × 6.894 = stress, MN/m².

No very good overall correlation exists between fatigue properties and any other mechanical property of a material. The best correlation is between the fatigue limit under completely reversed bending stress and the ordinary tensile strength. For many ferrous metals, the fatigue limit is approximately 0.40 to 0.60 times the tensile strength if the latter is below 200,000 lb/in^2. Low-alloy high-yield-strength steels often show higher values than this. The fatigue limit for nonferrous metals is approximately to 0.20 to 0.50 times the tensile strength. The fatigue limit in reversed shear is approximately 0.57 times that in reversed bending.

In some very important engineering situations components are cyclically stressed into the plastic range. Examples are thermal strains resulting from temperature oscillations and notched regions subjected to secondary stresses. Fatigue life in the plastic or **"low-cycle" fatigue** range has been found to be a function of plastic strain, and low-cycle fatigue testing is done with strain as the controlled variable rather than stress. Fatigue life N and cyclic plastic strain ε_p tend to follow the relationship

$$N\varepsilon_p^2 = C$$

where C is a constant for a material when $N < 10^5$. (See Coffin, A Study of Cyclic-Thermal Stresses in a Ductile Material, *Trans. ASME,* **76,** 1954, p. 947.)

The type of physical change occurring inside a material as it is repeatedly loaded to failure varies as the life is consumed, and a number of stages in fatigue can be distinguished on this basis. The early stages comprise the events causing **nucleation of a crack** or flaw. This is most likely to appear on the surface of the material; **fatigue failures** generally originate at a surface. This stage is well-known as **crack-initiation** stage. Following nucleation of the crack, it grows during the **crack-propagation** stage. Eventually the crack becomes large enough for some rapid terminal mode of failure to take over such as ductile rupture or brittle fracture. The rate of crack growth in the crack-propagation stage can be accurately quantified by fracture mechanics methods. Assuming an initial flaw and a loading situation as shown in Fig. 3.1.15, the rate of crack growth per cycle can generally be expressed as

$$da/dN = C_0\left(\Delta K_I\right)^n \tag{3.1.3}$$

where C_0 and n are constants for a particular material and ΔK_I is the range of stress intensity per cycle. K_I is given by (3.1.1). Using (3.1.3), it is possible to predict the number of cycles for the crack to grow to a size at which some other mode of failure can take over. Values of the constants C_0 and n are determined from specimens of the same type as those used for determination of K_{Ic} but are instrumented for accurate measurement of slow crack growth.

Constant-amplitude fatigue-test data are relevant to many rotary-machinery situations where constant cyclic loads are encountered. There are important situations where the component undergoes variable loads and where it may be advisable to use **random-load testing.** In this method, the load spectrum which the component will experience in service is determined and is applied to the test specimen artificially.

CREEP

Experience has shown that, for the design of equipment subjected to sustained loading at elevated temperatures, little reliance can be placed on the usual short-time tensile properties of metals at those temperatures. Under the application of a constant load it has been found that materials, both metallic and nonmetallic, show a gradual flow or **creep** even for stresses below the proportional limit at elevated temperatures. Similar effects are present in low-melting metals such as lead at room temperature. The deformation which can be permitted in the satisfactory operation of most high-temperature equipment is limited.

In metals, creep is a plastic deformation caused by slip occurring along crystallographic directions in the individual crystals, together with some flow of the grain-boundary material. After complete release of load, a small fraction of this plastic deformation is recovered with time. Most of the flow is nonrecoverable for metals and the resulting deformation is permanent.

The most common creep test is the **long-time creep test** under constant tensile load and the **stress-rupture test.** Other special forms are the **stress-relaxation test** and the **constant-strain-rate test.**

The **long-time creep test** is conducted by applying a dead weight to one end of a lever system, the other end being attached to the specimen surrounded by a furnace and held at constant temperature. The axial deformation is read periodically throughout the test and a curve is plotted of the strain ε_0 as a function of time t (Fig. 3.1.19). This is repeated for various loads at the same testing temperature. The portion of the curve OA in Fig. 3.1.19 is the region of **primary creep,** AB the region of **secondary creep,** and BC that of **tertiary creep.** The strain rates, or the slopes of the curve, are decreasing, constant, and increasing in these three regions, respectively. Since the period of the creep test is usually much shorter than the duration of the part in service, various extrapolation procedures are followed (see Gittus, "Creep, Viscoelasticity and Creep Fracture in Solids," Wiley, 1975). See Table 3.1.6.

In practical applications the region of constant-strain rate (secondary creep) is often used to estimate the probable deformation throughout the life of the part. It is thus assumed that this rate will remain constant during periods beyond the range of the test data. The working stress is chosen so that this total deformation will not be excessive. An **arbitrary creep strength,** which is defined as the stress at which a given temperature will result in 1 percent deformation in 100,000 h, has received a certain amount of recognition, but it is advisable to determine the proper stress for each individual case from diagrams of stress vs. creep rate (Fig. 3.1.20) (see "Creep Data," ASTM and ASME).

Additional temperatures (°F) and stresses (in 1,000 lb/in^2 or 1 ksi) for stated creep rates (percent per 1,000 h) for wrought nonferrous metals are as follows:

60-40 Brass: Rate 0.1, temp. 350°F (400°F), stress 8 ksi (2 ksi); rate 0.01, temp 300°F (350°F) [400°F], stress 10 ksi (3 ksi) [1 ksi].

Phosphor bronze: Rate 0.1, temp 400 (550) [700] [800], stress 15 (6) [4] [4]; rate 0.01, temp 400 (550) [700], stress 8 (4) [2].

Nickel: Rate 0.1, temp 800 (1,000), stress 20 (10).

70 Cu, 30 NI. Rate 0.1, temp 600 (750), stress 28 (13-18); rate 0.01, temp 600 (750), stress 14 (8-9).

Aluminum alloy 17 *S* (Duralumin): Rate 0.1, temp 300 (500) [600], stress 22 (5) [1.5].

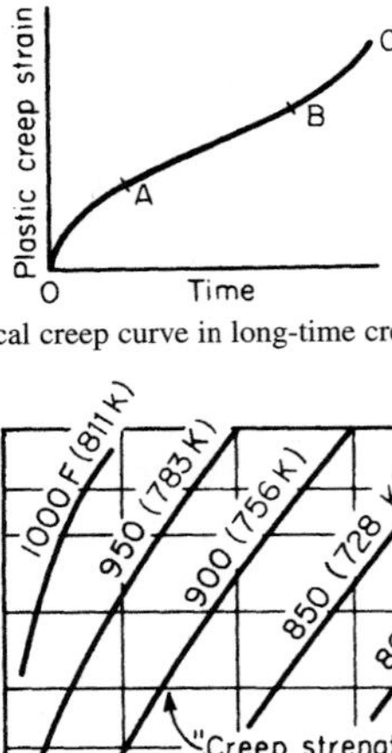

Figure 3.1.19 Typical creep curve in long-time creep test.

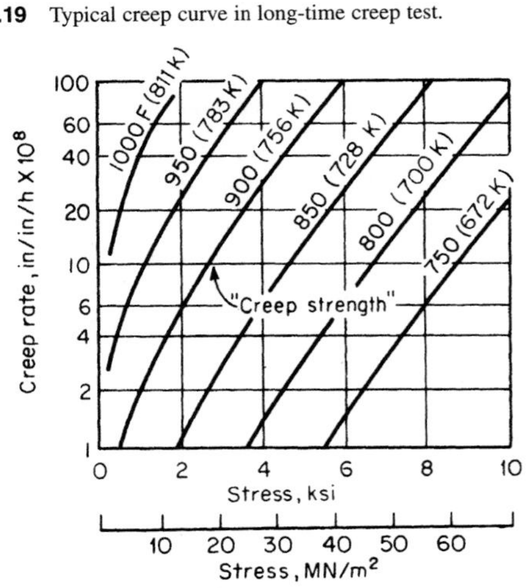

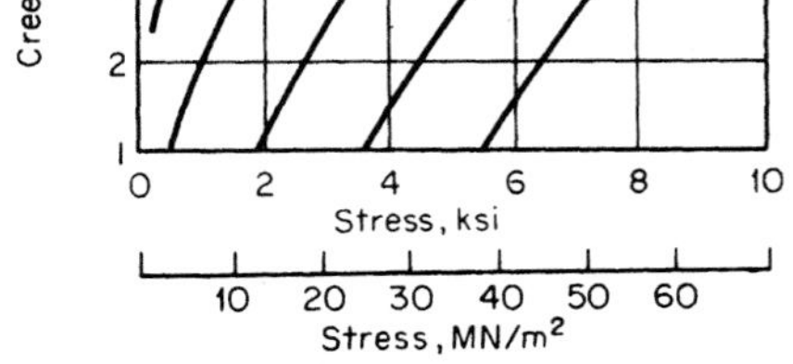

Figure 3.1.20 Creep rates for 0.35% C steel.

Table 3.1.6 Stresses for Given Creep Rates and Temperatures*

	Creep rate 0.1% per 1,000 h					Creep rate 0.01% per 1,000 h				
Material temp, °F	800	900	1,000	1,100	1,200	800	900	1,000	1,100	1,200
Wrought steels:										
SAE 1015	17-27	11-18	3-12	2-7	1	10-18	6-14	3-8	1	
0.20 C, 0.50 Mo	26-33	18-25	9-16	2-6	1-2	16-24	11-22	4-12	2	1
0.10-0.25 C, 4-6 Cr + Mo	22	15-18	9-11	3-6	2-3	14-17	11-15	4-7	2-3	1-2
SAE 4140	27-33	20-25	7-15	4-7	1-2	19-28	12-19	3-8	2-4	1
SAE 1030-1045	8-25	5-15	5	2	1	5-15	3-7	2-4	1	
Commercially pure iron	7		4		3	5		2		
0.15 C, 1-2.5 Cr, 0.50 Mo	25-35	18-28	8-20	6-8	3-4	20-30	12-18	3-12	2-5	1-2
SAE 4340	20-40	15-30	2-12	1-3		8-20		1-6		
SAE X3140	7-10		5-4			3-8		1-2		
0.20 C, 4-6 Cr	30	10-20	7-10	1				3-5		
0.25 C, 4-6 Cr + W	30	10-15	4-10	2-8			6-11	2-7		
0.16 C, 1.2 Cu		18	10-15	3	1		10-18	7-12		
0.20 C, 1 Mo	35	27	12			25	12	6		
0.10-0.40 C, 0.2-0.5 Mo, 1-2 Mn	30-40	12-20	4-14			25-28	8-15	2-8		0.5
SAE 2340	7-12	5	2							
SAE 6140	30	12	4			7	6	1		
SAE 7240	30	21	6-15	2		30	11	3-9	1	
Cr + Va + W, various	20-70	14-30	5-15			18-50	8-18	2-13		
Temp, °F	1,100	1,200	1,300	1,400	1,500	1,000	1,100	1,200	1,300	1,400
Wrought chrome-nickel steels:										
18-8†	10-18	5-11	3-10	2-5	2.5	11-16	5-12	2-10		1-2
10-25 Cr, 10-30 Ni‡	10-20	5-15	3-10	2-5			6-15	3-10	2-8	1-3
Temp, °F	800	900	1,000	1,100	1,200	800	900	1,000	1,100	1,200
Cast steels:										
0.20-0.40 C	10-20	5-10	3			8-15		1		
0.10-0.30 C, 0.5-1 Mo	28	20-30	6-12	2		20	10-15	2-5		
0.15-0.30 C, 4-6 Cr + Mo	25-30	15-25	8-15	8		20-25	9-15	2-7	2	
18-8§			20-25	15	10			20	15	8
Cast iron	20	8	4			10		2		
Cr Ni cast iron			9					3		

* Based on 1000-h tests. Stresses in 1,000 lb/in² or 1 ksi.
† Additional data. At creep rate 0.1 percent and 1,000 (1,600)°F the stress is 18-25 (1); at creep rate 0.01 percent at 1,500°F, the stress is 0.5.
‡ Additional data. At creep rate 0.1 percent and 1,000 (1,600)°F the stress is 10-30 (1).
§ Additional data. At creep rate 0.1 percent and 1,600°F the stress is 3; at creep rate 0.01 and 1,500°F, the stress is 2-3.

Lead pure (commercial) (0.03 percent Ca): At 110°F, for rate 0.1 percent the stress range, lb/in², is 150-180 (60-140) [200-220]; for rate of 0.01 percent, 50-90 (10-50) [110-150].

Stress of 1,000 lb/in² or 1 ksi = 6.894 MN/m² or MPa, $t_k = \frac{5}{9}(t_F + 459.67)$.

Structural changes may occur during a creep test, thus altering the metallurgical condition of the metal. In some cases, premature rupture appears at a low fracture strain in a normally ductile metal, indicating that the material has embrittled. This is a very insidious condition and difficult to predict. The **stress-rupture test** is well adapted to study this effect. It is conducted by applying a constant load to a specimen, in the same manner as for the long-time creep test, and record the time at which the specimen ruptures. This will give a single data point and the experiment should be repeated for different loads to obtain nominal stresses of rupture. The nominal stresses are then plotted vs. the times for fracture at constant temperature on a log-log scale (Fig. 3.1.21).

The stress reaction is measured in the **constant-strain-rate test** while the specimen is deformed at a constant strain rate. In the **relaxation test,** the decrease of stress with time is measured while the total strain (elastic + plastic) is maintained constant. The latter test

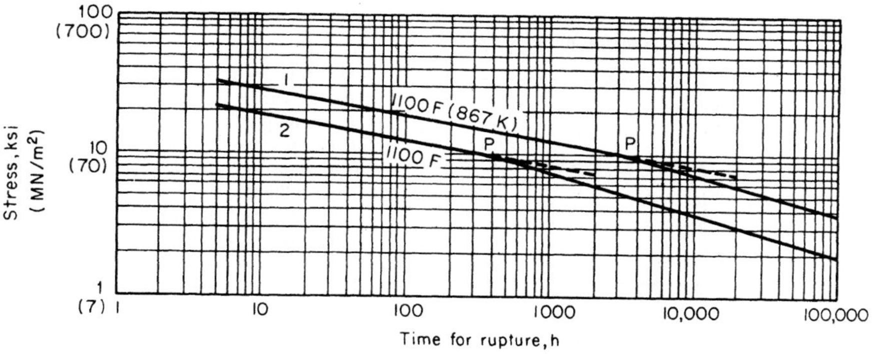

Figure 3.1.21 Relation between time to failure and stress for a 3% chromium steel. (1) Heat treated 2 h at 1,740°F (950°C) and furnace cooled; (2) hot rolled and annealed 1,580°F (860°C).

has direct application to the loosening of turbine bolts and to similar problems. Although some correlation has been indicated between the results of these various types of tests, no general correlation is yet available, and it has been found necessary to make tests under each of these special conditions to obtain satisfactory results.

The interrelationship between strain rate and temperature in the form of a velocity-modified temperature (see MacGregor and Fisher, A Velocity-modified Temperature for the Plastic Flow of Metals, *Jour. Appl. Mech.,* Mar. 1945) simplifies the creep problem in reducing the number of variables.

Superplasticity Superplasticity is the property of some metals and alloys which permits extremely large, uniform deformation at elevated temperature, in contrast to conventional metals which neck down and subsequently fracture after relatively small amounts of plastic deformation. Superplastic behavior requires a metal with small equiaxed grains, a slow and steady rate of deformation (strain rate), and a temperature elevated to somewhat more than half the melting point. With such metals, large plastic deformation can be brought about with lower external loads; ultimately, that allows the use of lighter fabricating equipment and facilitates production of finished parts to near-net shape.

Stress and strain rates are related for a metal exhibiting superplasticity. A factor in this behavior stems from the relationship between the applied stress and strain rates. This factor m—the strain rate sensitivity index—is evaluated from the equation $\sigma = K\dot{\varepsilon}^m$, where σ is the applied stress, K is a constant, and $\dot{\varepsilon}$ is the strain rate. Figure 3.1.22*a* plots a stress/strain rate curve for a superplastic alloy on log-log coordinates. The slope of the curve defines m, which is maximum at the point of inflection. Figure 3.1.22*b* shows the variation of m vs. ln $\dot{\varepsilon}$. Ordinary metals exhibit low values of m such as 0.2 or less; for those behaving superplastically, $m = 0.6$ to 0.8 or more. As m approaches 1, the behavior of the metal will be quite similar to that of a Newtonian viscous solid, which elongates plastically without necking down.

In Fig. 3.1.22*a*, in region I, the stress and strain rates are low and creep is predominantly a result of diffusion. In region III, the stress and strain rates are highest and creep is mainly the result of dislocation and slip mechanisms. In region II, where superplasticity is observed, creep is governed predominantly by grain boundary sliding.

Shape Memory

Some particular alloys have a remarkable property which may be quite useful. The **shape memory effect** inherent in an alloy is a thermoelastic property that results from a reversible martensitic transformation when heated or cooled within a range of temperatures below the melting point. In essence, a given geometry at elevated temperature is cooled and further deformed. Upon being heated to the previous elevated temperature, the deformation imparted at the lower temperature is negated, and the former elevated-temperature geometry is restored. A number of useful applications have been found for those alloys which have this property.

HARDNESS

Hardness has been variously defined as resistance to local penetration, to scratching, to machining, to wear or abrasion, and to yielding. The multiplicity of definitions, and corresponding multiplicity of hardness-measuring instruments, together with the lack of a fundamental definition, indicates that hardness may not be a fundamental property of a material but rather a composite one including yield strength, work hardening, true tensile strength, modulus of elasticity, and others.

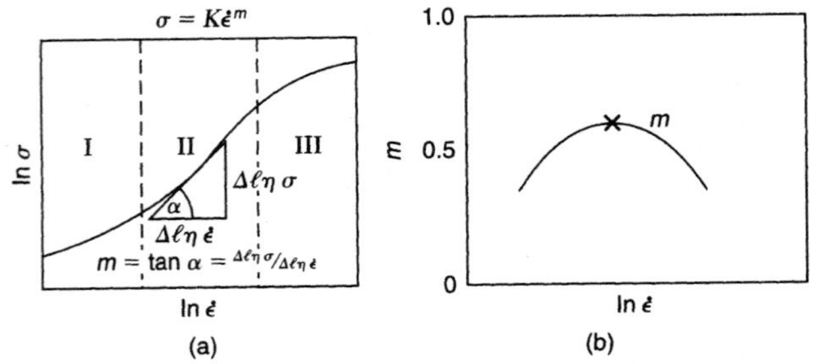

Figure 3.1.22 Stress and strain rate relations for superplastic alloys. (*a*) Log-log plot of $\sigma = K\dot{\varepsilon}^m$; (*b*) m as a function of strain rate.

Scratch hardness is measured by **Mohs scale** of minerals which is arranged so that each mineral will scratch the mineral of the next lower number. In recent mineralogical work and in certain microscopic metallurgical work, jeweled scratching points either with a set load or else loaded to give a set width of scratch have been used. Hardness in its relation to machinability and to wear and abrasion is generally dealt with in direct machining or wear tests, and little attempt is made to separate hardness itself, as a numerically expressed quantity, from the results of such tests.

The resistance to localized penetration, or **indentation hardness,** is widely used industrially as a measure of hardness, and indirectly as an indicator of other desired properties in a manufactured product. The indentation tests described below are essentially nondestructive, and in most applications may be considered non-marring, so that they may be applied to each piece produced; and through the empirical relationships of hardness to such properties as tensile strength, fatigue strength, and impact strength, pieces likely to be deficient in the latter properties may be detected and rejected.

Brinell hardness is determined by forcing a hardened sphere under a known load into the surface of a material and measuring the diameter of the indentation left after the test. The **Brinell hardness number,** or simply the **Brinell number,** is obtained by dividing the load used, in kilograms, by the actual surface area of the indentation, in square millimetres. The result is a pressure, but the units are rarely stated.

$$\text{BHN} = P\Big/\left[\frac{\pi D}{2}\left(D - \sqrt{D^2 - d^2}\right)\right]$$

where BHN is the Brinell hardness number; P the imposed load, kg; D the diameter of the spherical indenter, mm; and d the diameter of the resulting impression, mm.

Hardened-steel bearing balls may be used for hardness up to 450, but beyond this hardness specially treated steel balls or jewels should be used to avoid flattening the indenter. The standard-size ball is 10 mm and the standard loads are 3,000, 1,500, and 500 kg, along with 100, 125, and 250 kg for softer materials. If any other size of ball is used for special reasons, the load should be adjusted approximately as follows: for iron and steel, $P = 30D^2$; for brass, bronze, and other soft metals, $P = 5D^2$; for extremely soft metals, $P = D^2$ (see "Methods of Brinell Hardness Testing," ASTM). Readings obtained with other than the standard ball and loadings should have the load and ball size appended, as such readings are only approximately equal to those obtained under standard conditions.

The size of the specimen should be sufficient to ensure that no part of the plastic flow around the impression reaches a free surface, and in no case should the thickness be less than 10 times the depth of the impression. The load should be applied steadily and should remain on for at least 15 s in the case of ferrous materials and 30 s in the case of most nonferrous materials. Longer periods may be necessary on certain soft materials that exhibit creep at room temperature. In testing thin materials, it is not permissible to pile up several thicknesses of material under the indenter, as the readings so obtained will invariably be lower than the true readings. With such materials, smaller indenters and loads, or different methods of hardness testing, are necessary.

In the standard Brinell test, the diameter of the impression is measured with a low-power hand microscope, but for production work several testing machines are available which automatically measure the depth of the impression and from this give readings of hardness. Such machines should be calibrated frequently on test blocks of known hardness.

In the **Rockwell method** of hardness measurement, the depth of penetration of an indenter under certain arbitrary conditions of test is determined. The indenter may be either a steel ball of some specified diameter or a spherical-tipped conical diamond of 120° angle and 0.2-mm tip radius, called a "Brale." A *minor load* of 10 kg is first applied which causes an initial penetration and holds the indenter in place. Under this condition, the dial is set to zero and the major load applied. The values of the latter are 60, 100, or 150 kg. Upon removal of the major load, the reading is taken while the minor load is still on. The hardness number may then be read directly from the scale which measures penetration, and this scale is arranged so that soft materials with deep penetration give low hardness numbers.

A variety of combinations of indenter and major load are possible; the most commonly used are R_B using as indenter a 1/16-in ball and a major load of 100 kg and R_C using a Brale as indenter and a major load of 150 kg (see "Rockwell Hardness and Rockwell Superficial Hardness of Metallic Materials," ASTM).

Compared with the Brinell test, the Rockwell method makes a smaller indentation, may be used on thinner material, and is more rapid since hardness numbers are read directly and need not be calculated. However, the Brinell test may be made without special apparatus and is somewhat more widely recognized for laboratory use. There is also a **Rockwell superficial hardness test** similar to the regular Rockwell, except that the indentation is much shallower.

The **Vickers** method of hardness testing is similar in principle to the Brinell in that it expresses the result in terms of the pressure under the indenter and uses the same units, kilograms per square millimetre. The indenter is a diamond in the form of a square pyramid with an apical angle of 136°, the loads are much lighter, varying between 1 and 120 kg, and the impression is measured by means of a medium-power compound microscope.

$$V = P/(0.5393d^2)$$

where V is the Vickers hardness number, sometimes called the **diamond-pyramid hardness** (DPH); P the imposed load, kg; and d the diagonal of indentation, mm. The Vickers method is more flexible and is considered to be more accurate than either the Brinell or the Rockwell, but the equipment is more expensive than either of the others and the Rockwell is somewhat faster in production work.

Among the other hardness methods may be mentioned the **Scleroscope,** in which a diamond-tipped "hammer" is dropped on the surface and the rebound taken as an index of hardness. This type of apparatus is seriously affected by the resilience as well as the hardness of the material and has largely been superseded by other methods. In the **Monotron** method, a penetrator is forced into the material to a predetermined depth and the load required is taken as the indirect measure of the hardness. This is the reverse of the Rockwell method in principle, but the loads and indentations are smaller than those of the latter. In the **Herbert pendulum,** a 1-mm steel or jewel ball resting on the surface to be tested acts as the fulcrum for a 4-kg compound pendulum of 10-s period. The swinging of the pendulum causes a rolling indentation in the material, and from the behavior of the pendulum several factors in hardness, such as **work hardenability,** may be determined which are not revealed by other methods. Although the Herbert results are of considerable significance, the instrument is suitable for laboratory use only (see Herbert, The Pendulum Hardness Tester, and Some Recent Developments in Hardness Testing, *Engineer,* **135,** 1923, pp. 390, 686). In the **Herbert cloudburst** test, a shower of steel balls, dropped from a predetermined height, dulls the surface of a hardened part in proportion to its softness and thus reveals defective areas. A variety of **mutual indentation methods,** in which crossed cylinders or prisms of the material to be tested are forced together, give results comparable with the Brinell test. These are particularly useful on wires and on materials at high temperatures.

The relation among the scales of the various hardness methods is not exact, since no two measure exactly the same sort of hardness, and a relationship determined on steels of different hardnesses will be found only approximately true with other materials. The **Vickers-Brinell relation** is nearly linear up to at least 400, with the Vickers approximately 5 percent higher than the Brinell (actual values run from +2 to +11 percent) and nearly independent of the material. Beyond 500, the values become more widely divergent owing to the flattening of the Brinell ball. The **Brinell-Rockwell relation** is fairly satisfactory and is shown in Fig. 3.1.23. Approximate relations for the **Shore Scleroscope** are also given on the same plot.

The **hardness of wood** is defined by the ASTM as the load in pounds required to force a ball 0.444 in in diameter into the wood to a depth of 0.222 in, the speed of penetration being 1/4 in/min. For a summary of the work in hardness see Williams, "Hardness and Hardness Measurements," ASM.

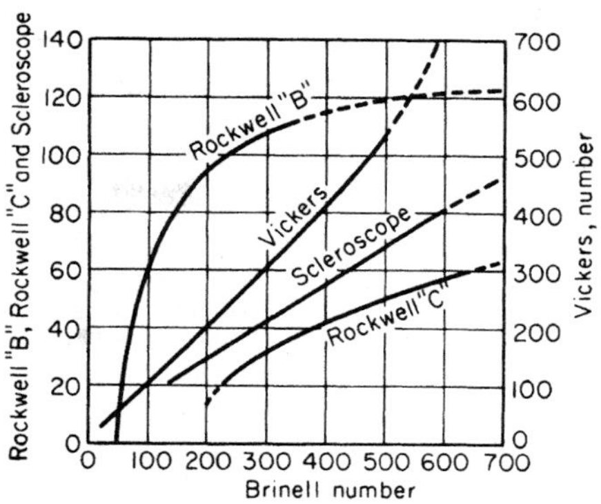

Figure 3.1.23 Hardness scales.

TESTING OF MATERIALS

Testing Machines Machines for the mechanical testing of materials usually contain elements (1) for gripping the specimen, (2) for deforming it, and (3) for measuring the load required in performing the deformation. Some machines (ductility testers) omit the measurement of load and substitute a measurement of deformation, whereas other machines include the measurement of both load and deformation through apparatus either integral with the testing machine (stress-strain recorders) or auxiliary to it (strain gages). In most general-purpose testing machines, the deformation is controlled as the independent variable and the resulting load measured, and in many special-purpose machines, particularly those for light loads, the load is controlled and the resulting deformation is measured. Special features may include those for constant rate of loading (pacing disks), for constant rate of straining, for constant load maintenance, and for cyclical load variation (fatigue).

In modern testing systems, the load and deformation measurements are made with load-and-deformation-sensitive transducers which generate electrical outputs. These outputs are converted to load and deformation readings by means of appropriate electronic circuitry. The readings are commonly displayed automatically on a recorder chart or digital meter, or they are read into a computer. The transducer outputs are typically used also as feedback signals to control the test mode (constant loading, constant extension, or constant strain rate). The load transducer is usually a load cell attached to the test machine frame, with electrical output to a bridge circuit and amplifier. The load cell operation depends on change of electrical resistivity with deformation (and load) in the transducer element. The deformation transducer is generally an extensometer clipped on to the test specimen gage length, and operates on the same principle as the load cell transducer: the change in electrical resistance in the specimen gage length is sensed as the specimen deforms. Optical extensometers are also available which do not make physical contact with the specimen. Verification and classification of extensometers is controlled by ASTM Standards. The application of load and deformation to the specimen is usually by means of a screw-driven mechanism, but it may also be applied by means of hydraulic and servohydraulic systems. In each case the load application system responds to control inputs from the load and deformation transducers. Important features in test machine design are the methods used for reducing friction, wear, and backlash. In older testing machines, test loads were determined from the machine itself (e.g., a pressure reading from the machine hydraulic pressure) so that machine friction made an important contribution to inaccuracy. The use of machine-independent transducers in modern testing has eliminated much of this source of error.

Grips should not only hold the test specimen against slippage but should also apply the load in the desired manner. Centering of the load is of great importance in compression testing, and should not be neglected in tension testing if the material is brittle. Figure 3.1.24 shows the theoretical errors due to off-center loading; the results are directly applicable to compression tests using swivel loading blocks. Swivel (ball-and-socket) holders or compression blocks should be used with all except the most ductile materials, and in compression testing of brittle materials (concrete, stone, brick), any rough faces should be smoothly capped with plaster of paris and one-third Portland cement. Serrated grips may be used to hold ductile materials or the shanks of other holders in tension; a taper of 1 in 6 on the wedge faces gives a self-tightening action without excessive jamming. Ropes are ordinarily held by wet eye splices, but braided ropes or small cords may be given several turns over a fixed pin and then clamped. Wire ropes should be zinced into forged sockets (solder and lead have insufficient strength). Grip selection for tensile testing is described in ASTM standards.

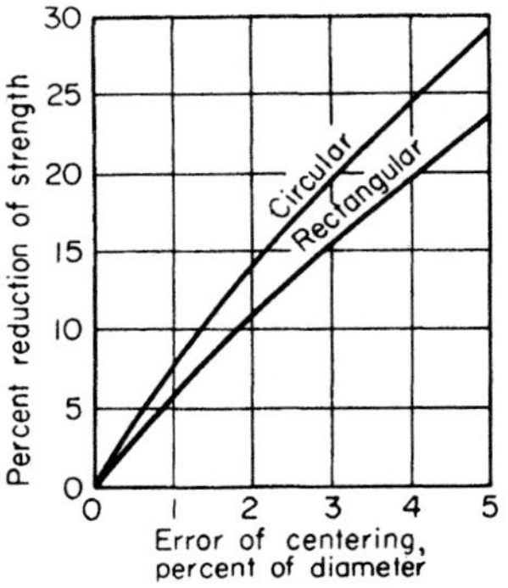

Figure 3.1.24 Effect of centering errors on brittle test specimens.

Accuracy and Calibration ASTM standards require that commercial machines have errors of less than 1 percent within the "loading range" when checked against acceptable standards of comparison at minimum five suitably spaced loads. The "loading range" may be any range through which the preceding requirements for accuracy are satisfied, except that it shall not extend below 100 times the least load to which the machine will respond or which can be read on the indicator. The use of calibration plots or tables to correct the results of an otherwise inaccurate machine is not permitted under any circumstances. Machines with errors less than 0.1 percent are commercially available (Tate-Emery and others), and somewhat greater accuracy is possible in the most refined research apparatus.

Dead loads may be used to check machines of low capacity; accurately calibrated proving levers may be used to extend the range of available weights. Various elastic devices (such as the Morehouse proving ring) made of specially treated steel, with sensitive distortion-measuring devices, and calibrated by dead weights at the NIST (formerly Bureau of Standards) are among the most satisfactory means of checking the higher loads.

Two **standard forms of test specimens** (ASTM) are shown in Figs. 3.1.25 and 3.1.26. In wrought materials, and particularly in those which have been cold-worked, different properties may be expected in different directions with respect to the direction of the applied work, and the test specimens, often called **coupons,** should be cut out from the parent material in such a way to give the strength in the desired direction. With the exception of fatigue specimens and specimens of extremely brittle materials, surface finish is of little practical importance, although extreme roughness tends to decrease the ultimate elongation.

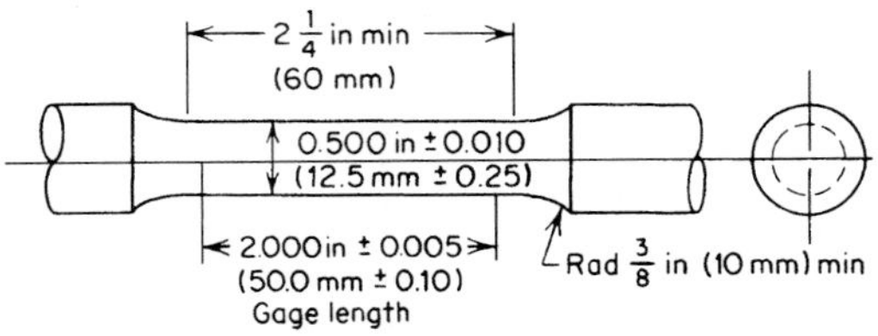

Figure 3.1.25 Test specimen, 2-in (50-mm) gage length, ½-in (12.5-mm) diameter. Others available for 0.35-in (8.75-mm) and 0.25-in (6.25-mm) diameters. *(ASTM.)*

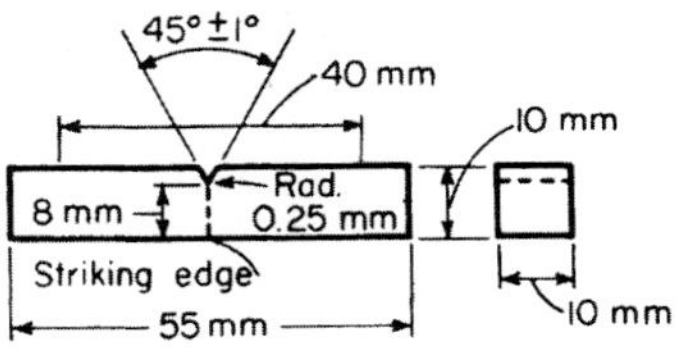

Figure 3.1.26 Charpy V-notch impact specimens. *(ASTM.)*

3.2 MECHANICS OF MATERIALS

by Ali M. Sadegh

REFERENCES: Timoshenko and MacCullough, "Elements of Strength of Materials," Van Nostrand. Seeley, "Advanced Mechanics of Materials," Wiley. Timoshenko and Goodier, "Theory of Elasticity," McGraw-Hill. Phillips, "Introduction to Plasticity," Ronald. Van Den Broek, "Theory of Limit Design," Wiley. Hetényi, "Handbook of Experimental Stress Analysis," Wiley. Dean and Douglas, "Semi-Conductor and Conventional Strain Gages," Academic. Robertson and Harvey, "The Engineering Uses of Holography," University Printing House, London. Sellers, "Basic Training Guide to the New Metrics and SI Units," National Tool, Die and Precision Machining Association. Roark and Young, "Formulas for Stress and Strain," McGraw-Hill. Perry and Lissner, "The Strain Gage Primer," McGraw-Hill. Donnell, "Beams, Plates, and Sheets," Engineering Societies Monographs, McGraw-Hill. Griffel, "Beam Formulas" and "Plate Formulas," Ungar. Durelli et al., "Introduction to the Theoretical and Experimental Analysis of Stress and Strain," McGraw-Hill. "Stress Analysis Manual," Department of Commerce, Pub. no. AD 759 199, 1969. Blodgett, "Welded Structures," Lincoln Arc Welding Foundation. "Characteristics and Applications of Resistance Strain Gages," Department of Commerce, *NBS Circ.* 528, 1954.

NOTE: The almost universal availability and utilization of computers in engineering practice has led to the development of many forms of software individually tailored to the solution of specific design problems in the area of mechanics of materials. Their use will permit the reader to amplify and supplement a good portion of the formulary and tabular collection in this section, as well as utilize those powerful computational tools in newer and more powerful techniques to facilitate solutions to problems. Many of the approximate methods, involving laborious iterative mathematical schemes, have been supplanted by the computer. Developments along those lines continue apace and bid fair to expand the types of problems handled, all with greater confidence in the results obtained thereby.

Main Symbols

UNIT STRESS

S = apparent stress

S_v or S_s = pure shearing

T = true (ideal) stress

S_p = proportional elastic limit

S_y = yield point

S_M = ultimate strength, tension

S_c = ultimate compression

S_v = vertical shear in beams
S_R = modulus of rupture

MOMENT

M = bending
M_t = torsion

EXTERNAL ACTION

P = force
G = weight of body
W = weight of load
V = external shear

MODULUS OF ELASTICITY

E = longitudinal
G = shearing
K = bulk
U_p = modulus of resilience
U_R = ultimate resilience

GEOMETRIC

l = length
A = area
V = volume
v = velocity
r = radius of gyration
I = rectangular moment of inertia
I_P or J = polar moment of inertia

DEFORMATION

e, e' = gross deformation
ε, ε' = unit deformation; strain
d or α = unit, angular
s' = unit, lateral
μ = Poisson's ratio
n = reciprocal of Poisson's ratio
r = radius
f = deflection

SIMPLE STRESSES AND STRAINS

Deformations are changes in form produced by external forces or loads that act on nonrigid bodies. Deformations are **longitudinal,** e, a lengthening (+) or shortening (−) of the body; and **angular,** α, a change of angle between the faces.

Unit deformation (strain) (dimensionless number) is the deformation per unit distance. Unit longitudinal deformation (longitudinal strain), $\varepsilon = e/l$ (Fig. 3.2.1). Unit angular-deformation tan α approximately equals alpha (Fig. 3.2.2).

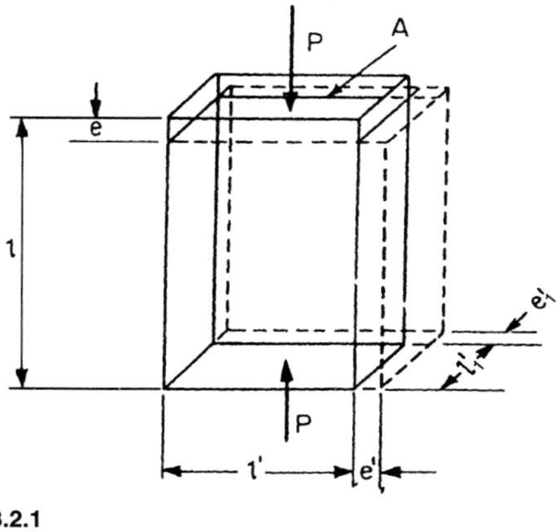

Figure 3.2.1

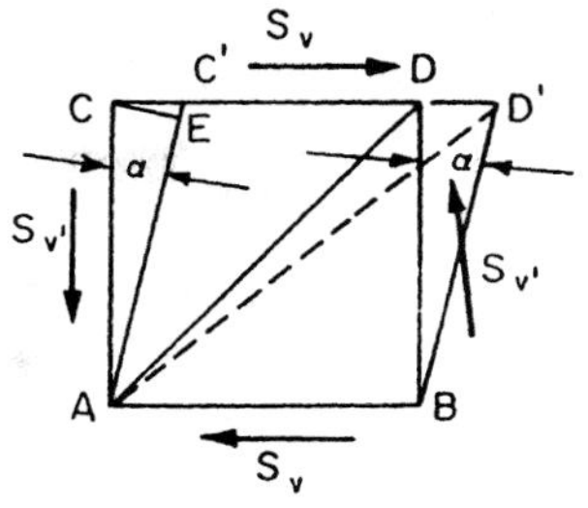

Figure 3.2.2

The accompanying lateral deformation results in unit lateral deformation (lateral strain) $\varepsilon' = e'/l'$ (Fig. 3.2.1). For homogeneous, isotropic material operating in the elastic region, the ratio $-\varepsilon'/\varepsilon$ is a constant and is a definite property of the material; this ratio is called **Poisson's ratio** μ.

A fundamental relation among the **three interdependent constants** E, G, and μ for a given material is $E = 2G(1 + \mu)$. Note that μ is always less than 0.5; thus the shearing modulus G is always smaller than the elastic modulus E. At the extremes, for example, $\mu \approx 0.5$ for rubber and paraffin; $\mu \approx 0$ for cork. For concrete, μ varies from 0.10 to 0.20 at working stresses and can reach 0.25 at higher stresses; μ for ordinary glass is about 0.25. In the absence of definitive data, μ for most structural metals can be taken to lie between 0.25 and 0.35. Extensive listings of Poisson's ratio are found in other sections; see Tables 3.1.3 and 4.1.9.

Stress is an internal distributed force, or, force per unit area; it is the internal mechanical reaction of the material accompanying deformation. Stresses always occur in pairs. Stresses are **normal** [tensile stress (+) and compressive stress (−)]; and **tangential,** or **shearing.**

Intensity of stress, or **unit stress,** S, lb/in^2 (kgf/cm^2), is the amount of force per unit of area (Fig. 3.2.3). P is the load acting through the center of gravity of the area. The uniformly distributed normal stress is

$$S = P/A$$

When the stress is not uniformly distributed, $S = dP/dA$.

A **long rod** will stretch under its own weight Q and a terminal load P (see Fig. 3.2.4). The total elongation e is that due to the terminal load plus that due to one-half the weight of the rod considered as acting at the end.

$$e = (Pl + Ql/2)/(AE)$$

The maximum stress is at the upper end.

When a **load** is **carried by several paths** to a support, the different paths take portions of the load in proportion to their stiffness, which is controlled by material (E) and by design.

EXAMPLE. Two pairs of bars rigidly connected (with the same elongation) carry a load P_0 (Fig. 3.2.5). A_1, A_2 and E_1, E_2 and P_1, P_2 and S_1, S_2 are cross

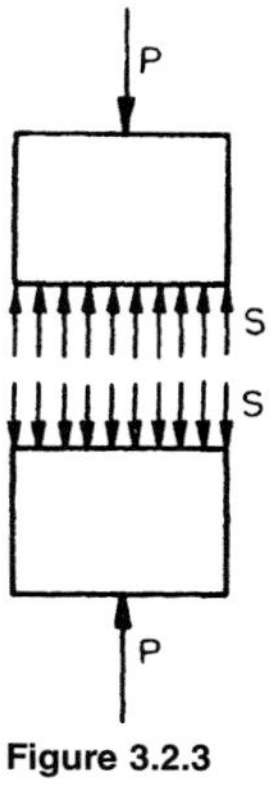

Figure 3.2.3

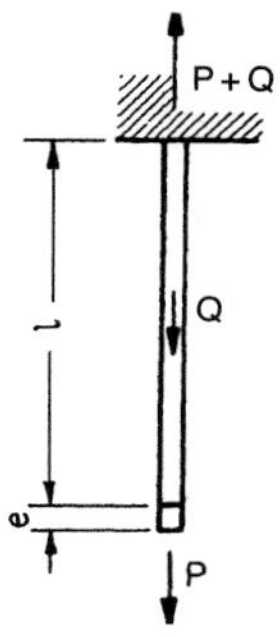

Figure 3.2.4

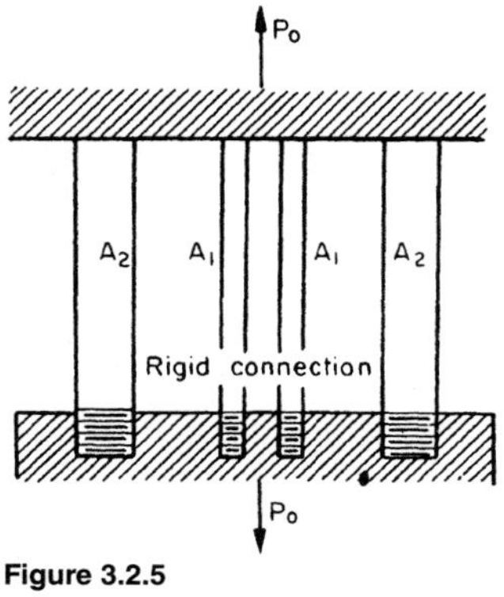

Figure 3.2.5

sections, moduli of elasticity, loads, and stresses of the bars, respectively; e = elongation.

$$e = P_1 l/(E_1 A_1) = P_2 l/(E_2 A_2)$$

$$P_0 = 2P_1 + 2P_2$$

$$S_2 = P_2/A_2 = \tfrac{1}{2}[P_0 E_2/(E_1 A_1 + E_2 A_2)]$$

$$S_1 = \tfrac{1}{2}[P_0 E_1/(E_1 A_1 + E_2 A_2)]$$

Shearing stresses (Fig. 3.2.2) act tangentially to surface of contact and do not change length of sides of elementary volume; they **change the angle between faces** and the **length of diagonal.** Two pairs of shearing stresses must act together. **Shearing stress intensities are of equal magnitude on all four faces of an element.** $S_v = S_v'$ (Fig. 3.2.6).

In the presence of **pure shear** on external faces (Fig. 3.2.6), the **resultant stress** S on one diagonal plane at 45° is pure tension and on the other diagonal plane pure compression; $S = S_v = S_v'$. Failure under pure shear is difficult to produce experimentally, except under torsion and in certain special cases. Figure 3.2.7 shows an ideal case, and Fig. 3.2.8 a common form of test piece that introduces bending stresses.

Let Fig. 3.2.9 represent the symmetric section of area A with a shearing force V acting through its centroid. If pure shear exists, $S_v = V/A$, and this shear would be uniformly distributed over the area A. When this **shear** is **accompanied by bending (transverse shear in beams),** the unit shear S_v increases from the extreme fiber to its maximum, which may or may not be at the neutral axis OZ. The unit shear parallel to OZ at a point d distant from the neutral axis (Fig. 3.2.9) is

$$S_v = \frac{V}{Ib}\int_d^e yz\,dy$$

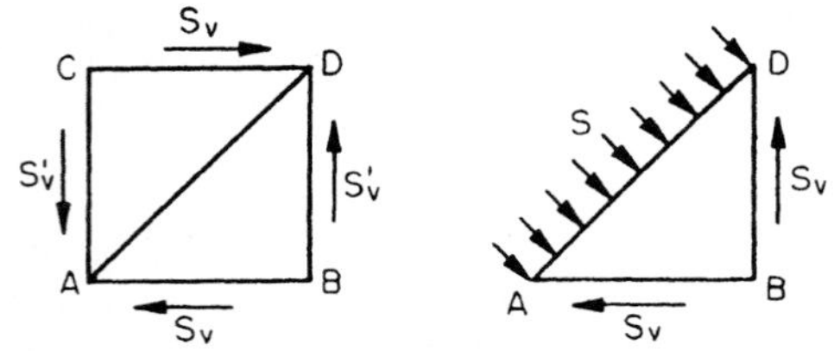

Figure 3.2.6

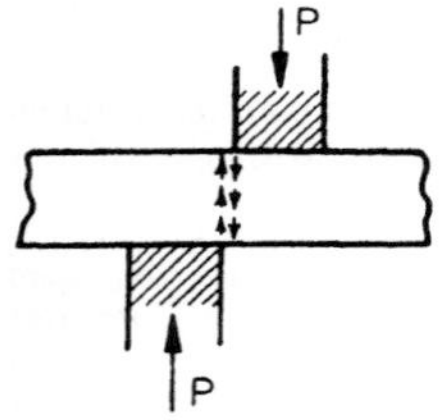

Figure 3.2.7

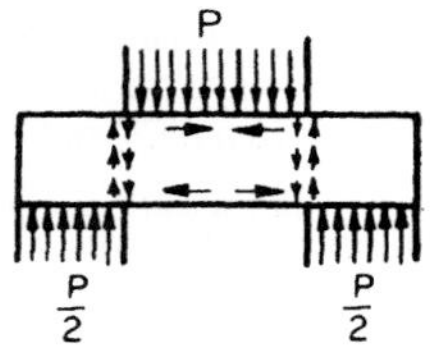

Figure 3.2.8

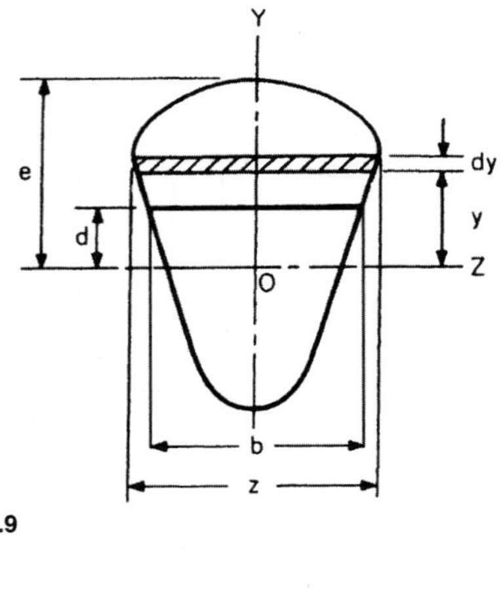

Figure 3.2.9

where z = the section width at distance y; and I is the moment of inertia of the **entire** section about the neutral axis OZ. Note that $\int_d^e yz\,dy$ is the first moment of the area above d with respect to axis OZ. For a **rectangular cross section** (Fig. 3.2.10*a*),

$$S_v = \frac{3}{2}\frac{V}{bh}\left[1 - \left(\frac{2y}{h}\right)^2\right]$$

$$S_v(\max) = \frac{3}{2}\frac{V}{bh} = \frac{3}{2}\frac{V}{A} \qquad \text{for } y = 0$$

For a **circular cross section** (Fig. 3.2.10*b*),

$$S_v = \frac{4}{3}\frac{V}{\pi r^2}\left[1 - \left(\frac{y}{r}\right)^2\right]$$

$$S_v(\max) = \frac{4}{3}\frac{V}{\pi r^2} = \frac{4}{3}\frac{V}{A} \qquad \text{for } y = 0$$

For a **circular ring** (thickness small in comparison with the major diameter), $S_v(\max) = 2V/A$, for $y = 0$.

For a **square cross section** (diagonal vertical, Fig. 3.2.10*c*),

$$S_v = \frac{V\sqrt{2}}{a^2}\left[1 + \frac{y\sqrt{2}}{a} - 4\left(\frac{y}{2}\right)^2\right]$$

$$S_v(\max) = 1.591\frac{V}{A} \qquad \text{for } y = \frac{e}{4}$$

For an **I-shaped cross section** (Fig. 3.2.10*d*),

$$S_v(\max) = \frac{3}{4}\frac{V}{a}\left[\frac{be^2 - (b-a)f^2}{be^3 - (b-a)f^3}\right] \qquad \text{for } y = 0$$

Consider a beam of uniform cross section that is subjected to a bending moment M. **Bending stress,** or the flexure formula, of the beam is $S = -My/I$, where y is the distance from the neutral axis and I is the moment of inertia of the cross section. For a circular cross section beam that is subjected to **torsion** M_t, pure shear stress in the beam is $S_s = M_t r/J$, where r is the radius of the cross section.

Elasticity is the ability of a material to return to its original dimensions after the removal of stresses. The **elastic limit** S_p is the limit of stress within which the deformation completely disappears after the removal of stress; that is, no set remains.

Hooke's law states that, within the elastic limit, deformation produced is proportional to the stress. Unless modified, the deduced formulas of mechanics apply only within the elastic limit. Beyond this, they are modified by experimental coefficients, as, for instance, the modulus of rupture.

The modulus of elasticity, lb/in^2 (kgf/cm^2), is the ratio of the increment of unit stress to increment of unit deformation within the elastic limit.

The modulus of elasticity in tension, or **Young's modulus,**

$$E = \text{unit stress/unit deformation} = Pl/(Ae)$$

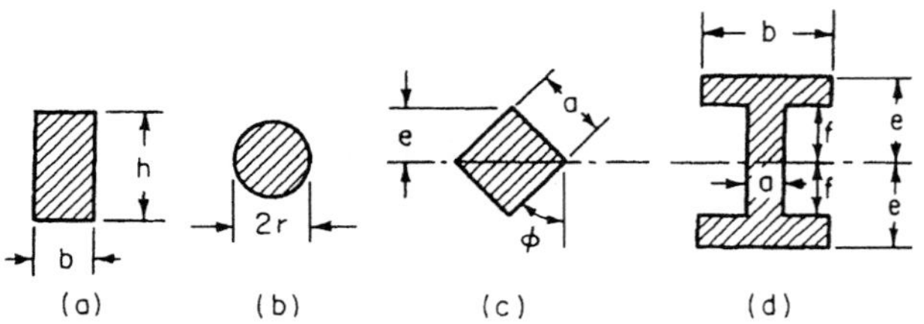

Figure 3.2.10

The modulus of elasticity in compression is similarly measured.

The **modulus of elasticity in shear** or **coefficient of rigidity,** $G = S_v/\alpha$ where α is expressed in radians (see Fig. 3.2.2).

The **bulk modulus of elasticity** K is the ratio of normal stress, applied to all six faces of a cube, to the change of volume.

Change of volume under normal stress is so small that it is rarely of significance. For example, given a body with length l, width b, thickness d, Poisson's ratio μ, and longitudinal strain ε, $V = lbd$ = original volume. The deformed volume = $(1 + \varepsilon)l(1 - \mu\varepsilon)b(1 - \mu\varepsilon)d$. Neglecting powers of ε, the deformed volume = $(1 + \varepsilon - 2\mu\varepsilon)V$. The change in volume is $\varepsilon(1 - 2\mu)V$; the **unit volumetric strain** is $\varepsilon(1 - 2\mu)$. Thus, a steel rod ($\mu = 0.3$, $E = 30 \times 10^6$ lb/in^2) compressed to a stress of 30,000 lb/in^2 will experience $\varepsilon = 0.001$ and a unit volumetric strain of 0.0004, or 1 part in 2,500.

The following relationships exist between the modulus of elasticity in tension or compression E, modulus of elasticity in shear G, bulk modulus of elasticity K, and Poisson's ratio μ:

$$E = 2G(1+\mu)$$
$$G = E/[2(1+\mu)]$$
$$\mu = (E-2G)/(2G)$$
$$K = E/[3(1-2\mu)]$$
$$\mu = (3K-E)/(6K)$$

Resilience U (in · lb)[(cm · kgf)] is the potential energy stored up in a deformed body. The amount of resilience is equal to the work required to deform the body from zero stress to stress S, when S does not exceed the elastic limit. For normal stress, resilience = work of deformation = average force times deformation $= \frac{1}{2} Pe = \frac{1}{2} AS \times Sl/E = \frac{1}{2}\, S^2V/E$.

Modulus of resilience U_p (in · lb/in^3) [(cm · kgf/cm^3)], or **unit resilience,** is the **elastic energy** stored up in a cubic inch of material at the elastic limit. For normal stress,

$$U_p = \tfrac{1}{2} S_p^2/E$$

The unit resilience for any other kind of stress, as shearing, bending, torsion, is a constant times one-half the square of the stress divided by the appropriate modulus of elasticity. For values, see Table 3.2.1.

Unit rupture work U_R, sometimes called **ultimate resilience,** is measured by the area of the stress-deformation diagram to rupture.

$$U_R = \tfrac{1}{3} e_u (S_y + 2S_M) \quad \text{(approx)}$$

where e_u is the total deformation at rupture.

For structural steel, $U_R = \frac{1}{3} \times \frac{27}{100} \times [35{,}000 + (2 \times 60{,}000)] = 13{,}950$ in · lb/in^3 (982 cm · kgf/cm^3).

EXAMPLE 1. A load P = 40,000 lb compresses a wooden block of cross-sectional area A = 10 in^2 and length = 10 in, an amount $e = \frac{4}{100}$ in. Stress $S = \frac{1}{10} \times 40{,}000 = 4{,}000$ lb/in^2. Unit elongation $s = \frac{4}{100} \times 10 = \frac{1}{250}$. Modulus of elasticity $E = 4{,}000 \times \frac{1}{250} = 1{,}000{,}000$ lb/in^2. Unit resilience $U_p = \frac{1}{2} \times 4{,}000 \times 4{,}000/1{,}000{,}000 = 8$ in · lb/in^3 (0.563 cm · kgf/cm^3).

EXAMPLE 2. A weight G = 5,000 lb falls through a height h = 2 ft; V = number of cubic inches required to absorb the shock without exceeding a stress of 4,000 lb/in^2. Neglect compression of block. Work done by falling weight = Gh = 5,000 × 2 × 12 in · lb (2,271 × 61 cm · kgf) Resilience of block = V × 8 in · lb = 5,000 × 2 × 12. Therefore, V = 15,000 in^3 (245,850 cm^3).

Thermal Stresses A bar will change its length when its temperature is raised (or lowered) by the amount $\Delta l_0 = \alpha l_0(t_2 - 32)$. The linear **coefficient of thermal expansion** α is assumed constant at normal temperatures and l_0 is the length at 32°F (273.2 K, 0°C). If this expansion (or contraction) is prevented, a **thermal-time stress** is developed, equal to $S = E\alpha(t_2 - t_1)$, as the temperature goes from t_1 to t_2. In thin flat plates the stress becomes $S = E\alpha(t_2 - t_1)/(1 - \mu)$; μ is Poisson's ratio. Such stresses can occur in castings containing large and small sections. Similar stresses also occur when heat flows through members because of the difference in temperature between one point and another. The heat flowing across a length b as a result of a linear drop in temperature Δt equals $Q = kA\ \Delta t/b$ Btu/h (cal/h). The thermal conductivity k is in Btu/(h)(ft^2)(°F)/(in of thickness) [cal/(h)(m^2)(k)/(m)]. The **thermal-flow stress** is then $S = E\alpha Qb/(kA)$. Note, when Q is substituted the stress becomes $S = E\alpha\ \Delta t$ as above, only t is now a function of distance rather than time. For coefficient of thermal expansion α, see Sec. 2.

EXAMPLE. A cast-iron plate 3 ft square and 2 in thick is used as a fire wall. The temperature is 330°F on the hot side and 160°F on the other. What is the thermal-flow stress developed across the plate?

$$S = E\alpha\ \Delta t = 13\times10^6 \times 6.5\times10^{-6} \times 170$$
$$= 14{,}360 \text{ lb/in}^2 \ (1{,}010 \text{ kgf/cm}^2)$$

or $$Q = 2.3\times9\times170/2 = 1{,}760 \text{ Btu/h}$$

and $$S = 13\times10^6 \times 6.5\times10^{-6} \times 1{,}760\times2/2.3\times9$$
$$= 14{,}360 \text{ lb/in}^2 \ (1{,}010 \text{ kgf/cm}^2)$$

COMBINED STRESSES

In the discussion that follows, the element is subjected to stresses lying in one plane; this is the case of **plane stress,** or **two-dimensional stress.**

Simple stresses, defined as such by the flexure and torsion theories, lie in planes normal or parallel to the line of action of the forces. Normal, as well as shearing, stresses may, however, exist in other directions. A particle out of a loaded member will contain normal and

Table 3.2.1 Resilience per Unit of Volume U_p

Tension or compression	$\frac{1}{2} S^2/E$	Torsion	
Shear	$\frac{1}{2} S_v^2/G$	Solid circular	$\frac{1}{4} S_v^2/G$
Beams (free ends)			
Rectangular section, bent in arc of circle; no shear	$\frac{1}{6} S^2/E$	Hollow, radii R_1 and R_2	$\frac{R_1^2+R_2^2}{R_1^2}\frac{1}{4}\frac{S_v^2}{G}$
Ditto, circular section	$\frac{1}{8} S^2/E$	Springs	
Concentrated center load; rectangular cross section	$\frac{1}{18} S^2/E$	Carriage	$\frac{1}{6} S^2/E$
		Flat spiral, rectangular section	$\frac{1}{24} S^2/E$
Ditto, circular cross section	$\frac{1}{24} S^2/E$	Helical: axial load, circular wire	$\frac{1}{4} S_v^2/G$
Uniform load, rectangular cross section	$\frac{5}{36} S^2/E$	Helical: axial twist	$\frac{1}{8} S^2/E$
1-beam section, concentrated center load	$\frac{3}{32} S^2/E$	Helical: axial twist, rectangular section	$\frac{1}{6} S^2/E$

(S = longitudinal stress; S_v = shearing stress; E = tension modulus of elasticity; G = shearing modulus of elasticity)

shearing stresses as shown in Fig. 3.2.11. Note that the four shearing stresses must be of the same magnitude, if equilibrium is to be satisfied.

If the particle is "cut" along the plane *AA*, equilibrium will reveal that, in general, normal as well as shearing stresses act upon the plane *AC* (Fig. 3.2.12). The normal stress on plane *AC* is labeled S_n, and shearing S_s. The application of equilibrium yields

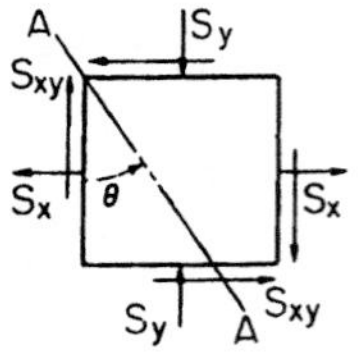

Figure 3.2.11

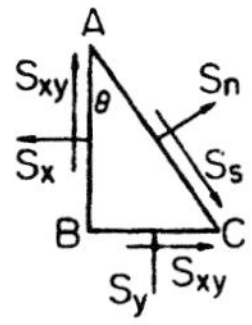

Figure 3.2.12

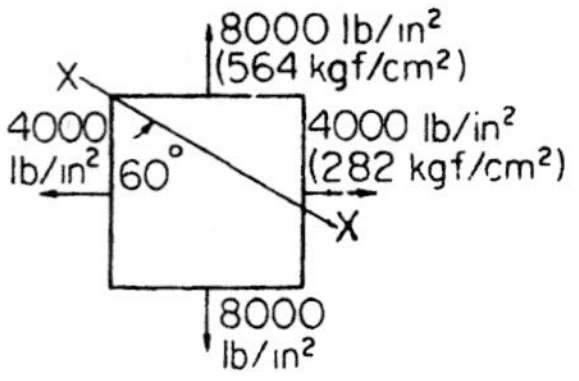

Figure 3.2.13

$$S_n = \frac{S_x + S_y}{2} + \frac{S_x - S_y}{2} \cos 2\theta + S_{xy} \sin 2\theta$$

and
$$S_s = \frac{S_x - S_y}{2} \sin 2\theta - S_{xy} \cos 2\theta$$

A *sign convention* must be used. A tensile stress is positive while compression is negative. A shearing stress is positive when directed as on plane *AB* of Fig. 3.2.12; that is, when the shearing stresses on the vertical planes form a clockwise couple, the stress is positive.

The planes defined by $\tan 2\theta = 2S_{xy}/S_x - S_y$, the *principal planes,* contain the **principal stresses**—the maximum and minimum normal stresses. These stresses are

$$S_M, S_m = \frac{S_x + S_y}{2} \pm \sqrt{\left(\frac{S_x - S_y}{2}\right)^2 + S_{xy}^2}$$

The maximum and minimum shearing stresses are represented by the quantity

$$S_{s\,M,m} = \pm \sqrt{\left(\frac{S_x - S_y}{2}\right)^2 + S_{xy}^2}$$

and they act on the planes defined by

$$\tan 2\theta = -\frac{S_x - S_y}{2S_{xy}}$$

EXAMPLE. The steam in a boiler subjects a particular particle on the outer surface of the boiler shell to a circumferential stress of 8,000 lb/in² and a longitudinal stress of 4,000 lb/in² as shown in Fig. 3.2.13. Find the stresses acting on the plane *XX*, making an angle of 60° with the direction of the 8,000 lb/in² stress. Find the principal stresses and locate the principal planes. Also find the maximum and minimum shearing stresses.

$$60^\circ\ S_n = \frac{4{,}000 + 8{,}000}{2} + \frac{4{,}000 - 8{,}000}{2}(-0.5000) + 0$$
$$= 7{,}000 \text{ lb/in}^2$$

$$60^\circ\ S_s = \frac{4{,}000 - 8{,}000}{2}(0.8660) - 0 = -1{,}732 \text{ lb/in}^2$$

$$S_{M,m} = \frac{4{,}000 + 8{,}000}{2} \pm \sqrt{\left(\frac{4{,}000 - 8{,}000}{2}\right)^2 + 0}$$
$$= 6{,}000 \pm 2{,}000$$
$$= 8{,}000 \text{ and } 4{,}000 \text{ lb/in}^2 \ (564 \text{ and } 282 \text{ kgf/cm}^2)$$

$$\text{at } \tan 2\theta = \frac{2 \times 0}{4{,}000 - 8{,}000} = 0 \quad \text{or} \quad \theta = 90^\circ \text{ and } 0^\circ$$

$$S_{s\,M,m} = \pm \sqrt{\left(\frac{4{,}000 - 8{,}000}{2}\right)^2 + 0}$$
$$= \pm 2{,}000 \text{ lb/in}^2 \ (\pm 141 \text{ kgf/cm}^2)$$

Mohr's Stress Circle The biaxial stress field with its combined stresses can be represented graphically by the Mohr stress circle. For instance, for the particle given in Fig. 3.2.11, Mohr's circle is as shown in Fig. 3.2.14. The stress *sign convention* previously defined must be adhered to. Furthermore, in order to locate the point (on Mohr's circle) that yields the stresses on a plane θ° from the vertical side of the particle (such as plane *AA* in Fig. 3.2.11), 2θ° must be laid off in the same direction from the radius to (S_x, S_{xy}). For the previous example, Mohr's circle becomes Fig. 3.2.15.

Eight special stress fields are shown in Figs. 3.2.16 to 3.2.23, along with Mohr's circle for each.

Combined Loading **Combined flexure and torsion** arise, for instance, when a shaft twisted by a torque M_t is bent by forces produced by belts or gears. An element on the surface, such as *ABCD* on the shaft of Fig. 3.2.24, is subjected to a **flexure stress** $S_x = Mc/I = 8Fl(\pi d^3)$ and a **torsional shearing stress** $S_{xy} = M_t c/J = 16M_t(\pi d^3)$. These stresses will induce combined stresses. The maximum combined stresses will be

$$S_n = \frac{1}{2}(S_x \pm \sqrt{S_x^2 + 4S_{xy}^2})$$

and
$$S_s = \pm\frac{1}{2}\sqrt{S_x^2 + 4S_{xy}^2}$$

The above situation applies to any case of normal stress with shear, as when a bolt is under both tension and shear. A beam subjected to both flexure and transverse shear is another case.

Combined torsion and longitudinal loads exist on a propeller shaft. An element on this shaft will contain a tensile stress computed using $S = F/A$ and a torsion shearing stress equal to $S_s = M_t c/J$. The free body of an element on the surface of a vertical turbine shaft is subjected to direct compression and torsion.

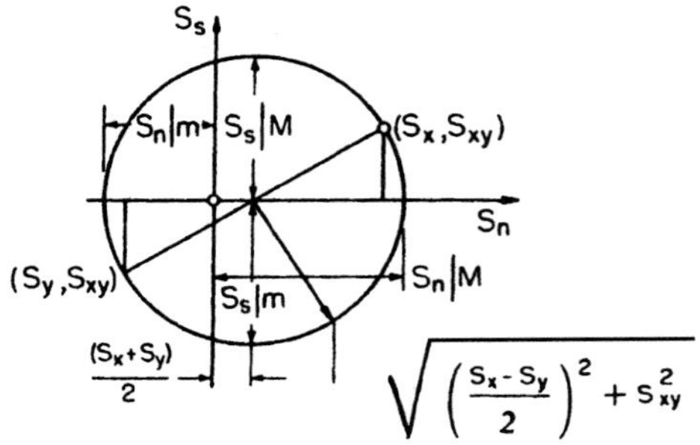

Figure 3.2.14

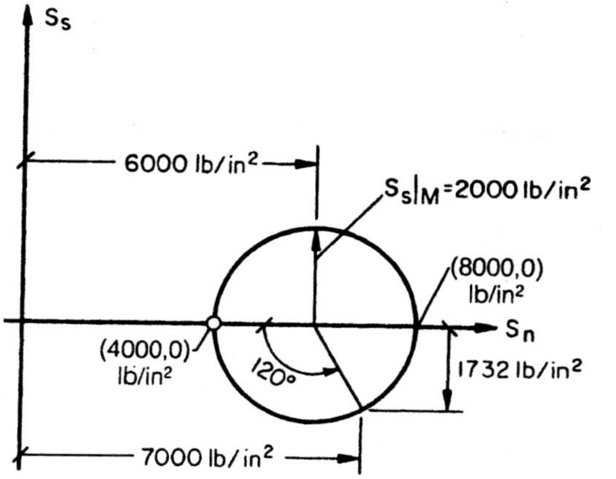

Figure 3.2.15

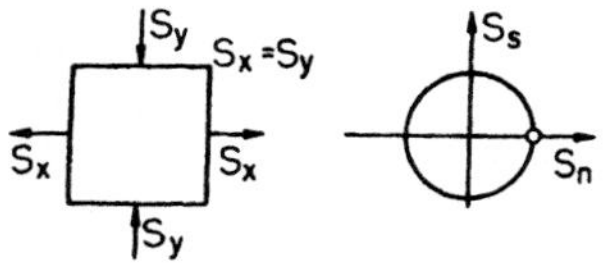

Figure 3.2.16

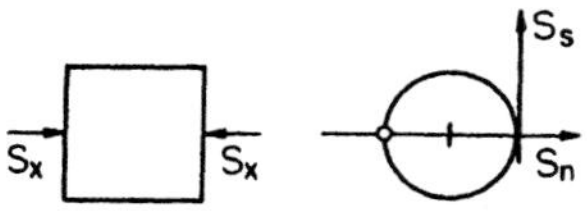

Figure 3.2.17

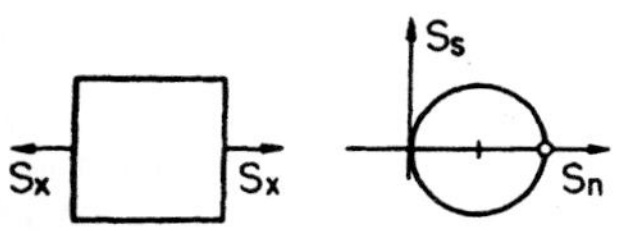

Figure 3.2.18

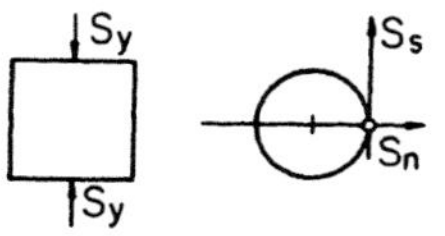

Figure 3.2.19

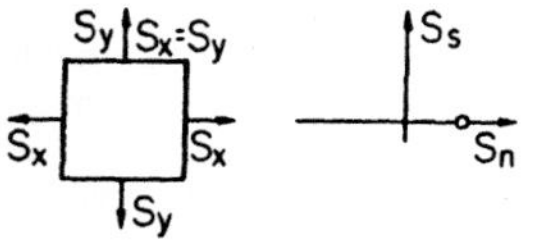

Figure 3.2.20

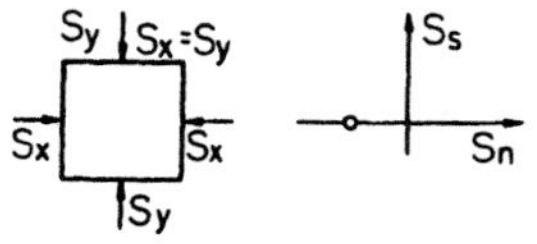

Figure 3.2.21

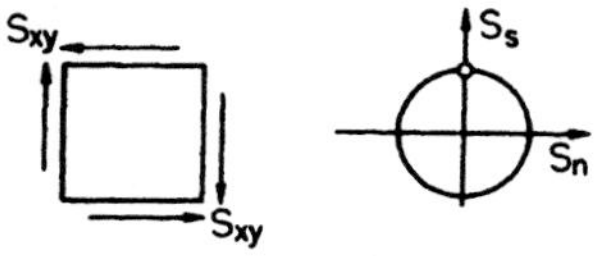

Figure 3.2.22

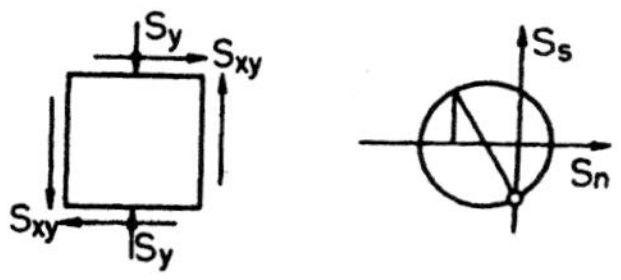

Figure 3.2.23

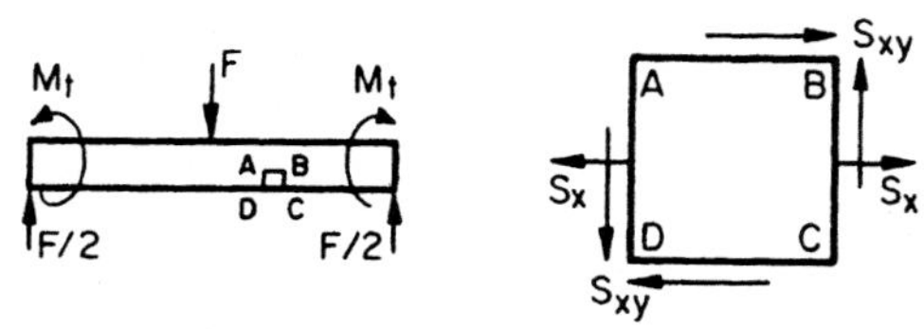

Figure 3.2.24

When combined loading results in stresses of the same type and direction, the addition is algebraic. Such a situation exists on an offset link like that of Fig. 3.2.25.

Mohr's Strain Circle Strain equations can also be derived for plane-strain fields. Strains e_x and e_y are the extensional strains (tension or compression) occurring at a point in two right-angle directions, and the change of the angle between them is γ_{xy}. The strain e at the point in any direction a at an angle θ with the x direction derives as

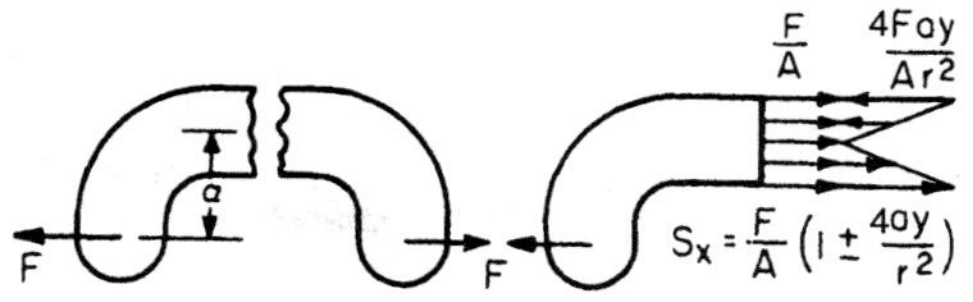

Figure 3.2.25

$$e_a = \frac{e_x + e_y}{2} + \frac{e_x - e_y}{2} \cos 2\theta + \frac{\gamma_{xy}}{2} \sin 2\theta$$

Similarly, the shearing strain γ_{ab} (change in the original right angle between directions a and b) is defined by

$$\gamma_{ab} = (e_x - e_y) \sin 2\theta + \gamma_{xy} \cos 2\theta$$

Inspection easily reveals that the above equations for e_a and γ_{ab} are mathematically identical to those for S_n and S_s. Thus, once a sign convention is established, a Mohr circle for strain can be constructed and used as the stress circle is used. The strain e is positive when an extension and negative when a contraction. If the direction associated with the first subscript a rotates counterclockwise during straining with respect to the direction indicated by the second subscript b, the shearing strain is positive; if clockwise, it is negative. In constructing the circle, positive extensional strains will be plotted to the right as abscissas and positive **half-shearing** strains will be plotted upward as ordinates.

For the strains shown in Fig. 3.2.26*a*, Mohr's strain circle becomes that shown in Fig. 3.2.26*b*. The extensional strain in the direction a, making an angle of θ_a with the x direction, is e_a, and the shearing strain is γ_{ab} counterclockwise. The strain 90° away is e_b. The maximum principal strain is e_M at an angle θ_M clockwise from the x direction. The other principal or minimum strain is e_m 90° away.

PLASTIC DESIGN

Early efforts in stress analysis were based on limit loads, that is, loads which stress a member "wholly" to the yield strength. Euler's famous paper on column action ("Sur la Force des Colonnes," Academie des Sciences de Berlin, 1757) deals with the column problem this way. More recently, the concept of limit loads, referred to as **limit,** or **plastic, design,** has found strong application in the design of certain structures. The theory presupposes a ductile material, absence of stress raisers, and fabrication free of embrittlement. Local load overstress is allowed, provided the structure does not deform appreciably.

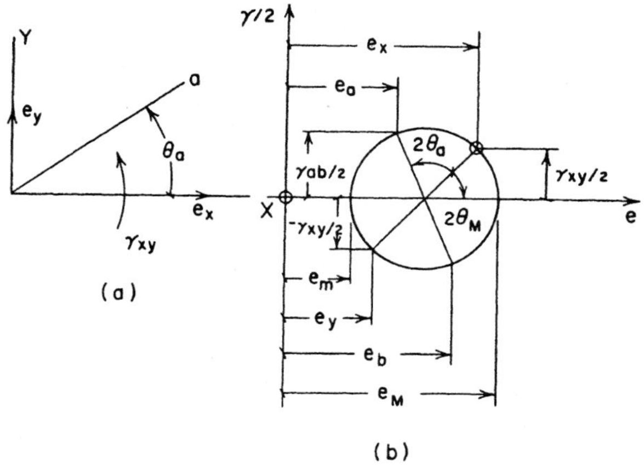

Figure 3.2.26

To visualize the limit-load approach, consider a simple beam of uniform section subjected to a concentrated load of midspan, as depicted in Fig. 3.2.27*a*. According to elastic theory, the outermost fiber on each side and at midspan—the section of maximum bending moment—will first reach the yield-strength value, across the depth of the beam, the stress distribution will, of course, follow the triangular pattern, becoming zero at the neutral axis. If the material is ductile, the stress in the outermost fibers will remain at the yield value until every other fiber reaches the same value as the load increases. Thus the stress distribution assumes the rectangular pattern before the *plastic hinge* forms and failure ensues.

The problem is that of finding the final limit load. Elastic-flexure theory gives the maximum load—triangular distribution—as

$$F_y = \frac{2S_y bh^2}{3l}$$

For the rectangular stress distribution, the limit load becomes

$$F_L = \frac{S_y bh^2}{l}$$

The ratio $F_L/F_y = 1.50$—an increase of 50 percent in load capability. The ratio F_L/F_y has been named **shape factor** (Jenssen, Plastic Design in Welded Structures Promises New Economy and Safety, *Welding Jour.,* Mar. 1959). See Fig. 3.2.27*b* for shape factors for some other sections. The shape factor may also be determined by dividing the first moment of area about the neutral axis by the section modulus.

A constant-section beam with both ends fixed, supporting a uniformly distributed load, illustrates another application of the plastic-load approach. The bending-moment diagram based on the elastic theory drawn in Fig. 3.2.28 (broken line) shows a moment at the center equal to one-half the moment at either end. A preferable situation, it might be argued, is one in which the moments are the same at the three stations—solid line. Thus, applying equilibrium to, say, the left half of the beam yields a bending moment at each of the three plastic hinges of

$$M_L = \frac{wl^2}{16}$$

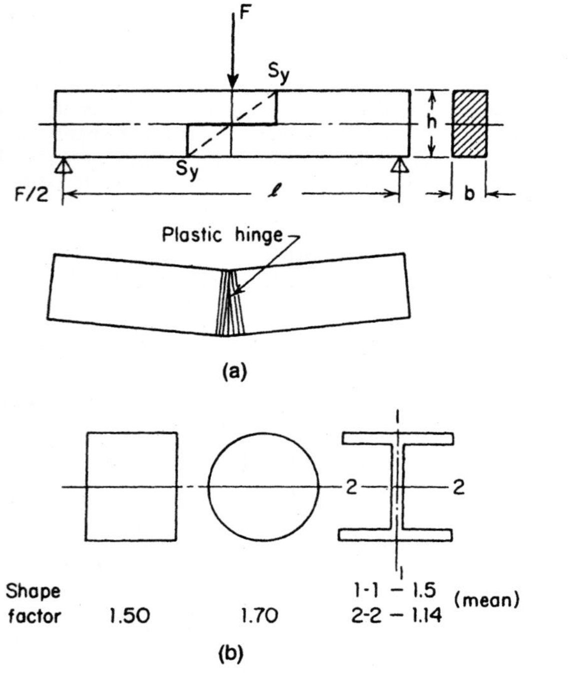

Figure 3.2.27

DESIGN STRESSES

If a machine part is to safely transmit loads acting upon it, a permissible maximum stress must be established and used in the design. This is the allowable stress, the working stress, or preferably, the **design stress.** The design stress should not waste material, yet should be large enough to prevent failure in case loads exceed expected values, or other uncertainties react unfavorably.

The design stress is determined by dividing the applicable material property—yield strength, ultimate strength, fatigue strength—by a **factor of safety.** The factor should be selected only after all **uncertainties** have been thoroughly considered. Among these are the uncertainty with respect to the magnitude and kind of operating load, the reliability of the material from which the component is made, the assumptions involved in the theories used, the environment in which the equipment might operate, the extent to which localized and fabrication stresses might develop, the uncertainty concerning causes of possible failure, and the endangering of human life in case of failure. Factors of safety vary from industry to industry, being the result of accumulated experience with a class of machines or a kind of environment. Many codes, such as the ASME code for power shafting, recommend design stresses found safe in practice.

In general, the **ductility** of the material determines the property upon which the factor should be based. Materials having an elongation of over 5 percent are considered ductile. In such cases, the factor of safety is based upon the yield strength or the endurance limit. For materials with an elongation under 5 percent, the ultimate strength must be used because these materials are **brittle** and fracture without yielding.

Factors of safety based on yield are often taken **between 1.5 and 4.0.** For more reliable materials or well-defined design and operating conditions, the lower factors are appropriate. In the case of untried materials or otherwise uncertain conditions, the larger factors are safer. The same values can be used when loads vary, but in such cases they are applied to the fatigue or endurance strength. When the ultimate strength determines the design stress (in the case of brittle materials), the factors of safety can be doubled.

Thus, under static loading, the design stress for, say, SAE 1020, which has a yield strength of 45,000 lb/in^2 (3,170 kgf/cm^2) may be taken at 45,000/2, or 22,500 lb/in^2 (1,585 kgf/cm^2), if a reasonably certain design condition exists. A Class 30 cast-iron part might be designed at 30,000/5 or 6,000 lb/in^2 (423 kgf/cm^2). A 2017S-0 aluminum-alloy component (13,000 lb/in^2 endurance strength) could be computed at a design stress of 13,000/2.5 or 5,200 lb/in^2 (366 kgf/cm^2) in the usual fatigue-load application.

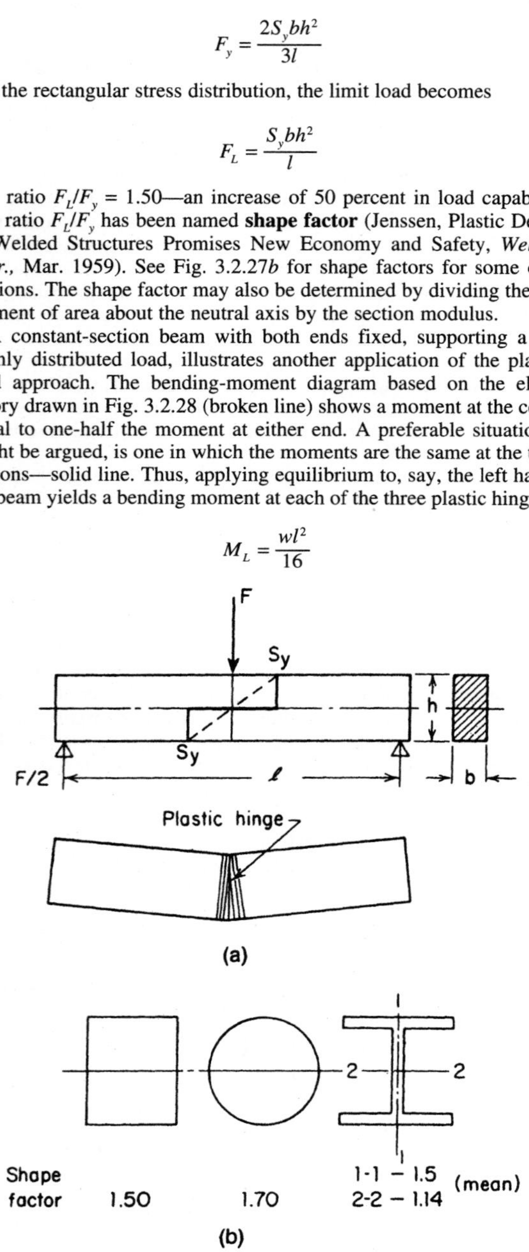

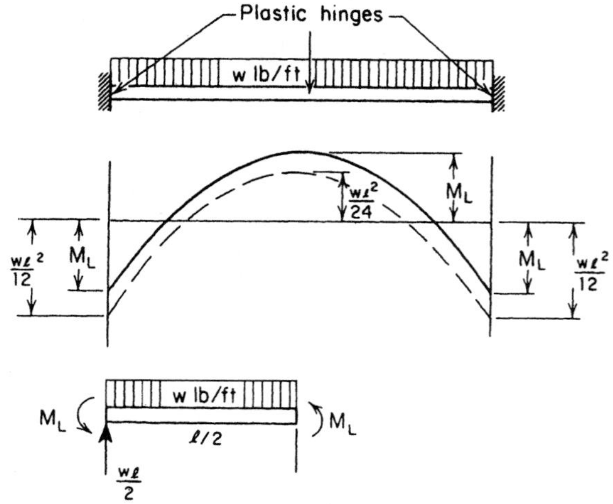

Figure 3.2.28

BEAMS

For properties of structural steel and wooden beams, see "Chapters Wood and Structural Design of Sections 4.2 and 4.6."

Notation

I = rectangular moment of inertia
I_p = polar moment of inertia
I/c = section modulus
M = bending moment
P, W' = concentrated load
Q or V = total vertical shear
R = reaction
S = unit normal stress
S_s or S_v = transverse shearing stress
W = total distributed load
f = deflection
i = slope
l = distance between supports
r = radius of gyration
r_c = radius of curvature
w = distributed load per longitudinal unit

A **simple beam** rests on supports at its ends that permit rotation. A **cantilever beam** is fixed (no rotation) at one end. When computing reactions and moments, distributed loads may be replaced by their resultants acting at the center of gravity of the distributed-load area.

Reactions are the forces and/or couples acting at the supports and holding the beam in place. In general, the weight of the beam should be accounted for.

The **bending moment** (pound-feet or pound-inches) (kgf · m) at any section is the algebraic sum of the external forces and moments acting on the beam on one side of the section. It is also equal to the moment of the internal-stress forces at the section, $M = \int s\, dA / y$. A bending moment that bends a beam convex downward (tensile stress on bottom fiber) is considered **positive,** while convex upward (compression on bottom) is **negative.**

The **vertical shear** V (lb) (kgf) effective on a section is the algebraic sum of all the forces acting parallel to and on one side of the section, $V = \Sigma F$. It is also equal to the sum of the transverse shear stresses acting on the section, $V = \int S_s\, dA$.

Moment and **shear diagram** may be constructed by plotting to scale the particular entity as the ordinate for each section of the beam. Such diagrams show in continuous form the variation along the length of the beam.

Moment-Shear Relation The shear V is the first derivative of moment with respect to distance along the beam, $V = dM/dx$. This relationship does not, however, account for any sudden changes in moment.

EXAMPLES. Figure 3.2.29 illustrates a simple beam subjected to a uniform load. $M = R_1 x - wx \times \frac{x}{2} = \frac{wlx}{2} - \frac{wx^2}{2}$ and $V = R_1 - wx = \frac{wl}{2} - wx$. Note also that $V = \frac{d}{dx}\left(\frac{wlx}{2} - \frac{wx^2}{2}\right) = \frac{wl}{2} - wx$.

Figure 3.2.30 is a simple beam carrying a uniformly varying load; $M = R_1 x - h\frac{x}{l} \times \frac{x}{2} \times \frac{x}{3} = \frac{hlx}{6} - \frac{hx^3}{6l}$, if h is in pounds per foot and weight of beam is neglected. The vertical shear $V = R_1 - \frac{hx}{l} \times \frac{x}{2} = \frac{hl}{6} - \frac{hx^2}{2l}$. Note again that $V = \frac{d}{dx}\left(\frac{hlx}{6} - \frac{hx^3}{6l}\right) = \frac{hl}{6} - \frac{hx^2}{2l}$.

Table 3.2.2 gives the reactions, bending-moment equations, vertical shear equations, and the deflection of some of the more common types of beams.

Maximum Safe Load on Steel Beams See Table 3.2.3 To obtain maximum safe load (or maximum deflection under maximum safe load) for any of the conditions of loading given in Table 3.2.5, multiply the corresponding coefficient in that table by the greatest safe load (or deflection) for distributed load for the particular section under consideration as given in Table 3.2.4.

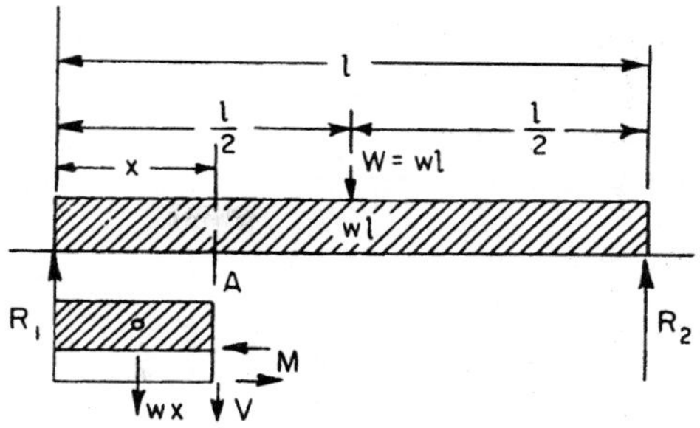

Figure 3.2.29

The following approximate factors for reducing the load should be used when beams are long in comparison with their breadth:

Ratio of unsupported (lateral) length to flange width or breadth	20	30	40	50	60	70
Ratio of greatest safe load to calculated load	1	0.9	0.8	0.7	0.6	0.5

Theory of Flexure A bent beam is shown in Fig. 3.2.31. The concave side is in compression and the convex side in tension. These are divided by the neutral plane of zero stress $A'B'BA$. The intersection of the neutral plane with the face of the beam is in the **neutral line or elastic curve** AB. The intersection of the neutral plane with the cross section is the neutral axis NN'.

It is assumed that a beam is prismatic, of a length at least 10 times its depth, and that the external forces are all at right angles to the axis of the beam and in a plane of symmetry, and that flexure is slight. Other assumptions are: (1) That the material is homogeneous, and obeys Hooke's law. (2) That **stresses are within the elastic limit.** (3) That every layer of material is free to expand and contract longitudinally and laterally under stress as if separate from other layers. (4) That the tensile and compressive moduli of elasticity are equal. (5) That the

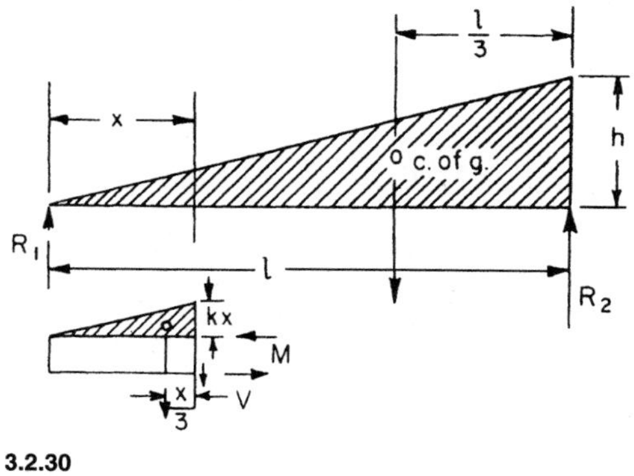

Figure 3.2.30

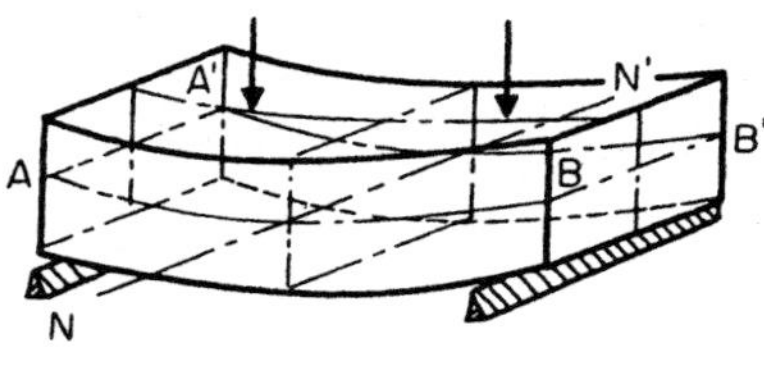

Figure 3.2.31

Table 3.2.2 Beams of Uniform Cross Section, Loaded Transversely

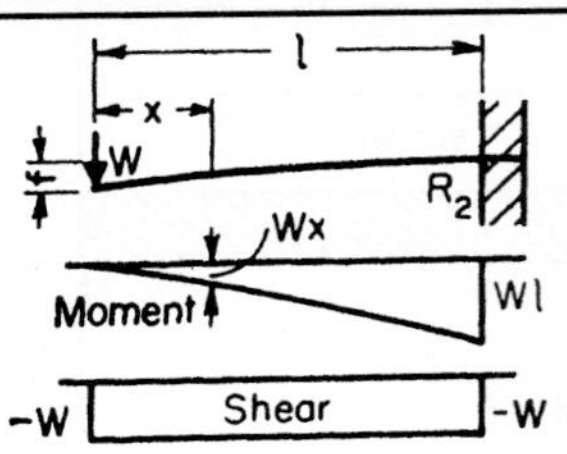

$$R_2 = W$$
$$M_x = -Wx$$
$$M_{max} = -Wl, (x = 1)$$
$$Q_x = -W$$
$$f = \frac{Wl^3}{3EI} \text{(max)}$$

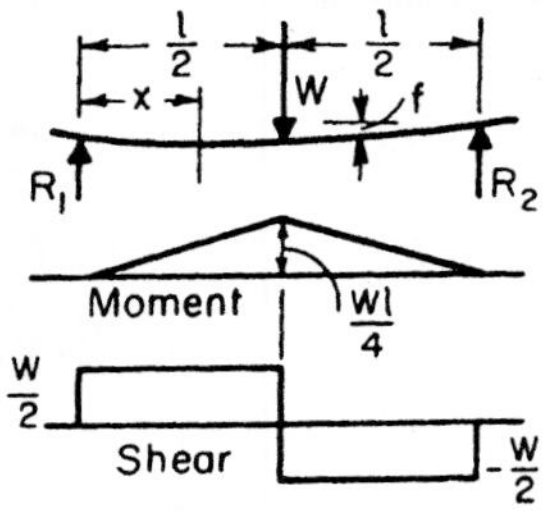

$$R_1 = \frac{W}{2}, R_2 = \frac{W}{2}$$
$$M_x = \frac{Wx}{2}$$
$$M_{max} = \frac{Wl}{4}, \left(x = \frac{l}{2}\right)$$
$$Q_x = \pm\frac{W}{2}$$
$$f = \frac{W}{EI}\frac{l^3}{48} \text{(max)}$$

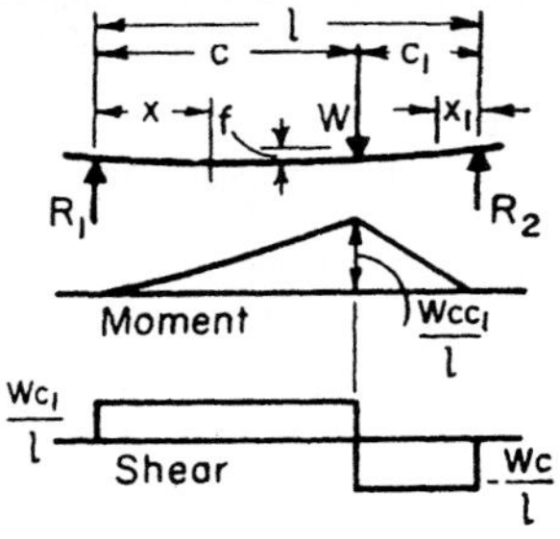

$$R_1 = \frac{Wc_1}{l}, R_2 = \frac{Wc}{l}$$
$$M_x = \frac{Wc_1 x}{l}, Mx' = \frac{Wcx_1}{l}$$
$$M_{max} = \frac{Wcc_1}{l}, (x_1 = c_1 \text{ or } x = c)$$
$$Q_x = \frac{Wc_1}{l}, Q_{x1} = \frac{Wc}{l}$$
$$f = \frac{Wc_1}{3EIl}\left[\frac{c(l + c_1)}{3}\right]^{3/2} \text{(max)}$$

Max f occurs at $x = \sqrt{c(l - c_1)/3}$

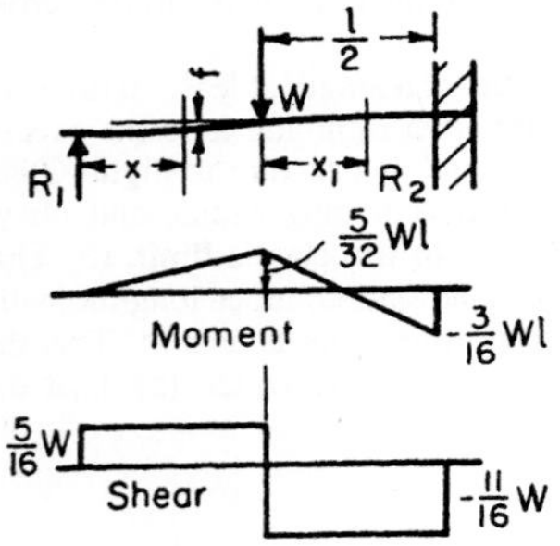

$$R_1 = \frac{5}{16}W, R_2 = \frac{11}{16}W$$
$$M_x = \frac{5}{16}Wx$$
$$M_{x1} = Wl\left(\frac{5}{32} - \frac{11}{16}\frac{x_1}{l}\right)$$
$$M_{max} = -\frac{3}{16}Wl, \left(x_1 = \frac{l}{2}\right)$$
$$Q_x = +\frac{5}{16}W, Q_{x1} = -\frac{11}{16}W$$
$$Q_{max} = -\frac{11}{16}W, \left(x = \frac{l}{2} \text{ to } x = l\right)$$
$$f = \frac{W}{EI}\frac{7l^3}{768}$$

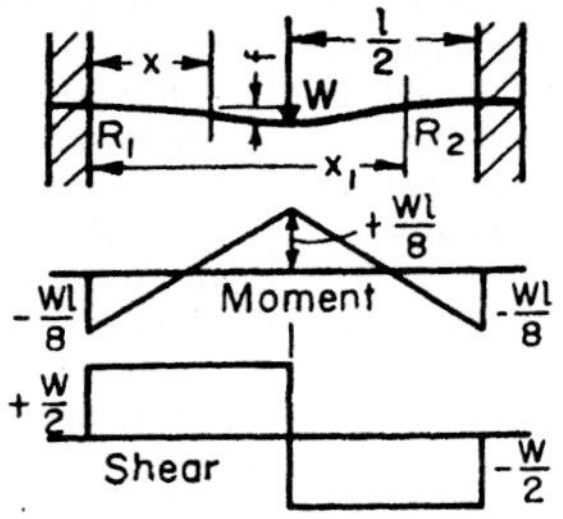

$$R_1 = \frac{W}{2}, R_2 = \frac{W}{2}$$
$$M_x = \frac{Wl}{2}\left(\frac{x}{l} - \frac{1}{4}\right)$$
$$M_{x1} = \frac{-Wl}{2}\left(\frac{x}{l} - \frac{3}{4}\right)$$
$$M_{max} = \frac{Wl}{8}, \left(x = \frac{l}{2}\right)$$
$$Q_x = \frac{W}{2}, Q_{x1} = -\frac{W}{2}$$
$$f = \frac{W}{EI}\frac{l^3}{192} \text{(max)}$$

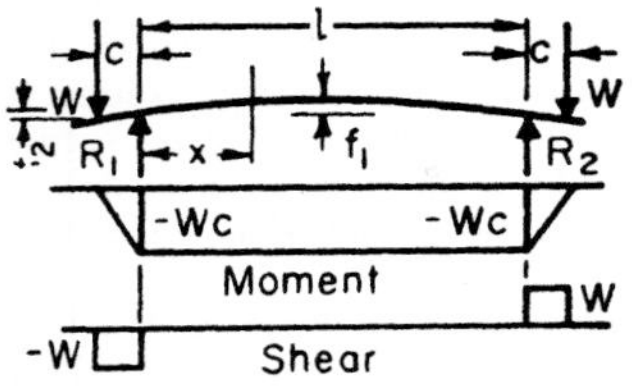

$$R_1 = W$$
$$R_2 = W$$
$$M_x = -W_c = \text{const}$$
$$Q_{W \text{ to } R_1} = -W$$
$$Q_{R_1 \text{ to } R_2} = 0$$
$$Q_{R_2 \text{ to } W} = +W$$
$$f_1 = \frac{Wcl^2}{EI8} \text{(max)}$$
$$f_2 = \frac{Wc^2}{EI3}\left(c + \frac{3l}{2}\right) \text{(max)}$$

Table 3.2.2 Beams of Uniform Cross Section, Loaded Transversely *(Continued)*

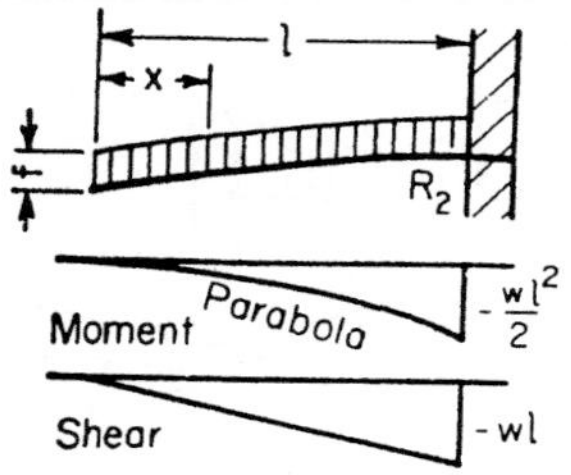

$R_2 = W = wl$

$M_x = -\frac{wx^2}{2}$

$M_{max} = -\frac{wl^2}{2}, (x = l)$

$Q_x = -wx$

$Q_{max} = -wl, (x = l)$

$f = \frac{W}{EI}\ \frac{l^3}{8}$ (max)

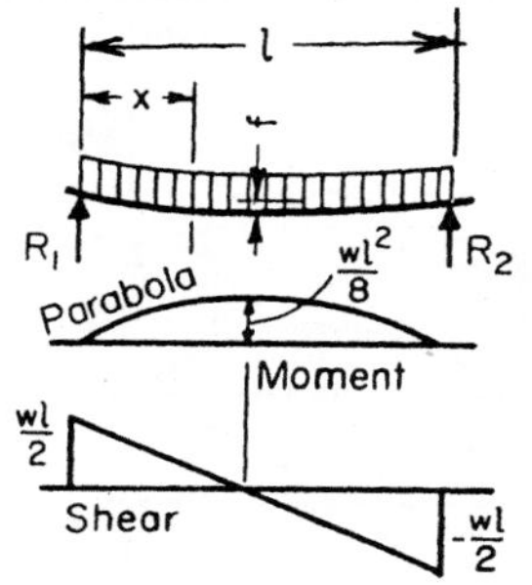

$R_1 = \frac{W}{2} = \frac{wl}{2}$

$R_2 = \frac{W}{2} = \frac{wl}{2}$

$M_x = \frac{wx}{2}(l - x)$

$M_{max} = \frac{wl^2}{8}, (x = ½l)$

$Q_x = \frac{wl}{2} - wx$

$Q_{max} = \frac{wl}{2}, (x = 0)$

$f = \frac{W}{EI}\ \frac{5l^3}{384}$ (max)

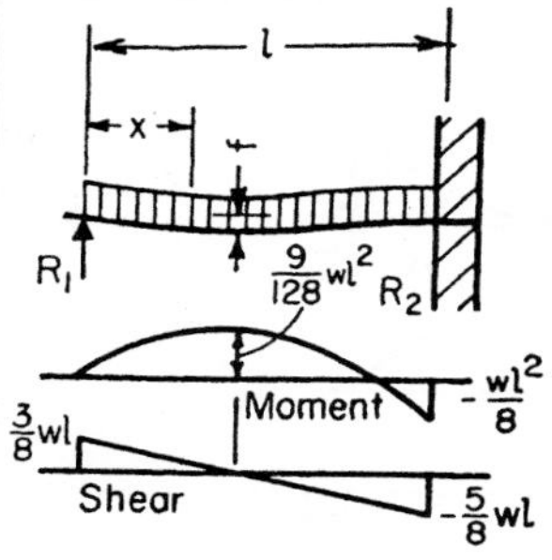

$R_1 = \frac{3}{8}\ W = \frac{3}{8}\ wl$

$R_2 = \frac{5}{8}\ W = \frac{5}{8}\ wl$

$M_x = \frac{wx}{2}\left(\frac{3}{4}\ l - x\right)$

$M_{max} = \frac{9}{128}\ wl^2, \left(x = \frac{3}{8}\ l\right)$

$M_{max} = -\frac{wl^2}{8}, (x = l)$

$Q_x = \frac{3}{8}\ wl - wx$

$Q_{max} = -\frac{5}{8}\ wl$

$f = \frac{W}{EI}\ \frac{l^3}{185}$ (max)

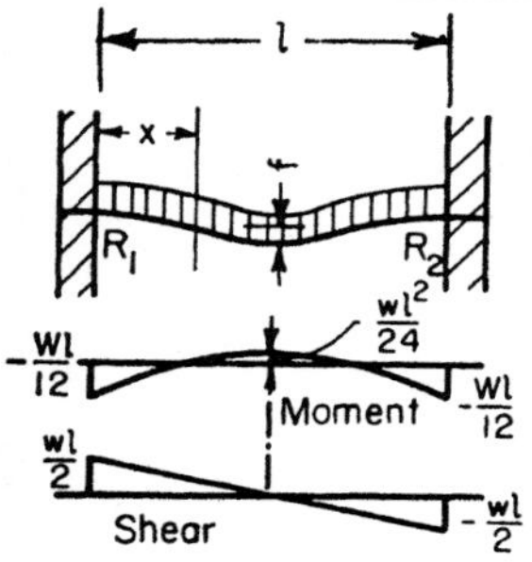

$R_1 = \frac{W}{2} = \frac{wl}{2},\ R_2 = \frac{W}{2} = \frac{wl}{2}$

$M_x = -\frac{wl^2}{2}\left(\frac{1}{6} - \frac{x}{l} + \frac{x^2}{l^2}\right)$

$M_{max} = -\frac{1}{12}wl^2, (x = 0, \text{or}\ x = l)$

$Q_x = \frac{wl}{2} - wx$

$Q_{max} = \pm\frac{wl}{2}$

$f = \frac{W}{EI}\ \frac{l^3}{384}$ (max)

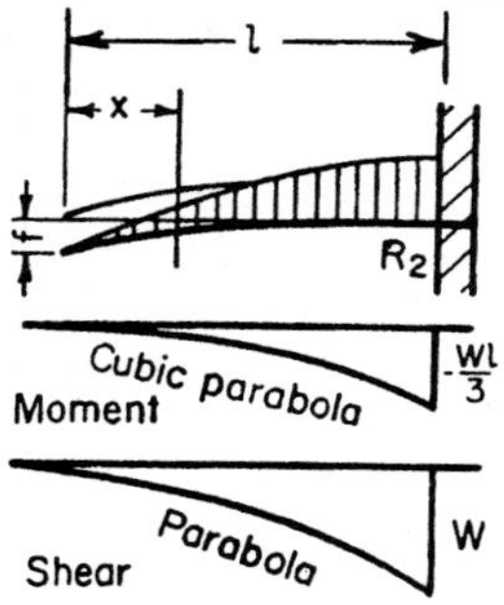

$R_2 = W =$ total load

$M_x = -\frac{W}{3}\ \frac{x^3}{l^2}$

$M_{max} = -\frac{Wl}{3}$

$Q_x = -\frac{Wx^2}{l^2}$

$Q_{max} = -W$

$f = \frac{W}{EI}\ \frac{l^3}{15}$ (max)

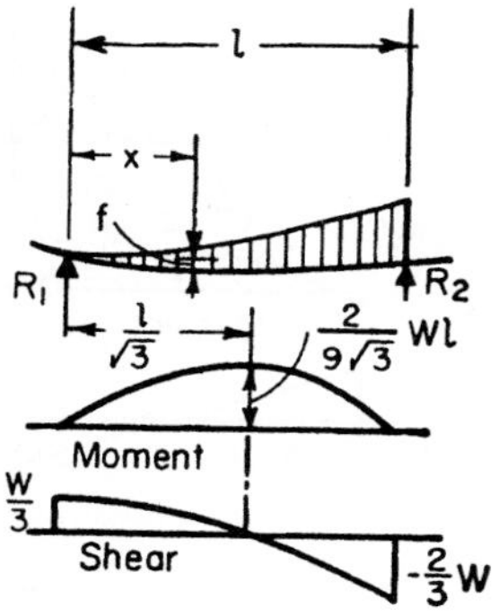

$R_1 = \frac{1}{3}W, R_2 = \frac{2}{3}W$

$M_x = \frac{Wx}{3}\left(1 - \frac{x^2}{l^2}\right)$

$M_{max} = \frac{2}{9\sqrt{3}}Wl, \left(x = \frac{1}{\sqrt{3}}\right)$

$Q_x = W\left(\frac{1}{3} - \frac{x^2}{l^2}\right)$

$Q_{max} = -\frac{2}{3}W, (x = l)$

$f = 0.01304\frac{Wl^3}{EI}$ (max)

(Continued)

Table 3.2.2 Beams of Uniform Cross Section, Loaded Transversely *(Continued)*

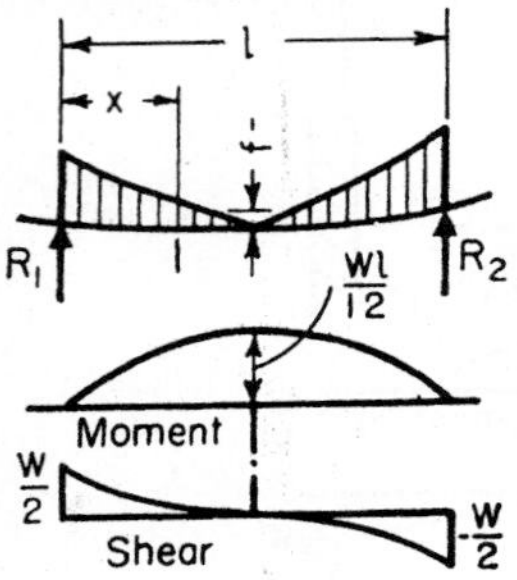

$$R_1 = \frac{W}{2}, R_2 = \frac{W}{2}$$

$$M_x = Wx\left(\frac{1}{2} - \frac{x}{l} + \frac{2x^2}{3l^2}\right)$$

$$M_{\max} = \frac{Wl}{12}, \left(x = \frac{1}{2}l\right)$$

$$Q_x = W\left(\frac{1}{2} - \frac{2x}{l} + \frac{2x^2}{l^2}\right)$$

$$Q_{\max} = \pm\frac{W}{2}, (x = 0)$$

$$f = \frac{W}{EI}\frac{3l^3}{320}(\max)$$

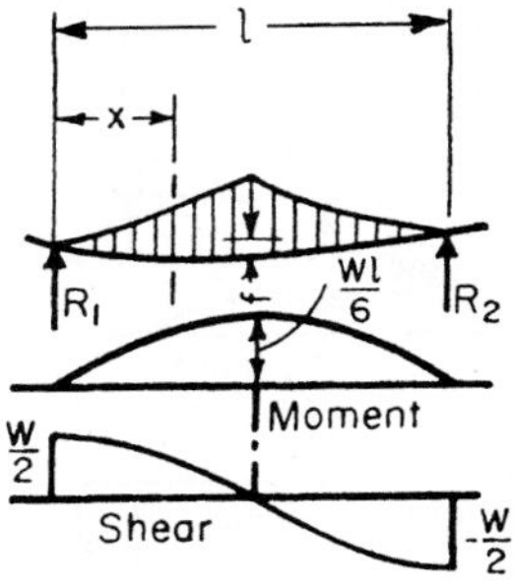

$$R_1 = \frac{W}{2}, R_2 = \frac{W}{2}$$

$$M_x = Wx\left(\frac{1}{2} - \frac{2}{3}\frac{x^2}{l^2}\right)$$

$$M_{\max} = \frac{Wl}{6}, \left(x = \frac{1}{2}l\right)$$

$$Q_x = W\left(\frac{1}{2} - \frac{2x^2}{l^2}\right)$$

$$Q_{\max} = \pm\frac{W}{2}, (x = 0)$$

$$f = \frac{W}{EI}\frac{l^3}{60}(\max)$$

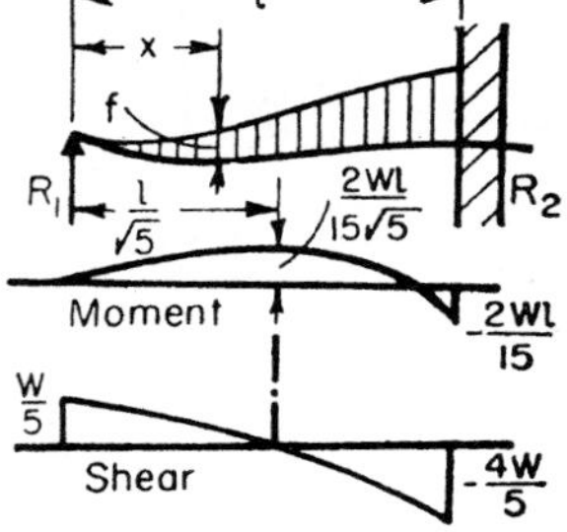

$$R_1 = \frac{W}{5}, R_2 = \frac{4W}{5}$$

$$M_x = Wx\left(\frac{1}{5} - \frac{x^2}{3l^2}\right)$$

$$M_{\max} = -\frac{2}{15}Wl \text{ at support 2}$$

$$Q_x = W\left(\frac{1}{5} - \frac{x^2}{l^2}\right)$$

$$Q_{\max} = -\frac{4W}{5}$$

$$f = \frac{16Wl^3}{1{,}500\sqrt{5EI}} = \frac{0.00477Wl^3}{EI}(\max)$$

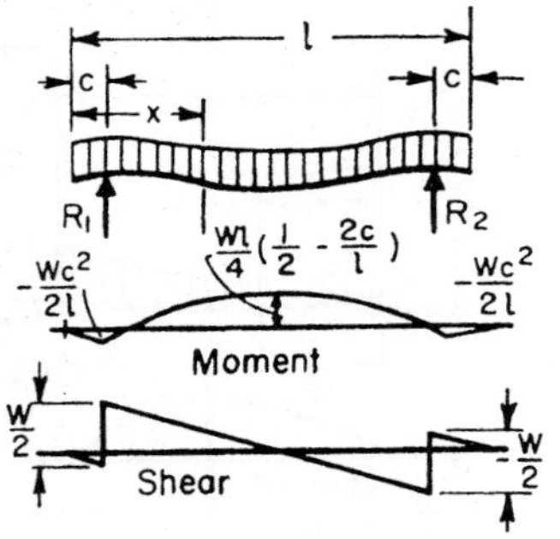

$$R_1 = \frac{W}{2} = \frac{wl}{2},\ R_2 = \frac{W}{2} = \frac{wl}{2}$$

$$M_x = \frac{Wx}{2}\left(1 - \frac{c}{x} - \frac{x}{l}\right), (x > c)$$

$$M_x = -\frac{Wx^2}{2l},\ (x \le c)$$

$$M_{\max} = \frac{Wl}{4}\left(\frac{1}{2} - \frac{2c}{l}\right), c \le \left(\frac{\sqrt{2} - 1}{2}\right)l$$

$$Q_x = \frac{W}{2} - wx(x > c)$$

$$Q_x = -wx(x \le c)$$

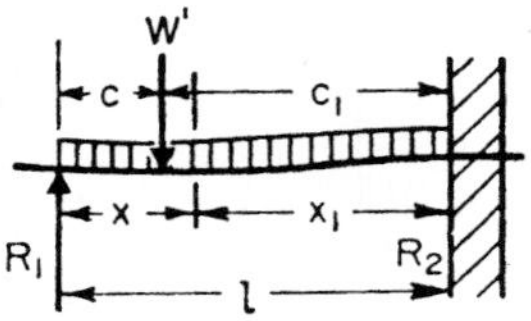

Concentrated load W'
Uniformly dist. load $W = wl$

$$R_1 = W'\frac{c_1^2(3c + 2c_1)}{2l^3} + \frac{3}{8}W$$

$$R_2 = W'\frac{(2c^2 + 6cc_1 + 3c_1^2)c}{2l^3} + \frac{5}{8}W$$

$$M_2 = W'\frac{cc_1(2c + c_1)}{2l^2} + W\left(\frac{1}{8}\right)$$

$$M_{W'} = W'\frac{cc_1^2(3c + 2c_1)}{2l^3} + W\frac{(3c_1 - c)c}{8l}$$

(a) $\frac{W'}{W} < \frac{l^2}{4c_1^2}\ \frac{5c - 3c_1}{3c + 2c_1}$

$$M_{c\max} = \frac{R_1^2}{2W}l,\ \left(x = \frac{R_1 l}{W}\right)$$

(b) $\frac{W'}{W} < \frac{l^2(3c_1 - 5c)}{4c(2c^2 + 6cc_1 + 3c_1^2)}$

$$M_{c_1\max} = W'c + \frac{(R_1 - W')^2}{2W}l, \left(x = \frac{R_1 - W'}{W}l\right)$$

Deflection under W'

$$f = \frac{W'}{EI}\frac{c^2c_1^3(4c + 3c_1)}{12l^3} + \frac{W}{EI}\frac{cc_1^2(3c + c_1)}{48l}$$

Table 3.2.2 Beams of Uniform Cross Section, Loaded Transversely *(Continued)*

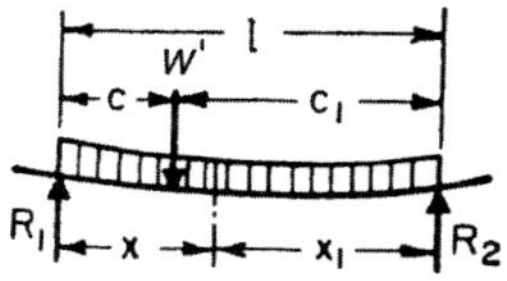

Concentrated load W'

Uniformly dist. load $W = wl;\ c < c_1$

$$R_1 = W'\frac{c_1}{l} + \frac{W_2}{2}$$

$$R_2 = W'\frac{c}{l} + \frac{W}{2}$$

$(a)\ \frac{W'}{W} < \frac{c_1 - c}{2c}$

$$M_{max} = R_2\frac{x_1}{2} = \frac{R_2^2 l}{2W},\left(x_1 = \frac{R_2 l}{W}\right)$$

$(b)\ \frac{W'}{W} > \frac{c_1 - c}{2c}$

$$M_{max} = \left(W' + \frac{W}{2}\right)\frac{cc_1}{l},(x_1 = c_1)$$

Deflection of beam under W':

$$f = \left(W' + \frac{l^2 + cc_1}{8cc_1}W\right)\frac{c^2c_1^2}{3EIl}$$

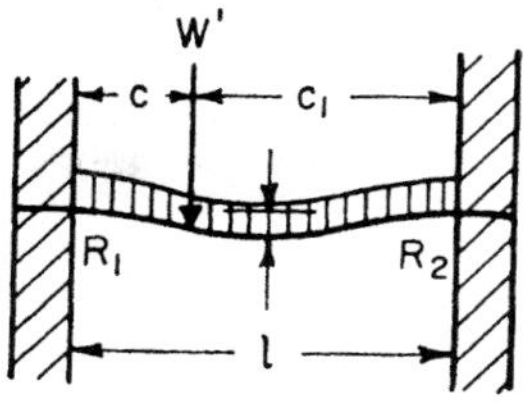

$c < c_1$

$$R_1 = W'\frac{(3c + c_1)c_1^2}{l^3} + \frac{W}{2}$$

$$R_2 = W'\frac{(c + 3c_1)c^2}{l^3} + \frac{W}{2}$$

$$M_{max} = M_1 = W'\frac{cc_1^2}{l^2} + \frac{Wl}{12}$$

Deflection under W'

$$f = \frac{1}{EI}\left(W'\frac{c^3c_1^3}{3l^3} + W\frac{c^2c_1^2}{24l}\right)$$

Table 3.2.3 Uniformally Distributed Loads on Simply Supported Rectangular Beams 1-in Wide*
(Laterally Supported Sufficiently to Prevent Buckling)
[Calculated for unit fiber stress at 1,000 lb/in² (70 kgf/cm²): nominal size]
Total load in pounds (kgf)† including the weight of beam

Span, ft (m)‡	Depth of beam, in (cm)§										
	6	7	8	9	10	11	12	13	14	15	16
5	800	1,090	1,420	1,800	2,220	2,690	3,200	3,750	4,350	5,000	5,690
6	670	910	1,180	1,500	1,850	2,240	2,670	3,130	3,630	4,170	4,740
7	570	780	1,010	1,290	1,590	1,920	2,280	2,680	3,110	3,570	4,060
8	500	680	890	1,120	1,390	1,680	2,000	2,350	2,720	3,130	3,560
9	440	600	790	1,000	1,230	1,490	1,780	2,090	2,420	2,780	3,160
10	400	540	710	900	1,110	1,340	1,600	1,880	2,180	2,500	2,840
11	360	490	650	820	1,010	1,220	1,450	1,710	1,980	2,270	2,590
12	330	450	590	750	930	1,120	1,330	1,560	1,810	2,080	2,370
13	310	420	550	690	850	1,030	1,230	1,440	1,680	1,920	2,190
14	290	390	510	640	790	960	1,140	1,340	1,560	1,790	2,030
15	270	360	470	600	740	900	1,070	1,250	1,450	1,670	1,900
16	250	340	440	560	690	840	1,000	1,170	1,360	1,560	1,780
17	230	320	420	530	650	790	940	1,100	1,280	1,470	1,670
18	220	300	400	500	620	750	890	1,040	1,210	1,390	1,580
19	210	290	380	470	590	710	840	990	1,150	1,320	1,500
20	200	270	360	450	560	670	800	940	1,090	1,250	1,420
22	180	250	320	410	500	610	730	850	990	1,140	1,290
24	160	230	290	370	460	560	670	780	910	1,040	1,180
26	150	210	270	340	420	520	610	720	840	960	1,090
28	140	190	250	320	390	480	570	670	780	890	1,010
30	130	180	240	300	370	450	530	630	730	830	950

* This table is convenient for wooden beams. For any other fiber stress S', multiply the values in table by $S'/1{,}000$. See Sec. 4.12. for properties of wooden beams of commercial sizes.
† To change to kgf, multiply by 0.454.
‡ To change to m, multiply by 0.305.
§ To change to cm, multiply by 2.54.

Table 3.2.4 Approximate Safe Loads in Pounds (kgf) on Steel Beams, Simply Supported, Single Span

Allowable fiber stress for steel, 16,000 lb/in² (1,127 kgf/cm²) (basis of table)

Beams simply supported at both ends.

L = distance between supports, ft (m)
A = sectional area of beam, in² (cm²)
D = depth of beam, in (cm)
a = interior area, in² (cm²)
d = interior depth, in (cm)
w = total working load, net tons (kgf)

	Greatest safe load, lb		Deflection, in	
Shape of section	Load in middle	Load distributed	Load in middle	Load distributed
Solid rectangle	$\frac{890AD}{L}$	$\frac{1,780AD}{L}$	$\frac{wL^3}{32AD^2}$	$\frac{wL^3}{52AD^2}$
Hollow rectangle	$\frac{890(AD-ad)}{L}$	$\frac{1,780(AD-ad)}{L}$	$\frac{wL^3}{32(AD^2-ad^2)}$	$\frac{wL^3}{52(AD^2-ad^2)}$
Solid cylinder	$\frac{667AD}{L}$	$\frac{1,333AD}{L}$	$\frac{wL^3}{24AD^2}$	$\frac{wL^3}{38AD^2}$
Hollow cylinder	$\frac{667(AD-ad)}{L}$	$\frac{1,333(AD-ad)}{L}$	$\frac{wL^3}{24(AD^2-ad^2)}$	$\frac{wL^3}{38(AD^2-ad^2)}$
I beam	$\frac{1,795AD}{L}$	$\frac{3,390AD}{L}$	$\frac{wL^3}{58AD^2}$	$\frac{wL^3}{93AD^2}$

cross section remains a plane surface. (The assumption of plane cross sections is strictly true only when the shear is constant or zero over the cross section, and when the shear is constant throughout the length of the beam.)

It follows then that: (1) The internal forces are in horizontal balance. (2) The **neutral axis contains the center of gravity** of the cross section, where there is no resultant axial stress. (3) The stress intensity varies directly with the distance from the neutral axis.

The moment of the elastic forces about the neutral axis, that is, the **stress moment** or **moment of resistance**, is $M = SI/c$, where S is an elastic unit stress at outer fiber whose distance from the neutral axis is c; and I is the rectangular moment of inertia about the neutral axis. I/c is the **section modulus.** This formula is for the **strength of beams.** For rectangular beams, $M = 1/6\ Sbh^2$, where b = breadth and h = depth; that is, the elastic **strength of beam sections** varies as follows: (1) for equal width, as the square of the depth; (2) for equal depth, directly as the width; (3) for equal depth and width, directly as the strength of the material; (4) if span varies, then for equal depth, width, and material, inversely as the span.

If a beam is cut in halves vertically, the two halves laid side by side each will carry only one-half as much as the original beam.

Tables 3.2.6 to 3.2.8 give the properties of various beam cross sections. For properties of structural-steel shapes, see Secs. 4.2 and 4.12.

Oblique Loading It should be noted that Table 3.2.6 includes certain cases for which the horizontal axis is not a neutral axis, assuming the common case of vertical loading. The rectangular section with the diagonal as a horizontal axis (Table 3.2.6) is such a case. These cases must be handled by the principles of oblique loading.

Table 3.2.5 Coefficients for Correcting Values in Table 3.2.4 for Various Methods of Support and of Loading, Single Span

Conditions of loading	Max relative safe load	Max relative deflection under max relative safe load
Beam supported at ends:		
Load uniformly distributed over span	1.0	1.0
Load concentrated at center of span	½	0.80
Two equal loads symmetrically concentrated	$l/4c$	
Load increasing uniformly to one end	0.974	0.976
Load increasing uniformly to center	¾	0.96
Load decreasing uniformly to center	3⁄2	1.08
Beam fixed at one end, cantilever:		
Load uniformly distributed over span	¼	2.40
Load concentrated at end	⅛	3.20
Load increasing uniformly to fixed end	⅜	1.92
Beam continuous over two supports equidistant from ends:		
Load uniformly distributed over span		
1. If distance $a > 0.2071l$	$l^2/(4a^2)$	
2. If distance $a < 0.2071l$	$\frac{l}{l-4a}$	
3. If distance $a = 0.2071l$	5.83	
Two equal loads concentrated at ends	$l/(4a)$	

NOTE: l = length of beam; c = distance from support to nearest concentrated load; a = distance from support to end of beam.

Every section of a beam has two principal axes passing through the center of gravity, and these two axes are always at right angles to each other. The principal axes are axes with respect to which the moment of inertia is, respectively, a maximum and a minimum, and for which the product of inertia is zero. For symmetrical sections, axes of symmetry are always principal axes. For unsymmetrical sections, like a **rolled angle** section (Fig. 3.2.32), the inclination of the principal axis with the X axis may be found from the formula $\tan 2\theta = 2I_{xy}/(I_y - I_x)$, in which θ = angle of inclination of the principal axis to the X axis, I_{xy} = the product of inertia of the section with respect to the X and Y axes, I_y = moment of inertia of the section with respect to the Y axis, I_x = moment of inertia of the section with respect to the X axis. When this principal axis has been found, the other principal axis is at right angles to it.

Calling the moments of inertia with respect to the principal axes I'_x and I'_y, the unit stress existing anywhere in the section at a point whose coordinates are x and y (Fig. 3.2.33) is $S = My\cos\alpha/I'_x + Mx\sin\alpha/I'_y$,

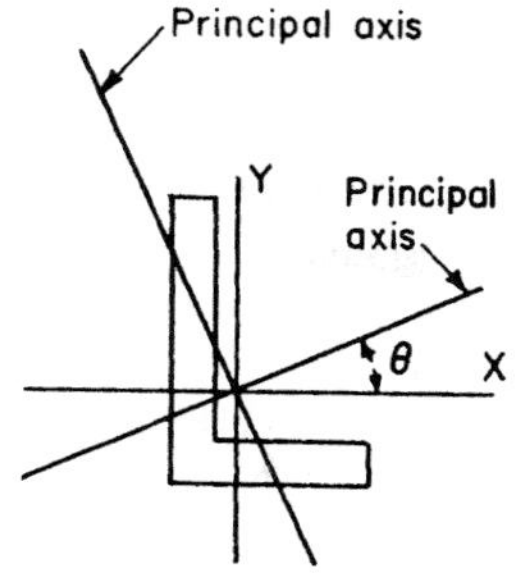

Figure 3.2.32

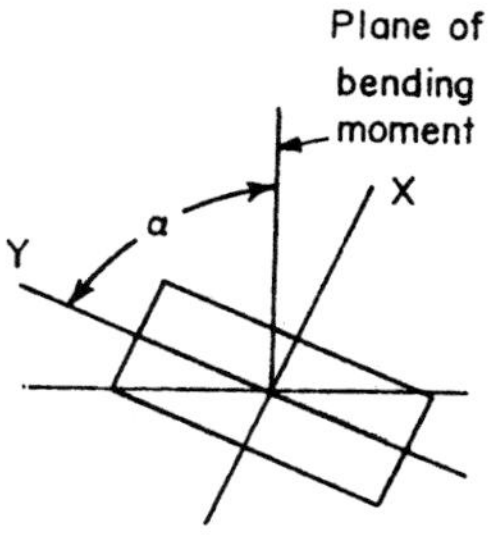

Figure 3.2.33

Table 3.2.6 Properties of Various Cross Sections*

(I = moment of inertia; I/c = section modulus; $r = \sqrt{I/A}$ = radius of gyration)

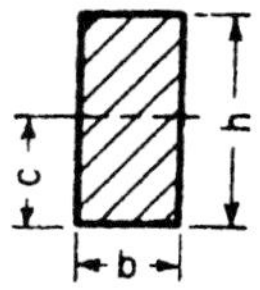	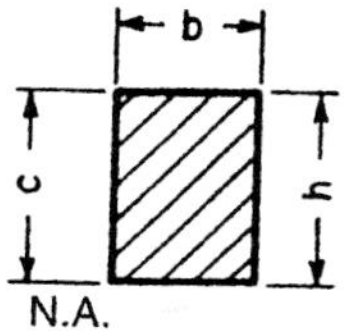	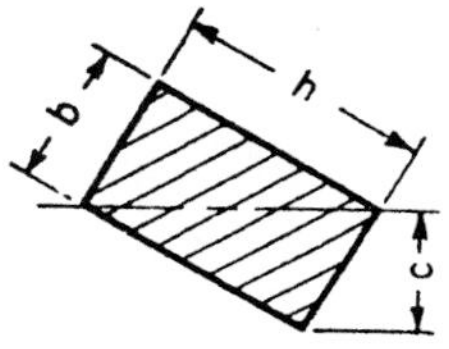	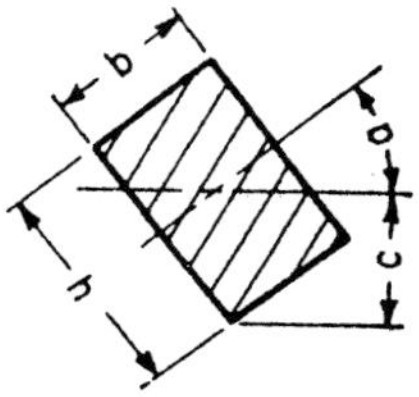
$I = \frac{bh^3}{12}$	$\frac{bh^3}{3}$	$\frac{b^3h^3}{6(b^2+h^2)}$	$\frac{bh}{12}(h^2\cos^2 a + b^2\sin^2 a)$
$\frac{I}{c} = \frac{bh^2}{6}$	$\frac{bh^2}{3}$	$\frac{b^2h^2}{6\sqrt{b^2+h^2}}$	$\frac{bh}{6}\left(\frac{h^2\cos^2 a + b^2\sin^2 a}{h\cos a + b\sin a}\right)$
$r = \frac{h}{\sqrt{12}} = 0.289h$	$\frac{h}{\sqrt{3}} = 0.577h$	$\frac{bh}{\sqrt{6(b^2+h^2)}}$	$\sqrt{\frac{h^2\cos^2 a + b^2\sin^2 a}{12}}$
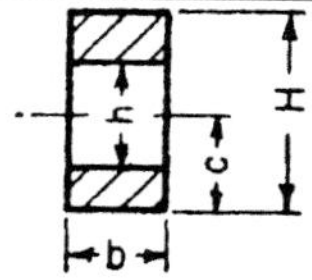	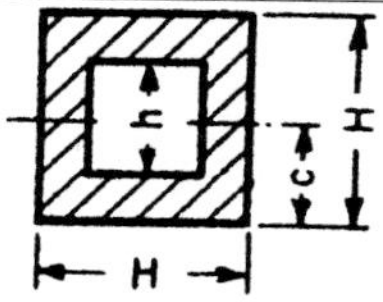	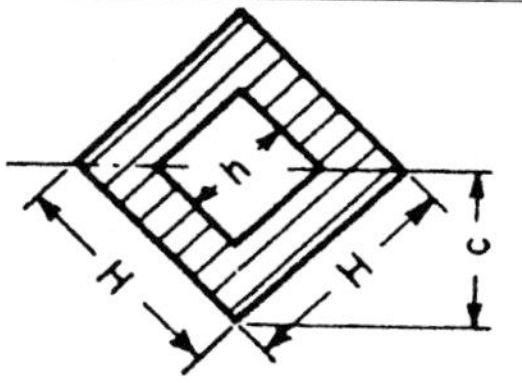	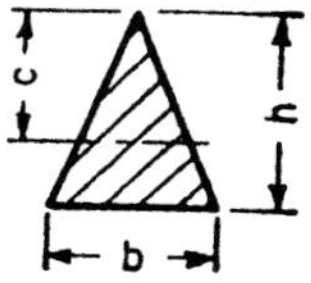
$I = \frac{b}{12}(H^3 - h^3)$	$\frac{H^4 - h^4}{12}$	$\frac{H^4 - h^4}{12}$	$\frac{bh^3}{36}; c = \frac{2}{3}h$
$\frac{I}{c} = \frac{b}{6}\,\frac{H^3 - h^3}{H}$	$\frac{1}{6}\,\frac{H^4 - h^4}{H}$	$\frac{\sqrt{2}}{12}\,\frac{H^4 - h^4}{H}$	$\frac{bh^2}{24}$
$r = \sqrt{\frac{H^3 - h^3}{12(H-h)}}$	$\sqrt{\frac{H^2 + h^2}{12}}$	$\sqrt{\frac{H^2 + h^2}{12}}$	$\frac{h}{\sqrt{18}}$
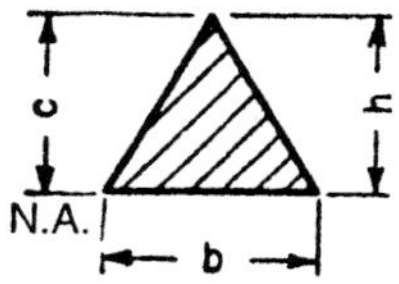	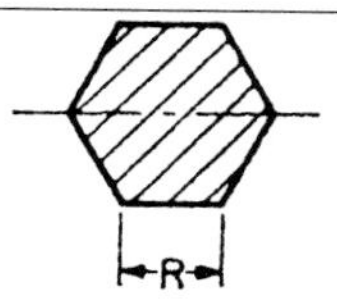	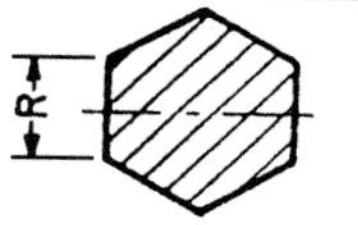	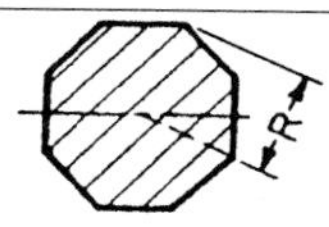
$I = \frac{bh^3}{12}$	$\frac{5\sqrt{3}}{16}R^4$		$\frac{1+2\sqrt{2}}{6}R^4$
$\frac{I}{c} = \frac{bh^2}{12}$	$\frac{5}{8}R^3$	$\frac{5\sqrt{3}}{16}R^3$	$0.6906R^3$
$r = \frac{h}{\sqrt{6}}$	$\sqrt{\frac{5}{24}}\,R$		$0.475R$

NOTE: Square, axis same as first rectangle, side = h; $I = h^4/12$; $I/c = h^3/6$; $r = 0.289h$.

(Continued)

Table 3.2.6 Properties of Various Cross Sections* (*Continued*)

Square, diagonal taken as axis: $I = h^4/12$; $I/c = 0.1179h^3$; $r = 0.289h$.

Equilateral Polygon A = area R = rad circumscribed circle r = rad inscribed circle n = no sides a = length of side Axis as in preceding section of octagon	$I = \frac{A}{24}(6R^2 - a^2)$ $= \frac{A}{48}(12r^2 + a^2)$ $= \frac{AR^2}{4}$ (approx)	$\frac{I}{c} = \frac{I}{r}$ $= \frac{I}{R\cos\frac{180°}{n}}$ $= \frac{AR}{4}$ (approx)	$\sqrt{\frac{I}{A}} = \sqrt{\frac{6R^2 - a^2}{24}} \approx \frac{R}{2}$ $= \sqrt{\frac{12r^2 + a^2}{48}}$
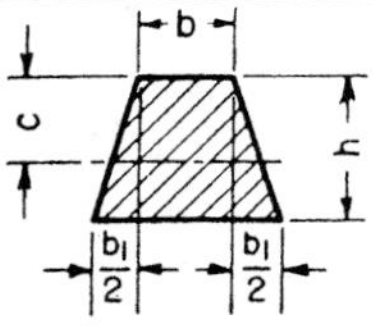	$I = \frac{6b^2 + 6bb_1 + b_1^2}{36(2b + b_1)} h^3$ $c = \frac{1}{3}\,\frac{3b + 2b_1}{2b + b_1} h$	$\frac{I}{c} = \frac{6b^2 + 6bb_1 + b_1^2}{12(3b + 2b_1)} h^2$	$r = \frac{h\sqrt{12b^2 + 12bb_1 + 2b_1^2}}{6(2b + b_1)}$
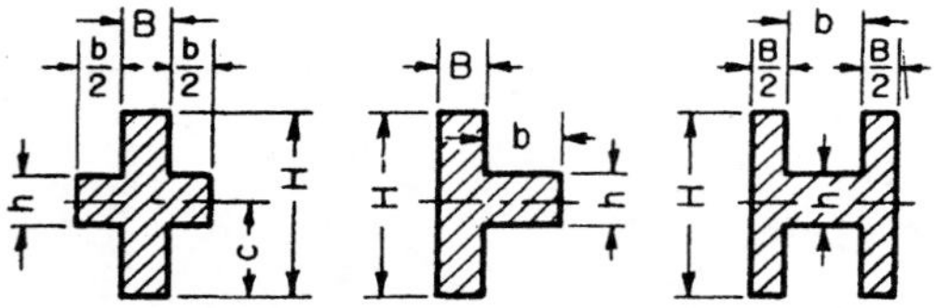		$I = \frac{BH^3 + bh^3}{12}$ $\frac{I}{c} = \frac{BH^3 + bh^3}{6H}$	$r = \sqrt{\frac{BH^3 + bh^3}{12(BH + bh)}}$
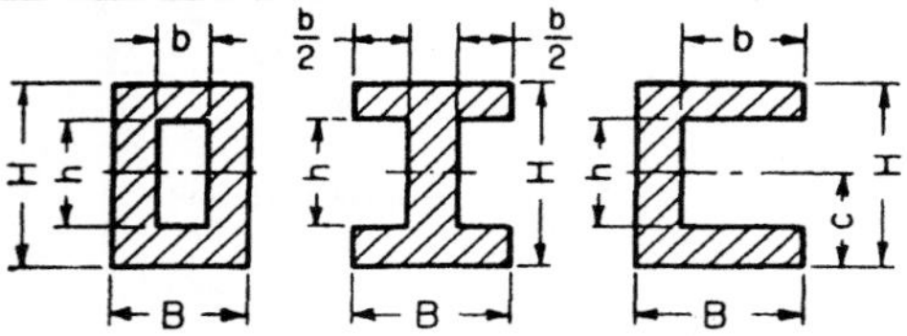		$I = \frac{BH^3 - bh^3}{12}$ $\frac{I}{c} = \frac{BH^3 - bh^3}{6h}$	$r = \sqrt{\frac{BH^3 - bh^3}{12(BH - bh)}}$
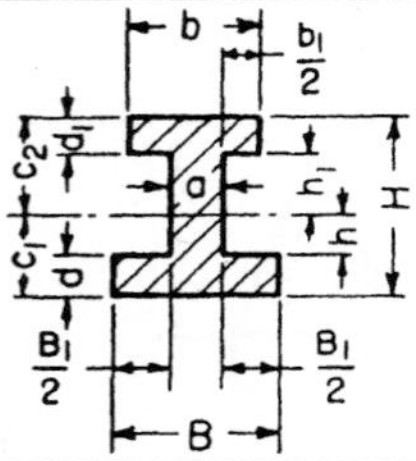	$I = ⅓(BC_1^3 - B_1h^3 + bc_3^3 - b_1h_1^3)$ $c_1 = \frac{1}{2}\,\frac{aH^2 + B_1d^2 + b_1d_1(2H - d_1)}{aH + B_1d + b_1d_1}$		$r = \sqrt{\frac{I}{Bd + bd_1 + a(h + h_1)}}$
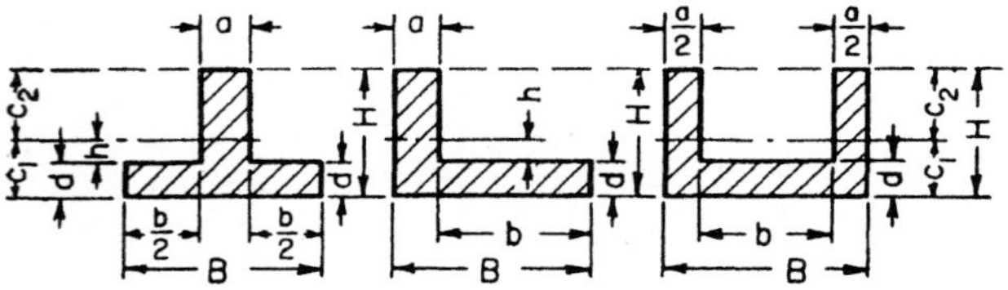			$I = ⅓(Bc_1^3 - bh^3 + ac_2^3)$ $c_1 = \frac{1}{2}\,\frac{aH^2 + bd^2}{aH + bd}$ $c_2 = H - c_1$ $r = \sqrt{\frac{I}{Bd + a(H - d)}}$
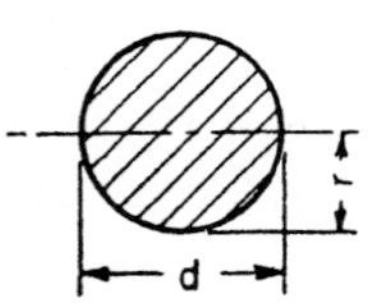	$I = \frac{\pi d^4}{64} = \frac{\pi r^4}{4} = \frac{A}{4}r^2$ $= 0.05d^4$ (approx)	$\frac{I}{c} = \frac{\pi d^3}{32} = \frac{\pi r^3}{4} = \frac{A}{4}r$ $= 0.1d^3$ (approx)	$\sqrt{\frac{I}{A}} = \frac{r}{2} = \frac{d}{4}$

Table 3.2.6 Properties of Various Cross Sections* *(Continued)*

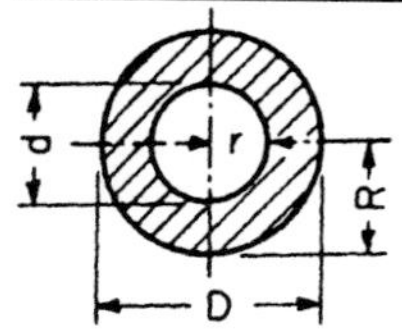 $d_m = \frac{1}{2}(D+d)$ $S = \frac{1}{2}(D-d)$	$I = \frac{\pi}{64}(D^4 - d^4)$ $= \frac{\pi}{4}(R^4 - r^4)$ $= \frac{1}{4}A(R^2 + r^2)$ $= 0.05(D^4 - d^4)$ (approx)	$\frac{I}{c} = \frac{\pi}{32}\frac{D^4 - d^4}{D}$ $= \frac{\pi}{4}\frac{R^4 - r^4}{R}$ $= 0.8d_m^2 s$ (approx) when $\frac{s}{d_m}$ is very small	$\sqrt{\frac{I}{A}} = \frac{\sqrt{R^2 + r^2}}{2} = \frac{\sqrt{D^2 + d^2}}{4}$
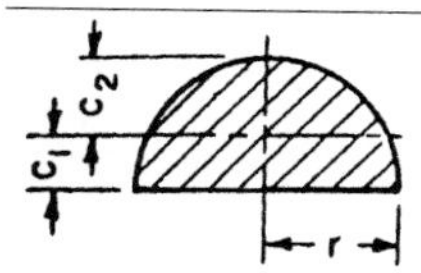	$I = r^4\left(\frac{\pi}{8} - \frac{8}{9\pi}\right)$ $= 0.1098r^4$	$\frac{I}{c_2} = 0.1908r^3$ $\frac{I}{c_1} = 0.2587r^2$ $c_1 = 0.4244r$	$\sqrt{\frac{I}{A}} = \frac{\sqrt{9\pi^2 - 64}}{6\pi}r = 0.264r$
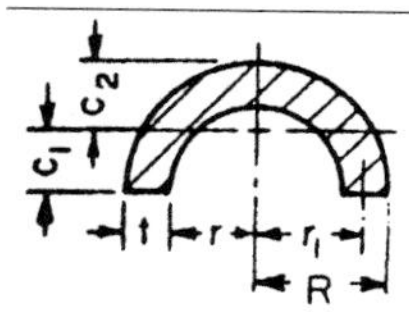	$I = 0.1098(R^4 - r^4)$ $- \frac{0.283R^2r^2(R-r)}{R+r}$ $= 0.3tr_1^3$ (approx) when $\frac{t}{r_1}$ is very small	$c_1 = \frac{4}{3\pi}\frac{R^2 + Rr + r^2}{R+r}$ $c_2 = R - c_1$	$\sqrt{\frac{I}{A}} = \sqrt{\frac{2I}{\pi(R^2 - r^2)}}$ $= 0.31r_1$ (approx)
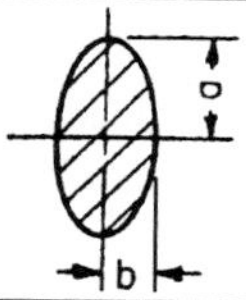	$I = \frac{\pi a^3 b}{4} = 0.7854a^3b$	$\frac{I}{c} = \frac{\pi a^2 b}{4} = 0.7854a^2b$	$r = \frac{a}{2}$
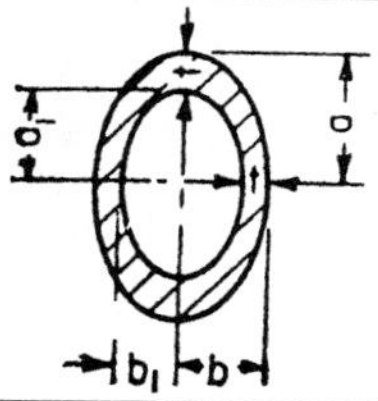	$I = \frac{\pi}{4}(a^3b - a_1^3b_1)$ $= \frac{\pi}{4}a^2(a+3b)t$ (approx)	$\frac{I}{c} = \frac{\pi}{4}a(a+3b)t$ (approx)	$r = \sqrt{\frac{I}{(\pi ab - a_1b_1)}}$ $= \frac{a}{2}\sqrt{\frac{a+3b}{a+b}}$ (approx)
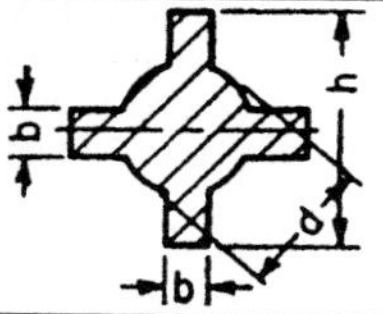	$I = \frac{1}{12}\left[\frac{3\pi}{16}d^4 + b(h^3 - d^3) + b^3(h-d)\right]$ $\frac{I}{c} = \frac{1}{6h}\left[\frac{3\pi}{16}d^4 + b(h^3 + d^3) + b^3(h-d)\right]$		$r = \sqrt{\frac{I}{\pi\frac{d^2}{4} + 2b(h-d)}}$ (approx)
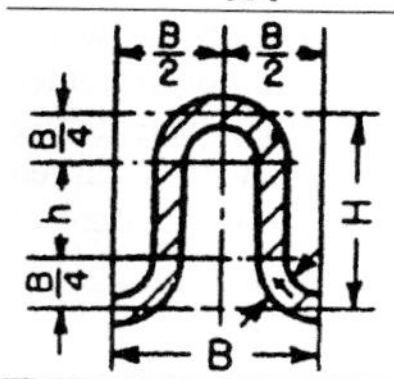	$I = \frac{t}{4}\left(\frac{\pi B^3}{16} + B^2h + \frac{\pi Bh^2}{2} + \frac{2}{3}h^3\right)$ $h = H - \frac{1}{2}B$ $\frac{I}{c} = \frac{2I}{H+t}$		$r = \sqrt{\frac{I}{2\left(\frac{\pi B}{4} + h\right)t}}$

(Continued)

Table 3.2.6 Properties of Various Cross Sections* (*Continued*)

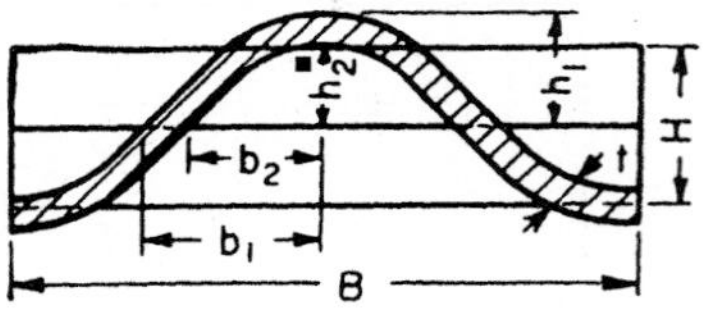

Corrugated sheet iron, parabolically curved

$$I = \frac{64}{105}\left(b_1h_1^2 - b_2h_2^3\right), \text{ where}$$

$$h_1 = \tfrac{1}{2}(H+t) \quad b_1 = \frac{1}{4}(B+2.6t)$$

$$h_2 = \tfrac{1}{2}(H-t) \quad b_2 = \tfrac{1}{4}(B-2.6t)$$

$$\frac{I}{c} = \frac{2I}{H+t}$$

$$r = \sqrt{\frac{3I}{t(2B+5.2H)}}$$

Approximate values of *least* radius of gyration r

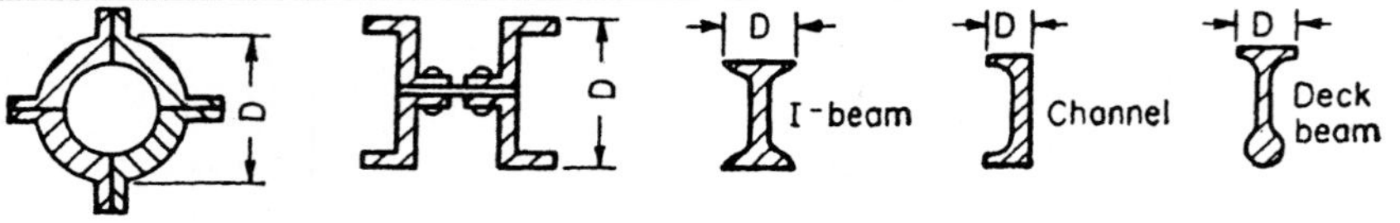

$r =$	$0.36336D$	$0.295D$	$D/4.58$	$D/3.54$	$D/6$

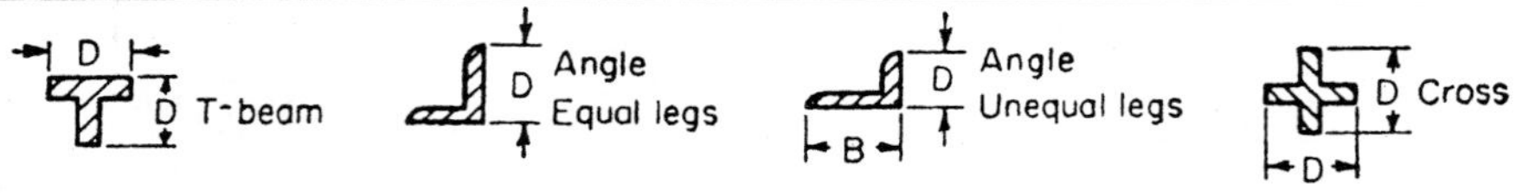

$r =$	$D/4.74$	$D/5$	$BD/[2.6(B+D)]$	$D/4.74$

* Some of the cross sections depicted in this table will be encountered most often in machinery as castings, forgings, or individual sections assembled and joined mechanically (or welded). A number of the sections shown are obsolete and will be encountered mainly in older equipment and/or building structures.

in which M = bending moment with respect to the section in question, α = the angle which the plane of bending moment or the plane of the loads makes with the y axis, $M \cos \alpha$ = the component of bending moment causing bending about the principal axis which has been designated as the X axis, $M \sin \alpha$ = the component of bending moment causing bending about the principal axis which has been designated as the Y axis. The sign of the two terms for unit stress may be determined by inspection in the usual way, and the result will be tension or compression as determined by the algebraic sum of the two terms.

In general, it may be stated that when the plane of the bending moment coincides with one of the principal axes, the other principal axis is the neutral axis. This is the ordinary case, in which the ordinary formula for unit stress may be applied. When the plane of the bending moment does not coincide with one of the principal axes, the above formula for oblique loading may be applied.

Internal Moment Beyond the Elastic Limit

Ordinarily, the expression $M = SI/c$ is used for stresses above the elastic limit, in which case S becomes an experimental coefficient S_R, the **modulus of rupture,** and the formula is empirical. The true relation is obtained by applying a stress-strain diagram from a tension and compression test to the cross section from a tension and compression test, as in Fig. 3.2.34. Figure 3.2.34 shows the side of a beam of depth d under flexure beyond its elastic limit; line 1-1 shows the distorted cross section; line 3-3, the usual rectilinear relation of stress to strain; and line 2-2, an actual stress-strain diagram, applied to the cross section of the beam, compression above and tension below. The neutral axis is then below the gravity axis. The **outer material** may be expected to develop **greater ultimate strength** than in simple stress, because of the reinforcing action of material nearer the neutral axis that is not yet overstrained. This leads to an **equalization of stress** over the cross section. S_R exceeds the ultimate strength S_M in tension as follows: for cast iron, $S_R = 2S_M$; for sandstone, $S_R = 3S_M$; for concrete, $S_R = 2.2S_M$; for wood (green), $S_R = 2.3S_M$.

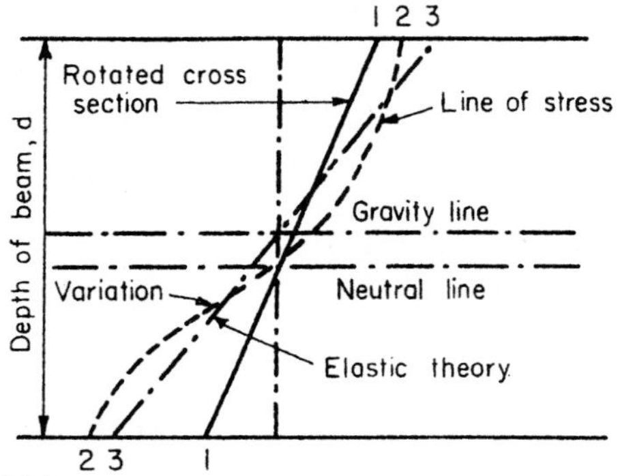

Figure 3.2.34

In the case of steel I-beams, failure begins practically when the elastic limit in the compression flange is reached.

Because of the support of adjoining material, the **elastic limit in flexure** S_p is also greater than in tension, depending upon the relation of breadth to depth of section. For the same breadth, the difference decreases with increase of height. No difference will occur in the case of an I-beam, or with hard materials.

Wide plates will not expand and contract freely, and the value of E will be increased on account of side constraint. As a consequence of lateral contraction of the fibers of the tension side of a beam and lateral swelling of fibers at the compression side, the cross section becomes distorted to a trapezoidal shape, and the neutral axis is at the center of gravity of the trapezoid. Strictly, this shape is one with a curved perimeter, the radius being r_c/μ, where r_c is the radius of curvature of the neutral line of the beam, and μ is Poisson's ratio.

Deflection of Beams

When a beam is subjected to bending, the fibers on one side elongate, while the fibers on the other side shorten (Fig. 3.2.35). These changes in length cause the beam to deflect. All points on the beam except those directly over the support fall below their original position, as shown in Figs. 3.2.31 and 3.2.35.

The **elastic curve** is the curve taken by the neutral axis. The radius of curvature at any point is

$$r_c = EI/M$$

A beam bent to a **circular curve** of constant radius has a constant bending moment.

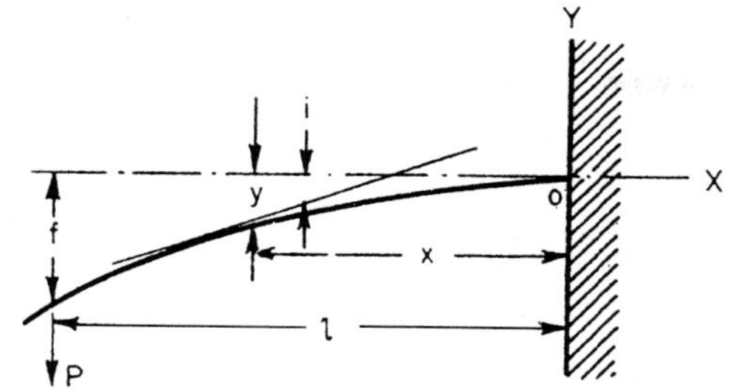

Figure 3.2.35

Replacing r_c in the equation by its approximate geometrical value, $1/r_c = d^2y/dx^2$, the fundamental equation from which the elastic curve of a bent beam can be developed and the deflection of any beam obtained is

$$M = EI\ d^2y/dx^2 \quad \text{(approx)}$$

Substituting the value of M, in terms of x, and integrating once, gives the slope of the tangent to the elastic curve of the beam at point x; $\tan i = dy/dx = \int_0^x M\,dx/(EI)$. Since i is usually small, $\tan i = i$, expressed in radians. A second integration gives the vertical deflection of any point of the elastic curve from its original position.

EXAMPLE. In the cantilever beam shown in Fig. 3.2.35, the bending moment at any section $= -P(l - x) = EI\ d^2y/(dx)^2$. Integrate and determine constant by the condition that when $x = 0$, $dy/dx = 0$. Then $EI\ dy/dx = -\ Plx + \frac{1}{2}Px^2$. Integrate again; and determine constant by the condition that when $x = 0$, $y = 0$. Then $EIy = -\frac{1}{2}Plx^2 + Px^3/6$. This is the equation of the elastic curve. When $x = l$, $y = f = -Pl^3/(3EI)$. In general, the two constants of integration must be determined simultaneously.

Deflection in general, f, may be expressed by the equation $f = Pl^3/(mEI)$, where m is a coefficient. See Tables 3.2.2 and 3.2.4 for values of f for beams of various sections and loadings. For coefficients of deflection of wooden beams and structural steel shapes, see Secs. 4.2 and 4.12.

Since I varies as the cube of the depth, the **stiffness,** or inverse deflection, of various **beams** varies, other factors remaining constant, inversely as the load, inversely as the cube of the span, and directly as the cube of the depth. This deflection is due to bending moment only. In general, however, the bending of beams involves transverse shearing stresses which cause **shearing strains** and thus **add to the total deflection.** The contribution of shearing strain to overall deflection becomes significant only when the beam span is very short. These strains may affect substantially the strength as well as the deflection of beams. When deflection due to transverse shear is to be accounted for, the differential equation of the elastic curve takes the form

$$EI\frac{d^2y}{dx^2} = EI\left(\frac{d^2y_b}{dx^2} + \frac{d^2y_s}{dx^2}\right) = M - \frac{kEI}{AG} \times \frac{d^2M}{dx^2}$$

where k is a factor dependent upon the beam cross section. Sergius Sergev, in "The Effect of Shearing Forces on the Deflection and Strength of Beams" (1947) gives $k = 1.2$ for rectangular sections, 10/9 for circular sections, and 2.4 for I-beams. He also points out that in the case of a deep, rectangular-section cantilever, carrying a concentrated load at the free end, the deflection due to shear may be up to 3.1 percent of that due to bending moment; if this beam supports a uniformly distributed load, it may be up to 4.1 percent. A deep, simple beam deflection may increase up to 15.6 percent when carrying a uniformly distributed load and up to 12.5 percent when the load is concentrated at midspan.

Design of beams may be based on **strength** (stress) or on **stiffness** if deflection must be limited. Deflection rather than stress becomes the criterion for design. For example in machine tools for which the relative positions of tool and workpiece must be maintained while the cutting loads are applied during operation. Similarly, large steam-turbine shafts supported on two end bearings must maintain alignment and tight critical clearances between the rotating blade assemblies and the stationary stator blades during operation. When more than one beam shares a load, each beam will assume a portion of the load that is proportional to its stiffness. **Superposition** may be used in connection with both stresses and deflections.

EXAMPLE. (Fig. 3.2.36). Two wooden stringers—one (A) 8 × 16 in in cross section and 20 ft in span, the other (B) 8 in × 8 in × 16 ft—carrying the center load P_0 = 22,000 lb are required, the load carried by each stringer. The deflections f of the two stringers must be equal. Load on $A = P_1$, and on $B = P_2$. $f = P_1l_1^3/(48EI_1) = P_2l_2^3/(48EI_2)$. Then $P_1/P_2 = l_2^3I_1/(l_1^3I_2) = 4$. $P_0 = P_1 + P_2 = 4P_2 + P_2$, whence P_2 = 22,000/5 = 4,400 lb (1,998 kgf) and P_1 = 4 × 4,400 = 17,600 lb (7,990 kgf).

Relation between Deflection and Stress

Combine the formula $M = SI/c = Pl/n$, where n is a constant, P = load, and l = span, with formula $f = Pl^3/(mEI)$, where m is a constant. Then

$$f = C''Sl^2/(Ec)$$

where C'' is a new constant $= n/m$. Other factors remaining the same, the **deflection varies directly as the stress and inversely as** E. If the span is constant, a shallow beam will submit to greater deformations than a deeper beam without exceeding a safe stress. If depth is constant, a beam of double span will attain a given deflection with only one-quarter the stress. Values of n, m, and C'' are given in Table 3.2.7 (for other values, see Table 3.2.2).

Graphical Relations

Referring to Fig. 3.2.37, the shear V acting at any section is equal to the total load on the right of the section, or

$$V = \int w\,dx$$

Since $w\,dx$ is the product of w, a loading intensity (which is expressed as a vertical height in the load diagram), and dx, an elementary length along the horizontal, evidently $w\,dx$ is the area of a small vertical strip of the **load diagram.** Then $\int w\,dx$ is the summation of all such vertical strips between two indefinite points. Thus, to obtain the shear in any section mn, find the area of the load diagram up to that section, and draw a second diagram called the **shear diagram,** any ordinate of which is proportional to the shear, or to the area in the load diagram to the right of mn. Since $V = dM/dx$,

$$\int V\,dx = M$$

By similar reasoning, a **moment diagram** may be drawn, such that the ordinate at any point is proportional to the area of the shear diagram to the right of that point. Since $M = EI\ d^2f/dx^2$,

$$\int M\,dx = EI\,(df/dx + C) = EI(i + C)$$

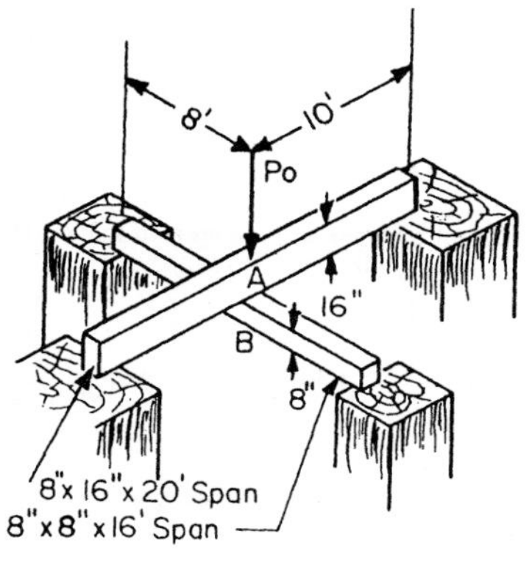

Figure 3.2.36

Table 3.2.7 Constants for the Relation between Deflection and Stress for Various Beams and Loading

Beam	Load	n	m	C''
Cantilever	Concentrated at end	1	3	1/3
Cantilever	Uniform	2	8	1/4
Simple	Concentrated at center	4	48	1/12
Simple	Uniform	8	384/5	5/48
Fixed ends	Concentrated at center	8	192	1/24
Fixed ends	Uniform	12	384	1/32
One end fixed, one end supported	Concentrated at center	16/3	768/7	7/144
One end fixed, one end supported	Uniform	128/9	185	1/13
Simple	Uniformly varying, maximum at center	6	60	1/10

if I is constant. Here C is a constant of integration. Thus i, the slope or grade of the elastic curve at any point, is proportional to the area of the moment diagram $\int M\,dx$ up to that point; and a **slope diagram** may be derived from the moment diagram in the same manner as the moment diagram was derived from the shear diagram.

If I is not constant, draw a new curve whose ordinates are M/I and use these M/I ordinates just as the M ordinates were used in the case where I was constant; that is, $\int (M/I)\,dx = E(i + C)$. The ordinate at any point of the slope curve is thus proportional to the area of the M/I curve to the right of that point. Again, since $iE = Edf/dx$.

$$\int iE\,dx = \int E\,df = E(f + C)$$

and thus the ordinate f to the elastic curve at any point is proportional to the area of the slope diagram $\int i\,dx$ up to that point. The equilibrium polygon may be used in drawing the **deflection curve** directly from the M/I diagram.

Thus, the five curves of load, shear, moment, slope, and deflection are so related that each curve is derived from the previous one by a process of graphical integration, and with proper regard to scales the deflection is thereby obtained.

The vertical distance from any point A (Fig. 3.2.38) on the elastic curve of a beam to the tangent at any other point B equals the moment of the area of the $M/(EI)$ diagram from A to B about A. This distance, the **tangential deviation** t_{AB}, may be used with the slope-area relation and the geometry of the elastic curve to obtain deflections. These theorems, together with the equilibrium equations, can be used to compute reactions in the case of statically **indeterminate beams.**

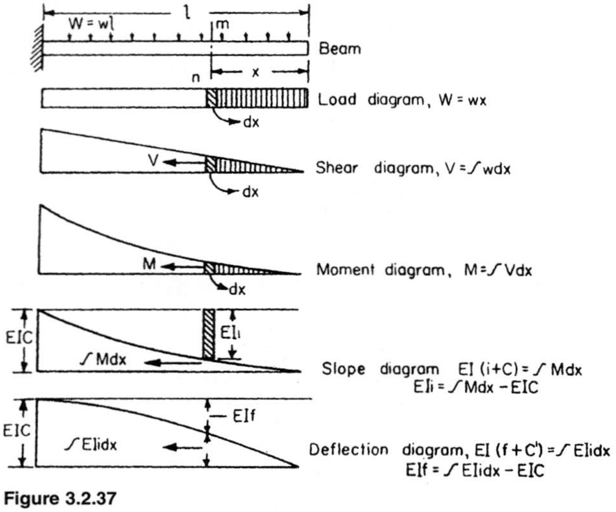

Figure 3.2.37

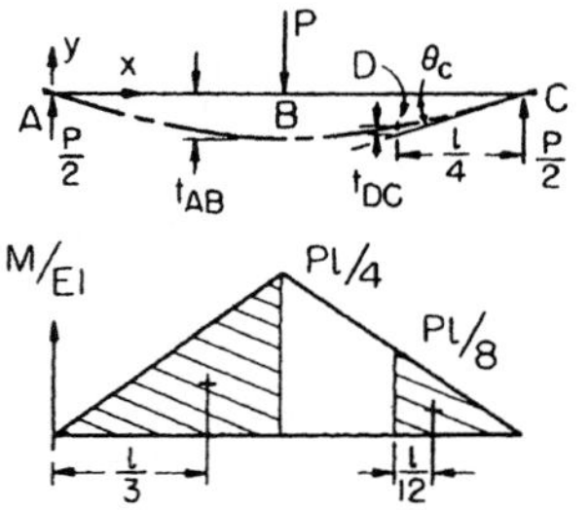

Figure 3.2.38

EXAMPLE. The deflections of points B and D (Fig. 3.2.38) are

$$y_B = -t_{AB} = \text{moment area } \frac{M}{EI}\bigg|_A^B A$$

$$= -\frac{1}{EI} \times \frac{Pl}{4} \times \frac{l}{4} \times \frac{l}{3} = -\frac{Pl^3}{48EI}$$

$$\theta_c = \nabla\theta\bigg|_B^C = \text{area } \frac{M}{EI}\bigg|_B^C = \frac{1}{EI} \times \frac{Pl}{4} \times \frac{l}{4} = \frac{Pl^2}{16EI}$$

$$y_D = -\left(\theta_c \times \frac{l}{4} - t_{DC}\right)$$

$$= -\frac{Pl^2}{16EI} \times \frac{l}{4} \times \frac{l}{EI} \times \frac{Pl}{8} \times \frac{l}{8} \times \frac{l}{12} = -\frac{11Pl^3}{768EI}$$

Resilience of Beams

The external work of a load gradually applied to a beam, and which increases from zero to P, is $\frac{1}{2}Pf$ and equals the **resilience** U. But, from the formulas $P = nSI/(cl)$ and $f = nSl^2/(mcE)$, where n and m are constants that depend upon loading and supports, S = fiber stress, c = distance from neutral axis to outer fiber, and l = length of span. Substitute for P and f, and

$$U = \frac{n^2}{m}\left(\frac{k}{c}\right)^2 \frac{S^2V}{2E}$$

where k is the radius of gyration and V the volume of the beam. For values of U, see Table 3.2.1.

The resilience of beams of similar cross section at a given stress is proportional to their volumes. The **internal resilience,** or the elastic deformation energy in the material of a beam in a length x is dU, and

$$U = \tfrac{1}{2}\int M^2\,dx/(EI) = \tfrac{1}{2}\int M\,di$$

where M is the moment at any point x, and di is the angle between the tangents to the elastic curve at the ends of dx. The values of resilience and deflection in special cases are easily developed from this equation.

Rolling Loads

Rolling or **moving loads** are those loads which may change their position on a beam. Figure 3.2.39 represents a beam with two equal concentrated moving loads, such as two wheels on a crane girder, or the wheels of a truck on a bridge. Since the maximum moment occurs where the shear is zero, it is evident from the shear diagram that the maximum moment will occur under a wheel. $x < a/2$:

$$R_1 = P\left(1 - \frac{2x}{l} + \frac{a}{l}\right)$$

$$M_2 = \frac{Pl}{2}\left(1 - \frac{a}{l} + \frac{2x}{l}\frac{a}{l} - \frac{4x^2}{l^2}\right)$$

$$R_2 = P\left(1 + \frac{2x}{l} - \frac{a}{l}\right)$$

$$M_1 = \frac{Pl}{2}\left(1 - \frac{a}{l} - \frac{2a^2}{l^2} + \frac{2x}{l}\frac{3a}{l} - \frac{4x^2}{l^2}\right)$$

$$M_2 \text{ max when } x = \tfrac{1}{4a}$$

$$M_1 \text{ max when } x = \tfrac{3}{4a}$$

$$M_{max} = \frac{Pl}{2}\left(1 - \frac{a}{2l}\right)^2 = \frac{P}{2l}\left(l - \frac{a}{2}\right)^2$$

EXAMPLE. Two wheel loads of 3,000 lb each, spaced on 5-ft centers, move on a span of $l = 15$ ft, $x = 1.25$ ft, and $R_2 = 2{,}500$ lb. $\therefore M_{max} = M_2 = 2{,}500 \times 6.25$ $(1{,}135 \times 1.90) = 15{,}600$ lb · ft (2,159 kgf · m).

Figure 3.2.40 shows the condition when two equal loads are equally distant on opposite sides of the center. The moment is equal under the two loads.

If the **two moving loads** are **of unequal weight,** the condition for **maximum moment** is that the maximum moment will occur under the heavy wheel when the center of the beam bisects the distance between the resultant of the loads and the heavy wheel. Figure 3.2.41 shows the position and the shear and moment diagrams.

When **several wheel loads** constituting a system occur, the several suspected wheels must be examined in turn to determine which will cause the greatest moment. The **position for the greatest moment** that can occur under a given wheel is, as stated, when the center of the span bisects the distance between the wheel in question and the resultant of all the loads then on the span. The **position for maximum shear** at the support will be when one wheel is passing off the span.

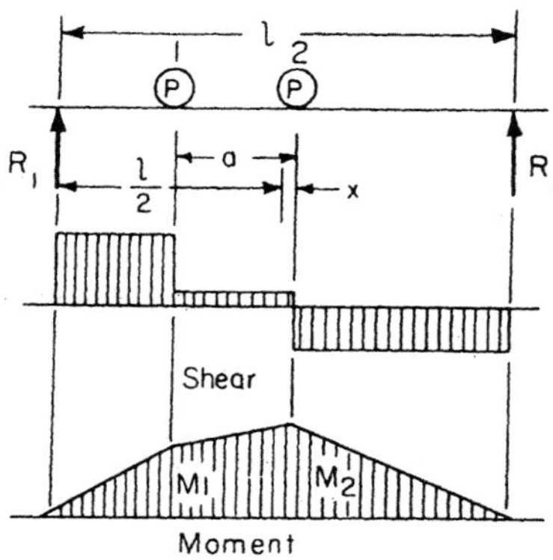

Figure 3.2.39

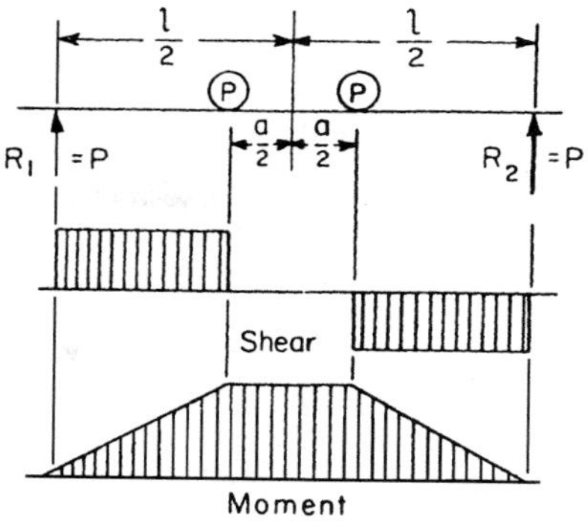

Figure 3.2.40

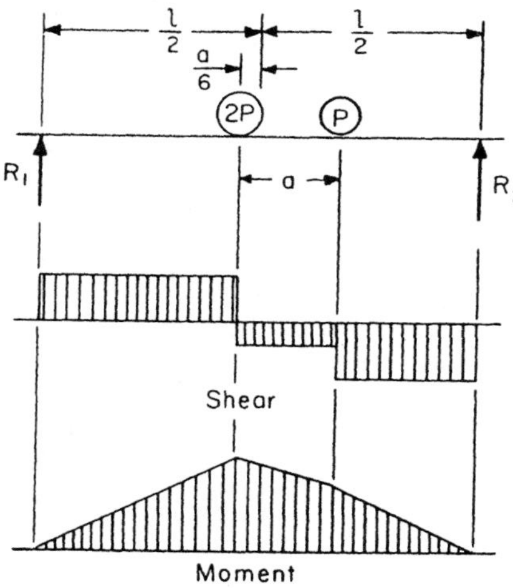

Figure 3.2.41

Constrained Beams

Constrained beams are those held or "built in" at one or both ends so that the tangent to the elastic curve remains fixed in direction. These beams are held at the ends in such a manner as to allow free horizontal motion, as illustrated by Fig. 3.2.42. A constrained beam is stiffer than a simple beam of the same material, because of the modification of the moment by an end resisting moment. Figure 3.2.43 shows the two most common cases of constrained beams. Also see Table 3.2.2.

Continuous Beams

A continuous beam is one resting upon several supports which may or may not be in the same horizontal plane. The general discussion for beams holds for continuous beams. $S_V A = V$, $SI/c = M$, and $d^2f/dx^2 = M/(EI)$. The **shear** at any section is equal to the algebraic sum of the

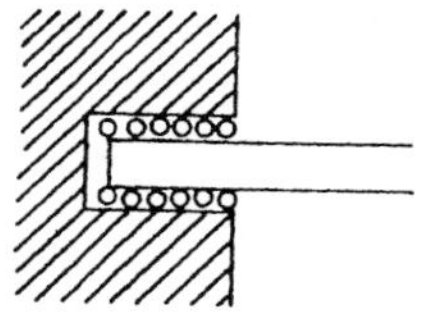

Figure 3.2.42

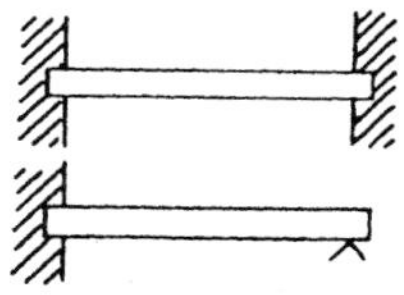

Figure 3.2.43

components parallel to the section of all external forces on either side of the section. The bending moment at any section is equal to the moment of all external forces on either side of the section. The relations stated above between shear and moment diagrams hold true for continuous beams. The bending moment at any section is equal to the bending moment at any other section, plus the shear at that section times its arm, plus the product of all the intervening external forces times their respective arms. To illustrate (Fig. 3.2.44):

$$V_x = R_1 + R_2 + R_3 - P_1 - P_2 - P_3$$

$$M_x = R_1(l_1 + l_2 + x) + R_2(l_2 + x) + R_3 x$$
$$- P_1(l_2 + c + x) - P_2(b + x) - P_3 a$$

$$M_x = M_3 + V_3 x - P_3 a$$

Table 3.2.8 gives the value of the moment at the various supports of a uniformly loaded continuous beam over equal spans, and it also gives the values of the shears on each side of the supports. Note that the shear is of opposite sign on either side of the supports and that the sum of the two shears is equal to the reaction.

Figure 3.2.45 shows the relation between the moment and shear diagrams for a uniformly loaded continuous beam of four equal spans (see Table 3.2.8). Table 3.2.8 also gives the **maximum bending moment** which will occur **between supports,** and in addition the position of this moment and the points of inflection (see Fig. 3.2.46).

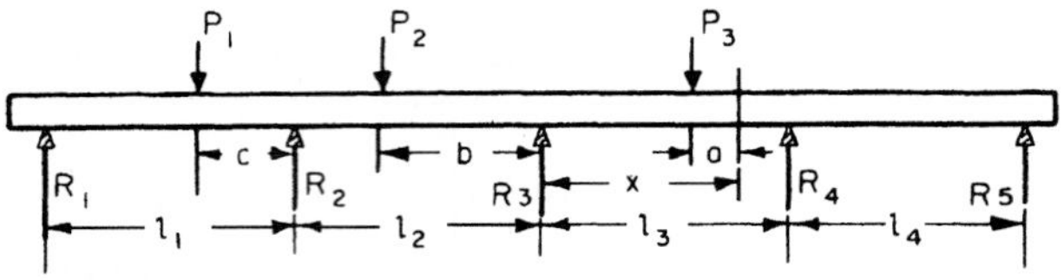

Figure 3.2.44

Figure 3.2.46 shows the values of the functions for a uniformly loaded continuous beam resting on three equal spans with four supports.

Continuous beams are **stronger and much stiffer than simple beams.** However, a small, unequal subsidence of piers will cause serious changes in sign and magnitude of the bending stresses, reactions, and shears.

Maxwell's Theorem When a number of loads rest upon a beam, the deflection at any point is equal to the sum of the deflections at this point due to each of the loads taken separately. Maxwell's theorem states that if unit loads rest upon a beam at two points *A* and *B*, the deflection at *A* due to the unit load at *B* equals the deflection at *B* due to the unit load at *A*.

Castigliano's theorem states that the deflection of the point of application of an external force acting on a beam is equal to the partial derivative of the work of deformation with respect to this force. Thus, if *P* is the force, *f* the deflection, and *U* the work of deformation, which equals the resilience,

$$dU/dP = f$$

Table 3.2.8 Uniformly Loaded Continuous Beams over Equal Spans
(Uniform load per unit length = *w*; length of equal span = *l*)

Number of supports	Notation of supports of span	Shear on each side of support. *L* = left, *R* = right. Reaction at any support is *L* + *R*		Moment over each support	Max moment in each span	Distance to point of max moment, measured to right from support	Distance to point of inflection, measured to right from support
		L	*R*				
2	1 or 2	0	1/2	0	0.125	0.500	None
3	1	0	3/8	0	0.0703	0.375	0.750
	2	5/8	5/8	1/8	0.0703	0.625	0.250
4	1	0	4/10	0	0.080	0.400	0.800
	2	6/10	5/10	1/10	0.025	0.500	0.276, 0.724
5	1	0	11/28	0	0.0772	0.393	0.786
	2	17/28	15/28	3/28	0.0364	0.536	0.266, 0.806
	3	13/28	13/28	2/28	0.0364	0.464	0.194, 0.734
6	1	0	15/38	0	0.0779	0.395	0.789
	2	23/38	20/38	4/38	0.0332	0.526	0.268, 0.783
	3	18/38	19/38	3/38	0.0461	0.500	0.196, 0.804
7	1	0	41/104	0	0.0777	0.394	0.788
	2	63/104	55/104	11/104	0.0340	0.533	0.268, 0.790
	3	49/104	51/104	8/104	0.0433	0.490	0.196, 0.785
	4	53/104	53/104	9/104	0.0433	0.510	0.215, 0.804
8	1	0	56/142	0	0.0778	0.394	0.789
	2	86/142	75/142	15/142	0.0338	0.528	0.268, 0.788
	3	67/142	70/142	11/142	0.0440	0.493	0.196, 0.790
	4	72/142	71/142	12/142	0.0405	0.500	0.215, 0.785
Values apply to:		wl	wl	wl^2	wl^2	l	l

NOTE: The numerical values given are coefficients of the expressions at the foot of each column.

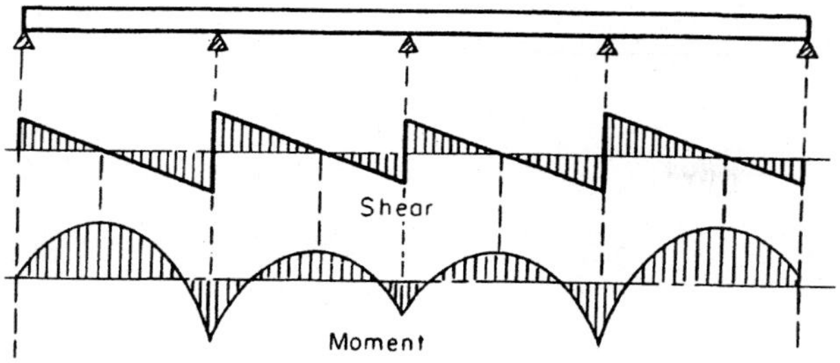

Figure 3.2.45

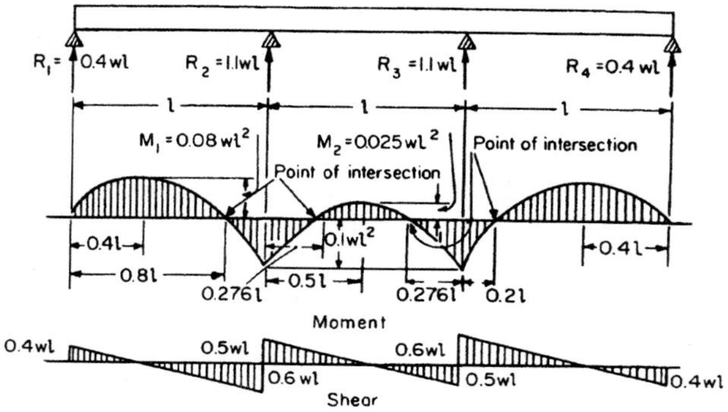

Figure 3.2.46

According to the **principle of least work,** the deformation of any structure takes place in such a manner that the work of deformation is a minimum.

Similarly, angular rotation of a point in a structure θ can be determined by $dU/dM_t = \theta$, where M_t is the bending or torsional moment.

In applying Castigliano's theorem, the strain energy must be expressed as a function of the load. If the structure is subjected to combination of loads, for example, axial force N, bending moment M, shearing force V, and torsional moment M_t, the strain energy is

$$U = \int \frac{N^2\,dx}{2EA} + \int \frac{M^2\,dx}{2EI} + \int \frac{f_s V^2\,dx}{2GA} + \int \frac{M_t^2\,dx}{2GJ}$$

Note that the last integral is valid only for a circular cross section. Using Castigliano's theorem, the displacement f is given by

$$f = \frac{1}{EA}\int N\frac{\partial N}{\partial P}dx + \frac{1}{EI}\int M\frac{\partial M}{\partial P}dx + \frac{1}{GA}\int f_s V\frac{\partial V}{\partial P}dx + \frac{1}{GJ}\int M_t\frac{\partial M_t}{\partial P}dx$$

Beams of Uniform Strength

Beams of uniform strength vary in section so that the unit stress S remains constant, the I/c varies as M. For **rectangular beams,** of breadth b and depth d, $I/c = bd^2/6$; and $M = Sbd^2/6$. Thus, for a cantilever beam of rectangular cross section, under a load P, $Px = Sbd^2/6$. If b is constant, d^2 varies with x, and the profile of the shape of the beam will be a parabola, as Fig. 3.2.47. If d is constant, b will vary as x and the beam will be triangular in plan, as shown in Fig. 3.2.48.

Shear at the end of a beam necessitates a modification of the forms determined above. The area required to resist shear will be P/S_v in a cantilever and R/S_v in a simple beam. The dotted extensions in Figs. 3.2.47 and 3.2.48 show the changes necessary to enable these cantilevers to resist shear. The extra material and cost of fabrication, however, make many of the forms impractical.

Table 3.2.9 shows some of the simple **sections of uniform strength.** In none of these, however, is shear taken into account.

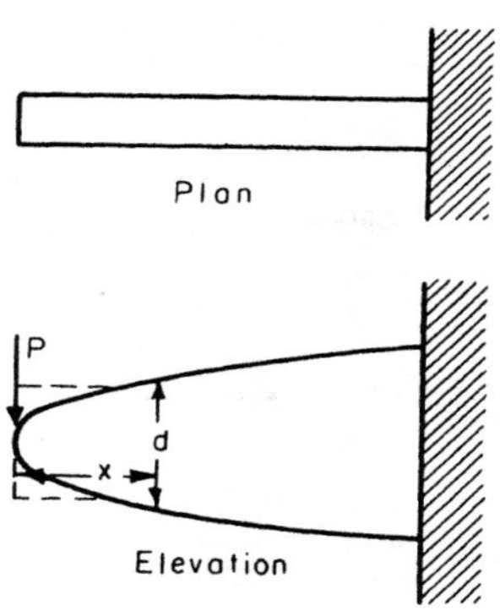

Figure 3.2.47

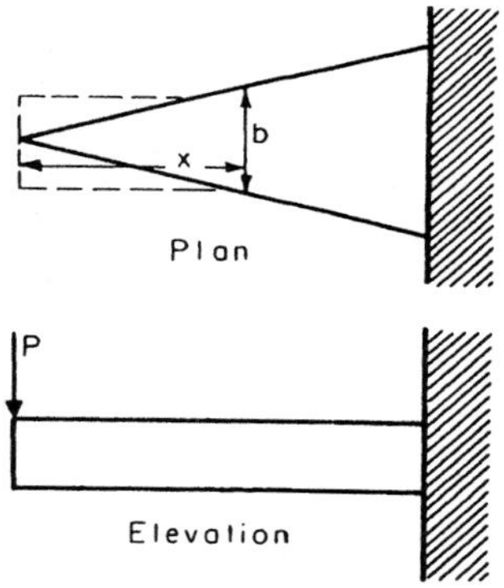

Figure 3.2.48

TORSION

Under torsion, a bar (Fig. 3.2.49) is twisted by a couple of magnitude Pp. Elements of the surface becomes helices of angle d, and a radius rotates through an angle θ in a length l, both d and θ being expressed in radians. S_v = shearing unit stress at distance r from center; I_p = *polar moment of inertia*; G = shearing modulus of elasticity. It is assumed that the cross sections remain planar surfaces. The strain on the cross section is wholly tangential, and is zero at the center of the section. Note that $ld = r\theta$.

In the case of a **circular cross section,** the stress S_v increases directly as the distance of the strained element from the center.

The polar moment of inertia I_p for any section may be obtained from $I_p = I_1 + I_2$, where I_1 and I_2 are the rectangular moments of inertia of the section about any two lines at right angles to each other, through the center of gravity.

The **external twisting moment** M_t is balanced by the internal resisting moment.

For strength, $M_t = S_v I_p / r$.

For stiffness, $M_t = aGI_p/l$.

The torsional resilience $U = \frac{1}{2}Ppa = S_v^2 I_p l/(2r^2G) = a^2GI_p/(2l)$.

The state of stress on an element taken from the surface of the shaft, as in Fig. 3.2.50, is pure shear. Pure tension exists at right angles to one 45° helix and pure compression at right angles to the opposite helix.

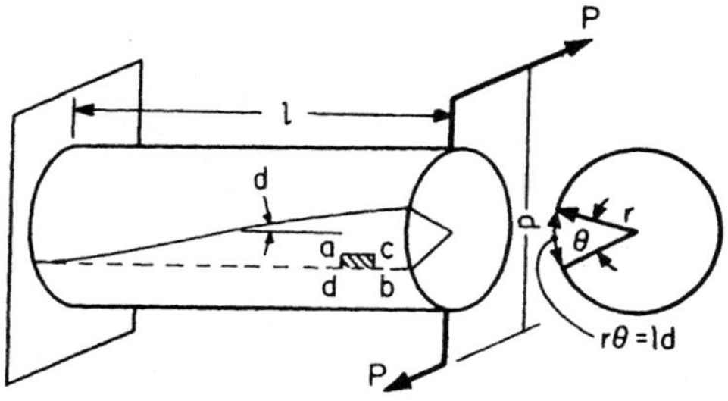

Figure 3.2.49

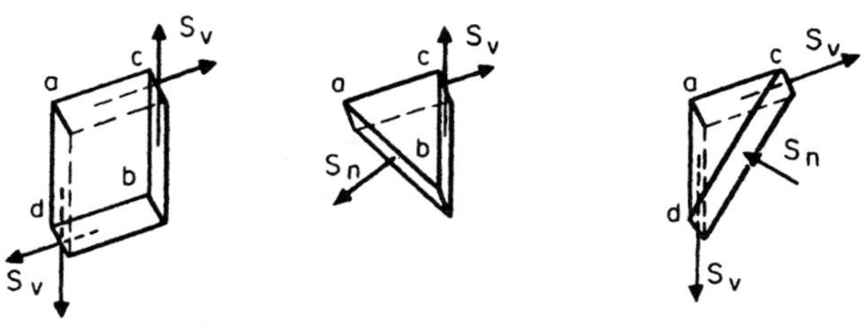

Figure 3.2.50

Table 3.2.9 Beams of Uniform Strength (in Bending)

Beam	Cross section	Elevation and plan	Formulas
1. Fixed at one end, load P concentrated at other end			
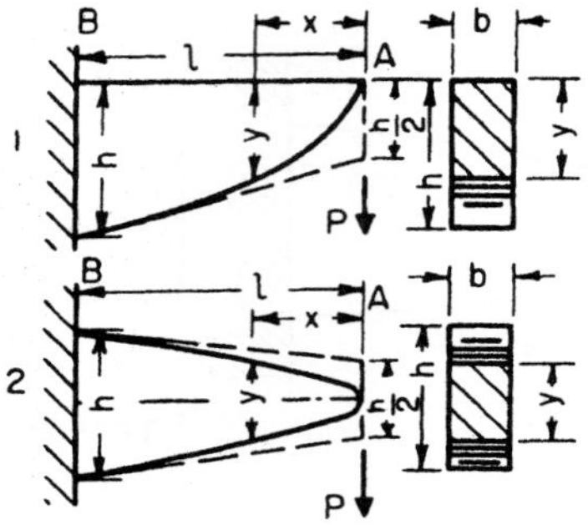	Rectangle: width (b) constant, depth (g) variable	Elevation: 1, top, straight line; bottom, parabola. 2, complete parabola Plan: rectangle	$y^2 = \frac{6P}{bS_s}x$ $h = \sqrt{\frac{6Pl}{bS_s}}$ Deflection at A: $f = \frac{8P}{bE}\left(\frac{l}{h}\right)^3$
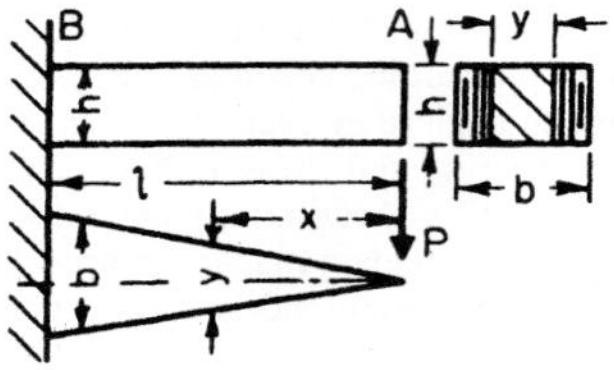	Rectangle: width (y) variable, depth (h) constant	Elevation: rectangle Plan: triangle	$y = \frac{6P}{h^2S_s}x$ $b = \frac{6Pl}{h^2S_s}$ Deflection at A: $f = \frac{6P}{bE}\left(\frac{l}{h}\right)^3$
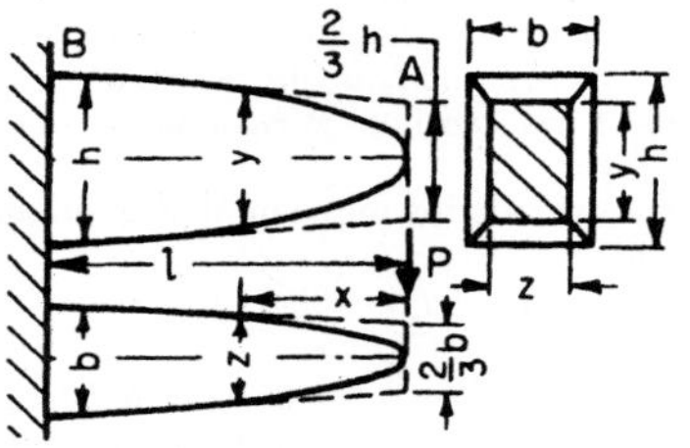	Rectangle: width (z) variable, depth (y) variable $\frac{z}{y} = k(\text{const.})$	Elevation: cubic parabola Plan: cubic parabola	$y^3 = \frac{6P}{kS_s}x$ $z = ky$ $h = \sqrt[3]{\frac{6Pl}{kS_s}}$ $b = kh$
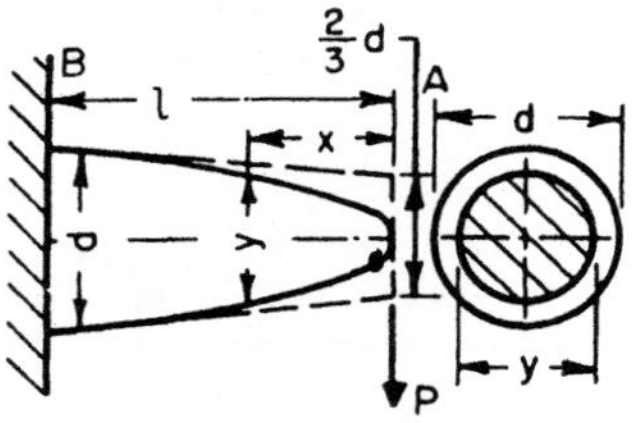	Circle: diameter (y) variable	Elevation: cubic parabola Plan: cubic parabola	$y^3 = \frac{32P}{\pi S_s}x$ $d = \sqrt[3]{\frac{32Pl}{\pi S_s}}$
2. Fixed at one end, load of total magnitude P uniformly distributed			
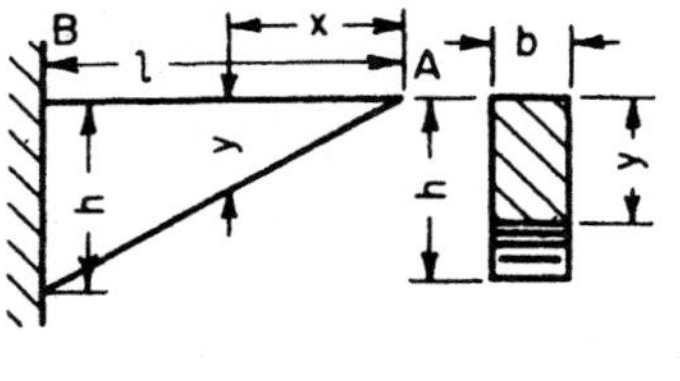	Rectangle: width (b) constant, depth (y) variable	Elevation: triangle Plan: rectangle	$y = x\sqrt{\frac{3P}{b/S}}$ $h = \sqrt{\frac{3Pl}{bS_s}}$ $f = 6\frac{P}{bE}\left(\frac{l}{h}\right)^3$
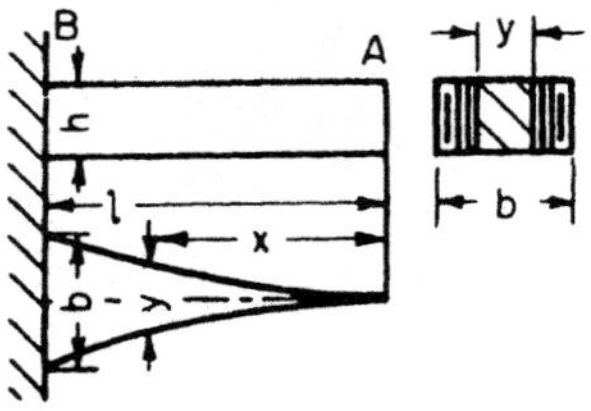	Rectangle: width (y) variable, depth (h) constant	Elevation: rectangle Plan: two parabolic curves with vertices at free end	$y = \frac{3Px^2}{lS_sh^2}$ $b = \frac{3Pl}{S_sh^2}$ Deflection at A: $f = \frac{3P}{bE}\left(\frac{l}{h}\right)^3$

Table 3.2.9 Beams of Uniform Strength (in Bending) (*Continued*)

Beam	Cross section	Elevation and plan	Formulas
2. Fixed at one end, load of total magnitude P uniformly distributed			
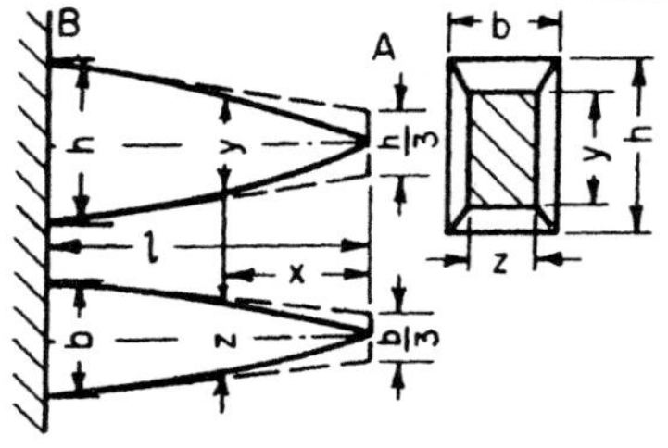	Rectangle: width (z) variable, depth (y) variable, $\frac{z}{y} = k$	Elevation: semicubic parabola Plan: semicubic parabola	$y^3 = \frac{3Px^2}{kS_s l}$ $z = ky$ $h = \sqrt[3]{\frac{3Pl}{kS_s}}$ $b = kh$
3. Supported at both ends, load P concentrated at point C			
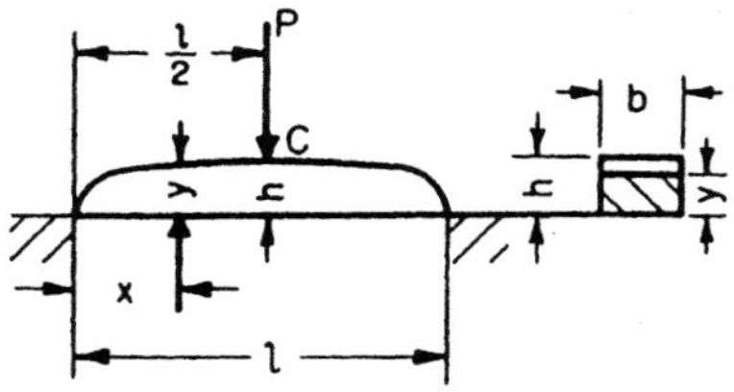	Rectangle: width (b) constant, depth (y) variable	Elevation: two parabolas, vertices at points of support Plan: rectangle	$y = \sqrt{\frac{3P}{S_s b}x}$ $h = \sqrt{\frac{3Pl}{2bS_s}}$ $f = \frac{P}{2Eb}\left(\frac{l}{h}\right)^3$
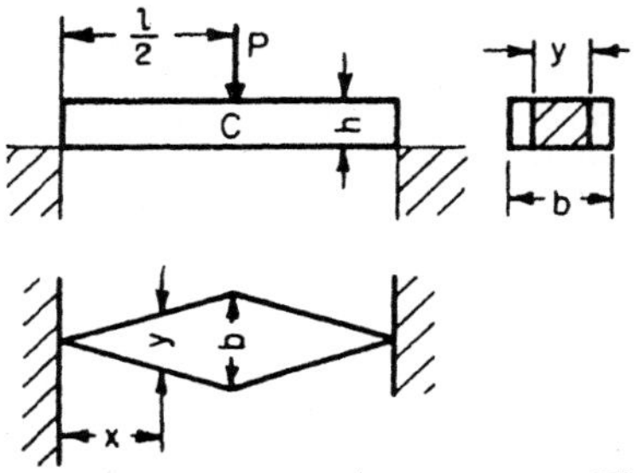	Rectangle: width (y) variable, depth (h) constant	Elevation: rectangle Plan: two triangles, vertices at points of support	$y = \frac{3P}{S_s h^2}x$ $b = \frac{3Pl}{2S_s h^2}$ $f = \frac{3Pl^3}{8Ebh^3}$
Load P moving across span			
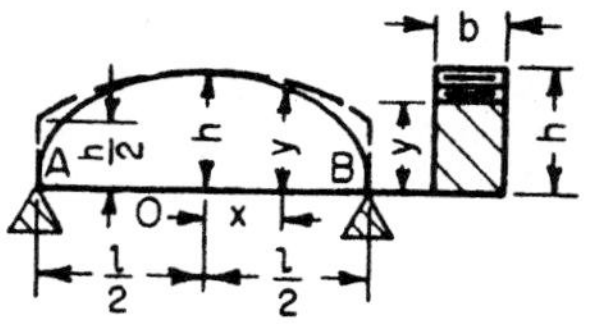	Rectangle: width (b) constant, depth (y) variable	Elevation: ellipse Major axis = l Minor axis = $2h$ Plan: rectangle	$\frac{x^2}{\left(\frac{l}{2}\right)^2} + \frac{y^2}{\frac{3Pl}{2bS_s}} = 1$ $h = \sqrt{\frac{2Pl}{2bS_s}}$
4. Supported at both ends, load of total magnitude P uniformly distributed			
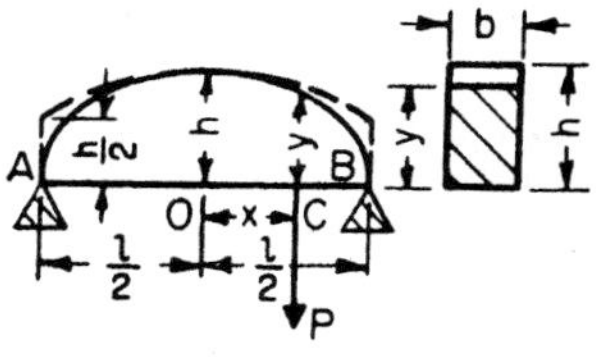	Rectangle: width (b) constant, depth (y) variable	Elevation: ellipse Plan: rectangle	$\frac{x^2}{\left(\frac{l}{2}\right)^2} + \frac{y^2}{\frac{3Pl}{4bS_s}} = 1$ $h = \sqrt{\frac{3Pl}{4bS_s}}$ Deflection at O: $f = \frac{1}{64}\frac{Pl^3}{EI}$ $= \frac{3}{16}\frac{P}{bE}\left(\frac{l}{h}\right)^3$

Reduced **formulas for shafts of various sections** are given in Table 3.2.11. When the ratio of shaft length to the largest lateral dimension in the cross section is less than approximately 2, the **end effects** may drastically affect the torsional stresses calculated.

Failure under torsion in brittle materials is a tensile failure at right angles to a helical element on the surface. Plastic materials twist off squarely. Fibrous materials separate in long strips.

Torsion of Noncircular Sections When a section is not circular, the unit stress no longer varies directly as the distance from the center. Cross sections become warped, and the greatest unit stress usually occurs at a point on the perimeter of the cross section *nearest* the axis of twist; thus, there is no stress at the corners of square and rectangular sections. The analyses become complex for noncircular sections, and the methods for solution of design problems using them most often admit only of approximations.

Torsion problems have been solved for many different noncircular cross sections by utilizing the **membrane analogy,** due to Prandtl, which makes use of the fact that the mathematical treatment of a twisted bar is governed by the same equations as for a membrane stretched over a hole which has the same shape as the bar and then inflated. The resulting three-dimensional surface provides the following: (1) The torque transmitted is proportional to twice the volume under the inflated membrane, and (2) the shear stress at any point is proportional to the slope of the curve measured perpendicular to that slope. In recent years, several other mathematical techniques have become widely used, especially with the aid of faster computational methods available from electronic computers.

By using **finite-difference methods,** the differential operators of the governing equations are replaced with difference operators which are related to the desired unknown values at a gridwork of points in the outline of the cross section being investigated.

The **finite-element method,** commonly referred to as **FEM,** deals with a spatial approximation of a complex shape which is then analyzed to determine deformations, stresses, etc. By using FEM, the exact structure is replaced with a set of simple structural elements interconnected at a finite number of nodes. The governing equations for the approximate structure can be solved exactly. Note that inasmuch as there is an exact solution for an approximate structure, the end result must be viewed, and the results thereof used, as approximate solutions to the real structure.

Using a finite-difference approach to Poisson's partial differential equation, which defines the stress functions for solid and hollow shafts with generalized contours, along with Prandl's membrane analogy, Isakower has developed a series of practical design charts (ARRAD-COMMISD Manual UN 80-5, January 1981, Department of the Army). Dimensionless charts and tables for transmitted torque and maximum shearing stress have been generated. Information for circular shafts with rectangular and circular keyways, external splines and milled flats, as well as rectangular and X-shaped torsion bars, is presented.

Assuming the stress distribution from the point of maximum stress to the corner to be parabolic, Bach derived the approximate expression, $S_sM = 9M_t/(2b^2h)$ for a rectangular section, b by h, where $h > b$. For closer results, the shearing stresses for a **rectangular section** (Fig. 3.2.51) may be expressed $S_A = M_t/(\alpha_A b^2 h)$ and $S_B = M_t/(\alpha_B b^2 h)$. The angle of twist for these shafts is $\theta = M_t l/(\beta G b^3 h)$. The factors α_A, α_B, and β are functions of the ratio h/b and are given in Table 3.2.10.

In the case of **composite sections,** such as a tee or angle, the torque that can be resisted is $M_t = G\ \theta \Sigma \beta h b^3$; the summation applies to each of the rectangles into which the section can be divided. The maximum stress occurs on the component rectangle having the largest b value. It is computed from

$$S_A = M_t \beta_A b_A / (\alpha_A \Sigma \beta h b^3)$$

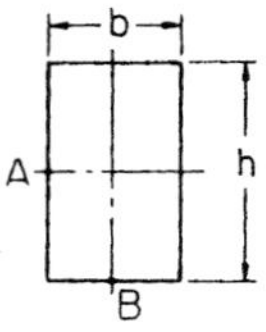

Figure 3.2.51

Torque, deflection, and work relations for some additional sections are given in Table 3.2.11.

COLUMNS

Members subjected to direct compression can be grouped into three classes. **Compression blocks** are so short (slenderness ratios below 30) that bending of member is unlikely. At the other limit, columns so slender that bending is primary, are the **long columns** defined by Euler's theory. The intermediate columns, quite common in practice, are called **short columns.**

Long columns and the more slender short columns usually fail by **buckling** when the **critical load** is reached. This is a matter of **instability;** that is, the column may continue to yield and deflect even though the load is not increased above critical. The **slenderness ratio** is the unsupported length divided by the least radius of gyration, parallel to which it can bend.

Long columns are handled by **Euler's column formula,**

$$P_{cr} = n\pi^2 EI/l^2 = n\pi^2 EA/(l/r)^2$$

The **coefficient** n accounts for **end conditions.** When the column is pivoted at both ends, $n = 1$; when one end is fixed and other rounded, $n = 2$; when both are fixed, $n = 4$; and when one end is fixed with the other free, $n = ¼$. The slenderness ratio that separates long columns from short ones depends upon the modulus of elasticity and the yield strength of the column material. When Euler's formula results in $(P_{cr}/A) > S_y$, strength rather than buckling causes failure, and the column ceases to be long. In round numbers, this **critical slenderness ratio** falls between 120 and 150. Table 3.2.12 gives additional facts concerning long columns.

Short Columns The stress in a short column may be considered partly due to compression and partly due to bending. A theoretical equation has not been derived. Empirical, though rational, expressions are, in general, based on the assumption that the permissible stress must be reduced below that which could be permitted were it due to compression only. The manner in which this reduction is made determines the type of equation as well as the slenderness ratio beyond which the equation does not apply. Figure 3.2.52 illustrates the situation. Some typical formulas are given in Table 3.2.13.

EAMPLE. A machine member unsupported for a length of 15 in has a square cross section 0.5 in on a side. It is to be subjected to compression. What maximum

Table 3.2.10 Factors for Torsion of Rectangular Shafts (Fig. 3.2.51)

$\frac{h}{b}$	1.00	1.50	1.75	2.00	2.50	3.00	4.00	5.00	6.00	8.00	10.0	∞
α_A	0.208	0.231	0.239	0.246	0.258	0.267	0.282	0.291	0.299	0.307	0.312	0.333
α_B	0.208	0.269	0.291	0.309	0.336	0.355	0.378	0.392	0.402	0.414	0.421	
β	0.141	0.196	0.214	0.229	0.249	0.263	0.281	0.291	0.299	0.307	0.312	0.333

Table 3.2.11 Torsion of Shafts of Various Cross Sections
(For strength and stiffness of shafts, see Sec. 6.2)

Cross section	Torsional resisting moment M_t	Angular twist, θ_1 (length = 1 in, radius = 1 in)		Work of torsion (V = volume)
		In terms of torsional moment	In terms of max shear	
	$\frac{\pi}{16} d^3 S_v$	$\frac{M_t}{GI_P} = \frac{32}{\pi d^4}\frac{M_t}{G}$	$2\frac{S_{v_{max}}}{G}\frac{1}{d}$	$\frac{1}{4}\frac{S^2_{v_{max}}}{G}V$ (Note 1)
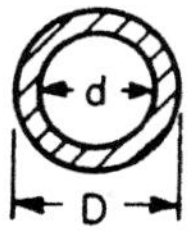	$\frac{\pi}{16}\frac{D^4 - d^4}{D}S_v$	$\frac{32}{\pi(D^4 - d^4)}\frac{M_t}{G}$	$2\frac{S_{v_{max}}}{G}\frac{1}{D}$	$\frac{1}{4}\frac{S^2_{v_{max}}}{G}\frac{D^2 + d^2}{D^2}V$ (Note 2)
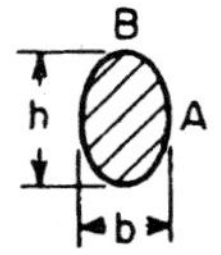	$\frac{\pi}{16} b^2 h S_v$ $(h > b)$	$\frac{16}{\pi}\frac{b^2 + h^2}{b^3 h^3}\frac{M_t}{G}$	$\frac{S_{v_{max}}}{G}\frac{b^2 + h^2}{bh^2}$	$\frac{1}{8}\frac{S^2_{v_{max}}}{G}\frac{b^2 + h^2}{h^2}V$ (Note 3)
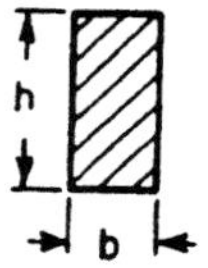	$\frac{2}{9} b^2 h S_v$ $(h > b)$	$3.6 * \frac{b^2 + h^2}{b^3 h^3}\frac{M_t}{G}$	$0.8 * \frac{S_{v_{max}}}{G}\frac{b^2 + h^2}{bh^2}$	$\frac{4}{45}\frac{S^2_{v_{max}}}{G}\frac{b^2 + h^2}{h^2}V$ (Note 4)
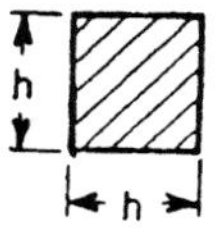	$\frac{2}{9} h^3 S_v$	$7.2\frac{1}{h^4}\frac{M_t}{G}$	$1.6\frac{S_{v_{max}}}{G}\frac{1}{h}$	$\frac{8}{45}\frac{S^2_{v_{max}}}{G}V$ (Note 5)
	$\frac{b^2}{20} S_v$	$46.2\frac{1}{b^4}\frac{M_t}{G}$	$2.31\frac{S_{v_{max}}}{G}\frac{1}{b}$	
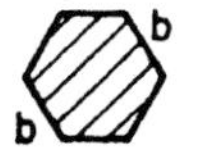	$\frac{b^3}{1.09} S_v$	$0.967\frac{1}{b^4}\frac{M_t}{G}$	$0.9\frac{S_{v_{max}}}{G}\frac{1}{b}$	

* When h/b =	1	2	4	8
Coefficient 3.6 becomes =	3.56	3.50	3.35	3.21
Coefficient 0.8 becomes =	0.79	0.78	0.74	0.71

NOTES: (1) $S_{v_{max}}$ at circumference. (2) $S_{v_{max}}$ at outer circumference. (3) $S_{v_{max}}$ at A; $S_{vB} = 16M_t/\pi bh^2$. (4) $S_{v_{max}}$ at middle of side h; in middle of b, $S_v = 9M_t/2bh^2$. (5) $S_{v_{max}}$ at middle of side.

Table 3.2.12 Strength of Round-ended Columns According to Euler's Formula*

Material	Cast iron	Wrought iron†	Low-carbon steel	Medium-carbon steel
Ultimate compressive strength, lb/in²	107,000	53,400	62,600	89,000
Allowable compressive stress, lb/in² (maximum)	7,100	15,400	17,000	20,000
Modulus of elasticity	14,200,000	28,400,000	30,600,000	31,300,000
Factor of safety	8	5	5	5
Smallest I allowable at worst section, in⁴	$\frac{Pl^2}{17{,}500{,}000}$	$\frac{Pl^2}{56{,}000{,}000}$	$\frac{Pl^2}{60{,}300{,}000}$	$\frac{Pl^2}{61{,}700{,}000}$
Limit of ratio, l/r	50.0	60.6	59.4	55.6
Rectangle $(r = b\sqrt{1/12})$, $l/b >$	14.4	17.5	17.2	16.0
Circle $(r = \frac{1}{4}d)$, $l/d >$	12.5	15.2	14.9	13.9
Circular ring of small thickness $(r = b\sqrt{1/8})$, $l/d >$	17.6	21.4	21.1	19.7

* P = allowable load, lb; l = length of column, in; b = smallest dimension of a rectangular section, in; d = diameter of a circular section, in; r = least radius of gyration of section.
† This material is no longer manufactured but may be encountered in existing structures and machinery.

Table 3.2.13 Typical Short-Column Formulas

Formula	Material	Code	Slenderness ratio
$S_w = 17{,}000 - 0.48\left(\frac{l}{r}\right)^2$	Carbon steels	AISC	$l/r < 120$
$S_w = 16{,}000 - 70(l/r)$	Carbon steels	Chicago	$l/r < 120$
$S_w = 15{,}000 - 5\left(\frac{l}{r}\right)$	Carbon steels	AREA	$l/r < 150$
$S_w = 19{,}000 - 100(l/r)$	Carbon steels	Am. Br. Co.	$60 < \frac{l}{r} < 120$
$S_{cr}^* = 135{,}000 - \frac{15.9}{c}\left(\frac{l}{r}\right)^2$	Alloy-steel tubing	ANC	$\frac{1}{\sqrt{c}r} < 65$
$S_w = 9{,}000 - 40\left(\frac{l}{r}\right)$	Cast iron	NYC	$\frac{l}{r} < 70$
$S_{cr}^* = 34{,}500 - \frac{245}{\sqrt{c}}\left(\frac{l}{r}\right)$	2017ST Aluminum	ANC	$\frac{1}{\sqrt{c}r} < 94$
$S_{cr}^* = 5{,}000 - \frac{0.5}{c}\left(\frac{l}{r}\right)^2$	Spruce	ANC	$\frac{1}{\sqrt{c}r} < 72$
$S_{cr}^* = S_y\left[1 - \frac{S_y}{4n\pi^2 E}\left(\frac{l}{r}\right)^2\right]$	Steels	Johnson	$\frac{l}{r} < \sqrt{\frac{2n\pi^2 E}{S_y}}$
$S_{cr}^*\dagger = \frac{S_y}{1 + \frac{ec}{r^2}\sec\left(\frac{l}{r}\sqrt{\frac{P}{4AE}}\right)}$	Steels	Secant	$\frac{l}{r} <$ critical

* S_{cr} = theoretical maximum, c = end fixity coefficient,
$c = 2$, both ends pivoted; $c = 2.86$, one pivoted, other fixed;
$c = 4$, both ends fixed; $c = 1$, one end fixed, one end free.
† e is initial eccentricity at which load is applied to center of column cross section.

safe load can be applied centrally, according to the AISC formula? At the computed load, what size section (also square) would be needed, if it were to be designed according to the AREA formula?

$$l/r = 15/0.5/\sqrt{12} = 104 \qquad \therefore \text{ short column}$$

$$P/A = 17{,}000 - 0.485(104)^2 = 11{,}730$$

or

$$P = 0.25 \times 11{,}730 = 2{,}940 \text{ lb } (1{,}335 \text{ kgf})$$

and

$$\frac{2{,}940}{a^2} \le 15{,}000 - 50\left(\frac{15la}{\sqrt{12}}\right) = 15{,}000 - \frac{2{,}600}{a}$$

thus

$$a^2 - 0.173a - 0.196 = 0 \quad \text{or} \quad a = 0.536 \text{ in } (1.36 \text{ cm})$$

Combined Flexure and Longitudinal Force Figure 3.2.53 shows a bar under flexure due to transverse and longitudinal loads. The maximum fiber stress S is made up of S_0, due to the direct action of load P, and S_b, due to the entire bending moment M. M is the algebraic sum of two bending moments, M_1 due to longitudinal load (+ for compression and − for tension), and M_2 due to transverse load. $M = M_2 \pm M_1$. Here $M_1 = Pf$ and $f = CS_b l^2/(Ec)$.

FOR THE CASE OF LONGITUDINAL COMPRESSION. $S_b I/c = M_2 + CPS_b l^2/(Eo)$, or $S_b = M_2 c(I - CPl^2/E)$. The maximum stress is $S = S_b + S_0$ compression. The constant C for the case of Fig. 3.2.53 is derived from the equations $P'l/4 = S_b I/c$ and $f = P'l^3/(48EI)$. Solving for f, $f = \frac{1}{12} S_b l^2/(Ec)$, or $C = \frac{1}{12}$. For a beam supported at the ends and uniformly loaded, $C = \frac{5}{48}$. Other cases can be similarly calculated.

FOR THE CASE OF LONGITUDINAL TENSION. $M = M_2 - Pf$, and $S_b = M_2 c/(I + CPl^2/E)$. The maximum stress is $S = S_b + S_0$, tension.

ECCENTRIC LOADS

When **short blocks** are **loaded eccentrically** in compression or in tension, that is, not through the center of gravity (cg), a combination of axial and bending stress results. The maximum unit stress S_M is the algebraic sum of these two unit stresses.

In Fig. 3.2.54 a load P acts in a line of symmetry at the distance e from cg; r = radius of gyration. The unit stresses are (1) S_c, due to P,

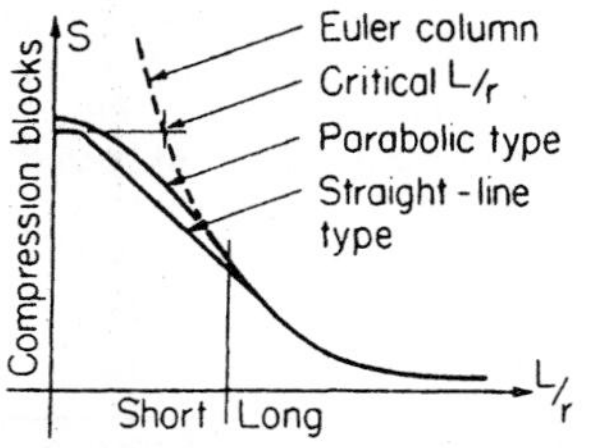

Figure 3.2.52

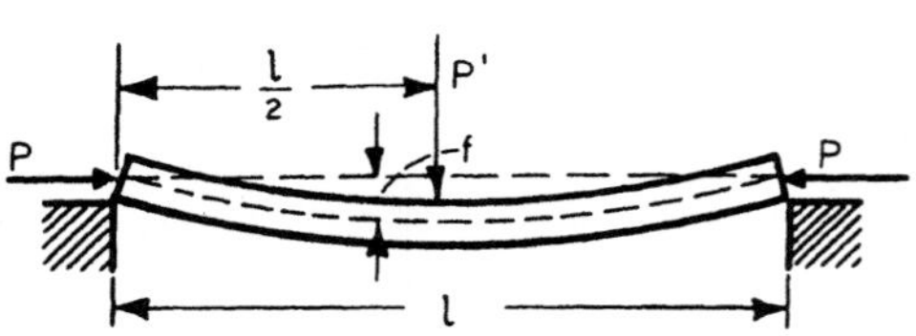

Figure 3.2.53

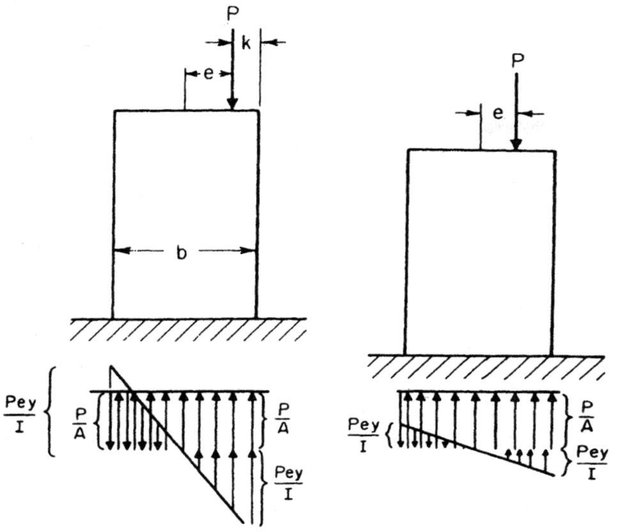

Figure 3.2.54

as if it acted through cg, and (2) S_b, due to the bending moment of P acting with a leverage of e about cg. Thus unit stress S at any point y is

$$S = S_c \pm S_b$$

$$= P/A \pm Pey/I$$

$$= S_c\left(1 \pm et/r^2\right)$$

positive for points on the same side of cg as P, and negative on the opposite side. For a **rectangular cross section** of width b, the maximum stress $S_M = S_c\,(1 + 6e/b)$. When P is outside the middle third of width b and is a compressive load, tensile stresses occur.

For **a circular cross section** of diameter d, $S_M = S_c(1 + 8e/d)$. The stress due to the weight of the solid will modify these relations.

NOTE. In these formulas e is measured from the gravity axis, and gives tension when e is greater than one-sixth the width (measured in the same direction as e), for rectangular sections; and when greater than one-eighth the diameter for solid circular sections.

If, as in certain classes of masonry construction, the **material cannot withstand tensile stress** and thus no tension can occur, the center of moments (Fig. 3.2.55) is taken at the center of stress. For a **rectangular section,** P acts at distance k from the nearest edge. Length under compression = $3k$, and $S_M = ⅔\,P/(hk)$. For a **circular section,** $S_M = [0.372 + 0.056(k/r)]\ P/k\sqrt{rk}$, where r = radius and k = distance of P from

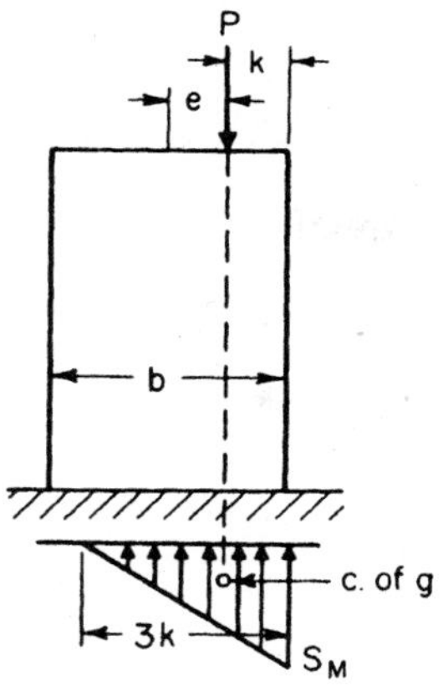

Figure 3.2.55

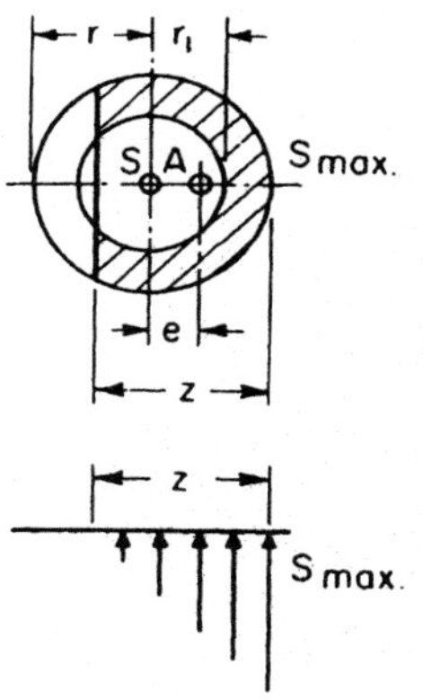

Figure 3.2.56

circumference. For a **circular ring,** S = average compressive stress on cross section produced by P; e = eccentricity of P; z = length of diameter under compression (Fig. 3.2.56). Values of z/r and of the ratio of S_{max} to average S are given in Tables 3.2.14 and 3.2.15.

CHIMNEY PROBLEM. Weight of chimney = 563,000 lb; e = 1.56 ft; OD of chimney = 10 ft 8 in; ID = 6 ft 6 ½ in. Overturning moment = Pe = 878,000 ft · lb, r_1/r = 0.6. e/r = 0.29. This gives $z/r > 2$. Therefore, the entire area of the base is under compression. Area under compression = 55.8 ft²; I = 546; S = 563,000/55.8 ± (878,000 × 5.33)/546 = 18,700 (max) and 1,500 (min) lb compression per ft². From Table 3.2.15, by interpolation, S_{max}/S_{av} = 1.85. ∴ S_{max} = (563,000/55.8) × 1.85 = 18,685 lb/ft² (91,313 kgf/m²).

The **kern** is the area around the center of gravity of a cross section within which any load applied will produce stress of only one sign throughout the entire cross section. Outside the kern, a load produces stresses of different sign. Figure 3.2.57 shows kerns (shaded) for various sections.

For **a circular ring,** the radius of the kern $r = D[1 + (d/D)^2]/8$.

For a **hollow square** (H and h = lengths of outer and inner sides), the kern is a square similar to Fig. 3.2.57a, where

$$r_{min} = \frac{H}{6}\ \frac{1}{\sqrt{2}}\left[1+\left(\frac{h}{H}\right)^2\right] = 0.1179H\left[1+\left(\frac{h}{H}\right)^2\right]$$

For a **hollow octagon** R_a and R_i = radii of circles circumscribing the outer and inner sides; thickness of wall = $0.9239(R_a - R_i)$, the kern is an octagon similar to Fig. 3.2.57c, where $0.2256R$ becomes $0.2256R_a\,[1 + (R_i/R_a)^2]$.

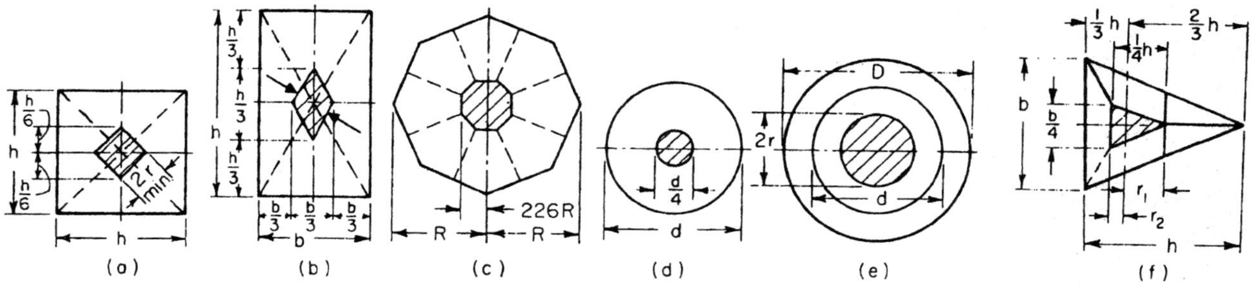

Figure 3.2.57

Table 3.2.14 Values of the Ratio z/r (Fig. 3.2.56)

$\frac{e}{r}$	$\frac{r_1}{r}$							$\frac{e}{r}$
	0.0	0.5	0.6	0.7	0.8	0.9	1.0	
0.25	2.00							0.25
0.30	1.82							0.30
0.35	1.66	1.89	1.98					0.35
0.40	1.51	1.75	1.84	1.93				0.40
0.45	1.37	1.61	1.71	1.81	1.90			0.45
0.50	1.23	1.46	1.56	1.66	1.78	1.89	2.00	0.50
0.55	1.10	1.29	1.39	1.50	1.62	1.74	1.87	0.55
0.60	0.97	1.12	1.21	1.32	1.45	1.58	1.71	0.60
0.65	0.84	0.94	1.02	1.13	1.25	1.40	1.54	0.65
0.70	0.72	0.75	0.82	0.93	1.05	1.20	1.35	0.70
0.75	0.59	0.60	0.64	0.72	0.85	0.99	1.15	0.75
0.80	0.47	0.47	0.48	0.52	0.61	0.77	0.94	0.80
0.85	0.35	0.35	0.35	0.36	0.42	0.55	0.72	0.85
0.90	0.24	0.24	0.24	0.24	0.24	0.32	0.49	0.90
0.95	0.12	0.12	0.12	0.12	0.12	0.12	0.25	0.95

Table 3.2.15 Values of the Ratio S_{max}/S_{avg} (Fig. 3.2.56)
(In determining S average, use load P divided by total area of cross section)

$\frac{e}{r}$	$\frac{r_1}{r}$							$\frac{e}{r}$
	0.0	0.5	0.6	0.7	0.8	0.9	1.0	
0.00	1.00	1.00	1.00	1.00	1.00	1.00	1.00	0.00
0.05	1.20	1.16	1.15	1.13	1.12	1.11	1.10	0.05
0.10	1.40	1.32	1.29	1.27	1.24	1.22	1.20	0.10
0.15	1.60	1.48	1.44	1.40	1.37	1.33	1.30	0.15
0.20	1.80	1.64	1.59	1.54	1.49	1.44	1.40	0.20
0.25	2.00	1.80	1.73	1.67	1.61	1.55	1.50	0.25
0.30	2.23	1.96	1.88	1.81	1.73	1.66	1.60	0.30
0.35	2.48	2.12	2.04	1.94	1.85	1.77	1.70	0.35
0.40	2.76	2.29	2.20	2.07	1.98	1.88	1.80	0.40
0.45	3.11	2.51	2.39	2.23	2.10	1.99	1.90	0.45
0.50	3.55	2.80	2.61	2.42	2.26	2.10	2.00	0.50
0.55	4.15	3.14	2.89	2.67	2.42	2.26	2.17	0.55
0.60	4.96	3.58	3.24	2.92	2.64	2.42	2.26	0.60
0.65	6.00	4.34	3.80	3.30	2.92	2.64	2.42	0.65
0.70	7.48	5.40	4.65	3.86	3.33	2.95	2.64	0.70
0.75	9.93	7.26	5.97	4.81	3.93	3.33	2.89	0.75
0.80	13.87	10.05	8.80	6.53	4.93	3.96	3.27	0.80
0.85	21.08	15.55	13.32	10.43	7.16	4.50	3.77	0.85
0.90	38.25	30.80	25.80	19.85	14.60	7.13	4.71	0.90
0.95	96.10	72.20	62.20	50.20	34.60	19.80	6.72	0.95
1.00	∞	∞	∞	∞	∞	∞	∞	1.00

CURVED BEAMS

The application of the flexure formula for a straight beam to the case of a curved beam results in error. When all "fibers" of a member have the same center of curvature, the **concentric** or common type of curved beam exists (see Fig. 3.2.58). Such a beam is defined by the Winkler-Bach theory. The stress at a point y units from the centroidal axis is

$$S = \frac{M}{AR}\left[1 + \frac{y}{Z(R+y)}\right]$$

M is the bending moment, positive when it increases curvature; Y is positive when measured toward the convex side; A is the cross-sectional area; R is the radius of the centroidal axis; Z **is a cross-section property** defined by

$$Z = -\frac{1}{A}\int \frac{y}{R+y}\, dA$$

Analytical expressions for Z of certain sections are given in Table 3.2.16. Also Z can be found by **graphical** integration methods (see any advanced strength book). The **neutral surface** shifts toward the center of curvature, or inside fiber, an amount equal to $e = ZR/(Z+1)$. The Winkler-Bach theory, though practically satisfactory, disregards radial stresses as well as lateral deformations and assumes pure bending. The **maximum stress** occurring on the inside fiber is $S = Mh_i/(AeR_i)$, while that on the outside fiber is $S = Mh_0/(AeR_0)$.

EXAMPLE. A split steel ring of rectangular cross section is subjected to a diametral force of 1,000 lb as shown in Fig. 3.2.59*a*. Compute the stress at the point 0.5 in from the outside fiber on plane *mm*. Also compute the maximum stress.

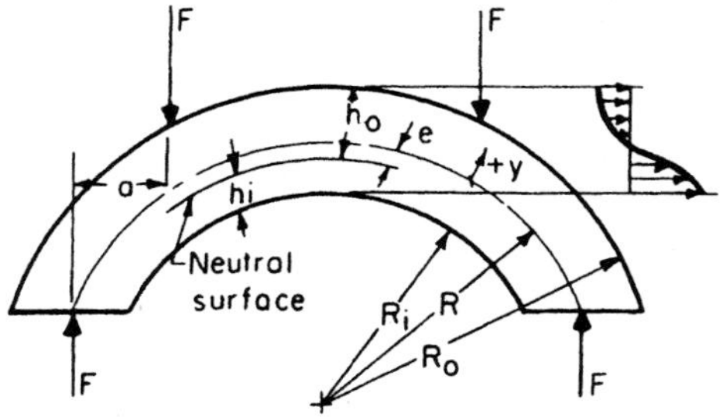

Figure 3.2.58

$$Z = -1 + \frac{R}{h}\ \ln\ \frac{R+C}{R-C}$$

$$= -1 + \frac{10}{4}\ \ln\ \frac{10+2}{10-2} = 0.0133$$

$$S_{1.5} = \frac{M}{AR}\left[1 + \frac{y}{Z(R+y)}\right] + \frac{F}{A}$$

$$= \frac{-1{,}000\times 10}{8\times 10}\left[1 + \frac{1.5}{0.0133(10+1.5)}\right] + \frac{1{,}000}{8}$$

$$= -1{,}250 + 125 = -1{,}125 \text{ lb/in}^2 \text{ (compr.)(79 kgf/cm}^2\text{)}$$

$$S_M = \frac{-1{,}000}{8}\left[1 + \frac{-2}{0.0133(10-2)}\right] + \frac{1{,}000}{8}$$

$$= 2{,}230 + 125 = 2{,}335 \text{ lb/in}^2 \text{ (166 kfg/cm}^2\text{)}$$

or

$$e = \frac{ZR}{Z+1} = \frac{0.0133\times 10}{0.0133+1} = 0.131$$

and

$$S_M = \frac{Mh_i}{AeR_i} + \frac{F}{A} = \frac{1{,}000\times 10\times 1.87}{8\times 0.131\times 8} + \frac{1{,}000}{8}$$

$$= 2{,}355 \text{ lb/in}^2 \text{ (166 kgf/cm}^2\text{)}$$

The **deflection** in curved beams can be computed by means of the moment-area theory. If the origin of axes is taken at the point whose deflection is wanted, it can be shown that the component displacements in the x and y directions are

$$\Delta_x = \int_0^s \frac{My\, ds}{EI} \qquad \text{and} \qquad \Delta_y = \int_0^s \frac{Mx\, ds}{EI}$$

The resultant deflection is then equal to $\Delta_0 = \sqrt{\Delta_x^2 + \Delta_y^2}$ in the direction defined by $\tan\theta = \Delta_y/\Delta_x$. Deflections can also be found conveniently by use of **Castigliano's theorem.** It states that in an elastic system the displacement in the direction of a force (or couple) and due to that force (or couple) is the partial derivative of the strain energy with respect to the force (or couple). Stated mathematically, $\Delta_z = \partial U/\partial F_z$. If a force does not exist at the point and/or in the direction desired, a dummy force may be applied. This force must then be eliminated by equating it to zero at the end.

EXAMPLE. A quadrant of radius R is fixed at one end as shown in Fig. 3.2.59*b*. The force F is applied in the radial direction at the free end B. Find the deflection of B.

Table 3.2.16 Analytical Expressions for Z

Section	Expression	Section	Expression
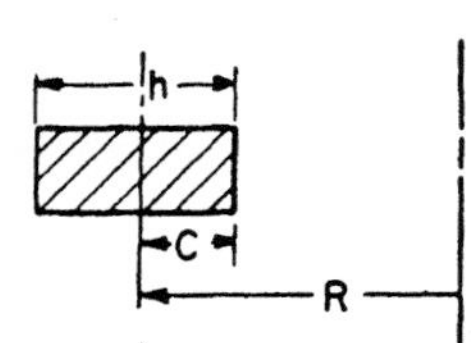	$Z = -1 + \frac{R}{h}\ln\frac{R+C}{R-C}$	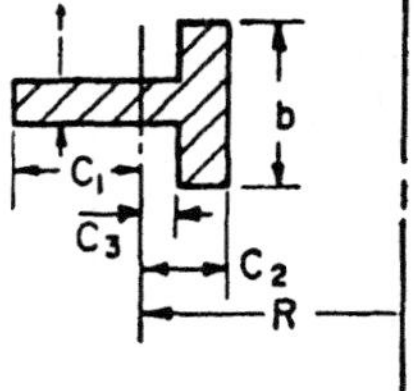	$Z = -1 + \frac{R}{A}\left[t\ln(R+C_1) + (b-t)\ln(R-C_3) - b\ln(R-C_3)\right]$ $A = tC_1 - (b-t)C_3 + bC_2$
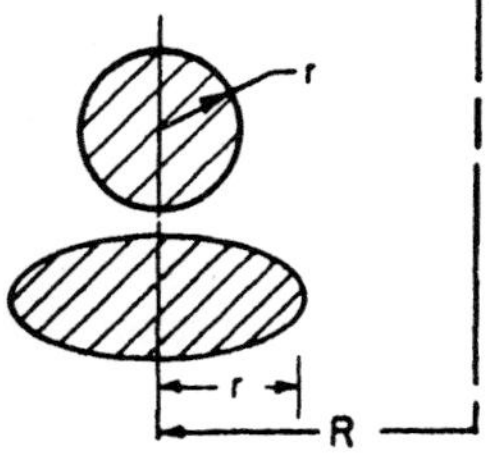	$Z = -1 + 2\left(\frac{R}{r}\right)\left[\frac{R}{r} - \sqrt{\left(\frac{R}{r}\right)^2 - 1}\right]$	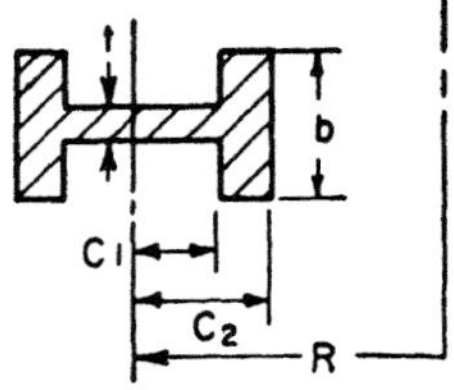	$Z = -1 + \frac{R}{A}\left[b\ln\frac{R+C_2}{R-C_2} + (t-b)\ln\frac{R+C_1}{R-C_1}\right]$ $A = 2[(t-b)C_1 + bC_2]$

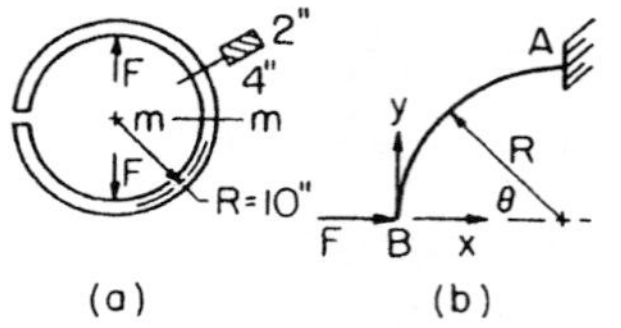

Figure 3.2.59

By moment area:

$$y = R\sin\theta \qquad x = R(1-\cos\theta)$$
$$ds = R\,d\theta \qquad m = FR\sin\theta$$
$${}_B\Delta_x = \frac{FR^3}{EI}\int_0^{\pi/2}\sin^2 d\theta = \frac{\pi FR^3}{4EI}$$
$${}_B\Delta_y = \frac{FR^3}{EI}\int_0^{\pi/2}\sin\theta(1-\cos\theta)d\theta = \frac{FR^3}{2EI}$$

and
$$\Delta_B = \frac{FR}{2EI}\sqrt{1+\frac{\pi^2}{4}}$$

at
$$\theta_x = \tan^{-1}\left(\frac{FR^3}{2EI}\times\frac{4EI}{\pi FR^3}\right) = \tan^{-1}\frac{2}{\pi} = 32.5^\circ$$

By Castigliano:

$${}_B\Delta_x = \frac{\partial U}{\partial F} = \frac{\partial}{\partial F_x}\int_0^{\pi/2}\frac{F^2R^3}{2EI}\sin^2\theta d\theta = \frac{\pi FR^3}{4EI}$$
$${}_B\Delta_y = \frac{\partial U}{\partial F_y} = \frac{\partial}{\partial F_y}\int_0^{\pi/2}\frac{[FR\sin\theta - F_yR(1-\cos\theta)]^2 R\,d\theta}{2EI}$$
$$= -\frac{FR^3}{2EI}$$

The F_y, assumed downward, is equated to zero, after the integration and differentiation are performed to find ${}_B\Delta_y$. The remainder of the computation is exactly as in the moment-area method.

Eccentrically Curved Beams These beams (Fig. 3.2.60) are bounded by arcs having different centers of curvature. In addition, it is possible for either radius to be the larger one. The one in which the section depth shortens as the central section is approached may be called the **arch beam**. When the central section is the largest, the beam is of the crescent type. **Crescent I** denotes the beam of larger outside radius and **crescent II** of larger inside radius. The stress at the **central section** of such beams may be found from $S = KMC/I$. In the case of rectangular cross section, the equation becomes $S = 6KM/(bh^2)$ where M is the bending moment, b is the width of the beam section, and h its height. The **stress factors** K for the **inner boundary**, established from photoelastic data, are given in Table 3.2.17. The outside radius is denoted by R_o and the inside by R_i. The geometry of crescent beams is such that the stress can be larger in **off-center sections**. The stress at the central section determined above must then be multiplied by the **position factor** k, given in Table 3.2.18. As in the concentric beam, the **neutral surface** shifts slightly toward the inner boundary (see Vidosic, Curved Beams with Eccentric Boundaries, *Trans. ASME,* 79, pp. 1317-1321).

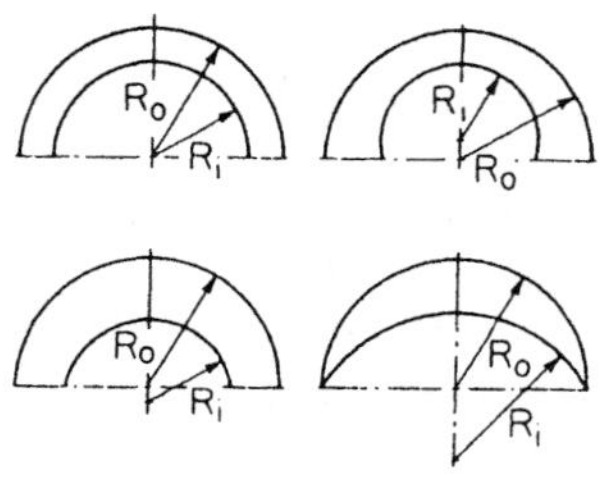

Figure 3.2.60

Table 3.2.17 Stress Factors for Inner Boundary at Central Section (See Fig. 3.2.60)

1. For the arch-type beams

(a) $K = 0.834 + 1.504\dfrac{h}{R_o+R_i}$ if $\dfrac{R_o+R_i}{h} < 5.$

(b) $K = 0.899 + 1.181\dfrac{h}{R_o+R_i}$ if $5 < \dfrac{R_o+R_i}{h} < 10.$

(c) In the case of larger section ratios use the equivalent beam solution.

2. For the crescent I-type beams

(a) $K = 0.570 + 1.536\dfrac{h}{R_o+R_i}$ if $\dfrac{R_o+R_i}{h} < 2.$

(b) $K = 0.959 + 0.769\dfrac{h}{R_o+R_i}$ if $2 < \dfrac{R_o+R_i}{h} < 20.$

(c) $K = 1.092\left(\dfrac{h}{R_o+R_i}\right)^{0.0298}$ if $\dfrac{R_o+R_i}{h} > 20.$

3. For the crescent II-type beams

(a) $K = 0.897 + 1.098\dfrac{h}{R_o+R_i}$ if $\dfrac{R_o+R_i}{h} < 8.$

(b) $K = 1.119\left(\dfrac{h}{R_o+R_i}\right)^{0.0378}$ if $8 < \dfrac{R_o+R_i}{h} > 20.$

(c) $K = 1.081\left(\dfrac{h}{R_o+R_i}\right)^{0.0270}$ if $\dfrac{R_o+R_i}{h} > 20.$

Table 3.2.18 Crescent-Beam Position Stress Factors (See Fig. 3.2.60)

Angle θ, deg	k Inner	k Outer
10	$1 + 0.055H/h$	$1 + 0.03H/h$
20	$1 + 0.164H/h$	$1 + 0.10H/h$
30	$1 + 0.365H/h$	$1 + 0.25H/h$
40	$1 + 0.567H/h$	$1 + 0.467H/h$
50	$1.521 - \dfrac{(0.5171 - 1.382H/h)^{1/2}}{1.382}$	$1 + 0.733H/h$
60	$1.756 - \dfrac{(0.2416 - 0.6506H/h)^{1/2}}{0.6506}$	$1 + 1.123H/h$
70	$2.070 - \dfrac{(0.4817 - 1.298H/h)^{1/2}}{0.6492}$	$1 + 1.70H/h$
80	$2.531 - \dfrac{(0.2939 - 0.7084H/h)^{1/2}}{0.3542}$	$1 + 2.383H/h$
90		$1 + 3.933H/h$

NOTE: All formulas are valid for $0 < H/h \le 0.325$. Formulas for the inner boundary, except for 40 deg, may be used to $H/h \le 0.36$. H = distance between centers.

IMPACT

A force or stress is considered **suddenly applied** when the duration of load application is less than one-half the **fundamental natural period** of vibration of the member upon which the force acts. Under impact, a compression wave propagates through the member at a velocity $c = \sqrt{E/\rho}$, where ρ is the mass density. As this compression wave travels back and forth by reflection from one end of the bar to the other, a maximum stress is produced which is many times larger than what it would be statically. An exact determination of this stress is most difficult. However, if conservation of kinetic and strain energies is applied, the **impact stress** is found to be

$$S' = S\sqrt{\frac{W}{W_b}\left(\frac{3W}{3W + W_b}\right)}$$

The weight of the striking mass is here denoted by W, that of the struck bar by W_b, while S is the static stress, W/A (A is the cross-sectional area of the bar). Above is the case of **sudden impact.** When the ratio W/W_b is small, the stress computed by the above equation may be erroneous. A better solution of this problem may result from

$$S' = S + S\sqrt{\frac{W}{W_b} + \frac{2}{3}}$$

If a weight W falls a distance h before striking a bar of mass W_b, energy conservation will yield the relation

$$S' = S\left(1 + \sqrt{1 + \frac{2h}{e} \times \frac{3W}{3W + W_b}}\right)$$

The elongation $e = \varepsilon l = Sl/E$. When the striking mass W is assumed rigid, the elasticity factor is taken equal to 1. Thus the equation becomes

$$S' = S(1 + \sqrt{1 + 2h/e})$$

If, in addition, h is taken equal to zero (sudden impact), the radical equals 1, and so the stress becomes $S' = 2S$. Since Hooke's law is applicable, the relations

$$e' = e(1 + \sqrt{1 + 2h/e}) \qquad \text{and} \qquad e' = 2e$$

are also true for the same conditions.

The expression may be converted, by using $v^2 = 2gh$, to

$$S' = S[1 + \sqrt{1 + v^2/(eg)}]$$

This might be called the **energy impact** form. If the **natural frequency** f_n of the bar is used, the stress equation is

$$S' = S(1 + \sqrt{1 + 0.204hf_n^2})$$

In general, the **maximum impact stress** in a **beam** and a **shaft** can be approximated from the simplified falling-weight equation. It is necessary, though, to substitute the maximum **deflection** y for e, in the case of beams, and for the **angle of twist** θ in the case of shafts. Of course $S = Mc/I$ and $M_t c/J$, respectively. Thus

$$S' = S\left[1 + \sqrt{1 + \frac{2h}{y(\text{or }\theta)}}\right]$$

For a more exact solution, elastic yield in each member must be considered. The theory then yields

$$S' = S\left[1 + \sqrt{1 + \frac{2h}{y}\left(\frac{35W}{35W + 17W_b}\right)}\right]$$

for a **simply supported beam** struck in the middle by a weight W.

THEORY OF ELASTICITY

For loaded members in which the stress distribution cannot be estimated by elementary strength-of-material methods, the more advanced mathematical principles of the **theory of elasticity** must be applied. When this is not possible, experimental stress analysis has to be used. Because of the complexity of solution, only some of the more practical problems have been solved by the theory of elasticity. The more general concepts and methods are presented.

Two kinds of forces may act on a body. **Surface forces** are distributed over the surface as the result of, for instance, the pressure of one body on another. Forces due to gravity, inertia, magnetism, etc., which act over the entire volume of a body, are called **body forces.** Both surface and body forces can be best handled if resolved into three orthogonal components. Surface forces are thus designated $\overline{X}$, $\overline{Y}$, and $\overline{Z}$, while body forces are labeled X, Y, and Z.

In general, there exists a normal stress σ and a shearing stress τ at each point of a loaded member. It is convenient to deal with components of each of these stresses on each of six orthogonal planes that bound the point element. Thus there are at each point **six stress components,** σ_x, σ_y, σ_z, $\tau_{yx} = \tau_{xy}$, $\tau_{xz} = \tau_{zx}$, and $\tau_{yz} = \tau_{zy}$. Similarly, if the normal unit strain is designated by the letter ε and shearing unit strain by γ, the **six components of strain** are defined by

$$\varepsilon_x = \partial u/\partial x \qquad \varepsilon_y = \partial v/\partial y \qquad \varepsilon_z = \partial w/\partial z$$

$$\gamma_{xy} = \partial u/\partial y + \partial v/\partial x \qquad \gamma_{yz} = \partial v/\partial z + \partial w/\partial y$$

and $\quad \gamma_{xz} = \partial u/\partial z + \partial w/\partial x$

The elastic **displacements** of particles on the body in the x, y, and z directions are identified as the u, v, and w **components,** respectively.

Since metals have the usually assumed elastic as well as isotropic properties, Hooke's law holds. Therefore, the interrelationships between stress and strain can easily be obtained.

$$\varepsilon_x = \frac{1}{E}[\sigma_x - \mu(\sigma_y + \sigma_z)]$$

$$\varepsilon_y = \frac{1}{E}[\sigma_y - \mu(\sigma_x + \sigma_z)]$$

$$\varepsilon_z = \frac{1}{E}[\sigma_z - \mu(\sigma_x + \sigma_y)]$$

$$\gamma_{xy} = \tau_{xy}/G \qquad \gamma_{xz} = \tau_{xz}/G$$

and $\quad \gamma_{yz} = \tau_{yz}/G$

The general case of strain can be obtained by superposing the elongation strains upon the shearing strains.

Problems depending upon theories of elasticity are considerably simplified if the stresses are all parallel to one plane or if all deformations occur in planes perpendicular to the length of the member. The first case is one of **plane stress,** as when a thin plate of uniform thickness is subjected to central, boundary forces parallel to the plane of the plate. The second is a case of **plane strain,** such as a gate subjected to hydrostatic pressure, the intensity of which does not vary along the gate's length. All particles therefore displace at right angles to the length, and so cross sections remain plane. In plane-stress problems, three of the six stress components vanish, thus leaving only σ_x, σ_y, and τ_{xy}. Similarly, in plane strain, only ε_x, ε_y, and γ_{xy} will not equal zero; thus the same three stresses σ_x, σ_y, and τ_{xy} remain to be considered. Plane problems can thus be represented by the element shown in Fig. 3.2.61. Equilibrium considerations applied to this particle result in the **differential equations of equilibrium** which reduce to

$$\frac{\partial \sigma_x}{\partial x} + \frac{\partial \tau_{xy}}{\partial y} + X = 0$$

and

$$\frac{\partial \sigma_y}{\partial y} + \frac{\partial \tau_{xy}}{\partial x} + Y = 0$$

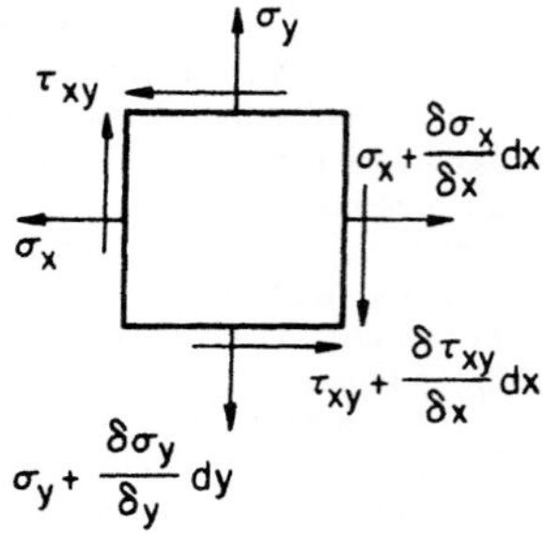

Figure 3.2.61

Since the two differential equations of equilibrium are insufficient to find the three stresses, a third equation must be used. This is the **compatibility equation** relating the three strain components. It is

$$\frac{\partial^2 \varepsilon_x}{\partial y^2} + \frac{\partial^2 \varepsilon_y}{\partial x^2} = \frac{\partial^2 \gamma_{xy}}{\partial x\, \partial y}$$

If strains are expressed in terms of the stresses, the compatibility equation becomes

$$\left(\frac{\partial^2}{\partial x^2} + \frac{\partial^2}{\partial y^2}\right)(\sigma_x + \sigma_y) = 0$$

Now, in any **two-dimensional** problem, the compatibility equation along with the differential equilibrium equations must be simultaneously solved for the three unknown stresses. This is accomplished using **stress functions,** which permit the integration and satisfy boundary conditions in each particular situation.

In **three-dimensional** problems, the third dimension must be considered. This results in three differential equations of equilibrium, as well as three compatibility equations. The six stress components can thus be found. The complexity involved in the solution of these equations is such, however, that only a few special cases have been solved.

In certain problems, such as rotating circular disks, **polar coordinates** become more convenient. In such cases, the stress components in a two-dimensional field are the **radial stress** σ_r, the **tangential stress** σ_θ, and the **shearing stress** $\tau_{r\theta}$. In terms of these stresses the **polar differential equations** become

$$\frac{\partial \sigma_r}{\partial r} + \frac{1}{r}\frac{\partial \tau_{r\theta}}{\partial \theta} + \frac{\sigma_r - \sigma_\theta}{r} + R = 0$$

and

$$\frac{1}{r}\frac{\partial \sigma_\theta}{\partial \theta} + \frac{\partial \tau_{r\theta}}{\partial r} + \frac{2\tau_{r\theta}}{r} = 0$$

The body force per unit volume is represented by R. The **compatibility equation** in polar coordinates is

$$\left(\frac{\partial^2}{\partial r^2} + \frac{1}{r}\frac{\partial}{\partial r} + \frac{1}{r^2}\frac{\partial^2}{\partial \theta^2}\right)\left(\frac{\partial^2 \phi}{\partial r^2} + \frac{1}{r}\frac{\partial \phi}{\partial r} + \frac{1}{r^2}\frac{\partial^2 \phi}{\partial \theta^2}\right) = 0$$

ϕ is again a stress function of r and θ that will provide a solution of the differential equations and satisfy boundary conditions. As an example, the exact solution of a **simply supported beam** carrying a uniformly distributed load w yields

$$\sigma_x = \frac{w}{2I}\,(l^2 - x^2)y + \frac{w}{2I}\left(\frac{2y^3}{3} - \frac{2c^2 y}{5}\right)$$

The origin of coordinates is at the center of the beam, $2c$ is the beam depth, and $2l$ is the span length. Thus the maximum stress at $x = 0$ and $y = c$ is $\sigma_x = \frac{wl^2c}{2I} + \frac{2}{15}\frac{wc^3}{I}$. The first term represents the stress as obtained by the elementary flexure theory; the second is a correction. The second term becomes negligible when c is small compared to l.

The important case of a **flat plate** of unit width with a **circular hole** of diameter $2a$ at its center, subjected to a uniform tensile load, has been solved using polar coordinates. If S is the uniform stress at some distance from the hole, r is measured from the center of the hole, and θ is the angle of r with respect to the longitudinal axis of the member, the stresses are

$$\sigma_r = \frac{S}{2}\left(1 - \frac{a^2}{r^2}\right) + \frac{S}{2}\left(1 + \frac{3a^4}{r^4} - \frac{4a^2}{r^2}\right)\cos 2\theta$$

$$\sigma_\theta = \frac{S}{2}\left(1 + \frac{a^2}{r^2}\right) - \frac{S}{2}\left(1 + \frac{3a^4}{r^4}\right)\cos 2\theta$$

$$\tau_r\theta = -\frac{S}{2}\left(1 - \frac{3a^4}{r^4} + \frac{2a^2}{r^2}\right)\sin 2\theta$$

CYLINDERS AND SPHERES

A thin-wall cylinder has a wall thickness such that the assumption of constant stress across the wall results in negligible error. Cylinders having internal-diameter-to-thickness (D/t) ratios greater than 10 are usually considered thin-walled. Boilers, drums, tanks, and pipes are often treated as such. Equilibrium equations reveal the circumferential, or hoop, stress to be $S = pr/t$ under an internal pressure p (see Fig. 3.2.62). If the cylinder is closed at the ends, a longitudinal stress of $pr/(2t)$ is developed. The tensile stress developed in a thin hollow sphere subjected to internal pressure is also $pr/(2t)$.

When thin-walled cylinders, such as vacuum tanks and submarines, are subjected to **external pressure, collapse** becomes the mode of failure. The shell is assumed perfectly round and of uniform thickness, the material obeys Hooke's law, the radial stress is negligible, and the normal stress distribution is linear. Other, lesser assumptions are also made. Using the theory of elasticity, R. G. Sturm (*Univ. Ill. Eng. Exp. Stn. Bull.*, no 12, Nov. 11, 1941) derived the **collapsing pressure** as

$$W_c = KE\left(\frac{t}{D}\right)^3 \qquad \text{lb/in}^2$$

The factor K, a numerical coefficient, depends upon the L/R and D/t ratios (D is outside-shell diameter), the kind of end support, and whether pressure is applied radially only, or at the ends as well. Figures 3.2.63 to 3.2.66, reproduced from the bulletin, supply the K values. N on these charts indicates the number of lobes into which the shell collapses. These values are for materials having Poisson's ratio $\mu = 0.3$. It may also be pointed out that in the case of long cylinders (infinitely long, theoretically) the value of K approaches $2/(l - \mu^2)$.

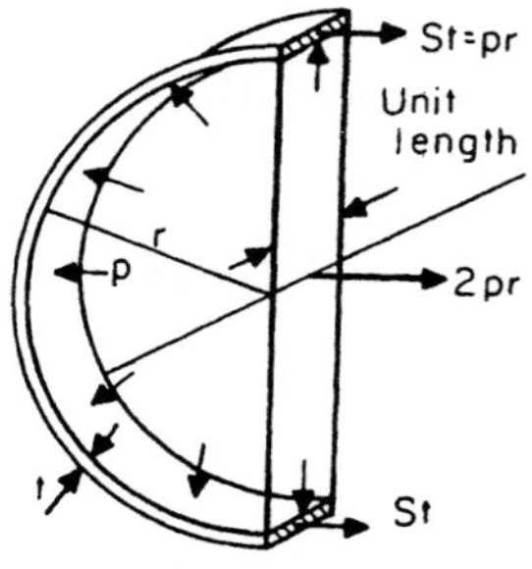

Figure 3.2.62

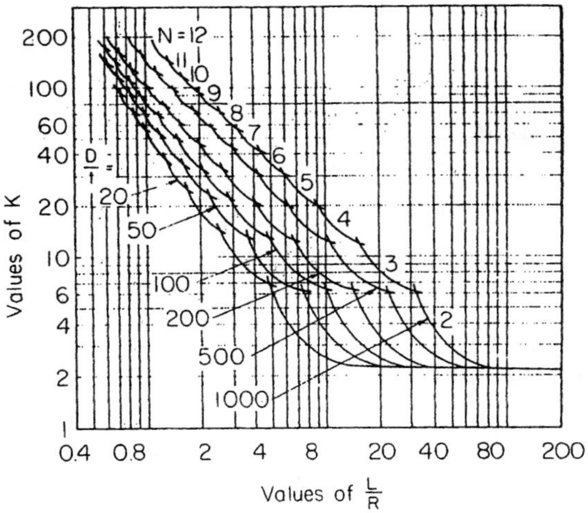

Figure 3.2.63

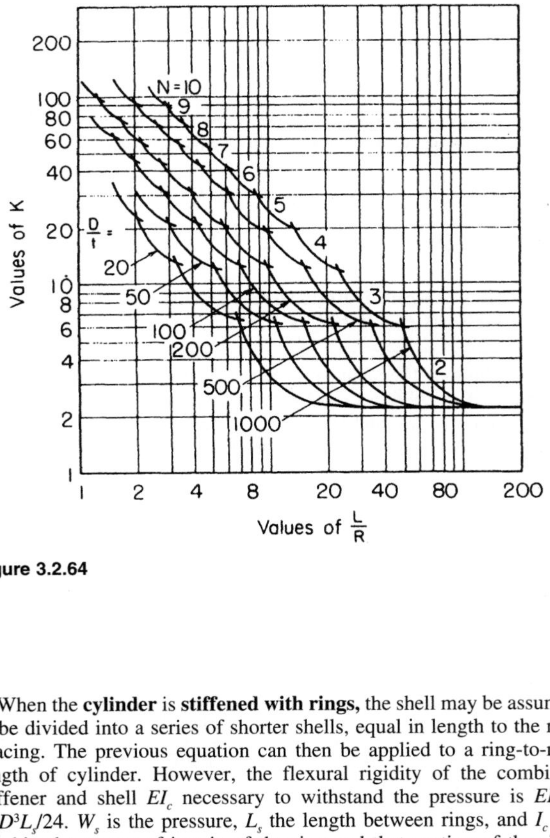

Figure 3.2.64

When the **cylinder** is **stiffened with rings,** the shell may be assumed to be divided into a series of shorter shells, equal in length to the ring spacing. The previous equation can then be applied to a ring-to-ring length of cylinder. However, the flexural rigidity of the combined stiffener and shell EI_c necessary to withstand the pressure is $EI_c = W_s D^3 L_s/24$. W_s is the pressure, L_s the length between rings, and I_c the combined moment of inertia of the ring and that portion of the shell assumed acting with the ring.

In some instances, cylinders collapse only after a stress in excess of the elastic limit has been reached; that is, **plastic range** stresses are present. In such cases the same equation applies, but the modulus of elasticity must be modified.

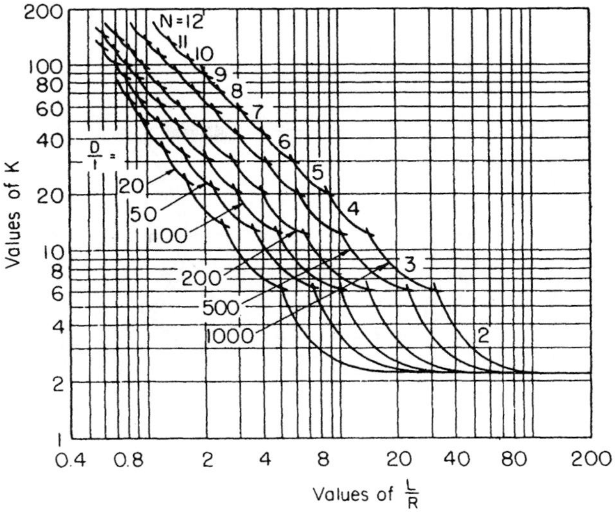

Figure 3.2.65

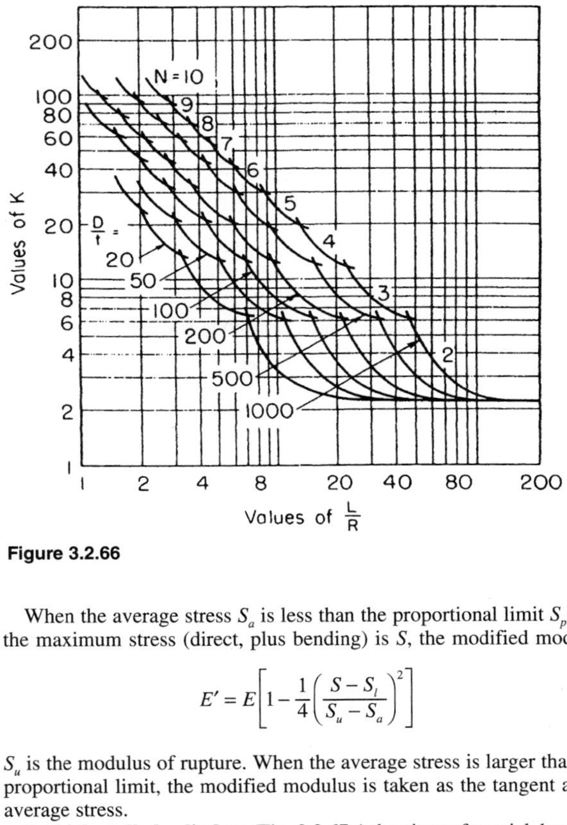

Figure 3.2.66

When the average stress S_a is less than the proportional limit S_p, and the maximum stress (direct, plus bending) is S, the modified modulus

$$E' = E\left[1 - \frac{1}{4}\left(\frac{S - S_l}{S_u - S_a}\right)^2\right]$$

S_u is the modulus of rupture. When the average stress is larger than the proportional limit, the modified modulus is taken as the tangent at the average stress.

In **thick-walled cylinders** (Fig. 3.2.67*a*) the circumferential, hoop, or tangential stress S_t is not uniform. In addition a radial stress S_r is present. When equilibrium is applied to the annulus taken out of Fig. 3.2.67*a* and shown in Fig. 3.2.67*b* and the equation is integrated, the general tangential and radial stress relations, called the **Lamé** equations, are derived.

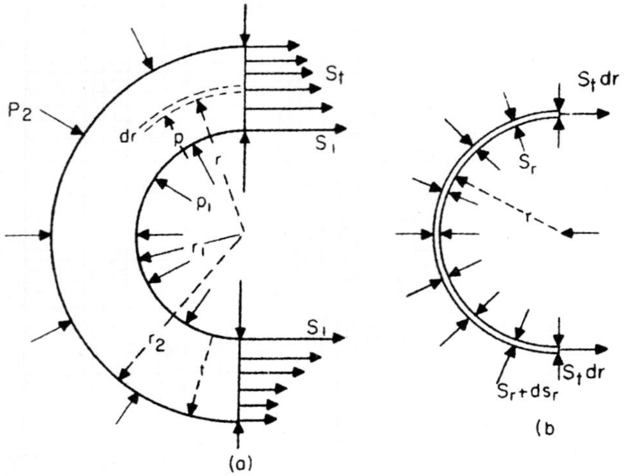

Figure 3.2.67

$$S_t = \frac{r_1^2 p_1 - r_2^2 p_2 + (p_1 - p_2) r_1^2 r_2^2 / r^2}{r_2^2 - r_1^2}$$

and $$S_r = \frac{r_1^2 p_1 - r_2^2 p_2 - (p_1 - p_2) r_1^2 r_2^2 / r^2}{r_2^2 - r_1^2}$$

When the external pressure $p_2 = 0$, the equations reduce to

$$S_t = \frac{r_1^2 p_1}{r_2^2 - r_1^2}\left(1 + \frac{r_2^2}{r^2}\right)$$

and $$S_r = \frac{r_1^2 p_1}{r_2^2 - r_1^2}\left(1 + \frac{r_2^2}{r^2}\right)$$

At the inner boundary the tangential elongation ε_t is equal to

$$\varepsilon_t = (S_t - \mu S_r)/E$$

The increase in the bore radius Δr_1 resulting therefrom is

$$\Delta r_1 = \frac{r_1 p_1}{E_h}\left(\frac{1 + r_1^2/r_2^2}{1 - r_1^2/r_2^2} + \mu\right)$$

Similarly a solid shaft of r radius under external pressure p_2 will have its radius decreased by the amount

$$\Delta r = -\frac{r p_2}{E_s}(1 - \mu)$$

In the case of a **press or shrink fit,** $p_1 = p_2 = p$. The sum of Δr_1 and Δr_1 absolute is the radial interference; twice this sum is the **diametral interference** Δ or

$$\Delta = 2 r_1 p\left[\frac{1}{E_h}\left(\frac{1 + r_1^2/r_2^2}{1 - r_1^2/r_2^2} + \mu\right) + \frac{1 - \mu}{E_s}\right]$$

If the hub and shaft materials are the same, $E_h = E_s = E$, and

$$\Delta = \frac{4 r_1 r_2^2 p}{E\left(r_2^2 - r_1^2\right)}$$

If the equation is solved for p and this value is substituted in Lamé's equation, the maximum tangential stress on the inner surface of the hub is found to be

$$S_t = \frac{E\Delta}{4 r_1 r_2^2}\left(r_2^2 + r_1^2\right)$$

EXAMPLE. The barrel of a field gun has an outside diameter of 9 in and a bore of 4.7 in. An internal pressure of 16,000 lb/in² is developed during firing. What maximum stress occurs in the barrel? An investigation of Lamé's equations for internal pressure reveals the maximum stress to be the tangential one on the inner surface. Thus,

$$S_t = p_1 \frac{r_1^2 + r_2^2}{r_2^2 - r_1^2} = \frac{16{,}000(2.35^2 + 4.5^2)}{4.5^2 - 2.35^2}$$

$$= 28{,}000 \text{ lb/in}^2 \ (1{,}972 \text{ kgf/cm}^2)$$

Oval Hollow Cylinders In Fig. 3.2.68, let a and b be the semiminor and semimajor axes. The bending moments at A and C will then be

$$M_0 = pa^2/2 - pI_x/(2S) - pI_y/(2S)$$

$$M_1 = M_0 - p(a^2 - b^2)/2$$

where I_x and I_y are the moments of inertia of the arc AC about the x and y axes, respectively. The **bending moment at any point** will be

$$M = M_0 - pa^2/2 + px^2/2 + py^2/2$$

Thick Hollow Spheres With an internal pressure p, where $p < T/0.65$.

$$r_2 = r_1\left[(T + 0.4p)/(T - 0.65p)\right]^{1/3}$$

The maximum tensile stress is on the inner surface, in the direction of the circumference. With an **external pressure** p, where $p < T/1.05$,

$$r_2 = r_1\left[T/(T - 1.05p)\right]^{1/3}$$

In both cases T is the true stress.

PRESSURE BETWEEN BODIES WITH CURVED SURFACES (CONTACT STRESSES)

(See Hertz, "Gesammelte Werke," vol. 1, pp. 159 *et seq.*, Barth.)

Two Spheres The radius A of the compressed area is obtained from the formula $A^3 = 0.68P(c_1 + c_2)/(1/r_1 + 1/r_2)$, in which P is the compressing force, c_1 and c_2 (= $1/E_1$ and $1/E_2$) are reciprocals of the respective moduli of elasticity, and r_1 and r_2 are the radii. (Reciprocal of Poisson's ratio is

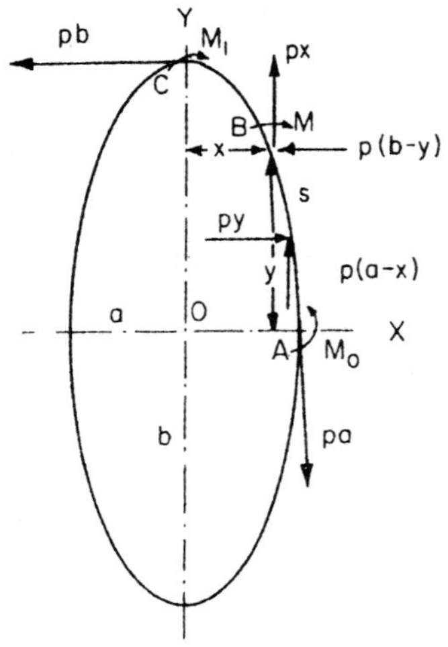

Figure 3.2.68

assumed to be $n = 10/3$.) The **greatest contact pressure** in the middle of the compressed surface will be $S_{max} = 1.5(P/\pi A^2)$, and

$$S^3_{max} = 0.235P(1/r_1 + 1/r_2)^2/(c_1 + c_2)^2$$

The **total deformation** of the two spheres will be Y, which is obtained from

$$Y^3 = 0.46P^2(c_1 + c_2)^2(1/r_1 + 1/r_2)$$

For $c_1 = c_2 = 1/E$, that is, **two spheres with the same modulus of elasticity,** it follows that $A^3 = 1.36P/E(1/r_1 + 1/r_2)$, $S^3_{max} = 0.059PE^2(1/r_1 + 1/r_2)^2$, and $Y^3 = 1.84P^2(1/r_1 + 1/r_2)/E^2$. If the radii of these spheres are also equal, $A^3 = 0.68Pr/E = 0.34Pd/E$; $S^3_{max} = 0.235PE^2/r^2 = 0.94PE^2/d^2$; and $Y^3 = 3.68P^2/(E^2r) = 7.36P^2/(E^2d)$.

Sphere and Flat Plate In this case $r_1 = r$ and $r_2 = \infty$, and the above formulas become $A^3 = 0.68Pr(c_1 + c_2) = 1.36Pr/E$, and

$$S^3_{max} = 0.235P/[r^2(c_1 + c_2)^2] = 0.059PE^2/r^2$$

$$Y^3 = 0.46P^2(c_1 + c_2)^2/r = 1.84P^2/(E^2r)$$

Two Cylinders The width b of the rectangular pressure surface is obtained from $(b/4)^2 = 0.29P(c_1 + c_2)/l[(1/r_1 + (1/r_2)]$, where r_1 and r_2 are the radii, and l the length

$$S^2_{max} = [4P/(\pi bl)]^2 = 0.35P(1/r_1 + 1/r_2)/l(c_1 + c_2)]$$

For cylinders with the same moduli of elasticity, $c_1 = c_2 = 1/E$, and $(b/4)^2 = 0.58P/El[(1/r_1) + (1/r_2)]$; and $S^2_{max} = 0.175PE(1/r_1 + 1/r_2)/l$. When $r_1 = r_2 = r$, $(b/4)^2 = 0.29Pr/(El)$, and $S^2_{max} = 0.35PE/(lr)$.

Cylinder and Flat Plate Here $r_1 = r$, $r_2 = \infty$, and the above formulas reduce to $(b/4)^2 = 0.29Pr(c_1 + c_2)/l = 0.58Pr/(El)$, and

$$S^2_{max} = 0.35P/[lr(c_1 + c_2)] = 0.175PE/(lr)$$

For application to ball and roller bearings and to gear teeth, see Sec. 6.

FLAT PLATES

The analysis of flat plates subjected to **lateral loads** is very involved because plates bend in all vertical planes. Strict mathematical derivations have therefore been accomplished only in some special cases. Most of the available formulas contain some amount of rational empiricism. Plates may be classified as (1) **thick plates,** in which transverse shear is important; (2) **average-thickness plates,** in which flexure stress predominates; (3) **thin plates,** which depend in part upon direct tension; and (4) **membranes,** which are subject to direct tension only. However, exact lines of demarcation do not exist.

The flat-plate formulas given apply primarily to symmetrically loaded **average-thickness plates** of constant thickness. They are valid only if the maximum deflection is small relative to the plate thickness; usually, $y \le 0.4t$. In the mathematical analyses, allowance for stress redistribution, because of slight local yielding, is usually not made. Since this yielding, especially in ductile materials, is beneficial, the formulas generally err on the side of safety. Certain cases of symmetrically loaded **circular and rectangular plates** are presented in Figs. 3.2.69 and 3.2.70. The maximum stresses are calculated from

$$S_M = k\frac{wR^2}{t^2} \qquad S_M = k\frac{P}{t^2} \qquad \text{or} \qquad S_M = k\frac{C}{t^2}$$

The first equation is for a uniformly distributed load w, lb/in²; the second supports a concentrated load P, lb; and the third a couple C, per unit length, uniformly distributed along the edge. Combinations of these loadings may be treated by superposition. The factors k and k_1 are given in Tables 3.2.19 and 3.2.20; R is the radius of circular plates or one side of rectangular plates, and t is the plate thickness. (In Figs. 3.2.69 and 3.2.70, r = radius of hole or piston for circular plates and r = smaller side of rectangular plates.)

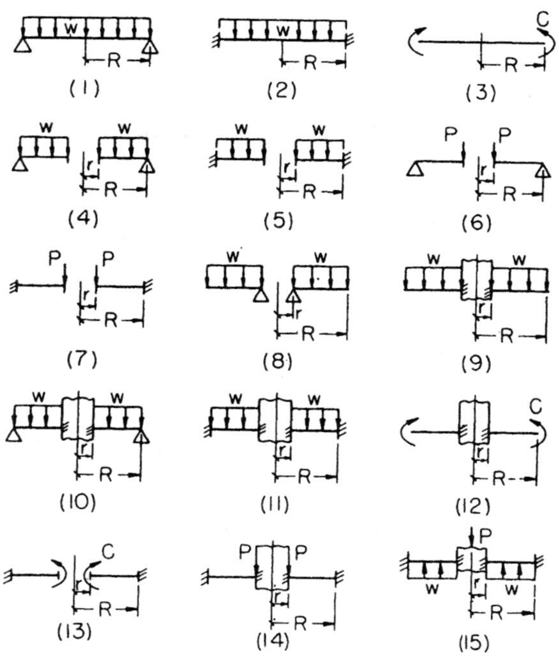

Figure 3.2.69

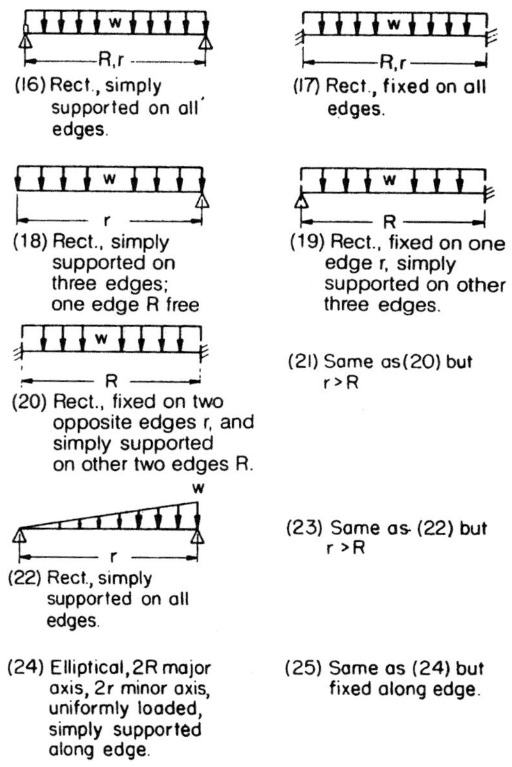

Figure 3.2.70

Table 3.2.19 Coefficients k and k_1 for Circular Plates
($\mu = 0.3$)

Case	k	k_1
1	1.24	0.696
2	0.75	0.171
3	6.0	4.2

	R/r											
	1.25		1.5		2		3		4		5	
Case	k	k_1	k	k_1	k	k_1	k	k_1	k	k_1	k	k_1
4	0.592	0.184	0.976	0.414	1.440	0.664	1.880	0.824	2.08	0.830	2.19	0.813
5	0.105	0.0025	0.259	0.0129	0.481	0.057	0.654	0.130	0.708	0.163	0.730	0.176
6	1.10	0.341	1.26	0.519	1.48	0.672	1.88	0.734	2.17	0.724	2.34	0.704
7	0.195	0.0036	0.320	0.024	0.455	0.081	0.670	0.171	1.00	0.218	1.30	0.238
8	0.660	0.202	1.19	0.491	2.04	0.902	3.34	1.220	4.30	1.300	5.10	1.310
9	0.135	0.0023	0.410	0.0183	1.04	0.0938	2.15	0.293	2.99	0.448	3.69	0.564
10	0.122	0.00343	0.336	0.0313	0.740	0.1250	1.21	0.291	1.45	0.417	1.59	0.492
11	0.072	0.00068	0.1825	0.005	0.361	0.023	0.546	0.064	0.627	0.092	0.668	0.112
12	6.865	0.2323	7.448	0.6613	8.136	1.493	8.71	2.555	8.930	3.105	9.036	3.418
13	6.0	0.196	6.0	0.485	6.0	0.847	6.0	0.940	6.0	0.801	6.0	0.658
14	0.115	0.00129	0.220	0.0064	0.405	0.0237	0.703	0.062	0.933	0.092	1.13	0.114
15	0.090	0.00077	0.273	0.0062	0.710	0.0329	1.54	0.110	2.23	0.179	2.80	0.234

Table 3.2.20 Coefficients k and k_1 for Rectangular and Elliptical Plates
($\mu = 0.3$)

	R/r									
	1.0		1.5		2.0		3.0		4.0	
Case	k	k_1	k	k_1	k	k_1	k	k_1	k	k_1
16	0.287	0.0443	0.487	0.0843	0.610	0.1106	0.713	0.1336	0.741	0.1400
17	0.308	0.0138	0.454	0.0240	0.497	0.0277	0.500	0.028	0.500	0.028
18	0.672	0.140	0.768	0.160	0.792	0.165	0.798	0.166	0.800	0.166
19	0.500	0.030	0.670	0.070	0.730	0.101	0.750	0.132	0.750	0.139
20	0.418	0.0209	0.626	0.0582	0.715	0.0987	0.750	0.1276	0.750	
21*	0.418	0.0216	0.490	0.0270	0.497	0.0284	0.500	0.0284	0.500	0.0284
22	0.160	0.0221	0.260	0.0421	0.320	0.0553	0.370	0.0668	0.380	0.0700
23*	0.160	0.0220	0.260	0.0436	0.340	0.0592	0.430	0.0772	0.490	0.0908
24	1.24	0.70	1.92	1.26	2.26	1.58	2.60	1.88	2.78	2.02
25	0.75	0.171	1.34	0.304	1.63	0.379	1.84	0.419	1.90	0.431

* Length ratio is r/R in cases 21 and 23.

The **maximum deflection** for the same cases is given by

$$y_M = k_1 \frac{wR^4}{Et^3} \qquad y_M = k_1 \frac{PR^2}{Et^3} \qquad \text{and} \qquad y_M = k_1 \frac{CR^2}{Et^3}$$

The factors k_1 are also given in the tables. For additional information, including shells, refer to ASME Handbook, "Metals Engineering: Design," McGraw-Hill.

THEORIES OF FAILURE

Material properties are usually determined from tests in which specimens are subjected to **simple stresses** under static or fluctuating loads. The attempt to apply these data to **bi- or triaxial stress fields** has resulted in the proposal of various theories of failure. Figure 3.2.71 shows the principal stresses on a triaxially stressed element. It is assumed, for simplicity, that $S_1 > S_2 > S_3$. Compressive stresses are negative.

1. **Maximum-stress theory** (Rankine) assumes failure occurs when the largest principal stress reaches the yield stress in a tension (or compression) specimen. That is, $S_1 = \pm S_y$.
2. **Maximum-shear theory** (Coulomb) assumes yielding (failure) occurs when the maximum shearing stress equals that in a simple tension (or compression) specimen at yield. Mathematically, $S_1 - S_3 = \pm S_y$.
3. **Maximum-strain-energy theory** (Beltrami) assumes failure occurs when the energy absorbed per unit volume equals the strain energy per unit

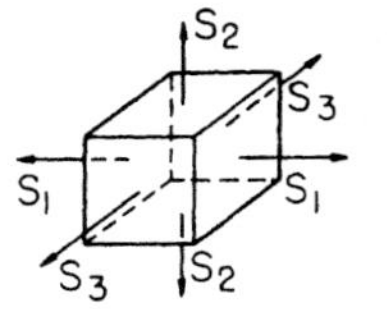

Figure 3.2.71

volume in a tension (or compression) specimen at yield. Mathematically, $S_1^2 + S_2^2 + S_3^2 - 2\mu(S_1S_2 + S_2S_3 + S_3S_1) = S_y^2$.

4. **Maximum-distortion-energy theory** (Huber, von Mises, Hencky) assumes yielding occurs when the distortion energy equals that in simple tension at yield. The distortion energy, that portion of the total energy which causes distortion rather than volume change, is

$$U_d = \frac{1+\mu}{3E}(S_1^2 + S_2^2 + S_3^2 - S_1S_2 - S_2S_3 - S_3S_1)$$

Thus failure is defined by

$$S_1^2 + S_2^2 + S_3^2 - (S_1S_2 + S_2S_3 + S_3S_1) = S_y^2$$

5. **Maximum-strain theory** (Saint-Venant) claims failure occurs when the maximum strain equals the strain in simple tension at yield or $S_1 - \mu(S_2 + S_3) = S_y$.

6. **Internal-friction theory** (Mohr). When the ultimate strengths in tension and compression are the same, this theory reduces to that of maximum shear. For principal stresses of opposite sign, failure is defined by $S_1 - (S_{uc}/S_u)\,S_2 = -S_{uc}$; if the signs are the same $S_1 = S_u$ or $-S_{uc}$, where S_{uc} is the ultimate strength in compression. If the principal stresses are both either tension or compression, then the larger one, say S_1, must equal S_u when S_1 is tension or S_{uc} when S_1 is compression.

A **graphical representation** of the first four theories applied to a biaxial stress field is presented in Fig. 3.2.72. Stresses outside the bounding lines in the case of each theory mean failure (yield or fracture). A comparison with experimental data proves the distortion-energy theory (4) best for ductile materials of equal tension-compression properties. When these properties are unequal, the internal friction theory (6) appears best. In practice, judging by some accepted codes, the maximum-shear theory (2) is generally used for ductile materials, and the maximum-stress theory (1) for brittle materials.

Fatigue failures cannot be related, theoretically, to elastic strength and thus to the theories described. However, experimental results justify this, at least to a limited extent. Therefore, the theory evaluation given above holds for **fluctuating stresses,** provided that principal stresses at the maximum load are used and the **endurance strength** in simple bending is substituted for the yield strength.

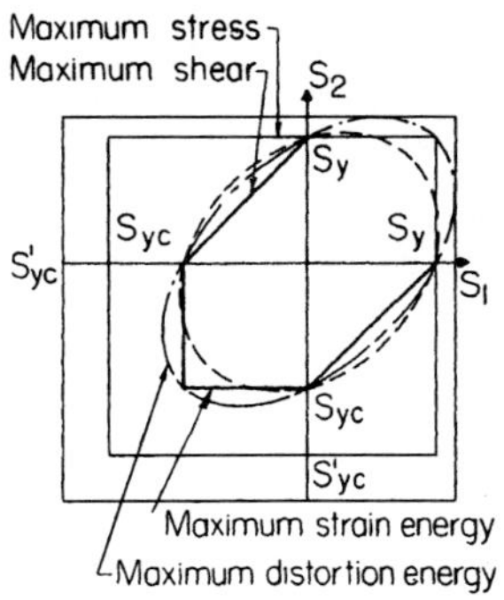

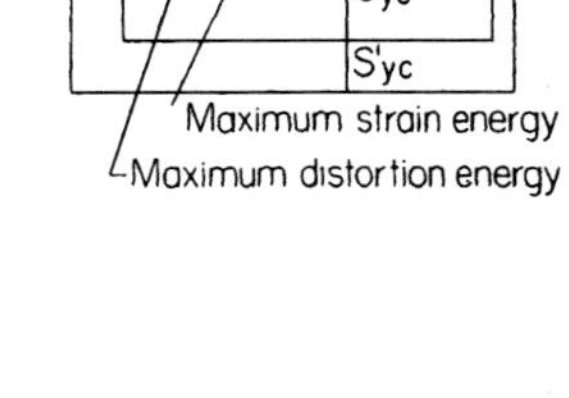

Figure 3.2.72

EXAMPLE. A steel shaft, 4 in in diameter, is subjected to a bending moment of 120,000 in · lb, as well as a torque. If the yield strength in tension is 40,000 lb/in², what maximum torque can be applied under the (1) maximum-shear theory and (2) the distortion-energy theory?

$$S_x = \frac{M_c}{I} = \frac{120{,}000 \times 2}{12.55}$$

$$= 19{,}100 \text{ lb/in}^2 \qquad S_{xy} = \frac{TC}{J} = \frac{T \times 2}{25.1} = 0.0798T$$

and

$$S_{M,m} = \frac{S_x}{2} \pm \sqrt{\left(\frac{S_x}{2}\right)^2 + S_{xy}^2}$$

$$S_M - S_m = S_y \quad \text{or} \quad 2\sqrt{\left(\frac{19{,}100}{2}\right)^2 + (0.0798T)^2} \tag{1}$$

$$= 40{,}000$$

or

$$T = 221{,}000 \text{ in} \cdot \text{lb}(254{,}150 \text{ cm} \cdot \text{kgf})$$

$$S_M^2 + S_m^2 - S_M S_m = S_y^2 \tag{2}$$

substituting and simplifying,

$$(9{,}550)^2 + 3\sqrt{\left(\frac{19{,}100}{2}\right)^2 + (0.0798T)^2} = (40{,}000)^2$$

or

$$T = 255{,}000 \text{ in} \cdot \text{lb}(293{,}250 \text{ cm} \cdot \text{kgf})$$

PLASTICITY

The reaction of materials to stress and strain in the plastic range is not fully defined. However, some concepts and theories have been proposed.

Ideally, a **purely elastic material** is one complying explicitly with Hooke's law. In a **viscous material,** the shearing stress is proportional to the shearing strain. The **purely plastic material** yields indefinitely, but only after reaching a certain stress. Combinations of these are the **elastoviscous** and the **elastoplastic** materials.

Engineering materials are not ideal, and usually contain some of the elastoplastic characteristics. The **total strain** ε_t is the sum of the **elastic strain** ε_o plus the **plastic strain** ε_p, as shown in Fig. 3.2.73, where the stress-strain curve is approximated by two straight lines. The **natural strain,** which is at the same time the total strain, is $\varepsilon = \int_{l}^{l} dl/l = \ln(l/l_o)$. In this equation, l is the instantaneous length, while l_o is the original length. In terms of the normal strain, the natural strain becomes $\bar{\varepsilon} = \ln(1 + \varepsilon_o)$. Since it is assumed that the volume remains constant, $l/l_o = A_o/A$, and so the natural stress becomes $\bar{S} = P/A = (P/A_o)(1 + \varepsilon_o)$. A_o is the original cross-sectional area. If the natural stress is plotted against strain on log-log paper, the graph is very nearly a straight line. The plastic-range relation is thus approximated by $\bar{S} = K\bar{\varepsilon}^n$, where the proportionality factor K and the **strain-hardening coefficient** n are determined from best fits to experimental data. Values of K and n determined by Low and Garofalo (*Proc. Soc. Exp. Stress Anal.,* vol. IV, no. 2, 1947) are given in Table 3.2.21.

The geometry of Fig. 3.2.73 can be used to arrive at a second approximate relation

$$\bar{S} = S_o + (\varepsilon_p - \varepsilon_o)\tan\theta = S_o\left(1 - \frac{H}{E}\right) + \varepsilon_p H$$

where $H = \tan\theta$ is a kind of **plastic modulus.**

The **deformation theory of plastic flow** for the general case of combined stress is developed using the above concepts. Certain additional assumptions involved include: principal plastic-strain directions are the same as principal stress directions; the elastic strain is negligible compared to plastic strain; and the ratios of the three principal shearing strains—$(\bar{\varepsilon}_1 - \bar{\varepsilon}_2)$, $(\bar{\varepsilon}_2 - \bar{\varepsilon}_3)$, $(\bar{\varepsilon}_3 - \bar{\varepsilon}_1)$—to

Table 3.2.21 Constants *K* and *n* for Sheet Materials

Material	Treatment	K, lb/in^2	n
0.05%C rimmed steel	Annealed	77,100	0.261
0.05%C killed steel	Annealed and tempered	73,100	0.234
Decarburized 0.05%C steel	Annealed in wet H_2	75,500	0.284
0.05/0.07% phos. low C	Annealed	93,330	0.156
SAE 4130	Annealed	169,400	0.118
SAE 4130	Normalized and tempered	154,500	0.156
Type 430 stainless	Annealed	143,000	0.229
Alcoa 24-S	Annealed	55,900	0.211
Reynolds R-301	Annealed	48,450	0.211

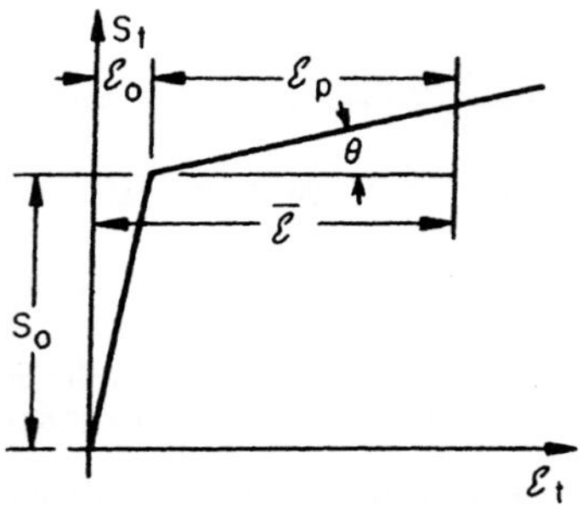

Figure 3.2.73

the principal shearing stresses—$(\bar{S}_1-\bar{S}_2)/2, (\bar{S}_2-\bar{S}_3)/2, (\bar{S}_3-\bar{S}_1)/2$—are equal. The relations between the principal strains and stresses in terms of the simple tension quantities become

$$\bar{\varepsilon}_1 = \bar{\varepsilon}/\bar{S}[\bar{S}_1 - (\bar{S}_2 + \bar{S}_3)/2]$$
$$\bar{\varepsilon}_2 = \bar{\varepsilon}/\bar{S}[\bar{S}_2 - (\bar{S}_3 + \bar{S}_1)/2]$$
$$\bar{\varepsilon}_3 = \bar{\varepsilon}/\bar{S}[\bar{S}_3 - (\bar{S}_1 + \bar{S}_2)/2]$$

If these equations are added, the plastic-flow theory is expressed:

$$\frac{\bar{S}}{\bar{\varepsilon}} = \sqrt{\frac{[(\bar{S}_1-\bar{S}_2)^2+(\bar{S}_2-\bar{S}_3)^2+(\bar{S}_3-\bar{S}_1)^2]/2}{2(\bar{\varepsilon}_1^2+\bar{\varepsilon}_2^2+\bar{\varepsilon}_3^2)/3}}$$

In the above equation

$$\sqrt{[(\bar{S}_1-\bar{S}_2)^2+(\bar{S}_2-\bar{S}_3)^2+(\bar{S}_3-\bar{S}_1)^2]/2} = \bar{S}_e$$

and $$\sqrt{2(\bar{\varepsilon}_1^2+\bar{\varepsilon}_2^2+\bar{\varepsilon}_3^2)/3} = \varepsilon_e$$

are the effective, or significant, stress and strain, respectively,

EXAMPLE. An annealed, stainless-steel type 430 tank has a 41-in inside diameter and has a wall 0.375 in thick. The ultimate strength of the stainless steel is 85,000 lb/in^2. Compute the maximum strain as well as the pressure at fracture.

The tank constitutes a biaxial stress field where $S_1 = pd/(2t)$, $S_2 = pd/(4t)$, and $S_3 = 0$. Taking the power stress-strain relation

$$\bar{S}_e = K\,\bar{\varepsilon}_e^{\,n} \qquad \text{or} \qquad \bar{\varepsilon}/\bar{S} = \bar{S}_e^{(1-n)/n}/K^{1/n}$$

thus $$\bar{\varepsilon}_1 = \frac{\bar{S}_e^{(1-n)/n}}{K^{1/n}}\left(\frac{3}{4}\,\bar{S}_1\right)\bar{\varepsilon} = 0$$

and $$\bar{\varepsilon}_3 = \frac{\bar{S}_e^{(1-n)/n}}{K^{1/n}}\left(-\frac{3}{4}\,\bar{S}_t\right) = -\bar{\varepsilon}_1$$

The maximum-shear theory, which is applicable to a ductile material under combined stress, is acceptable here. Thus rupture will occur at

$\bar{S}_1 - \bar{S}_3 = \bar{S}_u$, and

$$\bar{S}_e = \sqrt{\frac{1}{2}\left[\left(\frac{\bar{S}_1}{2}\right)^2 + \left(\frac{\bar{S}_1}{2}\right)^2 + \bar{S}_1^2\right]} = \sqrt{\frac{3}{4}\,\bar{S}_1^2} = \left(\frac{3}{4}\right)^{1/2}\bar{S}_u$$

$$\bar{\varepsilon}_1 = \frac{[(3/4)]^{1/2}\bar{S}_u]^{(1-n)/n}}{K^{1/n}}\left(\frac{3}{4}\,\bar{S}_u\right) = \left(\frac{3}{4}\right)^{1+0.229/0.458}\left(\frac{85{,}000}{143{,}000}\right)^{1/0.225}$$

$$= 0.0475 \text{ in/in } (0.0475 \text{ cm/cm})$$

since $$\bar{S}_u = S_1 = \frac{pd}{2t}, \text{ then } p = \frac{2t\bar{S}_u}{d}$$

or $$p = \frac{2\times 0.375\times 85{,}000}{41} = 1{,}550 \text{ lb/in}^2 \ (109 \text{ kgf/cm}^2)$$

ROTATING DISKS

Rotating circular disks may be of various profiles, of constant or variable thickness, with or without centrally and noncentrally located holes, and with radial, tangential, and shearing stresses.

The solution starts with the differential equations of equilibrium and compatibility and the subsequent application of appropriate boundary conditions for the derivation of working-stress equations.

If the disk thickness is small compared with the diameter, the variation of stress with thickness can be assumed to be negligible, and symmetry eliminates the shearing stress. In the rotating case, the disk weight is neglected, but its inertia force becomes the body-force term in the equilibrium equations.

Thus solved, the stress components in a solid disk become

$$\sigma_r = \frac{3+\mu}{8}\rho\omega^2(R^2 - r^2)$$

$$\sigma_\theta = \frac{3+\mu}{8}\rho\omega^2 R^2 - \frac{1+3\mu}{8}\rho\omega^2 r^2$$

where μ = Poisson's ratio; ρ = mass density, lb · s²/in⁴; ω = angular speed, rad/s; R = outside disk radius; and r = radius to point in question.

The largest stresses occur at the center of the solid disk and are

$$\sigma_r = \sigma_\theta = \frac{3+\mu}{8}\rho\omega^2 R^2$$

A disk with a central hole of radius r_h (no external forces) is subjected to the following stresses:

$$\sigma_r = \frac{3+\mu}{8}\rho\omega^2\left(R^2 + r_h^2 - \frac{R^2 r_h^2}{r^2} - r^2\right)$$

$$\sigma_\theta = \frac{3+\mu}{8}\rho\omega^2\left(R^2 + r_h^2 + \frac{R^2 r_h^2}{r^2} - \frac{1+3\mu}{3+\mu}r^2\right)$$

The maximum radial stress $\sigma_r|_M$ occurs at $r = \sqrt{Rr_h}$, and

$$\sigma_r|_M = \frac{3+\mu}{8}\rho\omega^2(R - r_h)^2$$

The largest tangential stress $\sigma_\theta|_M$ exists at the inner boundary, and

$$\sigma_\theta|_M = \frac{3+\mu}{4}\rho\omega^2\left(R^2 + \frac{1-\mu}{3+\mu}r_h^2\right)$$

As the hole radius r_h approaches zero, the tangential stress assumes a value twice that at the center of a rotating solid disk, given above.

3.3 PIPELINE FLEXURE STRESSES

Revised by Lawrence M. Matta

REFERENCES: Shipman, Design of Steam Piping to Care for Expansion, *Trans. ASME,* 1929. Wahl, Stresses and Reactions in Expansion Pipe Bends, *Trans. ASME,* 1927. Hovgaard, The Elastic Deformation of Pipe Bends, *Jour. Math. Phys.,* Nov. 1926, Oct. 1928, and Dec. 1929. M. W. Kellog Co., "The Design of Piping Systems," Wiley.

For details of pipe and pipe fittings see Sec. 6.7.

An initial piping design is primarily based on delivering fluids to the required locations at required rates while satisfying the fluid and gravity loads that occur during steady operation at design conditions. However, it is also important to provide adequate flexibility to avoid overstressing the piping due to transient loads. Thermal stresses, for example, have the potential to generating large forces and stresses in a piping system. If the piping is flexible enough, the expansion can occur without generating excessive stresses. Piping designed to be more flexible than necessary may waste materials and space. Thermal stresses are one source of concern, but design load cases may also include stresses due to earthquakes, wind, and other occasional loads.

While some simple geometries can be effectively analyzed using hand calculations, for anything more than the most basic designs, this quickly becomes impractical. Many high-quality software packages are available on the market for performing piping stress analysis. Such programs typically take into account static and dynamic conditions, restraint conditions, aboveground and buried configurations, etc. The iterative process of piping layout and analysis can be performed fairly efficiently using such software. The reader is referred to the technical literature for the most suitable and current software available for use in solving the immediate problems at hand.

The brief discussion in this section addresses the fundamental concepts entailed and sets forth the solution of simple systems as an exercise in application of the principles.

Nomenclature (see Figs. 3.3.1 and 3.3.2)

M_0 = end moment at origin, in · lb (N · m)
M = max moment, in · lb (N · m)
F_x = end reaction at origin in x direction, lb (N)
F_y = end reaction at origin in y direction, lb (N)
$S_l = (Mr/I)\alpha$ = max unit longitudinal flexure stress, lb/in² (N/m²)
$S_t = (Mr/I)\beta$ = max unit transverse flexure stress, lb/in² (N/m²)

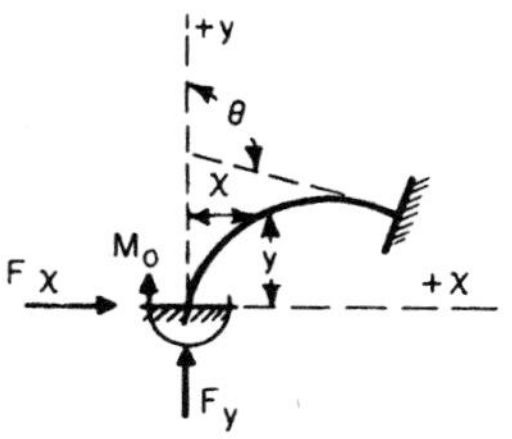

Figure 3.3.1

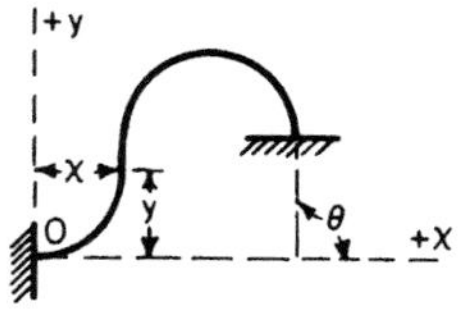

Figure 3.3.2

$S_s = (Mr/I)\gamma$ = max unit shearing stress, lb/in² (N/m²)
Δx = relative deflection of ends of pipe parallel to x direction caused by either temperature change or support movement, or both, in (m)
Δy = same as Δx but parallel to y direction, in. Note that Δx and Δy are positive if under the change in temperature the end opposite the origin tends to move in a positive x or y direction, respectively
t = wall thickness of pipe, in (m)
r = mean radius of pipe cross section, in (m)
λ = constant = tR/r^2
I = moment of inertia of pipe cross section about pipe centerline, in⁴ (m⁴)
E = modulus of elasticity of pipe at *actual working* temperature, lb/in² (N/m²)
K = flexibility index of pipe. K = 1 for all straight pipe sections, $K = (10 + 12\lambda^2)/(1 + 12\lambda^2)$ for all curved pipe sections where $\lambda > 0.335$ (see Fig. 3.3.3)
a, β, γ = ratios of actual max longitudinal flexure, transverse flexure, and shearing stresses to Mr/I for curved sections of pipe (see Fig. 3.3.3)
A, B, C, F, G, H = constants given by Table 3.3.2
θ = angle of intersection between tangents to direction of pipe at reactions
$\Delta\theta$ = change in θ caused by movements of supports, or by temperature change, or both, rad
ds = an infinitesimal element of length of pipe
s = length of a particular curved section of pipe, in (m)
R = radius of curvature of pipe centerline, in (m)

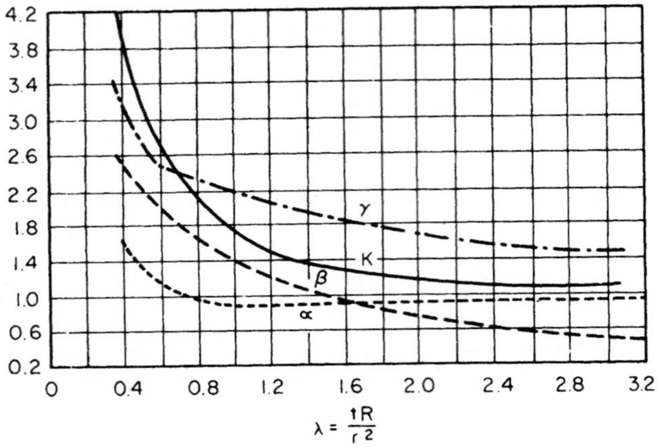

Figure 3.3.3 Flexure constants of initially curved pipes.

GENERAL DISCUSSION

Under the effect of changes in temperature of the pipeline, or of movement of support reactions (either translation or rotation), or both, the determination of stress distribution in a pipe becomes a statically indeterminate problem. In general the problem may be solved by a slight modification of the standard arch theory: $\Delta x = -K\int My\ ds(EI)$, $\Delta y = K\int Mx\ ds/(EI)$, and $\Delta\theta = K\int M\ ds/(EI)$ where the constant K is introduced to correct for the increased flexibility of a curved pipe, and where the integration is over the entire length of pipe between supports. In Table 3.3.1 are given equations derived by this method for moment and thrust at one reaction point for pipes in one plane that are fully fixed, hinged at both ends, hinged at one end and fixed at the other, or partly fixed. If the reactions at one end of the pipe are known, the moment distribution in the entire pipe then can be obtained by simple statics.

Since an initially curved pipe is more flexible than indicated by its moment of inertia, the constant K is introduced. Its value may be taken from Fig. 3.3.3, or computed from the equation given below. $K = 1$ for all straight pipe sections, since they act according to the simple flexure theory.

In Fig. 3.3.3 are given the flexure constants K, α, β, and γ for initially curved pipes as functions of the quantity $\lambda = tR/r^2$. The flexure constants are derived from the equations.

$$K = (10 + 12\lambda^2)/(1 + 12\lambda^2) \qquad \text{when } \lambda > 0.335$$

$$\alpha = \tfrac{2}{3}K\sqrt{(5 + 6\lambda^2)/18} \qquad \lambda \le 1.472$$

$$\alpha = K(6\lambda^2 - 1)/(6\lambda^2 + 5) \qquad \lambda > 1.472$$

Table 3.3.1 General Equations for Pipelines in One Plane (See Figs. 3.3.1 and 3.3.2)

Type of supports	Unsymmetric	Symmetric about y-axis
Both ends fully fixed	$M_0 = \dfrac{EI\,\Delta x(CF - AB) + EI\,\Delta y(BF - AG)}{2ABF + CGH - B^2H - A^2G - CF^2}$ $F_x = \dfrac{EI\,\Delta x(CH - A^2) + EI\,\Delta y(BH - AF)}{2ABF + CGH - B^2H - A^2G - CF^2}$ $F_y = \dfrac{EI\,\Delta x(BH - AF) + EI\,\Delta y(GH - F^2)}{2ABF + CGH - B^2H - A^2G - CF^2}$ $\Delta\theta = 0$	$M_0 = \dfrac{EI\,\Delta x F}{GH - F^2}$ $F_x = \dfrac{EI\,\Delta x H}{GH - F^2}$ $F_y = 0$ $\Delta\theta = 0$
Both ends hinged	$M_0 = 0$ $F_x = \dfrac{EI\,\Delta x C + EI\,\Delta y B}{CG - B^2}$ $F_y = \dfrac{EI\,\Delta x B + EI\,\Delta y G}{CG - B^2}$ $\Delta\theta = \dfrac{\Delta x(AB - CF) + \Delta y(AG - BF)}{CG - B^2}$	$M_0 = 0$ $F_x = \dfrac{EI\,\Delta x}{G}$ $F_y = 0$ $\Delta\theta = \dfrac{-\Delta x F}{G}$
Origin end only hinged, other end fully fixed	$M_0 = 0$ $F_x = \dfrac{EI\,\Delta x C + EI\,\Delta y B}{CG - B^2}$ $F_y = \dfrac{EI\,\Delta x B + EI\,\Delta y G}{CG + B^2}$ $\Delta\theta = \dfrac{\theta x(AB - CF) + \Delta y(AG - BF)}{CG - B^2}$	
In general for any specific rotation $\Delta\theta$ and movement Δx and Δy …	$M_0 = \dfrac{EI\,\Delta x(CF - AB) + EI\,\Delta y(BF - AG) + EI\,\Delta\theta(CG - B^2)}{2ABF + CGH - A^3G - CF^2 - B^2H}$ $F_x = \dfrac{EI\,\Delta x(CH - A^2) + EI\,\Delta y(BH - AF) + EI\,\Delta\theta(CF - AB)}{2ABF + CGH - A^2G - CF^2 - B^2H}$ $F_y = \dfrac{EI\,\Delta x(BH - AF) + EI\,\Delta y(GH - F^2) + EI\,\Delta\theta(BF - AG)}{2ABF + CGH - A^2G - CF^2 - B^2H}$	

$\beta = 18\lambda/(1+12\lambda^2)$

$\gamma = [8\lambda - 36\lambda^3 + (32\lambda^2 + 20/3)$

$\times \sqrt{(4/3)\lambda^2 + 5/18}\,] \div (1+12\lambda^2)$ when $\lambda < 0.58$

$= (12\lambda^2 + 18\lambda - 2)/(1+12\lambda^2)$ when $\lambda > 0.58$

The increased flexibility of the curved pipe is brought about by the tendency of its cross section to flatten. This flattening causes a transverse flexure stress whose maximum is S_r. Because the maximum longitudinal and maximum transverse stresses do not occur at the same point in the pipe's cross section, the resulting maximum shear is not one-half the difference of S_l and S_t; it is S_s. In the straight sections of the pipe, $\alpha = 1$, the transverse stress disappears, and $\lambda = \frac{1}{2}$. This discussion of S_s does not include the uniform transverse or longitudinal tension stresses induced by the internal pressure in the pipe; their effects should be added if appreciable.

Table 3.3.2 gives values of the constants *A*, *B*, *C*, *F*, *G*, and *H* for use in equations listed in Table 3.3.1. The values may be used (1) for the solution of any pipeline or (2) for the derivation of equations for standard shapes composed of straight sections and arcs of circles as of Fig. 3.3.5. Equations for shapes not given may be obtained by algebraic addition of those given. All measurements are from the left-hand end of the pipeline. Reactions and stresses are greatly influenced by end conditions. Formulas are given to cover the extreme conditions. The following suggestions and comments should be considered when laying out a pipeline:

Avoid expansion bends, and design the entire pipeline to take care of its own expansion.

The movement of the equipment to which the ends of the pipeline are attached must be included in the Δx and Δy of the equations.

Maximum flexibility is obtained by placing **supports** and **anchors** so that they will not interfere with the natural movement of the pipe.

That shape is most efficient in which the **maximum length of pipe is working** at the maximum safe stress.

Excessive **bending moment at joints** is more likely to cause trouble than excessive stresses in pipe walls. Hence, keep pipe joints away from points of high moment.

Reactions and stresses are greatly influenced by **flattening of the cross section** of the curved portions of the pipeline.

It is recommended that **cold springing** allowances be discounted in stress calculations.

Application to Two- and Three-Plane Pipelines Pipelines in more than one plane may be solved by the successive application of the preceding data, dividing the pipeline into two or more one-plane lines.

EXAMPLE 1. The unsymmetric pipeline of Fig. 3.3.4 has fully fixed ends. From Table 3.3.2 use $K = 1$ for all sections, since only straight segments are involved.

Upon introduction of $a = 120$ in (3.05 m), $b = 60$ in (1.52 m), and $c = 180$ in (4.57 m), into the preceding relations (Table 3.3.3) for *A*, *B*, *C*, *F*, *G*, *H*, the equations for the reactions at 0 from Table 3.3.1 become

$$M_0 = EI\,\Delta x(-7.1608\times10^{-5}) + EI\,\Delta y(-8.3681\times10^{-5})$$

$$F_x = EI\,\Delta x(+1.0993\times10^{-5}) + EI\,\Delta y(+3.1488\times10^{-6})$$

$$F_y = EI\,\Delta x(+3.1488\times10^{-6}) + EI\,\Delta y(+1.33717\times10^{-6})$$

Also it follows that

$$M_1 = M_0 + F_y a = EI\,\Delta x(+3.0625\times10^{-4}) + EI\,\Delta y(+7.6799\times10^{-5})$$

$$M_2 = M_1 - F_x b = EI\,\Delta x(-3.5333\times10^{-4}) + EI\,\Delta y(-1.1215\times10^{-4})$$

$$M_3 = M_2 + F_y c = EI\,\Delta x(+2.1345\times10^{-4}) + EI\,\Delta y(+1.2854\times10^{-4})$$

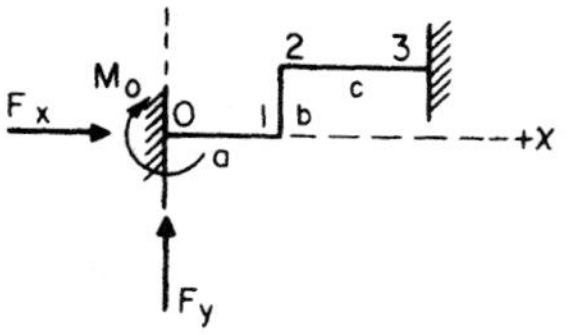

Figure 3.3.4

Thus the maximum moment *M* occurs at 3.

The total maximum longitudinal fiber stress ($a = 1$ for straight pipe)

$$S_l = \frac{F_x}{2\pi rt} \pm \frac{M_3 r}{I}$$

There is no transverse flexure stress since all sections are straight. The maximum shearing stress is either (1) one-half of the maximum longitudinal fiber stress as given above, (2) one-half of the hoop-tension stress caused by an internal radial pressure that might exist in the pipe, or (3) one-half the difference of the maximum longitudinal fiber stress and hoop-tension stress, whichever of these three possibilities is numerically greatest.

EXAMPLE 2. The equations of Table 3.3.1 may be employed to develop the solution of generalized types of pipe configurations for which Fig. 3.3.5 is a typical example. If only temperature changes are considered, the reactions for the right-angle pipeline (Fig. 3.3.5) may be determined from the following equations:

$$M_0 = C_1 EI\,\Delta x / R^2$$

$$F_x = C_2 EI\,\Delta x / R^3$$

$$F_y = C_3 EI\,\Delta x / R^3$$

In these equations, Δx is the *x* component of the deflection between reaction points caused by temperature change only. The values of C_1, C_2, and C_3 are given in Fig. 3.3.6 for $K = 1$ and $K = 2$. For other values of *K*, interpolation may be employed.

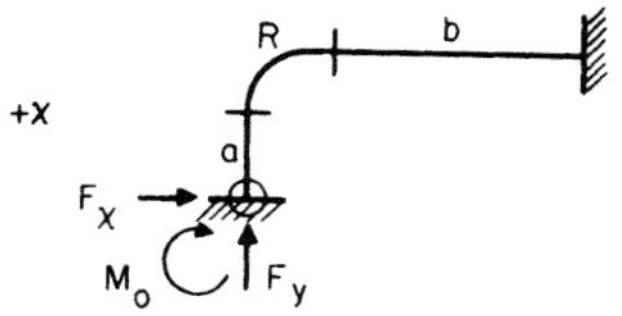

Figure 3.3.5 Right angle pipeline.

EXAMPLE 3. With $a/R = 20$ and $b/R = 3$, the value of C_1 is 0.185 for $K = 1$ and 0.165 for $K = 2$. If $K = 1.75$, the interpolated value of C_1 is 0.175.

Elimination of Flexure Stresses Pipeline flexure stresses that normally would result from movement of supports or from the tendency of the pipes to expand under temperature change often may be avoided entirely through the use of expansion joints (Sec. 6.7). Their use may simplify both the design of the pipeline and the support structure. When using expansion joints, the following suggestions should be considered: (1) select the expansion joint carefully for maximum temperature range (and deflection) expected so as to prevent damage to expansion fitting; (2) provide guides to limit movement at the expansion joint to direction permitted by joint; (3) provide adequate anchors at one end of each straight section or along its midlength, forcing movement to occur at the expansion joint yet providing adequate support for the pipeline; (4) mount expansion joints adjacent to an anchor point to prevent sagging of the pipeline under its own weight and do not depend upon the expansion joint for stiffness—it is intended to be flexible; (5) give consideration to effects of corrosion, since the corrugated character of expansion joints makes cleaning difficult.

Table 3.3.2 Values of *A*, *B*, *C*, *F*, *G*, and *H* for Various Piping Elements

	$A = K\int x ds$	$B = K\int xy ds$	$C = K\int x^2 ds$	$F = K\int y ds$	$G = K\int y^2 ds$	$H = K\int ds$
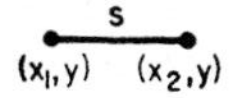	$\frac{s}{2}(x_1 + x_2)$	Ay	$s\left(\frac{s^2}{3} + x_1x_2\right)$	sy	Fy	s
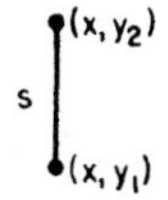	sx	$\frac{A}{2}(y_1 + y_2)$	Ax	$\frac{s}{2}(y_1 + y_2)$	$s\left(\frac{s^2}{3} + y_1y_2\right)$	s
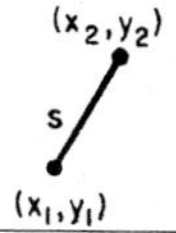	$\frac{s}{2}(x_1 + x_2)$	$\frac{A}{3}(y_1 + y_2) + \frac{s}{6}(x_1y_1 + x_2y_2)$	$\frac{s}{3}(x_1^2 + x_1x_2 + x_2^2)$	$\frac{s}{2}(y_1 + y_2)$	$\frac{s}{3}(y_1^2 + y_1y_2 + y_2^2)$	s
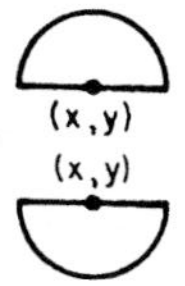	πKRx	$A\left(y + \frac{2R}{\pi}\right)$	$A\left(x + \frac{R^2}{2x}\right)$	$(\pi y + 2R)KR$	$Fy + \left(2y + \frac{\pi}{2}R\right)KR^2$	πKR
		$A\left(y - \frac{2R}{\pi}\right)$		$(\pi y - 2R)KR$	$Fy - \left(2y - \frac{\pi}{2}R\right)KR^2$	
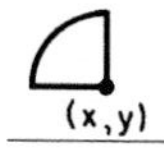	$\left(\frac{\pi x}{2} - R\right)KR$	$Ay + \left(x - \frac{R}{2}\right)KR^2$	$Ax + \left(\frac{\pi R}{4} - x\right)KR^2$	$\left(\frac{\pi y}{2} + R\right)KR$	$Fy + \left(\frac{\pi R}{4} + y\right)KR^2$	$\frac{\pi KR}{2}$

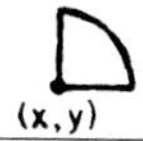	$\left(\frac{\pi y}{2}+R\right)KR$	$Ay-\left(x+\frac{R}{2}\right)KR^2$	$Ay+\left(\frac{\pi R}{4}+x\right)KR^2$	$\left(\frac{\pi y}{2}+R\right)KR$	$Fy+\left(\frac{\pi R}{4}+y\right)KR^2$	$\frac{\pi KR}{2}$
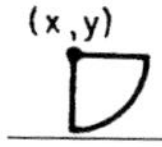		$Ay-\left(x+\frac{R}{2}\right)KR^2$		$\left(\frac{\pi y}{2}-R\right)KR$	$Fy+\left(\frac{\pi R}{4}-y\right)KR^2$	
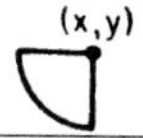	$\left(\frac{\pi x}{2}-2\right)KR$	$Ay-\left(x-\frac{R}{2}\right)KR^2$	$Ax+\left(\frac{\pi R}{4}-x\right)KR^2$	$\left(\frac{\pi R}{2}-R\right)KR$	$Fy+\left(\frac{\pi R}{4}-y\right)KR^2$	
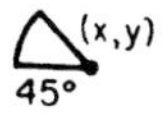	$\left(\frac{\pi x}{4}-\frac{R}{\sqrt{2}}\right)KR$	$Ay+\left[\left(1-\frac{\sqrt{2}}{2}\right)x-\frac{R}{4}\right]KR^2$	$Ax+\left[\left(\frac{\pi}{8}+\frac{1}{4}\right)R-\frac{x\sqrt{2}}{2}\right]KR^2$	$\left[\frac{\pi y}{4}+\left(1-\frac{\sqrt{2}}{2}\right)R\right]KR$	$Fy+\left[\left(1-\frac{\sqrt{2}}{2}\right)y+\left(\frac{\pi}{8}-\frac{1}{4}\right)R\right]KR^2$	$\frac{\pi KR}{4}$
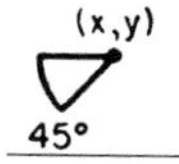		$Ay-\left[\left(1-\frac{\sqrt{2}}{2}\right)x-\frac{R}{4}\right]KR^2$		$\left[\frac{\pi y}{4}-\left(1-\frac{\sqrt{2}}{2}\right)R\right]KR$	$Fy-\left[\left(1-\frac{\sqrt{2}}{2}\right)y-\left(\frac{\pi}{8}-\frac{1}{4}\right)R\right]KR^2$	
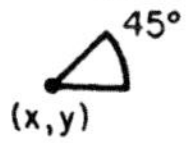	$\left(\frac{\pi x}{4}-\frac{R}{\sqrt{2}}\right)KR$	$Ay+\left[\left(1-\frac{\sqrt{2}}{2}\right)x+\frac{R}{4}\right]KR^2$	$Ax+\left[\left(\frac{\pi}{8}+\frac{1}{4}\right)R-\frac{x\sqrt{2}}{2}\right]KR^2$	$\left[\frac{\pi y}{4}+\left(1-\frac{\sqrt{2}}{2}\right)R\right]KR$	$Fy+\left[\left(1-\frac{\sqrt{2}}{2}\right)y+\left(\frac{\pi}{8}-\frac{1}{4}\right)R\right]KR^2$	
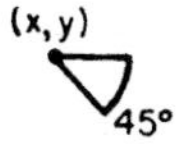		$Ay-\left[\left(1-\frac{\sqrt{2}}{2}\right)x+\frac{R}{4}\right]KR^2$		$\left[\frac{\pi y}{4}-\left(1-\frac{\sqrt{2}}{2}\right)R\right]KR$	$Fy-\left[\left(1-\frac{\sqrt{2}}{2}\right)y-\left(\frac{\pi}{8}-\frac{1}{4}\right)R\right]KR^2$	
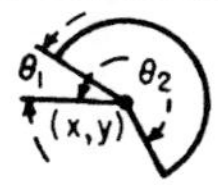	$[r(\theta_2-\theta_1)$ $-R(\sin\theta_2-\sin\theta_1)]KR$	$Ay-\left[x(\cos\theta_2-\cos\theta_1)\right.$ $\left.+\frac{R}{2}(\sin^2\theta_2-\sin^2\theta_1)\right]KR^2$	$Ay-\left[x(\sin\theta_2-\sin\theta_1)\right.$ $-\frac{R}{4}(\sin 2\theta_2-\sin 2\theta_1)$ $\left.-\frac{R}{2}(\theta_2-\theta_1)\right]KR^2$	$[y(\theta_2-\theta_1)$ $-R(\cos\theta_2-\cos\theta_1)]KR$	$Fy-\left[y(\cos\theta_2-\cos\theta_1)\right.$ $+\frac{R}{4}(\sin 2\theta_2-\sin 2\theta_1)$ $\left.-\frac{R}{2}(\theta_2-\theta_1)\right]KR^2$	$(\theta_2-\theta_1)KR$

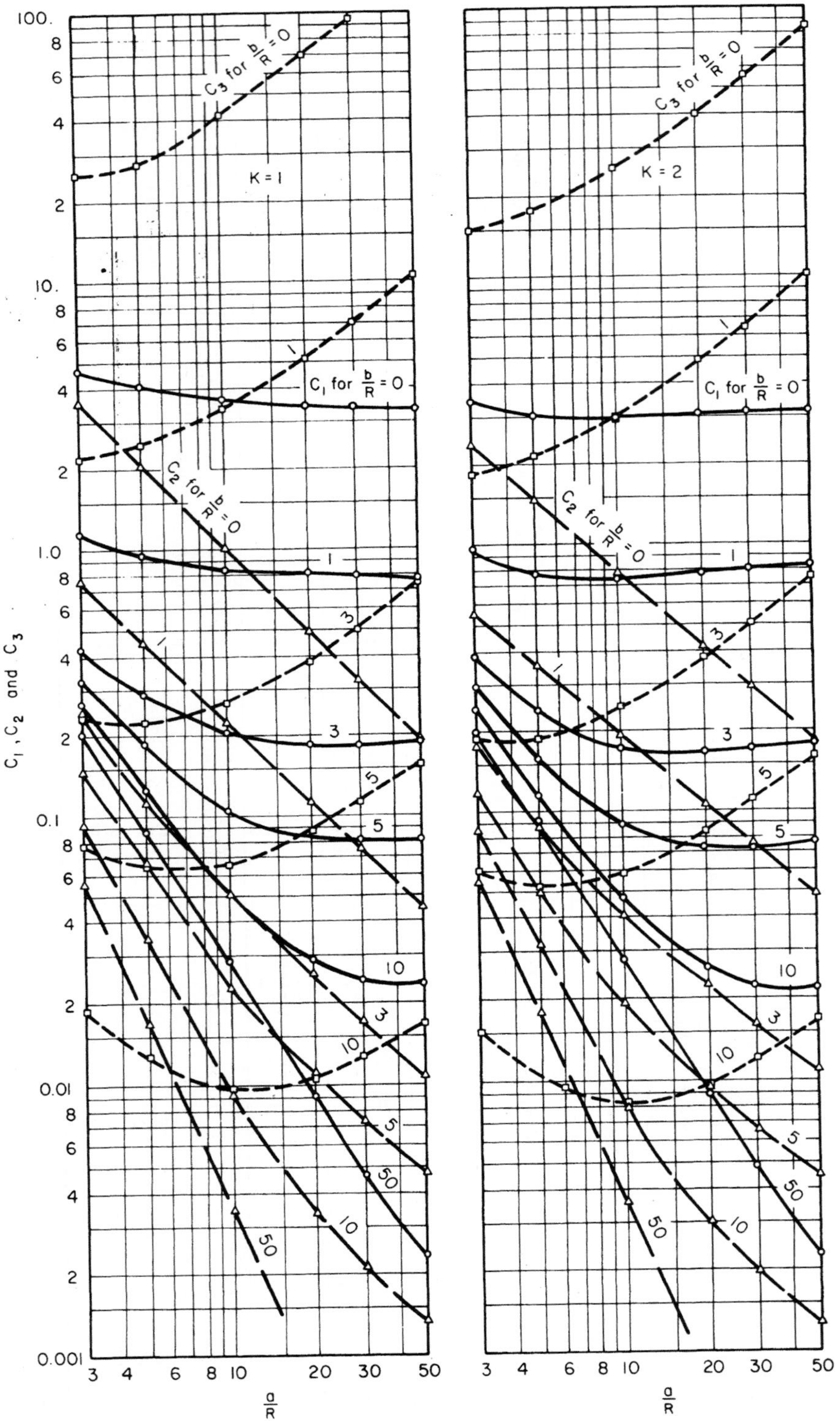

Figure 3.3.6 Reactions for right-angle pipelines.

Table 3.3.3 Example 1 Showing Determination of Integrals

Part of pipe	Values of integrals					
	A	B	C	F	G	H
0-1	$\frac{a^2}{2}$	0	$\frac{a^3}{3}$	0	0	a
1-2	ab	$\frac{ab^2}{2}$	a^2b	$\frac{b^2}{2}$	$\frac{b^3}{3}$	b
2-3	$\frac{c}{2}(2a+c)$	$\frac{bc}{2}(2a+c)$	$\frac{c^3}{3}+ac(a+c)$	bc	b^2c	c
Total 0-3	$\frac{a^2}{2}+ab+\frac{c}{2}(2a+c)$	$\frac{ab^2}{2}+\frac{bc}{2}(2a+c)$	$\frac{a^3}{3}+a^2b+\frac{c^3}{3}+ac(a+c)$	$\frac{b^2}{2}+bc$	$\frac{b^3}{3}+b^2c$	$a+b+c$

3.4 NONDESTRUCTIVE TESTING

by Francesco Lanza di Scalea and Gary L. Cloud

REFERENCES: Various authors, "Nondestructive Testing Handbook," American Society of Nondestructive Testing (ASNT), 3rd ed., 2007. Shull, ed., "Nondestructive Evaluation: Theory, Techniques and Applications," Marcel Dekker, Inc., 2002. Bray and Stanley, "Nondestructive Evaluation: A Tool in Design, Manufacturing and Service," CRC Press, 1997. Mix, "Introduction to Nondestructive Testing," 2nd ed., John Wiley & Sons, 2005. Malhorta and Carino, "Handbook of Nondestructive Testing of Concrete," CRC Press, 1991. Dodge, "Nondestructive Testing," Section 5.4 of Marks' Standard Handbook for Mechanical Engineers, 11th ed., McGraw Hill, 2007. Following journals, various authors: ASNT's Materials Evaluation, ASNT's Research in Nondestructive Evaluation, NDT & E International Journal, Structural Health Monitoring International Journal, Journal of Intelligent Materials Systems and Structures, Smart Materials and Structures, SEM's Experimental Mechanics, SEM's Experimental Techniques.

Nondestructive testing (NDT), also known as **Nondestructive evaluation (NDE)** or **Nondestructive inspection (NDI)**, is the use of noninvasive techniques to determine the integrity of a material, component or structure, or to quantitatively characterize geometrical, composition or mechanical properties of an object. A similar definition of NDT is provided by the American Society for Nondestructive Testing (ASNT) as "The examination of an object with technology that does not affect the object's future usefulness."

NDT can be used throughout the lifecycle of the material, component, or structure, all the way from the optimization of manufacturing processes, through the proper and safe operational deployment, to the end-of-life and proper disposal.

The typical implementation of an NDT test requires the use of transducers to extract relevant properties from the test object, followed by the interpretation of the data to determine the state of health or property of the object. A variation of NDT is **structural health monitoring (SHM)** that implies the use of transducers integrated with the test object (e.g., permanently attached to it) in an on-line, continuous monitoring mode, as opposed to an off-line, scheduled NDT inspection. However, the underlying fundamental physical and engineering principles of NDT and SHM monitoring are often the same.

Many NDT methods exist, each having particular characteristics. The practitioner must have knowledge of the advantages and drawbacks, including cost of implementation, of most techniques in order that the correct one may be chosen for a particular application. In many cases, more than one and maybe several approaches should be employed to obtain enough information to decide whether a component is in serviceable condition, if it is damaged but repairable, or if it should be scrapped. Further, repaired objects should be subjected to NDT analysis in order to find out if the repair is viable. Often times it is wise to begin inspection using a simple, quick, and inexpensive technique, such as visual examination, and then progress through a series of more sophisticated procedures.

Among the most common methods for NDT are visual inspection, liquid penetrant methods, magnetic particle methods, ultrasonic testing, impact-echo testing, acoustic emission testing, X-ray radiography, eddy current methods, microwave inspection, infrared thermography, digital speckle shearography, and digital speckle pattern interferometry. Some useful techniques that are in various stages of development include digital holographic interferometry, digital image correlation, optical transmittance scanning, low-coherence path matching interferometry, optical and radiological computed tomography, and the use of laser Doppler interferometry to identify damage through changes in vibrational response. The following sections discuss the essential details of each of the methods that are most widely implemented, generally progressing from the simplest to the more sophisticated.

VISUAL INSPECTION

The visual inspection method is the most basic and the most common NDT technique. It is the primary inspection mode for civil infrastructure, for example. Besides the obvious naked-eye observation by skilled operators, more sophisticated implementations of this method include the use of fiberscopes and borescopes to inspect hard-to-access locations such as the interiors of tanks and vessels, railroad tank cars and sewer lines. Autonomous or semi-autonomous robotic crawlers can also be used in hazardous environments that extend beyond the reach of fiberscopes. Visual inspection relies on the strong capability for pattern recognition of the human brain. However, the obvious drawback is the subjectivity of the operator that can often lead to false or inconsistent interpretations of the visual data. For this reason, proper training of the inspectors is a critical requirement of visual inspection.

LIQUID PENETRANT

Liquid penetrant tests are widely used in industry for the detection of surface cracks or surface porosity in various materials. The liquid penetrant technique consists of applying a thin film of a special liquid to the test part to highlight the images of surface features, specifically surface-breaking cracks. The important underlying physical principles of the liquid penetrant technique are those of surface tension, contact angle, and capillarity. The first requirement for a suitable liquid penetrant product is the ability to "wet" the underlying solid, that is, to have a contact angle less than 90 degrees. This depends on the specific liquid-solid combination, and it is therefore critical to choose the appropriate penetrant for a given test material. Once the penetrant is applied to the test object, the liquid penetrates the cracks through the capillarity effect (independent of gravity), that allows a deep penetration even into a closed crack. This fact allows the visual highlighting of the cracks.

Implementation of the liquid penetrant test requires the following steps: (a) cleaning the surface and applying the penetrant; (b) waiting for the "dwell time" to allow the capillarity action to take place (typical dwell time is 10-30 minutes for metals); (c) removing the excess penetrant with emulsifiers or solvents; (d) applying a developer (usually a powder) to draw the penetrant from the cracks to the surface where it can be seen; (e) visually interpreting the surface to detect cracks. This last step can be done under normal visible light (case of Type II penetrants that are usually dyed), or under ultraviolet (UV) illumination under dark conditions to increase the test sensitivity (case of Type I or fluorescent penetrants).

Drawbacks of the liquid penetrant technique include: (a) the need to apply the penetrant on the entire surface to be inspected; (b) the difficulty in testing porous materials; (c) the limitation of only detecting surface-breaking cracks; and (d) the influence of the operator's judgment in the evaluation of results. Standards exist where liquid penetrant results are collected on known defects and then compared to the actual tests.

MAGNETIC PARTICLE METHODS

Magnetic particle testing is used to detect discontinuities at or near the surface of ferromagnetic materials. One important advantage over liquid penetrant testing is the ability to detect subsurface cracks, beyond surface-breaking cracks. The test object is first magnetized. Finely milled iron particles coated with a dye pigment are then applied to the object. These particles are attracted to magnetic flux leakage fields caused by the discontinuities, and cluster to form an indication directly over the discontinuity. This indication can be visually detected under proper lighting conditions. A defect creates a leakage magnetic field only if it is perpendicular to the field. Hence, only certain defect orientations are detectable by this technique.

There are several ways to **create the magnetization**, including solenoids, yokes, threaded cables, etc. Either longitudinal magnetization, circular magnetization, or a combination of longitudinal and circular magnetizations can be employed. Magnetic-field direction and character are dependent upon how the magnetizing force is applied and upon the type of current that is used. For best sensitivity, the magnetizing current must flow in a direction parallel to the principal direction of the expected defect. Circular fields, produced by passing current through the object, are almost completely contained within the test object. Longitudinal fields, produced by coils or yokes, create external poles and a general-leakage field. Alternating, direct or half-wave direct currents may be used for the location of surface defects. Half-wave direct current is most effective for locating subsurface defects.

There are two **types of magnetic particles**, namely the "dry" type (dyed, usually red) and the "wet" type (fluorescent). The "dry" type can be viewed under visible light, whereas the fluorescent "wet" type requires illumination from a black light in a darkened environment.

Magnetic particle testing requires that a material is easy to magnetize (i.e., high relative permeability of 100 or more), and it cannot be used on nonferrous materials such as copper, brass, aluminum, titanium, or austenitic steel.

ULTRASONIC TESTING

Ultrasonic testing is a quite popular class of NDT methods that uses ultrasonic waves (defined here as transient vibrations at frequencies above the audible range, i.e., >20 kHz) to probe the test object in order to detect the presence of defects or characterize geometrical, composition or mechanical properties. **Transducers** are utilized to convert electrical energy into mechanical deformations to excite the wave in the object and, vice-versa, to convert mechanical deformations into electrical energy to detect the wave from the object. These transducers are most commonly piezoelectric devices that are coupled to the test object through an acoustic couplant (e.g., glycerin or similar gel) in order to minimize the acoustic impedance mismatch with the test material and ensure smooth transmission of the ultrasound energy. Besides piezoelectric transducers, other transduction means, including water, capacitive, magnetic or laser-based, can be used in those applications requiring non-contact testing such as, for example, in scanning or in-motion operations. Also, the ultrasonic transducers are often mounted on wedges to properly direct the waves into the test part according to Snell's law of refraction, which is the acoustic equivalent of the homonymous law of optics. The common **operating frequencies** of ultrasonic transducers range from 20 kHz to 5 MHz. Higher frequencies are possible (e.g., GHz) in applications requiring the resolution of very small features, such as in acoustic microscopy. The lower frequency range, below 500 kHz, is used in highly attenuating materials, such as porous media (e.g., ceramics) and heterogeneous media (e.g., concrete). The ~ MHz range is commonly used in more homogeneous materials such as metals.

There are several **wave modes** that can be utilized in an ultrasonic test of a solid object, including longitudinal and shear waves propagating in the bulk of the solid, Rayleigh waves propagating at the object's surface, and guided waves propagating in waveguide geometries such as plates, rods, and pipes. Each of these mode types has specific vibrational patterns and velocities. For this reason, proper identification of the mode(s) being used becomes necessary for an accurate interpretation of the ultrasonic signals.

The ultrasonic waves are affected by the presence of defects and, generally, discontinuities in the material. When the size of the defect is equal to or larger than the ultrasonic wavelength (calculated from the velocity divided by the frequency), the wave undergoes a reflection (echo) due to the mismatch in acoustic impedances between the base material and the discontinuity. When, instead, the size of the defect is smaller than the ultrasonic wavelength, the wave undergoes amplitude scattering that macroscopically manifests itself as an attenuation (damping) during propagation. In addition, the wave velocity is a function of basic material properties including density and elastic constants (stiffnesses). Consequently, material properties of this kind can be nondestructively characterized by the measurement of an ultrasonic wave velocity. Alternatively, when the wave velocity is known, the wave arrival time can indicate the spatial position of a reflector that, in the case of a free boundary, is directly related to the object's geometry, thus facilitating thickness gauging in pipes, pressure vessels and containment facilities.

The three **main modalities of ultrasonic testing** are: the Pulse-Echo Mode (or reflection mode), where the same transducer is used as the transmitter and the receiver, and a defect can be detected and located by monitoring a wave reflection; the Through-Transmission mode, where two separate transducers are used as transmitter and receiver, and the defect is detected by an attenuation measurement of the transmitted wave; and the Pitch-Catch mode that is a hybrid of the two previous modes where a separate transmitter and receiver are used. The pitch-catch mode is, for example, used in SHM applications that utilize sparse arrays of transducers over large areas.

The results of an ultrasonic test can be displayed as an **A-scan**, a **B-scan**, or a **C-scan**. An A-scan ("point" scan) displays voltage (the detected wave) as a function of time for a fixed point in the object. As such, it can provide information about the presence of a defect or other properties only at the point of location of the transducer. A B-scan (linear scan) displays wave amplitude or arrival time along a line of scanning of the transducer. As such, it can provide cross-sectional information on the test object. A C-scan (planar scan) displays wave amplitude or arrival time along a 2D grid of scanning, to produce the equivalent of a 2D picture of the test object.

A more sophisticated implementation of ultrasonic testing uses multielement **ultrasonic arrays** for **phased-array imaging** or **synthetic aperture focus imaging**. These techniques are also used in biomedical imaging. Phased-array imaging allows dynamic steering and focusing of the ultrasonic beam by appropriately time-delaying the excitation signals at the multiple elements of the array. This allows the concentration of more ultrasonic energy at critical locations in the test object. The synthetic aperture focusing mode is a related concept whereby synthetic focus can be achieved both in transmission and in reception by applying specific delays to the received waves so as to sum them coherently. Typical ultrasonic arrays for these imaging techniques use 16 or 32 elements, arranged in a linear fashion, for 2D imaging. These can be scanned in the third dimension to generate 3D images. Bi-dimensional ("matrix") arrays are also available to provide 3D images directly. The frequency of operation of the elements in a transducer array follows the same general rules outlined above for the traditional ultrasonic testing transducers. The drawback of the phased-array and synthetic focusing approaches is the increased complexity of the hardware and the signal processing algorithms, compared to a traditional ultrasonic testing that only uses one or two transducers. A relatively new area of ultrasonic testing is that of **nonlinear ultrasonics**, where nonlinear properties of the propagating waves, including higher-harmonic generation, wave mixing, etc., are monitored for increased sensitivity to structural conditions. For example, the onset of failure (e.g., dislocations or microcracks, before the development of macrocracks), that is challenging to detect with ultrasonic testing in the linear regime, can be easily detected in the nonlinear ultrasonic regime. The challenge of nonlinear ultrasonic testing remains the quantitative correlation of the nonlinear features to the properties of the test object. For this reason, nonlinear ultrasonics is very effective in monitoring damage growth in a test object, as opposed to determining the extent of damage in an absolute manner from one test.

Calibration standards, such as those established by the ASNT, are often required to compare the ultrasonic test signals to measurements on known object properties. Examples of calibration standards are specimens of known thickness and specimens with known reflectors such as flat-bottom holes.

IMPACT-ECHO TESTING

Impact-echo testing is a technique originally developed to detect internal cracks in concrete structural components or to determine the concrete components' thickness. Since concrete is a highly-attenuating material because of its heterogeneous nature, low frequency waves must be used in order to penetrate to sufficient depths. Accordingly, the impact-echo technique utilizes an impactor to generate waves in the ~ kHz frequency range, and an accelerometer to detect the waves at the surface of the concrete slab.

The mode of operation is similar to that of ultrasonic pulse-echo. The impact-generated wave undergoes multiple reflections from an internal crack or from the back surface of the concrete slab due to thickness reverberations. A "thickness resonance" is then established when the wavelength is equal to twice the slab thickness. This resonant wavelength then produces a resonant frequency equal to the wave velocity divided by the resonance wavelength. The resonant frequency is easily measured through a simple Fourier Transform of the time-domain signal received by the accelerometer. In the absence of an internal crack, the resonant frequency allows determination of the thickness of the slab for a known wave velocity. Alternatively, in the presence of an internal crack, a higher resonant frequency is detected and its value allows the determination of the depth of the crack.

Impact-echo can also be used to detect small voids in the grout of prestressing strands used in post-tensioned concrete structures. In this case, only a portion of the propagating wave energy is reflected from the void. Consequently, two resonant peaks are typically detected, a lower frequency associated with the slab thickness resonance, and a higher frequency associated with the grout voids.

The typical **frequency range** of impact-echo waves is 2-20 kHz. Depending on the specific test material and required inspection depth, the impact impulse can be tailored by using impactors of different weights and shapes. Generally, a heavier and larger impactor will generate lower frequency waves suitable for highly attenuating materials or large inspection depths, whereas a lighter and smaller impactor will generate higher frequency waves suitable for less heterogeneous materials or smaller inspection depths.

ACOUSTIC EMISSION TESTING

Acoustic emission testing is a variation of ultrasonic testing that uses the same propagating wave modes, but it relies on the test object itself to excite such waves. Therefore, acoustic emission is a method that requires only receiving transducers on the test object with the necessary excitation being provided only through mechanically or thermally stressing the object. The method is based on the fact that growing cracks, moving dislocations and similar transient events within the test object release mechanical energy in the form of an acoustic or ultrasonic wave that can then be detected by appropriately spaced transducers. The phenomenon of acoustic emission is very similar in principle to that of an earthquake event, generating seismic waves that can be detected by seismographs. Just as earthquakes can be detected and their epicenter located through triangulation of seismic records, a growing damage in a structure can be detected and located through ultrasonic records. Acoustic emission can be also used to detect and locate an impact event caused by a foreign object based on the waves generated by the impact event.

The frequencies of the acoustic emission waves follow the same general rules of ultrasonic testing frequencies, with low frequencies (e.g., below 500 kHz) typical of highly attenuating materials such as concrete or ceramics and high frequencies (500 kHz-2 MHz) typical of homogenous materials such as metals.

The acoustic emission signals are typically analyzed in terms of specific **signal features**. These include: number of acoustic emission "hits," signal peak amplitude, signal energy (area under the signal envelope), number of signal counts (threshold crossings), signal time duration, signal rise time (from start to peak amplitude), and the "cumulative" version of each of these features. Damage or an impact is detected by a sudden increase in the value of the features or, equivalently, by a sudden increase in the slope of the cumulative features.

The most basic **implementation of acoustic emission testing** for damage detection in a structure is the so-called "curve shape analysis," whereby the cumulative number of acoustic emission hits is plotted against time or applied loads. Damage is detected when the curve exhibits a "knee," or an abrupt increase in slope indicating increased number of emissions caused by the growth of new damage.

An important application of acoustic emission testing for periodic testing of structures (e.g., proof-testing of pressure vessels) utilizes the so-called **Kaiser effect** that refers to a multiple loading scenario. According to this effect, if the structure is undamaged, no new acoustic emission event should be detected if a new load does not exceed the previously applied maximum load. Alternatively, if new acoustic emission events are detected before the previous maximum load is exceeded (the so-called **Dunegan corollary** of the Kaiser effect), the structure has undergone some kind of damage since the last inspection. One exception to the Kaiser effect occurs in composite materials and concrete where pre-existing cracks and viscoelasticity may cause premature acoustic emission events without new damage occurring (the so-called Felicity Ratio principle). The Kaiser effect is used for proof testing of pressure vessels (e.g., ASTM standard E1419) and other structures. The Felicity Ratio principle is used for periodic inspection of composite or concrete materials in a slightly different modality.

An important capability of acoustic emission testing is that it yields the **location** of the defect or impact, beyond the simple detection of these. Acoustic emission source location is accomplished by triangulation techniques that use multiple ultrasonic receivers, in analogy with the location of the epicenter of an earthquake in seismology. Accordingly, to locate the source in a planar 2D structure, a minimum of three receivers are needed. By tracking the difference in arrival time of the acoustic emission waves at two different receivers, a hyperbola is identified as the locus of possible source locations. By using three receivers, therefore, the intersection of the two hyperbolae unequivocally locates the position of the acoustic emission source (damage or impact). This triangulation can be extended to locate the source in 3D (e.g., an embedded source in a 3D medium) by appropriately increasing the number of acoustic emission receivers.

Drawbacks of acoustic emission testing include (a) the requirement that the defect be "active" (i.e., growing), and (b) the difficulty to quantitative associate a certain acoustic emission to the specific type and size of damage (the so-called "source inversion" step).

X-RAY RADIOGRAPHY

X-rays are high energy electromagnetic waves with short wavelengths ranging between 0.01 and 10 nanometers. They were discovered by German physicist Wilhelm Rontgen in the late 1800s. Because of their high-energy, X-rays penetrate into most materials and they attenuate depending on the material composition and density. The most common implementation of X-ray radiography in medicine and NDT is the **projection radiography**, where a collimated X-ray beam is sent through the test part and detected, in transmission, by an appropriate X-ray detector. Changes in material composition or density cause changes in attenuation of the travelling radiation, causing a negative image of the part to be impressed on the detector plate.

X-ray radiation sources can be configured to act as either a point, line or an area source. They can be either electrically powered, generating photons with energy ranging from a few keV up to several MeV, or radioisotropic, radioactive materials that spontaneously emit X-rays from the decay of unstable atoms. **Radiation detectors** convert the incident X-rays into an optical or electrical signal that can then be converted into an image. Analog detectors and digital detectors exist. Analog detectors are in the form of films. Digital detectors can be in the form of photomultiplier tubes, charge-coupled devices (CCD), photostimulable phosphor devices, or thin film transistors. Many radiation detectors incorporate a device that converts the primary radiation into secondary lower energy electrons (screens) or light (scintillators or fluorescent screens). Generally, a collimator is also used at either the radiation source or the radiation detector to control the direction of propagation of the X-rays. A collimator is a material of a specific shape with specific ray absorption properties.

Physically, as an X-ray travels on a straight line through the object, it is attenuated in a manner that is proportional to the travelling distance through a material having a so-called **linear attenuation or absorption coefficient**. Generally, this coefficient is proportional to the density of the test material. X-ray imaging basically consists of creating a map of this attenuation coefficient for the test object, revealing its internal structure and features including defects. More specifically, the X-ray information that is recorded at a pixel (point) of the detector is related to the line integral of the attenuation coefficient along the straight propagation path. A series of line integrals are therefore collected at the detector's plate for a given X-ray propagation angle. The spatial Fourier Transform of the information collected at the detector's plate is related to the object's image through the Radon Transform. Inversion of the Radon Transform using filtered back propagation reconstructs the object's image (attenuation field) for the given X-ray projection angle.

In order to reconstruct 3D images of the test object, **computed tomography** (CT) is used, whereby the X-ray projection angles are changed to obtain a complete image of the object. The various projection angles can be generated by either rotating the test object with fixed positions of the X-ray source and detector, or by rotating the positions of the X-ray source and detector for a fixed orientation of the test object.

Penetrometers are used to indicate the contrast and definition existing in a radiograph. The type generally used in the United States is a small rectangular plate of the same material as the object being X-rayed. It is uniform in thickness (usually 2% of the object thickness) and has holes drilled through it. ASTM specifies hole diameters 1, 2, and 4 times the thickness of the penetrometer. Step, wire, and bead penetrometers are also used (see ASTM Materials Specification E94). Because of the variety of factors that affect the production and measurements of an X-ray image, operating factors are generally selected from reference tables or graphs which have been prepared from test data obtained for a range of operating conditions.

Four factors contribute to a good **picture quality**: (1) the effective focal-spot size of the radiation source should be as small as possible; (2) the source-to-object distance should be adequate for proper definition of the area of the object farthest from the film; (3) the film should be as close as possible to the object; and (4) the area of interest should be in the center of and perpendicular to the X-ray beams and parallel to the X-ray detector plane.

All materials may be inspected by radiographic means, but there are limitations to the configurations of materials. With optimum techniques, wires 0.0001 in (0.003 mm) in diameter can be resolved in small electrical components. At the other extreme, welded steel pressure vessels with 20-in (500-mm) wall thickness can be inspected by use of high-energy accelerators as a source of radiation.

Radiographic standards are published by ASTM, ASME, AWS, and API, primarily for detecting lack of penetration or lack of fusion in welded objects. Cast-metal objects are radiographed to detect conditions such as shrink, porosity, hot tears, cold shuts, inclusions, coarse structure, and cracks. The dangers connected with exposure of the human body to X-rays should be fully understood by any person responsible for the use of radiation equipment. NIST is a prime source of information concerning radiation safety. NRC specifies maximum permissible exposure to be a 1.25 R/year.

EDDY CURRENT TESTING

Eddy current testing is an NDT technique applied to inspection of electrically conductive materials. Currents in the test object are induced in response to a changing (AC) electromagnetic field that is generated by a changing current in a conductor that is placed in close proximity to the material. The magnetic fields are then detected either by electromagnetic induction in a coil or system of coils, or by other sensors such as the **Hall Effect** devices. The presence of discontinuities or material variations alters the eddy currents, thus changing the apparent impedance of the inducing coil or of a detection coil.

Typical applications of eddy current testing include the measurement of internal or external diameter of tubes and pipes, the detection of corrosion or cracking defects, and the inspection of coatings. A particularly successful application is the thickness measurement of metallic and nonmetallic coatings on metals. Coating thicknesses measured typically range from 0.0001 to 0.100 in (0.00025 to 0.25 cm). For these measurements to be possible, the coating conductivity must differ from that of the base metal.

Implementation of eddy current testing requires the **generation and the detection of eddy currents**. On the generation side, an oscillator (generally a coil) is required to produce a changing magnetic field close to the test. The oscillation frequency can range from 5 to ~500 Hz for the remote field method, or from 1 kHz to 2 MHz for testing nonferrous materials. On the detection side, a coil (which can be the same generation coil or a separate coil) or a Hall sensor can be used to detect the currents. The detection devices can be split into differential or cross-axis configurations. Test coils are of three general types: the circular coil, which surrounds an object; the bobbin coil, which is inserted within an object; and the probe coil, which is placed on the surface of an object. Coils are further classified as absolute, when testing is conducted without direct comparison with a reference object in another coil; or differential, when comparison is made through use of two coils connected in series opposition. Many variations of these coil types are utilized.

The **test frequency** determines the depth of current penetration into the test object, owing to the AC phenomenon of "skin effect." One "standard depth of penetration" is the depth at which the eddy currents are equal to 37 percent of their values at the surface. In a plane conductor, the depth of penetration varies inversely as the square root of the product of conductivity, permeability, and frequency. High frequency eddy currents are more sensitive to surface flaws or conditions, while low-frequency eddy currents are sensitive also to deeper internal flaws or conditions.

Correlation data must usually be obtained to determine whether test conditions for desired characteristics of a particular test object can be established. Because of the many factors which cause variations in electromagnetic properties of metals, care must be taken that the instrument response to the condition of interest is not nullified or duplicated by variations due to other conditions.

The instrumentation for the analysis and presentation of electric signals resulting from eddy current testing includes a variety of means, ranging from meters to oscilloscopes to computers. Instrument meter or alarm circuits are adjusted to be sensitive only to signals of a certain electrical phase or amplitude, so that selected conditions are indicated while others are ignored. Automatic and automated testing are among the principal advantages of the method.

MICROWAVE TESTING

Microwaves are electromagnetic radiations with wavelengths lying between 1 mm and 1 m. The operating frequency can be selected to maximize the interaction with dielectric layers, voids, inclusions, and surface flaws in the test object. Microwave inspection generally consists of measuring various properties of the electromagnetic wave scattered by, or transmitted through, the test object. The signal strength into a dielectric material is dictated by the incident power level, the loss factor (or absorption) of the material, and the wave frequency. In NDT, the waves are usually polarized to allow for selective interaction with an elongated anomaly.

Microwaves generated in a test instrument are transmitted by a waveguide through air to the test object. Analysis of the scattered (reflected) or transmitted energy indicates certain material characteristics, such as moisture content, composition, structure, density, degree of cure, aging, and presence of flaws. **Microwave equipment** include oscillators that produce a microwave signal at a prescribed frequency and power, network analyzers to detect amplitude and phase of the microwave, waveguides that are hollow tubes with rectangular or circular cross-section that act as a high-pass filter to allow the propagation of specific microwave oscillation modes, antennas in the form of horns to direct the microwave beams from the source to the test object, and detectors that are nonlinear devices incorporating a solid state detector such as a Schottky barrier diode.

Microwave NDT can be used for the internal inspection of dielectric materials, examples of which include ceramics, mica, polymers, glass, and the oxides of various metals. It can also be used for the surface inspection of conducting objects. Examples of **specific applications** include testing for delaminations in rocket casings, detection of defects in rocket propellants, thickness measurements of coatings such as ablative shields for aerospace reentry vehicles, detection of voids in composite honeycomb panels, detection of inclusions and porosity in ceramics and molded rubber, detection of surface cracks in metals even in the presence of dielectric materials such as paint, rust, dirt, etc., and the measurement of epoxy curing rates, moisture content in dielectric materials, and density variations in lumber.

INFRARED THERMOGRAPHY

Infrared thermography is an NDT technique that relies on the disruption of the heat conduction by an object's discontinuities (e.g., internal defects). The thermal diffusion equation governs the conduction of heat in a solid. An important material property governing this process is the **thermal diffusivity**. When an object has an internal discontinuity (e.g., a defect), the difference in thermal diffusivities between the base material and the discontinuity causes a disruption in the heat flow that generally slows down the heat conduction through the discontinuity, revealing the presence of the subsurface defect as a "hot spot" on the temperature profile of the object's surface.

An **infrared thermographic test** involves a heat source (e.g., flashlamps) and an infra-red (IR) camera. The IR camera measures surface temperature by tracking the object's emitted IR wavelength (i.e., wavelength larger than 0.68 micrometers) that is emitted by any object above absolute zero according to Planck's law. There are several **types of IR cameras**, ranging from low cost, uncooled microbolometers having a maximum frame rate of 30 Hz, sensitivity of 20-50 millikelvin, and fixed spectral range of 7.5-14 micrometers, to more expensive cooled detectors that possess frame rates of 1,000 Hz or higher, sensitivity better than 20 millikelvin, and choices of spectral ranges of 1-5 micrometers or 7.5-10.5 micrometers depending on test conditions. For quantitative temperature measurements, it is important to account for the emissivity of the test surface either in the instrument calibration constants or in post-processing.

In a typical **flashed thermography test**, the object's surface is heated by flashlamps, and the IR camera monitors the surface temperature evolution on the same side as the heating. While the pristine portions of the object cool normally according to the 3D heat diffusion equation, the areas having defects cool more slowly because of 2D or lateral diffusion, thereby producing "hot spots" in the surface temperature map. Hence, the defects produce discontinuities in the temperature distribution of the object's surface. In some cases, the test can be performed in transmission, whereby the heating is applied to one side of the object and the IR camera observation is made on the other side. In the transmission scheme, that can be advantageous with materials with low thermal diffusivity, defects produce "cool" spots at the object's surface. An interesting application of transmission thermography is in the detection of defects in containment structures (e.g., pressure vessels) or hollow structures in general (e.g., pipes and tubes) that can be easily heated by injection of a hot gas from the inside and viewed by the IR camera at the outside.

A more sophisticated implementation of infrared thermography requires the examination of the evolution of the object's temperature at specific pixels over time. In this **temporal mode** of operation, the time-varying temperature traces are recorded at the selected pixels. Proper analysis of these traces allows a more quantitative assessment of the underlying defect, including the determination of the depth of the defect. Referring to active flashed thermography, the identification of the precise time associated with a deviation in the cooling trend from a "normal" cooling trend ("deviation time") can be associated to the depth of the discontinuity within the test object. Specifically, deeper flaws will produce longer deviation times and vice-versa. The deviation time versus defect depth relationship can be previously calibrated on known defect depths or calculated from models, and then applied to the test at hand.

Infrared thermography can be also used in a **passive mode** without an active excitation. This implies that the thermal excitation is ambient. A suitable thermal excitation, for example, can be imparted by the sunlight shining on buildings at sunrise or by air-structure interaction in a flying object. In these cases, proper observation of the temperature evolution by the IR camera may suffice to detect internal defects in the test object.

Because of the nature of heat diffusion, infrared thermography works best as an NDT tool on defects that are close to the surface (subsurface features), for example, delaminations in composite panels or disbonds in adhesively-bonded joints. The technique is less accurate in detecting defects that lie deep inside an object because of the decreased effect of such deep features on the temperature map of the object's surface.

OPTICAL INTERFEROMETRY TECHNIQUES

Several valuable NDT techniques utilize, in one way or another, the physical phenomenon of interference of electromagnetic waves. The fundamental principles of optical interferometry are presented in considerable detail in Sec. 3.5 of this handbook and need not be duplicated here. The range of wavelengths employed ranges from the deep ultraviolet through the far infrared and into the microwave regime. Most of these methods, however, utilize radiation that falls within or near the fringes of the visible spectrum, so they fall within the category of **optical methods of NDT**.

A general advantage of a broad category of optical methods is that they yield **whole-field results**, which is to say that the analysis is presented in the form of a picture of the entire object under inspection. Defects and damage are automatically highlighted in the picture, although informed interpretation of the visual data is usually required. Another important characteristic of optical interferometric methods is that they are very sensitive, being capable of measuring deformations to a small fraction of the wavelength of the radiation used and so allowing the detection of very small subsurface defects.

Most of the various optical interference techniques share a common mode of operation when used for NDT. The fundamental idea is that a subsurface flaw will cause a slight perturbation in the surface displacement field that occurs as the test object is loaded. The optical measurement system is sensitive enough to detect and display this perturbation, usually as an anomaly in a **fringe pattern** or in a so-called **phase map**.

The fringe pattern or phase map is obtained by first recording on photographic film or digitally using a CCD device an image of the test object as it is illuminated with coherent light as from a laser. The object is then stressed a small amount by a mechanical load, pressure, vacuum, or temperature change. A second image is recorded. The intensity map of the first image is then subtracted from the second via analog processing if photographic film is used or, more usually, pixel-by-pixel in a computer. Appearing on the resulting difference image will be a set of **correlation fringes** that are not interference fringes at all but, rather, are loci of some constant value, such as of displacement or slope that is caused by the loading. Actual displacements can be determined by obtaining additional images and processing the image data arrays in the computer, but these steps are usually not executed for NDT applications. However, rudimentary processing of the digital data is often implemented to reduce noise, delineate the boundaries of the fringes, and facilitate interpretation of the optical data.

If no damage or defect is present in the test object, the rules governing the behavior of deformable solids dictate that the object surface displacements will be smoothly varying, and, so, the correlation fringe pattern will be regular. On the other hand, if a defect exists within the object, than there will appear in the correlation fringe pattern a perturbation such as a "bull's eye," a local change of fringe spacing, or a local warping of the fringes. If a surface crack is present, the fringes will be broken or discontinuous, and the phase map will show the location of the **tip of the crack** and can be used for **crack length measurement**.

The following sections present very briefly two alternative approaches to practical implementation of the principles described above.

DIGITAL SPECKLE PATTERN SHEAROGRAPHY

Digital speckle pattern shearography (DSPS), also named **electronic speckle pattern shearography (ESPS)** utilizes the **speckle pattern** that is observed in the image of an object that is illuminated with coherent light as from a laser. Section 3.5 of this handbook describes briefly how the speckle pattern is formed. Further, Sec. 3.5 outlines how the speckle effect is employed to implement speckle shearography to obtain a map of the change of slope of the surface of the test object as it is loaded. A defect or damage in the object will appear as a perturbation in the otherwise smoothly regular fringe pattern, as outlined above.

DSPS has developed into a valuable standard technique of NDT. It has supplanted hologram interferometry in being mandated by safety agencies for 100 percent inspection of tires in the commercial aircraft sector. Insurance companies require its use to document the integrity of costly durable goods such as the hulls of yachts. Inspections of structures of composite structures at manufacture, in the field, and after repair are other examples of important applications of this versatile technique. It is quite easy to use, and both portable and fixed proprietary turnkey devices and software are available from vendors. One reason for its convenience, especially in the portable forms applied in the field is that it functions as a **common path interferometer**, meaning that it is not affected to an appreciable level by disturbances such as vibrations or convection currents.

DIGITAL SPECKLE PATTERN INTERFEROMETRY

Digital speckle pattern interferometry (DSPI) historically called **Electronic speckle pattern interferometry (ESPI)**, like DSPS, uses the **speckle pattern** that appears in the recorded image of an object that is illuminated by coherent light. This technique is described in considerable detail within this handbook in Sec. 3.5 on experimental stress and strain analysis. While similar in procedure to DSPS, DSPI yields different information, namely, in its usual NDT configuration, a map of surface displacement parallel to the axis of observation. In other words, a **map of change of surface contours** of the specimen after loading is obtained, and such data might be more amenable to quantitative interpretation than the data obtained from DSPS. A drawback with DSPI is that it is not, in its usual forms, a common-path interferometer; so it is affected by vibrations and other ambient perturbations. However, a very simple and inexpensive **scatter-plate configuration** for implementation of DSPI, easily constructed in-house, converts it to a near-common-path form, greatly reducing environmental effects.

An example of an application of DSPI appears in Fig. 3.4.1, which shows a portion of a smoothed fringe pattern that was obtained from optical NDT of a cylindrical tank such as is used in vehicles fueled by high-pressure natural gas. The tanks are made by winding several layers of carbon and glass composite over a flask. DSPI was use to examine the tank to locate and characterize local subsurface damage caused by projectile impacts. Such defects might not be evident from visual inspection of the sacrificial exterior layers, but they would likely affect the integrity of the structure. Loading between the recording of the speckle patterns was only 40 psi. The fringes from undamaged areas are straight and parallel. As seen in the figure, the fringes from the small damaged area form a "bulls-eye" that proved to be centered on the impact point.

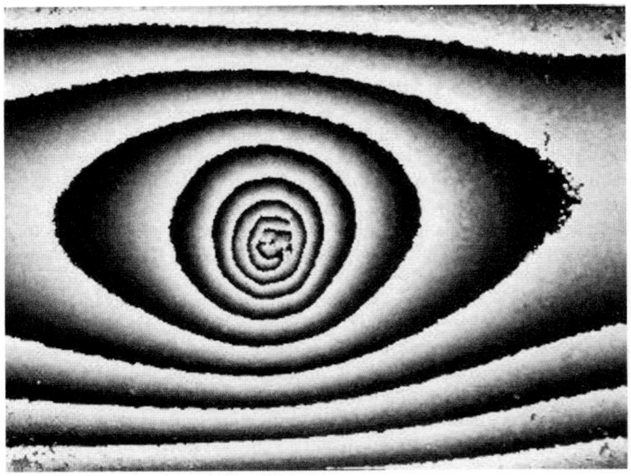

Figure 3.4.1 Fringe pattern from digital speckle pattern interferometry inspection of a composite natural gas fuel tank showing subsurface damage caused by impact.

OTHER OPTICAL TECHNIQUES

The powerful once-popular whole-field three-dimensional film-based measurement technique, **hologram interferometry (HI)**, is being succeeded by its electronic analog, **digital holographic interferometry (DHI)**. These methods are described in Sec. 3.5 of this book in the context of strain measurement. The digital version will certainly gain in acceptance as a favored NDT approach for critical applications. Likewise, **optical computed tomography** is being implemented for NDT. Pointwise scanning methods, including **optical transmittance scanning** and **low-coherence path length matching interferometry** are finding important applications, particularly in 3D NDT of transparent or translucent objects such as transparent armor and protective windows. The non-interferometric **Digital image correlation (DIC)** technique, used widely for displacement measurement and described in Sec. 3.5, is being brought into the NDT domain. These latter techniques, not broadly used as yet, are not discussed here; but awareness of their attributes would be of advantage to the NDT engineer.

3.5 EXPERIMENTAL STRESS AND STRAIN ANALYSIS

by Gary L. Cloud

REFERENCES: Sharpe (ed.), Handbook of Experimental Solid Mechanics, Springer, New York, 2008. Cloud, Optical Methods of Engineering Analysis, revised second printing, Cambridge University Press, New York, 1998. Shukla and Dally, Experimental Solid Mechanics, College House Enterprises, Knoxville, Tennessee, 2010.

Experimental stress and strain analysis, otherwise called **experimental mechanics**, is the quantitative measurement of the response of bodies and structures to stimuli, often in the form of applied loads. Applications of experimental mechanics are numerous and include, for example, biomechanics, geomechanics, fracture mechanics, infrastructure, materials characterization, manufacturing processes, transducer design, motion analysis, mechanical design, and nondestructive testing.

Many problems faced by analysts and designers stretch the limitations of analytical and numerical techniques, so experimental solutions are needed. Hybrid approaches wherein experimental data are used as input to a numerical solution that is later subjected to experimental validation are becoming more commonplace. The major problem that must be faced in measurement of strain in structures is that, for most engineering materials, the strain induced by loads is very small, the change in length per unit gage length being on the order of parts per million (microstrain), so a sensitivity of 1-10 microstrain is required. Photoelasticity is the only technique that offers a direct way to measure stress. Otherwise, to determine stress, the strain must be measured and converted to stress by use of the stress-strain relations for the material.

Many techniques are available to attack problems in experimental mechanics, and engineers should be familiar with a spectrum of methods in order that the one most appropriate for a given situation might be chosen. Some techniques have evolved, through long usage, into well-known standard approaches, while others are evolving or have not yet been thoroughly accepted by industrial practitioners. The former class includes resistance strain gages, photoelasticity, geometric moiré, and brittle coatings. Those less commonly implemented or still in evolution include holographic interferometry, speckle interferometry, speckle shearography, acoustic emission, thermography, digital photoelasticity, digital image correlation, digital holography, and moiré interferometry, among others. This article describes in some detail three of the standard methods, and the basic functions of some of the more advanced methods are also presented. Systems, supplies, information, and software for implementing these techniques are available from vendors.

RESISTANCE STRAIN GAGES

An electrical resistance strain gage consists of a length of a fine conducting wire or foil, often configured as a grid, that is intimately attached to the specimen. As the specimen is deformed, the gage is deformed by the same amount, so that its resistance changes. If the relationship between resistance change and strain is known for the gage material, then the observed resistance change can be converted to specimen strain at the gage location.

The relationship between strain and resistance change was discovered by Lord Kelvin and may be written,

$$\varepsilon=\frac{1}{F}\left(\frac{\Delta R}{R}\right)$$

Where ε is the strain along the gage axis, R is the gage resistance, ΔR is the change of gage resistance with strain, and F is the **gage factor** for the given gage. F is determined by the gage manufacturer through batch testing in a known uniaxial stress field and is provided to the gage user, as is the gage resistance.

Strain gages are employed in two distinct ways. First, they are used for direct measurement of strain. Second, they are commonly used in transducing devices in which strain measurements are used as indicators of some other physical parameter such as load, pressure, or acceleration.

Strain gages of many types, sizes, and configurations are available. The most common type is the **bonded metallic gage** in which the sensing element is a thin foil that is mounted on a plastic substrate. Strain gages must be carefully selected for an application in accordance with manufacturer's recommendations. Perfect bonding between gage and specimen is necessary, and rigorous but simple cleaning, bonding, wiring, and protective coating practices must be followed.

Gage factors for common metallic gages are about 2; so, since the strain is small, the resistance change per unit gage resistance is also small. An important part of a strain gage installation is the interface circuit that converts the small resistance change to a voltage that can be amplified and measured. Two basic interface circuits are commonly used, usually with additions that add capability and flexibility.

Wheatstone Bridge The Wheatstone bridge interface circuit is shown in Fig. 3.5.1.

In the usual unbalanced operational mode, the variable resistor R_2 is first adjusted to give zero initial voltage signal output E_0. Then, as the resistance of the gage R_1 is changed by ΔR_1, the signal voltage changes proportionately. Often, more than one and perhaps all of the resistors in the circuit are strain gages. The relationship between output voltage and gage strains is, for four identical active gages and where E is the supply voltage and F is the gage factor.

$$E_0=\frac{EF}{4}(-\varepsilon_1+\varepsilon_2-\varepsilon_3+\varepsilon_4)$$

This voltage is usually amplified and recorded by a voltmeter, an oscilloscope, or a digital DAQ system. Dedicated strain gage readout devices and/or the use of circuit calibration allow direct indication in strain units. The output signal is proportional to the excitation voltage applied to the bridge, so a stable power supply must be used to limit noise and to maintain constant circuit sensitivity. DC or AC excitation can be used; the latter facilitates the use of filtering and carrier amplification to reduce noise.

An advantage of the bridge circuit is that more than one active strain gage can be used to multiply output or to cancel unwanted inputs such as temperature effects. For example, consider the case where surface strain in a beam is to be measured. Two gages can be used. The one on the top of the beam might be subjected to tensile strain and the one on the bottom will experience compression. The gages are wired into the bridge circuit in adjacent arms, for example, R_1 and R_4. The bridge output will then be twice what will be seen if only one gage is used. Any effect of axial loads on the beam will be canceled as will temperature effects, because they have the same sign. The unbalanced bridge circuit can be used to measure either **static or dynamic strain**. It is the most widely used basic circuit, being implemented with many extra features in most commercial strain indicators and DAQ systems.

Potentiometer Circuit The potentiometer circuit, also called the ballast circuit or the voltage divider circuit, is shown in its basic form in Fig. 3.5.2.

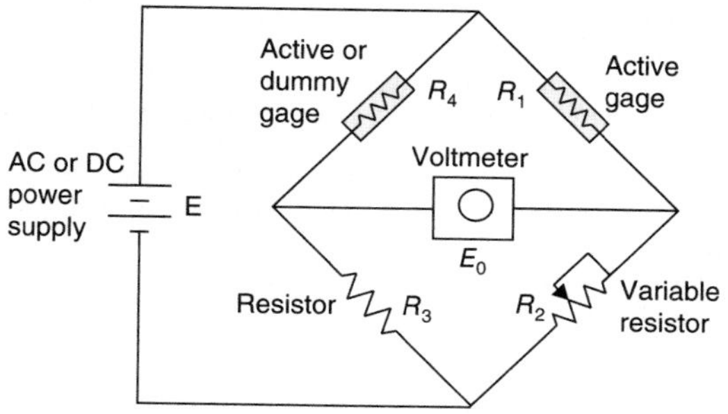

Figure 3.5.1 Wheatstone bridge circuit for resistance strain gages.

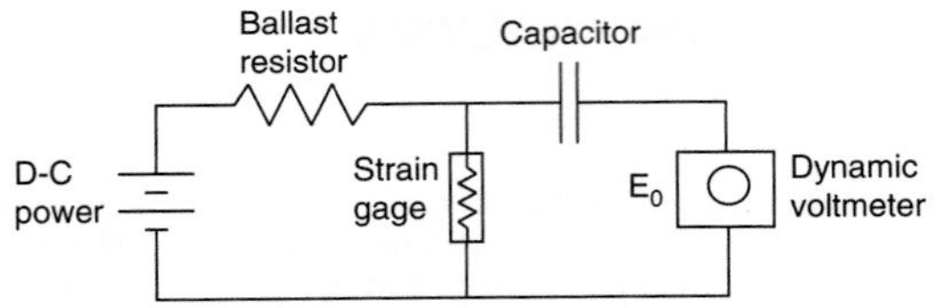

Figure 3.5.2 Potentiometer circuit for resistance strain gages.

This circuit is particularly simple to fabricate. The function of the ballast resistor in the circuit is to limit the current through the strain gage while still giving usable output. Usually, this circuit is used with only one strain gage, so care must be taken that unwanted inputs, such as temperature effects, are eliminated in other ways.

A disadvantage of the potentiometer circuit is that there is no way to initially balance it so that the starting output voltage is zero. The small change in output voltage that is caused by the strain gage will be superimposed on a relatively large bias voltage, meaning that one is faced with the problem of measuring accurately the small change in a large quantity. This fact limits the use of the circuit to problems involving rapidly varying or **dynamic strains**. A capacitor placed in the circuit acts as a high-pass filter so as to block the steady bias voltage and pass the time-varying component of interest to an amplifier. The circuit is DC powered, and the power supply must be stable and noise-free. In spite of these limitations, this circuit is exceptionally useful and easy to use in measuring strains in machinery, for example.

Circuit Calibration The output equations for the interface circuits are useful for basic design of the measuring system, but not for accurate calibration of the entire system for measuring strain. This calibration is best accomplished by creating a known resistance change at the input. Implementation involves placing a large shunt calibration resistor R_{cal} in parallel with the strain gage R_g. The equivalent calibration strain ε_{cal} is given by,

$$\varepsilon_{cal} = \frac{1}{F}\left(\frac{R_g}{R_g + R_{cal}}\right)$$

The relationship between circuit observable output (e.g., oscilloscope trace deflection) and input strain is thus established.

Temperature Effects Strain gages exhibit a resistance change with temperature that cannot be distinguished from the effects of strain. Large errors can result if gage temperature changes during an experiment and steps are not taken to eliminate these errors. Two approaches are used singly or together.

Manufacturers of strain gages provide gages that are self-compensated for broad classes of materials (e.g., brass, steel). These gages can be used without any other temperature compensation with good but not perfect results. They are especially useful for single-gage applications over short periods of time where mechanical strains are relatively large.

The output voltage equation for the Wheatstone bridge can be exploited to eliminate temperature effects in two related ways. In the first, an inactive **dummy gage** that is the same as the active gage is mounted on an unstressed coupon of the same material. It is wired into an arm of the circuit adjacent to the measuring gage. False strains from temperature are canceled. Another approach is to design the experiment so that two active measuring gages are placed in adjacent bridge arms. The desired strain signal is multiplied and the false thermal strain is eliminated. Similar approaches are used to eliminate temperature effects on gage lead wires, leading, for example, to the familiar "3-wire hookup" for single gage installations.

Transverse Sensitivity Resistance strain gages exhibit sensitivity to the strain component that is perpendicular to the gage axis. If the stress field is uniaxial, this transverse sensitivity does not cause errors because the gage factor is determined in such a field. In any other biaxial stress field, ignoring the effects of transverse sensitivity can lead to errors that range from negligible to very large, depending on the nature of the field and the strain component that is being measured. The **transverse sensitivity** K_T is determined by the gage manufacture by batch testing and is provided to the gage user. For typical foil gages, K_T is less than 2%.

The errors caused by transverse sensitivity are eliminated by using two gages in a biaxial stress field, even if only one strain component is sought. That is, for accurate determination of one strain in an unknown biaxial stress field, two gages are required. The two gages are mounted perpendicular to one another. Usually the gages are identical. The common gage factor F as given by the manufacturer is set on the strain indicator. The strain indicator gives an apparent strain for each gage, but the readings are not true strains. The readings are corrected as follows. Let the apparent indicated strain for gage A be Q_A and the reading for the second gage B be Q_B. The true or actual strain ε_A along the axis of gage A is,

$$\varepsilon_A = \frac{(1-\mu_0 K_T)}{(1-K_T{}^2)}(Q_A - K_T Q_B)$$

Note that if $Q_B >> Q_A$, the error caused by ignoring K_T will be large. The μ_0 in the equation is the Poisson ratio for the material that was used by the manufacturer in determining the gage factor and the transverse sensitivity factor. It is not always available, but may be taken to be 0.285 with reasonable safety. The true strain for gage B is found by permuting the subscripts in the above equation.

Strain Rosettes In a general biaxial strain field, there are three unknowns, the two principal strains and the orientation of the principal strain axes. If complete knowledge of the field is sought, then three strain measurements are required. Strain gages do not respond to shear strain, so three determinations of normal strain are used. Three ordinary gages can be employed, but gage manufacturers provide standard combinations of gages, called strain rosettes, that contain two or more gage elements in a single package that can be applied just like a single gage. The two most common configurations of gage elements in rosettes are (a) the **rectangular rosette**, having gage elements at 0°, 45°, and 90°; and (b) the **delta (equiangular) rosette**, having elements at 0°, 60°, and 120°.

The analysis of rosette data involves the strain transformation equations and the transverse sensitivity correction that is described above. Consider the case where a strain indicating system is used, so the output is in strain units. Set the usual manufacturer's gage factor F on the instrument. Take Q_n to be the apparent strain reading from gage n. The true strain for gage n is ε_n. Recall that the transverse sensitivity of each gage is K_T. The true strains for each gage element in the rectangular rosette are then found to be, $\varepsilon_1 = Q_1 - K_T Q_3$; $\varepsilon_2 = (1+K_T) Q_2 - K_T (Q_1 + Q_3)$; $\varepsilon_3 = Q_3 - K_T Q_1$. If gage elements 1 and 3 establish an x-y coordinate system, then the strain transformation equations yield $\varepsilon_x = \varepsilon_1$, $\varepsilon_y = \varepsilon_3$, and shear strain $\varepsilon_{xy} = \varepsilon_1 + \varepsilon_3 - 2\varepsilon_2$. With strains in the x-y coordinate system known, the principal strains and the principal axes can be found from the transformation equations or Mohr's circle. Similar data reduction equations for other types of rosettes are easily found in basic strain gage literature.

Strain Gage Transducers Resistance strain gages are used extensively in constructing transducers to measure physical quantities such as force, torque, pressure, and acceleration. Simple devices of this type can be easily fabricated in the lab for specific applications, or they may be purchased. As an example, consider the situation mentioned above where two gages are mounted on the top and bottom of a beam to measure the strain in the beam. If the output of the circuitry is calibrated in terms of the load on the beam, then the device becomes a load measuring transducer. Another instance is where a common wrench is instrumented with strain gages and calibrated so that the output is indicative of the torque transmitted by the wrench.

Critical factors in designing such transducers involve the design of the elastic member so that the physical quantity to be measured is converted to a detectable strain, the proper mounting of the gages to detect the strain, the design of the circuit so that it responds to the quantity of interest and rejects unwanted inputs, and the calibration of the device.

Because the strain in the elastic member is not of quantitative interest in a transducer, circuit calibration is not useful or appropriate. Instead, **direct calibration** is accomplished by applying known values of the desired physical quantity to the transducer and relating these known values to the output. For example, if a beam is used as a load transducer, then known weights are applied to the beam. The circuit output is then adjusted to indicate directly the values of the applied known load.

Fiber Optic Strain Gages Strain gages that are built into optical fibers have, in recent years, been used with great success in applications for which electrical resistance gages are not well suited. Two types are available from vendors, both utilizing one or another form of optical interferometry as described later in this chapter. The first type carries a tiny **Fabry-Perot interferometer** (2 mirrors) inside the optical fiber. The second relies on the interference and reflection of selected wavelengths of light by a **Bragg grating** that is imprinted inside the fiber. These strain gages share the advantages of small gage length, small diameter, low activating force, and inherent resistance to electrical and magnetic interference. Further, more than one gage can be implemented within a single fiber, which allows for multiplexing of the strain data to create, for example, a strain rosette. Thus, they are exceptionally useful for embedment inside specimens, particularly of composites, to measure internal strains. Their main disadvantage, at this writing, is the price of the gages and the equipment needed to interrogate them. The cost can be reduced somewhat by attaching them with thermoplastic or soluble adhesives so that the gages can be reused.

GEOMETRIC MOIRÉ

The geometric moiré or **mechanical moiré** effect is the mechanical occlusion of light by superimposed patterns to produce another pattern that is called a **moiré fringe pattern**. It is not an optical wave interference effect. Geometrical moiré provides a basis for a family of methods for measuring deformation, strain, shape, and contour. Advanced moiré methods, including moiré interferometry, are easily understood if the principles of geometric moiré are known.

This section describes the basic moiré phenomenon and its implementation for measurement of displacement and strain in a deformable body. Then, two techniques for using moiré to perform measurements of out-of-plane shape and deformation are outlined.

In-Plane Moiré Figure 3.5.3 illustrates the basic moiré phenomenon that appears when one system of closely spaced straight lines, a grill, is lain over a similar set. The two sets have different line spacings, and one set is also rotated slightly with respect to the other. At certain locations, the lines from one set block the gaps in the other set so that no light is allowed through. If the line spacings are sufficiently small, then the areas where the light is blocked appear to join up to form another system of broad lines, called a moiré fringe pattern. Simple experimentation with such a system shows that the moiré fringes undergo large movement when one of the grills is shifted only a small amount relative to the other grill. Clearly, the moiré pattern is a motion magnifier that can be exploited to give sensitive measurements of relative motion.

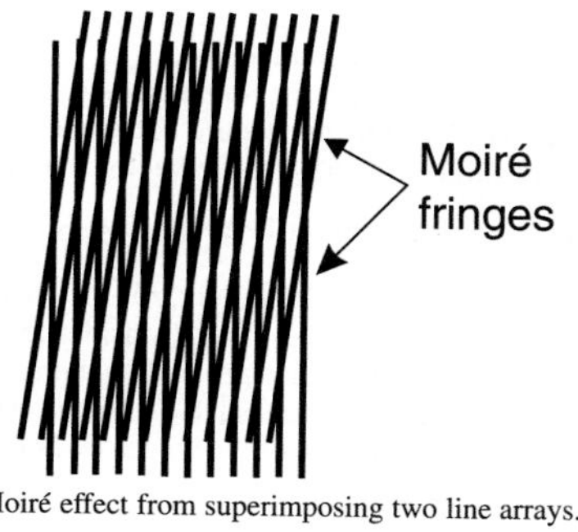

Figure 3.5.3 Moiré effect from superimposing two line arrays.

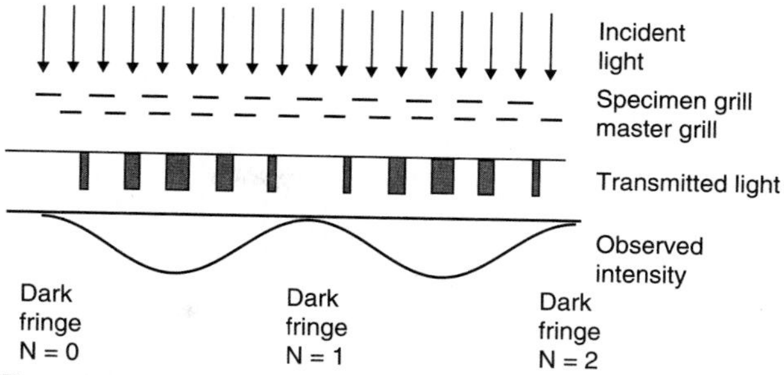

Figure 3.5.4 Moiré effect as used for displacement measurement.

Figure 3.5.4 shows a cross-sectional view of how the moiré effect is used to measure displacement and strain.

Light is projected through two grills that were originally identical but that now have slightly different spacings because one grill has been stretched relative to the other. Consider that this grill, called the specimen grill, has been stretched so that 10 lines fill the space occupied by 12 lines on the second grill, called the master grill. The widths of the cracks between the lines of the gratings vary slowly from zero to a maximum. Recall that the grill lines are actually close together, so the eye or camera does not see the light passing through the individual cracks. Rather, local averages are seen. The local average seems to oscillate twice from dark to light in the space shown, meaning two orders of moiré fringes are present. The extension or stretch in the specimen grill is $u = Np$, where N is the moiré fringe order and p is the line spacing or pitch of the master grill. The strain-displacement relationship is applied with the following result,

$$\varepsilon = \frac{\partial u}{\partial x} = \frac{\partial (Np)}{\partial x} = p\frac{\partial N}{\partial x}$$

which means that the strain along an axis perpendicular to the grill lines is the grill pitch times the gradient of the moiré fringe order (or fringe spacing) along the same axis. Similar relationships can be derived for strain in other directions and for the shear strain, so the complete strain map over the extent of the specimen can be determined. Fringes caused by small relative rotations of the grills are oriented perpendicular to the grating lines, so they do not affect the measurement of normal strain.

Specimen and Master Gratings Grills or gratings must be attached to the surface of the specimen and superimposed with a master grating using one of several observing techniques. For measurement of strain in metals, the gratings must be very fine, having ca. 10 to 100 lines per mm. A considerable body of art and technology has developed for creating and attaching specimen gratings. Superb photographically reproduced gratings for moiré work are available from vendors. Another good approach is to obtain fine photomechanical screens that are used by the printing industry. These screens are available as photographic negatives or as fine metallic mesh. They may be cut up and glued to the specimen using techniques similar to those used for attaching strain gages. More commonly, the purchased gratings are kept as intact masters, and are copied onto the specimen and as submaster gratings by photographic contact printing, by stencil techniques, or by vacuum deposition. Contact printing into a thin coating of the photoresist that is used for making printed circuit boards has been a method of choice for many practitioners.

Observing Techniques A direct and simple method of measuring displacements and strains in flat specimens involves the use of a transparent model. A grill or grating is attached or printed onto the specimen surface. The specimen is mounted in the loading frame, and a master grill is put into direct contact with the specimen grill and initially oriented to bring the moiré fringe orders to a minimum. The master grating is held in position with, for example, a spring clothes-pin. A camera set up at some distance from the specimen-master combination is used to photograph the moiré fringes as the specimen is loaded.

If a non-transparent specimen is used, the specimen grating must be made to have high visibility in reflected light. The direct superposition method can still be used, but fringe visibility will suffer. An alternative viewing method is to image the specimen grating in the back of a view camera. The master grating replaces the ground glass in the camera back, and the moiré pattern will be visible there. Direct photography of the specimen grating provides two other approaches. One is to use double-exposure, one exposure before loading and the other after loading. The moiré fringe pattern is then contained in the photograph. The second approach is to photograph the specimen grating for each loading state. These photographic replicas can be superimposed directly with a submaster or with each other at any later time, which is often an advantage in dealing with non-reversible processes, for example, those involving plastic deformation. A disadvantage of any procedure that involves imaging of the specimen grating is that the optical system must by capable of resolving finely spaced lines and very accurate focus must be established. Even so, the moiré fringes might not be of good contrast. In these cases, optical Fourier processing can be implemented to improve fringe visibility and reduce noise, while offering the possibility of enhancing strain sensitivity. Figure 3.5.5 shows a moiré fringe pattern obtained with these techniques.

Data Reduction The objective is to create a **map of strain distribution** from a moiré fringe pattern, so the derivative of the moiré fringe order with respect to position coordinates must be found. Conceptually, this is easily done by drawing an array of axes across the fringe pattern with the axes oriented perpendicular to the grill direction. The location of each moiré fringe relative to some fiducial origin along a chosen axis is measured and plotted as fringe order versus distance in specimen space. This graph is differentiated to obtain the fringe gradient, which is then multiplied by the grill pitch to obtain normal strain as a function of position along the axis. The process is repeated for each of the axes to complete the strain map. Fringe digitization and computer processing are useful, but manual processing serves well for small jobs, since only local fringe spacings are required.

Shadow Moiré The shadow method of geometric moiré utilizes the superimposition of a master grating with its own shadow. The fringes are loci of constant out-of-plane elevation, so they form an absolute **contour map** of the object being examined. This and related laser scanning methods are used in manufacturing and medical diagnosis, for example.

Figure 3.5.5 Portion of moiré pattern for measuring strain near a plastically deformed hole in aluminum.

To create a shadow moiré pattern, a master grill or grating of pitch p is placed in front of the specimen object. For large objects, the master grating can be created by stretching strings across a frame. The combination is illuminated with a collimated beam at incidence angle α. Observation is at normal incidence by means of a field lens that focuses the light to a point, at which is placed a camera that is focused on the object. The incident illumination creates a shadow of the grating on the surface of the specimen. The grating shadows on the specimen are elongated or foreshortened by a factor that depends on the inclination of the surface, and they are shifted laterally by an amount that depends on the incidence angle and the distance w from the master grating to the specimen. The camera superimposes the image of the master grating with its own distorted shadow, thereby forming a moiré fringe pattern. The fringes are numbered from some origin. If N is the moiré fringe order, then the following relationship can be derived for the distance w.

$$w = \frac{N_p}{\tan\alpha}$$

Projection Moiré Another useful method for using moiré for **out-of-plane displacement** measurement or for determining **changes of contour**, as opposed to absolute contour, involves projecting the reference grating onto the specimen by means of, for example, a slide projector. This method is especially useful when the shapes of two objects must be compared, a situation that arises in, for example, examining sheet metal stampings. An advantage is that only a small transparency of the specimen grating is required. Alternatively, a projected grating can be created by oblique interference of collimated beams of coherent light. Another advantage is that changes of contour are determined directly, and subtraction of absolute contours is not required. A disadvantage is that the imaging system must be able to resolve the individual grating lines, as with double-exposure in-plane moiré.

To create the fringe pattern, the grating is projected at incidence angle α onto the specimen surface in its initial state, or onto a master reference surface. The projected grating image is photographed by a camera along the normal to the surface. The specimen is then distorted or else the second surface is inserted if contour difference is sought. This grating is photographed over the first image. The developed double-exposure image will contain moiré fringes that are numbered as usual. At any point in the image where the moiré fringe order is N, the elevation difference w between the two states of the specimen is given by,

$$w = \frac{N_p}{\sin\alpha}$$

OPTICAL INTERFEROMETRY

Many important methods of experimental mechanics are based on the phenomenon of optical interference. Approaches in this class include, for example, photoelasticity, hologram interferometry, speckle interferometry, and moiré interferometry. Brief study of interference, one of the cornerstones of optics along with diffraction, facilitates quick understanding and use of all such methods.

Collinear Interference Light is considered to be electromagnetic waves that are characterized by an electric vector that oscillates in time and space. The magnitude of the electric vector may be taken to be a simple harmonic plane wave described as $E = A\cos\frac{2\pi}{\lambda}(z - vt)$, where A is amplitude, λ is wavelength of the light, z is distance along the optical path, v is wave velocity, and t is time. Consider two waves that are of equal amplitude, identically polarized and "coherent," meaning that they originate from the same source, such as a laser. These two waves travel along the same path, but one of the waves leads the other by some distance or phase difference r. The two waves will interfere with one

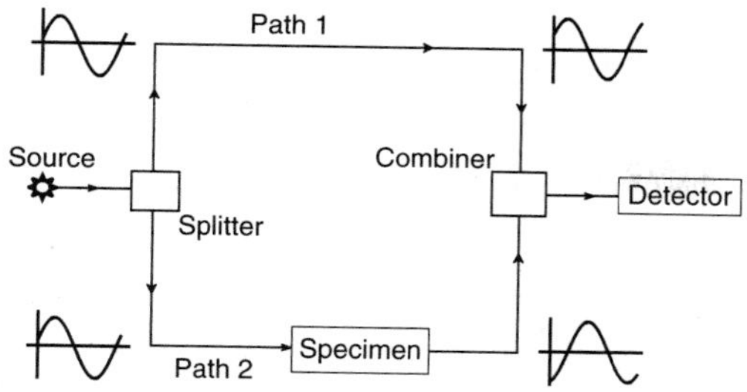

Figure 3.5.6 The generic interferometer model.

another to produce a resultant wave whose intensity I is a function of the phase difference,

$$I = 4A^2 \cos^2 \frac{\pi r}{\lambda}$$

The phase lag between two light waves, which is not visible to the eye or detectable by any other means, is converted, through the interference process, to an intensity difference, which can be measured. This transformation of invisible phase difference to visible intensity difference is the basis of all interferometry methods of measurement. It allows determination of physical quantities such as motion to within a fraction of the wavelength of light, giving sensitivities on the order of a few nanometers.

Generic Interferometer Figure 3.5.6 shows a conceptual model that contains all the important elements of an interferometric system for measurement. Light comes from a coherent source such as a laser and is split into two parts by a **beam splitter**. The two waves have identical phases as they leave this component, but they travel along different optical paths, one or both of which might contain optical elements or specimens. The two waves are out of phase as they are directed into a **combiner**, where they interfere to produce an intensity that depends on the **phase difference**. A **detector**, such as a photocell, an eye, or a camera records the intensity, from which the phase difference can be inferred. The phase difference is indicative of the "optical path length difference" between the two paths.

The conceptual model seems to imply that only pointwise measurements can be made with one wave. But, the process can be executed for every point in a broad field simultaneously if a beam of light is used and appropriate optical elements such as lenses, mirrors, and imaging devices are employed. The interference process produces a map of light and dark patches, called **interference fringes**, that are superimposed on the image of the object being studied. This whole-field capability is a great strength of optical methods in general.

Interferometry can be employed in various modes. The generic model suggests that the difference between the two paths can be obtained directly with one observation. More commonly, one path is held constant and two observations are taken for two different states of the specimen path. When subtracted, the change in only the specimen path is determined.

PHOTOELASTIC STRESS ANALYSIS

Photoelastic stress analysis is a method whereby **polarized light** is used to obtain, through interferometry, the stress state in a loaded transparent model or in a coating that is glued to the surface of a prototype. It is the only experimental technique that is capable of directly measuring stress, as opposed to strain, in a loaded engineering structure.

Stress Birefringence The speed of light in a vacuum is divided by the speed of light in a material to give the **refractive index** of the material. Optical path length is the physical path length (distance) traveled times the index of refraction. Changes or differences in optical path length cause phase lags or retardations in optical waves as they traverse a material.

In certain materials, including many plastics, the index of refraction at any point is closely related to the state of stress at a point, and such materials are said to exhibit stress-birefringence. Experiments on these materials show that:

(a) the index of refraction depends on the plane of polarization of light that passes through the material, that is, the index is direction-dependent,
(b) the state of index of refraction can be described in terms of principal values that are associated with principal directions,
(c) the principal values of refractive index are proportional to the principal stresses,
(d) the principal axes of refractive index coincide with the principal stress axes.

A conclusion is that, if the directions and magnitudes of the principal refractive indexes can be ascertained, then the principal stresses and principal directions can be determined, provided that the relation between stress and refractive index is known for the material. Determination of stress by interferometric observation of refractive index changes is accomplished by the method of photoelasticity.

Photoelasticity Figure 3.5.7 illustrates the interactions of light with a birefringent slab. Experiment and theory show that birefringent materials can transmit light only if the light is polarized in the principal directions. If a polarized wave falls on a birefringent material, it is immediately divided into two polarized components, one in each principal direction, so the beam splitter function that is necessary for interferometry is served by the surface of the birefringent material. The two waves created traverse the material at differing speeds, depending on the two principal values of refractive index. As these two waves exit the material, one will lag behind the other by an amount which is called the **relative retardation** R that depends on the local principal stresses and a material **stress-optic coefficient** C_σ according to the equation $R = C_\sigma(\sigma_1 - \sigma_2)d$, where σ_1 and σ_2 are the principal stresses and d is the thickness of the specimen. To determine R, the two wave components must be recombined interferometrically. This task is accomplished by a second polarizer in the system, known as the analyzer, which extracts the portion of each wave that lies parallel to the transmission axis of the analyzer to create the conditions for collinear interference. The intensity produced through interference is directly related to the relative retardation as described above.

If this interferometric process occurs simultaneously over the entire field, and a photograph is made of the result, interference fringes will be observed. Attention for the moment is limited to the case where the polarization axis for the light entering the specimen is perpendicular to the analyzer axis. For linear polarization, two separate families of fringes are produced, as shown in Fig. 3.5.8. These families are:

(a) isochromatic fringes that are loci of points having constant relative retardation R, meaning they are lines of constant $(\sigma_1 - \sigma_2)$. From the isochromatic family, the magnitudes of stresses throughout the specimen can be found.

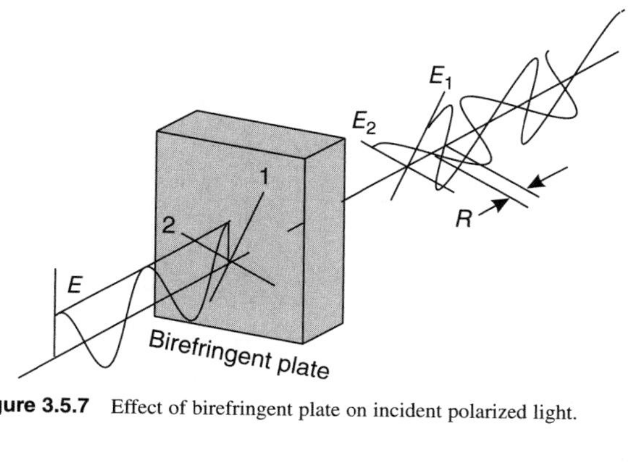

Figure 3.5.7 Effect of birefringent plate on incident polarized light.

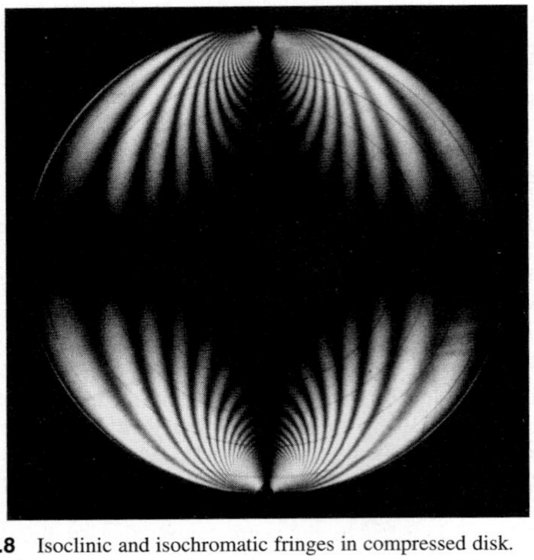

Figure 3.5.8 Isoclinic and isochromatic fringes in compressed disk.

(b) isoclinic fringes that are loci of points where the principal stress axes are parallel to the axes of polarizer and analyzer. This family is used to establish the principal stress directions for the entire specimen.

Circular Polarization A problem with the photoelastic measurements described above is that an isoclinic fringe always masks some of the isochromatics. The isoclinics can be eliminated by inserting two quarter-wave plates into the system, one on either side of the specimen.

A quarter-wave plate is a birefringent slab that produces a quarter-wavelength of relative retardation over its extent. The plates are inserted into the photoelastic measuring system with their principal axes oriented at ±45° to the crossed axes of polarizer and analyzer. The entering plane-polarized wave is divided into two equal orthogonal components, and, upon exit from the quarter-wave plate, one component will lag behind the other by $\lambda/4$. The resultant of the two out-of-phase waves will describe a circular helix in space, hence the term **circularly polarized**. The physical effect of using circular polarization is that the directional dependence of the interference is removed. The second $\lambda/4$ plate essentially cancels the effect of the first one, but does not restore the directional dependence.

Photoelastic Polariscope A combination of optical devices that is used for photoelasticity is called a polariscope. Many polariscope configurations are used. One setup, called the **diffused-light polariscope**, is shown in Fig. 3.5.9. In this system, a diffuser scatters light in all directions, only some of which passes through the system parallel to the optical axis. Polarizer, quarter-wave plates, specimen, and analyzer are arranged as described above. These elements are usually mounted in calibrated rings so that they can be rotated relative to one another and to the specimen. A field lens is used to direct the light to a point, where a camera lens is placed to create an image of the specimen on film or a CCD array. The distance from the field lens to the camera aperture must be made equal to the focal length of the field lens, otherwise the image is made with light that has not traveled parallel to the optical axis, and errors will be introduced. For quantitative measurements, a narrow-band color filter is used to create nearly monochromatic light (single wavelength), and/or a spectral lamp is used as the light source.

Material Calibration Accurate determination of the stress-optic coefficient C_σ for the photoelastic model material is prerequisite to good quantitative results. Calibration is accomplished by testing a model in which the stress field is known, such as a tensile specimen, a beam in pure bending, or a disc in diametral compression. The simple tension test is the most generally valid, since the stress state in tension does not depend on any assumptions of elastic linearity. The coefficient is dependent not only on the material, but also on time and wavelength, and these factors must be taken into account. Several different calibration procedures are valid; only a basic one is described here.

A "dog-bone" specimen of the model material is placed in a loading frame within the polariscope. A load is applied and the isochromatic order in the central section is carefully estimated by direct observation at a certain time t_1, say 10 minutes, after load. The load is increased by an increment, and the isochromatic order again estimated at t_1 after the load increase. The process is repeated for several loads, with the fringe order being determined always at the same time after changing the load. A plot of fringe order versus stress in the tensile specimen is constructed. The slope of this plot gives the stress-optic coefficient for time t_1 after load.

When the unknown specimen of the same material is tested to determine its stress state, the isochromatic orders are recorded for the same time t_1 after loading that was used in the calibration. Sometimes this is not possible. To avoid repeating the calibration testing, the isochromatic orders are often observed at a sequence of times after load so that a family of calibration results is obtained. Note that the definition of "stress-optic coefficient" is not standardized. Care is required in using calibration data provided by others.

Recording and Interpreting Isochromatics The specimen and polariscope are arranged as described above, first with the polarizer and analyzer axes crossed. The quarter-wave plates are also crossed and their axes are set to produce circularly polarized light. No light transits the system, so this setup is known as the **dark field** arrangement. The specimen is loaded, and, after waiting the prescribed time, the isochromatics are photographed. Immediately, the analyzer is rotated by 90° so as to produce a **light-field** arrangement, meaning that the background around the specimen will be bright. The new light-field isochromatic pattern is photographed. The reason that both light-field and dark field patterns are recorded is that twice as much data are obtained, which aids in interpretation. A typical dark-field pattern will look like that in Fig. 3.5.10.

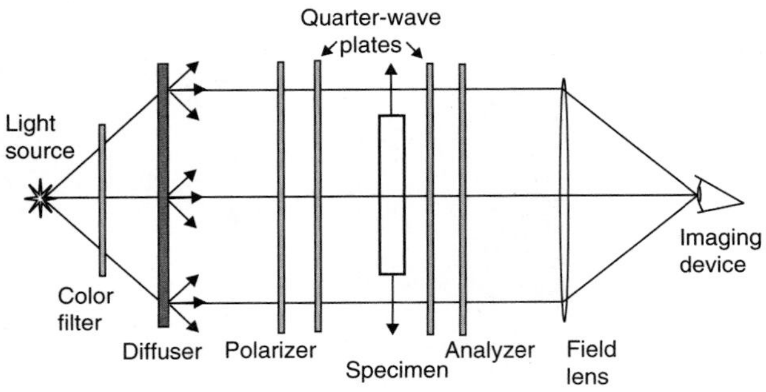

Figure 3.5.9 Diffused-light photoelastic polariscope.

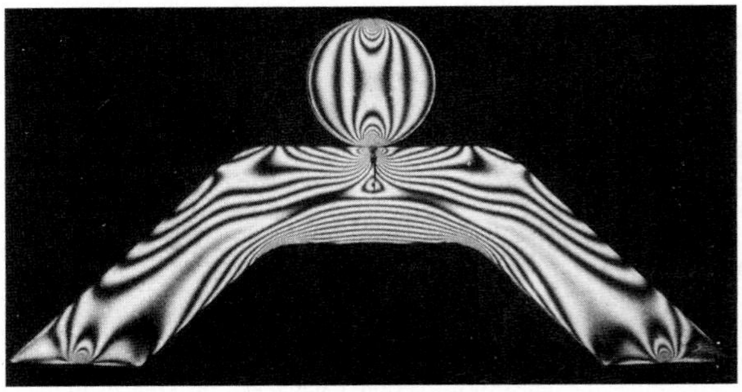

Figure 3.5.10 Isochromatic pattern indicating stresses in compressed disk and arch.

The dark fringes in the dark-field picture are whole-order isochromatics along which the relative retardation is an integral multiple of the wavelength. The isochromatics are numbered in order. Assigning correct orders is not a trivial task. One good technique is to keep track of them as the load is applied. Along these whole-order fringes, the relationship between stress and fringe order is found by combining equations already presented to give,

$$\sigma_1 - \sigma_2 = 2\tau_{\max} = \frac{m\lambda}{C_\sigma d}$$

where m is the fringe order. The dark fringes in the light-field photograph are half-order isochromatics that lie between the whole orders.

The **isochromatic fringes** give for every point inside the specimen only the difference between the principal stresses, which is the diameter of Mohr's circle and equals twice the maximum shear stress. Many methods to obtain the individual stresses have been devised. The problem is simplified along a boundary where there is no load and $\sigma_2 = 0$. For practical purposes, a **boundary stress plot** can be constructed to show stress concentrations. Normals to the boundary are erected with their lengths proportional to the local σ_1, and the tips of the normals are connected with a smooth curve.

Often, interpolation between whole- and half-orders of isochromatics does not give sufficient precision for critical areas. One of several precise methods of determining **fractional fringe orders** must then be used.

Recording and Interpreting Isoclinics For observation of isoclinics, a second model of a material that is optically less sensitive than that used for isochromatic study is often used in order that the isochromatic orders will be small and not mixed with the isoclinics. The model is placed in the polariscope, the quarter-wave plates are removed, and the polarizer and analyzer are set so their axes are horizontal and vertical, as are the axes of the specimen. The 0° isoclinic fringe appears, and it is recorded and labeled. Recording of isoclinics can be accomplished by direct tracing from a view camera back or a video monitor. They can also be photographed if heavy exposures are used to narrow the fringes.

The polarizer and analyzer are then rotated by an increment of, say, 10° in the same direction, and the isoclinic for that setting is recorded and labeled. This process is repeated incrementally until 90° is reached, where the isoclinic will duplicate the first one. The direction of rotation of the polarizers is arbitrary, but the direction used must be noted, usually on the isoclinic tracing. The result will be a complete **pattern of isoclinic fringes** that show the principal directions everywhere in the specimen.

Visualizing the stress direction map from the isoclinics is rather difficult, so an overlay orthogonal network of lines that are everywhere tangent to the principal directions, called the **stress trajectory map**, is often constructed from the isoclinic pattern.

Scaling stresses from model to prototype In order to use stresses obtained from a photoelasticity model, they must be transferred to the prototype component or structure, which likely is of a different material, of a different size, and subjected to different loads.

Two aspects of the problem must be given attention, the first being the difference of material properties and the second being differences in size and load. If the component being studied is elastic, simply connected (no holes), or multiply connected (with holes) but with no unequilibrated forces on a single boundary, and if the body forces are negligible, and if no displacement boundary conditions apply, then the stress distribution is completely independent of material properties. If displacement conditions are specified, then the modulus of elasticity is important. For the other cases, only the Poisson ratio is a factor, and the error in ignoring this effect will not exceed about 7%. In other words, for most problems, material properties are not a factor in transferring measured stresses from model to prototype.

The laws of dimensional similitude, developed from the **Buckingham theorem**, are used to attend to size and load differences. The resulting laws of **dimensional similarity** are almost intuitively obvious. Geometric similarity requires that the model be only scaled up or down relative to the prototype, meaning its shape and proportions cannot change. A similar rule applies for load similarity; the ratios of like loads between model and prototype must be constant. If displacement boundary conditions are specified, then the absolute load magnitudes must be scaled also, but this case rarely arises. If these similarity conditions are satisfied, then the stresses in model and prototype satisfy the following relationship, which shows how the stresses are scaled,

$$\sigma_p = \sigma_m \left(\frac{P_p}{P_m}\right)\left(\frac{a_m}{a_p}\right)\left(\frac{d_m}{d_p}\right)$$

where subscript m means model and p means prototype. P_p/P_m is the **load scale factor**, a_m/a_p is the **dimension scale factor**, and d_m/d_p is the **thickness scale factor**. This relationship applies only to two-dimensional (2D) problems, and it allows the thickness to be conveniently scaled differently from the lateral dimensions.

Reflection Photoelasticity The strain distribution in non-transparent prototype components and structures can be determined by an extension of the photoelasticity methods described above. The method requires that a thin layer of birefringent material be attached to the surface of the component using a cement that contains a light scattering medium such as aluminum powder. The polariscope is then folded in the middle. The polarized light (circular or linear) falls on the photoelastic coating and undergoes a relative retardation as it traverses the coating twice. The scattered (reflected) light travels through the second quarter-wave plate and the analyzer to the eye or a camera. As the prototype is loaded, isoclinic and isochromatic fringes will be obtained.

Interpretation of isochromatic fringes differs from before in that the surface strains, not stresses, are transferred to the birefringent coating by the cement. The material is calibrated in terms of principal strains, and the isochromatics are so interpreted. If stress data are required, then the prototype material properties and the stress-strain relations are used.

Reflection photoelasticity is highly developed because of its practicality, especially in the industrial setting.

Three-Dimensional (3D) Photoelasticity Photoelasticity can be employed to determine stress states in 3D objects by several techniques. All such methods, basically, slice the 3D model into a stack of 2D cases. Two such approaches are described briefly here.

The **embedded polariscope technique** uses a thin slice cut from the birefringent model. The slice is then glued back into the model, with polarizer and quarter-wave sheets replacing the saw kerfs. The result is a polariscope encased within the object with what is essentially a 2D model between the polarizers. The model is loaded and the fringe patterns for the slice obtained by passing unpolarized light through the model. Studies have shown that cutting the model and gluing it back together with the polarizers in place do not significantly affect the stress state if care is taken to use a matching cement. This method works very well for obtaining the stresses in the plane of a single slice of the object.

The **stress freezing and slicing method** requires that the model be brought to its second-order transition point by careful heating. A small deformation is then applied by loading, and this deformation is maintained while the model is cooled. The model is permanently deformed and the birefringence, which is related to the strains in this case, is "locked in." The model can be sliced into 2D cases for study in an ordinary polariscope without affecting the strain state. Again, the principal stress components in the plane of each slice are obtained. Additional observations, as by oblique incidence, are needed to completely characterize the 3D state of stress.

Digital Photoelasticity Extensive development of digital image acquisition and computer-based picture processing has resulted in the rise of several techniques that enhance the power of photoelasticity even as it has reduced the effort required to apply the technique. In the most basic instance, dark-field and light-field photographs are combined to multiply the visible fringe orders to increase data density. In another, gray levels are used to determine fractional fringe orders. **Phase-shifting and unwrapping techniques**, mentioned below in connection with speckle interferometry, are used to determine precisely the relative retardation map (phase map) for the entire extent of the model. **RGB (three-color) photoelasticity** uses the fringe colors obtained using white-light illumination to determine quickly and accurately the stress state over the entire field. Automated polariscopes have been invented that, along with appropriate software, allow a complete photoelastic analysis to be executed with little effort.

HOLOGRAPHIC INTERFEROMETRY

Holography is a method of storing and regenerating all the amplitude and phase information contained in the light that is scattered from a body that has been illuminated by a laser. Because all the information is reproduced, the reconstructed object beam is, ideally, indistinguishable from the original, so a full 3D image is created. Since it is possible to record the exact shape and position of the body in two different states, then it is possible to compare the two records interferometrically to obtain precise measurement of movement or deformation. This measurement technique is called holographic interferometry.

The making of a hologram is by oblique interference of two beams. The setup for creating this interference follows the model of the generic interferometer described above. A laser beam is split into two parts. One part, called the reference beam is expanded and made to fall on a high-resolution photographic plate or film. The other part, called the object beam, is expanded and made to illuminate the object being studied. The object scatters this illumination, so some of it falls on the same photographic film. The reference and object films interfere at the film to create a very fine complex grating, which is stored in the film emulsion. It is easy to show that this grating contains all information about the object shape, size, and position, although there will be no recognizable image stored.

Reconstruction of the holographic image is through diffraction by the exposed and developed film. The film is placed back into its original position and illuminated by only the reference beam. Grating diffraction will divide the beam into three parts, one of which, in the usual setup, recreates the object beam. If the eye is placed to receive this beam, a perfect virtual image of the object will appear in its original state.

Holographic interferometry is implemented in three different ways. In the double-exposure method, also called the **frozen-fringe technique**, a record of the initial state of the specimen is recorded as outlined above, but the film is not developed at this point. The object is then loaded or moved, and a second exposure of the object is recorded on the same film. The film is developed and replaced in the setup so as to reconstruct the images. The two superimposed images will be seen as one, and it will contain interferometric fringes that indicate the deformation or motion of the specimen. Alternative approaches are the **real-time (live fringe) technique** and the **time-average technique** for vibrating objects, neither of which are treated here.

Holographic interferometry has been primarily used for nondestructive testing of, for example, tires and turbine components. In this application, subsurface voids, disbonds, and damage cause local perturbations in the otherwise smooth holographic fringes.

Digital Hologram Interferometry For many years it was thought that a hologram could not be adequately recorded using a digital camera, because the spatial frequency range of the resulting diffraction grating was too great to be resolved by an electronic sensor array. However, both apparatus and technique have improved to the point where useful electronic holograms can be made; and, so, digital holographic interferometry can be implemented with success. The limited spatial resolution of digital sensor arrays gives rise to serious limitations in field size and viewing angles, but the technique shows promise for simplified studies of 3D deformations and vibrations as well as for nondestructive evaluation.

SPECKLE INTERFEROMETRY

This technique offers a way to achieve interferometric sensitivities for very rapid, whole-field, noncontacting measurements of displacement and strain. No special specimen preparation is required, and video imaging can be utilized.

The basic phenomenon employed in this class of techniques is that an object viewed or photographed when illuminated with laser light will exhibit a pattern of bright and dark granular spots that is fine and dense. This pattern is called a **speckle pattern**, and it is caused by local interference of the many waves that are scattered from every point on the illuminated object and that subsequently fall on the receiving film or sensor array. In order that the speckles have correct properties for measurement of deformation, two beams must be mixed at the sensor. In the first case, a reference beam is used as in holography. In the second instance, two object beams are used.

To utilize these speckles in electronic or digital speckle interferometry, a first digital image is captured to computer memory. The specimen is then moved or deformed, and a second image recorded. The second digital image is subtracted from the first pixel-by-pixel. This subtraction of two intensity patterns can be done in real-time on a desk-top computer. The resulting video image will exhibit fringes that indicate the local displacements of the surface of the specimen.

If a reference beam is used, if the illuminated object is photographed at near-normal incidence, and if the object beam is nearly parallel with the reference beam at the sensor, then the fringes indicate only the out-of-plane motion of the object and are analogous to Newton's rings. If two object beams are used, then they are set up to illuminate the object at equal and opposite angles of incidence and viewing is along the normal. In this case, the fringes correspond to in-plane motion or deformation. The measured in-plane displacements are then locally differentiated to obtain strain distribution.

The speckle methods just described produce **speckle correlation fringes** that are most commonly used in qualitative applications such as nondestructive testing. Phase-shifting techniques extend the method for quantitative measurement. A common implementation is to incorporate into one of the optical paths a mirror that can be moved by small amounts so as to change that path length. At least three images are captured for a matching number of imposed path length differences. Processing yields a whole-field **phase change map** that indicates both the sign and magnitude of displacement. The displacement map is locally differentiated in the computer. Thus, whole-field strain measurements with high sensitivity and small gage lengths can be accomplished in only a few seconds.

SPECKLE SHEAROGRAPHY

This approach uses the speckle effect described above in a different way but with much of the same equipment. An extra optical element, such as a Fresnel biprism or a split lens creates on the camera sensor two images of the specimen, one very slightly displaced from the other. A first image-pair is captured, the specimen is loaded, and the second image-pair is captured and subtracted from the first, typically in real time. The resulting difference of image-pairs exhibits fringes that indicate the in-plane derivative of the out-of-plane displacement (i.e., the change in slope), at each point of the object surface.

Electronic speckle shearography may be made highly quantitative by use of phase-shifting as outlined above, but it is primarily used without phase shifting as a very important nondestructive evaluation tool that is simple, quick, and easy to implement. As with holographic interferometry, the presence of flaws is indicated by local perturbations of the otherwise smooth fringe pattern.

SPECKLE PHOTOGRAPHY

The fine laser speckle patterns can be used as surface markers to directly measure specimen motion and deformation. The key phenomenon is that, for a limited range of motion, speckles move with the surface because they are signatures of local fine surface roughness features. The technique for using them is exceptionally simple. A high-resolution photograph is recorded for the specimen in its base state.

The specimen is then moved or loaded and a second record is made, usually as a double-exposure, although separate single exposures may be used. Microscopic examination shows that the doubly exposed film contains a multitude of very fine speckle pairs, with the separation between the elements of each of the pairs being equal to the local specimen displacement. The separations are usually measured using one or both of two techniques. The first involves interrogating the film point-by-point with a laser beam to cause **Young's fringes** to be cast on a screen, the spacing between the fringes being related to the pair spacing, the wavelength of the laser, and the distance between film and screen. The second approach involves **optical Fourier spatial filtering** to create a fringe pattern that is whole-field map of specimen in-plane displacement.

White Light Speckle Photography This technique differs from the laser speckle method just described in that the speckles are created by oblique lighting of the specimen surface so as to produce fine shadow patterns or else by directly applying them through coarse spraying or "splattering" of paint. The dynamic range of the white light approach is not as great as that demonstrated with laser speckles, but it is a useful technique that is easily implemented in the field and on very large structures including, for example, mine shafts and glaciers. More important, perhaps, is that this analysis approach is the precursor to the powerful but more complex digital image correlation method that is described later in this chapter.

MOIRÉ INTERFEROMETRY

To implement this extension of the moiré method, a grating having spatial frequencies on the order of several hundreds of lines per mm is created by **oblique interference** of two collimated laser beams. This grating is transferred to the surface of the specimen by one of several techniques. The specimen is then replaced in the same setup that was used to make the grating, and the specimen is loaded. **Reflective diffraction** of the incident beams from the specimen grating creates two slightly diverging beams, where the local divergence angle is related to the local change of spatial frequency of the specimen grating that is caused by strain. The two diffracted beams are collected by a camera lens and brought together to form two coincident images that appear as one. This image will contain interference fringes resulting from local oblique interference. Interpretation of the fringe pattern is the same as for geometric moiré. This important method overcomes the sensitivity limitation of geometric moiré. Its main disadvantage is that significant specimen preparation is required.

DIGITAL IMAGE CORRELATION

Digital Image Correlation (DIC) is a non-contacting measurement method that uses **digital signal processing algorithms** to track the motion in the camera image plane of subsets of either a coherent or a white-light **random speckle pattern**. This technique has developed over the last decade into a powerful tool for mapping displacements and, to a lesser extent, strain distributions in structural elements that undergo motions and distortions. It has become very useful in its commercial turn-key versions even as it continues to undergo refinements that enhance its sophistication, and it is displacing some older optical techniques in the experimentalist's kit.

Two-dimensional DIC (2D-DIC) uses a single camera to convert images of a nominally planar object that deforms predominantly in-plane into a full field map of in-plane displacements. **Three-dimensional DIC** (3D-DIC) converts images obtained from cameras located at multiple viewpoints into a full field map of the 3D displacements of all points in a planar or non-planar object. If the displacement data are of sufficient quality, not always the case, then they may be differentiated to obtain normal and shear strain distributions. The strain sensitivity varies according to the specimen size and other parameters, but good practice will yield strains to within ±100 microstrain in a typical application.

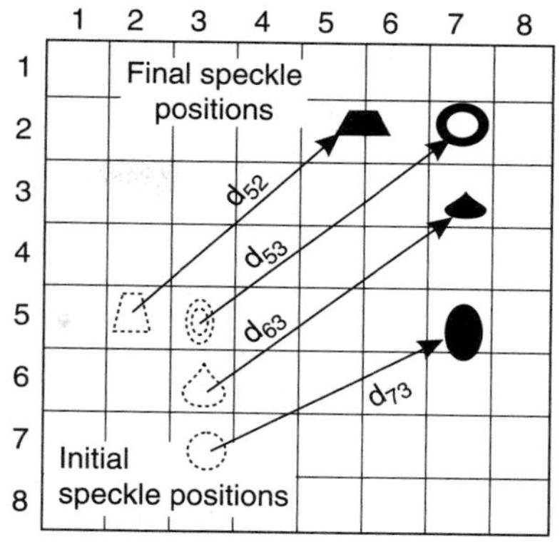

Figure 3.5.11 Before- and after-deformation states of an imaginary speckle pattern showing how the speckles move and are deformed.

Conceptual Model As is the case with any technique, understanding its fundamental basis of operation is important in order to avoid errors derived from misapplication or faulty procedure. This requirement is, perhaps, exceptionally critical for DIC, because almost all users of the technique simply purchase the very fine apparatus and user-friendly software that are available from vendors. Here, the necessary understanding of DIC is facilitated through a simple thought experiment that establishes the basis of 2D-DIC. The extension to 3D can then be visualized, but treatment of it is beyond the scope of this article.

Figure 3.5.11 shows a much-enlarged portion of a somewhat whimsical pattern of speckles that have been created on a planar specimen that undergoes deformation. Both the initial (before deformation) and final (after deformation) speckle patterns are shown, as would be the case in a double-exposure photograph, with the original speckles outlined with dotted lines. Overlain on this pattern is an 8 × 8 portion of the array of sensors that convert light intensity to voltage within the digital camera that is recording images of this small section of the specimen. The camera has remained stationary, so the speckles have moved to take up new positions within the sensor array. During the deformation process, the individual speckles have also undergone changes of size, shape, and spacing that, for reasons based on physics, are not too drastic unless the specimen begins to break. Note that some speckles might overlap more than one sensor. If the individual speckles could be photographed with sufficient resolution and somehow identified before and after deformation, their displacements (e.g., vector $\boldsymbol{d}_{52}$) could be determined with a high degree of precision, thus yielding, when extended to all the speckles on the specimen, a complete displacement map. This process is analogous to the speckle photography technique described above.

Unfortunately, a digital camera sensor array cannot track individual speckles with useful precision. It delivers only the intensity averaged over each sensor area, commonly called the **gray level**, for rendering into a photograph that is comprised of blocks called **pixels**. Each pixel corresponds to one of the sensors and displays only the gray level recorded by that sensor. Speckle size, shape, and, usually, color data are lost, as are their exact spacings. Following a single speckle between photographs recorded before and after specimen distortion cannot be done unless that speckle covers many pixels so as to form a recognizable image, a condition that compromises measurement precision. Otherwise, if a search is made for single pixel having a particular gray level, it is likely to turn up several hundred matches within the image of the deformed specimen.

To circumvent this limitation, use is made of the **local *patterns* of intensity information** characteristic of the collection of pixels inside **windows or subsets** of the images. The key to the method is the justifiable expectation that the gray-level pattern within a subset will not be seriously altered by the deformation of the specimen; meaning that, within the deformed image, only one matching collection will exist, and the problem is to find it. The probability of finding a single best match depends on the size of the subset and some characteristics of the speckle pattern. So, a search of the image of the deformed specimen is executed to locate a pattern of gray levels within a subset of pixels that is a good match to the pattern in the original undeformed subset. That is, the entire deformed image is scanned to find that one particular window in which the gray level distribution most closely matches that in the original image window. The match is rarely perfect owing to optical noise and the local distortions of the speckles that cause changes in the pixel intensity levels and distributions within the window.

Figure 3.5.12 illustrates the patterns of gray levels that the camera would record for the speckle pattern of Fig. 3.5.11, with the before- and after-deformation gray levels combined into one picture to save space. The **initial subset** is designated by the 4 × 4-pixel dotted square. The best **post-deformation match** can be found visually for this simple case, and it is shown within the square having the solid outline. This best match shows that the center of the original subset area has been displaced by 4 pixels to the right and 3 pixels upwards, which, when converted to specimen spatial dimensions, yields the components of the local displacement vector ***d*** of the center of the window. The uncertainty in this displacement is related to the size of the subset unless additional processing is implemented. Now, if this same search-and-match process is executed for all 4 × 4-pixel subsets in both the extended undeformed and deformed dense speckle fields characteristic of a structural component, a complete map of the displacement field will be obtained. Further, if the displacement data are dense enough and if the uncertainties are minimized, then the displacement components can be differentiated to obtain maps of the normal and shear strains, thus completing the experiment.

Implementation A real-life experiment would use a specimen containing, perhaps, thousands of speckles and a camera whose sensor array contains thousands of elements. Further, the gray level in each pixel would be assigned a numerical value, often running from 0 to 255. The subset matching process requires an immense number of computations of some metric that indicates the degree of match. This task is a

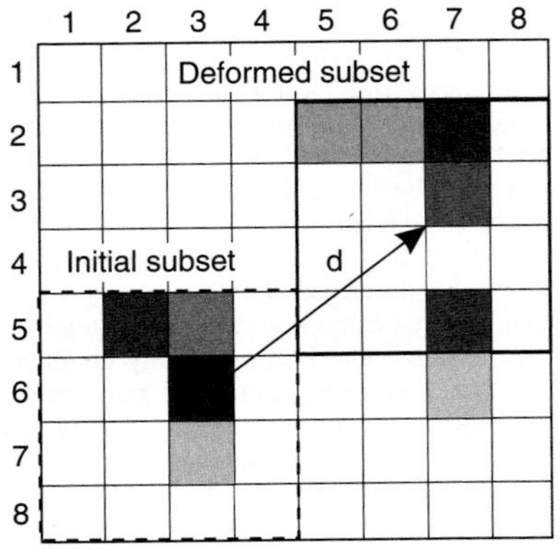

Figure 3.5.12 The speckle patterns of Fig. 3.5.11 as recorded in gray levels by a digital camera for before- and after-deformation states of the specimen, showing the best-match pixel subsets that result from DIC.

reasonable matter for even a small computer, and several standard algorithms are available to do such searching and matching. Consideration of the process of scanning the entire deformed image with the contents of all possible windows of the chosen size in the initial reference image suggests correctly that one is repetitiously evaluating a **standard correlation function** for two arrays of numbers to find the best matching subset pairs. This function is contained in the standard toolkits of software such as MatLab®, MathCad®, and Microsoft Excel® to name but three. The scanning and matching processes are expedited by optimizing the scanning paths, using an initial coarse search, and through use of estimates that are based on knowledge of material behavior and the results of earlier scans.

Optimized processing approaches, which are implemented in dedicated commercially available software, make use of an estimate of the deformation of the subset window and accommodate non-integer (pixel fraction) speckle displacements through interpolation within an image. A Taylor expansion of the displacement map to an affine transformation or a quadratic function is used to establish a displacement function. These steps require smoothly varying intensity distributions rather than the stepwise ones suggested by the example above, so the intensity arrays from each of the digital photographs must be smoothed using spline functions before the matching algorithm is implemented. This sophisticated data processing reduces displacement measurement uncertainties to a small fraction of the pixel size.

Much remains to be said about the practical execution of DIC. The necessary speckle pattern can be created using the laser speckle phenomenon described above in the section on speckle interferometry, through coarse spray-painting, powder adhesion, or by pressing on appliques. Speckle size relative to specimen and image sizes, speckle contrast and randomness, camera lens choices, and subset size and shape are among the many important factors to be considered.

Both 2D- and 3D-DIC require that **spatial calibration** be performed so that the image motions are converted into displacements with optimal accuracy. When photographic quality lens are used to acquire images, calibration in 2D-DIC is relatively simple, requiring only an image of a ruler or similar metric object. For 3D-DIC, pairs of images of a calibration object undergoing a set of 3D rotations are required to obtain the relative and positions of the cameras as well as to establish the mapping relationships between image and specimen spaces.

THERMOELASTICITY

Thermoelasticity is a noncontacting analysis technique that utilizes the minute local changes of temperature that are caused by stress. When a body is subjected to tensile stress the temperature of the body reduces slightly, and, conversely, when the stress is compressive, a slight increase in temperature occurs. In the elastic regime, this temperature change is directly proportional to the change in the sum of principal stresses on the component surface. The only surface preparation required is to ensure uniform **emissivity** of the component surface at the **infrared wavelengths** used by the detector, and this requirement can usually be achieved by the application of a thin layer of matte black paint. Since the detectors are measuring photon emission from a surface, thermoelasticity images can be recorded obliquely, with satisfactory results obtainable at up to 60° from the normal to the surface. This technique is especially useful in locating and evaluating stress concentrations as well as for detecting the initiation of cracks and measuring their progressions. Sophisticated equipment and software for thermoelasticity investigations are readily available from vendors. Additional details about this technique and apparatus may be found in Sec. 3.4 on Nondestructive Inspection.

3.6 MECHANICS OF COMPOSITE MATERIALS

by Michael W. Hyer and Scott W. Case

This section deals with some of the basic issues that must be addressed when applying the classic methods of mechanics of materials to predict the mechanical behavior of fiber-reinforced composite materials. The section deals exclusively with the mechanical behavior of materials that consist of stiff and strong fibers, such as glass, carbon, or aramid, embedded in a softer material, such as a thermosetting or thermoplastic polymer, known as the matrix material, or simply, the *matrix*. The role of the fiber is to carry the load, while the role of the matrix is to keep the fibers aligned, transfer the load from fiber to fiber, and protect the fibers from direct mechanical contact and exposure to environmental factors. Other fibers, such as boron, have been employed, and composite materials that use metal, such as aluminum or titanium, for the matrix have been developed. Composite materials which utilize boron fibers or a metal matrix are much less common and are quite expensive. The mechanical behavior of metal matrix composite materials can differ significantly from that of polymer matrix composite materials in that metals yield before failing, whereas the polymers used in composite materials are brittle by comparison and generally fail by cracking. This difference in the two matrix materials can lead to different failure criteria and therefore different analysis techniques to predict whether a particular structural component will fail or not. Here emphasis will be on polymer matrix composite materials. However, some of the assumptions and calculations are applicable to metal matrix composite materials if the stress levels in the matrix are below the yield level.

The application of the classical methods of the mechanics of materials to the analysis and design of structural components such as beams, plates, curved panels, and cylinders fabricated from fiber-reinforced composite materials requires considerably more steps than the analysis and design of a similar components made of conventional materials such as aluminum or steel. The primary reasons for this are: (1) the layered nature of components fabricated of fiber-reinforced composite materials; and (2) the fibers in the layers result in stiffness and strength properties that are directionally dependent, not only for that layer, but for the overall structural component. When material behavior is not directionally dependent, the material behavior is said to be *isotropic*. Except for slight rolling or other effects due to fabrication or consolidation, which are generally ignored, conventional materials such as aluminum or steel exhibit isotropic behavior. When material properties are directionally dependent, material behavior is said to be *anisotropic*, or non-isotropic. There are special classes of anisotropic material behavior, and one which will be defined shortly is *orthotropic* material behavior. To reach a point where the only two differences between a component fabricated from a fiber-reinforced composite material and one fabricated from a conventional material are the layered nature and anisotropic behavior, several steps and assumptions have to be considered. Specifically, the fiber and matrix materials each have their own stiffness and strength properties. When these two constituents are combined, the resulting mixture, or composite, has its own stiffness and strength properties, so-called *composite properties*. Composite properties depend not only on the properties of each constituent, but also on the quantity of each constituent in the composite material. A key measure of the quantity of a constituent is the *volume fraction* of that constituent in a given volume of composite material. There are models for predicting the properties of a composite material based on the properties and volume fraction of each constituent, in particular, the volume fraction of fibers and the volume fraction of matrix material. These models are referred to as *micromechanical models*. Several micromechanical models will be introduced here for computing such properties as overall stiffness or overall thermal expansion properties of a layer of composite material. These models are based on assumptions regarding the mechanical behavior of the individual fibers, the matrix surrounding them, and importantly, assumptions regarding the interaction between the fibers and matrix. The result is a collection of models that represent the combined stiffness and thermal expansion effects of all the fibers and the accompanying matrix material. A layer of composite material is assumed to exhibit these combined effects, and it is assumed the combined effects are uniformly distributed throughout the volume of the layer. The assumption of having a uniform distribution of composite properties, despite the fibers and matrix having individual properties, is the concept of *homogenization*, or *smearing*, of constituent properties. Once the smeared composite properties are predicted, the individual fibers and the matrix material lose their identities and a layer is treated as a single material with a single set of properties. Each layer may have different properties, because one layer may use glass fibers and another layer may use carbon fibers, or there may be a mixture of glass and carbon fibers within one layer, but within a layer, the properties are assumed to be uniform. Homogenization is an important concept. If this concept was not accepted, it would be impossible to study the mechanical behavior of fiber-reinforced composite materials, as it would be an insurmountable task to keep track of the behavior of each fiber and the matrix material surrounding each fiber.

After the effects of the fibers and matrix have been homogenized, before design and analysis can proceed, the concept of homogenization must be utilized a second time to represent the combined effects of all the layers on the overall behavior of a structural component. This second step is necessary because a component could be constructed of many, many layers of material. The assemblage of all the layers is referred to as a *laminate*. It becomes a large chore to keep track of the behavior of each layer in a laminate when interest may be on the overall behavior of the laminated component, for example, such as its lowest natural frequency or its buckling load. The steps for homogenization at this level also depend on assumptions regarding the behavior of the individual layers and assumptions regarding how each layer interacts with the surrounding layers. The study of the individual layers in a laminate and how they interact is often referred to as *macromechanics*. The behavior of a laminate when loading, support, and other conditions are considered is often referred to as *composite structural mechanics*. Many of the tools of mechanics apply to composite materials and structures because the tools are based on principles that do not depend on material behavior. Ultimately, however, material properties must be defined, and it is at this step that the design and analysis of composite structures and isotropic structures diverge.

MICROMECHANICS OF A LAYER

The practice of using fiber and matrix properties to estimate the homogenized response of a composite material has received much attention and some treatment is given in almost all introductory composite materials texts. Practical reasons for such an investigation include the ability to perform trade-off studies, such as varying the amount and type of fiber used, and the ability to calculate theoretically during the initial design process quantities that are difficult to determine experimentally. There are three principal approaches that have been employed in developing these models for homogenization: (1) mechanics of materials techniques; (2) elasticity solutions; and (3) empirical or semiempirical techniques. In the mechanics of materials approaches, simplified geometries and simplified material behavior models are usually employed. In general, these models do not capture the details of the stress and strain distributions within the fiber and matrix, and as a result, details of the actual fiber arrangement within the composite are unimportant. In the elasticity solutions, an attempt is made to determine the actual stresses and strains in the fiber and matrix materials. As a result, the geometric arrangement of the fibers, or *fiber packing*, within the composite can become a factor. Often, however, the actual fiber packing is represented by idealized packings, as shown in Fig. 3.6.1.

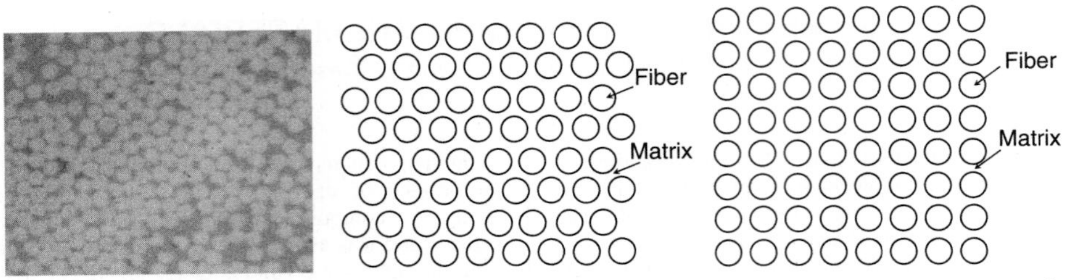

Figure 3.6.1 Photomicrograph of actual composite cross-section (left), and idealized representations of fiber packing: hexagonal (center) and square (right).

It is important to keep in mind, however, that these models are estimates of the actual material behavior, and that there is no substitute for direct experimental measurement of the desired properties. As a result, the models presented in this section are selected for their balance between ease of use and their accuracy.

Before discussing the micromechanics models in detail, it is necessary to begin with a description of the thermal-mechanical response of a single layer of unidirectional composite material. In discussing this response, it is convenient to use an orthogonal coordinate system that has one axis aligned with the direction of the fiber reinforcement, a second axis that lies in the plane of the layer, and the third axis through the thickness of the layer, as illustrated in Fig. 3.6.2. Such a coordinate system is referred to as the *principal material coordinate system* (or often the 1-2-3 coordinate system). In this case, the 1-direction is the fiber direction, while the 2- and 3-directions are transverse directions, often referred to as matrix directions. At this point, it is clear that the material properties will not be the same in all directions. For example, it is reasonable to expect the composite to be stronger and stiffer in the 1-direction than in either the 2- or 3-directions. It might also be expected that the 2- and 3-directions have similar—but not necessarily identical—properties because they are both directions perpendicular to the fiber direction. Accordingly, the response of an element of material at a point within layer that has different properties in the different directions is to be described. Such an element can be acted on by six stress components, as shown in Fig. 3.6.3.

In Fig. 3.6.3, σ_1, σ_2, and σ_3 are the normal stresses, in the principal material directions, τ_{12} is the in-plane shear stress, and τ_{13} and τ_{23} are through-thickness shear stresses. The corresponding normal strains are given as ε_1, ε_2, and ε_3; the engineering shear strains are given as γ_{12}, γ_{13}, and γ_{23}. The strains can then be related to the stresses as

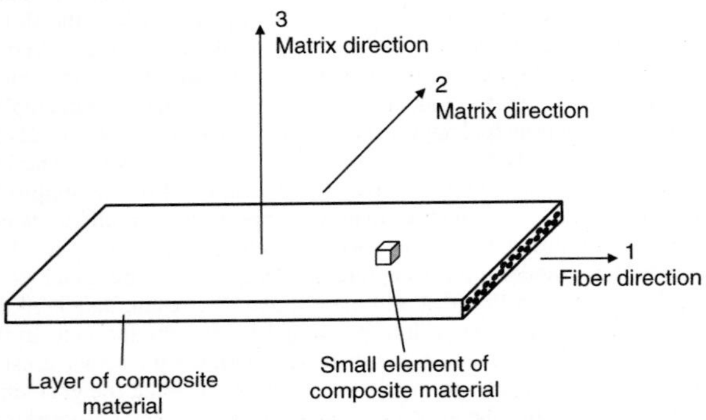

Figure 3.6.2 Layer of composite material illustrating principal material coordinate system.

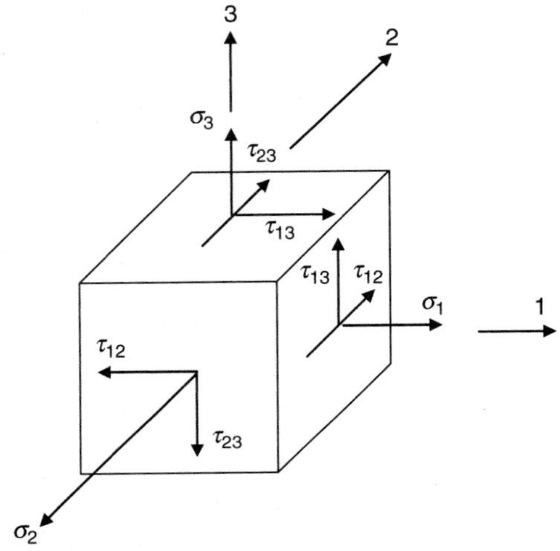

Figure 3.6.3 Illustration of stresses acting on the small element of composite material from Fig. 3.6.2.

$$\begin{Bmatrix} \varepsilon_1 \\ \varepsilon_2 \\ \varepsilon_3 \\ \gamma_{23} \\ \gamma_{13} \\ \gamma_{12} \end{Bmatrix} = \begin{bmatrix} \frac{1}{E_1} & -\frac{\nu_{21}}{E_2} & -\frac{\nu_{31}}{E_3} & 0 & 0 & 0 \\ -\frac{\nu_{12}}{E_1} & \frac{1}{E_2} & -\frac{\nu_{32}}{E_3} & 0 & 0 & 0 \\ -\frac{\nu_{13}}{E_1} & -\frac{\nu_{23}}{E_2} & \frac{1}{E_3} & 0 & 0 & 0 \\ 0 & 0 & 0 & \frac{1}{G_{23}} & 0 & 0 \\ 0 & 0 & 0 & 0 & \frac{1}{G_{13}} & 0 \\ 0 & 0 & 0 & 0 & 0 & \frac{1}{G_{12}} \end{bmatrix} \begin{Bmatrix} \sigma_1 \\ \sigma_2 \\ \sigma_3 \\ \tau_{23} \\ \tau_{13} \\ \tau_{12} \end{Bmatrix} \tag{3.6.1}$$

In Eq. (3.6.1), E_1 is the extensional modulus in the fiber direction, E_2 is the extensional modulus in the transverse 2-direction, and E_3 is the extensional modulus in the transverse 3-direction. The Poisson's ratios ν_{ij} are defined so $-\frac{\nu_{ij}}{E_i}\sigma_i$ is the strain in the j-direction due to a stress applied in the i-direction. The three shear moduli are G_{23}, G_{13}, and G_{12}. The extensional moduli, the shear moduli, and the Poisson's ratios are collectively referred to as *engineering properties*. The square six-by-six matrix in Eq. (1) is referred to as the *compliance matrix*, and is normally denoted by S. The stress-strain relations, in terms of S, are then given as

$$\begin{Bmatrix} \varepsilon_1 \\ \varepsilon_2 \\ \varepsilon_3 \\ \gamma_{23} \\ \gamma_{13} \\ \gamma_{12} \end{Bmatrix} = \begin{bmatrix} S_{11} & S_{12} & S_{13} & 0 & 0 & 0 \\ S_{21} & S_{22} & S_{23} & 0 & 0 & 0 \\ S_{31} & S_{32} & S_{33} & 0 & 0 & 0 \\ 0 & 0 & 0 & S_{44} & 0 & 0 \\ 0 & 0 & 0 & 0 & S_{55} & 0 \\ 0 & 0 & 0 & 0 & 0 & S_{66} \end{bmatrix} \begin{Bmatrix} \sigma_1 \\ \sigma_2 \\ \sigma_3 \\ \tau_{23} \\ \tau_{13} \\ \tau_{12} \end{Bmatrix} \tag{3.6.2}$$

A direct comparison of Eqs. (1) and (2) shows that the compliances may be calculated in terms of the engineering properties as

$$\begin{aligned} S_{11} &= \frac{1}{E_1} & S_{12} &= -\frac{\nu_{21}}{E_2} & S_{13} &= -\frac{\nu_{31}}{E_3} \\ S_{21} &= -\frac{\nu_{12}}{E_1} & S_{22} &= \frac{1}{E_2} & S_{23} &= -\frac{\nu_{32}}{E_3} \\ S_{31} &= -\frac{\nu_{13}}{E_1} & S_{32} &= -\frac{\nu_{23}}{E_2} & S_{33} &= \frac{1}{E_3} \\ S_{44} &= \frac{1}{G_{23}} & S_{55} &= \frac{1}{G_{13}} & S_{66} &= \frac{1}{G_{12}} \end{aligned} \tag{3.6.3}$$

The inverse of the compliance matrix is called the *stiffness matrix*, and is normally denoted by C. Once this inverse has been defined, the stress-strain relations can be written as

$$\begin{Bmatrix} \sigma_1 \\ \sigma_2 \\ \sigma_3 \\ \tau_{23} \\ \tau_{13} \\ \tau_{12} \end{Bmatrix} = \begin{bmatrix} C_{11} & C_{12} & C_{13} & 0 & 0 & 0 \\ C_{21} & C_{22} & C_{23} & 0 & 0 & 0 \\ C_{31} & C_{32} & C_{33} & 0 & 0 & 0 \\ 0 & 0 & 0 & C_{44} & 0 & 0 \\ 0 & 0 & 0 & 0 & C_{55} & 0 \\ 0 & 0 & 0 & 0 & 0 & C_{66} \end{bmatrix} \begin{Bmatrix} \varepsilon_1 \\ \varepsilon_2 \\ \varepsilon_3 \\ \gamma_{23} \\ \gamma_{13} \\ \gamma_{12} \end{Bmatrix} \tag{3.6.4}$$

The forms of Eqs. (2) and (4) suggest that 12 compliances or 12 stiffnesses are required to describe the stress-strain response of an element of composite material. However, it can be shown that the compliance matrix is symmetric, i.e., $S_{ij} = S_{ji}$. For that to be true, from Eq. (3) the following relations must hold:

$$\frac{\nu_{12}}{E_1} = \frac{\nu_{21}}{E_2} \qquad \frac{\nu_{13}}{E_1} = \frac{\nu_{31}}{E_3} \qquad \frac{\nu_{23}}{E_2} = \frac{\nu_{32}}{E_3} \tag{3.6.5}$$

The relations in Eq. (5) are referred to as the *reciprocity relations*. Thus, only nine independent engineering properties are needed to describe the elastic response of a layer of composite material.

In addition to the strains due to the stresses, expansional strains due to temperature changes and moisture absorption are often present. For such a case, the strains are related to the stresses by a modification of Eq. (1) as

$$\begin{Bmatrix} \varepsilon_1 \\ \varepsilon_2 \\ \varepsilon_3 \\ \gamma_{23} \\ \gamma_{13} \\ \gamma_{12} \end{Bmatrix} = \begin{bmatrix} \frac{1}{E_1} & -\frac{\nu_{21}}{E_2} & -\frac{\nu_{31}}{E_3} & 0 & 0 & 0 \\ -\frac{\nu_{12}}{E_1} & \frac{1}{E_2} & -\frac{\nu_{32}}{E_3} & 0 & 0 & 0 \\ -\frac{\nu_{13}}{E_1} & -\frac{\nu_{23}}{E_2} & \frac{1}{E_3} & 0 & 0 & 0 \\ 0 & 0 & 0 & \frac{1}{G_{23}} & 0 & 0 \\ 0 & 0 & 0 & 0 & \frac{1}{G_{13}} & 0 \\ 0 & 0 & 0 & 0 & 0 & \frac{1}{G_{12}} \end{bmatrix} \begin{Bmatrix} \sigma_1 \\ \sigma_2 \\ \sigma_3 \\ \tau_{23} \\ \tau_{13} \\ \tau_{12} \end{Bmatrix} + \begin{Bmatrix} \alpha_1 \\ \alpha_2 \\ \alpha_3 \\ 0 \\ 0 \\ 0 \end{Bmatrix} \Delta T + \begin{Bmatrix} \beta_1 \\ \beta_2 \\ \beta_3 \\ 0 \\ 0 \\ 0 \end{Bmatrix} \Delta M \tag{3.6.6}$$

In the above equation, α_1, α_2, and α_3 are coefficients of thermal expansion, β_1, β_2, and β_3 are coefficients of moisture expansion, ΔT is the change in temperature from a reference temperature, and ΔM is the change in moisture content within the material from a reference moisture content. (This reference moisture content is normally taken as zero.) It is important to note that there are no thermally-induced or moisture-induced shear strains in the principal material coordinate system. The form of the stress-strain relations in Eq. (6), and Eqs. (1), (2), and (4), defines *orthotropic* material behavior. Such behavior is characterized by: (1) a decoupling of shear and normal responses, resulting in many of the off-diagonal terms being zero; and (2) different material properties in the three mutually-perpendicular directions. Specifically, the zeros represent the fact that with orthotropic material behavior, normal strains and normal stresses are related, each particular shear strain is related to the corresponding shear stress, but there are no relations between normal strains and shear stresses, nor between shear strains and normal stresses. Also, as noted, there are no thermal- or moisture-induced shear strains. The form of the stress-strain relations for isotropic materials is the same as the form of Eq. (6), the difference being that the material properties are the same in all directions, and only two engineering properties are needed (e.g., the extensional modulus E and Poisson's ratio ν) to fully describe the response. That is, for the engineering properties $E_1 = E_2 = E_3 = E$, $\nu_{23} = \nu_{13} = \nu_{12} = \nu$, $G_{23} = G_{13} = G_{12} = G = E/2(1 + \nu)$, while for the expansional properties $\alpha_1 = \alpha_2 = \alpha_3 = \alpha$, and $\beta_1 = \beta_2 = \beta_3 = \beta$.

Equation (6) represents the general response of a layer of composite material subjected to a three-dimensional stress state. In many practical situations involving the use of composite materials, it may be assumed that the composite is in a state of *plane stress*: a state where three of the stress components are so much smaller than the other three they may be assumed to be equal to zero for subsequent stress analysis. Some situations where the plane-stress assumption may be employed is when considering beams, plates, and cylinders in which one dimension is at least an order of magnitude greater than the other dimensions. As an example, consider a plate in which the thickness direction, which is aligned with the 3-direction, is much less than the in-plane dimensions. The out-of-plane stress components σ_3, τ_{23}, and τ_{13} will then be much smaller than the in-plane stress components σ_1, σ_2, and τ_{12}. As a result, Eq. (6) can be simplified considerably.

Before examining the simplified stress-strain relations for the case of a plane-stress state, a number of cautions are in order. First, the plane-stress assumption can lead to inaccuracies. For example, delamination of laminates often initiates at free-edges due to the presence of out-of-plane stresses—exactly the stress components ignored in a plane-stress analysis. Second, locations of discontinuities in structures—such as at bonded joints, stiffeners, or ply drops—result in three-dimensional stress states. Finally, in many cases the through-thickness stresses are quite small in comparison to the in-plane stresses (particularly σ_1), but may still be large enough in comparison to the corresponding through-thickness strength values that failure may occur.

To see why the plane-stress assumption is important, the effects of this assumption on the stress-strain relations given in Eq. (6) can be examined. For the case of plane stress in which σ_3, τ_{23}, and τ_{13} are taken to be identically zero, the stress-strain relationship reduces to

$$\begin{Bmatrix} \varepsilon_1 \\ \varepsilon_2 \\ \gamma_{12} \end{Bmatrix} = \begin{bmatrix} \frac{1}{E_1} & -\frac{\nu_{12}}{E_1} & 0 \\ -\frac{\nu_{12}}{E_1} & \frac{1}{E_2} & 0 \\ 0 & 0 & \frac{1}{G_{12}} \end{bmatrix} \begin{Bmatrix} \sigma_1 \\ \sigma_2 \\ \tau_{12} \end{Bmatrix} + \begin{Bmatrix} \alpha_1 \\ \alpha_2 \\ 0 \end{Bmatrix} \Delta T + \begin{Bmatrix} \beta_1 \\ \beta_2 \\ 0 \end{Bmatrix} \Delta M \tag{3.6.7}$$

The three-by-three matrix is referred to as the *reduced compliance matrix*, although the expressions for the elements of the reduced compliance matrix in terms of the engineering properties are identical to those in the full six-by-six compliance matrix. In addition, it is important to remember that while the normal stress σ_3 is equal to zero for the plane-stress condition, the corresponding normal strain ε_3 is non-zero and is given by

$$\varepsilon_3 = -\frac{\nu_{13}}{E_1}\sigma_1 - \frac{\nu_{23}}{E_2}\sigma_2 + \alpha_3 \Delta T + \beta_3 \Delta M \tag{3.6.8}$$

Thus, in order to analyze the in-plane stress response of a layer of composite material to mechanical, thermal, and moisture loads, Eq. (3.6.7), it is necessary to know four engineering properties, plus the in-plane coefficients of thermal and moisture expansion. To use Eq. (3.6.8), four additional material properties are needed. To determine these many needed properties, the approach taken is to make use models based primarily on the mechanics of materials, and use other models when the mechanics of materials models do not yield results of sufficient accuracy, or when more accurate models result in expressions nearly as simple as the mechanics of materials results.

The basic starting point for the mechanics of materials model is the representative volume element shown in Fig. 3.6.4. This volume element consists of enough fiber and matrix so the volume fraction of fiber within the volume is the same as the average value in the composite. A model for the fiber-direction extensional modulus, E_1, can be developed by assuming that when a load is applied in the 1-direction, the strains in the fiber and matrix are the same. This *iso-strain* assumption leads to the result that

$$E_1 = V_f E_1^f + (1 - V_f) E_1^m \tag{3.6.9}$$

where V_f is the fiber volume fraction, E_1^f is the extensional modulus of the fiber along its length (the 1-direction), and E_1^m is the matrix extensional modulus in the 1-direction. Equation (9) is often referred to as the *rule-of-mixtures* (ROM) model for E_1 because the result is simply the fractional amount of fiber multiplied by the fiber extensional modulus added to the fractional amount of matrix multiplied by its extensional modulus. Likewise, if an iso-strain model is used to estimate the Poisson's ratio of the composite material, then the resulting expression is

$$\nu_{12} = V_f \nu_{12}^f + (1 - V_f) \nu_{12}^m \tag{3.6.10}$$

where ν_{12}^f is the fiber Poisson's ratio, and ν_{12}^m is the Poisson's ratio of the matrix. Despite representing a round fiber with a square cross section, numerical results to be presented later in this section will justify recommending the use of rule-of-mixtures estimates for E_1 and ν_{12}.

An inspection of the representative volume shown in Fig. 3.6.4 reveals that the iso-strain assumption is not appropriate for estimating the extensional modulus E_2. Rather, a first estimate would be that the stress σ_2 is the same in the fiber and the matrix, and is uniform, when a transverse load is applied. This *iso-stress assumption* results in an estimate for E_2 of the composite given by

$$\frac{1}{E_2} = \frac{V_f}{E_2^f} + \frac{(1 - V_f)}{E_2^m} \tag{3.6.11}$$

where E_2^f and E_2^m are the fiber and matrix transverse extensional modulus values, respectively. While this result based on the iso-stress assumption is often presented in introductory composite materials textbooks, it produces relatively inaccurate results, as will be shown later. An improved estimate may be developed by considering the representative volume element shown in Fig. 3.6.5. The analysis of this unit element is conducted by assuming that the central fiber/matrix element has an effective extensional modulus that may be calculated using the rule-of-mixtures. From there, an analysis of the deformation of the total element leads to the result that

$$\frac{1}{E_2} = \frac{1 - \sqrt{V_f}}{E_2^m} + \frac{\sqrt{V_f}}{E_2^f \sqrt{V_f} + E_2^m \left(1 - \sqrt{V_f}\right)}. \tag{3.6.12}$$

This *improved mechanics of materials model* yields results that in much better agreement with exact elasticity analyses than those given in Eq. (11), and is the recommended model for calculating E_2.

For the case of shear loading, the simple mechanics of materials analysis with an iso-stress assumption gives the result that

$$\frac{1}{G_{12}} = \frac{V_f}{G_{12}^f} + \frac{(1 - V_f)}{G_{12}^m} \tag{3.6.13}$$

As with the composite extensional modulus in the 2-direction, often referred to as the transverse extensional modulus, this iso-stress model gives poor estimates of the composite shear modulus G_{12}. It is possible to develop an analogous expression to Eq. (12) for the shear modulus. However, there is an *approximate* elasticity solution for a simplified geometry (concentric cylinders of fiber and matrix) available in

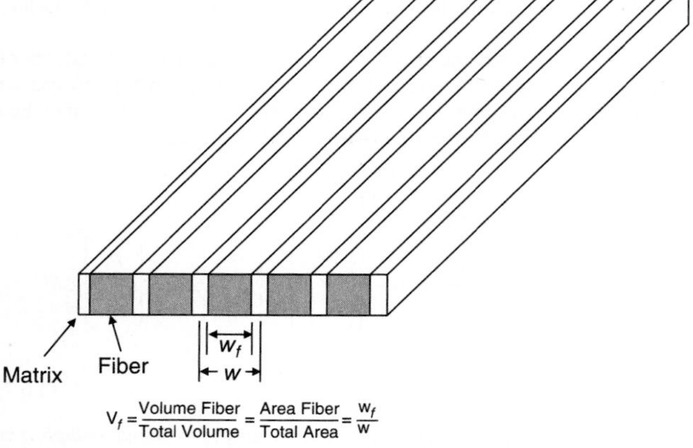

Figure 3.6.4 Representative volume element for micromechanics analysis.

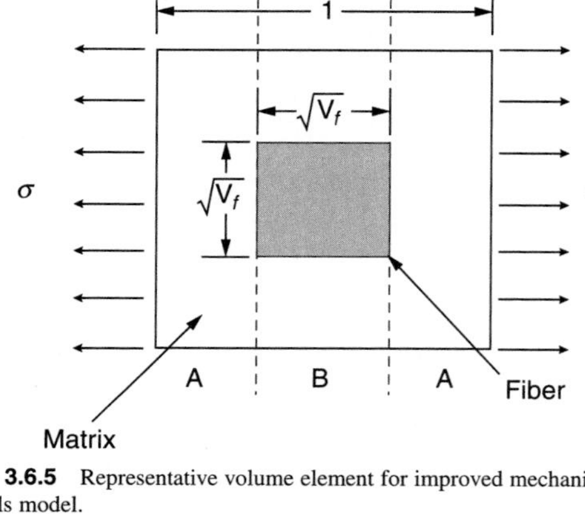

Figure 3.6.5 Representative volume element for improved mechanics of materials model.

closed-form. The resulting expression for the composite shear modulus is that

$$G_{12} = G_{12}^m \left\{ \frac{\left(G_{12}^m + G_{12}^f\right) - V_f\left(G_{12}^m - G_{12}^f\right)}{\left(G_{12}^m + G_{12}^f\right) + V_f\left(G_{12}^m - G_{12}^f\right)} \right\} \tag{3.6.14}$$

This model gives results that are extremely accurate in comparison to exact results for hexagonally- and square-packed arrays of fibers.

In addition to these models that describe the response of a layer of composite materials to mechanical loading conditions, it is also often useful to have micromechanics models for the expansional coefficients [such as the thermal expansion coefficients and moisture expansion coefficients that appear in Eq. (7)]. Using the mechanics of materials representative volume element of Fig. 3.6.4, it is possible to develop expressions for the coefficient of thermal expansion in the 1-direction, α_1, as

$$\alpha_1 = \frac{\alpha_1^f E_1^f V_f + \alpha_1^m E_1^m \left(1 - V_f\right)}{E_1} \tag{3.6.15}$$

where the denominator is the rule of mixtures expression for the fiber direction extensional modulus E_1 given by Eq. (3.6.9). The expression for the coefficient of moisture expansion, β_1, is developed by analogy. The coefficient of thermal expansion in the 2-direction, α_2, is given by

$$\begin{aligned} \alpha_2 = &\left[\alpha_2^f - \left(\frac{E_1^m}{E_1}\right) \nu_{12}^f (\alpha_1^m - \alpha_1^f)(1 - V_f)\right] V_f \\ &+ \left[\alpha_2^m + \left(\frac{E_1^f}{E_1}\right) \nu_{12}^m (\alpha_1^m - \alpha_1^f) V_f\right] (1 - V_f) \end{aligned} \tag{3.6.16}$$

Once again the coefficient of moisture expansion in the 2-direction may be developed by analogy.

In addition to the mechanics of materials, improved mechanics of materials, and elasticity models described above, semi-empirical models have been developed for calculating engineering properties of a layer based upon fiber and matrix properties. The most widely used are the Halpin-Tsai equations

$$\frac{M}{M_m} = \frac{1 + \eta \xi V_f}{1 - \eta V_f} \tag{3.6.17}$$

in which

$$\eta = \frac{\left(M_f / M_m\right) - 1}{\left(M_f / M_m\right) + \xi} \tag{3.6.18}$$

where

M = composite property E_2 or G_{12}

M_f = corresponding fiber property E_2^f or G_{12}^f

M_m = corresponding matrix property E_1^m, G_{12}^m, or ν_{12}^m

and ξ is a parameter having admissible values $0 \le \xi < \infty$, which may be different for E_2 and G_{12}. A simplified form of Eqs. (17) and (18), obtained by substitution of Eq. (18) into Eq. (17), is given by

$$M = M_m \frac{\left(M_f + \xi M_m\right) - \xi V_f \left(M_m - M_f\right)}{\left(M_f + \xi M_m\right) + V_f \left(M_m - M_f\right)} \tag{3.6.19}$$

A value of ξ = 1 gives excellent results for G_{12}—an identical result to that given by Eq. (14). A value off ξ = 2 has been found to give good results for E_2. If the matrix is isotropic—as is nearly always the case—then $E_1^m = E_2^m = E^m$, $\nu_{12}^m = \nu_{21}^m = \nu^m$, $G_{12}^m = G^m = E^m/2(1+\nu^m)$, $\alpha_1^m = \alpha_2^m = \alpha^m$.

An assessment of the accuracy of each of the models presented in this section may be obtained by comparing the results from these simple models to results from exact elasticity solutions. Such a comparison requires properties for the fiber and matrix materials. Representative properties for glass fibers, graphite fibers, and polymer matrix materials are taken as the values given in Table 3.6.1. Tables 3.6.2 and 3.6.3 provide a comparison of composite properties as a function of fiber volume fraction using the various relations discussed.

EXAMPLE A: Evaluate the engineering and thermal expansion properties of a layer of glass-polymer composite material having a fiber volume fraction of 60%.

APPROACH: Make use of the properties given in Table 3.6.1 along with the micromechanics models described above.

SOLUTION: The fiber properties are obtained from Table 3.6.1 as

$$E_1^f = E_2^f = 73.1 \text{ GPa}$$

$$G_{12}^f = 30.0 \text{ GPa}$$

$$\nu_{12}^f = 0.22$$

$$\alpha_1^f = \alpha_2^f = 5.04x10^{-6}/°C$$

The matrix properties are obtained from Table 3.6.1 as

$$E_1^m = E_2^m = 4.62 \text{ GPa}$$

$$G_{12}^m = 1.699 \text{ GPa}$$

$$\nu_{12}^m = 0.36$$

$$\alpha_1^m = \alpha_2^m = 41.4x10^{-6}/°C$$

The fiber volume fraction is given as $V_f = 0.60$.

For E_1, using Eq. (9):

$$\begin{aligned} E_1 &= V_f E_1^f + \left(1 - V_f\right) E_1^m \\ &= (0.60)(73.1) + (1 - 0.60)(4.62) \\ &= 45.7 \text{ GPa} \end{aligned}$$

This value compares favorably with the exact value of 45.7 GPa found using an elasticity solution (see Table 3.6.3).

For ν_{12}, using Eq. (10):

$$\begin{aligned} \nu_{12} &= V_f \nu_{12}^f + \left(1 - V_f\right) \nu_{12}^m \\ &= (0.60)(0.22) + (1 - 0.60)(0.36) \\ &= 0.276 \end{aligned}$$

This value compares favorably with the exact value of 0.269 found using an elasticity solution (see Table 3.6.3).

Table 3.6.1 Representative Properties for Graphite Fiber, Glass Fiber, and Polymeric Matrix Materials

	Graphite Fiber	Glass Fiber*	Polymer Matrix*
E_1 (GPa)	233	73.1	4.62
E_2 (GPa)	23.1	73.1	4.62
G_{12} (GPa)	8.96	30.0	1.699
ν_{12}	0.200	0.22	0.36
α_1 (/°C)	-0.540×10^{-6}	5.04	41.4×10^{-6}
α_2 (/°C)	10.10×10^{-6}	5.04	41.4×10^{-6}

* Assumed to be isotropic.

Table 3.6.2 Micromechanics Model Results for Graphite-polymer Composite Material

	Fiber Volume Fraction							
	0.0	0.1	0.2	0.3	0.4	0.5	0.6	0.7
E_1 (GPa), Eq. (9)	4.62	27.5	50.3	73.1	96.0	118.8	141.6	164.5
E_1 (GPa), Elasticity	4.62	27.5	50.3	73.2	96.0	118.9	141.7	164.5
ν_{12}, Eq. (10)	0.360	0.344	0.328	0.312	0.296	0.280	0.264	0.248
ν_{12}, Elasticity	0.360	0.343	0.327	0.311	0.297	0.282	0.269	0.256
E_2 (GPa), Eq. (11)	4.62	5.02	5.50	6.08	6.79	7.70	8.88	10.50
E_2 (GPa), Eq. (12)	4.62	5.61	6.48	7.40	8.45	9.67	11.15	12.98
E_2 (GPa), Eq. (19)	4.62	5.46	6.41	7.49	8.73	10.16	11.85	13.86
E_2 (GPa), Elasticity	4.62	5.78	6.54	7.40	8.43	9.70	11.29	13.27
G_{12} (GPa), Eq. (13)	1.699	1.848	2.03	2.24	2.51	2.86	3.31	3.93
G_{12} (GPa), Eq. (14)	1.699	1.947	2.23	2.57	2.97	3.45	4.05	4.80
G_{12} (GPa), Elasticity	1.699	1.947	2.23	2.57	2.97	3.46	4.05	4.82
α_1 (/°C) $\times 10^6$, Eq. (15)	41.4	5.81	2.54	1.315	0.671	0.275	7.17×10^{-3}	−0.1866
α_1 (/°C) $\times 10^6$, Elasticity	41.4	5.96	2.68	1.440	0.779	0.365	7.87×10^{-2}	−0.1323
α_2 (/°C) $\times 10^6$, Eq. (16)	41.4	49.7	46.2	42.0	37.6	33.1	28.5	23.9
α_2 (/°C) $\times 10^6$, Elasticity	41.4	48.9	44.9	40.3	35.7	31.2	26.7	22.4

Table 3.6.3 Micromechanics Model Results for Glass-polymer Composite Material

	Fiber Volume Fraction							
	0.0	0.1	0.2	0.3	0.4	0.5	0.6	0.7
E_1 (GPa), Eq. (9)	4.62	11.47	18.32	25.2	32.0	38.9	45.7	52.6
E_1 (GPa), Elasticity	4.62	11.48	18.33	25.2	32.0	38.9	45.7	52.6
ν_{12}, Eq. (10)	0.360	0.346	0.332	0.318	0.304	0.290	0.276	0.262
ν_{12}, Elasticity	0.360	0.343	0.327	0.311	0.297	0.282	0.269	0.256
E_2 (GPa), Eq. (11)	4.62	5.10	5.69	6.43	7.39	8.69	10.55	13.42
E_2 (GPa), Eq. (12)	4.62	6.25	7.56	9.02	10.78	13.03	16.07	20.5
E_2 (GPa), Eq. (19)	4.62	5.88	7.39	9.23	11.53	14.49	18.42	23.9
E_2 (GPa), Elasticity	4.62	5.77	6.84	8.14	9.82	12.16	15.56	20.7
G_{12} (GPa), Eq. (13)	1.699	1.875	2.09	2.37	2.73	3.21	3.91	5.00
G_{12} (GPa), Eq. (14)	1.699	2.03	2.44	2.94	3.59	4.44	5.62	7.36
G_{12} (GPa), Elasticity	1.699	2.03	2.44	2.94	3.58	4.44	5.64	7.47
α_1 (/°C) $\times 10^6$, Eq. (15)	41.4	18.22	12.38	9.71	8.19	7.20	6.51	6.00
α_1 (/°C) $\times 10^6$, Elasticity	41.4	18.56	12.75	10.06	8.50	7.46	6.72	6.16
α_2 (/°C) $\times 10^6$, Eq. (16)	41.4	45.0	42.2	38.2	33.8	29.1	24.4	19.62
α_2 (/°C) $\times 10^6$, Elasticity	41.4	43.9	40.3	35.8	31.1	26.4	21.8	17.38

For E_2, using Eq. (11):

$$\frac{1}{E_2} = \frac{V_f}{E_2^f} + \frac{(1-V_f)}{E_2^m}$$
$$= \frac{0.60}{73.1} + \frac{(1-0.60)}{4.62}$$
$$= 0.09478 \text{ GPa}^{-1}$$
$$E_2 = \frac{1}{0.09478 \text{ GPa}^{-1}} = 10.55 \text{ GPa}$$

This value is in relatively poor agreement with the exact value of 15.56 GPa found using the elasticity solution. For the improved mechanics of materials model, using Eq. (12):

$$\frac{1}{E_2} = \frac{1-\sqrt{V_f}}{E_2^m} + \frac{\sqrt{V_f}}{E_2^f\sqrt{V_f} + E_2^m\left(1-\sqrt{V_f}\right)}$$
$$= \frac{1-\sqrt{0.60}}{4.62} + \frac{\sqrt{0.60}}{(73.1)\sqrt{0.60} + (4.62)\left(1-\sqrt{0.60}\right)}$$
$$= 0.0622 \text{ GPa}^{-1}$$
$$E_2 = 16.07 \text{ GPa}$$

This model gives better agreement with the exact value for this case (and others as well) and is the recommended model for E_2. The semi-empirical Halpin-Tsai expression, Eq. (19), with $\xi = 2$ results in:

$$E_2 = E_2^m \frac{(E_2^f + \xi E_2^m) - \xi V_f(E_2^m - E_2^f)}{(E_2^f + \xi E_2^m) + V_f(E_2^m - E_2^f)}$$
$$= (4.62)\frac{(73.1 + (2)(4.62)) - (2)(0.60)(4.62 - 73.1)}{(73.1 + (2)(4.62)) + (0.60)(4.62 - 73.1)}$$
$$= 18.42 \text{ GPa}$$

For G_{12}, using Eq. (13):

$$\frac{1}{G_{12}} = \frac{V_f}{G_{12}^f} + \frac{(1-V_f)}{G_{12}^m}$$
$$= \frac{0.60}{30.0} + \frac{(1-0.60)}{1.699}$$
$$= 0.255 \text{ GPa}^{-1}$$
$$G_{12} = 3.91 \text{ GPa}$$

This value compares poorly with the exact value of 5.64 GPa obtained from the elasticity solution. The concentric cylinders approximate elasticity model, Eq. (14), gives:

$$G_{12} = G_{12}^m \left\{ \frac{(G_{12}^m + G_{12}^f) - V_f(G_{12}^m - G_{12}^f)}{(G_{12}^m + G_{12}^f) + V_f(G_{12}^m - G_{12}^f)} \right\}$$
$$= 1.699 \left\{ \frac{(1.699 + 30.0) - (0.60)(1.699 - 30.0)}{(1.699 + 30.0) + (0.60)(1.699 - 30.0)} \right\}$$
$$= 5.62 \text{ GPa}$$

This model gives good agreement with that of the exact solution, and is the recommended model for G_{12}.

For α_1, using Eq. (15):

$$\alpha_1 = \frac{\alpha_1^f E_1^f V_f + \alpha_1^m E_1^m (1-V_f)}{E_1^f V_f + E_1^m (1-V_f)}$$
$$= \frac{(5.04\times10^{-6})(73.1)(0.60) + (41.4\times10^{-6})(4.62)(1-0.60)}{(73.1)(0.60) + (4.62)(1-0.60)}$$
$$= 6.51\times10^{-6}/°\text{C}$$

This value compares favorably with the exact result of 6.72 × 10^{-6}/°C.

For α_2, using Eq. (16):

$$\alpha_2 = \left[\alpha_2^f - \left(\frac{E_1^m}{E_1}\right)\nu_{12}^f(\alpha_1^m - \alpha_1^f)(1-V_f)\right]V_f$$
$$+ \left[\alpha_2^m + \left(\frac{E_1^f}{E_1}\right)\nu_{12}^m(\alpha_1^m - \alpha_1^f)V_f\right](1-V_f)$$
$$= \left[5.04\times10^{-6} - \left(\frac{4.62}{45.7}\right)(0.22)(41.4\times10^{-6} - 5.04\times10^{-6})(1-0.60)\right](0.60)$$
$$+ \left[41.4\times10^{-6} + \left(\frac{73.1}{45.7}\right)(0.36)(41.4\times10^{-6} - 5.04\times10^{-6})(0.60)\right](1-0.60)$$
$$= 24.4\times10^{-6}/°\text{C}$$

This value compares favorably with the exact result of 21.8 × 10^{-6}/°C.

NOTE: It is important to note that in the above examples and all to follow it will be assumed that the fiber and matrix properties are quantified exactly by using three significant figures, as in Table 3.6.1. All subsequent numerical calculations are carried out starting with these properties and these subsequent numerical calculations will be reported rounded to three significant figures. It is also important to note that in all numerical examples to follow, a glass-polymer composite material with 60% fiber volume fraction is used. Because of their simplicity and accuracy, the engineering properties are computed by employing Eqs. (9), (10), (12), (14), (15), and (16).

Additional example results may be constructed by examining different fiber volume fractions or changing the fiber type, and comparing the results with the values given in Table 3.6.2 or Table 3.6.3, as appropriate for the fiber type chosen. For the graphite fibers, care must be taken to use of the appropriate extensional moduli and expansion coefficients, as these fibers are orthotropic.

Though it has not been necessary to this point to describe the thickness of a layer, in the subsequent numerical examples, a layer thickness, h, of 0.000125 m, or 0.125 mm, will be used.

DESCRIBING A LAMINATE

With details of the properties of a layer considered, attention turns to describing the behavior of a number of layers, bonded together, one next to the other, to form a laminate. An important part of the description of a laminate is the specification of its *stacking sequence*. The stacking sequence is a list that defines the through-thickness locations and fiber orientations of all the layers in a laminate. Referring to Fig. 3.6.6, which shows the exploded view of a laminate and the *laminate coordinate system*, the x-y-z coordinate system, the stacking sequence lists the layer fiber orientations relative to the +x-axis of the laminate coordinate system, starting with the layer at the negative-most z-position. The stacking sequence [±30/0]$_S$ describes the six-layer laminate shown in Fig. 3.6.6 which has +30° layers on the outside, −30° layers inside of those, and two 0° layers in the center. The stacking sequence could have been written [+30/−30/0/0/−30/+30]$_T$, where "T" signifies that the total laminate has been described, but the shorthand notation with the subscript "S" is more convenient and implies that the fiber orientation arrangement is symmetric with respect to the x–y plane, i.e., the z = 0 location. The formal definition of a symmetric

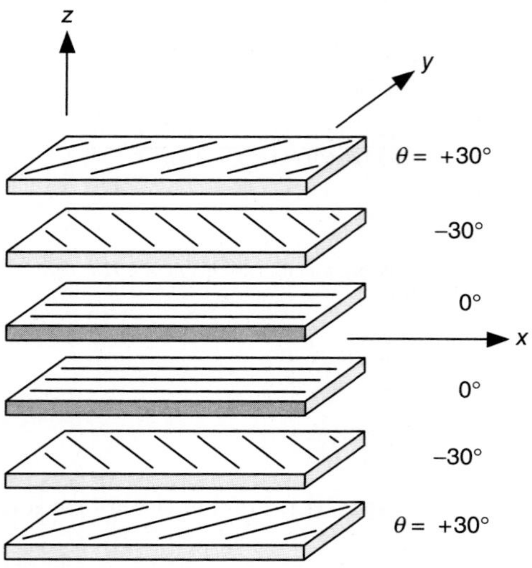

Figure 3.6.6 Exploded view of a $[\pm30/0]_S$ laminate and laminate coordinate system.

laminate will be given shortly. Other shorthand notation can be used, such as $[(\pm30/0)_3/(90/0)_2]_S$ to describe a 26-layer symmetric laminate which has two sub-sequences that repeat, one three times, the other twice. The notation $[\pm30]_{2S}$ is interpreted to mean $[(\pm30)_2]_S$. Of course it is necessary to specify the material properties and thickness of each layer to completely describe a laminate.

STRESS-STRAIN BEHAVIOR OF A LAYER: PRINCIPAL MATERIAL COORDINATE SYSTEM

Within the context of the macromechanical theory presented here to study laminate deformations and stresses, referred to as *classical lamination theory*, or *CLT*, a layer is assumed to be in a state of plane stress. Such a condition was previously described. Accordingly, from Eq. (7), the stress-strain behavior of a layer written in terms of the stresses and strains in the principal material, or 1-2-3, coordinate system, including the effects of thermal expansion, can be written as

$$\begin{Bmatrix} \sigma_1 \\ \sigma_2 \\ \tau_{12} \end{Bmatrix} = \begin{pmatrix} Q_{11} & Q_{12} & 0 \\ Q_{12} & Q_{22} & 0 \\ 0 & 0 & Q_{66} \end{pmatrix} \begin{Bmatrix} \varepsilon_1 - \alpha_1 \Delta T \\ \varepsilon_2 - \alpha_2 \Delta T \\ \gamma_{12} \end{Bmatrix} \qquad (3.6.20)$$

where

$$Q_{11} = \frac{E_1}{1-\nu_{12}\nu_{21}} \quad Q_{12} = \frac{\nu_{12}E_2}{1-\nu_{12}\nu_{21}} \quad Q_{22} = \frac{E_2}{1-\nu_{12}\nu_{21}} \quad Q_{66} = G_{12} \qquad (3.6.21)$$

The Q matrix is the *reduced stiffness matrix*. The terminology *reduced* is used to distinguish the three-by-three stiffness matrix of Eq. (20) from the full six-by-six stiffness matrix of Eq. (4) that must be used when all six components of stress and strain are considered. It is important to recognize that while the numerical values in the reduced compliance matrix are identical to those in the full six-by-six compliance matrix, the numerical values in the reduced stiffness matrix are different than those in the full six-by-six stiffness matrix. Again, the form of the reduced stiffness matrix in Eq. (20) is said to represent orthotropic material behavior. For simplicity, moisture content effects have not been included in Eq. (20). They can be included in the same manner as thermal expansion effects.

EXAMPLE B: Compute the numerical values of the reduced stiffness matrix for a glass-polymer composite.

APPROACH: Substitute the engineering properties directly into Eq. (21).

(As noted, all examples are based on using a 60% volume fraction of glass fibers, and because of their simplicity and accuracy, employing Eqs. (9), (10), (12), (14), (15), and (16) to calculate the material properties, as presented in Table 3.6.3.)

SOLUTION:

$$Q_{11} = \frac{E_1}{1-\nu_{12}\nu_{21}} = \frac{45.7\times10^9}{1-(0.276)(0.0970)} = 47.0\times10^9 = 47.0 \text{ GPa}$$

$$Q_{12} = \frac{\nu_{12}E_2}{1-\nu_{12}\nu_{21}} = \frac{(0.276)(16.07\times10^9)}{1-(0.276)(0.0970)} = 4.56 \text{ GPa}$$

$$Q_{22} = \frac{E_2}{1-\nu_{12}\nu_{21}} = \frac{16.07\times10^9}{1-(0.276)(0.0970)} = 16.51 \text{ GPa}$$

$$[Q] = \begin{pmatrix} Q_{11} & Q_{12} & 0 \\ Q_{12} & Q_{22} & 0 \\ 0 & 0 & Q_{66} \end{pmatrix} = \begin{pmatrix} 47.0 & 4.56 & 0 \\ 4.56 & 16.51 & 0 \\ 0 & 0 & 5.62 \end{pmatrix} \times 10^9$$

$$= \begin{pmatrix} 47.0 & 4.56 & 0 \\ 4.56 & 16.51 & 0 \\ 0 & 0 & 5.62 \end{pmatrix} \text{GPa}$$

The reciprocity relation of Eq. (5) was used to compute ν_{21}. We emphasize that in nearly all cases $\nu_{12} \neq \nu_{21}$. Of course, for an isotropic material $\nu_{12} = \nu_{21} = \nu$.

STRESS-STRAIN BEHAVIOR OF A LAYER: LAMINATE COORDINATE SYSTEM

Since there can be many fiber orientations within a laminate, each orientation having its own principal material coordinate system, it is more convenient to describe laminate behavior in terms of one coordinate system, namely the laminate, or *x-y-z*, coordinate system shown in Fig. 3.6.6. The stresses and strains in the laminate coordinate system, including thermal strains, can be expressed in terms of those in the principal material coordinate system by *transformation relations*, namely,

$$\begin{Bmatrix} \sigma_x \\ \sigma_y \\ \tau_{xy} \end{Bmatrix} = [T]^{-1} \begin{Bmatrix} \sigma_1 \\ \sigma_2 \\ \tau_{12} \end{Bmatrix} \quad \begin{Bmatrix} \varepsilon_x \\ \varepsilon_y \\ \frac{1}{2}\gamma_{xy} \end{Bmatrix} = [T]^{-1} \begin{Bmatrix} \varepsilon_1 \\ \varepsilon_2 \\ \frac{1}{2}\gamma_{12} \end{Bmatrix}$$
$$\begin{Bmatrix} \alpha_x \\ \alpha_y \\ \frac{1}{2}\alpha_{xy} \end{Bmatrix} \Delta T = [T]^{-1} \begin{Bmatrix} \alpha_1 \\ \alpha_2 \\ 0 \end{Bmatrix} \Delta T \qquad (3.6.22)$$

where

$$[T]^{-1} = \begin{bmatrix} T_{11} & T_{12} & T_{13} \\ T_{21} & T_{22} & T_{23} \\ T_{31} & T_{32} & T_{33} \end{bmatrix}^{-1} = \begin{bmatrix} m^2 & n^2 & -2mn \\ n^2 & m^2 & 2mn \\ mn & -mn & m^2 - n^2 \end{bmatrix} \qquad (3.6.23)$$

and where $m = \cos\theta$ and $n = \sin\theta$, θ being the angle measured from the + *x*-axis to the + 1-axis of the layer. In the above equation, the three-by-three matrix, denoted as T^{-1}, is referred to as the *inverse transformation matrix*. The word *inverse* is used because the stresses and strains of importance within a laminate are those parallel and perpendicular to the fibers, namely those in the principal material coordinate system. These stresses and strains can be used to determine at what load level a layer will fail. The relationship in Eq. (22) transforms those important stresses and strains into stresses and strains in the *x-y-z* coordinate system. The latter quantities are neither parallel nor perpendicular to the fibers, so it is more difficult to evaluate whether or not the stresses and strains are high enough to cause, for example, failure of a layer due to excess stress in the fiber direction.

Note well in the above relations the factor of one-half associated with the shear strain.

The above relations can be inverted and rewritten as

$$\begin{Bmatrix} \sigma_1 \\ \sigma_2 \\ \tau_{12} \end{Bmatrix} = [T] \begin{Bmatrix} \sigma_x \\ \sigma_y \\ \tau_{xy} \end{Bmatrix} \quad \begin{Bmatrix} \varepsilon_1 \\ \varepsilon_2 \\ \frac{1}{2}\gamma_{12} \end{Bmatrix} = [T] \begin{Bmatrix} \varepsilon_x \\ \varepsilon_y \\ \frac{1}{2}\gamma_{xy} \end{Bmatrix} \quad \begin{Bmatrix} \alpha_1 \\ \alpha_2 \\ 0 \end{Bmatrix} \Delta T = [T] \begin{Bmatrix} \alpha_x \\ \alpha_y \\ \frac{1}{2}\alpha_{xy} \end{Bmatrix} \Delta T \tag{3.6.24}$$

where

$$[T] = \begin{bmatrix} T_{11} & T_{12} & T_{13} \\ T_{21} & T_{22} & T_{23} \\ T_{31} & T_{32} & T_{33} \end{bmatrix} = \begin{bmatrix} m^2 & n^2 & 2mn \\ n^2 & m^2 & -2mn \\ -mn & mn & m^2 - n^2 \end{bmatrix} \tag{3.6.25}$$

The above three-by-three matrix, denoted as T, is referred to as the *transformation matrix.*

As a result of the fibers being at an angle relative to the laminate coordinate system, stresses and strains measured in the laminate coordinate system are quite different than those measured in the principal material coordinate system. The following example demonstrates the interesting strain response of a single layer caused by a temperature change, but measured in the laminate coordinate system.

EXAMPLE C: The temperature of an element cut from a single layer of glass-polymer composite with its fibers at +30° is decreased by –100°C. What shear strain is measured in the laminate coordinate system due to this temperature decrease? Note this temperature decrease is similar in magnitude to the decrease from the curing temperature of the material to room temperature.

APPROACH: The thermally-induced strains in the laminate coordinate system can be computed from the known values of the thermally-induced strains in the principal material coordinate system from Table 3.6.3 and the transformation relations of Eqs. (22) and (23).

SOLUTION: Since $\theta = 30°$, $m = \sqrt{3}/2$, $n = 1/2$, and

$$\begin{Bmatrix} \alpha_x \\ \alpha_y \\ \frac{1}{2}\alpha_{xy} \end{Bmatrix} \Delta T = [T(30°)]^{-1} \begin{Bmatrix} \alpha_1 \\ \alpha_2 \\ 0 \end{Bmatrix} \Delta T = \begin{bmatrix} \frac{3}{4} & \frac{1}{4} & \frac{-\sqrt{3}}{2} \\ \frac{1}{4} & \frac{3}{4} & \frac{\sqrt{3}}{2} \\ \frac{\sqrt{3}}{4} & \frac{-\sqrt{3}}{4} & \frac{1}{2} \end{bmatrix} \begin{Bmatrix} 6.51 \times 10^{-6} \\ 24.4 \times 10^{-6} \\ 0 \end{Bmatrix} (-100)$$

or

$$\alpha_x \Delta T = -1{,}099 \times 10^{-6} \quad \alpha_y \Delta T = -1{,}994 \times 10^{-6} \quad \alpha_{xy} \Delta T = 1{,}551 \times 10^{-6}$$

DISCUSSION: What is important to note is that although there is no shear strain in the principal material coordinate system, i.e., $\alpha_{12}\,\Delta T = 0$, a shear strain as well as contraction strains in the x- and y-directions are measured in the laminate coordinate system. Since polymer matrix composites are cured at an elevated temperature and then cooled, when a layer is part of a laminate, the thermally-induced strains in a layer caused by cooling the laminate from its curing temperature can produce residual thermal stresses. If the element is square and 0.1 m by 0.1 m in dimensions before the temperature change, after cooling the length in the x- and y-directions would be less. In addition, the original right angles would be changed by 1,551 micro-radians, denoted as 1,551 μrad, or 0.0888 deg, as shown in exaggerated fashion in Fig. 3.6.7, with the right angles of two opposite corners increasing, and the right angles of the other two opposite corners decreasing. This may not seem like much of an angle change, but it can cause significant stresses if resisted. The strains in the x- and y-directions would be denoted as -1,099 micro-strain and -1994 micro-strain, denoted, respectively, as –1,099 $\mu\varepsilon$ and –1994 $\mu\varepsilon$. The deformed element would have dimensions 0.0999 m and 0.0998 m in the x- and y-directions, respectively.

As it is convenient to transform the stresses and strains from each of the principal material coordinate systems of the various layers to the single laminate coordinate system, it is convenient to transform the stress-strain relation of each layer to the laminate coordinate system.

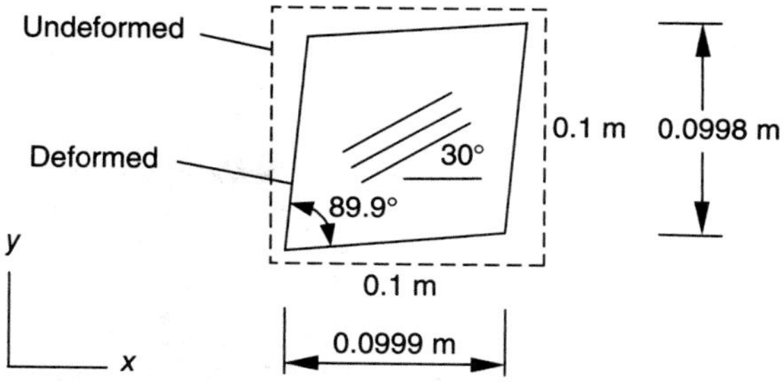

Figure 3.6.7 Cooling deformations of an element with fibers at oriented at +30°.

This can be accomplished by starting with the stress-strain relation of Eq. (20) and applying the transformation relations of Eq. (22). As a result, the stress-strain relation expressed in the laminate coordinate system is written as

$$\begin{Bmatrix} \sigma_x \\ \sigma_y \\ \tau_{xy} \end{Bmatrix} = \begin{pmatrix} \bar{Q}_{11} & \bar{Q}_{12} & \bar{Q}_{16} \\ \bar{Q}_{12} & \bar{Q}_{22} & \bar{Q}_{26} \\ \bar{Q}_{16} & \bar{Q}_{26} & \bar{Q}_{66} \end{pmatrix} \begin{Bmatrix} \varepsilon_x - \alpha_x \Delta T \\ \varepsilon_y - \alpha_x \Delta T \\ \gamma_{xy} - \alpha_{xy} \Delta T \end{Bmatrix} \tag{3.6.26}$$

where

$$\begin{aligned} \bar{Q}_{11} &= Q_{11}m^4 + 2(Q_{12} + 2Q_{66})m^2n^2 + Q_{22}n^4 \\ \bar{Q}_{12} &= (Q_{11} + Q_{22} - 4Q_{66})n^2m^2 + Q_{12}(n^4 + m^4) \\ \bar{Q}_{16} &= (Q_{11} - Q_{12} - 2Q_{66})nm^3 + (Q_{12} - Q_{22} + 2Q_{66})n^3m \\ \bar{Q}_{22} &= Q_{11}n^4 + 2(Q_{12} + 2Q_{66})m^2n^2 + Q_{22}m^4 \\ \bar{Q}_{26} &= (Q_{11} - Q_{12} - 2Q_{66})n^3m + (Q_{12} - Q_{22} + 2Q_{66})nm^3 \\ \bar{Q}_{66} &= (Q_{11} + Q_{22} - 2Q_{12} - 2Q_{66})n^2m^2 + Q_{66}(n^4 + m^4) \end{aligned} \tag{3.6.27}$$

As mentioned, $m = \cos\theta$ and $n = \sin\theta$, so the $\bar{Q}$ terms are functions of θ. The three-by-three $\bar{Q}$ matrix is referred to as the *transformed reduced stiffness matrix.* As can be observed, because of fiber orientation, compared to conventional isotropic materials or the orthotropic behavior of Eq. (20), unusual relations between stresses and strains occur. Specifically, the shear stress is related to the two normal strains, and the two normal stresses are related to the shear strain. There are no zeros in the $\bar{Q}$ matrix as there are with the Q matrix in Eq. (20). The behavior represented by Eq. (27) is referred to as *anisotropic* behavior. A word of caution: Note carefully the factors of one-half and two that occur in some relations but not in others. This is due to the use of engineering shear strains, γ_{12} and γ_{xy}, as opposed to tensor shear strains, which are given by $\frac{1}{2}\gamma_{12}$ and $\frac{1}{2}\gamma_{xy}$.

EXAMPLE D: A normal stress of 50 MPa is applied in the x-direction to a square element of glass-polymer composite with its fibers at 30° relative to the $+x$-axis while the temperature is kept constant, i.e., $\Delta T = 0$. No other stresses are present. What strains result from this applied stress?

APPROACH and SOLUTION: Substituting directly into the stress-strain relation expressed in the laminate coordinate system, Eq. (26), gives the answer, i.e.,

$$\begin{Bmatrix} \sigma_x \\ \sigma_y \\ \tau_{xy} \end{Bmatrix} = \begin{Bmatrix} 50 \times 10^6 \\ 0 \\ 0 \end{Bmatrix} = \begin{pmatrix} \bar{Q}_{11}(30°) & \bar{Q}_{12}(30°) & \bar{Q}_{16}(30°) \\ \bar{Q}_{12}(30°) & \bar{Q}_{22}(30°) & \bar{Q}_{26}(30°) \\ \bar{Q}_{16}(30°) & \bar{Q}_{26}(30°) & \bar{Q}_{66}(30°) \end{pmatrix} \begin{Bmatrix} \varepsilon_x \\ \varepsilon_y \\ \gamma_{xy} \end{Bmatrix}$$

where the $\bar{Q}$'s are computed using Eq. (27) with $\theta = 30°$, namely

$$\begin{pmatrix} \bar{Q}_{11}(30°) & \bar{Q}_{12}(30°) & \bar{Q}_{16}(30°) \\ \bar{Q}_{12}(30°) & \bar{Q}_{22}(30°) & \bar{Q}_{26}(30°) \\ \bar{Q}_{16}(30°) & \bar{Q}_{26}(30°) & \bar{Q}_{66}(30°) \end{pmatrix} = \begin{pmatrix} 33.4 & 10.54 & 10.05 \\ 10.54 & 18.15 & 3.14 \\ 10.05 & 3.14 & 11.60 \end{pmatrix} \times 10^9 \text{ N/m}^2$$

resulting in

$$\begin{pmatrix} 33.4 & 10.54 & 10.05 \\ 10.54 & 18.15 & 3.14 \\ 10.05 & 3.14 & 11.60 \end{pmatrix} \times 10^9 \begin{Bmatrix} \varepsilon_x \\ \varepsilon_y \\ \gamma_{xy} \end{Bmatrix} = \begin{Bmatrix} 50\times 10^6 \\ 0 \\ 0 \end{Bmatrix}$$

which leads to

$$\begin{Bmatrix} \varepsilon_x \\ \varepsilon_y \\ \gamma_{xy} \end{Bmatrix} = \begin{Bmatrix} 2,370 \\ -1,069 \\ -1,760 \end{Bmatrix} \times 10^{-6}$$

As shown in the Fig. 3.6.8, the element of material stretches in the x-direction and contracts in the y-direction, like an element of conventional material such as aluminum or steel would, the contraction being due to Poisson effects. Unlike an element of conventional material, however, the element experiences a shear strain. This results in corner angles that are not right angles when the square element deforms.

To be noted is the fact that there is distinct difference between stating that the element of material is stressed in the x-direction vs. stating that the element of material is strained in the x-direction. If the element is strained in the x-direction by, for example, 1,000 $\mu\varepsilon$, with the element experiencing no other strains, the stresses required are given by again using Eq. (26), namely,

$$\begin{Bmatrix} \sigma_x \\ \sigma_y \\ \tau_{xy} \end{Bmatrix} = \begin{pmatrix} \bar{Q}_{11}(30°) & \bar{Q}_{12}(30°) & \bar{Q}_{16}(30°) \\ \bar{Q}_{12}(30°) & \bar{Q}_{22}(30°) & \bar{Q}_{26}(30°) \\ \bar{Q}_{16}(30°) & \bar{Q}_{26}(30°) & \bar{Q}_{66}(30°) \end{pmatrix} \begin{Bmatrix} 1,000 \\ 0 \\ 0 \end{Bmatrix} \times 10^{-6}$$

or

$$\begin{Bmatrix} \sigma_x \\ \sigma_y \\ \tau_{xy} \end{Bmatrix} = \begin{Bmatrix} 33.4 \\ 10.54 \\ 10.05 \end{Bmatrix} \times 10^6 \text{ Pa} = \begin{Bmatrix} 33.4 \\ 10.54 \\ 10.05 \end{Bmatrix} \text{ MPa}$$

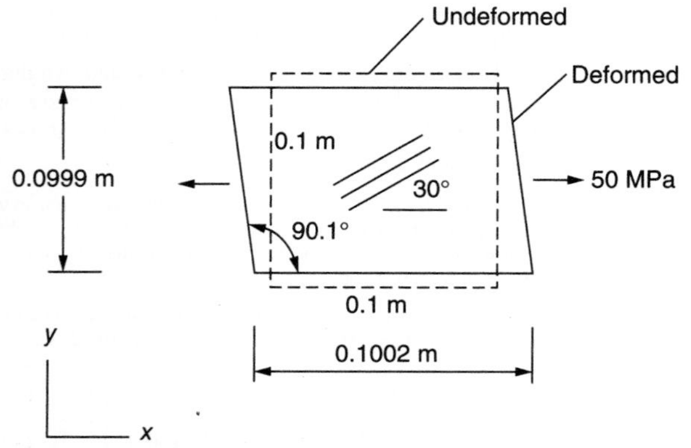

Figure 3.6.8 Deformations due to application of 50 MPa normal stress in x-direction.

It can be concluded that to produce this rather simple state of deformation, a tensile stress in the y-direction and a positive shear stress, in addition to a tensile stress in the x-direction, are required. These stresses are illustrated in Fig. 3.6.9. For a conventional material, although a tensile stress in the y-direction would be required, the shear stress would not be required because there is no tendency for the corner right angle to change. If the fiber angle was –30°, the same tensile stresses in the x- and y-directions would be required, but the shear stress required would be –10.05 MPa, a negative shear stress.

MACROMECHANICS OF A LAMINATE

When layers are combined to form a laminate, the unusual behavior in the various layers can interact to produce results that are even more unusual than the behavior of a single layer. To study the behavior of a laminate, it is assumed that within a laminate the layers are perfectly bonded together and do not slip relative to one another. Only small deformations are considered, much the same as when studying the mechanics of conventional materials. Like a layer, a laminate occupies a volume in three-dimensional space, so analysis and design of laminates is truly a three-dimensional problem. However, the plane-stress assumption eliminates some of the three-dimensionality of the problem, as does one other key assumption, namely the *Kirchhoff hypothesis*. As a direct result of the Kirchhoff hypothesis, the strains ε_x, ε_y, and γ_{xy} vary linearly through the thickness of the laminate. If the no-slip conditions were not true, the Kirchhoff hypothesis certainly would not be valid. The Kirchhoff hypothesis assumes that when the laminate deflects out of plane in the z-direction, all points through the thickness of the laminate at a given x- and y-location deflect out of plane the same amount. That is, the out-of-plane displacement of a point within a laminate is independent of z-location of the point. The assumed deformation characteristics of a laminate are further illustrated in Fig. 3.6.7, where the deformation is viewed in the x-z plane. The figure shows that straight line AA' through the thickness of the laminate before deformation remains straight after deformation, a result that leads to a linear variation of strains through the thickness of the laminate. Specifically, the displacements in the x-, y-, and z-directions of a generic point P at location x, y, and z within the laminate are assumed to have the following forms:

$$\begin{aligned} u(x,y,z) &= u^o(x,y) - z\frac{\partial w^o(x,y)}{\partial x} \\ v(x,y,z) &= v^o(x,y) - z\frac{\partial w^o(x,y)}{\partial y} \qquad (3.6.28) \\ w(x,y,z) &= w^o(x,y) \end{aligned}$$

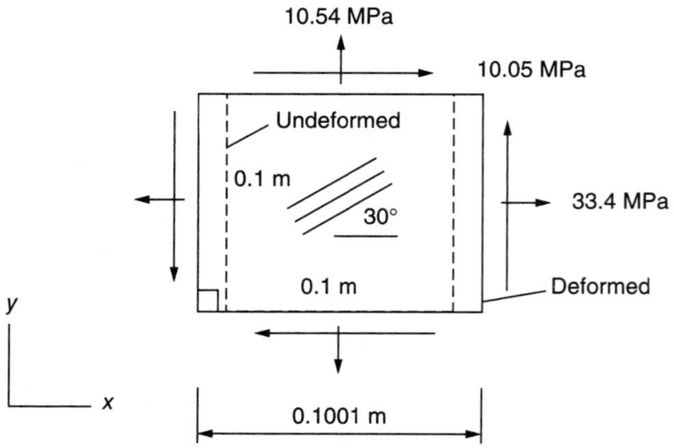

Figure 3.6.9 Stresses required to produce strain of 1,000 me in x-direction.

where the superscript "*o*" denotes the displacements of corresponding point P^o on the *geometric midplane* of the laminate, which is also referred to as the *reference surface*. Only small displacements and rotations are considered in the development of Eq. (28). As will be seen, if the deformation of the reference surface is known, i.e., if the strains and curvatures of the reference surface are known, then the strains, and subsequently the stresses, in every layer can be determined. The reference surface is somewhat like the neutral axis in simple beam analysis, except with composite materials the stresses are not necessarily zero at the reference surface location.

To be noted in Fig. 3.6.10 is the notation z_0, z_1, z_2, which describes the z-locations of the interfaces between layers, or alternatively, the z-locations of the lower and upper surface of the layers. Accordingly, $z_0 = -H/2$, H being the total laminate thickness, and the thickness of the k-th layer, h_k, is given by $z_k - z_{z-1}$.

The strains at any location within the laminate can be written in terms of the displacements as

$$
\begin{aligned}
\varepsilon_x &= \frac{\partial u}{\partial x} = \frac{\partial u^o}{\partial x} - z\frac{\partial^2 w^o}{\partial x^2} = \varepsilon_x^o + z\kappa_x^o \\
\varepsilon_y &= \frac{\partial v}{\partial y} = \frac{\partial v^o}{\partial y} - z\frac{\partial^2 w^o}{\partial y^2} = \varepsilon_y^o + z\kappa_y^o \\
\gamma_{xy} &= \frac{\partial u}{\partial y} + \frac{\partial v}{\partial x} = \frac{\partial u^o}{\partial y} + \frac{\partial v^o}{\partial x} - 2z\frac{\partial^2 w^o}{\partial x \partial y} = \gamma_{xy}^o + z\kappa_{xy}^o
\end{aligned}
\qquad (3.6.29)
$$

where

$$
\begin{Bmatrix} \varepsilon_x^o \\ \varepsilon_y^o \\ \gamma_{xy}^o \end{Bmatrix} = \begin{Bmatrix} \dfrac{\partial u^o}{\partial x} \\ \dfrac{\partial v^o}{\partial y} \\ \dfrac{\partial u^o}{\partial y} + \dfrac{\partial v^o}{\partial x} \end{Bmatrix} \qquad \begin{Bmatrix} \kappa_x^o \\ \kappa_y^o \\ \kappa_{xy}^o \end{Bmatrix} = \begin{Bmatrix} -\dfrac{\partial^2 w^o}{\partial x^2} \\ -\dfrac{\partial^2 w^o}{\partial y^2} \\ -2\dfrac{\partial^2 w^o}{\partial x \partial y} \end{Bmatrix} \qquad (3.6.30)
$$

The quantities ε_x^o, ε_y^o, and γ_{xy}^o are referred to as the *midplane* or *reference surface strains*. The quantities κ_x^o, κ_y^o, and κ_{xy}^o are the *midplane* or *reference surface curvatures* and are the bending curvatures the x- and y-directions, and the twist curvature, respectively, of the reference surface. By knowing the midplane strains and curvatures, the state of strain at any z-location within any layer can be computed though the

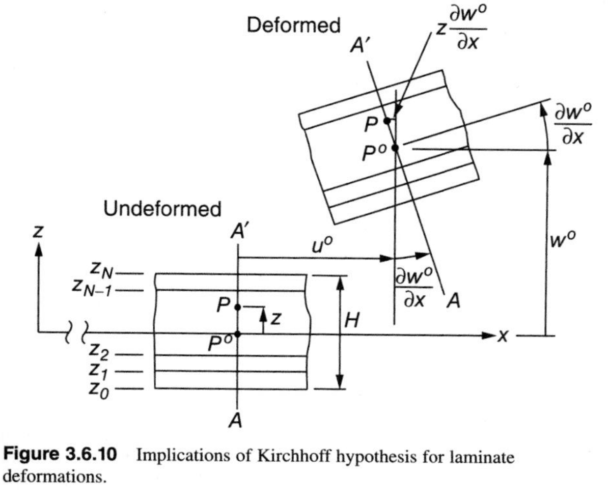

Figure 3.6.10 Implications of Kirchhoff hypothesis for laminate deformations.

use of Eq. (29). These strains, in turn, can be transformed to determine the strains in the principal material coordinate system of the layer at this z-location using the second equation of Eq. (24). The stresses in the principal material coordinate system at this z-location can then be computed using Eq. (20). As can be seen, then, knowing the midplane strains and curvatures is key to determining the stresses anywhere through the thickness the laminate. The plane-stress assumption and the Kirchhoff hypotheses are keys to the analysis simplifying to this degree. Without these assumptions, design and analysis of composite laminates would be intractable.

Since the midplane strains and curvatures are the keys to determining the stresses within a laminate, it is important to be able to compute these midplane quantities in terms of the loads applied to a laminate. In the application of composite materials, most often loads, in the form of pressures, forces, or bending moments, are specified rather than midplane strains and curvatures, or rather than the stresses. A relationship between the specified loads and the midplane strains and curvatures is therefore necessary. So far all that has been discussed are strains and stresses. Therefore, relationships between *loads* and the stresses or between *loads* and the midplane strains and curvatures need to be developed to be able to proceed with studying the behavior of laminates in practical applications. To develop these relations, the loads are defined in terms of the stresses as follows:

$$
\begin{aligned}
N_x &= \int_{-H/2}^{H/2} \sigma_x \, dz \quad N_y = \int_{-H/2}^{H/2} \sigma_y \, dz \quad N_{xy} = \int_{-H/2}^{H/2} \tau_{xy} \, dz \\
M_x &= \int_{-H/2}^{H/2} \sigma_x z \, dz \quad M_y = \int_{-H/2}^{H/2} \sigma_y z \, dz \quad M_{xy} = \int_{-H/2}^{H/2} \tau_{xy} z \, dz
\end{aligned}
\qquad (3.6.31)
$$

The Ns are the *force resultants* and the Ms are the *moment resultants*. The units of the force and moment resultants are force *per unit length* and moment *per unit length*, respectively. Collectively the six quantities are the *stress resultants*. The introduction of the stress resultants further reduces the three-dimensional nature of the problem. The *per unit length* part of the definition of the stress resultants is very important. The resultant N_x is the force in the x-direction per unit length of the laminate in the y-direction, N_y is force in the y-direction per unit length of laminate in the x-direction. Similar interpretations are associated with the moment resultants M_x and M_y. Since it is based on stress component τ_{xy}, the force resultant N_{xy} is the force in the x-direction per unit length of laminate in the y-direction. However, it is also the force in the y-direction per unit length of laminate in the x-direction, depending on which edge of the laminate is being discussed. A similar dual definition is associated with M_{xy}. It should be emphasized that it is impossible from equilibrium considerations to have N_{xy} and M_{xy} acting on one edge of an element of material and not have the same values of N_{xy} and M_{xy} acting on the adjacent edge. It is important to be aware of the positive sense of the stress resultants, which are illustrated in Fig. 3.6.11.

If the three components of stress from Eq. (26) are substituted into the integrands of the definitions of the six stress resultants, the integration with respect to z can be carried out. Considering N_x as an example, substitution of the first equation of stress-strain relation from Eq. (26) into the first equation of Eq. (31) results in

$$
\begin{aligned}
N_x = \int_{-H/2}^{H/2} \{ & \bar{Q}_{11}(\varepsilon_x^o + z\kappa_x^o - \alpha_x \Delta T) + \bar{Q}_{12}(\varepsilon_y^o + z\kappa_y^o - \alpha_y \Delta T) \\
& + \bar{Q}_{16}(\gamma_{xy}^o + z\kappa_{xy}^o - \alpha_{xy}\Delta T)\} \, dz
\end{aligned}
\qquad (3.6.32)
$$

Regrouping,

$$
\begin{aligned}
N_x = & \int_{-H/2}^{H/2} \{\bar{Q}_{11}(\varepsilon_x^o + z\kappa_x^o) + \bar{Q}_{12}(\varepsilon_y^o + z\kappa_y^o) + \bar{Q}_{16}(\gamma_{xy}^o + z\kappa_{xy}^o)\} \, dz \\
& - \int_{-H/2}^{H/2} (\bar{Q}_{11}\alpha_x + \bar{Q}_{12}\alpha_y + \bar{Q}_{16}\alpha_{xy}) \Delta T \, dz
\end{aligned}
\qquad (3.6.33)
$$

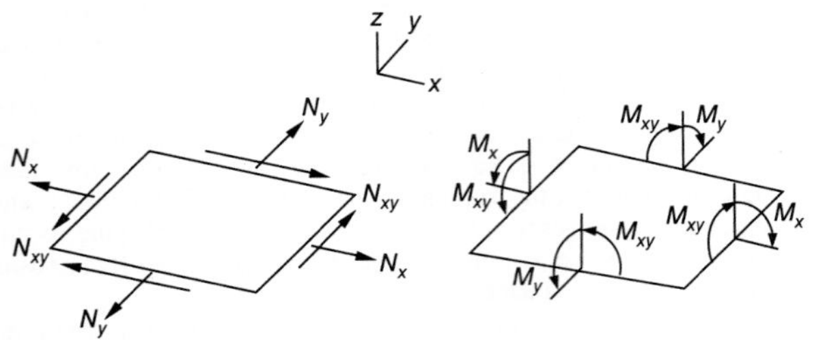

Figure 3.6.11 Illustration of positive sense of force and moment resultants.

The second integral is special. It is defined as the *effective thermal force resultant* in the x-direction, i.e.,

$$N_x^T = \int_{-H/2}^{H/2} \{\bar{Q}_{11}\alpha_x + \bar{Q}_{12}\alpha_y + \bar{Q}_{16}\alpha_{xy}\} \Delta T\, dz \qquad (3.6.34)$$

where the superscript T identifies that the resultant is due to thermal effects. If the temperature change is not a function of z, i.e., if the temperature change is uniform through the thickness of the laminate, then ΔT can be removed from the integral to yield

$$N_x^T = \left\{\int_{-H/2}^{H/2} \{\bar{Q}_{11}\alpha_x + \bar{Q}_{12}\alpha_y + \bar{Q}_{16}\alpha_{xy}\} dz\right\} \Delta T \qquad (3.6.35)$$

The integral within the brackets is called the *unit effective thermal force resultant*. It is only a function of material properties and the laminate geometry and is defined as

$$\hat{N}_x^T = \int_{-H/2}^{H/2} \{\bar{Q}_{11}\alpha_x + \bar{Q}_{12}\alpha_y + \bar{Q}_{16}\alpha_{xy}\} dz \qquad (3.6.36)$$

Note that in the definition of unit effective thermal force resultant the coefficients of thermal expansion combine with the reduced stiffnesses. This indicates that the magnitude of the thermal effects depends not only on the thermal expansion or contraction of the individual layers due to temperature changes, but also on how resistant the layers are to deformation, i.e., how stiff they are. Using the nomenclature of Eq. (36), Eq. (33) can be written as

$$N_x = \left\{\int_{-H/2}^{H/2} \bar{Q}_{11}\, dz\right\} \varepsilon_x^o + \left\{\int_{-H/2}^{H/2} \bar{Q}_{12}\, dz\right\} \varepsilon_y^o + \left\{\int_{-H/2}^{H/2} \bar{Q}_{16}\, dz\right\} \gamma_{xy}^o$$
$$+ \left\{\int_{-H/2}^{H/2} \bar{Q}_{11} z\, dz\right\} \kappa_x^o + \left\{\int_{-H/2}^{H/2} \bar{Q}_{12} z\, dz\right\} \kappa_y^o + \left\{\int_{-H/2}^{H/2} \bar{Q}_{16} z\, dz\right\} \kappa_{xy}^o - \hat{N}_x^T \Delta T \qquad (3.6.37)$$

where the reference surface strains and curvatures have been removed from the integral because they are not functions of z and are therefore not involved in the integration with respect to z. This is a very important point, and it is a direct result of the Kirchhoff hypothesis. As a final step, Eq. (37) is written as

$$N_x = A_{11}\varepsilon_x^o + A_{12}\varepsilon_y^o + A_{16}\gamma_{xy}^o + B_{11}\kappa_x^o + B_{12}\kappa_y^o + B_{16}\kappa_{xy}^o - \hat{N}_x^T \Delta T \qquad (3.6.38)$$

The As and Bs are defined as the integrals of the reduced stiffnesses through the thickness of the laminate, and hence represent *smeared, integrated,* or *overall stiffnesses* of the laminate. The reduced stiffnesses of every layer contribute to the smeared stiffnesses of the laminate. In the same spirit, the effective thermal force resultant represents smeared, integrated, or overall effects. Note that Eq. (38) relates the three reference surface strains and three reference surface curvatures to the force resultant and to thermal effects. This relationship and the other five to follow are quite important and absolutely necessary if the stresses within the various layers are to be computed from the loads. Before proceeding, it should be noted, as was stated at the onset, that smearing of properties is used for a second time with the definition of the As, Bs, and the effective thermal force resultant. The smearing of properties, be it at the micromechanics level or at the macromechanics, or laminate, level, is essential if progress is to be made in modeling and predicting behavior.

If the other five integrals in Eq. (31) are treated in a similar fashion and the results combined with Eq. (38), the following relationship results:

$$\begin{Bmatrix} N_x \\ N_y \\ N_{xy} \\ M_x \\ M_y \\ M_{xy} \end{Bmatrix} = \begin{bmatrix} A_{11} & A_{12} & A_{16} & B_{11} & B_{12} & B_{16} \\ A_{12} & A_{22} & A_{26} & B_{12} & B_{22} & B_{26} \\ A_{16} & A_{26} & A_{66} & B_{16} & B_{26} & B_{66} \\ B_{11} & B_{12} & B_{16} & D_{11} & D_{12} & D_{16} \\ B_{12} & B_{22} & B_{26} & D_{12} & D_{22} & D_{26} \\ B_{16} & B_{26} & B_{66} & D_{16} & D_{26} & D_{66} \end{bmatrix} \begin{Bmatrix} \varepsilon_x^o \\ \varepsilon_y^o \\ \gamma_{xy}^o \\ \kappa_x^o \\ \kappa_y^o \\ \kappa_{xy}^o \end{Bmatrix} - \begin{Bmatrix} \hat{N}_x^T \\ \hat{N}_y^T \\ \hat{N}_{xy}^T \\ \hat{M}_x^T \\ \hat{M}_y^T \\ \hat{M}_{xy}^T \end{Bmatrix} \Delta T \qquad (3.6.39)$$

The six-by-six matrix is referred to as the *laminate stiffness matrix* or, more simply, the *ABD matrix*. This matrix relates the stress resultants, or loads, to the deformations, or strains and curvatures, of the reference surface. Since the reduced stiffnesses generally do not vary through the thickness of any particular layer, the integrals that define the elements of the *ABD* matrix reduce to summations, namely,

$$A_{ij} = \sum_{k=1}^{N} \bar{Q}_{ij_k}(z_k - z_{k-1}) \quad B_{ij} = \tfrac{1}{2}\sum_{k=1}^{N} \bar{Q}_{ij_k}(z_k^2 - z_{k-1}^2) \quad D_{ij} = \tfrac{1}{3}\sum_{k=1}^{N} \bar{Q}_{ij_k}(z_k^3 - z_{k-1}^3) \qquad (3.6.40)$$

where the z_k are defined in Fig. 3.6.10 and N is the number of layers in the laminate. Since the coefficients of thermal expansion also generally do not vary through the thickness of any particular layer, the unit effective thermal stress resultants become summations given by

$$\begin{aligned}
\hat{N}_x^T &= \sum_{k=1}^{N} (\bar{Q}_{11_k}\alpha_{x_k} + \bar{Q}_{12_k}\alpha_{y_k} + \bar{Q}_{16_k}\alpha_{xy_k})(z_k - z_{k-1}) \\
\hat{N}_y^T &= \sum_{k=1}^{N} (\bar{Q}_{12_k}\alpha_{x_k} + \bar{Q}_{22_k}\alpha_{y_k} + \bar{Q}_{26_k}\alpha_{xy_k})(z_k - z_{k-1}) \\
\hat{N}_{xy}^T &= \sum_{k=1}^{N} (\bar{Q}_{16_k}\alpha_{x_k} + \bar{Q}_{26_k}\alpha_{y_k} + \bar{Q}_{66_k}\alpha_{xy_k})(z_k - z_{k-1}) \\
\hat{M}_x^T &= \tfrac{1}{2}\sum_{k=1}^{N} (\bar{Q}_{11_k}\alpha_{x_k} + \bar{Q}_{12_k}\alpha_{y_k} + \bar{Q}_{16_k}\alpha_{xy_k})(z_k^2 - z_{k-1}^2) \\
\hat{M}_y^T &= \tfrac{1}{2}\sum_{k=1}^{N} (\bar{Q}_{12_k}\alpha_{x_k} + \bar{Q}_{22_k}\alpha_{y_k} + \bar{Q}_{26_k}\alpha_{xy_k})(z_k^2 - z_{k-1}^2) \\
\hat{M}_{xy}^T &= \tfrac{1}{2}\sum_{k=1}^{N} (\bar{Q}_{16_k}\alpha_{x_k} + \bar{Q}_{26_k}\alpha_{y_k} + \bar{Q}_{66_k}\alpha_{xy_k})(z_k^2 - z_{k-1}^2)
\end{aligned} \qquad (3.6.41)$$

An inspection of the various elements of the *ABD* matrix reveals that not only do their definitions depend on the reduced stiffnesses of every layer, but their definitions also include the location within the laminate of every layer by virtue of z_k and z_{k-1}. The elements of the *ABD* matrix are thus weighted and smeared stiffness quantities, the weighting factor being the thickness and location of the various layers within the laminate. Strictly speaking, the values of the *A*s involve only the layer thicknesses, i.e., $z_k - z_{k-1} = h_k$, the thickness of the kth layer. On the other hand, the values of the *B*s and *D*s involve layer thicknesses and locations. Physically, the *A*s are the in-plane stiffnesses of the laminate, the *D*s are the bending stiffnesses, and the *B*s are unique to composite materials and are referred to as the *bending-stretching stiffnesses*. This nomenclature for the *B*s refers to the fact that through these terms, bending effects, namely, the curvatures, κ_x^o, κ_y^o, and κ_{xy}^o, are coupled to in-plane force resultants, N_x, N_y, and N_{xy}, and conversely, the in-plane strains, ε_x^o, ε_y^o, and γ_{xy}^o, are coupled to the bending moment resultants, M_x, M_y, and M_{xy}. The unit effective thermal stress resultants are also functions of stiffnesses and coefficients of thermal expansion weighted by geometry. With the elements of the *ABD* matrix and the unit effective thermal stress resultants depending on layer thickness and location, a layer can have more or less influence on the overall behavior of a laminate, depending on its location within the stacking sequence and it thickness. It is important to realize that once the location, material properties, and fiber orientation of each layer are known, the sums defining the *ABD* matrix and the unit effective thermal stress resultants can be computed and numerical values obtained.

Special cases of laminate stacking sequences

For so-called *symmetric laminates*, all elements of the *B* matrix are zero, as are the three unit effective thermal moment resultants if the temperature change ΔT is uniform through the thickness of the laminate. A symmetric laminate is defined as follows: If for every layer at a specific location and with a specific thickness, material properties, and fiber orientation located on one side of the geometric midplane, there is a layer with identical thickness, material properties, and fiber orientation at the mirror-image location on the other side of the geometric midplane, then the laminate is said to be symmetric. For so-called *balanced laminates*, A_{16}, A_{26}, and $\hat{N}_{xy}^T$ are zero. A balanced laminate is defined as follows: If for every layer with a specific thickness, material properties, and fiber orientation, there is a layer *somewhere* within the laminate with the same specific thickness and material properties, but with the *opposite* fiber orientation, then the laminate is said to be balanced. Symmetric and balanced laminates are very common and the relation of Eq. (39) simplifies considerably to

$$\begin{Bmatrix} N_x \\ N_y \\ N_{xy} \\ M_x \\ M_y \\ M_{xy} \end{Bmatrix} = \begin{bmatrix} A_{11} & A_{12} & 0 & 0 & 0 & 0 \\ A_{12} & A_{22} & 0 & 0 & 0 & 0 \\ 0 & 0 & A_{66} & 0 & 0 & 0 \\ 0 & 0 & 0 & D_{11} & D_{12} & D_{16} \\ 0 & 0 & 0 & D_{12} & D_{22} & D_{26} \\ 0 & 0 & 0 & D_{16} & D_{26} & D_{66} \end{bmatrix} \begin{Bmatrix} \varepsilon_x^o \\ \varepsilon_y^o \\ \gamma_{xy}^o \\ \kappa_x^o \\ \kappa_y^o \\ \kappa_{xy}^o \end{Bmatrix} - \begin{Bmatrix} \hat{N}_x^T \\ \hat{N}_y^T \\ 0 \\ 0 \\ 0 \\ 0 \end{Bmatrix} \Delta T \tag{3.6.42}$$

It is important to note that D_{16} and D_{26} are not zero. It is also important to remember that if the temperature is not uniform through the thickness of the laminate, even though the laminate is symmetric, Eq. (42) does not apply. Because of the simplifications, Eq. (42) can be written as two separate equations, namely,

$$\begin{Bmatrix} N_x \\ N_y \\ N_{xy} \end{Bmatrix} = \begin{bmatrix} A_{11} & A_{12} & 0 \\ A_{12} & A_{22} & 0 \\ 0 & 0 & A_{66} \end{bmatrix} \begin{Bmatrix} \varepsilon_x^o \\ \varepsilon_y^o \\ \gamma_{xy}^o \end{Bmatrix} - \begin{Bmatrix} \hat{N}_x^T \\ \hat{N}_y^T \\ 0 \end{Bmatrix} \Delta T = [A] \begin{Bmatrix} \varepsilon_x^o \\ \varepsilon_y^o \\ \gamma_{xy}^o \end{Bmatrix} - \begin{Bmatrix} \hat{N}_x^T \\ \hat{N}_y^T \\ 0 \end{Bmatrix} \Delta T \tag{3.6.43}$$

$$\begin{Bmatrix} M_x \\ M_y \\ M_{xy} \end{Bmatrix} = \begin{bmatrix} D_{11} & D_{12} & D_{16} \\ D_{12} & D_{22} & D_{26} \\ D_{16} & D_{26} & D_{66} \end{bmatrix} \begin{Bmatrix} \kappa_x^o \\ \kappa_y^o \\ \kappa_{xy}^o \end{Bmatrix} = [D] \begin{Bmatrix} \kappa_x^o \\ \kappa_y^o \\ \kappa_{xy}^o \end{Bmatrix} \tag{3.6.44}$$

where it is seen that for symmetric laminates, bending effects (M_x, M_y, M_{xy} and κ_x^o, κ_y^o, κ_{xy}^o) and in-plane effects (N_x, N_y, N_{xy} and ε_x^o, ε_y^o, γ_{xy}^o) decouple. Actually, Eq. (43) can be further simplified to two other equations as

$$\begin{Bmatrix} N_x \\ N_y \end{Bmatrix} = \begin{bmatrix} A_{11} & A_{12} \\ A_{12} & A_{22} \end{bmatrix} \begin{Bmatrix} \varepsilon_x^o \\ \varepsilon_y^o \end{Bmatrix} - \begin{Bmatrix} \hat{N}_x^T \\ \hat{N}_y^T \end{Bmatrix} \Delta T \tag{3.6.45}$$

$$N_{xy} = A_{66}\gamma_{xy}^o \tag{3.6.46}$$

Cross-ply laminates are laminates constructed of layers with fiber orientations of only 0° or 90°. For such laminates, because by Eq. (27) $\bar{Q}_{16}$ and $\bar{Q}_{26}$ are zero for every layer, all 16 and 26 terms and $\hat{N}_{xy}^T$ and $\hat{M}_{xy}^T$ in Eq. (39) are zero. For a symmetric cross-ply laminate, the relations of Eq. (39) simplify even more to give what can be written as four separate equations:

$$\begin{Bmatrix} N_x \\ N_y \end{Bmatrix} = \begin{bmatrix} A_{11} & A_{12} \\ A_{12} & A_{22} \end{bmatrix} \begin{Bmatrix} \varepsilon_x^o \\ \varepsilon_y^o \end{Bmatrix} - \begin{Bmatrix} \hat{N}_x^T \\ \hat{N}_y^T \end{Bmatrix} \Delta T \tag{3.6.47}$$

$$N_{xy} = A_{66}\gamma_{xy}^o \tag{3.6.48}$$

$$\begin{Bmatrix} M_x \\ M_y \end{Bmatrix} = \begin{bmatrix} D_{11} & D_{12} \\ D_{12} & D_{22} \end{bmatrix} \begin{Bmatrix} \kappa_x^o \\ \kappa_y^o \end{Bmatrix} \tag{3.6.49}$$

$$M_{xy} = D_{66}\kappa_{xy}^o \tag{3.6.50}$$

The existence of the four separate relations signifies a high degree of uncoupling of various stress resultants and reference surface deformations. Again as a reminder, if the temperature is not uniform through the thickness of the laminate, then Eqs. (47)-(50), do not apply. If the layers are all isotropic, or for a single isotropic layer, the form of Eq. (39) reduces to the form of Eqs. (47)-(50).

To follow are numerical examples of the *A*, *B*, and *D* matrices for a number of common laminates. The numerical examples are based upon the properties for a glass-polymer composite having a fiber volume fraction of 60%, as given in Example B.

EXAMPLE E.1: $[\pm 45/0/90]_S$

The subscript "S" signifies that the laminate is symmetric, and because there is a −45° layer for each +45° layer, the laminate is balanced. Neither the 0° nor 90° layers contribute to any 16 terms or the unit effective thermal stress resultants.

$$\begin{bmatrix} A_{11} & A_{12} & A_{16} \\ A_{12} & A_{22} & A_{26} \\ A_{16} & A_{26} & A_{66} \end{bmatrix} = \begin{bmatrix} 27.8 & 8.55 & 0 \\ 8.55 & 27.8 & 0 \\ 0 & 0 & 9.60 \end{bmatrix} \frac{\text{MN}}{\text{m}}$$

$$\begin{bmatrix} D_{11} & D_{12} & D_{16} \\ D_{12} & D_{22} & D_{26} \\ D_{16} & D_{26} & D_{66} \end{bmatrix} = \begin{bmatrix} 2.18 & 0.961 & 0.1784 \\ 0.961 & 1.945 & 0.1784 \\ 0.1784 & 0.1784 & 1.050 \end{bmatrix} \text{N-m}$$

$$\begin{Bmatrix} \hat{N}_x^T \\ \hat{N}_y^T \end{Bmatrix} = \begin{Bmatrix} 425 \\ 425 \end{Bmatrix} \frac{\text{N/m}}{^\circ\text{C}}$$

This eight-layer laminate is called a *quasi-isotropic* laminate because the in-plane extensional stiffnesses A_{11} and A_{22} are equal, but the in-plane shear stiffness A_{66} is not related to A_{11} ($= A_{22}$) as it would be for a truly isotropic material. In addition, the bending stiffnesses D_{11} and D_{22} are not equal, nor are D_{16} and D_{26} zero, as they would be for a truly isotropic material, but they are equal. Note also that $\hat{N}_x^T$ and $\hat{N}_y^T$ are equal. Herein, as stated before, the thickness of a layer is taken to be 0.000125 m (0.125 mm).

EXAMPLE E.2: $[\pm 60/0]_S$

$$\begin{bmatrix} A_{11} & A_{12} & A_{16} \\ A_{12} & A_{22} & A_{26} \\ A_{16} & A_{26} & A_{66} \end{bmatrix} = \begin{bmatrix} 20.8 & 6.41 & 0 \\ 6.41 & 20.8 & 0 \\ 0 & 0 & 7.20 \end{bmatrix} \frac{\text{MN}}{\text{m}}$$

$$\begin{bmatrix} D_{11} & D_{12} & D_{16} \\ D_{12} & D_{22} & D_{26} \\ D_{16} & D_{26} & D_{66} \end{bmatrix} = \begin{bmatrix} 0.675 & 0.363 & 0.0491 \\ 0.363 & 1.151 & 0.1570 \\ 0.0491 & 0.1570 & 0.400 \end{bmatrix} \text{N-m}$$

$$\begin{Bmatrix} \hat{N}_x^T \\ \hat{N}_y^T \end{Bmatrix} = \begin{Bmatrix} 319 \\ 319 \end{Bmatrix} \frac{\text{N/m}}{^\circ\text{C}}$$

This six-layer laminate is also quasi-isotropic. The bending stiffnesses D_{16} and D_{26} are not zero.

EXAMPLE E.3: $[\pm 30/0]_S$

$$\begin{bmatrix} A_{11} & A_{12} & A_{16} \\ A_{12} & A_{22} & A_{26} \\ A_{16} & A_{26} & A_{66} \end{bmatrix} = \begin{bmatrix} 28.4 & 6.41 & 0 \\ 6.41 & 13.20 & 0 \\ 0 & 0 & 7.20 \end{bmatrix} \frac{\text{MN}}{\text{m}}$$

$$\begin{bmatrix} D_{11} & D_{12} & D_{16} \\ D_{12} & D_{22} & D_{26} \\ D_{16} & D_{26} & D_{66} \end{bmatrix} = \begin{bmatrix} 1.191 & 0.363 & 0.1570 \\ 0.363 & 0.636 & 0.0491 \\ 0.1570 & 0.0491 & 0.400 \end{bmatrix} \text{N-m}$$

$$\begin{Bmatrix} \hat{N}_x^T \\ \hat{N}_y^T \end{Bmatrix} = \begin{Bmatrix} 315 \\ 323 \end{Bmatrix} \frac{\text{N/m}}{^\circ\text{C}}$$

It is interesting to note the equality of some of the terms in this laminate and the $[\pm 60/0]_S$ laminate above.

EXAMPLE E.4: $[\pm 30]_{2S}$

This eight-layer laminate is a symmetric and balanced *angle-ply* laminate.

$$\begin{bmatrix} A_{11} & A_{12} & A_{16} \\ A_{12} & A_{22} & A_{26} \\ A_{16} & A_{26} & A_{66} \end{bmatrix} = \begin{bmatrix} 33.4 & 10.54 & 0 \\ 10.54 & 18.15 & 0 \\ 0 & 0 & 11.60 \end{bmatrix} \frac{\text{MN}}{\text{m}}$$

$$\begin{bmatrix} D_{11} & D_{12} & D_{16} \\ D_{12} & D_{22} & D_{26} \\ D_{16} & D_{26} & D_{66} \end{bmatrix} = \begin{bmatrix} 2.78 & 0.878 & 0.314 \\ 0.878 & 1.512 & 0.0981 \\ 0.314 & 0.0981 & 0.966 \end{bmatrix} \text{N-m}$$

$$\begin{Bmatrix} \hat{N}_x^T \\ \hat{N}_y^T \end{Bmatrix} = \begin{Bmatrix} 421 \\ 429 \end{Bmatrix} \frac{\text{N/m}}{^\circ\text{C}}$$

It is interesting to compare the properties of this laminate with the $[\pm 30/0]_S$ above, which differs in thickness and the presence of 0° layers. The primary difference in the elements in the D matrix is due to differences in thickness, as z_k^3 is involved with the definition of the Ds, implying that doubling the thicknesses would increase the bending stiffnesses by a factor of eight.

EXAMPLE E.5: $[\pm 60/0/90]_S$

$$\begin{bmatrix} A_{11} & A_{12} & A_{16} \\ A_{12} & A_{22} & A_{26} \\ A_{16} & A_{26} & A_{66} \end{bmatrix} = \begin{bmatrix} 24.9 & 7.55 & 0 \\ 7.55 & 32.6 & 0 \\ 0 & 0 & 8.61 \end{bmatrix} \frac{\text{MN}}{\text{m}}$$

$$\begin{bmatrix} D_{11} & D_{12} & D_{16} \\ D_{12} & D_{22} & D_{26} \\ D_{16} & D_{26} & D_{66} \end{bmatrix} = \begin{bmatrix} 1.773 & 0.816 & 0.0736 \\ 0.816 & 2.65 & 0.235 \\ 0.0736 & 0.235 & 0.904 \end{bmatrix} \text{N-n}$$

$$\begin{Bmatrix} \hat{N}_x^T \\ \hat{N}_y^T \end{Bmatrix} = \begin{Bmatrix} 427 \\ 423 \end{Bmatrix} \frac{\text{N/m}}{^\circ\text{C}}$$

EXAMPLE E.6: $[0/90]_{2S}$

This is an eight-layer symmetric cross-ply laminate.

$$\begin{bmatrix} A_{11} & A_{12} & A_{16} \\ A_{12} & A_{22} & A_{26} \\ A_{16} & A_{26} & A_{66} \end{bmatrix} = \begin{bmatrix} 31.7 & 4.56 & 0 \\ 4.56 & 31.7 & 0 \\ 0 & 0 & 5.62 \end{bmatrix} \frac{\text{MN}}{\text{m}}$$

$$\begin{bmatrix} D_{11} & D_{12} & D_{16} \\ D_{12} & D_{22} & D_{26} \\ D_{16} & D_{26} & D_{66} \end{bmatrix} = \begin{bmatrix} 3.12 & 0.380 & 0 \\ 0.380 & 2.17 & 0 \\ 0 & 0 & 0.468 \end{bmatrix} \text{N-m}$$

$$\begin{Bmatrix} \hat{N}_x^T \\ \hat{N}_y^T \end{Bmatrix} = \begin{Bmatrix} 425 \\ 425 \end{Bmatrix} \frac{\text{N/m}}{^\circ\text{C}}$$

This laminate could be considered quasi-isotropic also. It is interesting to note the difference in D_{11} and D_{22}. The outer 0° layers provide most of the contribution to D_{11}, while the 90° layers provide most of the contribution to D_{22}. Since the 90° layers are closer to the reference surface than the 0° layers, the bending stiffness in the y-direction is less than the bending stiffness in the x-direction. On the other hand, the four 0° layers and the four 90° layers provide the same resistance to extension in the x- and y-directions, respectively, so $A_{22} = A_{11}$.

EXAMPLE E.7: $[0_4/90_4]_T$

This is an eight-layer *anti-symmetric* cross-ply laminate.

$$\begin{bmatrix} A_{11} & A_{12} & A_{16} & B_{11} & B_{12} & B_{16} \\ A_{12} & A_{22} & A_{26} & B_{12} & B_{22} & B_{26} \\ A_{16} & A_{26} & A_{66} & B_{16} & B_{26} & B_{66} \\ B_{11} & B_{12} & B_{16} & D_{11} & D_{12} & D_{16} \\ B_{12} & B_{22} & B_{26} & D_{12} & D_{22} & D_{26} \\ B_{16} & B_{26} & B_{66} & D_{16} & D_{26} & D_{66} \end{bmatrix}$$

$$= \begin{bmatrix} 31.7\frac{\text{MN}}{\text{m}} & 4.56\frac{\text{MN}}{\text{m}} & 0 & -3{,}810\text{ N} & 0 & 0 \\ 4.56\frac{\text{MN}}{\text{m}} & 31.7\frac{\text{MN}}{\text{m}} & 0 & 0 & 3{,}810\text{ N} & 0 \\ 0 & 0 & 5.62\frac{\text{MN}}{\text{m}} & 0 & 0 & 0 \\ -3{,}810\text{ N} & 0 & 0 & 2.65\frac{\text{N}}{\text{m}} & 0.380\frac{\text{N}}{\text{m}} & 0 \\ 0 & 3{,}810\text{ N} & 0 & 0.380\frac{\text{N}}{\text{m}} & 2.65\frac{\text{N}}{\text{m}} & 0 \\ 0 & 0 & 0 & 0 & 0 & 0.468\frac{\text{N}}{\text{m}} \end{bmatrix}$$

$$\begin{Bmatrix} \hat{N}_x^T \\ \hat{N}_y^T \\ \hat{N}_{xy}^T \\ \hat{M}_x^T \\ \hat{M}_y^T \\ \hat{M}_{xy}^T \end{Bmatrix} = \begin{Bmatrix} 425\frac{\text{N/m}}{^\circ\text{C}} \\ 425\frac{\text{N/m}}{^\circ\text{C}} \\ 0 \\ 19.77\times 10^{-4}\frac{\text{N}}{^\circ\text{C}} \\ -19.77\times 10^{-4}\frac{\text{N}}{^\circ\text{C}} \\ 0 \end{Bmatrix}$$

The only terms in the B matrix are B_{11}, which equals $-B_{22}$, but this single coupling term can cause interesting effects. There are also thermally-induced moments, which cause this laminate to curl out of plane when cooled from its elevated curing temperature. The two unit equivalent thermal moment resultants are equal in magnitude and opposite in sign. It is interesting to note that the A matrix is the same as for the previous cross-ply laminate.

INVERSE RELATIONS

Generally the force and moment resultants and temperature change are known for an element of laminate, and it is of interest to compute the strains and stresses in individual layers, or throughout the entire thickness of the laminate if a complete stress analysis is desired. It is convenient to invert the relations previously discussed so the reference surface strains and curvatures can be computed more easily. For a general laminate, from Eq. (39), the inverse relation is

$$\begin{bmatrix} \varepsilon_x^o \\ \varepsilon_y^o \\ \gamma_{xy}^o \\ \kappa_x^o \\ \kappa_y^o \\ \kappa_{xy}^o \end{bmatrix} = \begin{bmatrix} a_{11} & a_{12} & a_{16} & b_{11} & b_{12} & b_{16} \\ a_{12} & a_{22} & a_{26} & b_{21} & b_{22} & b_{26} \\ a_{16} & a_{26} & a_{66} & b_{61} & b_{62} & b_{66} \\ b_{11} & b_{21} & b_{61} & d_{11} & d_{12} & d_{16} \\ b_{12} & b_{22} & b_{62} & d_{12} & d_{22} & d_{26} \\ b_{16} & b_{26} & b_{66} & d_{16} & d_{26} & d_{66} \end{bmatrix} \left(\begin{Bmatrix} N_x \\ N_y \\ N_{xy} \\ M_x \\ M_y \\ M_{xy} \end{Bmatrix} + \begin{Bmatrix} \hat{N}_x^T \\ \hat{N}_y^T \\ \hat{N}_{xy}^T \\ \hat{M}_x^T \\ \hat{M}_y^T \\ \hat{M}_{xy}^T \end{Bmatrix} \Delta T \right) \tag{3.6.51}$$

where

$$\begin{bmatrix} a_{11} & a_{12} & a_{16} & b_{11} & b_{12} & b_{16} \\ a_{12} & a_{22} & a_{26} & b_{21} & b_{22} & b_{26} \\ a_{16} & a_{26} & a_{66} & b_{61} & b_{62} & b_{66} \\ b_{11} & b_{21} & b_{61} & d_{11} & d_{12} & d_{16} \\ b_{12} & b_{22} & b_{62} & d_{12} & d_{22} & d_{26} \\ b_{16} & b_{26} & b_{66} & d_{16} & d_{26} & d_{66} \end{bmatrix} = \begin{bmatrix} A_{11} & A_{12} & A_{16} & B_{11} & B_{12} & B_{16} \\ A_{12} & A_{22} & A_{26} & B_{12} & B_{22} & B_{26} \\ A_{16} & A_{26} & A_{66} & B_{16} & B_{26} & B_{66} \\ B_{11} & B_{12} & B_{16} & D_{11} & D_{12} & D_{16} \\ B_{12} & B_{22} & B_{26} & D_{12} & D_{22} & D_{26} \\ B_{16} & B_{26} & B_{66} & D_{16} & D_{26} & D_{66} \end{bmatrix}^{-1} \tag{3.6.52}$$

is the *laminate compliance matrix* or *inverse ABD matrix.*

For a symmetric balanced laminate, the inverses of Eqs. (43)-(46) are

$$\begin{Bmatrix} \varepsilon_x^o \\ \varepsilon_y^o \end{Bmatrix} = \begin{bmatrix} a_{11} & a_{12} \\ a_{12} & a_{22} \end{bmatrix} \begin{Bmatrix} N_x + \hat{N}_x^T \Delta T \\ N_y + \hat{N}_y^T \Delta T \end{Bmatrix} \tag{3.6.53}$$

$$\gamma_{xy}^o = a_{66} N_{xy} \tag{3.6.54}$$

$$\begin{Bmatrix} \kappa_x^o \\ \kappa_y^o \\ \kappa_{xy}^o \end{Bmatrix} = \begin{bmatrix} d_{11} & d_{12} & d_{16} \\ d_{12} & d_{22} & d_{26} \\ d_{16} & d_{26} & d_{66} \end{bmatrix} \begin{Bmatrix} M_x \\ M_y \\ M_{xy} \end{Bmatrix} = [d] \begin{Bmatrix} M_x \\ M_y \\ M_{xy} \end{Bmatrix} \tag{3.6.55}$$

where

$$\begin{bmatrix} a_{11} & a_{12} & 0 \\ a_{12} & a_{22} & 0 \\ 0 & 0 & a_{66} \end{bmatrix} = \begin{bmatrix} A_{11} & A_{12} & 0 \\ A_{12} & A_{22} & 0 \\ 0 & 0 & A_{66} \end{bmatrix}^{-1}$$
$$\begin{bmatrix} d_{11} & d_{12} & d_{16} \\ d_{12} & d_{22} & d_{26} \\ d_{16} & d_{26} & d_{66} \end{bmatrix} = \begin{bmatrix} D_{11} & D_{12} & D_{16} \\ D_{12} & D_{22} & D_{26} \\ D_{16} & D_{26} & D_{66} \end{bmatrix}^{-1} \tag{3.6.56}$$

The inverse matrices, namely the compliance matrices, are actually more useful for computing the reference surface strains and curvatures from the force and moment resultants. For a symmetric cross-ply laminate, the inverse relations are determined from Eqs. (47)-(50) as

$$\begin{Bmatrix} \varepsilon_x^o \\ \varepsilon_y^o \end{Bmatrix} = \begin{bmatrix} a_{11} & a_{12} \\ a_{12} & a_{22} \end{bmatrix} \begin{Bmatrix} N_x + \hat{N}_x^T \Delta T \\ N_y + \hat{N}_y^T \Delta T \end{Bmatrix} \tag{3.6.57}$$

$$\gamma_{xy}^o = a_{66} N_{xy} \tag{3.6.58}$$

$$\begin{Bmatrix} \kappa_x^o \\ \kappa_y^o \end{Bmatrix} = \begin{bmatrix} d_{11} & d_{12} \\ d_{12} & d_{22} \end{bmatrix} \begin{Bmatrix} M_x \\ M_y \end{Bmatrix} \tag{3.6.59}$$

$$\kappa_{xy}^o = d_{66} M_{xy} \tag{3.6.60}$$

where the compliance matrices are given by

$$\begin{bmatrix} a_{11} & a_{12} & 0 \\ a_{12} & a_{22} & 0 \\ 0 & 0 & a_{66} \end{bmatrix} = \begin{bmatrix} A_{11} & A_{12} & 0 \\ A_{12} & A_{22} & 0 \\ 0 & 0 & A_{66} \end{bmatrix}^{-1}$$
$$\begin{bmatrix} d_{11} & d_{12} & 0 \\ d_{12} & d_{22} & 0 \\ 0 & 0 & d_{66} \end{bmatrix} = \begin{bmatrix} D_{11} & D_{12} & 0 \\ D_{12} & D_{22} & 0 \\ 0 & 0 & D_{66} \end{bmatrix}^{-1} \tag{3.6.61}$$

The numerical values of the compliance matrices for the laminates discussed in Example E above are as follows:

EXAMPLE F.1: $[\pm 45/0/90]_S$

$$\begin{bmatrix} a_{11} & a_{12} & a_{16} \\ a_{12} & a_{22} & a_{26} \\ a_{16} & a_{26} & a_{66} \end{bmatrix} = \begin{bmatrix} 39.8 & -12.26 & 0 \\ -12.26 & 39.8 & 0 \\ 0 & 0 & 104.1 \end{bmatrix} \frac{\text{m}}{\text{GN}}$$

$$\begin{bmatrix} d_{11} & d_{12} & d_{16} \\ d_{12} & d_{22} & d_{26} \\ d_{16} & d_{26} & d_{66} \end{bmatrix} = \begin{bmatrix} 0.588 & -0.286 & -0.0514 \\ -0.286 & 0.662 & -0.0638 \\ -0.0514 & -0.0638 & 0.972 \end{bmatrix} \frac{1}{\text{N-m}}$$

where, for example,

$$39.8 \frac{\text{m}}{\text{GN}} = 39.8 \frac{\text{m}}{10^9 \text{N}} = 39.8 \times 10^{-9} \frac{\text{m}}{\text{N}}$$

EXAMPLE F.2: $[\pm 60/0]_S$

$$\begin{bmatrix} a_{11} & a_{12} & a_{16} \\ a_{12} & a_{22} & a_{26} \\ a_{16} & a_{26} & a_{66} \end{bmatrix} = \begin{bmatrix} 53.1 & -16.34 & 0 \\ -16.34 & 53.1 & 0 \\ 0 & 0 & 138.8 \end{bmatrix} \frac{\text{m}}{\text{GN}}$$

$$\begin{bmatrix} d_{11} & d_{12} & d_{16} \\ d_{12} & d_{22} & d_{26} \\ d_{16} & d_{26} & d_{66} \end{bmatrix} = \begin{bmatrix} 1.782 & -0.562 & 0.001835 \\ -0.562 & 1.095 & -0.361 \\ 0.001835 & -0.361 & 2.64 \end{bmatrix} \frac{1}{\text{N-m}}$$

EXAMPLE F.3: $[\pm 30/0]_S$

$$\begin{bmatrix} a_{11} & a_{12} & a_{16} \\ a_{12} & a_{22} & a_{26} \\ a_{16} & a_{26} & a_{66} \end{bmatrix} = \begin{bmatrix} 39.5 & -19.18 & 0 \\ -19.18 & 85.1 & 0 \\ 0 & 0 & 138.8 \end{bmatrix} \frac{\text{m}}{\text{GN}}$$

$$\begin{bmatrix} d_{11} & d_{12} & d_{16} \\ d_{12} & d_{22} & d_{26} \\ d_{16} & d_{26} & d_{66} \end{bmatrix} = \begin{bmatrix} 1.062 & -0.579 & -0.346 \\ -0.579 & 1.903 & -0.00631 \\ -0.346 & -0.00631 & 2.64 \end{bmatrix} \frac{1}{\text{N-m}}$$

EXAMPLE F.4: $[\pm 30]_{2S}$

$$\begin{bmatrix} a_{11} & a_{12} & a_{16} \\ a_{12} & a_{22} & a_{26} \\ a_{16} & a_{26} & a_{66} \end{bmatrix} = \begin{bmatrix} 36.7 & -21.3 & 0 \\ -21.3 & 67.5 & 0 \\ 0 & 0 & 86.2 \end{bmatrix} \frac{\text{m}}{\text{GN}}$$

$$\begin{bmatrix} d_{11} & d_{12} & d_{16} \\ d_{12} & d_{22} & d_{26} \\ d_{16} & d_{26} & d_{66} \end{bmatrix} = \begin{bmatrix} 0.454 & -0.256 & -0.1215 \\ -0.256 & 0.810 & 0.000881 \\ -0.1215 & 0.000881 & 1.074 \end{bmatrix} \frac{1}{\text{N-m}}$$

EXAMPLE F.5: $[\pm 60/0/90]_S$

$$\begin{bmatrix} a_{11} & a_{12} & a_{16} \\ a_{12} & a_{22} & a_{26} \\ a_{16} & a_{26} & a_{66} \end{bmatrix} = \begin{bmatrix} 43.1 & -10.00 & 0 \\ -10.00 & 33.0 & 0 \\ 0 & 0 & 116.2 \end{bmatrix} \frac{\text{m}}{\text{GN}}$$

$$\begin{bmatrix} d_{11} & d_{12} & d_{16} \\ d_{12} & d_{22} & d_{26} \\ d_{16} & d_{26} & d_{66} \end{bmatrix} = \begin{bmatrix} 0.657 & -0.203 & -0.000718 \\ -0.203 & 0.450 & -0.1006 \\ -0.000718 & -0.1006 & 1.132 \end{bmatrix} \frac{1}{\text{N-m}}$$

EXAMPLE F.6: $[0/90]_{2S}$

$$\begin{bmatrix} a_{11} & a_{12} & a_{16} \\ a_{12} & a_{22} & a_{26} \\ a_{16} & a_{26} & a_{66} \end{bmatrix} = \begin{bmatrix} 32.2 & -4.62 & 0 \\ -4.62 & 32.2 & 0 \\ 0 & 0 & 178.0 \end{bmatrix} \frac{\text{m}}{\text{GN}}$$

$$\begin{bmatrix} d_{11} & d_{12} & d_{16} \\ d_{12} & d_{22} & d_{26} \\ d_{16} & d_{26} & d_{66} \end{bmatrix} = \begin{bmatrix} 0.327 & -0.0573 & 0 \\ -0.0573 & 0.471 & 0 \\ 0 & 0 & 2.14 \end{bmatrix} \frac{1}{\text{N-m}}$$

EXAMPLE F.7: $[0_4/90_4]_T$

$$\begin{bmatrix} a_{11} & a_{12} & a_{16} & b_{11} & b_{12} & b_{16} \\ a_{12} & a_{22} & a_{26} & b_{21} & b_{22} & b_{26} \\ a_{16} & a_{26} & a_{66} & b_{61} & b_{62} & b_{66} \\ b_{11} & b_{21} & b_{61} & d_{11} & d_{12} & d_{16} \\ b_{12} & b_{22} & b_{62} & d_{12} & d_{22} & d_{26} \\ b_{16} & b_{26} & b_{66} & d_{16} & d_{26} & d_{66} \end{bmatrix}$$

$$= \begin{bmatrix} 39.1\frac{\text{m}}{\text{GN}} & -5.61\frac{\text{m}}{\text{GN}} & 0 & 5.62\times10^{-5}\,\frac{1}{\text{N}} & 0 & 0 \\ -5.61\frac{\text{m}}{\text{GN}} & 39.1\frac{\text{m}}{\text{GN}} & 0 & 0 & -5.62\times10^{-5}\,\frac{1}{\text{N}} & 0 \\ 0 & 0 & 178.0\frac{\text{m}}{\text{GN}} & 0 & 0 & 0 \\ 5.62\times10^{-5}\,\frac{1}{\text{N}} & 0 & 0 & 0.469\frac{\text{m}}{\text{N}} & -0.0673\frac{\text{m}}{\text{N}} & 0 \\ 0 & -5.62\times10^{-5}\,\frac{1}{\text{N}} & 0 & -0.0673\frac{\text{m}}{\text{N}} & 0.469\frac{\text{m}}{\text{N}} & 0 \\ 0 & 0 & 0 & 0 & 0 & 2.14\frac{\text{m}}{\text{N}} \end{bmatrix}$$

It is interesting to note that while a_{16} and a_{26} are always negative, the sign, and existence, of d_{16} and d_{26} vary with laminate. This has implications regarding the directions of the curvatures produced by applied moments, particularly the twist quantities.

LAMINATE STRESS ANALYSIS

With what has been presented, it is now possible to compute the stresses in an element of a laminate that is loaded by known applied force and moment resultants, and is subjected to a temperature change. The material properties, fiber orientation, and z-coordinates of each layer lead to numerical values for the elements of the A, B, and D matrices and unit effective thermal stress resultants. Knowing the magnitude and direction of the applied loads and the dimensions of the element, the force and moment resultants can be determined. The strains and curvatures of the reference surface can be computed using the various inverse relations of the previous section, depending on the particular laminate. The strains at every z-location though the thickness of the laminate can be determined from Eq. (29). These strains, in turn, can be transformed using Eq. (24) to determine the strains in the principal material coordinate system at every z-location. These strains can then be used in Eq. (20) to compute the stresses in the principal material coordinate system at every z-location. The approach is very systematic, the procedures to employ at each step are straight-forward and well-defined, and numerical values can be determined. It goes without saying that these steps are best executed by programming a calculator or computer to deal with the many algebraic relations. Executing the many algebraic steps by hand is very error-prone. However, any program should be thoroughly checked to be sure it is computing what is intended before using the results of the program for analysis and design. Once checked for accuracy, the program can be very useful.

Below are several examples that demonstrate the use of the steps described above to compute the stresses within a laminate.

EXAMPLE G: Consider a flat six-layer $[\pm 30/0]_S$ laminate with in-plane dimensions 0.25 m by 0.35 m subjected to a force of 15,000 N in the x-direction uniformly distributed along opposite edges, as shown in Fig. 3.6.12. The temperature change from the cure condition is $DT = -100°\text{C}$. Compute the stresses in each layer.

APPROACH: The relations between stress resultants and deformations and the deformations and stresses, and the transformation relations, all outlined above, are applied to the problem to determine the desired answers. It should be noted that the laminate is symmetric and balanced, so Eqs. (53)-(55) apply, and only a single force is applied. This considerably simplifies the calculations.

SOLUTION: Except for the effective thermal stress resultants, N_x is the only stress resultant, and since the force is uniformly distributed, it is given by

$$N_x = \frac{15{,}000\ \text{N}}{0.25\ \text{m}} = 60{,}000\ \text{N/m}$$

The unit effective thermal stress resultants consist of just the components given by Example E.3, namely,

$$\begin{Bmatrix} \hat{N}_x^T \\ \hat{N}_y^T \\ \hat{N}_{xy}^T \end{Bmatrix} = \begin{Bmatrix} 315 \\ 323 \\ 0 \end{Bmatrix} \text{N/m/°C}$$

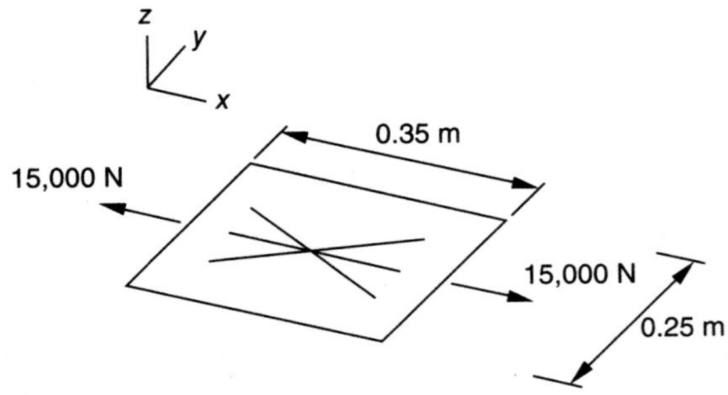

Figure 3.6.12 Laminate loaded in x-direction.

Since there are no moment resultants, the reference surface curvatures are zero. The two reference surface strains are given by direct application of Eqs. (53) and (54), with numerical values from Example F.3, to yield

$$
\left\{ \begin{array}{c} \varepsilon_x^o \\ \varepsilon_y^o \\ \gamma_{xy}^o \end{array} \right\} = \left[\begin{array}{ccc} a_{11} & a_{12} & 0 \\ a_{12} & a_{22} & 0 \\ 0 & 0 & a_{66} \end{array} \right] \left\{ \begin{array}{c} N_x + \hat{N}_x^T \Delta T \\ N_y + \hat{N}_y^T \Delta T \\ N_{xy} + \hat{N}_{xy}^T \Delta T \end{array} \right\}
$$

$$
= \left[\begin{array}{ccc} 39.5 & -19.18 & 0 \\ -19.18 & 85.1 & 0 \\ 0 & 0 & 138.8 \end{array} \right] \times 10^{-9} \left\{ \begin{array}{c} 60,000 + 315(-100) \\ 0 + 323(-100) \\ 0 + 0 \end{array} \right\}
$$

or

$$
\left\{ \begin{array}{c} \varepsilon_x^o \\ \varepsilon_y^o \\ \gamma_{xy}^o \end{array} \right\} = \left\{ \begin{array}{c} 2,370 \\ -1,151 \\ 0 \end{array} \right\} \times 10^{-6} + \left\{ \begin{array}{c} -625 \\ -2,140 \\ 0 \end{array} \right\} \times 10^{-6} = \left\{ \begin{array}{c} 1,745 \\ -3,290 \\ 0 \end{array} \right\} \times 10^{-6}
$$

The first term after the first equal sign is the strains due to the applied force and the second term is the strains due to the decrease in temperature. Due to the applied force, the element of laminate stretches in the direction of the applied force, contracts perpendicular to that direction, and experiences no shear strain. Due to thermal effects, the element of laminate contracts in both directions, but more in the y-direction. The net effect is stretching in the direction of the applied force and considerable contraction perpendicular to that direction, mostly due to thermal effects.

Transforming the strains to the principal material coordinate system for each fiber orientation from Eq. (24) results in

$$
\left\{ \begin{array}{c} \varepsilon_1^o(30°) \\ \varepsilon_2^o(30°) \\ \frac{1}{2}\gamma_{12}^o(30°) \end{array} \right\} = [T(30°)] \left\{ \begin{array}{c} \varepsilon_x^o \\ \varepsilon_y^o \\ \frac{1}{2}\gamma_{xy}^o \end{array} \right\} = \left[\begin{array}{ccc} \frac{3}{4} & \frac{1}{4} & \frac{\sqrt{3}}{2} \\ \frac{1}{4} & \frac{3}{4} & -\frac{\sqrt{3}}{2} \\ -\frac{\sqrt{3}}{4} & \frac{\sqrt{3}}{4} & \frac{1}{2} \end{array} \right] \left\{ \begin{array}{c} 1745 \\ -3290 \\ 0 \end{array} \right\} \times 10^{-6}
$$

$$
= \left\{ \begin{array}{c} 486 \\ -2030 \\ -2180 \end{array} \right\} \times 10^{-6}
$$

$$
\left\{ \begin{array}{c} \varepsilon_1^o(-30°) \\ \varepsilon_2^o(-30°) \\ \frac{1}{2}\gamma_{12}^o(-30°) \end{array} \right\} = [T(-30°)] \left\{ \begin{array}{c} \varepsilon_x^o \\ \varepsilon_y^o \\ \frac{1}{2}\gamma_{xy}^o \end{array} \right\} = \left[\begin{array}{ccc} \frac{3}{4} & \frac{1}{4} & -\frac{\sqrt{3}}{2} \\ \frac{1}{4} & \frac{3}{4} & \frac{\sqrt{3}}{2} \\ \frac{\sqrt{3}}{4} & \frac{\sqrt{3}}{4} & \frac{1}{2} \end{array} \right] \left\{ \begin{array}{c} 1,745 \\ -3,290 \\ 0 \end{array} \right\} \times 10^{-6}
$$

$$
= \left\{ \begin{array}{c} 486 \\ -2,030 \\ 2,180 \end{array} \right\} \times 10^{-6}
$$

$$
\left\{ \begin{array}{c} \varepsilon_1^o(0°) \\ \varepsilon_2^o(0°) \\ \frac{1}{2}\gamma_{12}^o(0°) \end{array} \right\} = [T(0°)] \left\{ \begin{array}{c} \varepsilon_x^o \\ \varepsilon_y^o \\ \frac{1}{2}\gamma_{xy}^o \end{array} \right\} = \left[\begin{array}{ccc} 1 & 0 & 0 \\ 0 & 1 & 0 \\ 0 & 0 & 1 \end{array} \right] \left\{ \begin{array}{c} 1745 \\ -3,290 \\ 0 \end{array} \right\} \times 10^{-6}
$$

$$
= \left\{ \begin{array}{c} 1745 \\ -3,290 \\ 0 \end{array} \right\} \times 10^{-6}
$$

In the principal material coordinate system there are substantial shear strains in the ±30° layers, and they are of opposite sign. The extensional strains are identical. Of course, no transformation is necessary for the 0° layers. Substituting the strains for each layer into the stress-strain relation of Eq. (20) provides the desired stresses:

$$
\left\{ \begin{array}{c} \sigma_1(+30°) \\ \sigma_2(+30°) \\ \tau_{12}(+30°) \end{array} \right\} = \left(\begin{array}{ccc} Q_{11} & Q_{12} & 0 \\ Q_{12} & Q_{22} & 0 \\ 0 & 0 & Q_{66} \end{array} \right) \left\{ \begin{array}{c} \varepsilon_1(+30°) - \alpha_1 \Delta T \\ \varepsilon_2(+30°) - \alpha_2 \Delta T \\ \gamma_{12}(+30°) \end{array} \right\}
$$

$$
\left\{ \begin{array}{c} \sigma_1(+30°) \\ \sigma_2(+30°) \\ \tau_{12}(+30°) \end{array} \right\} = \left(\begin{array}{ccc} 47.0 & 4.56 & 0 \\ 4.56 & 16.51 & 0 \\ 0 & 0 & 5.62 \end{array} \right) \times 10^9 \left\{ \begin{array}{c} 486 - 6.51(-100) \\ -2,030 - 24.4(-100) \\ -4,360 \end{array} \right\} \times 10^{-6}
$$

$$
= \left\{ \begin{array}{c} 55.3 \\ 11.94 \\ -24.5 \end{array} \right\} \text{MPa}
$$

$$
\left\{ \begin{array}{c} \sigma_1(-30°) \\ \sigma_2(-30°) \\ \tau_{12}(-30°) \end{array} \right\} = \left(\begin{array}{ccc} Q_{11} & Q_{12} & 0 \\ Q_{12} & Q_{22} & 0 \\ 0 & 0 & Q_{66} \end{array} \right) \left\{ \begin{array}{c} \varepsilon_1(-30°) - \alpha_1 \Delta T \\ \varepsilon_2(-30°) - \alpha_2 \Delta T \\ \gamma_{12}(-30°) \end{array} \right\}
$$

$$
\left\{ \begin{array}{c} \sigma_1(-30°) \\ \sigma_2(-30°) \\ \tau_{12}(-30°) \end{array} \right\} = \left(\begin{array}{ccc} 47.0 & 4.56 & 0 \\ 4.56 & 16.51 & 0 \\ 0 & 0 & 5.62 \end{array} \right) \times 10^9 \left\{ \begin{array}{c} 486 - 6.51(-100) \\ -2,030 - 24.4(-100) \\ 4,360 \end{array} \right\} \times 10^{-6}
$$

$$
= \left\{ \begin{array}{c} 55.3 \\ 11.94 \\ 24.5 \end{array} \right\} \text{MPa}
$$

$$
\left\{ \begin{array}{c} \sigma_1(0°) \\ \sigma_2(0°) \\ \tau_{12}(0°) \end{array} \right\} = \left(\begin{array}{ccc} Q_{11} & Q_{12} & 0 \\ Q_{12} & Q_{22} & 0 \\ 0 & 0 & Q_{66} \end{array} \right) \left\{ \begin{array}{c} \varepsilon_1(0°) - \alpha_1 \Delta T \\ \varepsilon_2(0°) - \alpha_2 \Delta T \\ \gamma_{12}(0°) \end{array} \right\}
$$

$$
\left\{ \begin{array}{c} \sigma_1(0°) \\ \sigma_2(0°) \\ \tau_{12}(0°) \end{array} \right\} = \left(\begin{array}{ccc} 47.0 & 4.56 & 0 \\ 4.56 & 16.51 & 0 \\ 0 & 0 & 5.62 \end{array} \right) \times 10^9 \left\{ \begin{array}{c} 1,745 - 6.51(-100) \\ -3,290 - 24.4(-100) \\ 0 \end{array} \right\} \times 10^{-6}
$$

$$
= \left\{ \begin{array}{c} 108.7 \\ -3.11 \\ 0 \end{array} \right\} \text{MPa}
$$

As might be expected, the stress in the fiber direction in the 0° layers is the highest. The ±30° layers experience a shear stress. If thermal effects are ignored, ΔT would be taken to be zero, resulting in

$$
\left\{ \begin{array}{c} \sigma_1(+30°) \\ \sigma_2(+30°) \\ \tau_{12}(+30°) \end{array} \right\} = \left\{ \begin{array}{c} 68.7 \\ 2.33 \\ -17.12 \end{array} \right\} \text{MPa} \qquad \left\{ \begin{array}{c} \sigma_1(-30°) \\ \sigma_2(-30°) \\ \tau_{12}(-30°) \end{array} \right\} = \left\{ \begin{array}{c} 68.7 \\ 2.33 \\ 17.12 \end{array} \right\} \text{MPa}
$$

$$
\left\{ \begin{array}{c} \sigma_1(0°) \\ \sigma_2(0°) \\ \tau_{12}(0°) \end{array} \right\} = \left\{ \begin{array}{c} 106.1 \\ -8.20 \\ 0 \end{array} \right\} \text{MPa}
$$

It can be seen that residual thermal effects increase some stress components and decrease others. Even though some stress components are small compared to others, for example σ_2 compared to σ_1, the material is much weaker perpendicular to the fibers than in the fiber direction. The small stress may be approaching levels to cause failure perpendicular to the fibers. The variations through the thickness of the three components of stress in the principal material coordinate system are shown in Fig. 3.6.13. The stresses are constant within each layer, but the overall distributions through the thickness are discontinuous, a characteristics which makes composite materials different than conventional materials such as aluminum or steel. For aluminum or steel, for example, the stress distributions through the thickness would be continuous and there would be no shear stress. It should be noted that there is no net shear strain for the laminate, though there are shear strains, but of opposite sign, for the ±30° layers. This characteristic is due to the balanced nature of the laminate.

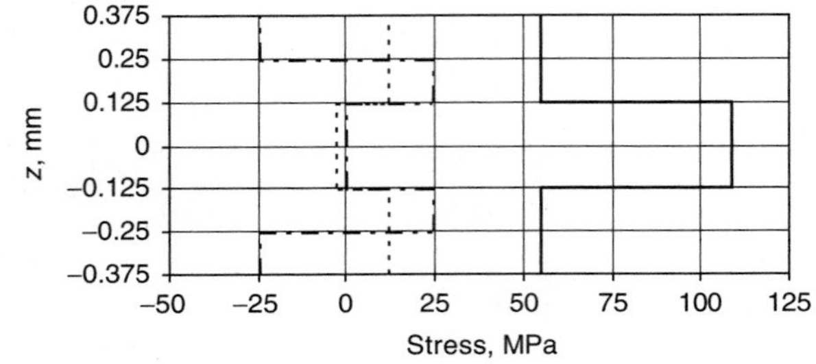

Figure 3.6.13 Through-thickness distribution of stresses: σ_1, solid; σ_2, dotted; τ_{12}, dashed.

EXAMPLE H: A flat eight-layer $[\pm 45/0/90]_S$ quasi-isotropic laminate 0.1 by 0.3 m is loaded by a 0.7 N · m moment, as shown in Fig. 3.6.14. The through-thickness distribution of the stresses in the principal material coordinate system is of interest. The temperature change from the cure temperature condition is –100°C.

APPROACH: The symmetric and balanced nature of the laminate, along with the fact that only a single moment is applied, simplifies the computations considerably. Equations (53)-(55) and the material properties from Examples E.1 and F.1 apply.

SOLUTION: If it is assumed that the applied moment is uniformly distributed along the opposite edges, the moment resultant is given by

$$M_x = \frac{-0.7}{0.1} = -7 \text{ N}\cdot\text{m/m}$$

The values of M_y and M_{xy} are zero. There are no unit equivalent thermal moment resultants, and the unit equivalent thermal force resultants are given by Example E.1 as

$$\begin{Bmatrix} \hat{N}_x^T \\ \hat{N}_y^T \\ \hat{N}_{xy}^T \end{Bmatrix} = \begin{Bmatrix} 425 \\ 425 \\ 0 \end{Bmatrix} \text{N/m/°C}$$

The reference surface strains are due only to the equivalent thermal force resultants and are computed as follows by way of Eqs. (53) and (54), with numerical values from Example F.1, as

$$\begin{Bmatrix} \varepsilon_x^o \\ \varepsilon_y^o \\ \gamma_{xy}^o \end{Bmatrix} = \begin{bmatrix} a_{11} & a_{12} & 0 \\ a_{12} & a_{22} & 0 \\ 0 & 0 & a_{66} \end{bmatrix} \begin{Bmatrix} \hat{N}_x^T \Delta T \\ \hat{N}_y^T \Delta T \\ \hat{N}_{xy}^T \Delta T \end{Bmatrix}$$

$$= \begin{bmatrix} 39.8 & -12.26 & 0 \\ -12.26 & 39.8 & 0 \\ 0 & 0 & 104.1 \end{bmatrix} \times 10^{-9} \begin{Bmatrix} 425(-100) \\ 425(-100) \\ 0 \end{Bmatrix} = \begin{Bmatrix} -1{,}171 \\ -1{,}171 \\ 0 \end{Bmatrix} \times 10^{-6}$$

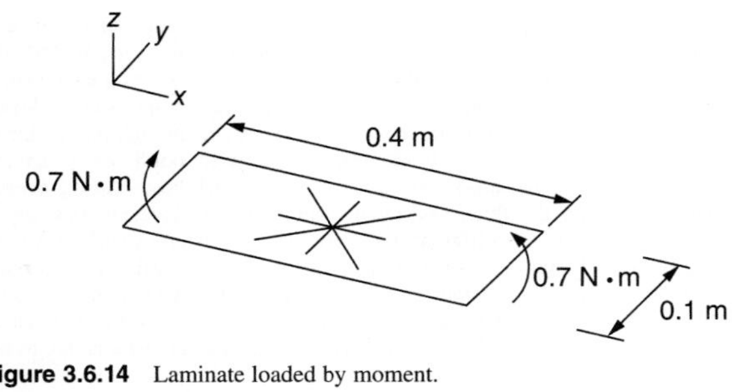

Figure 3.6.14 Laminate loaded by moment.

The reference surface curvatures are due to the applied moment and are computed by way of Eq. (55) and Example F.1 as

$$\begin{Bmatrix} \kappa_x^o \\ \kappa_y^o \\ \kappa_{xy}^o \end{Bmatrix} = \begin{bmatrix} d_{11} & d_{12} & d_{16} \\ d_{12} & d_{22} & d_{26} \\ d_{16} & d_{26} & d_{66} \end{bmatrix} \begin{Bmatrix} M_x \\ M_y \\ M_{xy} \end{Bmatrix}$$

$$= \begin{bmatrix} 0.588 & -0.286 & -0.0514 \\ -0.286 & 0.662 & -0.0638 \\ -0.0514 & -0.0638 & 0.972 \end{bmatrix} \begin{Bmatrix} -7 \\ 0 \\ 0 \end{Bmatrix} \frac{1}{\text{m}} = \begin{Bmatrix} -4.12 \\ 2.00 \\ 0.360 \end{Bmatrix} \frac{1}{\text{m}}$$

Note that in addition producing curvature component κ_x^o, as expected, the single moment M_x produces curvature components κ_y^o and κ_{xy}^o. The former curvature would exist if this was a conventional material such as aluminum or steel, and is called *anticlastic* curvature. It is due to Poisson's ratio, which is reflected in the value of d_{12}, and is related to $\bar{Q}_{12}$. The twist curvature κ_{xy}^o is due to the existence of d_{16}, which can be related to the existence of non-zero values of $\bar{Q}_{16}$ in the ±45° layers. Twist curvature would not exist if this were a conventional material.

The strain as a function of z in the laminate coordinate system is given by Eq. (29) as

$$\begin{Bmatrix} \varepsilon_x \\ \varepsilon_y \\ \gamma_{xy} \end{Bmatrix} = \begin{Bmatrix} \varepsilon_x^o \\ \varepsilon_y^o \\ \gamma_{xy}^o \end{Bmatrix} + z \begin{Bmatrix} \kappa_x^o \\ \kappa_y^o \\ \kappa_{xy}^o \end{Bmatrix} = \begin{Bmatrix} -1{,}171 \\ -1{,}171 \\ 0 \end{Bmatrix} \times 10^{-6} + z \begin{Bmatrix} -4.12 \\ 2.00 \\ 0.360 \end{Bmatrix}$$

These relations are valid for the full range of z, $-0.000500 \text{ m} \le z \le +0.000500 \text{ m}$. Transforming these strains to the principal material coordinate system by way of Eq. (24) for the various fiber orientations gives,

for $-0.000500 \text{ m} \le z \le -0.000375 \text{ m}$ and $0.000375 \le z \le -0.000500 \text{ m}$

$$\begin{Bmatrix} \varepsilon_1(45°) \\ \varepsilon_2(45°) \\ \tfrac{1}{2}\gamma_{12}(45°) \end{Bmatrix} = [T(45°)] \begin{Bmatrix} \varepsilon_x \\ \varepsilon_y \\ \tfrac{1}{2}\gamma_{xy} \end{Bmatrix} = \begin{bmatrix} \frac{1}{2} & \frac{1}{2} & 1 \\ \frac{1}{2} & \frac{1}{2} & -1 \\ -\frac{1}{2} & \frac{1}{2} & 0 \end{bmatrix}$$

$$\begin{Bmatrix} -1{,}171 \times 10^{-6} - 4.12z \\ -1{,}171 \times 10^{-6} + 2.00z \\ 0.180z \end{Bmatrix} = \begin{Bmatrix} -1{,}171 \times 10^{-6} - 0.878z \\ -1{,}171 \times 10^{-6} - 1.238z \\ 3.06z \end{Bmatrix}$$

for $-0.000375 \text{ m} \le z \le -0.000250 \text{ m}$ and $0.000250 \text{ m} \le z \le 0.000375 \text{ m}$

$$\begin{Bmatrix} \varepsilon_1(-45°) \\ \varepsilon_2(-45°) \\ \tfrac{1}{2}\gamma_{12}(-45°) \end{Bmatrix} = \begin{Bmatrix} -1{,}171 \times 10^{-6} - 1.238z \\ -1{,}171 \times 10^{-6} - 0.878z \\ -3.06z \end{Bmatrix}$$

for $-0.000250 \text{ m} \le z \le -0.000125 \text{ m}$ and $0.000125 \text{ m} \le z \le 0.000250 \text{ m}$

$$\begin{Bmatrix} \varepsilon_1(0°) \\ \varepsilon_2(0°) \\ \tfrac{1}{2}\gamma_{12}(0°) \end{Bmatrix} = \begin{Bmatrix} -1{,}171 \times 10^{-6} - 4.12z \\ -1{,}171 \times 10^{-6} + 2.00z \\ 0.1799z \end{Bmatrix}$$

for $-0.000125 \text{ m} \le z \le 0.000125 \text{ m}$

$$\begin{Bmatrix} \varepsilon_1(90°) \\ \varepsilon_2(90°) \\ \tfrac{1}{2}\gamma_{12}(90°) \end{Bmatrix} = \begin{Bmatrix} -1{,}171 \times 10^{-6} + 2.00z \\ -1{,}171 \times 10^{-6} - 4.12z \\ -0.1799z \end{Bmatrix}$$

Note each equation is valid for two ranges of z, the ranges depending on the z-location of the layers with the different fiber orientations. The stresses in the various layers in the principal material coordinate system can be computed directly from Eq. (20) using the above strains. Again there is a range of z for each relation, depending on the layer. The stresses in the principal material coordinate system are

for $-0.000500 \text{ m} \le z \le -0.000375 \text{ m}$ and $0.000375 \text{ m} \le z \le 0.000500 \text{ m}$

$$\begin{Bmatrix} \sigma_1(45^\circ) \\ \sigma_2(45^\circ) \\ \tau_{12}(45^\circ) \end{Bmatrix} = \begin{bmatrix} Q_{11} & Q_{12} & 0 \\ Q_{12} & Q_{22} & 0 \\ 0 & 0 & Q_{66} \end{bmatrix} \begin{Bmatrix} \varepsilon_1(45^\circ) - \alpha_1 \Delta T \\ \varepsilon_2(45^\circ) - \alpha_2 \Delta T \\ \gamma_{12}(45^\circ) \end{Bmatrix}$$

$$\begin{Bmatrix} \sigma_1(45^\circ) \\ \sigma_2(45^\circ) \\ \tau_{12}(45^\circ) \end{Bmatrix} = \begin{bmatrix} 47.0 & 4.56 & 0 \\ 4.56 & 16.51 & 0 \\ 0 & 0 & 5.62 \end{bmatrix}$$

$$\times 10^9 \begin{Bmatrix} -1{,}171 \times 10^{-6} - 0.878z - 6.51 \times 10^{-6}(-100) \\ -1{,}171 \times 10^{-6} - 1.238z - 24.4 \times 10^{-6}(-100) \\ 6.12z \end{Bmatrix}$$

$$\begin{Bmatrix} \sigma_1(45^\circ) \\ \sigma_2(45^\circ) \\ \tau_{12}(45^\circ) \end{Bmatrix} = \begin{Bmatrix} -18.62 - 46{,}900z \\ 18.62 - 24{,}400z \\ 34{,}400z \end{Bmatrix} \text{ MPa}$$

for $-0.000375 \text{ m} \le z \le -0.000250 \text{ m}$ and $0.000250 \text{ m} \le z \le 0.000375 \text{ m}$

$$\begin{Bmatrix} \sigma_1(-45^\circ) \\ \sigma_2(-45^\circ) \\ \tau_{12}(-45^\circ) \end{Bmatrix} = \begin{Bmatrix} -18.62 - 62{,}100z \\ 18.62 - 20{,}100z \\ -34{,}400z \end{Bmatrix} \text{ MPa}$$

for $-0.000250 \text{ m} \le z \le -0.000125 \text{ m}$ and $0.000125 \text{ m} \le z \le 0.000250 \text{ m}$

$$\begin{Bmatrix} \sigma_1(0^\circ) \\ \sigma_2(0^\circ) \\ \tau_{12}(0^\circ) \end{Bmatrix} = \begin{Bmatrix} -18.62 - 184{,}300z \\ 18.62 + 14{,}310z \\ 2{,}020z \end{Bmatrix} \text{ MPa}$$

for $-0.000125 \text{ m} \le z \le 0.000125 \text{ m}$

$$\begin{Bmatrix} \sigma_1(90^\circ) \\ \sigma_2(90^\circ) \\ \tau_{12}(90^\circ) \end{Bmatrix} = \begin{Bmatrix} -18.62 + 75{,}300z \\ 18.62 - 58{,}900z \\ -2{,}020z \end{Bmatrix} \text{ MPa}$$

The first terms in the expressions for the stresses are the stresses due to thermal effects. Because of the quasi-isotropic nature of the laminate, every layer experiences identical thermally-induced stresses, and the stress perpendicular to the fiber direction, σ_2, is equal but opposite in value to the stress in the fiber direction, σ_1. There are no thermally-induced shear stresses. The second terms, the terms linear in z, are the stresses due to the applied moment. These stresses vary linearly with z within each layer, but are discontinuous from layer to layer. The through-thickness distributions of the thermally-induced and moment-induced stresses are plotted separately in the Fig. 3.6.15 and Fig. 3.6.16, respectively.

It is seen that the largest tensile and compressive stresses in the principal material coordinate system due to the applied moment occur in layers 3 and 6, respectively, not in the outer layers. This is very much unlike the stress distributions in a conventional material such as aluminum or steel, where the stresses with the largest magnitude occur at $z = \pm H/2$. This characteristic of layered materials is often overlooked.

ENGINEERING PROPERTIES OF A LAMINATE

Often there is interest in what can be considered smeared engineering properties for a laminate. For example, it may be desirable to treat a laminate as a bar or beam made of a homogeneous material with an equivalent extensional modulus. Since a bar is usually associated with tension, or compression, and a beam is usually associated with bending, the equivalent modulus for a bar could be different than the equivalent modulus for a beam. This difference make sense, as the As depend differently on the thicknesses and locations of the layers than the Ds. Considering a bar and a one-dimensional state of stress, the stress-strain behavior can be described by

$$\sigma_x = E_x \varepsilon_x \tag{3.6.62}$$

where E_x is the extensional modulus of the material. If a material is homogeneous, then the stress would be uniform through the thickness. However, it has been shown in the example problems that, in general, for a laminated material the stress is not uniform through the thickness. Therefore, consider the stress in Eq. (62) to be the *average* stress, $\bar{\sigma}_x$. Then, considering that under tension or compression the strain in a bar constructed of a symmetric balanced laminate would be that given by the geometric midsurface, or reference surface, strain, Eq. (62) can be re-written as

$$\bar{\sigma}_x = E_x \varepsilon_x^o \tag{3.6.63}$$

The integral definition of the stress resultant N_x in Eq. (31) would reduce to

$$N_x = \bar{\sigma}_x H \tag{3.6.64}$$

where, recall, H is the thickness of the laminate. For a bar N_y and N_{xy} would be zero, resulting in, from Eq. (53),

$$\varepsilon_x^o = a_{11} N_x \tag{3.6.65}$$

Combining Eqs. (63)-(65) leads to

$$\varepsilon_x^o = \frac{\bar{\sigma}_x}{E_x} = a_{11} N_x = a_{11} \bar{\sigma}_x H \tag{3.6.66}$$

Equating the second and fourth parts of this relationship, the equivalent extensional modulus can thus be defined as

$$\left(E_x\right)_{\text{extension}} = \frac{1}{a_{11} H} \tag{3.6.67}$$

In a similar fashion, if instead of tension, bending is considered, then for a homogenous material the well-known moment-curvature relation for bending of a beam in the x-direction can be written as

$$M_x b = -E_x I \frac{d^2 w^o}{dx^2} = E_x I \kappa_x^o \tag{3.6.68}$$

where I is the second moment of area of the rectangular cross-section beam and is given by

$$I = \frac{1}{12} b H^3 \tag{3.6.69}$$

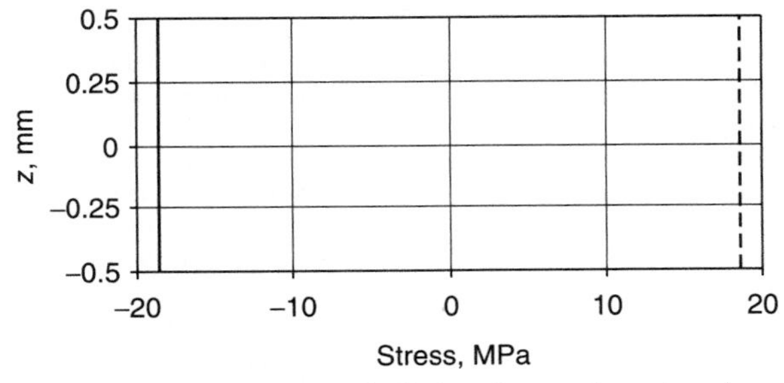

Figure 3.6.15 Through-thickness distribution of stresses due to thermal effects: σ_1, solid; σ_2, dotted; $\tau_{12} = 0$.

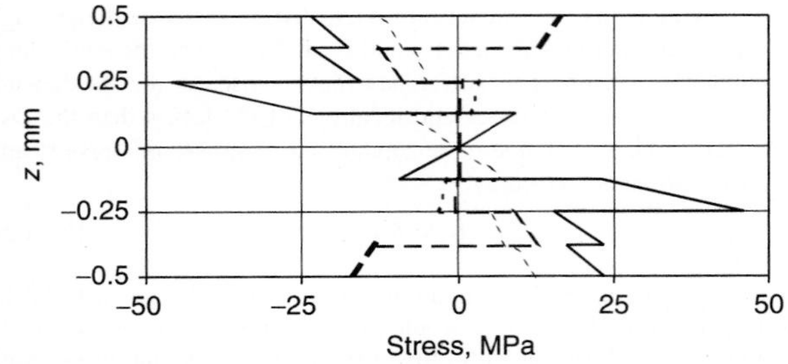

Figure 3.6.16 Through-thickness distribution of stresses due to applied moment: σ_1, solid; σ_2, dotted; τ_{12}, dashed.

The quantity b is the width of the beam (dimension in the y-direction in Fig. 3.6.11) that must be included with the moment since the definition of moment M_x as used herein is moment per unit in-plane dimension in the y-direction. For bending of a beam, M_y and M_{xy} would be zero, so the curvature in the x-direction for a symmetric balanced laminate is given by, from Eq. (55),

$$\kappa_x^o = d_{11} M_x \tag{3.6.70}$$

Combining Eqs. (68)-(70) results in

$$\kappa_x^o = \frac{M_x b}{E_x I} = \frac{M_x b}{E_x \frac{1}{12} b H^3} = d_{11} M_x \tag{3.6.71}$$

Using the third and fourth parts to the relationship provides the definition of the bending modulus as

$$\left(E_x\right)_{\text{bending}} = \frac{12}{d_{11} H^3} \tag{3.6.72}$$

Example I: Compute the extensional and bending moduli in the x-direction for an eight-layer glass-polymer $[\pm 45/0/90]_S$ laminate.

Approach: Use the results of Example F.1 directly in Eqs. (67) and (72).

Solution:

$$\left(E_x\right)_{\text{extension}} = \frac{1}{a_{11} H} = \frac{1}{39.8 \times 10^{-9} (8 \times 0.000125)} = 25.1 \text{ GPa}$$

$$\left(E_x\right)_{\text{bending}} = \frac{12}{d_{11} H^3} = \frac{12}{0.588 (8 \times 0.000125)^3} = 20.4 \text{ GPa}$$

Note the bending modulus is lower than the extensional modulus because the 0° layers are near the center of the laminate and have less influence on resistance to bending than on resistance to extension.

FAILURE OF COMPOSITE MATERIALS

Generally, the calculation of the stresses within a laminate is carried out so the values of the stresses can be compared with values known to cause failure. Knowing the levels of the stresses without having a metric to compare the stresses to is generally not too useful. Most often it is of interest to know what value of applied load for a particular laminate in a particular loading situation causes failure. As might be expected, failure of fiber-reinforced composite materials is a complex process, and historically no aspect of the behavior of composite materials has been studied and debated more than the issue of failure. The topic is complex because of the multiple modes of failure and because the strength of a composite material in tension is considerably different than the strength in compression. A composite can fail due to excess stress in the fiber direction, and the strength in the fiber direction in tension is greater than the strength in compression. Alternatively, a composite material can fail because of excess stress perpendicular to the fibers. The strength perpendicular to the fibers in tension is considerably less than the strength in compression. And finally, a composite can fail due to excess shear stress. The strength in shear is considerable less than the strength in tension in the fiber direction, though the strength in shear does not depend on the sign of the shear stress. These issues are further complicated by the fact that when the fibers are oriented at an angle relative to the loading direction, through the transformation relations, there is a stress component in the fiber direction, a component perpendicular to the fiber direction, and a shear stress. Which component of stress leads to failure? Or is a combination of the components? If it is a combination, how should the stresses be combined to predict failure? There are many issues related to failure of composite materials and there are a number of failure theories. Some theories are quite straightforward, while others are complicated and require a considerable number of experiments to determine all the parameters necessary to implement the theory. No one theory, or criterion, works perfectly for all materials in all situations. It is better to think of failure theories as *indicators* of failure rather than *predictors* of failure. Also, failure in one layer due to excess stress perpendicular to the fibers, for example, does not necessarily mean catastrophic failure of the entire laminate. The other layers may be able to assume a greater portion of load to compensate for the failed layer. Thus the issue of failure really becomes one of *damage accumulation*. With this viewpoint, failure of a laminate will occur when there is sufficient damage accumulation that not enough undamaged material remains to support the applied load. The level of damage accumulation that can be accepted depends on the application. For example, excess tensile stress perpendicular to the fibers generally causes cracking of a composite material, the cracks being parallel with the fibers. If some cracking is acceptable, then the prediction of cracking does not equate to failure of the laminate. However, the existence of any cracking at all is equivalent to the accumulation of some degree of damage. How much cracking can be tolerated very much depends on the application.

One failure theory that is easy to implement is the *maximum stress failure criterion*. The maximum stress failure criterion independently treats tension failure in the fiber direction, compression failure in the fiber direction, tension failure perpendicular to the fiber direction, compression failure perpendicular to the fiber direction, and failure in shear. There are thus five possible modes of failure. A known level of failure stress for each mode is required. Failure in tension in the fiber direction is a direct result of fibers fracturing or otherwise breaking. Failure in compression in the fiber direction is due to the kinking of fibers, generally due to the lack of support of the matrix material surrounding the fibers and the development of shear, or kink, bands within the fiber. Failure in tension perpendicular to the fibers is due to failure of the matrix material between the fibers, failure of the fibers, failure of the bond between the matrix material and the fibers, or a combination of all three failures. Failure in compression perpendicular to the fibers is generally due to crushing of the matrix material. Finally, failure in shear is also due to failure of the matrix material between the fibers, failure of the fibers, failure of the bond between the matrix material and the fibers, or a combination of all three. It is important to realize there is a variability to failure strengths, so the approach to use is to employ allowable stress levels for each mode that adequately account for the variability. If the following definitions for failure stresses in each of the modes are used,

σ_1^T = tensile failure stress in the 1-direction

σ_1^C = compression failure stress in the 1-direction (a negative number)

σ_2^T = tensile failure stress in the 2-direction

σ_2^C = compression failure stress in the 2-direction (a negative number)

τ_{12}^F = shear failure stress in 1-2 plane (a positive number)

then the maximum stress failure criterion states that failure occurs if any one of the following six equalities is satisfied:

$$\sigma_1 = \sigma_1^T \quad \sigma_1 = \sigma_1^C \quad \sigma_2 = \sigma_2^T \quad \sigma_2 = \sigma_2^C \quad \tau_{12} = \tau_{12}^F \quad \tau_{12} = -\tau_{12}^F \qquad (3.6.73)$$

The last two equations can be replaced by one, namely,

$$|\tau_{12}| = \tau_{12}^F \qquad (3.6.74)$$

EXAMPLE J: A thin-walled cylindrical pressure vessel with radius R = 0.25 m is constructed using eight-layers of glass-polymer composite and a stacking sequence of $[\pm 60/0/90]_S$. The fiber angles are specified relative to the axial direction of the cylinder. The end caps of the vessel are suitably designed and reinforced, so the concern is with failure away from the end caps. The change in temperature due to curing is –100°C. The level of internal pressure is increased from zero. In what layer, or layers, does the first failure occur, what is the mode of failure, and what is the pressure level? The failure stresses for the various modes of failure for the glass-polymer composite considered here are given in Table 3.6.4.

APPROACH: Since interest centers on the stresses away from the end caps, the stress resultants acting within the cylinder wall due to the internal pressure can be determined by considering equilibrium of a section of cylinder away from the end caps, just as is done with conventional materials. These stress resultants will be a function of cylinder radius as well as the pressure. Then, using the stress resultants, a stress analysis of the laminate can be conducted.

SOLUTION: As shown in Fig. 3.6.17, by considering equilibrium in the Y-direction of the section of cylinder of length ΔL, and equilibrium in the X-direction of the flat end cap, the axial and circumferential stress resultants acting on the cylinder wall can be computed as a function of pressure and cylinder radius. Summing forces in the X-direction on the end cap and in the Y-direction of the section of cylinder results in

$$\sum F_X = 0 : p\pi R^2 - N_x 2\pi R = 0 \rightarrow N_x = \tfrac{1}{2} pR$$

$$\sum F_Y = 0 : 2pR\Delta L - 2N_\theta \Delta L = 0 \rightarrow N_\theta = pR$$

or

$$\begin{Bmatrix} N_x \\ N_\theta \\ N_{x\theta} \end{Bmatrix} = \begin{Bmatrix} \tfrac{1}{2} \\ 1 \\ 0 \end{Bmatrix} l$$

where, for generality, the results are left in terms of p and R. These are the only two non-zero stress resultants acting within the cylinder wall, away from the end caps, due to the internal pressure. There are no moment resultants. There are, of course, equivalent thermal force resultants from Example E.5 given by

Table 3.6.4 Failure Stresses for Glass–Polymer Composite (MPa)

σ_1^T	σ_1^C	σ_2^T	σ_2^C	τ_{12}^F
1,000	–600	30	–120	70

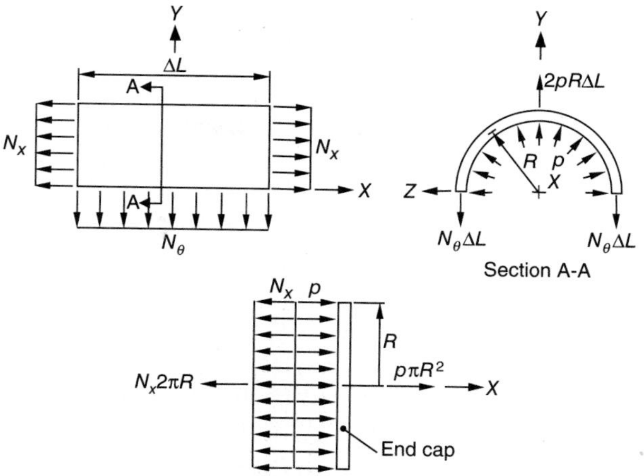

Figure 3.6.17 Forces acting on section of cylinder away from end caps and on end caps.

$$\begin{Bmatrix} \hat{N}_x^T \\ \hat{N}_\theta^T \\ 0 \end{Bmatrix} = \begin{Bmatrix} 427 \\ 423 \\ 0 \end{Bmatrix} \text{N/m/°C}$$

As can be seen, the subscript "y" in the nomenclature has been replaced by the subscript "θ" for this particular problem. This problem is like Example G as there are no moment resultants, and therefore no reference surface curvatures, only reference surface strains. However, the problem is unlike Example G in that the applied load level, in this case the internal pressure level, is not known. This is not a problem. The unknown pressure can be carried along symbolically in the calculations. Accordingly, from Eqs. (53) and (54), with numerical values from Examples E.5 and F.5, the reference surface strains are given by

$$\begin{Bmatrix} \varepsilon_x^o \\ \varepsilon_y^o \\ \gamma_{xy}^o \end{Bmatrix} = [a] \begin{Bmatrix} N_x + \hat{N}_x^T \Delta T \\ N_y + \hat{N}_y^T \Delta T \\ N_{xy} + \hat{N}_{xy}^T \Delta T \end{Bmatrix} = \begin{bmatrix} 43.1 & -10.00 & 0 \\ -10.00 & 33.0 & 0 \\ 0 & 0 & 116.2 \end{bmatrix}$$

$$\times 10^{-9} \begin{Bmatrix} \tfrac{1}{2} pR + 427(-100) \\ pR + 423(-100) \\ 0pR \end{Bmatrix}$$

For convenience, the pressure p in Pascals is converted to atmospheres using the relation

$$p = 101{,}400 p_a$$

where p_a is the internal pressure in atmospheres. The reference surface strains become

$$\begin{Bmatrix} \varepsilon_x^o \\ \varepsilon_y^o \\ \gamma_{xy}^o \end{Bmatrix} = \begin{Bmatrix} -1{,}418 + 1{,}172 p_a R \\ -970 + 2{,}840 p_a R \\ 0 p_a R \end{Bmatrix} \times 10^{-6}$$

The strains in the principal material coordinate system for the layers with 60° fiber orientation are computed from the transformation relation of Eq. (24) as

$$\begin{Bmatrix} \varepsilon_1(60^0) \\ \varepsilon_2(60^0) \\ \frac{1}{2}\gamma_{12}(60^0) \end{Bmatrix} = [T(60°)] \begin{Bmatrix} \varepsilon_x^o \\ \varepsilon_y^o \\ \frac{1}{2}\gamma_{xy}^0 \end{Bmatrix}$$

$$= \begin{bmatrix} \frac{1}{4} & \frac{3}{4} & \frac{\sqrt{3}}{2} \\ \frac{3}{4} & \frac{1}{4} & -\frac{\sqrt{3}}{2} \\ -\frac{\sqrt{3}}{4} & \frac{\sqrt{3}}{4} & -\frac{1}{4} \end{bmatrix} \begin{Bmatrix} -1{,}418+1{,}172p_aR \\ -970+2{,}840p_aR \\ 0p_aR \end{Bmatrix} \times 10^{-6}$$

Carrying out the algebra and doing the same for the other layer orientations gives

$$\begin{Bmatrix} \varepsilon_1(60^0) \\ \varepsilon_2(60^0) \\ \gamma_{12}(60^0) \end{Bmatrix} = \begin{Bmatrix} -1{,}082+2{,}420p_aR \\ -1{,}306+1{,}589p_aR \\ 388+1{,}446p_aR \end{Bmatrix} \times 10^{-6}$$

$$\begin{Bmatrix} \varepsilon_1(-60^0) \\ \varepsilon_2(-60^0) \\ \gamma_{12}(-60^0) \end{Bmatrix} = \begin{Bmatrix} -1{,}082+2{,}420p_aR \\ -1{,}306+1{,}589p_aR \\ -388-1{,}446p_aR \end{Bmatrix} \times 10^{-6}$$

$$\begin{Bmatrix} \varepsilon_1(0^0) \\ \varepsilon_2(0^0) \\ \gamma_{12}(0^0) \end{Bmatrix} = \begin{Bmatrix} -1{,}418+1{,}172p_aR \\ -970+2{,}840p_aR \\ 0 \end{Bmatrix} \times 10^{-6}$$

$$\begin{Bmatrix} \varepsilon_1(90^0) \\ \varepsilon_2(90^0) \\ \gamma_{12}(90^0) \end{Bmatrix} = \begin{Bmatrix} -970+2{,}840p_aR \\ -1{,}418+1{,}172p_aR \\ 0p_aR \end{Bmatrix} \times 10^{-6}$$

The stresses in the various layers can be computed using Eq. (20) and the above strains as follows:

$$\begin{Bmatrix} \sigma_1(60^0) \\ \sigma_2(60^0) \\ \tau_{12}(60^0) \end{Bmatrix} = [Q] \begin{Bmatrix} \varepsilon_1(60^0)-\alpha_1\Delta T \\ \varepsilon_2(60^0)-\alpha_2\Delta T \\ \gamma_{12}(60^0) \end{Bmatrix}$$

$$= \begin{bmatrix} 47.0 & 4.56 & 0 \\ 4.56 & 16.51 & 0 \\ 0 & 0 & 5.62 \end{bmatrix}$$

$$\times 10^9 \begin{Bmatrix} -1{,}082+2{,}420p_aR-6.51(-100) \\ -1{,}306+1{,}589p_aR-24.4(-100) \\ 388+1{,}446p_aR \end{Bmatrix} \times 10^{-6}$$

Carrying out the algebra, substituting for the radius with $R = 0.25$ m, and doing similar calculations for the other three layer angles results in

$$\begin{Bmatrix} \sigma_1(60^0) \\ \sigma_2(60^0) \\ \tau_{12}(60^0) \end{Bmatrix} = \begin{Bmatrix} -15.08+30.3p_a \\ 16.78+9.32p_a \\ 2.18+2.03p_a \end{Bmatrix} \text{ MPa}$$

$$\begin{Bmatrix} \sigma_1(-60^0) \\ \sigma_2(-60^0) \\ \tau_{12}(-60^0) \end{Bmatrix} = \begin{Bmatrix} -15.08+30.3p_a \\ 16.78+9.32p_a \\ -2.18-2.03p_a \end{Bmatrix} \text{ MPa}$$

$$\begin{Bmatrix} \sigma_1(0^0) \\ \sigma_2(0^0) \\ \tau_{12}(0^0) \end{Bmatrix} = \begin{Bmatrix} -29.3+17.00p_a \\ 20.8+13.07p_a \\ 0p_a \end{Bmatrix} \text{ MPa}$$

$$\begin{Bmatrix} \sigma_1(90^0) \\ \sigma_2(90^0) \\ \tau_{12}(90^0) \end{Bmatrix} = \begin{Bmatrix} -10.34+34.7p_a \\ 15.45+8.08p_a \\ 0p_a \end{Bmatrix} \text{ MPa}$$

Using Eq. (73), the stresses in each layer are equated to the failure levels for each failure mode and the value of p_a solved for. As the problem is formulated here, interest is in positive values of p_a, corresponding to internal pressure. However, for the sake of completeness, all six failure equations will be considered for the four fiber angles, resulting in 24 values of p_a. The smallest positive value is the value of internal pressure that first causes failure as the pressure is increased from zero. The corresponding layer and the corresponding mode of failure can be determined from the particular equation that leads to this lowest positive value of p_a. For convenience, the six equations of Eq. (73) are expressed as

$$\begin{Bmatrix} \sigma_1 \\ \sigma_2 \\ \tau_{12} \end{Bmatrix} = \begin{Bmatrix} \sigma_1^T \\ \sigma_2^T \\ \tau_{12}^F \end{Bmatrix} = \begin{Bmatrix} 1{,}000 \\ 30 \\ 70 \end{Bmatrix} \text{ MPa and } \begin{Bmatrix} \sigma_1 \\ \sigma_2 \\ \tau_{12} \end{Bmatrix} = \begin{Bmatrix} \sigma_1^C \\ \sigma_2^C \\ -\tau_{12}^F \end{Bmatrix} = \begin{Bmatrix} -600 \\ -120 \\ -70 \end{Bmatrix} \text{ MPa}$$

where failure stress levels from Table 3.6.4 have been used. For the 60° layers, using MPa, the above six failure equations are

$$\begin{Bmatrix} -15.08+30.3p_a \\ 16.78+9.32p_a \\ 2.18+2.03p_a \end{Bmatrix} = \begin{Bmatrix} 1{,}000 \\ 30 \\ 70 \end{Bmatrix} \text{ and } \begin{Bmatrix} -15.08+30.3p_a \\ 16.78+9.32p_a \\ 2.18+2.03p_a \end{Bmatrix} = \begin{Bmatrix} -600 \\ -120 \\ -70 \end{Bmatrix}$$

For the –60° layers, the six failure equations are

$$\begin{Bmatrix} -15.08+30.3p_a \\ 16.78+9.32p_a \\ -2.18-2.03p_a \end{Bmatrix} = \begin{Bmatrix} 1{,}000 \\ 30 \\ 70 \end{Bmatrix} \text{ and } \begin{Bmatrix} -15.08+30.3p_a \\ 16.78+9.32p_a \\ -2.18-2.03p_a \end{Bmatrix} = \begin{Bmatrix} -600 \\ -120 \\ -70 \end{Bmatrix}$$

For the 0° layers, the six failure equations are

$$\begin{Bmatrix} -29.3+17.00p_a \\ 20.8+13.07p_a \\ 0p_a \end{Bmatrix} = \begin{Bmatrix} 1{,}000 \\ 30 \\ 70 \end{Bmatrix} \text{ and } \begin{Bmatrix} -29.3+17.00p_a \\ 20.8+13.07p_a \\ 0p_a \end{Bmatrix} = \begin{Bmatrix} -600 \\ -120 \\ -70 \end{Bmatrix}$$

For the 90° layers, the six failure equations are

$$\begin{Bmatrix} -10.34+34.7p_a \\ 15.45+8.08p_a \\ 0p_a \end{Bmatrix} = \begin{Bmatrix} 1{,}000 \\ 30 \\ 70 \end{Bmatrix} \text{ and } \begin{Bmatrix} -10.34+34.7p_a \\ 15.45+8.08p_a \\ 0p_a \end{Bmatrix} = \begin{Bmatrix} -600 \\ -120 \\ -70 \end{Bmatrix}$$

From these 24 equations there are 10 positive values of p_a, 10 negative values of p_a, and four values of infinity. The 24 values of p_a are given in Table 3.6.5. (The numbers in parentheses in Table 3.6.5 will be discussed shortly.) The infinite values come from the two shear equations for the 0° layers and the two shear equations for the 90° layers. From these multiple values of p_a in the Table 3.6.5, the lowest level of internal pressure is $p_a = 0.704$ atmospheres. This value is obtained from the solution of the second equation for the positive failure stress levels for the 0° layer, i.e.,

$$20.8+13.07p_a = 30$$

This is interpreted to mean that at $p_a = 0.704$ atmospheres, cracks parallel to the fibers occur in the 0° layers. If matrix cracking is to be avoided, only pressures below this level are allowed. The answer to the original question is: Failure occurs when $p_a = 0.704$ atmospheres, the mode of failure is excess tensile stress perpendicular to the fibers, and it occurs in the 0° layers. It should be mentioned that for

Table 3.6.5 Summary of Failure Pressures (in Atmospheres) for $[\pm 60/0/90]_S$ Glass-polymer Cylindrical Pressure Vessel, R = 0.25 m, ΔT = −100 deg. C (Assuming ΔT = 0 in Parentheses)

Layer	Failure mode					
	σ_1^T	σ_1^C	σ_2^T	σ_2^C	$+\tau_{12}^F$	$-\tau_{12}^F$
+60°	33.5 (33.0)	−19.31 (−19.81)	1.417 (3.22)	−14.66 (−12.87)	33.4 (34.5)	−35.5 (−34.5)
−60°	33.5 (33.0)	−19.31 (−19.81)	1.417 (3.22)	−14.66 (−12.87)	−35.5 (−34.5)	33.4 (34.5)
0°	60.5 (58.8)	−33.6 (−35.3)	0.704 (2.29)	−10.77 (−9.18)	∞ (∞)	∞ (∞)
90°	29.1 (28.8)	−16.99 (−17.28)	1.801 (3.71)	−16.76 (−14.85)	∞ (∞)	∞ (∞)

internal pressure $p_a = 1.801$ atmospheres, all layers may exhibit cracks parallel to the fibers due to excessive values of tensile stress perpendicular to the fibers. It is important to realize that $p_a = 1.801$ atmospheres is an approximation because at $p_a = 0.704$ atmospheres the 0° layers lose stiffness perpendicular to their fibers, due to cracking, and the stresses in the other layers likely increase. However, $p_a = 1.801$ atmospheres provides some guidance as to the pressure to cause all layers to crack and thereby possibly lead to leakage of the pressure.

To go one step beyond what was asked for, a negative pressure of magnitude 10.77 atmospheres is obtained from the solution of the second equation for negative failure stress levels for the 0° layers, i.e.,

$$20.8 + 13.07 p_a = -120$$

This value is the negative value of p_a with the lowest magnitude. It is interpreted to mean that an *external* pressure of 10.77 atmospheres will cause failure of the 0° layers due to excessive compressive stress perpendicular to the fibers, assuming that the cylinder has not failed by buckling due to the external pressure.

The pressures in parentheses in Table 3.6.5 are the failure levels computed when ignoring thermal effects, i.e., assuming $\Delta T = 0$. As observed, the degree of influence of thermally-induced stresses varies with failure mode. However, for the tensile stresses perpendicular to the fibers the influence is the greatest, but the predicted failure mode remains the same, namely, excessive tensile stress in the 0° layers perpendicular to the fibers. However, thermal effects reduce the predicted failure pressure by more than a factor of 3 (0.704 vs. 2.29), an important conclusion.

Additional Reading

The following are general-purpose textbooks:

Hyer M.W., Stress Analysis of Fiber-Reinforced Composite Materials, DEStech Publications, Inc., Lancaster, PA, 2009.

Jones R.M., Mechanics of Composite Materials, 2nd ed., Taylor & Francis, Philadelphia, PA, 1999.

Herakovich C.T., Mechanics of Fibrous Composites, John Wiley and Sons, New York, NY, 1998.

Daniel I.M. and Ishai O., Engineering Mechanics of Composite Materials, 2nd ed., Oxford University Press, New York, NY, 2005.

Gibson R.F., Principles of Composite Material Mechanics, 3rd ed., CRC Press, Boca Raton, FL, 2011.

Reddy J.N., Mechanics of Laminated Composite Plates and Shells, Theory and Analysis, 2nd ed., CRC Press, Boca Raton, FL, 2004.

Swanson S.R., Introduction to Design and Analysis with Advanced Composite Materials, Prentice-Hall, Upper Saddle River, NJ, 1997.

Vinson J.R. and Sierakowski R.L., The Behavior of Structures Composed of Composite Materials, 2nd ed., Martinus Nijhoff Publishers, Boston, MA, 2002.

Christensen R.M., Mechanics of Composite Materials, Dover Publications, Mineola, NY, 2005.

Wolff E.G., Introduction to the Dimensional Stability of Composite Materials, DEStech Publications, Inc., Lancaster, PA, 2004.

The following books deal primarily with damage tolerance and failure:

Reifsnider K.L. and Case S.W., Damage Tolerance and Durability of Material Systems, John Wiley & Sons, New York, NY, 2002.

Hinton M.J., Soden P.D., Kaddour A.S., Failure Criteria in Fibre-Reinforced-Polymer Composites, Elsevier Publishers, 2004.

The following is a good reference:

Anonymous, The Composite Materials Handbook MIL-17, ASTM, West Conshohocken, PA.

Section 4
Materials of Engineering

BY

SIMON SIMMS *Professor of Chemistry, The City College of the City University of New York*
GLEN KOWACH *Professor of Chemistry, The City College of the City University of New York*
HAROLD W. PAXTON *United States Steel Professor Emeritus, Carnegie Mellon University*
JAMES D. REDMOND *Principal, Technical Marketing Resources, Inc.*
MALCOLM BLAIR *Technical and Research Director, Steel Founders' Society of America*
ROBERT E. EPPICH *Vice President, Technology, American Foundry Society*
L. D. KUNSMAN *Late Fellow Engineer, Research Labs, Westinghouse Electric Corp.*
C. L. CARLSON *Late Fellow Engineer, Research Labs, Westinghouse Electric Corp.*
J. RANDOLPH KISSELL *President, The TGB Partnership*
LARRY F. WIESERMAN *Retired Senior Technical Supervisor, ALCOA*
RICHARD L. BRAZILL *Technology Specialist, ALCOA*
FRANK E. GOODWIN *Executive Vice President, ILZRO, Inc.*
DON GRAHAM *Manager, Turning Products, Carboloy, Inc.*
JAMES H. MICHEL *Copper Development Association, Inc.*
JOHN H. TUNDERMANN *Former Vice President, Research and Technology, INCO Intl.*
FRANCIS H. (SAM) FROES *Retired and Consultant to the Light Metals Industry*
PETER K. JOHNSON *Consultant*
JOHN R. SCHLEY *Consultant*
ROBERT D. BARTHOLOMEW *Associate, Sheppard T. Powell Associates, LLC*
DAVID A. SHIFLER *Director of Cellular Materials Program, Office of Naval Research*
ROBERT J. ROSS *Project Leader, USDA Forest Service, and Research Professor, Michigan Technological University*
DAVID KRETSCHMANN *President, American Lumber Standard Committee, Incorporated*
C. ADAM SENALIK *Research General Engineer, USDA Forest Service*
XIPING WANG *Research Forest Products Technologist, USDA Forest Service*
ZHIYONG CAI *Project Leader, USDA Forest Service*
STAN LEBOW *Research Forest Products Technologist, USDA Forest Service*
LAURA HASBURGH *General Engineer, USDA Forest Service*
ALI M. SADEGH *Professor of Mechanical Engineering, The City College of the City University of New York*
GLENN E. ASAUSKAS *Lubrication Engineer, Chevron Corp.*
THOMAS A. FREDERICK *Retired Senior Scientist, Lyondell Basell; Principal Scientist, TF Solutions LLC*
STEPHEN R. SWANSON *Professor of Mechanical Engineering, University of Utah*
ERIK T. THOSTENSON *Professor of Mechanical Engineering, University of Delaware*
PARISA SABOORI *Professor of Mechanical Engineering, Manhattan College*
RAJIV SHIVPURI *Professor of Integrated Systems Engineering, Ohio State University*
DUANE K. MILLER *Engineering Services and Welding Design Consultant, Lincoln Electric Co.*
MICHAEL D. FLORCZYKOWSKI *Lincoln Electric Co.*
STEVEN R. SCHMID *Professor of Aerospace & Mechanical Engineering, University of Notre Dame*
SEROPE KALPAKJIAN *Professor of Mechanical, Materials, and Aerospace Engineering, Illinois Institute of Technology*
JACQUELINE J. LI *Professor of Mechanical Engineering, The City College of the City University of New York*

4.1 GENERAL PROPERTIES OF MATERIALS
Revised by Simon Simms and Glen Kowach

4.2 IRON AND STEEL
by Harold W. Paxton

4.3 IRON AND STEEL CASTINGS
by Malcolm Blair and Robert E. Eppich

4.4 NONFERROUS METALS AND ALLOYS; METALLIC SPECIALTIES
Contributors are shown at the head of each category.

4.5 CORROSION
by Robert D. Bartholomew and David A. Shifler

4.6 WOOD
by Staff, Forest Products Laboratory, USDA Forest Service, and several contributors

4.7 NONMETALLIC MATERIALS
Revised by Ali M. Sadegh

4.8 LUBRICANTS AND LUBRICATION
by Glenn E. Asauskas

4.9 PLASTICS
Revised by Thomas A. Frederick

4.10 FIBER COMPOSITE MATERIALS
by Stephen R. Swanson, Revised by Erik T. Thostenson

4.11 TRUSS DESIGN
Revised by Parisa Saboori

4.12 PLASTIC WORKING OF METALS
by Rajiv Shivpuri

4.13 WELDING AND CUTTING
by Duane K. Miller and Michael D. Florczykowski

4.14 MACHINING PROCESSES AND MACHINE TOOLS
by Steven R. Schmid and Serope Kalpakjian

4.15 SURFACE TEXTURE DESIGNATION, PRODUCTION, AND QUALITY CONTROL
by Ali M. Sadegh

4.16 FERROELECTRICS/PIEZOELECTRICS, MULTIFERROICS, AND SHAPE MEMORY ALLOYS
by Jacqueline J. Li

4.1 GENERAL PROPERTIES OF MATERIALS

Revised by Simon Simms and Glen Kowach

REFERENCES: "International Critical Tables," McGraw-Hill. "Smithsonian Physical Tables," Smithsonian Institution. Landolt, "Landolt-Börnstein, Zahlenwerte und Funktionen aus Physik, Chemie, Astronomie, Geophysik und Technik," Springer. "Handbook of Chemistry and Physics," Chemical Rubber Co. "Book of ASTM Standards," ASTM. "ASHRAE Refrigeration Data Book," ASHRAE. Brady, "Materials Handbook," McGraw-Hill. Mantell, "Engineering Materials Handbook," McGraw-Hill. International Union of Pure and Applied Chemistry, Butterworth Scientific Publications. "U.S. Standard Atmosphere," Government Printing Office. Tables of Thermodynamic Properties of Gases, *NIST Circ.* 564, ASME Steam Tables.

Thermodynamic properties of a variety of other specific materials are listed also in Secs. 2.1, 2.2, and 7.8. Sonic properties of several materials are listed in Sec. 10.6.

CHEMISTRY

Chemistry is the study of matter and its transformations. Matter is ubiquitous; it is anything that has mass and occupies space. Matter consists of one or more substances, and is present in several physical forms such as gas, liquid, and solid. The simplest form of a substance (Table 4.1.1.) is called an element and it may combine with other similar or different elements to form molecules or compounds. An element, in turn, is composed of very small units called atoms that are all alike and cannot be broken down by chemical processes. Most gases exist as molecules: oxygen consists of two atoms of the element oxygen, nitrogen dioxide has one atom of the element nitrogen and two atoms of oxygen, and methane has one atom of the element carbon and four atoms of the element hydrogen. Compounds consist of at least two different elements and a few are listed in Table 4.1.2. The atoms in molecules and compounds are combined in definite and fixed ratio and this determines the character of the substance.

The view generally accepted at present is that the atoms of all the chemical elements consist of several kinds of still smaller particles, three of which are known as protons, neutrons, and electrons, and others are

Table 4.1.1 Chemical Elements[a]

Element	Symbol	Atomic no.	Atomic weight[b]	Valence
Actinium	Ac	89		
Aluminum	Al	13	26.9815	3
Americium	Am	95		
Antimony	Sb	51	121.75	3, 5
Argon[c]	Ar	18	39.948	0
Arsenic	As	33	74.9216	3, 5
Astatine	At	85		
Barium	Ba	56	137.34	2
Berkelium	Bk	97		
Beryllium	Be	4	9.0122	2
Bismuth	Bi	83	208.980	3, 5
Boron	B	5	10.811[k]	3
Bromine[d]	Br	35	79.904[l]	1, 3, 5
Cadmium	Cd	48	112.40	2
Calcium	Ca	20	40.08	2
Californium	Cf	98		
Carbon	C	6	12.01115[k]	2, 4
Cerium	Ce	58	140.12	3, 4
Cesium[j]	Cs	55	132.905	1
Chlorine[e]	Cl	17	35.453[l]	1, 3, 5, 7
Chromium	Cr	24	51.996[l]	2, 3, 6
Cobalt	Co	27	58.9332	2, 3
Columbium (see Niobium)				
Copper	Cu	29	63.546[l]	1, 2
Curium	Cm	96		
Dysprosium	Dy	66	162.50	3
Einsteinium	Es	99		
Erbium	Er	68	167.26	3
Europium	Eu	63	151.96	2, 3
Fermium	Fm	100		
Fluorine[f]	F	9	18.9984	1
Francium	Fr	87		
Gadolinium	Gd	64	157.25	3
Gallium[j]	Ga	31	69.72	2, 3
Germanium	Ge	32	72.59	2, 4
Gold	Au	79	196.967	1, 3
Hafnium	Hf	72	178.49	4
Helium[c]	He	2	4.0026	0
Holmium	Ho	67	164.930	3
Hydrogen[g]	H	1	1.00797[h]	1
Indium	In	49	114.82	1, 2, 3
Iodine	I	53	126.9044	1, 3, 5, 7

Table 4.1.1 Chemical Elements[a] *(Continued)*

Element	Symbol	Atomic no.	Atomic weight[b]	Valence
Iridium	Ir	77	192.2	2, 3, 4, 6
Iron	Fe	26	55.847[l]	2, 3
Krypton[c]	Kr	36	83.80	0
Lanthanum	La	57	138.91	3
Lead	Pb	82	207.19	2, 4
Lithium[h]	Li	3	6.939	1
Lutetium	Lu	71	174.97	3
Magnesium	Mg	12	24.312	2
Manganese	Mn	25	54.9380	2, 3, 4, 6, 7
Mendelevium	Md	101		
Mercury[d]	Hg	80	200.59	1, 2
Molybdenum	Mo	42	95.94	3, 4, 5, 6
Neodymium	Nd	60	144.24	3
Neon[c]	Ne	10	20.183	0
Neptunium	Np	93		
Nickel	Ni	28	58.71	2, 3, 4
Niobium	Nb	41	92.906	2, 3, 4, 5
Nitrogen[e]	N	7	14.0067	3, 5
Nobelium	No	102		
Osmium	Os	76	190.2	2, 3, 4, 6, 8
Oxygen[e]	O	8	15.9994[k]	2
Palladium	Pd	46	106.4	2, 4
Phosphorus	P	15	30.9738	3, 5
Platinum	Pt	78	195.09	2, 4
Plutonium	Pu	94		
Polonium	Po	84		2, 4
Potassium	K	19	39.102	1
Praseodymium	Pr	59	140.907	3
Promethium	Pm	61		5
Protactinium	Pa	91		
Radium	Ra	88		2
Radon[l]	Rn	86		0
Rhenium	Re	75	186.2	1, 4, 7
Rhodium	Rh	45	102.905	3, 4
Rubidium	Rb	37	85.47	1
Ruthenium	Ru	44	101.07	3, 4, 6, 8
Samarium	Sm	62	150.35	3
Scandium	Sc	21	44.956	3
Selenium	Se	34	78.96	2, 4, 6
Silicon	Si	14	28.086[k]	4
Silver	Ag	47	107.868[l]	1
Sodium	Na	11	22.9898	1
Strontium	Sr	38	87.62	2
Sulfur	S	16	32.064[k]	2, 4, 6
Tantalum	Ta	73	180.948	4, 5
Technetium	Tc	43		
Tellurium	Te	52	127.60	2, 4, 6
Terbium	Tb	65	158.924	3
Thallium	T1	81	204.37	1, 3
Thorium	Th	90	232.038	3
Thulium	Tm	69	168.934	3
Tin	Sn	50	118.69	2, 4
Titanium	Ti	22	47.90	3, 4
Tungsten	W	74	183.85	3, 4, 5, 6
Uranium	U	92	238.03	4, 6
Vanadium	V	23	50.942	1, 2, 3, 4, 5
Xenon[c]	Xe	54	131.30	0
Ytterbium	Yb	70	173.04	2, 3
Yttrium	Y	39	88.905	3
Zinc	Zn	30	65.37	2
Zirconium	Zr	40	91.22	4

[a] All the elements for which atomic weights listed are metals, except as otherwise indicated. No atomic weights are listed for most radioactive elements, as these elements have no fixed value.

[b] The atomic weights are based upon nuclidic mass of $C^{12} = 12$.

[c] Inert gas. [d] Liquid. [e] Gas. [f] Most active gas. [g] Lightest gas. [h] Lightest metal. [j] Liquid at 25°C.

[k] The atomic weight varies because of natural variations in the isotopic composition of the element. The observed ranges are boron, ± 0.003; carbon, ± 0.00005; hydrogen, ± 0.00001; oxygen, ± 0.0001; silicon, ± 0.001; sulfur, ± 0.003.

[l] The atomic weight is believed to have an experimental uncertainty of the following magnitude: bromine, ± 0.001; chlorine, ± 0.001; chromium, ± 0.001; copper, ± 0.001; iron, ± 0.003; silver, ± 0.001. For other elements, the last digit given is believed to be reliable to ± 0.5.

SOURCE: Table courtesy IUPAC and Butterworth Scientific Publications.

Table 4.1.2 Solubility of Inorganic Substances in Water
(Number of grams of the substance soluble in 1,000 g of water. The common name of the substance is given in parentheses)

Substance	Composition	Temperature, °F (°C)		
		32 (0)	122 (50)	212 (100)
Aluminum sulfate	$Al_2(SO_4)_3$	313	521	891
Aluminum potassium sulfate (potassium alum)	$Al_2K_2(SO_4)_4 \cdot 24H_2O$	30	170	1,540
Ammonium bicarbonate	NH_4HCO_3	119		
Ammonium chloride (sal ammoniac)	NH_4Cl	297	504	760
Ammonium nitrate	NH_4NO_3	1,183	3,440	8,710
Ammonium sulfate	$(NH_4)_2SO_4$	706	847	1,033
Barium chloride	$BaCl_2 \cdot 2H_2O$	317	436	587
Barium nitrate	$Ba(NO_3)_2$	50	172	345
Calcium carbonate (calcite)	$CaCO_3$	0.018*		0.88
Calcium chloride	$CaCl_2$	594		1,576
Calcium hydroxide (hydrated lime)	$Ca(OH)_2$	1.77		0.67
Calcium nitrate	$Ca(NO_3)_2 \cdot 4H_2O$	931	3,561	3,626
Calcium sulfate (gypsum)	$CaSO_4 \cdot 2H_2O$	1.76	2.06	1.69
Copper sulfate (blue vitriol)	$CuSO_4 \cdot 5H_2O$	140	334	753
Ferrous chloride	$FeCl_2 \cdot 4H_2O$	644§	820	1,060
Ferrous hydroxide	$Fe(OH)_2$	0.0067‡		
Ferrous sulfate (green vitriol or copperas)	$FeSO_4 \cdot 7H_2O$	156	482	
Ferric chloride	$FeCl_3$	730	3,160	5,369
Lead chloride	$PbCl_2$	6.73	16.7	33.3
Lead nitrate	$Pb(NO_3)_2$	403		1,255
Lead sulfate	$PbSO_4$	0.042†		
Magnesium carbonate	$MgCO_3$	0.13‡		
Magnesium chloride	$MgCl_2 \cdot 6H_2O$	524		723
Magnesium hydroxide (milk of magnesia)	$Mg(OH)_2$	0.009‡		
Magnesium nitrate	$Mg(NO_3)_2 \cdot 6H_2O$	665	903	
Magnesium sulfate (Epsom salts)	$MgSO_4 \cdot 7H_2O$	269	500	710
Potassium carbonate (potash)	K_2CO_3	893	1,216	1,562
Potassium chloride	KCl	284	435	566
Potassium hydroxide (caustic potash)	KOH	971	1,414	1,773
Potassium nitrate (saltpeter or niter)	KNO_3	131	851	2,477
Potassium sulfate	K_2SO_4	74	165	241
Sodium bicarbonate (baking soda)	$NaHCO_3$	69	145	
Sodium carbonate (sal soda or soda ash)	$NaCO_3 \cdot 10H_2O$	204	475	452
Sodium chloride (common salt)	$NaCl$	357	366	392
Sodium hydroxide (caustic soda)	$NaOH$	420	1,448	3,388
Sodium nitrate (Chile saltpeter)	$NaNO_3$	733	1,148	1,755
Sodium sulfate (Glauber salts)	$Na_2SO_4 \cdot 10H_2O$	49	466	422
Zinc chloride	$ZnCl_2$	2,044	4,702	6,147
Zinc nitrate	$Zn(NO_3)_2 \cdot 6H_2O$	947		
Zinc sulfate	$ZnSO_4 \cdot 7H_2O$	419	768	807

* 59°F.
† 68°F.
‡ In cold water.
§ 50°F.

the subject of intense investigations. The protons, which are positively charged, neutrons, which are uncharged, along with other subatomic particles are located at the center or nucleus of the atom. The neutrons are particles having approximately the mass of a proton, and the mass of an electron is quite small, about 0.0544 percent of a proton. The electrons are negatively charged particles, all alike, external to the nucleus, and sufficient in number to neutralize the nuclear charge in an atom. The electrons are found outside the nucleus in predicted orbital levels and pathways. The electronic configuration, initially proposed by Bohr and substantially revamped by Schrodinger can be broadly envisioned as a system in which the electrons are revolving in orbits about the nucleus somewhat similar to the planets of the solar system around the sun.

The designation of each element is based on the number of protons, named the atomic number, in the nucleus of the atom. In a hydrogen atom, atomic number 1, there is one proton and one electron; in a radium atom, atomic number 88, there are 88 protons, 88 electrons, and 138 neutrons surrounding a nucleus 226 times as massive as the hydrogen nucleus. The essential feature that distinguishes one element from another is the atomic number. It also determines the position of the element in the periodic Table 4.1.3. Modern research has shown the existence of isotopes, that is, two or more species of atoms having the same atomic number and thus occupying the same place in the periodic system, but differing number of neutrons. These isotopes are chemically identical and are merely different species of the same chemical element. Most elements have been shown to consist of a mixture of isotopes.

Chemical reactions involve electrons that are furthest, often called valence electrons, from the nucleus. These outermost electrons are held lest tightly by the nucleus. Reactions occur when substances lose or gain or share the outer electrons.

A few elements have been shown to be radioactive. These have unstable nuclei which many believe to be due to the proton/neutron ratio. They undergo nuclear reactions to become more stable, which

Table 4.1.3 Periodic Table of the Elements

Periodic Table of Elements

Key: Atomic # / Symbol / Name / Atomic Mass (electron shells K, L, M, N, O, P, Q listed at the side of each cell)

C Solid · Hg Liquid · H Gas · Rf Unknown

Metals: Alkali metals · Alkaline earth metals · Lanthanoids · Actinoids · Transition metals · Poor metals
Nonmetals: Other nonmetals · Noble gases

Period	1	2	3	4	5	6	7	8	9	10	11	12	13	14	15	16	17	18	Shells
1	1 H Hydrogen 1.00794 [1]																	2 He Helium 4.002602 [2]	K
2	3 Li Lithium 6.941 [2 1]	4 Be Beryllium 9.012182 [2 2]											5 B Boron 10.811 [2 3]	6 C Carbon 12.0107 [2 4]	7 N Nitrogen 14.0067 [2 5]	8 O Oxygen 15.9994 [2 6]	9 F Fluorine 18.9984032 [2 7]	10 Ne Neon 20.1797 [2 8]	K L
3	11 Na Sodium 22.98976928 [2 8 1]	12 Mg Magnesium 24.3050 [2 8 2]											13 Al Aluminium 26.9815386 [2 8 3]	14 Si Silicon 28.0855 [2 8 4]	15 P Phosphorus 30.973762 [2 8 5]	16 S Sulfur 32.065 [2 8 6]	17 Cl Chlorine 35.453 [2 8 7]	18 Ar Argon 39.948 [2 8 8]	K L M
4	19 K Potassium 39.0983 [2 8 8 1]	20 Ca Calcium 40.078 [2 8 8 2]	21 Sc Scandium 44.955912 [2 8 9 2]	22 Ti Titanium 47.867 [2 8 10 2]	23 V Vanadium 50.9415 [2 8 11 2]	24 Cr Chromium 51.9961 [2 8 13 1]	25 Mn Manganese 54.938045 [2 8 13 2]	26 Fe Iron 55.845 [2 8 14 2]	27 Co Cobalt 58.933195 [2 8 15 2]	28 Ni Nickel 58.6934 [2 8 16 2]	29 Cu Copper 63.546 [2 8 18 1]	30 Zn Zinc 65.38 [2 8 18 2]	31 Ga Gallium 69.723 [2 8 18 3]	32 Ge Germanium 72.64 [2 8 18 4]	33 As Arsenic 74.92160 [2 8 18 5]	34 Se Selenium 78.96 [2 8 18 6]	35 Br Bromine 79.904 [2 8 18 7]	36 Kr Krypton 83.798 [2 8 18 8]	K L M N
5	37 Rb Rubidium 85.4678 [2 8 18 8 1]	38 Sr Strontium 87.62 [2 8 18 8 2]	39 Y Yttrium 88.90585 [2 8 18 9 2]	40 Zr Zirconium 91.224 [2 8 18 10 2]	41 Nb Niobium 92.90638 [2 8 18 12 1]	42 Mo Molybdenum 95.96 [2 8 18 13 1]	43 Tc Technetium (97.9072) [2 8 18 13 2]	44 Ru Ruthenium 101.07 [2 8 18 15 1]	45 Rh Rhodium 102.90550 [2 8 18 16 1]	46 Pd Palladium 106.42 [2 8 18 18 0]	47 Ag Silver 107.8682 [2 8 18 18 1]	48 Cd Cadmium 112.411 [2 8 18 18 2]	49 In Indium 114.818 [2 8 18 18 3]	50 Sn Tin 118.710 [2 8 18 18 4]	51 Sb Antimony 121.760 [2 8 18 18 5]	52 Te Tellurium 127.60 [2 8 18 18 6]	53 I Iodine 126.90447 [2 8 18 18 7]	54 Xe Xenon 131.293 [2 8 18 18 8]	K L M N O
6	55 Cs Caesium 132.9054519 [2 8 18 18 8 1]	56 Ba Barium 137.327 [2 8 18 18 8 2]	57–71	72 Hf Hafnium 178.49 [2 8 18 32 10 2]	73 Ta Tantalum 180.94788 [2 8 18 32 11 2]	74 W Tungsten 183.84 [2 8 18 32 12 2]	75 Re Rhenium 186.207 [2 8 18 32 13 2]	76 Os Osmium 190.23 [2 8 18 32 14 2]	77 Ir Iridium 192.217 [2 8 18 32 15 2]	78 Pt Platinum 195.084 [2 8 18 32 17 1]	79 Au Gold 196.966569 [2 8 18 32 18 1]	80 Hg Mercury 200.59 [2 8 18 32 18 2]	81 Tl Thallium 204.3833 [2 8 18 32 18 3]	82 Pb Lead 207.2 [2 8 18 32 18 4]	83 Bi Bismuth 208.98040 [2 8 18 32 18 5]	84 Po Polonium (208.9824) [2 8 18 32 18 6]	85 At Astatine (209.9871) [2 8 18 32 18 7]	86 Rn Radon (222.0176) [2 8 18 32 18 8]	K L M N O P
7	87 Fr Francium (223) [2 8 18 32 18 8 1]	88 Ra Radium (226) [2 8 18 32 18 8 2]	89–103	104 Rf Rutherfordium (261) [2 8 18 32 32 10 2]	105 Db Dubnium (262) [2 8 18 32 32 11 2]	106 Sg Seaborgium (266) [2 8 18 32 32 12 2]	107 Bh Bohrium (264) [2 8 18 32 32 13 2]	108 Hs Hassium (277) [2 8 18 32 32 14 2]	109 Mt Meitnerium (268) [2 8 18 32 32 15 2]	110 Ds Darmstadtium (271) [2 8 18 32 32 17 1]	111 Rg Roentgenium (272) [2 8 18 32 32 18 1]	112 Uub Ununbium (285) [2 8 18 32 32 18 2]	113 Uut Ununtrium (284) [2 8 18 32 32 18 3]	114 Uuq Ununquadium (289) [2 8 18 32 32 18 4]	115 Uup Ununpentium (288) [2 8 18 32 32 18 5]	116 Uuh Ununhexium (292) [2 8 18 32 32 18 6]	117 Uus Ununseptium	118 Uuo Ununoctium (294) [2 8 18 32 32 18 8]	K L M N O P Q

For elements with no stable isotopes, the mass number of the isotope with the longest half-life is in parentheses.

57 La Lanthanum 138.90547 [2 8 18 18 9 2]	58 Ce Cerium 140.116 [2 8 18 19 9 2]	59 Pr Praseodymium 140.90765 [2 8 18 21 8 2]	60 Nd Neodymium 144.242 [2 8 18 22 8 2]	61 Pm Promethium (145) [2 8 18 23 8 2]	62 Sm Samarium 150.36 [2 8 18 24 8 2]	63 Eu Europium 151.964 [2 8 18 25 8 2]	64 Gd Gadolinium 157.25 [2 8 18 25 9 2]	65 Tb Terbium 158.92535 [2 8 18 27 8 2]	66 Dy Dysprosium 162.500 [2 8 18 28 8 2]	67 Ho Holmium 164.93032 [2 8 18 29 8 2]	68 Er Erbium 167.259 [2 8 18 30 8 2]	69 Tm Thulium 168.93421 [2 8 18 31 8 2]	70 Yb Ytterbium 173.054 [2 8 18 32 8 2]	71 Lu Lutetium 174.9668 [2 8 18 32 9 2]
89 Ac Actinium (227) [2 8 18 32 18 9 2]	90 Th Thorium 232.03806 [2 8 18 32 18 10 2]	91 Pa Protactinium 231.03588 [2 8 18 32 20 9 2]	92 U Uranium 238.02891 [2 8 18 32 21 9 2]	93 Np Neptunium (237) [2 8 18 32 22 9 2]	94 Pu Plutonium (244) [2 8 18 32 24 8 2]	95 Am Americium (243) [2 8 18 32 25 8 2]	96 Cm Curium (247) [2 8 18 32 25 9 2]	97 Bk Berkelium (247) [2 8 18 32 27 8 2]	98 Cf Californium (251) [2 8 18 32 28 8 2]	99 Es Einsteinium (252) [2 8 18 32 29 8 2]	100 Fm Fermium (257) [2 8 18 32 30 8 2]	101 Md Mendelevium (258) [2 8 18 32 31 8 2]	102 No Nobelium (259) [2 8 18 32 32 8 2]	103 Lr Lawrencium (262) [2 8 18 32 32 9 2]

Ptable.com

may include ejection of neutrons, protons, and other particles, rearrangement of the nuclei to give smaller elements, and conversion of atomic mass to energy in the form of rays and heat.

Calculation of the Percentage Composition of Substances Add the atomic weights of the elements in the compound to obtain its molecular weight. Multiply the atomic weight of the element to be calculated by the number of atoms present (indicated in the formula by a subscript number) and by 100, and divide by the molecular weight of the compound. For example, hematite iron ore (Fe_2O_3) contains 69.94 percent of iron by weight, determined as follows: Molecular weight of Fe_2O_3 = (55.84 × 2) + (16.00 × 3) = 159.68. Percentage of iron in compound = (55.84 × 2) × 100/159.68 = 69.94. Some of the physical properties of commonly known matter are reported in Tables 4.1.4 to 4.1.11 below.

Table 4.1.4 Solubility of Gases in Water
(By volume at atmospheric pressure)

	t, °F (°C)		
	32 (0)	68 (20)	212 (100)
Air	0.032	0.020	0.012
Acetylene	1.89	1.12	
Ammonia	1,250	700	
Carbon dioxide	1.87	0.96	0.26
Carbon monoxide	0.039	0.025	
Chlorine	5.0	2.5	0.00
Hydrogen	0.023	0.020	0.018
Hydrogen sulfide	5.0	2.8	0.87
Hydrochloric acid	560	480	
Nitrogen	0.026	0.017	0.0105
Oxygen	0.053	0.034	0.185
Sulfuric acid	87	43	

SPECIFIC GRAVITIES AND DENSITIES AND OTHER PHYSICAL DATA

Table 4.1.5 Approximate Specific Gravities and Densities
(Water at 39°F and normal atmospheric pressure taken as unity) For more detailed data on any material, see the section dealing with the properties of that material. Data given are for usual room temperatures.

Substance	Specific gravity	Avg. density lb/ft³	Avg. density kg/m³
	Metals, alloys, ores*		
Aluminum, cast-hammered	2.55-2.80	165	2,643
Brass, cast-rolled	8.4-8.7	534	8,553
Bronze, aluminum	7.7	481	7,702
Bronze, 7.9-14% Sn	7.4-8.9	509	8,153
Bronze, phosphor	8.88	554	8,874
Copper, cast-rolled	8.80-8.95	556	8,906
Copper ore, pyrites	4.1-4.3	262	4,197
German silver	8.58	536	8,586
Gold, cast-hammered	19.25-19.35	1,205	19,300
Gold coin (U.S.)	17.18-17.20	1,073	17,190
Iridium	21.78-22.42	1,383	22,160
Iron, gray cast	7.03-7.13	442	7,079
Iron, cast, pig	7.2	450	7,207
Iron, wrought	7.6-7.9	485	7,658
Iron, spiegeleisen	7.5	468	7,496
Iron, ferrosilicon	6.7-7.3	437	6,984
Iron ore, hematite	5.2	325	5,206
Iron ore, limonite	3.6-4.0	237	3,796
Iron ore, magnetite	4.9-5.2	315	5,046
Iron slag	2.5-3.0	172	2,755
Lead	11.34	710	11,370
Lead ore, galena	7.3-7.6	465	7,449
Manganese	7.42	475	7,608
Manganese ore, pyrolusite	3.7-4.6	259	4,149
Mercury	13.546	847	13,570
Monel metal, rolled	8.97	555	8,688
Nickel	8.9	537	8,602
Platinum, cast-hammered	21.5	1,330	21,300
Silver, cast-hammered	10.4-10.6	656	10,510
Steel, cold-drawn	7.83	489	7,832
Steel, machine	7.80	487	7,800
Steel, tool	7.70-7.73	481	7,703
Tin, cast-hammered	7.2-7.5	459	7,352
Tin ore, cassiterite	6.4-7.0	418	6,695
Tungsten	19.22	1,200	18,820
Uranium	18.7	1,170	18,740
Zinc, cast-rolled	6.9-7.2	440	7,049
Zinc, ore, blende	3.9-4.2	253	4,052

Table 4.1.5 Approximate Specific Gravities and Densities (*Continued*)

Substance	Specific gravity	Avg. density lb/ft³	Avg. density kg/m³
	Various solids		
Cereals, oats, bulk	0.41	26	417
Cereals, barley, bulk	0.62	39	625
Cereals, corn, rye, bulk	0.73	45	721
Cereals, wheat, bulk	0.77	48	769
Cordage (natural fiber)	1.2-1.5	85	1,360
Cordage (plastic)	0.9-1.3	69	1,104
Cork	0.22-0.26	15	240
Cotton, flax, hemp	1.47-1.50	93	1,491
Fats	0.90-0.97	58	925
Flour, loose	0.40-0.50	28	448
Flour, pressed	0.70-0.80	47	753
Glass, common	2.40-2.80	162	2,595
Glass, plate or crown	2.45-2.72	161	2,580
Glass, crystal	2.90-3.00	184	1,950
Glass, flint	3.2-4.7	247	3,960
Hay and straw, bales	0.32	20	320
Leather	0.86-1.02	59	945
Paper	0.70-1.15	58	929
Plastics (see Sec. 4.9)			
Potatoes, piled	0.67	44	705
Rubber, caoutchouc	0.92-0.96	59	946
Rubber goods	1.0-2.0	94	1,506
Salt, granulated, piled	0.77	48	769
Saltpeter	2.11	132	2,115
Starch	1.53	96	1,539
Sulfur	1.93-2.07	125	2,001
Wool	1.32	82	1,315
	Timber, air-dry		
Apple	0.66-0.74	44	705
Ash, black	0.55	34	545
Ash, white	0.64-0.71	42	973
Birch, sweet, yellow	0.71-0.72	44	705
Cedar, white, red	0.35	22	352
Cherry, wild red	0.43	27	433
Chestnut	0.48	30	481
Cypress	0.45-0.48	29	465
Fir, Douglas	0.48-0.55	32	513
Fir, balsam	0.40	25	401
Elm, white	0.56	35	561

(Continued)

Table 4.1.5 Approximate Specific Gravities and Densities *(Continued)*

Substance	Specific gravity	Avg. density lb/ft³	Avg. density kg/m³
Hemlock	0.45-0.50	29	465
Hickory	0.74-0.80	48	769
Locust	0.67-0.77	45	722
Mahogany	0.56-0.85	44	705
Maple, sugar	0.68	43	689
Maple, white	0.53	33	529
Oak, chestnut	0.74	46	737
Oak, live	0.87	54	866
Oak, red, black	0.64-0.71	42	673
Oak, white	0.77	48	770
Pine, Oregon	0.51	32	513
Pine, red	0.48	30	481
Pine, white	0.43	27	433
Pine, Southern	0.61-0.67	38-42	610-673
Pine, Norway	0.55	34	541
Poplar	0.43	27	433
Redwood, California	0.42	26	417
Spruce, white, red	0.45	28	449
Teak, African	0.99	62	994
Teak, Indian	0.66-0.88	48	769
Walnut, black	0.59	37	593
Willow	0.42-0.50	28	449
Various liquids			
Alcohol, ethyl (100%)	0.789	49	802
Alcohol, methyl (100%)	0.796	50	809
Acid, muriatic (40%)	1.20	75	1,201
Acid, nitric (91%)	1.50	94	1,506
Acid, sulfuric (87%)	1.80	112	1,795
Chloroform	1.500	95	1,532
Ether	0.736	46	738
Lye, soda (66%)	1.70	106	1,699
Oils, vegetable	0.91-0.94	58	930
Oils, mineral, lubricants	0.88-0.94	57	914
Turpentine	0.861-0.867	54	866
Water			
Water, 4°C, max density	1.0	62.426	999.97
Water, 25°C	0.9970	62.24	997.0
Water, 100°C	0.9584	59.812	958.10
Water, ice	0.88-0.92	56	897
Water, snow, fresh fallen	0.125	8	128
Water, seawater	1.02-1.03	64	1,025
Ashlar masonry			
Granite, syenite, gneiss	2.4-2.7	159	2,549
Limestone	2.1-2.8	153	2,450
Marble	2.4-2.8	162	2,597
Sandstone	2.0-2.6	143	2,290
Bluestone	2.3-2.6	153	2,451
Rubble masonry			
Granite, syenite, gneiss	2.3-2.6	153	2,451
Limestone	2.0-2.7	147	2,355
Sandstone	1.9-2.5	137	2,194
Bluestone	2.2-2.5	147	2,355
Marble	2.3-2.7	156	2,500
Dry rubble masonry			
Granite, syenite, gneiss	1.9-2.3	130	2,082
Limestone, marble	1.9-2.1	125	2,001
Sandstone, bluestone	1.8-1.9	110	1,762
Brick masonry			
Hard brick	1.8-2.3	128	2,051
Medium brick	1.6-2.0	112	1,794
Soft brick	1.4-1.9	103	1,650
Sand-lime brick	1.4-2.2	112	1,794
Concrete masonry			
Cement, stone, sand	2.2-2.4	144	2,309
Cement, slag, etc.	1.9-2.3	130	2,082
Cement, cinder, etc.	1.5-1.7	100	1,602

Table 4.1.5 Approximate Specific Gravities and Densities *(Continued)*

Substance	Specific gravity	Avg. density lb/ft³	Avg. density kg/m³
Various building materials			
Ashes, cinders	0.64-0.72	40-45	640-721
Cement, portland, loose	1.5	94	1,505
Portland cement	3.1-3.2	196	3,140
Lime, gypsum, loose	0.85-1.00	53-64	849-1,025
Mortar, lime, set	1.4-1.9	103	1,650
		94	1,505
Mortar, portland cement	2.08-2.25	135	2,163
Slags, bank slag	1.1-1.2	67-72	1,074-1,153
Slags, bank screenings	1.5-1.9	98-117	1,570-1,874
Slags, machine slag	1.5	96	1,538
Slags, slag sand	0.8-0.9	49-55	785-849
Earth, etc., excavated			
Clay, dry	1.0	63	1,009
Clay, damp, plastic	1.76	110	1,761
Clay and gravel, dry	1.6	100	1,602
Earth, dry, loose	1.2	76	1,217
Earth, dry, packed	1.5	95	1,521
Earth, moist, loose	1.3	78	1,250
Earth, moist, packed	1.6	96	1,538
Earth, mud, flowing	1.7	108	1,730
Earth, mud, packed	1.8	115	1,841
Riprap, limestone	1.3-1.4	80-85	1,282-1,361
Riprap, sandstone	1.4	90	1,441
Riprap, shale	1.7	105	1,681
Sand, gravel, dry, loose	1.4-1.7	90-105	1,441-1,681
Sand, gravel, dry, packed	1.6-1.9	100-120	1,602-1,922
Sand, gravel, wet	1.89-2.16	126	2,019
Excavations in water			
Sand or gravel	0.96	60	951
Sand or gravel and clay	1.00	65	1,041
Clay	1.28	80	1,281
River mud	1.44	90	1,432
Soil	1.12	70	1,122
Stone riprap	1.00	65	1,041
Minerals			
Asbestos	2.1-2.8	153	2,451
Barytes	4.50	281	4,504
Basalt	2.7-3.2	184	2,950
Bauxite	2.55	159	2,549
Bluestone	2.5-2.6	159	2,549
Borax	1.7-1.8	109	1,746
Chalk	1.8-2.8	143	2,291
Clay, marl	1.8-2.6	137	2,196
Dolomite	2.9	181	2,901
Feldspar, orthoclase	2.5-2.7	162	2,596
Gneiss	2.7-2.9	175	2,805
Granite	2.6-2.7	165	2,644
Greenstone, trap	2.8-3.2	187	2,998
Gypsum, alabaster	2.3-2.8	159	2,549
Hornblende	3.0	187	2,998
Limestone	2.10-2.86	155	2,484
Marble	2.60-2.86	170	2,725
Magnesite	3.0	187	2,998
Phosphate rock, apatite	3.2	200	3,204
Porphyry	2.6-2.9	172	2,758
Pumice, natural	0.37-0.90	40	641
Quartz, flint	2.5-2.8	165	2,645
Sandstone	2.0-2.6	143	2,291
Serpentine	2.7-2.8	171	2,740
Shale, slate	2.6-2.9	172	2,758
Soapstone, talc	2.6-2.8	169	2,709
Syenite	2.6-2.7	165	2,645
Stone, quarried, piled			
Basalt, granite, gneiss	1.5	96	1,579
Limestone, marble, quartz	1.5	95	1,572
Sandstone	1.3	82	1,314

Table 4.1.5 Approximate Specific Gravities and Densities (*Continued*)

Substance	Specific gravity	Avg. density lb/ft³	Avg. density kg/m³
Shale	1.5	92	1,474
Greenstone, hornblend	1.7	107	1,715
Bituminous substances			
Asphaltum	1.1-1.5	81	1,298
Coal, anthracite	1.4-1.8	97	1,554
Coal, bituminous	1.2-1.5	84	1,346
Coal, lignite	1.1-1.4	78	1,250
Coal, peat, turf, dry	0.65-0.85	47	753
Coal, charcoal, pine	0.28-0.44	23	369
Coal, charcoal, oak	0.47-0.57	33	481
Coal, coke	1.0-1.4	75	1,201
Graphite	1.64-2.7	135	2,163
Paraffin	0.87-0.91	56	898
Petroleum	0.87	54	856
Petroleum, refined (kerosene)	0.78-0.82	50	801
Petroleum, benzine	0.73-0.75	46	737
Petroleum, gasoline	0.70-0.75	45	721
Pitch	1.07-1.15	69	1,105
Tar, bituminous	1.20	75	1,201
Coal and coke, piled			
Coal, anthracite	0.75-0.93	47-58	753-930
Coal, bituminous, lignite	0.64-0.87	40-54	641-866
Coal, peat, turf	0.32-0.42	20-26	320-417
Coal, charcoal	0.16-0.23	10-14	160-224
Coal, coke	0.37-0.51	23-32	369-513
Gases (see Sec. 2)			

* See also Sec. 4.4.

Compressibility of Liquids

If v_1 and v_2 are the volumes of the liquids at pressures of p_1 and p_2 atm, respectively, at any temperature, the coefficient of compressibility b is given by the equation

$$b = \frac{1}{v_1} \frac{v_1 - v_2}{p_2 - p_1}$$

The value of $b \times 10^6$ for oils at low pressures at about 70°F varies from about 55 to 80; for mercury at 32°F, it is 3.9; for chloroform at 32°F, it is 100 and increases with the temperature to 200 at 140°F; for ethyl alcohol, it increases from about 100 at 32°F and low pressures to 125 at 104°F; for glycerin, it is about 24 at room temperature and low pressure.

Table 4.1.6 Average Composition of Dry Air between Sea Level and 90-km (295,000-ft) Altitude

Element	Formula	% by Vol.	% by Mass	Molecular weight
Nitrogen	N_2	78.084	75.55	28.0134
Oxygen	O_2	20.948	23.15	31.9988
Argon	Ar	0.934	1.325	39.948
Carbon Dioxide	CO_2	0.0314	0.0477	44.00995
Neon	Ne	0.00182	0.00127	20.183
Helium	He	0.00052	0.000072	4.0026
Krypton	Kr	0.000114	0.000409	83.80
Methane	CH_4	0.0002	0.000111	16.043

From 0.0 to 0.00005 percent by volume of nine other gases.
Average composite molecular weight of air 28.9644.
SOURCE: "U.S. Standard Atmosphere," Government Printing Office.

Table 4.1.7 Specific Gravity and Density of Water at Atmospheric Pressure*
(Weights are in vacuo)

Temp, °C	Specific gravity	Density lb/ft³	Density kg/m³	Temp, °C	Specific gravity	Density lb/ft³	Density kg/m³
0	0.99987	62.4183	999.845	40	0.99224	61.9428	992.228
2	0.99997	62.4246	999.946	42	0.99147	61.894	991.447
4	1.00000	62.4266	999.955	44	0.99066	61.844	990.647
6	0.99997	62.4246	999.946	46	0.98982	61.791	989.797
8	0.99988	62.4189	999.854	48	0.98896	61.737	988.931
10	0.99973	62.4096	999.706	50	0.98807	61.682	988.050
12	0.99952	62.3969	999.502	52	0.98715	61.624	987.121
14	0.99927	62.3811	999.272	54	0.98621	61.566	986.192
16	0.99897	62.3623	998.948	56	0.98524	61.505	985.215
18	0.99862	62.3407	998.602	58	0.98425	61.443	984.222
20	0.99823	62.3164	998.213	60	0.98324	61.380	983.213
22	0.99780	62.2894	997.780	62	0.98220	61.315	982.172
24	0.99732	62.2598	997.304	64	0.98113	61.249	981.113
26	0.99681	62.2278	996.793	66	0.98005	61.181	980.025
28	0.99626	62.1934	996.242	68	0.97894	61.112	978.920
30	0.99567	62.1568	995.656	70	0.97781	61.041	977.783
32	0.99505	62.1179	995.033	72	0.97666	60.970	976.645
34	0.99440	62.0770	994.378	74	0.97548	60.896	975.460
36	0.99371	62.0341	993.691	76	0.97428	60.821	974.259
38	0.99299	61.9893	992.973	78	0.97307	60.745	973.041

* See also Sec. 2.2.

Table 4.1.8 Volume of Water as a Function of Pressure and Temperature

Temp, °F (°C)	Pressure, atm 0	500	1,000	2,000	3,000	4,000	5,000	6,500	8,000
32 (0)	1.0000	0.9769	0.9566	0.9223	0.8954	0.8739	0.8565	0.8361	
68 (20)	1.0016	0.9804	0.9619	0.9312	0.9065	0.8855	0.8675	0.8444	0.8244
122 (50)	1.0128	0.9915	0.9732	0.9428	0.9183	0.8974	0.8792	0.8562	0.8369
176 (80)	1.0287	1.0071	0.9884	0.9568	0.9315	0.9097	0.8913	0.8679	0.8481

SOURCE: "International Critical Tables."

Table 4.1.9 Basic Properties of Several Metals
(Staff contribution)*

Material	Density,† g/cm³	Coefficient of linear thermal expansion,‡ in/(in · °F) × 10⁻⁶	Thermal conductivity, Btu/(h · ft · °F)	Specific heat,‡ Btu/(lb · °F)	Approx melting temp, °F	Modulus of elasticity, lb/in² × 10⁶	Poisson's ratio	Ultimate Yield stress, lb/in² × 10³	stress, lb/in² × 10³	Elongation, %
Aluminum 2024-T3	2.77	12.6	110	0.23	940	10.6	0.33	50	70	18
Aluminum 6061-T6	2.70	13.5	90	0.23	1,080	10.6	0.33	40	45	17
Aluminum 7079-T6	2.74	13.7	70	0.23	900	10.4	0.33	68	78	14
Beryllium, QMV	1.85	6.4-10.2	85	0.45	2,340	40-44	0.024-0.030	27-38	33-51	1.0-3.5
Copper, pure	8.90	9.2	227	0.092	1,980	17.0	0.32		See "Metals Handbook"	
Gold, pure	19.32		172	0.031	1,950	10.8	0.42		18	30
Lead, pure	11.34	29.3	21.4	0.031	620	2.0	0.40-0.45	1.3	2.6	20-50
Magnesium AZ31B-H24 (sheet)	1.77	14.5	55	0.25	1,100	6.5	0.35	22	37	15
Magnesium HK31A-H24	1.79	14.0	66	0.13	1,100	6.4	0.35	29	37	8
Molybdenum, wrought	10.3	3.0	83	0.07	4,730	40.0	0.32	80	120-200	Small
Nickle, pure	8.9	7.2	53	0.11	2,650	32.0	0.31§		See "Metals Handbook"	
Platinum	21.45	5.0	40	0.031	3,217	21.3	0.39		20-24	35-40
Plutonium, alpha phase	19.0-19.7	30.0	4.8	0.034	1,184	14.0	0.15-0.21	40	60	Small
Silver, pure	10.5	11.0	241	0.056	1,760	10-11	0.37		18	48
Steel, AISI C1020 (hot-worked)	7.85	6.3	27	0.10	2,750	29-30	0.29	48	65	36
Steel, AISI 304 (sheet)	8.03	9.9	9.4	0.12	2,600	28	0.29	39	87	65
Tantalum	16.6	3.6	31	0.03	5,425	27.0	0.35		50-145	1-40
Thorium, induction melt	11.6	6.95	21.7	0.03	3,200	7-10	0.27	21	32	34
Titanium, B 120VCA (aged)	4.85	5.2	4.3	0.13	3,100	14.8	0.3	190	200	9
Tungesten	19.3	2.5	95	0.033	6,200	50	0.28		18-600	1-3
Uranium D-38	18.97	4.0-8.0	17	0.028	2,100	24	0.21	28	56	4

Room-temperature properties are given. For further information, consult the "Metals Handbook" or a manufacturer's publication.
* Compiled by Anders Lundberg, University of California, and reproduced by permission.
† To obtain the preferred density units, kg/m³, multiply these values by 1,000.
‡ See also Tables 4.1.10 and 4.1.11.
§ At 25°C.

Table 4.1.10 Coefficient of Linear Thermal Expansion for Various Materials
[Mean values between 32 and 212°F except as noted; in/(in · °F) × 10^{-4}]

Metals		Other Materials		Paraffin:	
Aluminum bronze	0.094	Bakelite, bleached	0.122	32-61°F	0.592
Brass, cast	0.104	Brick	0.053	61-100°F	0.724
Brass, wire	0.107	Carbon—coke	0.030	100-120°F	2.612
Bronze	0.100	Cement, neat	0.060	Porcelain	0.02
Constantan (60 Cu, 40 Ni)	0.095	Concrete	0.060	Quartz:	
German silver	0.102	Ebonite	0.468	Parallel to axis	0.044
Iron:		Glass:		Perpend. to axis	0.074
Cast	0.059	Thermometer	0.045	Quarts, fused	0.0028
Soft forged	0.063	Hard	0.033	Rubber	0.428
Wire	0.080	Plate and crown	0.050	Vulcanite	0.400
Magnalium (85 Al, 15 Mg)	0.133	Flint	0.044	Wood (‖ to fiber):	
Phosphor bronze	0.094	Pyrex	0.018	Ash	0.053
Solder	0.134	Granite	0.04-0.05	Chestnut and maple	0.036
Speculum metal	0.107	Graphite	0.044	Oak	0.027
Steel:		Gutta percha	0.875	Pine	0.030
Bessemer, rolled hard	0.056	Ice	0.283	Across the fiber:	
Bessemer, rolled soft	0.063	Limestone	0.023-0.050	Chestnut and pine	0.019
Nickel (10% Ni)	0.073	Marble	0.02-0.09	Maple	0.027
Type metal	0.108	Masonry	0.025-0.050	Oak	0.030
				Zirconium tungstate	−0.057

Table 4.1.11 Specific Heat of Various Materials
[Mean values between 32 and 212°F; Btu/(lb · °F)]

Solids		Granite	0.195	Wood:	
Alloys:		Graphite	0.201	Fir	0.65
Bismuth-tin	0.040-0.045	Gypsum	0.259	Oak	0.57
Bell metal	0.086	Hornblende	0.195	Pine	0.67
Brass, yellow	0.0883	Humus (soil)	0.44	Liquids	
Brass, red	0.090	Ice:		Acetic acid	0.51
Bronze	0.104	−4°F	0.465	Acetone	0.544
Constantan	0.098	32°F	0.487	Alcohol (absolute)	0.58
German silver	0.095	India rubber (Para)	0.27-0.48	Aniline	0.49
Lipowits's metal	0.040	Kaolin	0.224	Bensol	0.40
Nickel steel	0.109	Limestone	0.217	Chloroform	0.23
Rose's metal	0.050	Marble	0.210	Ether	0.54
Solders (Pb and Sn)	0.040-0.045	Oxides:		Ethyl acetate	0.478
Type metal	0.0388	Alumina (Al_3O_2)	0.183	Ethylene glycol	0.602
Wood's metal	0.040	Cu_2O	0.111	Fusel oil	0.56
40 Pb + 60 Bi	0.0317	Lead oxide (PbO)	0.055	Gasoline	0.50
25 Pb + 75 Bi	0.030	Lodestone	0.156	Glycerin	0.58
Asbestos	0.20	Magnesia	0.222	Hydrochloric acid	0.60
Ashes	0.20	Magnetite (Fe_3O_4)	0.168	Kerosene	0.50
Bakelite	0.3-0.4	Silica	0.191	Naphthalene	0.31
Basalt (lava)	0.20	Soda	0.231	Machine oil	0.40
Borax	0.229	Zinc oxide (ZnO)	0.125	Mercury	0.033
Brick	0.22	Paraffin wax	0.69	Olive oil	0.40
Carbon-coke	0.203	Porcelain	0.22	Paraffin oil	0.52
Chalk	0.215	Quarts	0.17-0.28	Petroleum	0.50
Charcoal	0.20	Quicklime	0.21	Sulfuric acid	0.336
Cinders	0.18	Salt, rock	0.21	Sea water	0.94
Coal	0.3	Sand	0.195	Toluene	0.40
Concrete	0.156	Sandstone	0.22	Turpentine	0.42
Cork	0.485	Serpentine	0.25	Molten metals:	
Corundum	0.198	Sulfur	0.180	Bismuth (535-725°F)	0.036
Dolomite	0.222	Talc	0.209	Lead (590-680°F)	0.041
Ebonite	0.33	Tufa	0.33	Sulfur (246-297°F)	0.235
Glass:		Vulcanite	0.331	Tin (460-660°F)	0.058
Normal	0.199				
Crown	0.16				
Flint	0.12				

Table 4.1.12 Index of Refraction of Various Solids, Liquids, and Gasses
[Note: Index of refraction of a vacuum is defined as 1.00]

Acetic acid	1.37
Acetone	1.36
Air	1.00029
Benzene	1.50
Carbon disulfide	1.63
Carbon tetrachloride	1.46
Chloroform	1.44
Diamond	2.42
Ethanol	1.36
Ether	1.35
Ethylene glycol	1.43
Hexane	1.37
Ice	1.31
Methanol	1.33
Nitrogen	1.00030
Polystyrene	1.59
Silica (amorphous)	1.46
Sodium chloride	1.52
Toluene	1.497
Water	1.33

4.2 IRON AND STEEL

by Harold W. Paxton

REFERENCES: "Metals Handbook," ASM International, latest ed., ASTM Standards, pt. 1. SAE Handbook. "Steel Products Manual," AISI. "Making, Shaping and Treating of Steel," AISE, latest ed.

CLASSIFICATION OF IRON AND STEEL

Iron (Fe) is not a high-purity metal commercially but contains other chemical elements which have a large effect on its physical and mechanical properties. The amount and distribution of these elements are dependent upon the method of manufacture. The most important commercial forms of iron are listed below.

Pig iron is the product of the blast furnace and is made by the reduction of iron ore.

Cast iron is an alloy of iron containing enough carbon to have a low melting temperature and which can be cast to close to final shape. It is not generally capable of being deformed before entering service.

Gray cast iron is an iron which, as cast, has combined carbon (in the form of cementite, Fe_3C) not in excess of a eutectoid percentage—the balance of the carbon occurring as graphite flakes. The term "gray iron" is derived from the characteristic gray fracture of this metal.

White cast iron contains carbon in the combined form. The presence of cementite or iron carbide (Fe_3C) makes this metal hard and brittle, and the absence of graphite gives the fracture a white color.

Malleable cast iron is an alloy in which all the combined carbon in a special white cast iron has been changed to free or temper carbon by suitable heat treatment.

Nodular (ductile) cast iron is produced by adding alloys of magnesium or cerium to molten iron. These additions cause the graphite to form into small nodules, resulting in a higher-strength, ductile iron.

Ingot iron, electrolytic iron (an iron-hydrogen alloy), and **wrought iron** are terms for low-carbon materials which are no longer serious items of commerce but do have considerable historical interest.

Steel is an alloy predominantly of iron and carbon, usually containing measurable amounts of manganese, and often readily formable.

Carbon steel is steel that owes its distinctive properties chiefly to the carbon it contains.

Alloy steel is steel that owes its distinctive properties chiefly to some element or elements other than carbon, or jointly to such other elements and carbon. Some alloy steels necessarily contain an important percentage of carbon, even as much as 1.25 percent. There is no complete agreement about where to draw the line between the alloy steels and the carbon steels.

Basic oxygen steel and **electric-furnace steel** are steels made by the basic oxygen furnace and electric furnace processes, irrespective of carbon content; the effective individual alloy content in engineering steels can range from 0.05 percent up to 3 percent, with a total usually less than 5 percent. Open-hearth and Bessemer steelmaking are no longer practiced in the United States.

Iron ore is reduced in a blast furnace to form **pig iron,** which is the raw material for practically all iron and steel products. Formerly, nearly 90 percent of the iron ore used in the United States came from the Lake Superior district; the ore had the advantages of high quality and the cheapness with which it could be mined and transported by way of the Great Lakes. With the rise of global steelmaking and the availability of high-grade ores and pellets (made on a large scale from low-grade ores) from many sources, the choice of feedstock becomes an economic decision.

The modern **blast furnace** consists of a vertical shaft up to 10 m or 40 ft in diameter and over 30 m (100 ft) high containing a descending column of iron ore, coke, and limestone and a large volume of ascending hot gas. The gas is produced by the burning of coke in the hearth of the furnace and contains about 34 percent carbon monoxide. This gas reduces the iron ore to metallic iron, which melts and picks up considerable quantities of carbon, manganese, phosphorus, sulfur, and

Table 4.2.1 Types of Pig Iron for Steelmaking and Foundry Use

Designation	Chemical composition, %*				Principal use
	Si	P	Mn	C†	
Basic pig, northern	1.50 max	0.400 max	1.01-2.00	3.5-4.4	Basic oxygen steel
In steps of	0.25		0.50		
Foundry, northern	3.50 max	0.301-0.700	0.50-1.25	3.0-4.5	A wide variety of castings
In steps of	0.25		0.25		
Foundry, southern	3.50 max	0.700-0.900	0.40-0.75	3.0-4.5	Cast-iron pipe
In steps of	0.25		0.25		
Ferromanganese (3 grades)	1.2 max	0.35 max	74-82	7.4 max	Addition of manganese to steel or cast iron
Ferrosilicon (silvery pig)	5.00-17.00	0.300 max	1.00-2.00	1.5 max	Addition of silicon to steel or cast iron

* Excerpted from "The Making, Shaping and Treating of Steel," AISE, 1984; further information in "Steel Products Manual," AISI, and ASTM Standards, Pt. 1.
† Carbon content not specified—for information only. Usually S is 0.05 max (0.06 for ferrosilicon) but S and P for basic oxygen steel are typically much lower today.

silicon. The **gangue** (mostly silica) of the iron ore and the ash in the coke combine with the limestone to form the blast-furnace slag. The pig iron and slag are drawn off at intervals from the hearth through the iron notch and cinder notch, respectively. Some of the larger blast furnaces produce around 10,000 tons of pig iron per day. The blast furnace produces a liquid product for one of three applications: (1) the huge majority passes to the steelmaking process for refining; (2) pig iron is used in foundries for making castings; and (3) ferroalloys, which contain a considerable percentage of another metallic element, are used as addition agents in steelmaking. Compositions of commercial pig irons and two ferroalloys (ferromanganese and ferrosilicon) are listed in Table 4.2.1.

Physical Constants of Unalloyed Iron Some physical properties of iron and even its dilute alloys are sensitive to small changes in composition, grain size, or degree of cold work. The following are reasonably accurate for "pure" iron at or near room temperature; those with an asterisk are sensitive to these variables perhaps by 10 percent or more. Those with a dagger (†) depend measurably on temperature; more extended tables should be consulted.

Specific gravity, 7.866; melting point, 1,536°C (2,797°F); heat of fusion 277 kJ/kg (119 Btu/lbm); thermal conductivity 80.2 W/(m·C) [557 Btu/(h·ft^2·in·°F)*†; thermal coefficient of expansion 12×10^{-6}/°C (6.7×10^{-6}/°F)†; electrical resistivity 9.7 $\mu\Omega \cdot$cm*†; and temperature coefficient of electrical resistance 0.0065/°C (0.0036/°F).†

Mechanical Properties Representative mechanical properties of annealed low-carbon steel (often similar to the former ingot iron) are as follows: yield strength 130 to 150 MPa (20 to 25 ksi); tensile strength 260 to 300 MPa (40 to 50 ksi); elongation 20 to 45 percent in 2 in; reduction in area of 60 to 75 percent; Brinell hardness 65 to 100. These figures are at best approximate and depend on composition (especially trace additives) and processing variables. For more precise data, suppliers or broader databases should be consulted.

Young's modulus for ingot iron is 202,000 MPa (29,300,000 lb/in^2) in both tension and compression, and the shear modulus is 81,400 MPa (11,800,000 lb/in^2). Poisson's ratio is 0.28. The **effect of cold rolling** on the tensile strength, yield strength, elongation, and shape of the stress-strain curve is shown in Fig. 4.2.1, which is for Armco ingot iron but would not be substantially different for other low-carbon steels.

Uses Low-carbon materials weld evenly and easily in all processes, can be tailored to be readily paintable and to be enameled, and with other treatments make an excellent low-cost soft magnetic material with high permeability and low coercive force for mass-produced motors and transformers. Other uses, usually after galvanizing, include culverts, flumes, roofing, siding, and housing frames; thin plates can be used in oil and water tanks, boilers, gas holders, and various nondemanding pipes; enameled sheet retains a strong market in ranges, refrigerators, and other household goods, in spite of challenges from plastics.

World Production From about 1970 to 1995, the annual world production of steel was remarkably steady at some 800,000,000 tons. Reductions in some major producers (United States, Japan, some European countries, and the collapsing U.S.S.R.) were balanced by the many

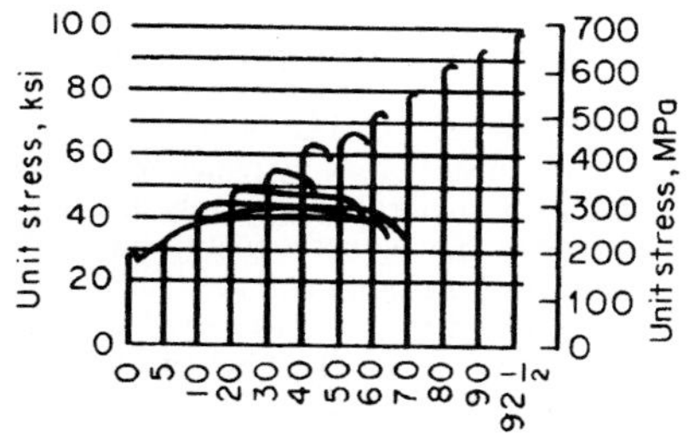

Figure 4.2.1 Effect of cold rolling on the stress-strain relationship of Armco ingot iron. *(Kenyon and Burns.)*

smaller countries which began production. The struggle for markets led to pricing that did not cover production costs and led to international political strains. A major impact began in the 1990s as China pushed its own production from a few tens of million tons toward 300,000,000 tons in 2004, pushing world production to over 1 billion tons.

The world raw material infrastructure (iron ore, coke, and scrap) was not able to react quickly to support this increase, and prices soared; steel prices followed. Recently (2006) long-term ore contracts reflected large price increases, so it appears that for some time **steel prices** will remain high by historical comparisons. The effects of material substitution are not yet clear, but as end users examine their options more closely, the marketplace will most likely settle the matter.

STEEL

Steel Manufacturing

Steel is produced by the removal of impurities from pig iron in a basic oxygen furnace or an electric furnace.

Basic Oxygen Steel This steel is produced by blowing pure (99 percent) oxygen either vertically under high pressure (1.2 MPa or 175 lb/in^2) onto the surface of molten pig iron (BOP) or through tuyeres in the base of the vessel (the Q-BOP process). Some facilities use a combination depending on local circumstances and product mix. This is an autogenous process that requires no external heat to be supplied. The furnaces are similar in shape to the former Bessemer converters but range in capacity to 275 metric tons (t) (300 net tons) or more. The barrel-shaped furnace or vessel may or may not be closed on the bottom, is open at the top, and can rotate in a vertical plane about a horizontal axis for charging and for pouring the finished steel. Selected scrap is charged into the vessel first, up to 30 percent by weight of the total charge. Molten pig iron (often purified from the raw blast-furnace hot metal to give lower sulfur, phosphorus, and sometimes silicon) is poured into the vessel. In the Q-BOP, oxygen must be flowing through the bottom tuyeres at this time to prevent clogging; further flow serves to refine the charge and carries in fluxes as powders. In the BOP process, oxygen is introduced through a water-cooled lance introduced through the top of the vessel. Within seconds after the oxygen is turned on, some iron in the charge is converted to ferrous oxide, which reacts rapidly with the impurities of the charge to remove them from the metal. As soon as reaction starts, limestone is added as a flux. Blowing is continued until the desired degree of purification is attained. The reactions take place very rapidly, and blowing of a heat is completed in about 20 min in a 200-net-ton furnace. Because of the speed of the process, a computer is used to calculate the charge required for making a given heat of steel, the rate and duration of oxygen blowing, and to regulate the quantity and timing of additions during the blow and for finishing the steel. Production rates of well over 270 t per furnace hour (300 net tons) can be attained. The comparatively low investment cost and low cost of operation have already made the basic oxygen process the largest producer of steel in the world, and along with electric furnaces, it almost completely replaces the basic open hearth as the major steelmaking process. No open hearths operate in the United States today.

Electric Steel The biggest change in steelmaking over the last 20 years is the fraction of steel made by remelting scrap in an electric furnace (EF), originally to serve a relatively nondemanding local market, but increasingly moving up in quality and products to compete with mills using the blast-furnace/oxygen steelmaking route. The economic competition is fierce and has served to improve choices for customers. In the last decade the fraction of steel made in the EF in the United States has gone above 50 percent by a combination of contraction in blast-furnace (BF) production (no new ones built, retirement of some, and a shortage of suitable coke) and an increase of total production, which has been met with EFs. At one time, there was concern that some **undesirable elements in scrap** that are not eliminated in steelmaking, notably copper and zinc, would increase with remelting cycles. However, the development of alternative **iron units** (direct reduced iron, iron carbide, and even pig iron) to dilute scrap additions has at least postponed this as an issue.

Early processes used three-phase alternating current, but increasingly the movement is to a single dc electrode with a conducting hearth. The high-power densities necessitate water cooling and improved basic refractory linings. Scrap is charged into the furnace, which usually contains some of the last heat to improve efficiency. Older practices often had a second slag made after the first meltdown and refining by oxygen blowing, but today, final refining takes place outside the melting unit in a ladle furnace, which allows refining, temperature control, and alloying additions to be made without interfering with the next heat. The materials are **continuously cast** including slabs only 50 mm (2 in) thick and casting directly to sheet in the 1- to 2-mm (0.04- to 0.08-in) thickness range is passing through the pilot stage.

The degree to which electric melting can replace more conventional methods is of great interest and depends in large part on the availability of sufficiently pure scrap at an attractive price and some improvements in surface quality to be able to make the highest-value products. Advances in EF technology are countered aggressively by new developments and cost control in traditional steelmaking; it may well be a decade or more before the pattern clarifies.

The **induction furnace** is simply a fairly small melting furnace to which the various metals are added to make the desired alloy, usually quite specialized. When steel scrap is used as a charge, it will be a high-grade scrap the composition of which is well known (see also Sec. 5).

Ladle Metallurgy One of the biggest contributors to quality in steel products is the concept of refining liquid steel outside the first melting unit—BOP, Q-BOP, or EF, none of which is well designed to perform the final refining function. In this separate unit, gases in solution (oxygen, hydrogen, and, to a lesser extent, nitrogen) can be reduced by vacuum treatment, carbon can be adjusted to desirable very low levels by reaction with oxygen in solution, alloy elements can be added, the temperature can be adjusted, and the liquid steel can be stirred by inert gases to float out inclusions and provide a homogeneous charge to the continuous casters which are now virtually ubiquitous. Reducing oxygen in solution means a "cleaner" steel (fewer nonmetallic inclusions) and a more efficient recovery of active alloying elements added with a purpose and which otherwise might end up as oxides.

Steel Ingots With the advent of continuous casters, ingot casting is now generally reserved for the production of relatively small volumes of material such as heavy plates and forgings which are too big for current casters. Ingot casting, apart from being inefficient in that the large volume change from liquid to solid must be handled by discarding the large void space usually at the top of the ingot (the **pipe**), also has several other undesirable features caused by the solidification pattern in a large volume, most notably significant differences in composition throughout the piece (**segregation**) leading to different properties, **inclusions** formed during solidification, and surface flaws from poor mold surfaces, splashing and other practices, which if not properly removed lead to defects in finished products (**seams, scabs, scale,** etc.). Some defects can be removed or attenuated, but others cannot; in general, with the exception of some very specialized tool and bearing steels, products from ingots are no longer state-of-the-art unless they are needed for size.

Continuous Casting This concept, which began with Bessemer in the 1850s, began to be a reliable production tool around 1970 and since then has replaced basically all ingot casting. Industrialized countries all continuously cast close to 100 percent of their production. Sizes cast range from 2-m (80-in)—or more—by 0.3-m (12-in) **slabs** down to 0.1-m (4-in) square or round **billets.** Multiple strands are common where production volume is important. Many heats of steel can be cast in a continuous string with changes of width possible during operation. Changes of composition are possible in succeeding ladles with a discard of the short length of mixed composition.

By intensive process control, it is often possible to avoid cooling the cast slabs to room temperature for inspection, enabling energy savings since the slabs require less reheating before hot rolling. If for some reason the slabs are cooled to room temperature, any surface defects which might lead to quality problems can be removed—usually by **scarfing** with an oxyacetylene torch or by grinding. Since this represents a yield loss, there is a real economic incentive to avoid the formation of such defects by paying attention to casting practices.

Mechanical Treatment of Steel

Cast steel, in the form of slabs, billets, or bars (these latter two differ somewhat arbitrarily in size) is treated further by various combinations of hot and cold deformation to produce a finished product for sale from the mill. Further treatments by fabricators usually occur before delivery to the final customer. These treatments have three purposes: (1) to change the shape by deformation or metal removal to desired tolerances; (2) to break up—at least partially—the segregation and large grain sizes inevitably formed during the solidification process and to redistribute the nonmetallic inclusions which are present; and (3) to change the properties. For example, these may be functional—strength or toughness—or largely aesthetic, such as reflectivity.

These purposes may be separable or in many cases may be acting simultaneously. An example is hot-rolled sheet or plate in which often the rolling schedule (reductions and temperature of each pass, and the cooling rate after the last reduction) is a critical path to obtain the properties and sizes desired and is often known as "heat treatment on the mill." The development of controls to do this has allowed much higher tonnages at attractive prices, and increasing robustness of rolls has allowed steel to be hot-rolled down to the 1 mm (0.04-in) range, where it can compete with the more expensive "cold-rolled" material.

Most steels are reduced after appropriate heating (to above 1,000°C) in various multistand hot rolling mills to produce **sheet, strip, plate, tubes, shaped sections,** or **bars.** More specialized deformation, e.g., by **hammer forging,** can result in working in more than one direction, with a distribution of inclusions which is not extended in one direction. Rolling, e.g., more readily imparts anisotropic properties. **Press forging** at slow strain rates changes the worked structure to greater depths and is preferred for high-quality products. The degree of reduction required to eliminate the cast structure varies from 4:1 to

10:1; clearly smaller reductions would be desirable but are currently not usual.

The slabs, blooms, and billets from the caster must be reheated in an atmosphere-controlled furnace to the working temperature, often from room temperature, but if practices permit, they may be charged hot to save energy. Coupling the hot deformation process directly to slabs at the continuous caster exit is potentially more efficient, but practical difficulties currently limit this to a small fraction of total production.

The steel is oxidized during heating to some degree, and this oxidation is removed by a combination of light deformation and high-pressure water sprays before the principal deformation is applied. There are differences in detail between processes, but as a representative example, the conventional production of wide "hot-rolled sheet" [>1.5 m (60 in)] will be discussed.

The slab, about 0.3 m (12 in) thick at about 1,200°C is passed through a *scale breaker* and high-pressure water sprays to remove the oxide film. It then passes through a set of *roughing passes* (possibly with some modest width reduction) to reduce the thickness to just over 25 mm (1 in), the ends are sheared perpendicular to the length to remove irregularities, and finally they are fed into a series of up to seven roll stands each of which creates a reduction of 50 to 10 percent passing along the train. Process controls allow each mill stand to run sufficiently faster than the previous one to maintain tension and avoid pileups between stands. The temperature of the sheet is a balance between heat added by deformation and that lost by heat transfer, sometimes with interstand water sprays. Ideally the temperature should not vary between head and tail of the sheet, but this is hard to accomplish.

The deformation encourages recrystallization and even some grain growth between stands; even though the time is short, temperatures are high. Emerging from the last stand between 815 and 950°C, the **austenite*** may or may not recrystallize, depending on the temperature. At higher temperatures, when austenite does recrystallize, the grain size is usually small (often in the 10- to 20-μm range). At lower exit temperatures austenite grains are rolled into "pancakes" with the short dimension often less than 10 μm. Since several **ferrite*** grains nucleate from each austenite grain during subsequent cooling, the ferrite grain size can be as low as 3 to 6 μm (ASTM 14 to 12). We shall see later that small ferrite grain sizes are a major contributor to the superior properties of today's carbon steels, which provide good strength and superior toughness simultaneously and economically.

Some of these steels also incorporate strong **carbide*** and **nitride*** formers in small amounts to provide extra strength from precipitation hardening; the degree to which these are undissolved in austenite during hot rolling affects recrystallization significantly. The subject is too complex to treat briefly here; the interested reader is referred to the ASM "Metals Handbook."

After the last pass, the strip may be cooled by programmed water sprays to between 510 and 730°C so that during coiling, any desired precipitation processes may take place in the coiler. The finished coil, usually 2 to 3 mm (0.080 to 0.120 in) thick and sometimes 1.3 to 1.5 mm (0.052 to 0.060 in) thick, which by now has a light oxide coating, is taken off line and either shipped directly or retained for further processing to make higher value-added products. Depending on composition, typical values of yield strength are from 210 up to 380 MPa (30 to 55 ksi), UTS in the range of 400 to 550 MPa (58 to 80 ksi), with an elongation in 200 mm (8 in) of about 20 percent. The higher strengths correspond to low-alloy steels.

About half the sheet produced is sold directly as **hot-rolled sheet.** The remainder is further cold-worked after scale removal by **pickling** and either is sold as cold-worked to various **tempers** or is recrystallized to form a very formable product known as **cold-rolled and annealed,** or more usually as **cold-rolled,** sheet. Strengthening by cold work is common in sheet, strip, wire, or bars. It provides an inexpensive addition to strength but at the cost of a serious loss of ductility, often a better surface finish, and a finished product held to tighter tolerances. It improves springiness by increasing the yield strength, but does not change the elastic moduli. Examples of the effect of cold working on carbon-steel drawn wires are shown in Figs. 4.2.2 and 4.2.3.

To make the highest class of formable sheet is a very sophisticated operation. After pickling, the sheet is again reduced in a multistand (three, four, or five) mill with great attention paid to tolerances and

* See Fig. 4.2.4 and the discussion thereof.

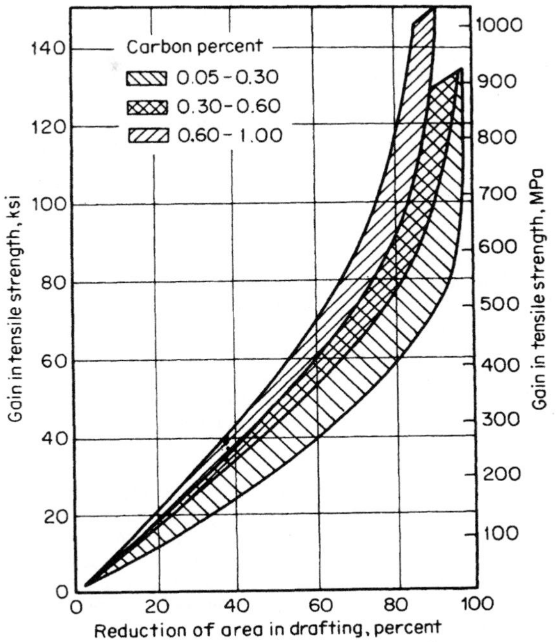

Figure 4.2.2 Increase of tensile strength of plain carbon steel with increasing amounts of cold working by drawing through a wire-drawing die.

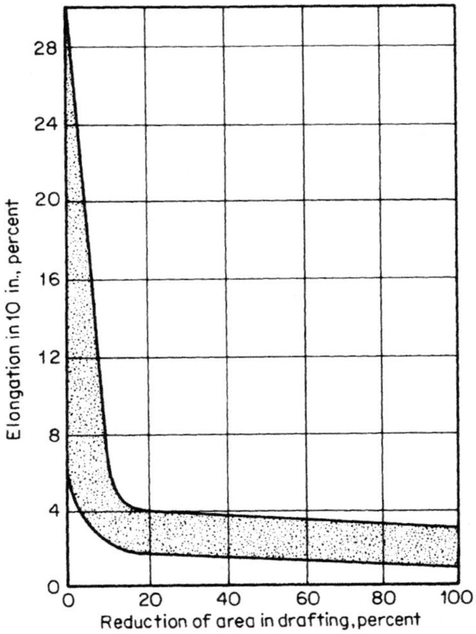

Figure 4.2.3 Reduction in ductility of plain carbon steel with increasing amounts of cold working by drawing through a wire-drawing die.

surface finish. Reductions per pass range from 25 to 45 percent in early passes to 10 to 30 percent in the last pass. The considerable heat generated necessitates an oil-water mixture to cool and to provide the necessary lubrication. The finished coil is degreased prior to annealing.

The purpose of **annealing** is to provide, for the most demanding applications, pancake-shaped grains after recrystallization of the cold-worked ferrite, in a matrix with a very sharp crystal texture containing little or no carbon or nitrogen in solution. The exact metallurgy is complex but well understood. Two types of annealing are possible: slow heating, holding, and cooling of coils in a hydrogen atmosphere (box annealing) lasting several days, or continuous feeding through a furnace with a computer-controlled time-temperature cycle. The latter is much quicker but very capital-intensive and requires careful and complex process control.

As requirements for formability are reduced, production controls can be relaxed. In order of increasing cost, the series is commercial quality (CQ), drawing quality (DQ), deep drawing quality (DDQ), and extra deep drawing quality (EDDQ). Even more formable steels are possible, but often are not commercially necessary.

Some other deformation processes are occasionally of interest, such as wire drawing, usually done cold, and extrusion, either hot or cold. **Hot extrusion** for materials that are difficult to work became practical through the employment of a glass lubricant. This method allows the hot extrusion of highly alloyed steels and other exotic alloys subjected to service at high loads and/or high temperatures.

Constitution and Structure of Steel

As a result of the methods of production, the following elements are always present in steel: carbon, manganese, phosphorus, sulfur, silicon, and traces of oxygen, nitrogen, and aluminum. Various alloying elements are frequently deliberately added, such as nickel, chromium, copper, molybdenum, niobium (columbium), and vanadium. The most important of the above elements in steel is carbon, and it is necessary to understand the effect of carbon on the internal structure of steel to understand the heat treatment of carbon and low-alloy steels.

The iron–iron carbide equilibrium diagram in Fig. 4.2.4 shows the phases that are present in steels of various carbon contents over a range of temperatures under equilibrium conditions. Pure iron when heated to 910°C (1,670°F) changes its internal crystalline structure from a body-centered cubic arrangement of atoms, **alpha iron,** to a face-centered cubic structure, **gamma iron.** At 1,390°C (2,535°F), it changes back to

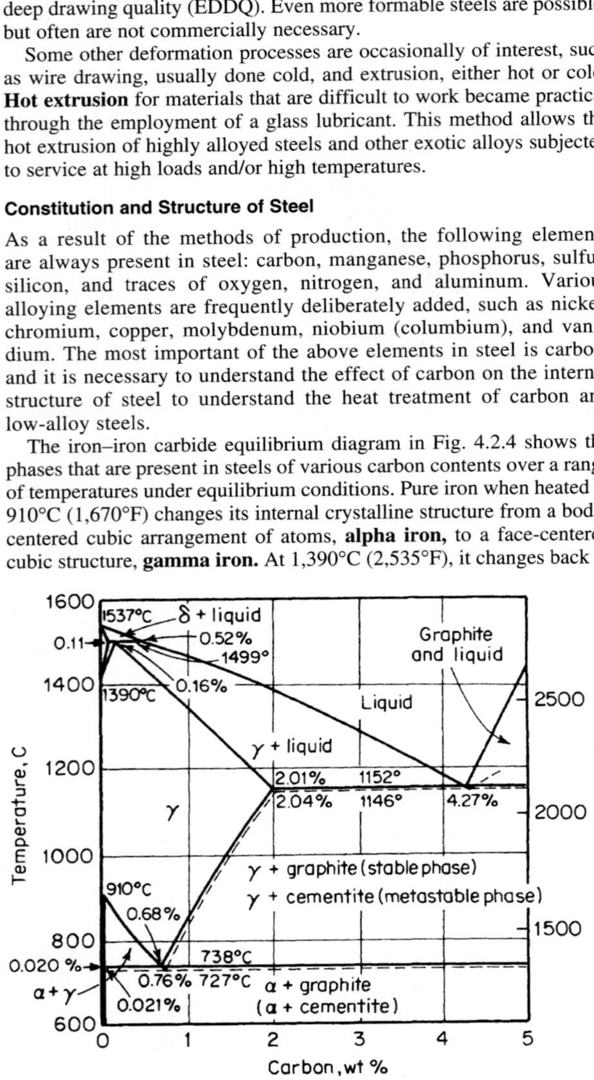

Figure 4.2.4 Iron–iron carbide equilibrium diagram, for carbon content up to 5 percent. (Dashed lines represent equilibrium with cementite, or iron carbide; adjacent solid lines indicate equilibrium with graphite.)

the body-centered cubic structure, **delta iron,** and at 1,539°C (2,802°F) the iron melts. When carbon is added to iron, it is found that it has only slight solid solubility in alpha iron (much less than 0.001 percent at room temperature at equilibrium). These small amounts of carbon, however, are critically important in many high-tonnage applications where formability is required. On the other hand, gamma iron will hold up to 2.0 percent carbon in solution at 1,130°C (2,066°F). The alpha iron containing carbon or any other element in solid solution is called **ferrite,** and the gamma iron containing elements in solid solution is called **austenite.** Usually when not in solution in the iron, the carbon forms a compound Fe_3C (iron carbide) which is extremely hard and brittle and is known as **cementite.** Other carbides of iron exist but are only of interest in rather specialized instances.

The temperatures at which the phase changes occur are called **critical points** (or temperatures) and, in the diagram, represent equilibrium conditions. In practice there is a lag in the attainment of equilibrium, and the critical points are found at lower temperatures on cooling and at higher temperatures on heating than those given, the difference increasing with the rate of cooling or heating.

The various critical points have been designated by the letter *A*; when obtained on cooling, they are referred to as *Ar,* on the heating as *Ac.* The subscripts *r* and *c* refer to *refroidissement* and *chauffage,* respectively, and reflect the early French contributions to heat treatment. The various critical points are distinguished from each other by numbers after the letters, being numbered in the order in which they occur as the temperature increases. Ac_1 represents the beginning of transformation of ferrite to austenite on heating; Ac_3 the end of transformation of ferrite to austenite on heating, and Ac_4 the change from austenite to delta iron on heating. On cooling, the critical points would be referred to as Ar_4, Ar_3, and Ar_1, respectively. The subscript 2, not mentioned here, refers to a magnetic transformation. The error came about through a misunderstanding of what rules of thermodynamics apply in phase diagrams. It must be remembered that the diagram represents the pure iron-iron carbide system at equilibrium. The varying amounts of impurities in commercial steels affect to a considerable extent the position of the curves and especially the lateral position of the eutectoid point.

Carbon steel in equilibrium at room temperature will have present both ferrite and cementite. The physical properties of ferrite are approximately those of pure iron and are characteristic of the metal. Cementite is itself hard and brittle; its shape, amount, and distribution control many of the mechanical properties of steel, as discussed later. The fact that the carbides can be dissolved in austenite is the basis of the heat treatment of steel, since the steel can be heated above the A_1 critical temperature to dissolve all the carbides, and then suitable cooling through the appropriate range will produce a wide and predictable range of the desired size and distribution of carbides in the ferrite.

If austenite with the **eutectoid** composition at 0.76 percent carbon (Fig. 4.2.4) is cooled slowly through the critical temperature, ferrite and cementite are rejected simultaneously, forming alternate plates or lamellae. This microstructure is called **pearlite,** since when polished and etched it has a pearly luster. When examined under a high-power optical microscope, however, the individual plates of cementite often can be distinguished easily. If the austenite contains less carbon than the eutectoid composition (i.e., **hypoeutectoid** compositions), free ferrite will first be rejected on slow cooling through the critical temperature until the remaining austenite reaches eutectoid composition, when the simultaneous rejection of both ferrite and carbide will again occur, producing pearlite. A hypoeutectoid steel at room temperature will be composed of areas of free ferrite and areas of pearlite; the higher the carbon percentage, the greater the amount of pearlite present in the steel. If the austenite contains more carbon than the eutectoid composition (i.e., **hypereutectoid** composition) and is cooled slowly through the critical temperature, then cementite is rejected and appears at the austenitic grain boundaries, forming a continuous cementite network until the remaining austenite reaches eutectoid composition, at which time pearlite is formed. A hypereutectoid steel, when slowly cooled, will exhibit areas of pearlite surrounded by a thin network of cementite, or iron carbide.

As the cooling rate is increased, the spacing between the pearlite lamellae becomes smaller; with the resulting greater dispersion of carbide preventing slip in the iron crystals, the steel becomes harder. Also, with an increase in the rate of cooling, there is less time for the separation of excess ferrite or cementite, and the equilibrium amount of these constituents will not be precipitated before the austenite transforms to pearlite. Thus with a fast rate of cooling, pearlite may contain less or more carbon than given by the eutectoid composition. When the cooling rate becomes very rapid (as obtained by quenching), the carbon does not have sufficient time to separate out in the form of carbide, and the austenite transforms to a highly elastically stressed structure super-saturated with carbon called **martensite.** This structure is exceedingly hard but brittle and requires **tempering** to increase the ductility. Tempering consists of heating martensite to some temperature below the critical temperature, causing the carbide to precipitate in the form of small spheroids, or especially in alloy steels, as needles or platelets. The higher the tempering temperature, the larger the carbide particle size, the greater the ductility of the steel, and the lower the hardness.

In a carbon steel, it is possible to have a structure consisting either of parallel plates of carbide in a ferrite matrix, the distance between the plates depending upon the rate of cooling, or of carbide spheroids in a ferrite matrix, the size of the spheroids depending upon the temperature to which the hardened steel was heated. (Some spheroidization occurs when pearlite is heated, but only at high temperatures close to the critical temperature range.)

Heat-Treating Operations

The following definitions of terms have been adopted by the ASTM, SAE, and ASM in substantially identical form.

Heat Treatment An operation, or combination of operations, involving the heating and cooling of a metal or an alloy in the solid state, for the purpose of obtaining certain desirable conditions or properties.

Quenching Rapid cooling by immersion in liquids or gases or by contact with metal.

Hardening Heating and quenching certain iron-base alloys from a temperature either within or above the critical range for the purpose of producing a hardness superior to that obtained when the alloy is not quenched. Usually restricted to the formation of martensite.

Annealing A heating and cooling operation implying usually a relatively slow cooling. The purpose of such a heat treatment may be (1) to remove stresses; (2) to induce softness; (3) to alter ductility, toughness, electrical, magnetic, or other physical properties; (4) to refine the crystalline structure; (5) to remove gases; or (6) to produce a definite microstructure. The temperature of the operation and the rate of cooling depend upon the material being heat-treated and the purpose of the treatment. Certain specific heat treatments coming under the comprehensive term annealing are as follows:

Full Annealing Heating iron base alloys above the critical temperature range, holding above that range for a proper period of time, followed by slow cooling to below that range. The annealing temperature is usually about 50°C (≈100°F) above the upper limit of the critical temperature range, and the time of holding is usually not less than 1 h for each 1-in section of the heaviest objects being treated. The objects being treated are ordinarily allowed to cool slowly in the furnace. They may, however, be removed from the furnace and cooled in some medium that will prolong the time of cooling as compared with unrestricted cooling in the air.

Process Annealing Heating iron-base alloys to a temperature below or close to the lower limit of the critical temperature range followed by cooling as desired. This heat treatment is commonly applied in the sheet and wire industries, and the temperatures generally used are from 540 to 705°C (about 1,000 to 1,300°F).

Normalizing Heating iron base alloys to approximately 40°C (about 100°F) above the critical temperature range followed by cooling to below that range in still air at ordinary temperature.

Patenting Heating iron base alloys above the critical temperature range followed by cooling below that range in air or a molten mixture of nitrates or nitrites maintained at a temperature usually between 425 and 565°C (about 800 to 1,050°F), depending on the carbon content of the steel and the properties required of the finished product. This treatment is applied in the wire industry to medium- or high-carbon steel as a treatment to precede further wire drawing.

Spheroidizing Any process of heating and cooling steel that produces a rounded or globular form of carbide. The following spheroidizing methods are used: (1) Prolonged heating at a temperature just below the lower critical temperature, usually followed by relatively slow cooling. (2) In the case of small objects of high-carbon steels, the spheroidizing result is achieved more rapidly by prolonged heating to temperatures alternately within and slightly below the critical temperature range. (3) Tool steel is generally spheroidized by heating to a temperature of 750 to 805°C (about 1,380 to 1,480°F) for carbon steels and higher for many alloy tool steels, holding at heat from 1 to 4 h, and cooling slowly in the furnace.

Tempering (also termed **Drawing)** Reheating hardened steel to some temperature below the lower critical temperature, followed by any desired rate of cooling. Although the terms *tempering* and *drawing* are practically synonymous as used in commercial practice, the term *tempering* is preferred.

Transformation Reactions in Carbon Steels Much of, but not all, the heat treatment of steel involves heating into the region above Ac_3 to form austenite, followed by cooling at a preselected rate. If the parts are large, heat flow may limit the available cooling rates. As an example selected for simplicity rather than volume of products, we may follow the possible transformations in a eutectoid steel over a range of temperature. (The reactions to produce ferrite in hypoeutectoid steels, which are by far the most common, do not differ in principle; the products are, of course, softer.) The curve is a derivative of the TTT (time-temperature-transformation) curves produced by a systematic study of austenite transformation rates isothermally on specimens thin enough to avoid heat flow complications. The data collected for many steels are found in the literature.

Figure 4.2.5 summarizes the rates of decomposition of a eutectoid carbon steel over a range of temperatures. Various cooling rates are shown diagrammatically, and it will be seen that the faster the rate of cooling, the lower the temperature of transformation, and the harder the product formed. At around 540°C (1,000°F), the austenite transforms rapidly to fine pearlite; to form martensite it is necessary to cool very rapidly through this temperature range to avoid the formation of pearlite before the specimen reaches the temperature at which the formation of martensite begins (M_s). The minimum rate of cooling that is required to form a fully martensite

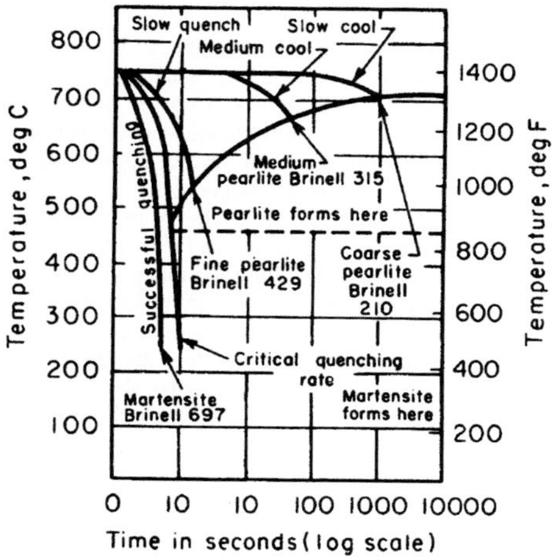

Figure 4.2.5 Influence of cooling rate on the product of transformation in a eutectoid carbon steel.

structure is called the **critical cooling rate.** No matter at what rate the steel is cooled, the only products of transformation of this steel will be pearlite or martensite. However, if the steel can be given an interrupted quench in a molten bath at some temperature between 205 and 540°C (about 400 and 1,000°F), an acicular structure, called **bainite,** of considerable toughness, combining high strength with high ductility, is obtained, and this heat treatment is known as austempering. A somewhat similar heat treatment called martempering can be utilized to produce a fully martensitic structure of high hardness, but free of the cracking, distortion, and residual stresses often associated with such a structure. Instead of quenching to room temperature, the steel is quenched to just above the martensitic transformation temperature and held for a short time to permit equalization of the temperature gradient throughout the piece. Then the steel may be cooled relatively slowly through the martensitic transformation range without superimposing thermal stresses on those introduced during transformation. The limitation of austempering and martempering for carbon steels is that these two heat treatments can be applied only to articles of small cross section, since the rate of cooling in salt baths is not sufficient to prevent the formation of pearlite in samples with diameter of more than ½ in.

The maximum hardness obtainable in a high-carbon steel with a fine pearlite structure is approximately 400 Brinell, although a martensitic structure would have a hardness of approximately 700 Brinell. Besides being able to obtain structures of greater hardness by forming martensite, a spheroidal structure will have considerably higher proof stress (i.e., stress to cause a permanent deformation of 0.01 percent) and ductility than a lamellar structure of the same tensile strength and hardness. It is essential, therefore, to form martensite when optimum properties are desired in the steel. This can be done with a piece of steel having a small cross section by heating the steel above the critical and quenching in water; but when the cross section is large, the cooling rate at the center of the section will not be sufficiently rapid to prevent the formation of pearlite. The characteristic of steel that determines its capacity to harden throughout the section when quenched is called **hardenability.** In the discussion that follows, it is pointed out that hardenability is significantly affected by most alloying elements. This term should not be confused with the ability of a steel to attain a certain hardness. The intensity of hardening, i.e., the maximum hardness of the martensite formed, is very largely dependent upon the carbon content of the steel.

Determination of Hardenability A long-established test for hardenability is the **Jominy test** which performs controlled water cooling on one end of a standard bar. Since the thermal conductivity of steel does not vary significantly, each distance from the quenched end (DQE) corresponds to a substantially unique cooling rate, and the structure obtained is a surrogate for the TTT curve on cooling. For a detailed account of the procedure, see the SAE Handbook. The figures extracted from the Jominy test can be extended to many shapes and quenching media. In today's practices, the factors discussed below can often be used to calculate hardenability from chemical composition with considerable confidence, leaving the actual test as a referee or for circumstances where a practice is being developed.

Three main factors affect the hardenability of steel: (1) austenite composition; (2) austenite grain size; and (3) amount, nature, and distribution of undissolved or insoluble particles in the austenite. All three determine the rate of decomposition, in the range of 540°C (about 1,000°F). The slower the rate of decomposition, the larger the section that can be hardened throughout, and therefore the greater the hardenability of the steel. Everything else being equal, the higher the carbon content, the greater the hardenability; this approach, however, is often counterproductive in that other strength properties may be affected in undesirable ways. The question of **austenitic grain size** is of considerable importance in any steel that is to be heat-treated, since it affects the properties of the steel to a considerable extent. When a steel is heated to just above the critical temperature, small polyhedral grains of austenite are formed. With increase in temperature, there is an increase in the size of grains, until at temperatures close to the melting point the grains are very large. Since the transformation of austenite to ferrite and pearlite usually starts at grain boundaries, a fine-grained steel will transform more rapidly than a coarse-grained steel because the latter has much less surface area bounding the grains than a steel with a fine grain size.

The grain size of austenite at a particular temperature depends primarily on the "pinning" of the boundaries by undissolved particles. These particles, which can be aluminum nitride from the deoxidation or various carbides and/or nitrides (added for their effect on final properties), dissolve as temperature increases, allowing grain growth. While hardenability is increased by large austenite grain size, this is not usually favored since some properties of the finished product can be seriously downgraded. Small particles in the austenite will act as nuclei for the beginning of transformation in a manner similar to grain boundaries, and therefore the presence of a large number of small particles (sometimes submicroscopic in size) will also result in low hardenability.

Determination of Austenitic Grain Size The subject of austenite grain size is of considerable interest because of the fact noted above that the grain size developed during heat treatment has a large effect on the physical properties of the steel. In steels of similar chemical analysis, the steel developing the finer austenitic grain size will have a lower hardenability but will, in general, have greater toughness, show less tendency to crack or warp on quenching, be less susceptible to grinding cracks, have lower internal stresses, and retain less austenite than coarse-grained steel. There are several methods of determining the grain-size characteristics of a steel.

The **McQuaid-Ehn test** (ASTM E112), which involves the outlining of austenite grains by cementite after a specific carburizing treatment, is still valid. It has been largely replaced, however, by quenching from the austenitizing treatment under investigation and observing the grain size of the resulting martensite after light tempering and etching with Vilella's reagent. There are several ways to report the grain size observed under the microscope, the one used most extensively being the ASTM index numbers. In fps or English units, the numbers are based on the formula: number of grains per square inch at $100x = 2^{N-1}$, in which N is the grain-size index. The usual range in steels will be from 1 to 128 grains/in^2 at $100x$, and the corresponding ASTM numbers will be 1 to 8, although today grain sizes up to 12 to 14 are common. Whereas at one time "coarse" grain sizes were 1 to 4, and "fine" grain sizes 5 to 8, these would not serve modern requirements. Most materials in service would be no coarser than 7 or 8, and the ferritic low-alloy high-strength steels routinely approach 12 or higher. Grain-size relationships in SI units are covered in detail in Designation E112 of ASTM Standards. For further information on grain size, refer to the ASM "Metals Handbook."

EFFECT OF ALLOYING ELEMENTS ON THE PROPERTIES OF STEEL

When relatively large amounts of alloying elements are added to steel, the characteristic behavior of carbon steels is not lost. Most **alloy steel** is medium- or high-carbon steel to which various elements have been added to modify its properties to an appreciable extent; the alloys as a minimum allow the properties characteristic of the carbon content to be fully realized even in larger sections, and in some cases may provide additional benefits. The percentage of alloy element required for a given purpose ranges from a few hundredths of 1 percent to possibly as high as 5 percent.

When ready for service, these steels will usually contain only two constituents, ferrite and carbide. The only way that an alloying element can affect the properties of the steel is to change the dispersion of carbide in the ferrite, change the properties of the ferrite, or change the characteristics of the carbide. The effect on the distribution of carbide is the most important factor, since in sections amenable to close control of structure, carbon steel is only moderately inferior to alloy steel. However, in large sections where carbon steels will fail to harden throughout the section even on a water quench, the hardenability of the steel can be increased at a cost by the addition of any alloying element (except

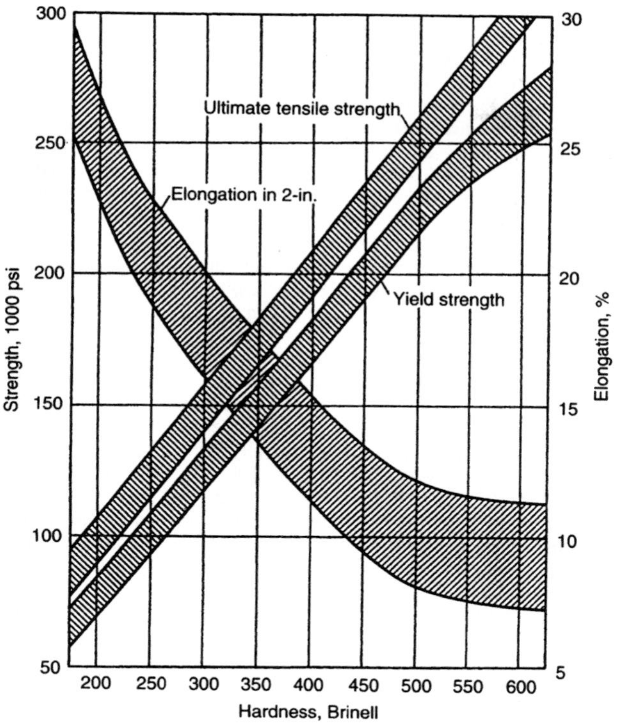

Figure 4.2.6 Range of tensile properties in several quenched and tempered steels at the same hardness values. *(Janitzky and Baeyertz. Source: ASM; reproduced by permission.)*

possibly cobalt). The increase in hardenability permits the hardening of a larger section of alloy steel than of plain carbon steel. The quenching operation does not have to be so drastic. Consequently, there is a smaller difference in temperature between the surface and center during quenching, and cracking and warping resulting from sharp temperature gradients in a steel during hardening can be avoided. The elements most effective in increasing the hardenability of steel are manganese, silicon, and chromium, or combinations of small amounts of several elements such as chromium, nickel, and molybdenum in SAE 4340, where the joint effects are greater than alloys acting singly.

Elements such as molybdenum, tungsten, and vanadium are effective in increasing the hardenability when dissolved in the austenite, but not when present in the austenite in the form of carbides. When dissolved in austenite, and thus contained in solution in the resulting martensite, they can modify considerably the rate of coarsening of carbides in tempered martensite. Tempering relieves the internal stresses in the hardened steel in part by precipitating various carbides of iron at fairly low temperature, which coarsen as the tempering temperature is increased. The increasing particle separation results in a loss of hardness and strength accompanied by increased ductility. See Fig. 4.2.6. Alloying elements can cause slower coarsening rates or, in some cases at temperatures from 500 to 600°C, can cause dissolution of cementite and the precipitation of a new set of small, and thus closely spaced, alloy carbides which in some cases can cause the hardness to actually rise again with no loss in toughness or ductility. This is especially important in tool steels. The presence of these stable carbide-forming elements enables higher tempering temperatures to be used without sacrificing strength. This permits these alloy steels to have a greater ductility for a given strength, or, conversely, greater strength for a given ductility, than plain carbon steels.

The third factor which contributes to the strength of alloy steel is the presence of the alloying element in the ferrite. Any element in solid solution in a metal will increase the strength of the metal, so that these elements will materially contribute to the strength of hardened and tempered steels. The elements most effective in strengthening the ferrite are phosphorus, silicon, manganese, and, to a lesser extent, nickel, molybdenum, and chromium. Carbon and nitrogen are very strong ferrite strengtheners but generally are not present in interstitial solution in significant amounts, and there are other processing reasons to actively keep the amount in solution small by adding strong carbide and/or nitride formers to give interstitial-free (IF) steels.

A final important effect of alloying elements discussed above is their influence on the austenitic grain size. Martensite formed from a fine-grained austenite has considerably greater resistance to shock than when formed from a coarse-grained austenite.

In Table 4.2.2, a summary of the effects of various alloying elements is given. Remember that this table indicates only the trends of the various elements, and the fact that one element has an important influence on one factor does not prevent it from having a completely different influence on another one.

PRINCIPLES OF HEAT TREATMENT OF IRON AND STEEL

When heat-treating a steel for a given part, certain precautions have to be taken to develop optimum mechanical properties in the steel.

Table 4.2.2 Trends of Influence of Some Alloying Elements

Element	Strengthening as dissolved in ferrite	Hardenability effects if dissolved in austenite	Effect on grain coarsening in austenite if undissolved as compound	Effects on tempered hardness, strength, and toughness
Al	*	*	†	None
Cr	*	†	†	*
Co	†	Negative	None	None
Cu	†	*	None	None
Mn	†	*	*	*
Mo	*	†	†	None
Nb (Cb)	None	†‡	†	†
Ni	*	*	None	None
P	†	*	None	None
Si	*	*	None	None
Ta	‡	†‡	†	†
Ti	†	†	†	‡
W	*	†	†	†
V	*	†	†	†

* Effects are moderate at best.
† Effects are strong to very strong.
‡ Effects not clear, or not used significantly.

Some of the major factors that have to be taken into consideration are outlined below.

Heating The first step in the heat treatment of steel is the heating of the material to above the critical temperature to make it fully austenitic. The **heating rate** should be sufficiently slow to avoid injury to the material through excessive thermal and transformational stresses. In general, hardened steel should be heated more slowly and uniformly than is necessary for soft stress-free materials. Large sections should not be placed in a hot furnace, the allowable size depending upon the carbon and alloy content. For high-carbon steels, care should be taken in heating sections as small as 50-mm (2-in) diameter, and in medium-carbon steels precautions are required for sizes over 150-mm (6-in) diameter. The **maximum temperature** selected will be determined by the chemical composition of the steel and its grain-size characteristics. In hypoeutectoid steel, a temperature about 25 to 50°C above the upper critical range is used, and in hypereutectoid steels, a temperature between the lower and the upper critical temperature is generally used to retain enough carbides to keep the austenite grain size small and preserve what is often limited toughness. Quenching temperatures are usually a little closer to the critical temperature for hypoeutectoid steels than to those for normalizing; annealing for softening is carried out just below Ac_1 for steels up to 0.3 percent C and just above for higher-carbon steels. Tables of suggested temperatures can be found in the ASM "Metals Handbook," or a professional heat treater may be consulted.

The **time** at maximum temperature should be such that a uniform temperature is obtained throughout the cross section of the steel. Care should be taken to avoid undue length of time at temperature, since this will result in undesirable grain growth, scaling, or decarburization of the surface. A practical figure often given for the total time in the hot furnace is 12 min/cm (about ½ h/in) of cross-sectional thickness. When the steel has attained a uniform temperature, the **cooling rate** must be such as to develop the desired structure; slow cooling rates (furnace or air cooling) to develop the softer pearlitic structures and high cooling rates (quenching) to form the hard martensitic structures. In selecting a **quenching medium** (see ASM "Metals Handbook"), it is important to select the quenching medium for a particular job on the basis of size, shape, and allowable distortion before choosing the steel composition. It is convenient to classify steels in two groups on the basis of depth of hardening: shallow hardening and deep hardening. **Shallow-hardening steels** may be defined as those which, in the form of 25-mm- (1-in-) diameter rounds, have, after brine quenching, a completely martensitic shell not deeper than 6.4 mm (¼ in). The shallow-hardening steels are those of low or no alloy content, whereas the deep-hardening steels have a substantial content of those alloying elements that increase penetration of hardening, notably chromium, manganese, and nickel. The high cooling rates required to harden shallow-hardening steel produce severe distortion and sometimes quench cracking in all but simple, symmetric shapes having a low ratio of length to diameter or thickness. Plain carbon steels cannot be used for complicated shapes where distortion must be avoided. In this case, water quenching must be abandoned and a less active quench used which materially reduces the temperature gradient during quenching. Certain oils are satisfactory but are incapable of hardening shallow-hardening steels of substantial size. A change in steel composition is required with a change from water to an oil quench. Quenching in oil does not entirely prevent distortion. When the degree of distortion produced by oil quenching is objectionable, recourse is taken to air hardening. The cooling rate in air is very much slower than in oil or water; so an exceptionally high alloy content is required. This means that a high price is paid for the advantage gained, in terms of both metal cost and loss in machinability, though it may be well justified when applied to expensive tools or dies. In this case, danger of cracking is negligible.

Liquids for Quenching Shallow-Hardening Steels Shallow-hardening steels require extremely rapid surface cooling in the quench, particularly in the temperature range around 550°C (1,020°F). A sub- merged water spray will give the fastest and most reproducible quench practicable. Such a quench is limited in application to simple short objects which are not likely to warp. Because of difficulty in obtaining symmetric flow of the water relative to the work, the spray quench is conducive to warping. The ideal practical quench is one that will give the required surface cooling without agitation of the bath. The addition of ordinary salt, sodium chloride, greatly improves the performance of water in this respect, the best concentration being around 10 percent. Most inorganic salts are effective in suppressing the formation of vapor at the surface of the steel and thus aid in cooling steel uniformly and eliminating the formation of soft spots. To minimize the formation of vapor, water-base quenching liquids must be kept cool, preferably under 20°C (about 70°F). The addition of some other soluble materials to water such as soap is extremely detrimental because of increased formation of vapor.

Liquids for Quenching Deeper-Hardening Steels When oil quenching is necessary, use a steel of sufficient alloy content to produce a completely martensitic structure at the surface over the heaviest section of the work. To minimize the possibility of cracking, especially when hardening tool steels, keep the quenching oil warm, preferably between 40 and 65°C (about 100 and 150°F). If this expedient is insufficient to prevent cracking, the work may be removed just before the start of the hardening transformation and cooled in air. Whether or not transformation has started can be determined with a permanent magnet, the work being completely nonmagnetic before transformation.

The cooling characteristics of quenching oils are difficult to evaluate and have not been satisfactorily correlated with the physical properties of the oils as determined by the usual tests. The standard tests are important with regard to secondary requirements of quenching oils. Low viscosity assures free draining of oil from the work and therefore low oil loss. A high flash and fire point assures a high boiling point and reduces the fire hazard which is increased by keeping the oil warm. A low carbon residue indicates stability of properties with continued use and little sludging. The steam-emulsion number should be low to ensure low water content, water being objectionable because of its vapor-filmforming tendency and high cooling power. A low saponification number assures that the oil is of mineral base and not subject to organic deterioration of fatty oils which give rise to offensive odors. Viscosity index is a valuable property for maintenance of composition.

In recent years, polymer-water mixtures have found application because of their combination of range of heat abstraction rates and relative freedom from fire hazards and environmental pollution. In all cases, the balance among productivity, danger of distortion and cracking, and minimum cost to give adequate hardenability is not simple; even though some general guides are available (e.g., "Metals Handbook," vol. 1, 10th ed.), consultation with experienced professionals is recommended.

Effect of the Condition of Surface The factors that affect the depth of hardening are the hardenability of the steel, the size of specimen, the quenching medium, and finally the condition of the surface of the steel before quenching. Steel that carries a heavy coating of scale will not cool so rapidly as a steel that is comparatively scale-free, and soft spots may be produced; or, in extreme cases, complete lack of hardening may result. It is therefore essential to minimize scaling as much as possible. Decarburization can also produce undesirable results such as nonuniform hardening and thus lowers the resistance of the material to alternating stresses (i.e., fatigue).

Tempering, as noted above, relieves quenching stresses and offers the ability to obtain useful combinations of properties through selection of tempering temperature. The ability of alloying elements to slow tempering compared to carbon steel allows higher temperatures to be used to reach a particular strength. This is accompanied by some usually modest increases in ductility and toughness. Certain high-hardenability steels are subject to delayed cracking after quenching and should be tempered without delay. Data on tempering behavior are available from many sources, such as ASM "Metals Handbook" or Bain and

Paxton, "Functions of the Alloying Elements in Steel," 3d ed., ASM International.

Relation of Design to Heat Treatment

Care must be taken in the design of a machine part to prevent cracking or distortion during heat treatment. With proper design the entire piece may be heated and cooled at approximately the same rate during the heat-treating operation. A light section should never be joined to a heavy section. Sharp reentrant angles should be avoided. Sharp corners and inadequate fillets produce serious stress concentration, causing the actual service stresses to build up to a point where they amount to two to five times the normal working stress calculated by the engineer in the original layout. The use of generous fillets is especially desirable with all high-strength alloy steels.

The modulus of elasticity of all commercial steels, either carbon or alloy, is the same so far as practical designing is concerned. The deflection under load of a given part is, therefore, entirely a function of the section of the part and is not affected by the composition or heat treatment of the steel. Consequently if a part deflects excessively, a change in design is necessary; either a heavier section must be used or the points of support must be increased.

COMPOSITE MATERIALS

For some applications, it is not necessary or even desirable that the part have the same composition throughout. The oldest method of utilizing this concept is to produce a high-carbon surface on low-carbon steel by **carburizing,** a high-temperature diffusion treatment, which after quenching gives a wear-resistant case 1 or 2 mm (0.04 or 0.08 in) thick on a fairly shock-resistant core. Clearly, this is an attractive process for gears and other complex machined parts. Other processes which produce a similar product are **nitriding** (which can be carried out at lower temperatures) and **carbonitriding** (a hybrid), or where corrosion resistance is important, by **chromizing.** Hard surfaces can also be obtained on a softer core by using selective heating to produce surface austenite (induction, flames, etc.) before quenching, or by depositing various hard materials on the surface by welding.

Many other types of surface treatment which provide corrosion protection and sometimes aesthetic values are common, beginning with paint or other polymeric films and ranging through enamels (a glass film); films such as zinc and its alloys which can be applied by dipping in molten baths or can be deposited electrolytically on one or both sides (galvanizing); tinplate for cans and other containers; chromium plating; or a light coherent scale of iron oxide. These processes are continually being improved, and they may be used in combinations, e.g., paint on a galvanized surface for exposed areas of automobiles.

Carburizing Various methods are available depending on the production volume. Pack carburizing can handle diverse feedstock by enclosing the parts in a sealed heat-resistant alloy box with carbonates and carbonaceous material, and by heating for several hours at about 925°C. The carbon dioxide evolved reacts with carbon to form carbon monoxide as the carrier gas, which does the actual carburizing. While the process can be continuous, much of it is done as a batch process, with consequent high labor costs and uncertain quality to be balanced against the flexibility of custom carburizing.

For higher production rates, furnaces with controlled atmospheres involving hydrocarbons and carbon monoxide provide better controls, low labor costs, shorter times to produce a given case depth [4 h versus 9 h for a 1-mm (0.04-in) case], and automatic quenching. Liquid carburizing using cyanide mixtures is even quicker for thin cases, but often it is not as economical for thicker ones; its great advantage is flexibility and control in small lots.

To provide the inherent value in this material of variable composition, it must be treated to optimize the properties of case and core by a double heat treatment. In many cases, to avoid distortion of precision parts, the material is first annealed at a temperature above the carburizing temperature and is cooled at a rate to provide good machinability. The part is machined and then carburized; through careful steel selection, the austenite grain size is not large at this point, and the material can be quenched directly without danger of cracking or distortion. Next the part is tempered; any retained austenite from the low M_s of the high-carbon case has a chance to precipitate carbides, raise its M_s, and transform during cooling after tempering. Care is necessary to avoid internal stresses if the part is to be ground afterward. An alternative approach is to cool the part reasonably slowly after carburizing and then to heat-treat the case primarily by suitable quenching from a temperature typical for hypereutectoid steels between A_1 and A_{cm}. This avoids some retained austenite and is helpful in high-production operations. The core is often carbon steel, but if alloys are needed for hardenability, this must be recognized in the heat treatment.

Nitriding A very hard, thin case can be produced by exposing an already quenched and tempered steel to an ammonia atmosphere at about 510 to 540°C, but unfortunately for periods of 50 to 90 h. The nitrogen diffuses into the steel and combines with strong nitride formers such as aluminum and chromium, which are characteristically present in steels where this process is to be used. The nitrides are small and finely dispersed; since quenching is not necessary after nitriding, dimensional control is excellent and cracking is not an issue. The core properties do not change since tempering at 550°C or higher temperature has already taken place. A typical steel composition is as follows: C, 0.2 to 0.3 percent; Mn, 0.04 to 0.6 percent; Al, 0.9 to 1.4 percent; Cr, 0.9 to 1.4 percent; and Mo, 0.15 to 0.25 percent.

Carbonitriding (Cyaniding) An interesting intermediate which rapidly adds both carbon and nitrogen to steels can be obtained by immersing parts in a cyanide bath just above the critical temperature of the core followed by direct quenching. A layer of about 0.25 mm (0.010 in) can be obtained in 1 h. The nitrides add to the wear resistance.

Local Surface Hardening For some parts which do not readily fit in a furnace, the surface can be hardened preferentially by local heating using flames, induction coils, electron beams, or lasers. The operation requires skill and experience, but in proper hands it can result in very good local control of structure, including the development of favorable surface compressive stresses to improve fatigue resistance.

Clad steels can be produced by one of several methods, including simple cladding by rolling a sandwich out of contact with air at a temperature high enough to bond (1,200°C); by explosive cladding where the geometry is such that the energy of the explosive causes a narrow molten zone to traverse along the interface and provide a good fusion bond; and by various casting and welding processes which can deposit a wide variety of materials (ranging from economical, tough, corrosion-resistant, or high-thermal-conductivity materials to hard and stable carbides in a suitable matrix). Many different product shapes lend themselves to these practices.

Chromizing Chromizing of low-carbon steel is effective in improving corrosion resistance by developing a surface containing up to 40 percent chromium. Some forming operations can be carried out on chromized material. Most chromizing is accomplished by packing the steel to be treated in a powdered mixture of chromium and alumina and then heating to above 1,260°C (2,300°F) for 3 or 4 h in a reducing atmosphere. Another method is to expose the parts to be treated to gaseous chromium compounds at temperatures above 845°C (about 1,550°F). Flat rolled sheets for corrosive applications such as auto mufflers can thus be chromized in open-coil annealing facilities.

THERMOMECHANICAL TREATMENT

The effects of mechanical treatment and heat treatment on the mechanical properties of steel have been discussed earlier in this section. Thermomechanical treatment consists of combining controlled (sometimes large) amounts of plastic deformation with the heat-treatment cycle to achieve improvements in yield strength beyond those attainable by the usual rolling practices alone or rolling following by a separate heat treatment. The tensile strength, of course, is increased at the same time as the yield strength (not necessarily to the same degree), and other properties such as ductility, toughness, creep resistance, and fatigue

Table 4.2.3 Classification of Thermomechanical Treatments

Class I. Deformation before austenite transformation
- *a.* Normal hot-working processes (hot/cold working)
- *b.* Deformation before transformation to martensite (ausforming, austforming, austenrolling, hot-cold working, marworking, warm working)
- *c.* Deformation before transformation to ferrite-carbide aggregates (austentempering)

Class II. Deformation during austenite transformation
- *a.* Deformation during transformation to martensite (Zerolling and Ardeform processes)
- *b.* Deformation during transformation to ferrite-carbide aggregates (flow tempering of bainite and isoforming)

Class III. Deformation after austenite transformation
- *a.* Deformation of martensite followed by tempering
- *b.* Deformation of tempered martensite followed by aging (flow tempering, marstraining, strain tempering, tempforming, warm working)
- *c.* Deformation of isothermal transformation products (patenting, flow tempering, warm working)

SOURCE: Radcliffe Kula, Syracuse University Press, 1964.

life can be improved. However, the high strength and hardness of thermomechanically treated steels limit their usefulness to the fabrication of components that require very little cold forming or machining, or very simple shapes such as strip and wire that can be used as part of a composite structure. Although the same yield strength may be achieved in a given steel by different thermomechanical treatments, the other mechanical properties (particularly the toughness) are not necessarily the same.

There are many possible combinations of deformation schedules and time-temperature relationships in heat treatment that can be used for thermomechanical treatment, and individual treatments will not be discussed here. Table 4.2.3 classifies broadly thermomechanical treatments into three principal groups related to the time-temperature dependence of the transformation of austenite discussed earlier under heat treatment. The names in parentheses following the subclasses in the table are those of some types of thermomechanical treatments that have been used commercially or have been discussed in the literature. At one time it was thought that these procedures would grow in importance, but in fact they are still used in a very minor way with the important exception of class 1*a* and class 1*c*, which are critically important in large tonnages.

COMMERCIAL STEELS

The wide variety of applications of steel for engineering purposes is due to the range of mechanical properties obtainable by changes in carbon content and heat treatment. Some typical applications of carbon steels are given in Table 4.2.4. Carbon steels can be subdivided roughly into three groups: (1) low-carbon steel, 0.01 to 0.25 percent carbon, for use where only moderate strength is required together with considerable plasticity; (2) machinery steels, 0.30 to 0.55 percent carbon, which can be heat-treated to develop high strength; and (3) tool steels, containing from 0.60 to 1.30 percent carbon (this range also includes rail and spring steels).

Table 4.2.4 Some Typical Applications of Carbon Steels

Percent C	Uses
0.01-0.10	Sheet, strip, tubing, wire nails
0.10-0.20	Rivets, screws, parts to be case-hardened
0.20-0.35	Structural steel, plate, forgings such as camshafts
0.35-0.45	Machinery steel—shafts, axles, connecting rods, etc.
0.45-0.55	Large forgings—crankshafts, heavy-duty gears, etc.
0.60-0.70	Bolt-heading and drop-forging dies, rails, setscrews
0.70-0.80	Shear blades, cold chisels, hammers, pickaxes, band saws
0.80-0.90	Cutting and blanking punches and dies, rock drills, hand chisels
0.90-1.00	Springs, reamers, broaches, small punches, dies
1.00-1.10	Small springs and lathe, planer, shaper, and slotter tools
1.10-1.20	Twist drills, small taps, threading dies, cutlery, small lathe tools
1.20-1.30	Files, ball races, mandrels, drawing dies, razors

The **product mix** has changed noticeably over the last 20 years as consumption rose from about 100 million to 120 to 130 million tpy, some of which were imports, both semifinished (slabs, billets, etc.) and finished. Rails, in Carnegie's time, " the" product, dropped enormously as track maintenance slipped; tool steels fell, reflecting competition from carbides and oxide tools. Sheet was the main gainer with a large contribution from coated sheet (largely galvanized, reflecting cosmetic influences in the auto industry). Hot-rolled sheet displaced some cold-rolled applications as noted above. As an indication of current (2006) volumes, both of these are approaching 20 million tpy in the United States, from 10 to 12 million tpy in the early 1990s; cold-rolled has remained virtually constant at 12 million tpy. However, some of the specialized forms of steel products remain very important, and their capabilities are sometimes unique. The combination of property, performance, and price must be evaluated for each case.

The chemical compositions and mechanical and physical properties of many of the steels whose uses are listed in Table 4.2.4 are covered by specifications adopted by the American Society for Testing and Materials (ASTM), the Society of Automotive Engineers (SAE), and the American Society of Mechanical Engineers (ASME); other sources include government specifications for military procurement, national specifications in other industrial countries, and smaller specialized groups with their own interests in mind. The situation is more complex than necessary because of mixtures of chemical composition ranges and property and geometric ranges. Simplification will take a great deal of time to develop, and understanding among users, when and if it is reached, will require a major educational endeavor. For the time being, most of the important specifications and equivalents are shown in the ASM "Metals Handbook," 10th ed., vol. 1.

Understanding Some Mechanical Properties of Steels

Before we give brief outlines of some of the more important classes of steels, a discussion of the important factors influencing selection of steels and an appreciation of those for relevant competitive materials may be helpful. The easy things to measure for a material, steel or otherwise, are chemical composition and a stress-strain curve, from which one can extract such familiar quantities as yield strength, tensile strength, and ductility expressed as elongation and reduction of area. Where the application is familiar and the requirements are not particularly crucial, this is still appropriate and, in fact, may be overkill; if the necessary properties are provided, the chemical composition can vary outside the specification with few or no ill effects. However, today's designs emphasize total life-cycle cost and performance and often will require consideration of several other factors such as manufacturing and assembly methods; costs involving weldability (or more broadly joinability), formability, and machinability; corrosion resistance in various environments; fatigue, creep, and fracture; dimensional tolerances; and acceptable disposability at the end of the life cycle. These factors interact with each other in such complex ways that merely using tabulated data can invite disaster in extreme cases. It is not possible to cover all these issues in detail, but a survey of important factors for some of the most common ones may be helpful.

Yield Strength In technical terms the yield stress, which measures the onset of plastic deformation (e.g., 0.1 or 0.2 percent permanent extension), occurs when significant numbers of "dislocations" move in the crystal lattice of the major phase present, usually some form of ferrite. Dislocations are line discontinuities in the lattice which allow crystal planes to slide over each other; they are always present at levels of 10^4 to 10^6 per square centimeter in undeformed metals, and this dislocation density can increase steadily to 10^{10} to 10^{12} per square centimeter after large deformations. Any feature which interferes with dislocation movement increases the yield stress. Such features include the following:

1. The lattice itself provides a lower limit (the **Peierls' stress**) by exhibiting an equivalent viscosity for movement. As a practical matter,

Table 4.2.5 Approximate Mechanical Properties for Various Tempers of Cold-Rolled Carbon Strip Steel

Temper	Tensile strength MPa	Tensile strength 1,000 lb/in²	Elongation in 50 mm or 2 in for 1.27-mm (0.050-in) thickness of strip, %	Remarks
No. 1 (hard)	621 ± 69	90 ± 10		A very stiff cold-rolled strip intended for flat blanking only, and not requiring ability to withstand cold forming
No. 2 (half-hard)	448 ± 69	65 ± 10	10 ± 6	A moderately stiff cold-rolled strip intended for limited bending
No. 3 (quarter-hard)	379 ± 69	55 ± 10	20 ± 7	A medium-soft cold-rolled strip intended for limited bending, shallow drawing, and stamping
No. 4 (skin-rolled)	331 ± 41	48 ± 6	32 ± 8	A soft ductile cold-rolled strip intended for deep drawing where no stretcher strains or fluting are permissible
No. 5 (dead-soft)	303 ± 41	44 ± 6	39 ± 6	A soft ductile cold-rolled strip intended for deep drawing where stretcher strains or fluting are permissible. Also for extrusions

SOURCE: ASTM A109. Complete specification should be consulted in ASTM Standards (latest edition).

this may be up to about 20 ksi (135 MPa) and is the same for all iron-based materials.

2. Dislocations need energy to cut through each other; thus, the stress necessary to continue deformation rises continuously as the dislocation density increases (work hardening). This is an important, inexpensive source of strength as various tempers are produced by rolling or drawing (see Table 4.2.5); it is accompanied by a significant loss of ductility.

3. Precipitates interfere with dislocation movement. The magnitude of interference depends sensitively on precipitate spacing, rising to a maximum at a spacing from 10 to 30 nm for practical additions. If a given heat treatment cannot develop these fine spacings (e.g., because the piece is so large that transformations take place at high temperatures), the strengthening is more limited. Precipitates are an important source of strength; martensite is used as an intermediate structure for carbon and alloy steels precisely because the precipitate spacing can be accurately controlled by tempering.

4. Small grain sizes, by interfering with the passage of dislocations across the boundaries, result in important increases in yield strength. To a very good approximation, the increase in yield is proportional to (grain diameter)$^{-1/2}$. A change from ASTM 8 to 12, for example, increases the yield strength by about 100 MPa.

5. Elements in solid solution also cause local lattice strains and make dislocation motion more difficult. The effect depends on the element; 1 percent P, for example, can double the hardness of iron or low-carbon steel. It is not used to this degree because of deleterious effects on toughness.

To a first approximation, these effects are additive; Fig. 4.2.7 depicts the effects of the iron lattice itself, solid solution strengthening, and grain size. Working the material would move all curves up the y axis by adding a work-hardening term. Figure 4.2.8 shows typical effects in an LAHS steel (see later) on the separated effects of manganese on yield strength. Manganese provides a little solid solution hardening and by its effect on ferrite grain size provides nonlinear strengthening. Figure 4.2.9 is a schematic of some combinations to obtain desired yield strengths and toughness simultaneously.

Tensile Properties Materials fail in tension when the increment of nominal stress from work hardening can no longer support the applied load on a decreasing diameter. The load passes through a maximum [the **ultimate tensile stress** (UTS)], and an instability **(necking)** sets in at that point. The triaxial stresses thus induced encourage the formation of internal voids nucleated at inclusions or, less commonly, at other particles such as precipitates. With increasing strain, these voids grow until they join and ultimately lead to a ductile failure.

Some plastic behavior of steels is sensitive to the number and type of inclusions. In a tensile test the UTS and **elongation to failure** are not affected much by increasing the number of inclusions (although the **uniform ductility** prior to necking is), but the **reduction in area** is; of greater importance, the energy absorbed to propagate a crack (related to the **fracture toughness**) is very sensitive to inclusion content. This relates directly to steelmaking practices, and to ladle metallurgy in particular, which has proved extremely effective in control of deleterious inclusions (not all inclusions are equally bad).

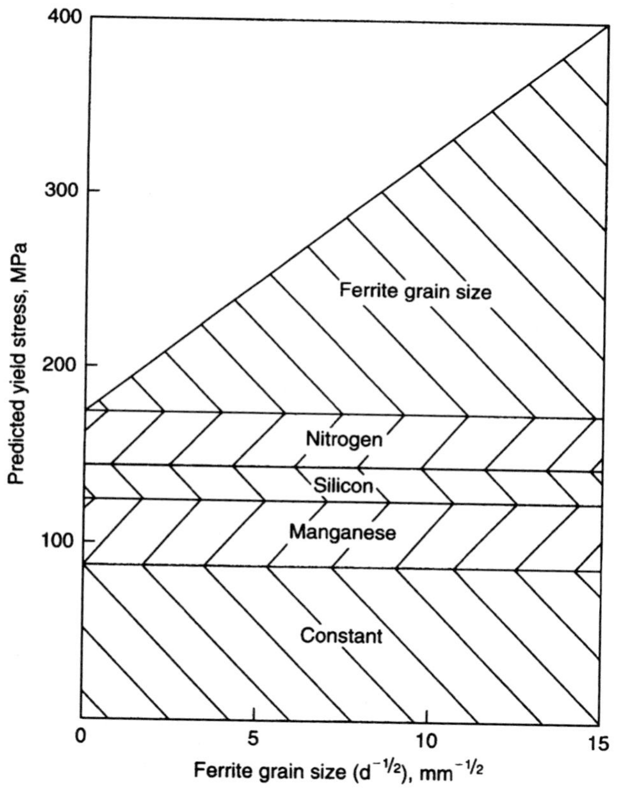

Figure 4.2.7 The components of yield stress predicted for an air-cooled carbon-manganese steel containing 1.0 percent manganese, 0.25 percent silicon, and 0.01 percent nitrogen. *(Source: Union Carbide Corp.; reproduced by permission.)*

Toughness A full treatment of this topic is not possible here, but we note the following:

1. As strength increases, toughness falls in all cases except where strengthening arises from grain-size reduction. Fine grain size thus is a double blessing; it will increase strength and toughness simultaneously.

2. The energy involved in crack growth depends on carbon content (Fig. 4.2.10). High-carbon materials have not only much lower

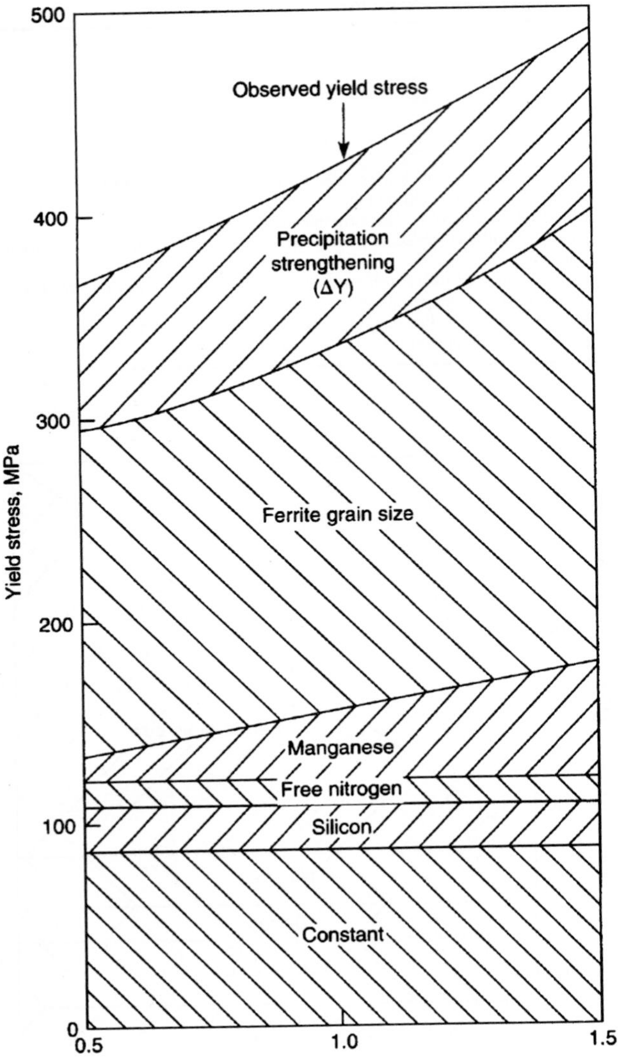

Figure 4.2.8 The effect of increasing manganese content on the components of the yield stress of steels containing 0.2 percent carbon, 0.2 percent silicon, 0.15 percent vanadium, and 0.015 percent nitrogen, normalized from 900°C (1650°F). *(Source: Union Carbide Corp., reproduced by permission.)*

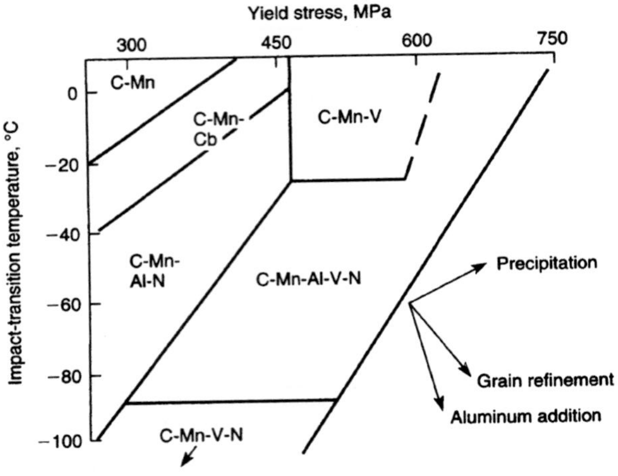

Figure 4.2.9 Combinations of yield strength and impact transition temperature available in normalized high strength, low alloy (HSLA) steels. *(Source: Union Carbide Corp.; reproduced by permission.)*

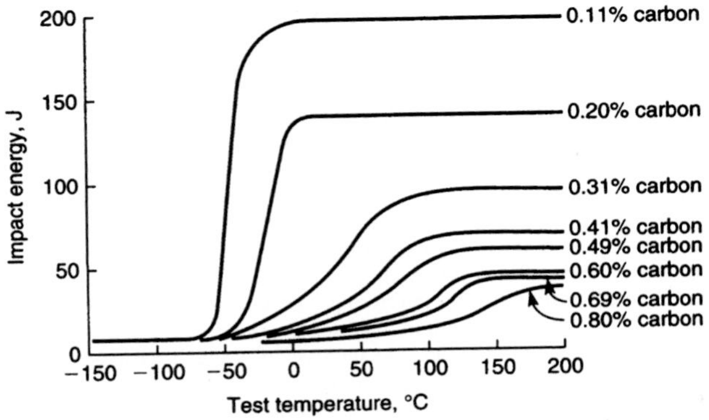

Figure 4.2.10 The effect of carbon, and hence the pearlite content on impact transition temperature curves of ferrite-pearlite steels. *(Source: Union Carbide Corp.; reproduced by permission.)*

propagation energies (e.g., the "shelf" in a Charpy test, or the value of K_c), but also higher **impact transition temperatures** (ITTs). Below the ITT the energy for crack propagation becomes very small, and we may loosely describe the steel as "brittle." There is, then, a real incentive to keep carbon in steel as low as possible, especially because pearlite does not increase the yield strength, but does increase the ITT, often to well above room temperature.

3. **Welding** is an important assembly technique. Since the hardenability of the steel can lead to generally undesirable martensite in the **heat-affected zone** (HAZ) and possibly to cracks in this region, the toughness of welds necessitates using steels with the lowest possible carbon equivalent. **Carbon equivalent** is based on a formula which includes C and strong hardenability agents such as Mn, Ni, and Cr (see Sec. 4.3). Furthermore, the natural tendency of austenite to grow to a very coarse grain size near the weld metal may be desirably restrained by adding elements such as Ti, which as undissolved carbides, interfere with grain growth. In critical applications, finer grain sizes after transformation during cooling increase toughness in the HAZ.

Other mechanical properties such as fatigue strength, corrosion resistance, and formability are frequently important enough to warrant consideration, but too short a description here may be misleading; standard texts and/or professional journals and literature should be consulted when necessary.

Steel is not necessarily the only material of choice in today's competitive world. The issue revolves on the method to make a legitimate and rational choice between two or more materials with many different properties and characteristics. Consider a very simple example (see Ashby, "Materials Selection in Mechanical Design," Pergamon Press, Oxford).

A furniture designer conceives of a lightweight table—a flat sheet of toughened glass supported on slender unbraced cylindrical legs. The legs must be solid, as light as possible, and must support the tabletop and whatever is placed on it without buckling. This involves consideration of the minimum weight and maximum aspect ratio of the legs. Examination of the mechanics shows that for minimum weight, we need a maximum value of the quantity $M_1 = E^{1/2}/\rho$, where E is Young's modulus and ρ is the density. For resistance to buckling, we need the maximum value of $M_2 = E$. If we plot E versus ρ of several materials, we can evaluate regions where both M_1 and M_2 are large. In doing so, we find attractive candidates are carbon-fiber-reinforced polymers (CFRPs) and certain engineering ceramics. They win out over wood, steel, and aluminum by their combination of high modulus and light weight. If we had other constraints such as cost, the CFRPs would be eliminated; as for toughness, the engineering ceramics would be unsuitable. Whenever many constraints must be satisfied simultaneously, their priority must be established. The net result is a compromise—but one which is better than an off-the-cuff guess!

This simple example can be generalized and used to develop a short list of very different alternatives in a reasonably quantitative and nonjudgmental manner for a wide variety of necessary properties. Ashby (op. cit.) provides a large collection of data.

Low-Carbon Steels Of the many low-carbon-steel products, sheet and strip steels are becoming increasingly important. The consumption of steel in the sheet and tinplate industry has accounted for approximately 60 percent of the total steel production in the United States. This large production has been made possible by refinement of continuous sheet and strip rolling mills. Applications in which large quantities of sheet are employed are tinplate for food containers; black, galvanized, and terne-coated sheets for building purposes; and high-quality sheets for automobiles, furniture, refrigerators, and countless other stamped, formed, and welded products. The difference between **sheet** and **strip** is based on width and is arbitrary. Cold working produces a better surface finish, improves the mechanical properties, and permits the rolling of thinner-gage material than hot rolling. Some hot-rolled products are sold as sheet or strip directly after coiling from the hot mill. Coils may be pickled and oiled for surface protection. Thicker coiled material up to 2 m (80 in) or more wide may be cut into plates, while thicker plates are produced directly from mills which may involve rolling a slab in two dimensions to produce widths approaching 5 m (200 in). Some plates are subject to heat treatment, occasionally elaborate, to produce particular combinations of properties for demanding applications. Structurals (beams, angles, etc.) are also produced directly by hot rolling, normally without further heat treating. Bars with various cross sections are also hot-rolled; some may undergo further heat treatment depending on service requirements. Wire rods for further cold drawing are also hot-rolled.

Much of the total hot-rolled sheet produced is cold-rolled further; the most demanding forming applications (automobiles and trucks, appliances, containers) require careful annealing, as discussed earlier. See Table 4.2.5.

Steels for deep-drawing applications must have a low yield strength and sufficient ductility for the intended purpose. This ductility arises from having a very small amount of interstitial atoms (carbon and nitrogen) in solution in the ferrite (**IF** or **interstitial-free** steels) and a sharp preferred orientation of the pancake-shaped ferrite grains. The procedures to obtain these are complex and go all the way back to steelmaking practices; they will not be discussed further here. They must also have a relatively fine grain size, since a large grain size will cause a rough finish, an "orange-peel" effect, on the deep drawn article. The sharp yield point characteristic of conventional low-carbon steel must be eliminated to prevent sudden local elongations in the sheet during forming, which result in strain marks called **stretcher strains** or **Lüders lines.** This can be done by cold working (Fig. 4.2.1), a reduction of only 1 percent in thickness usually being sufficient. This cold reduction is usually done by cold rolling, known as **temper rolling,** followed by alternate bending and reverse bending in a roller leveler. Temper rolling must always precede roller leveling because soft annealed sheets will "break" (yield locally) in the roller leveler. An important phenomenon in these temper-rolled low-carbon sheets is the return, partial or complete, of the sharp yield point after a period of time, by interstitial diffusion back to dislocations. This is known as **aging** in steel. The return of the yield point is accompanied by an increase in hardness and a loss in ductility. While a little more expensive, IF steels do not display a yield point and would not really require temper rolling, although they normally receive about 0.25 percent to improve surface finish. (The 1 percent typically used for non-IF steels would cause harmful changes in the stress-strain curve.) With these options available, aging is no longer the problem it used to be.

High-strength hot-rolled, cold-rolled, and galvanized sheets are now available with specified yield strengths. By the use of small alloy additions of niobium, vanadium, and sometimes copper, it is possible to meet the requirements of ASTM A607-70 for hot-rolled and cold-rolled sheets—345 MPa (50,000 lb/in^2) yield point, 483 MPa (70,000 lb/in^2) tensile strength, and 22 percent elongation in 2 in. By further alloy additions, sheets are produced to a minimum of 448 MPa (65,000 lb/in^2) yield point, 552 MPa (80,000 lb/in^2) tensile strength, and 16 percent elongation in 2 in. The sheets are available in coil form and are used extensively for metallic buildings and for welding into tubes for construction of furniture, etc.

Structural Carbon Steels Bridges and buildings historically have been constructed with structural carbon steel meeting the requirements of ASTM A36 (Table 4.2.6). This steel has a minimum yield point of 248 MPa (36,000 lb/in^2) and was developed to fill the need for a higher-strength structural carbon steel than the steels formerly covered by ASTM A7 and A373. The controlled composition of A36 steel provides good weldability and furnishes a significant improvement in the economics of steel construction. The structural carbon steels are available in the form of plates, shapes, sheet piling, and bars, all in the hot-rolled condition. A uniform strength over a range of section thickness is provided by adjusting the amount of carbon, manganese, and silicon in the A36 steel.

High-Strength Low-Alloy Steels While A36 still finds wide use, requirements for weight reduction, energy use in manufacture, and environmental savings have led to a steady, if slow, use of **higher-strength steels.** Slow adoption comes from several sources—the time necessary for specification changes, the education and general confidence level of designers, and inertia due to unfamiliarity with new materials.

The original development of inexpensive steels with higher yield stress began around 1930 when lighter railcars allowing higher payload were an issue. U.S. Steel with COR-TEN® and Bethlehem with Mayari-R™ used solid solution hardening by Cu, Ni, and P to achieve 350-MPa (50-ksi) yield, and a modest market developed. Some 20 years later, architects noted that under proper conditions, an attractive patina developed and new uses of "weathering steels" developed. The general corrosion resistance was improved by Cu, although care was (and is!) necessary where localized corrosion may be a concern.

In the 1960s, a new concept arose of using alloy carbonitrides to produce precipitation hardening. In the days before ladle metallurgy, reproducibility was not always adequate and usage was slow to develop. However, with ladle metallurgy and the contemporary development of controlled rolling and austenite recrystallization on the hot mill giving fine ferrite grain size, the growth of this class of materials has been encouraging. Much remains, to be done in this regard. These steels have in the past been referred to as "high-tensile steels" and "low-alloy steels," but the name **high-strength low-alloy steels,** abbreviated **HSLA steels,** is now the generally accepted designation.

HSLA steels are a group of steels, intended for general structural and miscellaneous applications, that have minimum yield strengths from 280 to 550 MPa (40 to 80 ksi) with greatest usage around 350 to 420 MPa (50 to 60 ksi). Many opportunities exist for increased usage of the readily weldable, reasonably priced, and tough material in the range of 500 MPa (70 ksi) and up; however, there are not many current prototypes which would encourage a movement to this strength level. These steels typically contain small amounts of alloying elements to achieve their strength in the hot-rolled or normalized condition. Among the elements used in small amounts, singly or in combination, are niobium, titanium, vanadium, manganese, copper, and phosphorus. A complete listing of HSLA steels available from producers in the United States and Canada shows many brands or variations, often not covered by ASTM or other specifications. These steels generally are available as sheet, strip, plates, bars, and shapes and often are sold as proprietary grades.

Some of the steels have improved resistance to corrosion, so that the necessary equal service life in a thinner section or longer life in the same section is obtained in comparison with that of a structural carbon steel member. Good resistance to repeated loading and good abrasion

Table 4.2.6 Mechanical Properties of Some Constructional Steels*

ASTM designation	Thickness range, mm (in)	Yield point, min MPa	Yield point, min 1,000 lb/in²	Tensile strength MPa	Tensile strength 1,000 lb/in²	Elongation in 200 mm (8 in) min, %	Suitable for welding?
			Structural carbon-steel plates				
ASTM A36	To 100 mm (4 in), incl.	248	36	400-552	58-80	20	Yes
			Low- and intermediate-tensile-strength carbon-steel plates				
ASTM A283	(structural quality)						
Grade A	All thicknesses	165	24	310	45	28	Yes
Grade B	All thicknesses	186	27	345	50	25	Yes
Grade C	All thicknesses	207	30	379	55	22	Yes
Grade D	All thicknesses	228	33	414	60	20	Yes
			Carbon-silicon steel plates for machine parts and general construction				
ASTM A284							
Grade A	To 305 mm (12 in)	172	25	345	50	25	Yes
Grade B	To 305 mm (12 in)	159	23	379	55	23	Yes
Grade C	To 305 mm (12 in)	145	21	414	60	21	Yes
Grade D	To 200 mm (8 in)	145	21	414	60	21	Yes
			Carbon-steel pressure-vessel plates				
ASTM A285							
Grade A	To 50 mm (2 in)	165	24	303-379	44-55	27	Yes
Grade B	To 50 mm (2 in)	186	27	345-414	50-60	25	Yes
Grade C	To 50 mm (2 in)	207	30	379-448	55-65	23	Yes
			Structural steel for locomotives and railcars				
ASTM A113							
Grade A	All thicknesses	228	33	414-496	60-72	21	No
Grade B	All thicknesses	186	27	345-427	50-62	24	No
Grade C	All thicknesses	179	26	331-400	48-58	26	No
			Structural steel for ships				
ASTM A131 (all grades)		221	32	400-490	58-71	21	No
			Heat-treated constructional alloy-steel plates				
ASTM A514	To 64 mm (2 in), incl.	700	100	800-950	115-135	18†	Yes
	Over 64 to 102 mm (2 to 4 in), incl.	650	90	750-950	105-135	17†	Yes

* See appropriate ASTM documents for properties of other plate steels, shapes, bars, wire, tubing, etc.
† Elongation in 50 mm (2 in), min.

resistance in service may be other characteristics of some of the steels. While high strength is a common characteristic of all HSLA steels, the other properties mentioned above may or may not be, singly or in combination, exhibited by any particular steel.

HSLA steels have found increasing, but still perhaps too limited, acceptance in many fields, among which are the construction of railroad cars, trucks, automobiles, trailers, and buses; welded steel bridges; television and power-transmission towers and lighting standards; columns in highrise buildings; portable liquefied petroleum gas containers; ship construction; oil storage tanks; air conditioning equipment; agricultural and earthmoving equipment.

Dual-Phase Steels The good properties of HSLA steels do not provide sufficient cold formability for automotive components which involve stretch forming. **Dual-phase steels** generate a microstructure of ferrite plus islands of austenite-martensite by quenching from a temperature between A_1 and A_3. This leads to continuous yielding (rather than a sharp yield point) and a high work-hardening rate with a larger elongation to fracture. Modest forming strains can give a yield strength in the deformed product of 350 MPa (about 50 ksi), comparable to HSLA steels.

Quenched and Tempered Low-Carbon Constructional Alloy Steels These steels, having yield strengths at the 689-MPa (100,000 lb/in²) level, are covered by ASTM A514, by military specifications, and for pressure-vessel applications, by ASME Code Case 1204. They are available in plates, shapes, and bars and are readily welded. Since they are heat-treated to a tempered martensitic structure, they retain excellent toughness at temperatures as low as −45°C (−50°F). Major cost savings have been effected by using these steels in the construction of pressure vessels, in mining and earthmoving equipment, and for major members of large steel structures.

Ultraservice Low-Carbon Alloy Steels (Quenched and Tempered) The need for high-performance materials with higher strength-to-weight ratios for critical military needs, for hydrospace explorations, and for aerospace applications has led to the development of quenched and tempered ultraservice alloy steels. Although these steels are similar in many respects to the quenched and tempered low-carbon constructional alloy steels described above, their significantly higher notch toughness at yield strengths up to 965 MPa (140,000 lb/in²) distinguishes the ultraservice alloy steels from the constructional alloy steels. These steels are not included in the AISI-SAE classification of alloy steels. There are numerous proprietary grades of ultraservice steels in addition to those covered by ASTM designations A543 and A579. Ultraservice steels may be used in large welded structures subjected to unusually high loads, and must exhibit excellent weldability and toughness. In some applications, such as hydrospace operations, the steels must have high resistance to fatigue and corrosion (especially stress corrosion) as well.

Maraging Steels For performance requiring high strength and toughness and where cost is secondary, age-hardening low-carbon martensites have been developed based on the essentially carbon-free iron-nickel system. The as-quenched martensite is soft and can be shaped before an aging treatment. Two classes exist: (1) a nominal 18 percent nickel steel containing cobalt, molybdenum, and titanium with yield strengths of 1,380 to 2,070 MPa (200 to 300 ksi) and an outstanding resistance to stress corrosion cracking (normally a major problem at these strengths) and (2) a nominal 12 percent nickel steel containing chromium, molybdenum, titanium, and aluminum adjusted to yield strengths of 1,034 to 1,379 MPa (150 to 200 ksi). The toughness of this series is the best available of any steel at these yield strengths.

Cryogenic-Service Steels For the economical construction of cryogenic vessels operating from room temperature down to the temperature of liquid nitrogen (−195°C or −320°F) a 9 percent nickel alloy steel has been developed. The mechanical properties as specified by ASTM A353 are 517 MPa (75,000 lb/in^2) minimum yield strength and 689 to 827 MPa (100,000 to 120,000 lb/in^2) minimum tensile strength. The minimum Charpy impact requirement is 20.3 J (15 ft·1 bf) at −195°C (−320°F). For lower temperatures, it is necessary to use austenitic stainless steel.

Machinery Steels A large variety of carbon and alloy steels is used in the automotive and allied industries. Specifications are published by SAE on all types of steel, and these specifications should be referred to for detailed information. A note on **specifications** may be in order. Many bodies set specifications on materials in their sphere of interest although efforts continue to rationalize the complexities. In a perfect world, a customer would specify expectations of properties. These would typically involve the range of mechanical properties, but also other matters of interest such as size, shape, surface finish, and weldability. However, composition often serves as the surrogate, and current specifications have considerable overlap. Relations between compositions and properties, preferably seen through the microstructure created by processing, are reasonably well known but not simple—many customers and suppliers arrive at suitable choices by mutual consultation, and by trial and error.

AISI, which for a long time coordinated its composition specifications with SAE, has discontinued its practice of designating steels, and the new arbiter is vol. 1 ("Materials") of "SAE Handbook." ASTM has a mixture of compositions, sizes, and properties required. The Unified Numbering System (UNS), which has been developed by a number of groups, is a designation of chemical composition and not a specification. It is described in SAE J1086 and ASTM E527.

A numerical index is used to identify the compositions of AISI (and SAE) steels. Most SAE alloy steels are made by the basic oxygen or basic electric furnace processes; a few steels that at one time were made in the electric furnace carry the prefix E before their number, i.e., E52100. However, with the almost complete use of ladle furnaces, this distinction is rarely necessary today. Steels are "melted" in the BOP or EF and "made" in the ladle. A series of four numerals designates the composition of the AISI steels; the first two indicate the steel type, and the last two indicate, as far as feasible, the average carbon content in "points" or hundredths of 1 percent. Thus 1,020 is a carbon steel with a carbon range of 0.18 to 0.23 percent, probably made in the basic oxygen furnace, and E4340 is a nickel-chromium molybdenum steel with 0.38 to 0.43 percent carbon made in the electric-arc furnace. A group of steels known as **H steels,** which are similar to the standard SAE steels, are being produced with a specified Jominy hardenability; these steels are identified by a suffix H added to the conventional series number. In general, these steels have a somewhat greater allowable variation in chemical composition but a smaller variation in hardenability than would be normal for a given grade of steel. This smaller variation in hardenability results in greater reproducibility of the mechanical properties of the steels on heat treatment; therefore, H steels have become increasingly important in machinery steels.

Boron steels are designated by the letter B inserted between the second and third digits, e.g., 50B44. The effectiveness of boron in increasing hardenability was a discovery of the late thirties, when it was noticed that heats treated with complex deoxidizers (containing boron) showed exceptionally good hardenability, high strength, and ductility after heat treatment. It was found that as little as 0.0005 percent of boron increased the hardenability of steels with 0.15 to 0.60 carbon, whereas boron contents of over 0.005 percent had an adverse effect on hot workability. Boron steels achieve special importance in times of alloy shortages, for they can replace such critical alloying elements as nickel, molybdenum, chromium, and manganese and, when properly heat-treated, possess physical properties comparable to the alloy grades they replace. Additional advantages for the use of boron in steels are a decrease in susceptibility to flaking, formation of less adherent scale, greater softness in the unhardened condition, and better machinability. It is also useful in low-carbon bainite and acicular ferrite steels.

Specific applications of these steels cannot be given, since the selection of a steel for a given part must depend upon an intimate knowledge of factors such as the availability and cost of the material, the detailed design of the part, and the severity of the service to be imposed. However, the mechanical properties desired in the part to be heat-treated will determine to a large extent the carbon and alloy content of the steel. Table 4.2.7 gives a résumé of mechanical properties that can be expected

Table 4.2.7 Mechanical Properties of Certain SAE Steels with Various Heat Treatments
Sections up to 40 mm (or 1½ diam or thickness)

Tempering temp		Tensile strength		Yield point				
°C	°F	MPa	1,000 lb/in^2	MPa	1,000 lb/in^2	Reduction of area, %	Elongation in 50 mm (2 in), %	Brinell hardness
SAE 1040 quenched in water from 815°C (1,500°F)								
315	600	862	125	717	104	46	11	260
425	800	821	119	627	91	53	13	250
540	1,000	758	110	538	78	58	15	220
595	1,100	745	108	490	71	60	17	216
650	1,200	717	104	455	66	62	20	210
705	1,300	676	98	414	60	64	22	205
SAE 1340 normalized at 865°C (1,585°F), quenched in oil from 845°C (1,550°F)								
315	600	1,565	227	1,420	206	43	11	448
425	800	1,248	181	1,145	166	51	13	372
540	1,000	966	140	834	121	58	17.5	297
595	1,100	862	125	710	103	62	20	270
650	1,200	793	115	607	88	65	23	250
705	1,300	758	110	538	78	68	25.5	234
SAE 4042 normalized at 870°C (1,600°F), quenched in oil from 815°C (1,500°F)								
315	600	1,593	231	1,448	210	41	12	448
425	800	1,207	175	1,089	158	50	14	372
540	1,000	966	140	862	125	58	19	297
595	1,100	862	125	758	110	62	23	260
650	1,200	779	113	683	99	65	26	234
705	1,800	724	105	634	92	68	30	210

Table 4.2.8 Effect of Size of Specimen on the Mechanical Properties of Some SAE Steels

Diam of section		Tensile strength		Yield point			Elongation	
in	mm	MPa	1,000 lb/in²	MPa	1,000 lb/in²	Reduction of area, %	in 50 mm (2 in), %	Brinell hardness
SAE 1040, water-quenched, tempered at 540°C (1,000°F)								
1	25	758	110	538	78	58	15	230
2	50	676	98	448	65	49	20	194
3	75	641	93	407	59	48	23	185
4	100	621	90	393	57	47	24.5	180
5	125	614	89	372	54	46	25	180
SAE 4140, oil-quenched, tempered at 540°C (1,000°F)								
1	25	1,000	145	883	128	56	18	297
2	50	986	143	802	125	58	19	297
3	75	945	137	814	118	59	20	283
4	100	869	126	758	110	60	18	270
5	125	841	122	724	105	59	17	260
SAE 8640, oil-quenched, tempered at 540°C (1,000°F)								
1	25	1,158	168	1,000	145	44	16	332
2	50	1,055	153	910	132	45	20	313
3	75	951	138	807	117	46	22	283
4	100	889	129	745	108	46	23	270
5	125	869	126	724	105	45	23	260

on heat-treating SAE steels, and Table 4.2.8 gives an indication of the effect of size on the mechanical properties of heat-treated steels.

The low-carbon SAE steels are used for carburized parts, cold-headed bolts and rivets, and for similar applications where high quality is required. The SAE 1100 series are low-carbon **free-cutting** steels for high-speed screw-machine stock and other machining purposes. These steels have high sulfur present in the steel in the form of manganese sulfide inclusions causing the chips to break short on machining. Manganese and phosphorus harden and embrittle the steel, which also contributes toward free machining. The high manganese contents are intended to ensure that all the sulfur is present as manganese sulfide. Lead was a common additive, but there are environmental problems; experiments have been conducted with selenium, tellurium, tin, and bismuth as replacements.

Cold-finished carbon-steel bars are used for bolts, nuts, typewriter and cash register parts, motor and transmission power shafting, piston pins, bushings, oil-pump shafts and gears, etc. Representative mechanical properties of cold-drawn steel are given in Table 4.2.9. Besides improved mechanical properties, cold-finished steel has better machining properties than hot-rolled products. The surface finish and dimensional accuracy are also greatly improved by cold finishing.

Forging steels, at one time between 0.30 and 0.40 percent carbon and used for axles, bolts, pins, connecting rods, and similar applications, can now contain up to 0.7 percent C with microalloying additions to refine the structure (for better toughness) and to deliver precipitation strengthening. In some cases, air cooling can be used, thereby saving the cost of alloy steels. These steels are readily forged and, after heat treatment, develop considerably better mechanical properties than low-carbon steels. For heavy sections where high strength is required, such as in crankshafts and heavy-duty gears, the carbon may be increased and sufficient alloy content may be necessary to obtain the desired hardenability.

TOOL STEELS

The application of tool steels can generally be fitted into one of the following categories or types of operations: cutting, shearing, forming, drawing, extruding, rolling, and battering. Each of these operations

Table 4.2.9 Representative Average Mechanical Properties of Cold-Drawn Steel

	Tensile strength		Yield strength		Elongation		
SAE no.	MPa	1,000 lb/in²	MPa	1,000 lb/in²	in 50 mm (2 in), %	Reduction of area, %	Brinell hardness
1,010	462	67	379	55.0	25.0	57	137
1,015	490	71	416	60.3	22.0	55	149
1,020	517	75	439	63.7	20.0	52	156
1,025	552	80	469	68.0	18.5	50	163
1,030	600	87	509	73.9	17.5	48	179
1,035	634	92	539	78.2	17.0	45	187
1,040	669	97	568	82.4	16.0	40	197
1,045	703	102	598	86.7	15.0	35	207
1,117	552	80	469	68.0	19.0	51	163
1,118	569	82.5	483	70.1	18.5	50	167
1,137	724	105	615	89.2	16.0	35	217
1,141	772	112	656	95.2	14.0	30	223

Sizes 16 to 50 mm (% to 2 in) diam, test specimens 50 × 13 mm (2 × 0.505 in).
SOURCE: ASM "Metals Handbook."

Table 4.2.10 Simplified Tool-Steel Classification*

Major grouping	Symbol	Types
Water-hardening tool steels	W	
Shock-resisting tool steels	S	
Cold-work tool steels	O	Oil hardening
	A	Medium-alloy air hardening
	D	High-carbon, high-chromium
Hot-work tool steels	H	H10-H19 chromium base H20-H39 tungsten base H40-H59 molybdenum base
High-speed tool steels	T	Tungsten base
	M	Molybdenum base
Special-purpose tool steels	F	Carbon-tungsten
	L	Low-alloy
	P	Mold steels P1-P19 low carbon P20-P39 other types

* Each subdivision is further identified as to type by a suffix number which follows the letter symbol.

requires in the tool steel a particular physical property or a combination of such metallurgical characteristics as hardness, strength, toughness, wear resistance, and resistance to heat softening, before optimum performance can be realized. These considerations are of prime importance in tool selection; but hardenability, permissible distortion, surface decarburization during heat treatment, and machinability of the tool steel are a few of the additional factors to be weighed in reaching a final decision. In actual practice, the final selection of a tool steel represents a compromise of the most desirable physical properties with the best overall economic performance. Tool steels have been identified and classified by the SAE into six major groups, based upon quenching methods, applications, special characteristics, and use in specific industries. These six classes are water-hardening, shock-resisting, cold-work, hot-work, high-speed, and special-purpose tool steels. A simplified classification of these six basic types and their subdivisions is given in Table 4.2.10.

Water-hardening tool steels, containing 0.60 to 1.40 percent carbon, are widely used because of their low cost, good toughness, and excellent machinability. They are shallow-hardening steels, unsuitable for nondeforming applications because of high warpage, and possess poor resistance to softening at elevated temperatures. Water-hardening tool steels have the widest applications of all major groups and are used for files, twist drills, shear knives, chisels, hammers, and forging dies.

Shock-resisting tool steels, with chromium-tungsten, silicon-molybdenum, or silicon-manganese as the dominant alloys, combine good hardenability with outstanding toughness. A tendency to distort easily is their greatest disadvantage. However, oil quenching can minimize this characteristic.

Cold-work tool steels are divided into three groups: oil-hardening, medium-alloy air-hardening, and high-carbon, high-chromium. In general, this class possesses high wear resistance and hardenability, develops little distortion, but at best is only average in toughness and in resistance to heat softening. Machinability ranges from good in the oil-hardening grade to poor in the high-carbon, high-chromium steels.

Hot-work tool steels are either chromium- or tungsten-based alloys possessing good nondeforming, hardenability, toughness, and resistance to heat-softening characteristics, with fair machinability and wear resistance. Either air or oil hardening can be employed. Applications are blanking, forming, extrusion, and casting dies where temperatures may rise to 540°C (1,000°F).

High-speed tool steels, the best-known tool steels, possess the best combination of all properties except toughness, which is not critical for high-speed cutting operations, and are either tungsten or molybdenum base types. Cobalt is added in some cases to improve the cutting qualities in roughing operations. They retain considerable hardness at a red heat. Very high heating temperatures are required for the heat treatment of high-speed steel and, in general, the tungsten-cobalt high-speed steels require higher quenching temperatures than the molybdenum steels. High-speed steel should be tempered at about 595°C (1,100°F) to increase the toughness; owing to a secondary hardening effect, the hardness of the tempered steels may be higher than as quenched.

Special-purpose tool steels are composed of the low-carbon, low-alloy, carbon-tungsten, mold, and other miscellaneous types.

SPRING STEEL

For small springs, steel is often supplied to spring manufacturers in a form that requires no heat treatment except perhaps a low-temperature anneal to relieve forming strains. Types of previously treated steel wire for small helical springs are **music wire** which has been given a special heat treatment called patenting and then cold-drawn to develop a high yield strength, **hard-drawn wire** which is of lower quality than music wire since it is usually made of lower-grade material and is seldom patented, and **oil-tempered wire** which has been quenched and tempered. The wire usually has a Brinell hardness between 352 and 415, although this will depend on the application of the spring and the severity of the forming operation. Steel for small flat springs has either been cold-rolled or quenched and tempered to a similar hardness.

Steel for both helical and flat springs which is hardened and tempered after forming is usually supplied in an annealed condition. Plain carbon steel is satisfactory for small springs; for large springs it is necessary to use alloy steels such as chrome-vanadium or silicon-manganese steel in order to obtain a uniform structure throughout the cross section. It is especially important for springs that the surface of the steel be free from all defects and decarburization, which lowers fatigue strength.

SPECIAL ALLOY STEELS

Many steel alloys with compositions tailored to specific requirements are reported periodically. Usually, the compositions and/or treatments are patented, and they are most likely to have registered trademarks and trade names. They are too numerous to be dealt with here in any great detail, but a few of the useful properties exhibited by some of those special alloys are mentioned.

Iron-silicon alloys with minimum amounts of both carbon and other alloying elements have been used by the electrical equipment industry for a long time; often these alloys are known as **electrical sheet steel.** Iron-nickel alloys with high proportions of nickel, and often with other alloying elements, elicit properties such as nonmagnetic behavior, high permeability, low hysteresis loss, low coefficient of expansion, etc.; iron-cobalt alloys combined with other alloying elements can result in materials with low resistivity and high hysteresis loss. The reader interested in the use of materials with some of these or other desired properties is directed to the extensive literature and data available.

STAINLESS STEELS

by James D. Redmond

REFERENCES: "Design Guidelines for the Selection and Use of Stainless Steel," Specialty Steel Institute of North America (SSINA), Washington, DC. Publications of the Nickel Development Institute, Toronto, Ontario, Canada. "Metals Handbook," 10th ed., ASM International. ASTM Standards.

When the chromium content is increased to about 11 percent in an iron-chromium alloy, the resulting material is generally classified as a stainless steel. With that minimum quantity of chromium, a thin, protective, passive film forms spontaneously on the steel. This passive film acts as a barrier to prevent corrosion. Further increases in chromium content strengthen the passive film and enable it to repair itself if it is damaged in a corrosive environment. Stainless steels are also heat-resistant because exposure to high temperatures (red heat and above) causes the formation of a tough oxide layer which retards further oxidation.

Other alloying elements are added to stainless steels to improve corrosion resistance in specific environments or to modify or optimize mechanical properties or characteristics. Nickel changes the crystal structure to improve ductility, toughness, and weldability. Nickel improves corrosion resistance in reducing environments such as sulfuric acid. Molybdenum increases pitting and crevice corrosion resistance in chloride environments. Carbon and nitrogen increase strength. Aluminum and silicon improve oxidation resistance. Sulfur and selenium are added to improve machinability. Titanium and niobium (columbium) are added to prevent **sensitization** by preferentially combining with carbon and nitrogen, thereby preventing intergranular corrosion.

Stainless Steel Grades There are more than 200 different **grades** of stainless steel. Each is alloyed to provide a combination of corrosion resistance, heat resistance, mechanical properties, or other specific characteristics. There are five **families** of stainless steels: austenitic, ferritic, duplex, martensitic, and precipitation hardening. Table 4.2.11 is a representative list of stainless steel grades by family and chemical composition as typically specified in ASTM Standards. The mechanical properties of many of these grades are summarized in Table 4.2.12. Note that the number designation of a stainless steel does not describe its performance when utilized to resist corrosion.

Austenitic stainless steels are the most widely used, and while most are designated in the 300 series, some of the highly alloyed grades, though austenitic, have other identifying grade designations, e.g., alloy 20, 904L, and the 6 percent molybdenum grades. These latter are often known by their proprietary designation. These stainless steel grades are available in virtually all wrought product forms, and many are employed as castings. Some austenitic grades in which manganese and nitrogen are substituted partially for nickel are also designated in the 200 series. Types 304 (18 Cr, 8 Ni) and 316 (17 Cr, 10 Ni, 2 Mo) are the workhorse grades, and they are utilized for a broad range of equipment and structures. Austenitic grades can be hardened by cold work but not by heat treatment; cold work increases strength with an accompanying decrease in ductility. They provide excellent corrosion resistance, respond very well to forming operations, and are readily welded. When fully annealed, they are not magnetic, but may become slightly magnetic when cold-worked. Compared to carbon steel, they have higher coefficients of thermal expansion and lower thermal conductivities. When austenitic grades are employed to resist corrosion, their performance may vary over a wide range, depending on the particular corrosive media encountered. As a general rule, increased levels of chromium, molybdenum, and nitrogen result in increased resistance to pitting and crevice corrosion in chloride environments. Type 304 is routinely used for atmospheric corrosion resistance and to handle lowchloride potable water. At the other end of the spectrum are the 6 percent Mo austenitic stainless steels, which have accumulated many years of service experience handling seawater in utility steam condensers and in piping systems on offshore oil and gas platforms. Highly alloyed austenitic stainless steels are subject to precipitation reactions in the range of 1,500 to 1,800°F (815 to 980°C), resulting in a reduction of their corrosion resistance and ambient-temperature impact toughness.

Low-carbon grades, or **L grades,** are restricted to very low levels of carbon (usually 0.030 wt % maximum) to reduce the possibility of sensitization due to chromium carbide formation either during welding or when exposed to a high-temperature thermal cycle. When a sensitized stainless steel is exposed subsequently to a corrosive environment, intergranular corrosion may occur. Other than improved resistance to intergranular corrosion, the low-carbon grades have the same resistance to pitting, crevice corrosion, and chloride stress corrosion cracking as the corresponding grade with the higher level of carbon (usually 0.080 wt % maximum).

There is a trend to dual-certify some pairs of stainless-steel grades. For example, an 18 Cr–8 Ni stainless steel low enough in carbon to meet the requirements of an L grade and high enough in strength to qualify as the standard carbon version may be **dual-certified.** ASTM specifications allow such a material to be certified and marked, for example, 304/304L and S30400/S30403.

Ferritic stainless steels are iron-chromium alloys with 11 to 30 percent chromium. They can be strengthened slightly by cold working. They are magnetic. They are difficult to produce in plate thicknesses, but are readily available in sheet and bar form. Some contain molybdenum for improved corrosion resistance in chloride environments. They exhibit excellent resistance to chloride stress corrosion cracking. Resistance to pitting and crevice corrosion is a function of the total chromium and molybdenum content. The coefficient of thermal expansion is similar to that of carbon steel. The largest quantity produced is Type 409 (11 Cr), used extensively for automobile catalytic converters, mufflers, and exhaust system components. The most highly alloyed ferritic stainless steels have a long service history in seawater-cooled utility condensers.

Duplex stainless steels combine some of the best features of the austenitic and the ferritic grades. Duplex stainless steels have excellent chloride stress corrosion cracking resistance and can be produced in the full range of product forms typical of the austenitic grades. Duplex stainless steels are comprised of approximately 50 percent ferrite and 50 percent austenite. They are magnetic. Their yield strength is about double that of the 300-series austenitic grades, so economies may be achieved from reduced piping and vessel wall thicknesses. The coefficient of thermal expansion of duplex stainless steels is similar to that of carbon steel and about 30 to 40 percent less than that of the austenitic grades. The general-purpose duplex grade is 2,205 (22 Cr, 5 Ni, 3 Mo, 0.15 N). Because they suffer embrittlement with prolonged high-temperature exposure, ASME constructions using the duplex grades are limited to a maximum 600°F (315°C) service temperature. Duplex stainless steel names often reflect their chemical composition or are proprietary designations.

Martensitic stainless steels are straight chromium grades with relatively high levels of carbon. The martensitic grades can be strengthened by heat treatment. To achieve the best combination of strength, corrosion resistance, ductility, and impact toughness, they are tempered in the range of 300 to 700°F (150 to 370°C). They have a 400-series designation; 410 is the general-purpose grade. They are magnetic. The martensitic grades containing up to about 0.15 wt % carbon—e.g., grades 403, 410, and 416 (a free-machining version) can be hardened to about 45 RC. The high-carbon martensitics, such as Types 440A, B, and C, can be hardened to about 60 RC. The martensitics typically exhibit excellent wear or abrasion resistance but limited corrosion resistance. In most environments, the martensitic grades have less corrosion resistance than Type 304.

Precipitation-hardening stainless steels can be strengthened by a relatively low-temperature heat treatment. The low-temperature heat treatment minimizes distortion and oxidation associated with higher-temperature heat treatments. They can be heat-treated to strengths greater than can the martensitic grades. Most exhibit corrosion resistance superior to the martensitics and approach that of Type 304. While some precipitation-hardening stainless steels have a 600-series designation, they are most frequently known by names which suggest their chemical composition, for example, 17–4PH, or by proprietary names.

The commonly used stainless steels have been approved for use in ASME boiler and pressure vessel construction. Increasing the carbon content increases the maximum **allowable stress values.** Nitrogen additions are an even more powerful strengthening agent. This may be seen by comparing the allowable stresses for 304L (0.030 C maximum), 304H (0.040 C minimum), and 304N (0.10 N minimum) in Table 4.2.13. In general, increasing the total alloy content—especially chromium, molybdenum, and nitrogen—increases the allowable design stresses. However, precipitation reactions which can occur in the most highly alloyed grades limit their use to a maximum of about 750°F (400°C). The duplex grade 2,205 has higher allowable stress values than even the most highly alloyed of the austenitic grades, up to a maximum service temperature of 600°F (315°C). The precipitation-hardening grades exhibit some of the highest allowable stress values but also are limited to about 650°F (345°C).

Copious amounts of information and other technical data regarding physical and mechanical properties, examples of applications and histories of service lives in specific corrosive environments, current costs, etc. are available in the references and elsewhere in the professional and trade literature.

Table 4.2.11 ASTM/AWS/AMS Chemical Composition (Wt. Pct)[a]

UNS number	Grade	C	Cr	NI	Mo	Cu	N	Other
					Austenitic stainless steels			
S20100	201	0.15	16.00-18.00	3.50-5.50	—[b]	—	0.25	Mn: 5.50-7.50
S20200	202	0.15	17.00-19.00	4.00-6.00	—	—	0.25	Mn: 7.50-10.00
S30100	301	0.15	16.00-18.00	6.00-8.00	—	—	0.50	—
S30200	302	0.15	17.00-19.00	8.00-10.00	—	—	0.10	—
S30300	303	0.15	17.00-19.00	8.00-10.00	—	—	—	S: 0.15 min
S30323	303Se	0.15	17.00-19.00	8.00-10.00	—	—	—	Se: 0.15 min
S30400	304	0.08	18.00-20.00	8.00-10.50	—	—	0.10	—
S30403	304L	0.030	18.00-20.00	8.00-12.00	—	—	0.10	—
S30409	304H	0.04-0.10	18.00-20.00	8.00-10.50	—	—	—	—
S30451	304N	0.08	18.00-20.00	8.00-10.50	—	—	0.10-0.16	—
S30500	305	0.12	17.00-19.00	10.50-13.00	—	—	—	—
S30800	308	0.08	19.00-21.00	10.00-12.00	—	—	—	—
S30883	308L	0.03	19.50-22.00	9.00-11.00	0.05	0.75		Si: 0.39-0.65
S30908	309S	0.08	22.00-24.00	12.00-15.00	—	—	—	—
S31008	310S	0.08	24.00-26.00	19.00-22.00	—	—	—	—
S31600	316	0.08	16.00-18.00	10.00-14.00	2.00-3.00	—	0.10	—
S31603	316L	0.030	16.00-18.00	10.00-14.00	2.00-3.00	—	0.10	—
S31609	316H	0.04-0.10	16.00-18.00	10.00-14.00	2.00-3.00	—	0.10	—
S31703	317L	0.030	18.00-20.00	11.00-15.00	3.00-4.00	—	0.10	—
S31726	317LMN	0.030	17.00-20.00	13.50-17.50	4.0-5.0	—	0.10-0.20	—
S32100	321	0.080	17.00-19.00	9.00-12.00	—	—	0.10	Ti: 5(C + N) min; 0.70 max
S34700	347	0.080	17.00-19.00	9.00-13.00	—	—	—	Nb: 10C min; 1.00 max
N08020	Alloy 20	0.07	19.00-21.00	32.00-38.00	2.00-3.00	3.00-4.00	—	Nb + Ta: 8C min; 1.00 max
N08904	904L	0.020	19.00-23.00	23.00-28.00	4.0-5.0	1.0-2.0	0.10	—
S31254	254 SMQ[c]	0.020	19.50-20.50	17.50-18.50	6.00-6.50	0.50-1.00	0.18-0.22	—
N08367	AL-6XN[d]	0.030	20.00-22.00	23.50-25.50	6.00-7.00	0.75	0.18-0.25	—
N08926	25-6MO[e]/1925hMo[f]	0.020	19.00-21.00	24.00-26.00	6.0-7.0	0.5-1.0	0.15-0.25	—
S32654	654 SMO[c]	0.020	24.00-25.00	21.00-23.00	7.00-8.00	0.30-0.60	0.45-0.55	Mn: 2.00-4.00
					Duplex stainless steels			
S31260	DP-3[g]	0.030	24.0-26.0	5.50-7.50	2.50-3.50	0.20-0.80	0.10-0.30	W 0.10-0.50
S31803	—	0.030	21.0-23.0	4.50-6.50	2.50-3.50	—	0.08-0.20	—
S32003	AL-2003[d]	0.030	19.5-22.5	3.0-4.0	1.50-2.00	—	0.14-0.20	—
S32101	LDX 2101[c]	0.040	21.0-22.0	1.35-1.70	0.10-0.80	—	0.20-0.25	—
S32205	2205	0.030	21.0-23.0	4.50-6.50	2.50-3.50	—	0.14-0.20	—
S32304	2304	0.030	21.5-24.5	3.00-5.00	0.05-0.60	0.05-0.50	0.05-0.20	—
S32550	Ferralium 255[h]	0.040	24.0-27.0	4.50-6.50	2.90-3.90	1.5-2.5	0.10-0.25	—
S32750	2507	0.030	24.0-26.0	6.00-8.00	3.00-5.00	0.50	0.24-0.32	—
S32900	329	0.08	23.00-28.00	2.50-5.00	1.00-2.00	—	—	—

Table 4.2.11 ASTM/AWS/AMS Chemical Composition (Wt. Pct)[a] (*Continued*)

UNS number	Grade	C	Cr	NI	Mo	Cu	N	Other
					Ferritic stainless steels			
S40500	405	0.08	11.50-14.50	0.60	—	—	—	Al: 0.10-0.30
S40900	409	0.08	10.50-11.75	0.50	—	—	—	Ti: 6xC Min; 0.75 max
S43000	430	0.12	16.00-18.00	—	—	—	—	—
S43020	430F	0.12	16.00-18.00	—	—	—	—	S: 0.15 min
S43035	439	0.07	17.00-19.00	0.5	—	—	0.04	Ti: 0.20 + 4(C + N) min; 1.10 max Al: 0.15 max
S43400	434	0.12	16.00-18.00	—	0.75-1.25	—	—	—
S44400	444 (18 Cr–2 Mo)	0.025	17.5-19.5	1.00	1.75-2.50	—	0.035	Ti + Nb: 0.20 + 4(C + N) min; 0.80 max
S44600	446	0.20	23.00-27.50	—	—	—	0.25	—
S44627	E-BRITE 26-1[d]	0.01	25.00-27.00	—	0.75—1.50	0.20	0.015	Nb: 0.05-0.20
S44660	SEA-CURE[i]	0.030	25.00-28.00	1.0-3.50	3.00-4.00	—	0.040	Ti + Nb: 0.20-1.00 and 6(C + N) min
S44735	AL 29-4C[c]	0.030	28.00-30.00	1.00	3.60-4.20	—	0.045	Ti + Nb: 0.20-1.00 and 6(C + N) min
					Martensitic stainless steels			
S40300	403	0.15	11.50-13.50	—	—	—	—	—
S41000	410	0.15	11.50-13.50	0.75	—	—	—	—
S41008	410S	0.08	11.50-13.50	0.60	—	—	—	—
S41600	416	0.15	12.00-14.00	—	—	—	—	—
S41623	416Se	0.15	12.00-14.00	—	—	—	—	Se: 0.15 min
S42000	420	0.15 min	12.00-14.00	—	—	—	—	—
S42020	420F	0.08	12.00-14.00	0.50	—	0.60	—	S: 0.15 min
S43100	431	0.20	15.00-17.00	1.25-2.50	—	—	—	—
S44002	440A	0.60-0.75	16.00-18.00	—	0.75	—	—	—
S44003	440B	0.75-0.95	16.00-18.00	—	0.75	—	—	—
S44004	440C	0.95-1.20	16.00-18.00	—	0.75	—	—	—
S44020	440F	0.95-1.20	16.00-18.00	0.50	—	0.6	—	S: 0.15 min
					Precipitation-hardening stainless steels			
S13800	XM-13/13-8Mo PH	0.05	12.25-13.25	7.50-8.50	2.00-2.50	—	0.01	Al: 0.90-1.35
S15700	632/15-7PH	0.09	14.00-16.00	6.50-7.75	2.00-3.00	—	—	Al: 0.75-1.00
S17400	630/17-4PH	0.07	15.00-17.50	3.00-5.00	—	3.00-5.00	—	Nb + Ta: 0.15-0.45
S17700	631/17-7PH	0.09	16.00-18.00	6.50-7.75	—	—	—	Al: 0.75-1.00
S35000	AMS 350	0.07-0.11	16.00-17.00	4.00-5.00	2.50-3.25	—	0.07-0.13	—
S35500	634/AMS 355	0.10-0.15	15.00-16.00	4.00-5.00	2.50-3.25	—	0.07-0.13	Mn: 0.50-1.25
S45000	XM-25/Custom 450[j]	0.05	14.00-16.00	5.00-7.00	0.50-1.00	1.25-1.75	—	Nb: 8C min
S45500	XM-16/Custom 455[j]	0.03	11.00-12.50	7.50-9.50	0.5	1.50-2.50	—	Ti: 0.90-1.40

[a] Maximum unless range or minimum is indicated.
[b] None required in the specification.
[c] Trademark of Outokumu.
[d] Trademark of Allegheny Ludlum Corp.
[e] Trademark of Special Metals.
[f] Trademark of Thyssen Krupp VDM.
[g] Trademark of Sumitomo Metals.
[h] Trademark of Langley Alloys Ltd.
[i] Trademark of Crucible Materials Corp.
[j] Trademark of Carpenter Technology Corp.

Table 4.2.12 ASTM Mechanical Properties (A 240, A 276, A 479, A 564, A 582)

UNS no.	Grade	Condition*	Tensile strength, min		Yield strength, min		Elongation in 2 in or 50 mm min, %	Hardness, max	
			ksi	MPa	ksi	MPa		Brinell	Rockwell B†
				Austenitic stainless steels					
S20100	201	Annealed	95	655	38	260	40.0	—‡	95
S20200	202	Annealed	90	620	38	260	40.0	241	—
S30100	301	Annealed	75	515	30	205	40.0	217	95
S30200	302	Annealed	75	515	30	205	40.0	201	92
S30300	303	Annealed	—	—	—	—	—	262	—
S30323	303Se	Annealed	—	—	—	—	—	262	—
S30400	304	Annealed	75	515	30	205	40.0	201	92
S30403	304L	Annealed	70	485	25	170	40.0	201	92
S30409	304H	Annealed	75	515	30	205	40.0	201	92
S30451	304N	Annealed	80	550	35	240	30.0	201	92
S30500	305	Annealed	75	515	30	205	40.0	183	88
S30908	309S	Annealed	75	515	30	205	40.0	217	95
S31008	310S	Annealed	75	515	30	205	40.0	217	95
S31600	316	Annealed	75	515	30	205	40.0	217	95
S31603	316L	Annealed	70	485	25	170	40.0	217	95
S31609	316H	Annealed	75	515	30	205	40.0	217	95
S31703	317L	Annealed	75	515	30	205	40.0	217	95
S31726	317LMN	Annealed	80	550	35	240	40.0	223	96
S32100	321	Annealed	75	515	30	205	40.0	217	95
S34700	347	Annealed	75	515	30	205	40.0	201	92
N08020	Alloy 20	Annealed	80	551	35	241	30.0	217	95
N08904	904L	Annealed	71	490	31	215	35.0	—	—
S31254	254 SMO	Annealed	94	650	44	300		223	96
N08367	AL-6XN	Annealed	95	655	45	310	30.0	233	—
N08926	25-6MO/1925hMo	Annealed	94	650	43	295	35.0	—	—
S32654	654 SMO	Annealed	109	750	62	430	40.0	250	—
				Duplex stainless steels					
S31260	DP-3	Annealed	100	690	70	485	20.0	290	—
S31803	—	Annealed	90	620	65	450	25.0	293	31HRC
S32003	AL 2003	Annealed	90	620	65	450	25.0	293	31HRC
S32101	LDX 2101	Annealed	94	650	65	450	30.0	290	—
S32205	2205	Annealed	95	655	65	450	25.0	293	31HRC
S32304	2304	Annealed	87	600	58	400	25.0	290	32HRC
S32550	Ferralium 255	Annealed	110	760	80	550	15.0	302	32HRC
S32750	2507	Annealed	116	795	80	550	15.0	310	32HRC
S32900	329	Annealed	90	620	70	485	15.0	269	28HRC
				Ferritic stainless steels					
S40500	405	Annealed	60	415	25	170	20.0	179	88
S40900	409	Annealed	55	380	25	205	20.0	179	88
S43000	430	Annealed	65	450	30	205	22.0	183	89
S43020	430F	Annealed	—	—	—	—	—	262	—
S43035	439	Annealed	60	415	30	205	22.0	183	89
S43400	434	Annealed	65	450	35	240	22.0	—	—
S44400	444(18 Cr–2 Mo)	Annealed	60	415	40	275	20.0	217	96
S44600	446	Annealed	65	515	40	275	20.0	217	96
S44627	E-BRITE 26-1	Annealed	65	450	40	275	22.0	187	90
S44660	SEA-CURE	Annealed	85	585	65	450	18.0	241	100
S44735	AL 29-4C	Annealed	80	550	60	415	18.0	255	25HRC
				Martensitie stainless steels					
S40300	403	Annealed	70	485	40	275	20.0	223	—
S41000	410	Annealed	65	450	30	205	20.0	217	96
S41008	410S	Annealed	60	415	30	205	22.0	183	89
S41600	416	Annealed	—	—	—	—	—	262	—
S41623	416Se	Annealed	—	—	—	—	—	262	—
S42000	420	Annealed	—	—	—	—	—	241	—
S42020	420F	Annealed	—	—	—	—	—	262	—
S43100	431	Annealed	—	—	—	—	—	285	—
S44002	440A	Annealed	—	—	—	—	—	269	—
S44003	440B	Annealed	—	—	—	—	—	269	—
S44004	440C	Annealed	—	—	—	—	—	269	—
S44020	440F	Annealed	—	—	—	—	—	285	—

Table 4.2.12 ASTM Mechanical Properties (A 240, A 276, A 479, A 564, A 582) *(Continued)*

UNS no.	Grade	Condition*	Tensile strength, min ksi	Tensile strength, min MPa	Yield strength, min ksi	Yield strength, min MPa	Elongation in 2 in or 50 mm min, %	Hardness, max Brinell	Hardness, max Rockwell B†
			Precipitation-hardening stainless steels						
S13800	XM-13/13-8Mo PH	Annealed	—	—	—	—	—	363	38HRC
		H950	220	1,520	205	1,410	10.0	430	45HRC
		H1000	205	1,410	190	1,310	10.0	400	43HRC
		H1050	175	1,210	165	1,140	12.0	372	40HRC
		H1150	135	930	90	620	14.0	283	30HRC
S15700	632/15-7PH	Annealed	—	—	—	—	—	269	100
		RH950	200	1,380	175	1,210	7.0	415	—
S17400	630/17-4PH	Annealed	—	—	—	—	—	363	38HRC
		H900	190	1,310	170	1,170	10.0	388	40HRC
		H925	170	1,170	155	1,070	10.0	375	38HRC
		H1025	155	1,070	145	1,000	12.0	331	35HRC
		H1100	140	965	115	795	14.0	302	31HRC
S17700	631/17-7PH	Annealed	—	—	—	—	—	229	98
		RH950	185	1,275	150	1,035	6.0	388	41HRC
S35500	634/AMS 355	Annealed	—	—	—	—	—	363	—
		H1000	170	1,170	155	1,070	12.0	341	37HRC
S45000	XM-25/Custom 450	Annealed	130	895	95	655	10.0	321	32HRC
		H900	180	1,240	170	1,170	10.0	363	39HRC
		H950	170	1,170	160	1,100	10.0	341	37HRC
		H1000	160	1,100	150	1,030	12.0	331	36HRC
		H1100	130	895	105	725	16.0	285	30HRC
S45500	XM-16/Custom 455	Annealed	—	—	—	—	—	331	36HRC
		H900	235	1,620	220	1,520	8.0	444	47HRC
		H950	220	1,520	205	1,410	10.0	415	44HRC
		H1000	205	1,410	185	1,280	10.0	363	40HRC

* Condition defined in applicable ASTM specification.
† Rockwell B unless hardness Rockwell C (HRC) is indicated.
‡ None required in ASTM specifications.

Table 4.2.13 Maximum Allowable Stress Values (ksi), Plate: ASME Section I; Section III, Classes 2 and 3; Section VIII, Division 1

UNS no.	Grade	−20 to 100°F	300°F	400°F	500°F	600°F	650°F	700°F	750°F	800°F	850°F	900°F
					Austenitic stainless steels							
S30403	304L	16.7	12.8	11.7	10.9	10.3	10.5	10.0	9.8	9.7	—*	—
S30409	304H	18.8	14.1	12.9	12.1	11.4	11.2	11.1	10.8	10.6	10.4	10.2
S30451	304N	20.0	16.7	15.0	13.9	13.2	13.0	12.7	12.5	12.3	12.1	11.8
S31603	316L	16.7	12.7	11.7	10.9	10.4	10.2	10.0	9.8	9.6	9.4	—
N08904	904L	17.8	15.1	13.8	12.7	12.0	11.7	11.4	—	—	—	—
S31254	254 SMO	23.5	21.4	19.9	18.5	17.9	17.7	17.5	17.3	—	—	—
					Duplex stainless steel							
S31803	—	22.5	21.7	20.9	20.4	20.2	—	—	—	—	—	—
S32101	LDX 2101	26.9	25.6	24.7	24.7	24.7	—	—	—	—	—	—
S32205	2205	25.7	24.8	23.9	23.3	23.1	—	—	—	—	—	—
					Ferritic stainless steels							
S40900	409	13.8	12.7	12.2	11.8	11.4	11.3	11.1	10.7	10.2	—	—
S43000	430	16.3	15.0	14.4	13.9	13.5	13.3	13.1	12.7	12.0	11.3	10.5
					Precipitation-hardening stainless steel							
S17400	630/17-4	35.0	35.0	34.1	33.3	32.8	32.6	—	—	—	—	—

* No allowable stress values at this temperature; material not recommended for service at this temperature.

Note:

1. The unprecedented expansion of supply by China (now about one half of world capacity), with the resulting economic dislocations, means that material selection could benefit from study of additional sources of information.

4.3 IRON AND STEEL CASTINGS

by Malcolm Blair and Robert E. Eppich

REFERENCES: "Metals Handbook," ASM. "Iron Casting Handbook," Iron Casting Society. "Malleable Iron Castings," Malleable Founders' Society. ASTM Specifications. "Ductile Iron Handbook," American Foundry Society. "Modern Casting," American Foundry Society. "Ductile Iron Data for Design Engineers," Quebec Iron and Titanium (QIT). "History Cast in Metal," American Foundry Society. "Steel Castings Handbook," Steel Founders' Society of America.

CLASSIFICATION OF CASTINGS

Cast-Iron Castings The term **cast iron** covers a wide range of iron-carbon-silicon alloys containing from 2.0 to 4.0 percent carbon and from 0.5 to 3.0 percent silicon. The alloy typically also contains varying percentages of manganese, sulfur, and phosphorus along with other alloying elements such as chromium, molybdenum, copper, and titanium. For a type and grade of cast iron, these elements are specifically and closely controlled. Cast iron is classified into five basic types; each type is generally based on graphite morphology as follows:

Gray iron
Ductile iron
Malleable iron
Compacted graphite iron
White iron

Even though chemistry is important to achieve the type of cast iron, the molten metal processing and cooling rates play major roles in developing each type. Different mechanical properties are generally associated with each of the five basic types of cast iron.

Figure 4.3.1 illustrates the range of carbon and silicon for each type. The chemistry for a specific grade within each type is usually up to the foundry producing the casting. There are situations where the chemistry for particular elements is specified by ASTM, SAE, and others; this should always be fully understood prior to procuring the casting.

Steel Castings There are two main classes of steel castings: carbon and low-alloy steel castings, and high-alloy steel castings. These classes may be broken down into the following groups: (1) **low-carbon steels** (C < 0.20 percent), (2) **medium-carbon steels** (C between 0.20 and 0.50 percent), (3) **high-carbon steels** (C > 0.50 percent), (4) **low-alloy steels** (total alloy ≤ 8 percent), and (5) **high-alloy steels** (total alloy > 8 percent). The tensile strength of cast steel varies from 60,000 to 250,000 lb/in^2 (400 to 1,700 MPa) depending on the composition and heat treatment. Steel castings are produced weighing from ounces to over 200 tons. They find universal application where strength, toughness, and reliability are essential.

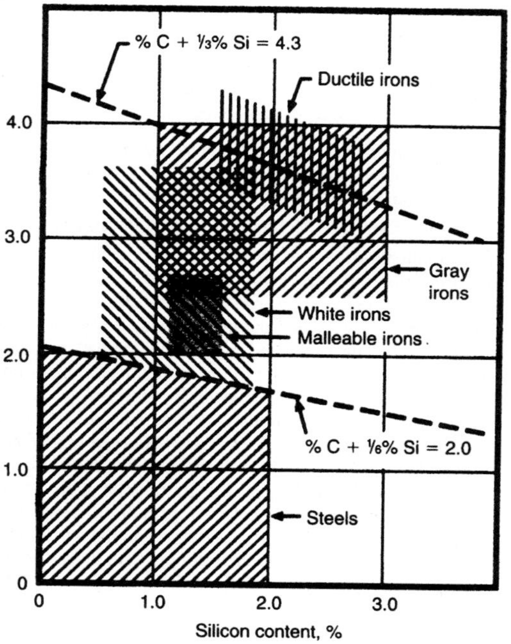

Figure 4.3.1 Approximate ranges of carbon and silicon for steels and various cast irons. *(QIT-Fer Titane, Inc.)*

CAST IRON

The engineering and physical properties of cast iron vary with the type of iron; the designer must match the engineering requirements with all the properties of the specific type of cast iron being considered. Machinability, for example, is significantly affected by the type of cast iron specified. Gray cast iron is the most machinable; the white cast irons are the least machinable. It should be noted that the machinability of gray cast iron can significantly change (improve) with time even when the hardness remains the same. This is primarily caused by nitrogen. The greatest improvement in machinability will occur in the first 30 days after the casting is produced.

Composition The properties of cast iron are controlled primarily by the graphite morphology and the quantity of graphite. Also to be considered is the matrix microstructure, which may be established either during cooling from the molten state (as-cast condition) or as a result of heat treatment. Except where previous engineering evaluations have established the need for specific chemistry or the cast iron is to be produced to a specific ASTM specification, such that chemistry is a part of that specification, the foundry normally chooses the composition that will meet the specified mechanical properties and/or graphite morphology.

Types of Cast Iron

Gray Iron Gray iron castings have been produced since the sixth century B.C. During solidification, when the composition and cooling rate are appropriate, carbon will precipitate in the form of graphite flakes that are interconnected within each eutectic cell. Figure 4.3.2 is

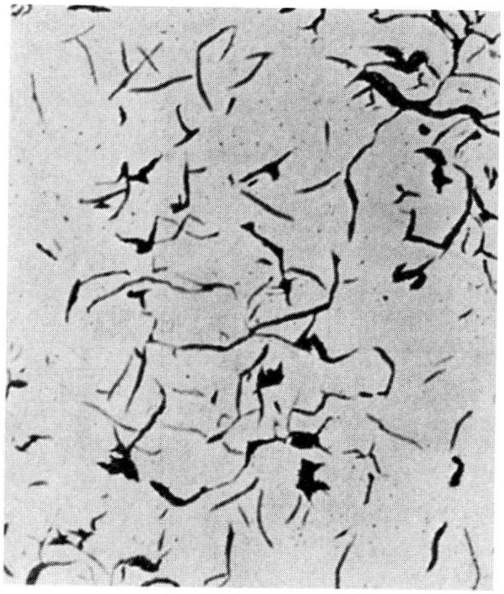

Figure 4.3.2 Gray iron with flake graphite. Unetched, 100 ×.

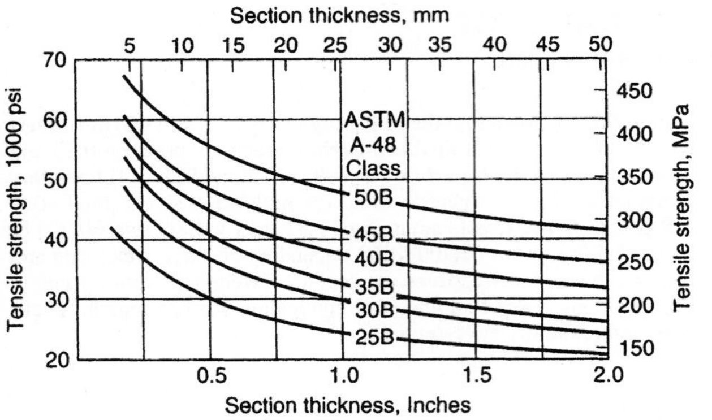

Figure 4.3.3 The influence of casting section thickness on the tensile strength and hardness for a series of gray irons classified by their strength as cast in 1.2-in-(30-mm) diameter, B bars. *(Steel Founders' Society of America.)*

a photomicrograph of a typical gray iron structure. The flakes appear black in the photograph. The sample is unetched; therefore the matrix microstructure does not show. Gray iron fractures primarily along the graphite. The fracture appears gray; hence the name gray iron. This graphite morphology establishes the mechanical properties. The cooling rate of the casting while it solidifies plays a major role in the size and shape of the graphite flakes. Thus the mechanical properties in a single casting can vary by as much as 10,000 lb/in^2 with thinner sections having higher strength (Fig. 4.3.3).

The flake graphite imparts unique characteristics to gray iron, such as excellent machinability, ability to resist galling, excellent wear resistance (provided the matrix microstructure is primarily pearlite), and excellent damping capacity. This same flake graphite exhibits essentially zero ductility and very low impact strength.

Mechanical Properties Gray iron **tensile strength** normally ranges from 20,000 to 60,000 lb/in^2 and is a function of chemistry, graphite morphology, and matrix microstructure. ASTM A48 recognizes nine grades based on the tensile strength of a separately cast test bar whose dimensions are selected to be compatible with the critical controlling section size of the casting. These dimensions are summarized in Table 4.3.1. Test bars are machined to the specified size prior to testing. Table 4.3.2 summarizes the various specifications under ASTM A48. Most engineering applications utilize gray iron with tensile strengths of 30,000 to 35,000 lb/in^2 (typically a B bar). An alternate method of specifying gray iron is to utilize SAE J431. This specification, shown in Table 4.3.3, is based on hardness (BHN) rather than tensile test bars.

Yield strength is not normally shown and is not specified because it is not a well-defined property of the material. There is no abrupt change in the stress-strain relation as is normally observed in other material. A proof stress or load resulting in 0.1 percent permanent strain will range from 65 to 80 percent of the tensile strength.

Table 4.3.1 ASTM A48: Separately Cast Test Bars for Use when a Specific Correlation Has Not Been Established between Test Bar and Casting

Thickness of wall of controlling section of casting, in (mm)	Test bar diameter (See Table 4.3.2)
Under 0.25 (6)	S
0.25 to 0.50 (6 to 12)	A
0.51 to 1.00 (13 to 25)	B
1.01 to 2 (26 to 50)	C
Over 2 (50)	S

SOURCE: Abstracted from ASTM data, by permission.

Table 4.3.2 ASTM A48: Requirements for Tensile Strength of Gray Cast Irons in Separately Cast Test Bars

Class no.	Tensile strength, min, ksi (MPa)	Nominal test bar diameter, in (mm)
20 A	20 (138)	0.88 (22.4)
20 B		1.2 (30.5)
20 C		2.0 (50.8)
20 S		Bars S*
25 A	25 (172)	0.88 (22.4)
25 B		1.2 (30.5)
25 C		2.0 (50.8)
25 S		Bars S*
30 A	30 (207)	0.88 (22.4)
30 B		1.2 (30.5)
30 C		2.0 (50.8)
30 S		Bars S*
35 A	35 (241)	0.88 (22.4)
35 B		1.2 (30.5)
35 C		2.0 (50.8)
35 S		Bars S*
40 A	40 (276)	0.88 (22.4)
40 B		1.2 (30.5)
40 C		2.0 (50.8)
40 S		Bars S*
45 A	45 (310)	0.88 (22.4)
45 B		1.2 (30.5)
45 C		2.0 (50.8)
45 S		Bars S*
50 A	50 (345)	0.88 (22.4)
50 B		1.2 (30.5)
50 C		2.0 (50.8)
50 S		Bars S*
55 A	55 (379)	0.88 (22.4)
55 B		1.2 (30.5)
55 C		2.0 (50.8)
55 S		Bars S*
60 A	60 (414)	0.88 (22.4)
60 B		1.2 (30.5)
60 C		2.0 (50.8)
60 S		Bars S*

* All dimensions of test bar S shall be as agreed upon between the manufacturer and the purchaser.

SOURCE: Abstracted from ASTM data, by permission.

Table 4.3.4 summarizes the mechanical properties of gray iron as they are affected by a term called the **carbon equivalent.** Carbon equivalent is often represented by the equation

$$CE = \text{Carbon (c)} + \tfrac{1}{3}\ \text{Silicon (Si)} + \tfrac{1}{3}\ \text{Phosphorus (P)}$$

Typical applications for gray iron are engine blocks, compressor housings, transmission cases, and valve bodies.

Ductile Iron Whereas gray iron has been used as an engineering material for hundreds of years, ductile iron was developed recently. It was invented in the late 1940s and patented by the International Nickel Company. By adding magnesium to the molten metal under controlled conditions, the graphite structure becomes spherical or nodular. The material resulting has been variously named *spheroidal graphite iron, sg iron, nodular iron,* and *ductile iron.* **Ductile iron** is preferred since it also describes one of its characteristics. The structure is shown in Fig. 4.3.4. As would be expected, this nodular graphite structure, when compared to the graphite flakes in gray iron, shows significant improvement in mechanical properties. However, the machinability and thermal conductivity of ductile iron are lower than those for gray iron.

The generally accepted grades of ductile iron are shown in Table 4.3.5, which summarizes ASTM A-536. In all grades the nodularity of the graphite remains the same. Thus the properties are controlled by the casting microstructure, which is influenced by both the chemistry and the cooling rate after solidification. The grades exhibiting high

Table 4.3.3 SAE J431: Grades of Automotive Gray Iron Castings Designed by Brinell Hardness

SAE grade	Specified hardness BHN*	Minimum tensile strength (for design purposes) lb/in²	MPa	Other requirements
G1800	187 max	18,000	124	
G2500	170-229	25,000	173	
G2500a†	170-229	25,000	173	3.4% min C and microstructure specified
G3000	187-241	30,000	207	
G3500	207-255	35,000	241	
G3500b†	207-255	35,000	241	3.4% min C and microstructure specified
G3500c†	207-255	35,000	241	3.5% min C and microstructure specified
G4000	217-269	40,000	276	

* Hardness at a designated location on the castings.
† For applications such as brake drums, disks, and clutch plates to resist thermal shock.
SOURCE: Abstracted from SAE data, by permission.

Table 4.3.4 Effect of Carbon Equivalent on Mechanical Properties of Gray Cast Irons

Carbon equivalent	Tensile strength, lb/in²*	Modulus of elasticity in tension at ½ load, lb/in² × 10⁶†	Modulus of rapture, lb/in²*	Deflection in 18-in span, in‡	Shear strength, lb/in²*	Endurance limit, lb/in²*	Compressive strength, lb/in²*	Brinell hardness, 3,000-kg load	Izod impact, unnotched, ft·lb
4.8	20,400	8.0	48,300	0.370	29,600	10,000	72,800	146	3.6
4.6	22,400	8.7	49,200	0.251	33,000	11,400	91,000	163	3.6
4.5	25,000	9.7	58,700	0.341	35,500	11,800	95,000	163	4.9
4.3	29,300	10.5	63,300	0.141	37,000	12,300	90,900	179	2.2
4.1	32,500	13.6	73,200	0.301	44,600	16,500	128,800	192	4.2
4.0	35,100	13.3	77,000	0.326	47,600	17,400	120,800	196	4.4
3.7	40,900	14.8	84,200	0.308	47,300	19,600	119,100	215	3.9
3.3	47,700	20.0	92,000	0.230	60,800	25,200	159,000	266	4.4

* × 6.89 = kPa. † × 0.00689 = MPa. ‡ × 25.4 = mm.

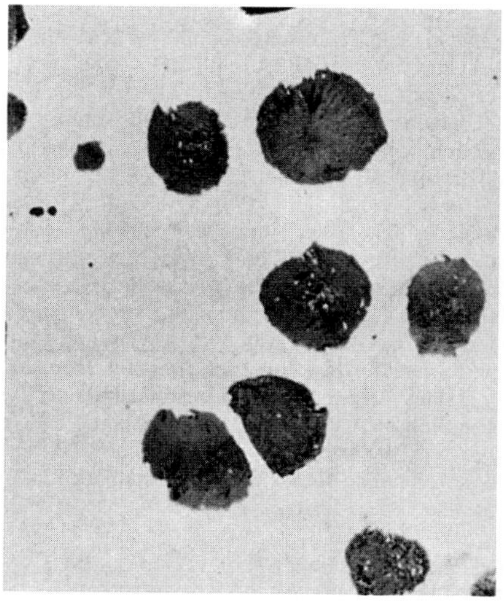

Figure 4.3.4 Ductile iron with spherulitic graphite. Unetched, 100 ×.

elongation with associated lower tensile and yield properties can be produced as cast, but annealing is sometimes used to enhance these properties. Grades 60-40-18 and 60-45-12 have a ferrite microstructure. Both major elements (C and S) and minor elements (Mn, P, Cr, and others) are controlled at the foundry in order to achieve the specified properties. Figure 4.3.5 illustrates the mechanical properties and impact strength as they are affected by temperature and test conditions.

An important new material within the nodular iron family has been developed within the last 20 years. This material, **austempered ductile iron (ADI),** exhibits tensile strengths up to 230,000 lb/in², but of greater significance is higher elongation, approximately 10 percent, compared to 2 percent for conventional ductile iron at equivalent tensile strength. The material retains the spherical graphite nodules characteristic of ductile iron, but the matrix microstructure of the heat-treated material now consists of acicular ferrite in a high-carbon austenite. Since 1990, ADI has been specified per ASTM 897. Properties associated with this specification are shown in Table 4.3.6 and are summarized in Fig. 4.3.6. Significant research has resulted in excellent documentation of other important properties such as fatigue resistance (Fig. 4.3.7), abrasion resistance (Fig. 4.3.8), and heat-treating response to specific compositions.

Malleable Iron Whiteheart malleable iron was invented in Europe in 1804. It was cast as white iron and heat-treated for many days, resulting in a material with many properties of steel. Blackheart malleable

Table 4.3.5 ASTM A536: Ductile Irons. Tensile Requirements

	Grade 60-40-18	Grade 65-45-12	Grade 80-55-06	Grade 100-70-03	Grade 120-90-02
Tensile strength, min, lb/in²	60,000	65,000	80,000	100,000	120,000
Tensile strength, min, MPa	414	448	552	689	827
Yield strength, min, lb/in²	40,000	45,000	55,000	70,000	90,000
Yield strength, min, MPa	276	310	379	483	621
Elongation in 2 in or 50 mm, min, %	18	12	6.0	3.0	2.0

SOURCE: Abstracted from ASTM data, by permission.

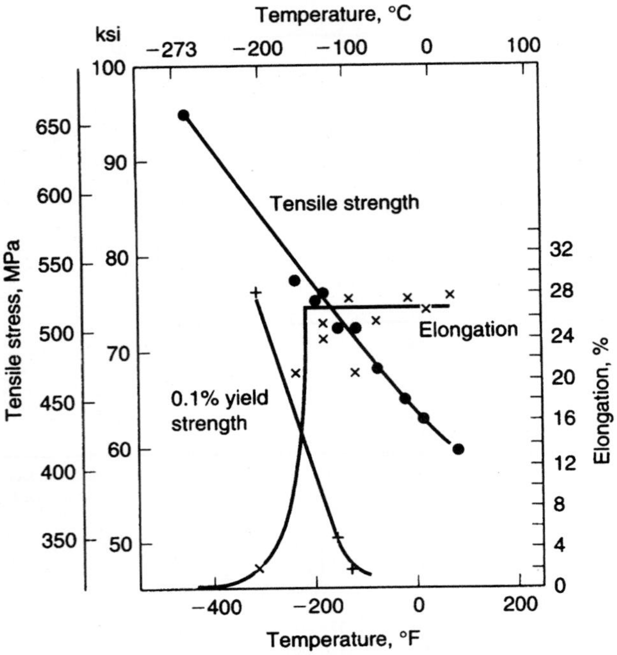

Figure 4.3.5 Effect of temperature on low-temperature properties of ferritic ductile iron. *(QIT-Fer Titane, Inc.)*

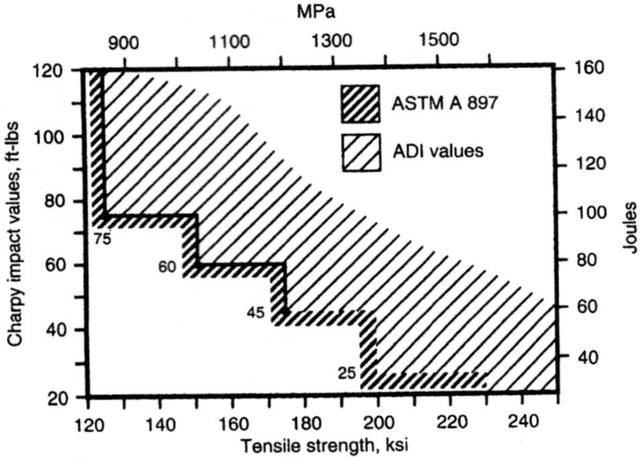

Figure 4.3.6 Austempered ductile iron mechanical properties. *(QIT-Fer Titane, Inc.)*

iron was invented in the United States by Seth Boyden. This material contains no free graphite in the as-cast condition, but when subjected to an extended annealing cycle (6 to 10 days), the graphite is present as irregularly shaped nodules called **temper carbon** (Fig. 4.3.9). All malleable iron castings produced currently are of blackheart malleable iron; the terms blackheart and whiteheart are obsolete.

The properties of several grades of malleable iron are shown in Table 4.3.7. The properties are similar to those of ductile iron, but the annealing cycles still range from 16 to 30 h. Production costs are higher than for ductile iron, so that virtually all new engineered products are designed to be of ductile iron. Malleable iron accounts for only 3 percent of total cast-iron production and is used mainly for pipe and electrical fittings and replacements for existing malleable-iron castings.

Compacted Graphite Iron Compacted graphite iron was developed in the mid-1970s. It is a controlled class of iron in which the graphite takes a form between that in gray iron and that in ductile iron. Figure 4.3.10 shows the typical graphite structure that is developed during casting. It was desired to produce a material that, without extensive alloying, would be stronger than the highest-strength grades of gray iron but would be more machinable than ductile iron. These objectives were met. The properties of the various grades of compacted iron are summarized in Table 4.3.8, which summarizes ASTM A842. Several other properties of this material should be considered. For example, its galling resistance is superior to that of ductile iron, and it is attractive for hydraulic applications; its thermal conductivity is superior to that of ductile iron but inferior to that of gray iron. Compacted graphite iron properties have resulted in its applications to satisfy a number of unique requirements.

White Iron When white iron solidifies, virtually all the carbon appears in the form of carbides. White irons are hard and brittle, and they break with a white fracture. These irons are usually alloyed with chromium and nickel; Table 4.3.9 summarizes ASTM A532. Their hardness is in the range of 500 to 600 BHN. The specific alloying that is required is a function of section size and application; there must be coordination between designer and foundry. These irons exhibit outstanding wear resistance and are used extensively in the mining industry for ball mill shell liners, balls, impellers, and slurry pumps.

Specialty Irons There are a number of highly alloyed gray and ductile iron grades of cast iron. These grades typically contain between 20 and 30 percent nickel along with other elements such as chromium and molybdenum. They have the graphite morphology characteristic of conventional gray or ductile iron, but alloying imparts significantly enhanced performance when wear resistance or corrosion resistance is required. Tables 4.3.10 and 4.3.11 summarize ASTM A436 for austenitic gray iron and ASTM A439 for austenitic ductile iron. This family of alloys is best known by its trade name *Ni-Resist.* Corrosion resistance of several of *Ni-Resist* alloys is summarized in Tables 4.3.12

Table 4.3.6 ASTM A897: Austempered Ductile Irons. Mechanical Property Requirements

	Grade 125/80/10	Grade 150/100/7	Grade 175/125/4	Grade 200/155/1	Grade 230/185/—
Tensile strength, min, ksi	125	150	175	200	230
Yield strength, min, ksi	80	100	125	155	185
Elongation in 2 in, min %	10	7	4	1	*
Impact energy, ft·lb†	75	60	45	25	*
Typical hardness, BHN, kg/mm²‡	269-321	302-363	341-444	388-477	444-555

* Elongation and impact requirements are not specified. Although grades 200/155/1 and 230/185/— are both primarily used for gear and wear resistance applications, Grade 200/155/1 has applications where some sacrifice in wear resistance is acceptable in order to provide a limited amount of ductility and toughness.

† Unnotched charpy bars tested at 72 ± 7°F. The values in the table are a minimum for the average of the highest three test values of the four tested samples.

‡ Hardness is not mandatory and is shown for information only.

SOURCE: Abstracted from ASTM data, by permission.

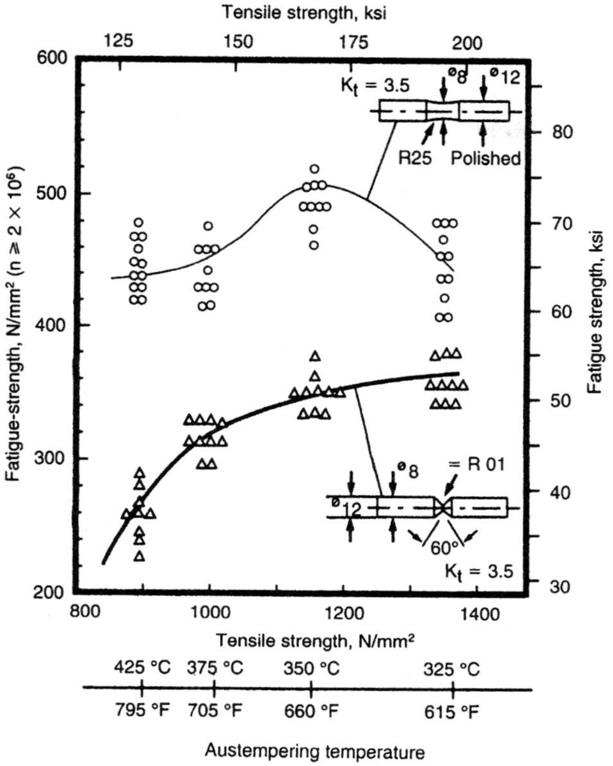

Figure 4.3.7 Fatigue properties of notched and unnotched ADI. *(QIT-Fer Titane, Inc.)*

and 4.3.13. Castings made from these alloys exhibit performance comparable to many stainless steels.

Several other specialty irons utilize silicon or aluminum as the major alloying elements. These also find specific application in corrosion or elevated-temperature environments.

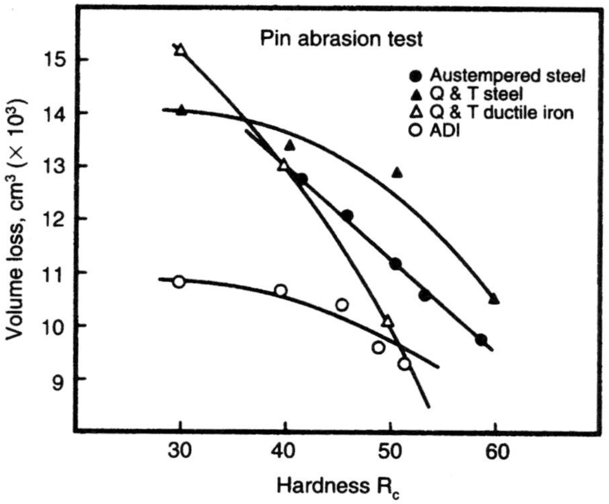

Figure 4.3.8 Comparison of pin abrasion test results of ADI, ductile iron, and two abrasion-resistant steels. *(QIT-Fer Titane, Inc.)*

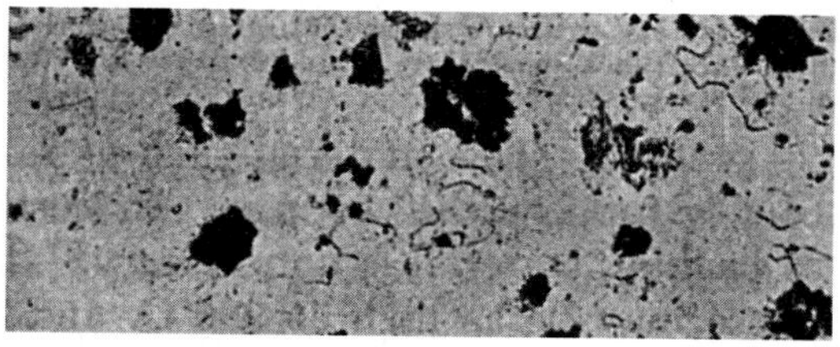

Figure 4.3.9 Temper carbon form of graphite in malleable iron. Unetched, 100 ×.

Table 4.3.7 SAE J158a: Malleable Irons. Mechanical Property Requirements

Grade	Casting hardness	Heat treatment
M 3210	156 BHN max.	Annealed
M 4504	163-217 BHN	Air-quenched and tempered
M 5003	187-241 BHN	Air-quenched and tempered
M 5503	187-241 BHN	Liquid-quenched and tempered
M 7002	229-269 BHN	Liquid-quenched and tempered
M 8501	269-302 BHN	Liquid-quenched and tempered

Source: Abstracted from SAE data, by permission.

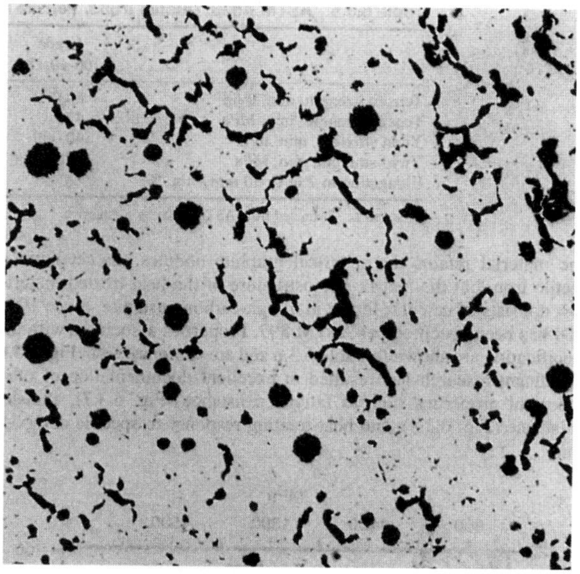

Figure 4.3.10 Compacted graphite iron. Unetched, 100 ×.

Table 4.3.8 ASTM A842: Compacted Graphite Irons. Mechanical Property Requirements

	Grade* 250	Grade 300	Grade 350	Grade 400	Grade† 450
Tensile strength, min, MPa	250	300	350	400	450
Yield strength, min, MPa	175	210	245	280	315
Elongation in 50 mm, min, %	3.0	1.5	1.0	1.0	1.0

* The 250 grade is a ferritic grade. Heat treatment to attain required mechanical properties and microstructure shall be the option of the manufacturer.

† The 450 grade is a pearlitic grade usually produced without heat treatment with addition of certain alloys to promote pearlite as a major part of the matrix.

Source: Abstracted from ASTM data, by permission.

Table 4.3.9 ASTM A532: White Irons. Chemical Analysis Requirements

Class	Type	Designation	Carbon	Manganese	Silicon	Nickel	Chromium	Molybdenum	Copper	Phosphorus	Sulfur
I	A	Ni-Cr-Hc	2.8-3.6	2.0 max	0.8 max	3.3-5.0	1.4-4.0	1.0 max	—	0.3 max	0.15 max
I	B	Ni-Cr-Lc	2.4-3.0	2.0 max	0.8 max	3.3-5.0	1.4-4.0	1.0 max	—	0.3 max	0.15 max
I	C	Ni-Cr-GB	2.5-3.7	2.0 max	0.8 max	4.0 max	1.0-2.5	1.0 max	—	0.3 max	0.15 max
I	D	Ni-HiCr	2.5-3.6	2.0 max	2.0 max	4.5-7.0	7.0-11.0	1.5 max	—	0.10 max	0.15 max
II	A	12% Cr	2.0-3.3	2.0 max	1.5 max	2.5 max	11.0-14.0	3.0 max	1.2 max	0.10 max	0.06 max
II	B	15% Cr-Mo	2.0-3.3	2.0 max	1.5 max	2.5 max	14.0-18.0	3.0 max	1.2 max	0.10 max	0.06 max
II	D	20% Cr-Mo	2.0-3.3	2.0 max	1.0-2.2	2.5 max	18.0-23.0	3.0 max	1.2 max	0.10 max	0.06 max
III	A	25% Cr	2.0-3.3	2.0 max	1.5 max	2.5 max	23.0-30.0	3.0 max	1.2 max	0.10 max	0.06 max

SOURCE: Abstracted from ASTM data, by permission.

Table 4.3.10 ASTM A436: Austenitic Gray Irons. Chemical Analysis and Mechanical Property Requirements

	Composition, %							
Element	Type 1	Type 1b	Type 2	Type 2b	Type 3	Type 4	Type 5	Type 6
Carbon, total, max	3.00	3.00	3.00	3.00	2.60	2.60	2.40	3.00
Silicon	1.00-2.80	1.00-2.80	1.00-2.80	1.00-2.80	1.00-2.00	5.00-6.00	1.00-2.00	1.50-2.50
Manganese	0.5-1.5	0.5-1.5	0.5-1.5	0.5-1.5	0.5-1.5	0.5-1.5	0.5-1.5	0.5-1.5
Nickel	13.50-17.50	13.50-17.50	18.00-22.00	18.00-22.00	28.00-32.00	29.00-32.00	34.00-36.00	18.00-22.00
Copper	5.50-7.50	5.50-7.50	0.50 max	0.50 max	0.50 max	0.50 max	0.50 max	3.50-5.50
Chromium	1.5-2.5	2.50-3.50	1.5-2.5	3.00-6.00*	2.50-3.50	4.50-5.50	0.10 max	1.00-2.00
Sulfur, max	0.12	0.12	0.12	0.12	0.12	0.12	0.12	0.12
Molybdenum, max	—	—	—	—	—	—	—	1.00
	Type 1	Type 1b	Type 2	Type 2b	Type 3	Type 4	Type 5	Type 6
Tensile strength, min, ksi (MPa)	25 (172)	30 (207)	25 (172)	30 (207)	25 (172)	25 (172)	20 (138)	25 (172)
Brinell hardness (3,000 kg)	131-183	149-212	118-174	171-248	118-159	149-212	99-124	124-174

* Where some matching is required, the 3.00 to 4.00 percent chromium range is recommended.
SOURCE: Abstracted from ASTM data, by permission.

Table 4.3.11 ASTM A439: Austenitic Ductile Irons. Chemical Analysis and Mechanical Property Requirements

	Type								
	D-2*	D-2B	D-2C	D-3*	D-3A	D-4	D-5	D-5B	D-5S
Element	Composition, %								
Total carbon, max	3.00	3.00	2.90	2.60	2.60	2.60	2.40	2.40	2.30
Silicon	1.50-3.00	1.50-3.00	1.00-3.00	1.00-2.80	1.00-2.80	5.00-6.00	1.00-2.80	1.00-2.80	4.90-5.50
Manganese	0.70-1.25	0.70-1.25	1.80-2.40	1.00 max†	1.00 max†	1.00 max†	1.00 max†	1.00 max†	1.00 max
Phosphorus, max	0.08	0.08	0.08	0.08	0.08	0.08	0.08	0.08	0.08
Nickel	18.00-22.00	18.00-22.00	21.00-24.00	28.00-32.00	28.00-32.00	28.00-32.00	34.00-36.00	34.00-36.00	34.00-37.00
Chromium	1.75-2.75	2.75-4.00	0.50 max†	2.50-3.50	1.00-1.50	4.50-5.50	0.10 max	2.00-3.00	1.75-2.25

	Type								
	D-2	D-2B	D-2C	D-3	D-3A	D-4	D-5	D-5B	D-5S
Element	Properties								
Tensile strength, min, ksi (MPa)	58 (400)	58 (400)	58 (400)	55 (379)	55 (379)	60 (414)	55 (379)	55 (379)	65 (449)
Yield strength (0.2% offset), min, ksi (MPa)	30 (207)	30 (207)	28 (193)	30 (207)	30 (207)	—	30 (207)	30 (207)	30 (207)
Elongation in 2 in or 50 mm, min, %	8.0	7.0	20.0	6.0	10.0	—	20.0	6.0	10
Brinell hardness (3,000 kg)	139-202	148-211	121-171	139-202	131-193	202-273	131-185	139-193	131-193

* Additions of 0.7 to 1.0% of molybdenum will increase the mechanical properties above 800°F (425°C).
† Not intentionally added.
SOURCE: Abstracted from ASTM data, by permission.

Allowances for Iron Castings

Dimension allowances that must be applied in the production of castings arise from the fact that different metals contract at different rates during solidification and cooling. The shape of the part has a major influence on the as-cast dimensions of the casting. Some of the principal allowances are discussed briefly here.

Shrinkage allowances for iron castings are of the order of in/ft (0.1 mm/cm). The indiscriminate adoption of this shrinkage allowance can lead to problems in iron casting dimensions. It is wise to discuss shrinkage allowances with the foundry that makes the castings before patterns are made.

Casting finish allowances (unmachined) and **machine finish allowances** are based on the longest dimension of the casting, although the weight of the casting may also influence these allowances. Other factors which affect allowances include pattern materials and the method of pattern construction. A useful guide is the international

Table 4.3.12 Corrosion Resistance of Ductile Ni-Resist Irons D-2 and D-2C

Corrosive media	Penetration, in/yr	
	Type D-2C	Type D-2
Ammonium chloride solution: 10% NH_4Cl, pH 5.15, 13 days at 30°C (86°F), 6.25 ft/min	0.0280	0.0168
Ammonium sulfate solution: 10% $(NH_4)2SO_4$, pH 5.7, 15 days at 30°C (86°F), 6.25 ft/min	0.0128	0.0111
Ethylene vapors & splash: 38% ethylene glycol, 50% diethylene glycol, 4.5% H_2O, 4% Na_2SO_4, 2.7% NaCl, 0.8% Na_2CO_3 + trace NaOH, pH 8-9, 85 days at 275-300°F	0.0023	0.0019*
Fertilizer: commercial "5-10-5", damp, 290 days at atmospheric temp	—	0.0012
Nickel chloride solution: 15% $NiCl_2$, pH 5.3, 7 days at 30°C (86°F), 6.25 ft/min	0.0062	0.0040
Phosphoric acid, 86%, aerated at 30°C (86°F), velocity 16 ft/min, 12 days	0.213	0.235
Raw sodium chloride brine, 300 gpl of chlorides, 2.7 gpl CaO, 0.06 gpl NaOH, traces of NH3 & H_2S, pH 6-6.5, 61 days at 50°F, 0.1-0.2 fps	0.0023	0.0020*
Seawater at 26.6°C (80°F), velocity 27 ft/s, 60-day test	0.039	0.018
Soda & brine: 15.5% NaCl, 9.0% NaOH, 1.0% Na_2SO_4, 32 days at 180°F	0.0028	0.0015
Sodium bisulfate solution: 10% $NaHSO_4$, pH 1.3, 13 days at 30°C (86°F), 6.25 ft/min	0.0431	0.0444
Sodium chloride solution: 5% NaCl, pH 5.6, 7 days at 30°C (86°F), 6.25 ft/min	0.0028	0.0019
Sodium hydroxide:		
50% NaOH + heavy conc. of suspended NaCl, 173 days at 55°C (131°F), 40 gal/min	0.0002	0.0002
50% NaOH saturated with salt, 67 days at 95°C (203°F), 40 gal/min	0.0009	0.0006
50% NaOH, 10 days at 260°F, 4 days at 70°F	0.0048	0.0049
30% NaOH + heavy conc. of suspended NaCl, 82 days at 85°C (185°F)	0.0004	0.0005
74% NaOH, 19³/4 days, at 260°F	0.005	0.0056
Sodium sulfate solution: 10% Na_2SO_4, pH 4.0, 7 days at 30°C (86°F), 6.25 ft/min	0.0136	0.0130
Sulfuric acid: 5%, at 30°C (86°F) aerated, velocity 14 ft/min, 4 days	0.120	0.104
Synthesis of sodium bicarbonate by Solvay process: 44% solid $NaHCO_3$ slurry plus 200 gpl NH_4Cl, 100 gpl NH_4HCO_3, 80 gpl NaCl, 8 gpl $NaHCO_3$, 40 gpl CO_2, 64 days at 30°C (86°F)	0.0009	0.0003
Tap water aerated at 30°F, velocity 16 ft/min, 28 days	0.0015	0.0023
Vapor above ammonia liquor: 40% NH_3, 9% CO_2, 51% H_2O, 109 days at 85°C (185°F), low velocity	0.011	0.025
Zinc chloride solution: 20% $ZnCl_2$, pH 5.25, 13 days at 30°C (86°F), 6.25 ft/min	0.0125	0.0064

* Contains 1 % chromium.
SOURCE: QIT-Fer Titane, Inc.

standard ISO 8,062 (Castings—System of Dimensional Tolerances and Machining Allowances). It is strongly recommended that designers discuss dimensional and machine finish allowance requirements with the foundry to avoid unnecessary costs.

STEEL CASTINGS

Composition There are five classes of commercial steel castings: low-carbon steels (C < 0.20 percent), medium-carbon steels (C between 0.20 and 0.50 percent), high-carbon steels (C > 0.50 percent), low-alloy steels (total alloy ≤ 8 percent), and high-alloy steels (total alloy > 8 percent).

Carbon Steel Castings Carbon steel castings contain less than 1.00 percent C, along with other elements which may include Mn (0.50 to 1.00 percent), Si (0.20 to 0.70 percent), max P (0.05 percent), and max S (0.06 percent). In addition, carbon steels, regardless of whether they are cast or wrought, may contain small percentages of other elements as residuals from raw materials or additives incorporated in the steelmaking process.

Table 4.3.13 Oxidation Resistance of Ductile Ni-Resist, Ni-Resist, Conventional and High-Silicon Ductile Irons, and Type 309 Stainless Steel

Oxidation resistance	Penetration (in/yr)	
	Test 1	Test 2
Ductile iron (2.5 Si)	0.042	0.50
Ductile iron (5.5 Si)	0.004	0.051
Ductile Ni-Resist type D-2	0.042	0.175
Ductile Ni-Resist type D-2C	0.07	—
Ductile Ni-Resist type D-4	0.004	0.0
Conventional Ni-Resist type 2	0.098	0.30
Type 309 stainless steel	0.0	0.0

Test 1—Furnace atmosphere—air, 4,000 h at 1,300°F.
Test 2—Furnace atmosphere—air, 600 h at 1,600-1,700°F, 600 h between 1,600-1,700°F, and 800-900°F, 600 h at 800-900°F.
SOURCE: QIT-Fer Titane, Inc.

Low-Alloy Steel Castings In a low-alloy steel casting, the alloying elements are present in percentages greater than the following: Mn, 1.00; Si, 0.70; Cu, 0.50; Cr, 0.25; Mo, 0.10; V, 0.05; W, 0.05; and Ti, 0.05. Limitations on phosphorus and sulfur contents apply to low-alloy steels as well as carbon steels unless they are specified to be different for the purpose of producing some desired effect, for example, free machining. Carbon and low-alloy steels account for approximately 85 percent of the steel castings produced in the United States.

High-Alloy Steel Castings Steel castings with total alloy content greater than 8 percent are generally considered to be high-alloy, and usually the steel castings industry requires that the composition contain greater than 11 percent Cr. This Cr content requirement eliminates the potential confusion arising from including austenitic manganese steels (Hadfield) in this group. High-alloy steels include corrosion-resistant, heat-resistant, and duplex stainless steels. Many of the cast corrosion-resistant and duplex stainless steels are similar to the wrought stainless steels, but their performance may not be the same. Table 4.3.14 lists cast high-alloy steels and similar wrought grades.

Methods of Manufacture of Steel Castings The manufacturing methods for steel castings include sand molding, investment casting, and centrifugal methods. Sand molding is the most widely used process; it requires the use of a wood, metal or plastic pattern which is a replica of the steel casting shape. This pattern is placed in a flask (box) and sand is packed around it. The pattern is removed and molten steel is introduced into the cavity producing a steel casting. The investment casting process is similar to the sand process but differs in that it utilizes a wax pattern which is encased in a ceramic shell. The wax is removed by steam in an autoclave. The shell is preheated in a furnace and metal is poured into the preheated shell. The investment casting process produces castings which have smoother surfaces and generally are much smaller than the sand molded product. Centrifugally cast products predominantly have a circular shape but are not necessarily limited to circular shapes. In terms of being an alternative casting method the

Table 4.3.14 ACI Designations for Heat- and Corrosion-Resistant High Alloy Castings

Cast designation.	UNS	Wrought alloy type*	Cast designation	UNS	Wrought alloy type*
CA15	J91150	410	CW2M	N26455	C4
CA15M	J91151	—	CW6M	N30107	—
CA28MWV	J91422	422	CW6MC	N26625	625
CA40	J91153	420	CW12MW	N30002	C
CA6NM	J91540	F6NM	CX2M	N26059	59
CA6N	J91650	—	CX2MW	N26022	C22
CB30	J91803	431	CY5SnBiM	N26055	—
CB7Cu-1	J92180	17-4	HA	—	T9
CB7Cu-2	J92110	15-5	HC	J92605	446
CC50	J92615	446	HD	J93005	327
CD3MWCuN	J93380	Zeron 100	HE	J93403	312
CD3MN	J92205	2205	HF	J92603	302B
CD4MCu	J93370	255	HH	J93503	309
CD6MN	J93371	—	HI	J94003	—
CE3MN	J93404	—	HK	J94224	310
CE30	J93423	—	HL	N08604	—
CE8MN	J93345	—	HN	J94213	—
CF3	J92500	304L	HP	N08705	—
CF8	J92600	304	HPNb	N28701	—
CF10SMnN	J92972	Nitronic 60	HPNbS	N28702	—
CF20	J92602	302	HT	N08605	330
CF3M	J92800	316L	HU	N08004	—
CF3MN	J92804	316LN	HW	N08001	—
CF8M	J92900	316	HX	N06006	—
CF8C	J92710	347	M25S	N24025	—
CF16F	J92701	303	M30C	N24130	—
CG6MMN	J93790	Nitronic 50	M30H	N24030	400
CG12	J93001	308	M35-1	N24135	400
CG8M	J93000	317	M35-2	N04020	—
CH20	J93402	309	N3M	N30003	B2
CK20	J94202	310	N7M	N30007	B
CK3MCuN	J93254	254SMO	N12MV	N30012	—
CN3MN	J94651	AL6XN			
CN7M	N08007	Alloy 20			
CN7MS	N02100	—			
CT15C	N08151	800			
CU5MCuC	N08826	825			

* Wrought alloy type references are listed only for the convenience of those who want corresponding wrought and cast grades. This table does not imply that the performance of the corresponding cast and wrought grades will be the same. Because the cast alloy chemical composition ranges are not the same as the wrought composition ranges, buyers should use cast alloy designations for proper identification of castings.

NOTE: Most of the standard grades listed are covered for general applications by ASTM specifications.

SOURCE: *Steel Founders' Society of America Handbook,* 5th ed.

centrifugal casting process does not generally compete with the sand or investment casting processes, they serve different markets.

Weight Range Depending on the manufacturing process steel castings can weigh from a few ounces to 200 tons. The thickness of steel castings can vary from 0.25 in. (6 mm) to the order of feet. In some special processes thicknesses 0.080 in (2 mm) has been achieved. In the investment casting process wall thicknesses of 0.060 in. (1.5 mm) can be achieved with sections tapering down to 0.030 in. (0.75 mm) are not uncommon.

Mechanical Properties The outstanding mechanical properties of cast steel are high strength, ductility, resistance to impact, stiffness, endurance strength, and resistance to both high and low temperatures. It is also weldable. The mechanical properties of carbon steel castings are shown in Figs. 4.3.11 and 4.3.12; those of low-alloy cast steels are shown in Figs. 4.3.13 and 4.3.14.

Tensile and Yield Strength Ferritic steels of a given hardness or hardenability have the same tensile strength whether cast, wrought, or forged regardless of alloy content. For design purposes involving tensile and yield properties, cast, wrought, and forged properties are considered the same.

Ductility If the ductility values of steels are compared with the hardness values, the cast, wrought, and forged values are almost identical. The longitudinal properties of wrought and forged steels are slightly higher than those for castings. The transverse properties of the wrought and forged steels are lower, by an amount that depends on the degree of working. Since most service conditions involve several directions of loading, the isotropic behavior of cast steels is particularly advantageous (Fig. 4.3.15).

Impact The notched-bar impact test is used often as a measure of the toughness of materials. Cast steels have excellent impact properties at normal and low temperatures. Generally wrought steels are tested in the direction of rolling, show higher impact values than cast steels of similar composition. Transverse impact values will be 50 to 70 percent lower than these values. Cast steels do not show directional properties. If the directional properties are averaged for wrought steels, the values obtained are comparable to the values obtained for cast steels of similar composition. The **hardenability** of cast steels is influenced by composition and other variables in the same manner as the hardenability of wrought steels. The ratio of the **endurance limit** to the tensile strength for cast steel varies from 0.42 to 0.50, depending somewhat on the composition and heat treatment of the steel. The notch-fatigue ratio varies from 0.28 to 0.32 for cast steels and is the same for wrought steels (Fig. 4.3.16).

In **wear testing,** cast steels react similarly to wrought steels and give corresponding values depending on composition, structure, and hardness.

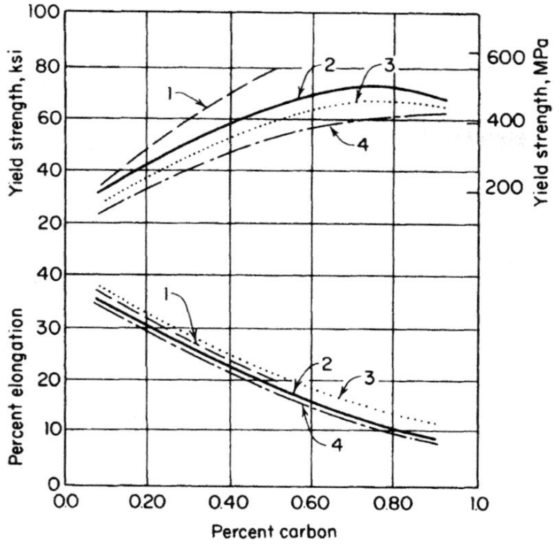

Figure 4.3.11 Yield strength and elongation versus carbon content for cast carbon steels. (1) Water quenched and tempered at 1,200°F; (2) normalized; (3) normalized and tempered at 1,200°F; (4) annealed.

Carbon cast steels of approximately 0.50 percent C and low-alloy cast steels of the chromium, chromium-molybdenum, nickel-chromium, chromium-vanadium, and medium-manganese types, all of which contain more than 0.40 percent C, exhibit excellent resistance to wear in service.

Corrosion Resistance Cast and wrought steels of similar composition and heat treatment appear to be equally resistant to corrosion in the same environments. Small amounts of copper in cast steels increase the resistance of the steel to atmospheric corrosion. High-alloy steels of chromium, chromium-nickel, and chromium-nickel-molybdenum types are normally used for corrosion service.

Heat Resistance Although not comparable to the high-alloy steels of the nickel-chromium types developed especially for heat resistance, the 4.0 to 6.5 percent Cr cast steels, particularly with additions of 0.75 to 1.25 percent W, or 0.40 to 0.70 percent Mo, and 0.75 to 1.00 percent Ti, show good strength and considerable resistance to scaling up to 1,000°F (550°C).

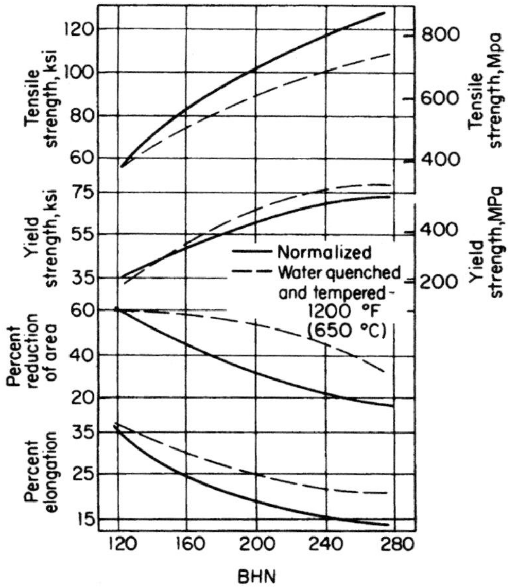

Figure 4.3.12 Tensile properties of carbon steel as a function of hardness.

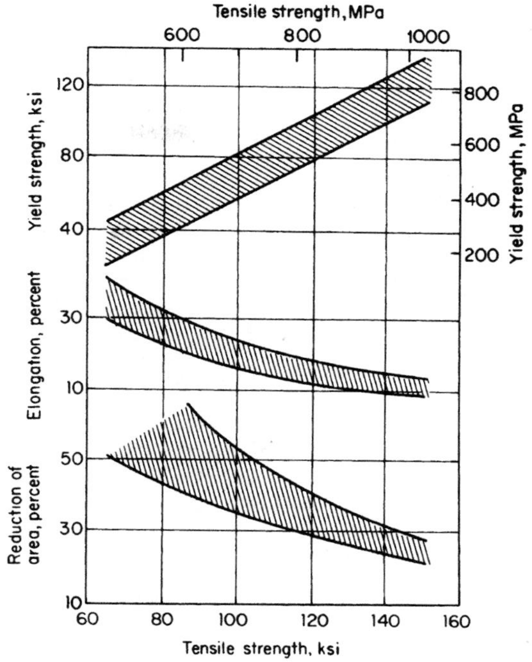

Figure 4.3.13 Properties of normalized and tempered low-alloy cast steels.

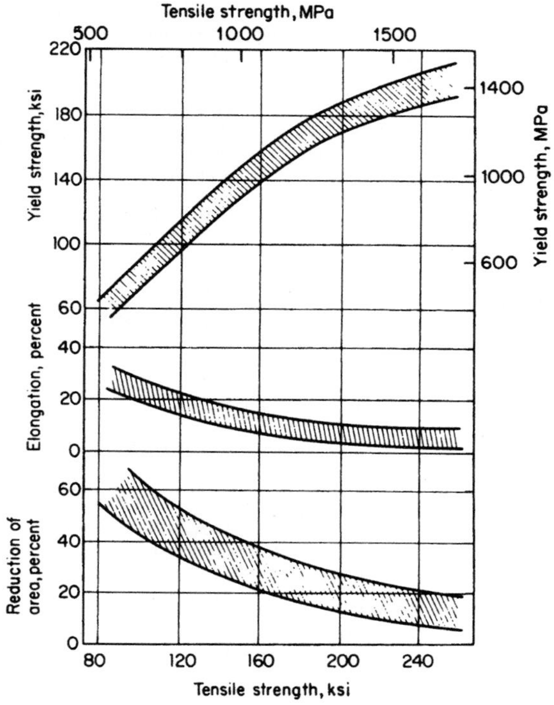

Figure 4.3.14 Properties of quenched and tempered low-alloy cast steels.

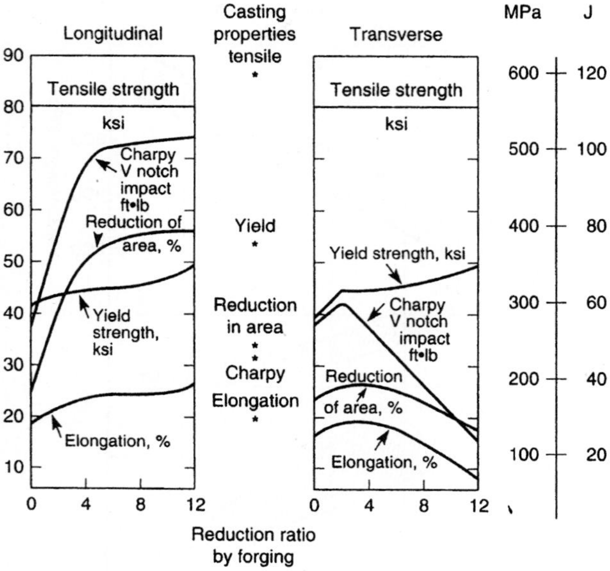

Figure 4.3.15 Influence of forging reduction on anisotropy for a 0.35 pecent carbon wrought steel. Properties for a cast of 0.35 percent carbon steel are shown with an asterisk for comparison purposes. (*Steel Founders' Society of America Handbook, 5th ed.*)

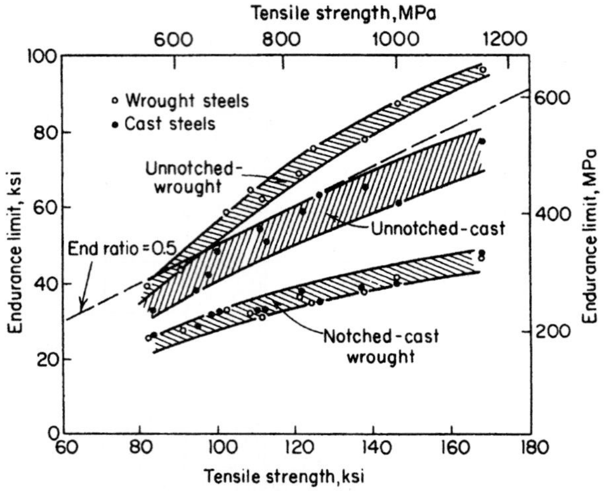

Figure 4.3.16 Variations of endurance limit with tensile strength for comparable carbon and low-alloy cast and wrought steels.

Many cast high-alloy nickel-chromium steels have **creep** and **rupture** properties superior to those of wrought materials. In many cases wrought versions of the cast grades are not available due to hot-working difficulties. These cast materials may be used in service conditions up to 2,000°F (1,100°C), with some grades capable of withstanding temperatures of 2,200°F (1,200°C).

Machinability of carbon and alloy cast steels is comparable to that of wrought steels having equivalent strength, ductility, hardness, and similar microstructure. Factors influencing the machinability of cast steels are as follows: (1) Microstructure has a definite effect on machinability of cast steels. In some cases it is possible to improve machinability by as much as 100 to 200 percent through heat treatments which alter the microstructure. (2) Generally speaking, hardness alone cannot be taken as the criterion for predicting tool life in cutting cast steels. (3) In general, for a given structure, plain carbon steels machine better than alloy steels. (4) When machining cast carbon steel (1,040) with carbides, tool life varies with the ratio of ferrite to pearlite in the microstructure of the casting, the 60 : 40 ratio resulting in optimum tool life. For equal tool life, the skin of a steel casting should be machined at approximately one-half the cutting speed recommended for the base metal. The machinability of various carbon and low-alloy cast steels is given in Table 4.3.15.

Welding steel castings presents the same problems as welding wrought steels. It is interesting to note that cast low-alloy steels show greater resistance to underbead cracking than their wrought counterparts.

Purchase Specifications for Steel Castings Many steel castings are purchased according to mechanical property requirements. The most reliable method employed for the purchase of steel castings is through the use of published specifications; the most common standards used are those published by ASTM. These standards are separated into two categories—(1) General Industrial Use and (2) Pressure Containing Parts. This separation is designed to identify the different testing requirements required by each broad market application. The ASTM standards development process requires input and approval by users and producers of steel castings. The ASTM standards are well understood by steel foundries and purchasers of steel castings. The property levels specified in ASTM specifications A27 and A148

Table 4.3.15 Machinability Index for Cast Steels

		Conventional		Metcut*	
Steel	BHN	Carbide	HSS	HSS	Carbide
B1112 Free machining steel (wrought)	179	. . .	100		
1,020 Annealed	122	10	90	160	400
1,020 Normalized	134	6	75	135	230
1,040 Double normalized	185	11	70	130	400
1,040 Normalized and annealed	175	10	75	135	380
1,040 Normalized	190	6	65	120	325
1,040 Normalized and oil-quenched	225	6	45	80	310
1,330 Normalized	187	2	40	75	140
1,330 Normalized and tempered	160	3	65	120	230
4,130 Annealed	175	4	55	95	260
4,130 Normalized and spheroidized	175	3	50	90	200
4,340 Normalized and annealed	200	3	35	60	210
4,340 Normalized and spheroidized	210	6	55	95	290
4,340 Quenched and tempered	300	2	25	45	200
4,340 Quenched and tempered	400		20	35	180
8,430 Normalized and tempered at 1,200°F (660°C)	200	3	50	90	200
8,430 Normalized and tempered at 1,275°F (702°C)	180	4	60	110	240
8,630 Normalized	240	2	40	75	180
8,630 Annealed	175	5	65	120	290

* The Metcut speed index number is the actual cutting speed (surface ft/min) which will give 1 h tool life in turning.

are shown in Table 4.3.16, these ASTM specifications are unusual in that they are what is known as mechanical property standards all other ASTM steel castings standards will, in addition to mechanical property requirements, specify the chemical composition limits of the steel grades. Table 4.3.17 lists current ASTM steel castings standards.

Steel Melting Practice Steel for steel castings is produced commercially by processes similar to those used for wrought steel production, but electric arc and electric induction melting furnaces predominate. Some steel foundries also use secondary refining processes such as **argon oxygen decarburization** (AOD) and/or **vacuum oxygen degassing** (VOD). Wire injection is also used selectively depending on service and process requirements for the steel casting. In addition to the materials listed in the ASTM standards, the steel casting industry will produce compositions tailored especially for particular service applications, and even in small quantity.

Allowances for Steel Castings

Dimensional allowances that must be applied in the production of castings are due to the fact that different metals contract at different rates during solidification and cooling. The shape of the part has a major influence on the as-cast dimensions of the casting. Some of the principal allowances are discussed briefly here.

Shrinkage allowances for steel castings vary from $^9/_{32}$ to $^1/_{16}$ in/ft (0.03 to 0.005 cm/cm). The amount most often used is ¼ in/ft (0.021 cm/cm), but if it is applied indiscriminately, it can lead to problems in steel casting dimensions. The best policy is to discuss shrinkage allowances with the foundry that makes the castings before patterns are made.

Minimum Section Thickness The fluidity of steel when compared to other metals is low. In order that all sections may be completely filled, a minimum value of section thickness must be adopted that is a function of the largest dimension of the casting. These values are suggested for design use:

Min section thickness		Max length of section	
in	mm	in	cm
¼	6	12	30
½	13	50	125

Casting finish allowances (unmachined) are based on the longest dimension of the casting; however, there are two other interconnected factors which influence these allowances. These are the length of the production run and the method of pattern construction. Often the purchaser will not be aware that there is a natural variation in the dimensions of a casting produced from a pattern. For example, the casting tolerance for the production of one casting is likely to be greater than that where a number of castings are required. The production of one casting does not allow for the adjustment of the pattern to ensure that the size or tolerance can be minimized. The other contributing factor is the method of construction and materials used for the pattern—soft wood patterns will wear faster than metal patterns. Information related to these issues is included in the SFSA "Steel Castings Handbook, Supplement 3, Dimensional Capabilities of Steel Castings."

Machining finish allowances added to the casting dimensions for machining purposes will depend entirely on the casting design. Definite values are not established for all designs, but Table 4.3.18 presents a brief guide to machining allowances on gears, wheels, and circular, flat castings.

Nondestructive Examination of Steel Castings

In addition to the mechanical tests such as tension, impact, and bend tests which may be used to verify mechanical properties of a heat or qualify weld procedures and welders, various nondestructive examinations may be carried out on steel castings. These nondestructive tests are used to refine the quality requirements of steel castings. These tests include:

Visual examination This requirement is universal and is referenced in the ASTM General Requirement Standards A703, A781, A957, and A985. More specific requirements related to surface appearance are addressed in A802 and A997.

Magnetic particle and dye penetrant examination These techniques provide a more discriminating method of visual inspection in that they highlight indications in the casting surface by fluorescence in the magnetic particle technique or colored dye with the dye penetrant method. Magnetic particle examination is limited to the examination of magnetic grades of steel. Magnetic particle inspection can be used to detect sub-surface indications at limited depths. Dye penetrant examination can be used on magnetic and non-magnetic grades of steel but is not capable of detecting sub-surface indications. The allowable indications for both techniques

Table 4.3.16 ASTM Requirements for Steel Castings in A27 and A148

Grade	Tensile strength, min, ksi (MPa)	Yield point min, ksi (MPa)	Elongation, min, %	Reduction of area, min, %
		ASTM A27		
U-60-30	60 (415)	30 (205)	22	30
60-30	60 (415)	30 (205)	24	35
65-35	65 (450)	35 (240)	24	35
70-36	70 (485)	36 (250)	22	30
70-40	70 (485)	40 (275)	22	30
		ASTM A148		
80-40	80 (550)	40 (275)	18	30
80-50	80 (550)	50 (345)	22	35
90-60	90 (620)	60 (415)	20	40
105-85	105 (725)	85 (585)	17	35
115-95	115 (795)	95 (655)	14	30
130-115	130 (895)	115 (795)	11	25
135-125	135 (930)	125 (860)	9	22
150-135	150 (1,035)	135 (930)	7	18
160-145	160 (1,105)	145 (1,000)	6	12
165-150	165 (1,140)	150 (1,035)	5	20
165-150L	165 (1,140)	150 (1,035)	5	20
210-180	210 (1,450)	180 (1,240)	4	15
210-180L	210 (1,450)	180 (1,240)	4	15
260-210	260 (1,795)	210 (1,450)	3	6
260-210L	260 (1,795)	210 (1,450)	3	6

SOURCE: Abstracted from ASTM data, by permission.

Table 4.3.17 A Summary of ASTM Specifications for Steel and Alloy Castings

A27	Steel Castings, Carbon for General Application
A128	Steel Castings, Austenitic Manganese
A148	Steel Castings, High Strength, for Structural Applications
A216	Steel Castings, Carbon, Suitable for Fusion Welding, for High-Temperature Service
A217	Steel Castings, Martensitic, Stainless and Alloy for Pressure Containing Parts, Suitable for High-Temperature Service
A297	Steel Castings, Iron-Chromium and Iron-Chromium-Nickel, Heat Resistant for General Applications
A351	Steel Castings, Austenitic, Austenitic-Ferritic (Duplex), for Pressure Containing Parts
A352	Steel Castings, Ferritic and Martensitic, for Pressure-Containing Parts, Suitable for Low-Temperature Service
A356	Steel Castings, Carbon, Low-Alloy and Stainless Steel, Heavy Walled for Steam Turbines
A389	Steel Castings, Alloy, Specially Heat-Treated, for Pressure-Containing Parts, Suitable for High-Temperature Service
A426	Centrifugally Cast Ferritic Alloy Steel Pipe for High-Temperature Service
A447	Steel Castings, Chromium-Nickel-Iron Alloy (25-12 Class) for High Temperature Service
A451	Centrifugally Cast Austenitic Steel Pipe for High-Temperature Service
A487	Steel Castings, Suitable for Pressure Service
A494	Castings, Nickel and Nickel Alloy
A560	Castings, Chromium-Nickel Alloy
A608	Centrifugally Cast Iron-Chromium-Nickel High-Alloy Tubing, for Pressure Applications at High Temperatures
A660	Centrifugally Cast Carbon Steel Pipe for High-Temperature Service
A703	Steel Castings, General Requirements, for Pressure-Containing Parts
A732	Casting, Investment, Carbon and Low-Alloy for General Applications
A743	Castings, Iron-Chromium, Iron-Chromium-Nickel, Corrosion-Resistant, for General Application
A744	Castings, Iron-Chromium, Iron-Chromium-Nickel, Corrosion-Resistant, for Severe Service
A747	Steel Castings, Stainless, Precipitation Hardening
A757	Steel Castings, Ferritic and Martensitic, for Pressure-Containing and Other Applications, for Low-Temperature Service
A781	Castings, Steel and Alloy, General Requirements, for General Industrial Use
A872	Centrifugally Cast Ferritic/Austenitic Stainless Steel Pipe for Corrosive Environments
A890	Castings, Iron-Chromium-Nickel-Molybdenum Corrosion Resistant, Duplex (Austenitic/Ferritic) for General Application
A915	Steel Castings, Carbon and Alloy, Chemical Requirements Similar to Standard Wrought Grades
A957	Investment Castings, Steel and Alloy, Common Requirements, for General Industrial Use
A958	Steel Castings, Carbon and Alloy, with Tensile Requirements, Chemical Requirements Similar to Standard Wrought Grades
A985	Steel Investment Castings, General requirements for Pressure Containing Parts
A990	Castings, Iron-Nickel-Chromium and Nickel Alloys, Specially Controlled for Pressure Retaining Parts for Corrosion Service
A995	Castings, Austenitic, Austenitic-Ferritic (Duplex), for Pressure Containing Parts
A1,001	High Strength Steel Castings in Heavy Sections
A1,002	Castings, Nickel-Aluminum Ordered Alloy

can be classified using ASTM A903 Steel Castings, Surface Acceptance Standards, Magnetic Particle and Liquid Penetrant Inspection.

Ultrasonic examination This technique is limited in application to the steels covered by the scope of ASTM A609 Standard Practice for Castings, Carbon, Low-Alloy, and Martensitic Stainless Steel, Ultrasonic Examination. It is not uncommon for ultrasonic testing to be used where heavy sections are to be examined, where radiographic examination would be impracticable due to long exposure times associated with the power available in the radiation source. Facilities with instruments such as linear accelerators provide shorter exposure times.

Radiographic examination Is commonly used where an assessment of the interior soundness of the casting is required. As steel contracts on solidifying by approximately 6 percent by volume, it is very difficult to produce a casting that does not have any shrinkage voids. The problem of shrinkage in castings is due to their complexity of design. To gain a more complete understanding of issues associated with shrinkage formation in steel castings it is recommended that contact be made with Steel Founders' Society of America for a more detailed and complete explanation of shrinkage formation.

A number of ASTM standards have been developed that classify the size of the indications using them to set quality levels, these standards utilize Reference Radiographs and are shown in Table 4.3.19.

A recent development in radiographic examination has been the use of digital imaging. This has a number of beneficial effects the first being that the exposure time to produce an image on a phosphor plate is reduced compared to a film image. The status of the development of these ASTM standards are shown in Table 4.3.20. The digital image standards ensure that the digital images are secure.

Table 4.3.18 Machining Allowances for Steel Castings

Specific dimension or reference line (plane) distance, in		Greatest dimension of casting							
		10 in		10-20 in		20-100 in		Over 100 in*	
	But not	0.xx in*	X/32 in*	0.xx in*	X/32 in*	0.xx in*	X/32 in*	0.xx in*	X/32 in*
Greater than	exceeding	+	+	+	+	+	+	+	+
0	2	0.187	6	0.218	7	0.25	8	0.312	10
2	5	0.25	8	0.281	9	0.312	10	0.437	14
5	10	0.312	10	0.344	11	0.375	12	0.531	17
10	20			0.406	13	0.468	15	0.625	20
20	50					0.562	18	0.75	24
50	75					0.656	21	0.875	28
75	100					0.75	24	1.00	32
100	500							1.25	40

		Machine tolerances on section thickness	
		0.XXX in*	X/32 in*
		+	+
0		½	0.062 2
½	½	0.092	3
1½	4	0.187	6
4		7	0.25 8
7		10	0.344 11
10			0.50 16

* × 2.54 = cm.

Table 4.3.19 ASTM Reference Radiograph Standards

ASTM ref	Title
E186	Standard Reference Radiographs for Heavy-Walled (2 to 4 ½ - in [51 to 114 mm]) Steel Castings
E192	Standard Reference Radiographs for Investment Steel Castings for Aerospace Applications
E446	Standard Reference Radiographs for Steel Castings up to 2 in (51 mm) in Thickness
E280	Standard Reference Radiographs Heavy-Walled (4 ½ to 12 in [114 to 305 mm]) Steel Castings

Table 4.3.20 ASTM Reference Radiograph Digital Versions

ASTM ref	Title
—	Standard Digital Reference Images for Heavy-Walled (2 to 4 ½ - in [51 to 114 mm]) Steel Castings – not available at this time
E2660	Standard Digital Reference Images for Investment Steel Castings for Aerospace Applications
E2868	Standard Digital Reference Images for Steel Castings up to 2 in (50.8 mm) in Thickness
—	Standard Digital Reference Images for Heavy-Walled (4 ½ to 12 in [114 to 305 mm]) Steel Castings – not available at this time, currently being balloted at ASTM

Use of Steel Castings

All industries use steel castings. They are used as wheels, sideframes, bolsters, and couplings in the railroad industry; rock crushers in the mining and cement industries; components in construction vehicles; suspension parts and fifth wheels for trucks; valves; pumps; armor. High-alloy steel castings are used in highly corrosive environments in the chemical, petrochemical, and paper industries. Generally, steel castings are applied widely when strength, fatigue, and impact properties are required, as well as weldability, corrosion resistance, and high-temperature strength. Often, overall economy of fabrication will dictate a steel casting as opposed to a competing product.

Casting Design Maximum service and properties can be obtained from castings only when they are properly designed. Design rules for castings have been prepared in detail to aid design engineers in preparing efficient designs. Engineers are referred to the following organizations for the latest publications available:

Steel Founders' Society of America, Crystal Lake, IL 60014.
American Foundry Society, Schaumburg, IL 60173.
Investment Casting Institute, Dallas, TX 75206.

4.4 NONFERROUS METALS AND ALLOYS; METALLIC SPECIALTIES

Contributors are shown at the head of each category.

References: Current edition of "Metals Handbook," American Society for Materials. Current listing of applicable ASTM, AISI, and SAE Standards. Publications of the various metal-producing companies. Publications of the professional and trade associations which contain educational material and physical property data for the materials promoted.

INTRODUCTION

by L. D. Kunsman and C. L. Carlson, Amended by Staff

Seven **nonferrous metals** are of primary commercial importance: copper, zinc, lead, tin, aluminum, nickel, and magnesium. Some 40 other elements are frequently alloyed with these to make the commercially important alloys. There are also about 15 minor metals that have important specific uses. The properties of these elements are given in Table 4.4.1.

Metallic Properties Metals are substances that characteristically are opaque crystalline solids of high reflectivity having good electrical and thermal conductivities, a positive chemical valence, and, usually, the important combination of considerable strength and the ability to flow before fracture. These characteristics are exhibited by the metallic elements (e.g., iron, copper, aluminum, etc.), or **pure metals,** and by combinations of elements (e.g., steel, brass, dural), or **alloys.** Metals are composed of many small **crystals,** which grow individually until they fill the intervening spaces by abutting neighboring crystals. Although the external shape of these crystals and their orientation with respect to each other are usually random, within each such **grain,** or crystal, the atoms are arranged on a regular three-dimensional lattice. Most metals are arranged according to one of the three common types of **lattice,** or **crystal structure:** face-centered cubic, body-centered cubic, or hexagonal close-packed. (See Table 4.4.1.)

The several properties of metals are influenced to varying degrees by the testing or service **environment** (temperature, surrounding medium) and the **internal structure** of the metal, which is a result of its chemical composition and previous history such as casting, hot rolling, cold extrusion, annealing, and heat treatment. These relationships, and the discussions below, are best understood in the framework of the several phenomena that may occur in metals processing and service and their general effect on metallic structure and properties.

Ores, Extractions, Refining All metals begin with the mining of ores and are successively brought through suitable physical and chemical processes to arrive at commercially useful degrees of purity. At the higher-purity end of this process sequence, scrap metal is frequently combined with that derived from ore. Availability of suitable ores and the extracting and refining processes used are specific for each metal and largely determine the price of the metal and the impurities that are usually present in commercial metals.

Melting Once a metal arrives at a useful degree of purity, it is brought to the desired combination of shape and properties by a series of physical processes, each of which influences the internal structure of the metal. The first of these is usually melting, during which several elements can be combined to produce an alloy of the desired composition. Depending upon the metal, its container, the surrounding atmosphere, the addition of alloy-forming materials, or exposure to vacuum, various chemical reactions may be utilized in the melting stage to achieve optimum results.

Casting The molten metal of desired composition is poured into some type of mold in which the heat of fusion is dissipated and the melt becomes a solid of suitable shape for the next stage of manufacture. Such castings either are used directly (e.g., sand casting, die casting, permanent-mold casting) or are transferred to subsequent operations (e.g., ingots to rolling, billets to forging or extrusion). The principal advantage of castings is their design flexibility and low cost. The typical casting has, to a greater or lesser degree, a large crystal size **(grain size),** extraneous **inclusions** from slag or mold, and **porosity** caused

Table 4.4.1 Physical Constants of the Principal Alloy-Forming Elements

	Atomic no.	Atomic weight	Density, lb/in³	Melting point, °F	Boiling point, °F	Specific heat*	Latent heat of fusion, Btu/lb	Linear coef of thermal exp. per °F × 10^{-6}	Thermal conductivity (near 68°F), Btu/(ft² · h · °F/in)	Electrical resistivity, $\mu\Omega \cdot$cm	Modulus of elasticity (tension), lb/in² × 10^6	Crystal structure†	Transition temp., °F	Symbol
Aluminum	13	26.97	0.09751	1,220.4	4,520	0.215	170	13.3	1,540	2.655	10	FCC		Al
Antimony	51	121.76	0.239	1,166.9	2,620	0.049	68.9	4.7-7.0	131	39.0	11.3	Rhom		Sb
Arsenic	33	74.91	0.207	1,497	1,130	0.082	159	2.6		35	11	Rhom		As
Barium	56	137.36	0.13	1,300	2,980	0.068		(10)		50	1.8	BCC		Ba
Beryllium	4	9.02	0.0658	2,340	5,020	0.52	470	6.9	1,100	5.9	37	HCP		Be
Bismuth	83	209.00	0.354	520.3	2,590	0.029	22.5	7.4	58	106.8	4.6	Rhom		Bi
Boron	5	10.82	0.083	3,812	4,620	0.309		4.6		1.8×10^{22}		O		B
Cadmium	48	112.41	0.313	609.6	1,409	0.055	23.8	16.6	639	6.83	8	HCP		Cd
Calcium	20	40.08	0.056	1,560	2,625	0.149	100	12	871	3.43	3	FCC/BCC	867	Ca
Carbon	6	12.010	0.0802	6,700	6,730	0.165		0.3-2.4	165	1,375	0.7	Hex/D		C
Cerium	58	140.13	0.25	1,460	4,380	0.042	27.2			78		HCP/FCC	572/1328	Ce
Chromium	24	52.01	0.260	3,350	4,500	0.11	146	3.4	464	13	36	BCC/FCC	3344	Cr
Cobalt	27	58.94	0.32	2,723	6,420	0.099	112	6.8	479	6.24	30	HCP/FCC/HCP	783/2048	Co
Columbium (niobium)	41	92.91	0.310	4,380	5,970	0.065		4.0		13.1	15	BCC		Cb
Copper	29	63.54	0.324	1,981.4	4,700	0.092	91.1	9.2	2,730	1.673	16	FCC		Cu
Gadolinium	64	156.9	0.287									HCP		Gd
Gallium	31	69.72	0.216	85.5	3,600	0.0977	34.5	10.1	232	56.8	1	O		Ga
Germanium	32	72.60	0.192	1,756	4,890	0.086	205.7	(3.3)		60×10^{6}	11.4	D		Ge
Gold	79	197.2	0.698	1,945.4	5,380	0.031	29.0	7.9	2,060	2.19	12	FCC		Au
Hafnium	72	178.6	0.473	3,865	9,700	0.0351		(3.3)		32.4	20	HCP/BCC	3540	Hf
Hydrogen	1	1.0030	3.026×10^{-6}	−434.6	−422.9	3.45	27.0		1.18			Hex		H
Indium	49	114.76	0.264	313.5	3,630	0.057	12.2	18	175	8.37	1.57	FCT		In
Iridium	77	193.1	0.813	4,449	9,600	0.031	47	3.8	406	5.3	75	FCC		Ir
Iron	26	55.85	0.284	2,802	4,960	0.108	117	6.50	523	9.71	28.5	BCC/FCC/BCC	1663/2554	Fe
Lanthanum	57	138.92	0.223	1,535	8,000	0.0448				59	5	HCP/FCC/?	662/1427	La
Lead	82	207.21	0.4097	621.3	3,160	0.031	11.3	16.3	241	20.65	2.6	FCC		Pb
Lithium	3	6.940	0.019	367	2,500	0.79	286	31	494	11.7	1.7	BCC		Li
Magnesium	12	24.32	0.0628	1,202	2,030	0.25	160	14	1,100	4.46	6.5	HCP		Mg
Manganese	25	54.93	0.268	2,273	3,900	0.115	115	12		185	23	CCX/CCX/FCT	1340/2010/2080	Mn

Mercury	80	200.61	0.4896	−37.97	675	0.033	4.9	33.8	58	94.1		Rhom		Hg
Molybdenum	42	95.95	0.369	4,750	8,670	0.061	126	3.0	1,020	5.17	50	BCC		Mo
Nickel	28	58.69	0.322	2,651	4,950	0.105	133	7.4	639	6.84	30	FCC		Ni
Niobium (see Columbium)														Nb
Nitrogen	7	14.008	0.042×10^{-3}	−346	−320.4	0.247	11.2		0.147			Hex		N
Osmium	76	190.2	0.813	4,900	9,900	0.031		2.6		9.5	80	HCP		Os
Oxygen	8	16.000	0.048×10^{-3}	−361.8	−297.4	0.218	5.9		0.171			C		O
Palladium	46	106.7	0.434	2,829	7,200	0.058	69.5	6.6	494	10.8	17	FCC		Pd
Phosphorus	15	30.98	0.0658	111.4	536	0.177	9.0	70		10^{17}		C		P
Platinum	78	195.23	0.7750	3,224.3	7,970	0.032	49	4.9	494	9.83	21	FCC		Pt
Plutonium	94	239	0.686	1,229				50-65		150		MC	6 forms	Pu
Potassium	19	39.096	0.031	145	1,420	0.177	26.1	46	697	6.15	0.5	BCC		K
Radium	88	226.05	0.18	1,300								?		Ra
Rhenium	75	186.31	0.765	5,733	10,700	0.0326	76			21	75	HCP		Re
Rhodium	45	102.91	0.4495	3,571	8,100	0.059		4.6	610	4.5	54	FCC		Rh
Selenium	34	78.96	0.174	428	1,260	0.084	29.6	21	3	12	8.4	MC/Hex	248	Se
Silicon	14	28.06	0.084	2,605	4,200	0.162	607	1.6-4.1	581	10^5	16	D		Si
Silver	47	107.88	0.379	1,760.9	4,010	0.056	45.0	10.9	2,900	1.59	11	FCC		Ag
Sodium	11	22.997	0.035	207.9	1,638	0.295	49.5	39	929	4.2	1.3	BCC		Na
Sulfur	16	32.066	0.0748	246.2	832.3	0.175	16.7	36	1.83	2×10^{-23}		Rhom/FCC	204	S
Tantalum	73	180.88	0.600	5,420	9,570	0.036		3.6	377	12.4	27	BCC		Ta
Tellurium	52	127.61	0.225	840	2,530	0.047	13.1	9.3	41	2×10^{-5}	6	Hex		Te
Thallium	81	204.39	0.428	577	2,655	0.031	9.1	16.6	2,700	18	1.2	HCP/BCC	446	Tl
Thorium	90	232.12	0.422	3,348	8,100	0.0355	35.6	6.2		18.6	11.4	FCC		Th
Tin	50	118.70	0.264	449.4	4,120	0.054	26.1	13	464	11.5	6	D/BCT	55	Sn
Titanium	22	47.90	0.164	3,074	6,395	0.139	187	4.7	119	47.8	16.8	HCP/BCC	1650	Ti
Tungsten	74	183.92	0.697	6,150	10,700	0.032	79	2.4	900	5.5	50	BCC		W
Uranium	92	238.07	0.687	2,065	7,100	0.028	19.8	11.4	186	29	29.7	O/Tet/BCC	1229/1427	U
Vanadium	23	50.95	0.217	3,452	5,430	0.120		4.3	215	26	18.4	BCC	2822	V
Zinc	30	65.38	0.258	787	1,663	0.092	43.3	9.4-22	784	5.92	12	HCP		Zn
Zirconium	40	91.22	0.23	3,326	9,030	0.066		3.1	116	41.0	11	HCP/BCC	1585	Zr

*Cal/(g · °C) at room temperature equals Btu/(IB · °F) at room temperature.

†FCC = face-centered cubic; BCC = body centered cubic; C = cubic; HCP = hexagonal closet packing; Rhomb = rhombohedral; Hex = hexagonal; FCT = face-centered tetragonal; O = orthorhombic; FCO = face-centered orthorhombic; CCX = cubic complex; D = diamond cubic; BCT = body-centered tetragonal; MC = monoclinic.

Source: ASM "Metals Handbook," revised and supplemented where necessary from Hampel's "Rare Metals Handbook" and elsewhere.

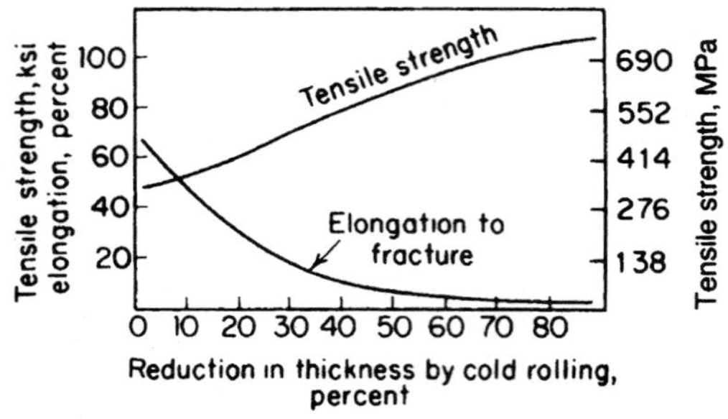

Figure 4.4.1 Effect of cold rolling on annealed brass (Cu 72, Zn 28).

by gas evolution and/or shrinking during solidification. The foundryman's art lies in minimizing the possible defects in castings while maximizing the economies of the process.

Metalworking The major portion of all metals is subjected to additional shape and size changes in the solid state. These metalworking operations substantially alter the internal structure and eliminate many of the defects typical of castings. The usual sequence involves first **hot working** and then, quite often, **cold working.** Some metals are only hot-worked, some only cold-worked; most are both hot and cold-worked. These terms have a special meaning in metallurgical usage.

A cast metal consists of an aggregate of variously oriented grains, each one a single crystal. Upon deformation, the grains flow by a process involving the slip of blocks of atoms over each other, along definite crystallographic planes. The metal is hardened, strengthened, and rendered less ductile, and further deformation becomes more difficult. This is an important method of increasing the strength of nonferrous metals. The effect of progressive **cold rolling** on brass (Fig. 4.4.1) is typical. Terms such as "soft," "quarter hard," "half hard," and "full hard" are frequently used to indicate the degree of hardening produced by such working. For most metals, the hardness resulting from cold working is stable at room temperature, but with lead, zinc, or tin, it will decrease with time.

Upon **annealing, or heating the cold-worked metal,** the first effect is to relieve macrostresses in the object without loss of strength; indeed, the strength is often increased slightly. Above a certain temperature, softening commences and proceeds rapidly with increase in temperature (Fig. 4.4.2). The cold-worked distorted metal undergoes a change called **recrystallization.** New grains form and grow until they have consumed the old, distorted ones. The temperature at which this occurs in a given time, called the **recrystallization temperature,** is lower and the resulting grain size finer, the more severe the working of the original piece. If the original working is carried out above this temperature, it does not harden the piece, and the operation is termed **hot working.** If the annealing temperature is increased beyond the recrystallization temperature or if the piece is held at that temperature for a long time, the average grain size increases. This generally softens and decreases the strength of the piece still further.

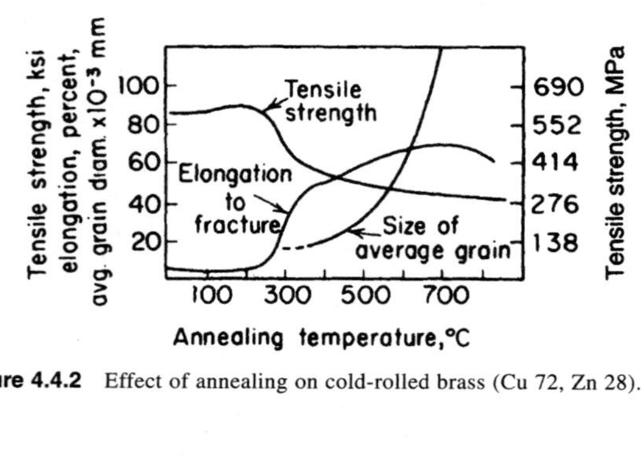

Figure 4.4.2 Effect of annealing on cold-rolled brass (Cu 72, Zn 28).

Hot working occurs when deformation and annealing proceed simultaneously, so that the resulting piece of new size and shape emerges in a soft condition roughly similar to the annealed condition.

Alloys The above remarks concerning structure and property control through casting, hot working, cold working, and annealing generally apply both to pure metals and to alloys. The addition of alloying elements makes possible other means for controlling properties. In some cases, the addition to a metal A of a second element B simply results in the appearance of some new crystals of B as a **mixture** with crystals of A; the resulting properties tend to be an average of A and B. In other cases, an entirely new substance will form—the **intermetallic compound AB,** having its own set of distinctive properties (usually hard and brittle). In still other cases, element B will simply dissolve in element A to form the **solid solution A(B)**. Such solid solutions have the characteristics of the solvent A modified by the presence of the **solute B,** usually causing increased hardness, strength, electrical resistance, and recrystallization temperature. The most interesting case involves the combination of solid solution A(B) and the precipitation of a second constituent, either B or AB, brought about by the precipitation-hardening heat treatment, which is particularly important in the major nonferrous alloys.

Precipitation Hardening Many alloys, especially those of aluminum but also some alloys of copper, nickel, magnesium, and other metals, can be hardened and strengthened by heat treatment. The heat treatment is usually a two-step process which involves (1) a solution heat treatment followed by rapid quenching and (2) a precipitation or aging treatment to cause separation of a second phase from solid solution and thereby cause hardening. After a solution treatment, these alloys are comparatively soft and consist of homogeneous grains of solid solution generally indistinguishable microscopically from a pure metal. If very slowly cooled from the solution treatment temperature, the alloy will deposit crystals of a second constituent, the amount of which increases as the temperature decreases. Rapid cooling after a solution treatment will retain the supersaturated solution at room temperature, but if the alloy is subsequently reheated to a suitable temperature, fine particles of a new phase will form and in time will grow to a microscopically resolvable size. At some stage in this precipitation process the hardness, the tensile strength, and particularly the yield strength of the alloy, will be increased considerably. If the reheating treatment is carried out too long, the alloy will overage and soften. The temperature and time for both solution and precipitation heat treatments must be closely controlled to obtain the best results. To some extent, precipitation hardening may be superimposed upon hardening resulting from cold working. Precipitation-hardened alloys have an unusually high ratio of proportional limit to tensile strength, but the endurance limit in fatigue is not increased to nearly the same extent.

Effect of Environment The properties of metals under "normal" conditions of 70°F and 50 percent relative humidity in clean air can be markedly changed under other conditions, with the various metals differing greatly in their degree of response to such changing conditions. The topic of **corrosion** is both important and highly specific (see Sec. 4.5). **Oxidation** is a chemical reaction specific to each metal and, wherever important, is so treated. The **effect of temperature,** however, can be profitably considered in general terms. The primary consequence of increase in temperature is increased atom movement, or **diffusion.** In the discussion above, the effects of recrystallization, hot working, solution treatment, and precipitation were all made possible by increased diffusion at elevated temperatures. The temperature at which such atom movements become appreciable is roughly proportional to the melting point of the metal. If their melting points are expressed on an absolute temperature scale, then various metals can be expected to exhibit comparable effects at about the same fraction of their melting points. Thus the recrystallization temperatures of lead, zinc, aluminum, copper, nickel, molybdenum, and tungsten are successively higher. The consequent rule of thumb is that alloys do not have useful structural strength at temperatures above about 0.5 of their absolute melting temperatures.

ALUMINUM AND ITS ALLOYS

by J. Randolph Kissell, Larry F. Wieserman, and Richard L. Brazill

REFERENCES: Kissell and Ferry, "Aluminum Structures," Wiley, 2002. Aluminum Association publications: "Aluminum Standards and Data 2017." "Aluminum Design Manual 2015." "Aluminum with Food and Chemicals." "Designation System for Aluminum Finishes," 2003. "Care of Aluminum," 5th ed. "Guidelines for In-Plant Handling of Aluminum Plate, Flat Sheet, and Coil." "Guidelines for Minimizing Water Staining of Aluminum," 2009. "Aluminum Technology, Applications, and Environment—A Profile of a Modern Metal," 6th ed. "Standards for Aluminum Sand and Permanent Mold Castings," 2008. "Welding Aluminum: Theory and Practice." ASM International publications: "Aluminum and Aluminum Alloys." "Corrosion of Aluminum and Aluminum Alloys." "The Surface Treatment and Finishing of Aluminum and Its Alloys." "Properties of Aluminum Alloys: Tensile, Creep, and Fatigue Data at High and Low Temperatures." "Fracture Resistance of Aluminum Alloys: Notch Toughness, Tear Resistance, and Fracture Toughness." "Aluminum Handbook," vol. 1. "Fundamentals and Materials." "Aluminum Handbook," vol. 2. "Forming Casting, Surface Treatment, Recycling, and Ecology." Minford, "Handbook of Aluminum Bonding Technology and Data," Marcel Dekker, 1993. "Metal Foams: A Design Guide," Butterworth Heinmann, 2000.

Introduction

Aluminum owes most of its applications to its low fabrication costs, low density [about 0.1 lb/in^3 (2,700 kg/m^3)] and the high strength to weight ratio of its alloys. Other uses rely on its corrosion resistance, electrical and thermal conductivity, reflectivity (infrared, visible, and ultraviolet), low heat emissivity, and toughness at low temperatures. Designs utilizing aluminum should take into account its relatively low modulus of elasticity [10,000 ksi (70,000 MPa)] and high coefficient of thermal expansion [13×10^{-6}/°F (23×10^{-6}/°C)]. Commercially pure aluminum contains at least 99 percent aluminum and is a ductile metal used where strength is not required. Aluminum alloys, on the other hand, possess better casting and machining characteristics and better mechanical properties than the pure metal and, therefore, are used for applications such as transportation (aircraft, trains, boats, cars), building products (roofing and siding, window and door frames, curtain walls, canopy and space frames), food and pharmaceutical packaging (cans and foil), sports equipment, heat exchangers, cryogenic containers, and optical components. A great variety of aluminum alloys and products are available from producers. A useful guide to those that are commonly available is the Aluminum Association publication "Aluminum Standards and Data."

Aluminum alloys are divided into two groups: wrought alloys and cast alloys, each with its own designation system specified in ANSI H35.1, "Alloy and Temper Designation Systems for Aluminum," maintained by the Aluminum Association.

Aluminum Alloy Designation Systems

WROUGHT ALLOYS

Wrought aluminum alloys are products produced by mechanical working such as rolling, extrusion, drawing, and forging to improve mechanical properties and are designated by a four-digit number. The first digit indicates the alloy group according to the main alloying element as follows:

	Designation no.
Aluminum, 99.00% and greater	1xxx
Aluminum alloys grouped by major alloying elements:	
Copper	2xxx
Manganese	3xxx
Silicon	4xxx
Magnesium	5xxx
Magnesium and silicon	6xxx
Zinc	7xxx
Other element	8xxx
Unused series	9xxx

The main alloying element is a good indicator of the properties of the alloy. For example, aluminum-copper alloys have greater strength but less corrosion resistance than pure aluminum. The second digit indicates modifications of the original alloy or impurity limits. For example, 2,319 is identical to 2,219 except for additional titanium to improve welding characteristics. The last two digits of the 1xxx series alloys indicate the aluminum purity. "Aluminum Standards and Data" provides the chemical compositions of commonly used alloys.

Some common wrought alloys (by major alloying elements) and applications are as follows:

1xxx Series: Aluminum of 99 percent or greater purity is used in the electrical and chemical industries for its electrical high conductivity, corrosion resistance, and workability. It is available extruded or rolled and is strengthened by cold working but not by heat treatment (see tempers below).

2xxx Series: These alloys are heat-treatable and some attain strengths comparable to those of steel alloys. They are less corrosion resistant than other aluminum alloys and thus are often clad with pure aluminum for corrosion resistance (see Alclad aluminum below). Alloys 2,014 and 2,024 are popular, 2,024 being perhaps the most widely used aircraft alloy. Many of the 2xxx alloys, including 2,014, have poor arc welding characteristics.

3xxx Series: Alloys in this series are strengthened only by cold working. Alloys 3,003, 3,004, and 3,105 are popular general-purpose alloys with moderate strengths and good workability, and are often used for sheet metal applications such as gutters, downspouts, roofing, and siding.

4xxx Series: Silicon added to alloys of this group lowers the melting point, making these alloys suitable for use as weld filler wire (such as 4,043) and brazing alloys.

5xxx Series: These alloys attain moderate to high strength by cold working and the addition of magnesium. They usually have the highest welded strengths of aluminum alloys and good corrosion resistance. Alloys 5,083, 5,086, 5,154, 5,454, and 5,456 are used in welded structures, including pressure vessels and ships. Alloy 5,052 is a popular, low-cost sheet metal alloy. For marine and other corrosive environments, 5xxx alloys of the -H116 and -H321 tempers are used for their corrosion resistance in freshwater and saltwater.

6xxx Series: Although these alloys usually are not as strong as those in the other heat-treatable series (2xxx and 7xxx), they offer a good combination of strength and corrosion resistance. Alloys 6,005A, 6,061, and 6,063 are used widely in structural applications, with 6,005A and 6,061 providing greater strength at a slightly greater cost. All three have good weldability, although welding significantly reduces their strength.

7xxx Series: Heat-treating alloys in this series produces some of the highest-strength aluminum alloys. They are frequently used in aircraft, and include 7,050, 7,075, and 7,475. 7,075-T6 sheet has a minimum tensile strength of 80 ksi (550MPa), the greatest of any commonly used aluminum alloy. Most alloys of this group contain copper and so are not readily welded, with 7,005 being an exception.

Aluminum-lithium alloys have higher strength and modulus of elasticity than those of any of the other alloys, but cost about 4 times as much. Aluminum-lithium alloys can be found in many aerospace structures such as wings, fuselage skins, and spacecraft to reduce the weight of the vehicle.

CAST ALUMINUM ALLOYS

Cast aluminum alloys are also designated by a four-digit number, but with a decimal point between the last two digits. The first digit indicates the alloy group according to the main alloying element as follows:

	Designation no.
Aluminum, 99.00% and greater	1xx.x
Aluminum alloys grouped by major alloying elements:	
Copper	2xx.x
Silicon, with copper and/or magnesium	3xx.x
Silicon	4xx.x
Magnesium	5xx.x
Zinc	7xx.x
Tin	8xx.x
Other element	9xx.x
Unused series	6xx.x

For the 1xx.x series the second two digits indicate the aluminum purity. The last digit indicates whether the product form is a casting (designated by 0) or ingot (designated by 1 or 2) and is often dropped in common usage for castings. A modification of the original alloy or impurity limits is indicated by a serial letter prefix before the numerical designation. An example is A356, a modification of 356 with slightly tighter controls on the minor alloying elements iron, copper, manganese, and zinc. The prefix letters are assigned in alphabetical order but omitting I, O, Q, and X.

Temper Designation System

The **temper designation system** applies to all aluminum and aluminum alloys, wrought or cast, except ingot. The temper designation follows the alloy designation, separated by a hyphen. The basic tempers are designated by letters as follows:

- F as fabricated (no control over thermal conditions or strain hardening)
- O annealed (the lowest strength temper of an alloy)
- H strain-hardened (applies to wrought products only)
- W solution heat-treated
- T thermally treated to produce stable tempers other than F, O, or H.

The H and T tempers are followed by one or more numbers, e.g., T6 or H14, indicating specific sequences of treatments. The properties of alloys of H and T tempers are discussed further below.

The alloys listed in Table 4.4.2 are divided into two classes: those which may be strengthened by only cold working (non-heat-treatable, designated by the H temper) and those which may be strengthened by thermal treatment (heat-treatable, designated by the T temper).

Heat treatment (see Table 4.4.3) begins with a solution heat treatment to put the soluble alloying elements into solid solution. For example, alloy 6,061 is heated to 990°F (530°C). This is followed by rapidly dropping the temperature, usually by quenching in water. At this point, the alloy is very workable; but if it is held at room temperature or above, strength gradually increases and ductility decreases

Table 4.4.2 Composition and Typical Room-Temperature Properties of Wrought Aluminum Alloys

Aluminum Assoc. alloy designation*	Nominal composition, % (balance aluminum)					Temper†	Density, lb/in^3	Electrical conductivity, % IACS	Brinell hardness, 500-kg load, 10-mm ball	Mechanical properties		
	Cu	Si	Mn	Mg	Other elements					Tensile yield strength,‡ ksi	Tensile strength, ksi	Elong, %, in 2 in§
Work-hardenable alloys												
1,100	0.12					O	0.098	59	23	5	13	35
						H14			32	17	18	9
						H18		57	44	22	24	5
3,003	0.12		1.2			O	0.099	50	28	6	16	30
						H14		41	40	21	22	8
						H18		40	55	27	29	4
5,052				2.5	0.25 Cr	O	0.097	35	47	13	28	25
						H34			68	31	38	10
						H38		35	77	37	42	7
5,056			0.12	5.0	0.12 Cr	O	0.095	29	65	22	42	35
						H18		27	105	59	63	10
Heat-treatable alloys												
2,011	5.5				0.4 Pb	T3	0.102	39	95	43	55	15
					0.4 Bi	T8		45	100	45	59	12
2,014	4.4	0.8	0.8	0.5		O	0.101	50	45	14	27	18
						T4, 451		34	105	42	62	20
						T6, 651		40	135	60	70	13
Alclad 2,014						O	0.101	40		10	25	21
						T4, 451				37	61	22
						T6, 651				60	68	10
2,017	4.0	0.5	0.7	0.6		O	0.101	50	45	10	26	22
						T4, 451		34	105	40	62	22
2,024	4.4		0.6	1.5		O	0.100	50	47	11	27	20
						T3		30	120	50	70	18
						T4, 351		30	120	47	68	20
Alclad 2,024						O	0.100			11	26	20
						T3				45	65	18
						T4, 351				42	64	19
2,025	4.4	0.8	0.8			T6	0.101	40	110	37	58	19
2,219	6.3		0.3		0.18 Zr	O		44		11	25	18
					0.10 V	T351				36	52	17
					0.06 Ti	T851				51	66	10
6,053		0.7	1.2		0.25 Cr	O	0.097	45	26	8	16	35
						T6		42	80	32	37	13
6,061	0.28	0.6		1.0	0.20 Cr	O	0.098	47	30	8	18	25
						T6, 651		43	95	40	45	12
6,063		0.4		0.7		O	0.097	58	25	7	13	25
						T6		53	73	31	35	12
7,075	1.6			2.5	5.6 Zn	O	0.101		60	15	33	17
					0.23 Cr	T6, 651		33	150	73	83	11
Alclad 7,075						O	0.101			14	32	17
						T6, 651				67	76	11

* Aluminum Association Standardized System of Alloy Designation adopted October 1954.
† Standard temper designations: O = fully annealed. H14 and H34 correspond to half-hard and H18 to H38 to hard strain-hardened tempers. T3 = solution treated and then cold-worked; T4 = solution heat-treated; T6 = solution treated and then artificially aged; T-51 = stretcher stress-relieved; T8 = solution treated, cold-worked, and artificially aged.
‡ At 0.2% offset.
§ Percent in 2 in, $\frac{1}{16}$-in thick specimen.

Table 4.4.3 Conditions for Heat Treatment of Aluminum Alloys

		Solution heat treatment		Precipitation heat treatment		
Alloy	Product	Temp, °F	Temper designation	Temp, °F	Time of aging	Temper designation
2,014	Extrusions	925-945	T4	315-325	18 h	T6
2,017	Rolled or cold finished wire, rod and bar	925-945	T4	Room	4 days	
2,024	Coiled Sheet	910-930	T4	Room	4 days	
6,063	Extrusions	940-960	T4	340-360	8 h	T6
6,061	Sheet	980-1,000	T4	310-330	18 h	T6
7,075	Forgings	860-880	W	215-235 / 340-360 *	6-8 h / 8-10 h	T73

°C = (°F − 32)/1.8.
* This is a two-stage treatment.

as precipitation of constituents of the supersaturated solution occurs. This gradual change in properties is called natural aging. Alloys that have received solution heat treatment only are designated by a W or T temper and can be more readily cold-worked in this condition. Alloys that naturally age to a substantially stable condition are designated with a T1 through T4 temper, depending on the shaping process and degree of cold work imparted. Some materials that are to receive severe forming operations, such as cold-driven rivets, are held in this workable condition by storing below 32°F (0°C). By applying a second heat treatment, called precipitation heat treatment or artificial aging, at slightly elevated temperatures for a controlled time, further strengthening is achieved and properties are stabilized. For example, 6,061 sheet is artificially aged when held at 320°F (160°C) for 18 h. Material so treated is designated T6 temper. Artificial aging changes the shape of the stress-strain curve above yield, thus affecting the tangent modulus of elasticity. For this reason, inelastic buckling strengths are determined differently for artificially aged alloys than for alloys that have not received such treatment.

Strain hardening, also called cold working, is used to strengthen non-heat-treatable alloys, which cannot be strengthened by heat treatment. Non-heat-treatable alloys may be given a heat treatment to stabilize mechanical properties.

Annealing reduces an alloy's strength to its lowest level but increases ductility. All wrought alloys may be annealed by heating to a prescribed temperature. For example, 6,061 is annealed when held at 775°F (415°C) for approximately 2 to 3 h. Strengths are degraded whenever strain-hardened or artificially aged material is heated above about 250°F (120°C). The loss in strength is proportional to the time at elevated temperature. For example, 6,061-T6 heated to 400°F (200°C) for no longer than 30 min or 300°F (150°C) for 1,000 h exhibits less than a 5 percent decrease in strength. Table 4.4.4 shows the effect of elevated temperatures on the strength of some alloy tempers.

Minimum **mechanical properties** (usually yield and ultimate tensile strengths and elongation) are specified for most wrought alloy tempers by the Aluminum Association in "Aluminum Standards and Data." These minimum properties are also listed in ASTM specifications, but are grouped by ASTM by product (such as sheet and plate) rather than by alloy. The minimum properties are established at the levels at which 99 percent of the material is expected to exceed at a confidence level of 95 percent. The strengths of aluminum members subjected to axial force, bending, shear, and torsion under static and fatigue loads given in the Aluminum Association's "Aluminum Design Manual" are calculated using these minimum properties.

Wrought Product Forms

Flat-rolled products include foil, sheet, and plate. **Foil** is defined as rolled product less than or equal to 0.0079 in (0.20 mm) thick. **Sheet** is less than 0.25 in (through 6.3 mm) thick but not less than 0.006 in (0.15 mm) thick; in the overlap region between 0.006 and 0.0079 in (0.15 and 0.2 mm), sheet and foil products are supplied to separate specifications. Decimal thicknesses to the nearest 0.001 in are preferred when ordering. Sheet is available in flat panels (solid, perforated, or expanded metal) and coiled. Commercial roofing sheet and siding sheet are available in a number of profiles, including corrugated, ribbed, and V-beam. **Plate** is 0.25 in (6.3 mm) thick and greater, up to about 16 in (400 mm) thick. Minimum bend radius for sheet and plate depends on alloy, temper, and thickness, but is generally greater than the bend radius for mild carbon steel.

Extrusions are among the most useful aluminum products and are produced by pushing heated metal through a die opening. A great variety of custom and standard shapes are extruded, such as I beams, angles, channels, pipe, rectangular and round tube, tees, and zees. Extrusion cross sections may be as large as those that fit with a 31-in (790-mm) circle, but commonly available shapes are limited to about an 18-in (450-mm) circle. Alloys extruded include 1,100, 1,350, 2,014, 2,024, 2,219, 3,003, 5,083, 5,086, 5,154, 5,454, 6,005, 6,005A, 6,061, 6,063, 6,082, 6,101, 6,105, 6,351, 6,360, 6,463, 7,005, 7,050, 7,075. The most common are the 6xxx series, especially 6,005A, 6,061, and 6,063. Rod, bar, and wire are also produced rolled or cold-finished as well as extruded. Tubes may be drawn or extruded. **Aluminum Conductors** are manufactured from extruded rod that is drawn into wire. On a weight basis, aluminum has twice the electrical conductivity of copper; on a volume basis, the conductivity of aluminum is about 62 percent that of copper. For electrical applications, a special, high-purity grade of aluminum is often used (1,350, also referred to as EC). Aluminum used as conductors is sometimes alloyed to improve strength with minimum decrease in conductivity. Table 4.4.8 lists common conductor alloys and gives their strengths and conductivities in various forms and treatments. In power transmission lines, the necessary strength for long spans is obtained by stranding aluminum wires about a core wire of steel (ACSR) or a higher-strength aluminum alloy. Since many soils are corrosive and electrically conductive, aluminum electrical cables are given nonpermeable coatings when used underground.

Forgings are produced by open-die and closed-die methods. Minimum mechanical properties are not published for open-die forgings, so structural applications usually require closed-die forgings. Like castings, forgings may be produced in complex shapes, but have more uniform properties and better ductility than castings and are used for products such as wheels and aircraft components requiring high strength-to-weight ratios. For some forging alloys, minimum mechanical properties are slightly lower in directions other than parallel to the grain flow. Common forging alloys are 2,014, 2,219, 5,083, 6,061, 7,050, and 7,175.

Cast Product Forms

Castings are used for parts of complex shapes and are produced by sand casting, permanent mold casting, die casting, investment casting, and other processes. The compositions and minimum mechanical properties of some cast aluminum alloys are given in Tables 4.4.5*a* to 4.4.5*d*. Castings generally exhibit greater variation in strength and less ductility than wrought products. Tolerances, draft requirements, heat treatments, and quality standards are given in the Aluminum Association's "Standards for Aluminum Sand and Permanent Mold castings." Tolerance and the quality and frequency of inspection must be specified by the user if inspection is desired.

Table 4.4.4 Typical Tensile Properties of Aluminum Alloys at Elevated Temperatures

Alloy and temper	Property*	Temp, °F* 75	300	400	500	700
		Wrought alloys				
1,100-H18	T.S.	24	18	6	4	2.1
	Y.S.	22	14	3.5	2.6	1.6
	El.	15	20	65	75	85
2,017-T4	T.S.	62	40	16	9	4.3
	Y.S.	40	30	13	7.5	3.5
	El.	22	15	35	45	70
2,024-T4	T.S.	68	45	26	11	5
	Y.S.	47	36	19	9	4
	El.	19	17	27	55	100
3,003-H18	T.S.	29	23	14	7.5	2.8
	Y.S.	27	16	9	4	1.8
	El.	10	11	18	60	70
3,004-H38	T.S.	41	31	22	12	5
	Y.S.	36	27	15	7.5	3
	El.	6	15	30	50	90
5,052-H34	T.S.	38	30	24	12	5
	Y.S.	31	27	15	7.5	3.1
	El.	16	27	45	80	130
6,061-T6	T.S.	45	34	19	7.5	3
	Y.S.	40	31	15	5	1.8
	El.	17	20	28	60	95
6,063-T6	T.S.	35	21	9	4.5	2.3
	Y.S.	31	20	6.5	3.5	2
	El.	18	20	40	75	105
7,075-T6	T.S.	83	31	16	11	6
	Y.S.	73	27	13	9	4.6
	El.	11	30	55	65	70
		Temp, °F* 75	300	400	500	600
		Sand castings				
355-T51	T.S.	28	24	14	10	6
	Y.S.	23	19	10	5	3
	El.	2	3	8	16	36
356-T6	T.S.	33	23	12	8	4
	Y.S.	24	20	9	5	3
	El.	4	6	18	35	60
		Permanent-mold castings				
355-T51	T.S.	30	23	15	10	6
	Y.S.	24	20	10	5	3
	El.	2	4	9	33	98
356-T6	T.S.	38	21	12	8	4
	Y.S.	27	17	9	5	3
	El.	5	10	30	55	70

*T.S., tensile strength ksi. Y.S., yield strength, ksi, 0.2 percent offset. El , elongation in 2 in, percent.

ksi × 6.895 = MPa °C = (°F − 32)/1.8.

Tensile tests made on pieces maintained at elevated temperatures for 10,000 h under no load. Stress applied at 0.05 in/(in/min) strain rate.

Sand castings (ASTM B26) are for larger parts—up to 7,000 lb (3,200 kg)—in relatively small quantities at slow solidification rates. The sand mold is used only once. Tolerance and minimum thicknesses for sand and castings are greater than those for other castings processes. **Permanent mold castings** (ASTM B108) are produced by pouring molten metal into a reusable mold, sometimes with a vacuum to assist the flow. While more expensive than sand casting, permanent mold casting can be used to make parts with wall thicknesses as thin as about 0.09 in (2.3 mm).

In **die casting** (ASTM B85), molten aluminum is injected into a reusable steel mold, or die, at high velocity; fast solidification rates are achieved.

A low-cost, general-purpose casting alloy is 356-T6. Alloy A356-T6 is common in more structurally demanding applications. Alloy A444-T4 provides excellent ductility, exceeding that of many wrought alloys. These three alloys rank among the most weldable of the casting alloys. The cast alloys with copper (2xx.x) generally offer higher strengths at elevated temperatures. In addition to its end use, alloy selection should account for fluidity, resistance to hot cracking, and pressure tightness.

Additive manufacturing is a general term for a new class of manufacturing where parts are built up layer-by-layer, using lasers or electron beams to locally melt metal powder having a cast structure. Powders made from aluminum-silicon-magnesium alloys similar to casting alloys are in common use in the direct metal laser sintering process.

Aluminum Foam

Aluminum foam is a new product that was developed for a variety of applications. The properties of aluminum foam can vary over a wide range, depending on the preparation methods and process conditions. Property selection is dictated by the application: low density (relative density 0.02 to 0.4), acoustic suppression, energy absorption, or thermal insulation. The aluminum foam structure may have open or closed porosity. Applications already developed include laminated (sheet-foam-sheet) car body structures, integrally molded foam parts, aluminum foam filling for tubular energy absorbers, highway sound management systems, specific pore size filters, and heat exchangers. The potential for new applications of aluminum foam is great.

Fabrication

Machining (see Sec. 9.4) Many aluminum alloys are easily machined without special techniques at faster cutting speeds than for other metals. Pure aluminum and alloys of aluminum-magnesium (5xxx) are harder to machine than alloys of aluminum-copper (2xxx) and aluminum-zinc (7xxx). The most machinable wrought alloy is 2,011 in the T3, T4, T6, and T8 tempers, producing small broken chips and excellent finish; it is used where physical properties are subordinate to high machinability, such as for screw machine products. Castings of aluminum-copper (2xx.x), aluminum-magnesium (5xx.x), aluminum-zinc (7xx.x), and aluminum-tin (8xx.x) alloys are among the best choices for good machinability.

Stamping Aluminum sheet metal is often formed into net shape parts such as automotive body parts and beverage cans with mass production stamping operations, which include blanking, bending, drawing, ironing, and hemming.

Joining

Mechanical fasteners (including rivets, bolts, and screws) are the most common method of joining aluminum parts, because the application of heat during welding decreases the strength of all but annealed material. Aluminum **bolts** (2,024-T4, 6,061-T6, or 7,075-T73) are available in diameters from ¼ in (6.4 mm) to 1 in (25 mm), with properties meeting ASTM F468. Mechanical properties are given in Table 4.4.6. Aluminum nuts (2,024-T4, 6,061-T6, or 6,262-T9) are also available (ASTM F467). Galvanized steel and 300 series stainless steel bolts are also used to join aluminum. Hole diameter usually exceeds bolt diameter by 0.063 in (1.6 mm) or less. **Rivets** are used to resist shear loads only. They do not develop sufficient clamping force between the parts joined and so cannot reliably resist tensile loads. In general, rivets of composition similar to the base metal are used. Table 4.4.7 lists common aluminum rivet alloys and their minimum ultimate shear strengths. Hole diameter for cold-driven rivets should be no more than 4 percent larger than the nominal rivet diameter. **Screws** of 2,024-T4, 7,075-T73, or stainless steel are often used to fasten aluminum sheet. Holes for fasteners may be drilled, reamed, or punched. Punching is not recommended if the metal thickness is greater than the hole diameter.

Table 4.4.5a Composition Limits of Aluminum Casting Alloys (Percent)

Alloy	Product*	Silicon	Iron	Copper	Manganese	Magnesium	Chromium	Nickel	Zinc	Titanium	Tin	Others	
												Each	Total
201.0	S	0.10	0.15	4.0-5.2	0.20-0.50	0.15-0.55	—	—	—	0.15-0.35	—	0.05	0.10
204.0	S&P	0.20	0.35	4.2-5.0	0.10	0.15-0.35	—	0.05	0.10	0.15-0.30	—	0.05	0.15
208.0	S&P	2.5-3.5	1.2	3.5-4.5	0.50	0.10	—	0.35	1.0	0.25	—	—	0.50
222.0	S&P	2.0	1.5	9.2-10.7	0.50	0.15-0.35	—	0.50	0.8	0.25	—	—	0.35
242.0	S&P	0.7	1.0	3.5-4.5	0.35	1.2-1.8	0.25	1.7-2.3	0.35	0.25	—	0.05	0.15
295.0	S	0.7-1.5	1.0	4.0-5.0	0.35	0.03	—	—	0.35	0.25	—	0.05	0.15
296.0	P	2.0-3.0	1.2	4.0-5.0	0.35	0.05	—	0.35	0.50	0.25	—	—	0.35
308.0	P	5.0-6.0	1.0	4.0-5.0	0.50	0.10	—	—	1.0	0.25	—	—	0.50
319.0	S&P	5.5-6.5	1.0	3.0-4.0	0.50	0.10	—	0.35	1.0	0.25	—	—	0.50
328.0	S	7.5-8.5	1.0	1.0-2.0	0.20-0.6	0.20-0.6	0.35	0.25	1.5	0.25	—	—	0.50
332.0	P	8.5-10.5	1.2	2.0-4.0	0.50	0.50-1.5	—	0.50	1.0	0.25	—	—	0.50
333.0	P	8.0-10.0	1.0	3.0-4.0	0.50	0.05-0.50	—	0.50	1.0	0.25	—	—	0.50
336.0	P	11.0-13.0	1.2	0.50-1.5	0.35	0.7-1.3	—	2.0-3.0	0.35	0.25	—	0.05	—
354.0	S&P	8.6-9.4	0.20	1.6-2.0	0.10	0.40-0.6	—	—	0.10	0.20	—	0.05	0.15
355.0	S&P	4.5-5.5	0.6	1.0-1.5	0.50	0.40-0.6	0.25	—	0.35	0.25	—	0.05	0.15
C355.0	S&P	4.5-5.5	0.20	1.0-1.5	0.10	0.40-0.6	—	—	0.10	0.20	—	0.05	0.15
356.0	S&P	6.5-7.5	0.6	0.25	0.35	0.20-0.45	—	—	0.35	0.25	—	0.05	0.15
A356.0	S&P	6.5-7.5	0.20	0.20	0.10	0.25-0.45	—	—	0.10	0.20	—	0.05	0.15
357.0	S&P	6.5-7.5	0.15	0.05	0.03	0.45-0.6	—	—	0.05	0.20	—	0.05	0.15
A357.0	S&P	6.5-7.5	0.20	0.20	0.10	0.40-0.7	—	—	0.10	0.04-0.20	—	0.05	0.15
359.0	S&P	8.5-9.5	0.20	0.20	0.10	0.50-0.7	—	—	0.10	0.20	—	0.05	0.15
360.0	D	9.0-10.0	2.0	0.6	0.35	0.40-0.6	—	0.50	0.50	—	0.15	—	0.25
A360.0	D	9.0-10.0	1.3	0.6	0.35	0.40-0.6	—	0.50	0.50	—	0.15	—	0.25
380.0	D	7.5-9.5	2.0	3.0-4.0	0.50	0.10	—	0.50	3.0	—	0.35	—	0.50
A380.0	D	7.5-9,5	1.3	3.0-4.0	0.50	0.10	—	0.50	3.0	—	0.35	—	0.50
383.0	D	9.5-11.5	1.3	2.0-3.0	0.50	0.10	—	0.30	3.0	—	0.15	—	0.50
384.0	D	10.5-12.0	1.3	3.0-4.5	0.50	0.10	—	0.50	3.0	—	0.35	—	0.50
390.0	D	16.0-18.0	1.3	4.0-5.0	0.10	0.45-0.65	—	—	0.10	0.20	—	—	0.20
B390.0	D	16.0-18.0	1.3	4.0-5.0	0.50	0.45-0.65	—	0.10	1.5	0.10	—	—	0.20
392.0	D	18.0-20.0	1.5	0.40-0.80	0.20-0.60	0.80-1.20	—	0.50	0.50	0.20	0.30	—	0.50
413.0	D	11.0-13.0	2.0	1.0	0.35	0.10	—	0.50	0.50	—	0.15	—	0.25
A413.0	D	11.0-13.0	1.3	1.0	0.35	0.10	—	0.50	0.50	—	0.15	—	0.25
C433.0	D	4.5-6.0	2.0	0.6	0.35	0.10	—	0.50	0.50	—	0.15	—	0.25
443.0	S&P	4.5-6.0	0.8	0.6	0.50	0.05	0.25	—	0.50	0.25	—	—	0.35
B443.0	S&P	4.5-6.0	0.8	0.15	0.35	0.05	—	—	0.35	0.25	—	0.05	0.15
A444.0	P	6.5-7.5	0.20	0.10	0.10	0.05	—	—	0.10	0.20	—	0.05	0.15
512.0	S	1.4-2.2	0.6	0.35	0.8	3.5-4.5	0.25	—	0.35	0.25	—	0.05	0.15
513.0	P	0.30	0.40	0.10	0.30	3.5-4.5	—	—	1.4-2.2	0.20	—	0.05	0.15
514.0	S	0.35	0.50	0.15	0.35	3.5-4.5	—	—	0.15	0.25	—	0.05	0.15
518.0	D	0.35	1.8	0.25	0.35	7.5-8,5	—	0.15	0.15	—	0.15	—	0.25
520.0	S	0.25	0.30	0.25	0.15	9.5-10.6	—	—	0.15	0.25	—	0.05	0.15
535.0	S&P	0.15	0.15	0.05	0.10-0.25	6.2-7.5	—	—	—	0.10-0.25	—	0.05	0.15
705.0	S&P	0.20	0.8	0.20	0.40-0.6	1.4-1.8	0.20-0.40	—	2.7-3.3	0.25	—	0.05	0.15
707.0	S&P	0.20	0.8	0.20	0.40-0.6	1.8-2.4	0.20-0.40	—	4.0-4.5	0.25	—	0.05	0.15
710.0	S	0.15	0.50	0.35-0.65	0.05	0.6-0.8	—	—	6.0-7.0	0.25	—	0.05	0.15
711.0	P	0.30	0.7-1.4	0.35-0.65	0.05	0.25-0.45	—	—	6.0-7.0	0.20	—	0.05	0.15
712.0	S	0.30	0.50	0.25	0.10	0.50-0.65	0.40-0.6	—	5.0-6.5	0.15-0.25	—	0.05	0.20
713.0	S&P	0.25	1.1	0.40-1.0	0.6	0.20-0.50	0.35	0.15	7.0-8.0	0.25	—	0.10	0.25
771.0	S	0.15	0.15	0.10	0.10	0.8-1.0	0.06-0.20	—	6.5-7.5	0.10-0.20	—	0.05	0.15
850.0	S&P	0.7	0.7	0.7-1.3	0.10	0.10	—	0.7-1.3	—	0.20	—	—	0.30
851.0	S&P	2.0-3.0	0.7	0.7-1.3	0.10	0.10	—	0.30-0.7	—	0.20	—	—	0.30
852.0	S&P	0.40	0.7	1.7-2.3	0.10	0.6-0.9	—	0.9-1.5	—	0.20	—	—	0.30

* S = sand casting, P = permanent mold casting, D = die casting.

Table 4.4.5*b* Mechanical Properties of Aluminum Sand Castings

Alloy	Temper	Minimum tensile ultimate strength, ksi	Minimum tensile yield strength, ksi	Minimum percent elongation in 2 in or 4*D*	Typical Brinnell hardness (500-kgf load, 10-mm ball)
201.0	T7	60.0	50.0	3.0	—
204.0	T4	45.0	28.0	6.0	—
242.0	O	23.0	A	A	70
242.0	T61	32.0	20.0	A	105
A242.0	T75	29.0	A	1.0	75
295.0	T4	29.0	13.0	6.0	60
295.0	T6	32.0	20.0	3.0	75
295.0	T62	36.0	28.0	A	95
295.0	T7	29.0	16.0	3.0	70
319.0	F	23.0	13.0	1.5	70
319.0	T5	25.0	A	A	80
319.0	T6	31.0	20.0	1.5	80
355.0	T51	25.0	18.0	A	65
355.0	T6	32.0	20.0	2.0	80
355.0	T71	30.0	22.0	A	75
C355.0	T6	36.0	25.0	2.5	—
356.0	F	19.0	9.5	2.0	55
356.0	T51	23.0	16.0	A	60
356.0	T6	30.0	20.0	3.0	70
356.0	T7	31.0	A	A	75
356.0	T71	25.0	18.0	3.0	60
A356.0	T6	34.0	24.0	3.5	80
A356.0	T61	35.0	26.0	1.0	—
443.0	F	17.0	7.0	3.0	40
B443.0	F	17.0	6.0	3.0	40
512.0	F	17.0	10.0	—	50
514.0	F	22.0	9.0	6.0	50
520.0	T4	42.0	22.0	12.0	75
535.0	F	35.0	18.0	9.0	70
705.0	T5	30.0	17.0 B	5.0	65
707.0	T7	37.0	30.0 B	1.0	80
710.0	T5	32.0	20.0	2.0	75
712.0	T5	34.0	25.0 B	4.0	75
713.0	T5	32.0	22.0	3.0	75
771.0	T5	42.0	38.0	1.5	100
771.0	T51	32.0	27.0	3.0	85
771.0	T52	36.0	30.0	1.5	85
771.0	T6	42.0	35.0	5.0	90
771.0	T71	48.0	45.0	2.0	120
850.0	T5	16.0	A	5.0	45
851.0	T5	17.0	A	3.0	45
852.0	T5	24.0	18.0	A	60

A = not required; B = to be determined only when specified by the purchaser. These properties are for separately cast test specimens, not the castings themselves.

Welding (see Sec. 9.3) Most wrought aluminum alloys are weldable by qualified welders using either fusion or resistance methods. Fusion welding is typically by **gas tungsten arc welding** (GTAW), commonly called TIG (for tungsten inert gas) welding, or **gas metal arc welding** (GMAW), referred to as MIG (metal inert gas) welding. TIG welding is usually used to join parts from about 1/32 to 1/8 in (0.8 to 3.2 mm) thick; MIG welding is usually used to weld thicker parts. The American Welding Society publication D1.2 "Structural Welding Code—Aluminum" provides requirements for structural applications of aluminum welding methods. Filler alloys are chosen for strength, corrosion resistance, and compatibility with the parent alloys to be welded. Cast alloys may also be welded, but are more susceptible to cracking than wrought alloys. All aluminum tempers other than annealed tempers suffer a reduction in strength in the heat-affected weld zone (known as the HAZ); this reduction is less in some of the aluminum-magnesium (5xxx) alloys. **Postweld heat-treatment** may be used to counter the reduction in strength caused by welding, but care must be taken to avoid embrittling or warping the weldment. **Resistance welding** includes spot welding, often used for lap joints, and seam welding. Friction stir welding and laser beam welding are newer welding methods that usually reduce the extent of heat-affected zones. Methods of **nondestructive testing** of aluminum welds include dye-penetrant and eddy current methods to detect flaws at the surface. Ultrasonic and radiographic (X-ray) inspections are used for detecting flaws on the interior of the part.

Brazing is also used to join aluminum alloys with relatively high melting points, such as 1,100, 3,003, and 6,063, using aluminum-silicon alloys such as 4,047 and 4,145. Special brazing alloys dominate the heat exchanger market, especially for air conditioners and radiators for trucks and cars. Aluminum may also be **soldered**, but corrosion resistance of soldered joints is inferior to that of welded, brazed, or mechanically fastened joints.

Adhesive Bonding Adhesive bonding uses special, proprietary polymeric resins to bond aluminum parts to other materials. The adhesive is typically classified as thermosetting or thermoplastic resin. Some advantages of adhesive bonding include: the contour of the part is not changed (unlike riveting or welding); the metallurgical properties remain unchanged in the bonded zone (unlike welding); and bonding can be accomplished between aluminum and expanded aluminum (ex. Honeycomb). The bonding of dissimilar metal parts (with little concern

Table 4.4.5c Mechanical Properties of Aluminum Permanent Mold Castings

Alloy	Temper	Minimum tensile ultimate strength, ksi	Minimum tensile yield strength, ksi	Minimum percent elongation in 2 in or 4*D*	Typical Brinnell hardness (500-kgf load, 10-mm ball)
204.0	T4 separately cast specimens	48.0	29.0	8.0	—
242.0	T571	34.0	—	—	105
242.0	T61	40.0	—	—	110
319.0	F	27.0	14.0	2.5	95
332.0	T5	31.0	—	—	105
333.0	F	28.0	—	—	90
333.0	T5	30.0	—	—	100
333.0	T6	35.0	—	—	105
333.0	T7	31.0	—	—	90
336.0	T551	31.0	—	—	105
336.0	T65	40.0	—	—	125
354.0	T61 separately cast specimens	48.0	37.0	3.0	
354.0	T61 castings, designated area	47.0	36.0	3.0	
354.0	T61 castings, no location designated	43.0	33.0	2.0	
354.0	T62 separately cast specimens	52.0	42.0	2.0	
354.0	T62 castings, designated area	50.0	42.0	2.0	
354.0	T62 castings, no location designated	43.0	33.0	2.0	
355.0	T51	27.0	—	—	75
355.0	T62	42.0	—	—	105
355.0	T7	36.0	—	—	90
355.0	T71	34.0	27.0	—	80
C355.0	T61 separately cast specimens	40.0	30.0	3.0	85-90
C355.0	T61 castings, designated area	40.0	30.0	3.0	
C355.0	T61 castings, no location designated	37.0	30.0	1.0	85
356.0	F	21.0	10.0	3.0	
356.0	T6	33.0	22.0	3.0	85
356.0	T71	25.0	—	3.0	70
A356.0	T61 separately cast specimens	38.0	26.0	5.0	80-90
A356.0	T61 castings, designated area	33.0	26.0	5.0	
A356.0	T61 castings, no location designated	28.0	26.0	3.0	
357.0	T6	45.0	—	3.0	—
A357.0	T61 separately cast specimens	45.0	36.0	3.0	100
A357.0	T61 castings, designated area	46.0	36.0	3.0	—
A357.0	T61 castings, no location designated	41.0	31.0	3.0	—
359.0	T61 separately cast specimens	45.0	34.0	4.0	90
359.0	T61 castings, designated area	45.0	34.0	4.0	
359.0	T61 castings, no location designated	40.0	30.0	3.0	
359.0	T62 separately cast specimens	47.0	38.0	3.0	100
359.0	T62 castings, designated area	47.0	38.0	3.0	
359.0	T62 castings, no location designated	40.0	30.0	3.0	
443.0	F	21.0	7.0	2.0	45
B443.0	F	21.0	6.0	2.5	45
A444.0	T4 separately cast specimens	20.0	—	20	—
A444.0	T4 castings, designated area	20.0	—	20	—
513.0	F	22.0	12.0	2.5	60
535.0	F	35.0	18.0	8.0	—
705.0	T1 or T5	37.0	17.0	10.0	
707.0	T1	42.0	25.0	4.0	
707.0	T7	45.0	35.0	3.0	
711.0	T1	28.0	18.0	7.0	70
713.0	T1 or T5	32.0	22.0	4.0	
850.0	T5	18.0	—	8.0	
851.0	T5	17.0	—	3.0	
851.0	T6	18.0	—	8.0	
852.0	T5	27.0	—	3.0	

When not stated otherwise, properties are for separately cast test specimens, not the castings themselves.

about galvanic corrosion), solid polymers, fiberglass blankets, woven fabrics (e.g., graphite, aramid, or glass fibers), and even wood (e.g., particle board) is possible. Additional benefits include: component design simplification, improved resistance to fatigue and vibration degradation, more uniform stress distribution during loading, and electrical, thermal, and sound insulation within an assembled component. The disadvantages of adhesively bonded joints include: use temperatures must remain below 500°F (260°C) or even lower for some adhesives; aluminum surfaces must be cleaned and sometimes primed, anodized, or conversion coated before applying the adhesive; thermosetting adhesives require curing above room temperature; and bond quality is difficult to evaluate without destructive testing. There are a number of detailed process specifications and standard practices established for critical applications, especially for the military and the aerospace industry

Table 4.4.5*d* Mechanical Properties of Aluminum Die Castings

Alloy	Typical tensile ultimate strength,* ksi	Typical tensile yield strength,* ksi	Typical elongation in 2 in (%)
360.0	44	25	2.5
A360.0	46	24	3.5
380.0	46	23	2.5
A380.0	47	23	3.5
383.0	45	22	3.5
384.0	48	24	2.5
390.0	40.5	35.0	< 1
B390.0	46.0	36.0	< 1
392.0	42.0	39.0	< 1
413.0	43	21	2.5
A413.0	42	19	3.5
C443.0	33	14	9.0
518.0	45	28	5

* These are properties of separately cast test bars, not the castings themselves.

Corrosion Resistance

Although aluminum is chemically active, the presence of a rapidly forming and firmly adherent self-healing oxide surface coating inhibits corrosive action except under conditions that tend to remove this surface film. Concentrated nitric and acetic acids are handled in aluminum not only because of its resistance to attack but also because any resulting corrosion products are colorless. For the same reason, aluminum is employed in the preparation and storage of foods and beverages. Hydrochloric acid and most alkalies dissolve the protective surface film and result in fairly rapid attack. Moderately alkaline soaps and the like can be used with aluminum if a small amount of sodium silicate is added. Aluminum is very resistant to sulfur and most of its gaseous products. Except in special applications, aluminum is not recommended in solutions of pH of less than 4 or greater than 8. Aluminum reacts with many organic compounds and should be used with appropriate corrosion inhibitors in such exposure. Never use glass cleaners or detergents containing ammonia on bare aluminum components. If used, the bright aluminum surface will be covered with a non-uniform white haze that can be removed only with an abrasive aluminum polish and much effort.

Table 4.4.6 Mechanical Properties of Aluminum Bolts

Alloy temper	Minimum tensile strength, ksi	Minimum shear strength, ksi
2024-T4	62	37
6061-T6	42	25
7075-T73	68	41

ksi × 6.895 = MPa

Table 4.4.7 Minimum Shear Strength of Aluminum Rivets

Alloy temper before driving	Minimum shear strength, ksi
1,100-H14	10
2,017-T4	33
2,024-T4	37
2,117-T4	26
5,056-H32	25
6,053-T61	20
6,061-T6	25
7,050-T7	39

ksi × 6.895 = MPa

Galvanic corrosion may occur when aluminum is electrically connected by an electrolyte to another metal. Aluminum is more anodic than most metals and will be sacrificed while protecting the other metal, a principle called cathodic protection. Consequently, aluminum is usually isolated from other metals such as steel (but not stainless steel) where moisture is present. Large castings of special aluminum alloys are used to protect steel in offshore oil platforms against corrosion.

Another form of corrosion is **exfoliation**, a delamination or peeling of layers of metal in planes approximately parallel to the metal surface, caused by the formation of corrosion product. Alloys with more than 3 percent magnesium (such as 5,083, 5,086, 5,154, and 5,456) and held at temperatures above 150°F (65°C) for extended periods are susceptible to this form of attack.

Care should be taken to store aluminum in a manner to avoid trapping water between adjacent flat surfaces, which causes **water stains.** These stains, which vary in color from dark gray to white, do not compromise strength but are difficult to remove and may be cosmetically unacceptable.

Ordinary atmospheric corrosion is resisted by aluminum and most of its alloys, and they may be used outdoors without any protective coating. (An exception is 2014, which is usually painted when exposed to the elements.) The pure metal is most resistant to attack, and additions of alloying elements decrease corrosion resistance, particularly after heat treatment. Under severe conditions of exposure such as may prevail in marine environments or where the metal is continually in contact with wood or other absorbent material in the presence of moisture, a protective coat of paint on the aluminum is advisable. Contact with treated wood is especially corrosive to aluminum. Pay particular attention to place a continuous polymeric barrier between aluminum components and the treated lumber. The zinc, copper, or other treatments for treated wood can greatly accelerate corrosion and form unsightly corrosion holes in the aluminum.

Clad products are wrought aluminum alloy products with a metallurgically bonded coating for the purposes of finishing (reflector sheet), brazing (brazing sheet), or corrosion protection. The cladding becomes an integral part of the material. In brazing sheet, a low melting point 4xxx (Al-Si) alloy is clad to a 3xxx or 6xxx core. Clad products include sheet, plate, wire, tube, and pipe. When the cladding is anodic to the core and intended to provide corrosion protection, the product is referred to as **Alclad.** Because the cladding usually has lower strength than the base metal, clad products have slightly lower strength than unclad products. Cladding thickness varies from 1.5 to 10 percent and can be on one or both sides of the product. Alclad products are available as 3,003 tube and 2,014, 2,024, 2,219, 3,003, 3,004, 6,061, 7,050, 7,075, and 7,475 sheet and plate. Another application for clad sheet is for architectural curtain wall panels. Wire, rod, and bar may also be clad to improve finished appearance and corrosion resistance.

Finishes

Anodizing is used to improve the corrosion resistance, wear resistance, and appearance of aluminum. It is also used to improve the thermal emissivity of heat sinks. Anodizing is achieved by making the parts to be treated the anode in an electrolytic bath such as sulfuric acid. This process produces a hard, adherent coating of aluminum oxide, usually 0.4 mil (0.01 mm) thick or greater. "The Surface Treatment and Finishing of Aluminum Alloys," by P. G. Sheasby and R. Pinner, provides useful information on anodizing. Any welding should be performed before anodizing, and filler alloys should be chosen judiciously for good anodized color match with the base alloy. The film is colorless on pure aluminum and tends to be gray or colored on alloys containing silicon, copper, or other elements. To provide a consistent color appearance after anodizing, anodizing quality (AQ) grade may be specified in certain alloys. Where appearance is the overriding concern, 5,005 sheet and 6,063 extrusions are preferred for anodizing. If a colored finish is

Table 4.4.8 Aluminum Electrical Conductors

Product, alloy, temper (size)	Tensile ultimate strength, ksi*	Tensile yield strength, ksi*	Min. electrical conductivity, % IACS†
Drawing Stock (Rod)			
1,350-O (0.375- to 1.000-in diameter)	8.5-14.0		61.8
1,350-H12, H22 (0.375- to 1.000-in diameter)	12.0-17.0		61.5
1,350-H14, H24 (0.375- to 1.000-in diameter)	15.0-20.0		61.4
1,350-H16, H26 (0.375- to 1.000-in diameter)	17.0-22.0		61.3
5,005-O (0.375-in diameter)	14.0-20.0		54.3
5,005-H12, H22 (0.375-in diameter)	17.0-23.0		54.0
5,005-H14, H24 (0.375-in diameter)	20.0-26.0		53.9
5,005-H16, H26 (0.375-in diameter)	24.0-30.0		53.8
8,017-H12, H22 (0.375-in diameter)	16.0-22.0		58.0
8,030-H12 (0.375-in diameter)	16.0-20.5		60.0
8,176-H14 (0.375-in diameter)	16.0-20.0		59.0
Wire			
1,350-H19 (0.0801 to 0.0900 in diameter)	26.0-27.5		61.0
5,005-H19 (0.0801 to 0.0900 in diameter)	37.0-39.0		53.5
6,201-T81 (0.0612 to 0.1327 in diameter)	46.0-48.0		52.5
8,176-H24 (0.0500 to 0.2040 in diameter)	15.0-17.0		61.0
Extrusions			
1,350-H111 (all)	8.5 min	3.5 min	61.0
6,101-H111 (0.250 to 2.000 in)	12.0 min	8.0 min	59.0
6,101-T6 (0.125 to 0.500 in)	29.0 min	25.0 min	55.0
6,101-T61 (0.125 to 0.749 in)	20.0 min	15.0 min	57.0
6,101-T61 (0.750 to 1.499 in)	18.0 min	11.0 min	57.0
6,101-T61 (1.500 to 2.000 in)	15.0 min	8.0 min	57.0
Rolled Bar			
1,350-H12 (0.125 to 1.000 in)	12.0 min	8.0 min	61.0
Sawed Plate			
1,350-H112 (0.125 to 0.499 in)	11.0 min	6.0 min	61.0
1,350-H112 (0.500 to 1.000 in)	10.0 min	4.0 min	61.0
1,350-H112 (1.001 to 1.500 in)	9.0 min	3.5 min	61.0
Sheet			
1,350-O (0.006 to 0.125 in)	8.0-14.0		61.8

*ksi × 6.895 = MPa.
† % IACS × 0.581 = MS/m.

desired, the electrolytically oxidized article may be treated with a dye solution. Care must be taken in selecting the dye when the part will be exposed to sunlight and weather, for not all have proven to be colorfast. Two-step electrolytic coloring, produced by first clear anodizing and then electrolytically depositing another metal or metal compound, can produce shades of bronze, burgundy, and blue. Anodized aluminum is used for lighting fixtures, automotive trim, cosmetic cases, and aluminum wheels.

Painting It is important to properly prepare the surface prior to the application of paint or lacquers. A thin anodic film makes an excellent paint base; alternatively, the aluminum surface may be chemically treated with a dilute phosphoric acid solution. Abrasion blasting may be used on parts thick enough to resist perforation or warping. Zinc chromate is frequently used as a primer, especially for corrosive environments, or another conversion coating or special primer may be used. Most paints are baked on; this can affect the strength of the metal if the bake temperature is high enough or applied long enough to cause annealing. Sometimes the paint baking process is used as the artificial aging heat treatment. Aluminum sheet is available with factory-baked paint finish; minimum mechanical properties must be obtained from the supplier. High-, medium-, and low-gloss paint finishes are available and are determined in accordance with ASTM D523.

Porcelain enameled finishes are ceramic coatings fired to coat the surface of aluminum. They are particularly useful for high-temperature applications such as appliances or cookware or for increased corrosion resistance to acids and alkalies.

The Aluminum Association's publication "Designation System for Aluminum Finishes" describes mechanical, chemical, and anodic finishes; "Care of Aluminum" describes aluminum finishes, cleaning, handling, storage, cleaners, and coatings.

BEARING METALS
by Frank E. Goodwin

REFERENCES: Current edition of ASM "Metals Handbook." Publications of the various metal producers. Trade association literature containing material properties. Applicable current ASTM and SAE Standards.

Babbitt metal is a general term used for soft tin and lead-base alloys which are cast as bearing surfaces on steel, bronze, or cast-iron shells. Babbitts have excellent **embeddability** (ability to embed foreign particles in itself) and **conformability** (ability to deform plastically to compensate for irregularities in bearing assembly) characteristics. These alloys may be run satisfactorily against a soft-steel shaft. The limitations of Babbitt alloys are the tendency to spread under high, steady loads and to fatigue under high, fluctuating loads. These limitations apply more particularly at higher temperatures; e.g., increase in temperature between 68 and 212°F (20 and 100°C) reduces the metal's strength by 50 percent. These limitations can be overcome by properly designing the thickness and rigidity of the backing material, properly choosing the Babbitt alloy for good mechanical characteristics, and ensuring a good bond between backing and bearing materials.

The important tin- and lead-base (Babbitt) bearing alloys are listed in Table 4.4.9. Alloys 1 and 15 have been used in internal-combustion engines, but the desire of automobile manufacturers to eliminate lead has resulted in the substitution of aluminum-based alloys. Alloy 1 performs satisfactorily at low temperatures, but alloy 15, an arsenic alloy, provides superior performance at elevated temperatures by virtue of its better high-temperature hardness, ability to support higher loads, and longer fatigue life. Alloys 2 and 3 contain more antimony, are harder, and are less likely to pound out. Alloys 7 and 8 are lead-base Babbitts

Table 4.4.9 Compositions and Properties of Some Babbitt Alloys

ASTM B23-00 grade	Composition, %					Compressive ultimate strength		Brinell hardness	
	Sn	Sb	Pb	Cu	As	68°F, lb/in²	20°C, MPa	68°F (20°C)	212°F (100°C)
1	91.0	4.5	0.35	4.5	—	12.9	88.6	17	8
2	89.0	7.5	0.35	3.5	—	14.9	103	24.5	12
3	84.0	8.0	0.35	8.0	—	17.6	121	27.0	14.5
7	10.0	15.0	74.5	0.5	0.45	15.7	108	22.5	10.5
8	5.0	15.0	79.5	0.5	0.45	15.6	108	20.0	9.5
15	1.0	16.0	82.5	0.5	1.10	—	—	21.0	13.0
Other alloys									
B	0.8	12.5	83.3	0.1	3.05	—	—		
ASTM B102, alloy Py 1815A	65	15	18	2	0.15	15	103	23	10

SOURCE: ASTM, reprinted with permission.

which will function satisfactorily under moderate conditions of load and speed. Alloy B has been used for diesel engine bearings. In general, increasing the lead content in tin-based Babbitt provides higher hardness, greater ease of casting, but lower strength values.

Silver lined bearings have an excellent record in heavy-duty applications in aircraft engines and diesels. For reciprocating engines, silver bearings normally consist of electrodeposited silver on a steel backing with an overlay of 0.001 to 0.005 in of lead. An indium flash on top of the lead overlay is used to increase corrosion resistance of the material.

Aluminum and **zinc alloys** are used for high-load, low-speed applications but have not replaced Babbits for equipment operating under a steady high-speed load. The Al 20 to 30, Sn 3 copper alloy is bonded to a steel bearing shell. The Al 6.25, Sn 1, Ni 1, copper alloy can be either used as-cast or bonded to a bearing shell. The Al 3 cadmium alloy with varying amounts of Si, Cu, and Ni can also be used in either of these two ways. The Zn 11, Al 1, Cu 0.02, magnesium (ZA-12) and Zn 27, Al 2, Cu 0.01 magnesium (ZA-27) alloys are used in cast form, especially in continuous-cast hollow-bar form.

Mention should be made of **cast iron** as a bearing material. The flake graphite in cast iron develops a glazed surface which is useful at surface speeds up to 130 ft/min and at loads up to 150 lb/in² approx. Because of the poor conformability of cast iron, good alignment and freedom from dirt are essential.

Copper-base bearing alloys have a wide range of bearing properties that fit them for many applications. Used alone or in combination with steel, Babbitt (white metal), and graphite, the bronzes and copper-leads meet the conditions of load and speed given in Table 4.4.10. **Copper-lead alloys** are cast onto steel backing strips in very thin layers (0.02 in) to provide bearing surfaces. Lead-based bearing alloys are still used in diesel engines; no lead-free alloy can withstand diesel firing loads.

Three families of copper alloys are used for bearing and wear-resistant alloys in cast form: **phosphor bronzes** (Cu-Sn), Cu-Sn-Pb alloys, and **manganese bronze, aluminum bronze,** and **silicon bronze.** Typical compositions and applications are listed in Table 4.4.10. Phosphor bronzes have residual phosphorus ranging from 0.1 to 1 percent. Hardness increases with phosphorus content. Cu-Sn-Pb alloys have high resistance to wear, high hardness, and moderate strength. The high lead compositions are well suited for applications where lubricant may be deficient. The Mn, Al, and Si bronzes have high tensile strength, high hardness, and good resistance to shock. They are suitable for a wide range of bearing applications.

Porous bearing materials are used in light- and medium-duty applications as small-sized bearings and bushings. Since they can operate for long periods without an additional supply of lubricant, such bearings are useful in inaccessible or inconvenient places where lubrication would be difficult. Porous bearings are made by pressing mixtures of copper and tin (bronze), and often graphite, Teflon, or iron and graphite, and sintering these in a reducing atmosphere without melting. Iron-based, oil-impregnated sintered bearings are often used. By controlling the conditions under which the bearings are made, porosity may be adjusted so that interconnecting voids of up to 35 percent of the total volume may be available for impregnation by lubricants. Applicable specifications for these bearings are given in Tables 4.4.11 and 4.4.12.

Miscellaneous A great variety of materials, e.g., rubber, wood, phenolic, carbon-graphite, cermets, ceramics, and plastics, are in use for special applications. Carbon-graphite is used where contamination by oil or grease lubricants is undesirable (e.g., textile machinery, pharmaceutical equipment, milk and food processing) and for elevated-temperature applications. Composite bushings of carbon graphite with metals, including copper and iron, are capable of service temperatures

Table 4.4.10 Compositions, Properties, and Applications of Some Copper-Base Bearing Metals

Specifications	Composition, %						Minimum tensile strength		Applications
	Cu	Sn	Pb	Zn	P	Other	ksi	MPa	
C86100, SAE J462	64	—	—	24	—	3 Fe, 5 Al, 4 Mn	119	820	Extra-heavy-duty bearings
C87610, ASTM B30	90	—	0.2	5	—	4 Si	66	455	High-strength hearings
C90700, ASTM B505	89	11	0.3	—	0.2	—	44	305	Worm gears and wheels; high-speed low-load bearings
C91100	84	16	—	—	—	—	35	240	Heavy load, low-speed bearings
C91300	81	19	—	—	—	—	35	240	Heavy load, low-speed bearings
C93200, SAE J462	83	7	7	3	—	—	35	240	General utility bearings, automobile fittings
C93700, ASTM B22	80	10	10	—	—	—	35	240	High-speed, high-pressure bearings, good corrosion resistant
C93800, ASTM B534	78	7	15	—	—	—	30	205	General utility bearings, backing for Babbit bearings
C94300, ASTM B584	70	5	25	—	—	—	27	185	Low-load, high-speed bearings

SOURCE: ASTM, reprinted with permission.

Table 4.4.11 Oil-Impregnated Iron-Base Sintered Bearings (ASTM Standard B439-83)

	Composition, %			
Element	Grade 1	Grade 2	Grade 3	Grade 4
Copper			7.0-11.0	18.0-22.0
Iron	96.25 min	95.9 min	Remainder*	Remainder*
Total other elements by difference, max	3.0	3.0	3.0	3.0
Combined carbon† (on basis of iron only)	0.25 max	0.25-0.60	—	—
Silicon, max	0.3	0.3	—	—
Aluminum, max	0.2	0.2	—	—

* Total of iron plus copper shall be 97% min.
† The combined carbon may be a metallographic estimate of the carbon in the iron.
SOURCE: ASTM, reprinted with permission.

Table 4.4.12 Oil-Impregnated Sintered Bronze Bearings (ASTM Standard B438-83)

	Composition, %			
	Grade 1		Grade 2	
Element	Class A	Class B	Class A	Class B
Copper	87.5-90.5	87.5-90.5	82.6-88.5	82.6-88.5
Tin	9.5-10.5	9.5-10.5	9.5-10.5	9.5-10.5
Graphite, max	0.1	1.75	0.1	1.75
Lead	*	*	2.0-4.0	2.0-4.0
Iron, max	1.0	1.0	1.0	1.0
Total other elements by difference, max	0.5	0.5	1.0	1.0

* Included in other elements.
SOURCE: ASTM, reprinted with permission.

up to 1,000°F (537°C). Notable among plastic materials are Teflon and nylon, the polycarbonate Lexan, and the acetal Delrin. Since Lexan and Delrin can be injection-molded easily, bearings can be formed quite economically from these materials.

CEMENTED CARBIDES
by Don Graham

REFERENCES: Schwartzkopf and Kieffer, "Refractory Hard Metals," Macmillan. German, "Powder Metallurgy Science," 2nd ed., Metal Powder Industries Federation. "Powder Metallurgy Design Manual," 2nd ed., Metal Powder Industries Federation, Princeton, NJ. Goetzal, "Treatise on Powder Metallurgy," Interscience. Schwartzkopf, "Powder Metallurgy," Macmillan. Davis (ed.), Tool Materials, "ASM Specialty Handbook," ASM, Metals Park, OH, 1995. Graham and Hale, Coated Cutting Tools, *Carbide and Tool J.*, May–June 1982, p. 34. Graham, Trends in Coatings to the Year 2000, "Cutting Tool Technology Conference Proceedings," SME, 1997.

Cemented carbides are a commercially and technically important class of composite material. Comprised primarily of hard tungsten carbide (WC) grains cemented together with a cobalt (Co) binder, they can contain major alloying additions of titanium carbide (TiC), titanium carbonitride (TiCN), tantalum carbide (TaC), niobium carbide (NbC), chromium carbide (CrC), etc., and minor additions of other elements. Because of the extremely high hardness, stiffness, and wear resistance of these materials, their primary use is in cutting tools for the material (usually metal, wood, composite, or rock) removal industry. Secondary applications are as varied as pen balls, food processing equipment, fuel pump parts, wear surfaces, fishing line guide rings, large high pressure components, and even wedding rings. Since their introduction in the mid-1920s a very large variety of grades of carbide have been developed and put into use in a diverse number of applications. They are ubiquitous throughout industry, particularly so in the high production metal working sector.

Most cemented carbide parts are produced by powder metallurgical processes. Powders of WC, Co, and sometimes TiC and TaC combined with NbC are blended by either ball or attritor milling. The mixed powder is them compacted under high pressure in appropriately shaped molds. Pressed parts are then sintered at temperatures between 1,250 and 1,500°C depending on the composition of the powder. During the sintering process, cobalt melts and wets the carbide particles. On cooling, cobalt "cements" the carbide grains together. Essentially complete densification takes place during sintering. A unique feature of cemented carbides is the fact that, although the part shrinks by as much as 40 percent by volume when the porosity is eliminated by the molten cobalt, the pressed part retains it geometric shape very accurately. This is an important feature since further shaping of this extremely hard material can be done accurately only by grinding with diamond wheels or electrodischarge machining. Lasers can be used to cut cemented carbide but not to tight tolerances. While binder metals such as iron or nickel are sometimes used, cobalt is preferred because of its ability to wet the tungsten carbide.

The metal removal industry consumes most of the cemented carbide that is produced. Tips having unique geometries, called inserts, are pressed and sintered. Many of these pressed and sintered parts are ready for use directly after sintering. Others are finish ground to either improve flatness on the seating surface or to produce very tight dimensional tolerances. Although the cutting edges on some inserts are left in the sharp condition, most are honed slightly to improve chippage resistance and reliability. Typically sharp edges are used for machining such materials as non-ferrous metals and alloys, high temperature alloys, alloys used for medical (hip and knee replacement parts for example) components, and non-metals. Honed inserts are used for all other applications, including the commercially important iron and steels industries.

The first cemented carbides developed consisted only of WC and Co. As cutting speeds were increased to take advantage of the high hardness of this material, temperatures at the tool/workpiece interface increased. In the presence of iron, steel, or titanium at high temperatures, carbide tools interact chemically with the work materials and result in "cratering," which, somewhat simplistically, is the dissolution of WC at the cutting edge. This results in a dimple on the surface of the insert that resembles a crater. Additions of secondary carbides like TiC and TaC minimize cratering—and thereby improve tool life—by virtue of their greater chemical stability. Grades containing these secondary carbides were introduced during the 1940s and were able to machine steel at speeds up to 600 surface feet per minute (sfpm) as opposed to 400 sfpm for uncoated carbide.

With the ever increasing desire for even greater metal removal rates, higher cutting speeds (and accompanying higher temperatures) were used. Chemical vapor deposition (CVD) overlay coatings were developed which were far more effective at preventing cratering than TiC and TaC additions to the carbide. These hard coatings have the additional benefit of increased abrasion resistance.

Titanium carbide coatings became available in 1969; titanium nitride (TiN) coatings followed shortly after that. Aluminum oxide (Al_2O_3) coatings for even higher speed applications were introduced in 1973. Today the large majority of coated carbide grades incorporate all three materials in distinct layers. Over the years dramatic improvements in deposition techniques and structural control have led to equally dramatic improvements in performance. Coatings typically result in tool life increases of at least 50 percent although 10-fold improvements are common. Extreme cases of 10,000 percent tool live improvements over uncoated carbide have been reported.

Each coating composition has its own specific advantages. For example, TiC and TiCN are very hard at lower temperatures and are ideal for very abrasive applications. Hence carbide grades designed for machining ductile irons, austempered ductile iron, compacted graphite iron, malleable iron, and alloy steels etc., incorporate relatively thick layers of these coatings.

TiN, along with other commercially available nitride coatings such as hafnium nitride (HfN) and zirconium nitride (ZrN), exhibits excellent metal build-up resistance. This feature is particularly important when machining softer, gummy alloys such as stainless steel, low carbon steel, and many high temperature alloys. Metal build-up occurs when fragments of the work material become welded to the cutting edge. This change in the geometry of the cutting edge results in a poor surface finish and, eventually, chippage of the cutting edge. Even grades designed for higher speeds often use the gold TiN for a top coat to both prevent metal build-up and to allow for easy determination of which insert edges have been used. Recently a very thin sputtered chrome coating has also been used to allow users to avoid the costly practice discarding unused cutting edges. Such detection is more difficult if the insert is black, particularly in poorly lit shops.

Aluminum oxide is very stable chemically and retains its hardness at elevated temperatures. Consequently, this coating is especially effective at elevated speeds. Grades designed for high productivity ferrous applications almost always feature a thick layer of Al_2O_3. Although clear, Al_2O_3 appears black due to the adhesive that bonds this coating to whatever sits below.

A more recent development in Al_2O_3 coatings is "texturing." Al_2O_3 is anisotropic. By preferentially aligning the oxide coating grains a tougher and harder coating can be produced. Several commercially available grades feature this textured coating and more will be available as time goes on.

Another commercially important coating technique is physical vapor deposition (PVD). For certain applications, this coating technique provides advantages over the CVD technique. First of all, PVD coatings are thinner and therefore replicate the cutting edge condition more accurately. This means that very sharp edges can be preserved with these coatings. Secondly, there is no loss in strength associated with the deposition of a PVD coating. The presence of a CVD coating by comparison usually reduces the edge toughness of a carbide insert by about 10 percent. As a consequence, PVD coated carbides are most often used for the turning of high temperature alloys, stainless steels at low to medium cutting speeds, and for a wide variety of drilling, threading, and even milling applications of a broad range of materials. PVD coated carbide is also used in a number of mold and die, punch, and wear applications.

The primary disadvantage of PVD coatings is their lack of thickness and consequent wear and crater resistance. Typical CVD coated inserts feature coating thickness ranging from 10 to 20 microns; PVD coatings are usually 2-4 microns thick.

The most common PVD coatings are TiN, titanium aluminum nitride (TiAlN), aluminum titanium nitride (AlTiN), and TiCN. Of these, TiAlN is the hardest and most stable and useful at relatively high speeds, though not as high as CVD coatings. Other coatings that can be obtained include ZrN, CrN, aluminum chromium oxygen nitrogen (AlCrON), titanium diboride (TiB_2), various hard carbon coatings, etc. The PVD area is very dynamic and improved versions of the listed coatings are always being introduced as are brand new coatings.

Design Considerations

Cemented carbide, in common with all brittle materials, may fragment in service, particularly under conditions of impact or upon release from high compressive loading. Additionally, if other failure modes—chipping, cratering, deformation, wear, thermal cracking—encountered during use are allowed to progress too far, fracture can occur. Precautionary measures must be taken to ensure that personnel and equipment are protected from flying fragments and sharp edges when working with brittle tools.

Failure of brittle materials frequently occurs as a result of tensile stress at or near the surface; thus the strength of brittle materials is very dependent on the surface condition of that material. Chips, scratches, thermal cracks, and grinding marks, which decrease the strength and breakage resistance of cemented carbide parts, should be avoided. Even electrodischarge machined (EDM) surfaces should be ground or lapped to a depth of 0.05 mm to remove damaged material.

Special care must be taken when one is grinding cemented carbides. Adequate ventilation of grinding spaces must comply with existing government regulations applicable to the health and safety of the workplace environment. During the fabrication of cemented carbide parts, particularly during grinding of tools and components, adequate provisions must be made for the removal and collection of generated dusts and cutting fluid mists that contain microscopic metal particles, even though those concentrations may be very low. Although the elements contained in these alloys are not radioactive, note that they may become radioactive when exposed to a sufficiently strong radiation source. The cobalt in the cemented carbide alloys, in particular, could be made radioactive.

Physical and Mechanical Properties

Typical property data are shown in Table 4.4.13. In addition to being an effective "cement," cobalt is the primary component that determines mechanical properties. For example, as cobalt content increases, toughness increases but hardness and wear resistance decrease. This tradeoff is illustrated in Table 4.4.13. A secondary factor that affects mechanical properties is WC grain size. Finer grain size leads to increased hardness and wear resistance but lower toughness.

Micrograin Carbides

An important segment of the cemented carbides is micrograin carbide. As their name implies, these grades feature grains that are less than 1 micron in diameter. As mentioned previously, as carbide grain size decreases, hardness increases but toughness decreases. However, once the WC grain size diameter drops below approximately 1 micron, this tradeoff becomes much more favorable. Further decreases in grain size result in continued increases in hardness but are accompanied by a much smaller drop in toughness. As a result, micrograin cemented carbides have outstanding hardness, wear resistance, and compressive strength combined with surprising toughness relatively to conventional carbides. This combination of properties makes these grades particularly useful in machining nickel-, cobalt-, and iron-based superalloys and other difficult to machine alloys, such as refractories and titanium alloys. They are also very useful in machining non-ferrous materials.

Mechanical Properties

The outstanding feature of cemented carbides is their hardness, high compressive strength, stiffness, and wear resistance. Unfortunately, but

Table 4.4.13 Properties of Cemented Carbides

Composition, wt. %	Grain size, μm	Hardness, *Ra*	Abrasion resistance, l/vol. loss cm^3	Density, g/cm^3	Transverse rupture strength, 1,000 lb/in^2	Ultimate compressive strength, 1,000 lb/in^2	Ultimate tensile strength, 1,000 lb/in^2	Modulus of elasticity, 10^6 lb/in^2	Proportional limit, 1,000 lb/in^2	Ductility, % elong	Fracture toughness, $(lb/in^2)(\sqrt{in})$	Thermal expansion 75 to 400°F, $in/(in \cdot °F) \times 10^{-6}$	Thermal conductivity, $cal/(s \cdot °C \cdot cm)$	Electrical resistivity, $\mu\Omega \cdot cm$
WC-3% Co	1.7	93	60	15.3	290	850		98	350			2.2	0.3	17
WC-6% Co	0.8	93.5	62	15										
WC-6% Co	1.1	92.8	60	15	335	860	160	92	370	0.2		2.9		
WC-6% Co	2.1	92	35	15	380	790		92	280		9,200	2.5	0.25	17
WC-6% Co	3	91	15	15	400	750	220	92	210		11,500	2.4	0.25	17
WC-9% Co	4	89.5	10	14.7	425	660		87	140		13,000	2.7	0.25	
WC-10% Co	1.8	91	13	14.6	440	750		85	230		11,500			
WC-10% Co	5.2	89	7	14.5	460	630		85	130			2.5		17
WC-13% Co	4	88.2	4	14.2	500	600		81	140	0.3	14,500	3	0.25	17
WC-16% Co	4	86.8	3	13.9	500	560	270	77	100	0.4	15,800	3.2	0.02	
WC-25% Co	3.7	84	2	13	440	450		67	60		21,000	3.5	0.2	18
Other materials*														
Tool steel (T-g)		85 (66 R_c)	2	8.4	575	600		34				6.5		
Carbon steel (1095)		79 (55 R_c)		7.8			300	30					0.12	20
Cast iron				7.3	105			15-30						
Copper				8.9								9.2	0.94	1.67

*Data for other materials are included to aid in comparing carbide properties with those of referenced materials.

as might be expected of hard materials, toughness and ductility of carbides are low. Table 4.4.13 lists some mechanical and physical property data for typical and popular cemented compositions.

Hardness and Wear Resistance

Most applications of cemented carbide alloys involve wear and abrasive conditions. In general, the wear resistance of these materials can be estimated based on hardness. Note that wear is a complex process and can occur by means other than normal abrasion. Should wear occur by microchipping, fracture, chemical attack, or metallurgical interactions, it will be obvious that performance does not correlate directly with hardness.

The data presented in Table 4.4.13 under the heading Abrasion Resistance were determined by using an abrasive wheel. Results from this simplified test should be used only as a rough guide in the selection of these alloys, since abrasive wear that occurs under actual service conditions is usually complex and varies greatly with particular circumstances. The performance characteristics of a given grade are, of course, best determined under actual service conditions.

In many applications, cemented carbides are subjected to relatively high operating temperatures. For example, the temperatures of the interface between the chip and the cutting edge of a carbide cutting tool during the machining process may reach between 500 and 1,200°C. Wear parts may have point contact temperatures in this range. The extent to which the cemented carbide maintains its hardness at the elevated temperatures encountered in use is an important consideration. The cemented carbides not only have high room temperature hardness, but also maintain hardness at elevated temperatures better than do steels and cast alloys.

Corrosion Resistance

The corrosion resistance of the various cemented carbide alloys is fairly good when compared with other materials, and they may be employed very advantageously in some corrosive environments where outstanding wear resistance is required. Generally they are not employed where corrosion resistance alone is the requirement because other materials usually can be found which are either cheaper or more corrosion resistant.

Corrosion may cause strength deterioration due to preferential attack on the binder phase. This results in the formation of microscopic surface defects to which cemented carbides, in common with other brittle materials, are sensitive.

Special corrosion resistant grades have been developed in which the cobalt binder has been modified or replaced by other metals or alloys, resulting in improved resistance to acid attack. Usually these grades feature a nickel (Ni) binder, perhaps alloyed with small amounts of chromium (Cr). Other factors such as temperature, pressure, and surface condition may significantly influence corrosive behavior, so that specific tests under actual operating conditions should be performed when possible to select a cemented carbide grade for service under corrosive conditions.

Cermets

Another class of composite material that fits in this general category is cermet. While not widely used, these alloys are more complex both in terms of their chemistry and structure but usually consist of TiCN or TiC grains and a nickel (Ni) or nickel-molybdenum (Ni-Mo) alloy binder. Other alloying elements such as Co and W are usually present. On occasion as many as twelve alloying elements will be present.

Cermets usually find application in the high speed finish machining of steels, stainless steels, and, occasionally, cast irons and high temperature alloys. They are also very useful for the machining powder metallurgical materials. As a rule, cermets provide better surface finishes, are more wear resistant but less tough than traditional cemented tungsten carbides. Recent processing improvements, such as sinter HIP (hot isostatic pressing) and a better understanding of the alloying characteristics of cermets, have helped improve their toughness immensely, so that cermets tough enough for milling are now available.

COPPER AND COPPER ALLOYS

by James H. Michel

REFERENCES: ASM International "Copper Specialty Metals Handbook," applicable current ASTM and SAE Standards and publications and website of the Copper Development Association www.copper.org and UNS alloy designations www.UNScopperalloys.org.

Copper and **copper alloys** are man's oldest metal systems and constitute one of the contemporary major groups of commercial metals. They are widely used because of their excellent electrical and thermal conductivities, outstanding corrosion resistance, and ease of fabrication. They are generally nonmagnetic and can be readily joined by conventional soldering, brazing, and welding processes (gas and resistance). From sustainability point of view of the copper currently in use today, 95 percent is totally recyclable. Fifty percent of the copper used is in wire form and an additional 30 percent is found in tubing. Sheet, strip, and plate product forms are found in architectural, electrical, and electronics.

Primary Copper

Copper used in the manufacture of fabricated brass mill or wire and cable products may originate in the ore body of open-pit or underground mines or from rigidly controlled recycled high-grade copper scrap.

Ores are of two types: oxide and sulfide. Processing to pure copper takes different paths for each ore type. Oxide ores (nominally 1/3 of the ore mined) are generally worked using a crushing and leaching process that results in a copper-rich process liquor which is then refined in an electrolytic cell that plates out copper directly. The cathode produced is 99.99 percent pure copper.

Sulfide ores (about 2/3 of the available and mined ores) require conventional crushing, milling, concentration by flotation, smelting to form a matte, conversion to blister copper, and final refining by either the **electrolytic** or the **electrowinning** processes to produce copper cathode, which is fully described in ASTM B115 Electrolytic Cathode Copper. This **electrochemical refining** step results in the deposition of virtually pure copper, with the major impurity being oxygen.

New Developments

Continuous Cast Wire Rod One of the most important production innovations in the copper industry in the past 30+ years has been the introduction of continuous cast wire rod technology. Within this period, the traditional 250-lb wire bar has been completely replaced with continuous cast wire rod product produced to meet the requirements of ASTM B49 Copper Redraw Rod for Electrical Purposes.

The Antimicrobial Property of Copper As evidenced by its historic practical application as one of the first disinfectants in water treatment, copper is bacteriostatic to many bacteria, viruses and cysts and microorganisms. Recent serendipitous and purposeful discovery of copper-control of pathogen growth for *Legionella pneumophila* in water systems and general microbial growth on surfaces are numerous (e.g., Rogers 1991; Kuhn 1983), and have given rise to registration of antimicrobial properties with the United States Environmental Protection Agency (US EPA 2008). At present registration includes over 280 copper alloys as antimicrobial materials, with human health claims that the materials kill >99.9 percent of six harmful human pathogens within 2 hours contact including MRSA (*Methicillin-resistant staphylococcus aureus*), VRE (*Vancomycin-resistant enterococcus*), *Enterobacter aerogenes*, *E. coli* O157:H7, *Staphylococcus aureus*, and *Pseudomonas aeruginosa*. Many of these benefits of copper are derived from the slow and steady environmental degradation (i.e., corrosion) of the underlying metal during its interaction with the environment, raising the ironic reality that for copper, at least some environmental degradation is highly desirable in some instances.

Low-Lead/No-Lead Alloys The passage of the Clean Water Act in the 1980s required that lower quantities of lead be used in the potable water system. This further led to requirements that the water path delivering potable water to the user/consumer contain 0.25 percent lead or less that went into effect in 2014. Many new alloys have been registered and utilized to provide alternatives to leaded alloys. Bismuth, silicon and sulfur are the elements generally being added to brasses (and other alloys) to produce an alloy that is both castable, formable, and machinable.

Classification of Copper Alloys

Wrought and cast copper and copper alloys are commonly categorized into six families: coppers, high-copper alloys, brasses, bronzes, copper nickels, and nickel silvers. They are systemically identified in **Unified Numbering System** (UNS) for metals and alloys.

The internationally recognized UNS designation system (begun in 1974) is an orderly method of defining and identifying the alloys. The copper section is administered by the Copper Development Association, but it is not a specification system. It eliminates the limitations and conflicts of alloy designations previously used and at the same time provides a workable method for the identification of mill and foundry products.

In the designation system, numbers from C10000 through C79999 denote **wrought alloys. Cast alloys** are numbered from C80000 through C99999. Within these two categories, compositions are further grouped into families of coppers and copper alloys shown in Table 4.4.14.

ASTM International (American Society for Testing Material) is a recognized source for product and performance standards. B01 Committee has the responsibility for Electrical Conductors and the B05 Committee covers copper and copper alloys products in general.

Coppers These metals have a designated minimum copper content of 99.3 percent or higher. They include the **oxygen-free coppers** (C10100 and C10200), made by melting prime-quality cathode copper under non-oxidizing conditions. These alloys are particularly suitable for applications requiring high electrical and thermal conductivities coupled with exceptional ductility and freedom from hydrogen embrittlement.

The most common copper is C11000—**electrolytic tough pitch**. Its electrical conductivity exceeds 100 percent IACS and is invariably selected for most wire and cable applications. Selective properties of wire produced to ASTM B1 Hard Drawn Copper Wire, B2 Medium Hard Drawn Copper Wire, and B3 Soft or Annealed Copper Wire are shown in Table 4.4.15. Relative electrical characteristics of copper an aluminum are shown in table 4.4.20.

If superior **machinability** is required, C14500 (tellurium copper), C14700 (sulfur copper), or C18700 (leaded copper) can be selected. With these coppers, superior machinability is gained at a modest sacrifice in electrical conductivity.

High-Copper Alloys In wrought form, these are alloys with designated copper contents less than 99.3 percent but more than 96 percent which do not fall into any other copper alloy group. The cast high-copper alloys have designated copper contents in excess of 94 percent. Mechanical properties of various tempers of C26000 are shown in Table 4.4.16.

Chromium and zirconium are added to increase elevated temperature strength with little decrease in conductivity.

Copper-beryllium alloys are **precipitation-hardening alloys** that combine moderate conductivity with very high strengths. To achieve these properties, a typical heat treatment would involve a solution heat treatment for 1 h at 1,450°F (788°C) followed by water quenching, then a precipitation heat treatment at 600°F (316°C) for 2 h.

Wrought Copper Alloys (Brasses and Bronzes)

There are approximately 230 wrought brass and bronze compositions. The most widely used is alloy C26000, which corresponds to a 70:30 copper-zinc composition and is most frequently specified unless high corrosion resistance or special properties of other alloys are required. For example, alloy C36000—free-cutting brass—is selected when extensive

Table 4.4.14 General Classification of Copper Alloys

Generic name	UNS Nos.	Composition
Wrought alloys		
Coppers	C10100-C15999	>99% Cu
High-copper alloys	C16000-C19999	>96% Cu
Brasses	C20000-C29999	Cu-Zn
Leaded brasses	C30000-C39999	Cu-Zn-Pb
Tin brasses	C40000-C49999	Cu-Zn-Sn-Pb
Phosphor bronzes	C50000-C52999	Cu-Sn-P
Leaded phosphor bronzes	C53000-C54999	Cu-Sn-Pb-P
Copper-phosphorus and copper-silver-phosphorus alloys	C55000-C59999	Cu-P-Ag
Aluminum bronzes	C60000-C64699	Cu-Al-Ni-Fe-Si-Sn
Silicon bronzes/brasses	C64700-C66199	Cu-Si-Sn
Other copper-zinc alloys	C66200-C69999	—
Copper nickels	C70000-C73499	Cu-Ni-Fe
Nickel silvers	C73500-C79999	Cu-Ni-Zn
Cast alloys		
Coppers	C80100-C81399	>99% Cu
High-copper alloys	C81400-C83299	>94% Cu
Red and leaded red brasses	C83300-C84999	Cu-Zn-Sn-Pb (75-89% Cu)
Yellow and leaded yellow brasses	C85000-C85999	Cu-Zn-Sn-Pb (57-74% Cu)
Manganese bronzes and leaded manganese bronzes	C86000-C87299	Cu-Zn-Mn-Fe-Pb
Silicon bronzes, silicon brasses	C87300-C88999	Cu-Zn-Si
Tin bronzes and leaded tin bronzes	C90000-C94699	Cu-Sn-Zn-Pb
Nickel-tin bronzes	C94700-C94999	Cu-Ni-Sn-Zn-Pb
Aluminum bronzes	C95000-C95999	Cu-Al-Fe-Ni
Copper nickels	C96000-C96999	Cu-Ni-Fe
Nickel silvers	C97000-C97999	Cu-Ni-Zn-Pb-Sn
Leaded coppers	C98000-C98999	Cu-Pb
Special alloys	C99000-C99999	—

Table 4.4.15 Tensile Properties of Various Brass Wire Alloys in the Half-Hard (HO2) Temper

Alloy UNS No.	Nominal Zinc Conten T, WL	Tensile Strength Ksi (MPa)	0.2% Yield Strength Ksi/MPa	Elongation in 2.0 in
C11000	0%	42 (283)	37 (255)	20%
C21000	5	47 (324)	44 (303)	17
C22000	10	52 (358)	47 (324)	12
C23000	15	56 (386)	48 (331)	14
C24000	20	60 (414)	43 (296)	18
C26000	30	62 (427)	51 (352)	23
C28000	40	64 (483)	50 (345)	10

Table 4.4.16 Tensile Properties of Various Rolled Tempers of Alloy C26000 Flat Products at 0.040 Inch Thickness

Rolled Temper	Nominal Cold Reduction	Tensile Strength Ksi (MPa)	0.2% Yield Strength Ksi (MPa)	Elongation in 2.0 in
OS040 (Annealed)	0%	48 (331)	16 (110)	59%
H01 (1/4 Hard)	11	54 (379)	33 (228)	46
H02 (1/2 Hard)	21	62 (427)	51 (352)	30
H04 (Hard)	37	76 (524)	72 (496)	10
H06 (Extra Hard)	50	88 (607)	83 (572)	3
H08 (Spring)	60	96 (648)	89 (614)	2
H10 (Extra Spring)	68	100 (682)	92 (634)	1

machining must be done, particularly on automatic screw machines. Other alloys containing aluminum, silicon, or nickel are specified for their outstanding corrosion resistance. The other properties of greatest importance include mechanical strength, fatigue resistance, ability to take a good finish, corrosion resistance, electrical and thermal conductivities, color, ease of fabrication, and machinability. Relative properties of various copper alloys compared to steel and aluminum are shown in 4.4.17.

The bronzes are divided into five alloy families: phosphor bronzes, aluminum bronzes, silicon bronzes, copper nickels, and nickel silvers.

Brasses These alloys contain zinc as the principal alloying element with or without other designated alloying elements such as iron, aluminum, nickel, and silicon. The wrought alloys comprise three main families of brasses: copper-zinc alloys; copper-zinc-lead alloys (leaded brasses); and copper-zinc-tin alloys (tin brasses). The cast alloys comprise four main families of brasses: copper-tin-zinc alloys (red, semi-red, and yellow brasses); **manganese bronze** alloys (high-strength yellow brasses); leaded manganese bronze alloys (leaded high-strength yellow brasses); and copper-zinc-silicon alloys (silicon brasses and bronzes). Ingot for remelting for the manufacture of castings may vary slightly from the ranges shown.

Bronzes are copper alloys in which the major element is not zinc or nickel. Originally *bronze* described alloys with tin as the only or principal alloying element. Today, the term generally is used not by itself, but with a modifying adjective. For wrought alloys, there are four main families of bronzes: copper-tin-phosphorus alloys (phosphor bronzes); copper-tin-lead-phosphorus alloys (leaded phosphor bronzes); copper-aluminum alloys (aluminum bronzes); and copper-silicon alloys (silicon bronzes).

The cast alloys comprise four main families of bronzes: copper-tin alloys (tin bronzes); copper-tin-lead alloys (leaded and high leaded tin bronzes); copper-tin-nickel alloys (nickel-tin bronzes); and copper-aluminum alloys (aluminum bronzes).

Table 4.4.17 Tensile Properties of Commercial Copper Alloys in the Annealed & Extra Hard (Nominally CR 50%) Tempers Compared to Steels & Aluminum

Alloy UNS No.	Tensile Strength Ksi (MPa)	0.2% Yield Strength Ksi (MPa)	Elongation in 2.0 in
C11000 Copper Annealed H06 (Extra Hard)	34 (235) 52 (358)	11 (76) 47 (324)	45% 5
C26000 Brass Annealed H06 (Extra Hard)	53 (365) 88 (607)	22 (150) 83 (572)	54 3
C51000 Phosphor Bronze Annealed H06 (Extra Hard)	50 (345) 96 (635)	22 (150) 80 (550)	50 6
C70600 Copper 10No. Annealed H06 (Extra Hard)	51 (350) 79 (545)	13 (90) 76 (525)	35 4
C75200 Nichol Silver Annealed H06 (Extra Hard)	58 (400) 92 (614)	25 (170) 83 (572)	41 1
1,008 Low-Carbon Steel Annealed (Hard)	44 (303) 70 (483)	25 (170) 60 (413)	41 18
304 Stainless Steel Annealed Cold Worked 50%	87 (600) 158 (1,089)	36 (245) 135 (931)	52 8
3,004 Aluminum (Soft)	26 (180)	10 (69)	22

The family of alloys known as **manganese bronzes**, in which zinc is the major alloying element, is included in the brasses.

Phosphor Bronzes Three tin bronzes, commonly referred to as **phosphor bronzes**, are the dominant alloys in this family. They contain 5, 8, and 10 percent tin and are identified, respectively, as alloys C51000, C52100, and C52400. Containing up to 0.4 percent phosphorus, which improves the casting qualities and hardens them somewhat, these alloys have excellent elastic properties.

Aluminum Bronzes These alloys with 5 and 8 percent aluminum find application because of their high strength and corrosion resistance, and sometimes because of their golden color. Those with 10 percent aluminum content or higher are very plastic when hot and have exceptionally high strength, particularly after heat treatment.

Silicon Bronzes There are three dominant alloys in this family in which silicon is the primary alloying agent but which also contain appreciable amounts of zinc, iron, tin, or manganese. These alloys are as corrosion-resistant as copper (slightly more so in some solutions) and possess excellent hot workability with high strengths. Their outstanding characteristic is that of ready weldability by all methods. The alloys are extensively fabricated by arc or acetylene welding into tanks and vessels for hot-water storage and chemical processing.

Copper-Nickels are alloys with nickel as the principal alloying element, with or without other designated alloying elements. They are extremely malleable and may be worked extensively without annealing. Because of their excellent corrosion resistance, they are used for condenser tubes for the most severe service. Alloys containing nickel have the best high-temperature properties of any copper alloy.

Nickel Silvers Nickel silvers are white and are often applied because of this property. They are tarnish resistant under atmospheric conditions. Nickel silver is the base for most silver-plated ware.

Copper-Nickel-Zinc Alloys, known commonly as **nickel silvers,** are alloys which contain zinc and nickel as the principal and secondary alloying elements, with or without other designated elements. The addition of nickel makes the alloys silver in color.

Special Alloys Alloys whose chemical compositions do not fall into any of previously described categories are combined under "miscellaneous or special alloys."

Properties and Processing of Copper Alloys

Basic physical properties of copper are listed in Table 4.4.18 and 4.4.19.

Mechanical Properties **Cold working** copper and copper alloys increase both tensile and yield strengths, with the more pronounced increase imparted to yield strength (see Figs. 4.4.3 and 4.4.4). For most alloys, tensile strength of the hardest cold-worked temper is approximately twice the tensile strength of the annealed temper,

Table 4.4.18 Physical Properties of Copper

Property	Value	Units	Value	Units
Atomic Number	29			
Atomic Weight	63.54			
Lattice Structure: Face Centered Cubic				
Density				
IEC Standard Value (1,913)	8.89	g/cm³	0.321	lb/in³
Typical Value at 20°C	8.92	g/cm³	0.322	lb/in³
Typical Value at 1,083°C (Solid)	8.32	g/cm³	0.300	lb/in³
Typical Value at 1,083°C (Liquid)	7.99	g/cm³	0.288	lb/in³
Melting Point	1,083	°C	1,981	°F
Boiling Point	2,595	°C	4,703	°F
Linear Coefficient of Thermal Expansion at:				
−253°C/−423°F	0.3 × 10-6	°C	0.17 × 10-6	°F
−183°C/−297°F	9.5 × 10-6	°C	5.28 × 10-6	°F
−191°C to 16°C/−312 to 61°F	14.1 × 10-6	°C	7.83 × 10	°F
25°C to 100°C/77 to 212°F	16.8 × 10-6	°C	9.33 × 10-6	°F
20°C to 200°C/68 to 392°F	17.3 × 10-6	°C	9.61 × 10-6	°F
20°C to 300°C/68 to 572°F	17.7 × 10-6	°C	9.83 × 10-6	°F
Specific Heat (Thermal Capacity) at:				
−253°C/−425°F	0.013	J/g°C	0.0031	Btu/lb°F
−150°C/−238°F	0.282	J/g°C	0.0674	Btu/lb°F
−50°C/−58°F	0.361	J/g°C	0.0862	Btu/lb°F
20°C/−68°F	0.386	J/g°C	0.0921	Btu/lb°F
100°C/−212°F	0.393	J/g°C	0.0939	Btu/lb°F
200°C/−392°F	0.403	J/g°C	0.0963	Btu/lb°F
Thermal Conductivity at:				
−253°C/−425°F	12.98	W-cm/cm²°C	750	Btu/ft/ft2h°F
−200°C/−328°F	5.74	W-cm/cm²°C	330	Btu/ft/ft2h°F
−183°C/−297°F	4.73	W-cm/cm²°C	270	Btu/ft/ft2h°F
−100°C/−148°F	4.35	W-cm/cm²°C	252	Btu/ft/ft2h°F
20°C/68°F	3.94	W-cm/cm²°C	227	Btu/ft/ft2h°F
100°C/212°F	3.85	W-cm/cm²°C	223	Btu/ft/ft2h°F
200°C/392°F	3.81	W-cm/cm²°C	220	Btu/ft/ft2h°F
300°C/572°F	3.77	W-cm/cm²°C	217	Btu/ft/ft2h°F

(Continued)

Table 4.4.18 Physical Properties of Copper (*Continued*)

Property	Value	Units	Value	Units
Electrical Conductivity (Volume) at:				
20°C/68°F (Annealed)	58.0-58.9	MS/m(mΩmm^2)	100.0-101.5	%IACS
20°C/68°F (Fully Cold Worked)	56.3	MS/m(mΩ·mm^2)	97.0	%IACS
Electrical Resistivity (Volume) at:				
20°C/68°F (Annealed)	0.017241-0.0170	Ω·mm^2/m	10.371-10.2	Ω(circmil/ft)
20°C/68°F (Annealed)	1.7241-1.70	μΩ·cm	0.6788-0.669	micrΩ-in
20°C/68°F (Fully Cold Worked)	0.0178	Ω·mm^2/m	10.7	Ω(circmil/ft)
20°C/68°F (Fully Cold Worked)	1.78	μΩ·cm	0.700	μΩ-in

Table 4.4.19 Physical Properties of Copper (*Continued*)

Property	Value	Units	Value	Units
Electrical Resistivity (Mass) at 20°C/68°F (Annealed):				
Mandatory Maximum	0.15328	Ω·g/m^2	875.4	Ω·lb/mile^2
Temperature Coefficient of Electrical Resistance (a) at 20°C °F: Annealed Copper of 100% IACS (Applicable from −100°C to 200°C/212°F to 392°F)	0.00393	°C	0.00218	°F
Fully Cold Worked Copper of 97% IACS (Applicable from 0°C to 100°C/68-212°F)	0.00381	°C	0.00238	°F
Modulus of Elasticity (Tension) at 20°C/68°F:				
Annealed	118,000	MPa	17 × 103	KSi
Cold Worked	118,000-132,000	MPa	17-19 × 103	KSi
Modulus of Rigidity (Torsion) at 20°C/68°F:				
Annealed	44,000	Mpa	6.4 × 103	KSi
Cold Worked	44,000-49,000	MPa	6.4-7 × 103	KSi
Latent Heat of Fusion	205	J/g		
Electro Chemical Equivalent For:				
Cu++	0.329	Mg/C		
Cu+	0.659	Mg/C		
Normal Electrode Potential (Hydrogen Electrode) For:				
Cu++	−0.344	V		
Cu+	−0.470	V		

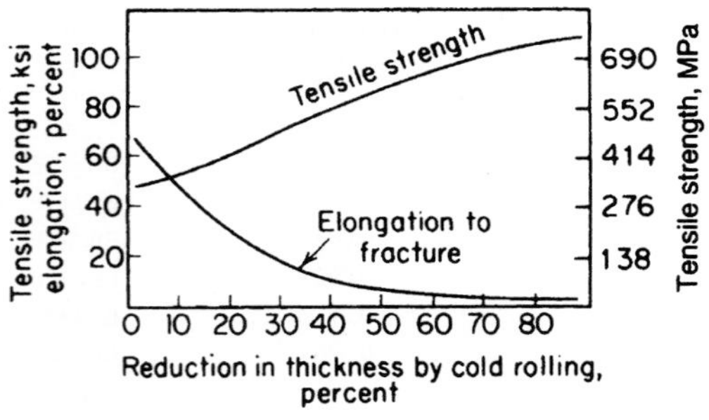

Figure 4.4.3 Effect of Cold Work on annealed C26000 Brass.

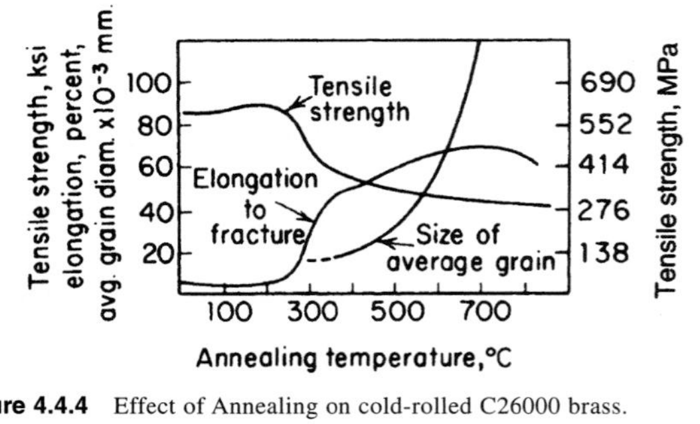

Figure 4.4.4 Effect of Annealing on cold-rolled C26000 brass.

whereas the yield strength of the hardest cold-worked temper can be up to 5 to 6 times that of the annealed temper. While hardness is a measure of temper, it is not an accurate one, because the determination of hardness is dependent upon the alloy, its strength level, and the method used to test for hardness.

All the brasses may be **hot-worked** irrespective of their lead content. Even for alloys having less than 60 percent copper and containing the beta phase in the microstructure, the process permits more extensive changes in shape than cold working because of the plastic (ductile) nature of the beta phase at elevated temperatures, even in the presence of lead. Hence a single hot-working operation can often replace a sequence of forming and annealing operations. Alloys for **extrusion, forging, or hot pressing contain the beta phase in varying amounts.**

Extruded sections of many copper alloys are made in a wide variety of shapes. In addition to architectural applications of extrusions, extrusion is an important production process since many objects such as hinges, pinions, brackets, and lock barrels can be extruded directly from bars.

Table 4.4.20 Physical & Mechanical Properties of Electrical Grades of Copper & Aluminum (Type 1350) Compared

Property	Units	Copper (C11000)	Aluminum (1,350)
Electrical Conductivity (Annealed)	%IACS	101	61
Electrical Resistivity (Annealed)	μΩ-cm	1.72	2.83
Thermal Conductivity at 20°C	W/m·K	397	230
Coefficient			

While the copper-zinc alloys (brasses) may contain up to 40 percent zinc, those with 30 to 35 percent zinc find the greatest application for they exhibit high ductility and can be readily cold-worked. With decreasing zinc content, the mechanical properties and corrosion resistance of the alloys approach those of copper. The properties of these alloys are listed in Table 4.4.20.

Heat Treating Tables 4 and 5 show the progressive effects of cold rolling and annealing of alloy C26000 flat products. **Cold rolling** clearly increases the hardness and the tensile and yield strengths while concurrently decreasing the ductility.

Annealing below a certain minimum temperature has practically no effect, but when the temperature is in the recrystallization range, a rapid decrease in strength and an increase in ductility occur. With a proper anneal, the effects of cold working are almost entirely removed. Heating beyond this point results in grain growth and comparatively little further increase in ductility. Table 4.4.17 shows the variation of properties of various copper alloys after annealing at the temperatures indicated. A comparison to properties of other materials is also provided.

Work-hardened metals are restored to a soft state by annealing. Single-phase alloys are transferred into unstressed crystals by **recovery, recrystallization,** and **grain growth.** In severely deformed metal, recrystallization occurs at lower temperatures than in lightly deformed metal. Also, the grains are smaller and more uniform in size when severely deformed metal is recrystallized.

Grain size can be controlled by proper selection of cold-working and annealing practices. Large amounts of prior cold work, rapid heating to annealing temperature, and short annealing times foster formation of fine grains. Larger grains are normally produced by a combination of limited deformation and long annealing times.

Practically all wrought copper alloys are used in the cold-worked condition to gain additional strength. Articles are often made from annealed stock and depend on the cold work of the forming operation to shape and harden them. When the cold work involved is too small to do this, brass rolled to a degree of temper consistent with final requirements should be used.

Brass for springs should be rolled as hard as is consistent with the subsequent forming operations. For articles requiring sharp bends, or for deep-drawing operations, annealed brass must be used. In general, smaller grain sizes are preferred.

Effect of Temperature Copper and all its alloys increase in strength slightly and uniformly as temperature decreases from room temperature. No low-temperature brittleness is encountered. Copper is useless for prolonged-stress service much above 400°F (204°C), but some of its alloys may be used up to 550°F (287°C). For restricted service above this temperature, only copper-nickel and copper-aluminum alloys have satisfactory properties.

For specific data, see Elevated-Temperature Properties of Copper Base Alloys, and Low-Temperature Properties of Copper and Selected Copper Alloys both on www.copper.org found under Resources; and the ASM International Copper Specialty Handbook.

Electrical and Thermal Conductivities The brasses (essentially copper-zinc alloys) have relatively good electrical and thermal conductivities, although the levels are somewhat lower than those of the coppers because of the effect of solute atoms on the copper lattice. For this reason, copper and high-copper alloys are preferred over the brasses containing more than a few percent total alloy content when high electrical or thermal conductivity is required for the application.

Machinability Machinability of copper alloys is governed by their metallurgical structure and related properties. In that regard, they are divided into three subgroups. All leaded brasses are yellow, have moderate electrical conductivity and fair hot-working qualities, but are relatively poor with respect to fabrication when compared with C26000.

While they all have excellent to satisfactory machinability ratings, the most important alloy for machined products is C36000, **free-cutting brass,** which is the standard material for automatic screw machine work where the very highest machinability is necessary.

ASTM B16 Free-Cutting Brass Rod, Bar, and Shapes for Use in Screw Machines is the dominant specification. Use of this material will often result in considerable savings over the use of steel. Most copper alloys are readily machined by usual methods using standard tools designed for steel, but machining is done at higher cutting speeds. Consideration of the wide range of characteristics presented by various types of copper alloys and the adaptation of machining practice to the particular material concerned will improve the end results to a marked degree.

Group A is composed of alloys of homogeneous structure: copper, wrought bronzes up to 10 percent tin, brasses and nickel silvers up to 37 percent zinc, aluminum bronzes up to 8 percent aluminum, silicon bronzes, and copper nickel. These alloys are all tough and ductile and form long, continuous chips. When they are severely cold-worked, they approach the group B classification in their characteristics.

Group B includes lead-free alloys of duplex structure, some cast bronzes, and most of the high-strength copper alloys. They form continuous but brittle chips by a process of intermittent shearing against the tool edge. Chatter will result unless work and tool are rigid.

Many of the basic **Group C** brasses and bronzes are rendered particularly adaptable for machining operations by the addition of 0.5 to 3.0 percent lead, which resides in the structure as minute, uniformly distributed droplets. They serve to break the chip and lubricate the tool cutting edge. Chips are fine, almost needlelike, and are readily removed. Very little heat is evolved, but the tendency to chatter is greater than for Group A alloys. Lead additions may be made to most copper alloys, but its low melting point makes hot working impossible. Tellurium and sulfur additions are used in place of lead when the combination of hot workability and good machinability is desired.

For **turning operations,** the tough alloys of Group A need a sharp top rake angle (208 to 308 for copper and copper nickel; 128 to 168 for the brasses, bronzes, and silicon bronzes with high-speed tools; 88 to 128 with carbide tools, except for copper, for which 168 is recommended). Type C (leaded) materials require a much smaller rake angle to minimize chatter, a maximum of 88 with high-speed steels and 38 to 68 with carbide. Type B materials are intermediate, working best with 68 to 128 rake angle with high-speed tools and 38 to 88 with carbide, the higher angle being used for the tougher materials. Side clearance angle should be 58 to 78 except for tough, "sticky" materials like copper and copper nickel, where a side rake of 108 to 158 is better.

Many copper alloys will **drill** satisfactorily with standard helix angle drills. Straight fluted tools (helix angle 08) are preferable for the Group C leaded alloys. Feed rates generally are 2 to 3 times faster than those used for steel. With Type B alloy, a fairly course feed helps to break the chip, and with Type A, a fine feed and high speed give best results, provided that sufficient feed is used to prevent rubbing and work hardening.

Corrosion Resistance For most of the last century, copper has been the preferred material for tubing used to convey potable water in domestic plumbing systems because of its outstanding corrosion resistance. In outdoor exposure, however, copper alloys containing less than 20 percent zinc are preferred because of their resistance to stress corrosion cracking, a form of corrosion originally called **season cracking.** For this type of corrosion to occur, a combination of tensile stress (either residual or applied) and the presence of a specific chemical reagent such as ammonia, mercury, mercury compounds, and cyanides is necessary. Stress relief annealing after forming alleviates the tendency for stress corrosion cracking. Alloys that are either zinc-free or contain less than 15 percent zinc generally are not susceptible to stress corrosion cracking.

Condenser tube alloys for freshwater power plant applications have been successfully replaced largely with alloy C70600 (90:10 copper-nickel). Likewise, in seawater or brackish water applications the copper nickel alloys are used because of their ability to perform in these harsh environments and in flowing seawater conditions. Also, aluminum bronze and nickel aluminum bronze have seen increased usage in seawater applications. Nickel aluminum bronze in particular is used for casting of most ship propellers of any size due to its seawater corrosion resistance and cavitation resistance.

Caution must be exercised, for copper and its alloys are not suitable for use in oxidizing acidic solutions or in the presence of moist ammonia and its compounds. Specific alloys are also limited to maximum flow velocities.

Fabrication Many copper alloys can be obtained in sheet, rod, and wire form, and some can be obtained as tubes. Most, in the annealed condition, will withstand extensive amounts of cold work and may be shaped to the desired form by deep drawing, flanging, forming, bending, and similar operations. If extensive cold work is planned, the material should be purchased in the annealed condition, and it may need intermediate annealing either to avoid metal failure or to minimize power consumption. Annealing is done at 900 to 1,300°F (482 to 704°C), depending on the alloy, and is usually followed by air cooling. Because of the ready workability of brass, it is often less costly to use than steel. Brass may be drawn at higher speeds than ferrous metals and with less wear on the tools. In cupping operations, a take-in of 45 percent is usual and on some jobs, it may be larger. Brass hardened by cold working is softened by annealing at about 1,100°F (593°C).

Temper Designations

Temper designations for wrought copper and copper alloys were originally specified on the basis of cold reduction imparted by the rolling of sheet or drawing of wire. Designations for rod, seamless tube, welded tube, extrusions, castings, and heat-treated products were not covered. In 1974, ASTM B601 Standard Practice for Temper Designations for Copper and Copper Alloys—Wrought and Cast, based on an alphanumeric code, was created to accommodate this deficiency. The general temper designation codes are listed in ASTM B601. Table 4.4.21 provides a general listing of copper alloy temper designations.

Cast Copper-Base Alloys

Casting makes it possible to produce parts with shapes that cannot be achieved easily by fabrication methods such as forming or machining. Often it is more economical to produce a part as a casting than to fabricate it by other means.

Copper alloy castings serve in applications that require superior corrosion resistance, high thermal or electrical conductivity, good bearing surface qualities, or other special properties.

All copper alloys can be successfully **sand-cast,** for this process allows the greatest flexibility in casting size and shape and is the most economical and widely used casting method, especially for limited production quantities. Permanent mold casting is best suited for tin, silicon, aluminum, and manganese bronzes, as well as yellow brasses. Most copper alloys can be cast by the centrifugal process.

Table 4.4.21 Examples of Temper Designations for Copper Alloys, From ASTM B601

Temper Designation	Temper Name or Condition
Annealed Condition	
010	Cast & Annealed
020	Hot Forged & Annealed
060	Soft Annealed
061	Annealed
081	Annealed to Temper: 1/4 Hard
0S015	Average Grain Size: 0.015mm
Cold Worked Tempers	
H01	1/4 Hard
H02	1/2 Hard
H04	Hard
H08	Spring
Cold Worked & Stress Relieved	
HR01	H01 & Stress Relieved
HR04	H04 & Stress Relieved
Precipitation Hardened Tempers	
TB00	Solution Heat Treated
TF00	TB00 & Age Hardened
TH02	TB00 Cold Worked & Aged
TM00/TM02/T008	Mill Hardened Tempers
Manufactured Tempers	
M01	As Sand Cast
M04	As Pressure Die Cast
M06	As Investment Cast

Brass die castings are made when great dimensional accuracy and/or a better surface finish is desired. While inferior in properties to hot-pressed parts, die castings are adaptable to a wider range of designs, for they can be made with intricate coring and with considerable variation in section thickness.

The dominant cast copper alloys with their composition and physical properties are shown in Tables 4.4.22 and 4.4.23.

Test results obtained on **standard test bars** (either attached to the casting or separately poured) indicate the quality of the metal used but not the specific properties of the casting itself because of variations of thickness, soundness, and other factors. The ideal casting is one with a fairly uniform metal section with ample fillets and a gradual transition from thin to thick parts.

Joining by Welding, Soldering, and Brazing

Welding Deoxidized copper will **weld** satisfactorily by the oxyacetylene method. Sufficient heat input to overcome its high thermal conductivity must be maintained by the use of torches considerably more powerful than those customary for steel, and preferably by additional preheating. The filler rod must be deoxidized. Gas-shielded

Table 4.4.22 Physical Properties of Five Common Wrought Copper Alloys

Alloy UNS No	Density lb/in^3 (g/cm^3)	Melting Point (or Solid US) °F (°C)	Electrical Conductivity %IACS (MS/m)	Thermal Conductivity Btu ft/ft^2 hr °F (Wcm/cm^2 °C)	Thermal Expansion Coefficient (Linear) $\times 10^{-6}$in/in°F ($\times 10^{-6}$cm/cm°C)
C11000	0.322 (8.92)	1,949°F (1,065°C)	101 (58)	226 (3.94)	9.33 (16.8)
C26000	0.308 (8.53)	1,680°F (915°C)	28 (16)	70 (1.21)	11.1 (19.9)
C51000	0.320 (8.86)	1,750°F (950°C)	15 (8.7)	40 (0.71)	– 9.9 (17.8)
C70600	0.323 (8.94)	2,010°F (1,100°C)	9 (5.2)	26 (0.46)	– 9.5 (17.1)
C75200	0.316 (8.73)	1,960°F (1,070°C)	6 (3.5)	19 (0.33)	– 9.0 (16.2)

Table 4.4.23 Composition and Properties of Selected Cast Copper Alloys

UNS Alloy Number	Nominal Composition	Melting Point Solidus °F (°C)	Electrical Conductivity %IACS (MS/M)	Thermal Conductivity Btu ft/ft^2 hr °F (cal. cm/s.cm^2.°C)	Thermal Expansion Coefficient (Linear) $\times 10^{-6}$in/in °F ($\times 10^{-6}$ cm/cm°C)
C83470	93 Cu, 4 Sn, 2 Zn	1,800 (982)	16.5	41.0 (0.170)	10.4 (18.7)
C83600	85 Cu, 5 Pb, 5 Sn, 5 Zn	1,570 (855)	15	41.6 (0.172)	10.0 (18.0)
C83800	82.9 Cu, 6 Pb, 3.8 Sn, 6.5 Zn	1,550 (845)	15	41.9 (0.173)	10.0 (18.0)
C84400	80 Cu, 7 Pb, 2.9 Sn, 8.5 Zn	1,540 (840)	16.4	41.9 (0.173)	10.0 (18.0)
C84800	76 Cu, 6.2 Pb, 2.5 Sn, 15 Zn	1,530 (832)	16.4	41.6 (0.172)	10.4 (18.7)
C85200	72 Cu, 2.6 Pb, 1.3 Sn, 23.5 Zn	1,700 (925)	18.6	48.5 (0.20)	11.5 (21)
C85400	67.5 Cu, 2.6 Pb, 1 Sn, 28 Zn	1,700 (925)	19.6	51 (0.21)	11.2 (20.2)
C86300	63 Cu, 3 Fe, 6.2 Al, 3.7 Mn, 25 Zn	1,625 (885)	9	21 (0.085)	12 (22)
C86500	57.5 Cu, 1.2 Fe, 1.5 Al, 1.5 Mn, 39 Zn	1,583 (862)	22	49.6 (0.089)	11.3 (20.3)
C87300	94 Cu, 1.1 Mn, 4 Si	1,680 (916)	6.7	28 (0.113)	10.9 (19.6)
C87610	92 Cu, 4 Si, 4 Zn	1,780 (971)	6	16.4 (0.067)	10.2 (17.6)
C89833	88.5 Cu, 5 Sn, 4 Zn, 2.2 Bi	1,454 (790)	17.8	41 (0.166)	13 (23.4)
C89836	89 Cu, 5.5 Sn, 3 Zn, 2.5 Bi	1,580 (860)	17.8	41 (0.166)	13 (23.4)
C90300	87.5 Cu, 8.2 Sn, 4 Zn	1,570 (855)	12	43 (0.18)	10.0 (18.0)
C92200	87.5 Cu, 2.1 Pb, 5 Sn, 3.7 Zn, .8 Ni	1,520 (825)	14.3	40 (0.17)	10.0 (18.0)
C93200	83 Cu, 7 Pb, 6.9 Sn, 2.5 Zn	1,570 (855)	12	34 (0.14)	10.0 (18.0)
C93700	80 Cu, 9.5 Pb, 10 SN	1,403 (762)	10	27 (0.11)	10.3 (18.5)
C95400	83 Cu, 4 Fe, 10.7 Al	1,570 (855)	13	34 (0.14)	9.0 (16.2)
C95400	83 Cu, 4 Fe, 10.7 Al	1,880 (1,025)	13	34 (0.14)	9.0 (16.2)
C95500	78 Cu, 4 Fe, 10.7 Al, 4.2 Ni	1,880 (1,025)	8.5	24 (0.10)	9.0 (16.2)
C95500	78 Cu, 4 Fe, 10.7 Al, 4.2 Ni	1,880 (1,025)	8.5	24 (0.10)	9.0 (16.2)
C95800	79 Cu, 4 Fe, 9 Al, 4.5 Ni, Mn 1.2	1,880 (1,025)	7.1	21 (0.085)	9.0 (16.2)
C96200	87 Cu, 1.4 Fe, 10 Ni	2,010 (1,100)	11	26 (0.11)	9.5 (17.3)
C96400	67 Cu, .8 Fe, 30 Ni,1 Nb	2,140 (1,170)	5	17 (0.068)	9.0 (16.2)
C97600	65 Cu, 4 Pb, 4 Sn, Zn 6, 20.2 Ni	2,027 (1,108)	5	13 (0.075)	9.3 (17)
C99500	Cu, 4 Fe, 1.2 Zn, 1.2 Al, 4.5 Ni, Si 1.2	2,050 (1121)	10	37.0 (0.153)	8.3 (14.9)
C99700	54 Cu, 1.7 Al, 13 Mn, 5 Ni, 22 Zn	1,615 (879)	3	37.0 (0.153)	11.8 (20.4)

arc welding is preferred. Tough-pitch copper will not result in high-strength welds because of embrittlement due to the oxygen content. Copper may be arc-welded, using shielded metal arc, gas metal arc (MIG), or gas tungsten arc (TIG) welding procedures using experienced operators. Filler rods of phosphor bronze or silicon bronze will give strong welds more consistently and are used where the presence of a weld of different composition and corrosion-resistance characteristics is not harmful. Brass may be welded by the oxyacetylene process but not by arc welding. A filler rod of about the same composition is used, although silicon is frequently added to prevent zinc fumes.

Copper-silicon alloys respond remarkably well to welding by all methods. The conductivity is not too high, and the alloy is, to a large extent, self-fluxing.

Applicable specifications for joining copper and copper alloys by welding include:

ANSI/AWS A5.6 Covered Copper and Copper Alloy Arc Welding Electrodes

ANSI/AWS A5.7 Copper and Copper Alloy Bare Welding Rods and Electrodes

ANSI/AWS A5.27 Copper and Copper Alloy Rods for Oxyfuel Gas Welding

The **fluxes** used for brazing copper joints are water-based and function by dissolving and removing residual oxides from the metal surface; they protect the metal from reoxidation during heating and promote wetting of the joined surfaces by the brazing filler metal.

Soldering Soldered joints with **capillary fittings** are used in plumbing for water and sanitary drainage lines. Such joints depend on capillary action drawing free-flowing molten solder into the gap between the fitting and the tube.

Flux acts as a cleaning and wetting agent which permits uniform spreading of the molten solder over the surfaces to be joined.

Selection of a solder depends primarily on the operating pressure and temperature of the system. Lead-free solders required for joining copper tube and fittings in potable water systems are covered by ASTM B32 Solder Metal.

As in brazing, the functions of soldering flux are to remove residual oxides, promote wetting, and protect the surfaces being soldered from oxidation during heating.

Fluxes best suited for soldering copper and copper alloy tube should meet the requirements of ASTM B813 Liquid and Paste Fluxes for Soldering Applications of Copper and Copper Alloy Tube with the joining accomplished per ASTM B828 Making Capillary Joints by Soldering of Copper and Copper Alloy Tube and Fittings.

JEWELRY METALS

Staff Contribution

References: ASM "Metals Handbook." Publications of the metal-producing companies. Standards applicable to classes of metals available from manufacturers' associations. Trade literature pertinent to each specific metal.

Gold is used primarily as a monetary standard; the small amount put to metallurgical use is for jewelry or decorative purposes, in dental work, for fountain-pen nibs, and as an electrodeposited protecting coating. Electroplated gold is widely used for electronic junction points (transistors and the like) and at the mating ends of telephone junction wires. An alloy with palladium has been used as a platinum substitute for laboratory vessels, but its present price is so high that it does not compete with platinum.

Silver has the highest electrical conductivity of any metal, and it found some use in bus bars during World War II, when copper was in short supply. Since its density is higher than that of copper and since government silver frequently has deleterious impurities, it offers no advantage over copper as an electrical conductor. Heavy-duty electrical contacts are usually made of silver. It is used in aircraft bearings and solders. Its largest commercial use is in tableware as sterling silver, which contains 92.5 percent silver (the remainder is usually copper). United States coinage used to contain 90 percent silver, 10 percent copper, but coinage is now manufactured from a debased alloy consisting of copper sandwiched between thin sheets of nickel.

Platinum has many uses because of its high melting point, chemical inertness, and catalytic activity. It is the standard catalyst for the oxidation of sulfur dioxide in the manufacture of sulfuric acid. Because it is inert toward most chemicals, even at elevated temperatures, it can be used for laboratory apparatus. It is the only metal that can be used for an electric heating element about 2,300°F without a protective atmosphere. Thermocouples of platinum with platinum-rhodium alloy are standard for high temperatures. Platinum and platinum alloys are used in large amounts in feeding mechanisms of glass-working equipment to ensure constancy of the orifice dimensions that fix the size of glass products. They are also used for electrical contacts, in dental work, in aircraft spark-plug electrodes, and as jewelry. It has been used for a long time as a catalyst in refining of gasoline and is in widespread use in automotive catalytic converters as a pollution control device.

Palladium follows platinum in importance and abundance among the platinum metals and resembles platinum in most of its properties. Its density and melting point are the lowest of the platinum metals, and it forms an oxide coating at a dull-red heat so that it cannot be heated in air above 800°F (426°C) approx. In the finely divided form, it is an excellent hydrogenation catalyst. It is as ductile as gold and is beaten into leaf as thin as gold leaf. Its hardened alloys find some use in dentistry, jewelry, and electrical contacts.

Iridium is one of the platinum metals. Its chief use are as a hardener for platinum jewelry alloys and as platinum contacts. Its alloys with osmium are used for tipping fountain-pen nibs. Isotope Ir 192 is one of the basic materials used in radiation therapy and is widely employed as a radioactive implant in oncological surgical procedures. It is the most corrosion-resistant element known.

Rhodium is used mainly as an alloying addition to platinum. It is a component of many of the pen-tipping alloys. Because of its high reflectivity and freedom from oxidation films, it is frequently used as an electroplate for jewelry and for reflectors for motion-picture projectors, aircraft searchlights, and the like.

LOW-MELTING-POINT METALS AND ALLOYS

by Frank E. Goodwin

REFERENCES: Current edition of ASM "Metals Handbook." Current listing of applicable ASTM and AWS standards. Publications of the various metal-producing companies.

Metals with low-melting temperatures offer a diversity of industrial applications. In this field, much use is made of the **eutectic**-type alloy, in which two or more elements are combined in proper proportion so as to have a minimum melting temperature. Such alloys melt at a single, fixed temperature, as does a pure metal, rather than over a range of temperatures, as with most alloys.

Liquid Metals A few metals are used in their liquid state. **Mercury** [mp, −39.37°F (−40°C)] is the only metal that is liquid below room temperature. In addition to its use in thermometers, scientific instruments, and electrical contacts, it is a constituent of some very low-melting alloys. Its application in dental amalgams is unique. Regulatory restrictions on mercury have resulted in substitution of alternatives because of health concerns. It has been used as a heat-exchange fluid, as have **sodium** and the sodium-potassium alloy **NaK** (see Sec. 7.8).

Moving up the temperature scale, **tin** melts at 449.4°F (232°C) and finds its largest single use in coating steel to make tinplate. Tin may be applied to steel, copper, or cast iron by hot-dipping, although for steel this method has been replaced largely by electrodeposition onto continuous strips of rolled steel. Typical tin coating thicknesses are 40 μ in (1 μm). **Solders** constitute an important use of tin. Sn-Pb solders are widely used, although Sn-Sb and Sn-Ag solders can be used in applications requiring lead-free compositions. Tin is alloyed in bronzes and battery grid material. Alloys of 12 to 25 percent Sn, with the balance lead, are applied to steel by hot-dipping and are known as **terneplate.**

Modern **pewter** is a tarnish-resistant alloy used only for ornamental ware and is composed of 91 to 93 percent Sn, 1 to 8 percent Sb, and 0.25 to 3 percent Cu. Alloys used for casting are lower in copper than those used for spinning hollowware. Pewter does not require intermediate annealing during fabrication. Lead was a traditional constituent of pewter but has been eliminated in modern compositions because of the propensity for lead to leach out into fluids contained in pewter vessels.

Pure **lead** melts at 620°F (327°C), and four grades of purities are recognized as standard; see Table 4.4.24. Corroding lead and common lead are 99.94 percent pure, while chemical lead and copper-bearing lead are 99.9 percent pure. Corroding lead is used primarily in the chemical industry, with most used to manufacture white lead or litharge (lead monoxide, PbO). Chemical lead contains residual copper that improves both its corrosion resistance and stiffness, allowing its extensive employment in chemical plants to withstand corrosion, particularly from sulfuric acid. Copper-bearing lead also has high corrosion resistance because of its copper content. Common lead is used for alloying and for battery oxides. Lead is strengthened by alloying with antimony at levels between 0 and 15 percent. Lead-tin alloys with tin levels over the entire composition range are widely used.

The largest use of lead is in **lead-acid batteries.** Deep cycling batteries typically use Pb 6-7 antimony, while Pb 0.1 Ca 0.3 tin alloys are used for maintenance-free batteries found in automobiles. Battery alloys usually have a minimum tensile strength of 6,000 lb/in^2 (41 MPa) and 20 percent elongation. Pb 0.85 percent antimony alloy is also used for sheathing of high-voltage power cables. Other cable sheathing alloys are based on Sn-Cd, Sn-Sb, or Te-Cu combined with lead. The lead content in cable sheathing alloys always exceeds 99 percent.

Table 4.4.24 Composition Specifications for Lead, According to ASTM B-29–03

	Composition, wt %			
Element	Corroding lead	Common lead	Chemical lead	Copper-bearing lead
Silver, max	0.015	0.005	0.020	0.020
Silver, min	—	—	0.002	—
Copper, max	0.0015	0.0015	0.080	0.080
Copper, min	—	—	0.040	0.040
Silver and copper together, max	0.025	—	—	—
Arsenic, antimony, and tin together, max	0.002	0.002	0.002	0.002
Zinc, max	0.001	0.001	0.001	0.001
Iron, max	0.002	0.002	0.002	0.002
Bismuth, max	0.050	0.050	0.005	0.025
Lead (by difference), min	99.94	99.94	99.90	99.90

SOURCE: ASTM, reprinted with permission.

Lead sheet for construction applications, primarily in the chemical industries, is based on either pure lead or Pb 6 antimony. When cold-rolled, this alloy has a tensile strength of 4,100 lb/in^2 (28.3 MPa) and an elongation of 47 percent. Lead sheet is usually in (1.2 mm) thick and weighs 3 lb/ft^2 (15 kg/m^2). **Architectural uses for lead** are numerous, primarily directed to applications where water is to be contained (shower and tiled bathing areas overlaid by ceramic tile). Often, copper is lead-coated to serve as roof flashing or valleys; in those applications, the otherwise offending verdigris coating which would develop on pure copper sheet is inhibited.

Ammunition for both sports and military purposes uses alloys containing percentages up to 8 Sb and 2 As. **Bullets** have somewhat higher alloy content than shot.

Type metals are the hardest and strongest lead alloys. Type metals are obsolescent by virtue of being superseded by newer printing techniques. Their use is restricted to existing printing machinery. Typical compositions are shown in Table 4.4.25. A small amount of copper may be added to increase hardness. Electrotype metal contains the lowest alloy content because it serves only as a backing to the shell and need not be hard. Linotype, or slug casting metal, must be fluid and capable of rapid solidification for its use in composition of newspaper type; thus metal of nearly eutectic composition is used. It is rarely used as the actual printing surface and therefore need not be as hard as stereotype and monotype materials.

Fusible Alloys (See Table 4.4.26.) These alloys are used typically as fusible links in sprinkler heads, as electric cutouts, as fire-door links, for making castings, for patterns in making match plates, for making electroforming molds, for setting punches in multiple dies, and for dyeing cloth. Some fusible alloys can be cast or sprayed on wood, paper, and other materials without damaging the base materials, and many of these alloys can be used for making hermetic seals. Since some of these alloys melt below the boiling point of water, they can be used in bending tubing. The properly prepared tubing is filled with the molten alloy and allowed to solidify, and after bending, the alloy is melted out by immersing the tube in boiling water. The volume changes during the solidification of a fusible alloy are, to a large extent, governed by the bismuth content of the alloy. As a general rule, alloys containing more than about 55 percent bismuth expand and those containing less than about 48 percent bismuth contract during solidification; those containing 48 to 55 percent bismuth exhibit little change in volume. The change in volume due to cooling of the solid metal is a simple linear shrinkage, but some of the fusible alloys owe much of their industrial importance to other volume changes, caused by change in structure of the solid alloy, which permit the production of castings having dimensions equal to, or greater than, those of the mold in which the metal was cast.

For **fire-sprinkler heads,** with a rating of 160°F (71°C) **Wood's metal** is used for the fusible-solder-alloy link. Wood's metal gives the most suitable degree of sensitivity at this temperature, but in tropical countries and in situations where industrial processes create a hot atmosphere (e.g., baking ovens, foundries), solders having a higher melting point must be used. Alloys of eutectic compositions are used since they melt sharply at a specific temperature.

Fusible alloys are also used as molds for thermoplastics, for the production of artificial jewelry in pastes and plastic materials, in foundry patterns, chucking glass lenses, as hold-down bolts, and inserts in plastics and wood.

Solders are filler metals that produce coalescence of metal parts by melting the solder while heating the base metals below their melting point. To be termed *soldering* (versus brazing), the operating temperature must not exceed 840°F (450°C). The filler metal is distributed between the closely fitting faying surfaces of the joint by capillary action. Compositions of solder alloys are shown in Table 4.4.27. The Pb 63 tin eutectic composition has the lowest melting point—361°F (183°C)—of the binary tin-lead solders. For joints in copper pipe and cables, a wide melting range is needed. Solders containing less than 5 percent Sn are used to seal precoated containers and in applications where the surface temperatures exceed 250°F (120°C). Solders with 10 to 20 percent Sn are used for repairing automobile radiators and bodies, while compositions containing 40 to 50 percent Sn are used

Table 4.4.25 Compositions and Properties of Type Metal*

	Composition, %			Liquidus		Solidus		Brinell
Service	Sn	Sb	Pb	°F	°C	°F	°C	hardness†
Electrotype	4	3	93	561	294	473	245	12.5
Linotype	5	11	84	475	246	462	239	22
Stereotype	6.5	13	80.5	485	252	462	239	22
Monotype	8	17	75	520	271	462	239	27

* The use of type metals is obsolescent in the printing industry.
† 0.39-in (10-mm) ball, 550-lb (250-kg) load
SOURCE: "Metals Handbook," ASM International, 10th ed.

Table 4.4.26 Compositions, Properties, and Applications of Fusible Alloys

Composition, %					Solidus		Liquidus		
Sn	Bi	Pb	Cd	In	op	°C	°F	°C	Typical application
8.3	44.7	22.6	5.3	19.10	117	47	117	47	Dental models, part anchoring, lens chucking
13.3	50	26.7	10.0	—	158	70	158	70	Bushings and locators in jigs and fixtures, lens chucking, reentrant tooling, founding cores and patterns, light sheet-metal embossing dies, tube bending
—	55.5	44.5	—	—	255	124	255	124	Inserts in wood, plastics, bolt anchors, founding cores and patterns, embossing dies, press-form blocks, duplicating plaster patterns, tube bending, hobbyist parts
12.4	50.5	27.8	9.3	—	163	73	158	70	Wood's metal-sprinkler heads
14.5	48.0	28.5	(9% Sb)	—	440	227	217	103	Punch and die assemblies, small bearings, anchoring for machinery, tooling, forming blocks, stripper plates in stamping dies
60	40	—	—	—	338	170	281	138	Locator members in tools and fixtures, electroforming cores, dies for lost-wax patterns, plastic casting molds, prosthetic development work, encapsulating avionic components, spray metallizing, pantograph tracer molds

SOURCE: "Metals Handbook," ASM International, 10th ed.

Table 4.4.27 Compositions and Properties of Selected Solder Alloys

Composition, %		Solidus temperature		Liquidus temperature		
Tin	Lead	°C	°F	°C	°F	Uses
2	98	316	601	322	611	Side seams for can manufacturing
5	95	305	581	312	594	Coating and joining metals
10	90	268	514	302	576	Sealing cellular automobile radiators, filling seams or dents
15	85	227	440	288	550	Sealing cellular automobile radiators, filling seams or dents
20	80	183	361	277	531	Coating and joining metals, or filling dents or seams in automobile bodies
25	75	183	361	266	511	Machine and torch soldering
30	70	183	361	255	491	
35	65	183	361	247	477	General-purpose and wiping solder
40	60	183	361	238	460	Wiping solder for joining lead pipes and cable sheaths; also for automobile radiator cores and heating units
45	55	183	361	227	441	Automobile radiator cores and roofing seams
50	50	183	361	216	421	Most popular general-purpose solder
60	40	183	361	190	374	Primarily for electronic soldering applications where low soldering temperatures are required
63	37	183	361	183	361	Lowest-melting (eutectic) solder for electronic applications
40	58 (2% Sb)	185	365	231	448	General-purpose, not recommended on zinc-containing materials
95	0 (5% Sb)	232	450	240	464	Joints on copper, electrical plumbing, heating, not recommended on zinc-containing metals
0	97.5 (2.5% Ag)	304	580	308	580	Copper, brass, not recommended in humid environments
95.5	0 (4% Cu, 0.5% Ag)	226	440	260	500	Potable water plumbing

SOURCE: "Metals Handbook," vol. 2, 10th ed., p. 553.

for manufacture of automobile radiators, electrical and electronic connections, roofing seams, and heating units. Silver is added to tin-lead solders for electronics applications to reduce the dissolution of silver from the substrate. Sb-containing solders should not be used to join base metals containing zinc, including galvanized iron and brass. A 95.5 percent Sn 4 Cu 0.5 Ag solder has supplanted lead-containing solders in potable water plumbing.

Brazing is similar to soldering but is defined as using filler metals which melt at or above 840°F (450°C). As in soldering, the base metal is not melted. Typical applications of brazing filler metals classified by AWS and ASTM are listed in Table 4.4.28; their compositions and melting points are given in Table 4.4.29.

METALS AND ALLOYS FOR USE AT ELEVATED TEMPERATURES

by John H. Tundermann

REFERENCES: "High Temperature High Strength Alloys," AISI. Simmons and Krivobok, Compilation of Chemical Compositions and Rupture Strength of Super-strength Alloys, *ASTM Tech. Pub.* 170-A. "Metals Handbook," ASM. Smith, "Properties of Metals at Elevated Temperatures," McGraw-Hill. Clark, "High-Temperature Alloys," Pittman. Cross, Materials for Gas Turbine Engines, *Metal Progress,* March 1965. "Heat Resistant Materials," ASM Handbook, vol. 3, 9th ed., pp. 187-350. Lambert, "Refractory Metals and Alloys," ASM Handbook, vol. 2, 10th ed., pp. 556-585. Watson et al., "Electrical Resistance Alloys," ASM Handbook, vol. 2, 10th ed., pp. 822-839.

Table 4.4.28 Brazing Filler Metals

AWS-ASTM filler-metal classification	Base metals joined
BAlSi (aluminum-silicon)	Aluminum and aluminum alloys
BCuP (copper-phosphorus)	Copper and copper alloys; limited use on tungsten and molybdenum; should not be used on ferrous or nickel-base metals
BAg (silver)	Ferrous and non-ferrous metals except aluminum and magnesium; iron, nickel, cobaltbase alloys; thin-base metals
BCu (copper)	Ferrous and non-ferrous metals except aluminum and magnesium
RBCuZn (copper-zinc)	Ferrous and non-ferrous metals except aluminum and magnesium; corrosion resistance generally inadequate for joining copper, silicon, bronze, copper, nickel, or stainless steel
BMg (magnesium)	Magnesium-base metals
BNi (nickel)	AISI 300 and 400 stainless steels; nickel- and cobalt-base alloys; also carbon steel, low-alloy steels, and copper where specific properties are desired

Some of data presented here refer to materials with names which are proprietary and registered trademarks. They include Inconel, Incoloy, and Nimonic (Inco Alloys International); Hastelloy (Haynes International); MAR M (Martin Marietta Corp.); and René (General Electric Co.).

Metals are used for an increasing variety of applications at elevated temperatures, "elevated" being a relative term that depends upon the specific metal and the specific service environment. Elevated-temperature properties of the common metals and alloys are cited in the several subsections contained within this section. This subsection deals with metals and alloys whose prime use is in **high-temperature applications.** [Typical operating temperatures are compressors, 750°F (399°C); steam turbines, 1,100°F (593°C); gas turbines, 2,000°F (1,093°C); resistance heating elements, 2,400°F (1,316°C); electronic vacuum tubes, 3,500°F (1,926°C), and lamps, 4,500°F (2,482°C).] In general, alloys for

Table 4.4.29 Compositions and Melting Ranges of Brazing Alloys

AWS designation	Nominal composition, %	Melting temp, °F
BAg-1	45 Ag, 15 Cu, 16 Zn, 24 Cd	1,125-1,145
BNI-7	13 Cr, 10 P, bal. Ni	1,630
BAu-4	81.5 Au, bal. Ni	1,740
BAlSi-2	7.5 Si, bal. Al	1,070-1,135
BAlSi-5	10 Si, 4 Cu, 10 Zn, bal. Al	960-1,040
BCuP-5	80 Cu, 15 Ag, 5 P	1,190-1,475
BCuP-2	93 Cu, 7 P	1,310-1,460
BAg-1a	50 Ag, 15.5 Cu, 16.5 Zn, 18 Cd	1,160-1,175
BAg-7	56 Ag, 22 Cu, 17 Zn, 5 Sn	1,145-1,205
RBCuZn-A	59 Cu, 40 Zn, 0.6 Sn	1,630-1,650
RBCuZn-D	48 Cu, 41 Zn, 10 Ni, 0.15 Si, 0.25 P	1,690-1,715
BCu-1	99.90 Cu, min	1,980
BCu-1a	99.9 Cu, min	1,980
BCu-2	86.5 Cu, min	1,980

°C = (°F − 32)/1.8.
SOURCE: "Metals Handbook," ASM.

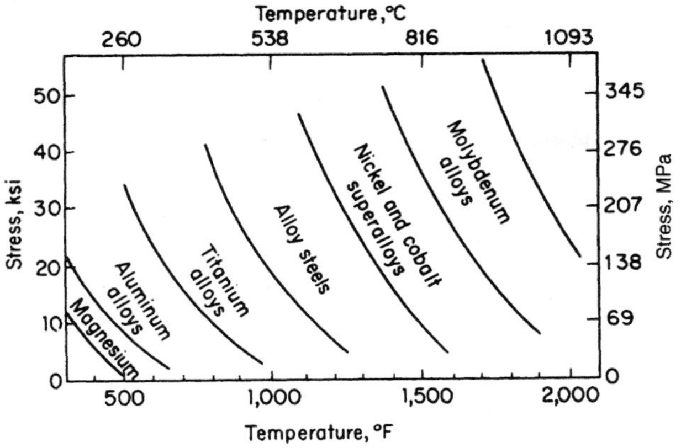

Figure 4.4.5 Stress-temperature application ranges for several alloy types; stress to produce rupture in 1,000 h.

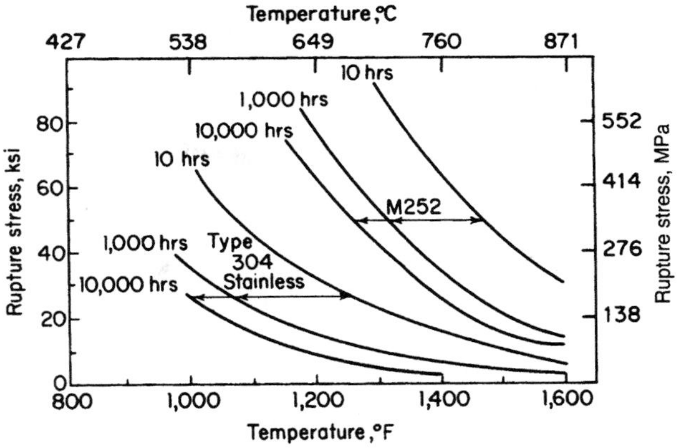

Figure 4.4.6 Effect of time on rupture strength of type 304 stainless steel and alloy M252.

high-temperature service must have melting points above the operating temperature, low vapor pressures at that temperature, resistance to attack (oxidation, sulfidation, corrosion) in the operating environment, and sufficient strength to withstand the applied load for the service life without deforming beyond permissible limits. At high temperatures, atomic diffusion becomes appreciable, so that time is an important factor with respect to surface chemical reactions, to **creep** (slow deformation under constant load), and to internal changes within the alloy during service. The effects of time and temperature are conveniently combined by the empirical **Larson-Miller parameter** $P = T(C + \log t) \times 10^{-3}$, where T = test temp in °R (°F + 460) and t = test time, h. The constant C depends upon the material but is frequently taken to be 20.

Many alloys have been developed specifically for such applications. The selection of an alloy for a specific high-temperature application is strongly influenced by service conditions (stress, stress fluctuations, temperature, heat shock, atmosphere, service life), and there are hundreds of alloys from which to choose. The following illustrations, data, and discussion should be regarded as examples. Vendor literature and more extensive references should be consulted.

Figure 4.4.5 indicates the general stress-temperature range in which various alloy types find application in elevated-temperature service. Figure 4.4.6 indicates the important effect of time on the strength of alloys at high temperatures, comparing a familiar stainless steel (type 304) with **superalloy** (M252).

Common Heat-Resisting Alloys A number of alloys containing large amounts of **chromium and nickel** are available. These have excellent oxidation resistance at elevated temperatures. Several of them have been developed as electrical-resistance heating elements; others are modifications of stainless steels, developed for general corrosion resistance. Selected data on these alloys are summarized in Table 4.4.30. The maximum temperature value given is for resistance to oxidation with a reasonable life. At higher temperatures, failure will be rapid because of scaling. At lower temperatures, much longer life will be obtained. At the maximum useful temperature, the metal may be very weak and frequently must be supported to prevent sagging. Under load, these alloys are generally useful only at considerably lower temperatures, say up to 1,200°F (649°C) max, depending upon permissible creep rate and the load.

Superalloys were developed largely to meet the needs of aircraft gas turbines, but they have also been used in other applications demanding

Table 4.4.30 Properties of Common Heat-Resisting Alloys

Alloy		Max. temp for oxidation resistance		Nominal chemical composition, %					Specific gravity	Coef of thermal expansion per °F × 10^{-6} (0–1,200°F)	Stress to rupture in 1,000 h, ksi			0.2% offset yield strength, ksi		
UNS no.	Name	°F	°C	C	Cr	Ni	Fe	Other			1,200°F (649°C)	1,500°F (816°C)	1,800°F (982°C)	Room temp	1,000°F (538°C)	1,200°F (649°C)
N 06004	62 Ni, 15 Cr	1,700	926		15	62	Bal		8.19	9.35						
N 06003	80 Ni, 20 Cr	2,100	1,148		20	80			8.4	9.8				63		
	Kanthal	2,450	1,371		25		Bal	3 Co, 5 Al	7.15							
N 06600	Inconel alloy 600	2,050	1,121		13	79	Bal		8.4		14.5	3.7		36	22	22
G 10150	1,015	1,000	537	0.15			Bal		7.8	8.36	2.7			42	20	10
S 50200	502	1,150	621	0.12	5		Bal	0.5 Mo	7.8	7.31	6	1.5		27	18	12
S 44600	446	2,000	1,093	0.12	26	0.3	Bal		7.6	6.67	4	1.2				
S 30400	304	1,650	871	0.06	18	9	Bal		7.9	10.4	11	3.5		32	14	11
S 34700	347	1,650	871	0.08	18.5	11.5	Bal	0.8 Nb	8.0	10.7	20			41	31	26
S 31600	316	1,650	873	0.07	18	13	Bal	2.5 Mo	8.0	10.3	25	7		41	22	21
S 31000	310	2,000	1,093	0.12	25	20	Bal		7.9	9.8	13.2	3.0	2.7	34	28	25
S 32100	321	1,650	899	0.06	18	10	Bal	0.5 Ti	8.0	10.7	17.5	3.7		39	27	25
	NA 22H	2,200	1,204	0.5	28	48	Bal	5W		8.6	30	18	3.6			
N 06617	Inconel alloy 617	2,050	1,121	0.1	22	Bal	3	9 Mo, 12.5 Co, 1.2 Al, 0.6 Ti	8.36	6.4	52	14	4	43	29	25
N 08800	Inconel alloy 800	1,800	982	0.1	21	32	Bal		7.94	7.9	24	7	2	36	26	26
N 08330	330	1,850	1,010	0.05	19	36	Bal		8.08	8.3		5	1.3	40	26	21

Table 4.4.31 Wrought Superalloys, Compositions*

	Chemical composition, %												
Common designation	C	Mn	Si	Cr	Ni	Co	Mo	W	Nb	Ti	Al	Fe	Other
19-9D L	0.2	0.5	0.6	19	9	....	1.2	1.25	0.3	0.3	....	Bal	
Timken	0.1	0.5	0.5	16	25	....	6	....	...	...	....	Bal	0.15 N
L.C. N-155	0.1	0.5	0.5	20	20	20	3	2	1	...	....	Bal	0.15 N
S-590	0.4	0.5	0.5	20	20	20	4	4	4	...	....	Bal	
S-816	0.4	0.5	0.5	20	20	Bal	4	4	4	...	....	4	
Nimonic alloy 80A	0.05	0.7	0.5	20	76	....	....	....	...	2.3	1.0	0.5	
K-42-B	0.05	0.7	0.7	18	42	22	....	....	...	2.0	0.2	14	
Hastelloy alloy B	0.1	0.5	0.5	..	65	....	28	....	...	...	....	6	0.4 V
M252†	0.10	1.0	0.7	19	53.5	10	10	....	...	2.5	0.75	2	
J1570	0.20	0.1‡	0.2‡	20	29	37.5	....	7	...	4.1	....	2	
HS-R235	0.10	....	....	16	Bal	1.5	6		...	3.0	1.8	8	
Hastelloy alloy X	0.10	....	....	21	48	2.0	9	1.0	...	...	....	18	
HS 25 (L605)	0.05	1.5	1.0‡	20	10	53	....	15	...	...	....	1.0	
A-286	0.05	....		15	26	....	1.3	....	...	2.0	0.35	55	0.3 V
Inconel alloy 718	0.05	0.2	0.2	19	52.5	....	3	....	5	0.9	0.5	18.5	
Inconel alloy X-750	0.1	1.0	0.5	15	72	1.0	....	....	...	2.5	0.7	7	

* For a complete list of superalloys, both wrought and cast, experimental and under development, see *ASTM Spec. Pub.* 170, "Compilation of Chemical Compositions and Rupture Strengths of Superstrength Alloys."

† Waspaloy has a similar composition (except for lower Mo content), and similar properties.

‡ Max.

high strength at high temperatures. These alloys are based on nickel and/or cobalt, to which are added (typically) chromium for oxidation resistance and a complex of other elements which contribute to hot strength, both by solid solution hardening and by forming relatively stable dispersions of fine particles. Hardening by cold work, hardening by precipitation-hardening heat treatments, and hardening by deliberately arranging for slow precipitation during service are all methods used to enhance the properties of these alloys. **Fabrication** of these alloys is difficult since they are designed to resist distortion even at elevated temperatures. Forging temperatures of about 2,300°F (1,260°C) are used with small reductions and slow rates of working. Many of these alloys are fabricated by precision casting. Cast alloys that are given a strengthening heat treatment often have better properties than wrought alloys, but the shapes that can be made are limited.

Vacuum melting is an important factor in the production of superalloys. The advantages which result from its application include the ability to melt higher percentages of reactive metals, improved mechanical properties (particularly fatigue strength), decreased scatter in mechanical properties, and improved billet-to-bar stock-conversion ratios in wrought alloys.

Table 4.4.31 to 4.4.34 list compositions and properties of some superalloys, and Fig. 4.4.7 indicates the temperature dependence of the rupture strength of a number of such alloys.

Cast tool alloys are another important group of materials having high-temperature strength and wear resistance. They are principally alloys of cobalt, chromium, and tungsten. They are hard and brittle, and they must be cast and ground to shape. Their most important application is for hard facing, but they compete with high-speed steels and

Table 4.4.32 Wrought Superalloys, Properties

Alloy		Max temp under load		Coef of thermal expansion, in/(in · °F) × 10^{-6}(70-1,200°F)	Specific gravity	Stress to rupture, ksi			Short-time tensile properties, ksi							
						At 1,200°F	At 1,500°F		Yield strength 0.2% offset				Tensile strength			
UNS	Name	°F	°C			1,000 h	100 h	1,000 h	Room	1,200°F	1,500°F	1,800°F	Room	1,200°F	1,500°F	1,800°F
K63198	19-9 DL	1,200	649	9.7	7.75	38	17	10	115	39	30		140	75	33	13
	Timken	1,350	732	9.25	8.06	36	13.5	9	96*	70*	16*		134	90	40	18
R30155	L.C. N-155	1,400	760	9.4	8.2	48	20.0	15	53	40	33	12	115	53	35	19
	S-590	1,450	788	8.0	8.34	40	19.0	15	78*	70	47	20	140	82	60	22
R30816	S-816	1,500	816	8.3	8.66	53	29.0	18	63*	45	41		140	112	73	25
N07080	Nimonic alloy 80A	1,450	788	7.56		56	24	15	80*	73	47		150	103	69	
—	K-42-B	1,400	760	8.5	8.23	38	22	15	105	84	52		158	117	54	
N10001	Hastelloy alloy B	1,450	788	6.9	9.24	36	17	10	58	42			135	94	66	24
N07252	M 252	1,500	816	7.55	8.25	70	26	18	90	75	65		160	140	80	20
—	J 1570	1,600	871	8.42	8.66	84	34	24	81	70	71	11	152	135	82	20
—	HS-R235	1,600	871	8.34	7.88	70	35	23	100	90	80	22	170	145	83	25
N06002	Hastelloy alloy X	1,350	732			30	14	10	56	41	37		113	83	52	15
R30605	HS 25 (L605)	1,500	816			54	22	18	70	35				75	50	23
K66286	A-286	1,350	732			45	13.8	7.7								
N07718	Inconel alloy 718	1,400	760	7.2	8.19	54	19	13	152	126			186	149		
N07750	Inconel alloy X-750	1,300	704	7.0	8.25				92	82	45		162	120	52	

* 0.02% offset.

Table 4.4.33 Cast Superalloys, Nominal Composition

	Chemical composition, %										
Common designation	C	Mn	Si	Cr	Ni	Co	Mo	W	Nb	Fe	Other*
Vitallium (Haynes 21)	0.25	0.6	0.6	27	2	Bal	6	—	—	1	—
61 (Haynes 23)	0.4	0.6	0.6	26	1.5	Bal	—	5	—	1	—
422-19 (Haynes 30)	0.4	0.6	0.6	26	16	Bal	6	—	—	1	—
X-40 (Haynes 31)	0.4	0.6	0.6	25	10	Bal	—	7	—	1	—
S-816	0.4	0.6	0.6	20	20	Bal	4	4	4	5	—
HE 1,049	0.45	0.7	0.7	25	10	45	—	15	—	1.5	0.4B
Hastelloy alloy C	0.10	0.8	0.7	16	56	1	17	4	—	5.0	0.3V
B 1,900 + Hf	0.1	—	—	8	Bal	10	6	—	—	—	6 Al 4 Ta 1 Ti 1 5 Hf
MAR-M 247	0.15	—	—	8.5	Bal	10	0.5	10	—	—	5.5 Al 3Ta 1 Ti 1.5 Hf
René 80	0.15	—	—	14	Bal	9.5	4	4	—	—	3 Al 5 Ti
MO-RE 2	0.2	0.3	0.3	32	48	—	—	15	—	Bal	1 Al

* Trace elements not shown.

cemented carbides for many applications, in certain instances being superior to both. (See Cemented Carbides, Sec. 4.4.) Typical compositions are given in Table 4.4.35; typical properties are: elastic modulus = 35×10^6 lb/in^2 (242 GPa); specific gravity = 8.8; hardness, BHN at room temp = 660; 1,500°F (815°C), 435, and 2,000°F (1,093°C) 340.

For still higher-temperature applications, the possible choices are limited to a few metals with high melting points, all characterized by limited availability and by difficulty of extraction and fabrication. Advanced alloys of metals such as chromium, niobium, molybdenum, tantalum, and tungsten are still being developed. Table 4.4.36 cites properties of some of these alloys. Wrought tool steel materials are described in "Tool Steels," ASM Handbook, vol. 3, 9th ed., pp. 421-447.

Chromium with room-temperature ductility has been prepared experimentally as a pure metal and as strong high-temperature alloys. It is not generally available on a commercial basis. Large quantities of chromium are used as melt additions to steels, stainless steels, and nickel-base alloys.

Molybdenum is similar to tungsten in most of its properties. It can be prepared in the massive form by powder metallurgy techniques, by inert-atmosphere or vacuum-arc melting. Its most serious limitation is its ready formation of a volatile oxide at temperatures of 1,400°F approx. In the worked form, it is inferior to tungsten in melting point, tensile strength, vapor pressure, and hardness, but in the recrystallized condition, the ultimate strength and elongation are higher. Tensile strengths up to 350,000 lb/in^2 (2,413 MPa) have been reported for hard drawn wire, and to 170,000 lb/in^2 (1,172 MPa) for soft wire. In the hard-drawn condition, molybdenum has an elongation of 2 to 5 percent, but after recrystallizing, this increases to 10 to 25 percent. Young's modulus is 50×10^6 lb/in^2 (345 GPa). It costs about the same as tungsten, per pound, but its density is much less. It has considerably better forming properties than tungsten and is extensively used for anodes, grids, and supports in vacuum tubes, lamps, and X-ray tubes. Molybdenum is generally used for winding electric furnaces for temperatures up to 3,000°F (1,649°C). As it must be protected against oxidation, such furnaces are usually operated in hydrogen. It is the common material for cathodes for radar devices, heat radiation shields, rocket nozzles and other missile components, and hot-working tools and die-casting cores in the metalworking industry. Its principal use is still as an alloying addition to steels, especially tool steels and high-temperature steels.

Molybdenum alloys have found increased use in aerospace and commercial structural applications, where their high stiffness, microstructural stability, and creep strength at high temperatures are required. They are generally stronger than niobium alloys but are not ductile in the welded condition. Molybdenum alloys are resistant to alkali metal corrosion.

Niobium (also known as **columbium**) is available as a pure metal and as the base of several niobium alloys. The alloys are used in nuclear applications at temperatures of 1,800 to 2,200°F (980 to 1,205°C) because of their low thermal neutron cross section, high strength, and good corrosion resistance in liquid or gaseous alkali metal atmospheres. Niobium alloys are used in leading edges, rocket nozzles, and guidance structures for hypersonic flight vehicles. A niobium-46.5 percent titanium alloy is used as a low-temperature superconductor material for magnets in magnetic resonance imaging machines.

Tantalum has a melting point that is surpassed only by tungsten. Its early use was as an electric-lamp filament material. It is more ductile than molybdenum or tungsten; the elongation for annealed material may be as high as 40 percent. The tensile strength of annealed sheet is about 50,000 lb/in^2 (345 MPa). In this form, it is used primarily in electronic capacitors. It is also used in chemical-processing equipment, where its high rate of heat transfer (compared with glass or ceramics) is particularly important, although it is equivalent to glass in corrosion resistance. Its corrosion resistance also makes it attractive for surgical implants. Like tungsten and molybdenum, it is prepared by powder metallurgy techniques; thus, the size of the piece that can be fabricated is limited by the size of the original pressed compact. Tantalum carbide is used in cemented-carbide tools, where it decreases the tendency to seize and crater. The stability of the anodic oxide film on tantalum leads to rectifier and capacitor applications. Tantalum and its **alloys** become competitive with niobium at temperatures above 2,700°F (1,482°C). As in the case of niobium, these materials are resistant to liquid or vapor metal corrosion and have excellent ductility even in the welded condition.

Tungsten has a high melting point, which makes it useful for high temperature structural applications. The massive metal is usually prepared by powder metallurgy from hydrogen-reduced powder. As a metal, its chief use is as filaments in incandescent lamps and electronic tubes, since its vapor pressure is low at high temperatures. Tensile strengths over 600,000 lb/in^2 (4,136 MPa) have been reported for fine tungsten wires; in larger sizes (0.040 in), tensile strength is only 200,000 lb/in^2 (1,379 MPa). The hard-drawn wire has an elongation of 2 to 4 percent, but the recrystallized wire is brittle. Young's modulus is 60 million lb/in^2 (415 GPa). It can be sealed directly to hard glass and so is used for lead-in wires. A considerable proportion of tungsten rod and sheet is used for electrical contacts in the form of disks cut from rod or sheet and brazed to supporting elements. The major part of the tungsten used is made into ferroalloys for addition to steels or into tungsten carbide for cutting tools. Other applications include elements of electronic tubes, X-ray tube anodes, and arc-welding electrodes. **Alloys** of tungsten that are commercially available are W-3 Re, W-25 Re, and thoriated tungsten. The rhenium-bearing alloys are more ductile than unalloyed tungsten at room temperature. Thoriated tungsten is stronger than unalloyed tungsten at temperatures up to the recrystallization temperature.

Table 4.4.34 Cast Superalloys, Properties

Common designation	Max temp under load		Coef of Thermal expansion in/(in · °F) $3\ 10^{-6}$ (70-1,200°F)	Specific gravity	Stress to rupture in 1,000 h, ksi			Short-time tensile properties, ksi							
								Yield strength 0.2% offset				Tensile strength			
	°F	°C			1,200°F	1,500°F	1,800°F	Room 68°F (20°C)	1,200°) (650°C)	1,500°F (816°C)	1,800°F (982°C)	Room 68°F (20°C)	1,200°F (650°C)	1,500°F (816°C)	1,800°F (982°C)
Vitallium (Haynes 21)	1,500	816	8.35	8.3	44	14	7	82	71	49		101	89	59	33
61(Haynes 23)	1,500	816	8.5	8.53	47	22	5.5	58	74	40	33	105	97	58	45
422-19 (Haynes 30)	1,500	816	8.07	8.31		21	7	55	37*	48		98	59*	64	37
N-40 (Haynes 31)	1,500	816	8.18	8.60	46	23	9.8	74	37*	44		101	77*	59	29
S-816	1,500	816	8.27	8.66	46	21	9.8					112			
HE 1049	1,650	899		8.9	75	35	7	80	72	62	36	90	82	81	51
Hastelloy alloy C			7.73	8.91	42.5	14.5	1.4	54	50			130	87	51	19
B 1,900 + Hf	1,800	982	7.7	8.22		55	15	120	134	109	60	140	146	126	80
MAR-M 247	1,700	927		8.53			18	117	117	108	48	140	152	130	76
René 80	1,700	927		8.16			15	124	105	90		149	149	123	

* Specimen not aged.

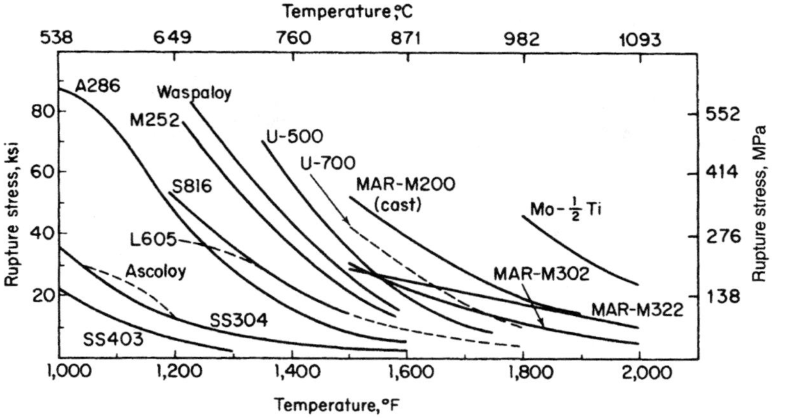

Figure 4.4.7 Temperature dependence of strengths of some high-temperature alloys; stress to produce rupture in 1,000 h.

METALS AND ALLOYS FOR NUCLEAR ENERGY APPLICATIONS

by L. D. Kunsman and C. L. Carlson, Amended by Staff

REFERENCES: Glasstone, "Principles of Nuclear Reactor Engineering," Van Nostrand. Hausner and Roboff, "Materials for Nuclear Power Reactors," Reinhold. Publications of the builders of reactors (General Electric Co., Westinghouse Corp., General Atomic Co.) and suppliers of metal used in reactor components. See also Sec. 7.8 commentary and other portions of Sec. 4.4.

The advent of atomic energy not only created a demand for new metals and alloys but also focused attention on certain properties and combinations of properties which theretofore had been of little consequence. Reactor technology requires special materials for fuels, fuel cladding, moderators, reflectors, controls, heat-transfer mediums, operating mechanisms, and auxiliary structures. Some of the properties pertinent to such applications are given in a general way in Table 4.4.37. In general, outside the reflector, only normal engineering requirements need be considered.

Nuclear Properties A most important consideration in the design of a nuclear reactor is the control of the number and speed of the neutrons resulting from fission of the fuel. The designer must have knowledge of the effectiveness of various materials in slowing down neutrons or in capturing them. The slowing-down power depends not only on the relative energy loss per atomic collision but also on the number of collisions per second per unit volume. The former will be larger, the lower the atomic weight, and the latter larger, the greater the atomic density and the higher the probability of a scattering collision. The effectiveness of a moderator is frequently expressed in terms of the moderating ratio, the ratio of the slowing-down power to the capture cross section. The capturing and scattering tendencies are measured in terms of nuclear cross section in barns (10^{-24} cm^2). Data for some of the materials of interest are given in Tables 4.4.38 to 4.4.40. The absorption data apply to slow, or thermal, neutrons; entirely different cross sections obtain for fast neutrons, for which few materials have significantly high capture affinity. Fast neutrons have energies of about 10^6 eV, while slow, or thermal, neutrons have energies of about

Table 4.4.35 Typical Compositions of Cast Tool Alloys

Name	C	Cr	W	Co	Fe	V	Ta	B	Other
Rexalloy	3	32	20	45					
Stellite 98 M2	3	28	18	35	9	4	0.1	0.1	3 Ni
Tantung G-2	3	15	21	40	. . .	. . .	19	0.2	
Borcoloy no. 6	0.7	5	18	20	Bal	1.3	. . . .	0.7	6 Mo
Colmonoy WCR 100	—	10	15	. . .	Bal	. . .	. . . .	3	

Table 4.4.36 Properties of Selected Refractory Metals

Alloy	Melting temp, °F	Density, lb/in^3 at 75°F	Elastic modulus at 75°F	Ductile-to-brittle transition temp, °F	Tensile strength, ksi (recrystallized condition)				
					75°F	1,800°F	2,200°F	3,000°F	3,500°F
Chromium	3,450	0.760	42	625		12	8		
Columbium	4,474	0.310	16	−185	45	12	10		
F48 (15 W, 5 Mo, 1 Zr, 0.05 C)			25		121	75	50		
Cb74 or Cb752 (10 W, 2 Zr)		0.326			84	60	36		
Molybdenum	4,730	0.369	47	85	75	34	22	6	3
Mo (1/2 Ti)					110	70	48	8	4
TZM (0.5 Ti, 0.08 Zr)					130	85	70	14	5
Tantalum	5,425	0.600	27	< −320	30	22	15	8	4
Ta (10 W)						65	35	12	8
Rhenium	5,460	0.759	68	< 75	170	85	60	25	11
Tungsten	6,170	0.697	58	645	85	36	32	19	10
W (10 Mo)								28	10
W (2 ThO_2)							40	30	25

Table 4.4.37 Requirements of Materials for Nuclear Reactor Components

Component	Neutron absorption cross section	Effect in slowing neutrons	Strength	Resistance to radiation damage	Thermal conductivity	Corrosion resistance	Cost	Other	Typical material
Moderator and reflector	Low	High	Adequate	—	—	High	Low	Low atomic wt	H_2O graphite, Be
Fuel*	Low	—	Adequate	High	High	High	Low	—	U, Th, Pu
Control rod	High	—	Adequate	Adequate	High	—	—	—	Cd, B_4C
Shield	High	High	High	—	—	—	—	High γ radiation absorption	Concrete
Cladding	Low	—	Adequate	—	High	High	Low	—	Al, Zr, stainless steel
Structural	Low	—	High	Adequate	—	High	—	—	Zr, stainless steel
Coolant	Low	—	—	—	High	—	—	Low corrosion rate, high heat capacity	H_2O, Na, NaK, CO_2 He

* U, Th, and Pu are used as fuels in the forms of metals, oxides, and carbides.

Table 4.4.38 Moderating Properties of Materials

Moderator	Slowing-down power, cm^{-1}	Moderating ratio
H_2O	1.53	70
D_2O	0.177	21,000
He	1.6×10^{-5}	83
Be	0.16	150
BeO	0.11	180
C (graphite)	0.063	170

SOURCE: Adapted from Glass, "Principles of Nuclear Reactor Engineering, Van Nostrand.

Table 4.4.39 Slow-Neutron Absorption by Structural Materials

Material	Relative neutron absorption per $cm^3 \times 10^3$	Relative neutron absorption for pipes of equal strength, 68°F (20°C)	Melting point °F	Melting point °C
Magnesium	3.5	10	1,200	649
Aluminum	13	102	1,230	666
Stainless steel	226	234	2,730	1,499
Zirconium	12.6	16	3,330	1,816

SOURCE: Leeser, *Materials & Methods,* **41,** 1955, p. 98.

Table 4.4.40 Slow-Neutron Absorption Cross Sections

Low		Intermediate		High	
Element	Cross section, barns	Element	Cross section, barns	Element	Cross section, barns
Oxygen	0.0016	Zinc	1.0	Manganese	12
Carbon	0.0045	Columbium	1.2	Tungsten	18
Beryllium	0.009	Barium	1.2	Tantalum	21
Fluorine	0.01	Strontium	1.3	Chlorine	32
Bismuth	0.015	Nitrogen	1.7	Cobalt	35
Magnesium	0.07	Potassium	2.0	Silver	60
Silicon	0.1	Germanium	2.3	Lithium	67
Phosphorus	0.15	Iron	2.4	Gold	95
Zirconium	0.18	Molybdenum	2.4	Hafnium	100
Lead	0.18	Gallium	2.8	Mercury	340
Aluminum	0.22	Chromium	2.9	Iridium	470
Hydrogen	0.32	Thallium	3.3	Boron	715
Calcium	0.42	Copper	3.6	Cadmium	3,000
Sodium	0.48	Nickel	4.5	Samarium	8,000
Sulfur	0.49	Tellurium	4.5	Gadolinium	36,000
Tin	0.6	Vanadium	4.8		
		Antimony	5.3		
		Titanium	5.8		

SOURCE: Leeser, *Materials & Methods,* **41,** 1955, p. 98.

Table 4.4.41 Resistance of Materials to Liquid Sodium and NaK*

	Good	Limited	Poor
< 1,000°F	Carbon steels, low-alloy steels, alloy steels, stainless steels, nickel alloys, cobalt alloys, refractory metals, beryllium, aluminum oxide, magnesium oxide, aluminum bronze	Gray cast iron, copper, aluminum alloys, magnesium alloys, glasses	Sb, Bi, Cd, Ca, Au, Pb, Se, Ag, S, Sn, Teflon
1,000-1,600°F	Armco iron, stainless steels, nickel alloys,† cobalt alloys, refractory metals‡	Carbon steels, alloy steels, Monel, titanium, zirconium, beryllium, aluminum oxide, magnesium oxide	Gray cast iron, copper alloys, Teflon, Sb, Bi, Cd, Ca, Au, Pb, Se, Ag, S, Sn, Pt, Si, Magnesium alloys

* For more complete details, see "Liquid Metals Handbook."
† Except Monel.
‡ Except titanium and zirconium.

2.5×10^{-2} eV. Special consideration must frequently be given to the presence of small amounts of high cross-section elements such as Co or W which are either normal incidental impurities in nickel alloys and steels or important components of high-temperature alloys.

Effects of Radiation Irradiation affects the properties of solids in a number of ways: dimensional changes; decrease in density; increase in hardness, yield, and tensile strengths; decrease in ductility; decrease in electrical conductivity; change in magnetic susceptibility. Another consideration is the activation of certain alloying elements by irradiation. Tantalum^{181} and Co^{60} have moderate radioactivity but long half-lives; isotopes having short lives but high-activity levels include Cr^{51}, Mn^{56}, and Fe^{59}.

Metallic Coolants The need for the efficient transfer of large quantities of heat in a reactor has led to use of several metallic coolants. These have raised new problems of pumping, valving, and corrosion. In addition to their thermal and flow properties, consideration must also be given the nuclear properties of prospective coolants. Extensive thermal, flow, and corrosion data on metallic coolants are given in the "Liquid Metals Handbook," published by USNRC. The resistances of common materials to liquid sodium and NaK are given qualitatively in Table 4.4.41. Water, however, remains the most-used coolant; it is used under pressure as a single-phase liquid, as a boiling two-phase coolant, or as steam in a superheat reactor.

Fuels There are, at present, only three fissionable materials, U^{233}, U^{235}, and Pu^{239}. Of these, only U^{235} occurs naturally as an isotopic "impurity" with natural uranium U^{238}. Uranium^{233} may be prepared from natural thorium, and Pu^{239} from U^{238} by neutron bombardment. Both uranium and thorium are prepared by conversion of the oxide to the tetrafluoride and subsequent reduction to the metal.

The properties of **uranium** are considerably affected by the three allotropic changes which it undergoes, the low-temperature forms being highly anisotropic. The strength of the metal is low (see Table 4.4.42) and decreases rapidly with increasing temperature. The corrosion resistance is also poor. Aluminum-base alloys containing uranium-aluminum intermetallic compounds have been used to achieve improved properties. **Thorium** is even softer than uranium but is very ductile. Like uranium it corrodes readily, particularly at elevated temperatures. Owing to its crystal structure, its properties are isotropic. **Plutonium** is unusual in possessing six allotropic forms, many of which have anomalous physical properties. The electrical resistivity of the alpha phase is greater than that for any other metallic element. Because of the poor corrosion resistance of nuclear-fuel metals, most fuels in power reactors today are in the form of oxide UO_2, ThO_2, or PuO_2, or mixtures of these. The carbides UC and UC_2 are also of interest in reactors using liquid-metal coolants.

Control-Rod Materials Considerations of neutron absorption cross sections and melting points limit the possible control-rod materials to a very small group of elements, of which only four—boron, cadmium, hafnium, and gadolinium—have thus far been prominent. **Boron** is a very light metal of high hardness (~3,000 Knoop) prepared by thermal or hydrogen reduction of BCl_3. Because of its high melting point, solid shapes of boron are prepared by powder-metallurgy techniques. Boron has a very high electrical resistivity at room temperature but becomes conductive at high temperatures. The metal in bulk form is oxidation

Table 4.4.42 Mechanical Properties of Metals for Nuclear Reactors

	Room temp						Elevated temp*		
	Longitudinal			Transverse					
Material	Yield strength, ksi	Ult strength, ksi	Elong,† %	Ult str, ksi	Elong,† %	Charpy impact ft . lb	°F	Ult str, ksi	Elong. %
Beryllium:									
Cast, extracted and annealed		40	1.82	16.6	0.18		392	62	23.5
Flake, extruded and annealed		63.7	5.0	25.5	0.30		752	43	29
Powder, hot-extruded	39.5	81.8	15.8	45.2	2.3	4.1	1,112	23	8.5
Powder, vacuum hot-pressed	32.1	45.2	2.3	45.2	2.3	0.8	1,472	5.2	10.5
Zirconium:									
Kroll-50% CW		82.6				14.8	250	32	
Kroil-annealed		49.0					500	23	
Iodide	15.9	35.9	31			2.5-6.0	700	17	
							900	12	
							1,500	3	
Uranium	25	53	<10			15	302	27	
							1,112	12	
Thorium	27	37.5	40				570	22	
							900	17.5	

* All elevated-temperature data on beryllium for hot-extruded powder; on zirconium for iodide material.
† Beryllium is extremely notch-sensitive. The tabulated data have been obtained under very carefully controlled conditions, but ductility values in practice will be found in general to be much lower and essentially zero in the transverse direction.

resistant below 1,800°F (982°C) but reacts readily with most halogens at only moderate temperatures. Rather than the elemental form, boron is generally used as boron steel or as the carbide, oxide, or nitride. **Cadmium** is a highly ductile metal of moderate hardness which is recovered as a by-product in zinc smelting. Its properties greatly resemble those of zinc. The relatively low melting point renders it least attractive as a control rod of the four metals cited. **Hafnium** metal is reduced from hafnium tetrachloride by sodium and subsequently purified by the iodide hot-wire process. It is harder and less readily worked than zirconium, to which it is otherwise very similar in both chemical and physical properties. Hafnium reacts easily with oxygen, and its properties are sensitive to traces of most gases. The very high absorption cross section of **gadolinium** renders it advantageous for fast-acting control rods. This metal is one of the rare earths and as yet is of very limited availability. It is most frequently employed as the oxide.

Beryllium Great interest attaches to beryllium because it is unique among the metals with respect to its very low neutron-absorption cross section and high neutron-slowing power. It may also serve as a source of neutrons when subjected to alpha-particle bombardment. Beryllium is currently prepared almost entirely by magnesium reduction of the fluoride, although fused-salt electrolysis is also practicable and has been used. The high affinity of the metal for oxygen and nitrogen renders its processing and fabrication especially difficult. At one time, beryllium was regarded as almost hopelessly brittle. Special techniques, such as vacuum hot pressing, or vacuum casting followed by hot extrusion of clad slugs, led to material having marginally acceptable, though highly directional, mechanical properties. The extreme toxicity of beryllium powder necessitates special precautions in all operations.

Zirconium Zirconium's importance in nuclear technology derives from its low neutron-absorption cross section, excellent corrosion resistance, and high strength at moderate temperatures. The metal is produced by magnesium reduction of the tetrachloride (Kroll process). The Kroll product is usually converted to ingot form by consumable-electrode-arc melting. An important step in the processing for many applications is the difficult chemical separation of the 1 to 3 percent hafnium with which zirconium is contaminated. Unless removed, this small hafnium content results in a prohibitive increase in absorption cross section from 0.18 to 3.5 barns. The mechanical properties of zirconium are particularly sensitive to impurity content and fabrication technique. In spite of the high melting point, the mechanical properties are poor at high temperatures, principally because of the allotropic transformation at 1,585°F (863°C). A satisfactory annealing temperature is 1,100°F (593°C). The low-temperature corrosion behavior is excellent but is seriously affected by impurities. The oxidation resistance at high temperatures is poor. Special alloys (Zircoloys) have been developed having greatly improved oxidation resistance in the intermediate-temperature range. These alloys have a nominal content of 1.5 percent Sn and minor additions of iron-group elements.

Stainless Steel Although stainless steel has a higher thermal neutron-absorption cross section than do the zirconium alloys, its good corrosion resistance, high strength, low cost, and ease of fabrication make it a strong competitor with zirconium alloys as a fuel-cladding material for water-cooled power-reactor applications. Types 348 and 304 stainless are used in major reactors.

MAGNESIUM AND MAGNESIUM ALLOYS

by Francis H. (Sam) Froes

REFERENCES: Annual Book of Standards, ASTM. Publications of the Dow Chemical Co, G.V. Raynor, The Physical Metallurgy of Magnesium and Its Alloys, Pergamon, 1959, C.S. Roberts Magnesium and Its Alloys, Wiloey, 1960, E.F. Emley, Principles of Magnesium Technology, Pergamon, 1966, Ian Polmear et al. "Light Alloys" 4th ed., Elsevier, 2017, Chapter 5, pp. 237-297. ASM Handbook. Vol. 2, 10th ed., "Properties and Selection: Nonferrous Alloys and Special-Purpose Materials," 1990, pp. 455-516. Karl U. Kainer "Magnesium Alloys and Technology," Wiley, 2003. Horst E. Friedrich and Barry L. Mordike "Magnesium Technology," Springer, 2006. Mihriban O. Pekguleryuz, Karl U. Kainer and Ali Kaya "Fundamentals of Magnesium Alloy Metallurgy," Woodhead Publishing Ltd, 2013.

Magnesium is used in many chemical compounds and biological applications, including use as a laxative and an antacid. However, this text will concentrate on the use as a metal in structural applications. Magnesium is the lightest metal of structural importance [108 lb/ft^3 (1.740 g/cm^3)]. However, a disadvantage of Magnesium is that it has a hcp structure resulting in low room temperature ductility, so that processing must be performed at greater than 200°F (93°C). The principal uses for **pure magnesium** are in aluminum alloys, steel desulfurization, and production of nodular iron. Principal uses for **magnesium alloys** are die castings (for automotive and computer applications), wrought products, and, to a lesser extent, gravity castings (usually for aircraft and aerospace components). Because of its chemical reactivity, magnesium can be used in pyrotechnic material and for sacrificial galvanic protection of other metals. Since magnesium in its molten state reacts with the oxygen in air, a protective atmosphere containing sulfur hexafluoride is employed as a controlled atmosphere.

Commercially pure magnesium contains 99.8 percent magnesium and is produced by extraction from seawater or by reduction from magnesite and dolomite ores. Prior to the mid-1990s, the world magnesium production was dominated by the electrolytic processes with the United States as the dominant supplier. As of 2005, there is a single US producer (Utah, US Magnesium). The US produced about 45 out of a 615 kton/yr (7%), compared to 140 out of 311 kton/yr (45%) in 1995. In the same year, China produced 400 out of the 615 kton/yr (65%) using the Pidgeon Process, compared to 12 out of 311 kton/yr (4%) in 1995. The basic chemical equation of the Pidgeon silicothermic reduction process, in which dolomite and ferrosilico are reacted together, is:

$$Si(s) + 2\,MgO(s) \leftrightarrow SiO_2(s) + 2\,Mg(g) \text{ (high temperature, distillation boiling zone)}$$

Chief impurities are iron, silicon, manganese, and aluminum.

The major use of magnesium, consuming about 50 percent of total magnesium production is as a component of aluminum alloys for beverage can stock. Another use for primary magnesium is in steel desulfurization. Magnesium, usually mixed with lime or calcium carbide, is injected beneath the surface of molten steel to reduce the sulfur content.

A growing application for magnesium alloys is die cast automotive components. Their light weight can be used to help achieve reduced fuel consumption, while high-ductility alloys are used for interior components which must absorb impact energy during collisions. These alloys can be cast by hot or cold chamber die-casting methods, producing a diverse group of components ranging from gear cases to steering wheel frames. Some vehicles use magnesium die castings as structural members to serve as instrument panel support beams and steering columns. Designs employing magnesium must account for the relatively low value of the modulus of elasticity (6.5×10^6 lb/in^2) and high thermal coefficient of expansion [14×10^{-6} per °F at 32°F (0°C) and 16×10^{-6} for 68 to 752°F (20 to 400°C)]. (See Tables 4.4.43 and 4.4.44.)

The development of jet engines and high-velocity aircraft, missiles, and spacecraft led to the development of magnesium alloys with improved **elevated-temperature properties**. These alloys employ the addition of some combination of rare earth elements, yttrium, manganese, zirconium, or, in the past, thorium (undesirable because it is radioactive). Such alloys have extended the useful temperatures at which magnesium can serve in structural applications to as high as 700 to 800°F.

Magnesium alloy forgings are used for applications requiring properties superior to those obtainable in castings. They are generally **press-forged.** Alloy AZ61 is a general-purpose alloy, while alloys AZ80A-T5 and ZK60A-T5 are used for the highest-strength press forgings of simple design. These alloys are aged to increase strength.

A wide range of extruded shapes are available in a number of compositions. Alloys AZ31B, AZ61A, AZ80A-T5, and ZK60A-T5 increase in cost and strength in the order named. **Impact extrusion** produces smaller, symmetric tubular parts.

Sheet is available in several alloys (see Table 4.4.44), in both the soft (annealed) and the hard (cold-rolled) tempers. Magnesium alloy sheet usually is hot-formed at temperatures between 400 and 650°F, although simple bends of large radius are made cold. With appropriate processing Magnesium alloys can be Superplastically Formed, Rapidly Solidified, Additively Manufactured, semi-solid processed, and injection molded.

Table 4.4.43 Typical Mechanical Properties of Magnesium Casting Alloys

Alloy	Condition or temper*	Nominal composition, % (balance Mg pius trace elements) Al	Zn	Mn, min	Zr	Other	Tensile strength, ksi	Tensile yield strength, ksi	Elongation, %, in 2 in	Shear strength, ksi	Strength, ksi Tensile	Yield	Hardness BHN	Electrical conductivity, % IACS‡
Permanent mold and sand casting alloys:														
AM100A	–T6	10.0		0.10			40	22	1	22			70	14
AZ63A	–F	6.0	3.0	0.15			29	14	6	18	60	40	50	14
	–T4						40	13	12	17	60	44	55	12
	–T5						30	14	4	17	60	40	55	
	–T6						40	19	5	20	75	52	73	15
AZ81A	–T4	7.5	0.7	0.13			40	12	15	17	60	44	55	12
AZ91E	– F	8.7	0.7	0.17			24	14	2	18	60	40	52	13
	–T4						40	12	14	17	60	44	53	11
	–T5						26	17	3					
	–T6						40	19	5	20	75	52	66	13
AZ92A	–F	9.0	2.0	0.10			24	14	2	18	50	46	65	12
	–T4						40	14	9	20	68	46	63	10
	–T5						26	16	2	19	50	46	70	
	–T6						40	21	2	22	80	65	84	14
EZ33A	–T5		2.5		0.8	3.5 RE†	23	15	3	22	57	40	50	25
K1A	–F				0.7		25	7	19	8				31
QE22A	–T6				0.7	2.0 RE,† 2.5 Ag	40	30	4	23				25
ZE41A	–T5		4.0		0.7	1.3 RE†	30	20	3.5	22	70	51	62	31
ZK51A	–T5		4.5		0.8		40	24	8	22	72	47	65	27
ZK61A	–T6		6.0		0.8		45	28	10	26			70	27
Die casting alloys:														
A291D	–F	9.0	0.7	0.15			34	23	3	20			75	
AM60B	–F	6.0	0.22 max	0.24			32	19	6-8				62	
AM50A	–F	5.0	0.22 max	0.26			32	18	8-10				57	
AS41B	–F	4.2	0.12 max	0.35		1.0 Si	31	18	6				75	
AE42X1	–F	4.0	0.22 max	0.25		2.4 RE†	33	20	8-10				57	

ksi × 6.895 = MPa
* –F = as cast; –T4 = artificially aged; –T5 = solution heat-treated; –T6 = solution heat-treated.
† RE = rare-earth mixture.
‡ Percent electrical conductivity/100 approximately equals thermal conductivity in cgs units.

Joining Magnesium alloys are joined by **riveting** or **welding**. Riveting is most common. Aluminum alloy rivets are used; alloy 5,052 is preferred, to minimize contact corrosion. Other aluminum alloy rivets can be used, but they are not as effective in inhibiting contact corrosion. Rivets should be anodized to prevent contact corrosion. Adhesive bonding is another accepted method for joining magnesium. It saves weight and improves fatigue strength and corrosion resistance.

Arc welding with inert-gas (helium or argon) shielding of the molten metal produces satisfactory joints. Butt joints are preferred, but any type of welded joint permissible for mild steel can be used. After welding, a stress relief anneal is necessary. Typical stress relief anneal times are 15 min at 500°F (260°C) for annealed alloys and 1 h at 400°F (204°C) for cold-rolled alloys.

Machining Magnesium in all forms is a free-machining metal, which is considerably easier to machine than competing metals (see Table below).

Comparative machinability of metals (from *Machining*, Magnesium Elektron Limited Handbook)

Metal	Relative power required*	Rough turning speeds (m s^{-1})	Drilling speeds (5-10 mm drill (m s^{-1})
Magnesium	1	up to 20	2.5-8.5
Aluminium	1.8	1.25-12.5	1-6.5
Cast iron	3.5	0.5-1.5	0.2-0.65
Mild steel	6.3	0.65-3.3	0.25-0.5
Stainless steel	10.0	0.3-1.5	0.1-0.35

* 1 = lowest

Standard tools such as those used for brass and steel can be used with slight modification. Relief angles should be from 78 to 128; rake angles from 08 to 158. High-speed steel is satisfactory and is used for most drills, taps, and reamers. Suitable grades of cemented carbides are better for production work and should be used where the tool design permits it.

Finely divided magnesium constitutes a fire hazard, and good housekeeping in production areas is essential. (See also Sec. 9.4.)

Corrosion Resistance and Surface Protection All magnesium alloys display good resistance to ordinary inland atmospheric exposure, to most alkalies, and to many organic chemicals. High-purity alloys introduced in the mid-1980s demonstrate excellent corrosion resistance (see Fig. 4.4.8) and are used in under-vehicle applications without coatings. However, galvanic couples formed by contact with most other metals, or by contamination of the surface with other metals during fabrication, can cause rapid attack of the magnesium when exposed to salt water. Protective treatments or coated fasteners can be used to isolate galvanic couples and prevent this type of corrosion.

Recently Magnesium Elektron have developed a number of rare earth containing alloys with attractive mechanical properties. WE54 (Mg-1.75 Nd-1.5%RE-5.1Y 0.5Zr) exhibits a YS of 205 MPa and UTS of 280 MPa in the T6 condition, and elevated temperature behavior is superior to many other Mg alloys. Better ductility was found in an alloy with reduced Y and increased Nd, cast WE43B (Mg-4.1Y-2.25Nd-1 Heavy RE-0.4 min Zr), exhibiting a YS of 185 MPa, UTS of 265 MPa, elongation of 7 percent and is useful to 300°C, which has seen applications in racing cars and helicopter transmission castings. However, this alloy has largely been superseded by Elektron® 21 (Mg-0.4Zn-2.9Nd-1.4Gd) which in the T6 condition exhibits a YS of 170 MPa, UTS of

Table 4.4.44 Properties of Wrought Magnesium Alloys

Alloy and temper*	Nominal composition, % (balance Mg plus trace elements)				Physical properties			Room-temperature mechanical properties (typical)							
					Density, lb/in^3	Thermal conductivity, cgs units, 68°F (20°C)	Electrical resistivity, $\mu\Omega \cdot$cm, 68°F (20°C)	Tensile strength, ksi	Tensile yield strength, ksi	Elongation in 2 in, %	Compressive yield strength, ksi	Shear strength, ksi	Bearing strength, ksi		Hardness BHN
	Al	Mn	Zn	Zr									Tensile	Yield	
Extruded bars, rods, shapes															
AZ31B-F	3.0		1.0		0.0639	0.18	9.2	38	29	15	14	19	56	33	49
AZ61A-F	6.5		1.0		0.0647	0.14	12.5	45	33	16	19	22	68	40	60
AZ80A-T5	8.5		0.5		0.0649	0.12	14.5	55	40	7	35	24	60	58	82
ZK60A-F			5.7	0.5	0.0659	0.28	6.0	49	38	14	33	24	76	56	75
-T5			5.7	0.5	0.0659	0.29	5.7	53	44	11	36	26	79	59	82
Extruded tube															
AZ31B-F	3.0		1.0		0.0639	0.18	9.2	36	24	16	12				46
AZ61A-F	6.5		1.0		0.0647	0.14	12.5	41	24	14	16				50
ZK60A-F			5.7	0.5	0.0659	0.28	6.0	47	35	13	25				75
-T5			5.7	0.5	0.0659	0.29	5.7	50	40	11	30				82
M1A		1.2			0.0635	0.31	5.4	37	26	11	12				
Sheet and plate															
AZ31B-H24	3.0		1.0		0.0639	0.18	9.2	42	32	15	26	29	77	47	73
								40	29	17	23	28	72	45	
								39	27	19	19	27	70	40	
AZ31B-0	3.0		1.0		0.0639	0.18	9.2	37	22	21	16	26	66	37	56
M1A		1.2			0.0635	0.31	5.4	37	26	11	12				
Tooling plate															
AZ31B	3.0		1.0		0.0639	0.18	9.2	35	19	12	10				
Tread plate															
AZ31B	3.0		1.0		0.0639	0.18	9.2	35	19	14	11				52

NOTE: For all above alloys: coefficient of thermal expansion = 0.0000145; modulus of elasticity = 6,500,000 lb/in^2; Position's ratio = 0.35.

* Temper: –F = as fabricated; –H24 = strain-hardened, then partially annealed; –O = fully annealed; –T5 = artificially aged.

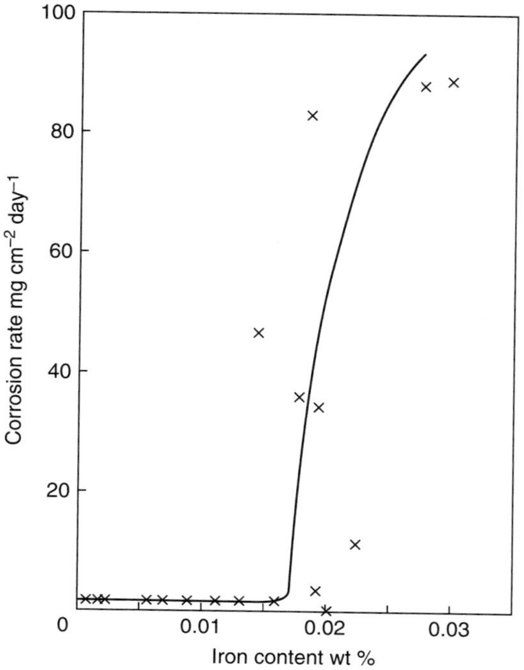

Figure 4.4.8 Effect of iron on corrosion of pure magnesium (alternate test in 3%NaCl) (*from Hanawalt, J. D., Nelson, C. E. and Peloubet, J. A., Trans. AIME, **147**, 273, 1942*).

280 MPa, a ductility of 5 percent and is useful to 200°C whilst demonstrating excellent corrosion resistance and castability. Elektron® 21 has achieved Aerospace Material Specification AMS 4,429 and is the first magnesium alloy ever to achieve full MMPDS MIL-HNDBK-V) design handbook entry, including statistical A & B values and elevated temperature information. Elektron® 21 is specified on aircraft such as the Boeing Apache and the Agusta Westland AW169.

Both Elektron® 21 (also known as ASTM EV31A), a sand casting, and Elektron® 43 (Mg-4.0Y-2.9RE—0.2Zr) (also known as ASTM WE43C), a wrought plate or extrusion alloy, have undergone extensive flammability testing by the Federal Aviation Administration (FAA). The FAA has shown that the use of these Elektron® 21 and Elektron®43 alloys in aircraft interior components (such as seat frames) does not reduce the level of safety of the aircraft when compared to heavier aluminium seat components, that is, a significant and major break-through for Mg alloys.

POWDERED METALS

by Peter K. Johnson

REFERENCES: German, "Powder Metallurgy Science," 2nd ed., Metal Powder Industries Federation, Princeton, NJ. "Powder Metallurgy Design Manual," 2nd ed., Metal Powder Industries Federation.

Powder metallurgy (PM) is an automated manufacturing process to make precision metal parts from metal powders or particulate materials. Basically a net or near-net shape metalworking process, a PM operation usually results in a finished part containing more than 97 percent of the starting raw material. Most PM parts weigh less than 5 1b (2.26 kg), although parts weighing as much as 35 1b (15.89 kg) can be fabricated in conventional PM equipment.

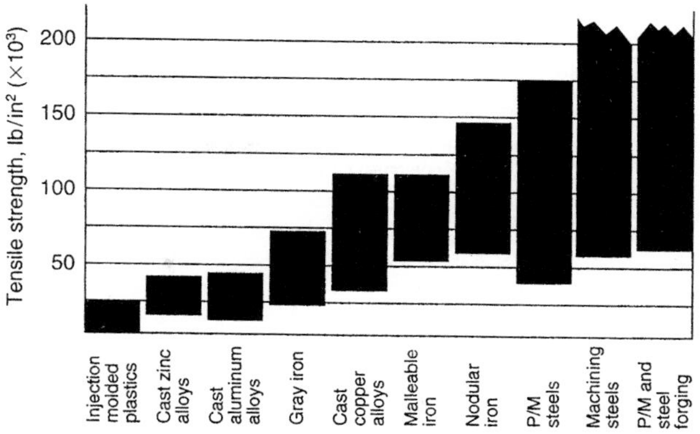

Figure 4.4.9 Comparison of material strengths.

In contrast to other metal-forming techniques, PM parts are shaped directly from powders, whereas castings originate from molten metal, and wrought parts are shaped by deformation of hot or cold metal, or by machining (Fig. 4.4.9). The PM process is cost-effective in manufacturing simple or complex shapes at, or very close to, final dimensions in production rates which can range from a few hundred to several thousand parts per hour. Normally, only a minimum amount of machining is required.

Powder metallurgy predates melting and casting of iron and other metals. Egyptians made iron tools, using PM techniques from at least 3,000 B.C. Ancient Inca Indians made jewelry and artifacts from precious metal powders. The first modern PM product was the tungsten filament for electric light bulbs, developed in the early 1900s. Oil-impregnated PM bearings were introduced for automotive use in the 1920s. This was followed by tungsten carbide cutting tool materials in the 1930s, automobile parts in the 1960s and 1970s, aircraft gas-turbine engine parts in the 1980s, and **powder forged** (PF) and **metal injection molding** (MIM) parts in the 1990s.

PM parts are used in a variety of end products such as lock hardware, garden tractors, snowmobiles, automobile automatic transmissions and engines, auto antilock brake systems (ABS), washing machines, power tools, hardware, firearms, copiers, off-road equipment, hunting knives, hydraulic assemblies, X-ray shielding, oil and gas drilling well-head components, and medical equipment. The typical five- or six-passenger car contains about 44 1b of PM parts, a figure that could increase within the next several years. Iron powder is used as a carrier for toner in electrostatic copying machines. People in the United States consume about 2 million 1b of iron powder annually in iron-enriched cereals and breads. Copper powder is used in electromagnetic imaging (EMI) shielding, environmentally friendly green bullets replacing lead, and metallic pigmented inks for packaging and printing. Aluminum powder is used in solid-fuel booster rockets for the space shuttle program. The spectrum of applications of powdered metals is very wide indeed.

There are five major processes that consolidate metal powders into precision shapes: conventional powder metallurgy (PM) (the dominant sector of the industry), metal injection molding (MIM), powder forging (PF), warm compaction, hot isostatic pressing (HIP), and cold isostatic pressing (CIP).

PM processes use a variety of alloys, giving designers a wide range of material properties (Tables 4.4.45 to 4.4.47). Major alloy groups include powdered iron and alloy steels, stainless steel, bronze, and brass. Powders of aluminum, copper, tungsten carbide, tungsten, and heavy metal alloys, tantalum, molybdenum, superalloys, titanium, high-speed tool steels, and precious metals are successfully fabricated by PM techniques.

Table 4.4.45 Properties of Powdered Ferrous Metals

Material	Density, g/cm³	Tensile strength*		Comments
		MPa	lb/in²	
Carbon steel				
MPIF F-0008	6.9	370†	54,000†	Often cost-effective
0.8% combined carbon (c.c.)	6.9	585‡	85,000‡	
Copper steel				
MPIF FC-0208	6.7	415†	60,000†	Good sintered strength
2% Cu, 0.8% c.c.	6.7	585‡	85,000‡	
Nickel steel				
MPIF FN-0205	6.9	345†	50,000†	Good heat-treated strength, impact energy
2% Ni, 0.5% c.c.	6.9	825‡	120,000‡	
MPIF FN-0405	6.9	385†	56,000†	
4% Ni, 0.5% c.c.	6.9	845‡	123,000‡	
Infiltrated steel				
MPIF FX-1005	7.3	530†	77,000†	Good strength, closed-off internal porosity
10% Cu, 0.5% c.c.	7.3	825‡	120,000‡	
MPIF FX-2008	7.3	550†	80,000†	
20% Cu, 0.8% c.c.	7.3	690‡	100,000‡	
Low-alloy steel, prealloyed Ni, Mo, Mn				
MPIF FL-4,605 HT	6.95	895‡	130,000‡	Good hardenability, consistency in heat treatment
MPIF FL-4,405 HT	7.10	1,100‡	160,000‡	
Stainless steels				
MPIF SS-316 N2 (316 stainless)	6.5	415†	60,000†	Good corrosion resistance, appearance
MPIF SS-410 HT (410 stainless)	6.5	725‡	105,000‡	

* Reference: MPIF Standard 35, Materials Standards for P/M Structural Parts. Strength and density given as typical values.
† As sintered.
‡ Heat-treated.

A PM microstructure can be designed to have controlled microporosity. This inherent advantage of the PM process can often provide special useful product properties. Sound and vibration damping can be enhanced. Components are impregnated with oil to function as self-lubricating bearings, resin impregnated to seal interconnecting microporosity, infiltrated with a lower-melting-point metal to increase strength and impact resistance, and steam-treated to increase corrosion resistance and seal microporosity. The amount and characteristics of the microporosity can be controlled within limits through powder characteristics, powder composition, and the compaction and sintering processes.

Powder metallurgy offers the following advantages to the designer:

1. PM eliminates or minimizes machining.
2. It eliminates or minimizes material losses.
3. It maintains close dimensional tolerances.
4. It offers the possibility of utilizing a wide variety of alloyed materials.
5. PM produces good surface finishes.
6. It provides components that may be heat-treated for increased strength or wear resistance.
7. It provides part-to-part reproducibility.
8. It provides controlled microporosity for self-lubrication or filtration.
9. PM facilitates the manufacture of complex or unique shapes that would be impractical or impossible with other metalworking processes.
10. It is suitable for moderate- to high-volume production.
11. It offers long-term performance reliability for parts in critical applications.

Table 4.4.46 Properties of Powdered Nonferrous Metals

Material	Use and characteristics
Copper	Electrical components
Bronze MPIF CTG-1001 10% tin, 1% graphite	Self-lubricating bearings: @ 6.6 g/cm³ oiled density, 160 MPa (23,000 lb/in²) K strength,* 17% oil content by volume
Brass	Electrical components, applications requiring good corrosion resistance, appearance, and ductility
MPIF CZ-1000 10% zinc	@ 7.9 g/cm³, yield strength 75 MPa (11,000 lb/in²)
MPIF CZ-3000 30% zinc	@7.9 g/cm³, yield strength 125 MPa (18,300 lb/in²)
Nickel silver MPIF CNZ-1818 18% nickel, 18% zinc	Improved corrosion resistance, toughness @ 7.9 g/cm³, yield strength 140 MPa (20,000 lb/in²)
Aluminum alloy	Good corrosion resistance, lightweight, good electrical and thermal conductivity
Titanium	Good strength/mass ratio, corrosion resistance

* Radial crashing constant. See MPIF Standard 35, Materials Standards for P/M Self-Lubricating Bearings.

Metal powders are materials precisely engineered to meet a wide range of performance requirements. Major metal powder production processes include atomization (water, gas, centrifugal); chemical (reduction of oxides, precipitation from a liquid, precipitation from a gas) and thermal; and electrolytic. **Particle shape,** size, and size distribution strongly influence the characteristics of powders, particularly their behavior during die filling, compaction, and sintering. The range of shapes covers highly spherical, highly irregular, flake, dendritic (needlelike), and sponge (porous). Metal powders are classified as elemental, partially alloyed, or prealloyed.

The three basic steps for producing conventional-density PM parts are mixing, compacting, and sintering.

Elemental or prealloyed metal powders are mixed with lubricants and/or other alloy additions to produce a homogeneous mixture.

Table 4.4.47 Properties of Powdered Soft Magnetic Metals

Material	Density, g/cm³	Maximum magnetic induction, kG*	Coercive field, Oe
Iron, low-density	6.6	9.5	1.8
Iron, high-density	7.2	12.0	1.8
Phosphorous iron, 0.45% P	7.0	11.5	1.5
Silicon iron, 3% Si	7.0	12.0	1.2
Nickel iron, 50% Ni	7.0	9.0	0.3
410 Stainless steel	7.1	10.0	2.0
430 Stainless steel	7.1	10.0	2.0

* Magnetic field—15 oersteds (Oe).

In compacting, a controlled amount of mixed powder is automatically gravity-fed into a precision die and is compacted, usually at room temperatures at pressures as low as 10 or as high as 60 or more tons/in^2 (138 to 827 MPa), depending on the density requirements of the part. Normally compacting pressures in the range of 30 to 50 tons/in^2 (414 to 690 MPa) are used. Special mechanical or hydraulic presses are equipped with very rigid dies designed to withstand the extremely high loads experienced during compaction. Compacting loose powder produces a *green compact* which, with conventional pressing techniques, has the size and shape of the finished part when ejected from the die and is sufficiently rigid to permit in-process handling and transport to a sintering furnace. Other specialized compacting and alternate forming methods can be used, such as powder forging, isostatic pressing, extrusion, injection molding, and spray forming.

During **sintering**, the green part is placed on a wide mesh belt and moves slowly through a controlled-atmosphere furnace. The parts are heated to below the melting point of the base metal, held at the sintering temperature, and then cooled. Basically a solid-state diffusion process, sintering transforms compacted mechanical bonds between the powder particles to metallurgical bonds. Sintering gives the PM part its primary functional properties. PM parts generally are ready for use after sintering. If required, parts can also be repressed, impregnated, machined, tumbled, plated, or heat-treated. Depending on the material and processing technique, PM parts demonstrate tensile strengths exceeding 200,000 lb/in^2 (1,379 MPa).

Warm compaction and **surface densification** are growing in use to improve the density and mechanical properties of PM parts. In warm compaction, powders and tooling members are preheated above room temperature. Surface densification by transverse rolling improves the performance of PM parts and highly loaded gears.

Powder forging (P/F) has become a standard technique for forming automotive connecting rods. More than 500 million P/F rods have been made since 1986. The use of MIM parts is growing rapidly. Applications include firearms, surgical tools, orthodontic appliances, electronics, and telecommunications. The **MIM process,** similar to plastic injection molding, injects a mixture of fine powder (less than 20 μm) and a binder into highly complex parts weighing less than 400 g. Rapid prototyping or desktop manufacturing is another PM technique that combines a three-dimensional computer model of the product and a laser sintering system to produce prototype and production parts. Powder is deposited in layers that form three-dimensional shapes. Two major processes are laser free-form manufacturing and selective laser sintering.

Development continues on PM and particulate materials covering conventional steels, stainless steels, high-speed steels, titanium, aluminum, nickel- and cobalt-based superalloys, refractory metals, composites, aluminides, and nanoscale materials.

NICKEL AND NICKEL ALLOYS

by John H. Tundermann

REFERENCES: Mankins and Lamb, Nickel and Nickel Alloys, "ASM Handbook," vol. 2, 10th ed., pp. 428-445. Tundermann et al., Nickel and Nickel Alloys, "Kirk Othmer Encyclopedia of Chemical Technology," vol. 17. Frantz, Low Expansion Alloys, "ASM Handbook," vol. 2, 10th ed., pp. 889-896.

Nickel is refined from sulfide and lateritic (oxide) ores using pyrometallurgical, hydrometallurgical, and other specialized processes. Sulfide ores are used to produce just over 50 percent of the nickel presently used in the world. High-purity nickel (99.7$^+$ percent) products are produced via electrolytic, carbonyl, and powder processes.

Electroplating accounts for about 10 percent of the total annual consumption of nickel. Normal nickel electroplate has properties approximating those of wrought nickel, but special baths and techniques can give much harder plates. Bonds between nickel and the base metal are usually strong. The largest use of nickel in plating is for corrosion protection of iron and steel parts and zinc-base die castings for automotive use. A 0.04- to 0.08-mm (1.5- to 3-mil) nickel plate is covered with a chromium plate only about one hundredth as thick to give a bright, tarnish-resistant, hard surface. Nickel electrodeposits are also used to facilitate brazing of chromium-containing alloys, to reclaim worn parts, and for electroformed parts.

Wrought nickel and nickel alloys are made by several melting techniques followed by hot and cold working to produce a wide range of product forms such as plate, tube, sheet, strip, rod, and wire.

The nominal composition and typical properties of selected commercial nickel and various nickel alloys are summarized in Tables 4.4.48 to 4.4.51.

Commercially pure nickels, known as Nickel 200 and 201, are available as sheet, rod, wire, tubing, and other fabricated forms. They are used where the thermal or electrical properties of nickel are required and where corrosion resistance is needed in parts that have to be worked extensively.

Commercial nickel may be forged or rolled at 871 to 1,260°C (1,600 to 2,300°F). It becomes increasingly harder below 871°C (1,600°F) but has no brittle range. The recrystallization temperature of cold-worked pure nickel is about 349°C (660°F), but commercial nickel recrystallizes at about 538°C (1,100° F) and is usually annealed at temperatures between 538 and 954°C (1,100 and 1,750°F).

The addition of certain elements to nickel renders it responsive to precipitation or aging treatments to increase its strength and hardness. In the unhardened or quenched state, Duranickel alloy 301 fabricates almost as easily as pure nickel, and when finished, it may be hardened by heating for about 8 h at 538 to 593°C (1,000 to 1,100°F). Intermediate anneals during fabrication are at 900 to 954°C (1,650 to 1,750°F). The increase in strength due to aging may be superimposed on that due to cold work.

Nickel castings are usually made in sand molds by using special techniques because of the high temperatures involved. The addition of 1.5 percent silicon and lesser amounts of carbon and manganese is necessary to obtain good casting properties. Data on nickel and nickel alloy castings can be found in Spear, Corrosion-Resistant Nickel Alloy Castings, "ASM Handbook," vol. 3, 9th ed, pp. 175-178.

Nickel has good to excellent resistance to corrosion in caustics and nonoxidizing acids such as hydrochloric, sulfuric, and organic acids, but poor resistance to strongly oxidizing acids such as nitric acid. Nickel is resistant to corrosion by chlorine, fluorine, and molten salts.

Nickel is used as a catalyst for the hydrogenation of organic fats and for many industrial processes. Porous nickel electrodes are used for battery and fuel cell applications.

Nickel-Copper Alloys Alloys containing less than 50 percent nickel are discussed under "Copper and Copper Alloys." **Monel** alloy 400 (see Tables 4.4.48 to 4.4.51) is a nickel-rich alloy that combines high strength, high ductility, and excellent resistance to corrosion. It is a homogeneous solid-solution alloy; hence its strength can be increased by cold working alone. In the annealed state, its tensile strength is about 480 MPa (70 ksi), and this may be increased to 1,170 MPa (170 ksi) in hard-drawn wire. It is available in practically all fabricated forms. Alloy 400 is hot-worked in the range 871 to 1,177°C (1,600 to 2,150°F) after rapid heating in a reducing, sulfur-free atmosphere. It can be cold-worked in the same manner as mild steel, but requires more power. Very heavily cold-worked alloy 400 may commence to recrystallize at 427°C (800°F), but in normal practice no softening will occur below 649°C (1,200°F). Annealing can be done for 2 to 5 h at about 760°C (1,400°F) or for 2 to 5 min at about 940°C (1,725°F). Nonscaling, sulfur-free atmospheres are required.

Because of its toughness, alloy 400 must be machined with high-speed tools with slower cutting speeds and lighter cuts than mild steel. A special grade containing sulfur, Monel alloy 405, should be used where high cutting speeds must be maintained. This alloy is essentially the same as the sulfur-free alloy in mechanical properties and corrosion resistance, and it can be hot-forged.

The short-time tensile strengths of alloy 400 at elevated temperatures are summarized in Table 4.4.51.

The fatigue endurance limit of alloy 400 is about 240 MPa (34 ksi) when annealed and 325 MPa (47 ksi) when hard-drawn. The action of corrosion during fatigue is much less drastic on alloy 400 than on steels of equal or higher endurance limit.

Table 4.4.48 Nominal Compositions, %, of Selected Nickel Alloys

Alloy	Ni	Cu	Fe	Cr	Mo	Al	Si	Mn	W	C	Co	Nb	Ti	Y_2O_3
Nickel 200	99.5	0.05	0.1				0.05	0.25		0.05				
Duranickel alloy 301	93.5	0.2	0.2			4.4	0.5	0.3		0.2			0.5	
Monel alloy 400	65.5	31.5	1.5				0.25	1.0		0.15				
Monel alloy K-500	65.5	29.5	1.0			3.0	0.15	0.5		0.15				
Hastelloy alloy C-276	56		5	15.5	16		0.05	1.0	4	0.02	2.5			
Inconel alloy 600	75.5	0.5	8	15.5			0.2	0.5		0.08				
Inconel alloy 625	62		2.5	22	9	0.2	0.2	0.2		0.05		3.5	0.2	
Inconel alloy 718	52.5		18.5	19	3	0.5	0.2	0.2		0.04		5	0.9	
Inconel alloy X-750	71	0.5	7	15.5		0.7	0.5	1.0		0.08	1.0		2.5	
Inconel alloy MA 754	78.5			20		0.3				0.05			0.5	0.6
Incoloy alloy 825	42	2.2	30	21	3	0.1	0.25	0.5		0.03			0.9	
Incoloy alloy 909	38		42				0.4			0.1	13.0	4.7	1.5	
Hastelloy alloy X	45		19.5	22	9		1.0	1.0	0.6	0.1	1.5			

Table 4.4.49 Typical Physical Properties of Selected Alloys

Alloy	UNS no.	Density, g/cm^3	Elastic modulus, GPa	Specific heat (20°C), J/(kg·K)	Thermal expansion (20°-93°C), μm/(m·K)	Thermal conductivity (20°C), W/(m·K)	Electrical resistivity (annealed), μΩ·m
Nickel 200	N02200	8.89	204	456	13.3	70	0.096
Duranickel alloy 301	N03301	8.25	207	435	13.0	23.8	0.424
Monel alloy 400	N04400	8.80	180	427	13.9	21.8	0.547
Monel alloy K-500	N05500	8.44	180	419	13.9	17.5	0.615
Hastelloy alloy C-276	N10276	8.89	205	427	11.2	9.8	1.29
Inconel alloy 600	N06600	8.47	207	444	13.3	14.9	1.19
Inconel alloy 625	N06625	8.44	207	410	12.8	9.8	1.29
Inconel alloy 718	N07718	8.19	211	450	13.0	11.4	1.25
Inconel alloy X-750	N07750	8.25	207	431	12.6	12.0	1.22
Inconel alloy MA 754	N07754	8.30	160	440	12.2	14.3	1.075
Incoloy alloy 825	N08825	8.14	206	440	14.0	11.1	1.13
Incoloy alloy 909	N19909	8.30	159	427	7.7	14.8	0.728
Hastelloy alloy X	N06002	8.23	205	461	13.3	11.6	1.16

Alloy 400 is highly resistant to atmospheric action, seawater, steam, foodstuffs, and many industrial chemicals. It deteriorates rapidly in the presence of moist chlorine and ferric, stannic, or mercuric salts in acid solutions. It must not be exposed when hot to molten metals, sulfur, or gaseous products of combustion containing sulfur.

Small additions of aluminum and titanium to the alloy 400 base nickel-copper alloy produces precipitation-hardenable Monel alloy K-500. This alloy is sufficiently ductile in the annealed state to permit drawing, forming, bending, or other cold-working operations but work-hardens rapidly and requires more power than mild steel. It is hotworked at 927 to 1,177°C (1,700 to 2,150°F) and should be quenched from 871°C (1,600°F) if the metal is to be further worked or hardened. Heat treatment consists of quenching from 871°C (1,600°F), cold working if desired, and reheating for 10 to 16 h at 538 to 593°C (1,000 to 1,100°F). If no cold working is intended, the quench may be omitted on sections less than 50 mm (2 in) thick and the alloy hardened at 593°C (1,100°F). The properties of the heat-treated alloy remain quite stable, at least up to 538°C (1,000°F). It is nonmagnetic down to −101°C (−150°F).

Table 4.4.50 Typical Mechanical Properties of Selected Nickel Alloys

Alloy	Yield strength, MPa	Tensile strength, MPa	Elongation, % in 2 in	Hardness
Nickel 201	150	462	47	110 HB
Duranickel alloy 301	862	1,170	25	35 HRC
Monel alloy 400	240	550	40	130 HB
Monel alloy K-500	790	1,100	20	300 HB
Hastelloy alloy C-276	355	790	60	90 HRB
Inconel alloy 600	310	655	40	36 HRC
Inconel alloy 625	520	930	42	190 HB
Inconel alloy 718	1,036	1,240	12	45 HRC
Inconel alloy X-750	690	1,137	20	330 HB
Inconel alloy MA 754	585	965	22	25 HRC
Incoloy alloy 825	310	690	45	75 HRB
Incoloy alloy 909	1,035	1,275	15	38 HRC
Hasteloy alloy X	355	793	45	90 HRB

Heat-Resistant Nickel-Chromium and Nickel-Chromium-Iron Alloys See "Metals and Alloys for Use at Elevated Temperatures" for further information on high-temperature properties of nickel alloys.

The addition of chromium to nickel improves strength and corrosion resistance at elevated temperature. Nickel chromium alloys such as 80 Ni 20 Cr are extensively used in electrical resistance heating applications. Typically, additions of up to 4 percent aluminum and 1 percent yttrium are made to these alloys to increase hot oxidation and corrosion resistance. Incorporating fine dispersions of inert oxides to this system significantly enhances high-temperature properties and microstructural stability. **Inconel** alloy MA 754, which contains 0.6 percent Y_2O_3, exhibits good fatigue strength and corrosion resistance at 1,100°C (2,010°F) and is used in gas-turbine engines and other extreme service applications.

Inconel alloy 600 is a nickel-chromium-iron alloy. Alloy 600 is a high-strength nonmagnetic [at −40°C (−40°F)] alloy which is used widely for corrosion- and heat-resisting applications at temperatures up to 1,204°C (2,200°F) in sulfidizing atmospheres. In sulfidizing atmospheres, the maximum recommended temperature is 816°C (1,500°F). Inconel alloy X-750 is an age-hardenable modification suitable for stressed applications at temperatures up to 649 to 816°C (1,200 to 1,500°F). The addition of molybdenum to this system provides alloys,

Table 4.4.51 Short-Time High-Temperature Properties of Hot-Rolled Nickel and Its Alloys*

	Temperature, °C							
	21	316	427	540	650	816	982	1,093
Nickel 200†								
Tensile strength, MPa	505	575	525	315	235	170	55	
Yield strength 0.2% offset, MPa	165	150	145	115	105			
Elongation, % in 2 in	40	50	52	55	57	65	91	
Monel alloy 400								
Tensile strength, MPa	560	540	490	350	205	110	55	
Yield strength 0.2% offset, MPa	220	195	200	160	125	60		
Elongation, % in 2 in	46	51	52	29	34	58	45	
Inconel alloy 600								
Tensile strength, MPa	585	545	570	545	490	220	105	75
Yield strength 0.2% offset, MPa	385	185	195	150	150			
Elongation, % in 2 in	50	51	50	21	5	23	51	67
Inconel alloy 718								
Tensile strength, MPa	1,280			1,140	1,030			
Yield strength 0.2% offset, MPa	1,050			945	870			
Elongation, % in 2 in	22			26	15			

* See also data on metals and alloys for use at elevated temperature elsewhere in Sec. 4.4.
† Nickel 200 is not recommended for use above 316°C; low-carbon nickel 201 is the preferred substitute.

e.g., Hastelloy alloy X, with additional solid-solution strengthening and good oxidation resistance up to 1,200°C (2,200°F).

The combination of useful strength and oxidation resistance makes nickel alloys frequent choices for high-temperature service. Table 4.4.51 indicates the changes in mechanical properties with temperature.

Corrosion-Resistant Nickel-Molybdenum and Nickel-Iron-Chromium Alloys A series of solid-solution nickel alloys containing molybdenum exhibit superior resistance to corrosion by hot concentrated acids such as hydrochloric, sulfuric, and nitric acids. Hastelloy alloy C-276, which also contains chromium, tungsten, and cobalt, is resistant to a wide range of chemical process environments including strong oxidizing acids, chloride solutions, and other acids and salts. Incoloy alloy 825, a nickel-iron-chromium alloy with additions of molybdenum, is especially resistant to sulfuric and phosphoric acids. These alloys are used in pollution control, chemical processing, acid production, pulp and paper production, waste treatment, and oil and gas applications. Although these types of alloys are used extensively in aqueous environments, they also have good elevated-temperature properties.

Nickel-Iron Alloys Nickel is slightly ferromagnetic but loses its magnetism at a temperature of 368°C (695°F) when pure. For commercial nickel, this temperature is about 343°C (650°F). Monel alloy 400 is lightly magnetic and loses all ferromagnetism above 93°C (200°F). The degree of ferromagnetism and the temperature at which ferromagnetism is lost are very sensitive to variations in composition and mechanical and thermal treatments.

An important group of soft magnetic alloys is the nickel irons and their modifications, which exhibit high initial permeability, high maximum magnetization, and low residual magnetization. Nickel-iron alloys also exhibit low coefficients of thermal expansion that closely match those of many glasses and are therefore often used for glass-sealing applications. Nickel-iron alloys such as Incoloy alloy 909, which contains cobalt, niobium, and titanium, have found wide applications in gas-turbine and rocket engines, springs and instruments, and as controlled-expansion alloys designed to provide high strength and low coefficients of thermal expansion up to 650°C (1,200°F).

Low-Temperature Properties of Nickel Alloys Several nickel-base alloys have very good properties at low temperatures, in contrast to ferrous alloys whose impact strength (the index of brittleness) falls off very rapidly with decreasing temperature. The impact strength remains nearly constant with most nickel-rich alloys, while the tensile and yield strengths increase as they do in the other alloys. Specific data on low-temperature properties may be found in White and Siebert, "Literature Survey of Low-Temperature Properties of Metals," Edwards, and "Mechanical Properties of Metals at Low Temperatures," *NBS Circ.* 520, 1952.

Welding Alloys Nickel alloys can be welded with similar-composition welding materials and basic welding processes such as gas tungsten arc welding (GTAW), gas metal arc welding (GMAW), shielded metal arc welding (SMAW), brazing, and soldering. The procedures used are similar to those for stainless steels. Nickel-base welding products are also used to weld dissimilar materials. Nickel-base filler metals, especially with high molybdenum contents, are typically used to weld other alloys to ensure adequate pitting and crevice corrosion resistance in final weld deposits. Nickel and nickel iron welding electrodes are used to weld cast irons.

(*Note:* Inconel, Incoloy, and Monel are trademarks of the Inco family of companies. Hastelloy is a trademark of Haynes International.)

TITANIUM AND ZIRCONIUM

by Francis H. (Sam) Froes and John R. Schley

REFERENCES: F.H. (Sam) Froes, Book Editor, Titanium Physical Metallurgy, Processing and Applications, ASM, Materials Park, Ohio, February 2015. F.H. Froes, D. Eylon and H.B. Bomberger Titanium Technology, International Titanium Association, 1985, B. Dutta and F.H. Froes, Titanium Additive Manufacturing, Elsevier, 2016,: Donachie, "Titanium, A Technical Guide," ASM International, Metals Park, OH. "Titanium, The Choice," rev. 1990, Titanium Development Association, Boulder, CO. ASM Metals Handbook, "Corrosion," vol. 13, 1987 edition, ASM International, Metals Park, OH.

Titanium

The metal titanium is the ninth most abundant element in the earth's surface and the fourth most abundant structural metal after aluminum, iron, and magnesium. It is a soft, ductile, silvery-gray metal with a density about 60 percent of that of steel and a melting point of 1,675°C (3,047°F). A combination of low density and high strength makes titanium attractive for structural applications, particularly in the aerospace industry and is leading to a plethoria of uses in Automobiles (see Table below). These characteristics, coupled with excellent corrosion resistance, have led to the widespread use of titanium in the chemical processing industries and in oil and gas production and refining.

Some Applications of Titanium Alloys in Automobiles

Component	Material	Manufacturer	Model
Connecting rods	Ti-3Al-2V-rare earth	Honda	Acura NSX
Connecting rods	Ti-6Al-4V	Ferrari	All 12-cyl.
Wheel rim screws	Ti-6Al-4V	Porsche	Sport wheel option
Brake pad guide pins	Ti grade 2	Daimler	S-Ciass
Brake sealing washers	Ti grade 1s	Volkswagen	All
Gearshift knob	Ti grade l	Honda	S2000 Roadster
Connecting rods	Ti-6Al-4V	Porsche	GT3
Valves	Ti-6Al-4V & PM-Ti	Toyota	Altezza 6-cyl.
Turbo charger wheel	Ti-6Al-4V	Daimler	Truck diesel
Suspension springs	TIMETAL LCB	Volkswagen	Lupo FSI
Wheel rim screws	β-6Al-4V	BMW	M-Techn. option
Valve spring retainers	β-titanium alloys	Mitsubishi	All 1.8 I – 4-cyl.
Turbo charger wheel	γ-TiAl	Mitsubishi	Lancer
Exhaust system	Ti grade 2	General Motors	Corvette Z06
Wheel rim screws	Ti-6Al-4V	Volkswagen	Sport package GTI
Valves	Ti-6Al-4V & PM-Ti	Nissan	Infiniti Q45
Suspension Springs	TIMETAL LCB	Ferrari	360 Stradale

Like iron, titanium can exist in two crystalline forms: hexagonal close-packed (hcp) below 883°C (1,621°F) and body-centered cubic (bcc) at higher temperatures up to the melting point. Titanium combines readily with many common metals such as aluminum, vanadium, zirconium, tin, and iron and is available in a variety of alloy compositions offering a broad range of mechanical and corrosion properties. Certain of these compositions are heat-treatable in the same manner as steels. In the case of Titanium the precipitation which occurs is a dispersion of the alpha phase in a beta matrix.

Production Processes

The commercial production of titanium begins with an ore, either ilmenite ($FeTiO_3$) or rutile (TiO_2), more often the latter, which passes through a series of chemical reduction steps leading to elemental titanium particles, referred to as **titanium sponge**. These processing steps involve chlorination of the oxide, in the presence of C, to form an intermediate chloride which is then reduced with Mg (Kroll Process) or Na (Hunter Process) to produce the titanium sponge. The sponge raw material then follows a sequence of processing operations resembling those employed in steelmaking operations. Beginning with the melting of titanium or titanium alloy ingots, the process follows a hot-working sequence of ingot breakdown by forging, then rolling to sheet, plate, or bar product. Titanium is commercially available in all common mill product forms such as billet and bar, sheet and plate, as well as tubing and pipe. It also is rendered into castings.

Mechanical Properties

Commercial titanium products fall into three structural categories corresponding to the allotropic forms of the metal. The alpha structure is present in **commercially pure** (CP) products and in certain alloys containing alloying elements that promote the alpha structure. The alpha-beta form occurs in a series of alloys that have a mixed structure at room temperature, and the beta class of alloys contains elements that stabilize the beta structure down to room temperature. Strength characteristics of the alpha-beta and beta alloys can be varied by heat treatment. Typical alloys in each class are shown in Table 4.4.52, together with their nominal compositions and tensile properties in the annealed condition. Certain of these alloys can be solution-heat-treated and aged to higher strengths.

Corrosion Properties

Titanium possesses outstanding **corrosion resistance** to a wide variety of environments especially oxidizing environments (see Fig. 4.4.10). This characteristic is due to the presence of a protective, strongly adherent oxide film on the metal surface. This film is normally transparent, and titanium is capable of healing the film almost instantly in any environment where a trace of moisture or oxygen is present. It is very stable and is only attacked by a few substances, most notably hydrofluoric acid. Titanium is unique in its ability to handle certain chemicals such as chlorine, chlorine chemicals, and chlorides. It is essentially inert in seawater, for example. The unalloyed CP grades typically are used for corrosion applications in industrial service, although the alloyed varieties are used increasingly for service where higher strength is required.

Fabrication

Titanium is cast and forged by conventional practices. Both means of fabrication are extensively used, particularly for the aerospace industry. Wrought titanium is readily fabricated in the same manner as and on the same equipment used to fabricate stainless steels and nickel alloys. For example, cold forming and hot forming of sheet and plate are done on press brakes and hydraulic presses with practices modified to

Table 4.4.52 Typical Commercial Titanium Alloys

	Tensile strength		0.2% Yield strength			Nominal composition, wt %						Impurity limits, wt %				
Designation	MPa	ksi	MPa	ksi	Elongation, %	Al	V	Sn	Zr	Mo	Other	N (max)	C (max)	H (max)	Fe (max)	O (max)
Unalloyed (CP grades:																
ASTM grade, 1	240	35	170	25	24	—	—	—	—	—	—	0.03	0.10	0.015	0.20	0.18
ASTM grade 2	340	50	280	40	20	—		—	—	—	—	0.03	0.10	0.015	0.30	0.25
ASTM grade 7	340	50	280	40	20	—	—		—	—	0.2 Pd	0.03	0.10	0.015	0.30	0.25
Alpha and near-alpha alloys:																
Ti-6 Al-2 Sn-4 Zr-2 Mo	900	130	830	120	10	6.0	—	2.0	4.0	2.0	—	0.05	0.05	0.0125	0.25	0.15
Ti-8 Al-1 Mo-l V	930	135	830	120	10	8.0	1.0	—	—	1.0	—	0.05	0.08	0.012	0.30	0.12
Ti-5 Al-2 5 Sn	830	120	790	115	10	5.0	—	2.5	—	—	—	0.05	0.08	0.02	0.50	0.20
Alpha-beta alloys:																
Ti-3 Al-2 5V.	620	90	520	75	15	3.0	2.5	—	—	—	—	0.015	0.05	0.015	0.30	0.12
Ti-6 Al-4 V*	900	130	830	120	10	6.0	4.0	—		—	—	0.05	0.10	0.0125	0:30	0.20
Ti-6 Al-6 V-2 Sn*	1,030	150	.970	140	8	6.0	6.0	2.0	—	—	0.75 Cu	0.04	0.05	0.015	1.0	0.20
Ti-6 Al-2 Sn-4 Zr-6 Mo†	1,170	170	1,100	160	8	6.0	—	2.0	4.0	6.0	—	0.04	0,04	0.0125	0.15	0.15
Beta alloys:																
Ti-10 V-2 Fe-3 Al†	1,170	180	1,100	170	8	3.0	10.0	—	—			0.05	0.05	0.015	2.2	0.13
Ti-15V-3 Al-3 Sn-3 Cr†	1,240	180	1,170	170	10	3.0	15.0	3.0	—	—	3.0 Cr	0.03	0.03	0.015	0.30	0.13
Ti-3 Al-8 V-6 Cr-4 Mo-4. Zr*	900	130	860	125	16	3.0	8.9		4.0	4.0	6.0 Cr	0.03	0.05	0.020	0.25	0.12

* Mechanical properties given for annealed condition; may be solution-treated and aged to increase strength.
† Mechanical properties given for solution-treated and aged condition.

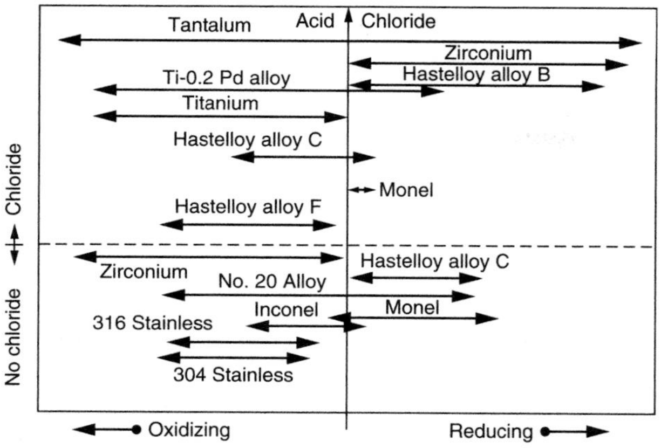

Figure 4.4.10 General corrosion behavior of commercial purity titanium and Ti-Pd alloys compared to other metals and alloys in oxidizing and reducing acids; with and without chloride ions. In general, each metal or alloy can be used for those environments below its respective solid lines.[17]

accommodate the forming characteristics of titanium. Recently Titanium has proven to be fabricable by Additive Manufacturing in a cost effective fashion. Titanium is machinable by all customary methods, again using practices recommended for titanium. Welding is the most common method used for joining titanium, for the metal is readily weldable. The standard welding process is **gas tungsten arc welding** (GTAW), **TIG or electron beam**. Two other welding processes, **plasma arc welding** (PAW) and **gas metal arc welding** (GMAW), or MIG, are used to a lesser extent. Since titanium reacts readily with the atmospheric gases oxygen and nitrogen, precautions must be taken to use an inert-gas shield to keep air away from the hot weld metal during welding. Usual standard practices apply to accomplish this. Titanium can be torch-cut by **oxyacetylene flame** or by **plasma torch,** but care must be taken to remove contaminated surface metal. For all titanium fabricating practices cited above, instructional handbooks and other reference materials are available. Inexperienced individuals are well advised to consult a titanium supplier or fabricator.

Applications

Titanium as a material of construction offers the unique combination of wide-ranging corrosion resistance, low density, and high strength. For this reason, titanium is applied broadly for applications that generally are separated into aerospace and nonaerospace uses, the latter termed *industrial*. The **aerospace applications** consume about 70 percent of the annual production of titanium, and the titanium alloys rather than the CP grades predominate since these applications depend primarily on titanium's high strength/weight ratio. This category is best represented by **aircraft gas-turbine engines,** the largest single consumer of titanium, followed by **airframe structural members** ranging from fuselage frames, wing beams, and landing gear to springs and fasteners. **Spacecraft** also take a growing share of titanium. Recently the intermetallic TiAl has been used in the hot sections of a GE gas turbine engine as it provides reduced density compared to nickel based superalloys. Nonaerospace or **industrial applications** more often use titanium for its corrosion resistance, sometimes coupled with a requirement for high strength. The unalloyed, commercially pure grades predominate in most corrosion applications and are typically represented by equipment such as heat exchangers, vessels, and piping.

A recent and growing trend is the application of titanium in the offshore oil and gas industries for service on and around offshore platforms. As applied there, titanium uses are similar to those onshore, but with a growing trend toward large-diameter pipe for subsea service. Strength is a major consideration in these applications, and titanium alloys are utilized increasingly.

In the automotive field (see Table above), the irreversible mandate for lightweight, fuel efficient Vehicles (see the effect of weight reduction on fuel economy in Fig. 4.4.11) appears to portend an increasing use of titanium and its alloys.

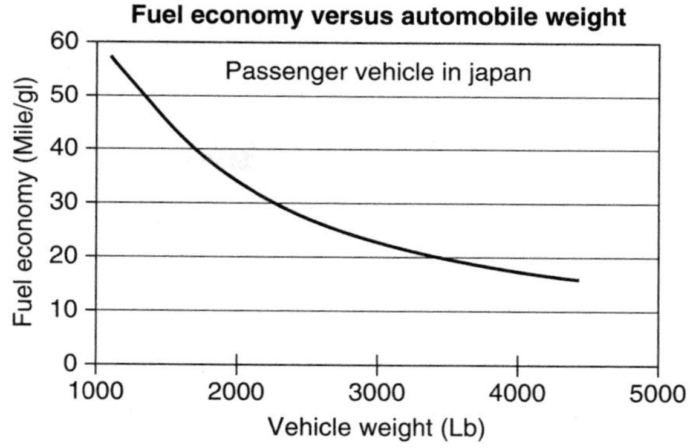

Figure 4.4.11 Fuel Economy versus Automobile Weight.

Zirconium

Zirconium was isolated as a metal in 1824 but was only developed commercially in the late 1940s, concurrently with titanium. Its first use was in nuclear reactor cores, and this is still a major application. Zirconium bears a close relationship to titanium in physical and mechanical properties, but has a higher density, closer to that of steel. Its melting point is 1,852°C (3,365°F) and, like titanium, it exhibits two allotropic crystalline forms. The pure metal is body-centered cubic (bcc) above 865°C (1,590°F) and hexagonal close-packed (hcp) below this temperature. Zirconium is available commercially in unalloyed form and in several alloyed versions, combined most often with small percentages of niobium or tin. It is corrosion-resistant in a wide range of environments.

Zirconium is derived from naturally occurring zircon sand processed in a manner similar to titanium ores and yielding a **zirconium sponge.** The sponge is converted to mill products in the same manner as titanium sponge. The principal commercial compositions intended for corrosion applications together with corresponding mechanical properties are listed in Tables 4.4.53 and 4.4.54.

Table 4.4.53 Chemical Compositions of Zirconium Alloys (Percent)

Grade (ASTM designation)	R60702	R60704	R60705	R60706
Zr + Hf, min	99.2	97.5	95.5	95.5
Hafnium, max	4.5	4.5	4.5	4.5
Fe + Cr	max 0.20	0.2-0.4	max 0.2	max 0.2
Tin	—	1.0-2.0	—	—
Hydrogen, max	0.005	0.005	0.005	0.005
Nitrogen, max	0.025	0.025	0.025	0.025
Carbon, max	0.05	0.05	0.05	0.05
Niobium	—	—	2.0-3.0	2.0-3.0
Oxygen, max	0.16	0.18	0.18	0.16

Table 4.4.54 Minimum ASTM Requirements for the Mechanical Properties of Zirconium at Room Temperature (Cold-Worked and Annealed)

Grade (ASTM designation)	R60702	R60704	R60705	R60706
Tensile strength, min, ksi (MPa)	55 (379)	60 (413)	80 (552)	74 (510)
Yield strength, min, ksi (MPa)	30 (207)	35 (241)	55 (379)	50 (345)
Elongation (0.2% offset), min, %	16	14	16	20
Bend test radius*	5*T*	5*T*	3*T*	2.5*T*

* Bend tests are not applicable to material over 0.187 in (4.75 mm) thick, and *T* equals the thickness of the bend test sample.

Zirconium is a reactive metal that owes its **corrosion resistance** to the formation of a dense, stable, and highly adherent oxide on the metal surface. It is resistant to most organic and mineral acids, strong alkalies, and some molten salts. A notable corrosion characteristic is its resistance to both strongly reducing hydrochloric acid and highly oxidizing nitric acid, relatively unique among metallic corrosion-resistant materials.

Zirconium and its alloys can be machined, formed, and welded by conventional means and practices much like those for stainless steels and identical to those for titanium. Cast shapes are available for equipment such as pumps and valves, and the wrought metal is produced in all standard mill product forms. Persons undertaking the fabrication of zirconium for the first time are well advised to consult a zirconium supplier.

The traditional application of zirconium in the nuclear power industry takes the form of fuel element cladding for reactor fuel bundles. The zirconium material used here is an alloy designated **Zircaloy,** containing about 1.5 percent tin and small additions of nickel, chromium, and iron. It is selected because of its low neutron cross section coupled with the necessary strength and resistance to water corrosion at reactor operating temperatures. Most zirconium is employed here. An interesting noncorrosion zirconium application in photography had been its use as a foil in flash bulbs, but the advent of the electronic flash has curtailed this. An active and growing corrosion application centers on fabricated chemical processing equipment, notably heat exchangers, but also a variety of other standard equipment.

ZINC AND ZINC ALLOYS

by Frank E. Goodwin

REFERENCES: Porter, "Zinc Handbook," Marcel Dekker, New York. ASM "Metals Handbook," 10th ed., vol. 2. ASTM, "Annual Book of Standards." "NADCA Product Specification Standards for Die Casting," Die Casting Development Council.

Zinc, one of the least expensive nonferrous metals, is produced from sulfide, silicate, or carbonate ores by a process involving concentration and roasting followed either by thermal reduction in a zinc-lead blast furnace or by leaching out the oxide with sulfuric acid and electrolyzing the solution after purification. Zinc distilled from blast-furnace production typically contains Pb, Cd, and Fe impurities that may be eliminated by fractional redistillation to produce zinc of 99.99 + percent purity. Metal of equal purity can be directly produced by the electrolytic process. Zinc ingot range from cast balls weighing a fraction of a pound to a 1.1-ton [1 metric ton (t)] "jumbo" blocks. Over 20 percent of zinc metal produced each year is from recycled scrap.

The three standard grades of zinc available in the United States are described in ASTM specification B-6 (see Table 4.4.55). **Special high grade** is overwhelmingly the most commercially important and is used in all applications except as a coating for steel articles galvanized after fabrication. In other applications, notably die casting, the higher levels of impurities present in grades other than special high grade can have harmful effects on corrosion resistance, dimensional stability, and formability. Special high grade is also used as the starting point for brasses containing zinc.

Table 4.4.55 ASTM Specification B6-03 for Slab Zinc

	Composition, %			
Grade	Lead, max	Iron, max	Cadmium, max	Zinc, min, by difference
Special high grade*	0.003	0.003	0.003	99.990
High grade	0.03	0.02	0.02	99.90
Prime western†	1.4	0.05	0.20	98.0

* Tin in special high grade shall not exceed 0.001%.
† Aluminum in prime western zinc shall not exceed 0.05%.
SOURCE: ASTM, reprinted with permission.

Galvanized Coatings

Protective coatings for steel constitute the largest use of zinc and rely upon the galvanic or sacrificial property of zinc relative to steel. Lead, antimony, or zinc can be added to produce the solidified surface pattern called **spangle** preferred on unpainted articles. The addition of aluminum in amounts of 5 and 55 percent results in coatings with improved corrosion resistance termed **Galfan** and **Galvalume,** respectively.

Zinc Die Castings

Zinc alloys are particularly well suited for making die castings since the melting point is reasonably low, resulting in long die life even with ordinary steels. High accuracies and good surface finish are possible. Alloys currently used for die castings in the United States are covered by ASTM Specifications B-86 and B-240. Nominal compositions and typical properties of these compositions are given in Table 4.4.56. The low limits of impurities are necessary to avoid disintegration of the casting by intergranular corrosion under moist atmospheric conditions. The presence of magnesium or nickel prevents this if the impurities are not higher than the specification values. The mechanical properties shown in Table 4.4.56 are from die-cast test pieces of 0.25-in (6.4-mm) section thickness. Zinc die castings can be produced with section thicknesses as low as 0.03 in (0.75 mm) so that considerable variations in properties from those listed must be expected. These die-casting alloys generally increase in strength with increasing aluminum content. The 8, 12, and 27 percent alloys can also be gravity-cast by other means.

Increasing copper content results in growth of dimensions at elevated temperatures. A measurement of the expansion of the die casting after exposure to water vapor at 203°F (95°C) for 10 days is a suitable index of not only stability but also freedom from susceptibility to intergranular corrosion. When held at room temperature, the copper-containing alloys begin to shrink immediately after removal from the dies; total **shrinkage** will be approximately two-thirds complete in 5 weeks. The maximum extent of this **shrinkage** is about 0.001 in/in (10 μm/cm). The copper-free alloys do not exhibit this effect. Stabilization may be effected by heating the alloys for 3 to 6 h at 203°F (95°C), followed by air cooling to room temperature.

Wrought Zinc

Zinc rolled in the form of sheet, strip, or plate of various thicknesses is used extensively for automobile trim, dry-cell battery cans, fuses, and plumbing applications. Compositions of standard alloys are

Table 4.4.56 Composition and Typical Properties of Zinc-Base Die Casting Alloys

	Composition, %					
Element	UNS Z33521 (AG40A) Alloy 3	UNS Z33522 (AG40B) Alloy 7	UNS Z35530 (AC41A) Alloy 5	UNS Z25630 ZA-8	UNS Z35630 ZA-12	UNS Z35840 ZA-27
Copper	0.25 max	0.25 max	0.75-1.25	0.8-1.3	0.5-1.25	2.0-2.5
Aluminum	3.5-4.3	3.5-4.3	3.5-4.3	8.0-8.8	10.5-11.5	25.0-28.0
Magnesium	0.020-0.05	0.005-0.020	0.03-0.08	0.015-0.030	0.015-0.030	0.010-0.020
Iron, max	0.100	0.075	0.100	0.10	0.075	0.10
Lead, max	0.005	0.0030	0.005	0.004	0.004	0.004
Cadmium, max	0.004	0.0020	0.004	0.003	0.003	0.003
Tin, max	0.003	0.0010	0.003	0.002	0.002	0.002
Nickel	—	0.005-0.020	—	—	—	
Zinc	Rest	Rest	Rest	Rest	Rest	Rest
Yield strength						
lb/in^2	32	32	39	41-43	45-48	52-55
MPa	221	221	269	283-296	310-331	359-379
Tensile strength						
lb/in^2	41.0	41.0	47.7	54	59	62
MPa	283	283	329	372	400	426
Elongation in 2 in (5 cm), %	10	14	7	6-10	4-7	2.0-3.5
Charpy impact on square specimens						
ft/lb	43	43	48	24-35	15-27	7-12
J	58	58	65	32-48	20-37	9-16
Brinell Hardness number on square specimens, 500-kg load, 10-mm ball, 30 s	82	80	91	100-106	95-105	116-122

SOURCE: ASTM, reprinted with permission.
Compositions from ASTM Specifications B-86–04 and B-240–04. Properties from NADCA Product Specification Standards for Die Castings.

Table 4.4.57 Typical Composition of Rolled Zinc (ASTM Specification B69–01), %

Lead	Iron, max	Cadmium	Copper	Magnesium	Zinc
0.05 max	0.010	0.005 max	0.001 max	—	Remainder
0.05-0.12	0.012	0.005 max	0.001 max	—	Remainder
0.30-0.65	0.020	0.20-0.35	0.005 max	—	Remainder
0.05-0.12	0.012	0.005 max	0.65-1.25	—	Remainder
0.05-0.12	0.015	0.005 max	0.75-1.25	0.007-0.02	Remainder

SOURCE: ASTM, reprinted with permission.

shown in Table 4.4.57. In addition, a comparatively new series of alloys containing titanium, exhibiting increased strength and creep resistance together with low thermal expansion, has become popular in Europe. A typical analysis is 0.5 to 0.8 percent Cu, 0.08 to 0.16 percent Ti, and maximum values of 0.2 percent Pb, 0.015 percent Fe, 0.01 percent Cd, 0.01 percent Mn, and 0.02 percent Cr. All alloys are produced by hot rolling followed by cold rolling when some stiffness and temper are required. Deep-drawing or forming operations are carried out on the softer grades, while limited forming to produce architectural items, plumbing, and automotive trim can be carried out using the harder grades.

Alloys containing 0.65 to 1.025 percent Cu are significantly stronger than unalloyed zinc and can be work-hardened. Since 1982, the United States one-cent coin has been made from a zinc-copper alloy that is copper-plated before coining. The addition of 0.01 percent Mg allows design stresses up to 10,000 lb/in^2 (69 MPa). The Zn-Cu-Ti alloy is much stronger, with a typical tensile strength of 25,000 lb/in^2 (172 MPa).

Rolled zinc may be easily formed by all standard techniques. The deformation behavior of rolled zinc and its alloys varies with direction; its crystal structure renders it nonisotropic. In spite of this, it can be formed into parts similar to those made of copper, aluminum, and brass, and usually with the same tools, provided the temperature is not below 70°F (21°C). More severe operations can best be performed at temperatures up to 125°F (52°C). When a cupping operation is performed, a take-in of 40 percent on the first draw is usual. Warm, soapy water is widely used as a lubricant. The soft grades are self-annealing at room temperature, but harder grades respond to deformation better if they are annealed between operations. The copper-free zincs are annealed at 212°F (100°C) and the zinc-copper alloys at 440°C (220°C). Welding is possible with a wire of composition similar to the base metals. Soldering with typical tin-lead alloys is exceptionally easy. The impact extrusion process is widely used for producing battery cups and similar articles.

Effect of Temperature

Properties of zinc and zinc alloys are very sensitive to temperature. **Creep resistance** decreases with increasing temperature, and this must be considered in designing articles to withstand continuous load. When steel screws are used to fasten zinc die castings, maximum long-term clamping load will be reached if an engagement length 4 times the diameter of the screw is used along with cut (rather than rolled) threads on the fasteners.

4.5 CORROSION

by Robert D. Bartholomew and David A. Shifler

REFERENCES: H.H. Uhlig, *The Corrosion Handbook*, Wiley, New York (1948). *Uhlig's Corrosion Handbook,* 2nd ed., R.Winston Revie, ed., Wiley Interscience, New York (2000). *Uhlig's Corrosion Handbook,* 3rd ed., R.Winston Revie, ed., Wiley Interscience, New York (2011). U.R. Evans, *An Introduction to Metallic Corrosion*, 3rd ed., Edward Arnold & ASM International, Metals Park, OH (1981). *Corrosion Basics - An Introduction*, 2nd ed., L.S. van Delinder, ed., NACE International, Houston (1984). G. Wranglen, *An Introduction to Corrosion and Protection of Metals*, Chapman and Hall, London (1986). M.G. Fontana, *Corrosion Engineering,* 3rd ed., McGraw-Hill, New York (1986). P.R. Roberge, Corrosion Engineering: Principles and Practice, McGraw-Hill, New York (2008). D.A. Jones, *Principles and Prevention of Corrosion*, MacMillan, New York (1992). D.A. Jones, *Principles and Prevention of Corrosion*, 2nd ed., Prentice Hall, Upper Saddle River, NJ (1996). E. McCafferty, Introduction to Corrosion Science, Springer, New York (2010). J.C. Scully, *The Fundamentals of Corrosion*, 3rd ed., Pergamon, New York (1990). H. Kaesche, *Metallic Corrosion*, 2nd ed., R.A. Rapp trans., NACE International, Houston (1985). *Corrosion*, 2nd ed., L.L. Sheir, ed., Newnes-Butterworths, London (1976). *Corrosion,* 3rd ed., L.L. Sheir, R.A. Jarman, eds., Butterworth-Heinemann, Oxford (1994). Shreir's Corrosion, J.A. Richardson, Ed., Elsevier, B.V. (2010). *Corrosion,* v. 13, Metals Handbook, 9th ed., ASM International, Metals Park, OH (1987). *Corrosion: Fundamentals, Testing, and Protection*, S.D. Cramer, B.S. Covino, Jr., eds., ASM International, Materials Park, OH (2003). P.J. Gellings, *Introduction to Corrosion Prevention and Control*, Delft University Press, Deflt, The Netherlands, (1985). A.J. Sedriks, *Corrosion of Stainless Steels*, 2nd ed., Wiley Interscience, New York (1996). M. Pourbaix, *Atlas of Electrochemical Equilibria in Aqueous Solutions*, NACE International, Houston (1974). G. Prentice, *Electrochemical Engineering Principles*, Prentice Hall, Englewoods Cliffs, NJ (1991). J. Newman, K.E. Thomas-Alyea, *Electrochemical Systems*, 3rd ed., Wiley Interscience, New York (2004). J. O'M. Bockris, A.K.N. Reddy, *Modern Electrochemistry*, Plenum, New York (1970). J. O'M. Bockris, S.U.M. Khan, *Surface Electrochemistry - A Molecular Level Approach*, Plenum, New York (1993). Corrosion and Corrosion Control, R.W. Revie and HH. Uhlig, eds., Wiley-Interscience, Hoboken, NJ, (2008). Corrosion Mechanisms in Theory and Practice, P. Marcus, Ed., 3rd ed., CRC Press (2011).

INTRODUCTION

Corrosion is the deterioration of a material or its properties because of its reaction with the environment. Materials may include metals and alloys, nonmetals, woods, ceramics, plastics, composites, etc. Although corrosion as a science is barely 150 years old, its effects have impacted mankind for thousands of years. The importance of understanding the causes, initiation, and propagation of corrosion and the methods for controlling its degradation is threefold. First, corrosion generates an **economical impact** through materials losses and failures of various structures and components. A recent study estimated that the combined annualized economic losses from 1999 to 2001 in the United States from corrosion was 276 billion dollars [J.H. Payer, Y.P. Virmani, G.H. Koch, M.P.H. Brongers, N.G. Thompson, "Corrosion Costs and Preventative Strategies in the United States," Publication No. FHWA-RD-01-156, U.S. Federal Highway Administration, (2001); "Cost of Corrosion Study Unveiled," *Materials Performance*, supplement, NACE International (July 2002)]. Studies of corrosion costs in other countries have determined that 3 to 4 percent of GNP is related to economic losses from corrosion. The indirect cost of corrosion is estimated to be at least equal to the direct cost. In that case, the total cost of corrosion is **993** billion dollars in March 2013 and estimated to exceed **1 trillion** dollars by June 2013 (based on estimates of GDP from http://forecasts.org/gdp.htm). These estimates consider only the direct economic costs. Indirect costs are difficult to assess but can include plant downtime, loss of products or services, loss of efficiency, contamination, and overdesign of structures and components. Application of current corrosion control technology (discussed later) could recover or avoid 15 to 20 percent of the costs due to corrosion.

Second, an important consideration of corrosion prevention or control is improved **safety**. Catastrophic degradation and failures of pressure vessels, petrochemical plants, boilers, aircraft sections, tanks containing toxic materials, and automotive parts have led to thousands of personal injuries and deaths, which often result in subsequent litigation. If the safety of individual workers or the public is endangered in some manner by selection of a material or use of a corrosion control method, that approach should be abandoned. Once assurance is given that this criterion is met through proper materials selection and improved design and corrosion control methods, other factors can be evaluated to provide the optimum solution.

Third, understanding factors leading to corrosion can **conserve resources**. The world's supply of easily extractable raw materials is limited and corrosion control can minimize use of precious resources. Corrosion also imparts wastage of energy, water resources, and raw or processed materials.

Corrosion can occur through chemical or electrochemical reactions and is usually an interfacial process between a material and its environment. Deterioration solely by physical means such as erosion, galling, or fretting is not generally considered corrosion. **Chemical corrosion** may be a heterogeneous reaction which occurs at the metal/environment interface and involves the material (metal) itself as one of the reactants. Such occurrences may include metals in strong acidic or alkaline solutions, high temperature oxidation metal/gas reactions where the oxide or compound is volatile, the breakdown of chemical bonds in polymers, dissolution of solid metals or alloys in a liquid metal (e.g., aluminum in mercury), or the dissolution of metals in fused halides or nonaqueous solutions.

However, even in acidic or alkaline solutions and in most high temperature environments, corrosion by electrochemical processes consisting of two or more partial reactions involves the transfer of electrons or charges. An electrochemical reaction requires: (1) an anode; (2) a cathode; (3) an electrolyte; and (4) a complete electrical circuit. In the reaction of a metal (M) in hydrochloric acid, the metal is oxidized and the electrons are generated at the anode [see (Eq. 4.5.1)]. At the same time, hydrogen cations are reduced and electrons consumed to generate hydrogen gas at the cathode, the **hydrogen evolution reaction** (HER) as described in Eq. (4.5.2).

$$\text{Anodic (oxidation) reaction: } M \longrightarrow M^{+n} + ne^- \quad (4.5.1)$$

$$\text{Cathodic (reduction) reaction: } 2H^+ + 2e^- \longrightarrow H_2 \quad (4.5.2)$$

Generally, oxidation occurs at the anode, while reduction occurs at the cathode. Corrosion is usually involved at the anode where the metal or alloy is oxidized; this causes dissolution, wastage, and penetration. Cathodic reactions significant to corrosion are few. These may include, in addition to the HER, the following:

$$\text{Oxygen reduction (acidic solutions): } O_2 + 4H^+ + 4e^- \longrightarrow 2H_2O \quad (4.5.3)$$

$$\text{Oxygen reduction (neutral or basic): } O_2 + 2H_2O + 4e^- \longrightarrow 4OH^- \quad (4.5.4)$$

$$\text{Metal ion reduction: } Fe^{+3} + e^- \longrightarrow Fe^{+2} \quad (4.5.5)$$

$$\text{Metal deposition: } Cu^{+2} + 2e^- \longrightarrow Cu \quad (4.5.6)$$

FORMS OF CORROSION

It is convenient to classify corrosion by the various forms in which it exists, preferably by visual examination-although in many cases analysis may require more sophisticated methods. An arbitrary list of **corrosion types** is: (1) uniform corrosion; (2) galvanic corrosion; (3) crevice corrosion; (4) pitting; (5) intergranular corrosion; (6) dealloying or selective leaching; (7) erosion-corrosion; (8) environmentally-induced cracking; and (9) high temperature corrosion.

Uniform corrosion is the most common form of corrosion. It is typified by an electrochemical or chemical attack that affects the entire exposed surface. The metal becomes thinner and eventually fails. Unlike many of the other forms of corrosion, uniform corrosion can be easily measured by weight loss tests and by using equations. The life of structures and components suffering from uniform corrosion then can be accurately estimated.

Stray-current corrosion or **stray-current electrolysis** is caused by externally induced electrical currents and is usually independent of environmental factors such as oxygen concentration or pH. Stray currents are escaped currents that follow paths other than their intended circuit via a low resistance path through soil, water, or any suitable electrolyte. Direct currents from electric mass transit systems, welding machines, and implied cathodic protection systems (discussed later) are major sources of stray-current corrosion. At the point where stray currents enter the unintended structure, the sites will become cathodic because of changes in potential. Areas where the stray currents leave the metal structure become anodic and become sites where serious corrosion can occur. Alternating currents (AC) generally cause less severe damage than direct currents. Stray-current corrosion may vary over short time periods and its appearance is sometimes similar to galvanic corrosion [Stray Current Corrosion: The Past, Present, and Future of Rail Transit Systems, M.J. Szeliga, ed., NACE International, Houston, TX (1994)].

Galvanic corrosion occurs when a metal or alloy is electrically coupled to another metal, alloy, or to a conductive nonmetal either directly or through an external path in a common, conductive medium. A potential difference usually exists between dissimilar metals, which causes a flow of electrons between them. The corrosion rate of the less corrosion resistant (active) anodic metal is increased, while the corrosion rate of the more corrosion resistant (noble) cathodic metal or alloy is decreased. Galvanic corrosion may be affected by many factors including: (1) potential difference between two or more alloys; (2) reaction kinetics; (3) alloy composition; (4) protective film characteristics in the given environment; (5) mass transport (migration, diffusion, and/or convection); (6) bulk solution properties (oxygen content, pH, conductivity, pollutant/contamination level); (7) cathode/anode ratio; (8) polarization; (9) volume of solution; (10) temperature; (11) type of joint; (12) flow rate; and (13) geometry of coupling members [J.W. Oldfield, "Electrochemical Theory of Galvanic Corrosion," *Galvanic Corrosion*, STP 978, H.P. Hack, ed., ASTM, Philadelphia, p. 5 (1988); R. Francis, *British Corrosion Journal*, **29**, 53 (1994); *R.* Francis, *Galvanic Corrosion: A Practical Guide for Engineers*, NACE International, Houston, TX (2001).].

A driving force for corrosion or current flow is the potential difference formed between dissimilar metals. The extent of accelerated corrosion resulting from galvanic corrosion is related to potential differences between dissimilar metals or alloys, the nature of the environment, the polarization behavior of metals involved in the galvanic couple, and the geometric relationship of the component metals. A **galvanic series of metals and alloys** is useful for predicting the possibility for galvanic corrosion in seawater (see Fig. 4.5.1) [R. Baboian, *Galvanic and Pitting Corrosion - Field and Laboratory Studies*, STP 576, ASTM, Philadelphia (1976)]. A galvanic series must be arranged according to the potentials of the metals or alloys in a specific electrolyte at a given temperature and conditions. Separation of two metals in the galvanic series may indicate the possible magnitude of the galvanic corrosion. Since corrosion product films and other changes in the surface composition may occur in different environments, no singular value may be assigned for a particular metal or alloy. This is important since polarity reversal may occur. For example, at ambient temperatures, zinc is more active than steel and is used to cathodically protect steel; however, above 60°C (such as found in hot water heaters), corrosion products formed on zinc may cause the steel to become anodic and corrode preferentially to zinc.

The magnitude of the potential difference between the dissimilar materials does not indicate the degree of galvanic corrosion because potentials are a function of the thermodynamics and not of the reaction kinetics that may occur. The relative surface kinetics of the galvanic

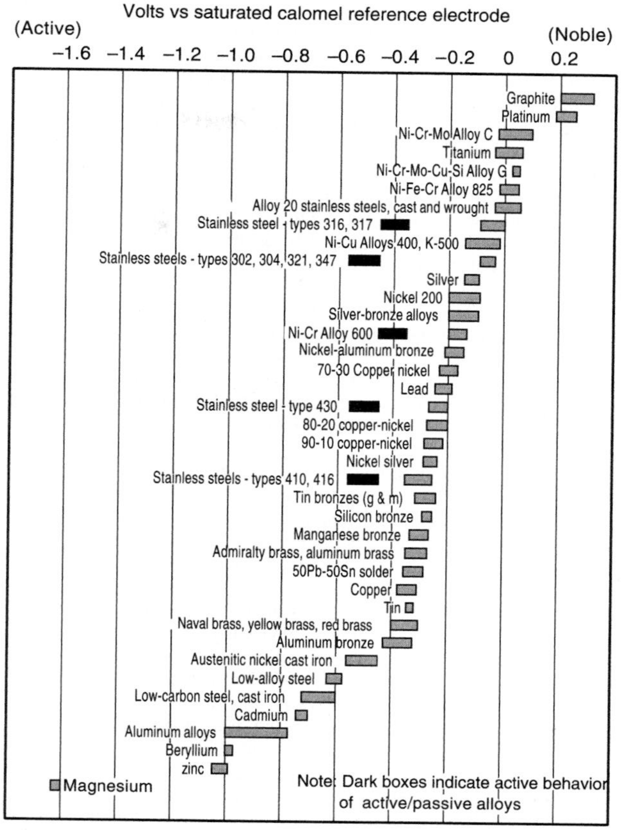

Figure 4.5.1 Galvanic Series in seawater. (F.L. LaQue, "*Marine Corrosion*", John Wiley & Sons, New York, p.179 (1975)].

members ultimately will determine the severity of galvanic corrosion. A difference of 50 mV between select dissimilar materials can lead to severe corrosion while a potential difference of 800 mV has been successfully coupled [*R.* Francis, *Galvanic Corrosion: A Practical Guide for Engineers*, NACE International, Houston, TX (2001)].

Corrosion from galvanic effects is usually worse near the couple junction and decreases with distance from the junction. The distance influenced by galvanic corrosion will be affected by solution conductivity. In low conductivity solutions, galvanic attack may be observed as a sharp localized groove at the junction, while in high conductivity solutions, galvanic corrosion may be shallow and spread over a relatively long distance. Another important factor in galvanic corrosion is the cathodic-to-anodic area ratio. Since anodic and cathodic currents (μA) must be equal in a corrosion reaction, an unfavorable area ratio is composed of a large cathode and a small anode. The current density ($\mu A/cm^2$) and corrosion rate are greater for the small anode than for the large cathode. Galvanic corrosion rates for an unfavorable cathode-to-anode ration may be 100-1,000 times greater than if the cathode and anode areas were equal. The method by which dissimilar metals or alloys are joined may affect the degree of galvanic corrosion. Welding, where a gradual transition from one material to another often exists, could react differently in a system where two materials are insulated by a gasket but electrically connected elsewhere through the solution, or differently still in a system connected by fasteners.

Galvanic corrosion may be minimized by: (1) selecting combinations of metals or alloys near each other in the galvanic series; (2) avoiding unfavorably large cathode/anode area ratios; (3) completely insulating dissimilar metals whenever possible to disrupt the electrical circuit; (4) adding inhibitors to decrease the aggressiveness of the environment; (5) avoiding threaded joints connecting materials far apart in the galvanic series; (6) designing for readily replaceable anodic parts; and (7) adding a third metal that is anodic to both metals or alloys in the galvanic couple [*Galvanic Corrosion*, STP 978, H.P. Hack, ed., ASTM, Philadelphia (1988)].

Crevice corrosion is the intensive localized corrosion that may occur because of the presence of narrow openings or gaps between metal-to-metal (under bolts, lap joints, or rivet heads) or nonmetal-to-metal (under gaskets, surface deposits, or dirt) components and small volumes of stagnant solution. Crevice corrosion sometimes is referred to as **underdeposit corrosion** or **gasket corrosion**. Resistance to crevice corrosion can vary from one alloy/environment system to another. Crevice corrosion may range from near uniform attack to severe localized dissolution at the metal surface within the crevice and may occur in a variety of metals and alloys ranging from relatively noble metals such as silver and copper to active metals such as aluminum and titanium. Stainless steels are susceptible to crevice corrosion. Fibrous gasket materials that draw solution into a crevice by capillary action are particularly effective in promoting crevice corrosion. The environment can be any neutral or acidic aggressive solution, but solutions containing chlorides are particularly conducive to crevice corrosion. In cases where the bulk environment is particularly aggressive, general corrosion may deter localized corrosion at a crevice site.

A condition for crevice corrosion is a crevice or gap that is wide enough to permit solution entry, but is sufficiently narrow to maintain a stagnant zone of solution within the crevice to restrict entry of cathodic reactants and removal of corrosion products through diffusion and migration. Crevice gaps are usually 0.025 to 0.1 mm (0.001 to 0.004 in.) wide. The critical depth for crevice corrosion is dependent on the gap width, and may be only 15 to 40 μm deep. Most cases of crevice corrosion occur in near-neutral solutions in which dissolved oxygen is the cathodic reactant. In seawater, localized corrosion of copper and its alloys (due to differences in Cu^{+2} concentration) is different from that of the stainless steel group because the attack occurs outside of the crevice rather than within. In acidic solutions where hydrogen ion is the cathodic reactant, crevice corrosion actually occurs at the exposed surface near the crevice. Decreasing crevice width or increasing crevice depth, bulk chloride concentration, and/or acidity will increase the potential for crevice corrosion. In some cases, deep crevices may restrict propagation because of the voltage drop through the crevice solution [T.S. Lee, R.M. Kain, J.W. Oldfield, *Corrosion/83*, paper 69, NACE International, Houston (1983)]. Survey of Literature on Crevice Corrosion (1979-1998), F.P. I Jsseling, ed., European Federation of Corrosion Publications, No. 30 (2000).

A condition common to crevice corrosion is the development of an occluded localized environment that differs considerably from the bulk solution. The basic overall mechanism of crevice corrosion is the dissolution of a metal [Eq. (4.5.1)] and the reduction of oxygen to hydroxide ions [Eq. (4.5.4)]. Initially, these reactions take place uniformly, both inside and outside the crevice. However, after a short time, oxygen within the crevice is depleted because of restricted convection and oxygen reduction ceases in this area. Because the area within the crevice is much smaller than the external area, oxygen reduction remains virtually unchanged. Once oxygen reduction ceases within the crevice, the continuation of metal dissolution tends to produce an excess of positive charge within the crevice. To maintain charge neutrality, chloride or possibly hydroxide ions migrate into the crevice. This results in an increased concentration of metal chlorides within the crevice, which hydrolyzes in water according to Eq. (4.5.7) to form an insoluble hydroxide and a free acid.

$$M^+Cl^- + H_2O \longrightarrow MOH\downarrow + H^+Cl^- \qquad (4.5.7)$$

The increased acidity boosts the dissolution rates of most metals and alloys, which increases migration and results in a rapidly accelerating and autocatalytic process. The chloride concentration within crevices exposed to neutral chloride solutions is typically three to ten times higher than in the bulk solution. The pH within the crevice is 2-3. The pH drop with time and critical crevice solution that will cause breakdown in stainless steels can be calculated [J.W. Oldfield, W.H. Sutton, *Brit. Corrosion J.*, **13**, 13 (1978); *Survey of Literature on Crevice Corrosion (1979-1998)*, European Federation of Corrosion Publications, Number 30, F.P. I Jsseling, ed., IOM Communications, London (2000).].

Optimum crevice corrosion resistance can be achieved with an active/passive metal that possesses: (1) a narrow active/passive transition; (2) a small critical current density; (3) an extended passive region. Crevice corrosion may be minimized by avoiding riveted, lap, or bolted joints in favor of properly welded joints, designing vessels for complete drainage and removal of sharp corners and stagnant areas, removing deposits frequently, and using solid, **nonabsorbent gaskets** (such as Teflon) wherever possible.

Pitting is a form of extremely localized attack that results in a cavity or hole in the metal or alloy. Deterioration by pitting is one of the most dangerous types of localized corrosion, but its unanticipated occurrences and propagation rates are difficult to consider in practical engineering designs. Pits are often covered by corrosion products. Depth of pitting is sometimes expressed by the **pitting factor**, which is the ratio of deepest metal penetration to average metal penetration as determined by weight loss measurements. A pitting factor of one denotes uniform corrosion.

Pitting is usually associated with the breakdown of a passive metal. Breakdown involves the existence of a critical potential, E_b, induction time at a potential > E_b, presence of aggressive species (Cl^-, Br^-, ClO_4^-, etc.), and discrete sites of attack. Pitting is associated with local imperfections in the passive layer. Nonmetallic inclusions, second phase precipitates, grain boundaries, scratch lines, dislocations, and other surface inhomogeneities can become initiation sites for pitting [Z. Szklarska-Smialowska, *Pitting of Metals*, NACE International, Houston (1986); Z. Szklarska-Smialowska, *Pitting and Crevice Corrosion*, NACE International, Houston (2005).]. The induction time before pits initiate may range from days to years. Induction time depends on the metal, aggressiveness of the environment, and the potential, and tends to decrease if either the potential or the concentration of aggressive species increases.

Once pits are initiated, they continue to grow through an autocatalytic process (i.e., the corrosion conditions within the pit are both stimulating and necessary for continuing pit growth). Pit growth is controlled by the rate of depolarization at the cathodic areas. Oxygen in aqueous environments and ferric chloride in various industrial environments are effective depolarizers. The propagation of pits is virtually identical to crevice corrosion. Factors contributing to localized corrosion such as crevice corrosion and pitting of various materials are found in other publications [*Localized Corrosion-Cause of Metal Failure*, STP 516, ASTM, Philadelphia, (1972)], [*Localized Corrosion*, NACE-3, B.F. Brown, J. Kruger, R.W. Staehle, eds., NACE International, Houston (1974)], [*Advances in Localized Corrosion*, NACE-9, H.S. Isaacs, U. Bertocci, J. Kruger, S. Smialowska, eds., NACE International, Houston (1990)], [*Critical Factors in Localized Corrosion*, PV 92-9, G.S. Frankel, R.C. Newman, eds., The Electrochemical Society, Pennington, NJ (1992); *Critical Factors in Localized Corrosion II*, PV 95-15, P.M. Natishan, R.G. Kelly, G.S. Frankel and R. Newman, eds., The Electrochemical Society, Pennington, NJ (1996); *Critical Factors in Localized Corrosion III, A Symposium in Honor of the 70th Birthday of Jerome Kruger*, PV 98-17, R.G. Kelly, P.M. Natishan, G.S. Frankel, and R.C. Newman, eds., The Electrochemical Society, Pennington, NJ (1999); *Critical Factors in Localized Corrosion IV: A Symposium in Honor of the 65th Birthday of Hans Bohni,* PV2002-24, S. Virtanen, P. Schmuki, and G.S. Frankel, eds., The Electrochemical Society, Pennington, NJ (2002).].

Copper and its alloys suffer localized corrosion by nodular pitting in which the attacked areas are covered with small mounds or nodules composed of corrosion product and $CaCO_3$ precipitated from the water. The pit interior is covered with CuCl which prevents formation of a protective Cu_2O layer. Pitting occurs most often in cold, moderately hard to hard waters, but also happens in soft waters above 60°C. Pitting is associated with the presence of very thin carbon (cathodic) films from lubricants used during tube manufacture. Copper pitting is favored by a high sulfate/chloride ratio. Copper-nickel tubing suffers from accelerated corrosion in seawater due to the presence of both sulfides and oxygen. Sulfides prevent the formation of a protective oxide layer, which allows the anodic reaction to proceed unabated supported by the oxygen reduction reaction.

Caustic corrosion, frequently referred to as **caustic gouging** or **ductile gouging,** is a form of localized corrosion that occurs as a result of fouled heat-transfer surfaces of water-cooled tubes and the presence of an active corrodent in the water of high pressure boilers. Porous deposits that form and grow in these high heat input areas produce conditions that allow sodium hydroxide to permeate the deposits by a process called **wick boiling**, and concentrate to extremely high levels (e.g., 100 ppm bulk water concentrated to 220,000 ppm NaOH in water film under these deposits) which dissolves the protective magnetite and then reacts directly with the steel to cause rapid corrosion. A smooth, irregular thinning occurs, usually downstream of flow disruptions or in horizontal or inclined tubing [R.D. Port, H.M. Herro, *The Nalco Guide to Boiler Tube Analysis*, McGraw-Hill, New York, 57-70 (1991); D.A. Shifler, J.F. Wilkes, "Historical Perspectives of Intergranular Corrosion of Boiler Metal by Water," **CORROSION/93**, paper 44, NACE International, Houston, TX (1993)].

Intergranular corrosion is localized attack that follows a narrow path along the grain boundaries of a metal or alloy. Intergranular corrosion is caused by the presence of impurities, precipitation of one of the alloying elements, or depletion of one of these elements at or along the grain boundaries. The driving force of intergranular corrosion is the difference in corrosion potential that exists between a thin grain boundary zone and the bulk of immediately adjacent grains. Intergranular corrosion most commonly occurs in aluminum or copper alloys and austenitic stainless steels.

When austenitic stainless steels such as 304 are heated or placed in service at temperatures of 950-1,450°C, they become sensitized and susceptible to intergranular corrosion, because chromium carbide ($Cr_{26}C_6$), is formed. This effectively removes chromium from the vicinity of the grain boundaries, thereby approaching the corrosion resistance of carbon steel in these chromium-depleted areas. Methods to avoid **sensitization** and control intergranular corrosion of austenitic stainless steels are: (1) employing high temperature heat treatment (solution quenching) at 1,950-2,050°C; (2) adding elements that are stronger carbide formers (stabilizers such as columbium (niobium), niobium plus tantalum, or titanium) than chromium; and (3) lowering the carbon content below 0.03 percent. Stabilized austenitic stainless steels can experience intergranular corrosion if heated above 2,250°C, where columbium carbides dissolve as may occur during welding. Intergranular corrosion of stabilized stainless steels is called **knife-line** attack. This attack transpires in a narrow band in the parent metal adjacent to a weld. Reheating the steel to around 1,950-2,250°C reforms the stabilized carbides.

Surface carburization also can cause intergranular corrosion in cast austenitic stainless steels. Overaging of aluminum alloys can form precipitates such as $CuAl_2$, $FeAl_3$, $MgAl_3$, $MgZn_2$, and $MnAl_6$ along grain boundaries and promote intergranular corrosion. Some nickel alloys can form intergranular precipitates of carbides and intermetallic phases during heat treatment or welding. Alloys with these intergranular precipitates are subject to intergranular corrosion.

Dealloying involves the selective removal of the most electrochemically active constituent metal in the alloy. Selective removal of one metal can result in either localized attack leading to possible perforation (plug dealloying) or in a more uniform attack (layer dealloying) resulting in loss of component strength. Although selective removal of metals such as Al, Fe, Si, Co, Ni, and Cr from their alloys has occurred, the most common dealloying is the selective removal of zinc from brasses called **dezincification**. Dezincification of brasses (containing 15% or more zinc) takes place in soft waters, especially if the carbon dioxide content is high. Other factors that increase the susceptibility of brasses to dealloying are high temperatures, waters with high chloride content, low water velocity, crevices, and deposits on the metal surface. Dezincification is experienced over a wide pH range. Layer dezincification is favored when the environment is acidic and the brass has a high zinc content. Plug dezincification tends to develop when the environment is slightly acidic, neutral, or alkaline and the zinc content is relatively low. It has been proposed that either zinc is selectively dissolved from the alloy leaving a porous structure of metallic copper *in situ* within the brass lattice, or that both copper and zinc are dissolved initially, but copper immediately redeposits at sites close to where the brass dissolved.

Dezincification can be minimized by reducing oxygen in the environment or by cathodic protection. Dezincification of α-brass (70% Cu, 30% Zn) can be minimized by alloy additions of 1 percent tin. Small amounts (~0.05%) of arsenic, antimony, or phosphorus promote effective inhibitor of dezincification for 70/30 α-brass (inhibited Admiralty brass). Brasses containing less than 15 percent zinc have not been reported to suffer dezincification.

Gray cast iron selectively leaches iron in mild environments, particularly in underground piping systems. Graphite is cathodic to iron, which sets up a galvanic cell. The dealloying of iron from gray cast iron leaves a porous network of rust, voids and graphite termed **graphitization** or **graphitic corrosion**. The cast iron loses both its strength and its metallic properties, but experiences no apparent dimensional changes. Graphitization does not occur in nodular or malleable cast irons because a continuous graphite network is not present. White cast iron also is immune to graphitization since it has essentially no available free carbon.

Dealloying also can occur at high temperatures. Exposure of stainless steels to low-oxygen atmospheres at 1,800°F (980°C) results in the selective oxidation of chromium and the creation of a more protective scale, and the depletion of chromium under the scale in the substrate metal. Decarburization can occur in carbon steel tubing if heated above 1,600°F (870°C). Denickelification of austenitic stainless steels occurs in liquid sodium.

Corrosion rates are often dependent on fluid flow and the availability of appropriate species required to drive electrochemical reactions. An increase in fluid flow may increase corrosion rates by removing protective films or increasing the diffusion or migration of deleterious species. An increase in fluid flow may also decrease corrosion rates by eliminating aggressive ion concentration or enhancing passivation or inhibition by transporting the protective species to the fluid/metal interface. There are a number of mechanisms caused by the conjoint action of flow and corrosion that result in several types of flow-influenced corrosion: (1) mass transport-controlled corrosion; (2) phase transport controlled corrosion; (3) erosion corrosion; and (4) cavitation [E. Heitz, *Corrosion*, v. 47, 135 (1991); also *Corrosion/90,* paper 1, NACE International, Houston, TX (1990)]. Mass transport-controlled corrosion implies that the rate of corrosion is dependent on the convective mass transfer processes at the metal/fluid interface. Corrosion by mass transport will often be streamlined and smooth. Phase transport-controlled corrosion suggests that the wetting of the metal surface by a corrosive phase is flow dependent. This may occur from the separation of one liquid phase from another or by the formation of second phase from a liquid. An example of the second mechanism is the formation of discrete bubbles or a vapor phase from boiler water in horizontal or inclined tubes in high heat-flux areas under low flow conditions (departure from nucleate boiling). The distribution and morphology of the corrosive attack will be related to the distribution of the corrosive phase — the corroded sites will frequently display rough, irregular surfaces and coated with thick, porous corrosion deposits. [D.A. Shifler, "Flow-Influenced Corrosion in a Waste Energy Sparge Line," *Materials Performance*, v. 37, 38-44 (1998).]

Erosion-corrosion is the acceleration or increase of attack because of the relative movement between a corrosive fluid and the metal surface. Mechanically, the conjoint action of erosion and corrosion not only damages the protective film but also the metal or alloy surface itself by the abrasive action of a fluid (gas or liquid) at high velocity. The inability to maintain a protective passive film on the metal or alloy surface raises the corrosion rate. Erosion-corrosion is characterized by grooves, gullies, waves, rounded holes, and valleys—usually exhibiting in a directional manner. Components exposed to moving fluids and subject to erosion-corrosion include bends, elbows, tees, valves, pumps, blowers, propellers, impellers, agitators, condenser tubing, orifices, turbine blades, nozzles, wear plates, baffles, boiler tubes, and ducts.

The hardness of the alloy or metal and/or the mechanical or protective nature of the passive film will affect the resistance to erosion-corrosion in a particular environment. Copper and brasses are particularly susceptible to erosion-corrosion. Models and experimental results have shown that flow stream velocity and increases of particle size, sharpness, density, and concentration increase the erosion corrosion rate. Increases in fluid viscosity, density, target material hardness, and/or pipe diameter decrease the corrosion rate. Often, an escalation in the fluid velocity increases the attack by erosion-corrosion. The effect may be nil until a critical velocity is reached, above which attack may increase very rapidly. The critical velocity is dependent on the alloy, passive layer, temperature, and nature of the fluid environment. The increased attack may be attributed, in part, to increasing cathodic reactants to the metal surface. Increased velocity may decrease attack by increasing the effectiveness of inhibitors or by preventing silt or dirt from depositing and causing crevice corrosion.

Erosion-corrosion includes impingement, cavitation, and fretting corrosion. **Impingement attack** occurs when a fluid strikes the metal surface at high velocity at directional change sections such as bends, tees, turbine blades, and inlets of pipes or tubes. In the majority of cases involving impingement attack, a geometrical feature of the system results in turbulence. Air bubbles or solids present in turbulent flow will accentuate attack.

Cavitation damage is caused by the formation and collapse of vapor bubbles in a liquid near a metal surface. Cavitation occurs where high velocity water and hydrodynamic pressure changes are encountered, such as ship propellers or pump impellers. If pressure on a liquid is lowered drastically by flow divergence, the water vaporizes and forms bubbles. These bubbles generally collapse rapidly, producing shock waves with pressures as high as 60,000 lb/in^2 (410 MPa), which destroys the passive film. The newly exposed metal surface corrodes, reforms a passive film which is destroyed again by another cavitation bubble, and so forth. As many as 2,000,000 bubbles may collapse over a small area in one second. Cavitation damage causes both corrosion and mechanical effects.

Fretting corrosion is a form of damage caused at the interface of two closely fitting surfaces under load when subjected to vibration and slip. Fretting is explained as: (1) the surface oxide is ruptured at localized sites prompting reoxidation; or (2) material wear and friction cause local oxidation. The degree of plastic deformation is greater for softer materials than for hard; seizing and galling may often occur. Lubrication, increased hardness, increased friction, and the use of gaskets can reduce fretting corrosion.

Environmentally-induced cracking is a brittle fracture of an otherwise ductile material due to the conjoint action of tensile stresses in a specific environment over a period of time. Environmentally-induced cracking includes: (1) stress corrosion cracking; (2) hydrogen damage; (3) corrosion fatigue; (4) liquid metal embrittlement; and (5) solid metal induced embrittlement.

Stress corrosion cracking (SCC) refers to service failures due to the joint interaction of tensile stress with a specific corrodent and a susceptible normally ductile material. The observed crack propagation is the result of mechanical stress and corrosion reactions. Stress can be externally applied, but most often is the result of residual tensile stresses. SCC can initiate and propagate with little outside evidence of corrosion or macroscopic deformation and usually provides no warning of impending failure. Cracks often initiate at surface flaws that were preexisting or formed during service. Branching is frequently associated with SCC; the cracks tend to be very fine and tight and visible only with special techniques or by microscopic examination. SCC can be intergranular or transgranular. Intergranular SCC is often associated with grain boundary precipitation or grain boundary segregation of alloy constituents such as found in sensitized austenitic stainless steels. Transgranular SCC is related to crystal structure, anisotropy, grain size and shape, dislocation density and geometry, phase composition, yield strength, ordering, and stacking fault energies. The specific corrodents of SCC for different alloys include: (1) carbon steels - sodium hydroxide; (2) stainless steels - NaOH or chlorides; (3) α-brass - ammoniacal solutions; (4) aluminum alloys - aqueous Cl^-, Br^-, I^- solutions; (5) titanium alloys - aqueous Cl^-, Br^-, I^- solutions, organic liquids; (6) high nickel alloys - high-purity steam [*Environment-Induced Cracking of Metals*, NACE-10, R.P. Gangloff, M.B. Ives, eds., NACE International, Houston (1990)], [*Parkins Symposium on Fundamental Aspects of Stress Corrosion Cracking*, S.M. Bruemmer *et al.*, eds., The Minerals, Metals, & Materials Society, Warrendale, PA (1992); R.L. Jones, *Stress Corrosion Cracking: Materials Performance and Evaluation*, ASM International, Materials Park, OH (1992); *Chemistry and Electrochemistry of Corrosion and Stress Corrosion Cracking: A Symposium Honoring the Contributions of R.W. Staehle*, R.H. Jones ed., The Minerals, Metals, & Materials Society, Warrendale, PA (2001).].

Hydrogen interactions can affect most engineering metals and alloys. Terms to describe these interactions have not been universally agreed upon. Hydrogen damage or hydrogen attack is caused by the diffusion of hydrogen through carbon and low alloy steels. Hydrogen reacts with carbon, either in elemental form or as carbides, to form methane gas. Methane accumulates at grain boundaries and can cause local pressures to exceed the yield strength of the material. Hydrogen damage causes intergranular, discontinuous cracks and decarburization of the steel. Once cracking occurs, the damage is irreversible. In utility boilers, hydrogen damage has been associated with fouled heat transfer surfaces. Hydrogen damage is temperature dependent with a threshold temperature of about 400°F (204°C). Hydrogen damage has been observed in boilers operating at pressures of 450-2,700 psi (3.1-18.6 MPa) and tube metal temperatures of 600-950°F (316-510°C). Hydrogen damage may occur in either acidic or alkaline conditions, although this contention is not universally accepted.

Hydrogen embrittlement results from penetration and adsorption of atomic hydrogen into an alloy matrix. This process results in a decrease of toughness and ductility of alloys due to hydrogen ingress. Hydrogen may be removed by baking at 200-300°F (93-149°C), which returns the alloy mechanical properties very near to those existing before hydrogen entry. Hydrogen embrittlement can affect most metals and alloys. Hydrogen blistering results from atomic hydrogen penetration at internal defects near the surface such as laminations or nonmetallic inclusions where molecular hydrogen forms. Pressure from H_2 can cause local plastic deformation or surface exfoliation [*Hydrogen Effects on Material Behavior*, N.R. Moody, A.W. Thompson, eds., The Minerals, Metals, & Materials Society, Warrendale, PA (1990); D.A. Shifler, J.F. Wilkes, "Historical Perspectives of Intergranular Corrosion of Boiler Metal by Water," **CORROSION/93**, paper 44, NACE International, Houston, TX (1993)].

For higher alloyed steels, the higher the strength of the alloy, the more susceptible the alloy will be to internal hydrogen embrittlement and to hydrogen environmental embrittlement. Hydrogen solubility in an ultra high strength steel (UHHS)/environment system is controlled by trapping and the binding energies of the particular trap sites. Understanding hydrogen interactions and transport in UHSS microstructures is necessary to control hydrogen embrittlement [D. Li, R.P. Gangloff, J.R. Scully, "Hydrogen Trap States in Ultrahigh-Strength AERMET 100 Steel," *Metallurgical and Materials Transactions A*, v. 35A, 849 (2004); R.L.S. Thomas, J.R. Scully, R.P. Gangloff, "Internal Hydrogen Embrittlement of Ultrahigh-Strength AERMET 100 Steel," *Metallurgical and Materials Transactions A*, v. 34A, 327 (2003).].

When a metal or alloy is subjected to cyclic, changing stresses, cracking and fracture can occur at stresses lower than the yield stress through fatigue. A clear relationship exists between stress amplitude and number of cyclic loads; an endurance limit is the stress level below which fatigue will not occur indefinitely. Fracture solely by fatigue is not viewed as an example of environmental cracking. However, the conjoint action of a corrosive medium and cyclic stresses on a material is termed **corrosion fatigue**. Corrosion fatigue failures are usually "thick-walled" brittle ruptures showing little local deformation. An endurance limit often does not exist in corrosion fatigue failures. Cracking often begins at pits, notches, surface irregularities, welding defects, or sites of intergranular corrosion. Surface features of corrosion fatigue cracking vary with alloys and specific environmental conditions. Crack paths can be transgranular (carbon steels, aluminum alloys, etc.) or intergranular (copper, copper alloys, etc.); exceptions to the norm are found in each alloy system. "Oyster shell" markings may denote corrosion fatigue failures, but corrosion products and corrosion often cover and obscure this feature. Cracks usually propagate in the direction perpendicular to the principal tensile stress; multiple, parallel cracks are usually present near the principal crack which led to failure. Applied cyclic stresses are caused by mechanical restraints or thermal fluctuations; susceptible areas may also include structures that possess residual stresses from fabrication or heat treatment. Low pH, and high stresses, stress cycles, and oxygen content or corrosive species concentrations will increase the probability of corrosion fatigue cracking [*Corrosion Fatigue-Mechanics, Metallurgy, Electrochemistry, and Engineering*, STP 801, T.W. Crooker, B.N. Leis, eds., ASTM, Philadelphia (1983)], [*Corrosion Fatigue-Chemistry, Mechanics, and Microstructure,* A.J. McEvily, R.W. Staehle, eds., NACE International, Houston (1986)].

Liquid metal embrittlement (LME) or liquid metal induced embrittlement is the catastrophic brittle fracture of a normally ductile metal when coated by a thin film of liquid metal and subsequently stressed in tension. Cracking may either be intergranular or transgranular. Usually, a solid that has little or no solubility in the liquid and forms no intermetallic compounds with the liquid constitutes a couple with the liquid. Fracture can occur well below the yield stress of the metal or alloy [D.R.H. Jones, *Engineering Failure Analysis*, 1, 51 (1994)]. A partial list of examples of LME is: (1) aluminum (by mercury, gallium, indium, tin-zinc, lead-tin, sodium, lithium, and inclusions of lead, cadmium, or bismuth in aluminum alloys); (2) copper and select copper alloys (mercury, antimony, cadmium, lead, sodium, lithium, and bismuth); (3) zinc (mercury, gallium, indium, Pb-Sn solder); (4) titanium or titanium alloys (mercury, zinc, and cadmium); (5) iron and carbon steels (aluminum, antimony, cadmium, copper, gallium, indium, lead, lithium, zinc, and mercury); and (6) austenitic and nickel-chromium steels (tin, and zinc).

Solid metal induced embrittlement (SMIE) occurs below the melting temperature, T_m, of the solid in various LME couples. Many instances of loss of ductility or strength, and brittle fracture have taken place with electroplated metals or coatings and inclusions of low-melting alloys below T_m. Cadmium-plated steel, leaded steels, and titanium embrittled by cadmium, silver, or gold are materials affected by SMIE. The prerequisites for SMIE and LME are similar, although multiple cracks are formed in SMIE and SMIE crack propagation is 100-1,000 times slower than LME cracking. More detailed descriptions of LME and SMIE are discussed elsewhere [*Embrittlement of Metals*, M.H. Kambar, ed., American Institute of Mining, Metallurgical and Petroleum Engineers, St. Louis (1984)].

High temperature corrosion is a phenomenon that occurs in components that operate at high temperatures (>250°C), such as in power generation (nuclear and fossil fuel boilers), pulp and paper boilers, waste incinerators, solid-oxide fuel cells, mineral and metallurgical processing, industrial, aviation and marine gas turbines, and petrochemical and chemical processing.

High-temperature corrosion may occur in numerous environments and is affected by numerous factors such as temperature, alloy or protective coating composition, time, and gas composition. Alloys often rely upon the oxidation reaction to develop a protective scale to resist corrosion attack such as oxidation, sulfidation, carburization, and other forms of high temperature attack. However, nearly all metals, alloys and materials of technological interest will oxidize and corrode at high temperatures. The nature of the corrosion products, the rate of corrosion and the mechanism can vary wildly depending on the environment and the temperature. High temperature corrosion is not restricted to the gaseous phase—solid ash and salt deposits contribute to the corrosive effect, with associated erosion and removal of scale. In the liquid phase, molten metals and molten salts pose their own unique variety of challenges, causing highly complex and environmental-dependent corrosion.

The rate by which a material may degrade in a high temperature environment will depend on the reaction or range of operative reactions, the reaction thermodynamics, kinetics, and chemical physical properties of the material exposed to this environment. A material can degrade at high temperatures upon exposure to gaseous, liquid (molten ash, salts, or metal/alloys), or solid state reaction processes such as diffusion and interdiffusion, precipitation, and creep.

High temperature corrosion/degradation of materials may occur through a number of potential processes: (1) oxidation, (2) carburization and metal dusting, (3) nitridation, (4) sulfidation, (5) hot corrosion, (6) chloridation and other halogenization reactions, (7) hydrogen interactions, (8) molten metals, (9) molten salts, (10) aging reactions such as sensitization, (11) creep, (12) erosion corrosion and wear. Various materials (alloys, high temperature coatings, refractories, glasses) are exposed to high temperature environments and experience varying degrees of degradation [R.B. Pond, Jr., D.A. Shifler, "High Temperature Corrosion Failures," in Volume 11: Failure_Analysis, Metals Handbook, ASM International, Materials Park, OH, (2002); D.A. Shifler, "High Temperature Gaseous Corrosion Testing," Volume 13A, Corrosion Principles, ASM Handbook, ASM International, Materials Park, OH, (2003)].

Oxide growth advances with time by parabolic, linear, cubic, or logarithmic rates. The corrosion rate can be influenced considerably by the presence of contaminants, particularly if they are adsorbed on the metal surface. Each alloy has an upper temperature limit where the oxide scale, based on its chemical composition and mechanical properties, cannot protect the underlying alloy. Catastrophic oxidation refers to metal-oxygen reactions that occur at continuously increasing rates or break away from protective behavior very rapidly. Molybdenum, tungsten, vanadium, osmium, rhenium, and alloys containing Mo or V may oxidize catastrophically. Internal oxidation may take place in copper or silver-based alloys containing small levels of Al, Zn, Cd, or Be when more stable oxides are possible below the surface of the metal/scale interface. Sulfidation forms sulfide phases with less stable base metals, rather than oxide phases, when H_2S, S_2, SO_2, and other gaseous sulfur species have a sufficiently high concentration. In corrosion resistant alloys, sulfidation usually occurs when Al or Cr is tied up with sulfides which interferes with the process of developing a protective oxide. Carburization of high-temperature alloys is possible in reducing carbon-containing environments. Hot corrosion involves the high-temperature, liquid phase attack of alloys by eutectic alkali-metal sulfates that solubilize an protective alloy oxide, and exposes the bare metal, usually iron, to oxygen which forms more oxide and promotes subsequent metal loss at temperatures of 1,000-1,300°F (538-704°C). Vanadium pentoxide and sodium oxide can also form low melting point (~1,000°F [538°C]) liquids capable of causing alloy corrosion. These are usually contaminants in fuels, either coal or oil.

Alloy strength decreases rapidly when metal temperatures approach 900-1,000°F and above. Creep entails a time-dependent deformation involving grain boundary sliding and atom movements at high temperatures. When sufficient strain has developed at the grain boundaries, voids and microcracks develop, and grow and coalesce to form larger cracks until failure occurs. Creep rate will increase and the projected time to failure decrease when stress and/or tube metal temperature is increased [*High Temperature Corrosion*, NACE-10, R.A. Rapp, ed., NACE International, Houston (1983)], [G.Y. Lai, *High Temperature Corrosion of Engineering Alloys*, ASM International, Materials Park, OH (1990); B. Gleeson, "High-Temperature Corrosion of Metallic Alloys and Coatings," in: *Corrosion and Environmental Degradation of Materials*, M. Schütze (ed.), Volume 19 of the Series: *Materials Science and Technology*, R.W. Cahn, P. Haasen, E.J. Kramer (Series Editors), Wiley-VCH,

Germany, pp. 173-228 (2000); A.S. Khanna, *Introduction to High Temperature Oxidation and Corrosion*, ASM International, Materials Park, OH (2002); John Stringer Symposium on High Temperature Corrosion, P.F. Tortorelli, I.G. Wright, P.Y. Hou, Eds., ASM International, Materials Park, OH, (2003); High Temperature Corrosion and Materials Chemistry IV, E.J. Opila, P.Y. Hou, T. Maruyama, B. Pieraggi, M.J. McNallan, E. Wuchina, D. A. Shifler, PV 2003-12, The Electrochemical Society, Pennington, NJ (2003); Proceedings of the Workshop on Materials and Practice to Improve Resistance to Fuel Derived Environmental Damage in Land and Sea Based Turbines, G.R. Edwards, D.L. Olson, R. Visawantha, T. Stringer, P. Bollinger, October 22-23, 2002, EPRI Publication, Electric Power Research Institute, 3412 Hillview Avenue, Palo Alto, CA, 94304 (2003)].

CORROSION PROTECTION METHODS

The basis of **corrosion protection** methods involves restricting or controlling anodic and/or cathodic portions of corrosion reactions, changing the environmental variables, or breaking the electrical contact between anodes and cathodes. Corrosion protection methods include: (1) proper materials selection; (2) design; (3) coatings; (4) use of inhibitors; (5) anodic protection; and (6) cathodic protection [*NACE Corrosion Engineer's Reference Book,* 3rd ed., Baboian, ed., NACE International, Houston (2002)].

The most common method to reduce corrosion is selecting the right material for the environmental conditions and applications. Ferrous and nonferrous metals and alloys, thermoplastics, nonmetallic linings, resin coatings, composites, glass, concrete, and nonmetallic elements are a few of the potential materials available for selection [P.L. Mangonon, *The Principles of Materials Selection for Engineering Design*, Prentice Hall, New York (1999)]. Experience and data, either in-house or from outside vendors and fabricators, may assist in the materials selection process. Many materials can be eliminated by service conditions (temperature, pressure, strength, chemical compatibility). Corrosion data for a particular chemical or environment may be obtained through a literature survey [*Corrosion Abstracts,* NACE International], [*Corrosion Data Survey-Metals Section*, 6th ed. (1985); *Handbook of Corrosion Data*, 2nd ed., ASM International, Materials Park, OH, (1994); *Corrosion Data Survey-Nonmetals Section*, NACE International, Houston 5th ed., (1975)] or from expert systems or databases. Different organizations (ASTM, ASM International, ASME, NACE International) may offer references to help solve problems and make predictions about the corrosion behavior of candidate materials. The material should be cost-effective and resistant to the various forms of corrosion (discussed earlier) that may be encountered in its application. General rules may be applied to determine resistance of metals and alloys. For reducing or nonoxidizing environments (such as air-free acids and aqueous solutions) nickel, copper, and their alloys are usually satisfactory. For oxidizing conditions, chromium alloys are used, while in extremely powerful oxidizing environments, titanium and its alloys have shown superior resistance. The corrosion resistance of chromium and titanium can be enhanced in hot concentrated oxidizer-free acid by small additions of platinum, palladium, or rhodium.

Many costs related to corrosion could be eliminated by **proper design**. The designer should have a credible knowledge of corrosion or should work in cooperation with a materials and corrosion engineer. The design should avoid gaps or structures where dirt or deposit could easily form crevices; horizontal faces should slope for easy drainage. Tanks should be properly supported and employ the use of drip skirts and dished, fatigue-resistant bottoms [*Guidelines for the Welded Fabrication of Nickel Alloys for Corrosion-Resistant* Service, part 3, Nickel Development Institute, Toronto, Canada (1994)]. Connections should avoid dissimilar metals or alloys widely separated in the galvanic series for lap joints or fasteners, if possible. Welded joints should avoid crevices, microstructural segregation, and high residual or applied stresses. Positioning of parts should account for prevailing environmental conditions. The design should avoid local differences in concentration, temperature, and velocity or turbulence. All parts requiring maintenance or replacement must be easily accessible. All parts to be coated must be accessible for the coating application.

Metals and alloys degrade by several known corrosion mechanisms. Relatively thin coatings of metallic, organic, and inorganic materials can provide an effective barrier between metals/alloys and its environment. **Protective coatings** include metallic, chemical conversion, inorganic nonmetallic, and organic coatings. Before the application of an effective coating on metals and alloys, it is necessary to clean the surface carefully to remove, dirt, grease, salts, and oxides such mill scale and rust. Metallic coatings may be applied by cladding, electrodeposition (copper, cadmium, nickel, tin, chromium, silver, zinc, gold), hot dipping (tin, zinc/galvanizing, lead, and aluminum), diffusion (chromium/chromizing, zinc/sherardizing, boron, silicon, aluminum/calorizing or aluminizing, tin, titanium, molybdenum), metallizing or flame spraying (zinc, aluminum, lead, tin, stainless steels), vacuum evaporation (aluminum), ion implantation, and other methods. All commercially prepared coatings are porous to a degree. Corrosion or galvanic action may influence the performance of the metal coating. Noble (determined by the galvanic series) coatings must be thicker and have a minimum of pores and small pore size to delay entry of a deleterious fluid. Some pores of noble metal coatings are filled with an organic lacquer or a second lower melting metal is diffused into the initial coating. Sacrificial or cathodic coatings, such as cadmium zinc, and in some environments, tin, and aluminum, cathodically protect the base metal. Porosity of sacrificial coatings is not critical as long as cathodic protection of the base metal continues. Higher solution conductivity allows larger defects in the sacrificial coating; thicker coatings provide longer times of effective cathodic protection.

Conversion coatings are produced by electrochemical reaction of the metal surface to form adherent, protective corrosion products. Anodizing aluminum forms a protective film of Al_2O_3. Phosphate and chromate treatments provide temporary corrosion resistance and usually provide a basis for painting. Chromates are extremely effective and widely used as corrosion inhibitors for high strength Al alloys in aerospace applications. The combined properties of storage, release, migration, and irreversible reduction provided by chromate coatings underlie their outstanding corrosion protection. Of special note is the action of Cr^{VI} as a "suicidal inhibitor" in which reduction to Cr^{III} locks in the inhibiting film by converting mobile Cr^{VI} to insoluble and substitution inert Cr^{III} as discovered in the university research program for the USAF. (G. Frankel, "Mechanism of Al Alloy Corrosion and the Role of Chromate Inhibitors," AFOSR contract No. F49620-96-1-0479, Final Report, 1 December 2001). Inorganic nonmetallic coatings include vitreous enamels, glass linings, cement, or porcelain enamels bonded on metals. Susceptibility to mechanical damage or cracking by thermal shock are major disadvantages of these coatings. An inorganic coating must be perfect and defect-free unless cathodic protection is applied.

Paints, lacquers, coal-tar or asphalt enamels, waxes, and varnishes are typical organic coatings. Converted epoxy, epoxy polyester, moisture-cured polyurethane, vinyl, chlorinated rubber, epoxy ester, oil-modified phenolics, and zinc-rich organic coatings are other barriers used to control corrosion. Paints, which are a mixture of insoluble particles of pigments (e.g., TiO_2, Pb_3O_4, Fe_2O_3, $ZnCrO_4$) in a continuous organic or aqueous vehicle such as linseed or tung oil, are the most common organic coatings. All paints are permeable to water and oxygen to a degree and are subject to mechanical damage and eventual breakdown. Inhibitor and antifouling agents are added to improve corrosion protection. Underground pipelines and tanks should be covered with thicker coatings of asphalt or bituminous paints, in conjunction with cloth or plastic wrapping. Polyester and polyethylene have been used for tank linings and as coatings for tank bottoms [*Corrosion Prevention by Protective Coatings*, C.G. Munger, NACE International, Houston (1984); Corrosion Prevention by Protective Coatings, 2nd ed., C.C. Munger, L.D. Vincent, NACE International, Houston, TX (1999); Organic and Inorganic Coatings for Corrosion Prevention–Research and Experiences (European Federation of Corrosion Report No. 20, L. Fedrizzi, P.L. Bonora, eds., Institute of

Materials, London (1997)]. Coatings Technology Handbook, 3rd ed., A.A. Tracton, ed., CRC Press (2005).

Polymer technology and nanotechnology allow engineers to design functionality and aesthetics into coatings. AS higher demands are placed on coatings, requiring them to environmentally friendly, self-cleaning. self-healing, capable of indicating corrosion, superdurable, etc. has lead to research and development of intelligent coatings [Intelligent Coatings for Corrosion Control, A. Tiwari, J. Rawlins, L.H. Hihara, eds., Butterworth-Heinemann, London (2015)].

Inhibitors, when added above a threshold concentration, decrease the corrosion rate of a material in an environment. If the level of an inhibitor is below the threshold concentration, corrosion could be accelerated than if the inhibitor was completely absent. Inhibitors may be organic or inorganic and fall into several different classes: (1) passivators or oxidizers; (2) precipitators; (3) cathodic or anodic; (4) organic adsorbents; (5) vapor phase; and (6) slushing compounds. Any one inhibitor may be grouped into one or more classification. Factors such as temperature, fluid velocity, pH, salinity, cost, solubility, interfering species, and metal or alloy may determine both the effectiveness and the choice of inhibitor. Federal EPA regulations may limit inhibitor choices (such as chromate use) based on its toxicity and its disposal requirements.

Passivators in contact with the metal surface act as depolarizers, initiating high anodic current densities and shifting the potential into the passivation range of active/passive metal and alloys. However, if present in concentrations below 10^{-3} to 10^{-4} M, these inhibitors can cause pitting. Passivating inhibitors are anodic inhibitors and may include oxidizing anions such as chromate, nitrite, and nitrate. Phosphate, molybdate, and tungstate require oxygen to passivate steel and are considered nonoxidizing. Nonoxidizing sodium benzoate, cinnamate, and polyphosphate compounds effectively passive iron in the near-neutral range by facilitating oxygen adsorption. Alkaline compounds (NaOH, Na_3PO_4, $Na_2B_4O_7$, and Na_2O-$nSiO_4$) indirectly assist iron passivation by enhancing oxygen adsorption.

Precipitators are film-forming compounds which create a general action over the metal surface and subsequently indirectly interfere with both anodes and cathodes. Silicates and phosphates in conjunction with oxygen provide effective inhibition by forming deposits. Calcium and magnesium interfere with inhibition by silicates; 2-3 ppm of polyphosphates is added to overcome this. The levels of calcium and phosphate must be balanced for effective calcium phosphate inhibition. Addition of a zinc salt often improves polyphosphate inhibition.

Cathodic inhibitors (cathodic poisons, cathodic precipitates, oxygen scavengers) are generally cations which migrate toward cathode surfaces where they are selectively precipitated either chemically or electrochemically to increase circuit resistance and restrict diffusion of reducible species to the cathodes. Cathodic poisons interfere with and slow the HER reaction, thereby slowing the corrosion process. Depending on pH, arsenic, bismuth, antimony, sulfides, and selenides are useful cathodic poisons. Sulfides and arsenic can cause hydrogen blistering and hydrogen embrittlement. Cathodic precipitation-type inhibitors such as calcium and magnesium carbonates and zinc sulfate in natural waters requires adjustment of the pH to be effective. Oxygen scavengers help inhibit corrosion, either alone or with another inhibitor, to prevent cathodic depolarization. Sodium sulfite, hydrazine, carbohydrazide, hydroquinone, methylethylketoxime, and diethylhydroxylamine are oxygen scavengers.

Organic inhibitors, in general, affect the entire surface of a corroding metal and affect both the cathode and anode to differing degrees, depending on the potential and the structure or size of the molecule. Cationic, positively charged inhibitors such as amines and anionic, negatively charged inhibitors such as sulfonates will be adsorbed preferentially depending on whether the metal surface is either negatively or positively charged. Soluble organic inhibitors form a protective layer only a few molecules thick; if an insoluble organic inhibitor is added, the film may become about 0.003″ (76 μm) thick. Thick films show good persistence by continuing to inhibit even when the inhibitor is no longer being injected into the system.

Vapor phase inhibitors consist of volatile aliphatic and cyclic amines and nitrites that possess high vapor pressures. Such inhibitors are placed in the vicinity of the metal to be protected (e.g., inhibitor-impregnated paper) which transfers the inhibiting species to the metal by sublimation or condensation where they adsorb on the surface and retard the cathodic, anodic, or both corrosion processes. This inhibitor protects against water and/or oxygen. Vapor phase inhibitors are usually only effective if used in closed spaces and are used primarily to retard atmospheric corrosion. Slushing compounds are polar inorganic or organic additives in oil, grease, or wax that adsorb on the metal surface to form a continuous protective film. Suitable additives include organic amines, alkali and alkaline-earth metal salts of sulfonated oils, sodium nitrite, and organic chromates [*Corrosion Inhibitors*, C.C. Nathan, ed., NACE International, Houston (1973); *Reviews of Corrosion Inhibitor Science and Technology*, v. 1, A. Raman, P. Labine, eds., NACE International, Houston, TX (1993); *Reviews of Corrosion Inhibitor Science and Technology*, v. 2, A. Raman, P. Labine, eds., NACE International, Houston, TX (1996)].

Environmental concerns and regulations over using some traditional corrosion control materials has led to research and implementation of alternative corrosion control technologies for (1) volatile organic compounds (VOCs), (2) chromates, and (3) cadmium. By far, the most important environmental impact from paints and coatings is the release of VOCs from drying. Virtually everything but the solids in a typical coating formulation is released to the air around the surface being coated. In an enclosed system, such as a paint booth, some of this emission may be captured before release to the atmosphere. New coating formulations with higher proportions of solids meet current environmental requirements; eventually, 100 percent solids coating formulations will be common.

Hexavalent chromium is a cancer hazard that can damage lungs, skin, eyes, and nasal passages. Substitutes for hard chromium coatings are trivalent chromium, nickel-base binary and ternary plating chemicals. Alternatives to electroplating are pressure controlled atomization processes (Thermal Spray), HVOF (High Velocity Oxygen Fuel) processes, amorphous nanocrystalline composite depositions, flexible preceramic coatings deposition, organic sealants application, inductive-coupled radio frequency plasma torch, and cold spray. Cadmium plating has been used to protect steel components in marine environments. Cadmium has been used because of its hardness, close dimensional tolerance, lubricity, and its ability to restrict hydrogen permeation into or out of high-strength steels. Cadmium plating is frequently used for fasteners and other very tight tolerance parts because of the dual qualities of lubricity at minimal thickness and superior sacrificial corrosion protection. However, cadmium is highly toxic to personnel and the environment, thereby necessitating the search for alternative corrosion control coating systems. Research is vigorous to develop cadmium replacement coatings with mechanical and sacrificial properties and corrosion performance equivalent or exceeding cadmium. Coatings that show some promise are Zn-Ni, Zn-Sn, and Ni-Zn-P, Al-Co-Ce alloys, but further research and testing will be required.

The degradation of materials in marine environments depends on the expected materials demands and service requirements for the component or systems. Corrosion is dependent on a number of factors and understanding these factors can optimize the ability of modelling to predict galvanic interactions in various environments and under various operating conditions.

The computational, storage, and analytical instrumentation are readily available to organize, compile, and verify the many **corrosion models** developed from previous materials-based research and empirical investigations related to materials performance in the various environments. This is an arduous task for there are many parameters and factors, which affect passivity and the breakdown of passivity of materials. There are limited corrosion models that can describe, formulate, and improve our understanding of materials interactions leading to protection or corrosion in marine environments. Yet, small changes, additions, or subtractions of various environmental parameters may alter:

(1) the corrosion resistance of materials; (2) the precise mechanism(s) of breakdown; and (3) the accuracy of corrosion models to explain the mode of breakdown. It is important that computational or mechanistic models not be used beyond their capabilities. In the future, fundamental research on the atomic level will help clarify many aspects of current corrosion models, modify earlier empirically-based models, or develop and verify new corrosion models. Fundamentally-based research will also improve our understanding of critical materials properties and the effects that various marine environmental interactions have on materials. Better physical and computational models will improve our understanding of corrosion and corrosion control that can lead to predictive and prognostic capabilities to better structural design [D.A. Shifler, "*Galvanic Corrosion Modeling in Flowing Systems*," **Proceedings** ELECTROCOR 2005: Simulation of Electrochemical Processes, C.A. Brebbia, V.G. DeGiorgi, R.A. Adey, WITPress, Southampton, UK, 59 (2005); D.A. Shifler, "Understanding Material Interactions in Marine Environments to Promote Extended Structural Life," *Corrosion Science* (2005)].

Anodic protection is based on formation of a protective film on metals by externally applied anodic currents. Figure 4.5.2 illustrates the effect. The applied anodic current density is equal to the difference between total oxidation and reduction rates of the system, or $i_{app} = i_{oxid} - i_{red}$. The potential range in which anodic protection is achieved is the protection range or passive region. At the optimum potential, E_A, the applied current is approximately 1 μA/cm². Anodic protection is limited to active/passive metals and alloys and can be utilized in environments ranging from weak to very aggressive. The applied current is usually equal to the corrosion of the protected system which can be used to monitor the instantaneous corrosion rate. Operating conditions in the field usually can be accurately and quickly determined by electrochemical laboratory tests [O.L. Riggs, C.E. Locke, *Anodic Protection-Theory and Practice in the Prevention of Corrosion*, Plunum Press, New York, (1981)].

Cathodic protection (CP) controls corrosion by supplying electrons to a metal structure, thereby suppressing metal dissolution and increasing hydrogen evolution. Figure 4.5.2 shows that when an applied cathodic current density ($i_{app} = i_{red} - i_{oxid}$) of 10,000 μA/cm² on a bare metal surface shifts the potential in the negative or active direction to E_c, the corrosion rate has been reduced to 1 μA/cm². CP is applicable to all metals and alloys and is common for use in aqueous and soil environments. CP can be accomplished by (1) impressed current from an external power source through an inert anode; or (2) use of galvanic couplings by sacrificial anodes. Sacrificial anodes include zinc, aluminum, magnesium, and alloys containing these metals.

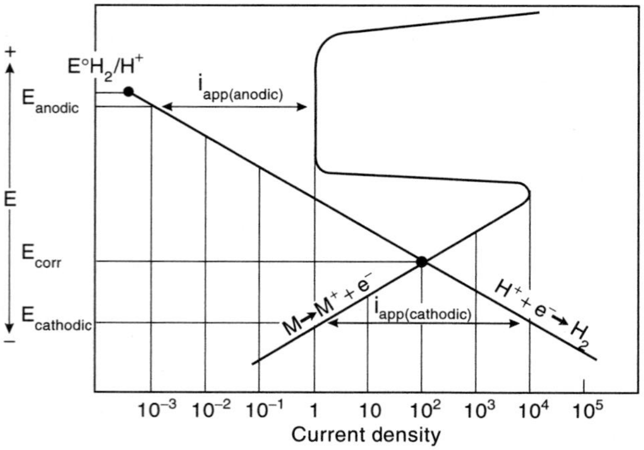

Figure 4.5.2 Idealized polarization common for most metals, at E_{corr}, where anodic ($i_{corr} - i_a$) and cathodic currents are equal.

Design of cathodic protection systems can significantly affect the usable lifetime of a structure. Poor designs can be far-more costly to implement than optimal designs. Both sacrificial and impressed cathodic protection systems must be immersed in a conductive liquid medium. Improper design can cause overprotection, with resulting paint blistering and accelerated corrosion of some alloys, underprotection, with resultant structure corrosion, or stray current corrosion of nearby Structures [*Designing Cathodic Protection Systems for Marine Structures and Vehicles, ,4STM SIP 1370,* H.P. Hack, ed., American Society for Testing and Materials, West Conshohocken, PA, (1999)].

The total protective current is directly proportional to the surface area of the metal being protected; hence, CP is combined with surface coatings where only the coating pores or holidays and damaged spots need be cathodically protected (e.g., for 10 miles of a bare pipe, a current of 500 A would be required, while for 10 miles of a superior coated pipe (5 × 10⁶ ohm/ft²) 0.03 A would be required). The polarization potential can be measured versus a reference, commonly saturated $Cu/CuSO_4$. The criterion for proper CP is −0.85 V versus this reference electrode. Overprotection by CP (denoted by potentials < −0.85 V) can lead to blistering and debonding of e pipe coatings and hydrogen embrittlement of steel from hydrogen gas evolution. Another problem with CP involves stray currents to unintended structures. The proper CP for a system is determined empirically and must resolve a number of factors to be effective; [Peabody's Control of Pipeline Corrosion, 2nd ed., R/L/ Bianchetti, ed., NACE International, Houston TX (2001)], [J. Morgan, *Cathodic Protection*, 2nd ed., NACE International, Houston (1987)], *Handbook of Cathodic Corrosion Protection: Theory and Practice of Electrochemical Protection Processes*, 3rd ed., W. von Baeckmann, W. Schwenk, W. Prinz, Gulf Publishing (1997)], [Dr. R. Johnsen, "Cathodic Protection," Institute of Engineering Design and Materials, Norwegian University of Science and Technology, Trondheim, Norway, Rev 2, (April 10, 2006)].

CORROSION IN INDUSTRIAL AND POWER PLANT STEAM GENERATING SYSTEMS

Types of Steam Generators

The type of steam generator will dictate the types of corrosion related damage experienced. Most steam generators can be grouped into the following categories: heat recovery steam generators (**HRSG**), direct-fired boilers (i.e., a fuel is burned in the furnace), process waste heat boilers, once through steam generators (OTSG) in thermal enhanced oil recovery (EOR) systems, nuclear steam generators, and electric boilers. This section focuses on the corrosion of two categories of steam generators: direct fired boilers and HRSG units. Generally the term 'steam generator' is used herein when referring to both. Details about the design of steam generators are discussed in Section 7 and the references. [Steam: Its Generation and Use, 40th ed., S.C. Stultz, J.B. Kitto, eds., Babcock and Wilcox, (1992)].

Steam/Water Cycle Components

Figure 4.5.3 presents an example of a steam/water cycle for a power plant with a direct fired boiler. The flow is basically in a clockwise direction in the Figure. Major components in the cycle includes the condenser, **deaerator**, boiler (including the economizer, steam generator, superheater, and reheater tubes), and the turbine.

Heat Recovery Steam Generator (**HRSG**) units have a simpler condensate/feedwater systems (often with no external feedwater heaters or **deaerators**) and more complicated drum and steam systems (often 2 or 3 drums operating at different pressures with superheated steam from the intermediate pressure drum feeding the high pressure cold reheat lines and superheated steam from the low pressure drum feeding the LP turbine). For industrial and cogeneration facilities, a portion of the steam produced would be used in the process and process condensate normally would be treated in a condensate polisher and returned to the condensate/feedwater system.

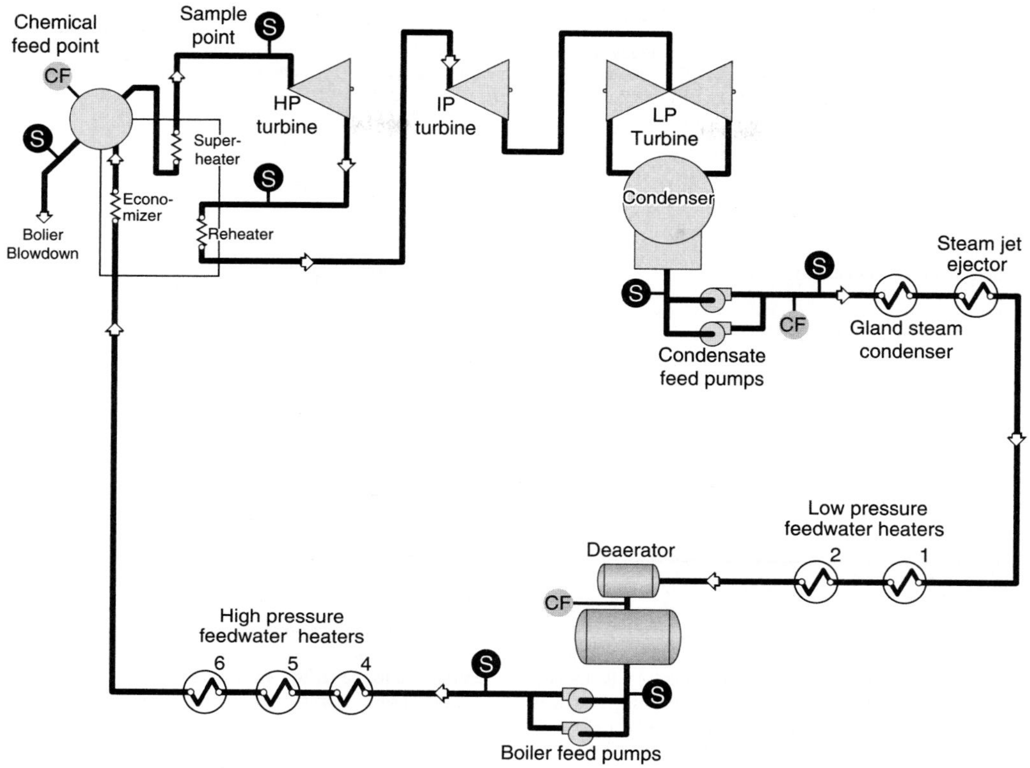

Figure 4.5.3 Example of a steam/water cycle for a direct fired boiler.

Steam/Water Cycle Corrosion

Table 4.5.1 provides a summary of the different corrosion mechanisms experienced in the steam/water cycle components indicated. Corrosion can occur on the internal or external surfaces of the components. In most components, internal pitting refers to the inner diameter (ID) of the tube and the external pitting refers to the outer diameter (OD) of heat transfer tubes. Underdeposit corrosion includes caustic gouging and phosphate gouging (commonly called acid phosphate corrosion) as well as concentration of other salts under deposits (e.g., in steam generator tubes during condenser leaks or other contamination events). A detailed (370 pages) discussion of corrosion of steam/water cycle materials was presented in the "ASME Handbook On Water Technology For Thermal Power Systems," Chap. 9, ASME, New York (1989). It is cautioned that "Chapter 17 Management of Cycle Chemistry" of this ASME handbook would need to be rewritten to reflect the changes in understanding regarding corrosion, carryover, monitoring, and chemistry control in steam/water cycle components. There are a number of other types of failure mechanisms experienced by steam/water cycle components—including thermal and/or vibration fatigue, erosion, overheating, manufacturing defects, and maintenance problems. Some of the actual failure mechanisms listed are combinations of mechanical and chemical mechanisms, such as corrosion fatigue and stress corrosion cracking (SCC). The identification, cause(s), and corrective action(s) for most types of failure mechanism was presented in "The Boiler Tube Failure Metallurgical Guide, Volume 1: Technical Report" TR-102433-V1, EPRI, Palo Alto, CA (1993). A significantly expanded and updated version of this document was developed by EPRI [and a later version of this document "Boiler and Heat Recovery Steam Generator Failures: Theory and Practice," Volumes 1-3, EPRI, Palo Alto, 10,23,063 (2011)]. In addition to the mechanisms recognized in that document, stress assisted corrosion can be treated as an additional mechanism rather than just a subset of corrosion fatigue [E.M. Labuda and R.D. Bartholomew, "Stress Assisted Corrosion: Case Histories," Power Plant Chemistry, March, 2005].

Deaerators can experience oxygen pitting and corrosion fatigue. NACE Standard RP0590-96, "Recommended Practice for Prevention, Detection, and Correction of Deaerator Cracking," 1996) was largely written in response to deaerator failures from corrosion fatigue cracking. The interconnecting piping in the feedwater system can experience **FAC**—sometimes with catastrophic consequences. For information on corrosion in nuclear steam generators, a detailed government document on corrosion mechanisms and corrosion predictions is available via the Internet. ["Incorporating Aging Effects into Probabilistic Risk Assessment – A Feasibility Study Utilizing Reliability Physics Models," NUREG/CR-5632, Office of Nuclear Regulatory Research, Washington, D.C., August, 2001.]

Control of Waterside/Steamside Corrosion in the Steam/Water Cycle

In order to minimize corrosion associated with the watersides and steam sides of steam generators and other steam/water cycle components, a comprehensive chemistry program should be instituted. There is a wide variety of chemical treatment programs which are utilized for these systems depending on the materials of construction, operating temperatures, pressures, heat fluxes, contaminant levels, and purity criteria for components using the steam.

Consensus documents are available from the ASME which provide useful information related to chemistry and corrosion control for various types of steam/water cycles and situations: industrial boiler chemistry "Consensus On Operating Practices For The Control Of Feedwater And

Table 4.5.1 Summary of Corrosion Mechanisms for Steam/Water Cycle Components

Corrosion mechanism	Condenser	Feedwater heaters	Economizer	Steam generator	Superheater	Reheater	Steam turbine
Internal pitting	X	X	X	X	X	X	X
External pitting	X	X	X	X			
Stress corrosion cracking	X	X			X	X	X
Stress-assisted corrosion			X	X	X		
Corrosion fatigue			X	X			X
Dealloying	X	X					
Ammonia grooving	X						
Ennoblement	X						
Microbiologically influenced corrosion (MIC)	X	While rare, MIC can occur due to improper handling and storage before components are fully assembled or installed.					
Flow-accelerated corrosion (FAC)		X	X	X (HRSG)			
Erosion corrosion	X	X	X				
Dew point corrosion	*		X				
Under deposit corrosion				X			
Hydrogen damage	X (Rare)			X			
Galvanic corrosion	X						
Intergranular corrosion	X				X	X	
Sulfide corrosion	X			X			
Liquid ash corrosion				X	X	X	
Halogenation					X	X	
Crevice corrosion	X						

* Air-cooled condenser.

Boiler Water Chemistry In Modern Industrial Boilers," CRTD-Vol. 34, ASME, New York (1998); layup: ["Consensus for the Layup of Boilers, Turbines, Turbine Condensers and Auxiliary Equipment," CRTD-Vol. 66, 2002]; sampling and monitoring: ["Consensus on Operating Practices for the Sampling and Monitoring of Feedwater and Boiler Water Chemistry in Modern Industrial Boilers," CRTD-Vol. 81, 2006]; HRSG chemistry control: ["Consensus on Operating Practices for Control of Water and Steam Chemistry in Combined Cycle and Cogeneration Power Plants," ASME, (2012)]; tube sampling and analyses: [Consensus on Best Tube Sampling Practices for Boilers and Nonnuclear Steam Generators. CRTD-Vol. 103, (2014)]; and [Consensus on Pre-Commissioning Stages for Cogeneration and Combined Cycle Power Plants, 2017].

For the electric power industry, a series of detailed chemistry guidelines are available from the Electric Power Research Institute (EPRI). EPRI is a research organization funded by its utility members. Their first chemistry guides were published in the 1982 and 1986 for PWR nuclear steam generators and fossil-fired steam generators (respectively) in the electric utility industry. A number of facilities experienced failures in boilers and other cycle chemistry components despite adherence to these guidelines. EPRI continues to update both documents to reflect the changes in understanding regarding corrosion and corrosion product transport. For example, as of this document, the latest chemistry guideline document for the secondary water system of PWR nuclear steam generators was in its 6th version (2004). EPRI currently has a number of chemistry guidelines tailored for different types of steam generators (fossil fired boilers, **HRSG**, fluidized bed boilers, etc.). For EPRI member organizations, it is suggested to obtain the latest guideline for your type of steam generator.

A good chemistry control program requires a high purity makeup water to the steam/water cycle. For high pressure electric utility boilers, two stage demineralization (e.g., cation/anion/mixed bed ion exchange; reverse osmosis/membrane decarbonation/continuous electrodeionization, and combinations of these) is generally utilized. Consult the cited reference for information on design of makeup treatment systems. ["Guidelines For Makeup Water Treatment, Conventional Fossil and Heat Recovery Steam GeneratorWater/Steam Cycle Makeup," 10,19,635, EPRI, Palo Alto (2010)]. For industrial boilers, one or two stage demineralization or softening (sometimes with dealkalization) are utilized for **makeup water treatment**—depending on the boiler pressure. Unsoftened makeup has been utilized in small, low pressure (e.g., ≤15 psig, ≤100 kPa) boiler systems which receive large amounts of condensate return, but this practice increases the potential for corrosion and deposition. Some makeup systems also include vacuum degasifiers for the removal of dissolved oxygen from the makeup water prior to its entrance into the steam/water cycle.

Condensate returns often require purification or polishing. **Condensate polishing** can be achieved with specially designed filters, softeners, or mixed bed ion exchangers. Industrial boilers typically use softeners and/or filters (e.g., electromagnetic or cartridge filtration). Once-through boilers typically use full-flow mixed bed polishers (preferred) or precoat (with crushed ion exchange resin) filters. Mixed bed condensate polishers also are strongly advised for drum type steam generators with condenser cooling water having a high content of chloride and/or a high ratio of hardness to alkalinity. These systems are more subject to underdeposit corrosion of steam generator tubes in the event of a condenser tube leak.

Minimizing the transport of corrosion products to the steam generator is essential for corrosion control in the steam generator. Corrosion products from the feedwater deposit on steam generator tubes, inhibit cooling of tube surfaces, and provide an evaporative-type concentration mechanism of dissolved salts under boiler tube deposits. Therefore, the chemistry of water in contact with the tube surfaces is often much worse than in the bulk water.

There are two basic philosophies of **feedwater treatment**. The first type of feedwater treatment, which is used for industrial boilers and many high pressure power boilers, relies primarily on raising the feedwater pH to 8.5-10 (each plant will have a smaller control range) and maintaining a reducing environment by eliminating dissolved oxygen (to <5-10 ppb) and applying a reducing agent. In this environment, the predominant oxide formed on steel is a mixed metal oxide composed primarily of magnetite (Fe_3O_4). The mixed oxide layer on steel formed in deoxygenated solutions resists dissolution or erosion and thus controls iron transport to the boiler and corrosion of the underlying metal. Copper oxides are maintained in a reduced state (Cu_2O) which is less subject to transport to the steam generator (or carry over into the steam). However, excessive application of reducing agents can result in low dissolved oxygen levels (<2 ppb) and low oxidizing reduction potentials (ORP) which render steel susceptible to flow accelerated corrosion (**FAC**).

FAC is the gradual dissolution of magnetite layers from steel surfaces in flowing systems. It only occurs in areas of high flow or turbulence and the metal surface often has the scalloped surface morphology – which is sometimes referred to as erosion corrosion. FAC occurs between 100°C (212°F) and 250°C (482°F) with a maximum attack around 150°C (302°F) for single phase FAC and 140-180°C (284-356°F) for two phase (i.e., steam/water mixtures) FAC. FAC is decreased by raising the pH, raising the dissolved oxygen level (as little as 1-2 ppb sometimes can have a significant effect), reducing or eliminating the feed of oxygen scavengers (reducing agents), changing to an alloy with greater chromium content (e.g., T22 or P22 provide substantial resistance), and reducing the flow-induced erosive forces (e.g., reducing turbulence by decreasing fluid velocity, minimizing sharp bends and avoiding pressure reductions or other sources of flashing and two phase flow).

FAC has caused gradual thinning and catastrophic failures of feedwater lines (several resulting in fatalities), economizer header tube stubs, economizer tubes, evaporator tubes (particularly in low pressure **HRSG** evaporator sections), steam drum internals, riser tubes, feeder tubes, riser headers, deaerator floors, deaerator drop-legs, feed pump minimum recirculation lines, heater drain lines, heater division plates, air cooled condensers, and attemperation water lines in power plants. Facilities feeding oxygen scavengers should evaluate their system for areas likely to have FAC and should check the thickness of metal in those locations. Modeling programs are available for selecting areas for routine evaluation of FAC as discussed in "Flow-Accelerated Corrosion in Power Plants," TR-106611, EPRI, Palo Alto, CA, 1996.

An adapted version of the deoxygenated (<5-10 ppb) feedwater treatment program for systems with high purity makeup and without any copper present relies solely on mechanical deaeration (no reducing agent is applied). Currently, 5-10 ppb is targeted to provide a slightly more oxidizing solution and a more oxidized oxide layer on steel (in the feedwater system) which has been found to be more resistant to **FAC** than steel with primarily a magnetite layer. The only chemical applied for condensate/feedwater chemistry control is ammonia [and/or organic amines) and the program is called all volatile treatment - oxidizing (AVT(O)).

The second basic type of **feedwater treatment** program is **oxygenated treatment** (OT). While used for many years in Germany and Russia, OT was not generally utilized in the United States until the 1990's. It involves a completely different approach than the other treatment programs since oxygen is fed and it relies on the formation of a gamma ferric oxide (Fe_2O_3) or gamma ferric oxide hydrate (FeOOH) layer on the steel surfaces in the preboiler cycle.

Originally, **oxygenated treatment** was based on neutral pH values (7-8) and oxygen levels of 50-250 ppb in very pure feedwater (cation conductivities <0.15 uS/cm) Later, a program based on slightly alkaline pH values (8.0-8.5) and 30-150 ppb of dissolved oxygen was developed. This treatment program was originally referred to as combined water treatment, but it is now commonly referred to as Oxygenated Treatment (OT) in the United States. A higher pH range is used by a number of facilities. OT is the standard treatment for once-through boilers with high purity makeup and all-steel steam/water cycles (if full-flow **condensate polishing** is provided, as recommended, copper alloys are permitted in the condensers). While much less common, there are some once through boilers with nickel/copper high pressure feedwater heaters which require reducing [AVT(R)] feedwater treatment rather than OT to avoid excessive copper transport. A primary benefit of OT has been to minimize deposit-induced pressure drop through the boiler during service. The reduction in iron oxide transport into and deposition in the boiler can also significantly reduce the frequency of chemical cleanings. OT also has been applied (with oxygen levels of about 30-50 ppb in the feedwater) for the condensate and feedwater systems of drum type boilers and HRSG units to minimize FAC and corrosion product transport. However, OT requires extremely pure feedwater and pitting can result from severely contaminated feedwater. The details of Oxygenated Treatment Programs for once through boilers was presented in "Cycle Chemistry Guidelines For Fossil Plants: Oxygenated Treatment," TR-102285, EPRI, Palo Alto, CA (1994). This document also presented guidance for utilizing OT for drum-type boilers and **HRSG** units, although OT in drum type units incurs a greater potential risk because feedwater contaminants are concentrated in the drums. Updated guidance for the most common treatment programs for high pressure boiler and HRSG units are available in EPRI guidelines (Comprehensive Cycle Chemistry Guidelines for Power Plants, 10,21,767, 2011).

As indicated in Figure 4.5.3, a **deaerator** is included in many steam/water cycles for the removal of dissolved oxygen and other non-condensable gases. Deaerators are direct contact steam heaters which spray feedwater into fine droplets and strip the dissolved oxygen from the water with steam. This is often referred to as mechanical deaeration. Deaerators normally operate at pressures of 5-180 psig (35-1,241 kPa), although some old deaerators operated at 1-2 psig. (6.9-13.8 kPa). Most modern deaerators reliably remove dissolved oxygen to ≤5-7 ppb, provided they are properly maintained and operated. While sometimes rated for 7 ppb, the performance of deaerating condensers currently is more variable in industry (e.g., 2-40 ppb) and involves maintaining proper vacuum, controlling air inleakage, condensate subcooling, and deaerating or installing spray headers for the makeup.

Following mechanical deaeration, chemical reducing agents called oxygen scavengers often are applied to minimize residual dissolved oxygen and/or to inhibit pitting (enhance passivation). This is recommended for all systems containing copper in feedwater heaters or steam-fed heat exchangers (copper corrosion can be minimized at ≤2 ppb) and for steel systems with elevated levels of contamination.

A wide variety of oxygen scavengers are utilized. For boilers below 700-900 psig (4.8-6.2 MPa) sodium sulfite (often catalyzed with cobalt) is commonly used although other non-volatile or volatile oxygen scavengers also may be effective. In industrial boiler systems, severe oxygen tuberculation and pitting and failures of economizer tubes has occurred due to excessive oxygen and inadequate feed of sodium sulfite. A slight excess of sodium sulfite is applied to the feedwater resulting in a residual of sulfite in the boiler water. The exact level of sulfite maintained depends primarily on the boiler pressure and oxygen concentrations in the feedwater - although other factors such as boiler water pH, steam/condensate purity, feedwater temperatures etc., are involved. Sodium sulfite should not be used in boilers operating above 900 psig (6.2 MPa) - as a significant amount of the sulfite decomposes to volatile sulfur species at the elevated temperatures. The resultant acidic species decrease steam and condensate pH values - contributing to increased corrosion in steam/water cycle components. Sulfur species in steam may contribute to SCC. Also, sulfite cannot be used in systems that directly inject feedwater into steam for attemperation (desuperheating).

Commonly used volatile oxygen scavengers (reducing agents) include hydrazine, carbohydrazide, hydroquinone, diethylhydroxylamine (DEHA), and methylethylketoxime. These are normally applied to maintain a free residual of the scavenger in the feedwater at the economizer inlet to the boiler. Due to the slower reaction rates, routine dissolved oxygen monitoring usually is more crucial for boilers using volatile oxygen scavengers than boilers using sulfite. It is possible for dissolved oxygen and these volatile reducing agents to coexist due to slow reaction rates. A benefit and a disadvantage of some of the volatile oxygen scavengers (e.g., hydrazine, carbohydrazide, DEHA, etc.,) is their ability to enhance metal passivation. This has been useful in systems with poorly operating **deaerator**s because, they maintain a magnetite layer on the steel and appear to inhibit oxygen pitting – despite the presence of elevated dissolved oxygen levels. However, excessive feed of these volatile reducing agents in systems with demineralized makeup must be avoided as they can lead **FAC**.

Sulfite is an effective oxygen scavenger, but has little effect on passivation beyond the effect of oxygen removal. Therefore, it will not convert oxide films on steel and does not cause **FAC**. However, coexistence of sulfite and oxygen does not protect the steel against operational oxygen pitting. Additional information on oxygen scavengers can be found in the NACE technical committee report 3A194 ("Oxygen Scavengers in Steam Generating Systems and in Oil Production," NACE, Houston, TX, 1994).

Aqueous solutions of ammonia or volatile neutralizing amines are usually applied to the feedwater to control feedwater, steam and condensate pH. For all steel cycles these pH levels can range from 9.0 to 10.0 with the higher pH values resulting in lower iron oxide transport. The higher pH and amine levels results in higher baseline conductivities which obscures the ability to detect system contamination (which may increase the potential for contamination causing deposits or scale in the steam generator). Condensate/feedwater/steam pH values of about 10 with ammonia are used to minimize FAC and iron transport in all steel cycles for HRSG units, especially those with air cooled condensers. In systems containing both steel and copper alloys, pH levels in the feedwater, steam and condensate are typically controlled in the 8.5-9.4 range. The exact pH control limits usually are a much tighter range depending on the water purity, amine used, temperatures, and type of copper alloys present.

Neutralizing amines commonly utilized in steam/water cycles include cyclohexylamine, morpholine, ethanolamine, diethanolamine, diethylaminoethanol, dimethyl isopropanolamine, and methoxypropylamine. For large steam distribution systems, a blend of 2-4 amines with different distribution coefficients (ratio of amine in steam to condensate) is often applied to provide pH control throughout the steam/condensate system. Amines tend to decompose in high pressure, high temperature systems and produce organic acids which may contribute to corrosion.

Filming amines are high molecular weight amines applied to develop thin films on surfaces to serve as a barrier against corrosion. Applications date at least back to the middle of the twentieth century. Various types of filming amines used up through the 1990's in the United States sometimes resulted in fouling of equipment (e.g., steam traps, condensate polishers) when the waxy films and associated iron oxides sloughed off surfaces. Subsequent, high molecular weight amines (currently called polyamines) continue to be introduced and applied. This is an area of ongoing research and caution is advised in the application of this technology. However, while results vary there are a few reports of potentially beneficial use of filming amines to protect steam/water cycle components during out-of-service periods [Alternative Layup Techniques for Major Components in Fossil Steam/Water Cycles: Film Forming Amines, 3,002,000,713. (2014)].

In addition to the condensate/**feedwater treatment** program, chemistry must be controlled in the steam generator (boiler water) to minimize corrosion. For once-through boilers, the steam generator treatment program is identical to the feedwater treatment program (normally OT). Currently, the most common types of drum type steam generator treatment programs in use in the United States are coordinated **phosphate treatment**, congruent phosphate treatment, equilibrium phosphate treatment, phosphate continuum treatment, phosphate treatment with caustic, caustic treatment, and all volatile treatment (AVT).

The variety of phosphate and caustic treatments are designed to provide a reserve of alkalinity to avoid acid attack and minimize pH fluctuations. Caustic or acidic materials concentrating underneath deposits in boiler tubing contributes to most types of waterside corrosion. It is generally desired to control the pH in a range which controls the level of free caustic alkalinity and acidic constituents within acceptable values. Phosphate also precipitates a portion of the hardness contamination entering with the feedwater as a hardness phosphate sludge in the steam drum (or generation bank downcomer tubes) before reaching the higher heat evaporator sections.

The type of **phosphate treatment** mainly depends on the operating pressure and water purity, low pressure boilers on softened makeup generally operate with substantial amounts of free caustic alkalinity in the boiler water. Industrial boilers operating up to about 900 psig (6.2 MPa) receiving demineralized makeup generally limit free sodium hydroxide (caustic) in the steam generator to $\leq$1-10 ppm and intermediate and high pressure boilers and steam generators usually limit caustic to $\leq$0-1 ppm. These limits are primarily set to avoid caustic corrosion in the boiler. Facilities with free caustic and carryover also can be susceptible to stress corrosion cracking of stainless steel steam system components. These facilities favor the use of all volatile treatment or of coordinated phosphate treatment with a sodium-to phosphate mole ratio (actually moles of sodium alkalinity divided by moles of phosphate) in the boiler water up to a maximum of 3.0, which corresponds to trisodium phosphate. The minimum ratio varied but typically sodium- to-phosphate ratios were controlled at 2.4-3.0 for boilers operating less than 1,000 psig (6.9 MPa) and 2.6-3.0 for higher pressure boilers. The minimum ratio for higher pressure boilers has been raised as indicated in the next paragraph.

Another series of captive alkalinity programs called congruent treatment used lower sodium to phosphate ratios (e.g., range of 2.2-2.6), but these were found to lead to acid phosphate corrosion in high pressure boilers (mainly >1,800 psig, >12.4 MPa). Acid phosphate corrosion results from reaction of sodium phosphates with magnetite to form acidic compounds on tube surfaces with the simultaneous generation of caustic in solution. During load reductions, these deposits can redissolve and contribute to corrosion. Research studies determined the minimum sodium-to-phosphate ratio to avoid acid phosphate corrosion as a function of temperature or drum pressure [P.R. Tremaine et al., "Phosphate Interactions With Metal Oxides Under High-Performance Boiler Hideout Conditions," Paper No. 35, International Water Conference, Pittsburgh (1993), and P.R. Tremaine et al., "Solubility and Thermodynamics of Sodium Phosphate Reaction Products Under Hideout Conditions in High-Pressure Boilers," IWC-96-21, International Water Conference, Pittsburgh, (1996). For high pressure boilers (especially $\geq$2,000 psig or $\geq$13.8 MPa), equilibrium phosphate (which restricts the maximum level of phosphate applied and maintains sodium alkalinity to phosphate ratios above 3.0 with up to 1 ppm of caustic) currently is considered to be one of the best forms of **phosphate treatment** for boilers and **HRSG** units. [Jan Stodola, "Fifteen Years of Equilibrium Phosphate Treatment (Correct Use of Phosphates in Drum Boilers)," Power Plant Chemistry, Volume 5, No. 2, 2003]. While they may not be optimal for the boiler, slightly higher phosphate levels and lower minimum ratios (to 2.7 for at 1,500-2,000 psig or 10.3-13.8 MPa, and down to 2.4 <1,000 psig or <6.9 MPa) has been safely used in some lower pressure boilers with lower heat flux, minimal phosphate hideout, and more variable chemistry control. With phosphate or caustic treatment the steam generator pH should always be greater than the feedwater pH to ensure nonvolatile alkali is present.

AVT is an extension of the **feedwater treatment** program. Steam generator water pH control is maintained by the portion of amines or ammonia from the feedwater which remains in the steam generator water. Since these materials are volatile, the steam generator pH normally is 0.2-0.3 pH units less than the feedwater pH. Due to the low concentrations and poor buffering capacity of ammonia or amines in steam generator water, strict purity limits are maintained on the feedwater and steam generator water. AVT is generally only utilized in high pressure steam generators (>1,800 psig, 12.4 kPa) with condensate polishers, although some lower pressure boilers (850 psig, 5.9 kPa) boilers also utilize this treatment. Many all steel **HRSG** units also rely solely on AVT due to problems with phosphate hideout in the steam generator and carryover of water from the drums into the steam.

Control of Fireside Corrosion of Steam Generators

Corrosive constituents in fuel at appropriate metal temperatures may promote fireside corrosion in boiler tube steel. For coal, lignite, oil, or refuse, the corrosive ingredients can form liquids (liquid ash corrosion) that solubilize the oxide film on tubing and react with the underlying metal to reduce the tube wall thickness. This can occur in the combustion zone (e.g., waterwall tubes) or convection pass (e.g., superheaters). Technologies to reduce NOx emissions from the boiler also result in more reducing conditions in the furnace, which greatly increases the potential for liquid ash corrosion. This has been alleviated by installing air ports along the walls to create an air blanket and installing stainless steel and high alloy (e.g., nickel) weld overlays. For black liquor recovery boilers, sulfur or carbon (char) induced corrosion can occur through the direct reaction with the metal oxide layer. To combat the high corrosion rates of carbon steel in the lower furnace of black liquor recovery boilers, stainless steel and nickel alloy overlays and composite tubes have been used. Cracking of these composite tubes around air ports and in floor tubes currently is the focus of much investigation. High nickel alloys also have been used in flue gas desulfurization systems to combat the severe corrosion resulting from the hot, wet environment with low pH values and high dissolved solids levels.

Fluidized bed boilers can experience severe fireside wastage of boiler tubes due to erosion. Fireside erosion related mechanisms also can cause severe damage to tubes in the convection pass. Dew-point corrosion may occur on the fireside surfaces of the economizer and other lower temperature surfaces (e.g., near drum corrosion in generation bank tubes for black liquor recovery boilers) when condensation can form acidic products [W.T. Reid "External Corrosion And Deposits—Boiler And Gas Turbine" American Elsevier, New York (1971), and James Labossiere and Jeff Henry, "Chromizing for near-drum corrosion protection," TAPPI Journal, September, (1999)].

In the preceding discussion, we indicated a few examples of corrective measures for controlling fireside corrosion. To summarize, fireside corrosion often can be mitigated by: (1) material selection (e.g., stainless or nickel alloys); (2) purchase of fuels with low impurity levels (limit levels of sulfur, chloride, potassium, sodium, vanadium, etc.,); (3) purification of fuels (e.g., coal washing); (4) application of fuel additives (e.g., magnesium oxide); (5) adjustment of operating conditions (e.g., increasing percent excess air, % solids in recovery boilers, etc.,); and (6) modification of lay-up practices (e.g., dehumidification systems).

Control of Steam Turbine Corrosion

Steam turbine corrosion is primarily controlled by maintaining steam purity during operation and proper layup during out-of-service periods. As the pressure and temperature decrease through the turbine, contaminants in the steam deposit on the turbine. Fortunately, current researchers indicate that no more than 1 percent of materials in excess of their steam solubility actually deposit in turbines due to the high velocities. Nevertheless, the deposits which happen to form can lead to pitting (particularly in the case of sodium chloride) during a subsequent shutdown or out-of-service period where oxygen levels will be elevated in the condensate in contact with turbine surfaces. Once initiated, corrosion sites can lead to cracking and stress corrosion cracking during subsequent service. EPRI has various publications on turbine damage ["Turbine Steam Path Damage: Theory and Practice, Volumes 1 and 2," EPRI, Palo Alto, CA, 1999 and "Turbine Steam Chemistry and Corrosion," 1,006,283, EPRI, Palo Alto, CA, September, 2001] and has worked on development of a model to predict turbine damage based on damage function analysis (DFA) and the point defect model (PDM) of corrosion [Engelhardt, G.R., Macdonald, D.D., Zhang, Y., Dooley, R.B., "Deterministic Prediction of Corrosion Damage in Low Pressure Steam Turbines," Power Plant Chemistry, Volume. 6, No. 11, November, 2004]. Steam turbine corrosion during out-of-service periods can be minimized by removing moisture (e.g., applying dehumidified air), and/or removing oxygen (e.g., applying a nitrogen purge). Information on layup is available from the reference [Cycling, Startup, Shutdown, and Layup Fossil Plant Cycle Chemistry Guidelines for Operators and Chemists. EPRI, Palo Alto, CA: 2009. 1,015,657].

The source of steam contamination for once through steam generators is the feedwater and for drum type and other subcritical steam generators is a combination of carryover into the saturated steam and contamination in the attemperation water (which is added to reduce the temperature of the steam). There are several methods of desuperheating to control steam temperatures. The most common in the power industry is the spray of feedwater directly into the steam. A common and the most preferred means of desuperheating for industrial boilers is to use a 'sweet water condenser' which condenses saturated steam (with feedwater) for injection into the steam. Some older boilers have drum-type attemperators where the steam is condensed in tube bundles immersed in boiler water in the mud drum. Unfortunately, these tubes can leak and significantly contaminate the steam.

CORROSION IN HEATING AND COOLING WATER SYSTEMS AND COOLING TOWERS

Control of corrosion in the watersides of cooling and heating water systems begins with proper preconditioning of the system. This usually involves alkaline detergent cleaning, extensive flushing and initial application of elevated corrosion inhibitor levels (2-3 times normal). Galvanized towers need to be treated differently as discussed in "White Rust: An Industry Update and Guide Paper," Association of Water Technologies, 2002. Objectives of the subsequent cooling water treatment program are as follows.

- Prevent corrosion of metals in the system
- Prevent the formation of scale (e.g., calcium carbonate, calcium sulfate, calcium phosphate, etc.)
- Minimize the amount and deposition of suspended solids
- Minimize biological growths

Substantial emphasis of a cooling tower treatment program is placed on preventing deposits of scale, suspended solids, or biological matter from forming on system components. The primary reason for this emphasis is to maintain the performance of the cooling system. However, minimizing deposition is integral to corrosion control to avoid problems with underdeposit corrosion or MIC attack. Stainless steels are particularly susceptible to MIC: stainless steel condensers have failed in less than a year of operation due to MIC related to stagnant flow conditions. Automatic ball cleaning or brush cleaning systems are usually recommended for stainless steel condenser tubes. Suspended particles can lead to erosion of tubing and piping - particularly copper tubing. In HVAC systems, sidestream filters or full-flow strainers have been installed to minimize suspended matter in the cooling water. Sidestream softening also has been used to reduce scale forming constituents in cooling tower water system at zero liquid discharge facilities.

Depending on the inherent corrosivity of the cooling water, corrosion control in open-recirculating cooling water systems may consist merely of controlling pH, conductivity and alkalinity levels. However, in systems with extensive piping networks, corrosion inhibitors and deposit inhibitors (dispersants) are often utilized to protect materials in the system. The corrosion inhibitors can include the following compounds: zinc salts, molybdate salts, triazoles, orthophosphates, polyphosphates, phosphonates and polysilicates. Often, two or more of these corrosion inhibitors are used together. Deposit inhibitors usually are called dispersants and consist of water soluble organic compounds such as polyacrylates, phosphonates, and various copolymers and terpolymers. More complete discussions of cooling water corrosion inhibitors and deposit control agents are available in the cited references [Robert Cavanaugh, Developing Cooling Water Treatment, Parts 1-4, The Analyst, 2008; Colin Frayne, Organic Water Treatment Inhibitors, Parts I & II, The Analyst, 2009].

Oxidizing or non-oxidizing biocides also are required in cooling tower systems to inhibit biological activity. Oxidizing biocides result in increased general corrosion rates for copper and copper alloys and to a lesser extent for ferrous metals. However, oxidizing biocides are sometimes required to provide adequate control of slime forming bacteria and/or Legionella bacteria. Control of Legionella, applications of oxidizing biocides, and monitoring requirements for open cooling water systems are covered in the references [ANSI/ASHRAE Standard 188-2015, Legionellosis: Risk Management for Building Water Systems, Jun 26, 2015; Legionellosis Guideline: Best Practices for Control of Legionella, WTB-148 (08), Cooling Technologies Institute, 2008; "Legionella 2003: An Update and Statement by the Association of Water Technologies (AWT)," McLean, VA, 2003; "Application of Oxidizing Biocides," WTP-141, Cooling Technology Institute, 1994; Robert D. Bartholomew, "Bromine Based Biocides for Cooling Water Systems: A Literature Review," International Water Conference, Pittsburgh, PA, 1998; Standard Recommended Practice "Online Monitoring of Cooling Waters," NACE, Houston, TX. RP0189-2002 (2002)].

For closed cooling and heating water systems, corrosion inhibitors are usually applied and pH levels are controlled at 8-11. Corrosion inhibitors utilized in closed cooling and heating water systems include the inhibitors listed for open, recirculating cooling water systems except for zinc salts. Also, nitrite salts are often used in closed water systems.

The definition of a closed system is the loss of 5 percent of the system volume per year. Corrosion control in closed cooling systems is greatly facilitated by minimizing water loss from the system. Makeup water introduced to replace leakage introduces oxygen, organic matter to support biological activity, and scale constituents. Utilization of pumps with mechanical seals (with filtered seal water) rather than packing seals can greatly reduce water losses from closed systems. Control of oxygen levels by minimizing water leakage and aeration is probably the single most important factor for reducing corrosion in these systems. Glycol, intermittently used for freeze protection of out-of-service components, often is not able to be adequately purged from the system. The glycol is converted to organic acids, which contributes to low pH values and elevated corrosion. Whenever possible other means of freeze protection are suggested.

In heating and cooling water systems, concerns regarding corrosion control generally focus on the waterside or interior of piping and components. However, substantial corrosion can occur on the exterior of these piping systems underneath insulation. External corrosion of piping generally requires a source of external moisture. In chill water and secondary water systems, moisture can be introduced through exposure to the air and subsequent condensation of moisture on the pipe surface.

While chill water piping systems generally are installed with vapor barriers, breaks in the vapor barrier can be present due to inadequate installation or subsequent maintenance activities. Secondary water system piping is sometimes not installed with vapor barriers because the cooling water temperature is designed to be above the expected dew point. However, humidity levels in the building can sometimes be elevated and lead to condensation at higher temperatures than initially expected.

Leaks from piping at fittings or joints can saturate the insulation with moisture and lead to corrosion. This type of moisture source can be particularly troublesome in hot water piping as the moisture can be evaporated to dryness and wet/dry corrosion can result. Corrosion rates during this transition from wetness to dryness can be many times higher than corrosion of a fully wetted surface. [Corrosion Under Wet Thermal Insulation, W.I. Pollock, C.N. Steely, eds. NACE International, Houston (1990).] If the moisture can be eliminated, corrosion underneath insulation should be negligible. However, if the moisture cannot be eliminated, coatings may need to be applied to the metal surface. Inhibitors are also available to mitigate corrosion underneath insulation due to condensation, but they may not be effective at controlling corrosion due to leakage because the inhibitors can be washed away. Preferably, insulated systems would have all welded joints in the piping system to prevent joint leakage and corrosion under insulation.

4.6 WOOD

by Staff, Forest Products Laboratory, USDA Forest Service, and several contributors

REFERENCES: Freas and Selbo, Fabrication and Design of Glued Laminated Wood Structural Members, *U.S. Dept. Agr. Tech. Bull.*, no. 1069, 1954. Hunt and Garratt, "Wood Preservation," McGraw-Hill. "Wood Handbook: Wood as an Engineering Material," Gen. Tech. Rep. FPL-GTR-190, rev. 2010. "Design Properties of Round, Sawn and Laminated Preservatively Treated Construction Poles as Posts," ASAE Eng. Practice EP 388.2. American Society of Agricultural Engineering Standards, ANSI/AWC National Design Specification (NDS) National Design Specification for Wood Construction 2015 Edition. American Wood Council, 222 Catoctin Circle, SE, Suite 201, Leesburg, VA 20,175. ANSI/AWC National Design Specification (NDS) Supplement: Design Values for Wood Construction 2015 Edition. American Wood Council, 222 Catoctin Circle, SE, Suite 201, Leesburg, VA 20,175. AITC 111-05 Recommended Practice for Protection of Structural Glued Laminated Timber During Transit, Storage and Erection, American Institute for Timber Construction, www.aitc-glulam.org "Laminated Timber Design Guide," American Institute of Timber Construction, www.aitc-glulam.com, 2001. "Timber Construction Manual, 6th ed., Wiley, New York, 2012. American National Standard, Structural Glued Laminated Timber. ANSI/AITC A190.1-2007, www.aitc-glulam.org. ASTM International, "Annual Book of Standards," Vol. 04.10, "Wood," ASTM International, W. Conshohocken, PA, 2015: "Standard Terminology Relating to Wood," ASTM D 9; "Standard Practice for Establishing Structural Grades and Related Allowable Properties for Visually Graded Dimension Lumber," ASTM D 245; "Standard Definitions of Terms Relating to Veneer and Plywood," ASTM D 1,038; "Standard Nomenclature of Domestic Hardwoods and Softwoods," ASTM D 1,165; "Standard Practice for Establishing Allowable Properties for Visually Graded Dimension Lumber from In-Grade Tests of Full Size Specimens," ASTM D, 1990; "Standard Methods for Establishing Clear Wood Strength Values," ASTM D 2,555; "Standard Practice for Evaluating Allowable Properties of Grades of Structural Lumber," ASTM D 2,915; "Standard Test Methods for Mechanical Properties of Lumber and Wood Based Structural Material," ASTM D 4,761; "Standard Method for Establishing Design Stresses for Round Timber Piles," ASTM D 2,899; "Standard Practice for Establishing Stress Grades for Structural Members Used in Log Buildings," ASTM D 3,957; "Standard Practice for Establishing Stresses for Structural Glued Laminated Timber," ASTM D 3,737; "Standard Specification for Evaluation for Structural Composite Lumber Products," ASTM D 5,456. "Standard Test Method for Surface Burning Characteristics of Building Materials," ASTM E 84; "Standard Test Methods for Fire Tests of Building Construction and Materials," ASTM E 119.] AWPA "Book of Standards," American Wood Preserver's Association, Selma, AL, 2015. C415 Gluelam Floor Beams APA-EWA, Tacoma, WA, 2006. Cassens and Feist, "Exterior Wood in the South—Selection, Applications, and Finishes," Gen. Tech. Rep. FPL-GTR-69, USDA Forest Service, Forest Products Laboratory, Madison, WI, 1991. Chudnoff, Tropical Timbers of the World, *Agric. Handb.* no. 607, 1984. Green, "Moisture Content and the Shrinkage of Lumber," Res. Pap. FPL-RP-489, USDA Forest Service, Forest Products Laboratory, Madison WI, 1989. Green and Kretschmann, "Moisture Content and the Properties of Clear Southern Pine," Res. Pap. FPL-RP-531, USDA Forest Service, Forest Products Laboratory, Madison, WI, 1995. Green, Evans, and Craig, Durability of Structural Lumber Products at High Temperatures: Part I, 60°C at 75% RH and 82°C at 30% RH, *Wood Fiber Sci.* 35, no. 4, 2003, pp. 499-523. James, "Dielectric Properties of Wood and Hardboard: Variation with Temperature, Frequency, Moisture Content, and Grain Orientation," Res. Pap. FPL-RP-245, USDA Forest Service, Forest Products Laboratory, Madison, WI, 1975. James, "Electric Moisture Meters for Wood," Gen. Tech. Rep. FPL-GTR-6, USDA Forest Service, Forest Products Laboratory, Madison, WI, 1988. Kohara and Okamoto, Studies of Japanese Old Timbers, *Sci. Rpts. Saikyo. Univ.* no. 7, 1995, pp. 9-20. MacLean, Thermal Conductivity of Wood, *Heat, Piping, Air Cond.* **13,** no. 6, 1941, pp. 380-391. Ross, "Nondestructive Evaluation of Wood," General Technical Report FPL-GTR-238, USDA Forest Service, Forest Products Laboratory, Madison, WI, 2015. TenWolde et al., "Thermal Properties of Wood and Wood Panel Products for use in Buildings," ORNL/Sub/87-21,697/1, Oak Ridge National Laboratory, Oak Ridge, TN, 1988. Weatherwax and Stamm, The Coefficients of Thermal Expansion of Wood and Wood Products, *Trans. ASME*, **69,** no. 44, 1947, pp. 421-432. Wilkes, Thermo-Physical Properties Data Base Activities at Owens-Corning Fiberglas, in "Thermal Performance of the Exterior Envelopes of Buildings," ASHRAE SP 28, American Society of Heating, Refrigerating, and Air-Conditioning Engineers, Atlanta, 1981.

The authors acknowledge the following individuals for their contributions to the previous edition: David W. Green, Robert White, Anton Tenwolde, William Simpson, Joseph Murphy, and Roland Hernandez.

COMPOSITION, STRUCTURE, AND NOMENCLATURE

by Robert J. Ross

Wood is a naturally formed organic material consisting essentially of elongated tubular elements called cells arranged in a parallel manner for the most part. These cells vary in dimensions and wall thickness with position in the tree, age, conditions of growth, and kind of tree. The walls of the cells are formed principally of chain molecules of cellulose, polymerized from glucose residues and oriented as a partly crystalline material. These chains are aggregated in the cell wall at a variable angle, roughly parallel to the axis of the cell. The cells are cemented by an amorphous material called lignin. The complex structure of the gross wood approximates a rhombic system. The direction parallel to the grain and the axis of the stem is longitudinal (L), the two axes across the grain are radial (R) and tangential (T) with respect to the cylinder of the tree stem. This anisotropy and the molecular orientation account for the major differences in physical and mechanical properties with respect to direction which are present in wood.

Natural variability of any given physical measurement in wood approximates the normal probability curve. It is traceable to the difference in the growth of individual samples and at present can be controlled with only limited success. For engineering purposes, statistical evaluation is employed for determination of safe working limits.

Lumber is classified as **hardwood,** which is produced by the broad-leafed trees (angiosperms), such as oak, maple, and ash; and **softwood,** the product of coniferous trees (gymnosperms), such as pines, larch, spruce, and hemlock. The terms *hard* and *soft* have no relation to the actual hardness of the wood. **Sapwood** is the living wood of pale color on the outside of the stem. **Heartwood** is the inner core of physiologically inactive wood in a tree and is usually darker, somewhat heavier, due to infiltrated material, and more decay-resistant than the sapwood. Other terms relating to wood, veneer, and plywood are defined in ASTM D 9, D 1,038, and the "Wood Handbook."

Standard nomenclature of lumber is based on commercial practice which groups woods of similar technical qualities but separate botanical identities under a single name. For listings of domestic hardwoods and softwoods, see ASTM D 1,165 and the "Wood Handbook."

The chemical composition of woody cell walls is generally about 40 to 50 percent cellulose, 15 to 35 percent lignin, less than 1 percent mineral, 20 to 35 percent hemicellulose, and the remainder extractable matter of a variety of sorts. Softwoods and hardwoods have about the same cellulose content.

PHYSICAL AND MECHANICAL PROPERTIES OF CLEAR WOOD

by David Kretschmann, C. Adam Senalik, and Xiping Wang

Moisture Relations

Wood is a hygroscopic material which contains water in varying amounts, depending upon the relative humidity and temperature of the surrounding atmosphere. The standard reference condition for wood is **oven-dry** weight, which is determined by drying at 100 to 105°C until there is no significant change in weight.

Mechanical Properties

Average **mechanical properties** determined from tests on clear, straight-grained wood at 12 percent moisture content are given in Table 4.6.1. Approximate standard deviation(s) can be estimated from

$$s = CX$$

where X is average value for species and

$$c = \begin{cases} 0.10 & \text{for specific gravity} \\ 0.22 & \text{for modulus of elasticity} \\ 0.16 & \text{for modulus of rupture} \\ 0.18 & \text{for maximum crushing strength parallel to grain} \\ 0.14 & \text{for compression strength perpendicular to grain} \\ 0.25 & \text{for tensile strength perpendicular to grain} \\ 0.25 & \text{for impact bending strength} \\ 0.14 & \text{for shear strength parallel to grain} \end{cases}$$

Relatively few data are available on tensile strength parallel to the grain. The modulus of rupture is considered to be a conservative estimate for the tensile strength of clear wood.

Mechanical properties remain constant as long as the moisture content is above the fiber saturation point. Below the fiber saturation point, properties generally increase with decreasing moisture content down to about 8 percent. Below about 8 percent moisture content, some properties, principally tensile strength parallel to the grain and shear strength, may decrease with further drying. An approximate adjustment for clear wood properties between about 8 percent moisture and green can be obtained by using an annual compound-interest type of formula

$$P_2 = P_1\left(1 + \frac{C}{100}\right)^{-(M_2 - M_1)}$$

where P_1 is the known property at moisture content M_1, P_2 is the property to be calculated at moisture content M_2, and C is the assumed percentage change in property per percentage change in moisture content. Values of P_1 at 12 percent moisture content are given in Table 4.6.1

PROPERTIES OF LUMBER PRODUCTS

by David Kretschmann

Visually Graded Structural Lumber

Stress-graded structural lumber is produced under two systems: visual grading and machine grading. Visual structural grading is the oldest stress grading system. It is based on the premise that the mechanical properties of lumber differ from those of clear wood because many growth characteristics of lumber affect its properties; these characteristics can be seen and judged by eye (ASTM D 245). The principal growth feature affecting lumber properties are the size and location of knots, sloping grain, and density.

Grading rules for lumber nominally 2 to 4 in (standard 38 to 89 mm) thick (dimension lumber) are published by grading agencies (listing and addresses are given in "National Design Specification," American Wood Council, 2015). For most species, allowable properties are based on test results from full-size specimens graded by agency rules, sampled according to ASTM D 2,915, and tested according to ASTM D 4,761. Procedures for deriving allowable properties from these tests are given in ASTM D 1990. Allowable properties for visually graded hardwoods and a few softwoods are derived from clear-wood data following principles given in ASTM D 2,555. Derivation of the allowable strength properties accounts for within-species variability by starting with a nonparametric estimate of the 5th percentile of the data. Thus, 95 of 100 pieces would be expected to be stronger than the assigned property. The allowable strength properties are based on an assumed normal duration of load of 10 years. Tables 4.6.2 and 4.6.3 show the grades and allowable properties for the four most commonly used species groupings sold in the United States. The allowable strength values in bending, tension, shear, and compression parallel to the grain can be multiplied by factors for other load durations. Some commonly used factors are 0.90 for permanent (50-years) loading, 1.15 for snow loads (2 months), and 1.6 for wind/earthquake loading (10 min). The most recent edition of "National Design Specification" should be consulted for updated property values and for property values for other species and size classifications.

Allowable properties are assigned to visually graded lumber at two moisture content levels: green and 19 percent maximum moisture content (assumed 15 percent average moisture content). Because of the influence of knots and other growth characteristics on lumber properties, the effect of moisture content on lumber properties is generally less than its effect on clear wood. The factors of Table 4.6.4 are for adjusting the properties in Tables 4.6.2 and 4.6.3 from 15 percent moisture content to green. The Annex of ASTM D 1990 provides formulas that can be used to adjust lumber properties to any moisture content between green and 10 percent. Below about 8 percent moisture content, some properties may decrease with decreasing values, and care should be exercised in these situations (Green and Kretschmann, 1995).

Shrinkage in commercial lumber differs from that in clear wood primarily because the grain in lumber is seldom oriented in purely radial and tangential directions. Approximate formulas used to estimate shrinkage of lumber for most species are

$$S_w = 6.031 - 0.215M$$
$$S_t = 5.062 - 0.181M$$

where S_w is the shrinkage across the wide [8-in (203-mm)] face of the lumber in a 2 × 8 (standard 38 × 184 mm), S_t is the shrinkage across the narrow [2-in(51-mm)] face of the lumber, and M is moisture content (percent). As with clear wood, shrinkage is assumed to occur below a moisture content of 28 percent. Because extractives make wood less hygroscopic, less shrinkage is expected in redwood, western redcedar, and northern white cedar (Green, 1989).

The reversible effect of **temperature** on lumber properties appears to be similar to that on clear wood. For simplicity, "National Design Specification" uses conservative factors to account for reversible reductions in properties as a result of heating to 150°F(65°C) or less

Table 4.6.1 Strength and Related Properties of Wood at 12% Moisture Content (Average Values from Tests on Clear Pieces of 2 in × 2 in Cross Section per ASTM D 143)

Kind of wood	Specific gravity, oven-dry volume	Density at 12% m.c. lb/ft³	Shrinkage, % from green to oven-dry condition based on dimension when green		Static bending		Max crushing strength parallel to grain, lb/in²	Compression perpendicular to grain at proportional limit, lb/in²	Tensile strength perpendicular to grain, lb/in²	Impact bending, height of drop in inches for failure with 50-lb hammer	Shear strength parallel to grain, lb/in², avg of *R* and *T*	Hardness perpendicular to grain, lb, avg of *R* and *T*
			Rad.	Tan.	Modulus of rupture, lb/in²	Modulus of elasticity, ×10³ lb/in²						
Hardwoods												
Ash, white	0.60	42	4.9	7.8	15,400	1,740	7,410	1,160	940	43	1,910	1,320
Basswood	0.37	26	6.6	9.3	8,700	1,460	4,730	370	350	16	990	410
Beech	0.64	45	5.5	11.9	14,900	1,720	7,300	1,010	1,010	41	2,010	1,300
Birch, yellow	0.62	43	7.3	9.5	16,600	2,010	8,170	970	920	55	1,880	1,260
Cherry, black	0.50	35	3.7	7.1	12,300	1,490	7,110	690	560	29	1,700	950
Cottonwood, eastern	0.40	28	3.9	9.2	8,500	1,370	4,910	380	580	20	930	430
Elm, American	0.50	35	4.2	9.5	11,800	1,340	5,520	690	660	39	1,510	830
Elm, rock	0.63	44	4.8	8.1	14,800	1,540	7,050	1,230		56	1,920	1,320
Sweetgum	0.52	36	5.4	10.2	12,500	1,640	6,320	620	760	32	1,600	850
Hickory, shagbark	0.72	50	7.0	10.5	20,200	2,160	9,210	1,760		67	2,430	
Maple, sugar	0.63	44	4.8	9.9	15,800	1,830	7,830	1,470		39	2,330	1,450
Oak, red, northern	0.63	44	4.0	8.6	14,300	1,820	6,760	1,010	800	43	1,780	1,290
Oak, white	0.68	48	5.6	10.5	15,200	1,780	7,440	1,070	800	37	2,000	1,360
Poplar, yellow	0.42	29	4.6	8.2	10,100	1,580	5,540	500	540	24	1,190	540
Tupelo, black	0.50	35	5.1	8.7	9,600	1,200	5,520	930	500	22	1,340	810
Walnut, black	0.55	38	5.5	7.8	14,600	1,680	7,580	1,010	690	34	1,370	1,010
Softwoods												
Cedar, western red	0.32	22	2.4	5.0	7,500	1,110	4,560	460	220	17	990	350
Cypress, bald	0.46	32	3.8	6.2	10,600	1,440	6,360	730	270	24	1,000	510
Douglas-fir, coast	0.48	34	4.8	7.6	12,400	1,950	7,230	800	340	31	1,130	710
Hemlock, eastern	0.40	28	3.0	6.8	8,900	1,200	5,410	650		21	1,060	500
Hemlock, western	0.45	31	4.2	7.8	11,300	1,630	7,200	550	340	23	1,290	540
Larch, western	0.52	36	4.5	9.1	13,000	1,870	7,620	930	430	35	1,360	830
Pine, red	0.46	32	3.8	7.2	11,000	1,630	6,070	600	460	26	1,210	560
Pine, pondersona	0.40	28	3.9	6.2	9,400	1,290	5,320	580	420	19	1,130	460
Pine, eastern white	0.35	24	2.1	6.1	8,600	1,240	4,800	440	310	18	900	380
Pine, western white	0.38	27	4.1	7.4	9,700	1,460	5,040	470		23	1,040	420
Pine, shortleaf	0.51	36	4.6	7.7	13,100	1,750	7,270	820	470	33	1,390	690
Redwood, old growth	0.40	28	2.6	4.4	10,000	1,340	6,150	700	240	19	940	480
Spruce, sitka	0.40	28	4.3	7.5	10,200	1,570	5,610	580	370	25	1,150	510
Spruce, black	0.42	29	4.1	6.8	10,800	1,610	5,960	550		23	1,230	520

SOURCE: Tabulated from "Wood Handbook," Tropical Woods no. 95, and unpublished data from the USDA Forest Service, Forest Products Laboratory.

Table 4.6.2 Base Design Values for Visually Graded Dimension Lumber*
(Tabulated design values are for normal load duration and dry service conditions)

Species and Commercial grade	Size classification, in	Bending F_b	Tension parallel to grain F_t	Shear parallel to grain F_v	Compression perpendicular to grain $F_{c\perp}$	Compression parallel to grain F_c	Modulus of elasticity E	Grading rules agency
		Design values, lb/in^2						
				Douglas-fir–Larch				
Select structural		1,500	1,000	180	625	1,700	1,900,000	WCLIB
No. 1 and better	2-4 thick	1,200	800	180	625	1,550	1,800,000	WWPA
No. 1		1,000	675	180	625	1,500	1,700,000	
No. 2	2 and wider	900	575	180	625	1,350	1,600,000	
No. 3		525	325	180	625	775	1,400,000	
Stud		700	450	180	625	850	1,400,000	
Construction	2-4 thick	1,000	650	180	625	1,650	1,500,000	
Standard		575	375	180	625	1,400	1,400,000	
Utility	2-4 wide	275	175	180	625	900	1,300,000	
				Hem–Fir				
Select structural		1,400	925	150	405	1,500	1,600,000	WCLIB
No. 1 and better	2-4 thick	1,100	725	150	405	1,350	1,500,000	WWPA
No. 1		975	625	150	405	1,350	1,500,000	
No. 2	2 and wider	850	525	150	405	1,300	1,300,000	
No. 3		500	300	150	405	725	1,200,000	
Stud		675	400	150	405	800	1,200,000	
Construction	2-4 thick	975	600	150	405	1,550	1,300,000	
Standard		550	325	150	405	1,300	1,200,000	
Utility	2-4 wide	250	150	150	405	850	1,100,000	
				Spruce–Pine–Fir				
Select structural	2-4 thick	1,250	700	135	425	1,400	1,500,000	NLGA
No. 1 – No. 2		875	450	135	425	1,150	1,400,000	
No. 3	2 and wider	500	250	135	425	650	1,200,000	
Stud		675	350	135	425	725	1,200,000	
Construction	2-4 thick	1,000	500	135	425	1,400	1,300,000	
Standard		550	275	135	425	1,150	1,200,000	
Utility	2-4 wide	275	125	135	425	750	1,100,000	

*Lumber dimensions: Tabulated design values are applicable to lumber that will be used under dry conditions such as in most covered structures. For 2- to 4-in-thick lumber, the dry dressed sizes shall be used regardless of the moisture content at the time of manufacture or use. In calculating design values, the natural gain in strength and stiffness that occurs as lumber dries has been taken into consideration as well as the reduction in size that occurs when unseasoned lumber shrinks. The gain in load-carrying capacity due to increased strength and stiffness resulting from drying more than offsets the design effect of size reductions due to shrinkage. Size factor C_F, flat-use factor C_{fu}, and wet-use factor C_M are given in Table 4.6.4.
SOURCE: Table used by permission of the American Wood Council 2015, NDS Supplement.

Table 4.6.3 Design values for Visually Graded Southern Pine Dimension Lumber*
(Tabulated design values are for normal load duration and dry services conditions)

Commercial grade	Size classification, in	Bending F_b	Tension parallel to grain F_t	Shear parallel to grain F_v	Compression perpendicular to grain $F_{c\perp}$	Compression parallel to grain F_c	Modulus of elasticity E	Grading rules agency
		Design values, lb/in^2						
Dense select structural	2-4 thick	2,700	1,900	175	660	2,050	1,900,000	SPIB
Select structural		2,350	1,650	175	565	1,900	1,800,000	
Nondense select structural	2-4 wide	2,050	1,450	175	480	1,800	1,600,000	
No. 1 dense		1,650	1,100	175	660	1,750	1,800,000	
No. 1		1,500	1,000	175	565	1,650	1,600,000	
No. 1 nondense		1,300	875	175	480	1,550	1,400,000	
No. 2 dense		1,200	750	175	660	1,500	1,600,000	
No. 2		1,100	675	175	656	1,450	1,400,000	
No. 2 nondense		1,050	600	175	480	1,450	1,300,000	
No. 3 and stud		650	400	175	565	850	1,300,000	
Construction	2-4 thick	875	500	175	565	1,600	1,400,000	
Standard		475	275	175	565	1,300	1,200,000	
Utility	4 wide	225	125	175	565	850	1,200,000	

(Continued)

Table 4.6.3 Design values for Visually Graded Southern Pine Dimension Lumber* (*Continued*)
(Tabulated design values are for normal load duration and dry services conditions)

		Design values, lb/in^2						
Commercial grade	Size classification, in	Bending F_b	Tension parallel to grain F_t	Shear parallel to grain F_v	Compression perpendicular to grain $F_{c\perp}$	Compression parallel to grain F_c	Modulus of elasticity E	Grading rules agency
Dense select structural	2-4 thick	2,400	1,650	175	660	1,900	1,900,000	
Select structural		2,100	1,450	175	565	1,800	1,800,000	
Nondense select structural	5-6 wide	1,850	1,300	175	480	1,700	1,600,000	
No. 1 dense		1,500	1,000	175	660	1,650	1,800,000	
No. 1		1,350	875	175	565	1,550	1,600,000	
No. 1 nondense		1,200	775	175	480	1,450	1,400,000	
No. 2 dense		1,050	650	175	660	1,450	1,600,000	
No. 2		1,000	600	175	565	1,400	1,400,000	
No. 2 nondense		950	525	175	480	1,350	1,300,000	
No. 3 and stud		575	350	175	565	800	1,300,000	
Dense select structural	2-4 thick	2,200	1,550	175	660	1,850	1,900,000	
Select structural		1,700	1,350	175	565	1,700	1,800,000	
Nondense select structural	8 wide	2,100	1,200	175	480	1,650	1,600,000	
No. 1 dense		1,350	900	175	660	1,600	1,800,000	
No. 1		1,250	800	175	565	1,500	1,600,000	
No. 1 nondense		1,100	700	175	480	1,400	1,400,000	
No. 2 dense		975	600	175	660	1,400	1,600,000	
No. 2		925	550	175	565	1,350	1,400,000	
No. 2 nondense		875	500	175	480	1,300	1,300,000	
No. 3 and stud		525	325	175	565	775	1,300,000	
Dense select structural	2-4 thick	1,950	1,300	175	660	1,800	1,900,000	
Select structural		1,700	1,150	175	565	1,650	1,800,000	
Nondense select structural	10 wide	1,500	1,050	175	480	1,600	1,600,000	
No. 1 dense		1,200	800	175	660	1,550	1,800,000	
No. 1		1,050	700	175	565	1,450	1,600,000	
No. 1 nondense		950	625	175	480	1,400	1,400,000	
No. 2 dense		850	525	175	660	1,350	1,600,000	
No. 2		800	475	175	565	1,300	1,400,000	
No. 2 nondense		750	425	175	480	1,250	1,300,000	
No. 3 and stud		475	275	175	565	750	1,300,000	
Dense select structural	2-4 thick	1,800	1,250	175	660	1,750	1,900,000	
Select structural		1,600	1,100	175	565	1,650	1,800,000	
Nondense select structural	12 wide	1,400	975	175	480	1,550	1,600,000	
No. 1 dense		1,100	750	175	660	1,500	1,800,000	
No. 1		1,000	650	175	565	1,400	1,600,000	
No. 1 nondense		900	575	90	480	1,350	1,400,000	
No. 2 dense		800	500	90	660	1,300	1,600,000	
No. 2		750	450	90	565	1,250	1,400,000	
No. 2 nondense		700	400	90	480	1,350	1,300,000	
No. 3 and stud		450	250	90	565	825	1,300,000	

*For size factor C_F appropriate size adjustment factors have already been incorporated in the tabulated design values for most thicknesses of southern pine dimension lumber. For dimension lumber 4 in thick, 8 in and wider, tabulated bending design values F_b shall be permitted to be multiplied by the size factor $C_F = 1.1$. For dimension lumber wider than 12 in, tabulated bending, tension, and compression parallel-to-grain design values for 12-in-wide lumber shall be multiplied by the size factor $C_F = 0.9$. Repetitive member factor C_r, flat-use factor C_{fu}, and wet-service factor C_M are given in Table 4.6.4.

SOURCE: Table used by permission of the American Wood Council, Washington, DC.

(Table 4.6.5). No increase in properties is taken for temperatures colder than normal because in practice it is difficult to ensure that the wood temperature remains consistently low. Permanent effects of temperature are not given in "National Design Specification" (see Green et al., 2003).

Mechanically Graded Structural Lumber

Machine-stress-rated (MSR) lumber and machine-evaluated lumber (MEL) are two types of mechanically graded lumber. The three basic components of both mechanical grading systems are (1) sorting and prediction of strength through machine-measured nondestructive determination of properties coupled with visual assessment of growth characteristics, (2) assignment of allowable properties based upon strength prediction, and (3) quality control to ensure that assigned properties are being obtained. Grade names for MEL start with an M designation. Grade "names" for MSR lumber are a combination of the allowable bending stress and the average modulus of elasticity; e.g., 1650f-1.4E means an allowable bending stress of 1,650 lb/in^2(11.4 MPa) and modulus of elasticity of 1.4×10^6 lb/in^2 (9.7 GPa). Selected grades of mechanically graded lumber and their allowable properties are given in Table 4.6.6. Additional grades are available (see AWC, 2015).

Structural Composite Lumber

Types of Structural Composite Lumber Structural composite lumber (SCL) refers to several types of reconstituted products that have been developed to meet the demand for high-quality material for the manufacture of engineered wood products and structures.

SCL products are characterized by smaller pieces of wood glued together into sizes common for solid-sawn lumber. There are two broad categories. One type of SCL product manufactured by laminating specially graded veneer with all plies parallel to the length is called

Table 4.6.4 Adjustment Factors

Size factor C_F for Table 4.6.2 (Douglas-fir–Larch, Hem–Fir, Spruce–Pine–Fir)

Tabulated bending, tension, and compression parallel to grain design values for dimension lumber 2 to 4 in thick shall be multiplied by the following size factors:

Grades	Width, in	F_b thickness, in: 2 & 3	F_b thickness, in: 4	F_t	F_c
Select structural	2, 3, and 4	1.5	1.5	1.5	1.15
No. 1 and better	5	1.4	1.4	1.4	1.1
No. 1, no. 2, no. 3	6	1.3	1.3	1.3	1.1
	8	1.2	1.3	1.2	1.05
	10	1.1	1.2	1.1	1.0
	12	1.0	1.1	1.0	1.0
	14 and wider	0.9	1.0	0.9	0.9
Stud	2, 3, and 4	1.1	1.1	1.1	1.05
	5 and 6	1.0	1.0	1.0	1.0
Construction and standard	2, 3, and 4	1.0	1.0	1.0	1.0
Utility	4	1.0	1.0	1.0	1.0
	2 and 3	0.4	—	0.4	0.6

Repetitive-member factor C_r for Tables 6.7.2 and 6.7.3

Bending design values F_b for dimension lumber 2 to 4 in thick shall be multiplied by the repetitive factor C_r = 1.15, when such members are used as joists, truss chords, rafters, studs, planks, decking, or similar members which are in contact or spaced not more than 24 in on centers, are not less than 3 in number, and are joined by floor, roof, or other load-distributing elements adequate to support the design load.

Flat-use factor C_{fu} for Tables 6.7.2 and 6.7.3

Bending design values adjusted by size factors are based on edgewise use (load applied to narrow face). When dimension lumber is used flatwise (load applied to wide face), the bending design value F_b shall also be multiplied by the following flat-use factors:

Width, in	Thickness, in: 2 and 3	Thickness, in: 4
2 and 3	1.0	—
4	1.1	1.0
5	1.1	1.05
6	1.15	1.05
8	1.15	1.05
10 and wider	1.2	1.1

Wet-use factor C_M for Tables 6.7.2 and 6.7.3

When dimension lumber is used where moisture content will exceed 19 percent for an extended period, design values shall be multiplied by the appropriate wet-service factors from the following table:

F_b	F_t	F_v	$F_{c\perp}$	F_c	E
0.85*	1.0	0.97	0.67	0.8†	0.9

*When $F_bC_F \leq 1{,}150$ lb/in², $C_M = 1.0$.
†When $F_cC_F \leq 750$ lb/in², $C_M = 1.0$.
SOURCE: Used by permission of the American Wood Council, Washington, DC.

laminated veneer lumber (LVL). Another type of SCL product consists of strands of wood or strips of veneer glued together under high pressures and temperatures. Depending upon the component material, this product is called laminated strand lumber (LSL), parallel strand lumber (PSL), or oriented strand lumber (OSL). SCL is used as a replacement for lumber in various applications and in the manufacture of other engineered wood products, such as prefabricated wood I-joists, which take advantage of engineering design values that can be greater than those commonly assigned to sawn lumber.

Laminated veneer lumber is manufactured from layers of veneer with the grain of all the layers parallel. This contrasts with plywood, which consists of adjacent layers with the grain perpendicular. Most

Table 4.6.5 Temperature Factors C_t for Short-Term Exposure (Reversible Effect)

Design values	In-service moisture conditions	C_t: $T \leq 100°F$	C_t: $100°F < T \leq 125°F$	C_t: $125°F < T \leq 150°F$
F_t, E	Green or dry	1.0	0.9	0.9
F_b, F_v, F_c, and $F_{c\perp}$	≤ 19% green	1.0	0.8	0.7
		1.0	0.7	0.5

SOURCE: Table used by permission of the American Wood Council, Washington, DC.

Table 4.6.6 Design Values for Mechanically Graded Dimension Lumber
(Tabulated design values are for normal load duration and dry service conditions.)

		Design values, lb/in²				
Species and commercial grade	Size classification, in	Bending	Tension parallel to grain	Compression parallel to grain	Modulus of elasticity	Grading rules agency
		Machine-stress-rated lumber				
750f-1.4E		750	425	925	1,400,000	SPIB
850f-1.4E		850	475	975	1,400,000	SPIB
900f-1.0E		900	350	1,050	1,000,000	WCLIB, WWPA, NELMA, NSLB
975f-1.6E		975	550	1,450	1,600,000	SPIB
1,050f-1.2E		1,050	450	1,225	1,200,000	SPIB
1,050f-1.6E		1,050	575	1,500	1,600,000	SPIB
1,200f-1.2E		1,200	600	1,400	1,200,000	NLGA, WCLIB, WWPA, NELMA, NSLB
1,200f-1.3E		1,200	600	1,400	1,300,000	SPIB
1,200f-1.6E		1,200	650	1,550	1,600,000	SPIB
1,250f-1.4E		1,250	800	1,475	1,400,000	WCLIB
1,250f-1.6E		1,250	725	1,600	1,600,000	SPIB
1,350f-1.3E		1,350	750	1,600	1,300,000	NLGA, WCLIB, WWPA, NELMA, NSLB
1,350f-1.4E		1,350	750	1,600	1,600,000	SPIB
1,400f-1.2E		1,400	800	1,600	1,200,000	NLGA
1,450f-1.3E		1,450	800	1,625	1,300,000	NLGA, WCLIB, WWPA, NELMA, NSLB
1,450f-1.5E		1,450	875	1,625	1,500,000	WCLIB
1,500f-1.4E		1,500	900	1,650	1,400,000	NLGA, WCLIB, WWPA, NELMA, NSLB
1,500f-1.5E		1,500	900	1,650	1,500,000	SPIB
1,500f-1.6E		1,500	900	1,650	1,600,000	SPIB
1,500f-1.7E		1,500	900	1,650	1,700,000	SPIB
1,600f-1.4E		1,600	950	1,675	1,400,000	NLGA, WWPA
1,650f-1.3E		1,650	1,020	1,700	1,300,000	NLGA
1,650f-1.5E		1,650	1,020	1,700	1,500,000	NLGA, SPIB, WCLIB, WWPA,NELMA, NSLB
1,650f-1.6E		1,650	1,175	1,700	1,600,000	WCLIB
1,650f-1.7E		1,650	1,020	1,750	1,700,000	SPIB
1,700f-1.6E		1,700	1,175	1,725	1,600,000	WCLIB
1,750f-2.0E		1,750	1,125	1,725	2,000,000	NLGA
1,800f-1.5E		1,800	1,300	1,750	1,500,000	NLGA, SPIB, WCLIB, WWPA, NELMA, NSLB
1,800f-1.6E		1,800	1,175	1,750	1,600,000	WCLIB
1,800f-1.8E		1,800	1,200	1,750	1,800,000	SPIB, WWPA
1,800f-2.0E		1,800	1,175	1,750	2,000,000	WCLIB
1,850f-1.7E		1,850	1,175	1,750	1,700,000	SPIB
1,950f-1.5E	2 & less in thickness	1,950	1,375	1,800	1,500,000	SPIB
1,950f-1.7E		1,950	1,375	1,800	1,700,000	NLGA, SPIB, WCLIB, WWPA, NELMA, NSLB
2,000f-1.6E		2,000	1,300	1,825	1,600,000	NLGA
2,100f-1.8E	2 & wider	2,100	1,575	1,875	1,800,000	NLGA, SPIB, WCLIB, WWPA, NELMA, NSLB
2,250f-1.7E		2,250	1,750	1,925	1,700,000	NLGA
2,250f-1.8E		2,250	1,750	1,925	1,800,000	NLGA, WCLIB
2,250f-1.9E		2,250	1,750	1,925	1,900,000	NLGA, SPIB, WCLIB, WWPA, NELMA, NSLB
2,400f-1.8E		2,400	1,925	1,975	1,800,000	NLGA
2,400f-2.0E		2,400	1,925	1,975	2,000,000	NLGA, SPIB, WCLIB, WWPA, NELMA, NSLB
2,500f-2.2E		2,500	1,750	2,000	2,200,000	WCLIB
2,550f-1.8E		2,550	1,400	2,000	1,800,000	SPIB
2,550f-2.1E		2,550	2,050	2,025	2,100,000	NLGA, SPIB, WCLIB, WWPA, NELMA, NSLB
2,700f-2.0E		2,700	1,800	2,100	2,000,000	WCLIB
2,700f-2.2E		2,700	2,150	2,100	2,200,000	NLGA, SPIB, WCLIB, WWPA, NELMA, NSLB
2,850f-1.8E		2,850	1,600	2,100	1,800,000	SPIB
2,850f-2.3E		2,850	2,300	2,150	2,300,000	NLGA, SPIB, WCLIB, WWPA, NELMA, NSLB
3,000f-2.4E		3,000	2,400	2,200	2,400,000	NLGA, SPIB
		Machine-evaluated lumber				
M-5		900	500	1,050	1,100,000	SPIB
M-6		1,100	600	1,300	1,000,000	SPIB
M-7		1,200	650	1,400	1,100,000	SPIB
M-8		1,300	700	1,500	1,300,000	SPIB
M-9		1,400	800	1,600	1,400,000	SPIB
M-10		1,400	800	1,600	1,200,000	NLGA, SPIB
M-11		1,550	850	1,675	1,500,000	NLGA, SPIB
M-12		1,600	850	1,675	1,600,000	NLGA, SPIB
M-13		1,600	950	1,675	1,400,000	NLGA, SPIB
M-14		1,800	1,000	1,750	1,700,000	NLGA, SPIB
M-15		1,800	1,100	1,750	1,500,000	NLGA, SPIB

Table 4.6.6 Design Values for Mechanically Graded Dimension Lumber (*Continued*)
(Tabulated design values are for normal load duration and dry service conditions.)

Species and commercial grade	Size classification, in	Design values, lb/in^2				Grading rules agency
		Bending	Tension parallel to grain	Compression parallel to grain	Modulus of elasticity	
M-16		1,800	1,300	1,750	1,500,000	SPIB
M-17	2 & less in thickness	1,950	1,300	2,050	1,700,000	SPIB
M-18		2,000	1,200	1,825	1,800,000	NLGA,SPIB
M-19	2 & wider	2,000	1,300	1,825	1,600,000	NLGA, SPIB
M-20		2,000	1,600	2,100	1,900,000	SPIB
M-21		2,300	1,400	1,950	1,900,000	NLGA, SPIB
M-22		2,350	1,500	1,950	1,700,000	NLGA, SPIB
M-23		2,400	1,900	1,975	1,800,000	NLGA, SPIB
M-24		2,700	1,800	2,100	1,900,000	NLGA, SPIB
M-25		2,750	2,000	2,100	2,200,000	NLGA, SPIB
M-26		2,800	1,800	2,150	2,000,000	NLGA, SPIB
M-27		3,000	2,000	2,400	2,100,000	SPIB
M-28		2,200	1,600	1,900	1,700,000	SPIB
M-29		1,550	850	1,650	1,700,000	SPIB
M-30		2,050	1,050	1,850	1,700,000	SPIB
M-31		2,850	1,600	2,150	1,900,000	SPIB
M-32		750	425	925	1,400,000	SPIB
M-33		850	475	975	1,600,000	SPIB
M-34		975	550	1,450	1,600,000	SPIB
M-36		1,050	575	1,500	1,600,000	SPIB
M-36		1,200	650	1,550	1,600,000	SPIB
M-37		1,250	725	1,600	1,600,000	SPIB
M-38		1,500	900	1,650	1,600,000	SPIB
M-39		1,650	1,020	1,750	1,700,000	SPIB
M-40		1,850	2,000	1,850	1,700,000	SPIB

Lumber dimensions: Tabulated design values are applicable to lumber that will be used under dry conditions such as in most covered structures. For 2- to 4-in-thick lumber, the *dry* dressed sizes shall be used regardless of the moisture content at the time of manufacture or use. In calculating design values, natural gain in strength and stiffness that occurs as lumber dries has been taken into consideration as well as reduction in size that occurs when unseasoned lumber shrinks. The gain in load-carrying capacity due to increased strength and stiffness resulting from drying more than offsets the design of size reductions due to shrinkage. **Shear parallel to grain F_v and compression perpendicular to grain F_c:** Design values for shear parallel to grain $F_{v\parallel}$ and compression perpendicular to grain $F_{c\perp}$ are identical to the design values given in Tables 6.7.2 and 6.7.3 for No. 2 visually graded lumber of the appropriate species. When the F_v or $F_{c\perp}$ values shown on the grade stamp differ from the values shown in the tables, the values shown on the grade stamp shall be used for design. **Modulus of elasticity E and tension parallel to grain F_t:** For any given bending design value F_b, the average modulus of elasticity E and tension parallel to grain F_t design value may vary depending upon species, timber source, or other variables. The E and F_t values included in the F_b and E grade designations are those usually associated with each F_b level. Grade stamps may show higher or lower values if machine rating indicates the assignment is appropriate. When the E or F_t values shown on a grade stamp differ from the values in Table 4.6.6, the values shown on the grade stamp shall be used for design. The tabulated F_b and F_c values associated with the designated F_b value shall be used for design. **Additional grades are available.**

SOURCE: Table used by permission of the American Wood Council, Washington, DC.

manufacturers use sheets of 1/10- to 1/6-in-(2.5- to 4.2-mm-) thick veneer. These veneers are stacked up to the required thickness and may be laid end to end to the desired length with staggered end joints in the veneer. Waterproof adhesives are generally used to bond the veneer under pressure. The resulting product is a billet of lumber that may be up to in (44 mm) thick, 4 ft (1.2 m) wide, and 80 ft (24.4 m) long. The billets are then ripped to the desired width and cut to the desired length. The common sizes of LVL closely resemble those of sawn dimension lumber.

Parallel-strand lumber is manufactured from strands or elongated flakes of wood. One North American product is made from veneer clipped to in (13 mm) wide and up to 8 ft (2.4 m) long. Another product is made from elongated flakes and technology similar to that used to produce oriented strandboard. A third product is made from mats of interconnected strands crushed from small logs that are assembled into the desired configuration. All the products use waterproof adhesive that is cured under pressure. The size of the product is controlled during manufacture through adjustments in the amount of material and pressure applied. Parallel-strand lumber is commonly available in the same sizes as structural timbers or lumber.

Laminated strand lumber (LSL) and oriented strand lumber (OSL) products are an extension of the technology used to produce oriented strandboard (OSB) structural panels. The products have more similarities than differences. The main difference is that the aspect ratio of strands used in LSL is higher than for OSL (AF&PA 2006). One type of LSL uses strands that are about 0.3 m (12 in.) long, which is somewhat longer than the strands commonly used for OSB. Waterproof adhesives are used in the manufacture of LSL. One type of product uses an isocyanate type of adhesive that is sprayed on the strands and cured by steam injection. This product needs a greater degree of alignment of the strands than does OSB and higher pressures, which result in increased densification. Both LSL and OSL are proprietary products; LSL is sold as TimberStrand. Applications such as studs and millwork are common.

Properties of Structural Composite Lumber Standard design values have not been established for either LVL or PSL. Rather, standard procedures are available for developing these design values (ASTM D 5,456). Commonly, each manufacturer follows these procedures and submits supporting data to the appropriate regulatory authority to establish design properties for the product. Thus, design information for LVL and PSL varies among manufacturers and is given in their product literature. Generally the engineering design properties compare favorably with or exceed those of high-quality solid dimension lumber. Example design values accepted by U.S. building codes are given in Table 4.6.7.

Glued-Laminated Timber

Structural **glued-laminated timber (glulam)** is an engineered, stress-rated product of a timber laminating plant. It is comprised of assemblies

Table 4.6.7 Example Design Values for Structural Composite Lumber

	Bending stress		Modulus of elasticity		Horizontal shear	
Product	lb/in^2	MPa	×10^3 lb/in^2	GPa	lb/in^2	MPa
LVL	2,800	19.2	2,000	13.8	190	1.31
PSL, type A	2,900	20.0	2,000	13.8	210	1.45
PSL, type B	1,500	10.3	1,200	8.3	150	1.03

of suitably selected and prepared wood laminations bonded together with adhesives. The grain of all laminations is approximately parallel longitudinally. Individual laminations are typically made from nominal 2-in- (standard 38-mm-) thick lumber when used for straight or slightly cambered members, and nominal 1-in- (standard 19-mm-) thick lumber when used for curved members. Individual lamination pieces may be joined end to end to produce laminated timbers much longer than the laminating stock itself. Pieces may also be placed or glued edge to edge to make members wider than the input stock. Straight members up to 140 ft (42 m) long and more than 7 ft (2.1 m) deep have been manufactured, with size limitations generally resulting from transportation constraints. Curved members have been used in domed structures spanning over 500 ft (152 m).

The arrangement of lumber grades in the laminated cross section depends on the anticipated loading parallel or perpendicular to the wide faces of the laminations. When loads are applied perpendicular to the wide faces of the laminations, referred to as horizontally laminated, glulam cross sections are typically designed as bending combinations. **Bending combinations** have higher-grade laminations in the outer top and bottom layers to carry the highest stresses and lower-grade laminations in the core layers where stresses are minimal. In situations where all laminations are subjected to the same load, such as compression, tension, or in bending with loads applied parallel to the wide faces of the laminations (vertically laminated), glulam cross sections are typically designed as axial combinations. **Axial combinations** typically have the same lumber grade throughout the cross section.

Manufacture Glulam members used in structural applications in the United States must be manufactured by a laminating plant that meets the requirements of the national standard, ANSI A190.1. This standard contains requirements for production, testing, and certification of the product. Plants meeting these requirements can place their product quality mark on the glulam, which contains vital information regarding the type, species, and properties of the glulam combination. Figure 4.6.1 shows examples of glulam product quality marks and describes the features required in the mark.

Manufacturing standards cover many softwoods and hardwoods; Douglas-fir and southern pine are the most commonly used softwood species. Glulam members can be used in either dry- or wet-use conditions. Dry use, which is defined as a condition resulting in a moisture content of 16 percent or less, permits manufacturing with nonwaterproof adhesives; however, nearly all manufacturers in North America use waterproof adhesives exclusively. For wet-use conditions, these waterproof adhesives are required. For wet-use conditions in which the moisture content is expected to exceed 20 percent, pressure preservative treatment is recommended (AWPA C28). Lumber can be pressure-treated with water-based preservatives prior to gluing, provided that special procedures are followed in the manufacture. For treatment after gluing, oil-based preservatives are generally recommended.

Glulam is generally manufactured at moisture content below 16 percent. For most dry-use applications, it is important to protect the glulam timber from increases in moisture content. End sealers, surface sealers, primer coats, and wrapping may be applied at the manufacturing plant to provide protection from changes in moisture content. Protection will depend upon the final use and finish of the timber.

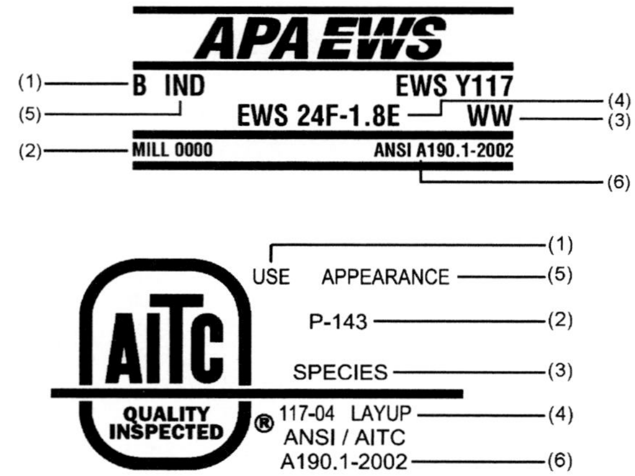

Figure 4.6.1 Typical product quality stamps showing vital information for glulam combination. (1) structural use of member (B, simple span bending member; C, compression member; T, tension member; CB, continuous or cantilevered span bending member); (2) mill number; (3) species; (4) structural grade designation, indicating species, design bending strength and stiffness; (5) appearance grade designation (framing, industrial, architectural, or premium); (6) identification of conformance to ANSI A190.1 standard.

Special precautions are necessary during handling, storage, and erection to prevent structural damage to glulam members. Padded or nonmarring slings are recommended; cable slings or chokers should be avoided unless proper blocking protects the members. Technical manuals on transit, storage, and erection of glulam members are provided by both AITC and APA-EWS.

Design Glulam members are available in standard sizes with standardized design properties. The following standard widths are established to match the widths of standard sizes of lumber, less an allowable amount for finishing the edges of the manufactured beams:

3 or 3⅛ in (76 or 79 mm)
5 or 5⅛ in (127 or 130 mm)
6¾ in (171 mm)
8½ or 8¾ in (216 or 222 mm)
10½ or 10¾ in (267 or 273 mm)

Standard beam depths are common multiples of lamination thickness of either 1⅜ or 1½ in (35 or 38 mm). There are no standard beam lengths, although most uses will be on spans where the length is from 10 to 20 times the depth. Technical publications containing allowable spans for various loadings of standard sizes of beams are available from either AITC or APA-EWS.

The design stresses for beams in bending for dry-use applications are standardized in multiples of 200 lb/in^2 (1.4 MPa) with the range of 2,000 to 3,000 lb/in^2 (13.8 to 20.7 MPa). Modulus of elasticity values associated with these design stresses in bending vary from 1.6 to 2.0×10^6 lb/in^2. A bending stress of 2,400 lb/in^2 (16.5 MPa) and a modulus of elasticity of 1.8×10^6 lb/in^2 (12.4 GPa) are most commonly specified, and the designer needs to verify the availability of beams with higher values. Design properties must be adjusted for wet-use applications. Detailed information on other design properties for beams as well as design properties and procedures for arches and other uses is given in "National Design Specification" (AWC, 2015).

Round Timbers

Round timbers in the form of poles, piles, or construction logs represent some of the most efficient uses of forest products because of the minimum of processing required. Poles and piles are generally debarked or peeled, seasoned, graded, and treated with a preservative

Table 4.6.8 Design Stresses for Selected Species of Round Timbers for Building Construction

Type of timber and species	Bending lb/in²	Shear lb/in²	Comp ⊥ lb/in²	Comp// lb/in²	Modulus of elasticity × 10⁶ lb/in² E	Modulus of elasticity × 10⁶ lb/in² E_{min}	Specific Gravity
Poles*							
Pacific Coast Douglas-fir	2,050	160	490	1,300	1.7	0.69	0.50
Lodgepole Pine	1,275	125	265	825	1.1	0.43	0.42
Ponderosa Pine	1,200	175	295	775	1.0	0.40	0.43
Red Pine	1,350	125	270	850	1.3	0.52	0.42
Southern Pine (Grouped)	1,950	160	440	1,250	1.5	0.60	0.55
Western Hemlock	1,550	165	275	1,050	1.3	0.56	0.47
Western Larch	1,900	170	405	1,250	1.5	0.66	0.49
Western Red Cedar	1,250	140	260	875	1.0	0.36	0.34
Piles†							
Pacific Coast Douglas-fir	2,050	160	490	1,300	1.7	0.69	0.40
Red Pine	1,350	125	270	850	1.3	0.52	0.40
Southern Pine (Grouped)	1,950	160	440	1,250	1.5	0.60	0.40

*From "Timber Construction Manual" (AITC, 1994).
†From "National Design Specification" (AWC, 2015).

prior to use. Construction logs are often shaped to facilitate their use. See Table 4.6.8.

Poles The primary use of wood poles is to support utility and transmission lines. An additional use is for building construction. Each of these uses requires that the poles be pressure-treated with preservatives following the applicable AWPA standard (C1). For utility structures, pole length may vary from 30 to 125 ft (9.1 to 38.1 m). Poles for building construction rarely exceed 30 ft (9.1 m). Southern pines account for the highest percentage of poles used in the United States because of their favorable strength properties, excellent form, ease of treatment, and availability. Douglas-fir and western redcedar are used for longer lengths; other species are also included in the ANSI O5.1 standard (ANSI 1992) that forms the basis for most pole purchases in the United States.

Design procedures for the use of ANSI O5.1 poles in utility structures are described in the "National Electric Safety Code" (NESC). For building construction, design properties developed based on ASTM D 2,899 (see ASTM, 2015) are provided in "National Design Specification Supplement" (AWC, 2015).

Piles Most piles used for foundations in the United States utilize either southern pine or Douglas-fir. Material requirements for timber piles are given in ASTM D 25, and preservative treatment should follow the applicable AWPA standard (C1 or C3). Design stress and procedures are provided in "National Design Specifications."

Construction Logs Log buildings continue to be a popular form of construction because nearly any available species of wood can be used. Logs are commonly peeled prior to fabrication into a variety of shapes. There are no standardized design properties for construction logs, and when they are required, log home suppliers may develop design properties by following an ASTM standard (ASTM D 3,957).

PROPERTIES OF STRUCTURAL PANEL PRODUCTS

by Zhiyong Cai

Structural panel products are a family of wood products made by bonding veneer, strands, particles, or fibers of wood into flat sheets. The members of this family are (1) plywood, which consists of products made completely or in part from wood veneer; (2) flakeboard, made from strands, wafers, or flakes; (3) structural particleboard, made from particles; and (4) composite panels. Plywood and flakeboard make up a large percentage of the panels used in structural applications such as roof, wall, and floor sheathing; thus, only those two types will be described here.

Plywood

Plywood is the name given to a wood panel composed of relatively thin layers or plies of veneer with the wood grain of adjacent layers at right angles. The outside plies are called faces or face and back plies, the inner plies with grain parallel to that of the face and back are called cores or centers, and the plies with grain perpendicular to that of the face and back are called crossbands. In four-ply plywood, the two center plies are glued with the grain direction parallel to each ply, making one center layer. Total panel thickness is typically not less than in (1.6 mm) or more than 3 in (76 mm). Veneer plies may vary as to number, thickness, species, and grade. Stock plywood sheets usually measure 4 by 8 ft (1.2 by 2.4 m), with the 8-ft (2.4-m) dimension parallel to the grain of the face veneers.

The alternation of grain direction in adjacent plies provides plywood panels with dimensional stability across their width. It also results in fairly similar axial strength and stiffness properties in perpendicular directions within the panel plane. The laminated construction results in a distribution of defects and markedly reduces splitting (compared to solid wood) when the plywood is penetrated by fasteners.

Two general classes of plywood, covered by separate standards, are available: construction and industrial plywood, and hardwood and decorative plywood. Construction and industrial plywood is covered by Product Standard PS 1 or PS 2; and hardwood and decorative plywood is covered by ANSI/HPVA HP-1. Each standard recognizes different exposure durability classifications, which are primarily based on the moisture resistance of the glue used, but sometimes also address the grade of veneer used.

The exposure durability classifications for construction and industrial plywood specified in PS-1 are: exterior and exposure 1. Exterior plywood is bonded with exterior (waterproof) glue and is suitable for repeated wetting and redrying or long-term exposure to weather or other conditions of similar severity. Exposure 1 plywood is bonded with exterior glue and is suitable for uses not permanently exposed to the weather. Exposure 1 panels are intended to resist the effects of moisture on structural performance due to construction delays or other conditions of similar severity.

The exposure durability classifications for hardwood and decorative plywood specified in ANSI/HPVA HP-1 are, in decreasing order of moisture resistance, as follows: technical (exterior), type I (exterior), type II (interior), and type III (interior). Hardwood and decorative plywood are not typically used in applications where structural performance is a prominent concern. Therefore, most of the remaining discussion of plywood performance will concern construction and industrial plywood.

A very significant portion of the market for construction and industrial plywood is in residential construction. This market reality has resulted in the development of performance standards for sheathing and single-layer subfloor or underlayment for residential construction by APA–The Engineered Wood Association (APA). Plywood panels conforming to these performance standards for sheathing are marked with grade stamps (grade stamps are different for different agencies). Typically, grade stamps must show (1) conformance to the plywood product standards; (2) recognition as a quality assurance agency by the National Evaluation Services (NES), which is affiliated with the Council of American Building Officials; (3) exposure durability classification; (4) thickness of panel; (5) span rating, 32/16, which refers to the maximum allowable roof support spacing of 32 in (813 mm) and maximum floor joist spacing of 16 in (406 mm); (6) conformance to the performance-rated standard of the agency; (7) manufacturer's name or mill number; and (8) grades of face and core veneers.

The span-rating system for plywood was established to simplify specification of plywood without resorting to specific structural engineering design. This system indicates performance without the need to refer to species group or panel thickness. It gives the allowable span when the face grain is placed across supports.

If design calculations are desired, a design guide is provided by APA in "Plywood Design Specifications" (PDS, 2004). The design guide contains tables of grade stamp references, section properties, and allowable stresses for plywood used in construction of buildings and similar related structures. For the design and fabrication of curved panels, plywood-lumber beams, plywood stress-skin panels, plywood sandwich panels, and all-plywood beams, information is available in supplements to the PDS standard from APA.

If calculations for the actual physical and mechanical properties of plywood are desired, formulas relating the properties of the particular wood species in the component plies to the laminated panel are provided in "Wood Handbook." These formulas could be applied to plywood of any species, provided the basic mechanical properties of the species are known. Note, however, that the formulas yield predicted actual properties (not design values) of plywood made of defect-free veneers.

Structural Flakeboards

Structural flakeboards are wood panels made from specially produced flakes—typically from relatively low-density species, such as aspen or pine—and bonded with an exterior-type water-resistant adhesive. Two major types of flakeboards are recognized, **oriented strandboard** (OSB) and **waferboard**. OSB is a flakeboard product made from wood strands (long and narrow flakes) that are formed into a mat of three to five layers. The outer layers are aligned in the long panel direction, while the inner layers may be aligned at right angles to the outer layers or may be randomly aligned. In waferboard, a product made almost exclusively from aspen wafers (wide flakes), the flakes are not usually oriented in any direction, and they are bonded with an exterior-type resin. Because flakes are aligned in OSB, the bending properties (in the aligned direction) of this type of flakeboard are generally superior to those of waferboard. For this reason, OSB is the predominant form of structural flakeboard. Panels commonly range from 0.25 to 0.75 in (6 to 19 mm) thick and are 4 by 8 ft (1 by 2 m) in surface dimension. However, thicknesses up to 1.125 in (28.58 mm) and surface dimensions up to 8 by 24 ft (2 by 7 m) are available by special order.

A substantial portion of the market for structural flakeboard is in residential construction. For this reason, structural flakeboards are usually marketed as conforming to a product standard for sheathing or single-layer subfloor or underlayment and are graded as a performance-rated product (PRP-108) similar to that for construction plywood. The Voluntary Product Standard PS 2 is the performance standard for wood-based structural-use panels, which includes such products as plywood, composites, OSB, and waferboard. The PS 2 is not a replacement for PS 1, which contains necessary veneer grade and glue bond requirements as well as prescriptive layup provisions and includes many plywood grades not covered under PS 2.

Design capacities of the APA performance-rated products, which include OSB and waferboard, can be determined by using procedures outlined in the APA Technical Note N375A. In this reference, allowable design strength and stiffness properties, as well as nominal thicknesses and section properties, are specified based on the span rating of the panel. Additional adjustment factors based on panel grade and construction are provided.

Because of the complex nature of structural flakeboards, formulas for determining actual strength and stiffness properties, as a function of the component material, are not available.

DURABILITY OF WOOD IN CONSTRUCTION

by Stan Lebow and Laura Hasburgh

Biological Challenge

In the natural ecosystem, wood residues are recycled into the nutrient web through the action of wood-degrading fungi, insects, and other organisms. In some circumstances, these same natural recyclers have the potential to degrade wood used in construction. Termites and decay fungi are the most destructive, but other organisms, such as wood boring beetles and carpenter ants, can be important in some regions. Several types of marine organisms can attack wood used in brackish water and saltwater. If conditions are favorable for the survival of one or more of these wood-destroying organisms, wood that has natural durability or that has been treated with preservatives should be employed to ensure the integrity of the structure.

Conditions Favorable for Biodeterioration

Three environmental components govern the development of wood-degrading organisms within terrestrial wood construction: moisture, oxygen supply, and temperature. Of these, moisture seems most directly influenced by design and construction practices. The oxygen supply isusually adequate except for materials submerged below water or deep within the soil, and wood-degrading organisms have evolved to survive in a wide range of temperatures. For these reasons, most of the following discussion will focus on relationships between wood moisture and potential for wood deterioration.

Decay fungi require free water within the wood cells (above 20 to 25 percent moisture content) to survive. Soil provides a continuing source of moisture. Nondurable wood in contact with the ground will decay most rapidly at the ground line where moisture from soil and the supply of oxygen within the wood support growth of decay fungi. Deep within the soil, a limited supply of oxygen prevents growth of most decay fungi. As the distance above ground increases, the decreasing moisture content limits fungal growth unless the wood is wetted through exposure to rain or from water entrapped as a consequence of design or construction practices. Wood absorbs water through cut ends about 11 times faster than through lateral surfaces. Consequently, decay develops first at the joints in above ground construction.

Extremely low oxygen content in wood submerged below water will prevent growth of decay fungi, but other microorganisms can slowly colonize submerged wood over decades or centuries of exposure. Thus, properties of such woods need to be reconfirmed when old, submerged structures are retrofitted.

Most wood-degrading insects also prefer moist wood, but their moisture requirements vary somewhat by species. Subterranean termites native to the U.S. mainland establish a continuous connection with soilto maintain adequate moisture in wood that is being attacked above ground. Formosan termites have the capacity to use sources of aboveground moisture without establishing a direct connection with the soil. Dry-wood termites survive only in tropical or neotropical areas with sufficiently high relative humidity to elevate the wood moisture content to levels high enough to provide moisture for insect metabolism. Sometypes of wood-boring beetles, such as powder post beetles, can colonize wood with low moisture contents. Carpenter ants initially invade wood that is moist or decayed, but may then expand into adjacent dry wood.

Methods for Protecting Wood

Protection with Good Design The most important aspect of protecting wood in construction is use of designs that minimize moisture accumulation. Many wood structures have endured for several hundred years because of sound design and construction practices. In nearly all those old buildings the wood has been kept dry by a barrier over the structure (roof plus overhang), by maintaining a separation between the ground and the wood elements (foundation), and by preventing accumulation of moisture in the structure (ventilation). Where Formosan subterranean termites occur, good designs and construction practices that eliminate sources of above ground moisture are particularly important. Channeling of water from roof drains, collection of condensate from air conditioners, and proper installation of flashing are examples of important considerations. Other design and construction practices also help to protect wood from insect attack. Chemical or physical barriers are often used to prevent subterrane an termites from moving from the soil into a building. Many beetles attack only certain species or groups of woods that may be used in speciality items such as joinery. When these species are used, it is important to use wood that is not preinfested at the time of construction. Today's engineered wood products will last for centuries if good design practices are used. However, there are many structures where it is impractical to prevent decay and insect attack simply through design. In these situations part of or all the structure may need to be constructed with naturally durable wood species or with wood that has been treated with preservatives.

Naturally Durable Wood The heartwood of old-growth trees of certain species, such as bald cypress, redwood, cedars, and several white oaks, is naturally resistant or very resistant to decay fungi. Heartwood of several other species, such as Douglas-fir, longleaf pine, eastern white pine, and western larch, is moderately resistant to wood decay fungi. Some tropical species, such as ipe, are also imported because of their natural durability. Although decay resistance and resistance to termite attack are correlated, not all species that are decay-resistant are also resistant to termite attack. A more complete listing of naturally durable woods is given in the "Wood Handbook." One limitation on the use of these species is that the durability is variable between trees and even between pieces cut from a single tree. In addition, the supply of naturally durable species is limited relative to the demand for durable wood products. Consequently, other forms of protection, such as preservative treatment, are more frequently used in current construction. Naturally durable species are not effective in preventing marine borer attack, and pressure treatment with wood preservatives is required to protect wood used in marine or brackish waters.

Preservative Treatments Wood preservatives have been used for over a hundred years. They are broadly classified as either water type or oil type based on the chemical composition of the preservative and the solvent used during the treating process. Each type of preservative has advantages and disadvantages, depending on the application. The most common oil-type preservatives are creosote, pentachlorophenol, and copper naphthenate. Because wood treated with oil-type preservatives may be oily to the touch or have a noticeable odor, it is not usually used for applications involving frequent human contact or for inhabited structures. Oil-type preservatives may offer improved water repellency and typically do not promote corrosion of fasteners. Water-based preservatives leave a dry, paintable surface and are commonly used to treat wood for residential applications such as decks and fences. They are primarily used to treat softwoods because hardwoods treated with these preservatives may not be well protected from soft-rot attack. Until recently, the most widely used water-type preservative was chromated copper arsenate (CCA). However, CCA has now been voluntarily phased out for most applications around residential areas and where human contact is expected. More recent water-based preservatives, such as alkaline copper quat or copper azole, rely primarily on copper to protect the wood. Water-based wood preservatives containing copper can increase susceptibility to corrosion, and metal fasteners used with the treated wood should be hot-dipped galvanized or stainless steel. Copper-free formulations are also available for treatment of wood that is not placed in direct contact with the ground.

Borates are another type of waterborne preservative that may be used to treat interior building components for protection against insect attack. Borates should not be used where they are exposed to soil or rainfall because they are readily leached from the wood. In non-load-bearing applications such as exterior millwork around windows and doors, wood is usually protected with water-repellent preservative treatments that are applied by nonpressure processes. Standards for preservative treatment are published by the American Wood Protection Association, and detailed information on wood preservation is given in the "Wood Handbook." It is important to ensure that wood is being treated to standard specifications. The U.S. Department of Commerce American Lumber Standard Committee (ALSC) accredits third-party inspection agencies for treated wood products. Updated lists of accredited agencies can be found on the ALSC Website (www.alsc.org). If the wood has been treated to these specifications, it will have the quality mark or symbol of an ALSC-accredited inspection agency.

Protection from Weathering The combination of sunlight and other weathering agents will slowly remove the surface fibers of wood products. Removal of fibers can be greatly reduced by providing a wood finish; if the finish is properly maintained, the removal of fibers can benearly eliminated. Information on wood finishes is available in Cassens and Feist, "Exterior Wood in the South."

Protection from Fire In general, proper design for fire safety allows the use of untreated wood. When there is a need to reduce the potential for heat contribution or flame spread, fire-retardant treatments are available. Although fire-retardant coatings or dip treatments are available, effective treatment often requires that the wood be pressure-impregnated with the fire-retardant chemicals. These chemicals include inorganic salts such as monoammonium and diammonium phosphate, ammonium sulfate, zinc chlorate, sodium tetraborate, and boric acid. Resin polymerized after impregnation into wood is used to obtain aleach-resistant treatment. Such amino resin systems are based on urea, melamine, dicyandiamide, and related compounds. An effective treatment can reduce the ASTM E 84 flame spread to less than 25 and show no evidence of significant progressive combustion when the ASTM E 84 test is continued for an additional 20-min period. When the external source of heat is removed, the flames from fire-retardant-treated woodwill generally self-extinguish. Many fire-retardant treatments reducethe generation of combustible gases by lowering the thermal degradation temperature. ASTM standards have been developed to address other performance requirements such as initial strength loss due to treatment and kiln drying, strength loss when exposed to elevated temperatures in use, excessive hygroscopicity in areas of high humidity, and loss of fire performance due to outdoor exposure to rainfall.

4.7 NONMETALLIC MATERIALS

Revised by Ali M. Sadegh

ABRASIVES

References: "Abrasives: Their History and Development," The Norton Co. Searle, "Manufacture and Use of Abrasive Materials," Pitman. Heywood, "Grinding Wheels and Their Uses," Penton. "Boron Carbide," The Norton Co. "Abrasive Materials," annual review in "Minerals Yearbook," U.S. Bureau of Mines. "Abrasive Engineering," Hitchcock Publishing Co. Coated Abrasives Manufacturer's Institute, "Coated Abrasives—Modern Tool of Industry," McGraw-Hill. Wick, Abrasives; Where They Stand Today, *Manufacturing Engineering and Management,* **69,** no. 4, Oct. 1972. Burls, "Diamond Grinding; Recent Research and Development," Mills and Boon, Ltd., London. Coes, Jr., "Abrasives," Springer-Verlag, New York. *Proceedings of the American Society for Abrasives Methods.* 1971, ANSI B74.1 to B74.3 1977 to 1993. Washington Mills Abrasive Co., Technical Data, *Mechanical Engineering,* Mar. 1984. "Modern Abrasive Recipes," *Cutting Tool Engineering,* April 1994. "Diamond Wheels Fashion Carbide Tools," *Cutting Tool Engineering,* August 1994. Synthetic Gemstones, *Compressed Air Magazine,* June 1993. Krar and Ratterman, "Superabrasives," McGraw-Hill. Current ASTM and ANSI standards.

Manufactured (Artificial) Abrasives

Manufactured abrasives dominate the scene for commercial and industrial use, because of the greater control over their chemical composition and crystal structure, and their greater uniformity in size, hardness, and cutting qualities, as compared with natural abrasives. Technical advances have resulted in abrasive "machining tools" that are economically competitive with many traditional machining methods, and in some cases replace and surpass them in terms of productivity. Fused electrominerals such as silicon carbide, aluminum oxide, alumina zirconia, alumina chromium oxide, and alumina titania zirconia are the most popular conventional **manufactured abrasives.**

Grains

Manufactured (industrial or synthetic) **diamonds** are produced from graphite at pressures 29.5 to 66.5 × 10^6 N/m2 (8 to 18 × 10^5 lb/in^2) and temperatures from 1,090 to 2,420°C (2,000 to 4,400°F), with the aid of metal catalysts. **Crystalline alumina,** as chemically purified $A1_2O_3$, is very adaptable in operations such as precision grinding of sensitive steels. **Silicon carbide (SiC)** corresponds to the mineral moissanite, and has a hardness of about 9.5 (Mohs scale) or 2,500 (Knoop), and specific gravity 3.2. It is insoluble in acid and is infusible but decomposes above 2,230°C (4,060°F).

Natural Abrasives

Diamond of the **bort** variety, crushed and graded into usable sizes and bonded with synthetic resin, metal powder, or vitrified-type bond, is used extensively for grinding tungsten- and tantalum-carbide cutting tools, and glass, stone, and ceramics.

Garnet Certain deposits of garnet having a hardness between quartz and corundum are used in the manufacture of abrasive paper.

Abrasive Waterjets

Such tough-to-cut materials as titanium, ceramics, metallic honeycomb structures, glass, graphite, and bonding compounds can be successfully cut with abrasive waterjets. Abrasive waterjets use pressurized water, up to 60 lb/in2, to capture solid abrasives by flow momentum, after which the mixture is expelled through a sapphire nozzle to form a highly focused, high-velocity cutting "tool."

ADHESIVES

References: Cagle (ed.), "Handbook of Adhesive Bonding," McGraw-Hill. Bikerman, "The Science of Adhesive Joints," Academic. Shields, "Adhesives Handbook," CRC Press (Division of The Chemical Rubber Co.). Cook, "Construction Sealants and Adhesives," Wiley. Patrick, "Treatise on Adhesives," Marcel Dekker. NASA SP-5,961 (01) Technology Utilization, "Chemistry Technology: Adhesives and Plastics," National Technical Information Services, Virginia. Simonds and Church, "A Concise Guide to Plastics," Reinhold. Lerner, Kotsher, and Sheckman, "Adhesives Red Book," Palmerton Publishing Co., New York. *Machine Design,* June 1976. ISO 6354–1982. "1994 Annual Book of ASTM Standards," vol. 15.06 (Adhesives). ANSI/ASTM Standards D896–D3,808. Current ASTM and ANSI Standards.

Adhesives are substances capable of holding materials together in a useful manner by surface attachment. Some of the advantages and disadvantages of adhesive bonding are as follows:

Advantages Ability to bond similar or dissimilar materials of different thicknesses; fabrication of complex shapes not feasible by other fastening means; smooth external joint surface; economic and rapid assembly; uniform distribution of stresses; weight reduction; vibration damping; prevention or reduction of galvanic corrosion; insulating properties.

Disadvantages Surface preparation; long cure times; optimum bond strength not realized instantaneously, service-temperature limitations; service deterioration; assembly fire or toxicity; tendency to creep under sustained load.

A broad scheme of classification is given in Table 4.7.1.

For the vocabulary of adhesives the reader should refer to ISO 6,354–1982, and for standard definitions refer to ASTM D907–82. Procedures for the testing of adhesive strength viscosity, storage life, fatigue properties, etc. can be found in ANSI/ASTM D950-D3,808.

Thermoplastic adhesives are a general class of adhesives based upon long-chained polymeric structure, and are capable of being softened by the application of heat.

Thermosetting adhesives are a general class of adhesives based upon *cross-linked* polymeric structure, and are incapable of being softened once solidified. A recent development (1994) in cross-linked aromatic polyesters has yielded a very durable adhesive capable of withstanding temperatures of 700°F before failing. It retains full strength through 400°F. It is likely that widespread applications for it will be found first in the automotive and aircraft industries.

Thermoplastic and thermosetting adhesives are cured (set, polymerized, solidified) by heat, catalysis, chemical reaction, free-radical activity, radiation, loss of solvent, etc., as governed by the particular adhesive's chemical nature.

Elastomers are a special class of thermoplastic adhesive possessing the common quality of substantial flexibility or elasticity.

Anaerobic adhesives are a special class of thermoplastic adhesive (polyacrylates) that set *only* in the *absence* of air (oxygen). The two basic types are: (1) *machinery*—possessing shear strength only, and (2) *structural*—possessing both tensile and shear strength.

Pressure-sensitive adhesives are permanently (and aggressively) tacky (sticky) solids which form immediate bonds when two parts are brought together under pressure. They are available as films and tapes as well as hot-melt solids.

For high-performance adhesive applications (engineering or machine parts) the following grouping is convenient.

Thread locking	Anaerobic acrylic
Hub mounting	Anaerobic acrylic—compatible materials or flow migration unimportant.
	Modified acrylic—large gaps or migration must be avoided.
	Epoxy—maximum strength at high temperatures
Bearing mounting	Anaerobic acrylic—compatible materials necessary and flow into bearing area to be prevented.
	Modified acrylic—for lowest cost
Structural joining	Epoxies and modified epoxies—for maximum strength (highest cost)
	Acrylics—anaerobic or modified cyanacrylates
Gasketing	Silicones—primarily anaerobic

Table 4.7.1 Classification of Adhesives

Origin and basic type		Adhesive material
Natural	Animal	Albumen, animal glue (including fish), casein, shellac, beeswax
	Vegetable	Natural resins (gum arabic, tragacanth, colophony, Canada balsam, etc.); oils and waxes (carnauba wax, linseed oils); proteins (soybean); carbohydrates (starch, dextrins)
	Mineral	Inorganic materials (silicates, magnesia, phosphates, litharge, sulfur, etc.); mineral waxes (paraffin); mineral resins (copal, amber); bitumen (including asphalt)
Synthetic	Elastomers	Natural rubber (and derivatives, chlorinated rubber, cyclized rubber, rubber hydrochloride) Synthetic rubbers and derivatives (butyl, polyisobutylene, polybutadiene blends (including styrene and acrylonitrile), polyisoprenes, polychloroprene, polyurethane, silicone, polysulfide, polyolefins (ethylene vinyl chloride, ethylene polypropylene) Reclaim rubbers
	Thermoplastic	Cellulose derivatives (acetate, acetate-butyrate, caprate, nitrate, methyl cellulose, hydroxy ethyl cellulose, ethyl cellulose, carboxy methyl cellulose) Vinyl polymers and copolymers (polyvinyl acetate, alcohol, acetal, chloride, polyvinylidene chloride, polyvinyl alkyl ethers Polyesters (saturated) [polystyrene, polyamides (nylons and modifications)] Polyacrylates (methacrylate and acrylate polymers, cyanoacrylates, acrylamide) Polyethers (polyhydroxy ether, polyphenolic ethers) Polysulfones
	Thermosetting	Amino plastics (urea and melamine formaldehydes and modifications) Epoxides and modifications (epoxy polyamide, epoxy bitumen, epoxy polysulfide, epoxy nylon) Phenolic resins and modifications (phenol and resorcinol formaldehydes, phenolic-nitrile, phenolic-neoprene, phenolic-epoxy) Polyesters (unsaturated) Polyaromatics (polyimide, polybenzimidazole, polybenzothiazole, polyphenylene) Furanes (phenol furfural)

SOURCE: Shields, "Adhesives Handbook," CRC Press (Division of the Chemical Rubber Co.).

CERAMICS

REFERENCES: Kingery, "Introduction to Ceramics," Wiley. Norton, "Elements of Ceramics," Addison-Wesley. "Carbon Encyclopedia of Chemical Technology," Interscience. Humenik, "High-Temperature Inorganic Coatings," Reinhold. Kingery, "Ceramic Fabrication Processes," MIT. McCreight, Rauch, Sr., Sutton, "Ceramic and Graphite Fibers and Whiskers; A Survey of the Technology," Academic. Rauch, Sr., Sutton, McCreight, "Ceramic Fibers and Fibrous Composite Materials," Academic. Hague, Lynch, Rudnick, Holden, Duckworth (compilers and editors), "Refractory Ceramics for Aerospace; A Material Selection Handbook," The American Ceramic Society. Waye, "Introduction to Technical Ceramics," Maclaren and Sons, Ltd., London. Hove and Riley, "Ceramics for Advanced Technologies," Wiley. McMillan, "Glass-Ceramics," Academic. *Machine Design,* May 1983. "1986 Annual Book of ASTM Standards," vols. 15.02, 15.04 (Ceramics). Current ASTM and ANSI Standards.

Ceramic materials are a diverse group of nonmetallic, inorganic solids with a wide range of compositions and properties. Their structure may be either crystalline or glassy. The desired properties are often achieved by high-temperature treatment (firing or burning).

Traditional ceramics are products based on the silicate industries, where the chief raw materials are naturally occurring minerals such as the clays, silica, feldspar, and talc. While silicate ceramics dominate the industry, newer ceramics, sometimes referred to as **electronic** or **technical** ceramics, are playing a major role in many applications. **Glass ceramics** are important for electrical, electronic, and laboratory-ware uses. Glass ceramics are melted and formed as glasses, then converted, by controlled nucleation and crystal growth, to polycrystalline ceramic materials.

Manufacture Typically, the manufacture of traditional-ceramic products involves blending of the finely divided starting materials with water to form a plastic mass which can be formed into the desired shape. The plasticity of clay constituents in water leads to excellent forming properties. Formation processes include extrusion, pressing, and ramming. Unsymmetrical articles can be formed by "slip-casting" techniques, where much of the water is taken up by a porous mold. After the water content of formed articles has been reduced by drying, the ware is **fired** at high temperature for fusion and/or reaction of the components and for attainment of the desired properties. Firing temperatures can usually be considerably below the fusion point of the pure components through the use of a **flux**, often the mineral feldspar. Following burning, a vitreous ceramic coating, or **glaze**, may be applied to render the surface smooth and impermeable.

Quite different is the **fusion casting** of some refractories and refractory blocks and most glasses, where formation of the shaped article is carried out after fusion of the starting materials.

Properties The physical properties of ceramic materials are strongly dependent on composition, microstructure (phases present and their distribution), and the history of manufacture. Volume pore concentration can vary widely (0 to 30%) and can influence shock resistance, strength, and permeability. Most traditional ceramics have a glassy phase, a crystalline phase, and some porosity. The last can be eliminated at the surface by glazing. Most ceramic materials are resistant to large compressive stresses but fail readily in tension. Resistance to abrasion, heat, and stains, chemical stability, rigidity, good weatherability, and brittleness characterize many common ceramic materials.

Products Many traditional-ceramic products are referred to as **whiteware** and include pottery, semivitreous wares, electrical porcelains, sanitary ware, and dental porcelains. Building products which are ceramic include brick and structural tile and conduit, while refractory blocks, as well as many abrasives, are also ceramic in nature. Porcelain **enamels** for metals are opacified, complex glasses which are designed to match the thermal-expansion properties of the substrate. Important **technical ceramics** include magnetic ceramics, with magnetic properties but relatively high electrical resistance; nuclear ceramics, including uranium dioxide fuel elements; barium titanate as a material with very high dielectric constant. Several of the pure oxide ceramics with superior physical properties are being used in electrical and missile applications where high melting and deformation temperatures and stability in oxygen are important. Fibrous ceramic composed of ziconium oxide fibers, such as Zircar, provide optimum combination of strength, low thermal conductivity, and high temperature resistance to about 2,490°C (4,500°F). Partially stabilized zirconia, because of its steel-strong and crack-resisting properties, is finding high-temperature applications, for example, in diesel engines.

Various forms of ceramics are available from manufacturers such as paper, "board," blankets, tapes, and gaskets.

A fair amount of development is under way dealing with **reinforcing ceramics** to yield **ceramic matrix composite** materials. The driving interest here is potential applications in high-temperature, high-stress environments, such as those encountered in aircraft and aerospace structural components. None of the end products are available yet "off the shelf," and applications are likely to be quite circumscribed and relegated to solution of very specialized design problems.

Ceramic inserts suitably shaped and mechanically clamped in tool holders are applied as cutting tools, and they find favor especially for cutting abrasive materials of the type encountered in sand castings or forgings with abrasive oxide crusts. They offer the advantage of resisting abrasion at high tool/chip interface temperatures, and consequently they exhibit relatively long tool life between sharpenings. They tend to be somewhat brittle and see limited service in operations requiring interrupted cuts; when they are so applied, tool life between sharpenings is very short.

ELASTOMERS

The trend toward irregular roof surfaces—folded plates, hyperbolic paraboloids, domes, barrel shells—has brought the increased use of plastics or synthetic rubber elastomers (applied as fluid or sheets) as roofing membranes. Such membranes offer light weight, adaptability to any roof slope, good heat reflectivity, and high elasticity at moderate temperatures. Negatively, elastomeric membranes have a more limited range of satisfactory substrate materials than conventional ones. Table 4.7.2 presents several such membranes, and the method of use.

RUBBER AND RUBBERLIKE MATERIALS (ELASTOMERS)

REFERENCES: Dawson and Porritt, "Rubber: Physical and Chemical Properties," Research Association of British Rubber Manufacturers. Davis and Blake, "The Chemistry and Technology of Rubber," Reinhold. ASTM Standards on Rubber Products. "Rubber Red Book Directory of the Rubber Industry," *The Rubber Age*. Flint, "The Chemistry and Technology of Rubber Latex," Van Nostrand. "The Vanderbilt Handbook," R.T. Vanderbilt Co. Whitby, Davis, and Dunbrook, "Synthetic Rubber," Wiley. "Rubber Bibliography," Rubber Division, American Chemical Society. Noble, "Latex in Industry," *The Rubber Age*. "Annual Report on the Progress of Rubber Technology," Institution of the Rubber Industry. Morton (ed.). "Rubber Technology," Van Nostrand Reinhold. ISO 188-1982. "1986 Annual Book of ASTM Standards," vols. 09.01, 09.02 (Rubber). Current ASTM and ANSI Standards.

To avoid confusion by the use of the word **rubber** for a variety of natural and synthetic products, the term **elastomer** has come into use, particularly in scientific and technical literature, as a name for both natural and synthetic materials which are elastic or resilient and in general resemble natural rubber in feeling and appearance.

The utility of rubber and synthetic elastomers is increased by compounding. In the raw state, elastomers are soft and sticky when hot, and hard or brittle when cold. **Vulcanization** extends the temperature range within which they are flexible and elastic. In addition to vulcanizing agents, ingredients are added to make elastomers stronger, tougher, and harder, to make them age better, to color them, and in general to modify them to meet the needs of service conditions. Few rubber products today are made from rubber or other elastomers alone.

The elastomers of greatest commercial and technical importance today are natural rubber, GR-S, Neoprene, nitrile rubbers, and butyl.

Natural rubber of the best quality is prepared by coagulating the latex of the *Hevea brasilensis* tree, cultivated chiefly in the Far East. This represents nearly all of the natural rubber on the market today.

Unloaded vulcanized rubber will stretch to approximately 10 times its length and at this point will bear a **load** of 13.8×10^6 N/m^2 (10 tons/in^2). It can be compressed to one-third its thickness thousands of times without injury. When most types of vulcanized rubber are stretched, their resistance increases in greater proportion than the extension. Even when stretched almost to the point of rupture, they recover very nearly their original dimensions on being released, and then gradually recover a part of the residual distortion.

Freshly cut or torn raw rubber possesses the power of **self-adhesion** which is practically absent in vulcanized rubber. Cold water preserves rubber, but if exposed to the air, particularly to the sun, rubber goods tend to become hard and brittle. Dry heat up to 49°C (120°F) has little deteriorating effect; at temperatures of 181 to 204°C (360 to 400°F) rubber begins to melt and becomes sticky; at higher temperatures, it becomes entirely carbonized. Unvulcanized rubber is **soluble** in gasoline, naphtha, carbon bisulfide, benzene, petroleum ether, turpentine, and other liquids.

Most rubber is **vulcanized,** i.e., made to combine with sulfur or sulfur-bearing organic compounds or with other chemical cross-linking agents. Vulcanization, if properly carried out, improves mechanical properties, eliminates tackiness, renders the rubber less susceptible to temperature

Table 4.7.2 Elastomeric Membranes

	Material	Method of application	Number of coats or sheets
Fluid-applied	Neoprene-Hypalon*	Roller, brush, or spray	2 + 2
	Silicone	Roller, brush, or spray	2
	Polyurethane foam, Hypalon* coating	Spray	2
	Clay-type asphalt emulsion reinforced with chopped glass fibers†	Spray	1
Sheet-applied	Chlorinated polyethylene on foam	Adhesive	1
	Hypalon* on asbestos felt	Adhesive	1
	Neoprene-Hypalon*	Adhesive	1 + surface paint
	Tedlar‡ on asbestos felt	Adhesive	1
	Butyl rubber	Adhesive	1
Traffic decks	Silicone plus sand	Trowel	1 + surface coat
	Neoprene with aggregate§	Trowel	1 + surface coat

* Registered trademark of E. I. du Pont de Nemours & Co., for chlorosulfonated polyethylene.
† Frequently used with coated base sheet.
‡ Registered trademark of E. I. du Pont de Nemours & Co. for polyvinyl fluoride.
§ Aggregate may be flint, sand, or crushed walnut shells.
SOURCE: C. W. Griffin, Jr., "Manual of Built-up Roof Systems," McGraw-Hill, 1970, with permission of the publisher.

changes, and makes it insoluble in all known solvents. It is impossible to dissolve vulcanized rubber unless it is first decomposed. Other ingredients are added for general effects as follows:

To increase tensile strength and resistance to abrasion: carbon black, precipitated pigments, as well as organic vulcanization accelerators.

To cheapen and stiffen: whiting, barytes, talc, silica, silicates, clays, fibrous materials.

To soften (for purposes of processing or for final properties): bituminous substances, coal tar and its products, vegetable and mineral oils, paraffin, petrolatum, petroleum oils, asphalt.

Vulcanization accessories, dispersion and wetting mediums, etc.: magnesium oxide, zinc oxide, litharge, lime, stearic and other organic acids, degras, pine tar.

Protective agents (natural aging, sunlight, heat, flexing): condensation amines, waxes.

Coloring pigments: iron oxides, especially the red grades, lithopone, titanium oxide, chromium oxide, ultramarine blue, carbon and lampblacks, organic pigments of various shades.

Specifications should state suitable physical tests. Tensile strength and extensibility tests are of importance and differ widely with different compounds.

GR-S is an outgrowth and improvement of German Buna S. The quantity now produced far exceeds all other synthetic elastomers. It is made from butadiene and styrene, which are produced from petroleum. These two materials are copolymerized directly to GR-S, which is known as a butadiene-styrene copolymer. GR-S has recently been improved, and now gives excellent results in tires.

Neoprene is made from acetylene, which is converted to vinylacetylene, which in turn combines with hydrogen chloride to form chloroprene. The latter is then polymerized to Neoprene.

Nitrile rubbers, an outgrowth of German Buna N or Perbunan, are made by a process similar to that for GR-S, except that acrylonitrile is used instead of styrene. This type of elastomer is a butadiene-acrylonitrile copolymer.

Butyl, one of the most important of the synthetic elastomers, is made from petroleum raw materials, the final process being the copolymerization of isobutylene with a very small proportion of butadiene or isoprene.

Polysulfide rubbers having unique resistance to oxidation and to softening by solvents are commercially available and are sold under the trademark Thiokol.

Ethylene-propylene rubbers are notable in their oxidation resistance. **Polyurethane** elastomers can have a tensile strength up to twice that of conventional rubber, and solid articles as well as foamed shapes can be cast into the desired form using prepolymer shapes as starting materials. **Silicone** rubbers have the advantages of a wide range of service temperatures and room-temperature curing. **Fluorocarbon** elastomers are available for high-temperature service. Polyester elastomers have excellent impact and abrasion resistance.

No one of these elastomers is satisfactory for all kinds of service conditions, but rubber products can be made to meet a large variety of service conditions.

The following examples show some of the important properties required of rubber products and some typical services where these properties are of major importance:

Resistance to abrasive wear: auto-tire treads, conveyor-belt covers, soles and heels, cables, hose covers, V belts.

Resistance to tearing: auto inner tubes, tire treads, footwear, hotwater bags, hose covers, belt covers, V belts.

Resistance to flexing: auto tires, transmission belts, V belts (see Sec. 4.2 for information concerning types, sizes, strengths, etc., of V belts), mountings, footwear.

Resistance to high temperatures: auto tires, auto inner tubes, belts conveying hot materials, steam hose, steam packing.

Resistance to cold: airplane parts, automotive parts, auto tires, refrigeration hose.

Minimum heat buildup: auto tires, transmission belts, V belts, mountings.

High resilience: auto inner tubes, sponge rubber, mountings, elastic bands, thread, sandblast hose, jar rings, V belts.

High rigidity: packing, soles and heels, valve cups, suction hose, battery boxes.

Long life: fire hose, transmission belts, tubing, V belts.

Electrical resistivity: electricians' tape, switchboard mats, electricians' gloves.

Electrical conductivity: hospital flooring, nonstatic hose, matting.

Impermeability to gases: balloons, life rafts, gasoline hose, special diaphragms.

Resistance to ozone: ignition distributor gaskets, ignition cables, windshield wipers.

Resistance to sunlight: wearing apparel, hose covers, bathing caps.

Resistance to chemicals: tank linings, hose for chemicals.

Resistance to oils: gasoline hose, oil-suction hose, paint hose, creamery hose, packinghouse hose, special belts, tank linings, special footwear.

Stickiness: cements, electricians' tapes, adhesive tapes, pressure-sensitive tapes.

Low specific gravity: airplane parts, forestry hose, balloons.

No odor or taste: milk tubing, brewery and wine hose, nipples, jar rings.

Special colors: ponchos, life rafts, welding hose.

Table 4.7.3 gives a comparison of some important characteristics of the most important elastomers when vulcanized. The lower part of the table indicates, for a few representative rubber products, preferences in the use of different elastomers for different service conditions without consideration of cost.

Specifications for rubber goods may cover the chemical, physical, and mechanical properties, such as elongation, tensile strength, permanent set, and oven tests, minimum rubber content, exclusion of reclaimed rubber, maximum free and combined sulfur contents, maximum acetone and chloroform extracts, ash content, and many construction requirements. It is preferable, however, to specify properties such as resilience, hysteresis, static or dynamic shear and compression modulus, flex fatigue and cracking, creep, electrical properties, stiffening, heat generation, compression set, resistance to oils and chemicals, permeability, brittle point, etc., in the temperature range prevailing in service, and to leave the selection of the elastomer to a competent manufacturer.

Latex, imported in stable form from the Far East, is used for various rubber products. In the manufacture of such products, the latex must be compounded for vulcanizing and otherwise modifying properties of the rubber itself. Important products made directly from compounded latex include surgeons' and household gloves, thread, bathing caps, rubberized textiles, balloons, and sponge. A recent important use of latex is for "foam sponge," which may be several inches thick and used for cushions, mattresses, etc.

Gutta-percha and **balata,** also natural products, are akin to rubber chemically but more leathery and thermoplastic, and are used for some special purposes, principally for submarine cables, golf balls, and various minor products.

Rubber Derivatives

Rubber derivatives are chemical compounds and modifications of rubber, some of which have become of commercial importance.

Chlorinated rubber, produced by the action of chlorine on rubber in solution, is nonrubbery, incombustible, and extremely resistant to many chemicals. As commercial **Parlon,** it finds use in corrosion-resistant paints and varnishes, in inks, and in adhesives.

Rubber hydrochloride, produced by the action of hydrogen chloride on rubber in solution, is a strong, extensible, tear-resistant, moisture-resistant, oil-resistant material, marketed as **Pliofilm** in the form of tough transparent films for wrappers, packaging material, etc.

Cyclized rubber is formed by the action of certain agents, e.g., sulfonic acids and chlorostannic acid, on rubber, and is a thermoplastic, nonrubbery, tough or hard product. One form, **Thermoprene,** is used in the **Vulcalock process** for adhering rubber to metal, wood, and concrete, and in chemical-resistant paints. **Pliolite,** which has high resistance

Table 4.7.3 Comparative Properties of Elastomers

	Natural rubber	GR-S	Neoprene	Nitrile rubbers	Butyl	Thiokol
Tensile properties	Excellent	Good	Very good	Good	Good	Fair
Resistance to abrasive wear	Excellent	Good	Very good	Good	Good	Poor
Resistance to tearing	Very good	Poor	Good	Fair	Very good	Poor
Resilience	Excellent	Good	Good	Fair	Poor	Poor
Resistance to heat	Good	Fair	Good	Excellent	Good	Poor
Resistance to cold	Excellent	Good	Good	Good	Excellent	Poor
Resistance to flexing	Excellent	Good	Very good	Good	Excellent	Poor
Aging properties	Excellent	Excellent	Good	Good	Excellent	Good
Cold flow (creep)	Very low	Low	Low	Very low	Fairly low	High
Resistance to sunlight	Fair	Fair	Excellent	Good	Excellent	Excellent
Resistance to oils and solvents	Poor	Poor	Good	Excellent	Fair	Excellent
Permeability to gases	Fairly low	Fairly low	Low	Fairly low	Very low	Very low
Electrical insulation	Fair	Excellent	Fair	Poor	Excellent	Good
Flame resistance	Poor	Poor	Good	Poor	Poor	Poor
Auto tire tread	Preferred	Alternate				
Inner tube	Alternate				Preferred	
Conveyor-belt cover	Preferred		Alternate			
Tire sidewall	Alternate	Preferred				
Transmission belting	Preferred		Alternate			
Druggist sundries	Preferred					
Gasoline and oil hose				Preferred		
Lacquer and paint hose						Preferred
Oil-resistant footwear			Preferred			
Balloons	Alternate		Preferred			
Jar rings	Alternate	Preferred				
Wire and cable insulation	Alternate	Preferred				

to many chemicals and has low permeability, is used in special paints, paper, and fabric coatings. **Marbon-B** has exceptional electrical properties and is valuable for insulation. **Hypalon** (chlorosulphonated polyethylene) is highly resistant to many important chemicals, notably ozone and concentrated sulfuric acid, for which other rubbers are unsuitable.

SILICONES

Silicones are organosilicon oxide polymers characterized by remarkable temperature stability, chemical inertness, waterproofness, and excellent dielectric properties. Among related products are the following:

Water repellents in the form of extremely thin films which can be formed on paper, cloth, ceramics, glass, plastics, leather, powders, or other surfaces. These have great value for the protection of delicate electrical equipment in moist atmospheres.

Oils with high flash points [above 315°C (600°F)], low pour points, −84.3°C (−120°F), and with a constancy of viscosity notably superior to petroleum products in the range from 260 to −73.3°C (500 to −100°F). These oil products are practically incombustible. They are in use for hydraulic servomotor fluids, damping fluids, dielectric liquids for transformers, heat-transfer mediums, etc., and are of special value in aircraft because of the rapid and extreme temperature variations to which aircraft are exposed. At present they are not available as lubricants except under light loads.

Greases and compounds for plug cocks, spark plugs, and ball bearings which must operate at extreme temperatures and speeds.

Varnishes and resins for use in electrical insulation where temperatures are high. Layers of glass cloth impregnated with or bounded by silicone resins withstand prolonged exposure to temperatures up to 260°C (500°F). They form paint finishes of great resistance to chemical agents and to moisture. They have many other industrial uses.

Silicone rubbers which retain their resiliency for −45 to 270°C (−50 to 520°F) but with much lower strength than some of the synthetic rubbers. They are used for shaft seals, oven gaskets, refrigerator gaskets, and vacuum gaskets.

4.8 LUBRICANTS AND LUBRICATION

by Glenn E. Asauskas

(For general discussion of friction, viscosity, bearings, and coefficients of friction, see Secs. 1 and 6.)

REFERENCES: STLE, "Handbook of Lubrication," 3 vols., CRC Press. ASME, "Wear Control Handbook." ASM, "ASM Handbook," vol. 18, "Friction, Lubrication, and Wear Technology." ASTM, "Annual Book of ASTM Standards," 3 vols., "Petroleum Products and Lubricants." NLGI, "Lubricating Grease Guide." Fuller, "Theory and Practice of Lubrication for Engineers," Wiley. Shigley and Mischke, "Bearings and Lubrication: A Mechanical Designers' Workbook," McGraw-Hill. Stipina and Vesely, "Lubricants and Special Fluids," Elsevier. Miller, "Lubricants and Their Applications," McGraw-Hill. Shubkin (ed.), "Synthetic Lubricants and High Performance Functional Fluids," Marcel Dekker. Bartz (ed.), "Engine Oils and Automotive Lubrication," Dekker. Hamrock, "Fundamentals of Fluid Film Lubrication," McGraw-Hill. Neale, "Lubrication: A Tribology Handbook," Butterworth-Heinemann. Ramsey, "Elastohydrodynamics," Wiley. Arnell, "Tribology, Principles and Design Applications," Springer-Verlag. Czichos, "Tribology, A Systems Approach to . . . Friction, Lubrication and Wear," Elsevier. Bhushan and Gupta, "Handbook of Tribology: Materials, Coatings, and Surface Treatments," McGraw-Hill. Yamaguchi, "Tribology of Plastic Materials," Elsevier.

Tribology, a term introduced in the 1960s, is the science of rubbing surfaces and their interactions, including friction, wear, and lubrication phenomena. **Lubrication** primarily concerns modifying friction and reducing wear and damage at the surface contacts of solids rubbing against one another. Anything introduced between the surfaces to accomplish this is a **lubricant.**

LUBRICANTS

Lubricants can be liquids, solids, or even gases, and they are most often oils or greases. Liquid lubricants often provide many functions in addition to controlling friction and wear, such as scavenging heat, dirt, and wear debris; preventing rust and corrosion; transferring force; and acting as a sealing medium.

Engineers are called upon to select and evaluate lubricants, to follow their performance in service, and to use them to best advantage in the design of equipment. Lubricant manufacturers and distributors may have hundreds of lubricants in their product line, each described separately in the product literature as to intended applications, properties, and benefits, as well as performance in selected standard tests. Lubricants are selected according to the needs of the particular application. Careful lubricant selection helps obtain improved performance, lower operating cost, and longer service life, for both the lubricant and the equipment involved. Industry's demands for efficient, competitive equipment and operations, which meet the latest environmental regulations, create continued demand for new and improved lubricants. Equipment manufacturers and suppliers specify lubricants that suit their particular equipment and its intended operating conditions, and their recommendations should be followed.

LIQUID LUBRICANTS

A liquid lubricant consists of (1) a mixture of selected base oils and additives, (2) blended to a specific viscosity, with (3) the blend designed to meet the performance needs of a particular type of service. A lubricant may contain several base oils of different viscosities and types, blended to meet viscosity requirements and to help meet performance requirements, or to minimize cost. Additives may be used to help meet viscosity as well as various performance needs.

Petroleum Oils Lubricants made with petroleum base oils are widely used because of their general suitability to much of existing equipment and availability at moderate cost. Most petroleum base oils, often called mineral oils, are prepared by conventional refining processes from naturally occurring hydrocarbons in crude oils. The main crude oil types are paraffinic- and naphthenic-based, terms referring to the type of chemical structure of most molecules. Paraffinic oils are most often preferred as lubricants, although naphthenic oils are important in certain applications.

Table 4.8.1 API Base Stock Categories

Group	Sulfur, wt %		Saturates, wt %	VI*
I	>0.03	and/or	<90	80-119
II	<0.03	and	>90	80-119
III	<0.03	and	>90	>120
IV	All Polyalphaolefins (PAO)			
V	All Stocks not included in Groups I-IV (Pale Oils and non-PAO Synthetics)			

* All commercial paraffinic base oils have VI > 95.

The American Petroleum Institute (API) base stock categories were established in the early 1990s to make it easier for lubricant blenders to interchange one base stock for another (Table 4.8.1). The original five categories were patterned on the base oils which were available at the time. Since that time, the lines between the categories have become less distinct, but these categories are still useful in describing base oils.

In general, older-technology base oils, produced by solvent refining and solvent dewaxing technology, fall into group I. More modern base oils, using hydroprocessing technology, fall into groups II and III (group III if the VI is 120 or greater). Polyalphaolefins (PAOs) are group IV, and everything else is group V. Thus, group V can include quite low-quality base stocks such as naphthenic base oils, as well as very high-quality base stocks such as esters.

Synthetic oils are artificially made, as opposed to naturally occurring petroleum fluids. Synthetic oil types include polyalphaolefins (PAOs), diesters, polyol esters, alkylbenzenes, polyalkylene glycols, phosphate esters, silicones, and halogenated hydrocarbons. Synthetic oils include diverse types of chemical compounds, so few generalities apply to all synthetic oils. Generally, synthetic oils are organic chemicals. Synthetic oils cost much more than petroleum oils, and usually they have some lubricant-related properties that are good or excellent, but they are not necessarily strong in all properties. Their stronger properties often include greater high-temperature stability and oxidation resistance, wider temperature operating range, and improved energy efficiency. As with petroleum base oils, some of their weaknesses are correctable by additives. Some synthetic esters, such as polyol esters, have good biodegradability, a property of increasing need for environmental protection.

Synthetic lubricants, because of their high cost, often are used only where some particular property is essential. There are instances where the higher cost of synthetics is justified, even though petroleum lubricants perform reasonably well. This may occur, e.g., where use of synthetics reduces the oxidation rate in high-temperature service and as a result extends the oil life substantially. Properties of some synthetic lubricants are shown in Table 4.8.2, the assessments based on formulated lubricants, not necessarily the base oils. Synthetic oil uses include

Table 4.8.2 Relative Properties of Synthetic Lubricants*

	Viscosity index	High-temperature stability	Low-temperature Lubricity	Hydrolytic properties	Fire stability	resistance	Volatility
Polyalphaolefins	Good	Good	Good	Good	Excellent	Poor	Good
Diesters	Varies	Excellent	Good	Excellent	Fair	Fair	Average
Polyol esters	Good	Excellent	Good	Good	Good	Poor	Average
Alkylbenzenes	Poor	Fair	Good	Good	Excellent	Poor	Average
Polyalkylene glycols	Excellent	Good	Good	Good	Good	Poor	Good
Phosphate esters	Poor	Excellent	Good	Varies	Fair	Excellent	Average
Silicones	Excellent	Excellent	Poor	Excellent	Fair	—	Good
Fluorinated lubes	Excellent	Excellent	Varies	Fair	Excellent	Excellent	Average

* Comparisons made for typical formulated products with additive packages, not for the base oils alone.
SOURCE: *Courtesy of Texaco's magazine Lubrication,* **78**, no. 4, 1992.

lubricants for gears, bearings, engines, hydraulic systems, greases, compressors, and refrigeration systems, although petroleum oils tend to dominate these applications. Lubricants blended with both petroleum and synthetic base oils are called **semisynthetic lubricants.**

Vegetable oils have very limited use as base stocks for lubricants, but are used in special applications. Naturally occurring esters, such as from rapeseed or castor plants, are used as the base oils in some hydraulic fluids, and are used where there are environmental concerns of possibility of an accidental fluid spill or leakage. These oils have a shorter life and lower maximum service temperature, but have a high degree of biodegradability.

Fatty oils, extracted from vegetable, animal, and fish sources, have excellent lubricity because of their glyceride structure, but are seldom used as base oils because they oxidize rapidly. A few percent of fatty oils are sometimes added to petroleum oils to improve lubricity (as in worm-gear applications), or to repel moisture (as in steam engines), and this use is often referred to as **compounding.** They also are used widely in forming soaps which, in combination with other oils, set up grease structures.

Additives are chemical compounds added to base oils to modify or enhance certain lubricant performance characteristics. The amount of additives used varies with the type of lubricant, from a few parts per million in some types of lubricants to over 20 percent in some engine oils. The types of additives used include antioxidants, antiwear agents, extreme-pressure agents, viscosity index improvers, dispersants, detergents, pour point depressants, friction modifiers, corrosion inhibitors, rust inhibitors, metal deactivators, tackiness agents, antifoamants, air-release agents, demulsifiers, emulsifiers, odor control agents, and biocides.

LUBRICATION REGIMES

The viscosity of a lubricant and its additive content are to a large extent related to the lubrication regimes expected in its intended application. It is desirable, but not always possible, to design mechanisms in which the rubbing surfaces are totally separated by lubricant films. **Hydrostatic** lubrication concerns equipment designed such that the lubricant is supplied under sufficient pressure to separate the rubbing surfaces. **Hydrodynamic** lubrication is a more often used regime, and it occurs when the motion of one surface over lubricant on the other surface causes sufficient film pressure and thickness to build up to separate the surfaces. Achieving this requires certain design details, as well as operation within a specific range of speed, load, and lubricant viscosity. Increasing speed or viscosity increases the film thickness, and increasing load reduces film thickness. Plain journal bearings, widely used in automotive engines, operate in this regime. For concentrated nonconforming contacts, such as rolling-element bearings, cams, and certain type of gears, the surface shape and high local load squeeze the oil film toward surface contact. Under such conditions, **elastohydrodynamic** lubrication (EHL) may occur, where the high pressures in the film cause it to become very viscous and to elastically deform and separate the surfaces.

Boundary lubrication occurs when the load is carried almost entirely by the rubbing surfaces, separated by films of only molecular dimensions. **Mixed-film lubrication** occurs when the load is carried partly by the oil film and partly by contact between the surfaces' roughness peaks (asperities) rubbing one another. In mixed-film and boundary lubrication, lubricant viscosity becomes less important and chemical composition of the lubricant film more important. The films, although very thin, are needed to prevent surface damage. At loads up to a certain level, antiwear additives, the most common one being zinc dithiophosphate (ZDTP), are used in the oil to form films to help support the load and reduce wear. ZDTP films can have thicknesses from molecular to much larger dimensions. In very highly loaded contacts, such as hypoid gears, **extreme-pressure** (EP) additives are used to minimize wear and prevent scuffing. EP additives may contain sulfur, phosphorus, or chlorine, and they react with the high-temperature contact surfaces to form molecular-dimension films. Antiwear, EP, and friction modifier additives function only in the mixed and boundary regimes.

LUBRICANT TESTING

Lubricants are manufactured to have specific characteristics, defined by physical and chemical properties, and performance characteristics. Most of the tests used by the lubricants industry are described in ASTM (American Society for Testing and Materials) Standard Methods. **Physical tests** are frequently used to characterize petroleum oils, since lubricant performance often depends upon or is related to physical properties. Common physical tests include measurements of viscosity, density, pour point, API gravity, flash point, fire point, odor, and color. **Chemical tests** measure such things as the amount of certain elements or compounds in the additives, or the acidity of the oil, or the carbon residue after heating the oil at high temperature. **Performance tests** evaluate particular aspects of in-service behavior, such as oxidation stability, rust protection, ease of separation from water, resistance to foaming, and antiwear and extreme-pressure properties. Many of the tests used are small-size or bench tests, for practical reasons. To more closely simulate service conditions, full-scale standardized performance tests are required for certain types of lubricants, such as oils for gasoline and diesel engines. Beyond these industry standard tests, many equipment builders require tests in specific machines or in the field. The full scope of lubricant testing is rather complex, as simple physical and chemical tests are incapable of fully defining in-service behavior.

VISCOSITY TESTS

Viscosity is perhaps the single most important property of a lubricant. Viscosity needs to be sufficient to maintain oil films thick enough to minimize friction and wear, but not so viscous as to cause excessive efficiency losses. Viscosity reduces as oil temperature increases. It increases at high pressure and, for some fluids, reduces from shear, during movement in small clearances.

Kinematic viscosity is the most common and fundamental viscosity measurement, and it is obtained by oil flow by gravity through capillary-type instruments at low-shear-rate flow conditions. It is measured by ASTM Standard Method D445. A large number of commercially available capillary designs are acceptable. Automated multiple-capillary machines also are sold. Kinematic viscosity is typically measured at 40 and 100°C and is expressed in centistokes, cSt (mm^2/s).

Saybolt viscosity, with units of Saybolt universal seconds (SUS), was once widely used. There is no longer a standard method for direct measurement of SUS. Some industrial consumers continue to use SUS, and it can be determined from kinematic viscosity using ASTM D2,161.

Dynamic viscosity is measured for some lubricants at low temperatures (ASTM D4,684, D5,293, and D2,983); it is expressed in centipoise cP (mPa · s). Kinematic and dynamic viscosities are related by the equation cSt = cP/density, where density is in g/cm^3.

Viscosity grades, not specific viscosities, are used to identify viscosities of lubricants in the marketplace, each viscosity grade step in a viscosity grade system having minimum and maximum viscosity limits. There are different viscosity grade systems for automotive engine oils, gear oils, and industrial oils. Grade numbers in any one system are independent of grade numbers in other systems.

ISO Viscosity Grades The International Standards Organization (ISO) has a viscosity grade system (ASTM D2,422) for industrial oils. Each grade is identified by ISO VG followed by a number representing the nominal 40°C (104°F) kinematic viscosity in centistokes for that grade. This system contains 18 viscosity grades, covering the range from 2 to 1,500 cSt at 40°C, each grade being about 50 percent higher in viscosity than the previous one. The other viscosity grade systems are described later.

The **variation of viscosity with temperature** for petroleum oils can be determined from the viscosities at any two temperatures, such as 40 and 100°C kinematic viscosity, using a graphical method or calculation (ASTM D341). The method is accurate only for temperatures above the wax point and below the initial boiling point, and for oils not containing certain polymeric additives.

The **viscosity index** (VI) is an empirical system expressing an oil's rate of change in viscosity with change in temperature. The system, developed some years ago, originally had a 0 to 100 VI scale, based on two oils thought to have the maximum and minimum limits of viscosity-temperature sensitivity. Subsequent experience identified oils far outside the VI scale in both directions. The VI system is still used, and the VI of an oil can be calculated from its 40 and 100°C kinematic viscosity using ASTM D2,270, or by easy-to-use tables (ASTM Data Series, DS 39B). Lubricants with high VI have less change in viscosity with temperature change, desirable for a lubricant in service where temperatures vary widely.

Base oils with more than 100 VI can be made from a wide variety of crude oil distillates by solvent refining, by hydrogenation, by selective blending of paraffinic base oils, by adding a few percent of highmolecular-weight polymeric additives called **viscosity index improvers,** or by combinations of these methods. A few highly processed petroleum base oils and many synthetic oils have high VIs.

Effect of Pressure on Viscosity When lubricating oils are subjected to high pressures, thousands of lb/in^2, their viscosity increases. When oil-film pressures are in this order of magnitude, their influence on viscosity becomes significant. Empirical equations are available to estimate the effect of pressure on viscosity. In rolling-contact bearings, gears, and other machine elements, the high film pressures will influence viscosity with an accompanying increase in frictional forces and load-carrying capacity.

Effect of Shear on Viscosity The viscosity of lubricating oils not containing polymeric additives is independent of **shear rate**, except at low temperatures where wax is present. Where polymeric additives are used, as in multigrade engine oils, high shear rates in lubricated machine elements cause **temporary viscosity loss**, as the large polymer molecules temporarily align in the direction of flow. Engine oils consider this effect by high-temperature/high-shear-rate measurements (ASTM D4,683 or ASTM D4,741). Such oils can suffer **permanent viscosity loss** in service, as polymer molecules are split by shear stresses into smaller molecules with less thickening power. The amount of the permanent viscosity loss depends on the type of polymeric additive used, and it is measurable by comparing the viscosity of the new oil to that of the used oil, although other factors need considering, such as any oxidative thickening of the oils.

OTHER PHYSICAL AND CHEMICAL TESTS

Cloud Point Petroleum oils, when cooled, may become plastic solids, either from wax formation or from the fluid congealing. With some oils, the initial wax crystal formation becomes visible at temperatures slightly above the solidification point. When that temperature is reached at specific test conditions, it is known as the cloud point (ASTM D2,500). The cloud point cannot be determined for those oils in which wax does not separate prior to solidification or in which the separation is invisible.

The cloud point indicates the temperature below which clogging of filters may be expected in service. Also, it identifies a temperature below which viscosity cannot be predicted by ASTM D2,270, as the wax causes a higher viscosity than is predicted.

Pour point is the temperature at which cooled oil will just flow under specific test conditions (ASTM D97). The pour point indicates the lowest temperature at which a lubricant can readily flow from its container. It provides only a rough guide to its flow in machines. Pour point depressant additives are often used to reduce the pour point, but they do not affect the cloud point. Other tests, described later, are used to more accurately estimate oil flow properties at low temperatures in automotive service.

Density and Gravity ASTM D1,298 may be used for determining density (mass), specific gravity, and the API (American Petroleum Institute) gravity of lubricating oils. Of these three, petroleum lubricants tend to be described by using API gravity. The specific gravity of an oil is the ratio of its weight to that of an equal volume of water, both measured at 60°F (16°C), and the API gravity of an oil can be calculated by

$$\text{API gravity, deg} = (141.5/\text{sp. gr. @ } 60°/60°\text{F}) - 131.5$$

The gravity of lubricating oils is of no value in predicting their performance. Low-viscosity oils have higher API gravities than higherviscosity oils of the same crude-oil series. Paraffinic oils have the lowest densities or highest API gravities, naphthenic are intermediate, and animal and vegetable oils have the heaviest densities or lowest API gravities.

Flash and Fire Points The flash point of an oil is the temperature to which an oil has to be heated until sufficient flammable vapor is driven off to flash when brought into momentary contact with a flame. The fire point is the temperature at which the oil vapors will continue to burn when ignited. ASTM D92, which uses the Cleveland open-cup (COC) tester, is the flash and fire point method used for lubricating oils. In general, for petroleum lubricants, the open flash point is 30°F or (17°C) higher than the closed flash (ASTM D93), and the fire point is some 50 to 70°F or (28 to 39°C) above the open flash point.

Flash and fire points may vary with the nature of the original crude oil, the narrowness of the distillation cut, the viscosity, and the method of refining. For the same viscosities and degree of refinement, paraffinic oils have higher flash and fire points than naphthenic oils. While these values give some indication of fire hazard, they should be taken as only one element in fire risk assessment.

Color The color of a lubricating oil is obtained by reference to transmitted light; the color by reflected light is referred to as **bloom**. The color of an oil is not a measure of oil quality, but indicates the uniformity of a particular grade or brand. The color of an oil normally will darken with use. ASTM D1,500 is for visual determination of the color of lubricating oils, heating oils, diesel fuels, and petroleum waxes, using a standardized colorimeter. Results are reported in terms of the ASTM color scale and can be compared with the former ASTM union color. The color scale ranges from 0.5 to 8; oils darker than color 8 may be diluted with kerosene by a prescribed method and then observed. Very often oils are simply described by visual assessment of color: brown, black, etc. For determining the color of petroleum products lighter than 0.5, the ASTM D156 Test for Saybolt Color of Petroleum Products can be used.

Carbon residue, the material left after heating an oil under specified conditions at high temperature, is useful as a quality control tool in the refining of viscous oils, particularly residual oils. It does not correlate with carbon-forming tendencies of oils in internal-combustion engines. Determination is most often made by the Conradson procedure (ASTM D189). Values obtained by the more complex Ramsbottom procedure (ASTM D524) are sometimes quoted. The correlation of the two values is given in both methods.

Ash Although it is unlikely that well-refined oils that do not contain metallic additives will yield any appreciable ash from impurities or contaminants, measurement can be made by ASTM D482. A more useful determination is sulfated ash by ASTM D874, as applied to lubricants containing metallic additives.

Neutralization number and **total acid number** are interchangeable terms indicating a measure of acidic components in oils, as determined by ASTM D664 or D974. The original intent was to indicate the degree of refining in new oils, and to follow the development of oxidation in service, with its effects on deposit formation and corrosion. However, many modern oils contain additives which, in these tests, act similar to undesirable acids and are indistinguishable from them. Caution must be exercised in interpreting results without knowledge of additive behavior. Change in acid number is more significant than the absolute value in assessing the condition of an oil in service. Oxidation of an oil is usually accompanied by an increase in acid number.

Total base number (TBN) is a measure of alkaline components in oils, especially those additives used in engine oils to neutralize acids formed during fuel combustion. Some of these acids get in the crankcase with gases that blow by the rings. Today, TBN is generally measured by ASTM D2,896 or D4,739, as it provides a better indication of an oil's ability to protect engines from corrosive wear than ASTM D664, which was previously used for TBN measurement.

Antifoam The foaming characteristic of crankcase, turbine, or circulating oils is checked by a foaming test apparatus (ASTM D892), which blows air through a diffuser in the oil. The test measures the amount of foam generated and the rate of settling. Certain additive oils tend to foam excessively in service. Many types of lubricants contain a minute quantity of antifoam inhibitor to prevent excessive foaming in service. With the advent of the HEUI (hydraulically actuated, electronically controlled unit injector) fuel injection system, in diesel trucks, the Navistar 7.3 L oil aeration test has become standard in API classification CG-4, CH-4, Cl-4, and Cl-4 plus.

Oxidation Tests Lubricating oils may operate at relatively high temperatures in the presence of air and catalytically active metals or metallic compounds. Oxidation becomes significant when oil is operated above 150°F (66°C). Some lubricating-oil sump temperatures may exceed 250°F (121°C). The oxidation rate doubles for about every 18°F (10°C) rise in oil temperature above 150°F (66°C). The resultant oil oxidation increases viscosity, acids, sludge, and lacquer.

To accelerate oxidation to obtain practical test length, oxidation tests generally use temperatures above normal service, large amounts of catalytic metals, and sufficient oxygen. Two standard oxidation tests, ASTM D943 and D2,272, are often applied to steam turbine and similar industrial oils. There are many other standard oxidation tests for other specific lubricants. Because test conditions are somewhat removed from those encountered in service, care should be exercised in extrapolating results to expected performance in service.

Rust Prevention Lubricating oils are often expected to protect ferrous surfaces from rusting when modest amounts of water enter the lubricating system. Many oils contain additives specifically designed for that purpose. Steam turbine and similar industrial oils often are evaluated for rust prevention by ASTM D665 and D3,603. Where more protection is required, as when a thin oil film must protect against a moisture-laden atmosphere, ASTM D1,748 is a useful method.

Water Separation In many applications, it is desirable for the oil to have good water-separating properties, to enable water removal before excessive amounts accumulate and lead to rust and lubrication problems. The demulsibility of an oil can be measured by ASTM D1,401 for light- to moderate-viscosity oils and by ASTM D2,711 for heavier-viscosity oils.

Wear and EP Tests A variety of test equipment and methods have been developed for evaluating the wear protection and EP properties (**load-carrying capacity before scuffing**) of lubricant fluids and greases under different types of heavy-duty conditions. The tests are simplified, accelerated tests, and correlations to service conditions are seldom available, but they are important in comparing different manufacturers' products and are sometimes used in lubricant specifications. Each apparatus tends to emphasize a particular characteristic, and a given lubricant will not necessarily show the same EP properties when tested on different machines. ASTM methods D2,783 (fluids) and D2,509 (grease) use the **Timken EP tester** to characterize the EP properties of industrialtype gear lubricants, in tests with incremented load steps. ASTM D3,233 (fluids) uses the **Falex pin** and **vee block** tester for evaluating the EP properties of oils and gear lubricants, using either continuous (method A) or incremented (method B) load increase during the test. Evaluation of wear-prevention properties also can be conducted using this apparatus, by maintaining constant load. ASTM methods D2,783 (fluids) and D2,596 (grease) use the **four-ball EP tester** for evaluating the EP properties of oils and greases. ASTM methods D4,172 (fluids) and D2,266 (grease) use the **four-ball wear tester** to evaluate the wear-prevention properties of oils and grease.

GREASES

Lubricating greases consist of lubricating liquid dispersed in a thickening agent. The lubricating liquid is about 70 to 95 wt % of the finished grease and provides the principal lubrication, while the thickener holds the oil in place. Additives are often added to the grease to impart special properties. Grease helps seal the lubricating fluid in and the contaminants out. Grease also can help reduce lubrication frequency, particularly in intermittent operation.

The lubricating liquid is usually a napthenic or paraffinic petroleum oil or a mixture of the two. Synthetic oils are used also, but because of their higher cost, they tend to be used only for specialty greases, such as greases for very low or high temperature use. Synthetic oils used in greases include polyalphaolefins, dialkylbenzenes, dibasic acid esters, polyalkylene glycols, silicones, and fluorinated hydrocarbons. The viscosity of the lubricating liquid in the grease should be the same as would be selected if the lubricant were an oil.

Soaps are the most common thickeners used in greases. Complex soaps, pigments, modified clays, chemicals (such as polyurea), and polymers are also used, alone or in combination. Soaps are formed by reaction of fatty material with strong alkalies, such as calcium hydroxide. In this saponification reaction, water, alcohol, and/or glycerin may be formed as by-products. Soaps of weak alkalies, such as aluminum, are formed indirectly through further reactions. Metallic soaps, such as sodium, calcium, and lithium, are the most widely used thickeners today.

Additives may be incorporated in the grease to provide or enhance tackiness, to provide load-carrying capability, to improve resistance to oxidation and rusting, to provide antiwear or EP properties, and to lessen sensitivity to water. The additives sometimes are solid lubricants such as graphite, molybdenum disulfide, metallic powders, or polymers, which grease can suspend better than oil. Solid fillers may be used to improve grease performance under extremely high loads or shock loads.

The **consistency** or **firmness** is the characteristic which may cause grease to be chosen over oil in some applications. Consistency is determined by the depth to which a cone penetrates a grease sample under specific conditions of weight, time, and temperature. In ASTM D217, a standardized cone is allowed to drop into the grease for 5 s at 77°F (25°C). The resulting depth of fall, or **penetration,** is measured in tenths of millimeters. If the test is made on a sample simply transferred to a standard container, the results are called **unworked penetration.** To obtain a more uniform sample, before the penetration test, the grease is worked for 60 strokes in a mechanical worker, a churnlike device. Variations on unworked and worked penetrations, such as prolonged working, undisturbed penetration, or block penetration, also are described in ASTM D217. The worked 60-stroke penetration is the most widely used indication of grease consistency. The National Lubricating Grease Institute (NLGI) used this penetration method to establish a system of **consistency numbers** to classify differences in grease consistency or firmness. The consistency numbers range from 000 to 6, each representing a range of 30 penetration units, and with a 15 penetration unit gap between each grade, as shown in Table 4.8.3.

As grease is heated, it may change gradually from a semisolid to a liquid state, or its structure may weaken until a significant amount of oil is lost. Grease does not exhibit a true melting point, which implies a sharp change in state. The weakening or softening behavior of grease when heated to a specific temperature can be defined in a carefully controlled heating program with well-defined conditions. This test gives a temperature called the **dropping point,** described in ASTM D566 or D2,265. Typical dropping points are given in Table 4.8.4. At its **dropping point,** a grease already has become fluid, so the maximum application temperature must be well below the dropping point shown in the table.

Table 4.8.3 Grease Consistency Ranges

Consistency no.	Appearance	Worked penetration
000	Semifluid	445-475
00	Semifluid	400-430
0	Semifluid	355-385
1	Soft	310-340
2	Medium	265-295
3	Medium hard	220-250
4	Hard	175-205
5	Very hard	130-160
6	Block type	85-115

The oil constituent of a grease is loosely held. This is necessary for some lubrication requirements, such as those of ball bearings. As a result, on opening a container of grease, free oil is often seen. The tendency of a grease to **bleed** oil may be measured by ASTM D1,742. Test results are related to bleeding in storage, not in service or at elevated temperatures.

Texture of a grease is determined by formulation and processing. Typical textures are shown in Table 4.8.4. They are useful in identification of some products, but are of only limited assistance for predicting behavior in service. In general, smooth, buttery, short-fiber greases are preferred for rolling-contact bearings and stringy, fibrous products for sliding service. Stringiness or tackiness imparted by polymeric additives helps control leakage, but this property may be diminished or eliminated by the shearing action encountered in service.

Grease **performance** depends only moderately on physical test characteristics, such as penetration and dropping point. The likelihood of suitable performance may be indicated by these and other tests. However, the final determination of how a lubricating grease will perform in service should be based on observations made in actual service.

SOLID LUBRICANTS

Solid lubricants are materials with low coefficients of friction compared to metals, and they are used to reduce friction and wear in a variety of applications. There are a large number of such materials, and they include graphite, molybdenum disulfide, polytetrafluoroethylene, talc, graphite fluoride, polymers, and certain metal salts. The many diverse types have a variety of different properties, operating ranges, and methods of application. The need for lubricants to operate at extremes of temperature and environment beyond the range of organic fluids, such as in the space programs, helped foster development of solid lubricants.

One method of using solid lubricants is to apply them as thin films on the bearing materials. The film thicknesses used range from 0.0002 to 0.0005 in (0.005 to 0.013 mm). There are many ways to form the films. Surface preparation is very important in all of them. The simplest method is to apply them as **unbonded solid lubricants,** where granular or powdered lubricant is applied by brushing, dipping, or spraying, or in a liquid or gas carrier for ease of application. Burnishing the surfaces is beneficial. The solid lubricant adheres to the surface to some degree by mechanical or molecular action. More durable films can be made by using **bonded solid lubricants,** where the lubricant powders are mixed with binders before being applied to the surface. Organic binders, if used, can be either room-temperature air-cured type or those requiring thermal setting. The latter tend to be more durable. Air-cured films are generally limited to operating temperatures below 300°F (260°C), while some thermoset films may be satisfactory to 700°F (371°C). Inorganic ceramic adhesives combined with certain powdered metallic solid lubricants permit film use at temperatures in excess of 1,200°F (649°C). The performance of solid-lubricant films is influenced by the solid lubricant used, the method of application, the bearing-surface-material finish and its hardness, the binder-solid lubricant mix, the film and surface pretreatment, and the application's operating conditions. Solid-lubricant films can be evaluated in certain standard tests, including ASTM D2,510 (adhesion), D2,511 (thermal shock sensitivity), D2,625 (wear endurance and load capacity), D2,649 (corrosion characteristics), and D2,981 (wear life by oscillating motion).

Many plastics are also solid lubricants compared to the friction coefficients typical of metals. Plastics are used without lubrication in many applications. Strong plastics can be compounded with a variety of solid lubricants to make plastics having both strength and low friction. Solid-lubricant powders and plastics can be mixed, compacted, and sintered to form a lubricating solid, such as for a bearing. Such materials also can be made by combining solid-lubricant fibers with other stronger plastic fibers, either woven together or chopped, and used in compression-molded plastics. Bushings made of low-friction plastic materials may be press-fitted into metal sleeves. The varieties of low-friction plastic materials are too numerous to mention. Bearings made with solid low-friction plastic materials are commercially available.

Solid lubricants may also be embedded in metal matrices to form solid-lubricant composites of various desirable properties. Graphite is the most common solid lubricant used, and others include molybdenum disulfide and other metal-dichalcogenides. The base metals used include copper, aluminum, magnesium, cadmium, and others. These materials are made by a variety of processes, and their uses range from electrical contacts to bearings in heavy equipment.

LUBRICATION SYSTEMS

There are many positive methods of applying lubricating oils and greases to ensure proper lubrication. Bath and circulating systems are common. There are constant-level lubricators, gravity-feed oilers, multiple-sight feeds, grease cups, forced-feed lubricators, centralized lubrication systems, and air-mist lubricators, to name a few. Selecting the method of lubricant application is as important as the lubricant itself. The choice

Table 4.8.4 General Characteristics of Greases

Base	Texture	Typical dropping point, °F (°C)	Max usable temperature,* °F (°C)	Water resistance	Primary uses
Sodium	Fibrous	325-350 (163-177)	200 (93)	P–F	Older, slower bearings with no water
Calcium	Smooth	260-290 (127-143)	200 (93)	E	Moderate temperature, water present
Lithium 12-hydroxystearate	Smooth	380-395 (193-202)	250 (121)	G	General bearing lubrication
Polyurea	Smooth	450+ (232+)	350 (177)	G–E	Sealed-for-life bearings
Calcium complex	Smooth	500+ (260+)	350 (177)	F–E	Used at high temperatures, with frequent relubrication
Lithium complex	Smooth	450+ (232+)	325 (163)	G–E	Used at high temperatures, with frequent relubrication
Aluminum complex	Smooth	500+ (260+)	350 (177)	G–E	Used at high temperatures, with frequent relubrication
Organoclay	Smooth	500+ (260+)	250 (121)	F–E	Used at high temperatures, with frequent relubrication

* Continuous operation with relatively infrequent lubrication.

Table 4.8.5 SAE Viscosity Grades for Engine Oil[a] (SAE J300 DEC 95)

SAE viscosity grade	Low temperature viscosity, cP, max, at temp, °C, max		Viscosity,[d] cSt, at 100°C		High-shear-rate viscosity,[e] cP, min, at 150°C and 10^6 s^{-1}
	Cranking[b]	Pumping[c]	Min	Max	
0W	3,250 at −30	60,000 at−40	3.8	—	—
5W	3,500 at −25	60,000 at−35	3.8	—	—
10W	3,500 at −20	60,000 at−30	4.1	—	—
15W	3,500 at −15	60,000 at−25	5.6	—	—
20W	4,500 at −10	60,000 at−20	5.6	—	—
25W	6,000 at −5	60,000 at−15	9.3	—	—
20	—	—	5.6	< 9.3	2.6
30	—	—	9.3	<12.5	2.9
40	—	—	12.5	<16.3	2.9[f]
40	—	—	12.5	<16.3	3.7[g]
50	—	—	16.3	<21.9	3.7
60	—	—	21.9	<26.1	3.7

[a] All values are critical specifications as defined in ASTM D3,244.
[b] ASTM D5,293.
[c] ASTM D4,684. Note that the presence of any yield stress detectable by this method constitutes a failure, regardless of viscosity.
[d] ASTM D445 Note: 1 cP = 1 mPa · s 1 cSt = 1 mm²/s.
[e] ASTM D4,683, CEC L-36-A-90 (ASTM D4,741).
[f] 0W-40, 5W-40, and 10W-40 grade oils.
[g] 15W-40, 20W-40, 25W-40, and 40 grade oils.

of the device and the complexity of the system depend upon many factors, including the type and quantity of the lubricant, reliability and value of the machine elements, maintenance schedules, accessibility of the lubrication points, labor costs, and other economic considerations, as well as the operating conditions.

The importance of removing foreign particles from circulating oil has been increasingly recognized as a means to minimize wear and avoid formation of potentially harmful deposits. Adequate **filtration** must be provided to accomplish this. In critical systems involving closely fitting parts, monitoring of particles on a regular basis may be undertaken.

LUBRICATION OF SPECIFIC EQUIPMENT

Internal-combustion-engine oils are required to carry out numerous functions to provide adequate lubrication. Crankcase oils, in addition to reducing friction and wear, must keep the engine clean and free from rust and corrosion, must act as a coolant and sealant, and must serve as a hydraulic oil in engines with hydraulic valve lifters. The lubricant may function over a wide temperature range and in the presence of atmospheric dirt and water, as well as with combustion products that blow by the rings into the crankcase. It must be resistant to oxidation, sludge, and varnish formation in a wide range of service conditions.

Engine oils are heavily fortified with additives, such as **detergents,** to prevent or reduce deposits and corrosion by neutralizing combustion by-product acids; **dispersants,** to help keep the engine clean by solubilizing and dispersing sludge, soot, and deposit precursors; **oxidation inhibitors,** to minimize oil oxidation, particularly at high temperatures; **corrosion inhibitors,** to prevent attack on sensitive bearing metals; **rust inhibitors,** to prevent attack on iron and steel surfaces by condensed moisture and acidic corrosion products, aggravated by low-temperature stop-and-go operation; **pour point depressants,** to prevent wax gelation and improve low-temperature flow properties; **viscosity-index improvers,** to help enable adequate low-temperature flow, along with sufficient viscosity at high temperatures; **antiwear** additives, to minimize wear under boundary lubrication conditions, such as cam and lifter, and cylinder-wall and piston-ring surfaces; **defoamants,** to allow air to break away easily from the oil; and **friction modifiers,** to improve fuel efficiency by reducing friction at rubbing surfaces.

The **SAE viscosity grade system for engine oils** (SAE J300), shown in Table 4.8.5 provides W (winter) viscosity grade steps (e.g., 5W) for oils meeting certain primarily low-temperature viscosity limits, and non-W grade steps (e.g., 30) for oils meeting high-temperature viscosity limits. The system covers both single-grade oils (e.g., SAE 30), not designed for winter use, and multigrade oils (e.g., SAE 5 W-30), formulated for year-round service. This system is subject to revision as improvements are made, and the table shows the December 1995 version. Each W grade step has maximum low-temperature viscosity limits for cranking (D5,293) and pumping (D4,684) bench tests, and a minimum 100°C kinematic viscosity limit. The lower the W grade, the colder the potential service use. Pumping viscosity limits are based on temperatures 10°C below those used for cranking limits, to provide a safety margin so that an oil that will allow engine starting at some low temperature also will pump adequately at that temperature. Each non-W grade step has a specific 100°C kinematic viscosity range and a minimum 150°C high-shear-rate viscosity limit. High-temperature high-shear-rate viscosity limits are a recent addition to J300, added in recognition that critical lubrication in an engine occurs largely at these conditions. Two 150°C high-shear-rate viscosity limits are shown for 40 grade oils, each for a specific set of W grades, the limit being higher for viscosity grades used in heavy-duty diesel engine service.

Gasoline Engines Two systems are used at this time to define gasoline-engine-oil performance levels, not counting systems outside the United States. The API service category system has been used for many years and is the result of joint efforts of many organizations. Since 1970, gasoline engine performance categories in this system have used the letter S (spark ignition) followed by a second letter indicating the specific performance level. Performance category requirements are based primarily on performance in various standard engine tests. The categories available change periodically as higher performance categories are needed for newer engines, and as older categories become obsolete when tests on which they are based are no longer available. At this writing, SM is the highest performance level for gasoline engines.

The newer ILSAC system is based primarily on the same tests, but has a few additional requirements. This system was developed by the International Lubricant Standardization and Approval Committee (ILSAC), a joint organization of primarily U.S. and Japanese automobile and engine manufacturers. The first category is in use, ILSAC GF-1.

API Service Symbol API Certification Mark

Figure 4.8.1 API marks for engine oil containers. (*From API Publication 1,509, 12th ed., 1993. Reprinted by courtesy of the American Petroleum Institute.*)

There are similarities as well as distinct differences between the two systems. Both are displayed on oil containers using API licensed marks. The API service category system uses the API service symbol, and ILSAC system uses the API certification mark, both shown in Fig. 4.8.1. Either or both marks can be displayed on an oil container. In the API service symbol, the performance level (SM), viscosity grade (SAE 5W-30), and energy conserving for 5W-20, 5W-30, and 10W-30 viscosity grades are shown. None of these are shown in the API certification mark of the ILSAC system. The API certification mark remains unchanged as new performance levels are developed and regardless of the viscosity grade of the oil in the container. ILSAC categories require the highest energy-conserving level, and only allow a few viscosity grades, those recommended for newer engines. The viscosity grades allowed are SAE OW-20, 5W-20, 5W-30, and 10W-30. The viscosity grade may be displayed separately. In the ILSAC system, to ensure high-performance-level oils, the API certification mark is licensed annually and only for oils meeting the latest ILSAC performance level.

Diesel Engines API performance-level categories for commercial heavy-duty diesel engine oils, using the API service category system, begin with the letter C (compression ignition), followed by additional letters and numbers indicating the performance level and engine type, the performance letter changing as category improvements are needed. At this writing, Cl-4 plus is the highest performance category for four-cycle turbocharged heavy-duty diesel engines, and CF-2 is the highest for two-cycle turbocharged heavy-duty diesel engines. Viscosity grades used depend on the engine manufacturer's recommendations for the ambient temperature range. SAE 15W-40 oils are very popular for most four-cycle diesel engines, and single-grade oils, such as SAE 40, are used primarily in hotter climates and for two-cycle diesel engines.

The engine oil is increasingly becoming a major part in the total design, affecting performance of critical components as well as engine life. Emission control restrictions are tightening up on diesel engines, and this is affecting the engine oils. Designs of pistons are changing, and rings are being located higher on the pistons and are more subject to carbon deposits. Deposits are being controlled partly by use of proper engine oils. One new engine design with reduced emissions uses the engine oil as hydraulic fluid to operate and improve control of the fuel injectors, and this has created a need for improved oil deaeration performance. Most engine manufacturers maintain lists of oils that perform well in their engines.

Larger diesel engines found in marine, railroad, and stationary service use crankcase oils that are not standardized as to performance levels, unlike the oils discussed above. They are products developed by reputable oil suppliers working with major engine builders. In general, they are formulated along the same lines as their automotive counterparts. However, particularly in marine service where fuels often contain relatively high levels of sulfur, the detergent additive may be formulated with a high degree of alkalinity to neutralize sulfuric acid resulting from combustion.

Gas engines pose somewhat different lubrication requirements. They burn a clean fuel which gives rise to little soot, but conditions of operation are such that nitrogen oxides formed during combustion can have a detrimental effect on the oil. Suitable lubricants may contain less additive than those which must operate in a sootier environment. However, the quality of the base oil itself assumes greater importance. Special selection of crude source and a high degree of refining must be observed to obtain good performance. The most common viscosity is SAE 40, although SAE 30 is also used. Where cold starting is a factor, SAE 15W-40 oils are available.

Steam Turbines Although they do not impose especially severe lubrication requirements, steam turbines are expected to run for very long periods, often measured in many years, on the same oil charge. Beyond that, the oil must be able to cope with ingress of substantial amounts of water. Satisfactory products consist of highly refined base oils with rust and oxidation (R&O) inhibitors, and they must show good water-separating properties (demulsifiability). There is demand for longer-life turbine oil in the power generation industry, such as 6,000 h or more of life in the ASTM D943 oxidation bench test. This demand is being met by using more severely processed base oils and with improved R&O inhibitors. Most large utility turbines operate with oil viscosity ISO VG 32. Where gearing is involved, ISO VG 100 is preferred. Smaller industrial turbines, which do not have circulating systems typical of large units, may operate with ISO VG 68. Circulating systems should be designed with removal of water in mind, whether by centrifuge, coalescer, settling tank, or a combination of these methods. Where the turbine governor hydraulic system is separate from the central lubrication system, fire-resistant phosphate ester fluids are often used in the governor hydraulic system.

Gas turbines in industrial service are lubricated with oils similar to the ISO VG 32 product recommended for steam turbines; but to obtain acceptable service life under higher-temperature conditions, they may be inhibited against oxidation to a greater extent. There is demand for longer-life oils for gas turbines, and, as for steam turbines, the demand is being met by using more severely processed base oils and improved R&O inhibitors. Gas turbines aboard aircraft operate at still higher bearing temperatures which are beyond the capability of petroleum oils. Synthetic organic esters are used, generally complying with military specifications.

Gears API System of Lubricant Service Designations for Automotive Manual Transmissions and Axles (SAE J308B) defines service levels for automotive gear oils. The service levels include API GL-1 to GL-6 and API PL-1 to PL-2, related to gear types and service conditions. Also, there are SAE Viscosity Grades for Axle and Transmission Lubricants (SAE J306C) for gear oils in automotive service. In this system, there are winter grades from SAE 70W to 85W, based on low-temperature viscosity (ASTM D2,983) limits and 100°C kinematic viscosity (ASTM D445) limits, and grades from SAE 90 to 250, based on 100°C kinematic viscosity limits. Many multigrade oils are marketed based on it, capable of year-round operation.

The American Gear Manufacturers Association (AGMA) has classification systems for industrial gear lubricants, defined in AGMA 250.04 and 251.02. These systems specify lubricant performance grades by AGMA lubricant numbers, each grade linked to a specific ISO viscosity grade. There are three systems for enclosed gear drives, and they cover (1) rust- and oxidation-inhibited, (2) extreme pressure, and (3) compounded lubricants. Three systems for open gear drives cover (1) rustand oxidation-inhibited, (2) extreme pressure, and (3) residual lubricants.

The various gear types in use differ in severity of lubrication requirements. For spur, bevel, helical, and herringbone gears, elastohydrodynamic lubrication occurs at rated speeds and loads. Oils used for these types of gears need not contain additives that contribute to oil film strength, but may contain antioxidants, rust inhibitors, defoamants, etc., depending on the application. Worm gears need oils with film strength additives, such as compounded lubricants (containing fatty oils in a petroleum base), and use oils of relatively high viscosity, such as ISO VG 460. Other gear types, such as spur, use oils of widely varying viscosities, from ISO VG 46 to 460 or higher. The choice relates to operating conditions, as well as any need to lubricate other mechanisms with the same oil. Hypoid gears, widely used in automobiles, need oils containing effective EP additives. There is increased use of synthetic gear oils where wide operating temperature ranges and need for long life are involved. For example, polyalkylene glycol gear oils, which have high viscosity indices and excellent load capacities, are being used in industrial applications involving heavily loaded or high-temperature worm gears. Synthetic gear oils based on PAO base stocks have performed well in closed gearbox applications. With open gears, the lubricant often is applied sporadically and must adhere to the gear for some time, resisting being scraped off by meshing teeth. Greases, and viscous asphaltic-base lubricants, often containing EP additives, are among the gear lubricants used in this service. Asphaltic products may be diluted with a solvent to ease application, the solvent evaporating relatively quickly from the gear teeth.

Rolling-Contact Bearings These include ball bearings of various configurations, as well as roller bearings of the cylindrical, spherical, and tapered varieties and needle bearings. They may range in size from a few millimeters to several meters. They are basically designed to carry load under EHL conditions, but considerable sliding motion may exist as well as rolling motion at points of contact between rolling elements and raceways. In addition, sliding occurs between rolling elements and separators and, in roller bearings, between the ends of rollers and raceway flanges. In many instances, it is desirable to incorporate antiwear additives to deal with these conditions.

A choice can frequently be made between oil and grease. Where the lubricant must carry heat away from the bearings, oil is the obvious selection, especially where large quantities can be readily circulated. Also, where contamination by water and solids is difficult to avoid, oil is preferred, provided an adequate purification system is furnished. If circulation is not practical, then grease may be the better choice. Where access to bearings is limited, grease may be indicated. Grease is also preferred in instances where leakage of lubricant might be a problem.

Suitable oils generally contain rust and oxidation inhibitors, and antiwear additives. Viscosity is selected on the basis of EHL considerations or builder recommendations for the particular equipment, and it may range anywhere from ISO VG 32 to 460, typically perhaps ISO VG 68.

Where grease is employed, the selection of type will be governed by operating conditions including temperature, loading, possibility of water contamination, and frequency and method of application. Lithium 12-hydroxystearate grease, particularly of no. 2 consistency grade, is widely used. Where equipment is lubricated for life at the time of manufacture, such as electric motors in household appliances, polyurea greases are often selected.

Air Compressors **Reciprocating** compressors require lubrication of the cylinder walls, packing, and bearings. Temperatures at the cylinder wall are fairly high, and sufficient viscosity must be provided. Oils of ISO VG from about 68 to 460 may be specified. Since water condensation may be encountered, rust inhibitor is needed in addition to oxidation inhibitor. A principal operating problem concerning the lubricant is development of carbonaceous deposits on the valves and in the piping. This can seriously interfere with valve operation and can lead to disastrous fires and explosions. It is essential to choose an oil with a minimum tendency to form such deposits. Conradson and Ramsbottom carbon residue tests are of little value in predicting this behavior, and builders and reputable oil suppliers should be consulted. Maintenance procedures need to ensure that any deposits are cleaned on a regular basis and not be allowed to accumulate. In units which call for all-loss lubrication to the cylinders, feed rates should be reduced to the minimum recommended levels to minimize deposit-forming tendencies. Increasingly, certain synthetic oils, such as fully formulated diesters, polyalkylene glycols, and PAOs, are being recommended for longer, trouble-free operation.

Screw compressors present a unique lubrication problem in that large quantities of oil are sprayed into the air during compression. Exposure of large surfaces of oil droplets to hot air is an ideal environment for oxidation to occur. This can cause lacquer deposition to interfere with oil separator operation which is essential to good performance. Although general-purpose rust- and oxidation-inhibited oils, as well as crankcase oils, are often used, they are not the best choice. Specially formulated petroleum oils, in ISO VG 32 to 68, are available for these severe conditions. For enhanced service life, synthetic organic ester fluids of comparable viscosity are often used.

Refrigeration compressors vary in their lubricant requirements depending on the refrigerant gas involved, particularly as some lubricant may be carried downstream of the compressor into the refrigeration system. If any wax from the lubricant deposits on evaporator surfaces, performance is seriously impaired. Ammonia systems can function with petroleum oils, even though ammonia is only poorly miscible with such oils. Many lubricants of ISO VG 15 to 100 are suitable, provided that the pour point is somewhat below evaporator temperature and that it does not contain additives that react with ammonia. Miscibility of refrigerant and lubricant is important to lubrication of some other types of systems. Chlorinated refrigerants (CFCs and HCFCs) are miscible with oil, and highly refined, low-pour, wax-free naphthenic oils of ISO VG 32 to 68 are often used, or certain wax-free synthetic lubricants. CFC refrigerants, which deplete the ozone layer, will cease being produced after 1995. Chlorine-free hydrofluorocarbon (HFC) refrigerants are now widely used, as they are non-ozone-depleting. HFCs are largely immisible in petroleum-based lubricants, but partial miscibility of refrigerant and lubricant is necessary for adequate lubrication in these systems. Synthetic lubricants based on polyol ester or polyalkylene glycol base oils have miscibility with HFCs and are being used in HFC air conditioning and refrigeration systems.

Hydraulic Systems Critical lubricated parts include pumps, motors, and valves. When operated at rated load, certain pumps and motors are very sensitive to the lubricating quality of the hydraulic fluid. When the fluid is inadequate in this respect, premature wear occurs, leading to erratic operation of the hydraulic system. For indoor use, it is best to select specially formulated antiwear hydraulic oils. These contain rust and oxidation inhibitors as well as antiwear additives, and they provide good overall performance in ISO VG 32 to 68 for many applications. Newer hydraulic systems have higher operating temperatures and pressures and longer drain intervals. These systems require oils with improved oxidation stability and antiwear protection. Fluids with good filterability are being demanded to allow use of fine filters to protect the critical clearances of the system. Hydraulic fluids may contain VI improver and/or pour point depressant for use in low-temperature outside service, and antifoamants and demulsifiers for rapid release of entrained air and water. Hydraulic equipment on mobile systems has greater access to automotive crankcase oils and automatic transmission fluids and manufacturers may recommend use of these oils in their equipment. Where biodegradability is of concern, fluids made of vegetable oil (rapeseed) are available.

Systems Needing Fire-Resistant Fluids Where accidental rupture of an oil line may cause fluid to splash on a surface above about 600°F (310°C), a degree of fire resistance above that of petroleum oil is desirable. Four classes of fluids, generally used in hydraulic systems operating in such an environment, are available. **Phosphate esters** offer the advantages of a good lubricant requiring little attention in service, but they require special seals and paints and are quite expensive. **Waterglycol** fluids contain some 40 to 50 percent water, in a uniform solution of diethylene glycol, or glycol and polyglycol, to achieve acceptable fire resistance. They require monitoring in service to ensure proper content of water and rust inhibitor. **Invert emulsions** contain 40 to 50 percent water dispersed in petroleum oil. With oil as the outside phase of the emulsion, lubrication properties are fairly good, but the fluid must be monitored to maintain water content and to ensure that the water remains adequately dispersed. Finally, there are **conventional emulsions** which contain 5 to 10 percent petroleum oil dispersed in water. With the water as the outside phase, the fluid is a rather poor lubricant, and equipment requiring lubrication must be designed and selected to operate with these emulsions. Steps must also be taken to ensure that problems such as rusting, spoilage, and microbial growth are controlled.

Note that *fire-resistant* is a relative term, and it does not mean nonflammable. Approvals of the requisite degree of fire resistance are issued by Factory Mutual Insurance Company and by the U.S. Bureau of Mines where underground mining operations are involved.

Metal Forming The functions of fluids in machining operation are (1) to cool and (2) to lubricate. Fluids remove the heat generated by the chip/tool rubbing contact and/or the heat resulting from the plastic deformation of the work. Cooling aids tool life, preserves tool hardness, and helps to maintain the dimensions of the machined parts. Fluids lubricate the chip/tool interface to reduce tool wear, frictional heat, and power consumption. Lubricants aid in the reduction of metal welding and adhesion to improve surface finish. The fluids may also serve to carry away chips and debris from the work as well as to protect machined surfaces, tools, and equipment from rust and corrosion.

Many types of fluids are used. Most frequently they are (1) mineral oils, (2) soluble oil emulsions, or (3) chemicals or synthetics. These are often compounded with additives to impart specific properties. Some metalworking operations are conducted in a controlled gaseous atmosphere (air, nitrogen, carbon dioxide).

The choice of a **metalworking fluid** is very complicated. Factors to be considered are (1) the metal to be machined, (2) the tools, and (3) the type of operation. Tools are usually steel, carbide, or ceramic. In operations where chips are formed, the relative motion between the tool and chip is high-speed under high load and often at elevated temperature. In chipless metal forming—drawing, rolling, stamping, extruding, spinning—the function of the fluid is to (1) lubricate and cool the die and work material and (2) reduce adhesion and welding on dies. In addition to fluids, solids such as talc, clay, and soft metals may be used in drawing operations.

In addition to the primary function to lubricate and cool, the cutting fluids should not (1) corrode, discolor, or form deposits on the work; (2) produce undesirable fumes, smoke, or odors; (3) have detrimental physical effects on operators. Fluids should also be stable, resist bacterial growth, and be foam-resistant.

In the machining operations, many combinations of tools, workpieces, and operating conditions may be encountered. In some instances, straight mineral oils or oils with small amounts of additives will suffice, while in more severe conditions, highly compounded oils are required. The effectiveness of the additives depends upon their chemical activity with newly formed, highly reactive surfaces; these combined with the high temperatures and pressures at the contact points are ideal for chemical reactions. Additives include fatty oils, sulfur, sulfurized fatty oils, and sulfurized, chlorinated, and phosphorus additives. These agents react with the metals to form compounds which have a lower shear strength and may possess EP properties. Oils containing additives, either dark or transparent, are most widely used in the industry.

Oils with sulfur- and chlorine-based additives raise some issues of increasing concern, as the more active sulfurized types stain copper and the chlorinated types have environmental concerns and are not suitable for titanium. New cutting oils have been developed which perform well and avoid these concerns.

Water is probably the most effective coolant available but can seldom be used as an effective cutting fluid. It has little value as a lubricant and will promote rusting. One way of combining the cooling properties of water with the lubricating properties of oil is through the use of soluble oils. These oils are compounded so they form a stable emulsion with water. The main component, water, provides effective cooling while the oil and compounds impart desirable lubricating, EP, and corrosion-resistance properties.

The ratio of water to oil will influence the relative lubrication and cooling properties of the emulsion. For these nonmiscible liquids to form a stable emulsion, an emulsifier must be added. Water hardness is an important consideration in the forming of an emulsion, and protection against bacterial growth should be provided. Care in preparing and handling the emulsion will ensure more satisfactory performance and longer life. Overheating, freezing, water evaporation, contamination, and excessive air mixing will adversely affect the emulsion. When one is discarding the used emulsion, it is often necessary to "break" the emulsion into its oil and water components for proper disposal.

Newer-technology soluble oils, using nonionic emulsifiers instead of anionic emulsifiers, have improved properties over earlier-technology oils, and have less tendency to form insoluble soaps with hard water, which results in fewer filter changes, less machine downtime, and lower maintenance costs. They also have longer emulsion life, from greater bacterial and fungal resistance.

Water-soluble chemicals, and synthetics, are essentially solutions or microdispersions of a number of ingredients in water. These materials are used for the same purposes described above for mineral and soluble oils. Water-soluble synthetic fluids are seeing increased use and have greater bioresistance than soluble oils and provide longer coolant life and lower maintenance.

Health Considerations Increased attention is being focused on hazards associated with manufacture, handling, and use of all types of industrial and consumer materials, and lubricants are no exception. Much progress has been made in removing substances suspected of creating adverse effects on health. Suppliers can provide information on general composition and potential hazards requiring special handling of their products, in the form of Material Safety Data Sheets (MSDSs). It is seldom necessary to go further to ascertain health risks. It is necessary that personnel working with lubricants observe basic hygienic practices. They should avoid wearing oil-soaked clothing, minimize unnecessary exposure of skin, and wash exposed skin frequently with approved soaps.

Used-Lubricant Disposal The nonpolluting disposal of used lubricants is becoming increasingly important and requires continual attention. More and more legislation and control are being enacted, at local, state, and national levels, to regulate the disposal of wastes. Alternative methods of handling specific used lubricants may be recommended, such as in situ purification and refortifying the oil and returning it to service, often in mobile units provided by the lubricant distributor. Also, waste lubricating oils may be reprocessed into base oils for rerefined lubricant manufacture. The use of waste lubricating oils as fuels is being increasingly regulated because of air pollution dangers. If wastes are dumped on the ground or directly into sewers, they may eventually be washed into streams and water supplies and become water pollutants. They may also interfere with proper operation of sewage plants. Improper burning may contribute to air pollution. Wastes must be handled in such a way as to ensure nonpolluting disposal. Reputable lubricant suppliers are helpful in suggesting general disposal methods, although they cannot be expected to be knowledgeable about all local regulations.

Lubricant management programs are being used increasingly by industry to minimize operating cost, including lubricant, labor, and waste disposal costs. These programs can be managed by lubricant user/supplier teams. Fluid management programs include the following activities: lubrication selection; lubricant monitoring during service; reclamation and refortification; and disposal. The supplier and/or lubricant service firms often have the major role in performing the last three activities. These programs can be beneficial to both lubricant user and supplier.

4.9 PLASTICS

Revised by Thomas A. Frederick

REFERENCES: Billmyer, "Textbook of Polymer Science," 3rd ed. Brydson, "Plastics Materials," 4th ed. Birley, Heath, and Scott, "Plastics Materials: Properties and Applications." Current publications of and sponsored by the Society of the Plastics Industry (SPI) and similar plastics industry organizations. Manufacturers' specifi- cations, data, and test reports. "Modern Plastics Encyclopedia," McGraw-Hill.

GENERAL OVERVIEW OF PLASTICS

Plastics are ubiquitous engineering materials which are wholly or in part composed of long, chainlike molecules called **polymers.** These polymer chains are made from the chemical reaction (**polymerization**) of many small molecules, known as monomers. The result is a polymer material made from repeating units of one or more monomers. The structure and properties of each polymer material is largely the result of the chemical structure of these monomers. With sufficiently long chains, resulting in high enough molecular weights, these polymers have useful, engineering properties. While carbon is the element common to all commercial polymers; hydrogen, oxygen, nitrogen, sulfur, halogens, and silicon can be present in varying proportions. Polymers may be divided into two classes: **thermoplastic** and **thermosetting.** The former reversibly melt to become highly viscous liquids and solidify upon cooling. The resultant solids will be elastic, ductile, tough, or brittle, depending on the structure of the solid as evolved from the molten state and the use temperature (Fig. 4.9.1). Thermosetting polymers are infusible without thermal or mechanical degradation. Thermosetting polymers (often termed **thermosets**) cure by a chain- linking chemical reaction usually initiated at elevated temperature and pressure, although there are types which cure at room temperature through the use of catalysts. The more highly cured the polymer, the higher its heat distortion temperature and the harder and more brittle it becomes.

As a class, plastics possess a combination of physical and mechanical properties which are attractive to the designer. There are certain properties of plastics which cannot be replicated in metals, including light weight and density (specific gravity rarely greater than 2, with the normal value in the range of 0.9 to 1.7—compare this with magnesium, the lightest structural metal of significance, whose specific gravity is 1.75); optical properties which may range from near glass-like clarity to complete opacity, with the ability to be compounded with colors in an almost limitless range; low thermal conductivity and good electrical resistance; the ability to impart excellent surface finish to parts made by many of the primary production processes. Conversely, when compared to metals, plastics generally have lower elastic modulus—and thus, are inherently more flexible; have lower flexural and impact strength, and inferior toughness; and have lower dimensional stability. Some engineering plastics, when reinforced with graphite or other fibers, can achieve similar properties to metals with lower densities. Some properties of plastics which are most desirable are high strength/weight ratios and the ability to process raw materials through to a finished size and shape in one of several basic operations; this last item has a large impact on the overall costs of finished products by virtue of eliminating secondary operations.

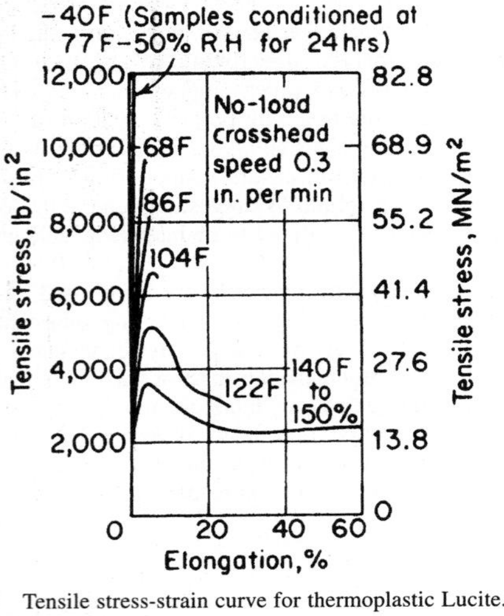

Figure 4.9.1 Tensile stress-strain curve for thermoplastic Lucite. (*The Du Pont Co.*)

The seemingly endless variety of all types of plastics which are available in the marketplace is continually augmented by new compositions; if an end product basically lends itself to the use of plastics, there is most likely to be some existing composition available to satisfy the design requirements. See Table 4.9.1.

The properties of a given plastic are generally protected by the addition of the appropriate stabilizers and can be modified by additives such as colorants, processing aids, impact modifiers, and fillers and reinforcements. Likewise, where local strength requirements are beyond the capacity of the plastic, ingenious configurations of **metallic inserts** can be incorporated into the manufacture of the plastic part and become structurally integral with the part itself; female threaded inserts and male threaded studs are the most common type of inserts found in plastic part production.

Plastic resins may themselves be used as **adhesives** to join other plastics or other paired materials including wood/wood, wood/plastic, metal/plastic, metal/metal, and wood/metal. Structural **sandwich panels,** with the inner layer of sheet or foamed plastic, are an adaptation of composite construction.

Foamed plastic has found widespread use as thermal insulation, a volume filler, cushioning material, and many lightweight consumer items.

The ability to be recycled is an important attribute of thermoplastics, especially those produced in high volumes, whereas thermosets are generally not able to be reused. In view of the massive amounts of plastic in the consumer stream, **recycling** has become more the norm than the exception; indeed, in some jurisdictions, recycling of spent plastic consumables is mandated by law, while in others, the incentive to recycle manifests itself as a built-in cost of the product—a tax, as it were, imposed at the manufacturing phase of production to spur recycling efforts. (See Recycling and Sustainability on pages following.)

The raw materials cost for plastics will vary. The economies of quantity production are self-evident from a study of applicable statistics of plastics production. Consider that in 2015, global production of plastic raw materials was in excess of 270 million metric tons (t). The final overall cost of a finished plastic part is impacted by the cost of raw materials, of course, but further, the use of plastic itself affects the design, manufacture, and shipping components of the final cost; these latter are not inconsequential in arriving at cost comparisons of production with plastics vis-à-vis metals. Suffice it to say that in general, when all costs are accounted for and when design requirements can be met with plastics, the end cost of a plastic item is often decidedly favorable.

RAW MATERIALS

The source of virgin raw materials, or **feedstock,** is generally petroleumor natural gas. The feedstock is converted to monomers, which thenbecome the basis for plastic resins. As recycling efforts increase, conversion of postconsumer plastic items (i.e., scrap) may result in increasing amounts of plastics converted back to feedstock by application of hydrolytic and pyrolytic processes.

The plastic is supplied to primary processers usually in the form of powder, solid pellets, or plastisols (liquid or semisolid dispersions of finely powdered polymer in a nonmigrating liquid).

Compounding, mixing, and blending result in plastic materials ready to be fed into any of the various fabrication processes.

COMMON TYPES OF PLASTICS

Polyolefins are represented by polyethylene (high density, low density and linear low density) and polypropylene, the two most commonly used plastics families. They are easily injection or blow molded, processed into film and fibers, and used in many other processes. They are highly crystalline polymers with excellent chemical resistance and low density. Their relatively low cost and wide versatility as to product types allow them to be used in a wide range of high volume applications. High density polyethylene (HDPE) is the stiffest polyethylene and is used for injection molding, blow molding, and rotational molding. The low density PE and linear low density (LLPE) are softer and tougher and are primarily used for various types of film. Ultrahigh molecular weight polyethylene (UHMWPE) is a specialty grade that is usually compression molded or solution processed. Sheet made from UHMWPE has excellent abrasion resistance and very low coefficient of friction. Polypropylene has higher stiffness and heat resistance than polyethylene and is very versatile. Toughness can be improved via rubber modification either during manufacture or in subsequent compounding. Polybutene also has some useful niche applications.

Styrenics include polystyrene which is more rigid than polyolefins, can be very clear, and can be made into foamed materials. Impact can be greatly improved by adding butadiene rubber (HIPS). Heat and chemical resistance can be improved with styrene-acrylonitrile (SAN) copolymer. The combination of both gives ABS, a very versatile material with improved engineering properties such as toughness, and aesthetics such as gloss.

Polyvinyl chloride is a very rigid plastic and is relatively inert and can provide a very wide range of rigidity and flexibility by adding plasticizers to levels as high as 40 percent or more. PVC is commonly extruded into rigid profiles for construction applications and pipe. Flexible (plasticized) PVC is used for simulated leather, soft overlays, soft laminates, etc.

Polyamides include nylon 66 and nylon 6 as well as other nylon types. Polyamides are semicrystalline and have relatively high melting temperatures. They have good wear properties, high stiffness and strength, making them useful for many injection molded components. They also make durable fibers that are easily dyed with a wide range of colors.

Polyesters include a wide range of plastics where each contains ester functional groups. These include thermoplastics such as polyethylene terephthalate (PET) that makes excellent fibers for apparel and injection stretch blow molded bottles. Polybutylene terephthalate (PBT) is tougher and more durable than PET and used unfilled, glass reinforced and impact modified. Other Polyesters can also be used to make a wide range of thermosetting resins for coatings, castings, and layered with glass to make larger structures such as boats. Polycarbonate is a very tough, stiff and high clarity engineering plastic.

Other frequently used thermoplastics include polyvinyl acetates, polyacetals, and acrylics such as polymethyl methacrylate (PMMA).

PRIMARY FABRICATION PROCESSES

Plastics have a wide range of ways in which they can be processed which can vary considerably across different plastics types, and in some cases many different processes can be used on the same plastics type. Some of the highest volume processed products are extruded films and fibers. Most of the high volume film production is cast or biaxially oriented film made from polyethylenes and polypropylene, but there are specialty films made from many types of plastics. Most fibers are made from semicrystalline plastics (polymers that exhibit crystallizability under typical processing conditions such as polyethylene, polypropylene, polyamides, polyesters, etc.). These fibers can go into nonwovens, carpeting, apparel, ropes, etc. Profile extrusion including sheet for thermoforming or larger thicknesses for structural applications is an important process for many types of plastics. Injection molding is very commonly used to make large numbers of uniquely designed parts in an economical way. Various types of blow molding are used especially for liquid and other containers. Rotational molding is used to make large vessels. Other processes include compression molding, expanded-bead molding to make foams, filament winding over a form, press laminating, vacuumforming, and open molding (hand layup).

In those cases where secondary operations are required to be performed on plastic parts, the usual chip-producing material-removal processes are employed. In those regards, due diligence must be paid to the nature of the material being cut. Thermoplastics will soften from heat generated during cutting, and their low flexural modulus may require backing to prevent excessive deflection in response to cutting forces; thermosets may prove abrasive (even without fillers) and cause rapid wear of cutting edges. Cutting operations on some soft thermoplastics will experience elastic flow of the material beyond the cutting region, necessitating multiple cuts, each of smaller magnitude.

Additive manufacturing, sometimes referred to as 3D printing, is another processing technique that is becoming mainstream. There are several types of additive manufacturing techniques and many different plastics can be used. The common feature is the ability to make intricate, highly detailed parts that cannot be made in any other way. At this time these processes are most useful for making prototypes, highly personalized products (such as medical prostheses), or other low volume applications.

ADHESIVES, ASSEMBLY, AND FINISHES

Most adhesives operate to affect solvent-based **bonding**, and they are either one- or two-part materials. While these adhesives cure, the evaporation of volatile organic compounds (VOCs) constitutes an environmental hazard which is subject to increasing regulation. Water-based bonding agents, on the other hand, emit no VOCs and effectively are environmentally benign. Adhesives which are constituted of only solid material contain no solvents or water, hence no evaporants. A small quantity of bonding is performed via **radiation-curable adhesives.** The source of energy is ultraviolet radiation or electron beam radiation.

The adhesives cited above, and others, display a range of properties which must be assessed as to suitability for the application intended.

Welding of plastics most often implies bonding or sealing of plastic sheet through the agent of ultrasonic energy, which, when applied locally, trans- forms high-frequency ultrasonic energy to heat, which, in turn, fuses the plastic locally to achieve the bond. The design of the joint is critical to the successful application of ultrasonic welding when the pieces are other than thin sheets. **Solvent welding** consists of localized softening of the pieces to be joined and the intimate mixture of the plastic surface material of both parts. The solvent is usually applied with the parts fixed or clamped in place, and sufficient time is allowed for the solvent to react locally with the plastic surfaces and then evaporate.

Most plastic pieces are amenable to the application of **decorative finishes** either during fabrication (e.g., in the mold) or as a secondary operation. Suitable paints are applied as required, often via silk screening. Electrolytic plating, vacuum metallizing, hot stamping, printing, and application of labels via pressure-sensitive adhesives are well-established methods in wide use.

ADDITIVES

The inherent properties of most plastics can be tailored to impart other desired properties or to enhance existing ones by the introduction of additives. A wide variety of additives allow compounding a specific

Table 4.9.1 Properties of Plastic Resins and Compounds

Materials	Properties	ASTM test method	ABS[a]						
				Injection molding grades					
			Extrusion grade	ABS/Nylon	Heat-resistant	Medium-impact	High-impact	Platable grade	20% Glass fiber-reinforced
Processing	1a. Melt flow, g/10 min	D1,238	0.85-1.0		1.1-1.8	1.1-1.8	1.1-18	1.1	
	1. Melting temperature, 8°C. T_m (crystalline)								
	T_g (amorphous)		88-120		110-125	102-115	91-110	100-110	100-110
	2. Processing temperature range, °F. (C = compression; T = transfer, I = injection; E = extrusion)		E:350-500	I:460-520	C:325-500 I:475-550	C:325-350 I:390-525	C:325-350 I:380-525	C:325-400 I:350-500	C:350-500 I:350-500
	3. Molding pressure range, 10^3 lb/in^2			8-25	8-25	8-25	8-25	8-25	15-30
	4. Compression ratio		2.5-2.7	1.1-2.0	1.1-2.0	1.1-2.0	1.1-2.0	1.1-2.0	
	5. Mold (linear) shrinkage, in/in	D955	0.004-0.007	0.003-0.010	0.004-0.009	0.004-0.009	0.004-0.009	0.005-0.008	0.001-0.002
Mechanical	6. Tensile strength at break, lb/in^2	D638[b]	2,500-8,000	4,000-6,000	4,800-7,500	5,500-7,500	4,400-6,300	5,200-6,400	10,500-13,000
	7. Elongation at break, %	D638[b]	20-100	40-300	3-45	5-60	5-75		2-3
	8. Tensile yield strength, lb/in^2	D638[b]	4,300-6,400	4,300-6,300	4,300-7,000	5,000-7,200	2,600-5,900	6,700	
	9. Compressive strength (rupture or yield), lb/in^2	D695	5,200-10,000		7,200-10,000	1,800-12,500	4,500-8,000		13,000-14,000
	10. Flexural strength (rupture or yield), lb/in^2	D790	4,000-14,000	8,800-10,900	9,000-13,000	7,100-13,000	5,400-11,000	10,500-11,500	14,000-17,500
	11. Tensile modulus, 10^3 lb/in^2	D638[b]	130-420	260-320	285-360	300-400	150-350	320-380	740-880
	12. Compressive modulus, 10^3 lb/in^2	D695	150-390		190-440	200-450	140-300		800
	13. Flexural modulus, 10^3 lb/in^2 73°F	D790	130-440	250-310	300-400	310-400	179-375	340-390	650-800
	200°F	D790							
	250°F	D790							
	300°F	D790							
	14. Izod impact, ft·lb/in of notch ($\frac{1}{8}$-in-thick specimen)	D256A	1.5-12	15-20	2.0-6.5	3.0-6.0	6.0-9.3	4.0-8.3	1.1-1.4
	15. Hardness Rockwell	D785	R75-115	R93-105	R100-115	R102-115	R85-106	R103-109	M85-98, R107
	Shore/Barcol	D2,240/ D2,583							
Thermal	16. Coef. of linear thermal expansion, 10^6 in/(in·°C)	D696	60-130	90-110	60-93	80-100	95-110	47-53	20-21
	17. Deflection temperature under flexural load, °F 264 lb/in^2	D648	170-220	130-150	220-240 annealed 181-193[g]	200-220 annealed	205-215 annealed	190-222 annealed	210-220
	66 lb/in^2	D648	170-235	180-195	230-245 annealed	215-225 annealed	210-225 annealed	215-222 annealed	220-230
	18. Thermal conductivity, 10^{-4} cal·cm/(s·cm^2·°C)	C177			4.5-8.0				4.8
Physical	19. Specific gravity	D792	1.02-1.08	1.06-1.07	1.05-1.08	1.03-1.06	1.01-1.05	1.04-1.07	1.18-1.22
	20. Water absorption ($\frac{1}{8}$-in-thick specimen), % 24 h	D570	0.20-0.45		0.20-0.45	0.20-0.45	0.20-0.45		0.18-0.20
	Saturation	D570							
	21. Dielectric strength ($\frac{1}{8}$-in-thick specimen), short time, V/mil)	D149	350-500		350-500	350-500	350-500	420-550	450-460

NOTE: Footnotes [a] to [f] are at end of table.
SOURCE: Abstracted from "Modern Plastics Encyclopedia," 1995.

Acetal						
Homopolymer	Copolymer	Extrusion and blow molding grade (terpolymer)	20% Glass-reinforced homopolymer	25% Glass-coupled copolymer	2%-20% PTFE-filled copolymer	Chemically lubricated homopolymer
1-20	1-90	1.0	6.0			6
172-184	160-175	160-170	175-181	160-180	160-175	175
I:380-470	C:340-400 I:360-450	E:360-400	I:350-480	I:365-480	I:350-445 I:325-500 E:360-500	I:400-440
10-20	8-20		10-20	8-20	8-20	
3.0-4.5	3.0-4.5	3.0-4.0		3.0-4.5	3.0-4.5	
0.018-0.025	0.020 (Avg.)	0.02	0.009-0.012	0.004 (flow) 0.018 (trans.)	0.018-0.029	
9,700-10,000			8,500-9,000	16,000-18,500	8,300	9,500
10-75	15-75	67	6-12	2-3	30	40
9,500-12,000	8,300-10,400	8,700	7,500-8,250	16,000	8,300	9,500
15,600-18,000 @ 10%	16,000 @ 10%	16,000	18,000 @ 10%	17,000 @ 10%	11,000-12,600	
13,600-16,000	13,000	12,800	10,700-16,000	18,000-28,000	11,500	13,000
400-520	377-464		900-1,000	1,250-1,400	250-280	450
670	450					
380-490	370-450	350	600-730	1,100	310-360	400
120-135			300-360			130
75-90			250-270			80
1.1-2.3	0.8-1.5	1.7	0.5-1.0	1.0-1.8	0.5-1.0	1.4
M92-94, R120	M75-90	M84	M90	M79-90, R110	M79	M90
50-112	61-110		33-81	17-44	52-68	
253-277	185-250	205	315	320-325	198-225	257
324-342	311-330	318	345	327-331	280-325	329
5.5	5.5				4.7	
1.42	1.40	1.41	1.54-1.56	1.58-1.61	1.40	1.42
0.25-1	0.20-0.22	0.22	0.25	0.22-0.29	0.15-0.26	0.27
0.90-1	0.65-0.80	0.8	1.0	0.8-1.0	0.5	1.00
400-500 (90 mil)	500 (90 mil)		490 (125 mil)	480-580	400-410	400 (125 mil)

Table 4.9.1 Properties of Plastic Resins and Compounds *(Continued)*

Materials	Properties	ASTM test method	Acrylic				Acrylonitrile	
			Sheet	Molding and extrusion compounds			Molding and extrusion	Injection
			Cast	PMMA	Impact-modified	Heat-resistant		
Processing	1a. Melt flow, g/10 min	D1,238		1.4-27	1-11	1.6-8.0		12
	1. Melting temperature, °C. T_m (crystalline)						135	
	T_g (amorphous)		90-105	85-105	80-103	100-165	95	
	2. Processing temperature range, °F. (C = compression; T = transfer, I = injection; E = extrusion)			C:300-425 I:325-500 E:360-500	C:300-400 I:400-500 E:380-480	C:350-500 I:400-625 E:360-550	C:320-345 I:410 E:350-410	380-420
	3. Molding pressure range, 10^3 lb/in^2			5-20	5-20	5-30	20	20
	4. Compression ratio			1.6-3.0		1.2-2.0	2	2-2.5
	5. Mold (linear) shrinkage, in/in	D955	1.7	0.001-0.004 (flow) 0.002-0.008 (trans.)	0.002-0.008	0.002-0.008	0.002-0.005	0.002-0.005
Mechanical	6. Tensile strength at break, lb/in^2	D638[b]	66-11,000	7,000-10,500	5,000-9,000	9,300-11,500	9,000	
	7. Elongation at break, %	D638[b]	2-7	2-5.5	4.6-7.0	2-10	3-4	3-4
	8. Tensile yield strength, lb/in^2	D638[b]		7,800-10,600	5,500-8,470	10,000	7,500	9,500
	9. Compressive strength (rupture or yield), lb/in^2	D695	11,000-19,000	10,500-18,000	4,000-14,000	15,000-17,000	12,000	12,000
	10. Flexural strength (rupture or yield), lb/in^2	D790	12,000-17,000	10,500-19,000	7,000-14,000	12,000-18,000	14,000	14,000
	11. Tensile modulus, 10^3 lb/in^2	D638[b]	450-3,100	325-470	200-500	350-650	510-580	500-550
	12. Compressive modulus, 10^3 lb/in^2	D695	390-475	370-460	240-370	450		
	13. Flexural modulus, 10^3 lb/in^2 73°F	D790	390-3,210	325-460	200-430	450-620	500-590	480
	200°F	D790				150-440		
	250°F	D790				350-420		
	300°F	D790						
	14. Izod impact, ft·lb/in of notch (⅛-in-thick specimen)	D256A	03-0.4	0.2-0.4	0.40-2.5	0.2-0.4	2.5-6.5	2.5
	15. Hardness Rockwell	D785	M80-102	M68-105	M35-78	M94-100	M72-78	M60
	Shore/Barcol	D2,240/ D2,583						
Thermal	16. Coef. of linear thermal expansion, 10^6 in/(in·°C)	D696	50-90	50-90	48-80	40-71	66	66
	17. Deflection temperature under flexural load, °F 264 lb/in^2	D648	98-215	155-212	165-209	190-310	164	151
	66 lb/in^2	D648	165-235	165-225	180-205	200-315	172	166
	18. Thermal conductivity, 10^{-4} cal·cm/(s·cm^2·°C)	C177	4.0-6.0	4.0-6.0	4.0-5.0	2.0-4.5	6.2	6.1
Physical	19. Specific gravity	D792	1.17-1.20	1.17-1.20	1.11-1.18	1.16-1.22	1.15	1.15
	20. Water absorption (⅛-in-thick specimen), % 24 h	D570	0.2-0.4	0.1-0.4	0.19-0.8	0.2-0.3	0.28	
	Saturation	D570						
	21. Dielectric strength (⅛-in-thick specimen), short time, V/mil	D149	450-550	400-500	380-500	400-500	220-240	220-240

Cellulosic				Epoxy				
Ethyl cellulose molding compound and sheet	Cellulose acetate		Cellulose acetate butyrate	Bisphenol molding compounds		Casting resins and compounds		
	Sheet	Molding compound	Molding compound	Glass fiber-reinforced	Mineral-filled	Unfilled	Aluminum-filled	Flexibilized
135	230	230	140	Thermoset	Thermoset	Thermoset	Thermoset	Thermoset
C:250-390 I:350-500		C:260-420 I:335-490	C:265-390 I:335-480	C:300-330 T:280-380	C:250-330 T:250-380			
8-32		8-32	8-32	1-5	0.1-3			
1.8-2.4		1.8-2.6	1.8-2.4	3.0-7.0	2.0-3.0			
0.005-0.009		0.003-0.010	0.003-0.009	0.001-0.008	0.002-0.010	0.001-0.010	0.001-0.005	0.001-0.010
2,000-8,000	4,500-8,000	1,900-9,000	2,600-6,900	5,000-20,000	4,000-10,800	4,000-13,000	7,000-12,000	2,000-10,000
5-40	20-50	6-70	40-88	4		3-6	0.5-3	20-85
		2,500-6,300						
		3,000-8,000	2,100-7,500	18,000-40,000	18,000-40,000	15,000-25,000	15,000-33,000	1,000-14,000
4,000-12,000	6,000-10,000	2,000-16,000	1,800-9,300	8,000-30,000	6,000-18,000	13,000-21,000	8,500-24,000	1,000-13,000
			50-200	3,000	350	350		
					650			1-350
		1,200-4,000	90-300	2,000-4,500	1,400-2,000			
0.4	2.0-8.5	1.0-7.8	1.0-10.9	0.3-10.0	0.3-0.5	0.2-1.0	0.4-1.6	2.3-5.0
R50-115	R85-120	R17-125	R31-116	M100-112	M100-M112	M80-110	M55-85	
								Shore D65-89
100-200	100-150	80-180	110-170	11-50	20-60	45-65	5.5	20-100
115-190		111-195	113-202	225-500	225-500	115-550	190-600	73-250
		120-209	130-227					
3.8-7.0	4-8	4-8	4-8	4.0-10.0	4-35	4.5	15-25	
1.09-1.17	1.28-1.32	1.22-1.34	1.15-1.22	1.6-2.0	1.6-2.1	1.11-1.40	1.4-1.8	0.96-1.35
0.8-1.8	2.0-7.0	1.7-6.5	0.9-2.2	0.04-0.20	0.03-0.20	0.08-0.15	0.1-4.0	0.27-0.5
350-500	250-600	250-600	250-400	250-400	250-420	300-500		235-400

Table 4.9.1 Properties of Plastic Resins and Compounds (*Continued*)

Materials	Properties	ASTM test method	Fluoroplastics: Polychlorotrifluoroethylene	Fluoroplastics: Polytetrafluoroethylene: Granular	Fluoroplastics: Polytetrafluoroethylene: 25% Class fiber-reinforced	Fluoroplastics: Polyvinylidene fluoride: Molding and extrusion	Fluoroplastics: Polyvinylidene fluoride: EMI shielding (conductive); 30% PAN carbon fiber	Fluoroplastics: Modified PE-TFE: 25% Glass fiber-reinforced
Processing	1a. Melt flow, g/10 min	D1,238						
	1. Melting temperature, °C. T_m (crystalline)			327	327	141-178		270
	T_g (amorphous)		220			−60 to −20		
	2. Processing temperature range, °F. (C = compression; T = transfer, I = injection; E = extrusion)		C:460-580 I:500-600 E:360-590			C:360-550 I:375-550 E:375-550	I:430-500	C:575-625 I:570-650
	3. Molding pressure range, 10^3 lb/in^2		1-6	2-5	3-8	2-5		2-20
	4. Compression ratio		2.6	2.5-4.5		3		
	5. Mold (linear) shrinkage, in/in	D955	0.010-0.015	0.030-0.060	0.018-0.020	0.020-0.035	0.001	0.002-0.030
Mechanical	6. Tensile strength at break, lb/in^2	D638^b	4,500-6,000	3,000-5,000	2,000-2,700	3,500-7,250	14,000	12,000
	7. Elongation at break, %	D638^b	80-250	200-400	200-300	12-600	0.8	8
	8. Tensile yield strength, lb/in^2	D638^b	5,300			2,900-8,250		
	9. Compressive strength (rupture or yield), lb/in^2	D695	4,600-7,400	1,700	1,000-1,400 @ 1% strain	8,000-16,000		10,000
	10. Flexural strength (rupture, or yield), lb/in^2	D790	7,400-11,000		2,000	9,700-13,650	19,800	10,700
	11. Tensile modulus, 10^3 lb/in^2	D638^b		58-80	200-240	200-80,000	2,800	1,200
	12. Compressive modulus, 10^3 lb/in^2	D695	150-300	60		304-420		
	13. Flexural modulus, 10^3 lb/in^2 73°F	D790	170-200	80	190-235	170-120,000	2,100	950
	200°F	D790	180-260					450
	250°F	D790						310
	300°F	D790						200
	14. Izod impact, ft·lb/in of notch (⅛-in-thick specimen)	D256A	2.5-5	3	2.7	2.5-80	1.5	9.0
	15. Hardness Rockwell	D785	R75-112			R79-83, 85		R74
	Shore/Barcol	D2,240/ D2,583	Shore D75-80	Shore D50-65	Shore D60-70	Shore D80, 82 65-70		
Thermal	16. Coef. of linear thermal expansion, 10^6 in/(in·°C)	D696	36-70	70-120	77-100	70-142		10-32
	17. Deflection temperature under flexural load, °F 264 lb/in^2	D648		115		183-244	318	410
	66 lb/in^2	D648	258	160-250		280-284		510
	18. Thermal conductivity, 10^{-4} cal·cm/(s·cm^2·°C)	C177	4.7-5.3	6.0	8-10	2.4-3.1		
Physical	19. Specific gravity	D792	2.0.8-2.2	2.14-2.20	2.2-2.3	1.77-1.78	1.74	1.8
	20. Water absorption (⅛-in-thick specimen), % 24 h	D570	0	<0.01		0.03-0.06	0.12	0.02
	Saturation	D570						
	21. Dielectric strength (⅛-in-thick specimen), short time, V/mil	D149	500-600	480	320	260-280		425

Furan	Phenolic				
	Molding compounds, phenol-formaldehyde				Casting resins
			Impact-modified		
Asbestos-filled	Wood flour-filled	High-strength glass fiber-reinforced	Fabric- and rag-filled	Cellulose-filled	Unfilled
		0.5-10			
Thermoset	Thermoset	Thermoset	Thermoset	Thermoset	Thermoset
C:275-300	C:290-380 I:330-400	C:300-380 I:330-390 T:300-350	C:290-380 I:330-400 T:300-350	C:290-380 I:330-400	
0.1-0.5	2-20	1-20	2-20	2-20	
	1.0-1.5	2.0-10.0	1.0-1.5	1.0-1.5	
	0.004-0.009	0.001-0.004	0.003-0.009	0.004-0.009	
3,000-4,500	5,000-9,000	7,000-18,000	6,000-8,000	3,500-6,500	5,000-9,000
	0.4-0.8	0.2	1-4	1-2	1.5-2.0
10,000-13,000	25,000-31,000	16,000-70,000	20,000-28,000	22,000-31,000	12,000-15,000
600-9,000	7,000-14,000	12,000-60,000	10,000-14,000	5,500-11,000	11,000-17,000
1,580	800-1,700	1,900-3,300	900-1,100		400-700
		2,740-3,500			
	1,000-1,200	1,150-3,300	700-1,300	900-1,300	
	0.2-0.6	0.5-18.0	0.8-3.5	0.4-1.1	0.24-0.4
R110	M100-115	E54-101	M105-115	M95-115	M93-120
		Barcol 72			
	30-45	8-34	18-24	20-31	68
	300-370	350-600	325-400	300-350	165-175
	4-8	8-14	9-12	6-9	3.5
1.75	1.37-1.46	1.69-2.0	1.37-1.45	1.38-1.42	1.24-1.32
0.01-0.02	0.3-1.2	0.03-1.2	0.6-0.8	05-0.9	0.1-0.36
		0.12-1.5			
	260-400	140-400	200-370	300-380	250-400

Table 4.9.1 Properties of Plastic Resins and Compounds *(Continued)*

Materials	Properties	ASTM test method	Polyamide					
			Nylon, Type 6					
					Toughened			
			Molding and extrusion compound	30%-35% Glass fiber-reinforced	33% Glass fiber-reinforced	High-impact copolymers and rubber-modified compounds	Impact-modified; 30% glass fiber-reinforced	Cast
Processing	1a. Melt flow, g/10 min	D1,238	0.5-10			1.5-5.0		
	1. Melting temperature, °C. T_m (crystalline)		210-220	210-220	210-220	210-220	220	227-238
	T_g (amorphous)							
	2. Processing temperature range, °F. (C = compression; T = transfer; I = injection; E = extrusion)		I:440-550 E:440-525	I:460-550	I:520-550	I:450-580 E:450-550	I:480-550	
	3. Molding pressure range, 10^3 lb/in^2		1-20	2-20		1-20	3-20	
	4. Compression ratio		3.0-4.0	3.0-4.0		3.0-4.0	3.0-4.0	
	5. Mold (linear) shrinkage, in/in	D955	0.003-0.015	0.001-0.005	0.001-0.003	0.008-0.026	0.003-0.005	
Mechanical	6. Tensile strength at break, lb/in^2	D638[b]	6,000-24,000	24-27,600[c]; 18,900[d]	17,800[c]	6,300-11,000[c]	21,000[c]; 14,500[d]	12,500
	7. Elongation at break, %	D638[b]	30-100[c]; 300[d]	2.2-3.6[c]	4.0[c]	150-270[c]	5[c]-8[d]	20-30
	8. Tensile yield strength, lb/in^2	D638[b]	13,100[c], 7,400[d]			9,000[c]-9,500		
	9. Compressive strength (rupture or yield), lb/in^2	D695	13,000-16,000[c]	19,000-24,000[c]		3,900[c]		17,000
	10. Flexural strength (rupture or yield), lb/in^2	D790	15,700[c]; 5,800[d]	34-3,600[c]; 21,000[d]	25,800[c]	5,000-12,000[c]		16,500
	11. Tensile modulus, 10^3 lb/in^2	D638[b]	380-464[c]; 100-247[d]	1,250-1,600[c]; 1,090[d]			1,220[c]-754[d]	500
	12. Compressive modulus, 10^3 lb/in^2	D695	250[d]					325
	13. Flexural modulus, 10^3 lb/in^2 73°F	D790	390-410[c]; 140[d]	1,250-1,400[c]; 800-950[d]	1,110[c]	110-320[c]; 130[d]	1,160[c]-600[d]	430
	200° F	D790				60-130[c]		
	250°F	D790						
	300°F	D790						
	14. Izod impact, ft·lb/in of notch (⅛-in-thick specimen)	D256A	0.6-2.2[c]; 30[d]	2.1-3.4[c]; 3.7-5.5[d]	3.5[c]	1.8–No break[c] 1.8–No break[d]	2.2[c]–6[d]	0.7-0.9
	15. Hardness Rockwell	D785	R119[c]; M100-105[c]	M93–96c; M78[d]		R81-113[c]; M50		R115-125
	Shore/Barcol	D2,240/ D2,583						D78.83
Thermal	16. Coef. of linear thermal expansion, 10^6 in/(in·°C)	D696	80-83	16-80		72-120	20-25	50
	17. Deflection temperature under flexural load, °F 264 lb/in^2	D648	155-185[c]	392-420[c]	400[c]	113-140[c]	410[c]	330-400
	66 lb/in^2	D648	347-375[c]	420-430[c]	430[c]	260-367[c]	428[c]	400-430
	18. Thermal conductivity, 10^{-4} cal·cm/(s·cm^2·°C)	C177	5.8	5.8-11.4				
Physical	19. Specific gravity	D792	1.12-1.14	1.35-1.42	1.33	1.07-1.17	1.33	1.15-1.17
	20. Water absorption (⅛-in-thick specimen), % 24 h	D570	1.3-1.9	0.90-1.2	0.86	1.3-1.7	2.0	0.3-0.4
	Saturation	D570	8.5-10.0	6.4-7.0		8.5	6.2	5-6
	21. Dielectric strength (⅛-in-thick specimen), short time, V/mil	D149	400[c]	400-450[c]		450-470[c]		500-600

Polyamide (*Continued*)						
Nylon, Type 66						
		Toughened		Lubricated		
Molding compound	30%-33% Glass fiber-reinforced	15%-33% Glass fiber-reinforced	Antifriction molybdenum disulfide-filled	5% Silicone	30% PTFE	5% Molybdenum disulfide and 30% PTFE
255-265	260-265	256-265	249-265	260-265	260-265	260-265
I:500-620	I:510-580	I:530-575	I:500-600	I:530-570	I:530-570	I:530-570
1-25	5-20		5-25			
3.0-4.0	3.0-4.0					
0.007-0.018	0.002-0.006	0.0025-0.0045[c]	0.007-0.018	0.015	0.007	0.01
13,700[c]; 11,000[d]	27,600[c]; 20,300[d]	10,900-20,300[c]; 14,500[d]	10,500-13,700[c]	8,500[c]	5,500[c]	7,500[c]
15-80[c]; 150-300[d]	2.0-3.4[c]; 3-7[d]	4.7[c]; 8[d]	4.4-40[c]			
8,000-12,000[c]; 6,500-8,500[d]	25,000[c]					
12,500-15,000[c] (yld.)	24,000-40,000[c]	15,000[c]-20,000	12,000-12,500[c]			
17,900-1,700[c]; 6,100[d]	40,000[c]; 29,000[d]	17,400-29,900[c]	15,000-20,300[c]	15,000[c]	8,000[c]	1,200[c]
230-550[c]; 230-500[d]	1,380[c]; 1,090[d]	1,230[c]; 943[d]	350-550[c]			
410-470[c]; 185[d]	1,200-1,450[c]; 800[d]; 900	479-1,100[c]	420-495[c]	300[c]	460[c]	400[c]
0.55-1.0[c]; 0.85-2.1[d]	1.6-4.5[c]; 2.6-3.0[d]	>3.2-5.0	0.9-4.5[c]	1.0[c]	0.5[c]	0.6[c]
R120[c]; M83[c]; M95-105[d]	R101-119[c]; M101-102[c]; M96[d]	R107[c]; R115; R116; M86[c]; M70[d]	R119[c]			
80	15-54	43	54	63.0	45.0	
158-212[c]	252-490[c]	446-470	190-260[c]	170	180	185
425-474[c]	260-500[c]	480-495	395-430			
5.8	5.1-11.7					
1.13-1.15	1.15-1.40	1.2-1.34	1.15-1.18	1.16	1.34	1.37
1.0-2.8	0.7-1.1	0.7-1.5	0.8-1.1	1.0	0.55	0.55
8.5	5.5-6.5	5	8.0			
600[c]	360-500		360[c]			

Table 4.9.1 Properties of Plastic Resins and Compounds ***(Continued)***

Materials	Properties	ASTM test method	Polycarbonate: Unfilled molding and extrusion resins	Polycarbonate: Glass fiber-reinforced	Polycarbonate: Impact-modified polycarbonate/polyester blends	Polycarbonate: Lubricated
			High viscosity	10% glass		10%-15% PTFE, 20% glass fiber-reinforced
Processing	1a. Melt flow, g/10 min	D1,238	3-10	7.0		
	1. Melting temperature, °C. T_m (crystalline)					
	T_g (amorphous)		150	150		150
	2. Processing temperature range, °F. (C = compression; T = transfer, I = injection; E = extrusion)		I:560	I:520-650	I:475-560	I:590-650
	3. Molding pressure range, 10^3 lb/in²		10-20	10-20	15-20	
	4. Compression ratio		1.74-5.5		2-2.5	
	5. Mold (linear) shrinkage, in/in	D955	0.005-0.007	0.002-0.005	0.006-0.009	0.002
Mechanical	6. Tensile strength at break, lb/in²	D638[b]	9,100-10,500	7,000-10,000	7,600-8,500	12,000-15,000
	7. Elongation at break, %	D638[b]	110-120	4-10	120-165	2
	8. Tensile yield strength, lb/in²		9,000	8,500-11,600	7,400-8,300	
	9. Compressive strength (rupture or yield), lb/in²	D695	10,000-12,500	12,000-14,000	7,000	11,000
	10. Flexural strength-(rupture or yield), lb/in²	D790	12,500-13,500	13,700-16,000	10,900-12,500	18,000-23,000
	11. Tensile modulus, 10^3 lb/in²	D638[b]	315	450-600		1,200
	12. Compressive modulus, 10^3 lb/in²	D695	350	520		
	13. Flexural modulus, 10^3 lb/in² 73°F	D790	330-340	460-580	280-325	850-900
	200°F	D790	275	440		
	250°F	D790	245	420		
	300°F	D790				
	14. Izod impact, ft·lb/in of notch (⅛-in-thick specimen)	D256A	12-18 @ ⅛ in 2.3 @ ¼ in	2-4	2-18	1.8-3.5
	15. Hardness Rockwell	D785	M70-M75 R118-122	M62-75	R114-122	
	Shore/Barcol	D2,240/ D2,583				
Thermal	16. Coef. of linear thermal expansion, 10^6 in/(in·°C)	D696	68	32-38	80-95	21.6-23.4
	17. Deflection temperature under flexural load, °F 264 lb/in²	D648	250-270	280-288	190-250	280-290
	66 lb/in²	D648	280-287	295	223-265	290
	18. Thermal conductivity, 10^{-4} cal·cm/(s·cm²·°C)	C177	4.7	4.6-5.2	4.3	
Physical	19. Specific gravity	D792	1.2	1.27-1.28	1.20-1.22	1.43-1.5
	20. Water absorption (⅛-in-thick specimen), % 24 h	D570	0.15	0.12-0.15	0.12-0.16	0.11
	Saturation	D570	0.32-0.35	0.25-0.32	0.35-0.60	
	21. Dielectric strength (⅛-in-thick specimen), short time, V/mil	D149	380–>400	470-530	440-500	

Polyester, thermoplastic				Polyester, thermoset and alkyd		
Polybutylene terephthalate		Polyethylene terephthalate		Cast		Glass fiber-reinforced
Unfilled	30% Glass fiber-reinforced	Unfilled	30% Glass fiber-reinforced	Rigid	Flexible	Preformed, chopped roving
220-267	220-267	212-265	245-265	Thermoset	Thermoset	Thermoset
		68-80				
I:435-525	I:440-530	I:440-660 E:520-580	I:510-590			C:170-320
4-10	5-15	2-7	4-20			0.25-2
		3.1	2-3			1.0
0.009-0.022	0.002-0.008	0.002-0.030	0.002-0.009			0.0002-0.002
8,200-8,700	14,000-19,000	7,000-10,500	20,000-24,000	600-13,000	500-3,000	15,000-30,000
50-300	2-4	30-300	2-7	<2.6	40-310	1-5
8,200-8,700		8,600	23,000			
8,600-14,500	18,000-23,500	11,000-15,000	25,000	13,000-30,000		15,000-30,000
12,000-16,700	22,000-29,000	12,000-18,000	30,000-36,000	8,500-23,000		10,000-40,000
280-435	1,300-1,450	400-600	1,300-1,440	300-640		800-2,000
375	700					
330-400	850-1,200	350-450	1,200-1,590	490-610		1,000-3,000
			520			
			390			
0.7-1.0	0.9-2.0	0.25-0.7	1.5-2.2	0.2-0.4	>7	2-20
M68-78	M90	M94-101; R111	M90; M100			
				Barcol 35-75	Shore D84-94	Barcol 50-80
60-95	15-25	65×10^{-6}	18-30	55-100		20-50
122-185	385-437	70-150	410-440	140-400		>400
240-375	421-500	167	470-480			
4.2-6.9	7.0	3.3-3.6	6.0-7.6			
1.30-1.38	1.48-1.53	1.29-1.40	1.55-1.70	1.04-1.46	1.01-1.20	1.35-2.30
0.08-0.09	0.06-0.08	0.1-0.2	0.05	0.15-0.6	0.5-2.5	0.01-1.0
0.4-0.5	0.35	0.2-0.3				
420-5,500	460-560	420-550	430-650	380-500	250-400	350-500

Table 4.9.1 Properties of Plastic Resins and Compounds (*Continued*)

	Materials		Polyethylene and ethylene copolymers					
			Low and medium density		High density			Cross-linked
				LDPE copolymers				
	Properties	ASTM test method	Linear copolymer	Ethylene-vinyl acetate	Poly-ethylene homopolymer	Ultrahigh molecular weight	30% Class fiber-reinforced	Molding grade
Processing	1a. Melt flow, g/10 min	D1,238		1.4-2.0	5-18			
	1. Melting temperature, °C. T_m (crystalline)		122-124	103-110	130-137	125-138	120-140	
	T_g (amorphous)							
	2. Processing temperature range, °F. (C = compression; T = transfer; I = injection; E = extrusion)		I:350-500 E:450-600	C:200-300 I:350-430 E:300-380	I:350-500 E:350-525	C:400-500	I:350-600	C:240-450 I:250-300
	3. Molding pressure range, 10^3 lb/in^2		5-15	1-20	12-15	1-2	10-20	
	4. Compression ratio		3		2			
	5. Mold (linear) shrinkage, in/in	D955	0.020-0.022	0.007-0.035	0.015-0.040	0.040	0.002-0.006	0.007-0.090
Mechanical	6. Tensile strength at break, lb/in^2	D638[b]	1,900-4,000	2,200-4,000	3,200-4,500	5,600-7,000	7,500-9,000	1,600-4,600
	7. Elongation at break, %	D638[b]	100-965	200-750	10-1,200	420-525	1.5-2.5	10-440
	8. Tensile yield strength, lb/in^2	D638[b]	1,400-2,800	1,200-6,000	3,800-4,800	3,100-4,000		
	9. Compressive strength (rupture or yield), lb/in^2	D695			2,700-3,600		6,000-7,000	2,000-5,500
	10. Flexural strength (rupture or yield), lb/in^2	D790					11,000-12,000	2,000-6,500
	11. Tensile modulus, 10^3 lb/in^2	D638[b]	38-75	7-29	155-158		700-900	50-500
	12. Compressive modulus, 10^3 lb/in^2	D695						50-150
	13. Flexural modulus, 10^3 lb/in^2 73°F	D790	40-105	7.7	145-225	130-140	700-800	70-350
	200°F	D790						
	250°F	D790						
	300°F	D790						
	14. Izod impact, ft · lb/in of notch (⅛-in-thick specimen)	D256A	1.0-No break	No break	0.4-4.0	No break	1.1-1.5	1-20
	15. Hardness Rockwell	D785				R50	R75-90	
	Shore/Barcol	D2,240/ D2,583	Shore D55-56	Shore D17-45	Shore D66-73	Shore D61-63		Shore D55-80
Thermal	16. Coef. of linear thermal expansion, 10^6 in/(in · °C)	D696		160-200	59-110	130-200	48	100
	17. Deflection temperature under flexural load, °F 264 lb/in^2	D648				110-120	250	105-145
	66 lb/in^2	D648			175-196	155-180	260-265	130-225
	18. Thermal conductivity, 10^{-4} cal · cm/(s · cm^2 · °C)	C177			11-12		8.6-11	
Physical	19. Specific gravity	D792	0.918-0.940	0.922-0.943	0.952-0.965	0.94	1.18-1.28	0.95-1.45
	20. Water absorption (⅛-in-thick specimen), % 24 h	D570		0.005-0.13	<0.01	<0.01	0.02-0.06	0.01-0.06
	Saturation	D570						
	21. Dielectric strength (⅛-in-thick specimen), short time, V/mil	D149		620-760	450-500	710	500-550	230-550

Polyimide				Polypropylene				
Thermoplastic		Thermoset		Homopolymer			Copolymer	
Unfilled	30% Glass fiber-reinforced	Unfilled	50% Glass fiber-reinforced	Unfilled	10%-30% Glass fiber-reinforced	Impact-modified, 40% mica-filled	Unfilled	10%-20% Glass fiber-reinforced
4.5-7.5				0.4-38.0	1-20		0.6-44.0	0.1-20
388	388	Thermoset	Thermoset	160-175	168	168	150-175	160-168
250-365	250			−20			−20	
C:625-690 I:734-740 E:734-740	I:734-788	460-485	C:460 I:390 T:390	I:375-550 E:400-500	I:425-475	I:350-470	I:375-550 E:400-500	I:350-480
3-20	10-30	7-29	3-10	10-20			10-20	
1.7-4	1.7-2.3	1-1.2		2.0-2.4			2-2.4	
0.0083	0.0044	0.001-0.01	0.002	0.010-0.025	0.002-0.008	0.007-0.008	0.010-0.025	0.003-0.01
10,500-17,100	24,000	4,300-22,900	6,400	4,500-6,000	6,500-13,000	4,500	4,000-5,500	5,000-8,000
7.5-90	3	1		100-600	1.8-7	4	200-500	3.0-4.0
12,500-13,000		4,300-22,900		4,500-5,400	7,000-10,000		3,000-4,300	
17,500-40,000	27,500	19,300-32,900	34,000	5,500-8,000	6,500-8,400		3,500-8,000	5,500-5,600
10,000-28,800	35,200	6,500-50,000	21,300	6,000-8,000	7,000-20,000	7,000	5,000-7,000	7,000-11,000
300-400	1,720	460-4,650		165-225	700-1,000	700	130-180	
315-350	458	421		150-300				
360-500	1,390	422-3,000	1,980	170-250	310-780	600	130-200	355-510
				50			40	
				35			30	
210	1,175	1,030-2,690						
1.5-1.7	2.2	0.65-15	5.6	0.4-1.4	1.0-2.2	0.7	1.1-14.0	0.95-2.7
E52-99, R129, M95	R128, M104	110-120M	M118	R80-102	R92-115		R65-96	R100-103
							Shore D70-73	
45-56	17-53	15-50	13	81-100	21-62		68-95	
460-680	469	572->575	660	120-140	253-288	205	130-140	260-280
				225-250	290-320		185-220	305
2.3-4.2	8.9	5.5-12	8.5	2.8	5.5-6.2		3.5-4.0	
1.33-1.43	1.56	1.41-1.9	1.6-1.7	0.900-0.910	0.97-1.14	1.23	0.890-0.905	0.98-1.04
0.24-0.34	0.23	0.45-125	0.7	0.01-0.03	0.01-0.05		0.03	0.01
1.2								
415-560	528	480-508	450	600			600	

Table 4.9.1 Properties of Plastic Resins and Compounds ***(Continued)***

Materials	Properties	ASTM test method	Polystyrene and styrene copolymers: Polystyrene homopolymers: High and medium flow	Polystyrene homopolymers: Heat-resistant	Polystyrene homopolymers: 20% Long and short glass fiber-reinforced	Rubber-modified: High-impact	Styrene copolymers: Styrene-acrylonitrile (SAN): Molding and extrusion	Styrene copolymers: High heat-resistant copolymers: 20% Glass fiber-reinforced
Processing	1a. Melt flow, g/10 min	D1,238				5.8		
	1. Melting temperature, °C. T_m (crystalline)							
	T_g (amorphous)		74-105	100-110	115	9.3-105	100-200	
	2. Processing temperature range, °F. (C = compression; T = transfer; I = injection; E = extrusion)		C:300-400 I:350-500 E:350-500	C:300-400 I:350-500 E:350-500	I:400-550	I:350-525 E:375-500	C:300-400 I:360-550 E:360-450	I:425-550
	3. Molding pressure range, 10^3 lb/in^2		5-20	5-20	10-20	10-20	5-20	
	4. Compression ratio		3	3-5		4	3	
	5. Mold (linear) shrinkage, in/in	D955	0.004-0.007	0.004-0.007	0.001-0.003	0.004-0.007	0.003-0.005	0.003-0.004
Mechanical	6. Tensile strength at break, lb/in^2	D638^b	5,200-7,500	6,440-8,200	10,000-12,000	1,900-6,200	10,000-11,900	10,000-14,000
	7. Elongation at break, %	D638^b	1.2-2.5	2.0-3.6	1.0-1.3	20-65	2-3	1.4-3.5
	8. Tensile yield strength, lb/in^2	D638^b		6,440-8,150		2,100-6,000	9,920-12,000	
	9. Compressive strength (rapture or yield), lb/in^2	D695	12,000-13,000	13,000-14,000	16,000-17,000		14,000-15,000	
	10. Flexural strength (rupture or yield), lb/in^2	D790	10,000-14,600	13,000-14,000	14,000-18,000	3,300-10,000	11,000-19,000	16,300-22,000
	11. Tensile modulus, 10^3 lb/in^2	D638^b	330-475	450-485	900-1,200	160-370	475-560	850-900
	12. Compressive modulus, 10^3 lb/in^2	D695	480-490	495-500			530-580	
	13. Flexural modulus, 10^3 lb/in^2 73°F	D790	380-490	450-500	950-1,100	160-390	500-610	800-1,050
	200°F	D790						
	250°F	D790						
	300°F	D790						
	14. Izod impact, ft·lb/in of notch (⅛-in-thick specimen)	D256A	0.35-0.45	0.4-0.45	0.9-2.5	0.95-7.0	0.4-0.6	2.1-2.6
	15. Hardness Rockwell	D785	M60-75	M75-84	M80-95, R119	R50-82; L–60	M80, R83	
	Shore/Barcol	D2,240/ D2,583						
Thermal	16. Coef. of linear thermal expansion, 10^6 in/(in·°C)	D696	50-83	68-85	39.6-40	44.2	65-68	20
	17. Deflection temperature under flexural load, °F 264 lb/in^2	D648	169-202	194-217	200-220	170-205	214-220	231-247
	66 lb/in^2	D648	155-204	200-224	220-230	165-200	220-224	
	18. Thermal conductivity, 10^{-4} cal·cm/(s·cm^2·°C)	C177	3.0	3.0	5.9		3.0	
Physical	19. Specific gravity	D792	1.04-1.05	1.04-1.05	1.20	1.03-1.06	1.06-1.08	1.20-1.22
	20. Water absorption (⅛-in-thick specimen), % 24 h	D570	0.01-0.03	0.01	0.07-0.01	0.05-0.07	0.15-0.25	0.1
	Saturation	D570	0.01-0.03	0.01	0.3		0.5	
	21. Dielectric strength (⅛-in-thick specimen), short time, V/mil	D149	500-575	500-525	425		425	

Polyurethane					Silicone		
Thermoset			Thermoplastic		Casting resins	Liquid injection molding	Molding and encapsulating compounds
Casting resins		55%-65% Mineral-filled potting and casting compounds	Unreinforced molding	10%-20% Glass fiber-reinforced molding compounds	Flexible (including RTV)	Liquid silicone rubber	Mineral- and/or glass-filled
Liquid	Unsaturated						
Thermoset	Thermoset	Thermoset	75-137		Thermoset	Thermoset	Thermoset
				120-160			
C:43-250		25 (casting)	I:430-500 E:430-510	I:360-410		I:360-420	C:280-360 I:330-370 T:330-370
0.1-5				8-11		1-2	0.3-6
							2.0-8.0
0.020		0.001-0.002	0.004-0.006	0.004-0.010	0.0-0.006	0.0-0.005	0.0-0.005
175-10,000	10,000-11,000	1,000-7,000	7,200-9,000	4,800-7,500	350-1,000	725-1,305	500-1,500
100-1,000	3-6	5-55	60-180	3-70	20-700	300-1,000	80-800
			7,800-11,000				
20,000				5,000			
700-4,500	19,000		10,200-15,000	1,700-6,200			
10-100			190-300	0.6-1.40			
10-100							
10-100	610		235-310	40-90			
25 to flexible	0.4		1.5-1.8	10-14-No break			
			R>100; M48	R45-55			
Shore A10-13, D90	Barcol 30-35	Shore A90, D52-85			Shore A10-70	Shore A20-70	Shore A10-80
100-200		71-100		34	10-19	10-20	20-50
Varies over wide range	190-200		158-260	115-130			>500
			176-275	140-145			
5		6.8-10			3.5-7.5		7.18
1.03-1.5	1.05	1.37-2.1	1.2	1.22-1.36	0.97-2.5	1.08-1.14	1.80-2.05
0.2-1.5	0.1-0.2	0.06-0.52	0.17-0.19	0.4-0.55	0.1		0.15
			0.5-0.6	1.5			0.15-0.40
300-500		500-750 @ $\frac{1}{16}$ in.	400	600	400-550		200-550

Table 4.9.1 Properties of Plastic Resins and Compounds (*Continued*)

Materials	Properties	ASTM test method	Vinyl polymers and copolymers				
			PVC molding compound, 20% glass fiber-reinforced	PVC and PVC-acetate MC, sheets, rods, and tubes			Molding and extrusion compounds
				Rigid	Flexible, unfilled	Flexible, filled	Chlorinated polyvinyl chloride
Processing	1a. Melt flow, g/10 min	D1,238					
	1. Melting temperature, °C. T_m (crystalline)						
	T_g (amorphous)		75-105	75-105	75-105	75-105	110
	2. Processing temperature range, °F. (C = compression; T = transfer; I = injection; E = extrusion)		I:380-400 E:390-400	C:285-400 I:300-415	C:285-350 I:320-385	C:285-350 I:320-385	C:350-400 I:395-440 E:360-420
	3. Molding pressure range, 10^3 lb/in^2		5-15	10-40	8-25	1-2	15-40
	4. Compression ratio		1.5-2.5	2.0-2.3	2.0-2.3	2.0-2.3	1.5-2.5
	5. Mold (linear) shrinkage, in/in	D955	0.001	0.002-0.006	0.010-0.050 0.002-0.008	0.008-0.035	0.003-0.007
Mechanical	6. Tensile strength at break, lb/in^2	D638[b]	8,600-12,800	5,900-7,500	1,500-3,500	1,000-3,500	6,800-9,000
	7. Elongation at break, %	D638[b]	2-5	40-80	200-450	200-400	4-100
	8. Tensile yield strength, lb/in^2	D638[b]		5,900-6,500			6,000-8,000
	9. Compressive strength (rupture or yield), lb/in^2	D695		8,000-13,000	900-1,700	1,000-1,800	9,000-22,000
	10. Flexural strength (rupture or yield), lb/in^2	D790	14,200-22,500	10,000-16,000			14,500-17,000
	11. Tensile modulus, 10^3 lb/in^2	D638[b]	680-970	350-600			341-475
	12. Compressive modulus, 10^3 lb/in^2	D695					335-600
	13. Flexural modulus, 10^3 lb/in^2 73°F	D790	680-970	300-500			380-450
	200°F	D790					
	250°F	D790					
	300°F	D790					
	14. Izod impact, ft·lb/in of notch (⅛-in-thick specimen)	D256A	1.0-1.9	0.4-22	Varies over wide range	Varies over wide range	1.0-5.6
	15. Hardness Rockwell	D785	R108-119				R117-112
	Shore/Barcol	D2,240/ D2,583	Shore D85-89	Shore D65-85	Shore A50-100	Shore A50-100	
Thermal	16. Coef. of linear thermal expansion, 10^6 in/(in·°C)	D696	24-36	50-100	70-250		62-78
	17. Deflection temperature under flexural load, °F 264 lb/in^2	D648	165-174	140-170			202-234
	66 lb/in^2	D648		135-180			215-247
	18. Thermal conductivity, 10^{-4} cal·cm/(s·cm^2·°C)	C177		3.5-5.0	3-4	3-4	3.3
Physical	19. Specific gravity	D792	1.43-1.50	1.30-1.58	1.16-1.35	1.3-1.7	1.49-1.58
	20. Water absorption (⅛-in-thick specimen), % 24 h	D570	0.01	0.04-0.4	0.15-0.75	0.5-1.0	0.02-0.15
	Saturation	D570					
	21. Dielectric strength (⅛-in-thick specimen), short time, V/mil	D149		350-500	300-400	250-300	600-625

[a] Acrylonitrile-butadiene-styrene. [b] Tensile test method varies with material: D638 is standard for thermoplastics; D651 for rigid thermoset plastics; D412 for elastomeric plastics; D882 for thin plastics sheeting. [c] Dry, as molded (approximately 0.2% moisture content). [d] As conditioned to equilibrium with 50% relative humidity. [e] Test method is ASTM D4,092. [f] *Pseudo* indicates that the thermoset and thermoplastic components were mixed in the form of pellets or powder prior to fabrication.

type of plastic to meet some desired end result; a few groups of additive types are listed below.

Stabilizers

Antioxidants are substances compounded with plastics resins to prevent oxidative degeneration which may be brought about by exposure to heat during processing or during use in applications at elevated temperatures, during exposure to radiation such as during sterilization, or to certain impurities or high shear loads. Plastics will degrade when subjected to high temperatures; thermoplastics will soften and eventually melt; thermosets will char and burn. Both thermoplastic and thermosets parts are subjected to heat during processing and may see service above room temperature during its useful life. The judicious addition of an appropriate **heat stabilizer** will prevent thermal degradation under those circumstances. Some of the more effective heat-stabilizing agents contain lead, antimony, or cadmium and are subject to increasingly stringent environmental limits. **Ultraviolet stabilizers** retard or prevent the degradation of some strength properties as well as color, which result from exposure to sunlight. **Antacids** counteract the degradative effects of catalyst residues and other acid residuals.

Process Aids

Various additives are used to allow plastics to be more easily processed. **Mold release compounds** facilitate the removal of molded parts from mold cavities with retention of surface finish on the finished parts and elimination of pickup of molding material on mold surfaces. The use of **lubricants** serve to impart lubricity to plastic parts and make many types of fabrication easier. **Antistatic compounds** are used to reduce the static buildup which is very common in plastics. The inherent good electrical insulation property of plastic often leads to buildup of static electric charges. The static charge enhances dust pickup and retention to the plastic surface. At the very least, this may inhibit the action of a mold-release agent; at worst, the leakage of static electric charge may be an explosion or fire hazard. **Antiblocks** are used to reduce stickiness of polymer to itself in film and other applications where large amounts of surface area are in contact before fully cooling. These are usually inorganic particulates. **Blowing into foaming agents** are compressed gas or a liquid which will evolve gas when heated, resulting in a network ofinterstices in expanded foam.

Property Modifiers

A wide range of additives or materials may be added to polymers to subtly or significantly modify one or many properties. These plastics are usually compounded together in a separate compounding step, but when used in small quantities, can sometime be used in the primary melt processing step such as at the injection molding machine or extruder. These can include other polymers in blends that give added characteristics. Polymers can be modified with functional groups and added to act as **compatibilizers** (with other polymers) or **coupling agents** (with inorganic materials such as filers and reinforcements). Elastomeric **impact modifiers** can be added to greatly improve the impact property performance. Some common impact modifiers include ethylene copolymerized with propylene, butene or octene; styrene-butadiene copolymers such as SBS and SEBS, and EPDM. Some plastics which have sufficient ability to crystallize can have molding cycle times reduced by adding appropriate **nucleators**, which also can increase stiffness related properties.

Fillers and reinforcements can be added in large percentages (10%-50+%) to greatly modify a wide range of properties. These include talc, calcium carbonate, mica, clay, wollastonite, cellulose, glass spheres and fibers, natural synthetic fibers, etc. These compounds reduce the amount of plastic used but almost always increase the density in the final product. Most of these also increase flexural modulus and heat resistance performance while the reinforcing types such as fibers and other forms with aspect ratios greater than one can also increase tensile strength and some other properties. One relatively new area for significant property improvement is to add "nano" particles (**nano materials**) that when reduced to very small dimensions, can offer very significant property improvements with relatively small levels of added material. For example, certain clays can be delaminated (termed exfoliation) into very thin platelets with the proper compounding equipment and conditions. Relatively low levels (<5%) can significantly decrease permeability of oxygen or other gases, alter electrical properties, increase stiffness and other properties; so the plastics can reach new performance realms.

Antibacterial agents (biocides) act to inhibit the growth of bacteria, especially when the type of filler used can be an attractive host to bacterial action (wood flour, for example), or when the final plastic part will be used in certain environments (exterior siding). **Flame retardants** are introduced primarily for safety, and they work by increasing the ignition temperature and lowering the rate of combustion. **Plasticizers** increase the softness and flexibility of the plastic resin, but often there is an accompanying loss of tensile strength, increased flammability, etc. A common example is when PVC, a very rigid material, is heavily modified to make soft, pliable materials.

Aesthetics

Colorants are dyes or pigments that impart color to plastics. Dyes allow a wider range of color. Colorants enhance the utility of plastic products where cosmetic and/or aesthetic effects are required. Ultraviolet (UV)radiation from sunlight degrades the color unless a UV-inhibiting agent is incorporated into the plastic resin. **Clarifiers** are typically used in plastics such as polypropylene to improve the clarity of injection molded parts.

RECYCLING AND SUSTAINABILITY

The feedstock for all plastics is almost always petroleum or natural gas. These materials constitute valuable and irreplaceable natural resources. Plastic materials may degrade; plastic parts may break or no longer be of service; certainly, very little plastic per se "wears away." The waste of a valuable resource implicit in discarding scrap plastic in landfills or the often environmentally hazardous incineration of some plastics has led to very strong efforts toward recycling plastic scrap. Separation of plastic by types, aided by unique identification symbols molded into consumer items, is mandated by law in many localities which operate an active recycling program. Different plastic types present a problem when comingled during recycling, but some post-recycling consumer goods have found their way to the market. Efforts will continue as recycling of plastics becomes universal; the costs associated with the overall recycling effort will become more attractive as economies of large-scale operations come into play.

Ideally, plastic recycling is best when clean scrap segregated as to type. In reality, this is generally not the case, and for those reasons, a number of promising scrap processing directions are being investigated. Among those is depolymerization, whereby the stream of plastic scrap is subjected to pyrolytic or hydrolytic processing which results in reversion ofthe plastic to feedstock components. This regenerated feedstock can then be processed once more into the monomers, then to virgin polymeric resins. Other concepts will come under consideration in the near future.

Beyond recycling, one way to promote sustainability of plastics is through the use of "bio-based" monomers. This can involve the development of new polymers based on naturally available monomers, or the production of existing monomers from natural sources other than oil and natural gas; for examples, ethylene and propylene from cane sugar.

In the future, there will be ongoing, evolving discussions about many approaches for improving the sustainability of plastics. For each plastics type, a balance among monomer sourcing, recycling, biodegradation, useful, clean incineration for energy reclamation, and other aspects will need to be addressed.

4.10 FIBER COMPOSITE MATERIALS

by Stephen R. Swanson, Revised by Erik T. Thostenson

INTRODUCTION

Composite materials are composed of two or more physically discrete constituents where one constituent—referred to as the reinforcement—is distributed within an interconnected phase—referred to as the matrix. Examples of composite materials can be found widely in nature. For example, wood is primarily composed of cellulose fibers bound together with a naturally occurring polymer known as lignin. Traditional engineering materials, such as metals, ceramics and most polymers, are considered to be homogeneous and isotropic—in that the mechanical and physical properties do not depend on direction in the material. Wood is known to have very different properties longitudinal and transverse to the grain direction. Most synthetic composite materials draw on this inspiration from nature. Like wood, the properties of composite materials are anisotropic where the properties depend strongly on the arrangement of the reinforcement in the matrix. Through the use of advanced manufacturing processes, the engineer can control the distribution of the reinforcement and design the material properties for specific applications.

Composite materials are typically classified based on their type of matrix as either polymer matrix composites (PMCs), metal matrix composites (MMCs), or ceramic matrix composites (CMCs). The properties of composite materials depend on both the constituent materials (reinforcement and matrix) as well as their interaction (bonding at the fiber/matrix interface). Most polymers typically have low Young's modulus and strength. The properties of PMCs are dominated by the much higher stiffness and strength of the reinforcing fiber. There are many ways to increase the strength of metals through alloying, work hardening, and heat treatment. While materials formed with various aluminum alloys and temper designations may have yield strengths that can vary by an order or magnitude all aluminum alloys have essentially the same Young's modulus. As a result, the primary motivation behind MMCs is for increasing the stiffness of the metal matrix. Monolithic ceramics are extremely brittle and, while they have excellent high-temperature stability, they are rarely used in structural applications due to their low fracture toughness. The motivation for CMCs is mainly to increase the fracture toughness. Often CMC systems are composed of the same reinforcement and matrix, such as silicon carbide reinforced silicon carbide (SiC/SiC). Since ceramics are processed and often used at high temperatures having the same reinforcement and matrix reduces residual internal stresses due to thermal expansion. The reinforcement increases the fracture toughness due to the introduction of different fracture mechanisms such as crack deflection, interface debonding, and fiber pullout. These different fracture mechanisms increase the overall work of fracture.

Examples of engineering use of composites date back to the use of straw in clay by the Egyptians. Advanced composites using fiber-reinforced polymer matrices have created a revolution in high-performance structures in recent years and have found much broader applications than metal and ceramic-based composites. Advanced composite materials offer significant advantages in strength and stiffness coupled with low density relative to conventional metallic materials. Along with this structural performance comes the freedom to select the constituent materials and the orientation of the fibers for optimum performance. Modern composites have been described as being revolutionary in the sense that the material can be designed as well as the structure.

Initial applications of composites started with **glass fibers** and over the past 50 years there have been major advances in producing a variety of high-performance fibers such as carbon, aramid, boron, and silicon carbide. The **stiffness** and **strength-to-weight** properties make materials such as carbon fiber composites attractive for applications in aerospace, automotive, and sporting goods. For example, depending on the grade of carbon fiber the tensile strength varies, but a typical range of values is 3.1 to 5.5 GPa (450 to 800 ksi) for fiber tensile strength and a Young's modulus on the order of 240 GPa (35 Msi), combined with a specific gravity of 1.7. Thus the fiber itself is stronger than 7,075 T6 aluminum by a factor of 5 to 10, is stiffer by a factor of 3.5, and weighs approximately 60 percent as much. In addition, composites often have superior resistance to environmental attack, and glass fiber composites are used extensively in the chemical industries and in marine applications because of this advantage. Both glass and carbon fiber composites have been investigated for infrastructure applications, such as bridges and reinforced concrete because of this environmental resistance.

The potential advantages of high-performance fiber composites in mechanical design are obvious. On the other hand, material costs can considerably higher than for conventional materials, such as aluminum. The cost competitiveness of composites often depends on how important weight reduction or environmental resistance is to the overall application. While glass fibers usually cost less than aluminum on a weight basis, carbon fibers are still considerably more expensive. In addition to the material cost, the cost of manufacturing can be significant and depends on the manufacturing process used to form the composite. There is considerable attention being given to driving-down the cost of composites manufacturing through automation and process optimization. In some applications composite structures can achieve significant cost savings in manufacturing, often by reducing the number of parts involved in a complex assembly.

REINFORCEMENTS

Advanced composite materials typically utilize continuous high-performance fibers to reinforce the polymer matrix. Depending on the performance requirements discontinuous, or chopped, fibers are sometimes used. Discontinuous fibers are described in terms of their non-dimensional aspect ratio (length/diameter), and a continuous fiber has an infinite aspect ratio. Composites manufactured with discontinuous fibers will have lower strength and stiffness relative to continuous fiber composites but can have advantages in terms of cost or manufacturability. In recent decades, nanoscale materials have emerged as important reinforcements in polymer composites. Composites reinforced with nanoscale constituents are typically referred to as nanocomposites.

Continuous-fiber materials are available in a number of different forms, with the specific form utilized depending on the manufacturing process and performance requirements. The individual fibers themselves have small diameters, with sizes of 5 to 7 μm (0.0002 to 0.0003 in) typical for carbon fibers. Fibers are generally spun into bundles of several thousand fibers, often referred to as a **tow** or **roving**. For example, carbon fibers are available in bundles ranging from 2,000 to 12,000 fibers per bundle. Depending on the manufacturing process these fiber bundles can be utilized directly in their bundle form. The individual tows may be further combined using a variety of textile processes such as **weaving**, **braiding,** or **knitting**.

Common reinforcements for composites are described later and Table 4.10.1 lists the properties of some fibers commonly used in composites.

Glass Fibers Glass fibers are widely used in a variety of commercial products and is probably the most commonly used fiber for composites. Glass fibers are chemically composed of amorphous silica (SiO_2). Different glass compositions have modifier ions that alter the properties or manufacturability of the fibers. There are two grades of glass fibers that are primarily used in structural composites are E-glass and S-glass. E-glass is the most frequently used in composites and the

Table 4.10.1 Mechanical Properties of Typical Fibers

Fiber	s	Fiber Density		Tensile Strength		Tensile Modulus	
		lb/in³	g/cm³	ksi	GPa	Msi	GPa
E-glass	8-14	0.092	2.54	500	3.45	10.5	72.4
S-glass	8-14	0.090	2.49	665	4.58	12.5	86.2
Polyethylene	10-12	0.035	0.97	392	2.70	12.6	87.0
Aramid (Kevlar 49)	12	0.052	1.44	525	3.62	19.0	130.0
HS carbon, T300	7	0.063	1.74	514	3.54	33.6	230
AS4 carbon	7	0.065	1.80	580	4.00	33.0	228
IM7 carbon	5	0.065	1.80	785	5.41	40.0	276
Boron	50-203	0.094	2.60	500	3.44	59.0	407

E refers to its original use in electrical applications. S-glass has a higher silica content and has both higher Young's modulus and strength. While glass fibers are relatively low-cost their performance is limited in applications where high stiffness and strength to weight is required. Compared to other high-performance fibers the stiffness and strength tend to be lower with higher density. Glass fibers have been used extensively where corrosion resistance is important such as piping for the chemical industry and in marine applications.

Carbon Fiber A wide range of commercial carbon fibers have been developed for composites. Carbon fibers are broadly classified based on their precursor materials. These precursors undergo a variety of heat treatment processes to form the graphitic structure. Initially, carbon fibers were produced using a Rayon-based precursor but today commercial carbon fibers are typically based on either **polyacrylonitrile (PAN)** or petroleum **pitch** precursors. PAN-based carbon fibers are the most commonly used fibers used widely in aerospace and for some sporting goods because of their relatively high stiffness and high strength/weight ratios. Carbon fibers vary in strength and stiffness depending on their precursor materials as well as the processing variables in manufacturing the fiber. As a result, different grades are available (high modulus or intermediate modulus), with the tradeoff being between high modulus and high strength. Most PAN-based carbon fibers have an intermediate modulus, typically ranging between 200 and 300 GPa. Fibers produced from a mesophase pitch precursor have higher Young's modulus but much lower strength. Pitch-based fibers have higher thermal and electrical conductivities. Because of the high-cost of PAN-based precursors there have been recent efforts to develop alternative precursors. The most commonly studied alternative precursors include lignin and Lyocell (reconstituted cellulose).

Aramid Fiber Aramid fibers sold under the trade name **Kevlar** were introduced by the DuPont company in the 1970s and offer higher strength and stiffness relative to glass, coupled with light weight, high tensile strength but lower compressive strength. The compressive strength of Kevlar is approximately 1/8th of its tensile strength. Several different grades of Kevlar have been developed and the most commonly used for composites is Kevlar 49 because of its higher tensile strength and Young's modulus. Kevlar fibers are not as brittle as either glass or carbon fibers. Aramid fiber is often used as a higher-performance replacement for glass fiber in industrial and sporting goods and in protective clothing.

Other Fiber Reinforcements Ultra-high molecular weight polyethylene (UHMWPE) fibers have been developed and are attractive for composites because of their high strength and extremely low density, which gives them high strength-to-weight. These fibers are often referred to by their trade names of Spectra and Dyneema, which are manufactured by different companies. UHMWPE is limited to a very low temperature range, typically below 150°C and it is difficult to bond with to common matrix materials, which limits its potential application in structural composites. **Boron fibers** offer very high stiffness but at very high cost. These fibers have been used in specialized applications in both aluminum and polymeric matrices. A number of other fibers, including silicon carbide (SiC) and aluminum oxide (Al_2O_3), have been developed for ceramic matrices to enable use in very high-temperature applications such as engine components.

Nanoscale Reinforcements A new class of composites has been developed based on nanoscale reinforcements. These reinforcements are often in the form of particles, platelets, and fibers/tubes. Although these reinforcements are discontinuous, certain nanoplatelet and nanofiber materials can have extremely large aspect ratios (length/thickness for platelets and length/diameter for fibers). Among the most important nanomaterials for reinforcement are layered materials, such as montmorillonite clay and graphene, and carbon nanotubes. The first nanocomposites to gain wide attention were based on montmorillonite clay. Under the right processing conditions, the layers of the clay are able to exfoliate, or break apart from one another. These layers have high aspect ratio and often result in improved mechanical properties at low reinforcement contents. These nanocomposites were also demonstrated to have better thermal stability, flammability resistance, and barrier (liquid/gas permeability) properties. Carbon nanomaterials have much higher Young's modulus and strength than clay and, similar to carbon fiber, are potentially more desirable. Graphene is a single layer of graphite and is a platelet like clay. Carbon nanotubes, first observed in the early 1990s, have remarkable stiffness and strength with low density. With carbon fiber there is always a trade-off between stiffness and strength but carbon nanotubes have high Young's modulus, near 1,000 GPa, along with high tensile strengths, often reported in the range of 10-100 GPa. One unique aspect is the ability of nanotubes to create electrically conductive networks in polymers at very low volume fractions. Forms of nanotubes being developed include producing them in continuous macroscopic assemblies, such as bundles or sheets. There has also been considerable research aimed at hybridizing nanoscale reinforcements with traditional fibers. Nanotubes have diameters on the order of a few nanometers (10^{-9} meters) as compared to traditional fibers on the micron scale (10^{-6} meters). This three order of magnitude difference enables nanotubes and other nanoscale reinforcements to reinforce regions in between individual fibers and layers in a composite.

MATRICES

The selection of a polymer matrix depends on a variety of factors including performance requirements, such as use temperature or chemical resistance, fiber/matrix bonding and compatibility, and processing consideration. There are two classes of polymer matrices—thermosets and thermoplastics. **Thermosetting polymers** (such as **epoxies** and **polyesters**) are widely utilized in advanced composites, and a large amount of characterization data are available for these materials. Epoxies generally provide superior performance but are more costly as compared to polyesters. Thermosetting materials are initially liquids and require a chemical reaction to occur in order to solidify in a process known as curing. Typical cure temperatures for epoxies are in the range of 121 to 177°C (250 to 350°F). An important development has been high-toughness epoxies, which offer significantly improved resistance to damage from accidental impact but at higher cost. Polyester and vinyl ester are often used in less demanding applications. **Thermoplastic polymers** are able to soften upon heating and solidify upon cooling. A major potential advantage of thermoplastic composites is reduced processing times since there is no cure cycle involved. Thermoplastics are often used with discontinuous fibers in injection molding and there has been significant developments over the past two decades in advancing the materials and manufacturing technology for continuous fiber thermoplastic composites. High-performance thermoplastics such as polyetheretherketone (PEEK) and polyetherimide (PEI) have gained attention due to their excellent thermal stability, enabling higher use temperatures. Thermoplastic matrices also tend to have high fracture toughness. Another advantage is the potential recyclability of thermoplastics since the lifecycle of materials is being considered more as a design factor.

Polymeric matrix materials have a limited temperature range for many practical applications, with epoxies usually limited to 150°C (300°F) or less, depending on the material formulation. Higher-temperature thermosetting polymers available, such as polyimides, usually show increased brittleness. They are used for cowlings and ducts for jet engines. MMCs can be utilized in higher-temperature applications. Aluminum matrix has been utilized with boron and carbon fibers. Ceramic matrix materials are being developed for service at still higher temperatures. As discussed earlier, the fibers in CMCs are not necessarily higher in strength and stiffness than the matrix but are used primarily to increase the fracture toughness.

FIBER/MATRIX INTERFACE

In addition to the constituents—the reinforcement and the matrix—the properties of composites are also strongly influenced by the adhesion at the fiber/matrix interface. The load is transferred between the matrix and the fibers through shear stresses at the interface. For polymer-based composites fibers often have specific surface treatments, referred to as sizings, that make the fibers more compatible with the matrix. These sizings increase compatibility by promoting wetting of the fiber with the polymer. In addition, the sizings often contain chemical groups called coupling agents that promote the formation of chemical bonds at the fiber surface. In the region of the interface the properties of the polymer can be different than the bulk polymer matrix. Often referred to as the "interphase" this region can be formed by the sizing applied to the fiber.

COMPOSITES MANUFACTURING

There is a fundamental link between composites manufacturing and the final properties of the finished composite, since the manufacturing process forms both the final shape as well as the microstructure. The commonly utilized manufacturing processes are highlighted here and many variations have been developed.

Hand Layup The hand layup approach is probably the oldest and most common technique for manufacturing composites. As the name implies the reinforcement and matrix is laid-up manually onto a mold and then consolidated into a composite part. In **wet layup** the manufacturing technician manually places the dry reinforcement, typically in the form of a woven fabric or random mat, and the thermosetting polymer matrix is worked into the fabric typically with brushes and serrated rollers to fully impregnate the fibers and remove entrapped air. A variation on the wet layup is the **spray-up** process, where the reinforcement and the thermosetting resin is sprayed onto a mold surface using a spray gun that chops the fiber while simultaneously spraying the polymer. These processes are labor-intensive and the quality of the final composite part is influenced by the skill of the operator.

To provide better repeatability in the manufacturing process the reinforcement and matrix can be combined together in a process known as **prepregging**. This process is widely used with thermosetting polymeric resins. After combining the fibers together with the polymer matrix the resin is partially cured and the resulting **ply** is placed on a paper backing. The prepregged material is available commercially in continuous rolls. These rolls must be kept refrigerated until they are utilized and the assembled product is cured. Typical volume fractions of fiber are on the order of 60 percent. Prepregs based on thermoplastic matrices have also been developed.

With either the wet layup or prepreg approaches the plies are laid down in the desired pattern and then they are typically layered with several additional materials used in the curing and consolidation process. The objective is to remove volatiles and excess air to facilitate consolidation of the laminate. To this end, the laminate is covered with a **peel ply,** for removal of the other curing materials, and a **breather ply,** which is often a nonwoven polyester fabric. Optionally, a bleeder may be used to absorb excess resin. Finally, the assembly is covered with a vacuum bag and is sealed at the edges. A vacuum is drawn, and after inspection, heat is applied often applied to cure the polymer matrix. The vacuum bag results in the application of atmospheric pressure to the laminate.

In many cases an autoclave, a specialized oven that is also a pressure vessel, is used to apply higher pressures, typically on the order of 0.1 to 0.7 MPa (20 to 100 lb/in^2), to ensure final consolidation. Autoclave processing ensures good lamination but is an expensive piece of machinery and also costly to operate. For thermoplastic matrix composites higher pressures are required to achieve consolidation. The lamination process is often performed by hand, but automated machinery can be used to layup prepreg tapes and fibers (tows). These machines can be large enough to construct wing panels or other large structures. Because of the high cost of owning and operating an autoclave there have been significant advances in developing prepreg systems that are able to be consolidated under vacuum pressure only. These specialized prepregs are often referred to as out-of-the-autoclave prepregs.

Filament Winding The filament winding process consists of winding the fiber around a tool, typically referred to as a mandrel, to form the structure. Usually the mandrel rotates while fiber placement is synchronized to wrap the mandrel in a helical pattern. The polymer matrix may be added to the fiber by passing the dry fiber tows through a bath at the time of placement, a process called **wet winding**, or the tows may be prepregged prior to winding. Filament winding typically creates geometries that are a surface of revolution where the shape is axisymmetric and is widely used to manufacture pressure vessels, pipes, rocket motor cases, and other similar products. Although the process lends itself most readily to axisymmetric shapes advances in machine design and automation enables the production of non-axisymmetric shapes. Filament winding is a more automated manufacturing process compared to hand layup. After winding pressure consolidation and cure is typically performed in either an oven or autoclave.

Pultrusion The continuous pultrusion process is utilized to produce composites that have a constant cross-section such as rods or other structural shapes (C-channels, I-beams). In pultrusion fibers or fabrics are typically pulled through a matrix bath and then a heated die of the desired cross-section. As the fibers and resin are pulled through the die the polymer matrix is cured. The final composite is then cut to the desired length after coming out of the die. Although the pultrusion process is most typically used with thermoset polymer matrices, it has also been adapted for processing of thermoplastic composites.

Liquid Molding Processes A variety of manufacturing processes have been developed where a liquid resin is infiltrated into fibers during the manufacturing process. The most commonly utilized liquid molding processes are **resin transfer molding (RTM)** and **vacuum-assisted resin transfer molding (VARTM)**. Both of these processes involve first creating a fibrous structure that is the approximate near-net-shape as the final part, referred to as a **preform**. This preform is then placed into a mold and then the polymer matrix is infused into the fibers prior to curing. In RTM the mold is typically a two-sided mold in the shape of the final composite part and the resin is injected under positive pressure into the mold to infiltrate and wet-out the fibers. The molds are designed so that vents allow air to escape as the resin is injected.

Although vacuum is sometimes drawn on the mold during traditional RTM, the process commonly referred to as VARTM usually refers to placing the textile preform onto a one-sided tooling and sealing in a vacuum bag. Vacuum is drawn on the preform and the resin infiltrates the fibers with atmospheric pressure pushing the liquid resin into the preform. Because of less complexity in tooling the VARTM process has been utilized to produce large structural composites. Another liquid molding process is **resin film infusion (RFI)**. Instead of injecting the liquid resin into the fiber preform a film of resin is either placed adjacent to the preform or interleaved between layers of the preform. Upon heating the resin film reduces in viscosity and then infiltrates into the fibers. The RFI process is typically performed either under vacuum or within an autoclave.

DESIGN AND ANALYSIS

The fundamental way in which fiber composites, and in particular continuous-fiber composites, differ from conventional engineering materials such as metals is that their properties are **highly directional**. The stiffness and strength in the fiber direction are often higher than in the direction transverse to the fibers by orders of magnitude. Thus a basic principle is to align the fibers in directions where stiffness and strength are needed. A well-developed theoretical basis called **classical lamination theory** has been established to predict the overall composite stiffness and to calculate stresses within the layers of a composite laminate. This theory can also be applied to filament wound structures and to textile preform structures if allowances are made for the fiber waviness produced by the undulating fiber paths. The basic assumption is that fiber composites are orthotropic materials. Thin composite layers require four independent elastic constants to characterize the stiffness; those are the longitudinal fiber direction modulus (E_1), transverse fiber direction modulus (E_2), in-plane shear modulus (G_{12}), and one of the two Poisson ratios (V_{12} or V_{21}), with the other related through symmetry of the orthotropic stress-strain matrix. The material constants are routinely provided by material suppliers but can also be predicted using micromechanics techniques or measured in the laboratory. Lamination theory is then used to predict the overall stiffness of the laminate and to calculate stresses within the individual layers under mechanical and thermal loads.

Transverse properties are much lower than those in the fiber direction, so that fibers must usually be oriented in more than one direction, even if only to take care of secondary loads. Laminates are described in terms of the fiber orientations of the individual layers. For example, a laminate may consist of fibers in an axial direction combined with fibers oriented at ±60° to this direction, commonly designated as an $[0_m/\pm60_n]_s$ laminate, where m and n refer to the number of plies in the axial and ±60° directions and s stands for symmetry with respect to the midplane of the laminate. Midplane symmetry is often desirable since a laminate that is asymmetric about the midplane of the laminate exhibits coupling between extension and curvature. Although not all laminates are designed to be symmetric, residual stresses form upon consolidation due to differential thermal expansion of the layers and laminates with extension/bending coupling will cause flat panels to curve when cooled from elevated-temperatures during processing. Laminate constructions of [0/±60] or the [0/±45/90] are of practical importance since the in-plane extensional stiffness is equal in all directions and are referred to as **quasi-isotropic**.

Note that the **stiffness of composite laminates** is lower than that of the fibers themselves for two reasons. First, the individual plies contain fibers and often a much less stiff matrix. With high-stiffness fibers and polymeric matrices, the contribution of the matrix can be neglected, so that the stiffness in the fiber direction is essentially the **fiber volume fraction** V_F times the fiber stiffness. Fiber volume fractions are often in the range of 50 to 60 percent and depend on the manufacturing process. Second, laminates contain fibers oriented in more than one direction, so that the stiffness in any one direction is less than it would be in the fiber direction of a unidirectional laminate. The load is shared between the piles based on their relative stiffness in that direction, and the ply stiffness is strongly influenced by fiber orientation. As an example, the in-plane stiffness of a quasi-isotropic carbon fiber/polymeric matrix laminate is about 45 percent of the stiffness of a unidirectional composite in the fiber direction, which in turn is approximately 60 percent of the fiber modulus for a 60 percent fiber volume fraction material.

Fiber composites typically have excellent strength-to-weight properties. Like stiffness, strength is reduced as compared to the fiber strength because of the fiber volume fraction and because not all the fibers can be oriented in one direction. The overall reduction in apparent strength because of these two factors is similar to that for stiffness, although this is a very rough guide. As an example, the strength of a quasi-isotropic carbon/epoxy laminate under uniaxial tensile load has been reported to be about 35 to 40 percent of that provided by the same material in a unidirectional form.

Fiber composite materials can fail in several different modes. Damage in composite laminates is a progressive process where initial damage is initiated in the polymer matrix or at the fiber/matrix interface. The design goal is usually to have the failure mode be dominated by in-plane failure due to fiber fracture, as this takes advantage of the strong fibers. Unanticipated loads, or poorly designed laminates, can lead to ultimate failure of the matrix, which is much weaker. For example, an in-plane shear loads on a [0/90] layup would be resisted primarily by the matrix and not the fiber, with the potential result of failure at low applied load. Adding fibers in the ±45 directions will resist shear loading more effectively.

Fiber composites respond differently to compressive loads than they do to tensile loads. As discussed earlier, some fibers, such as aramid and the very high-modulus carbon fibers, are inherently weaker in compression than in tension. Regardless of the fiber properties in compression the strength of unidirectional composites in compression are lower than in tension. The mechanism of compressive failure in unidirectional composites involve local bending or buckling of the very small-diameter fibers against the support provided by the matrix. This phenomenon is known as **microbuckling**. The compressive failure often involves the formation of localized kink bands in the fiber bundles or plies.

Matrix cracking in polymer composites usually occurs in the transverse plies of laminates, often at an applied load less than half that for ultimate fiber failure. This is caused by the lower failure strain of the matrix and also the stress concentration effect of the fibers. Matrix cracking is also influenced by residual thermal stresses caused by the differences in coefficients of thermal expansion in the fibers and matrix. In some cases these matrix cracks are tolerated, while in other cases they produce undesirable effects. An example of the latter is that matrix cracking increases the permeability in unlined pressure vessels, which can result in leakage of gasses or liquids.

Another potential failure mode is **delamination**. Delamination occurs when cracks form between the individual layers of the composite and is often caused by high interlaminar shear and through-the-thickness tensile stresses, perhaps combined with manufacturing deficiencies such as inadequate cure or matrix porosity. Delamination can often be initiated by unexpected out-of-plane loads, such as local impact. In addition, delamination can start at the edges of laminates due to "free edge stresses" which have no counterpart in isotropic materials such as metals.

Fiber composites usually have excellent fatigue properties. However, stress concentrations or accidental damage can result in localized damage areas that can lead to failure under fatigue loading.

Suggested Reading

General Theory and Mechanics of Composites

Agarwal BD, Broutman LJ, Chandrashekhara K. (2006). *Analysis and performance of fiber composites*. John Wiley & Sons.

Gibson RF. (2016). *Principles of composite material mechanics*. CRC press. Chicago.

Hull D, Clyne TW. (1996). *An introduction to composite materials*. Cambridge University Press.

TEXTILE COMPOSITES

Chou TW. (1992). *Microstructural design of fiber composites.* Cambridge University Press.

COMPOSITES MANUFACTURING

Strong AB. (2008). *Fundamentals of composites manufacturing: materials, methods and applications.* Society of Manufacturing Engineers.

Advani SG, Sozer EM (eds.). (2010). *Process modeling in composites manufacturing.* CRC Press.

NANOCOMPOSITES

Thostenson ET, Li C, Chou TW. (2005). Nanocomposites in context. *Composites Science and Technology, 65*(3), 491-516.

Pandey G, Thostenson ET. (2012). Carbon nanotube-based multifunctional polymer nanocomposites. *Polymer Reviews, 52*(3), 355-416.

Qian H, Greenhalgh ES, Shaffer MS, Bismarck A. (2010). Carbon nanotube-based hierarchical composites: a review. *Journal of Materials Chemistry, 20*(23), 4,751-4,762.

4.11 TRUSS DESIGN

Revised by Parisa Saboori

REFERENCES: "Manual of Steel Construction—Allowable Stress Design," American Institute of Steel Construction. "National Design Specification for Wood Construction," American Forest and Paper Association. "Design Values for Wood Construction," American Forest and Paper Association. "Uniform Building Code" (as applicable), International Conference of Building Officials. SEI/ASCE 7-05, "Minimum Design Loads for Buildings and Other Structures," American Society of Civil Engineers. Blodgett, "Design of Welded Structures," J. F. Lincoln Arc Welding Foundation. "Masonry Designers Guide," The Masonry Society. "Building Code Requirements for Structural Concrete and Commentary," American Concrete Institute. "Wood Handbook," Forest Products. Laboratory, USDA. "International Building Code" (IBC), International Code Council. "Cold Formed Steel Framing Design Guide," AISI. *Note:* The editions of the references that currently apply and that are cited in the applicable codes must be used.

LOADS AND FORCES

Buildings and other structures should be designed and constructed to support safely all loads of both a permanent and transient nature, without exceeding the allowable stresses for the specified materials of construction. Usually, there are governing local codes, either unique to the jurisdiction or incorporating accepted codes (International Building Code, Building Code Requirements for Structural Concrete, Uniform Building Code, etc.,—sometimes amended locally), that will guide the selection of appropriate loads. In general, the following will enter into the design considerations: **dead loads,** generally agreed to include the weight of all permanent construction; **live loads,** produced by the use or occupancy of the building; **external** (or **environmental**) **loads** resulting from wind, snow, rain, ice, and earthquakes. The individual types of loading may be used singly or in combination, depending upon the use and location of the building and its inherent design features. For instance, the building may be designed to impound water on the roof for the purpose of keeping the roof deck (and the building below) cooler. In that case the collected and retained water has a major impact on the design of the roof structure. The governing code will address the matter of loading per se, or the use of combined loads by reference to other definitive sources whose information has standing. Absent specific guidance in the local governing code, for areas not so regulated, the data cited in the paragraphs below have been culled from many years of successful application.

Live loads on floors are generally regulated by the building codes in cities or states. For areas not regulated, the following values will serve as a guide for live loads in lb/ft^2 (kPa): rooms for habitation, 40 (1.92); offices, 50 (2.39); halls with fixed seats, 60 (2.87); corridors, halls, and other spaces where a crowd may assemble, 100 (4.76); light manufacturing or storage, 125 (5.98); heavy manufacturing or storage, 250 (11.95); foundries, warehouses 200 to 300 (9.58 to 14.37). Floor decks and beams that support only a small floor area must also be designed for any local concentrations of load that may be placed upon them. Girders, columns, and members that support large floor areas, except in buildings such as warehouses where the full load may extend over the whole area, may often be designed for live loads progressively reduced as the supported area becomes greater. Where live loads, such as cranes and machinery, produce **impact** or **vibration;** static loads should be increased as follows: elevator machinery, 100 percent; reciprocating machinery, 50 percent; others, 25 percent.

Roof live loads should be taken as a minimum of 20 lb per horizontal ft^2 (957.6 kPa) for essentially flat roofs (rise less than 4 in/ft), varying linearly with increasing slope to 12 lb/ft^2 for steep slopes (rise greater than 12 inches per foot). Reductions may also be made on the basis of tributary area greater than 200 ft^2 (18.58 m^2) of the member under consideration, to a maximum of 40 percent for tributary areas greater than 600 ft^2. The minimum roof live load shall be 12 lb/ft^2 after all reduction factors have been applied. Where **snow loads** occur, appropriate design values should be based on the local building codes.

Dead loads are due to the weight of the structure, partitions, finishes, and all permanent equipment not included in the live load. The weights of common building materials used in floors and roofs are given below (see also Sec. 4).

Material	Weight, lb/ft^2
Asphalt and felt, 4-ply	3
Corrugated asbestos board	5
Glass, corrugated wire	5-6
Glass, sheet, ⅛ in thick	2
Lead, ⅛ in thick	8
Plaster ceiling (suspended)	10
Acoustical tile	1-2
Sheet metal	1-2
Shingles, wood	3
Light-weight concrete over metal deck	30-45
Sheathing, 1 in wood	3
Skylight, 3⁄16 to ¼ in, glass and frame	6-8
Slate, 3⁄16 to ½ in thick	8-20
Tar and gravel, 5-ply	6
Tar and slag, 5-ply	5
Roof tiles, plain, ⅜ in thick	20

NOTE: lb/ft^2 × 0.04788 = kPa.

Earthquake Effects Two distinct methods of designing structures for seismic loads are currently accepted. The more familiar method, employed by the Uniform Building Code (UBC), yields equivalent loading for use with the allowable stress design approach. The National Earthquake Hazard Reduction Program (NEHRP) has developed Recommended Provisions for the Development of Seismic Regulations for New Buildings, which is based on the ultimate strength design approach. The NEHRP approach is the basis for model codes, such as the IBC. SEI/ASCE 7-05 has an extended and detailed presentation applicable to determination of rational earthquake loads.

Wind Pressures on Structures Every building and component of buildings should be designed to resist wind effects, determined by taking into consideration the geographic location, exposure, and both the shape and height of the structure. For structures sensitive to dynamic effects, such as buildings with a high height-to-width ratio, structures sensitive to wind-excited oscillations, such as vortex shedding or icing, and very tall buildings, special consideration should be given to design for wind effects and procedures used should be in accordance with approved national standards. Wind loads should not be reduced for the shading effects of adjacent buildings.

Wind pressure on walls of buildings should be assumed to be a minimum of 15 lb/ft^2 (0.72 kPa) on surfaces less than 50 ft above the ground. For buildings in exposed locations and in locations with high wind velocity, pressures should be calculated based on the procedures outlined in SEI/ASCE 7-05, or as dictated by the governing code. Boundary-layer wind tunnels, which simulate the variation of wind speed with height and gusting, are used to better estimate the design wind loading.

The effect of the sudden application of **gust loads** has sometimes been blamed for peculiar failures due to wind. In most cases, these failures can be explained from the pressure distributions in a steady wind. If a relatively flexible structure such as a radio tower, chimney, or skyscraper with a natural period of 1 to 5 s is set into vibration, the stresses in the structure may be increased over those calculated from a static-load analysis. Provision should be made for this effect by increasing the static design wind. A rational analysis should be performed to calculate the magnitude of this increase, which will depend on the flexibility of the structure.

Even a steady wind may give rise to periodic forces which may build up into large vibrations and lead to failure of the structure when the frequency of the exciting force coincides with one of the natural frequencies of vibration of the structure. The periodic exciting force may be due to the separation of a system of **Kármán vortices** (Fig. 4.11.1) in the wake of the body. The exciting frequency n in cycles per second is related to d, the dimension of the body normal to the wind velocity V. Dangerous vibrations related to the "flutter" of airplane wings may arise on bridges and similar flat bodies. These self-induced vibrations may be caused (1) by a negative slope of the curve of lift against angle of attack (Den Hartog, op. cit.) or (2) by a dynamic instability which arises when a body having two or more degrees of freedom (such as bending and torsion) moves in such a manner as to extract energy out of the air stream. The first of these vibrations will occur at one of the natural frequencies of the structure. The second type of vibration will occur at a frequency intermediate between the natural frequencies of the structure.

Figure 4.11.1 Von Kármán vortices.

The wind forces and the **pressure distribution** over a structure corresponding to a design wind V can be determined by **model testing** in a wind tunnel. Extrapolation from model to full scale is based on the fact that at every other point on the body, the pressure p is proportional to the stagnation pressure q (see Sec. 8) and thus the ratio p/q = constant for a fixed point on the body, as the scale of the model or the velocity of the wind is changed. Since the principal component of the wind force is due to the pressures, the force F acting on the surface S is

$$F \approx p_{avg}S = (p/q)_{avg}qS$$

and denoting $(p/q)_{avg}$ by a normal force coefficient or **shape factor** C_N

$$F = C_N \tfrac{1}{2}\rho V^2 S = C_N qS$$

The shape factors so obtained apply to full scale for structures with sharp edges whose principal resistance is due to the pressure forces. For bodies that do not have any sharp edges perpendicular to the flow, such as spheres or streamlined bodies, the factor C_N is not constant. It depends upon the Reynolds number (see Sec. 1). For such bodies, the law for variation of the shape factor C_N must be determined experimentally before safe predictions of full-scale forces can be made from model measurements.

Radio Towers and Other Framed Structures For open frame towers, by using round structural members instead of flat and angular sections, a substantial reduction in wind force can be effected.

Load Combinations Methods of combining types of loading vary with the governing local codes. Dead loads are usually considered to act all the time in combination with either full or reduced live, wind, earthquake, and temperature loads. The reductions in live loads are based on the improbability of fully loaded tributary areas, generally when the tributary area exceeds 150 ft^2. Most codes consider that wind and earthquake loads need not be taken to act simultaneously. There are two basic design methods in use: allowable stress or working stress design and limit states or ultimate strength design. Whereas allowable stress design compares the actual working stresses with an allowable value, the limit states approach determines the adequacy of each element by comparing the ultimate strength of the element with the factored design loads. For load combinations see SEI/ASCE 7-05, or as dictated by the governing code.

DESIGN OF STRUCTURAL MEMBERS

Members are usually proportioned so that stresses do not exceed allowable **working stresses** which are based on the strength of the material and, in the case of compressive stresses, on the stiffness of the element under compression. Internal forces and moments in **simple beams,** columns, and pin-connected truss bars are obtained by means of the equations of static equilibrium. **Continuous beams,** rigid frames, and other members characterized by practically rigid joints require for analysis additional equations derived from consideration of deflections and rotations.

Design may also be on the basis of the **ultimate strength** of members, the factor of safety being embodied in stipulated increases in the design loads. In steel-frame construction, the procedures of **plastic design** determine points where the material may be allowed to yield, forming *plastic hinges,* and the resulting redistribution of internal forces permits a more efficient use of the material.

Floors and Roofs

Except in reinforced-concrete flat-slab construction, floors and roofs generally consist of flat decks supported upon beams, girders, or trusses. The decks may usually be considered a series of beamlike strips spanning between beams and themselves designed as beams. The design of a beam consists chiefly in proportioning its cross section to resist the maximum bending and shear and providing adequate connections at its supports, without exceeding the unit stresses allowed in the materials (see Sec. 3) and limiting maximum live load deflection at midspan to $\frac{1}{360}$ of the span.

Up to spans of 20 to 30 ft (6.1 to 9.1 m), either wood or steel beams of uniform section are generally more economical than trusses, while for spans above 50 to 70 ft (15.2 to 21.3 m), trusses are usually more economical. Between these limits, the line of economy is not well-defined. Conditions that favor the use of trusses are as follows: (1) identical trusses are repeated many times, (2) the height of the building need not be increased for the greater depth of the truss, and (3) fire protection of wood or metal is not required.

Roof trusses often have their top chords sloped with the roof. Common trusses for steeply pitched roofs are shown in Figs. 4.11.2 to 4.11.6, with the top chord panels being equal in each truss. The members shown by heavy lines are in compression under ordinary loads, those in light lines in tension. The trusses of Figs. 4.11.6 to 4.11.9 are adapted for either steel or wood. In wooden trusses, the tensile web members may be steel rods with plates, nuts, and threaded ends. The truss of Fig. 4.11.6 is usually made of steel.

The forces in any member of these trusses under a vertical load uniformly distributed may be found by multiplying the coefficients in Tables 4.11.1 to 4.11.5 by the panel load P on the truss. For other slopes, types of trusses, or loads, see "General Procedure" below. Trusses for flat roofs are commonly of one of the types shown in Figs. 4.11.7 to 4.11.13 except when the top chords conform to the slope of the roof.

Floor trusses normally have parallel chords. Common types are shown in Figs. 4.11.7 to 4.11.13 in which heavy lines indicate members in compression, light lines in tension, and dash lines members with only nominal stress, under equal vertical panel loads. The panel lengths l in each truss are equal. The stress in each member is written next to the member in the figure, in terms of the panel load and the lengths of members. For a truss like one of the figures turned upside down, the stresses in the chords and diagonals remain the same in magnitude but reversed in sign. Forces in verticals must be computed (compare

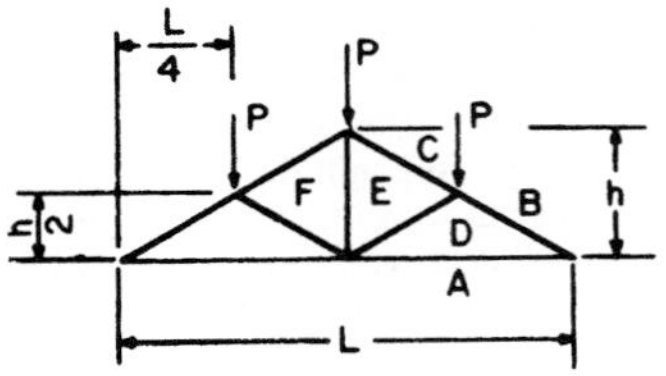

Figure 4.11.2

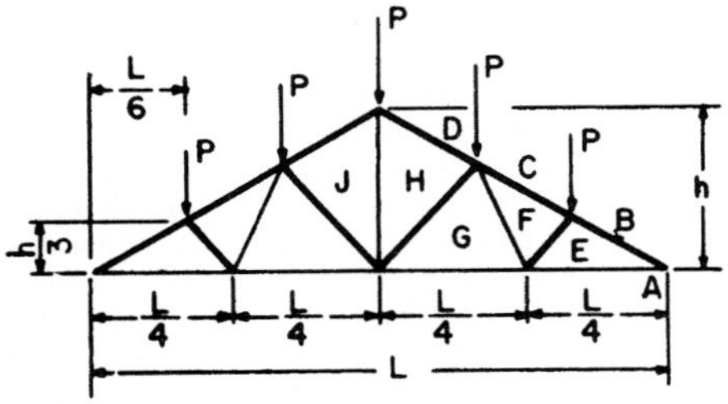

Figure 4.11.3

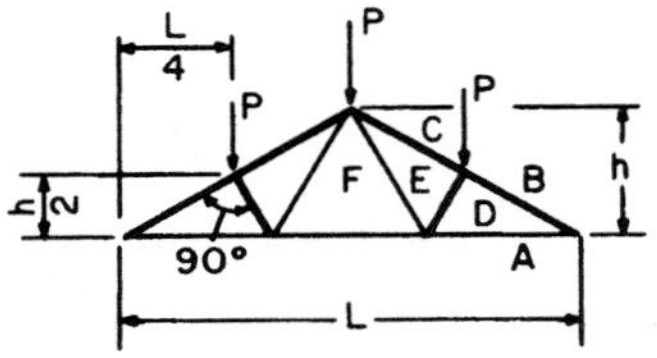

Figure 4.11.4

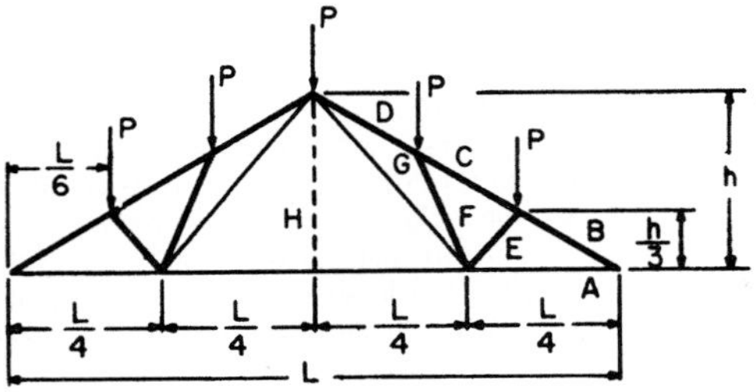

Figure 4.11.5

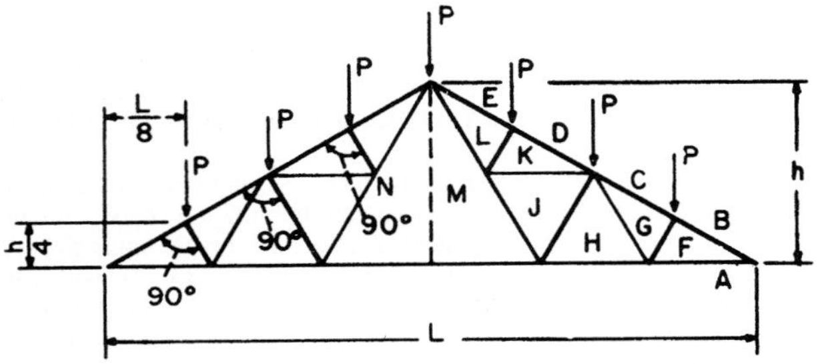

Figure 4.11.6

Figure 4.11.2 to 4.11.6 Types of steep roof trusses.

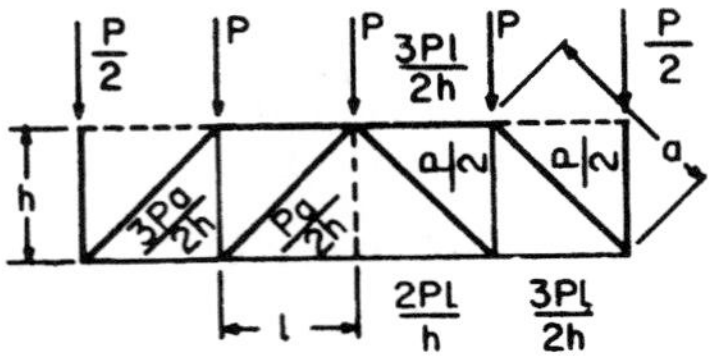

Figure 4.11.7

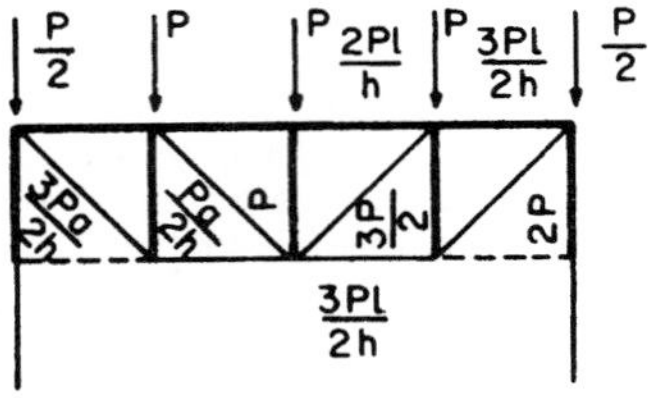

Figure 4.11.8

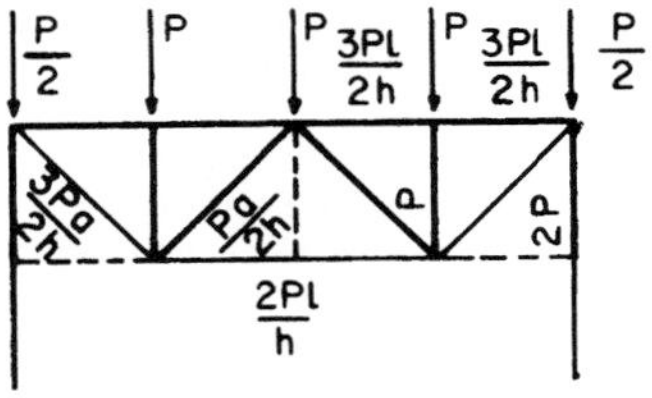

Figure 4.11.9

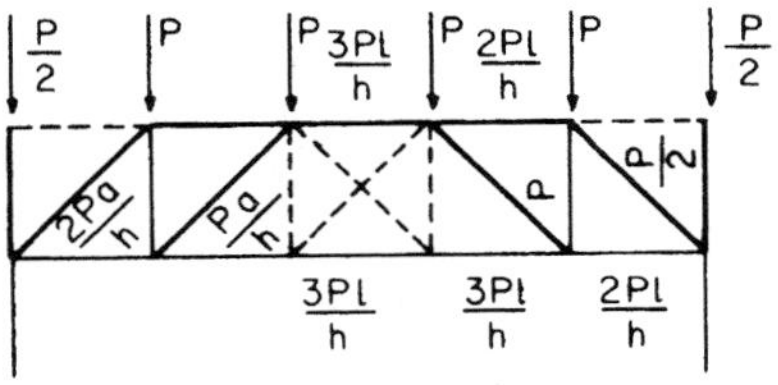

Figure 4.11.10

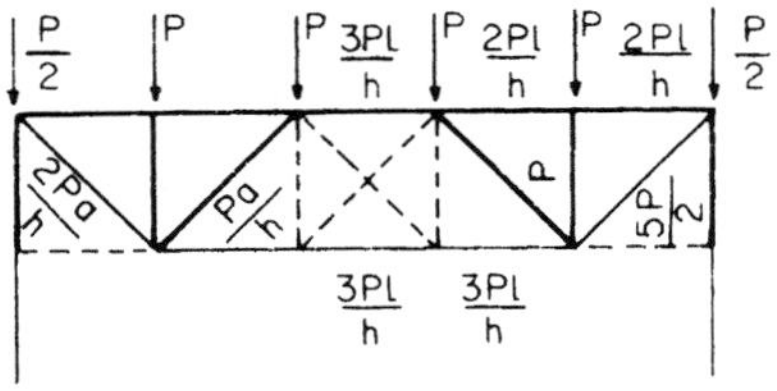

Figure 4.11.11

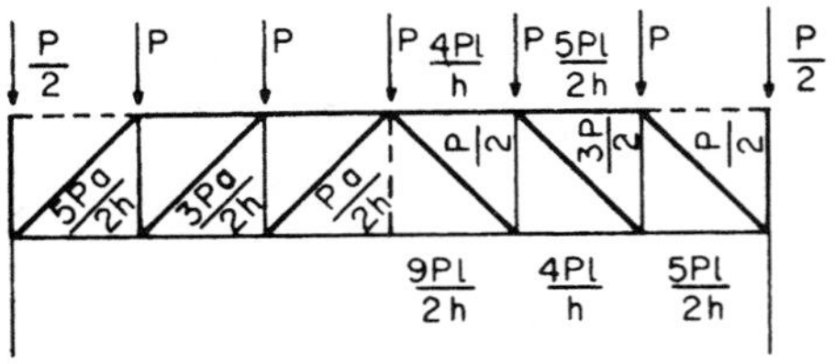

Figure 4.11.12

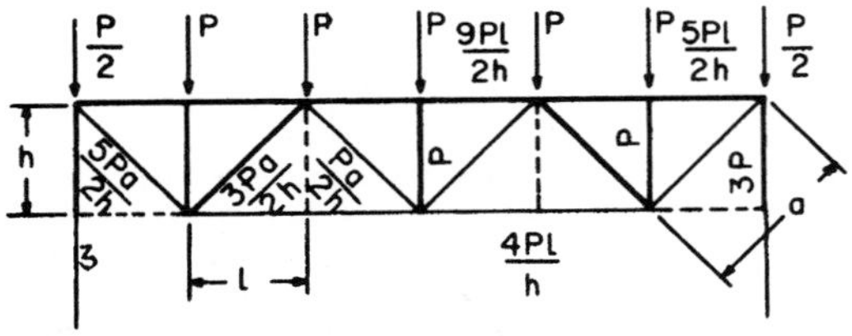

Figure 4.11.13

Figure 4.11.7 to 4.11.13 Types of floor and roof trusses.

Table 4.11.1 Coefficients for Truss Shown in Fig. 4.11.2

Pitch *h*/*L*	Coefficients of *P* for force in				
	AD	*BD*	*CE*	*DE*	*EF*
⅓	2.25	2.71	1.80	0.90	1.00
0.288*	2.60	3.00	2.00	1.00	1.00
¼	3.00	3.35	2.24	1.12	1.00
⅕	3.75	4.03	2.69	1.35	1.00

*30° Slope.

Table 4.11.2 Coefficients for Truss Shown in Fig. 4.11.3

Pitch *h*/*L*	Coefficients of *P* for force in								
	AE	*AG*	*BE*	*CF*	*DH*	*EF*	*FG*	*GH*	*HJ*
⅓	3.75	3.00	4.50	2.7	3.60	0.83	0.72	1.25	2.00
0.288*	4.33	3.46	5.00	3.0	4.00	0.88	0.73	1.32	2.00
¼	5.00	4.00	5.59	3.35	4.47	0.94	0.75	1.42	2.00
⅕	6.25	5.00	6.74	4.04	5.39	1.07	0.79	1.60	2.00

*30° slope.

Table 4.11.3 Coefficients for Truss Shown in Fig. 4.11.4

Pitch *h*/*L*	Coefficients of *P* for force in					
	AD	*AF*	*BD*	*CE*	*DE*	*EF*
⅓	2.25	1.50	2.70	2.15	0.83	0.75
0.288*	2.60	1.73	3.00	2.50	0.87	0.87
¼	3.00	2.00	3.35	2.91	0.89	1.00
⅕	3.75	2.50	4.04	3.67	0.93	1.25

*30° slope.

Table 4.11.4 Coefficients for Truss Shown in Fig. 4.11.5

Pitch *h*/*L*	Coefficients of *P* for force in							
	AE	*AH*	*BE*	*CF*	*DG*	*EF*	*FG*	*GH*
⅓	3.75	2.25	4.51	3.91	4.51	0.83	1.42	2.50
0.288*	4.33	2.60	5.00	4.33	5.00	0.88	1.45	2.65
¼	5.00	3.00	5.59	4.84	5.59	0.94	1.49	2.83
⅕	6.25	3.75	6.73	5.83	6.73	1.07	1.57	3.20

*30° slope.

Table 4.11.5 Coefficients for Truss Shown in Fig. 4.11.6

Pitch *h*/*L*	Coefficients of *P* for force in											
	AF	*AH*	*AM*	*BF*	*CG*	*DK*	*EL*	*FG* *KL*	*GH* *JK*	*HJ*	*JM*	*LM*
⅓	5.25	4.50	3.00	6.31	5.75	5.20	4.65	0.83	0.75	1.66	1.50	2.25
0.288*	6.06	5.20	3.46	7.00	6.50	6.00	5.50	0.87	0.87	1.73	1.73	2.60
¼	7.00	6.00	4.00	7.83	7.38	6.93	6.48	0.89	1.00	1.79	2.00	3.00
⅕	8.75	7.50	5.00	9.42	9.05	8.68	8.31	0.93	1.25	1.86	2.50	3.75

*30° slope.

Figs. 4.11.7 and 4.11.8). For other loads and other types of trusses see "General Procedure" below.

Weights of Trusses The approximate weight in pounds of a wooden roof truss may be taken as $W = LS(L/25 + L^2/6{,}000)$, where L is the span and S the spacing of trusses in feet. The approximate weight in pounds of a steel roof truss may be taken as $W = \frac{1}{3}LS(\sqrt{L} + \frac{1}{8}L)$.

Choice of Roof Trusses

Wooden Trusses. For pitched roofs with spans up to 20 ft (6.1 m) the simple king-post truss (Fig. 4.11.2) may be used. For spans up to 40 ft (12.2 m) the trusses of Fig. 4.11.2 and Fig. 4.11.4 are good. For spans up to 60 ft (18.3 m) the trusses of Figs. 4.11.3 and 4.11.5 are good. The number of panels rarely exceeds eight or the panel length, 10 ft (3 m). For flat roofs, the Howe truss (Figs. 4.11.7, 4.11.10, and 4.11.12) is built of wood with steel rods for verticals; the depth is one-eighth to one-twelfth the span. Wooden trusses are usually spaced 10 to 15 ft (3.0 to 4.6 m) on centers. Wood is rarely used in roof trusses with span over 60 ft (18.3 m) long. Steel trusses for pitched roofs may well take the form shown in Figs. 4.11.3 to 4.11.6 for spans up to 100 ft (30.5 m). For flat roofs, the Warren truss (Figs. 4.11.9, 4.11.11, and 4.11.13) is usually constructed in steel; the depth of the trusses ranges from one-eighth to one-twelfth the span, with trusses spaced from 15 to 25 ft (4.6 to 7.6 m) on centers.

Stresses in Trusses

An ideal truss is a framework consisting of straight bars or members connected at their ends by frictionless ball-and-socket joints. The external forces are applied only at these ball-and-socket joints. Internal forces and stresses in such straight bars are axial, either tension or compression, without bending. Since frictionless ball-and-socket joints are impossible, and the ends of bars are often bolted or welded, the ideal truss is never realized. For purposes of analysis, the primary stresses, which are always axial, are determined on the assumption that the truss under consideration conforms to the ideal. Secondary stresses are additional stresses, generally flexural or bending, brought about by all the factors that make the actual truss different from the ideal. In the following discussion, only primary forces and stresses will be considered.

Analytical Solution of Trusses

General Procedure After all external forces (loads and reactions) have been determined, the internal force or stress in any member is found (1) by taking a section, making an imaginary cut through the members of the truss, including the one whose stress is to be found, so as to separate the truss into two parts; (2) by isolating either of these parts; (3) by replacing each bar cut by a force, representing the force in the bar; and (4) by applying the equations of statics to the part isolated.

The various ways in which sections can be taken and the equations used to determine the forces are illustrated by a solution of the truss in Fig. 4.11.14.

A section 1–1 (Fig. 4.11.14) may be taken around a joint, L_0. Isolate the forces inside the section (Fig. 4.11.15*a*). Assume the unknown forces to be tension. Since the forces are concurrent and coplanar, two independent equations of statics will establish equilibrium. These may be either $\Sigma x = 0$ and $\Sigma y = 0$ or $\Sigma M = 0$ taken about two axes perpendicular to the plane of the forces and passing through two points selected so that neither point is the intersection of the forces, or so that the line joining the two points is not coincident with either of the unknown forces.

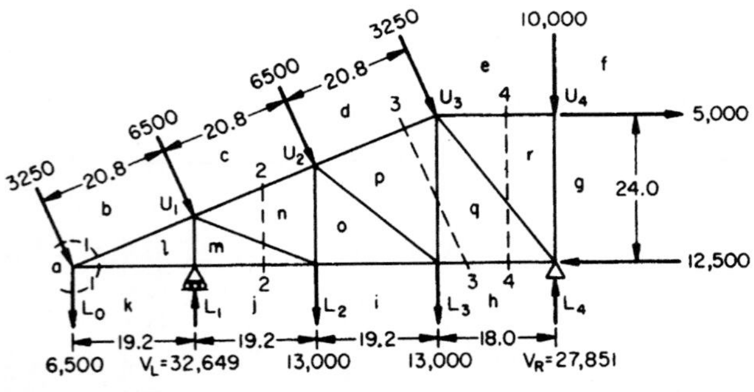

Figure 4.11.14

Using the first set of equations and taking components of all forces along horizontal and vertical axes; for example, the horizontal component of the 3,250-lb force is 1,250 and the vertical component is 3,000 lb.

$$\Sigma x = 0 = s_{lk} + s_{bl}(19.2/20.8) + 1{,}250$$
$$\Sigma y = 0 = s_{bl}(8/20.8) - 9{,}500$$

From these equations, $s_{lk} = -24{,}050$ and $s_{bl} = 24{,}700$.

The minus sign indicates that the force acts opposite to the assumed direction. If all the unknown forces are assumed to be in tension, then a plus sign in the result indicates that the force is tension and a minus sign indicates compression.

Hence, s_{lk} is 24,050 lb compression and s_{bl} is 24,700 lb tension.

Using the $\Sigma M = 0$ twice,

$$\Sigma M \text{ about } U_1 = 0$$
$$= -s_{lk} \times 8 - 1{,}250 \times 8 - 9{,}500 \times 19.2$$

from which $S_{lk} = 24{,}050$ lb compression.

$$\Sigma M \text{ about } L_1 = 0 = s_{bl}(8/20.8)19.2 - 9{,}500 \times 19.2$$

from which $s_{bl} = 24{,}700$ tension.

Instead of taking a section around a joint, a cut may be made vertically or inclined, cutting a number of bars such as Fig. 4.11.15*b* or Fig. 4.11.15*c*. If only three members are cut, and they are neither concurrent nor parallel, the forces can be found by taking moments of all forces on either side of the section about axes passing through the intersections of any two members.

Considering the part to the left of section 2–2, the forces in the three members (Fig. 4.11.15*b*) may be determined by taking moments about L_0, U_1, and L_2 of all forces acting on the part on either side of the section, for example, on the left of the section because it has the fewer forces:

$$\Sigma M \text{ about } L_0 = s_{mn}(8/20.8)38.4 + 2{,}500 \times 8 - (32{,}649 - 6{,}000)19.2$$
$$\Sigma M \text{ about } U_1 = 0 = -s_{mj} \times 8 - 1{,}250 \times 8 - 9{,}500 \times 19.2$$
$$\Sigma M \text{ about } L_2 = 0 = s_{cn}(19.2/20.8)16 + 2{,}500 \times 8 - 9{,}500 \times 38.4 + 26{,}649 \times 19.2$$

from which $s_{mn} = 33{,}290$ tension, $s_{mj} = 24{,}050$ compression, and $s_{cn} = 11{,}298$ compression.

This method is sometimes called the "method of moments."

Considering the part to the right of section 3–3, the forces in the three members (Fig. 4.11.15*c*) may be determined by taking moments about L_0, U_3, L_3 of all the forces acting on part or either side of the section, for example, on the right of the section because it has the fewer forces:

$$\Sigma M \text{ about } L_0 = 0 = s_{pq} \times 57.6 + 6{,}250 \times 24 + 3{,}000 \times 57.6 - (27{,}851 - 10{,}000) \times 75.6$$
$$\Sigma M \text{ about } L_3 = 0 = -s_{dp}(19.2/20.8)24 + 6{,}250 \times 24 - 17{,}851 \times 18$$
$$\Sigma M \text{ about } U_3 = 0 = s_{qh} \times 24 + 12{,}500 \times 24 - 17{,}851 \times 18$$

from which $s_{pq} = 17{,}825$ tension, $s_{dp} = 7{,}733$ compression, and $s_{qh} = 888$ tension.

Considering the part to the right of section 4–4, the forces in the three members (Fig. 4.11.15*d*) may be found by taking moments about axes where two unknowns intersect, for example, about U_3 and L_4. Since two unknown forces are parallel, their lines of action do not intersect.

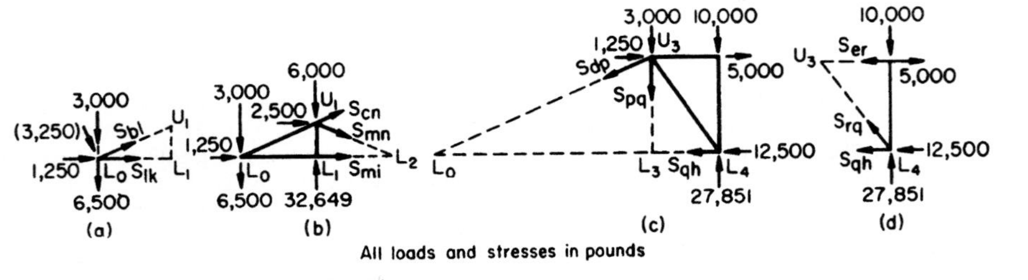

Figure 4.11.15 Resolution of forces for truss shown in Fig. 4.11.14.

However, the equation $\Sigma y = 0$ will enable one to find the force in the member which is not parallel to the other two:

$$\Sigma M \text{ about } U_3 = 0 = s_{qh} \times 24 - (27{,}851 - 10{,}000)18 + 12{,}500 \times 24$$
$$\Sigma M \text{ about } L_4 = 0 = -s_{er} \times 24 + 5{,}000 \times 24$$
$$\Sigma y = 0 = s_{rq}(24/30) + 27{,}851 - 10{,}000$$

from which s_{qh} = 888 tension, s_{er} = 5,000 tension, and s_{rq} = 22,314 compression.

STEEL CONSTRUCTION

(Note: In the design of steel structures, 1,000 lb is frequently designated as a kilopound or "kip," and a stress of 1 kip per square inch is designated as 1 ksi.)

Structural steel design was based only on the allowable stress design (ASD) approach until the introduction of the load and resistance factor design (LRFD) technique in the mid-1980s. The LRFD approach is an ultimate strength design approach, similar to that adopted by the American Concrete Institute for concrete design. Both ASD and LRFD are accepted in current codes. The ASD method is still the most commonly used for design.

Specifications The following are in part condensed excerpts from the Specifications of the American Institute of Steel Construction.

Material Ordinary steel for rolled shapes, plates, and bars is typically specified by ASTM A36, with a yield stress of 36,000 lb/in^2 (248.2 MPa). However, advances in mill production methods have resulted in most steels satisfying the higher strength requirements of ASTM A572, Grade 50, leading to increased use of higher-strength steel at little or no cost premium. Other higher-strength steels used in structures are A440, A441, A588, and A242. Steel materials for pipe and tube, specified by ASTM A53 (welded-seam pipe) and A500 (cold-formed), have yield strengths of 33,000 to 50,000 lb/in^2 (227.5 to 344.7 MPa).

Ordinary unfinished machine bolts are specified by A307. Bolts used for structural steel connections are typically high-strength bolts specified by A325 or A490. Riveting is now rarely used, but may often be encountered in older structures. The most common rivets are A502, Grade 1.

Allowable Stresses in A36 Steel

	lb/in^2	MPa
Tension F_t:		
On gross section	22,000	151.6
On net section, except at pinholes	29,000	200
On net section, at pinholes	16,000	110.2
Compression F_c: See Table 4.11.6		
Bending tension and compression on extreme fibers F_b:		
Basic stress, reduced in certain cases	22,000	151.6
Compact, adequately braced beams	24,000	165.3
Rectangular bearing plates	27,000	186.0
Shear F_v: Web of beams, gross section	14,500	99.9

Allowable Stresses* in Riveted and Bolted Connections

	lb/in^2	MPa
Bearing: A36 steel		
Pins in reamed, drilled, or bored holes	32,400	223.3
Bolts and rivets	69,000	475.7
Roller, lb/lin in (N/lin cm)	760 × diam (in)	1,131 × diam (cm)
Shear: bearing-type connections†		
A502, grade 1 hot-driven rivets	17,500	120.6
A307 bolts	10,000	68.9
A325 bolts when threading is excluded from shear planes (std. holes)	30,000	206.8
A325 bolts when threading is not excluded from shear planes (std. holes)	21,000	144.7
Shear: friction-type connections† (with threads included or excluded from shear plane)		
A325 bolts in standard holes	17,500	120.6
A325 bolts in oversized or short slotted holes	15,000	103.4
A325 bolts in long slotted holes	12,000	82.6
Tension:		
A502, grade 1, hot-driven rivets	23,000	158.5
A307 bolts	20,000	137.9
A325 bolts	44,000	303.3
Bending in pins of A36 steel	27,000	186.1

* Allowable stresses are based on nominal body area of fasteners unless indicated.

† Rivets or bolts may not share loads with welds on bearing-type connections but may do so in friction-type connections.

Proportion of Parts

Most simple beams, columns, and truss members are proportioned to limit the actual stresses to the allowable stresses stipulated above. Other stability or serviceability criteria may control the design. **Deflection** may govern in such members as cantilevers and lightly loaded roof beams. **Buckling,** rather than strength, may govern the design of compression members. The slenderness ratio Kl/r, where Kl is the effective length of the member and r is its radius of gyration, should be limited to 200 in compression members, and L/r limited to 300 in tension members. Kl should not be less than the actual unbraced length l in columns of a frame which depends on its bending stiffness for lateral stiffness. **Width-thickness ratios** are specified for projecting elements under compression. Repeated fluctuations in stress leading to fatigue may be a controlling factor. Rules are given for **combined stresses** of tension, compression, bending, and shear.

Tension members should be proportioned for the gross and net section, deducting for bolt or rivet holes in (0.3 cm) larger than the nominal diameter of the fastener.

Columns and other **compression members** subject to eccentric load or to axial load and bending are governed by special rules. A

Table 4.11.6 Allowable Stress, ksi, for Compression Members of A36 Steel

Main and secondary members, Kl/r not more than 120						Main members, Kl/r, 121-200			
$\frac{Kl}{r}$	F_a	$\frac{Kl}{r}$	F_a	$\frac{Kl}{r}$	F_a	$\frac{Kl}{r}$	F_a	$\frac{Kl}{r}$	F_a
1	21.56	41	19.11	81	15.24	121	10.14	161	5.76
5	21.39	45	18.78	85	14.79	125	9.55	165	5.49
10	21.16	50	18.35	90	14.20	130	8.84	170	5.17
15	20.89	55	17.90	95	13.60	135	8.19	175	4.88
20	20.60	60	17.43	100	12.98	140	7.62	180	4.61
25	20.28	65	16.94	105	12.33	145	7.10	185	4.36
30	19.94	70	16.43	110	11.67	150	6.64	190	4.14
35	19.58	75	15.90	115	10.99	155	6.22	195	3.93
40	19.19	80	15.36	120	10.28	160	5.83	200	3.73

NOTE: 1 ksi = 6.89 MPa.

long-established rule is that $f_a/F_a + f_b/F_b$ should be equal to or less than unity, where f_a is the axial stress, f_b the bending stress, and F_a and F_b are the corresponding allowable stresses if axial or bending stress alone exist. This is still considered valid when f_a/F_a is less than 0.15. Joints shall be fully spliced, except that where reversal of stress is not expected and the joint is laterally supported, the ends of the members may be milled to plane parallel surfaces normal to the stresses and abutted with sufficient splicing to hold the connected members accurately in place. Column bases should be milled on top for the column bearing, except for rolled steel bearing plates 4 in (10 cm) or less in thickness.

Beams and girders, of rolled section or built-up, should in general be sized such that the bending moment M divided by the section modulus S is less than the allowable bending stress F_b. For built-up sections a rule of thumb is, for A36 steel, flanges in compression should have a thickness of $\frac{1}{16}$ the projecting half width, and webs should have a thickness of $\frac{1}{320}$ the maximum clear distance between flanges. Web stiffeners should be provided at points of high concentrated loads; additional web stiffeners are required in plate girders. Splices in the webs of plate girders should be made by plates on both sides of the web. When two or more rolled beams or channels are used side by side to form a beam, they should be connected at separators spaced no more than 5 ft (1.52 m); beams deeper than 12 in (30 cm) are to have at least two bolts to each separator.

The **lateral force on crane runways** due to the effect of moving crane trolleys may be assumed as 20 percent of the sum of the weights of the lifted loads and of the crane trolley (but exclusive of the other parts of the crane) applied at the top of the rail, one-half on each side of the runway, and shall be considered as acting in either direction normal to the runway rail. The **longitudinal force** may be assumed as 10 percent of the maximum wheel reactions of the crane applied at the top of the rail.

Bolted or riveted connections carrying calculated stress, except lacing and sag bars, should be designed to support not less than 6,000 lb (27.0 kN). Rivets or high-strength bolts are preferred in all places, and both are implied in these paragraphs wherever "bolting" is mentioned; unfinished bolts, A307, may be used in the shop or in field connections of lightly loaded structures, or for secondary members, bracing, and the like.

Members in tension or compression, meeting at a joint, shall have their lines of center of gravity pass through a point, if practicable; if not, provision shall be made for the eccentricity. A group of bolts transmitting stress to a member shall have its center of gravity in the line of the stress, if practicable; if not, the group shall be designed for the resulting eccentricity. Where stress is transmitted from one member to another by bolts through a loose filler greater than ¼ in in thickness, except in slip critical connections using high-strength bolts, the filler shall be extended beyond the connected member and the extension secured by enough bolts or sufficient welding to distribute the total stress in the member uniformly over the combined sections of the member and the filler.

Most bolted connections transmit shearing forces by developing the shearing or bearing values of the bolts, but bolts in certain connections, such as shelf angles and brackets, are required to transmit tension forces.

Bolts shall be proportioned by the nominal diameter. Rivets and A307 bolts whose grip exceeds 5 diam shall be allowed 1 percent less safe stress for each $\frac{1}{16}$ in (0.16 cm) excess length. The minimum distance between centers of bolt holes shall be $2\frac{2}{3}$ diam of the bolt; but preferably not less than 3 diam.

The minimum distance from the center of any bolt hole to a sheared edge shall be $2\frac{1}{4}$ in (5.7 cm) for $1\frac{1}{4}$-in (32-mm) bolts, 2 in (5.1 cm) for $1\frac{1}{8}$-in (28-mm) bolts, $1\frac{3}{4}$ in (4.4 cm) for 1-in (25-mm) bolts, $1\frac{1}{2}$ in (3.8 cm) for $\frac{7}{8}$-in (22-mm) bolts, $1\frac{1}{4}$ in (3.2 cm) for $\frac{3}{4}$-in (19-mm) bolts, $1\frac{1}{8}$ in (2.8 cm) for $\frac{5}{8}$-in (16-mm) bolts, and $\frac{7}{8}$ (2.22 cm) for $\frac{1}{2}$-in (13-mm) bolts. The distance from any edge shall not exceed 12 times the thickness of the plate and shall not exceed 6 in (15 cm).

Design of Members

Properties of Standard Structural Shapes Detailed dimensions, geometric properties, and other data for standard rolled sections are listed in the AISC Manual of Steel Construction. This reference is usually the starting point when one is embarking on a design utilizing steel shapes. For sizes not standard, or for *unusual* (*generally much higher*) geometric properties required for special applications, an endless variety of beam and column shapes can be fabricated by welding the desired sections from plate stock.

A great variety of tees is produced by shearing or gas-cutting standard beams (S shapes) or wide-flange sections (W shapes) length-wise at midheight of the web, making two T sections.

Welding The main advantage of assembling and connecting steel frames by welding is the reduction in the amount of metal used. The saving in metal is achieved by (1) elimination of bolt holes which reduce the net section of tension members, (2) simplification of details, and (3) elimination of splice plate and gusset plate material. (See also Sec. 9.3.)

Allowable stresses in welds depend on the type of weld, the manner of loading, and the relative strengths of the weld metal and the base metal. For **complete-penetration groove welds** in which the full edge thickness of the thinner part to be joined is beveled in preparation for welding, allowable stresses due to tension or compression normal to the effective area or parallel to the axis of the weld are the same as the base metal, and the allowable shear stress on the effective area is 0.3 times the nominal tensile strength of the weld metal (limited by 0.4 times yield stress in the base metal). The effective area for a complete-penetration groove weld is the width of the part joined times the thickness of the thinner part.

For **partial-penetration groove welds,** allowable stresses due to compression normal to the effective area and tension or compression parallel to the axis of the weld are the same as the base metal. For tension normal to the effective area, the allowable stress is 0.3 times the nominal tensile strength of the weld metal (limited by 0.6 times the yield stress of the base metal); for shear parallel to the axis of the weld, the allowable stress on the effective area is 0.3 times the nominal tensile strength of weld metal (limited by 0.4 times yield stress of the base metal). The effective thickness of partial-penetration groove welds depends on the welding process, welding position, and the included angle of the groove but may be safely taken as the depth of the chamfer less $\frac{1}{8}$ in.

For **fillet welds,** allowable shear stresses on the effective area are taken as 0.3 times nominal tensile strength of the weld (limited by 0.4 times yield stress of the base metal), and for tension or compression parallel to the axis of the weld, allowable stresses are the same as the base metal. Fillet welds are not to be loaded in tension or compression normal to the effective area; they transfer loads between members in shear only. The effective area of a fillet weld is the overall length of the full-size weld times the shortest distance from the root to the face (normally leg size × sin 45°). Allowable shear in a fillet weld is taken as 930 lb/in of length (163 N/mm) for each $\frac{1}{16}$ in (1.59 mm) of leg size. (See Sec. 9.) Typical details of welded connections are indicated in Fig. 4.11.16.

Safe Loads for Steel Beams To determine the safe load uniformly distributed, as limited by bending, for a structural steel beam on a given span, apply the formula $W = 8F_bs/l$, where W is the total load, lb.; F_b is the allowable fiber stress (24,000 lb/in^2 or any other); S is the section modulus for the beam in question, and l is the span, in. (This formula may also be used with equivalent metric units.) The safe load concentrated at midspan is one-half this amount. For other safe loads, note that F_bS is the safe resistance to bending in inch-pounds (or newton-metres)

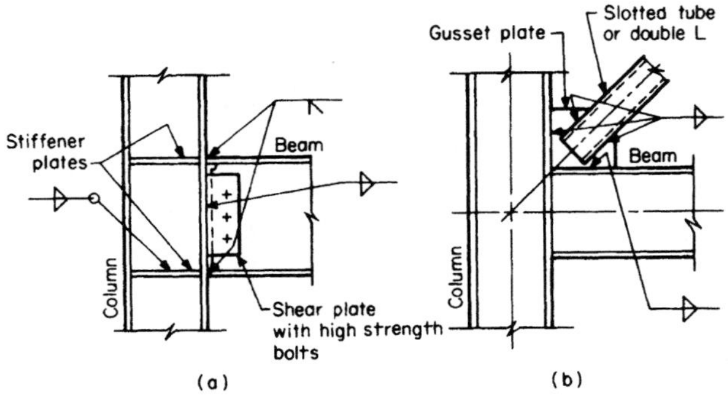

Figure 4.11.16 Welded connections. (*a*) Moment connection; (*b*) bracing connection.

Table 4.11.7 Coefficients of Deflection for Simply Supported Steel Beams under Uniformly Distributed Loads

Span, ft	Fiber stress of 24,000 lb/in²	Span, ft	Fiber stress of 24,000 lb/in²	Span, ft	Fiber stress of 24,000 lb/in²	Span, ft	Fiber stress of 24,000 lb/in²
1	0.026	14	4.87	27	18.1	39	37.7
2	0.098	15	5.59	28	19.5	40	39.8
3	0.223	16	6.36	29	20.9	41	41.8
4	0.398	17	7.18	30	22.4	42	43.9
5	0.621	18	8.04	31	23.9	43	45.8
6	0.892	19	8.97	32	25.4	44	48.0
7	1.23	20	9.93	33	27.0	45	50.4
8	1.59	21	10.9	34	28.7	46	52.6
9	2.01	22	12.1	35	30.5	47	54.7
10	2.48	23	13.1	36	32.2	48	57.1
11	3.00	24	14.3	37	34.1	49	59.5
12	3.58	25	15.6	38	35.8	50	62.2
13	4.20	26	16.8				

NOTE: For a load concentrated at midspan, use $\frac{4}{5}$ of the coefficient given. 1 ft = 0.305 m; 1 lb/in² = 6.89 kPa.

afforded by the beam. Compute the load, of whatever type or distribution, which will produce a maximum bending moment equal to safe moment of resistance (see Sec. 3 for bending-moment formulas).

To select a beam to support a given load, compute the maximum bending moment in inch-pounds, divide by the allowable fiber stress F_b, and refer to the table for a beam having a section modulus which is not smaller than the quotient.

Formulas for the safe loads and deflections of beams with various methods of support and of loading are given in Sec. 3.

Short beams should be investigated for crippling of the web. In the tables are given the safe end reactions for beams of A36 steel resting on a seat $3\frac{1}{2}$ in (9 cm) long along the axis of the beam. Short beams should also be investigated for shear, by dividing the maximum shear, in pounds, by the area of the web, excluding the flanges.

Single angles used as beams and loaded in the plane of axis *X-X* or *Y-Y* tend to deflect laterally as well as in the plane of the loads. Unless this is prevented, as by pairing the angles back to back and securing them together, the unit fiber stress due to bending may be as much as 40 percent above that computed by dividing the bending moment by S for the axis perpendicular to the plane of the loads. The relation $f = M/S$ does not hold for single angles and for *Z* bars, which are unsymmetrical about both axes.

Deflection of Beams Table 4.11.7 gives **coefficients of deflection** for steel shapes under uniformly distributed loads, and is based on the formula; deflection in inches = $30fL^2/Ed$, the table giving the values of $30fL^2/E$. (f = fiber stress, lb/in²; L = span, ft; d = depth of section, in; E = modulus of elasticity = 29,000,000 lb/in².)

To find the deflection in inches of a section symmetrical about the neutral axis, such as a beam, channel, etc., divide the coefficient in the table corresponding to given span and fiber stress by the depth of the section in inches.

To find the deflection in inches of a section which is not symmetrical about the neutral axis but which is symmetrical about an axis at right angles thereto, such as a tee or pair of angles, divide the coefficient corresponding to given span and fiber stress by twice the distance of extreme fiber from neutral axis obtained from table of elements of sections.

To find the deflection in inches of a section for any other fiber stress than that given, multiply this fiber stress by either of the coefficients in the table for the given span and divide by the fiber stress corresponding to the coefficients used.

I beams and channels loaded to a fiber stress of 24,000 lb/in² (165.5 MPa) will not deflect in excess of 1/360 in of the span (allowed for plastered ceilings) if the depth in inches (cm) is not less than 0.74 (6.21) times the span in feet (m) for uniform loads and 0.60 (5.00) times the span for a central concentration.

Beam Supports Steel beams are supported at the ends generally (1) by means of web connections to girders and columns, (2) by resting on structural-steel seats, or (3) by resting on masonry. Limiting values of end reactions of the second type, for seats $3\frac{1}{2}$ in (9 cm) long, are listed in the AISC Manual of Steel Construction. Standard AISC web connections of the first type are called **framed beam** connections and are designated by the number of rows of bolts. Examples of connections are given in Fig. 4.11.17. These connections may be specified as "Standard 3 row, 4 row, etc., connections." Connections must always be designated and detailed for the calculated design reaction.

The capacity of web connections is governed by the shearing of the fastener, or the bearing of the fastener on the web or on the material to which the beam is connected, or by the strength of the connecting angles. For example, the supporting values of standard framed beam connections, using $\frac{7}{8}$ in fasteners in members of A36 material, are given in Table 4.11.8. For fasteners in webs thicker than 0.34 in use the values in the column headed Double Shear; for thinner webs, bearing limits the value, and the coefficients for web bearing are to be used. For $\frac{3}{4}$-in fasteners, multiply tabular bearing values by $\frac{6}{7}$ and shear values by $\frac{36}{49}$. Fasteners connecting the outstanding legs to the

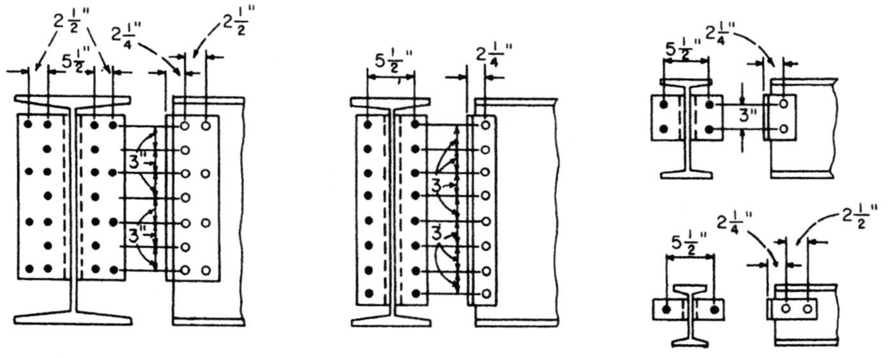

Figure 4.11.17 Framed beam connections.

Table 4.11.8 Values of Standard Framed-Beam Connections
(⅞-in A325 HS bolts in standard holes,* A36 members†)

AISC designation	Two angles thickness × length	Shear 1,000 lb	Bearing on beam web (t), 1,000 lb
10 rows	⁵⁄₁₆ × 2′5½″	204	609t
9 rows	⁵⁄₁₆ × 2′2½″	184	548t
8 rows	⁵⁄₁₆ × 1′11½″	164	487t
7 rows	⁵⁄₁₆ × 1′8½″	143	426t
6 rows	⁵⁄₁₆ × 1′5½″	123	365t
5 rows	⁵⁄₁₆ × 1′2½″	102	304t
4 rows	⁵⁄₁₆ × 0′11½″	81.8	243t
3 rows	⁵⁄₁₆ × 0′8½″	61.3	182t
2 rows	⁵⁄₁₆ × 0′5½″	40.9	121t

* Values indicated are for slip-critical connections or bearing type where threads are not excluded from the shear plane. For bearing-type connections where threads are excluded from the shear plane, shear values may be increased by 1.47.

† If the web of the supporting beam is thinner than 0.17 in (0.42 in if beams frame on both sides), bearing must also be investigated.

NOTE: 1 in = 2.54 cm; 1 lb = 4.45 N.

supporting metal are in double shear if two beams are framed opposite or in single shear if a beam is connected on one side only. If the supporting material of A36 steel is thinner than 0.34 in in double-shear connections or thinner than 0.17 in in single-shear connections, the capacity is limited by bearing loads. The value of any ⅞-in fastener in bearing on A36 material is 60,900t, where t is he thickness of the plate in inches. The value of ⅞ in A502, grade 1 rivets or A325 HS bolts (slip-critical connections) is 9,020 lb in single shear and 20,400 lb in double shear. The corresponding values for A307 unfinished bolts are 6,000 lb and 12,000 lb, respectively.

Cast-iron columns were often used in the past instead of wood, to save space, in the lower stories of heavy buildings, but are obsolete. Occasionally they are encountered in repair and alteration work in older buildings. The ratio of length to least radius of gyration l/r should not exceed 70, and the average unit stress under axial compression should not exceed $9{,}000 - 40l/r$ lb/in^2.

Steel joists consisting of lightweight rolled sections, thin for their height, or open-web trussed members fabricated by welding or otherwise, are used with economy in buildings where spans are long and loads are light, and where a plaster ceiling affords sufficient fire protection. They are rarely used in industrial buildings, except to support roof loads.

Steel pipe is often used for columns under light loads. Table 4.11.9 gives the safe loads on standard size pipes (ASTM A501 or A53, grade B) used as columns. For extra-strong and double extra-strong pipe used as columns, the safe loads will increase approximately in the same proportion as the weight per foot. (See Sec. 6.7.)

Structural steel tubing (ASTM A500, grade B), in square or rectangular cross section with F_y = 46 ksi (317.1 MPa), is also used for columns, bracing members, and unbraced members subjected to large torsional loads. The closed box shape makes tube sections especially suited for resisting torsional loads.

Corrugated metal deck and siding is used for roofs and walls, respectively, to span between purlins for roof loads or between girts for wind loads. The decking is sized to resist the bending caused by these loads. Roof decking is often also used as a diaphragm to transfer wind or seismic loads to the lateral bracing system below. Load tables specifying safe loads for different spans are available from metal deck manufacturers.

Lightweight, cold-rolled, galvanized steel sheet metal structural sections are used when the design will allow. A popular use is as a substitute for wood framing (wall studs and light beams). Assembly is commonly through use of self-drilling, self-tapping screws with corrosion-resistant plating, although the thicker product is amenable to welding. The most common cross sections rolled are shown in Fig. 4.11.18. Wall stud sections are manufactured with punch-outs along their length to facilitate easy installation of electric wiring, piping, etc. The sections are fabricated per ASTM 653 specifications and come in minimum yield strengths of 33 and 50 ksi. Some of the larger sections are used as beams and columns when used as-rolled or when assembled into stiffer cross sections. (See Steel Stud Manufacturers Association literature.)

Fire Resistance The resistance to fire of building materials has been tested extensively by various agencies.

Table 4.11.9 Safe Axial Loads for Standard (Schedule 40) Steel Pipe Columns, kips
(Stress according to AISC specification for A501 pipe*)

Nominal pipe size, in	Outside diam, in	Wall thickness, in	Effective length of column K1, ft										
			6	7	8	9	10	11	12	14	16	18	20
3	3.500	0.216	38	36	34	31	28	25	22	16	12	10	
3½	4.000	0.226	48	46	44	41	38	35	32	25	19	15	12
4	4.500	0.237	59	57	54	52	49	46	43	36	29	23	19
5	5.563	0.258	83	81	78	76	73	71	68	61	55	47	39
6	6.625	0.280	110	108	106	103	101	98	95	89	82	75	67
8	8.625	0.322	171	168	166	163	161	158	155	149	142	135	127
10	10.750	0.365	246	243	241	238	235	232	229	223	216	209	201
12	12.750	0.375	303	301	299	296	293	291	288	282	275	268	261

* Yield stress is 36 ksi (248.2 MPa). Pipe ordered to ASTM A53, type E or S grade B, or to API standard 5L grade B will have a yield point of 35 ksi (241.3 MPa) and may be designed at stresses allowed for A501 pipe.

NOTE: 1 in = 2.54 cm; 1 ft = 0.305 m. For dimensions of standard pipe see Sec. 6. Safe loads above underscore lines are for values of Kl/r more than 120 but not over 200.

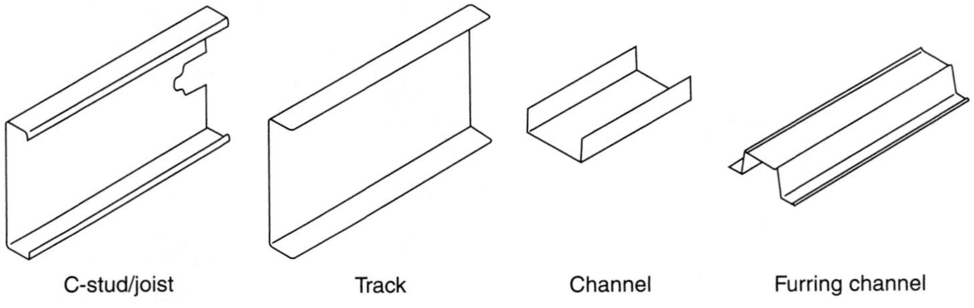

Figure 4.11.18 Light-gage sheet metal cross sections.

4.12 PLASTIC WORKING OF METALS

by Rajiv Shivpuri

(See also Secs. 3 and 4.)

REFERENCES: Crane,"Plastic Working of Metals and Power Press Operations," Wiley. Woodworth, "Punches, Dies and Tools for Manufacturing in Presses," Henley. Jones, "Die Design and Die Making Practice," Industrial Press. Stanley, "Punches and Dies," McGraw-Hill. DeGarmo, "Materials and Processes in Manufacturing," Macmillan. "Modern Plastics Encyclopedia and Engineers Handbook," Plastics Catalogue Corp., New York. "The Tool Engineers Hand-book," McGraw-Hill. Bridgman, "Large Plastic Flow and Fracture," McGraw-Hill. "Cold Working of Metals," ASM. Pearson, "The Extrusion of Metals," Wiley.

PLASTIC WORKING TECHNIQUES

In the **metalworking** operations, as distinguished from metal cutting, material is forced to move into new shapes by plastic flow. **Hot-working** is carried on above the recovery temperature, and spontaneous recovery, or annealing, occurs about as fast as the properties of the material are altered by the deformation. This process is limited by the chilling of the material in the tools, scaling of the material, and the life of the tools at the required temperatures. **Cold-working** is carried on at room temperature and may be applied to most of the common metals. Since, in most cases, no recovery occurs at this temperature, the properties of the metal are altered in the direction of increasing strength and brittleness throughout the working process, and there is consequently a limit to which cold-working may be carried without danger of fracture (Fig. 4.12.1).

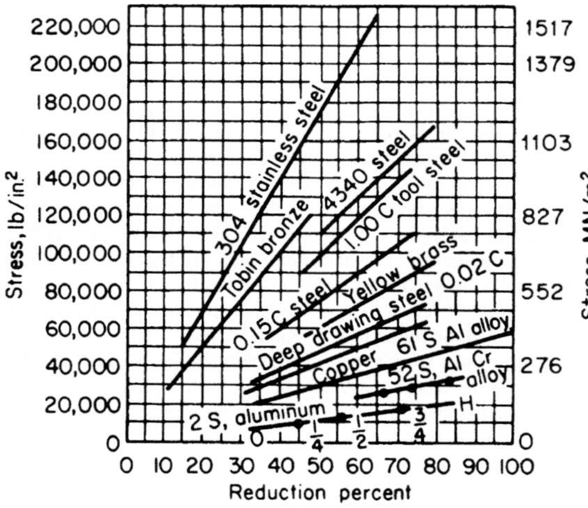

Figure 4.12.1 Plastic range chart of commonly worked metals.

ROLLING OPERATIONS

Rolling of sheets, coils, bars, and shapes is a primary process using plastic ranges both above and below recrystallization to prepare metals for further working or for fabrication. Metal squeezed in the bite area of the rolls moves out lengthwise with very little spreading in width. This compressive working above the yield point of the metal may be aided in some cases by maintaining a substantial tensile strain in the direction of rolling.

A cast or forged billet or slab is preheated for the preliminary breakdown stage of rolling, although considerable progress has been made in continuous casting, in which the molten metal is poured continuously into a mold in which the metal is cooled progressively until it solidifies (albeit still at high temperature), whence it is drawn off as a quasi-continuous billet and fed directly into the first roll pass of the rolling mill. The increased speed of operation and production and the increased efficiency of energy consumption are obvious. Most new mills, especially minimills, have incorporated continuous casting as the normal method of operation. A reversing hot mill may achieve 5,000 percent elongation of an original billet in a series of manual or automatic passes. Alternatively, the billet may pass progressively through, say, 10 hot mills in rapid succession. Such a production setup requires precise control so that each mill stand will run enough faster than the previous one to make up for the elongation of the metal that has taken place. Hot-rolled steel may be sold for many purposes with the black mill scale on it. Alternatively, it may be acid-pickled to remove the scale and treated with oil or lime for corrosion protection. To prevent scale from forming in hot-rolling, a nonoxidizing atmosphere may be maintained in the mill area, a highly special plant design (Fig. 4.12.2).

Pack rolling of a number of sheets stacked together provides means of retaining enough heat to hot-roll thin sheets, as for high-silicon electric steels.

Cold-rolling is practical in production of thin coil stock with the more ductile metals. The number of passes or amount of reduction between anneals is determined by the rate of work hardening of the metal. Successive stands of cold-rolling help to retain heat generated in working. Tension provided by mill reels and between stands helps to increase the practical reduction per step. Bright annealing in a controlled atmosphere avoids surface pockmarks, which are difficult to get out. For high-finish stock, the rolls must be maintained with equal finish.

Cold Rolling of Threads and Gears Threaded parts, mostly fasteners, are cold-rolled with special tooling to impart a typical helical thread geometry to the part. The thread profile is most often a standard 60° vee, although thread profiles (i.e., Acme) for power threads are possible and have been produced. In one form, the tooling consists of two tapered reciprocating dies with the desired thread profile cut thereon. A blank of diameter smaller than the OD of the screw is positioned between the dies when they are at maximum separation. Then, as the dies reciprocate and decrease the gap between them, the blank is gripped, rolled, and plastically deformed to the desired screw profile. The indenting dies displace metal upward to form the upper part of the thread profile. There is no waste metal, and the fastener suffers no tears at the root. The resulting cold work and plastic displacement of metal results in a superior product. Subsequent heat treatment of the fastener may follow, depending on the strength properties desired for a particular application.

Thread rolling is also accomplished by using a nest of three profiled rotating rollers. In that case, the blank is fed axially when the rollers are at maximum gap and then is plastically deformed as the roll gap closes during rotation.

Gear profiles also can be rolled. The action is similar to that described for rolling threads, except that the dies or rollers are profiled

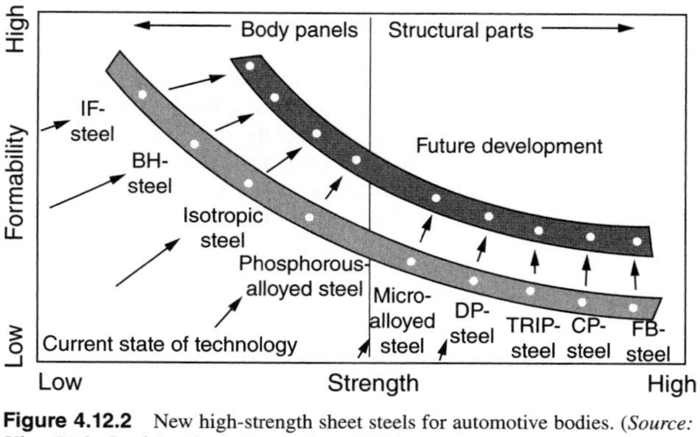

Figure 4.12.2 New high-strength sheet steels for automotive bodies. (*Source: Ultra Light Steel Autobody Consortium, USA.*)

to impart the desired involute gear profile. The gear blank is caught between the rollers or dies, and conjugate action ensues between the blank and tooling as the blank progresses through the operating cycle. In some applications, the rolled gear is produced slightly oversize to permit a finer finish by subsequent hobbing. The advantage lies in the reduction of metal cut by the hob, thereby increasing the production rate as well as maintaining the beneficial cold-worked properties imparted to the metal by cold rolling.

Thread and gear rolling enables high production rates; most threaded fasteners in production are of this type.

Protective coating is best exemplified by high-speed tinplate mills in which coil stock passes continuously through the necessary series of cleaning, plating, and heating steps. Zinc and other metals are also applied by plating but not on the same scale. **Clad** sheets (high-strength aluminum alloys with pure aluminum surface for protection against electrolytic oxidation) are produced by rolling together; an aluminum alloy billet is hot-rolled together with plates of pure aluminum above and below it through a series of reducing passes, with precautions to ensure clean adhesion.

On the other hand, prevention of adhesion, as by a separating film, is essential in the final stages of **foil rolling,** where two coils may have to be rolled together. Such foil may then be **laminated** with suitable adhesive to paper backing materials for wrapping purposes. (See also Sec. 4.)

Shape-rolling of structural shapes and rails is usually a hot operation with roll-pass contours designed to distribute the displacement of metal in a series of steps dictated largely by experience. **Contour rolling** of relatively thin stock into tubular, channel, interlocking, or varied special cross sections is usually done cold in a series of roll stands for lengthwise bending and setting operations. There is also a wide range of simple bead-rolling, flange-rolling, and seam-rolling operations in relatively thin materials, especially in connection with the production of barrels, drums, and other containers.

Oscillating or **segmental rolling** probably developed first in the manually fed contour rolling of agricultural implements. In some cases, the suitably contoured pair of roll inserts or roll dies oscillates before the operator, to form hot or cold metal. In other cases, the rolls rotate constantly, toward the operator. The working contour takes only a portion of the circumference, so that a substantial clearance angle leaves a space between the rolls. This permits the operator to insert the blank to the tong grip between the rolls and against a fixed gage at the back. Then, as rotation continues, the roll dies grip and form the blank, moving it back to the operator. This process is sometimes automated; such units as **tube-reducing mills** oscillate an entire rolling-mill assembly and feed the work over a mandrel and into the contoured rolls, advancing it and possibly turning it between reciprocating strokes of the roll stands for cold reduction, improved concentricity, and, if desired, the tapering or forming of special sections.

Spinning operations (Fig. 4.12.3) apply a rolling-point pressure to relatively limited-lot production of cup, cone, and disk shapes, from floor lamps and TV tube housings to car wheels and large tank ends. Where substantial metal thickness is required, powerful machines and hydraulic servo controls may be used. Some of the large, heavy sections and difficult metals are spun hot.

Rolling operations are distinguished by the relatively rapid and continuous application of working pressure along a limited line of contact. In determining the working area, consider the lineal dimension (width of coil), the bite (reduction in thickness), and the roll-face deflection, which tends to increase the contact area. Approximations of rolling-mill load and power requirements have been worked out in literature of the AISE and ASME.

SHEARING

The shearing group of operations includes such **power press operations** as blanking, piercing, perforating, shaving, broaching, trimming, slitting, and parting. Shearing operations traverse the entire plastic range of metals to the point of failure.

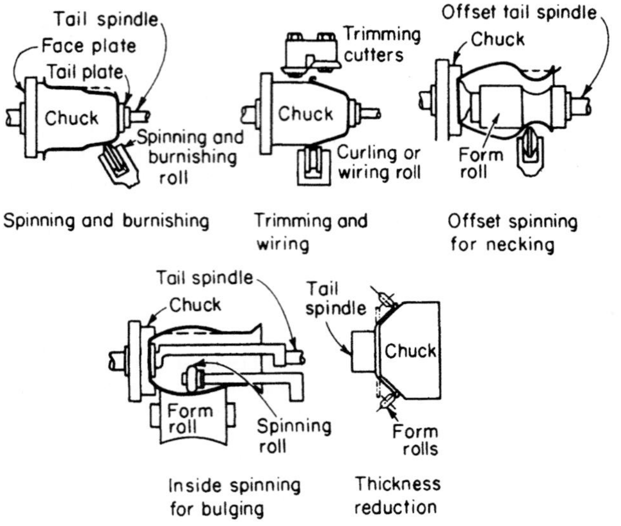

Figure 4.12.3 Spinning operations.

The maximum pressure *P*, in pounds, required in shearing operations is given by the equation $P = \pi DtS = Lts$, where *s* is the resistance of the material to shearing, lb/in^2; *t* is the thickness of the material, in; *L* is the length of cut, in, which is the circumference of a round blank πD or the periphery of a rectangular or irregular blank. Approximate values of *s* are given in Table 4.12.1.

Shear (Fig. 4.12.4) is the advance of that portion of the shearing edge which first comes in contact with the material to be sheared over the last portion to establish contact, measured in the direction of motion. It should be a function of the thickness *t*. Shear reduces the maximum pressure because, instead of shearing the whole length of cut at once, the shearing action takes place progressively, shearing at only a portion of the length at any instant. The maximum pressure for any case where the shear is equal to or greater than *t* is given by $P_{max} = P_{av}\ t/\text{shear}$, where P_{av} is the average value of the pressure on a punch, with shear = *t*, from the time it strikes the metal to the time it leaves.

Distortion results from shearing at an angle (Fig. 4.12.4) and accordingly, in blanking, where the blank should be flat, the punch should be flat, and the shear should be on the die. Conversely, in hole punching, where the scrap is punched out, the die should be flat and the shear should be on the punch. Where there are a number of punches, the effect of shear may be obtained by stepping the punches.

Crowding results during the plastic deformation period, before the fracture occurs, in any shearing operation. Accordingly, when small delicate punches are close to a large punch, they should be stepped shorter than the large punch by at least a third of the metal thickness.

Clearance between the punch and die is required for a clean cut and durability. An old rule of thumb places the clearance all around the punch at 8 to 10 percent of the metal thickness for soft metal and up to 12 percent for hard metal. Actually, hard metal requires less clearance for a clean fracture than soft, but it will stand more. In some cases, with delicate punches, clearance is as high as 25 percent. Where the hole diameter is important, the punch should be the desired diameter and the clearance should be added to the die diameter. Conversely, where the blank size is important, the die and blank dimensions are the same and clearance is deducted from the punch dimensions.

The **work per stroke** may be approximated as the product of the maximum pressure and the metal thickness, although it is only about 20 to 80 percent of that product, depending upon the clearance and ductility of the metal. Reducing the clearance causes secondary fractures and increases the work done. With sufficient clearance for a clean fracture, the work is a little less than the product of the maximum pressure, the

Table 4.12.1 Approximate Resistance to Shearing in Dies

Material	Annealed state		Hard, cold-worked	
	Resistance to shearing, lb/in²*	Penetration to fracture, percent	Resistance to shearing, lb/in²*	Penetration to fracture, percent
Lead	3,500	50	Anneals at room temperature	
Tin	5,000	40	Anneals at room temperature	
Aluminum 2S, 3S	9,000-11,000	60	13,000-16,000	30
Aluminum 52S, 61S, 62S	12,000-18,000	. . .	24,000-30,000	
Aluminum 75S	22,000	. . .	46,000	
Zinc	14,000	50	19,000	25
Copper	22,000	55	28,000	30
Brass	33,000-35,000	50-55	52,000	25-30
Bronze 90–10	. . .	. . .	40,000	
Tobin bronze	36,000	25	42,000	
Steel 0.10C	35,000	50	43,000	38
Steel 0.20C	44,000	40	55,000	28
Steel 0.30C	52,000	33	67,000	22
Steel 0.40C	62,000	27	78,000	17
Steel 0.60C	80,000	20	102,000	9
Steel 0.80C	97,000	15	127,000	5
Steel 1.00C	115,000	10	150,000	2
Stainless steel	57,000	39		
Silicon steel	65,000	30		
Nickel	35,000	55		

* 1,000 lb/in² = 6.895 MN/m².

NOTE: Available test data do not agree closely. The above table is subject to verification with closer control of metal analysis, rolling and annealing conditions, and die clearances. In dinking dies, steel-rule dies, hollow cutters, etc., cutting-edge resistance is substantially independent of thickness: cotton glove cloth (stack, 2 or 3 in thick), 240 lb/in; kraft paper (stack tested, 0.20 in thick), 385 lb/in, celluloid [in thick, warmed in water to 120 to 150°F (49 to 66°C)], 300 lb/in.

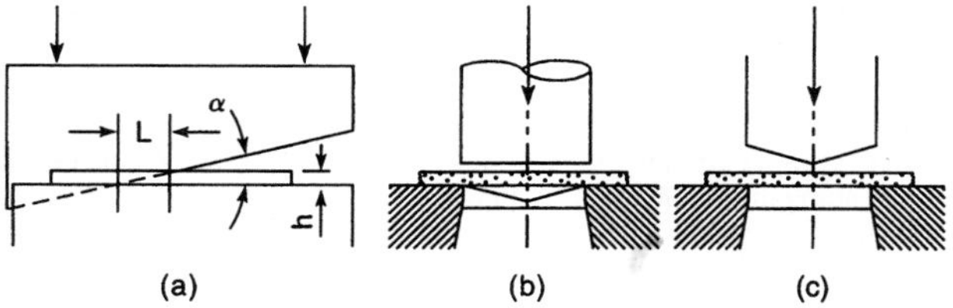

α = shearing angle
t = h = thickness of sheet
L = shearing length

Figure 4.12.4 Shearing forces can be reduced by providing a rake or shear on (*a*) the blades in a guillotine, (*b*) the die in blanking, (*c*) the punch in piercing. (*J. Schey, "Introduction to Manufacturing Processes," McGraw-Hill, 1987.*)

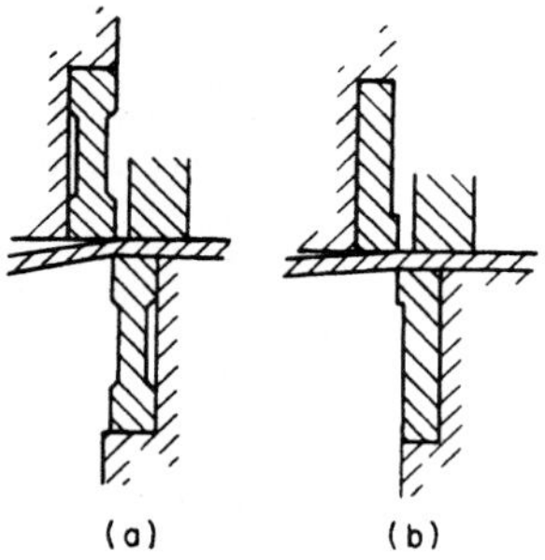

Figure 4.12.5 Squaring shears.

metal thickness, and the percentage reduction in thickness at which the fracture occurs. Approximate values for this are given in Table 4.12.1. The **power required** may be obtained from the work per stroke plus a 10 to 20 percent friction allowance.

Shaving A sheared edge may be squared up roughly by shaving once, allowing for the shaving of mild sheet steel about 10 percent of the metal thickness. This allowance may be increased somewhat for thinner material and should be decreased for thicker and softer material. In making several cuts, the amount removed is reduced each time. For extremely fine finish a round-edged burnishing die or punch, say 0.001 or 0.0015 in tight, may be used. Aluminum parts may be blanked (as for impact extrusion) with a fine finish by putting a 30° bevel, approx one-third the metal thickness on the die opening, with a near metal-to-metal fit on the punch and die, and pushing the blank through the highly polished die.

Squaring shears for sheet or plate may have their blades arranged in either of the ways shown in Fig. 4.12.5. The square-edged blades in Fig. 4.12.5*a* may be reversed to give four cutting edges before they are reground. Single-edged blades, as shown in Fig. 4.12.5*b*, may have a clearance angle on the side where the blades pass, to reduce the working friction. They may also be ground at an angle or rake, on the face which comes in contact with the metal. This reduces the bending and consequent distortion at the edge. Either type of blade distorts also in the other direction owing to the angle of shear on the length of the blades (see Fig. 4.12.4).

Circular cutters for slitters and circle shears may also be square-edged (on most slitters) or knife-edged (on circle shears). According to one rule, their diameter should be not less than 70 times the metal thickness.

Knife-edge hollow cutters working against end-grain maple blocks represent an old practice in cutting leather, rubber, and cloth in multiple thicknesses. **Steel-rule dies,** made up of knife-edge hard-steel strip economically mounted against a steel plate in a wood matrix with rubber strippers and cutting against hard saw-steel plates, extend the practice to corrugated-carton production and even some limited-lot metal cutting.

Higher precision is often required in **finish shearing** operations on sheet material. For ease of subsequent operations and assembly, the cut edges should be clean (acceptable burr heights and good surface finish) and perpendicular to the sheet surface. The processes include precision or fine blanking, negative clearance blanking, counterblanking, and

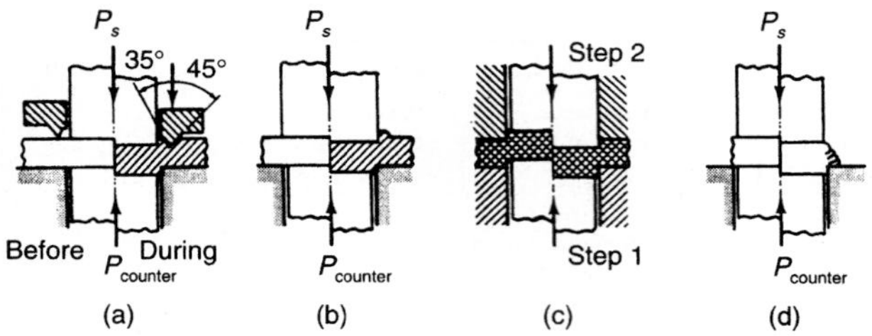

Figure 4.12.6 Parts with finished edges can be produced by (*a*) precision blanking, (*b*) negative-clearance blanking, (*c*) counterblanking, (*d*) shaving a previously sheared part. (*J. Schey, "Introduction to Manufacturing Processes," McGraw-Hill, 1987.*)

shaving, as shown in Fig. 4.12.6. By these methods either the plastic behavior of material is suppressed or the plastically deformed material is removed.

BENDING

The bending group of operations is performed in **presses** (variety), **brakes** (metal furniture, cornices, roofing), **bulldozers** (heavy rolled sections), **multiple-roll forming machines** (molding, etc.), **draw benches** (door trim, molding, etc.), **forming rolls** (cylinders), and **roll straighteners** (strips, sheets, plates).

Spring back, due to the elasticity of the metal and amount of the bend, may be compensated for by overbending or largely prevented by striking the metal at the radius with a **coining** (i.e., squeezing, as in production of coins) pressure sufficient to set up compressive stresses to counterbalance surface tensile stresses. A very narrow bead may be used to localize the pinch where needed and minimize danger to the press in squeezing on a large area. Under such conditions, good sharp bends in V dies have been obtained with two to four times the pressure required to shear the metal across the same section.

These are illustrated in Fig. 4.12.7, where P_b is the **bending load** on the press brake, W_b is the width of the die support, and $P_{counter}$ is the counterload. The bending load can be obtained from

$$P_b = wt^2(\text{UTS})/W_b$$

where t and w are the sheet thickness and width, respectively, and UTS is the ultimate tensile strength of the sheet material.

Bending Allowance The thickness of the metal over a small radius or a sharp corner is 10 or 15 percent less than before bending because the metal moves more easily in tension than in compression. For the same reason the neutral axis of the metal moves in toward the center of the corner radius. Therefore, in figuring the length of blank L to be allowed for the bend up to an inside radius r of two or three times the metal thickness t, the length may be figured closely as along a neutral line at $0.4t$ out from the inside radius. Thus, with reference to

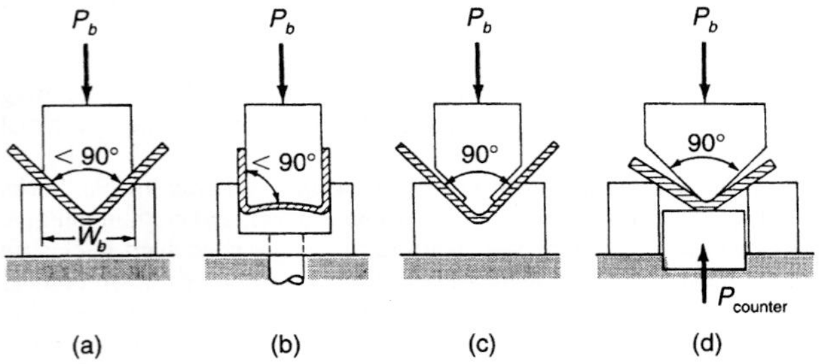

Figure 4.12.7 Springback may be neutralized or eliminated by (*a*), (*b*) overbending; (*c*) plastic deformation at the end of the stroke; (*d*) subjecting the bend zone to compression during bending. *[Part (d) after V. Kupka, T. Nakagawa, and H. Tyamoto, CIRP 22:73-74 (1973).] (Source: J. Schey, "Introduction to Manufacturing Processes," McGraw-Hill, 1987.)*

Fig. 4.12.8, for any angle a in deg and other dimensions in inches, $L = (r + 0.4t)2\pi a/360 = (r + 0.4t)a/57.3$.

The factor $0.4t$, which locates the neutral axis, is subject to some variation (say 0.35 to $0.45t$) according to radius, condition of metal, and angle. In figuring allowances for sharp bends, note that the metal builds up on the compression side of the corner. Therefore, in locating the neutral axis, consider an inside radius r of about $0.05t$ as a minimum.

Roll straighteners work on the principle of bending the metal beyond its elastic limit in one direction over rolls small enough in diameter, in proportion to the metal thickness, to give a permanent set, and then taking that bend out by repeatedly reversing it in direction and reducing it in amount. Metal is also straightened by gripping and stretching it beyond its elastic limit and by hammering; the results of the latter operation depend entirely upon the skill of the operator.

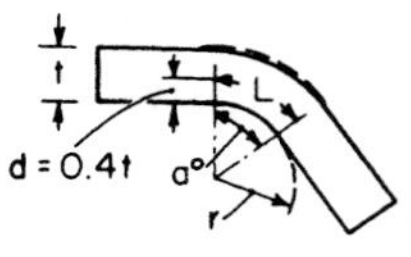

Figure 4.12.8 Bending allowance.

For approximating **bending loads,** the beam formula may be used but must be very materially increased because of the short spans. Thus, for a span of about 4 times the depth of section, the bending load is about 50 percent more than that indicated by the beam formula. It increases from this to nearly the shearing resistance of the section where some **ironing** (i.e., the thinning of the metal when clearance between punch and die is less than the metal thickness) occurs. Where hit-home dies do a little coining to "set" the bend, the pressure may range from two or three times the shearing resistance, and with striking beads and proper care, up to very much higher figures.

The work to roll-bend a sheet or plate t in thick with a volume of V in^3, into curved shape of radius r in, is given as $W = CS(t/r)V/48$ ft · lb, in which S is the tensile strength and C is an experience factor between 1.4 and 2.

The **equipment for bending** consists of mechanical presses for short bends, press brakes (mechanical and hydraulic) for long bends, and roll formers for continuous production of profiles. The bends are achieved by bending between tools, wiping motion around a die corner, or bending between a set of rolls. These bending actions are illustrated in Fig. 4.12.9. Complex shapes are formed by repeated bending in simple tooling or by passing the sheet through a series of rolls which progressively bend it into the desired profile. Roll forming is economical for continuous forming for large volume production. Press brakes can be computer-controlled with synchronized feeding and bending as well as spring-back compression.

DRAWING

Drawing includes operations in which metal is pulled or drawn, in suitable containing tools, from flat sheets or blanks into cylindrical cups or rectangular or irregular shapes, deep or shallow. It also includes reducing operations on shells, tube, wire, etc., in which the metal being drawn is pulled through dies to reduce the diameter or size of the shape. All drawing and reducing operations, by an applied tensile stress in the material, set up circumferential compressive stresses which crowd the metal into the desired shape. The relation of the shape or diameter before drawing to the shape or diameter after drawing determines the magnitude of the stresses. Excessive draws or reductions cause thinning or tearing out near the bottom of a shell. Severe cold-drawing operations require very ductile material and, in consequence of the amount of plastic deformation, harden the metal rapidly and necessitate annealing to restore the ductility for further working.

The **pressure used in drawing** is limited to the load to shear the bottom of the shell out, except in cases where the side wall is ironed thinner, when wall friction makes somewhat higher loads possible. It is less than this limit for round shells which are shorter than the limiting height and also for rectangular shells. Drawing occurs only around the corner radii of rectangular shells, the straight sides being merely free bending.

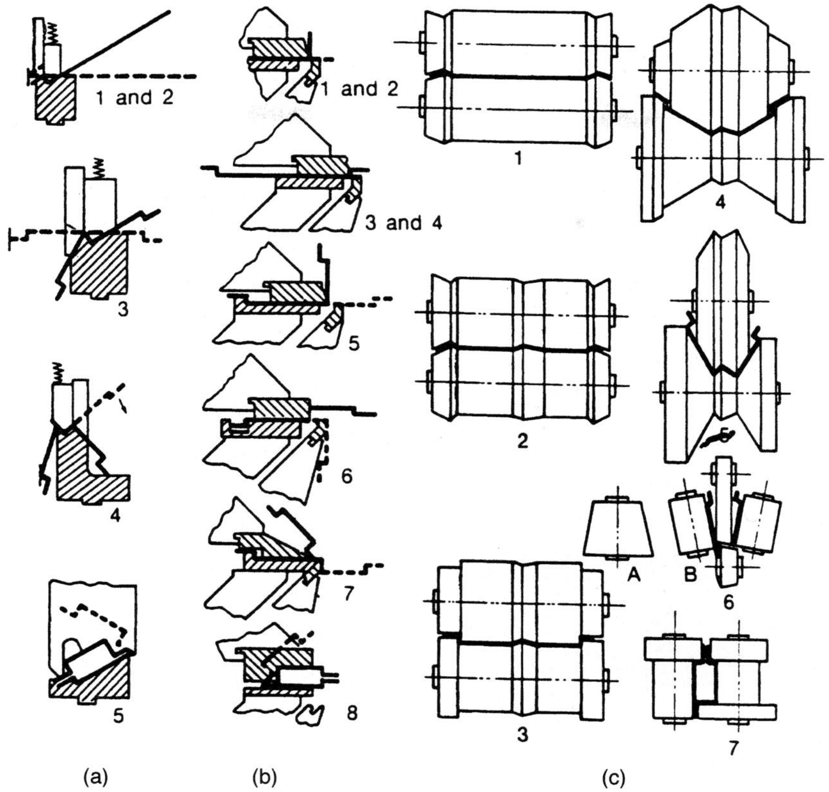

Figure 4.12.9 Complex profiles formed by a sequence of operations on (*a*) press brakes, (*b*) wiping dies, (*c*) profile rolling. *(After Oehler; Biegen, Hanser Verlag, Munich, 1963.) (Source: J. Schey, "Introduction to Manufacturing Processes," McGraw-Hill, 1987.)*

A holding pressure is required in most initial drawing and some redrawing, to prevent the formation of wrinkles due to the circumferential compressive stresses. Where the blank is relatively thin compared with its diameter, the blank-holding pressure for round work is likely to vary up to about one-third of the drawing pressure. For material heavy enough to provide sufficient internal resistance to wrinkling, no pressure is required. Where a drawn shape is very shallow, the metal must be stretched beyond its elastic limit in order to hold its shape, making it necessary to use higher blank-holding loads, often in excess of the drawing pressure. To grip the edges sufficiently to do this, it is often advisable to use **draw beads** on the blank-holding surfaces if sufficient pressure is available to form these beads.

In sheet/deep-drawing practice, the punch force P can be approximated by

$$P = \pi D_p t_0 (\text{UTS})(D_0/D_p - 0.7)$$

where t_0 is the blank thickness and, D_0 and D_p are the diameters of the blank and the punch.

The blank holder pressure for avoiding defects such as wrinkling of bottom/wall tear-out is kept at 0.7 to 1.0 percent of the sum of the yield and the UTS of the material. Punch/die clearances are chosen to be 7 to 14 percent greater than the sheet blank thickness t_0. The die corner radii are chosen to avoid fracture at the die corner from puckering or wrinkling. Recommended values of D_0/t_0 for deep-drawn cups are 6 to 15 for cups without flange and 12 to 30 for cups with flange. These values will be smaller for relatively thick sheets and larger for very thin blank thickness. For deeper-drawn cups, they may be redrawn or reverse-drawn, the latter process taking advantage of strain softening on reverse drawing. When the material has marked strain-hardening propensities, it may be necessary to subject it to an intermediate annealing process to restore some of its ductility and to allow progression of the draws to proceed.

Some shells, which are very thick or very shallow compared with their diameter, do not require a blank holder. Blank-holding pressure may be obtained through toggle, crank, or cam mechanisms built into the machine or by means of air cylinders, spring-pressure attachments, or rubber bumpers under the bolster plate. The length of car springs should be about 18 in/in (18 cm/cm) of draw to give a fairly uniform drawing pressure and long life. The use of car springs has been largely superseded by hydraulic and pneumatic cushions. Rubber bumpers may be figured on a basis of about 7.5 lb/in^2 (50 kN/m^2) of cross-sectional area per 1 percent of compression. In practice they should never be loaded beyond 20 percent compression, and as with springs, the greater the length relative to the working stroke, the more uniform is the pressure.

Deep Drawing and Hydroforming of Sheet Metal Parts Sheet metal parts are conventionally deep drawn using rigid steel punches and dies (Fig. 4.12.10). An alternative approach is to use flexible media (Fig. 4.12.11) such as water, gas, or rubber as either the male or female die, and to perform the hydroforming process in a closed die. In hydromechanical deep drawing, the die is replaced by a fluid (Fig. 4.12.11) while in high-pressure sheet metal hydroforming the punch is replaced by the fluid medium. The use of flexible media often permits greater drawability and the possibility of combining many steps in one operation, such as permitting joining and trimming simultaneously with forming (Fig. 4.12.12). Complex parts can be made in a single step by using thin sheets, thus reducing the cost and weight of the shell structure.

Dimensions of Drawn Shells The smallest and deepest round shell that can be drawn from any given blank has a diameter d of 65 to

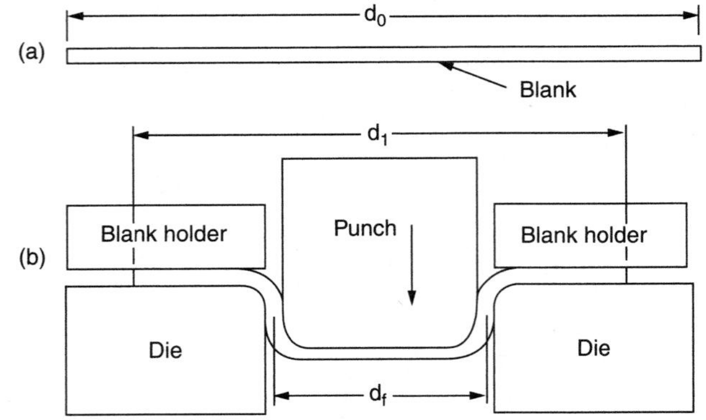

Figure 4.12.10 Schematic of the conventional deep drawing process for sheet metal parts. (*a*) Initial blank, (*b*) drawing in process.

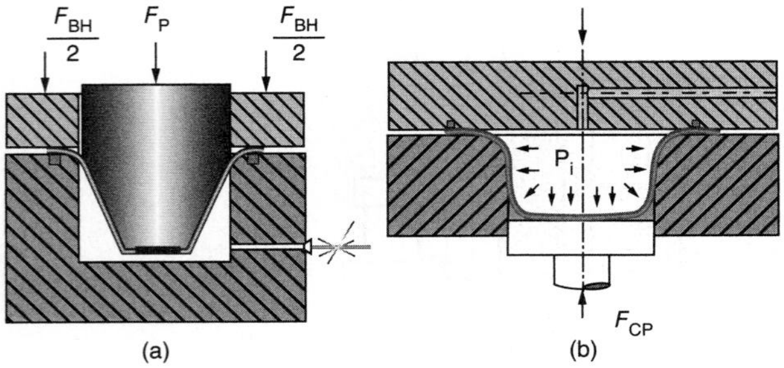

Figure 4.12.11 Forming by flexible media. (*a*) Hydromechanical deep drawing (*Source: K. Seigert and M. Aust, Hydromechanical Deep-Drawing, Production Engineering,* VII/2, *Annals of the German Academic Society for Production Engineering, pp. 7-12.)* (*b*) High-pressure hydroforming. *M. Kliener, W. Homberg, and A. Brosius, Processes and Control of Sheet Metal Hydroforming. International Conference on Advanced Technology of Plasticity, Germany, 2, 1999, pp. 1,243-1,252.)*

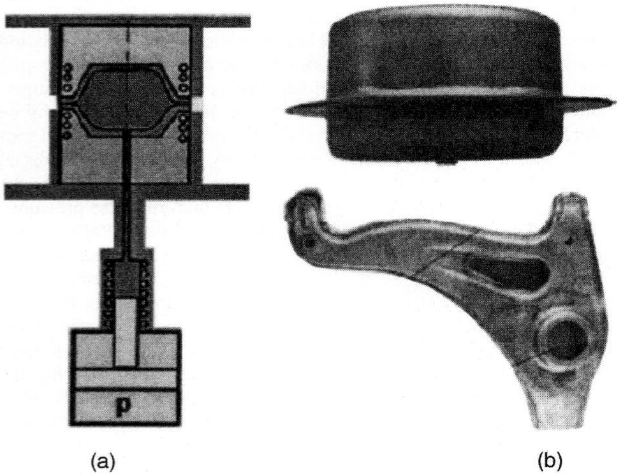

Figure 4.12.12 Combined hydroforming, joining, and trimming of sheet metal parts. (*a*) In process; (*b*) typical parts. (*Source: P. Hein and M. Geiger, Advanced Process Control Strategies for the Hydroforming of Sheet Metal Pairs, Int. Conf. Adv. Techn. Plasticity, Germany,* **2**, 1999, pp. 1,267-1,272.)

50 percent of the blank diameter D. The height of these shells is $h = 0.35d$ to $0.75d$, approximately. Higher shells have occasionally been drawn with ductile material and large punch and die radii. Greater thickness of material relative to the diameter also favors deeper drawing.

The area of the bottom and of the side walls added together may be considered as equal to the area of the blank for approximations. If the punch radius is appreciable, the area of a neutral surface about $0.4t$ out from the inside of the shell may be taken for approximations. Accurate blank sizes may be obtained only by trial, as the metal tends to thicken toward the top edge and to get thinner toward the bottom of the shell wall in drawing.

Approximate diameters of blanks for shells are given by the expression $\sqrt{d^2 + 4dh}$, where d is the diameter and h the height of the shell.

In **redrawing** to smaller diameters and greater depths the amount of reduction is usually decreased in each step. Thus in double-action redrawing with a blank holder, the successive reductions may be 25, 20, 16, 13, 10 percent, etc. This progression is modified by the relative thickness and ductility of the metal. Single-action redrawing without a blank holder necessitates smaller steps and depends upon the shape of the dies and punches. The steps may be 19, 15, 12, 10 percent, etc. Smaller reductions per operation seem to make possible greater total reductions between annealings.

Rectangular shells may be drawn to a depth of 4 to 6 times their corner radius. It is sometimes desirable, where the sheet is relatively thin, to use draw beads at the corners of the shell or near reverse bends in irregular shapes to hold back the metal and assist in the prevention of wrinkles.

Work in drawing is approximately the product of the length of the draw, and the maximum punch pressure, as the load rises quickly to the peak, remains fairly constant, and drops off sharply at the end of the draw unless there is stamping or wall friction. To this, add the work of blank holding which, in the case of cam and toggle pressure, is the product of the blank-holding pressure and the spring of the press at the pressure (which is small). For single-action presses with spring, rubber, or air-drawing attachments it is the product of the average blank-holding pressure and the length of draw.

Rubber-die forming, especially of the softer metals and for limited-lot production, uses one relatively hard member of metal, plaster, or plastic with a hard powder filler to control contour. The mating member may be a rubber or neoprene mattress or a hydraulically inflatable bag, confined and at 3,000 to 7,000 lb/in^2 (20 to 48 MN/m^2). Babbitt, oil, and water have also been used directly as the mobile member. A large hydraulic press is used, often with a sliding table or tables, and even static containers with adequate pumping systems.

Warm Forming of Aluminum and Magnesium Sheets Aluminum and magnesium are used to decrease the weight of automotive and aerospace parts. Aluminum and magnesium exhibit increased ductility at elevated temperatures (Fig. 4.12.13). Magnesium does have many limitations, but its use for structural parts is growing.

Tailor Welded Blanks (TWB) in Forming With the demand for reducing both automotive structure weight and manufacturing costs many new processes are being employed:

1. Different parts are formed separately and then joined by laser. Part forming is independently optimized, followed by trimming and weld assembly. Forming is easier but welding along curved lines is more complex (Fig. 4.12.14*a*).

2. The blanks are welded, and then the panel is formed in one die. Welding is simpler, but forming is considerably more complex. Dimensional tolerances are better controlled (Fig. 4.12.14*b*).

3. Blanks of varying thickness are tailor-welded using laser techniques to create a single blank that subsequently is formed into the required geometry (Fig. 4.12.15). The forming process of TWBs is very complex as the blank areas with different thicknesses flow differently during the drawing operation.

4. Friction stir welding is used to create a single blank made of different materials or different sheet thickness. This composite blank is then drawn to final geometry (Fig. 4.12.16).

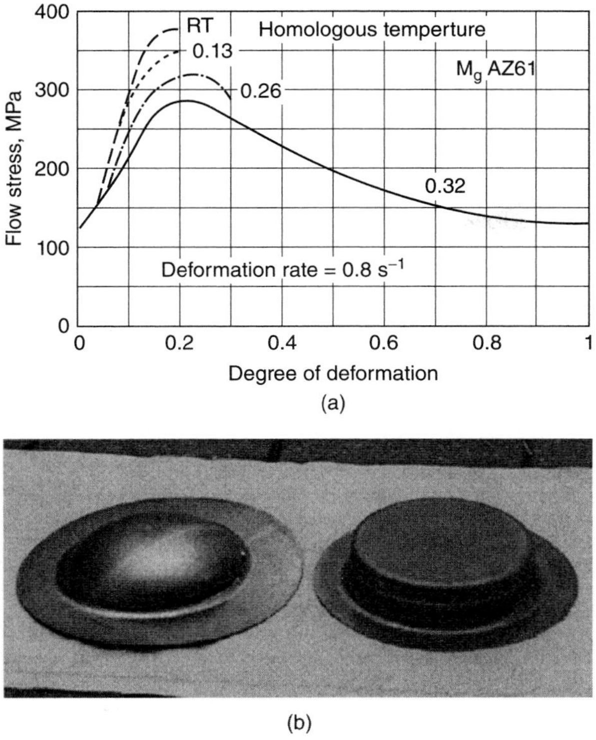

Figure 4.12.13 Drawing of magnesium sheets at elevated temperatures. (*a*) Flow stress of magnesium AZ61. *[Chabbi, et al., 2000, Hot and Cold Forming Behavior of Magnesium Alloys AZ31 and AZ61, in K. U. Kainer (ed.), "Magnesium Alloys and Their Applications," Wiley-VCH, D, pp. 621-627.]* (*b*) Drawn AZ31B cups at room temperature (left) and at 230°C (right). *(S. Novotni, Innenhochdruk-Umformen von Belchen aus Aluminium-und Magnesiumlegierungen bei erhohter Temperatur, Ph.D. thesis, Elrangen University, Germany.)*

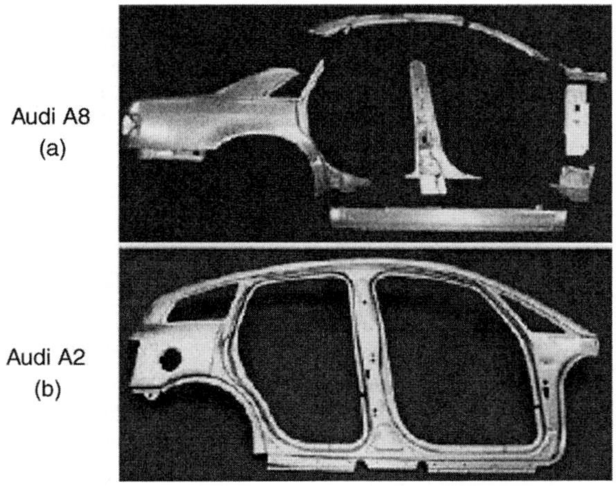

Figure 4.12.14 (*a*) Many parts are formed separately and joined to form side panels and pillars; (*b*) single blank is trimmed and then formed. *(Source: Ruch et al., Grobserienfertigung von Aluminumkarosserien, in K. Siegert, "Neuere Entwicklungen inder Blechumformung," MATINFO, Frankfurt, D, 2000.)*

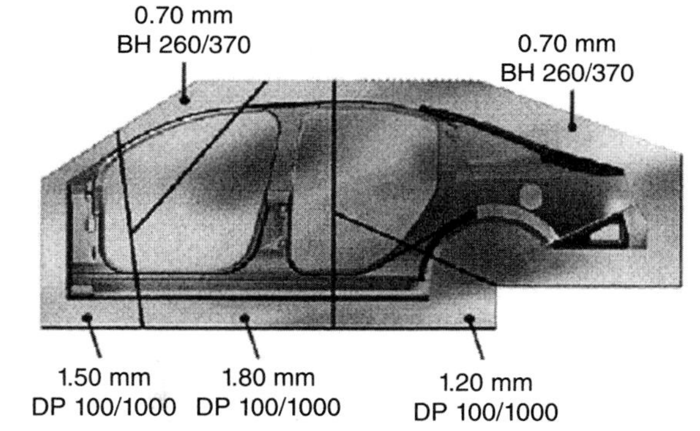

Figure 4.12.15 Use of tailor-welded blanks in automotive side panels using different materials of unequal thickness. *(Source: Ultra Light Steel Autobody Consortium, USA.)*

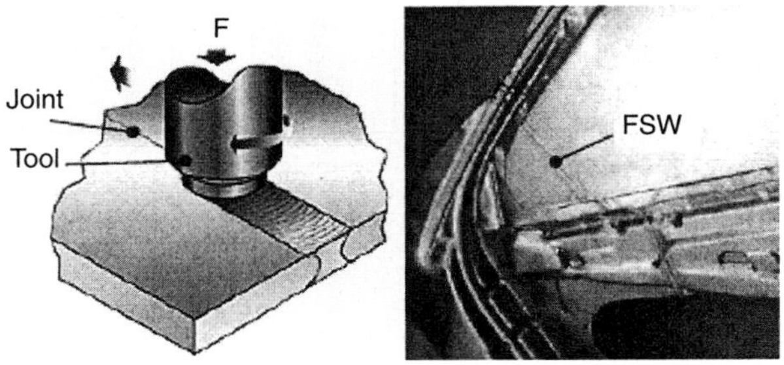

Figure 4.12.16 Use of friction stir welding to create a blank. *(Source: The Welding Institute.)*

Hot drawing above the recrystallization range applies single- and double-action drawing principles. For light gages of plastics, paper, and hexagonal-lattice metals such as magnesium, dies and punches may be heated by gas or electricity. For thick steel plate and heat-treatable alloys, the mass of the blank may be sufficient to hold the heat required.

High-Pressure Hydroforming of Tubes Tubes formed to various cross sections and bent to various shapes are widely used in automotive frames. There are a number of variants of this process (Fig. 4.12.17), including forming under (1) tensile and compressive conditions, (2) bending conditions, and (3) shear conditions. Each variant is intended to impart a particular deformation to the tube by a predetermined motion of the tools. Motions include axial compression due to tool motion and circumferential expansion due to internal pressure. In high-pressure forming, pressurized force can reach 35,000 tons. Lead times can be very high due to the slow pressurization and depressurization required for each forming cycle. These high pressures lead to metal flow-related defects, such as buckling, wrinkling, and bursting, and a safe working window has to be determined for success (Fig. 4.12.18).

Incremental Forming The incremental forming process employs local forming of the workpiece and then rotating the workpiece or the tools to form the entire surface. Since the contact area between the tools and the workpiece is small, the force required is small and the friction conditions are favorable. Consequently, high deformation levels are achieved that are not possible by conventional means. These processes include flow forming, radial forming, rotary forming, orbital forming, spinning, shear forming, and incremental sheet forming (single-point or two-point contact). Incremental sheet forming uses sheet blanks held in a fixture while rotating tools incrementally stretch the blank to the required shape (Fig. 4.12.19). In the flow forming process for thin axi-symmetric parts, a roller deforms the metal and changes the thickness of the formed piece (Fig. 4.12.20).

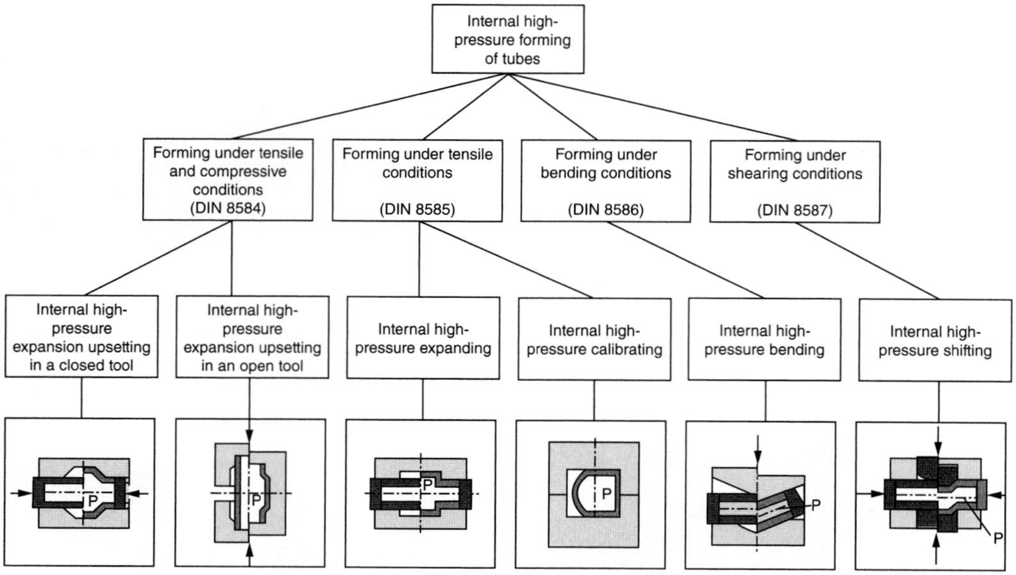

Figure 4.12.17 Process and tooling variations in the hydroforming of tubes with high-pressure media. *(Source: D. Schmoeckel, C. Hielscher, and R. Huber, Metal Forming of Tubes and Sheets with Liquid and Other Flexible Media, CIRP Annals,* **48**, *no. 2, 1999, pp. 497-513.)*

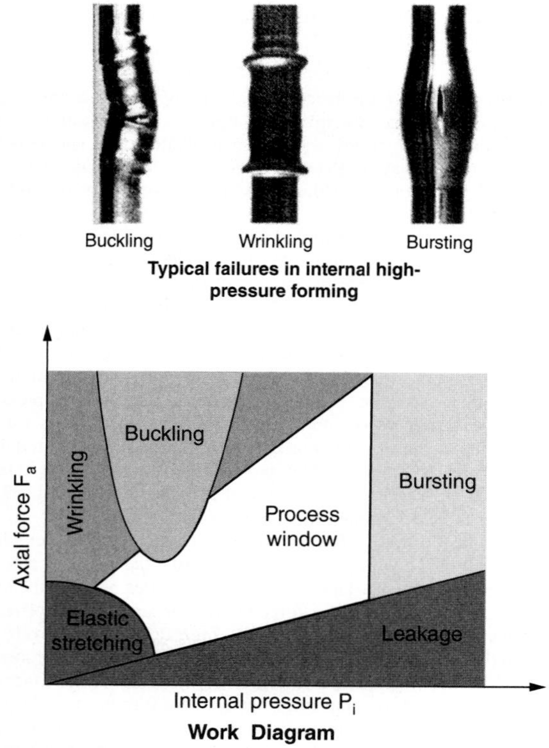

Figure 4.12.18 Typical failures in tube hydroforming and the safe working window for the process. *(Source: D. Schmoeckel, C. Hielscher, and R. Huber, Metal Forming of Tubes and Sheets with Liquid and Other Flexible Media, CIRP Annals,* **48**, *no. 2, 1999, pp. 497-513.)*

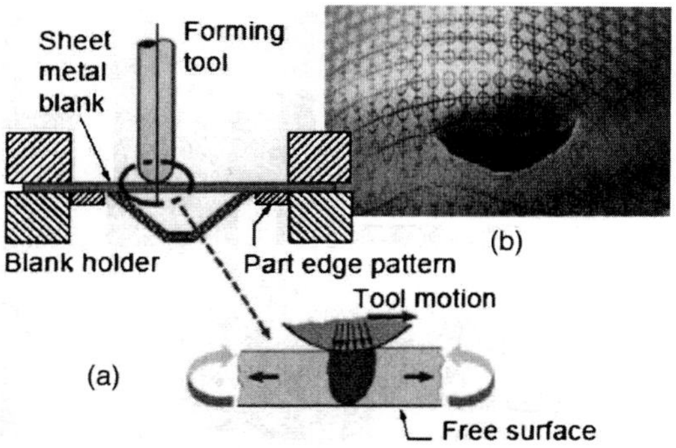

Figure 4.12.19 Incremental forming of thin sheet parts. (*a*) Process principle; (*b*) increased strains possible. *(Source: J. Jesweit and E. Hagan, Rapid Prototyping of a Headlight with Sheet Metal. Trans. NAMRI, Society of Manufacturing Engineers, May 2002.)*

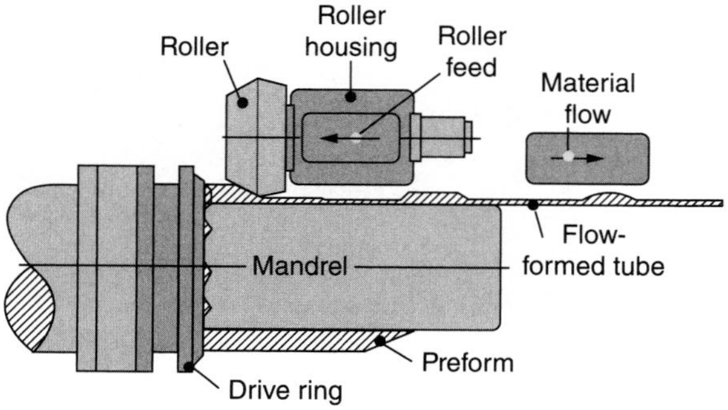

Figure 4.12.20 Flow forming a rim for an automotive wheel.

Lubricants for Presswork Many jobs may be done dry, but better results and longer life of dies are obtained by the use of a lubricant. Lard or sperm oil is used when punching iron, steel, or copper. Petroleum jelly is used for drawing aluminum. A soap solution is commonly used for drawing brass, copper, or steel. One manufacturer uses 90 percent mineral oil, 5 percent rosin, and 5 percent oleic acid for light work and an emulsion of a mineral oil, degras, and a pigment consisting of chalk, sulfur, or lithopone for heavy work. (See also Sec. 4.)

For heavy drawing operations and extrusion, steels may have a zinc phosphate coating bonded on, and a zinc or sodium stearate bonded to that, to withstand pressures over 300,000 lb/in^2 (2,070 MN/m^2). An anodized coating for aluminum may be used as a host for the lubricant. It is reported that such a chemical treatment plus a lacquer or plastic coating and a lubricant is effective for severe ironing operations.

The problem is to prevent local pressure welding from starting as galling or pickup, with resulting scratching, by maintaining a fluid film separation between metal surfaces. At moderate pressures, almost any viscous liquid lubricant will do the job. Rust protection and easy removal of the lubricant are often major factors in the choice.

Lubricants for metalworking are often classified based on their interface friction coefficient, which depends on the workpiece material and the lubricant being used. The **interface friction coefficient** μ is often defined as the ratio of the friction force to the normal force at the interface. Typical values of μ for different workpiece-lubricant pairs are included in Table 4.12.2.

Shock-wave Forming For short runs, it is applied in several ways. Explosive forming, especially for large-area drawn or formed shapes, usually requires one metal contour-control die immersed in a large container of fluid, or even in a lake or pond. Explosives manufacturers have developed means of computing the charge and the distance that it should be suspended above the blank to be formed. The space back of the blank in the die has to be evacuated. A blank-holding ring to minimize wrinkle formation in the flange area is bolted very tightly to the die, with an O-ring seal to prevent leakage.

Electrohydraulic forming is similar to explosive forming except that the shock wave is imparted electrically from a large battery of capacitors. **Magnetic forming** uses the same source of power but does not require a fluid medium. A flexible pancake coil delivers the magnetic shock pulse.

Electromagnetic Forming (EMF) In this process the energy of a pulsed magnetic field is used with a contact-free tool to join metals with good electrical conductivity and magnetic permeability, such as aluminum. The rapid discharge of a high-voltage capacitor through a coil in the tool generates an intense magnetic field, inducing eddy currents in the workpiece. The magnetic forces acting between the tool and the workpiece cause movement of the workpiece, thus permanently changing its shape (Fig. 4.12.21). This process can be used with tubular workpieces to join two tubes or to change tube shape.

BULK FORMING

The squeezing group of operations are those in which the metal is worked in compression. Resultant tensile strains occur, however; in cases where the metal is thin compared with its area and there is an appreciable movement of the metal, there results a pyramiding of pressure toward the center of the die which may prove serious. The metal is incompressible (beyond about 1 percent), and consequently, to reduce the thickness of any volume of metal in the center of the blank, its area must be increased, which involves spreading or stretching all the metal

Table 4.12.2 Typical Lubricants* and Friction Coefficients in Plastic Deformation

Workplace material	Working	Forging		Extrusion† lubricant	Wire drawing		Rolling		Sheet metalworking	
		Lubricant	μ		Lubricant	μ	Lubricant	μ	Lubricant	μ
Sn, Pb, Zn alloys		FO-MO	0.05	FO or soap	FO	0.05	FA-MO or MO-EM	0.05 0.1	FO-MO	0.05
Mg alloys	Hot or warm	GR and/or MoS_2	0.1-0.2	None			MO-FA-EM	0.2	GR in MO or dry soap	0.1-0.2
Al alloys	Hot	GR or MoS_2	0.1-0.2	None			MO-FA-EM	0.2		
	Cold	FA-MO or dry soap	0.1 0.1	Lanolin or soap on PH	FA-MO-EM, FA-MO	0.1 0.03	1-5% FA in MO (1-3)	0.03	FO, lanolin, or FA-MO-EM	0.05-0.1
Cu alloys	Hot	GR	0.1-0.2	None (or GR)			MO-EM	0.2		
	Cold	Dry-soap, wax, of tallow	0.1	Dry soap or wax or tallow	FO-soap-EM, MO	0.1	MO-EM	0.1	FO-soap-EM or FO-soap	0.05-0.1
Steels	Hot	GR	0.1-0.2	GL (100-300), GR			None or GR-EM	ST‡ 0.2	GR	0.2
	Cold	EP-MO or soap, on PH	0.1 0.05	Soap on PH	Dry soap or soap on PH	0.05 0.03	10% FO-EM	0.05	EP-MO, EM, soap, or polymer	0.05-0.1
Stainless steel, Ni and alloys	Hot	GR	0.1-0.2	GL (100-300)			None	ST‡	GR	0.2
	Cold	CL-MO or soap on PH	0.1 0.05	CL-MO or soap on PH	Soap on PH or CL-MO	0.03 0.05	FO-CL-EM or CL-MO	0.1 0.05	CL-MO, soap, or polymer	0.1
Ti alloys	Hot	GL or GR	0.2	GL (100-300)					GR, GL	0.2
	Cold	Soap or MO	0.1	Soap on PH	Polymer	0.1	MO	0.1	Soap or polymer	0.1

* Some more frequently used lubricants (hyphenation indicates that several components are used in the lubricant):
CL = chlorinated paraffin
EM = emulsion; listed lubricating ingredients are finely distributed in water
EP = "extreme-pressure" compounds (containing S, Cl, and P)
FA = fatty acids and alcohols, e.g., oleic acid, stearic acid, stearyl alcohol
FO = fatty oils, e.g., palm oil and synthetic palm oil
GL = glass (viscosity at working temperature in units of poise)
GR = graphite; usually in a water-base carrier fluid
MO = mineral oil (viscosity in parentheses, in units of centipoise at 40°C).
PH = phosphate (or similar) surface conversion, providing keying of lubricant
† Friction coefficients are misleading for extrusion and therefore are not quoted here.
‡ The symbol ST indicates sticking friction.
SOURCE: John A. Schey, *Introduction to Manufacturing Processes,* McGraw-Hill, New York, 1987.

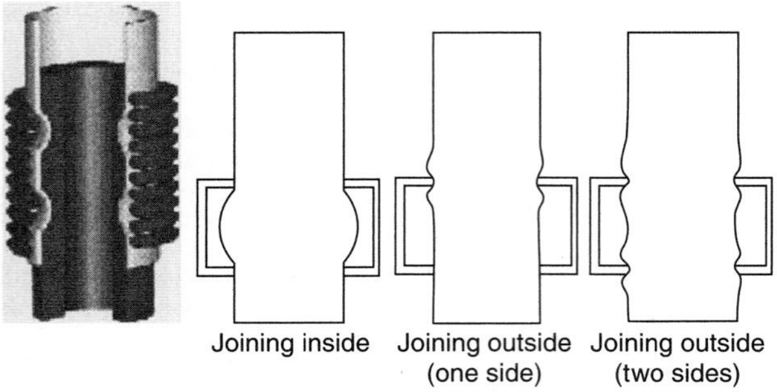

Figure 4.12.21 Electromagnetic forming and joining of tubes. *(Source: C. Beerwald, A. Brosius, and M. Kliener, Determination of Flow Stress at Very High Strain Rates by a Combination of Magnetic Forming and FEM Calculation, International Workshop on Friction and Flow Stress in Cutting and Forming, Paris, pp. 175-182.)*

around it. The surrounding metal acts like shrunk bands and offers a resistance increasing toward the center, and often many times the compressive resistance of the material.

Squeezing operations and particularly the squeezing of steel are practically the severest of all press operations. They may be divided into four general classifications according to severity, although in every group there will be found examples of working to the limit of what the die steels will stand, which may be taken at about 100 tons/in^2. The severer operations, such as cold bottom extrusion and wall extrusion, are limited to the softer metals. Squeezing operations ordinarily require pressure through a very short distance, the pressure starting at the compressive yield strength of the material over the surface being squeezed and rising to a maximum at bottom stroke. This maximum is greatest when the metal is thin compared with its area or when the die is entirely closed as for coining. Care must be taken, on all squeezing operations, in the setting of presses and avoidance of double blanks or extra-heavy blanks as the presses must be stiff. In squeezing solidly across bottom center the mechanical advantage is such that a small difference in thickness or setup can make a very large difference in pressure exerted. For this reason high-speed self-contained hydraulic presses, with automatic pressure-control and size blocks, are now finding favor for some of this work.

Sizing, or the flattening or surfacing of parts of forgings or castings, is usually the least severe of the squeezing group. Tolerances are ordinarily closer than for the milling operations which are supplanted. When extremely close tolerances are required, say plus or minus one-thousandth of an inch (0.025 mm), arrange substantial size blocks to take half or two-thirds of the total load. These take up uniformly the bearing-oil films and any slight deflection of the bed and bolster and minimize the error in springback due to variation in thickness, hardness, and area of the rough forging or casting. The usual amount left for squeezing is $\frac{1}{32}$ to $\frac{1}{16}$ in (0.8 to 1.6 mm). Presses may be selected for this service on a basis of 60 to 80 tons/in^2 (830 to 1,100 MN/m^2), although 100 tons/in^2 (1,380 MN/m^2) is more often used in the automobile trade for reserve capacities. When figuring from experimental results obtained in testing machines, the recorded loads are usually doubled in selecting a press, in order to allow for the difference among the speed of the machines, the positive action, and a safety margin.

Swaging or **cold forging** involves squeezing of the blank to an appreciably different shape. Success in performing such operations on steel usually depends upon squeezing a relatively small area with freedom to flow without restraint. Dies for this work must usually be substantially backed up with hardened steel plates. The edge of the blank after coining is usually ragged and must be trimmed for appearance.

Cold-Forging of Steel Parts More and more parts in machinery and automotive drive components are being cold-forged to net shape.

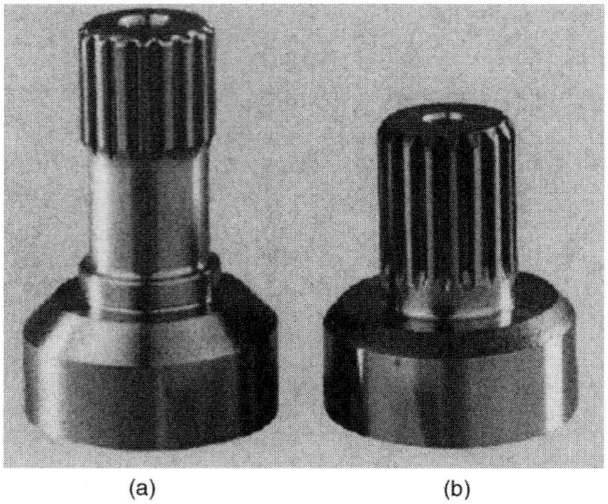

Figure 4.12.22 Comparison of (*a*) machined and (*b*) cold-forged gearbox endpiece. *(Source: D. Landgrebe, Precision Forming and Machining, "Recent Developments in Metalforming Technology," ERC/NSM, The Ohio State University, Columbus, 2002.)*

Due to the work-hardening characteristics of the cold metal, the tool pressures can be several times the flow stress of the material (700 to 2,100 MPa for steels). Consequently, precise control of metal flow, good lubrication, and specially designed and shrink-fitted carbide dies are required. An example of a cold-forged transmission part is shown in Fig. 4.12.22. A machined part is also included for comparison purposes. Note that the hobbed part on the left requires longer run-out clearance that increases the part length. No such clearance is required for the cold-forged spline.

Cold-Forging of Aluminum Parts Cold-forged aluminum parts (Fig. 4.12.23) are increasingly replacing machined steel parts, primarily due to excellent formability and the lower weight of aluminum. A cold-forged aluminum column part is backward can-extruded, followed by successive ironing operations, reducing the wall thickness to 1.5 mm and enabling final forming of the bellow.

Hot forging is similar in certain respects to the above but permits much greater movement of metal. When dies are used, hot forging may be done in drop hammers, percussion presses, power presses, or forging machines. Steam or pneumatic hammers, helve hammers, or hydraulic presses, most often employ plain anvils.

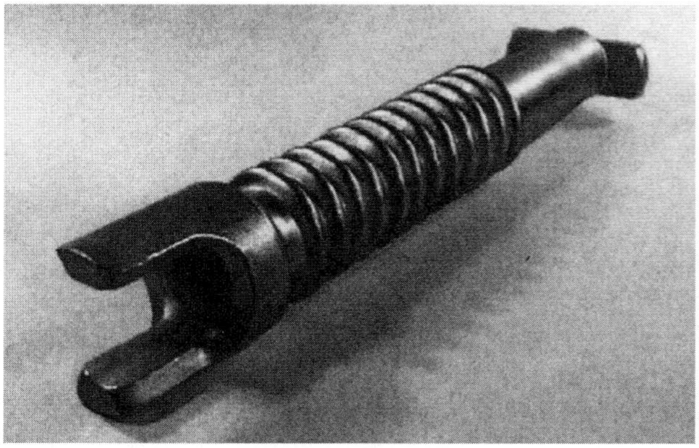

Figure 4.12.23 Automotive steering column manufactured by cold-forging aluminum alloy. *(Source: N. Bay, Cold Forming of Aluminum—State of Art. Jour. Mater. Processing Technol., **71**, 1997, pp. 76-90.)*

Drop hammers are rated according to weight of ram. For carbon steel they may be selected on a basis of 50 to 55 lb of ram weight per square inch of projected area (3.5 to 3.9 kg/cm^2) of the forging, including as much of the flash as is squeezed. This allowance should be increased to 60 lb/in^2 (4.2 kg/cm^2) for 0.20 carbon steel, 70 lb/in^2 (4.9 kg/cm^2) for 0.30 carbon steel, and up to about 130 lb/in^2 (9.2 kg/cm^2) for tungsten steel.

In figuring the **forging pressure,** multiply the projected area of the forging, including the portion of the flash that is squeezed, by approximately one-third of the cold compressive strength of the material. Another method gives the forging pressure at three to four times the compressive strength of the material at forging temperatures times the projected area, for presses; or at ten times the compressive strength at forging temperatures times the projected area, for hammers. The pressure builds up to a rather high figure at bottom stroke owing to the cooling of the metal particularly in the flash and to the small amount of relief for excess metal which the flash allows.

In **heading operations,** hot or cold, the length of wire or rod that can be gathered into a head, without side restraint, in a single operation, is limited to three times the diameter. In coining and then cold heading large heads, wire of about 0.08 carbon must be used to avoid excessive strain-hardening.

Forging Dies Drop-forge dies are usually of steel or steel castings. A good all-around grade of **steel** is a 0.60 percent carbon open-hearth. Dies of this steel will forge mild steel, copper, and tool steel satisfactorily if the number of forgings required is not too large. For a large number of tool-steel forgings, tool-steel dies of 0.80 to 0.90 percent carbon may be used and for extreme conditions, 3½ percent nickel steel.

Die blocks of alloy steels have special value for the production of drop forgings in large quantities. Widely used die materials and their recommended hardness are listed in Table 4.12.3. For a typical hot-working steel, the relationship between the hardness and the UTS is

HRC	*UTS, ksi (MPa)*
30	140 (960)
40	185 (1,250)
50	250 (1,700)
60	350 (2,400)

The pressure on dies can be kept as high as 80 percent of the above values. For higher tool pressures, carbide (tungsten carbide is the most common) die material is used. Carbides can withstand very high compressive pressures but have poor tensile properties. Consequently, carbide dies and inserts are always kept under compression, often by the use of shrink rings. These are some design guidelines for punches and dies:

1. *Long punch.* The punch pressure should be kept below the buckling stress σ_b

$$\sigma_b = \sigma_y\left[1-\left(\frac{4\sigma_y}{\pi^2 E}\right)\left(\frac{L_p}{D_p}\right)^2\right]$$

where σ_y is the yield stress of the punch material, E the elastic modulus 30,000 ksi (210 GPa) for steel and 50,000 ksi (350 GPa) for tungsten carbide, L_p is the punch length and D_p is the punch diameter.

2. *Short punch.* The failure mode is plastic upset. Therefore, the punch pressure should be kept less than the yield stress of the punch material [180 ksi (1,200 MPa) for steel and 500 ksi (3,000 MPa) for tungsten carbide].

3. *Flat platen.* Flat platens fail by plastic yielding. A common formula for calculating acceptable maximum platen pressure p is

$$p = \sigma_y \text{ (diameter of platen/diameter of workpiece)}$$

4. *Die cavity.* The die pressures in a deep cavity (impression) are much higher than in a flat platen (die), resulting in die burst-out. A more conservative value of 150 ksi (1,000 MPa) is used for acceptable die pressure. If shrink rings are used, higher values of 250 ksi (1,700 MPa) for steels and 400 ksi (2,700 MPa) for tungsten carbide can be used for the inner insert.

Table 4.12.3 Typical Die Materials for Deformation Processes*

Process	Die material† and HRC for working			
	Al, Mg, and Cu alloys		Steels and Ni alloys	
Hot forging	6G	30-40	6G	35-45
	H12	48-50	H12	40-56
Hot extrusion	H12	46-50	H12	43-47
Cold extrusion:				
Die	W1, A2	56-58	A2, D2	58-60
	D2	58-60	WC	
Punch	A2, D2	58-60	A2, M2	64-65
Shape drawing	O1	60-62	M2	62-65
	WC		WC	
Cold rolling	O1	55-65	O1, M2	56-65
Blanking	Zn alloy		As for Al, and	
	W1	62-66	M2	60-66
	O1	57-62	WC	
	A2	57-62		
	D2	58-64		
Deep drawing	W1	60-62	As for Al, and	
	O1	57-62	M2	60-65
	A2	57-62	WC	
	D2	58-64		
Press forming	Epoxy/metal powder		As for Al	
	Zn alloy			
	Mild steel			
	Cast iron			
	O1, A2, D2			

* Compiled from "Metals Handbook," 9th ed., vol. 3, American Society for Metals, Metals Park, Ohio, 1980.

† Die materials mentioned first are for lighter duties, shorter runs. Tool steel compositions, percent (representative members of classes):
6 G (prehardened die steel): 0.5 C, 0.8 Mn, 0.25 Si, 1 Cr, 0.45 Mo, 0.1 V
H12 (hot-working die steel): 0.35 C, 5 Cr, 1.5 Mo, 1.5 W, 0.4 V
W1 (water-hardening steel): 0.6-1.4 C
01 (oil-hardening steel): 0.9 C, 1 Mn, 0.5 Cr
A2 (air-hardening steel): 1 C, 5 Cr, 1 Mo
D2 (cold-working die steel): 1.5 C, 12 Cr, 1 Mo
M2 (Mo high-speed steel): 0.85 C, 4 Cr, 5 Mo, 6.25 W, 2 V
WC (tungsten carbide)

SOURCE: John A. Schey, *Introduction to Manufacturing Processes,* McGraw-Hill, New York, 1987.

For large massive dies or for intermittent service, chrome-nickel-molybdenum alloys are preferred. In closed die work, where the dies must dissipate considerable heat, the tungsten steels are preferred, with resulting increase in wear resistance but decrease in toughness.

For very large pieces with deep impressions, **cast-steel** dies are sometimes used. For large dies likely to spring in hardening, 0.85 **carbon steel** high in manganese is sometimes used unhardened.

Good **die-block proportions** for width and depth are as follows:

Width, in (cm)	8 (20)	10 (24)	12 (30)	14 (36)
Depth, in (cm)	6 (15)	7 (18)	7 (18)	7 or 8 (18 or 20)

For ordinary work, 1½ in (4 cm) of metal between impression and edge of block is sufficient.

Dimensions of **dovetailed die shanks:** for hammers up to 1,200 lb (550 kg) 4 in (10 cm) wide and 1⅛ in (3 cm) deep, with sides dovetailed at angles of 6° with the vertical; for hammers from 1,200 to 3,000 lb (550 to 1,360 kg) in size, 6 in (15 cm) wide and 1½ in (4 cm) deep, with 6° angles.

The **minimum draft for the impressions** is 7°, although for parts difficult to draw this may be increased up to 15°. It is not uncommon to have several drafts in the same impression.

Carbon steel and tool-steel **dies** are **hardened** by heating in a carbonizing box packed in charcoal and dipping face downward over a jet of brine. The jet is allowed to strike into the impression, thus freeing the face of steam and producing uniform hardness. After hardening, they are drawn in an oil bath to a temperature of 500 to 550°F (260 to 290°C).

The forging production per pair of dies is largely affected by the size and shape of the impression, the material forged, the material in the dies, the quality of heating of stock to be forged, and the care exercised in use. It may vary from a few hundred pieces to 50,000 or more.

Coining, Stamping, and Embossing The metal is well confined in closed dies in which it is forced to flow to fill the shape. The U.S. Mint gives the following pressures; silver quarter, 100 tons/in^2 (1,380 MN/m^2); nickel (0.25 Ni, 0.75 Cu), 90 tons/in^2 (1,240 MN/m^2); copper cent, 40 tons/in^2 (550 MN/m^2). In stamping designs, lettering, etc., in sheet metal the thickness is so little compared with the area that there is practically no relief for excess pressure. Where sharp designs are required, as in stamping panels, the dies should be arranged to strike on a narrow line [say $\frac{1}{32}$ in (0.8 mm)] around the outline. If a sharp design is not obtained, it is often best to correct deflection in the machine by shimming or more substantial backing. Increasing the pressure only aggravates the condition and may break the press. General practice for light overall stamping is to allow 5 to 10 tons/in^2 (68 to 132 MN/m^2) of area that is to be stamped, except in areas where the yield point must be exceeded.

Extrusion is the severest of the squeezing processes. The metal is forced to flow rapidly through an orifice, being otherwise confined and subject largely to the laws of hydraulics, with allowances for restraint of flow and for work hardening. Power-press **impact extrusion** began with tin and lead collapsible tubes. It has been extended to the backward and forward extrusion of aluminum, brass, and copper in pressure ranges of 30 to 60 or more tons/in^2 (413 to 825 MN/m^2), and mild steel at pressures up to 165 tons/in^2 (2,275 MN/m^2). Hot impact extrusion of steel, as in projectile piercing, ranges from about 25 to 50 tons/in^2 (345 to 690 MN/m^2). **Forward extrusion** of long tubes, rods, and shapes usually performed hot in hydraulic presses has been extended from the softer metals to the extrusion of steels. Most work is done horizontally because of the lengths of the extrusions. Some vertical mechanical-press equipment is used in hot extrusion of steel tubing.

EQUIPMENT FOR WORKING METALS

The **mass production** industries use an extremely wide variety of machines to force materials to flow plastically into desired shapes (as compared with the more gradual methods of obtaining shapes by cutting away surplus material on machine tools). The application of working pressure may use hydraulic, pneumatic, mechanical, or electric means to apply pressing, hammering, or rolling forces. Mechanically and hydraulically actuated devices cover much the same range. In general, the mechanical equipment is faster, easier to maintain, and more efficient to operate by reason of energy-storing flywheels. The hydraulic equipment is more flexible and more easily adjusted to limited lots in pressure, positions, and strokes. Mechanical handling or feeding devices incorporated in or serving many of these more or less specialized machines further extend their productivity.

Power presses consist of a frame or substantial construction with devices for holding the dies or tools and a moving member or slide for actuating one portion of the dies. This slide usually receives its movement from a crankshaft furnished with a clutch for intermittent operation and a flywheel to supply the sudden power requirement. **Hydraulic presses** have no crankshaft, clutch, or flywheel but employ rams actuated by pumps.

The crankshaft is ordinarily the limiting factor in the pressure capacity of the machine and accordingly is often taken as the basis for tonnage ratings. There is no uniform basis for this rating, owing to variations in shaft proportions and materials and in the different relative severity of various press operations. The following valuation is tentative and is based on the shaft diameter in the main bearings. The bending strength is figured at a section through the center of the crankpin and the combined bending and torsional strength at the inside ends of the main bearings, taking the bending fulcrum at a distance out from these points equal to one-third the length of the main bearings. In the case of double-crank presses and twin-drive arrangements, the relative proportion of the torsional load must be varied to suit, but, except in the cases of long strokes, it is usually small. The working strength is based upon a stress in the extreme fibers of 28,000 lb/in^2 (193 MN/m^2). The limit bearing capacities are taken approximately at 5,000 lb/in^2 (35 MN/m^2) on the crankpins and 2,500 lb/in^2 (18 MN/m^2) average over the main bearings for ordinary steel on cast-iron press bearings with proper grooving. On the knuckle-joint-type presses with hardened tool-steel bearing surfaces and flood lubrication, the bearings will take up to about 30,000 lb/in^2 (207 MN/m^2). On eccentric-type shafts where the main bearings support right up to the oversize pin on each side, the limiting factor is the bearing load. The shaft is practically in shear, so that it has a considerable overload capacity (about $7d^2$ tons). In Table 4.12.4, uniform-diameter single crankshafts are those in which for manufacturing reasons the diameter is the same at the crankpin and at the main bearings. Other crank shafts have an oversize crankpin to balance the bending load at the center with the combined bending and torsional load at the side. The strength of the shaft is figured at mid-stroke, and the stroke and tonnage capacity are given in terms of the diameter d at the main bearings. Where the working load comes on only near the bottom stroke, the shaft press capacity may be figured as if the stroke were shorter in proportion.

Table 4.12.4 gives the rated capacities of a series of power presses as a function of the shaft diameter.

The speed of operation of the press depends upon the energy requirement and the crankpin velocity. The latter determines the velocity of impact on the tools. In blanking, the blow varies directly with the contact speed and the thickness and hardness of the material. In drawing operations the variation depends upon contact speed, ductility of material, lubrication, etc.

The energy required per stroke is practically the product of the average load and the working distance, plus friction allowance, assumed at about 16 percent. On short-stroke operations, such as blanking, the working energy is supplied almost entirely by slowing down the flywheel; motor and belt pull serve merely to return the flywheel to speed during the large part of the cycle in which no work is done. In drawing operations, the working period is considerable, and in many cases the belt takes the largest part of the working load. In this case, add to the available flywheel energy, the work done by the belt. This amounts, for example, to the allowable belt loading (flat or Vee), multiplied by the ratio of the belt velocity to the crankpin velocity, multiplied by the length of the working stroke on the crank circle in feet. The maximum flywheel slowdown has been assumed to be up to 10 percent for continuous operation and up to 20 percent for intermittent operation. The following formula is based upon average press-flywheel proportions and a slowdown of 10 percent. The result may be doubled for 20 percent slowdown.

The flywheel capacity per stroke at 10 percent slowdown in inch-tons equals $WD^2N^2/5{,}260{,}000{,}000$, where W is the weight of the flywheel, lb, D is the diameter, in, and N is rotation speed, r/min. (See Sec. 6.2.)

The difference between nongeared and geared presses is only in speed of operation and the relatively greater flywheel capacity.

Table 4.12.4 Power-Press Shaft Capacities

Type of press crankshaft	Max stroke, in*	Capacity tons†
Single crank, single drive, uniform diameter	d‡	$2.8d^2$
Single crank, single drive, oversize crankpin	d	$3.5d^2$
Single crank, single drive, oversize crankpin	$2d$	$2.2d^2$
Single crank, single drive, oversize crankpin	$3d$	$1.6d^2$
Single crank, twin drive, oversize crankpin	$2d$	$3.5d^2$
Single crank, twin drive, oversize crankpin	$3d$	$2.7d^2$
Double crank, single drive, oversize crankpin	$0.75d$	$5.5d^2$
Double crank, single drive, oversize crankpin	d	$4.4d^2$
Double crank, single drive, oversize crankpin	$2d$	$2.5d^2$
Double crank, single drive, oversize crankpin	$3d$	$1.7d^2$
Double crank, twin drive, oversize crankpin	$1.5d$	$5.5d^2$
Double crank, twin drive, oversize crankpin	$3d$	$3.2d^2$
Single eccentric, single or twin drive	$0.5d$	$4.3d^2$

* 1 in = 2.54 cm.
† 1 ton = 8.9 kN.
‡ d = shaft diameter in main bearing.
SOURCE: F. W. Bliss Co.

Press frames are designed for stiffness and usually have considerable excess strength. Good practice is to figure cast-iron sections for a stress of about 2,000 to 3,000 lb/in^2 (13.8 to 20.6 MN/m^2). **C-frame presses** are subjected to an appreciable arc spring amount ordinarily of between 0.0005 and 0.002 in/ton (1.5 and 6 mm/MN), because the center of gravity of the frame section is a considerable distance back of the working centerline of the press. **Straight-sided presses** eliminate that portion of the spring or deflection which is on an arc. **Built-up frame presses** are held together with steel tie rods shrunk in under an initial tension in excess of the working load so that they minimize stretch in that portion of the press.

Power presses are built in a very wide variety of styles and sizes with shafts ranging from 1- to 21-in (2.5- to 53-cm) diam. Over a large part of this range they are built with C frames for convenience, straight-sided frames for heavier and thinner work, eccentric shafts for heavy forgings and stampings, double crankshafts for wide jobs, four-point presses for large panel work, underdrive presses in high-production plants where repairs to presses would interfere with flow of production, and knuckle-join presses for intensely high pressures at the very bottom of the stroke. All these are classified as single-action presses and are used for most of the operations previously discussed.

Double-action presses combine the functions of blank holding with drawing. In the smaller sizes, such presses have cams mounted on the cheeks of the crankshaft to actuate the outer or blank-holding slide. In larger machines, toggle mechanisms are provided to actuate the outer slide, with the advantage that the blank-holding load is taken on the frame instead of the crankshaft. Both of these types afford a considerable power saving over single-action presses equipped with drawing attachments, because the latter must add the blank-holding pressure to the working load for the full depth of the draw.

Types of presses include **foot presses,** in which the pendulum type has the lowest mechanical advantage and the longest stroke; the **lever type,** which has higher mechanical advantage and shorter strokes; **toggle** or **knuckle type,** which has the highest mechanical advantage and works through the shortest stroke with considerable advantage obtainable from the use of tie rods on fine stamping or embossing work; long-stroke rack and pinion-driven presses; triple-action drawing presses; cam-actuated presses; etc.

Screw presses consist of a conventional frame and a slide which is forced down by a steep pitch screw on the upper end of which is a flywheel or weight bar. Hand-operated machines are used for die testing and for small production stamping, embossing, forming, and other work requiring more power than foot presses. Power-driven screw presses are built with a friction drive for the flywheel and automatic control to limit the stroke. Such presses are built in comparatively large sizes and used to a considerable extent for press forging. They lack the accuracy and speed of power presses built for this work but have a safety factor which power presses have not, in that their action is not positive. In this they closely resemble a drop hammer, although their motion is slower. The energy available for work in these presses is $\frac{1}{2}I_f v^2 + \frac{1}{2}I_s v^2$, in which I_f is the moment of inertia of the flywheel, I_s is the moment of inertia of the spindle, and v is the angular velocity of both.

Self-contained fast-acting **hydraulic presses** are being increasingly used. Equipped with motor-driven variable-displacement oil hydraulic pumps, the speed and pressure of the operating ram or rams are under instant and automatic control; this is particularly advantageous for deep drawing operations. The punch can be brought into initial contact with the work without shock and moved with a uniform controlled velocity through the drawing portion of the cycle. The cold drawing of stainless steels and aluminum alloys (in which the control of drawing speed is vital), as well as the hot drawing of magnesium, is best done on hydraulic presses.

The hydraulic press is used on the rubber pad, or *Guerin,* process of blanking or forming metals, in which a laminated-rubber pad replaces one half—usually the female half—of a die. In forming aluminum the practice has developed of using inexpensive dies of soft metal, vulcanized fiber, plastic, wood, or plaster; and cast dies in industries which, like the aircraft industry, require short runs on many different sizes of shapes and parts.

The older accumulator type of hydraulic-press construction is still used for hot extrusion and some forging work.

Pneumatic Hammers A self-contained type of pneumatic forge hammer (the Bêché) has an air-operated ram with an air-compressing cylinder integral with the frame. The ram is raised by admitting compressed air beneath the ram piston; at the same time a partial vacuum is caused above it. The ram is forced down by a reversal of this action.

The **terminal velocity** (velocity at ram-workpiece contact) of the steam, pneumatic, or hydraulic assisted hammer can be calculated as follows:

$$v = \left[2hg\left(1 + \frac{Ap_m}{H}\right)\right]^{0.5}$$

where A is the area of the piston and p_m the mean pressure in the drive cylinder. The hammer energy is often calculated by dividing the energy required for plastic work by the mechanical efficiency of the hammer.

Hydraulic presses are energized by pressurized liquid, usually oil. They can deliver high tonnage but are slow. Due to large die chilling (heat loss to dies) present in hydraulic press forging, they are not usually used for hot forging, except in isothermal forging, where slow speed and large die-workpiece contact times are not a major limitation. They are especially suited to sheet metal forming operations, where slower ram speeds produce lower impact loads, speeds can be varied during the stroke, and multiple actions can be obtained for blank holder and die cushion operation.

Hammers and presses are often selected based on their characteristics such as energy, ram mass (t_{up}), force or tonnage, ram speed (stroking rate), stroke length, bed area, and mechanical efficiency. The characteristics for various hammers and presses are summarized in Table 4.12.5.

Rotary motion is used **for working sheet metal** in a variety of machines, including bending rolls (three rolls); rolling straighteners with five, seven, or more rolls; roll forming machines, in which a series of rolls in successive pairs are used to bend the strip material step by step to some desired shape; a series of two-spindle and multiple-spindle machines used for rolling beads, threads, knurls, flanges, and trimming or curling the edges of drawn shells of cylinders; seaming machines for double seaming, crimping, curling, and other operations in the production of tin cans, pieced tinware, etc.; and spinning machines for spinning, burnishing, trimming, curling, shape forming, and thickness reduction. Various production spinning operations and tool arrangements are shown in Fig. 4.12.3.

Plate-Straightening Machines The **horsepower require**d for plate-straightening machines operating on steel plate is shown in Table 4.12.6.

Power required for angle-iron-straightening machines: for 4-in angles, 12 hp; for 6-in, 18 hp; for 8-in, 25 hp.

Power or hydraulic presses are used to straighten **large rolled sections.** The presses make 20 to 30 strokes per min, and the amount of flexure is regulated by inserting wedges or pieces of flat iron. The beams are supported on rolls so they can be easily handled. The **power required for presses** of this kind is as follows:

Depth of girder, in*	4	6	8	10	12	16	20	24
Horsepower (approx)†	3	4	7	11	13	19	23	35

* 1 in = 2.54 cm.
† 1 hp = 0.746 kW.

Horizontal plate-bending machines consist of two stationary rolls and a third vertical adjustable upper roll which can be fitted obliquely for taper bending and is held in bearings with spherical seats. The diameter of the rolls can be determined approximately from the equation $r^2 = bt$, in which r is the radius of the roll, b the width of the plate or sheet, and t its thickness, all in inches.

The **power requirements** of horizontal plate and sheet bending machines are shown in Table 4.12.7.

Numerically Adaptive Bending of Tubes and Profiles (Fig. 4.12.24) Extrusions, profiles, and tubes find greater use in car and truck body structures due to their high stiffness under bending loads. The shapes are

Table 4.12.5 Characteristics of Hammers and Presses*

Equipment type	Energy,† kN · m	Ram mass, kg	Force,‡ kN	Speed, m/s	Strokes per min	Stroke, m	Bed area, m × m	Mechanical efficiency
Hammers								
Mechanical	0.5-40	30-5,000		4-5	350-35	0.1-1.6	0.1 × 0.1 to 0.4 × 0.6	0.2-0.5
Steam and air	20-600	75-17,000 (25,000)		3-8	300-20	0.5-1.2	0.3 × 0.4 to (1.2 × 1.8)	0.05-0.3
Counterblow	5-200 (1,250)			3-5	60-7		0.3 × 0.4 to (1.8 × 5)	0.2-0.7
Herf	15-750			8-20	<2			0.2-0.6
Presses								
Hydraulic, forging			100-80,000 (800,000)	<0.5	30-5	0.3-1 (3)	0.5 × 0.5 to (3.5 × 8)	0.1-0.6
Hydraulic, sheet metalworking			10-40,000	<0.5	130-20	0.1-1	0.2 × 0.2 to 2 × 6	0.5-0.7
Hydraulic, extrusion			1,000-50,000 (200,000)	<0.5	<2	0.8-5	0.06-0.6 diam. container	0.5-0.7
Mechanical, forging			10-80,000	<0.5	130-10	0.1-1	0.2 × 0.2 to 2 × 3	0.2-0.7
Horizontal upsetter			500-30,000 (1-9 in diam)	<1	90-15	0.05-0.4	0.2 × 0.2 to 0.8 × 1	0.2-0.7
Mechanical, sheet metalworking			10-20,000	<1	180-10	0.1-0.8	0.2 × 0.2 to 2 × 6	0.3-0.7
Screw			100-80,000	<1	35-6	0.2-0.8	0.2 × 0.3 to 0.8 × 1	0.2-0.7

* From a number of sources, chiefly A. Geleji, *Forge Equipment, Rolling Mills and Accessories*, Akademiai Kiado, Budapest, 1967.
† Multiply number in column by 100 to get m·kg, by 0.73 to get 10^3 lbf·ft.
‡ Divide number by -10 to get tons. Numbers in parentheses indicate the largest sizes, available in only a few places in the world.
SOURCE: John A. Schey, *Introduction to Manufacturing Processes*, McGraw-Hill, New York, 1987.

Table 4.12.6 Power Requirement for Plate-Straightening Machines (Steel)

Thickness of plate, in*	0.25	0.4	0.6	0.8	1.0	1.2	1.4	1.6
Width of plate, in*	48.0	52.0	60.0	72.0	88.0	102.0	120.0	140.0
Diameter of rolls, in*	5.0	8.0	10.0	12.0	13.0	14.0	15.0	16.0
Horsepower (approx)†	6.0	8.0	12.0	20.0	30.0	55.0	90.0	130.0

* 1 in = 2.54 cm.
† 1 hp = 0.746 kW.

Table 4.12.7 Power Requirement for Plate and Sheet Bending Machines (Steel)

Thickness of plate, in*	0.5	0.6	0.8	1	1.2
Horsepower for plate 120 in wide†	10.0	12.0	18.0	27.0	40.0
Horsepower for plate 240 in wide†	30.0	30.0	40.0	55.0	75.0

* 1 in = 2.54 cm.
† 1 hp = 0.746 kW.

bent or curved to conform to the geometric needs of design and assembly. Therefore, new fixtures and numerically controlled machines and tooling (Fig. 4.12.24) have been developed to stretch bend the structural shapes.

Vertical plate-bending machines have a hydraulically operated piston which moves an upper and a lower pair of rolls between inclined surfaces of the stationary upright and the crosshead. The bending is done piece by piece against a second stationary upright. Heavy ship plates are rigidly clamped down and bent by a roll operated by two hydraulic pistons. For angular bends or for the production of warped surfaces, the pistons can be operated independently or together. In vertical machines, angles and other rolled shapes are bent between suitably shaped rolls. Pipes are filled with sand to prevent flattening when being bent. For some work, pipes are bent hot between suitable forms operated by hydraulic pressure.

The **rotary swaging machines** for tapering, closing in, and reducing tubes, rods, and hollow articles is essentially a cage carrying a number of rollers and revolving at high speed; e.g., 14 rolls in a cage revolving at 600 r/min will strike 8,400 blows per min on the work.

A rapid succession of light blows is applied to a considerable variety of commercial **riveting** operations such as pneumatic riveting. Another method of riveting, described as spinning, involves rotating small rollers rapidly over the top of the rivet and at the same time applying pressure. Neither of these methods involves pressures as intense as those used in riveting by direct pressure, either hot or cold. Power presses and C-frame riveters, employing hydraulic pressure or air pressure of 80 to 100 lb/in^2 (550 to 690 kN/m^2), are designed to apply 150,000 lb/in^2 (1,035 MN/m^2) on the cross section of the body of the rivet for hot-working and 300,000 lb/in^2 (2,070 MN/m^2) for cold-working. The pieces joined should be pressed together by a pressure 0.3 to 0.4 times that used in riveting.

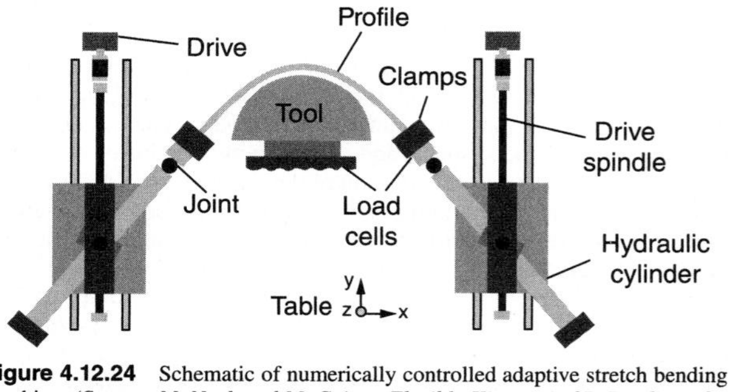

Figure 4.12.24 Schematic of numerically controlled adaptive stretch bending machine. *(Source: M. Nock and M. Geiger, Flexible Kinematic 3D-Bending of Tubes and Profiles, International Conference on the Advanced Technology of Plasticity, Japan, 1, 2002, pp. 643-648.)*

4.13 WELDING AND CUTTING

by Duane K. Miller and Michael D. Florczykowski

REFERENCES: American Welding Society (AWS): "Welding Handbook" (six volumes), "Structural Welding Code," "Filler Metal Specifications," "Specification for the Design of Welded Joints in Machinery and Equipment," "Welding Terms and Definitions," "Brazing Manual," "Thermal Spraying," "Welding of Chromium-Molybdenum Steels," "Standard Symbols for Welding, Brazing, and Nondestructive Examination," The James F. Lincoln Arc Welding Foundation: Blodgett, "Design of Weldments," "Design of Welded Structures," "Procedure Handbook of Welding," AISC: "Steel Construction Manual—ASD," "Steel Construction Manual—LRFD," latest editions, Miller, "Design Guide 21 Welded Connections—A Primer for Engineers," McGraw Hill: "Manufacturing Engineering Handbook 2nd ed.," ANSI: "Safety in Welding, Cutting, and Allied Processes," ASM: International "Metals Handbook Vol. 1, 10th ed.," ASTM: "ASTM A941," 2000a edition; Rolfe and Barsom: "Fracture and Fatigue Control in Structures."

INTRODUCTION

This section on welding and thermal cutting is part of the chapter that addresses Manufacturing Processes and appropriately so. However, the general topic of welding involves more than just manufacturing: welds are typically associated with connections and the field of welded connection design is broad. The ability to successfully weld various materials together is dependent on the chemical and mechanical properties of the materials being joined. Thermal welding and cutting processes typically alter the properties of the materials being joined. Accordingly, welding requires an extensive knowledge of materials engineering.

The prominent failure mode for connections, and for welded connections in particular, is fatigue resulting from cyclic loading. Unwelded products are also subject to fatigue, but welded connections add additional complexities. Some products and systems are subject to only static loading, that is, loads are sustained and rarely change. For many welded products such as earthmoving equipment, material handling systems, rotating equipment and presses, loads are applied and removed, resulting in cyclic loading that may lead to fatigue failure.

While Welding Engineering is a broad field of study, this section on Welding and Cutting will address the major themes of welding and thermal cutting processes, welded connection design, cyclically loaded welded connections and material considerations. This section is not comprehensive in its coverage of welding; other resources will be required to effectively use all the information that is introduced here. For further study, consult the References provided.

The primary welding and cutting focus of this section is connections of components made of steel that exceeds 1/8 in (3 mm). This concentration was selected in order to focus on welds expected to transfer loads through assemblies that carry significant loads. For many sheet metal applications, such as guards and ductwork, the welds need only be strong enough to hold the parts in place and are therefore not subject to formal design analysis. Such welds are often made with resistance welding methods and the materials are usually severed by shearing or laser cutting. Thicker materials are usually chosen when stresses are higher, and the welded connections must be designed to resist the transferred loads. In manufacturing, these thicker materials are often severed with thermal cutting processes, and subsequently joined with arc welding processes. All of the major welding processes are discussed, but the design-related portions deal exclusively with arc welds.

The focus on welding and thermal cutting of steel also has a practical basis: welded assemblies of thicker materials are typically made of steel base metals. Other materials are welded, with stainless steel and aluminum being popular alternatives. Welding considerations for steel applications are extensively considered, and an introduction to welding on non-steel materials is contained at the end of this section.

WELDING PROCESSES

The American Welding Society (AWS) recognizes over 60 different welding, brazing, and soldering processes that can be grouped into the following broad categories: arc welding, oxyfuel gas welding (OFW), high energy beam welding, resistance welding (RW), solid-state welding (SSW), other welding processes, and finally soldering and brazing. In each broad category, there are a variety of individual welding processes with both similar and also unique features. For example, all arc welding processes utilize an electric arc, while all brazing and soldering processes involve melting the filler metal, but not the base metal.

Each individual welding process has advantages and limitations, as well as unique features and applications; in some cases, their capabilities overlap considerably. Thus, two manufacturing engineers may select different processes for the same application; both decisions may be correct. Entire handbooks have been written on some of the individual processes discussed herein; of necessity, the coverage in this section is only introductory. It should, however, provide the design engineer with a sufficient background to facilitate knowledgeable interaction with the manufacturing department.

This section assumes that the design engineer is responsible for selecting material types, material thicknesses, and joint designs, calculating stresses in the part, and stresses across the joint, and determining the weld type and size. It further assumes that a manufacturing engineer is responsible for determining the means and methods by which a product can be produced at the most economical rate, with the required quality level, and in a safe manner for all concerned.

Obviously, the design engineer and the manufacturing engineer must interact. Understanding the constraints imposed upon the other party can help achieve the goals of low cost, quality, and safety. For example, a seemingly minor change in material composition can have significant welding implications. Welds that have been specified in design to be larger than needed will automatically increase production costs. Ongoing interaction between the design engineer and the manufacturing engineer can help optimize operations.

Please note: Unless otherwise noted, all of the definitions provided in quotation marks are from *AWS 3.0M/A3.0: Standard Welding Terms and Definitions*. They are used here with the permission of the American Welding Society.

Shielded Metal Arc Welding (SMAW)

SMAW is "an arc welding process with an arc between a covered electrode and the weld pool. The process is used with shielding from the decomposition of the electrode covering, without the application of pressure, and with filler metal coming from the electrode." Often called "stick welding" or "manual welding," SMAW is a common fusion welding process. SMAW, illustrated in Figs. 4.13.1 and 4.13.2, is used in the shop and in the field for fabrication, erection, maintenance, and repairs.

With SMAW, the welder manually propels the arc along the length of the joint, as well as manually feeds the electrode toward the weld

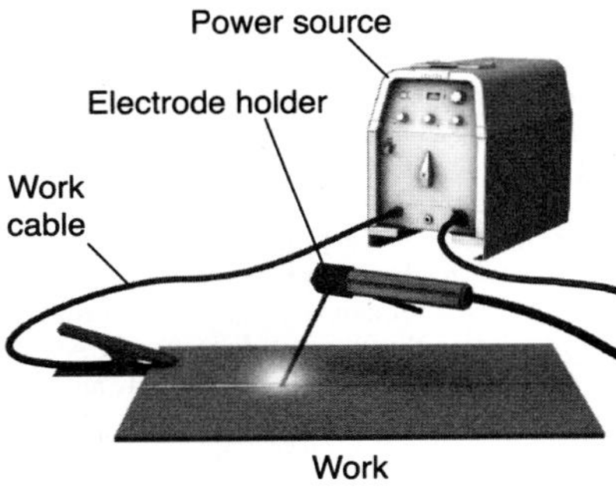

Figure 4.13.1 Shielded Metal Arc Welding (SMAW) process.

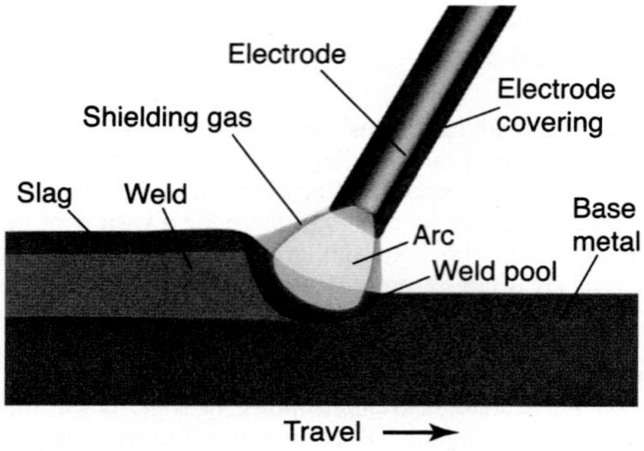

Figure 4.13.2 SMAW process details.

pool. Considerable skill is required to accomplish these two tasks. The welder must maintain a specific arc length, the distance from the end of the electrode to the weld pool. If the arc length is too long, the arc will go out; if too short, the electrode will fuse to the weld pool.

Versatility is perhaps the greatest advantage of SMAW. With one power supply, a wide range of materials can be welded by simply changing the electrode to be used, and adjusting the power source output setting. The same machine can weld thin or thick material, as well as steel, stainless steel, and cast iron. Welding can be performed in all positions. For these reasons, SMAW remains popular for maintenance and repair applications.

The greatest shortcoming of SMAW is its relative inefficiency. The electrodes are used in lengths of 9 to 18 in (200-400 mm), and approximately 2 in (50 mm) of each electrode is unusable and must be discarded. Once the electrode is consumed, the welder must stop welding, remove the stub, insert another electrode, and start welding again. These work interruptions are inherent to the process, and reduce production rates accordingly.

Additionally, there is a limited amount of current that can be conducted through the length of the electrode. This in turn limits the amount of metal that can be deposited, slowing productivity further. While SMAW was extensively used in production environments in the 1940s and 1950s, it is less commonly used today in developed countries where wages are high. On the other hand, in less developed countries where labor costs are low and capital for equipment may be scarce, SMAW continues to be popular.

Flux Cored Arc Welding (FCAW)

FCAW is "an arc welding process using an arc between a continuous filler metal electrode and the weld pool. The process is used with shielding from a flux contained within the tubular electrode, with or without additional shielding from an externally supplied gas, and without the application of pressure." FCAW electrodes are supplied on spools, coils, reels, and other devices that may contain as little as 1 lb (0.5 kg) or up to 1,000 lbs (500 kg) or more. As shown in Figs. 4.13.3 and 4.13.4, the long lengths of electrode permit welding to be performed without the interruptions inherent to SMAW.

In FCAW, the electrode is fed through a welding gun by a wire feeder, a mechanical device that delivers the electrode at a regulated rate, eliminating the need for the welder to manually deliver the electrode to the joint. Accordingly, FCAW is known as "semiautomatic" welding since the welder simply propels the welding gun along the length of the joint. Due to the nature of the power supplies used for FCAW, the arc length is automatically regulated, eliminating the need for the welder to maintain this critical distance. FCAW requires less skill than SMAW. FCAW can also be performed automatically with hard automation, or robotics.

Within the welding gun is a hollow copper tube called a contact tube or contact tip. Electrical energy is transferred from the gun to the electrode at this point. A short length of electrode, typically around 1 in

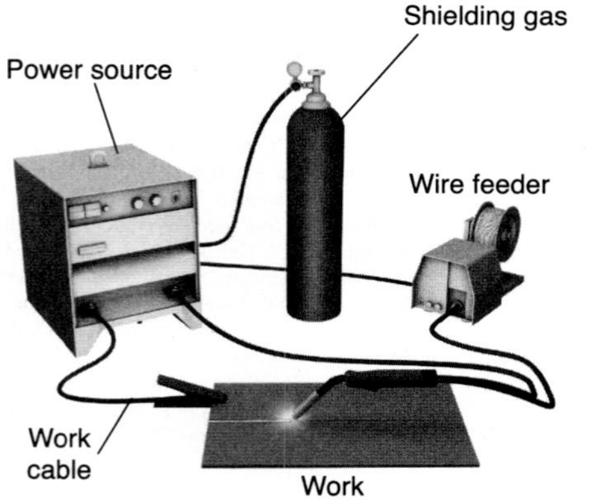

Figure 4.13.3 Flux Cored Arc Welding (FCAW) process.

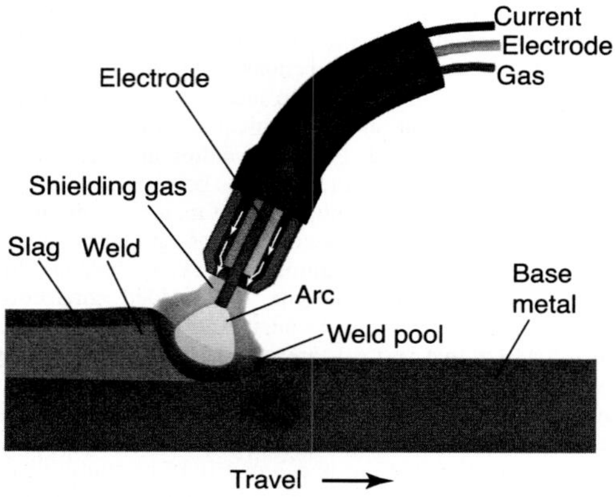

Figure 4.13.4 FCAW process details.

(25 mm) extends beyond the end of the contact tip during welding. This short length of electrode is capable of conducting considerable current, much more than is possible with SMAW electrodes. Thus, the combination of the "continuous" electrode with the higher welding currents makes FCAW more productive than SMAW.

Two versions of flux cored arc welding exist: gas shielded FCAW (or FCAW-G) and self-shielded FCAW (or FCAW-S). Shielding gas is required for FCAW-G. Electrodes are designed and classified based on the requirements specific to FCAW-G or FCAW-S. Shielding gas should not be used with self-shielded electrodes as this will cause poor weld quality. A variety of shielding gasses may be used with FCAW-G, with carbon dioxide or argon/carbon dioxide mixtures being common for steel applications. In general, the gas shielded version is more versatile with a greater range of available filler metals. The self-shielded version is ideal for welding outdoors where wind may disturb the gas shield associated with FCAW-G.

When a weld is made with FCAW, a slag covering remains behind on the surface of the completed weld. The slag must be removed upon completion of the weld, an activity that adds to the cost of making the weld.

FCAW began to replace SMAW in many high production situations in the 1960s and 1970s and is common in a variety of industries today. The flux contained in the electrode allows the process to handle moderate levels of scale and rust as well as other contaminants on the surface of the material being joined. With the proper electrode and procedure, all-position welding is possible. In general, FCAW may be viewed as a replacement for SMAW, albeit at the cost of more expensive equipment and less versatility.

Gas Metal Arc Welding (GMAW)

GMAW is "an arc welding process, using an arc between a continuous filler metal electrode and the weld pool." The process is used with shielding from an externally supplied gas and without the application of pressure. Often called "MIG" welding (metal inert gas), the process is similar in concept to FCAW except that the electrode contains no flux, and the finished weld does not have an extensive slag coating. GMAW, shown in Figs. 4.13.5 and 4.13.6, can be used semi-automatically or automatically, and is commonly used for robotic applications.

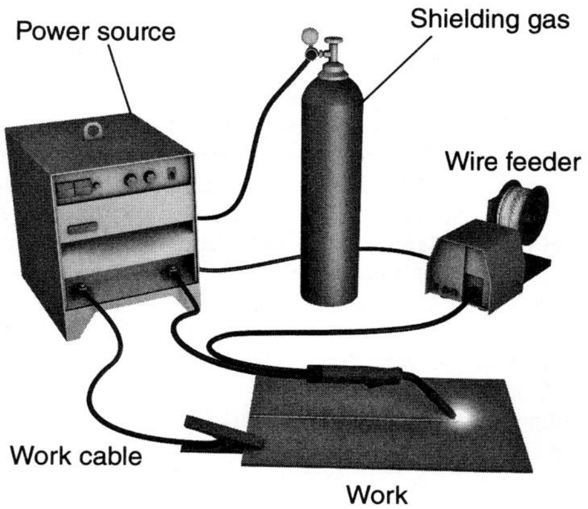

Figure 4.13.5 Gas Metal Arc Welding (GMAW) process.

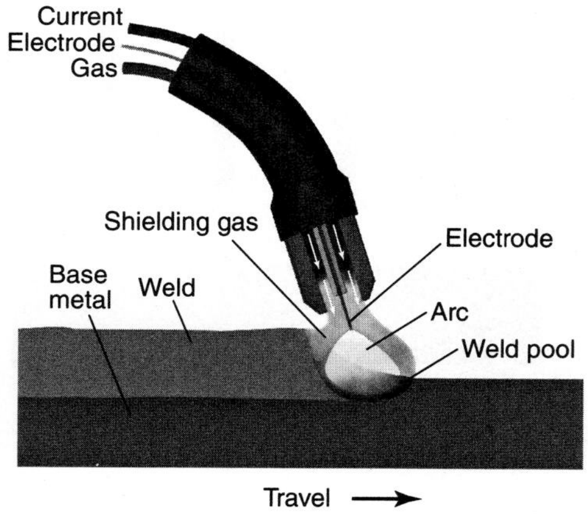

Figure 4.13.6 GMAW process details.

In its most basic form, GMAW can use the same type of equipment as FCAW. GMAW may use either solid or metal-cored electrodes. Solid electrodes are essentially wires of a specified composition and diameter. Metal-cored electrodes, also called "composite" electrodes, are tubular and contain metal powders in their core. While they are similar in construction to FCAW electrodes, the essential difference is that GMAW metal-cored electrodes do not contain flux and slag forming ingredients. A wide range of shielding gasses may be used with GMAW, depending on the material being welded and the mode of metal transfer. Carbon dioxide may be used for welding on steel with GMAW. GMAW that uses CO_2 gas may be called "MAG" welding (metal active gas). Purely inert gases such as argon and helium can be used, typically in conjunction with minor components of other gases, such as oxygen. Blends of two gasses are typical, and mixtures of three or more gases are sometimes used.

Since a finished GMAW weld contains little or no slag, post-welding slag removal is not necessary, making the process ideal for automatic and robotic welding, as well as for multiple pass welds. The absence of fluxing agents in the electrode causes the GMAW process to be less tolerant of surface contaminants such as scale and rust. For out-of-position welding, the slag associated with SMAW and FCAW can be used to support the liquid metal; without this slag, welding out-of-position with GMAW is more difficult.

The nature of the arc and the transfer of metal from the electrode to the puddle can take on a variety of forms with GMAW, depending on the welding parameters and shielding gas. Distinctions in the type of transfer are typically identified as modes of transfer. The five most common transfer modes are described below. Each mode has its own advantages and limitations.

GMAW Spray Transfer

Spray transfer, as applied to GMAW, is "metal transfer in which molten metal from a consumable electrode is propelled axially across the arc in small droplets." As shown in Fig. 4.13.7, the droplets are smaller in diameter than the electrode diameter, which itself is small. Spray transfer is associated with relatively high current and voltage levels. High quality welds with particularly good appearance are obtained. The shielding gas used in spray arc transfer for steel applications is composed of at least 80 percent argon, with the balance either CO_2 or O_2. Typical mixtures would include 90/10 argon/CO_2, and 95/5 argon/O_2. Due to the intensity of the arc and gravitational pull on the weld puddle, spray transfer for steel applications is restricted to welding in the flat and horizontal positions. Furthermore, spray transfer is restricted to welding on relatively thick materials.

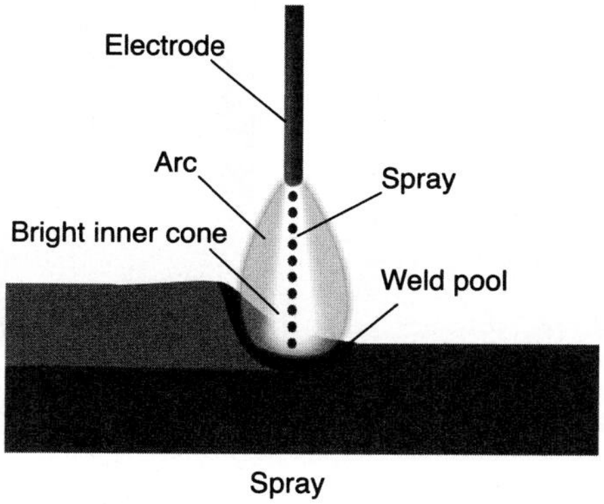

Figure 4.13.7 GMAW Spray Transfer process.

GMAW Globular Transfer

Globular transfer is the "transfer of molten metal in large drops from a consumable electrode across the arc," as illustrated in Fig. 4.13.8. The droplets are typically larger in diameter than the electrode. The large drops hitting the weld puddle can create considerable spatter, droplets from which may fuse to the surface of the base metal, away from the weld. The arc is relatively rough and the weld appearance is inferior to GMAW spray transfer. Two compelling factors, however, encourage the use of globular transfer in steel applications: first, lower cost CO_2 shielding gas can be used and secondly, the heating effects on the welding gun and the welding operator are less with globular transfer versus spray transfer.

GMAW Short Circuiting Transfer (GMAW-S)

Short circuiting transfer GMAW is a "gas metal arc welding process variation in which the consumable electrode is deposited during repeated short circuits." An electrical short circuit is achieved when two separate electrical conductors physically touch each other and current is conducted across the interface. For short circuit GMAW, the short occurs between the electrode and the workpiece: there is no arc at this point. High electrical currents flow through the short circuit,

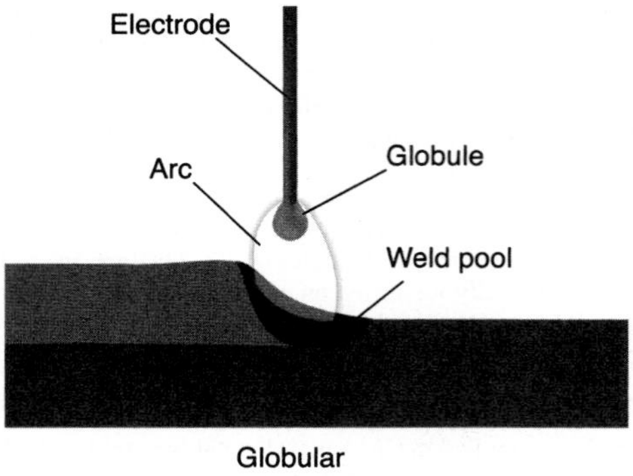

Figure 4.13.8 GMAW Globular Transfer process.

causing the electrode to overheat and essentially "explode," eliminating the short. Momentarily, the arc is re-established and molten metal is transferred to the weld pool. As shown in Fig. 4.13.9, the electrode is fed into the weld pool again, creating another short and repeating the sequence, which occurs hundreds of times per second.

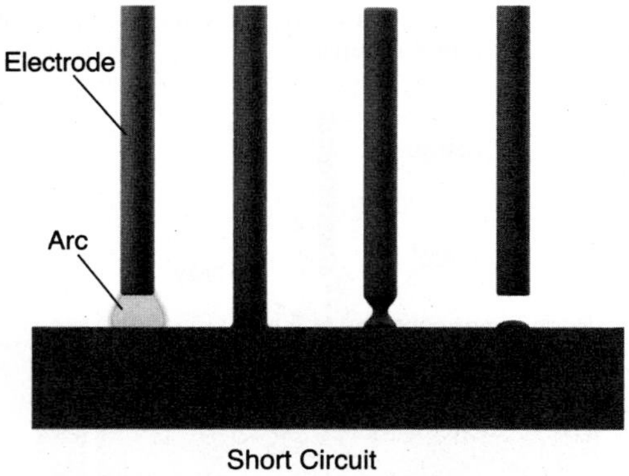

Figure 4.13.9 GMAW Short Circuiting Transfer process.

While the process mode is officially called "short circuit gas metal arc welding," it is often called "short arc." GMAW-S is ideally suited for welding on thin sections of material 1/8 in (3 mm) or less (although it can be used on thicker materials) because it is a relatively "cold" welding process. This low energy mode of transfer allows the process to be used when welding in the vertical and overhead position.

A major disadvantage of GMAW-S is directly related to the "cold" nature of the transfer mode: GMAW-S is notorious for a welding defect known as incomplete fusion, more commonly called "cold lap." The energy associated with a GMAW-S welding procedure may be sufficient to melt the filler metal, creating a weld bead, but insufficient to cause the weld metal to fuse to the base metal. The weld may have good visual appearance, but the connection may have limited or no overall strength, making this possibility a major concern.

The "explosion" that occurs during the electrical short associated with GMAW-S results in considerable weld spatter, fine droplets of liquid weld metal that solidify and fuse to the base metal surface, another limitation of the process.

GMAW Controlled Short Circuiting Transfer

There is a subset of GMAW-S that involves regulated control of the shorting process. As previously mentioned, the short of the welding electrode to the weld pool causes the electrode to "explode," releasing spatters of molten metal that often stick to the surface of the material being welded. Several proprietary and patented methods are used to regulate the shorting process, usually with high speed electronic controls. Rather than allowing the uncontrolled rise in welding current during the shorting process, the power supply regulates the amount of energy that is delivered, eliminating the "explosion" and the corresponding generation of spatter. This overcomes one of the shortcomings of conventional GMAW-S, but does not eliminate the possibility of incomplete fusion. However, for welding on thin materials, these controlled short circuiting transfer modes permit quality welding with almost no spatter.

GMAW Pulsed Spray Transfer (GMAW-P)

In gas metal arc welding, pulsed spray transfer is a "variation of spray transfer in which the welding power is cycled from a low level to a high level, at which point spray transfer is attained, resulting in a lower average voltage and current." The welding power supply automatically pulses between a higher energy setting where metal is transferred from the electrode to the weld pool, to a lower energy setting where the arc is maintained, but no metal is transferred. The cycle repeats itself hundreds of times per second. The high energy setting is similar to that used with spray transfer, thus the term "pulsed spray." The process may be called "pulsed arc" or "pulse welding."

The pulsing of the energy from high to low values reduces energy input into the weld joint. This enables GMAW-P, shown in Fig. 4.13.10, to be used out-of-position (unlike GMAW-spray transfer) when welding on steel. The lower energy input also causes less distortion of the base metal. The only drawback to GMAW-P is that more complex and expensive equipment is required.

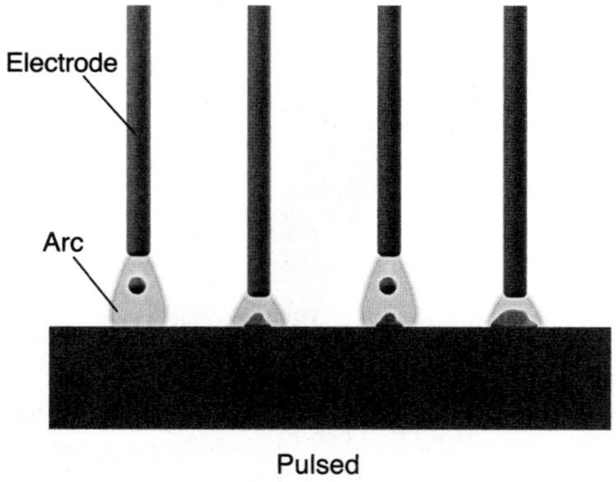

Figure 4.13.10 GMAW Pulsed Transfer process.

Gas Tungsten Arc Welding (GTAW)

GTAW is "an arc welding process using an arc between a tungsten electrode (nonconsumable) and the weld pool. The process is used with shielding gas and without the application of pressure." In this fusion welding process, often called "TIG" welding, the tungsten electrode conducts the welding current from the torch to the workpiece, as shown in Fig. 4.13.11. The tungsten electrode is not intended to deliver molten metal to the weld pool (although this can happen, leading to weld imperfections). Filler metal, if needed for the application, is supplied from a rod that is not electrically charged. Figure 4.13.12 illustrates that the filler metal is dipped into the weld pool and melted by the heat of the pool. Argon is the typical shielding gas, although helium or argon-helium mixtures may be used as well.

GTAW is ideally suited to welding nonferrous materials such as stainless steel and aluminum, and is very effective for joining thin sections. Highly skilled welders are required for GTAW, but the resulting weld quality can be excellent. The process is often used to weld exotic materials, such as titanium. Critical repair welds, as well as root passes in pressure piping, are typical applications. The completed GTAW weld, when properly made, has excellent appearance, and contains no

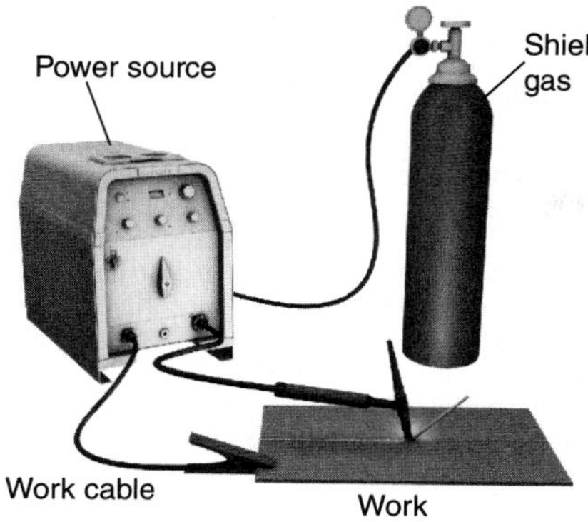

Figure 4.13.11 Gas Tungsten Arc Welding (GTAW) process.

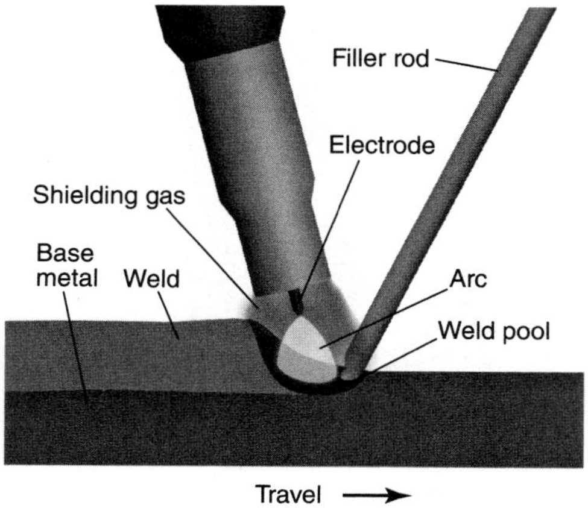

Figure 4.13.12 GTAW process details.

slag. GTAW requires the base metal to be relatively clean and free of surface contaminants.

GTAW is inherently slow, and thus, often used only in situations where no other process is viable. It is typically used in the manual mode, although automated and robotic applications are possible.

Submerged Arc Welding (SAW)

SAW is "an arc welding process using an arc, or arcs, between a bare metal electrode and the weld pool. The arc and molten metal are shielded by a blanket of granular flux on the workpieces. The process is used without pressure and with filler metal from the electrodes and sometimes from a supplemental source (welding rod, flux, or metal granules)." The process is illustrated in Figs. 4.13.13 and 4.13.14. The flux for SAW performs the same functions as flux for SMAW or FCAW welding: cleansing the weld from surface contaminants and then forming a slag to protect the molten weld pool.

Since the SAW arc is completely covered by the flux, welds are made without the arc flash, spatter, sparks or smoke that characterize the open-arc processes. The flux also covers the weld pool, making it impossible for the welder to observe its size and shape. As a result, SAW is typically used in the automatic mode but in some cases may be used semi-automatically.

SAW is capable of high productivity because it can use high welding currents, resulting in higher deposition rates and deeper penetration. For even higher deposition rates, a second or third electrode (or even more) can be added into the system to increase productivity further. Welds made under the protective layer of flux are excellent in appearance and spatter-free. Another benefit of the SAW process is freedom

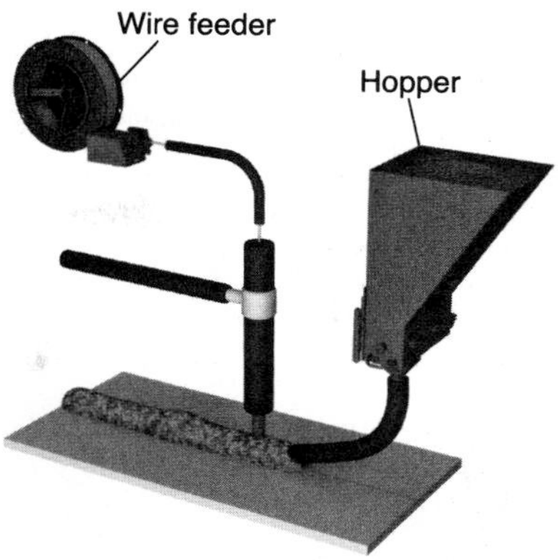

Figure 4.13.13 Submerged Arc Welding (SAW) process.

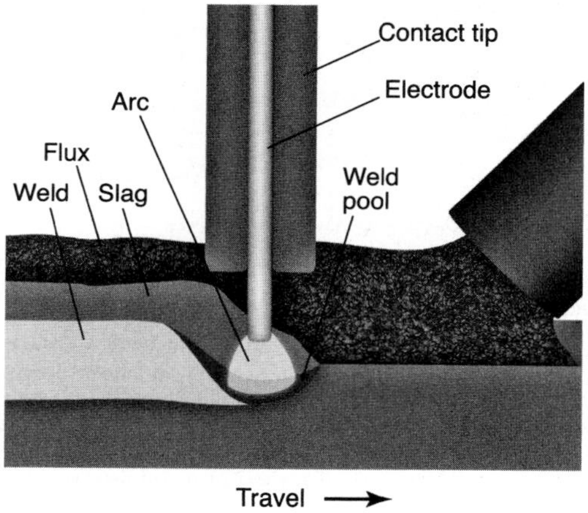

Figure 4.13.14 SAW process details.

from the open arc. This means that the welder is not required to use the standard protective helmet, and multiple welding operations can be conducted in small areas without the need for extensive shields to guard the operators from arc flash. The process produces very little fume, which is another production advantage, particularly in situations with limited ventilation.

SAW is restricted to welding in the flat and horizontal position. For shop fabrication, the use of positioners, or simple reorientation of the weldment, can facilitate in-position welding. However, field conditions may prohibit such opportunities, and thus restrict the suitability of SAW.

Plasma Arc Welding (PAW)

PAW is "an arc welding process employing a constricted arc between a nonconsumable electrode and the weld pool (transferred arc) or between the electrode and the constricting nozzle (nontransferred arc)." In the transferred arc mode, the arc occurs between the electrode and the workpiece as it does in GTAW, the primary difference being the constriction afforded by the secondary gases and torch design. With the nontransferred arc mode, arcing is contained within the torch between a tungsten electrode and a surrounding nozzle.

The constricted arc results in higher localized arc energies than are experienced with GTAW, resulting in faster welding speeds. Applications for PAW are similar to those for GTAW. The only significant disadvantage of PAW is the equipment cost, which is higher than that for GTAW. Most PAW is done with the transferred arc mode, although this mode utilizes a non-transferred arc for the first step of operation.

An arc and plasma are initially established between the electrode and the nozzle. When the torch is properly located, a switching system will redirect the arc toward the workpiece, as shown in Fig. 4.13.15. Since the arc and plasma are already established, the reliable transfer of the arc to the workpiece is easily accomplished, so PAW is often preferred for automated applications. PAW can be applied manually, automatically, or robotically.

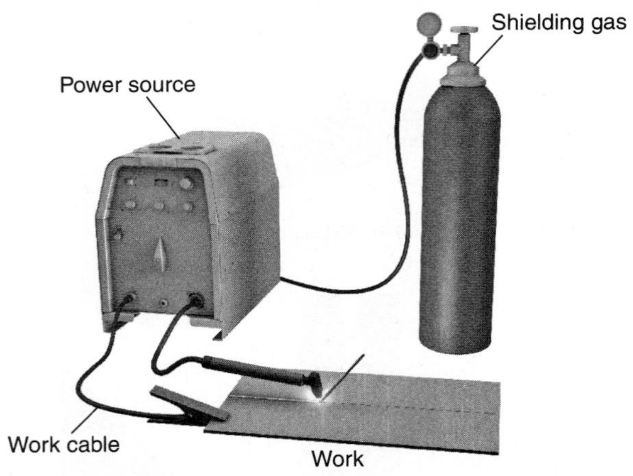

Figure 4.13.15 Plasma Arc Welding (PAW) process.

Arc Stud Welding (SW)

SW is "an arc welding process using an arc between a metal stud, or similar part, and the other workpiece. The process is used without filler metal, with or without shielding gas or flux, with or without partial shielding from a ceramic or graphite ferrule surrounding the stud, and with the application of pressure after the faying surfaces are sufficiently heated." Figure 4.13.16 shows that most of the molten metal and any contamination are expelled from the weld area as the stud is mechanically forced into the weld pool.

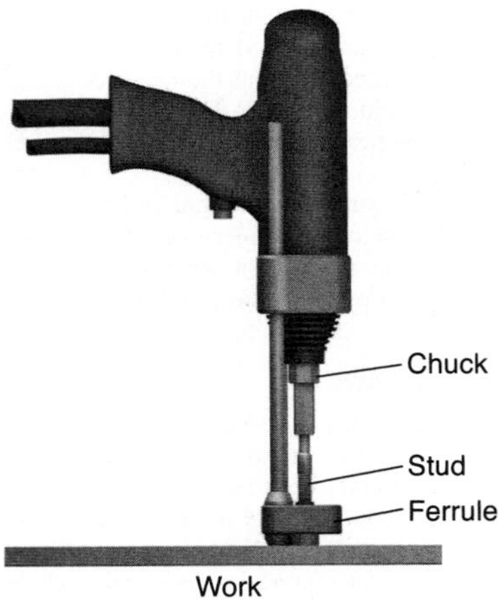

Figure 4.13.16 Arc Stud Welding (SW) process.

Stud welding can be used to attach threaded studs to the base metal, such as metal office furniture or electrical cabinets. SW is used to attach headed shear stud connectors to beams to facilitate composite action when the studs are embedded in concrete. The process is automated and fairly simple to use. The keys to obtaining a quality weld are to weld on relatively clean materials, use clean studs, and obtain the proper balance between welding current and arcing time.

Oxyfuel Gas Welding (OFW)

OFW is a "group of welding processes producing coalescence of workpieces by heating them with an oxyfuel gas flame." The heat for oxyfuel gas welding is supplied by burning a mixture of oxygen and a suitable combustible gas. The gases are mixed in a torch that controls the welding flame. The torch is moved along the juncture of two metals, and a filler metal, if used, is manually fed into the weld pool by the welder. Perhaps the biggest advantage of the oxyfuel gas welding process, illustrated in Figs. 4.13.17 and 4.13.18, is the increased ability of the operator to control the amount of heat that is introduced into the base metal, compared to other welding processes. However, OFW is relatively slow, is not easily automated, and is rarely used for production welding.

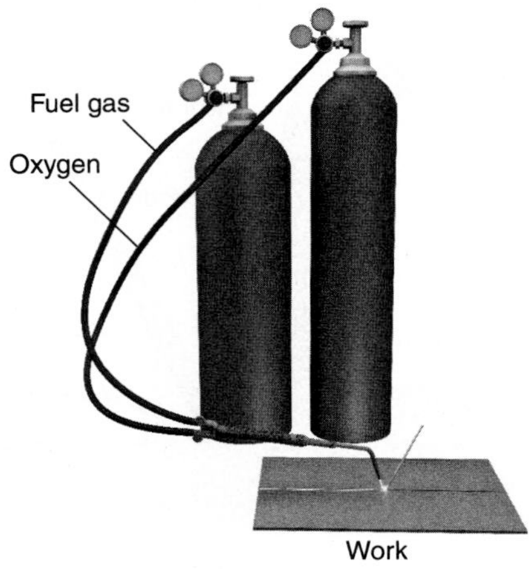

Figure 4.13.17 Oxyfuel Welding (OFW) process.

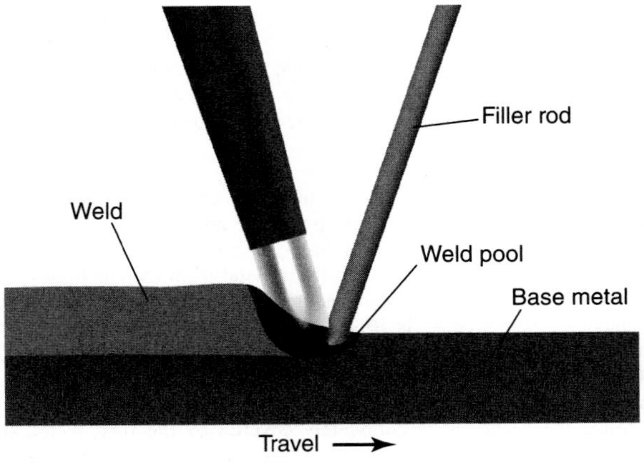

Figure 4.13.18 OFW process details.

Acetylene is typically used as the fuel gas for oxyfuel welding because of its high flame temperature. Other fuel gasses such as natural gas may be suitable for heating or brazing, but do not generate the intense energy needed for welding. The same torch used for oxyfuel welding can be used for brazing, soldering, and thermal cutting. It can be used to heat parts as well. This versatility makes oxyfuel welding sets a popular tool for maintenance, repair, and light fabrication.

High Energy Beam Welding

Electron Beam Welding (EBW) EBW is "a welding process producing coalescence with a concentrated beam, composed primarily of high-velocity electrons, impinging upon the joint." It is used without shielding gas and without the application of pressure. An electron beam gun, consisting of an emitter, a bias electrode, and an anode, is used to create and accelerate the beam of electrons. Auxiliary components such as beam alignment, focus, and deflection coils may be used with the electron beam gun; the entire assembly is referred to as the electron beam gun column.

The advantages of the process arise from the extremely high energy density in the focused beam which produces deep, narrow welds at high speed, minimizing distortion and other deleterious effects of high heat input. Major applications are with metals and alloys highly reactive to gases in the atmosphere or those volatilized from the base metal being welded.

EBW generally takes place in a vacuum chamber, restricting the size of parts that can be welded to those that can fit within the chamber. Although the welding may be relatively fast, the need to draw a vacuum each time a new part is to be welded slows the process cycle time down considerably. The need to provide precision parts and fixtures so that the beam can be precisely aligned with the joint to ensure complete fusion is another drawback. Gapped joints are not normally welded because of fixture complexity and the extreme difficulty of manipulating filler metal into the tiny, rapidly moving, weld puddle under high vacuum. Other disadvantages of the process arise from the cost, complexity, and skills required to operate and maintain the equipment, and the need to protect operating personnel from the X-rays generated during the operation.

Laser Beam Welding (LBW) LBW is a process that produces "coalescence with the heat from a laser beam impinging on the joint." In simple terms, the laser is an energy conversion device that transforms energy from a primary source into an electromagnetic beam of radiation of a particular frequency, producing intense localized heating. An inert gas shields the molten puddle. Filler metal may be added in some applications.

LBW is increasing in popularity, primarily because the cost of the equipment has been reduced. The total energy input can be carefully controlled, minimizing the problems associated with excessive energy input (such as distortion). Single pass welds in materials 1 in (25 mm) thick or more are possible, without the need for filler metal and joint preparation. The source of laser energy can be located away from the weld joint and the energy transmitted by optics (mirrors and lenses) or optical cables. Thus, the laser beam can be directed into areas that may be inaccessible with other processes. Many types of material have been successfully welded with LBW and extremely high productivity rates can be achieved. LBW requires careful alignment of the parts, clean surfaces, accurate fixturing, and is suitable only for automatic applications. Unlike EBW, LBW does not require a vacuum for welding, nor does it emit radiation during welding.

Resistance Welding (RW) RW is "a group of welding processes producing coalescence of the faying surfaces with the heat obtained from the resistance of the workpieces to the flow of the welding current in a circuit of which the workpieces form part and by the application of pressure." When welding current is passed through the part; the interface between the parts constitutes a point of high electrical resistance and thus heating and melting is concentrated in this area. Pressure is applied before, during, and after welding, creating the atomic closeness that is required in welding.

Since there is no means by which oxides can be reduced or removed with RW, surface cleanliness is a critical variable. The resistance between the materials being welded, as well as the resistance between the electrodes and the work, must be consistent for quality welding. Different base metals have unique cleanliness requirements. Resistance welding on coated steels can be challenging.

The electrical energy is introduced to the parts by means of electrodes, typically made of alloyed copper. For high productivity situations, the electrodes are water cooled to prevent overheating. In addition to introducing the electrical energy into the part, the electrodes apply pressure to the part.

Resistance Spot Welding (RSW) RSW is "a resistance welding process producing a spot weld." In RSW, the parts are lapped and held in place under pressure, as shown in Fig. 4.13.19. The size and shape of the electrodes control the size and shape of the welds, which are usually circular, illustrated in Fig. 4.13.20. Resistance spot welding machines vary from small, manually operated units to large, elaborate units designed to produce welds on complex assemblies. Portable gun-type machines are available for use where the assemblies are too large to be transported to a fixed machine. Spot welding guns are often mounted on robots that allow for rapid repositioning of the guns.

Figure 4.13.19 Resistance Spot Welding (RSW) process.

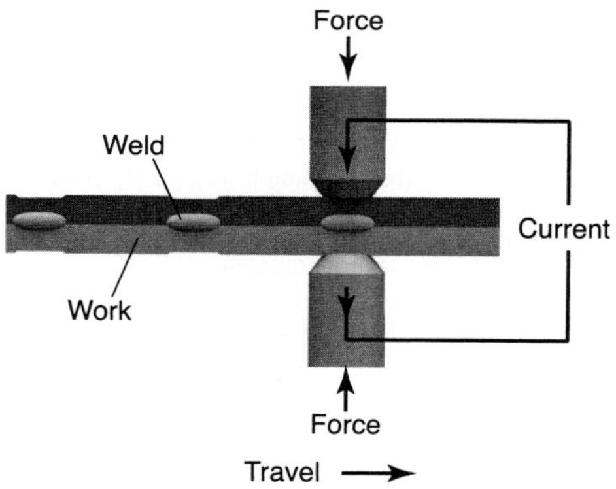

Figure 4.13.20 RSW process details.

RSW is often used as a replacement for screws and rivets on assemblies that do not need to be disassembled. Material up to 1/8 in (3 mm) thick can be readily spot welded. The process is fast and economical, although it relies on expensive equipment. RSW requires little skill and can be easily automated. Resistance spot welding is only possible on lap joints and has limited capacity in terms of strength.

Resistance Seam Welding (RSEW) RSEW produces "a weld at the faying surfaces of overlapped parts progressively along a length of a joint." The resistance seam welding process produces a series of spot welds made by circular or wheel type electrodes as shown in Fig. 4.13.21. The continuous weld is formed by a series of overlapping spot welds. The circular electrode rotates, maintaining constant pressure on the parts being welded. The welding current is pulsed, creating the overlapping spot welds. RSEW permits a seam to be sealed, which is useful in making tanks that are required to retain fluids.

The seam for RSEW must be straight or gently curving; abrupt changes in the seam create welding challenges. Welds can be made quickly and at a relatively low cost, but the equipment is expensive. Arc welded seams typically exhibit greater strength than RSEW seams.

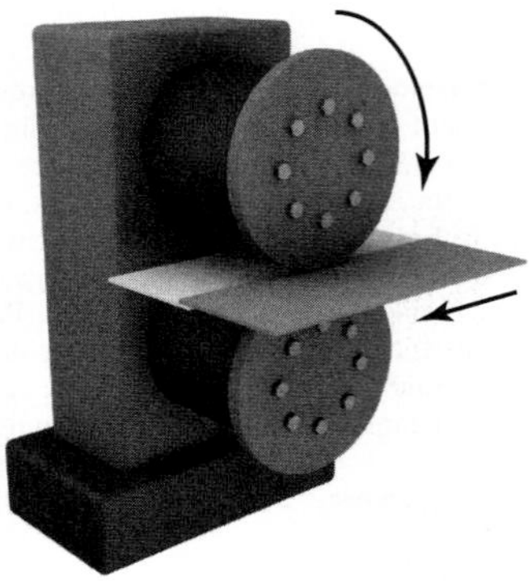

Figure 4.13.21 Resistance Seam Welding (RSEW) process.

Projection Welding (PW) PW is a "resistance welding process in which the weld size, shape, and placement is determined by the presence of a projection, embossment, or intersection in one overlapping member which serves to localize the applied heat and force." The heat for welding is derived from the localization of resistance at predetermined points. Single or multiple welds may be made somewhat more easily than with resistance spot welding. When made in multiples, all welds may be made simultaneously. The projection is the point of welding and thus the weld location is better controlled than with RSW.

Projection welding can be used to join mounting devices such as nuts, bolts, and brackets to other materials. The "dimple" effect associated with RSW is eliminated on the part that contains the projection. The use of the projection permits greater variations in thicknesses between the materials being joined, with ratios of 6:1 being possible. The process relies on the projection which must be created in a prior manufacturing operation.

Solid-State Welding (SSW) SSW is "a group of welding processes producing coalescence by the application of pressure without melting any of the joint components." These processes tend to be automatically applied by dedicated machines made for very specific applications. They are particularly well suited for joining materials or combinations of materials that, for metallurgical reasons, cannot be melted and re-solidified without major difficulties.

Friction Welding (FRW) FRW is "a solid-state welding process producing a weld under the compressive force contact of workpieces rotating or moving relative to one another to produce heat and plastically displace material from the faying surfaces." The process, shown in Fig. 4.13.22, is considered solid-state although some melting may occur. Two approaches are used to introduce the mechanical energy to the rotating parts. With the direct drive method, one part is attached to a motor while the other is fixed from rotation. The rotating part is pressed toward the fixed part, resulting in friction that generates intense local heat. Surface oxides and contaminants are mechanically removed from the interface. After a select amount of time, the rotating part is stopped, either by a brake or by the resistance of the formed weld.

The other method, called friction inertia welding (or, simply inertia welding) involves a flywheel that delivers the energy to one part, the other part being fixed. The two parts are pressed together and the kinetic energy stored in the rotating flywheel is converted to heat. As the flywheel loses energy and as the weld becomes stronger, rotation eventually stops.

Regardless of drive method, the applications for friction welding are the same, typically involving objects that are circular in cross section. No filler metal is needed, nor is flux or filler metal required. Surface cleanliness is not a factor as the oxidized material is expelled out of the joint during welding. The expelled material, called flash, is usually removed as a post weld operation. The process is automated and little operator skill is required. The two parts to be joined must be nearly symmetrical and circular. The equipment used for the process is specialized and expensive, although it is relatively inexpensive to make each individual weld.

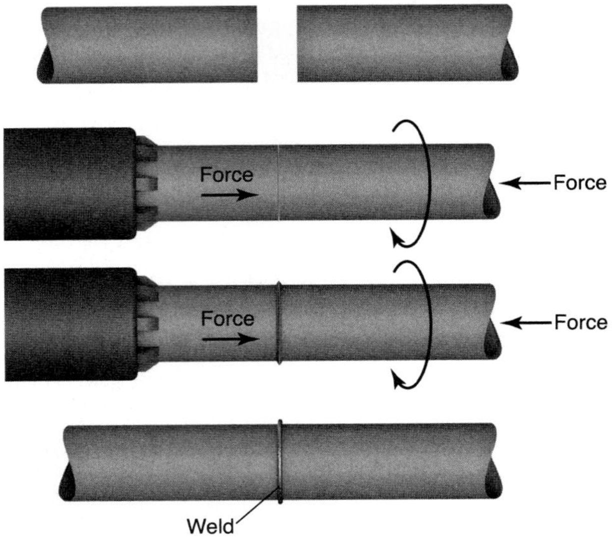

Figure 4.13.22 Friction Welding (FRW) process.

Ultrasonic Welding (USW) USW is "a solid-state welding process producing a weld by the local application of high-frequency vibratory energy as the workpieces are held together under pressure." As shown in Fig. 4.13.23, the two parts to be welded are clamped between a welding tool and an anvil. The welding tool is essentially a transducer that converts electric frequencies to ultra-high-frequency mechanical vibrations. The vibration and pressure cause the atoms between the two pieces to bond. USW is capable of welding metals to metals, plastics to plastics, and metals to nonmetals.

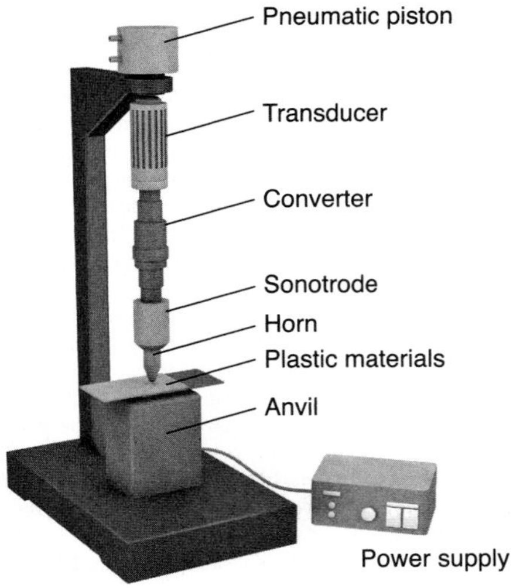

Figure 4.13.23 Ultrasonic Welding (USW) process.

The process is restricted to relatively thin materials and is used to weld electronic components. While one part must be thin, typically under 0.10 in (2.5 mm), the part against the anvil can be thicker. Ultrasonic welding does not generate high temperatures, enabling the process to be used to join encapsulating material for explosives. It is often used in applications that might otherwise be joined by resistance welding, but where the size of the part, or the need to avoid heating the part, makes USW preferable. Like resistance welding, ultrasonic welding is restricted to lap joint configurations.

Cold Welding (CW) CW is "a solid-state welding process in which pressure is used to produce a weld at room temperature with substantial deformation of the weld." The process relies on pressure, without heat, to create metallurgical bonds. To successfully cold weld materials, at least one piece to be welded must be highly ductile and must not be subject to work hardening. Soft aluminum and copper can be cold welded, as can gold, silver, platinum, and palladium. Dissimilar metals can be joined, such as copper and aluminum. Joints are restricted to butts and laps. Clean surfaces are a necessity. Pressure is applied by mechanical means: hydraulic or mechanical presses, rolls, or special tools.

Diffusion Welding (DFW) DFW is "a solid-state welding process producing a weld by the application of pressure at elevated temperature with no macroscopic deformation or relative motion of the workpieces. The process relies on pressure and temperature to create bonding although the temperatures are below the melting point. Similar and dissimilar materials can be joined by diffusion welding. Sometimes, an intermediate layer of yet another material is inserted between the materials being welded.

DFW is a three step process. First, the combination of temperature and pressure brings the two materials into intimate contact (by yielding and creep deformation). At this point for similar materials, the interface between the two pieces is essentially a grain boundary. Next, diffusion comes into play; and the boundary between the materials starts to disappear as diffusion occurs between the two materials on opposite sides of the previous interface. Finally, grain growth occurs where separate grains grow into one. With little heat involved, base metal properties are not significantly affected and distortion is minimized.

Explosion Welding (EXW) EXW is "a solid state welding process producing a weld by high velocity impact of the workpieces as the result of controlled deformation." The two sheets of material to be joined are stacked on top of each other, with a carefully controlled space between the two, called the standoff distance. An explosive is placed on the top of the pack. When detonated, the explosion accelerates the top part onto the bottom part, forcing atomic closeness. Some heating is involved, not from the explosion but from the collision. The explosion always starts at one edge of the part and progresses across the part. As the explosion progresses, the air between the prime component (upper piece) and the base component (lower piece) compresses, creating a high velocity jet stream which removes the surface oxides and contaminants.

As is true with all solid state processes, joining dissimilar metals is a common application of explosion welding. Specifically, cladding large sheets of dissimilar material is often done with EXW. The process is also used to make transitional joints. Transitional joints may be used in situations where fusion welding is impossible (such as dissimilar metal applications where there is a large difference in melting temperatures). Once a transition joint is made with EXW, arc welding may then be used to join each side of the transition joint to its respective material.

Friction Stir Welding (FSW) FSW is a variation of friction welding producing a weld by the friction heating and plastic material displacement caused by a rapidly rotating tool traversing the weld joint. The process uses a cylindrical, shouldered tool that is mounted in a machine with the overall configuration of a vertical mill, as shown in Fig. 4.13.24. Friction causes the metal being welded to heat and soften, but not melt. The plasticized material is moved from the leading edge of the tool to the trailing edge, creating a solid-state bond.

FSW is an automatic welding process, requiring application-specific tooling and equipment. Since no melting is involved and temperatures are low, it is commonly used to weld aluminum, usually for making butt seams.

Other Welding Processes

Electroslag Welding (ESW) ESW is "a welding process producing coalescence of metals with molten slag, melting the filler metal and the surfaces of the workpieces. The weld pool is shielded by this slag, which moves along the full cross section of the joint as welding progresses." Technically, it is not arc welding, but resistance welding, since the filler metal and base metal are heated by electrical resistance. ESW, illustrated in Fig. 4.13.25, uses a solid or tubular electrode that is fed through an electrically conductive hot slag that melts the electrode, adding metal to the weld pool. ESW is used for vertical-up welding, with the weld pool contained by copper dams (or shoes) on the sides of the weld; metallic backing that fuses to the weld can also be used. Groove welds in butt and T joints are the most common applications. The welds are completed in a single pass.

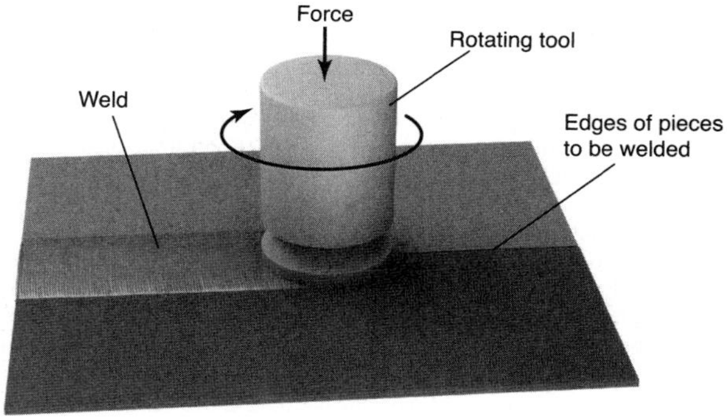

Figure 4.13.24 Friction Stir Welding (FSW) process.

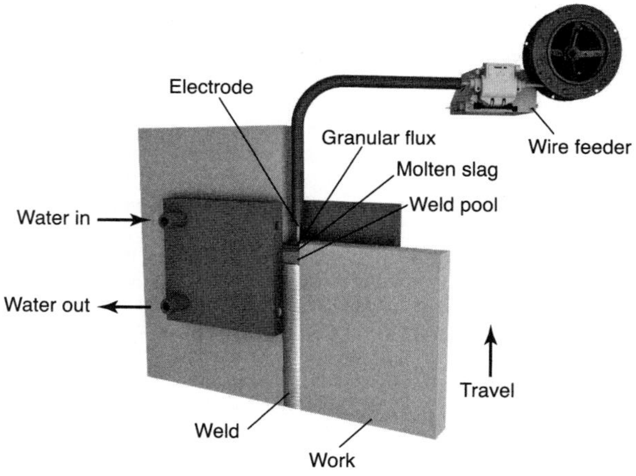

Figure 4.13.25 Electroslag Welding (ESW) process.

When electroslag welding is started, it functions like SAW; an arc buried under the flux melts the base metal, filler metal and flux, forming a slag. However, unlike SAW, the slag for ESW is electrically conductive. After a slag blanket is established, the electrical current is conducted from the electrode, through the slag, and into the workpiece. The high currents transferred through the slag keep it hot. As the electrode is fed through this hot slag, it melts, and molten metal drips from the electrode into the weld pool. No arc is involved, except when the process is started.

Very high deposition rates can be obtained with ESW, leading to productivity gains. Normally, the joint details involve square edge preparations, eliminating plate beveling costs. In some cases, material handling is reduced—plates do not need to be flipped as is the case for double-sided welds made with SAW, for example. Angular distortion can be reduced, as compared to single-sided welds in V and bevel grooves. ESW is ideal for thicker materials, and typical applications are 1 in (25 mm) thick or greater. Due to the high levels of heat input, often ten times or more higher than those typical of submerged arc welding, the heat-affected zone (HAZ) is large.

Thermite Welding (TW) TW produces "coalescence of metals by heating them with superheated molten metal from a chemical reaction between a metal oxide and aluminum." The process is commonly used to join sections of rail together, or large diameter reinforcing steel bars. A butt joint is created between the two parts to be joined and molds are applied to the sides. The joint is typically preheated to high temperatures and then the thermite mix or charge is positioned above the joint. The charge is ignited and the exothermic reaction begins, melting the material that flows into the joint cavity. The lower density slag flows to the top and out of the joint. Once the metal solidifies, the mold is removed.

The primary advantage of thermite welding is that most of the energy associated with it is generated by the thermite reaction, making the process highly portable. TW is quite specialized and requires the two parts to have similar cross sections. The materials used for each weld tend to be expensive, although no other major costs are involved. Properly done, thermite welds can be of high quality. Improperly done, however, TW is relatively unforgiving. While steel is the most common material to be joined by thermite welding, copper can also be welded with TW.

SOLDERING AND BRAZING

Soldering and brazing are similar joining processes that both involve melting of a filler metal, but no melting of the base metal(s). The thermal energy required for soldering and brazing is generated in a variety of ways.

Soldering

Soldering describes "a group of joining processes in which the workpiece(s) and solder are heated to the soldering temperature to form a soldered joint." The soldering filler metal has a melting point that does not exceed 840°F (450°C) and is below the melting point of the base metal. The solder is introduced into the closely fitting parts by capillary action. Typically, the bond between the solder and the base metal is a metallurgical bond, although for some material combinations it is an adhesive bond. Successful soldering operations rely on clean base metal surfaces and the proper fluxing to enable the solder to wet the base metal surface.

Many methods are used to deliver the thermal energy to the part to be soldered. These include the common examples of soldering irons and torches. Dip soldering involves a molten bath of solder into which the parts to be joined are dipped. Parts can be heated in ovens or furnaces or heated with induction coils or infrared sources. Resistance heating of the parts is possible as well. Wave soldering is a specialized process used in the manufacture of printed circuit boards.

A variety of filler metals are available for soldering, depending partly on the base metals being soldered. Common alloys in solders include lead, tin, antimony, copper, silver, cadmium, and zinc. Various combinations of these alloys are used, and the ratio of the two elements will affect the melting temperature, wetting capabilities, and solidification characteristics.

Fluxes serve to cleanse already clean surfaces, to protect the surface from further oxidation, and to protect the molten filler metal from oxidation. The selection of flux type depends on the type of material being soldered. Flux is provided in the core of the filler metal, or as a liquid, a paste, or in dry powder form.

Because of the relatively low temperatures involved, soldering does not have a major effect on the base metal properties. Many different materials can be joined by soldering, providing the proper flux and filler metal are selected. The diverse heating methods allow for considerable flexibility in manufacturing. The key to consistent solder joint quality is careful control of the process.

Brazing

Brazing is "a group of joining processes producing the bonding of materials by heating them to the brazing temperature in the presence of a brazing filler metal having a liquidus above 840°F (450°C) and below the solidus of the base metals. The brazing filler metal is distributed and retained between the closely fitted faying surfaces of the joint by capillary action." As with soldering, various methods of heating are also available for brazing, and a wide variety of fluxes and filler metals are available that allow many different types of base metals to be brazed.

It is possible to make a "braze weld." Brazing relies on capillary action to draw the melted filler metal into the joint. A braze weld fills a joint with brazing filler material in a manner similar to adding welding filler metal to the joint, except braze welding never melts the base metal. The process is often used for repairing broken cast iron parts. The reduced heat and the soft deposited metal make braze welding ideal for such repairs, albeit at a reduced strength.

THERMAL CUTTING AND GOUGING

Oxyfuel Gas Cutting (OFC)

OFC is "a group of oxygen cutting processes using heat from an oxyfuel gas flame." The process relies on the chemical reaction of oxygen with the metal at elevated temperatures. OFC is used to cut steels and to prepare bevel and vee grooves. In this process, shown in Fig. 4.13.26, the metal is heated to its ignition temperature, or kindling point, using a series of preheat flames. After the ignition temperature is reached, a high-velocity stream of pure oxygen is introduced, which causes rapid oxidation to occur. The force of the oxygen stream blows the oxides out of the joint, resulting in a clean cut. The oxidation process also generates additional thermal energy, which is radially conducted into the surrounding steel, increasing the temperature of the steel ahead of the cut. The next portion of the steel is raised to the kindling temperature, and the cut proceeds.

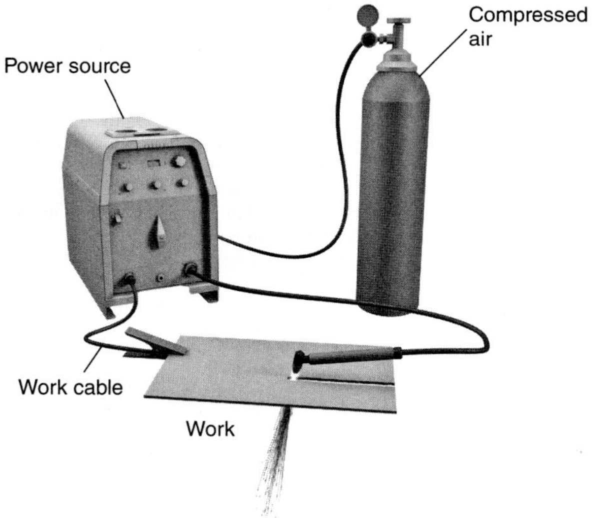

Figure 4.13.26 Oxyfuel Gas Cutting (OFC) process.

Because oxyfuel cutting is a combustion process that generates thermal energy, the process is well suited for cutting even very thick sections of steel, up to 12 in (300 mm) and greater. Oxidization resistant materials such as stainless steel and aluminum cannot be cut with conventional oxyfuel cutting methods.

Oxyfuel cutting of steel is done more rapidly than with mechanical cutting systems. The cut can be curved, beveled, and configured in ways that are difficult to achieve by mechanical means. Simple hand cutting systems are very economical, and the equipment is highly portable. Dimensional control is more difficult with oxyfuel cutting than with mechanical cutting systems. The process produces showers of sparks and molten metal, creating potential fire hazards. When

Figure 4.13.27 Air Carbon Arc Cutting (CAC-A) and Gouging process.

applied to hardenable steels, the edge may become extremely hard, brittle, and crack-sensitive. Localized heating and cooling can distort the cut parts.

Plasma Arc Cutting (PAC) and Plasma Arc Gouging (PAG)

PAC is "an arc cutting process employing a constricted arc and removing molten metal with a high-velocity jet of ionized gas issuing from the constricting orifice." PAC was developed initially to cut materials that do not permit the use of the oxyfuel process—stainless steel and aluminum. It was found, however, that plasma arc cutting offered economic advantages when applied to thinner sections of carbon steel, especially those less than 1 in (25 mm) thick. PAC can cut thin steels at higher speeds than oxyfuel cutting, providing the power supply is of ample capacity. The higher travel speeds have corollary advantages such as reduced distortion and reduced metallurgical changes on the cut surface.

Because the thermal energy generated by oxyfuel cutting during the oxidation process is not present in plasma cutting, PAC is not economical when thick sections of steel are cut. PAC is used to cut thick sections of stainless steel and aluminum where oxyfuel cutting is not viable. PAC may be used manually, or it can be automated. Larger parts can be cut on large cutting tables. The tables may be filled with water, and submerging the parts being cut in the water keeps noise, fumes and distortion to a minimum.

With minor changes to the plasma torch, the same process can be used for gouging, creating cavities necessary for U and J groove welds, or for removal of metal from joints that must be backgouged.

Laser Beam Cutting (LBC)

LBC is "a thermal cutting process severing metal by locally melting or vaporizing it with the heat from a laser beam. The process is used with or without assist gas to aid the removal of molten and vaporized material." While lasers can be used to cut thicker materials, the most common applications involve metals less than 3/8 in (10 mm) thick. For sheet steel application, very high travel speeds are possible.

Laser beam cutting has become a popular alternative to the stamping operations used in the past. With different programs, computer controlled cutting paths enable one machine to make an endless series of parts, eliminating costly dies and time consuming die changes. Cut surfaces of very high quality are possible in a variety of materials, both metals and nonmetals.

Air Carbon Arc Cutting (CAC-A) and Gouging

CAC-A is "a carbon arc cutting process variation removing molten metal with a jet of air." Air carbon arc gouging is the same process, but is used to produce cavities in material, for preparing for U or J groove profiles, for back-gouging joints, or for excavating material for a weld repair. The air carbon arc gouging system (Fig. 4.13.27) utilizes an electric arc to melt the base material; a high-velocity jet of compressed air subsequently blows the molten material away.

The process uses a standard welding power source and typically a copper coated carbon electrode. Gouging electrodes come in a range of sizes: larger electrodes require more energy but remove metal more rapidly. Air carbon arc gouging may be manually or automatically performed, the latter being more popular for the preparation of U and J groove weld profiles. Metal removal is fast and the equipment simple and relatively inexpensive. The process generates considerable noise, smoke, and a shower of sparks, all of which can create safety hazards.

Thermal Spraying (THSP)

THSP is "a group of processes in which finely divided metallic or nonmetallic surfacing materials are deposited in a molten or semi-molten condition on a substrate to form a thermal spray deposit. The surfacing material may be in the form of powder, rod, cord, or wire." The spray materials may be heated with an oxyfuel gas flame, electric arc, plasma or explosive gas mixture. The process is sometimes called metalizing, metal spraying or flame spraying.

Thermal spraying (Fig. 4.13.28) is used in maintenance operations to restore worn services. For new construction, it is used to apply special corrosion resistant or wear resistant properties to surfaces.

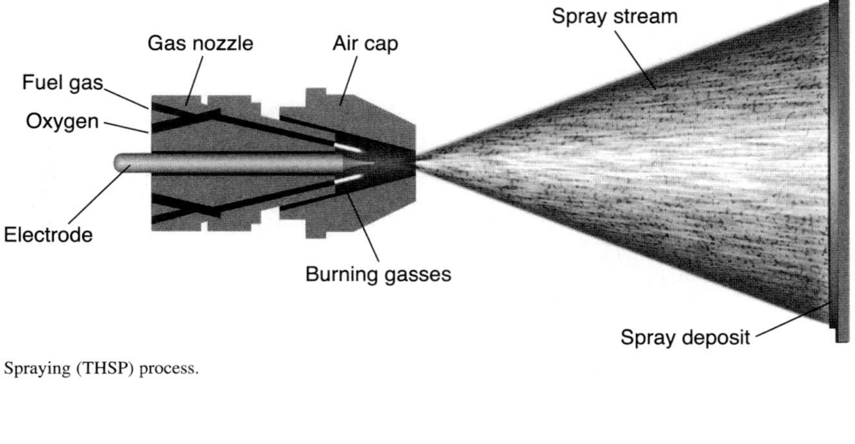

Figure 4.13.28 Thermal Spraying (THSP) process.

The part onto which the spray is applied is known as the substrate. Many substrate materials can be thermally sprayed. Regardless of the material, cleanliness is critical. In some cases, a bond coat is used between the substrate and the coating. Finally, there is the coating material, and it too may be composed of many different metals or nonmetals.

Sprayed materials typically bond to each other effectively, although the bond to the substrate may not be strong even when the process is performed correctly, depending on the materials involved. Incorrect procedures may result in poor bonding.

JOINTS AND WELDS

A **welded connection** consists of two or more pieces of base metal joined by weld metal. Design engineers determine joint type and generally specify weld type and the required throat dimension. Manufacturing engineers typically select the specific weld joint details to be used.

Joint Types

When pieces of steel are brought together to form a **joint,** they will assume one of the five configurations presented in Fig. 4.13.29. Joint types are descriptions of the relative positions of the materials to be joined and do not imply a specific type of weld. Butt joints are used in direct splices of material. T-joints are commonly used in many applications. Corner joints are often associated with box sections. Edge joints are usually reserved for applications involving thin materials and used where a joint must be sealed.

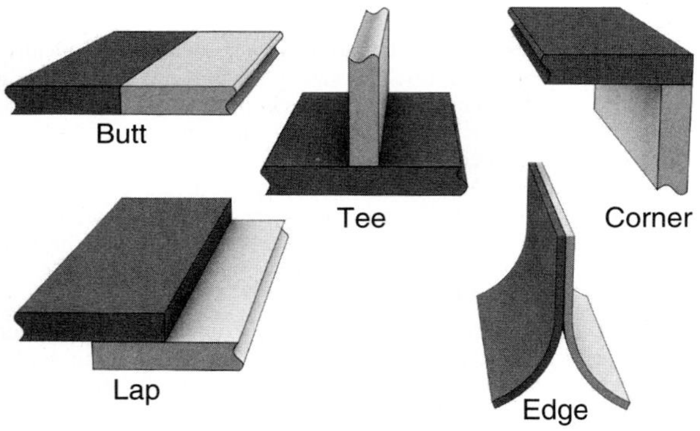

Figure 4.13.29 Joint types.

Weld Types

There are three major weld types: fillet welds, groove welds, and plug/slot welds, as shown in Fig. 4.13.30.

Fillet Welds

Fillet welds have a triangular cross section and are applied to the surfaces of the materials they join. By themselves, fillet welds do not fully fuse the cross-sectional areas of parts they join, although it is still possible to develop full-strength connections with fillet welds. The size of a fillet weld is usually determined by measuring the leg, even though the weld is designed by specifying the required throat. For equal-legged, flat-faced fillet welds applied to T-joints where the two plates intersect at a 90° angle, the throat dimension is found by multiplying the leg size by 0.707, or cosign (90°/2).

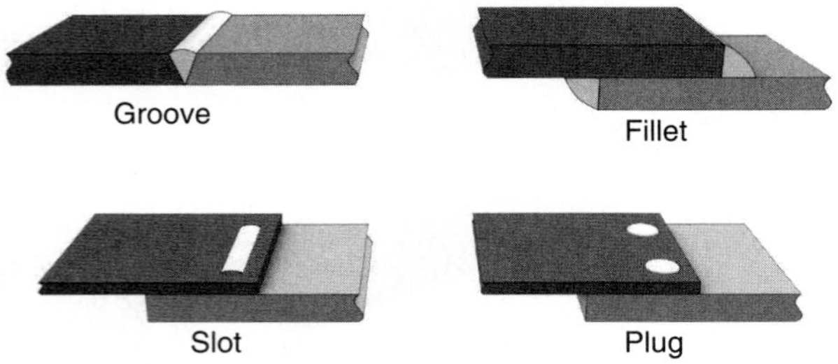

Figure 4.13.30 Weld types.

Groove Welds

Groove welds have two subcategories: complete joint penetration (CJP) groove welds and partial joint penetration (PJP) groove welds, as illustrated in Fig. 4.13.31. By definition, a CJP groove welds has a throat dimension equal to the thickness of the material it joins; a PJP groove weld is one with a throat dimension less than the thickness of the materials joined.

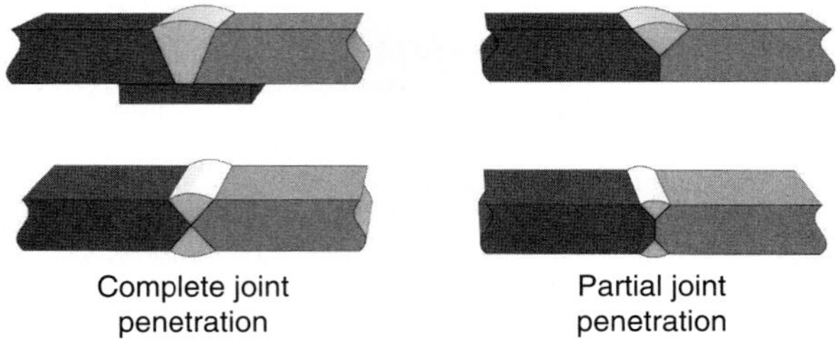

Figure 4.13.31 Types of groove welds.

CJP groove welds develop the full strength of the attached material when matching strength filler metal is used and the weld is full length, and when the applied loading is static (i.e., the loading is not cyclic). Accordingly, no design calculations are necessary when CJP groove welds are specified and these conditions are met. However, this convenience can lead to an abuse: CJP welds may be specified when the engineer simply wants to avoid determining what weld size is necessary. In some cases, this leads to welds that are much larger, and more expensive, than is justified.

PJP groove welds do not develop the full strength of the attached material. The engineer must determine the size of the PJP groove welds in order to provide a connection with the required strength. Associated with PJP groove welds is an **effective throat.** This term is used to differentiate between the depth of groove preparation and the probable depth of fusion that will be achieved. The **effective throat** on a PJP groove weld is abbreviated by *S*. The required **depth of groove preparation** is designated by a capital *D*. Since the designer may not know which welding process a fabricator will select, it is necessary only to specify the dimension for *S*. The manufacturing engineer then selects the welding process, determines the position of welding, and applies the appropriate *S* dimension. In most cases, both the *S* and *D* dimensions will appear on the welding symbols of shop drawings, with the effective throat dimension shown in parentheses.

Plug and Slot Welds

Plug welds are made by depositing weld metal in a prepared hole. Similarly, slot welds are made by depositing weld metal in a prepared slot. Two criteria determine proper hole or slot geometries: the fused area must yield a connection of sufficient strength, and the cavity must be configured in a way that allows for consistent fusion. Generally, plug and slot welds are designed to be at least as wide as the thickness of the member in which the hole is made, plus 5/16 in (8 mm).

Plug and slot welds are designed to resist shear. Generally, these welds are used in conjunction with other welds as well. When additional capacity from plug or slot welds is needed, additional plug or slot welds can be specified. These welds perform poorly in cyclically-loaded connections.

Selection of Weld Type

The performance and cost of welded connections is significantly affected by the selection of the proper weld type. Some weld types cannot be used in certain joint types. Groove welds can be applied to butt, tee, and corner joints, but cannot be used in lap and edge joints. Fillet welds can be used in tee, corner, and lap joints, but not in butt joints. Slot and plug welds are uniquely used in lap joints.

When the full tensile strength of the connected material of a butt joint is required, CJP groove welds must be used. If the same joint has limited tensile or shear loading across the joint, properly sized PJP groove welds will usually constitute a lower cost option. If loaded in compression, bearing of the un-fused root area can be considered as part of the load path.

For tee and corner joints, CJP and PJP groove welds are options, as are fillet welds. The loads transferred through these joints are often small, and fillet welds are ideal when this is the case. When fillet welds are required to be very large, with throat dimensions exceeding approximately 1 in (25 mm), PJP groove welds with an equivalent throat will generally constitute a lower cost option. For tee and corner joints, more expensive CJP groove welds should be used only when the loads transferred through the connection justify their use.

Lap joints are usually joined with fillet welds although lightly loaded connections may justify the use of plug or slot welds.

For cyclically loaded connections, the individual selecting a weld type must also consider the fatigue behavior of the various options.

Specification of Welding Details

A system of welding symbols has been developed by the AWS; the details are specified in AWS A2.4 *Standard Symbols for Welding, Brazing, and Nondestructive Examination*. Examples of these symbols are shown in Fig. 4.13.32.

DETERMINING WELD SIZE

The fundamental purpose of a weld is to hold two or more pieces of material together or, alternately stated, to permit forces to be transferred from one piece of material into another. Accordingly, welds must be of a size that permits the transfer of forces through the connection without overloading the connection. The transferred loads are resisted by the effective area of the weld, which is the product of the effective throat times the effective length of the weld. This localized stress should not exceed the allowable stress for the weld. Organizations such as the AWS and American Institute of Steel Construction (AISC), as well as other organizations, have established allowable stress levels for welds. Figure 4.13.33 shows allowable stress levels for various weld types.

There is no need to determine CJP groove weld sizes since, by definition, the weld throat is equal to the thickness of the material joined. For statically loaded, full length CJP groove welds made with matching strength filler metal, the strength of the connection is not governed by the weld but controlled by the capacity of the base metal.

The allowable stress on the effective area of PJP groove welds, fillet welds, and plug/slot welds is 30% of the minimum specified tensile strength of the filler metal used to make a given weld. For the common situation where weld metal with a 70 ksi (480 MPa) classification tensile strength is used (as is the case for filler metals such as E7,018, E70S-6, and E70T-1, for example), the permitted stress on the effective area is 0.30×70 or 21 ksi (0.30×480 or 145 MPa). The 30 percent factor accomplishes several tasks: it includes a conversion from tensile strength to shear strength and incorporates a factor of safety against rupture of 2.5. This factor, which was experimentally established in the late 1960s and has been used in a variety of structural and mechanical fields since then, is applicable to statically loaded connections.

Filler Metal Strength

The third column of Fig. 4.13.33 addresses the required filler metal strength level, which may be classified as matching, undermatching, or overmatching. **Matching filler metal** has the same, or slightly higher, minimum specified yield and tensile strength as the base metal. CJP groove welds in tension require the use of matching weld metal—otherwise, the strength of the welded connection will be lower than that of the base metal. **Undermatching filler metal** deposits welds of a strength lower than that of the base metal. Undermatching filler metal may be deposited in fillet welds and PJP groove welds as long as the designer specifies a throat size that will compensate for the reduction in weld metal strength. An **overmatching filler metal** deposits weld metal that is stronger than the base metal; this is undesirable unless, for practical reasons, lower-strength filler metal is unavailable for the application. When overmatching filler metal is used, if the weld is stressed to its maximum allowable level, the base metal can be overstressed, resulting in failure in the fusion zone. Designers must ensure that connection strength, including the fusion zone, meets the application requirements.

In welding high-strength steel, it is generally desirable to utilize undermatching filler metal for secondary welds. High-strength steel may require additional preheat and greater care in welding because there is an increased tendency to crack, especially if the joint is restrained. Undermatching filler metals such as E70 are the easiest to use and are preferred, provided the weld is sized to impart sufficient strength to the joint.

In many situations, the most difficult part of determining a weld size is quantifying the loads transferred through the connection. The stresses transferred through a connection are not always the same as the stresses in the member and the distribution of forces in a connection are often different than the forces in the member.

Consider the four illustrations in Figs. 4.13.34 to 4.13.37. In each situation, the welded portion has a box-like configuration. Each example involves four rectangular plate elements, joined at the corner with four longitudinal welds. The welded joints are essentially the same, but the loads transferred through the connections differ significantly.

Figure 4.13.34 assumes the box assembly functions as a hanger and the tensile forces are uniformly distributed across the whole cross section. Essentially no load is transferred across the connection and therefore relatively small welds are all that is needed to hold the pieces together. Shipping and handling loads will likely create the greatest stresses these welds will ever need to transfer.

The same assembly could be a column and loaded in compression as shown in Fig. 4.13.35. If totally unwelded, the slender plates making up the box could buckle in compression. Small welds will keep the four plates together in the box configuration which provides significant resistance to buckling. As compared to the same assembly loaded in tension, the compressive loading will result in more force transfer between the four elements, yet the transferred loads are still relatively small.

In Fig. 4.13.36, the same assembly is now a beam subject to bending. The load transferred between the four plates forming the box is horizontal (or longitudinal) shear, and the loading of concern is shear on the effective area of the weld. The shear force transferred from web to flange is typically small, and minimum-sized PJP welds—even intermittent PJP welds—are often adequate to resist the transferred load.

The example shown in Fig. 4.13.37 uses the same general configuration of material in the joint area, but full tensile loads in the outboard plates are transferred through the connections. This is a much different loading configuration as compared to the situations shown in Figs. 4.13.34 to 4.13.36, yet the welded joint detail is the same. Depending on the magnitude of the load and the length of the joint, a CJP groove weld may be required in this final example.

Welded connections do not need to develop the full capacity of the attached material, but rather must be of a size that is sufficient to safely transfer the loads that are applied to the connection. While this task is not always easy, it results in economical welded connections.

Directly and Indirectly Loaded Connections

It is often helpful to consider welded connections as either directly loaded, or indirectly loaded. In the case of directly loaded connections, the applied loads result in forces that are directly transferred through the connection. Several examples of directly loaded connections are

BASIC ARC WELDING SYMBOLS

LOCATION	Flash Or Upset	Groove					
	Square	Bevel	V	U	J	Flare Bevel	Flare V
Arrow-Side							
Other Side							
Both Sides							
No Arrow Side Or Other Side Significance	Not Used Except For Flash Or Upset Welds	Not Used	Not Used	Not Used	Not Used	Not Used	Not Used

(a)

LOCATION	Flange		Fillet	Plug or Slot	Back or Backing	Melt-Thru	Surfacing
	Edge	Corner					
Arrow Side					GROOVE WELD SYMBOL		Not Used
Other-Side					GROOVE WELD SYMBOL		Not Used
Both Sides	Not Used	Not Used		Not Used	Not Used	Not Used	Not Used
No Arrow Side Or Other Side Significance	Not Used	Not Used	Not Used	Not Used	Not Used	Not Used	

(b)

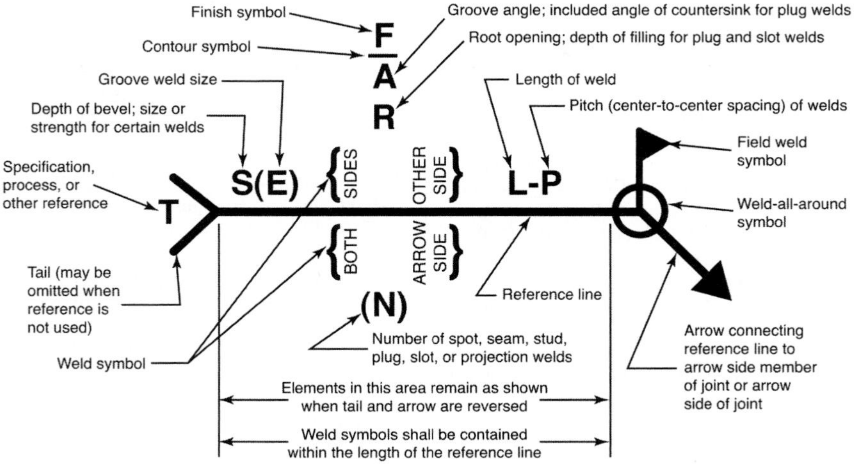

(c)

Figure 4.13.32 Basic weld symbols (adapted from AWS A2.4, used with permission from AWS).

Type of Applied Stress	Allowable Stress	Required Filler Metal Strength Level
	CJP Groove Welds	
Tension normal to the effective area[a]	Same as base metal	Matching filler metal shall be used[b]
Compression normal to effective area	Same as base metal	Filler metal with a strength level equal to or one classification (10 ksi [70 MPa]) less than matching filler metal may be used
Tension or compression parallel to axis of the weld[c]	Not a welded joint design consideration	Filler metal with a strength level equal to or less than matching filler metal may be used
Shear on effective area	0.30 × classification tensile strength of filler metal except shear on the base metal shall not exceed 0.40 × yield strength of the base metal	
	PJP Groove Welds	
Tension normal to the effective area	0.30 × classification tensile strength of filler metal	Filler metal with a strength level equal to or less than matching filler metal may be used
Compression normal to effective area of weld in joints designed to bear	0.90 × classification tensile strength of filler metal, but not more than 0.90 × yield strength of the connected base metal	
Compression normal to effective area of weld in joints not designed to bear	0.75 × classification tensile strength of filler metal	
Tension or compression parallel to axis of the weld[c]	Not a welded joint design consideration	
Shear parallel to axis of effective area	0.30 × classification tensile strength of filler except shear on the base metal shall not exceed 0.40 × yield strength of the base metal	
	Fillet Welds	
Shear on effective area or weld	0.30 × classification tensile strength of filler metal except that the base metal net section shear area stress shall not exceed 0.40 × yield strength of the base metal[d,e]	Filler metal with a strength level equal to or less than matching filler metal may be used
Tension or compression parallel to axis of the weld[c]	Not a welded joint design consideration	
	Plug and Slot Welds	
Shear parallel to the faying surface on the effective area[f]	0.30 × classification tensile strength of filler metal	Filler metal with a strengrh level equal to or less than matching filler metal may be used

[a] For definitions of effective areas, see 2.4.

[b] For matching filler metal to base metal strength for code approved steels, see Table 3.1, **Table 3.2**, and Table 4.9.

[c] Fillet welds and groove welds joining components of built-up members are allowed to be designed without regard to the tension and compression stresses in the connected components parallel to the weld axis although the area of the weld normal to the weld axis may be included in the cross- sectional area of the member.

[d] The limitation on stress in the base metal to 0.40 × yield point of base metal does not apply to stress on the diagrammatic weld leg; however, a check shall be made to assure that the strength of the connection is not limited by the thickness of the base metal on the net area around the connection. particularly in the case of a pair of fillet welds on opposite sides of a plate element.

[e] Alternatively, see 2.6.4.2, 2.6.4.3, and 2.6.4.4 Note d (above) applies.

[f] The strength of the connection shall also be limited by the tear-out load capacity of the thinner base metal on the perimeter area around the connection.

Figure 4.13.33 Table 2.3 from AWS D1.1:2015 (used with permission from AWS).

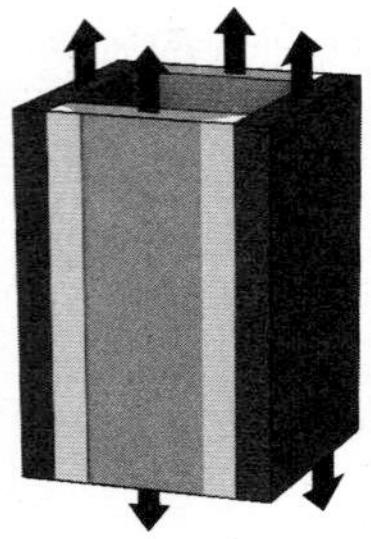

Figure 4.13.34 A welded box assembly with welds subject to "parallel tension."

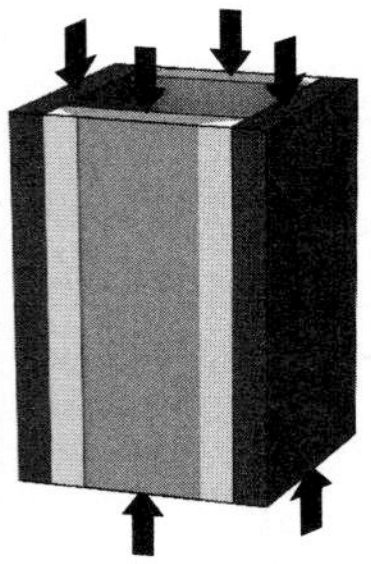

Figure 4.13.35 A welded box assembly with welds subject to "parallel compression."

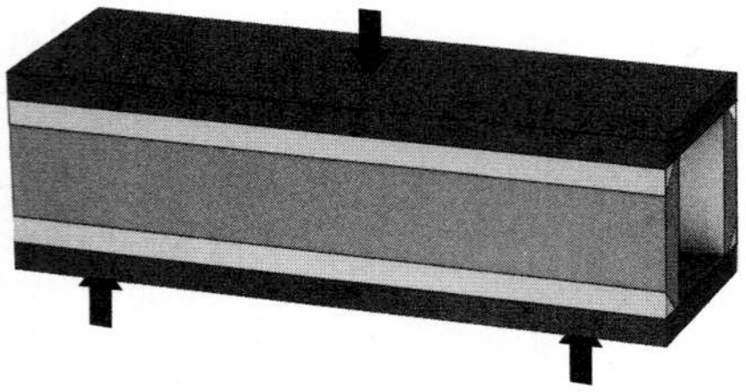

Figure 4.13.36 A welded box assembly with welds subject to shear, due to bending.

shown in Fig. 4.13.38. In such cases, the forces are usually more easily established, require larger welds, and involve welds that are more critical. The actual load may be tension, compression or shear. Welds that are directly loaded may be called primary welds.

Indirectly loaded welded connections are those where the applied force is not directly transferred through the connection, but rather creates secondary forces that are resisted by the weld. Consider the built-up beam shown in Fig. 4.13.39: the applied load results in horizontal (longitudinal) shear between the web and the flange. This

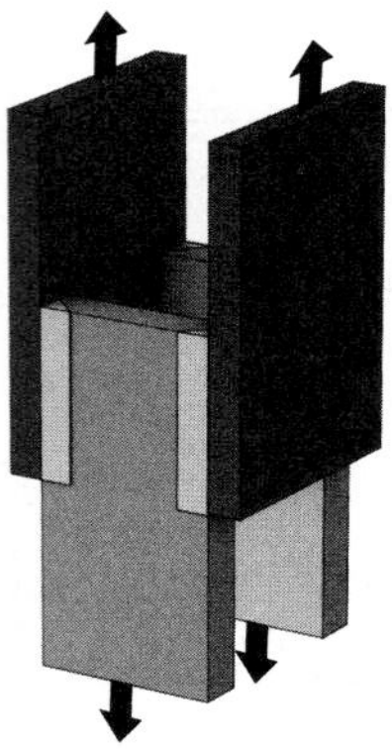

Figure 4.13.37 A welded box assembly with welds subject to shear, due to direct tensile loading.

force is usually small, and the welds joining the web to flange can be small as well. Also in Fig. 4.13.39, note the box section subject to torsion: the applied load results in shear between the four plate elements. Such shear forces are usually small and again, justify the use of smaller weld sizes. Indirectly loaded welds are often called secondary welds. Other examples of indirectly loaded connections are given in Fig. 4.13.40.

Treating the Weld as a Line

Treating the weld as a line allows for proper sizing of welds by modifying standard design formulas. Using this method, only the length of the weld must be known in advance. Without treating the weld as a line, one must know the throat dimension in addition to the weld length in order to calculate the capacity of the weld. The weld size, throat dimension and ultimately leg dimension, is what needs to be determined. The standard design formulas result in units of stress, ksi (MPa), but when the weld is treated as a line, the modified formulas result in units of force per unit length, lbs/in (N/m), as shown in Table 4.13.1. Section properties of welded connections also differ from the section properties

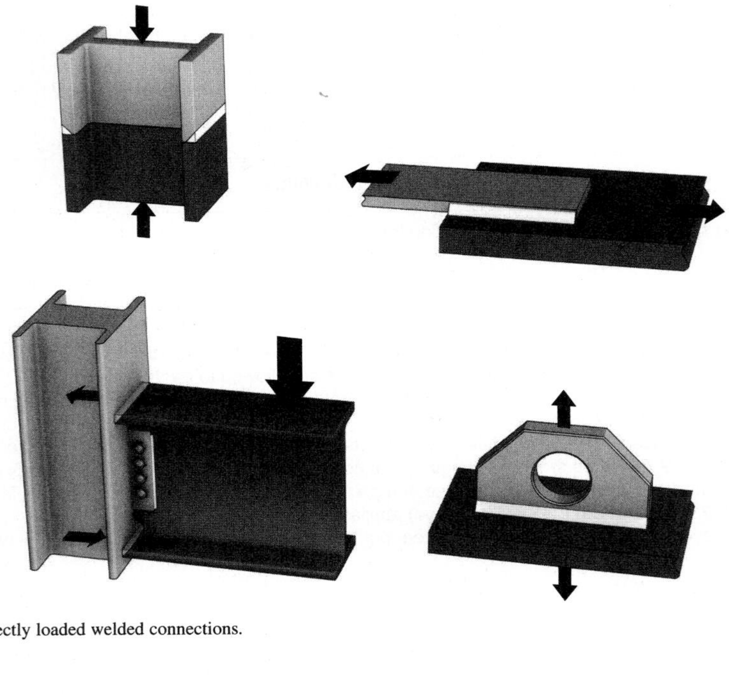

Figure 4.13.38 Examples of directly loaded welded connections.

Figure 4.13.39 Examples of indirectly loaded welded connections.

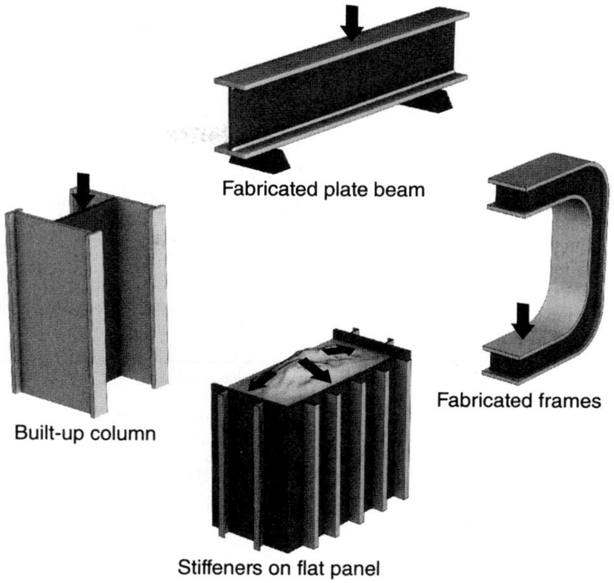

Figure 4.13.40 Additional examples of indirectly loaded welded connections.

of only a member such as a beam, channel, or round bar. Table 4.13.1 lists the section properties of common weld configurations for both bending, S_w, and twisting, J_w.

To determine the weld throat size, t_w, the resultant force applied to the weld is divided by the allowable strength of the weld. The leg size of the weld is then calculated by dividing the throat dimension by 0.707 (for 90° intersections), as follows:

$$t_w = \frac{f_{resultant}}{f_{allowable}}$$

$$\omega = \frac{t_w}{0.707}$$

where,

t_w is the effective throat size, in (mm)

$f_{resultant}$ is the force per length applied to the weld, lbs/in (N/m)

$f_{allowable}$ is the allowable strength of the weld, ksi (MPa)

and

ω is the fillet weld leg size

The resultant force is calculated using the equations in Tables 4.13.1 and 4.13.2. The allowable force is given as a percentage of the electrode classification. As discussed previously, for shear on the effective throat of PJP groove welds and fillet welds, AWS D1.1 and AISC specify

Table 4.13.1 Properties of a Weld Treated as a Line

Type of loading		Standard design formula	Treating the weld as a line
		Stress lb/in^2	Force, lb/in
Primary welds transmit entire load at this point			
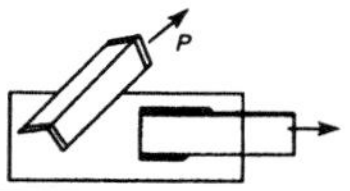	Tension or compression	$\sigma = \frac{P}{A}$	$f = \frac{P}{A_w}$
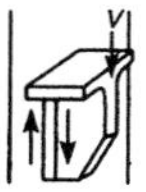	Vertical shear	$\sigma = \frac{V}{A}$	$f = \frac{V}{A_w}$
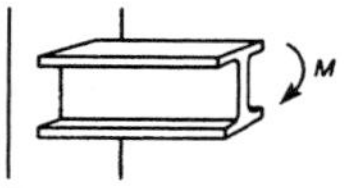	Bending	$\sigma = \frac{M}{S}$	$f = \frac{M}{S_w}$
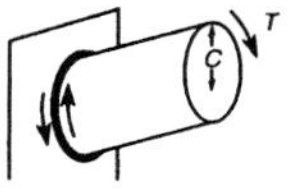	Twisting	$\sigma = \frac{TC}{J}$	$f = \frac{TC}{J_w}$

(Continued)

Table 4.13.1 Properties of a Weld Treated as a Line *(Continued)*

Type of loading		Standard design formula Stress lb/in^2	Treating the weld as a line Force, lb/in
Secondary welds hold section together—low stress			
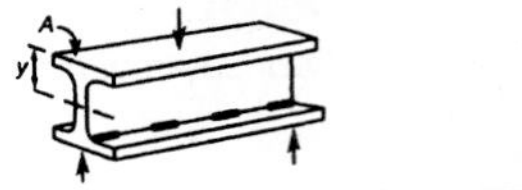	Horizontal shear	$\tau = \frac{VA_y}{It}$	$f = \frac{VA_y}{In}$
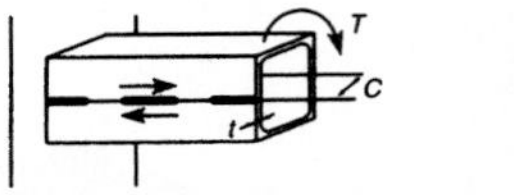	Torsion horizontal shear*	$\tau = \frac{T}{2At}$	$f = \frac{T}{2A}$

A = area contained within median line.
*Applies to closed tubular section only.

(*a*) Design formulas used to determine forces on a weld

b = width of connection, in
d = depth of connection, in
A = area of flange material held by welds in horizontal shear, in^2
y = distance between center of gravity of flange material and N.A. Of whole section, in
I = moment of inertia of whole section, in.^{4}
C = distance of outer fiber, in
t = thickness of plate, in
J = polar moment of inertia of section, in.^{4}
P = tensile or compressive load, lb
V = vertical shear load, lb

M = bending moment, in · lb
T = twisting moment, in · lb
A_w = length of weld, in
S_w = section modulus of weld, in^2
J_w = polar moment of inertia of weld, in^3
N_x = distance from x axis to face
N_y = distance from y axis to face
S = stress in standard design formula, lb/in^2
f = force in standard design formula when weld is treated as a line, lb/in
n = number of welds

(*b*) Definition of terms

Table 4.13.2 Properties of welded construction treated as a line

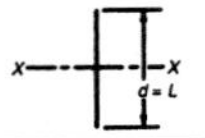	$I_X = \frac{d^3}{12}$	$S_X = \frac{d^2}{6}$	
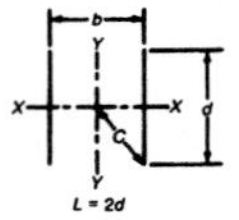	$I_X = \frac{d^3}{6}$ $I_Y = \frac{b^2 d}{2}$	$S_X = \frac{d^2}{3}$ $S_Y = bd$	$J_W = \frac{d}{6}(3b^2 + d^2)$ $C = \frac{(b^2 + d^2)^{1/2}}{2}$
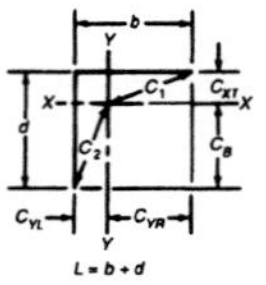	$I_X = \frac{d^3}{12}\left(\frac{4b+d}{b+d}\right)$ $I_Y = \frac{b^3}{12}\left(\frac{b+4d}{b+d}\right)$ $J_W = \frac{b^3 + d^3}{12} + \frac{bd(b^2 + d^2)}{4(b+d)}$ $C_T = \frac{d^2}{2(b+d)}$ $C_{YL} = \frac{b^2}{2(b+d)}$	$S_{XT} = \frac{d}{6}(4b+d)$ $S_{YL} = \frac{b}{6}(b+4d)$ $C_B = \frac{d}{2}\left(\frac{2b+d}{b+d}\right)$ $C_{YR} = \frac{b}{2}\left(\frac{b+2d}{b+d}\right)$	$S_{XB} = \frac{d^2}{6}\left(\frac{4b+d}{2b+d}\right)$ $S_{YR} = \frac{b^2}{6}\left(\frac{b+4d}{2d+b}\right)$ $C_1 = \left(C_T^2 + C_{YR}^2\right)^{1/2}$ $C_2 = \left(C_B^2 + C_{YL}^2\right)^{1/2}$
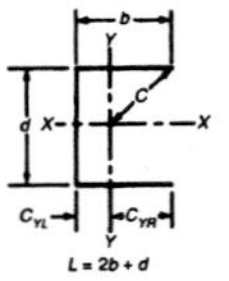	$I_X = \frac{d^2}{12}(6b+d)$ $I_Y = \frac{b^3}{3}\left(\frac{b+2d}{2b+d}\right)$ $C_{YL} = \frac{b^2}{2b+d}$ $C = \left[C_{YR}^2 + \left(\frac{d}{2}\right)^2\right]^{1/2}$	$S_X = \frac{d}{6}(6b+d)$ $S_{YL} = \frac{b}{3}(b+2d)$ $C_{YR} = \frac{b(b+d)}{2b+d}$ $J_W = \frac{b^3}{3}\left(\frac{b+2d}{2b+d}\right) + \frac{d^2}{12}(6b+d)$	 $S_{YR} = \frac{b^2}{3}\left(\frac{b+2d}{b+d}\right)$

Table 4.13.2   Properties of welded construction treated as a line

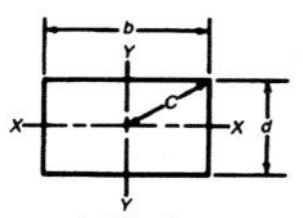

$$l_X = \frac{b^2}{6}(3b+d)$$

$$l_Y = \frac{b^2}{6}(b+3d)$$

$$J_W = \frac{(b+d)^3}{6}$$

$$S_X = \frac{d}{3}(3b+d)$$

$$S_Y = \frac{b}{3}(b+3d)$$

$$C = \frac{(b^2+d^2)^{1/2}}{2}$$

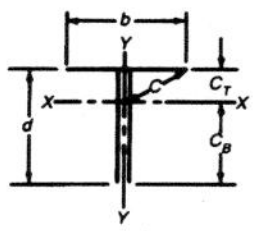

$$l_X = \frac{d^3}{3}\left(\frac{2b+d}{b+2d}\right)$$

$$l_Y = \frac{d^3}{12}$$

$$J_W = \frac{d^3}{3}\left(\frac{2b+d}{b+2d}\right)+\frac{b^3}{12}$$

$$S_{XT} = \frac{d}{3}(2b+d)$$

$$S_Y = \frac{d^2}{6}$$

$$C_B = d\left(\frac{b+d}{b+2d}\right)$$

$$C = \left[C_T^2 + \left(\frac{b}{2}\right)^2\right]^{1/2}$$

$$S_{XB} = \frac{d^2}{3}\left(\frac{2b+d}{b+d}\right)$$

$$C_T = \frac{d^2}{b+2d}$$

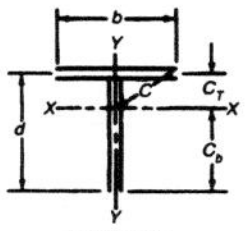

$$l_X = \frac{d^3}{6}\left(\frac{4b+d}{b+d}\right)$$

$$l_Y = \frac{d^3}{6}$$

$$J_W = \frac{d^3}{6}\left(\frac{4b+d}{b+d}\right)+\frac{b^2}{6}$$

$$S_{XT} = \frac{d}{3}(4b+d)$$

$$S_Y = \frac{b^2}{3}$$

$$C_B = \frac{d}{2}\left(\frac{2b+d}{b+d}\right)$$

$$C = \left[C_T^2 + \left(\frac{b}{2}\right)^2\right]^{1/2}$$

$$S_{XB} = \frac{d^2}{3}\left(\frac{4b+d}{b+d}\right)$$

$$C_T = \frac{d^2}{2(b+d)}$$

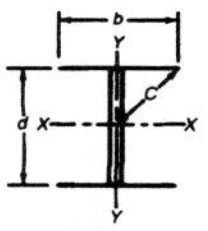

$$l_X = \frac{d^2}{6}(3b+d)$$

$$l_Y = \frac{b^3}{6}$$

$$J_W = \frac{d^2}{6}(3b+d)+\frac{b^3}{6}$$

$$S_X = \frac{d}{3}(3b+d)$$

$$S_Y = \frac{b^2}{3}$$

$$C = \frac{(b^2+d^2)^{1/2}}{2}$$

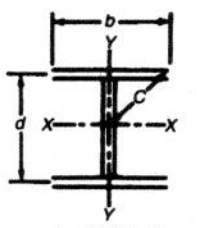

$$l_X = \frac{d^2}{6}(6b+d)$$

$$l_Y = \frac{b^3}{3}$$

$$J = \frac{d^2}{6}(6b+d)+\frac{b^3}{3}$$

$$S_X = \frac{d}{3}(6b+d)$$

$$S_Y = \frac{2}{3}b^2$$

$$C = \frac{(b^2+d^2)^{1/2}}{2}$$

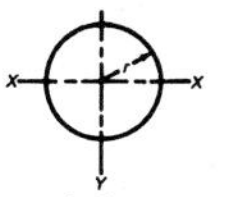

$$l = \pi r^3$$

$$S_W = \pi r^2$$

$$J_W = 2\pi r^3$$

Key:

π	= 3.1416	
b	= Length of the horizontal weld, in. (mm)	
C	= Distance from the center of gravity to the point of combined stress, in. (mm)	
C_B	= Distance from the X-axis to the bottom, in. (mm)	
C_T	= Distance from the X-axis to the top, in. (mm)	
C_{YR}	= Distance from the Y-axis to far right side, in. (mm)	
C_{YL}	= Distance from the Y-axis to the far left side, in. (mm)	
d	= Length of the vertical weld, in. (mm)	
J	= Polar moment of inertia of an area, in.[4] (mm[4])	
J_W	= Polar moment of inertia of a line weld, in.[3] (mm[3])	
L	= Total length of the weld, in. (mm)	

l	= Moment of inertia, in.[4] (mm[4])	
l_X	= Moment of inertia about the X-axis, in.[4] (mm[4])	
l_Y	= Moment of inertia about the Y-axis, in.[4] (mm[4])	
r	= Radius, in. (mm)	
S	= Section modulus of an area, in.[3] (mm[3])	
S_W	= Section modulus of a line weld, in.[2] (mm[2])	
S_X	= Section modulus about the X-axis	
S_{XB}	= Section modulus of the bottom piece about the X-axis	
S_{XT}	= Section modulus of the top piece about the X-axis	
S_Y	= Section modulus about the Y-axis	
S_{YL}	= Section modulus of the left piece about the Y-axis	
S_{YR}	= Section modulus of the right piece about the Y-axis	

SOURCE: Adapted from Blodgett, O. W., 1966, *Design of Welded Structures*, Cleveland: The James F. Lincoln Arc Welding Foundation, Section 7.4, Table 5.

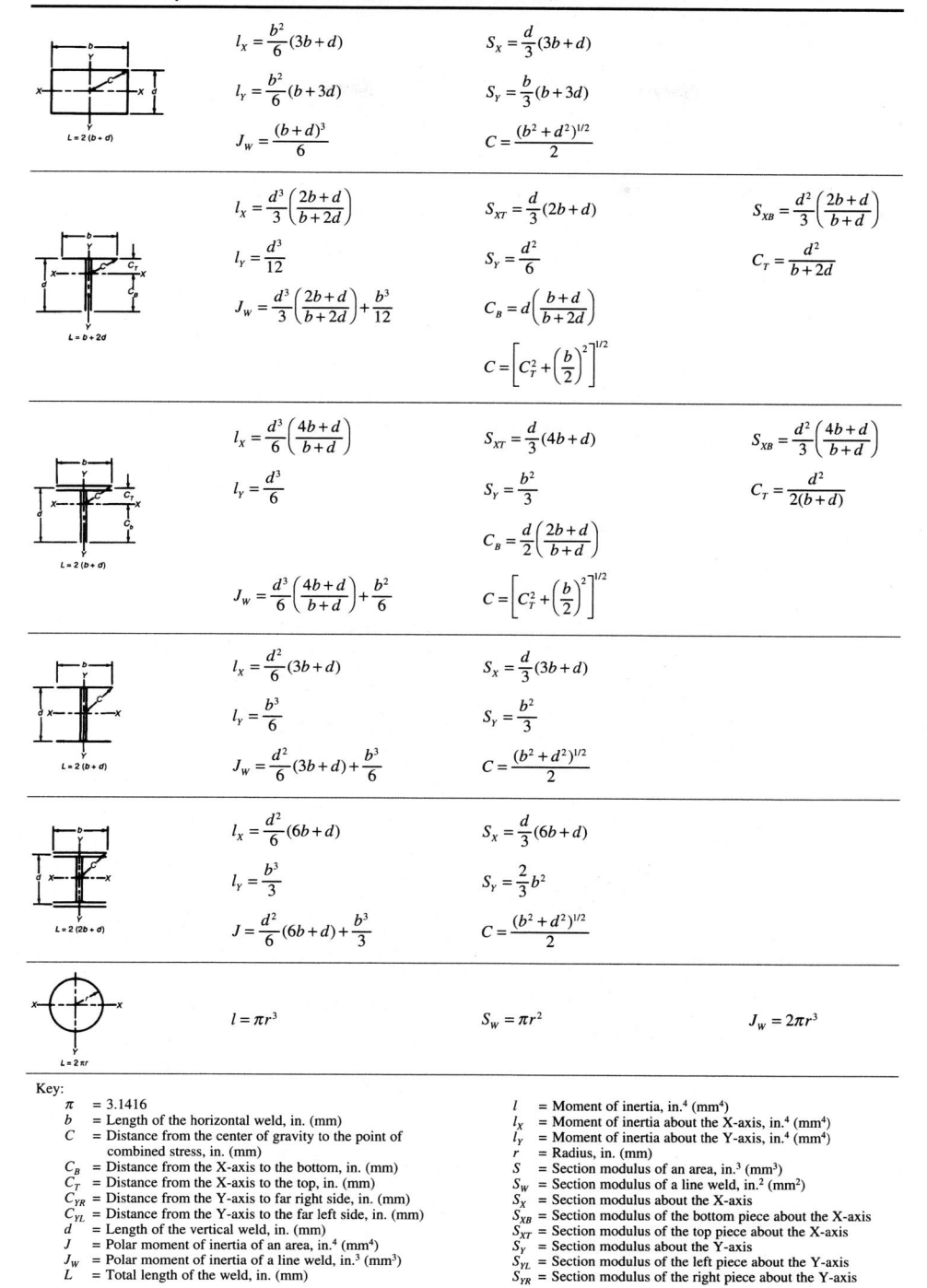

30 percent of the electrode classification tensile strength. The following steps outline how to calculate weld size by treating the weld as a line.

Steps for determining weld size: Find the position on the welded connection where the combination of forces will be maximum. There may be more than one location which must be considered. Find the value of each of the component forces on the welded connection at this point. Use Table 4.13.1 for the standard design formula to find the force on the weld and to find the properties of the weld treated as a line. Combine (vectorially) all the forces on the weld at this point. Determine the required weld size by dividing the value obtained by vectorially combined forces on the weld by the allowable force determined from Fig. 4.13.33 or Fig. 4.13.44.

Figure 4.13.41 presents a design example that demonstrates how to incorporate the design equations and electrode classifications.

Step 1. Identify the location where the combined forces are a maximum.

Where:

CG = Center of gravity
P = Applied load, lbf (N)
C_{YR} = Greatest distance from CG in the horizontal direction, in. (mm)
d = Length of the vertical weld, in. (mm)
f_h = Horizontal force component, lb/in. (N/mm)
f_r = Resultant force, lb/in. (N/mm)
f_s = Force component due to shear, lb/in. (N/mm)
f_v = Vertical twisting force component, lb/in. (N/mm)

Source: Adapted from Blodgett, O. W., 1966, Design of Welded Structures, *Cleveland: The James F. Lincoln Arc Welding Foundation, Figure 15.*

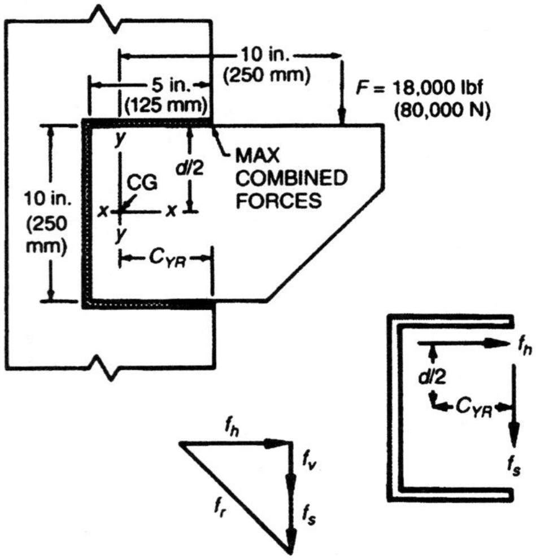

Where:

CG = Center of gravity

P = Applied load, lbf (N)

C_{YR} = Greatest distance from CG in the horizontal direction, in. (mm)

d = Length of the vertical weld, in. (mm)

f_h = Horizontal force component, lb/in. (N/mm);

f_r = Resultant force, lb/in. (N/mm)

f_s = Force component due to shear, lb/in. (N/mm)

f_v = Vertical twisting force component, lb/in. (N/mm)

Figure 4.13.41 Bracket joined to a column face plate.

The location of maximum combined forces is located at the right hand side of both the top and bottom welds (only one location needs to be considered).

Step 2. Determine the forces acting at the location of maximum combined forces.

The polar moment of inertia of the weld is determined from Table 4.13.2 as follows:

$$J_W = \frac{b^3}{3}\frac{(b+2d)}{(2b+d)} + \frac{d^2}{12}(6b+d)$$
$$= \frac{5^3}{3}\frac{(5+20)}{(10+10)} + \frac{10^2}{12}(30+10)$$
$$= 385 \text{ in.}^3\ (6.3\times 10^6 \text{ mm}^3)$$

where

J_w = Polar moment of inertia of a line weld, in.^{3} (mm^3);
b = Length of horizontal weld, in. (mm); and
d = Length of vertical weld, in. (mm).

The horizontal component of twisting, f_h, is determined from the "Twisting" equation in Table 4.13.1, as follows:

$$f_h = \frac{T\left(\frac{d}{2}\right)}{J_W} = \frac{(180{,}000)\left(\frac{10}{2}\right)}{385}$$
$$= 2{,}340 \text{ lb/in. } (410 \text{ N/mm})$$

where

f_h = Horizontal force component due to twisting, lb/in. (N/mm);
T = Torsional moment [= 18,000 × 10 = 180,000 in. lb (31.5 kN mm)];
d = Length of vertical weld, in (mm); and
J_w = Polar moment of inertia of a line weld, in.^{3} (mm^3).

The vertical twisting component is determined from the "Tension or Compression" equation in Table 4.13.1, as follows:

$$f_v = \frac{TC_{YR}}{J_W} = \frac{(180{,}000)(3.75)}{385}$$
$$= 1{,}750 \text{ lb/in. } (306 \text{ N/mm})$$

where

f_v = Vertical twisting force component, lb/in. (N/mm);
T = Torsional moment, in. lb (kN mm);
C_{YR} = Horizontal distance from the center of gravity to the point of combined stress.

$$C_{YR} = \frac{b(b+d)}{(2b+d)} = \frac{5(5+10)}{2(5)+10} = \frac{75}{20} = 3.75 \text{ in. } (95 \text{ mm});$$

J_w = Polar moment of inertia of a line weld, in.^{3} (mm^3).

The vertical shear force is determined from the equation for "Tension and Compression" in Table 4.13.1, as follows:

$$f_S = \frac{P}{L_W} = \frac{18{,}000}{20} = 900 \text{ lb/in. } (158 \text{ N/mm})$$

where

f_s = Vertical shear force, lb/in. (N/mm);
P = Applied load, lbf (N); and
L_w = Total length of the weld, in. (mm).

Step 3. Combine the forces to determine the resultant force, as follows:

$$f_r = \left[f_h^2 + (f_v + f_s)^2\right]^{1/2}$$

where

f_r = Resultant force, lb/in. (N/mm);
f_h = Horizontal force component due to twisting, lb/in. (N/mm);

f_v = Vertical twisting force component, lb/in. (N/mm); and
f_s = Vertical shear force, lb/in. (N/mm).

Step 4. Calculate the fillet weld leg size (assume E70 electrode is used).

$$\frac{f_r}{F_{allowable}} = \frac{f_r}{(0.707)(0.30)(F_{EXX})}$$

$$\omega = \frac{3{,}540}{(0.707)(0.30)(70{,}000)}$$

$$= 0.238 \text{ in}$$

Or a ¼-in (25-mm) leg fillet weld.

DESIGN OF WELDS SUBJECT TO CYCLIC LOADING

Fatigue is the process of cumulative damage caused by repeated fluctuating loads. Such damage occurs at regions of stress or strain concentrations that cause the localized stress to exceed the yield stress of the material. Fatigue damage can occur in any product when it is subjected to a sufficient number of loading cycles of enough magnitude, whether welded or not. However, welded connections have some extra challenges related to fatigue that must be considered.

Many welded products are subject to cyclic loading: agricultural equipment, earthmoving equipment, fans, compressors, material handling systems, cranes—virtually any product that moves in operation has the potential of fatigue damage. Even shipping loads may cause some products to experience fatigue damage. Non-moving structures like bridges and material storage racks can be subject to fatigue due to the cyclic loads applied to the structure.

Many fatigue models have been developed for the design of welded connections. This section will describe the model that is used in the AISC Specifications. This system has been adopted by the AWS and incorporated into AWS D1.1 Structural Welding Code—Steel, and AWS D14.4 Specification for the Design of Welded Joints in Machinery and Equipment. While the model was developed for structures such as bridges and cyclically loaded components in buildings, it has been successfully used in many other industries. This empirically derived model was developed from cyclic testing of large or full scale welded assemblies. The AISC Specifications which define the requirements for and use of the model described herein are available as a free download from www.aisc.org.

Fatigue loading can be divided into two categories: low cycle fatigue and high cycle fatigue. The AISC model and this section deal with the latter: high cycle loading. Typically, low cycle fatigue involves failures that occur at less than 1,000 cycles (and might even be fewer than a dozen) whereas high cycle fatigue deals with tens of thousands of loading cycles. For steel products, low cycle fatigue usually involves overall stress levels that result in inelastic (plastic) loading; high cycle fatigue involves overall stresses that are elastic (although the localized stresses at geometric irregularities will be inelastic).

The two key variables that predict the life of a welded connection are the stress range and the geometry of the connection. Thus, for a given number of loading cycles and connection geometry, the permissible stress range can be calculated and the connection designed so that the applied stress range is less than the permissible range.

One advantage of the AISC method is that stresses are determined in the conventional manner, and there is no need to assign any stress concentration factor to the computed stresses. Rather, based on statistically reliable experimental data, the actual stress concentrations are considered by identifying geometric details of connections that are the same as or similar to those that were experimentally evaluated.

Stress range is defined as the maximum stress minus the minimum stress. Such stresses may result from applied tension, bending or torsion, and the stresses are computed in the same manner as in the case of static loads. For applications involving cyclic loading, the stresses are first calculated with the dead load only. Next, the stresses due to the cyclic loads are determined. Finally, the stresses for the two loading conditions are combined. The difference between the maximum total stress and the minimum total stress is the stress range.

The stress range, not the maximum stress, is the best predictor of fatigue behavior for welded assemblies. As will be shown, the applied stress may be compressive or tensile, and the behavior of the welded component in terms of fatigue crack initiation will be the same, provided that the stress range is the same. This behavior is a result of the residual tensile stresses that are inherent to as-welded assemblies. In and around the weld, there are residual tensile stresses that are at or near the yield stress of the material.

The stress range ($\Delta\sigma$) is the difference between the maximum stress (σ_{max}) and the minimum stress (s_{min}), and is mathematically expressed as follows:

$$\Delta\sigma = \sigma_{max} - \sigma_{min}$$

Four conditions will be described to illustrate the concept of stress range. In each case, the welded assembly consists of a symmetrical beam with flanges that contain CJP groove welds, as shown in Fig. 4.13.42. The two conditions will consider a bare beam, shown in Fig. 4.13.42 part A; the latter two conditions will consider a beam with a non-composite concrete slab, as shown in Fig. 4.13.42 part B.

For the first example, consider the beam shown in Fig. 4.13.42 part A. Initially, no load is applied to the beam. Assuming the mass of the beam results in negligible stresses, then the stress across the welded connections in both top and bottom flanges is zero. Next, a vertical load is applied to the beam, resulting in tensile forces across the the bottom flange weld. The graph in Fig. 4.13.42 part C shows the changes in tensile stresses as a result of the application and removal of the load. The stresses vary from zero to a maximum level, assumed to be 15 ksi (105 MPa) in this example. The stress range (i.e., the maximum stress minus the minimum stress) is therefore 15 – 0 or 15 ksi (105 – 0 or 105 MPa).

For the second example, the top flange weld will be considered. Under the same loading conditions, the top flange experiences compression. With a symmetrical beam, the stresses across the CJP groove weld range from zero to a compressive stress of 15 ksi (105 MPa), or more precisely, –15 ksi (–105 MPa), as shown in Fig. 4.13.42 part D. The stress range is therefore 0 – (–15) or 15 ksi [0 – (–105) or 105 MPa]. Notice that the stress range in the second example is identical to the stress range in the first example.

For the final two examples, the same beam as was used in the first two examples is used again, but a concrete slab (non-composite) has been added as shown in Fig. 4.13.42 part B. This creates a constant load, and the bottom flange weld experiences a constant stress of +5 ksi (+35 MPa) (i.e., tension). The top flange weld experiences a constant stress of –5 ksi (–35 MPa) (i.e., compression).

For the third example, consider the bottom flange weld. The same load used in the previous examples is applied. This creates a total tensile stress in the bottom flange weld of 20 ksi (140 MPa). The stress range is therefore 20 – 5 or 15 ksi (140 – 35 or 105 MPa) as is shown in Fig. 4.13.42 part E. Again, the stress range is the same as the two previous cases.

Finally, for the fourth example, consider the top flange of the beam shown in Fig. 4.13.42 part B. The concrete slab results in a compressive stress in the top flange weld which is –5 ksi (–35 MPa). When the load is applied, the compressive stress is increased to –20 ksi (–140 MPa) as seen in Fig. 4.13.42 part F. The stress range is therefore –5 – (–20) or 15 ksi [–35– (–140) or 105 MPa]. Again, the stress range is the same as the previous cases.

These four examples, each with different stress states, all experienced the same stress range. The stress range in the examples is the same because the cyclic load was the same. And, since the welded flanges in the beam were the same in each example, the corresponding fatigue life is predicted to be the same. This is true, despite the fact that loading conditions varied significantly: zero to tension (Example 1); zero to compression (Example 2); tension to tension (Example 3); and compression to compression (Example 4). In addition to the consideration of cyclic loading, the static strength of the system must also be established, and the design of the members and connections in terms of the static strength may be different.

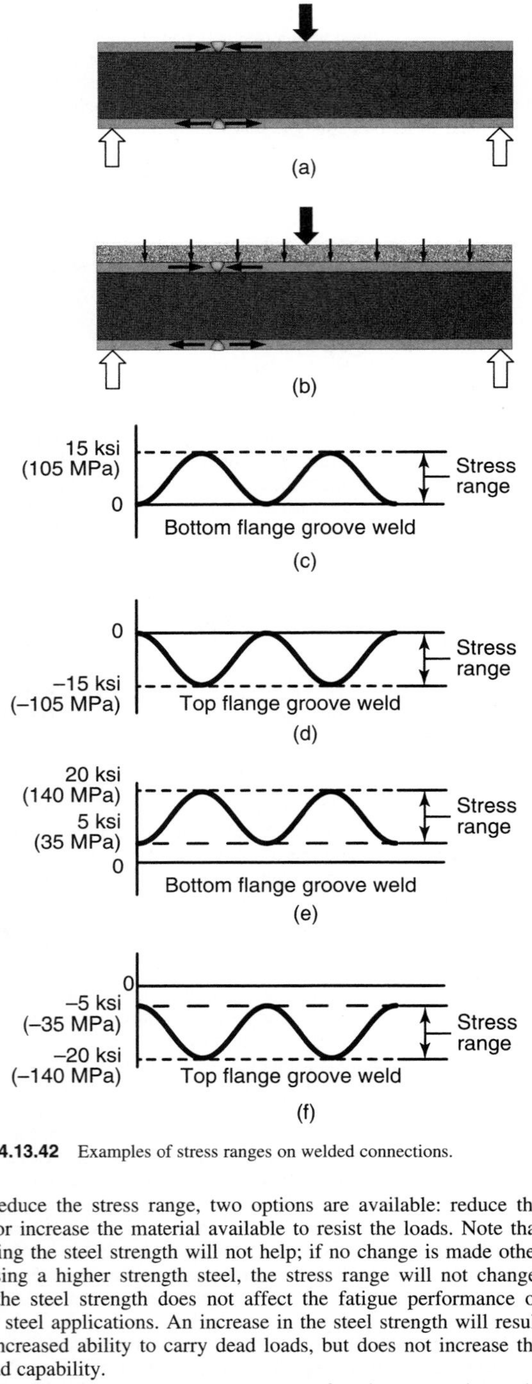

Figure 4.13.42 Examples of stress ranges on welded connections.

To reduce the stress range, two options are available: reduce the loads, or increase the material available to resist the loads. Note that increasing the steel strength will not help; if no change is made other than using a higher strength steel, the stress range will not change. Thus, the steel strength does not affect the fatigue performance of welded steel applications. An increase in the steel strength will result in an increased ability to carry dead loads, but does not increase the live load capability.

Ironically, using higher strength steel often increases the problems of fatigue, because when higher strength steels are specified, smaller members can be used to resist the steady load. These smaller members, however, will result in increased stresses and increased stress ranges. With the increase in stress ranges, the design with the higher strength steel (and smaller members) will have a reduced life expectancy. Thus, given that under most situations loads cannot be reduced, the only design option is to increase the material available to resist the loads.

The second of the three key variables involved in designing members subject to cyclic loading is the geometry of the connection. The principal influence of the geometry lies in the nature and extent of the stress raiser that is created. Abrupt, sharp changes create significant stress raisers, whereas smooth, flowing changes do not. Geometric factors that must be considered include the type of weld (whether CJP, PJP, fillet, or plug/slot), the orientation of the weld (parallel vs. transverse to the applied load), the length of the weld (continuous vs. intermittent), the treatment of the weld reinforcement (whether left-in-place or ground flush), and the quality of the weld.

This lengthy list of variables, and the myriad combinations that might exist, may seem daunting, but fortunately, all these factors have been distilled into eight categories (for welded connections) into which the various combinations can be placed. For example, the longitudinal fillet weld joining the web to the flange of a plate girder has essentially the same fatigue behavior as the transverse CJP groove weld that makes a flange splice on the same girder, providing the reinforcement on the groove weld has been removed and that the weld is nondestructively inspected to ensure internal soundness.

In the AISC model, the eight categories of welded details have been assigned alphabetical designations, from A to F. Categories B and E both have primed subsets (i.e., B' and E'). Category A details have the highest permissible stress range for a given number of cycles, or alternately, for a given stress range, will yield the longest life. Category B details have a reduced permissible stress range as compared to Category A.

Figure 4.13.43 is a plot of the Stress Range (F_{SR}) versus the number of loading cycles (N) for the various categories of details. The graph is a log-log plot, so minor changes in the stress range can have a major effect on the cycle life associated with a given detail. With the exception of Category F (discussed later), all the curves are parallel to each other, allowing for the use of a similar equation to compute the design stress range. Category A details are permitted to have the highest stress range for a given life expectancy (i.e., number of cycles). Moving from Category A to E', the permissible stress ranges progressively decrease for a given number of cycles of loading.

Associated with each detail is a threshold value, represented by the horizontal portion of the curves shown in Fig. 4.13.43. If applied stress ranges are held at or below this value, infinite life is expected from the structure. For Category A details, the threshold behavior is experienced after approximately two million cycles, while Category B details achieve this behavior at approximately three million cycles and Category E' details at about 20 million cycles. Category A details have a threshold value of 24 ksi (165 MPa), while Category B details have a reduced value of 16 ksi (110 MPa). Category E' details have the lowest value of 2.6 ksi (18 MPa).

The AISC Specification Appendix 3 contains descriptions and illustrations of dozens of details. Part of this Appendix is shown in Fig. 4.13.44. It is a fairly simple procedure to review the various examples and to identify a situation the same as, or similar to, the structural detail of interest. From the tables, the stress category can be obtained, as well as a constant C_f and a threshold value F_{TH}. The AISC Specification includes a complete listing of fatigue categories and the corresponding constants (C_f) and threshold values (F_{TH}).

Category A applies exclusively to unwelded base metal, but includes flame cut base metal. Categories B-E' generally involve details where fatigue cracking would be expected to occur in the base metal near a weld toe, or at the end of a weld. Category F governs fatigue failures that occur through the throat of a weld, whether a fillet, a PJP, or a plug/slot. Fortunately, Category F never needs to be a limiting factor in a design: the weld can be made larger to accommodate the stress range.

Fatigue Calculations

The basic equation for determining the maximum design stress range (F_{SR}) assumes the form of the following:

$$F_{SR} = (C_f / N)^{0.333} \geq F_{TH}$$

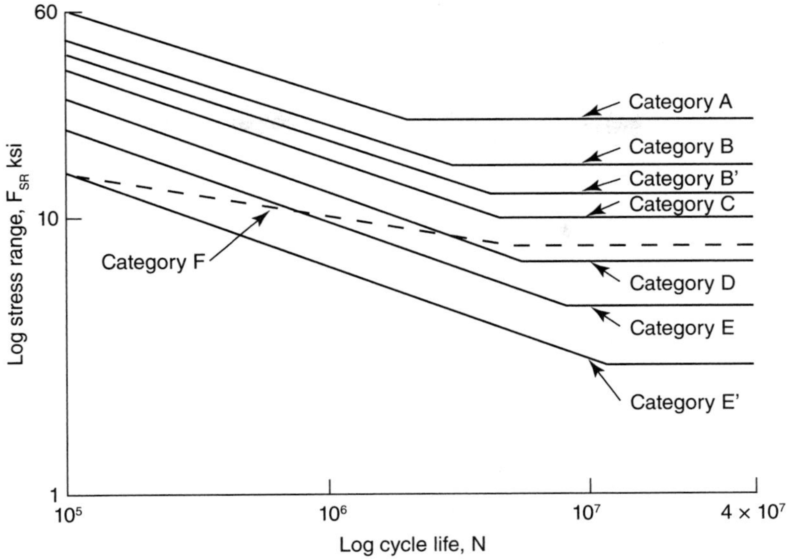

Figure 4.13.43 The AISC allowable stress range for various details.

where

F_{SR} = design stress range, ksi
C_f = constant for the geometric detail
N = number of stress range fluctuations (i.e., number of cycles)
F_{TH} = threshold fatigue stress range for infinite design life, ksi

For metric practice, the above equation is modified as follows:

$$F_{SR} = (329\ C_f/N)^{0.333} \geq F_{TH}$$

The coefficient C_f can be obtained from Table 3.1 of the AISC Specification. The number of stress range fluctuations can be determined from the life expectancy of the structure, multiplied by the frequency of loading. For infinite life, the design stress range (F_{SR}) must be held below the threshold fatigue stress range (F_{TH}).

For Category F, the equation above assumes a slightly different format, as follows:

$$F_{SR} = (C_f/N)^{0.167} \geq F_{TH}$$

For metric practice, the above equation is modified as follows:

$$F_{SR} = \{(11 \times 10^4\ C_f)/N\}^{0.167} \geq F_{TH}$$

The exponent reflects the difference in the slope of the curve for Category F material as can be seen in Fig. 4.13.43. This difference causes the constant C_f for Category F to be two magnitudes higher than that for the others.

Restrictions and Special Fabrication Requirements

The AISC model assumes some operational conditions and fabrication practices will be in place in order to make the model applicable. The constraints include:

- The maximum stress due to live and dead loads does not exceed 66 percent of yield.
- The steel members are subject to only mildly corrosive conditions, such as normal atmospheric conditions.
- The operational temperature does not exceed 300°F (150°C).

Special fabrication requirements include:

- Longitudinal steel backing (if used) is to be continuous for the length of the joint. Any splices in the backing are required to be joined with CJP grove welds, with reinforcement removed. Such splicing is required to take place prior to assembly in the joint.
- Transverse steel backing is required to be removed.
- Reinforcing, or contouring fillet welds are required on top of PJP or CJP groove welds in T and corner joints. AISC requires this to be no less than ¼ in (6 mm).
- Surface roughness for thermally cut edges does not exceed 1,000 min (25 mm).
- Reentrant corners at cuts, copes, weld access hole as other geometric contours have a minimum radius of 3/8 in (10 mm).

This overview of the AISC method of dealing with fatigue of welded connections does not provide the data needed for design purposes; the complete AISC appendix must be used. The reader should download the document and examine all the details to correctly apply this concept to actual applications.

BASE METALS FOR WELDING

General Considerations

The designer of a product will normally select the base material for a given application, considering variables such as strength, stiffness, density, ductility, toughness, and other factors that affect the performance of the product. Cost and availability are always factors that must be recognized as well. When welding is involved, the relative ease associated with welding the material should also be considered; this is sometimes called weldability. Weldability is highly dependent on the chemical composition of the base metal to be welded, a factor that is not normally of primary importance to the product designer, unless the product is required to resist corrosive elements. It is possible, and indeed common, that a material that satisfies all the designer's expectations in terms of product performance may be difficult or impossible to weld in a manner that is consistent, reliable, and cost effective.

In many cases, industry standards and codes provide listings of materials that can be easily welded, and such references should be reviewed when materials are selected. The relative weldability of a material is often described in the material standard. Suppliers of welding materials and base metals have literature that describes how their products are to be used for welded applications as well.

Before a material is selected for a welded application, it is desirable to know the general type of material (e.g., steel vs. aluminum), the specific type and grade, the chemical composition, the mechanical properties, and the method of manufacture (e.g., hot rolled vs. quench and tempered). Knowing the general type of material is essential—without

TABLE A-3.1 (continued)
Fatigue Design Parameters

Description	Stress Category	Constant C_f	Threshold F_{TH} ksi (MPa)	Potential Crack Initiation Point
SECTION 3 – WELDED JOINTS JOINING COMPONENTS OF BUILT-UP MEMBERS				
3.1 Base metal and weld metal in members without attachments built up of plates or shapes connected by continuous longitudinal complete-joint-penetration groove welds, back gouged and welded from second side, or by continuous fillet welds.	B	120 x 10^8	16 (110)	From surface or internal discontinuities in weld away from end of weld
3.2 Base metal and weld metal in members without attachments built up of plates or shapes, connected by continuous longitudinal complete-joint-penetration groove welds with backing bars not removed, or by continuous partial-joint-penetration groove welds.	B′	61 x 10^8	12 (83)	From surface or internal discontinuities in weld, including weld attaching backing bars
3.3 Base metal at weld metal terminations of longitudinal welds at weld access holes in connected built-up members.	D	22 x 10^8	7 (48)	From the weld termination into the web or flange
3.4 Base metal at ends of longitudinal intermittent fillet weld segments.	E	11 x 10^8	4.5 (31)	In connected material at start and stop locations of any weld deposit
3.5 Base metal at ends of partial length welded coverplates narrower than the flange having square or tapered ends, with or without welds across the ends; and coverplates wider than the flange with welds across the ends.				In flange at toe of end weld or in flange at termination of longitudinal weld or in edge of flange with wide coverplates
Flange thickness (t_f) ≤ 0.8 in. (20 mm)	E	11 x 10^8	4.5 (31)	
Flange thickness (t_f) > 0.8 in. (20 mm)	E′	3.9 x 10^8	2.6 (18)	
3.6 Base metal at ends of partial length welded coverplates wider than the flange without welds across the ends.	E′	3.9 x 10^8	2.6 (18)	In edge of flange at end of coverplate weld
SECTION 4 – LONGITUDINAL FILLET WELDED END CONNECTIONS				
4.1 Base metal at junction of axially loaded members with longitudinally welded end connections. Welds shall be on each side of the axis of the member to balance weld stresses.				Initiating from end of any weld termination extending into the base metal
t ≤ 0.5 in. (12 mm)	E	11 x 10^8	4.5 (31)	
t > 0.5 in. (12 mm)	E′	3.9 x 10^8	2.6 (18)	

TABLE A-3.1 (continued)
Fatigue Design Parameters

	Illustrative Typical Examples
SECTION 3 – WELDED JOINTS JOINING COMPONENTS OF BUILT-UP MEMBERS	
3.1	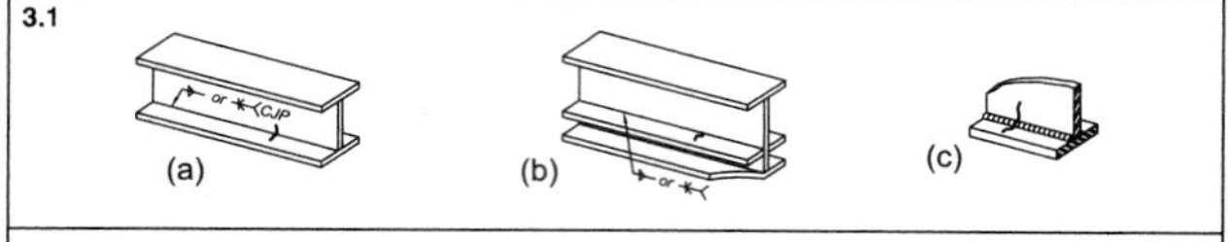
3.2	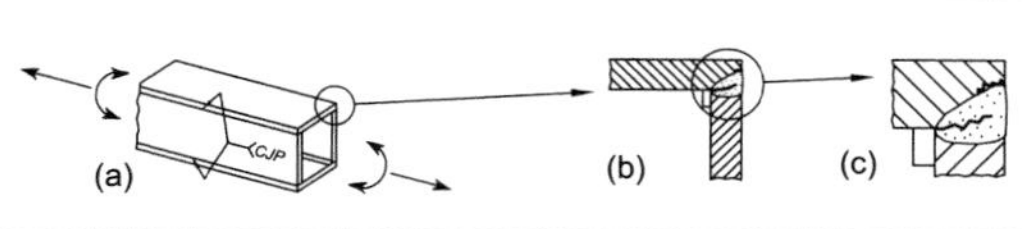
3.3	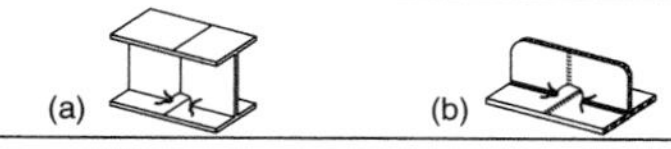
3.4	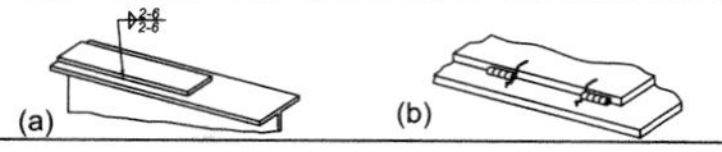
3.5	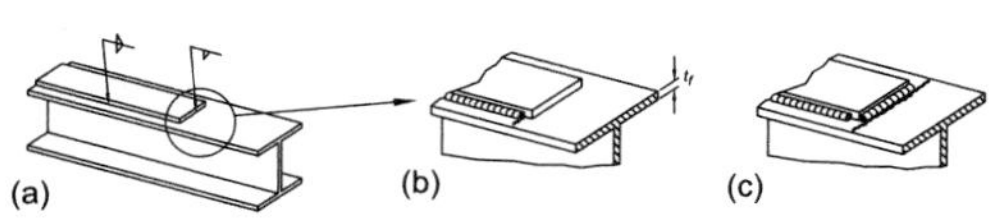
3.6	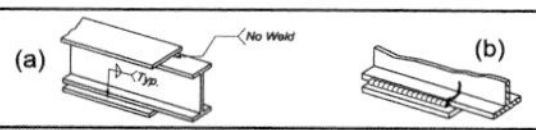
SECTION 4 – LONGITUDINAL FILLET WELDED END CONNECTIONS	
4.1	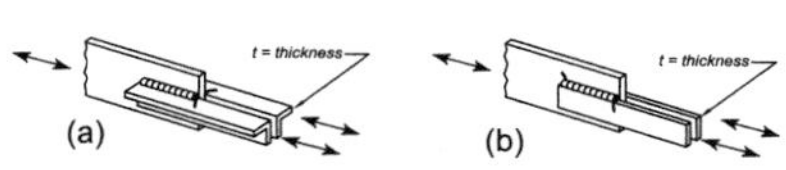

Figure 4.13.44 A portion of the fatigue category descriptions from AISC.

that knowledge, no responsible decisions regarding welding of the material can be made.

Knowledge of the chemical composition will directly affect the weldability and determine whether preheat, post heat, and other thermal controls are needed. The chemical composition may also be required to select the proper filler metal, particularly for applications involving corrosive environments. Chemical composition ranges are typically specified for required elements, often shown as minimum and maximum limits. For undesirable elements, a maximum content is usually provided. In some cases, residual elements may be present in the material, typically introduced from the raw materials used to make the final product. These elements are normally held to a low level and are typically not problematic. However, when present to a sufficient level, residual elements can be problematic in welded applications.

The mechanical properties of the base metal may need to be replicated in the weld metal, depending on the weld type and loading conditions. Mechanical properties may be specified in terms of minimum values (with no maximum limit) or in terms of a range. In many cases, the "extra" strength provided by materials supplied with no upper limit may be viewed as desirable by the product designer, but may be problematic when welding is performed.

The method by which the material was produced is an important welding consideration. Hot rolled steels, for example, are generally easily welded with proper welding procedures. Quenched and tempered steels, in contrast, require additional controls on the welding procedures in order to be successfully welded. Other steels may obtain their strength by cold working. The heating and cooling cycles associated with most welding processes may alter the metallurgical structure that was deliberately created by the materials supplier, and this should be considered as part of the base metal selection process; it may also affect the suitability of the proposed welding process. Aluminum alloys that may be softened by arc welding processes can be successfully welded with friction stir welding, for example.

Steel

Steel is the most commonly used metal and is accordingly the most frequently welded metal. Steels can be subdivided into the following categories: carbon steels, high-strength low-alloy steels, low alloy steels, and alloy steel. These terms are defined in standards such as ASTM A941 and are generally based upon the carbon and alloy content in the material.

Carbon steel is defined as a steel with carbon of not more than 2.0 percent, and manganese of not more than 1.65 percent, and according to ASTM A941 "does not prescribe a minimum limit for chromium, cobalt, molybdenum, nickel, niobium (columbium), tungsten (wolfram), vanadium, or zirconium." Importantly, the term "carbon steel" not only specifies a limit on carbon, but also identifies the material as one that has a limited amount of manganese, and no minimum limit on eight listed elements. Not present in the list of unspecified elements is silicon; some silicon is typically added to carbon steels.

High-strength low-alloy (HSLA) steels are defined both in terms of composition and strength. Low levels of specific alloys are required to be present, but the level is not as high as would be required for an "alloy steel," and the yield strength of the product is at least 36 ksi or 250 MPa. The carbon content is typically low, in the range of 0.05-0.25 percent. These steels are designed with a primary focus on mechanical properties, and a controlled limit on the alloys used to make the steel. HSLA steel may be supplied in the as-rolled condition, but may also receive after-rolling mechanical or thermal treatments.

Low-alloy steels are defined in a similar way to HSLA steels, albeit with no reference to the minimum yield strength. Specific alloys in limited quantities are added to impart specific properties. Included in this category of steels are many of the SAE-AISI designated steels.

Alloy steels contain one or more of the following elements, to a level greater than the value shown (by mass percent): 0.30 aluminum, 0.0008 boron, 0.30 chromium, 0.30 cobalt, 1.40 copper, 0.40 lead, 1.65 manganese, 0.08 molybdenum, 0.30 nickel, 0.06 niobium, 0.60 silicon, 0.10 vanadium, 0.05 zirconium, 0.10 for any other alloying element, except sulfur, phosphorous, carbon, or nitrogen.

Carbon steels can be subdivided, based on the carbon content: low-carbon steels with up to 0.30 percent carbon, medium-carbon steels with 0.30-0.60 percent, and high-carbon steels with 0.60-1.0 percent. As is the case for all carbon steels, these subdivisions contain some manganese, perhaps silicon, but no deliberate additions of other alloying ingredients.

Low-carbon steels are easily welded, particularly when the carbon content is less than 0.20 percent. Preheat is not normally required when welding is done at a room temperature of 70°F (20°C). Low-carbon steels with 0.20-0.30 percent carbon may require some preheat to preclude cracking. For sheet steel applications where formability is important, the carbon content may be very low (less than 0.10%), along with low manganese levels, making the steels easy to weld with resistance welding processes. Brazing of low-carbon steels is also common.

Medium-carbon steels can be successfully welded with a variety of processes, albeit with the need for better control of the thermal cycle that will occur during welding. Preheat in the range of 300-500°F (150-260°C) may be required, depending on the thickness of the material, restraint, and the hydrogen content of the weld metal.

High-carbon steels are rarely welded as they require specialized welding procedures. Preheat levels of 500°F (260°C) are typically required, along with post weld thermal treatments. When high-carbon steels are selected, welding procedures should be carefully developed and prototype assemblies carefully evaluated before material selection is finalized.

Low-alloy and high-alloy steels have widely varying weldability, depending on the alloy(s) involved, and the level of the alloy, as well as the carbon content. For those steels with limited levels of carbon and alloy, the welding techniques described for low- and medium-carbon steels will likely apply. For higher alloy and carbon contents, welding becomes more difficult, requiring techniques such as preheat, postheat, and post weld heat treatments.

High-strength low-alloy steels are commonly used in welded construction. HSLA steels are designed to achieve specific mechanical properties, but with the alloy content held to low limits. As the yield strength increases, the weldability decreases, requiring more attention to the welding procedures. Some HSLA steels are welded at room temperature without the need for any preheat, while others will need to be preheated to moderate levels 300°F (150°C), depending on the specific steel, thickness, restraint, and other factors.

In general, when steels are welded, the deposited weld metal is not required to have a chemical composition that is identical to that of the base metal. Typically, the goal is for the strength of the deposited weld metal to be appropriate, even if the composition is different. For example, steel weld metal often has lower carbon contents and higher levels of alloy than the base metal. The types of alloys used for weld metals may be different than for base metals. As a rule, if the strength level is acceptable, differences in the chemical composition are acceptable. A notable exception occurs when weathering steel is used: for such applications, the weld metal is expected to have similar atmospheric corrosion resisting properties. However, even in this case, the chemical composition of the weld metal and the weathering steel are not required to be identical.

Stainless Steel

Stainless steel is an iron-base alloy containing a minimum of 10.5 percent chromium, and a maximum carbon content of generally less than 2 percent. A thin, dense surface film of chromium oxide which forms on stainless steel imparts superior corrosion resistance; its passivated nature inhibits scaling and prevents further oxidation, hence the term "stainless." There are five types of stainless steels, and they are grouped by their metallurgical characteristics which range from fully austenitic to fully ferritic. Most stainless steels have good weldability and may be joined by many processes, including arc welding, resistance welding, electron and laser beam welding, and brazing. With any of these, the joint surfaces and any filler metal must be clean.

The coefficient of thermal expansion for the austenitic stainless steels is nearly 50 percent greater than that of carbon steel, while the coefficient of thermal conductivity is about one-third of that of carbon steel. These two factors make distortion control a significantly greater challenge when welding on stainless steel. Austenitic stainless steels also have lower electrical conductivity (i.e., higher electrical resistivity) as compared to carbon steel. Combined with the slower thermal conductivity, welding procedures need to be adjusted when a transition is made from carbon steel to austenitic stainless steel. Welding currents are generally lower in both arc welding and resistance welding because of the higher electric resistivity.

Ferritic Stainless Steels contain 11.5 to 30 percent Cr, up to 0.20 percent C, and small amounts of ferrite stabilizers, such as Al, Nb, Ti, and Mo. They are ferritic at all temperatures, do not transform to austenite, and are not hardenable by heat treatment. This group includes types 405, 409, 430, 442, and 446. To weld ferritic stainless steels, filler metals should match or exceed the Cr level of the base metal.

Martensitic Stainless Steels contain 11.5 to 18 percent Cr, up to 1.2 percent C, and small amounts of Mo and Ni. They will transform to austenite on heating and, therefore, can be hardened by formation of martensite on cooling. This group includes types 403, 410, 414, 416, 420, 422, 431, and 440. Weld cracks may appear on cooled welds as a result of martensite formation. The Cr and C content of the filler metal should generally match these elements in the base metal. Preheating and interpass temperature in the 400 to 600°F (200 to 315°C) range is recommended for welding most martensitic stainless steels. Steels with over 0.20 percent C often require a postweld heat treatment to avoid weld cracking.

Austenitic Stainless Steels contain 16 to 26 percent Cr, 10 to 24 percent Ni and Mn, up to 0.25 percent C, and small amounts of Mo, Ti, Nb, and Ta. The balance between Cr, Ni, and Mn is normally adjusted to provide a microstructure of 90 to 100 percent austenite. 80 percent of the stainless steels welded are austenitic. These alloys have good strength and high toughness over a wide temperature range, and they resist oxidation to over 1,000°F (540°C). This group includes types 302, 304, 310, 316, 321, and 347. Filler metals for these alloys should generally match the base metal, but for most alloys should also provide a microstructure with some ferrite to avoid hot cracking.

Two problems are associated with welding austenitic stainless steels: sensitization of the weld-heat-affected zone and hot cracking of weld metal. Sensitization is caused by chromium carbide precipitation at the austenitic grain boundaries in the heat-affected zone when the base metal is heated to 800 to 1,600°F (425 to 870°C). When chromium carbides precipitate, chromium is removed from solution in the vicinity of the grain boundaries, and this condition leads to intergranular corrosion. The problem can be alleviated by using low-carbon stainless-steel base metal (types 302L, 316L, etc.) and low-carbon filler metal. Alternately, stabilized stainless steel base metals and filler metals are available which contain alloying elements that react preferentially with carbon, thereby not depleting the chromium content in solid solution and keeping it available for corrosion resistance. Type 321 contains titanium and type 347 contains niobium and tantalum, all of which are stronger carbide formers than chromium.

The second problem associated with welding austenitic stainless steels is hot cracking, caused by low-melting-point metallic compounds of sulfur and phosphorus which penetrate grain boundaries. When present in the weld metal or heat-affected zone, these ingredients cause cracks to appear as the weld cools and shrinkage stresses develop. Hot cracking can be prevented by adjusting the composition of the base metal and filler metal to obtain a microstructure with a small amount of ferrite in the austenite matrix. The ferrite provides ferrite-austenite boundaries which control the sulfur and phosphorus compounds and thereby prevent hot cracking.

Precipitation-Hardening (PH) Stainless Steels contain alloying elements such as aluminum which permit hardening by a solution and aging heat treatment. There are three categories of PH stainless steels: martensitic, semiaustenitic, and austenitic. Martensitic PH stainless steels are hardened by quenching from the austenitizing temperature [around 1,900°F (1,340°C)] and then aging between 900 and 1,150°F (480 to 620°C). Semiaustenitic PH stainless steels do not transform to martensite when cooled from the austenitizing temperature because the martensite transformation temperature is below room temperature. Austenitic PH stainless steels remain austenitic after quenching from the solution temperature, even after substantial amounts of cold work. If maximum strength is required of martensitic PH and semiaustenitic PH stainless steels, matching, or nearly matching, filler metal should be used, and before welding, the work pieces should be in the annealed or solution-annealed condition.

After welding, a complete solution heat treatment plus an aging treatment is preferred. If postweld solution treatment is not feasible, the components should be solution-treated before welding and then aged after welding. Thick sections of highly restrained parts are sometimes welded in the overaged condition. These require a full heat treatment after welding to attain maximum strength properties. Austenitic PH stainless steels are the most difficult to weld because of hot cracking. Welding is preferably done with the parts in a solution-treated condition, under minimum restraint and with minimum heat input. Filler metals of the Ni-Cr-Fe type, or of conventional austenitic stainless steel, are preferred.

Duplex Stainless Steels are the most recently developed type of stainless steel, and they have a microstructure of approximately equal amounts of ferrite and austenite. Their advantages over conventional austenitic and ferritic stainless steels include higher yield strength and greater stress corrosion cracking resistance. The duplex microstructure is attained in steels containing 21 to 25 percent Cr and 5 to 7 percent Ni by hot-working at 1,800 to 1,900°F (1,000 to 1,050°C), followed by water quenching. Weld metal of this composition will be mainly ferritic because the deposit will solidify as ferrite and will transform only partly to austenite without hot working or annealing. Since hot working or annealing most weld deposits is not feasible, the metal composition filler is generally modified by adding Ni (up to 10%); this results in increased amounts of austenite in the as-welded microstructure. Welding electrode used for duplex stainless steel is primarily selected based on maintaining the ferrite austenite phase balance.

Cast Iron

Even though cast iron has a high-carbon content and is a relatively brittle material, welding can be performed successfully if proper precautions are taken. Generally, welds on cast iron are used to repair the casting not for joining the casting to itself or another material. Two approaches have been used to successfully repair cast iron: keep it cool, and keep it hot. The former seeks to limit the thermal expansion and contractions, and thus limit the accumulation of residual stresses that might cause cracking. Small welds are made and the part is allowed to cool before another small weld is made. The opposite approach is to use preheat. The preheat slows the cooling rate of the weld and the heat affected zone, mitigating cracking tendencies. In some cases, in addition to preheating the area to be welded, it may be necessary to elevate the temperature of the entire casting. Heating the entire casting will avoid differential thermal expansion and contraction. Gray iron castings may be enclosed in insulation or packed in dry sand to allow for slow cooling after welding. Other irons may require postheat treatment immediately after welding to restore mechanical properties. Common electrodes used for welding on cast iron include ESt (a steel electrode) and ENiFe (a nickel-iron alloy electrode). Many different welding processes have been used to weld cast iron, the most common being SMAW, gas welding, and braze welding.

Aluminum and Aluminum Alloys

Most aluminum alloys are readily weldable. Aluminum is divided into two main groups: heat treatable and non-heat treatable. The heat treatable alloys include 2xxx, 6xxx, and 7xxx series. While 2xxx and 7xxx are not readily weldable, 6xxx is easily welded. The most commonly used alloy and heat treatment is 6,061-T6. Non-heat treatable alloys (the 3xxx, 4xxx, and 5xxx series) are regularly welded with success. Welding practices for aluminum must be adjusted from those used for steel because of a lower melting temperature range of approximately 900 to 1,215°F (480 to 660°C), higher thermal conductivity (about 2 to 4 times that of mild steel), and higher rate of thermal expansion (about twice that of carbon steel). This means that, when compared with mild steel, much higher welding speeds are needed and greater care must be exercised to avoid distortion. For resistance welding, much higher current densities are required.

Aluminum forms a thin transparent surface oxide which gives the material its corrosion resistance. Because that oxide has a much higher melting temperature [3,700°F (2,000°C)] than the aluminum, it must be removed prior to welding by mechanical brushing, machining, or by chemical methods. Aluminum is typically fusion welded with GTAW or GMAW. For GTAW alternating current should be used. Alternating current provides for surface cleaning action with the positive cycle of the sine wave and penetration with the negative cycle of the sine wave. GMAW aluminum uses either a DCEP polarity or a controlled wave form to produce sound and visually appealing welds. GMAW also has the advantage of higher deposition rates than GTAW. Some specialized equipment may be necessary such as a push-pull gun, special contact tips, and a water cooled torch. Many solid state welding methods are effective in joining aluminum such as friction stir welding or friction welding.

Copper and Copper Alloys

In welding commercially pure copper, it is important to select the correct type. Electrolytic, or "tough-pitch," copper contains a small percentage of copper oxide, which at welding temperatures leads to oxide embrittlement. For welded assemblies it is recommended that deoxidized, or oxygen-free, copper be used and that welding rods, when needed, be of the same analysis. Copper and its alloys may be welded using GTAW, GMAW, high energy beam processes, oxyfuel welding, and resistance processes. SMAW electrodes are also commercially available. Brazing with brazing filler metals conforming to BAg, BCuP, and RBCuZn-A classifications is also employed. The high heat conductivity of copper requires special consideration in welding; generally higher welding currents are necessary along with concurrent supplementary heating.

SAFETY

Welding is safe when sufficient measures are taken to protect the welder from potential hazards. When these measures are overlooked or ignored, welders can be subject to electric shock; overexposure to radiation, fumes, and gases; and fires and explosion. Any of these can be fatal. Everyone associated with welding operations should be aware of the potential hazards and help ensure that safe practices are employed. Infractions must be reported to the appropriate responsible authority.

ANSI Z49.1:2012, *Safety in Welding, Cutting, and Allied Processes*, available as a free download from AWS (http://www.aws.org/library/doclib/AWS-Z49.pdf), should be consulted for information on welding safety. A printed copy is also available for purchase from AWS (http://pubs.aws.org/p/1,081/z491,2012-safety-in-welding-and-cutting-and-allied-processes-aws-z491). From the AWS website, a variety of Safety & Health Fact Sheets also can be downloaded. Additional information about welding safety is available from American Welding Society, 8,669 NW 36 Street, # 130, Miami, Florida 33,166-6,672. Phone: 800-443-9,353 or 305-443-9,353.

Credits

Substantial portions of this chapter have been incorporated from other handbooks the author(s) have written. The sections on welding processes were taken from the McGraw-Hill publication *Manufacturing Engineering Handbook,* 2nd edition. The section on fatigue was adapted from the AISC publication *Design Guide 21: Welded Connections, A Primer for Engineers.* Extensive quotations from AWS publications are included with the American Welding Society's permission. The cooperation of these organizations is appreciated.

4.14 MACHINING PROCESSES AND MACHINE TOOLS

by Steven R. Schmid and Serope Kalpakjian

REFERENCES: Altintas, Y., *Machining Automation: Metal Cutting Mechanics, Machine Tool Vibrations, and CNC Design*, Cambridge, 2012. Knight, W.A., and Boothroyd, G., *Fundamentals of Machining and Machine Tools*, 3rd ed., Dekker, 2006. *Machinery's Handbook*, Industrial Press, 2012. *Machining and Metalworking Handbook*, 3rd ed., McGraw-Hill, 2005. Malkin, S., and Guo, C., *Grinding Technology*, 2nd ed., Industrial Press, 2008. Overby, A., *CNC Machining Handbook*, McGraw-Hill, 2010. Shaw, M.C., *Metal Cutting Principles*, 2nd ed., Oxford, 2005. Stephenson, D., and Agapiou, J.S., *Metal Cutting: Theory and Practice*, 2nd ed., CRC Press, 2005.

INTRODUCTION

Machining processes, which include cutting, grinding, and various nonmechanical processes, are often necessary for the following reasons: (1) closer dimensional tolerances, surface roughness, or surface-finish characteristics may be required than are available by casting, forming, powder metallurgy, and other shaping processes, and (2) part geometries may be too complex or too costly to be manufactured by other processes. However, machining inevitably wastes material, production rates may be low, and the processes used may have detrimental effects on the workpiece surface.

Traditional machining generally includes turning, milling, boring, drilling, reaming, threading, shaping, planing, and broaching processes, as well as abrasive processes, such as grinding, ultrasonic machining, lapping, and honing. Non-abrasive machining processes include electrical and chemical means of material removal and the use of abrasive jets, water jets, laser beams, electron beams, or combinations of these methods.

BASIC MECHANICS OF METAL CUTTING

The basic mechanics of machining processes (Fig. 4.14.1) are shown, in simplest form, in Fig. 4.14.2. A cutting tool, with a certain **rake angle** α (positive as shown) and **relief angle,** moves along the surface

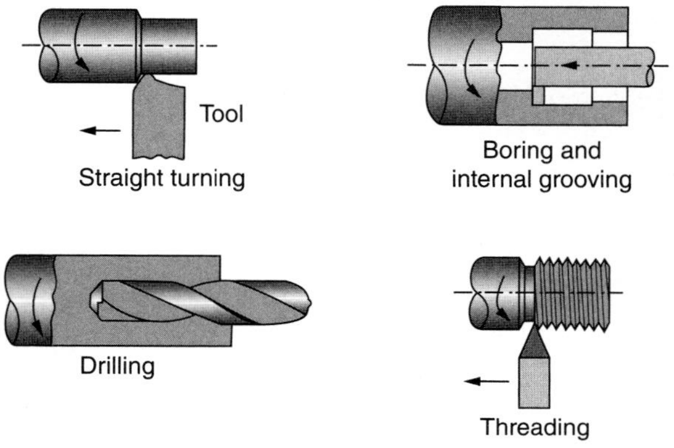

Figure 4.14.1 Examples of chip-type machining operations.

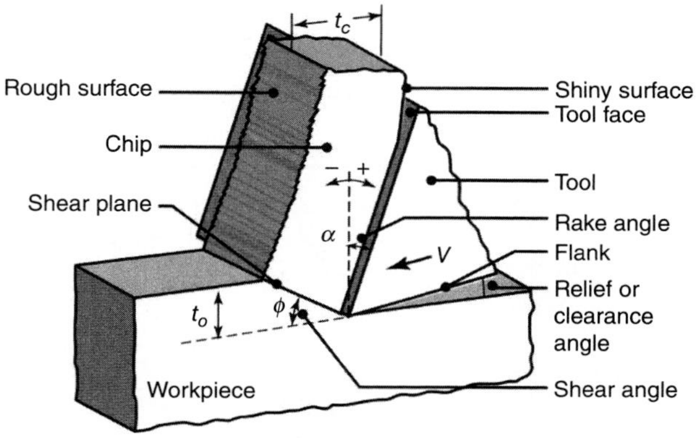

Figure 4.14.2 Basic mechanics of metal cutting process.

of the workpiece at a depth of cut t_o. The material ahead of the tool is sheared continuously along the **shear plane,** which makes an angle ϕ with the surface of the workpiece. This angle is called the **shear angle** and, together with the rake angle, determines the chip thickness t_c. The ratio of t_o to t_c is called the *cutting ratio r*. The relationship between the shear angle, the rake angle, and the cutting ratio is given by the equation $\tan \phi = r \cos \alpha/(1 - r \sin \alpha)$. It can readily be seen that the shear angle is important in that it controls the thickness of the chip which, in turn, has a major influence on cutting performance. The *shear strain* that the material undergoes during cutting is given by the equation $\gamma = \cot \phi + \tan (\phi - \alpha)$; in metal cutting they usually are less than 5.

Although simple models assume that the shear zone is a well-defined, narrow plane, experimental investigations have shown that this may not always be so; more elaborate models have been developed. Various formulas have been developed to define the shear angle in terms of such factors as the rake angle and the friction angle β. (See Fig. 4.14.3.)

Because of the large shear strains that the chip undergoes, it becomes hard and brittle. In most cases, the chip curls away from the tool; among possible factors contributing to chip curl are nonuniform normal stress distribution on the shear plane, strain hardening, and thermal effects in the cutting zone.

Regardless of the type of machining operation, some basic types of chips found in practice are shown in Fig. 4.14.4. **Continuous chips** are formed by continuous deformation of the workpiece material ahead of the tool, followed by smooth flow of the chip along the **rake face** of the tool. This type of chip is typically obtained in cutting ductile materials at high speeds.

Discontinuous chips consist of segments produced by fracture of the metal ahead of the tool; the segments may be either loosely connected to each other or unconnected. Such chips are most often found

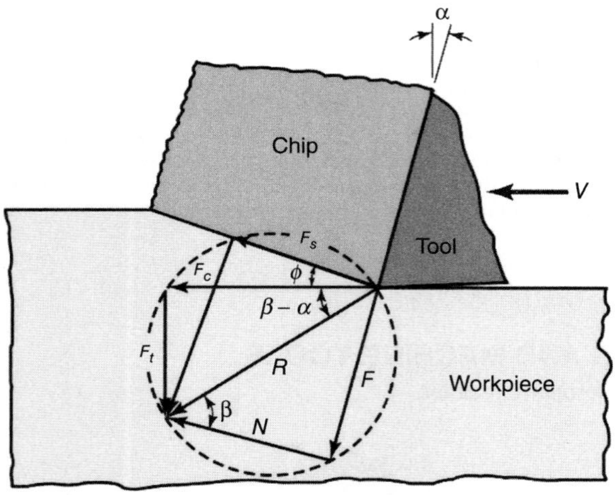

Figure 4.14.3 Force system in metal cutting process.

in machining brittle materials or ductile materials machined at very low speeds or negative or low rake angles.

Inhomogeneous (serrated) chips consist of regions of high and small strain in the chip. Such chips are characteristic of metals with low thermal conductivity or with metals in which yield strength decreases sharply with temperature, such as various titanium alloys.

Built-up edge chips consist of workpiece metal that adheres to the tool tip while the chip itself flows continuously along the rake face. This type of chip is often observed in machining at low speeds and is associated with high adhesion between chip and tool; built-up edge is undesirable because it adversely affects surface finish.

The **forces** acting on the cutting tool are shown in Fig. 4.14.3. The *resultant force R* has two components, F_c and F_t. The *cutting force* F_c in the direction of tool travel determines the amount of work done in cutting; the *thrust force* F_t does no work but, together with F_c, deflects the tool. The resultant force also has two components on the shear plane: force F_s shears the metal along the shear plane, and F_n is the normal force on this plane. Two other force components on the face of the tool are: the *friction force F* and the *normal force N*. Whereas F_c is always in the direction shown in Fig. 4.14.3, F_t may be in the opposite direction to that shown in the figure; this occurs when both the rake angle and the depth of cut are large, and friction is low.

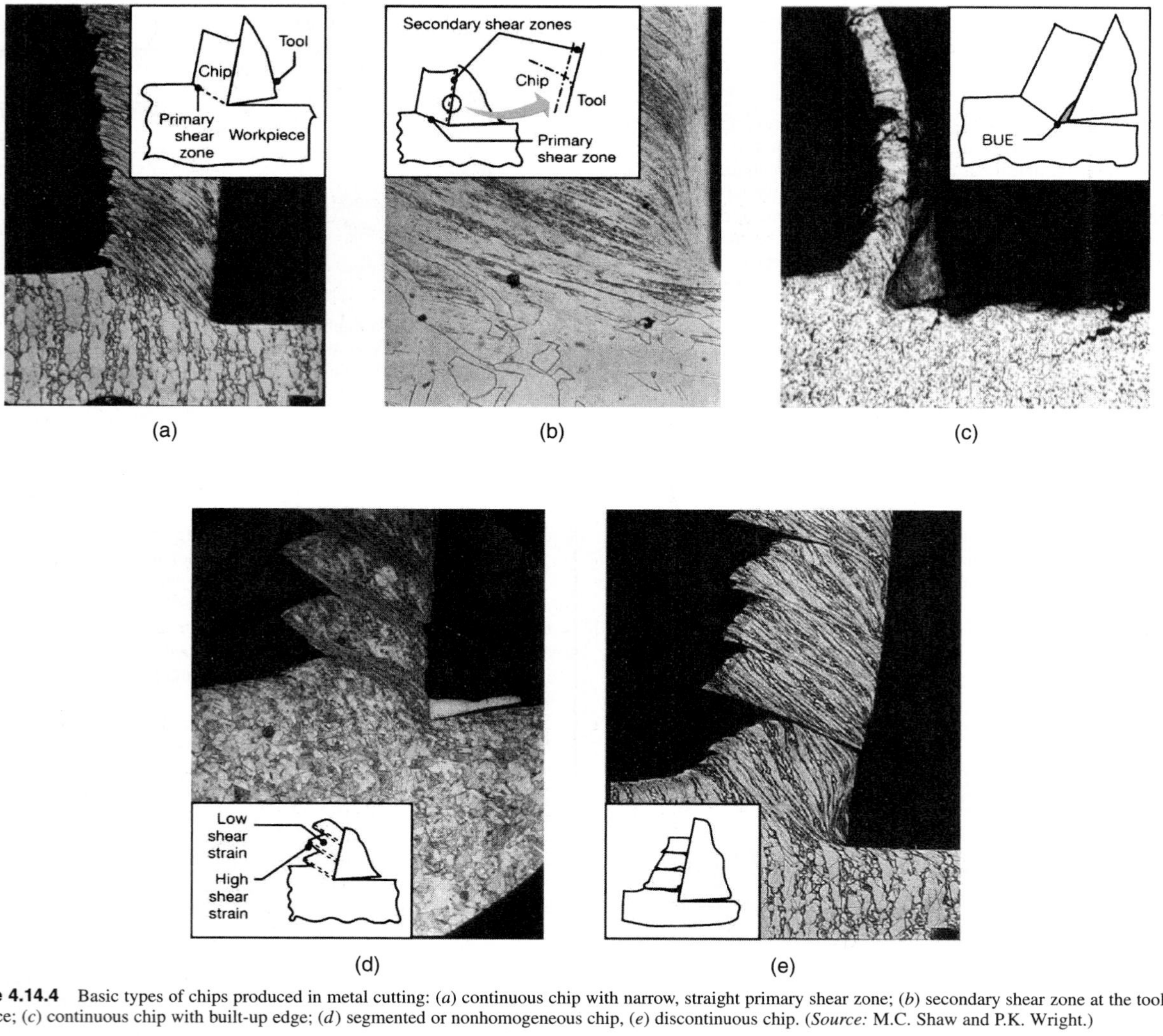

Figure 4.14.4 Basic types of chips produced in metal cutting: (*a*) continuous chip with narrow, straight primary shear zone; (*b*) secondary shear zone at the tool-chip interface; (*c*) continuous chip with built-up edge; (*d*) segmented or nonhomogeneous chip, (*e*) discontinuous chip. (*Source:* M.C. Shaw and P.K. Wright.)

From the geometry of Fig. 4.14.3, the following relationships can be derived: The **coefficient of friction** at the tool-chip interface is $\mu = (F_t + F_c \tan \alpha)/(F_c - F_t \tan \alpha)$. The **friction force** along the tool is $F = F_t \cos \alpha + F_c \sin \alpha$. The *shear stress* in the shear plane is $\tau = (F_c \sin \phi \cos \phi - F_t \sin^2 \phi) / A_0$, where A_0 is the cross-sectional being removed from the workpiece. The coefficient of friction, μ, on the tool face is a complex but important factor in cutting; it can be reduced by such means as using an effective cutting fluid, higher cutting speed, improved tool material, or chemical additives in the workpiece material.

The net **power** consumed is $P = F_cV$; since F_c is a function of tool geometry, workpiece material, and processing variables, it is difficult to reliably calculate its value in a particular machining operation. Depending on the material machined and the condition of the tool, **unit power** (power per unit volume of material removed) requirements in machining range from 0.2 hp-min/in^3 (0.55 W-s/mm^3) for aluminum and magnesium alloys to 3.5 for high-strength alloys.

The power consumed in cutting is transformed mostly to **heat,** most of which is usually carried away by the chip itself, and the remainder is dissipated to the tool, the workpiece, and the cutting fluid. An increase in cutting speed or feed will increase the proportion of the heat transferred to the chip. It has been observed that in turning, the average interface **temperature** between the tool and the chip increases with cutting speed and feed, while the influence of the depth of cut on temperature has been found to be limited. Interface temperatures up to 1,500 to 2,000°F (800 to 1,100°C) have been measured in metal cutting. The use of a cutting fluid removes heat and thus reduces temperatures on the cutting edge.

In cutting metals at high speeds, the chips typically become very hot; they can cause safety hazards because the long chips entangle with the tooling. Tools and tool inserts are equipped with **chip breaker;** this feature curls the chips, causing them to break into short sections.

A phenomenon of major significance in metal cutting is **tool wear.** Numerous factors determine the type and rate at which wear occurs on the tool surfaces. The critical variables that directly affect tool wear are tool temperature, type of tool material and its hardness, grade and condition of workpiece, abrasiveness of the microconstituents in the workpiece material, tool geometry, feed, speed, and type of cutting fluid. The wear patterns that develop depend on the relative significance of each of these variables.

Tool wear is generally classified as (1) **flank wear** (Fig. 4.14.5); (2) **crater wear;** (3) **localized wear,** such as the rounding of the cutting edge; (4) **chipping** and **thermal softening** (plastic flow) of the cutting edge; and (5) **wear notch,** a concentrated wear or a deep **groove** on the tool.

In general, wear on the flank or relief side of the tool is the most dependable guide for **tool life,** because it directly influences workpiece surface finish. A wear land of 0.060 in (1.5 mm) on high-speed steel tools and 0.015 in (0.4 mm) on carbide tools is usually used as the criterion for tool replacement. Cutting speed has the greatest influence on tool life; the relationship between tool life and cutting speed is given by the *Taylor equation* $VT^n = C$, where V is the cutting speed; T is the actual cutting time to develop a certain wear land, min; C is a constant where its value depends on workpiece material and process variables,

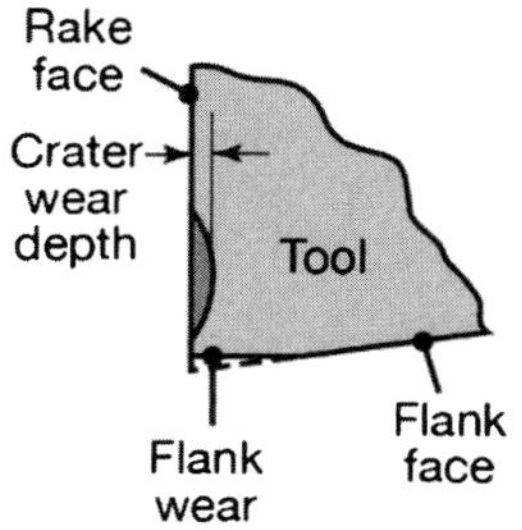

Figure 4.14.5 Types of tool wear in cutting.

numerically equal to the cutting speed that gives a tool life of 1 min; and n is an exponent, where its value depends on the workpiece material and the processing variables.

The recommended cutting speed for a high-speed steel tool is generally the one that produces a 60- to 120-min tool life. With carbide tools, a 30- to 60-min tool life may be satisfactory; for difficult-to-machine materials, lower values may be considered. The values of n typically range from 0.08 to 0.2 for high-speed steels, 0.1 to 0.15 for cast alloys, 0.2 to 0.5 for uncoated carbides, 0.4 to 0.6 for coated carbides, and 0.5 to 0.7 for ceramics.

When using tool-life equations caution should be exercised in extrapolating the curves beyond the operating region for which they were developed. Also, in a log-log plot, tool-life curves may be linear over a short cutting-speed range but are rarely linear over a wide range of cutting speeds. In spite of the extensive data developed to date, no simple formulas have been developed for quantitative relationships between tool life and various processing variables.

An aspect of machining important for computer-controlled equipment is **tool-condition monitoring,** while the machine tool is in continuous operation with little or no operator supervision. Most state-of-the-art machine tools are now equipped with tool-condition monitoring systems. Two common techniques involve the use of (1) transducers, installed on the tool holder, continually monitoring torque and forces; and (2) acoustic emission, through a piezoelectric transducer. In both methods the signals are automatically analyzed and interpreted for tool wear or chipping; corrective actions are then taken before any significant damage occurs to the workpiece.

A term commonly used in machining and comprising most of the topics above is **machinability,** generally defined in terms of (1) tool life, (2) power requirement, and (3) surface integrity. Thus, for example, a good machinability rating for a particular material would indicate a combination of long tool life, low power requirement, and a good surface. It is difficult to develop quantitative relationships between these variables, although tool life is considered as the important factor; in actual production, it is usually expressed as the number of pieces machined between tool changes. Various tables are available in the technical literature, summarizing the machinability rating for a variety of materials.

The major factors influencing **surface finish** in machining are (1) the local geometry of the cutting tool in contact with the workpiece, (2) fragments of built-up edge that are deposited on the workpiece surface, and (3) vibration and chatter. Improvements in surface finish may be obtained, to various degrees, generally by increasing the cutting speed and decreasing the feed and depth of cut. Changes in the cutting fluid, tool geometry, and tool material are also important, as are the microstructure and chemical composition of the workpiece material.

As a result of mechanical working and thermal effects in machining, **residual stresses** are generally developed on the surfaces of materials. These stresses may cause warping of the workpiece and also affect its resistance to fatigue and stress corrosion. To minimize residual stresses, sharp tools, medium feeds, and medium depths of cut are recommended.

Because of plastic deformation, thermal effects, and chemical reactions during machining, alterations of the machined surface may take place, significantly affecting the **surface integrity** of a part. Typical detrimental effects include burns, cracks, distortion, reduction in fatigue strength, compromise of stress-corrosion properties, and development of residual stresses. Improvements in surface integrity may be obtained by post-processing techniques, such as sanding, peening, finish machining, fine grinding, and polishing.

Vibration and **chatter** in machining are complex phenomenon, and are often the cause of premature tool failure, poor surface finish, damage to the workpiece, and if excessive, even the machine tool. Vibration may be **forced** or **self-excited.** The term chatter is commonly used to designate self-excited vibrations; the amplitudes are usually very high. Although there is no complete solution to all types of vibration problems in machining, certain measures may be taken. For example, if the vibration is being forced, it may be possible to remove or isolate the forcing element from the machine tool. In cases where the forcing frequency is at or near a natural frequency, either the forcing frequency or the natural frequency may be modified; damping will also greatly reduce the amplitude. Self-excited vibrations are generally controlled by increasing the stiffness and damping capacity of the machine tool. (See also Secs. 1 and 3.)

Good machining practice requires a rigid setup. The machine tool must be capable of providing the **stiffness** required for the particular machining conditions. If a rigid setup is not available, the depth of cut must be reduced. The length of end mills and drills, for example, should be kept to a minimum. Tools with a large nose radius or with a long, straight cutting edge increase the possibility of chatter.

TURNING AND LATHES

Turning is a common machining operation capable of producing circular parts with straight or various axisymmetric profiles. The cutting tools may be single point or form tools, and the most common machine used is a **lathe.** The basic turning operation is shown in Fig. 4.14.6, where the workpiece is held in a **chuck,** rotating at N (r/min). The tool moves along the length of the workpiece, at a feed f (in/r or mm/r), and removes material at a radial depth d, reducing its diameter from D_0 to D_f.

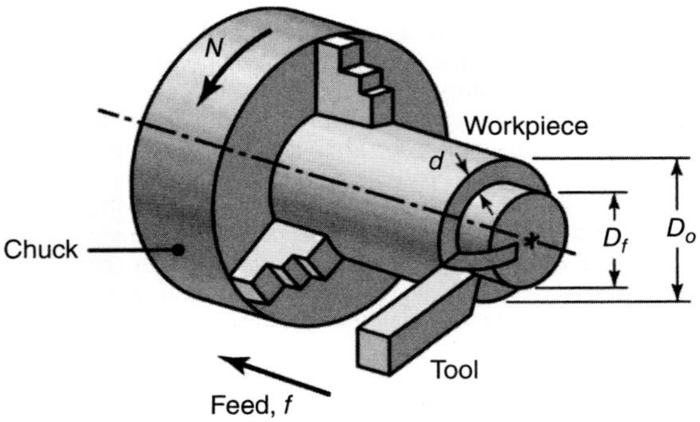

Figure 4.14.6 A turning operation on a round workpiece held in a three-jaw chuck.

Tool Shapes for Turning

The standard **nomenclature** for single-point tools is shown in Fig. 4.14.7. Traditionally, the point of a single-point tool is shaped by grinding on the end of the shank. Most tools now consist of an **insert** (Fig. 4.14.8); each insert has multiple cutting edges, and can easily be indexed within the tool mounting. The optimum tool shape for each workpiece material and each machining operation depends on several factors, some of which are summarized in Table 4.14.1.

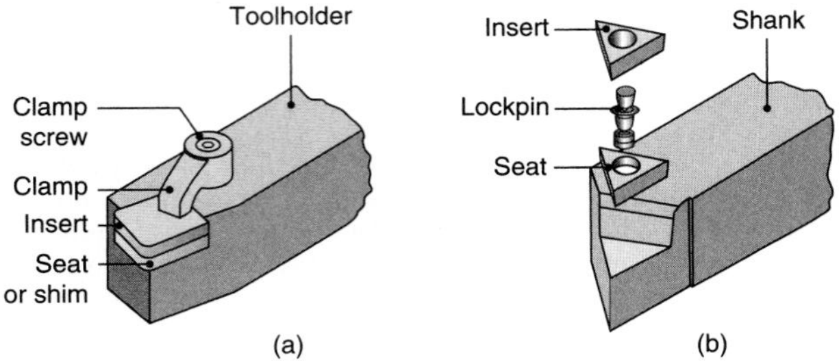

Figure 4.14.8 Inserts and methods of mounting them onto toolholders: (*a*) clamping; (*b*) using a lockpin.

Positive **rake angles** reduce cutting forces, hence power, and reduce deflection; however, a high positive rake angle may result in premature failure of the cutting edge. Positive angles are generally used for cutting lower-strength materials; for higher-strength or brittle materials, negative rake angles may be used. **Back rake angle** usually controls the direction of chip flow, and is of less importance than the **side rake angle.** The purpose of a **relief angle** is to avoid interference and rubbing between the workpiece and tool flank surfaces. In general, they should be small for materials with high stiffness and larger for compliant materials; excessive relief angles may weaken the tool tip. The **side cutting-edge angle** influences the length of tool-chip contact; large angles could cause tool chatter. The purpose of the **nose radius** is to produce a smooth machined surface finish and longer tool life, by increasing the strength of the cutting edge. A large nose radius makes the tool stronger and it may be used for roughing cuts. However, large nose radii may lead to tool chatter; a small radius reduces forces and is therefore preferred on thin or slender workpieces.

BORING

Boring is a process for producing internal straight cylindrical surfaces or profiles; the processing characteristics and tooling are similar to those for turning operations. The **horizontal** boring machine is particularly adapted for workpieces that cannot be conveniently revolved and for milling, slotting, drilling, tapping, boring, and reaming long holes. **Vertical boring mills** are used for a wide range of applications; their advantage lies in the ease of clamping a workpiece onto a horizontal table, resembling a four-jaw, independent chuck, with extra radial

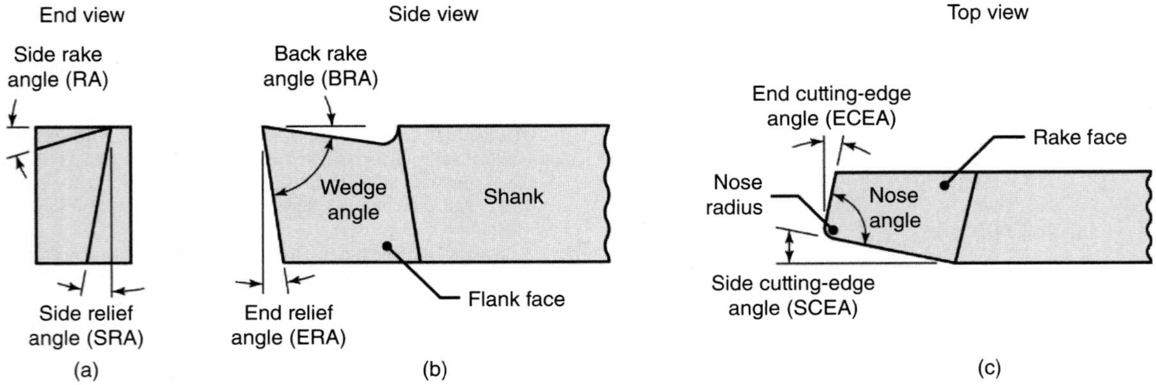

Figure 4.14.7 Standard nomenclature for single-point cutting tools.

Table 4.14.1 Recommended Tool Geometry for Turning, deg

Material	High-speed steel					Carbide inserts				
	Back rake	Side rake	End relief	Side relief	Side and end cutting edge	Back rake	Side rake	End relief	Side relief	Side and end cutting edge
Aluminum and magnesium alloys	20	15	12	10	5	0	5	5	5	15
Copper alloys	5	10	8	8	5	0	5	5	5	15
Steels	10	12	5	5	15	−5	−5	5	5	15
Stainless steels	5	8-10	5	5	15	−5–0	−5–5	5	5	15
High-temperature alloys	0	10	5	5	15	5	0	5	5	45
Refractory alloys	0	20	5	5	5	0	0	5	5	15
Titanium alloys	0	5	5	5	15	−5	−5	5	5	5
Cast irons	5	10	5	5	15	−5	−5	5	5	15
Thermoplastics	0	0	20-30	15-20	10	0	0	20-30	15-20	10
Thermosets	0	0	20-30	15-20	10	0	15	5	5	15

T-slots. The table revolves the workpiece, allowing for boring and associated operations.

Boring Recommendations

Depth of cut, feed, cutting speed, and tool materials are generally the same as for turning operations. Tool deflections, chatter, and dimensional accuracy can be concerns if the boring bar has to extend significantly inward for the machining operation to be completed, and also the radial space within the workpiece may be limited. Boring bars can be designed to dampen vibrations and reduce chatter that typically can occur in using long bars.

DRILLING

Drilling is a common hole-making process that uses a fluted **drill** as a cutting tool; drilled holes may be subjected to subsequent operations, such as by reaming and honing, for better surface finish and dimensional accuracy.

Twist drills (Fig. 4.14.9) are the most common and are available in a wide range of diameters and lengths. The shape of the groove in a drill is important; the best is one that gives a straight cutting edge and allows a full curl to develop in the chip. The **helix angles** of the flutes vary from 10 to 45°, while the standard **point angle** is 118°. The point of the drill may be ground either in the **standard** or the **crankshaft** geometry. Table 4.14.2 gives the drill geometry for high-speed steel twist drills for a variety of workpiece materials.

Among common **types of drills** (Fig. 4.14.10) are the combined drill and countersink or **center drill,** a short drill used to center shafts before turning them: the **step drill,** with two or more diameters; the **spade drill,** used for large and deep holes; the **gun drill,** used for relatively small-diameter long holes, as in a gun; the **core drill,** to drill cored holes; the **oil-hole drill,** with holes or tubes in its body through which oil is forced to the cutting lips; and three- and four-fluted drills, used to enlarge holes after a leader hole has been cored, punched, or drilled. **Trepanning tools** are used to cut a core from a piece of metal instead of reducing all the metal removed to chips.

The most common tool material for drills is **high-speed steel.** Drills are also made with **carbide inserts** (such as in masonry drills) to form the chisel edge and both cutting edges of the drill; they are used primarily for drilling abrasive or very hard materials.

Drilling Recommendations

General recommendations for speeds and feeds in drilling a variety of materials are given in Table 4.14.3; hole depth is also a factor in selecting drilling parameters. A general **troubleshooting guide** for drilling is given in Table 4.14.4.

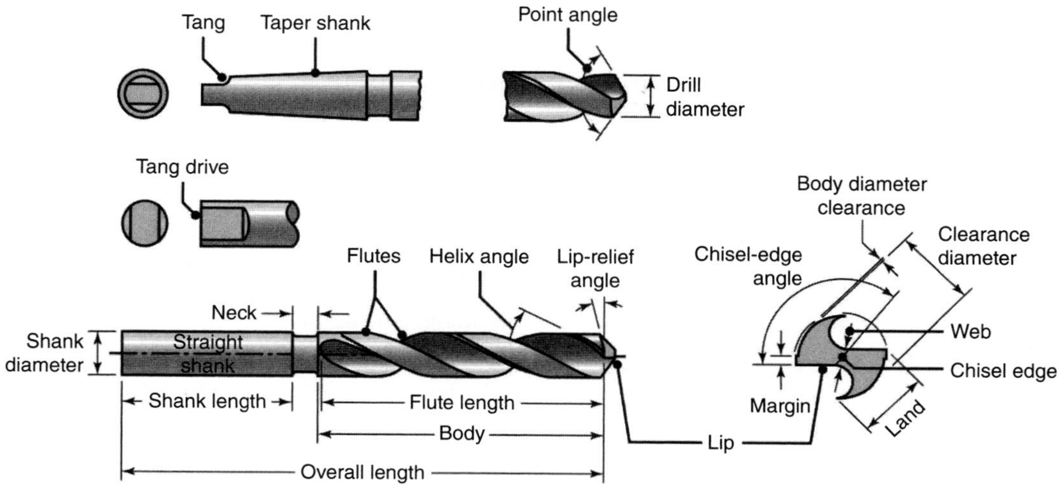

Figure 4.14.9 Straight shank twist drill.

Table 4.14.2 Recommended Drill Geometry for High-Speed Steel Twist Drillls

Material	Point angle, deg	Lip relief angle, deg	Chisel edge angle, deg	Helix angle, deg	Point grind
Aluminum alloys	90-118	12-15	125-135	24-48	Standard
Magnesium alloys	70-118	12-15	120-135	30-45	Standard
Copper alloys	118	12-15	125-135	10-30	Standard
Steels	118	10-15	125-135	24-32	Standard
High-strength steels	118-135	7-10	125-135	24-32	Crankshaft
Stainless steells, low-strength	118	10-12	125-135	24-32	Standard
Stainless steels, high-strength	118-135	7-10	120-130	24-32	Crankshaft
High-temperature alloys	118-135	9-12	125-135	15-30	Crankshaft
Refractory alloys	118	7-10	125-135	24-32	Standard
Titanium alloys	118-135	7-10	125-135	15-32	Crankshaft
Cast irons	118	8-12	125-135	24-32	Standard
Plastics	60-90	7	120-135	29	Standard

SOURCE: "Machining Data Jandbook," published by Machinability Data Center, Metcut Research Associates Inc.

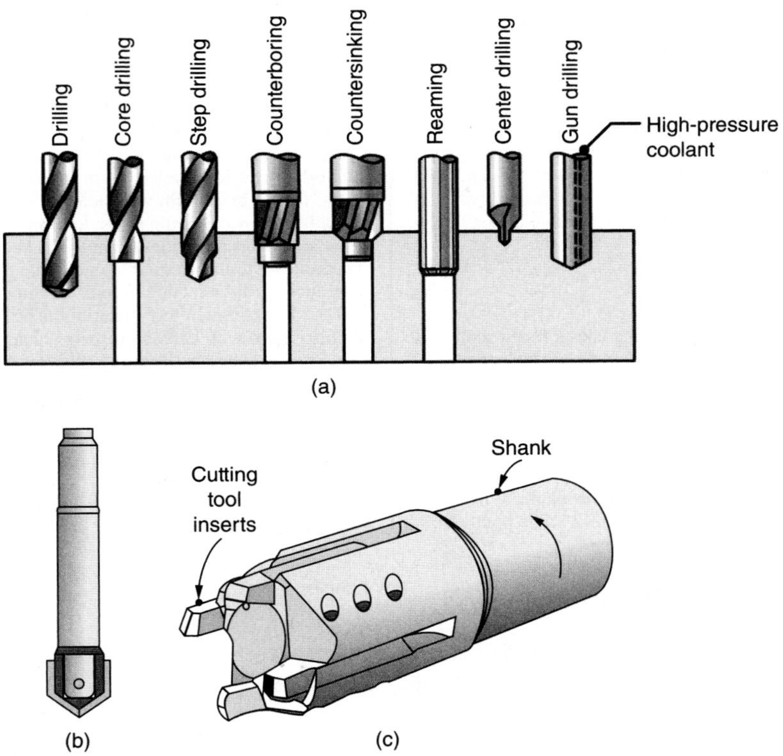

Figure 4.14.10 (*a*) Various types of drills and drilling and reaming operations; (*b*) spade drill; (*c*) trepanning tool with four cutting-tool inserts.

Vertical drilling machines are usually designated by a dimension that roughly indicates the diameter of the largest circle that can be drilled, ranging from around 6 to 50 in. Heavy-duty drill presses of the vertical type, with speed and feed drives, are constructed with a box-type column instead of the older cylindrical column. The size of a **radial drill** is designated by the length of the arm, which represents the radius that can be drilled.

REAMING

A **reamer** is a multiple-cutting-edge tool used to enlarge or finish holes, and for dimensional accuracy and good finish in a part. The **rose reamer** is heavy bodied, with end cutting edges; it is used to remove material and to true a hole (to improve its roundness) prior to reaming. It is similar to the three- and four-fluted drills. **Fluted reamers** cut principally on its periphery, removing only about 0.004 to 0.008 in (0.1 to 0.2 mm) on the bore.

A reamer may be straight or helically fluted; the latter provides smoother cutting and gives a better finish. **Expansion reamers** permit a slight expansion by the action of a wedge; they provide slight variations in size for job shop use. **Adjustable reamers** have means of mechanically adjusting inserted blades, so that a definite size can be maintained through several internal grinding cycles. **Shell reamers** constitute the cutting portion of the tool that fits interchangeably on arbors, to make

Table 4.14.3 General Recommendations for Drilling

	Surface Speed		Feed, mm/rev (in/rev) Drill diameter		Spindle speed (rpm) Drill diameter	
Workpiece material	m/min	ft/min	1.5 mm (0.060 in.)	12.5 mm (0.5 in.)	1.5 mm (0.060 in.)	12.5 mm (0.5 in.)
Aluminum alloys	30-120	100-400	0.025 (0.001)	0.30 (0.012)	6,400-25,000	800-3,000
Magnesium alloys	45-120	150-400	0.025 (0.001)	0.30 (0.012)	9,600-25,000	1,100-3,000
Copper alloys	15-60	50-200	0.025 (0.001)	0.25 (0.010)	3,200-12,000	400-1,500
Steels	20-30	60-100	0.025 (0.001)	0.30 (0.012)	4,300-6,400	500-800
Stainless steels	10-20	40-60	0.025 (0.001)	0.18 (0.007)	2,100-4,300	250-500
Titanium alloys	6-20	20-60	0.010 (0.0004)	0.15 (0.006)	1,300-4,300	150-500
Cast irons	20-60	60-200	0.025 (0.001)	0.30 (0.012)	4,300-12,000	500-1,500
Thermoplastics	30-60	100-200	0.025 (0.001)	0.13 (0.005)	6,400-12,000	800-1,500
Thermosets	20-60	60-200	0.025 (0.001)	0.10 (0.004)	4,300-12,000	500-1,500

Table 4.14.4 General Troubleshooting Guide for Drilling Operations

Problem	Probable causes
Drill breakage	Dull drill; drill seizing in hole because of chips clogging flutes; feed too high; lip relief angle too small.
Excessive drill wear	Cutting speed too high; ineffective cutting fluid; rake angle too high; drill burned and strength lost when being sharpened.
Tapered hole	Drill misaligned or bent; lips not equal; web not central.
Oversized hole	Same as above; machine spindle loose; chisel edge not central; side pressure on workpiece.
Poor hole surface finish	Dull drill; ineffective cutting fluid; welding of workpiece material on drill margin; improperly ground drill; improper alignment.

several sizes. The most common tool materials for reamers are high-speed steels and carbide.

THREADING

Threads can be produced on the outside or inside of a solid cylinder or a cone using (1) single-point cutting tools (see Fig. 4.14.1), (2) threading chasers, (3) taps, (4) dies, (5) thread milling, (6) thread rolling, and (7) grinding. There are numerous types of **taps,** such as hand, machine screw, pipe, and combined pipe tap and drill. The cutting speed depends upon workpiece hardness, hole depth, and the taper. Blind holes require lower speeds; effective cutting fluid allow for increased speeds. Taps are made of high-speed steels. **Threading dies** are used for external threads; they may be solid, adjustable, spring adjustable, or self-opening die heads.

MILLING

Milling is one of the most versatile machining processes and is capable of producing a variety of shapes, involving flat surfaces, slots, and a variety of contours (Fig. 4.14.11). **Milling machines** use **milling cutters** with multiple teeth, in contrast to the single-point tools used on a lathe or a planer. They are made in a wide variety of shapes and sizes, including those with carbide **inserts.** The nomenclature of tooth parts and angles is standardized as shown in Fig. 4.14.12.

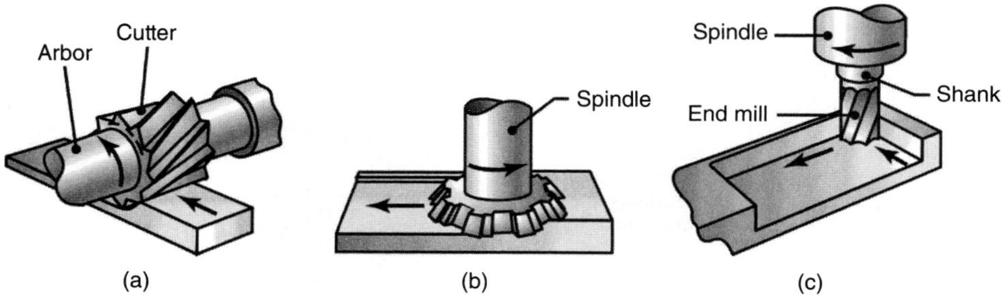

Figure 4.14.11 Basic types of milling cutters and operations. (*a*) Peripheral milling; (*b*) face milling; (*c*) end milling.

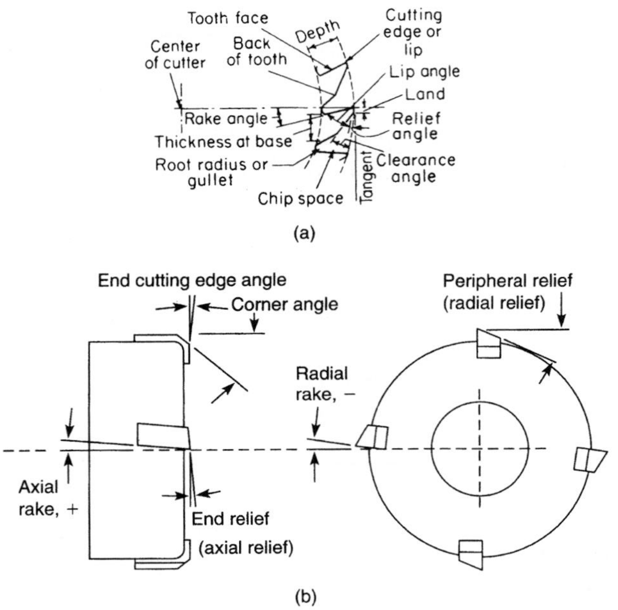

Figure 4.14.12 (*a*) Plain milling cutter teeth; (*b*) face milling cutter.

There are two basic types of milling operations: **peripheral (slab)** and **face** milling. In the **peripheral** or **slab milling** process, the axis of the cutter is parallel to the surface being milled, In **face milling,** the cutter axis is generally at a right angle to the surface. The slab-milling process consists of two types: **conventional (up) milling** (as in Fig. 4.14.11*a*) and **climb (down) milling.** Each has its advantages, and the choice depends on several factors, including the type and condition of the machine tool, tool life, surface finish, and processing parameters. **End milling** (Fig. 4.14.11*c*) is an important operation; using a variety of end mills, it is particularly capable of producing complex internal and external shapes, including curved, stepped, and pocketed, and at high speeds. One common application is in producing cavities in dies (die sinking).

MILLING MACHINES

The classification of **milling machines** is based on design, operation, or purpose. **Knee-and-column** type machines have the table and the saddle supported on the vertically-adjustable knee. The table is fed longitudinally on the saddle, and the latter transversely on the knee to give the operation three degrees of freedom. The machines have horizontal or vertical spindles. **Horizontal** universal machines have a swiveling table, for more complex workpiece geometries (i.e., additional degrees of freedom). **Vertical** milling machines, with fixed or sliding heads, are similar to the horizontal type, and are used for face or end milling and can be provided with a rotary table for machining cylindrical surfaces.

Fixed-bed machines have a spindle mounted in a head, dovetailed to and sliding on the face of the column; the table rests directly on the bed. They are simple and rigidly built and are used primarily for high-production work. These machines are typically provided with work-holding fixtures and may be constructed as plain or multiple-spindle machines, simple or duplex. **Planer-type** milling machines are used only on the heaviest work. They can machine a number of surfaces of a particular part or group of parts, arranged in series in fixtures on the table.

GEAR MANUFACTURING

This section describes the basic processes for manufacturing a variety for gears and the equipment used. Modern **computer-controlled** machine tools, with several features, can now perform these operations efficiently, with good dimensional control and high productivity.

Gear Cutting

Gear-cutting processes are generally classified as either **forming** or **generating** type. In forming, the shape of the cutter is reproduced on the workpiece; in generating, the shape produced on the workpiece depends on both the shape of the tool and the relative motion between the cutter and the workpiece during the machining operation. In general, a generating process is more accurate than in forming. In **form cutting,** the cutter has the shape of the space between the teeth. For this reason, form cutting will produce precise tooth profiles only with an accurate cutter and constant-width tooth space, such as on spur and helical gears. Single-space form milling with disk-type cutters is particularly suitable for gears with large teeth, because the cutting action of a milling cutter is more effective than that of the tools used for generating. Form milling of spur gears is done on machines that retract and index the gear blank automatically.

In **gear generating,** the cutter can be considered as one of the gears in a conjugate pair and the gear blank as the other. The correct relative motion between the tool arbor and the blank arbor is obtained by means of a train of indexing gears within the machine. On a **gear shaper,** the generating tool is a pinion-shaped cutter that rotates slowly at the proper speed as if meshing with the blank; the cutting action is produced by reciprocation of the cutter parallel to the workpiece axis. These machines can cut spur and helical gears, both internal and external; they can also cut continuous-tooth helical (herringbone) gears and are particularly suitable for cluster gears or gears that are close to a shoulder.

On a **rack shaper,** the generating tool is a segment of a rack that moves perpendicular to the axis of the blank while the blank rotates about a fixed axis, at the speed corresponding to conjugate action between the rack and the blank. The cutting action is produced by reciprocation of the cutter parallel to the axis of the blank. Since it is impracticable to have more than 6 to 12 teeth on a rack cutter, the cutter must be disengaged from the blank at suitable intervals and returned to the starting point; the blank meanwhile remaining fixed. These machines can cut both spur and helical external gears.

A **hob** (Fig. 4.14.13) is basically a worm, or screw, made into a generating tool by cutting a series of longitudinal slots or gashes

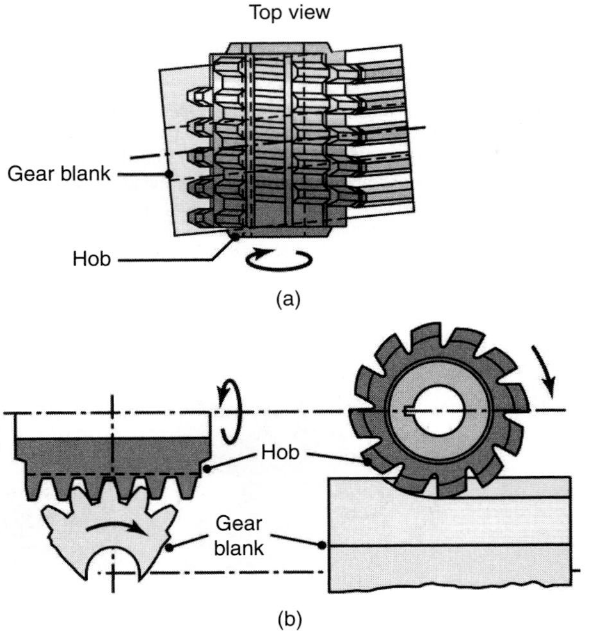

Figure 4.14.13 A gear-cutting hob.

to form gear teeth; a hob may have one, two, or three threads. On involute hobs with a single thread, the generating portion of the hob-tooth profile usually has straight sides (like an involute rack tooth) in a section taken at right angles to the thread. Gear cutting with a hob is sometimes referred to as **gashing.** In addition to the conjugate rotary motions of the hob and the workpiece, the hob must be fed parallel to the workpiece axis for a distance greater than the face width of the gear. The hobbing process is continuous until all the teeth are cut.

The same machines and the same hobs used for cutting spur gears also can be used for helical gears. The threads of worms are usually cut with a disk-type milling cutter, on a thread-milling machine, and finished by grinding. Worm gears are usually cut with a hob on the machines used for hobbing spur and helical gears. Except for the relief on the teeth and an allowance for grinding, the hob is a counterpart of the worm. The hob and workpiece axes are inclined to one another at the shaft angle of the worm and gear set, usually 90°. The hob may be fed in to full depth in a radial direction or parallel to the hob axis. The generating method for straight bevel gears is analogous to the rack-generating method used for spur gears. Instead of using a rack with several complete teeth, however, the cutter has only one straight cutting edge that moves in the plane of the tooth.

The machines for cutting spiral bevel gears operate on essentially the same principle as those for straight bevel gears, but the cutter is different. The spiral cutter is basically a disk with a number of straight-sided cutting blades protruding from its periphery on one side to form the rim of a cup. The machines have means for indexing, retracting, and producing a generating roll.

Gear Shaving

Shaving improves the surface finish and profile accuracy of machined spur and helical gears (internal and external); this is a free-cutting operation that removes small amounts of metal from the working surfaces of the gear teeth. The teeth on the shaving cutter have a series of sharp-edged rectangular grooves running from tip to root; the intersection of the grooves with the tooth profiles creates cutting edges. When the cutter and the workpiece are made to move relative to one another along the tooth depth direction, the cutting edges remove metal from the gear teeth. Shaving requires less time than grinding, but it cannot be used on gears harder than approximately 400 HB (42 HRC).

Gear Grinding

Machines for grinding spur and helical gears utilize either a forming or a generating process. For form grinding, a disk-type grinding wheel is first dressed to the proper shape with a diamond held on a special dressing attachment. For each number of teeth, a special index plate with V-type notches on its periphery is required. When grinding helical gears, a means for producing a helical motion of the blank must be provided.

For generating grinding, the grinding wheel may be a disk-type, double-conical wheel with an axial section equivalent to the basic rack of the gear system. A master gear, similar to the gear being ground, is mounted on the workpiece arbor and it meshes with a master rack; the generating roll is created by rolling the master gear in the stationary rack. Spiral bevel and hypoid gears can be ground on the machines on which they were generated. The grinding wheel has the shape of a flaring cup, with a double-conical rim having a cross section equivalent to the surface that is the envelope of the rotary cutter blades.

PLANING AND SHAPING

Planers are used for rough and finish machining of large, flat surfaces, although arcs and special forms also can be made with appropriate tools and attachments. Surfaces can be finished by scraping, such as for ways and long dovetails, and for parts of machine tools. **Shapers** are used for miscellaneous planing, surfacing, notching, key seating, and production of flat surfaces on small parts. The tool is held in a holder supported on a clapper on the end of a ram, that is reciprocated hydraulically or by crank and rocker arm, in a straight line.

BROACHING

Broaching is a process whereby a **broach** is used to finish machine internal or external surfaces, such as flat surfaces, holes of circular, square, or irregular section, keyways, teeth of internal gears, and splines. Broaching round holes gives higher accuracy and better surface finish than by reaming; however, because the broach may be guided only by the workpiece it is cutting, the hole may not be accurate with respect to previously machined surfaces. For higher accuracy, it is better to first broach and then to machine the surfaces.

A **broach** is typically long and is provided with several cutting teeth, so graded in size that each tooth removes a small layer of material when the broach is pulled or pushed, either through the previously prepared leader hole or the past surface. The main features of the broach are the pitch, degree of taper or increase in height of each successive tooth, relief, tooth depth, and rake. The **pitch** of the teeth (the distance from a corresponding point on one tooth profile to the next) depends on tooth strength, length of cut, and the shape and size of chips. The pitch should be as coarse as possible to provide ample chip clearance, but at least two teeth should be in contact with the workpiece at all times.

The degree of **taper,** or increase in size per tooth, depends largely on the hardness or toughness of the material to be broached and the surface finish desired. Usually the first few teeth coming in contact with the workpiece are undersized, but of uniform taper, for the greatest feed per tooth. As the finished size is approached, the teeth remove smaller and smaller layers, with several teeth at the finishing end of nearly zero taper. For medium-sized broaches, the taper per tooth is 0.001 to 0.003 in (0.025 to 0.076 mm); large broaches remove 0.005 to 0.010 in (0.127 to 0.254 mm) per tooth or even more. The teeth are given a **front rake angle** of 5 to 15° to give a curl to the chip, provide a cleaner cut surface, and reduce the power consumption. The **land** of the cutting edge has a relief varying from 1 to 3°, with a clearance of 30 to 45°.

Push broaches are usually shorter than pull broaches, being 6 to 14 in (150 to 350 mm) long, depending on their diameter and the volume of metal to be removed. In many cases, for accuracy, four to six broaches of the push type constitute a set, used in sequence to finish the surface being broached. Push broaches usually have a large cross-sectional area so as to be sufficiently rigid. **Pull broaches** are attached to the crosshead of the broaching machine by means of a key slot and key, by a threaded connection, or by a head that fits into an automatic broach puller. The threaded connection is used where the broach is not removed from the drawing head while the workpiece is placed over the cutter, as in cutting a keyway. In enlarging holes, however, the small end of the pull broach must first be extended through the reamed, drilled, or cored hole and then fixed in the drawing head before being pulled through the workpiece. With pull broaches, pulling tends to straighten the hole, whereas push broaches follow any irregularity of the leader hole.

Processing Parameters for Broaching Cutting speeds for broaching may range from 5 ft/min (1.5 m/min) for high-strength materials to as high as 30 to 50 ft/min (9 to 15 m/min) for aluminum and magnesium alloys. The most common tool materials are high speed steels; carbide tips are also widely used. An emulsion is often used as a cutting fluid for general work; oils may also be used.

Broaching Machines

Push broaching machines are usually vertical, while pull broaching machines may be either vertical or horizontal. In either case, the ram may be driven hydraulically or by screw, rack, or crank. Both are made in the duplex- and multiple-head type.

CUTTING OFF

Cutting off involves parting or slotting bars, tubes, plate, or sheet by various means. The machines used for cutting off include lathes (using a single-point cutting tool), hacksaws, band saws, circular saws, friction saws, and thin abrasive wheels. Cutting off may also be carried out by shearing and cropping, as well as using flame, electric arc, abrasive jets, and electron or laser beams.

Hacksaw blades are made 8 to 12 in long, 7/16 to 9/16 in wide, and 0.025 in thick. The number of teeth per inch for cutting soft steel or cast iron is typically 14; for tool steel and angle iron, 18; for brass, copper, and heavy tubing, 24; and for sheet metal and thin tube, 32. In **power hacksaws,** the blade reciprocates while translating through the workpiece, which is held in a vise on the bed. With high-speed steel blades, cutting speeds range from about 30 strokes per minute for high strength materials to 180 strokes per minute for carbon steels.

Band saws may be vertical, horizontal, or universal types. The **kerf** or width of cut is small with a consequently small loss in material. The teeth of band saws, like those of hacksaws, are set with the regular alternate type to prevent binding, one bent to the right and the next to the left; or with the alternate and center set, in which one tooth is bent to the right, the second to the left, and the third straight in the center. With high-speed steel saw blades, band speeds range from about 30 ft/min (10 m/min) for high-temperature alloys to as much as 1,300 ft/min (400 m/min) for aluminum and magnesium alloys; the speed should be decreased as the workpiece thickness increases.

Circular saws are made in a wide variety of styles and sizes. They may have teeth of various shapes, such as radial face teeth for small fine-tooth saws; radial face teeth with a land for fine-tooth saws; alternate bevel-edged teeth to break up the chips, with every other tooth beveled 45° on each side with the next tooth plain; alternate side-beveled teeth; and one tooth beveled on the right, the next on the left, and the third beveled on both sides, each leaving slightly overlapping flats on the periphery of the teeth. The most common high-speed steels for circular saws are M2 and M7; saws may also have brazed carbide teeth for harder materials.

Abrasive cutoff wheels are made of thin resinoid or rubber bonded abrasives, and they operate at surface speeds of up to 16,000 ft/min (4,800 m/min). These wheels are used for cutting off tubes, structural shapes, hardened steels, ceramics, and composite materials. **Friction sawing machines** are generally used for cutting off structural shapes; peripheral speeds are about 20,000 ft/min (6,000 m/min). Material removal takes place by the combination of high sliding speed and friction, whereby the temperature rise is sufficient to cause melting of the workpiece. The wheels may be plain on the periphery, V-notched, or with notches.

Grinders

Modern grinding machines are **computer controlled,** and have such features as automatic workpiece loading and unloading, part clamping, and automatic dressing and wheel shaping. Grinders can be equipped with probes and gages, for determining the relative position of the wheel and workpiece surfaces, as well as with tactile-sensing features, whereby diamond dressing-tool breakage, for example, can be rapidly detected during the dressing cycle.

Surface grinders are generally for flat surfaces. The movement of the grinding wheel may be along the surface of the workpiece (traverse grinding, through-feed grinding, or cross-feeding), or the wheel may move radially into the workpiece (plunge grinding). Surface grinders make up the largest percentage of grinders used in production, followed by cylindrical grinders, tool and cutter grinders, and internal grinders.

Centerless grinders are high-production machines used for grinding cylindrical surfaces; the workpiece is supported not by centers (hence the term "centerless") or chucks, but by a blade. Typical parts made are roller bearings, piston pins, engine valves, camshafts, and similar components; parts as small as 0.1-mm (0.004-in.) in diameter can be ground. The grinders are capable of wheel surface speeds on the order of 10,000 m/min (35,000 ft/min), and typically are made of cubic boron nitride.

The **cylindrical grinder** is a companion machine to a lathe; shafts, cylinders, rods, studs, and a wide variety of other cylindrical parts are first roughed out on the lathe, then finished accurately to size on a cylindrical grinder. The workpiece is carried on centers, rotated slowly and traversed past the face of the grinding wheel. **Internal grinders** are used for finishing holes in bushings, rolls, sleeves, cutters, and similar parts; spindle speeds in the range of 15,000 to 30,000 r/min are common.

Although grinding is generally regarded as a finishing operation, it is possible to increase the rate of stock removal, whereby the process becomes competitive with milling; this type of grinding operation is called **creep-feed grinding.** It uses such equipment as reciprocating table or vertical-spindle grinders, and rotary-table surface grinders, with capacities up to 300 hp (220 kW). Typical stock removal may range up to ¼ in (6.4 mm), with wheel speeds up to 5,000 surface ft/min (1,500 m/min).

MACHINING AND GRINDING OF PLASTICS

The low strength of thermoplastics permits high cutting speeds and feeds, but their low thermal conductivity and greater resilience require higher relief angles and lower rake angles, in order to avoid undersize cutting. Hard and sharp tools should be used. A cutting fluid, such as a blast of air or a stream of water, improves turning and cutting of plastics, as it prevents the heating of the tool and causes the chips to remain brittle and to break, rather than become sticky and gummy. A zero or slightly negative back rake and a relief angle of 8 to 12° should be used. For thermoplastics, cutting speeds generally range from 250 to 400 ft/min (75 to 120 m/min), and for thermosetting plastics from 400 to 1,000 ft/min (120 to 300 m/min). Recommended tool materials are high-speed steels and carbide.

In sawing plastics a pitch ranging from 3 to 14 teeth/in (1.2 to 5.5 teeth/cm) is typically used, with fewer teeth per length for thicker workpieces. Speeds for thermoplastics range up to 4,000 ft/min (1,200 m/min), and for thermosetting plastics up to 5,500 ft/min (1,700 m/min). High-carbon-steel blades are recommended, and an air blast is helpful in preventing the chips from sticking to the saw. Abrasive saws operating at 3,500 to 6,000 ft/min (1,000 to 1,800 m/min) are also used for cutting off operations.

Thermoplastics and thermosets can be ground with relative ease, usually using silicon carbide wheels; as in machining, temperature rise should be minimized. Polishing and buffing are done on several types of plastics, using special compounds containing wax or fine abrasives. Buffing wheels should have loose stitching. Vinyl plastics can be buffed and polished with fabric wheels of standard types, using light pressures.

MACHINING AND GRINDING OF CERAMICS

The technology of machining and grinding of ceramics, as well as composite materials, has advanced rapidly, with good surface characteristics and product integrity. Ceramics can be machined with carbide, high-speed steel, or diamond tools; care should be exercised because of the brittle nature of ceramics, resulting possible surface damage. *Machinable ceramics* have been developed that minimize machining difficulties. Grinding of ceramics is generally done with diamond wheels.

OTHER MACHINING PROCESSES

In addition to mechanical methods of material removal, there are several other important processes that may be preferred over conventional methods. Among important factors to be considered are the hardness of the workpiece material, the shape of the workpiece, its sensitivity to thermal damage, residual stresses, dimensional tolerances, and economics. Some of these processes, described below, produce a heat-affected layer on the surface; improvements in surface integrity may be achieved by post-processing techniques, such as polishing or peening. Almost all machines are now computer-controlled, and have various features to improve productivity.

Electric-discharge machining The principle of electrical-discharge machining (EDM), also called electro-discharge or spark-erosion

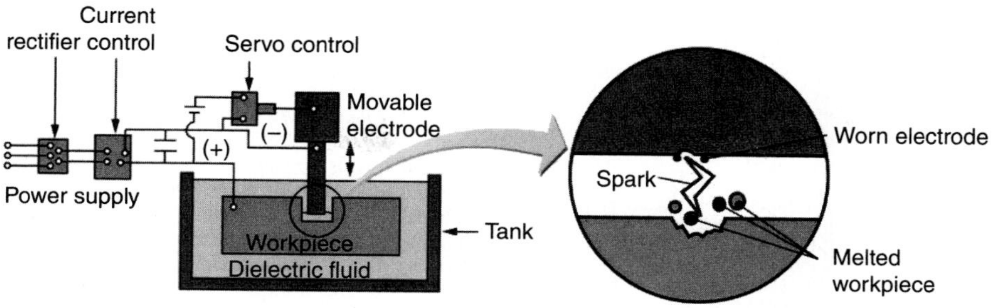

Figure 4.14.14 Schematic diagram of the electric-discharge machining process.

machining, is based on the erosion of metals by spark discharges. When two current-conducting wires are allowed to touch each other, an arc is produced; observing closely the point of contact between the two wires, it will be noted that a small portion of the metal has been eroded away, leaving a crater on the surface. The EDM process has become one of the most important and widely used production technologies in manufacturing.

The basic EDM system, shown schematically in Fig. 4.14.14, consists of a shaped tool (electrode) and the workpiece, connected to a DC power supply and placed in a dielectric (electrically nonconducting) fluid. When the potential between the tool and the workpiece is sufficiently high, the dielectric breaks down and a transient spark discharges through the fluid, removing a very small amount of metal from the workpiece surface. The capacitor discharge is repeated continuously, at rates between 200 and 500 kHz, with voltages usually ranging between 50 and 380 V, and currents from 0.1 to 500 A. Surface finish may range from 1,000 μin *R*q (25 μm) in roughing cuts, to less than 25 min (0.6 mm) in finishing cuts.

The response of materials to this process depends mostly on their thermal properties. Thermal conductivity, heat capacity, and latent heats of melting and vaporization are important; hardness and strength do not generally affect metal removal rates. The EDM process is applicable to all electrically conducting materials. The electrode is usually made of graphite, copper-tungsten, or copper alloys. Tool wear is an important consideration; in order to control dimensional tolerances and minimize cost, the ratio of tool material removed to workpiece material removed should be low. This ratio varies with different tool and workpiece material combinations and with operating conditions, and a particular tool material may not be best for all workpieces. In machining some steels, tool wear can be minimized by reversing the polarity and using copper tools, known as no-wear EDM.

The EDM process has numerous applications, such as machining cavities and dies, cutting small-diameter holes, blanking parts from sheets, cutting off rods of materials with poor machinability, and flat or form grinding. It is also applied to sharpening tools, cutters, and broaches, and the process can be used to generate almost any geometry with a suitable tool.

Thick plates may be cut with **wire EDM** (Fig. 4.14.15). A slowly moving wire travels a prescribed path along the workpiece and cuts the metal, with sparks acting like saw teeth. The wire, usually about 0.01 in (0.25 mm) in diameter, is made of brass, copper, or tungsten, and is generally used only once. This process is also used in making tools and dies from hard materials, provided that they are electrically conducting.

Electric-discharge grinding (EDG) is similar to the EDM process, with the exception that the electrode is in the shape of a grinding wheel, made of graphite or brass. Removal rates are up to 1.5 in^3/h (25 cm^3/h), with dimensional tolerances on the order of 0.001 in (0.025 mm). The wheel is operated around 100 to 600 surface ft/min (30 to 180 m/min) to minimize splashing of the dielectric fluid. Typical applications of this process are in grinding of carbide tools and dies, thin slots in hard materials, and production grinding of intricate forms.

The **electrochemical machining (ECM)** process (Fig. 4.14.16) uses electrolytes that dissolve the reaction products formed on the workpiece by electrochemical action; ECM is basically a reverse electroplating process. The electrolyte is pumped at high velocities through the tool, and a gap of 0.005 to 0.020 in (0.13 to 0.5 mm) is maintained. A DC power supply maintains very high current densities between the tool and the workpiece. In most applications, a current density of 1,000 to

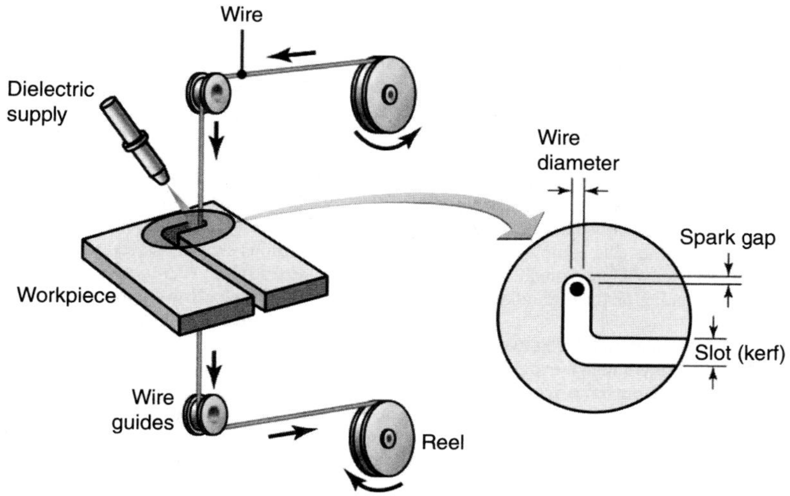

Figure 4.14.15 Schematic diagram of the wire EDM process.

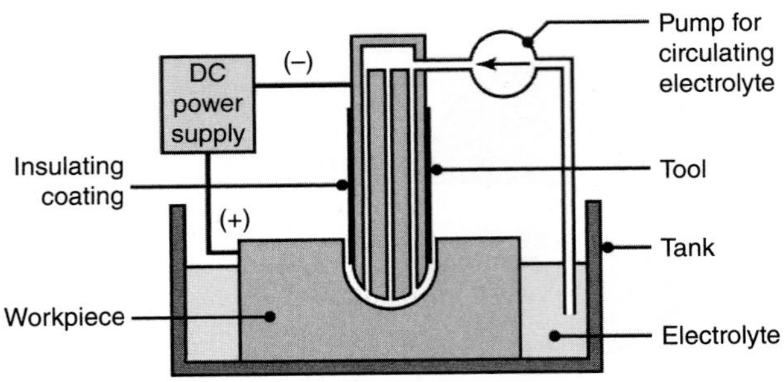

Figure 4.14.16 Schematic diagram of the electrochemical machining process.

5,000 A/in^2 is required. Removal rates up to 1 in^3/min (16 cm^3/min) can be achieved, with a 10,000-A power supply.

The **electrochemical grinding (ECG)** process (Fig. 4.14.17) is a combination of electrochemical machining and abrasive cutting, where most of the metal removal results from the electrolytic action. The process consists of a rotating cathode, a neutral electrolyte, and abrasive particles in contact with the workpiece. The equipment is similar to a conventional grinding machine except for the electrical accessories. The cathode usually consists of a metal-bonded diamond or aluminum oxide wheel. An important function of the abrasive grains is to maintain a space for the electrolyte between the wheel and workpiece.

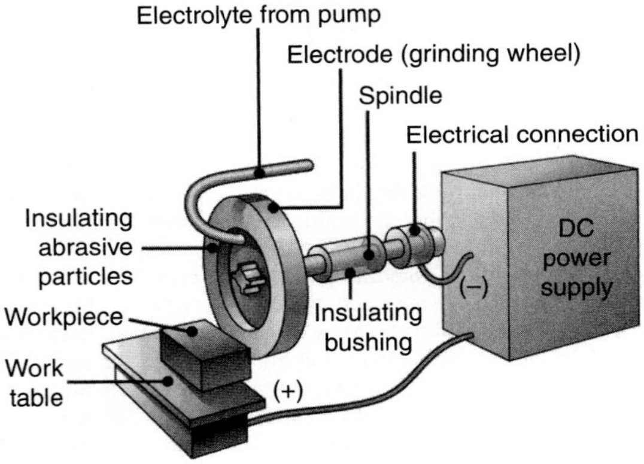

Figure 4.14.17 Schematic diagram of the electrochemical grinding process.

Surface finish, dimensional accuracy, and the metal-removal rate are influenced by the composition of the electrolyte. Aqueous solutions of sodium silicate, borax, sodium nitrate, and sodium nitrite are commonly used as electrolytes. The process is primarily used for machining high-strength materials and sharpening of cutters.

Electrochemical discharge grinding (ECDG) is a combination of electric-discharge and electrochemical methods of material removal. The electrode is graphite rotating wheel that electrochemically grinds the workpiece; the intermittent spark discharges remove the oxide films that form as a result of electrolytic action. The equipment is similar to that for electrochemical grinding. Typical applications include machining of fragile parts and for re-sharpening or form grinding of carbides and tools, such as milling cutters.

In **chemical machining (CM)** material is removed by chemical or electrochemical dissolution of preferentially exposed surfaces of the workpiece. Selective attack on different areas is controlled by masking or by partial immersion. There are two processes involved: chemical **milling** and chemical **blanking.** Milling applications produce shallow cavities for overall weight reduction, and are also used to make tapered sheets, plates, or extrusions. Masking with paint or tapes is common. Masking materials may be elastomers (such as butyl rubber, neoprene, and styrene-butadiene) or plastics (such as polyvinyl chloride, polystyrene, and polyethylene). Typical blanking applications are decorative panels, printed-circuit etching, and thin stampings. Etchants are solutions of sodium hydroxide for aluminum, and solutions of hydrochloric and nitric acids for steel.

Ultrasonic machining (USM) is a process in which a tool is given a high-frequency, low-amplitude oscillation, which, in turn, transmits a high velocity to fine abrasive particles that are present between the tool and the workpiece, usually suspended in a carrier fluid. Minute particles of the workpiece are chipped away with each impact with the abrasive particles. Aluminum oxide, boron carbide, or silicon carbide grains are used in a water slurry (usually 50% by volume), that also carries away the debris. Grain size ranges from 200 to 1,000 (see Sec. 4).

The equipment consists of an electronic oscillator, a transducer, a connecting cone or toolholder, and the tool. The oscillatory motion is obtained most conveniently by magnetostriction, at about 20 kHz and a stroke of 0.002 to 0.005 in (0.05 to 0.13 mm). The tool material is typically cold-rolled steel or stainless steel, and is brazed, soldered, or fastened mechanically to the transducer through a toolholder; the tool is about 0.003 to 0.004 in (0.075 to 0.1 mm) smaller than the cavity it produces. Dimensional tolerances of 0.0005 in (0.013 mm) or better can be obtained, using fine abrasives. The ultrasonic machining process is used in drilling, engraving, cavity sinking, slicing, and broaching. It is best suited for materials that are hard and brittle, such as ceramics, hardened steels, carbides, borides, ferrites, glass, and precious stones.

In **water jet machining (WJM),** water is ejected from a nozzle, at pressures as high as 200,000 lb/in^2 (1,400 MPa), and acts as a saw. The process is suitable for cutting and deburring of a variety of materials, including polymers, paper, and brick, in thicknesses ranging from 0.03 to 1 in (0.8 to 25 mm) or more. The cut can be started at any location, wetting is minimal, and no deformation of the rest of the piece takes place. Abrasives can be added to the water stream to increase material removal rate, known as **abrasive water jet machining (AWJM).** In **abrasive jet machining (AJM),** material is removed by fine particles of aluminum oxide or silicon carbide, carried in a high-velocity stream of air, nitrogen, or carbon dioxide. The gas pressure ranges up to 120 lb/in^2 (800 kPa), with a nozzle velocity of up to 1,000 ft/s (300 m/s). Nozzles are made of tungsten carbide or sapphire. Typical applications are in drilling, sawing, slotting, and deburring of hard, brittle materials such as glass.

In **laser-beam machining (LBM),** material is removed by converting electric energy into a narrow beam of light focused on the workpiece. The high energy density of the beam is capable of melting and vaporizing all materials, and consequently there is a thin heat affected zone. The most commonly used laser types are CO_2 (pulsed or continuous-wave) and Nd:YAG. Typical applications include cutting a variety of metallic and nonmetallic materials, drilling (as small as 0.0002 in or 0.005 mm in diameter), and marking. The efficiency of cutting increases with decreasing thermal conductivity and reflectivity of the material. Because of the inherent flexibility of the process, programmable and computer-controlled laser cutting has become an important process, particularly in cutting profiles and multiple holes of various shapes and sizes on large sheets. Cutting speeds may range up to 25 ft/min (7.5 m/min).

The **electron-beam machining (EBM)** process removes material by focusing high-velocity electrons on the workpiece. Unlike lasers, this process is carried out in a vacuum chamber and is used for drilling small holes, scribing, and cutting slots in variety of materials, including ceramics.

Several **hybrid processes** have been developed that combine power sources or mechanisms of material removal. An example is **laser microjet,** illustrated in Fig. 4.14.18, that uses a low-pressure laminar

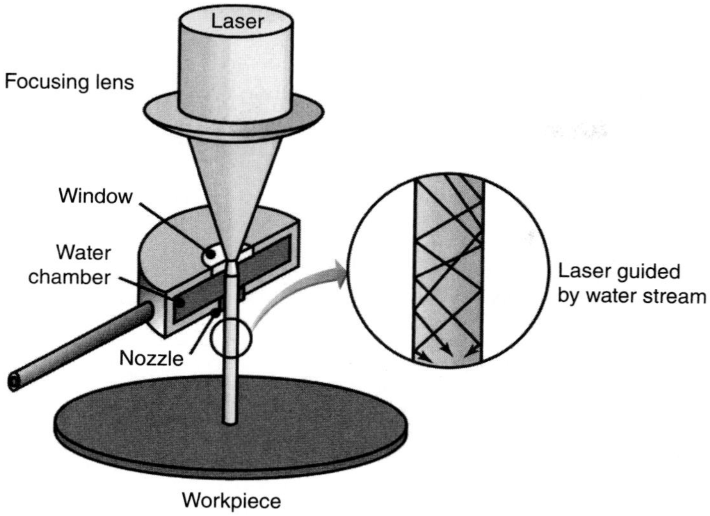

Figure 4.14.18 Schematic illustration of the laser microjet process.

water stream, to serve as a variable-length, fiber-optic cable to direct the laser and deliver laser power at the bottom of the kerf. This method has an advantage in that the laser focus depth is very large, and cuts at high aspect ratios cuts can be made. The water jet is produced by a sapphire or diamond nozzle, with an opening of 25 to 100 μm, and exerts a force less than 0.1 N. Material removal is due to the action of the laser, while the water provides cooling (thus reducing the heat-affected zone), and prevents splatter from bonding to the workpiece. **Hybrid machines**, where multiple operations are performed on a part without changing fixtures, are also commercially available.

MACHINE-TOOL STRUCTURES

The design of **machine-tool structures** requires analysis of such factors as form and materials of structures, stresses, weight, and manufacturing and performance considerations. The best approach to obtain the ultimate in machine-tool accuracy is to employ both improvements in structural stiffness and compensation of deflections by use of special controls. The C-frame structure has been used extensively in the past because it provides ready accessibility to the working area of the machine. With the advent of computer control, the box-type frame, with its considerably improved static stiffness, becomes practical since the need for manual access to the working area is greatly reduced. The use of a box-type structure with thin walls can provide low weight for a given stiffness.

The light-weight-design principle offers high dynamic stiffness by providing a high natural frequency of the structure through combining high static stiffness with low weight rather than through the use of large mass. Dynamic stiffness is the stiffness exhibited by the system when subjected to dynamic excitation where the elastic, the damping, and the inertia properties of the structure are involved; it is a frequency-dependent quantity.

AUTOMATION IN MACHINE TOOLS

Automation involves several activities, such as material handling, processing, assembly, inspection, and packaging. Its primary objective is to lower manufacturing costs through controlled production and product quality, lower labor cost, reduce damage to workpieces during handling, higher level safety for personnel, and economy of floor space. Automation may be partial, such as gaging in cylindrical grinding, or it may be total automation.

The conditions that play a role in decisions concerning automation are rising production costs, high percentage of rejects, lagging output, scarcity of skilled labor, hazardous working conditions, and work requiring repetitive operations. Factors that must be carefully studied prior to deciding on automation are high initial cost of equipment, maintenance problems, and the type of product. (See also Sec. 12.)

Mass production with modern machine tools has been achieved through the development of self-contained **power-head** production units and the development of **transfer mechanisms.** Power-head units, consisting of a frame, electric driving motor, gearbox, and tool spindles, are available for numerous types of machining operations. Transfer mechanisms move the workpieces from station to station by various methods, and transfer-type machines can be arranged in various configurations, such as a straight line or a U pattern. Various types of machine tools for mass production can be built from components, known as the **building-block** principle. Such a system combines flexibility and adaptability with high productivity. (See **machining centers.**)

Numerical control (NC) is a method of controlling the movements of machine components by numbers. First applied to machine tools in the 1950s, numerically controlled machine tools are classified according to the type of machining operation. For example, in drilling and boring machines, the positioning and the cutting take place sequentially (**point-to-point**). In die-sinking machines, on the other hand, positioning and cutting take place simultaneously, described as **continuous-path** machines; because they require more exacting specifications, they give rise to more complex problems. A large variety of programming systems has been developed, and machines now perform over a very wide range of machining conditions to improve dimensional accuracy without requiring adjustment to, for example, eliminate chatter. Complex contours can be machined that would be almost impossible by any other method.

In **computer numerical control (CNC),** a microcomputer is a part of the control panel of the machine tool. Its advantages are ease of operation, simpler programming, higher dimensional accuracy, versatility, and lower maintenance costs.

4.15 SURFACE TEXTURE DESIGNATION, PRODUCTION, AND QUALITY CONTROL

by Ali M. Sadegh

REFERENCES: American National Standards Institute, "Surface Texture," ANSI/ASME B 46.1-1985, and "Surface Texture Symbols," ANSI Y 14.36-1978. Broadston, "Control of Surface Quality," Surface Checking Gage Co., Hollywood, CA. ASME, "Metals Engineering Design Handbook," McGraw-Hill SME, "Tool and Manufacturing Engineers Handbook," McGraw-Hill.

DESIGN CRITERIA

Surfaces produced by various processes exhibit distinct differences in texture. These differences make it possible for honed, lapped, polished, turned, milled, or ground surfaces to be easily identified. As a result of its unique character, the surface texture produced by any given process can be readily compared with other surfaces produced by the same process through the simple means of comparing the average size of its irregularities, using applicable standards and modern measurement methods. It is then possible to predict and control its performance with considerable certainty by limiting the range of the average size of its characteristic surface irregularities. Surface texture standards make this control possible.

Variations in the texture of a critical surface of a part influence its ability to resist wear and fatigue; to assist or destroy effective lubrication; to increase or decrease its friction and/or abrasive action on other parts, and to resist corrosion, as well as affect many other properties that may be critical under certain conditions.

Clay has shown that the load-carrying capacity of nitrided shafts of varying degrees of roughness, all running at 1,500 r/min in diamond-turned lead-bronze bushings finished to 20 μin (0.50 μm), varies as shown in Fig. 4.15.1. The effects of roughness values on the friction between a flat slider on a well-lubricated rotating disk are shown in Fig. 4.15.2.

Surface texture control should be a normal design consideration under the following conditions:

1. For those parts whose roughness must be held within closely controlled limits for optimum performance. In such cases, even the process may have to be specified. Automobile engine cylinder walls, which should be finished to about 13 μin (0.32 μm) and have a circumferential (ground) or an angular (honed) lay, are an example. If too rough, excessive wear occurs, if too smooth, piston rings will not seat properly, lubrication is poor, and surfaces will seize or gall.

2. Some parts, such as antifriction bearings, cannot be made too smooth for their function. In these cases, the designer must optimize the tradeoff between the added costs of production and various benefits derived from added performance, such as higher reliability and market value.

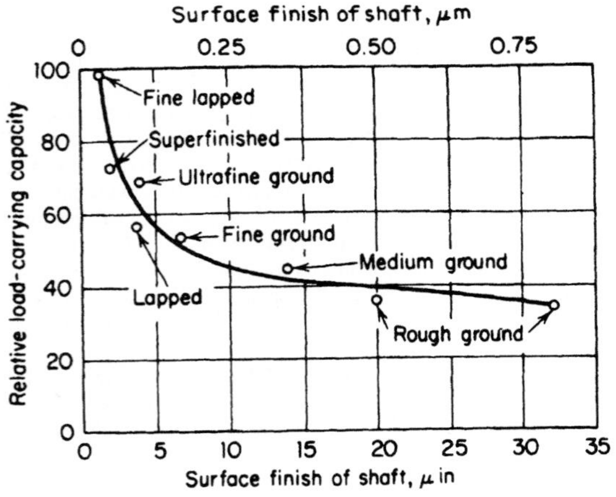

Figure 4.15.1 Load-carrying capacity of journal bearings related to the surface roughness of a shaft. *(Clay, ASM Metal Progress, Aug. 15, 1955.)*

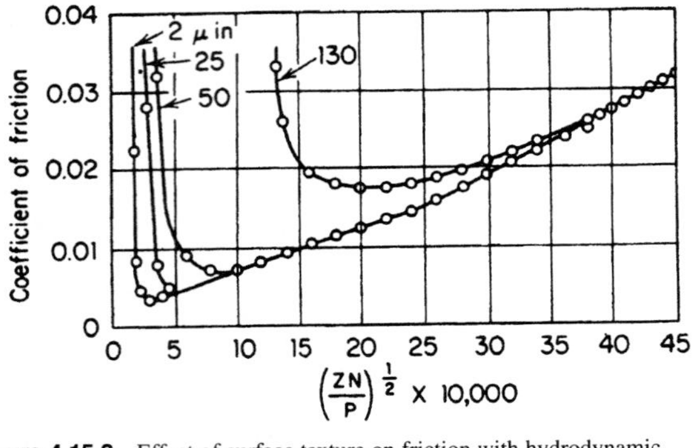

Figure 4.15.2 Effect of surface texture on friction with hydrodynamic lubrication using a flat slider on a rotating disk. Z = oil viscosity, cP; N = rubbing speed, ft/min; P = load, lb/in^2.

3. There are some parts where surfaces must be made as smooth as possible for optimum performance regardless of cost, such as gages, gage blocks, lenses, and carbon pressure seals.

4. In some cases, the nature of the most satisfactory finishing process may dictate the surface texture requirements to attain production efficiency, uniformity, and control even though the individual performance of the part itself may not be dependent on the quality of the controlled surface. Hardened steel bushings, for example, which must be ground to close tolerance for press fit into housings, could have outside surfaces well beyond the roughness range specified and still perform their function satisfactorily.

5. For parts which the shop, with unjustified pride, has traditionally finished to greater perfection than is necessary, the use of proper surface texture designations will encourage rougher surfaces on exterior and other surfaces that do not need to be finely finished. Significant cost reductions will accrue thereby.

It is the designer's responsibility to decide which surfaces of a given part are critical to its design function and which are not. This decision should be based upon a full knowledge of the part's function as well as of the performance of various surface textures that might be specified. From both a design and an economic standpoint, it may be just as unsound to specify too smooth a surface as to make it too rough—or to control it at all if not necessary. Wherever normal shop practice will produce acceptable surfaces, as in drilling, tapping, and threading, or in keyways, slots, and other purely functional surfaces, unnecessary surface texture control will add costs which should be avoided.

Whereas each specialized field of endeavor has its own traditional criteria for determining which surface finishes are optimum for adequate performance, Table 4.15.1 provides some common examples for design review.

DESIGNATION STANDARDS, SYMBOLS, AND CONVENTIONS

The precise definition and measurement of surface texture irregularities of machined surfaces are almost impossible because the irregularities are very complex in shape and character and, being so small, do not lend themselves to direct measurement. Although both their shape and length may affect their properties, control of their average height and direction usually provides sufficient control of their performance. The standards do not specify the surface texture suitable for any particular application, nor the means by which it may be produced or measured. Neither are the standards concerned with other surface qualities such as appearance, luster, color, hardness, microstructure, or corrosion and wear resistance, any of which may be a governing design consideration.

Table 4.15.1 Typical Surface Texture Design Requirements

(250 μin)	6.3	Clearance surfaces Rough machine parts
(125 μin)	3.2	Mating surfaces (static) Chased and cut threads Clutch-disk faces Surfaces for soft gaskets
(63 μin)	1.60	Piston-pin bores Brake drums Cylinder block, top Gear locating faces Gear shafts and bores Ratchet and pawl teeth Milled threads Rolling surfaces Gearbox faces Piston crowns Turbine-blade dovetails
(32 μin)	0.80	Broached holes Bronze journal bearings Gear teeth Slideways and gibs Press-fit parts Piston-rod bushings Antifriction bearing seats Sealing surfaces for hydraulic tube fittings
(16 μin)	0.40	Motor shafts Gear teeth (heavy loads) Spline shafts O-ring grooves (static) Antifriction bearing bores and faces Camshaft lobes Compressor-blade airfoils Journals for elastomer lip seals
(13 μin)	0.32	Engine cylinder bores Piston outside diameters Crankshaft bearings
(8 μin)	0.20	Jet-engine stator blades Valve-tappet cam faces Hydraulic-cylinder bores Lapped antifriction bearings
(4 μin)	0.10	Ball-bearing races Piston pins Hydraulic piston rods Carbon-seal mating surfaces
(2 μin)	0.050	Shop-gage faces Comparator anvils
(1 μin)	0.025	Bearing balls Gages and mirrors Micrometer anvils

The standards provide definitions of the terms used in delineating critical surface-texture qualities and a series of symbols and conventions suitable for their designation and control. In the discussion which follows, the reference standards used are "Surface Texture" (ANSI/ASME B46.1-1985) and "Surface Texture Symbols" (ANSI Y 14.36-1978).

The basic ANSI symbol for designating surface texture is the checkmark with horizontal extension shown in Fig. 4.15.3. The symbol with the triangle at the base indicates a requirement for a machining allowance, in preference to the old *f* symbol. Another, with the small circle in the base, prohibits machining; hence surfaces must be produced without the removal of material by processes such as cast, forged, hot- or cold-finished, die-cast, sintered- or injection-molded, to name a few. The surface-texture requirement may be shown at A; the machining allowance at B; the process may be indicated above the line at C; the roughness width cutoff (sampling length) at D, and the lay at E. The ANSI symbol provides places for the insertion of numbers to specify a wide variety of texture characteristics, as shown in Table 4.15.2.

Control of **roughness,** the finely spaced surface texture irregularities resulting from the manufacturing process or the cutting action of tools or abrasive grains, is the most important function accomplished through the use of these standards, because roughness, in general, has a greater effect on performance than any other surface quality. The roughness-height index value is a number which equals the arithmetic average deviation of the minute surface irregularities from a hypothetical perfect surface, expressed in either millionths of an inch (microinches, μin, 0.000001 in.) or in micrometers, μm, if drawing dimensions are in metric, SI units.

The term *roughness cutoff,* a characteristic of tracer-point measuring instruments, is used to limit the length of trace within which the asperities of the surface must lie for consideration as roughness. Asperity spacings greater than roughness cutoff are then considered as waviness.

Waviness refers to the secondary irregularities upon which roughness is superimposed, which are of significantly longer wavelength and are usually caused by machine or work deflections, tool or workpiece vibration, heat treatment, or warping. Waviness can be measured by a dial indicator or a profile recording instrument from which roughness has been filtered out. It is rated as maximum peak-to-valley distance. For fine waviness control, techniques involving

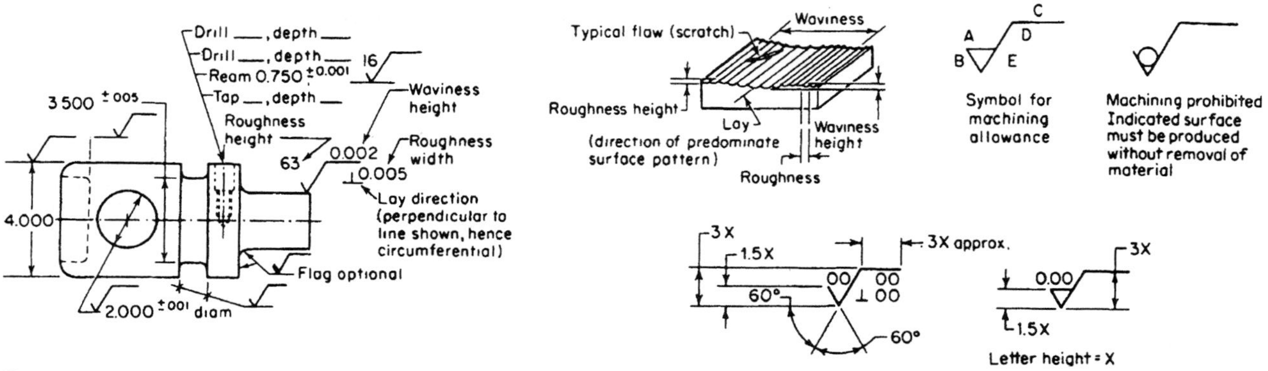

Figure 4.15.3 Application and use of surface texture symbols.

Table 4.15.2 Application of Surface Texture Values to Surface Symbols

(63) 1.6 √	Roughness average rating is placed at the left of the long leg. The specification of only one rating shall indicate the maximum value and any lesser value shall be acceptable. Specify in micrometers (microinches).
(63) 1.6 (32) 0.8 √	The specification of maximum value and minimum value roughness average ratings indicates permissible range of value rating. Specify in micrometers (microinches).
(32) 0.8 √ 0.05	Maximum waviness height rating is placed above the horizontal extension. Any lesser rating shall be acceptable. Specify in millimeters (inches).
(32) 0.8 √ 0.05 – 100	Maximum waviness spacing rating is placed above the horizontal extension and to the right of the waviness height rating. Any lesser rating shall be acceptable. Specify in millimeters (inches).
(63) 1.6 / 3.5 √	Machining is required to produce the surface. The basic amount of stock provided for machining is specified at the left of the short leg of the symbol. Specify in millimeters (inches).
(63) 1.6 √	Removal of material by machining is prohibited.
(32) 0.8 √ ⊥	Lay designation is indicated by the lay symbol placed at the right of the long leg.
(32) 0.8 √ 2.5 (0.100)	Roughness sampling length or cutoff rating is placed below the horizontal extension. When no value is shown, 0.80 mm is assumed. Specify in millimeters (inches).
(32) 0.8 √ ⊥0.5	Where required, maximum roughness spacing shall be placed at the right of the lay symbol. Any lesser rating shall be acceptable. Specify in millimeters (inches).

contact-area determination in percent (90, 75, 50% preferred) may be required. Waviness control by interferometric methods is also common, where notes, such as "Flat within XX helium light bands," may be used. Dimensions may be determined from the precision length table.

Lay refers to the direction of the predominant visible surface roughness pattern.

Flaws are imperfections in a surface that occur only at infrequent intervals. They are usually caused by nonuniformity of the material, or they result from damage to the surface subsequent to processing, such as scratches, dents, pits, and cracks. Flaws should not be considered in surface texture measurements, as the standards do not consider or classify them. Acceptance or rejection of parts having flaws is strictly a matter of judgment based upon whether the flaw will compromise the intended function of the part.

To call attention to the fact that surface texture values are specified on any given drawing, a note and typical symbol may be used as follows:

√ Surface texture per ANSI B46.1

Values for nondesignated surfaces can be limited by the note

XX√ All machined surfaces except as noted

QUALITY CONTROL (SIX SIGMA)

Quality control is a system that outlines the policies and procedures necessary to improve and control the various processes in manufacturing that will ultimately lead to improved business performance.

Six Sigma is a quality management program to achieve "six sigma" levels of quality. It was pioneered by Motorola in the mid-1980s and has spread to many other manufacturing companies. In statistics, sigma refers to the standard deviation of a set of data. Therefore, **six sigma** refers to six standard deviations. Likewise, **three sigma** refers to three standard deviations. In probability and statistics, the standard deviation is the most commonly used measure of statistical dispersion; that is, it measures the degree to which values in a data set are spread. The standard deviation is defined as the square root of the variance, that is, the root mean square (rms) deviation from the average. It is defined in this way to give us a measure of dispersion.

Assuming that defects occur according to a standard normal distribution, this corresponds to approximately 2 quality failures per million parts manufactured. In practical application of the six sigma methodology, however, the rate is taken to be 3.4 per million.

Initially, many believed that such high process reliability was impossible, and **three sigma** (67,000 defects per million opportunities, or DPMO) was considered acceptable. However, market leaders have measurably reached six sigma in numerous processes.

4.16 FERROELECTRICS/PIEZOELECTRICS, MULTIFERROICS, AND SHAPE MEMORY ALLOYS

by Jacqueline J. Li

FERROELECTRICS/PIEZOELECTRICS

Ferroelectrics are materials exhibiting a spontaneous dielectric polarization and **hysteresis effects** in the relation between dielectric displacement and electric field. This ferroelectric behavior is observed only below a transition temperature T_c—the so-called Curie temperature. Above the Curie temperature, the materials are no longer ferroelectric and show normal dielectric behavior.

Ferroelectric phenomena were first observed in 1920 by Valasek in Rochelle salt [304-59-6], $KNaC_4H_4O_6{\cdot}4H_2O$. In his description, Valasek called attention to the analogy with ferromagnetic phenomena. *Ferro* is a prefix associated with iron-containing materials. But the hysteresis nature in the relation between dielectric displacement and electric field is remarkably similar to the induction-magnetic field relation for ferromagnetic materials. Ferromagnetic materials generally contain iron. Ferroelectrics are named after the similar hysteresis loop and rarely contain iron as a significant constituent.

Table 4.16.1 (Table I-1 in "Ferroelectric Crystals," Franco Jona and G. Shirane, Dover, 1993) provides a partial list of ferroelectric crystals arranged approximately in chronological order of their discovery.

Most ferroelectric materials are ceramics; some polymers also exhibit ferroelectric behaviors. At present, the number of known pure compounds with ferroelectric properties is about 200; they are classified

Table 4.16.1 Partial List of Ferroelectric Crystals

Name and chemical formula	Curie temperature, °C	Spontaneous polarization, 10^{-6} C/cm^2	Year in which reported
Rochelle salt $NaKC_4H_4O_6 \cdot 4H_2O$	+23	0.25	1921
Lithium ammonium tartrate $Li(NH_4)C_4H_4O_6 \cdot H_2O$	−170	0.20	1951
Potassium dihydrogen phosphate KH_2PO_4	−150	4.0	1935
Potassium dideuterium phosphate KD_2PO_4	−60	5.5	1942
Potassium dihydrogen arsenate KH_2AsO_4	−177	5.0	1938
Barium titanate $BaTiO_3$	+120	26.0	1945
Lead titanate $PbTiO_3$	+490	>50	1950
Potassium niobate $KNbO_3$	+415	30	1951
Potassium tantalate $KTaO_3$	−260	?	1951
Cadmium (pyro) niobate $Cd_2Nb_2O_7$	−85	~10	1952
Lead (meta) niobate $PbNb_2O_6$	+570	?	1953
Guanidinium aluminium sulfate hexahydrate $C(NH_2)_3Al(SO_4)_2 \cdot 6H_2O$	None	0.35	1955
Methylammonium aluminum alum $CH_3\ NH_3Al(SO_4)_2 \cdot 12H_2O$	−96	1.0	1956
Ammonium sulfate $(NH_4)_2SO_4$	−50	0.25	1956
Trilglycine sulfate $(NH_2CH_2COOH)_2 \cdot H_2SO_4$	+49	2.8	1956
Colemanite $CaB_3O_4(OH)_3 \cdot H_2O$	−7	0.65	1956
Dicalcium strontium propionate $Ca_2Sr(CH_3CH_2COO)_6$	+8	0.3	1957
Lithium acid selenite $LiH_3(SeO_3)_2$	None	10.0	1959
Sodium nitrite $NaNO_2$	+160	7.0	1958

into about 40 families. Examples of typical ferroelectrics are: potassium dihydrogen phosphate, KH_2PO_4 (commonly abbreviated as KDP), and a number of isomorphous phosphates and arsenates; barium titanate, $BaTiO_3$, and other isomorphous double oxides; Rochelle salt (sodium potassium tartrate tetrahydrate, $NaKC_4H_4O_6{\cdot}4H_2O$); and some other isomorphous crystals. The most widely used ferroelectrics now are lead zirconate titanate solid solutions, Pb(Zr-Ti)O_3 (commonly known as PZT). Before we introduce the general properties and structures of ferroelectric materials, we focus on barium titanate, which is one of the simplest ferroelectric materials.

Barium Titanate

In order to better understand the ferroelectric properties, let's concentrate on barium titanate ($BaTiO_3$), which is the most extensively investigated ferroelectric material and whose structure is far simpler than that of any other ferroelectrics. Its crystal structure is of perovskite type. This structure is common to a large family of compounds with the general formula ABO_3, whose representative in nature is the mineral $CaTiO_3$, called perovskite. Figure 4.16.1*a* shows a unit cell of the cubic perovskite-type lattice of $BaTiO_3$. The Curie temperature of $BaTiO_3$ is at 120°C. Above 120°C, $BaTiO_3$ remains in the centrosymmetric cubic structure and nonpolar phase (nonpiezoelectric, or paraelectric). Upon cooling just below 120°C, $BaTiO_3$ undergoes a phase transformation to a tetragonal modification as shown in Fig. 4.16.1*b*. In this process, the positively charged titanium ion moves from its original central position to a position aligned along the polar axis (or *c* axis), and together with the negatively charged oxygen ions, forms a dipole moment. The total dipole moments in the material with a unit volume define the spontaneous polarization *P*. Between 5°C and 120°C, the crystal has a stable tetragonal structure and the lattice parameters are $a = 0.3992$ nm and $c = 0.4032$ nm. Such a poled crystal can undergo a 180° domain switch under a reversed electric field, or a 90° switch under a compressive stress or under an electric field perpendicular to its original polar direction, as shown in Fig. 4.16.1*c*.

Hysteresis Loop

The ferroelectric material can have zero overall **polarization** under zero applied electric field, due to a random orientation of microscopic-scale domains, regions in which the *c* axes of adjacent unit cells have a common direction. If we first apply a small electric field, we will have only linear relationship between polarization *P* and the electric field *E* because the field is not large enough to switch any of the domains and the crystal will behave like a normal paraelectric. Figure 4.16.2 shows schematic plot of *P* versus *E*. When the applied electric field increases, the dipole orientations are favored to the direction parallel to the applied field direction. In this case, domains with such orientations "grow" at the expense of other, less favorably oriented, ones. When the applied electric field increases more, more domains will switch over in the favored applied field direction and the polarization will increase rapidly, until it reaches a state in which all the domains are aligned in the applied electric field direction. This is a state of saturation and the crystal consists now of a single domain.

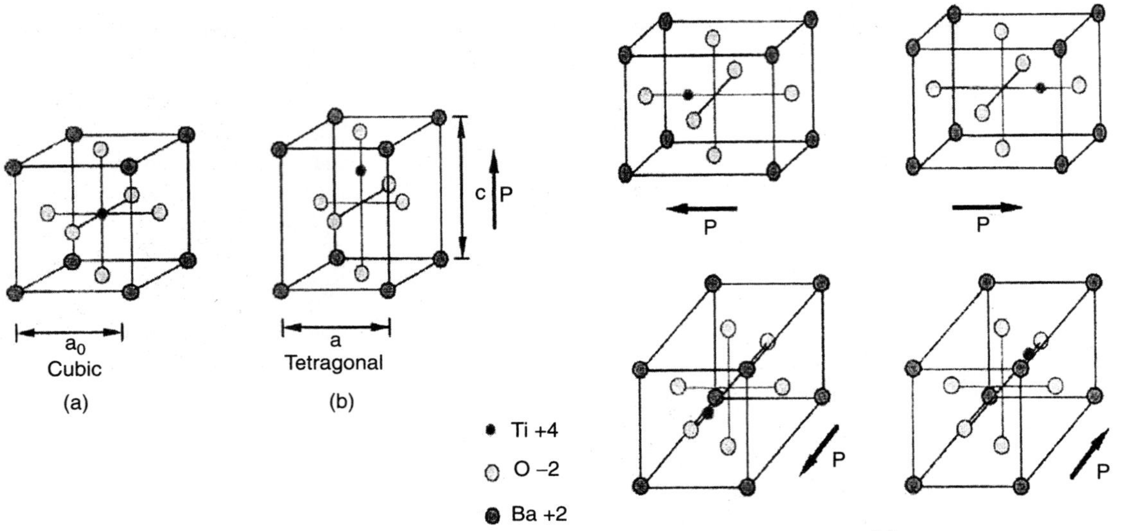

Figure 4.16.1 (*a*) Cubic structure of a $BaTiO_3$ unit cell above the Curie point of 120°C. (*b*) Tetragonal structure below 120°C. (*c*) 90° reorientation under a compressive stress or a perpendicular electric field.

If now we decrease the applied field to zero, the polarization will generally not return to zero and some of the domains will remain aligned in the applied field direction. The polarization at this point is called remanant polarization P_r. The extrapolation of the linear portion of the top curve back to the polarization axis represents the value of the spontaneous polarization P_s.

In order to observe the overall polarization of a ferroelectric material, an electric field in the opposite direction needs to be applied. The value of the electric field required to reduce the polarization P to zero is called the *coercive field* E_c. Further increase of the field in the opposite direction will, of course, cause complete alignment of the dipoles in this direction, and the cycle can be completed by reversing the field direction once again.

From Fig 4.16.2, the hysteresis loop of the polarization P versus the applied field E is obtained, which is the most important characteristic of a ferroelectric material. The essential feature of a ferroelectric is not that it has a spontaneous polarization, but rather that this spontaneous polarization can be reversed by means of an electric field.

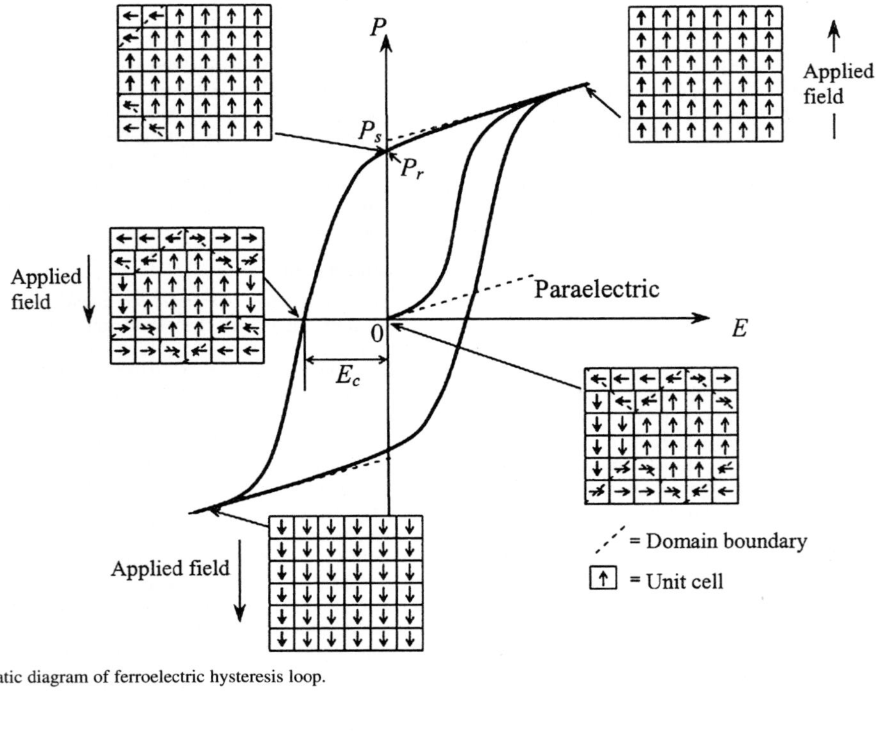

Figure 4.16.2 Schematic diagram of ferroelectric hysteresis loop.

Electromechanical Coupling Behavior

Because of the nature of polarization and the reorientation or domain switch of a ferroelectric under a specific electric and/or mechanical loading, a coupling behavior between electric and mechanical can be observed. When a ferroelectric material is subjected to an electric field, it can be deformed up to 0.2 percent of strain. Figure 4.16.3 shows the relation between the axial strain and the electric field of a PLZT, which gives a typical butterfly-shaped strain versus electric field curve. On the other hand, under the applied compressive stress loading, a ferroelectric produces a measurable voltage change.

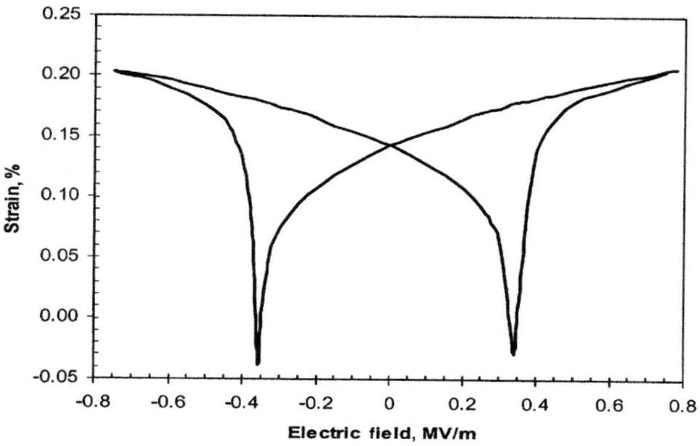

Figure 4.16.3 The butterfly-shaped strain versus electric field relations of a PLZT.

Piezoelectricity

Although ferroelectricity is an intriguing phenomenon, it does not have the practical importance that ferromagnetic materials display (in areas such as magnetic storage of information). The most common applications of ferroelectrics stem from a closely related phenomenon, piezoelectricity. The prefix *piezo* comes from the Greek word for pressure. Piezoelectricity is the ability of certain crystalline materials to develop an electric charge proportional to a mechanical stress. The piezoelectric phenomenon was first discovered by J. Curie and P. Curie in 1880. The direct piezoelectric effect is that electric polarization is produced by mechanical stress. Closely related to it is the converse effect, in which a crystal becomes strained when an electric field is applied. We can write the two effects that occur in piezoelectrics as

Direct piezoelectric effect: $$E = g\sigma \qquad (4.16.1)$$

Converse piezoelectric effect: $$e = dE \qquad (4.16.2)$$

where E is the electric field (V/m), σ is the applied stress (Pa), e is the strain, and g and d are **piezoelectric constants.** The g constant is related to the d constant by the permittivity:

$$g = d/\varepsilon' = d/K\varepsilon_0$$

where ε' is permittivity and K is the relative dielectric constant, which equals to the ratio between the charge stored on an electroded slab of material brought to a given voltage and the charge stored on a set of identical electrodes separated by vacuum ($K = \varepsilon'/\varepsilon_0$). Some piezoelectric materials and their typical values for d are listed in Table 4.16.2.

All ferroelectric materials exhibit piezoelectric behavior in limited elastic range. But not all pyroelectrics that possess a spontaneous polarization are ferroelectric in the sense of possessing a reorientable electric moment. While ferroelectric materials exhibit nonlinear coupling behavior between electrical and mechanical systems and hysteresis loops as discussed above, piezoelectrics have only a linear electrical-mechanical coupling effect. The principal reason that ferroelectrics were discovered so much later was because the formation of domains of differently oriented polarization within virgin single crystals led to a lack of any net polarization and very small pyroelectric and piezoelectric response.

Piezoelectric Polymers

So far we have seen plenty of examples of ceramics showing piezoelectricity. It is known that some natural organic substances, for example, rubber, wood, and hair, are piezoelectric. Bone also shows a piezoelectricity of the same order as wood. The symmetry of rubber and wood is the same as that for piezoelectric ceramics. In 1969, Kawai discovered that the piezoelectricity in polyvinylidene fluoride (PVDF or PVF_2) is several times larger than that of quartz. A thin PVDF film was extended and poled at 100°C under a high field, such as 300 kV/cm. Some improvements were made in the poling procedure and a high d constant, about 30 pC/N, was obtained.

Table 4.16.2 The Piezoelectric Constant *d* for Selected Materials

Material	Piezoelectric constant $d \times 10^{-12}$, C/Pa · m² = m/V
Piezoelectric crystals	
Quartz (SiO_2)	2.3
Rochelle salt ($NaKC_4H_4O_6 \cdot 4H_2O$)	11
EDT ($C_6H_{12}N_2O_6$)	10-20
DKT ($N_2C_4H_4O_6 \cdot H_2O$)	10-20
Amonium dihydrogen phosphate—ADP ($NH_4H_2PO_4$)	23
$BaTiO_3$	85.6
Lithium niobate ($LiNbO_3$)	6
Lithium tantalite ($LiTaO_3$)	8
Piezoelectric semiconductors	
BeO	0.24
ZnO	12.4
CdS	10.3
CdSe	7.8
AIN	5
Piezoelectric ceramics	
PZTs ($PbZr\ O_3$ and $PbTiO_3$)	250-365
PLZT	1,188
$BaTiO_3$	191
$PbNb_2O_6$	80

Table 4.16.3 Characteristics of Piezoelectric Polymers

	Polymer PVDF	Copolymer P(VDF–TrFE)*	Copolymer P(VDF–TeFE)†	Copolymer P(VDCN–VAc)‡	Composite PZT/PVDF
ρ [10^3 kg/m^3]	1.78	1.88	1.90	1.20	5.5
$\varepsilon_{33}/\varepsilon_0$	6.2	6.0	5.5	6.0	30
tan δ	0.25	0.15			
d_{31} [pC/N]	28	12.5	8	6	
e_{33} [C/m^2]	−0.16	−0.23		−0.14	0.18
g_{33} [10^{-3} V · m/N]	−320	−380			26
k_t	0.20	0.30	0.21	0.22	0.07
$c_{33}{}^D$ [10^9 N/m^2]	9.1	11.3			26.6
v_{31} [km/s]	2.26	2.40			2.2
Q_m	10	20			
Z_0 [10^6 kg/m^2 · s]	4.0	4.5	4.4	3.2	12.1
T_C [°C]	170	130	141		
P_r [C/m^2]	0.055	0.082	0.071	0.035	
E_c [Kv/mm]	45	38	39		
p [μC/m^2 · K]§	35	50	40	30	

*$VDF_{0.75}$ $TrFE_{0.25}$.
† $VDF_{0.8}$ $TeFE_{0.2}$.
‡ $VDCN_{0.5}$ $VAc_{0.5}$.
§ Pyroelectric coefficient.
VDF: vinylidene fluoride; TrFE: trifluoroethylene; TeFE: tetrafluoroethylene; VDCN: vinylidene cyanide; VAc: vinylacetate.

PVDF is a polymer that has crystallinity of 40 to 50 percent. The PVDF crystal is **dimorphic,** with two phases (α and β phases). The β phase is polar and piezoelectric. To improve the piezoelectricity, some copolymers have been developed. The copolymer VDF/TrFE consists mostly of the piezoelectric β phase of PVDF with high crystallinity, about 90 percent. The magnitude of polarization in TrEF is only half of that in VDF, but the excellent crystallinity improves the piezoelectric coupling in VDF polymers.

Because of the flexibility of the piezoelectric polymer, it usually is supplied in the form of a thin film. This feature makes it suitable for applications in headphones and speakers. Furthermore, the thin-film transducer is applicable to a high-frequency ultrasonic generator. The performance of these materials greatly depends on the thermal and poling conditions. Typical characteristics of piezoelectric polymers are given in Table 4.16.3 (Table 9.4 in Ikeda's book).

Application of Piezoelectrics

Because of the electric and mechanical coupling behavior in piezoelectric materials, they can be used as transducers between electrical and mechanical energy. They also are active materials (or "smart" materials), which have the ability to perform both sensing and responding functions. This ability traditionally is associated only with living systems, as opposed to inanimate matter. The distinguishing feature of smart materials is, then, their ability to imitate this fundamental aspect of life. In practice, piezoelectric transducers are limited to devices involving only very small mechanical displacements and small amounts of electric charge per cycle. They are also the ideal candidates as transducers, actuators, or sensors in microelectromechanical systems (MEMS). A list of representative applications of piezoelectric materials is given below.

1. **Phonograph cartridges.** Phonograph pickups have used more piezoelectric ceramic elements than all other applications combined. In the period from 1935 to 1950, the phonograph pickup field was dominated by the first ferroelectric single crystal, Rochelle salt. A schematic sketch of phonograph pickup elements in the conventional configuration of a cemented sandwich of flat plates of ceramic on a brass vane is shown in Fig. 4.16.4. The forces available from the phonograph record groove are only 1 to 5 g. Typical output voltage for the ceramic is from 0.1 to 0.5 V. Modified lead titanate zirconates (PZTs) have largely replaced barium titanate, as they give approximately twice as much output.

2. **Ultrasonic transducers.** A common application of piezoelectric materials is in ultrasonic transducers. Figure 4.16.5 shows a typical

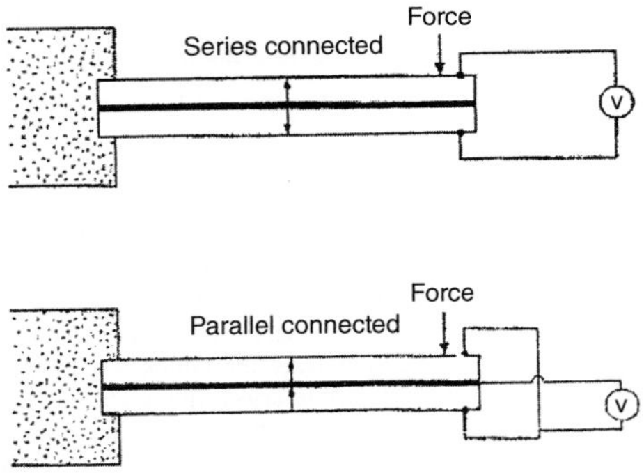

Figure 4.16.4 Ceramic phonograph pickup, series- and parallel-connected. *(From Jaffe, Cook, and Jaffe, Fig. 12.1.)*

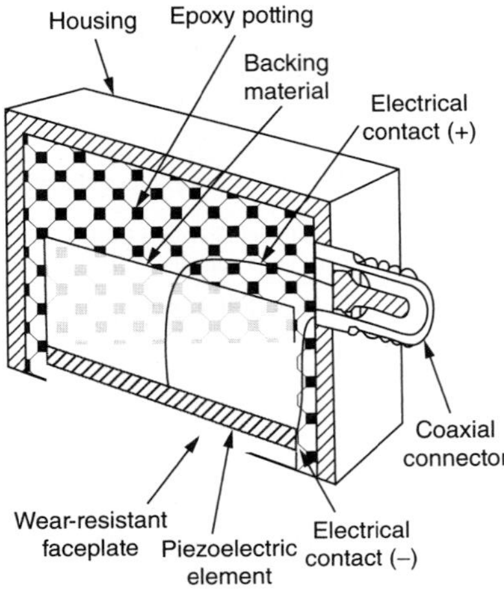

Figure 4.16.5 A cross section view of an ultrasonic transducer. (*From Metals Handbook, 8th ed. Vol. 11, American Society for Metals, Metals Park, Ohio,* 1976.)

cross section of an ultrasonic transducer. In this cutaway view, the piezoelectric crystal or element is encased in a convenient housing. The constraint of the backing material causes the reverse piezoelectric effect to generate a pressure when the faceplate is pressed against a solid material to be inspected. When the transducer is operated in this way (as an ultrasonic transmitter), an AC electrical signal produces an ultrasonic signal of the same frequency (usually in megahertz). When the transducer is operated as an ultrasonic receiver, the direct piezoelectric effect is employed. In that case, the high-frequency elastic wave striking the faceplate generates a measurable voltage oscillation of the same frequency.

3. **Instrument transducers.** PZT ceramics have become the dominant material for piezoelectric accelerometers. In addition to their use in generating ultrasonic waves of frequencies in the order of 1 to 5 MHz to detect flaws and other inhomogeneities in solids or in joints between solid bodies, miniature ceramic transducers have been inserted into blood vessels to record periodic pressure changes connected with the heartbeat. Special-purpose transducers such as blast gages and noise-sensing elements to detect boiling of cooling water in nuclear reactors have been served well by piezoelectrics. Much ingenuity has been expended on other schemes that utilize piezoelectrics for measurement of forces that can be related to input variables of aerospace, medical, and industrial instrumentation.

4. **Air transducers.** Air transducers have been widely used in microphones, especially in the hearing-aid field. But with the advent of the transistor, the market became dominated by electrodynamic microphones of low impedance. Progress in techniques for manufacturing thin ceramic bending elements has resulted in recouping some of the microphone market for piezoelectricity. Piezoelectrics also work well as tuned ultrasonic microphones in remote control of television sets.

5. **Actuators.** One example of piezoelectric materials as smart materials is the piezoelectric actuator used in an impact dot-matrix printer. A schematic illustration of such a ceramic actuator is shown in Fig. 4.16.6. In this case, of course, the signal is coming in the form of a computer text or image to be printed. Each character formed in a typical printer is a 24 × 24 dot matrix. A printing ribbon is impacted by the multiwire array. The printing element is composed of a multilayer piezoelectric device involving 100 thin ceramic sheets. Each sheet is 100 μm thick. The hinged lever provides a displacement magnification of 30 times, resulting in a net motion at the ink ribbon of 0.5 mm with an energy transfer efficiency of >50 percent. The advantages of the piezoelectric design application compared with conventional electromagnetic types are order-of-magnitude higher printing speed, order-of-magnitude lower energy consumption, and reduced printing noise. The reason for the last advantage is that the low level of heat generation by the piezoelectric allows greater sound shielding.

There are many other applications of piezoelectrics such as delay line transducers, wave filters, high-voltage sources, and sensors. Good sources of information on piezoelectric devices and their applications can be found in Katz's book (1959) which emphasizes circuit applications, Mason's book (1958), and some other materials science handbooks.

Glossary

Converse piezoelectric effect: A crystal becomes strained when an electric field is applied.

Curie temperature: The temperature at which such a transition from the polar into the nonpolar state occurs.

Direct piezoelectric effect: Electric polarization is produced by mechanical stress.

Domains: Regions with uniform alignment of the electric dipoles or uniform polarization.

Ferroelectric crystal: Pyroelectric crystal with reversible polarization.

Hysteresis loop: The loop traced out by the polarization as the electric field is cycled. A similar loop occurs in magnetic materials.

Paraelectric: A symmetric unit cell material; only a small polarization is possible, as the applied electric field causes a small induced dipole.

Piezoelectricity: A linear interaction between electrical and mechanical systems.

Polarization: Alignment of dipoles so that a charge can be permanently stored.

Pyroelectric: A crystal having a spontaneous polarization.

Pyroelectricity: An interaction between electrical and thermal systems.

Spontaneous polarization: Measured in terms of dipole moment per unit volume, or with reference to the charges induced on the surfaces perpendicular to the polarization, in terms of charge per unit area.

MULTIFERROICS

Multiferroics are materials possess at least two ferroic orders (ferroelectric, ferromagnetic, or ferroelastic) with both ferroelectric and ferromagnetic properties. There are very few multiferroic materials in nature.[16] Some of the multiferroic materials also possess the magnetoelactric (ME) coupling effect between the magnetic and electrical properties. Even though the single-phase multiferroics are rare, with the development of material design and processing, multiferroic composites containing both ferroelectric and ferromagnetic phase with high ME effect can be made. The ME effect has been widely investigated due to its potential applications as multifunctional devices.

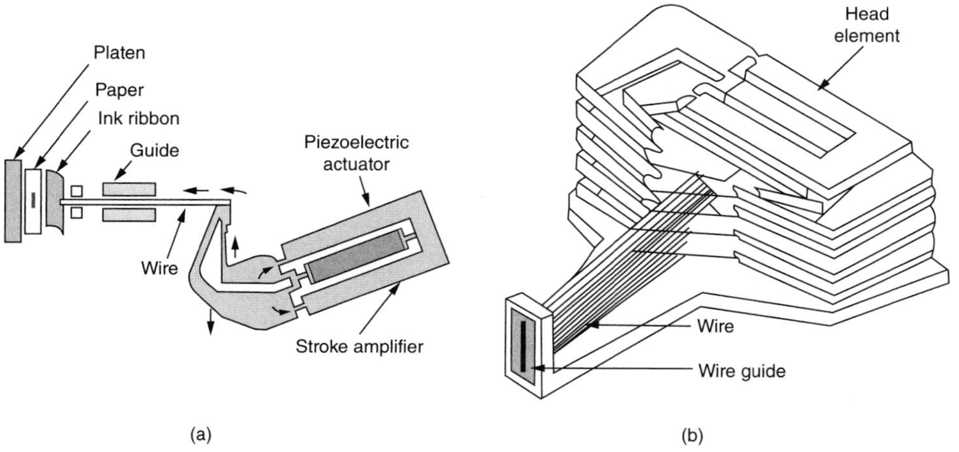

Figure 4.16.6 A schematic illustration of a piezoelectric actuator in an impact dot-matrix printer. (*a*) Overall structure of the printer head. (*b*) Close-up of the multilayer piezoelectric printer-head element. (*From K. Uchino, MRS Bulletin,* **18***; 42, 1993.*)

Megnetoelectric Effect Magnetoelectric effect was first discovered by Röntgen in 1888[17] when he observed a moving dielectric became magnetized under an electric field. Separately, in 1894, Curie indicated the possibility of intrinsic ME effect of crystals on the bases of symmetric considerations.[18,19] The term of "magnetoelectric" was named by Debye a few years later.[20] The direct magnetoelectric effect is defined by the electric polarization $\boldsymbol{P}$ response upon the applied magnetic field $\boldsymbol{H}$ while the converse magnetoelectric effect as the magnetization $\boldsymbol{M}$ versus the applied electric field $\boldsymbol{E}$. The linear relationship between them for isotropic materials can be summarized as:[21]

Direct magnetoelectric effect: $$P = \alpha H \quad (4.16.3)$$

Converse magnetoelectric effect: $$M = \alpha E \quad (4.16.4)$$

where α is the magnetoelectric coupling coefficient for isotropic materials.

From the consideration of the free energy of the system in Landau theory, the ME response in general is limited by

$$\alpha_{ij}^2 \leq \epsilon_{ii}\mu_{jj} \quad (4.16.5)$$

where α_{ij} is the ME coupling coefficient tensor, ϵ_{ii} and μ_{jj} are the diagonalized electric permittivity and magnetic permeability tensors, respectively. For isotropic materials, α_{ij} reduces to only single independent coefficient α as indicated in Eqs. (4.16.3) and (4.16.4). According to Eq. (4.16.5), the ME coupling coefficient can only be large for ferroelectric and/or ferromagnetic materials.

Single-Phase Multiferroics and Magnetoelectrics Until 1970s, only about 80 compounds have been identified to display magnetoelectric coupling effect in single-phase materials.[22] The peak value of ME coupling coefficient is about 30 ps m^{-1}. Table 4.16.4 shows the peak value of ME coupling coefficient of some single-phase materials. Due to the low value of ME coupling coefficient, the applications of single-phase ME materials have been limited or not promising.

Multiferroic Composites With greater flexibility in materials design, multiferroic composites by combining piezoelectric and magnetic materials have become more attractive and promising choice for future application devices using ME coupling effect. In general, the physical properties of a composite are determined by the properties of each constituent. There are three classifications of the composite properties:[19] a sum property, a scaling property, and a product property. The sum property of the composites usually is lying between the properties of each constituent, while the scaling property is that the amplitude for composite is higher than each individual phase. The product properties of a composite are only found in the composite but not in each individual compound. The ME effect in composite materials is basically due to the product property. The ME effect in single-phase materials belongs to the intrinsic property of the material while it is extrinsic property in composites.

Neither piezoelectric nor magnetic material exhibits the ME effect, but composites of these two phases can produce remarkable ME effect. Thus the ME effect is a result of the product of the magnetostrictive effect (magnetic/mechanical effect) in the magnetic phase and the piezoelectric effect (mechanical/electrical effect) in the piezoelectric one, the ME effect can be written as:[23]

$$\mathrm{ME}_H\ \text{Effect} = \frac{magnetic}{Mechanical} \times \frac{Mechanical}{electric}$$

$$\mathrm{ME}_E\ \text{Effect} = \frac{electric}{Mechanical} \times \frac{Mechanical}{magnetic}.$$

The magnetoelectric coupling is achieved via elastic interaction. For instance, under a magnetic field, the shape change in the magnetic phase passes on the piezoelectric phase which resulting the electric polarization. The ME coupling coefficients of some composites are given in Table 4.16.4.

Applications The most important property of multiferroics is the magnetoelectric coupling which can be used in the applications. Due to the coupling between the magnetic and electric field, the ME materials can be used as transducers, sensors, memory devices, etc. In reference [21], various applications have been discussed. Here we only give a brief summary as follows:

A. **Magnetic sensors:** Both AC and DC magnetic field sensors can be introduced by using ME composites. The working principle of magnetic sensing is simple and direct. When probing a magnetic field, the magnetic part in the composite deforms which producing a proportional charge in piezoelectric phase.

B. **Electric current sensors:** ME composites have also been suggested as good candidates for electric current sensing. Figure 4.16.7(*a*) schematically shows a design of the current sensor using ring-type ME laminate with magnetized Terfenol-D and a circumferentially poled piezoelectric PMN-PT layers [Ref. 21–Fig. 45(a)]. Figure 4.16.8(*b*) shows the ME voltage response to a square wave current passing through the wire [Ref. 24–Fig. 4].

C. **Transformers and gyrators:** Due to the coupling behavior between magnetic and electric, ME composites can also be easily applied in the transformer devices such as voltage gain and power conversion devices, and gyrators. Dong, et al.,[25] have reported the magnetoelectric gyration effect in Terfenol-D/PZT laminated composites at the electromechanical resonance. Figure 4.16.8[25] shows the illustrations of ME I-V conversion using a ME Terfenol-D/PZT laminate. The input current I_{in} produces an AC magnetic field H_{ac} that in turn excites

Table 4.16.4 Magnetoelectric Coupling for Some Materials

Multiferroic ME materials	ME effect
Single-phase*	Peak α (ps·m^{-1})
Cr_2O_3	4.13
$LiCoPO_4$	30.6
Yttrium iron garnet (YIG) films	30
$TbPO_4$	36.7
Composites**	Coupling constant (mV·cm^{-1}·Oe^{-1})
$BaTiO_3$ and $CoFe_2O_4$	50
Terfenol-D and PZT in polymer matrix	42
Terfenol-D in polymer matrix/PZT in polymer matrix Laminates	3,000
Terfenol-D and PZT laminates	4,800
La0.7Sr0.3MnO_3/PZT laminates	60
$NiFe_2O_4$/PZT laminates	1,400

* Data was collected from ref. [16].
** Data was collected from ref. [19].

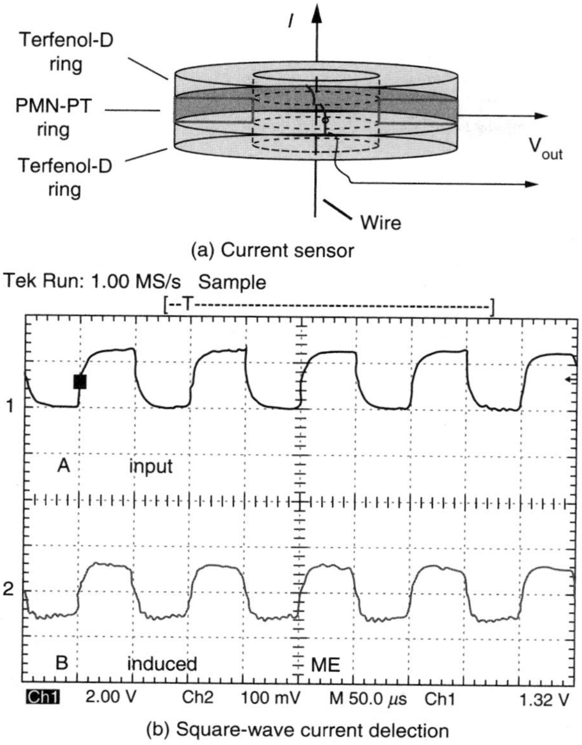

Figure 4.16.7 ME current sensor [Ref. 21–Fig. 45].

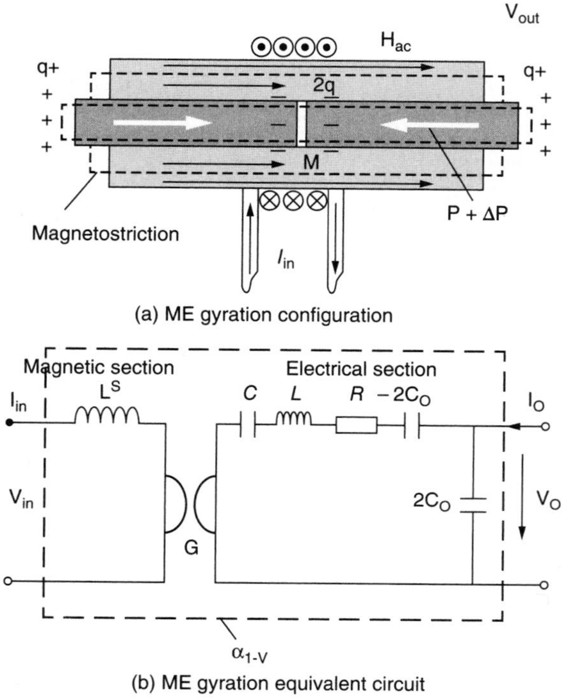

Figure 4.16.8 ME laminate gyrator [Ref. 25–Fig. 1].

magnetostrictive strain in Terfenol-D layers, then forcing the laminated piezoelectric layer to deform which further changes the polarization and produces an output voltage V_{out} across the PZT layer. The I-V conversion is shown in Fig. 4.16.9 [Ref. 41 - Fig. 2 insert].

D. **Microwave devices:** In current microwave devices such as tunable signal processing devices, resonators, filters and phase shifters, ferrites are the main and important materials to be used. But, to achieve such a tunability, a bias magnetic field and relative large power are necessary.[42] Since the ME composites can convert between the electric and magnetic fields through the mechanical straining without additional inductors, and the electric tuning process is faster than traditional magnetic tuning process with little power consumption, it is an ideal candidate for future tunable signal processing devices used in microwaves, battlefield radars and communication devices. Figure 4.16.10 shows a schematic diagram for a microstrip line resonator designed by Fetisov and Srinivasan.[42] PZT (piezoelectric) and YIG (Ferromagnetic) bilayer composite has been constructed in the resonator. It has been demonstrated in their device, for nominal electric fields, the tuning range is five to eight times frequency half-width of the FMR absorption. Such tunability could be potentially used in microwave resonators.

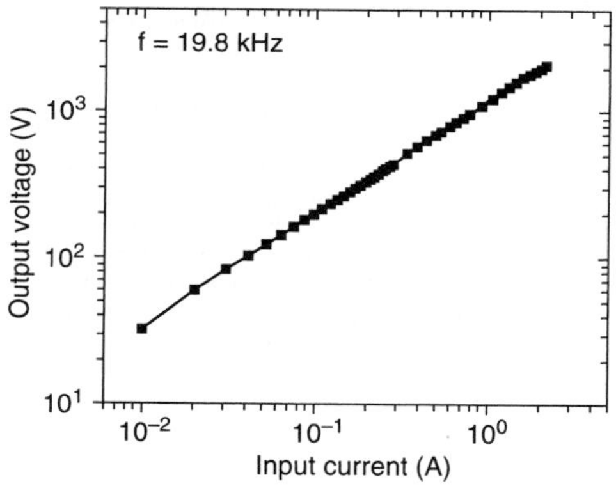

Figure 4.16.9 I-V conversion of an ME Terfenol-D/PZT laminate gyrator [Ref. 25–Fig. 2 insert].

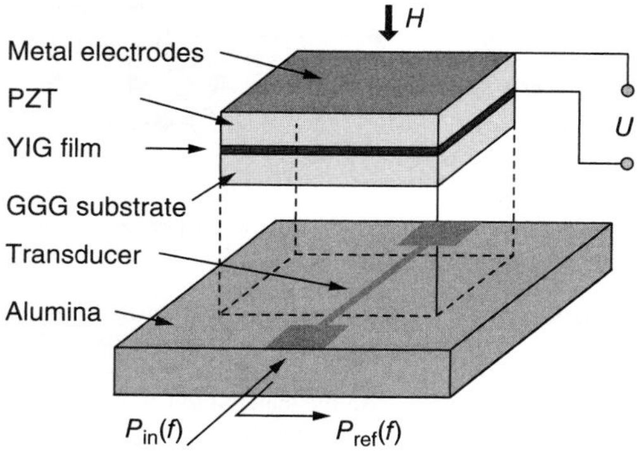

Figure 4.16.10 Schematic diagram of a PZT/YIG bilayer ME composite for microstripline resonator [Ref. 26–Fig. 1].

SHAPE MEMORY ALLOYS

Shape memory alloys (SMAs) are metals that, after being strained, at a certain temperature revert back to their original shape. A change in their crystal structure above their transformation temperature causes them to return to their original shape. They exhibit two very unique properties: pseudo-elasticity and the shape memory effect.

The shape memory effect was first discovered by the Swedish physicist Arne Olander in 1932, who used an alloy of gold and cadmium, but not until the 1960s were any serious research advances made in the field of shape memory alloys. In early 1960s, researchers at the U.S. Naval Ordnance Laboratory discovered the shape memory effect in nickel titanium alloy and named it Nitinol (an acronym for nickel titanium naval ordnance laboratory). Still, not until the early 1970s did several commercially available products appear that implemented shape memory alloys. The most effective and widely used SM alloys include NiTi (nickel-titanium), CuZnAl, and CuAlNi.

Most shape memory alloys also exhibit **pseudo-elastic behavior,** which is a generic name for both rubberlike behavior and superelasticity.

Both the shape memory effect and pseudo-elasticity are mainly due to the solid-state phase transformation of the material. There are two stable phases in shape memory alloys—martensite at low temperature and austenite at high temperature.

Martensite (after the German scientist Martens) was originally used to designate the hard microconstituent formed in quenched steels. Now it is used in a wider context to apply to products of all **martensitic transformations.** Such transformations take place in crystalline solids by coordinated displacements of atoms or molecules over distances smaller than interatomic distances in the parent phase. For this reason the transformations are described as being **diffusionless.** More descriptions of martensite can be found in other sections.

Austenite was the name originally given to the face-centered-cubic (FCC) crystal structure of iron. It is now used in a wider context for other metals, alloys with cubic structures.

In general, a shape memory alloy starts martensitic transformation at a temperature M_s during cooling; the transformation is complete when a lower temperature M_f is reached, where the material is said to be in the martensitic state. Above a certain temperature A_s, the material starts to transform back to austenite, and the transformation is completed at a higher temperature, A_f. A schematic illustration of two phases of nitinol, which is a typical shape memory alloy that contains a nearly equal mixture of nickel (55 wt.%) and titanium, is shown in Fig. 4.16.11. If no forces applied during heating or cooling, the microstructure of the alloy will change without macroscopically noticeable shape change.

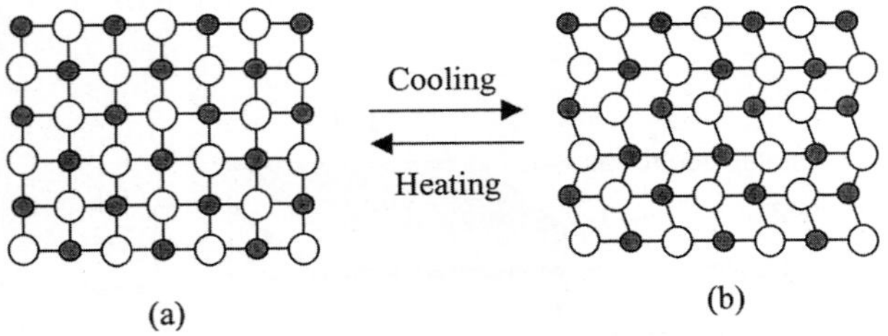

Figure 4.16.11 Schematic sketch of NiTi microstructure. (*a*) Austenite; (*b*) martensite (twined).

In the martensitic condition, an alloy can be easily deformed into a new shape by nonconventional atomistic mechanisms. The alloy will remain in the new deformed shape as long as the temperature is kept constant. Heating will cause the material to transform back to the austenitic phase and restore its original shape. This restoring effect is called **shape memory effect** (Fig. 4.16.12).

In most cases, the memory effect is one-way, which means that, upon cooling, a shape memory alloy does not undergo any shape change, even though the structure changes back to martensite. When the martensite is strained up to several percent, however, that strain is retained until the material is heated again, at which time shape recovery occurs. Upon recooling, the material does not spontaneously change shape, but must be deliberately strained if shape recovery is again desired.

It is possible in some of the shape memory alloys to cause two-way shape memory. That is, shape change occurs upon both heating and cooling. The amount of this shape change is always significantly less than obtained with the one-way shape memory effect, and very little stress can be exerted by the alloy as it tries to assume its low-temperature shape. The heating shape change can still exert very high forces, as with the one-way memory.

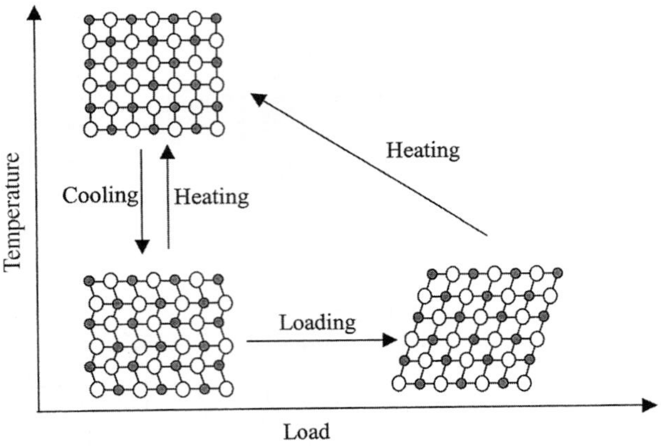

Figure 4.16.12 Schematic microstructure diagram of the shape memory effect.

Shape memory alloys also exhibit pseudo-elastic behavior. Unlike the shape memory effect, pseudo-elasticity occurs without a change of temperature. In the austenite condition, an SMA undergoes a phase transformation from austenite to deformed martensite (or stress-induced martensite) when a stress loading is applied. Once the stress is released, the material returns to the austenite state, which means its shape is restored. This is called **superelasticity.** A typical stress-strain curve is shown in Fig. 4.16.13, while the load-temperature diagram in Fig. 4.16.14 shows the transformation process during a loading and unloading cycle.

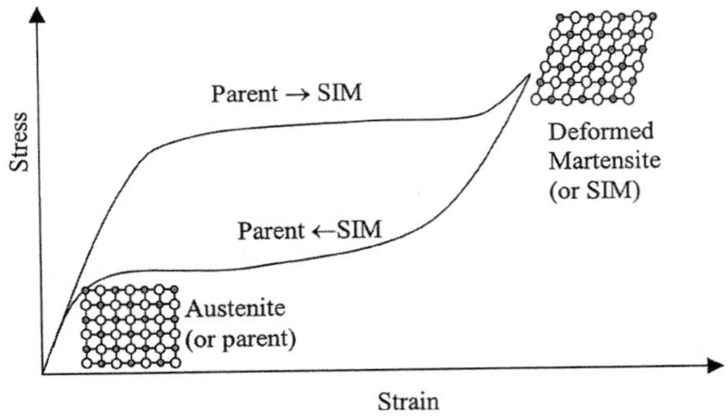

Figure 4.16.13 Schematic stress-strain curve of a NiTi alloy loaded above M_s temperature and then unloaded. Stress-induced martensite (SIM) is formed during loading, which disappears upon unloading.

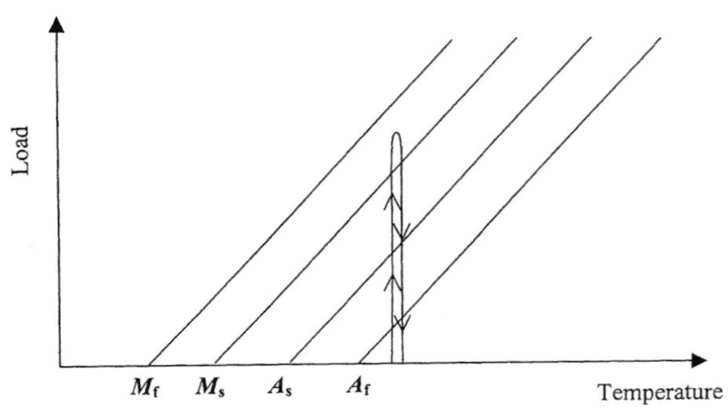

Figure 4.16.14 Load diagram of the superelastic effect above A_f temperature.

Table 4.16.5 Some Selected Properties of Shape Memory Alloys

Alloy		Composition, wt.%	Transformation temperature A_f, °C	Modulus of elasticity, GPa	Recovery strain, %
Nitinol*	SE508 tubing	Ni: 55.8 Ti: balance O (max): 0.05 C (max): 0.02	<15	41-75	10
	SE508 wiring	Same as tubing	5-18	41-75	10
	SM495 wire	Ni: 54.5 Ti: balance O (max): 0.05 C (max): 0.02	60	28-41	10
Ni-Ti		Ni: 49-51	–200 to 110	28-83	8.5
Cu-Al-Ni		Al: 14-14.5 Ni: 3-4.5	<200	80-85	4
Cu-Zn-Al		Zn: 38.5-41.5 Al: A few wt.%	<120	70-72	4

* From Nitinol.com.

Rubberlike behavior was first observed in 1932 by Olander, who studied an Au-Cd alloy. This phenomenon has also been found in several other SMA alloy systems, such as Cu-Al-Ni, Cu-Au-Zn, and Cu-Zn.

Considering the Au-Cd alloy, it is interesting to note that the rubberlike behavior is found only after the martensite is aged at room temperature for at least a few hours. Freshly transformed specimens exhibit the shape memory effect, and bent rods will not spring back to their original shape without aging.

Ruberlike behavior also constitutes a mechanical type of shape memory, as does the process of superelasticity, and, in fact, these two phenomena cannot be distinguished on the basis of stress-strain curves alone. But rubberlike behavior is characteristic of a fully martensitic structure, whereas supperelastic behavior is associated with formation of martensite from the parent phase under stress. These two types of behavior collectively fall in the category of **pseudo-elasticity.**

The most commercially available SMAs are the NiTi alloys and the copper-base alloys. Properties of the two systems (see Table 4.16.5) are quite different. The NiTi alloys have greater shape memory strain (up to 10 percent versus 4 to 5 percent for the copper-base alloys), tend to be much more thermally stable, have excellent corrosion resistance, and have much higher ductility. On the other hand, the copper-base alloys are much less expensive, can be melted and extruded in air with ease, and have a wider range of potential transformation temperatures. The two alloy systems thus have advantages and disadvantages that must be considered in a particular application.

Applications of Shape Memory Alloys

Some examples of applications using shape memory effect are listed below.

- **Coffeepots:** A nickel-titanium spring in coffeepots marketed in Japan (Fig. 4.16.15) is trained to open a valve and release hot water at the proper temperature to brew a perfect pot of coffee.

Figure 4.16.15 Coffeepot.

- **Linear motor for remote control:** Locks, valves, screens, and other devices can be remotely controlled by a **shape memory linear motor.** A self-limiting electric heating element (PTC) is fixed onto a **shape memory bending element.** Applied heat causes shape recovery of the bending element that is transformed into linear motion of the actuator. As the electric current drops, the temperature falls and the actuator returns to its starting position. Energy consumption is low, and heating elements to operate at any desired voltage are easily available. The main advantage of this application is the direct linear movement, without the need to convert a rotary movement into a linear one. Another advantage is its noiseless operation.
- **Hubble space telescope:** In April 1990 the Hubble space telescope was launched into space aboard the shuttle *Discovery*. The telescope was put into orbit around the earth for observing planets and stars and with the aim to obtain pictures 50 times sharper compared to observations from the earth. Several systems of the Hubble space telescope were driven by solar energy. The solar panels that were folded during the launching were unfolded and positioned after the Hubble space telescope was brought into position. The mechanism, designed by Dornier, is reliable and space- and energy-efficient. It used a specially bent Ni-Ti shape memory element to keep the solar panels in the locked position during launch. At a temperature of 115°F, the shape memory element remembered its original straight form and unlocked the panels.

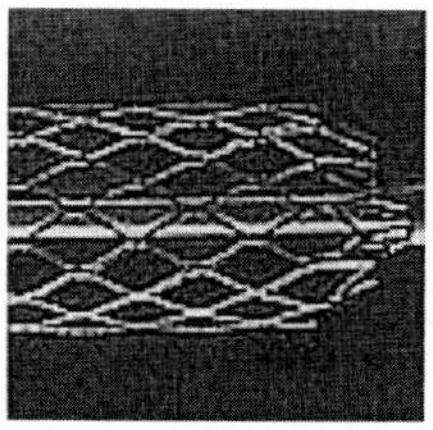

Figure 4.16.16 Stent.

- **Vascular stents:** A vascular stent is a device that is implanted in a blood vessel (vein or artery) to provide structural reinforcement of the vessel wall (Fig. 4.16.16). The introduction of SMA to stent manufacture allows greater recovery strains for use in wider vessels, or stronger recovery force in narrower vessels.
- **Antiscald devices:** Basically these are an extension that fits between a shower head and water pipe, containing a Ti-based SMA element that expands if water becomes too hot, choking off the flow. A very cheap, easy, and effective alternative to more expensive technologies.
- **Connecting rings:** Rings and cylinders made of SMA that is austenite at ambient temperature can provide significant constraining force around the inner circumference, ensuring a strong join that can be released by cooling the coupling. For instance, NiTi rings on electrical connectors (Fig. 4.16.17) provide a secure joint at ambient temperatures. Yet, they can easily be released by chilling.

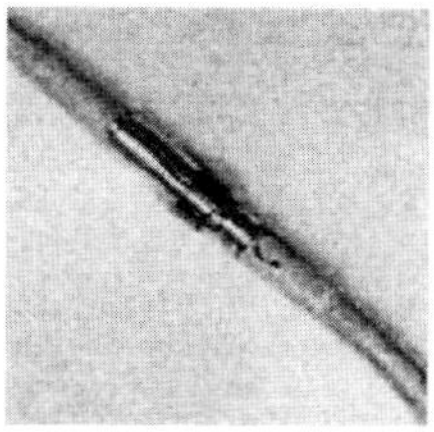

Figure 4.16.17 Electric connector.

Some examples of applications in which pseudo-elasticity is used are:

- **Eyeglass frames:** Nitinol is extensively used in the manufacture of eyeglass frames. It is used in the superelastic state or in the combined shape memory and superelastic states. The frames hold prescription lenses in the proper position so that the wearer gets the maximum benefit from them.
- **Endodontic wire—high-fatigue superelastic wire:** Nickel-titanium alloy has entered the world of endodontics. The flexibility of nickel titanium makes it an ideal material for use in the manufacture of endodontic instruments. Endodontic files using NiTi alloys, both hand and rotary types, have the potential to greatly improve a clinician's ability to negotiate curved root canal systems. Current research is under progress to better define the limitations and strengths of NiTi, and to determine the file designs and techniques to take best advantage of this unique material.
- **Orthodontic arches—low-deformation-resistance superelastic wire:** Superelastic shape memory wires have found wide use as orthodontic wires. The difference between these shape memory wires and a normal wire is a large elastic deformation combined with a low plateau stress obtainable in the shape memory wire. The advantages for the patient are two-fold: fewer visits to the orthodontist, because of the larger elastic stroke, and more comfort, because of lower stress levels.

Glossary

A_s: The temperature at which austenite starts to form.

A_f: The temperature at which a material is completely transformed into austenite.

M_s: The temperature at which the martensitic transformation starts upon cooling.

M_f: The temperature at which the martensitic transformation is completed, where the material is said to be in the martensitic state.

Austenite: The name originally given to the FCC crystal structure of iron. It is a stronger phase with a cubic structure in shape memory alloys at high temperature.

Martensite: The name originally used to designate the hard microconstituent formed in quenched steels. Now it is a phase that forms as the result of a diffusionless solid-state transformation. It is a more deformable phase in shape memory alloys at lower temperature.

Pseudo-elasticity: A generic name for both rubberlike flexibility and superelasticity.

Rubberlike behavior: Some SME alloys show rubberlike flexibility. When bars are bent, they spontaneously unbend upon release of the stress. In order to obtain this behavior, the martensite, after its initial formation, usually must be aged for a period of time; unaged martensites show typical SME behavior. The rubberlike behavior, unlike superelasticity, is a characteristic of the martensite phase—not the parent phase.

Shape memory effect (SME): The ability of certain materials to develop microstructures that, after being deformed, can return to the material to its initial shape when heated. Upon cooling, the specimen does not return to its deformed shape.

Superelasticity (SE): The result of stress-induced phase transformation from austenite directly into deformed martensite under external loading. When the deforming stress is released, the material will transform back to its stable austenitic state and shape. The martensite disappears (reverses) when the stress is released, giving rise to a superelastic stress-strain loop with some stress hysteresis.

REFERENCES

1. *Ferroelectrics/Piezoelectrics.*
2. Cady, "*Piezoelectricity,*" Dover, 1964.
3. Cady, (1964). "Piezoelectricity," Dover.
4. Ferroelectrics in "*Ullmann's Encyclopedia of Industrial Chemistry,*" Wiley, 2003.
5. "Ferroelectrics" Journal.
6. IEEE Standard on Piezoelectricity, 1987.
7. Ikeda, (1996). "*Fundamentals of Piezoelectricity,*" Oxford.
8. Jaffe, Cook, and Jaffe, "*Piezoelectric Ceramics,*" Academic Press, 1971.
9. Jona and Shirane, "*Ferroelectric Crystals,*" Dover, 1993.
10. Katz, "*Solid State Magnetic and Dielectric Devices,*" John Wiley, New York, 1959.
11. Lines and Glass, "*Principles and Applications of Ferroelectrics and Related Materials,*" Oxford, 1997.
12. Mason, "*Physical Acoustics and the Properties of Solids,*" Van Nostrand, Princeton, NJ, 1958.
13. Nye, "*Physical Properties of Crystals,*" Oxford, 1979.
14. Shackelford, "*Introduction to Materials Science for Engineers,*" Sec. 15.4, Prentice Hall, 2000.
15. Valasek, *Phys. Rev.*, **15**:537-538 (1920); **17**:475-481 (1921).
16. W. Eerenstein, N. D. Mathur, and J. F. Scott, *Nature*, **442**:759-765, 2006.
17. W. C. Röntgen, *Ann. Phys.*, **35**:264, 1888.
18. P. Curie, *J. Physique*, **3**:393, 1894 .
19. M. Fiebig, *J. Phys. D: Appl. Phys.*, **38**:R123, 2005.
20. P. Debye, *Z. Phys.*, **36**:300, 1926.
21. C. W. Nan, M. I. Bichurin, S. X. Dong, D. Viehland, and G. Srinivasan, *J. App. Phys.*, **103**:031101, 2008.
22. H. Schmid, *Int. J. Magn.*, **4**:337, 1973.
23. C. W. Nan, *Phys. Rev.*, **B 50**:6082, 1994.
24. S. X. Dong, J. G. Bai, J. Y. Zhai, J. F. Li, G. Q. Lu, D. Viehland, S. J. Zhang, and T. R. Shrout, *Appl. Phys. Lett.*, **86**:182506, 2005.
25. S. X. Dong, J. Y. Zhai, J. F. Li, and D. Viehland, *Appl. Phys. Lett.*, **89**:243512, 2006.
26. Y. K. Fetisov and G. Srinivasan, *Appl. Phys. Lett.*, **88**:143503, 2006.
27. Shape Memory Alloys
28. Bever, "*Encyclopedia of Materials Science and Engineering,*" Vol. 4, Pergamon, MIT Press, 1986.
29. Chang and Read, *Trans. AIME*, **191**(47), 1951.
30. http://www.nitinol.com, http://www.amtbe.com
31. Wayman, "Shape memory and related phenomena," *Progress in Materials Science,* **36**:203-224, 1992.

Section 5

Fuels and Furnaces

BY

JAMES G. SPEIGHT *Consultant*
THOMAS BUTCHER *Brookhaven National Laboratory*
CHARLES E. BAUKAL, JR. *Director, John Zink Institute*
CHARLES O. VELZY *Past President, American Society of Mechanical Engineers*
LEONARD M. GRILLO *Grillo Engineering Company*
ARVIND C. THEKDI *President, Energy and Environmental Efficiency Management (E3M), Inc.*

5.1 FUELS
by James G. Speight and Thomas Butcher

5.2 INDUSTRIAL COMBUSTION FURNACES
by Charles E. Baukal, Jr.

5.3 MUNICIPAL WASTE COMBUSTION
by Charles O. Velzy and Leonard M. Grillo

5.4 INDUSTRIAL ELECTRIC HEATING
by Arvind C. Thekdi

5.1 FUELS

by James G. Speight and Thomas Butcher

COAL

Coal is a sedimentary organic rock that has been formed from accumulations of organic matter, likely along the shores of shallow seas as well as along the edges of lakes or rivers. Flat swampy areas that are periodically flooded are the best candidates for coal formation. During non-flooding periods of time, thick accumulations of dead plant material pile up and, as the water levels rise, the organic debris is covered by water, sand, and soils. The water (often salty), sand, and soils can prevent the oxidative decay and transport of the organic debris to other locales. As a consequence, the buried organic debris begins to go evolve into the carbonaceous matter that eventually leads to the coal series as it is progressively buried under more sand and silt. The compressed and/or heated organic debris begins driving off volatiles, leaving behind primarily coal, or the carbonaceous mass that eventually become coal.

More generally, coal is a black or brownish-black combustible solid of varying types (ranks) (Table 5.1.1) formed by the decomposition of vegetation in the absence of air. Microscopy can identify plant tissues, resins, spores, etc. that existed in the original structure. It is composed principally of carbon, hydrogen, oxygen, and small amounts of sulfur and nitrogen. Associated with the organic matrix are water and as many as 65 other chemical elements. Coal is used directly as a fuel, a chemical reactant, and a source of organic chemicals.

Mining

Coal is mined by either underground or surface methods. In the underground mining method, the coal beds (coal seams) are made accessible through shaft, drift, or slope entries (vertical, horizontal, or inclined, respectively), depending on location of the bed relative to the terrain.

Table 5.1.1 General Description of Different Types of Coal Including Peat for Comparison

Rank (excluding peat)	Properties
Peat	A mass of recently accumulated to partially carbonized plant debris that is not classed as coal; an organic sediment that has a carbon content <60% w/w on a dry ash-free basis.
Lignite	The lowest rank of coal (also called *brown coal*) which may contain recognizable plant structures; the heating value is <8,300 Btu/lb on a mineral matter free basis and a carbon content from 60% to 70% w/w on a dry ash-free basis.
Subbituminous	Coal with a carbon content on the order of 71% to 77% w/w on a dry ash-free basis) and a heating value between 8,300 and 13,000 Btu/lb on a mineral matter free basis; on the basis of heating value, this coal is subdivided into subbituminous A, subbituminous B and subbituminous C coals.
Bituminous	Coal with a carbon content on the order of 77% and 87% w/w on a dry ash-free basis and a heating value that is much higher than lignite or subbituminous coal; on the basis of volatile matter production, bituminous coal is subdivided into low volatile bituminous coal, medium volatile bituminous coal, and high volatile bituminous coal; often referred to as *soft coal* (in relation to anthracite).
Anthracite	The highest rank of coal with a carbon content greater than 87% w/w on a dry ash-free basis and the highest heating value per pound on a mineral matter free basis; often subdivided into semi-anthracite, anthracite, and meta-anthracite on the basis of carbon content and *hard coal* (in relation to bituminous coal).

The most widely used methods of coal mining in the United States are termed conventional mining and continuous mining. In conventional mining the coal is usually broken from the face by means of blasting agents or by pressurized air or carbon dioxide devices. In preparation for breaking, the coal may be cut horizontally or vertically by cutting machines and holes drilled for charging explosives. The broken coal is then picked up by loaders and discharged to conveyors or cars. The latter, makes use of *continuous mining machines* which of the ripping, boring, or milling types or hybrid combinations of these. These miners cut and break the coal from the face and load it onto conveyors, shuttle cars, or railcars in a single operation. A method that has increased in use over the past several decades is termed *longwall mining*, which employs shearing or plowing machines to break coal from more extensive faces. Also, pillars to support the roof are not needed because the roof is caved under controlled conditions behind the working face.

An important requirement in all mining systems is roof support. When the roof rock consists of strong sandstone or limestone, relatively uncommon, little or no support may be required over large areas. Most mine roofs consist of shale formations and must be reinforced. Permanent supports may consist of arches, crossbars and legs, or single posts made of steel or wood. Screw or hydraulic jacks, with or without crossbars, often serve as temporary supports. Long roof bolts, driven into the roof and anchored in sound strata above, are used widely for support, permitting freedom of movement for machines and personnel.

Ventilation is another necessary factor in underground mining to provide a proper atmosphere for personnel and to dilute or remove dangerous concentrations of methane and coal dust. The ventilation system must be well-designed so that adequate air is supplied across the working faces without creating more dust.

When coal occurs near the surface, *strip mining* or *open-pit mining* is often more economical than underground mining. In the United States, this is especially true in states west of the Mississippi River, such as Wyoming, where coal seams are many feet thick and relatively low in sulfur.

In preparation for surface mining, core drilling is conducted to survey the underlying coal seams, usually with dry-type rotary drills. To commence the mining operations, the overburden (the material that covers the coal seams) must then be removed after being loosened by ripping or drilling and blasting. Ripping can be accomplished by bulldozers or scrapers and the overburden and coal are then removed by shovels, draglines, bulldozers, or wheel excavators. The first two may have a bucket capacity on the order of (or larger than) 200 yd^3 (153 m^3). Draglines are most useful for very thick cover or long dumping ranges. Once the coal is mined, it is transported by trucks or tractor-trailers with capacities up to 240 short tons (1 short ton = 2,000 lbs). Reclamation of stripped coal land is necessary and involves returning the land to near its original contour, replanting with ground cover or trees, and sometimes providing water basins and arable land.

Transportation

Transportation of coal from the mine to the point of consumption is primarily by rail (67%) followed by barges (11%) and trucks (10%); approximately 1 percent moves through pipelines. Approximately 11 percent of the coal consumed is burned at mine-mouth power plants, for it is cheaper and easier to transport electric power than bulk coal. Several long-distance coal-slurry pipelines are proposed but only one, 273 mi (439 km) long, is in commercial use.

Overall energy statistics for the United States show that coal accounts for approximately 30 percent of the total energy production, with the balance coming from petroleum and natural gas. The use of the coal generally falls into the following categories: (i) electric power, 87.2 percent, (2) industrial, 8.4 percent, (iii) coke production, 3.8 percent, as well as (iv) commercial and residential use, 0.6 percent. The reserves of coal in the United States are ample for several hundred years even allowing for the increased production of electric power and the synthesis of fuels and chemicals. The total reserve base is estimated to be almost 4 trillion tons, of which 1.7 trillion tons is identified resources.

Preparation

Once mined, the coal must be treated by mechanical means to remove impurities to supply a marketable product. Mechanical mining has increased the proportion of fine coal and noncoal minerals in the product. At the preparation plant run-of-mine coal (raw uncleaned coal from the mine) is usually given a preliminary size reduction with roll crushers or rotary breakers. Large or heavy impurities are then removed by hand picking or screening. Magnetic iron compounds are usually removed by magnets. Before washing, the coal may be given a preliminary size fractionation by screening. Typically, the preparation practice will be based on the density difference between the coal and the associated impurities. Heavy-medium separators using magnetite or sand suspensions in water come closest to ideal gravity separation conditions.

Mechanical devices that are used for coal cleaning include jigs, classifiers, washing tables, cyclones, and centrifuges. Fine coal, less than 1⁄4 in (6.3 mm) is usually treated separately, and may be cleaned by froth flotation. Dewatering of the washed and sized coal may be accomplished by screening, centrifuging, or filtering, and finally, the fine coal may be heated to complete the drying. Before shipment the coal may be dust-proofed and freezeproofed with oil or salt.

Removal of sulfur from coal is an important aspect of preparation because of the role of sulfur dioxide in air pollution. Pyrite (FeS_2), the main inorganic sulfur-containing mineral, is partly removed along with other minerals in conventional cleaning. Processes to improve pyrite removal are being developed. These include magnetic and electrostatic separation, chemical leaching, and specialized froth flotation.

Storage

Coal storage is an important aspect of coal production. In many cases, the coal user may stockpile several thousand tons of coal (in different stockpiles) to cover periods when coal delivery is slow. However, coal may heat spontaneously, with the likelihood of self-heating greatest among coals of lowest rank. The heating begins when freshly broken coal is exposed to air, typically during alternating periods of precipitation (rain, snow) followed by dry periods with higher air temperatures. The self-heating process accelerates with increase in temperature and active burning will result if the heat from oxidation is not dissipated. The finer (smaller) sizes of coal, having more surface area per unit weight than the larger sizes, are more susceptible to spontaneous heating.

The prevention of spontaneous heating in storage poses a problem but can be of minimized by reducing the tendency for oxidation of the coal dissipating any heat produced. Air may carry away heat, but it also brings oxygen to create more heat. Spontaneous heating can be prevented or lessened by (i) storing coal underwater, (ii) compressing the pile in layers, as with a road roller, to retard access of air, (iii) storing large-size coal, (iv) preventing any segregation of sizes in the pile, (v) storing in small piles, (vi) keeping the storage pile as low as possible—six feet is the limit for many coals, (vii) keeping storage away from any external sources of heat, (viii) avoiding any draft of air through the coal, and (ix) using older portions of the storage first and to avoid accumulation of old coal in places.

It is also desirable to monitor the temperature of the coal stock pile such as by the insertion of a thermometer (protected by an iron tube driven into the coal pile). When the coal reaches a temperature of 50°C (120°F), the coal should be moved. Using water to put out a fire, although effective for the immediate moment, may only delay the necessity of moving the coal. Furthermore, water treatment did not always a safe method since steam and coal can react at high temperatures to form carbon monoxide and hydrogen:

$$C_{coal} + H_2O \rightarrow CO + H_2$$

Classification and Properties

Coal may be classified (i) by rank, (ii) by variety, (iii) by size, and sometimes (iv) by use. Classification by rank is a common method of coal identification and used the degree of metamorphism or progressive alteration in the natural series from lignite (low-rank coal) to anthracite (high-rank coal) (Table 5.1.2). The basic scheme uses the fixed carbon value (FC) and heating value (HV) from a proximate

Table 5.1.2 Classification of Coals by Rank (ASTM D388)*

		Fixed carbon limits, (%, dry, mineral-matter-free basis)		Volatile matter limits (%, dry, mineral-matter-free basis)		Calorific value limits Btu/lb, (moist, mineral-matter-free basis)[b]		
Class	Group	Equal to or greater than	Less than	Greater than	Equal to or less than	Equal to or greater than	Less than	Agglomerating character
I. Anthracitic	1. Metaanthracite	98	—	—	2	—	—	Nonagglomerating
	2. Anthracite	92	98	2	8	—	—	
	3. Semianthracite[c]	86	92	8	14	—	—	
II. Bituminous	1. Low-volatile bituminous coal	78	86	14	22	—	—	Commonly agglomerating[e]
	2. Medium-volatile bituminous coal	69	78	22	31	—	—	
	3. High-volatile A bituminous coal	—	69	31	—	14,000[d]	—	
	4. High-volatile B bituminous coal	—	—	—	—	13,000	14,000	
	5. High-volatile C bituminous coal	—	—	—	—	11,500	13,000	
						10,500	11,500	Agglomerating
III. Subbituminous	1. Subbituminous A coal	—	—	—	—	10,500	11,500	Nonagglomerating
	2. Subbituminous B coal	—	—	—	—	9,500	10,500	
	3. Subbituminous C coal	—	—	—	—	8,300	9,500	
IV. Lignite	1. Lignite A	—	—	—	—	6,300	8,300	
	2. Lignite B	—	—	—	—	—	6,300	

SOURCE: ASTM (2004).

[a] This classification does not include a few coals, principally nonbanded varieties, that have unusual physical and chemical properties and that come within the limits of fixed-carbon or calorific value of the high-volatile bituminous and subbituminous ranks. All of these coals either contain less than 48% dry, mineral-matter-free fixed carbon or have more than 15,500 moist, mineral-matter-free British thermal units per pound.

[b] Moist refers to coal containing its natural inherent moisture but not including visible water on the surface of the coal.

[c] If agglomerating, classify in low-volatile group in the bituminous class.

[d] Coals having 69% or more fixed carbon on the dry, mineral-matter-free basis shall be classified according to fixed carbon, regardless of calorific value.

[e] It is recognized that there may be nonagglomerating varieties in these groups of the bituminous class, and there are notable exceptions in the high-volatile C bituminous group.

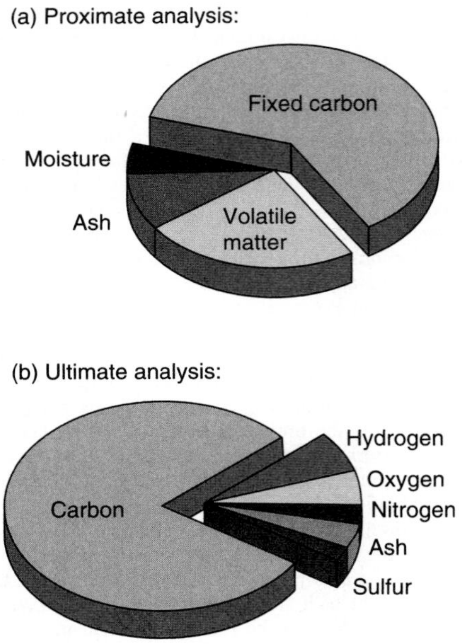

Figure 5.1.1 Comparison of data types from proximate analysis and ultimate analysis.

analysis, calculated on the mineral-matter-free (mmf) basis (Fig. 5.1.1). The higher-rank coals are classified according to the fixed carbon (dry basis) and the lower-rank coals according to the heating value (Btu/lb, moist basis) (Table 5.1.3).

To classify coal according to rank, the fixed carbon (FC) and the heating value (HV) can be calculated to a moisture-free basis by the Parr formulas:

$$\text{FC (dry, mmf)} = (\text{FC} - 0.15S)/[100 - (M + 1.08A + 0.55S)] \times 100$$

$$\text{VM (dry, mmf)} = 100 - FC$$

$$\text{HV (dry, mmf)} = (\text{Btu} - 50S)/[100 - (M + 1.08A + 0.55S)]$$

In these equations, *FC* is the % w/w fixed carbon, *VM* is the % w/w volatile matter, *M* is the % w/w moisture, *A* is the % w/w ash, *S* is the % w/w sulfur, all on a moist basis, HV is the heating value, Btu/lb.

Table 5.1.3 Proximate Analysis of Selected US Coals

Coal	Moisture (% fuel)	Ash (% dry fuel)	Volatile matter (% dry fuel)	Fixed carbon (% dry fuel)
Antelope	23	5.51	43.3	51.2
Belle Ayr	25	6.84	48.2	45.8
Beulah Lignite	26.89	13.86	42.78	43.36
Black Thunder	21.3	6.46	54.26	39.28
Blind Canyon	3.3	10.4	43.58	46.03
Decker	17.99	5.11	44.03	50.87
Eagle Butte	23.74	6.4	45.55	48.05
Eastern Kentucky	1.17	10.154	36.82	53.03
Hanna Basin	11	11.05	38.4	50.6
Illinois #6	11.83	10.13	40.95	48.91
Kentucky #11	3.97	22.21	34.97	42.82
Kentucky #9	8.3	14.64	37.88	47.47
Pittsburgh #8	1.02	10.68	40.16	49.16
Pocahontas #3	0.62	4.51	18.49	77
Roland	9.89	6.29	44.52	49.2
San Miguel Lignite	18.5	52.178	30.62	17.42
Upper Freeport	0.74	21.66	24.09	54.25

Table 5.1.4 Formulas Used for Reporting Coal Analysis (ASTM D-3180)

1. For conversion from the as-determined (*ad*) basis to the as-received (*ar*) basis (all values are expressed in weight percent). For moisture (A/):
$M_{ar} = M_{ad} \times (100 - ADL/100) + ADL$

ADL = air-dry loss in weight percent of as-received sample.

2. For hydrogen (H) and oxygen (O), including the hydrogen and oxygen in the moisture associated with the sample:
$H_{ar} = (H_{ad}, - 0.1119\ M_{ad}) \times (100 - ADL/100) + 0.1119\ M_{ar}$
$O_{ar} = (O_{ad} - 0.8881\ M_{ad}) \times (100 - ADL/100) + 0.8881\ M_{ar}$

3. For hydrogen and oxygen not including the hydrogen and oxygen in the moisture associated with the sample:
$H_{ar} = (H_{ad} - 0.1119\ M_{ad}) \times (100 - ADL/100)$
$O_{ar} = (O_{ad} - 0.8881\ M_{ad}) \times (100 - ADL/100)$

Note: The parameters must be expressed as a weight percentage, except gross calorific value, which is expressed as Btu/lb.

Other more recent formulas for interconverting the analytic data are also available (Table 5.1.4).

As a result of the classification system, various types of coal can be identified: (i) *meta-anthracite*, (ii) *anthracite*, (iii) *semi-anthracite*, (iv) *low-volatile bituminous coal*, (v) *medium-volatile bituminous coal*, (vi) *high-volatile A bituminous coal*, (vii) *high-volatile B bituminous coal*, (viii) *high-volatile C bituminous coal*, (ix) *subbituminous coal*, and (x) *lignite*.

Meta-anthracite is a high-carbon coal that approaches graphite in structure and composition. It usually is slow to ignite and difficult to burn. It has little commercial importance. *Anthracite*, sometimes called hard coal, is hard, compact, and shiny black, with a generally conchoidal fracture. It ignites with some difficulty and burns with a short, smokeless, blue flame. Anthracite is used primarily for space heating and as a source of carbon. It is also used in electric power generating plants in or close to the anthracite-producing area. The iron and steel industry uses some anthracite in blends with bituminous coal to make coke, for sintering iron-ore fines, for lining pots and molds, for heating, and as a substitute for coke in foundries. *Semi-anthracite* is dense, but softer than anthracite. It burns with a short, clean, bluish flame and is somewhat more easily ignited than anthracite. The uses are approximately the same as for anthracite.

Low-volatile bituminous coal is grayish black, granular in structure, and friable on handling. It cakes in a fire and burns with a short flame that is usually considered smokeless under all burning conditions. It is used for space heating and steam raising and as a constituent of blends for improving the coke strength of higher-volatile bituminous coals. Low-volatile bituminous coals cannot be carbonized alone in slot-type ovens because they expand on coking and damage the walls of the ovens. *Medium-volatile bituminous coal* is an intermediate stage between high-volatile and low-volatile bituminous coal and therefore has some of the characteristics of both. Some of these coal types are relatively soft and friable. *High-volatile A bituminous coal* has distinct bands of varying luster. It is hard and handles well with little breakage. It includes some of the best steam and coking coal. On burning in a fuel bed, it cakes and emits smoke when it is improperly fired. The coking property is often improved by blending with more strongly coking medium-volatile bituminous coal and low-volatile bituminous coal. *High-volatile B bituminous coal* is similar to *high-volatile A bituminous coal* but has slightly higher bed moisture and oxygen content and is less strongly coking. It is good coal for steam raising and space heating. Some of it is blended with more strongly coking coals for making metallurgical coke. *High-volatile C bituminous coal* is a stage lower in rank than the B bituminous coal and therefore has a progressively higher bed moisture and oxygen content. It is used primarily for steam raising and space heating.

Subbituminous coals usually show less evidence of banding than bituminous coals. They have a high-moisture content, and on exposure to air, they disintegrate or "slack" because of shrinkage from loss of moisture. They are non-caking and non-coking, and their primary use is for steam raising and space heating. *Lignite*, the lowest rank coal, is brown to black in color and have a bed moisture content of 30 percent

Table 5.1.5 Sampling and Analytical Methods Used for Coal Evaluation

Test/Property	Results/Comments
Sample information	
Sample history	Sampling date, sample type, sample origin (mine, location)
Sampling protocols	Assurance that sample represents gross consignment
Chemical properties	
Proximate analysis	Determination of the "approximate" overall composition (i.e., moisture, volatile matter, ash and fixed carbon content)
Ultimate analysis	Absolute measurement of the elemental composition (i.e., carbon, hydrogen, sulfur, nitrogen, and oxygen content)
Sulfur forms	Chemically bonded sulfur: organic, sulfide, or sulfate
Ash properties	
Elemental analysis	Major elements
Mineralogical analysis	Analysis of the mineral content
Trace element analysis	Analysis of trace elements; some enrichment in ash
Ash fusibility	Qualitative observation of temperature at which ash passes through defined stages of fusing and flow

to 45 percent w/w with a resulting lower heating value than higher-rank coals. Like subbituminous coal, lignite has a tendency to disintegrate during air drying. They are noncaking and non-coking. Lignite can be burned on traveling or spreader stokers and in pulverized form.

Except for anthracite, the values of fixed carbon and heating value increase from the lowest to the highest rank as the percent of volatile matter and moisture decrease. Other coals are hard and do not disintegrate on handling. They cake in a fuel bed and smoke when improperly fired. These coals make cokes of excellent strength and are either carbonized alone or blended with other bituminous coals. When carbonized alone, only those coals that do not expand appreciably can be used without damaging oven walls.

Sampling, Composition, and Characteristics

Sampling Because of its complexity of coal, the analysis of coal requires care in sampling, preparation, and selection of the method of analysis (Table 5.1.5) whether the sample is to be used for proximate analysis or ultimate analysis (Fig. 5.1.1). In fact, the key to determining the nature of a coal pile is in the sampling methods employed. Because coal is a heterogeneous material, collection and handling of samples that adequately represent the bulk lot of coal are required if the analytical and test data are to be meaningful. Coal is best sampled when in motion, as it is being loaded or unloaded from belt conveyors or other coal-handling equipment, by collecting increments of uniform weight evenly distributed over the entire lot. Each increment should be sufficiently large and so taken to represent properly the various sizes of the coal.

Two sampling procedures are recognized: (i) commercial sampling and (ii) special-purpose sampling, such as classification by rank or performance. The *commercial sampling procedure* is intended for an accuracy such that, if many samples are taken from a large lot of coal, the test results would fall within ±10 percent of the ash content of these samples. For commercial sampling of lots up to 1,000 tons, it is recommended that one gross sample be taken as long as the sample represents the lot.

However, for lots over 1,000 tons, the following alternatives may be used: (i) separate gross samples may be taken for each 1,000 tons of coal or fraction thereof, and a weighted average of the analytical determinations of these prepared samples may be used to represent the lot, (ii) separate gross samples may be taken for each 1,000 tons or fraction thereof, and the -20 mesh or -60 mesh samples taken from the gross samples may be mixed together in proportion to the tonnage represented by each sample and one analysis carried out on the composite sample, and (iii) one gross sample may be used to represent the lot, provided that at least four times the usual minimum number of increments are taken. In *special-purpose sampling*, the increment requirements used in the commercial sampling procedure are increased according to prescribed rules.

Analysis Proximate analysis, sulfur content, and calorific values are the analytical determination most commonly used for industrial characterization of coal (Table 5.1.6). The *proximate analysis* is the simplest means for determining the distribution of products obtained during heating. It separates the products into four groups: (i) water or moisture, (ii) volatile matter consisting of gases and vapors, (iii) fixed carbon consisting of the carbonized residue less ash, and (iv) ash derived from the mineral impurities in the coal.

Moisture is the loss in weight obtained by drying the coal at a temperature between 104 and 110°C (220 and 230°F) under prescribed conditions. Further heating at higher temperatures may remove more water, but this moisture usually is considered part of the coal substance. The moisture obtained by the standard method consists of (i) surface or extraneous moisture that may come from external sources such as percolating waters in the mine, rain, condensation from the air, or water from a coal washing operation, (ii) inherent moisture, sometimes referred to as *bed moisture*, which is so closely held by the coal substance that it does not separate these two types of moisture. Coal may be air-dried at room temperature or somewhat above, thereby determining an *air-drying loss*, but this result is not the extraneous moisture because part of the inherent moisture also vaporizes during the drying.

Mine samples taken at freshly exposed faces in the mine, which are free from visible surface moisture, give the best information as to inherent moisture content (bed moisture content). Such moisture content ranges in value from 2 to 4% w/w for anthracite and for bituminous coals of the eastern Appalachian field, such as the Pocahontas, Sewell, Pittsburgh, Freeport, and Kittanning coal seams. In the western part of this field, especially in Ohio, the inherent moisture ranges from 4 to 10% w/w. In the interior fields of Indiana, Illinois, western Kentucky, Iowa, and Missouri, the range is from 8 to 17% w/w. In subbituminous coals the

Table 5.1.6 Ranges of Composition for the Various Ranks of Coal

	Anthracite	Bituminous	Subbituminous	Lignite
Moisture (%)	3-6	2-15	10-25	25-45
Volatile matter (%)	2-12	15-45	28-45	24-32
Fixed carbon (%)	75-85	50-70	30-57	25-30
Ash (%)	4-15	4-15	3-10	3-15
Sulfur (%)	0.5-2.5	0.5-6	0.3-1.5	0.3-2.5
Hydrogen (%)	1.5-3.5	4.5-6	5.5-6.5	6-7.5
Carbon (%)	75-85	65-80	55-70	35-45
Nitrogen (%)	0.5-1	0.5-2.5	0.8-1.5	0.6-1.0
Oxygen (%)	5.5-9	4.5-10	15-30	38-48
Btu/lb	12,000-13,500	12,000-14,5000	7,500-10,000	6,000-75,000
Density (g/mL)	1.35-1.70	1.28-1.35	1.35-1.40	1.40-1.45

inherent moisture ranges from 15 to 30% w/w, and in lignite from 30 percent to 45 percent w/w. The total amount of moisture in commercial coal may be greater or less than that of the coal in the mine. Freshly mined subbituminous coal and lignite lose moisture rapidly when exposed to the air. The extraneous or surface moisture in coal is a function of the surface exposed, each surface being able to hold a film of moisture. Fine sizes hold more moisture than lump. Coal which in the mine does not contain more than 4% w/w moisture may in finer sizes hold as much as 15% w/w; the same coal in lump sizes, even after underwater storage, may contain little more moisture than originally in the mine.

Volatile matter does not exist in coal as such but is produced by decomposition of the coal when heated. In the standard method of analysis, the *volatile matter* is the loss in weight, less moisture, obtained by heating the coal for 7 minutes in a covered crucible at approximately 950°C (1,742°F) under specified conditions. Volatile matter consists chiefly of the combustible gases, hydrogen, carbon monoxide, methane and other hydrocarbons, tar vapors, volatile sulfur compounds, and some noncombustible gases such as carbon dioxide and water vapor. The composition of the volatile matter varies greatly with different coals and can also vary with the rate of heating. The inert or noncombustible gas may range from 4% of the total volatile matter in low-volatile coals to 40% w/w in subbituminous coals.

Fixed carbon is the carbonaceous residue that remains after coal is heated under standard conditions—as for example, using the test method for the determination of the volatile matter. The usual method of determining the fixed carbon yield is to subtract from 100 the sum of the % w/w of the moisture, the % w/w volatile matter, and the % w/w ash obtained as a result of the proximate analysis. The yields of fixed carbon not represent the total carbon in the coal because a considerable part of the carbon is expelled as volatile matter in combination with hydrogen as hydrocarbons and with oxygen as carbon monoxide and carbon dioxide. It also is not pure carbon because it may contain several tenths % w/w of hydrogen and oxygen, 0.4 to 1.0% of nitrogen, and approximately 50% w/w of the sulfur that was in the original coal.

In the standard method, *ash* is the inorganic residue that remains after burning the coal in a muffle furnace to a final temperature of 700 to 750°C (1,292 to 1,382°F). Coal ash is composed largely of compounds of silicon, aluminum, iron, and calcium, with smaller quantities of compounds of magnesium, titanium, sodium, and potassium. The ash as determined is usually less than the inorganic mineral matter originally present in the coal. During incineration, various weight changes take place, such as loss of water of constitution of the silicate minerals, loss of carbon dioxide from carbonate minerals, oxidation of iron pyrites to iron oxide, and fixation of a part of the oxides of sulfur by bases such as calcium and magnesium. The chemical composition of coal ash varies widely depending on the mineral constituents associated with the coal (Tables 5.1.7 and 5.1.8). The ash of subbituminous coals may have more calcium oxide (CaO) and magnesium oxide (MgO) than the ash of bituminous coals.

Ultimate analysis expresses the composition of coal as sampled in % w/w of carbon, hydrogen, nitrogen, sulfur, oxygen, and ash (Table 5.1.6). The carbon includes that present in the organic coal substance as well as a minor amount that may be present as mineral carbonates. In actual practice, the hydrogen and oxygen values include those of the organic coal substance as well as those present in the form of moisture and the water of constitution of the silicate minerals. In certain other countries, the values for hydrogen and oxygen are corrected for the moisture in the coal and are reported separately. The ash is the same as reported in the proximate analysis; the sulfur, carbon, hydrogen, and nitrogen are determined chemically. Oxygen in coal is usually estimated by subtracting the sum of carbon, hydrogen, nitrogen, sulfur, and ash from 100.

Table 5.1.7 Major Inorganic Constituents of Coal Ash

Constituents	Representative percentage
SiO_2	40-90
Al_2O_3	20-30
Fe_2O_3	5-25
CaO	1-15
MgO	0.5-4
Na_2O	0.5-3
K_2O	0.5-3
SO_3	0.5-10
P_2O_5	0-1
TiO_2	0-2

Table 5.1.8 The Pyrite Group of Minerals

Mineral	Formula (approx.)
Pyrite	FeS_2
Marcasite	FeS_2
Braoite	$(Ni \cdot Fe)S_2$
Laurite	RuS_2

Sulfur occurs in three forms in coal: (i) pyritic sulfur, which is sulfur combined with iron as pyrite or marcasite, FeS_2, (ii) organic sulfur, which is sulfur combined with coal substance as a heteroatom or as a bridge atom, (iii) sulfate sulfur, which is sulfur combined mainly with iron or calcium together with oxygen as iron sulfate, $FeSO_4$, or calcium sulfate, $CaSO_4$. Pyrite and marcasite are recognized by their metallic luster and pale brass-yellow color, although some marcasite is almost white. Organic sulfur may comprise from approximately 20 to 85% w/w of the total sulfur in the coal. Most freshly mined coal contains only very small quantities of sulfate sulfur; it increases in weathered coal. The total sulfur content of coal mined in the United States varies from approximately 0.4 to 5.5% w/w on a dry coal basis.

The *gross calorific value* of a fuel expressed in Btu/lb of fuel is the heat produced by complete combustion of a unit quantity, at constant volume, in an oxygen bomb calorimeter under standard conditions. The values includes the latent heat of the water vapor in the products of combustion. Since the latent heat is not available for making steam in actual operation of boilers, a net caloric value is sometimes determined by the following formula:

Net calorific value = gross calorific value – (92.70 × total hydrogen, % in coal)

In this equation, both the net calorific value and the gross calorific value are in Btu/lb. The gross calorific value may also be approximated by the Dulong formula:

$$\text{Btu/lb} = 14{,}544C\ 1\ 62{,}028(H - \underline{O/8) + }1\ 4{,}050S(\text{Btu/lb} \times 2.328\ 5\ \text{kJ/kg})$$

In this equation, C, H, O, and S are weight fractions from the ultimate analysis. For anthracite, semi-anthracite, and bituminous coal, the calculated values are usually within 1.5% w/w of those determined by the bomb calorimeter. For subbituminous coal and lignite, the calculated values show deviations often reaching 4 and 5% w/w.

Because coal ash is a mixture of various components, it does not have a definite melting point. The gradual softening and fusion of the ash is not merely the successive melting of the various ash constituents but is a more complicated process in which reactions involving the formation of new and more fusible compounds take place. The *fusibility of coal ash* is determined by heating a triangular pyramid (cone), 0.75 in high and 0.25 in wide at each side of the base, made up of the ash together with a small amount of organic binder. As the cone is heated, three temperatures are noted: (i) the initial deformation temperature, IDT, or the temperature at which the first rounding of the apex or the edges of the cone occurs, (ii) the softening temperature, ST, or the temperature at which the cone has fused down to a spherical lump, and (iii) the fluid temperature, FT, or the temperature at which the cone has spread out in a nearly flat layer. The softening interval is the degrees of temperature difference between (ii) and (i), the flowing interval the difference between (iii) and (ii), and the fluidity range the difference between (iii) and (i). Of the three, the softening temperature is most widely used.

Table 5.1.9 Physical Properties Often Used for Determining Coal Suitability for Use

Test/Property	Results/Comments
Physical properties	
Density	True density as measured by helium displacement
Specific gravity	Apparent density
Pore structure	Specification of the porosity or ultrafine structure of coals and nature of pore structure between macro, micro, and transitional pores
Surface area	Determination of total surface area by heat of absorption
Reflectivity	Useful in petrographic analyses
Mechanical properties	
Strength	Specification of compressibility strength
Hardness/abrasiveness	Specification of scratch and indentation hardness; also abrasive action of coal
Friability	Ability to withstand degradation in size on handling, tendency toward breakage
Grindability	Relative amount of work needed to pulverize coal
Dustiness index	Amount of dust produced when coal is handled
Thermal properties	
Calorific value	Indication of energy content
Heat capacity	Measurement of the heat required to raise the temperature of a unit amount of coal 1°
Thermal conductivity	Time rate of heat transfer through unit area, unit, thickness, unit temperature difference
Plastic/agglutinating	Changes in a coal upon heating; caking properties of coal
Agglomerating index	Grading on nature of residue from 1-g sample when heated at 950°C (1,550°F)
Free swelling index	Measure of the increase in volume when a coal is heated without restriction
Electrical properties	
Electrical resistivity	Electrical resistivity of coal measured in ohm-centimeters
Dielectic constant	Measure of electrostatic polarizability

Data on ash fusion characteristics are useful to the combustion engineer concerned with evaluation of the clinkering tendencies of coals used in combustion furnaces and with corrosion of metal surfaces in boilers due to slag deposits. The types of mineral matter occurring in different coals are not well related to rank or geographic location, although there is a tendency for midcontinent coals (Indiana to Oklahoma) to have low ash fusion temperatures. High significance is attached to all of the previously indicated fusion temperatures and the intervals between them. The initial deformation temperature is sometimes identified with surface stickiness, the softening temperature with plastic distortion or sluggish flow, and the fluid temperature with liquid mobility. Long fusion intervals often produce tough, dense, slags while short fusion intervals favor the production of porous, friable structures.

Properties The properties of coal are used to determine the specifications that play a role in the purchase of coal vary widely depending on the intended use (Table 5.1.9), whether for coke or for a combustion unit and whether it meets the standards imposed by the customer. Most of the properties must be taken into consideration, for example, the sulfur content and swelling properties of a coal used for metallurgical coke production, the heating value of a coal used for steam generation, and the slagging properties of its ash formed after combustion.

The *specific gravity* of coal is the ratio of the weight of solid coal to the weight of an equal volume of water. It is useful in calculating the weight of solid coal as it occurs in the ground for estimating the tonnage of coal per acre of surface. An increase in ash-forming mineral matter increases the specific gravity; for example, bituminous coals of Alabama, ranging from 2 to 15% w/w ash and from 2 to 4.5% w/w moisture, vary in specific gravity from 1.26 to 1.37.

Bulk density is the weight per cubic foot of broken coal. It varies according to the specific gravity of the coal, its size distribution, its moisture content, and the amount of orientation when piled. The range of weight from subbituminous coal to anthracite is from 44 to 59 lb/ft^3 when loosely piled; when piled in layers and compacted, the weight per cubic foot may increase as much as 25% w/w. The weight of fuel in a pile can usually be determined to within 10 to 15% w/w by measuring its volume. Typical weights of coal, as determined by shoveling it loosely into a box of 8 ft^3 capacity, are as follows: anthracite, 50 to 58 lb/ft^3; low- and medium-volatile bituminous coal, 49 to 57 lb/ft^3; high-volatile bituminous and subbituminous coal, 42 to 57 lb/ft^3.

The *grindability* of coal, or the ease with which it can be ground fine enough for use as a pulverized fuel, is a composite of several specific physical properties such as hardness, tensile strength, and fracture. A laboratory procedure adopted by ASTM (D409) for evaluating grindability (the Hardgrove machine method) uses a specially designed grinding apparatus to determine the relative grindability or ease of pulverizing coal in comparison with a standard coal, chosen as 100 grindability. Primarily, the ASTM Hardgrove grindability test is used for estimating how various coals affect the capacity of commercial pulverizing equipment.

A general relationship exists between grindability of coal and its rank. Coals that are easiest to grind (highest grindability index) are those of approximately 14 to 30% w/w volatile matter on a dry, ash-free basis. Coals of either lower or higher volatile matter content usually are more difficult to grind. The relationship of grindability and rank, however, is not sufficiently precise for grindability to be estimated from the chemical analysis, partly because of the variation in grindability of the various petrographic and mineral components. Grindability indexes of U.S. coals range from approximately 20 for an anthracite to 120 for a low-volatile bituminous coal.

The *agglomerating* character is used to differentiate between certain adjacent groups. Coals are considered agglomerating if, in the test to determine volatile matter, they produce either a coherent button that will support a 500-g weight or a button that shows swelling or cell structure.

Most bituminous coals, when heated at uniformly increasing temperatures in the absence or partial absence of air, fuse and become *plastic*. These coals may be designated as either caking or coking in different degrees. *Caking* usually refers to the fusion process in a boiler furnace. *Coking coals* are those that make good coke, suitable for metallurgical purposes where the coke must withstand the burden of the ore and flux above it. Coals that are caking in a fuel bed do not necessarily make good coke in a coke oven. Subbituminous coal, lignite, and anthracite are non-caking coals.

The test for the *free-swelling index* of coal is a test that measures the free-swelling properties of coal and gives an indication of the caking characteristics of the coal when burned on fuel beds. It is not intended to determine the expansion of coals in coke ovens. The test consists in heating 1 g of pulverized coal in a silica crucible over a gas flame under prescribed conditions to form a coke button, the size and shape of which are then compared with a series of standard profiles numbered 1 to 9 in increasing order of swelling.

Coal Conversion

Coal can also be converted to gaseous fuels, liquid fuels, and solid fuels. Coal-based electricity is well-established and highly reliable. Over 41% of global electricity is currently based on coal. The generation technologies are well-established and technical capacity and human expertise is widespread. Unlike gaseous, liquid or intermittent renewable sources, coal can be stockpiled at the power station and stocks drawn on to meet demand without depending on primary supply. Because of the geographical diversity of coal reserves, some power stations can be located at mine-mouth and still be close to the market, thereby minimizing transmission and distribution losses. In addition, environmental regulations related to coal use must be balanced and include improved designs for coke-making which have already entered the market.

Carbonization Carbonization is the destructive distillation of coal in the absence of air accompanied by the production of carbon and liquid and gaseous products. The coke produced by the carbonization of coal is used in the iron and steel industry and as a domestic smokeless fuel. The gradual depletion of indigenous good quality coking coal resources, research initiatives have provided the scientific basis and scope of utilizing different type inferior grade coals as blend components for metallurgical coke making. Application of the knowledge base backed by validation tests in pilot plants has helped in optimizing the blend components for coke production.

Three different approaches to coal carbonization may be adopted in the future: (i) improvement of coke production by introducing process modifications, (ii) the building of a new coking system, and (iii) the development of new carbon materials with different specifications and new functions. Improvement in existing technology includes the construction of new heating systems which reduce NOx formation and the systematic study of blended coal as the feedstock for coke.

Liquefaction The production of liquid fuels—gasoline and diesel—from coal is not a new process. The first patent was registered in 1913, with the more common Fischer-Tropsch indirect liquefaction process patented in 1925. *Direct coal liquefaction* converts coal to a liquid by dissolving coal in a solvent at high temperature and pressure. This process is highly efficient, but the liquid products require further refining ("hydrocracking" or adding hydrogen over a catalyst) to achieve high grade fuel characteristics. *Indirect coal liquefaction* first gasifies the coal with steam to form synthesis gas (syngas—a mixture of hydrogen and carbon monoxide). The sulfur is removed from this gas and the mixture adjusted according to the desired product. The synthesis gas is then condensed over a catalyst—the Fischer-Tropsch process—to produce high quality, ultra-clean products.

The development of a coal to liquids (CTL) industry can serve to hedge against oil-related energy security risks. Using domestic coal reserves, or accessing the relatively stable international coal market, can allow countries such as the United States to minimize their exposure to oil price volatility while providing the liquid fuels needed for economic stability and growth.

Gasification Clean coal technology using gasification is a promising alternative to meet the global energy demand. There are indications that the use of coal as a global energy source has caught up with the use of natural gas and will (at current rates of use) surpass natural gas use by 2030. Most existing coal gasification technologies perform best on high rank (bituminous) coal and petroleum refinery waste products but are inefficient, less reliable and expensive to operate when processing low rank coal. These low-grade coal reserves including low rank and high ash coal remain underutilized as a global energy sources despite being available in abundance. Coal gasification for electric power generation enables the use of a technology common in modern natural gas fired power plants, the use of *combined cycle* technology to recover more of the energy released by burning the fuel. In a combined cycle plant the gas is burned in a turbine that generates electricity, then steam is generated from the hot exhaust and used to power a second generator. This method achieves an efficiency of 45 to 50% compared to traditional power plants which only use one cycle to create electricity at only 35% efficiency.

The increased efficiency of the *combined cycle* for electrical power generation results in a 50% decrease in carbon dioxide emissions compared to conventional coal power plants. As the technology required to develop economical methods of *carbon sequestration*, the removal of carbon dioxide from combustion by-products to prevent its release to the atmosphere, coal gasification units could be modified to further reduce their climate change impact because a large part of the carbon dioxide generated can be separated from the syngas *before* combustion.

Clean Coal Technologies On a global macro level, coal will continue to form a significant portion of the world's energy supply for the next several decades and that the United States and the world will rely on clean coal technologies to be the bridge leading to future renewable energy solutions. *Clean coal technology* is an umbrella term used to describe technologies being developed that aim to reduce the environmental impact of coal energy generation. These include chemically washing minerals and impurities from the coal, gasification, treating the flue gases with steam to remove sulfur dioxide, carbon capture and storage (CCS) technologies to capture the carbon dioxide from the flue gas and dewatering lower rank coals (brown coals) to improve the calorific quality, and thus the efficiency of the conversion into electricity. Clean coal technology usually addresses atmospheric problems resulting from burning coal. Historically, the primary focus was on sulfur dioxide and particulates, because it is the most important gas which leads to acid rain.

Using clean coal technologies for energy will benefit many and as technology makes coal cleaner to burn and ways to protect the environment improve, coal will be a safe alternative. The more options that consumers have for energy needs, the quality and affordability of the energy will improve.

Gas Cleaning Due to the harmful impact of air pollutants, regulatory agencies have enacted strict regulations to limit their emission. Indeed, the introduction of the strict limiting values imposed by various governments during the twentieth century has seen improvements in gas cleaning technologies of which multistage flue gas cleaning concepts initially tended to predominate. These were characterized by selective separation stages for the individual pollutants. The captured pollutants, such as sulfur dioxide, were usually converted into recyclable products, in this case gypsum. More recently, a trend has been observed in many countries over the past decade which moves in the direction of less complex, integrated flue gas cleaning processes (Fig. 5.1.2).

Instead, mixed products are produced which can be disposed in secure landfill areas such as old mines or, after suitable physicochemical treatment together with adequate environmental compatibility.

Chemicals from Coal An array of products can be made via coal conversion processes. For example, refined coal tar is used in the manufacture of chemicals, such as creosote oil, naphthalene, phenol, and benzene. Ammonia gas recovered from coke ovens is used to manufacture ammonia salts, nitric acid, and agricultural fertilizers. Thousands of different products have coal or coal by-products as components: soap, aspirins, solvents, dyes, plastics, and fibers, such as rayon and nylon.

There are process schemes to convert syngas to ammonia, methanol, substitute natural gas (SNG), and Fischer-Tropsch derived transportation fuels. These novel and efficient schemes minimize plant energy consumption, water use and emissions while increasing the output of valuable products.

Environmental Aspects of Coal Use

The conversion of any feedstock to liquid fuels is an energy intensive one, and process emissions must be considered. While the CTL process

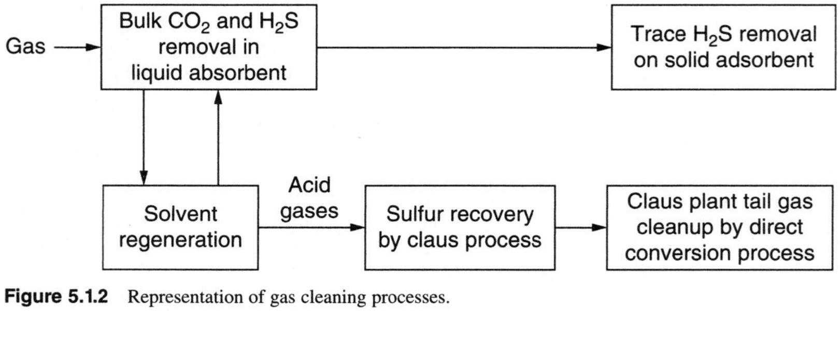

Figure 5.1.2 Representation of gas cleaning processes.

is more carbon dioxide intensive than conventional petroleum refining, there are options for preventing or mitigating emissions. Due to the broad global distribution of coal reserves, emissions may also be avoided through shorter fuel transport distances.

For CTL plants, CCS can be a low-cost method of addressing carbon dioxide concerns and may result in greenhouse gas emissions. Where co-processing of coal and biomass is undertaken, and combined with CCS, greenhouse gas emissions over the full fuel cycle may be as low as one-fifth of those from fuels provided by conventional oil. CCS involves the capture of carbon dioxide emissions from the source, followed by transportation to, and storage in, geological formations. Once the carbon has been captured there are several storage options available. The carbon dioxide can be stored in deep saline aquifers or be used to assist in enhanced oil recovery and subsequent carbon dioxide storage.

One of the major environmental advantages of coal gasification is the opportunity to remove impurities such as sulfur and mercury and soot *before* burning the fuel, using readily available chemical engineering processes. In addition, the ash produced is in a vitreous or glasslike state which can be recycled as concrete aggregate, unlike pulverized coal power plants which generate ash that must be landfilled, potentially contaminating groundwater.

Energy Security

As global demand for energy continues to rise—especially in rapidly industrializing and developing economies—energy security concerns become ever more important. To provide solid economic growth, and to maintain levels of economic performance, energy must be readily available, affordable, and able to provide a reliable source of power without vulnerability to long-term or short-term disruptions. Energy security considerations are potentially even more important in the transport sector. Oil products provide the majority (>90%) of the energy used in transportation. As the price of crude oil has fluctuated over recent years, the possibility of hedging risks using alternative fuels—CTL, gas to liquids (GTL), and biomass to liquids (BTL)—is being seriously considered.

Coal, which has a unique role to play in meeting the demand for a secure energy, is well-established and is a reliable, secure, and affordable fuel for both power generation and industrial applications. The production and utilization of coal is based on well-proven and widely used technologies and is built on a vast infrastructure and a strong base of expertise worldwide. Alternative liquid fuels can be made from solid (e.g., coal or biomass) or gaseous (e.g., natural gas) feedstocks, and can be used in existing vehicle fleets with no, or little, modification. Coal will have a significant role to play in the provision of these alternative fuels—it is the most affordable of the fossil fuels and is widely distributed around the world.

However, it cannot be ignored that coal does face environmental challenges that have implications for energy security and sustainable development. Nevertheless, the coal industry has a proven track record of developing technology pathways which have successfully addressed (and continue to address) environmental concerns at local and national levels. In terms of energy security for many countries, a key issue is resource availability—the actual physical amount of the coal resource available within the borders of the country. This leads to the security of a continuous supply of coal as a source of energy, particularly electricity, to meet consumer demand at any given time. In this respect, coal has attributes that make a positive contribution to energy security as part of a balanced energy mix.

While coal has furthered social development through electricity generation around the world, coal can also be transformed to liquid and gaseous fuels to guard against oil import dependence and price shocks. Coal gasification is a further example where indigenous (or imported) fuels can be transformed to address environmental concerns while enhancing energy security.

Energy from renewable resources (biomass) can also cause challenges to those who need to ensure a stable and reliable flow of electricity through the grid. Wind power has been the dominant option for integration to existing energy networks but wind power is expensive, and suffers from reliability and intermittency problems. It is now generally accepted by power industry engineers and energy analysts that the level of wind power capacity in a grid should not exceed around 10% without the grid operator incurring significant costs to deal with the intermittency issues.

Environmental issues should be addressed in a non-discriminatory manner—while recognizing the benefits that arise when coal is used as a source of energy. Clear, long-term environmental policies provide certainty, allowing investments to be made in advanced coal technologies that bring enhanced environmental performance. Unlimited pollution in the name of energy generation is not the answer nor is the passage of unrealistic environmental regulations that will cause mankind to freeze in the green darkness.

A balance must be struck and only then will the full potential of coal as a secure form of energy (for the United States) be realized.

References

1. ASTM International. Annual Book of Standards. ASTM International Conshohocken, Pennsylvania.

2. Lee S., Speight J.G., and Loyalka S. *Handbook of Alternative Fuel Technologies*, 2nd ed., CRC Press, Taylor & Francis Group, Boca Raton, Florida, 2014.

3. Speight J.G. *Synthetic Fuels Handbook: Properties, Processes, and Performance*, McGraw-Hill, New York, 2008.

4. Speight J.G. *The Refinery of the Future*, Gulf Professional Publishing, Elsevier, Oxford, United Kingdom, 2011.

5. Speight J.G. (Editor). *The Biofuels Handbook*, Royal Society of Chemistry, London, United Kingdom, 2011.

6. Speight J.G. *The Chemistry and Technology of Coal*, 3rd ed., CRC Press, Taylor & Francis Group, Boca Raton, Florida, 2013.

7. Speight J.G. *Coal-Fired Power Generation Handbook*, Scrivener Publishing, Beverly, Massachusetts, 2013.

8. Speight J.G. *Handbook of Coal Analysis*, 2nd ed., John Wiley & Sons Inc., Hoboken, New Jersey, 2015.

BIOMASS AND BIOFUELS

Biomass is the collective name for *renewable materials* which includes to (1) energy crops grown specifically to be used as fuel, such as wood or various types of grass, (2) agricultural residues and by-products, such as straw, sugarcane fiber, rice hulls animal waste, and (3) residues from forestry, construction, and other wood-processing industries. Biomass is a renewable resource that has received great attention due to environmental considerations and the increasing demands of energy worldwide. Biomass has a negligible content of sulfur, nitrogen, and ash which lead to lower emissions of sulfur dioxide, nitrogen oxides, and soot than conventional fossil fuels. Following form this, a *biofuel* is any fuel that is derived from biomass. A biofuel has also been defined as any fuel with an 80% v/v minimum content of materials derived from living organisms harvested within the 10 years preceding its manufacture.

In addition, a *primary biomass feedstock* is biomass that has been harvested or collected from the field or forest where it is grown. Examples of primary biomass feedstocks currently being used for bioenergy (specifically, bio-oil) include grains and oilseed crops used for transportation fuel production (Fig. 5.1.3), plus some crop residues (such as orchard trimmings and nut hulls) and some residues from logging and forest operations that are currently used for heat and power production. In the future, it is anticipated that a larger proportion of the residues inherently generated from food crop harvesting, as well as a larger proportion of the residues generated from ongoing logging and forest operations, will be used for bioenergy. Additionally, as the bioenergy industry develops, both woody and herbaceous perennial crops will be planted and harvested specifically for bioenergy and product end-uses.

Secondary biomass feedstocks are a by-product of processing of the primary feedstocks in which there has been substantial physical or chemical breakdown of the primary biomass and production of by-products. Specific examples of secondary biomass include sawdust

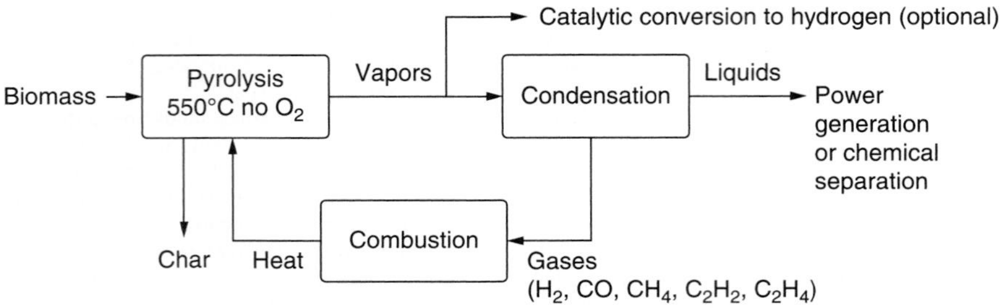

Figure 5.1.3 Production of bio-oil from biomass by pyrolysis.

from sawmills, black liquor (a byproduct of paper making), and whey cheese (which is a by-product of cheese making processes). Manures from concentrated animal feeding operations are collectable secondary biomass resources. Vegetable oil used for production of biodiesel that is derived directly from the processing of oilseeds for various uses is also a secondary biomass resource.

The term *tertiary biomass feedstock* includes (i) post-consumer residues and wastes, such as fat, grease, construction and (ii) demolition wood debris, other waste wood from the urban environments, as well as (iii) packaging wastes, municipal solid wastes, and landfill gases. A category *other wood waste from the urban environment* includes trimmings from urban trees, which technically fits the definition of primary biomass but, because this material is normally handled as a waste stream along with other post-consumer wastes from urban environments, this type of waster tends to be considered as part of the tertiary biomass stream.

Biomass conversion to energy continues to be a subject of intensive study for both developed and less developed countries. In the United States, combustion of biomass contributes only a relatively small percentage of the total energy supply, and it is mostly in the form of agricultural wastes and paper. Almost any plant material can be the raw material for gasification from which a variety of products can be created catalytically from the carbon monoxide, carbon dioxide, and hydrogen. Typical commercial products are methanol, ethanol, methyl acetate, acetic anhydride, and hydrocarbons. Alcohols are of interest as fuels for transportation and power generation. Producer gas has been introduced into small compression ignition engines, and the diesel oil feed has been reduced. Technical and environmental problems are still not solved in the large-scale use of biomass.

Plants and vegetables are another source of biomass-derived oils. Fatty acid esters have been used in diesel engines alone and in blends with diesel oil, and although the esters are effective, some redesign and changes in operation are required. In developing countries where fossil fuels are costly, deforestation has led to land erosion and its consequences. Useful biomass materials include most of the precoal organic vegetation such as wood, grass, the organic portion of domestic waste, and wood waste from demolition sites (Table 5.1.10) and food processing wastes like bagasse from sugar production. Digestible food processing wastes can be converted to biogas (Table 5.1.11).

Biodiesel

Biodiesel is an oxygenated biofuel which can be used to displace petroleum-sourced No. 2 heating oil and diesel fuel. The feedstocks for biodiesel production are plant-oil based and include soy oil, a wide variety of other crop oils, waste vegetable oils from cooking operations, and emerging sources such as algal oils. Animal fats can also be used to produce biodiesel. These oils are technically impractical to use directly as fuels due largely to their relatively high viscosity. Direct use of these oils would require changes to the equipment including fuel heating and changes to storage, filtering, and atomization systems.

The process of producing biodiesel breaks apart the large triglyceride molecule of the oil to produce the smaller esters and a lower viscosity fuel. The basics of biodiesel production are illustrated in Figs. 5.1.4 and 5.1.5. Methanol is combined with the plant-sourced oil in the presence of a catalyst for this relatively simple, low pressure transesterification reaction. The produced fuel is cleaned to remove residual methanol, water, and catalyst.

With an annual U.S. production capacity of 2.1 billion gallons, the U.S. biodiesel industry is well developed. The properties of biodiesel for use as a blendstock in diesel and heating oil are fully defined in ASTM Standard D6751. The properties of blends for use in home heating oil and diesel fuel are also formally defined in ASTM standards. A formal industry quality standard (BQ9000) is in place to promote good practices for fuel production. The U.S. Department of Energy publishes a comprehensive Handling and Use Guide which is periodically updated.

Table 5.1.10 Composition of Selected Biomass Feedstocks

Biomass type:	Clean wood	Verge grass**	Domestic waste***	Wood****
Moisture content, % w/w wet biomass	50	60	54	20
Mineral matter content*, % w/w dry biomass	1.3	8.4	18.9	0.9
LHV (as received), MJ/kg	7.7	5.4	6.4	13.9
HHV (as received), MJ/kg	9.6	7.4	8.3	15.4
Composition, % w/w (maf)				
Carbon	49.1	48.7	51.9	48.4
Hydrogen	6	6.4	6.7	5.2
Oxygen	44.3	42.5	38.7	45.2
Nitrogen	0.48	1.9	2.2	0.15
Sulfur	0.01	0.14	0.5	0.03
Chlorine	0.1	0.39	0.3	0.08

* Determined as ash by combustion.
** Taken from the side of a roadway.
*** Organic waste.
**** From demolition sites.

Table 5.1.11 Composition of Biogas (% v/v)

Methane (CH_4)	55-65
Carbon dioxide (CO_2)	35-45
Hydrogen sulphide (H_2S)	0-1
Nitrogen (N_2)	0-3
Hydrogen (H_2)	0-1
Oxygen (O_2)	0-2
Ammonia (NH_3)	0-1

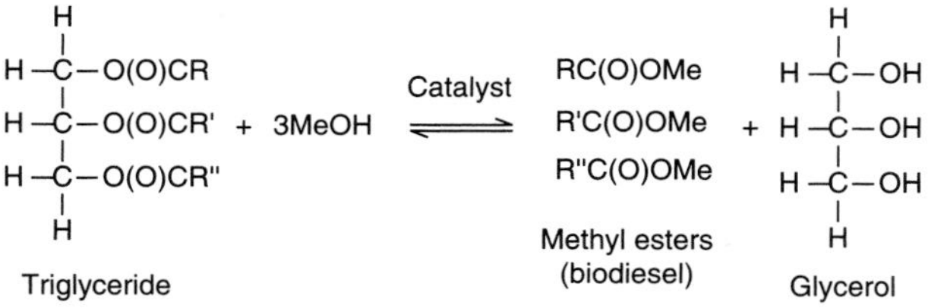

Figure 5.1.4 The chemistry of biodiesel production.

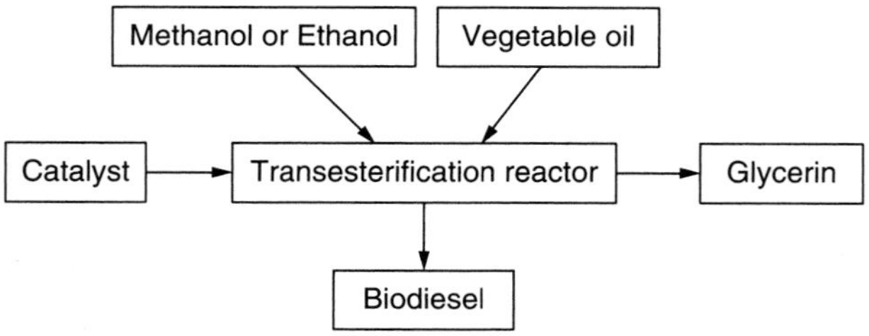

Figure 5.1.5 Biodiesel production catalytic by a catalytic process.

It is common practice to refer to biodiesel blends as BXX where XX represents the volume percent biodiesel in the mixture. For example B20 is a commonly used blend of 20% biodiesel in petroleum diesel. With ~11% oxygen by weight, the heating value of B100 is roughly 14% lower than that of petroleum diesel (Btu/gal basis). The flash point of biodiesel is slightly higher than that of petroleum diesel and the two fuels are fully miscible. Biodiesel is nearly sulfur-free (under 10 ppm S). Low temperature flow properties of biodiesel are different than those of petroleum-sourced distillate fuels and are strongly oil-source dependent. Biodiesel fuels gel at higher temperatures than petroleum-sourced fuels. At low blend levels, B20 and less, this has not been found to be a significant concern. There is interest in using biodiesel at high blend levels, up to B100 and, in this case, careful consideration must be given to the fuel delivery and storage temperature conditions and the properties of the specific biodiesel being used. At higher blend levels compatibility with elastomers used in pumps and other fuel system components should be evaluated for the application. At low blend levels this is not a strong concern. For example, blend levels to B5 in heating oil are formally recognized as equivalent to petroleum-based heating oil in the ASTM D396 standard.

Peat

Peat, a product of the early stages in the metamorphosis of vegetable matter into coal (Fig. 5.1.6), is the product of partial decomposition and disintegration of plant remains in water bogs, swamps or marshlands, and in the absence of air. Like all material of vegetable origin, peat is a complex mixture of carbon, hydrogen, and oxygen in a ratio similar to that of cellulose and lignin. Generally, peat is low in essential growing elements (K, S, Na, P) and ash. Trace elements, when found in the peat bed, are usually introduced by leaching from adjacent strata. The

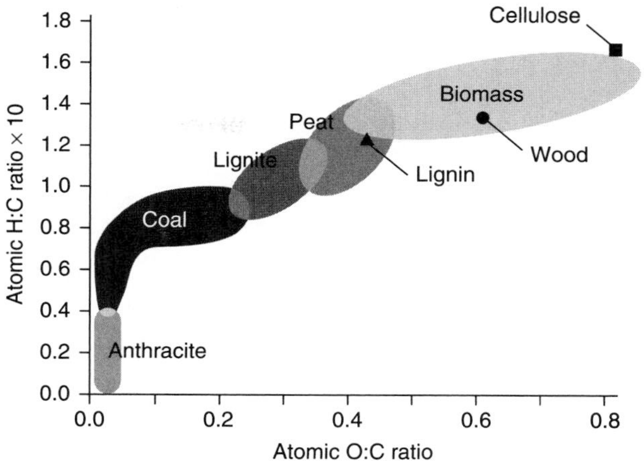

Figure 5.1.6 Representation of peat and biomass on the atomic H/C and atomic O/C diagram.

Federal Trade Commission specifies that to be so labeled, peat must contain at least 75 percent peat with the rest composed of normally related soil materials. The water content of undrained peat in a bog is 92 to 95 percent but it is reduced to 10 to 50 percent when peat is used as a fuel. Peat is harvested by large earth-moving equipment from a drained bog dried by exposure to wind and sun. The most popular method used in Ireland and the Soviet Union involves harrowing of drum-cut peat and allowing it to field dry before being picked up mechanically or pneumatically.

The chemical and physical properties of peat vary considerably, depending on the source and the method of processing (Table 5.1.12). The proximate analyses of samples (Table 5.1.13), calculated back to a dry basis, follow a similar broad pattern: 55 to 70 percent volatile matter, 30 to 40 percent fixed carbon and 2 to 10 percent ash. The dry, ash-free ultimate analysis ranges from 53 to 63 percent carbon, 5.5. to 7 percent hydrogen, 30 to 40% w/w oxygen, 0.3 to 0.5% w/w sulfur, and 1.2 to 1.5% w/w nitrogen.

Wood—Direct Combustion

The scope of direct biomass combustion is broad, reaching from residential wood stoves to automated wood chip combustion for power plants. Fuels used include cordwood, wood pellets, wood chips, fine wood dust, and wood waste materials. These fuels represented about 2% of the annual U.S. fuel use in 2015. Some 2.5 million homes use wood as the primary heating fuel and 9 million use wood as a secondary fuel. The direct use of wood rose sharply between 2005 and 2012, driven at least in part by high oil prices over this time period. Regionally, air quality has been impacted by the rapid rise in wood use, leading to increasing interest in low emission technology. Older wood combustion technologies can have very high emission factors for carbon monoxide, hydrocarbons, and particulates. The situation has been similar in Europe leading to the same pressures for improved technology and considerable movement of technology between Europe and the U.S.

Table 5.1.12 Properties of Different Types of Processed Peat

Property	Air-dried	Milled*	Briquettes
Moisture, % w/w	25-50	50-55	10-12
Density lb/ft^3	15-25	20-30	30-50
Calorific value, Btu/lb	6,200	3,700-5,300	8,000

* Peat in granulated or crumb form which has been mechanically harvested from the surface of a bog.

Table 5.1.13 Typical Analysis of Dry Wood Fuels*

Proximate analysis, % w/w	
Volatile matter	74-82
Fixed carbon	17-23
Mineral matter (as mineral ash)	0.5-2.2
Ultimate analysis, %, w/w	
Carbon	49.6-53.1
Hydrogen	5.8-6.7
Oxygen	39.8-43.8
Heating value	
Btu/lb	8,560-9,130
kJ/kg	19,900-21,250
*Ash-fusion temperature***	
Initial, °F	2,650-2,760
Initial, °C	14,545-1,515
Fluid, °F	2,730-2,830
Fluid, °C	1,490-1,550

* Typically, wood contains no sulfur and approximately 0.1% w/w nitrogen. By comparison, cellulose contains 44.5% w/w carbon, 6.2% w/w hydrogen, and 49.3% w/w oxygen.
** For convenience, rounded to the nearest 5°.

Fresh timber contains 30 to 50 percent moisture, mostly in the cell structure of the wood, and after air drying for a year, the moisture content reduces to 18 to 25 percent. Kiln-dried wood contains about 8 percent moisture (Table 5.1.13). Wood pellets, a convenient commercial fuel made from very fine waste streams, typically contain less than 8% moisture. The moisture content of wood obviously has a strong effect on heating value (Table 5.1.14). Residential cordwood stoves are regionally very attractive in part because they can operate during an electric power outage. Technologies to reduce emissions include two stage combustion and exhaust catalysts which are very effective. Residential pellet stoves offer the convenience of automatic feed and complex delivery/storage/ and feeding systems are now available. Pellet systems operate with much lower particulate emissions than cordwood systems generally.

For residential hydronic heating, cord wood, wood pellet, and wood chip boilers are available. With cord wood, two stage combustion systems with oxygen sensors and automatic excess air control are now common on the market. Thermal storage is very often used to reduce cycling and improve the performance of all biomass-fired hydronic heaters. Boilers with integrated emission control catalysts are also emerging on the market. Residential cordwood warm air furnaces, and dual fuel, cordwood/oil, heating appliances are available.

In larger sizes automatic feed systems are more common with wood chip or pellet fuels, although hand-fed biomass boilers are available even to 20 million Btu/hr firing rates. Biomass gasifiers which produce low Btu fuel gas are available with the possibility of refitting to existing oil- and gas-fired boilers. Wood chip gasifiers with gas cleanup and feeding of low Btu gas to engines for power production are also in use. Particulate emission controls, for example electrostatic precipitators, are more commonly used in these larger size systems.

Wood Charcoal

Wood charcoal is made by heating wood to a high temperature in the absence of air. Wood loses up to 75% w/w and 50% v/v of its volume owing to the elimination of moisture and volatile matter. Thus, charcoal has a higher heating value per cubic foot than the original wood, especially if the final product is compacted in the form of briquettes. Charcoal is marketed in the form of lumps, powder, or briquettes and finds some use as a fuel for curing, restaurant cooking, and a picnic fuel. Its nonfuel uses, particularly in the chemical industry, are as an adsorption medium for purifying gas and liquid streams and as a decolorizing agent.

Table 5.1.14 Effect of Moisture of the Heating Value of Wood

Moisture % w/w	Ratio, water/wood w/w	Heating value Btu/lb
0	0	8,750
20	0.25	7,000
50	1.00	4,375
80	4.00	1,750

Table 5.1.15 Properties of Byproduct Fuels

Fuel*	Heating value, Btu/lb (dry)	Moisture, % w/w	Ash, moisturefree
Black liquor (sulfate)	6,500	35	40-45
Cattle manure	7,400	50-75	17
Coffee grounds	10,000	65	1.5
Corncobs	9,300	10	1.5
Cottonseed cake	9,500	10	8
Municipal refuse	9,500	43	8
Pine hark	9,500	40-50	5-10
Rice straw or hulls	6,000	7	15
Scrap tires	16,400	0.5	6
Wheat straw	8,500	10	4

* Listed alphabetically rather than be preference.

In addition to peat and wood, several lesser-known fuels are in common use for the generation of industrial steam and power (Table 5.1.15). Aside from their value as a fuel, the burning of wastes minimizes a troublesome disposal problem that could have serious environmental impact. Nearly all these waste fuels are cellulosic in character, and the heating value is a function of the carbon content. Ash content is generally low, but much moisture could be present from processing, handling, and storage. On a moisture- and ash-free basis the heating values can be estimated at 8,000 Btu/lb; more resinous materials about 9,000 Btu/lb.

Typically, charcoal burns at temperatures, up to 2,700°C (4,890°F) and due to the porosity, it is sensitive to the flow of air and the heat generated can be moderated by controlling the air flow to the fire. Charcoal briquettes can burn up to approximately 1,260°C (2,300°F) with a forced air blower forges.

Bagasse

Bagasse is the fibrous material left after pressing the juice from sugarcane or harvesting the seeds from sunflowers. The chopped waste usually contains about 50% w/w moisture and is burned in much the same manner as wood waste. Spreader stokers or cyclone burners are used. Supplemental fuel is added sometimes to maintain steady combustion and to provide energy for the elimination of moisture. A typical analysis of dry bagasse from Puerto Rico is 44.47% w/w carbon, 6.3% w/w hydrogen, 49.7% w/w oxygen, and 1.4% w/w mineral matter (determined as mineral ash) with a heating value is on the order of 8,400 Btu/lb.

Bagasse is often used as a primary fuel source for sugar mills; when burned in quantity, it produces sufficient heat energy to supply all the needs of a typical sugar mill. Furnaces have been developed to burn wastes, and some preferences emerge by operating characteristics. Spreader stokers are preferred for wood waste and bagasse. Tangential firing seems to be used for coffee grounds, rice hulls, some wood waste, and chars from coal or lignite. Traveling-grate stokers are used for industrial wastes and coke breeze.

Biogas

Biogas, sometimes incorrectly known as swamp gas, typically refers to a biofuel gas produced by anaerobic digestion or fermentation of organic matter including manure, sewage sludge, municipal solid waste, biodegradable waste, or any other biodegradable feedstock, under anaerobic conditions. Depending on where it is produced biogas is also called swamp gas, marsh gas, landfill gas, and digester gas.

The composition of biogas varies depending upon the composition of the waste material in the landfill and the anaerobic digestion process. Landfill gas typically has methane concentrations around 50%. Advanced waste-treatment technologies can produce biogas with 55 to 75% v/v methane; often air is introduced (5% by volume) for microbiological desulfurization. For example, the constituents (by volume) of biogas generally are: methane (50 to 75%), carbon dioxide (25 to 50%), nitrogen (0 to 10%), hydrogen (0 to 1%), hydrogen sulfide (0 to 3%), and oxygen (0 to 2%).

Biogas is produced by methanogenic organisms in landfills, swamps and marshes, by the decomposition of organic (both animal and plant) matter. The anaerobic (low or no oxygen present in the system) decomposition produces methane, carbon dioxide, a small amount of hydrogen and a small amount of heat. Acidity, the carbon to nitrogen ratio of the feedstock, temperature and percentage of solids are factors that influence the quality and quantity of biogas produced. Normally, cow manure, enriched by vegetable matter is the ideal feedstock. The other principal type of biogas is wood gas, which is created by gasification of wood or another biomass. This type of biogas is comprised primarily of nitrogen, hydrogen, and carbon monoxide, with trace amounts of methane. Industrial anaerobic digesters and mechanical biological treatment systems are becoming increasingly popular. Methane is a potent greenhouse gas but is also a very clean, easy-burning fuel, and hence harnessing it from natural sources such as landfills is beneficial to the natural environment.

Landfill gas is another source of biogas generated from the decomposition of the biodegradable organic fraction of municipal solid waste by microorganisms. This generally occurs under semi-controlled conditions in a landfill; its constituents depend on the composition and age of the waste. Large variations may exist in the composition of landfill gas due to differences in sources of municipal solid waste and operating conditions at the landfill. The three main gas constituents (methane, CH_4, carbon dioxide, CO_2, and hydrogen sulfide, H_2S) are used to characterize landfill gas.

Typically, landfill gas is composed of 45 to 60% v/v methane, 40 to 60% v/v carbon dioxide, 0 to 1.0% v/v hydrogen sulfide, 0 to 0.2% v/v hydrogen (H_2), trace amounts of nitrogen (N_2), low molecular weight hydrocarbons (dry volume basis) and water vapor (saturated). The specific gravity of landfill gas is approximately 1.02 to 1.06. The complete combustion of 1 m^3 of methane releases approximately 38 MJ, in comparison combustion of 1 m^3 of landfill gas releases 20 to 26 MJ/m^3 (dependent on the methane content)—this represents a low-value to medium-value fuel in comparison to methane. In addition, landfill gas requires a pressure of 34,450 kPa for liquefaction, whereas methane must be successively cooled to below −160°C (−256°F) for liquefaction.

When biogas is sent to processing for impurity removal, the cleaned gas has characteristics of the biogas are close to the characteristics of as natural gas. In this instance the producer of the biogas can utilize the local gas distribution networks. The gas must be very clean to reach pipeline quality. Water (H_2O), hydrogen sulfide (H_2S), and particulates are removed if present at high levels or if the gas is to be completely cleaned. Carbon dioxide is less frequently removed, but it must also be separated to achieve pipeline quality gas. If the gas is to be used without extensively cleaning, it is sometimes cofired with natural gas to improve combustion. Biogas cleaned up to pipeline quality is called renewable natural gas. In this form the gas can be used in any application in which natural gas is used. The heating value of the raw biogas is on the order of 600 Btu/ft^3 (5,340 $calories/m^3$). The raw gas will burn in an engine, but corrosion can occur unless proper materials of construction are selected.

Bio-oil

Bio-oil (sometimes called biocrude) is the liquid condensate produced from biomass (forestry residues, crop residues, waste paper, and organic waste) by means of processes such as pyrolysis and thermochemical conversion. Depending on the feedstock, the bio-oil could be compatible with existing refinery technology and can be converted into fuels, such as green diesel. Alternatively, the bio-oil may be produced in a biorefinery, which is analogous to conventional oil refineries and petrochemical complexes that have evolved over many years to maximize process synergies, energy integration and feedstock utilization to drive down production costs. In a manner similar to the petroleum refinery, a biorefinery would integrate a variety of conversion processes to produce multiple product streams such as fuels and other chemicals from biomass. In short, a biorefinery would combine the essential technologies to transform biomass feedstocks and other feedstocks into a range of industrially useful intermediates. However, the type of biorefinery would have to be differentiated by the character of the feedstock. For example, the crop biorefinery would use raw materials such as cereals or maize and the lignocellulose biorefinery would use raw material with high cellulose content, such as straw, wood, and paper waste.

In the hydrothermal upgrading process, biomass is treated with water at high temperature (300 to 350°C, 570 to 660°F) and pressure 1,770 to 2,650 psi) to produce bio-oil. This can be separated by flashing or extraction to heavy oil (suitable for co-combustion in coal power stations) and light oil, which can be upgraded by hydrodeoxygenation to biofuels. On the other hand, rapid thermal processing occurs at a temperature of about 500°C (930°F), when a turbulent stream of hot sand flashes the biomass into a vapor which is then rapidly condensed into a liquid. This process produces high yields of bio-oil (typically on the order of 65 to 75% w/w yield of pyrolysis oil from dried lignocellulosic biomass).

Bio-oil is a dark brownish viscous liquid that bears some resemblance to certain types of crude oil. However, bio-oil is a complex oxygenated compound comprised of water, water-soluble compounds, such as acids, esters, etc., and water-insoluble compounds, usually called pyrolytic lignin because it comes from the lignin fraction of the biomass. The elemental composition of bio-oil is like that of the parent biomass. Because of its high oxygen content, the heating value (Btu per gallon) of bio-oil is lower than fossil fuel, typically only about half the heating value of fossil crude such as heavy fuel oil. However, bio-oil contains less nitrogen and only traces of metals or sulfur.

Bio-oil is acidic with a pH in the 2-4 range, making it highly unstable and corrosive—the acidity can be lessened by the addition of readily available base compounds. It therefore, presents transportation/piping and storage challenges including the tendency to corrode most metals. Hence, it is usually transported and stored in stainless steel containers.

Like other biofuels, the properties of bio-oil are variable and dependent on the feedstock. The specific gravity of the liquid is about 1.10 to 1.25, which means it is slightly heavier than water, heavier than fuel oil, and significantly heavier than the bulk density of the original biomass. The viscosity of the bio-oil varies from as low as 25 cP up to 1,000 cP, depending on the water content of the bio-oil and the original feedstock from which the bio-oil was produced.

Refuse-Derived Fuels (RDFs)

Refuse derived fuels usually refers to waste material that has been converted to a fuel of consistent composition for commercial application. The fuel consists predominantly of combustible components of municipal waste such as plastics and biodegradable waste. The fuel is extracted from refuse using a mix of mechanical and/or biological treatment methods, the last of which is the fuel refining and treatment to produce a saleable and usable product.

Refuse disposal (solid waste disposal) is one of the most serious environmental problems facing countries and it is now an accepted environmental philosophy that wastes have value and should be utilized based on principles of reduce, reuse, recover, and recycle. Through recovery, domestic, urban, agricultural, crop and forestry wastes can be converted into useful products as solid char, gases and liquid fuels using technologies such as Fischer-Tropsch reactions, anaerobic digestion of organic waste and/or sewage producing biogas, and pyrolysis and gasification of wood, paper, plastic, scrap tires, light industrial waste, producing synthetic diesel or synthesis gas. Technologies to produce high-quality fuels from a variety of refuse streams provide an important alternative to biofuel feedstocks that sometimes competes with use as food and animal fodder. The renewable gases and liquid fuels are also hydrogen rich and a possible source of low-carbon fuels.

Although several sources of refuse exist, the issues lie in gathering the raw material, processing it into a usable form and composition, and delivering it to the point of combustion. Almost any carbonaceous material and a suitable binder can be converted to a form that can be fired into a pulverized fuel or a stoker boiler. Each application should be evaluated to eliminate carryover of low-density material, such as loose paper, and where in the boiler that combustion takes place.

Some installations are cofired with fossil fuels and most of the byproduct fuels (Table 5.1.15) might be used alone or with a binder. An ideal situation would be nearby waste streams from a major production facility. After mixing and extrusion, the pellets would be storable and transported only a short distance. If the pellet density is close to the normal boiler feed, the problems of separation during transportation and firing are reduced or eliminated. The problem of hazardous emissions remains if the waste streams are contaminated; if they are clean-burning, hazardous emissions might reduce the apparent cleanliness of the effluent.

References

1. Alleman T. *Biodiesel Handling and Use Guide*, 5th ed., U.S. Department of Energy, Office of Energy Efficiency and Renewable Energy, 2016. Report DOE/GO-102016-4865.

2. Chadeesingh R. The Fischer-Tropsch Process. *The Biofuels Handbook*, In: J.G. Speight (ed.). Royal Society of Chemistry, London, United Kingdom. Part 3, Chapter 5. Page 476-517, 2011.

3. Gupta R.B., and Demirbas A. *Gasoline, Diesel, and Ethanol Biofuels from Grasses and Plants*. Cambridge University Press, Cambridge, United Kingdom, 2010.

4. John E., and Singh K. Production of Fuels from Domestic and Industrial Waste. *The Biofuels Handbook*, In: J.G. Speight (ed.). Royal Society of Chemistry, London, United Kingdom. Part 3, Chapter 1, Page 333-376, 2011.

5. John E., and Singh K. Properties of Fuels from Domestic and Industrial Waste. *The Biofuels Handbook*, In: J.G. Speight (ed.). Royal Society of Chemistry, London, United Kingdom. Part 3, Chapter 2. Page 377-407, 2011.

6. Lee S., and Shah Y.T. *Biofuels and Bioenergy: Processes and Technologies*. CRC Press, Taylor & Francis Group, Boca Raton, Florida, 2013.

7. Lee S., Speight J.G., and Loyalka S. *Handbook of Alternative Fuel Technologies*, 2nd ed., CRC Press, Taylor & Francis Group, Boca Raton, Florida, 2014.

8. Mokhatab S., Poe W.A., and Speight J.G. *Handbook of Natural Gas Transmission and Processing*, Elsevier, Amsterdam, Netherlands, 2006.

9. Singh R., and Biofuels N. *The Biofuels Handbook*, In: J.G. Speight (ed.). Royal Society of Chemistry, London, United Kingdom. Part 1, Chapter 5. Page 160-198, 2011.

10. Speight J.G. *Synthetic Fuels Handbook: Properties, Processes, and Performance*. McGraw-Hill, New York, 2008.

11. Speight J.G. A Biorefinery. *The Biofuels Handbook*. Royal Society of Chemistry, London, United Kingdom. Part 1, Chapter 4. Page 118-159, 2011.

12. Speight J.G. *Gasification of Unconventional Feedstocks*. Gulf Professional Publishing, Elsevier, Oxford, United Kingdom, 2014.

13. Speight J.G., and Singh K. *Environmental Management of Energy from Biofuels and Biofeedstocks*. Scrivener Publishing, Beverly, Massachusetts, 2014.

14. Speight J.G., and Islam M.R. *Peak Energy – Myth or Reality*. Scrivener Publishing, Beverly, Massachusetts, 2016.

15. U.S. Energy Information Administration. Monthly Biodiesel Production Report. Available: http://www.eia.gov/biofuels/biodiesel/production. Accessed Jan 27, 2017.

16. U.S. Energy Information Administration. Biomass Explained – Wood and Woodwaste. 2016. Available: http://www.eia.gov/energyexplained/?page=biomass_wood. Accessed Jan 27, 2017.

17. U.S. Environmental Protection Agency. Burn Wise. Available: https://www.epa.gov/burnwise. Accessed Jan 27, 2017.

18. Weiss L. et al. New York State Wood Heat Report: An Energy, Environmental, and Market Assessment. New York State Energy Research and Development Authority (NYSERDA) Report 15-26. Available: http://www.nyserda.ny.gov. Accessed Jan 27, 2017.

PETROLEUM AND PETROLEUM PRODUCTS

Petroleum (also called *crude oil*) occurs throughout the world, and commercial fields have been located on every continent. Petroleum resides together with water, and sometimes free gas (*gas cap*), in very small holes (pore spaces) and fractures. The size, shape, and degree of interconnection of the pores vary considerably from place to place in an individual reservoir. The anatomy of a reservoir is complex, both microscopically and macroscopically and, because of the various types of accumulations and the existence of wide ranges of both rock and fluid properties, reservoirs respond differently and must be treated individually.

The Petroleum Family

Petroleum may be qualitatively described as brownish green to black liquids of specific gravity from approximately 0.810 to 0.985 and having a boiling range from approximately 20°C (68°F) to above 350°C (660°F), above which active decomposition ensues when distillation is attempted. Petroleum contains up to 35% v/v or more of components boiling in the gasoline range, as well as varying proportions of kerosene hydrocarbons and higher-boiling-point constituents up to the viscous and nonvolatile compounds present in lubricants and the asphalts. The composition of the petroleum obtained from the well is variable and depends on both the original composition of the petroleum in situ and the manner of production and stage reached in the life of the well or reservoir.

The term *petroleum* covers a wide assortment of materials consisting of mixtures of hydrocarbons and other compounds containing variable amounts of sulfur, nitrogen, and oxygen, which may vary widely in API gravity and sulfur content (Table 5.1.16) as well as viscosity and the amount of residuum (the portion of crude oil boiling above 510°C, 950°F). Metal-containing constituents, notably those compounds that contain vanadium and nickel, usually occur in the more viscous crude oils in amounts up to several thousand parts per million and can have serious consequences during processing of these feedstocks. Thus, because petroleum is a mixture of widely varying constituents and proportions, its physical properties also vary widely and the color from colorless to black and there are different types of petroleum in the *petroleum family*.

Other members of the petroleum family that may invoke the need for enhanced recovery methods, including the application of hydraulic fracturing techniques (Speight, 2015a) and which are worthy of mention here include: (i) crude oil from tight formations, (ii) foamy oil, (iii) heavy oil, (iv) high acid crude oil, and (v) opportunity crude oil.

Crude Oil from Tight Formations Generally, unconventional *tight oil* resources are found at considerable depths in sedimentary rock formations that are characterized by very low permeability. While some of the tight oil plays produce oil directly from shales, tight oil resources are also produced from low-permeability siltstone formations, sandstone formations, and carbonate formations that occur in close association with a shale source rock. Tight formations scattered thought North America have the potential to produce crude oil (*tight oil*). Such formations might be composted of shale sediments or sandstone sediments.

In a conventional sandstone reservoir, the pores are interconnected so gas and oil can flow easily from the rock to a wellbore. In tight sandstones, the pores are smaller and are poorly connected by very narrow capillaries which results in low permeability—typically the effective permeability of less than 1 milliDarcy (<1 mD). A shale play is a geographic area containing an organic-rich fine-grained sedimentary rock

Table 5.1.16 Examples of the API Gravity and Sulfur Content of Crude Oils from Various Countries

Country	Crude oil	API	Sulfur % w/w
Abu Dhabi (UAE)	Abu Al Bukhoosh	31.6	2.00
Abu Dhabi (UAE)	Zakum (Lower)	40.6	1.05
Algeria	Zarzaitine	43.0	0.07
Angola	Cabinda	31.7	0.17
Angola	Palanca	40.1	0.11
Australia	Barrow Island	37.3	0.05
Australia	Saladin	48.2	0.02
Australia	Skua	41.9	0.06
Brazil	Garoupa	30.0	0.68
Brazil	Sergipano Platforma	38.4	0.19
Brunei	Champion Export	23.9	0.12
Brunei	Seria	40.5	0.06
Cameroon	Kole Marine	32.6	0.33
Cameroon	Lokele	20.7	0.46
Canada (Alberta)	Rainbow	40.7	0.50
Canada (Alberta)	Wainwright-Kinsella	23.1	2.58
China	Nanhai Light	40.6	0.06
China	Shengli	24.2	1.00
Colombia	Cano Limon	29.3	0.51
Congo (Brazzaville)	Emeraude	23.6	0.600
Dubai (UAE)	Fateh	31.1	2.000
Dubai (UAE)	Margham Light	50.3	0.040
Ecuador	Oriente	29.2	0.880
Egypt	Gulf of Suez	31.9	1.520
Egypt	Ras Gharib	21.5	3.640
Gabon	Gamba	31.4	0.090
Gabon	Rabi-Kounga	33.5	0.070
Ghana	Salt Pond	37.4	0.097
India	Bombay High	39.2	0.150
Indonesia	Bima	21.1	0.250
Indonesia	Kakap	51.5	0.050
Indonesia	Katapa	50.8	0.060
Iran	Aboozar (Ardeshir)	26.9	2.480
Iran	Rostam	35.9	1.550
Iran	Salmon (Sassan)	33.9	1.910
Iraq	Basrah Heavy	24.7	3.500
Iraq	Basrah Light	33.7	1.950
Ivory Coast	Espoir	32.3	0.340
Kazakhstan	Kumkol	42.5	0.07
Kuwait	Kuwait Export	31.4	2.52
Libya	Bu Attifel	43.3	0.04
Libya	Buri	26.2	1.76
Malaysia	Bintulu	28.1	0.08
Malaysia	Dulang	39.0	0.12
Mexico	Maya	22.2	3.30
Mexico	Olmeca	39.8	0.80
Neutral Zone	Eocene	18.6	4.55
Neutral Zone	Hout	32.8	1.91
Nigeria	Bonny Medium	25.2	0.23
Nigeria	Brass River	42.8	0.06
North Sea (Denmark)	Danish North Sea	34.5	0.260
North Sea (Norway)	Ekofisk	39.2	0.169
North Sea (Norway)	Emerald	22.0	0.750
North Sea (Norway)	Oseberg	33.7	0.310
North Sea (UK)	Alba	20.0	1.330
North Sea (UK)	Innes	45.7	0.130
North Yemen	Alif	40.3	0.10
Oman	Oman Export	34.7	0.94
Papua New Guinea	Kubutu	44.0	0.04
Peru	Loreto Peruvian	33.1	0.23
Qatar	Dukhan (Qatar Land)	40.9	1.27
Qatar	Qatar Marine	36.0	1.42
Ras Al Khaiman (UAE)	Ras Al Khaiman	44.3	0.15
Russia	Siberian Light	37.8	0.42
Saudi Arabia	Arab Extra Light (Berri)	37.2	1.15
Saudi Arabia	Arab Heavy (Safaniya)	27.4	2.80
Sharjah (UAE)	Mubarek	37.0	0.62
Sumatra	Duri	20.3	0.21
Syria	Souedie	24.9	3.82
Timor Sea (Indonesia)	Hydra	37.5	0.08
Trinidad Tobago	Galeota Mix	32.8	0.27
Tunisia	Ashtart	30.0	0.99
USA (Alaska)	Alaskan North Slope	27.5	1.11
USA (Alaska)	Drift River	35.3	0.09
USA (California)	Huntington Beach	20.7	1.38
USA (Florida)	Sunniland	24.9	3.25
USA (Louisiana)	Lake Arthur	41.9	0.06
USA (Louisiana)	Ostrica	32.0	0.30
USA (Michigan)	Lakehead Sweet	47.0	0.31
USA (New Mexico)	New Mexico Intermediate	37.6	0.17
USA (New Mexico)	New Mexico Light	43.3	0.07
USA (Oklahoma)	Basin-Cushing Composite	34.0	1.95
USA (Texas)	Coastal B-2	32.2	0.22
USA (Texas)	West Texas Intermediate	40.8	0.34
USA (Wyoming)	Tom Brown	38.2	0.100
USA (Wyoming)	Wyoming Sweet	37.2	0.330
Venezuela	Oficina	33.3	0.780
Venezuela	Temblador	21.0	0.830
Vietnam	Bach Ho (White Tiger)	38.6	0.030
Vietnam	Dai Hung (Big Bear)	36.9	0.080
Yemen	Masila	30.5	0.670
Zaire	Zaire	31.7	0.130

that underwent physical and chemical compaction during diagenesis to produce the following characteristics: (i) clay to silt sized particles, (ii) high percentage of silica, and sometimes carbonate minerals, (iii) thermally mature, (iv) hydrocarbon-filled porosity—on the order of 6 to 14%, (v) low permeability—on the order of <0.1 mD, (vi) large areal distribution, and (vii) fracture stimulation required for economic production. In addition, tight oil formations are heterogeneous and vary widely over relatively short distances. Thus, even in a single horizontal production well, the amount of oil recovered may vary as may recovery within a field or even between adjacent wells. This makes evaluation of *shale plays* and decisions regarding the profitability of wells on a lease difficult. In addition, tight reservoirs which contain only crude oil (without natural gas as the pressurizing agent) cannot be economically produced.

The most notable tight oil plays in North America include the Bakken shale, the Niobrara formation, the Barnett shale, the Eagle Ford shale, and the Miocene Monterey play of California's San Joaquin Basin (California) and the Cardium play (Alberta, Canada). In many of these tight formations, the existence of large quantities of crude oil has been known for decades and efforts to commercially produce those resources have occurred sporadically with typically disappointing results. However, starting in the mid-2000s, advancements in well drilling and stimulation technologies combined with high oil prices have turned tight oil resources into one of the most actively explored and produced targets in North America. Other known tight formations (on a worldwide basis) include the R'Mah Formation (Syria), the Sargelu Formation (northern Persian Gulf region), the Athel Formation (Oman), the Bazhenov Formation and Achimov Formation (West Siberia, Russia), the Coober Pedy Formation (Australia), the Chicontepex Formation (Mexico), and the Vaca Muerta field (Argentina).

Typical of the crude oil from tight formations (*tight oil, tight light oil, and tight shale oil* have been suggested as alternate terms) is the Bakken crude oil which is a light highly volatile crude oil. Bakken crude oil is a light sweet (low-sulfur) crude oil that has a relatively high proportion of volatile constituents. The production of the oil also yields a significant amount of volatile gases (including propane, C_3H_8, and butane, C_4H_{10}) and low-boiling liquids (such as pentane, C_5H_{12}, and natural gasoline—a low-boiling naphtha fraction), which are often referred to collectively as (low-boiling or light) naphtha. Natural gasoline (sometime also referred to as *gas condensate*) is a mixture of low-boiling liquid hydrocarbons that is suitable for blending with the various refinery fraction that make up sales gasoline. Because of the presence of low-boiling hydrocarbons from the Bakken formation, the Bakken crude oil can become extremely explosive, even at relatively low ambient temperatures. Crude oil from tight shale formation is also characterized by low-asphaltene content, low-sulfur content and a significant molecular weight distribution of the paraffinic wax content—in addition to the low-boiling paraffins, paraffins having C_{10} to C_{60} chains have been found, with some tights crude oils containing carbon chains up to C_{72}.

Finally, the properties of crude oils from tight formations are highly variable. Density and other properties can show wide variation, even within the same field. The Bakken crude is light and sweet with an API of 42° and a sulfur content of 0.19% w/w. Similarly, Eagle Ford is a light sweet feed, with a sulfur content of approximately 0.1% w/w and with published API gravity between 40° and 62° API.

Foamy Oil Foamy oil is oil-continuous foam that contains dispersed gas bubbles produced at the well head from heavy oil reservoirs under solution gas drive. The nature of the gas dispersions in oil distinguishes foamy oil behavior from conventional heavy oil. The gas that comes out of solution in the reservoir does not coalesce into large gas bubbles nor into a continuous flowing gas phase. Instead it remains as small bubbles entrained in the crude oil, keeping the effective oil viscosity low while providing expansive energy that helps drive the oil toward the producing. Foamy oil accounts for unusually high production in heavy oil reservoirs under solution-gas drive. Thus, in some heavy oil reservoirs, due to the properties of the oil and the sand and due to the production methods, the released gas forms foam with the oil and remains subdivided in the form of dispersed bubbles much longer.

Reservoirs that exhibit foamy oil behavior are typically characterized by the appearance of an oil-continuous foam at the wellhead. When oil is produced as this non-equilibrium mixture, reservoirs can perform with higher than expected rates of production: up to 30 times that predicted by Darcy's Law, and lower than expected production gas-oil-ratios. Moreover, foamy oil flow is often accompanied by sand production along with the oil and gas—the presence of sand at the wellhead leads to sand dilation and the presence of high porosity, high permeability zones (wormholes) in the reservoir. It is generally believed that in the field, the high rates and recoveries observed are the combination of the foamy oil mechanism and the presence of the wormholes.

Heavy Oil Heavy oil is a *type* of petroleum that is different from conventional petroleum insofar as they are much more difficult to recover from the subsurface reservoir because the oil is viscous and does not flow easily. The common characteristic properties (relative to conventional crude oil) are high specific gravity, low hydrogen to carbon ratios, high carbon residues, and high contents of asphaltenes, heavy metal, sulfur, and nitrogen. Additional refining processes are required to produce more useful fractions, such as: naphtha, kerosene, and gas oil (Fig. 5.1.7).

There are large resources of *heavy oil* in Canada, Venezuela, Russia, the United States and many other countries. The resources in North America alone provide a small percentage of current oil production (approximately 2%), existing commercial technologies could allow for significantly increased production. Under current economic conditions,

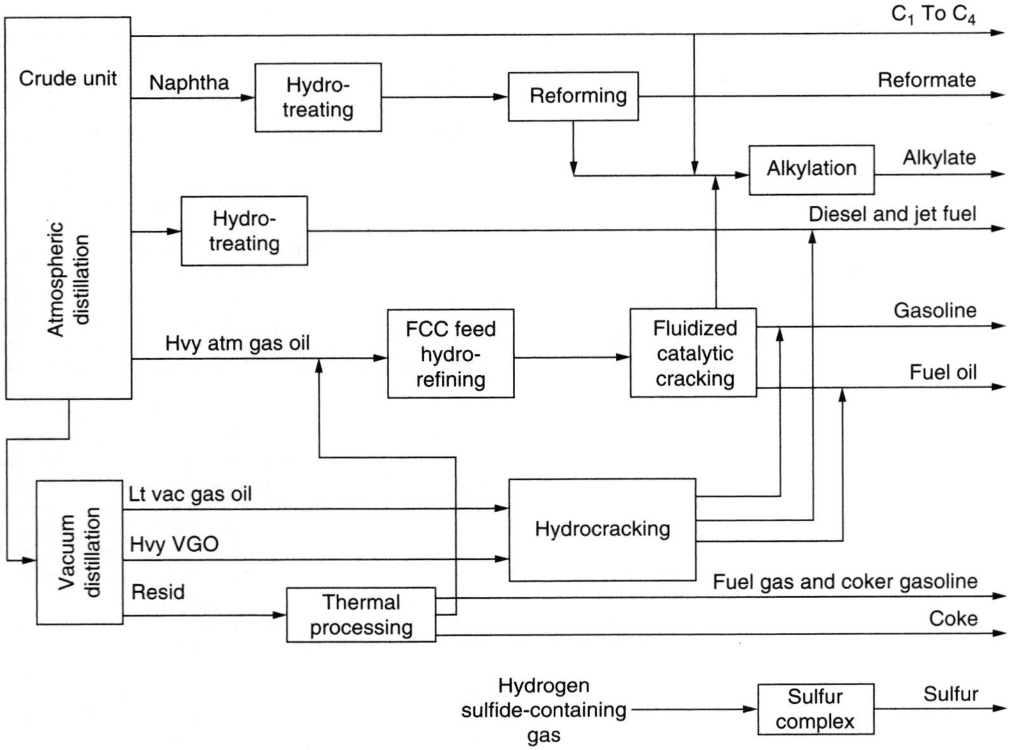

Figure 5.1.7 Schematic overview of a refinery.

Table 5.1.17 Simplified Differentiation between Conventional Crude Oil, Tight Oil, Heavy Crude Oil, Extra Heavy Crude Oil, and Tar Sand Bitumen*

Conventional Crude Oil
- Mobile in the reservoir; API gravity: >25°
- High-permeability reservoir
- Primary recovery
- Secondary recovery

Tight Oil
- Similar properties to the properties of conventional crude oil; API gravity: >25°
- Immobile in the reservoir
- Low-permeability reservoir
- Horizontal drilling into reservoir
- Fracturing (typically multifracturing) to release fluids/gases

Heavy Crude Oil
- More viscous than conventional crude oil; API gravity: 10-20°
- Mobile in the reservoir
- High-permeability reservoir
- Secondary recovery
- Tertiary recovery (enhanced oil recovery—EOR; e.g., steam stimulation)

Extra Heavy Crude Oil
- Similar properties to the properties of tar sand bitumen; API gravity: <10°
- Mobile in the reservoir
- High-permeability reservoir
- Secondary recovery
- Tertiary recovery (enhanced oil recovery—EOR; e.g., steam stimulation)

Tar Sand Bitumen
- Immobile in the deposit; API gravity: <10°
- High-permeability reservoir
- Mining (often preceded by explosive fracturing)
- Steam assisted gravity draining (SAGD)
- Solvent methods (VAPEX)
- Extreme heating methods
- Innovative methods**

* This list is not intended for use as a means of classification.

** Innovative methods exclude tertiary recovery methods and methods such as steam assisted gravity drainage (SAGD) and vapor assisted extraction (VAPEX) methods but do include variants or hybrids thereof.

heavy oil can be profitably produced, but at a smaller profit margin than for conventional oil, due to higher production costs and upgrading costs in conjunction with the lower market price for heavier crude oils. Furthermore, heavy oil accounts for more than double the resources of conventional oil in the world and heavy oil offers the potential to satisfy current and future oil demand.

The term *heavy oil* has also been arbitrarily (incorrectly) used to describe both the heavy oils that require thermal stimulation of recovery from the reservoir and the bitumen in bituminous sand (tar sand, called oil sand in Canada) formations from which the heavy bituminous material is recovered by a mining operation. *Extra heavy oil* is a non-descript term (related to viscosity) of little scientific meaning which is usually applied to tar sand bitumen, which is generally incapable of free flow under reservoir conditions. The general difference is that extra heavy oil, which may have properties similar to tar sand bitumen in the laboratory but, unlike tar sand bitumen in the deposit, has some degree of mobility in the reservoir or deposit (Table 5.1.17). Extra heavy oil can flow at the (typically) reservoir temperature and can be produced economically, without additional viscosity-reduction techniques, through variants of conventional processes such as long horizontal wells, or multilaterals. This is the case, for instance, in the Orinoco basin (Venezuela) or in offshore reservoirs of the coast of Brazil but, once outside of the influence of the high reservoir temperature, these oils are too viscous at surface to be transported through conventional pipelines and require heated pipelines for transportation. Alternatively, the oil must be partially upgraded or fully upgraded or diluted with a light hydrocarbon (such as aromatic naphtha) to create a mix that is suitable for transportation.

In the context of this section, the heavy oil has an API gravity less than 20 with a variable sulfur content (Table 5.1.18). However, it must be

Table 5.1.18 Examples of the API Gravity and Sulfur Content of Heavy Oils

Country	Crude oil	API	Sulfur % w/w
Brazil	Albacor Leste	18.9	0.66
Canada (Alberta)	Athabasca	8.0	4.8
Canada (Alberta)	Cold Lake	13.2	4.11
Canada (Alberta)	Lloydminster	16.0	2.60
Canada (Alberta)	Wabasca	19.6	3.90
Chad	Bolobo	16.8	0.14
Chad	Kome	18.5	0.20
China	Bozhong	16.7	0.30
China	Qinhuangdao	16.0	0.26
China	Zhao Dong	18.4	0.25
Colombia	Castilla	13.3	0.22
Colombia	Chichimene	19.8	1.12
Congo	Yombo	17.7	0.33
Ecuador	Ecuador Heavy	18.2	2.23
Ecuador	Napo	19.2	1.98
Guatemala	Xan-Coban	18.7	6.00
Indonesia	Kulin (South)	19.8	0.30
Iran	Soroosh (Cyrus)	18.1	3.30
Kuwait	Eocene	18.4	4.00
USA (California)	Arroyo Grande/Edna	14.9	2.03
USA (California)	Belridge (South)	13.7	1.00
USA (California)	Beta Offshore	15.9	3.60
USA (California)	Beta Offshore	16.9	3.30
USA (California)	Hondo Monterey	19.4	4.70
USA (California)	Hondo Monterey	17.2	4.70
USA (California)	Huntington Beach	19.4	2.00
USA (California)	Huntington Beach	14.4	0.90
USA (California)	Kern River	13.3	1.10
USA (California)	Kern River	14.4	1.02
USA (California	Lost Hills	18.4	1.00
USA (California)	Midway Sunset	12.6	1.60
USA (California)	Midway Sunset	11.0	1.55
USA (California)	Monterey	12.2	2.30
USA (California)	Mount Poso	16.0	0.70
USA (California)	Newport Beach	15.1	2.00
USA (California)	Point Arguello	19.5	3.50
USA (California)	Point Pedernales	15.9	5.10
USA (California)	San Ardo	12.2	2.30
USA (California)	San Joaquin Valley	15.7	1.20
USA (California)	Santa Maria	13.7	5.20
USA (California)	Sockeye	19.6	5.30
USA (California)	Sockeye	15.9	5.40
USA (California)	Torrance	18.2	1.80
USA (California)	Wilmington	18.6	1.59
USA (California)	Wilmington	16.9	1.70
USA (Mississippi)	Baxterville	16.3	3.02
Venezuela	Bachaquero	16.3	2.35
Venezuela	Bachaquero	12.2	2.80
Venezuela	Bachaquero	14.4	2.52
Venezuela	Bachaquero Heavy	10.7	2.78
Venezuela	Boscan	10.1	5.50
Venezuela	Hamaca	8.4	3.82
Venezuela	Jobo	9.2	4.10
Venezuela	Laguna	10.9	2.66
Venezuela	Lagunillas Heavy	17.0	2.19
Venezuela	Merey	18.0	2.28
Venezuela	Morichal	12.2	2.78
Venezuela	Pilon	14.1	1.91

Table 5.1.18 Examples of the API Gravity and Sulfur Content of Heavy Oils (*Continued*)

Country	Crude oil	API	Sulfur % w/w
Venezuela	Tia Juana Pesado	12.1	2.70
Venezuela	Tia Juana Heavy	18.2	2.24
Venezuela	Tia Juana Heavy	11.6	2.68
Venezuela	Tremblador	19.0	0.80
Venezuela	Zuata	15.7	2.69

For reference, Athabasca tar sand bitumen has API = 8° and sulfur content = 4.8-5.0% w/w (Table 5.1.19).

recognized that some of the heavy oil sufficiently liquid to be recovered by pumping operations and are already being recovered by this method. Refining depends on the characteristics (properties) since these heavy oils fall into a range of high viscosity (Fig. 5.1.8), which is the reason for the application of thermal conditions or dilution with a suitable solvent (such as aromatic naphtha) to enable heavy oil to flow in pipeline or within the refinery system.

High Acid Crude Oil High acid crude oils are crude oils that contain considerable proportions of naphthenic acids which, as commonly used in the petroleum industry, refers collectively to the organic acids present in the crude oil. A naphthenic acid is a monobasic carboxyl group attached to a saturated cycloaliphatic structure. However, it has been a convention accepted in the oil industry that all organic acids in crude oil are called naphthenic acids. Naphthenic acids in crude oils are now known to be mixtures of low to high molecular weight acids and the naphthenic acid fraction also contains other acidic species. In many instances, the high-acid crude oils are the heavier crude oils. The total acid matrix is therefore complex and it is unlikely that a simple titration, such as the traditional methods for measurement of the total acid number, can give meaningful results to use in predictions of problems.

High acid crude oils cause corrosion in the refinery—corrosion is predominant at temperatures more than 180°C (355°F) and occurs particularly in the atmospheric distillation unit (the first point of entry of the high-acid crude oil) and in the vacuum distillation units. Damage is in the form of unexpected high corrosion rates on alloys that would normally be expected to resist sulfidic corrosion (particularly steels with more than 9% Cr). In some cases, even very highly alloyed materials (i.e., 12% Cr, type 316 stainless steel (SS) and type 317 SS, and

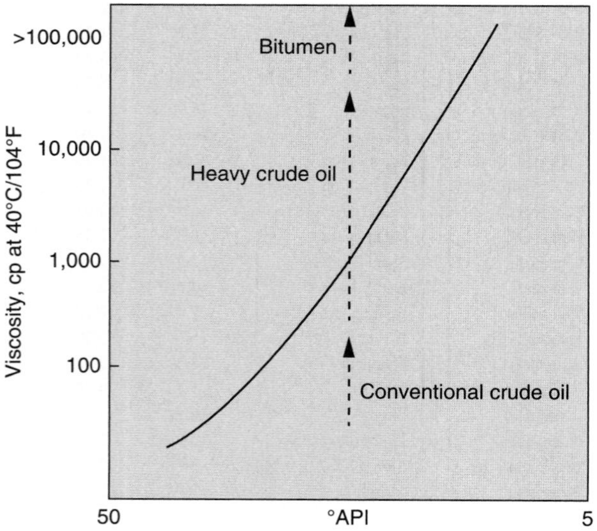

Figure 5.1.8 General relationship of viscosity to API gravity.

in severe cases even 6% Mo), stainless steel has been found to exhibit sensitivity to corrosion under these conditions.

The corrosion reaction processes involve the formation of iron naphthenates:

$$Fe + 2RCOOH \rightarrow Fe(RCOO)_2 + H_2$$

$$Fe(RCOO)_2 + H_2S \rightarrow FeS + 2RCOOH$$

The iron naphthenates are soluble in oil and the surface is relatively film free. In the presence of hydrogen sulfide, a sulfide film is formed which can offer some protection depending on the acid concentration. If the sulfur-containing compounds are reduced to hydrogen sulfide, the formation of a potentially protective layer of iron sulfide occurs on the unit walls and corrosion is reduced. When the reduction product is water, coming from the reduction of sulfoxides, the naphthenic acid corrosion is enhanced.

Thermal decarboxylation can occur during the distillation process during which the temperature of the crude oil in the distillation column can be as high as 400°C (750°F):

$$RCO_2H \rightarrow RH + CO_2$$

However, not all acidic species in petroleum are derivatives of carboxylic acids (–COOH) and some of the acidic species are resistant to high temperatures. For example, acidic species (such as phenol derivatives) appear in the vacuum residue after having been subjected to the inlet temperatures of an atmospheric distillation tower and a vacuum distillation tower. In addition, for the acid species that are volatile, naphthenic acids are most active at their boiling point and the most severe corrosion generally occurs on condensation from the vapor phase back to the liquid phase.

In addition, to taking preventative measure for the refinery to process these feedstocks without serious deleterious effects on the equipment, refiners need to develop programs for detailed and immediate feedstock evaluation so that they can understand the qualities of a crude oil very quickly and it can be valued appropriately and management of the crude processing can be planned meticulously.

Opportunity Crude Oils *Opportunity crude oils* are either new crude oils with unknown or poorly understood properties relating to processing issues or are existing crude oils with well-known properties and processing concerns. Opportunity crude oils are often, but not always, heavy crude oils but in either case are more difficult to process due to high levels of solids (and other contaminants) produced with the oil, high levels of acidity, and high viscosity. These crude oils may also be incompatible with other oils in the refinery feedstock blend and cause excessive equipment fouling when processed either in a blend or separately.

In addition, to taking preventative measure for the refinery to process these feedstocks without serious deleterious effects on the equipment, refiners need to develop programs for detailed and immediate feedstock evaluation so that they can understand the qualities of a crude oil very quickly and it can be valued appropriately and management of the crude processing can be planned meticulously. For example, the compatibility of opportunity crudes with other opportunity crudes and with conventional crude oil and heavy oil is a very important property to consider when making decisions regarding which crude to purchase. Blending crudes that are incompatible can lead to extensive fouling and processing difficulties due to unstable asphaltene constituents. These problems can quickly reduce the benefits of purchasing the opportunity crude in the first place. For example, extensive fouling in the crude preheat train may occur resulting in decreased energy efficiency, increased emissions of carbon dioxide, and increased frequency at which heat exchangers need to be cleaned. In a worst-case scenario, crude throughput may be reduced leading to significant financial losses.

Opportunity crude oils, while offering initial pricing advantages, may have composition problems which can cause severe problems at the refinery, harming infrastructure, yield, and profitability. Before refining, there is the need for comprehensive evaluations of and feedstocks composed of opportunity crude oils. This will assist the refiner to manage the ever-changing crude oil quality input to a refinery—including quality and quantity requirements and situations, crude oil variations, contractual specifications, and risks associated with such opportunity crudes.

Tar Sand Bitumen

In addition, to conventional petroleum and heavy crude oil, there remains an even more viscous material that offers some relief to the potential shortfalls in supply. This is the *bitumen* found in *tar sand* (called *oil sand* in Canada) deposits. Tar sands, also variously called oil sands or bituminous sands, are a loose-to-consolidated sandstone or a porous carbonate rock, impregnated with bitumen, a heavy asphaltic crude oil with an extremely high viscosity under reservoir conditions. Many of these reserves are only available with some difficulty and optional refinery scenarios will be necessary for conversion of these materials to liquid products because of the substantial differences in character between conventional petroleum and tar sand bitumen (Table 5.1.19).

The term *tar sand bitumen* (also, on occasion referred to as *extra heavy oil* (which is dependent upon mobility in the deposit) and *native asphalt*, (an incorrect term) includes a wide variety of reddish brown to black materials of near solid to solid character that exist in nature either with no mineral impurity or with mineral matter contents that exceed 50% by weight. Bitumen is frequently found filling pores and crevices of sandstone, limestone, or argillaceous sediments, in which case the organic and associated mineral matrix is known as *rock asphalt*. Tar sand deposits occur throughout the world and the largest deposits occur in Alberta, Canada (the Athabasca, Wabasca, Cold Lake, and Peace River areas), and in Venezuela. Smaller deposits occur in the United States, with the larger individual deposits in Utah, California, New Mexico, and Kentucky.

Table 5.1.19 Comparison of the Properties of Crude Oil and Tar Sand Bitumen

Property	Bitumen	Conventional
Gravity, °API	8.6	25-37
Distillation, °F		
Vol%		
IBP	—	
5	430	
10	560	
30	820	
50	1,010	< 650
Viscoity		
SUS, I 00°F (38°C)	35,000	< 30
SUS, 210°F (99°C)	513	
Pour point, °F	+50	
Elemental analysis, wt %		
Carbon	83.1	86
Hydroen	10.6	13.5
Sulfur	4.8	0.1-2.0
Nitrogen	0.4	0.2
Oxygen	1.1	
Hydrocarbon type, wt %		
Asphaltenes	19	<10
Resins	32	
Oils	49	>60
Metals, ppm		
Vanadium	250	
Nickel	100	
Iron	75	2-10
Copper	5	
Ash, wt %	0.75	0
Conradson carbon, wt %	13.5	1-2
Net heating value, Btu/lb	17,500	~19,500

The most appropriate definition of *tar sands* is found in the writings of the United States government, viz.:

> *Tar sands are the several rock types that contain an extremely viscous hydrocarbon which is not recoverable in its natural state by conventional oil well production methods including currently used enhanced recovery techniques. The hydrocarbon-bearing rocks are variously known as bitumen-rocks oil, impregnated rocks, oil sands, and rock asphalt.*

This definition presents an indication of the character of the bitumen by references to the method of recovery. Thus, the bitumen found in tar sand deposits is an extremely viscous material that is *immobile under reservoir conditions* and cannot be recovered through a well by the application of secondary or enhanced recovery techniques. Mining methods match the requirements of this definition (since mining is not one of the specified recovery methods) and the bitumen can be recovered by alteration of the natural state (such as thermal conversion) to a product that is then recovered. In this sense, changing the natural state (the chemical composition) as occurs during several thermal processes (such as some in-situ combustion processes) also matches the requirements of the definition. By inference and by omission, conventional petroleum and heavy oil are also included in this definition. Petroleum is the material that can be recovered by conventional production methods whereas heavy oil is the material that can be recovered by enhanced recovery methods.

It is incorrect to refer to native bituminous materials as *tar* or *pitch*. Although the word tar is descriptive of the black, heavy bituminous material, it is best to avoid its use with respect to natural materials and to restrict the meaning to the volatile or near-volatile products produced in the destructive distillation of such organic substances as coal (Speight, 2013a). In the simplest sense, pitch is the distillation residue of the various types of tar. Thus, alternative names, such as *bituminous sand* or *oil sand*, are gradually finding usage, with the former name (bituminous sands) more technically correct. The term *oil sand* is also used in the same way as the term *tar sand*, and these terms are used interchangeably throughout this text.

The properties and composition of the tar sand and the bitumen contained in the tar sand significantly influence the selection of recovery and the bitumen conversion processes and vary among the bitumen from different deposits. In the so-called wet sands or water-wet sands of the Athabasca deposit, a layer of water surrounds the sand grain, and the bitumen partially fills the voids between the wet grains. Utah tar sands lack the water layer; the bitumen is directly in contact with the sand grains without any intervening water; such tar sands are sometimes referred to as oil-wet sands. Typically, more than 99% w/w of mineral matter is composed of quartz and clays—the latter are detrimental to most refining processes. High concentrations of heteroatoms (nitrogen, oxygen, sulfur, and metals) tend to increase viscosity, increase the bonding of bitumen with minerals, reduce yields, and make processing more difficult.

The term *extra heavy oil* is a recently-evolved term (related to viscosity) of little scientific meaning. While this type of oil may resemble tar sand bitumen and does not flow easily, extra heavy oil is generally recognized as having mobility in the reservoir compared to tar sand bitumen, which is typically incapable of mobility (free flow) under reservoir conditions. For example, the tar sand bitumen located in Alberta Canada is not mobile in the deposit and requires extreme methods of recovery to recover the bitumen. On the other hand, much of the extra heavy oil located in the Orinoco basin of Venezuela requires recovery methods that are less extreme because of the mobility of the material in the reservoir. Whether the mobility of extra heavy oil is due to a high reservoir temperature (that is higher than the pour point of the extra heavy oil) or due to other factors is variable and subject to localized conditions in the reservoir.

Chemical and Physical Properties

The chemical and physical properties of petroleum (and the members of the petroleum family) vary considerably because of the variations in composition. The specific gravity of petroleum ranges from 0.8 (45.3 degrees API) for the lighter crude oils to over approximately 1 (ca. 10° API) for heavy oil. There is also considerable variation in

Table 5.1.20 Examples of Petroleum Products

Product	Lower carbon limit	Upper carbon limit	Lower boiling point, degrees C	Upper boiling point, degrees C	Lower boiling point, degrees F	Upper boiling point, degrees F
Refinery gas	C1	C4	–161	–1	–259	31
Liquefied petroleum gas	C3	C4	–42	–1	–44	31
Naphtha	C5	C17	36	302	97	575
Gasoline	C4	C12	–1	216	31	421
Kerosene/diesel fuel	C8	C18	126	258	302	575
Aviation turbine fuel	C8	C16	126	287	302	548
Fuel oil	C12	>C20	216	421	>343	>649
Lubricating oil	>C20		>343		>649	
Wax	C17	>C20	302	>343	575	>649
Asphalt	>C20		>343		>649	
Coke	>C50*		>1,000*		>1,832*	

* Carbon number and boiling point are estimates and used for illustrative purposes only.

viscosity; lighter crude oils range from 2 to 100 cSt, heavy oil typically has a viscosity on the order of several thousand centistokes. However, there is a tendency for the ultimate composition of petroleum to vary over narrower limits:

Carbon: 83.0 to 87.0% w/w;
Hydrogen: 10.0 to 14.0% w/w;
Nitrogen: 0.1 to 1.5% w/w;
Oxygen: 0.1 to 1.5% w/w;
Sulfur: 0.1 to 5.0% w/w;
Metals (nickel plus vanadium): 10 to 500 ppm w/w.

Physical Properties of Petroleum Products

Petroleum products in contrast to *petrochemicals*, are those bulk fractions that are derived from petroleum and have commercial value as a bulk product (Table 4.1.20). In the strictest sense, petrochemicals are also petroleum products but they are individual chemicals that are used as the basic building blocks of the chemical industry. Petroleum products are sold in the United States by barrels of 42 gal corrected to 60°F (Table 5.1.21) Their *specific gravity* is expressed on an arbitrary scale termed degrees API.

The *high heat value* (HHV) of petroleum products is determined by combustion in a bomb with oxygen under pressure. It may also be calculated, in products free from impurities, by the formula:

$$Q_v = 22{,}320 - 3{,}780d^2$$

Qv is the HHV at constant volume in Btu/lb and d is the specific gravity at 60/60°F from which the *low heat value* (LHV) at constant pressure Q_p may be calculated by the relation:

$$Q_p = Qv - 90.8H$$

H is the weight percentage of hydrogen and can be obtained from the equation:

$$H = 26 - 15d$$

Typical heats of combustion of petroleum and petroleum products that free from water, ash, and sulfur vary (within an estimated accuracy of 1 percent) with the API gravity (Table 5.1.22). The heat value should be corrected when the oil contains sulfur by using the HHV of 4,050 Btu/lb for sulfur.

The *specific heat*, c, of petroleum products of specific gravity d and at temperature t (°F) is given by the equation:

$$c = s0.388\ 1\ 0.00045td/2d$$

The *heat of evaporation*, L (Btu/lb) may be calculated from the equation:

$$L = (110.9 - 0.09t)/d$$

The heat of vaporization per gallon (measured at 60°F) is:

$$8.34Ld = 925 - 0.75t$$

This indicates that the heat of vaporization per gallon depends only on the temperature of vaporization t and varies over the range of 450 for the heavier products to 715 for gasoline (Table 5.1.23). These data have an estimated accuracy within ±10% when the vaporization is at constant temperature and at pressures below 50 lb/inch^2 and without any chemical change.

Properties and Specifications for Petroleum Products

Gasoline *Gasoline* is a complex mixture of hydrocarbons that distills within the range of 38 to 205°C (100 to 400°F). Commercial gasolines are blends of straight-run, cracked, reformed, and natural gasolines. On the other hand, *straight-run gasoline* is recovered from crude petroleum by distillation and contains a large proportion of normal hydrocarbons of the paraffin series. The octane number of straight run gasoline is too low for use in modern engines, and it is reformed and blended with other products to improve its combustion properties. *Cracked gasoline* is manufactured by heating crude-petroleum distillation fractions or residues under pressure, or by heating with or without pressure in the presence of a catalyst (fluid catalytic cracking), in which case the correct name is *FCC naphtha* or *cracked naphtha*. In the fluid catalytic cracking process, higher molecular weight (higher-boiling) hydrocarbons are decomposed into lower molecular weight (lower-boiling) products, some of which distill in the naphtha-gasoline range. The octane number of the cracked gasoline (cracked naphtha) is usually above the octane number of straight-run gasoline. *Reformed gasoline* is made by passing gasoline fractions over catalysts in such a manner that low-octane-number hydrocarbons are molecularly rearranged to high-octane-number components. Many of the catalysts use platinum and other metals deposited on a silica and/or alumina support. *Natural gasoline* is obtained from natural gas by liquefying those constituents which boil in the gasoline range either by compression and cooling or by absorption in oil. Natural gasoline is too volatile for general use, but proper characteristics can be secured by distillation or by blending. It is often blended with gasolines to adjust their volatility characteristics to meet climatic conditions. The specifications

Table 5.1.21 Coefficients of Expansion of Petroleum Products at 60°F

Deg API	<15	15-34.9	35-50.9	51-63.9	64-78.9	79-88.9	89-93.9	94-100
Coef of expansion	0.00035	0.0004	0.0005	0.0006	0.0007	0.0008	0.00085	0.0009

Table 5.1.22 Relationship of Heat Content to API Gravity

API at 60°F	Density, lb/gal*	High heat value (HHV) at constant volume Q_v, Btu Per lb	Per gal	Low heat value (LHV) at constant pressure Q_p, Btu Per lb	Per gal
10	8.337	18,540	154,600	17,540	146,200
20	7.787	19,020	148,100	17,930	139,600
30	7.305	19,420	141,800	18,250	133,300
40	6.879	19,750	135,800	18,510	127,300
50	6.5	20,020	130,100	18,720	121,700
60	6.16	20,260	124,800	18,900	116,400
70	5.855	20,460	119,800	19,020	112,500
80	5.578	20,630	115,100	19,180	107,000

* Btu/lb × 2.328 = kJ/kg; Btu/gal × 279 = kJM³.

for gasoline provide for various volatility classes, varying from low-volatility gasolines to minimize vapor lock to high-volatility gasoline that permits easier starting during cold weather.

The *antiknock character* of gasoline is an important property because engine power output and fuel economy are limited by the antiknock characteristics of the fuel. The antiknock index is currently defined as the average of the research octane number (RON) and motor octane number (MON). The *research octane number* is a measure of antiknock performance under mild operating conditions at low to medium engine speeds while the *motor octane number* is indicative of antiknock performance under more severe conditions, such as those encountered during power acceleration at relatively high engine speeds.

Reformulated motor gasoline is a product that is composed of various refinery streams that have been carefully blended to reduce the various environmental issues that arise from the use of automobiles. In fact, there has been a serious effort to produce reformulated gasoline components from a variety of processes (Table 5.1.24).

Diesel Fuel Diesel is a liquid product distilling over the range of 150 to 400°C (300 to 750°F). The carbon number ranges on average from approximately C_{13} to C_{21}. The chemical composition of a typical diesel fuel and how it applies to the individual specifications—API gravity, distillation range, freezing point, and flash point—are directly attributable to both the carbon number and the compound classes present in the finished fuel but these are subject to various country codes.

Aviation Gasoline Gasolines for aircraft piston engines have a narrower boiling range than motor gasolines; i.e., they have fewer low-boiling and fewer high-boiling components. *Jet fuels* or *aviation turbine fuels* are not limited by antiknock requirements, and they have wider boiling-point ranges to provide greater availability. Fuel JP-4 is a relatively wide-boiling-point range distillate that encompasses the boiling point range of gasoline and kerosene. The average initial boiling point is approximately 60°C (140°F), and the endpoint is approximately 235°C (455°F). Fuel JP-5 is a high-flash point kerosene type of fuel with an initial boiling point of approximately 175°C (346°F) and endpoint of approximately 255°C (490°F). There are a variety of grades and specifications for jet fuel because of its use as a commercial and a military fuel and the chemical composition of each fuel, API gravity, distillation range, freezing point, and flash point are attributable to both the carbon number and the compound types present in the finished fuel.

Table 5.1.23 Latent Heat of Vaporization of Petroleum Products

Product	Gravity, °API	Average boiling temp, °F	Heat of vaporization Btu/lb	Btu/gal
Gasoline	60	280	116	715
Naphtha	50	340	103	670
Kerosene	40	440	86	595
Fuel oil	30	580	67	490

Table 5.1.24 Production of Reformulated Gasoline Constituents

Process	Objective
Catalytic reformer prefractionation	Reduce benzene
Reformate fractionation	Reduce benzene
Isomerization	Increase octane
Aromatics saturation	Reduce total aromatics
Catalytic reforming	Oxygenate for octane enhancement
MTBE synthesis	Provide oxygenates
Isobutane dehydrogenation	Feedstock for oxygenate synthesis
Catalytic cracker naphtha fractionation	Increase alkylate
	Increase oxygenates
	Reduce olefins and sulfur
Feedstock hydrotreating	Reduce sulfur

Kerosene *Kerosene* is less volatile than gasoline and has a higher flash point, to provide greater safety in handling. Other quality tests are specific gravity, color, odor, distillation range, sulfur content, and burning quality. Most kerosene is used for heating, ranges, and illumination, so it is treated with sulfuric acid to reduce the content of aromatics, which burn with a smoky flame. Specification tests for quality control include flash point (minimum 46°C, 115°F), distillation endpoint (maximum 300°C, 572°F), sulfur, and color.

Diesel Fuel *Diesel fuel* for diesel engines requires variability in its performance since the engines range from small, high-speed engines used in trucks and buses to large, low-speed stationary engines for power plants. Thus, several grades of diesel fuel are needed for different classes of service (Table 5.1.25). The combustion characteristics of diesel fuels are expressed in terms of the cetane number, a measure of ignition delay. A short ignition delay, i.e., the time between injection and ignition, is desirable for a smooth-running engine. Some diesel fuels contain cetane improvers, which usually are alkyl nitrates. The cetane number is determined by an engine test or an approximate value (the *cetane index*) can be calculated for fuels that do not contain cetane improvers. In addition, to cetane improvers, the list includes antioxidants, corrosion inhibitors, and dispersants. The dispersants are added to prevent agglomeration of gum or sludge deposits so these deposits can pass through filters, injectors, and engine parts without plugging them.

Gas-Turbine Fuel Gas turbine fuel is a catch-all term used to describe a variety of duels that are designed for engines that accept most commercial fuels, such as natural gas, propane, diesel fuel, and kerosene as well as renewable fuels such as E85 (typically, an ethanol fuel blend of 85% v/v denatured ethanol fuel and 15% v/v gasoline), biodiesel (diesel fuel produced from biomass), and biogas (methane-rich gas produced from biomass). However, when running on kerosene or diesel, starting sometimes requires the assistance of a more volatile product such as propane gas.

Table 5.1.25 Illustration of the Various Grades and Grades of Diesel Fuel

Grade	Use
1-D	A volatile distillate fuel for engines in service requiring frequent speed and load changes
2-D	A distillate fuel of lower volatility for engines in industrial and heavy mobile service
4-D	A fuel for low- and medium-speed engines
Type*	
C-B	Diesel fuel oils for city bus and similar operations
T-T	Fuels for diesel engines in trucks, tractors, and similar service
R-R	Fuels for railroad diesel engines
S-M	Heavy-distillate and residual fuels for large stationary and marine diesel engines

* An additional guide to fuel selection is the grouping of fuels by the type of service.

Table 5.1.26 Grades of Turbine Fuels

Grade	Use
O-GT	Naphtha or other low-flash-point hydrocarbon liquid which also includes jet B fuel
1-GT	Volatile distillate for gas turbines that requires a fuel that burns cleaner than grade 2-GT; corresponds in physical properties to no. 1 burner fuel and grade 1-D diesel fuel
2-GT	Distillate fuel of low ash and medium volatility, suitable for turbines not requiring grade 1-GT; corresponds in physical properties to no. 2 burner fuel and grade 2-D diesel fuel
3-GT*	Low-volatility, low-ash fuel that may contain residual components
4-GT*	Low-volatility fuel containing residual components and having higher vanadium content than grade 3-GT

* The viscosity ranges of grades 3-GT and 4-GT bracket the viscosity ranges of no. 4, no. 5 light, no. 5 heavy, and no. 6 burner fuels.

The specification for a gas turbine fuel requires that the fuel should be a homogeneous mixture of hydrocarbons free of inorganic acid, and free of excessive amounts of solid or fibrous foreign matter likely to make frequent cleaning of suitable strainers necessary. All grades containing residual components shall remain homogeneous in normal storage and not separated by gravity into light and heavy oil components outside the viscosity limits for the grade which include properties such as: flash point test, pour point test, water and sediment test, carbon residue test, ash test, distillation test, viscosity test, density test, and sulfur test. Five grades of gas-turbine fuels are specified by the type of service and engine (Table 5.1.26).

Fuel Oil Fuel oil used for domestic purposes or for small heating installations will have lower viscosities and lower sulfur content (Table 5.1.27). In large-scale industrial boilers, heavier-grade fuel oil is used but the sulfur content is by local (or national) environmental regulations. The sulfur content of fuel oil is limited and must meet the legal requirements of the locality in which they are to be used. The additional refinery processing needed by some residual fuels to meet low-sulfur-content regulations may lower the viscosity enough to cause the fuels to change the grade classification. The *flash point* is also because of safety considerations. Asphaltene content, carbon residue value, mineral matter (measured as combustion as), water content, and metal content requirements are included in some specifications.

The *pour point*, which is the lowest temperature at which the fuel will retain its fluidity, is limited in the various specifications to the local requirements and fuel-handling facilities. The upper limit is often on the order of 10°C (50°F), in warm climates the pour point specification may be somewhat higher. Another important specification requirement is the heat of combustion which is usually on the order of 10,000 cal/kg (gross) or 9,400 cal/kg (net). Blending of fuel oils to obtain lower viscosity values and to mix fuel oils of different chemical characteristics is a source of deposit and sludge formation in the vessel fuel systems. This incompatibility is mainly observed when high-asphaltene fuel oils are blended with diluents or other fuel oils of a paraffinic nature.

Table 5.1.27 Grades and Heat Content of Fuel Oil

Grade	Heating value, Btu/gal	Comments
Fuel Oil No. 1	132,900-137,000	Distillate oil intended for vaporizing pot-type burners and other domestic burners
Fuel Oil No. 2	137,000-141,800	Distillate oil for general-purpose domestic heating in burners not requiring No. 1 fuel oil
Fuel Oil No. 4	143,100-148,100	Industrial burners; preheating not usually required for handling or burning
Fuel Oil No. 5 (Light)	146,800-150,000	Preheating may be required depending upon climate and type of equipment
Fuel Oil No. 5 (Heavy)	149,400-152,000	Preheating may be required for burning and, in cold climates, may be required for handling
Fuel Oil No. 6	151,300-155,900	Bunker C; Preheating required for burning and handling

Gas Oil *Gas oil* is a general term applied to distillates boiling between kerosene and lubricating oil. The middle distillates refer to products from the middle boiling range of petroleum and include kerosene, diesel fuel, distillate fuel oil, and light gas oil. Waxy distillate and lower boiling lubricating oils are sometimes included in the middle distillates. The remainder of the crude oil includes the higher boiling lubricating oils, gas oil, and residuum (the non-volatile fraction of the crude oil). The name *gas oil* was derived from the initial use for making illuminating gas but it is now used as burner fuel, diesel fuel, and feedstock to a catalytic cracking unit. Thus, after physical separation into such constituents as light and heavy naphtha, kerosene, and light and heavy gas oils, selected petroleum fractions may be subjected to conversion processes, such as thermal cracking (coking) and catalytic cracking.

References

1. ASTM International. *Annual Book of Standards*, ASTM International, West Conshohocken, Pennsylvania, 2016.
2. Hsu C.S., and Robinson P.R. (eds.). *Practical Advances in Petroleum Processing*, Volume 1 and Volume 2, Springer Science, New York, 2006.
3. Speight J.G. *Crude Oil Assay Database*, Knovel, Elsevier, New York, 2012. Online version available at: http://www.knovel.com/web/portal/browse/display?_EXT_KNOVEL_ DISPLAY_bookid=5485&VerticalID=0. Last accessed November 5, 2015.
4. Speight J.G. *Shale Oil Production Processes*, Gulf Professional Publishing, Elsevier, Oxford, United Kingdom, 2012.
5. Speight J.G. *The Chemistry and Technology of Coal*, 3rd ed., CRC Press, Taylor & Francis Group, Boca Raton, Florida, 2013.
6. Speight J.G. *Heavy Oil Production Processes*, Gulf Professional Publishing, Elsevier, Oxford, United Kingdom, 2013.
7. Speight J.G. *Oil Sand Production Processes*, Gulf Professional Publishing, Elsevier, Oxford, United Kingdom, 2013.
8. Speight J.G. *Heavy and Extra Heavy Oil Upgrading Technologies*, Gulf Professional Publishing, Elsevier, Oxford, United Kingdom, 2013.
9. Speight J.G. *The Chemistry and Technology of Petroleum*, 5th ed., CRC Press, Taylor and Francis Group, Boca Raton, Florida, 2014.
10. Speight J.G. *High Acid Crudes*, Gulf Professional Publishing, Elsevier, Oxford, United Kingdom, 2014.
11. Speight J.G. *Oil and Gas Corrosion Prevention*, Gulf Professional Publishing, Elsevier, Oxford, United Kingdom, 2014.
12. Speight J.G. *Handbook of Hydraulic Fracturing*, John Wiley & Sons Inc., Hoboken, New Jersey, 2015.
13. Speight J.G. *Handbook of Petroleum Product Analysis*, 2nd ed., John Wiley & Sons Inc., Hoboken, New Jersey, 2015.
14. Speight J.G. *Asphalt Materials Science and Technology*, Butterworth-Heinemann, Elsevier, Oxford, United Kingdom, 2105.
15. Speight J.G. *Introduction to Enhanced Recovery Methods for Heavy Oil and Tar Sands*, 2nd ed., Gulf Professional Publishing, Elsevier, Oxford, United Kingdom, 2016.
16. Speight J.G. *Handbook of Hydraulic Fracturing*, John Wiley & Sons Inc., Hoboken, New Jersey, 2016.
17. Speight J.G. *Deep Shale Oil and Gas*, Gulf Professional Publishing, Elsevier, Oxford, United Kingdom, 2017.
18. Speight J.G. *Handbook of Petroleum Refining*, CRC Press, Taylor and Francis Group, Boca Raton, Florida, 2017.
19. US EIA. *Review of Emerging Resources*, US Shale Gas and Shale Oil Plays. Energy Information Administration, United States Department of Energy, Washington, DC, 2011.
20. US EIA. *Technically Recoverable Shale Oil and Shale Gas Resources: An Assessment of 137 Shale Formations in 41 Countries outside the United States*, Energy Information Administration, United States Department of Energy, Washington, DC, 2013.

GASEOUS FUELS

Fuel gas is any one of several fuels that, at standard conditions of temperature and pressure, are gaseous. Fuel gases consist primarily of hydrocarbons with four or less carbon atoms (as well as lesser amounts of carbon monoxide and hydrogen) which have boiling points that are lower than room temperature and these products are gases at ambient temperature and pressure. For example, *natural gas* (predominantly *methane*, CH_4) is the lowest boiling and least complex of all hydrocarbons. Higher boiling hydrocarbons up to n-octane (C_8H_{18}) may also be present in which case the gas is referred to as *wet gas*, which is usually processed (refined) to remove the high-boiling hydrocarbons (when isolated, the higher boiling hydrocarbon mixture is often referred to as *natural gas condensate*. In fact, there are wells from which the predominant product is *condensate*.

There are several types of gases that are suitable for use (or designation as) fuel gases. These are (alphabetically): (i) coalbed methane, (ii) coal gas, (iii) liquefied petroleum gas, (iv) manufactured gas, (v) natural gas, (vi) shale gas, and (vii) synthesis gas.

Coalbed Methane

Coalbed methane (CBM) is predominantly methane extracted from coal beds (coal seams) and has become an important source of energy in United States, Canada, and other countries. The gas is adsorbed into the solid matrix of the coal and is often referred to as *sweet gas* because of its lack of hydrogen sulfide. Unlike natural gas from conventional reservoirs, coalbed methane contains very little higher molecular weight hydrocarbons such as propane or butane or condensate. It often contains up to a few percent carbon dioxide. Some coal seams contain little methane, with the predominant coal seam gas being carbon dioxide.

In fact, coalbed methane has been suggested with sufficient justification that the materials comprising a coalbed fall broadly into the following two categories: (i) volatile low-molecular-weight materials (components) that can be liberated from the coal by pressure reduction, mild heating, or solvent extraction and (ii) materials that will remain in the solid state after the separation of volatile components.

Techniques to detect and deal with coalbed methane in mines have ranged from (i) the classic canary in a cage to detect an oxygen-poor atmosphere, (ii) ventilation fans to force the replacement of a methane-rich environment with outside air, and (iii) drilling coalbed methane wells in front of the coal face to try to degas the coal prior to exposing the mine to the coalbed methane.

With the unique method of storage of coalbed methane, the preponderance of the gas is available only to very low coal-face pressures. The coal-face pressure is set by a combination of flowing wellhead pressure and the hydrostatic head exerted by standing liquid within the wellbore. Effective removal of higher molecular weight hydrocarbons and other constituents (*liquids removal* or *deliquefication*) techniques can reduce or remove the backpressure caused by accumulated liquid.

Coal Gas

Coal gas is any gaseous product that is produced by gasification of coal or by carbonization of coal. Most of these products have been replaced by natural gas (the city and town gas works are the industries of the past) but still find use in some parts of the world. The products of coal gasification are varied insofar as the gas composition varies with the system employed. It is emphasized that the gas product must be first freed from any pollutants such as particulate matter and sulfur compounds before further use, particularly when the intended use is a water gas shift or methanation.

Low heat-content gas (low-Btu gas) is produced during the gasification when the oxygen is not separated from the air and, thus, the gas product invariably has a low heat-content (ca. 150-300 Btu/ft^3). Low heat-content gas is also the usual product of in situ gasification of coal which is used essentially as a technique for obtaining energy from coal without the necessity of mining the coal. The process is a technique for utilization of coal which cannot be mined by other techniques. The nitrogen content of low heat-content gas ranges from somewhat less than 33% v/v to slightly more than 50% v/v and cannot be removed by any reasonable means; the presence of nitrogen at these levels renders the product gas to be low heat-content. The nitrogen also strongly limits the applicability of the gas to chemical synthesis. Two other noncombustible components (water, H_2O, and carbon dioxide, CO) further lower the heating value of the gas; water can be removed by condensation and carbon dioxide by relatively straightforward chemical means. The two major combustible components are hydrogen and carbon monoxide; the H_2/CO ratio varies from approximately 2:3 to approximately 3:2 but methane may also make an appreciable contribution to the heat content of the gas. Of the minor components hydrogen sulfide is the most significant and the amount produced is, in fact, proportional to the sulfur content of the feed coal. Any hydrogen sulfide present must be removed by one, or more, of several procedures.

Low heat-content gas is of interest to industry as a fuel gas or even, on occasion, as a raw material from which ammonia, methanol, and other compounds may be synthesized.

Medium heat-content gas (medium-Btu gas) has a heating value in the range 300-550 Btu/ft^3 and the composition is much like that of low heat-content gas, except that there is virtually no nitrogen. The primary combustible gases in medium heat-content gas are hydrogen and carbon monoxide.

Medium heat-content gas is considerably more versatile than low heat-content gas; like low heat-content gas, medium heat-content gas may be used directly as a fuel to raise steam, or used through a combined power cycle to drive a gas turbine, with the hot exhaust gases employed to raise steam, but medium heat-content gas is especially amenable to synthesize methane (by methanation), higher hydrocarbons (by Fischer-Tropsch synthesis), methanol, and a variety of synthetic chemicals. The reactions used to produce medium heat-content gas are the same as those employed for low heat-content gas synthesis, the major difference being the application of a nitrogen barrier (such as the use of pure oxygen) to keep diluent nitrogen out of the system.

In medium heat-content gas, the H_2/CO ratio varies from 2:3 to ca. 3:1 and the increased heating value correlates with higher methane and hydrogen contents as well as with lower carbon dioxide contents. Furthermore, the very nature of the gasification process used to produce the medium heat-content gas has a marked effect upon the ease of subsequent processing. For example, the CO_2-acceptor product is quite amenable to use for methane production because it has (i) the desired H_2/CO ratio just exceeding 3:1, (ii) an initially high methane content, and (iii) relatively low water and carbon dioxide contents. Other gases may require appreciable shift reaction and removal of large quantities of water and carbon dioxide prior to methanation.

High heat-content gas (high-Btu gas) is essentially pure methane and often referred to as *synthetic natural gas* or *substitute natural gas* (SNG). However, to qualify as SNG, a product must contain at least 95% methane; the energy content of synthetic natural gas is 980-1,080 Btu/ft^3. The commonly accepted approach to the synthesis of high heat-content gas is the catalytic reaction of hydrogen and carbon monoxide.

$$3H_2 + CO \rightarrow CH_4 + H_2O$$

To avoid catalyst poisoning, the feed gases for this reaction must be quite pure and, therefore, impurities in the product are rare. The large quantities of water produced are removed by condensation and recirculated as very pure water through the gasification system. The hydrogen is usually present in slight excess to ensure that the toxic carbon monoxide is reacted; this small quantity of hydrogen will lower the heat content to a small degree.

The carbon monoxide/hydrogen reaction is somewhat inefficient as a means of producing methane because the reaction liberates large quantities of heat. In addition, the methanation catalyst is troublesome and prone to poisoning by sulfur compounds and the decomposition of metals can destroy the catalyst. Thus, hydrogasification may be employed to minimize the need for methanation.

$$C_{coal} + 2H_2 \rightarrow CH_4$$

The product of hydrogasification is far from pure methane and additional methanation is required after hydrogen sulfide and other impurities are removed.

Liquefied Petroleum Gas

Liquefied petroleum gas (LPG) is the term applied to certain specific hydrocarbons and their mixtures, which exist in the gaseous state under atmospheric ambient conditions but can be converted to the liquid state under conditions of moderate pressure at ambient temperature. LPG is stored under pressure to keep the two hydrocarbons liquefied within the container. Before use, LPG passes through a pressure relief valve that causes a reduction in pressure and the liquid vaporizes (gasifies). Winter-grade LPG is mostly propane, the lower boiling of the two gases that is easier to vaporize at lower temperatures. Summer-grade LPG contains higher amounts of butane.

Thus, LPG is a hydrocarbon mixture usually containing propane ($CH_3CH_2CH_3$), *n*-butane ($CH_3CH_2CH_2CH_3$), isobutane [$CH_3CH(CH_3)CH_3$], and to a lesser extent propylene ($CH_3CH{=}CH_2$) or the butylene isomers ($CH_3CH_2CH{=}CH_2$, $CH_3CH{=}CH_2CH_3$). The most common commercial products are propane, butane, or some mixture of the two, and they are generally extracted from natural gas or crude petroleum. Propylene and the butylene isomers result from cracking other hydrocarbons in a petroleum refinery and are two important chemical feedstocks.

The presence of propylene and butylene isomers ($CH_3CH_2CH{=}CH_2$ and/or $CH_3CH{=}CHCH_3$) in LPG used as fuel gas is not critical. The vapor pressures of these olefins are slightly higher than those of propane and butane and the flame speed is substantially higher but this may be an advantage since the flame speeds of propane and butane are slow. However, one issue that often limits the proportion of olefins in LPG is the propensity of the olefins to form soot.

In addition, LPG is usually available in different grades (usually specified as: *commercial propane, commercial butane, commercial propane-butane* (P-B) *mixtures*, and *special duty propane*). During the use of LPG, the gas must vaporize completely and burn satisfactorily in the appliance without causing any corrosion or producing any deposits in the system.

Commercial propane consists predominantly of propane and/or propylene while *commercial butane* is mainly composed of butanes and/or butylenes. Both must be free from harmful amounts of toxic constituents and free from mechanically entrained water (that may be further limited by specifications). Analysis by gas chromatography is possible and often preferred.

Commercial Propane-Butane mixtures are produced to meet requirements such as volatility, vapor pressure, specific gravity, hydrocarbon composition, sulfur and its compounds, corrosion of copper, residues, and water content. These mixtures are used as fuels in areas and at times where low ambient temperatures are less frequently encountered. On the other hand, *special Duty Propane* is intended for use in spark-ignition engines and the specification includes a minimum *motor octane number* to ensure satisfactory antiknock performance. Propylene ($CH_3CH{=}CH_2$) has a significantly lower octane number than propane, so there is a limit to the amount of this component that can be tolerated in the mixture.

LPG and liquefied natural gas can share the facility of being stored and transported as a liquid and then vaporized and used as a gas. To achieve this, LPG must be maintained at a moderate pressure but at ambient temperature. The liquefied natural gas can be at ambient pressure but must be maintained at a temperature of roughly −1 to 60°C (30 to 140°F). In fact, in some applications it is economical and convenient to use LPG in the liquid phase. In such cases, certain aspects of gas composition (or quality such as the ratio of propane to butane and the presence of traces of heavier hydrocarbons, water and other extraneous materials) may be of lesser importance compared to the use of the gas in the vapor phase.

For normal use, the contaminants of LPG are controlled at a level at which they do not corrode fittings and appliances or impede the flow of the gas. For example, hydrogen sulfide (H_2S) and carbonyl sulfide (COS) should be absent. Organic sulfur to the level required for adequate odorization (*stenching*) is a normal requirement in LPG, dimethyl sulfide (CH_3SCH_3) and ethyl mercaptan (C_2H_5SH) are commonly used at a concentration of up to 50 ppm. Natural gas is similarly treated possibly with a wider range of volatile sulfur compounds. The presence of water in LPG (or in natural gas) is undesirable since it can produce hydrates that will cause, for example, line blockage due to the formation of hydrates under conditions where the water *dew point* is attained. If the amount of water is above acceptable levels, the addition of a small methanol will counteract any such effect.

In addition to other gases, LPG may also be contaminated by higher boiling constituents such as the constituents of middle distillates to lubricating oil. These contaminants become included in the gas during handling and must be prevented from reaching unacceptable levels. Olefins and especially diolefins are prone to polymerization and should be removed.

Manufactured Gas

Manufactured gas is a fuel-gas mixture made from other solid, liquid, or gaseous materials, such as coal, coke, oil, or natural gas. The principal types of manufactured gas are retort coal gas, coke oven gas, water gas, carbureted water gas, producer gas, oil gas, reformed natural gas, and reformed propane or LPG. Several processes for making SNG from coal have been developed.

Manufactured gas should not be confused with natural gas. Manufactured, or artificial, gas was produced from coal, coal and oil mixtures, or from petroleum. Almost without exception, in the U.S. manufactured gas was produced by means of three processes: (i) coal carbonization process, (ii) the carbureted water gas process, and (iii) oil gas process.

The *coal carbonization process* was the primary commercial mode of manufacturing gas from ca. 1816 to 1875. After 1875, newer processes and technologies gradually replaced coal carbonization. Coal gas was produced through the distillation of bituminous coal in heated, anaerobic vessels called retorts. In this process, coal was broken down into its volatile components through the action of heat in a nearly oxygen free environment. During the retorting phase, approximately two-fifths of the coal's weight was converted into volatile non-solids or gases. Most of the remaining amount of the coal was converted into solids, primarily coke. From the retort, the gases were drawn off into a device known as the hydraulic main where some of the vapors were converted to liquids and remained in gaseous state. The liquids, or liquors as they were often referred to, consisted of (contaminated) water and coal tar, and its associated wastes. The remaining vaporous material was coal gas. After exiting the hydraulic main, the gas ran through a condenser where the gas cooled and additional coal tar and other impurities were removed. Quite often, the gas was next funneled through a device known as an exhausted, which further cooled the gas. The coal gas, however, still contained impurities, primarily gaseous ammonia and sulfur compounds. These were removed by "washing" the gas in water and by running the gas through beds of moist lime or moist iron oxides. After this final purification process, the coal gas passed through the station meter, where it was measured, and on to gas storage holder. From the holder, the gas was distributed via street mains to the consumer.

The *carbureted water gas process* represented a technological advance that spurred on the growth of America's manufactured gas industry. At its core, the process consisted of enriching a form or coal gas, known as water gas (blue gas) and thus increasing its caloric/energy value. By injecting oil into a vessel containing heated water gas, the oil and vapor combined, forming a gaseous fuel with a thermal content of approximately 300-350 Btu/ft^3. Typically, a carbureted water gas plant consisted of a brick-lined, cylindrical, steel vessel, the generator, the carburetor (carbureted), and a super-heater. It was in the generator that the coal or coke was distilled to generate gaseous product. As the gas was drawn off from the generator and passed into the carburetor, the oil was introduced into the vapor.

The *oil gas process* is the only one of the three manufactured gas processes discussed here that did not use coal as a raw material. The first large-scale oil gas plant went on line in Oakland, CA, in 1902. The process as a whole was very similar to carbureted water gas process. The oil gas process consisted of thermo-cracking oil in a steam environment to produce the raw gas rather than distilling coal. The oil was heated and cracked in a vessel like the generator used in the carbureted water gas process. From there the gas passed to a vaporizer, where it was enriched with additional injections of oil and then routed through the super heater. After exiting the super heater, the gas was scrubbed and processed for distribution in much the same way as was carbureted water gas. Many of the same waste products associated with the production of coal gases, notably tars containing polynuclear aromatic hydrocarbons (PNAs, or polyaromatic hydrocarbons, PAHs) were also generated during gas manufacture.

Natural Gas

Natural gas is a combustible mixture of hydrocarbon gases. Within the natural gas family, the composition of *associated gas* (a by-product of oil production and the oil recovery process) is extremely variable, even within the gas from a petroleum reservoir. After the gas and fluids are brought to the surface, they are separated at a tank battery at or near the production lease into a hydrocarbon liquid stream (crude oil or natural gas condensate), a produced water stream (brine, which is water containing mineral salts), and a gas stream. Gas that occurs in reservoirs without the presence of crude oil is referred to as *nonassociated natural gas*. Typically, these reservoirs occur at greater depth. If the fluid gas brought to the surface remains gaseous, then it is referred to as *dry gas*. If the surface pressures cause some liquid hydrocarbons to evolve, it is referred to *wet gas*. However, while nonassociated gas reservoirs are likely to be found at greater depths, upward migration from the source rock over geologic time can result in shallow gas reservoirs.

Associated natural gas traditionally has high proportions of *natural gas liquids* (NGLs, such as ethane, propane, butane, and pentanes and higher molecular weight hydrocarbon constituents) and is referred to as *rich gas*. The higher molecular weight constituents (i.e., the C_5^+ product) are commonly referred to as *natural gasoline*. Rich gas will have a high heating value and a high hydrocarbon dew point. When referring to natural gas liquids in the gas stream, the term *gallon per thousand cubic feet* is used as a measure of high molecular weight hydrocarbon content. On the other hand, the composition of non-associated gas (sometimes called *well gas*) is deficient in natural gas liquids. The gas is produced from geological formations that typically do not contain much, if any, hydrocarbon liquids. As expected, the properties of associated gas and non-associated gas differ considerably when both gases are in the unrefined state but once refined to produce methane, the properties of the methane are evident.

The primary component of natural gas is methane (CH_4), the slowest molecular weight hydrocarbon molecule. Natural gas also contains higher molecular weight gaseous hydrocarbons such as ethane (C_{2H6}), propane (C_{3H8}), and butane (C_{4H10}), as well as higher molecular weight hydrocarbons and other sulfur-containing constituents in varying amounts (Table 5.1.28). The higher molecular weight hydrocarbons include pentane (C_5H_{12}) isomers, hexane (C_6H_{14}) isomers, heptane (C_7H_{16}) isomers, as well as trace amounts of octane (C_8H_{18}) isomers. Some aromatics [BTX – benzene (C_6H_6), toluene ($C_6H_5CH_3$), and the xylenes ($CH_3C_6H_4CH_3$)] can also be present, raising safety issues due to their toxicity.

The non-hydrocarbon gas portion of the natural gas contains nitrogen (N_2), carbon dioxide (CO_2), helium (He), hydrogen sulfide (H_2S), water vapor (H_2O), and other sulfur compounds (such as carbonyl sulfide (COS) and mercaptans (e.g., CH_3SH) and trace amounts of other gases). Carbon dioxide and hydrogen sulfide are commonly referred to as *acid gases* since they form corrosive compounds in the presence of water. Nitrogen, helium, and carbon dioxide are also referred to as *diluents* since none of these burn, and thus they have no heating value. Mercury can also be present either as the metal in vapor phase or as an organometallic compound in liquid fractions. Concentration levels are generally very small, but even at very small concentration levels, mercury can be detrimental due its toxicity and its corrosive properties (reaction with aluminum alloys).

Table 5.1.28 Variation of Composition of Natural Gas

	% v/v
Methane	70 to 90
Ethane	4 to 12
Propane	0 to 4
Butane	0.1 to 1.0
Pentane	0.1 to 0.2
Carbon dioxide	0 to 8
Oxygen	0 to 0.2
Nitrogen	0 to 5
Hydrogen sulfide	0 to 5
Rare gases	trace

Generally, the hydrocarbon constituents having a higher molecular weight than methane, carbon dioxide, and hydrogen sulfide are removed from natural gas prior to its use as a fuel. Organic sulfur compounds and hydrogen sulfide are common contaminants that must be removed prior to most uses. Gas with a significant amount of sulfur impurities, such as hydrogen sulfide, is termed *sour gas* and often referred to as *acid gas*. Processed natural gas that is available to end-users is tasteless and odorless. However, before gas is distributed to end-users, it is odorized by adding small amounts of thiols, to assist in leak detection.

Mixed gas is a gas prepared by adding natural gas or LPG to a manufactured gas, giving a product with better utility and higher heat content or Btu value.

The regular supply of gas during periods of peak demand or in emergencies is known as *peak shaving*. Most gas utilities, particularly natural-gas utilities at the end of long-distance transmission lines, maintain peak-shaving or standby equipment. Also, gas utilities in many cases have established natural-gas storage facilities close to their distribution systems. This allows gas to be stored underground near the point of consumption during periods of low demand, as in summer, and then withdrawn to meet peak or emergency demands, as may occur in winter. Propane-air mixtures are the major supplements to natural gas for peak shaving use. Gas manufactured by cracking various petroleum distillates supplies most of the remaining peak requirements.

Refinery Gas

The terms *refinery gas* or *petroleum gas* are often used to identify LPG or even gas that emanates from the top of a refinery distillation column. Refinery gas varies in composition and volume, depending on crude origin and on any additions to the crude made at the loading point—in this sense, it is probably the most difficult gas to analyze. Furthermore, it is not uncommon to reinject low molecular weight hydrocarbons such as propane and butane into the crude before dispatch by tanker or pipeline. This results in a higher vapor pressure of the crude, but it allows one to increase the quantity of light products obtained at the refinery. Since light ends in most petroleum markets command a premium, while in the oil field itself propane and butane may have to be reinjected or flared, the practice of *spiking* crude oil with LPG is becoming common.

Thus, *refinery gas* (*fuel gas*) is the non-condensable gas that is obtained during distillation of crude oil or treatment (cracking, thermal decomposition) of petroleum. Refinery gas is produced in considerable quantities during the different refining processes and is used as fuel for the refinery itself and as an important feedstock to produce petrochemicals. It consists mainly of hydrogen (H_2), methane (CH_4), ethane (C_2H_6), propane (C_3H_8), butane (C_4H_{10}), and olefins (RCH=CHR1, where R and R1 can be hydrogen or a methyl group) and may also include off-gases from petrochemical processes (Table 5.1.29). Olefins such as ethylene (CH_2=CH_2, boiling point: −104°C, −155°F), propene (propylene, CH_3CH=CH_2, boiling point: −47°C, −53°F), butene (butene-1, CH_3CH_2CH=CH_2, boiling point: −5°C, 23°F) iso-butylene ($(CH_3)_2C$=CH_2, −6°C, 21°F), cis- and trans-butene-2 (CH_3CH=$CHCH_3$, boiling point: ca. 1°C, 30°F) and butadiene (CH_2=$CHCH$=CH_2, boiling point: −4°C, 24°F) as well as higher boiling olefins are produced by various refining processes.

Still gas is broad terminology for low-boiling hydrocarbon mixtures and is the lowest boiling fraction isolated from a distillation (*still*) unit in the refinery. If the distillation unit is separating light hydrocarbon fractions, the still gas will be almost entirely methane with only traces of ethane (CH_3CH_3) and ethylene (CH_2=CH_2). If the distillation unit is handling higher boiling fractions, the still gas might also contain

Table 5.1.29 Possible Constituents of Refinery Gas

Gas	Molecular Weight	Boiling Point 1atm °C °F	Density at 60°F (15.6°C), 1atrn g/liter	Density at 60°F (15.6°C), 1atrn Relative to Air = 1
Methane	16.043	−161.5 (−258.7)	0.6786	0.5547
Ethylene	28.054	−103.7 (−154.7)	1.1949	0.9768
Ethane	30.068	−88.6 (−127.5)	1.2795	1.0460
Propylene	42.081	−47.7 (−53.9)	1.8052	1.4757
Propane	44.097	−42.1 (−43.8)	1.8917	1.5464
1,2-Butadiene	54.088	10.9 (51.6)	2.3451	1.9172
1,3-Butadiene	54.088	−4.4 (24.1)	2.3491	1.9203
1-Butene	56.108	−6.3 (20.7)	2.4442	1.9981
cis-2-Butene	56.108	3.7 (38.7)	2.4543	2.0063
trans-2-Butene	56.108	0.9 (33.6)	2.4543	2.0063
iso-Butene	56.104	−6.9 (19.6)	2.4442	1.9981
n-Butane	58.124	−0.5 (31.1)	2.5320	2.0698
iso-Butane	58.124	−11.7 (10.9)	2.5268	2.0656

propane ($CH_3CH_2CH_3$), butane ($CH_3CH_2CH_2CH_3$), and their respective isomers. *Fuel gas* and still gas are terms that are often used interchangeably but the term *fuel gas* is intended to denote the product's destination-to be used as a fuel for boilers, furnaces, or heaters.

In a series of processes commercialized under the names Platforming, Powerforming, Catforming, and Ultraforming, paraffinic and naphthenic (cyclic non-aromatic) hydrocarbons, in the presence of hydrogen and a catalyst are converted into aromatics, or isomerized to more highly branched hydrocarbons. Catalytic reforming processes thus not only result in the formation of a liquid product of higher octane number, but also produce substantial quantities of gases. The latter are rich in hydrogen, but also contain hydrocarbons from methane to butanes, with a preponderance of propane ($CH_3CH_2CH_3$), *n*-butane ($CH_3CH_2CH_2CH_3$) and isobutane [$(CH_3)_3CH$]. The composition of these gases varies in accordance with reforming severity and reformer feedstock. Since all catalytic reforming processes require substantial recycling of a hydrogen stream, it is normal to separate reformer gas into a propane ($CH_3CH_2CH_3$) and/or a butane [$CH_3CH_2CH_2CH_3/(CH_3)_3CH$] stream, which becomes part of the refinery LPG production, and a lighter gas fraction, part of which is recycled. In view of the excess of hydrogen in the gas, all products of catalytic reforming are saturated, and there are usually no olefin-type gases present in either gas stream.

A second group of refining operations that contributes to gas production is that of the catalytic cracking processes. Thermofor catalytic cracking, and other variants in which heavy gas oils are converted into cracked gas, LPG, catalytic naphtha, fuel oil, and coke by contacting the heavy hydrocarbon with the hot catalyst. Both catalytic and thermal cracking processes, the latter being now largely used to produce chemical raw materials, result in the formation of unsaturated hydrocarbons, particularly ethylene ($CH_2{=}CH_2$), but also propylene (propene, $CH_3CH{=}CH_2$), isobutylene [isobutene, $(CH_3)_2C{=}CH_2$] and the *n*-butenes ($CH_3CH_2CH{=}CH_2$, and $CH_3CH{=}CHCH_3$) in addition to hydrogen (H_2), methane (CH_4) and smaller quantities of ethane (CH_3CH_3), propane ($CH_3CH_2CH_3$), and butane isomers [$CH_3CH_2CH_2CH_3$, $(CH_3)_3CH$]. Diolefins, such as butadiene ($CH_2{=}CHCH{=}CH_2$), are also present.

Additional gases are produced in refineries with coking or visbreaking facilities for the processing of their heaviest crude fractions. In the visbreaking process, fuel oil is passed through externally fired tubes and undergoes liquid phase cracking reactions, which result in the formation of lighter fuel oil components. Oil viscosity is thereby reduced, and some gases, mainly hydrogen, methane, and ethane, are formed. Substantial quantities of both gas and carbon are also formed in coking (both fluid coking and delayed coking) in addition to the middle distillate and naphtha. When coking a residual fuel oil or heavy gas oil, the feedstock is preheated and contacted with hot carbon (coke) which causes extensive cracking of the feedstock constituents of higher molecular weight to produce lower molecular weight products ranging from methane, via liquefied petroleum gas(es) and naphtha, to gas oil and heating oil. Products from coking processes tend to be unsaturated and olefin-type components predominate in the tail gases from coking processes.

A further source of refinery gas is hydrocracking, a catalytic high-pressure pyrolysis process in the presence of fresh and recycled hydrogen. The feedstock is again heavy gas oil or residual fuel oil, and the process is mainly directed at the production of additional middle distillates and gasoline. Since hydrogen is to be recycled, the gases produced in this process again must be separated into lighter and heavier streams; any surplus recycle gas and the LPG from the hydrocracking process are both saturated.

Both hydrocracker and catalytic reformer tail gases are commonly used in catalytic desulfurization processes. In the latter, feedstocks ranging from light to vacuum gas oils are passed at pressures of 500-1,000 psi with hydrogen over a hydrofining catalyst. This results mainly in the conversion of organic sulfur compounds to hydrogen sulfide:

$$[S]_{feedstock} + H_2 \rightarrow H_2S + \text{hydrocarbons}$$

The process also produces some light hydrocarbons by hydrocracking.

The first and most important aspect of gaseous testing is the measurement of the volume of gas. However, in all cases, it is the composition of the gas in terms of *hydrocarbon type* that is more important in the context of the application. For example, in petrochemical applications, the presence of propylene and butylene above 10% v/v can have an adverse effect on hydrodesulfurization prior to steam reforming. On the other hand, petrochemical processes, such as in the production of *iso*-octane from *iso*-butane and butylene, can require the exclusion of the saturated hydrocarbons. Furthermore, refinery gas specifications will vary depending on the gas quality available and the end use. For fuel uses, gas as specified above presents little difficulty used as supplied. Alternatively, a gas of constant Wobbe Index, say for gas turbine use, could readily be produced by the user. Part of the combustion air would be diverted into the gas stream by a Wobbe Index controller. This would be set to supply gas at the lowest Wobbe Index of the undiluted gas.

Still gas is a term reserved for low-boiling hydrocarbon mixtures that represent the lowest boiling fraction isolated from an atmospheric distillation unit (*still*) in the refinery. Typically, the still gas will contain methane, ethane (CH_3CH_3), propane ($CH_3CH_2CH_3$), butane ($CH_3CH_2CH_2CH_3$), and their respective isomers. If cracked products have been recycled to the atmospheric distillation unit, the still gas may also contain volatile olefins such as ethylene ($CH_2{=}CH_2$).

Shale Gas

Shale gas is natural gas produced from shale formations (Table 5.1.30) that typically function as both the reservoir and source rocks for the natural gas. In terms of chemical makeup (Table 5.1.31), shale gas is typically a dry gas composed primarily of methane (60 to 95% v/v),

Table 5.1.30 Shale Gas Formations in the United States and Canada

Formation	Period	Location
Antrim Shale	Late Devonian	Michigan Basin, Michigan
Baxter Shale	Late Cretaceous	Vermillion Basin, Colorado, Wyoming
Barnett Shale	Mississippian	Fort Worth and Permian basins, Texas
Bend Shale	Pennsylvanian	Palo Duro Basin, Texas
Cane Creek Shale	Pennsylvanian	Paradox Basin, Utah
Caney Shale	Mississippian	Arkoma Basin, Oklahoma
Chattanooga Shale	Late Devonian	Alabama, Arkansas, Kentucky, Tennessee
Chimney Rock Shale	Pennsylvanian	Paradox Basin, Colorado, Utah
Cleveland Shale	Devonian	Eastern Kentucky
Clinton Shale	Early Silurian	Eastern Kentucky
Cody Shale	Cretaceous	Oklahoma, Texas
Colorado Shale	Cretaceous	Central Alberta, Saskatchewan
Conasauga Shale	Middle Cambrian	Black Warrior Basin, Alabama
Dunkirk Shale	Upper Devonian	Western New York
Duvernay Shale	Late Devonian	West Central Alberta
Eagle Ford Shale	Late Cretaceous	Maverick Basin, Texas
Ellsworth Shale	Late Devonian	Michigan Basin, Michigan
Excello Shale	Pennsylvanian	Kansas, Oklahoma
Exshaw Shale	Devonian-Mississippian	Alberta, Northeast British Columbia
Fayetteville Shale	Mississippian	Arkoma Basin, Arkansas
Fernie Shale	Jurassic	West central Alberta, Northeast British Columbia
Floyd/Neal Shale	Late Mississippian	Black Warrior Basin, Alabama, Mississippi
Frederick Brook Shale	Mississippian	New Brunswick, Nova Scotia
Gammon Shale	Late Cretaceous	Williston Basin, Montana
Gordondale Shale	Early Jurassic	Northeast British Columbia
Gothic Shale	Pennsylvanian	Paradox Basin, Colorado, Utah
Green River Shale	Eocene	Colorado, Utah
Haynesville/Bossier Shale	Late Jurassic	Louisiana, East Texas
Horn River Shale	Middle Devonian	Northeast British Columbia
Horton Bluff Shale	Early Mississippian	Nova Scotia
Hovenweep Shale	Pennsylvanian	Paradox Basin, Colorado, Utah
Huron Shale	Devonian	East Kentucky, Ohio, Virginia, West Virginia
Klua/Evie Shale	Middle Devonian	Northeast British Columbia
Lewis Shale	Late Cretaceous	Colorado, New Mexico
Mancos Shale	Cretaceous	San Juan Basin, New Mexico, Uinta Basin, Utah
Manning Canyon Shale	Mississippian	Central Utah
Marcellus Shale	Devonian	New York, Ohio, Pennsylvania, West Virginia
McClure Shale	Miocene	San Joaquin Basin, California
Monterey Shale	Miocene	Santa Maria Basin, California
Montney-Doig Shale	Triassic	Alberta, Northeast British Columbia
Moorefield Shale	Mississippian	Arkoma Basin, Arkansas
Mowry Shale	Cretaceous	Bighorn and Powder River basins, Wyoming
Muskwa Shale	Late Devonian	Northeast British Columbia
New Albany Shale	Devonian-Mississippian	Illinois Basin, Illinois, Indiana
Niobrara Shale	Late Cretaceous	Denver Basin, Colorado
Nordegg/Gordondale Shale	Late Jurassic	Alberta, Northeast British Columbia
Ohio Shale	Devonian	East Kentucky, Ohio, West Virginia
Pearsall Shale	Cretaceous	Maverick Basin, Texas
Percha Shale	Devonian-Mississippian	West Texas
Pierre Shale	Cretaceous	Raton Basin, Colorado
Poker Chip Shale	Jurassic	West Central Alberta, Northeast British Columbia
Queenston Shale	Ordovician	New York
Rhinestreet Shale	Devonian	Appalachian Basin
Second White Speckled Shale	Late Cretaceous	Southern Alberta
Sunbury Shale	Mississippian	Appalachian Basin
Utica Shale	Ordovician	New York, Ohio, Pennsylvania, West Virginia, Quebec
Wilrich/Buckinghorse/Garbutt/Moosebar Shale	Early Cretaceous	West Central Alberta, Northeast British Columbia
Woodford Shale	Devonian-Mississippian	Oklahoma, Texas

Table 5.1.31 Composition of Selected Unprocessed Shale Gas Samples

	(Volume %)			
Component	Marcellus	Appalachian	Haynesville	Eagle Ford
Methane	97.131	79.084	96.323	74.595
Ethane	2.441	17.705	1.084	13.824
Propane	0.095	0.566	0.205	5.425
C_4+	0.014	0.034	0.203	4.462
Hexanes+	0.001	0.000	0.061	0.478
Carbon Dioxide	0.040	0.073	1.816	1.536
Nitrogen	0.279	2.537	0.369	0.157
Totallnerts ($CO_2 + N_2$)	0.318	2.609	2.184	1.693
TOTAL	**100.0**	**100.0**	**100.1**	**100.5**
HHV (BTU/SCF)	1,031.6	1,133.2	1,009.8	1,307.1
Hydrocarbon Dew Point (°F)	–96.8	–41.3	9.7	119.6
Wobbe Number (BTU/SCF)	1,367.1	1,397.0	1,320.1	1,490.0

but some formations do produce wet gas. The Antrim and New Albany plays have typically produced water and gas. Gas shale formations are organic-rich shale formations that were previously regarded only as source rocks and seals for gas accumulating in the strata near sandstone and carbonate reservoirs of traditional onshore gas.

Gas production from unconventional shale gas reservoirs (such as tight shale formations) has become more common in the past decade. Produced shale gas observed to date has shown a broad variation in compositional makeup, with some having wider component ranges, a wider span of minimum and maximum heating values, and higher levels of water vapor and other substances than pipeline tariffs or purchase contracts may typically allow. Indeed, because of these variations in gas composition, each shale gas formation can have unique processing requirements for the produced shale gas to be marketable.

Ethane can be removed by cryogenic extraction while carbon dioxide can be removed through a scrubbing process. However, it is not always necessary (or practical) to process shale gas to make its composition identical to *conventional* transmission-quality gases. Instead, the gas should be interchangeable with other sources of natural gas now provided to end-users. The interchangeability of shale gas with conventional gases is crucial to its acceptability and eventual widespread use in the United States.

Although not highly sour in the usual sense of having high hydrogen sulfide content, and with considerable variation from play to resource to resource and even from well to well within the same resource (due to extremely low permeability of the shale even after fracturing), shale gas often contains varying amounts of hydrogen sulfide with wide variability in the carbon dioxide content. The gas is not ready for pipelining immediately after it has exited the shale formation.

The challenge in treating such gases is the low (or differing) hydrogen sulfide/carbon dioxide ratio and the need to meet pipeline specifications. In a traditional gas processing plant, the olamine of choice for content for hydrogen sulfide removal is N-methyldiethanolamine (MDEA) but whether this olamine will suffice to remove the hydrogen sulfide without removal of excessive amounts of carbon dioxide requires consistent monitoring.

Gas treatment may begin at the wellhead—condensates and free water usually are separated at the wellhead using mechanical separators, he observes. Gas, condensate and water are separated in the field separator and are directed to separate storage tanks and the gas flows to a gathering system. After the free water has been removed, the gas is still saturated with water vapor, and depending on the temperature and pressure of the gas stream, may need to be dehydrated or treated with methanol to prevent hydrates as the temperature drops. But this may not be always the case in actual practice.

Synthesis Gas

Synthesis gas is a mixture of carbon monoxide (CO) and hydrogen (H_2) that is used as a fuel gas but is also the beginning of a wide range of chemicals. The production of synthesis gas, that is, mixtures of carbon monoxide and hydrogen has been known for several centuries. But it is only with the commercialization of the Fischer-Tropsch reaction that the importance of synthesis gas has been realized.

Synthesis gas can be produced from a carbonaceous feedstock (such as a crude oil residuum, heavy oil, tar sand bitumen, and biomass) by gasification (partially oxidizing) the feedstock:

$$[2CH]_{feedstock} + O_2 \rightarrow 2CO + H_2$$

The initial partial oxidation step consists of the reaction of the feedstock with a quantity of oxygen insufficient to burn it completely, making a mixture consisting of carbon monoxide, carbon dioxide, hydrogen, and steam.

Success in partially oxidizing heavy feedstocks depends mainly on details of the burner design. The ratio of hydrogen to carbon monoxide in the product gas is a function of reaction temperature and stoichiometry and can be adjusted, if desired, by varying the ratio of carrier steam to oil fed to the unit.

Properties of Fuel Gases

Since the composition of natural, manufactured, and mixed gases can vary so widely, no single set of specifications could cover all situations. The requirements are usually based on performances in burners and equipment, on minimum heat content, and on maximum sulfur content. Gas utilities in most states come under the supervision of state commissions or regulatory bodies, and the utilities must provide a gas that is acceptable to all types of consumers and that will give satisfactory performance in all kinds of consuming equipment. However, particularly relevant are the heating values of the various fuel gases and their constituents (Table 5.1.32). For this reason, measurement of the properties of fuel gases is an important aspect of fuel gas technology.

Analysis The different methods for gas analysis include absorption, distillation, combustion, mass spectroscopy, infrared spectroscopy, and gas chromatography. Absorption methods involve absorbing individual constituents one at a time in suitable solvents and recording of contraction in volume measured. Distillation methods depend on the separation of constituents by fractional distillation and measurement of the volumes distilled. In combustion methods, certain combustible elements are caused to burn to carbon dioxide and water, and the volume changes are used to calculate composition. Infrared spectroscopy is useful applications and for the most accurate analyses, mass spectroscopy and gas chromatography are the preferred methods.

Flame Propagation The *rate of flame propagation* (also referred to as the *burning velocity*) in gas-air mixtures is of importance in utilization problems, including those dealing with burner design and rate of energy release. There are several methods that have been used

Table 5.1.32 Gross and Net Heating Values of Various Gases

	Gross Heating Value		Net Heating Value	
Gas*	(Btu/ft³)**	(Btu/lb)	(Btu/ft³)	(Btu/lb)
Butane	3,225	21,640	2,977	19,976
Butylene	3,077	20,780	2,876	19,420
Carbon monoxide	323	4,368	323	4,368
Coal gas	149	16,500		
Ethane	1,783	22,198	1,630	20,295
Ethylene	1,631	21,884	1,530	20,525
Hexane	4,667	20,526	4,315	18,976
Hydrogen	325	61,084	275	51,628
Hydrogen sulfide	672	7,479		
Methane	1,011	23,811	910	21,433
Natural Gas	950	19,500	850	17,500
Pentane	3,981	20,908	3,679	19,322
Propane	2,572	21,564	2,371	19,834
Propylene	2,336	21,042	2,185	19,683

* Listed alphabetically.
** 1 Btu/ft³ = 8.9 kcal/m.

Table 5.1.33: Flammability Limits of the Constituents of Fuel Gases

Gas*	LFL % v/v in air	UFL % v/v in air
n-Butane	1.6	8.4
Butylene (1-butene)	1.7	9.7
Carbon monoxide	12.5	74.2
Carbonyl sulfide	12.5	74
2,2-Dimethyl propane	1.4	7.5
Ethane	3.0	12.4
Ethylene	2.7	36
n-Heptane	1.0	7.0
n-Hexane	1.0	7.5
Hydrogen	4.0	75
Hydrogen sulfide	4.0	46
Isobutane	1.82	9.6
Methane	5.0	15
Methyl mercaptan	3.9	21.8
n-Octane	1.0	7.0
iso-Octane	0.6	6.0
n-Pentane	1.4	7.8
iso-Pentane	1.3	9.2
Propane	2.1	9.5
Propylene	2.0	11.1

* Listed alphabetically.

for measuring such burning velocities, in both laminar and turbulent flames. Results by the various methods do not agree, but any one method does give relative values of utility. Maximum burning velocities for turbulent flames are greater than those for laminar flames.

Flammability Before a fire or explosion can occur, three conditions must be met simultaneously: a fuel gas—a combustible gas—and oxygen (air) must exist in certain proportions, along with an ignition source, such as a spark or flame. The ratio of fuel and oxygen that is required varies with each combustible gas or vapor. The minimum concentration of a combustible gas (or vapor) necessary to support combustion of the gas in air is defined as the *lower explosive limit* (LEL) for that gas. Below this level, the mixture is too lean to burn. The maximum concentration of a gas (or vapor) that will burn in air is defined as the *upper explosive limit* (UEL). Above this level, the mixture is too rich to burn. The range between the LEL and UEL is known as the flammable range for that gas or vapor.

Thus, the lower and upper limits of flammability indicate the percentage of combustible gas in air below which and above which flame will not propagate (Table 5.1.33). When flame is initiated in mixtures having compositions within these limits, it will propagate and therefore the mixtures are flammable. A knowledge of flammable limits and their use in establishing safe practices in handling gaseous fuels is important; for example, when purging equipment used in gas service, in controlling factory or mine atmospheres, or in handling liquefied gases.

The calculation of flammable limits is accomplished by Le Chatelier's modification of the mixture law, which (expressed in the simplest form) is:

$$L = 100/(p_1/N_1 + p_2/N_2 \ldots + p_n N_n)$$

L is the volume percentage of fuel gas in a limited mixture of air and gas; $p_1, p_2, \ldots, p_n$ are the volume percentages of each combustible gas present in the fuel gas, calculated on an air- and inert-free basis so that $p_1 + p_2 + \ldots p_n = 100$; and $N_1, N_2, \ldots N_n$ are the volume percentages of each combustible gas in a limit mixture of the individual gas and air. The foregoing relation may be applied to gases with inert content of 10 percent or less without introducing an absolute error of more than 1 or 2 percent in the calculated limits.

Odorization Odorization is a technique that is applied for safety reasons to the largest consumer product—natural gas. Since natural gas as delivered to pipelines has practically no odor, the addition of an odorant is required by most regulations in order that the presence of the gas can be detected readily in case of accidents and leaks. This odorization is provided by the addition of trace amounts of some organic sulfur compounds to the gas before it reaches the consumer. The standard requirement is that a user will be able to detect the presence of the gas by odor when the concentration reaches 1 percent of gas in air. Since the lower limit of flammability of natural gas is approximately 5 percent, this 1 percent requirement is essentially equivalent to one-fifth the lower limit of flammability. The combustion of these trace amounts of odorant does not create any serious problems of sulfur content or toxicity.

Physical Constants The specific gravity of gases, including liquefied petroleum gases, may be determined conveniently by a number of methods and a variety of instruments. The heat value of gases is generally determined at constant pressure in a flow calorimeter in which the heat released by the combustion of a definite quantity of gas is absorbed by a measured quantity of water or air. A continuous recording calorimeter is available for measuring heat values of natural gas.

References

1. ASTM International. *Annual Book of Standards*, ASTM International, West Conshohocken, Pennsylvania, 2016.

2. Bahadori A. *Natural Gas Processing: Technology and Engineering Design*, Gulf Professional Publishing, Elsevier, Amsterdam, Netherlands, 2014.

3. Chadeesingh R. The Fischer-Tropsch Process. *The Biofuels Handbook*, In: J.G. Speight (ed.). The Royal Society of Chemistry, London, United Kingdom. Part 3, Chapter 5. Page 476-517, 2011.

4. Kidnay A.J., Parrish W.R., and McCartney D.G. *Fundamentals of Natural Gas Processing*, 2nd ed., CRC Press, Taylor & Francis Group, Boca Raton, Florida, 2011.

5. Mokhatab S., Poe W.A., and Speight J.G. *Handbook of Natural Gas Transmission and Processing*, Elsevier, Amsterdam, Netherlands, 2006.

6. Smil V. *Natural Gas: Fuel for the 21st Century*, John Wiley & Sons Inc., Hoboken, New Jersey, 2015.

7. Lee S., Speight J.G., and Loyalka S. *Handbook of Alternative Fuel Technologies*, 2nd ed., CRC Press, Taylor & Francis Group, Boca Raton, Florida, 2014.

8. Luque R., and Speight J.G. (ed.). *Gasification for Synthetic Fuel Production: Fundamentals, Processes, and Applications*, Woodhead Publishing, Elsevier, 2015.

9. NFPA. *National Fuel Gas Handbook*, 2015 ed., National Fire Protection Association, Quincy, Massachusetts, 2015.

10. Speight J.G. *Natural Gas: A Basic Handbook*, GPC Books, Gulf Publishing Company, Houston, Texas, 2007.

11. Speight J.G. *The Chemistry and Technology of Coal*, 3rd ed., CRC Press, Taylor & Francis Group, Boca Raton, Florida, 2013.

12. Speight J.G. *Shale Gas Production Processes*, Gulf Professional Publishing, Elsevier, Oxford, United Kingdom, 2103.

13. Speight J.G. *The Chemistry and Technology of Petroleum*, 5th ed., CRC Press, Taylor & Francis Group, Boca Raton, Florida, 2014.

14. Speight J.G. *Gasification of Unconventional Feedstocks*, Gulf Professional Publishing, Elsevier, Oxford, United Kingdom, 2014.

15. Speight J.G. *Handbook of Petroleum Product Analysis*, 2nd ed., John Wiley & Sons Inc., Hoboken, New Jersey, 2015.

16. Speight J.G. *Handbook of Hydraulic Fracturing*, John Wiley & Sons Inc., Hoboken, New Jersey, 2016.

17. Speight J.G. *Handbook of Petroleum Refining*, CRC Press, Taylor & Francis Group, Boca Raton, Florida, 2017.

SYNTHETIC FUELS

Synthetic (liquid and gaseous) fuels (alternative fuels) in commercial use can be produced from sources other than natural gas and crude oil. An alternative fuel or a synthetic fuel (*synfuel*) is defined by the context of the fuel use. In the context of substitutes for petroleum-based fuel, the term alternative fuel or synthetic fuel implies any available fuel or energy source and may also refer to a fuel derived from a renewable energy sources. Fuels such as natural gas and gasoline have an atomic hydrogen/carbon ratio on the order of 2.0 while the fuel sources have lower atomic hydrogen/carbon ratio (on the order of 1.0 to 1.5) (Table 5.1.34). The source of hydrogen to upgrade a synthetic fuel can be intramolecular in which a carbonaceous low-hydrogen residue (e.g., coke) is produced or intermolecular in which hydrogen is added from an external source.

The process of producing synfuels is often referred to as *coal-to-liquids* (CTL), *gas-to-liquids* (GTL) or *biomass-to-liquids* (BTL), depending on the initial feedstock.

Coal to liquids involves the thermal decomposition is one method to produce liquid fuels from coal, and it is the principal method used to convert oil shale and tar sand bitumen to liquid fuels. Moreover, as gasification and liquefaction are carried out at elevated temperatures pyrolysis may be considered a first stage in any conversion process. Thus, gaseous or liquid synthetic fuels are obtained by converting a carbonaceous material to a gaseous or liquid form, respectively. In the United States and many other countries, the most abundant naturally occurring materials suitable for this purpose are biomass, natural gas, coal, and oil shale. The bitumen from tar sand deposits are also suitable for synthetic fuel production and large deposits occur in Alberta (Canada).

Table 5.1.34 Elemental Composition and Ash Content (% w/w) (Dry Basis) (% w/w) of the Sources of Synthetic Fuels*

Fuel	C	H	O	N	S	Ash
Bark	47	6	47	0	0.1	3.9
Birch wood	49	6	44	1	0	0.5
Miscanthus	50	6	44	1	0	3.3
Pine wood	49	6	44	1	0	0.5
Reed grass	49	6	43	2	0	8.8
Sugar cane	50	6	44	1	0	3.7
Wheat straw	50	6	44	1	0	4.7
Peat**	53	6	38	1	0.2	5.6
Coal**	80	5	6.7	1	0.5	7

* Listed alphabetically rather than by preference.
** Included for comparison.

The production of synthetic fuels from biomass (*biomass to liquids*) usually (but not always) involves a degree of thermal conversion. In a very general sense, thermal decomposition is often used to mean liquid production by thermal decomposition but gaseous and solid product may also be produced. For example, *cracking* (*pyrolysis*) refers to the decomposition of organic matter by heat in the absence of air. Thermal decomposition is frequently used to mean the same, although it generally connotes the breakdown of inorganic compounds. The petroleum industry tends to use the words *cracking* and *coking* for the thermal decomposition of petroleum constituents.

Gas-to-liquids is a process for conversion of natural gas into longer-chain hydrocarbons. Thus, in the process, a methane-rich gas streams can be converted into liquid fuels either via direct conversion or via synthesis gas as an intermediate using the Fischer-Tropsch process. Using such processes, refineries can convert some of their gaseous waste products into valuable fuel oils, which can be sold as or blended only with diesel fuel. This process will be increasingly significant as crude oil resources are depleted, while natural gas supplies are projected to last into the twenty-second century.

Several factors do make synthetic fuels attractive relative to conventional petroleum-based fuels: (i) the raw material (coal and tar sand) is available in quantities sufficient to meet current demand for centuries, (ii) many of the raw materials can produce gasoline, diesel, or kerosene directly without the need for additional refining steps such as reforming or cracking, (iii) in some cases, there is no need to convert vehicle engines to use a different fuel, and (iv) the distribution network is already in place. However, the full potential of incineration is not being realized because of widespread fears regarding environmental pollution. Modern conversion plants with adequate environmental safeguards and careful monitoring have been shown to be a safe and cost effective technology that is likely to increase in importance during the next two-to-four decades.

Direct production of a gaseous fuel, a liquid fuel, or a solid fuel from the source is the obvious option. For example, gaseous and liquid fuels can be produced from (i) coal, (ii) biomass, (iii) oil shale, (iv) tar sand bitumen, and (v) the Fischer-Tropsch process. However, the gasification of various carbonaceous feedstocks can yield clean gases for combustion or synthesis gas with a controlled ratio of hydrogen to carbon monoxide and, in addition, liquid fuels can also be produced from these sources. The Fischer-Tropsch process is a different concept insofar as it is not the direct reduction of synthetic fuels from a source but involves the production of synthesis gas (a mixture of carbon monoxide and hydrogen) followed by the catalytic hydrogenation of carbon monoxide thereby providing a flexible method of synthetic fuel production.

Sources

Biomass The most important biomass energy sources are wood and wood wastes (Tables 5.1.35 and 5.1.36), agricultural crops and their waste byproducts, animal wastes, waste from food processing, and aquatic plants and algae. The average majority of biomass energy is produced from wood and wood wastes (64%), followed by municipal solid waste (24%), agricultural waste (5%), and landfill gases (5%). For countries where the cultivation of energy crops is difficult as synthetic fuel feedstocks, lignocellulosic materials are an attractive option to produce biofuels. Lignocellulosic materials serve as a cheap and abundant feedstock, which is required to produce fuel ethanol from renewable resources. Producing bioethanol from lignocellulosic materials may alleviate many of the environmental and food-versus-fuel concerns that are drawbacks of producing bioethanol from food crops like sugar or corn.

Biodiesel has been mainly produced from edible vegetable oils all over the world. More than 95% of global biodiesel production is made from edible vegetable oils. Biodiesel is also made from renewable oil crops such as soybean (U.S.), rapeseed and sunflower (Northern U.S., Canada, and Europe), oil palm (Malaysia and Indonesia), and jatropha (India). It can also be made from recycled vegetable oils and animal

Table 5.1.35 Major Categories of Biomass Feedstocks

Forest products	Wood; logging residues; trees; shrubs; and wood residues; sawdust; bark
Biorenewable wastes	Agricultural wastes, crop residues, mill wood wastes, urban wood wastes, urban organic wastes
Energy crops	Short rotation woody crops, herbaceous woody crops, grasses, starch crops, sugar crops, forage crops, oilseed crops, switchgrass, miscanthus
Aquatic plants	Algae, water weed, water hyacinth, reed and rushes
Food crops	Grains, oil crops
Sugar crops	Sugar cane, sugar beets, molasses, sorghum
Landfill	Hazardous waste, non-hazardous waste, inert waste, liquid waste
Organic wastes	Municipal solid waste, industrial organic wastes, municipal sewage and sludge
Algae	Prokaryotic algae, eukaryotic algae, kelps
Mosses	Bryophyta, polytrichales
Lichens	Crustose lichens, foliose lichens, fruticose lichen

fats. In the EU and the United States, the preparation of biodiesel is made from rapeseed oil, soybean oil, sunflower palm, and mustard seed oil. Rapeseed is the major feedstock and about 84% of global biodiesel is produced from rapeseed oil.

The benefits of biodiesel also depend on the type of oilseed used. Seeds of high oil content, such as sunflower (40 to 50% w/w oil), rapeseed (42 to 48% w/w oil), and soybean seeds (18 to 20% w/w oil) have gained much attention lately as renewable energy sources both because of their relatively high yield per hectare. Rapeseed produces about 435 liters of biodiesel per acre, but high-yield rapeseed can produce as much as 550 liters. Soybeans produce about 160 liters per acre. Sunflowers can produce 280 liters of biodiesel per acre, palm oil as much as 975 liters.

Microalgal biofuels are a viable alternative to biodiesel. Microalgae are photosynthetic microorganisms that convert sunlight, water and carbon dioxide to algal biomass. In fact, algae have the potential to dwarf all the other biodiesel feedstocks due to their efficiency in photosynthesizing solar energy into chemical energy in high yield (Table 5.1.37). Many algae are exceedingly rich in oil, which can be converted to biodiesel. The oil productivity of many microalgae exceeds the best producing oil crops. The oil content of some microalgae exceeds 80% of dry weight of algae biomass. Oil levels on the order of 20 to 50% w/w are quite common. Microalgae with high oil productivities are desired for producing biodiesel.

Coal Coal contains very little hydrogen (atomic ratio: approximately 0.5 to 0.8) and commercial transportation fuels contain 15 to 18% w/w hydrogen giving an atomic hydrogen-carbon ratio on the order of 1.6 to 2.0)—at the upper end of the atomic ratio scale is methane with a hydrogen/carbon atomic ratio of 4:1. In addition, coal also contains mineral matter and constituents such as oxygen, sulfur, and nitrogen, where hydrogen is consumed to form water, hydrogen sulfide, and ammonia, respectively. Some oxygen is removed as carbon dioxide.

The process objectives of coal liquefaction by direct methods are mainly focused on easing operating process severity, minimizing hydrogen requirement, and making the product liquid more environmentally acceptable. Due to the recent trends of high and fluctuating petroleum

Table 5.1.36 Composition of Different Biomass Types (% w/w, dry basis)

Type	Cellulose	Hemicellulose	Lignin	Others	Ash
Soft wood	41	24	28	2	0.4
Hard wood	39	35	20	3	0.3
Pine bark	34	16	34	14	2
Straw (wheat)	40	28	17	11	7
Rice husks	30	25	12	18	16
Peat*	10	32	44	11	6

* Included for comparison.

Table 5.1.37 Oil Yields from Common Energy Crops*

Energy crop	kg oil/hectare kg oil/2.47 acres
Coconut	2,260
Corn (maize)	145
Hemp	305
Jatropha	1,590
Microalgae	40,000-120,000
Mustard seed	481
Oil palm	5,000
Opium poppy	978
Peanuts	890
Rapeseed	1,000
Safflower	655
Soybean	375
Sunflowers	800

* Listed alphabetically rather than by preference.

price in the world market, the relative process economics of coal liquefaction is also much more favorable. Considering the vast amount of coal reserve throughout the world and its global distribution of major deposits, this alternative is even more attractive and also very practical.

The conversion of coal to liquid fuels (*liquefaction*) is accomplished by four principal methods: (1) direct catalytic hydrogenation, (2) solvent extraction, (3) indirect catalytic by hydrogenation (of carbon monoxide), and (4) pyrolysis. Several processes are available for each approach for adding hydrogen to the coal and removing undesirable constituents. The amount of hydrogen consumed is determined by the properties desired in the final product, whether it is (i) gases, (ii) naphtha, (iii) kerosene, and (iv) heavy fuel oil. Furthermore, upgrading a primary synthetic fuel (i.e., a fuel produced directly from the source) involves the addition of hydrogen and more hydrogen is required to upgrade coal than to process petroleum into finished products. The ranges are generalized to indicate the relative need for processing and the range of product distributions. Premium fuels are in the kerosene/gasoline range, which includes diesel and jet fuels.

In one approach to direct hydrogenation, coal and the catalyst are mixed with coal-derived recycle oil and the slurry is pumped into a high-pressure system where hydrogen is present. Operating conditions are generally on the order of 400 to 480°C and 1,500 to 5,000 psi. A high-boiling synthetic crude oil (also call *syncrude*) is produced at the milder conditions, and high yields of distillable oils are produced at the more severe conditions. Effective catalysts contain iron, molybdenum, cobalt, nickel, and tungsten. Many chemicals of commercial value are also found in the synthetic crude oil.

The *solvent extraction* process solubilizes and disperses coal in a hydroaromatic solvent that transfers some of its hydrogen to the coal. Ash and insoluble coal are separated from the liquid product to recover recycle oil and product oil. The carbonaceous residue is reacted with steam in a gasifier to produce the hydrogen needed. By recycling some of the mineral matter from the coal as a catalyst and increasing the severity of the operating conditions, a lower-boiling product (SRC II product) is formed. The SRC I product is a solid at room temperature.

Further improvement can be achieved by hydrogenation of the solvent to control the hydroaromatic content of the recycle stream and to improve the product quality. The Exxon *donor solvent process* (EDS Process) uses this latter technique, and no catalyst is required in the first contacting vessel. One version of the solvent extraction system is a two-stage hydrogenolysis. The integrated sequence starts with pulverized coal in a recycle solvent pumped into a reactor where the slurry is hydrogenated for a short time at 425 to 450°C (800 to 840°F) and 2,400 psi. The partially hydrogenated product is vacuum distilled to recover solvent and primary product. Solids are next separated by a critical solvent or an antisolvent process and the ash-free oil is hydrogenated at 400°C (750°F) and 2,700 psi to produce a final product boiling below 350°C (660°F) and any higher-boiling oil is recovered in the vacuum distillation step.

The *pyrolysis-based* process depends on the controlled thermal decomposition of coal without the addition of hydrogen. Most of the coal feedstock is recovered as solid product—gaseous and liquid products are volatilized from the reactor. The liquid product can be hydrotreated further for sulfur removal and upgrading to specification fuels. Byproduct gas and char must be utilized for the process to be economical. Yields of primary products depend on the coal source, the rate of heating, the ultimate temperature reached, and the atmosphere in which the reaction takes place. Both single and multistage processes have been developed but few are used commercially, except for the production of metallurgical coke and char.

Gasification of coal and petroleum residua is currently used on a wide scale to generate electricity. However, almost any type of organic material can be used as the raw material for gasification, such as biomass, wood or even waste plastic. Gasification processes are reasonably efficient and were used in the coal industry for many years to manufacture illuminating gas (coal gas) for gas lighting, before electric lighting became widely available. When synthesis gas contains a significant amount of nitrogen, the nitrogen must be removed. Cryogenic processing has great difficulty in recovering pure carbon monoxide when relatively large volumes of nitrogen are present due to carbon monoxide and nitrogen having very similar boiling points which are −191.5°C and −195.79°C, respectively. Instead process there is process technology selectively removes carbon monoxide by complexation/decomplexation of carbon monoxide in a proprietary solvent containing cuprous aluminum chloride ($CuAlCl_4$) dissolved in an organic liquid such as toluene. The purified carbon monoxide can have purity greater than 99%. The reject gas from the process can contain carbon dioxide, nitrogen, methane, ethane and hydrogen. The reject gas can be further processed on a pressure swing absorption system to remove hydrogen and the hydrogen and carbon dioxide recombined in the proper ratio for Fischer-Tropsch fuels and methanol.

The products from the gasification of coal may be of low, medium, or high heat-content (high-Btu) content as dictated by the process as well as by the ultimate use for the gas. Coal can also be converted into liquid fuels by indirect liquefaction which involves gasification of coal to mixtures of carbon monoxide and hydrogen (synthesis gas) followed by application of the Fischer-Tropsch process in which the synthesis gas is converted to hydrocarbons under catalytic conditions of temperature and pressure.

In the current context, the term *solid fuel* refers to various types of solid material that is used as fuel to produce energy and provide heating, usually released through combustion of the fuel. However, for the purposes of this section, coal itself is not included in this definition.

The most common coal-based solid fuel, coke, is a solid carbonaceous residue derived from low-ash, low-sulfur bituminous coal from which the volatile constituents are driven off by baking in an oven without oxygen at temperatures as high as 1,000°C (1,832°F) so that the fixed carbon and residual ash are fused together. Coke is used as a fuel and as a reducing agent in smelting iron ore in a blast furnace. Coke from coal is gray, hard, and porous and has a heating value of 24.8 million Btu/ton. Byproducts of this conversion of coal to coke include coal tar, ammonia, light oils, and coal gas. On the other hand, petroleum coke is the solid residue obtained during petroleum refining but at lower temperature, which resembles coal-based coke but contains too many impurities to be useful in metallurgical applications.

Next to combustion, carbonization to produce coke represents one of the most popular, and oldest, uses of coal. The thermal decomposition of coal on a commercial scale is often more commonly referred to as carbonization and is more usually achieved by the use of temperatures up to 1,500°C (2,730°F). The degradation of the coal is quite severe at these temperatures and produces (in addition to the desired coke) substantial amounts of gaseous products.

The original process of heating coal (in rounded heaps; the hearth process) remained the principal method of coke production for many centuries, although an improved oven in the form of a beehive shape was developed in the Newcastle area of England in about 1759. The method of coke production was initially the same as to produce charcoal, that is, stockpiling coal in round heaps (known as milers), igniting the piles, and then covering the sides with a clay-type soil. The smoke generated by the partial combustion of coal tars and gases was soon a major problem near residential areas, and the coking process itself could not be controlled because of climatic elements: rain, wind, and ice. It was this new modification of the charring-coking kiln which laid the foundation for the modern coke blast furnace: the beehive process.

Carbonization is essentially a process to produce a carbonaceous residue by thermal decomposition (with simultaneous removal of distillate) of organic substances.

$$[C]_{\text{organic carbon}} \rightarrow [C]_{\text{coke/char/carbon}} + \text{liquids} + \text{gases}$$

The process is a complex sequence of events which can be described in terms of several important physicochemical changes, such as the tendency of the coal to soften and flow when heated (plastic properties or the relationship to carbon-type in the coal). In fact, some coals become quite fluid at temperatures on the order of 400 to 500°C (750 to 930°F) and there is a considerable variation in the degree of maximum plasticity, the temperature of maximum plasticity, as well as the plasticity temperature range for various types of coal. The yields of tar and low molecular weight liquids are to some extent variable but are greatly dependent on the process parameters, especially temperature, as well as the type of coal.

Municipal Solid Waste *Municipal solid waste* (MSW) is a waste that includes predominantly household waste (domestic waste) with sometimes the addition of *commercial waste* collected by a municipality within a given area (Table 5.1.38). They are in either solid or semi-solid form and generally exclude industrial hazardous waste. The term *residual waste* relates to waste left from household sources containing materials that have not been separated out or sent for reprocessing.

The technologies used to recover energy from MSW include (i) waterwall incineration and (ii) modular incineration. *Waterwall incineration* involves the combustion of unprocessed municipal solid waste

Table 5.1.38 Various Sources and Types of Waste

Source	Facilities, Activities, or Locations	Types of Waste
Commercial	Stores, restaurants, markets, office buildings, hotels, motels, print shops, service stations, repair shops, etc.	Paper, cardboard, plastics, wood, waste and/or surplus lower grade biomass, food waste, glass, metals, waste edible oils, special wastes (consumer electronics, batteries, oil, and tyres), hazardous wastes, *etc.*
Construction and demolition	Construction sites, road repair/renovation sites, razing of buildings, broken pavement	Wood, steel, concrete, dirt, stones, bricks, plaster, shingles, plumbing parts, heating parts, electrical parts, asbestos, *etc.*
Institutional	Schools, hospitals, prisons, governmental centers	Same as Commercial
Municipal services	Street cleaning, landscaping, catch basin cleaning, parks, beaches, other recreational areas	Garbage, street sweepings, landscape and tree trimmings, catch basin debris, dead animals, abandoned vehicles, and general waste from parks, beaches, and recreational areas
Residential	Homes (single, multifamily), apartments (low, medium, high rise)	Food waste (animal, fruit, or vegetable residues), paper, cardboard, plastics, textiles, leather, rubber, furniture, crockery, yard wastes, wood, glass, tin cans, aluminum, other metals, waste edible oils, ashes, street leaves, special wastes (consumer electronics, batteries, oil, and tyres) household wastes, *etc.*
Treatment plant sites	Water, wastewater and industrial treatment processes, *etc.*	Plant waste, residual sludges

(mass burning) or processed municipal solid waste in a furnace with integral boiler tubes. Advantages of waterwall incineration are (i) pre-processing of waste is not required, except for large bulky items, (ii) the ash produced could be used for block manufacture, and (iii) auxiliary fuel is not required during normal steady state operation. Modular incineration is used in smaller plants and is an assembly of prefabricated major components assembled in the field to form a total operation.

Feedstock preparation plays an important role in any thermal waste conversion approach. *Refuse derived fuel* (RDF) is produced when municipal solid waste is processed by mechanical means picking, separation of inert materials (glass, minerals, ferrous and non-ferrous metals), flailing, screening, shredding, and conditioning to produce a more homogenous material prior to incineration. RDF systems utilize *pre-treated* municipal solid waste. There are several types of RDF: (i) coarse, (ii) fluff, (iii) powder, and (iv) densified which differ in complexity, power requirements, particle size and whether the material is compacted into pellets or briquettes. RDF can be used as the primary fuel source in a dedicated boiler or co-fired with natural gas or landfill gas. The heating value of nearly all solid waste fuels is a function of carbon content. The mineral matter content (measured as ash by combustion) is generally low, but the amount of moisture is highly variable and depends upon moisture generation plus the effects of processing, handling and storage.

Scrap tires can be processed into a waste derived fuel (WDF) and mixed with other fuels without major alteration to the existing combustion facility. Tire derived fuels (TDFs) are commonly used as supplemental fuel in cement kilns which are technologically and environmentally superior for selective waste disposal due to favorable internal kiln conditions as flame temperature (1,850°C, 3,360°F), residence time, alkaline environment, high gas turbulence and mixing which ensure complete combustion, and the added benefit of having no bottom ash disposal problem since the residual materials (e.g., steel from steel-belted tires) are incorporated into the clinker. The rationale behind the use of TDF is as follows: (i) excellent combustion characteristics and high heating value, (ii) lower cost and ready availability, (iii) relatively low sulfur content, and (iv) reducing the environmental burden and health effects of tire stockpiles.

Thus, agricultural, industrial and urban waste has considerable promise as a feedstock for gasification because it contains relatively more lignin, which biological processes cannot convert. Such waste is abundant in most developing countries and can be harnessed for production of fuels and petrochemical intermediates. Knowing the potential of the waste for gasification and subsequent fuel production is essential for reducing pressure on traditional energy sources.

Oil Shale *Oil shale* is a complex and intimate mixture of organic and inorganic materials that vary widely in composition and properties. In general terms, oil shale is a fine-grained sedimentary rock that is rich inorganic matter and yields oil when heated. Some oil shale is genuine shale but others have been mis-classified and are actually siltstones, impure limestone, or even impure coal. Oil shale does not contain oil and only produces oil when it is heated to about 500°C (about 932°F), when some of the organic material is transformed into a distillate similar to crude oil (Speight, 2008).

Oil shale is a fine-grained sedimentary rock containing relatively large amounts of organic matter (called *kerogen*), an organic sediment from which significant amounts of shale oil and combustible gas can be extracted by destructive distillation (Longwell, 1990; Scouten, 1990; Lee, 1991; Bartis et al., 2005).

Oil shale, or the kerogen contained therein, does not have definite geological definition or a specific chemical formula. Different types of oil shale vary by the chemical consist, type of kerogen, age, and depositional history, including the organisms from which they were derived. Furthermore, the term *oil shale* is a misnomer since the shale formation does not contain oil nor is it commonly shale. The organic material is chiefly *kerogen*—a naturally-occurring carbonaceous material of variable composition and indeterminate structure—and the *shale* is usually a relatively hard rock, called marl. Properly processed, kerogen can be converted into a substance somewhat similar to petroleum. However, the kerogen in oil shale has not gone through the *oil window* by which petroleum is produced and to be converted into a liquid hydrocarbon product, it must be heated to a high temperature. By this process the organic material is converted into a liquid, which must be further processed to produce oil which is said to be better than the lowest grade of oil produced from conventional oil deposits, but of lower quality than the upper grades of conventional oil.

Oil shale occurs in many parts of the world ranging from deposits of little or no economic value to those that occupy thousands of square miles and contain many billions of barrels of potentially extractable shale oil. Total world resources of oil shale are conservatively estimated at 2.6 trillion barrels (2.6×10^{12} bbls) of oil equivalent. With the continuing decline of petroleum supplies, accompanied by increasing costs of petroleum-based products, oil shale presents opportunities for supplying some of the fossil energy needs of the world in the years ahead.

In the United States, there are two principal types of oil shale types (i) Green River shale from the Green River Formation in Colorado, Utah, and Wyoming, and (ii) the Devonian-Mississippian black shale of the eastern and Midwestern states. The Green River shale is considerably richer in organic material, occurs in thicker seams, and therefore are the more likely to be exploited for synthetic fuel manufacture. In the Green River oil shale, the kerogen is not bound to the rock and the largest concentrations of kerogen are found in sedimentary non-reservoir rocks such as marlstone (a mix of carbonates, silicates, and clay). In contrast the black shale of the eastern and Midwestern states is true shale, insofar as it is composed predominantly of the illite.

The organic (kerogen) content of oil shale is much higher than those of normal and ordinary rocks, and typically range from 1 to 5 percent by mass (lean shale) to 15-20 percent by mass (rich shale). This natural resource is widely scattered in the entire world, and occurrences are scientifically closely linked to the history and geological evolution of the earth. Due to its abundance and wide distribution throughout the world, its utilization has a long history, both documented and undocumented. It is also obvious that the shale must have been relatively easy sources for domestic energy requirements for the ancient world, mainly due to the ease of handling and transportation, solid fuels were more convenient in the earlier human history and the examples are plentiful, including wood and coal.

There are two conventional approaches to oil shale processing. In one, the shale is fractured in situ and heated to obtain gases and liquids by wells. The second is by mining, transporting, and heating the shale to about 450°C (°F), adding hydrogen to the resulting product, and disposing of and stabilizing the waste. Both processes use considerable amounts of water. The total energy and water requirements together with environmental and monetary costs (to produce shale oil in significant quantities) have so far made production uneconomic. During and following the oil crisis of the 1970s, major oil companies, working on some of the richest oil shale deposits in the world in western United States, spent several billion dollars in various unsuccessful attempts to commercially extract shale oil.

Shale oil, sometimes termed *retort oil*, is the liquid oil condensed from the effluent in oil shale retorting and typically contains appreciable amounts of water and solids, as well as having an irrepressible tendency to form sediments. However, shale oils are sufficiently different from crude oil that processing shale oil presents some unusual problems. Just like the term *oil sand* (tar sand in the United States), the term *oil shale* is a misnomer since the mineral does not contain oil nor is it always *shale*. The organic material is chiefly *kerogen* and the *shale* is usually a relatively hard rock, called marl. Properly processed, kerogen can be converted into a substance somewhat similar to petroleum which is often better than the lowest grade of oil produced from conventional oil reservoirs but of lower quality than conventional light oil.

Retorting is the process of heating oil shale in order to recover the organic material, predominantly as a liquid. To achieve economically attractive recovery of product, temperatures of 400-600°C (750–1,100°F) are required. A retort is simply a vessel in which the oil shale is heated from which the product gases and vapors can escape to a collector. The amount of *shale oil* that can be recovered from a given deposit depends upon many factors. Some deposits or portions thereof, such as large areas of the Devonian black shale in eastern United States, may be too deeply buried to economically mine in the foreseeable future. Surface land uses may greatly restrict the availability of some oil shale deposits for development, especially those in the industrial western countries. The bottom line in developing a large oil shale industry will be governed by the price of petroleum. When the price of shale oil is comparable to that of crude oil because of diminishing resources of crude, then shale oil may find a place in the world fossil energy mix.

The active devolatilization of oil shale begins at about 350 to 400°C (660 to 750°F), with the peak rate of oil evolution at about 425°C (800°F), and with devolatilization essentially complete in the range of 470 to 500°C (880 to 930°F). At temperatures near 500°C (930°F), the mineral matter, consisting mainly of calcium/magnesium and calcium carbonates, begins to decompose yielding carbon dioxide as the principal product. The properties of crude shale oil are dependent on the retorting temperature, but more importantly on the temperature-time history because of the secondary reactions accompanying the evolution of the liquid and gaseous products. The produced shale oil is dark brown, odoriferous, and tending to waxy oil.

Tar Sand Tar sand is a common term for oil-impregnated sediments that can be found in almost every continent. High-grade tar sands have a porosity of 25 to 35 percent and contain about 18% w/w bitumen. The sand grains are wetted by about 2 percent of water, making them hydrophilic and thus more amenable to hot-water extraction. Solvent extraction, thermal retorting, in situ combustion, and steam injection methods have been tested.

Reserves of tar sands in the United States have been estimated to be on the order of 30 billion barrels (30×10^9 bbls) of petroleum equivalent. However, most of the deposits are too deep for surface mining and require in situ treatment before extraction. Some of the surface deposits are worked to produce asphalt for highway application and other minor uses. After mining (when surface mining is applied to recovery of the tar sand), the tar sand is transported to an extraction plant, where a hot water process separates the bitumen from sand, water, and minerals. The separation takes place in separation cells. Hot water is added to the sand, and the resulting slurry is piped to the extraction plant where it is agitated. The combination of hot water and agitation releases bitumen from the oil sand, and causes tiny air bubbles to attach to the bitumen droplets, that float to the top of the separation vessel, where the bitumen can be skimmed off. Further processing removes residual water and solids. The bitumen is then transported and converted to synthetic crude oil by thermal processes.

There are two coking processes that have been applied to the production of liquids from Athabasca bitumen. Delayed coking is practiced at the Suncor (formerly Great Canadian Oil Sands) plant, whereas Syncrude employs a fluid coking process which produces less coke than the delayed coking in exchange for more liquids and gases. In each case the charge is converted to distillate oils, coke, and light gases. The coke fraction and product gases can be used for plant fuel. The coker distillate is a partially upgraded material and is a suitable feed for hydrodesulfurization to produce a low-sulfur synthetic crude oil.

Sulfur is distributed throughout the boiling range of the delayed coker distillate, as with distillates from direct coking. Nitrogen is more heavily concentrated in the higher boiling fractions but is present in most of the distillate fractions. Raw coker naphtha contains significant quantities of olefins and diolefins that must be saturated by downstream hydrotreating. The gas oil has a high aromatic content typical of coker gas oils.

Catalytic hydrotreating is used for secondary upgrading to remove impurities and enhance the quality of the final synthetic crude oil product (Table 5.1.39). In a typical catalytic hydrotreating unit, the feedstock is mixed with hydrogen, preheated in a fired heater and then charged under high pressure to a fixed-bed catalytic reactor. Hydrotreating converts sulfur and nitrogen compounds present in the feedstock to hydrogen sulfide and ammonia. Sour gases from the hydrotreater(s) are treated for use as plant fuel. Hydrocracking may also be employed at this stage to improve product yields and quality. The synthetic crude oil is a blend of naphtha, distillate, and gas oil range materials, with no residuum (1,050°F+, 565°C+ material). The light, sweet synthetic crude marketed by Suncor today is called Suncor Oil Sands Blend A (OSA). Syncrude Canada Ltd. market a fully-hydrotreated blend utilizing fluidized-bed coking technology as the primary upgrading step. This product is referred to as Syncrude sweet blend (SSB).

Table 5.1.39 Properties of Tar Sand Bitumen, Synthetic Crude Oil, and Conventional Crude Oil

Property	Athabasca bitumen	Synthetic crude oil	Conventional crude oil
Specific gravity	1.03	0.85	0.85-0.90
Viscosity, centipoises:			
38°C/100°F	750,000	210	<200
100°C/212°F	11,300		
Pour point, °F	>50	−35	ca. −20
Elemental analysis (% w/w):			
Carbon	83.0	86.3	86.0
Hydrogen	10.6	13.4	13.5
Nitrogen	0.5	0.02	0.2
Oxygen	0.9	0.00	<0.5
Sulfur	4.9	0.03	<2.0
Ash	0.8	0.0	0.0
Nickel (ppm)	250	0.01	<10.0
Vanadium (ppm)	100	0.01	<10.0
Fractional composition (% w/w):			
Asphaltenes (pentane)	17.0	0.0	<10.0
Resins	34.0	0.0	<20.0
Aromatics	34.0	40.0	>30.0
Saturates	15.0	60.0	>30.0
Carbon residue (% w/w):			
Conradson	14.0	<0.5	<10.0

Process Parameters

When biomass, coal, oil shale, or tar sand bitumen are thermally decomposed, hydrogen-rich volatile matter is distilled and a carbon-rich solid residue is left behind. The carbon and mineral matter remaining behind is the residual char. In this regard, the term *carbonization* is sometimes used as a synonym for coal pyrolysis. However, carbonization has as its aim the production of a solid char, whereas in synthetic fuel production greatest interest centers on liquid and gaseous hydrocarbons. Assuming all process parameters are equal, the *composition of the raw material* is important in determining the yield of distillable products. The principal material property defining the yield is the atomic hydrogen-to-carbon ratio (derived from the elemental analysis). On the other hand, the composition of the volatile products evolved during thermal decomposition is largely determined by the raw organic material.

The *reaction temperature* affects both the amount and composition of the volatile yields. When a fuel produces volatile products, the residence time of the products within the hot zone and the temperature of the hot zone can markedly affect the distortion of the final products. Secondary and tertiary products will be formed form the primary products and to use the delayed coking process as an analogy, the products may undergo several thermal alternations before stabilization in the final molecular form.

Pressure will also affect the yield of distillable products since there is also a relationship between pressure and residence time. Generally, higher pressures favor cracking reactions and produce higher yields of lower molecular weight hydrocarbon gases, whereas lower pressure will lead to larger tar and oil fractions. When hydrogen pressure is employed (*hydrocracking* or *hydropyrolysis*), the yield of distillable product can be expected to increase because of the attendant hydrogenation diminishes the tendency for the formation of higher molecular weight products (tar and coke).

For the solid feedstocks, *particle size* is known to influence product yield. Larger particles heat up more slowly, so the average particle temperatures will be lower, and hence volatile yields may be expected to be less. In addition, the time taken for the thermal products to diffuse out of the larger particle (i.e., longer residence time in the hot zone) also contributes to product distribution. However, if the particle size is sufficiently small, the feedstock will be heated relatively uniformly and, with rapid diffusion of the products form the hot zone, a different product slate can be anticipated.

Fischer-Tropsch Process

The Fischer-Tropsch process is a different concept insofar as it is not the direct reduction of synthetic fuels from a source but involves the catalytic hydrogenation of carbon monoxide and the process is a flexible method of synthetic fuel production. The process commences with the gasification of a carbonaceous feedstock into a combustible gas (synthesis gas, a mixture of carbon monoxide and hydrogen) by partial combustion (in an oxygen-starved environment) (Table 5.1.40). The synthesis gas (syngas) is then processed into hydrocarbon products. Synthesis gas can also be converted to methanol (methyl alcohol, CH_3OH), which can be used as a fuel, fuel additive, or further processed into gasoline and diesel fuel.

In the process, operating conditions are on the order of 200 to 400°C (380 to 750°F) and 300 to 500 psi. The temperature of the exothermic reaction [−39.4 kcal/(g mol)] is controlled by carrying out the reaction in fixed beds, fluidized beds, slurry, and dilute phase systems with heat removal. The diesel fuel produced from this process has a high cetane number; the paraffin waxes can be of high quality. Generally, the gasoline produced by indirect synthesis has a low octane number because of its aliphatic nature. One method of producing a high-octane gasoline is to make methanol from a 2:1 hydrogen/carbon monoxide mixture in a first stage and then process the alcohol over a zeolite catalyst.

Catalysts can be prepared from Fe, Ni, Co, Ru, Zn, and Th either alone or promoted on a support. Each catalyst gives a different product distribution that is also a function of the method of preparation and pretreatment. In South Africa, SASOL (South African Synthetic Oil Limited) applies advanced gasification and liquefaction processes to achieve fuels the product mix consists of naphtha and kerosene/diesel.

Chemically, the Fischer-Tropsch process is represented as the conversion of carbon monoxide and hydrogen to hydrocarbons and water:

$$nCO + (2n + 1)H_2 \rightarrow C_nH_{(2n+2)} + nH_2O$$

The initial reactants in the above reaction (i.e., CO and H_2) can be produced by other reactions such as the partial combustion of a hydrocarbon:

$$C_nH_{(2n+2)} + \tfrac{1}{2}nO_2 \rightarrow (n + 1)H_2 + nCO$$

Or by the gasification of coal or biomass:

$$C + H_2O \rightarrow H_2 + CO$$

The energy needed for this endothermic reaction of coal or biomass and steam is usually provided by the (exothermic) combustion with air or oxygen. This leads to the following reaction:

$$2C + O_2 \rightarrow 2CO$$

The mixture of carbon monoxide and hydrogen is called synthesis gas (syngas). The resulting hydrocarbon products are refined to produce the desired synthetic fuel.

The carbon dioxide and carbon monoxide is generated by partial oxidation of coal and wood-based fuels. The utility of the process is primarily in its role in producing liquid hydrocarbons from a solid feedstock, such as coal or solid carbon-containing wastes of various types. Non-oxidative pyrolysis of the solid material produces syngas which can be used directly as a fuel without being taken through Fischer-Tropsch transformations. If a liquid fuel, lubricant, or wax is required, the Fischer-Tropsch process can be applied successfully in the manufacture.

Table 5.1.40 Composition of Synthesis Gas from Biomass Gasification

Constituents	% v/v (dry and nitrogen free)
Carbon monoxide (CO)	28-36
Hydrogen (H_2)	22-32
Carbon dioxide (CO_2)	21-30
Methane (CH_4)	8-11
Ethylene (C_2H_4)	2-4

References

1. Chadeesingh R. The Fischer-Tropsch Process. *The Biofuels Handbook*, In: J.G. Speight (ed.). The Royal Society of Chemistry, London, United Kingdom. Part 3, Chapter 5. Page 476-517, 2011.
2. Dahiya A. *Bioenergy: Biomass to Biofuels*. Academic Press, New York, 2014.
3. John E., and Singh K. Production of Fuels from Domestic and Industrial Waste. *The Biofuels Handbook*, In: J.G. Speight (ed). The Royal Society of Chemistry, London, United Kingdom. Part 3, Chapter 1. Page 333-376, 2011.
4. John E., and Singh K. Properties of Fuels from Domestic and Industrial Waste. *The Biofuels Handbook*, In: J.G. Speight (ed). The Royal Society of Chemistry, London, United Kingdom. Part 3, Chapter 2. Page 377-407, 2011.
5. Kreith K., and Goswami D.Y. *Handbook of Energy Efficiency and Renewable Energy, Florida*. CRC Press, Taylor & Francis Group, Boca raton, Florida, 2007.
6. Lee S., and Shah Y.T. *Biofuels and Bioenergy: Processes and Technologies*. CRC Press, Taylor & Francis Group, Boca Raton, Florida, 2012.
7. Lee S. *Oil Shale Technology*. CRC Press, Taylor & Francis Group, Boca Raton, Florida, 2013.
8. Lee S., Speight J.G., and Loyalka S. *Handbook of Alternative Fuel Technologies*, 2nd ed., CRC Press, Taylor & Francis Group, Boca Raton, Florida, 2014.
9. Ludwig C., Hellweg S., and Stucki S. (eds). *Municipal Solid Waste Management: Strategies and Technologies for Sustainable Solutions*. Springer Science, New York, 2003.
10. Luque R., Campelo J., and Clark J. (eds). *Handbook of Biofuels Production: Processes and Technologies*. Woodhead Publishing, Elsevier, Cambridge, United Kingdom, 2014.
11. Luque R., and Speight J.G. (eds). *Gasification for Synthetic Fuel Production: Fundamentals, Processes, and Applications*. Woodhead Publishing, Elsevier, Cambridge, United Kingdom, 2015.
12. Mousdale D.M. Biofuels: *Biotechnology, Chemistry, and Sustainable Development*. CRC Press, Taylor & Francis Group, Boca Raton, Florida, 2008.
13. Ramroop Singh N. Biofuels. *The Biofuels Handbook*. In: J.G. Speight (ed). The Royal Society of Chemistry, London, United Kingdom. Part 1, Chapter 5. Page 160-200, 2011.
14. Speight J.G. *Synthetic Fuels Handbook: Properties, Process, and Performance*. McGraw Hill, New York, 2008.
15. Speight J.G. *Shale Oil Production Processes*. Gulf Professional Publishing, Elsevier, Oxford, United Kingdom, 2012.
16. Speight J.G. *The Chemistry and Technology of Coal*, 3rd ed., CRC Press, Taylor & Francis Group, Boca Raton, Florida, 2013.
17. Speight J.G. *Oil Sand Production Processes*. Gulf Professional Publishing, Elsevier, Oxford, United Kingdom, 2013.
18. Speight J.G. *Coal-Fired Power Generation Handbook*. Scrivener Publishing, Beverly, Massachusetts, 2013.
19. Speight J.G. *Gasification of Unconventional Feedstocks*. Gulf Professional Publishing, Elsevier, Oxford, United Kingdom, 2014.
20. Speight J.G. *The Chemistry and Technology of Petroleum*, 5th ed., CRC Press, Taylor and Francis Group, Boca Raton, Florida, 2014.
21. Speight J.G. *Introduction to Enhanced Recovery Methods for Heavy Oil and Tar Sands*, 2nd ed. Gulf Professional Publishing, Elsevier, Oxford, United Kingdom, 2016.
22. Speight J.G. *Handbook of Petroleum Refining*. CRC Press, Taylor & Francis Group, Boca Raton, Florida, 2017.

EXPLOSIVES

by J. Edmund Hay

U.S. Department of the Interior

REFERENCES: Meyer, "Explosives," Verlag Chemie. Cook, "The Science of High Explosives," Reinhold. Johansson and Persson, "Detonics of High Explosives," Academic. Davis, "The Chemistry of Powder and Explosives," Wiley. "Manual on Rock Blasting." Aktiebolaget Atlas Diesel, Stockholm. Dick, Fletcher, and D'Andrea, Explosives and Blasting Procedures Manual, *BuMines Inf. Circ.* 8925, 1983.

The term *explosives* refers to any substance or article which is able to function by explosion (i.e., the extremely rapid release of gas and heat) by chemical reaction within itself.

Explosive substances are commonly divided into two types: (1) high or **detonating explosives** are those which normally function by detonation, in which the chemical reaction is propagated by a shock wave that in turn is driven by the energy released; and (2) low or **deflagrating explosives,** normally function by deflagration, in which the chemical reaction is propagated by convective, conductive, and/or radiative heat transfer. High explosives are further divided into two types: primary explosives are those for which detonation is the only mode of reaction, and secondary explosives may either detonate or deflagrate, depending on a variety of conditions. Low explosives are usually pyrotechnics or propellants for guns, rockets, or explosive-actuated devices.

It is important to note the word *normally.* Not only can many high explosives deflagrate rather than detonate under certain conditions, but also some explosives which are normally considered to be "low" explosives can be forced to detonate. Also note that the definition of *explosive* is itself somewhat elastic—the capability to react explosively is strongly dependent on the geometry, density, particle size, confinement, and initiating stimulus, as well as the chemical composition. Historically, immense grief has resulted from ignorance of these facts. In normal use, the term *explosive* refers to those substances or articles which have been classified as explosive by the test procedures recommended by the United Nations (UN) Committee on the Transport of Dangerous Goods.

According to the UN scheme of classification, most high explosives are designated class 1.1 (formerly called *class A*), and most low explosives are designated class 1.3 (formerly called *class B*). Explosive devices of minimal hazard (formerly called *class C*) are designated class 1.4.

However, a very important subclass of secondary high explosives is designated class 1.5 (formerly called *blasting agents* or *nitrocarbonitrates*). The distinction is based primarily on **sensitivity:** In simplified terms, the initiation of detonation of a class 1.5 explosive requires a stronger stimulus than that provided by a detonator with a 0.45-g PETN base charge which exhibits a low tendency to the deflagration-to-detonation transition.

The important physical properties of high explosives include their bulk density, detonation rate, critical and "ideal" diameters (or thickness), "sensitivity," and strength.

The detonation rate is the linear speed at which the detonation propagates through the explosive, and it ranges from about 6,000 ft/s (1.8 km/s) to about 28,000 ft/s (8.4 km/s), depending on the density and other properties of the explosive.

The *critical diameter* (or thickness, in the case of a sheet explosive) is the minimum diameter or thickness at which detonation can propagate through the material. For most class 1.1 explosives, this is between 1 and 30 mm; for class 1.5 explosives, it is usually greater than 50 mm. This value depends to some extent on the confinement provided by the material adjacent to the explosive. For explosives whose diameter (thickness) is only slightly above the critical value, the detonation rate increases with increasing diameter (thickness), finally attaining a value for which no increase in rate is observed for further increases in dimension. The diameter or thickness at which this occurs is called the *ideal diameter* or *thickness.*

Sensitivity and strength are two of the most misunderstood properties of explosives, in that there is a persistent belief that each of these terms refers to a property with a unique value. Each of these properties can be quantitatively determined by a particular test or procedure, but there are far more such procedures than can even be listed in this space, and there are large deviations from correlation between the values determined by these different tests.

Descriptions of some of the more common explosives or types of high explosive are given below.

Ammonium nitrate-fuel oil (ANFO) blasting agents are the most widely used type of explosive product, accounting for about 90 percent of explosives in the United States. They usually contain 5.5 to 6 percent fuel oil, typically no. 2 diesel fuel. If used underground, the oil content must be carefully regulated to minimize the production of toxic fumes. Some ANFO compositions contain aluminum or densifying agents. Premixed ANFO is usually shipped in 50- to 100-lb bags, although bulk shipment and storage are practiced in certain operations. ANFO has a density of 0.85 to 1.0 g/cm^3 and a detonation velocity in the range of 10,000 to 14,000 ft/s (3,000 to 4,300 m/s). ANFO may also incorporate densifying agents and other fuels such as aluminum, and it may be blended with **emulsions** (discussed later). The density of such compositions may run as high as 1.5 g/cm^3, and the detonation rate as high as 16,000 ft/s (4,800 m/s). ANFO and other blasting agents require the use of high-explosive **primers** to initiate detonation. Commonly used primers are cartridges of ordinary dynamite or specially cast charges of to lb of military-type explosives such as **composition B or pentolite**. The efficiency of ANFO lies in the method of loading, which fills the borehole completely and provides good coupling with the burden.

Water-based compositions fall into two types. **Water gels** are gelled solutions of ammonium nitrate containing other oxidizers, fuels, and sensitizers such as amine nitrates or finely milled aluminum, in solution or suspension. **Emulsions** are emulsions of ammonium nitrate solution with oil, and they may contain additional oxidizers, fuels, and sensitizers. These compositions are classified either as explosives or as blasting agents depending on their sensitivity. Both types contain a thickening agent to prevent segregation of suspended solids. For larger operations, mobile mixing trucks capable of high-speed mixing and loading of the composition directly into the large-diameter vertical boreholes have become popular. Another advantage of this on-site mixing and loading is the ability to change composition and strength between bottom and top loads. The density of the explosive-sensitized composition is usually about 1.4 g/cm^3 but may be as high as 1.7 g/cm^3 for the aluminum-sensitized type; detonation velocities vary from 10,000 to 17,000 ft/s (3,000 to 5,200 m/s).

Dynamite is a generic term covering a multitude of nitroglycerine-sensitized mixtures of carbonaceous materials (wood, flour, starch) and oxygen-supplying salts such as ammonium nitrate and sodium nitrate. The nitroglycerin contains ethylene glycol dinitrate or other nitrated compounds to lower its freezing point, and antacids, such as chalk or zinc oxide, are divided into nongelatinous or granular and gelatinous types, the latter containing nitrocellulose. All dynamites are cap sensitive.

Straight dynamites are graded by the percentage of explosive oil they contain; this may be as low as 15 percent and as high as 60 percent. A typical percentage formulation for a 40 percent straight dynamite is: nitroglycerin, 40; sodium nitrate, 44; antacid, 2; carbonaceous material, 14. The rate of detonation increases with grade from 9,000 to 19,000 ft/s (2,700 to 5,800 m/s). Straight dynamites now find common use only in ditching where propagation by influence is practiced.

Ammonia dynamites differ from straight dynamites in that some of the sodium nitrate and much of the explosive oil have been replaced by ammonium nitrate. Strength of ammonia dynamites ranges from 15 to 60 percent, each grade having the same weight strength as the corresponding straight dynamite when compared in the ballistic mortar. A typical percentage formula for a 40 percent ammonia dynamite is: explosive oil, 14; ammonium nitrate, 36; sodium nitrate, 33; antacid, 1; carbonaceous material, 16. The rate of detonation, 4,000 to 17,000 ft/s (1,200 to 5,200 m/s), again increases with grade. Low-density, high-weight-strength compositions are popular in many applications, but ANFO has displaced them in numerous operations.

Blasting gelatin is the strongest and highest-velocity explosive used in industrial operations. It consists essentially of explosive oil (nitroglycerin plus ethylene glycol dinitrate) colloided with about 7 percent nitrocellulose. It is completely water-resistant but has a poor fume rating and consequently finds only limited use.

Gelatin dynamites correspond to straight dynamites except that the explosive oil has been gelatinized by nitrocellulose; this results in a cohesive mixture having improved water resistance. Under confinement, the gelatins develop high velocity, ranging from 8,500 to 22,000 ft/s (2,600 to 6,700 m/s) and increasing between the grades of 20 and 90 percent. An approximate percentage composition for a 40 percent grade is: explosive oil, 32; nitrocellulose, 0.7; sulfur, 2; sodium nitrate, 52; antacid, 1.5; and carbonaceous material, 11. In the common grades of 40 and 60 percent, fume characteristics are good, making these types useful for underground hard-rock blasting.

Ammonia gelatin dynamites are similar to the ammonia dynamites except for their nitrocellulose content. These used to be popular in quarrying and hard-rock mining. Their excellent fume characteristics make them suitable for use underground, but again ANFO is widely used. The rates of detonation of 7,000 to 20,000 ft/s (2,000 to 6,000 m/s) are somewhat less than the straight gelatins. A typical percentage composition for the 40 percent grade is: gelatinized explosive oil, 21; ammonium nitrate, 14; sodium nitrate, 49; with antacid and combustible making up the remainder.

The **semigels** are important variants of the ammonia gels; these contain less explosive oil, sodium nitrate, and nitrocellulose and more ammonium nitrate than the corresponding grade of ammonia gel. Rates of detonation fall in the limited range of 10,000 to 13,000 ft/s (3,000 to 4,000 m/s). These powders are cohesive and have good water resistance and good fume characteristics.

Permissible explosives are powders especially designed for use in underground coal mines, which have passed a series of tests established by the Bureau of Mines. The most important of these tests concern incendivity of the explosives—their tendency to ignite methane-air or methane-coal dust-air mixtures. Permissible explosives are either granular or gelatinous; the granular type makes up the bulk of the powders used today. Typically, a granular permissible contains the following, in percent: explosive oil, 9; ammonium nitrate, 65; sodium nitrate, 5; sodium chloride, 10; carbonaceous material, 10; and antacid, 1. Gels contain nitrocellulose for improved water resistance and more explosive oil. Detonation velocities for the granular grades vary from 4,500 to 11,000 ft/s (1,400 to 3,400 m/s), and for the gels from 10,500 to 18,500 ft/s (3,200 to 5,600 m/s). Many water-based permissible formulations with comparable physical and safety properties are now marketed as well.

Liquid oxygen explosives (LOX) once saw considerable use in coal strip mines but have been completely displaced by ANFO or water-based compositions. LOX consisted of bags of pressed carbon black or specially processed char that were saturated with liquid oxygen just before loading into the borehole. The rate of detonation ranged from 12,000 to 18,000 ft/s (3,700 to 5,500 m/s).

Military explosives, originally developed for such uses as bomb, shell, and mine loads and demolition work, have been adapted to many industrial explosive applications. The more common military explosives are listed in Table 5.1.41, with their compositions, ballistic mortar strengths, and detonation velocities.

Amatol was used early in World War II, largely because of the short supply of TNT. Modifications of amatol have been used as industrial blasting agents. **Explosive D,** or ammonium picrate, by virtue of its extreme insensitivity, was used in explosive-filled armor-piercing shells and bombs.

RDX, a very powerful explosive compound, was widely used during World War II in many compositions, of which Compositions A, B, and C were typical. **Composition A** was used as a shell loading; **B** was used as a bomb and shaped charge filling; C, being plastic enough to allow molding to desired shapes, was developed for demolition work. Compositions that also contained aluminum powder were developed for improved underwater performance (**Torpex, HBX**). RDX has found limited commercial application as the base charge in some detonators, the filling for special-purpose detonating fuses or cordeau detonants, and the explosive in small shaped charges used as oil well perforators and tappers for open hearth steel furnaces.

PETN has never found wide military application because of its sensitivity and relative instability. It is used extensively, however, as the core of detonating fuses and in caps and, mixed with **TNT,** in boosters.

Tetryl, once widely used by the military as a booster loading and commercially as a base charge in detonators, has been displaced by other compositions. **Tetrytol** found limited application as a demolition charge.

TNT is a very widely used military explosive. Its stability, insensitivity, convenient melting point (81°C), and relatively low cost have made it the explosive of choice either alone or in admixture with other materials for loadings which are to be cast. A free-flowing pelletized form has found application in certain types of blasting requiring high loading density, where it is used to fill the cavity formed in sprung holes or the free space around the column of other explosives in the borehole.

DUST EXPLOSIONS

by Harry C. Verakis and John Nagy
Mine Safety and Health Administration

REFERENCES: Nagy and Verakis, "Development and Control of Dust Explosions," Dekker. "Classification of Dusts Relative to Electrical Equipment in Class II Hazardous Locations," NMAB 353-4, National Academy of Sciences, Washington. "Fire Protection Handbook," 17th ed., National Fire Protection Assoc. *BuMines Rept. Inv.* 4725, 5624, 5753, 5971, 7132, 7208, 7279, 7507. Eckhoff, "Dust Explosions in the Process Industries," Butterworth-Heinemann.

A dust explosion hazard exists where combustible dusts accumulate or are processed, handled, or stored. The possibility of a dust explosion may often be unrecognized because the material in bulk form presents little or no explosion hazard. However, if the material is dispersed in the atmosphere, the potential for a dust explosion is increased significantly. The first well-recorded dust explosion occurred in a flour mill in Italy in 1785. Dust explosions continue to plague industry and cause serious disasters with loss of life, injuries, and property damage. For example, there were about 100 reportable dust explosions (excluding grain dust) from 1970 to 1980 which caused about 25 deaths and a yearly property loss averaging about $20 million. A complete and accurate record of the number of dust explosions, deaths, injuries, and property damage is unavailable because reporting of each incident is not required unless a fatality occurs or more than five persons are seriously injured. In recent years, most dust explosions have involved wood, grain, resins and plastics, starch, and aluminum. Most of the incidents occurred during crushing or pulverizing, buffing or grinding, conveying, and at dust collectors.

Despite the well-recognized hazards inherent with explosible dusts, the vast amount of technical data accumulated and published, and standards for prevention of and protection from dust explosions, severe property damage and loss of life occur every year. As an example of the severity of dust explosions, there was a series of explosions in four grain elevators during December 1977, which caused 59 deaths, 47 injuries, and nearly $60 million in property damage.

A **dust explosion** is the rapid combustion of a cloud of particulate matter in a confined or partially confined space in which heat is generated at a higher rate than it is dissipated. In a confined space, the explosion is characterized by relatively rapid development of pressure with the evolution of large quantities of heat and reaction products. The condition necessary for a dust explosion to occur is the simultaneous presence of a dust cloud of proper concentration in air or gas that will support combustion and an ignition source.

Dust means particles of materials smaller than 0.016 in in diameter, or those particles passing a no. 40 U.S. standard sieve, 425 μm (this definition relates to the limiting size, not to the average particle size of

Table 5.1.41 Physical Characteristics of Military Explosives

Explosive	Composition, %	Ballistic mortar strength (TNT = 100)	Density, g/cm³	Rate of detonation, m/s
80/20 amatol	Ammonium nitrate, 80; TNT, 20	117	Cast	4,500
50/50 amatol	Ammonium nitrate, 50; TNT, 50	122	Cast	5,600
Composition A (pressed)	RDX, 91; wax, 9	134	0.80	4,560
			1.20	6,340
			1.50	7,680
			1.60	8,130
Composition B (cast)	RDX, 59.5; TNT, 39.5; wax, 1	130	1.65	7,660
Composition C-3 (plastic)	RDX, 77; tetryl, 3; mononitrotoluene, 5; dinitrotoluene, 10; TNT, 4; nitrocellulose, 1	145	1.55	8,460
Composition C-4 (plastic)	RDX, 91; dioctyl sebacate, 5.3; polyisobutylene, 2.1; oil, 1.6	—	1.59	8,000
Explosive D	Ammonium picrate	97	0.80	4,000
			1.20	5,520
			1.50	6,660
			1.60	7,040
HBX-1	RDX, 40; TNT, 38; aluminum, 17; desensitizer, 5	130	1.70	7,310
Lead azide*	Lead azide	—	2.0	4,070
			3.0	4,630
			4.0	5,180
PETN	Pentaerythritol tetranitrate	145	0.80	4,760
			1.20	6,340
			1.50	7,520
			1.60	7,920
50/50 Pentolite	PETN, 50; TNT, 50	120	1.20	5,410
			1.50	7,020
			1.60	7,360
			Cast	7,510
Picric acid	Trinitrophenol	108	1.20	5,840
			1.50	6,800
			Cast	7,350
RDX (cyclonite)	Cyclotrimethylene trinitramine	150	0.80	5,110
			1.20	6,550
			1.50	7,650
			1.60	8,000
			1.65	8,180
Tetryl	Trinitrophenylmethylnitramine	121	0.80	4,730
			1.20	6,110
			1.50	7,160
			1.60	7,510
75/25 Tetrytol	Tetryl, 75; TNT, 25	113	1.60	7,400
TNT	Trinitrotoluene	100	0.80	4,170
			1.20	5,560
			1.50	6,620
			1.60	6,970
			Cast	6,790

* Primary compound for blasting caps.

the material); and *explosible* dust means a dust which, when dispersed, is ignited by spark, flame, heated coil, or in the Godbert-Greenwald furnace at or below 730°C, when tested in accordance with the equipment and procedures described in *BuMines Rept. Inv.* 5624.

Explosibility Factors

Empirical methods and experimental data are the chief guides in evaluating relative dust explosion hazards. A mathematical model correlating some of the numerous interrelated factors affecting dust explosion development in closed vessels has been developed. Details of the model are presented in Nagy and Verakis, "Development and Control of Dust Explosions."

Dust Composition Many industrial dusts are not pure compounds. The severity of a dust explosion varies with the chemical constitution and certain physical properties of the dust. High percentages of noncombustible material, such as mineral matter or moisture, reduce the ease of oxidation, and oxygen requirements influence the explosibility of dusts. Volatile, combustible components in such materials as coals, asphalts, and pitches increase explosibility.

Dust composition also affects the amount and type of products produced in an explosion. Organic materials evolve new gaseous products, whereas most metals form solid oxides during combustion in an air atmosphere.

Particle Size and Surface Area Explosibility of dusts increases with a decrease in particle size. Fine dust particles have greater surface area, more readily disperse into a cloud, mix better with air, remain longer in suspension, and oxidize more rapidly and completely than coarse particles. Decrease of particle size generally results in lower ignition temperature, lower igniting energy, lower minimum explosive concentration, and higher pressure and rates of pressure rise. Some metals, such as chromium, become explosive only at very fine particle sizes (average particle diameter of 3 μm), and almost all metals become pyrophoric if reduced to very fine powder.

Range of Explosibility Most combustible dusts have a well-defined lower limit, but the upper limit is usually indefinite. The upper limit has been determined for only a few dusts, but these data have only limited importance in practice. The range of dust explosibility is normally 0.015 to greater than 10 oz/ft³ (10 kg/m³). The optimum concentration producing the strongest dust explosions is about 0.5 to 1.0 oz/ft³. Table 5.1.42 gives explosion characteristics for a number of dusts at a concentration of 0.5 oz/ft³. A typical example of the effect of dust concentration on maximum pressure and maximum rate of pressure rise

Table 5.1.42 Explosive Characteristics of Various Dusts*

Type of dust	Ignition temperature of dust cloud, °C	Minimum igniting energy, J	Minimum explosive concentration, oz/ft³	Maximum explosion pressure, lb/in² gage	Maximum rate of pressure rise, Ib/(m²)(s)	Terminal oxygen concentration, %†
Agricultural:						
Alfalfa	530	0.320	0.105	92	2,200	
Cereal grass	550	0.800	0.250	52	500	
Cinnamon	440	0.030	0.060	114	3,900	
Citrus peel	730	0.045	0.065	71	2,000	
Cocoa	500	0.120	0.065	55	900	
Coffee	720	0.160	0.085	53	300	
Com	400	0.040	0.055	95	6,000	
Corncob	480	0.080	0.040	110	3,100	
Com dextrine	410	0.040	0.040	105	7,000	
Cornstarch	390	0.030	0.040	115	9,000	
Cotton linters	520	1.920	0.500	48	150	
Cottonseed	530	0.120	0.055	96	3,000	
Egg white	610	0.640	0.140	58	500	
Flax shive	430	0.080	0.080	81	800	
Garlic	360	0.240	0.100	80	2,600	
Grain, mixed	430	0.030	0.055	115	5,500	
Grass seed	490	0.260	0.290	34	400	
Guar seed	500	0.060	0.040	98	2,400	
Gum, Manila (copal)	360	0.030	0.030	88	5,600	
Hemp hard	440	0.035	0.040	103	10,000	
Malt, brewers	400	0.035	0.055	92	4,400	
Milk, skim	490	0.050	0.050	83	2,100	
Pea flour	560	0.040	0.050	95	3,800	
Peanut hull	460	0.050	0.045	82	4,700	
Peat, sphagnum	460	0.050	0.045	84	2,200	
Pecan nutshell	440	0.050	0.030	106	4,400	
Pectin	410	0.035	0.075	112	8,000	
Potato starch	440	0.025	0.045	97	8,000	
Pyrethrum	460	0.080	0.100	82	1,500	
Rauwolfia vomitoria root	420	0.045	0.055	106	7,500	
Rice	440	0.050	0.050	93	2,600	
Safflower	460	0.025	0.055	84	2,900	
Soy flour	550	0.100	0.060	111	1,600	15
Sugar, powdered	370	0.030	0.045	91	1,700	
Walnut shell, black	450	0.050	0.030	97	3,300	
Wheat flour	440	0.060	0.050	104	4,400	
Wheat, untreated	500	0.060	0.065	98	4,400	
Wheat starch	430	0.025	0.045	100	6,500	
Wheat straw	470	0.050	0.055	99	6,000	
Yeast, torula	520	0.050	0.050	105	2,500	
Carbonaceous:						
Asphalt, resin, volatile content 57.5%	510	0.025	0.025	94	4,600	
Charcoal, hardwood mix, volatile content 27.1%	530	0.020	0.140	100	1,800	18
Coal, Colo., Brookside, volatile content 38.7%	530	0.060	0.045	88	3,200	
Coal, III., no. 7, volatile content 48.6%	600	0.050	0.040	84	1,800	15
Coal, Ky., Breek, volatile content 40.6%	610	0.030	0.050	88	4,000	
Coal, Pa., Pittsburgh, volatile content 37.0%	610	0.060	0.055	83	2,300	17
Coal, Pa., Thick Freeport, volatile content 35.6%	595	0.060	0.060	77	2,200	
Coal, W. Va., no. 2 Gas, volatile content 37.1%	600	0.060	0.060	82	1,600	
Coal, Wyo., Laramie no. 3, volatile content 43.3%	575	0.050	0.050	92	2,000	
Gilsonite, Utah, volatile content 86.5%	580	0.025	0.020	78	3,700	
Lignite, Calif., volatile content 60.4%	450	0.030	0.030	90	8,000	
Pitch, coal tar, volatile content 58.1%	710	0.020	0.035	88	6,000	
Chemical compounds:						
Benzoic acid, C_6H_5COOH	620	0.020	0.030	74	5,500	
Phosphorus pentasulfide, P_2S_5, slowly cooled to give single crystals	280	0.015	0.050	54	10,000+	
Phosphorus pentasulfide, P_2S_5, cooled quickly	290	0.015	0.050	58	7,500	13

Table 5.1.42 Explosive Characteristics of Various Dusts* (*Continued*)

Type of dust	Ignition temperature of dust cloud, °C	Minimum igniting energy, J	Minimum explosive concentration, oz/ft³	Maximum explosion pressure, lb/in² gage	Maximum rate of pressure rise, lb/(in²)(s)	Terminal oxygen concentration, %†
Chemical compounds (*Continued*):						
Phthalimide, $C6H_4(CO)_2NH$	630	0.050	0.030	79	4,500	
Potassium bitartrate, $KHC_4H_4O_6$	520					
Salicylanilide, *o*-$HOC_6H_4CONHC_6H_5$	610	0.020	0.040	61	4,400	
Sodium thiosulfate, anhydrous, $Na_2S_2O_3$	510					
Sorbic acid, $CH_3(CH{:}CH)_2$ COOH	470	0.015	0.020	88	10,000+	
Sucrose, $C_{12}H_{22}O_{11}$	420	0.040	0.045	82	4,200	14
Sulfur, S_8, 100% finer than 4 μm	210	0.020	0.045	56	3,100	
Sulfur, S_8, avg particle size 4 μm	190	0.015	0.035	78	4,700	12
Drugs:						
Aspirin (acetylsalicylic acid), *o*-$CH_3COOC_6H_4COOH$, fine	660	0.025	0.050	83	10,000+	
Mannitol (hexahydric alcohol), $CH_2OH(CHOH)_4CH_2OH$	460	0.040	0.065	82	2,800	
Secobarbital sodium, $C_{12}H_{17}N_2O_3Na$	520	0.960	0.100	54	500	
Vitamin C, ascorbic acid, $C_6H_8O_6$	460	0.060	0.070	88	4,800	15
Explosives and related compounds:						
Dinitrobenzamide	500	0.045	0.040	94	6,500	
Dinitrobenzoic acid	460	0.045	0.040	92	4,300	
Dinitro-sym-diphenyl-urea (dmitrocarbanilide)	550	0.060	0.095	87	2,500	
Dinitrotoluamide (3,5-dinitro-ortho-toluamide)	500	0.015	0.050	106	10,000–	13
Metals:						
Aluminum	650	0.015	0.045	100	10,000+	2
Antimony	420	1.920	0.420	8	100	16
Boron	470	0.060	0.100	90	2,400	
Cadmium	570	4.000				
Chromium	580	0.140	0.230	56	5,000	14
Cobalt	760					
Copper	900					
Iron	420	0.020	0.100	46	6,000	10
Lead	710					
Magnesium	520	0.020	0.020	94	10,000+	0
Molybdenum	720					
Nickel	950+					
Selenium	950+					
Silicon	780	0.080	0.100	106	10,000+	12
Tantalum	630	0.120	0.200	51	3,700	
Tellurium	550					
Thorium	270	0.005	0.075	48	3,300	0
Tin	630	0.080	0.190	37	1,300	15
Titanium	460	0.010	0.045	80	10,000+	0
Tungsten	950+					
Uranium	20	0.045	0.060	53	3,400	0
Vanadium, 86%	500	0.060	0.220	48	600	13
Zinc	600	0.640	0.480	48	1,800	9
Zirconium	20	0.005	0.045	65	9,000	0
Alloys and compounds:						
Aluminum-cobalt	950	0.100	0.180	78	8,500	
Aluminum-copper	930	1.920	0.280	27	500	
Aluminum-iron	550	0.720	0.500	21	100	
Aluminum-magnesium	430	0.020	0.020	90	10,000	0
Aluminum-nickel	940	0.080	0.190	79	10,000	14
Aluminum-silicon, 12% Si	670	0.060	0.040	74	7,500	
Calcium silicide	540	0.130	0.060	73	10,000+	8
Ferrochromium, high-carbon	790		2.000			19
Ferromanganese, medium-carbon	450	0.080	0.130	47	4,200	
Ferrosilicon, 75% Si	860	0.400	0.420	87	3,600	16
Ferrotitanium, low-carbon	370	0.080	0.140	53	9,500	13
Ferrovanadium	440	0.400	1.300			17
Thorium hydride	260	0.003	0.080	60	6,500	6
Titanium hydride	440	0.060	0.070	96	10,000+	13

Table 5.1.42 Explosive Characteristics of Various Dusts* *(Continued)*

Type of dust	Ignition temperature of dust cloud, °C	Minimum igniting energy, J	Minimum explosive concentration, oz/ft³	Maximum explosion pressure, lb/in² gage	Maximum rate of pressure rise, lb/(in²)(s)	Terminal oxygen concentration, %†
Alloys and compounds (*Continued*):						
Uranium hydride	20	0.005	0.060	43	6,500	0
Zirconium hydride	350	0.060	0.085	69	9,000	8
Plastics:						
Acetal resin (polyformaldehyde)	440	0.020	0.035	89	4,100	11
Acrylic polymer resin Methyl methacrylate-ethyl acrylate	480	0.010	0.030	85	6,000	11
Alkyd resin Alkyd molding compound	500	0.120	0.155	15	150	15
Allyl resin, allyl alcohol derivative, CR-39	500	0.020	0.035	106	10,000+	13
Amino resin, urea-formaldehyde molding compound	450	0.080	0.075	89	3,600	17
Cellulosic fillers, wood flour	430	0.020	0.035	110	5,500	17
Cellulosic resin, ethyl cellulose molding compound	320	0.010	0.025	102	6,000	11
Chlorinated polyether resin, chlorinated polyether alcohol	460	0.160	0.045	66	1,000	
Cold-molded resin, petroleum resin	510	0.030	0.025	94	4,600	
Coumarone-indene resin	520	0.010	0.015	93	10,000+	14
Epoxy resin	530	0.020	0.020	86	6,000	12
Fluorocarbon resin, fluoroethylene polymer	600					
Furan resin, phenol furfural	520	0.010	0.025	90	8,500	14
Ingredients, hexamethylenetetramine	410	0.010	0.015	98	10,000+	14
Miscellaneous resins, petrin acrylate monomer	220	0.020	0.045	104	10,000+	
Natural resin, rosin, DK	390	0.010	0.015	87	10,000+	14
Nylon polymer resin	500	0.020	0.030	89	7,000	13
Phenolic resin, phenol-formaldehyde molding compound	500	0.020	0.030	92	10,000+	14
Polycarbonate resin	710	0.020	0.025	78	4,700	15
Polyester resin, polyethylene terephthalate	500	0.040	0.040	91	5,500	13
Polyethylene resin	410	0.010	0.020	83	5,000	12
Polymethylene resin, carboxypolymethylene	520	0.640	0.115	70	5,500	
Polypropylene resin	420	0.030	0.020	76	5,000	
Polyurethane resin, polyurethane foam	510	0.020	0.025	88	3,700	
Rayon (viscose) flock	520	0.240	0.055	88	1,700	
Rubber, synthetic	320	0.030	0.030	93	3,100	15
Styrene polymer resin, polystyrene latex	500	0.020	0.020	91	7,000	13
Vinyl polymer resin, polyvinyl butyral	390	0.010	0.020	84	2,000	14

* Data taken from the following *Bureau of Mines Reports of Investigations:* RI 5753, "Explosibility of Agricultural Dusts"; RI 5971, "Explosibility of Dusts Used in the Plastics Industry"; RI 6516, "Explosibility of Metal Powders"; RI 7132, "Dust Explosibility of Chemicals, Drugs, Dyes and Pesticides"; RI 7208, "Explosibility of Miscellaneous Dusts." The data were obtained using the equipment described in RI 5624, "Laboratory Equipment and Test Procedures for Evaluating Explosibility of Dusts."

† The terminal oxygen concentration is the limiting oxygen concentration in air-CO_2 atmosphere required to prevent ignition of dust clouds by electric spark.

from explosions in closed vessels of various size and shape is shown in Figs. 5.1.9 and 5.1.10.

Ignition Source Ignition sources known to have initiated dust explosions in industry include electric sparks and arcs in fuses, faulty wiring, motors and other appliances, static electrical fuses, faulty wiring, motors and other appliances, static electrical discharges, open flames, frictional, or metallic sparks, glowing particles, overheated bearings and other machine parts; hot electric bulbs, overheated driers, and other hot surfaces; dust layers may also ignite by these sources as well as by spontaneous ignition. Ignition temperatures of many dust clouds are given in Table 5.1.42. Normally the ignition temperature of a dust layer is considerably less than for a dust cloud. The position and intensity of the ignition source affect dust-explosion development; detailed information on these factors is presented in *BuMines Rept. Inv.* 7507.

Turbulence Turbulence has a slight effect on maximum pressure, but a marked effect on the rates of pressure rise for dust explosions.

Experiments show the maximum rate of pressure rise in a highly turbulent dust-air mixture can be as much as 8 times higher than in a nonturbulent mixture (*BuMines Repts. Inv.* 5815 and 7507 and Nagy and Verakis).

Moisture and Other Inerts Moisture in a dust absorbs heat and tends to reduce the explosibility of a dust. A high concentration of moisture in the dust also tends to reduce the dispersibility of a dust. An increase in moisture content causes an increase in ignition temperature and a reduction in maximum pressure and rates of pressure rise. However, the amount of moisture required to produce a marked lowering of the explosibility parameters is higher than can ordinarily be tolerated in industrial processes. Most mineral inert dusts admixed with a combustible absorb heat during the combustion reaction and reduce explosibility similar to the action of water. Some chemical compounds, such as sodium and potassium carbonates, act as inhibitors and are more effective than mineral inerts; the limiting inert dust concentration required to prevent ignition and explosion depends on the strength of the igniting source.

Atmospheric Oxygen Concentration The pressure and rate of pressure development decrease as the oxygen concentration in the atmosphere decreases. The ignition sensitivity of dusts decreases with decrease in oxygen concentration and for most dusts, ignition and explosion can be prevented by reducing the oxygen concentration to a safe value. Carbon dioxide, nitrogen, argon, helium, and water vapor are effective diluents. For highly reactive metal powders, only argon and

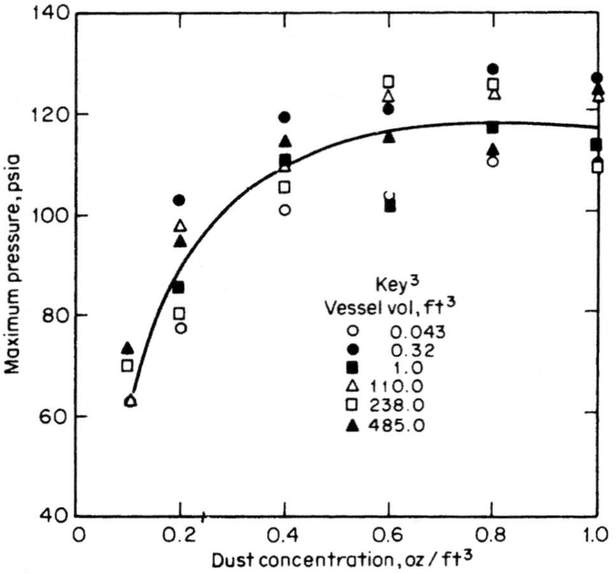

Figure 5.1.9 Electrical resistance heated furnace with resistance elements.

helium are chemically inert. Limiting oxygen concentrations using carbon dioxide as a diluent are given in Table 5.1.42 for many dusts. With carbon dioxide as a diluent, a reduction of oxygen in the atmosphere to 11 percent is sufficient to prevent ignition by sparks for all dusts tested except the metallic powders. With nitrogen as the diluent, ignition of nonmetallic dusts is prevented by diluting the atmosphere to 8 percent oxygen. Some metal dusts, such as magnesium, titanium, and zirconium, ignite by spark in a pure carbon dioxide atmosphere. Freons and halons

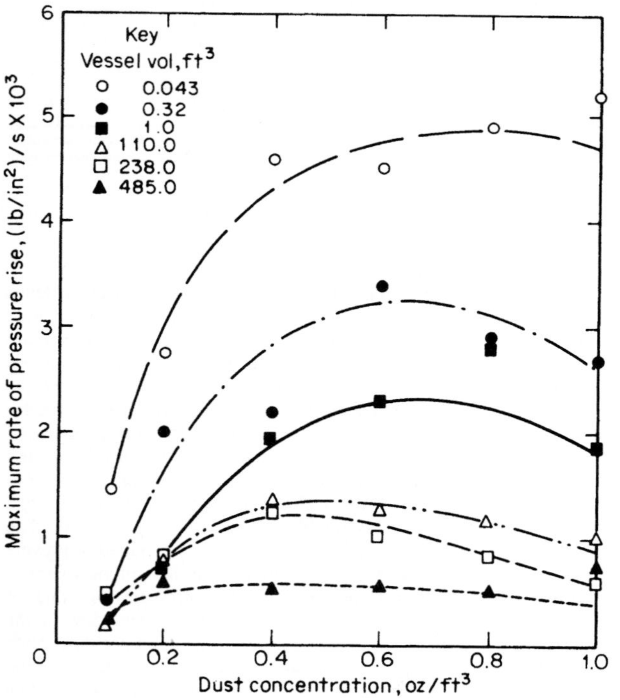

Figure 5.1.10 Effect of dust concentration on maximum rate of pressure rise by explosions of cellulose acetate dust in closed vessels.

are sometimes used as diluent gases, but if metal dusts are involved, they can intensify rather than suppress ignition. The limiting oxygen concentration decreases as the dust becomes finer in particle size; limiting oxygen concentration varies slightly with dust concentration and is lowest at concentrations two to five times the stoichiometric mixture.

Relative Dust Explosion Hazards

Table 5.1.42 gives test results of selected dusts whose explosive characteristics have been evaluated in the laboratory by the Bureau of Mines. The data were obtained for dusts passing a no. 200 sieve and represent the most hazardous of the specific materials tested. The values are relative rather than absolute since the test apparatus and experimental procedures affect the results to some degree. The samples were dried before testing only if the moisture content exceeded 5 percent. Detailed description of the equipment and procedures for the small-scale testing are given in *BuMines Rept. Inv.* 5624.

Ignition Temperature The ignition temperature of a dust cloud was determined by dispersing dust through a heated cylindrical furnace. The ignition temperature is the minimum furnace temperature at which flame appears at the bottom of the furnace in one or more trials in a group of four.

Minimum Energy The minimum electrical spark energy required to ignite a dust cloud was determined by dispersing the dust in a vertically mounted, $2^3/_4$-in-diameter, 12-in-long tube. The dust is dispersed by an air blast and then a condenser of known capacitance and voltage is discharged through a spark gap located within the dust dispersion. The top of the tube is enclosed with a paper diaphragm. The minimum energy for ignition of the dust cloud is the least amount of energy required to produce flame propagation 4 in or longer in the tube.

Minimum Concentration The minimum explosive concentration or the lower explosive limit of a dust cloud was determined in a vertically mounted, $2^3/_4$-in diameter, 12-in-long tube using a continuous sparkigniting source. A known weight of dust was dispersed within the tube by an air blast. The lowest weight of dust at which sufficient pressure develops to burst a paper diaphragm enclosing the tube or which causes flame to fill the tube is used to calculate the minimum explosive concentration; this calculation is made utilizing the tube volume.

Maximum Pressure and Rates of Pressure Rise The maximum pressure and rates of pressure rise developed by a dust explosion were determined by dispersing dust in a closed steel tube. A continuous spark is used for ignition. A transcribed pressure-time record is obtained during the test. The maximum rate is the steepest slope of the pressure-time curve. Normally, explosion tests are made at dust concentrations of 0.10 to 2.0 oz/ft³. Maximum pressure is primarily dependent on dust composition and independent of vessel size and shape. The maximum rate of pressure rise increases as vessel size decreases.

Explosibility Index The overall explosion hazard of a dust is related to the ignition sensitivity and to explosion severity and is characterized by empirical indexes. The ignition sensitivity of a dust cloud depends on the ignition temperature, minimum energy, and minimum concentration. The explosion severity of a dust depends on the maximum pressure and maximum rate of pressure rise. The indexes are not derived from theoretical considerations, but provide a numerical rating consistent with research observations and practical experience. Results obtained for a sample dust are compared with values for a standard Pittsburgh-seam coal dust. The indexes are defined as follows:

$$\text{Ignition sensitivity} = \frac{(\text{Ign temp} \times \text{min energy} \times \text{min conc})\ \text{Pittsburgh coal dust}}{(\text{ign temp} \times \text{min energy} \times \text{min conc})\ \text{sample dust}}$$

$$\text{Explosion severity} = \frac{(\text{max explosive pressure} \times \text{max rate of pressure rised})\ \text{sample dust}}{(\text{max explosive pressure} \times \text{max rate of pressure rise})\ \text{Pittsburgh coal dust}}$$

Explosibility index = ignition sensitivity × explosion severity

A dust having ignition and explosion characteristics equivalent to the standard Pittsburgh-seam coal has an explosibility index of unity. The relative hazard of dusts is further classified by the following adjective ratings: fire, weak, moderate, strong, or severe. The notation $\ll 0.1$ designates a combustible dust presenting primarily a fire hazard as ignition of the dust cloud is not obtained by a spark or flame source, but only by an intense, heated surface source. These ratings are correlated with the empirical indexes as follows:

Type of explosion	*Ignition sensitivity*	*Explosion severity*	*Index of explosibility*
Fire			<0.1
Weak	<0.2	<0.5	<0.1
Moderate	0.2-1.0	0.50-1.0	0.1-1.0
Strong	1.0-5.0	1.0-2.0	1.0-10
Severe	>5.0	>2.0	>10

Prevention of Dust Explosions

There are codes published by the National Fire Protection Association which contain recommendations for a number of dust-producing industries.

Additional sources of information may be found in the NFPA "Fire Protection Handbook," the Factory Mutual "Handbook of Industrial Loss Prevention," *BuMines Rept. Inv.* 6543, Nagy and Verakis, and Eckhoff.

Safeguards against explosions include, but are not limited to the following:

Good housekeeping: An excellent means to minimize the potential for and extent of an explosion is good housekeeping. Control of dust spillage or leakage and elimination of dust accumulations removes the fuel required for an explosion.

Limited personnel: Wherever a hazardous operation must be performed, the number of persons should be limited to the minimum required for safe operation.

Elimination of Ignition Sources All sources of ignition should be eliminated from equipment containing combustible dust and from adjacent areas. Open flames or lights and smoking should be prohibited. The use of electric or gas cutting and welding equipment for repairs should be avoided unless dust-producing machinery is shut down and all dust has been removed from the machines and from their vicinity. Proper control methods should be instituted for materials susceptible to spontaneous combustion. Additional safety measures to follow are ground and bonding of all equipment to prevent the accumulation of static electrical charges; strict adherence to the National Electric Code when installing electrical equipment and wiring in hazardous locations; use of magnetic separators to prevent entrance of ferrous materials into dust-grinding mills; use of nonferrous blades in fans through which dust passes; and avoidance of spark-producing tools in certain industries and of high-speed shafting and belts. Safeguards against ignition by lightning should also be considered. (See also NFPA no. 77, Static Electricity and NFPA no. 78, Lightning Protection.)

Building and Equipment Construction Buildings should be constructed to minimize the collection of dust on beams, ledges, and other surfaces, particularly overhead. Vacuum cleaning is preferable to other methods for dust removal, but soft push brooms may be used without serious hazard. Buildings, including inside partitions, where combustible dusts are handled or stored should be detached units of incombustible construction. Hazardous units within buildings should be separated by substantial fire walls.

Grinders, conveyors, elevators, collectors, and other equipment which may produce dust clouds should be as dust-tight as possible; they should have the smallest practical interior volume and should be constructed to withstand dust explosion pressures. The degree of turbulence within and around an enclosure should be kept to a minimum to prevent dust from being suspended.

Dust collectors should preferably be located outside of buildings or detached rooms and near the dust source. The choice of a suitable dust collector depends on the particle size, dryness, explosibility, dust concentration, gas velocity and temperature, efficiency and space requirements, and economic considerations. (See also Sec. 7.)

Inerted Atmosphere Equipment such as grinders, conveyors, pulverizers, mixers, dust collectors, and sacking machines can frequently be protected by using an inerted atmosphere or explosion suppression systems. The inert gas for this purpose may be obtained by dilution of air with flue gases from boilers, internal-combustion engines, or other sources, or by dilution with carbon dioxide, nitrogen, helium from high-pressure cylinders, or gas from inert-gas generators. The amount and rate of application of inert gas required depend upon the permissible oxygen concentration, leakage loss, atmospheric and operating conditions, equipment to be protected, and application method. Addition of inert dusts to the combustible dust may also prevent explosive dust-air mixtures from forming in and around equipment. (See NFPA no. 69, Explosion Prevention Systems.)

Relief Venting To reduce structural damage and to protect personnel from dust explosions, dust collectors and other equipment and the rooms in which dust-producing machinery is located should be provided with relief vents. Relief vents properly designed and located will sufficiently relieve explosion pressures in most instances and direct explosion gases away from occupied areas. The vents may be unrestricted or free openings; hinged or pivoted sash that swing outward at a low internal pressure; fixed sash with light wall anchorages; scored glass panes; light wall panels; monitors or skylights; paper, metal foil, or other diaphragms that burst at low pressures; poppet-type vent closures; pullout diaphragms; or other similar arrangements. Vents should be located near potential sources of ignition to keep explosion pressure at a minimum and to prevent a dust explosion from developing into a detonation in long ducts.

Empirical formulas and mathematical methods for calculating the vent area to limit pressure from an explosion have been developed. Unfortunately, because of the many factors involved in determining venting requirements, none of these methods can be considered entirely satisfactory to cover the complete range of situations confronting an equipment or building designer. For example, the maximum pressure that can develop in a vented enclosure during a dust explosion is affected by factors such as the chemical affinity of the combustible material with oxygen, heat of combustion, particle size distribution, degree of turbulence, uniformity of the dust cloud, size and energy of the igniting source, location of the igniting source relative to the vent, area of the vent opening, bursting strength or inertial resistance of the vent closure, initial pressure and initial temperature within an enclosure, and the oxygen concentration of the atmosphere. Because of the complexity of the phenomena during explosion in a vented vessel, information on the required vent area to limit the excess or explosion pressure to be a safe value for a given vessel or structure under specific conditions is still estimated from data obtained by physical tests usually made under severe test conditions. Extremely reactive dusts such as magnesium and aluminum are difficult or nearly impossible to vent successfully if an explosion occurs under optimum conditions. Agricultural dusts, most plastic-type dusts, and other metallic dusts can usually be vented successfully. Materials containing oxygen or a mixture with an oxidant should be subjected to tests before venting is attempted. Information and recommendations on venting are given in NFPA no. 68, "Guide for Venting of Deflagrations" (1994). A mathematical analysis showing the relationship of numerous factors affecting the venting of explosions is presented in "Development and Control of Dust Explosions" (Nagy and Verakis, Dekker).

Information published by others shows that higher values of maximum pressures and rates of pressure rise are normally obtained in vessels larger and differently shaped than the Hartmann apparatus described in *BuMines RI 5624*. Data on maximum pressures and rates obtained from the Hartmann apparatus are shown in Table 5.1.42. A comparison of test data from the Hartmann apparatus and a 1-m^3 vessel is shown in "Development and Control of Dust Explosions." NFPA no. 68, "Guide for Venting of Deflagrations" (1994), recommends using the rate of pressure rise data obtained from closed vessels, 1-m^3 or larger, in venting calculations.

Combating Dust Fires

The following points should be observed when one is dealing with dust fires, in addition to the usual recommendations for fire prevention and firefighting, including sprinkler protection.

1. Attention should be directed to the potential hazard of spontaneous heating of dust products, particularly when grinding or pulverizing processes are used.

2. First-aid and firefighting equipment should be installed. Small hoses with spray nozzles or automatic sprinkler systems fitted with spray or fog nozzles are particularly satisfactory. The fine spray wets the dust and is not so likely to raise a dust cloud as with a solid stream. Portable extinguishers used to combat dust fires should be provided with similar devices for safe discharge.

3. Large hose of fire department size giving solid water streams should be used with caution; a dust cloud may be formed with consequent risk of explosion. Plant employees and the fire department should be advised of this potential hazard in advance. Hose equipped with spray or fog nozzles should be provided and kept ready for an emergency.

4. Fires involving aluminum, magnesium, or some other metal powders are difficult to extinguish. Sand, talc, or other dry inert materials, and special proprietary powders designed for this purpose should be used. These materials should be applied gently to smother the fire. Materials such as hard pitch can completely seal the dust from oxygen and may be used. (See NFPA Code nos. 48, 65, and 651.)

ROCKET FUELS

by Randolph T. Johnson

Indian Head Division, Naval Surface Warfare Center

REFERENCES: Billig, Tactical Missile Design Concepts, Johns Hopkins Applied Physics Laboratory, *Technical Digest,* 1983, **4,** no. 3. Roth and Capener, Propellants, Solid, "Encyclopedia of Explosives and Related Items," Kave, ed., PATR 2700, vol. 8, U.S. Army Armament Research and Development Command, Dover, NJ, 1978. "Solid Propellant Selection and Characterization," NASA Design Criteria Guide, NASA SP 8064, 1971.

Until the development of the solid-fueled Polaris missile in the late 1950s and early 1960s, rocket motors tended to be divided into two distinct categories: rocket motors greater than 3 ft in diameter were liquid-fueled, while smaller rocket motors were solid-fueled. Since that time, a number of quite large rocket motors have been developed which use solid fuel; these include the Minuteman series and, most recently, the solid rocket boosters used on the space shuttle, each of which has over 1,000,000 lb of propellant. There are exceptions to the rule in the other direction, too; e.g., the air-launched Bullpup rocket motor started out as a liquid-fueled unit, changed to a solid fuel, then back to a liquid fuel.

Design Criteria

Before selecting the proper propellant system for a given rocket motor, the designer must consider the parameters by which the design is constrained. The following criteria generally need to be considered when choosing the propellant(s) for a given rocket motor:

Envelope Constraints The designer must consider the volume, mass and shape limits within which the rocket motor is constrained. For example, an air-launched rocket may be limited by the carrying capacity of the aircraft, the size of the launcher, and the size of the payload.

Performance Requirements In general, the requirements imposed on a rocket motor are expressed as the minimum required and/or the maximum allowed to complete the mission and the maximum allowed to prevent damage to the payload, launcher, etc. Such parameters include velocity, range, burn time, and acceleration.

Environmental Conditions The environments seen by various rocket motors differ dramatically, depending on the intended use. For example, a submarine-launched strategic missile lives a pampered life in near ideal conditions while a field-launched barrage rocket or an air-launched missile see near worse-case environments. Some of the parameters to consider in the choice of propellants include temperature limits of storage and operation, vibration and shock spectra to be experienced and survived, and the moisture and corrosion environments the rocket motor (and possibly the propellant) may be expected to encounter. It may be possible to reduce the effects of these environments on the propellant by rocket motor design, but the propellant chemist and the rocket motor designer must be willing to work together to come up with a viable combination.

Safety Requirements Safety considerations include toxic and explosion hazards in the manufacture of the rocket motor and in servicing, handling, and use. The use of the rocket must be considered in the safety margin built into its design. If it is to be used in a "human-rated" system, such as an ejection seat, it must be of more conservative design than if it were to be used in a barrage rocket, for instance.

Service Life The designer must consider the duration over which the rocket motor will be in service. It is, in general, less expensive to make a number of rocket motors at a time and then store them, than to make the same number in several small lots over a period of time with the accordingly increased start-up and shutdown costs. A very short service life leads to heightened costs for repetitive shipping to and from storage, as well as rework and replacement of overaged units. There is yet another hidden cost of a very short service life: the risk that an overaged unit will be inadvertently used with potentially catastrophic results.

Maintenance The availability of maintenance will affect the choice of propellants for the given unit. The unit with readily available maintenance facilities will have far less severe constraints than the unit which must function after years of storage and/or far from the reach of service facilities.

Smokiness The choice of propellants must be influenced by whether or not the mission will allow the use of a propellant which leaves a smoky trail. Propellants containing a metal fuel, as well as certain other ingredients, leave a large smoky trail which reveals the location of the launch point. This is particularly undesirable in tactical rocket systems for it leaves the user revealed to the enemy. The smoke from metal fuels is termed **primary** smoke as it is a product of the primary combustion, **secondary** smoke comes from the reaction of propellant combustion products with the atmosphere, such as the reaction of hydrogen chloride with moisture in the air to leave a hydrochloric acid cloud, or water vapor condensing in cold air to leave a contrail.

Cost The cost of a rocket motor must be divided into a number of categories including design, test, ingredients, processing, components, surveillance, maintenance, rework, and disposal. The hidden cost of nonfunction (reliability) should also be factored into the equation.

Liquid Propellants versus Solid Propellants

The first choice to be made in the selection of propellants is whether to use a liquid or a solid propellant. Each of these comes with its own set of advantages and disadvantages which must be weighed in the selection process.

Liquid propellants offer the possibility of extremely high impulse per unit weight of propellant. The thrust of a liquid-propellant rocket motor may be easily modulated by controlling the flow rate of the propellant into the combustion chamber.

Liquid propellants, however, tend to be of low density, which leads to large packages for a given total energy. Most liquid propellants have a very limited usable temperature range because of freezing or vaporization. Because they require the use of valves, pipes, pumps, and the like, liquid-propellant rocket motors are relatively complex and require a high level of maintenance.

The use of liquid rocket propellants is predominant in older strategic rocket motors, such as Titan, and in space flight, such as the space shuttle main engines and attitude-control motors.

Solid propellants offer the possibility of use over a wide range of environmental conditions. They offer a significantly higher density than liquid propellants. Solid-propellant rocket motors are much simpler than liquid rocket motors, as the propellant grain forms at least one wall of the combustion chamber. This simplicity of design leads to low maintenance and high reliability. Safety is somewhat higher than with liquid propellants, for there are no volatile and hazardous liquids to spill during handling and storage.

Solid propellants, however, do not attain the impulse levels of the more energetic liquid propellants. Furthermore, once ignited, the thrust-time profile will be as dictated by the propellant surface history and propellant burn rate for the nozzle given; this profile is not easily altered, unlike that of a liquid-propellant rocket motor suitably equipped.

Solid-propellant rocket motors are now used in every size from small thrusters on the Dragon antitank round to the boosters on the space shuttle. They are used when a preprogrammed thrust-time history is appropriate.

Liquid Propellants Once the decision has been made to use liquid propellant, the designer is confronted with the decision of whether to use a monopropellant or a bipropellant system.

A **monopropellant** is a fuel which requires no separate oxidizer, but provides its propulsive energy through its own decomposition. The advantage of a monopropellant is the inherent simplicity of having only one liquid to supply to the combustion chamber. The principal disadvantage of the commonly used monopropellants is their very low impulse compared to most bipropellant or solid-propellant systems. Two of the more commonly used monopropellants are ethylene oxide and hydrogen peroxide.

Bipropellant systems use two liquids, an oxidizer and a fuel, which are merged and burned in a combustion chamber. There is a much wider selection of fuel constituents for bipropellant systems than for monopropellant systems. Bipropellants offer much higher impulse values than do the monopropellants, and higher than those available with solid propellants. The primary disadvantage of bipropellant systems is the need for a far more complex piping and metering system than is required for monopropellants.

Bipropellants may be subdivided further into two categories, **hypergolic** (in which the two constituents ignite on contact) and **nonhypergolic.** Hypergolic systems eliminate the need for a separate igniter, thus decreasing system complexity; however, this increases the fire hazard if the fuel system should leak.

When **cryogenic liquids** such as liquid hydrogen and liquid oxygen are used, the problems of storage become of major significance. A significant penalty in weight and complexity must be paid to store these liquids and hold them at temperature. Furthermore, these liquids charge a significant penalty, because their very low density enlarges the packaging requirements even more. This penalty is in the parasitic weight of the packaging and the increased drag induced by the increased skin area. These problems have limited the use of these propellants to systems where immediate response is not required, such as space launches, as opposed to strategic or tactical systems.

Fuels for bipropellant systems include methyl alcohol, ethyl alcohol, aniline, turpentine, unsymmetrical dimethylhydrazine (UDMH), hydrazine, JP-4, kerosene, hydrogen, and ammonia. **Oxidizers** for bipropellant systems include nitric acid, hydrogen peroxide, oxygen, fluorine, and nitrogen tetroxide. In some cases, mixtures of these will yield superior properties to either ingredient used alone; e.g., 50:50 mixture of UDMH with hydrazine is sometimes used in lieu of either alone.

Solid Propellants Once the decision has been made to use a solid propellant, one must decide between a case-bonded and cartridge-loaded propellant grain. The decision then must be made among composite, double-base, and composite modified double-base propellants.

Case-Bonded versus Cartridge-Loaded Case bonding refers to the technique by which the propellant grain is mechanically (adhesively) linked to the motor case. Cartridge-loaded propellant grains are retained in the motor case by purely mechanical means.

Case bonding takes advantage of the strength of the motor case to support the propellant grain radially and longitudinally; this permits the use of low-modulus propellants. This technique also yields high volumetric efficiency.

Since the propellant-to-case bond effectively inhibits the outer surface of the propellant, the quality of this bond becomes critical to the proper function of the rocket motor. This outer inhibition also tends to limit the potential propellant grain surface configurations, thus limiting the interior ballistician's leeway in tailoring the rocket motor ballistics. Since the propellant grain is bonded to the motor case, the grain must be cast into the motor case or secondarily bonded to it. The former method leads to reduced flexibility in scheduling motor manufacture. The second technique can lead to quality control problems if a bare, uninhibited grain is bonded to the motor case, or to a loss in volumetric loading efficiency if a bare grain is inhibited or cast into a premolded form that is bonded to the motor case. Bonding the propellant grain to the motor case also makes it difficult to dispose of the rock motor when it reaches the end of its service life, especially in regard to making the motor case suitable for reuse.

Cartridge loading of the propellant grain offers flexibility of manufacture of the rocket motor, as the motor-case manufacture and propellant-grain manufacture may be pursued independently. This technique also allows the interior ballistician a free hand in configuring the propellant surface to obtain optimal ballistics. Since the propellant grain is held in the motor case mechanically, it is a simple matter to remove the propellant at the end of its service life and reuse the motor case and associated hardware.

The cartridge-loaded propellant grain must be inhibited in a separate operation, as opposed to the case-bonded grain. The free-standing grain must be supported so that it is not damaged by vibration and shock; correspondingly, the propellant must have substantial strength of its own to withstand the rigors of the support system and the vibrations and shocks that filter through the system. The presence of the mechanical support system and the inhibitor, and the space required to slide the propellant grain into the motor case, prevent the cartridge-loaded propellant grain from having the volumetric loading efficiency of a case bonded propellant grain.

Composite versus Double-Base versus Composite Modified Double-Base Composite propellants consist primarily of a binder material such as polybutadiene (artificial rubber) and finely ground solid fuels (such as aluminum) and oxidizers (such as ammonium perchlorate). Double-base propellants consist primarily of stabilized nitrocellulose and nitroglycerine. Composite modified double-base propellants use a double-base propellant for a binder, with the solid fillers commonly found in composite propellants.

Composite propellants offer moderately high impulse levels and widely tailorable physical properties. They may be designed to function over a wide range of temperatures and, depending upon the fillers, have high ignition temperatures and consequently very favorable safety features.

Composite propellants at the present time are limited to cast applications. The binders are petroleum-based, and subject to the vagaries of oil availability and price. Many of the formulations for their binders contain toxic ingredients as well as ingredients which are sensitive to moisture. The water sensitivity carries over to the filler materials which tend to dissolve or agglomerate in the presence of moisture in the air. Composite propellants also are notoriously difficult to adequately inhibit once the propellant has fully cured.

Historically, a number of binder materials have been used in the manufacture of composite propellants. Among these are asphalt, polysulfides, polystyrene-polyester, and polyurethanes. Propellants of recent development have been predominantly products of the polybutadiene family: carboxy-terminated polybutadiene (CTPB), hydroxy-terminated polybutadiene (HTPB), and carboxy-terminated polybutadiene-acrylonitrile (CTBN). It should be pointed out that the binder material also serves as a fuel, so a satisfactory propellant for many purposes may be manufactured without a separate fuel added to it. While the addition of metallic fuels to the propellant significantly increases the energy of the propellant, it also produces primary smoke in the form of metal oxides. The most commonly added metal is aluminum, but magnesium, beryllium, and other metals have been tried. The most commonly used oxidizer (which makes up the preponderance of the weight of the propellant) is ammonium perchlorate. Other oxidizers which have been used include ammonium nitrate, potassium nitrate, and potassium perchlorate.

Double-base propellants have been used in guns since Alfred Nobel's discovery of ballistite in the nineteenth century. Their first significant use in rocketry came during World War II in barrage rockets. All our modern double-base propellants are descended from JPN, of World War II vintage. These propellants may be cast or extruded, use relatively nontoxic ingredients which are relatively insensitive to water, do not use petroleum derivatives to any appreciable extent, and are easily inhibited by solvent bonding an inert material to the surface of the propellant. It is possible to attain very low temperature coefficients of burn rate with these propellants; it is also possible to attain a double-base propellant which declines in burn rate with pressure over a portion of its usable pressure range, as opposed to the more usual monotonic increase in burn rate. This "mesa" burning, as it is called, allows the interior ballistician to more easily stabilize the pressure and thrust of the rocket motor.

Double-base propellants tend to have low to moderate impulse levels and are limited in temperature range not only by a high glass transition temperature, but also by their tendency to soften, liberate nitroglycerine, and decompose at higher temperatures. These propellants tend to autoignite at low temperatures and tend to be explosive hazards. Their physical properties are essentially fixed by the properties of nitrocellulose and are not easily tailored to changing requirements.

Double-base propellants contain nitrocellulose of various nitration levels (usually 12.6 percent nitrogen), nitroglycerine, and a stabilizer. Various inert plasticizers are added to modify either the flame temperature or the physical properties of the propellant.

Composite modified double-base propellants offer the highest energy levels presently available with solid propellants. This energy comes at an extremely high price, however: the storage and operating temperature limits only differ from minimum to maximum by about 20°F. These propellants are also extremely shock-sensitive and have very low autoignition temperatures. They are limited in use to systems such as submarine-launched ballistic missiles, where the temperature and vibration conditions are closely controlled. Classically, these compositions have contained nitrocellulose, nitroglycerine, aluminum, ammonium perchlorate, and the explosive HMX, which serves both as an oxidizer and gas-producing additive.

Propellant Properties and Interior Ballistics

The thrust and pressure profiles of a rocket motor must be controlled in order to meet the design criteria noted earlier. With liquid and solid propellants, the nominal controls differ dramatically, but in the elimination of spurious pulses and the control of nozzle erosion, the two types of propellant are more similar than different.

Pressure and Thrust Control In design of a liquid-propellant rocket motor, the prime considerations are properly sizing the combustor and nozzle and metering the propellant flow to attain the desired thrust. If variations in thrust are desired, these may be carried out either through preprogramming the flow by orifice size or by pump pressure; similarly, the flow may be varied on command by the operator.

The situation with a solid rocket motor is less clear-cut. The rate of gas generation inside the rocket motor is controlled by the burn rate of the propellant and the amount of burn surface available.

The burn rate of the propellant may be varied by the addition of various chemical catalysts; iron compounds are sometimes added to composite propellants and lead compounds to double-base propellants in order to speed burning. Coolants such as oxamide are added to composite propellants and inert plasticizers are added to double-base propellants to slow burn rates. The burn rate of composite propellants may be changed by changing oxidizers or by modifying the size distribution of the oxidizer. In efforts to achieve ultra-high burn rates, silver or aluminum wires have been added to propellants to increase heat conduction and so speed burning.

The surface area of a propellant grain is controlled by its shape and by the amount and areas in which the propellant grain is inhibited. This surface area tends to change in configuration as the propellant burns. The propellant grain designer must evolve a grain configuration which yields the desired pressure-time and thrust-time history. Most often, the objective in grain design is to achieve a neutral to slightly regressive (constant to slightly decreasing) thrast-tirne history.

In some cases (as where a high-thrust phase is to be followed by a low- to moderate-thrust phase), it is necessary to develop a grain or grains with varying geometries and/or varying burn rates to achieve the desired thrust-time profile. These results may require such techniques as using tandem or coaxial grains with different geometries and/or different burn rates and perhaps different propellants.

The pressure within the combustion chamber depends not only on the rate of combustion, but also on the size of the nozzle throat. Throat size tends to decrease with heating, but the surface tends to wear away with the hot gases and embedded particles eroding material from the surface of the throat and nozzle exit cone. In some cases, this loss of material may serve to assist the interior ballistician in the quest for the ideal thrust-time profile.

Combustion Instability Rocket motors are sometimes given to sudden, erratic pressure excursions for a number of reasons. Liquid-propellant fuel may surge; solid propellant grains may crack; material may be ejected from the motor and temporarily block the nozzle. Sometimes these excursions cannot be explained by any of the above possibilities, but will be attributed to "unstable combustion." The causes of unstable or "resonant" combustion are still under investigation.

A number of techniques have been developed to reduce or eliminate the pressure excursions brought on by unstable combustion, but there is, as yet, no panacea.

Nozzle Erosion As the propellant gases pass out of the rocket motor through the nozzle, their heat is partially transferred to the nozzle. The heated nozzle material softens and tends to be eroded by the mechanical and chemical action of the propellant gases. The degree of erosion is heightened when a metal fuel is added to the propellant gases, particulate matter is contained in the gases, the gases are corrosive, or the propellant gases are oxidizing.

One technique widely used in reducing nozzle erosion is to add coolant to the propellant formulation. Ablatives are sometimes added to the nozzle or chamber ahead of the nozzle throat; gases generated by the ablating material form a boundary layer to protect the nozzle from the hot propellant gases in the core flow. In liquid propellant rockets, the fuel (where stability permits) may be used as a coolant fluid. Nozzle inserts are probably the most commonly used technique to limit nozzle erosion. The nozzle shell is usually made of aluminum or steel when inserts are used and the nozzle throat is made of heat-resistant material. For a small amount of permissible erosion, carbon inserts are used; when no erosion is acceptable, tungsten or molybdenum inserts are used.

5.2 INDUSTRIAL COMBUSTION FURNACES

by Charles E. Baukal, Jr.

REFERENCES: Mullinger and Jenkins, *Industrial and Process Furnaces* (2nd ed.), Butterworth-Heinemann, 2014. Trinks et al., *Industrial Furnaces* (6th ed.), Wiley, 2004.

There are many different names that are used for the combustion chamber containing the heating system and the material being heated, often referred to as the load. Ovens and dryers are combustors that typically operate at lower temperatures and are used to remove moisture from materials. Heaters are typically refractory-lined chambers usually considered to be lower temperature combustors used to heat solids and liquids. These lower temperature combustors are not considered here. Furnaces are refractory-lined chambers usually considered to be higher temperature combustors used for heating and melting solids. Kilns (pronounced "kills") are combustors used to bring about physical and chemical changes in materials and are a type of furnace that is usually associated with thermal processing of nonmetallic solids such as the ceramic, cement, and lime industries. The chemical changes are often through calcination that may involve water evaporation, volatile evolution, and even partial combustion of the load material.

Solid, liquid, and gaseous fuels are used in furnaces. These fuels are discussed in detail in other sections in this chapter. They are only briefly considered in this section as they pertain to specific furnaces. Due to space limitations there are many aspects of furnaces that will be considered here briefly if at all. For example, heat transfer calculations for sizing furnaces, furnace construction details, and fuel and load handling equipment are not included here. This section is intended to be representative but not comprehensive so not all industrial combustion furnaces will be discussed.

FURNACES

The proper choice of furnace type depends on the specific application, as there are many possibilities. One important consideration is whether materials will be processed in a batch or continuous mode. Another important consideration is the kind of fuel that will be used. Other considerations include the furnace geometry and required temperature range. The processing method and furnace geometry are briefly considered next.

The two volume set of books by Trinks and Mawhinney (1961, 1967) is considered by many to be the bible of furnaces and is a good reference that deals with all aspects of furnaces including heat transfer and heat recovery. However, Trinks and Mawhinney have specifically limited their definition of industrial furnaces to only those processes which raise the temperature of the material (mainly metals), but do not cause any phase changes (melting or vaporization) or any chemical changes (e.g., calcining). Even with that narrow definition of furnace, they list 61 applicable processes ranging in material temperatures from 300 to 3,250°F (150 to 1,790°C). Here, industrial furnace is more broadly defined to include processes that not only raise the material temperature, but which may also cause phase changes and chemical reactions.

Firing Method

There are two main methods used to heat materials that are usually referred to as direct and indirect firing. The method chosen is normally dictated by the process requirements. If the material being processed must not come into contact with the combustion products, then indirect firing is used. If there are no such requirements, direct firing is preferred as it is normally more efficient and straightforward.

Direct Firing In a direct-fired furnace, there is nothing between the combustion products generated by the burners and the load, as shown in Fig. 5.2.1. This is the predominant type used in most industrial heating applications and essentially all of the high temperature processes.

Indirect Firing In an indirect-fired furnace, there is some intermediate surface between the combustion products and the load, as shown in Fig. 5.2.2. That surface is commonly some type of ceramic due to the high temperatures, although metals are used in some cases. This surface is designed to prevent the combustion products from contacting the load and reducing the quality of the finished product. That quality can be reduced by chemically altering the product or physically changing the surface. An example of the first is in many metallurgical processes, especially heat treating, where the metal product must be heated in a protective atmosphere containing H_2, N_2, and CO, with only negligible quantities of O_2 and H_2O which are detrimental to the quality of the metal during heat treating. Another type of indirect-fired furnace is where the flame gases are separated from the load for either transport or safety reasons.

Two methods are commonly used to separate the combustion products from the load. One is to use open-flame burners but to have a

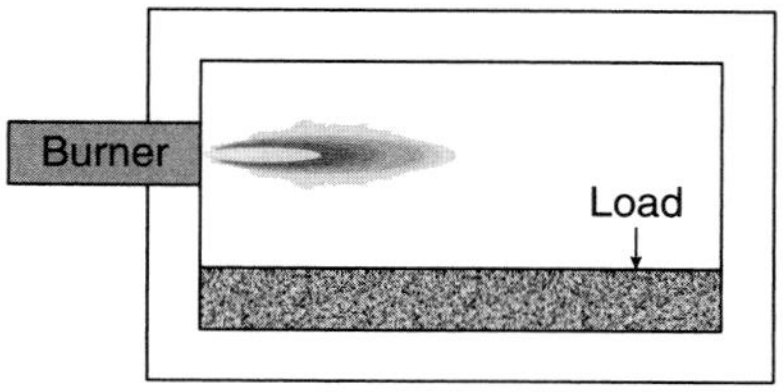

Figure 5.2.1 Electrical resistance heated furnace with resistance elements.

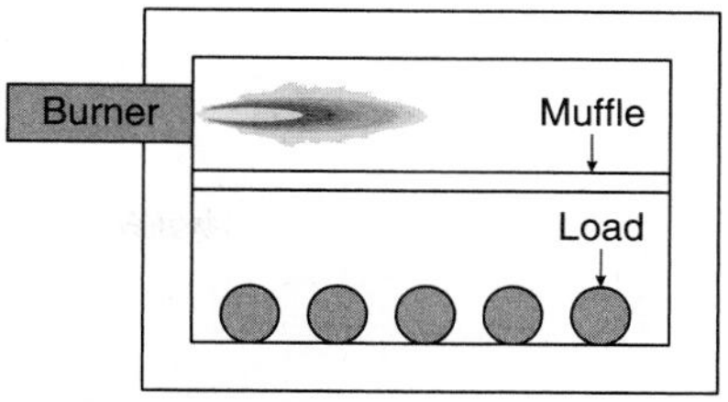

Figure 5.2.2 Elevation view of an indirect-fired furnace (CRC Press, with permission).

separator, sometimes referred to as a muffle, across the entire combustion space between the flames and the load. Some of the challenges of this method include supporting the separator because of the high temperatures, maximizing the heat transfer from the flames to the separator to optimize the thermal efficiency, and getting a good gas seal around the perimeter of the separator due to thermal expansion. The second method commonly used in industrial combustion applications to separate the exhaust products from the load is to use radiant tube burners (Fig. 5.2.3). In that method, the flame from each individual burner is contained in a metallic or ceramic tube. This often improves the overall heat transfer to the separator because of the improved forced convection

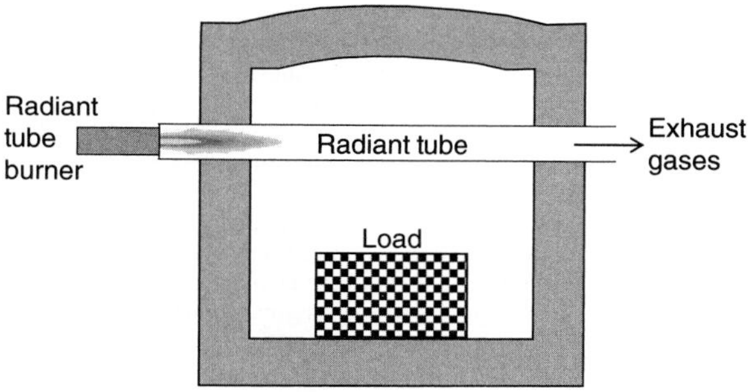

Figure 5.2.3 Elevation view of radiant-tube-fired furnace.

inside the tube. As with the muffle, supporting long radiant tubes can be a problem as well as the seal between the typically metal burner and ceramic tube. A different challenge compared to the muffle is getting uniform heat flux from the radiant tubes as the hottest gases are produced at the burner end with the coldest gases exiting from the tube. This can be mitigated through proper design of the radiant tube burner.

The analysis of an indirect-fired furnace may be more complicated than for a direct-fired furnace due to the presence of the heat exchange surface between the flames and the load. The heat transfer from the flames to that surface is normally by both radiation and forced convection. Then the heat must be conducted through that surface where it then radiates to the load. This simplifies the analysis of the heat transfer to the load that is essentially all by radiation.

Heat Distribution The distribution of burners within a furnace has a significant impact on how heat will be transferred. As will be discussed later in this chapter, the furnace geometry is a large determining factor on where burners will be placed. For example, in a rotary furnace (see Fig. 5.2.4), which is often long and relatively narrow in diameter, a single burner centered on one end of the furnace and firing toward the other end is often the preferred geometry to get proper heat distribution. The desired flame length is often a specific fraction (e.g., half) of the length of the cylinder. In a rectangular-shaped furnace, multiple burners are often used and distributed on one or more walls. In a side-port glass-melting furnace (see Fig. 5.2.5), burners are typically located on two opposite side walls and directly opposed to each other.

The furnace designer, in conjunction with the burner manufacturer, must decide how many burners will be used and where they will be located in the furnace. There are several factors that must be

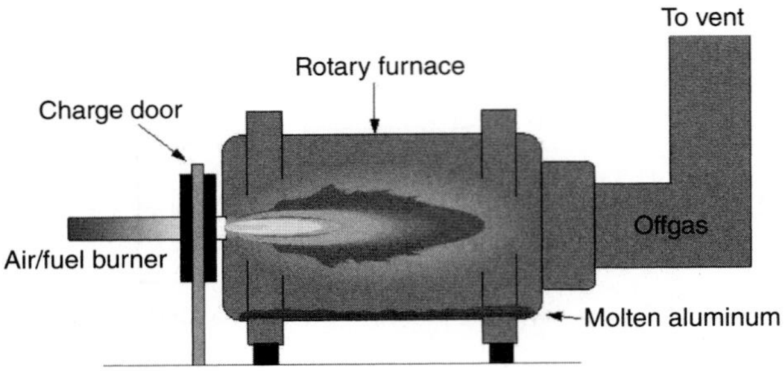

Figure 5.2.4 Conventional single pass, rotary aluminum melting furnace (CRC Press, with permission).

considered. As with most things, economics is an important consideration. It is generally more expensive to use two smaller burners than it is to use one larger burner, not only because of the cost of the burners which is incrementally higher, but also because of the associated extra piping and controls equipment. Another consideration is the physical layout of the furnace structural steel, which limits where burners can be placed in order to avoid steel support members. Still another consideration is the heat transfer distribution within the combustion chamber. Multiple smaller burners often have a more uniform heat distribution pattern than single larger burners, as shown in Fig. 5.2.6 where the size of the flame is proportional to the firing rate. In both cases, the average amount of energy released per unit volume of combustion space is the same, only the distribution of the energy release is different. As can be seen in this example, more uniform heating is usually achieved with a greater number of smaller burners, compared to fewer larger burners.

The actual burner layout is also a function of the burner design and the firing rate of each burner. In certain applications, it may be preferable to have different burners firing at different rates. As an example, Fig. 5.2.7 shows the burner closest to the incoming materials firing the hardest and the burner closest to the outgoing materials firing at the lowest rate. In some cases, the incoming material can absorb as much heat as is available, without overheating. It is also sometimes the case that too much heat applied to the outgoing materials can reduce the product quality by overheating. In some applications, different burner types may also be used in a single furnace. An example is shown in Fig. 5.2.8, which is referred to as oxy/fuel boosting. In that process, oxy/fuel burners are located at the feed end of the furnace to provide more heat than with conventional air/fuel burners. This is used in glass

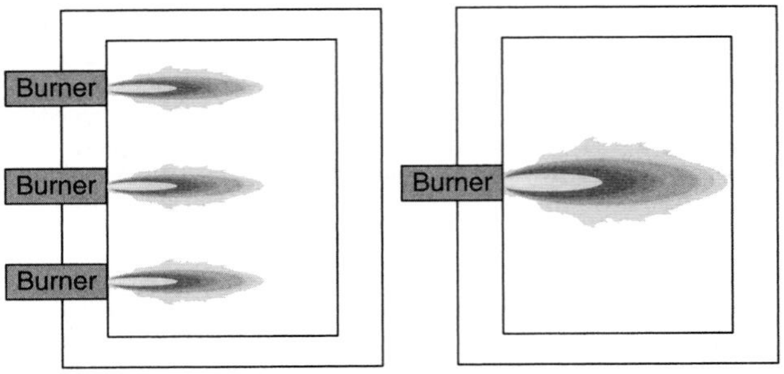

Figure 5.2.6 Burner distribution effects shown in a plan view of two furnaces with equal total firing rates: (*a*) 3 smaller burners vs. (*b*) 1 larger burner (CRC Press, with permission).

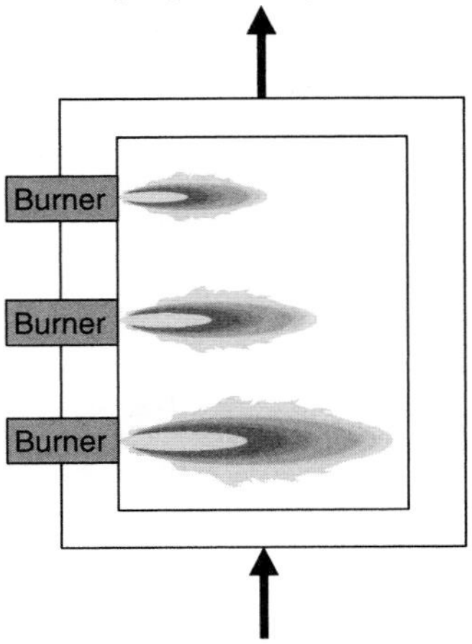

Figure 5.2.7 Plan view of burner profiling where burners are fired at different rates to get a certain heat flux profile (CRC Press, with permission).

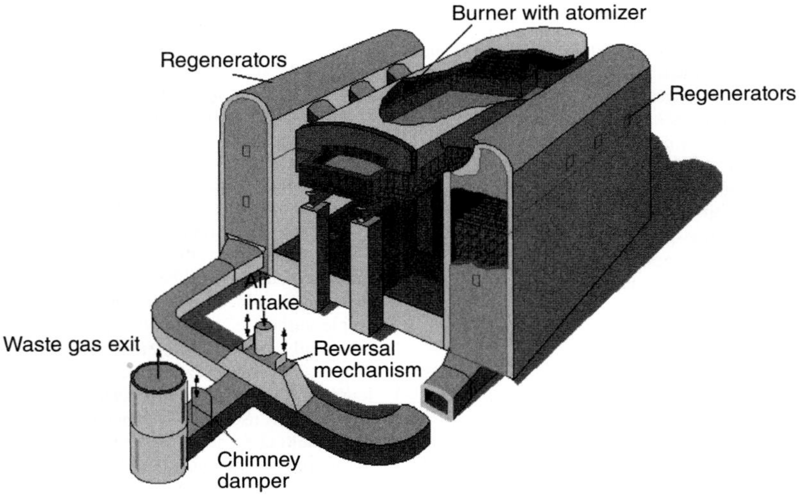

Figure 5.2.5 Typical side-port glass melting furnace (CRC Press, with permission).

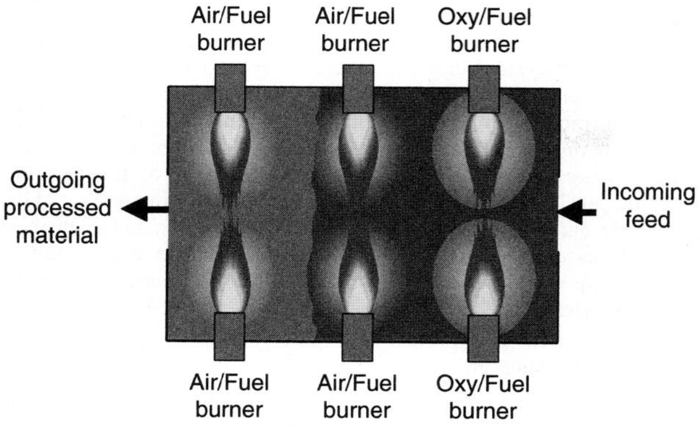

Figure 5.2.8 Plan view of an oxy/fuel boosting in a furnace (CRC Press, with permission).

melting furnaces to increase productivity. In that application, the heat transfer distribution in the furnace is even more complicated because the air/fuel burners fire on only one side at a time and cycle to the other side about every 15-30 minutes. Another example of a variation of the heat distribution in the furnace is shown in Fig. 5.2.9. The scrap aluminum is loaded through the top of the furnace. The burners do not fire perpendicular to the wall, but fire at an angle to create a specific

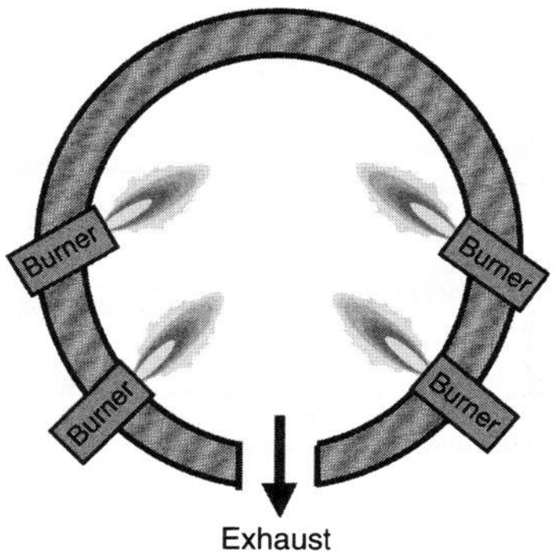

Figure 5.2.9 Plan view of a round-top aluminum melting furnace (CRC Press, with permission).

flow pattern around the scrap, instead of having all four burners fire directly toward the center of the furnace. This furnace relies heavily on both convection and radiation, depending on the portion of the melting cycle. Right after a new charge of scrap is added to the furnace that may be piled above the height of the burners, forced convection plays an important role by flowing through the gaps in the scrap pile. As the pile melts down to a flat bath, radiation becomes more dominant.

Load Processing Method

Industrial furnaces are commonly either batch or continuous processes. The specific method has a strong influence on the heat transfer mechanisms that will be important for each type of system. There are also some hybrid processes that are a combination of batch and continuous methods.

Batch Processing In a batch process, materials are loaded into a combustor, heat is then applied, and the materials are removed from the combustor at the completion of the heating cycle. An example is an electric arc furnace (EAF) where scrap metal is loaded into the furnace, melted, and then tapped out for casting. Batch processes are typically a more complicated system to analyze because of their time-dependent nature. Since the initially-loaded material is usually much colder than the combustor, convection and radiation are important modes of heat transfer at the beginning of the heating cycle. The temperature inside the combustor is also transient, where colder temperatures are present near the opened charging door and higher temperatures are more prevalent farther from the door as they have been less affected by the heat lost through the open door. During the heating cycle, the temperature distributions inside the combustor and in the load are both dynamically changing. Later in the heating cycle, as the load temperature comes closer to that of the combustor, radiation often becomes the predominant heat transfer mechanism.

Continuous In a continuous heat process, the material is constantly fed through the combustor, heated and withdrawn from the combustor. During normal operations the combustor is at steady-state conditions with little variation, for example, in the wall temperatures, except at start-up, shut-down, during product changes, or when there are process problems. An example of a continuous process is the production of glass where raw batch material is continuously fed into the furnace inlet and molten glass is removed at the furnace outlet. A continuous heating process is often more straightforward to analyze and model because of the steady-state nature of the system. However, this does not necessarily mean that analysis is simple, merely that time-dependence can be excluded. Continuous heating processes may have a more complicated temperature distribution in the combustor to accommodate the heating of the load. Where a batch combustion process may have fairly uniform furnace wall temperatures inside the combustor, continuous furnaces may have a wider range of wall temperatures corresponding to the wide spatial temperature differences in the load which is typically much colder at the feed end than at the discharge end.

An example of a continuous furnace is a tunnel kiln (see Fig. 5.2.10) which is used to make bricks, refractories, and ceramics. The materials to be heated, sometimes referred to as ware, are loaded onto cars and pushed through the kiln at regular intervals. The ware passes through three zones: preheating, firing, and cooling. In the preheating zone, the ware is preheated by the exhaust products from the firing zone. The primary function of the preheating zone is to remove moisture and burn out any organic material that might be present. In the firing zone, roof- and/or side-mounted burners are fired in the free space around the ware. The function of this zone is to promote high-temperature chemical transformations, such as the decomposition of carbonates, and to heat the ware to the sintering temperature. In the third zone, the cooling zone, the ware is cooled down by the incoming combustion air which

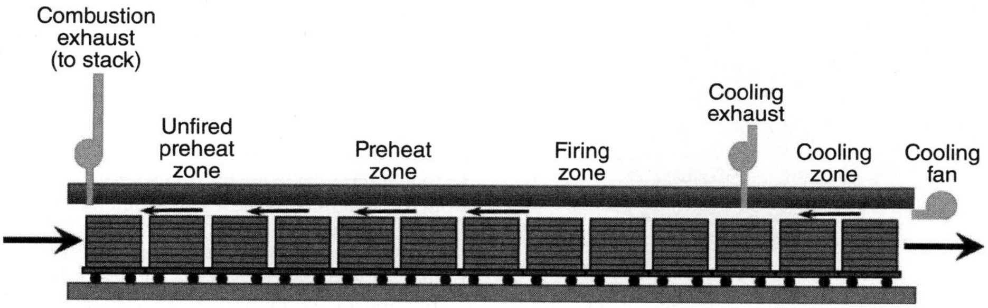

Figure 5.2.10 Elevation side view of a tunnel kiln.

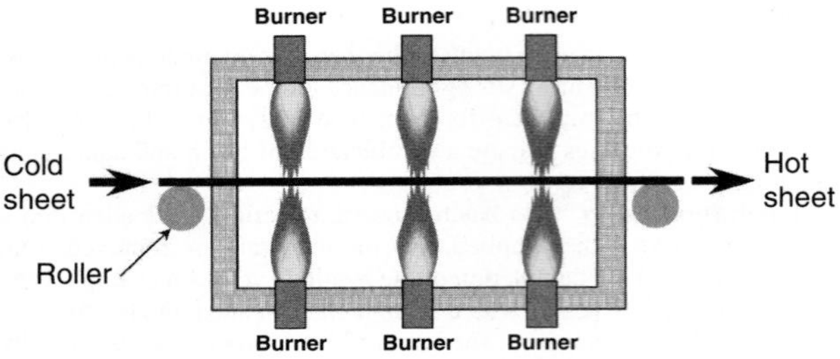

Figure 5.2.11 Elevation view of flame impingement inside a continuous reheat furnace (CRC Press, with permission).

is simultaneously preheated by the hot ware to improve the thermal efficiency of the system. Some of the preheated air is removed from this zone and used in other parts of the plant.

Another example of a continuous heating furnace is shown in Fig. 5.2.11. Continuous reheat furnaces are used to heat up sheet steel prior to forming. The increased metal temperatures make forming easier. Another version of this type of furnace is where there are burners firing on the side of metal sheets (see Fig. 5.2.12) to heat them back

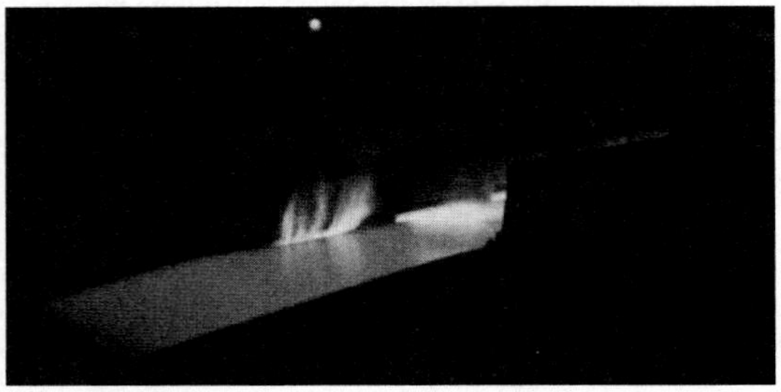

Figure 5.2.12 Flame impingement used to heat the edge of metal sheets (CRC Press, with permission).

up as the edges cool faster than the main body of the sheet. In general, the more non-uniform the sheet temperature, the lower the quality of the finished product or the more metal wasted by cutting off the edges.

Hybrid Some processes, like reverberatory furnaces (see Fig. 5.2.13), are hybrids because they have some features of both batch and continuous furnaces. Side-well reverbs often have a provision for continuously feeding new scrap into a well on the side of the furnace,

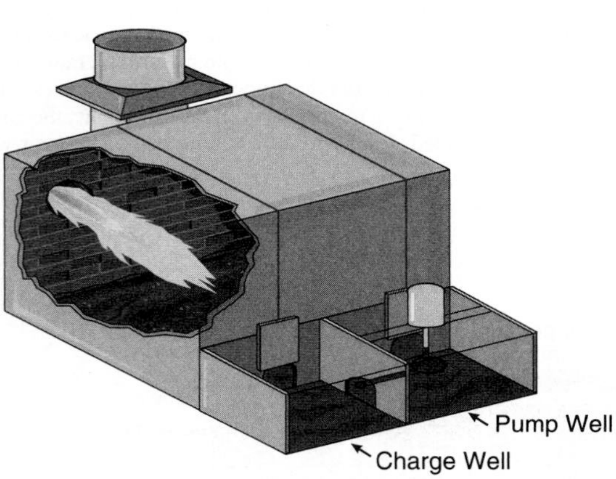

Figure 5.2.13 Sketch of a reverberatory furnace (CRC Press, with permission).

outside of the combustion chamber, so that the door to that chamber does not need to be opened to charge new materials. However, heats are done in discrete cycles which take on average from 4 to 10 hours depending on the alloy being made, the furnace conditions, and the operating procedures at the given plant. At the end of the heat, the molten aluminum is tapped out of the furnace through a tap hole and then is poured into molds to make ingots for use later, or directly into a specific part like an aluminum wheel. During the tapping portion of the cycle, the burners are normally on a low-fire mode to maintain the heat in the furnace without overheating the aluminum. The heating portion of the cycle is more continuous in nature. The heat transfer inside the furnace varies more than during certain parts of the cycle than in others.

Heat Transfer Medium

Another factor that has a significant influence on the heat transfer in the furnace is the medium used to transport the energy from the burners to the load. In industrial combustion, this medium is most commonly a gas, but liquids and solids are also used. In some heat treating processes, no medium at all is used (vacuum furnaces).

Gaseous Medium Gases are by far the most common medium used to transport convective heat from the burners to the load and to the furnace walls. In most direct-fired processes, the combustion products themselves transport heat to the load by forced convection. The actual rate of heat transfer depends on the flow geometry (gas velocity, how the gas flows over the load, and the shape of the load), the temperatures of both the gases and the load, and the gas composition. Processes are often specifically designed to maximize this convective heat transfer. For example, in a counter-flow rotary kiln (see Fig. 5.2.14), the combustion products flow in one direction while the load flows in the other

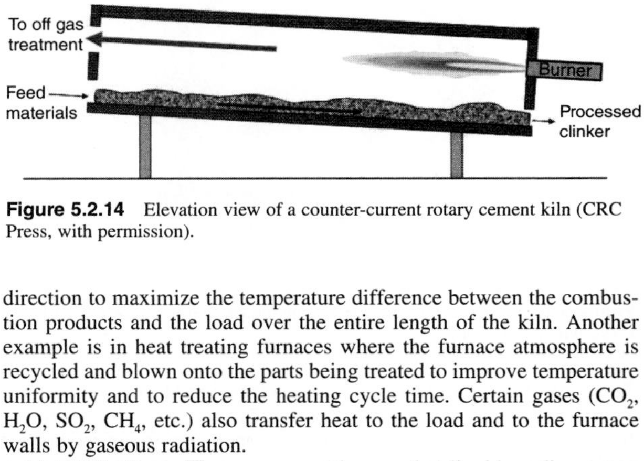

Figure 5.2.14 Elevation view of a counter-current rotary cement kiln (CRC Press, with permission).

direction to maximize the temperature difference between the combustion products and the load over the entire length of the kiln. Another example is in heat treating furnaces where the furnace atmosphere is recycled and blown onto the parts being treated to improve temperature uniformity and to reduce the heating cycle time. Certain gases (CO_2, H_2O, SO_2, CH_4, etc.) also transfer heat to the load and to the furnace walls by gaseous radiation.

Liquid Medium There are several ways that liquid mediums may transport heat in an industrial furnace. One way is by convection from molten materials. For example, in a glass-making furnace, raw cold batch materials are distributed over the surface of the molten glass at the feed end of the furnace. The batch materials are typically melted into the molten glass before they reach halfway across the furnace. Another example of heat transfer from liquids is in a scrap aluminum melting furnace. As the aluminum melts and flows down to the bottom of the furnace, it convectively heats the solid scrap that it flows over. In some metal production processes, there is a slag layer on the top of the molten metal. This layer is deliberately created by the steel maker to minimize oxidizing the desired metal. Then heat must be transferred to the liquid slag layer, which in turn transfers heat to the molten metal. Heat may also be conducted through a liquid medium by conduction. For example, in a reverberatory furnace used for producing aluminum, the energy is received at the surface of the molten liquid aluminum. Because of the high thermal conductivity of aluminum, the heat is quickly conducted to lower levels through the liquid aluminum.

Solid Medium There are several ways that heat may be transferred through a solid medium in an industrial furnace. One method is by luminous radiation produced by solid soot particles generated in the flame. Another method is by thermal conduction between contacting solid parts. For example, in a vacuum furnace, energy is received at the outer boundaries of the pile of parts. Then heat is conducted into the parts by thermal conduction between the parts in contact with each other. Solid parts may also transfer heat by radiating to each other. Although not considered here, in a fluidized bed combustor, heat is transferred by circulating suspended solid particles.

Geometry

The geometry refers to both the shape and the size of the combustion chamber. An important parameter that determines the heat loss from a combustor is the external surface area. Heat is lost by both convection and radiation from the outside surface walls. Combustor geometries with higher surface areas will normally have higher losses than those with lower surface areas, assuming everything else is equivalent. One way that combustors are designed is based on the heat release per unit of combustor volume (e.g., kW/m^3 or $Btu/hr\text{-}ft^3$). It is then possible to compare the ratio of the surface area to the volume for different geometries:

$$SV = \frac{\text{surface area}}{\text{volume}} \tag{5.2.1}$$

The higher the SV ratio, the higher the proportion of heat loss for that combustor design. A spherical combustor would have the following surface area to volume ratio:

$$SV_{\text{sphere}} = \frac{4\pi r^2}{\frac{4}{3}\pi r^3} = \frac{3}{r} \tag{5.2.2}$$

where r is the radius of the sphere. This shows that the larger the sphere, the lower the SV_{sphere}, which normally means proportionally lower heat losses. Similarly, a cubical combustor would have the following surface area to volume ratio:

$$SV_{\text{cube}} = \frac{6l^2}{l^3} = \frac{6}{l} \tag{5.2.3}$$

where l is the length of each side of the cube. As the cube size increases, SV_{cube} decreases, which again should result in proportionately lower heat losses (see Fig. 5.2.15). This implies that larger furnaces should have proportionally lower heat losses and therefore be more thermally efficient than smaller furnaces, all other factors being equivalent. This is true in most cases, although real industrial processes are often more complicated than strictly considering the size. For example, in a batch process, loading and unloading the furnace may take proportionately more or less time for a larger furnace, depending on the material handling methods and downstream processes.

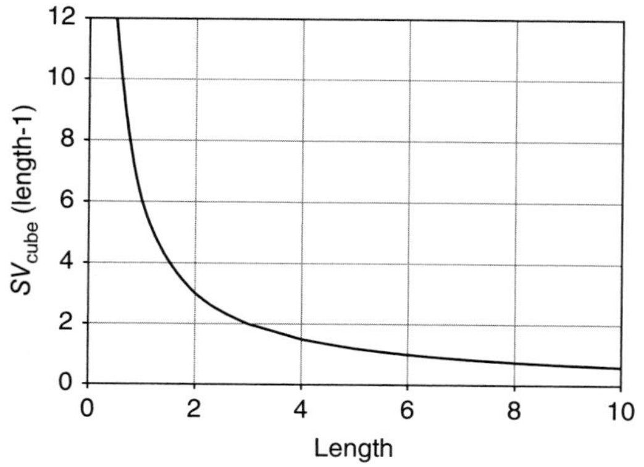

Figure 5.2.15 Surface area-volume ratio (SV) for a cubical furnace as a function of the length of each side of the cube (CRC Press, with permission).

It is also instructive to compare the relative shapes of combustors and their impact on heat losses. For the sake of comparison, the volume of the furnace will be held constant as the heat release per unit furnace volume will be assumed to be fixed. This would not be strictly true if the heat losses varied for different furnace geometries as more heat would be required for furnaces with higher heat losses. However, the heat losses for a well-insulated furnace are commonly relatively small, so fixing the furnace volume is reasonable to compare surface area effects. The base case will be a cubical furnace with the length of each side equal to 1 unit (l = 1). Then, the volume of that cube would be 1 unit^3 and the surface area would be 6 unit^2, with an SV_{cube} = 6 unit^{-1}. For a spherical combustor with an equal volume of 1 unit^3, the radius would be r = 0.62 units. Then the SV_{sphere} would be 4.8 unit^{-1}. The ratio of surface to volume ratios for the sphere to the cube ($SV_{\text{sphere}}/SV_{\text{cube}}$) equals 0.81. This means that a cubical furnace would normally have proportionally higher heat losses than a spherical furnace, for equal volumes. However, it is not as practical to build a spherical furnace, compared to a cubical furnace.

One can also compare a rectangular furnace to a cubical furnace to see how the heat loss would be affected. Again assume a base case of a cubical furnace with a unit volume. Choose a rectangular furnace that has two sides of equal length l_1 with the remaining side twice as long or $l_2 = 2l_1$, also with a volume of 1 unit^3. Then l_1 would be 0.79 units to get an equivalent volume to the cube and the $SV_{\text{rectangular}}$ = 6.3. The ratio of the surface to volume ratios for this rectangular shape and the cube of equal volume ($SV_{\text{rectangular}}/SV_{\text{cube}}$) equals 1.05. This indicates that the heat loss should be higher for a rectangular furnace compared to a cubical furnace of equal volume. It shows a comparison of the cubical, spherical, and rectangular furnaces that all have a unit volume of 1 unit^3. Figure 5.2.16 shows a scale drawing with a comparison of the three furnace geometries of equal volume.

There are numerous combustor geometries in industrial combustion processes. Only a few representative geometries are considered next including typical applications.

Cylindrical Rotary kilns are typically cylindrical, refractory-lined vessels, which are used to heat solid materials like cement, lime, and scrap aluminum. They typically have high length to diameter ratios, are slightly inclined from the horizontal, and rotate. The material is fed in at the higher end and exits at the lower end of the kiln. This geometry is normally characterized by a single burner located in or near the center of one end of the kiln and firing parallel to the axis of the kiln. In counter-current rotary kilns, the material is fed in on the opposite end of the burner. In co-current rotary kilns, the material inlet and the burner are at the same end. The flames are usually round to match the cylindrical furnace geometry and to transfer heat uniformly around the furnace. Longer flame lengths are often desired to get better heat transfer coverage over a longer length of the furnace.

The heat transfer in a rotary kiln is unique compared to other geometries because it transfers heat under the load, as well as on top. The flame transfers heat to the load and to the exposed furnace walls. Because the furnace rotates, the heated walls rotate under the load and transfer heat to the bottom of the load by thermal conduction. In addition, the solids typically tumble inside the kiln, which constantly exposes new material to the heat source. Tumbling the load and heating from below help increase the efficiency of this process.

An example of a rotary furnace is a sludge incinerator as shown in Fig. 5.2.17. The biggest challenge in sludge incineration is the large amount of energy required to evaporate the large quantity of water contained in the sludge. The heating value of the sludge is minimal. Therefore, large quantities of auxiliary fuel are required.

Another example of a cylindrical geometry is the EAF which was conceived as a unit to melt scrap steel and recycle the iron units back to useful service. It does not rely on molten iron from a blast furnace, but rather solid scrap is melted by energy input from electrical arcs

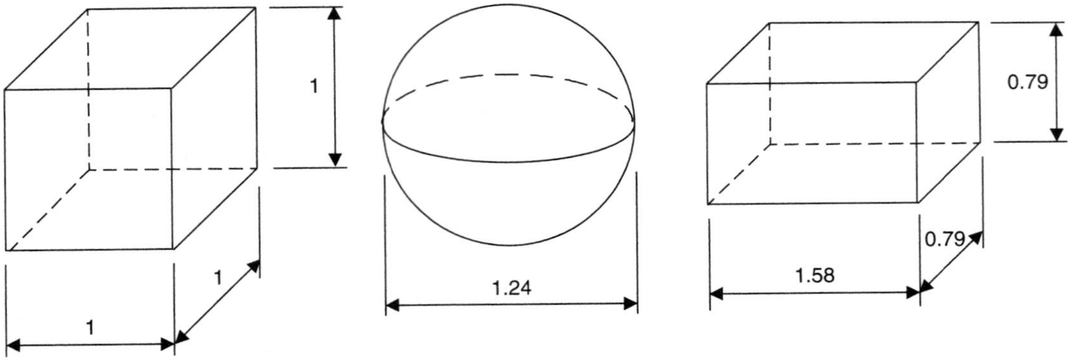

Figure 5.2.16 Cubical, spherical, and rectangular furnaces with volumes of 1 unit^3 (CRC Press, with permission).

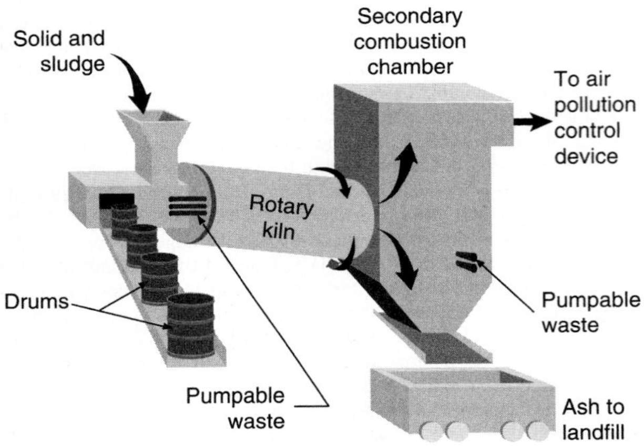

Figure 5.2.17 Sludge processing in a rotary kiln incinerator (CRC Press, with permission).

passed between three carbon electrodes. Originally this furnace was conceived as an all-electric melter. Freedom from the blast furnace and comparatively cheap electric power fueled the growth of the EAF, particularly in what came to be known as mini-mills. Due to the very high temperature requirements in most ferrous metal processes, oxy/fuel burners have become standard equipment on most EAFs. Initially, firing through the slag door, oxy/fuel burners mounted on a boom or a carriage were introduced to increase scrap melting. Later, to efficiently target the cold spots in the EAF, the burner location was moved to either the roof or sidewalls, to aim at the cold spots associated with the spaces between the electrodes, with one to four burners in a furnace, providing supplemental energy. Figure 5.2.18 shows a sketch of a three burner, wall-fired supplemental system, firing between the electrodes. In addition to productivity improvements of 5 to 20 percent, the burners provide economical energy for melting scrap at a lower cost.

Rectangular The rectangular shape is one of the workhorses in industrial combustion. One example of a horizontally-fired rectangular furnace is the reverb furnace, commonly used in aluminum melting, which is shown in Fig. 5.2.13.

As with other furnace geometries, much of the energy received by the load comes from surface radiation from the refractory-lined furnace walls. In most cases, there is very little heat transfer to or from the furnace below the level of the material loading. The objective is normally to make the furnace as adiabatic as possible to minimize heat losses and to maintain the material temperature.

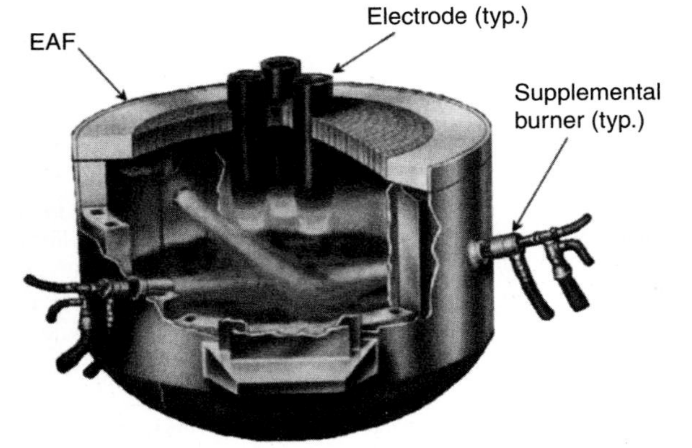

Figure 5.2.18 Electric Arc Furnace with three supplemental burners mounted in the sidewalls and firing between the three electrodes (CRC Press, with permission).

As discussed above, burners may be located in a variety of locations, depending on the process. In glass furnaces, burners are located in the side walls that fire parallel to the floor and roof. In reverberatory furnaces used to melt aluminum, burners are located in the side wall, but are typically angled down toward the load. Figure 5.2.19 shows examples of overfired and underfired furnaces.

Furnace Types

A few common types of furnaces are briefly discussed in this section. Although there are many types of furnaces, the ones chosen here are representative of a variety of furnaces which are often variations of those discussed here.

Reverberatory A reverberatory or "reverb" furnace (see Fig. 5.2.13) is a rectangular refractory-lined box that is typically used to process non-ferrous metals. Burners are located in one wall or in the roof.

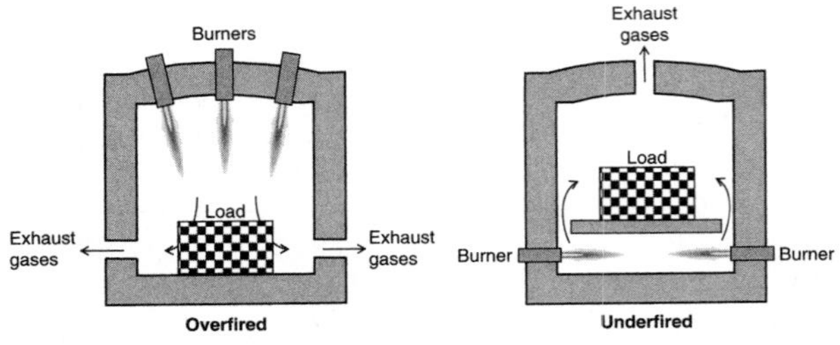

Figure 5.2.19 Elevation view of overfired and underfired furnaces.

The exhaust may be in the roof or on one of the walls, often in between the burners. Reverberatory furnaces are operated as semi-continuous processes as previously discussed. The burners are used primarily to heat the refractory walls of the reverb. The hot walls then re-radiate or reverberate the energy to the load inside the furnace. In the case of aluminum, the surface emissivity of molten aluminum ranges from 0.004 to 0.55. However, molten aluminum readily forms a slag layer on the surface containing aluminum oxide which has an emissivity of 0.11 to 0.19. This results in poor heat transfer qualities as the primary mode of heat transfer is radiation to the surface. Therefore, the dross layer thickness should be minimized. The furnace is designed to compensate for this by having a high bath surface area-to-melt ratio, compared to some other furnace designs.

There are two common methods for loading reverbs: direct charging and indirect charging. In direct charging, a charging door to the combustion chamber is opened and the load is directly placed into the combustion chamber. The burners in that chamber normally go to a low-fire mode during the charging and then go back to a high fire mode after the door has been closed. This method of operation is less efficient as the heat losses from the opened door are substantial. This is also a much more batch operation because of the discrete load charges.

Indirect charge reverbs are often more efficient and are closer to a continuous process because of the charging method, although they are still batch processes as there are discrete times when the load is tapped from the furnace. The load is commonly charged into a chamber attached to the side of the combustion chamber, often referred to as a side well. This type of reverb is called a side-well reverb. These are more thermally efficient as no doors need to be opened on the combustion chamber for charging materials. The solid load is placed into the side well and then carried into the combustion chamber by circulating molten metal.

Forced convection also plays an important role in reverbs, especially during the initial part of the heating cycle after a new charge has been added directly into the combustion chamber. High velocity burners direct hot combustion products toward the load to enhance convective heating. As the solids melt, then radiation becomes more and more important. Thermal conduction also plays an important role in reverbs since the heat transferred to the surface of the load must be conducted down into the load. In the case of aluminum which has a very high thermal conductivity, the heat is conducted rapidly through the load resulting in a more uniform temperature distribution.

Reverb furnaces also have large wall surface areas to maximize the heat transfer to the bath. This also increases the wall losses. For furnaces that do not have special submerged pumps to circulate the molten metal, the bath depth is often shallow to ensure good heat transfer through the bath. For furnaces with pumps, the bath depth is usually thicker because the increased mixing enhances the heat transfer through the bath. It is not uncommon for reverbs to have some type of heat recuperation for preheating the incoming combustion air, to improve the thermal efficiency.

Shaft Kiln A vertical shaft kiln is shown in Fig. 5.2.20. It acts like a counterflow heat exchanger where the hot exhaust products from the burners located at the bottom flow upward while the incoming cold materials flow downward. These vertical shaft kilns have been used for years for calcining limestone and for the manufacture of cement. The heat transfer is fairly unique compared to other industrial heating processes because of the vertical counterflow of materials going down and exhaust products going up. Convection plays a more important role and wall radiation is much less important, compared to many other processes. Thermal conduction is also more important because of the direct contact between moving layers of materials. Feeding solid materials from the top of the furnace is a unique aspect of this furnace configuration.

Rotary Furnace A rotary kiln is shown in Fig. 5.2.21. The raw materials are fed in at the top of the inclined, slowly rotating cylinder and exit at the opposite end. These are often continuous processes where raw material is continuously fed in one end and treated materials exit at the other end. The raw materials are solid materials that are often granular in composition so that they can easily flow through the kiln. These kilns also normally operate as counterflow heat exchangers with the burner at the discharge end and the incoming feed at the opposite end. However, there are also co-flow versions of this type of kiln. Rotary kilns are used for processing bulk materials like cement, lime, and other minerals. Large cement kilns can be up to 20 ft (6 m) in diameter and 500 ft (160 m) long.

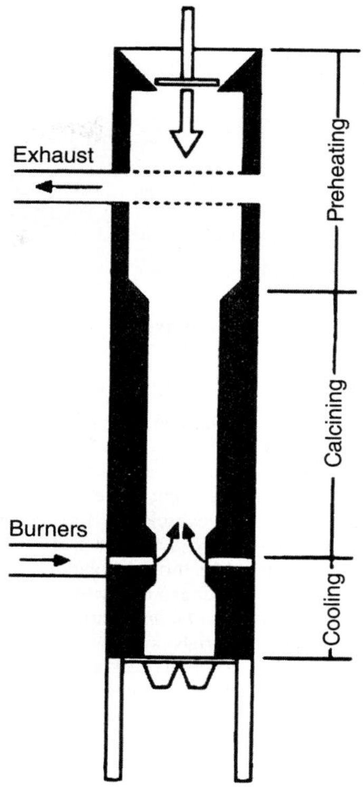

Figure 5.2.20 Elevation view of a vertical shaft kiln (CRC Press, with permission).

One important design feature of the rotary furnace is that the granular load is constantly agitated and new material is continuously being exposed to the heat. The material in contact with the furnace shell also gets heating from the bottom by thermal conduction. In some industrial processes, the load may melt and change from a solid to a liquid. An example is rotary kiln (see Fig. 5.2.22) furnaces used to melt scrap aluminum, especially used beverage containers (UBCs). In those kilns, the scrap aluminum is fed in one end and molten aluminum is tapped off the other end. That particular process is batch in nature as raw materials are fed into the kiln at discrete intervals. When the kiln is fully loaded,

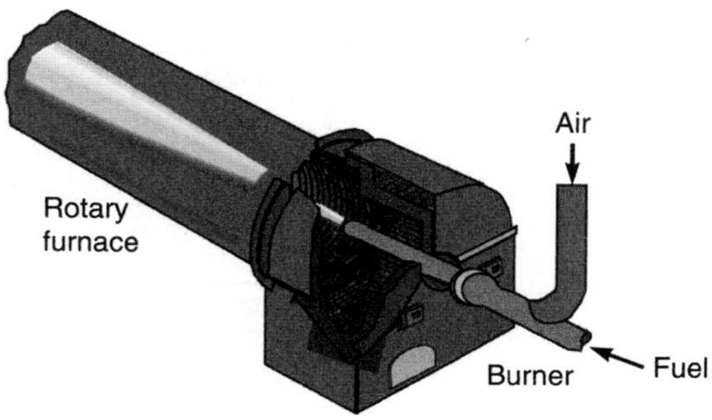

Figure 5.2.21 Elevation view of a rotary kiln (CRC Press, with permission).

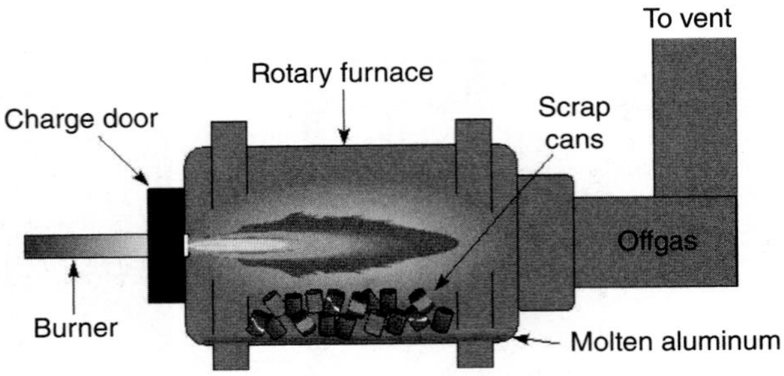

Figure 5.2.22 Elevation view of a rotary scrap aluminum melting furnace (CRC Press, with permission).

it will continue heating the aluminum until it is fully melted at which time the furnace is tapped.

There are some unique challenges associated with the heat transfer in rotary kilns. The kiln is at a slight angle to the horizontal and the load is at an angle to the kiln because of the furnace rotation. The load moves but not in as well-defined a manner as a liquid or a gas. The load is constantly mixed by the furnace rotation and is heated from above by the walls and the burner and below by the wall. Volatiles may evolve from the load which can be an additional source of energy in the heating process. Some materials processes in rotary kilns may have some endothermic chemical reactions such as calcining, pyrolysis, and sintering. Some processes can be three phase if the solid feed material melts and becomes liquid in the furnace, in addition to the combustion gases and possible gases evolving from the process.

HEAT RECOVERY

Heat recovery devices are often used to improve the efficiency of combustion systems. Some of these devices are incorporated into the burners, but more commonly they are another component in the combustion system, separate from the burners. These heat recovery devices incorporate some type of heat exchanger, depending on the application. The two most common types have been recuperators and regenerators which are discussed next. The classifications of these heat exchangers are shown in Fig. 5.2.23. Gas recirculation effects on heat transfer are also considered in this section.

Recuperators

A recuperator is a low to medium-temperature (up to about 1,300°F or 700°C), continuous heat exchanger that uses the sensible energy from hot combustion products to preheat something such as the incoming combustion air or water for generating steam. These heat exchangers are commonly counterflow where the highest temperatures for both the combustion products and the combustion air are at one end of the exchanger with the coldest temperatures at the other end. Lower temperature recuperators are normally made of metal, while higher temperature recuperators may be made of ceramics. Recuperators are typically used in lower temperature applications because of the limitations of the metals used to construct these heat exchangers.

A recuperative melter is a unit melter equipped with a recuperator. The furnace is fired with 4 to 20 individual burners. These furnaces range in size from as large as 280 tons (250 metric tons) per day of glass to as small as 20 tons (18 metric tons) per day of glass. These furnaces are common in fiberglass production but can also be used to produce frit. Some recuperative furnaces are used in the container industry, though this is not common. Furnace life is a function of glass type being produced. For example, a 6-year furnace life is typical for wool fiberglass. A typical recuperative melter is shown in Fig. 5.2.24.

An example of a waste heat boiler used for heat recovery is shown in Fig. 5.2.25. In some cases, the waste materials being destroyed in a thermal oxidizer (TO) may contain little if any heating value. Then, a large amount of energy must be combusted in the TO because of the endothermic reactions. This means a large amount of energy is available in the exhaust gases. If a plant can use additional steam capacity, a waste heat boiler is typically included in the TO system. If there is no need for additional steam, another common option is to use the flue gases to heat a heat transfer fluid (e.g., hot oil) used in the plant.

Another type of recuperative heat recovery system is to preheat the incoming load materials. Figure 5.2.26 shows an example of a waste gas preheater used in a thermal oxidation process. A counter-flow heat exchanger is used with the incoming cooler waste gases on one side and the hot exhaust gases on the other side.

Regenerators

A regenerator is a higher-temperature, transient heat exchanger that is used to improve the energy efficiency of higher temperature heating and melting processes, particularly in the high temperature processing industries like glass production (e.g., Fig. 5.2.27). Regenerators are

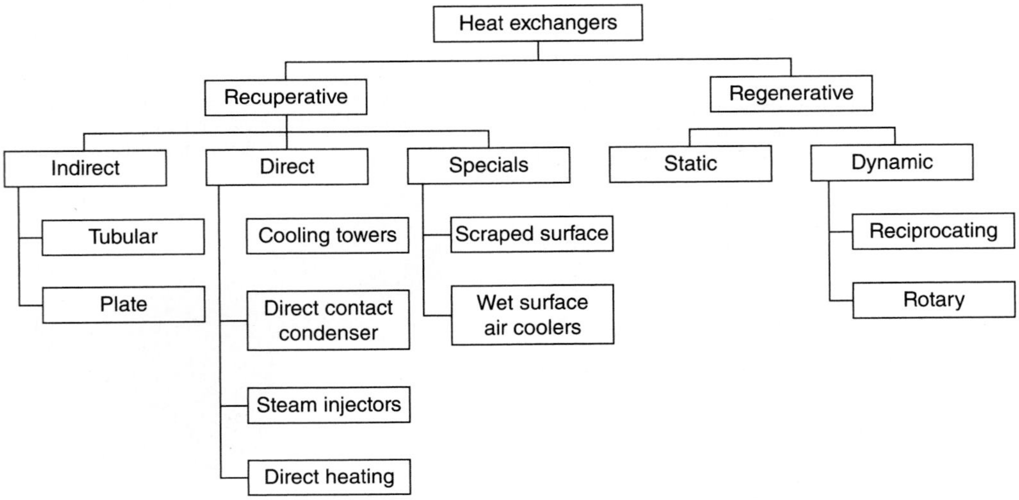

Figure 5.2.23 Heat exchanger classifications (CRC Press, with permission).

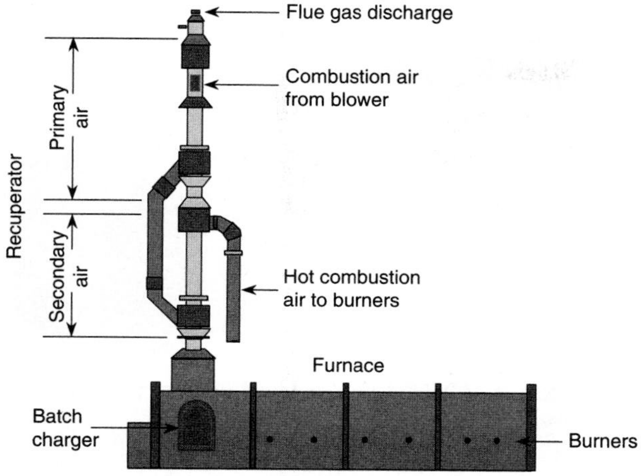

Figure 5.2.24 Typical recuperative glass melter (CRC Press, with permission).

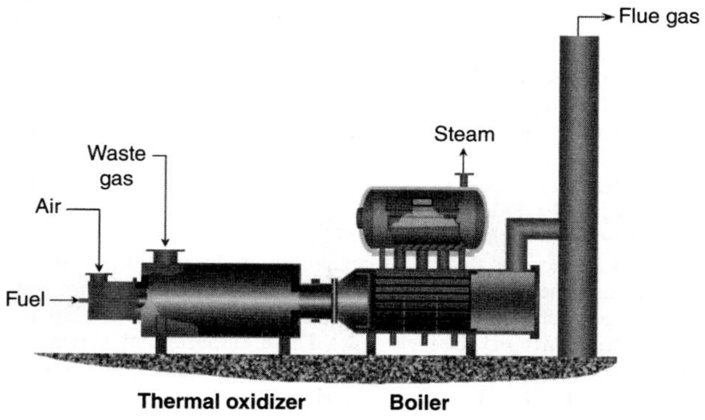

Figure 5.2.25 Thermal oxidizer with a waste heat boiler (CRC Press, with permission).

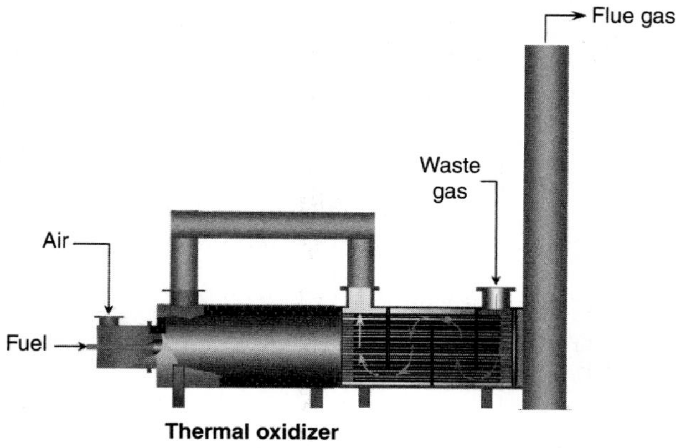

Figure 5.2.26 Thermal oxidizer with a waste gas preheater (CRC Press, with permission).

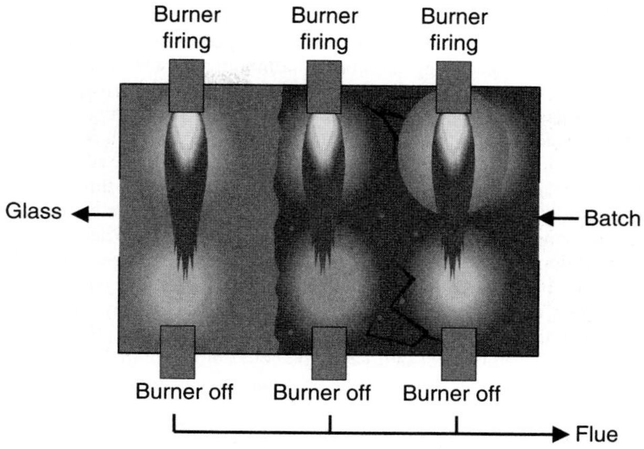

Figure 5.2.27 Multiple burners in a side-fired regenerative glass furnace (CRC Press, with permission).

sometimes referred to as "capacitive heat exchangers" and are mainly used in gas/gas heat recovery. In a regenerator, energy from the hot combustion products is temporarily stored in a unit constructed of some type of packing, such as firebricks. This energy is then used to heat the incoming combustion air during a given part of the firing cycle up to temperatures in excess of 2,400°F (1,300°C).

There are also burners that incorporate regenerators inside them (see Fig. 5.2.28). When these burners are used in a furnace, one set of burners fires for a time and the exhaust flows through another set of burners that are not firing at that time. The combustion air flowing through the burners that are firing gets preheated by the internal regenerative media normally made out of some type of ceramic. The regenerative media in the firing burners cools down with time as the combustion air convects away heat, while the media in the burners that are off heats up as the exhaust flows through them. After a fixed amount of time, typically on the order of 15 to 30 minutes, the firing cycle reverses and the burners that were off begin firing and that burners that were firing are turned off. This process captures some of the energy in the exhaust gases that would normally be lost and improves the overall thermal efficiency of the process.

Large regenerators that are constructed of a honeycomb of firebricks are often called a chequer work or simply a chequer. Other packing materials may also be used depending on the application. The packing material needs to be able to tolerate the thermal shock of constant thermal cycling and may need to withstand a potentially corrosive environment, such as in the manufacturing of glass. Many different patterns, such as the "square chimney" or "closed basket weave" are used in arranging the packing material, depending on the material and

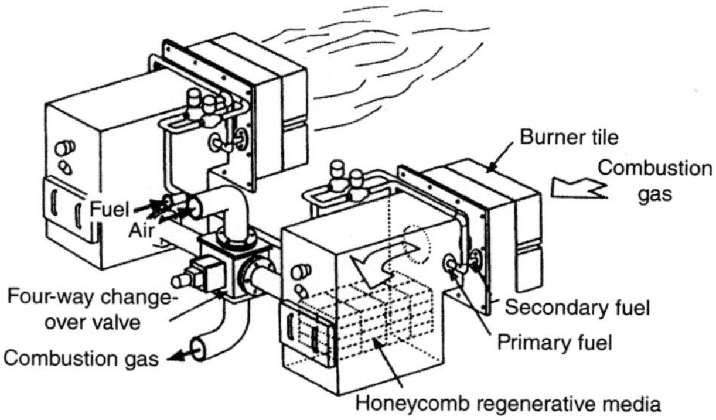

Figure 5.2.28 Schematic diagram for an alternating type of regenerative burner (CRC Press, with permission).

the pressure drop requirements. An additional requirement is that the regenerators should either not be easily plugged or should be easy to clean if they do get plugged, in order to minimize downtime.

Regenerators are normally operated in pairs because of the normal requirement for a continuous stream of preheated air. During one part of the cycle, the hot combustion gases are flowing through one of the regenerators and heating up the refractory bricks (known as the "hot blow" or "hot period"), while the combustion air is flowing through and cooling down the refractory bricks in the second regenerator (known as the "cold blow" or "cold period"). Both the exhaust gases and the combustion air directly contact the bricks in the regenerators, although not both at the same time since each is in a different regenerator at any given time. After a sufficient amount of time (usually from 5 to 30 min.), the cycle is reversed so that the cooler bricks in the second regenerator are then reheated while the hotter bricks in the first regenerator exchange their heat with the incoming combustion air. A reversing valve is used to change the flow from one gas to another in each regenerator. This is shown schematically in Fig. 5.2.29.

In the more efficient counter flow regenerator operation, the hot gases pass through the regenerator in the opposite direction that the cold gases go through the regenerator, while in the less efficient parallel flow operation, both the hot and cold gases pass through the regenerator in the same direction. There are also rotary regenerators which are single units that use the hot combustion products to preheat a rotating ring which is then used to preheat the incoming combustion air. The ring rotates so that it is continually heating and cooling. Then, only a single regenerator is needed to provide a continuous stream of preheated air. The basic problem with rotary regenerators is that they are not completely gas tight so that exhaust products leak into the combustion air stream and vice versa.

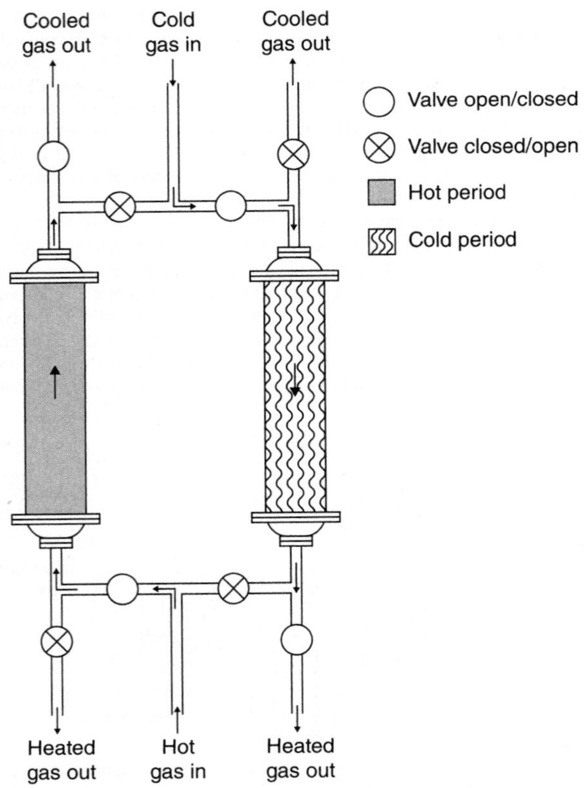

Figure 5.2.29 Schematic of a fixed bed regenerator (CRC Press, with permission).

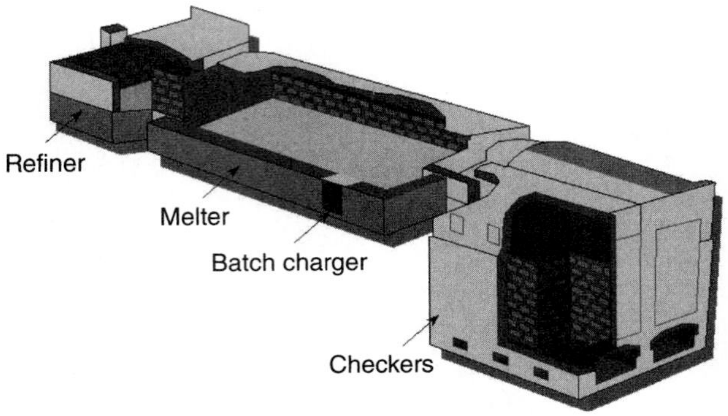

Figure 5.2.30 Typical end-port glass melting furnace (CRC Press, with permission).

End-Port Regenerative Glass Melting Furnace End-port regenerative furnaces are typically used for producing less than 250 tons (230 metric tons) of glass per day. In an end port furnace, the ports are located on the furnace back wall. Batch is charged into the furnace near the back wall on one or both of the side walls. Figure 5.2.30 shows the layout of a typical end port furnace. These furnaces are commonly used for producing container glass, but are also used for producing tableware and sodium silicates. For container production, a furnace campaign typically lasts 8 years.

Side-Port Regenerative Furnace Side-port regenerative furnaces have ports located on the furnace side walls. Batch is charged into the furnace from the back wall. Figure 5.2.5 shows the layout of a typical side port furnace. Side-port regenerative furnaces are typically used for producing greater than 250 tons (230 metric tons) of glass per day. A side port furnace for float glass commonly produces 500 to 700 tons (460 to 630 metric tons) of glass per day. For container glass, side port furnaces ordinarily produce between 250 and 350 tons (230 and 320 metric tons) of glass per day. These furnaces are commonly used in container and float glass production, but are also used for the production of tableware and sodium silicates.

Gas Recirculation

There are two common types of gas recirculation in industrial combustors: furnace and flue gas. These are briefly considered next.

Flue Gas Recirculation This type is schematically shown in Fig. 5.2.31. In this process, the exhaust products are taken out of the flue and recirculated back into the furnace, often through the burners. The recirculation requires an external, high temperature fan to move

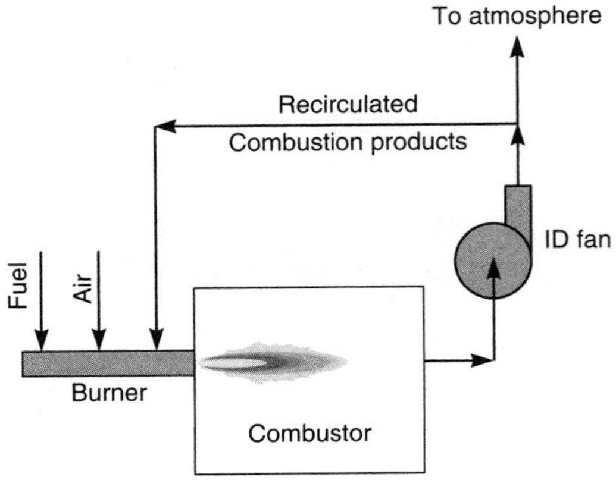

Figure 5.2.31 Schematic of flue gas recirculation (CRC Press, with permission).

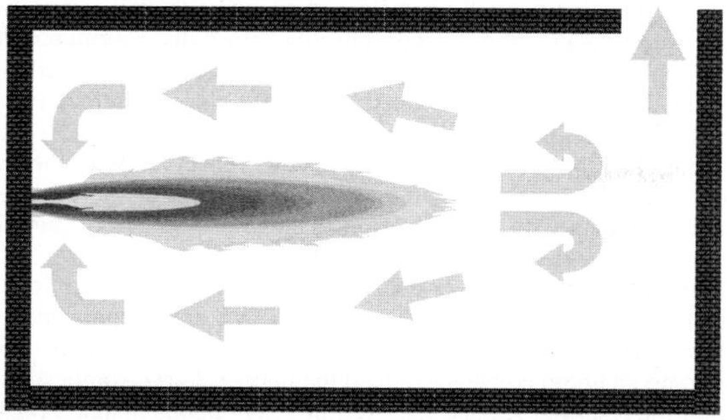

Figure 5.2.32 Schematic of furnace gas recirculation (CRC Press, with permission).

the gases and the associated high temperature duct work to transport the gases. There are two common reasons for employing this type of recirculation. In the past, the major reason was for heat recovery. The hot combustion exhaust products were given a second chance to transfer their energy to a lower temperature heat load. A more recent reason to use this technology is to reduce NOx emissions. Although the recirculated gases may be hot, they are not nearly as hot as the flame gases so the recirculated gases actually cool the flame. The reduced flame temperature reduces NOx emissions which are highly temperature dependent.

This type of recirculation can have a significant impact on the heat transfer in the furnace. One primary effect is that the forced convection heat transfer is often increased due to the added mass of gases flowing through the furnace and the higher overall gas velocities. However, the forced convection may be tempered by the lower overall gas temperature by diluting the hot flame products with colder flue gases. Since the thermal efficiency increases for nearly all processes employing flue gas recirculation, the overall effect on convection is normally to increase it.

Furnace Gas Recirculation This is a related process where combustion products are recirculated inside the combustor (see Fig. 5.2.32). An example of a burner designed to produce this type of recirculation is shown in Fig. 5.2.33. There, the high velocity core jet of pure oxygen induces furnace gases to flow into the burner to dilute the flame temperature. Other burner designs use the flow inside the furnace to recirculate the exhaust products back toward the burners. This technique also tends to improve thermal efficiencies and reduce NOx emissions, but to a lesser extent compared to flue gas recirculation. However, the capital and operating costs are also less since no external fan and ductwork are required. Therefore, the technique chosen is a balance of economics and desired benefits. The heat transfer in the furnace is similarly affected as with flue gas recirculation, but the magnitude of the change is usually smaller.

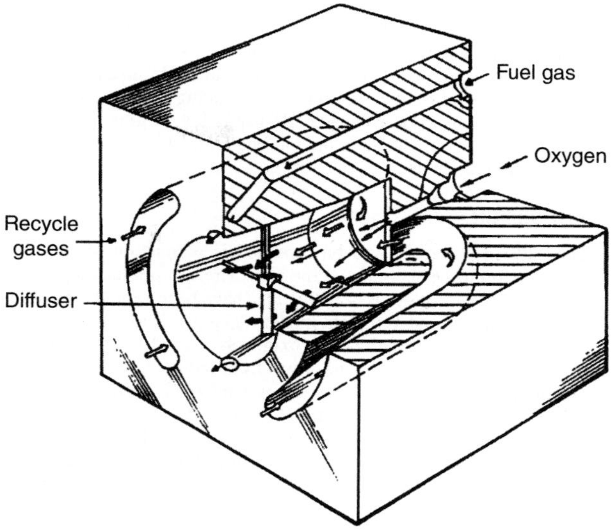

Figure 5.2.33 Oxy-fuel burner incorporating furnace gas recirculation (U.S. patent 4,954,076).

5.3 MUNICIPAL WASTE COMBUSTION

by Charles O. Velzy and Leonard M. Grillo

REFERENCES: *Proc. Biennial ASME National Waste Processing Conf.*, 1964-2005. "Design Considerations in Heat Recovery from Refuse," International Symposium on Energy Recovery from Refuse, 1975. Velzy et al., eds., "CRC Handbook on Energy Efficiency," chap. 24, Waste-to-Energy Combustion, 2015. "Steam, Its Generation and Use," 40th ed., Babcock and Wilcox Co. Kirklin et al., The Variability of Municipal Solid Waste and Its Relationship to the Determination of the Calorific Value of Refuse Derived Fuels, *Resources and Conservation*, **9**, 1982, pp. 281-300.

Waste combustion is a method for processing of solid wastes by the burning of the combustible portions. It reduces the volume of solid wastes and eliminates the possibility of pollution of groundwater from putrescible organic waste, and the residue may serve as a source of mineral constituents and as a fill. With the application of boilers, beneficial use of the energy generated from burning of the waste is achieved.

NATURE OF THE FUEL

The refuse which is received at a waste combustor today contains a high proportion of paper; plastics; some wood; vegetable and animal waste; and varying amounts of cloth, leather, and rubber—together with metal cans, glass, and other noncombustible matter. Collections may also include metal appliances, furniture, tree limbs, waste building material, broken concrete, and other coarse waste matter, commonly classified as rubbish. Yard waste, such as grass clippings, is generally not accepted at a waste combustion facility. There may be a wide variation in moisture content of refuse, depending on the weather. Thus, after a storm, the moisture content may be so high that the combustibility of the waste is adversely impacted. Industrial and hazardous waste should be specifically identified and combustion facilities designed for the particular waste.

TYPES OF FURNACES

The type of furnace for waste combustors is dictated largely by the type of grate around which the furnace is built. Except in small plants, the modern furnace is equipped with a mechanical grate.

The "Controlled Air" Furnace In the late 1970s, a type of combustion unit, batch fed, utilizing two chambers for staged combustion and followed by a waste heat boiler, was installed in smaller communities in the United States. Such units, while less efficient than larger furnaces, can be factory-assembled in large segments (or modular components) and therefore are less expensive than large, field-erected units. Thus they extended the economics of waste-to-energy plants to the smaller communities. In such plants, the refuse is normally dumped on a receiving floor and is then pushed by front end loader into a ram for feeding into the combustion units. In these units, combustion occurs on hearths, which are fitted with rams to move the material through the furnace from one hearth to the next as it is burned.

The Waterwall Furnace Waterwall furnaces are rectangular, with movement provided by travel of the grate or by a reciprocating or rocking action of the grate sections. Refuse is usually stored in a pit and loaded to a feed hopper by overhead cranes. In areas with high water tables, the refuse may be stored on a tipping floor and loaded to the feed hopper by apron conveyor. Refuse is fed by gravity from the hopper through a vertical chute onto a feed table, where a ram pushes the waste onto the grate.

PLANT DESIGN

Capacity The capacity to be provided is a function of the area and population to be served, and the rate of refuse production for the population served. If records of collections have been kept, the capacity can be determined and forecasts made; lacking records, the quantity of refuse may be estimated as approximately 4 lb (1.8 kg) per capita per day, when there is little or no waste from industry, to 5 lb (2.3 kg) per capita per day where there is some waste from industry. Nearly all facilities, regardless of size, operate 24 h/day, 7 days/week.

Location An isolated site may be preferred to avoid the possible objections of neighbors to the proximity of a waste disposal plant. However, well-designed and well-operated waste combustors which do not present a nuisance may also be installed in light industrial and commercial areas, thereby avoiding the economic burden of extended truck routes. Since considerable vertical distance is involved in passing refuse through a combustor, there is an advantage in a sloping or hillside site. Collection trucks can then deliver refuse at the higher elevation while the residue trucks operate at the lower elevation with a minimum of site grading.

Refuse-Handling Facilities Scales are provided for recording the weight of material delivered by collection trucks. Trucks then proceed to the tipping floor at the edge of the storage pit. This enclosed area must be large enough to permit several trucks at a time to maneuver to and from the dumping position. The number of tipping positions is determined by the size of the facility and the type of truck (packer or transfer trailer) delivering the waste.

Since collections usually are limited to one 8-h daily shift (with partial weekend operation) while burning is usually continuous over 24 h, ample **storage** must be provided. Seasonal and cyclic variations must also be considered in establishing the storage requirements.

Refuse storage in larger plants is normally in long, deep pits extending along the front of the furnaces. Waste is rehandled with the overhead cranes to mix and stack the waste prior to combustion. In waste combustor plants with a tipping floor instead of a pit, floor dumping, mixing, and stacking are done using a front end loader.

Feeding the Furnaces In a large waste combustor, refuse is transferred from storage pit to the furnace hopper by a crane equipped with a grapple. Large facilities with tipping floors use front end loaders to push waste onto an apron conveyor to feed the hopper. Large combustors are fed and the resulting residue removed in a continuous manner. Smaller controlled air furnaces are usually fed directly by front end loader to the feed hopper. Some controlled air combustors batch feed waste or remove the residue intermittently. Batch feeding or batch discharge of residue is undesirable because of the resulting variations in furnace temperatures, adverse impact on furnace side walls, and increased air emissions.

The charging hopper in a waterwall furnace leads to a vertical charging chute, 12 to 14 ft (3.6 to 4.2 m) high, 4 to 5 ft wide, and as long as the width of the furnace, sometimes as much as 35 ft. This chute is kept full of refuse to maintain an air seal and to prevent a back fire from the combustor. At the bottom of the charging chute is a feeding table, with hydraulic rams that push the waste into the furnace onto the first row of grates. (See Fig. 5.3.1.)

FURNACE DESIGN

The basic design factors which determine furnace capacity are grate area and furnace volume. Both provision for and quantity of underfire air, and provision for quantity and method of applying overfire air influence capacity. The required grate area depends upon the selected burning rate, which varies between 60 and 90 lb/(ft2 h) of refuse in

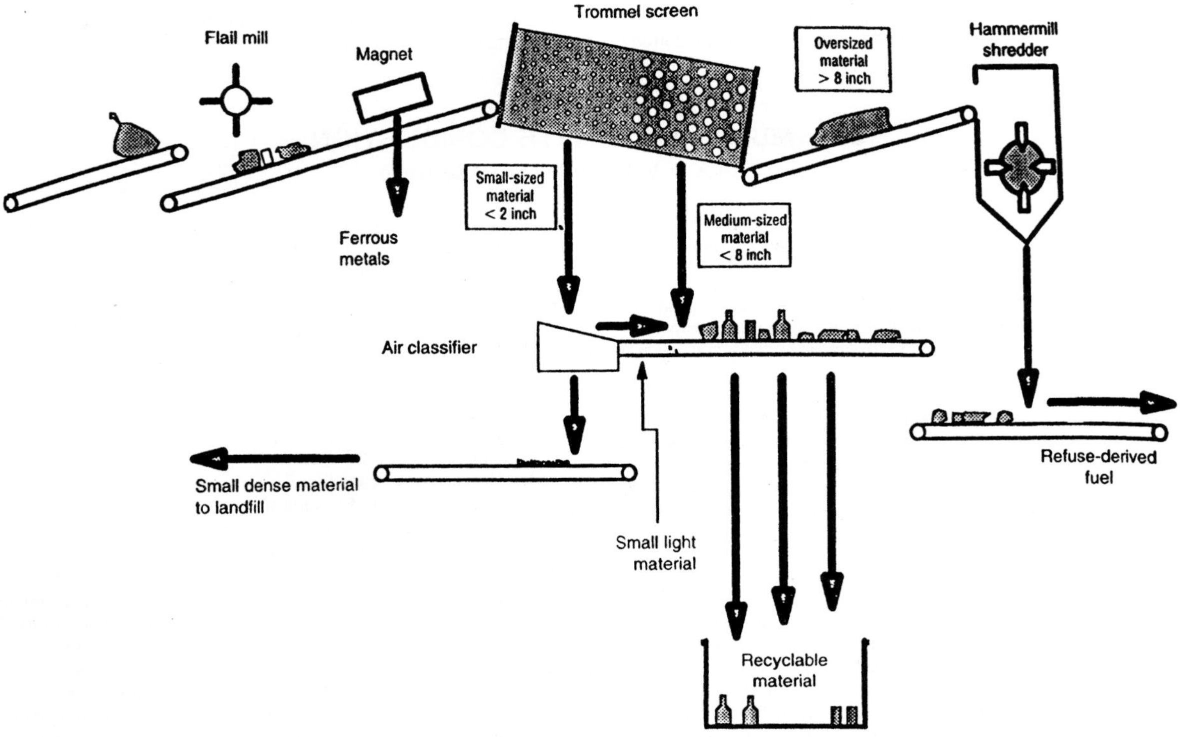

Figure 5.3.1 Process diagram for a typical RDF system. (*Roy F. Weston, Inc.*)

practice. Conservative design, with reasonable reserve capacity calls for a burning rate between 60 and 70 lb/ft2 of grate area.

Furnace volume is a function of the rate of heat release from the fuel. A commonly accepted minimum volume is that which results from a heat release of 20,000 Btu/(ft^3 h). Thus, at this rate, if the fuel has a heat content of 5,000 Btu/lb, the burning rate would be 4 lb/(h ft^3) of furnace volume. A conservative design, allowing for some overload and possible quantities of refuse of high heat content, would be from 20 to 30 ft^3/ton of rated capacity.

The primary objective of a mechanical **grate** is to convey the refuse automatically from the point of feed through the drying, combustion, and burnout zones to the point of residue discharge with a proper depth of fuel and in a period of time to accomplish complete combustion. The rate of movement of the grate or its parts is adjustable to meet varying conditions.

A secondary, but important, objective is to tumble the refuse to aid in completeness of combustion. In the United States, there are several types of mechanical grates: (1) reciprocating, (2) rocking, and (3) a proprietary water-cooled rotary combustor. In addition, traveling grates are generally used for combusting RDF. The reciprocating and rocking grates tumble the material by movement of the grate elements. The rotary combustor slowly rotates to tumble the material which is inside the cylinder. **Furnace configuration** is largely dictated by the type of grate used. When built with a mechanical grate, the furnace is rectangular in plan and the height is dependent upon the volume required by the limiting rate of heat release.

The total **air capacity** provided in waste combustors must be more than the theoretical amount required for combustion in order to obtain complete combustion and to control temperatures—particularly with dry, high-heat-content refuse. The total combustion air requirements may range to as high as 10 lb of air/lb of refuse. For the mechanical-grate furnace chamber, two blower systems are usually provided to supply combustion air to the furnace. Fan capacities can be divided, with half or more from the underfire air fan and somewhat less than half from the overfire air fan, with dampers on fan inlets and air distribution ducts for control. The pressure on the underfire air system for most U.S. grate systems approximates 3 in of water. The pressure on the overfire air should be high enough so that the jet effect on passage through properly proportioned and distributed nozzles in the furnace roof and walls produces sufficient turbulence and retains the gases in the primary furnace chamber long enough to ensure complete combustion.

Heat Recovery Perhaps the most potentially attractive form of recovery, or extraction, of resources from municipal solid wastes is recovery of energy from the combustion process.

Several options exist when one is considering a waste-to-energy system. These options include mass burning in a refractory-walled furnace with a waste-heat boiler inserted in the flue downstream; mass burning in a water-walled furnace with the convection surface immediately downstream; and refuse preprocessing and separation of the combustible fraction with combustion taking place in a spreader stoker boiler partially in suspension and partially on a grate. This latter option is generally termed combustion of a refuse-derived fuel. A partial list of waterwall and RDF plants in North America in operation as of the end of 2016 extracting energy from combustion of municipal-type waste is shown in Table 5.3.1. This list excludes plants built for developmental or experimental purposes and plants that utilize specialized industrial wastes.

Figure 5.3.2 illustrates a plant with heat recovery, while Fig. 5.3.3 shows a cross section through a typical RDF facility.

In considering the above options, one should take into account the overall energy balance in the various systems. These systems can be grouped under the following general categories: burning as-received refuse; burning mechanically processed refuse; and gasification. In mechanical processing systems, less heat will be available for use than there was prior to processing.

The tabulation below from published data regarding the production of a fuel gas from refuse will partially illustrate the net energy loss in converting the available energy to another form:

Composition	*Percent by volume*	*Percent by weight*	*Percent of total carbon*
CO	47	62.1	70
H_2	33	3.1	
CO_2	14	29.1	21
CH_4	4	3.0	6
C_2H_2	1	1.4	3
N_2	1	1.3	

Table 5.3.1 Energy-from-Waste Plants Larger than 400 Tons/Day Commissioned 1989-1995

Plant location	Plant size tons/d	Start-up year	T	Energy sold
Broward County, FL, south	2,250	1991	Mass burn water wall	Electricity
Camden, NJ	1,050	1991	Mass burn water wall	Electricity
Chester, PA	2,688	1991	Rotary kiln water wall	Electricity
Essex County, NJ	2,250	1990	Mass burn water wall	Electricity
Fort Myers, FL (Lee County)	1,200	1994	Mass burn water wall	Electricity
Gloucester County, NJ	575	1990	Mass burn water wall	Electricity
Honolulu, HI	2,000/900	1990/2012	RDF/Mass burn water wall	Electricity
Hudson Falls, NY	400	1992	Mass burn water wall	Electricity
Huntington, NY	750	1991	Mass burn water wall	Electricity
Huntsville, AL	690	1990	Mass burn water wall	Steam
Kent County, MI	625	1990	Mass burn water wall	Electricity/steam
Lake County, FL	528	1991	Mass burn water wall	Electricity
Lancaster County, PA	1,200	1991	Mass burn water wall	Electricity
Long Beach, CA	1,380	1990	Mass burn water wall	Electricity
Lorton, VA (Fairfax County)	3,000	1990	Mass burn water wall	Electricity
Montgomery County, PA	1,200	1991	Mass burn water wall	Electricity
Pasco County, FL	1,050	1991	Mass burn water wall	Electricity
Rochester, MA	2,700	1988/1993	RDF water wall	Electricity
Southeast CT (Preston, CT)	600	1992	Mass burn water wall	Electricity
Spokane, WA	800	1991	Mass burn water wall	Electricity
Union County, NJ	1,440	1994	Mass burn water wall	Electricity
West Palm Beach, FL	2,000/3,000	1989/2015	RDF/Mass burn	Electricity

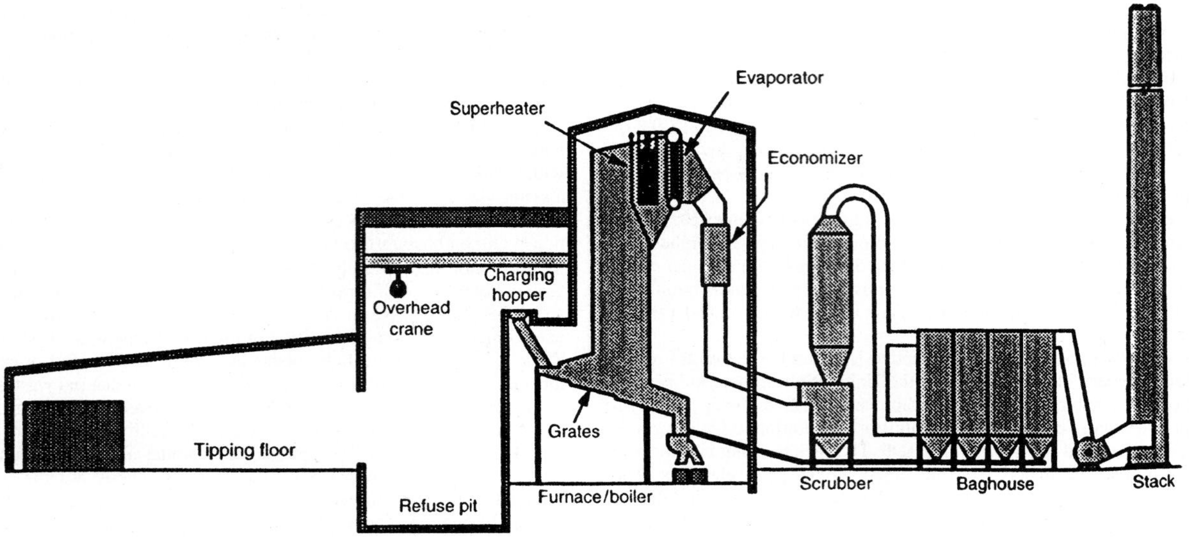

Figure 5.3.2 Typical cross section of a mass-burn facility. (*Roy F. Weston, Inc.*)

Note that 21 percent of the carbon is in CO_2 which will not burn, while 70 percent of the carbon is in CO where 30 percent of the elemental energy in carbon is no longer available. While this heat is not wasted, it is lost energy not available to do further work.

A tabulation of energy losses and total net available energy, for refuse with an initial heat content of 5,000 Btu/lb, is given in Table 5.3.2.

Of the 5,000 Btu/lb in the refuse as received, the tabulated data indicate the useful energy that may be made available through combustion. While the data are not absolute, the relative magnitudes are meaningful, provided similar degrees of design efficiency and sophistication of control are used for each process.

Other factors to consider in selection and design of heat-recovery facilities include efficiency of boiler facilities, furnace-chamberdesign, and combustion air supply. While in older plants with waste-heat boilers installed in downstream flues, steam production averaged 1.5 to 1.8 lb/lb of refuse, in today's water-walled furnaces and suspension-fired units, steam production is on the order of 3.0 to 3.5 lb/lb of refuse. The lower efficiency in waste-heat boiler units is due to higher heat losses in the plant stack gases, in turn caused by higher excess air levels required to control combustion temperatures properly in the refractory-lined primary furnace enclosure.

In most water-walled furnaces and furnaces in which shredded combustible refuse fractions are burned, the usual configuration is a tall primary chamber with the gases passing out the top and into the superheater and convection boiler surfaces after completion of combustion. It has been found necessary in mass-burning water-wall plants to coat a substantial height of the primary combustion chamber with Inconel overlay and to limit average gas velocities to under 15 ft/s. Gas velocity entering the boiler convection bank should be less than 30 ft/s.

In all combustion systems, combustion air is supplied from both under and over the grate. Overfire air is utilized to enhance turbulence and mixing of combustion gases with the combustion air, and for completion of combustion. Accordingly, this air is best introduced through numerous relatively small (1½- to 3-in-diameter) nozzles, at

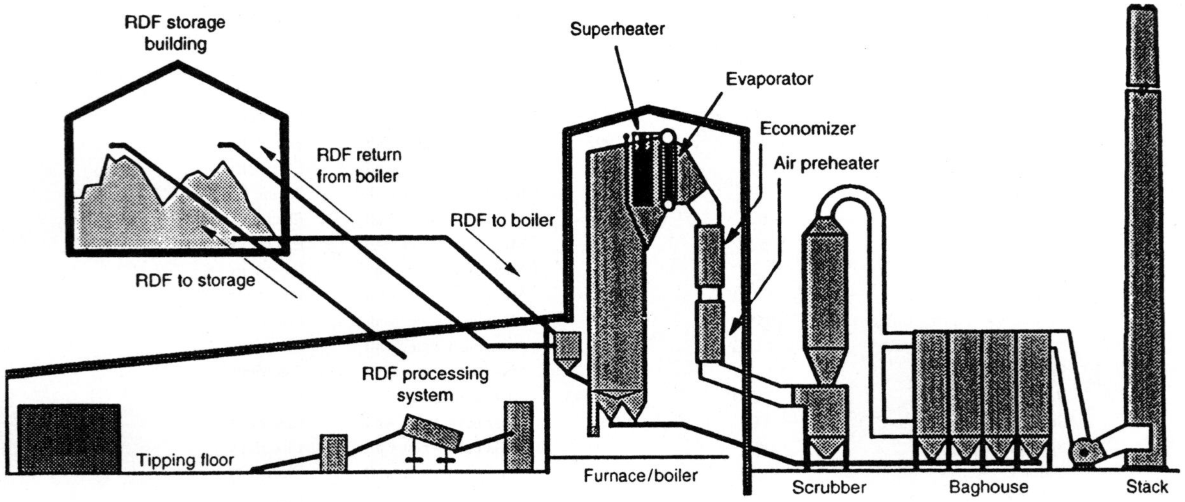

Figure 5.3.3 Typical cross section of an RDF facility. (*Roy F. Weston, Inc.*)

Table 5.3.2 Energy Losses and Total Net Available Energy for Refuse with Initial Heat Content of 5,000 Btu/lb

Process	Energy loss, %			Total net available energy, Btu/lb
	Processing	Combustion	Total	
As received	1	29	30	3,500
Dry shredding	18	25	43	2,850
Gasification	32	25	57	2,150

pressures of 20 in of water and higher. Ideally, provision should be made for the introduction of the overfire air at several different elevations in the furnace.

Flues and chambers beyond the furnace convey gases to the air pollution control equipment for removal of fly ash and other pollutants, and to the stack. The draft for a waste combustor is provided by an induced-draft fan. (See Secs. 2 and 10.) Most plants include extensive air pollution control equipment.

Air Pollution Control The Federal New Source Performance Standards as of 2009 for air pollution control regulate the 11 pollutants shown in Table 5.3.3. Some jurisdictions have promulgated emission requirements more stringent than the federal standards.

To meet current emission requirements, two basic techniques have been utilized on waste combustors: electrostatic precipitation and fabric filters preceded by lime addition in a slurry form into a spray dryer absorber, so-called "dry scrubbing." Electrostatic precipitation has performed reliably and has given predictable emission test results. The dry scrubber/fabric filter air pollution control system is considered by many to represent the most efficient combination of pollution control systems currently available for particulates and acid gases and is used on most facilities. Many older facilities that originally employed electrostatic precipitators have been retrofitted with fabric filters. Selective noncatalytic reduction using ammonia or urea has been applied at most plants for NO_x control. Activated carbon injection for control of mercury and dioxin emissions has also been installed on most facilities. Good combustion control is used to limit emission of organics.

Residue Discharge and Disposal The residue from refuse burning consists of relatively fine, light ash mixed with items such as burned tin cans, partly melted glass, and pieces of metal. Discharge from furnaces may be through a chute into a conveyor trough filled with water for quenching and then carried by flight conveyor to a storage building, or a truck loading facility. Alternatively, a ram discharger, in which ash is submerged in water, effects discharge onto a dry conveyor. Usually there are two residue removal conveyors arranged such that the residue can be discharged onto either conveyor, allowing one to be used as a standby.

The lower end of the discharge chute leading to the trough or extractor is submerged in a water seal to prevent entrance of cold air to the furnace. In design of the residue discharge system, the dimensions should be large because of the nature of the material handled, and the metal used should be selected to withstand severe abrasive service. Final disposal of the residue is by dumping at a suitable location, which for modern plants usually means a monofill. Volume required for disposal is 5 to 15 percent of that required for dumping raw refuse.

Miscellaneous Facilities Good working environment and reasonable comfort for the staff should be provided.

Table 5.3.3 Federal New Source Performance Standards

Pollutant	Emission limit*
Dioxin/furans	30 ng/dscm
Cadmium	35 ug/dscm
Lead	400 ug/dscm
Mercury	50 ug/dscm or 85% reduction
Opacity	10%
Particulate matter	25 mg/dscm
Hydrogen chloride	29 ppmdv or 95% reduction
Nitrogen oxides	165-230 ppmdv (depending on technology)
Sulfur dioxide	29 ppmdv or 75% removal
Fugitive ash	5% of observation period
Carbon monoxide	50-250 ppmdv (depending on technology)

* dscm = dry standard cubic meter; ppmdv = parts per million dry volume.

COMBUSTION CALCULATIONS

Among the factors directly affecting design are **moisture** and **combustible content** of refuse as received, heat released by combustion, temperature control, and water requirements. The design of furnaces, chambers, flues, and other plant elements should be based on characteristics which result in large sizes. Controls should provide satisfactory operation for loads below the maximum. The computations which follow are for relatively typical heat releases.

The prime factors in **heat calculations** are the moisture and combustible content of the refuse and heat released by burning the combustible portion of the refuse. The moisture content may vary from 20 to 50 percent by weight, and the combustible content may range from 25 to 70 percent. The combustible portion is composed largely of cellulose and similar materials, mixed with proteins, fats, oils, waxes, rubber, and plastics. The heat released by burning cellulose is approximately 8,000 Btu/lb, while that released by the plastics, fats, oils, etc., is approximately 17,000 Btu/lb. If cellulose, plastics, oil, and fat exist in the refuse in the ratio of 5:1, the heat content of the combustible matter will be 9,500 Btu/lb. The heat content per pound of refuse as received, for varying proportions of moisture and noncombustibles, is given in Table 5.3.4.

Determination of the **air requirement** is illustrated by computation with refuse of 5,000 Btu/lb heat content where the composition is: combustible, 52.5 percent; noncombustible, 22.5 percent.

Carbon and hydrogen are the essential fuel elements in combustion of refuse; sulfur and other elements which oxidize during combustion are present in trace amounts and do not contribute significantly to the heat of combustion. Carbon and hydrogen content can be determined from a complete analysis of the refuse, but such an analysis is of questionable value because of the variable character of refuse and the difficulty of obtaining representative samples. The carbon-to-hydrogen ratio in municipal solid waste is approximately 7.5:1.0. A typical ultimate analysis, which is assumed for the sample calculation, is shown in Table 5.3.5. It is probable that 1 to 3 lb of combustible material per 100 lb of refuse will escape unburned with the residue. For the sake of clarity in the illustrated computations, complete combustion is assumed.

Oxygen requirements and products of combustion can be determined from the reactions as follows:

Carbon $C + O_2 \rightarrow CO_2$
Atomic wt: $12 + 32 = 44$
Ratio: $1 + 2.667 = 3.667$

Hydrogen $2H_2 + O_2 \rightarrow 2H_2O$
Atomic wt: $4 + 32 = 36$
Ratio: $1 + 8 = 9$

Table 5.3.4 Heat Content of Refuse, as Received

	Noncombustible, %							
	10		25		20		25	
Moisture, %	Comb., %	Heat content*	Comb., %	Heat content*	Comb., %	Heat content*	Comb., %	Heat content*
50	40	3,800	35	3,325	30	2,850	25	2,375
40	50	4,750	45	4,275	40	3,800	35	3,325
30	60	5,700	55	5,225	50	4,750	45	4,275
20	70	6,650	65	6,175	60	5,700	55	5,225

* Btu/lb.

Table 5.3.5 Typical Ultimate Analysis

Element	Percentage
Carbon	28.0
Hydrogen	3.7
Oxygen	20.2
Nitrogen	0.4
Sulfur	0.2
Chlorine	0.0
Ash	22.5
Moisture	25.0
Total	**100.0**

The theoretical air required per 100 lb of refuse follows from these figures where air is considered to contain 23.15 percent oxygen.

$$\text{Air required} = (28 \times 2.667 + 3.7 \times 8 - 20.2)/0.2315$$
$$= 363 \text{ lb/100 lb refuse}$$

For waste combustion, furnace temperature must be controlled to minimize furnace maintenance. Normally, excess air in a waterwall furnace will range from 50 to 100 percent. With no other provision for heat absorption, such as in a refractory lined furnace, it is necessary to introduce excess air well beyond the needs for complete combustion, e.g., 140 percent of theoretical so that, in the example cited:

$$\text{Total air} = 1.4 \times 363 = 508 \text{ lb/100 lb refuse}$$

To summarize the quantities for a computation of furnace temperature, a materials balance is given in Table 5.3.6 equating the input to the furnace and output for 100 lb of refuse. In this tabulation, allowance is made for moisture in the air at a commonly accepted rate of 0.0132 lb/lb of dry air. Some residue quench water will be evaporated, and the moisture added to the flue gases is estimated at 5 lb for each 100 lb of refuse burned. Since the assumed analyses are not precise, an exact balance is not obtained, but the indicated computations are sufficiently accurate for combustor design.

In Table 5.3.6, total air is broken down into oxygen and nitrogen on the basis that 23.15 percent of the air is oxygen. To compute the air in the "output," or flue gas, the nitrogen is the same as the "input." Oxygen is diminished by the amount consumed in combustion. Since carbon and hydrogen unite with oxygen during combustion, the oxygen consumed per 100 lb refuse is

For carbon,	28 × 2.667 = 74.7 lb
For hydrogen,	3.7 × 8 = 29.6 lb
	104.3 lb
Less oxygen in fuel	−20.2 lb
	Total = 84.1 lb

Abiabatic flame temperature is the maximum theoretical temperature that can be reached by the products of combustion of a specific fuel-air combination. To calculate this temperature, the total heat input in the fuel is adjusted to subtract the heat input required to vaporize moisture in the fuel, moisture produced in combustion of hydrogen, and the residue quench water that is vaporized. A loss is assumed to account for incomplete combustion and other small losses. The remaining heat energy is the sensible heat available in the furnace gas. It can be calculated per 100 lb of refuse from the data of Table 5.3.6 and the enthalpy data of Fig. 5.3.4 as follows:

Input, 100 lb × 5,000 Btu/lb			500,000 Btu
Losses:			
Heat of vaporization deduction			
Moisture			
In refuse	25.0		
From burning hydrogen	33.3		
From residue quench	5.0		
	63.3		
63.3 × 1.050 Btu/lb		66,500	
Assumed loss due to incomplete combustion and other losses 2%		10,000	
Total deduction			−76,500
Sensible heat available in furnace gas			423,500 Btu
Enthalpy of gas, 418.810 Btu ÷ 766.3 lb			553 Btu/lb

Table 5.3.6 Materials Balance for Furnace lb/100 lb of Refuse

Input:			
Refuse			
Combustible material		52.5	
Moisture		25.0	
Noncombustible		22.5	100.0
Total air, at 140% excess air			
Oxygen	117.7		
Nitrogen	390.7		508.4
Moisture in air			6.7
Residue quench water			5.0
Total			620.1
Output:			
CO_2(28 × 3.3667)			102.7
Air			
Oxygen (117.7 − 84.1)	33.6		
Nitrogen	390.7		424.3
Moisture			
In refuse		25.0	
From burning hydrogen		33.3	
In air		6.7	
In residue quench water		5.0	70.0
Furnace gas subtotal			597.0
Noncombustible material			22.5
Sulfur and Nitrogen			0.6
Total			620.1

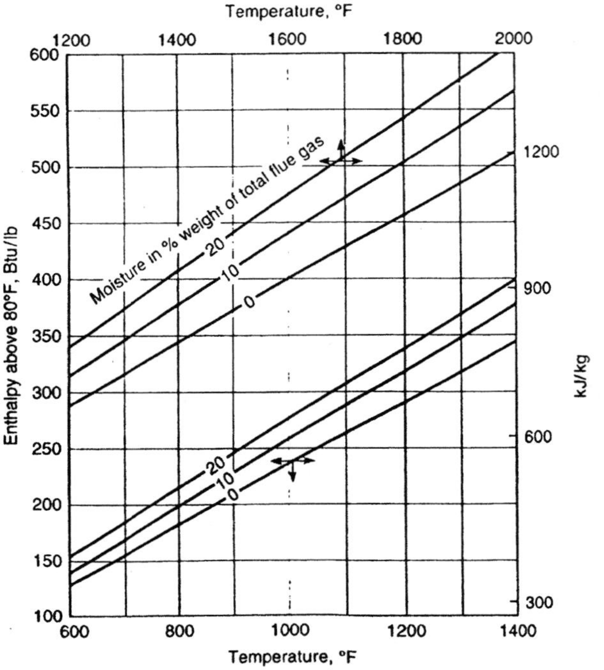

Figure 5.3.4 Enthalpy of flue gas above 80°F.

% Moisture in furnace gas		
Moisture vaporized		63.3
Moisture in air		6.7
		70.0
70.0 ÷ 766.3 = 9.1%		
Temperature of furnace gas at 553 Btu/lb and 9.1% moisture from Fig. 5.3.4	1,950°F	

RECOVERY

Many waste-to-energy (WTE) plants incorporate waste processing/materials recovery facilities. Such facilities are a natural adjunct at plants producing RDF. Relatively active stable markets have developed for newsprint, glass, metals, and certain specific plastics. Recent surveys of waste composition indicate that after recycling, paper and noncombustibles have decreased slightly while high-heat-value plastics have increased significantly as a percentage of municipal solid waste going to disposal. The impact of these changes in waste composition has been to increase slightly the heat content of waste available for processing in WTE plants and to remove some of the more troublesome materials from a materials handling standpoint.

Fly ash has been used to a limited extent as a concrete additive and as a road base. Waste combustor residue has been used for land reclamation in low areas and, in some cases, as a road-base material. Generally, before such use, the fly ash and residue must be tested to ensure that they are not classified as hazardous waste.

As this nation's energy needs become more critical in the future, this readily available source of energy should be tapped more frequently. The technology is available now for successful application of these techniques if provision is made for adequate funding and properly trained operating staffs.

5.4 INDUSTRIAL ELECTRIC HEATING

by Arvind C. Thekdi

INTRODUCTION

Use of electricity as a source of heat for furnaces and ovens has been used extensively in industry. Heat produced from electricity is applied for thermal processing of materials in heating equipment such as furnaces, ovens, dryers, boilers, heaters for processes such as heating, drying, curing, melting, and forming to manufacture or transform products.

Use of electricity for heating offers several advantages. They include: relatively high energy efficiency; electric heating can be used for higher temperature (>2,000°F) processes with relatively high energy efficiency; safe—little or no explosion hazards for heating systems using vacuum or air atmosphere; no flue Gases to deal with at the site of use; no Pollution or emissions of Green House Gases (GHG) such as CO_2 and NO_x; usually lower initial cost for furnace or oven equipment due to absence or lack of fuel firing system; easy to install and operate compared to fuel fired systems; automation could be simpler—easier, possibility of achieving shorter heating time and better temperature uniformity for the processed material with improved yield with consistent product quality; and easier to install a unit as a part of production cell since most electrical systems have smaller foot print and may not require exhaust gas collection, treatment, and discharge arrangement.

However, there are several issues that may go against the use of electric heating. They need to be considered before making final decision on use of electrical system to replace an existing or installing a new fuel-fired system. These issues are: higher operating cost, since the competing energy source, natural gas, usually is lower cost making electric per unit production 2-4 times higher; for most commonly used resistance heating system, the heating elements have relatively short life; for metals heating applications, there is a danger of heating elements shorting due to possibilities of metal parts dropping on the heating elements or touching the elements; the heating elements are prone to corrosion, soot deposit, oxidation or other type of deterioration if the furnace operating environment is not controlled properly; for an electrically heated system additional expenses may occur for adding electrical power supply such as a substation, transformer; soot deposits and corrosion of exposed heating elements in applications using process atmospheres or special gas mixture; longer furnace length for the same heat input, particularly for continuous high temperature furnaces.

The following list gives most commonly used electric heating technologies, often referred to as electrotechnology.

1. Resistance Heating can be applied as direct resistance heating or indirect resistance heating. Direct resistance heating involves using the material being processed as electrical resistor to generate heat when electric current is passed through the material itself. Indirect heating involves use of electrical heating elements to generate heat and transfer this heat to the material being heated by direct radiation or combination of radiation and convection. Direct resistance heating is used extensively in the glass industry. Resistance furnaces are also used for holding molten iron and aluminum. Direct resistance processing is also used for welding steel tubes and pipes. Indirect resistance heaters are used for a variety of applications, including heating water, sintering ceramics, heat pressing fabrics, brazing and preheating metal for forging, stress relieving, and sintering. This method is also used to heat liquids, including water, paraffin, acids, and caustic solutions. Applications in the food industry are also common, including keeping oils, fats, and other food products at the proper temperature. Standard electrical resistance furnaces are designed to operate at temperatures within the range 1,000-2,000°F (550-1,200°C).

2. Induction Heating is a non-contact heating technology that uses electromagnetic properties of electrically conductive materials. When the electrical current passes along a conductor, an electromagnetic field is produced around that wire. Induction heating involves use of this electromagnetic field to produce eddy currents in the metal workpiece being heated. The workpiece is heated primarily by the current in the same manner as resistance heating by passing current through the workpiece. Induction heating can be used to heat or melt metals such as copper, aluminum, steel, or brass. It can also be used for semiconductors such as graphite, carbon, or silicon carbide. Nonconductive materials such as plastics or glass can be heated by using an electrically conductive susceptor like graphite and transfer heat to the nonconductive material being heated.

3. Electric Arc Heating occurs when an electric arc struck between an electrode and metal is used to heat and melt metals or provide heat for other processes at very high temperatures. Most prominent industrial application of direct arc heating is for melting steel scrap. Other applications include use of submerged electric arc in the Ferro alloy industry. In direct-arc melting furnaces, one or more graphite electrodes are used to carry AC or DC current and an arc is struck between the electrodes and metal, usually in the form of pile of scrap metal. The arc temperature, in the range of 1,800°C (3,272°F), and 3,000°C (5,432°F).

4. Electric Infrared processing involves radiation heat transfer or electromagnetic energy transfer from a heat source that produces wavelength within the range of 0.75 to 10^3 μm. This requires emitting sources in the range of 500-4,200°F. Infrared processing systems are used for heating, drying, curing, thermal-bonding, sintering, and sterilizing applications.

5. Microwave Heating or Processing involves interaction of polar water molecules and charged ions. The friction resulting from molecule alignment and migration of charged ions in rapidly alternating electromagnetic field generates heat within materials such as foods. Microwave heating uses 915 and 2,450 MHz frequencies for heating applications. Typical applications of microwave heating in industries include food drying or processing, rubber and plastic heating, curing of adhesives in plywood and construction lumber.

6. Radio Frequency Heating is in the same class as microwave heating. While high-frequency microwaves have wavelengths from 300 MHz to 300 GHz, the radio frequency has a frequency spectrum from 3 to 300 MHz. Radio frequency waves are longer than microwaves, enabling them to penetrate larger objects than microwave energy. Typical applications of radio frequency heating include drying natural and synthetic textiles, drying water-based adhesives, emulsions and coatings, preheating fiber mat board, moisture removal from glass fibers, etc.

7. Electron Beam (EB) Processing involves use of kinetic energy of an accelerated stream of electrons to produce heat or chemical and physical changes in various substances at the location of impingement of the electrons. Use of EB in appropriate applications offer several advantages such as superior process control, tighter tolerances, elimination of limitations in polymer compositions, compounding and extrusion, short treatment times, etc. EB is primarily used by automotive and aerospace industry in welding of thick material to thin material, joining of dissimilar metals, and some combinations that are unweldable. Recently EB has been applied to surface heat treatment such as hardening of high wear components. Other applications include low temperature processing of materials, curing of multiple layers of web material with simultaneous curing of surface coatings.

8. Ultraviolet (UV) Processing utilizes electromagnetic energy within a wavelength of 4-400 nm. UV processing involves application of UV radiation to certain liquid polymeric substances to transform them into a solid coating. UV curing is fundamentally different from the conventional thermal drying process, since exposure to UV radiation results in change in molecular structure of coating in the form of solid without using a solvent. Typical application of UV processing include curing of thick layers such as used in floor coverings, in-depth drying for heavier ink coats as used in silk screen, surface hardening in organic coatings, curing of polymeric materials in electronic parts manufacturing.

9. Plasma Heating uses a special form of ionized gas that conducts electricity. It is generated by using two water cooled electrodes with a gap in between to create an arc that can be described as "white glow" and looks similar to lightning bolts. Plasmas can be generated in a wide range of temperatures reaching millions of degrees (e.g., for fusion reactors), typical industrial uses of thermal plasmas are at temperatures in the range of 5,000-10,000°C (9,000-18,000°F). Industrial applications of plasma heating system include municipal waste management, ladle and tundish heating, melting (ferrous and non-ferrous metals), vacuum melting, recovery processes, and hazardous waste treatments.

10. Laser Heating uses laser beam to heat material on which laser beam strikes. A laser is a coherent and focused beam of photons meaning that is all in one wavelength, unlike ordinary light that showers on us in many wavelengths. Laser heating is used for numerous heating applications including local heating and melting of material at the desired or selected area of a part to be processed. One of the many widely used of laser heating is heat treating of parts used in many industrial applications. Laser heat treating makes use of the laser energy to rapidly and efficiently heat a selected area of the surface of a metal component above the transformational temperature.

Figure 5.4.1 Electrical resistance heated furnace with resistance elements.

HYBRID SYSTEMS

A hybrid heating system is the combination of electric heating and heating by another type of heating source such as fuel firing or steam heating to optimize energy use and increase overall process thermal efficiency. Hybrid process heating systems utilize a combination of process heating technologies based on different energy sources and/or different heating methods of the same energy source. Hybrid process heating systems that combine multiple forms of heat transfer through radiant, conductive, and/or convective methods can reduce heating time, increase energy efficiency, and improve product quality. Examples of hybrid systems include electric infrared in combination with either an electric convection oven or a gas convection oven and a paper-drying process that combines a natural gas or electric-based infrared technology with a steam-based drum dryer.

RESISTANCE HEATING AND MELTING

Resistance heating is the simplest and oldest electric-based method of heating and melting metals and nonmetals. The heating chamber of a standard furnace is an enclosure with a refractory lining, a surrounding layer of heat insulation, and an outer casing of steel plate, or for large furnaces an outer layer of brick or tile, as indicated by Figs. 5.4.1 and 5.4.2. The hearth of a furnace often is constructed of a heat-resisting alloy, made in sections to prevent warping or refractory materials. In

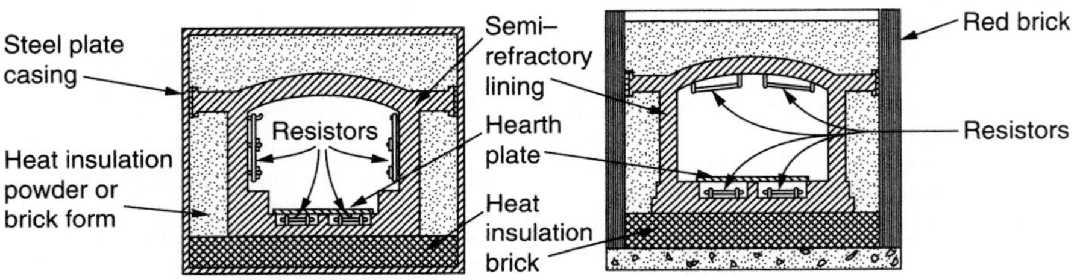

Figure 5.4.2 Different orientation of electrical resistance elements in a furnace.

some continuous furnaces, the conveyor forms the hearth while in others a separate hearth is required.

The inner lining of the heating chamber may use one or more layers of insulation. This insulation can be refractory material, such as fire brick or insulting brick, or fibrous insulation in the form of blanket or board. The selected material offers thermal insulating properties and heat storage. The current trend is to use low density fibrous material wherever possible to reduce heat storage and tolerance to thermal cycling. However, for higher temperature applications and where it is necessary, as in case of metals manufacturing industries, it is common to use refractory material in the shape of bricks or other shapes.

Furnace Atmospheres A mixture of air and the gases evolved from the charge constitutes a natural atmosphere in the heating chamber of a resistance heating furnace. The composition of such an atmosphere in a batch furnace is variable during a heating cycle. A natural atmosphere in the heating chamber of a continuous furnace is mainly air. Natural atmospheres are used where the extent of the action of oxygen on the material during the heating cycle is not objectionable and for processes where that chemical action is desired. In some cases, as in case of heat treating of metallic parts, it is necessary to replace air with a mixture of other gases such as N_2, H_2, CO, CH_4, etc.

Heating Elements or Resistors The performance and life of heating element depend on properties of the material used for heating element. The required properties for heating elements are: high melting point; free from oxidation in open atmosphere; high tensile strength; sufficient ductility to draw the metal or alloy in the form of wire; high resistivity; low temperature coefficient of resistance.

Resistance heating furnaces or ovens use one of the three classes of elements: (1) Metallic [nickel-chromium (Ni-Cr) and iron-chromium-aluminum (Fe-Cr-Al)], (2) Silicon carbide (SiC), and (3) Molybdenum disilicide ($MoSi_2$). Most of these elements have maximum operating temperature or maximum element temperature (MET). Values of MET for commonly used elements in dry air are shown in Fig. 5.4.3. When these elements are used in a furnace, an additional factor, the furnace atmosphere or gas composition within the furnace must be considered in placement of the elements.

Maximum allowable Furnace temperature (MFT) for different heating elements under commonly used operating conditions is given in Fig. 5.4.4. From this figure, it is clear that for all practical operations (exception being use of hydrogen in a furnace) the maximum allowable furnace temperature is possible when the elements are operated in air atmosphere. The element manufacturers are making continuous advancements in the materials; hence it is always

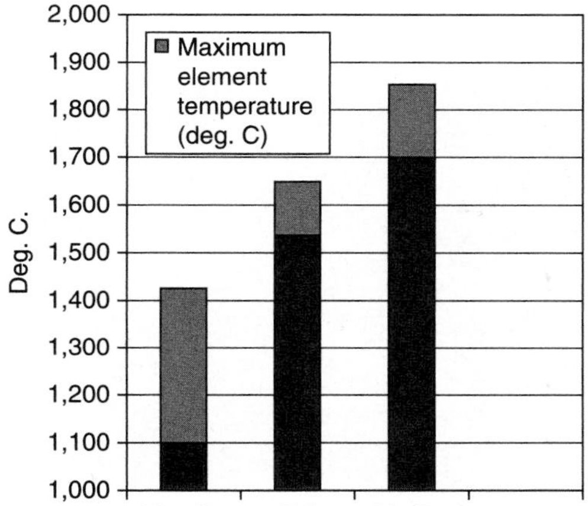

Figure 5.4.3 Maximum element temperature vs. element type.

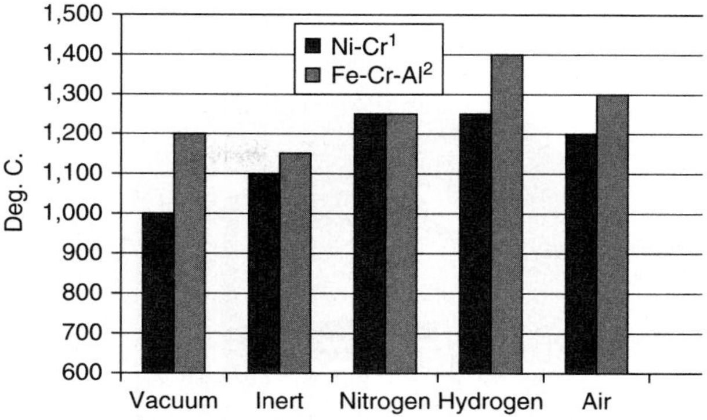

Figure 5.4.4 Maximum element temperatures for metallic elements in selected atmospheres.

recommended that the users contact the element suppliers to obtain the latest information.

Once the operating temperature for a class of element is established, the next important consideration is allowable watt loading or energy input for the element.

Electrical Resistance Furnace Configurations Electrically heated furnaces are built in many configurations to meet specific requirements. They can be broadly classified in the following categories.

1. Radiation heating furnaces (Fig. 5.4.5) where the electrical resistance elements are directly exposed to the material being heated.
2. Convection or indirectly heated furnaces (Fig. 5.4.6) where a set of baffles or a muffle or an inner cover is placed between the electrical heating elements and the work.
3. Use of radiant tubes or other enclosure that contains heating elements. In this system, electrical heating elements are placed in a high-temperature metal tube so they are not in direct contact with the furnace gases or atmosphere as shown in Fig. 5.4.7.
4. Direct resistance heating where electrical current flows directly through the material and the material becomes a type of electrical resistor is shown in Fig. 5.4.8.

Figure 5.4.5 A kiln with directly exposed heating elements (Courtesy L&L Kiln Mfg Inc., NJ).

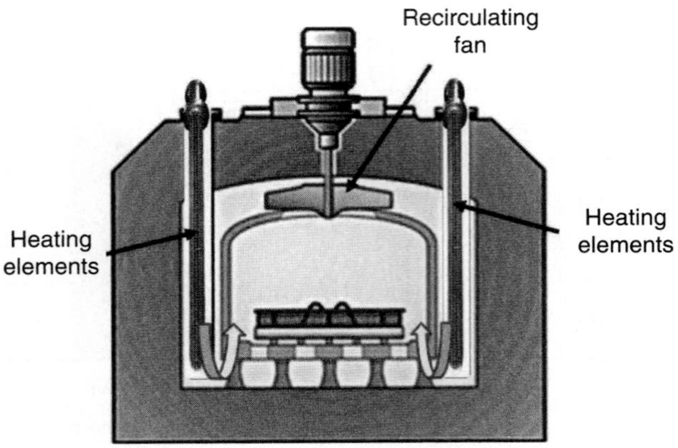

Figure 5.4.6 A convection heating furnace with electrical elements behind baffles or shields and a circulating fan for convection heating (Courtesy: Kanthal Corporation).

References

1. Kanthal A.B., "Kanthal Super Heating Elements Handbook," 1999.
2. Custom heating elements Inc.
3. *IMI-NFG Course on Processing of Glass—Lecture 3: Basics of industrial glass melting furnaces mathieu.hubert@celsian.nl* 31 Pictures from Sorg, report "Glass melting technology," available online *All-electric melting.* Mathieu Hubert, PhD CelSian Glass & Solar Eindhoven, The Netherlands mathieu.hubert@celsian.nl.

INDUCTION HEATING AND MELTING

Induction heating involves heating of a conductive material by placing it in a variable magnetic field. Induction heating can be achieved by two methods. In the first method, electrical current (I) induced by a variable magnetic field heats the material by Joule effect or I^2R losses. As shown in Fig. 5.4.9, this induces electromagnetic flux (EMF) resulting in eddy currents or induction currents which results in heat.

The second method, referred to as hysteretic heating, involves production of energy within the part by alternating magnetic field generated by the coil modifying the component's magnetic polarity. The heat generated by hysteresis is generally much lower (<10%) than that generated by induction current.

Induction heating finds applications in processes where temperatures are as low as 100°C (212°F) and as high as 3,000°C (5,432°F). It is also used in short heating processes lasting for less than half a second and in heating processes that extend over several months.

The depth at which induction heating provides heat (or depth of penetration) is referred to as skin effect. A range of penetration defined in terms of dimensions such as diameter for rounds or depth in case of other shapes, at different frequencies for commonly heated materials at frequency varying from 50 to 50,000 Hz is given in Table 5.4.1. This table shows that the depth of penetration reduces significantly as frequency increases. For example a 1-in diameter steel bar can be stress-relieved by heating it to 700°C (1,292°F) using a 500 Hz induction system. However, a 5,000 Hz system will be needed to harden the same bar by heating it to 870°C (1,600°F). Hence selection of lower frequency is preferred for heating of metals.

Application of Induction Heating Induction heating is used for a variety of processes in manufacturing operation. Some of the most commonly used applications include metal melting, metal heat treating

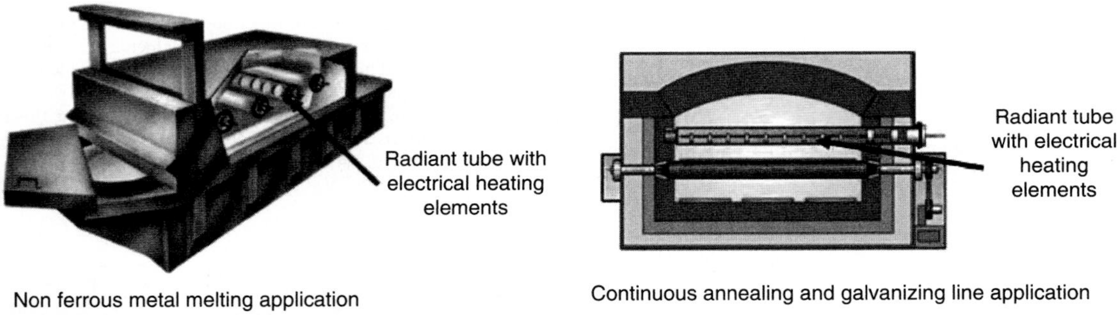

Figure 5.4.7 Heating systems using encased electrical elements in a metal or ceramic tube (radiant tube) for heat treating and melting applications.

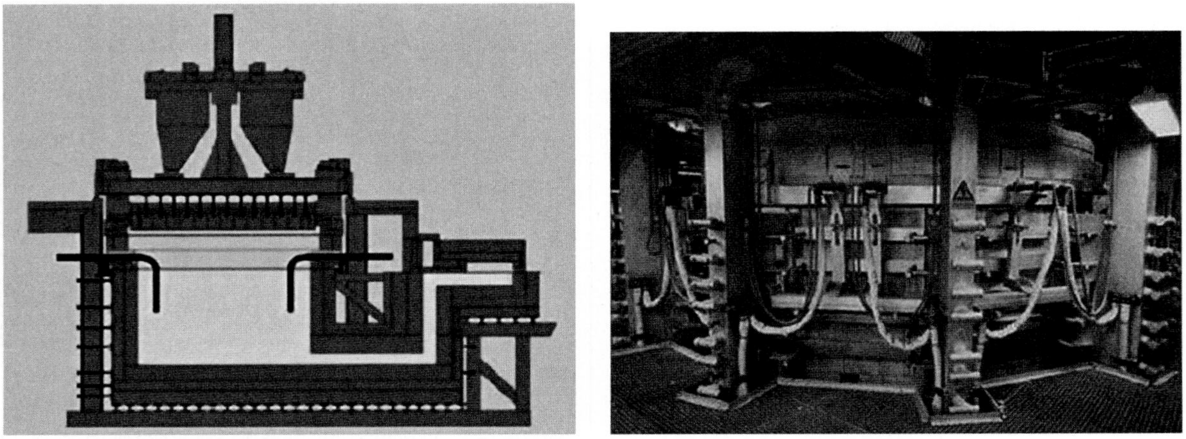

Figure 5.4.8 Direct resistance heating system for a glass melting furnace (Reference 3).

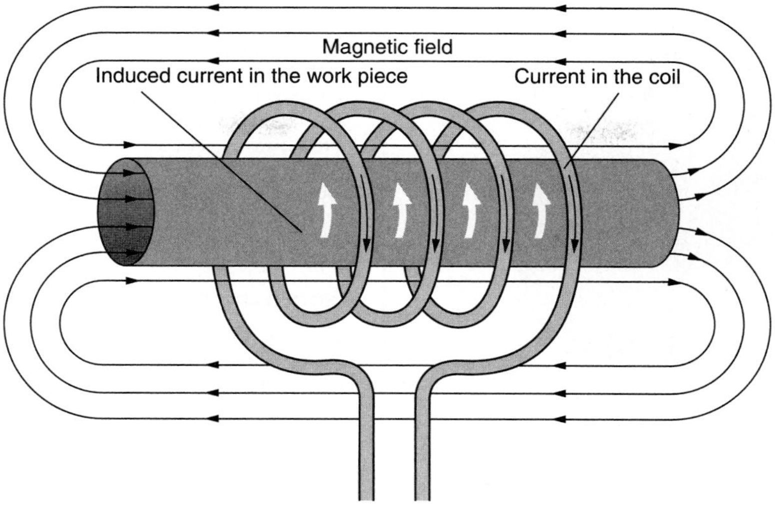

Figure 5.4.9 Induction heating method and components (Courtesy: Ambrell Precision Induction Heating).

including hardening, annealing, tempering, brazing, bonding of metals and nonmetals to cure bonding adhesives and sealants, welding, heating of metals for forging and forming, melting metals such as steel, aluminum, brass, and special applications such as shrink fitting, glass wool fiberizing, etc. Induction heating is also used for targeted heating of parts, to heat graphite crucibles (containing other materials) and in the semiconductor industry for the heating of silicon and other semiconductors.

Energy Use in Induction Melting Furnaces Energy use in induction melting furnaces depends on various factors such as design of the unit, size or melting capacity, selection of power supply system, operating practices, and type of feed material used. General areas of energy use are: heat supply to metal for heating and melting (melt), heat used for water cooling, electrical losses, radiation loss, and other miscellaneous losses. It is difficult to generalize heat usage for all types and sizes of induction melting furnaces. However, based on study of several units it is possible to present energy distribution as performance of a typical induction melting furnace. Figure 5.4.10 shows that approximately 66 percent of the energy is used for metal melting, 27-28 percent is used for cooling water, 4-5 percent is used as electrical losses in power supply system, 2.5-3 percent is wasted as radiation loss and the remaining, approximately 0.6 percent is waste as other losses.

Advantages of Induction Melting

- **Higher Yield:** Induction melting furnaces do not have combustion products that contain metal oxidizing components such as H_2O, O_2 and CO_2. Hence, the oxidation losses are almost nonexistent. This results in significantly higher yield and correspondingly higher productivity.
- **Faster Startup:** Full power from power supply is available instantaneously thus reducing the time to reach working temperature.
- **Flexibility:** No molten metal is necessary as heal to start medium frequency coreless induction melting equipment.
- **Natural Stirring:** Medium frequency units can give a strong stirring action resulting in a homogeneous melt.
- **Cleaner Melting:** No bi-product are produced due to cleaner melting environment.
- **Compact Installation:** High malting rates can be obtained from small furnaces and the unit can be located very close to the final point of use for molten metal.

Table 5.4.1 Approximate Frequencies for Through-heating Some Common Materials (Courtesy EFD Induction)

	Material									
	Steel non-magnetic		Steel magnetic		Brass		Copper		Aluminum and aluminum	
	Final temp.		Final temp.		Final temp.		Final temp.		Final temp.	
	Deg. C.	Deg. F.	Deg. C.	Deg. F.	Deg. C.	Deg. F.	Deg. C.	Deg. F.	Deg. C.	Deg. F.
Frequency	1,200	2192	700	1292	800	1472	850	1562	500	932
Hz	Ø mm	Inch	Ø mm	Inch	Ø mm	Inch	Ø mm	Inch	Ø mm	Inch
50	150-500	5.9-19.7	27-75	1.0-3.0	110-	4.33-	50-	2.0-	50-	2.0-
500	60-250	2.4-1.0	8-35	0.3-1.4	35-440	1.4-17.3	22-800	0.9-31.5	22-800	0.9-31.5
1,000	40-175	1.6-6.9	6-25	0.2-1.0	30-300	1.2-11.8	15-600	0.6-23.6	15-600	0.6-23.6
3,000	25-100	1.0-3.9	3.5-14	0.1-0.6	15-180	0.6-7.1	9-350	0.4-13.8	9-350	0.4-13.8
5,000	20-85	0.8-3.3	2.5-10.5	0.1-0.4	10-130	0.4-5.1	7-260	0.3-10.2	7-260	0.3-10.2
10,000	14-60	0.6-2.4	2-8.5	0.1-0.3	8-100	0.3-3.9	5-180	0.2-0.7	5-180	0.2-0.7
20,000	10-40	0.4-1.6	1.5-5.5	0.05-0.2	6-75	0.2-3.0	3-125	0.1-4.9	3-125	0.1-4.9
60,000	5-22	0.2-0.9	0.7-3.0	0.02-0.11	3.5-40	0.13-1.5	2-75	0.1-0.3	2-75	0.1-0.3
100,000	4-17	0.2-0.7	0.5-2.0	0.01-0.07	2.5-30	0.09-1.18	1.5-60	0.1-0.2	1.5-60	0.1-0.2
500,000	1.8-8	0.1-0.3	0.2-1.0	0.00-0.03	1.2-15	0.04-0.59	0.6-20	0.02-0.8	0.6-20	0.02-0.8

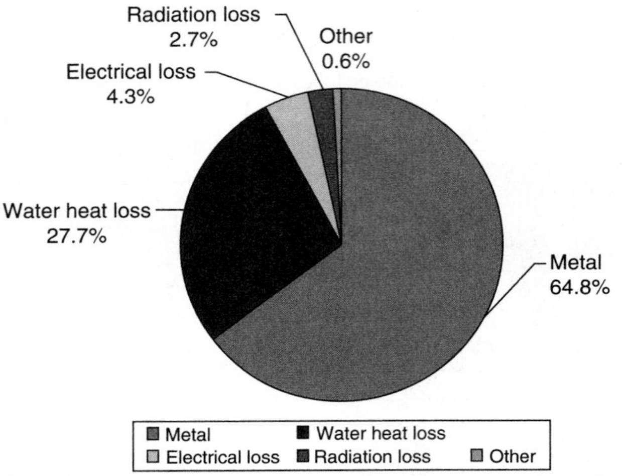

Figure 5.4.10 Heat balance for a typical induction melting system (Reference 2).

- **Energy Conservation:** Overall energy efficiency in induction melting ranges from 55 to 75 percent, and is significantly better than combustion processes.
- **No Flue Gases:** Induction furnaces do not use combustion as a source of heat hence there are no large volumes of flue gases to deal with. No need to install emission control equipment for large volumes of gases.

Disadvantages

- Charge materials must be clean of oxidation products and of known composition, and some alloying elements may be lost due to oxidation.
- Removal of sulfur and phosphonate is limited, so selection of charges with less impurity is required.
- Limited production capacity. Most commonly used induction melting furnaces have limited melting capacity compared to other options such as arc melting or cupola furnaces.

References

1. Rudnev, Valery, "Induction Heating Serves Today's Forging Inductry". Available at: www.Forge.com; Inductotherm Group. Available at: www.inductotherm.com

2. Semiatin S. L. and Stutz D. E., "Induction Heat Treatment of Steel", American Society for Metals, 1986.

3. Ambrell Precision Induction Heating. Available at: www.ambrell.com/ambrell-induction-heating

4. EFD Induction. Available at: http://www.efd-induction.com/

5. Atlas Foundry web site http://www.atlasfdry.com/

6. Biswas S., Peaslee K. D., and Lekakh S., "Increasing Melting Energy Efficiency in Steel Foundries". Missouri University of Science & Technology, Rolla, MO, USA.

Paper 12-040.pdf, Page 1 of 8, AFS Transactions 2012 © American Foundry Society, Schaumburg, IL USA.

ELECTRIC ARC FURNACES

Electric arc furnaces (EAFs) use an electric arc developed between two electrodes or electrically conducting material when high voltage is applied to the electrodes. As shown in Fig. 5.4.11, a three-phase furnace uses three electrodes which are connected to the power supply system and carry AC current.

A typical arrangement for injecting chemical energy for the EAF furnace is shown in Fig. 5.4.12, multiple burners of appropriate size of heat input capacity are installed at the required angle in the furnace shell. The burners are water cooled and are fired in the earlier part of the furnace heat cycle or time. Carbon fines and flux material fine are injected in the furnace by using specially designed injection nozzles.

Direct Current (DC) Arc Furnace DC arc furnaces have single graphite electrode (cathode) and the current flows down from this graphite electrode to an anode that is mounted in the bottom of the furnace. In these furnaces the arc established between the top electrode tip and the slag bath. The top of the electrode is connected as the cathode, and the conductive bottom system is connected as the anode located at the bottom of the molten metal bath. Typical DC arc furnace configuration is shown in Fig. 5.4.13.

Energy consumption for EAFs EAFs today typically use between 380 and 650 kWh of energy to produce 1 ton of steel. Theoretically, at 100 percent efficiency, electric energy consumption will be only 320 kWh/ton for iron metal charge. The factors that influence the specific consumption of electric energy are:

- The charge weight, tapping weight, weight of fluxes, and charge carbon added to the charge materials,
- Number of scrap charges,
- Tapping temperature, time between heats and time to produce the heat,

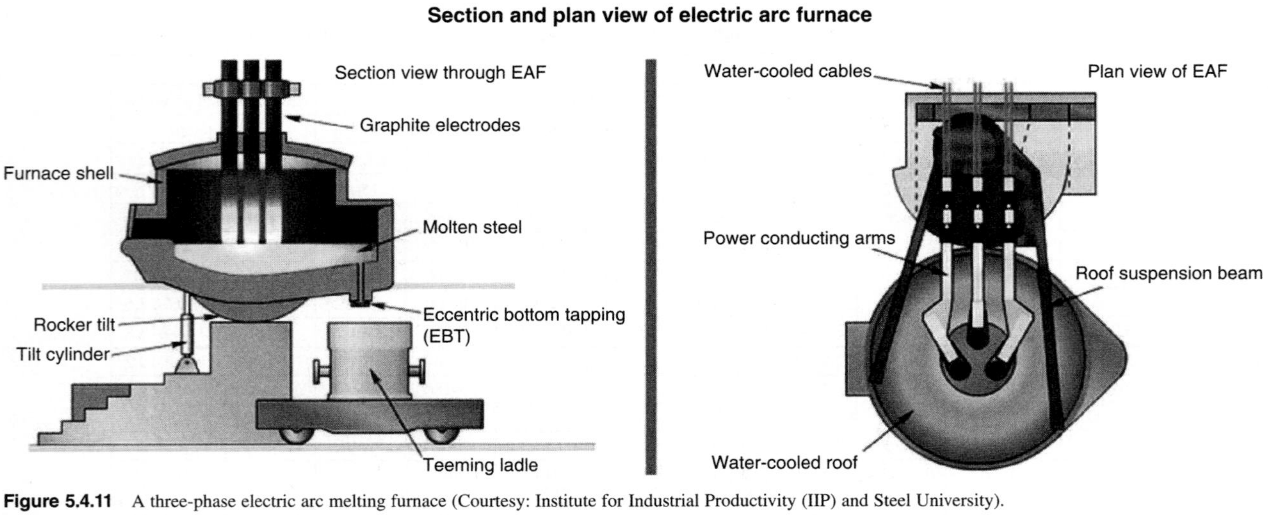

Figure 5.4.11 A three-phase electric arc melting furnace (Courtesy: Institute for Industrial Productivity (IIP) and Steel University).

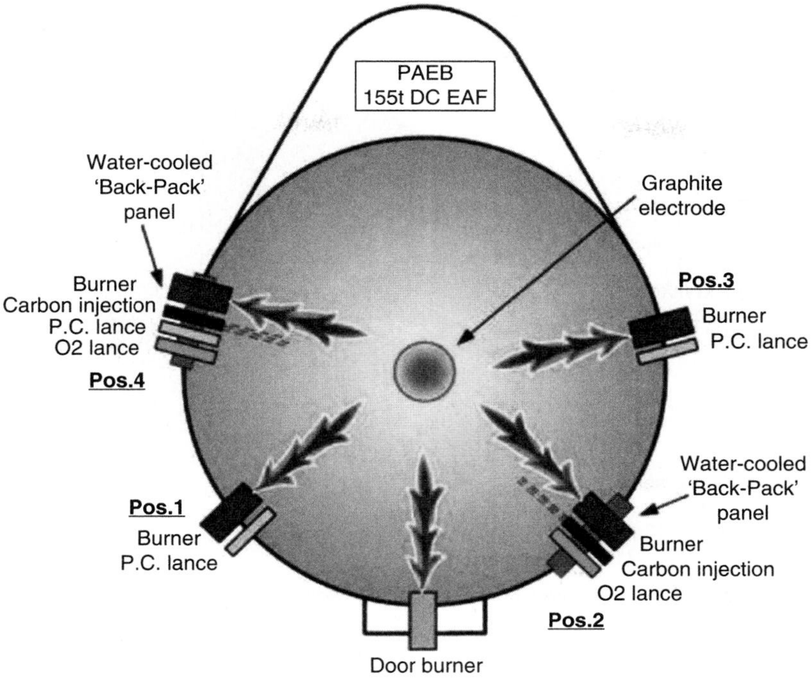

Figure 5.4.12 EAF injection system including oxy-fuel lances or burners (Reference 9).

• Amount of burner gas and distribution points of the gas, amount of blowing oxygen and distribution points of the blowing oxygen,

• Injection of carbon and flux materials into the slag during melting, and type of materials used for melting (including the use of hot metal).

Theoretical energy use for melting cold charge material of scrap and alternate ferrous materials is in the range of 320-350 kWh/ton (Reference 6). However, additional heat losses from water cooling, exhaust gases, etc. increases the energy consumption for all electric EAF in the range of 500-650 kWh/ton of molten metal.

A heat balance representing energy use distribution for a EAF is given in Fig. 5.4.14. This heat balance is based on typical energy and material input shown in Fig. 5.4.14. Note: Energy use values in this figure are based on "long ton" or "tonne," which is different from U.S. ton. The conversion factor is: 1 tonne = 1.12 U.S. tons. This figure shows that a large percentage of the total energy input is wasted as heat in exhaust gas (38.1%) as sensible and chemical energy, for water cooling (7.9%) and heat content of slag (3.1%).

References

1. IIP web site www.iipnetwork.org/

2. IIP website http://ietd.iipnetwork.org/content/direct-current-dc-arc-furnace

3. Energy Efficiency Improvement and Cost Saving Opportunities for the U.S. Iron and Steel Industry. Available at: http://www.energystar.gov/ia/business/industry/Iron_Steel_Guide.pdf?25eb-abc5 (link is external)

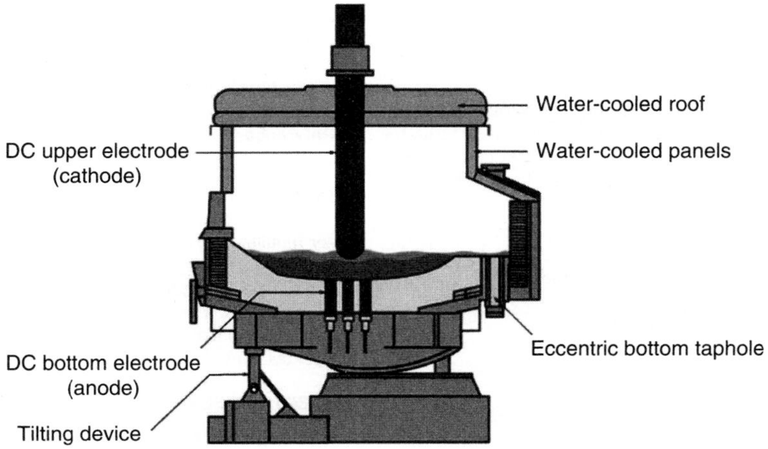

Figure 5.4.13 Single electrode DC electric arc melting furnace.

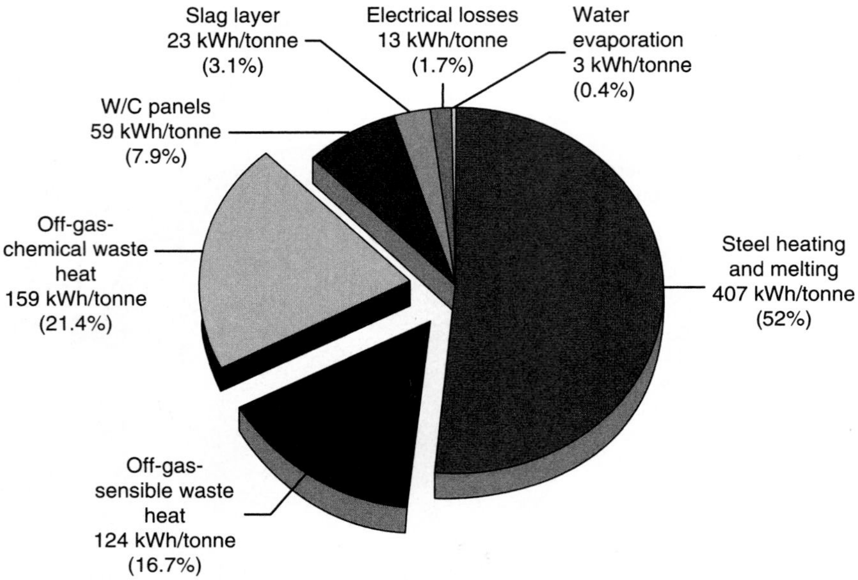

Figure 5.4.14 Heat balance for a typical electric arc melting furnace (Reference 6: Nimbalkar, Thekdi paper at AIST conference 2014).

4. JFE 21st Century Foundation. Available at: http://www.jfe-21st-cf.or.jp/chapter_2/2h_1.html

5. Steel University. Available at: https://steeluniversity.org/

6. Nimbalkar S., Thekdi A., Keiser J., Storey J., "Waste Heat Recovery from High Temperature Off-Gases". A paper presented at 2014 AIST conference, Pittsburgh, PA.

7. Electric Arc Furnace Steelmaking—By Jeremy A. T. Jones, Nupro Corporation, on AISI Steel Works web site Available at: http://www.steel.org/making-steel/how-its-made/processes/processes-info/electric-arc-furnace-steelmaking.aspx

8. Energetics report—Energy and Environmental Profile of Steel Industry.

9. Saxena R., "Developments in EAF technology". Metallurgical & Materials Engineering Department, MNIT, Jaipur, 2014. Available at: http://documentslide.com/download/link/developments-in-electric-arc-furnaces

10. Kleimt B. et al., "Dynamic Control of EAF Burners and Injectors for Carbon and Oxygen for Improved and Reproducible Furnace Operation and Slag Foaming". EU report EUR 23920 EN, 2009.

ELECTRIC INFRARED HEATING

Infrared heating is a class of radiation heating that provides thermal radiation within a certain wavelength. The range of wavelengths for infrared radiation ranges from 0.750 to 1,000 μm (750 to 1,000,000 nm).

Specific industrial applications for infrared include thermal processes such as:

- Adhesive drying
- Annealing and curing of rubber
- Drying of parts (coated with paints or varnishes)
- Profile drying textiles and paper
- Drying coatings on steel and aluminum coil
- Ink curing
- Thermoforming of plastics
- Powder coating curing
- Shrink wrapping
- Silk screening

Types of Electric-based Infrared Emitters Electric infrared heating systems are typically comprised of an emitter (temperature source), a reflector system that directs the radiant energy toward the part, and a control system designed for the application. An emitter and reflector supply the infrared energy and represent one of the most important components of the heating system. A typical system used for heating parts conveyed on a conveyor is shown in Fig. 5.4.15.

The emitter, as the heat source is called, is a key part of the system. Manufacturers typically classify emitters based on the wavelength of emitted energy. These classifications are described in Table 5.4.2 with further details of commonly used emitters.

The amount of energy given by an emitter depends on its operating temperature which, in turn, depends on the type of electrical resistance element used. In this discussion, the term radiant efficiency is defined as the total amount of energy that is emitted from the source as infrared radiation while the balance of heat energy from the source is transferred by convection and conduction.

A summary of characteristics and performance of commonly used emitters is given in Table 5.4.3.

Because infrared systems can heat a product in as little as seconds, accurate control is critical. Using infrared without giving thought to the controls can lead to quality issues. Figure 5.4.16 shows a schematic of a typical electric infrared system with a simple feedback control system where temperature is the control point.

References

1. Orfeuil M. and Robin A., "Electric Process Heating". Published by Battelle Press, 1987.

2. Improving Process Heating System Performance. A Sourcebook for Industry, 3rd ed., U.S. Department of Energy, Energy Efficiency and Renewable Energy, 2015.

3. Fostoria Electric Infrared Heating Manual in pdf format. Available at: https://www.infraredheaters.com/basic.html

4. Innovative Industries. Available at: web site www.innovativeindustries.com

ELECTRON BEAM PROCESSING

The basic components needed to produce a typical electron beam are an electron gun and a magnetic optic for controlling the beam to the target. Configuration of a typical EB system is given in Fig. 5.4.17.

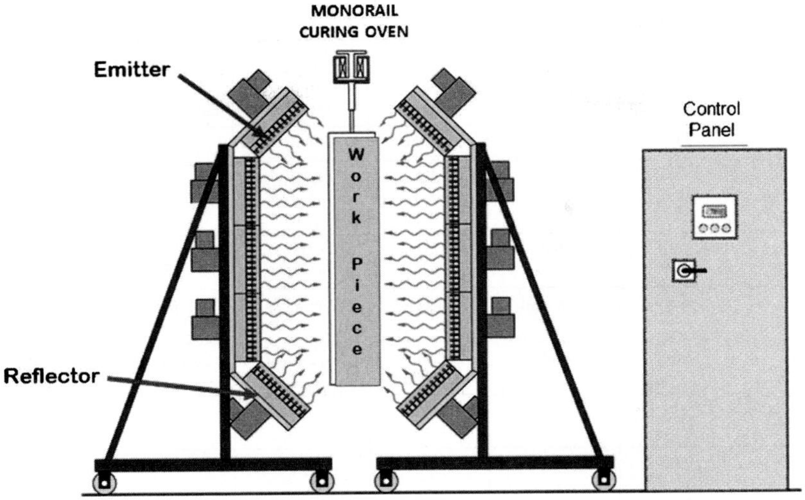

Figure 5.4.15 A typical electric infrared oven (Courtesy: Innovative Industries, Cleveland, Ohio).

Table 5.4.2 Infrared Wavelengths, Temperature, and Heating Characteristics. (Source: Process Heating Sourcebook Published by the U.S. Department of Energy)

Infrared wavelength category	Source temperature* range (typical)	Peak wavelength/ radiant output	General category of IR emitters (typical)
Short (near)	Up to 4,000°F	1.15 μm/90%	Quartz lamps having an inert gas to sustain the high source temperature. Often called T-3 for its 3/8 inch diameter lamp envelope. Considered very fast response
Medium (middle)	Up to 1,800°F	2.3 μm/60%	Embedded coils in ceramic, surface mounted coils/ribbons, carbon fiber, coils in quartz tubes Depending on emitter; medium to fast response
Long (far)	Up to 1,000°F	3-5 μm/50%	Embedded coils in ceramics and metal sheath Typically slow heat up time to temp; often because the heating element is of a heavier gauge metal

* Source: Temperature is the actual operating temperature of the infrared emitter. This is a determining factor as to the primary wavelength and radiant efficiency of the emitter. With controls, this source temperature can changed, within limits of the material, thereby changing both the radiant efficiency and wavelength.

The EB system uses an electron gun (consisting of a cathode, grid, and anode) to generate and accelerate the primary beam in a vacuum chamber. This beam consists of a stream of electronics that is focused on a workpiece using magnetic optical system.

In EB heating, metals are heated to intense temperatures when a directed beam of electrons is focused against the work surface. In EB curing, a liquid is chemically transformed to a solid on the work surface by a stream of directed electrons. EB processing can be done under vacuum, partial vacuum, and non-vacuum conditions.

Process, Applications, and Industries EB processing is used in many thermal process applications; welding metals, machining holes and slots, surface hardening of metals, and melting. In addition to thermal heating, EB competes with conventional drying methods to cure coatings, inks, and adhesives.

References

1. Improving Process Heating System Performance. A Sourcebook for Industry, 3rd ed., U.S. Department of Energy, Energy Efficiency and Renewable Energy, 2015.

2. eb industries. Available at: https://www.ebindustries.com/electron-beam-welding/

DIELECTRIC HEATING

Radio Frequency Heating and Microwave Heating

Dielectric heating is a method of volumetric heating of a non-conducting material by using electromagnetic waves in a selected wavelength range. Volumetric heating is heating of the entire volume or mass of material as opposed to heating by conventional method of heat conduction that progresses layer by layer in the material being heated. Examples of non-conducting materials include paper, plastics, fabrics, wood, chemicals, food items, etc.

A selected range of electromagnetic wavelength (Fig. 5.4.18) is known as microwave and radio frequency region. Microwaves and radio frequency waves are electromagnetic waves comparable to light or radar waves. The frequency spectrum most commonly used for dielectric heating is in the range of 3 MHz to 30 GHz, corresponding to a wavelength of 100 m to 1 cm. Within this range in USA, radio frequency process heating uses frequencies 6.78, 13.56, 27.12 and 40.68 MHz. Microwave heating is usually carried out at 915 MHz or 2,450 MHz frequency. (Reference 1).

Depending on their dielectric properties, the waves can penetrate finite distance into the material, resulting in a volumetric heating. The penetration depth is defined as the depth at which the incident wave intensity is reduced to 1/e (about 36.8%). This means that a certain amount of wave energy remains and results in heating below the penetration depth. The penetration depth of most industrially used materials as a minimum is in the range of 5-10 cm, allowing for a homogeneous heating of most products processed in industry.

Values of penetration depth of some materials are given in Table 5.4.4. These values are at 25°C and 2,450 MHz (2.45 GHz) (Reference 4).

A typical continuous unit used in drying application is shown in Fig. 5.4.19. This includes all the components described later in this section.

In addition, to the components such as power generator, applicator, etc., the system uses a conveyor to carry the material through the heating system and air cooling system for maintain safe temperature for the

Table 5.4.3 Characteristics and performance of infrared emitters (Source: Maurice Orfeuil and Albert Robin," Electric Process Heating," published by Battelle Press, 1987)

Radiation range	Short IR		Medium IR			Long IR	
Emitter types	Lamps	Quartz tubes	Quartz tubes	Silica radiating panels	Metallic radiating panels	Pyrex radiating panels	Ceramic elements
Operating temperature (°C)	2,000	2,200	1,050	650	750	350	300 to 700
Wavelength corresponding to maximum monochromatic emittance (μm)	1.4	1.2	2.2	3.5	2.8	4.6	3 to 5
Maximum installed power density (kW/m^2)	10	300	70	25	40	15	40
Thermal inertia corresponding to 90% reduction in emitted energy	1 second	1 second	30 seconds	2 minutes	2 minutes	5 minutes	5 minutes
Average service life (h)	3,000	5,000	5,000 to 10,000	Several years	Several years	Several years	Several years
Energy transferred by direct radiation (1) as percentage of emitted energy	75	80	60	50	55	45	50
Maximum product temperature (2) (°C)	300	600	500	450	400	250	500
Radiation reflected by products	High (requires insulation)		Low			Low	
Radiation penetration	Adequate		Average			Low	
Treatment quality	Homogeneous heating appropriate for thick products (a few millimeters)		Relatively superficial heating. Radiation well absorbed by wet products			Superficial heating appropriate for thin products	

(1) This percentage varies with the importance accorded to forced convection.
(2) In normal operation.

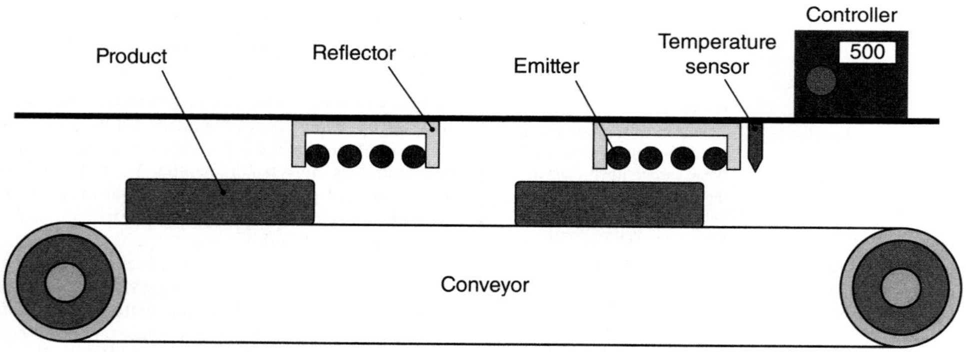

Figure 5.4.16 Simple schematic of an electric infrared heating system.
(SOURCE: Improving Process Heating System Performance. A Sourcebook for Industry, 3rd ed., U.S. Department of Energy, Energy Efficiency and Renewable Energy, 2015).

power supply system components. As may be required for some drying applications it is necessary, as shown in this figure, to use air exhaust system that is required to purge the vapors or gases driven off the material being processed.

Applications of Radio Frequency Heating

Radio frequency heating is used to achieve fast heating, volumetric heating, high efficiency, good control, and better product quality. Most conventional heating processes are limited by the heat conductivity and physical alterations of the product (i.e., evaporation of liquids, phase transformations). As high frequency heating is independent of the heat conductivity of the product and is a volumetric heating, the heat-up time is much shorter than for conventional heating. Some of the most commonly used industrial applications are listed in Table 5.4.5. Due to its inverse temperature profile, dielectric heating can also improve the drying speed of a product without risking damages to the material. The industry branches were dielectric heating is mostly used are ceramic, food, rubber, chemistry, and building materials.

Improve the Efficiency of Existing Radio Frequency Systems

• Verify that the correct frequency is being used. The amount of heat generated is a function not only of the output of the power supply, but also the frequency of the field.
• Use programmable logic controller to optimize your process. Good control systems allow for precise application of heat at the proper temperature for the correct amount of time.
• Consider a hybrid radio frequency/convection heating system. The efficiency of convection dryer drops significantly as the moisture level in the material decreases, and radio frequency energy couples directly

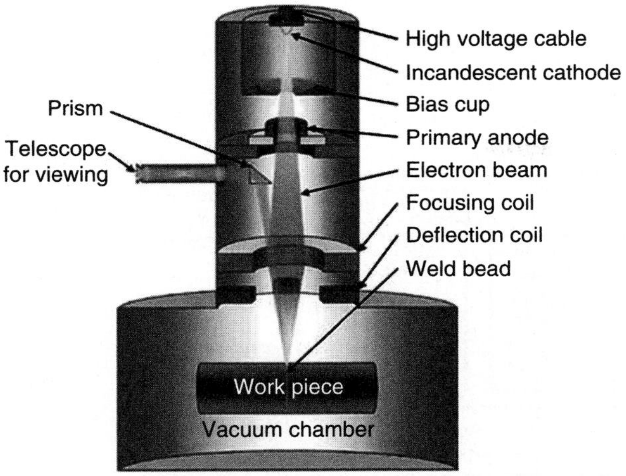

Figure 5.4.17 Configuration of a typical electron beam system (Courtesy: eb industries).

with the water. At this point, radio frequency energy is more efficient at removing the moisture. This technology is particularly useful for drying of heat sensitive materials, since efficient drying can occur without exceeding the boiling point of water.

Industrial Microwave Heating Microwave heating systems are used extensively in manufacturing. Microwave heating is a subsector of dielectric heating that includes radio frequency heating and microwave heating. As in case of radio frequency heating microwave offers volumetric heating, where heat is produced and used within the material itself. It is not necessary to transfer heat from the surface or outer layer of the material being heated to the center or core of the material. Difference between radio frequency heating and microwave heating is in the frequency of electrical power used for heating.

The microwave technology is used for a variety of heating processes such as dehydration, sterilization, tempering (thawing), boost heating, and cooking. The technology can be used either in conjunction with, or in place of, conventional heating methods to maximize energy efficiency and minimize process time and waste.

Application of MW Heating in Industries Microwave heating is being used for many applications for heating of materials such as food, rubber, plastic while several other applications are being explored and

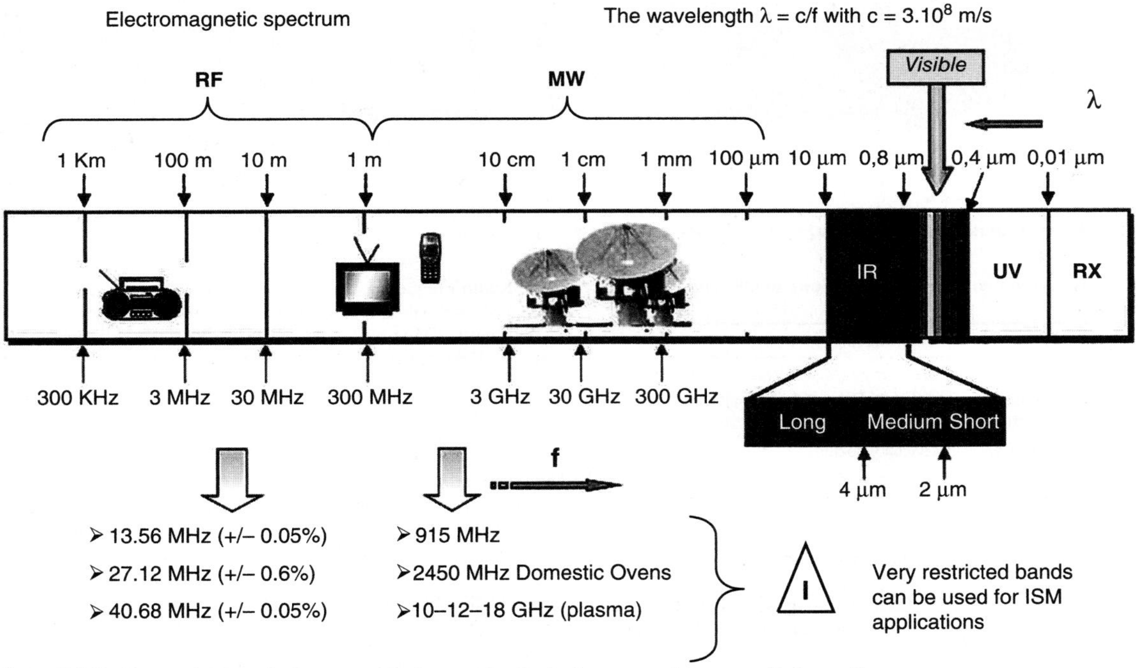

Figure 5.4.18 Frequencies for radio frequency and microwave heating in electromagnetic spectrum (Reference 7).

Table 5.4.4 Penetration Depth for Different Dielectric Materials Using Microwave Frequency (2.45 GHz at 25°C) Heating

Material	Penetration Depth (cm)
Aluminium Oxide	1,460
Silicon Carbide	7.0
PTFE	4,700
PVC	200
Wood (40% H_2O)	4.0
Water	1.8

experimented. A list of commonly used applications compiled from various sources (References 1 to 3) is given below.

1. Food tempering includes tempering of food such as meat, fish, fruit, butter, and other foodstuffs from freezer temperature to process temperature close to ambient temperature for further processing.
2. Preheating for rubber vulcanization of rubber extrusions that require rapid and uniform heating. In some cases, this is combined with conventional infrared heating or air heating.
3. Atmospheric pressure drying of various materials such as textiles, ceramics, coated paper, leather, food items. In some cases, it is efficient to use hybrid heating that includes microwave heating with air drying.

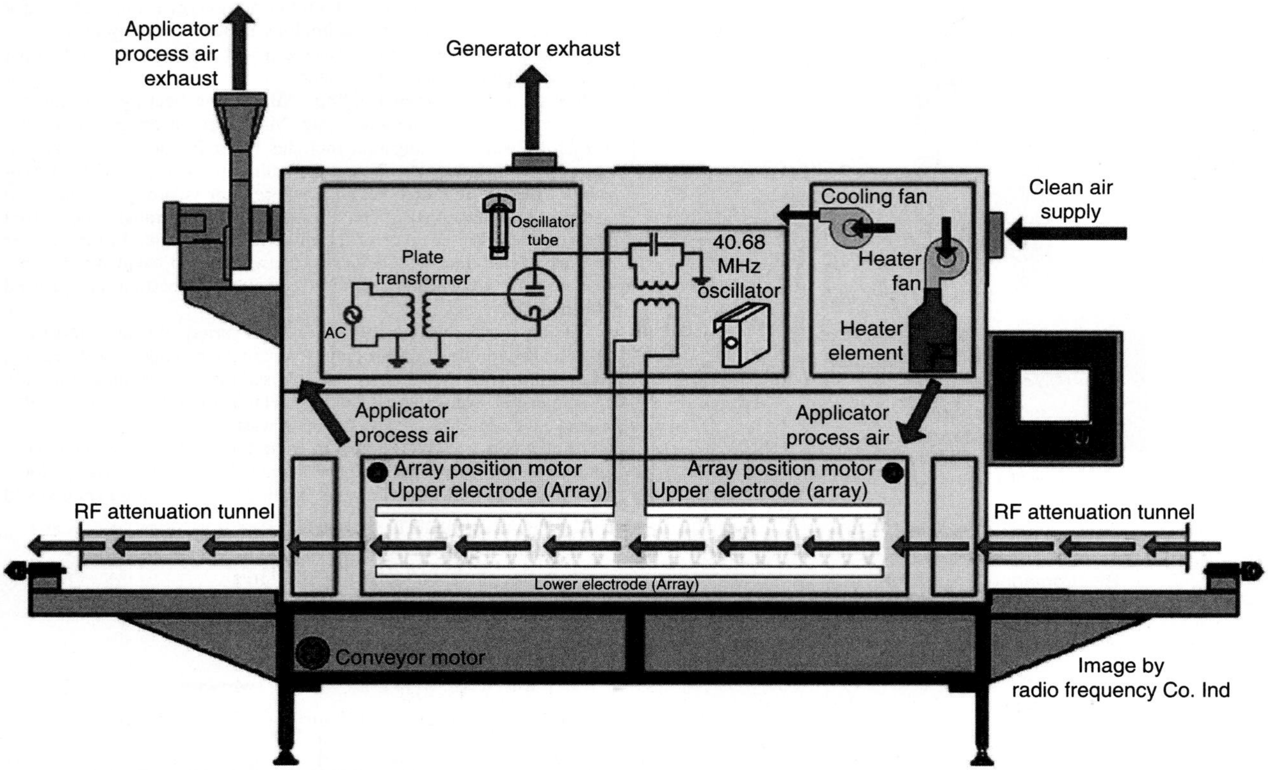

Figure 5.4.19 A continuous radio frequency heating system (Courtesy: Radio Frequency Corporation).

Table 5.4.5 Traditional Industrial Applications of MW and RF Technologies

Radio Frequency	Microwave
Drying (Textiles, Paper, Wood)	Tempering (Meats, Vegetables, Fruits)
Plastic Preform Heating	Vacuum Drying Pharmaceuticals
Post Baking Crackers/Biscuits	Ceramics Sintering
Resins & Adhesives Curing	Bacon Cooking
Plastics Welding	Curing Rubber

4. Vacuum drying: drying of fruit juices, beverages, drugs, and pharmaceutical pellets.

5. Ceramic processing that includes slip casting, sintering of a wide range of ceramics and composites, joining and calcining of superconductors or electroceramics.

6. Other potential applications promoted include food processing, asphalt hole patcher, vitrification of nuclear wastes, treatment of highly toxic substances, waster recover of plastics, pyrolysis, heating of resins, polymerization, heating of oil sands, and the processing of minerals.

References

1. Blaker, G of PSC – Litzler Company. Private communications.
2. Metaxas, A. C., “Foundations of Electroheat: A Unified Approach”. John Wiley and Sons, 1996.
3. Metaxas, A. C. and Meredith, R. J., “Industrial Microwave Heating” (IEE, 1983, reprinted 1988 and 1993).
4. Maurice O. and Albert R., “Electric Process Heating Technologies/Equipment/Applications.” Battelle Press, 1987.
5. Cresco J., US Department of Energy, EERE 2016 Private Communications.
6. Radio Frequency Company web site www.radiofrequency.com
7. PSC web site http://www.pscrfheat.com/
8. Tang J., “Microwave (and RF) Heating in Food Processing Applications.” Department of Biological Systems Engineering Washington State University, Pullman, WA.

ULTRAVIOLET (UV) PROCESSING

Ultraviolet processing of materials commonly known as UV curing is a photochemical process in which high-intensity ultraviolet light produced from a light source is used to cure or “dry” specially formulated inks, coatings or adhesives. Applications where conventional curing systems are used.

How the Technology Works UV radiation segment of the electromagnetic spectrum is represented by wavelength within 100-400 nanometer (nm). Applying UV radiation to certain liquid polymeric substances transforms (cures) them into a solid coating. UV curing technology uses specially formulated ink or coating that contains monomers, oligomers, and photoinitiator. This coating solidifies when it is exposed to UV radiation. Figure 5.4.20 shows simplified illustration of three-step process of UV curing.

The steps are:

1. Application of specially formulated ink containing monomers, oligomers, and photoreactive photoinitiators on the substrate such as paper, metal, plastic, glass, etc.;

2. Exposure of the coating to a specific range of UV light using one of the several different types of light sources such as LED lamps. This results in phot reaction at almost instantaneously;

3. Hardening of liquid due to chemical reaction and retrieval of the finished final product.

Note that this figure shows use of LED light source; however, as discussed alter the UV light source can be selected from several different types of UV light sources.

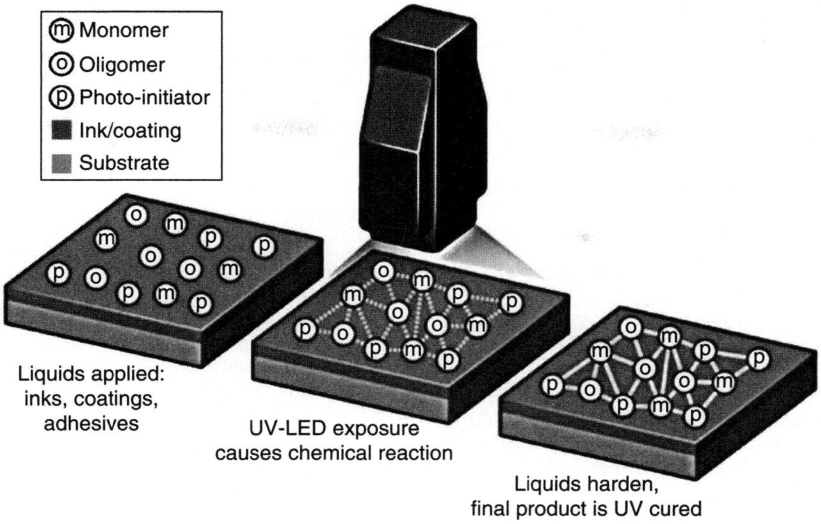

Figure 5.4.20 Illustration of simplified steps for UV curing of parts showing LED lamp as light source (Courtesy: LEDS Magazine and Phoseon Technology).

Industrial Equipment Used for UV Curing A typical UV curing system, shown in Figure 5.4.21, used in industries is designed as a batch processing unit or continuous system. In a batch unit, the coated material is placed in an enclosure or light chamber and the material is cured while in the enclosure in batches—one lot or unit at a time. Continuous systems are used for mass production where the material is conveyed continually through the enclosed unit, light chamber, by using a material handling mechanism such as a belt, rolls, or another method of transporting. The material is exposed to UV light source and the coating goes through chemical reaction to produce hardened coating.

Process, Applications, and Industries UV curing differs significantly from conventional heating where convection and radiant heating is used to vaporize solvent or water while the solid content of the coting mixture sets in. This process generates and emits organic or other types of vapors that need to be treated to meet environmental regulations. The process also requires finite amount of time for the entire process to take place. However, UV-based systems typically have more rapid curing speeds, produce fewer emissions, and can cure heat-sensitive substrates. The cross linking of molecules requires minimal or no solvents as part of the coating. These systems require special UV-curable coatings and generally a custom-made lamp system for a particular application. UV curing takes about 25 percent of the energy required by a thermal-based system using a fuel-fired oven. Use of UV system can increase output because of the nearly instantaneous curing time. Although UV coatings are more expensive on a cost-per-gallon basis, they do not require costly emission control system such as thermal oxidizers to destroy VOCs emitted by solvent-based coatings. In addition, there is no reduction in the cured coating thickness versus applied coating thickness.

Photoinitiators typically absorbs light in two wavelength regions, either 260 nm or 365 nm. Wavelength of intense absorption tends to favor coating surface cure while wavelength of lower absorption tends to favor coating through cure, which is why some coatings require a

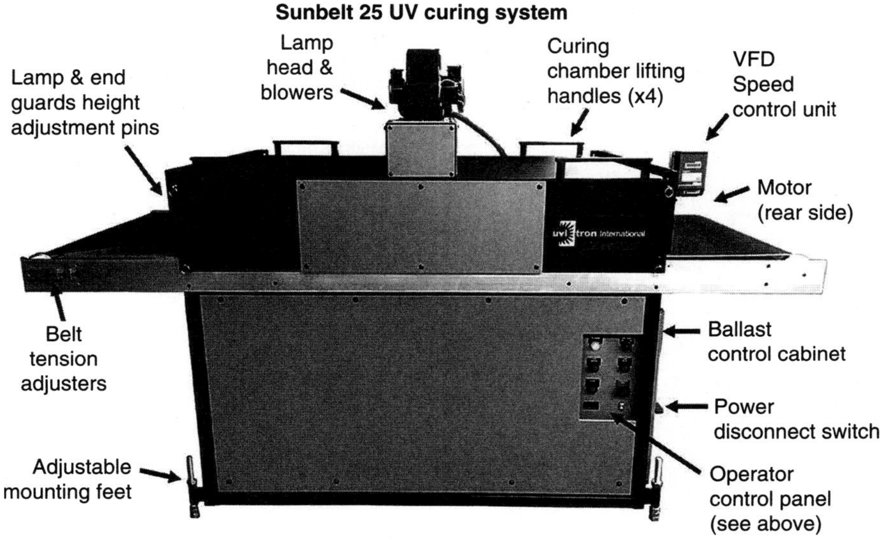

Figure 5.4.21 A typical continuous type UV curing system (Courtesy: Unitron International Inc.).

Table 5.4.6 Four Ranges of UV Wavelengths and Their Applications

UV Type	Range (nm)	Uses
UVA	315-400	For curing of thick layers, i.e., floor coverings w/wear layers
UVB	280-315	In-depth drying for heavier inks coats, i.e., silk screen
UVC	200-280	Maintain reaction for through curing, i.e., general ink surface
UVV	100-200	For surface cure/touch care; activates photoinitiators; germicidal UV

mixture of multiple photoinitiators. There are four ranges of UV wavelengths as listed in Table 5.4.6.

The four main applications for UV curing are coatings, printing, adhesives, and electronic parts.

Coatings Common industrial coatings cured with UV radiation include those applied to wood, metals, paper, plastics, vinyl flooring, and wires. The coating can be a liquid or a powder, with both having similar characteristics.

Printing Lithographic, silk screen, and flexographic printing operations can use UV curable inks instead of solvent-based, thermally cured inks.

Adhesives Adhesive materials processed with UV radiation are common in the structural and packaging markets.

Electronic parts UV processing is used throughout the electronics and communications parts manufacturing industry to cure polymeric materials, especially with printed circuit board lithography.

UV processing is also used in the wastewater industry to treat water and to purify indoor air. New applications of use of UV processing are being developed continuously by the researchers and equipment suppliers in collaboration with the industrial end users.

References

1. Cortelyou B., "UV-LED Advancements Extend the Promise in Curing." Published in UV LED Magazine, March 11, 2014.
2. Uvitron International Inc. Private Communications with Mr. Yevgeniy Mikhaylichenko.
3. Radtech Internationl North America, "UV/EB Curing Primer 1." Available at: *infohouse.p2ric.org/ref/29/28394.pdf*
4. Traction A. A., "Coatings Technology: Fundamentals, Testing, and Processing Techniques." Published by CRC press, 2006.

SUMMARY OF ELECTRICAL HEATING TECHNOLOGIES (ELECTROTECHNOLOGIES)

Table 5.4.7 summarizes the electrotechnologies that were covered in this section, with their applications, the primary industries that utilize them, and competing technologies.

Table 5.4.7 Summary of Electrotechnologies for Process Heating

Electrotechnology	Applications	Primary Industries	Competing Technologies
Arc Furnaces	Steel production, melting steel, iron and refractory metals, smelting	Primary metal production, especially steelmaking	Oxygen furnaces (coal), plasma processing
Electric infrared processing	Heating, drying, curing, thermal-bonding, sintering, sterilizing	Fabricated metal products, transportation equipment, plastics and rubber	Electron beam processing, ultraviolet curing, resistance heating, induction heating
Electron beam processing	Melting, drying, welding, machining, curing, crosslinking, grafting	Fabricated metal products, machinery, transportation equipment, plastics	Infrared processing, laser processing, ultraviolet curing, induction melting, arc welding
Induction heating and melting	Heating, heat treating, melting	Fabricated metal products, primary metals, transportation equipment, machinery	Resistance heating, infrared processing, electron beam processing, gas furnaces
Laser processing	Heat treating, hardening, trimming and cutting, welding	Transportation equipment, fabricated metal products, electronics	Electron beam processing, plasma processing, induction heating, arc welding
Microwave and radio frequency processing	Heating, tempering, drying, cooking, plasma production	Food/beverage, plastics and rubber, wood products, chemicals	Resistance heating, induction heating, ovens
Plasma Processing	Melting scrap, remelting for refining, reduction, surface hardening, welding, cutting	Primary metals (titanium, high-alloy steels, tungsten)	Arc furnaces, oxygen furnaces, electron beam processing, laser processing
Resistance heating and melting	Heating, brazing, sintering, melting	Glass production, metal products, plastics, chemicals, food	Induction heating and melting, infrared processing, microwave/radio frequency processing, gas furnaces
Ultraviolet curing	Curing, coatings, printing, adhesives, purifying	Computers, electronics, printing, fabricated metal products, transportation equipment, machinery	Electron beam processing, infrared processing

Section **6**

Machine Elements

BY

ALI M. SADEGH *Professor of Mechanical Engineering, The City College of the City University of New York*

VITTORIO (RINO) CASTELLI *Senior Research Fellow, Retired, Xerox Corp.; Engineering Consultant*

STEVEN V. BOYD *The Timken Corporation*

MICHAEL D. ALLEGA *The Timken Corporation*

SARAH A. MEYER *The Timken Corporation*

MATTHEW TONES *Sr. Applications Engineer, Garlock Sealing Technologies, an EnPro Industries Company*

JAMES DRAGO *Manager, Engineering, Garlock Sealing Technologies, an EnPro Industries Company*

6.1 MECHANISMS
by Ali M. Sadegh

6.2 MACHINE ELEMENTS
by Ali M. Sadegh

6.3 GEARING
Revised by Ali M. Sadegh

6.4 FLUID FILM BEARINGS
by Vittorio (Rino) Castelli

6.5 BEARINGS WITH ROLLING CONTACT
by Steven V. Boyd, Michael D. Allega, and Sarah A. Meyer

6.6 PACKINGS, GASKETS, AND SEALS
by Matthew Tones and James Drago

6.7 PIPE, PIPE FITTINGS, AND VALVES
Revised by Ali M. Sadegh

6.1 MECHANISMS

by Ali M. Sadegh

REFERENCES: Beggs, "Mechanism," McGraw-Hill. Hrones and Nelson, "Analysis of the Four Bag Linkage," Wiley. Jones, "Ingenious Mechanisms for Designers and Inventors," 4 vols., Industrial Press. Moliam, "The Design of Cam Mechanisms and Linkages," Elsevier. Chironis, "Gear Design and Application," McGraw-Hill.

NOTE: The reader is referred to the current and near-past professional literature for extensive material on linkage mechanisms. The vast number of combinations thereof has led to the development of computer software programs to aid in the design of specific linkages.

Definition A **mechanism** is the part of a machine that contains two or more pieces arranged so that the motion of one compels the motion of the other(s), as prescribed by the nature of the combination.

LINKAGES

Links may be of any form so long as they do not interfere with the desired motion. The simplest form is four bars *A, B, C,* and *D,* fastened together at their ends by cylindrical pins, and which are all movable in parallel planes. If the links are of different lengths and each is fixed in turn, there will be four possible combinations; but as two of these are similar there will be produced three mechanisms having distinctly different motions. Thus, in Fig. 6.1.1, if *D* is fixed, *A* can rotate and *C* oscillate, creating a **beam-and-crank** mechanism, as used on side-wheel steamers. If *B* is fixed, the same motion will result. If *A* is fixed (Fig. 6.1.2), links *B* and *D* can rotate, creating a **drag-link** mechanism

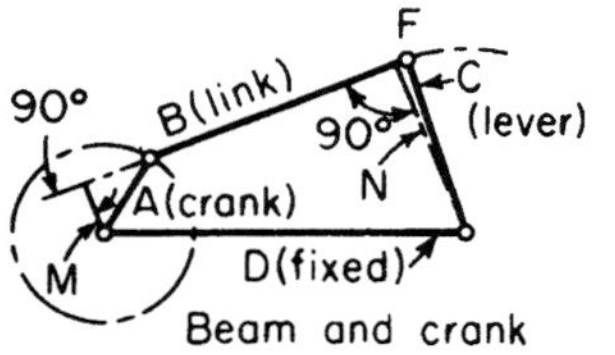

Figure 6.1.1 Beam-and-crank mechanism.

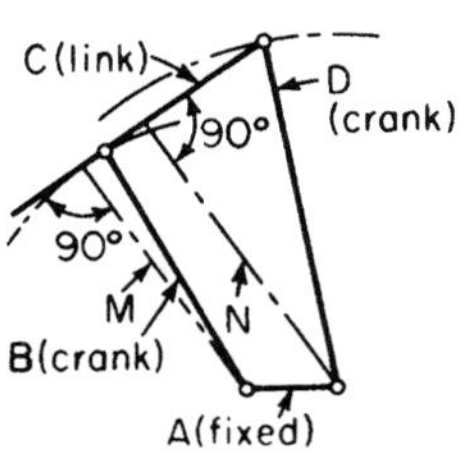

Figure 6.1.2 Drag-link mechanism.

that can be used to feather the floats on paddle wheels. If link *C is fixed* (Fig. 6.1.3), *D* and *B* can only oscillate, and a **rocker** mechanism sometimes used in straight-line motions is produced. It is customary to call a rotating link a **crank,** an oscillating link a beam or **lever,** and a connecting link a **connecting rod** or **coupler.** Discrete points on the coupler, crank, or lever can be pressed into service to provide a desired motion. The fixed link is often enlarged and used as the supporting frame.

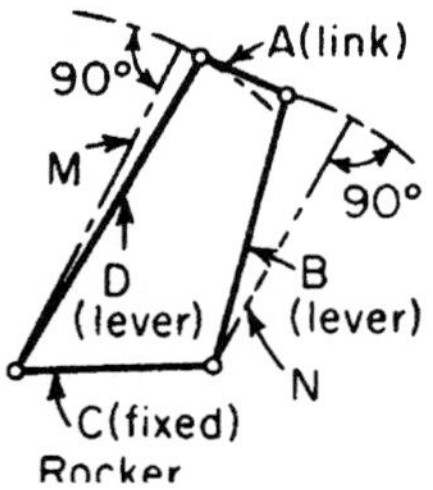

Figure 6.1.3 Rocker mechanism.

If the pin joint *F* in the linkage (Fig. 6.1.1) is replaced by a slotted piece *E* (Fig. 6.1.4), no change will be produced in the resulting motion, and if the length of links *C* and *D* is made infinite, the slotted piece *E* will become straight and the motion of the slide will be that of pure translation, thus obtaining the engine, or **sliding-block linkage** (Fig. 6.1.5).

In the sliding-block linkage (Fig. 6.1.5), if the long link *B* is fixed (Fig. 6.1.6), *A* will rotate and *E* will oscillate, and infinite links *C* and *D* may be indicated as shown. This gives the **swinging-block linkage.** When used for quick-return motion, the slotted piece and slide are usually interchanged (Fig. 6.1.7) which in no way changes the resulting motion. If the short link *A* is fixed (Fig. 6.1.8), *B* and *E* can both rotate, and the mechanism known as the **turning-block linkage** is obtained.

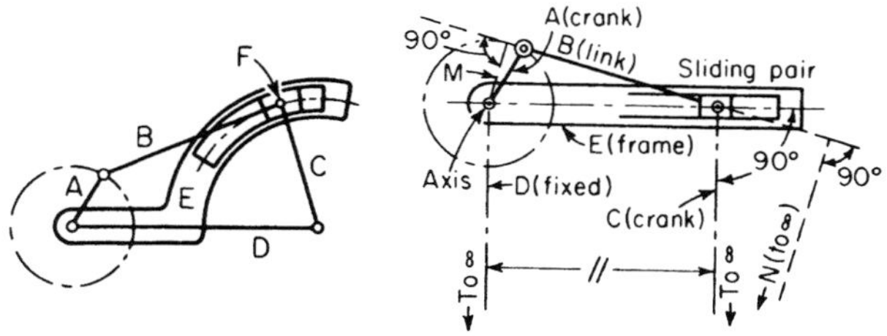

Figures 6.1.4 and 6.1.5 Sliding-block linkage.

This is better known under the name of the **Whitworth quick-return motion,** and is generally constructed as in Fig. 6.1.9. The **ratio of time of advance to time of return** *H*/*K* of the two quick-return motions (Figs. 6.1.7) and may be found by locating, in the case of the swinging

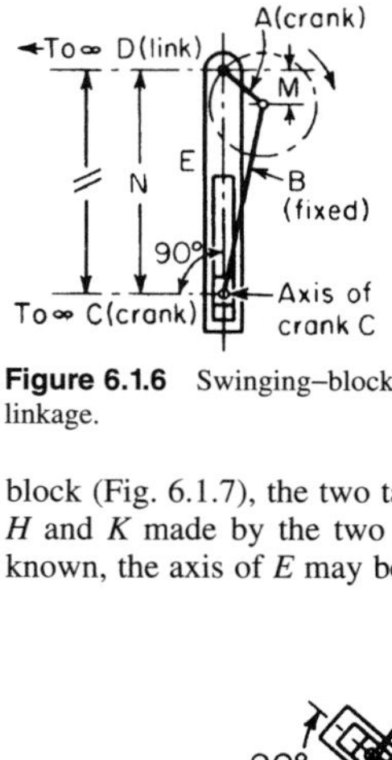

Figure 6.1.6 Swinging-block linkage.

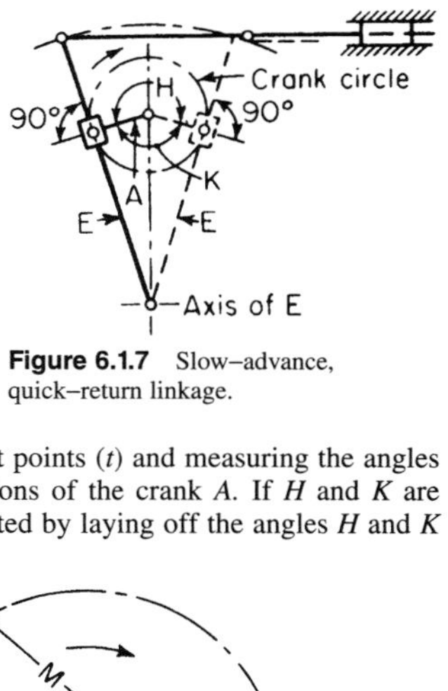

Figure 6.1.7 Slow-advance, quick-return linkage.

block (Fig. 6.1.7), the two tangent points (*t*) and measuring the angles *H* and *K* made by the two positions of the crank *A*. If *H* and *K* are known, the axis of *E* may be located by laying off the angles *H* and *K*

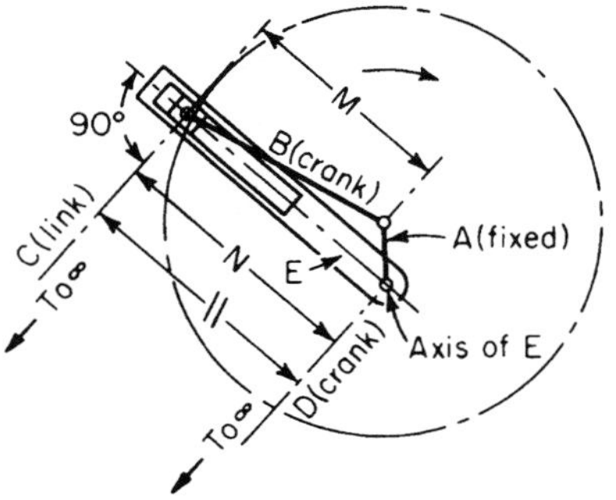

Figure 6.1.8 Turning-block linkage.

on the crank circle and drawing the tangents *E,* their intersection giving the desired point. For the turning-block linkage (Fig. 6.1.9), determine the angles *H* and *K* made by the crank *B* when *E* is in the horizontal position; or, if the angles are known, the axis of *E* may be determined by drawing a horizontal line through the two crankpin positions (S) for the given angle, and the point where a line through the axis of B cuts E perpendicularly will be the axis of E.

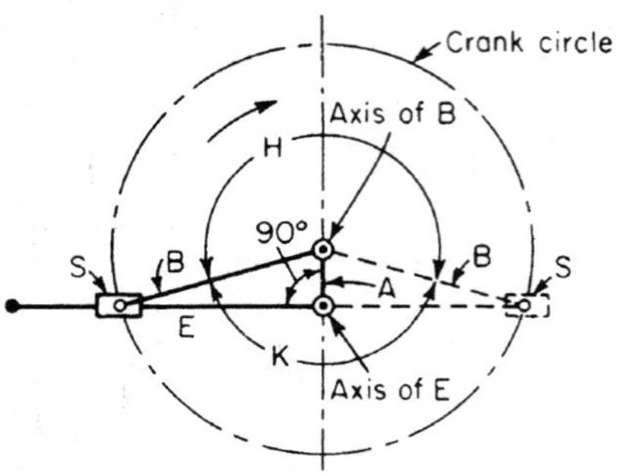

Figure 6.1.9 Whitworth quick–return motion.

Velocities of any two or more points on a link must fulfill the following conditions (see Sec. 1). (1) Components along the link must be equal and in the same direction (Fig. 6.1.10): $V_a = V_b = V_c$. (2) Perpendiculars to V_A, V_B, V_C from points A, B, and C must intersect at a common point d, the **instant center** (or instantaneous axis). (3) The velocities of points A, B, and C are directly proportional to their distances from this center (Fig. 6.1.11): $V_A/a = V_B/b = V_C/c$. For a straight link the tips of the vectors representing the velocities of any number of points on the link will be on a line (Fig. 6.1.12); abc = a straight line. To find the velocity of any point when the velocity and direction of any two other points are known, condition 2 may be used, or a combination of conditions 1 and 3. The **linear velocity ratio** of any two points on a linkage may be found by determining the distances e and f to the instant [center (Fig. 6.1.13); $V_c/V_b = e/f$]. This may be simplified by noting that a line drawn parallel to e and cutting B forms two similar triangles efB and sAy, which gives $V_c/V_b = e/f = s/A$. The **angular velocity ratio** for any position of two oscillating or rotating links A and C (Fig. 6.1.1), connected by a movable link B, may be determined by scaling the length of the perpendiculars M and N from the axes of rotation to the centerline of the movable link. The angular velocity ratio is inversely proportional to these perpendiculars, or $O_C/O_A = M/N$. This method may be applied directly to a linkage having a sliding pair if the two infinite links are redrawn perpendicular to the sliding pair, as indicated in Fig. 6.1.14. M and N are shown also in Figs. 6.1.1, 6.1.2, 6.1.3, 6.1.5, 6.1.6, and 6.1.8. In Fig. 6.1.5 one of the axes is at infinity; therefore, N is infinite, or the slide has pure translation.

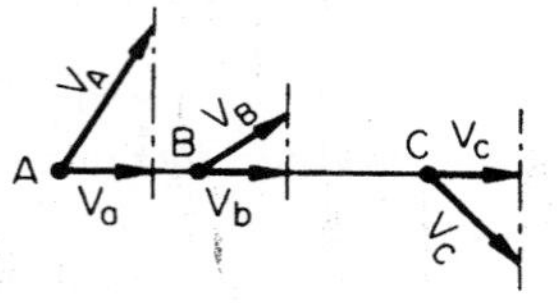

Figure 6.1.10

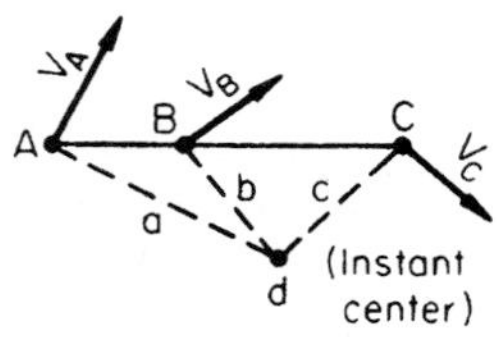

Figure 6.1.11

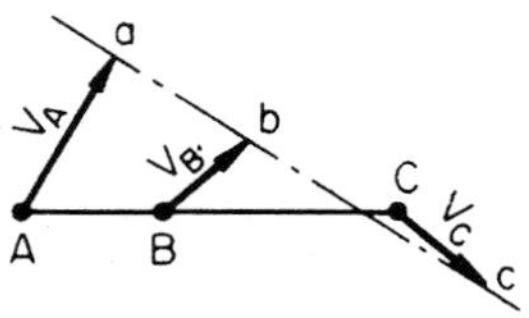

Figure 6.1.12

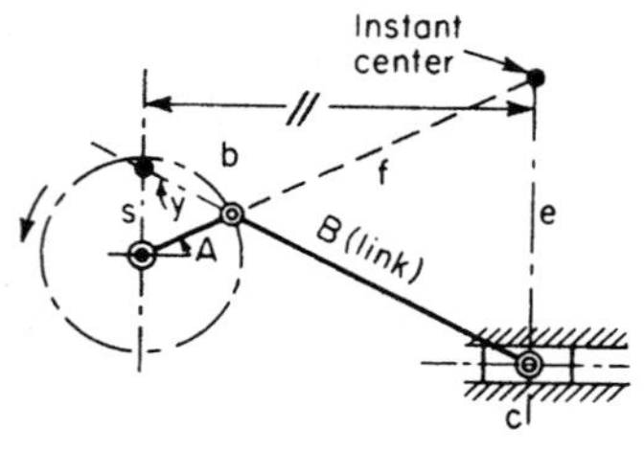

Figure 6.1.13

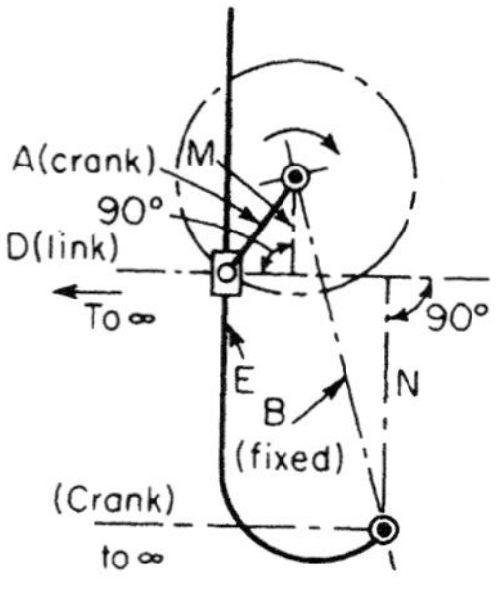

Figure 6.1.14

Forces A mechanism must deliver as much work as it receives, neglecting friction; therefore, the force at any point F multiplied by the velocity V_F in the direction of the force at that point must equal the force at some other point P multiplied by the velocity V_P at that point; or the forces are inversely as their velocities and $F/P = V_P/V_F$. It is at times more convenient to equate the moments of the forces acting around each axis of rotation (sometimes using the instant center) to determine the force acting at some other point. In Fig. 6.1.15, $F \times a \times c/(b \times d) = P$.

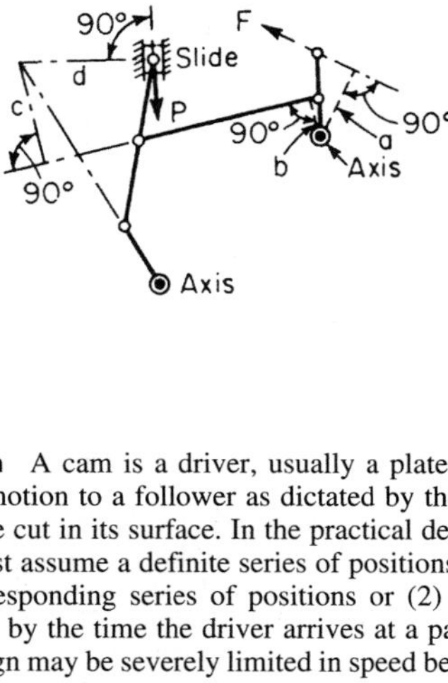

Figure 6.1.15

CAMS

Cam Diagram A cam is a driver, usually a plate or cylinder, that communicates motion to a follower as dictated by the geometry of its edge or a groove cut in its surface. In the practical design of cams, the follower (1) must assume a definite series of positions while the driver occupies a corresponding series of positions or (2) must arrive at a definite location by the time the driver arrives at a particular position. The former design may be severely limited in speed because the interrelationship between the follower and cam positions may yield a follower displacement versus time function that involves large values for the successive time derivatives, indicating large accelerations and forces, with concomitant large impacts and accompanying noise. The second design centers about finding that particular interrelationship between the follower and cam positions that results in the minimum forces and impacts so that the speed may be made quite large. In either case, the desired interrelationship must be put into hardware as discussed below. In the case of high-speed machines, small irregularities in the cam surface or geometry can be severely detrimental.

A stepwise displacement in time for the follower running on a cam driven at constant speed is, of course, impossible because the follower would require infinite velocities. A step in velocity for the follower would result in infinite accelerations; these in turn would introduce forces that approach infinite magnitudes, which would tend to destroy the machine. A step in acceleration causes a large jerk and large shock waves to be transmitted and reflected throughout the parts; this generates noise and would tend to limit the life of the machine. A step in jerk, the third derivative of the follower displacement with respect to time, seems altogether acceptable. In those designs requiring or exhibiting clearance between the follower and cam (usually at the bottom of the stroke), as gentle and slow a ramp portion as can be tolerated must be inserted on either side of the clearance region to limit the magnitude of the acceleration and jerk to a minimum. The tolerance on the clearance adjustment must be small enough to assure that the follower will be left behind and picked up gradually by the gentle ramp portions of the cam.

The three most common forms of motion used are uniform motion (Fig. 6.1.16), harmonic motion (Fig. 6.1.17), and uniformly accelerated and retarded motion (Fig. 6.1.18). In plotting the diagrams (Fig. 6.1.18) for this last motion, divide *ac* into an even number of equal parts and *bc*

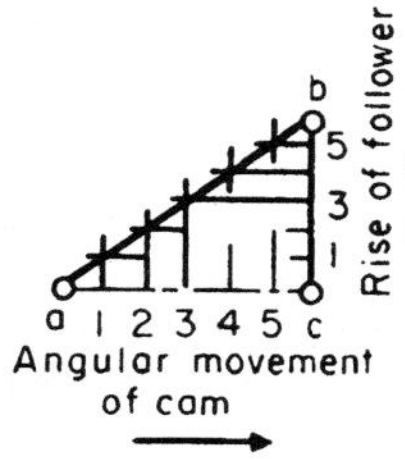

Figure 6.1.16 Uniform motion.

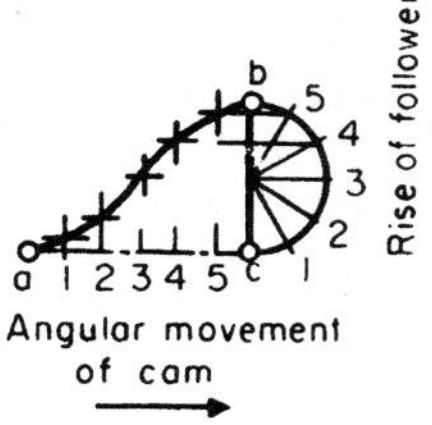

Figure 6.1.17

into the same number of parts with lengths increasing by a constant increment to a maximum and then decreasing by the same decrement, as, for example, 1, 3, 5, 5, 3, 1, or 1, 3, 5, 7, 9, 9, 7, 5, 3, 1. In order to prevent shock when the direction of motion changes, as at *a* and *b* in the uniform motion, the harmonic motion may be used; if the cam is to be operated at high speed, the uniformly accelerated and retarded motion should preferably be employed. In either case there is a very gradual change of velocity.

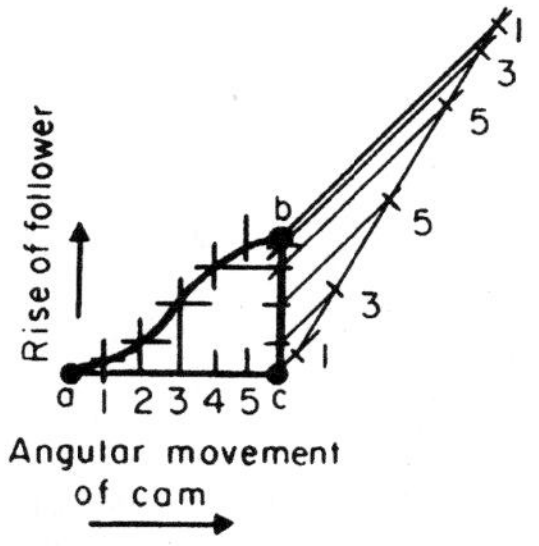

Figure 6.1.18

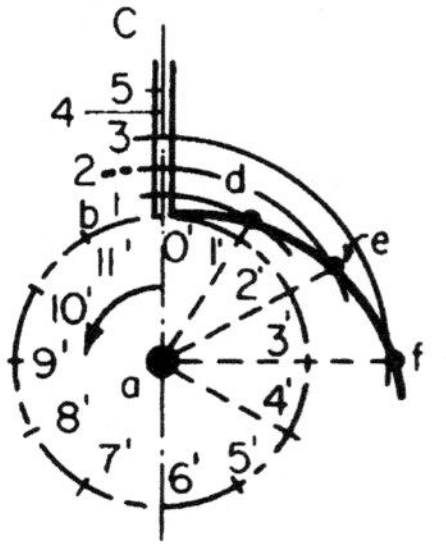

Figure 6.1.19

Pitch Line The actual pitch line of a cam varies with the type of motion and with the position of the follower relative to the cam's axis. Most cams are constructed by one of the following four cases.

Follower on Line of Axis (Fig. 6.1.19). To draw the pitch line, subdivide the motion *bc* of the follower in the manner indicated in Figs. 6.1.16, 6.1.17, and 6.1.18. Draw a circle with a radius equal to the smallest radius of the cam *a*0 and subdivide it into angles 0*a*1′, 0*a*2′, 0*a*3′, etc., corresponding with angular displacements of the cam for positions 1, 2, 3, etc., of the follower. With *a* as a center and radii *a*1, *a*2, *a*3, etc., strike arcs cutting radial lines at *d, e, f*, etc. Draw a smooth curve through points *d, e, f*, etc.

Offset Follower (Fig. 6.1.20). Divide *bc* as indicated in Figs. 6.1.16, 6.1.17, and 6.1.18. Draw a circle of radius *ac* (highest point of rise of follower) and one tangent to *cb* produced. Divide the outer circle into parts 1′, 2′, 3′, etc., corresponding with the angular displacement

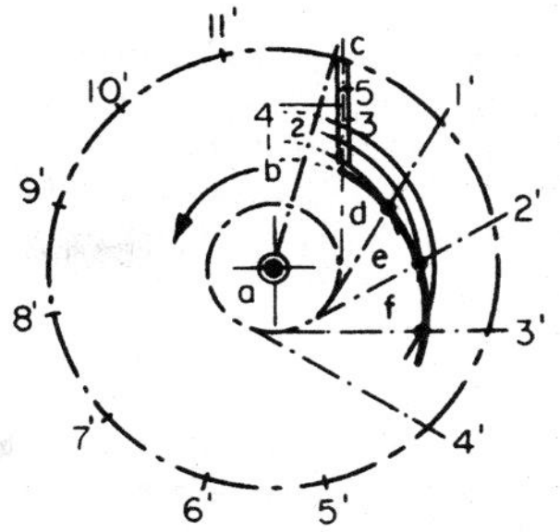

Figure 6.1.20

of the cam for positions 1, 2, 3, etc., of the follower, and draw tangents from points 1′, 2′, 3′, etc., to the small circle. With *a* as a center and radii *a*1, *a*2, *a*3, etc., strike arcs cutting tangents at *d, e, f*, etc. Draw a smooth curve through *d, e, f*, etc.

Rocker Follower (Fig. 6.1.21). Divide the stroke of the slide *S* in the manner indicated in Figs. 6.1.16, 6.1.17, and 6.1.18, and transfer these points to the arc *bc* as points 1, 2, 3, etc. Draw a circle of radius *ak* and divide it into parts 1′, 2′, 3′, etc., corresponding with angular

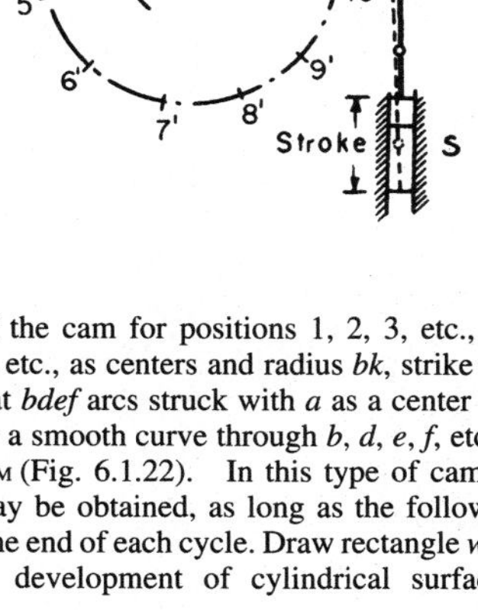

Figure 6.1.21

displacements of the cam for positions 1, 2, 3, etc., of the follower. With *k*, 1′, 2′, 3′, etc., as centers and radius *bk*, strike arcs *kb*, 1′*d*, 2′*e*, 3′*f*, etc., cutting at *bdef* arcs struck with *a* as a center and radii *ab*, *a*1, *a*2, *a*3, etc. Draw a smooth curve through *b, d, e, f*, etc.

Cylindrical Cam (Fig. 6.1.22). In this type of cam, more than one complete turn may be obtained, as long as the follower returns to its starting point at the end of each cycle. Draw rectangle *wxyz* (Fig. 6.1.22) representing the development of cylindrical surface of the cam.

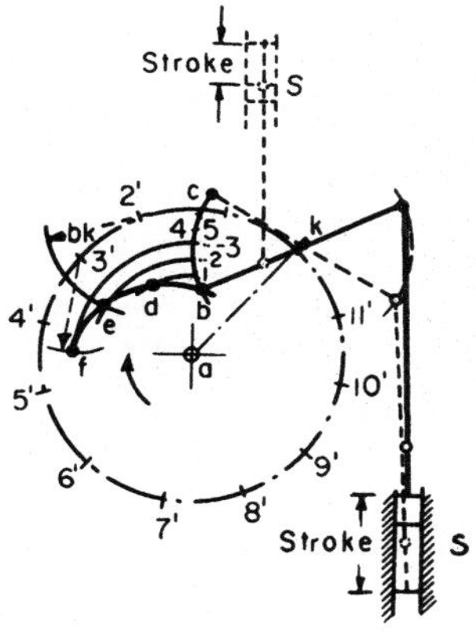

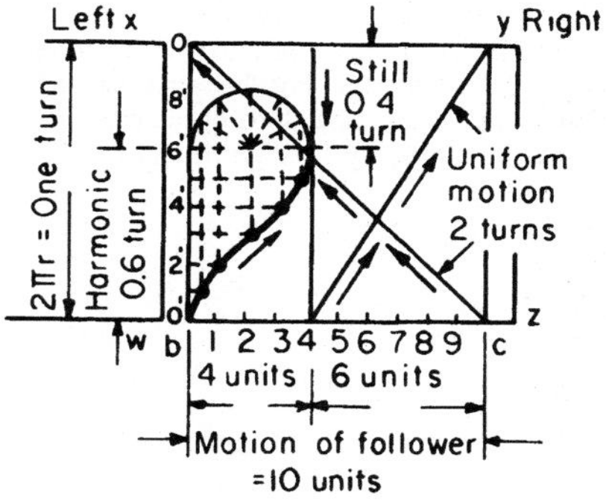

Figure 6.1.22 Cylindrical cam.

Subdivide the desired motion of the follower *bc* horizontally in the manner indicated in Figs. 6.1.16, 6.1.17, and 6.1.18, and plot the corresponding angular displacement 1′, 2′, 3′, etc., of the cam vertically. Through the intersection of lines from these points draw a smooth curve. This may best be shown by an example, assuming the following data for the diagram in Fig. 6.1.22: Total motion of follower = *bc*; circumference of cam = $2\pi r$. Follower moves harmonically 4 units to right in 0.6 turn, then rests (or "dwells") 0.4 turn, and finishes with uniform motion 6 units to right and 10 units to left in 2 turns.

Cam Design In the practical design of cams the following points must be noted. If only a small force is to be transmitted, a sliding contact may be used, otherwise use a **rolling contact.** For the latter the pitch line must be corrected in order to get the true slope of the cam. An approximate construction (Fig. 6.1.23) may be employed by using the pitch line as the center of a series of arcs with radii equal to those

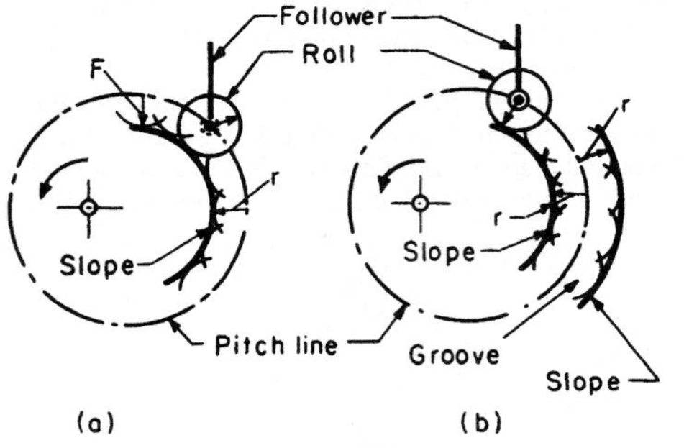

Figure 6.1.23

of the follower roll that will be used. A smooth curve drawn tangent to the arcs will then give the slope desired for a roll along periphery of the cam (Fig. 6.1.23*a*) or in a groove (Fig. 6.1.23*b*). For plate cams the roll should be a small cylinder, as in Fig. 6.1.24*a*. In cylindrical cams it is usually sufficiently accurate to make the roll conical, as in Fig. 6.1.24*b*,

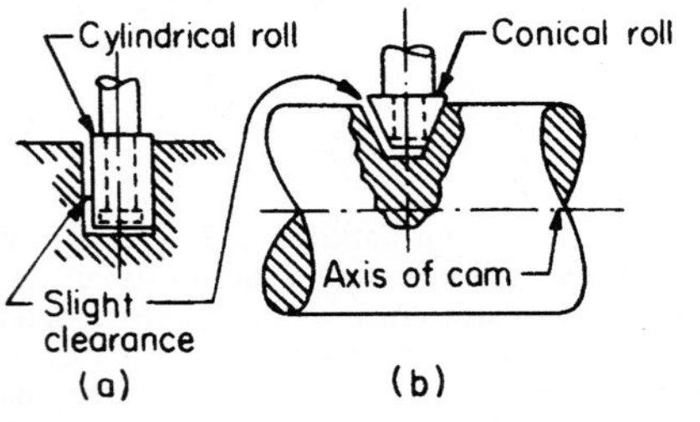

Figure 6.1.24 Plate cam.

in which case the taper of the roll produced should intersect the axis of the cam. If the pitch line *abc* is made too sharp (Fig. 6.1.25) the follower will not rise the full amount. In order to prevent this **loss of rise,** the pitch line should have a radius of curvature at all parts of not less than the roll's diameter plus ⅛ in. For the same rise of follower, *a*, the angular motion of the cam, *O*, the slope of the cam changes considerably, as indicated by the heavy lines *A*, *B*, and *C* (Fig. 6.1.26). Care

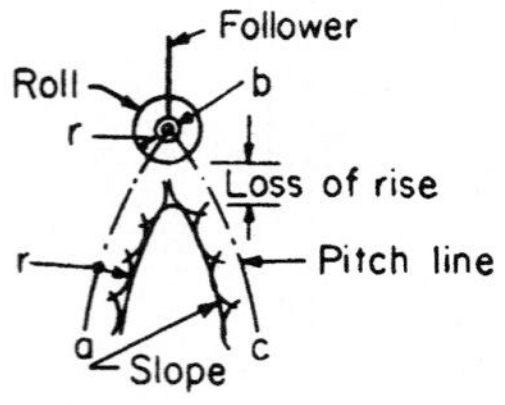

Figure 6.1.25

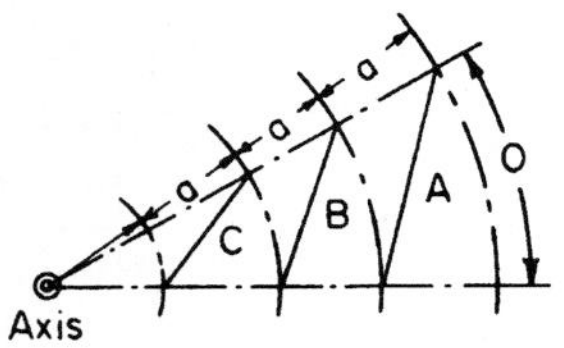

Figure 6.1.26

should be taken to keep a moderate slope and thereby keep down the side thrust on the follower, but this should not be carried too far, as the cam would become too large and the friction increase.

ROLLING SURFACES

In order to connect two shafts so that they shall have a definite angular velocity ratio, rolling surfaces are often used; and in order to have no slipping between the surfaces they must fulfill the following two conditions: the line of centers must pass through the point of contact,

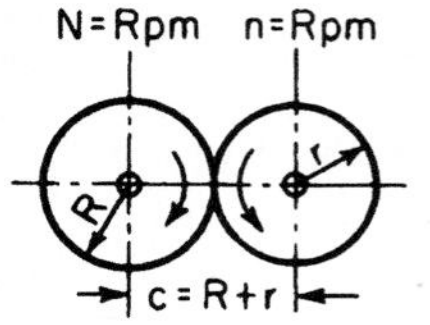

Figure 6.1.27

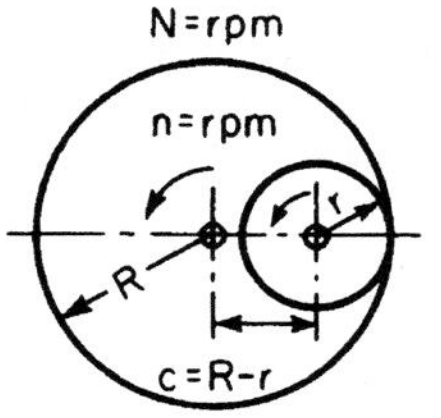

Figure 6.1.28

and the arcs of contact must be of equal length. The angular velocities, expressed usually in r/min, will be inversely proportional to the radii: $N/n = r/R$. The two surfaces most commonly used in practice, and the only ones having a constant angular velocity ratio, are cylinders where the shafts are parallel, and cones where the shafts (projected) intersect at an angle. In either case there are two possible directions of rotation, depending upon whether the surfaces roll in opposite directions (external contact) or in the same direction (internal contact). In Fig. 6.1.27, $R = nc/(N + n)$ and $r = Nc/(N + n)$; in Fig. 6.1.28, $R = nc/(N - n)$ and

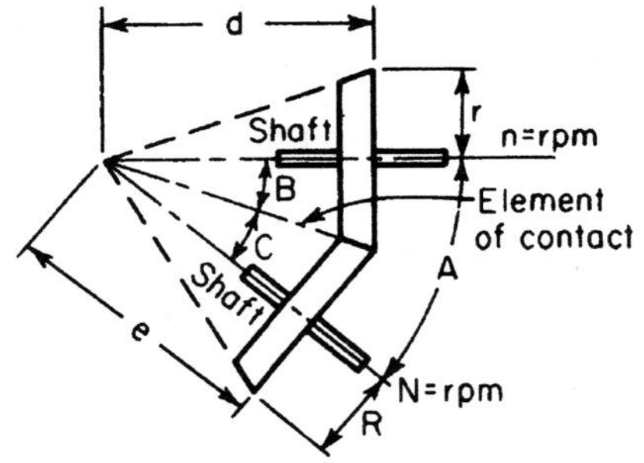

Figure 6.1.29

$r = Nc/(N - n)$. In Fig. 6.1.29, $\tan B = \sin A/(n/N + \cos A)$ and $\tan C = \sin A/(N/n + \cos A)$; in Fig. 6.1.30, $\tan B = \sin A/(N/n - \cos A)$, and $\tan C = \sin A/(n/N - \cos A)$. With the above values for the angles *B* and *C*, and the length *d* or *e* of one of the cones, *R* and *r* may be calculated.

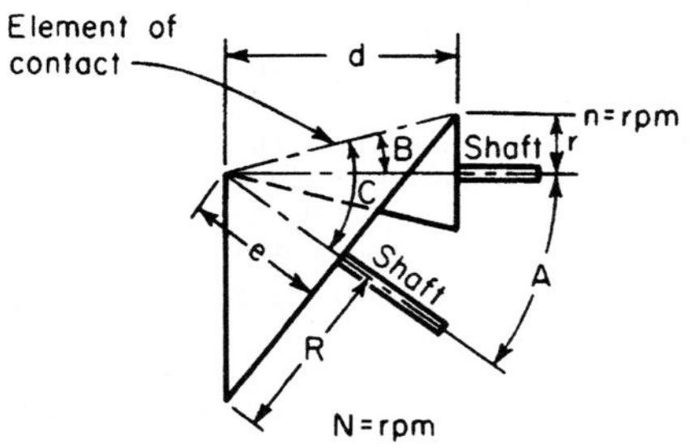

Figure 6.1.30

	A	B	C	D
Gear locked, Fig. 6.1.31	+1	+1	+1	+1
Arm locked, Fig. 6.1.31	0	−1	$+1 \times {}^{50}\!/_{20}$	$+1 \times {}^{50}\!/_{20} \times {}^{20}\!/_{40}$
Addition, Fig. 6.1.31	+1	0	$+3\frac{1}{2}$	$-\frac{1}{4}$
Gears locked, Fig. 6.1.32	+1	+1	+1	+1
Arm locked, Fig. 6.1.32	0	−1	$+1 \times {}^{30}\!/_{20}$	$+1 \times {}^{30}\!/_{20} \times {}^{20}\!/_{70}$
Addition, Fig. 6.1.32	+1	0	$+2\frac{1}{2}$	$+1\frac{3}{7}$

The natural limitations of **rolling without slip,** with the use of pure rolling surfaces limited to the transmission of very small amounts of torque historically led to the alteration of geometric surfaces to include teeth and tooth spaces, that is, **toothed wheels,** or simply **gears.** Modern gear tooth systems are described in greater detail in Sec. 6.3. This brief discussion is limited to the kinematic considerations of some common gear combinations.

EPICYCLIC TRAINS

Epicyclic trains are combinations of gears in which some of or all the gears have a compound motion made of a rotation about an axis and a translation or revolution of that axis. The gears are usually connected by a link called an arm, which often rotates about the axis of the first gear. Such trains may be calculated by first considering the case where all gears are locked while the arm is rotated and then the case where the gears are rotated while the arm is locked. The algebraic sum of the separate motions will give the desired result. The following examples and method of tabulation will illustrate this. The figures on each gear refer to the number of teeth for that gear.

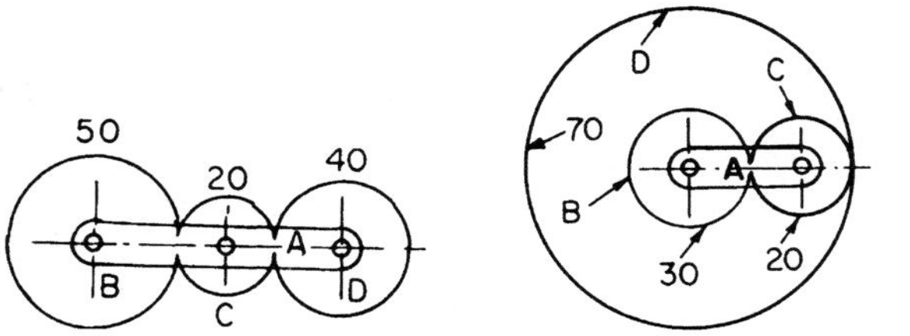

Figures 6.1.31 and 6.1.32 Epicyclic trains.

In Figs. 6.1.31 or 6.1.32 lock the gears and turn the arm A right-handed through 1 revolution (+1); then lock the arm and turn the gear B back to where it started (−1); gears C and D will have rotated the amount indicated in the tabulation. Then the algebraic sums will give the relative turns of each gear. That is, in Fig. 6.1.31, for one turn of the arm, B does not move and C turns in the same direction $3\frac{1}{2}$ r, and D in the opposite direction $\frac{1}{4}$ r; whereas in Fig. 6.1.32, for one turn of the arm, B does not turn, but C and D turn in the same direction as the arm, respectively, $2\frac{1}{2}$ and $1\frac{3}{7}$ r. (*Note:* The arm in the above case was turned +1 for convenience, but any other value might be used.)

Bevel epicyclic trains are epicyclic trains containing bevel gears and may be calculated by the preceding method, but it is usually simpler to use the general formula which applies to all cases of epicyclic trains:

$$\frac{\text{Turns of } C \text{ relative to arm}}{\text{Turns of } B \text{ relative to arm}} = \frac{\text{absolute turns of } C - \text{turns of arm}}{\text{absolute turns of } B - \text{turns of arm}}$$

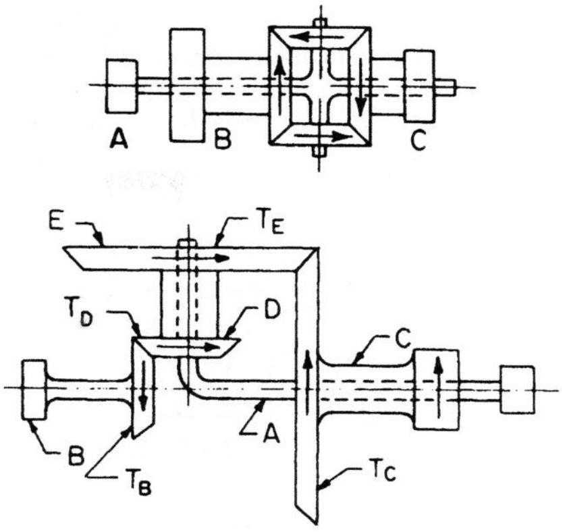

Figures 6.1.33 and 6.1.34 Bevel epicyclic trains.

The left-hand term gives the value of the train and can always be expressed in terms of the number of teeth (T) on the gears. Care must be used, however, to express it as either plus (+) or minus (−), depending upon whether the gears turn in the same or opposite directions.

$$\frac{\text{Relative turns of } C}{\text{Relative turns of } B} = \frac{C - A}{B - A} = -1 \qquad \text{(in Fig. 6.1.33)}$$

$$= +\frac{T_E}{T_C} \times \frac{T_B}{T_D} \qquad \text{(Fig. 6.1.34)}$$

OTHER MECHANISMS

Pulley Block (Fig. 6.1.35) Given the weight W to be raised, the force F necessary is $F = V_W W/V_F = W/n$ = load/number of ropes, V_W and V_F being the respective velocities of W and F.

Differential Chain Block (Fig. 6.1.36)

$$F = V_W W/V_F = W(D - d)/(2D)$$

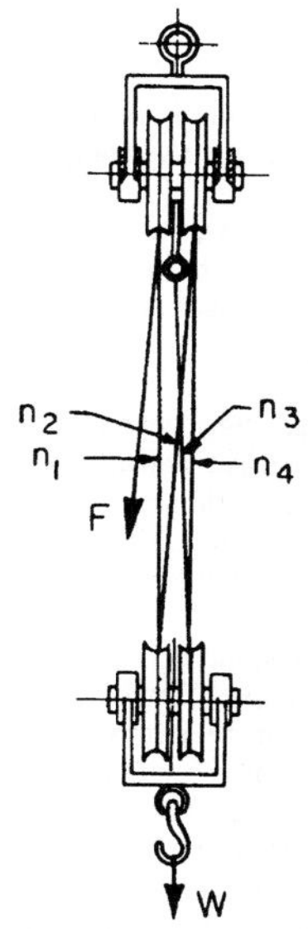

Figure 6.1.35 Pulley block.

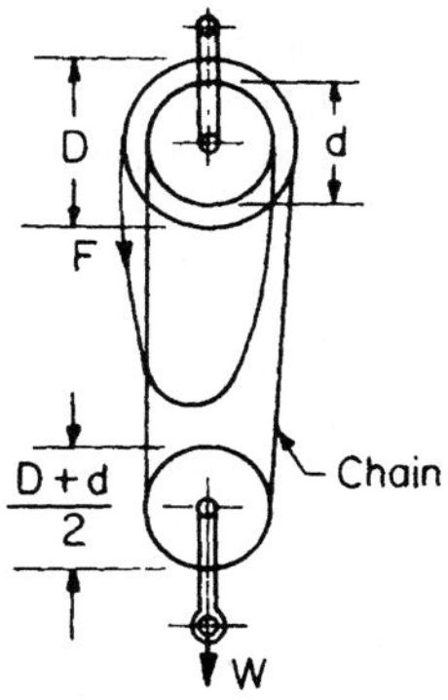

Figure 6.1.36 Differential chain block.

Worm and Wheel (Fig. 6.1.37) $F = \pi d(n/T)W/(2\pi R) = WP(d/D)/(2\pi R)$, where n = number of threads, single, double, triple, etc.

Triplex Chain Block (Fig. 6.1.38) This geared hoist makes use of the epicyclic train. $W = FL/\{M[1 + (T_D/T_C) \times (T_B/T_A)]\}$, where T = number of teeth on gears.

Toggle Joint (Fig. 6.1.39) $P = Fs(\cos A)/t$.

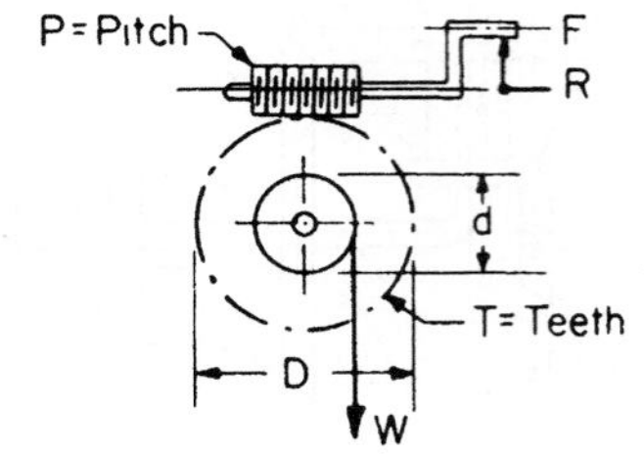

Figure 6.1.37 Worm and worm wheel.

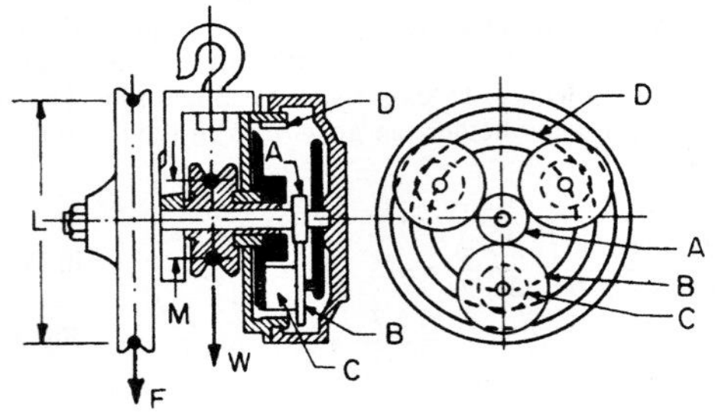

Figure 6.1.38 Triplex chain block.

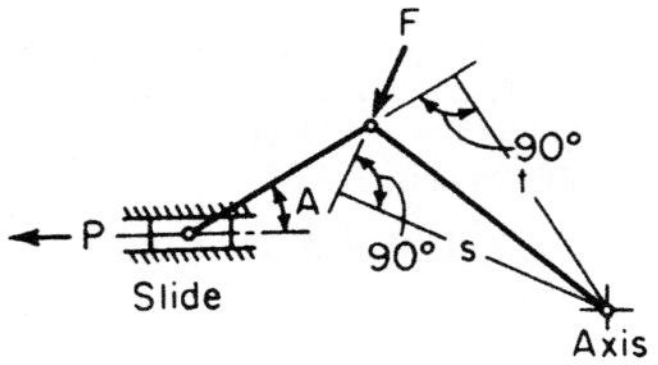

Figure 6.1.39 Toggle joint.

Geneva Wheel Mechanism In this mechanism the driven member makes one-fourth of a revolution for each revolution of the driver. The slots are positioned so that a pin enters and leaves the slots tangentially (Fig. 6.1.40).

Parallel Mechanism A parallel mechanism that is used to enlarge or to reduce movements is shown in Fig. 6.1.41. Another application of parallel mechanism that is used in drafting is shown in Fig. 6.1.42.

Harmonic Drive Mechanism

Invented in the 1950s, harmonic drive or a harmonic gear system is a mechanical **speed-changing device.** Harmonic drive transmissions are noted for their high reduction ratios with concentric shafts and their ability to reduce backlash and vibration in a motion control system. It operates on a different principle from, and has capabilities beyond the scope of, conventional speed changers.

As shown in Figs. 6.1.43 and 6.1.44, a harmonic drive consists of three basic elements: (1) an elastically deformable (flexible) metallic thin-walled ring, having splines or gear teeth on its external surface; (2) a circular thick-walled rigid ring having the same internal splines or gear teeth as in the flexible spline; and (3) an elliptical or similar-shape cam with rotating bearings. The external flexible spline has fewer teeth than the internal circular spline. The flexible spline ring is placed inside the rigid ring. The elliptical cam with rotating bearings is mounted inside the flexible ring and forces the flexible splines (teeth) to push

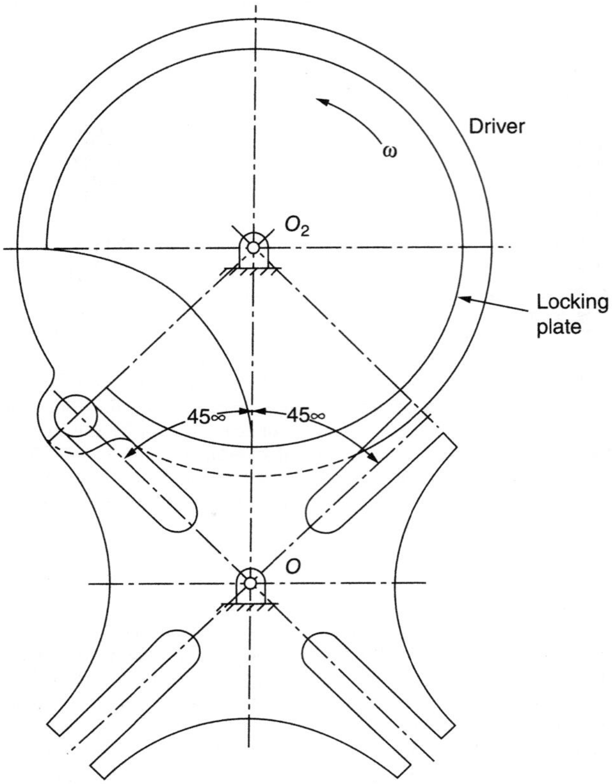

Figure 6.1.40

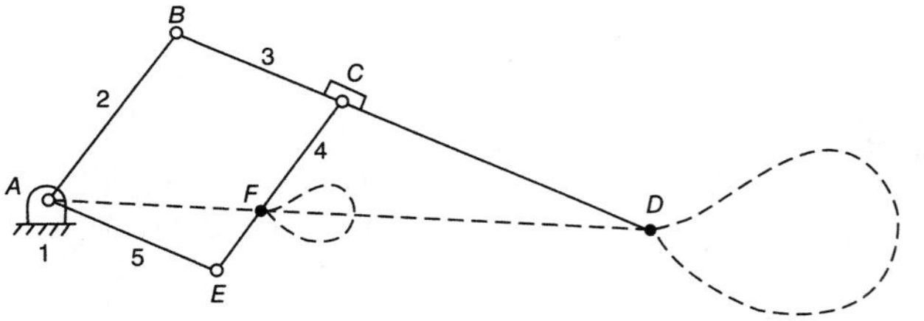

Figure 6.1.41 Parallel mechanism with magnification.

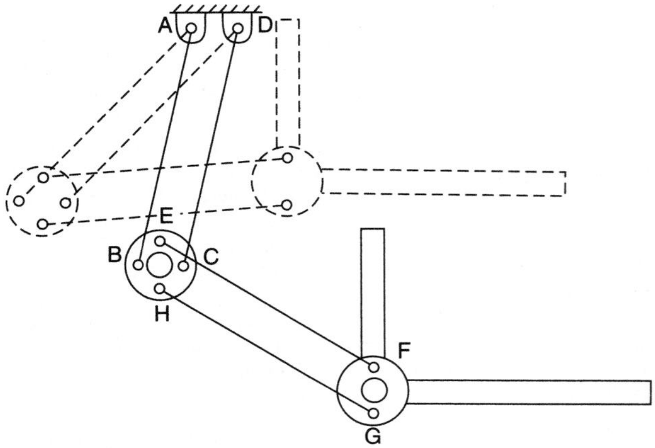

Figure 6.1.42 Parallel mechanism.

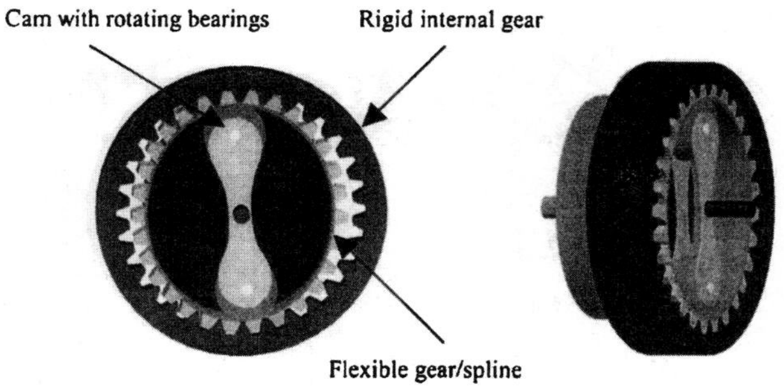

Figure 6.1.43 Harmonic drive mechanisms with two rotating bearings.

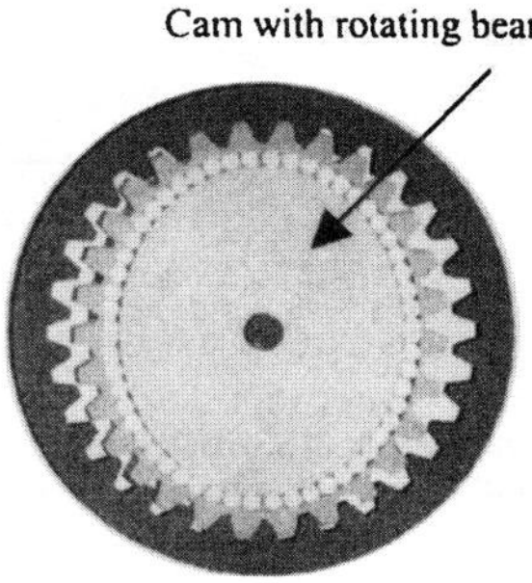

Figure 6.1.44 Harmonic drive mechanisms with many rotating bearings.

deeply into the rigid ring at two opposite points while rotating. The thin flexible ring deflects elastically as it rolls on the inside of the slightly larger rigid circular ring. The flexible ring rotates at a speed that is the difference in the number of teeth on the rigid ring and on the flexible ring (Fig. 6.1.43). This method basically preloads the teeth, which reduces backlash. The flexible spline is made of special steel with a high resistance to fatigue.

The input (motor) shaft is attached to the oval-shaped cam with thin ball bearings placed around the outer circumference of the oval cam. The gear's output shaft is attached to the flexible spline. The circular spline is attached to the gearbox along its outer circumference. If the elliptical cam or the motor shaft rotates in a cw direction, the flexible spline will rotate in a ccw direction and vice versa.

The reduction ratio is calculated by dividing the number of teeth on the rigid ring by the difference between the numbers of teeth on the rigid and flexible rings. For example, if a fixed spline has 80 teeth and its corresponding flexible spline has 78 teeth, then the reduction ratio is 80/(80 − 78) = 40:1.

Harmonic Drive Limitations Harmonic drives have single-stage speed reductions of 30:1 to 350:1. The lower ratios are limited because of fatigue considerations on the flexible spline ring. The harmonic drive is normally not suitable for high power levels above 6 kW.

Harmonic Drive Applications Harmonic drives offer many benefits including zero backlash, high positional accuracy, low vibration, small size and light weight, and compact design. The applications are numerous in robotics, machine tools, and medical and aerospace applications.

SLIDER CRANK MECHANISM

Kinematics The slider crank mechanism is widely used in automobile engines, punch presses, feeding mechanisms, etc. For such mechanisms, displacements, velocities, and accelerations of the parts are important design parameters. The basic mechanism is shown in Fig. 6.1.45. The slider displacement x is given by

$$x = r\cos\theta + l\sqrt{1-[(r/l)\sin\theta]^2}$$

where r = crank length, l = connecting-rod length, θ = crank angle measured from top dead center position. Slider velocity is given by

$$\dot{x} = V = -r\omega\left(\sin\theta + \frac{r\sin 2\theta}{2l\cos\beta}\right)$$

where ω = instantaneous angular velocity of the crank at position θ and

$$\cos\beta = \sqrt{1-(r/l)^2\sin^2\theta}$$

Slider acceleration is given by

$$a = \ddot{x} = -r\alpha\left(\sin\theta + \frac{r}{l}\frac{\sin 2\theta}{2\cos\beta}\right) - r\omega^2\left(\cos\theta + \frac{r}{l}\frac{\cos 2\theta}{\cos\beta} + \frac{r^3}{l^3}\frac{\sin^2 2\theta}{4\cos^3\beta}\right)$$

where α = instantaneous angular acceleration of the crank at position θ.

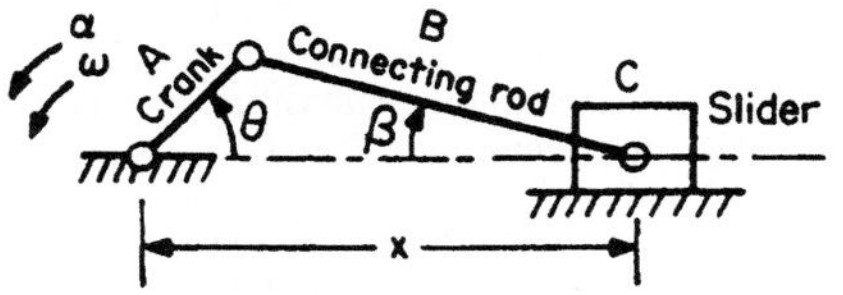

Figure 6.1.45 Basic slider–crank mechanism.

Forces Neglecting gravity effects, the forces in a mechanism arise from those produced by input and output forces or torques (herewith called static components). Such forces may be produced by driving motors, shaft loads, expanding cylinder gases, etc. The net forces on the various links cause accelerations of the mechanism's masses, and can be thought of as dynamic components. The static components must be borne by the various links, thus giving rise to internal stresses in those parts. The supporting bearings and slide surfaces also feel the effects of these components, as do the support frames. Stresses are also induced by the dynamic components in the links, and such components cause shaking forces and shaking moments in the support frame.

By building appropriately designed countermasses onto the basic mechanism, the support frame and its bearings can be relieved of a significant portion of the dynamic component effects, sometimes called inertia effects. The augmented mechanism is then considered to be "balanced." The static components cannot be relieved by any means, so that the support frame and its bearings must be designed to carry safely the static component forces and not be overstressed. Figure 6.1.46 illustrates a common, simple form of approximate balancing. Sizing

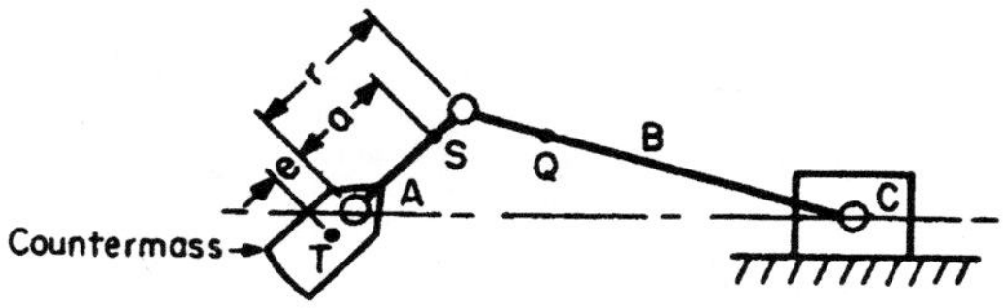

Figure 6.1.46 "Balanced" slider–crank mechanism where T = center of mass, S = center of mass of crank A, and Q = center of mass of connecting rod B.

of the countermass is somewhat complicated because the total inertia effect is the vector sum of the separate link inertias, which change in magnitude and direction at each position of the crank. The countermass D is sized to contain sufficient mass to completely balance the crank plus an additional mass (effective mass) to "balance" the other links (connecting rod and slider). In simple form, the satisfying condition is approximately

$$M_D e\omega^2 = M_A a\omega^2 + M_{\text{eff}} e\omega^2$$

where a = distance from crankshaft to center of mass of crank (note that most often a is approximately equal to the crank radius r); e = distance from crankshaft to center of mass of countermass; M_{eff} = additional mass of countermass to "balance" connecting rod and slider.

From one-half to two-thirds the slider mass (e.g., $\frac{1}{2}M_c \le M_{eff} \le \frac{2}{3}M_c$) is usually added to the countermass for a single-cylinder engine. For critical field work, single and multiple slider crank mechanisms are dynamically balanced by experimental means.

Forces and Torques Figure 6.1.47a shows an exploded view of the slider crank mechanism and the various forces and torques on the links (neglecting gravity and weight effects). Inertial effects are shown as broken-line vectors and are manipulated in the same manner as the actual or real forces. The inertial effects are also known as **D'Alembert forces.** The meanings are as follows: $F_1 = -M_D e\omega^2$, parallel to crank A; $F_2 = -M_a e\alpha$, perpendicular to crank A; $F_3 = (z/l)\,F_9$; F_4 = as found in Fig. 6.1.47b; $F_5 = F_2 - F_7$; $F_6 = F_1 - F_8$; $F_7 = -M_A r\omega^2$, parallel to crank A; $F_8 = -M_A r\alpha$, perpendicular to crank A; $F_9 = -M_B \times$ absolute acceleration of Q, where acceleration of point Q can be found graphically by constructing an acceleration polygon of the mechanism (see, e.g., Shigley, "Kinematic Analysis of Mechanisms," McGraw-Hill); $F_{10} = (y/l)F_9$; $F_{11} = -M_c \times$ absolute acceleration of slider $\ddot{x}$; F = normal wall force (neglecting friction); F = external force on slider's face, where the vector sum $F + F_4 + F_{10} + F_{11} + F_{12} = 0$; T = external crankshaft torque, where the algebraic sum $T + F_3E + F_4W + F_2e + F_7a = 0$ (note that signs must reflect direction of torque); $T_t = F_3E + F_wW$ = transmitted torque; K = the effective location of force F_9. The distance Δ of force F_9 from the center of mass of connecting rod B (see Fig. 6.1.47a) is given by

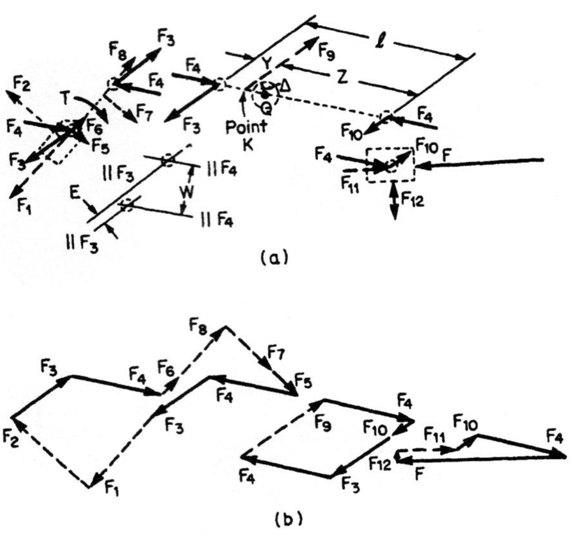

Figure 6.1.47 (a) Forces and torques; (b) force polygons.

$\Delta = J_{B(cm)} \times$ angular acceleration of link $B/M_B \times$ absolute acceleration of point Q. Figure 6.1.47b shows the force polygons of the separate links of the mechanism.

6.2 MACHINE ELEMENTS

by Ali M. Sadegh

REFERENCES: American National Standards Institute (ANSI) Standards. International Organization for Standardization (ISO) Standards. Morden, "Industrial Fasteners Handbook," Trade and Technical Press. Parmley, "Standard Handbook of Fastening and Joining," McGraw-Hill. Bickford, "An Introduction to the Design and Behavior of Bolted Joints," Marcel Dekker. Maleev, "Machine Design," International Textbook. Shigley, "Mechanical Engineering Design," McGraw-Hill. *Machine Design* magazine, Penton/IPC. ANSI/Rubber Manufacturers Assn. (ANSI/RMA) Standards. "Handbook of Power Transmission Flat Belting," Goodyear Rubber Products Co. "Industrial V-Belting," Goodyear Rubber Products Co. Carlson, "Spring Designer's Handbook," Marcel Dekker. American Chain Assn., "Chains for Power Transmission and Material Handling—Design and Applications Handbook," Marcel Dekker. "Power Transmission Handbook," DAYCO. "Wire Rope User's Manual," American Iron and Steel Institute. Blake, "Threaded Fasteners—Materials and Design," Marcel Dekker. Young, "Roark's Formulas for Stress and Strain," McGraw-Hill.

NOTE: At this writing, conversion to metric hardware and machine elements continues. SI units are introduced as appropriate, but the bulk of the material is still presented in the form in which the designer or reader will find it available.

SCREW FASTENINGS

At present there exist two major standards for screw threads, namely Unified inch screw threads and metric screw threads. Both systems enjoy a wide application globally, but movement toward a greater use of the metric system continues.

Unified Inch Screw Threads (or Unified Screw Threads)

The Unified Thread Standard originated by an accord of screw thread standardization committees of Canada, the United Kingdom, and the United States in 1984. The Unified Screw-Thread Standard was published by ANSI as American Unified and American Screw Thread Publication B1.1-1974, revised in 1982 and then again in 1989. Revisions did not tamper with the basic 1974 thread forms. In conjunction with Technical Committee No. 1 of the ISO, the Unified Standard was adopted as an ISO Inch Screw Standard (ISO 5864-1978).

Of the numerous and different screw thread forms, those of greatest consequence are

UN—unified (no mandatory radiused root)

UNR—unified (mandatory radiused root; minimum $0.108 = p$)

UNJ—unified (mandatory larger radiused root; recommended $0.150 = p$)

M—metric (inherently designed and manufactured with radiused root; has $0.125 = p$)

MJ—metric (mandatory larger radiused root; recommended $0.150 = p$)

The basic American screw thread profile was standardized in 1974, and it now carries the UN designations (UN = unified). ANSI publishes these standards and all subsequent revisions. At intervals these standards are published with a "reaffirmation date" (i.e., R1988). In 1969 an *international* basic thread profile standard was established, and it is designated as M. The ISO publishes these standards with yearly updates. The UN and M profiles are the same, but **UN screws** are manufactured to **inch** dimensions while **M screws** are manufactured to **metric** dimensions.

The **metric system** has only the two thread forms: **M,** standard for commercial uses, and **MJ,** standard for aerospace use and for aerospace-quality commercial use.

Certain groups of diameter and pitch combinations have evolved over time to become those most used commercially. Such groups are called **thread series.** Currently there are 11 UN series for inch products and 13 **M** series for metric products.

The Unified standard comprises the following two parts:

1. **Diameter-pitch combinations.** (See Tables 6.2.1 to 6.2.5.)

a. UN inch series:	
Coarse	UNC or UNRC
Fine	UNF or UNRF
Extra-fine	UNEF or UNREF
Constant-pitch	UN or UNR
b. Metric series:	
Coarse	M
Fine	M

NOTE: Radiused roots apply only to external threads. The preponderance of important commercial use leans to UNC, UNF, 8UN (eight-threaded), and metric coarse M. Aerospace and aerospace-quality applications use UNJ and MJ.

2. **Tolerance classes.** The amounts of tolerance and allowance distinguish one thread class from another. Classes are designated by one of three numbers (1, 2, 3), and either letter A for external threads or letter B for internal threads. Tolerance decreases as class number increases. Allowance is specified only for classes 1A and 2A. Tolerances are based on engagement length equal to nominal diameter. 1A/1B—liberal tolerance and allowance required to permit easy assembly even with dirty or nicked threads. 2A/2B—most commonly used for general applications, including production of bolts, screws, nuts, and similar threaded fasteners. Permits external threads to be plated. 3A/3B—for closeness of fit and/or accuracy of thread applications where zero allowance is needed. 2AG—allowance for rapid assembly where high-temperature expansion prevails or where lubrication problems are important.

   Unified screw threads are designated by a set of numbers and letter symbols signifying, in sequence, the nominal size, threads per inch, thread series, tolerance class, hand (only for left hand), and in some instances in parentheses a Thread Acceptability System Requirement of ANSI Bl.3.

   EXAMPLE. ¼-20 UNC-2A-LH (21), or optionally 0.250-20 UNC-2A-LH (21), where ¼ = nominal size (fractional diameter, in, or screw number, with decimal equivalent of either being optional); 20 = number of threads per inch, *n*; UNC = thread form and series; 2A = tolerance class; LH = left hand (no symbol required for right hand); (21) = thread gaging system per ANSI B1.3.

3. **Load considerations**

   a. Static loading. Only a slight increase in tensile strength in a screw fastener is realized with an increase in *root* rounding radius, because minor diameter (hence cross-sectional area at the root) growth is small. Thus the basic tensile stress area formula is used in stress calculations for all thread forms. See Tables 6.2.2, 6.2.3, and 6.2.4. The designer should take into account such factors as stress concentration as applicable.

   b. Dynamic loading. Few mechanical joints can remain absolutely free of some form of fluctuating stress, vibration, stress reversal, or impact. Metal-to-metal joints of very high-modulus materials or non-elastic-gasketed high-modulus joints plus preloading at assembly (preload to be greater than highest peak of the external fluctuating load) can realize absolute static conditions inside the screw fastener. For ordinary-modulus joints and elastic-gasketed joints, a fraction of the external fluctuating load will be transmitted to the interior of the screw fastener. Thus the fastener must be designed for fatigue according to a **static plus fluctuating load** model. See discussion under "Strength" later.

   Since **fatigue failures** generally occur at locations of high **stress concentration,** screw fasteners are especially vulnerable because of the abrupt change between head and body, notchlike conditions at the thread roots, surface scratches due to manufacturing, etc. The highest stress concentrations occur at the thread roots. The stress concentration factor can be very large for non-rounded roots, amounting to about 6 for sharp or flat roots, to less than 3 for UNJ and MJ threads which are generously rounded. This can effectively double the fatigue life. UNJ and MJ threads are especially well suited for dynamic loading conditions.

Screw Thread Profile

Basic Profile The basic profiles of UN and UNR are the same, and these in turn are identical to those of ISO metric threads. Basic thread shape (60° thread angle) and basic dimensions (major, pitch, and minor diameters; thread height; crest, and root flats) are defined. See Fig. 6.2.1.

Design Profile Design profiles define the maximum material (no allowance) for external and internal threads, and they are derived from the basic profile. UN threads (external) may have either flat or rounded crests and roots. UNR threads (external) must have rounded roots, but may have flat or rounded crests. UN threads (internal) *must* have rounded roots. Any rounding must clear the basic flat roots or crests.

Basic major diameter	Largest diameter of basic screw thread.
Basic minor diameter	Smallest diameter of basic screw thread.
Basic pitch diameter	Diameter to imaginary lines through thread profile and parallel to axis so that thread and groove widths are equal. These three definitions apply to both external and internal threads.
Maximum diameters (external threads)	Basic diameters minus allowance.
Minimum diameters (internal threads)	Basic diameters.
Pitch	$1/n$ (n = number of threads per inch).
Tolerance	Inward variation tolerated on maximum diameters of external threads and outward variation tolerated on minimum diameters of internal threads.

Multiple-threaded screw is a screw that has two or more threads cut beside each other (imagine two or more strings wound side by side around a shaft). **Pitch** is the axial distance between threads. Pitch is equal to the lead in a single-threaded screw. **Lead** is the axial distance the nut advances in one complete revolution of the screw. The lead is equal to the pitch times the number of threads.

Metric Screw Threads

Metric screw thread standardization has been under the aegis of the International Organization for Standardization (ISO). The ISO basic profile is essentially the same as the Unified screw thread basic form, and it is shown in Fig. 6.2.1. The ISO thread series (see Table 6.2.5) are those published in ISO 261-1973. Increased overseas business sparked U.S. interest in metric screw threads, and the ANSI, through its Special Committee to Study Development of an Optimum Metric Fastener System, in joint action with an ISO working group (ISO/TC 1/TC 2), established compromise recommendations regarding metric screw threads. The approved results appear in ANSI B1.13-1979 (Table 6.2.6). This ANSI metric thread series is essentially a selected subset (boxed-in portion of Table 6.2.5) of the larger ISO 261-1973 set. The M profiles of tolerance class 6H/6g are intended for metric applications where inch class 2A/2B has been used.

Metric Tolerance Classes for Threads Tolerance classes are a selected combination of tolerance grades and tolerance positions applied to length- of-engagement groups.

Tolerance grades are indicated as numbers for crest diameters of nut and bolt and for pitch diameters of nut and bolt. Tolerance is the acceptable variation permitted on any such diameter.

Tolerance positions are indicated as letters, and are allowances (fundamental deviations) as dictated by field usage or conditions. Capital letters are used for internal threads (nut) and lower case for external threads (bolt).

Table 6.2.1 Standard Series Threads (UN/UNR)*

Nominal size, in		Basic major diameter, in	Threads per inch											Nominal size, in
			Series with graded pitches			Series with constant pitches								
Primary	Secondary		Coarse UNC	Fine UNF	Extra-fine UNEF	4UN	6UN	8UN	12UN	16UN	20UN	28UN	32UN	
0		0.0600	—	80	—	—	—	—	—	—	—	—	—	0
	1	0.0730	64	72	—	—	—	—	—	—	—	—	—	1
2		0.0860	56	64	—	—	—	—	—	—	—	—	—	2
	3	0.0990	48	56	—	—	—	—	—	—	—	—	—	3
4		0.1120	40	48	—	—	—	—	—	—	—	—	—	4
5		0.1250	40	44	—	—	—	—	—	—	—	—	—	5
6		0.1380	32	40	—	—	—	—	—	—	—	—	UNC	6
8		0.1640	32	36	—	—	—	—	—	—	—	—	UNC	8
10		0.1900	24	32	—	—	—	—	—	—	—	—	UNF	10
	12	0.2160	24	28	32	—	—	—	—	—	—	UNF	UNEF	12
1/4		0.2500	20	28	32	—	—	—	—	—	UNC	UNF	UNEF	1/4
5/16		0.3125	18	24	32	—	—	—	—	—	20	28	UNEF	5/16
3/8		0.3750	16	24	32	—	—	—	—	UNC	20	28	UNEF	3/8
7/16		0.4375	14	20	28	—	—	—	—	16	UNF	UNEF	32	7/16
1/2		0.5000	13	20	28	—	—	—	—	16	UNF	UNEF	32	1/2
9/16		0.5625	12	18	24	—	—	—	UNC	16	20	28	32	9/16
5/8		0.6250	11	18	24	—	—	—	12	16	20	28	32	5/8
	11/16	0.6875	—	—	24	—	—	—	12	16	20	28	32	11/16
3/4		0.7500	10	16	20	—	—	—	12	UNF	UNEF	28	32	3/4
	13/16	0.8125	—	—	20	—	—	—	12	16	UNEF	28	32	13/16
7/8		0.8750	9	14	20	—	—	—	12	16	UNEF	28	32	7/8
	15/16	0.9275	—	—	20	—	—	—	12	16	UNEF	28	32	15/16
1		1.0000	8	12	20	—	—	UNC	UNF	16	UNEF	28	32	1
	1 1/16	1.0625	—	—	18	—	—	8	12	16	20	28	—	1 1/16
1 1/8		1.1250	7	12	18	—	—	8	UNF	16	20	28	—	1 1/8
	1 3/16	1.1875	—	—	18	—	—	8	12	16	20	28	—	1 3/16
1 1/4		1.2500	7	12	18	—	—	8	UNF	16	20	28	—	1 1/4
	1 5/16	1.3125	—	—	18	—	—	8	12	16	20	28	—	1 5/16
1 3/8		1.3750	6	12	18	—	UNC	8	UNF	16	20	28	—	1 3/8
	1 7/16	1.4375	—	—	18	—	6	8	12	16	20	28	—	1 7/16
1 1/2		1.5000	6	12	18	—	UNC	8	UNF	16	20	28	—	1 1/2
1	1 9/16	1.5625	—	—	18	—	6	8	12	16	20	—	—	1 9/16
1 5/8		1.6250	—	—	18	—	6	8	12	16	20	—	—	1 5/8
	1 11/16	1.6875	—	—	18	—	6	8	12	16	20	—	—	1 11/16
1 3/4		1.7500	5	—	—	—	6	8	12	16	20	—	—	1 3/4
	1 13/16	1.8125	—	—	—	—	6	8	12	16	20	—	—	1 13/16
1 7/8		1.8750	—	—	—	—	6	8	12	16	20	—	—	1 7/8
	1 15/16	1.9375	—	—	—	—	6	8	12	16	20	—	—	1 15/16
2		2.0000	4 1/2	—	—	—	6	8	12	16	20	—	—	2
	2 1/8	2.1250	—	—	—	—	6	8	12	16	20	—	—	2 1/8
2 1/4		2.2500	4 1/2	—	—	—	6	8	12	16	20	—	—	2 1/4
	2 3/8	2.3750	—	—	—	—	6	8	12	16	20	—	—	2 3/8
2 1/2		2.5000	4	—	—	UNC	6	8	12	16	20	—	—	2 1/2
	2 5/8	2.6250	—	—	—	4	6	8	12	16	20	—	—	2 5/8
2 3/4		2.7500	4	—	—	UNC	6	8	12	16	20	—	—	2 3/4
	2 7/8	2.8750	—	—	—	4	6	8	12	16	20	—	—	2 7/8
3		3.0000	4	—	—	UNC	6	8	12	16	20	—	—	3
	3 1/8	3.1250	—	—	—	4	6	8	12	16	—	—	—	3 1/8
3 1/4		3.2500	4	—	—	UNC	6	8	12	16	—	—	—	3 1/4
	3 3/8	3.3750	—	—		4	6	8	12	16	—	—	—	3 3/8
3 1/2		3.5000	4	—	—	UNC	6	8	12	16	—	—	—	3 1/2
	3 5/8	3.6250	—	—	—	4	6	8	12	16	—	—	—	3 5/8
3 3/4		3.7500	4	—	—	UNC	6	8	12	16	—	—	—	3 3/4
	3 7/8	3.8750	—	—	—	4	6	8	12	16	—	—	—	3 7/8
4		4.0000	4	—	—	UNC	6	8	12	16	—	—	—	4
	4 1/8	4.1250	—	—	—	4	6	8	12	16	—	—	—	4 1/8
4 1/4		4.2500	—	—	—	4	6	8	12	16	—	—	—	4 1/4
	4 3/8	4.3750	—	—	—	4	6	8	12	16	—	—	—	4 3/8
4 1/2		4.5000	—	—	—	4	6	8	12	16	—	—	—	4 1/2
	4 5/8	4.6250	—	—	—	4	6	8	12	16	—	—	—	4 5/8
4 3/4		4.7500	—	—	—	4	6	8	12	16	—	—	—	4 3/4
	4 7/8	4.8750	—	—	—	4	6	8	12	16	—	—	—	4 7/8
5		5.0000	—	—	—	4	6	8	12	16	—	—	—	5
	5 1/8	5.1250	—	—	—	4	6	8	12	16	—	—	—	5 1/8
5 1/4		5.2500	—	—	—	4	6	8	12	16	—	—	—	5 1/4
	5 3/8	5.3750	—	—	—	4	6	8	12	16	—	—	—	5 3/8
5 1/2		5.5000	—	—	—	4	6	8	12	16	—	—	—	5 1/2
	5 5/8	5.6250	—	—	—	4	6	8	12	16	—	—	—	5 5/8
5 3/4		5.7500	—	—	—	4	6	8	12	16	—	—	—	5 3/4
	5 7/8	5.8750	—	—	—	4	6	8	12	16	—	—	—	5 7/8
6		6.0000	—	—	—	4	6	8	12	16	—	—	—	6

* Series designation shown indicates the UN thread form; however, the UNR thread form may be specified by substituting UNR in place of UN in all designations for external use only.
SOURCE: ANSI B1.1-1982; reaffirmed in 1989, reproduced by permission.

Table 6.2.2 Basic Dimensions for Coarse Thread Series (UNC/UNRC)

Nominal size, in	Basic major diameter D, in	Threads per inch n	Basic pitch diameter* E, in	UNR design minor diameter external† K_s, in	Basic minor diameter internal K, in	Section at minor diameter at $D - 2h_b$, in²	Tensile stress area, ‡ in²
1 (0.073)§	0.0730	64	0.0629	0.0544	0.0561	0.00218	0.00263
2 (0.086)	0.0860	56	0.0744	0.0648	0.0667	0.00310	0.00370
3 (0.099)§	0.0990	48	0.0855	0.0741	0.0764	0.00406	0.00487
4 (0.112)	0.1120	40	0.0958	0.0822	0.0849	0.00496	0.00604
5 (0.125)	0.1250	40	0.1088	0.0952	0.0979	0.00672	0.00796
6 (0.138)	0.1380	32	0.1177	0.1008	0.1042	0.00745	0.00909
8 (0.164)	0.1640	32	0.1437	0.1268	0.1302	0.01196	0.0140
10 (0.190)	0.1900	24	0.1629	0.1404	0.1449	0.01450	0.0175
12 (0.216)§	0.2160	24	0.1889	0.1664	0.1709	0.0206	0.0242
1/4	0.2500	20	0.2175	0.1905	0.1959	0.0269	0.0318
5/16	0.3125	18	0.2764	0.2464	0.2524	0.0454	0.0524
3/8	0.3750	16	0.3344	0.3005	0.3073	0.0678	0.0775
7/16	0.4375	14	0.3911	0.3525	0.3602	0.0933	0.1063
1/2	0.5000	13	0.4500	0.3334	0.4167	0.1257	0.1419
9/16	0.5625	12	0.5084	0.4633	0.4723	0.162	0.182
5/8	0.6250	11	0.5660	0.5168	0.5266	0.202	0.226
3/4	0.7500	10	0.6850	0.6309	0.6417	0.302	0.334
7/8	0.8750	9	0.8028	0.7427	0.7547	0.419	0.462
1	1.0000	8	0.9188	0.8512	0.8647	0.551	0.606
1 1/8	1.1250	7	1.0322	0.9549	0.9704	0.693	0.763
1 1/4	1.2500	7	1.1572	1.0799	1.0954	0.890	0.969
1 3/8	1.3750	6	1.2667	1.1766	1.1946	1.054	1.155
1 1/2	1.5000	6	1.3917	1.3016	1.3196	1.294	1.405
1 3/4	1.7500	5	1.6201	1.5119	1.5335	1.74	1.90
2	2.0000	4 1/2	1.8557	1.7353	1.7594	2.30	2.50
2 1/4	2.2500	4 1/2	2.1057	1.9853	2.0094	3.02	3.25
2 1/2	2.5000	4	2.3376	2.2023	2.2294	3.72	4.00
2 3/4	2.7500	4	2.5876	2.4523	2.4794	4.62	4.93
3	3.0000	4	2.8376	2.7023	2.7294	5.62	5.97
3 1/4	3.2500	4	3.0876	2.9523	2.9794	6.72	7.10
3 1/2	3.5000	4	3.3376	3.2023	3.2294	7.92	8.33
3 3/4	3.7500	4	3.5876	3.4523	3.4794	9.21	9.66
4	4.0000	4	3.8376	3.7023	3.7294	10.61	11.08

* British: effective diameter.
† See formula under definition of tensile stress area in Appendix B of ANSI B1.1-1987.
‡ Design form. See Fig. 2B in ANSI B1.1-1982 or Fig. 1 in 1989 revision.
§ Secondary sizes.
SOURCE: ANSI B1.1-1982, revised 1989; reproduced by permission.

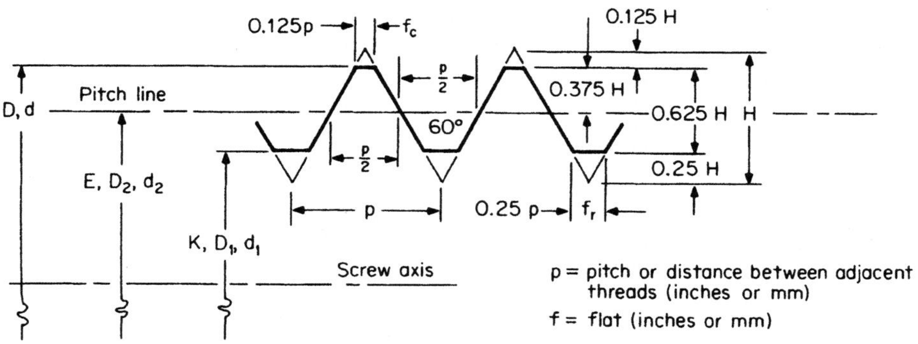

Name	Symbol: Inch bolt or nut	Symbol: Metric bolt	Symbol: Metric nut
Major diameter	D	d	D
Pitch diam. (inch); effective diam. (metric)	E	d_2	D_2
Minor diameter	K	d_1	D_1

Figure 6.2.1 Basic thread profile.

Table 6.2.3 Basic Dimensions for Fine Thread Series (UNF/UNRF)

Nominal size, in	Basic major diameter D, in	Threads per inch n	Basic pitch diameter* E, in	UNR design minor diameter external† K_s, in	Basic minor diameter internal K, in	Section at minor diameter at $D - 2h_b$, in²	Tensile stress area, ‡ in²
0 (0.060)	0.0600	80	0.0519	0.0451	0.0465	0.00151	0.00180
1 (0.073)§	0.0730	72	0.0640	0.0565	0.0580	0.00237	0.00278
2 (0.086)	0.0860	64	0.0759	0.0674	0.0691	0.00339	0.00394
3 (0.099)§	0.0990	56	0.0874	0.0778	0.0797	0.00451	0.00523
4 (0.112)	0.1120	48	0.0985	0.0871	0.0894	0.00566	0.00661
5 (0.125)	0.1250	44	0.1102	0.0979	0.1004	0.00716	0.00830
6 (0.138)	0.1380	40	0.1218	0.1082	0.1109	0.00874	0.01015
8 (0.164)	0.1640	36	0.1460	0.1309	0.1339	0.01285	0.01474
10 (0.190)	0.1900	32	0.1697	0.1528	0.1562	0.0175	0.0200
12 (0.216)§	0.2160	28	0.1928	0.1734	0.1773	0.0226	0.0258
1/4	0.2500	28	0.2268	0.2074	0.2113	0.0326	0.0364
5/16	0.3125	24	0.2854	0.2629	0.2674	0.0524	0.0580
3/8	0.3750	24	0.3479	0.3254	0.3299	0.0809	0.0878
7/16	0.4375	20	0.4050	0.3780	0.3834	0.1090	0.1187
1/2	0.5000	20	0.4675	0.4405	0.4459	0.1486	0.1599
9/16	0.5625	18	0.5264	0.4964	0.5024	0.189	0.203
5/8	0.6250	18	0.5889	0.5589	0.5649	0.240	0.256
3/4	0.7500	16	0.7094	0.6763	0.6823	0.351	0.373
7/8	0.8750	14	0.8286	0.7900	0.7977	0.480	0.509
1	1.0000	12	0.9459	0.9001	0.9098	0.625	0.663
1 1/8	1.1250	12	1.0709	1.0258	1.0348	0.812	0.856
1 1/4	1.2500	12	1.1959	1.1508	1.1598	1.024	1.073
1 3/8	1.3750	12	1.3209	1.2758	1.2848	1.260	1.315
1 1/2	1.5000	12	1.4459	1.4008	1.4098	1.521	1.581

* British: effective diameter.
† See formula under definition of tensile stress area in Appendix B of ANSI B1.1-1982.
‡ Design form. See Fig. 2B of ANSI B1.1-1982, or Fig. 1 in 1989 revision.
§ Secondary sizes.
SOURCE: ANSI B1.1-1982, revised 1989; reproduced by permission.

Table 6.2.4 Basic Dimensions for Extra-Fine Thread Series (UNEF/UNREF)

Nominal size, in		Basic major diameter D, in	Threads per inch n	Basic pitch diameter* E, in	UNR design minor diameter external† K_s, in	Basic minor diameter internal K, in	Section at minor diameter at $D - 2h_b$, in²	Tensile stress area, ‡ in²
Primary	Secondary							
	12 (0.216)	0.2160	32	0.1957	0.1788	0.1822	0.0242	0.0270
1/4		0.2500	32	0.2297	0.2128	0.2162	0.0344	0.0379
5/16		0.3125	32	0.2922	0.2753	0.2787	0.0581	0.0625
3/8		0.3750	32	0.3547	0.3378	0.3412	0.0878	0.0932
7/16		0.4375	28	0.4143	0.3949	0.3988	0.1201	0.1274
1/2		0.5000	28	0.4768	0.4573	0.4613	0.162	0.170
9/16		0.5625	24	0.5354	0.5129	0.5174	0.203	0.214
5/8		0.6250	24	0.5979	0.5754	0.5799	0.256	0.268
	11/16	0.6875	24	0.6604	0.6379	0.6424	0.315	0.329
3/4		0.7500	20	0.7175	0.6905	0.6959	0.369	0.386
	13/16	0.8125	20	0.7800	0.7530	0.7584	0.439	0.458
7/8		0.8750	20	0.8425	0.8155	0.8209	0.515	0.536
	15/16	0.9375	20	0.9050	0.8780	0.8834	0.598	0.620
1		1.0000	20	0.9675	0.9405	0.9459	0.687	0.711
	1 1/16	1.0625	18	1.0264	0.9964	1.0024	0.770	0.799
1 1/8		1.1250	18	1.0889	1.0589	1.0649	0.871	1.901
	1 3/16	1.1875	18	1.1514	1.1214	1.1274	0.977	1.009
1 1/4		1.2500	18	1.2139	1.1839	1.1899	1.090	1.123
	1 5/16	1.3125	18	1.2764	1.2464	1.2524	1.208	1.244
1 3/8		1.3750	18	1.3389	1.3089	1.3149	1.333	1.370
	1 7/16	1.4375	18	1.4014	1.3714	1.3774	1.464	1.503
1 1/2		1.5000	18	1.4639	1.4339	1.4399	1.60	1.64
	1 9/16	1.5625	18	1.5264	1.4964	1.5024	1.74	1.79
1 5/8		1.6250	18	1.5889	1.5589	1.5649	1.89	1.94
	1 11/16	1.6875	18	1.6514	1.6214	1.6274	2.05	2.10

* British: effective diameter.
† Design form. See Fig. 2B in ANSI B1.1-1982 or Fig. 1 in 1989 revision.
‡ See formula under definition of tensile stress area in Appendix B in ANSI B1.1-1982.
SOURCE: ANSI B1.1-1982 revised 1989; reproduced by permission.

Table 6.2.5 ISO Metric Screw Thread Standard Series

Nominal size diam, mm			Pitches, mm														Nominal size diam, mm
Column*			Series with graded pitches		Series with constant pitches												
1	2	3	Coarse	Fine	6	4	3	2	1.5	1.25	1	0.75	0.5	0.35	0.25	0.2	
0.25			0.075	—	—	—	—	—	—	—	—	—	—	—	—	—	0.25
0.3			0.8	—	—	—	—	—	—	—	—	—	—	—	—	—	0.3
	0.35		0.09	—	—	—	—	—	—	—	—	—	—	—	—	—	0.35
0.4			0.1	—	—	—	—	—	—	—	—	—	—	—	—	—	0.4
	0.45		0.1	—	—	—	—	—	—	—	—	—	—	—	—	—	0.45
0.5			0.125	—	—	—	—	—	—	—	—	—	—	—	—	—	0.5
	0.55		0.125	—	—	—	—	—	—	—	—	—	—	—	—	—	0.55
0.6			0.15	—	—	—	—	—	—	—	—	—	—	—	—	—	0.6
	0.7		0.175	—	—	—	—	—	—	—	—	—	—	—	—	—	0.7
0.8			0.2	—	—	—	—	—	—	—	—	—	—	—	—	—	0.8
	0.9		0.225	—	—	—	—	—	—	—	—	—	—	—	—	—	0.9
1			0.25	—	—	—	—	—	—	—	—	—	—	—	—	0.2	1
	1.1		0.25	—	—	—	—	—	—	—	—	—	—	—	—	0.2	1.1
1.2			0.25	—	—	—	—	—	—	—	—	—	—	—	—	0.2	1.2
	1.4		0.3	—	—	—	—	—	—	—	—	—	—	—	—	0.2	1.4
1.6			0.35	—	—	—	—	—	—	—	—	—	—	—	—	0.2	1.6
	1.8		0.35	—	—	—	—	—	—	—	—	—	—	—	—	0.2	1.8
2			0.4	—	—	—	—	—	—	—	—	—	—	—	0.25	—	2
	2.5		0.45	—	—	—	—	—	—	—	—	—	—	—	0.25	—	2.2
2.5			0.45	—	—	—	—	—	—	—	—	—	—	0.35	—	—	2.5
3			0.5	—	—	—	—	—	—	—	—	—	—	0.35	—	—	3
	3.5		0.6	—	—	—	—	—	—	—	—	—	—	0.35	—	—	3.5
4			0.7	—	—	—	—	—	—	—	—	—	0.5	—	—	—	4
	4.5		0.75	—	—	—	—	—	—	—	—	—	0.5	—	—	—	4.5
5			0.8	—	—	—	—	—	—	—	—	—	0.5	—	—	—	5
		5.5	—	—	—	—	—	—	—	—	—	—	0.5	—	—	—	5.5
6			1	—	—	—	—	—	—	—	—	0.75	—	—	—	—	6
		7	1	—	—	—	—	—	—	—	—	0.75	—	—	—	—	7
8			1.25	1	—	—	—	—	—	—	1	0.75	—	—	—	—	8
		9	1.25	—	—	—	—	—	—	—	1	0.75	—	—	—	—	9
10			1.5	1.25	—	—	—	—	—	1.25	1	0.75	—	—	—	—	10
		11	1.5	—	—	—	—	—	—	—	1	0.75	—	—	—	—	11
12			1.75	1.25	—	—	—	—	1.5	1.25	1	—	—	—	—	—	12
	14		2	1.5	—	—	—	—	1.5	1.25†	1	—	—	—	—	—	14
		15	—	—	—	—	—	—	1.5	—	1	—	—	—	—	—	15
16			2	1.5	—	—	—	—	1.5	—	1	—	—	—	—	—	16
		17	—	—	—	—	—	—	1.5	—	1	—	—	—	—	—	17
	18		2.5	1.5	—	—	—	2	1.5	—	1	—	—	—	—	—	18
20			2.5	1.5	—	—	—	2	1.5	—	1	—	—	—	—	—	20
	22		2.5	1.5	—	—	—	2	1.5	—	1	—	—	—	—	—	22
24			3	2	—	—	—	2	1.5	—	1	—	—	—	—	—	24
		25	—	—	—	—	—	2	1.5	—	1	—	—	—	—	—	25
		26	—	—	—	—	—	—	1.5	—	1	—	—	—	—	—	26
	27		3	2	—	—	—	2	1.5	—	1	—	—	—	—	—	27
		28	—	—	—	—	—	2	1.5	—	1	—	—	—	—	—	28
30			3.5	2	—	—	(3)	2	1.5	—	1	—	—	—	—	—	30
		32	—	—	—	—	—	2	1.5	—	—	—	—	—	—	—	32
	33		3.5	2	—	—	(3)	2	1.5	—	—	—	—	—	—	—	33
		35‡	—	—	—	—	—	—	1.5	—	—	—	—	—	—	—	35‡
36			4	3	—	—	—	2	1.5	—	—	—	—	—	—	—	36
		38	—	—	—	—	—	—	1.5	—	—	—	—	—	—	—	38
	39		4	3	—	—	—	2	1.5	—	—	—	—	—	—	—	39
		40	—	—	—	—	3	2	1.5	—	—	—	—	—	—	—	40
42			4.5	3	—	4	3	2	1.5	—	—	—	—	—	—	—	42
	45		4.5	3	—	4	3	2	1.5	—	—	—	—	—	—	—	45

* Thread diameter should be selected from column 1, 2 or 3; with preference being given in that order.
† Pitch 1.25 mm in combination with diameter 14 mm has been included for spark plug applications.
‡ Diameter 35 mm has been included for bearing locknut applications.
NOTE: The use of pitches shown in parentheses should be avoided wherever possible. The pitches enclosed in the bold frame, together with the corresponding nominal diameters in columns 1 and 2, are those combinations which have been established by ISO Recommendations as a selected "coarse" and "fine" series for commercial fasteners. Sizes 0.25 mm through 1.4 mm are covered in ISO Recommendation R68 and, except for the 0.25-mm size, in ANSI B1.10.
SOURCE: ISO 261-1973, reproduced by permission.

Table 6.2.6 Limiting Dimensions of Standard Series Threads for Commercial Screws, Bolts, and Nuts

			External thread (bolt), mm										Internal thread (nut), mm					
Nominal size diam, mm	Pitch p, mm	Basic thread designation	Tol. class	Allowance	Major diameter Max	Major diameter Min	Pitch diameter Max	Pitch diameter Min	Pitch diameter Tol.	Minor diameter Max*	Minor diameter Min†	Tol. class	Minor diameter Min	Minor diameter Max	Pitch diameter Min	Pitch diameter Max	Pitch diameter Tol.	Major diam, min
1.6	0.35	M1.6	6g	0.019	1.581	1.496	1.354	1.291	0.063	1.151	1.063	6H	1.221	1.321	1.373	1.458	0.085	1.600
1.8	0.35	M1.8	6g	0.019	1.781	1.696	1.554	1.491	0.063	1.351	1.263	6H	1.421	1.521	1.573	1.568	0.085	1.800
2	0.4	M2	6g	0.019	1.981	1.886	1.721	1.654	0.067	1.490	1.394	6H	1.567	1.679	1.740	1.830	0.090	2.000
2.2	0.45	M2.2	6g	0.020	2.180	2.080	1.888	1.817	0.071	1.628	1.525	6H	1.713	1.838	1.908	2.000	0.095	2.200
2.5	0.45	M2.5	6g	0.020	2.480	2.380	2.188	2.117	0.071	1.928	1.825	6H	2.013	2.138	2.208	2.303	0.095	2.500
3	0.5	M3	6g	0.020	2.980	2.874	2.655	2.580	0.075	2.367	2.256	6H	2.459	2.599	2.675	2.775	0.100	3.000
3.5	0.6	M3.5	6g	0.021	3.479	3.354	3.089	3.004	0.085	2.742	2.614	6H	2.850	3.010	3.110	3.222	0.112	3.500
4	0.7	M4	6g	0.022	3.978	3.838	3.523	3.433	0.090	3.119	2.979	6H	3.242	3.422	3.545	3.663	0.118	4.000
4.5	0.75	M4.5	6g	0.022	4.478	4.338	3.991	3.901	0.090	3.558	3.414	6H	3.688	3.878	4.013	4.131	0.118	4.500
5	0.8	M5	6g	0.024	4.976	4.826	4.456	4.361	0.095	3.994	3.841	6H	4.134	4.334	4.480	4.605	0.125	5.000
6	1	M6	6g	0.026	5.974	5.794	5.324	5.212	0.112	4.747	4.563	6H	4.917	5.153	5.350	5.500	0.150	6.000
7	1	M7	6g	0.026	6.974	6.794	6.234	6.212	0.112	5.747	5.563	6H	5.917	6.153	6.350	6.500	0.150	7.000
8	1.25	M8	6g	0.028	7.972	7.760	7.160	7.042	0.118	6.439	6.231	6H	6.647	6.912	7.188	7.348	0.160	8.000
	1	M8 × 1	6g	0.026	7.974	7.794	7.324	7.212	0.112	6.747	6.563	6H	6.918	7.154	7.350	7.500	0.150	8.000
10	1.5	M10	6g	0.032	9.968	9.732	8.994	8.862	0.132	8.127	7.879	6H	8.376	8.676	9.026	9.206	0.180	10.000
	1.25	M10 × 1.25	6g	0.028	9.972	9.760	9.160	9.042	0.118	8.439	8.231	6H	8.646	8.911	9.188	9.348	0.160	10.000
12	1.75	M12	6g	0.034	11.966	11.701	10.829	10.679	0.150	9.819	9.543	6H	10.106	10.441	10.863	11.063	0.200	12.000
	1.25	M12 × 1.25	6g	0.028	11.972	11.760	11.160	11.028	0.118	10.439	10.217	6H	10.646	10.911	11.188	11.368	0.180	12.000
14	2	M14	6g	0.038	13.962	13.682	12.663	12.503	0.160	11.508	11.204	6H	11.835	12.210	12.701	12.913	0.212	14.000
	1.5	M14 × 1.5	6g	0.032	13.968	13.732	12.994	12.854	0.140	12.127	11.879	6H	12.376	12.676	13.026	13.216	0.190	14.000
16	2	M16	6g	0.038	15.962	15.682	14.663	14.503	0.160	13.508	13.204	6H	13.385	14.210	14.701	14.913	0.212	16.000
	1.5	M16 × 1.5	6g	0.032	15.968	15.732	14.994	14.854	0.140	14.127	13.879	6H	14.376	14.676	15.026	15.216	0.190	16.000
18	2.5	M18	6g	0.038	17.958	17.623	16.334	16.164	0.170	14.891	14.541	6H	15.294	15.744	16.376	16.600	0.224	18.000
	1.5	M18 × 1.5	6g	0.032	17.968	17.732	16.994	15.854	0.140	16.127	15.879	6H	16.376	16.676	17.026	17.216	0.190	18.000
20	2.5	M20	6g	0.042	19.958	19.623	18.334	18.164	0.170	16.891	16.541	6H	17.294	17.744	18.376	18.600	0.224	20.000
	1.5	M20 × 1.5	6g	0.032	19.968	19.732	18.994	18.854	0.140	18.127	17.879	6H	18.376	18.676	19.026	19.216	0.190	20.000
22	2.5	M22	6g	0.042	21.958	21.623	20.334	20.164	0.170	18.891	18.541	6H	19.294	19.744	20.376	20.600	0.224	22.000
	1.5	M22 × 1.5	6g	0.032	21.968	21.732	20.994	20.854	0.140	20.127	19.879	6H	20.376	20.676	21.026	21.216	0.190	22.000
24	3	M24	6g	0.048	23.952	23.577	22.003	21.803	0.200	20.271	19.855	6H	20.752	21.252	22.051	22.316	0.265	24.000
	2	M24 × 2	6g	0.038	23.962	23.682	22.663	22.493	0.170	21.508	21.194	6H	21.835	22.210	22.701	22.925	0.224	24.000
27	3	M27	6g	0.048	26.952	26.577	25.003	24.803	0.200	23.271	22.855	6H	23.752	24.252	25.051	25.316	0.265	27.000
	2	M27 × 2	6g	0.038	26.962	26.682	25.663	25.493	0.170	24.508	24.194	6H	24.835	25.210	25.701	25.925	0.224	27.000
30	3.5	M30	6g	0.053	29.947	29.522	27.674	27.462	0.212	25.653	25.189	6H	26.211	26.771	27.727	28.007	0.280	30.000
	2	M30 × 2	6g	0.038	29.962	29.682	28.663	28.493	0.170	27.508	27.194	6H	27.835	28.210	28.701	28.925	0.224	30.000
33	3.5	M33	6g	0.053	32.947	32.522	30.674	30.462	0.212	28.653	28.189	6H	29.211	29.771	30.727	31.007	0.280	33.000
	2	M33 × 2	6g	0.038	32.962	32.682	31.663	31.493	0.170	30.508	30.194	6H	30.835	31.210	31.701	31.925	0.224	33.000
36	4	M36	6g	0.060	35.940	35.465	33.342	33.118	0.224	31.033	30.521	6H	31.670	32.270	33.402	33.702	0.300	36.000
	3	M36 × 3	6g	0.048	35.952	35.577	34.003	33.803	0.200	32.271	31.855	6H	32.752	33.252	34.051	34.316	0.265	36.000
39	4	M39	6g	0.060	38.940	38.465	36.342	36.118	0.224	34.033	33.521	6H	34.670	35.270	36.402	36.702	0.300	39.000
	3	M39 × 3	6g	0.048	38.952	38.577	37.003	36.803	0.200	35.271	34.855	6H	35.752	36.252	37.051	37.316	0.265	39.000

* Design form, see Figs. 2 and 5 of ANSI B1.13M-1979 (or Figs. 1 and 4 in 1983 revision).
† Required for high-strength applications where rounded root is specified.
SOURCE: [Appeared in ASME/SAE Interpretive document Metric Screw Threads, B1.13 (Nov. 3, 1966), pp. 9, 10.] ISO 261-1973, reproduced by permission.

There are three established groups of length of thread engagement, S (short), N (normal), and L (long), for various diameter-pitch combinations. Normal length of thread engagement is calculated from the formula $N = 4.5pd^{0.2}$, where p is pitch and d is the smallest nominal size within each of a series of groupings of nominal sizes.

In conformance with coating (or plating) requirements and demands of ease of assembly, the following tolerance positions have been established:

Bolt	*Nut*	
e		Large allowance
g	G	Small allowance
h	H	No allowance

See Table 6.2.7 for preferred tolerance classes.

ISO metric screw threads are designated by a set of number and letter symbols signifying, in sequence, metric symbol, nominal size, × (symbol), pitch, tolerance grade (on pitch diameter), tolerance position (for pitch diameter), tolerance grade (on crest diameter), and tolerance position (for crest diameter).

EXAMPLE. M6 × 0.75-5g6g, where M = metric symbol; 6 = nominal size, × = symbol; 0.75 = pitch-axial distance of adjacent threads measured between corresponding thread points (millimeters); 5 = tolerance grade (on pitch diameter); g = tolerance position (for pitch diameter); 6 = tolerance grade (on crest diameter); g = tolerance position (for crest diameter).

Power Transmission Screw Threads: Forms and Proportions

The **Acme** thread appears in four series [ANSI B1.8-1973 (revised 1988) and B1.5-1977]. Generalized dimensions for the series are given in Table 6.2.8.

The 29° general-purpose thread (Fig. 6.2.2) is used for all Acme thread applications outside of special design cases.

The 29° stub thread (Fig. 6.2.3) is used for heavy-loading designs and where space constraints or economic factors make a shallow thread advantageous.

The 60° stub thread (Fig. 6.2.4) finds special applications in the machine-tool industry.

The 10° modified square thread (Fig. 6.2.5) is, for all practical purposes, equivalent to a "square" thread.

For selected Acme diameter-pitch combinations, see Table 6.2.9.

Three classes (2G, 3G, 4G) of general-purpose threads have clearances on all diameters for free movement. A fourth class (5G)

Table 6.2.7 Preferred Tolerance Classes

	Length of engagement														
	External threads (bolts)									Internal threads (nuts)					
	Tolerance position e (large allowance)			Tolerance position g (small allowance)			Tolerance position h (no allowance)			Tolerance position G (small allowance)			Tolerance position H (no allowance)		
Quality	Group S	Group N	Group L	Group S	Group N	Group L	Group S	Group N	Group L	Group S	Group N	Group L	Group S	Group N	Group L
Fine							3h4h	4h	4h5h				4H	5H	6H
Medium		6e	7e6e	5g6g	6g	7g6g	5h6h	6h	7h6h	5G	6G	7G	5H	6H	7H
Coarse					8g	9g8g					7G	8G		7H	8H

NOTE: Fine quality applies to precision threads where little variation in fit character is permissible. Coarse quality applies to those threads which present manufacturing difficulties, such as the threading of hot-rolled bars or tapping deep blind holes.

SOURCE: ISO 261-1973, reproduced by permission.

Table 6.2.8 Acme Thread Series
(D = outside diam, p = pitch. All dimensions in inches.)
(See Figs. 6.2.2 to 6.2.5.)

	Thread dimensions			
Symbols	29° general purpose	29° stub	60° stub	10° modified
t = thickness of thread	$0.5p$	$0.5p$	$0.5p$	$0.5p$
R = basic depth of thread	$0.5p$	$0.3p$	$0.433p$	$0.5p$*
F = basic width of flat	$0.3707p$	$0.4224p$	$0.250p$	$0.4563p$†
G = (see Figs. 6.2.2, 6.2.3, 6.2.4)	$F - (0.52 \times \text{clearance})$	$F - (0.52 \times \text{clearance})$	$0.227p$	$F - (0.17 \times \text{clearance})$
E = basic pitch diam	$D - 0.5p$	$D - 0.3p$	$D - 0.433p$	$D - 0.5p$
K = basic minor diam	$D - p$	$D - 0.6p$	$D - 0.866p$	$D - p$
Range of threads, per inch	1-16	2-16	4-16	

* A clearance of at least 0.010 in is added to h on threads of 10-pitch and coarser, and 0.005 in on finer pitches, to produce extra depth, thus avoiding interference with threads of mating parts of a minor or major diameter.

† Measured at crest of screw thread.

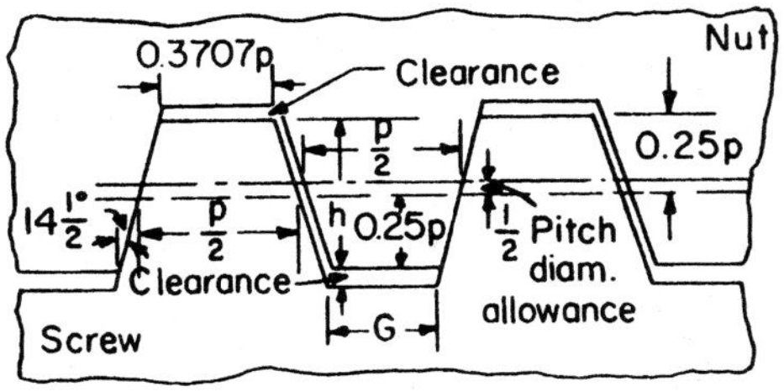

Figure 6.2.2 29° Acme general-purpose thread.

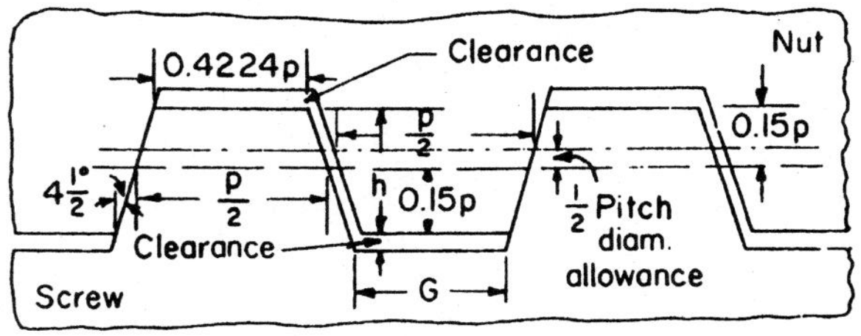

Figure 6.2.3 29° stub Acme thread.

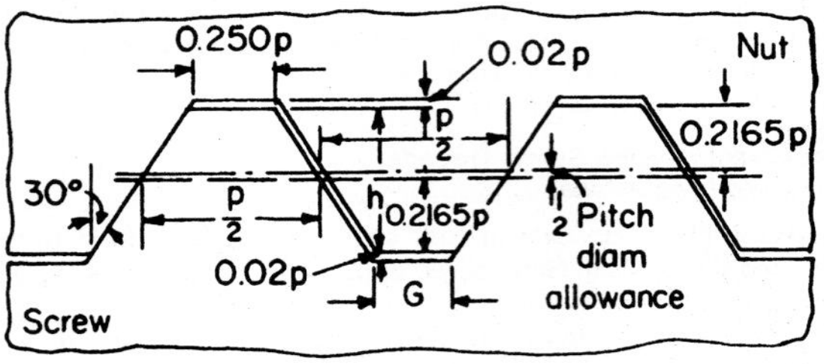

Figure 6.2.4 60° stub Acme thread.

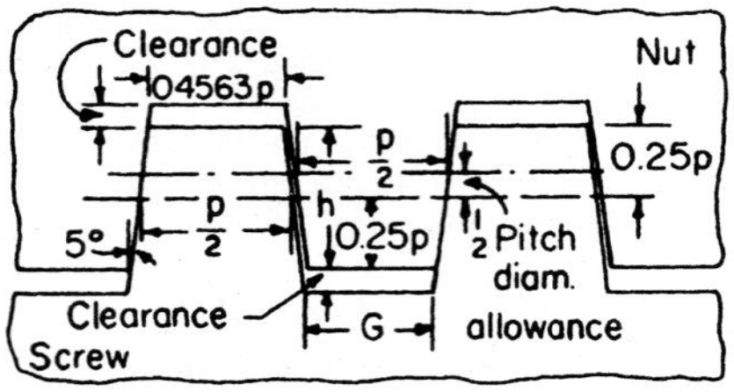

Figure 6.2.5 10° modified square thread.

Table 6.2.9 Acme Thread Diameter-Pitch Combinations
(See Figs. 6.2.2 to 6.2.5.)

Size	Threads per inch	Size	Threads per inch	Size	Threads per inch	Size	Threads per inch	Size	Threads per inch
1/4	16	5/8	8	1 1/4	5	2 1/4	3	4	2
5/16	14	3/4	6	1 3/8	4	2 1/2	3	4 1/2	2
3/8	12	7/8	6	1 1/2	4	2 3/4	3	5	2
7/16	12	1	5	1 3/4	4	3	2		
1/2	10	1 1/8	5	2	4	3 1/2	2		

of general-purpose threads has no allowance or clearance on the pitch diameter for purposes of minimum end play or backlash.

Raising and Lowering Torques In a single-threaded screw, the torque T_R required for raising load F, or tightening, is

$$T_R = \frac{Fd_m}{2}\left(\frac{l + \pi \mu d_m \sec\alpha}{\pi d_m - \mu l \sec\alpha}\right)$$

where d_m is the mean diameter of the screw, μ is the coefficient of friction between screw and nut threads, l is the lead, and α is one-half of Acme thread angle (2α). Note $\alpha = 0$ for square threads. The torque require to lower the load is

$$T_L = \frac{Fd_m}{2}\left(\frac{\pi \mu d_m \sec\alpha - l}{\pi d_m + \mu l \sec\alpha}\right)$$

When a screw has a collar for lifting a load, additional torque T_c should be added to the lifting torque, $T_c = F\mu_c d_c/2$, where the subscript c is for the collar.

High-Strength Bolting Screw Threads

High-strength bolting applications include pressure vessels, steel pipe flanges, fittings, valves, and other services. They can be used for either hot or cold surfaces where high tensile stresses are produced when the joints are made up. For sizes 1 in and smaller, the ANSI coarse-thread series is used. For larger sizes, the ANSI 8-pitch thread series is used (see Table 6.2.10).

Ball Screws

A ball screw is an improvement over an Acme screw, and it generally consists of a set of bearing balls within the ball nut thread and a ball screw shaft. The thread form in which the bearing balls ride is a semicircular shape or an oval shape formed from two arcs of the same radius with offset centers; see Figs. 6.2.6 and 6.2.7.

Bearing ball circuit is the closed path that the bearing balls follow through the ball nut. Ball nuts may have one or more circuits.

The primary load path in a ball screw assembly is through the circuit (or circuits) of bearing balls within the ball nut, where a ball enters the ball path between the nut and screw carrying the load one or more turns around the screw, as shown in Fig. 6.2.7.

Return guide is a path for returning the bearing balls to the beginning of the circuit. **Pitch** is defined as the axial distance between threads. Pitch is equal to the lead in a single-start screw and is equal to one-half of the lead in double-start screw.

The advantages of ball screws include a lead accuracy of up to 0.0005 in, quiet and smooth operation, minimal backlash, and high efficiency. Because of these features ball screws have long life and low wear. Ball screws are used in linear motion and linear actuator devices in which high accuracy and axial stiffness are required. Ball screws are available in many types and sizes. For further information the reader is referred to the manufacturers' specifications.

Design Considerations. (1) Flanges may be used with the nut assembly to carry axial and lateral loads. (2) When a ball screw assembly is used in an orientation other than vertical, it is important to orient the return tubes to optimize ball nut operation. As shown in Fig. 6.2.8, return guides down should be avoided.

Screw Threads for Pipes

American National Standard Taper Pipe Thread (ANSI/ASME B1.20.1-1983) This thread is shown in Fig. 6.2.9. It is made to the following specifications: The taper is 1 in 16 or 0.75 in/ft. The basic length of the external taper thread is determined by $L_2 = p(0.8D + 6.8)$, where D is the basic outside diameter of the pipe (see Table 6.2.11). Thread designation and notation is written as: nominal size, number of threads per inch, thread series. For example: 3/8-18 NPT, 1/8-27 NPSC, 1/2-14 NPTR, 1/8-27 NPSM, 1/8-27 NPSL, 1-11.5 NPSH, where N = National (American) Standard, T = taper, C = coupling, S = straight, M = mechanical, L = locknut, H = hose coupling, and R = rail fittings. Where pressure-tight joints are required, it is intended that taper pipe threads be made up wrench-tight with a sealant. Descriptions of thread series include: NPSM = free-fitting mechanical joints for fixtures, NPSL = loose-fitting mechanical joints with locknuts, NPSH = loosefitting mechanical joints for hose coupling.

American National Standard Straight Pipe Thread (ANSI/ASME B1.20.1-1983) This thread can be used to advantage for the following: (1) pressure-tight joints with sealer; (2) pressure-tight joints without sealer for drain plugs, filler plugs, etc.; (3) free-fitting mechanical joints for fixtures; (4) loose-fitting mechanical joints with locknuts; and (5) loose-fitting mechanical joints for hose couplings. Dimensions are shown in Table 6.2.12.

American National Standard Dry-Seal Pipe Threads (ANSI B1.20.3-1976 (inch), ANSI B1.20.4-1976 (metric translation) Thread

Table 6.2.10 Screw Threads for High-Strength Bolting
(All dimensions in inches)

Size	Threads per inch	Allowance (minus)	Major diam	Major diam tolerance	Max pitch diam*	Max pitch diam tolerance	Minor diam max	Nut max minor diam	Nut max minor diam tolerance	Nut max pitch diam*	Nut max pitch diam tolerance
5/4	20	0.0010	0.2490	0.0072	0.2165	0.0026	0.1877	0.2060	0.0101	0.2211	0.0036
5/16	18	0.0011	0.3114	0.0082	0.2753	0.0030	0.2432	0.2630	0.0106	0.2805	0.0041
3/8	16	0.0013	0.3737	0.0090	0.3331	0.0032	0.2990	0.3184	0.0111	0.3389	0.0045
7/16	14	0.0013	0.4362	0.0098	0.3898	0.0036	0.3486	0.3721	0.0119	0.3960	0.0049
1/2	13	0.0015	0.4985	0.0104	0.4485	0.0037	0.4041	0.4290	0.0123	0.4552	0.0052
9/16	12	0.0016	0.5609	0.0112	0.5068	0.0040	0.4587	0.4850	0.0127	0.5140	0.0056
5/8	11	0.0017	0.6233	0.0118	0.5643	0.0042	0.5118	0.5397	0.0131	0.5719	0.0059
3/4	10	0.0019	0.7481	0.0128	0.6831	0.0045	0.6254	0.6553	0.0136	0.6914	0.0064
7/8	9	0.0021	0.8729	0.0140	0.8007	0.0049	0.7366	0.7689	0.0142	0.8098	0.0070
1	8	0.0022	0.9978	0.0152	0.9166	0.0054	0.8444	0.8795	0.0148	0.9264	0.0076
1 1/8	8	0.0024	1.1226	0.0152	1.0414	0.0055	0.9692	1.0045	0.0148	1.0517	0.0079
1 1/4	8	0.0025	1.2475	0.0152	1.1663	0.0058	1.0941	1.1295	0.0148	1.1771	0.0083
1 3/8	8	0.0025	1.3725	0.0152	1.2913	0.0061	1.2191	1.2545	0.0148	1.3024	0.0086
1 1/2	8	0.0027	1.4973	0.0152	1.4161	0.0063	1.3439	1.3795	0.0148	1.4278	0.0090
1 5/8	8	0.0028	1.6222	0.0152	1.5410	0.0065	1.4688	1.5045	0.0148	1.5531	0.0093
1 3/4	8	0.0029	1.7471	0.0152	1.6659	0.0068	1.5937	1.6295	0.0148	1.6785	0.0097
1 7/8	8	0.0030	1.8720	0.0152	1.7908	0.0070	1.7186	1.7545	0.0148	1.8038	0.0100
2	8	0.0031	1.9969	0.0152	1.9157	0.0073	1.8435	1.8795	0.0148	1.9294	0.0104
2 1/8	8	0.0032	2.1218	0.0152	2.0406	0.0075	1.9682	2.0045	0.0148	2.0545	0.0107
2 1/4	8	0.0033	2.2467	0.0152	2.1655	0.0077	2.0933	2.1295	0.0148	2.1798	0.0110
2 1/2	8	0.0035	2.4965	0.0152	2.4153	0.0082	2.3431	2.3795	0.0148	2.4305	0.0117
2 3/4	8	0.0037	2.7463	0.0152	2.6651	0.0087	2.5929	2.6295	0.0148	2.6812	0.0124
3	8	0.0038	2.9962	0.0152	2.9150	0.0092	2.8428	2.8795	0.0148	2.9318	0.0130
3 1/4	8	0.0039	3.2461	0.0152	3.1649	0.0093	3.0927	3.1295	0.0148	3.1820	0.0132
3 1/2	8	0.0040	3.4960	0.0152	3.4148	0.0093	3.3426	3.3795	0.0148	3.4321	0.0133

The Unified form of thread shall be used. Pitch diameter tolerances include errors of lead and angle.
* The maximum pitch diameters of screws are smaller than the minimum pitch diameters of nuts by these amounts.

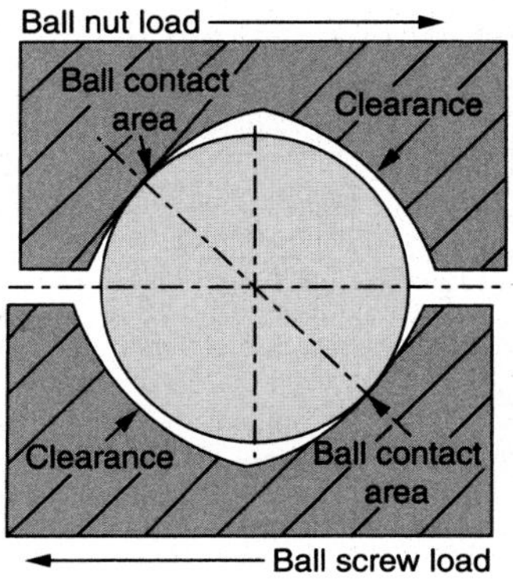

Figure 6.2.6 Cross section of ball screw and the nut.

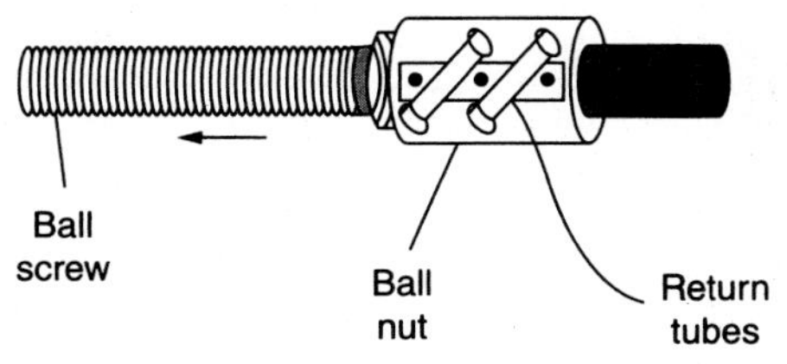

Figure 6.2.7 Ball screw. (*Courtesy Nook Industries, Inc., Cleveland, Ohio.*)

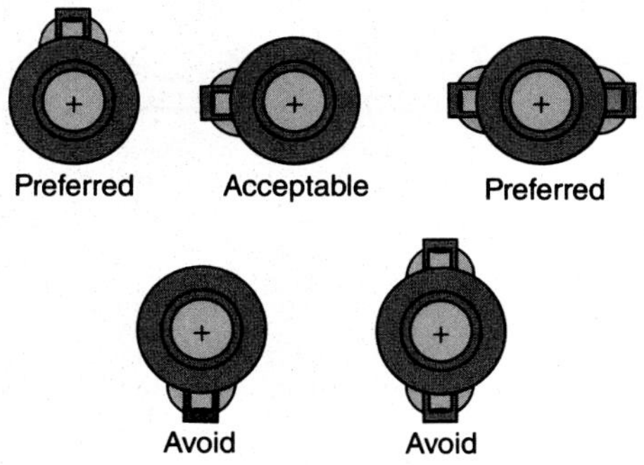

Figure 6.2.8 Design considerations of ball screw return tubes. (*Courtesy Nook Industries, Inc., Cleveland, Ohio.*)

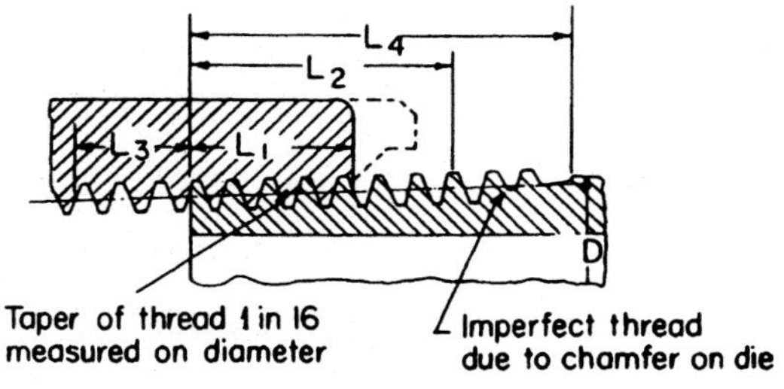

Figure 6.2.9 American National Standard taper pipe threads.

Table 6.2.11 ANSI Taper Pipe Thread
(All dimensions in inches)
(See Fig. 6.2.9.)

Nominal pipe size	OD of pipe	Threads per inch	Pitch of thread	Hand-tight engagement length L_1	Effective thread external length L_2	Wrench makeup length for internal thread length L_3	Overall length external thread L_4
1/16	0.3125	27	0.03704	0.160	0.2611	0.1111	0.3896
⅛	0.405	27	0.03704	0.180	0.2639	0.1111	0.3924
¼	0.540	18	0.05556	0.200	0.4018	0.1667	0.5946
⅜	0.675	18	0.05556	0.240	0.4078	0.1667	0.6006
½	0.840	14	0.07143	0.320	0.5337	0.2143	0.7815
¾	1.050	14	0.07143	0.339	0.5457	0.2143	0.7935
1	1.315	11½	0.08696	0.400	0.6828	0.2609	0.9845
1¼	1.660	11½	0.08696	0.420	0.7068	0.2609	1.0085
1½	1.900	11½	0.08696	0.420	0.7235	0.2609	1.0252
2	2.375	11½	0.08696	0.436	0.7565	0.2609	1.0582
2½	2.875	8	0.12500	0.682	1.1375	0.2500	1.5712
3	3.500	8	0.12500	0.766	1.2000	0.2500	1.6337
3½	4.000	8	0.12500	0.821	1.2500	0.2500	1.6837
4	4.500	8	0.12500	0.844	1.3000	0.2500	1.7337
5	5.563	8	0.12500	0.937	1.4063	0.2500	1.8400
6	6.625	8	0.12500	0.958	1.5125	0.2500	1.9462
8	8.625	8	0.12500	1.063	1.7125	0.2500	2.1462
10	10.750	8	0.12500	1.210	1.9250	0.2500	2.3587
12	12.750	8	0.12500	1.360	2.1250	0.2500	2.5587
14 OD	14.000	8	0.12500	1.562	2.2500	0.2500	2.6837
16 OD	16.000	8	0.12500	1.812	2.4500	0.2500	2.8837
18 OD	18.000	8	0.12500	2.000	2.6500	0.2500	3.0837
20 OD	20.000	8	0.12500	2.125	2.8500	0.2500	3.2837
24 OD	24.000	8	0.12500	2.375	3.2500	0.2500	3.6837

Table 6.2.12 ANSI Straight Pipe Threads
(All dimensions in inches)

		Pressure-tight with seals		Pressure-tight without seals		Free-fitting (NPSM)				Loose-fitting (NPSL)			
						External		Internal		External		Internal	
Nominal pipe size (1)	Threads per inch (2)	Pitch diam, max (3)	Minor diam, min (4)	Pitch diam, max (5)	Minor diam, min (6)	Pitch diam, max (7)	Major diam, max (8)	Pitch diam, max (9)	Minor diam, min (10)	Pich diam, max (11)	Major diam, max (12)	Pitch diam, max (13)	Minor diam, min (14)
⅛	27	0.3782	0.342	0.3736	0.3415	0.3748	0.399	0.3783	0.350	0.3840	0.409	0.3989	0.362
¼	18	0.4951	0.440	0.4916	0.4435	0.4899	0.527	0.4951	0.453	0.5038	0.541	0.5125	0.470
⅜	18	0.6322	0.577	0.6270	0.5789	0.6270	0.664	0.6322	0.590	0.6409	0.678	0.6496	0.607
½	14	0.7851	0.715	0.7784	0.7150	0.7784	0.826	0.7851	0.731	0.7963	0.844	0.8075	0.753
¾	14	0.9956	0.925	0.9889	0.9255	0.9889	1.036	0.9956	0.941	1.0067	1.054	1.0179	0.964
1	11½	1.2468	1.161	1.2386	1.1621	1.2386	1.296	1.2468	1.181	1.2604	1.318	1.2739	1.208
1¼	11½	1.5915	1.506	—	—	1.5834	1.641	1.5916	1.526	1.6051	1.663	1.6187	1.553
1½	11½	1.8305	1.745	—	—	1.8223	1.880	1.8305	1.764	1.8441	1.902	1.8576	1.792
2	11½	2.3044	2.219	—	—	2.2963	2.354	2.3044	2.238	2.3180	2.376	2.3315	2.265
2½	8	2.7739	2.650	—	—	2.7622	2.846	2.7739	2.679	2.7934	2.877	2.8129	2.718
3	8	3.4002	3.277	—	—	3.3885	3.472	3.4002	3.305	3.4198	3.503	3.4393	3.344
3½	8	3.9005	3.777	—	—	3.8888	3.972	3.9005	3.806	3.9201	4.003	3.9396	3.845
4	8	4.3988	4.275	—	—	4.3871	4.470	4.3988	4.304	4.4184	4.502	4.4379	4.343
5	8	—	—	—	—	5.4493	5.533	5.4610	5.366	5.4805	5.564	5.5001	5.405
6	8	—	—	—	—	6.5060	6.589	6.5177	6.423	6.5372	6.620	6.5567	6.462
8	8	—	—	—	—	—	—	—	—	8.5313	8.615	8.5508	8.456
10	8	—	—	—	—	—	—	—	—	10.6522	10.735	10.6717	10.577
12	8	—	—	—	—	—	—	—	—	12.6491	12.732	12.6686	12.574

designation and notation include nominal size, number of threads per inch, thread series, class. For example, 1/8-27 NPTF-1, 1/8-27 NPTF-2, 1/8-27 PTF-SAE short, 1/8-27 NPSI, where N = National (American) standard, P = pipe, T = taper, S = straight, F = fuel and oil, I = intermediate. NPTF has two classes: class 1 = specific inspection of root and crest truncation *not* required; class 2 = specific inspection of root and crest truncation *is* required. The series includes: NPTF for all types of service; PTF-SAE short where clearance is not sufficient for full thread length as NPTF; NPSF, nontapered, economical to produce, and used with soft or ductile materials; NPSI nontapered, thick sections with little expansion.

Dry-seal pipe threads resemble tapered pipe threads except the form is truncated (see Fig. 6.2.10), and $L_4 = L_2 + 1$ (see Fig. 6.2.9). Although these threads are designed for nonlubricated joints, as in automobile work, under certain conditions a lubricant is used to prevent galling. Table 6.2.13 lists truncation values.

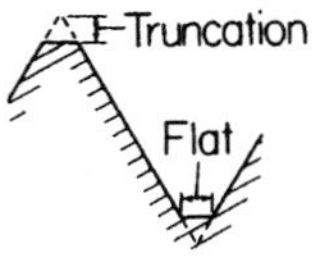

Figure 6.2.10 American National Standard dry-seal pipe thread.

Table 6.2.13 ANSI Dry-Seal Pipe Threads*
(See Fig. 6.2.10.)

Threads per inch n	Truncation, in Min	Truncation, in Max	Width of flat Min	Width of flat Max
27 Crest	0.047p	0.094p	0.054p	0.108p
Root	0.094p	0.140p	0.108p	0.162p
18 C	0.047p	0.078p	0.054p	0.090p
R	0.078p	0.109p	0.090p	0.126p
14 C	0.036p	0.060p	0.042p	0.070p
R	0.060p	0.085p	0.070p	0.098p
11½ C	0.040p	0.060p	0.046p	0.069p
R	0.060p	0.090p	0.069p	0.103p
8 C	0.042p	0.055p	0.048p	0.064p
R	0.055p	0.076p	0.064p	0.088p

* The *truncation* and *width-of-flat* proportions listed above are also valid in the metric system.

Tap drill sizes for tapered and straight pipe threads are listed in Table 6.2.14.

Wrench bolt heads, nuts, and wrench openings have been standardized (ANSI 18.2-1972). Bolt head and nut dimensions are given in Table 6.2.15.

Table 6.2.14 Suggested Tap Drill Sizes for Internal Pipe Threads

	Taper pipe thread							Straight pipe thread			
		Minor diameter at distance		Drill for use without reamer		Drill for use with reamer		Minor diameter			
Size	Probable drill oversize cut (mean)	L_1 from large end	$L_1 + L_3$ from large end	Theoretical drill size*	Suggested drill size†	Theoretical drill size‡	Suggested drill size†	NPSF	NPSI	Theoretical drill size§	Theoretical drill size†
	1	2	3	4	5	6	7	8	9	10	11
					Inch						
1/16-27	0.0038	0.2443	0.2374	0.2405	"C" (0.242)	0.2336	"A" (0.234)	0.2482	0.2505	0.2444	"D" (0.246)
1/32-27	0.0044	0.3367	0.3293	0.3323	"Q" (0.332)	0.3254	21/64 (0.328)	0.3406	0.3429	0.3362	"R" (0.339)
1/4-18	0.0047	0.4362	0.4258	0.4315	7/17 (0.438)	0.4211	27/64 (0.422)	0.4422	0.4457	0.4375	7/16 (0.438)
3/8-18	0.0049	0.5708	0.5604	0.5659	9/16 (0.562)	0.5555	9/16 (0.563)	0.5776	0.5811	0.5727	27/64 (0.578)
1/2-14	0.0051	0.7034	0.6901	0.6983	45/64 (0.703)	0.6850	11/16 (0.688)	0.7133	0.7180	0.7082	45/64 (0.703)
3/4-14	0.0060	0.9127	0.8993	0.9067	29/32 (0.906)	0.8933	57/64 (0.891)	0.9238	0.9283	0.9178	59/64 (0.922)
1-11½	0.0080	1.1470	1.1307	1.1390	1 9/64 (1.141)	1.1227	1 1/8 (1.125)	1.1600	1.1655	1.1520	1 5/32 (1.156)
1¼-11½	0.0100	1.4905	1.4742	1.4805	1 31/64 (1.484)	1.4642	1 11/32 (1.469)				
1½-11½	0.0120	1.7295	1.7132	1.7175	1 23/32 (1.719)	1.7012	1 45/64 (1.703)				
2-11½	0.0160	2.2024	2.1861	2.1864	2 3/16 (2.188)	2.1701	2 11/64 (2.172)				
2½-8	0.0180	2.6234	2.6000	2.6054	2 39/64 (2.609)	2.5820	2 27/64 (2.578)				
3-8	0.0200	3.2445	3.2211	3.2245	3 15/64 (3.234)	3.2011	3 13/64 (3.203)				
					Metric						
1/16-27	0.097	6.206	6.029	6.109	6.1	5.932	6.0	6.304	6.363	6.207	6.2
1/8-27	0.112	8.551	8.363	8.438	8.4	8.251	8.2	8.651	8.710	8.539	8.5
1/4-18	0.119	11.080	10.816	10.961	11.0	10.697	10.8	11.232	11.321	11.113	11.0
3/8-18	0.124	14.499	14.235	14.375	14.5	14.111	14.0	14.671	14.760	14.547	14.5
1/2-14	0.130	17.867	17.529	17.737	17.5	17.399	17.5	18.118	18.237	17.988	18.0
3/4-14	0.152	23.182	22.842	23.030	23.0	22.690	23.0	23.465	23.579	23.212	23.0
1-11½	0.203	29.134	28.720	28.931	29.0	28.517	28.5	29.464	29.604	29.261	29.0
1¼-11½	0.254	37.859	37.444	37.605	37.5	37.190	37.0				
1½-11½	0.305	43.929	43.514	43.624	43.5	43.209	43.5				
2-11½	0.406	55.941	55.527	55.535	56.0	55.121	55.0				
2½-8	0.457	66.634	66.029	66.177	66.0	65.572	65.0				
3-8	0.508	82.410	81.815	81.902	82.0	81.307	81.0				

* Column 4 values equal column 2 values minus column 1 values.
† Some drill sizes listed may not be standard drills, and in some cases, standard metric drill sizes may be closer to the theoretical inch drill size and standard inch drill sizes may be closer to the theoretical metric drill size.
‡ Column 6 values equal column 3 values minus column 1 values.
§ Column 10 values equal column 8 values minus column 1 values.
SOURCE: ANSI B1.20.3-1976 and ANSI B1.20.4-1976, reproduced by permission.

Table 6.2.15 Width Across Flats of Bolt Heads and Nuts
(All dimensions in inches)

Nominal site or basic major diam of thread	Dimensions of regular bolt heads unfinished, square, and hexagon		Dimensions of heavy bolt heads unfinished, square, and hexagon		Dimensions of cap-screw heads hexagon		Dimensions of setscrew heads		Dimensions of regular nuts and regular jam nuts, unfinished, square, and hexagon (jam nuts, hexagon only)		Dimensions of machine-screw and stove-bolt nuts, square and hexagon		Dimensions of heavy nuts and heavy jam nuts, unfinished, square, and hexagon (jam nuts, hexagon only)	
	Max	Min	Max	Min	Max	Min	Max	Min	Max	Min	Max	Min	Max	Min
No. 0	—	—	—	—	—	—	—	—	—	—	0.1562	0.150		
No. 1	—	—	—	—	—	—	—	—	—	—	0.1562	0.150		
No. 2	—	—	—	—	—	—	—	—	—	—	0.1875	0.180		
No. 3	—	—	—	—	—	—	—	—	—	—	0.1875	0.180		
No. 4	—	—	—	—	—	—	—	—	—	—	0.2500	0.241		
No. 5	—	—	—	—	—	—	—	—	—	—	0.3125	0.302		
No. 6	—	—	—	—	—	—	—	—	—	—	0.3125	0.302		
No. 8	—	—	—	—	—	—	—	—	—	—	0.3438	0.332		
No. 10	—	—	—	—	—	—	—	—	—	—	0.3750	0.362		
No. 12	—	—	—	—	—	—	—	—	—	—	0.4375	0.423		
1/4	0.3750	0.362	—	—	0.4375	0.428	0.2500	0.241	0.4375	0.425	0.4375	0.423	0.5000	0.488
5/16	0.5000	0.484	—	—	0.5000	0.489	0.3125	0.302	0.5625	0.547	0.5625	0.545	0.5938	0.578
3/8	0.5625	0.544	—	—	0.5625	0.551	0.3750	0.362	0.6250	0.606	0.6250	0.607	0.6875	0.669
7/16	0.6250	0.603	—	—	0.6250	0.612	0.4375	0.423	0.7500	0.728	—	—	0.7812	0.759
1/2	0.7500	0.725	0.8750	0.850	0.7500	0.736	0.5000	0.484	0.8125	0.788	—	—	0.8750	0.850
9/16	0.8750	0.847	0.9375	0.909	0.8125	0.798	0.5625	0.545	0.8750	0.847	—	—	0.9375	0.909
5/8	0.9375	0.906	1.0625	1.031	0.8750	0.860	0.6250	0.606	1.0000	0.969	—	—	1.0625	1.031
3/4	1.1250	1.088	1.2500	1.212	1.0000	0.983	0.7500	0.729	1.1250	1.088	—	—	1.2500	1.212
7/8	1.3125	1.269	1.4375	1.394	1.1250	1.106	0.8750	0.852	1.3125	1.269	—	—	1.4375	1.394
1	1.5000	1.450	1.6250	1.575	1.3125	1.292	1.0000	0.974	1.5000	1.450	—	—	1.6250	1.575
1 1/8	1.6875	1.631	1.8125	1.756	1.5000	1.477	1.1250	1.096	1.6875	1.631	—	—	1.8125	1.756
1 1/4	1.8750	1.812	2.0000	1.938	1.6875	1.663	1.2500	1.219	1.8750	1.812	—	—	2.0000	1.938
1 3/8	2.0625	1.994	2.1857	2.119	—	—	1.3750	1.342	2.0625	1.994	—	—	2.1875	2.119
1 1/2	2.2500	2.175	2.3750	2.300	—	—	1.5000	1.464	2.2500	2.175	—	—	2.3750	2.300
1 5/8	2.4375	2.356	2.5625	2.481	—	—	—	—	2.4375	2.356	—	—	2.5625	2.481
1 3/4	2.6250	2.538	2.7500	2.662	—	—	—	—	2.6250	2.538	—	—	2.7500	2.662
1 7/8	2.8125	2.719	2.9375	2.844	—	—	—	—	2.8125	2.719	—	—	2.9375	2.844
2	3.0000	2.900	3.1250	3.025	—	—	—	—	3.0000	2.900	—	—	3.1250	3.025
2 1/4	3.3750	3.262	3.5000	3.388	—	—	—	—	3.3750	3.262	—	—	3.5000	3.388
2 1/2	3.7500	3.625	3.8750	3.750	—	—	—	—	3.7500	3.625	—	—	3.8750	3.750
2 3/4	4.1250	3.988	4.2500	4.112	—	—	—	—	4.1250	3.988	—	—	4.2500	4.112
3	4.5000	4.350	4.6250	4.475	—	—	—	—	4.5000	4.350	—	—	4.6250	4.475
3 1/4	—	—	—	—	—	—	—	—	—	—	—	—	5.0000	4.838
3 1/2	—	—	—	—	—	—	—	—	—	—	—	—	5.3750	5.200
3 3/4	—	—	—	—	—	—	—	—	—	—	—	—	5.7500	5.562
4	—	—	—	—	—	—	—	—	—	—	—	—	6.1250	5.925

Regular bolt heads are for general use. Unfinished bolt heads are not finished on any surface. Semifinished bolt heads are finished under head.
Regular nuts are for general use. Semifinished nuts are finished on bearing surface and threaded. Unfinished nuts are not finished on any surface but are threaded.

Machine Screws

Machine screws are defined according to head types as follows:

Flat Head This screw has a flat surface for the top of the head with a countersink angle of 82°. It is standard for machine screws, cap screws, and wood screws.

Round Head This screw has a semielliptical head and is standard for machine screws, cap screws, and wood screws except that for the cap screw it is called *button head.*

Fillister Head This screw has a rounded surface for the top of the head, the remainder being cylindrical. The head is standard for machine screws and cap screws.

Oval Head This screw has a rounded surface for the top of the head and a countersink angle of 82°. It is standard for machine screws and wood screws.

Hexagon Head This screw has a hexagonal head for use with external wrenches. It is standard for machine screws.

Socket Head This screw has an internal hexagonal socket in the head for internal wrenching. It is standard for cap screws.

These screw heads are shown in Fig. 6.2.11; pertinent dimensions are in Table 6.2.16. There are many more machine screw head shapes available to the designer for special purposes, and many are found in the literature. In addition, many different screw head configurations have been developed to render fasteners "tamperproof"; these, too, are found in manufacturers' catalogs or the trade literature.

Eyebolts

Eyebolts are classified as rivet, nut, or screw and can be used on a swivel. See Fig. 6.2.12 and Table 6.2.17. The safe working load may be obtained for each application by applying an appropriate factor of safety.

Driving recesses come in many forms and types and can be found in company catalogs. Figure 6.2.13 shows a representative set.

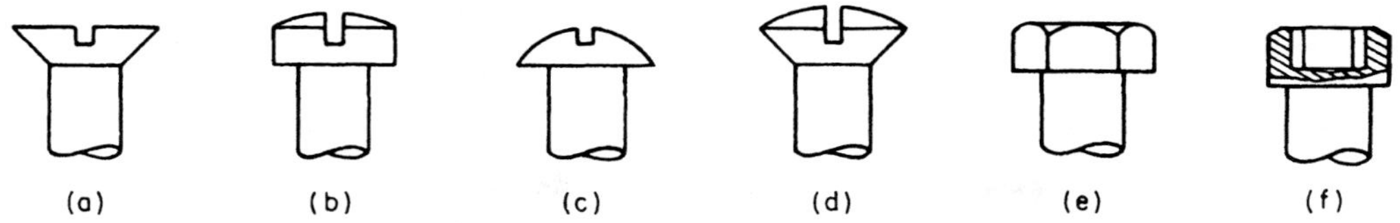

Figure 6.2.11 Machine screw heads. (*a*) Flat; (*b*) fillister; (*c*) round; (*d*) oval; (*e*) hexagonal; (*f*) socket.

Table 6.2.16 Head Diameters (Maximum), in

Machine screws						
Nominal size	Screw diam	Flat head	Round head	Fillister head	Oval head	Hexagonal head across flats
2	0.086	0.172	0.162	0.140	0.172	0.125
3	0.099	0.199	0.187	0.161	0.199	0.187
4	0.112	0.225	0.211	0.183	0.225	0.187
5	0.125	0.252	0.236	0.205	0.252	0.187
6	0.138	0.279	0.260	0.226	0.279	0.250
8	0.164	0.332	0.309	0.270	0.332	0.250
10	0.190	0.385	0.359	0.313	0.385	0.312
12	0.216	0.438	0.408	0.357	0.438	0.312
1/4	0.250	0.507	0.472	0.414	0.507	0.375
5/16	0.3125	0.636	0.591	0.519	0.636	0.500
3/8	0.375	0.762	0.708	0.622	0.762	0.562
Cap screws						
Nominal size	Screw diam	Flat head	Button head	Fillister head	Socket head	
1/4	0.250	1/2	7/16	3/8	3/8	
5/16	0.3125	5/8	9/16	7/16	7/16	
3/8	0.375	3/4	5/8	9/16	9/16	
7/16	0.4375	13/16	3/4	5/8	5/8	
1/2	0.500	7/8	13/16	3/4	3/4	
9/16	0.5625	1	15/16	13/16	13/16	
5/8	0.625	1 1/8	1	7/8	7/8	
3/4	0.750	1 3/8	1 1/4	1	1	
7/8	—	—	—	1 1/8	1 1/8	
1	—	—	—	1 5/16	1 5/16	

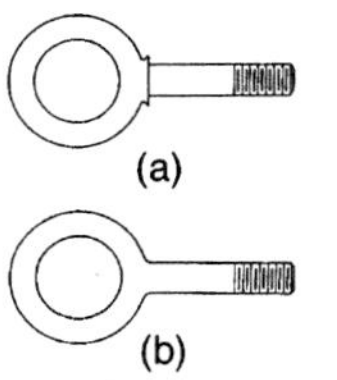

Figure 6.2.12 Eyebolts.

Setscrews are used for fastening collars, sheaves, gears, etc., to shafts to prevent relative rotation or translation. They are available in a variety of head and point styles, as shown in Fig. 6.2.14. A complete tabulation of dimensions is found in ANSI/ASME B18.3-1982 (R86), ANSI 18.6.2-1977 (R93), and ANSI 18.6.3-1977 (R91). Holding power for various sizes is given in Table 6.2.18.

Locking Fasteners

Locking fasteners are used to prevent loosening of a threaded fastener in service and are available in a wide variety differing vastly in design, performance, and function. Since each has special features which may make it of particular value in the solution of a given machine problem, it is important that great care be exercised in the selection of a particular design in order that its properties may be fully utilized. These fasteners may be divided into six groups, as follows: seating lock, spring stop nut, interference, wedge, blind, and

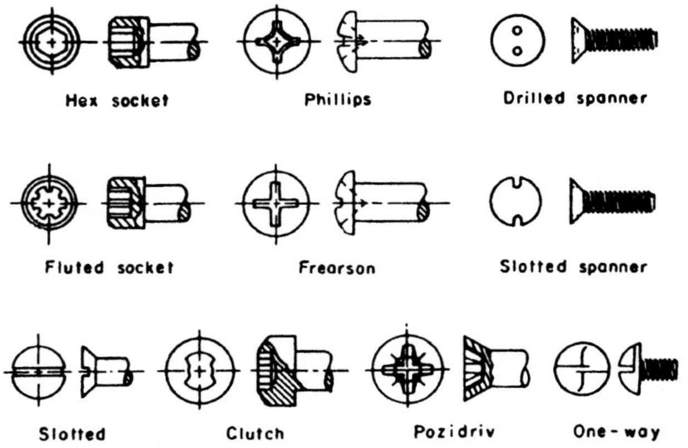

Figure 6.2.13 Driving recesses. (*Adapted, with permission, from Machine Design.*)

Table 6.2.17 Regular Nut Eyebolts—Selected Sizes
(Thomas Laughlin Co., Portland, Me.)
(All dimensions in inches)
(See Fig. 6.2.12.)

Diam and shank length	Thread length	Eye dimension		Approx breaking strength, lb	Diam and shank length	Thread length	Eye dimension		Approx breaking strength, lb
		ID	OD				ID	OD	
$\frac{1}{4} \times 2$	$1\frac{1}{2}$	$\frac{1}{2}$	1	2,200	$\frac{3}{4} \times 6$	3	$1\frac{1}{2}$	3	23,400
$\frac{5}{16} \times 2\frac{1}{4}$	$1\frac{1}{2}$	$\frac{5}{8}$	$1\frac{1}{4}$	3,600	$\frac{3}{4} \times 10$	3	$1\frac{1}{2}$	3	23,400
$\frac{3}{8} \times 4\frac{1}{2}$	$2\frac{1}{2}$	$\frac{3}{4}$	$1\frac{1}{2}$	5,200	$\frac{3}{4} \times 15$	5	$1\frac{1}{2}$	3	23,400
$\frac{1}{2} \times 3\frac{1}{4}$	$1\frac{1}{2}$	1	2	9,800	$\frac{7}{8} \times 8$	4	$1\frac{3}{4}$	$3\frac{1}{2}$	32,400
$\frac{1}{2} \times 6$	3	1	2	9,800	1×6	3	2	4	42,400
$\frac{1}{2} \times 10$	3	1	2	9,800	1×9	4	2	4	42,400
$\frac{5}{8} \times 4$	2	$1\frac{1}{4}$	$2\frac{1}{2}$	15,800	1×18	7	2	4	42,400
$\frac{5}{8} \times 6$	3	$1\frac{1}{4}$	$2\frac{1}{2}$	15,800	$1\frac{1}{4} \times 8$	4	$2\frac{1}{2}$	5	67,800
$\frac{5}{8} \times 10$	3	$1\frac{1}{4}$	$2\frac{1}{2}$	15,800	$1\frac{1}{4} \times 20$	6	$2\frac{1}{2}$	5	67,800

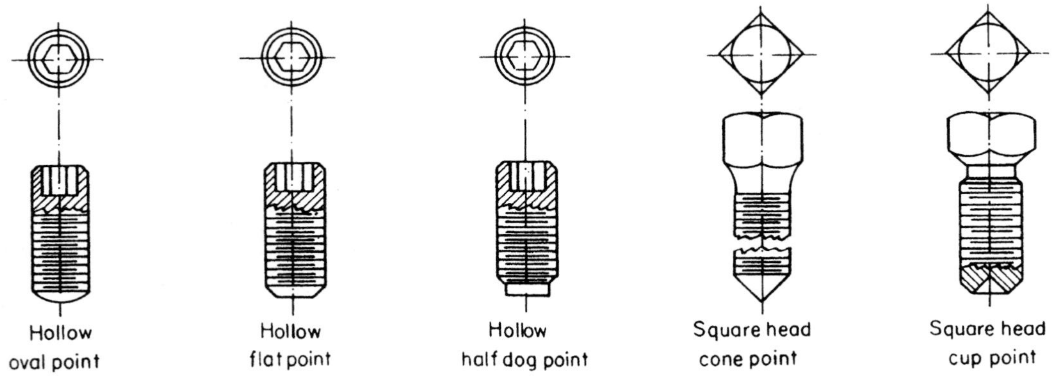

Figure 6.2.14 Setscrews.

Table 6.2.18 Cup-Point Setscrew Holding Power

Nominal screw size	Seating torque, lb · in	Axial holding power, lb
No. 0	0.5	50
No. 1	1.5	65
No. 2	1.5	85
No. 3	5	120
No. 4	5	160
No. 5	9	200
No. 6	9	250
No. 8	20	385
No. 10	33	540
$\frac{1}{4}$ in	87	1,000
$\frac{5}{16}$ in	165	1,500
$\frac{3}{8}$ in	290	2,000
$\frac{7}{16}$ in	430	2,500
$\frac{1}{2}$ in	620	3,000
$\frac{9}{16}$ in	620	3,500
$\frac{5}{8}$ in	1,225	4,000
$\frac{3}{4}$ in	2,125	5,000
$\frac{7}{8}$ in	5,000	6,000
1 in	7,000	7,000

NOTES: 1. Torsional holding power in inch-pounds is equal to one-half of the axial holding power times the shaft diameter in inches.
2. Experimental data were obtained by seating an alloy-steel cup-point setscrew against a steel shaft with a hardness of Rockwell C 15. Screw threads were class 3A, tapped holes were class 2B. Holding power was defined as the minimum load necessary to produce 0.01 in of relative movement between the shaft and the collar.
3. Cone points will develop a slightly greater holding power; flat, dog, and oval points, slightly less.
4. Shaft hardness should be at least 10 Rockwell C points less than the setscrew point.
5. Holding power is proportional to seating torque. Torsional holding power is increased about 6% by use of a flat on the shaft.
6. Data by F. R. Kull, Fasteners Book Issue, *Mach. Des.*, Mar. 11, 1965.

quick-release. The **seating-lock type** locks only when firmly seated and is therefore free-running on the bolt. The **spring stop-nut type** of fastener functions by a spring action clamping down upon the bolt. The **prevailing torque type** locks by elastic or plastic flow of a portion of the fastener material. A recent development employs an adhesive coating applied to the threads. The **wedge type** locks by relative wedging of either elements or nut and bolt. The **blind type** usually utilizes spring action of the fastener, and the **quick-release type** utilizes a quarter-turn release device. An example of each is shown in Fig. 6.2.15.

One such specification developed for prevailing torque fasteners by the Industrial Fasteners Institute is based on locking torque and may form a precedent for other types of fasteners as well.

Coach and lag screws find application in wood, or in masonry with an expansion anchor. Figure 6.2.16*a* shows two types, and Table 6.2.19 lists pertinent dimensions.

Wood screws [ANSI B18.22.1-1975 (R81)] are made in lengths from $\frac{1}{4}$ to 5 in for steel and from $\frac{1}{4}$ to $3\frac{1}{2}$ in for brass screws, increasing by $\frac{1}{8}$ in up to 1 in, by $\frac{1}{4}$ in up to 3 in, and by $\frac{1}{2}$ in up to 5 in. Sizes are given in Table 6.2.20. Screws are made with flat, round, or oval heads. Figure 6.2.16*b* shows several heads.

Washers [ANSI B18.22.1-1975 (R81)] for bolts and lag screws, either round or square, are made to the dimensions given in Table 6.2.21. For other types of lock washers, see Fig. 6.2.17*a* and *b*.

Self-tapping screws are available in three types. **Thread-forming** tapping screws plastically displace material adjacent to the pilot hole. **Thread-cutting** tapping screws have cutting edges and chip cavities (flutes) and form a mating thread by removing material adjacent to the pilot hole. Thread-cutting screws are generally used to join thicker and

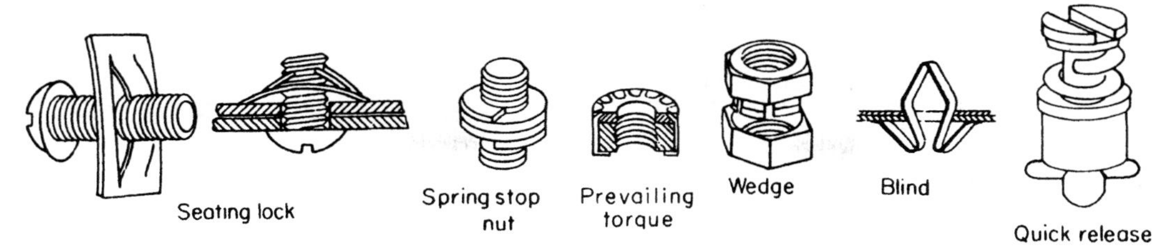

Figure 6.2.15 Locking fasteners.

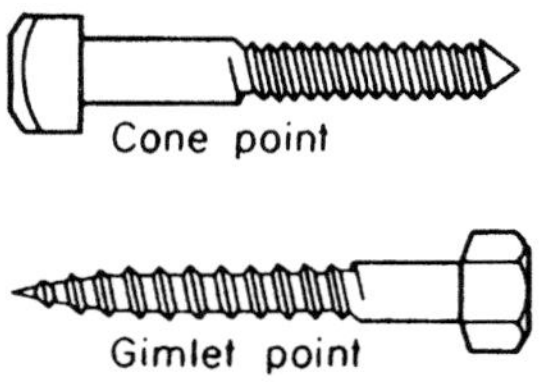

Figure 6.2.16*a* Coach and lag screws.

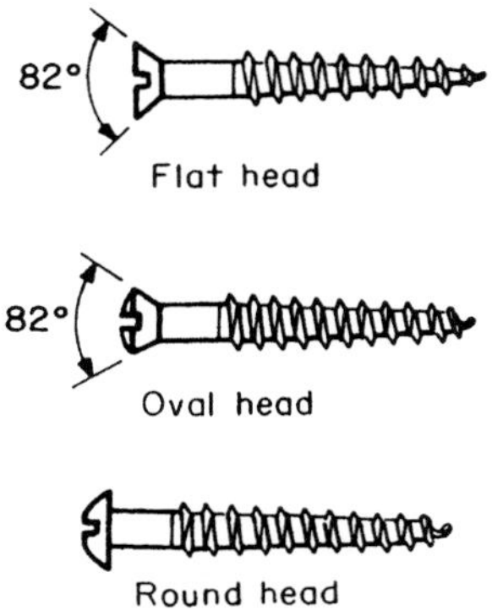

Figure 6.2.16*b* Wood screws.

harder materials and require a lower driving torque than thread-forming screws. **Metallic drive** screws are forced into the material by pressure and are intended for making permanent fastenings. These three types are further classified on the basis of thread and point form as shown in Table 6.2.22. In addition, to these body forms, a number of different head types are available. Basic dimensional data are given in Table 6.2.23.

Carriage bolts have been standardized in ANSI B18.5-1971, revised 1990. They come in styles shown in Fig. 6.2.18. The range of bolt diameters is no. 10 (= 0.19 in) to 1 in, no. 10 to ¾ in, no. 10 to ½ in, and no. 10 to ¾ in, respectively.

Materials, Strength, and Service Adaptability of Bolts and Screws Materials

Table 6.2.24 shows the relationship between selected metric bolt classes and SAE and ASTM grades. The first number of a metric bolt class equals the minimum tensile strength (ultimate) in megapascals (MPa) divided by 100, and the second number is the approximate ratio between minimum yield and minimum ultimate strengths.

A variety of specially configured heads is available to prevent vandalism; they go by the generic term **tamper-proof.**

EXAMPLE. Class 5.8 has a minimum ultimate strength of approximately 500 MPa and a minimum yield strength approximately 80 percent of minimum ultimate strength.

Strength The fillet between head and body, the thread runout point, and the first thread to engage the nut all create stress concentrations causing local stresses much greater than the average tensile stress in the bolt body. The complexity of the stress patterns renders ineffective the ordinary design calculations based on yield or ultimate stresses. Bolt strengths are therefore determined by laboratory tests on bolt-nut assemblies and published as proof loads. Fastener manufacturers are required to periodically repeat such tests to ensure that their products meet the original standards.

In order that a bolted joint remain firmly clamped while carrying its external load P, the bolt must be tightened first with sufficient torque to induce an initial tensile preload F_i. The total load F_B experienced by the bolt is then $F_B = F_i + \varepsilon P$. The fractional multiplier ε is given by $\varepsilon = K_B/(K_B + K_M)$, where K_B = elastic constant of the bolt and $1/K_M = 1/K_N + 1/K_W + 1/K_G + 1/K_J$. K_N = elastic constant of the nut; K_W = elastic constant of the washer; K_G = elastic constant of the gasket; K_J = elastic constant of the clamped surfaces or joint.

Table 6.2.19 Coach and Lag Screws

Diam of screw, in	¼	5/16	⅜	7/16	½	⅝	¾	⅞	1
No. of threads per inch	10	9	7	7	6	5	4½	4	3½
Across flats of hexagon and square heads, in	⅜	15/32	9/16	21/32	¾	15/16	1⅛	1 5/16	1½
Thickness of hexagon and square heads, in	3/16	¼	5/16	⅜	7/16	17/32	⅝	¾	⅞

Length of threads for screws of all diameters

Length of screw, in	1½	2	2½	3	3½	4	4½
Length of thread, in	To head	1½	2	2¼	2½	3	3½
Length of screw, in	5	5½	6	7	8	9	10-12
Length of thread, in	4	4	4½	5	6	6	7

Table 6.2.20 American National Standard Wood Screws

Number	0	1	2	3	4	5	6	7	8
Threads per inch	32	28	26	24	22	20	18	16	15
Diameter, in	0.060	0.073	0.086	0.099	0.112	0.125	0.138	0.151	0.164
Number	9	10	11	12	14	16	18	20	24
Threads per inch	14	13	12	11	10	9	8	8	7
Diameter, in	0.177	0.190	0.203	0.216	0.242	0.268	0.294	0.320	0.372

Table 6.2.21 Dimensions of Steel Washers, in

	Plain washer			Lock washer		
Bolt size	Hole diam	OD	Thickness	Hole diam	OD	Thickness
3/16	1/4	9/16	3/64	0.194	0.337	0.047
1/4	5/16	3/4	1/16	0.255	0.493	0.062
5/16	3/8	7/8	1/16	0.319	0.591	0.078
3/8	7/16	1	5/64	0.382	0.688	0.094
7/16	1/2	1 1/4	5/64	0.446	0.781	0.109
1/2	9/16	1 3/8	3/32	0.509	0.879	0.125
9/16	5/8	1 1/2	3/32	0.573	0.979	0.141
5/8	11/16	1 3/4	1/8	0.636	1.086	0.156
3/4	13/16	2	1/8	0.763	1.279	0.188
7/8	15/16	2 1/4	5/32	0.890	1.474	0.219
1	1 1/16	2 1/2	5/32	1.017	1.672	0.250
1 1/8	1 1/4	2 3/4	5/32	1.144	1.865	0.281
1 1/4	1 3/8	3	5/32	1.271	2.058	0.312
1 3/8	1 1/2	3 1/4	11/64	1.398	2.253	0.344
1 1/2	1 5/8	3 1/2	11/64	1.525	2.446	0.375
1 5/8	1 3/4	3 3/4	11/64			
1 3/4	1 7/8	4	11/64			
1 7/8	2	4 1/4	11/64			
2	2 1/8	4 1/2	11/64			
2 1/4	2 3/8	4 3/4	3/16			
2 1/2	2 5/8	5	7/32			

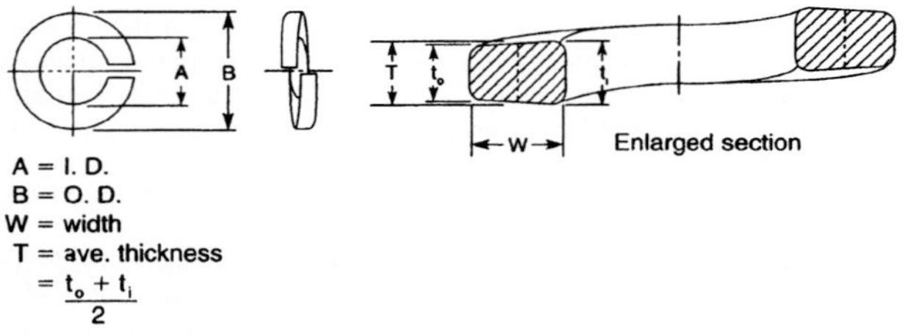

Figure 6.2.17a Regular helical spring lock washer.

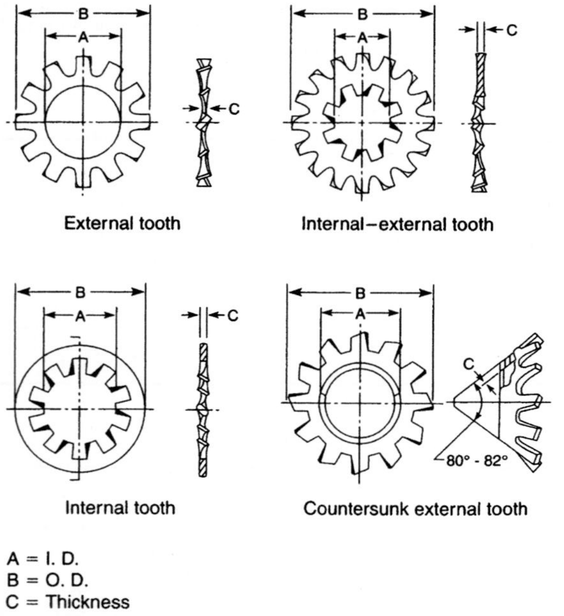

Figure 6.2.17b Toothed lock washers.

By manipulation, the fractional multiplier can be written $\varepsilon = 1/(1 + K_M/K_B)$. When K_M/K_B approaches 0, $\varepsilon \rightarrow 1$. When K_M/K_B approaches infinity, $\varepsilon \rightarrow 0$. Generally, K_N, K_W, and K_J are much stiffer than K_B, while K_G can vary from very soft to very stiff. In a metal-to-metal joint, K_G is effectively infinity, which causes K_M to approach infinity and ε to approach 0. On the other hand, for a very soft gasketed joint, $K_M \rightarrow 0$ and $\varepsilon \rightarrow 1$. For a metal-to-metal joint, then, $F_B = F_i + 0 \times P = F_i$; thus no fluctuating load component enters the bolt. In that case, the bolt remains at *static force* F_i at all times, and the static design will suffice. For a very soft gasketed joint, $F_B = F_i + 1 \times P = F_i + P$, which means that if P is a dynamically fluctuating load, it will be superimposed onto the static value of F_i. Accordingly, one must use the *fatigue design* for the bolt. Of course, for conditions between $\varepsilon = 0$ and $\varepsilon = 1$, the load within the bolt body is $F_B = F_i + \varepsilon P$, and again the *fatigue design* must be used.

In general, one wants as much preload as a bolt and joint will tolerate, without damaging the clamped parts, encouraging stress corrosion, or reducing fatigue life. For ungasketed, unpressurized joints under static loads using high-quality bolt materials, such as SAE 3 or better, the preload should be about 90 percent of proof load.

The **proof strength** is the stress obtained by dividing **proof load** by **stress area.** Stress area is somewhat larger than the root area and can be found in thread tables, or calculated approximately from a diameter which is the mean of the root and pitch diameters.

Initial sizing of bolts can be made by calculating area = (% × proof load)/(proof strength). See Table 6.2.25 for typical physical properties.

NOTE: In European practice, proof stress of a given grade is independent of diameter and is accomplished by varying chemical composition with diameters.

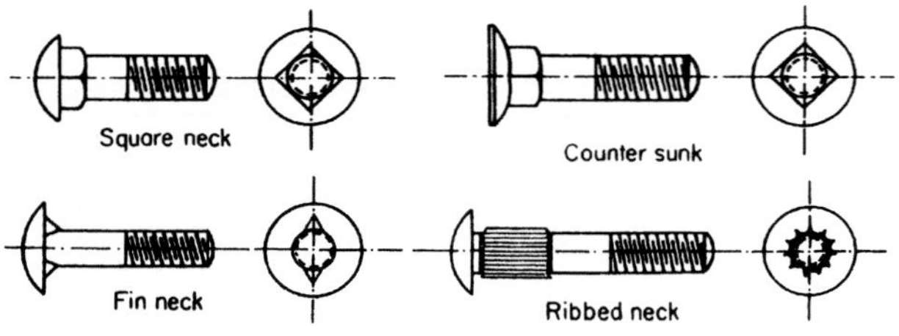

Figure 6.2.18 Carriage bolts.

Table 6.2.22 Tapping Screw Forms

ASA type and thread form	Description and recommendations
 AB	Spaced thread, with gimlet point, designed for use in sheet metal, resin-impregnated plywood, wood, and asbestos compositions. Used in pierced or punched holes where a sharp point for starting is needed.
 B	Type B is a blunt-point spaced-thread screw, used in heavy-gage sheet-metal and nonferrous castings.
 BP	Same as type B, but has a 45 deg included angle unthreaded cone point. Used for locating and aligning holes or piercing soft materials.
 C	Blunt point with threads approximating machine-screw threads. For applications where a machine-screw thread is preferable to the spaced-thread form. Unlike thread-cutting screws, type C makes a chip-free assembly.
 U	Multiple-threaded drive screw with steep helix angle and a blunt, unthreaded starting pilot. Intended for making permanent fastenings in metals and plastics. Hammered or mechanically forced into work. Should not be used in materials less than one screw diameter thick.
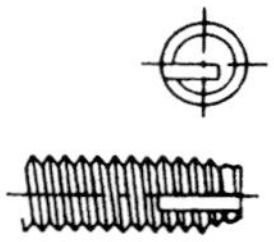 D	Blunt point with single narrow flute and threads approximating machine-screw threads. Flute is designed to produce a cutting edge which is radial to screw center. For low-strength metals and plastics; for high-strength brittle metals; and for rethreading clogged pretapped holes.
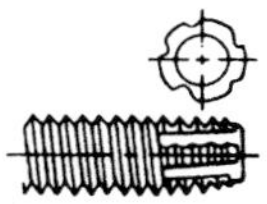 F	Approximate machine-screw thread and blunt point.
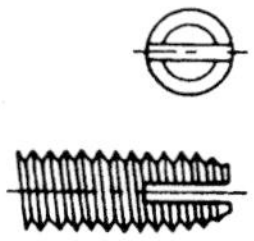 G	Approximate machine-screw thread with single through slot which forms two cutting edges. For low-strength metals and plastics.
 T	Same as type D with single wide flute for more chip clearance.
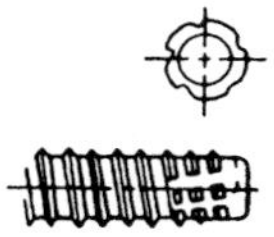 BF	Spaced thread with blunt point and five evenly spaced cutting grooves and chip cavities. Wall thickness should be 1½ times major diameter of screw. Reduces stripping in brittle plastics and die castings.
 BT	Same as type BF except for single wide flute which provides room for twisted, curly chips.
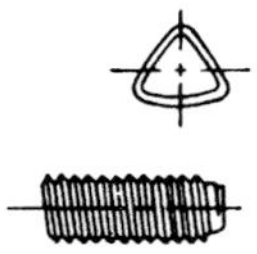 TT	Thread-rolling screws roll-form clean, screw threads. The plastic movement of the material it is driven into locks it in place. The Taptite form is shown here.

SOURCE: *Mach. Des.*, Mar. 11, 1965.

Table 6.2.23 Self-Tapping Screws

Screw size	Basic major diam, in	Threads per inch AB	B, BP	C	D, F, G, T	BF, BT	Type U Max outside diam, in	Type U Number of thread starts
00	—	—	—	—	—	—	0.060	6
0	0.060	40	48	—	—	48	0.075	6
1	0.073	32	42	—	—	42		
2	0.086	32	32	56 and 64	56 and 64	32	0.100	8
3	0.099	28	28	48 and 56	48 and 56	28		
4	0.112	24	24	40 and 48	40 and 48	24	0.116	7
5	0.125	20	20	40 and 44	40 and 44	20		
6	0.138	18	20	32 and 40	32 and 40	20	0.140	7
7	0.151	16	19	—	—	19	0.154	8
8	0.164	15	18	32 and 36	32 and 36	18	0.167	8
10	0.190	12	16	24 and 32	24 and 32	16	0.182	8
12	0.216	11	14	24 and 28	24 and 28	14	0.212	8
14	0.242	10	—	—	—	—	0.242	9
¼	0.250	—	14	20 and 28	20 and 28	14		
16	0.268	10						
18	0.294	9						
5/16	0.3125	—	12	18 and 24	18 and 24	12	0.315	14
20	0.320	9						
24	0.372	9						
⅜	0.375	—	12	16 and 24	16 and 24	12	0.378	12
7/16	0.4375	—	10	—	—	10		
½	0.500	—	10	—	—	10		

Table 6.2.24 ISO Metric Fastener Materials

Metric bolt	Metric nut class normally used	Roughly equivalent U.S. bolt materials: SAE J429 grades	ASTM grades
4.6	4 or 5	1	A193, B8; A307, grade A
4.8	4 or 5	2	
5.8	5	2	
8.8	8	5	A325; A449
9.9	9	5+	A193, B7 and B16
10.9	10 or 12	8	A490; A354, grade 8D
12.9	10 or 12		A540, B21 through B24

SOURCE: Bickford, "An Introduction to the Design and Behavior of Bolted Joints," Marcel Dekker, 1981; reproduced by permission. See Appendix G for additional metric materials.

General Notes on the Design of Bolted Joints

Bolts subjected to shock and sudden change in load are found to be more serviceable when the unthreaded portion of the bolt is turned down or drilled to the area of the root of the thread. The drilled bolt is stronger in torsion than the turned-down bolt.

When a number of bolts are employed in fastening together two parts of a machine, such as a cylinder and cylinder head, the load carried by each bolt depends on its relative tightness, the tighter bolts carrying the greater loads. When the conditions of assembly result in differences in tightness, lower working stresses must be used in designing the bolts than otherwise are necessary. On the other hand, it may be desirable to have the bolts the weakest part of the machine, since their breakage from overload in the machine may result in a minimum replacement cost. In such cases, the breaking load of the bolts may well be equal to the load which causes the weakest member of the machine connected to be stressed up to the elastic limit.

Bolts screwed up tight have an initial stress due to the tightening (preload) before any external load is applied to the machine member. The initial tensile load due to screwing up for a tight joint varies approximately as the diameter of the bolt, and may be estimated at 16,000 lb/in of diameter. The actual value depends upon the applied torque and the efficiency of the screw threads. Applying this rule to bolts of 1-in diam or less results in excessively high stresses, thus demonstrating why bolts of small diameter frequently fail during assembly.

It is advisable to use as large-diameter bolts as possible in pressure-tight joints requiring high tightening loads.

In pressure-tight joints without a gasket the force on the bolt under load is essentially never greater than the initial tightening load. When a gasket is used, the total bolt force is approximately equal to the initial tightening load plus the external load. In the first case, deviations from the rule are a result of elastic behavior of the joint faces without a gasket, and inelastic behavior of the gasket in the latter case. The following generalization will serve as a guide. If the bolt is more yielding than the connecting members, it should be designed simply to resist the initial tension or the external load, whichever is greater. If the probable yielding of the bolt is 50 to 100 percent of that of the connected members, take the resultant bolt load as the initial tension plus one-half the external load. If the yielding of the connected members is probably four to five times that of the bolt (as when certain packings are used), take the resultant bolt load as the initial tension plus three-fourths the external load.

In cases where bolts are subjected to cyclic loading, an increase in the initial tightening load decreases the operating stress range. In certain applications it is customary to fix the tightening load as a fraction of the yield-point load of the bolt.

In order to avoid the possibility of bolt failure in pressure-tight joints and to obtain uniformity in bolt loads, some means of determining initial bolt load **(preload)** is desirable. Calibrated torque wrenches are available for this purpose, reading directly in inch-pounds or inch-ounces. Inaccuracies in initial bolt load are possible even when using a torque wrench, owing to variations in the coefficient of friction between the nut and the bolt and, further, between the nut or bolthead and the abutting surface.

An exact method to determine the preload in a bolt requires that the bolt elongation be measured. For a through bolt in which both ends are accessible, the elongation is measured, and the preload force P is obtained from the relationship

$$P = AEe \div l$$

where E = modulus of elasticity, l = original length, A = cross-sectional area, e = elongation. In cases where both ends of the bolt are not accessible, strain-gage techniques may be employed to determine the strain in the bolt, and thence the preload.

Table 6.2.25a Specifications and Identification Markings for Bolts, Screws, Studs, Sems,[a] and U Bolts[b]
(Multiply the strengths in kpsi by 6.89 to get the strength in MPa.)

SAE grade	ASTM grade	Metric[c] grade	Nominal diameter, in	Proof strength, kpsi	Tensile strength, kpsi	Yield[d] strength, kpsi	Core hardness, Rockwell min/max	Products[e]
1	A307	4.6	¼ thru 1½	33	60	36	B70/B100	B, Sc, St
2		5.8	¼ thru ¾	55	74	57	B80/B100	B, Sc, St
		4.6	Over ¾ thru 1½	33	60	36	B70/B100	B, Sc, St
4		8.9	¼ thru 1½	65[f]	115	100	C22/C32	St
5	A449 or A325 type 1	8.8	¼ thru 1	85	120	92	C25/C34	B, Sc, St
		7.8	Over 1 thru 1½	74	105	81	C19/C30	B, Sc, St
		8.6	Over 1½ to 3	55	90	58		B, Sc, St
5.1		8.8	No. 6 thru ⅝	85	120		C25/C40	Se
		8.8	No. 6 thru ½	85	120		C25/C40	B, Sc, St
5.2	A325 type 2	8.8	¼ thru 1	85	120	92	C26/C36	B, Sc
7[g]		10.9	¾ thru 1½	105	133	115	C28/C34	B, Sc
8	A354 Grade BD	10.9	¼ thru 1½	120	150	130	C33/C39	B, Sc, St
8.1		10.9	¼ thru 1½	120	150	130	C32/C38	St
8.2		10.9	¼ thru 1	120	150	130	C35/C42	B, Sc
	A574	12.9	0 thru ½	140	180	160	C39/C45	SHCS
		12.9	⅝ thru 1½	135	170	160	C37/C45	SHCS

[a] Sems are screw and washer assemblies.
[b] Compiled from ANSI/SAE J429j; ANSI B18.3.1-1978; and ASTM A307, A325, A354, A449, and A574.
[c] Metric grade is *xx.x* where *xx* is approximately 0.01 S_{ut} in MPa and *.x* is the ratio of the minimum S_y to S_{ut}.
[d] Yield strength is stress at which a permanent set of 0.2% of gauge length occurs.
[e] B = bolt, Sc = Screws, St = studs, Se = sems, and SHCS = socket head cap screws.
[f] Entry appears to be in error but conforms to the standard, ANSI/SAE J429j.
[g] Grade 7 bolts and screws are roll-threaded after heat treatment.
NOTE: Company catalogs should be consulted regarding *proof loads*. However, approximate values of proof loads may be calculated from: proof load = proof strength × stress area.
SOURCE: Shigley and Mitchell, "Mechanical Engineering Design," 4th ed., McGraw-Hill, 1983, by permission.

Table 6.2.25b ASTM and SAE Grade Head Markings for Steel Bolts and Screws

Grade marking	Specification and material	Grade marking	Specification and material
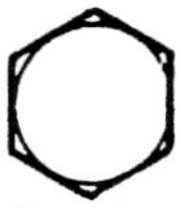 No marks	SAE grade 1 Low- or medium-carbon steel ASTM A307 Low-carbon steel SAE grade 2 Low- or medium-carbon steel	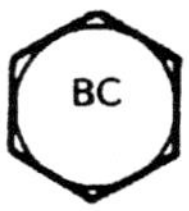	ASTM A354 grade BC Alloy steel, quenched and tempered
	SAE grade 5 ASTM A449 Medium-carbon steel, quenched and tempered		SAE grade 7 Medium-carbon alloy steel, quenched and tempered, roll-threaded after heat treatment
	SAE grade 5.2 Low-carbon martensite steel, quenched and tempered		SAE grade 8 Medium-carbon alloy steel, quenched and tempered ASTM A354 grade BD Alloy steel, quenched and tempered
	ASTM A325 type 1 Medium-carbon steel, quenched and tempered; radial dashes optional		SAE grade 8.2 Low-carbon martensite steel, quenched and tempered
	ASTM A325 type 2 Low-carbon martensite steel, quenched and tempered		ASTM A490 type 1 Alloy steel, quenched and tempered
	ASTM A325 type 3 Atmospheric corrosion (weathering) steel, quenched and tempered		ASTM A490 type 3 Atmospheric corrosion (weathering) steel, quenched and tempered

SOURCE: ANSI B18.2.1–1981 (R92), Appendix III, p. 41. By permission.

High-strength bolts designated ASTM A325 and A490 are almost exclusively employed in the assembly of structural steel members, but they are applied in mechanical assemblies such as flanged joints. The direct tension indicator (DTI) (Fig. 6.2.19), for use with high-strength bolts, allows **bolt preload** to be applied rapidly and simply. The device is a hardened washer with embossed protrusions (Fig. 6.2.19*a*). Tightening the bolt causes the protrusions to flatten and results in a decrease in the gap between washer and bolthead. The prescribed degree of bolt-tightening load, or preload, is obtained when the gap is reduced to a predetermined amount (Fig. 6.2.19*c*). A feeler gauge of a given thickness is used to determine when the gap has been closed to the prescribed amount (Fig. 6.2.19*b* and *c*). With a paired bolt and DTI, the degree of gap closure is proportional to bolt preload. The system is reported to provide bolt-preload force accuracy within +15 percent of that prescribed (Fig. 6.2.19*d*). The devices are available in both inch and metric series and are covered under ASTM F959 and F959M.

Preload-indicating bolts and nuts provide visual assurance of preload in that tightening to the desired preload causes the wavy flange to flatten flush with the clamped assembly (Fig. 6.2.20).

In drilling and tapping cast iron for steel studs, it is necessary to tap to a depth equal to 1½ times the stud diameter so that the strength of the cast-iron threads in shear may equal the tensile strength of the stud. Drill sizes and depths of hole and thread are given in Table 6.2.26.

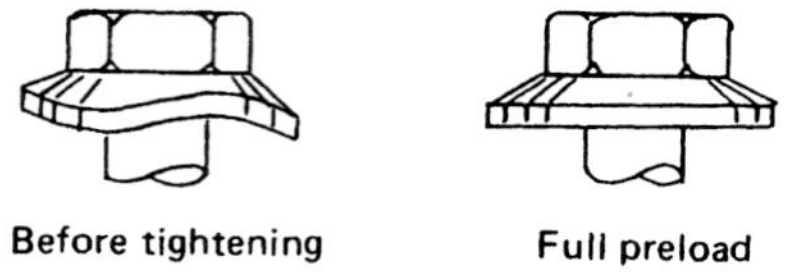

Figure 6.2.20 Load-indicating wavy-flange bolt (or nut).

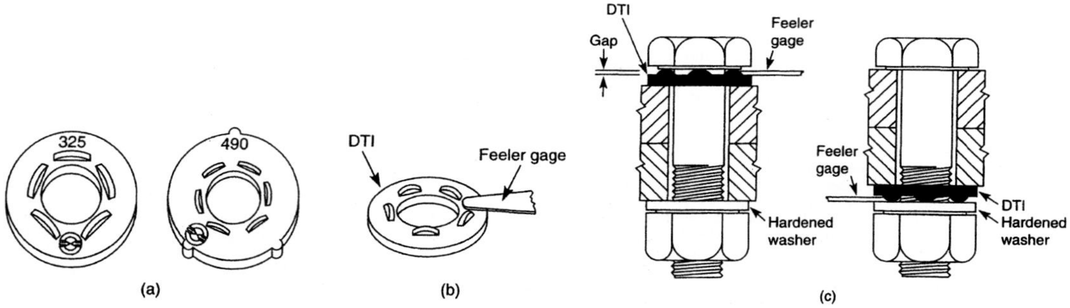

DTI Gaps To Give Required Minimum Bolt Tension

DTI Fitting	325	490
Under bolt head		
Plain finish DTIs	0.015″ (.40 mm)	0.015″
Mechanically galvanized DTIs	0.005″ (.125 mm)	—
Epoxy coated on mechanically galvanized DTIs	0.005″ (.125 mm)	—
Under turned Element		
With hardened washers (plain finish)	0.005″ (.125 mm)	0.005″

With average gaps equal or less than above, bolt tensions will be greater than in adjacent listing

Minimum Bolt Tensions
In thousands of pounds (Kips)

Bolt dia.	A325	A490
1/2″	12	—
5/8″	19	—
3/4″	28	35
7/8″	39	49
1″	51	64
1 1/8″	56	80
1 1/4″	71	102
1 3/8″	85	121
1 1/2″	103	—

(d)

Figure 6.2.19 Direct tension indicators; gap and tension data. *(Source: J & M Turner Inc.)*

Table 6.2.26 Depths to Drill and Tap Cast Iron for Studs

Diam of stud, in	1/4	5/16	3/8	7/16	1/2	9/16	5/8	3/4	7/8	1
Diam of drill, in	13/64	17/64	5/16	3/8	27/64	31/64	17/32	41/64	3/4	55/64
Depth of thread, in	3/8	15/32	9/16	21/32	3/4	27/32	15/16	1 1/8	1 5/16	1 1/2
Depth to drill, in	7/16	17/32	5/8	23/32	27/32	15/16	1 1/32	1 1/4	1 7/16	1 5/8

It is not good practice to drill holes to be tapped through the metal into pressure spaces, for even though the bolt fits tightly, leakage will result that is difficult to eliminate.

Self-sealing Screws and Nuts Screws and bolts used for vacuum or pressure chambers are sealed by placing elastic, soft metallic, or plastic materials in the shape of an O ring or a washer under the head of a screw or close to the threads of its nut. By tightening the screw or the bolt, the deformable O ring or the soft washer fills the space between the threads of the screw and the member and seals the chamber. For further information, refer to manufacturers.

Screw thread inserts made of high-strength material (Fig. 6.2.21) are useful in many cases to provide increased thread strength and life. Soft

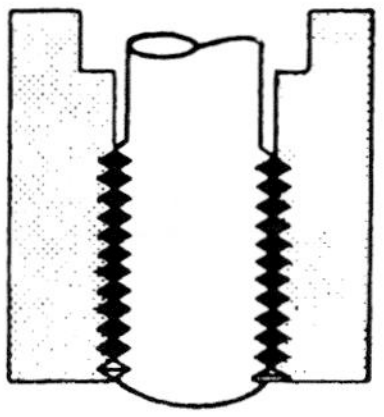

Figure 6.2.21 Screw thread insert.

or ductile materials tapped to receive thread inserts exhibit improved load-carrying capacity under static and dynamic loading conditions. Holes in which threads have been stripped or otherwise damaged can be restored through the use of thread inserts.

Holes for thread inserts are drilled oversize and specially tapped to receive the insert selected to mate with the threaded fastener used. The standard material for inserts is 18-8 stainless steel, but other materials are available, such as phosphor bronze and Inconel. Recommended insert lengths are given in Table 6.2.27.

Drill Sizes Unified thread taps are listed in Table 6.2.28.

Table 6.2.27 Screw-Thread Insert Lengths
(Heli-Coil Corp.)

Shear strength of parent material, lb/in²	Bolt material ultimate tensile strength, lb/in²				
	60,000	90,000	125,000	170,000	220,000
	Length in terms of nominal insert diameter				
15,000	$1\frac{1}{2}$	2	$2\frac{1}{2}$	3	
20,000	1	$1\frac{1}{2}$	2	$2\frac{1}{2}$	3
25,000	1	$1\frac{1}{2}$	2	2	$2\frac{1}{2}$
30,000	1	1	$1\frac{1}{2}$	2	2
40,000	1	1	$1\frac{1}{2}$	$1\frac{1}{2}$	2
50,000	1	1	1	$1\frac{1}{2}$	$1\frac{1}{2}$

Table 6.2.28 Tap-Drill Sizes for American National Standard Screw Threads
(The sizes listed are the commercial tap drills to produce approx 75% full thread)

Size	Coarse-thread series: Threads per inch	Coarse-thread series: Tap drill size	Fine-thread series: Threads per inch	Fine-thread series: Tap drill size	Size	Coarse-thread series: Threads per inch	Coarse-thread series: Tap drill size	Fine-thread series: Threads per inch	Fine-thread series: Tap drill size
No. 0	—	—	80	$\frac{3}{64}$	$\frac{3}{4}$	10	$\frac{21}{32}$	16	$\frac{11}{16}$
No. 1	64	No. 53	72	No. 53	$\frac{7}{8}$	9	$\frac{49}{64}$	14	$\frac{13}{16}$
No. 2	56	No. 50	64	No. 50	1	8	$\frac{7}{8}$	14	$\frac{15}{16}$
No. 3	48	No. 47	56	No. 45	$1\frac{1}{8}$	7	$\frac{63}{64}$	12	$1\frac{3}{64}$
No. 4	40	No. 43	48	No. 42	$1\frac{1}{4}$	7	$1\frac{7}{64}$	12	$1\frac{11}{64}$
No. 5	40	No. 38	44	No. 37	$1\frac{3}{8}$	6	$1\frac{7}{32}$	12	$1\frac{19}{64}$
No. 6	32	No. 36	40	No. 33	$1\frac{1}{2}$	6	$1\frac{21}{64}$	12	$1\frac{27}{64}$
No. 8	32	No. 29	36	No. 29	$1\frac{3}{4}$	5	$1\frac{35}{64}$		
No. 10	24	No. 25	32	No. 21	2	$4\frac{1}{2}$	$1\frac{25}{32}$		
No. 12	24	No. 16	28	No. 14	$2\frac{1}{4}$	$4\frac{1}{2}$	$2\frac{1}{32}$		
$\frac{1}{4}$	20	No. 7	28	No. 3	$2\frac{1}{2}$	4	$2\frac{1}{4}$		
$\frac{5}{16}$	18	F	24	I	$2\frac{3}{4}$	4	$2\frac{1}{2}$		
$\frac{3}{8}$	16	$\frac{5}{16}$	24	Q	3	4	$2\frac{3}{4}$		
$\frac{7}{16}$	14	U	20	$\frac{25}{64}$	$3\frac{1}{4}$	4	3		
$\frac{1}{2}$	13	$\frac{27}{64}$	20	$\frac{29}{64}$	$3\frac{1}{2}$	4	$3\frac{1}{4}$		
$\frac{9}{16}$	12	$\frac{31}{64}$	18	$\frac{33}{64}$	$3\frac{3}{4}$	4	$3\frac{1}{2}$		
$\frac{5}{8}$	11	$\frac{17}{32}$	18	$\frac{37}{64}$	4	4	$3\frac{3}{4}$		

RIVET FASTENINGS

Forms and Proportion of Rivets The forms and proportions of small and large rivets have been standardized and conform to ANSI B18.1.1-1972 (R89) and B18.1.2-1972 (R89) (Fig. 6.2.22*a* and *b*).

Materials Specifications for Rivets and Plates See Secs. 6 and 12.2.

Conventional signs to indicate the form of the head to be used and whether the rivet is to be driven in the shop or the field at the time of erection are given in Fig. 6.2.23. **Rivet lengths** and **grips** are shown in Fig. 6.2.22*b*.

For **structural riveting,** see Sec. 4.11.

Punched vs. Drilled Plates Holes in plates forming parts of riveted structures are punched, punched and reamed, or drilled. Punching, while cheaper, is objectionable. The holes in different plates cannot be spaced with sufficient accuracy to register perfectly on being assembled. If the hole is punched out, say $\frac{1}{16}$ in smaller than is required and then reamed to size, the metal injury by cold flow during punching will be removed. Drilling, while more expensive, is more accurate and does not injure the metal.

Tubular Rivets

In tubular rivets, the end opposite the head is made with an axial hole (partway) to form a thin-walled, easily upsettable end. As the material at the edge of the rivet hole is rolled over against the surface of the joint, a **clinch** is formed (Fig. 6.2.24*a*).

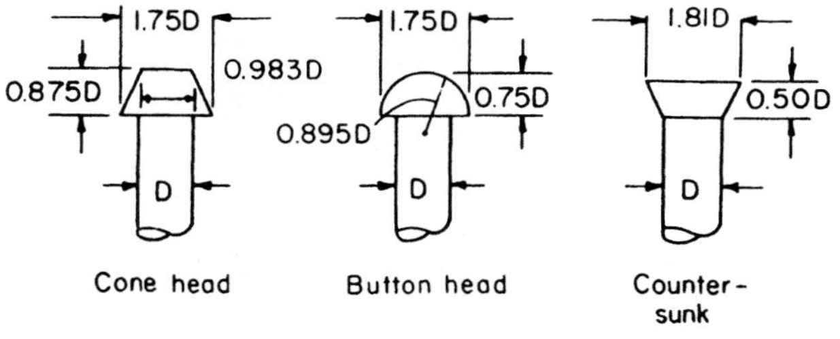

Figure 6.2.22a Rivet heads.

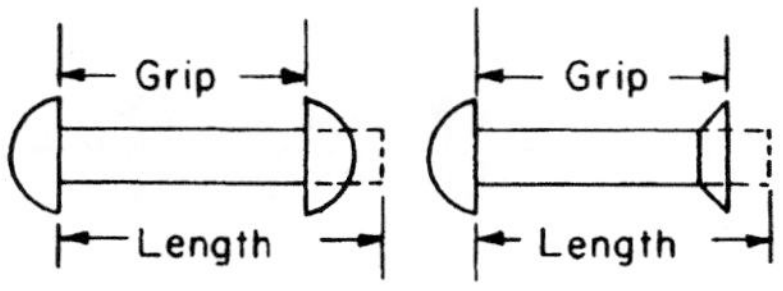

Figure 6.2.22b Rivet length and grip.

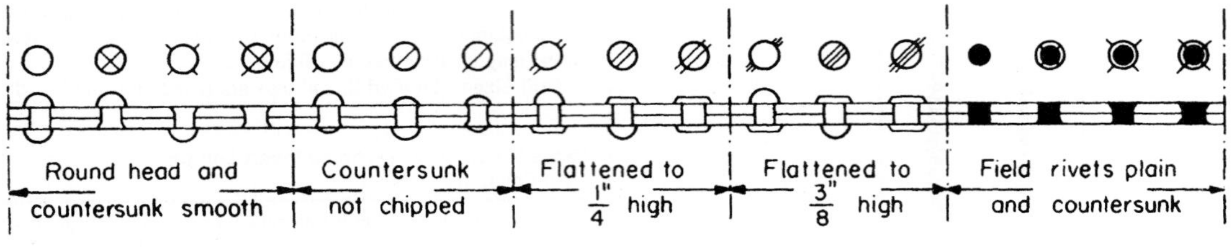

Figure 6.2.23 Conventional signs for rivets.

Two-part tubular rivets have a thin-walled head with attached thin-walled rivet body and a separate thin-walled expandable plug. The head-body is inserted through a hole in the joint from one side, and the plug from the other. By holding an anvil against the plug bottom and hammering on the head, the plug is caused to expand within the head, thus locking both parts together (Fig. 6.2.24*a*).

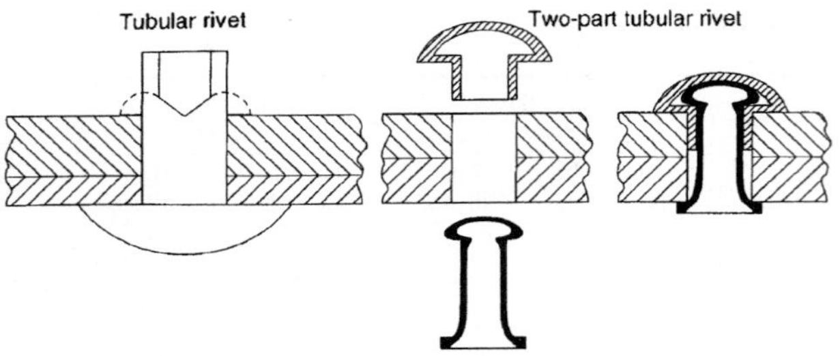

Figure 6.2.24*a* Tubular rivets.

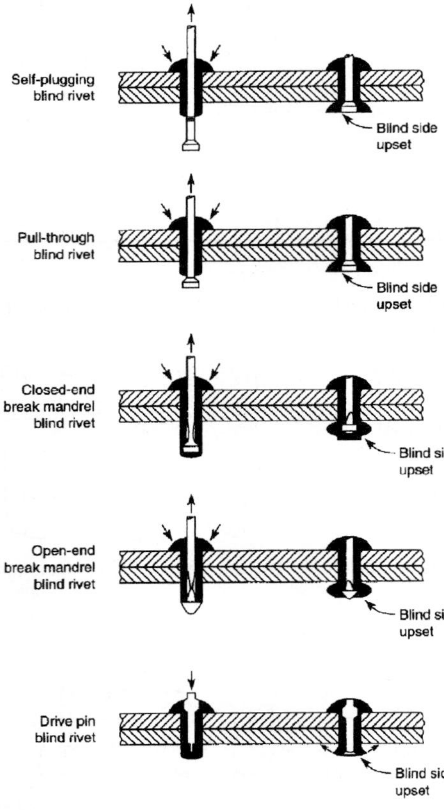

Figure 6.2.24*b* Blind rivets.

Blind Rivets

Blind rivets are inserted and set all from one side of a structure. This is accomplished by mechanically expanding, through the use of the rivet's built-in mandrel, the back (blind side) of the rivet into a bulb or upset head after insertion. Blind rivets include the *pull type* and *drive-pin type.*

The pull-type rivet is available in two configurations: a self-plugging type and a pull-through type. In the self-plugging type, part of the mandrel remains permanently in the rivet body after setting, contributing additional shear strength to the fastener. In the pull-through type, the entire mandrel is pulled through, leaving the installed rivet empty.

In a drive-pin rivet, the rivet body is slotted. A pin is driven forward into the rivet, causing both flaring of the rivet body and upset of the blind side (Fig. 6.2.24*b*).

KEYS, PINS, AND COTTERS

Keys and key seats have been standardized and are listed in ANSI B17.1-1967 (R89). Descriptions of the principal key types follow.

Woodruff keys [ANSI B17.2-1967 (R90)] are made to facilitate removal of pulleys from shafts. They should not be used as sliding keys. Cutters for milling out the key seats, as well as special machines for using the cutters, are to be had from the manufacturer. Where the hub of the gear or pulley is relatively long, two keys should be used. Slightly rounding the corners or ends of these keys will obviate any difficulty met with in removing pulleys from shafts. The key is shown in Fig. 6.2.25 and the dimensions in Table 6.2.29.

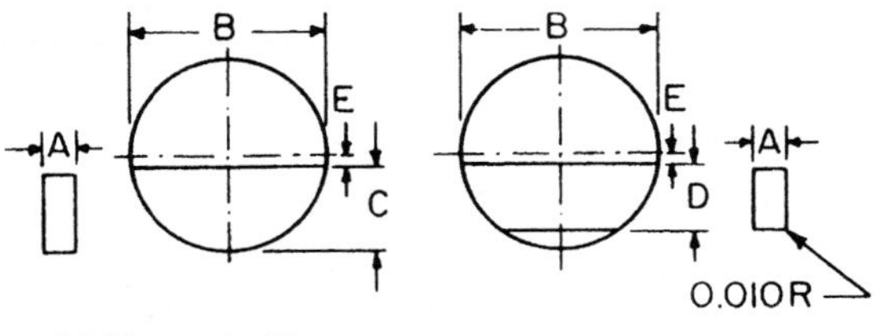

Figure 6.2.25 Woodruff key.

Square and flat plain taper keys have the same dimensions as gib-head keys (Table 6.2.30) up to the dotted line of Fig. 6.2.26. **Gib-head keys** (Fig. 6.2.26) are necessary when the smaller end is inaccessible for drifting out and the larger end is accessible. It can be used, with care, with all sizes of shafts. Its use is forbidden in certain jobs and places for safety reasons. Proportions are given in Table 6.2.30.

The minimum stock length of keys is 4 times the key width, and maximum stock length of keys is 16 times the key width. The increments of increase of length are 2 times the width.

Sunk keys are made to the form and dimensions given in Fig. 6.2.27 and Table 6.2.31. These keys are adapted particularly to the case of fitting adjacent parts with neither end of the key accessible. **Feather keys** prevent parts from turning on a shaft while allowing them to move in a lengthwise direction. They are of the forms shown in Fig. 6.2.28 with the same base dimensions as given in Table 6.2.31.

Table 6.2.29 Woodruff Key Dimensions [ANSI B17.2-1967 (R90)]
(All dimensions in inches)

Key no.	Nominal key size, $A \times B$	Width of key A Max	Width of key A Min	Diam of key B Max	Diam of key B Min	Height of key C Max	Height of key C Min	Height of key D Max	Height of key D Min	Distance below E
204	1/16 × 1/2	0.0635	0.0625	0.500	0.490	0.203	0.198	0.194	0.188	3/64
304	3/32 × 1/2	0.0948	0.0928	0.500	0.490	0.203	0.198	0.194	0.188	3/64
305	3/32 × 5/8	0.0948	0.0938	0.625	0.615	0.250	0.245	0.240	0.234	1/16
404	1/8 × 1/2	0.1260	0.1250	0.500	0.490	0.203	0.198	0.194	0.188	3/64
405	1/8 × 5/8	0.1260	0.1250	0.625	0.615	0.250	0.245	0.240	0.234	1/16
406	1/8 × 3/4	0.1260	0.1250	0.750	0.740	0.313	0.308	0.303	0.297	1/16
505	5/32 × 5/8	0.1573	0.1563	0.625	0.615	0.250	0.245	0.240	0.234	1/16
506	5/32 × 3/4	0.1573	0.1563	0.750	0.740	0.313	0.308	0.303	0.297	1/16
507	5/32 × 7/8	0.1573	0.1563	0.875	0.865	0.375	0.370	0.365	0.359	1/16
606	3/16 × 3/4	0.1885	0.1875	0.750	0.740	0.313	0.308	0.303	0.297	1/16
607	3/16 × 7/8	0.1885	0.1875	0.875	0.865	0.375	0.370	0.365	0.359	1/16
608	3/16 × 1	0.1885	0.1875	1.000	0.990	0.438	0.433	0.428	0.422	1/16
609	3/16 × 1 1/8	0.1885	0.1875	1.125	1.115	0.484	0.479	0.475	0.469	5/64
807	1/4 × 7/8	0.2510	0.2500	0.875	0.865	0.375	0.370	0.365	0.359	1/16
808	1/4 × 1	0.2510	0.2500	1.000	0.990	0.438	0.433	0.428	0.422	1/16
809	1/4 × 1 1/8	0.2510	0.2500	1.125	1.115	0.484	0.479	0.475	0.469	5/64
810	1/4 × 1 1/4	0.2510	0.2500	1.250	1.240	0.547	0.542	0.537	0.531	5/64
811	1/4 × 1 3/8	0.2510	0.2500	1.375	1.365	0.594	0.589	0.584	0.578	3/32
812	1/4 × 1 1/2	0.2510	0.2500	1.500	1.490	0.641	0.636	0.631	0.625	7/64
1,008	5/16 × 1	0.3135	0.3125	1.000	0.990	0.438	0.433	0.428	0.422	1/16
1,009	5/16 × 1 1/8	0.3135	0.3125	1.125	1.115	0.484	0.479	0.475	0.469	5/64
1,010	5/16 × 1 1/4	0.3135	0.3125	1.250	1.240	0.547	0.542	0.537	0.531	5/64
1,011	5/16 × 1 3/8	0.3135	0.3125	1.375	1.365	0.594	0.589	0.584	0.578	3/32
1,012	5/16 × 1 1/2	0.3135	0.3125	1.500	1.490	0.641	0.636	0.631	0.625	7/64
1,210	3/8 × 1 1/4	0.3760	0.3750	1.250	1.240	0.547	0.542	0.537	0.531	5/64
1,211	3/8 × 1 3/8	0.3760	0.3750	1.375	1.365	0.594	0.589	0.584	0.578	3/32
1,212	3/8 × 1 1/2	0.3760	0.3750	1.500	1.490	0.641	0.636	0.631	0.625	7/64

Numbers indicate the nominal key dimensions. The last two digits give the nominal diameter (B) in eighths of an inch, and the digits preceding the last two give the nominal width (A) in thirty-seconds of an inch. Thus, 204 indicates a key 2/32 × 4/8 or 1/16 × 1/2 in; 1210 indicates a key 12/32 × 10/8 or 3/8 × 1 1/4 in.

Table 6.2.30 Dimensions of Square and Flat Gib-Head Taper Stock Keys,* in

Shaft diam	Square type, Key: Max width W	Square type, Key: Height at large end,† H	Square type, Gib head: Height C	Square type, Gib head: Length D	Square type, Gib head: Height edge of chamfer E	Flat type, Key: Max width W	Flat type, Key: Height at large end,† H	Flat type, Gib head: Height C	Flat type, Gib head: Length D	Flat type, Gib head: Height edge of chamfer E	Tolerance: On width (−)	Tolerance: On height (+)
1/2-9/16	1/8	1/8	1/4	7/32	5/32	1/8	3/32	3/16	1/8	1/8	0.0020	0.0020
5/8-7/8	3/16	3/16	5/16	9/32	7/32	3/16	1/8	1/4	3/16	5/32	0.0020	0.0020
15/16-1 1/4	1/4	1/4	7/16	11/32	11/32	1/4	3/16	5/16	1/4	3/16	0.0020	0.0020
1 5/16-1 3/8	5/16	5/16	9/16	13/32	13/32	5/16	1/4	3/8	5/16	1/4	0.0020	0.0020
1 7/16-1 3/4	3/8	3/8	11/16	15/32	15/32	3/8	1/4	7/16	3/8	5/16	0.0020	0.0020
1 13/16-2 1/4	1/2	1/2	7/8	19/32	5/8	1/2	3/8	5/8	1/2	7/16	0.0025	0.0025
2 5/16-2 3/4	5/8	5/8	1 1/16	23/32	3/4	5/8	7/16	3/4	5/8	1/2	0.0025	0.0025
2 7/8-3 1/4	3/4	3/4	1 1/4	7/8	7/8	3/4	1/2	7/8	3/4	5/8	0.0025	0.0025
3 3/8-3 3/4	7/8	7/8	1 1/2	1	1	7/8	5/8	1 1/16	7/8	3/4	0.0030	0.0030
3 7/8-4 1/2	1	1	1 3/4	1 3/16	1 3/16	1	3/4	1 1/4	1	13/16	0.0030	0.0030
4 3/4-5 1/2	1 1/4	1 1/4	2	1 7/16	1 7/16	1 1/4	7/8	1 1/2	1 1/4	1	0.0030	0.0030
5 3/4-6	1 1/2	1 1/2	2 1/2	1 3/4	1 3/4	1 1/2	1	1 3/4	1 1/2	1 1/4	0.0030	0.0030

* Stock keys are applicable to the general run of work and the tolerances have been set accordingly. They are not intended to cover the finer applications where a closer fit may be required.
† This height of the key is measured at the distance W equal to the width of the key, from the gib head.

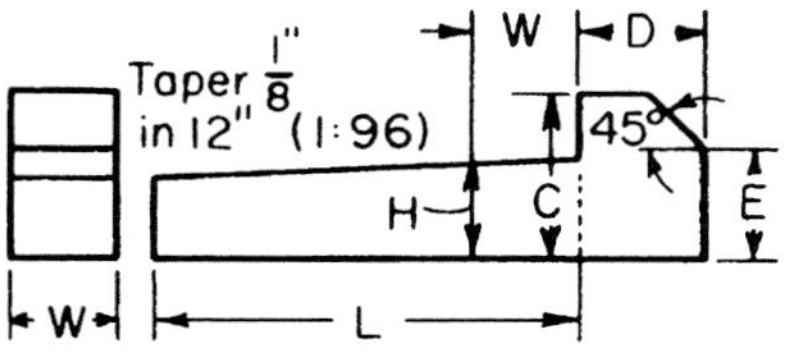

Figure 6.2.26 Gib-head taper stock key.

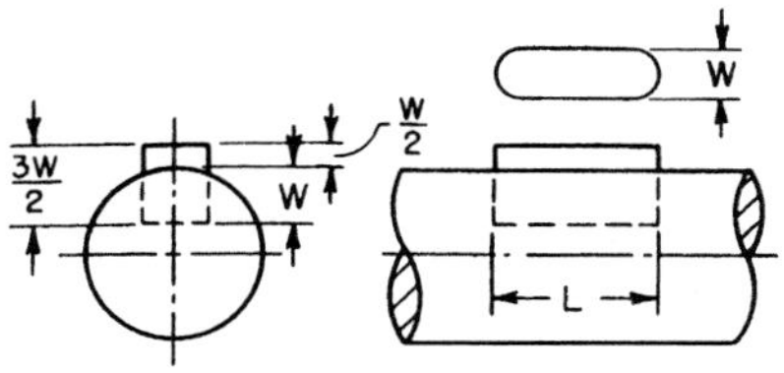

Figure 6.2.27 Sunk key.

Table 6.2.31 Dimensions of Sunk Keys
(All dimensions in inches. Letters refer to Fig. 6.2.24)

Key no.	*L*	*W*	Key no.	*L*	*W*	Key no.	*L*	*W*	Key no.	*L*	*W*
1	1/2	1/16	13	1	3/16	22	1 3/8	1/4	54	2 1/4	1/4
2	1/2	3/32	14	1	7/32	23	1 3/8	5/16	55	2 1/4	5/16
3	1/2	1/8	15	1	1/4	F	1 3/8	3/8	56	2 1/4	3/8
4	5/8	3/32	B	1	5/16	24	1 1/2	1/4	57	2 1/4	7/16
5	5/8	1/8	16	1 1/8	3/16	25	1 1/2	5/16	58	2 1/2	5/16
6	5/8	5/32	17	1 1/8	7/32	G	1 1/2	3/8	59	2 1/2	3/8
7	3/4	1/8	18	1 1/8	1/4	51	1 3/4	1/4	60	2 1/2	7/16
8	3/4	5/32	C	1 1/8	5/16	52	1 3/4	5/16	61	2 1/2	1/2
9	3/4	3/16	19	1 1/4	3/16	53	1 3/4	3/8	30	3	3/8
10	7/8	5/32	20	1 1/4	7/32	26	2	3/16	31	3	7/16
11	7/8	3/16	21	1 1/4	1/4	27	2	1/4	32	3	1/2
12	7/8	7/32	D	1 1/4	5/16	28	2	5/16	33	3	9/16
A	7/8	1/4	E	1 1/4	3/8	29	2	3/8	34	3	5/8

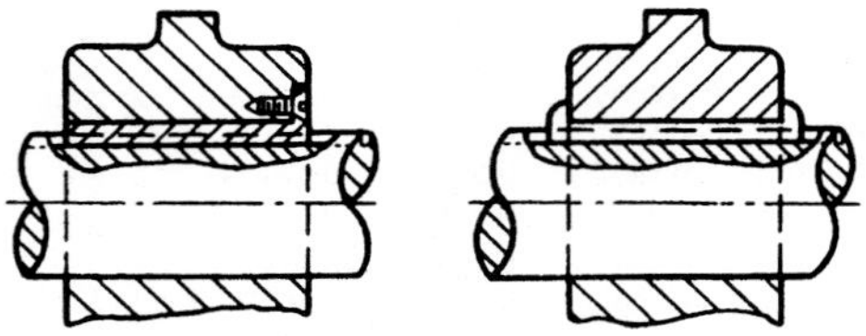

Figure 6.2.28 Feather key.

In **transmitting large torques,** it is customary to use two or more keys.

Another means for fastening gears, pulleys flanges, etc., to shafts is through the use of mating pairs of tapered sleeves known as **grip springs.** A set of sleeves is shown in Fig. 6.2.29. For further references see data issued by the Ringfeder Corp.

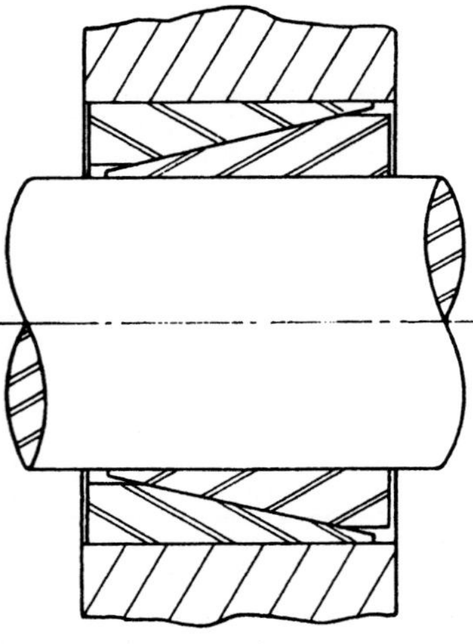

Figure 6.2.29 Grip springs.

Tapered pins (Fig. 6.2.30) can be used to transmit very small torques or for positioning. They should be fitted so that the parts are drawn together to prevent their working loose when the pin is driven home. Table 6.2.32 gives dimensions of Morse tapered pins.

The Groov-Pin Corp., New Jersey, has developed a special **grooved pin** (Fig. 6.2.31) which may be used instead of smooth taper pins in certain cases.

Straight pins, likewise, are used for transmission of light torques or for positioning. **Spring pins** have been used widely for many years. Two types shown in Figs. 6.2.32 and 6.2.33 deform elastically in the radial direction when driven; the resiliency of the pin material locks the pin in place. They can replace straight and taper pins and combine the advantages of both, i.e., simple tooling, ease of removal, reusability, ability to be driven from either side.

Cotter pins (Fig. 6.2.34) are used to secure or lock nuts, clevises, etc. Driven into holes in the shaft, the eye prevents complete passage, and the split ends, deformed after insertion, prevent withdrawal.

When two rods are to be joined so as to permit movement at the joint, a round pin is used in place of a cotter. In such cases, the proportions may be as shown in Fig. 6.2.35 (knuckle joint).

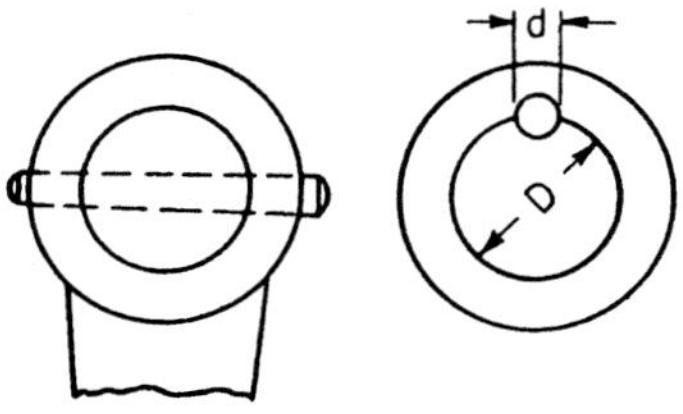

Figure 6.2.30 Taper pins.

Figure 6.2.31 Grooved pins.

Figure 6.2.32 Roll pin.

Figure 6.2.33 Spiral pins.

Figure 6.2.34 Cotter pin.

Table 6.2.32 Morse Standard Taper Pins
(Taper, 1/8 in/ft. Lengths increase by 1/4 in. Dimensions in inches)

Size no.	0	1	2	3	4	5	6	7	8	9	10
Diam at large end	0.156	0.172	0.193	0.219	0.250	0.289	0.341	0.409	0.492	0.591	0.706
Length	0.5-3	0.5-3	0.75-3.5	0.75-3.5	0.75-4	0.75-4	0.75-5	1-5	1.25-5	1.5-6	1.5-6

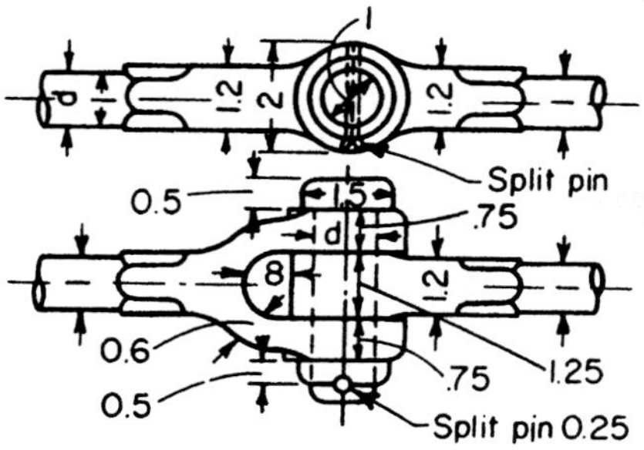

Figure 6.2.35 Knuckle joint.

SPLINES

Involute spline proportions, dimensions, fits, and tolerances are given in detail in ANSI B92.1-1970. External and internal involute splines (Fig. 6.2.36) have the same general form as involute gear teeth, except that the teeth are one-half the depth of standard gear teeth and the pressure angle is 30°. The spline is designated by a fraction in which the numerator is the diametral pitch and the denominator is always twice the numerator.

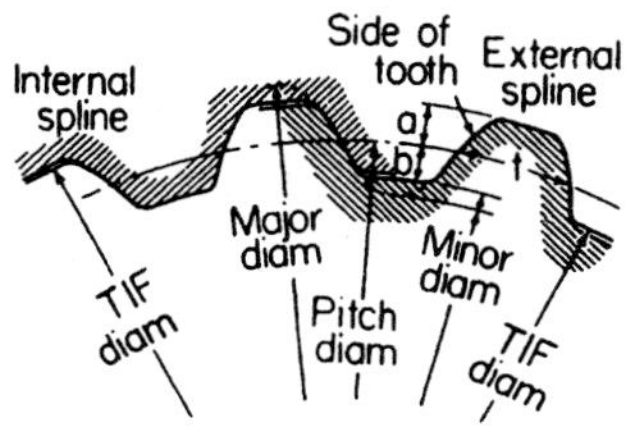

Figure 6.2.36 Involute spline.

There are 17 series, as follows: 2.5/5, 3/6, 4/8, 5/10, 6/12, 8/16, 10/20, 12/24, 16/32, 20/40, 24/48, 32/64, 40/80, 48/96, 64/128, 80/160, 128/256. The number of teeth within each series varies from 6 to 50. Both a flat-root and a fillet-root type are provided. There are three **types of fits:** (1) **major diameter**—fit controlled by varying the major diameter of the external spline; (2) **sides of teeth**—fit controlled by varying tooth thickness and customarily used for fillet-root splines; (3) **minor diameter**—fit controlled by varying the minor diameter of the internal spline. Each type of fit is further divided into three classes: (*a*) **sliding**—clearance at all points; (*b*) **close**—close on either major diameter, sides of teeth, or minor diameter; (*c*) **press**—interference on either the major diameter, sides of teeth, or minor diameter. Important basic formulas for tooth proportions are:

D = pitch diam
N = number of teeth
P = diametral pitch
p = circular pitch
t = circular tooth thickness
a = addendum
b = dedendum
D_O = major diam
TIF = true involute form diam
D_R = minor diam

Flat and Fillet Roots

$$D = N/P$$
$$p = \pi/P$$
$$t = p/2$$
$$a = 0.5000/P$$
$$D_O\ (\text{external}) = \frac{N+1}{P}$$
$$\text{TIF (internal)} = \frac{N+1}{P}$$
$$D_R = \frac{N+1}{P} \qquad \text{(minor-diameter fits only)}$$
$$\text{TIF (external)} = \frac{N-1}{P}$$

$b = 0.600/P + 0.002$ (For major-diameter fits, the internal spline dedendum is the same as the addendum; for minor-diameter fits, the dedendum of the external spline is the same as the addendum.)

Fillet Root Only

½ through ¹²⁄₂₄	¹⁶⁄₃₂ through ⁴⁸⁄₉₆
$D_O\,(\text{internal}) = \frac{N+1.8}{P}$	$D_O\,(\text{internal}) = \frac{N+1.8}{P}$
$D_R\,(\text{external}) = \frac{N-1.8}{P}$	$D_R\,(\text{external}) = \frac{N-2}{P}$
$b\,(\text{internal}) = 0.900/P$	$b\,(\text{internal}) = 0.900/P$
$b\,(\text{external}) = 0.900/P$	$b\,(\text{external}) = 1.000/P$

Internal spline dimensions are basic while external spline dimensions are varied to control fit.

The advantages of involute splines are: (1) maximum strength at the minor diameter, (2) self-centering equalizes bearing and stresses among all teeth, and (3) ease of manufacture through the use of standard gear-cutting tools and methods.

The design of involute splines is critical in shear. The torque capacity may be determined by the formula $T = LD^2S_s/1.2732$, where L = spline length, D = pitch diam, S_s = allowable shear stress.

Parallel-side splines have been standardized by the SAE for 4, 6, 10, and 16 spline fittings. They are shown in Fig. 6.2.37; pertinent data are in Tables 6.2.33 and 6.2.34.

DRY AND VISCOUS COUPLINGS

A **coupling** makes a semipermanent connection between two shafts. They are of three main types: **rigid, flexible,** and **fluid.**

Rigid Couplings

Rigid couplings are used only on shafts which are perfectly aligned. The **flanged-face coupling** (Fig. 6.2.38) is the simplest of these. The

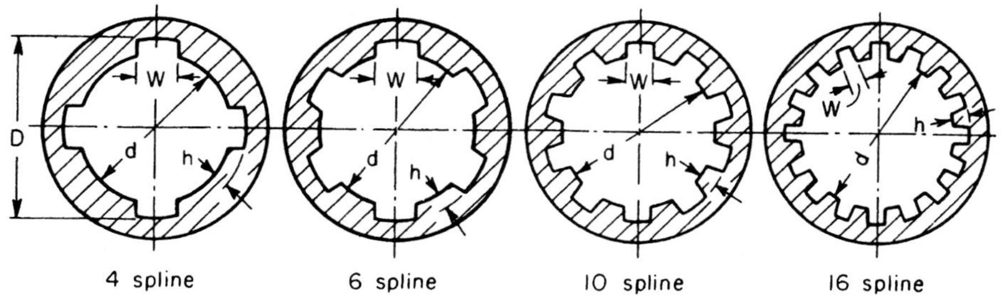

Figure 6.2.37 Parallel-sided splines.

Table 6.2.33 Dimensions of Spline Fittings, in
(SAE Standard)

Nominal diam	4-spline for all fits		6-spline for all fits		10-spline for all fits		16-spline for all fits	
	D max*	*W* max†	*D* max*	*W* max†	*D* max*	*W* max†	*D* max*	*W* max†
¾	0.750	0.181	0.750	0.188	0.750	0.117		
⅞	0.875	0.211	0.875	0.219	0.875	0.137		
1	1.000	0.241	1.000	0.250	1.000	0.156		
1⅛	1.125	0.271	1.125	0.281	1.125	0.176		
1¼	1.250	0.301	1.250	0.313	1.250	0.195		
1⅜	1.375	0.331	1.375	0.344	1.375	0.215		
1½	1.500	0.361	1.500	0.375	1.500	0.234		
1⅝	1.625	0.391	1.625	0.406	1.625	0.254		
1¾	1.750	0.422	1.750	0.438	1.750	0.273		
2	2.000	0.482	2.000	0.500	2.000	0.312	2.000	0.196
2¼	2.250	0.542	2.250	0.563	2.250	0.351		
2½	2.500	0.602	2.500	0.625	2.500	0.390	2.500	0.245
3	3.000	0.723	3.000	0.750	3.000	0.468	3.000	0.294
3½	—	—	—	—	3.500	0.546	3.500	0.343
4	—	—	—	—	4.000	0.624	4.000	0.392
4½	—	—	—	—	4.500	0.702	4.500	0.441
5	—	—	—	—	5.000	0.780	5.000	0.490
5½	—	—	—	—	5.500	0.858	5.500	0.539
6	—	—	—	—	6.000	0.936	6.000	0.588

* Tolerance allowed of −0.001 in for shafts ¾ to 1¾ in, inclusive; of −0.002 for shafts 2 to 3 in, inclusive; −0.003 in for shafts 3½ to 6 in, inclusive, for 4-, 6-, and 10-spline fittings; tolerance of −0.003 in allowed for all sizes of 16-spline fittings.
† Tolerance allowed of −0.002 in for shafts ¾ in to 1¾ in, inclusive; of −0.003 in for shafts 2 to 6 in, inclusive, for 4-, 6-, and 10-spline fittings; tolerance of −0.003 allowed for all sizes of 16-spline fittings.

Table 6.2.34 Spline Proportions

No. of splines	*W* for all fits	Permanent fit		To slide when not under load		To slide under load	
		h	*d*	*h*	*d*	*h*	*d*
4	0.241*D*	0.075*D*	0.850*D*	0.125*D*	0.750*D*		
6	0.250*D*	0.050*D*	0.900*D*	0.075*D*	0.850*D*	0.100*D*	0.800*D*
10	0.156*D*	0.045*D*	0.910*D*	0.070*D*	0.860*D*	0.095*D*	0.810*D*
16	0.098*D*	0.045*D*	0.910*D*	0.070*D*	0.860*D*	0.095*D*	0.810*D*

flanges must be keyed to the shafts. The **keyless compression coupling** (Fig. 6.2.39) affords a simple means for connecting abutting shafts without the necessity of key seats on the shafts. When drawn over the slotted tapered sleeve the two flanges automatically center the shafts

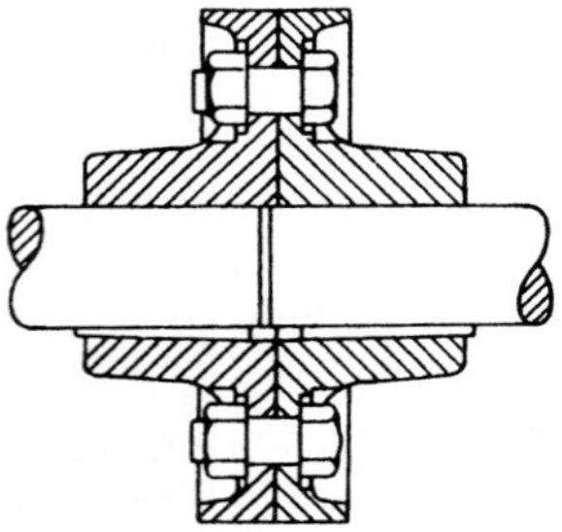

Figure 6.2.38 Flanged face coupling.

Figure 6.2.39 Keyless compression coupling.

and provide sufficient contact pressure to transmit medium or light loads. **Ribbed-clamp couplings** (Fig. 6.2.40) are split longitudinally and are bored to the shaft diameter with a shim separating the two halves. It is necessary to key the shafts to the coupling.

Flexible Couplings

Flexible couplings are designed to connect shafts which are misaligned either laterally or angularly. A secondary benefit is the absorption of impacts due to fluctuations in shaft torque or angular speed. The

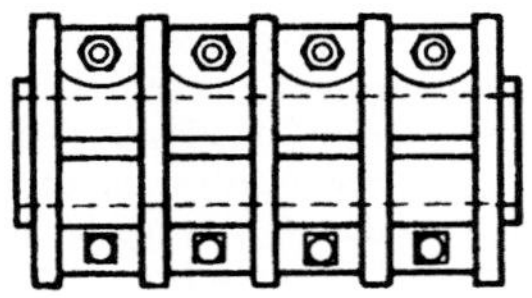

Figure 6.2.40 Ribbed-clamp coupling.

Oldham, or **double-slider, coupling** (Fig. 6.2.41) may be used to connect shafts which have only lateral misalignment. The **"Fast" flexible coupling** (Fig. 6.2.42) consists of two hubs each keyed to its respective shaft. Each hub has generated splines cut at the maximum possible distance from the shaft end. Surrounding the hubs is a casing or sleeve which is split transversely and bolted by means of flanges. Each

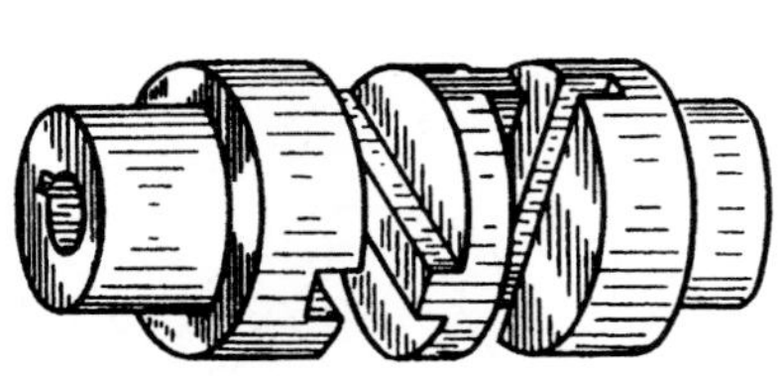

Figure 6.2.41 Double-slider coupling.

half of this sleeve has generated internal splines cut on its bore at the end opposite to the flange. These internal splines permit a definite error of alignment between the two shafts.

Another type, the Waldron coupling (Midland-Ross Corp.), is shown in Fig. 6.2.43.

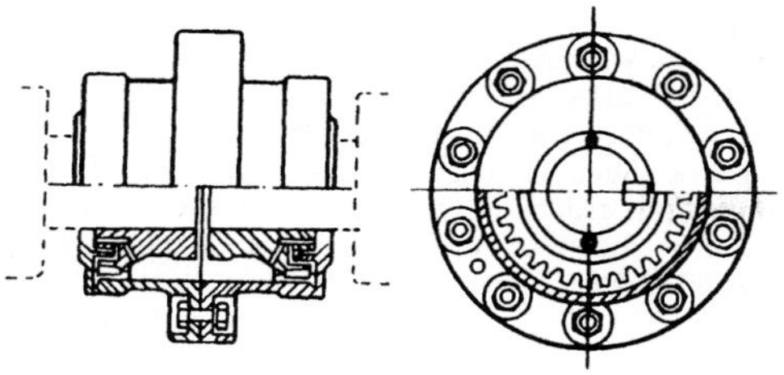

Figure 6.2.42 "Fast" flexible coupling.

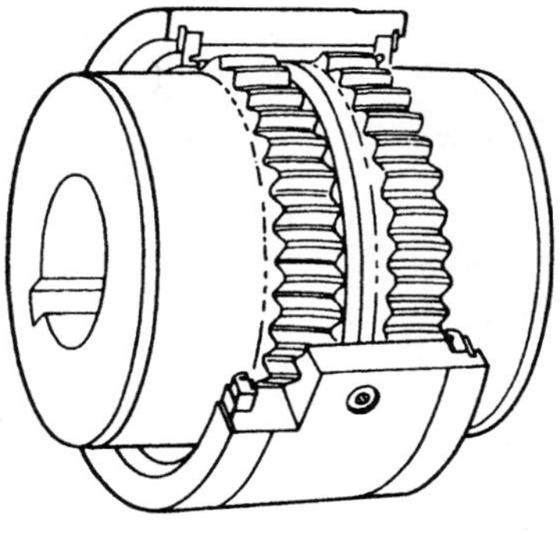

Figure 6.2.43 Waldron coupling.

The chain coupling shown in Fig. 6.2.44 uses silent chain, but standard roller chain can be used with the proper mating sprockets. Nylon links enveloping the sprockets are another variation of the chain coupling.

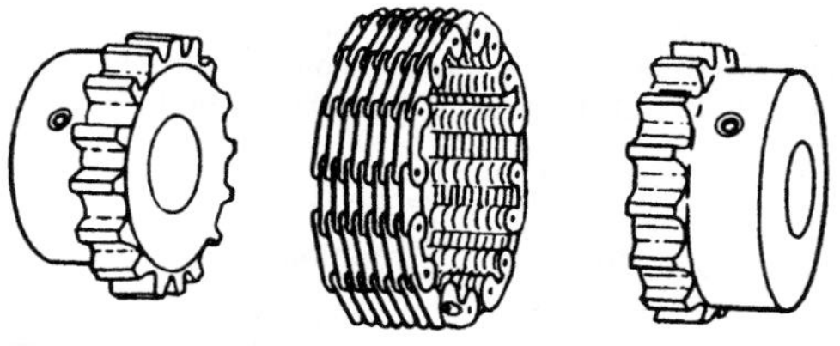

Figure 6.2.44 Chain coupling.

Steelflex couplings (Fig. 6.2.45) are made with two grooved steel hubs keyed to their respective shafts. Connection between the two halves is secured by a specially tempered alloy-steel member called the "grid."

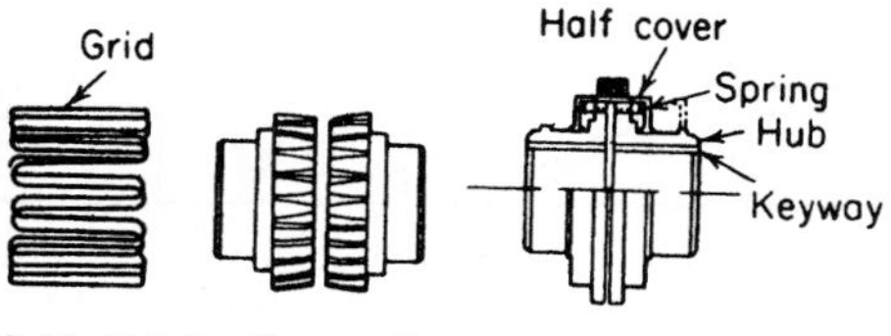

Figure 6.2.45 Falk Steelflex coupling.

In the rubber flexible coupling shown in Fig. 6.2.46, the torque is transmitted through a comparatively soft rubber section acting in shear. The type in Fig. 6.2.47 loads the intermediate rubber member in compression. Both types permit reasonable shaft misalignment and are recommended for light loads only.

Universal joints are used to connect shafts with much larger values of misalignment than can be tolerated by the other types of flexible couplings. Shaft angles up to 30° may be used. The Hooke's-universal joint (Fig. 6.2.48) suffers a loss in efficiency with increasing angle which may be approximated for angles up to 15° by the following relation: efficiency = $100(1-0.003\theta)$, where θ is the angle between the shafts. The velocity ratio between input and output shafts with a single universal joint is equal to

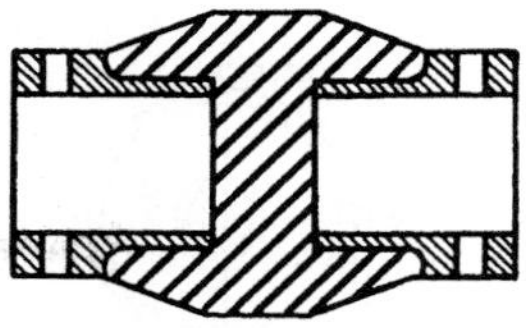

Figure 6.2.46 Rubber flexible coupling, shear type.

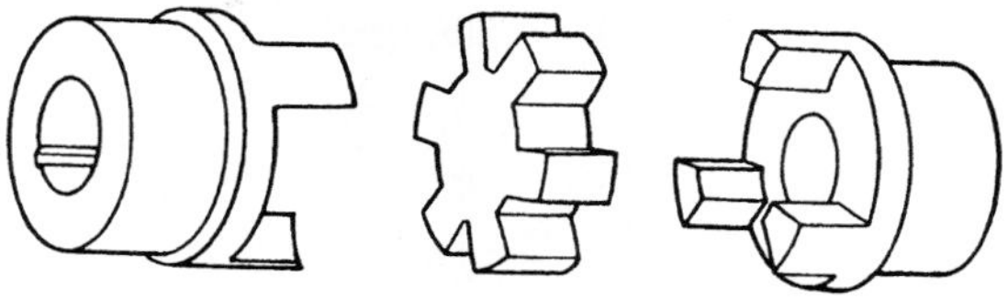

Figure 6.2.47 Rubber flexible coupling, compression type.

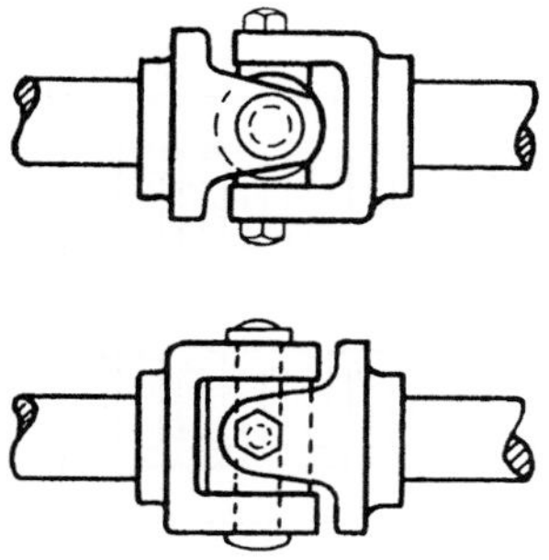

Figure 6.2.48 Hooke's universal joint.

$$\omega_2/\omega_1 = \cos\theta/1 - \sin^2\theta\sin^2(\alpha + 90°)$$

where ω_2 and ω_1 are the angular velocities of the driven and driving shafts, respectively, θ is the angle between the shafts, and α is the angular displacement of the driving shaft from the position where the pins on the drive-shaft yoke lie in the plane of the two shafts. A velocity ratio of 1 may be obtained at any angle using two Hooke's-type joints and an intermediate shaft. The intermediate shaft must make equal angles with the main shafts, and the driving pins on the yokes attached to the intermediate shaft must be set parallel to each other.

The Bendix-Weiss "rolling-ball" universal joint provides constant angular velocity. Torque is transmitted between two yokes through a set of four balls such that the centers of all four balls lie in a plane which bisects the angle between the shafts. Other variations of constant velocity universal joints are found in the Rzeppa, Tracta, and double Cardan types.

Fluid Couplings

(See also Sec. 8.)

Fluid couplings (Fig. 6.2.49*a*) have two basic parts—the input member, or impeller, and the output member, or runner. There is no mechanical connection between the two shafts, power being transmitted by kinetic energy in the operating fluid. The impeller *B* is fastened to the flywheel *A* and turns at engine speed. As this speed increases, fluid within the impeller moves toward the outer periphery because of centrifugal force. The circular shape of the impeller directs the fluid toward the runner *C*, where its kinetic energy is absorbed as torque delivered by shaft *D*. The positive pressure behind the fluid causes flow to continue toward

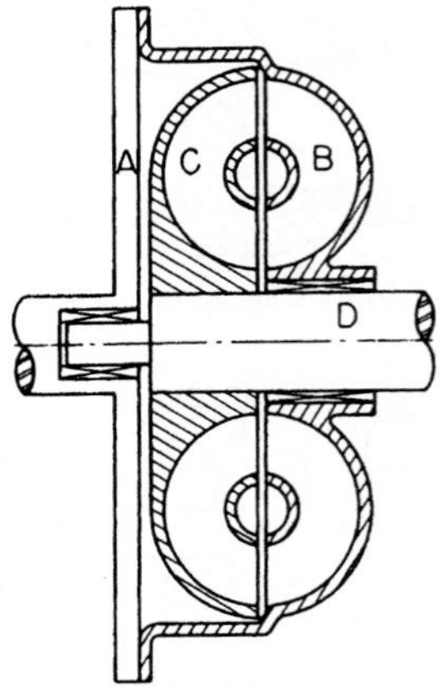

Figure 6.2.49*a* Fluid coupling. (*A*) Flywheel; (*B*) impeller; (*C*) runner; (*D*) output shaft.

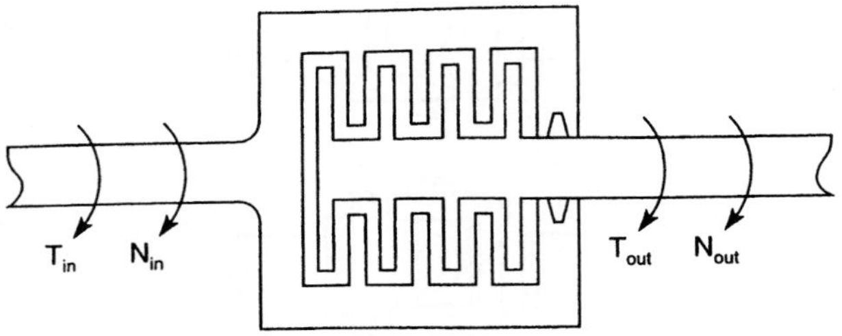

Figure 6.2.49*b* Schematic of viscous coupling.

the hub and back through the impeller. The toroidal space in both the impeller and runner is divided into compartments by a series of flat radial vanes.

The torque capacity of a fluid coupling with a full-load slip of about 2.5 percent is $T = 0.09n^2D^5$, where n is the impeller speed, hundreds of r/min, and D is the outside diameter, ft. The output torque is equal to the input torque over the entire range of input-output speed ratios. Thus the prime mover can be operated at its most effective speed regardless of the speed of the output shaft. Other advantages are that the prime mover cannot be stalled by application of load and that there is no transmission of shock loads or torsional vibration between the connected shafts.

Viscous couplings are becoming major players in mainstream frontwheel-drive applications and are already used in four-wheel-drive vehicles.

Torque transmission in a viscous coupling relies on shearing forces in an entrapped fluid between axially positioned disks rotating at different angular velocities (Fig. 6.2.49*b*), all encased in a lifetime leakproof housing. A hub carries the so-called inner disks while the housing carries the so-called outer disks. Generally Silicone is the working fluid.

Operation of the coupling is in normal (slipping) mode when torque is being generated by viscous shear. However, prolonged slipping under severe starting conditions causes heat-up, which in turn causes the fluid, which has a high coefficient of thermal expansion, to expand considerably with increasing temperature. It then fills the entire available space, causing a rapid pressure increase, which in turn forces the disks together into metal-to-metal frictional contact. Torque transmission now increases substantially. This self-induced torque amplification is known as the **hump effect.** The point at which the hump occurs can be set by the design and coupling setup. Under extreme conditions, 100 percent lockup occurs, thus providing a self-protecting relief from overheating as fluid shear vanishes. This effect is especially useful in autos using viscous couplings in their limited-slip differentials, when one wheel is on low-friction surfaces such as ice. The viscous coupling transfers torque to the other gripping wheel. This effect is also useful when one is driving up slopes with uneven surface conditions, such as rain or snow, or on very rough surfaces. Such viscous coupling differentials have allowed a weight and cost reduction of about 60 percent. A more complete account can be found in Barlage, Viscous Couplings Enter Main Stream Vehicles, *Mech. Eng.*, Oct. 1993.

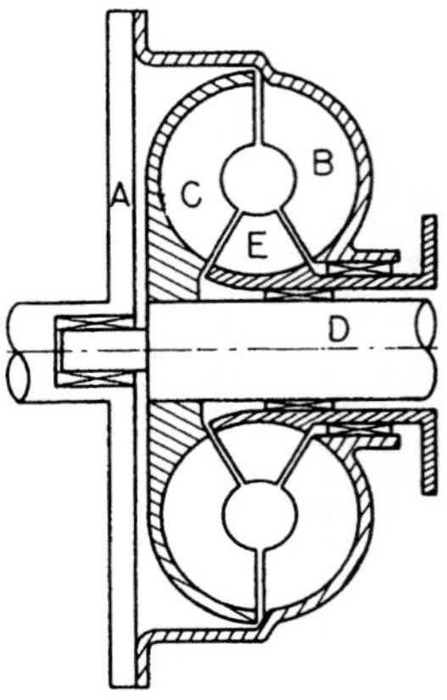

Figure 6.2.50 Hydraulic torque converter. (*A*) Flywheel; (*B*) impeller; (*C*) runner; (*D*) output shaft; (*E*) reactor.

A **hydraulic torque converter** (Fig. 6.2.50) is similar in form to the hydraulic coupling, with the addition of a set of stationary guide vans, the reactor, interposed between the runner and the impeller. All blades in a converter have compound curvature. This curvature is designed to control the direction of fluid flow. Kinetic energy is therefore transferred as a result of both a scalar and vectorial change in fluid velocity. The blades are designed such that the fluid will be moving in a direction parallel to the blade surface at the entrance (Fig. 6.2.51) to each section. With a design having fixed blading, this can be true at only one value of runner and impeller velocity, called the design point. Several design modifications are possible to overcome this difficulty. The angle of the blades can be made adjustable, and the elements can be divided into sections operating independently of each other according to the load requirements. Other refinements include the addition of multiple stages in the runner and reactor stages as in steam reaction turbines (see Sec. 7). The advantages of a torque converter are the ability to multiply starting torque 5 to 6 times and to serve as a stepless transmission. As in the coupling, torque varies as the square of speed and the fifth power of diameter.

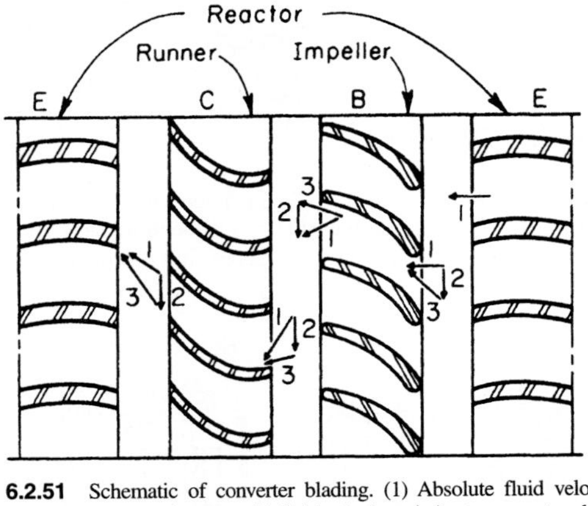

Figure 6.2.51 Schematic of converter blading. (1) Absolute fluid velocity; (2) velocity vector—converter elements; (3) fluid velocity relative to converter elements.

Optimum efficiency (Fig. 6.2.52) over the range of input-output speed ratios is obtained by a combination converter coupling. When the output speed rises to the point where the torque multiplication factor is 1.0, the clutch point, the torque reaction on the reactor element reverses direction. If the reactor is mounted to freewheel in this opposite direction, the unit will act as a coupling over the higher speed ranges. An automatic friction clutch (see "Clutches," below) set to engage at or near the clutch point will also eliminate the poor efficiency of the converter at high output speeds.

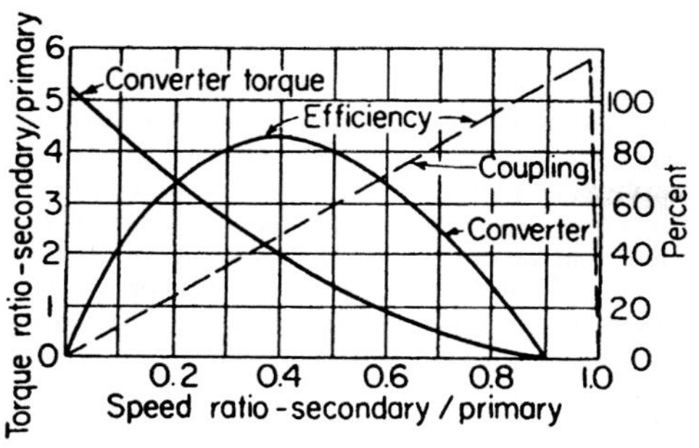

Figure 6.2.52 Hydraulic coupling characteristic curves. (*Heldt, "Torque Converters and Transmissions," Chilton.*)

CLUTCHES

Clutches are couplings which permit the disengagement of the coupled shafts during rotation.

Positive clutches are designed to transmit torque without slip. The **jaw clutch** is the most common type of positive clutch. These are made with **square jaws** (Fig. 6.2.53) for driving in both directions or **spiral jaws** (Fig. 6.2.54) for unidirectional drive. Engagement speed should be limited to 10 r/min for square jaws and 150 r/min for spiral jaws. If disengagement under load is required, the jaws should be finish-machined and lubricated.

Friction clutches are designed to reduce coupling shock by slipping during the engagement period. They also serve as safety devices by slipping when the torque exceeds their maximum rating. They may be divided into two main groups, **axial** and **rim clutches,** according to the direction of contact pressure.

The **cone clutch** (Fig. 6.2.55) and the **disk clutch** (Fig. 6.2.56) are examples of axial clutches. The disk clutch may consist of either a single plate or multiple disks. Table 6.2.35 lists typical friction coefficient for various materials and important design data. The torque capacity of a disk clutch is given by $T = 0.5ifF_aD_m$, where T is the torque, i the number of pairs of contact surfaces, f the applicable coefficient of friction, F_a the axial engaging force, and D_m the mean diameter of the clutch facing. The spring forces holding a disk clutch in engagement are usually of relatively high value, as given by the allowable contact pressures. In order to lower the force required at the operating lever, elaborate linkages are required, usually having lever ratios in the range of 10 to 12. As these linkages must rotate with the clutch, they must be adequately balanced and the effect of centrifugal forces must be

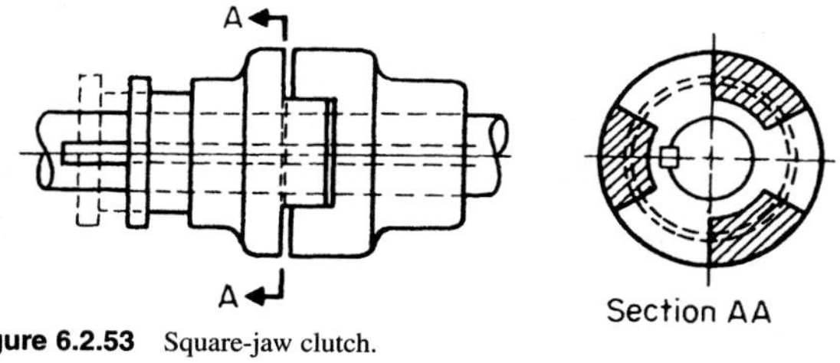

Figure 6.2.53 Square-jaw clutch.

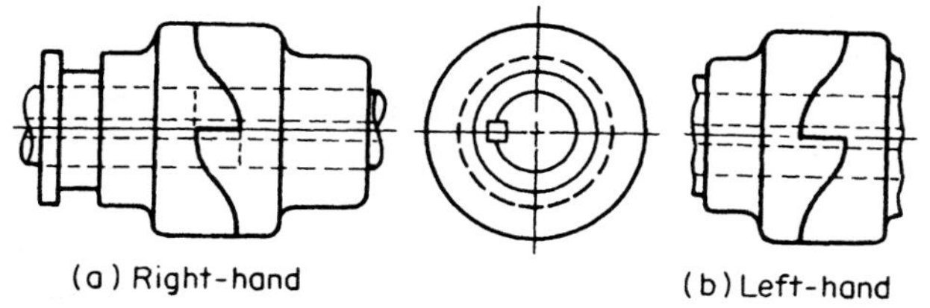

Figure 6.2.54 Spiral-jaw clutch.

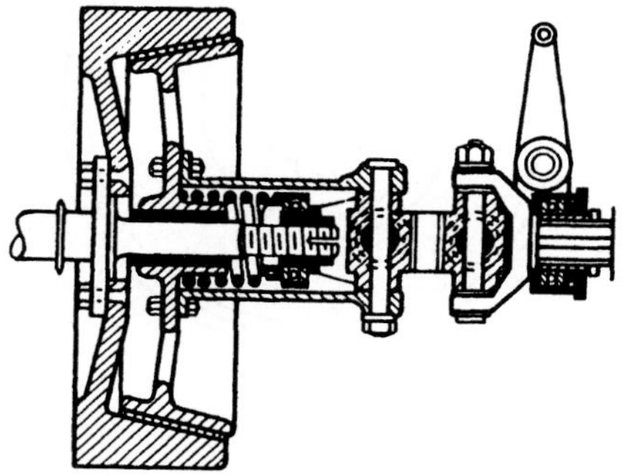

Figure 6.2.55 Cone clutch.

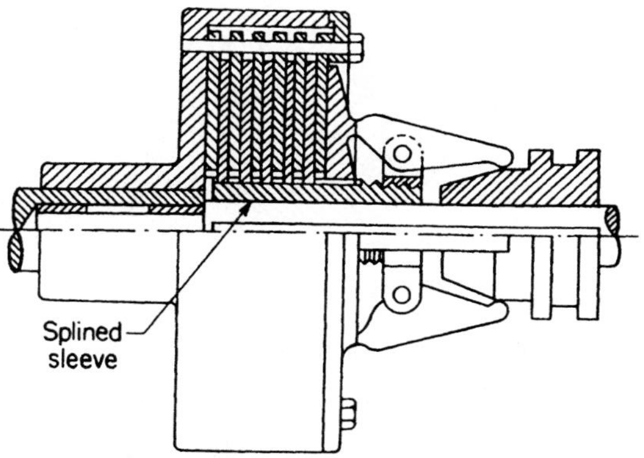

Figure 6.2.56 Multidisk clutch.

Table 6.2.35 Friction Coefficients and Allowable Pressures

Materials in contact	Friction coefficient f			Allowable pressure, lb/in²
	Dry	Greasy	Lubricated	
Cast iron on cast iron	0.2-0.15	0.10-0.06	0.10-0.05	150-250
Bronze on cast iron	—	0.10-0.05	0.10-0.05	80-120
Steel on cast iron	0.30-0.20	0.12-0.07	0.10-0.06	120-200
Wood on cast iron	0.25-0.20	0.12-0.08	—	60-90
Fiber on metal	—	0.20-0.10	—	10-30
Cork on metal	0.35	0.30-0.25	0.25-0.22	8-15
Leather on metal	0.5-0.3	0.20-0.15	0.15-0.12	10-30
Wire asbestos on metal	0.5-0.35	0.30-0.25	0.25-0.20	40-80
Asbestos blocks on metal	0.48-0.40	0.30-0.25	—	40-160
Asbestos on metal, short action	—	—	0.25-0.20	200-300
Metal on cast iron, short action	—	—	0.10-0.05	200-300

SOURCE: Maleev, Machine Design, International Textbook, by permission.

considered. Disk clutches are often run wet, either immersed in oil or in a spray. The advantages are reduced wear, smoother action, and lower operating temperatures. Disk clutches are often operated automatically by either air or hydraulic cylinders as, for examples, in automobile automatic transmissions.

Rim clutches may be subdivided into two groups: (1) those employing either a band or block (Fig. 6.2.57) in contact with the rim and (2) overrunning clutches (Fig. 6.2.58) employing the wedging action of a roller or sprag. Clutches in the second category will automatically engage in one direction and freewheel in the other.

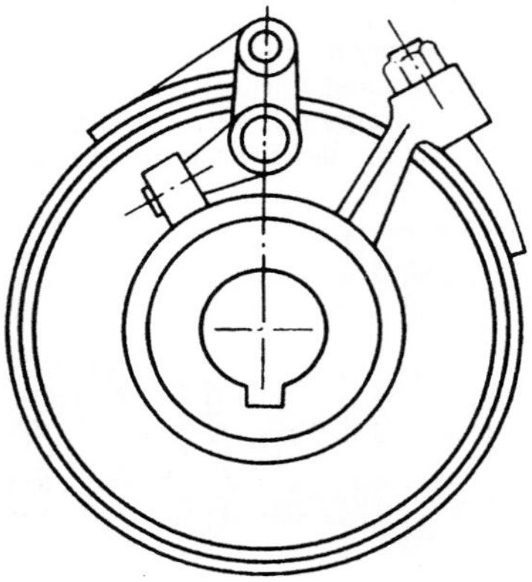

Figure 6.2.57 Band clutch.

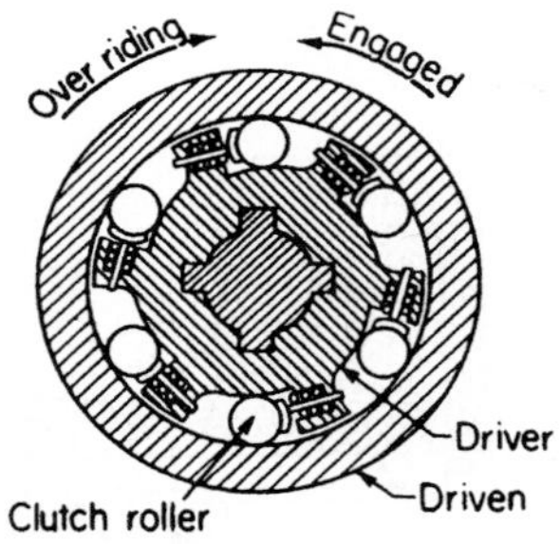

Figure 6.2.58 Overrunning clutch.

HYDRAULIC POWER TRANSMISSION

Hydraulic power transmission systems comprise machinery and auxiliary components which function to generate, transmit, control, and utilize hydraulic power. The **working fluid,** a pressurized incompressible liquid, is usually either a petroleum base or a fire-resistant type. The latter are water and oil emulsions, glycol-water mixtures, or synthetic liquids such as silicones or phosphate esters.

Liquid is pressurized in a **pump** by virtue of its resistance to flow; the pressure difference between pump inlet and outlet results in flow. Most hydraulic applications employ positive-displacement pumps of the gear, vane, screw, or piston type; piston pumps are axial, radial, or reciprocating (see Sec. 10).

Power is transmitted from pump to controls and point of application through a combination of **conduit and fittings** appropriate to the particular application. Flow characteristics of hydraulic circuits take into account fluid properties, pressure drop, flow rate, and pressure-surging tendencies. Conduit systems must be designed to minimize changes in flow velocity, velocity distribution, and random fluid eddies, all of which dissipate energy and result in pressure drops in the circuit. Pipe, tubing, and flexible hose are used as hydraulic power conduits; suitable fittings are available for all types and for transition from one type to another.

Controls are generally interposed along the conduit between the pump and point of application (i.e., an actuator or motor), and act to control pressure, volume, or flow direction.

Pressure control valves, of which an ordinary safety valve is a common type (normally closed), include relief and reducing valves and pressure switches (Figs. 6.2.59 and 6.2.60). Pressure valves, normally closed, can be used to control sequential operations in a hydraulic circuit. **Flow control valves** throttle flow to or bypass flow around the unit being controlled, resulting in pressure drop and temperature increase as pressure energy is dissipated. Figure 6.2.61 shows a simple needle valve with variable orifice usable as a flow control valve. **Directional control valves** serve primarily to direct hydraulic fluid to the point of application. Directional control valves with rotary and sliding spools are shown in Figs. 6.2.62 and 6.2.63.

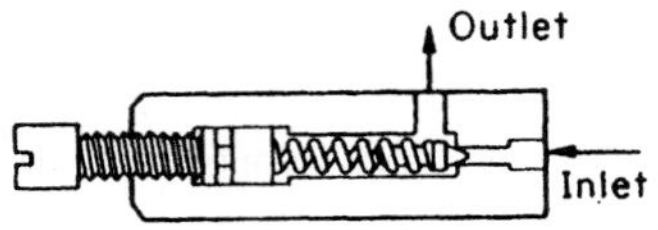

Figure 6.2.59 Relief valve.

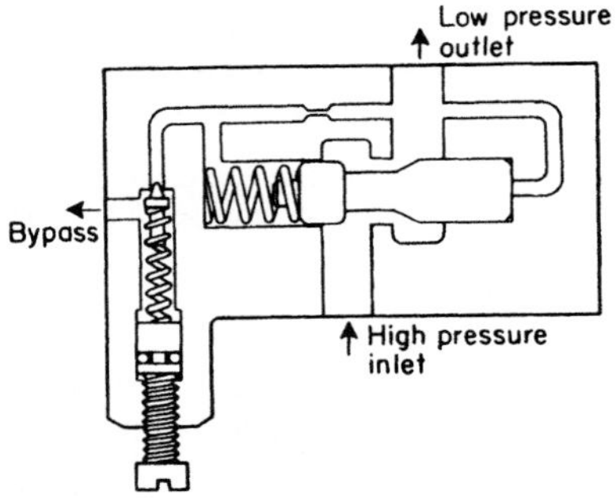

Figure 6.2.60 Reducing valve.

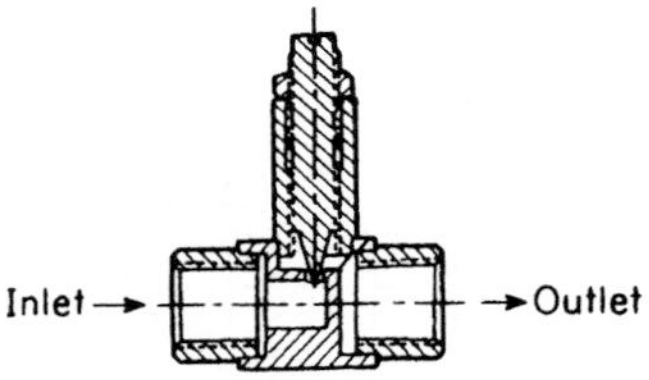

Figure 6.2.61 Needle valve.

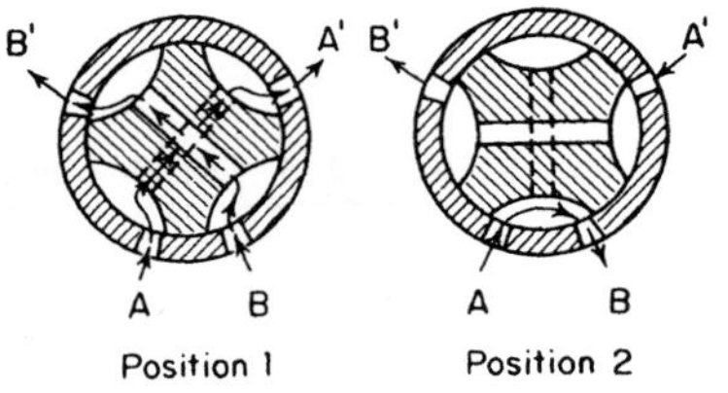

Figure 6.2.62 Rotary-spool directional flow valve.

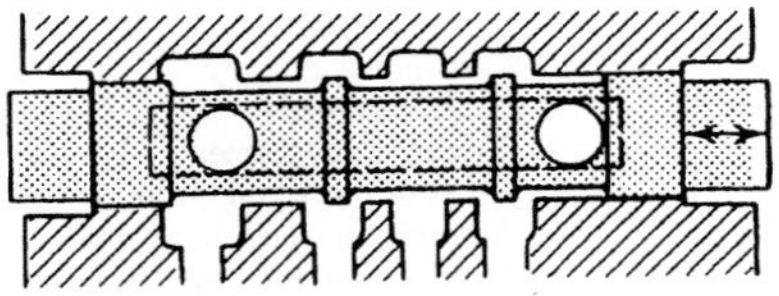

Figure 6.2.63 Sliding-spool directional flow valve.

A poppet (valve) mechanism is shown in Fig. 6.2.64, a diaphragm valve in Fig. 6.2.65, and a shear valve in Fig. 6.2.66.

Accumulators are effectively "hydraulic flywheels" which store potential energy by accumulating a quantity of pressurized hydraulic fluid in a suitable enclosed vessel. The bag type shown in Fig. 6.2.67 uses pressurized gas inside the bag working against the hydraulic fluid outside the bag. Figure 6.2.68 shows a piston accumulator.

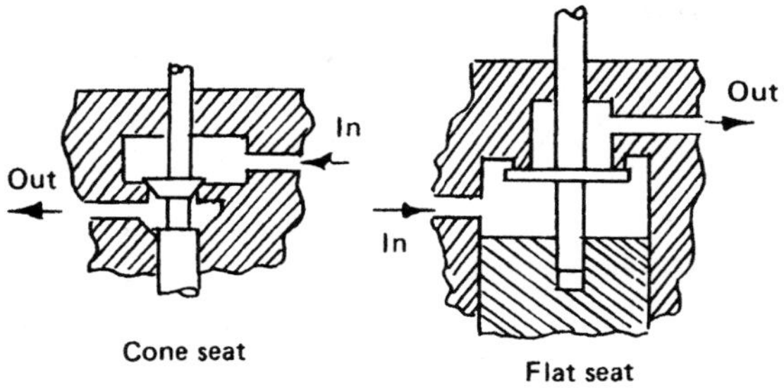

Figure 6.2.64 Poppet valve.

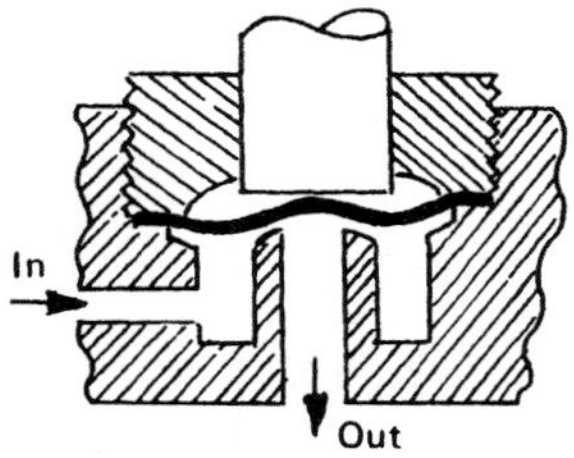

Figure 6.2.65 Diaphragm valve.

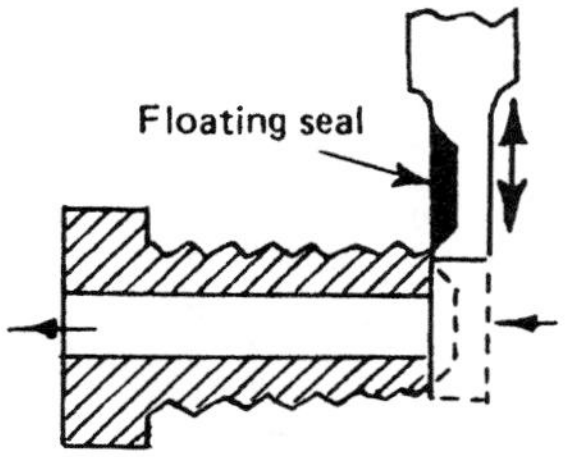

Figure 6.2.66 Shear valve.

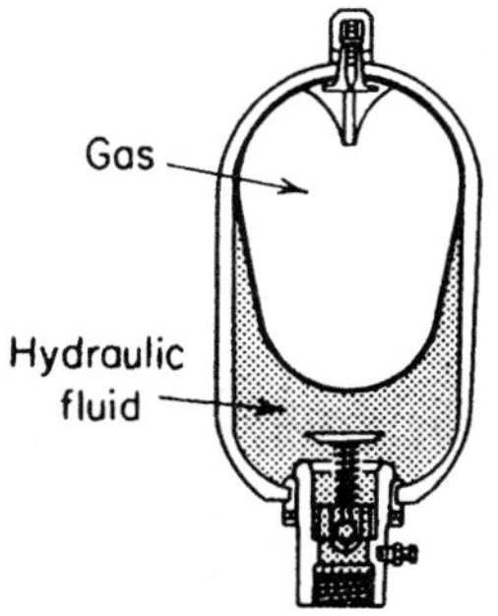

Figure 6.2.67 Bag accumulator.

Pressurized hydraulic fluid acting against an **actuator** or motor converts fluid pressure energy into mechanical energy. Motors providing continuous rotation have operating characteristics closely related to their pump counterparts. A linear actuator, or cylinder (Fig. 6.2.69), provides straight-line reciprocating motion; a rotary actuator (Fig. 6.2.70) provides arcuate oscillatory motion. Figure 6.2.71 shows a one-shot booster (linear motion) which can be used to deliver sprays through a nozzle.

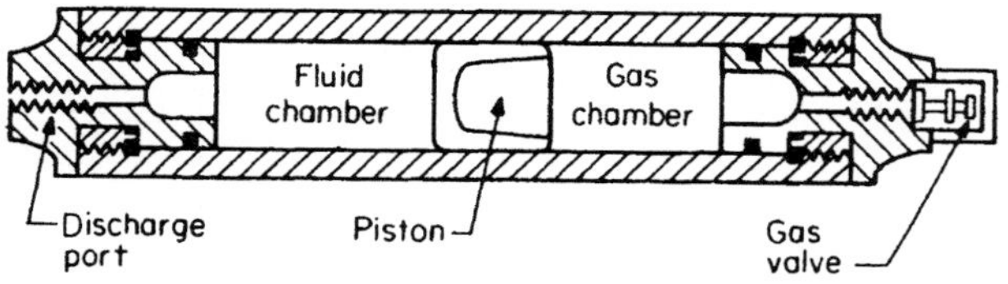

Figure 6.2.68 Piston accumulator.

Hydraulic fluids (liquids and air) are conducted in **pipe, tubing,** or **flexible hose.** Hose is used when the lines must flex or in applications in which fixed, rigid conduit is unsuitable. Table 6.2.36 lists SAE standard hoses. Maximum recommended operating pressure for a broad range of industrial applications is approximately 25 percent of rated bursting pressure. Due consideration must be given to the operating- temperature range; most applications fall in the range from −40 to 200°F (−40 to 95°C). Higher operating temperatures can be accommodated with appropriate materials.

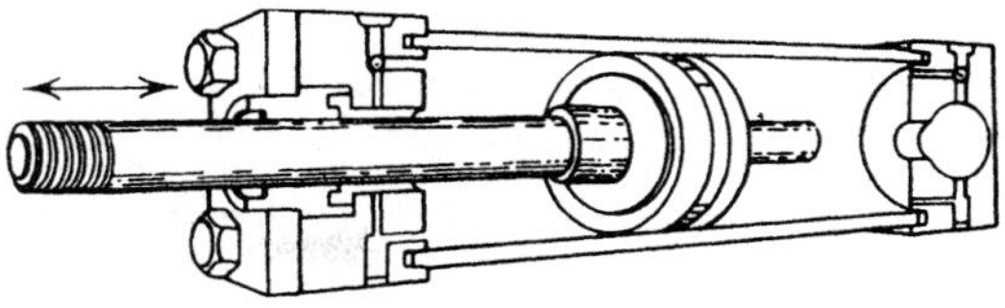

Figure 6.2.69 Linear actuator or hydraulic cylinder.

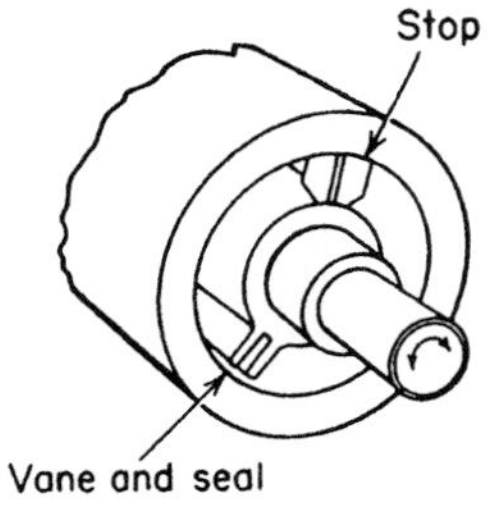

Figure 6.2.70 Rotary actuator.

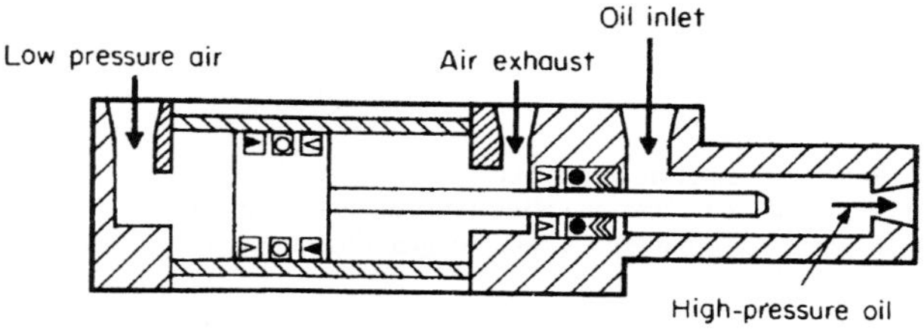

Figure 6.2.71 One-shot booster.

Table 6.2.36 SAE Standard Hoses

100R1A	One-wire-braid reinforcement, synthetic rubber cover
100R1T	Same as R1A except with a thin, nonskive cover
100R2A	Two-wire-braid reinforcement, synthetic rubber cover
100R2B	Two spiral wire plus one wire-braid reinforcement, synthetic rubber cover
100R2AT	Same as R2A except with a thin, nonskive cover
100R2BT	Same as R2B except with a thin, nonskive cover
100R3	Two rayon-braid reinforcement, synthetic rubber cover
100R5	One textile braid plus one wire-braid reinforcement, textile braid cover
100R7	Thermoplastic tube, synthetic fiber reinforcement, thermoplastic cover (thermoplastic equivalent to SAE 100R1A)
100R8	Thermoplastic tube, synthetic fiber reinforcement, thermoplastic cover (thermoplastic equivalent to SAE 100R2A)
100R9	Four-ply, light-spiral-wire reinforcement, synthetic rubber cover
100R9T	Same as R9 except with a thin, nonskive cover
100R10	Four-ply, heavy-spiral-wire reinforcement, synthetic rubber cover
100R11	Six-ply, heavy-spiral-wire reinforcement, synthetic rubber cover

Hose fittings are of the screw-type or swaged, depending on the particular application and operating pressure and temperature. A broad variety of hose-end fittings is available from the industry.

Pipe has the advantage of being relatively cheap, is applied mainly in straight runs, and is usually of steel. Fittings for pipe are either standard pipe fittings for fairly low pressures or more elaborate ones suited to leakproof high-pressure operation.

Tubing is more easily bent into neat forms to fit between inlet and outlet connections. Steel and stainless-steel tubing is used for the highest-pressure applications; aluminum, plastic, and copper tubing is also used as appropriate for the operating conditions of pressure and temperature. Copper tubing hastens the oxidation of oil-base hydraulic fluids; accordingly, its use should be restricted either to air lines or with liquids which will not be affected by copper in the operating range.

Tube fittings for permanent connections allow for brazed or welded joints. For temporary or separable applications, **flared** or **flareless fittings** are employed (Figs. 6.2.72 and 6.2.73). SAE J514, ANSI B116.1-1974 and B116.2-1974 pertain to tube fittings. The variety of fittings available is vast; the designer is advised to refer to manufacturers' literature for specifics.

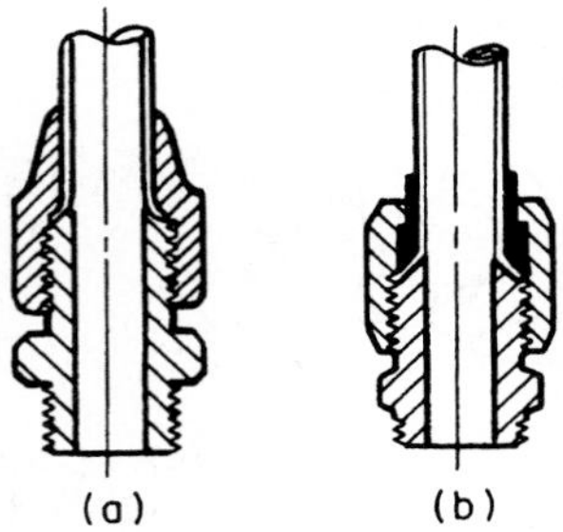

Figure 6.2.72 Flared tube fittings. (*a*) A 45° flared fitting; (*b*) triple-lok flared fitting. *(Parker-Hannafin Co.)*

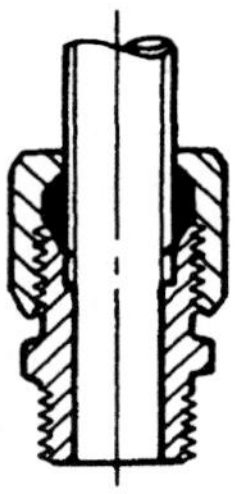

Figure 6.2.73 Ferulok flareless tube fitting. *(Parker-Hannafin Co.)*

Parameters entering into the design of a hydraulic system are volume of flow per unit time, operating pressure and temperature, viscosity characteristics of the fluid within the operating range, and compatibility of the fluid/conduit material.

Flow velocity in suction lines is generally in the range of 1 to 5 ft/s (0.3 to 1.5 m/s); in discharge lines it ranges from 10 to 25 ft/s (3 to 8 m/s).

The pipe or tubing is under internal pressure. Selection of material and wall thickness follows from suitable equations (see Sec. 3). Safety factors range from 6 to 10 or higher, depending on the severity of the application (i.e., vibration, shock, pressure surges, possibility of physical abuse, etc.). JIC specifications provide a guide to the designer of hydraulic systems.

BRAKES

Brakes may be classified as: (1) rim type—internally expanding or externally contracting shoes, (2) band type, (3) cone type, (4) disk or axial type, (5) miscellaneous.

Rim Type—Internal Shoe(s) (Fig. 6.2.74)

$$F = \begin{cases} \dfrac{M_N - M\mu}{d} & \text{clockwise rotation} \\[2ex] \dfrac{M_N + M\mu}{d} & \text{counterclockwise rotation} \end{cases}$$

where

$$M\mu = \frac{\mu P_a B r}{\sin\theta_a}\int_{\theta_1}^{\theta_2}(\sin\theta)(r - d\cos\theta)\,d\theta$$

B = face width of frictional material; P_a = maximum pressure; θ_a = angle to point of maximum pressure (if $\theta_2 > 90°$; then $\theta_a = 90°$; $\theta_2 < 90°$, then $\theta_a = \theta_2$); μ = coefficient of friction; r = radius of drum; d = distance from drum center to brake pivot;

$$M_N = \frac{P_a B r d}{\sin\theta_a}\int_{\theta_1}^{\theta_2}\sin^2\theta\,d\theta$$

and torque on drum is

$$T = \frac{\mu P_a B r^2(\cos\theta_1 - \cos\theta_2)}{\sin\theta_a}$$

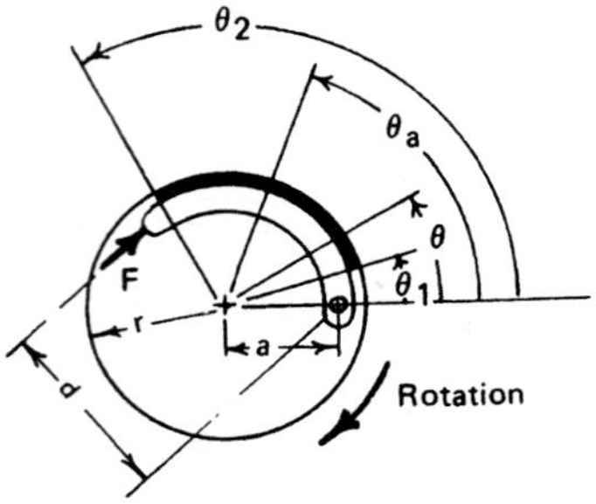

Figure 6.2.74 Rim brake: internal friction shoe.

Self-locking of the brake ($F = 0$) will occur for clockwise rotation when $M_N = M_\mu$. This self-energizing phenomenon can be used to advantage without actual locking if is replaced by a larger value μ' so that $1.25\mu \le \mu' \le 1.50$, from which the pivot position a can be solved.

In automative use there are two shoes made to pivot in opposition, so that self-energization is present and can be used to great advantage (Fig. 6.2.75).

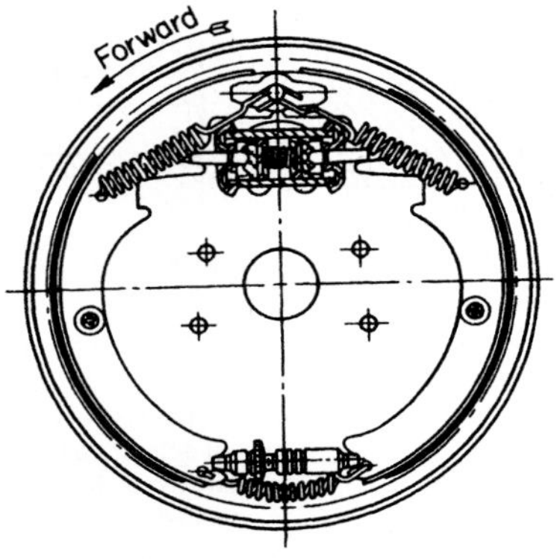

Figure 6.2.75 Internal brake.

Rim Type—External Shoe(s) (Fig. 6.2.76)

The equations for M_N and M_μ are the same as above:

$$F = \begin{cases} \dfrac{M_N + M\mu}{d} & \text{clockwise rotation} \\[2ex] \dfrac{M_N - M\mu}{d} & \text{counterclockwise rotation} \end{cases}$$

Self-locking ($F = 0$) can occur for counterclockwise rotation at $M_N = M_\mu$.

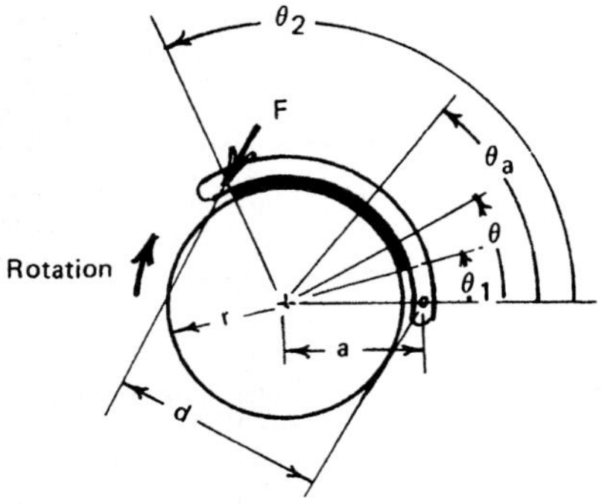

Figure 6.2.76 Rim brake: external friction shoe.

Band Type (Figs. 6.2.77*a*, *b*, and *c*)

Flexible band brakes are used in power excavators and in hoisting. The bands are usually of an asbestos fabric, sometimes reinforced with copper wire and impregnated with asphalt.

In Fig. 6.2.77*a*, F = force at end of brake handle; P = tangential force at rim of wheel; f = coefficient of friction of materials in contact; a = angle of wrap of band, deg; T_1 = total tension in band on tight side; T_2 = total tension in band on slack side. Then $T_1 - T_2 = P$ and $T_1/T_2 = 10^{0.0076fa} = 10^b$, where $b = 0.0076fa$. Also, $T_2 = P/(10^b - 1)$ and $T_1 = P \times 10^b/(10^b - 1)$. The values of $10^{0.0076fa}$ are given in Fig. 6.2.93 for a in radians.

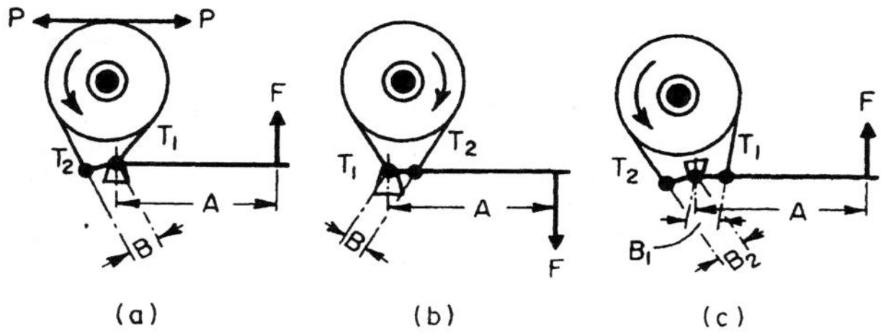

Figure 6.2.77 Band brakes.

For the arrangement shown in Fig. 6.2.77*a*,

$$FA = T_2B = PB/(10^b - 1)$$

and

$$F = PB/[A(10^b - 1)]$$

For the construction illustrated in Fig. 6.2.77*b*,

$$F = PB/\{A[10^b/(10^b - 1)]\}$$

For the differential brake shown in Fig. 6.2.77*c*,

$$F = (P/A)[(B_2 - 10^bB_1)/(10^b - 1)]$$

In this arrangement, the quantity 10^bB_1 must always be less than B_2, or the band will grip the wheel and the brake, or part of the mechanism to which it is attached, will rupture.

It is usual in practice to have the leverage ratio A/B for band brakes about 10:1.

If f for wood on iron is taken at 0.3 and the angle of wrap for the band is 270°, i.e., subtends three-fourths of the circumference, then $10^b = 4$ approx; the loads required for a given torque will be as follows for the cases just considered and for the leverage ratios stated above:

Band brake, Fig. 6.2.77*a*	$F = 0.033P$
Band brake, Fig. 6.2.77*b*	$F = 0.133P$
Band brake, Fig. 6.2.77*c*	$F = 0.016P$

In the case of Fig. 6.2.77*c*, the dimension B_2 must be greater than $B_1 \times 10^b$. Accordingly, B_1 is taken as ¼, A as 10, and, since $10^b = 4$, B_2 is taken as 1½.

The principal function of a brake is to absorb energy. This energy appears at the surface of the brake as heat, which must be carried away at a sufficiently rapid rate to prevent burning of the wooden blocks. Suitable proportions may be arrived at as follows:

Let p = unit pressure on brake surface, lb/in² = R (reaction against block)/area of block; v = velocity of brake rim surface, ft/s = $2\pi rn/60$, where n = speed of brake wheel, r/min. Then pv = work absorbed per in² of brake surface per second, and $pv \le 1{,}000$ for intermittent applications of load with comparatively long periods of rest and poor means for carrying away heat (wooden blocks); $pv \le 500$ for continuous application of load and poor means for carrying away heat (wooden blocks); $pv \le 1{,}400$ for continuous application of load with effective means for carrying away heat (oil bath).

Cone Brake (see Fig. 6.2.78)

Uniform Wear

$$F = \frac{\pi P_a d}{2}(D - d)$$

$$T = \frac{\pi \mu P_a d}{8 \sin \alpha}(D^2 - d)$$

where P_a = maximum pressure occurring at $d/2$.

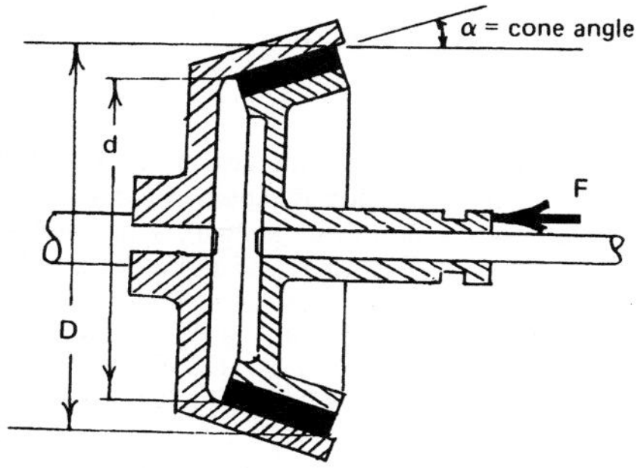

Figure 6.2.78 Cone brake.

Uniform Pressure

$$F = \frac{\pi P_a}{4}(D^2 - d^2)$$

$$T = \frac{\pi \mu P_a}{12 \sin \alpha}(D^3 - d^3) = \frac{F\mu}{3 \sin \alpha} \times \frac{D^3 - d^3}{D^2 - d^2}$$

Figure 6.2.79 shows a cone brake arrangement used for lowering heavy loads.

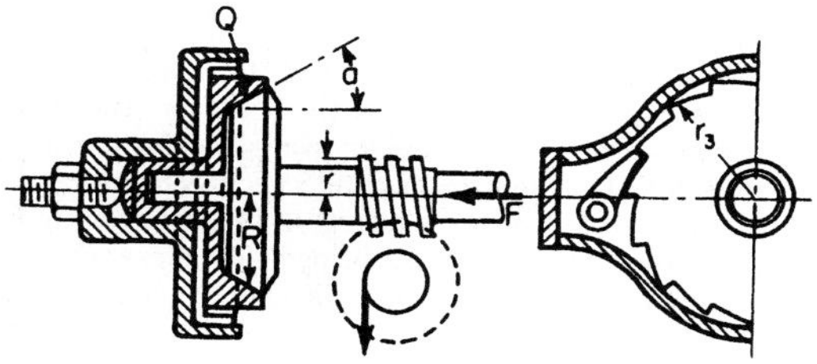

Figure 6.2.79 Cone brake for lowering loads.

Disk Brakes (see Fig. 6.2.80)

Disk brakes are free from "centrifugal" effects, can have large frictional areas contained in small space, have better heat dissipation qualities than the rim type, and enjoy a favorable pressure distribution.

Uniform Wear

$$F = \frac{\pi P_a d}{2}(D - d)$$

$$T = \frac{F\mu}{4}(D + d)$$

Uniform Pressure

$$F = \frac{\pi P_a}{4}(D^2 - d^2)$$

$$T = \frac{F\mu}{3} \times \frac{D^3 - d^3}{D^2 - d^2}$$

These relations apply to a single surface of contact. For caliper disk, or multidisk brakes, the above relations are applied for each surface of contact.

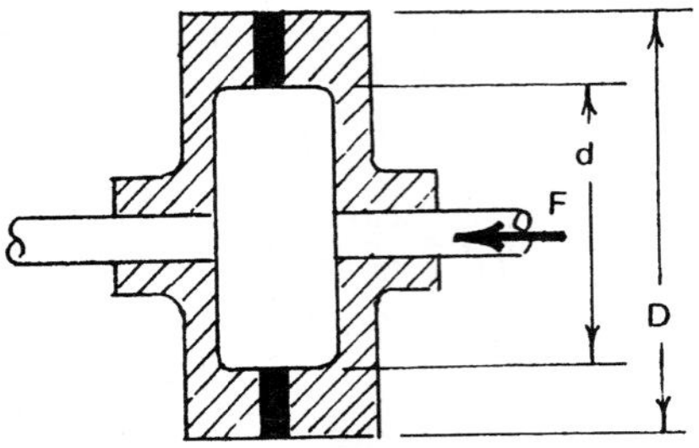

Figure 6.2.80 Disk brake.

Selected friction materials and properties are listed in Table 6.2.37.

Frequently disk brakes are made as shown in Fig. 6.2.81. The pinion *Q* engages the gear in the drum (not shown). When the load is to be raised, power is applied through the gear and the connection between *B* and *C* is accomplished by the advancing of *B* along *A* and the clamping of the friction disks *D* and *d* and the ratchet wheel *E*. The reversal

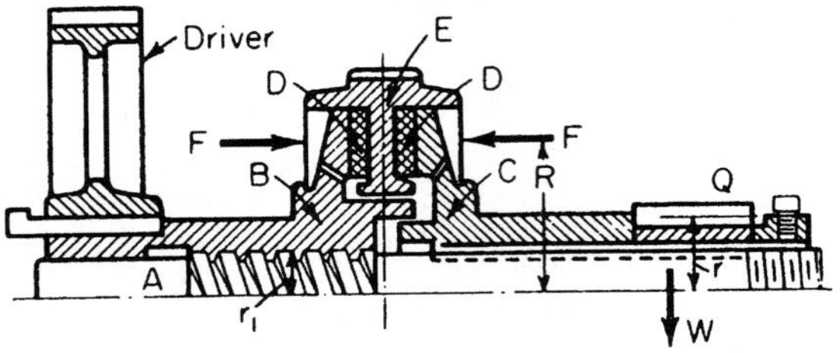

Figure 6.2.81 Disk brake.

of the motor disconnects *B* and *C*. In lowering the load, only as much reversal of rotation of the gear is given as is needed to reduce the force in the friction disks so that the load may be lowered under control.

A **multidisk brake** is shown in Fig. 6.2.82. This type of construction results in an increase in the number of friction faces. The drum shaft is geared to the pinion *A,* while the motive power for driving comes through the gear *G*. In raising the load, direct connection is between *G, B,* and *A*. In lowering, *B* moves relatively to *G* and forces the friction plates together, those plates fixed to *E* being held stationary by the pawl on *E*. In the figure, there are three plates fixed to *E,* one fixed to *G,* and one fixed to *C*.

Eddy-current brakes (Fig. 6.2.83) are used with flywheels where quick braking is essential, and where large kinetic energy of the rotating masses precludes the use of block brakes due to excessive heating, as in reversible rolling mills. A number of poles *a* are electrically excited (north and south in turn) and create a magnetic flux which permeates the gap and the iron of the rim, causing eddy current. The flywheel energy is converted through these currents into heat. The hand brake

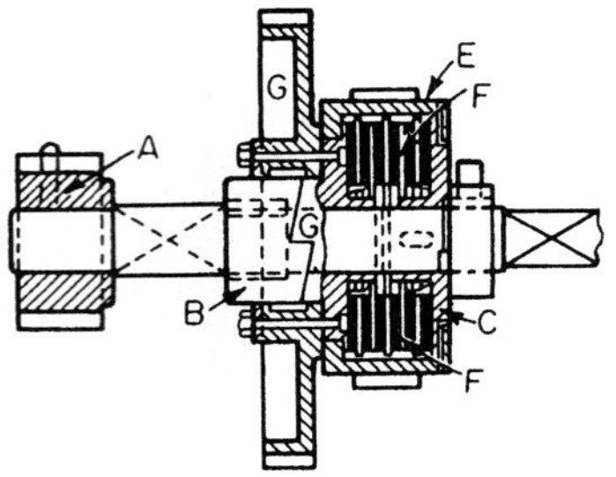

Figure 6.2.82 Multidisk brake.

b may be used for quicker stopping when the speed of the wheel is considerably decreased; i.e., when the eddy-current brake is inefficient. Two brakes are provided to avoid bending forces on the shaft.

Electric brakes are often used in cranes, bridges, turntables, and machine tools, where an automatic application of the brake is important as soon as power is cut off. The brake force is supplied by an adjustable spring which is counteracted by the force of a solenoid or a centrifugal thrustor. Interruption of current automatically applies the spring-activated brake shoes. Figures 6.2.84 and 6.2.85 show these types of electric brake.

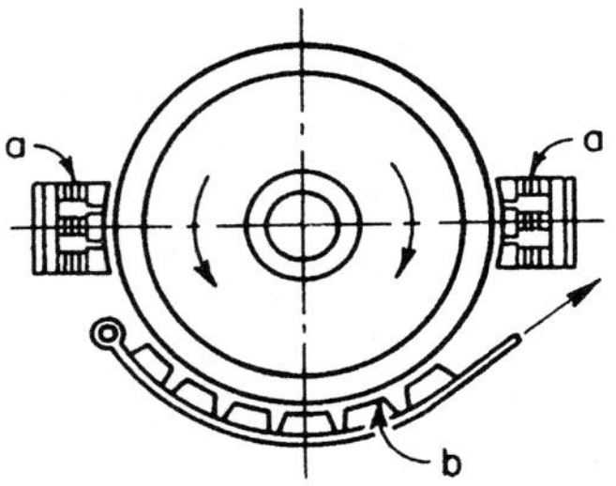

Figure 6.2.83 Eddy-current brake.

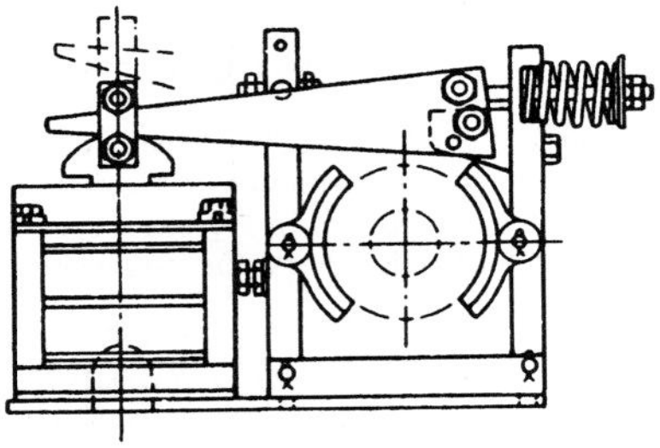

Figure 6.2.84 Solenoid-type electric brake.

Table 6.2.37 Selected Friction Materials and Properties

Material	Opposing material	Friction coefficient			Max pressure		Max temperature	
		Dry	Wet	In oil	lb/in²	kPa	°F	°C
Sintered metal	Cast iron or steel	0.1-0.4	0.05-0.01	0.05-0.08	150-250	1,000-1,720	450-1,250	232-677
Wood	Cast iron or steel	0.2-0.35	0.16	0.12-0.16	60-90	400-620	300	149
Leather	Cast iron or steel	0.3-0.5	0.12		10-40	70-280	200	93
Cork	Cast iron or steel	0.3-0.5	0.15-0.25	0.15-0.25	8-14	55-95	180	82
Felt	Cast iron or steel	0.22	0.18		5-10	35-70	280	138
Asbestos—woven	Cast iron or steel	0.3-0.6	0.1-0.2	0.08-0.10	50-100	350-700	400-500	204-260
Asbestos—molded	Cast iron or steel	0.2-0.5	0.08-0.12	0.06-0.09	50-150	350-1,000	400-500	204-260
Asbestos—impregnated	Cast iron or steel	0.32	0.12					
Cast iron	Cast iron	0.15-0.20	0.05	0.03-0.06	150-250	1,000-1,720	500	260
Cast iron	Steel			0.03-0.06	100-250	690-1,720	500	260
Graphite	Steel	0.25	0.05-0.1	0.12 (av)	300	2,100	370-540	188-282

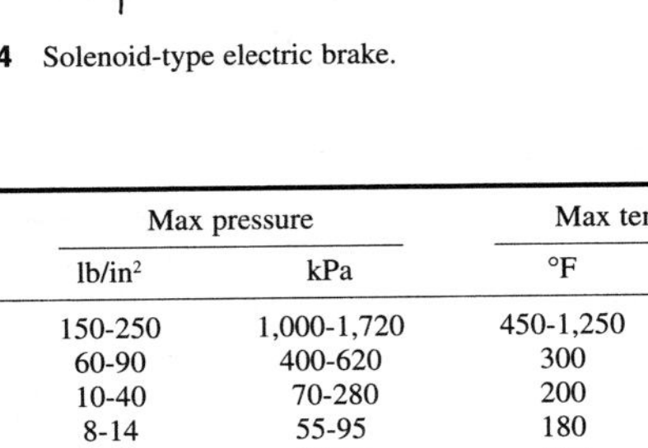

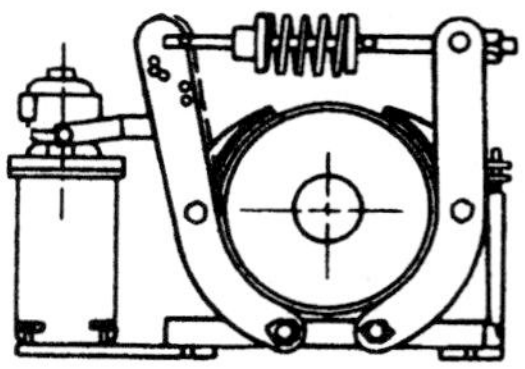

Figure 6.2.85 Thrustor-type electric brake.

SHRINK, PRESS, DRIVE, AND RUNNING FITS

Inch Systems ANSI/ASME B4.1-1967 (R2009) recommends preferred sizes, allowances, and tolerances for fits between plain cylindrical parts. Such fits include bearing, shrink and drive fits, etc. Terms used in describing fits are defined as follows: **Allowance:** minimum clearance (positive allowance) or maximum interference (negative allowance) between mating parts. **Tolerance:** total permissible variation of size. **Limits of size:** applicable maximum and minimum sizes. **Clearance fit:** one having limits of size so prescribed that a clearance always results when mating parts are assembled. **Interference fit:** In this case, limits are so prescribed that interference always results on assembly. **Transition fit:** This may have either a clearance or an interference on assembly. **Basic size:** one from which limits of size are derived by the application of allowances and tolerances. **Unilateral tolerance:** In this case a variation in size is permitted in only one direction from the basic size.

Fits are divided into the following general classifications: (1) running and sliding fits, (2) locational clearance fits, (3) transition fits, (4) locational interference fits, and (5) force or shrink fits.

1. **Running and sliding fits.** These are intended to provide similar running performance with suitable lubrication allowance throughout the range of sizes. These fits are further subdivided into the following classes:

Class RC1: close-sliding fits. Intended for accurate location of parts which must assemble without perceptible play.

Class RC2: sliding fits. Parts made with this fit move and turn easily but are not intended to run freely; also, in larger sizes they may seize under small temperature changes.

Class RC3: precision-running fits. These are intended for precision work at slow speeds and light journal pressures but are not suitable where appreciable temperature differences are encountered.

Class RC4: close-running fits. For running fits on accurate machinery with moderate surface speeds and journal pressures, where accurate location and minimum play is desired.

Classes RC5 and RC6: medium-running fits. For higher running speeds or heavy journal pressures.

Class RC7: free-running fits. For use where accuracy is not essential, or where large temperature variations are likely to be present, or both.

Classes RC8 and RC9: loose-running fits. For use with materials such as cold-rolled shafting or tubing made to commercial tolerances.

Limits of clearance given in ANSI B4.1-1967 (R87) for each of these classes are given in Table 6.2.38. Hole and shaft tolerances are listed on a unilateral tolerance basis in this reference to give the clearance limits of Table 6.2.38, the hole size being the basic size.

2. **Locational clearance fits.** These are intended for normally stationary parts which can, however, be freely assembled or disassembled. These are subdivided into various classes which run from snug fits for parts requiring accuracy of location, through medium clearance fits (spigots) to the looser fastener fits where freedom of assembly is of prime importance.

3. **Transition fits.** These are for applications where accuracy of location is important, but a small amount of either clearance or interference is permissible.

4. **Locational interference fits.** Used where accuracy of location is of prime importance and for parts requiring rigidity and alignment with no special requirements for bore pressure.

Data on clearance limits, interference limits, and hole and shaft diameter tolerances for locational clearance fits, transition fits, and locational interference fits are given in ANSI B4.1-1967 (R87).

5. **Force or shrink fits.** These are characterized by approximately constant bore pressures throughout the range of sizes; interference varies almost directly as the diameter, and the differences between maximum and minimum values of interference are small. These are divided into the following classes:

Class FN1: light-drive fits. For applications requiring light assembly pressures (thin sections, long fits, cast-iron external members).

Class FN2: medium-drive fits. Suitable for ordinary steel parts or for shrink fits on light sections. These are about the tightest fits that can be used on high-grade cast-iron external members.

Class FN3: heavy-drive fits. For heavier steel parts or shrink fits in medium sections.

Classes FN4 and FN5: force fits. These are suitable for parts which can be highly stressed. Shrink fits are used instead of press fits in cases where the heavy pressing forces required for mounting are impractical.

Table 6.2.38 Limits of Clearance for Running and Sliding Fits (Basic Hole)
(Limits are in thousandths of an inch on diameter)

Nominal size range, in	Class								
	RC1	RC2	RC3	RC4	RC5	RC6	RC7	RC8	RC9
0-0.12	0.1 0.45	0.1 0.55	0.3 0.95	0.3 1.3	0.6 1.6	0.6 2.2	1.0 2.6	2.5 5.1	4.0 8.1
0.12-0.24	1.5 0.5	0.15 0.65	0.4 1.2	0.4 1.6	0.8 2.0	0.8 2.7	1.2 3.1	2.8 5.8	4.5 9.0
0.24-0.40	0.2 0.6	0.2 0.85	0.5 1.5	0.5 2.0	1.0 2.5	1.0 3.3	1.6 3.9	3.0 6.6	5.0 10.7
0.40-0.71	0.25 0.75	0.25 0.95	0.6 1.7	0.6 2.3	1.2 2.9	1.2 3.8	2.0 4.6	3.5 7.9	6.0 12.8
0.71-1.19	0.3 0.95	0.3 1.2	0.8 2.1	0.8 2.8	1.6 3.6	1.6 4.8	2.5 5.7	4.5 10.0	7.0 15.5
1.19-1.97	0.4 1.1	0.4 1.4	1.0 2.6	1.0 3.6	2.0 4.6	2.0 6.1	3.0 7.1	5.0 11.5	8.0 18.0
1.97-3.15	0.4 1.2	0.4 1.6	1.2 3.1	1.2 4.2	2.5 5.5	2.5 7.3	4.0 8.8	6.0 13.5	9.0 20.5
3.15-4.73	0.5 1.5	0.5 2.0	1.4 3.7	1.4 5.0	3.0 6.6	3.0 8.7	5.0 10.7	7.0 15.5	10.0 24.0

In Table 6.2.39 are listed the limits of interference (maximum and minimum values) for the above classes of force or shrink fits for various diameters, as given in ANSI B4.1-1967 (R87). Hole and shaft tolerances to give these interference limits are also listed in this reference.

Metric System ANSI B4.2-1978 (R94) and ANSI B4.3-1978 (R94) define limits and fits for cylindrical parts, and provide tables listing preferred values.

The standard ANSI B4.2-1978 (R94) is essentially in accord with ISO R286.

ANSI B4.2 provides 22 basic deviations, each for the shaft (a to z plus js), and the hole (A to Z plus Js). International has 18 tolerance grades: IT 01, IT 0, and IT 1 through 16.

IT grades are roughly applied as follows: measuring tools, 01 to 7; fits, 5 to 11; material, 8 to 14; and large manufacturing tolerances, 12 to 16. See Table 6.2.41 for metric preferred fits.

Basic size—The basic size is the same for both members of a fit, and is the size to which limits or deviations are assigned. It is designated by 40 in 40H7.

Deviation—The algebraic difference between a size and the corresponding basic size.

Upper deviation—The algebraic difference between the maximum limit of size and the corresponding basic size.

Lower deviation—The algebraic difference between the minimum limit of size and the corresponding basic size.

Table 6.2.39 Limits of Interference for Force and Shrink Fits
(Limits are in thousandths of an inch on diameter)

Nominal size range, in	Class FN 1	FN 2	FN 3	FN 4	FN 5
0.04-0.12	0.05 0.5	0.2 0.85		0.3 0.95	0.3 1.3
0.12-0.24	0.1 0.6	0.2 1.0		0.95 1.2	1.3 1.7
0.24-0.40	0.1 0.75	0.4 1.4		0.6 1.6	0.5 2.0
0.40-0.56	0.1 0.8	0.5 1.6		0.7 1.8	0.6 2.3
0.56-0.71	0.2 0.9	0.5 1.6		0.7 1.8	0.8 2.5
0.71-0.95	0.2 1.1	0.6 1.9		0.8 2.1	1.0 3.0
0.95-1.19	0.3 1.2	0.6 1.9	0.8 2.1	1.0 2.3	1.3 3.3
1.19-1.58	0.3 1.3	0.8 2.4	1.0 2.6	1.5 3.1	1.4 4.0
1.58-1.97	0.4 1.4	0.8 2.4	1.2 2.8	1.8 3.4	2.4 5.0
1.97-2.56	0.6 1.8	0.8 2.7	1.3 3.2	2.3 4.2	3.2 6.2
2.56-3.15	0.7 1.9	1.0 2.9	1.8 3.7	2.8 4.7	4.2 7.2
3.15-3.94	0.9 2.4	1.4 3.7	2.1 4.4	3.6 5.9	4.8 8.4
3.94-4.73	1.1 2.6	1.6 3.9	2.6 4.9	4.6 6.9	5.8 9.4
4.73-5.52	1.2 2.9	1.9 4.5	3.4 6.0	5.4 8.0	7.5 11.6
5.52-6.30	1.5 3.2	2.4 5.0	3.4 6.0	5.4 8.0	9.5 13.6
6.30-7.09	1.8 3.5	2.9 5.5	4.4 7.0	6.4 9.0	9.5 13.6

Table 6.2.40 Preferred Sizes (Metric)

Basic size, mm First choice	Second choice	Basic size, mm First choice	Second choice	Basic size, mm First choice	Second choice
1		10		100	
	1.1		11		110
1.2		12		120	
	1.4		14		140
1.6		16		160	
	1.8		18		180
2		20		200	
	2.2		22		220
2.5		25		250	
	2.8		28		280
3		30		300	
	3.5		35		350
4		40		400	
	4.5		45		450
5		50		500	
	5.5		55		550
6		60		600	
	7		70		700
8		80		800	
	9		90		900
				1,000	

SOURCE: ANSI B4.2-1978 (R94), reproduced by permission.

Fundamental deviation—That one of the two deviations closest to the basic size. It is designated by the letter H in 40H7.

Tolerance—The difference between the maximum and minimum size limits on a part.

International tolerance grade (IT)—A group of tolerances which vary depending on the basic size, but which provide the same relative level of accuracy within a grade. It is designated by 7 in 40H7 (IT 7).

Tolerance zone—A zone representing the tolerance and its position in relation to the basic size. The symbol consists of the fundamental deviation letter and the tolerance grade number (i.e., H7).

Hole basis—The system of fits where the minimum hole size is basic. The fundamental deviation for a hole basis system is H.

Shaft basis—Maximum shaft size is basic in this system. Fundamental deviation is h. NOTE: Capital letters refer to the hole and lowercase letters to the shaft.

Clearance fit—A fit in which there is clearance in the assembly for all tolerance conditions.

Interference fit—A fit in which there is interference for all tolerance conditions. Table 6.2.40 lists preferred metric sizes. Table 6.2.41 lists preferred tolerance zone combinations for clearance, transition, and interference fits. Table 6.2.42 lists dimensions for the grades corresponding to preferred fits. Table 6.2.43*a* and *b* lists limits (numerical) of preferred hole-basis clearance, transitions, and interference fits.

Stresses Produced by Shrink or Press Fit

STEEL HUB ON STEEL SHAFT. The maximum equivalent stress, pounds per square inch, set up by a given press-fit allowance (in inches per inch of shaft diameter) is equal to $3x \times 10^7$, where x is the allowance per inch of shaft diameter (Baugher, *Trans. ASME*, 1931, p. 85). The press-fit pressures set up between a steel hub and shaft, for various ratios d/D between shaft and hub outside diameters, are given in Fig. 6.2.86. These curves are accurate to 5 percent even if the shaft is hollow, provided the inside shaft diameter is not over 25 percent of the outside. The equivalent stress given above is based on the maximum shear theory and is numerically equal to the radial-fit pressure added to the tangential tension in the hub. Where the shaft is hollow, with an inside diameter equal to more than about 25 percent of the outside diameter, the allowance in inches per inch to obtain an equivalent hub stress of 30,000 lb/in² may be determined by using Lamé's thick-cylinder formulas *(Jour. Appl. Mech.*, 1937, p. A-185). It should be noted that these curves hold only when the maximum equivalent stress is below the yield point; above the yield point, plastic flow occurs and the stresses are less than calculated.

Table 6.2.41 Description of Preferred Fits (Metric)

ISO symbol: Hole basis	ISO symbol: Shaft basis	Description
		Clearance fits
H11/c11	C11/h11	*Loose-running* fit for wide commercial tolerances or allowances on external members.
H9/d9	D9/h9	*Free-running* fit not for use where accuracy is essential, but good for large temperature variations, high running speeds, or heavy journal pressures.
H8/f7	F8/h7	*Close-running* fit for running on accurate machines and for accurate location at moderate speeds and journal pressures.
H7/g6	G7/h6	*Sliding fit* not intended to run freely, but to move and turn freely and locate accurately.
H7/h6	H7/h6	*Locational clearance* fit provides snug fit for locating stationary parts; but can be freely assembled and disassembled.
		Transition fits
H7/k6	K7/h6	*Locational transition* fit for accurate location, a compromise between clearance and interference.
H7/n6	N7/h6	*Locational transition* fit for more accurate location where greater interference is permissible.
		Interference fits
H7/p6*	P7/h6	*Locational interference* fit for parts requiring rigidity and alignment with prime accuracy of location but without special bore pressure requirements.
H7/s6	S7/h6	*Medium-drive* fit for ordinary steel parts or shrink fits on light sections, the tightest fit usable with cast iron.
H7/u6	U7/h6	*Force* fit suitable for parts which can be highly stressed or for shrink fits where the heavy pressing forces required are impractical.

More clearance ↑ — More interference ↓

* Transition fit for basic sizes in range from 0 through 3 mm.
SOURCE: ANSI B4.2-1978 (R94), reproduced by permission.

Table 6.2.42 International Tolerance Grades

Basic sizes: Over	Basic sizes: Up to and including	IT6	IT7	IT8	IT9	IT10	IT11
		Tolerance grades, mm					
0	3	0.006	0.010	0.014	0.025	0.040	0.060
3	6	0.008	0.012	0.018	0.030	0.048	0.075
6	10	0.009	0.015	0.022	0.036	0.058	0.090
10	18	0.011	0.018	0.027	0.043	0.070	0.110
18	30	0.013	0.021	0.033	0.052	0.084	0.130
30	50	0.016	0.025	0.039	0.062	0.100	0.160
50	80	0.019	0.030	0.046	0.074	0.120	0.190
80	120	0.022	0.035	0.054	0.087	0.140	0.220
120	180	0.025	0.040	0.063	0.100	0.160	0.250
180	250	0.029	0.046	0.072	0.115	0.185	0.290
250	315	0.032	0.052	0.081	0.130	0.210	0.320
315	400	0.036	0.057	0.089	0.140	0.230	0.360
400	500	0.040	0.063	0.097	0.155	0.250	0.400
500	630	0.044	0.070	0.110	0.175	0.280	0.440
630	800	0.050	0.080	0.125	0.200	0.320	0.500
800	1,000	0.056	0.090	0.140	0.230	0.360	0.560
1,000	1,250	0.066	0.105	0.165	0.260	0.420	0.660
1,250	1,600	0.078	0.125	0.195	0.310	0.500	0.780
1,600	2,000	0.092	0.150	0.230	0.370	0.600	0.920
2,000	2,500	0.110	0.175	0.280	0.440	0.700	1.100
2,500	3,150	0.135	0.210	0.330	0.540	0.860	1.350

SOURCE: ANSI B4.2-1978 (R94), reproduced by permission.

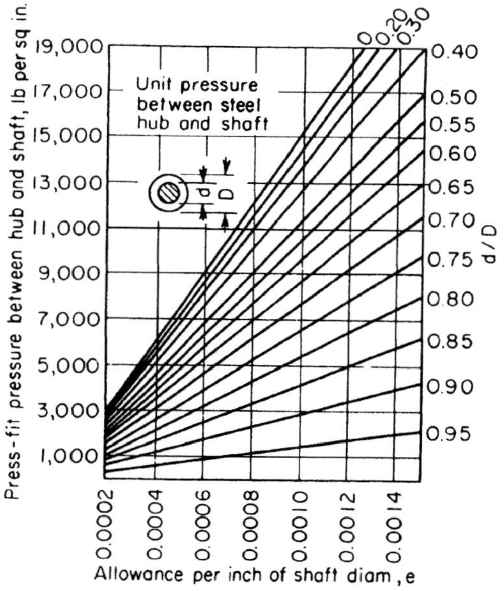

Figure 6.2.86 Press-fit pressures between steel hub and shaft.

Cast-Iron Hub on Steel Shaft Where the shaft is solid, or hollow with an inside diameter not over 25 percent of the outside diameter, Fig. 6.2.87 may be used to determine maximum tensile stresses in the cast-iron hub, resulting from the press-fit allowance; for various ratios d/D, Fig. 6.2.88 gives the press-fit pressures. These curves are based on a modulus of elasticity of 30×10^6 lb/in^2 for steel and 15×10^6 for cast iron. For a hollow shaft with an inside diameter more than about ¼ the outside, the Lamé formulas may be used.

Pressure Required in Making Press Fits The force required to press a hub on the shaft is given by $\pi fpdl$, where l is length of fit, p the unit press-fit pressure between shaft and hub, f the coefficient of friction, and d the shaft diameter. Values of f varying from 0.03 to 0.33 have been reported, the lower values being due to yielding of the hub as a consequence of too high a fit allowance; the average is around 0.10 to 0.15. (For additional data see Horger and Nelson, "Design Data and Methods," ASME, 1953, pp. 87-91.)

Table 6.2.43a Limits of Fits
(Dimensions in mm)

		Preferred hole basis clearance fits														
		Loose-running			Free-running			Close-running			Sliding			Locational clearance		
Basic size		Hole H11	Shaft c11	Fit	Hole H9	Shaft d9	Fit	Hole H8	Shaft f7	Fit	Hole H7	g6	Fit	Hole H7	Shaft h6	Fit
1	max	1.060	0.940	0.180	1.025	0.980	0.070	1.014	0.994	0.030	1.010	0.998	0.018	1.010	1.000	0.016
	min	1.000	0.880	0.060	1.000	0.955	0.020	1.000	0.984	0.006	1.000	0.992	0.002	1.000	0.994	0.000
1.2	max	1.260	1.140	0.180	1.225	1.180	0.070	1.214	1.194	0.030	1.210	1.198	0.018	1.210	1.200	0.016
	min	1.200	1.080	0.060	1.200	1.155	0.020	1.200	1.184	0.006	1.200	1.192	0.002	1.200	1.194	0.000
1.6	max	1.660	1.540	1.180	1.625	1.580	0.070	1.614	1.594	0.050	1.610	1.598	0.018	1.610	1.600	0.016
	min	1.600	1.480	0.060	1.600	1.555	0.020	1.600	1.584	0.006	1.600	1.592	0.002	1.600	1.594	0.000
2	max	2.060	1.940	0.180	2.025	1.980	0.070	2.014	1.994	0.030	2.010	1.998	0.018	2.010	2.000	0.016
	min	2.000	1.880	0.060	2.000	1.955	0.020	2.000	1.984	0.006	2.000	1.992	0.002	2.000	1.994	0.000
2.5	max	2.560	2.440	0.180	2.525	2.480	0.070	2.514	2.494	0.030	2.510	2.498	0.018	2.510	2.500	0.016
	min	2.500	2.380	0.060	2.500	2.455	0.020	2.500	2.484	0.006	2.500	2.492	0.002	2.500	2.494	0.000
3	max	3.060	2.940	0.180	3.025	2.980	0.070	3.014	2.994	0.030	3.010	2.998	0.018	3.010	3.000	0.016
	min	3.000	2.880	0.060	3.000	2.955	0.020	3.000	2.984	0.006	3.000	2.992	0.002	3.000	2.994	0.000
4	max	4.075	3.930	0.220	4.030	3.970	0.090	4.018	3.990	0.040	4.012	3.996	0.024	4.012	3.000	0.020
	min	4.000	3.855	0.070	4.000	3.940	0.030	4.000	3.978	0.010	4.000	3.988	0.004	4.000	3.992	0.000
5	max	5.075	4.930	0.220	5.030	4.970	0.090	5.018	4.990	0.040	5.012	4.996	0.024	5.012	5.000	0.020
	min	5.000	4.855	0.070	5.000	4.940	0.030	5.000	4.978	0.010	5.000	4.988	0.004	5.000	4.992	0.000
6	max	6.075	5.930	0.220	6.030	5.970	0.090	6.018	5.990	0.040	6.012	5.996	0.024	6.012	6.000	0.020
	min	6.000	5.855	0.070	6.000	5.940	0.030	6.000	5.978	0.010	6.000	5.988	0.004	6.000	5.992	0.000
8	max	8.090	7.920	0.260	8.036	7.960	0.112	8.022	7.987	0.050	8.015	7.995	0.029	8.015	8.000	0.024
	min	8.000	7.830	0.080	8.000	7.924	0.040	8.000	7.972	0.013	8.000	7.986	0.005	8.000	7.991	0.000
10	max	10.090	9.920	0.260	10.036	9.960	0.112	10.022	9.987	0.050	10.015	0.995	0.029	10.015	10.000	0.024
	min	10.000	9.830	0.080	10.000	9.924	0.040	10.000	9.972	0.013	10.000	9.986	0.005	10.000	9.991	0.000
12	max	12.110	11.905	0.315	12.043	11.950	0.136	12.027	11.984	0.061	12.018	11.994	0.035	12.018	12.000	0.029
	min	12.000	11.795	0.095	12.000	11.907	0.050	12.000	11.966	0.016	12.000	11.983	0.006	12.000	11.989	0.000
16	max	16.110	15.905	0.315	16.043	15.950	0.136	16.027	15.984	0.061	16.018	15.994	0.035	16.018	16.000	0.029
	min	16.000	15.795	0.095	16.000	15.907	0.050	16.000	15.966	0.016	16.000	15.983	0.006	16.000	15.989	0.000
20	max	20.130	19.890	0.370	20.052	19.935	0.169	20.033	19.980	0.074	20.021	19.993	0.041	20.021	20.000	0.034
	min	20.000	19.760	0.110	20.000	19.883	0.065	20.000	19.959	0.020	20.000	19.980	0.007	20.000	19.987	0.000
25	max	25.130	24.890	0.370	25.052	24.935	0.169	25.033	24.980	0.074	25.021	24.993	0.041	25.021	25.000	0.034
	min	25.000	24.760	0.110	25.000	24.883	0.065	25.000	24.959	0.020	25.000	24.980	0.007	25.000	24.987	0.000
30	max	30.130	29.890	0.370	30.052	29.935	0.169	30.033	29.980	0.074	30.021	29.993	0.041	30.021	30.000	0.034
	min	30.000	29.760	0.110	30.000	29.883	0.065	30.000	29.959	0.020	30.000	29.980	0.007	30.000	29.987	0.000
40	max	40.160	39.880	0.440	40.062	39.920	0.204	40.039	39.975	0.089	40.025	39.991	0.050	40.025	40.000	0.041
	min	40.000	39.720	0.120	40.000	39.858	0.080	40.000	39.950	0.025	40.000	39.975	0.009	40.000	39.984	0.000
50	max	50.160	49.870	0.450	50.062	49.920	0.204	50.039	49.975	0.089	50.025	49.991	0.050	50.025	50.000	0.041
	min	50.000	49.710	0.130	50.000	49.858	0.080	50.000	49.950	0.025	50.000	49.975	0.009	50.000	49.984	0.000
60	max	60.190	59.860	0.520	60.074	59.900	0.248	60.046	59.970	0.106	60.030	59.990	0.059	60.030	60.000	0.049
	min	60.000	59.670	0.140	60.000	59.826	0.100	60.000	59.940	0.030	60.000	59.971	0.010	60.000	59.981	0.000
80	max	80.190	79.850	0.530	80.074	79.900	0.248	80.046	79.970	0.106	80.030	79.990	0.059	80.030	80.000	0.049
	min	80.000	79.660	0.150	80.000	79.826	0.100	80.000	79.940	0.030	80.000	70.971	0.010	80.000	79.981	0.000
100	max	100.220	99.830	0.610	100.087	99.880	0.294	100.054	99.964	0.125	100.035	99.988	0.069	100.035	100.000	0.057
	min	100.000	99.610	0.170	100.000	99.793	0.120	100.000	99.929	0.036	100.000	99.966	0.012	100.000	99.978	0.000
120	max	120.220	119.820	0.620	120.087	119.880	0.294	120.054	119.964	0.125	120.035	119.988	0.069	120.035	120.000	0.057
	min	120.000	119.600	0.180	120.000	119.793	0.120	120.000	119.929	0.036	120.000	119.966	0.012	120.000	119.978	0.000
160	max	160.250	159.790	0.710	160.100	159.855	0.345	160.063	159.957	0.146	160.040	159.986	0.079	160.040	160.000	0.065
	min	160.000	159.540	0.210	160.000	159.755	0.145	160.000	159.917	0.043	160.000	159.961	0.014	160.000	159.975	0.000

Table 6.2.43b Limits of Fits
(Dimensions in mm)

		Preferred hole basis transition and interference fits														
		Locational transition			Locational transition			Locational interference			Medium drive			Force		
Basic size		Hole H7	Shaft k6	Fit	Hole H7	Shaft n6	Fit	Hole H7	Shaft p6	Fit	Hole H7	Shaft s6	Fit	Hole H7	Shaft u6	Fit
1	max	1.010	1.006	0.010	1.010	1.010	0.006	1.010	1.012	0.004	1.010	1.020	−0.004	1.010	1.024	−0.008
	min	1.000	1.000	−0.006	1.000	1.004	−0.010	1.000	1.006	−0.012	1.000	1.014	−0.020	1.000	1.018	−0.024
1.2	max	1.210	1.206	0.010	1.210	1.210	0.006	1.210	1.212	0.004	1.210	1.220	−0.004	1.210	1.224	−0.008
	min	1.200	1.200	−0.006	1.200	1.204	−0.010	1.200	1.206	−0.012	1.200	1.214	−0.020	1.200	1.218	−0.024
1.6	max	1.610	1.606	0.010	1.610	1.610	0.006	1.610	1.612	0.004	1.610	1.620	−0.004	1.610	1.624	−0.008
	min	1.610	1.600	−0.006	1.600	1.604	−0.010	1.600	1.606	−0.012	1.600	1.614	−0.020	1.600	1.618	−0.024
2	max	2.010	2.006	0.010	2.010	2.010	0.006	2.010	2.012	0.004	2.010	2.020	−0.004	2.010	2.024	−0.008
	min	2.000	2.000	−0.006	2.000	2.004	−0.010	2.000	2.006	−0.012	2.000	2.014	−0.020	2.000	2.018	−0.024
2.5	max	2.510	2.506	0.010	2.510	2.510	0.006	2.510	2.512	0.004	2.510	2.520	−0.004	2.510	2.524	−0.008
	min	2.500	2.500	−0.006	2.500	2.504	−0.010	2.500	2.506	−0.012	2.500	2.514	−0.020	2.500	2.518	−0.024
3	max	3.010	3.006	0.010	3.010	3.010	0.006	3.010	3.012	0.004	3.010	3.020	−0.004	3.010	3.024	−0.008
	min	3.000	3.000	−0.006	3.000	3.004	−0.010	3.000	3.006	−0.012	3.000	3.014	−0.020	3.000	3.018	−0.024
4	max	4.012	4.009	0.011	4.012	4.016	0.004	4.012	4.020	0.000	4.012	4.027	−0.007	4.012	4.031	−0.011
	min	4.000	4.001	−0.009	4.000	4.008	−0.016	4.000	4.012	−0.020	4.000	4.019	−0.027	4.000	4.023	−0.031
5	max	5.012	5.009	0.011	5.012	5.016	0.004	5.012	5.020	0.000	5.012	5.027	−0.007	5.012	5.031	−0.011
	min	5.000	5.001	−0.009	5.000	5.008	−0.016	5.000	5.012	−0.020	5.000	5.019	−0.027	5.000	5.023	−0.031
6	max	6.012	6.009	0.011	6.012	6.016	0.004	6.012	6.020	0.000	6.012	6.027	−0.007	6.012	6.031	−0.011
	min	6.000	6.001	−0.009	6.000	6.008	−0.016	6.000	6.012	−0.020	6.000	6.019	−0.027	6.000	6.023	−0.031
8	max	8.015	8.010	0.014	8.015	8.019	0.005	8.015	8.024	0.000	8.015	8.032	−0.008	8.015	8.037	−0.013
	min	8.000	3.001	−0.010	8.000	8.010	−0.019	8.000	8.015	−0.024	8.000	8.023	−0.032	8.000	8.028	−0.037
10	max	10.015	10.010	0.014	10.015	10.019	0.005	10.015	10.024	0.000	10.015	10.032	−0.008	10.015	10.037	−0.013
	min	10.000	10.031	−0.010	10.000	10.010	−0.019	10.000	10.015	−0.024	10.000	10.023	−0.032	10.000	10.028	−0.037
12	max	12.018	12.012	0.017	12.018	12.023	0.006	12.018	12.029	0.000	12.018	12.039	−0.010	12.018	12.044	−0.015
	min	12.000	12.001	−0.012	12.000	12.012	−0.023	12.000	12.018	−0.029	12.000	12.028	−0.039	12.000	12.033	−0.044
16	max	16.018	16.012	0.017	16.018	16.023	0.006	16.018	16.029	0.000	16.018	16.039	−0.010	16.018	16.044	−0.015
	min	16.000	16.001	−0.012	16.000	16.012	−0.023	16.000	16.018	−0.029	16.000	16.028	−0.039	16.000	16.033	−0.044
20	max	20.021	20.015	0.019	20.021	20.028	0.006	20.021	20.035	−0.001	20.021	20.048	−0.014	20.021	20.054	−0.020
	min	20.000	20.002	−0.015	20.000	20.015	−0.028	20.000	20.022	−0.035	20.000	20.035	−0.048	20.000	20.041	−0.054
25	max	25.021	25.015	0.019	25.021	25.028	0.006	25.021	25.035	−0.001	25.021	25.048	−0.014	25.021	25.061	−0.027
	min	25.000	25.002	−0.015	25.000	25.015	−0.028	25.000	25.022	−0.035	25.000	25.035	−0.048	25.000	25.048	−0.061
30	max	30.021	30.015	0.019	30.021	30.028	0.006	30.021	30.035	−0.001	30.021	30.048	−0.014	30.021	30.061	−0.027
	min	30.000	30.002	−0.015	30.000	30.015	−0.028	30.000	30.022	−0.035	30.000	30.035	−0.048	30.000	30.048	−0.061
40	max	40.025	40.018	0.023	40.025	40.033	0.008	40.025	40.042	−0.001	40.025	40.059	−0.018	40.025	40.076	−0.035
	min	40.000	40.002	−0.018	40.000	40.017	−0.033	40.000	40.026	−0.042	40.000	40.043	−0.059	40.000	40.060	−0.076
50	max	50.025	50.018	0.023	50.025	50.033	0.008	50.025	50.042	−0.001	50.025	50.059	−0.018	50.025	50.086	−0.045
	min	50.000	50.002	−0.018	50.000	50.017	−0.033	50.000	50.026	−0.042	50.000	50.043	−0.059	50.000	50.070	−0.086
60	max	60.030	60.021	0.028	60.030	60.039	0.010	60.030	60.051	−0.002	60.030	60.072	−0.023	60.030	60.106	−0.057
	min	60.000	60.002	−0.021	60.000	60.020	−0.039	60.000	60.032	−0.051	60.000	60.053	−0.072	60.000	60.087	−0.106
80	max	80.030	80.021	0.028	80.030	80.039	0.010	80.030	80.051	−0.002	80.030	80.078	−0.029	80.030	80.121	−0.072
	min	80.000	80.002	−0.021	80.000	80.020	−0.039	80.000	80.032	−0.051	80.000	80.059	−0.078	80.000	80.102	−0.121
100	max	100.035	100.025	0.032	100.035	100.045	0.012	100.035	100.059	−0.002	100.035	100.093	−0.036	100.035	100.146	−0.089
	min	100.000	100.003	−0.025	100.000	100.023	−0.045	100.000	100.037	−0.059	100.000	100.071	−0.093	100.000	100.124	−0.146
120	max	120.035	120.025	0.032	120.035	120.045	0.012	120.035	120.059	−0.002	120.035	120.101	−0.044	120.035	120.166	−0.109
	min	120.000	120.003	−0.025	120.000	120.023	−0.045	120.000	120.037	−0.059	120.000	120.079	−0.101	120.000	120.144	−0.166
160	max	160.040	160.028	0.037	160.040	160.052	0.013	160.040	160.068	−0.003	160.040	160.125	−0.060	160.040	160.215	−0.150
	min	160.000	160.003	−0.028	160.000	160.027	−0.052	160.000	160.043	−0.068	160.000	160.100	−0.125	160.000	160.190	−0.215

SOURCE: ANSI B4.2-1978 (R94), reprinted by permission.

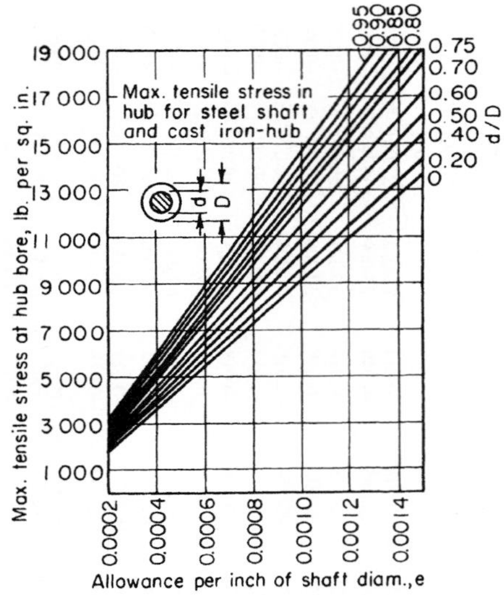

Figure 6.2.87 Variation of tensile stress in cast-iron hub in press-fit allowance.

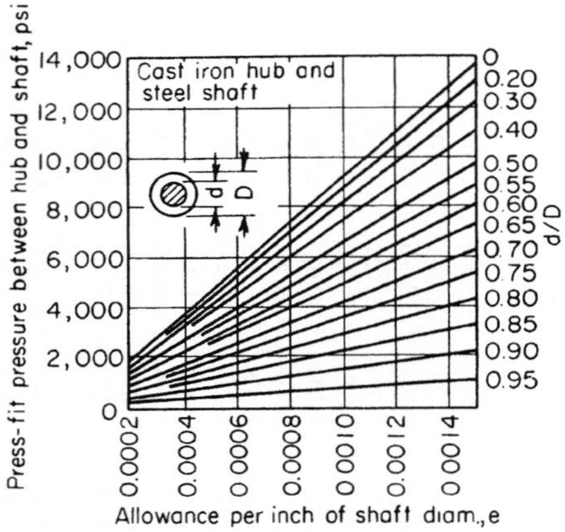

Figure 6.2.88 Press-fit pressures between cast-iron hub and shaft.

Torsional Holding Ability The torque required to cause complete slippage of a press fit is given by $T = \frac{1}{2}\pi fpld^2$. Local slippage will usually occur near the end of the fit at much lower torques. If the torque is alternating, stress concentration and rubbing corrosion will occur at the hub face so that, eventually, fatigue failure may occur at considerably lower torques. Only in cases of static torque application is it justifiable to use ultimate torque as a basis for design.

A designer can often improve shrink-, press-, and slip-fit cylindrical assemblies with adhesives. When applied, adhesives can achieve high frictional force with attendant greater torque transmission without extra bulk, and thus augment or even replace press fits, compensate for differential thermal expansion, make fits with leakproof seals, eliminate backlash and clearance, etc.

The adhesives used are the anaerobic (see Sec. 4.8, "Adhesives") variety, such as Loctite products. Such adhesives destabilize and tend to harden when deprived of oxygen. Design suggestions on the use of such adhesives appear in industrial catalogs.

SHAFTS, AXLES, AND CRANKS

Most shafts are subject to combined bending and torsion, either of which may be steady or variable. Impact conditions, such as sudden starting and stopping, will cause momentary peak stresses greater than those related to the steady or variable portions of operation.

Design of shafts requires a theory of failure to express a stress in terms of loads and shaft dimensions, and an allowable stress as fixed by material strength and safety factor. Maximum shear theory of failure and distortion energy theory of failure are the two most commonly used in shaft design. Material strengths can be estimated from any one of several analytic representations of combined-load fatigue test data, starting from the linear (Soderberg, modified Goodman) which tend to give conservative designs to the nonlinear (Gerber parabolic, quadratic, Kececioglu, Bagci) which tend to give less conservative designs.

When linear representations of material strengths are used, and where both bending and torsion stresses have steady and variable components, the maximum shear theory and the distortion energy theory lead to somewhat similar formulations:

$$d = \left\{ \varepsilon \frac{n}{\pi} \left[\left(\frac{T_a}{S_{se}} + \frac{T_m}{S_{sy}} \right)^2 + \left(\frac{M_a}{S_{se}} + \frac{M_m}{S_{sy}} \right)^2 \right]^{1/2} \right\}^{1/3}$$

where $\varepsilon = 32$ (maximum shear theory) or 48 (distortion energy theory); d = shaft diameter; n = safety factor; T_a = amplitude torque = $(T_{max} - T_{min})/2$; T_m = mean torque = $(T_{max} + T_{min})/2$; M_a = amplitude bending moment = $(M_{max} - M_{min})/2$; M_m = mean bending moment = $(M_{max} + M_{min})/2$; S_{sy} = yield point in shear; S_{se} = completely corrected shear endurance limit = $S'_{se}k_a k_b k_c k_d/K_f$; S'_{se} = statistical average endurance limit of mirror finish, standard size, laboratory test specimen at standard room temperature; k_a = surface factor, a decimal to adjust S'_{se} for other than mirror finish; k_b = size factor, a decimal to adjust S'_{se} for other than standard test size; k_c = reliability factor, a decimal to adjust S'_{se} to other than its implied statistical average of 50 percent safe, 50 percent fail rate (50 percent reliability); k_d = temperature factor, decimal, to adjust S'_{se} to other than room temperature; $K_f = 1 + q(K_t - 1)$ = actual or fatigue stress concentration factor, a number greater than unity to adjust the nominal stress implied by T_a and M_a to a peak stress as induced by stress-raising conditions such as holes, fillets, keyways, press fits, etc. (K_f for T_a need not necessarily be the same as K_f for M_a); q = notch sensitivity; K_t = theoretical or geometric stress concentration factor.

For specific values of endurance limits and various factors, the reader is referred to the technical literature (e.g., ASTM, NASA technical reports, ASME technical papers) or various books [e.g., "Machinery's Handbook" (Industrial Press, New York) and machine design textbooks].

If one allows for a variation of at least 15 percent, the following approximation is useful for the endurance limit in bending: $S'_{se} = 0.5S_{ut}$. This becomes for maximum shear theory $S'_{se} = 0.5(0.5S_{ut})$ and for distortion energy theory $S'_{se} = 0.577(0.5S_{ut})$.

Representative or approximate values for the various factors mentioned above were abstracted from Shigley, "Mechanical Engineering Design," McGraw-Hill, and appear below with permission.

Surface Factor k_a

	S_{ut}, kips			
Surface condition	60	120	180	240
Polished	1.00	1.00	1.00	1.00
Ground	0.89	0.89	0.89	0.89
Machined or cold-drawn	0.84	0.71	0.66	0.63
Hot-rolled	0.70	0.50	0.39	0.31
As forged	0.54	0.36	0.27	0.20

Size Factor k_b

$$k_b = \begin{cases} 0.869d^{-0.097} & 0.3 \text{ in} < d \le 10 \text{ in} \\ 1 & d \le 0.3 \text{ in or } d \le 8 \text{ mm} \\ 1.189d^{-0.097} & 8 \text{ mm} < d < 250 \text{ mm} \end{cases}$$

Reliability Factor k_c

Reliability, %	k_c
50	1.00
90	0.89
95	0.87
99	0.81

Temperature Factor k_d

Temperature		
°F	°C	k_d
840	450	1.00
940	482	0.71
1,020	550	0.42

Notch Sensitivity q

	Notch radius r, in			
S_{ut}, kips	0.02	0.06	0.10	0.14
60	0.56	0.70	0.74	0.78
100	0.68	0.79	0.83	0.85
150	0.80	0.90	0.91	0.92
200	0.90	0.95	0.96	0.96

See Table 6.2.44 for fatigue stress concentration factors for plain press fits.

A general representation of material strengths (Marin, Design for Fatigue Loading, *Mach. Des.* **29,** no. 4, Feb. 21, 1957, pp. 128-131, and series of the same title) is given as

$$\left(\frac{S_a}{S_e}\right)^m + \left(\frac{KS_m}{S_{ut}}\right)^P = 1$$

where S_a = variable portion of material strength; S_m = mean portion of material strength; S_e = adjusted endurance limit; S_{ut} = ultimate strength. Table 6.2.45 lists the constants m, K, and P for various failure criteria.

For purposes of design, safety factors are introduced into the equation resulting in:

$$\left(\frac{\sigma'_{a,p}}{S_e/n_{se}}\right)^m + \left(\frac{K\sigma'_{m,p}}{S_{ut}/n_{ut}}\right)^P = 1$$

where

$$\sigma'_{a,p} = \frac{1}{\pi d^3}\sqrt{(32n_{Ma}M_a)^2 + 3(16n_{\tau a}T_a)^2}$$

$$\sigma'_{m,p} = \frac{1}{\pi d^3}\sqrt{(32n_{Mm}M_m)^2 + 3(16n_{\tau m}T_m)^2}$$

and n_{ij} = safety factor pertaining to a particular stress (i.e., n_a = safety factor for amplitude shear stress).

Stiffness of shafting may become important where critical speeds, vibration, etc., may occur. Also, the lack of sufficient stiffness in shafts may give rise to bearing troubles. Critical speeds of shafts in torsion or bending and shaft deflections may be calculated using the methods of Sec. 3. For shafts of variable diameter see Spotts, "Design of Machine Elements," Prentice-Hall. In order to avoid trouble where sleeve bearings are used, the angular deflections at the bearings in general must be kept within certain limits. One rule is to make the shaft deflection over the bearing width equal to a small fraction of the oil-film thickness. Note that since stiffness is proportional to the modulus of elasticity, alloy-steel shafts are no stiffer than carbon-steel shafts of the same diameter.

Crankshafts For calculating the torsional stiffness of crankshafts, the formulas given in Sec. 3 may be used.

Marine-engine shafts and **diesel-engine crankshafts** should be designed not only for strength but for avoidance of critical speed. (See Applied Mechanics, *Trans. ASME*, **50,** no. 8, for methods of calculating critical speeds of diesel engines.)

Table 6.2.45 Constants for Use in $(S_a/S_e)^m + (KS_m/S_{ut})^P = 1$

Failure theory	K	P	m
Soderberg	S_{ut}/S_y	1	1
Bagci	S_{ut}/S_y	4	1
Modified Goodman	1	1	1
Gerber parabolic	1	2	1
Kececioglu	1	2	m†
Quadratic (elliptic)	1	2	2

$$^{\dagger}m = \begin{cases} 0.8914 \text{ UNS G 10180 } H_B = 130 \\ 0.9266 \text{ UNS G 10380 } H_B = 164 \\ 1.0176 \text{ UNS G 41300 } H_B = 207 \\ 0.9685 \text{ UNS G 43400 } H_B = 233 \end{cases}$$

PULLEYS, SHEAVES, AND FLYWHEELS

Arms of pulleys, sheaves, and flywheels are subjected to stresses due to condition of founding, to details of construction (such as split or solid), and to conditions of service, which are difficult to analyze. For these reasons, no accurate stress relations can be established, and the following formulas must be understood to be only approximately correct. It has been established experimentally by Benjamin (*Am. Mach.*, Sept. 22, 1898) that thin-rim pulleys do not distribute equal loads to the several pulley arms. For these, it will be safe to assume the tangential force on the pulley rim as acting on half of the number of arms. Pulleys with comparatively thick rims, such as engine band wheels, have all the arms taking the load. Furthermore, while the stress action in the arms is similar to that in a beam fixed at both ends, the amount of restraint at the rim

Table 6.2.44 K_f Values for Plain Press Fits
(Obtained from fatigue tests in bending)

Shaft material	Shaft diam, in	Collar or hub material	K_f	Remarks
0.42% carbon steel	1⅝	0.42% carbon steel	2.0	No external reaction through collar
0.45% carbon axle steel	2	Ni-Cr-Mo steel (case-hardened)	2.3	No external reaction through collar
0.45% carbon axle steel	2	Ni-Cr-Mo steel (case-hardened)	2.9	External reaction taken through collar
Cr-Ni-Mo steel (310 Brinell)	2	Ni-Cr-Mo steel (case-hardened)	3.9	External reaction taken through collar
2.6% Ni steel (57,000 lb/in² fatigue limit)	2	Ni-Cr-Mo steel (case-hardened)	3.3-3.8	External reaction taken through collar
Same, heat-treated to 253 Brinell	2	Ni-Cr-Mo steel (case-hardened)	3.0	External reaction taken through collar

depending on the rim's elasticity, it may nevertheless be assumed for purposes of design that cantilever action is predominant. The bending moment at the hub in arms of thin-rim pulleys will be $M = PL/(\frac{1}{2}N)$, where M = bending moment, in-lb; P = tangential load on the rim, lb; L = length of the arm, in; and N = number of arms. For thick-rim pulleys and flywheels, $M = PL/N$.

For arms of elliptical section having a width of two times the thickness, where E = **width of arm section** at the rim, in, and s_t = intensity of tensile stress, lb/in²

$$E = \sqrt[3]{40PL/(s_tN)} \text{ (thin rim)} = \sqrt[3]{20PL/(s_tN)} \text{ (thick rim)}$$

For single-thickness belts, P may be taken as $50B$ lb and for double-thickness belts $P = 75B$ lb, where B is the width of pulley face, in. Then $E = K \times \sqrt[3]{BL/(s_tN)}$, where k has the following values: for thin rim, single belt, 13; thin rim, double belt, 15; thick rim, single belt, 10; thick rim, double belt, 12. For cast iron of good quality, s_t due to bending may be taken at 1,500 to 2,000. The arm section at the rim may be made from $\frac{2}{3}$ to $\frac{3}{4}$ the dimensions at the hub.

For high-speed pulleys and flywheels, it becomes necessary to check the arm for tension due to rim expansion. It will be safe to assume that each arm is in tension due to one-half the centrifugal force of that portion of the rim which it supports. That is, $T = As_t = Wv^2/(2NgR)$, where T = tension in arm, lb; N = number of arms; v = speed of rim, ft/s; R = radius of pulley, ft; A = area of arm section, in²; W = weight of pulley rim, lb; and s_t = intensity of tensile stress in arm section, lb/in², whence $s_t = WRn^2/(6{,}800NA)$, where n = r/min of pulley.

Arms of flywheels having heavy rims may be subjected to severe stress action due to the inertia of the rim at sudden load changes. There being no means of predicting the probable maximum to which the inertia may rise, it will be safe to make the arms equal in strength to $\frac{3}{4}$ of the shaft strength in torsion. Accordingly, for elliptical arm sections,

$$N \times 0.5E^3 s_t = \tfrac{3}{4} \times 2s_s d^3 \quad \text{or} \quad E = 1.4d\sqrt[3]{s_s/(s_tN)}$$

For steel shafts with $s_s = 8{,}000$ and cast-iron arms with $s = 1{,}500$,

$$E = 2.4d/\sqrt[3]{N} = 1.3d \text{ (for 6 arms)} = 1.2d \text{ (for 8 arms)}$$

where $2E$ = width of elliptical arm section at hub, in (thickness = E), and d = shaft diameter, in.

Rims of belted pulleys cast whole may have the following proportions (see Fig. 6.2.89):

$$t_2 = \tfrac{3}{4}h + 0.005D \qquad t_1 = 2t_2 + C \qquad W = \tfrac{9}{8}B \text{ to } \tfrac{5}{4}B$$

where h = belt thickness, $C = \frac{1}{24}W$, and B = belt width, all in inches.

Engine band wheels, flywheels, and pulleys run at **high speeds** are subjected to the following stress actions in the rim:

Considering the rim as a free ring, i.e., without arm restraint, and made of cast iron or steel, $s_t = v^2/10$ (approx.), where s_t = intensity of tensile stress, lb/in², and v = rim speed, ft/s. For beam action between the arms of a solid rim, $M = Pl/12$ (approx.), where M = bending moment in rim, in · lb; P = centrifugal force of that portion of rim between arms, lb, and l = length of rim between arms, in; from which $s_t = WR^2n^2/(450N^2Z)$, where W = weight of entire rim, lb; R = radius of wheel, ft; n = r/min of wheel; and Z = section modulus of rim section, in³. In case the rim section is of the forms shown in Fig. 6.2.89, care must be taken that the flanges do not reduce the section modulus from that of the rectangular section. For **split rims** fastened with bolts the stress analysis is as follows:

Let w = weight of rim portion, lb (with length L, in) lb; w_1 = weight of lug, lb; L_1 = lever arm of lug, in; and s_t = intensity of tensile stress lb/in² in rim section joining arm. Then $s_t = 0.00034n^2R(w_1L_1 + wL/2)/Z$ where n = r/min of wheel; R = wheel radius, ft; and Z = section modulus of rim section, in³. The above equation gives the value of s_t for bending when the bolts are loose, which is the worst possible condition that may arise. On this basis of analysis, s_t should not be greater than 8,000 lb/in². The stress due to bending in addition to the stress due to rim expansion as analyzed previously will be the probable maximum intensity of stress for which the rim should be checked for strength. The flange bolts, because of their position, do not materially relieve the bending action. In case a tie rod leads from the flange to the hub, it will be *safe* to consider it as an additional factor of safety. When the tie rod is kept tight, it very materially strengthens the rim.

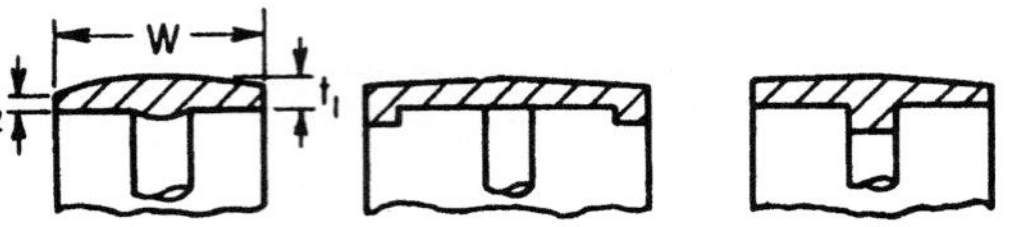

Figure 6.2.89 Rims for belted pulleys.

A more accurate method for calculating maximum stresses due to centrifugal force in flywheels with arms cast integral with the rim is given by Timoshenko, "Strength of Materials," Pt. II, 1941, p. 98. More exact equations for calculating stresses in the arms of flywheels and pulleys due to a combination of belt pull, centrifugal force, and changes in velocity are given by Heusinger, *Forschung*, 1938, p. 197 (http://link.springer.com/article/10.1007%2FBF02583230). In both treatments, shrinkage stresses in the arms due to casting are neglected.

Large flywheels for high rim speeds and severe working conditions (as for rolling-mill service) have been made from flat-rolled steel plates with holes bored for the shaft. A group of such plates may be welded together by circumferential welds to form a large flywheel. By this means, the welds do not carry direct centrifugal loads, but serve merely to hold the parts in position. Flywheels up to 15-ft diam have been constructed in this way.

BELT DRIVES

Flat-Belt Drives

The primary drawback of flat belts is their reliance on belt tension to produce frictional grip over the pulleys. Such high tension can shorten bearing life. Also, tracking may be a problem. However, flat belts, being thin, are not subject to centrifugal loads and so work well over small pulleys at high speeds in ranges exceeding 9,000 ft/min. In light service flat belts can make effective clutching drives. Flat-belt drives have efficiencies of about 90 percent, which compares favorably to geared drives. Flat belts are also quiet and can absorb torsional vibration readily.

Leather belting has an ultimate tensile strength ranging from 3,000 to 5,000 lb/in². Average values of breaking strength of good oak-tanned belting are as follows: single (double) in solid leather 900 (1,400); at riveted joint 600 (1,200); at laced joint 350 lb/in of width. Well-made cemented joints have strengths equal to the belt, leather-laced and riveted joints about one-third to two-thirds as strong, and wire-laced joints about 85 to 95 percent as strong.

Rubber belting is made from fabric or cord impregnated and bound together by vulcanized rubber compounds. The fabric or cord may be of cotton or rayon. Nylon cord and steel cord or cable are also available. Advantages are high tensile strength, strength to hold metal fasteners satisfactorily, and resistance to deterioration by moisture. The best rubber fabric construction for most types of service is made from hard or tight-woven fabric with a "skim coat" or thin layer of rubber between plies. The cord type of construction allows the use of smaller pulley diameters than the fabric type, and also develops less stretch in service. It must be used in the endless form, except in cases where the oil-field type of clamp may be used.

Initial tensions in rubber belts run from 15 to 25 lb/ply/in width. A common rule is to cut belts 1 percent less than the minimum tape-line measurement around the pulleys. For heavy loads, a $1\frac{1}{2}$ percent allowance is usually required, although, because of shrinkage, less initial tension is required for wet or damp conditions. Initial tensions of 25 lb/ply/in may overload shafts or bearings. Maximum safe tight-side tensions for rubber belts are as follows:

Duck weight, oz	28	32	32.66	34.66	36
Tension, lb/ply/in width	25	28	30	32	35

Table 6.2.46a Service Factors *S*

Application	Squirrel-cage ac motor: Normal torque, line start	Squirrel-cage ac motor: High torque	Wound rotor ac motor (slip ring)	Single-phase capacitor motor	DC shunt-wound motor	Diesel engine 4 or more cyl, above 700 r/min
Agitators	1.0-1.2	1.2-1.4	1.2			
Compressors	1.2-1.4		1.4	1.2	1.2	1.2
Belt conveyors (ore, coal, sand)	—	1.4	—	—	1.2	
Screw conveyors	—	1.8	—	—	1.6	
Crushing machinery	—	1.6	1.4	—	—	1.4-1.6
Fans, centrifugal	1.2	—	1.4	—	1.4	1.4
Fans, propeller	1.4	2.0	1.6	—	1.6	1.6
Generators and exciters	1.2	—	—	—	1.2	2.0
Line shafts	1.4	—	1.4	1.4	1.4	1.6
Machine tools	1.0-1.2	—	1.2-1.4	1.0	1.0-1.2	
Pumps, centrifugal	1.2	1.4	1.4	1.2	1.2	
Pumps, reciprocating	1.2-1.4	—	1.4-1.6	—	—	1.8-2.0

Centrifugal forces at high speeds require higher tight-side tensions to carry rated horsepower.

Rubber belting may be bought in endless form or made endless in the field by means of a vulcanized splice produced by a portable electric vulcanizer. For endless belts the drive should provide take-up of 2 to 4 percent to allow for length variation as received and for stretch in service. The amount of take-up will vary with the type of belt used. For certain drives, it is possible to use endless belts with no provision for take-up, but this involves a heavier belt and a higher initial unit tension than would be the case otherwise. Ultimate tensile strength of rubber belting varies from 280 to 600 lb or more per inch width per ply. The weight varies from 0.02 to 0.03 or more lb/in width/ply. Belts with steel reinforcement are considerably heavier. For horsepower ratings of rubber belts, see Table 6.2.46*c*.

Arrangements for Belt Drives In belt drives, the centerline of the belt advancing on the pulley should lie in a plane passing through the midsection of the pulley at right angles to the shaft. Shafts inclined to each other require connections as shown in Fig. 6.2.90*a*. In case guide pulleys are needed their positions can be determined as shown in Fig. 6.2.90*a*, *b*, and *c*. In Fig. 6.2.90*d* the center circles of the two pulleys to be connected are set in correct relative position in two planes, being the angle between the planes (= supplement of angle between shafts). If any two points as *E* and *F* are assumed on the line of intersection *MN* of the planes, and tangents *EG*, *EH*, *FJ*, and *FK* are drawn from them to the circles, the center circles of the guide pulleys must be so arranged that these tangents are also tangents to them, as shown. In other words, the middle planes of the guide pulleys must lie in the planes *GEH* and *JFK*.
When these conditions are met, the belts will run in either direction on the pulleys.

To avoid the necessity of taking up the **slack** in belts which have become stretched and permanently lengthened, a **belt tightener** such as shown in Fig. 6.2.91 may be employed. It should be placed on the slack side of the belt and nearer the driving pulley than the driven pulley. Pivoted motor drives may also be used to maintain belt tightness with minimum initial tension.

Length of Belt for a Given Drive The length of an **open belt** for a given drive is equal to $L = 2C + 1.57(D + d) + (D - d)^2/(4C)$,

Table 6.2.46b Arc of Contact Factor *K*—Rubber Belts

Arc of contact, deg	140	160	180	200	220
Factor *K*	0.82	0.93	1.00	1.06	1.12

Table 6.2.46c Horsepower Ratings of Rubber Belts
(hp/in of belt width for 180° wrap)

	Ply	Belt speed, ft/min: 500	1,000	1,500	2,000	2,500	3,000	4,000	5,000	6,000	7,000	8,000
32-oz fabric	3	0.7	1.4	2.1	2.7	3.3	3.9	4.9	5.6	6.0		
	4	0.9	1.9	2.8	3.6	4.4	5.2	6.5	7.4	7.9		
	5	1.2	2.3	3.4	4.5	5.5	6.5	8.1	9.2	9.8		
	6	1.4	2.8	4.1	5.4	6.6	7.8	9.6	11.0	11.7		
	7	1.6	3.2	4.7	6.2	7.7	9.0	11.2	12.8	13.6		
	8	1.8	3.6	5.3	7.0	8.7	10.2	12.7	14.6	15.5		
32-oz hard fabric	3	0.7	1.5	2.2	2.9	3.5	4.1	5.1	5.8	6.2	6.1	5.5
	4	1.0	2.0	3.0	3.9	4.7	5.5	6.8	7.8	8.3	8.1	7.3
	5	1.3	2.5	3.7	4.9	5.9	6.9	8.5	9.8	10.3	9.1	9.0
	6	1.5	3.0	4.5	5.9	7.1	8.3	10.2	11.7	12.3	12.1	10.7
	7	1.7	3.5	5.2	6.9	8.3	9.7	11.9	13.6	14.3	14.1	12.4
	8	1.9	4.0	5.9	7.9	9.5	11.1	13.6	15.5	16.3	16.0	14.1
	9	2.1	4.5	6.6	8.9	10.6	12.4	15.3	17.4	18.3	17.9	15.8
	10	2.3	5.0	7.3	9.8	11.7	13.7	17.0	19.3	20.3	19.8	17.5
No. 70 rayon cord	3	1.6	3.1	4.6	6.0	7.3	8.6	10.6	12.0	12.7	12.3	10.7
	4	2.1	4.1	6.1	8.0	9.8	11.5	14.5	16.6	17.8	17.8	16.4
	5	2.6	5.1	7.6	10.1	12.3	14.5	18.3	21.1	23.0	23.5	22.2
	6	3.1	6.2	9.2	12.1	14.8	17.5	22.1	25.7	28.1	28.9	27.9
	7	3.6	7.2	10.7	14.1	17.4	20.4	26.0	30.3	33.2	34.5	33.7
	8	4.1	8.2	12.2	16.2	19.9	23.4	29.8	34.8	38.4	40.0	39.4

Table 6.2.46*d* Minimum Pulley Diameters—Rubber Belts, in

		Belt speed, ft/min										
	Ply	500	1,000	1,500	2,000	2,500	3,000	4,000	5,000	6,000	7,000	8,000
32-oz fabric	3	4	4	4	4	5	5	5	6	6		
	4	4	5	6	6	7	7	8	9	10		
	5	6	7	9	10	10	11	12	13	14		
	6	9	10	11	13	14	14	16	18	19		
	7	13	14	16	17	18	19	21	22	24		
	8	18	19	21	22	23	24	25	27	29		
32-oz hard fabric	3	3	3	3	4	4	4	4	5	5	6	7
	4	4	4	5	5	6	6	7	7	8	9	12
	5	5	6	7	8	8	9	10	11	12	13	16
	6	6	8	10	11	11	12	13	15	16	18	21
	7	10	12	14	15	15	16	17	19	20	22	26
	8	14	16	17	18	19	20	21	23	24	27	31
	9	18	20	21	22	23	24	25	27	28	31	36
	10	22	24	25	26	27	28	29	31	33	35	41
No. 70 rayon cord	3	5	6	7	7	8	8	9	10	11	12	13
	4	7	8	9	9	10	11	12	12	14	15	17
	5	9	10	11	12	13	13	15	16	17	19	21
	6	13	14	15	16	16	17	18	19	21	23	25
	7	16	17	18	19	20	21	22	23	24	26	29
	8	19	20	22	23	23	24	25	26	28	30	33

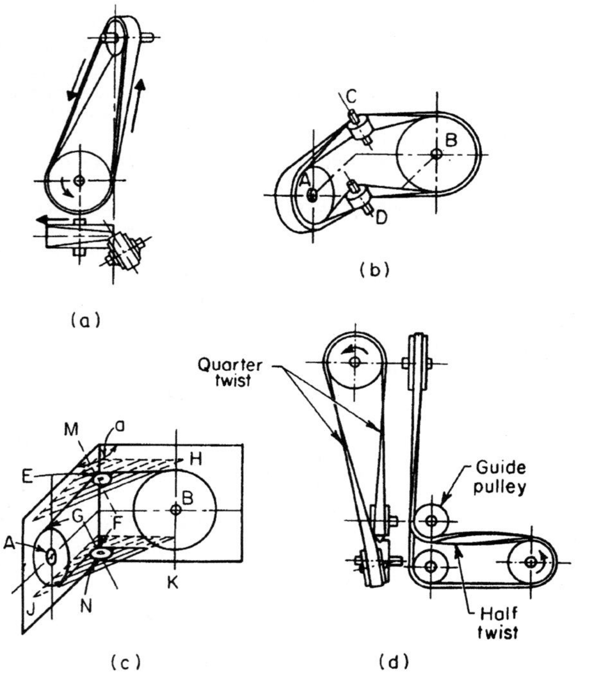

Figure 6.2.90 Arrangements for flat-belt drives.

where L = length of belt, in; D = diam of large pulley, in; d = diam of small pulley, in; and C = distance between pulley centers, in. Center distance C is given by $C = 0.25b + 0.25\sqrt{b^2 - 2(D-d)^2}$, where $b = L - 1.57(D + d)$. When a **crossed belt** is used, the length in $L = 2C + 1.57(D + d) + (D + d)^2/(4C)$.

Step or Cone Pulleys For belts operating on step pulleys, the pulley diameters must be such that the belt will fit over any pair with equal tightness. With **crossed belts**, it will be apparent from the equation for length of belt that the sum of the pulley diameters need only be constant in order that the belt may fit with equal tightness on each pair of pulleys. With open belts, the length is a function of both the sum and

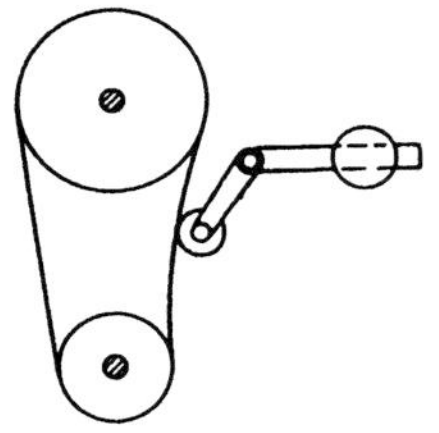

Fig. 6.2.91 Belt tightener.

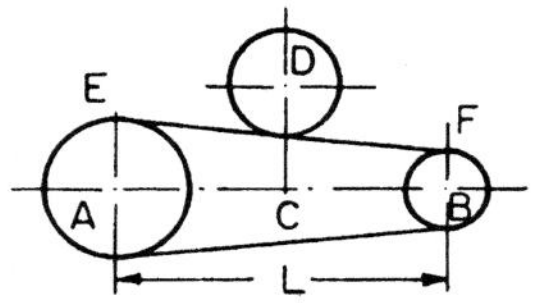

Fig. 6.2.92 Symbols for cone pulley graphical method.

the difference of the pulley diameters; hence no direct solution of the problem is possible, but a graphical approach can be of use.

A graphical method devised by Smith (*Trans. ASME*, 10) is shown in Fig. 6.2.92. Let A and B be the centers of any pair of pulleys in the set, the diameters of which are known or assumed. Bisect AB in C, and draw CD at right angles to AB. Take CD = 0.314 times the center distance L, and draw a circle tangent to the belt line EF. The belt line of any other pair of pulleys in the set will then be tangent to this circle. If the angle EF makes with AB is greater than 18°, draw a tangent to the circle D, making an angle of 18° with AB; and from a center on CD distant 0.298L above C, draw an arc tangent to this 18° line. All belt lines with angles greater than 18° will be tangent to this last drawn arc.

A very slight error in a graphical solution drawn to any scale much under full size will introduce an error seriously affecting the equality of belt tensions on the various pairs of pulleys in the set, and where much power is to be transmitted it is advisable to calculate the pulley diameters from the following **formulas** derived from Burmester's graphical method ("Lehrbuch der Mechanik").

Let D_1 and D_2 be, respectively, the diameters of the smaller and larger pulleys of a pair, $n = D_2/D_1$, and l = distance between shaft centers, all in inches. Also let $m = 1.58114l - D_0$, where D_0 = diam of both pulleys for a speed ratio $n = 1$. Then $(D_1 + m)^2 + (nD_1 + m)^2 = 5l^2$. First settle on values of D_0, l, and n, and then substitute in the equation and solve for D_1. The diameter D_2 of the other pulley of the pair will then be nD_1. The values are correct to the fourth decimal place.

The speeds given by cone pulleys should increase in a **geometric ratio;** i.e., each speed should be multiplied by a constant a in order to obtain the next higher speed. Let n_1 and n_2 be, respectively, the lowest and highest speeds (r/min) desired and k the number of speed changes. Then $a = \sqrt[k-1]{n_2/n_1}$. In practice, a ranges from 1.25 up to 1.75 and even 2.

The ideal value for a in machine-tool practice, according to Carl G. Barth, would be 1.189. In the example below, this would mean the use of 16 speeds instead of 8.

EXAMPLE. Let $n_1 = 16$, $n_2 = 400$, and $k = 8$, to be obtained with four pairs of pulleys and a back gear. From formula $a = \sqrt[7]{25} = 1.584$, hence speeds will be 16, ($16 \times 1.584 =$) 25.34, ($25.34 \times 1.584 =$) 40.14, and similarly 63.57, 100.7, 159.5, 252.6, and 400. The first four speeds are with the back gear in; hence the back-gear ratio must be $100.7 \div 16 = 6.29$.

Transmission of Power by Flat Belts The theory of flat-belt drives takes into account changes in belt tension caused by friction forces between belt and pulley, which, in turn, cause belt elongation or contraction, thus inducing relative movement between belt and pulley. The transmission of power is a complex affair. A lengthy mathematical presentation can be found in Firbank, "Mechanics of the Flat Belt Drive," ASME Paper 72-PTG-21. A simpler, more conventional analysis used for many years yields highly serviceable designs.

The turning force (tangential) on the rim of a pulley driven by a flat belt is equal to $T_1 - T_2$, where T_1 and T_2 are, respectively, the tensions in the driving (tight) side and following (slack) side of the belt. (For the relations of T_1 and T_2 at low peripheral speeds, see Sec. 1.) Log $(T_1/T_2) = 0.0076fa$ when the effect of centrifugal force is neglected and $T_1/T_2 = 10^{0.0076fa}$. Figure 6.2.93 gives values of this function. When the speeds are high, however, the relations of T_1 to T_2 are modified by centrifugal stresses in the belt, in which case log $(T_1/T_2) = 0.0076f(1 - x)a$, where f = coefficient of friction between the belt and pulley surface, a = angle of wrap, and $x = 12wv^2/(gt)$ in which w = weight of 1 in^3 of belt material, lb; v = belt speed, ft/s; $g = 32.2$ ft/s^2; and t = allowable working tension, lb/in^2. Values of x for leather belting (with $w = 0.035$ and $t = 300$) are as follows:

u	30	40	50	60	70	
x	0.039	0.070	0.118	0.157	0.214	
uv	80	90	100	110	120	130
x	0.279	0.352	0.435	0.526	0.626	0.735

Barth et al. (*Trans. ASME*, 1909) seem to show that f is a function of the belt velocity, varying according to the formula $f = 0.54 - 140/(500 + V)$ for leather belts on iron pulleys, where V = belt velocity for ft/min. For practical design, however, the following values of f may be used: for leather belts on cast-iron pulleys, $f = 0.30$; on wooden pulleys, $f = 0.45$; on paper pulleys, $f = 0.55$. The treatment of belts with belt dressing, pulleys with cork inserts, and dampness are all factors which greatly modify these values, tending to make them higher.

The **arc of contact** on the smaller of two pulleys connected by an open belt, in degrees, is approximately equal to $180 - 60(D - d)/l$, where D and d are the larger and smaller pulley diameters and l the distance between their shaft centers, all in inches. This formula gives an error not exceeding 0.5 percent.

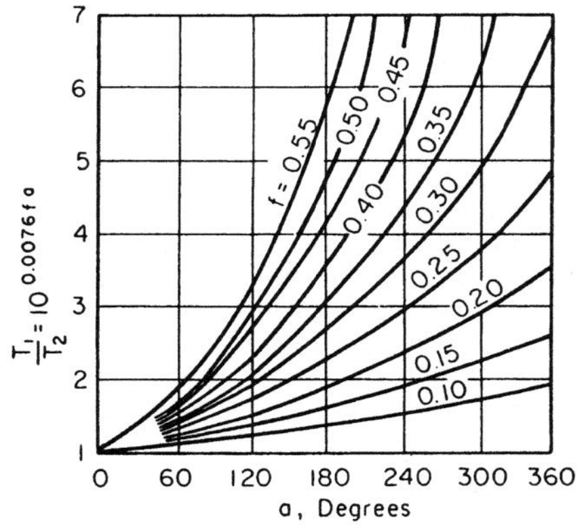

Figure 6.2.93 Values of $10^{0.0076fa}$.

Selecting a Belt Selecting an appropriate belt involves calculating horsepower per inch of belt width as follows:

$$\text{hp/in} = (\text{demanded hp} \times S)/(K \times W)$$

where demanded hp = horsepower required by the job at hand; S = service factor; K = arc factor; W = proposed belt width (determined from pulley width). One enters a belt manufacturer's catalog with *hp/in*, *belt speed*, and *small pulley diameter*, then selects that belt which has a matching maximum hp/in rating. See Table 6.2.46*a*, *b*, *c*, and *d* for typical values of S, K, hp/in ratings, and minimum pulley diameters.

V-Belt Drives

V-belt drives are widely used in power transmission, despite the fact that they may range in efficiency from about 70 to 96 percent. Such drives consist essentially of endless belts of trapezoidal cross section which ride in V-shaped pulley grooves (see Fig. 6.2.95*a*). The belts are formed of cord and fabric, impregnated with rubber, the cord material being cotton, rayon, synthetic, or steel. V-belt drives are quiet, able to absorb shock and operate at low bearing pressures. A V-belt should ride with the top surface approximately flush with the top of the pulley groove; clearance should be present between the belt base and the base of the groove so that the belt rides on the groove walls. The friction between belt and groove walls is greatly enhanced beyond normal values because sheave groove angles are made somewhat less than beltsection angles, causing the belt to wedge itself into the groove. See Table 6.2.55*a* for standard groove dimensions of sheaves.

The cross section and lengths of V belts have been standardized by ANSI in both inch and SI (metric) units, while ANSI and SAE have standardized the special category of automobile belts, again in both units. Standard designations are shown in Table 6.2.47, which also

Table 6.2.47 V-Belt Standard Designations—A Selection

	Inch standard		Metric standard	
Type	Section	Minimum sheave diameter, in	Section	Minimum sheave diameter, mm
Heavy-duty	A	3.0	13 C	80
	B	5.4	16 C	140
	C	9.0	22 C	214
	D	13.0	32 C	355
	E	21.04		
Automotive	0.25	2.25	6 A	57
	0.315	2.25	8 A	57
	0.380	2.40	10 A	61
	0.440	2.75	11 A	70
	0.500	3.00	13 A	76
	11/16	3.00	15 A	76
	3/4	3.00	17 A	76
	7/8	3.50	20 A	89
	1.0	4.00	23 A	102
Heavy-duty narrow	3 V	2.65		
	5 V	7.1		
	8 V	12.3		
Notched narrow	3 VX	2.2		
	5 VX	4.4		
Light-duty	2 L	0.8		
	3 L	1.5		
	4 L	2.5		
	5 L	3.5		
Synchronous belts	MXL			
	XL			
	L			
	H			
	XH			
	XXH			

NOTE: The use of smaller sheaves than minimum will tend to shorten belt life.
SOURCE: Compiled from ANSI/RMA IP-20, 21, 22, 23, 24, 25, 26; ANSI/SAE J636C.

includes minimum sheave diameters. V belts are specified by combining a standard designation (from Table 6.2.47) and a belt length; inside length for the inch system, and pitch (effective) length for metric system. Sheaves are specified by their pitch diameters, which are used for velocity ratio calculations in which case inside belt lengths must be converted to pitch lengths for computational purposes. Pitch lengths are calculated by adding a conversion factor to inside length (i.e., $L_p = L_s + \Delta$). See Table 6.2.48 for conversion factors. Table 6.2.49 lists standard inside inch lengths L_s and Table 6.2.50 lists standard metric pitch (effective) lengths L_p.

Table 6.2.48 Length Conversion Factors Δ

Belt section	Size interval	Conversion factor	Belt section	Size interval	Conversion factor
A	26-128	1.3	D	120-210	3.3
B	35-210	1.8	D	≥240	0.8
B	≥240	0.3	E	180-210	4.5
C	51-210	2.9	E	≥240	1.0
C	≥240	0.9			

SOURCE: Adapted from ANSI/RMA IP-20-1977 (R88) by permission.

Table 6.2.49 Standard Lengths L_s, in, and Length Correction Factors K_2: Conventional Heavy-Duty V Belts

	Cross section				
L_s	A	B	C	D	E
26	0.78				
31	0.82				
35	0.85	0.80			
38	0.87	0.82			
42	0.89	0.84			
46	0.91	0.86			
51	0.93	0.88	0.80		
55	0.95	0.89			
60	0.97	0.91	0.83		
68	1.00	0.94	0.85		
75	1.02	0.96	0.87		
80	1.04				
81		0.98	0.89		
85	1.05	0.99	0.90		
90	1.07	1.00	0.91		
96	1.08		0.92		
97		1.02			
105	1.10	1.03	0.94		
112	1.12	1.05	0.95		
120	1.13	1.06	0.96	0.88	
128	1.15	1.08	0.98	0.89	
144		1.10	1.00	0.91	
158		1.12	1.02	0.93	
173		1.14	1.04	0.94	
180		1.15	1.05	0.95	0.92
195		1.17	1.06	0.96	0.93
210		1.18	1.07	0.98	0.95
240		1.22	1.10	1.00	0.97
270		1.24	1.13	1.02	0.99
300		1.27	1.15	1.04	1.01
330			1.17	1.06	1.03
360			1.18	1.07	1.04
390			1.20	1.09	1.06
420			1.21	1.10	1.07
480				1.13	1.09
540				1.15	1.11
600				1.17	1.13
660				1.18	1.15

SOURCE: ANSI/RMA IP-20-1977 (R88), reproduced by permission.

Table 6.2.50 Standard Pitch Lengths L_p (Metric Units) and Length Correction Factors K_2

13C		16C		22C		32C	
L_p	K_2	L_p	K_2	L_p	K_2	L_p	K_2
710	0.83	960	0.81	1,400	0.83	3,190	0.89
750	0.84	1,040	0.83	1,500	0.85	3,390	0.90
800	0.86	1,090	0.84	1,630	0.86	3,800	0.92
850	0.88	1,120	0.85	1,830	0.89	4,160	0.94
900	0.89	1,190	0.86	1,900	0.90	4,250	0.94
950	0.90	1,250	0.87	2,000	0.91	4,540	0.95
1,000	0.92	1,320	0.88	2,160	0.92	4,720	0.96
1,075	0.93	1,400	0.90	2,260	0.93	5,100	0.98
1,120	0.94	1,500	0.91	2,390	0.94	5,480	0.99
1,150	0.95	1,600	0.92	2,540	0.96	5,800	1.00
1,230	0.97	1,700	0.94	2,650	0.96	6,180	1.01
1,300	0.98	1,800	0.95	2,800	0.98	6,560	1.02
1,400	1.00	1,900	0.96	3,030	0.99	6,940	1.03
1,500	1.02	1,980	0.97	3,150	1.00	7,330	1.04
1,585	1.03	2,110	0.99	3,350	1.01	8,090	1.06
1,710	1.05	2,240	1.00	3,550	1.02	8,470	1.07
1,790	1.06	2,360	1.01	3,760	1.04	8,850	1.08
1,865	1.07	2,500	1.02	4,120	1.06	9,240	1.09
1,965	1.08	2,620	1.03	4,220	1.06	10,000	1.10
2,120	1.10	2,820	1.05	4,500	1.07	10,760	1.11
2,220	1.11	2,920	1.06	4,680	1.08	11,530	1.13
2,350	1.13	3,130	1.07	5,060	1.10	12,290	1.14
2,500	1.14	3,330	1.09	5,440	1.11		
2,600	1.15	3,530	1.10	5,770	1.13		
2,730	1.17	3,740	1.11	6,150	1.14		
2,910	1.18	4,090	1.13	6,540	1.15		
3,110	1.20	4,200	1.14	6,920	1.16		
3,310	1.21	4,480	1.15	7,300	1.17		
		4,650	1.16	7,680	1.18		
		5,040	1.18	8,060	1.19		
		5,300	1.19	8,440	1.20		
		5,760	1.21	8,820	1.21		
		6,140	1.23	9,200	1.22		
		6,520	1.24				
		6,910	1.25				
		7,290	1.26				
		7,670	1.27				

SOURCE: ANSI/RMA IP-20-1988 revised, reproduced by permission.

For given large and small sheave diameters and center-to-center distance, the needed V-belt length can be computed from

$$L_p = 2C + 1.57(D + d) + \frac{(D - d)^2}{4C}$$

$$C = \frac{K + \sqrt{K^2 - 32(D - d)^2}}{16} \qquad K = 4L_p - 6.28(D + d)$$

where C = center-to-center distance; D = pitch diameter of large sheave; d = pitch diameter of small sheave; L_p = pitch (effective) length.

Arc of contact on the smaller sheave (degrees) is approximately

$$\theta = 180 - \frac{60(D - d)}{C}$$

Transmission of Power by V Belts Unfortunately there is no theory or mathematical analysis that is able to explain all experimental results reliably. Empirical formulations based on experimental results, however, do provide very serviceable design procedures, and together with data published in V-belt manufacturers' catalogs provide the engineer with the necessary V-belt selection tools.

For satisfactory performance under most conditions, ANSI provides the following empirical single V-belt power-rating formulation (inch and metric units) for 180° arc of contact and average belt length:

$$H_r = \left(C_1 - \frac{C_2}{d} - C_3(r - d)^2 - C_4 \log rd\right) rd + C_2 r\left(1 - \frac{1}{K_A}\right)$$

where H_r = rated horsepower for inch units (rated power kW for metric units); C_1, C_2, C_3, C_4 = constants from Table 6.2.52; r = r/min of high-speed shaft times 10^{-3}; K_A = speed ratio factor from Table 6.2.51; d = pitch diameter of small sheave, in (mm).

Selecting a Belt Selecting an appropriate belt involves calculating horsepower per belt as follows:

$$NH_r = (\text{demanded hp} \times K_s)/(K_1K_2)$$

where H_r = hp/belt rating, either from ANSI formulation above, or from manufacturer's catalog (see Table 6.2.54); demanded hp = horsepower required by the job at hand; K_s = service factor accounting for driver and driven machine characteristics regarding such things as shock, torque level, and torque uniformity (see Table 6.2.53); K_1 = angle of contact correction factor (see Fig. 6.2.94*a*); K_2 = length correction factor (see Tables 6.2.49 and 6.2.50); N = number of belts.

Table 6.2.51 Approximate Speed-Ratio Factor K_A for Use in Power-Rating Formulation

D/*d* range	K_A	*D*/*d* range	K_A
1.00-1.01	1.0000	1.15-1.20	1.0586
1.02-1.04	1.0112	1.21-1.27	1.0711
1.05-1.07	1.0226	1.28-1.39	1.0840
1.08-1.10	1.0344	1.40-1.64	1.0972
1.11-1.14	1.0463	Over 1.64	1.1106

SOURCE: Adapted from ANSI/RMA IP-20-1977 (R88), by permission.

Table 6.2.52 Constants C_1, C_2, C_3, C_4 for Use in Power-Rating Formulation

Belt section	C_1	C_2	C_3	C_4
		Inch		
A	0.8542	1.342	2.436×10^{-4}	0.1703
B	1.506	3.520	4.193×10^{-4}	0.2931
C	2.786	9.788	7.460×10^{-4}	0.5214
D	5.922	34.72	1.522×10^{-3}	1.064
E	8.642	66.32	2.192×10^{-3}	1.532
		Metric		
13C	0.03316	1.088	1.161×10^{-8}	5.238×10^{-3}
16C	0.05185	2.273	1.759×10^{-8}	7.934×10^{-3}
22C	0.1002	7.040	3.326×10^{-8}	1.500×10^{-2}
32C	0.2205	26.62	7.037×10^{-8}	3.174×10^{-2}

SOURCE: Compiled from ANSI/RMA IP-20-1977 (R88), by permission.

Table 6.2.53 Approximate Service Factor K_s for V-Belt Drives

	Load			
Power source torque	Uniform	Light shock	Medium shock	Heavy shock
Average or normal	1.0-1.2	1.1-1.3	1.2-1.4	1.3-1.5
Nonuniform or heavy	1.1-1.3	1.2-1.4	1.4-1.6	1.5-1.8

SOURCE: Adapted from ANSI/RMA IP-20-1977 (R88), by permission.

See Fig. 6.2.94*b* for selection of V-belt cross section.

V band belts, effectively joined V belts, serve the function of multiple single V belts (see Fig. 6.2.95*b*).

Long center distances are not recommended for V belts because excess slack-side vibration shortens belt life. In general $D \leqq C \leqq 3(D + d)$. If longer center distances are needed, then link-type V belts can be used effectively.

Since belt-drive capacity is normally limited by slippage of the smaller sheave, V-belt drives can sometimes be used with a flat, larger pulley rather than with a grooved sheave, with little loss in capacity. For instance the flat surface of the flywheel in a large punch press can serve such a purpose. The practical range of application is when speed ratio is over 3 : 1, and center distance is equal to or slightly less than the diameter of the large pulley.

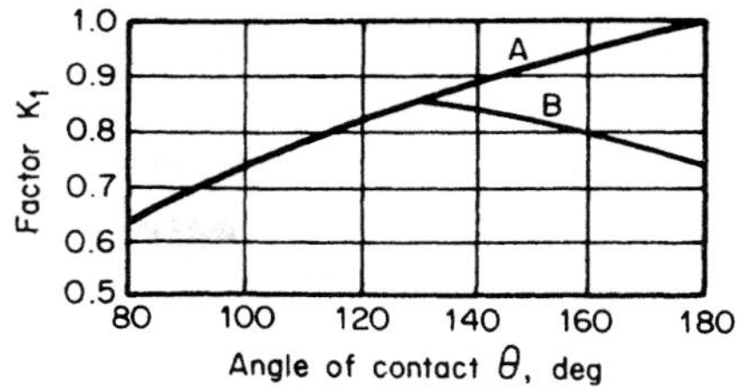

Figure 6.2.94a Angle-of-contact correction factor, where A = grooved sheave to grooved pulley distance (V to V) and B = grooved sheave to flat-face pulley distance (V to flat).

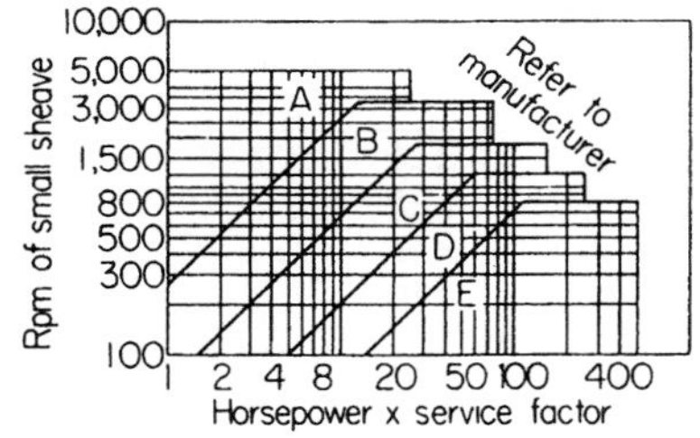

Figure 6.2.94b V-belt section for required horsepower ratings. Letters A, B, C, D, E refer to belt cross section. (See Table 6.2.53 for service factor.)

Figure 6.2.95 V- and V-band belt cross section.

In general the V-belting of skew shafts is discouraged because of the decrease in life of the belts, but where design demands such arrangements, special deep-groove sheaves are used. In such cases center distances should comply with the following:

For 90° turn	$C_{min} = 5.5(D + W)$
For 45° turn	$C_{min} = 4.0(D + W)$
For 30° turn	$C_{min} = 3.0(D + W)$

where W = width of group of individual belts. Selected values of W are shown in Table 6.2.55*d*.

Figure 6.2.96 shows a 90° turn arrangement, from which it can be seen that the horizontal shaft should lie some distance Z higher than the center of the vertical-shaft sheave. Table 6.2.55*c* lists the values of Z for various center distances in a 90° turn arrangement.

Cogged V belts have cogs molded integrally on the underside of the belt (Fig. 6.2.97*a*). Sheaves can be up to 25 percent smaller in diameter with cogged belts because of the greater flexibility inherent in the cogged construction. An extension of the cogged belt mating with a sheave or pulley notched at the same pitch as the cogs leads to a drive particularly useful for timing purposes.

Table 6.2.54 Horsepower Ratings of V Belts

Belt section	Speed of faster shaft, r/min	Rated horsepower per belt for small sheave pitch diameter, in							Additional horsepower per belt for speed ratio				
		2.60	3.00	3.40	3.80	4.20	4.60	5.00	1.02-1.04	1.08-1.10	1.15-1.20	1.28-1.39	1.65-over
A	200	0.20	0.27	0.33	0.40	0.46	0.52	0.59	0.00	0.01	0.01	0.02	0.03
	800	0.59	0.82	1.04	1.27	1.49	1.70	1.92	0.01	0.04	0.06	0.08	0.11
	1,400	0.87	1.25	1.61	1.97	2.32	2.67	3.01	0.02	0.06	0.10	0.15	0.19
	2,000	1.09	1.59	2.08	2.56	3.02	3.47	3.91	0.03	0.09	0.15	0.21	0.27
	2,600	1.25	1.87	2.47	3.04	3.59	4.12	4.61	0.04	0.12	0.19	0.27	0.35
	3,200	1.37	2.08	2.76	3.41	4.01	4.57	5.09	0.05	0.14	0.24	0.33	0.43
	3,800	1.43	2.23	2.97	3.65	4.27	4.83	5.32	0.06	0.17	0.28	0.40	0.51
	4,400	1.44	2.29	3.07	3.76	4.36	4.86*	5.26*	0.07	0.20	0.33	0.46	0.59
	5,000	1.39	2.28	3.05	3.71	4.24*	4.48*	4.64*	0.07	0.22	0.37	0.52	0.65
	5,600	1.29	2.17	2.92	3.50*				0.08	0.25	0.42	0.58	0.75
	6,200	1.11	1.98	2.65*					0.09	0.28	0.46	0.64	0.83
	6,800	0.87	1.68*	2.24*					0.10	0.30	0.51	0.71	0.91
	7,400	0.56	1.28*						0.11	0.33	0.55	0.77	0.99
	7,800	0.31*							0.12	0.35	0.58	0.81	1.04
		4.60	5.20	5.80	6.40	7.00	7.60	8.00					
B	200	0.69	0.86	1.02	1.18	1.34	1.50	1.61	0.01	0.02	0.04	0.05	0.07
	600	1.68	2.12	2.56	2.99	3.41	3.83	4.11	0.02	0.07	0.12	0.16	0.21
	1,000	2.47	3.16	3.84	4.50	5.14	5.78	6.20	0.04	0.12	0.19	0.27	0.35
	1,400	3.13	4.03	4.91	5.76	6.59	7.39	7.91	0.05	0.16	0.27	0.38	0.49
	1,800	3.67	4.75	5.79	6.79	7.74	8.64	9.21	0.07	0.21	0.35	0.49	0.63
	2,200	4.08	5.31	6.47	7.55	8.56	9.48	10.05	0.09	0.26	0.43	0.60	0.77
	2,600	4.36	5.69	6.91	8.01	9.01	9.87	10.36	0.10	0.30	0.51	0.71	0.91
	3,000	4.51	5.89	7.11	8.17	9.04	9.73*		0.12	0.35	0.58	0.82	1.05
	3,400	4.51	5.88	7.03	7.95*				0.13	0.40	0.66	0.93	1.19
	3,800	4.34	5.64	6.65*					0.15	0.44	0.74	1.04	1.33
	4,200	4.01	5.17*						0.16	0.49	0.82	1.15	1.47
	4,600	3.48							0.18	0.54	0.90	1.25	1.61
	5,000	2.76*							0.19	0.59	0.97	1.36	1.75
		7.00	8.00	9.00	10.00	11.00	12.00	13.00					
C	100	1.03	1.29	1.55	1.81	2.06	2.31	2.56	0.01	0.03	0.05	0.08	0.10
	400	3.22	4.13	5.04	5.93	6.80	7.67	8.53	0.04	0.13	0.22	0.30	0.39
	800	5.46	7.11	8.73	10.31	11.84	13.34	14.79	0.09	0.26	0.43	0.61	0.78
	1,000	6.37	8.35	10.26	12.11	13.89	15.60	17.24	0.11	0.33	0.54	0.76	0.97
	1,400	7.83	10.32	12.68	14.89	16.94	18.83	20.55	0.15	0.46	0.76	1.06	1.36
	1,800	8.76	11.58	14.13	16.40	18.37	20.01	21.31*	0.20	0.59	0.98	1.37	1.75
	2,000	9.01	11.90	14.44	16.61	18.37	19.70*		0.22	0.65	1.08	1.52	1.95
	2,400	9.04	11.87	14.14					0.26	0.78	1.30	1.82	2.34
	2,800	8.38	10.85*						0.30	0.91	1.52	2.12	2.73
	3,000	7.76							0.33	0.98	1.63	2.28	2.92
	3,200	6.44*							0.36	1.07	1.79	2.50	3.22

* For footnote see end of table on next page.

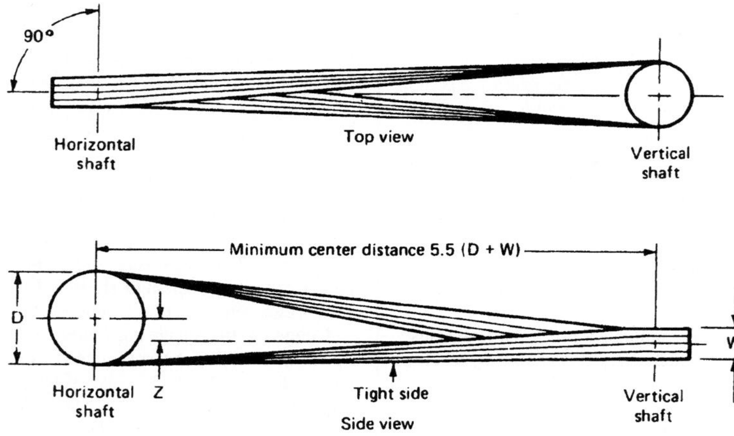

Figure 6.2.96 Quarter-turn drive for V belts.

Table 6.2.54 Horsepower Ratings of V Belts *(Continued)*

Belt section	Speed of faster shaft, r/min	Rated horsepower per belt for small sheave pitch diameter, in							Additional horsepower per belt for speed ratio				
		12.00	14.00	16.00	18.00	20.00	22.00	24.00	1.02-1.04	1.08-1.10	1.15-1.20	1.28-1.39	1.65-over
D	50	1.96	2.52	3.08	3.64	4.18	4.73	5.27	0.02	0.06	0.10	0.13	0.17
	200	6.28	8.27	10.24	12.17	14.08	15.97	17.83	0.08	0.23	0.38	0.54	0.69
	400	10.89	14.55	18.12	21.61	25.02	28.35	31.58	0.15	0.46	0.77	1.08	1.38
	600	14.67	19.75	24.64	29.33	33.82	38.10	42.15	0.23	0.69	1.15	1.61	2.07
	800	17.70	23.91	29.75	35.21	40.24	44.83	48.94	0.31	0.92	1.54	2.15	2.77
	1,000	19.93	26.94	33.30	38.96	43.86	47.93	51.12	0.38	1.15	1.92	2.69	3.46
	1,200	21.32	28.71	35.05	40.24	44.18*			0.46	1.39	2.31	3.23	4.15
	1,400	21.76	29.05	34.76*					0.54	1.62	2.69	3.77	4.84
	1,600	21.16	27.81*						0.62	1.85	3.08	4.30	5.53
	1,800	19.41							0.69	2.08	3.46	4.84	6.22
	1,950	17.28*							0.75	2.25	3.75	5.25	6.74
		18.00	21.00	24.00	27.00	30.00	33.00	36.00					
E	50	4.52	5.72	6.91	8.08	9.23	10.38	11.52	0.04	0.11	0.18	0.26	0.33
	100	8.21	10.46	12.68	14.87	17.04	19.19	21.31	0.07	0.22	0.37	0.51	0.66
	200	14.68	18.86	22.97	27.00	30.96	34.86	38.68	0.15	0.44	0.73	1.03	1.32
	300	20.37	26.29	32.05	37.67	43.13	48.43	53.58	0.22	0.66	1.10	1.54	1.98
	400	25.42	32.87	40.05	46.95	53.55	59.84	65.82	0.29	0.88	1.47	2.06	2.64
	500	29.86	38.62	46.92	54.74	62.05	68.81	75.00	0.37	1.10	1.84	2.57	3.30
	600	33.68	43.47	52.55	60.87	68.36	74.97	80.63	0.44	1.32	2.20	3.08	3.96
	700	36.84	47.36	56.83	65.15	72.22	77.93*		0.51	1.54	2.57	3.60	4.62
	800	39.29	50.20	59.61	67.36	73.30*			0.59	1.76	2.94	4.11	5.28
	900	40.97	51.89	60.73					0.66	1.98	3.30	4.63	5.94
	1,000	41.84	52.32	60.04*					0.73	2.21	3.67	5.14	6.60
	1,100	41.81	51.40*						0.81	2.43	4.04	5.65	7.26
	1,200	40.83							0.88	2.65	4.41	6.17	7.93
	1,300	38.84*							0.95	2.87	4.77	6.68	8.59

* Rim speed above 6,000 ft/min. Special sheaves may be necessary.
SOURCE: Compiled from ANSI/RMA IP-20-1998 revised, by permission.

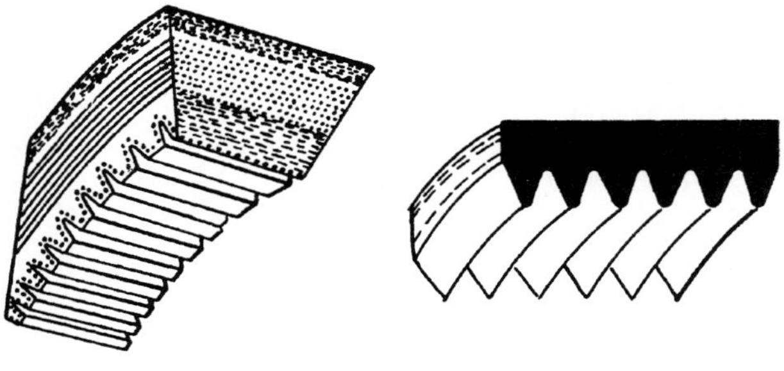

(a) (b)

Figure 6.2.97 Special V belts. (*a*) Cogged V belt; (*b*) ribbed V belt.

Ribbed V belts are really flat belts molded integrally with longitudinal ribbing on the underside (Fig. 6.2.97*b*). Traction is provided principally by friction between the ribs and sheave grooves rather than by wedging action between the two, as in conventional V-belt operation. The flat upper portion transmits the tensile belt loads. Ribbed belts serve well when substituted for multiple V-belt drives and for all practical purposes eliminate the necessity for belt-matching in multiple V-belt drives.

Adjustable Motor Bases

To maintain proper belt tensions on short center distances, an adjustable motor base is often used. Figure 6.2.98 shows an embodiment of such a base made by the Automatic Motor Base Co., in which adjustment for proper belt tension is made by turning a screw which opens or closes the center distance between pulleys, as required. The carriage portion of the base is spring loaded so that after the initial adjustment for belt tension has been made by the screw, the spring will compensate for a normal amount of stretch in the belts. When there is more stretch than can be accommodated by the spring, the screw is turned to provide the necessary belt tensions. The carriage can be moved while the unit is in operation, and the motor base is provided for vertical as well as horizontal mounting.

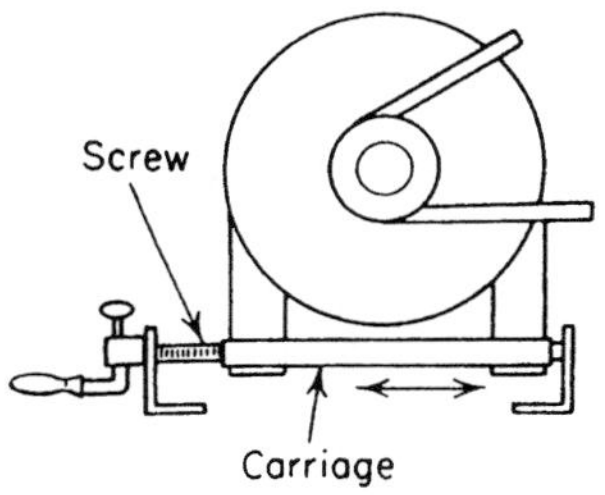

Figure 6.2.98 Adjustable motor base.

CHAIN DRIVES

Roller-Chain Drives

The advantages of finished steel roller chains are high efficiency (around 98 to 99 percent), no slippage, no initial tension required, and chains may travel in either direction. The basic construction of roller chains is shown in Fig. 6.2.99 and Table 6.2.56.

The shorter the pitch, the higher the permissible operating speed of roller chains. Horsepower capacity in excess of that provided by a single chain may be had by the use of multiple chains, which are essentially parallel single chains assembled on pins common to all strands. Because of its lightness in relation to tensile strength, the effect of centrifugal pull does not need to be considered; even at the unusual speed of 6,000 ft/min, this pull is only 3 percent of the ultimate tensile strength.

Table 6.2.55*a* Standard Sheave Groove Dimensions, in

Face width of standard and deep groove sheaves
Face width = $S_g (N_g - 1) + 2 S_e$
where:
N_g = number of grooves

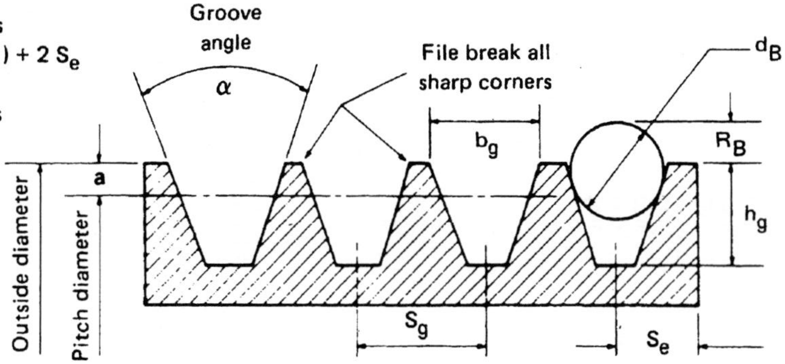

Cross section	Outside diameter range	Groove angle $\alpha \pm 0.33°$	b_g	h_g min	R_B min	$d_B \pm 0.0005$	$S_g \pm 0.025$	S_e	Pitch diameter range	Minimum recommended pitch diameter	$2a$
	Standard groove dimensions						Drive design factors				
A	Up through 5.65 Over 5.65	34 38	0.494 0.504 ± 0.005	0.460	0.148 0.149	0.4375 (7/16)	0.625	$0.375^{+0.090}_{-0.062}$	Up through 5.40 Over 5.40	3.0	0.250
B	Up through 7.35 Over 7.35	34 38	0.637 0.650 ± 0.006	0.550	0.189 0.190	0.5625 (9/16)	0.750	$0.500^{+0.120}_{-0.065}$	Up through 7.00 Over 7.00	5.4	0.350
A/B	Up through 7.35 Over 7.35	34 38	0.612 0.625 ± 0.006	0.612	0.230 0.226	0.5625 (9/16)	0.750	$0.500^{+0.120}_{+0.065}$		A = 3.0 B = 5.4	A = 0.620* B = 0.280
C	Up through 8.39 Over 8.39 to and incl. 12.40 Over 12.40	34 36 38	0.879 0.877 ± 0.007 0.895	0.750	0.274 0.276 0.277	0.7812 (25/32)	1.000	$0.688^{+0.160}_{-0.70}$	Up through 7.99 Over 7.99 to and incl. 12.00 Over 12.00	9.0	0.400
D	Up through 13.59 Over 13.59 to and incl. 17.60 Over 17.60	34 36 38	1.256 1.271 ± 0.008 1.283	1.020	0.410 0.410 0.411	1.1250 (1⅛)	1.438	+0.220 −0.80	Up through 12.99 Over 12.99 to and incl. 17.00 Over 17.00	13.0	0.600
E	Up through 24.80 Over 24.80	36 38	1.527 1.542 ± 0.010	1.270	0.476 0.477	1.3438 (1 11/32)	1.750	$1.125^{+0.280}_{-0.090}$	Up through 24.00 Over 24.00	21.0	0.800

* The *a* values shown for the A/B combination sheaves are the geometrically derived values. These values may be different than those shown in manufacturers' catalogs.
SOURCE: "Dayco Engineering Guide for V-Belt Drives," Dayco Corp., Dayton, OH, 1981, reprinted by permission.

Table 6.2.55*b* Classical Deep Groove Sheave Dimensions, in

Cross section	Deep groove dimensions: Outside diameter range	Groove angle $\alpha \pm 0.33°$	b_g	h_g min	$2a$	$S_g \pm 0.025$	S_e
A	Up through 5.96 Over 5.96	34 38	0.539 0.611 ± 0.005	0.615	0.560	0.750	$0.438^{+0.090}_{-0.062}$
B	Up through 7.71 Over 7.71	34 38	0.747 0.774 ± 0.006	0.730	0.710	0.875	$0.562^{+0.120}_{-0.065}$
C	Up through 9.00 Over 9.00 to and incl. 13.01 Over 13.01	34 36 38	1.066 1.085 + 0.007 1.105	1.055	1.010	1.250	$0.812^{+0.160}_{-0.070}$
D	Up through 14.42 Over 14.42 to and incl. 18.43 Over 18.43	34 36 38	1.513 1.541 ± 0.008 1.569	1.435	1.430	1.750	$1.062^{+0.220}_{-0.080}$
E	Up through 25.69 Over 25.69	36 38	1.816 1.849 + 0.010	1.715	1.690	2.062	$1.312^{+0.280}_{-0.090}$

SOURCE: "Dayco Engineering Guide for V-Belt Drives," Dayco Corp., Dayton, OH, 1981, reprinted by permission.

Table 6.2.55c *Z* Dimensions for Quarter-Turn Drives, in

Center distance	3V, 5V, 8V Z dimension	A B, C, D Z dimension
20		0.2
30		0.2
40		0.4
50		0.4
60	0.2	0.5
80	0.3	0.5
100	0.4	1.0
120	0.6	1.5
140	0.9	2.0
160	1.2	2.5
180	1.5	3.5
200	1.8	4.0
220	2.2	5.0
240	2.6	6.0

SOURCE: "Dayco Engineering Guide for V-Belt Drives," Dayco Corp., Dayton, OH, 1981, reprinted by permission.

Table 6.2.55*d* Width *W* of Set of Belts Using Deep-Groove Sheaves, in

No. of belts	V-belt cross section: 3V	5V	8V	A	B	C	D
1	0.4	0.6	1.0	0.5	0.7	0.9	1.3
2	0.9	1.4	2.3	1.3	1.6	2.2	3.1
3	1.4	2.2	3.6	2.0	2.5	3.4	4.8
4	1.9	3.0	4.0	2.8	3.3	4.7	6.6
5	2.4	3.8	6.2	3.5	4.2	5.9	8.3
6	2.9	4.7	7.6	4.3	5.1	7.2	10.1
7	3.4	5.5	8.9	5.0	6.0	8.4	11.8
8	3.9	6.3	10.2	5.8	6.8	9.7	13.6
9	4.4	7.1	11.5	6.5	7.7	10.9	15.3
10	4.9	7.9	12.8	7.3	8.6	12.2	17.1

SOURCE: "Dayco Engineering Guide for V-Belt Drives," Dayco Corp., Dayton, OH, 1981, reprinted by permission.

Sprocket wheels with fewer than 16 teeth may be used for relatively slow speeds, but 18 to 24 teeth are desirable for high-speed service. Sprockets with fewer than 25 teeth, running at speeds above 500 or 600 r/min, should be heat-treated to give a tough wear-resistant surface testing between 35 and 45 on the Rockwell C hardness scale.

If the speed ratio requires the larger sprocket to have as many as 128 teeth, or more than eight times the number on the smaller sprocket, it is usually better, with few exceptions, to make the desired reduction in two or more steps. The ANSI tooth form ASME B29.1 2011 allows roller chain to adjust itself to a larger pitch circle as the pitch of the chain elongates owing to natural wear in the pin-bushing joints. The greater the number of teeth, the sooner the chain will ride out too near the ends of the teeth.

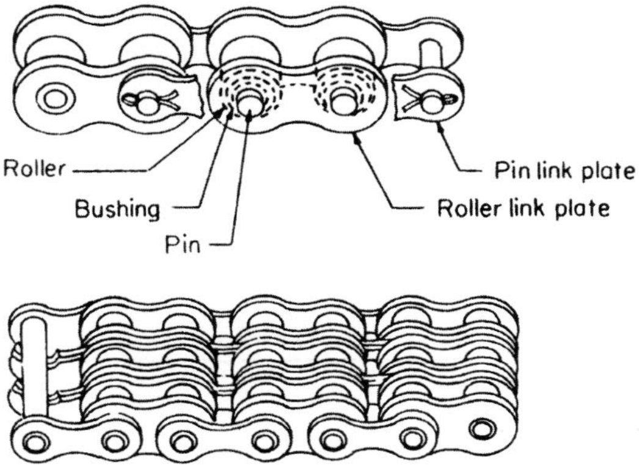

Figure 6.2.99 Roller chain construction.

Idler sprockets may be used on either side of the standard roller chain, to take up slack, to guide the chain around obstructions, to change the direction of rotation of a driven shaft, or to provide more wrap on another sprocket. Idlers should not run faster than the speeds recommended as maximum for other sprockets with the same number of teeth. It is desirable that idlers have at least two teeth in mesh with the chain, and it is advisable, though not necessary, to have an idler contact the idle span of chain.

Horsepower ratings are based upon the number of teeth and the rotative speed of the smaller sprocket, either driver or follower. The pin-bushing bearing area, as it affects allowable working load, is the important factor for medium and higher speeds. For extremely slow speeds, the chain selection may be based upon the ultimate tensile strength of the chain. For chain speeds of 25 ft/min and less, the chain pull should be not more than $1/5$ of the ultimate tensile strength; for 50 ft/min, $1/6$; for 100 ft/min, $1/7$; for 150 ft/min, $1/8$; for 200 ft/min, $1/9$; and for 250 ft/min, $1/10$ of the ultimate tensile strength.

Ratings for **multiple-strand chains** are proportional to the number of strands. The recommended numbers of strands for multiple chains are 2, 3, 4, 6, 8, 10, 12, 16, 20, and 24, with the maximum overall width in any case limited to 24 in.

The **horsepower ratings** in Table 6.2.57 are modified by the **service factors** in Table 6.2.58. Thus for a drive having a nominal rating of 3 hp,

Table 6.2.56 Roller-Chain Data and Dimensions, in*

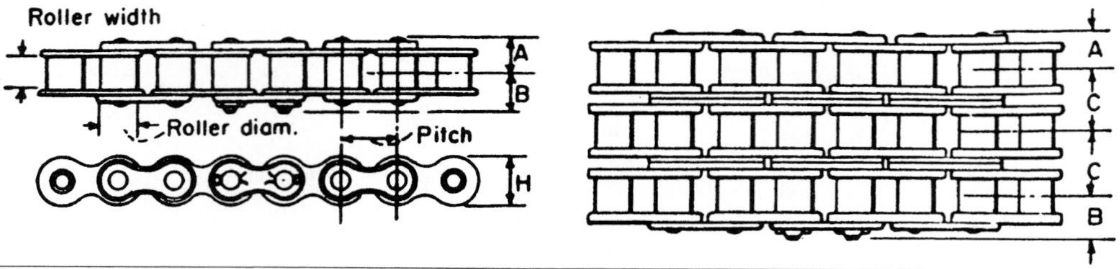

ANSI chain no.	ISO chain no.	Roller Pitch	Roller Width	Roller Diam	Pin diam	Roller link plate Thickness	Roller link plate Height H	Dimension A	Dimension B	Dimension C	Tensile strength per strand, lb	Recommended max speed r/min 12 teeth	18 teeth	24 teeth
25	04C-1	¼	⅛	0.130	0.091	0.030	0.230	0.150	0.190	0.252	925	5,000	7,000	7,000
35	06C-1	⅜	3/16	0.200	0.141	0.050	0.344	0.224	0.290	0.399	2,100	2,380	3,780	4,200
41	085	½	¼	0.306	0.141	0.050	0.383	0.256	0.315	—	2,000	1,750	2,725	2,850
40	08A-1	½	5/16	0.312	0.156	0.060	0.452	0.313	0.358	0.566	3,700	1,800	2,830	3,000
50	10A-1	⅝	⅜	0.400	0.200	0.080	0.594	0.384	0.462	0.713	6,100	1,300	2,030	2,200
60	12A-1	¾	½	0.469	0.234	0.094	0.679	0.493	0.567	0.897	8,500	1,025	1,615	1,700
80	16A-1	1	⅝	0.625	0.312	0.125	0.903	0.643	0.762	1.153	14,500	650	1,015	1,100
100	20A-1	1¼	¾	0.750	0.375	0.156	1.128	0.780	0.910	1.408	24,000	450	730	850
120	24A-1	1½	1	0.875	0.437	0.187	1.354	0.977	1.123	1.789	34,000	350	565	650
140	28A-1	1¾	1	1.000	0.500	0.218	1.647	1.054	1.219	1.924	46,000	260	415	500
160	32A-1	2	1¼	1.125	0.562	0.250	1.900	1.250	1.433	2.305	58,000	225	360	420
180		2¼	1 13/32	1.406	0.687	0.281	2.140	1.421	1.770	2.592	76,000	180	290	330
200	40A-1	2½	1½	1.562	0.781	0.312	2.275	1.533	1.850	2.817	95,000	170	260	300
240	48A-1	3	1⅞	1.875	0.937	0.375	2.850	1.722	2.200	3.458	135,000	120	190	210

* For conversion to metric units (mm) multiply table values by 25.4.

Table 6.2.57 Selected Values of Horsepower Ratings of Roller Chains

ANSI no. and pitch, in	Number of teeth in small socket	Small sprocket, r/min 50	500	1,200	1,800	2,500	3,000	4,000	5,000	6,000	8,000	10,000
25 ¼	11	0.03	0.23	0.50	0.73	0.98	1.15	1.38	0.99	0.75	0.49	0.35
	15	0.04	0.32	0.70	1.01	1.36	1.61	2.08	1.57	1.20	0.78	0.56
	20	0.06	0.44	0.96	1.38	1.86	2.19	2.84	2.42	1.84	1.201	0.86
	25	0.07	0.56	1.22	1.76	2.37	2.79	3.61	3.38	2.57	1.67	1.20
	30	0.08	0.68	1.49	2.15	2.88	3.40	4.40	4.45	3.38	2.20	1.57
	40	0.12	.092	2.03	2.93	3.93	4.64	6.00	6.85	5.21	3.38	2.42
35 ⅜	11	0.10	0.77	1.70	2.45	3.30	2.94	1.91	1.37	1.04	0.67	0.48
	15	0.14	1.08	2.38	3.43	4.61	4.68	3.04	2.17	1.65	1.07	0.77
	20	0.19	1.48	3.25	4.68	6.29	7.20	4.68	3.35	2.55	1.65	1.18
	25	0.24	1.88	4.13	5.95	8.00	9.43	6.54	4.68	3.56	2.31	1.65
	30	0.29	2.29	5.03	7.25	9.74	11.5	8.59	6.15	4.68	3.04	2.17
	40	0.39	3.12	6.87	9.89	13.3	15.7	13.2	9.47	7.20	4.68	—
41 ½	11	0.13	1.01	1.71	0.93	(0.58)	0.43	0.28	0.20	0.15	0.10	
	15	0.18	1.41	2.73	1.49	(0.76)	0.69	0.45	0.32	0.24	0.16	
	20	0.24	1.92	4.20	2.29	(1.41)	1.06	0.69	0.49	0.38		
	25	0.31	2.45	5.38	3.20	(1.97)	1.49	0.96	0.69	0.53		
	30	0.38	2.98	6.55	4.20	(2.58)	1.95	1.27	0.91	0.69		
	40	0.51	4.07	8.94	6.47	(3.97)	3.01	1.95	1.40			
40 ½	11	0.23	1.83	4.03	4.66	(3.56)	2.17	1.41	1.01	0.77	0.50	
	15	0.32	2.56	5.64	7.43	(4.56)	3.45	2.24	1.60	1.22	0.79	
	20	0.44	3.50	7.69	11.1	(7.03)	5.31	3.45	2.47	1.88		
	25	0.56	4.45	9.78	14.1	(9.83)	7.43	4.82	3.45	2.63		
	30	0.68	5.42	11.9	17.2	(12.9)	9.76	6.34	4.54	3.45		
	40	0.93	7.39	16.3	23.4	(19.9)	15.0	9.76	6.99			
50 ⅝	11	0.45	3.57	7.85	5.58	(3.43)	2.59	1.68	1.41	1.20	0.92	
	15	0.63	4.99	11.0	8.88	(5.46)	4.13	2.68	2.25	1.92		
	20	0.86	6.80	15.0	13.7	(8.40)	6.35	4.13	3.46	2.95		
	25	1.09	8.66	19.0	19.1	(11.7)	8.88	5.77	4.83			
	30	1.33	10.5	23.2	25.1	(15.4)	11.7	7.58				
	40	1.81	14.4	31.6	38.7	(23.7)	18.0					

Table 6.2.57 Selected Values of Horsepower Ratings of Roller Chains *(Continued)*

ANSI no. and pitch, in	Number of teeth in small sprocket	Small sprocket, r/min										
		10	50	100	200	500	700	1,000	1,400	2,000	2,700	4,000
60	11	0.18	0.77	1.44	2.69	6.13	8.30	11.4	9.41	5.51	(3.75)	1.95
¾	15	0.25	1.08	2.01	3.76	8.57	11.6	16.0	15.0	8.77	(6.18)	3.10
	20	0.35	1.47	2.75	5.13	11.7	15.8	21.8	23.1	13.5	(9.20)	
	25	0.44	1.87	3.50	6.52	14.9	20.1	27.8	32.2	18.9	(12.9)	
	30	0.54	2.28	4.26	7.94	18.1	24.5	33.8	42.4	24.8	(16.9)	
	40	0.73	3.11	5.81	10.8	24.7	33.5	46.1	62.5	38.2		
80	11	0.42	1.80	3.36	6.28	14.3	19.4	19.6	12.8	6.93	4.42	
1	15	0.59	2.52	4.70	8.77	20.0	27.1	31.2	18.9	11.0	7.04	
	20	0.81	3.44	6.41	12.0	27.3	37.0	48.1	29.0	17.0		
	25	1.03	4.37	8.16	15.2	34.7	47.0	64.8	40.6	23.8		
	30	1.25	5.33	9.94	18.5	42.3	57.3	78.9	53.3	31.2		
	40	1.71	7.27	13.6	25.3	57.7	78.1	108	82.1	48.1		
100	11	0.81	3.45	6.44	12.0	27.4	37.1	23.4	14.2	8.29		
1¼	15	1.13	4.83	9.01	16.8	38.3	51.9	37.3	22.5	13.2		
	20	1.55	6.58	12.3	22.9	52.3	70.8	57.5	34.7	20.3		
	25	1.97	8.38	15.5	29.2	66.6	90.1	80.3	48.5	28.4		
	30	2.40	10.2	19.0	35.5	81.0	110	106	63.7	10.0		
	40	3.27	13.9	26.0	48.5	111	150	163	98.1			
120	11	1.37	5.83	10.9	20.3	46.3	46.3	27.1	16.4	9.59		
1½	15	1.91	8.15	15.2	28.4	64.7	73.8	43.2	26.1			
	20	2.61	11.1	20.7	38.7	88.3	114	66.5	40.1			
	25	3.32	14.1	26.4	49.3	112	152	92.9	56.1			
	30	4.05	17.2	32.1	60.0	137	185	122	73.8			
	40	5.52	23.5	43.9	81.8	187	253	188	59.5			
140	11	2.12	9.02	16.8	31.4	71.6	52.4	30.7	18.5			
1¾	15	2.96	12.6	23.5	43.9	100	83.4	48.9	29.5			
	20	4.04	17.2	32.1	59.9	137	128	75.2	45.4			
	25	5.14	21.9	40.8	76.2	174	180	105	63.5			
	30	6.26	26.7	49.7	92.8	212	236	138				
	40	8.54	36.4	67.9	127	289	363	213				
160	11	3.07	13.1	24.4	45.6	96.6	58.3	34.1				
2	15	4.30	18.3	34.1	63.7	145	92.8	54.4				
	20	5.86	25.0	46.4	86.9	198	143	83.7				
	25	7.40	31.8	59.3	111	252	200	117				
	30	9.08	38.7	72.2	125	307	263	154				
	40	12.4	52.8	98.5	184	419	404					
180	11	4.24	18.1	33.7	62.9	106	64.1	37.5				
2¼	15	5.93	25.3	47.1	88.0	169	102	59.7				
	20	8.10	34.5	64.3	120	260	157	92.0				
	25	10.3	43.9	81.8	153	348	220					
	30	12.5	53.4	99.6	186	424	289					
	40	17.1	72.9	136	254	579	398					
200	11	5.64	24.0	44.8	83.5	115						
2½	15	7.88	33.5	62.6	117	184						
	20	10.7	45.8	85.4	159	283						
	25	13.7	58.2	109	203	396						
240	11	9.08	38.6	72.1	135							
3	15	12.7	54.0	101	188							
	20	17.3	73.7	138	257							
	25	22.0	93.8	175	327							

NOTE: The sections separated by heavy lines denote the method of lubrication as follows: type A (left section), manual; type B (middle section), bath or disk; type C (right section), oil stream.
SOURCE: Supplementary action of ANSI B29.1-1975 (R93), adapted by permission.

Table 6.2.58 Service Factors for Roller Chains

Power source	Load		
	Smooth	Moderate shock	Heavy shock
Internal combustion engine with hydraulic drive	1.0	1.2	1.4
Electric motor or turbine	1.0	1.3	1.5
Internal combustion engine with mechanical drive	1.2	1.4	1.7

SOURCE: ANSI B29.1-1975, adapted by permission.

subject to heavy shock, abnormal conditions, 24-h/day operation, the chain rating obtained from Table 6.2.57 should be at least $3 \times 1.7 = 5.1$ hp.

Chain-Length Calculations Referring to Fig. 6.2.100, L = length of chain, in; P = pitch of chain, in; R and r = pitch radii of large and small sprockets, respectively, in; D = center distance, in; A = tangent length, in; a = angle between tangent and centerline; N and n = number of teeth on larger and smaller sprockets, respectively; $180 + 2a$ and $180 - 2a$ are angles of contact on larger and smaller sprockets, respectively, deg.

$$a = \sin^{-1}[(R - r)/D] \qquad A = D\cos a$$

$$L = NP(180 + 2a)/360 + nP(180 - 2a)/360 + 2D\cos a$$

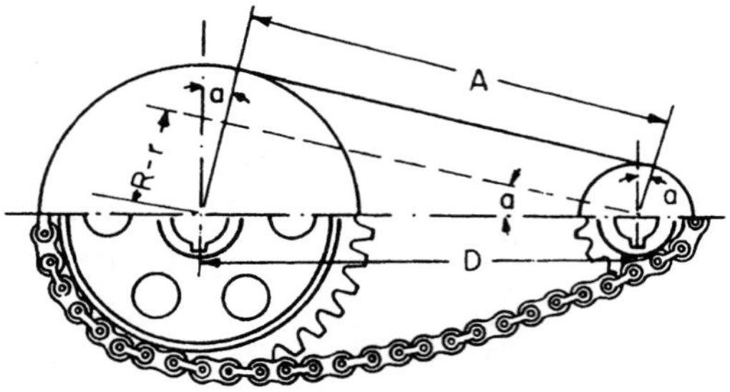

Figure 6.2.100 Symbols for chain length calculations.

If L_p = length of chain in pitches and D_p = center distance in pitches,

$$L_p = (N + n)/2 + a(N - n)/180 + 2D_p \cos a$$

Avoiding the use of trigonometric tables,

$$L_p = 2C + (N + n)/2 + K(N - n)^2/C$$

where C is the center distance in pitches and K is a variable depending upon the value of $(N - n)/C$. Values of K are as follows:

$(N - n)/C$	0.1	1.0	2.0	3.0
K	0.02533	0.02538	0.02555	0.02584
$(N - n)/C$	4.0	5.0	6.0	
K	0.02631	0.02704	0.02828	

Formulas for chain length on multisprocket drives are cumbersome except when all sprockets are the same size and on the same side of the chain. For this condition, the chain length in pitches is equal to the sum of the consecutive center distances in pitches plus the number of teeth on one sprocket.

Actual chain lengths should be in even numbers of pitches. When necessary, an odd number of pitches may be secured by the use of an offset link, but such links should be avoided if possible. An offset link is one pitch; half roller link at one end and half pin link at the other end. If center distances are to be nonadjustable, they should be selected to give an initial snug fit for an even number of pitches of chain. For the average application, a center distance equal to 40 ± 10 pitches of chain represents good practice.

There should be at least 120° of wrap in the arc of contact on a power sprocket. For ratios of 3:1 or less, the wrap will be 120° or more for any center distance or number of teeth. To secure a wrap of 120° or more, for ratios greater than 3:1, the center distance must be not less than the difference between the pitch diameters of the two sprockets.

Sprocket Diameters N = number of teeth; P = pitch of chain, in; D = diameter of roller, in. The pitch of a standard roller chain is measured from the center of a pin to the center of an adjacent pin.

$$\text{Pitch diam} = p / \sin\frac{180}{N}$$

$$\text{Bottom diam} = \text{pitch diam} - D$$

$$\text{Outside diam} = P\left(0.6 + \cot\frac{180}{N}\right)$$

$$\text{Caliper diam} = \left(\text{pitch diam} \times \cos\frac{90}{N}\right) - D$$

The exact bottom diameter cannot be measured for an odd number of teeth, but it can be checked by measuring the distance (caliper diameter) between bottoms of the two tooth spaces nearest opposite to each other. Bottom and caliper diameters must not be oversize—all tolerances must be negative. ANSI negative tolerance $= 0.003 + 0.001P\sqrt{N}$ in.

Design of Sprocket Teeth for Roller Chains The **section profile** for the teeth of roller chain sprockets, recommended by ANSI, has the proportions shown in Fig. 6.2.101. Let P = chain pitch; W = chain width (length of roller); n = number of strands of multiple chain; M = overall width of tooth profile section; H = nominal thickness of the link plate, all in inches. Referring to Fig. 6.2.101, $T = 0.93W - 0.006$, for single-strand chain; $= 0.90W - 0.006$, for double- and triple-strand chains; $= 0.88W - 0.006$, for quadruple- or quintuple-strand chains; and $= 0.86W - 0.006$, for sextuple-strand chain and over. $C = 0.5P$. $E = \frac{1}{8}P$. $R(\min) = 1.063P$. $Q = 0.5P$. $A = W + 4.15H + 0.003$. $M = A(n - 1) + T$.

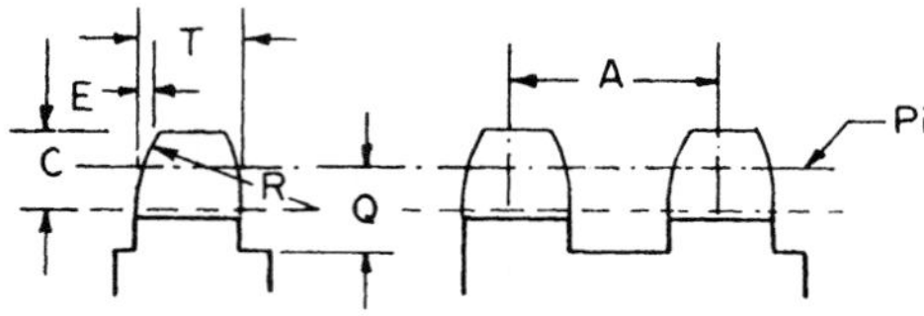

Figure 6.2.101 Sprocket tooth sections.

Inverted-tooth (silent) chain drives have a typical tooth form shown in Fig. 6.2.102. Such chains should be operated in an oil-retaining casing with provisions for lubrication. The use of offset links and chains with an uneven number of pitches should be avoided.

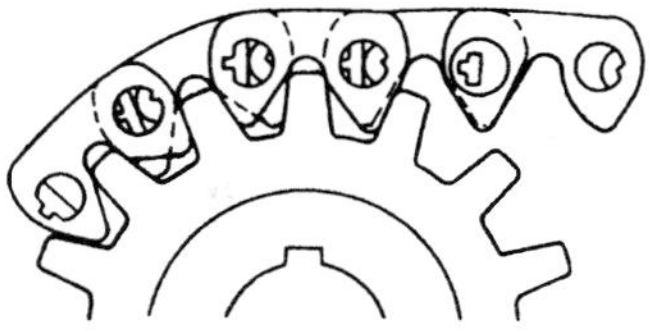

Figure 6.2.102 Inverted tooth (silent-chain) drive.

Horsepower ratings per inch of silent chain width, given in ANSI B29.2-1957 (R1971), for various chain pitches and speeds, are shown in Tables 6.2.59 and 6.2.60. These ratings are based on ideal drive conditions with relatively little shock or load variation, an average life of 20,000 h being assumed. In utilizing the horsepower ratings of the tables, the nominal horsepower of the drive should be multiplied by a service factor depending on the application. Maximum, or worst-case scenario, service factors are listed in Table 6.2.61.

For details on lubrication, sprocket dimensions, etc., see ANSI B29.2M-1957(R71). In utilizing Tables 6.2.59 and 6.2.60 (for a complete set of values, see ANSI B29.2M-1982) the required chain width is obtained by dividing the design horsepower by the horsepower ratings given. For calculating silent-chain lengths, the procedure for roller-chain drives may be used.

Flywheels

In slider crank mechanisms (see Sec. 6.1 Mechanism), one can surmise that both the slider force F and the external crank shaft torque may be functions of crank angle θ. Even if one or the other were deliberately kept constant, the remaining one would still be a function of θ. If a steady-speed crank is desired (ω = constant and α = zero), then external crankshaft torque T must be constantly adjusted to equal transmitted torque T_i. In such a situation a motor at the crankshaft would suffer

Table 6.2.59 Horsepower Rating per Inch of Chain Width, Silent-Chain Drive (Small Pitch)

Pitch, in	No. of teeth in small sprocket	Small sprocket, r/min											
		500	600	700	800	900	1,200	1,800	2,000	3,500	5,000	7,000	9,000
3/16	21	0.41	0.48	0.55	0.62	0.68	0.87	1.22	1.33	2.03	2.58	3.12	3.35
	25	0.49	0.58	0.66	0.74	0.82	1.05	1.47	1.60	2.45	3.13	3.80	4.10
	29	0.57	0.67	0.76	0.86	0.95	1.21	1.70	1.85	2.83	3.61	4.40	4.72
	33	0.64	0.75	0.86	0.97	1.07	1.37	1.90	2.08	3.17	4.02	4.85	—
	37	0.71	0.84	0.96	1.08	1.19	1.52	2.11	2.30	3.48	4.39	5.24	—
	45	0.86	1.02	1.15	1.30	1.43	1.83	2.53	2.75	4.15	5.21	—	—
	Type*				I				II			III	

Pitch, in	No. of teeth in small sprocket	Small sprocket, r/min												
		100	500	1,000	1,200	1,500	1,800	2,000	2,500	3,000	3,500	4,000	5,000	6,000
3/8	21	0.58	2.8	5.1	6.0	7.3	8.3	9.0	10	11	12	12	12	10
	25	0.69	3.3	6.1	7.3	8.8	10	11	13	14	15	5	15	14
	29	0.80	3.8	7.3	8.5	10	12	13	15	16	18	19	19	18
	33	0.90	4.4	8.3	9.8	12	14	15	18	19	21	21	21	20
	37	1.0	4.9	9.1	11	14	15	16	20	21	24	24	24	—
	45	1.3	6.0	11	13	16	19	20	24	26	28	29	—	—
	Type*		I				II					III		

Pitch, in	No. of teeth in small sprocket	Small sprocket, r/min										
		100	500	700	1,000	1,200	1,800	2,000	2,500	3,000	3,500	4,000
1/2	21	1.0	5.0	6.3	8.8	10	14	14	15	16	16	—
	25	1.2	5.0	7.5	10	13	16	18	20	21	21	20
	29	1.4	6.3	8.8	13	14	19	21	24	25	25	25
	33	1.6	7.5	10	14	16	23	24	28	29	30	29
	37	1.9	8.8	11	16	19	25	26	30	33	33	—
	45	2.5	10	14	19	23	30	30	36	39	—	—
	Type*		I		II					III		

Pitch, in	No. of teeth in small sprocket	Small sprocket, r/min									
		100	500	700	1,000	1,200	1,800	2,000	2,500	3,000	3,500
5/8	21	1.6	7.5	10	13	15	19	20	20	20	—
	25	1.9	8.8	11	16	19	24	25	26	26	24
	29	2.1	10	14	19	21	28	30	31	31	29
	33	2.5	11	16	21	25	33	34	36	36	34
	37	2.8	13	18	24	28	36	39	43	41	—
	45	3.4	16	21	29	34	44	46	—	—	—
	Type*		I		II			III			

Pitch, in	No. of teeth in small sprocket	Small sprocket, r/min								
		100	500	700	1,000	1,200	1,500	1,800	2,000	2,500
3/4	21	2.3	10	14	18	20	23	24	25	24
	25	2.8	13	16	21	25	29	31	31	30
	29	3.1	15	20	26	30	34	36	38	38
	33	3.6	16	23	30	34	39	43	44	44
	37	4.0	19	25	34	39	44	48	49	49
	45	4.9	23	30	40	46	53	56	58	—
	Type*		I		II			III		

* Type I: manual, brush, or oil cup. Type II: bath or disk. Type III: circulating pump.
NOTE: For best results, smaller sprocket should have at least 21 teeth.
SOURCE: Adapted from ANSI B29.2M-1982.

fatigue effects. In a combustion engine the crankshaft would deliver a fluctuating torque to its load.

Inserting a flywheel at the crankshaft allows the peak and valley excursions of ω to be considerably reduced because of the flywheel's ability to absorb energy over periods when $T > T_t$ and to deliver back into the system such excess energy when $T < T_t$. Figure 6.2.103*a* illustrates the above concepts, also showing that over one cycle of a repeated event the excess (+) energies and the deficient (–) energies are equal. The greatest crank speed change tends to occur across a single large positive loop, as illustrated in Fig. 6.2.103*a*.

Sizing the Flywheel For the single largest energy change we can write

$$\Delta E_{ab} \int_a^b (T - T_t)\,d\theta = \frac{J_0}{2}(\omega_2^2 - \omega_1^2)$$

$$= \frac{J_0}{2}(\omega_2 - \omega_1)(\omega_2 + \omega_1)$$

where J_0 = flywheel moment of inertia plus effective mechanism moment of inertia.

Table 6.2.60 Horsepower Rating per Inch of Chain Width, Silent-Chain Drive (Large Pitch)

Pitch, in	No. of teeth in small sprocket	Small sprocket, r/min										
		100	200	300	400	500	700	1,000	1,200	1,500	1,800	2,000
1	21	3.8	7.5	11	15	18	23	29	31	33	33	—
	25	5.0	8.8	14	18	21	28	35	39	41	41	41
	29	5.0	11	16	20	25	33	11	46	50	51	50
	33	6.3	13	18	24	29	38	49	54	59	59	58
	37	6.8	14	20	26	33	43	54	60	65	66	—
	45	8.8	16	25	31	39	51	65	71	76	—	—
	Type*		I				II				III	

Pitch, in	No. of teeth in small sprocket	Small sprocket, r/min										
		100	200	300	400	500	600	700	800	1,000	1,200	1,500
$1\frac{1}{4}$	21	6.3	11	18	23	26	30	33	36	40	41	—
	25	7.5	14	20	26	31	36	40	44	50	53	53
	29	8.6	16	24	31	38	43	48	53	59	63	64
	33	9.9	19	28	35	43	49	55	60	69	73	74
	37	11	21	30	40	48	55	63	68	76	81	—
	45	13	26	38	49	59	68	75	81	91	—	—
	Type*		I			II				III		

Pitch, in	No. of teeth in small sprocket	Small sprocket, r/min										
		100	200	300	400	500	600	700	800	900	1,000	1,200
$1\frac{1}{2}$	21	8.8	16	24	30	36	40	44	46	49	49	—
	25	10	20	29	38	44	50	55	59	61	65	64
	29	13	24	34	44	51	59	65	70	74	75	76
	33	14	28	39	50	59	68	75	80	85	88	89
	37	16	30	44	59	66	76	84	90	96	99	—
	45	19	38	54	68	81	93	101	108	113	—	—
	Type*	I			II				III			

Pitch, in	No. of teeth in small sprocket	Small sprocket, r/min								
		100	200	300	400	500	600	700	800	900
2	21	16	29	40	50	53	63	65	—	—
	25	18	35	49	61	70	78	83	85	85
	29	21	41	58	73	84	93	99	103	103
	33	25	46	66	83	96	106	114	118	118
	37	28	53	75	72	110	124	128	131	—
	45	34	64	90	113	131	144	151	—	—
	Type*	I		II				III		

* Type I: manual, brush, or oil cup. Type II: bath or disk. Type III: circulating pump.
NOTE: For best results, small sprocket should have at least 21 teeth.
SOURCE: Adapted from ANSI B29.2-1982.

Table 6.2.61 Service Factors* for Silent-Chain Drives

Application	Fluid-coupled engine or electric motor		Engine with straight mechanical drive		Torque converter drives	
	10 h	24 h	10 h	24 h	10 h	24 h
Agitators	1.1	1.4	1.3	1.6	1.5	1.8
Brick and clay machinery	1.4	1.7	1.6	1.9	1.8	2.1
Centrifuges	1.4	1.7	1.6	1.9	1.8	2.1
Compressors	1.6	1.9	1.8	2.1	2.0	2.3
Conveyors	1.6	1.9	1.8	2.1	2.0	2.3
Cranes and hoists	1.4	1.7	1.6	1.9	1.8	2.1
Crushing machinery	1.6	1.9	1.8	2.1	2.0	2.3
Dredges	1.6	1.9	1.8	2.1	2.0	2.3
Elevators	1.4	1.7	1.6	1.9	1.8	2.1
Fans and blowers	1.5	1.8	1.7	2.0	1.9	2.2
Mills—flour, feed, cereal	1.4	1.7	1.6	1.9	1.8	2.1
Generator and excitors	1.2	1.5	1.4	1.7	1.6	1.9
Laundry machinery	1.2	1.5	1.4	1.7	1.6	1.9
Line shafts	1.6	1.9	1.8	2.1	2.0	2.3
Machine tools	1.4	1.7	—	—	—	—
Mills—ball, hardinge, roller, etc.	1.6	1.9	1.8	2.1	—	—
Mixer—concrete, liquid	1.6	1.9	1.8	2.1	2.0	2.3
Oil field machinery	1.6	1.9	1.8	2.1	2.0	2.3
Oil refinery equipment	1.5	1.8	1.7	2.0	1.9	2.2
Paper machinery	1.5	1.8	1.7	2.0	1.9	2.2
Printing machinery	1.5	2.0	1.4	1.7	1.6	1.9
Pumps	1.6	1.9	1.8	2.1	2.0	2.3
Rubber plant machinery	1.5	1.8	1.7	2.0	1.9	2.2
Rubber mill equipment	1.6	1.9	1.8	2.1	2.0	2.3
Screens	1.5	1.8	1.7	2.0	1.9	2.2
Steel plants	1.3	1.6	1.5	1.8	1.7	2.0
Textile machinery	1.1	1.4	—	—	—	—

* The values shown are for the maximum worst-case scenario for each application. The table was assembled from ANSI B29.2M-1982, by permission. Because the listings above are maximum values, overdesign may result when they are used. Consult the ANSI tables for specific design values.

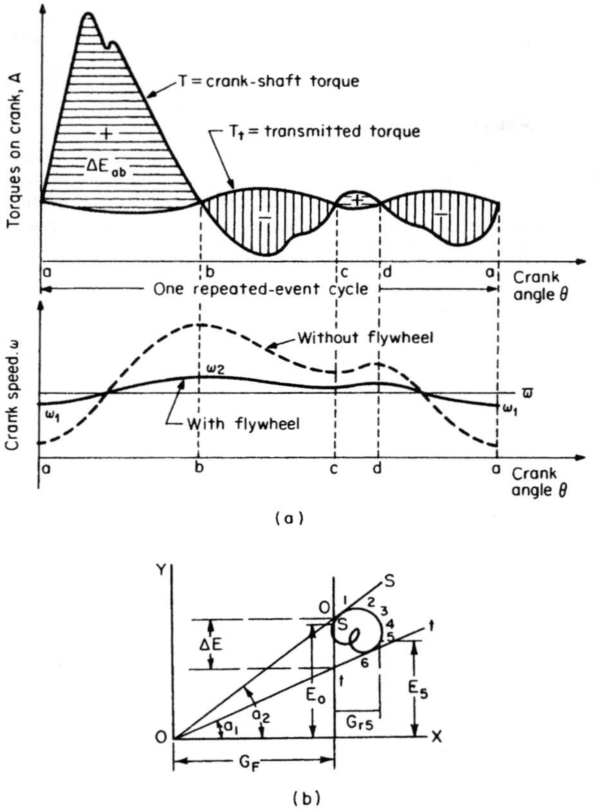

Figure 6.2.103 Sizing the flywheel. (*a*) Variation of torque and crank speed vs. crank angle, showing ΔE_{ab}; (*b*) graphics for Wittenbauer's analysis.

Define $\bar{\omega} \doteq (\omega_2 + \omega_1)/2$ and C_s = coefficient of speed fluctuation = $(\omega_2 - \omega_1)/\bar{\omega}$. Hence $\Delta E_{ab} = J_0 C_s \bar{\omega}^2$ Acceptable values of C_s are:

Pumps	0.03-0.05
Machine tools	0.025-0.03
Looms	0.025
Paper mills	0.025
Spinning mills	0.015
Crusher	0.02
Electric generators, ac	0.003
Electric generators, dc	0.002

Evaluating ΔE_{ab} involves finding the integral

$$\int_a^b (T - T_1)\, d\theta$$

which can be done graphically or by a numerical technique such as Simpson's rule.

Wittenbauer's Analysis for Flywheel Performance This method does not involve more computation work than the one described above, but it is more accurate where the reciprocating parts are comparatively heavy. Wittenbauer's method avoids the inaccuracy resulting from the evaluation of the inertia forces on the reciprocating parts on the basis of the uniform nominal speed of rotation for the engine.

Let the crankpin velocity be represented by v_r and the velocity of any moving masses (m_1, m_2, m_3, etc.,) at any instant of phase be represented, respectively, by v_1, v_2, v_3, etc. The kinetic energy of the entire engine system of moving masses may then be expressed as

$$E = \tfrac{1}{2}(m_1 v_1^2 + m_2 v_2^2 + m_3 v_3^2 + \cdots) = \tfrac{1}{2} M_r v_r^2$$

or, the single reduced mass M_r at the crankpin which possesses the equivalent kinetic energy is

$$M_r = [m_1(v_1/v_r)^2 + m_2(v_2/v_r)^2 + m_3(v_3/v_r)^2 + \cdots]$$

In an engine mechanism, sufficiently accurate values of M_r can be obtained if the weight of the connecting rod is divided between the crankpin and the wrist pin so as to retain the center of gravity of the rod in its true position; usually 0.55 to 0.65 of the weight of the connecting rod should be placed on the crankpin, and 0.45 to 0.35 of the weight on the wrist pin. M_r is a variable in engine mechanisms on account of the reciprocating parts and should be found for a number of crank positions. It should include all moving masses except the flywheel.

The total energy E used in accelerating reciprocating parts from the beginning of the forward stroke up to any crank position can be obtained by finding from the indicator cards the total work done in the cylinder (on both sides of the piston) up to that time and subtracting from it the work done in overcoming the resisting torque, which may usually be assumed constant. The mean energy of the moving masses is $E_0 = \tfrac{1}{2} M_r v_r^2$.

In Fig. 6.2.103*b*, the reduced weights of the moving masses $G_F + G_{r5}$ are plotted on the X axis corresponding to different crank positions. $G_F = gM_F$ is the reduced flywheel weight and $G_{r5} = gM_{r5}$ is the sum of the other reduced weights. Against each of these abscissas is plotted the energy E available for acceleration measured from the beginning of the forward stroke. The curve $O123456$ is the locus of these plotted points.

The diagram possesses the following property: Any straight line drawn from the origin O to any point in the curve is a measure of the velocity of the moving masses; tangents bounding the diagram measure the limits of velocity between which the crankpin will operate. The maximum linear velocity of the crankpin in feet per second is $v_2 = \sqrt{2g}$ tan a_2, and the minimum velocity is $v_1 = \sqrt{2g}$ tan a_1. Any desired change in v_1 and v_2 may be accomplished by changing the value of G_F, which means a change in the flywheel weight or a change in the flywheel weight reduced to the crankpin. As G_F is very large compared with G_r and the point 0 cannot be located on the diagram unless a very large drawing is made, the tangents are best formed by direct calculation:

$$\tan\alpha_2 = \frac{v_r^2}{2g}(1+k) \qquad \tan\alpha_1 = \frac{v_r^2}{2g}(1-k)$$

where k is the coefficient of velocity fluctuation. The two tangents *ss* and *tt* to the curve $O123456$, thus drawn, cut a distance ΔE and on the ordinate E_0. The reduced flywheel weight is then found to be

$$G_F(\Delta E)g/(v_r^2 k)$$

SPRINGS

It is assumed in the following formulas that the springs are in no case stressed beyond the elastic limit (i.e., that they are perfectly elastic) and that they are subject to Hooke's law.

Notation

P = safe load, lb
f = deflection for a given load P, in
l = length of spring, in
V = volume of spring, in^3
S_s = safe stress (due to bending), lb/in^2
S_v = safe shearing stress, lb/in^2
U = resilience, in · lb

For sheet metal and wire gauge, ferrous and nonferrous, see Table 6.2.75 and metal suppliers' catalogs.

The **work** in inch-pounds performed in **deflecting a spring** from 0 to f (spring duty) is $U = Pf/2 = S_s^2 V/(CE)$. This is based upon the assumption that the deflection is proportional to the load, and C is a constant dependent upon the shape of the springs and E is modulus of elasticity.

The **time of vibration** T (in seconds) **of a spring** (weight not considered) is equal to that of a simple circular pendulum whose length l_0 equals the deflection f (in feet that is produced in the spring by the load P). $T = \pi\sqrt{l_0/g}$, where g = acceleration of gravity, ft/s^2.

Springs Subjected to Bending

1. **Rectangular plate spring** (Fig. 6.2.104).

$$P = bh^2S_s/(6l) \quad I = bh^3/12 \quad U = Pf/2 = VS_s^2/(18E)$$
$$f = Pl^3/(3EI) = 4Pl^3/(bh^3E) = 2l^2S_s/(3hE)$$

2. **Triangular plate spring** (Fig. 6.2.105). The elastic curve is a circular arc.

$$P = bh^2S_s/(6l) \quad I = bh^3/12 \quad U = Pf/2 = S_s^2V/(6E)$$
$$f = Pl^3/(2EI) = 6Pl^3/(bh^3E) = l^2S_s/(hE)$$

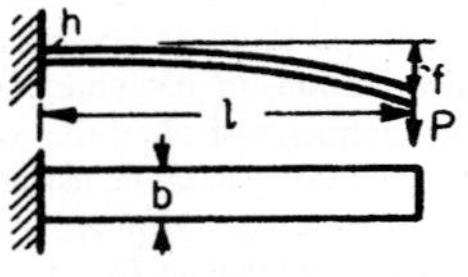

Figure 6.2.104 Rectangular plate spring.

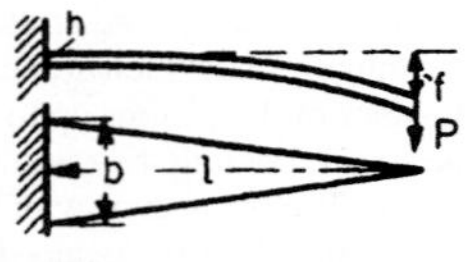

Figure 6.2.105 Triangular plate spring.

3. **Rectangular plate spring with end tapered** in the form of a cubic parabola (Fig. 6.2.106). The elastic curve is a circular arc; P, I, and f same as for triangular plate spring (Fig. 6.2.105); $U = Pf/2 = S_s^2V/(9E)$.

The strength and deflection of **single-leaf flat springs** of various forms are given (Bruce, *Am. Mach.*, July 19, 1900) by the formulas $h = al^2/f$ and $b = cPl/h^2$. The volume of the spring is given by $V = vlbh$. The values of constants a and c and the resilience in inch-pounds per cubic inch are given in Table 6.2.62, in terms of the safe stress S_s. Values of v are given also.

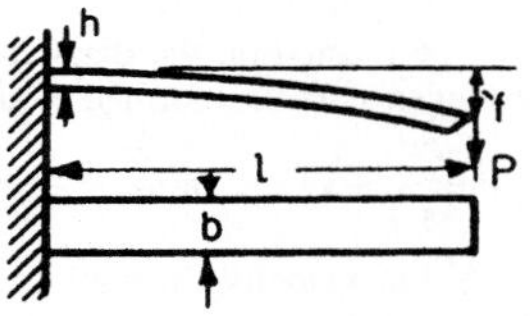

Figure 6.2.106 Rectangular plate spring: tapered end.

4. **Compound (leaf or laminated) springs.** If several springs of rectangular section are combined, the resulting compound spring should (1) form a beam of uniform strength that (2) does not open between the joints while bending (i.e., elastic curve must be a circular arc). Only the type immediately following meets both requirements, the others meeting only the second requirement.

5. **Laminated triangular plate spring** (Fig. 6.2.107). If the triangular plate spring shown at I is cut into an even number (= $2n$) of strips of equal width (in this case eight strips of width $b/2$), and these strips are combined, a laminated spring will be formed whose carrying capacity will equal that of the original uncut spring; or $P = nbh^2S_s/(6l)$; $n = 6Pl/(bh^2S_s)$.

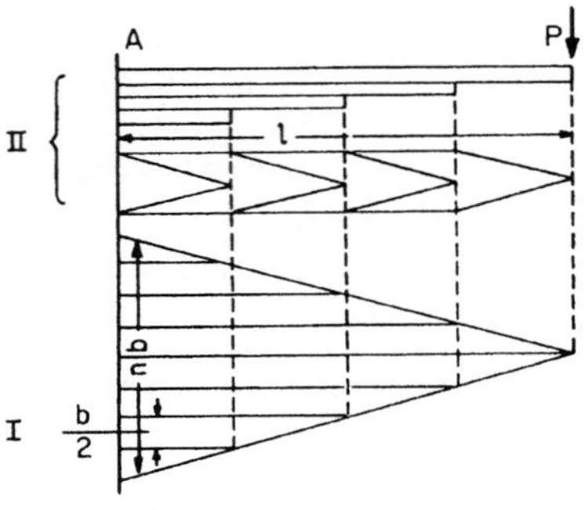

Figure 6.2.107 Laminated triangular plate spring.

6. **Laminated rectangular plate spring with lead ends tapered** in the form of a cubical parabola (Fig. 6.2.108); see case 3 above.

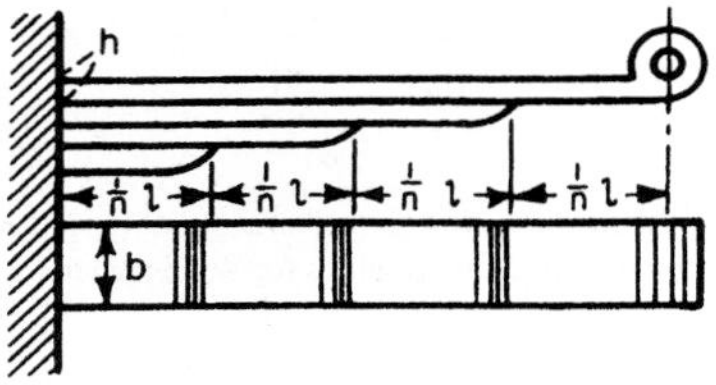

Figure 6.2.108 Laminated rectangular plate spring with leaf end tapered.

Table 6.2.62 Strength and Deflection of Single-Leaf Flat Springs

Load applied at end of spring; $c = 6/S_s$				Load applied at center of spring; $c = 6/4S_s$			
Plans and elevations of springs	a	U	v	Plans and elevations of springs	a	U	v
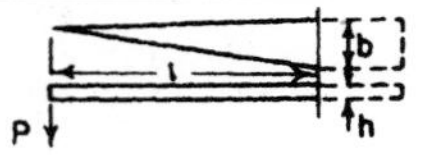	$\frac{S_s}{E}$	$\frac{S_s^2}{6E}$	$\frac{1}{2}$	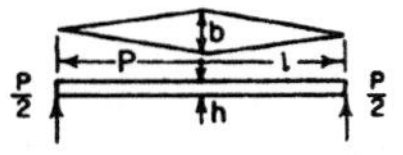	$\frac{S_s}{4E}$	$\frac{S_s^2}{6E}$	$\frac{1}{2}$
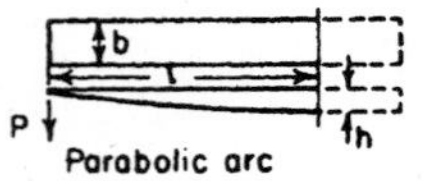 Parabolic arc	$\frac{4S_s}{3E}$	$\frac{S_s^2}{6E}$	$\frac{2}{3}$	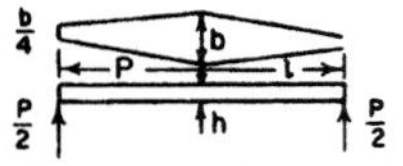	$\frac{0.87S_s}{4E}$	$\frac{0.70S_s^2}{6E}$	$\frac{5}{8}$
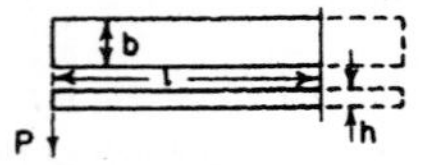	$\frac{2S_s}{3E}$	$\frac{0.33S_s^2}{6E}$	1	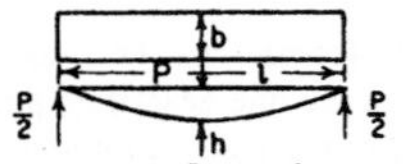 Parabolic arcs	$\frac{S_s}{3E}$	$\frac{S_s^2}{6E}$	$\frac{2}{3}$
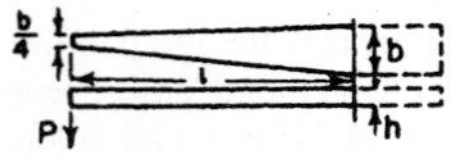	$\frac{0.87S_s}{E}$	$\frac{0.70S_s^2}{6E}$	$\frac{5}{8}$	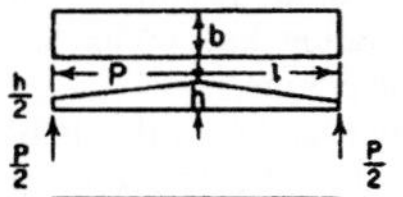	$\frac{1.09S_s}{4E}$	$\frac{0.725S_s^2}{6E}$	$\frac{3}{4}$
	$\frac{1.09S_s}{E}$	$\frac{0.725S_s^2}{6E}$	$\frac{3}{4}$	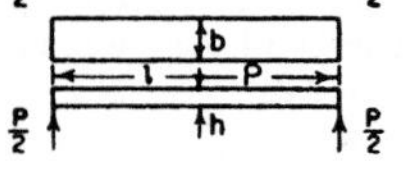	$\frac{S_s}{6E}$	$\frac{0.33S_s^2}{6E}$	I

7. **Laminated trapezoidal plate spring with leaf ends tapered** (Fig. 6.2.109). The ends of the leaves are trapezoidal and are tapered according to the formula

$$z = \frac{h}{\sqrt[3]{1 + (b_1/b)(a/x - 1)}}$$

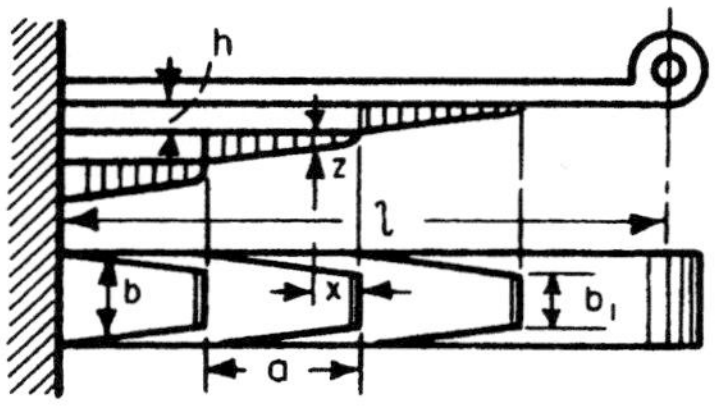

Figure 6.2.109 Laminated trapezoidal plate spring with leaf ends tapered.

8. **Semielliptic springs** (for locomotives, trucks, etc.). Referring to Fig. 6.2.110, the load $2P$ (lb) acting on the spring center band produces a tensional stress $P/\cos a$ in each of the inclined shackle links. This is resolved into the vertical force P and the horizontal force $P \tan a$, which together produce a bending moment $M = P(l + p \tan a)$. The equations given in (1), (2), and (3) apply to curved as well as straight springs. The bearing force $= 2P = (2nbh^2/6)[S_s/(l + p \tan a)]$, and the deflection $= [6l^2/(nbh^3)]P(l + p \tan a)/E = l^2S_2/(hE)$.

In addition, to the bending moment, the leaves are subjected to the tension force $P \tan a$ and the transverse force P, which produce in the upper leaf an additional stress $S = P \tan a/(bh)$, as well as a transverse shearing stress.

In determining the number of leaves n in a given spring, allowance should be made for an excess load on the spring caused by the vibration. This is usually done by decreasing the allowable stress about 15 percent.

The foregoing does not take account of initial stresses caused by the band. For more detailed information, see Wahl, "Mechanical Springs," Penton.

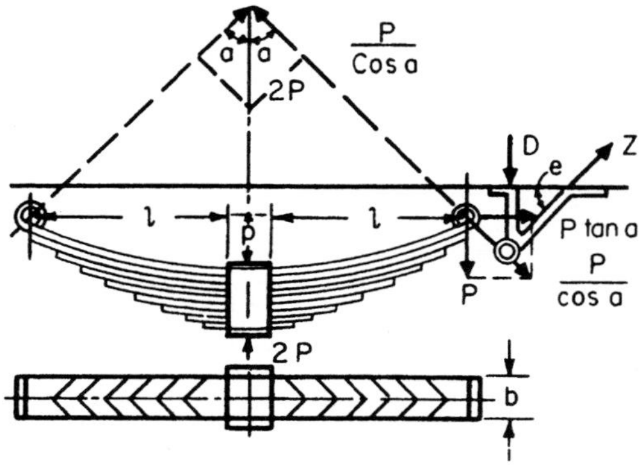

Figure 6.2.110 Semielliptic springs.

9. **Elliptic springs.** Safe load $P = nbh^2S_s/(6l)$, where $l = \frac{1}{2}$ distance between bolt eyes (less $\frac{1}{2}$ length of center band, where used); deflection $f = 4l^2S_sK/(hE)$, where

$$K = \frac{1}{(1-r)^3}\left[\frac{1-r^2}{2} - 2r(1-r) - r^2 \ln r\right]$$

r being the number of full-length leaves ÷ total number (n) of leaves in the spring. All dimensions in inches. For semielliptic springs, the deflection is only half as great. Safe load $= nbh^2S_s/(3l)$. (Peddle, *Am. Mach.*, Apr. 17, 1913.)

Coiled Springs In these, the load is applied as a couple Pr which turns the spring while winding or holds it in place when wound up. If the spindle is not to be subjected to bending moment, P must be replaced by two equal and opposite forces ($P/2$) acting at the circumference of a circle of radius r. The formulas are the same in both cases. The springs are assumed to be fixed at one end and free at the other. The bending moment acting on the section of least resistance is always Pr. The length of the straightened spring $= l$. See Benjamin and French, Experiments on Helical Springs, *Trans. ASME*, **23**, p. 298.

For **heavy closely coiled helical springs** the usual formulas are inaccurate and result in stresses greatly in excess of those assumed. See Wahl, Stresses in Heavy Closely-Coiled Helical Springs, *Trans. ASME*, 1929. In springs 10 to 12 and 15 to 18, the quantity k is unity for lighter springs and has the stated values (supplied by Wahl) for heavy closely coiled springs.

10. **Spiral coiled springs of rectangular cross section** (Fig. 6.2.111).

$$P = bh^2S_s/(6rk) \quad I = bh^3/12 \quad U = Pf/2 = S_s^2V/(6Ek^2)$$
$$f = ra = Plr^2/(EI) = 12Plr^2/(Ebh^3) = 2rlS_s/(hEk)$$

For heavy closely coiled springs, $k = (3c - 1)/(3c - 3)$, where $c = 2R/h$ and R is the minimum radius of curvature at the center of the spiral.

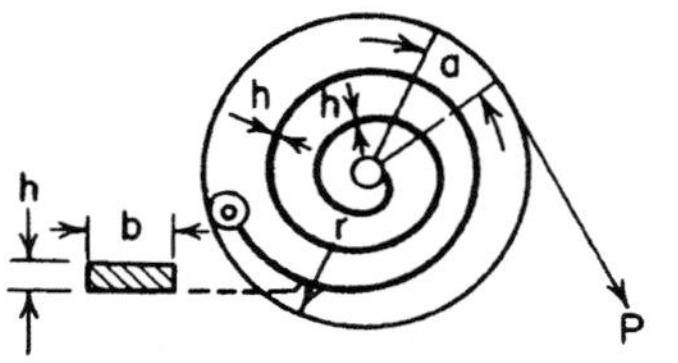

Figure 6.2.111 Spiral coiled spring: rectangular cross section.

11. **Cylindrical helical spring of circular cross section** (Fig. 6.2.112).

$$P = \pi d^3S_s/(32rk) \quad I = \pi d^4/64 \quad U = Pf/2 = S_s^2V/(8Ek^2)$$
$$f = ra = Plr^2/(EI) = 64Plr^2/(\pi Ed^4) = 2rlS_s/(dEk)$$

For heavy closely coiled springs, $k = (4c - 1)/(4c - 4)$, where $c = 2r/d$.

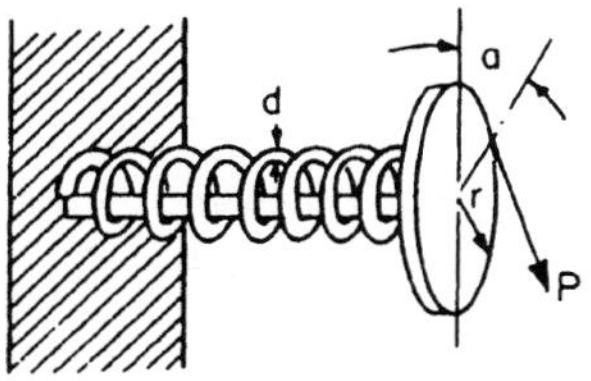

Figure 6.2.112 Cylindrical helical spring: circular cross section.

12. **Cylindrical helical spring of rectangular cross section** (Fig. 6.2.113).

$$P = bh^2S_s/(6rk) \quad I = bh^3/12 \quad U = Pf/2 = S_s^2V/(8Ek^2)$$
$$f = ra = Plr^2/(EI) = 12Plr^2/(Ebh^3) = 2rlS_s/(hEk)$$

For heavy closely coiled springs, $k = (3c - 1)/(3c - 3)$, where $c = 2r/h$.

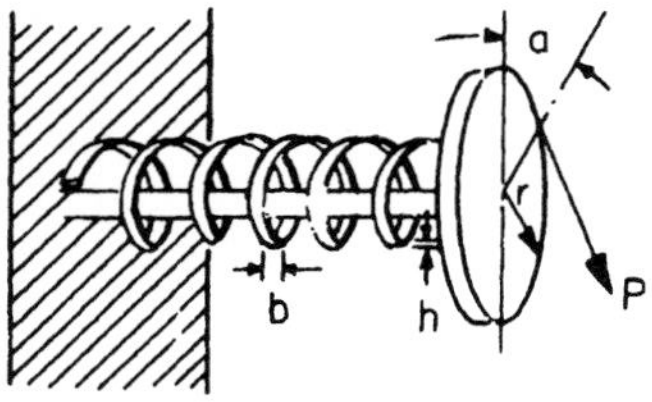

Figure 6.2.113 Cylindrical helical spring: rectangular cross section.

Springs Subjected to Torsion

The statements made concerning coiled springs subjected to bending apply also to springs 13 and 14.

13. **Straight bar spring of circular cross section** (Fig. 6.2.114).

$$P = \pi d^3 S_v/(16r) = 0.1963 d^3 S_v/r \qquad U = Pf/2 = S_v^2 V/(4G)$$
$$f = ra = 32r^2 lP/(\pi d^4 G) = 2rl\,S_v/(dG)$$

14. **Straight bar spring of rectangular cross section** (Fig. 6.2.115).

$$P = 2b^2 hS_v/(9r) \qquad K = b/h$$
$$U = Pf/2 = 4S_v^2 V(K^2+1)/(45G) \quad \text{max when } K = 1$$
$$f = ra = 3.6r^2 lP(b^2+h^2)/(b^3h^3G) = 0.8rlS_v(b^2+h^2)/(bh^2G)$$

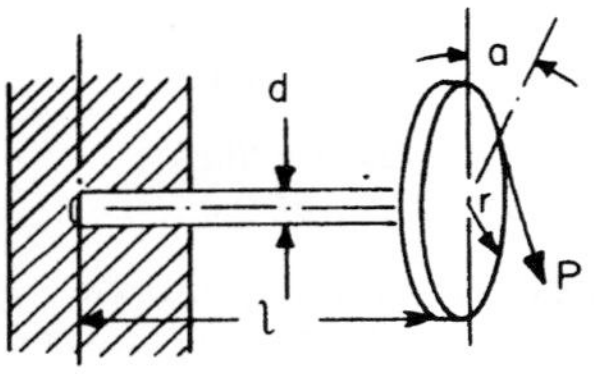

Figure 6.2.114 Straight bar spring: circular cross section.

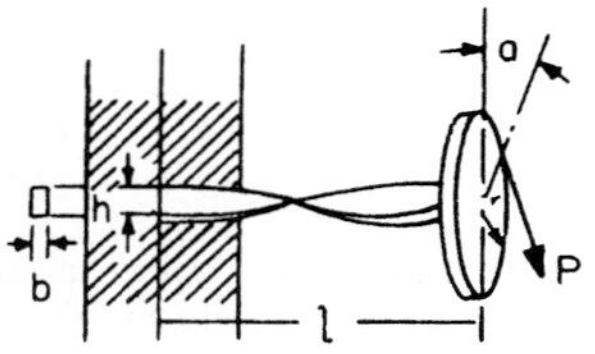

Figure 6.2.115 Straight bar spring: rectangular cross section.

Springs Loaded Axially in Tension

Note: For springs 15 to 18, r = mean radius of coil; n = number of coils.

15. **Cylindrical helical spring of circular cross section** (Fig. 6.2.116).

$$P = \pi d3S_v/(16rk) = 0.1963d^3S_v/(rk)$$
$$U = Pf/2 = S_v^2 V/(4Gk^2)$$
$$f = 64nr^3P/(d^4G) = 4\pi nr^2 S_v/(dGk)$$

For heavy closely coiled springs, $k = (4c - 1)/(4c - 4) + 0.615/c$, where $c = 2r/d$.

16. **Cylindrical helical spring of rectangular cross section** (Fig. 6.2.117).

$$P = 2b^2hS_v/(9rk) \qquad K = b/h$$
$$U = Pf/2 = 4S_v^2 V(K^2+1)/(45Gk^2) \quad \text{max when } K = 1$$
$$f = 7.2\pi nr^3 P(b^2+h^2)/(b^3h^3G) = 1.6\pi nr^2 S_v(b^2+h^2)/(bh^2Gk)$$

For heavy closely coiled springs, $k = (4c - 1)/(4c - 4) + 0.615/c$, where $c = 2r/b$. b and h are the dimensions of the rectangular cross-section of the spring.

17. **Conical helical spring of circular cross section** (Fig. 6.2.118*a*).

l = length of developed spring
d = diameter of wire
r = maximum mean radius of coal

$$P = \pi d^3 S_v/(16rk) = 0.1963d^3S_v/(rk)$$
$$U = Pf/2 = S_v^2 V/(8Gk^2)$$
$$f = 16r^2 lP/(\pi d^4 G) = 16\pi r^3 P/(d^4G)$$
$$= rlS_v/(dGk) = \pi nr^2 S_v/(dGk)$$

Figure 6.2.116 Cylindrical helical spring: circular cross section.

Figure 6.2.117 Cylindrical helical spring: rectangular cross section.

For heavy closely coiled springs, $k = (4c - 1)/(4c - 4) + 0.615/c$, where $c = 2r/d$.

18. **Conical helical spring of rectangular cross section** (Fig. 6.2.118*b*).

b = small dimension of section
d = large dimension of section
r = maximum mean radius of coil

$$K = b/h\,(\le 1) \qquad P = 2b^2hS_v/(9rk)$$
$$U = Pf/2 = 2S_v^2 V(K^2+1)/(45Gk^2) \qquad \text{max when } K = 1$$
$$f = 1.8r^2 lP(b^2+h^2)/(b^3h^3G) = 1.8\pi nr^3 P(b^2+h^2)/(b^3h^3G)$$
$$= 0.4rlS_v(b^2+h^2)/(bh^2Gk) = 0.4\pi nr^2 S_v(b^2+h^2)(bh^2Gk)$$

For heavy closed coiled springs, $k = (4c - 1)/(4c - 4) + 0.615c$, where $c = 2r/r_o - r_i$.

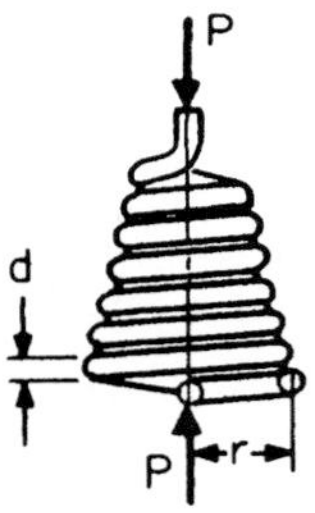

Figure 6.2.118*a* Conical helical spring: circular cross section.

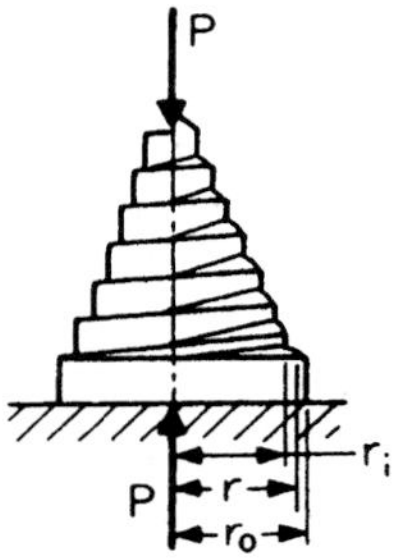

Figure 6.2.118*b* Conical helical spring: rectangular cross section.

19. **Truncated conical springs** (17 and 18). The formulas under 17 and 18 apply for truncated springs. In calculating deflection f, however, it is necessary to substitute $r_1^2 + r_2^2$ for r^2, and $\pi n(r_1 + r_2)$ for πnr, r_1 and r_2 being, respectively, the greatest and least mean radii of the coils.

Note: The preceding formulas for various forms of coiled springs are sufficiently accurate when the cross-section dimensions are small in comparison with the radius of the coil, and for small pitch. Springs 15 to 19 are for either tension or compression but formulas for springs 17 and 18 are good for compression only until the largest coil flattens out; then r becomes a variable, depending on the load.

Design of Helical Springs

When sizing a new spring, one must consider the spring's available working space and the loads and deflections the spring must experience. Refinements dictated by temperature, corrosion, reliability, cost, etc. may also enter design considerations. The two basic formulas of load and deflection (see item 15, Fig. 6.2.116) contain eight variables (f, P, d, S, r, k, n, G) which prevent one from being able to use a one-step solution. For instance, if f and P are known and S and G are chosen, there still remain d, r, k, and n to be found.

A variety of solution approaches are available, including: (1) slide-rule-like devices available from spring manufacturers, (2) nomographic methods (Chironis, "Spring Design and Application," McGraw-Hill; Tsai, Speedy Design of Helical Compression Springs by Nomography Method, *J. of Eng. for Industry*, Feb. 1975), (3) table methods (Carlson, "Spring Designer's Handbook," Marcel Dekker), (4) formula method (ibid.), and (5) computer programs and subroutines.

Design by Tables

Safe working loads and deflections of cylindrical helical springs of round steel wire in tension or compression are given in Table 6.2.63. The table is based on the formulas given for spring 15. d = diameter of steel wire, in; D = pitch diameter (center to center of wire), in; P = safe working load for given unit stress, lb; f = deflection of 1 coil for safe working load, in.

Table 6.2.63 Representative of Safe Working Loads *P* and Deflections *f* of Cylindrical Helical Steel Springs of Circular Cross Section

(For closely coiled springs, divide given load and deflection values by the curvature factor *k*.)

Allowable unit stress, lb/in²	Diam, in	Pitch diameter *D*, in											
		D	5/32	3/16	1/4	5/16	3/8	7/16	1/2	5/8	3/4	7/8	1
150,000	0.035	*P*	16.2	13.4	10.0	8.10	6.66	5.75	4.96	4.05	3.39		
		f	.026	.037	.067	.105	.149	.200	.276	.420	.608		
	0.041	*P*	26.2	21.6	16.2	13.0	10.8	9.27	8.10	6.52	5.35	4.57	
		f	.023	.032	.057	.089	.128	.175	.229	.362	.512	.697	
	0.047	*P*	39.1	32.6	24.5	19.6	16.4	13.9	12.3	9.80	8.10	6.92	6.14
		f	.019	.028	.049	.078	.112	.153	.200	.311	.449	.610	.800
	0.054	*P*	59.4	49.6	37.2	29.7	24.6	21.2	18.5	14.7	12.4	10.5	9.25
		f	.016	.024	.043	.067	.098	.133	.174	.273	.390	.532	.695
	0.062	*P*		74.9	56.1	44.9	37.3	32.0	28.0	22.4	18.6	16.1	13.9
		f		.021	.037	.058	.084	.115	.151	.235	.340	.460	.605
	0.063	*P*		78.2	58.7	46.9	39.2	33.9	29.4	23.5	19.6	16.8	14.7
		f		.020	.037	.057	.083	.113	.148	.233	.335	.445	.591
	0.072	*P*		117.	80.7	70.0	58.7	50.2	43.6	35.2	29.0	25.0	21.9
		f		.018	.032	.050	.077	.100	.130	.203	.294	.405	.521
	0.080	*P*			121	96.6	80.5	69.1	60.4	48.2	48.2	34.6	30.1
		f			.029	.045	.065	.090	.117	.183	.262	.359	.470
140,000	0.092	*P*			171	136	113	97.6	85.5	68.9	57.3	48.8	42.6
		f			.023	.037	.053	.072	.098	.148	.214	.291	.388
	0.093	*P*			178	142	118	99.5	89.0	71.2	59.1	50.9	44.3
		f			.023	.036	.052	.071	.093	.146	.211	.286	.376
	0.105	*P*				204	170	147	127	102	85.4	73.0	63.4
		f				.032	.047	.064	.083	.122	.188	.256	.336
	0.120	*P*				303	253	217	190	152	126	108	95.2
		f				.028	.041	.055	.073	.114	.174	.223	.296
	0.125	*P*					286	245	214	171	143	121	107
		f					.039	.053	.069	.109	.169	.213	.278
	0.135	*P*					359	309	270	217	171	154	135
		f					.036	.049	.064	.106	.145	.198	.260
	0.148	*P*						408	356	285	237	207	178
		f						.045	.059	.092	.132	.180	.236

Allowable unit stress, lb/in²	Diam, in	Pitch diameter *D*, in														
		D	7/16	1/2	5/8	3/4	7/8	1	1 1/8	1 1/4	1 3/8	1 1/2	1 5/8	1 3/4	1 7/8	2
140,000	0.156	*P*	480	418	330	270	234	208	185	167	152	139	128	119	111	104
		f	0.42	.056	.087	.125	.171	.223	.282	.350	.422	.509	.588	.685	.785	.896
	0.162	*P*		468	376	311	276	234	207	187	170	156	143	134	125	117
		f		.054	.085	.121	.165	.216	.273	.338	.409	.488	.566	.663	.757	.863
	0.177	*P*		608	487	406	347	305	270	243	223	205	187	174	163	152
		f		.049	.077	.115	.151	.198	.251	.311	.375	.447	.522	.606	.695	.793
125,000	0.187	*P*		642	522	426	367	320	284	256	233	213	197	183	170	160
		f		.041	.065	.093	.127	.166	.210	.260	.314	.373	.487	.510	.584	.665
	0.192	*P*		696	556	465	396	348	309	278	254	233	214	199	186	174
		f		.040	.063	.091	.124	.160	.205	.252	.308	.366	.428	.499	.571	.652
	0.207	*P*			694	579	495	432	385	346	315	288	266	247	232	216
		f			.059	.085	.115	.151	.191	.236	.286	.342	.396	.462	.570	.607
	0.218	*P*			812	678	580	509	452	408	360	339	310	291	270	255
		f			.055	.080	.109	.142	.180	.223	.269	.321	.374	.437	.488	.570
	0.225	*P*			895	746	640	560	498	447	407	372	345	320	299	280
		f			.054	.078	.106	.138	.175	.213	.262	.312	.372	.425	.486	.565
	0.244	*P*			1,120	950	811	711	632	570	517	475	438	406	381	356
		f			.049	.071	.098	.138	.161	.200	.240	.287	.336	.391	.449	.537
	0.250	*P*				1,027	880	760	685	617	560	513	476	440	410	385
		f				.070	.095	.131	.157	.191	.236	.281	.328	.385	.439	.524

	0.263	*P*	1,195	1,125	895	795	717	652	598	551	501	478	448
		f	.066	.089	.118	.149	.183	.224	.266	.312	.363	.416	.475
	0.281	*P*	1,450	1,240	1,087	969	863	794	724	665	620	580	543
		f	.062	.085	.111	.140	.172	.209	.250	.292	.340	.390	.443
	0.283	*P*		1,264	1,110	985	886	805	740	682	634	592	564
		f		.084	.111	.139	.169	.207	.246	.289	.338	.356	.440
115,000	0.312	*P*		1,575	1,376	1,220	1,100	1,000	915	845	775	733	687
		f		.070	.092	.116	.144	.174	.207	.242	.283	.322	.368
	0.331	*P*			1,636	1,455	1,316	1,187	1,090	1,000	932	870	818
		f			.088	.109	.135	.163	.194	.227	.264	.343	.346
	0.341	*P*			1,820	1,620	1,452	1,325	1,214	1,120	1,040	970	910
		f			.082	.105	.127	.156	.186	.218	.256	.293	.330
	0.362	*P*			2,140	1,910	1,714	1,560	1,430	1,318	1,220	1,147	1,070
		f			.079	.100	.123	.149	.177	.207	.243	.273	.317
	0.375	*P*				2,110	1,940	1,780	1,580	1,458	1,354	1,265	1,185
		f				.079	.117	.144	.172	.201	.234	.268	.308
	0.393	*P*				2,430	2,180	1,984	1,820	1,680	1,560	1,458	1,365
		f				.092	.114	.137	.164	.195	.223	.256	.292
	0.406	*P*					2,400	2,170	2,000	1,840	1,710	1,600	1,500
		f					.108	.134	.159	.168	.217	.248	.284
	0.430	*P*					2,875	2,610	2,400	2,210	2,050	1,918	1,798
		f					.104	.126	.150	.175	.204	.234	.267
	0.437	*P*					3,000	2,730	2,500	2,310	2,140	2,000	1,800
		f					.100	.124	.148	.173	.201	.231	.264

Allowable unit stress, lb/in²	Diam, in	Pitch diameter *D*, in											
		D	1¼	1⅜	1½	1⅝	1¾	1⅞	2	2¼	2½	2¾	3
110,000	0.460	*P*		3,065	2,800	2,580	2,400	2,230	2,100	1,865	1,680	1,530	1,400
		f		.112	.134	.157	.183	.209	.239	.303	.374	.447	.536
	0.468	*P*		3,265	2,940	2,725	2,530	2,375	2,210	1,970	1,770	1,610	1,472
		f		.111	.132	.154	.182	.206	.235	.295	.368	.444	.530
	0.490	*P*		3,675	3,270	3,115	2,890	2,710	2,535	2,245	2,025	1,840	1,690
		f		.106	.126	.148	.172	.196	.225	.284	.351	.424	.506
	0.500	*P*			3,610	3,320	3,090	2,890	2,710	2,410	2,160	1,970	1,810
		f			.123	.144	.168	.192	.220	.274	.347	.415	.495
	0.562	*P*				4,700	4,390	4,090	3,830	3,420	3,080	2,790	2,565
		f				.128	.149	.175	.195	.248	.306	.372	.440
	0.625	*P*					6,100	5,600	5,260	4,660	4,210	3,825	3,505
		f					.134	.154	.176	.218	.275	.328	.397
100,000	0.687	*P*							6,325	5,660	5,090	4,630	4,250
		f							.145	.183	.228	.274	.3278
	0.750	*P*								7,400	6,640	6,030	5,540
		f								.178	.218	.252	.299
	0.812	*P*									8,420	7,660	7,000
		f									.192	.232	.276
	0.875	*P*									1,0830	9,550	8,700
		f									.179	.218	.257
90,0.00	0.937	*P*										1,0600	9,700
		f										.179	.217
	1.000	*P*											1,1780
		f											.206
	1.125	*P*											
		f											
	1.250	*P*											
		f											
80,000	1.375	*P*											
		f											

The table is based on the values of unit stress indicated, and G = 12,500,000. For any other value of unit stress, divide the tabular value by the unit stress used in the table and multiply by the unit stress to be used in the design. For any other value of G, multiply the value of f in the table by 12,500,000 and divide by the value of G chosen. For **square steel wire,** multiply values of P by 1.06, and values of f by 0.75. For **round brass wire,** take S_s=10,000 to 20,000, and multiply values of f by 2 (Howe).

EXAMPLES OF USE OF TABLE 6.2.63. 1. Required the safe load (P) for a spring of -in round steel with a pitch diameter (D) of 3½ in. In the line headed D, under 3½, is given the value of P, or 678 lb. This is for a unit stress of 115,000 lb/in². The load P for any other unit stress may be found by dividing the 678 by 115,000 and multiplying by the unit stress to be used in the design. To determine the number of coils this spring would need to compress (say) 6 in under a load of (say) 678 lb, take the value of f under 678, or 0.938, which is the deflection of one coil under the given load. Therefore, 6/0.938 = 6.4, say 7, equals the number of coils required. The spring will therefore be 2⅝ in long when closed (7 × ⅜), counting the working coils only, and must be 8⅝ in long when unloaded. Whether there is an extra coil at one end which does not deflect will depend upon the details of the particular design. The deflection in the above example is for a unit stress of 115,000 lb/in². The rule is, divide the deflection by 115,000 and multiply by the unit stress to be used in the design.

2. A 7⁄16-in steel spring of 3½-in OD has its coils in close contact. How much can it be extended without exceeding the limit of safety? The maximum safe load for this spring is found to be 1,074 lb, and the deflection of one coil under this load is 0.810 in. This is for a unit stress of 115,000 lb/in². Therefore, 0.810 is the greatest admissible opening between any two coils. In this way, it is possible to ascertain whether or not a spring is overloaded, without knowledge of the load carried.

Design by Formula

A design formula can be constructed by equating calculated stress s_v (from load formula, Fig. 6.2.119) to an allowable working stress in torsion:

$$s_v = \sigma_{\max} = \frac{16rP}{\pi d^3}\left(\frac{4c-1}{4c-4} + \frac{0.615}{c}\right) = \frac{16rPk}{\pi d^3} = \frac{S_v}{K_{sf}}$$

where P = axial load on spring, lb; $r = D/2$ = mean radius of coil, in; D = mean diameter of coil, in (outside diameter minus wire diameter); d = wire diameter, in; $\sigma_{\max}$ = torsional stress, lb/in²; K_{sf} = safety factor; and $c = D/d$. Note that the expression in parentheses $[(4c - 1)(4c - 4) + 0.615/c]$ is the **Wahl correction factor** k, which accounts for the added stresses in the coils due to curvature and direct shear. See Fig. 6.2.119. Values of S_v, yield point in shear from standard tests, are strongly dependent on d, hence the availability of S_v in the literature is limited. However, an empirical relationship between S_{uT} and d is available (see Shigley, "Mechanical Engineering Design," McGraw-Hill, 4th ed., p. 452). Using also the approximate relations $S_y = 0.75S_{ut}$ and $S_v = 0.577S_y$ results in the following relationship:

$$S_v = \frac{0.43A}{d^m}$$

where A and m are constants (see Table 6.2.64).

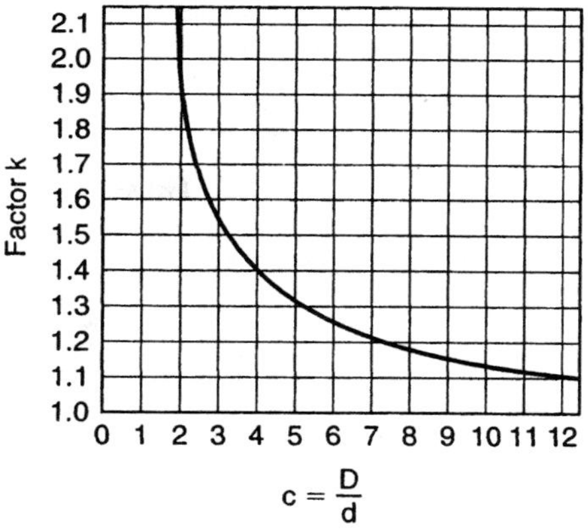

Figure 6.2.119 Wahl correction factor.

Substituting S_v and rearranging yield the following useful formula:

$$d^{3-m} = \frac{K_{sf}16rPk}{\pi 0.43A}$$

EXAMPLE. A slow-speed follower is kept in contact with its cam by means of a helical compression spring, in which the minimum spring force desired is 20 lb to assure firm contact, while maximum spring force is not to exceed 60 lb to prevent excessive surface wear on the cam. The follower rod is ¼ inch in diameter, and the rod enclosure where the spring is located is ⅞ inch in diameter. Maximum displacement is 1.5 in.

Choose $r = 0.5 \times 0.75 = 0.375$. From Table 6.2.64 choose $m = 0.167$ and A = 169,000. Choose $K_{sf} = 2$. Assume $k = 1$ to start.

$$d^{3-0.167} = \frac{(2)(0.375)(60)(16)}{(\pi)(0.43)(169{,}000)} = 0.00315$$

$$d = (0.00315)^{0.35298} = 0.131 \text{ in}$$

$$\text{Spring constant} = \frac{\Delta P}{\Delta f} = \frac{40}{1.5} = 26.667$$

and from the deflection formula, the number of active turns

$$n = \frac{(0.131)^4(11{,}500{,}000)}{(26.667)(64)(0.375)^3} = 37.6$$

For squared and ground ends add two dead coils, so that

$$n_{\text{total}} = 40$$

$$H = \text{solid height} = (40)(0.131) = 5.24 \text{ in}$$

$$f_0 = \text{displacement from zero to maximum load} = 60/26.667 = 2.249$$

$$L_0 = \text{approximate free length} = 5.24 + 2.249 = 7.49 \text{ in}$$

NOTE: Some clearance should be added between coils so that at maximum load the spring is not closed to its solid height. Also, the nearest commercial stock size should be selected, and recalculations made on this stock size for S_v,

Table 6.2.64 Constants for Use in $S_v = 0.43A/d^m$

Material	Size range, in	Size range, mm	Exponent m	Constant A kips	Constant A MPa
Music wire*	0.004-0.250	0.10-6.55	0.146	196	2,170
Oil-tempered wire†	0.020-0.500	0.50-12	0.186	149	1,880
Hard-drawn wire‡	0.028-0.500	0.70-12	0.192	136	1,750
Chrome vanadium§	0.032-0.437	0.80-12	0.167	169	2,000
Chrome silicon¶	0.063-0.375	1.6-10	0.112	202	2,000

* Surface is smooth, free from defects, and has a bright, lustrous finish.
† Has a slight heat-treating scale which must be removed before plating.
‡ Surface is smooth and bright, with no visible marks.
§ Aircraft-quality tempered wire; can also be obtained annealed.
¶ Tempered to Rockwell C49 but may also be obtained untempered.
SOURCE: Adapted from "Mechanical Engineering Design," Shigley, McGraw-Hill, 1983 by permission.

remembering to include k at this juncture. If recalculated S_v is satisfactory as compared to published values, the design is retained, otherwise enough iterations are performed to arrive at a satisfactory result. Figure 6.2.120 shows a plot of S_{uT} versus d. To convert S_{uT} to S_v, multiply S_{uT} by 0.43.

$$c = 2(0.375/0.131) = 5.73$$

$$k = \frac{4(5.73)-1}{4(5.73)-4} + \frac{0.615}{5.73} = 1.257$$

$$S_v = \frac{(K_{sf})(16)(0.375)(1.257)(60)}{\pi(0.131)^3} = (K_{sf})(64,072)$$

Now S_{uT} = 235,000 lb/in² (from Fig. 6.2.120) and S_v (tabulated) = (0.43)(235,000) = 101,696 lb/in² so that K_{sf} = 101,696/64,072 = 1.59, a satisfactory value.

NOTE: The original choice of a generous K_{sf} = 2 was made to hedge against the large statistical variations implied in the empirical formula $S_v = 0.43A/d^m$.

The basis for design of springs in parallel or in series is shown in Fig. 6.2.121.

Belleville Springs Often called dished, or conical spring, washers, Belleville springs occupy a very small space. They are stressed in a very complex manner, and provide unusual spring rate curves (Fig. 6.2.122a). These springs are nonlinear, but for some proportions, they behave with approximately linear characteristics in a limited range. Likewise, for some proportions they can be used through a spectrum of spring rates, from positive to flat and then through a negative region. The snap-through action, shown at point A in Fig. 6.2.122b, can be useful in particular applications requiring reversal of spring rates. These springs are used for very high and special spring rates. They are extremely sensitive to slight variations in their geometry. A wide range is available commercially.

Parallel	Series
$P_T = P_I + P_{II}$	$P_T = P_I = P_{II}$
$\Delta_T = \Delta_I = \Delta_{II}$	$\Delta_T = \Delta_I + \Delta_{II}$
$K_T = K_I + K_{II}$	$K_T = \dfrac{K_I K_{II}}{K_I + K_{II}}$

where P = load, lb
Δ = deflection, in.
K = spring rate, lb/in.

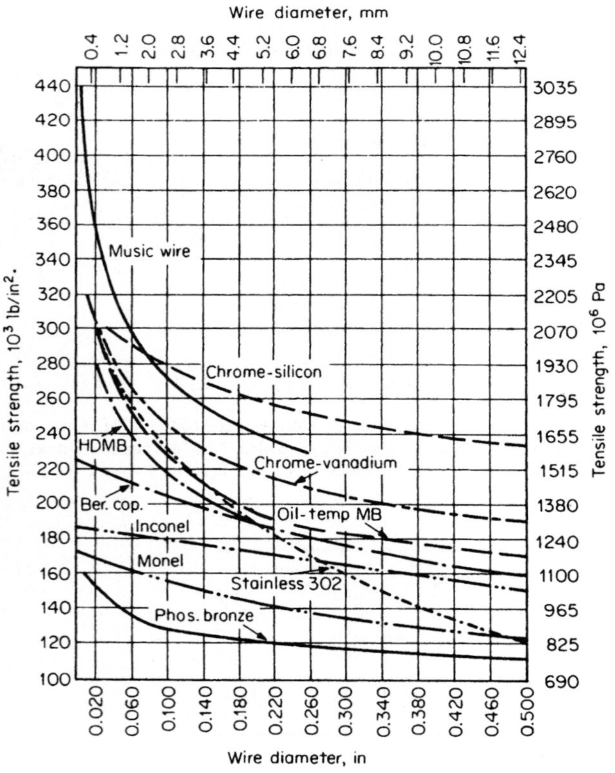

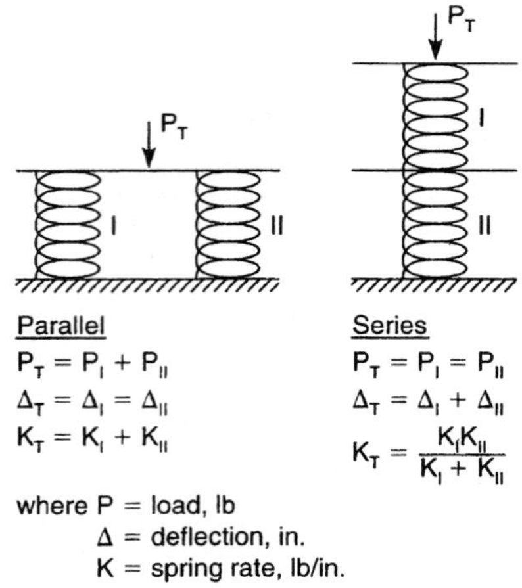

Figure 6.2.121 Springs in parallel and in series.

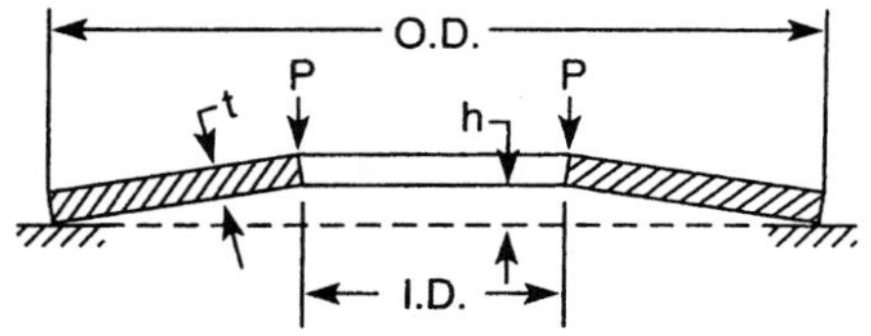

Figure 6.2.122a Sectional view of Belleville spring.

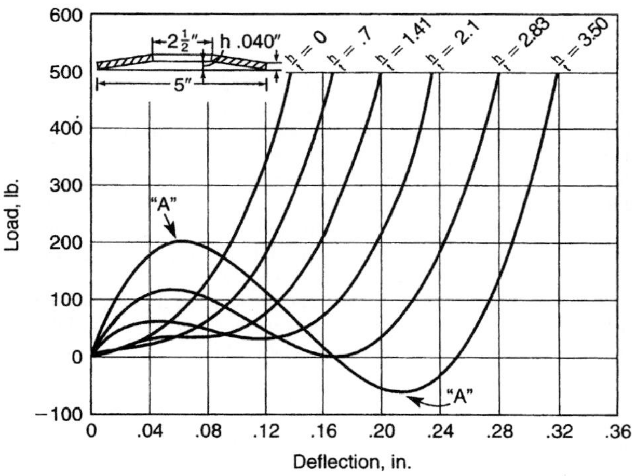

Figure 6.2.122b Load deflection curves for a family of Belleville springs. *(Associated Spring Corp.)*

Figure 6.2.120 Minimum tensile strength for the most popular spring materials, spring-quality wire. *(Reproduced from Carlson, "Spring Designer's Handbook," Marcel Dekker, by permission.)*

WIRE ROPE

When power source and load are located at extreme distances from one another, or loads are very large, the use of wire rope is suggested. Design and use decisions pertaining to wire ropes rest with the user, but manufacturers generally will help users toward appropriate choices. The following material, based on the Committee of Wire Rope Producers, "Wire Rope User's Manual," current edition, may be used as an initial guide in selecting a rope.

Wire rope is composed of (1) wires to form a strand, (2) strands wound helically around a core, and (3) a core. Classification of wire

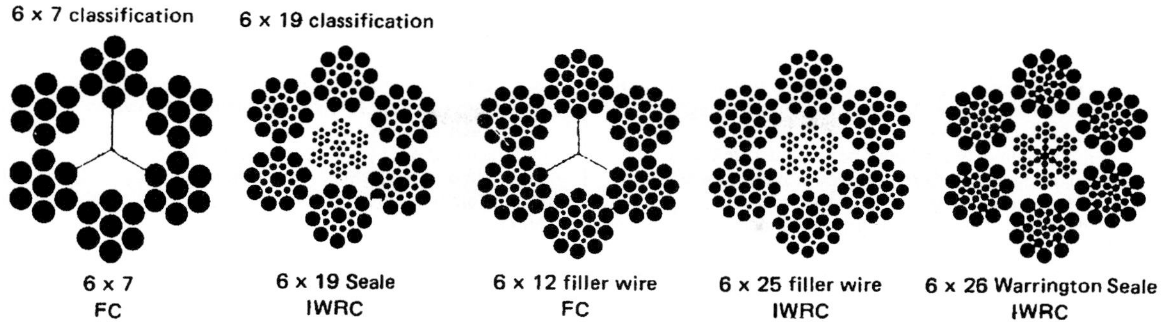

Figure 6.2.123 Cross sections of some commonly used wire rope construction. *(Reproduced from "Wire Rope User's Manual," AISI, by permission.)*

ropes is made by giving the number of strands, number of minor strands in a major strand (if any), and nominal number of wires per strand. For example 6 × 7 rope means 6 strands with a nominal 7 wires per strand (in this case no minor strands, hence no middle number). A nominal value simply represents a range. A nominal value of 7 can mean anywhere from 3 to 14, of which no more than 9 are outside wires. A full rope description will also include length, size (diameter), whether wire is preformed or not prior to winding, direction of lay (right or left, indicating the direction in which strands are laid around the core), grade of rope (which reflects wire strength), and core. The most widely used classifications are: 6 × 7, 6 × 19, 6 × 37, 6 × 61, 6 × 91, 6 × 127, 8 × 19, 18 × 7, 19 × 7. Some special constructions are: 3 × 7 (guardrail rope); 3 × 19 (slusher); 6 × 12 (running rope); 6 × 24 and 6 × 30 (hawsers); 6 × 42 and 6 × 6 × 7 (tiller rope); 6 × 3 × 19 (spring lay); 5 × 19 and 6 × 19 (marlin clad); 6 × 25B, 6 × 27H, and 6 × 30G (flattened strand). The diameter of a rope is the circle which just contains the rope. The right-regular lay (in which the wire is twisted in one direction to form the strands and the strands are twisted in the opposite direction to form the rope) is most common. Regular-lay ropes do not kink or untwist and handle easily. Lang-lay ropes (in which wires and strands are twisted in the same direction) are more resistant to abrasive wear and fatigue failure.

Cross sections of some commonly used wire rope are shown in Fig. 6.2.123. Figure 6.2.124 shows rotation-resistant ropes, and Fig. 6.2.125 shows some special-purpose constructions.

The core provides support for the strands under normal bending and loading. Core materials include fibers (hard vegetable or synthetic) or steel (either a strand or an independent wire rope). Most common core designations are: fiber core (FC), independent wire-rope core (IWRC), and wire-strand core (WSC). Lubricated fiber cores can provide lubrication to the wire, but add no real strength and cannot be used in high temperature environments. Wire-strand or wire-rope cores add from 7 to 10 percent to strength, but under nonstationary usage tend to wear from interface friction with the outside strands. Great flexibility can be achieved when wire rope is used as strands. Such construction is very pliable and friction resistant. Some manufacturers will provide plastic coatings (nylon, Teflon, vinyl, etc.) upon request. Such coatings help provide resistance to abrasion, corrosion, and loss of lubricant. *Crushing* refers to rope damage caused by excessive pressures against drum or sheave, improper groove size, and multiple layers on drum or sheave. Consult wire rope manufacturers in doubtful situations.

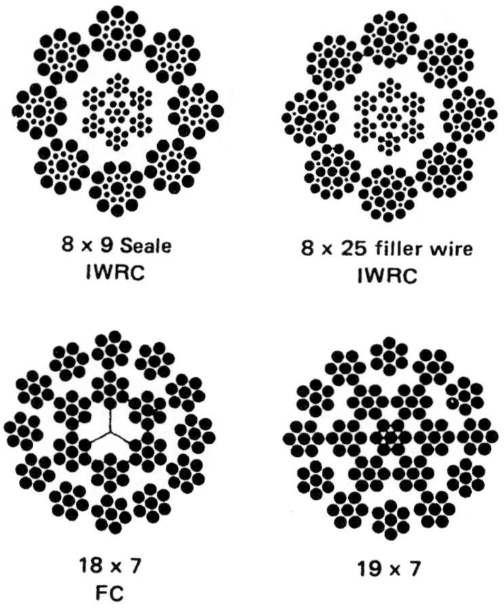

Figure 6.2.124 Cross section of some rotation-resistant wire ropes. *(Reproduced from "Wire Rope User's Manual," AISI, by permission.)*

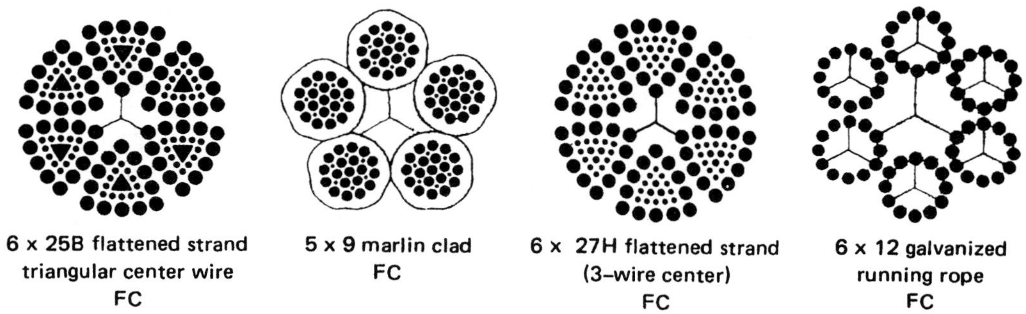

Figure 6.2.125 Some special constructions. *(Reproduced from "Wire Rope User's Manual," AISI, by permission.)*

Wire-rope materials and their strengths are reflected as grades. These are: traction steel (TS), mild plow steel (MPS), plow steel (PS), improved plow steel (IPS), and extra improved plow (EIP). The plow steel strength curve forms the basis for calculating the strength of all steel rope wires. American manufacturers use color coding on their ropes to identify particular grades.

The grades most commonly available and tabulated are IPS and EIP. Two specialized categories, where selection requires extraordinary attention, are elevator and rotation-resistant ropes.

Elevator rope can be obtained in four principal grades: iron, traction steel, high-strength steel, and extra-high-strength steel.

Bronze rope has limited use; iron rope is used mostly for older existing equipment.

Selection of Wire Rope

Appraisal of the following is the key to choosing the rope best suited to the job: resistance to breaking, resistance to bending fatigue, resistance to vibrational fatigue, resistance to abrasion, resistance to crushing, and reserve strength. Along with these must be an appropriate choice of safety factor, which in turn requires careful consideration of all loads, acceleration-deceleration, shocks, rope speed, rope attachments, sheave arrangements as well as their number and size, corrosive and/or abrasive environment, length of rope, etc. An approximate selection formula can be written as:

$$\text{DSL} = \frac{(\text{NS})K_b}{K_{sf}}$$

where DSL (demanded static load) = known or dead load **plus** additional loads caused by sudden starts or stops, shocks, bearing friction, etc., tons; NS (nominal strength) = published test strengths, tons (see Table 6.2.65); K_b = a factor to account for the reduction in nominal strength due to bending when a rope passes over a curved surface such as a stationary sheave or pin (see Fig. 6.2.126); K_{sf} = safety factor. (For average operation use $K_{sf} = 5$. If there is danger to human life or other critical situations, use $8 \le K_{sf} \le 12$. For instance, for elevators moving at 50 ft/min, $K_{sf} = 8$, while for those moving at 1,500 ft/min, $K_{sf} = 12$.)

Having made a tentative selection of a rope based on the demanded static load, one considers next the wear life of the rope. A loaded rope

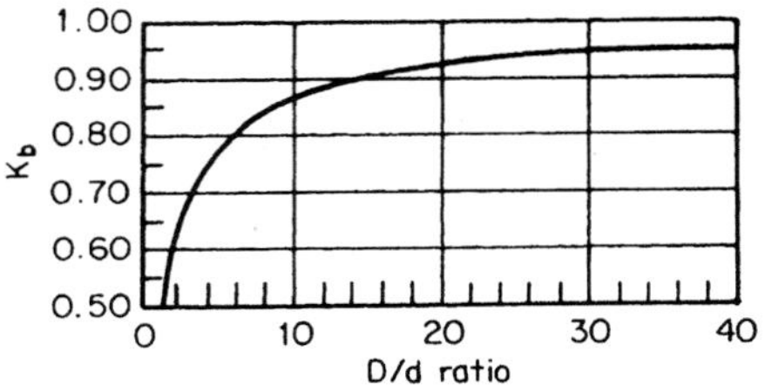

Figure 6.2.126 Values of K_{bend} vs. D/d ratios (D = sheave diameter, d = rope diameter), based on standard test data for 6 × 9 and 6 × 17 class ropes. *(Compiled from "Wire Rope User's Manual," AISI, by permission.)*

Table 6.2.65 Selected Values of Nominal Strengths of Wire Rope

			Fiber core				IWRC					
									Nominal strength			
	Nominal diameter		Approximate mass		Nominal strength IPS		Approximate mass		IPS		EIP	
Classification	in	mm	lb/ft	kg/m	tons	t	lb/ft	kg/m	tons	t	tons	t
6 × 7 Bright (uncoated)	1/4	6.4	0.09	0.14	2.64	2.4	0.10	0.15	2.84	2.58		
	3/8	9.5	0.21	0.31	5.86	5.32	0.23	0.34	6.30	5.72		
	1/2	13	0.38	0.57	10.3	9.35	0.42	0.63	11.1	10.1		
	5/8	16	0.59	0.88	15.9	14.4	0.65	0.97	17.1	15.5		
	7/8	22	1.15	1.71	30.7	27.9	1.27	1.89	33.0	29.9		
	1 1/8	29	1.90	2.83	49.8	45.2	2.09	3.11	53.5	48.5		
	1 3/8	35	2.82	4.23	73.1	66.3	3.12	4.64	78.6	71.3		
6 × 19 Bright (uncoated)	1/4	6.4	0.11	0.16	2.74	2.49	0.12	0.17	2.94	2.67	3.40	3.08
	3/8	9.5	0.24	0.35	6.10	5.53	0.26	0.39	6.56	5.95	7.55	6.85
	1/2	13	0.42	0.63	10.7	9.71	0.46	0.68	11.5	10.4	13.3	12.1
	5/8	16	0.66	0.98	16.7	15.1	0.72	1.07	17.7	16.2	20.6	18.7
	7/8	22	1.29	1.92	32.2	29.2	1.42	2.11	34.6	31.4	39.8	36.1
	1 1/8	29	2.13	3.17	52.6	47.7	2.34	3.48	56.5	51.3	65.0	59.0
	1 3/8	35	3.18	4.73	77.7	70.5	3.5	5.21	83.5	75.7	96.0	87.1
	1 5/8	42	4.44	6.61	107	97.1	4.88	7.26	115	104	132	120
	1 7/8	48	5.91	8.8	141	128	6.5	9.67	152	138	174	158
	2 1/8	54	7.59	11.3	179	162	8.35	12.4	192	174	221	200
	2 3/8	60	9.48	14.1	222	201	10.4	15.5	239	217	274	249
	2 5/8	67	11.6	17.3	268	243	12.8	19.0	288	261	331	300
6 × 37 Bright (uncoated)	1/4	6.4	0.11	0.16	2.74	2.49	0.12	0.17	2.94	2.67	3.4	3.08
	3/8	9.5	0.24	0.35	6.10	5.53	0.26	0.39	6.56	5.95	7.55	6.85
	1/2	13	0.42	0.63	10.7	9.71	0.46	0.68	11.5	10.4	13.3	12.1
	5/8	16	0.66	0.98	16.7	15.1	0.72	1.07	17.9	16.2	20.6	18.7
	7/8	22	1.29	1.92	32.2	29.2	1.42	2.11	34.6	31.4	39.5	36.1
	1 1/8	29	2.13	3.17	52.6	47.7	2.34	3.48	56.5	51.3	65.0	59.0
	1 3/8	35	3.18	4.73	77.7	70.5	3.50	5.21	83.5	75.7	96.0	87.1
	1 5/8	42	4.44	6.61	107	97.1	4.88	7.26	115	104	132	120
	1 7/8	48	5.91	8.8	141	128	6.5	9.67	152	138	174	158
	2 1/8	54	7.59	11.3	179	162	8.35	12.4	192	174	221	200
	2 3/8	60	9.48	14.1	222	201	10.4	15.5	239	217	274	249
	2 7/8	67	11.6	17.3	268	243	12.8	19.0	288	261	331	300
	3 1/8	74	13.9	20.7	317	287	15.3	22.8	341	309	392	356
		80	16.4	24.4	371	336	18.0	26.8	399	362	458	415

SOURCE: "Wire Rope User's Manual," AISI, adapted by permission.

Table 6.2.66 Suggested Allowable Radial Bearing Pressures of Ropes on Various Sheave Materials

Material	Regular lay rope, lb/in²				Lang lay rope, lb/in²			Flattened strand lang lay, lb/in²	Remarks
	6 × 7	6 × 19	6 × 37	8 × 19	6 × 7	6 × 19	6 × 37		
Wood	150	250	300	350	165	275	330	400	On end grain of beech, hickory, gum.
Cast iron	300	480	585	680	350	550	660	800	Based on minimum Brinell hardness of 125.
Carbon-steel casting	550	900	1,075	1,260	600	1,000	1,180	1,450	30-40 carbon. Based on minimum Brinell hardness of 160.
Chilled cast iron	650	1,100	1,325	1,550	715	1,210	1,450	1,780	Not advised unless surface is uniform in hardness.
Manganese steel	1,470	2,400	3,000	3,500	1,650	2,750	3,300	4,000	Grooves must be ground and sheaves balanced for high-speed service.

SOURCE: "Wire Rope User's Manual," AISI, reproduced by permission.

bent over a sheave stretches elastically and so rubs against the sheave, causing wear of both members. Drum or sheave size is of paramount importance at this point.

Sizing of Drums or Sheaves

Diameters of drums or sheaves in wire rope applications are controlled by two main considerations: (1) the radial pressure between rope and groove and (2) degree of curvature imposed on the rope by the drum or sheave size.

Radial pressures can be calculated from $p = 2T/(Dd)$, where p = unit radial pressure, lb/in²; T = rope load, lb; D = tread diameter of drum or sheave, in; d = nominal diameter of rope, in. Table 6.2.66 lists suggested allowable radial bearing pressures of ropes on various sheave materials.

All wire ropes operating over drums or sheaves are subjected to cyclical stresses, causing shortened rope life because of fatigue. Fatigue resistance or relative service life is a function of the ratio D/d. Adverse effects also arise out of relative motion between strands during passage around the drum or sheave. Additional adverse effects can be traced to poor match between rope and groove size, and to lack of rope lubrication. Table 6.2.67 lists suggested and minimum sheave and drum ratios for various rope construction. Table 6.2.68 lists relative bending life factors; Figure 6.2.127 shows a plot of relative rope service life versus D/d.

Table 6.2.67 Sheave and Drum Ratios

Construction* †	Suggested	Minimum
6 × 7	72	42
19 × 7 or 18 × 7 Rotation-resistant	51	34
6 × 19 S	51	34
6 × 25 B flattened strand	45	30
6 × 27 H flattened strand	45	30
6 × 30 G flattened strand	45	30
6 × 21 FW	45	30
6 × 26 WS	45	30
6 × 25 FW	39	26
6 × 31 WS	39	26
6 × 37 SFW	39	26
6 × 36 WS	35	23
6 × 43 FWS	35	23
6 × 41 WS	32	21
6 × 41 SFW	32	21
6 × 49 SWS	32	21
6 × 46 SFW	28	18
6 × 46 WS	28	18
8 × 19 S	41	27
8 × 25 FW	32	21
6 × 42 Tiller	21	14

* WS—Warrington Seale; FWS—Filler Wire Seale; SFW—Seale Filler Wire; SWS—Seale Warrington Seale; S—Seale; FW—Filler Wire.

† D = tread diameter of sheave; d = nominal diameter of rope. To find any tread diameter from this table, the diameter for the rope construction to be used is multiplied by its nominal diameter d. For example, the minimum sheave tread diameter for a ½-in 6 × 21 FW rope would be ½ in (nominal diameter) × 30 (minimum ratio), or 15 in.

NOTE: These values are for reasonable service. Other values are permitted by various standards such as ANSI, API, PCSA, HMI, CMAA, etc. Similar values affect rope life.

SOURCE: "Wire Rope User's Manual," AISI, reproduced by permission.

Table 6.2.68 Relative Bending Life Factors

Rope construction	Factor	Rope construction	Factor
6 × 7	0.61	6 × 36 WS	1.16
19 × 7 or 18 × 7	0.67	6 × 43 FWS	1.16
Rotation-resistant	0.81	6 × 41 WS	1.30
6 × 19 S	0.90	6 × 41 SFW	1.30
6 × 25 B flattened strand	0.90	6 × 49 SWS	1.30
6 × 27 H flattened strand	0.90	6 × 43 FW (2 op)	1.41
6 × 30 G flattened strand	0.89	6 × 46 SFW	1.41
6 × 21 FW	0.89	6 × 46 WS	1.41
6 × 26 WS	1.00	8 × 19 S	1.00
6 × 25 FW	1.00	8 × 25 FW	1.25
6 × 31 WS	1.00	6 × 42 Tiller	2.00
6 × 37 SFW			

SOURCE: "Wire Rope User's Manual," AISI, reproduced by permission.

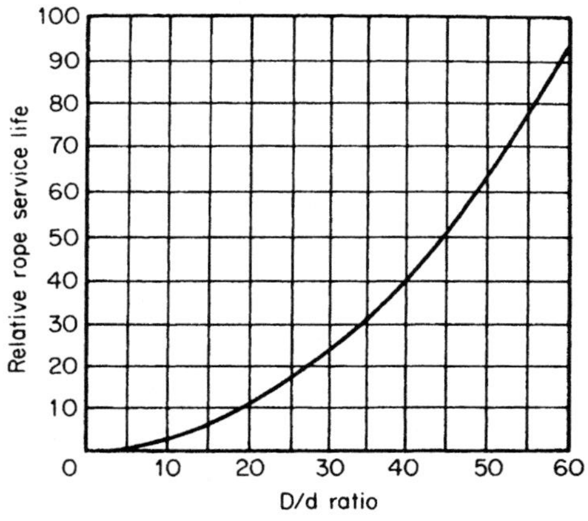

Figure 6.2.127 Service life curves for various D/d ratios. Note that this curve takes into account only bending and tensile stresses. *(Reproduced from "Wire Rope User's Manual," AISI, by permission.)*

Table 6.2.69 Minimum Sheave- and Drum-Groove Dimensions*

Nominal rope diameter		Groove radius			
		New		Worn	
in	mm	in	mm	in	mm
1/4	6.4	0.135	3.43	.129	3.28
5/16	8.0	0.167	4.24	.160	4.06
3/8	9.5	0.201	5.11	.190	4.83
7/16	11	0.234	5.94	.220	5.59
1/2	13	0.271	6.88	.256	6.50
9/16	14.5	0.303	7.70	.288	7.32
5/8	16	0.334	8.48	.320	8.13
3/4	19	0.401	10.19	.380	9.65
7/8	22	0.468	11.89	.440	11.18
1	26	0.543	13.79	.513	13.03
1 1/8	29	0.605	15.37	.577	14.66
1 1/4	32	0.669	16.99	.639	16.23
1 3/8	35	0.736	18.69	.699	17.75
1 1/2	38	0.803	20.40	.759	19.28
1 5/8	42	0.876	22.25	.833	21.16
1 3/4	45	0.939	23.85	.897	22.78
1 7/8	48	1.003	25.48	.959	24.36
2	52	1.085	27.56	1.025	26.04
2 1/8	54	1.137	28.88	1.079	27.41
2 1/4	58	1.210	30.73	1.153	29.29
2 3/8	60	1.271	32.28	1.199	30.45
2 1/2	64	1.338	33.99	1.279	32.49
2 5/8	67	1.404	35.66	1.339	34.01
2 3/4	71	1.481	37.62	1.409	35.79
2 7/8	74	1.544	39.22	1.473	37.41
3	77	1.607	40.82	1.538	39.07
3 1/8	80	1.664	42.27	1.598	40.59
3 1/4	83	1.731	43.97	1.658	42.11
3 3/8	87	1.807	45.90	1.730	43.94
3 1/2	90	1.869	47.47	1.794	45.57
3 3/4	96	1.997	50.72	1.918	48.72
4	103	2.139	54.33	2.050	52.07
4 1/4	109	2.264	57.51	2.178	55.32
4 1/2	115	2.396	60.86	2.298	58.37
4 3/4	122	2.534	64.36	2.434	61.82
5	128	2.663	67.64	2.557	64.95
5 1/4	135	2.804	71.22	2.691	68.35
5 1/2	141	2.929	74.40	2.817	71.55
5 3/4	148	3.074	78.08	2.947	74.85
6	154	3.198	81.23	3.075	78.11

* Values given are applicable to grooves in sheaves and drums; they are not generally suitable for pitch design since this may involve other factors. Further, the dimensions do not apply to traction-type elevators; in this circumstance, drum- and sheave-groove tolerances should conform to the elevator manufacturer's specifications. Modern drum design embraces extensive considerations beyond the scope of this publication. It should also be noted that dram grooves are now produced with a number of oversize dimensions and pitches applicable to certain service requirements.

SOURCE: "Wire Rope User's Manual," AISI, reproduced by permission.

Table 6.2.69 lists minimum drum (sheave) groove dimensions. Periodic groove inspection is recommended, and worn or corrugated grooves should be re-machined or the drum replaced, depending on severity of damage.

Seizing and Cutting Wire Rope Before a wire rope is cut, seizings (bindings) must be applied on either side of the cut to prevent rope distortion and flattening or loosened strands. Normally, for preformed ropes, one seizing on each side of the cut is sufficient, but for ropes that are not preformed a minimum of two seizings on each side is recommended, and these should be spaced six rope diameters apart (see Fig. 6.2.128). Seizings should be made of soft or annealed wire or strand, and the width of the seizing should never be less than the diameter of the rope being seized. Table 6.2.70 lists suggested seizing wire diameters.

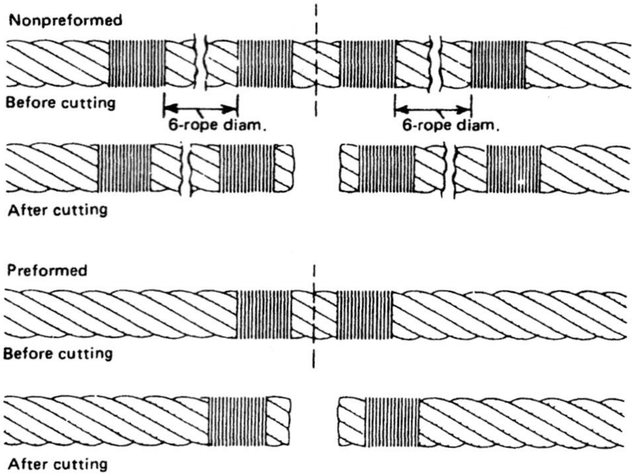

Figure 6.2.128 Seizings. *(Reproduced from "Wire Rope User's Manual," AISI, by permission.)*

Table 6.2.70 Seizing*

Rope diameter		Suggested seizing wire diameter†	
in	mm	in	mm
1/8-5/16	3.5-8.0	0.032	0.813
3/8-9/16	9.4-14.5	0.048	1.21
5/8-15/16	16.0-24.0	0.063	1.60
1-1 5/16	26.0-33.0	0.080	2.03
1 3/8-1 11/16	35.0-43.0	0.104	2.64
1 3/4 and larger	45.0 and larger	0.124	3.15

* Length of the seizing should not be less than the rope diameter.

† The diameter of seizing wire for elevator ropes is usually somewhat smaller than that shown in this table. Consult the wire rope manufacturer for specific size recommendations. Soft annealed seizing strand may also be used.

SOURCE: "Wire Rope User's Manual," AISI, reproduced by permission.

Wire Rope Fittings or Terminations End terminations allow forces to be transferred from rope to machine, or load to rope, etc. Figure 6.2.129 illustrates the most commonly used end fittings or terminations. Not all terminations will develop full strength. In fact, if all of the rope elements are not held securely, the individual strands will sustain unequal loads causing unequal wear among them, thus shortening the effective rope service life. Socketing allows an end fitting which reduces the chances of unequal strand loading.

Wire rope manufacturers have developed a recommended procedure for socketing. A tight wire serving band is placed where the socket base will be, and the wires are unlaid, straightened, and "broomed" out. Fiber core is cut close to the serving band and removed, wires are cleaned with a solvent such as SC-methyl chloroform, and brushed to remove dirt and grease. If additional cleaning is done with muriatic acid this must be followed by a neutralizing rinse (if possible, ultrasonic cleaning is preferred). The wires are dipped in flux, the socket is positioned, zinc (spelter) is poured and allowed to set, the serving band is removed, and the rope lubricated.

A somewhat similar procedure is used in thermoset resin socketing.

Socketed terminations generally are able to develop 100 percent of nominal strength.

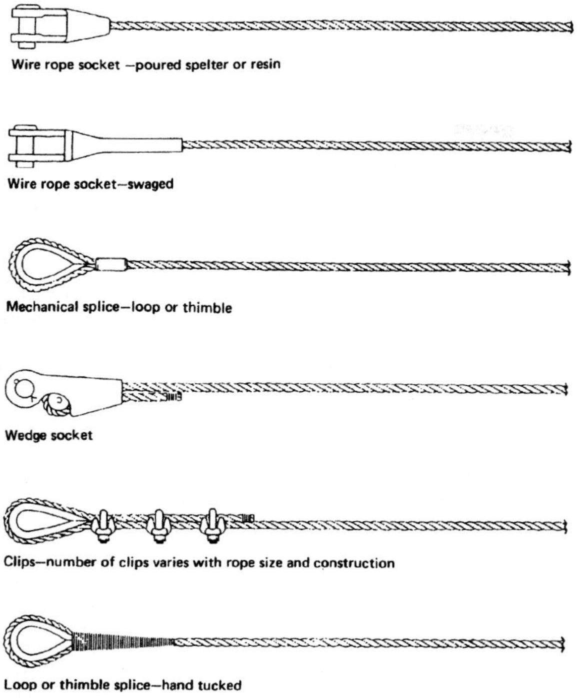

Figure 6.2.129 End fittings, or terminations, showing the six most commonly used. *(Reproduced from "Wire Rope User's Manual," AISI, by permission.)*

FIBER LINES

The breaking strength of various fiber lines is given in Table 6.2.71.

Knots, Hitches, and Bends

No two parts of a knot which would move in the same direction if the rope were to slip should lie alongside of and touching each other. The knots shown in Fig. 6.2.130 are known by the following names:

A, bight of a rope; *B*, simple or overhand knot; *C*, figure 8 knot; *D*, double knot; *E*, boat knot; *F*, bowline, first step; *G*, bowline, second step; *H*, bowline, completed; *I*, square or reef knot; *J*, sheet bend or weaver's knot; *K*, sheet bend with a toggle; *L*, carrick bend; *M*, "stevedore" knot completed; *N*, "stevedore" knot commenced; *O*, slip knot; *P*, Flemish loop; *Q*, chain knot with toggle; *R*, half hitch; *S*, timber hitch; *T*, clove hitch; *U*, rolling hitch; *V*, timber hitch and half hitch; *W*, blackwall hitch; *X*, fisherman's bend; *Y*, round turn and half hitch; *Z*, wall knot commenced; *AA*, wall knot completed; *BB*, wall-knot crown commenced; *CC*, wall-knot crown completed.

The bowline *H*, one of the most useful knots, will not slip, and after being strained is easily untied. Knots *H*, *K*, and *M* are easily untied after being under strain. The knot *M* is useful when the rope passes through an eye and is held by the knot, as it will not slip, and is easily untied after being strained. The wall knot is made as follows: Form a bight with strand 1 and pass the strand 2 around the end of it, and the strand 3 around the end of 2, and then through the bight of 1, as shown at *Z* in the figure. Haul the ends taut when the appearance is as shown in *AA*. The end of the strand 1 is now laid over the center of the knot, strand 2 laid over 1, and 3 over 2, when the end of 3 is passed through the bight of 1, as shown at *BB*. Haul all the strands taut, as shown at *CC*. The "stevedore" knot (*M*, *N*) is used to hold the end of a rope from passing through a hole. When the rope is strained, the knot draws up

Table 6.2.71 Breaking Strength of Fiber Lines, lb

Size, in: Diam	Size, in: Cir	Manila	Composite	Sisal	Sisal mixed	Sisal hemp	Agave or jute	Nylon	Dacron	Poly-ethylene	Poly-propylene (mono-filament)	Esterlon (polyester)
3/16	5/8	450	—	360	340	310	270	1,000	850	700	800	720
1/4	3/4	600	—	480	450	420	360	1,500	1,380	1,200	1,200	1,150
5/16	1	1,000	—	800	750	700	600	2,500	2,150	1,750	2,100	1,750
3/8	1 1/8	1,350	—	1,080	1,010	950	810	3,500	3,000	2,500	3,100	2,450
7/16	1 1/4	1,750	—	1,400	1,310	1,230	1,050	4,800	4,500	3,400	3,700	3,400
1/2	1 1/2	2,650	—	2,120	1,990	1,850	1,590	6,200	5,500	4,100	4,200	4,400
9/16	1 3/4	3,450	—	2,760	2,590	2,410	2,070	8,300	7,300	4,600	5,100	5,700
5/8	2	4,400	—	3,520	3,300	3,080	2,640	10,500	9,500	5,200	5,800	7,300
3/4	2 1/4	5,400	—	4,320	4,050	3,780	3,240	14,000	12,500	7,400	8,200	9,500
13/16	2 1/2	6,500	—	5,200	4,880	4,550	3,900	17,000	15,000	8,900	9,800	11,500
7/8	2 3/4	7,700	—	—	—	—	—	20,000	17,500	10,400	11,500	13,500
1	3	9,000	—	7,200	6,750	6,300	5,400	24,000	20,000	12,600	14,000	16,500
1 1/16	3 1/4	10,500	—	8,400	7,870	7,350	6,300	28,000	22,500	14,500	16,100	19,000
1 1/8	3 1/2	12,000	—	9,600	9,000	8,400	7,200	32,000	25,000	16,500	18,300	21,500
1 1/4	3 3/4	13,500	—	10,800	10,120	9,450	8,100	36,500	28,500	18,600	21,000	24,300
1 5/16	4	15,000	—	12,000	11,250	10,500	9,000	42,000	32,000	21,200	24,000	28,000
1 1/2	4 1/2	18,500	16,600	14,800	13,900	12,950	11,100	51,000	41,000	26,700	30,000	34,500
1 5/8	5	22,500	20,300	18,000	16,900	15,800	13,500	62,000	50,000	32,700	36,500	41,500
1 3/4	5 1/2	26,500	23,800	21,200	19,900	18,500	15,900	77,500	61,000	39,500	44,000	51,000
2	6	31,000	27,900	24,800	23,200	21,700	18,600	90,000	72,000	47,700	53,000	61,000
2 1/8	6 1/2	36,000	—	—	—	—	—	105,000	81,000	55,800	62,000	70,200
2 1/4	7	41,000	36,900	32,800	30,800	28,700	—	125,000	96,000	63,000	70,000	81,000
2 1/2	7 1/2	46,500	—	—	—	—	—	138,000	110,000	72,500	80,500	92,000
2 5/8	8	52,000	46,800	41,600	39,000	36,400	—	154,000	125,000	81,000	90,000	103,000
2 7/8	8 1/2	58,000	—	—	—	—	—	173,000	140,000	92,000	100,000	116,000
3	9	64,000	57,500	51,200	48,000	44,800	—	195,000	155,000	103,000	116,000	130,000
3 1/4	10	77,000	69,300	61,600	57,800	53,900	—	238,000	190,000	123,000	137,000	160,000
3 1/2	11	91,000	—	—	—	—	—	288,000	230,000	146,000	162,000	195,000
4	12	105,000	94,500	84,000	78,800	73,500	—	342,000	275,000	171,000	190,000	230,000

Breaking strength is the maximum load the line will hold at the time of breaking, The working load of a line is one-fourth to one-fifth of the breaking strength.
SOURCE: Adapted, by permission of the U.S. Naval Institute, Annapolis, MD, and Wall Rope Works, Inc., New York, NY.

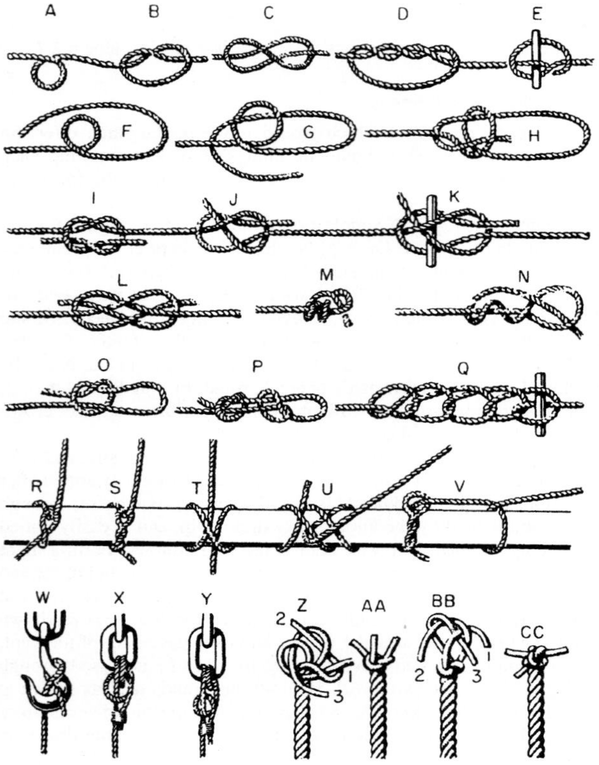

Figure 6.2.130 Knots, hitches, and bends.

tight, but it can be easily untied when the strain is removed. If a knot or hitch of any kind is tied in a rope, its failure under stress is sure to occur at that place. The shorter the bend in the standing rope, the weaker is the knot. The approximate strength of knots compared with the full strength of (dry) rope (= 100), based on Miller's experiments (*Mach.*, 1900, p. 198), is as follows: eye splice over iron thimble, 90; short splice in rope, 80; *S* and *Y*, 65; *H*, *O*, and *T*, 60; *I* and *J*, 50; *B* and *P*, 45.

NAILS AND SPIKES

Nails are either **wire nails** of circular cross section and constant diameter or **cut nails** of rectangular cross section with taper from head to point. The larger sizes are called **spikes.** The length of the nail is expressed in the **"penny"** system, the equivalents in inches being given in Tables 6.2.72 to 6.2.74. The letter d is the accepted symbol for penny. A keg of nails weighs 100 lb. **Heavy hinge nails** or **track nails** with countersunk heads have chisel points unless diamond points are specified. **Plasterboard nails** are smooth with circumferential grooves and have diamond points. **Spikes** are made either with flat heads and diamond points or with oval heads and chisel points.

WIRE AND SHEET-METAL GAUGES

In the metal industries, the word *gauge* has been used in various systems, or scales, for expressing the thickness or weight per unit area of thin plates, sheet, and strip, or the diameters of rods and wire. Specific diameters, thicknesses, or weights per square foot have been or are denoted in gauge systems by certain numerals followed by the word *gauge*, for example, no. 12 gauge, or simply 12 gauge. Gauge numbers for flat rolled products have been used only in connection with thin materials (Table 6.2.75). Heavier and thicker, flat rolled materials are usually designated by thickness in English or metric units.

There is considerable danger of confusion in the use of gauge number in both foreign and domestic trade, which can be avoided by specifying thickness or diameter in inches or millimeters.

DRILL SIZES

See Table 6.2.76.

Table 6.2.72 Wire Nails for Special Purposes
(Steel wire gauge)

	Barrel nails		Barbed roofing nails		Barbed dowel nails	
Length, in	Gauge	No. per lb	Gauge	No. per lb	Gauge	No. per lb
⅝	15½	1,570	—	—	8	394
¾	15½	1,315	13	729	8	306
⅞	14½	854	12	478	8	250
1	14½	750	12	416	8	212
1⅛	14½	607	12	368	8	183
1¼	14	539	11	250	8	16
1⅜	13	386	11	228	8	145
1½	13	355	10	167	8	131
1¾	—	—	10	143		
2	—	—	9	104		

	Clout nails		Slating nails		Fine nails		
Length, in	Gauge	No. per lb	Gauge	No. per lb	Length, in	Gauge	No. per lb
¾	15	999					
⅞	14	733					
1	14	648	12	425	1	16½	1,280
1⅛	14	580	10½	229			
1¼	13	398	—	—	1	17	1,492
1⅜	13	365					
1½	13	336	10½	190	1⅛	15	757
1¾	—	—	10	144	1⅛	16	984
2	—	—	9	104			

Table 6.2.73 Wire Nails and Spikes
(Steel wire gauge)

		Casing nails		Finishing nails		Clinch nails		Shingle nails	
Size of nail	Length, in	Gauge	No. per lb	Gauge	No. per lb	Gauge	No. per lb	Gauge	No. per lb
2d	1	15½	940	16½	1.473	14	723	13	434
3d	1¼	14½	588	15½	880	13	432	12	271
4d	1½	14	453	15	634	12	273	12	233
5d	1¾	14	389	15	535	12	234	12	203
6d	2	12½	223	13	288	11	158		
7d	2¼	12½	200	13	254	11	140		
8d	2½	11½	136	12½	196	10	101		
9d	2¾	11½	124	12½	178	10	91.4		
10d	3	10½	90	11½	124	9	70		
12d	3¼	10½	83	11½	113	9	64.1		
16d	3½	10	69	11	93	8	50		
20d	4	9	51	10	65	7	36.4		
30d	4¼	9	45						
40d	5	8	37						

		Boat nails				Hinge nails				Flooring nails	
		Heavy		Light		Heavy		Light			
Size of nail	Length, in	Diam, in	No. per lb	Diam, in	No. per lb	Diam, in	No. per lb	Diam, in	No. per lb	Gauge	No. per lb
4d	1½	¼	47	3/16	82	¼	53	3/16	90		
6d	2	¼	36	3/16	62	¼	39	3/16	66	11	168
8d	2½	¼	29	3/16	50	¼	31	3/16	53	10	105
10d	3	⅜	11	¼	24	⅜	12	¼	25	9	72
12d	3¼	⅜	10.4	¼	22	⅜	11	¼	23	8	56
16d	3½	⅜	9.6	¼	20	⅜	10	¼	22	7	44
20d	4	⅜	8	¼	18	⅜	8	¼	19	6	32

		Common wire nails and brads		Barbed car nails				Spikes			
				Heavy		Light					
Size of nail	Length, in	Gauge	No. per lb	Gauge	No. per lb	Gauge	No. per lb	Size	Length, in	Gauge	Approx no. per lb
2d	1	15	847	—	—	—	—	10d	3	6	43
3d	1¼	14	548	—	—	—	—	12d	3¼	6	39
4d	1½	12½	294	10	179	12	284	16d	3½	5	31
5d	1¾	12½	254	9	124	10	152	20d	4	4	23
6d	2	11½	167	9	108	10	132	30d	4½	3	18
7d	2¼	11½	150	8	80	9	95	40d	5	2	14
8d	2½	10¼	101	8	72	9	88	50d	5½	1	11
9d	2¾	10¼	92	7	55	8	65	50d	6	1	10
10d	3	9	66	7	50	8	59	—	7	5/16 in	7
12d	3¼	9	61	6	39	7	46	—	8	⅜	4.1
16d	3½	8	47	6	36	7	43	—	9	⅜	3.7
20d	4	6	30	5	27	6	32	—	10	⅜	3.3
30d	4½	5	23	5	24	6	28	—	12	⅜	2.7
40d	5	4	18	4	18	5	22				
50d	5½	3	14	3	14	4	17				
60d	6	2	11	3	13	4	15				

Table 6.2.74 Cut Steel Nails and Spikes
(Sizes, lengths, and approximate number per lb)

Size	Length, in	Common	Clinch	Finishing	Casing and box	Fencing	Spikes	Barrel	Slating	Tobacco	Brads	Shingle
2d	1	740	400	1,100	—	—	—	450	340			
3d	1¼	460	260	880	—	—	—	280	280			
4d	1½	280	180	530	420	—	—	190	220			
5d	1¾	210	125	350	300	100	—	—	180	130		
6d	2	160	100	300	210	80	—	—	—	97	120	
7d	2¼	120	80	210	180	60	—	—	—	85	94	
8d	2½	88	68	168	130	52	—	—	—	68	74	90
9d	2¾	73	52	130	107	38	—	—	—	58	62	72
10d	3	60	48	104	88	26	—	—	—	48	50	60
12d	3¼	46	40	96	70	20	—	—	—	—	40	
16d	3½	33	34	86	52	18	17	—	—	—	27	
20d	4	23	24	76	38	16	14					
25d	4¼	20	—	—	—	—	—					
30d	4½	16½	—	—	30	—	11					
40d	5	12	—	—	26	—	9					
50d	5½	10	—	—	20	—	7½					
60d	6	8	—	—	16	—	6					
—	6½	—	—	—	—	—	5½					
—	7	—	—	—	—	—						5

Table 6.2.75 Comparison of Standard Gauges*
Thickness of diameter, in

Gauge no.	BWG; Stubs Iron Wire	AWG; B&S	U.S. Steel Wire; Am. Steel & Wire; Washburn & Moen; Steel Wire	Galv. sheet steel	Manufacturers' standard
0000000	—	—	0.4900	—	—
000000	—	0.580000	0.4615	—	—
00000	—	0.516500	0.4305	—	—
0000	0.454	0.460000	0.3938	—	—
000	0.425	0.409642	0.3625	—	—
00	0.380	0.364796	0.3310	—	—
0	0.340	0.324861	0.3065	—	—
1	0.300	0.289297	0.2830	—	—
2	0.284	0.257627	0.2625	—	—
3	0.259	0.229423	0.2437	—	0.2391
4	0.238	0.204307	0.2253	—	0.2242
5	0.220	0.181940	0.2070	—	0.2092
6	0.203	0.162023	0.1920	—	0.1943
7	0.180	0.144285	0.1770	—	0.1793
8	0.165	0.128490	0.1620	0.1681	0.1644
9	0.148	0.114423	0.1483	0.1532	0.1495
10	0.134	0.101897	0.1350	0.1382	0.1345
11	0.120	0.090742	0.1205	0.1233	0.1196
12	0.109	0.080808	0.1055	0.1084	0.1046
13	0.095	0.071962	0.0915	0.0934	0.0897
14	0.083	0.064084	0.0800	0.0785	0.0747
15	0.072	0.057068	0.0720	0.0710	0.0673
16	0.065	0.050821	0.0625	0.0635	0.0598
17	0.058	0.045257	0.0540	0.0575	0.0538
18	0.049	0.040303	0.0475	0.0516	0.0478
19	0.042	0.035890	0.0410	0.0456	0.0418
20	0.035	0.031961	0.0348	0.0396	0.0359
21	0.032	0.028462	0.03175	0.0366	0.0329
22	0.028	0.025346	0.0286	0.0336	0.0299
23	0.025	0.022572	0.0258	0.0306	0.0269
24	0.022	0.020101	0.0230	0.0276	0.0239
25	0.020	0.017900	0.0204	0.0247	0.0209
26	0.018	0.015941	0.0181	0.0217	0.0179
27	0.016	0.014195	0.0173	0.0202	0.0164
28	0.014	0.012641	0.0162	0.0187	0.0149

Table 6.2.75 Comparison of Standard Gauges* (*Continued*)

Gauge no.	BWG; Stubs Iron Wire	AWG; B&S	U.S. Steel Wire; Am. Steel & Wire; Washburn & Moen; Steel Wire	Galv. sheet steel	Manufacturers' standard
29	0.013	0.011257	0.0150	0.0172	0.0135
30	0.012	0.010025	0.0140	0.0157	0.0120
31	0.010	0.008928	0.0132	0.0142	0.0105
32	0.009	0.007950	0.0128	0.0134	0.0097
33	0.008	0.007080	0.0118	—	0.0090
34	0.007	0.006305	0.0104	—	0.0082
35	0.005	0.005615	0.0095	—	0.0075
36	0.004	0.005000	0.0090	—	0.0067
37	—	0.004453	0.0085	—	0.0064
38	—	0.003965	0.0080	—	0.0060
39	—	0.003531	0.0075	—	—
40	—	0.003144	0.0070	—	—

* Principal uses—BWG: strips, bands, hoops, and wire; AWG or B&S: nonferrous sheets, rod, and wire; U.S. Steel Wire: steel wire except music wire; manufacturers' standard: uncoated steel sheets.

Table 6.2.76 Diameters of Small Drills

Number, letter, metric, and fractional drills in order of size (rounded to 4 decimal places)

No.	Ltr	mm	in	Diam, in	No.	Ltr	mm	in	Diam, in	No.	Ltr	mm	in	Diam, in
		0.10		0.0039			1.10		0.0433			2.40		0.0945
		0.15		0.0059			1.15		0.0452	41				0.0960
		0.20		0.0079	56				0.0465			2.45		0.0964
		0.25		0.0098				3/64	0.0468	40				0.0980
		0.30		0.0118			1.20		0.0472			2.50		0.0984
80				0.0135			1.25		0.0492	39				0.0995
		0.35		0.0137			1.30		0.0512	38				0.1015
79				0.0145	55				0.0520			2.60		0.1024
			1/64	0.0156			1.35		0.0531	37				0.1040
		0.40		0.0157	54				0.0550			2.70		0.1063
78				0.0160			1.40		0.0551	36				0.1065
		0.45		0.0177			1.45		0.0570			2.75		0.1082
77				0.0180			1.50		0.0590				7/64	0.1093
		0.50		0.0197	53				0.0595	35				0.1100
76				0.0200			1.55		0.0610			2.80		0.1102
75				0.0210				1/16	0.0625	34				0.1110
		0.55		0.0216			1.60		0.0630	33				0.1130
74				0.0225	52				0.0635			2.90		0.1142
		0.60		0.0236			1.65		0.0649	32				0.1160
73				0.0240			1.70		0.0669			3.00		0.1181
72				0.0250	51				0.0670	31				0.1200
		0.65		0.0255			1.75		0.0688			3.10		0.1220
71				0.0260	50				0.0700				1/8	0.1250
		0.70		0.0275			1.80		0.0709			3.20		0.1260
70				0.0280			1.85		0.0728			3.25		0.1279
69				0.0292	49				0.0730	30				0.1285
		0.75		0.0295			1.90		0.0748			3.30		0.1299
68				0.0310	48				0.0760			3.40		0.1339
			1/32	0.0312			1.95		0.0767	29				0.1360
		0.80		0.0314				5/64	0.0781			3.50		0.1378
67				0.0320	47				0.0785	28				0.1405
66				0.0330			2.00		0.0787				9/64	0.1406
		0.85		0.0334			2.05		0.0807			3.60		0.1417
65				0.0350	46				0.0810	27				0.1440
		0.90		0.0354	45				0.0820			3.70		0.1457
64				0.0360			2.10		0.0827	26				0.1470
63				0.0370			2.15		0.0846			3.75		0.1476
		0.95		0.0374	44				0.0860	25				0.1495
62				0.0380			2.20		0.0866			3.80		0.1496
61				0.0390			2.25		0.0885	24				0.1520
		1.00		0.0393	43				0.0890			3.90		0.1535
60				0.0400			2.30		0.0906	23				0.1540
59				0.0410			2.35		0.0925				5/32	0.1562
		1.05		0.0413	42				0.0935	22				0.1570
58				0.0420				3/32	0.0937			4.00		0.1575
57				0.0430										

Table 6.2.76 Diameters of Small Drills ***(Continued)***

No.	Ltr	mm	in	Diam, in
21				0.1590
20				0.1610
		4.10		0.1614
		4.2		0.1654
19				0.1660
		4.25		0.1673
		4.30		0.1693
18				0.1695
			11/64	0.1718
17				0.1730
		4.40		0.1732
16				0.1770
		4.50		0.1772
15				0.1800
		4.60		0.1811
14				0.1820
13				0.1850
		4.70		0.1850
		4.75		0.1850
			3/16	0.1875
		4.80		0.1890
12				0.1890
11				0.1910
		4.90		0.1920
10				0.1935
9				0.1960
		5.00		0.1968
8				0.1990
		5.10		0.2008
7				0.2010
			13/64	0.2031
6				0.2040
		5.20		0.2047
5				0.2055
		5.25		0.2066
		5.30		0.2087
4				0.2090
		5.40		0.2126
3				0.2130
		5.50		0.2165
			7/32	0.2187
		5.60		0.2205
2				0.2210
		5.70		0.2244
		5.75		0.2263
1				0.2280
		5.80		0.2283
		5.90		0.2323
	A			0.2340
			15/64	0.2340
		6.00		0.2362
	B			0.2380
		6.10		0.2402
	C			0.2420
		6.20		0.2441
	D			0.2460
		6.25		0.2460
		6.30		0.2480
	E		1/4	0.2500
		6.40		0.2519
		6.50		0.2559
	F			0.2570
		6.60		0.2598
	G			0.2610
		6.70		0.2637
			17/64	0.2656
		6.75		0.2657
	H			0.2660
		6.80		0.2677
		6.90		0.2717
	I			0.2720
		7.00		0.2756
	J			0.2770
		7.10		0.2795
	K			0.2810
			9/32	0.2812
		7.20		0.2835
		7.25		0.2854
		7.30		0.2874
	L			0.2900
		7.40		0.2913
	M			0.2950
		7.50		0.2953
			19/64	0.2968
		7.60		0.2992
	N			0.3020
		7.70		0.3031
		7.75		0.3051
		7.80		0.3071
		7.90		0.3110
			5/16	0.3125
		8.00		0.3150
	O			0.3160
		8.10		0.3189
		8.20		0.3228
	P			0.3230
		8.25		0.3248
		8.30		0.3268
			21/64	0.3281
		8.40		0.3307
	Q			0.3320
		8.50		0.3346
		8.60		0.3386
	R			0.3390
		8.70		0.3425
			11/32	0.3437
		8.75		0.3444
		8.80		0.3464
	S			0.3480
		8.90		0.3504
		9.00		0.3543
	T			0.3580
		9.10		0.3583
			23/64	0.3593
		9.20		0.3622
		9.25		0.3641
		9.30		0.3661
	U			0.3680
		9.40		0.3701
		9.50		0.3740
			3/8	0.3750
	V			0.3770
		9.60		0.3780
		9.70		0.3819
		9.75		0.3838
		9.80		0.3858
	W			0.3860
		9.90		0.3898
			25/64	0.3906
		10.00		0.3937
	X			0.3970
	Y			0.4040
			11/32	0.4062
	Z			0.4130
		10.50		0.4134
			27/64	0.4218
		11.00		0.4331
			7/16	0.4375
		11.50		0.4528
			29/64	0.4531
			15/32	0.4687
		12.00		0.4724
			31/64	0.4843
		12.50		0.4921
			1/2	0.5000
		13.00		0.5118
			33/64	0.5156
			17/32	0.5312
		13.50		0.5315
			35/64	0.5468
		14.00		0.5512
			9/16	0.5625
		14.50		0.5708
			37/64	0.5781
		15.00		0.5905
			19/32	0.5937
			39/64	0.6093
		15.50		0.6102
			5/8	0.6250
		16.00		0.6299
			41/64	0.6406
		16.50		0.6496
			21/32	0.6562
		17.00		0.6693
			43/64	0.6718
			11/16	0.6875
		17.50		0.6890
			45/64	0.7031
		18.00		0.7087
			23/32	0.7187
		18.50		0.7283
			47/64	0.7374
		19.00		0.7480
			3/4	0.7500
			49/64	0.7656
		19.50		0.7677
			25/32	0.7812
		20.00		0.7874
			51/64	0.7968
		20.50		0.8070
			13/16	0.8125
		21.00		0.8267
			53/64	0.8281
			27/32	0.8437
		21.50		0.8464
			55/64	0.8593
		22.00		0.8661
			7/8	0.8750
		22.50		0.8858
			57/64	0.8906
		23.00		0.9055
			29/32	0.9062
			59/64	0.9218
		23.50		0.9251
			15/16	0.9375
		24.00		0.9448
			61/64	0.9531
		24.50		0.9646
			31/32	0.9687
		25.00		0.9842
			63/64	0.9843
			1.0	1.0000
		25.50		1.0039

SOURCE: Adapted from Colvin and Stanley, "American Machinists' Handbook," 8th ed., McGraw-Hill, New York, 1945.

6.3 GEARING

Revised by Ali M. Sadegh

REFERENCES: Buckingham, "Manual of Gear Design," Industrial Press. Cunningham, Noncircular Gears, *Mach. Des.*, Feb. 19, 1957. Cunningham and Cunningham, Rediscovering the Noncircular Gear, *Mach. Des.*, Nov. 1, 1973. Dudley, "Gear Handbook," McGraw-Hill. Dudley, "Handbook of Practical Gear Design," McGraw-Hill. Michalec, "Precision Gearing: Theory and Practice," Wiley. Shigely, "Engineering Design," McGraw-Hill. AGMA Standards. "Gleason Bevel and Hypoid Gear Design," Gleason Works, Rochester. "Handbook of Gears: Inch and Metric" and "Elements of Metric Gear Technology," Designatronics, New Hyde Park, NY. Adams, "Plastics Gearing: Selection and Application," Marcel Dekker.

Notation

a = addendum
b = dedendum
B = backlash, linear measure along pitch circle
c = clearance
C = center distance
d = pitch diam of pinion
d_b = base circle diam of pinion
d_o = outside diam of pinion
d_r = root diam of pinion
D = pitch diameter of gear
D_P = pitch diam of pinion
D_G = pitch diam of gear
D_o = outside diam of gear
D_b = base circle diam of gear
D_t = throat diam of wormgear
F = face width
h_k = working depth
h_t = whole depth
inv ϕ = involute function ($\tan \phi - \phi$)
l = lead (advance of worm or helical gear in 1 rev)
$l_p(l_G)$ = lead of pinion (gear) in helical gears
L = lead of worm in one revolution
m = module
m_G = gear ratio ($m_G = N_G/N_P$)
m_p = contact ratio (of profiles)
M = measurement of over pins
$n_p(n_G)$ = speed of pinion (gear), r/min
$N_P(N_G)$ = number of teeth in pinion (gear)
n_w = number of threads in worm
p = circular pitch
p_b = base pitch
p_n = normal circular pitch of helical gear
P_d = diametral pitch
P_{dn} = normal diametral pitch
R = pitch radius
R_c = radial distance from center of gear to center of measuring pin
$R_P(R_G)$ = pitch radius of pinion (gear)
R_T = testing radius when rolled on a variable-center-distance inspection fixture
s = stress
t = tooth thickness
t_n = normal circular tooth thickness
$T_P(T_G)$ = formative number of teeth in pinion (gear) (in bevel gears)
v = pitch line velocity
X = correction factor for profile shift
α = addendum angle of bevel gear
γ = pitch angle of bevel pinion
γ_R = face angle at root of bevel pinion tooth
γ_o = face angle at tip of bevel pinion tooth
Γ = pitch angle of bevel gear
Γ_R = face angle at root of bevel gear tooth
Γg_o = face angle at tip of bevel gear tooth
δ = dedendum angle of bevel gear
$\overline{\Delta C}$ = relatively small change in center distance C
ϕ = pressure angle
ϕ_n = normal pressure angle
ψ = helix or spiral angle
$\psi_P(\psi_G)$ = helix angle of teeth in pinion (gear)
Σ = shaft angle of meshed bevel pair

BASIC GEAR DATA

Gear Types Gears are grouped in accordance with tooth forms, shaft arrangement, pitch, and quality. Tooth forms and shaft arrangements are:

Tooth form	*Shaft arrangement*
Spur	Parallel
Helical	Parallel or under an angle
Worm	under an angle
Bevel	Intersecting
Hypoid	Bevel with offset between shafts

Pitch definitions (see Fig. 6.3.1). **Diametral pitch** P_d is the ratio of number of teeth in the gear to the diameter of the pitch circle D measured

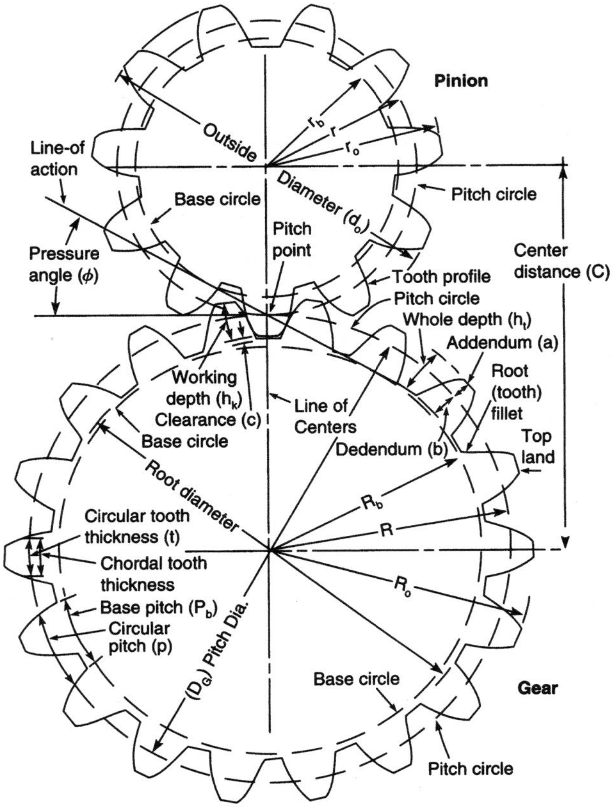

Figure 6.3.1*a* Basic gear geometry and nomenclature.

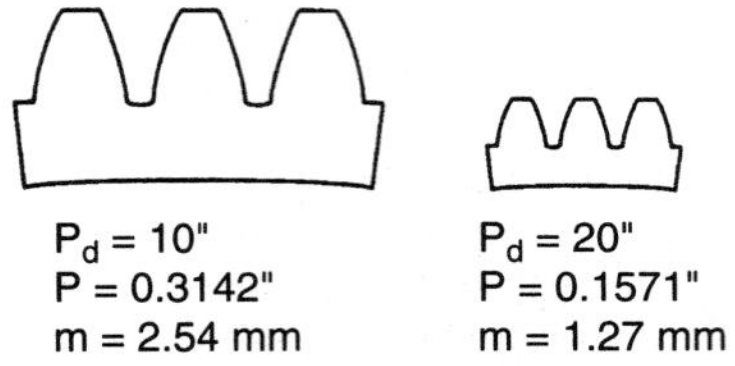

Figure 6.3.1*b* Comparison of pitch, module, and tooth size.

in inches, $P_d = N/D$. **Circular pitch** p is the linear measure in inches along the pitch circle between corresponding points of adjacent teeth. From these definitions, $P_d\, p = \pi$. The **base pitch** p_b is the distance along the line of action between successive involute tooth surfaces. The base and circular pitches are related as $p_b = p \cos \phi$, where ϕ = the pressure angle.

Pitch circle is the imaginary circle that rolls without slippage with a pitch circle of a mating gear. The pitch (circle) diameter equals $D = N/P_d = Np/\pi$. The basic relation between P_d and p is $P_d\, p = \pi$.

Tooth size is related to pitch. In terms of diametral pitch P_d, the relationship is inverse; that is, large P_d implies a small tooth, and small P_d implies a large tooth. Conversely, there is a direct relationship between tooth size and circular pitch p. A small tooth has a small p, but a large tooth has a large p. (See Fig. 6.3.1*b*.) In terms of P_d, coarse teeth comprise P_d less than 20; fine teeth comprise P_d of 20 and higher. (See Fig. 6.3.1*b*.) **Quality** of gear teeth is classified as commercial, precision, and ultraprecision.

Pressure angle ϕ for all gear types is the acute angle between the common normal to the profiles at the contact point and the common pitch plane. For spur gears it is simply the acute angle formed by the common tangent between base circles of mating gears and a normal to the line of centers. For standard gears, pressure angles of 14½°, 20°, and 25° have been adopted by ANSI and the gear industry (see Fig. 6.3.1*a*). The 20° pressure angle is most widely used because of its versatility. The higher pressure angle 25° provides higher strength for highly loaded gears. Although 14½° appears in standards, and in past decades was extensively used, it is used much less than 20°. The 14½° standard is still used for replacement gears in old design equipment, in applications where backlash is critical, and where advantage can be taken of lower backlash with change in center distance.

The **base circle** (or **base cylinder**) is the circle from which the involute tooth profiles are generated. The relationship between the base-circle and pitch-circle diameter is $D_b = D \cos \phi$.

Tooth proportions are established by the addendum, dedendum, working depth, clearance, tooth circular thickness, and pressure angle (see Fig. 6.3.1). In addition, gear face width F establishes thickness of the gear measured parallel to the gear axis.

For involute teeth, proportions have been standardized by ANSI and AGMA into a limited number of systems using a basic rack for specification (see Fig. 6.3.2 and Table 6.3.1). Dimensions for the basic rack

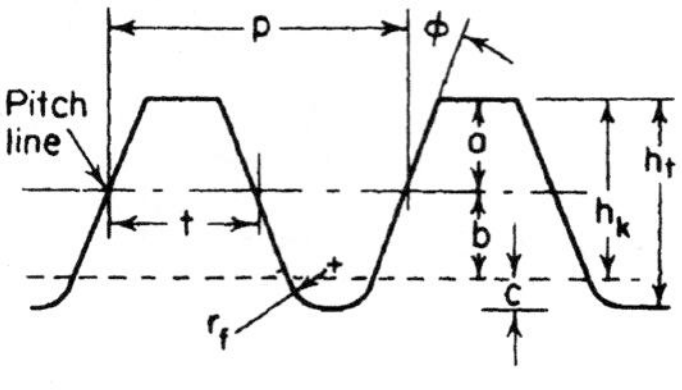

Figure 6.3.2*a* Basic rack for involute gear systems. a = addendum; b = dedendum; c = clearance; h_k = working depth; h_t = whole depth; p = circular pitch; r_f = fillet radius; t = tooth thickness; ϕ = pressure angle.

are normalized for diametral pitch = 1. Dimensions for a specific pitch are obtained by dividing by the pitch. Standards for basic involute spur, helical, bevel, and wormgear designs are listed in Table 6.3.2.

Gear ratio (or **mesh ratio**) m_G is the ratio of number of teeth in a meshed pair, expressed as a number greater than 1; $m_G = N_G/N_P$, where the pinion is the member having the lesser number of teeth. For spur

Table 6.3.2 Gear System Standards

Gear type	ANSI/AGMA/ISO	Title
Spur and helical	1003-H07 1103-H07(metric)	Tooth Proportions for Fine-Pitch Spur and Helical Gearing
Spur and helical	917-B97	Design Manual for Fine-Pitch Gearing
Bevel gears	23509	Design Manual for Bevel Gears (Straight, Zerol, Spiral, and Hypoid)
Worm gearing	6022-C93	Design Manual for Coarse-Pitch Cylindrical Worm Gearing
Worm gearing	6035-A02	Design, Rating and Application of Industrial Globoidal Wormgearing

Table 6.3.1 Tooth Proportions of Basic Rack for Standard Involute Gear Systems

		Tooth proportions for various standard systems					
		1	2	3	4	5	6
Tooth parameter (of basic rack)	Symbol, Figs. 6.3.1*a* and 6.3.2	Full-depth involute, 14½°	Full-depth involute, 20°	Stub involute, 20°	Coarse-pitch involute spur gears, 20°	Coarse-pitch involute spur gears, 25°	Fine-pitch involute, 20°
1. System sponsors		ANSI and AGMA	ANSI	ANSI and AGMA	AGMA	AGMA	ANSI and AGMA
2. Pressure angle	ϕ	14½°	20°	20°	20°	25°	20°
3. Addendum	a	$1/P_d$	$1/P_d$	$0.8/P_d$	$1.000/P_d$	$1.000/P_d$	$1.000/P_d$
4. Min dedendum	b	$1.157/P_d$	$1.157/P_d$	$1/P_d$	$1.250/P_d$	$1.250/P_d$	$1.200/P_d + 0.002$
5. Min whole depth	h_t	$2.157/P_d$	$2.157/P_d$	$1.8/P_d$	$2.250/P_d$	$2.250/P_d$	$2.2002/P_d$ 0.002 in
6. Working depth	h_k	$2/P_d$	$2/P_d$	$1.6/P_d$	$2.000/P_d$	$2.000/P_d$	$2.000/P_d$
7. Min clearance	h_c	$0.157/P_d$	$0.157/P_d$	$0.200/P_d$	$0.250/P_d$	$0.250/P_d$	$0.200/P_d + 0.002$ in
8. Basic circular tooth thickness on pitch line	t	$1.5708/P_d$	$1.5708/P_d$	$1.5708/P_d$	$\pi/(2P_d)$	$\pi/(2P_d)$	$1.5708/P_d$
9. Fillet radius in basic rack	r_f	1⅓ × clearance	1½ × clearance	Not standardized	$0.300/P_d$	$0.300/P_d$	Not standardized
10. Diametral pitch range		Not specified	Not specified	Not specified	19.99 and coarser	19.99 and coarser	20 and finer
11. Governing standard:							
ANSI		B6.1	B6.1	B6.1			
AGMA		201.02 (withdrawn)	(withdrawn)	201.02 (withdrawn)	201.02 (withdrawn)	201.02 (withdrawn)	1,003-H07

and parallel-shaft helical gears, the base circle ratio must be identical to the gear ratio. The speed ratio of gears is inversely proportionate to their numbers of teeth. Only for standard spur and parallel-shaft helical gears is the pitch diameter ratio equal to the gear ratio and inversely proportionate to the speed ratio. Usually, speed (velocity) ratio $V_r = N_{driver}/N_{driven}$. Thus, V_r can be less than or greater than 1.

Metric Gears—Tooth Proportions and Standards

Metric gearing not only is based upon different units of length measure but also involves its own unique design standard. This means that metric gears and American-standard-inch diametral-pitch gears are not interchangeable.

In the metric system the *module m* is analogous to pitch and is defined as

$$m = \frac{D}{N} = \text{mm of pitch diameter per tooth}$$

Note that, for the module to have proper units, the pitch diameter must be in millimeters.

The **metric module** was developed in a number of versions that differ in minor ways. The German DIN standard is widely used in Europe and other parts of the world. The Japanese have their own version defined in JIS/JGMA standards. Deviations among these and other national standards are minor, differing only as to dedendum size and root radii. The differences have been resolved by the new unified module standard promoted by the International Standards Organization (ISO). This unified version (Fig. 6.3.3) conforms to the new SI in all respects. All major industrial countries on the metric system have shifted to this ISO standard, which also is the basis for American metric gearing. Table 6.3.3 lists pertinent current ISO metric standards.

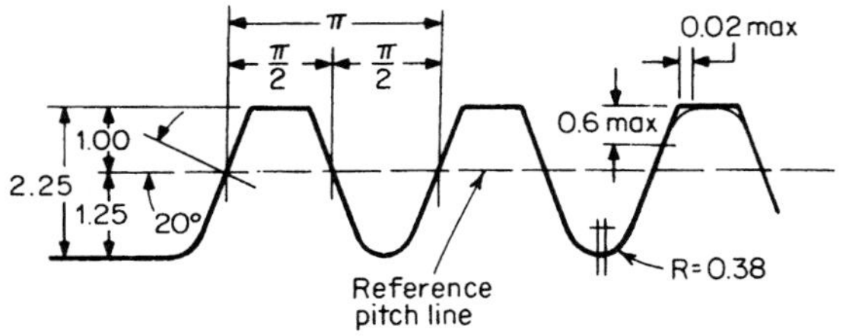

Figure 6.3.3 The ISO basic rack for metric module gears.

Table 6.3.3 ISO Metric Gearing Standards

ISO 53:1998	Cylindrical gears for general and heavy engineering—Basic rack
ISO 54:1996	Cylindrical gears for general and heavy engineering—Modules
ISO 677:1976	Straight bevel gears for general and heavy engineering—Basic rack
ISO 678:1976	Straight bevel gears for general and heavy engineering—Modules and diametral pitches
ISO 701:1998	International gear notation—symbols for geometric data
ISO 1122-1:1998	Glossary of gear terms—Part 1: Geometric definitions
ISO 1328-1:2013	Cylindrical gears—ISO system of flank tolerance classification
ISO 1340:1976	Cylindrical gears—Information to be given to the manufacturer by the purchaser in order to obtain the gear required
ISO 1341:1976	Straight bevel gears—Information to be given to the manufacturer by the purchaser in order to obtain the gear required
ISO 2203:1973	Technical drawings—Conventional representation of gears

Tooth proportions for standard spur and helical gears are given in terms of the basic rack. Dimensions, in millimeters, are normalized for module $m = 1$. Corresponding values for other modules are obtained by multiplying each dimension by the value of the specific module *m*. Major tooth parameters are described by this standard:

Tooth form: Straight-sided and full-depth, forming the basis of a family of full-depth interchangeable gears.

Pressure angle: 20°, conforming to worldwide acceptance.

Addendum: Equal to module *m*, which corresponds to the American practice of $1/P_d$ = addendum.

Dedendum: Equal to 12.5 *m*, which corresponds to the American practice of $1.25/P_d$ = dedendum.

Root radius: Slightly greater than American standards specifications.

Tip radius: A maximum is specified, whereas American standards do not specify. In practice, U.S. manufacturers can specify a tip radius as near zero as possible.

Note that the basic racks for metric and American inch gears are essentially identical, but metric and American standard gears are not interchangeable.

The preferred standard gears of the metric system are not interchangeable with the preferred diametral-pitch sizes. Table 6.3.4 lists commonly used pitches and modules of both systems (preferred values are boldface).

FUNDAMENTAL RELATIONSHIPS OF SPUR AND HELICAL GEARS

Center distance is the distance between axes of mating gears and is determined from $C = (n_G + N_p)/(2P_d)$, or $C = (D_G + D_P)/2$. Deviation from ideal center distance of involute gears is not detrimental to proper (conjugate) gear action which is one of the prime superiority features of the involute tooth form.

Contact Ratio Referring to the top part of Fig. 6.3.4 and assuming no tip relief, the pinion engages in the gear at *a*, where the outside circle of the gear tooth intersects the line of action *ac*. For the usual spur gear and pinion combinations there will be two pairs of teeth theoretically in contact at engagement (a gear tooth and its mating pinion tooth considered as a pair). This will continue until the pair ahead (bottom part of Fig. 6.3.4) disengages at *c*, where the outside circle of the pinion intersects the line of action *ac*, the movement along the line of action being *ab*. After disengagement the pair behind will be the only pair in contact until another pair engages, the movement along the line of action for single-pair contact being *bd*. Two pairs are theoretically in contact during the remaining intervals, *ab* + *dc*.

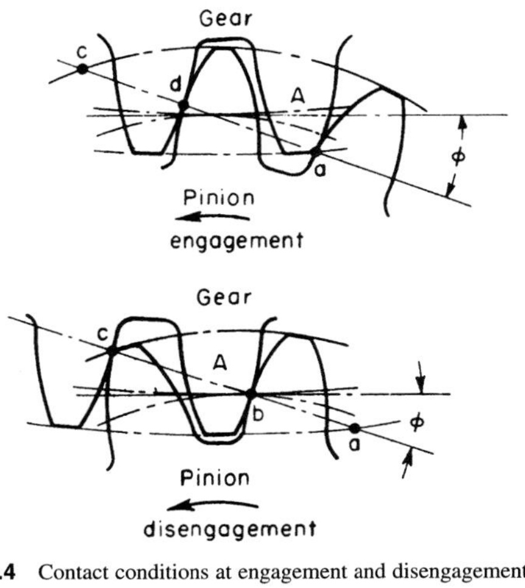

Figure 6.3.4 Contact conditions at engagement and disengagement.

Table 6.3.4 Metric and American Gear Equivalents

Diametral pitch P_d	Module m	Circular pitch		Circular tooth thickness		Addendum	
		in	mm	in	mm	in	mm
1/2	50.8000	6.2832	159.593	3.1416	79.7965	2.0000	50.8000
0.5080	**50**	6.1842	157.080	3.0921	78.5398	1.9685	50
0.5644	45	5.5658	141.372	2.7850	70.6858	1.7730	45
0.6048	**42**	5.1948	131.947	2.5964	65.9734	1.6529	42
0.6513	39	4.8237	122.522	2.4129	61.2610	1.5361	39
0.7056	**36**	4.4527	113.097	2.2249	56.5487	1.4164	36
3/4	33.8667	4.1888	106.396	2.0943	53.1977	1.3333	33.8667
0.7697	33	4.0816	103.673	2.0400	51.8363	1.2987	33
0.8467	**30**	3.7105	94.248	1.8545	47.1239	1.1806	30
0.9407	27	3.3395	84.823	1.6693	42.4115	1.0627	27
1	25.4000	3.1416	79.800	1.5708	39.8984	1.0000	25.4001
1.0583	**24**	2.9685	75.398	1.4847	37.6991	0.9452	24
1.1546	22	2.7210	69.115	1.3600	34.5575	0.8658	22
1.2700	**20**	2.4737	62.832	1.2368	31.4159	0.7874	20
1.4111	**18**	2.2263	56.548	1.1132	28.2743	0.7087	18
1.5	16.9333	2.0944	53.198	1.0472	26.5988	0.6667	16.933
1.5875	**16**	1.9790	50.267	0.9894	25.1327	0.6299	16
1.8143	**14**	1.7316	43.983	0.8658	21.9911	0.5512	14
2	12.7000	1.5708	39.898	0.7854	19.949	0.5000	12.7000
2.1167	**12**	1.4842	37.699	0.7420	18.8496	0.4724	12
2.5	10.1600	1.2566	31.918	0.6283	15.9593	0.4000	10.1600
2.5400	**10**	1.2368	31.415	0.6184	15.7080	0.3937	10
2.8222	9	1.1132	28.275	0.5565	14.1372	0.3543	9
3	8.4667	1.0472	26.599	0.5235	13.2995	0.3333	8.4667
3.1416	8.0851	1.0000	25.400	0.5000	12.7000	0.3183	0.0851
3.1750	**8**	0.9895	25.133	0.4948	12.5664	0.3150	8.00
3.5	7.2571	0.8976	22.799	0.4488	11.3994	0.2857	7.2571
3.6286	**7**	0.8658	21.991	0.4329	10.9956	0.2756	7.000
3.9078	**6.5**	0.8039	20.420	0.4020	10.2101	0.2559	6.5
4	6.3500	0.7854	19.949	0.3927	9.9746	0.2500	6.3500
4.2333	**6**	0.7421	18.850	0.3710	9.4248	0.2362	6.0000
4.6182	**5.5**	0.6803	17.279	0.3401	8.6394	0.2165	5.5
5	5.0801	0.6283	15.959	0.3142	7.9794	0.2000	5.080
5.0802	**5**	0.6184	15.707	0.3092	7.8537	0.1968	5.000
5.3474	**4.75**	0.5875	14.923	0.2938	7.4612	0.1870	4.750
5.6444	**4.5**	0.5566	14.138	0.2783	7.0688	0.1772	4.500
6	4.2333	0.5236	13.299	0.2618	6.6497	0.1667	4.233
6.3500	**4**	0.4947	12.565	0.2473	6.2827	0.1575	4.000
6.7733	**3.75**	0.4638	11.781	0.2319	5.8903	0.1476	3.750
7	3.6286	0.4488	11.399	0.2244	5.6998	0.1429	3.629
7.2571	**3.5**	0.4329	10.996	0.2164	5.4979	0.1378	3.500
7.8154	**3.25**	0.4020	10.211	0.2010	5.1054	0.1279	3.250
8	3.1750	0.3927	9.974	0.1964	4.9886	0.1250	3.175
8.4667	**3**	0.3711	9.426	0.1855	4.7130	0.1181	3.000
9	2.8222	0.3491	8.867	0.1745	4.4323	0.1111	2.822
9.2364	**2.75**	0.3401	8.639	0.1700	4.3193	0.1082	2.750
10	2.5400	0.3142	7.981	0.1571	3.9903	0.1000	2.540
10.1600	**2.50**	0.3092	7.854	0.1546	3.9268	0.0984	2.500
11	2.3091	0.2856	7.254	0.1428	3.6271	0.0909	2.309
11.2889	**2.25**	0.2783	7.069	0.1391	3.5344	0.0886	2.250
12	2.1167	0.2618	6.646	0.1309	3.3325	0.0833	2.117
12.7000	**2**	0.2474	6.284	0.1236	3.1420	0.0787	2.000
13	1.9538	0.2417	6.139	0.1208	3.0696	0.0769	1.954
14	1.8143	0.2244	5.700	0.1122	2.8500	0.0714	1.814
14.5143	**1.75**	0.2164	5.497	0.1082	2.7489	0.0689	1.750
15	1.6933	0.2094	5.319	0.1047	2.6599	0.0667	1.693
16	1.5875	0.1964	4.986	0.0982	2.4936	0.0625	1.587
16.9333	**1.5**	0.1855	4.712	0.0927	2.3562	0.0591	1.500
18	1.4111	0.1745	4.432	0.0873	2.2166	0.0556	1.411
20	1.2700	0.1571	3.990	0.0785	1.9949	0.0500	1.270
20.3200	**1.25**	0.1546	3.927	0.0773	1.9635	0.0492	1.250
22	1.1545	0.1428	3.627	0.0714	1.8136	0.0455	1.155
24	1.0583	0.1309	3.325	0.0655	1.6624	0.0417	1.058
25.4000	**1**	0.1237	3.142	0.0618	1.5708	0.0394	1.000
28	0.90701	0.1122	2.850	0.0561	1.4249	0.0357	0.9071
28.2222	**0.9**	0.1113	2.827	0.0556	1.4137	0.0354	0.9000
30	0.84667	0.1047	2.659	0.0524	1.3329	0.0333	0.8467
31.7500	**0.8**	0.0989	2.513	0.04945	1.2566	0.0315	0.8000
32	0.79375	0.0982	2.494	0.04909	1.2468	0.0313	0.7937
33.8667	**0.75**	0.0928	2.357	0.04638	1.1781	0.0295	0.7500

Table 6.3.4 Metric and American Gear Equivalents *(Continued)*

Diametral pitch P_d	Module m	Circular pitch, in	Circular pitch, mm	Circular tooth thickness, in	Circular tooth thickness, mm	Addendum, in	Addendum, mm
36	0.70556	0.0873	2.217	0.04363	1.1083	0.0278	0.7056
36.2857	**0.7**	0.0865	2.199	0.04325	1.0996	0.0276	0.7000
40	0.63500	0.0785	1.994	0.03927	0.9975	0.0250	0.6350
42.3333	**0.6**	0.0742	1.885	0.03710	0.9423	0.0236	0.6000
44	0.57727	0.0714	1.814	0.03570	0.9068	0.0227	0.5773
48	0.52917	0.0655	1.661	0.03272	0.8311	0.0208	0.5292
50	0.50800	0.0628	1.595	0.03141	0.7976	0.0200	0.5080
50.8000	**0.5**	0.06184	1.5707	0.03092	0.7854	0.0197	0.5000
63.5000	**0.4**	0.04947	1.2565	0.02473	0.6283	0.0157	0.4000
64	0.39688	0.04909	1.2469	0.02454	0.6234	0.0156	0.3969
67.7333	0.375	0.04638	1.1781	0.02319	0.5890	0.0148	0.3750
72	0.35278	0.04363	1.1082	0.02182	0.5541	0.0139	0.3528
72.5714	0.35	0.04329	1.0996	0.02164	0.5498	0.0138	0.3500
78.1538	0.325	0.04020	1.0211	0.02010	0.5105	0.0128	0.3250
80	0.31750	0.03927	0.9975	0.01964	0.4987	0.0125	0.3175
84.6667	**0.3**	0.03711	0.9426	0.01856	0.4713	0.0118	0.3000
92.3636	0.275	0.03401	0.8639	0.01700	0.4319	0.0108	0.2750
96	0.26458	0.03272	0.8311	0.01636	0.4156	0.0104	0.2646
101.6000	0.25	0.03092	0.7854	0.01546	0.3927	0.00984	0.2500
120	0.21167	0.02618	0.6650	0.01309	0.3325	0.00833	0.2117
125	0.20320	0.02513	0.6383	0.01256	0.3192	0.00800	0.2032
127.0000	**0.2**	0.02474	0.6284	0.01237	0.3142	0.00787	0.2000
150	0.16933	0.02094	0.5319	0.01047	0.2659	0.00667	0.1693
169.3333	0.15	0.01855	0.4712	0.00928	0.2356	0.00591	0.1500
180	0.14111	0.01745	0.4432	0.00873	0.2216	0.00555	0.1411
200	0.12700	0.01571	0.3990	0.00786	0.1995	0.00500	0.1270
203.2000	0.125	0.01546	0.3927	0.00773	0.1963	0.00492	0.1250

Contact ratio expresses the average number of pairs of teeth theoretically in contact and is obtained numerically by dividing the length of the line of action by the normal pitch. For full-depth teeth, without undercutting, the contact ratio is $m_p = (\sqrt{D_0^2 - D_b^2} + \sqrt{d_0^2 - d_b^2} - 2C \sin\phi)/(2p \cos\phi)$. The result will be a mixed number with the integer portion the number of pairs of teeth always in contact and carrying load, and the decimal portion the amount of time an additional pair of teeth are engaged and share load. As an example, for m_p between 1 and 2:

Load is carried by one pair, $(2 - m_p)/m_p$ of the time.

Load is carried by two pairs, $2(m_p - 1)/m_p$ of the time.

In Figs. 6.3.5 to 6.3.7, contact ratios are given for standard generated gears, the lower part of Figs. 6.3.5 and 6.3.6 representing the effect of undercutting.

These charts are applicable to both standard diametral pitch gears made in accordance with American standards and also standard metric gears that have an addendum of one module.

Tooth Thickness For standard gears, the tooth thickness t of mating gears is equal, where $t = p/2 = \pi/(2P_d)$ measured linearly along the arc of the pitch circle. The tooth thickness t_1 at any radial point of the tooth (at diameter D_1) can be calculated from the known thickness t at the pitch radius $D/2$ by the relationship $t_1 = t(D_1/D) - D_1\,(\text{inv}\,\phi_1 - \text{inv}\,\phi)$,

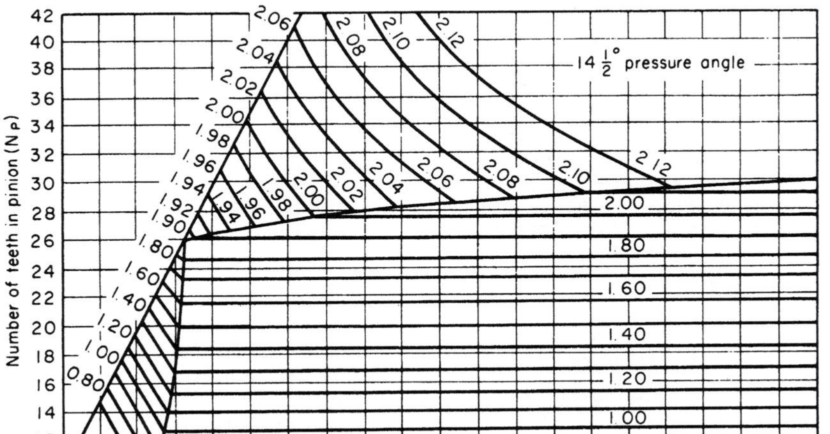

Figure 6.3.5 Contact ratio, spur gear pairs—full depth, standard generated teeth, 14½° pressure angle.

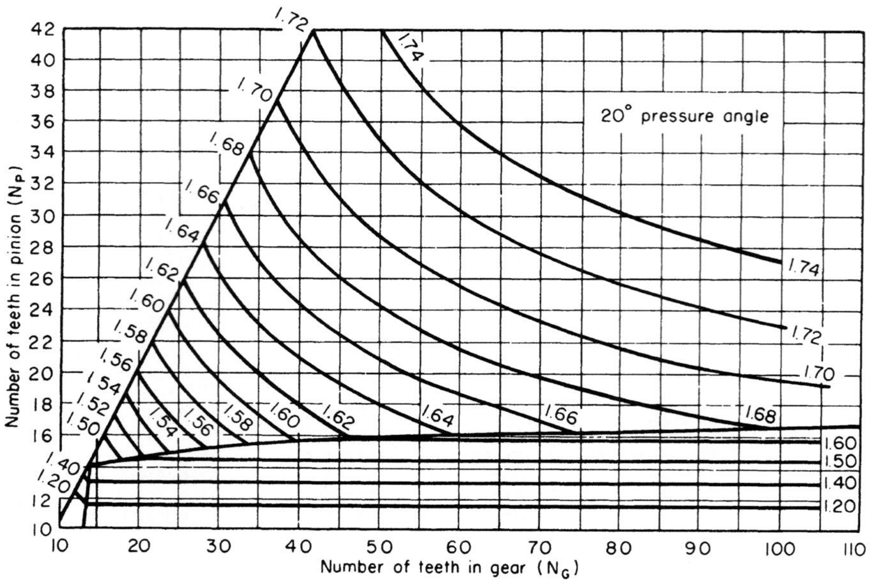

Figure 6.3.6 Contact ratio, spur gear pairs—full-depth standard generated teeth, 20° pressure angle.

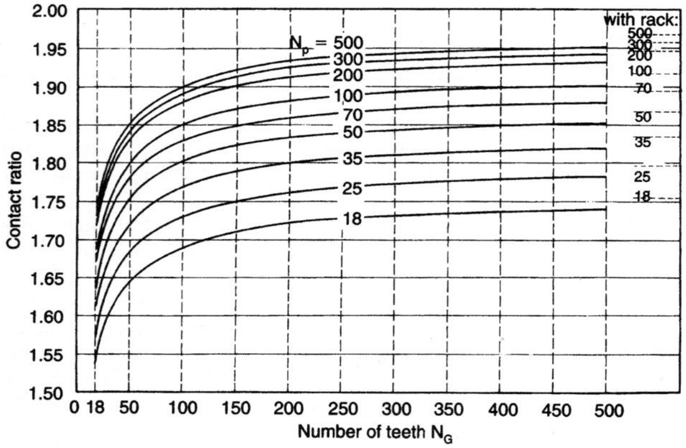

Figure 6.3.7 Contact ratio for large numbers of teeth—spur gear pairs, full-depth standard teeth, 20° pressure angle. (*Data by R. Feeney and T. Wall.*)

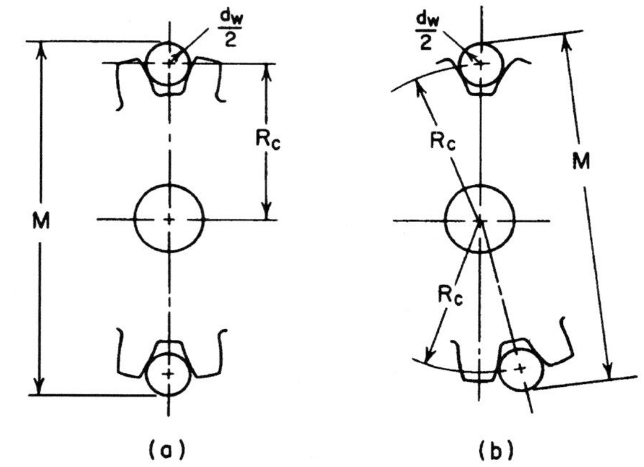

Figure 6.3.8 Geometry of over-pins measurements (*a*) for an even number of teeth and (*b*) for an odd number of teeth.

where inv $\phi = \tan\phi - \phi$ = involute function. Units for ϕ must be radians. Tables of values for inv ϕ from 0 to 45° can be found in the references (Buckingham and Dudley).

Over-plus measurements (spur gears) are another means of deriving tooth thickness. If cylindrical pins are inserted in tooth spaces diametrically opposite one another (or nearest space for an odd number of teeth) (Fig. 6.3.8), the tooth thickness can be derived from the measurement M as follows:

$$t = D(\pi/N + \text{inv}\ \phi_1 - \text{inn}\ \phi - d_w/D_b)$$

$$\cos\phi_1 = (D\cos\phi)/2R_c$$

$$R_c = (M - d_w)/2 \qquad \text{for even number of teeth}$$

$$R_c = (M - d_w)/[2\cos(90/N)] \qquad \text{for odd number of teeth}$$

where d_w = pin diameter, R_c = distance from gear center to center of pin, and M = measurement over pins.

For the reverse situation, the over-pins measurement M can be found for a given tooth thickness t at diameter D and pressure angle ϕ by the following: inv $\phi_1 = t/D + \text{inv}\ \phi + d_w/(D\cos\phi) - \pi/N$, $M = D\cos\phi/\cos\phi_1 + d_w$ (for even number of teeth), $M = (D\cos\phi/\cos\phi_1)\cos(90°/N) + d_w$ (for odd number of teeth).

Table values of over-pins measures (see Dudley and Van Keuren) facilitate measurements for all standard gears including those with slight departures from standard. (For correlation with tooth thickness and testing radius, see Michalec, *Product Eng.*, May 1957, and "Precision Gearing: Theory and Practice," Wiley.)

Testing radius R_T is another means of determining tooth thickness and refers to the effective pitch radius of the gear when rolled intimately with a master gear of known size calibration. (See Michalec, *Product Eng.*, Nov. 1956, and "Precision Gearing: Theory and Practice," op. cit.) For standard design gears the testing radius equals the pitch radius. The testing radius may be corrected for small departures Δt from ideal tooth thickness by the relationship, $R_T = R + \Delta t/2 \tan\phi$, where $\Delta t = t_1 - t$ and is positive and negative, respectively, for thicker and thinner tooth thicknesses than standard value t.

Backlash B is the amount by which the width of a tooth space exceeds the thickness of the engaging tooth measured on the pitch circle. Backlash does not adversely affect proper gear function except for lost motion upon reversal of gear rotation. Backlash inevitably occurs because of necessary fabrication tolerances on tooth thickness

Table 6.3.5 Metric Spur Gear Design Formulas

To obtain:	From known	Use this formula*
Pitch diameter D	Module; diametral pitch	$D = mN$
Circular pitch p_c	Module; diametral pitch	$p_c = m\pi = \frac{D}{N}\pi = \frac{\pi}{P}$
Module m	Diametral pitch	$m = \frac{25.4}{P}$
No. of teeth N	Module and pitch diameter	$N = \frac{D}{m}$
Addendum a	Module	$a = m$
Dedendum b	Module	$b = 1.25m$
Outside diameter D_o	Module and pitch diameter or number of teeth	$D_o = D + 2m = m(N + 2)$
Root diameter D_r	Pitch diameter and module	$D_r = D - 2.5m$
Base circle diameter D_b	Pitch diameter and pressure angle ϕ	$D_b = D \cos \phi$
Base pitch p_b	Module and pressure angle	$p_b = m\pi \cos \phi$
Tooth thickness at standard pitch diameter T_{std}	Module	$T_{std} = \frac{\pi}{2}m$
Center distance C	Module and number of teeth	$C = \frac{m(N_1 + N_2)}{2}$
Contact ratio m_p	Outside radii, base-circle radii, center distance, pressure angle	$m_p = \frac{\sqrt{{}_1R_o^2 - {}_1R_b^2} + \sqrt{{}_2R_o^2 - {}_2R_b^2} - C\sin\phi}{m\pi\cos\phi}$
Backlash (linear) B (along pitch circle)	Change in center distance	$B = 2(\Delta C) \tan \phi$
Backlash (linear) B (along pitch circle)	Change in tooth thickness, T	$B = \Delta T$
Backlash (linear) (along line of action) B_{LA}	Linear backlash (along pitch circle)	$B_{LA} = B \cos\phi$
Backlash (angular) B_a	Linear backlash (along pitch circle)	$B_a = 6{,}880 \frac{B}{D}$ (arc minutes)
Min. number teeth for no undercutting, N_c	Pressure angle	$N_c = \frac{2}{\sin^2 \phi}$

* All linear dimensions in millimeters.

and center distance plus need for clearance to accommodate lubricant and thermal expansion. Proper backlash can be introduced by a specified amount of tooth thinning or slight increase in center distance. The relationship between small change in center distance $\overline{\Delta C}$ and backlash is $B = 2\overline{\Delta C}$ tan ϕ (see Michalec, "Precision Gearing: Theory and Practice").

Total composite error (tolerance) is a measure of gear quality in terms of the net sum of irregularity of its testing radius R_T due to pitch-circle runout and tooth-to-tooth variations (see Michalec, op. cit.).

Tooth-to-tooth composite error (tolerance) is the variation of testing radius R_T between adjacent teeth caused by tooth spacing, thickness, and profile deviations (see Michalec, op. cit.).

Profile shifted gears have tooth thicknesses that are significantly different from nominal standard value; excluded are deviations caused by normal allowances and tolerances. They are also known as **modified gears, long and short addendum gears,** and **enlarged gears.** They are produced by cutting the teeth with standard cutters at enlarged or reduced outside diameters. The result is a relative shift of the two families of involutes forming the tooth profiles, simultaneously with a shift of the tooth radially outward or inward (see Fig. 6.3.9). Calculation of operating conditions and tooth parameters are

$$C_1 = \frac{C\cos\phi}{\cos\phi_1}$$

$$\text{inv } \phi_1 = \text{inv } \phi + \frac{N_P(t'_G + t'_P) - \pi D_P}{D_P(N_P + N_G)}$$

$$t'_G = t + 2X_G \tan\phi$$

$$t'_P = t + 2X_P \tan\phi$$

$$D'_G = (N_G/P_d) + 2X_G$$

$$D'_P = (N_P/P_d) + 2X_P$$

$$D'_o = D'_G + (2/P_d)$$

$$d'_o = D'_P + (2/P_d)$$

where ϕ = standard pressure angle, ϕ_1 = operating pressure angle, C = standard center distance = $(N_G + N_P)/2P_d$, C_1 = operating center distance, X_G = profile shift correction of gear, and X_P = profile shift correction of pinion. The quantity X is positive for enlarged gears and negative for thinned gears.

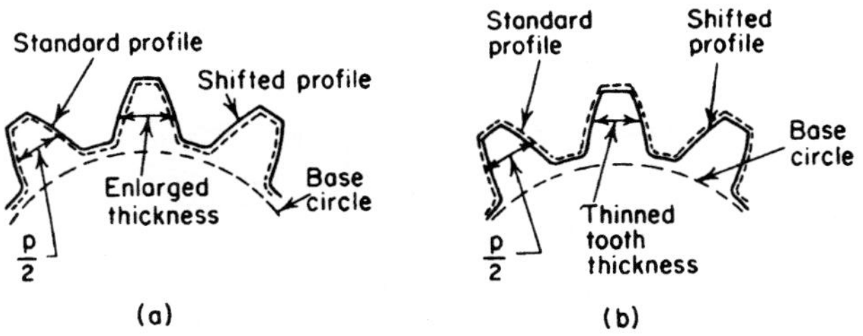

Figure 6.3.9 Geometry of profile-shifted teeth. (*a*) Enlarged case; (*b*) thinned tooth thickness case.

Metric Module Gear Design Equations Basic design equations for spur gearing utilizing the metric module are listed in Table 6.3.5. (See Designatronics, "Elements of Metric Gear Technology.")

HELICAL GEARS

Helical gears divide into two general applications: for driving parallel shafts and for driving skew shafts (mostly at right angles), the latter often referred to as *crossed-axis* helical gears. The helical tooth form may be imagined as consisting of an infinite number of staggered laminar spur gears, resulting in the curved cylindrical helix.

Pitch of helical gears is definable in two planes. The diametral and circular pitches measured in the plane of rotation (transverse) are defined as for spur gears. However, pitches measured normal to the tooth are related by the cosine of the helix angle; thus normal diametral pitch = $P_{dn} = P_d$ /cos ψ, normal circular pitch = $p_n = p$ cos ψ, and $P_{dn}\, p_n = \pi$. **Axial pitch** is the distance between corresponding sides of adjacent teeth measured parallel to the gear axis and is calculated as $p_a = p$ cot ψ.

Pressure angle of helical gears is definable in the normal and transverse planes by $\tan \phi_n = \tan \phi \cos \psi$. The transverse pressure angle, which is effectively the real pressure angle, is always greater than the normal pressure angle.

Tooth thickness t of helical gears can be measured in the plane of rotation, as with spur gears, or normal to the tooth surface t_n. The relationship of the two thicknesses is $t_n = t \cos \psi$.

Over-Pins Measurement of Helical Gears Tooth thicknesses t at diameter d can be found from a known over-pins measurement M at known pressure angle ϕ, corresponding to diameter D as follows:

$$R_c = (M - d_w)/2 \qquad \text{for even number of teeth}$$

$$R_c = (M - d_w)/[2 \cos (180/2N)] \qquad \text{for odd number of teeth}$$

$$\cos \phi_1 = (D \cos \phi)/2R_c$$

$$\tan \phi_n = \tan \phi \cos \psi$$

$$\cos \psi_b = \sin \phi_n / \sin \phi$$

$$t = D[\pi/N + \text{inv } \phi_1 - \text{inv } \phi - d_w/(D \cos \phi \cos \psi_b)]$$

Parallel-shaft helical gears must conform to the same conditions and requirements as spur gears with parameters (pressure angle and pitch) consistently defined in the transverse plane. Since standard spur gear cutting tools are usually used, normal plane values are standard, resulting in nonstandard transverse pitches and nonstandard pitch diameters and center distances. For parallel shafts, helical gears must have identical helix angles, but must be of opposite hand (left and right helix directions). The commonly used helix angles range from 15 to 35°. To make most advantage of the helical form, the advance of a tooth should be greater than the circular pitch; recommended ratio is 1.5 to 2 with 1.1 minimum. This overlap provides two or more teeth in continual contact with resulting greater smoothness and quietness than spur gears. Because of the helix, the normal component of the tangential pressure on the teeth produces end thrust of the shafts. To remove this objection, gears are made with helixes of opposite hand on each half of the face and are then known as herringbone gears (see Fig. 6.3.10).

Crossed-axis helical gears, also called *spiral* or *screw gears* (Fig. 6.3.11), are a simple type of involute gear used for connecting

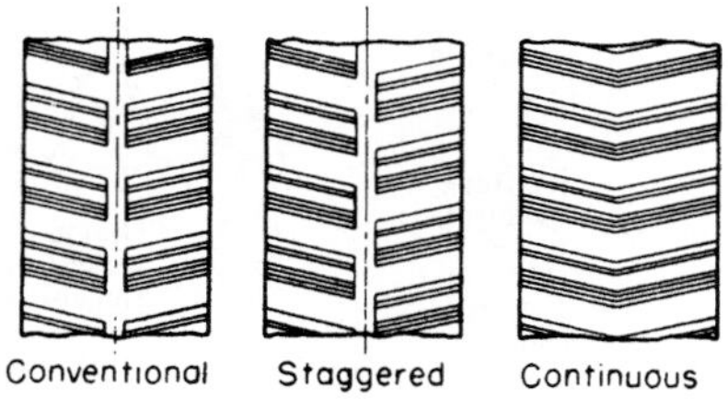

Figure 6.3.10 Herringbone gears.

non-parallel, nonintersecting shafts. Contact is the point and there is considerably more sliding than with parallel-axis helicals, which limits the load capacity. The individual gear of this mesh is identical in form and specification to a parallel-shaft helical gear. Crossed-axis helicals can connect any shaft angle Σ, although 90° is prevalent. Usually, the helix angles will be of the same hand, although for some extreme cases it is possible to have opposite hands, particularly if the shaft angle is small.

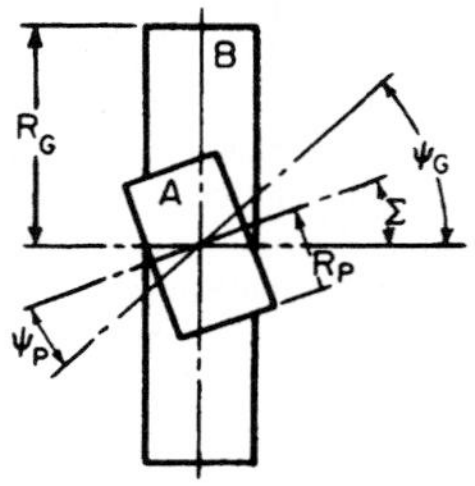

Figure 6.3.11 Crossed-axis helical gears.

Helical Gear Calculations For parallel shafts the center distance is a function of the helix angle as well as the number of teeth, that is, $C = (N_G + N_P)/(2P_{dn} \cos \psi)$. This offers a powerful method of gearing shafts at any specified center distance to a specified velocity ratio. For crossed-axis helicals the problem of connecting a pair of shafts for any velocity ratio admits of a number of solutions, since both the pitch radii and the helix angles contribute to establishing the velocity ratio. The formulas given in Tables 6.3.6 and 6.3.7 are of assistance in calculations. The notation used in these tables is as follows:

$N_P(N_G)$ = number of teeth in pinion (gear)
$D_P(D_G)$ = pitch diam of pinion (gear)
$p_P(p_G)$ = circular pitch of pinion (gear)
p = circular pitch in plane of rotation for both gears
P_d = diametral pitch in plane of rotation for both gears
p_n = normal circular pitch for both gears
P_{dn} = normal diametral pitch for both gears
ψ_G = tooth helix angle of gear
ψ_P = tooth helix angle of pinion
$l_P(l_G)$ = lead of pinion (gear)
= lead of tooth helix
$n_P(n_G)$ = r/min of pinion (gear)
Σ = angle between shafts in plan
C = center distance

Table 6.3.6 Helical Gears on Parallel Shafts

To find:	Formula
Center distance C	$\dfrac{N_G + N_P}{2P_{dn} \cos \psi}$
Pitch diameter D	$\dfrac{N}{P_d} = \dfrac{N}{P_{dn} \cos \psi}$
Normal diametral pitch P_{dn}	$\dfrac{P_d}{\cos \psi}$
Normal circular pitch p_n	$p \cos \psi$
Pressure angle ϕ	$\tan^{-1} \dfrac{\tan \phi_n}{\cos \psi}$
Contact ratio m_p	$\dfrac{\sqrt{{}_G D_o^2 - {}_G D_b^2} + \sqrt{{}_P D_o^2 - {}_P D_b^2} + 2C \sin\phi}{2p \cos \phi} + \dfrac{F \sin\psi}{p_n}$
Gear ratio m_G	$\dfrac{N_G}{N_P} = \dfrac{D_G}{D_P}$

Table 6.3.7 Crossed Helical Gears on Skew Shafts

To find:	Formula
Center distance C	$\dfrac{P_n}{2\pi}\left(\dfrac{N_G}{\cos \psi_G} + \dfrac{N_P}{\cos \psi_P}\right)$
Pitch diameter D_G, D_P	$D_G = \dfrac{N_G}{P_{dG}} = \dfrac{N_G}{P_{dn} \cos \psi_G} = \dfrac{N_G P_n}{\pi \cos \psi_G}$
	$D_P = \dfrac{N_P}{P_{dP}} = \dfrac{N_P}{P_{dn} \cos \psi_P} = \dfrac{n_P P_n}{\pi \cos \psi_P}$
Gear ratio m_G	$\dfrac{N_G}{N_P} = \dfrac{D_G \cos \psi_G}{D_P \cos \psi_P}$
Shaft angle Σ	$\psi_G + \psi_P$

NONSPUR GEAR TYPES*

Bevel gears are used to connect two intersecting shafts in any given speed ratio. The tooth shapes may be designed in any of the shapes shown in Fig. 6.3.12. A special type of gear known as a **hypoid** was

* In the following text relating to bevel gearing, all tables and figures have been extracted from Gleason Works publications, with permission.

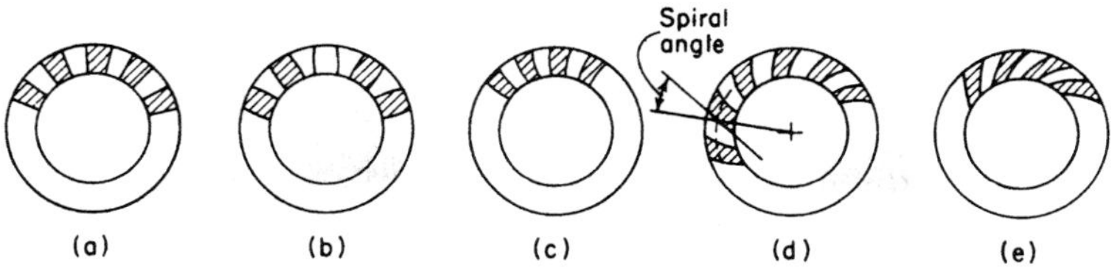

Figure 6.3.12 Bevel gear types. (*a*) Old-type straight teeth; (*b*) modern Coniflex straight teeth (exaggerated crowning); (*c*) Zerol teeth; (*d*) spiral teeth; (*e*) hypoid teeth.

developed by Gleason Works for the automotive industry (see *Jour. SAE*, **18,** no. 6). Although similar in appearance to a spiral bevel, it is not a true bevel gear. The basic pitch rolling surfaces are hyperbolas of revolution. Because a "spherical involute" tooth form has a curved crown tooth (the basic tool for generating all bevel gears), Gleason used a straight-sided crown tooth which resulted in bevel gears differing slightly from involute form. Because of the figure 8 shape of the complete theoretical tooth contact path, the tooth form has been called "octoid." Straight-sided bevel gears made by reciprocating cutters are of this type. Later, when curved teeth became widely used (spiral and Zerol), practical limitations of such cutters resulted in introduction of the "spherical" tooth form which is now the basis of all curved tooth bevel gears. (For details see Gleason's publication, "Guide to Bevel Gears.") Gleason Works also developed the generated tooth form Revacycle and the nongenerated tooth forms Formate and Helixform, used principally for mass production of hypoid gears for the automotive industry.

Referring to Fig. 6.3.13, we see that the pitch surfaces of bevel gears are frustums of cones whose vertices are at the intersection of the axes; the essential elements and definitions follow.

Addendum angle α**:** The angle between elements of the face cone and pitch cone.

Back angle: The angle between an element of the back cone and a plane of rotation. It is equal to the pitch angle.

Back cone: The angle of a cone whose elements are tangent to a sphere containing a trace of the pitch circle.

Back-cone distance: The distance along an element of the back cone from the apex to the pitch circle.

Cone distance A_o**:** The distance from the end of the tooth (heel) to the pitch apex.

Crown: The sharp corner forming the outside diameter.

Crown-to-back: The distance from the outside diameter edge (crown) to the rear of the gear.

Dedendum angle δ**:** The angle between elements of the root cone and pitch cone.

Face angle γ_o**:** The angle between an element of the face cone and its axis.

Face width F**:** The length of teeth along the cone distance.

Front angle: The angle between an element of the front cone and a plane of rotation.

Generating mounting surface, GMS: The diameter and/or plane of rotation surface or shaft center which is used for locating the gear blank during fabrication of the gear teeth.

Heel: The portion of a bevel gear tooth near the outer end.

Mounting distance, MD: For assembled bevel gears, the distance from the crossing point of the axes to the registering surface, measured along the gear axis. Ideally, it should be identical to the pitch apex to back.

Mounting surface, MS: The diameter and/or plane of rotation surface which is used for locating the gear in the application assembly.

Octoid: The mathematical form of the bevel tooth profile. Closely resembles a spherical involute but is fundamentally different.

Pitch angle: The angle formed between an element of the pitch cone and the bevel gear axis. It is the half angle of the pitch.

Pitch apex to back: The distance along the axis from apex of pitch cone to a locating registering surface on back.

Registering surface, RS: The surface in the plane of rotation which locates the gear blank axially in the generating machine and the gear in application. These are usually identical surfaces, but not necessarily so.

Root angle γ_R: The angle formed between a tooth root element and the axis of the bevel gear.

Shaft angle Σ: The angle between mating bevel-gear axes; also, the sum of the two pitch angles.

Spiral angle ψ: The angle between the tooth trace and an element of the pitch cone, corresponding to helix angle in helical gears. The spiral angle is understood to be at the mean cone distance.

Toe: The portion of a bevel tooth near the inner end.

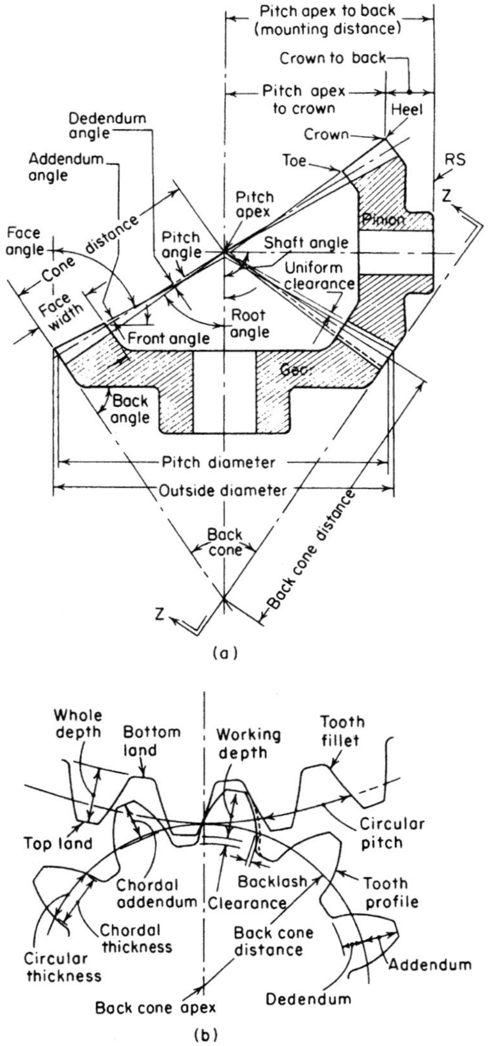

Figure 6.3.13 Geometry of bevel gear nomenclature. (*a*) Section through axes; (*b*) view along axis Z/Z.

Bevel gears are described by the parameter dimensions at the large end (heel) of the teeth. Pitch, pitch diameter, and tooth dimensions, such as addendum are measurements at this point. At the large end of the gear, the tooth profiles will approximate those generated on a spur gear pitch circle of radius equal to the back cone distance. The formative number of teeth is equal to that contained by a complete spur gear. For pinion and gear, respectively, this is $T_P = N_P/\cos\ \gamma$; $T_G = N_G/\cos\ \Gamma$, where T_P and T_G = formative number teeth and N_P and N_G = actual number teeth.

Although bevel gears can connect intersecting shafts at any angle, most applications are for right angles. When such bevels are in a 1:1 ratio, they are called **miter gears.** Bevels connecting shafts other than 90° are called **angular bevel gears.** The speeds of the shafts of bevel gears are determined by $n_P/n_G = \sin\ \Gamma/\sin\ \gamma$, where $n_P(n_G)$ = r/min of pinion (gear), and $\gamma(\Gamma)$ = pitch angle of pinion (gear).

All standard bevel gear designs in the United States are in accordance with the **Gleason bevel gear system.** This employs a basic pressure angle of 20° with long and short addendums for ratios other than 1:1 to avoid undercut pinions and to increase strength.

20° Straight Bevel Gears for 90° Shaft Angle Since straight bevel gears are the easiest to produce and offer maximum precision, they are frequently a first choice. Modern straight-bevel-gears generators produce a tooth with localized tooth bearing designated by the Gleason registered tradename Coniflex. These gears, produced with a circular cutter, have a slightly crowned tooth form (see Fig. 6.3.12*b*). Because of the superiority of Coniflex bevel gears over the earlier reciprocating cutter produced straight bevels and because of their faster production, they are the standards for all bevel gears. The design parameters of Fig. 6.3.13 are calculated by the formulas of Table 6.3.8. Backlash data are given in Table 6.3.9.

Table 6.3.8 Straight Bevel Gear Dimensions*
(All linear dimensions in inches)

1. Number of pinion teeth†	n	5. Working depth	$h_k = \frac{2.000}{P_d}$
2. Number of gear teeth†	N	6. Whole depth	$h_t = \frac{2.188}{P_d} + 0.002$
3. Diametral pitch	P_d	7. Pressure angle	ϕ
4. Face with	F	8. Shaft angle	Σ

	Pinion	Gear
9. Pitch diameter	$d = \frac{n}{P_d}$	$D = \frac{N}{P_d}$
10. Pitch angle	$\gamma = \tan^{-1}\frac{n}{N}$	$\Gamma = 90° - \gamma$
11. Outer cone distance	$A_O = \frac{D}{2\ \sin\ \Gamma}$	
12. Circular pitch	$p = \frac{3.1416}{P_d}$	
13. Addendum	$a_{OP} = h_k - a_{OG}$	$a_{OG} = \frac{0.540}{P_d} + \frac{0.460}{P_d(N/n)^2}$
14. Dedendum‡	$b_{OP} = \frac{2.188}{P_d} - a_{OP}$	$b_{OG} = \frac{2.188}{P_d} - a_{OG}$
15. Clearance	$c = h_t - h_k$	
16. Dedendum angle	$\delta_P = \tan^{-1}\frac{b_{OP}}{A_O}$	$\delta_G = \tan^{-1}\frac{b_{OG}}{A_O}$
17. Face angle of blank	$\gamma_O = \gamma + \delta_G$	$\Gamma_O = \Gamma + \delta_P$
18. Root angle	$\gamma_R = \gamma - \delta_P$	$\Gamma_R = \Gamma - \delta_G$
19. Outside diameter	$d_O = d + 2a_{OP}\ \cos\ \gamma$	$D_O = D + 2a_{OG}\ \cos\ \Gamma$
20. Pitch apex to crown	$x_O = \frac{D}{2} - a_{OP}\sin\gamma$	$X_O = \frac{d}{2} - a_{OG}\ \sin\ \Gamma$
21. Circular thickness	$t = p - T$	$T = \frac{p}{2} - (a_{OP} - a_{OG})\ \tan\phi - \frac{K}{P_d}$
22. Backlash	B See Table 6.3.9	(see Fig. 6.3.14)
23. Chordal thickness	$t_C = t - \frac{t^2}{6d^2} - \frac{B}{2}$	$T_C = T - \frac{T^2}{6D^2} - \frac{B}{2}$
24. Chordal addendum	$a_{CP} = a_{OP} + \frac{t^2\ \cos\ \gamma}{4d}$	$a_{CG} = a_{OG} + \frac{T^2\ \cos\ \Gamma}{4D}$
25. Tooth angle	$\frac{3{,}438}{A_O}\left(\frac{t}{2} + b_{OP}\ \tan\phi\right)$ minutes	$\frac{3{,}438}{A_O}\left(\frac{T}{2} + b_{OG}\ \tan\phi\right)$ minutes
26. Limit-point width (L.F.)	$W_{LOP} = (T - 2b_{OP}\ \tan\phi) - 0.0015$	$W_{LOG} = (t - 2b_{OG}\ \tan\phi) - 0.0015$
27. Limit-point width (S.E.)	$W_{LiP} = \frac{A_O - F}{A_O}(T - 2b_{OP}\ \tan\phi) - 0.0015$	$W_{LiG} = \frac{A_O - F}{A_O}(t - 2b_{OG}\ \tan\phi) - 0.0015$
28. Tool-point width	$W = W_{LiP}$ − stock allowance	$W = W_{LiG}$ − stock allowance

* Abstracted from "Gleason Straight Bevel Gear Design," Tables 6.3.8 and 6.3.9 and Fig. 6.3.4. Gleason Works, Inc.

† Numbers of teeth; ratios with 16 or more teeth in pinion: 15/17 and higher; 14/20 and higher; 13/31 and higher. These can be cut with 20° pressure angle without undercut.

‡ The actual dedendum will be 0.002 in greater than calculated.

Table 6.3.9 Recommended Normal Backlash for Bevel Gear Meshes*

P_d	Backlash range	P_d	Backlash range
1.00-1.25	0.020-0.030	3.50-4.00	0.007-0.009
1.25-1.50	0.018-0.026	4-5	0.006-0.008
1.50-1.75	0.016-0.022	5-6	0.005-0.007
1.75-2.00	0.014-0.018	6-8	0.004-0.006
2.00-2.50	0.012-0.016	8-10	0.003-0.005
2.50-3.00	0.010-0.013	10-12	0.002-0.004
3.00-3.50	0.008-0.011	Finer than 12	0.001-0.003

* The table gives the recommended normal backlash for gears assembled ready to run. Because of manufacturing tolerances and changes resulting from heat treatment, it is frequently necessary to reduce the theoretical tooth thickness by slightly more than the tabulated backlash in order to obtain the correct backlash in assembly. In case of choice, use the smaller backlash tolerances.

Angular straight bevel gears connect shaft angles other than 90° (larger or smaller), and the formulas of Table 6.3.8 are not entirely applicable, as shown in the following:

Item 8, shaft angle, is the specified non-90° shaft angle.

Item 10, pitch angles. Shaft angle Σ less than 90°, $\tan \gamma = \sin \Sigma/(N/n + \cos \Sigma)$; shaft angle Σ greater than 90°, $\tan \gamma = \sin (180 - \Sigma)/[N/n \cos (180° - \Sigma)]$.

For all shaft angles, $\sin \gamma/\sin \Gamma = n/N$; $\Gamma = \Sigma - \gamma$.

Item 13, addendum, requires calculation of the equivalent 90° bevel gear ratio m_{90}, $m_{90} = [N \cos \gamma/(n \cos \Gamma)]^{1/2}$. The value m_{90} is used as the ratio N/n when applying the formula for addendum. The quantity under the radical is always the absolute value and is therefore always positive.

Item 20, pitch apex to crown, $x_o = A_0 \cos \gamma - a_{op} \sin \gamma$, $X_o = A_0 \cos \Gamma - a_{oG} \sin \Gamma$.

Item 21, circular thickness, except for high ratios, K may be zero.

Spiral Bevel Gears for 90° Shaft Angle The spiral curved teeth produce additional overlapping tooth action which results in smoother gear action, lower noise, and higher load capacity. The spiral angle also influences the efficiency. In new developments this often leads to lower spiral angles for higher efficiency. Design parameters are calculated by formulas of Table 6.3.10.

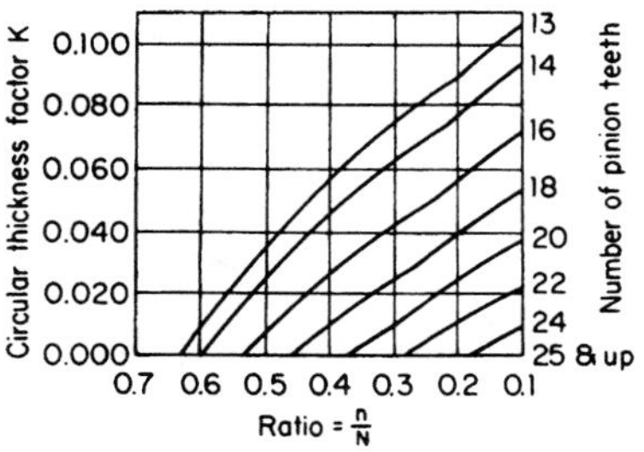

Figure 6.3.14 Circular thickness factor for straight bevel gears.

Table 6.3.10 Spiral Bevel Gear Dimensions
(All linear dimensions in inches)

1. Number of pinion teeth	n	5. Working depth	$h_k = \dfrac{1.700}{P_d}$
2. Number of gear teeth	N	6. Whole depth	$h_t = \dfrac{1.888}{P_d}$
3. Diametral pitch	P_d	7. Pressure angle	ϕ
4. Face width	F	8. Shaft angle	Σ

	Pinion	Gear
9. Pitch diameter	$d = \dfrac{n}{P_d}$	$D = \dfrac{N}{P_d}$
10. Pitch angle	$\gamma = \tan^{-1} \dfrac{n}{N}$	$1 = 90° - \gamma$
11. Outer cone distance	$A_O = \dfrac{D}{2 \sin \Gamma}$	
12. Circular pitch	$p = \dfrac{3.1416}{P_d}$	
13. Addendum	$A_{OP} = h_k - a_{OG}$	$a_{OG} = \dfrac{0.460}{P_d} + \dfrac{0.390}{P_d(N/n)^2}$
14. Dedendum	$b_{OP} = h_t - a_{OP}$	$b_{OG} = h_t - a_{OG}$
15. Clearance	$c = h_t - h_k$	
16. Dedendum angle	$\delta_p = \tan^{-1} \dfrac{b_{OP}}{A_O}$	$\delta_G = \tan^{-1} \dfrac{b_{OG}}{A_O}$
17. Face angle of blank	$\gamma_O = \gamma + \delta_G$	$\Gamma_O = \Gamma + \delta_P$
18. Root angle	$\gamma_R = \gamma - \delta_P$	$\Gamma_R = \Gamma - \delta_G$
19. Outside diameter	$d_O = d + 2a_{OP} \cos \gamma$	$D_O = D + 2a_{OG} \cos \Gamma$
20. Pitch apex to crown	$x_O = \dfrac{D}{2} - a_{OP} \sin \gamma$	$X_O = \dfrac{d}{2} - a_{OG} \sin \Gamma$
21. Circular thickness	$t = p - T$	$T = \dfrac{p}{2} (a_{OP} - a_{OG}) \dfrac{\tan \phi}{\cos \phi} - \dfrac{K}{P_d}$ (see Fig. 6.3.15)
22. Backlash	See Table 6.3.9	
23. Hand of spiral	Left or right	Right or left
24. Spiral angle	35°	
25. Driving member	Pinion or gear	
26. Direction of rotaion	Clockwise or counterclockwise	

SOURCE: Gleason, "Spiral Bevel Gear System."

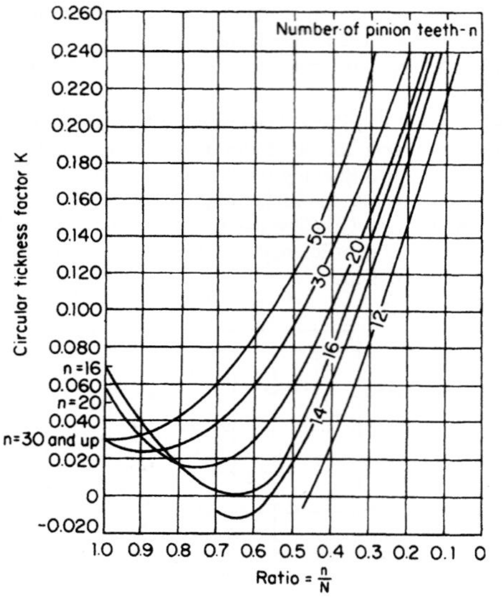

Figure 6.3.15 Circular thickness factors for spiral bevel gears with 20° pressure angle and 35° spiral angle. Left-hand pinion driving clockwise or right-hand pin-ion driving counterclockwise.

Angular Spiral Bevel Gears Several items deviate from the formulas of Table 6.3.10 in the same manner as angular straight bevel gears. Therefore, the same formulas apply for the deviating items with only the following exception:

Item 21, circular thickness, the value of K in Fig. 6.3.15 must be determined from the equivalent 90° bevel ratio (m_{90}) and the equivalent 90° bevel pinion. The latter is computed as $n_{90} = n \sin \Gamma_{90}/\cos \gamma$, where $\tan \Gamma_{90} = m_{90}$.

The Zerol bevel gear is a special case of a spiral bevel gear with a mean spiral angle of zero degree. Its invention was based on the fact that in former times straight bevel gears could not be ground, but aircraft applications always require a grinding finishing process for quality reasons. Details can be obtained from Gleason Works.

Hypoid gears are special and are often used in to automotive applications. In former times lapping after hardening was the hard finishing operation of choice leaving residual distortions due to the hardening process in the finished gearset. Due to the increased requirements in NVH, ease of manufacture, efficiency, strength and gear quality, grinding becomes increasingly the operation of choice. This process allows predetermined higher order flank form optimizations and influences on the surface structure that are realized within a closed loop correction.

WORMGEARS AND WORMS

Worm gearing is used for obtaining large speed reductions between non-intersecting shafts making an angle of 90° with each other. If a wormgear such as shown in Fig. 6.3.16 engages a straight worm, as shown in Fig. 6.3.17, the combination is known as **single enveloping worm gearing.** If a wormgear of the kind shown in Fig. 6.3.16 engages a worm as shown in Fig. 6.3.18, the combination is known as **double enveloping worm gearing.**

With worm gearing, the **velocity ratio** is the ratio between the number of teeth on the wormgear and the number of threads on the worm. Thus, a 30-tooth wormgear meshing with a single threaded worm will have a velocity ratio of 1:30; that is, the worm must make 30 rv in order to revolve the wormgear once. For a double threaded worm, there will be 15 rv of the worm to one of the wormgear, etc. High-velocity ratios are thus obtained with relatively small wormgears.

Tooth proportions of the worm in the central section (Fig. 6.3.17) follow standard rack designs, such as 14½, 20, and 25°. The mating wormgear is cut conjugate for a unique worm size and center distance. The

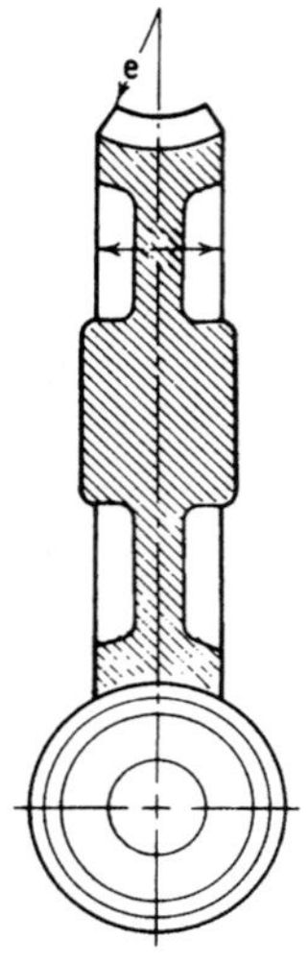

Figure 6.3.16 Single enveloping worm gearing.

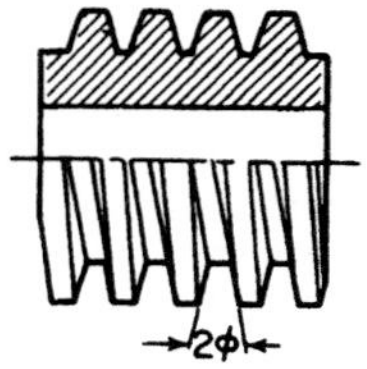

Figure 6.3.17 Straight worm.

geometry and related design equations for a straight-sided cylindrical worm are best seen from a development of the pitch plane (Fig. 6.3.19).

$$D_w = \text{ptich diameter of down} = \frac{n_w p_n}{\pi \sin \lambda}$$

$$p_n = p \cos \lambda = \frac{\pi D_w}{n_w} \sin \lambda$$

$$L = \text{load or worm} = n_w p$$

$$D_g = \text{pitch diameter of wormgear}$$

$$= \frac{N_g}{P_d} = \frac{pN_g}{\pi} = \frac{P_n N_g}{\pi \cos \lambda}$$

$$C = \text{center distance}$$

$$= \frac{D_w + D_g}{2} = \frac{p_n}{2\pi}\left(\frac{N_g}{\cos \lambda} + \frac{n_w}{\sin \lambda}\right)$$

where n_w = number of threads in worm; N_g = number of teeth in wormgear; Z = velocity ratio = N_g/n_w.

The pitch diameter of the wormgear is established by the number of teeth, which in turn comes from the desired gear ratio. The pitch

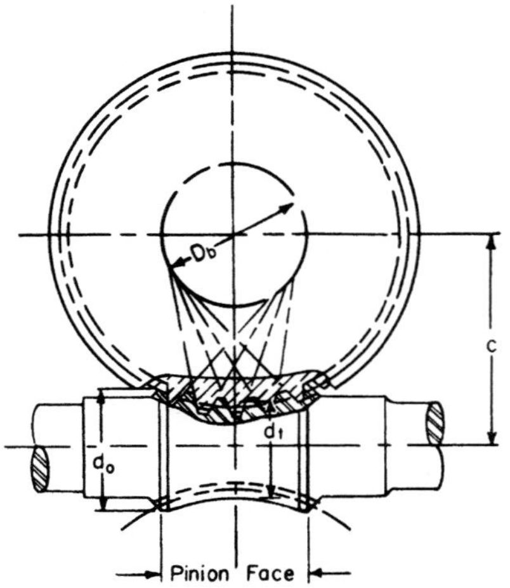

Figure 6.3.18 Double enveloping worm gearing.

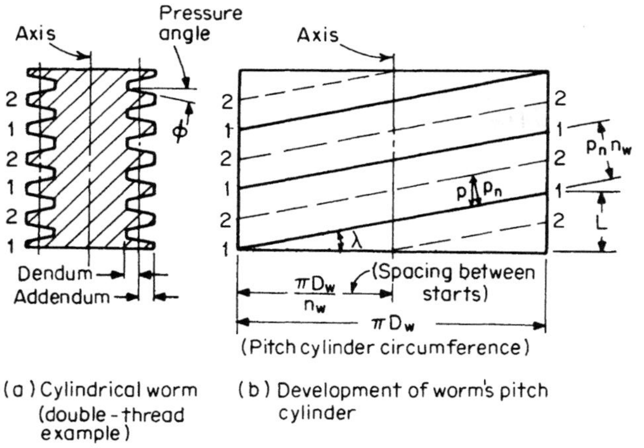

Figure 6.3.19 Cylindrical worm geometry and design parameters.

diameter of the worm is somewhat arbitrary. The lead must match the wormgear's circular pitch, which can be satisfied by an infinite number of worm diameters; but for a fixed lead value, each worm diameter has a unique lead angle. AGMA offers a design formula that provides near optimized geometry:

$$D_w = \frac{C^{0.875}}{2.2}$$

where C = center distance. Wormgear face width is also somewhat arbitrary. Generally it will be $\frac{3}{5}$ to $\frac{2}{3}$ of the worm's outside diameter.

Worm mesh nonreversibility, a unique feature of some designs, occurs because of the large amount of sliding in this type of gearing. For a given coefficient of friction there is a critical value of lead angle below which the mesh is nonreversible. This is generally 10° and lower but is related to the materials and lubricant. Most single thread worm meshes are in this category. This locking feature can be a disadvantage or in some designs can be put to advantage.

Double enveloping worm gearing is special in both design and fabrication. Application is primarily where a high load capacity in small space is desired. Currently, there is only one source of manufacture in the United States: Cone Drive Division of Ex-Cello Corp. For design details and load ratings consult publications of Cone Drive and AGMA Standards.

Other Gear Types

Gears for special purposes include the following (details are to be found in the references):

SRH gears (Gleason Works) are super reduction hypoids based on hypoid gear sets replacing worm gears with ratios of up to 1:100 while delivering all features of modern bevel gear technology including improvement in NVH and efficiency. These sets are usually ground for hard finishing.

Spiroid (Illinois Tool Works) gears, used to connect skew shafts, resemble a hypoid-type bevel gear but in performance are more like worm meshes. They offer very high ratios and a large contact ratio resulting in high strength. The **Helicon** (Illinois Tool Works) gear is a variation in which the pinion is not tapered, and ratios under 10:1 are feasible.

Beveloid (Vinco Corp.) gears are tapered involute gears which can couple intersecting shafts, skew shafts, and parallel shafts.

Face gears have teeth cut on the rotating face plane of the gear and mate with standard involute spur gears. They can connect intersecting or nonparallel, nonintersecting shafts.

Noncircular gears or **function gears** are used for special motions or as elements of analog computers. They can be made with elliptical, log-arithmic, spiral, and other functions. See Cunningham references; also, Cunningham Industries, Inc., Stamford, CT.

DESIGN STANDARDS

In addition to the ANSI and AGMA standards on basic tooth proportions, the AGMA sponsors a large number of national standards dealing with gear design, specification, and inspection. (Consult AGMA, 1500 King St., Arlington, VA 22314, for details.) Helpful general references are AGMA, "Gear Handbook," 390.03 (replaced by AGMA 2009-B01 and 2011-B14) and ANSI/AGMA, "Gear Classification and Inspection Handbook," 2000-A88 (replaced by ANSI/AGMA 2015-1, 2015-2), which establish a system of quality classes for all gear sizes and pitches, ranging from crude coarse commercial gears to the highest orders of fine and coarse ultra-precision gears. In the old but still often used quality system there are 13 quality classes, numbered from 3 through 15 in ascending quality. Today's system shows the highest quality at the lowest number correlating to other international systems. Tolerances are given for key functional parameters: runout, pitch, profile, lead, total composite error, tooth-to-tooth composite error, and tooth thickness. Also, tooth thickness tolerances and recommended mesh backlash are included. These are related to diametral pitch and pitch diameter in recognition of fabrication achievability. Data are available for spur, helical, herringbone, bevel, and worm gearing; and spur and helical racks. Special sections cover gear applications and suggested quality number; gear materials and treatments; and standard procedure for identifying quality, material, and other pertinent parameters. These data are too extensive for inclusion in this handbook, and the reader is referred to the cited AGMA references.

STRENGTH AND DURABILITY

Gear teeth fail in two classical manners: tooth breakage and surface fatigue pitting. Instrument gears and other small, lightly loaded gears are designed primarily for tooth-bending beam strength since minimizing size is the priority. Power gears, usually larger, are designed for both strengths, with surface durability often more critical. Expressions for calculating the beam and surface stresses started with the Lewis-Buckingham formulas and now extend to the latest AGMA formulas.

The Lewis formula for analysis of beam strength, now relegated to historical reference, serves to illustrate the fundamentals that current formulas utilize. In the Lewis formula, a tooth layout shows the load assumed to be at the tip (Fig. 6.3.20). From this Lewis demonstrated that the beam strength $W_b = FSY/P_d$, where F = face width; S = allowable stress; Y = Lewis form factor; P_d = diametral pitch. The form factor Y is derived from the layout as $Y = 2P_d/3$. The value of Y varies with tooth design (form and pressure angle) and number of teeth. In the case of a helical gear tooth, there is a thrust force W_{th} in the axial direction that arises and must be considered as a component of bearing load. See Fig. 6.3.21*b*. Buckingham modified the Lewis formula to include dynamic effects on beam strength and developed equations for evaluating surface stresses. Further modifications were made by other investigators, and have resulted in the most recent AGMA rating formulas which are the basis of most gear designs in the United States.

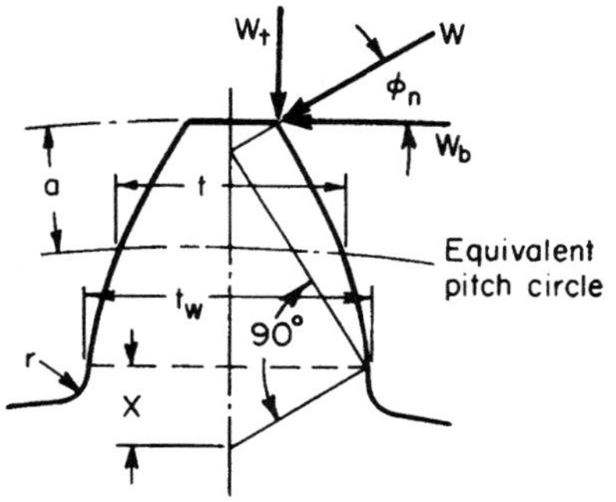

Figure 6.3.20 Layout for beam strength (Lewis formula).

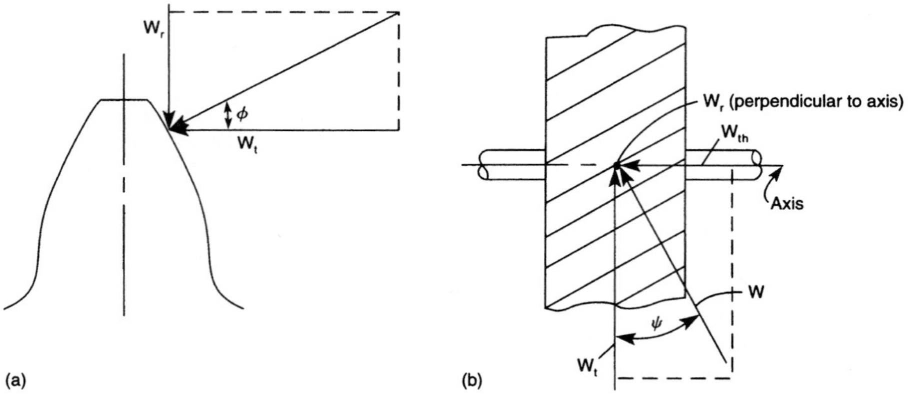

Figure 6.3.21 Forces on spur and helical teeth. (a) Spur gear; (b) helical gear.

AGMA Strength and Durability Rating Formulas

For many decades the AGMA Gear Rating Committee has developed and provided tooth beam strength and surface durability (pitting resistance) formulas suitable for modern gear design. Over the years, the formulas have gone through a continual evolution of revision and improvement. The intent is to provide a common basis for rating various gear types for differing applications and thus have a uniformity of practice within the gear industry. This has been accomplished via a series of standards, many of which have been adopted by ANSI.

The latest standards for rating bending beam strength and pitting resistance are ANSI/AGMA 2001-D04, "Fundamental Rating Factors and Calculation Methods for Involute, Spur, and Helical Gear Teeth" (available in English and metric units) and AGMA 908-B89, "Geometry Factors for Determining the Pitting Resistance and Bending Strength of Spur, Helical, and Herringbone Gear Teeth." These standards have replaced AGMA 218.01 with improved formulas and details.

The rating formulas in Tables 6.3.11 and 6.3.12 are abstracted from ANSI/AGMA 2001-D04, "Fundamental Rating Factors and Calculation Methods for Involute, Spur and Helical Gear Teeth," with permission.

Overload factor K_o is intended to account for an occasional load in excess of the nominal design load W_t. It can be established from experience with the particular application. Otherwise use $K_o = 1$.

Size factor K_s is intended to factor in material nonuniformity due to tooth size, diameter, face width, etc. AGMA has not established factors for general gearing; use $K_s = 1$ unless there is information to warrant using a larger value.

Load distribution factor K_m reflects the nonuniform loading along the lines of contact due to gear errors, installation errors, and deflections. Analytical and empirical methods for evaluating this factor are presented in ANSI/AGMA 2001-D04 but are too extensive to include here. Alternately, if appropriate for the application, K_m can be extrapolated from values given in Table 6.3.14.

Surface condition factor C_f is affected by the manufacturing method (cutting, shaving, grinding, shotpeening, etc.). Standard factors have not been established by AGMA. Use $C_f = 1$ unless experience can establish confidence for a larger value.

Geometry factors I and J relate to the shape of the tooth at the point of contact, the most heavily loaded point. AGMA 908-B89 (Information Sheet, Geometry Factors for Determining the Pitting Resistance and Bending Strength for Spur, Helical and Herringbone Gear Teeth) presents detailed procedures for calculating these factors. The standard also includes a collection of tabular values for a wide range of gear tooth designs, but they are too voluminous to be reproduced here in their entirety. Earlier compact graphs of I and J values in AGMA 218.01 are still valid. They are presented here along with several curves from AGMA 6010-E88, which are based upon 218.01. See Figs. 6.3.23 to 6.3.31.

Table 6.3.11 AGMA Pitting Resistance Formula for Spur and Helical Gears
(See Note 1 below.)

$$s_c = C_p\sqrt{W_t K_o K_v K_s \frac{K_m}{dF}\frac{C_f}{I}}$$

where s_c = contact stress number, lb/in²
C_p = elastic coefficient,* $(\text{lb/in}^2)^{0.5}$ (see text and Table 6.3.13)
W_t = transmitted tangential load, lb
K_o = overload factor (see text)
K_v = dynamic factor (see Fig. 6.3.22)
K_s = size factor (see text)
K_m = load distribution factor (see text and Table 6.3.14)
C_f = surface condition factor for pitting resistance (see text)
F = net face width of narrowest member, in
I = geometry factor for pitting resistance (see text and Figs. 8.2.23 and 6.3.24)
d = operating pitch diameter of pinion, in

$$= \frac{2C}{m_G + 1} \quad \text{for external gears}$$

$$= \frac{2C}{m_G - 1} \quad \text{for internal gears}$$

where C = operating center distance, in
m_G = gear ratio (never less than 1.0)

Allowable contact stress umber s_{ac}

$$s_c \le \frac{s_{ac} Z_N C_H}{S_H K_T K_R}$$

where s_{ac} = allowable contact stress number, lb/in² (see Tables 6.3.15 and 6.3.16; Fig. 6.3.34)
Z_N = stress cycle factor for pitting resistance (see Fig. 6.3.35)
C_H = hardness ratio for pitting resistance (see text and Figs. 6.3.36 and 6.3.37)
S_H = safety factor for pitting (see text)
K_T = temperature factor (see text)
K_R = reliability factor (see Table 6.3.19)

* **Elastic coefficient C_p** can be calculated from the following equation when the paired materials in the pinion-gear set are not listed in Table 6.3.13:

$$C_p = \sqrt{\frac{1}{\pi[(1-\mu_P^2)/E_P + (1-\mu_G^2)/E_G]}}$$

where μ_P (μ_G) = Poisson's ratio for pinion (gear)
E_P (E_G) = modulus of elasticity for pinion (gear), lb/in²

Note 1: If the rating is calculated on the basis of uniform load, the transmitted tangential load is

$$W_t = \frac{33{,}000P}{v_t} = \frac{2T}{d} = \frac{126{,}000P}{n_p d}$$

where P = transmitted power, hp
T = transmitted pinion torque, lb · in
v_t = pitch line velocity at operating pitch diameter, ft/min = $\dfrac{\pi n_p d}{12}$

Table 6.3.12 AGMA Bending Strength Fundamental Formula for Spur and Helical Gears
(See Note 1 in Table 6.3.11.)

$$s_t = W_t K_o K_v K_s \frac{P_d}{F} \frac{K_m K_B}{J}$$

where s_t = bending stress number, lb/in²
K_B = rim thickness factor (see Fig. 6.3.38)
J = geometry factor for bending strength (see text and Figs. 6.3.25 to 6.3.31)
P_d = transverse diametral pitch, in⁻¹*;
P_{dn} for helical gears

$$P_d = \frac{\pi}{p_x \tan \psi_s} = P_{dn} \cos \psi_s \qquad \text{for helical gears}$$

where P_{dn} = normal diametral pitch, in⁻¹
p_x = axial pitch, in
ψ_s = helix angle at standard pitch diameter

$$\psi_s = \arcsin \frac{\pi}{p_x P_{dn}}$$

Allowable bending stress numbers s_{at}

$$s_t \le \frac{s_{at} Y_N}{S_F K_T K_R}$$

where s_{at} = allowable bending stress number, lb/in² (see Tables 6.3.17 and 6.3.18 and Figs. 6.3.39 to 6.3.41)
Y_N = stress cycle factor for bending strength (see Fig. 6.3.42)
S_F = safety factor for bending strength (see text)

Allowable contact stress s_{ac} and **allowable bending stress S_{at}** are obtainable from Tables 6.3.15 to 6.3.18. Contact stress hardness specification applies to the start of active profile at the center of the face width, and for bending stress at the root diameter in the center of the tooth space and face width. The lower stress values are for general design purposes; upper values are for high-quality materials and high-quality control. (See ANSI/AGMA 2001-D04, tables 7 through 10, regarding detailed metallurgical specifications; stress grades 1, 2, and 3; and type A and B hardness patterns.)

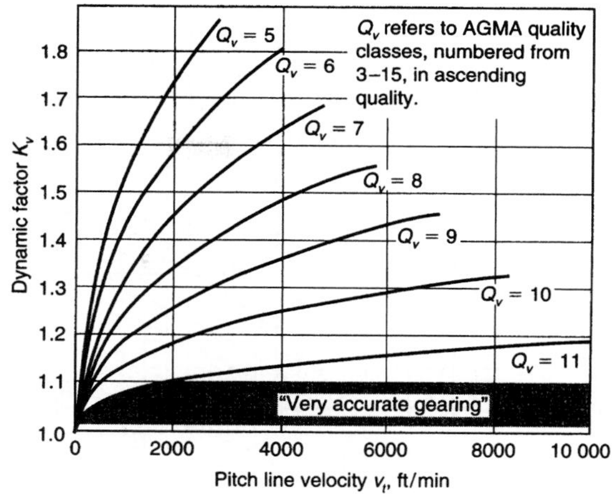

Figure 6.3.22 Dynamic factor K_v. *(Source: ANSI/AGMA 2001-D04, with permission, in accordance with former quality assignment.)*

For **reversing loads,** allowable bending stress values, s_{at} are to be reduced to 70 percent. If the rim thickness cannot adequately support the load, an additional derating factor K_B is to be applied. See Fig. 6.3.38.

Hardness ratio factor C_H applies when the pinion is substantially harder than the gear, and it results in work hardening of the gear and increasing its capacity. Factor C_H applies to only the gear, not the pinion. See Figs. 6.3.36 and 6.3.37.

Safety factors S_H and S_F are defined by AGMA as factors beyond K_O and K_R; they are used in connection with extraordinary risks, human or economic. The values of these factors are left to the designer's judgment as she or he assesses all design inputs and the consequences of possible failure.

Table 6.3.13 Elastic Coefficient C_p

Pinion material	Pinion modulus of elasticity E_P, lb/in² (MPa)	Gear material and modulus of elasticity E_G, lb/in² (MPa): Steel 30 × 10⁶ (2 × 10⁵)	Malleable iron 25 × 10⁶ (1.7 × 10⁵)	Nodular iron 24 × 10⁶ (1.7 × 10⁵)	Cast iron 22 × 10⁶ (1.5 × 10⁵)	Aluminum bronze 17.5 × 10⁶ (1.2 × 10⁵)	Tin bronze 16 × 10⁶ (1.1 × 10⁵)
Steel	30 × 10⁶ (2 × 10⁵)	2,300 (191)	2,180 (181)	2,160 (179)	2,100 (174)	1,950 (162)	1,900 (158)
Malleable iron	25 × 10⁶ (1.7 × 10⁵)	2,180 (181)	2,090 (174)	2,070 (172)	2,020 (168)	1,900 (158)	1,850 (154)
Nodular iron	24 × 10⁶ (1.7 × 10⁵)	2,160 (179)	2,070 (172)	2,050 (170)	2,000 (166)	1,880 (156)	1,830 (152)
Cast iron	22 × 10⁶ (1.5 × 10⁵)	2,100 (174)	2,020 (168)	2,000 (166)	1,960 (163)	1,850 (154)	1,800 (149)
Aluminum bronze	17.5 × 10⁶ (1.2 × 10⁵)	1,950 (162)	1,900 (158)	1,880 (156)	1,850 (154)	1,750 (145)	1,700 (141)
Tin bronze	16 × 10⁶ (1.1 × 10⁵)	1,900 (158)	1,850 (154)	1,830 (152)	1,800 (149)	1,700 (141)	1,650 (137)

Poisson's ratio = 0.30.
SOURCE: ANSI/AGMA 2005-B88, with permission.

Table 6.3.14 Load-Distribution Factor K_m for Spur Gears*

Characteristics of support	Face width, in: 0-2	6	9	16 up
Accurate mountings, small bearing clearances, minimum deflection, precision gears	1.3	1.4	1.5	1.8
Less rigid mountings, less accurate gears, contact across full face	1.6	1.7	1.8	2.2
Accuracy and mounting such that less than full face contact exists	Over 2.2			

* An approximate guide only. See ANSI/AGMA 2001-C95 for derivation of more exact values.
SOURCE: Darle W. Dudley, "Gear Handbook," McGraw-Hill, New York, 1962.

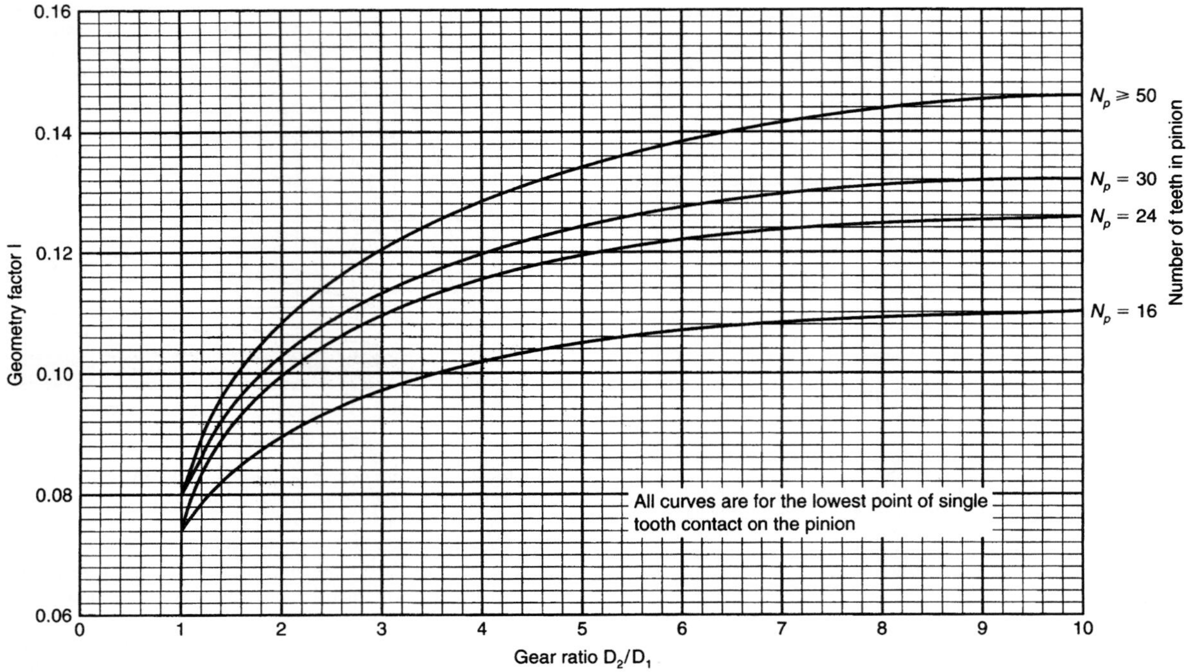

Figure 6.3.23 Geometry factor *I* for 20° full-depth standard spur gears. (*Source: ANSI/AGMA 218.01, with permission.*)

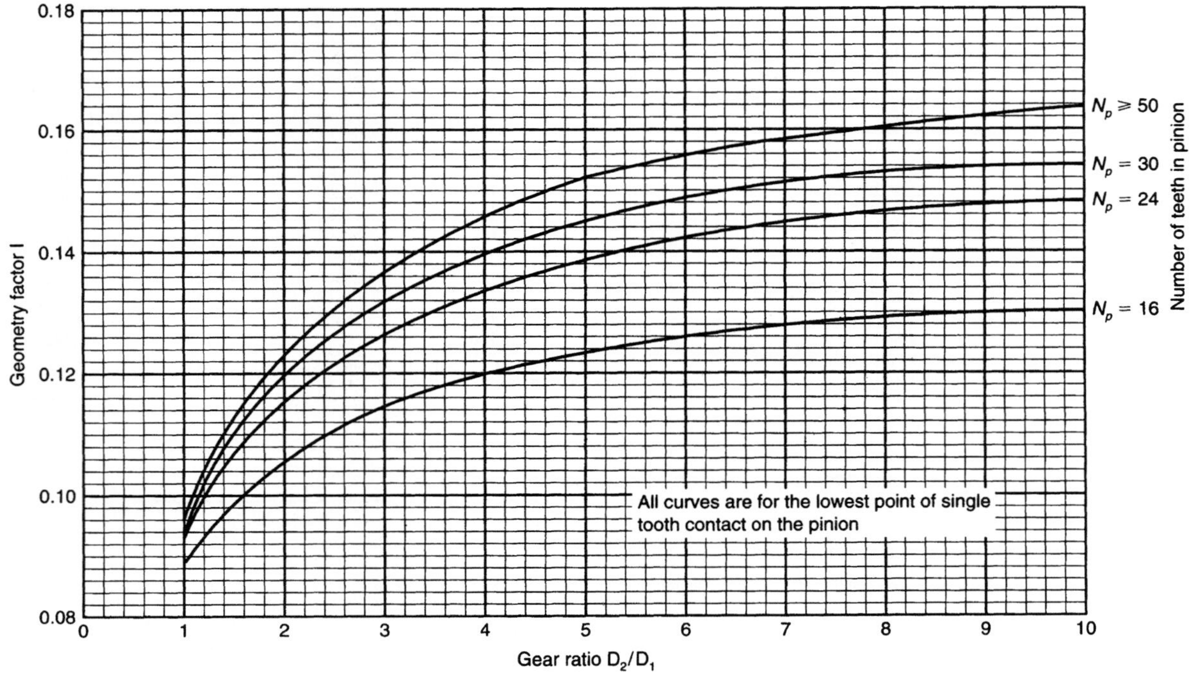

Figure 6.3.24 Geometry factor *I* for 25° full-depth standard spur gears. (*Source: ANSI/AGMA 218.01, with permission.*)

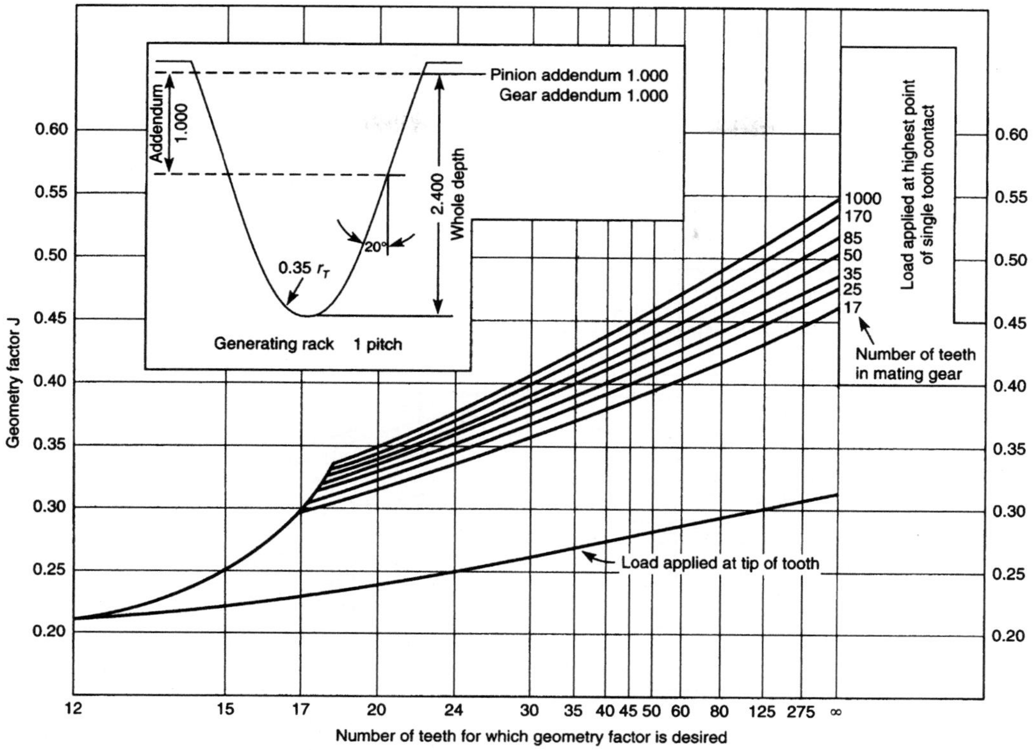

Figure 6.3.25 Geometry factor J for 20° standard addendum spur gears. (*Source: ANSI/AGMA 218.01, with permission.*)

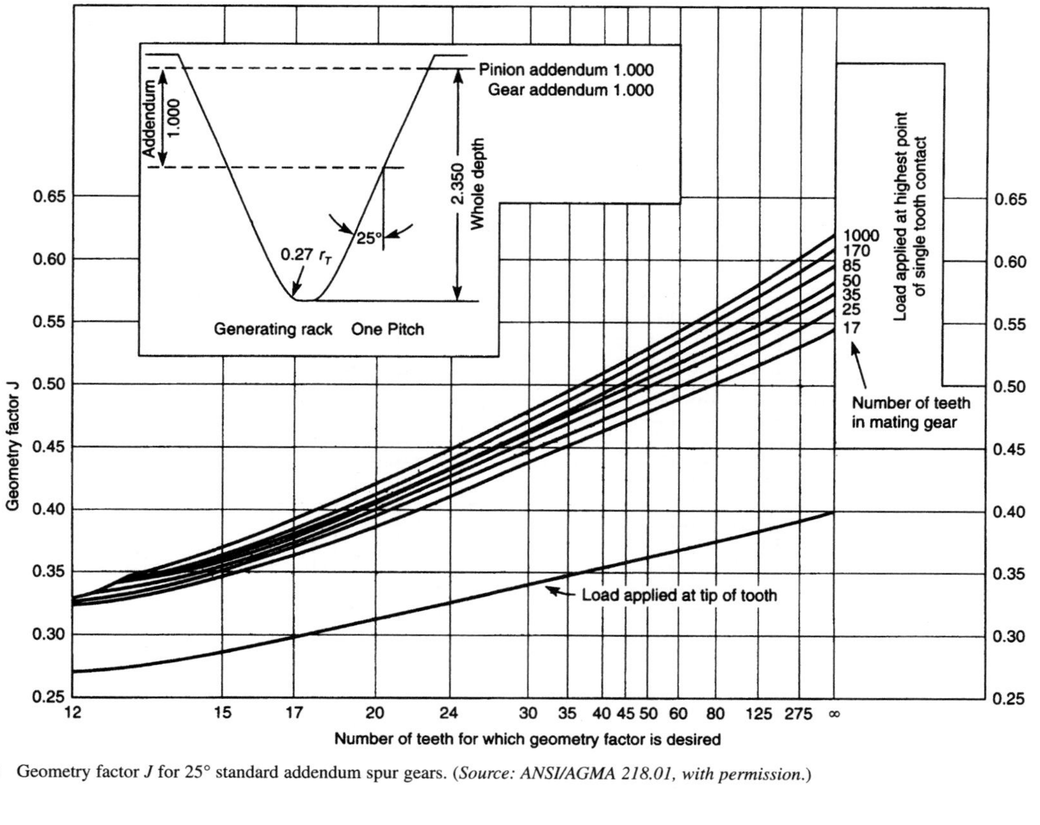

Figure 6.3.26 Geometry factor J for 25° standard addendum spur gears. (*Source: ANSI/AGMA 218.01, with permission.*)

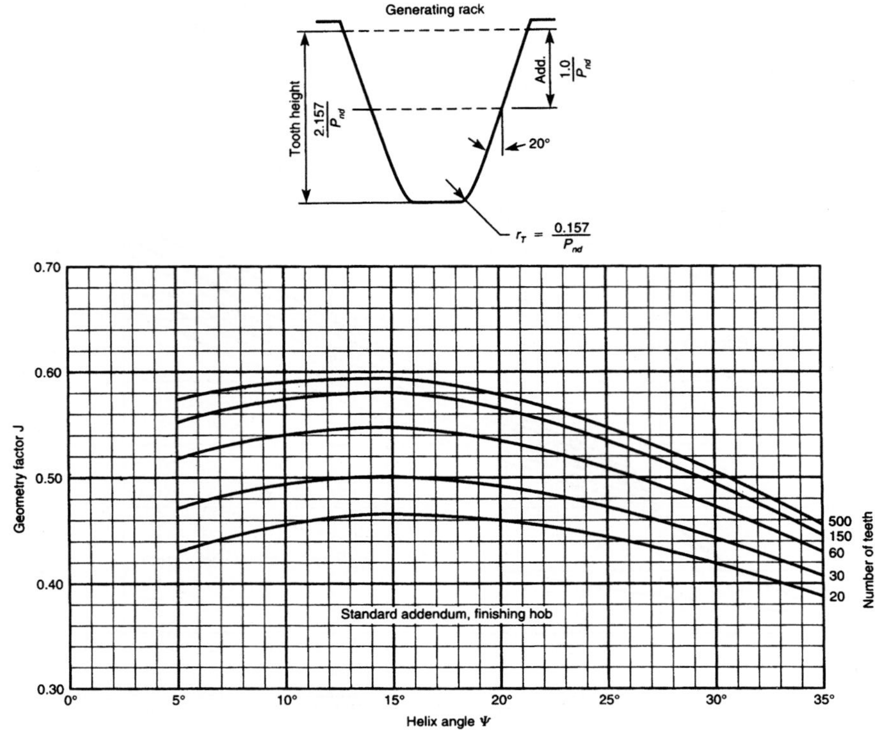

Figure 6.3.27 Geometry factor J for 20° normal pressure angle helical gears. (Standard addendum, finishing hob.) (*Source: ANSI/AGMA 218.01, with permission.*)

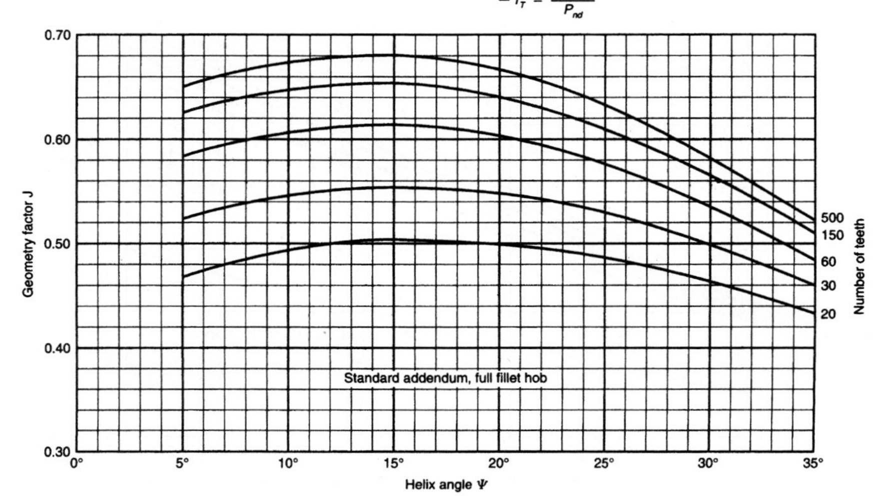

Figure 6.3.28 Geometry factor J for 20° normal pressure angle helical gears. (Standard addendum, full fillet hob.) (*Source: ANSI/AGMA 218.01, with permission.*)

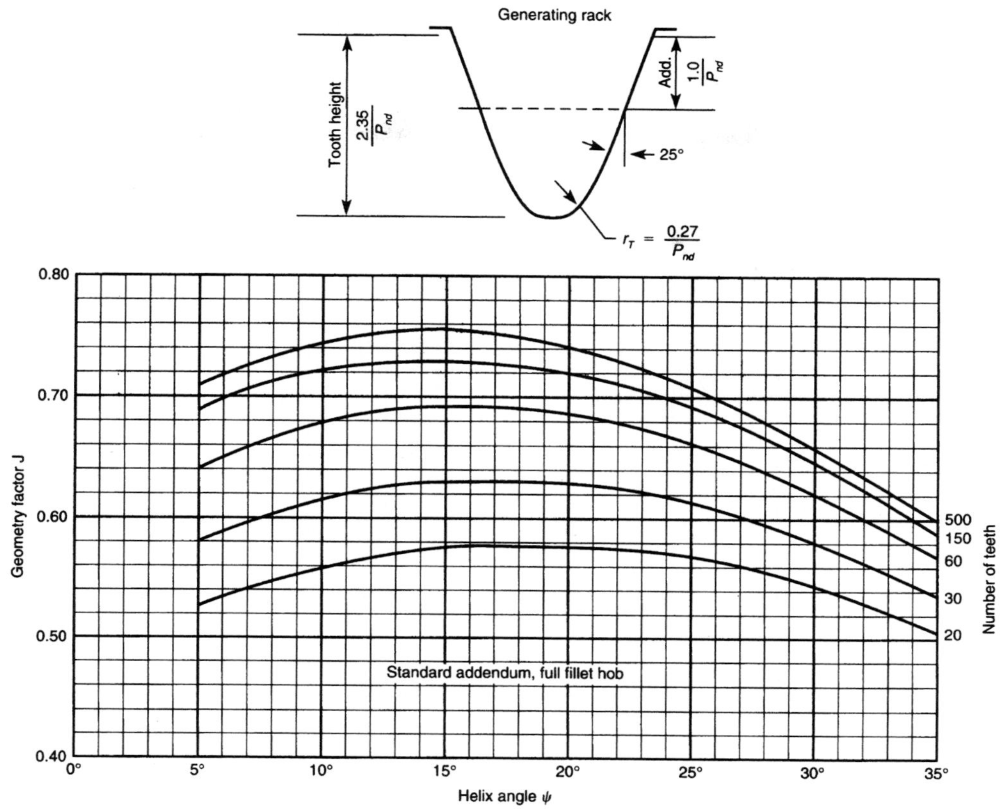

Figure 6.3.29 Geometry factor J for 25° normal pressure angle helical gears. (Standard addendum, full fillet hob.) (*Source: ANSI/AGMA 218.01, with permission.*)

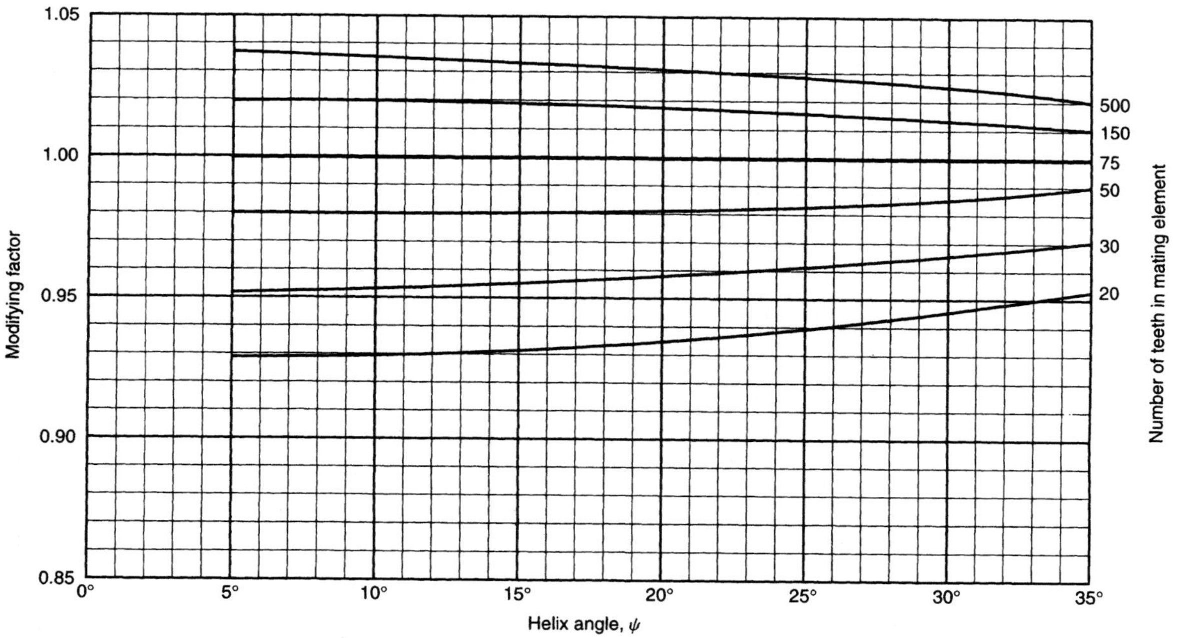

Figure 6.3.30 Factor J multipliers for 20° normal pressure angle helical gears. The modifying factor can be applied to the J factor when other than 75 teeth are used in the mating element. (*Source: ANSI/AGMA 218.01, with permission.*)

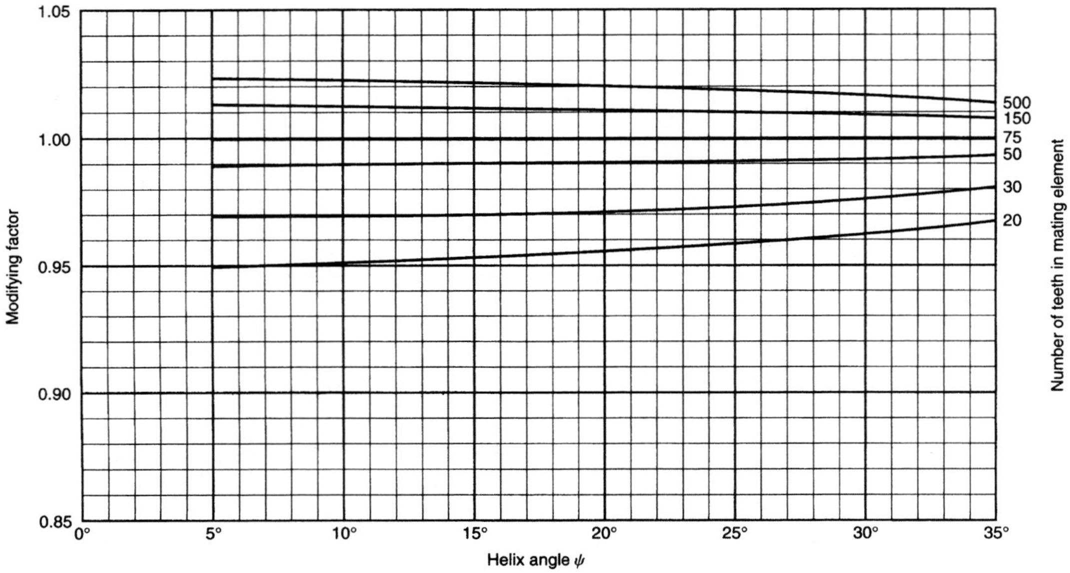

Figure 6.3.31 Factor J multipliers for 25° normal pressure angle helical gears. The modifying factor can be applied to the J factor when other than 75 teeth are used in the mating element. (*Source: ANSI/AGMA 6010-E88, with permission.*)

Table 6.3.15 Allowable Contact Stress Number s_{ac} for Steel Gears

Material designation	Heart treatment	Minimum surface hardness*	Allowable contact stress number s_{ac}, lb/in^2 Grade 1	Grade 2	Grade 3
Steel	Through-hardened*	Fig. 6.3.34	Fig. 6.3.34	Fig. 6.3.34	—
	Flame-* or induction-hardened*	50 HRC	170,000	190,000	—
		54 HRC	175,000	195,000	—
	Carburized and hardened*	Table 9 Note 1	180,000	225,000	275,000
	Nitrided* (through-hardened steels)	83.5 HR15N	150,000	163,000	175,000
		84.5 HR15N	155,000	168,000	180,000
2.5% Chrome (no aluminum)	Nitrided*	87.5 HR15N	155,000	172,000	189,000
Nitralloy 135M	Nitrided*	90.0 HR15N	170,000	183,000	195,000
Nitralloy N	Nitrided*	90.0 HR15N	172,000	188,000	205,000
2.5% Chrome (no aluminum)	Nitrided*	90.0 HR15N	176,000	196,000	216,000

Note 1: Table 9 and Tables 7, 8, and 10 cited in Tables 6.3.16 to 6.3.18 are in ANSI/AGMA 2001-D04.
* The allowable-stress numbers indicated may be used with the case depths shown in Figs. 6.3.32 and 6.3.33.
SOURCE: Abstracted from ANSI/AGMA 2001-D04, with permission.

Table 6.3.16 Allowable Contact Stress Number s_{ac} for Iron and Bronze Gears

Material	Material designation*	Heat treatment	Typical minimum surface hardness	Allowable contact stress number s_{ac}, lb/in^2
ASTM A48 gray cast iron	Class 20	As cast	—	50,000-60,000
	Class 30	As cast	174 HB	65,000-75,000
	Class 40	As cast	201 HB	75,000-85,000
ASTM A536 ductile (nodular) iron	Grade 60-40-18	Annealed	140 HB	77,000-92,000
	Grade 80-55-06	Quenched & tempered	179 HB	77,000-92,000
	Grade 100-70-03	Quenched & tempered	229 HB	92,000-112,000
	Grade 120-90-02	Quenched & tempered	269 HB	103,000-126,000
Bronze	—	Sand-cast	Minimum tensile strength 40,000 lb/in^2	30,000
	ASTM B-148 Alloy 954	Heat-treated	Minimum tensile strength 90,000 lb/in^2	65,000

* See ANSI/AGMA 2004-C08, "Gear Materials and Heat Treatment Manual."
SOURCE: Abstracted from ANSI/AGMA 2001-D04, with permission.

Table 6.3.17 Allowable Bending Stress Number s_{at} for Steel Gears

Material designation	Heat treatment	Minimum surface hardness	Allowable bending stress number s_{at}, lb/in²		
			Grade 1	Grade 2	Grade 3
Steel	Through-hardened	Fig. 6.3.39	Fig. 6.3.39	Fig. 6.3.39	—
	Flame* or induction-hardened* with type A pattern	Table 8§	45,000	55,000	—
	Flame* or induction hardened* with type B pattern	Table 8§ Note 1	22,000	22,000	—
	Carburized and hardened*	Table 9§ Note 2	55,000	65,000 or 70,000†	75,000
	Nitrided*‡ (through-hardened steels)	83.5 HR15N	Fig. 6.3.40	Fig. 6.3.40	—
Nitralloy 135M, nitralloy N, and 2.5% Chrome (no aluminum)	Nitrided‡	87.5 HR15N	Fig. 6.3.41	Fig. 6.3.41	Fig. 6.3.41

Note 1: See Table 8 in ANSI/AGMA 2001-D04.
Note 2: See Table 9 in ANSI/AGMA 2001-D04.
* The allowable-stress numbers indicated may be used with the case depths shown in Figs. 6.3.32 and 6.3.33.
† If bainite and microcracks are limited to grade 3 levels, 70,000 lb/in² may be used.
‡ The overload capacity of nitrided gears is low. Since the shape of the effective S–N curve is flat, the sensitivity to shock should be investigated before one proceeds with the design.
§ The tabular material is too extensive to record here. Refer to ANSI/AGMA 2001-D04, Tables 7 to 10.
SOURCE: Abstracted from ANSI/AGMA 2001-D04, with permission.

Table 6.3.18 Allowable Bending Stress Number s_{at} for Iron and Bronze Gears

Material	Material designation*	Heat treatment	Typical minimum surface hardness	Allowable bending stress number s_{at}, lb/in²
ASTM A48 gray cast iron	Class 20	As cast	—	5,000
	Class 30	As cast	174 HB	8,500
	Class 40	As cast	201 HB	13,000
ASTM A536 ductile (nodular) iron	Grade 60-40-18	Annealed	140 HB	22,000-33,000
	Grade 80-55-06	Quenched & tempered	179 HB	22,000-33,000
	Grade 100-70-03	Quenched & tempered	229 HB	27,000-40,000
	Grade 120-90-02	Quenched & tempered	269 HB	31,000-44,000
Bronze		Sand-cast	Minimum tensile strength 40,000 lb/in²	5,700
	ASTM B-148 Alloy 954	Heat-treated	Minimum tensile strength 90,000 lb/in²	23,600

* See ANSI/AGMA 2004-B89, "Gear Materials and Heat Treatment Manual."
SOURCE: Abstracted from ANSI/AGMA 2001-C95, with permission.

Table 6.3.19 Reliability Factors K_R

Requirements of application	K_R*
Fewer than one failure in 10,000	1.50
Fewer than one failure in 1,000	1.25
Fewer than one failure in 100	1.00
Fewer than one failure in 10	0.85†
Fewer than one failure in 2	0.70†,‡

* Tooth breakage is sometimes considered a greater hazard than pitting. In such cases a greater value of K_R is selected for bending.
† At this value, plastic flow might occur rather than pitting.
‡ From test data extrapolation.
SOURCE: Abstracted from ANSI/AGMA 2001-D04, with permission.

Temperature factor $K_T = 1$ when gears operate with oil temperature not exceeding 250 = F.

Reliability factor K_R accounts for statistical distribution of material failures. Typically, material strength ratings (Tables 6.3.13 and 6.3.15 to 6.3.18; Figs. 6.3.34 and 6.3.42) are based on probability of one failure in 100 at 10^7 cycles. Table 6.3.19 lists reliability factors that may be used to modify the allowable stresses and the probability of failure.

Strength and Durability of Bevel, Worm, and Other Gear Types

For bevel gears, consult the referenced Gleason publications; for worm gearing refer to AGMA standards; for other special types, refer to Dudley, "Gear Handbook."

GEAR MATERIALS

(See Tables 6.3.15 to 6.3.18.)

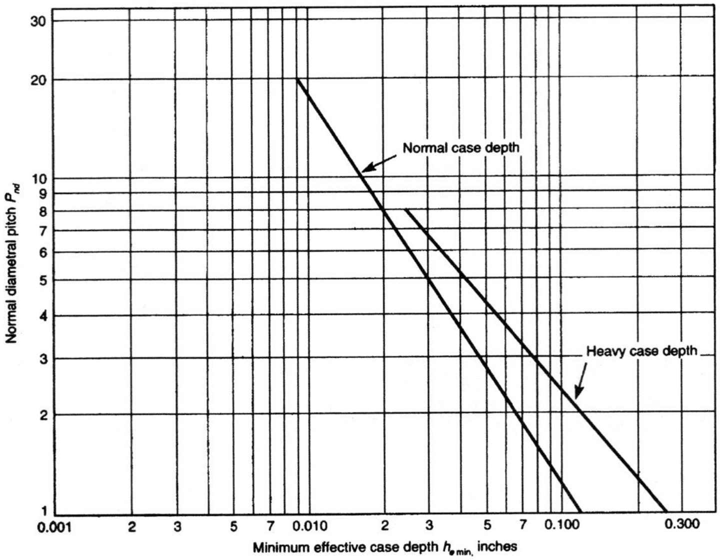

Figure 6.3.32 Minimum effective case depth for carburized gears $h_{e\,min}$. Effective case depth is defined as depth of case with minimum hardness of 50 RC. Total case depth to core carbon is approximately 1.5 times the effective case depth. (*Source: ANSI/AGMA 2001-D04, with permission.*)

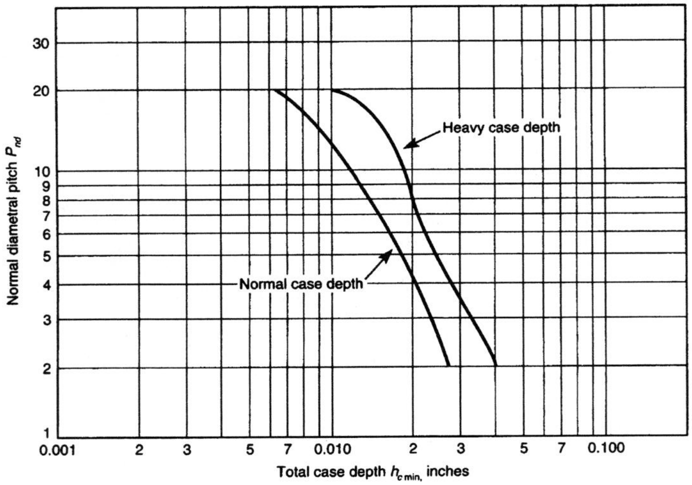

Figure 6.3.33 Minimum total case depth for nitrided gears $h_{c\,min}$. (*Source: ANSI/AGMA 2001-D04, with permission.*)

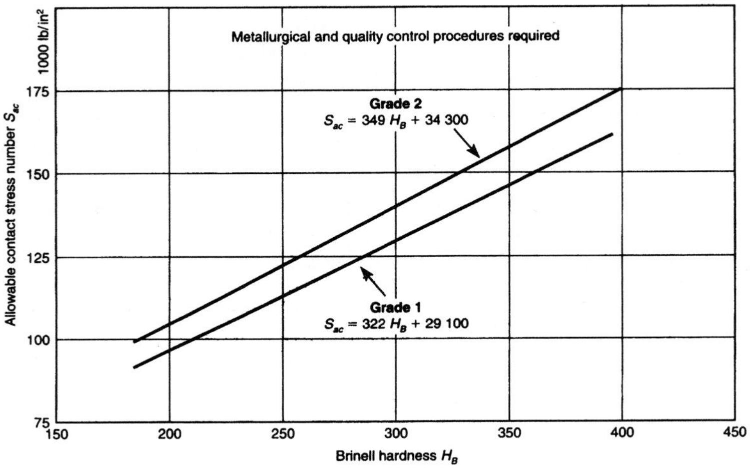

Figure 6.3.34 Allowable contact stress number for through-hardened steel gears s_{ac}. (*Source: ANSI/AGMA 2001-D04, with permission.*)

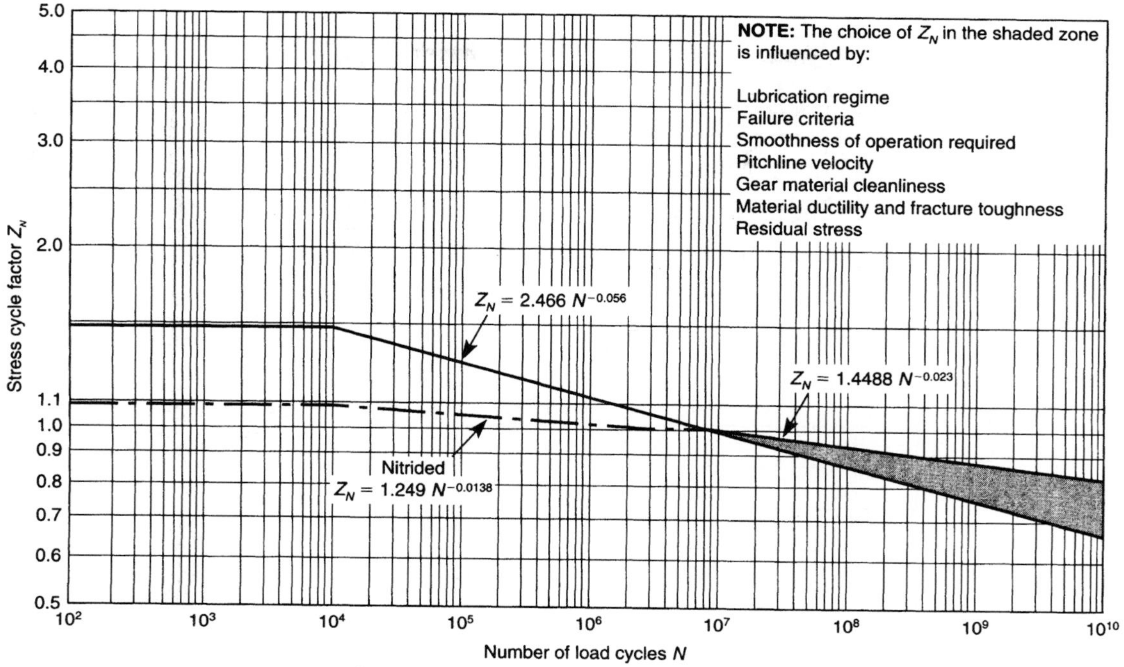

Figure 6.3.35 Pitting resistance stress cycle factor Z_N. (*Source: ANSI/AGMA 2001-D04, with permission.*)

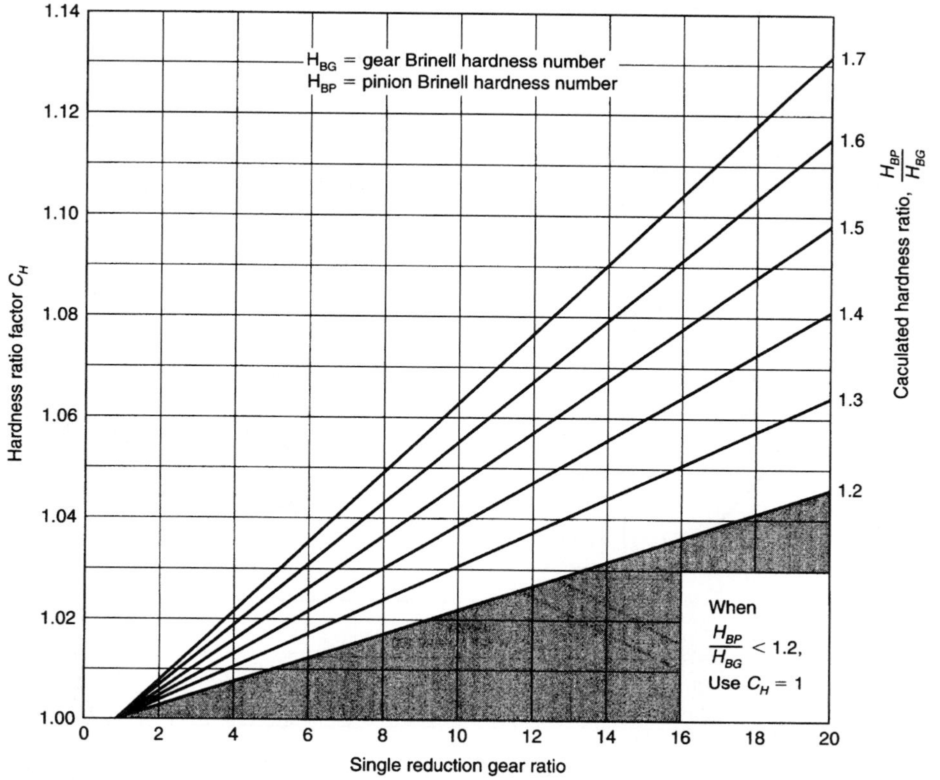

Figure 6.3.36 Hardness ratio factor C_H (through-hardened). (*Source: ANSI/AGMA 2001-D04, with permission.*)

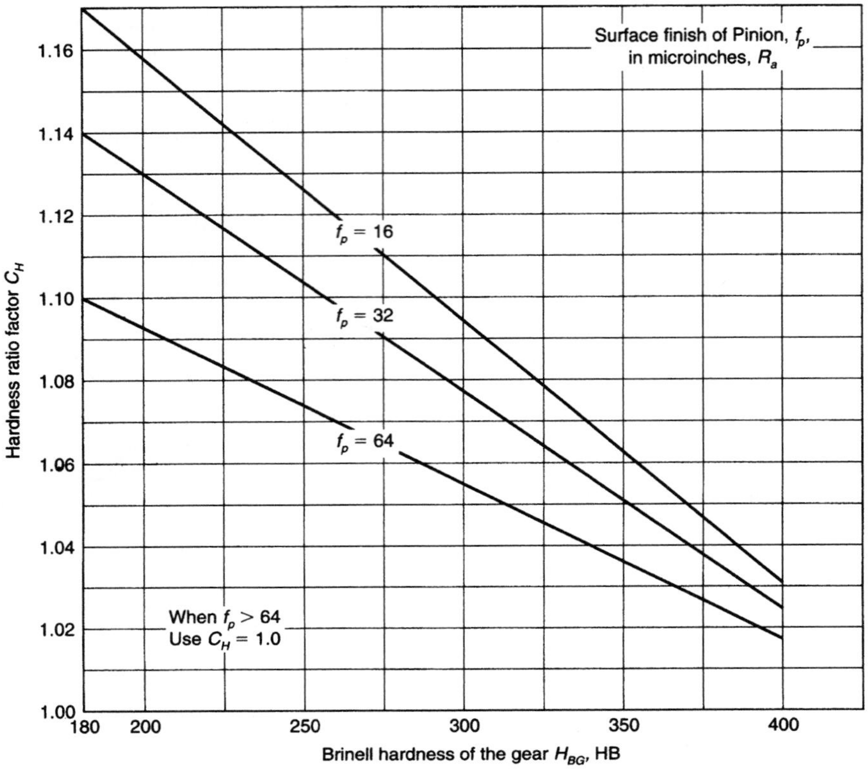

Figure 6.3.37 Hardness ratio factor C_H (surface-hardened pinions). (*Source: ANSI/AGMA 2001-D04, with permission.*)

Metals

Plain carbon steels are most widely used as the most economical; similarly, cast iron is used for large units or intricate body shapes. Heat-treated carbon and alloy steels are used for the more severe load- and wear-resistant applications. Pinions are usually made harder to equalize wear. Strongest and most wear-resistant gears are a combination of heat-treated high-alloy steel cores with case-hardened teeth. (See Dudley, "Gear Handbook," chap. 10.) Bronze is particularly recommended for wormgears and crossed helical gears. Stainless steels are limited to special corrosion-resistant environment applications. Aluminum alloys are used for light-duty instrument gears and airborne lightweight requirements.

Sintered powdered metals technology offers commercial high-quality gearing of high strength at very economical production costs. Die-cast gears for light-duty special applications are suitable for many products.

For power gear applications, heat treatment is an important part of complete and proper design and specification. Heat treatment descriptions and specification tolerances are given in the reference cited below.

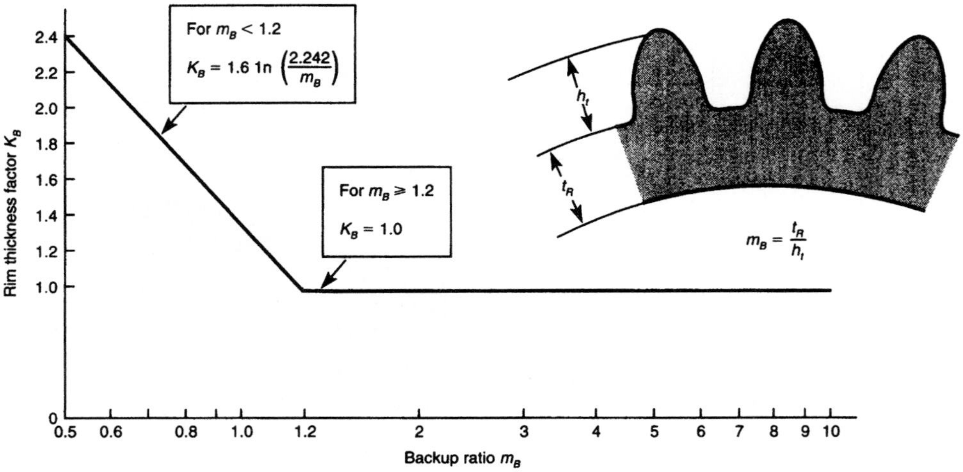

Figure 6.3.38 Rim thickness factor K_B. (*Source: ANSI/AGMA 2001-D04, with permission.*)

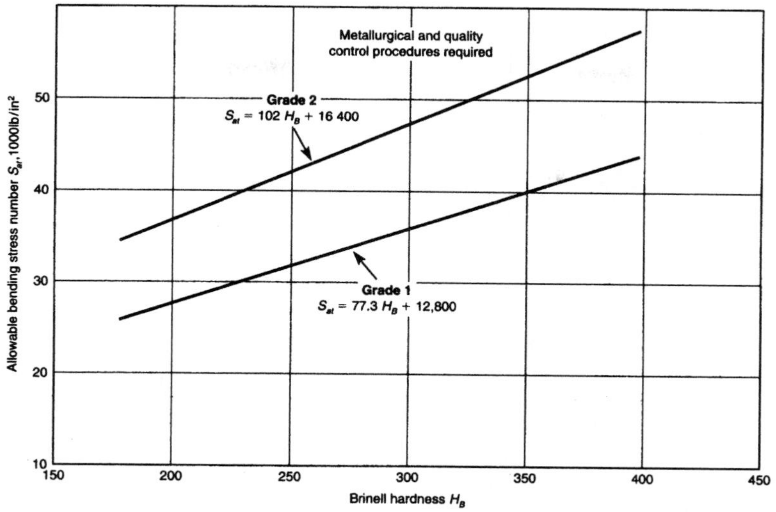

Figure 6.3.39 Allowable bending stress numbers for through-hardened steel gears s_{at}. *(Source: ANSI/AGMA 2001-D04, with permission.)*

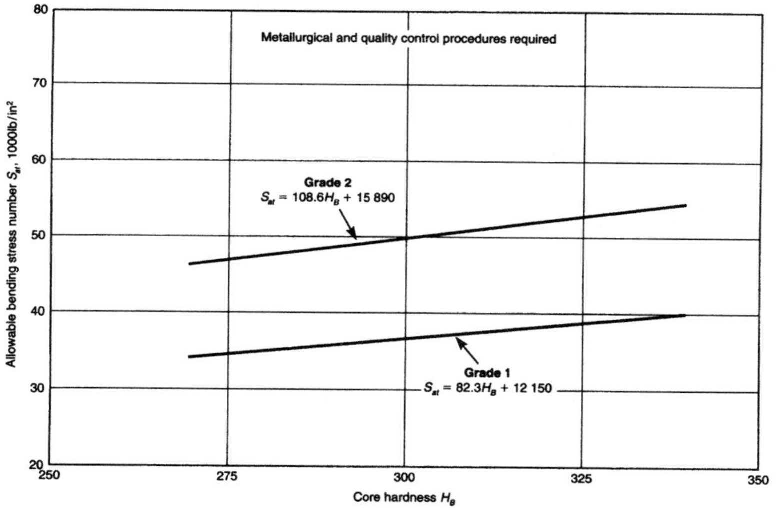

Figure 6.3.40 Allowable bending stress numbers s_{at} for nitrided through-hardened steel gears (i.e., AISI 4140 and 4340). *(Source: ANSI/AGMA 2001-D04, with permission.)*

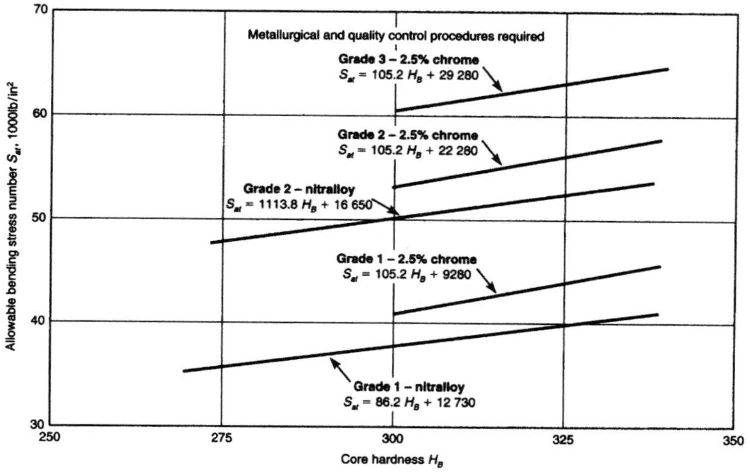

Figure 6.3.41 Allowable bending stress numbers for nitrided steel gears s_{at}. *(Source: ANSI/AGMA 2001-D04, with permission.)*

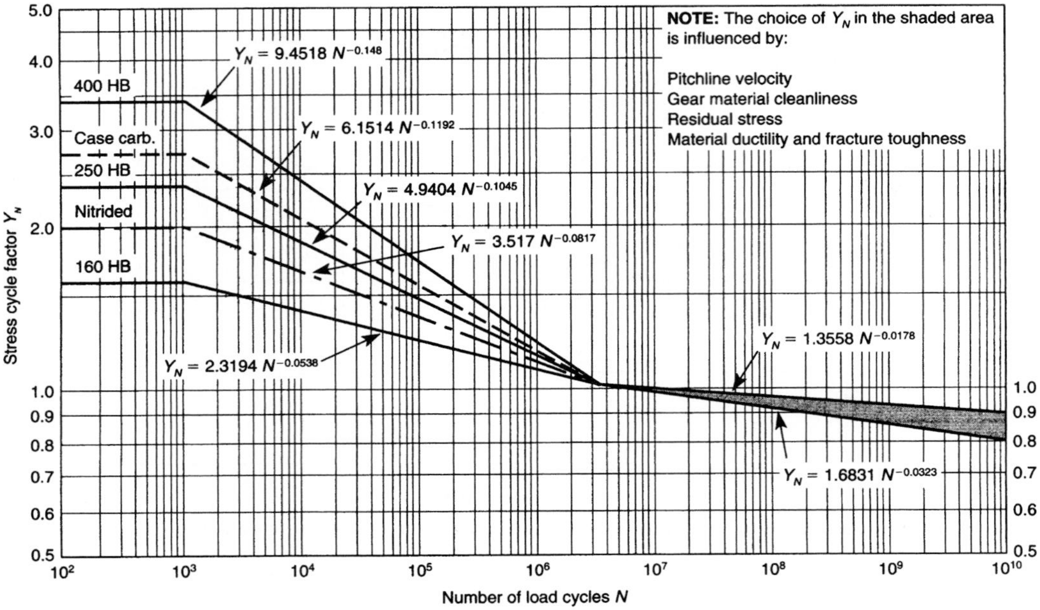

Figure 6.3.42 Bending strength stress cycle factor Y_N. *(Source: ANSI/AGMA 2001-D04, with permission.)*

Precision gears of the small device and instrument types often require protective coatings, particularly for aircraft, marine, space, and military applications. There is a wide choice of chemical and electroplate coatings offering a variety of properties and protection.

For pertinent properties and details of the above special materials and protective coatings see Michalec, "Precision Gearing," chap. 9, Wiley.

Plastics

In recent decades, various forms of nonmetallic gears have displaced metal gears in particular applications. Most plastics can be hobbed or shaped by the same methods used for metallic gears. However, high-strength composite plastics suitable for good-quality gear molding have become available, along with the development of economical high-speed injection molding machines and improved methods for producing accurate gear molds.

The most significant features and advantages of plastic gear materials are:

Cost-effectiveness of injection molding process

Wide choice of characteristics: mechanical strength, density, friction, corrosion resistance, etc.

One-step production; no preliminary or secondary operations Uniformity of parts

Ability to integrate special shapes, etc., into gear body Elimination of machining operations

Capability to mold with metallic insert hubs, if required, for more precise bore diameter or body stability

Capability to mold with internal solid lubricants Ability to operate without lubrication

Quietness of operation

Consistent with trend toward greater use of plastic housings and components

Plastic gears do have limitations relative to metal gears. The most significant are:

Less load-carrying capacity

Cannot be molded to as high accuracy as machined metal gears Much larger coefficient of expansion compared to metals

Less environmentally stable with regard to temperature and water absorption

Can be negatively affected by some chemicals and lubricants

Initial high cost in mold manufacture to achieve proper tooth geometry accuracies

Narrower range of temperature operation, generally less than 250°F and not lower than 0°F

For further information about plastic gear materials and achievable precision, consult the cited reference: Michalec, "Precision Gearing: Theory and Practice."

For a comprehensive presentation of gear molding practices, design, plastic materials, and strength and durability of plastic gears, consult the cited reference: Designatronics, "Handbook of Gears: Inch and Metric," pp. T131–T158.

GEAR LUBRICATION

Proper lubrication is important to prevention of premature wear of tooth surfaces. In the basic action of involute tooth profiles there is a significant sliding component along with rolling action. In worm gearing sliding is the predominant consideration. Thus, a lubricant is essential for all gearing subject to measurable loadings, and even for lightly or negligibly loaded instrument gearing it is needed to reduce friction. Excellent oils and greases are available for high unit load, high speed gearing. See Secs. 6.11 and 8.4; also consult lubricant suppliers for recommendations and latest available high-quality lubricants with special-purpose additives.

General and specific information for lubrication of gearing is found in ANSI/AGMA 9005-E02, "Industrial Gear Lubrication," which covers open and enclosed gearing of all types. AGMA lists a family of lubricants in accordance with viscosities, numbered 1 through 13, with a cross-reference to equivalent ISO grades. See Table 6.3.20. For AGMA's lubrication recommendations for open and closed gearing related to pitch line speed and various types of lubrication systems, refer to Tables 6.3.21 to 6.3.23. For worm gearing and other information, see ANSI/AGMA 9005-E02.

Information in Table 6.3.24 may be used as a quick guide to gear lubricants and their sources for general-purpose instrument and medium-size gearing. Lubricant suppliers should be consulted for specific high-demand applications.

Often gear performance can be enhanced by special additives to the oil. For this purpose, colloidal additives of graphite, molybdenum disulfide

Table 6.3.20 Viscosity Ranges for AGMA Lubricants

Rust- and oxidation-inhibited gear oils, AGMA lubricant no.	Viscosity range[a] mm²/s (cSt) at 40°C	Equivalent ISO grade[a]	Extreme pressure gear lubricants,[b] AGMA lubricant no.	Synthetic gear oils,[c] AGMA lubricant no.
0	28.8-35.2	32		0 S
1	41.4-50.6	46		1 S
2	61.2-74.8	68	2 EP	2 S
3	90-110	100	3 EP	3 S
4	135-165	150	4 EP	4 S
5	198-242	220	5 EP	5 S
6	288-352	320	6 EP	6 S
7, 7 Comp[d]	414-506	460	7 EP	7 S
8, 8 Comp[d]	612-748	680	8 EP	8 S
8A Comp[d]	900-1,100	1,000	8A EP	—
9	1,350-1,650	1,500	9 EP	9 S
10	2,880-3,520	—	10 EP	10 S
11	4,140-5,060	—	11 EP	11 S
12	6,120-7,480	—	12 EP	12 S
13	190-220 cSt at 100°C (212°F)[e]	—	13 EP	13 S
Residual compounds,[f] AGMA lubricant no.	**Viscosity ranges[e] cSt at 100°C (212°F)**			
14R	428.5-857.0			
15R	857.0-1,714.0			

[a] Per ISO 3448, "Industrial Liquid Lubricants—ISO Viscosity Classification," also ASTM D2422 and British Standards Institution B.S. 4231.
[b] Extreme-pressure lubricants should be used only when recommended by the gear manufacturer.
[c] Synthetic gear oils 9S to 13S are available but not yet in wide use.
[d] Oils marked *Comp* are compounded with 3% to 10% fatty or synthetic fatty oils.
[e] Viscosities of AGMA lubricant no. 13 and above are specified at 100°C (210°F) since measurement of viscosities of these heavy lubricants at 40°C (100°F) would not be practical.
[f] Residual compounds—diluent type, commonly known as solvent cutbacks—are heavy oils containing a volatile, nonflammable diluent for ease of application. The diluent evaporates, leaving a thick film of lubricant on the gear teeth. Viscosities listed are for the base compound without diluent.
CAUTION: These lubricants may require special handling and storage procedures. Diluent can be toxic or irritating to the skin. Do not use these lubricants without proper ventilation. Consult lubricant supplier's instructions.
SOURCE: Abstracted from ANSI/AGMA 9005-E02, with permission.

Table 6.3.21 AGMA Lubricant Number Guidelines for Open Gearing (Continuous Method of Application)[a,b]

Ambient temperature[c] °C (°F)	Character of operation	Pressure lubrication: Pitch line velocity: Under 5 m/s (1,000 ft/min)	Pressure lubrication: Pitch line velocity: Over 5 m/s (1,000 ft/min)	Splash lubrication: Pitch line velocity: Under 5 m/s (1,000 ft/min)	Splash lubrication: Pitch line velocity: 5-10 m/s (1,000-3,000 ft/min)	Idler immersion: Pitch line velocity: Up to 1.5 m/s (300 ft/min)
−10 to 15[d] (15-60)	Continuous	5 or 5 EP	4 or 4 EP	5 or 5 EP	4 or 4 EP	8-9 8 EP-9 EP
	Reversing or frequent "start/stop"	5 or 5 EP	4 or 4 EP	7 or 7 EP	6 or 6 EP	8-9 8 EP-9 EP
10 to 50[d] (50-125)	Continuous	7 or 7 EP	6 or 6 EP	7 or 7 EP	8-9[f] 8 EP-9 EP	11 or 11 EP
	Reversing or frequent "start/stop"	7 or 7 EP	6 or 6 EP	9-10[e] 9 EP-10 EP		

[a] AGMA lubricant numbers listed above refer to gear lubricants shown in Table 6.3.20. Physical and performance specifications are shown in Tables 1 and 2 of ANSI/ASMA 9005-D94. Although both R & O and EP oils are listed, the EP is preferred. Synthetic oils in the corresponding viscosity grades may be substituted where deemed acceptable by the gear manufacturer.
[b] Does not apply to worm gearing.
[c] Temperature in vicinity of the operating gears.
[d] When ambient temperatures approach the lower end of the given range, lubrication systems must be equipped with suitable heating units for proper circulation of lubricant and prevention of channeling. Check with lubricant and pump suppliers.
[e] When ambient temperature remains between 30°C (90°F) and 50°C (125°F) at all times, use 10 or 10 EP.
[f] When ambient temperature remains between 30°C (90°F) and 50°C (125°F) at all times, use 9 or 9 EP.
SOURCE: Abstracted from ANSI/AGMA 9005-E02, with permission.

Table 6.3.22 AGMA Lubricant Number Guidelines for Open Gearing Intermittent Applications[a,b,c]
Gear pitch line velocity does not exceed 7.5 m/s (1,500 ft/min)

Ambient temperature,[d] °C (°F)	Intermittent spray systems[e]: R&O or EP lubricant	Intermittent spray systems[e]: Synthetic lubricant	Intermittent spray systems[e]: Residual compound[f]	Gravity feed or forced-drip method[g]: R&O or EP lubricant	Gravity feed or forced-drip method[g]: Synthetic lubricant
−10-15 (15-60)	11 or 11 EP	11 S	14 R	11 or 11 EP	11 S
5-40 (40-100)	12 or 12 EP	12 S	15 R	12 or 12 EP	12 S
20-50 (70-125)	13 or 13 EP	13 S	15 R	13 or 13 EP	13 S

[a] AGMA viscosity number guidelines listed above refer to gear oils shown in Table 6.3.20.
[b] Does not apply to worm gearing.
[c] Feeder must be capable of handling lubricant selected.
[d] Ambient temperature is temperature in vicinity of gears.
[e] Special compounds and certain greases are sometimes used in mechanical spray systems to lubricate open gearing. Consult gear manufacturer and spray system manufacturer before proceeding.
[f] Diluents must be used to facilitate flow through applicators.
[g] EP oils are preferred, but may not be available in some grades.
SOURCE: Abstracted from ANSI/AGMA 9005-E02, with permission.

Table 6.3.23 AGMA Lubricant Number Guidelines for Enclosed Helical, Herringbone, Straight Bevel, Spiral Bevel, and Spur Gear Drives[a]

Pitch line velocity[b,c] of final reduction stage	AGMA lubricant numbers,[a,d,e] ambient temperature °C (°F)[f,g] −40 to −10 (−40 to +14)	−10 to +10 (14 to 50)	10 to 35 (50 to 95)	35 to 55 (95 to 131)
Less than 5 m/s (1,000 ft/min)[h]	3 S	4	6	8
5-15 m/s (1,000-3,000 ft/min)	3 S	3	5	7
15-25 m/s (3,000-5,000 ft/min)	2 S	2	4	6
Above 25 m/s (5,000 ft/min)[h]	0 S	0	2	3

[a] AGMA lubricant numbers listed above refer to R&O and synthetic gear oil shown in Table 6.3.20. Physical and performance specifications are shown in Tables 1 and 3 of ANSI/AGMA 9005-E02. EP or synthetic gear lubricants in the corresponding viscosity grades may be substituted where deemed acceptable by the gear drive manufacturer.
[b] Special considerations may be necessary at speeds above 40 m/s (8,000 ft/min). Consult gear drive manufacturer for specific recommendations.
[c] Pitch line velocity replaces previous standards' center distance as the gear drive parameter for lubricant selection.
[d] Variations in operating conditions such as surface roughness, temperature rise, loading, speed, etc., may necessitate use of a lubricant of one grade higher or lower. Contact gear drive manufacturer for specific recommendations.
[e] Drives incorporating wet clutches or overrunning clutches as backstopping devices should be referred to the gear manufacturer, as certain types of lubricants may adversely affect clutch performance.
[f] For ambient temperatures outside the ranges shown, consult the gear manufacturer.
[g] Pour point of lubricant selected should be at least 5°C (9°F) lower than the expected minimum ambient starting temperature. If the ambient starting temperature approaches lubricant pour point, oil sump heaters may be required to facilitate starting and ensure proper lubrication (see 5.1.6 in ANSI/AGMA 9005-E02).
[h] At the extreme upper and lower pitch line velocity ranges, special consideration should be given to all drive components, including bearing and seals, to ensure their proper performance.
SOURCE: Abstracted from ANSI/AGMA 9005-E02, with permission.

Table 6.3.24 Typical Gear Lubricants

Lubricant type	Military specification	Useful temp range, °F	Commercial source and specification (a partial listing) Source	Identification	Remarks and applications
Oils					
Petroleum	MIL-L-644B	−10 to 250	Exxon Corporation Franklin Oil and Gas Co. Royal Lubricants Co. Texaco	#4035 or Unvis P-48 L-499B Royco 380 1692 Low Temp. Oil	Good general-purpose lubricant for all quality gears having a narrow range of operating temperature
Diester	MIL-L-6085A	−67 to 350	Anderson Oil Co. Eclipse Pioneer Div., Bendix Shell Oil Co. E. F. Houghton and Co.	Windsor Lube I-245X Pioneer P-10 AeroShell Fluid 12 Cosmolubric 270	General-purpose, low-starting torque, and stable over a wide temperature range. Particularly suited for precision instrument gears and small machinery gears
Diester	MIL-L-7808C	−67 to 400	Sinclair Refining Co. Socony Mobil Oil Co. Bray Oil Co. Exxon Corporation	Aircraft Turbo S Oil Avrex S Turbo 251 Brayco 880 Exxon Turbine Oil 15	Suitable for oil spray or mist system at high temperature. Particularly suitable for high-speed power gears
Silicone		−75 to 350	Dow-Corning Corp.	DC200	Rated for low-starting torque and lightly loaded instrument gears
Silicone		−100 to 600	General Electric Co.	Versilube 81644	Best load carrier of silicone oils with widest temperature range. Applicable to power gears requiring wide temperature ranges
Greases					
Diester oil-lithium soap	MIL-G-7421A	−100 to 200	Royal Lubricants Co. Texaco	Royco 21 Low Temp. No. 1888	For moderately loaded gears requiring starting torques at low temperatures
Diester oil-lithium soap	MIL-G-3278A	−67 to 250	Exxon Corporation Shell Oil Co. Sinclair Refining Co. Bray Oil Co.	Beacon 325 AeroShell Grease II Sinclair 3278 Grease Braycote 678	General-purpose light grease for precision instrument gears, and generally lightly loaded gears
Petroleum oil-sodium soap	MIL-L-3545	−20 to 300	Exxon Corporation Standard Oil Co. of Calif.	Andok 260 RPM Aviation Grease #2	A high-temperature lubricant for high speed and high loads
Mineral oil-sodium soap	—	−25 to 250	Exxon Corporation	Andok C	Stiff grease that channels readily. Suitable for high speeds and highly loaded gears

Table 6.3.25 Solid Oil Additives

Lubricant type	Temperature range, °F	Source	Identification	Remarks
Colloidal graphite	Up to 1,000	Acheson Colloids Co.	SLA 1275	Good load capacity, excellent temperature resistance
Colloidal MoS_2	Up to 750	Acheson Colloids Co.	SLA 1286	Good antiwear
Colloidal Teflon	Up to 575	Acheson Colloids Co.	SLA 1612	Low coefficient of friction

(MoS_2), and Teflon are very effective. These additives are particularly helpful to reduce friction and prevent wear; they are also very beneficial to reduce the rate of wear once it has begun, and thus they prolong gear life. The colloidal additive combines chemically with the metal surface material, resulting in a tenacious layer of combined material interposed between the base metals of the meshing teeth. The size of the colloidal additives is on the order of 2 mm, sufficiently fine not to interfere with the proper operation of the lubricant system filters. Table 6.3.25 lists some commercial colloidal additives and their sources of supply.

Plastic gears are often operated without any external lubrication, and they will provide long service life if the plastic chosen is correct for the application. Plastics manufacturers and their publications can be consulted for guidance. Alternatively, many plastic gear materials can be molded with internal solid lubricants, such as MoS_2, Teflon, and graphite.

GEAR INSPECTION AND QUALITY CONTROL

Gear performance is not only related to the design, but also depends upon obtaining the specified quality. Details of gear inspection and control of subtle problems relating to quality are given in Michalec, "Precision Gearing," Chap. 11.

COMPUTER MODELING AND CALCULATIONS

A feature of the latest AGMA rating standards is that the graphs, including those presented here, are accompanied by equations which allow application of computer-aided design. Gear design equations and strength and durability rating equations have been computer modeled by many gear manufacturers, users, and university researchers. Numerous software programs, including integrated CAD/CAM, are available from these places, and from computer system suppliers and specialty software houses. It is not necessary for gear designers, purchasers, and fabricators to create their own computer programs.

With regard to gear tooth strength and durability ratings, many custom gear house designers and fabricators offer their own computer modeling which incorporates modifications of AGMA formulas based upon experiences from a wide range of applications.

The following organizations offer software programs for design and gear ratings according to methods outlined in AGMA publications: Fairfield Manufacturing Company Gear Software; The Gleason Works, Geartech Software, Inc.; PC Gears; Universal Technical Systems, Inc. For details and current listings, refer to AGMA's latest "Catalog of Technical Publications."

6.4 FLUID FILM BEARINGS

by Vittorio (Rino) Castelli

REFERENCES: "General Conference on Lubrication and Lubricants," ASME. Fuller, "Theory and Practice of Lubrication for Engineers," 2d ed., Wiley. Booser, "Handbook of Lubrication, Theory and Design," vol. 2, CRC Press. Barwell, "Bearing Systems, Principles and Practice," Oxford Univ. Press. Cameron, "Principles of Lubrication," Longmans Greene. "Proceedings," Second International Symposium on Gas Lubrication, ASME. Gross, "Fluid-Film Lubrication," Wiley. Gunter, "Dynamic Stability of Rotor-Bearing Systems," NASA SP-113, Government Printing Office. Khonsari and Booser, "Applied Tribology," Wiley, 2001. Loeb and Rippel, Determination of Optimum Properties for Hydrostatic Bearings, *ASLE Tribology Trans.* 1958.

Plain bearings, according to their function, may be:

Journal bearings, cylindrical, carrying a rotating shaft and a radial load.

Thrust bearings, the function of which is to prevent axial motion of a rotating shaft.

Guide bearings, to guide a machine element in its translational motion, usually without rotation of the element.

In exceptional cases of design, or with a complete **failure of lubrication,** a bearing may run dry. The coefficient of friction is then between 0.25 and 0.40, depending on the materials of the rubbing surfaces. With the **bearing barely greasy,** or when the bearing is well lubricated but the speed of rotation is very slow, boundary lubrication takes place. The coefficient of friction may vary from 0.08 to 0.14. This condition occurs also in any bearing when the shaft is starting from rest if the bearing is not equipped with an oil lift.

Semifluid, or **mixed,** lubrication exists between the journal and bearing when the conditions are not such as to form a load-carrying fluid film and thus separate the surfaces. Semifluid lubrication takes place at comparatively low speed, with intermittent or oscillating motion, heavy load, insufficient oil supply to the bearing (wick or waste-lubrication, drop-feed lubrication). Semifluid lubrication may also exist in thrust bearings with fixed parallel-thrust collars, in guide bearings of machine tools, in bearings with copious lubrication where the shaft is bent or the bearing is misaligned, or where the bearing surface is interrupted by improperly arranged oil grooves. The coefficient of friction in such bearings may range from 0.02 to 0.08 (Fuller, Mixed Friction Conditions in Lubrication, *Lubrication Eng.*, 1954).

Fluid or **complete lubrication,** when the rubbing surfaces are completely separated by a fluid film, provides the lowest friction losses and prevents wear. A certain amount of oil must be fed to the oil film in order to compensate for end leakage and maintain its carrying capacity. Such lubrication can be provided under pressure from a pump or gravity tank, by automatic lubricating devices in self-contained bearings (oil rings or oil disks), or by submersion in an oil bath (thrust bearings for vertical shafts).

Notation

R = radius of bearing, length
r = radius of journal, length
$c = mr = R - r$ = radial clearance, length
W = bearing load, force
μ = viscosity = force × time/length^2
Z = viscosity, centipoise (cP); 1 cP = 1.45×10^{-7} lb · s/in^2 (0.001 N · s/m^2)
β = angle between load and entering edge of oil film
η = coefficient for side leakage of oil
ν = kinematic viscosity = μ/ρ, length^2/time
R_e = Reynolds number = umr/ν
P_a = absolute ambient pressure, force/area

$P = W/(ld)$ = unit pressure, lb/in²
N = speed of journal, r/min
m = clearance ratio (diametral clearance/diameter)
F = friction force, force
A = operating characteristic of plain cylindrical bearing
P' = alternate operating characteristic of plain cylindrical bearing
h_0 = minimum film thickness, length
ε = eccentricity ratio, or ratio of eccentricity to radial clearance
e = eccentricity = distance between journal and bearing centers, length
f = coefficient of friction
f' = friction factor = $F/(\pi rlpu^2)$
l = length of bearing, length
$d = 2r$ = diameter of journal, length
K_f = friction factor of plain cylindrical bearing
t_w = temperature of bearing wall
t_0 = temperature of air
t_1 = temperature of oil film
u = surface speed, length/time
ω = angular velocity, rad/time
ρ = mass density, mass/length³
Λ = bearing compressibility parameter = $6\mu\omega r^2/(P_a c^2)$

INCOMPRESSIBLE AND COMPRESSIBLE LUBRICATION

Depending on the fluid employed and the pressure regime, the fluid density may or may not vary appreciably from the ambient value in the load-carrying film. Typically, oils, water, and liquid metals can be considered incompressible, while gases exhibit compressibility effects even at modest loads. The difference comes from the fact that, in incompressible lubricants, fluid flow rates are linearly proportional to pressure differences, whereas for compressible lubricants the mass flow rates are proportional to the difference of some power of the pressure. This is because the pressure affects the fluid density. The bearing behavior is somewhat dissimilar. In incompressible lubrication, gage pressures can be used and the value of the ambient pressure has no effect on the load-carrying capacity, which is linearly related to viscosity and speed. This is not true in compressible lubrication, where the value of ambient pressure has a direct effect on the load-carrying capacity which, in turn, increases with viscosity and speed, but only up to a limit dependent on the bearing geometry. In what follows, incompressible lubrication is treated first and compressible lubrication second.

Incompressible (Plain Cylindrical Journal Bearings)

Fluid lubrication in plain cylindrical bearings depends on the viscosity of the lubricant, the speed of the bearing components, the geometry of the film, and possible external sources of pressurized lubricant. The oil is entrained by the journal into the film by the action of the viscosity which, if the passage is convergent, causes the creation of a pressure field, resulting in a force sufficient to float the journal and carry the load applied to it.

The **minimum film thickness** h_0 determines the closest approach of the journal and bearing surfaces (Fig. 6.4.1). The allowable closest approach depends on the finish of these surfaces and on the rigidity of the journal and bearing structures. In practice, h_0 = 0.00075 in (0.019 mm) is common in electric motors and generators of medium speed, with steel shafts in babbitted bearings; h_0 = 0.003 in (0.076 mm) to 0.005 in (0.127 mm) for large steel shafts running at high speed in babbitted bearings (turbogenerators, fans), with pressure oil-supply for lubrication; h_0 = 0.0001 in (0.0025 mm) to 0.0002 in (0.005 mm) in automotive and aviation engines, with very fine finish of the surfaces.

Figure 6.4.2 gives the relationship between ε and the load-carrying coefficient A for a plain cylindrical journal. The operating characteristic of the bearing is

$$A = (132/\eta)(1{,}000\,m)^2[P/(ZN)]$$

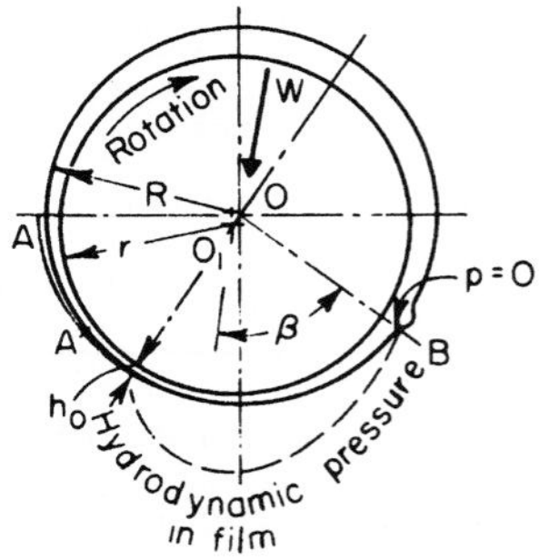

Figure 6.4.1 Journal bearing with perfect lubrication.

In Fig. 6.4.1, β is the angle between the direction of the load W and the entering edge of the load-carrying oil film, in degrees. The entering edge is at the place where the hydrodynamic pressure is equal or nearly equal to the atmospheric pressure and may be at the location of the oil-distributing groove B, or at the end of the machined recess pocket as at AA. For **complete bearings,** that is, when the inner surface of the bearing is not interrupted by grooves, β may be taken as 90°. The reason for this assumption is the fact that, where the film diverges, the bearing pumping action tends to generate negative pressure, which liquids cannot sustain. The film **cavitates,** that is, it breaks up in regions of fluid intermixed with either air or fluid vapor, while the pressure does not deviate substantially from ambient. For a 120° bearing with a central load, β may be taken as 60°.

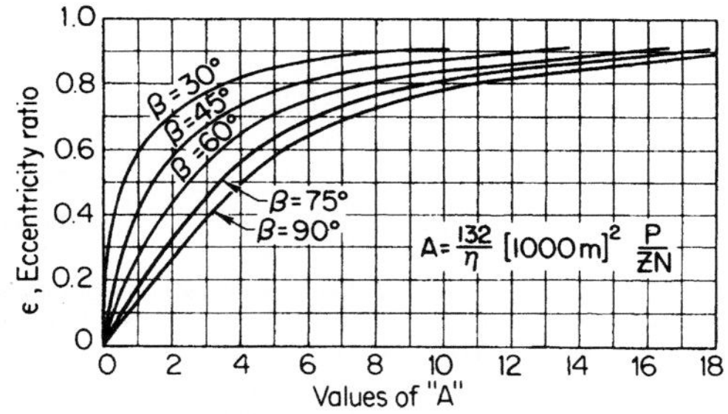

Figure 6.4.2 Eccentricity ratio for a plain cylindrical journal.

The coefficient η corrects for side leakage. There is a loss of load-carrying capacity caused by the drop in the hydrodynamic pressure p in the oil film from the midsection of the bearing toward its ends; $p = 0$ at the ends. The value of η depends on the length-diameter ratio l/d and ε, the eccentricity ratio. Values of η are given in Fig. 6.4.3.

Example 1. A generator bearing, 6 in diam by 9 in long, carries a vertical downward load of 8,650 lb; N = 720 r/min. The diametral clearance of the bearing is 0.012 in; the bearing is split on its horizontal diameter, and the lower half is relieved 40° down on each side, for oil distribution along journal; the bearing arc is therefore 100°; with the load vertical, β = 50°; bearing temperature 160°F. The absolute viscosity of the oil in the film is 12 centipoises (medium turbine oil). $P = W/ld$ = 160 lb/in²; $\mu = 12 \times 1.45 \times 10^{-7} = 17.4 \times 10^{-7}$ lb·s/in². The solution is one of trial and error. By using Fig. 6.4.3 in conjunction with Fig. 6.4.2, only a few trials are necessary to obtain the answer. As a first trial assume ε = 0.85. For an l/d ratio of 1.5 in Fig. 6.4.3, η, the end-leakage factor, will be 0.77. Compute A using this value of η. m = 0.012/6 = 0.002.

$$A = \frac{132}{0.77}(2)^2\frac{160}{12 \times 720} = 12.7$$

Enter Fig. 6.4.2 with this value of a and at β = 50°, and find that ε = 0.9. This value is larger than the initial assumption for ε. As a second trial, ε = 0.88. Then η = 0.8, A = 12.2, and ε = 0.89. This is a sufficiently close check. The minimum film thickness is

$$h_0 = mr(1 - \varepsilon) = 0.002 \times 3 \times 0.12 = 0.0007 \text{ in } (0.01778 \text{ mm})$$

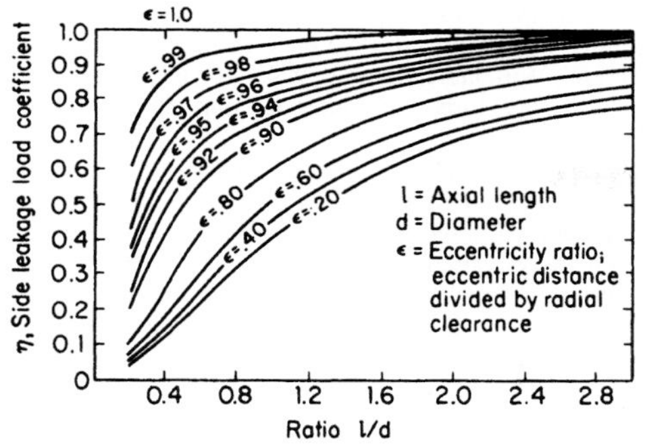

Figure 6.4.3 The slide leakage load coefficient "eta" in terms of length-diameter ratio l/d.

For severe operating conditions the value of A may exceed 18, the limit of Fig. 6.4.2. For complete journal bearings under extreme operating conditions, Fig. 6.4.4 should be used. The ordinate is P', defined as shown. The curves are drawn for various values of l/d instead of values of β as in Fig. 6.4.2. Values of ε may thus be obtained directly (Dennison, Film-Lubrication Theory and Engine-Bearing Design, *Trans. ASME*, **58,** 1936).

EXAMPLE 2. A 360° journal bearing $2\frac{1}{2}$ in diam and $3\frac{7}{8}$ in long carries a steady load of 3,875 lb. Speed N = 500 r/min; diametral clearance, 0.0064 in; average viscosity of the oil in the film, 23.4 centipoises (SAE 20 light motor oil at 105°F). $P = 3{,}875/(2.5 \times 3.875) = 400$ lb/in^2. Value of $m = 0.0064/2.5 = 0.00256$. Value of l/d = 1.55. First, attempt to use Figs. 6.4.2 and 6.4.3 in this solution. Assume eccentricity ratio ε is 0.9. Then, in Fig. 6.4.3, with l/d = 1.55, value of η is determined as 0.8. A is calculated as 37. This is completely off scale in Fig. 6.4.2. Consider instead Fig. 6.4.4. Value of P' is computed as

$$P' = 6.9(2.56)^2 \frac{400}{23.4 \times 500} = 1.54$$

In Fig. 6.4.4, enter the curves with P' = 1.54, and move left to intersect the curve for l/d = 1.5. Drop downward to read a value for $1/(1 - \varepsilon)$ of 16. Then $\frac{1}{16} = 1 - \varepsilon$, or the eccentricity ratio $\varepsilon = \frac{15}{16}$, or 0.94. The minimum film thickness, as in Example 1 = $h_0 = mr(1 - \varepsilon)$, or

$$h_0 = 0.000256 \times 1.25\,(1 - 0.94) = 0.0002 \text{ in } (0.0051 \text{ mm})$$

Allowable mean bearing pressures in bearings with fluid film lubrication are given in Table 6.4.1. If the load maintains the same magnitude and direction when the journal is at rest (heavily loaded shafts, heavy gears), the mean bearing pressure should be somewhat less than when bearings are loaded only when running.

For internal-combustion-engine bearing design, Etchells and Underwood (*Mach. Des.*, Sept. 1942) list the following maximum design pressures for bearing alloys, pounds per square inch of projected area: lead-base babbitt (75 to 85 percent lead, 4 to 10 percent tin, 9 to 15 percent antimony) 600 to 800; tin-base babbitt (0.35 to 0.6 percent lead, 86 to 90 percent tin, 4 to 9 percent antimony, 4 to 6 percent copper) 800 to 1,000; cadmium-base alloy (0.4 to 0.75 percent copper, 97 percent cadmium, 1 to 1.5 percent nickel, 0.5 to 1.0 percent silver) 1,200 to 1,500; copper-lead alloy (45 percent lead, 55 percent copper) 2,000 to 3,000; copper-lead (25 percent lead, 3 percent tin, 72 percent copper) 3,000 to 4,000; silver (0.5 to 1.0 percent lead on surface, 99 percent silver) 5,000 up. The above pressures are based on fatigue life of 500 h at 300°F bearing temperature, and a bearing metal thickness 0.01 to 0.015 in for lead-, tin-, and cadmium-base metals and 0.25 in for copper, lead, and silver. At lower temperatures the life will be greatly extended.

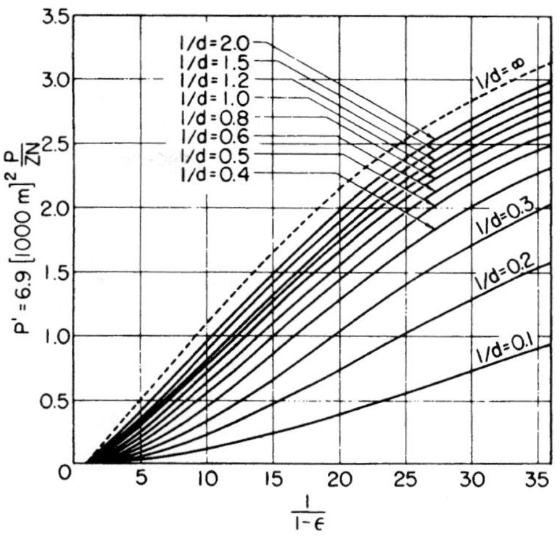

Figure 6.4.4 Load-carrying parameter in terms of eccentricity.

Much higher pressures are encountered in rolling element bearings, such as ball and roller bearings, and gears. In these situations, the formation of fluid films capable of preventing contact between surface asperities is aided by the increase of viscosity with pressure, as exhibited by most lubricating oils. The relation is typically exponential, $\mu = \mu_0 e^{\alpha p}$, where α is the so-called pressure coefficient of viscosity.

Length-diameter ratios are usually chosen between l/d = 1 and l/d = 2, although many engine bearings are designed with l/d = 0.5, or even less. In shorter bearings, the carrying capacity of the oil film is greatly impaired by the effect of side leakage. Longer bearings are used to restrain the shaft from vibration, as in line shafts, or to position the shaft accurately, as in machine tools. In power machines, the tendency is toward shorter bearings. Typical values are as follows:

Table 6.4.1 Current Practice in Mean Bearing Pressures

Type of bearing	Permissible pressure, lb/in^2, of projected area	Type of bearing	Permissible pressure, lb/in^2, of projected area
Diesel engines, main bearings	800-1,500	Automotive gasoline engines, main bearings	500-1,000
Crankpin	1,000-2,000	Crankpin	1,500-2,500
Wrist pin	1,800-2,000	Air compressors, main bearings	120-240
Electric motor bearings	100-200	Crankpin	240-400
Marine diesel engines, main bearings	400-600	Crosshead pin	400-800
Crankpin	1,000-1,400	Aircraft engine crankpin	700-2,000
Marine line-shaft bearings	25-35	Centrifugal pumps	80-100
Steam engines, main bearings	150-500	Generators, low or medium speed	90-140
Crankpin	800-1,500	Roll-neck bearings	1,500-2,500
Crosshead pin	1,000-1,800	Locomotive crankpins	1,500-1,900
Flywheel bearings	200-250	Railway-car axle bearings	300-350
Marine steam engine, main bearings	275-500	Miscellaneous ordinary bearings	80-150
Crankpin	400-600	Light line shaft	15-25
Steam turbines and reduction gears	100-220	Heavy line shaft	100-150

turbogenerators, 0.8 to 1.5; gasoline and diesel engines for main and crankpin bearings, 0.4 to 1.0, with most values between 0.5 and 0.8; generators and motors, 1.5 to 2.0; ordinary shafting, heavy, with fixed bearings, 2 to 3; light, with self-aligning bearings, 3 to 4; machine-tool bearings, 2 to 4; railroad journal bearings, 1.2 to 1.8.

For the **clearance between journal and bearing** see Fits in Sec. 6. Medium fits may be used for journals running at speeds under 600 r/min, and free fits for speeds over 600 r/min. Kingsbury suggests for these journals a diametral clearance = 0.002 + 0.001*d* in. In journals running at high speed, diametral clearance = 0.002*d* should be used in order to lower the friction losses in the bearing. All units are in inches. The most satisfactory clearance should, of course, be based on a complete bearing analysis which includes both load-carrying capacity and heat generation due to friction. For example, a bearing designed to run at the extremely high speed of 50,000 r/min uses a diametral clearance of 0.0025 in for a journal with 0.8-in diameter, giving a clearance ratio, clearance/diameter, of 0.00316.

For high-speed internal-combustion-engine bearings using forced-feed lubrication, medium fits are used. Federal-Mogul recommends the following diametral clearances in inches per inch of shaft diameter for insert-type bearings: tin-base and high-lead babbitts, 0.0005; cadmium-silver-copper, 0.0008; copper-lead, 0.001.

The dependence of the **coefficient of friction** for journal bearings on the bearing clearance, lubricant viscosity, rotational speed, and loading pressure, as reported by McKee and others, is shown in Sec. 1. A plot of the coefficient of friction against the parameter *ZN/P* is a convenient method for showing this relationship. *ZN/P* is a parameter based on mixed units. *Z* is the viscosity in centipoise, *N* is r/min, *P* is the mean pressure on the bearing due to the load, pounds per square inch of projected area, and *m* is the clearance ratio. Values of *ZN/P* greater than about 30 indicate fluid film conditions in the bearings. If the viscosity of the lubricant becomes lower or if there is a reduction in rotational speed or an increase in load, the value of *ZN/P* will become smaller until the coefficient of friction reaches a minimum value. Any further reduction in *ZN/P* will produce breakdown of the oil film, marking the transition from fluid film lubrication with complete separation of the moving surfaces to semifluid or mixed lubrication, where there is partial contact. As soon as semifluid conditions are initiated, there will be a sharp increase in the coefficient of friction. The critical value of *ZN/P*, where this transition takes place, will be lowest for a rigid bearing and shaft with finely finished surfaces.

Figure 6.4.5 shows a generalization of the relationship between the coefficient of friction for a journal bearing and the parameter *ZN/P*, indicating the various possible lubrication regimes that may be expected. For optimum design, a value of *ZN/P* somewhere between 30 and 300 would be recommended, but, in any case, the determination of minimum film thickness h_0 should be the deciding parameter. For extremely large values of *ZN/P*, resulting from high speeds and low loads, whirl instability may be developed. (See material on gas-lubricated bearings in this section.) With large values of *ZN/P* and a lubricant having a low kinematic viscosity, turbulent conditions may develop in the bearing clearance.

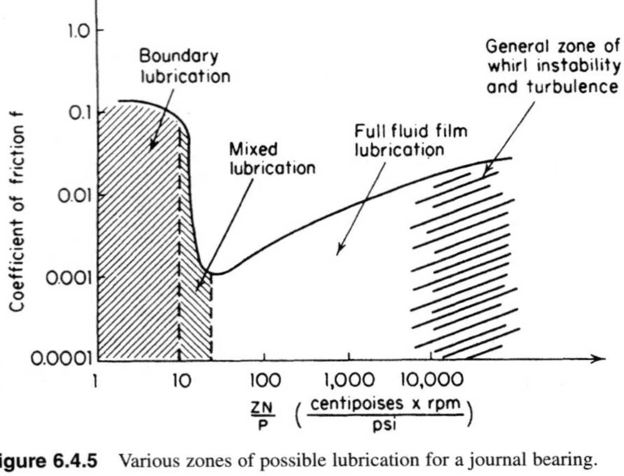

Figure 6.4.5 Various zones of possible lubrication for a journal bearing.

The friction force in plain journal bearings may be estimated by the use of the expression $F = K_f \mu Nrl/m$, where μ is in lb·s/in² units. The value of K_f depends upon the magnitude of ε and the type of bearing. Figure 6.4.6 shows values of K_f for a complete bearing, a 150° partial bearing, and a 120° partial bearing, assuming that the clearance space is at all times filled with lubricant. Note that *F* is the friction force at the surface of the bearing. Consequently, the friction torque is obtained by multiplying *F* by the bearing radius.

EXAMPLE 3. As an illustration of the use of Fig. 6.4.6, determine the friction force in the bearing of Example 2. This is a complete journal bearing 2½-in diam by 3⅞ in. The value of ε was determined as 0.94. From Fig. 6.4.6, $K_f = 2.8$. Then

$$F = \frac{2.8 \times 23.4 \times 1.45 \times 10^{-7} \times 500 \times 1.25 \times 3.875}{0.00256}$$
$$= 0.97 \text{ lb}(4.08 \text{ kg})$$

The coefficient of friction *F/W* = 8.97/3,875 = 0.00231. The mechanical loss in the bearing is *FV*/33,000 hp, where *V* is the peripheral velocity of the journal, ft/min.

$$\text{Friction hp} = (8.97 \times 500 \times \pi \times 2.5)/(33{,}000 \times 12)$$
$$= 0.089 \text{ hp}(66.37 \text{ W})$$

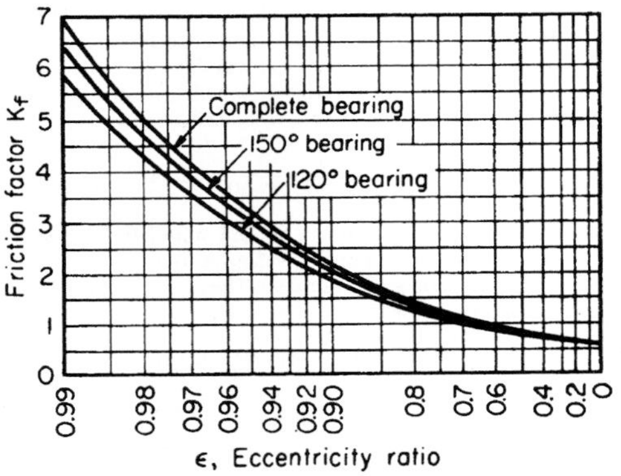

Figure 6.4.6 Variation of the friction factor of a bearing with eccentricity ratio.

Departure from laminarity in the fluid film of a journal bearing will increase the friction loss. Figure 6.4.7 (Smith and Fuller, Journal Bearing Operation at Super-laminar Speeds, *Trans. ASME*, **78,** 1956) shows test results for such bearings, expressed in terms of a Reynolds number for the fluid film, $R_e = umr/\nu$. Laminar conditions hold up to an R_e of about 1,000. Friction may be calculated for laminar flow by using Fig. 6.4.6 or the left branch of the curve in Fig. 6.4.7, where $f' = 2/R_e$, and which applies to low values of the eccentricity ratio ($K_f = 0.66$). The values from Fig. 6.4.7 may be converted to friction torque *T* by the use of the expression $T = f'\pi p u^2 r^2 l$, where ρ is the mass density of the lubricant. In Fig. 6.4.7, a transition region spans values of the Reynolds number from 1,000 to 1,600. Here, two types of flow instability can occur. Usually, the first is due to **Taylor vortices** which are wrapped in regular circumferential structures, each of which occupies the entire clearance. The onset of this phenomenon takes place at a value of the Reynolds number exceeding the threshold $R_e = 41.1(r/c)^{1/2}$. The second instability is due to turbulence, occurring at $R_e > 2{,}000$.

EXAMPLE 4. A journal bearing is 4.5 in diameter by 4.5 in long. Speed 22,000 r/min. *mr* = 0.002 in. Viscosity μ, 1 cP (water) = 1.45 × 10⁻⁷ lb·s/in²; mass density $p = 62.4/1{,}728 \times 386 = 9.35 \times 10^{-5}$ lb·s²/in⁴; $\nu = \mu/p = 1.45 \times 10^{-7}/9.35 \times 10^{-5} = 0.155 \times 10^{-2}$ in²/s; $u = 22{,}000 \times 2\pi \times 2.25/60 = 5{,}180$ in/s; $R_e = 5{,}180 \times 0.002/0.155 \times 10^{-2} = 6{,}680$. This would indicate turbulence in the film. Value of f'

is then $0.078/6{,}680^{0.43} = 0.078/44.2 = 1.765 \times 10^{-3}$. Friction torque $T = 1.765 \times 10^{-3} \times \pi \times 9.35 \times 10^{-5} \times 5{,}180^2 \times 2.25^2 \times 4.5$, $T = 317.5$ in·lb. Friction horsepower $= 2\pi TN/12 \times 33{,}000 = 2\pi \times 317.5 \times 22{,}000/12 \times 33{,}000$, FHP = 111 (82.77 kW).

In self-contained bearings (electric motor, line shaft, etc.) without external oil or water cooling, the **heat dissipation** is equal to the heat generated by friction in the bearing.

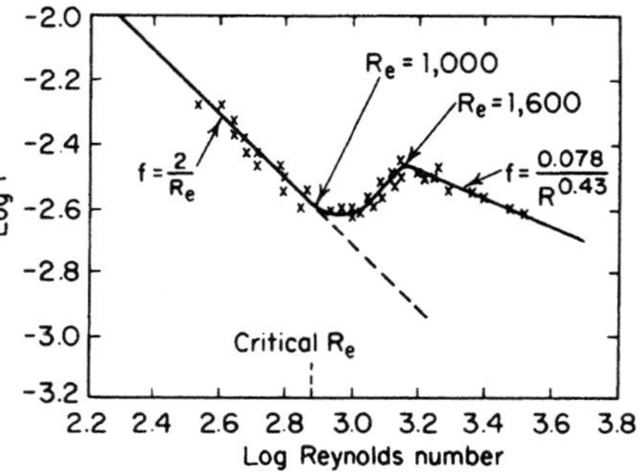

Figure 6.4.7 Friction f' as a function of the Reynolds number for an unloaded journal bearing with $l/d = 1$. *(Smith and Fuller.)*

The heat dissipated from the outside bearing wall to the surrounding air is governed by the laws of heat transfer $Q = hS(t_w - t_0)$, where S is the surface area from which the heat is convected, Q is the rate of energy flow; t_w and t_0 are the temperatures of the wall and ambient air, respectively; and h is the heat convection coefficient, which has values from 2.2 Btu/(h·ft²·°F) for still air to 6.5 Btu/(h·ft²·°F) for air moving at 500 ft/min. Calculations of heat loss are extremely important due to the strong temperature dependence of the viscosity of most oils.

The temperature of the oil film will be higher than the temperature of the bearing wall. Typical ranges of values according to Karelitz (*Trans. ASME*, **64,** 1942), Pearce (*Trans. ASME*, **62,** 1940), and Needs (*Trans. ASME*, 1948) for self-contained bearings with oil bath, oil ring, and waste-packed lubrication are shown in Fig. 6.4.8.

EXAMPLE 5. The frictional loss for the generator bearing of Example 1, computed by the method outlined in Example 3, is 0.925 hp with $\varepsilon = 0.88$, $K_f = 1.6$, and $F = 27$ lb. Operating in moving air the heat dissipated by the bearing housing will be $L = 6.5S(t_w - t_0)$. Since this is a self-contained bearing, the heat dissipated is also equal to the heat generated by friction in the oil film, or $L = 0.925 \times 2{,}545 = 2{,}355$ Btu/h. With $S = 25 \times 6 \times 9/144 = 9.4$ ft², $t_w = t_0 = 2{,}355/6.5 \times 9.4 = 38.5$°F. This is the temperature rise of the bearing wall above the ambient room temperature. For an 80°F room, the wall temperature of the bearing would be about 118°F. In Fig. 6.4.8 an oil-ring bearing in moving air with a temperature rise

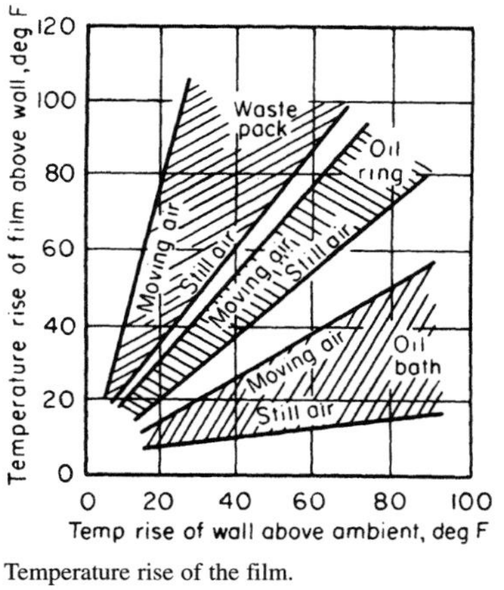

Figure 6.4.8 Temperature rise of the film.

of wall over ambient of 38°F should have a film temperature 50°F higher than that of the wall. The film temperature on the basis of Fig. 6.4.8 will then be 80 +38 + 50, or 168°F. This is close enough to the value of the film temperature of 160°F from Example 1, with which the friction loss in the bearing was computed, to indicate that this bearing can operate without the need for external cooling.

To predict the operating temperature of a self-contained bearing, the cut-and-try method shown above may be used. First, an oil-film temperature is assumed. Viscosity and friction losses are calculated. Then the temperature rise of the wall over ambient is computed so as to dissipate to the atmosphere an amount of heat equal to the friction loss. Lastly from Fig. 6.4.8 the corresponding oil-film temperature is estimated and compared to the value that was originally assumed. A few adjustments of the assumed film temperature will produce satisfactory agreement and indicate the leveling-off temperature of the bearing. Self-contained bearings have been built with diameters of 3, 8, and 24 in (7.62, 20.32, and 60.96 cm) to operate at shaft speeds of 3,600, 1,000, and 200 r/min, respectively. These designs indicate a rough limit for bearings with no external cooling. The highest bearing temperature permissible with normal lubricants is about 210°F (100°C).

The temperature of automotive-type bearings is held within safe limits by using a **pressure-feed oil supply.** Sufficient lubricant is forced through the bearing to act as a coolant and prevent overheating. One widely used practice is to place a circumferential groove at the center of the bearing to which the oil supply is fed. This is effective as far as cooling is concerned but has the disadvantage of interrupting the active length of the bearing and lowering its l/d ratio (see Fig. 6.4.9). The axial flow through each side of the bearing is given by

$$Q_1 = \frac{\Delta P n^3 e^4 \pi}{6\mu b}\left(1 + \frac{3}{2}\varepsilon^2\right)$$

where b is the effective axial length of the half bearing and ΔP is the difference between the oil pressure in the circumferential groove and the pressure at the ends of the bearing. The value of the last term in this equation will vary from 1.0 for a concentric shaft and bearing indicated by $\varepsilon = 0$ to a value of 2.5 for the extreme case of the shaft touching the bearing wall, indicated when $\varepsilon = 1$. Most of the heat caused by friction in the bearing is carried away by the circulating oil. Permissible temperature rises for this type of bearing may range from 15 to 50°F (8 to 28°C). In extreme cases a rise of 100°F (55°C) can be tolerated for high-strength bearing materials. The lower values of temperature rise usually indicate needlessly large oil flow. Such a condition will result in an excessive friction loss in the bearing.

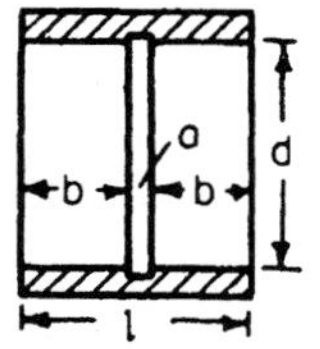

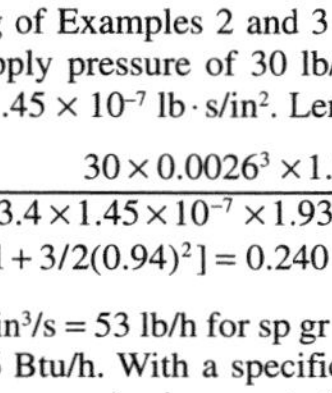

Figure 6.4.9 Bearing with central circumferential groove.

EXAMPLE 6. The bearing of Examples 2 and 3 is lubricated by a circumferential groove with an oil supply pressure of 30 lb/in² and, as before, $\varepsilon = 0.94$, $m = 0.0026$, and $\mu = 23.4 \times 1.45 \times 10^{-7}$ lb·s/in². Length b is about 1.93 in.

$$Q_1 \text{ flow out one side} = \frac{30 \times 0.0026^3 \times 1.25^4 \times \pi}{6 \times 23.4 \times 1.45 \times 10^{-7} \times 1.93} \times [1 + 3/2(0.94)^2] = 0.240 \text{ in}^3/\text{s } (3.93 \text{ cm}^3/\text{s})$$

Total flow (two sides) = 0.48 in³/s = 53 lb/h for sp gr = 0.85. The friction loss from Example 3 = 0.089 hp = 226 Btu/h. With a specific heat of 0.5 Btu/(lb·°F) and assuming that all the friction energy is given up to the oil in the form of heat, the temperature rise $\Delta t = 226/0.5 \times 53 = 8.5$°F (4.72°C).

A definite **minimum rate of oil feed** is required to maintain a fluid film in journal bearings. This makes no allowance for the additional flow that may be needed to cool the bearings. However, many industrial bearings run at relatively low speeds with light loads and, as a consequence, additional oil flow to provide cooling is not necessary. But if a fluid film is desired, a definite minimum amount of lubricant is required. If the volume of lubricant fed to the bearing is less than this minimum requirement, there will not be a complete fluid film in the bearing. Friction will rise, wear will become greater, and the satisfactory service life of such a bearing will be reduced. This minimum lubricant supply can be evaluated by using the equation

$$Q_M = K_M urml$$

where Q_M is the flow rate and K_M is approximately 0.006.

EXAMPLE 7. The minimum feed rate for a journal bearing $2\frac{1}{2}$-in diam by $2\frac{1}{8}$ in long will be determined. Diametral clearance is 0.0045 in; speed, 1,230 r/min; load, 40 lb/in² based on projected area. $u = 1{,}230 \times \pi \times 2.125 = 10{,}220$ in/min, $r = 1.062$ in, $m = 0.0045/2.125 = 0.00212$, $l = 2.125$ in. Substituting,

$$Q_M = 0.006 \times 10{,}220 \times 1.062 \times 0.00212 \times 2.125$$
$$= 0.28 \text{ in}^3/\text{min}$$

(Fuller and Sternlicht, Preliminary Investigation of Minimum Lubricant Requirements of Journal Bearings, *Trans. ASME*, **78,** 1956.)

Many bearings are supplied with oil at low rates of feed by **felts, wicks,** and **drop-feed oilers.** Wicks can supply substantial rates of feed if they are properly designed. The two basic types of wick feed are siphon wicks, as shown in Fig. 6.4.10, and bottom wicks, as shown in Fig. 6.4.11. Data on oil delivery for these wicks are shown in Figs. 6.4.12 and 6.4.13. The data, from the American Felt Co., are for SAE Fl felts, based on a cross-sectional area of 0.1 in². The flow rate is indicated in drops per minute. One drop equals 0.0026 in³ or 0.043 cm³.

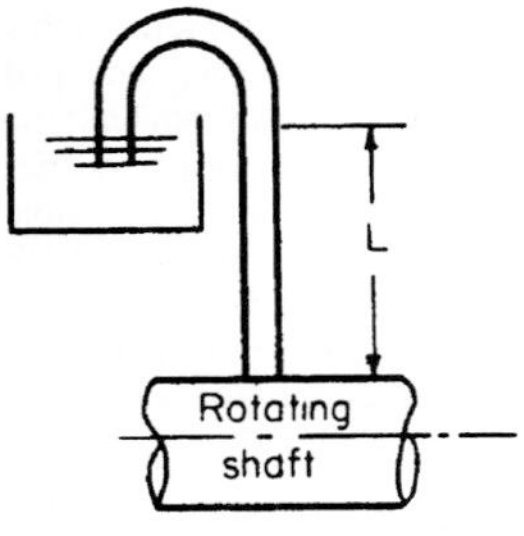

Figure 6.4.10 Siphon wick.

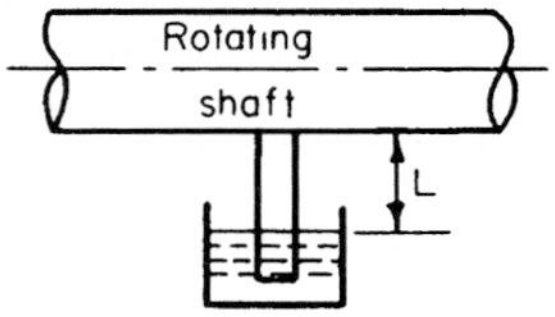

Figure 6.4.11 Bottom wick.

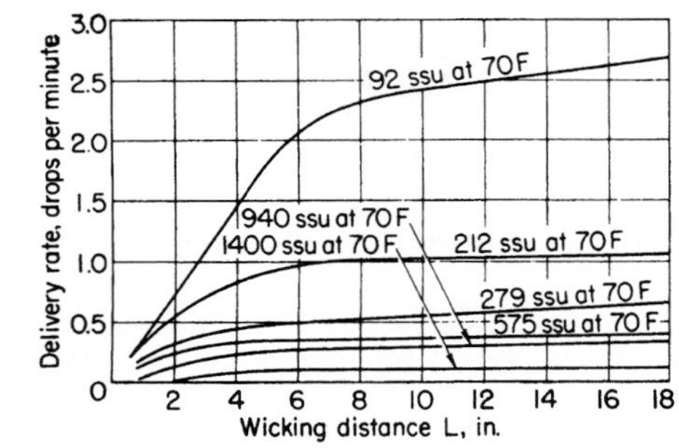

Figure 6.4.12 Oil delivery with siphon wick (Fig. 6.4.10).

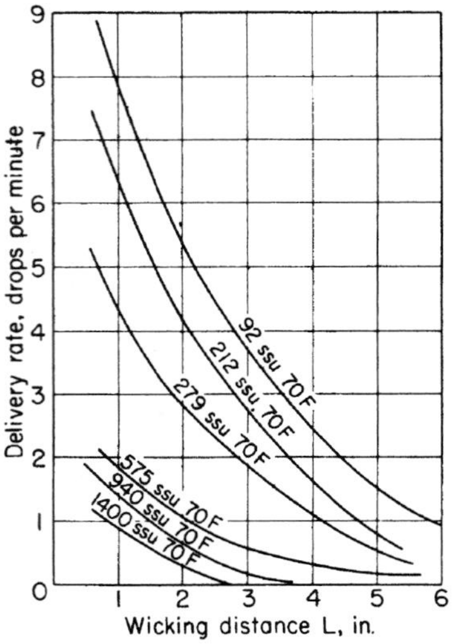

Figure 6.4.13 Oil delivery with bottom wick (Fig. 6.4.11).

EXAMPLE 8. If it is desired to deliver 12.5 drops/min to a journal bearing, and if the viscosity of the oil is 212 s Saybolt Universal at 70°F, and if *L*, Fig. 6.4.10, is 5 in, what size of round wick would be required? From Fig. 6.4.12, for the stated conditions the delivery rate would be 0.9 drop/min for an area of 0.1 in². If 12.5 drops/min is needed, this would mean an area of 12.5 divided by 0.9 and multiplied by 0.1, or 1.4 in². For a round wick this would mean a diameter of 1 in (3.49 cm).

If a **bottom wick** is considered with $L = 4$ in, Fig. 6.4.11, then in Fig. 6.4.13 the delivery rate using the same oil would be 1.6 drops/min; and if 12.5 drops/min is required, the area would be 12.5 divided by 1.6 and multiplied by 0.1, or 0.78 in². This would mean a bottom wick of 1 in diam if it is round (2.54 cm).

When journal **bearings** are **started, stopped,** or **reversed,** or whenever conditions are such that the operating value of *ZN/P* falls below the critical value for that bearing, the oil film will be ruptured and metal-to-metal contact will increase friction and cause wear. This condition can be eliminated by using a **hydrostatic oil lift.** High-pressure oil is introduced to the area between the bottom of the journal and the bearing (Fig. 6.4.14). If the pressure and quantity of flow are great enough, the shaft, whether it is rotating or not, will be raised and supported by an oil film. Neglecting axial flow, which is small, the flow up one side is

$$Q_1 = \frac{Wrm^3}{A\mu} \text{ in}^2/\text{s}$$

and the inlet pressure required, $P_o = \mu Q_1 B/(br^2m^3)$ where b is the axial length of the high-pressure recess. Values of A and B are dimensionless factors which represent geometric effects and are given in the following table as a function of ε:

Current practice is to make the total area of the high-pressure recess in a bearing $2\frac{1}{2}$ to 5 percent of the projected area *ld* of the bearing. It is generally desirable to use a check valve in the supply line to the oil lift so that, when the journal builds up a hydrodynamic oil-film pressure, reverse flow of oil in the supply line will be prevented.

EXAMPLE 9. A 4,000-in-diam journal rests in a bearing of 4.012-in-diam. SAE 30 oil at 100°F (105 cP) is supplied under pressure to a groove at the lowest point in the bearing. Length of bearing, 6 in, length of groove, 3 in, load on

ε	0	0.1	0.2	0.3	0.4	0.5	0.6	0.7	0.75	0.8	0.85	0.9
A	24.0	28.1	33.8	41.6	53.3	72.0	105	173	237	360	613	1,320
B	18.9	23.2	29.0	38.2	52.7	77.9	128	246	344	634	1,260	3,360

ε	0.91	0.92	0.93	0.94	0.95	0.96	0.97	0.98	0.99
A	1,620	2,070	2,620	3,530	5,040	7,800	13,700	30,600	121,000
B	4,340	5,810	8,040	11,800	18,400	32,100	65,300	179,000	348,000

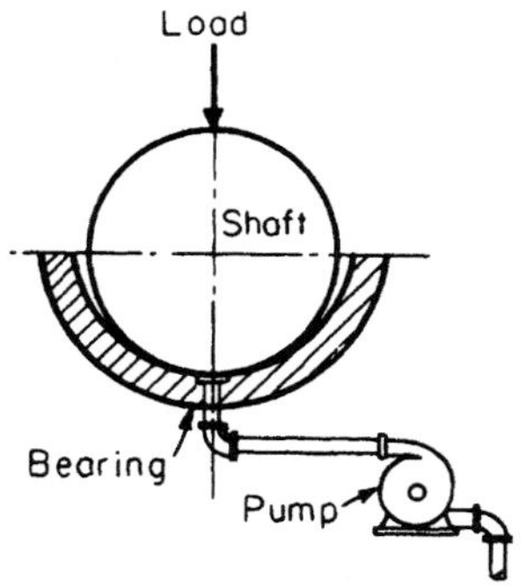

Figure 6.4.14 Diagram of oil lift.

bearing, 3,600 lb. What inlet pressure and oil flow are needed to raise the journal 0.004 in?

$$h_0 = mr(1-\varepsilon)$$
$$0.004 = 0.006(1-\varepsilon)$$
$$\varepsilon = 0.333$$

From the table, $A = 44.5$, $B = 42$.

$$Q_1 = \frac{3{,}600\times 2}{44.5\times 105\times 1.45\times 10^{-7}}(0.003)^3$$
$$= 0.287\ \text{in}^3/\text{s, one side } (4.70\ \text{cm}^3/\text{s})$$

Flow from both sides = $(0.287 \times 2) \times {}^{60}\!/_{231}$ = 0.149 gal/min (0.564 1/min).
Oil supply pressure is

$$P_o = \frac{105\times 1.45\times 10^{-7}\times 0.287\times 42}{3\times 4}\times\frac{1}{0.003^3} = 566\ \text{lb/in}^2$$

An adjustable constant-volume pump or a spur-gear pump with a capacity of about 1,000 lb/in² (6.894 kN/m²) should be used to allow for pressure that may be built up in the line before the journal begins to rise.

Other configurations for hydrostatically lubricated journal bearings are shown in Fig. 6.4.15. These were obtained by means of electric analog solutions (Loeb, Determination of Flow, Film Thickness and Load-Carrying Capacity of Hydrostatic Bearings through the Use of the Electric Analog Field Plotter, *Trans. ASLE*, **1**, 1958). The data from Fig. 6.4.15 are exact for a uniform film thickness corresponding to $\varepsilon = 0$ but may be used with discretion for other values of ε. More recently these solutions are achieved by numerical approximations with the help of digital computers.

Multiple recesses are used in externally pressurized bearings in order to provide local **stiffness.** This term indicates that the bearing resists shaft motions in any direction, and it is achieved by properly arranging the feeding network according to a strategy called **compensation.** Three main types are employed: orifice (and its variant, inherent), capillary, and fixed flow rates. In the first two, the idea is to insert a hydraulic resistance in each of the recess feeding lines and to use a single pump to feed all recesses. The flow rate q through orifices varies with the square root of the pressure drop Δp

$$q \propto \sqrt{\Delta p}$$

while for capillary tubes the relation is linear:

$$q = \frac{\pi \Delta p d^4}{64 l_1 \mu}$$

The general rule of thumb in designing orifices or capillary restrictors is to generate a pressure drop approximately equal to that taking place through the bearing, that is, from the recesses to the ambient. The recess geometry and distribution, on the other hand, are designed so that $W = 0.5p_{\text{recess}}\,DL$. Thus, the pump supply pressure is 4 times the average bearing pressure. The bearing stiffness is usually equal to $K = 0.5p_{\text{recess}}\,DL/c$.

The third method of compensation consists of forcing the same amount of flow to reach each recess regardless of clearance distribution. This can be achieved either by using separate pumps for each recess or by using a hydraulic device called a *flow divider*. With recess distributions as indicated above, the pump pressure need only be double

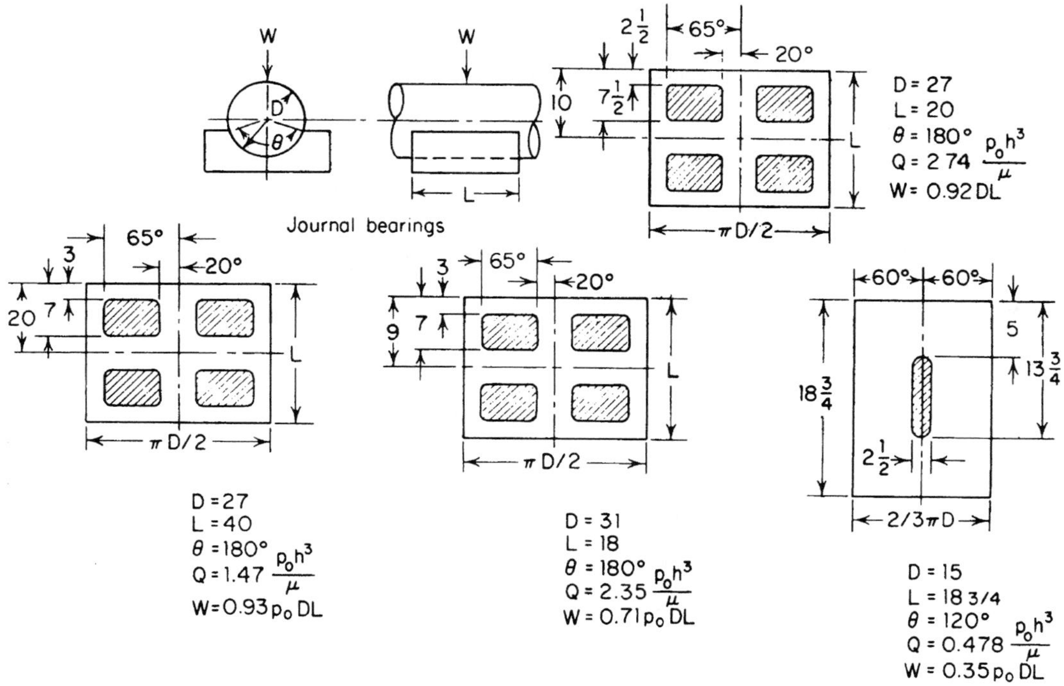

Figure 6.4.15 Load-carrying capacity and flow for journal bearings *(Loeb)*. Lengths in inches.

the average bearing pressure; thus, this method of compensation leads to half the power dissipation of the other two. It is commonly used in large machinery, where power consumption must be limited. The polar axis bearings of the 200-in Hale telescope on Mount Palomar were the first large-scale demonstration of this technique. The azimuth axis thrust bearing of the 270-ft-diameter Goldstone radio telescope is probably the largest example of this type of bearing.

ELEMENTS OF JOURNAL BEARINGS

Typical dimensions of solid and split **bronze bushings** are given in Table 6.4.2.

Bronze bushings made from hard-drawn sheets and rolled into cylindrical shape are made with a wall thickness of only 1/32 in for bearings up to 1/2 in diam and with a wall thickness of 1/16 in for bearings from 1 in diam up. The wall thickness of these bearings depends chiefly upon the strength of the material which supports them. Bushings of this type are pressed into place, and the bearing surface is finished by burnishing with a slightly tapered bar to a mirror finish. The allowable bearing pressures may exceed those of cast bronze shown in Table 6.4.1 by 10 to 20 percent.

Babbitt linings in larger bearings are generally employed in thickness of 1/8 in or over and must be provided with sufficient anchorage in the supporting shell. The anchors take the form of dovetailed grooves or holes drilled in the shell and counterbored from the outside.

Improved conditions are obtained by sweating or bonding the babbitt to the shell by tinning the latter, using potassium chlorate as flux. Tin-base babbitts and other low-strength materials evidence some yielding when subjected to heavy pressures. This tendency may be alleviated by the use of a thinner layer of the bearing material, fused either to a bronze or to a steel shell. This improves the fatigue life of the bearing material. Standard bearing inserts of this type are available in tin-base babbitts, high-lead babbitts, cadmium alloys, and copper-lead mixtures in diameters up to about 6 in (15.24 cm) (Fig. 6.4.16). A few materials can be obtained in sizes up to 8 in (20.32 cm). Some types are available with flanges or with other special features. The bearing lining may vary from about 0.001 in (0.025 mm) to 0.1 in (2.5 mm) in thickness depending upon the size of the bearing.

Figure 6.4.17 shows the principal types of bonded babbitt linings. Figure 6.4.17*a* is for normal operating conditions. Figure 6.4.17*b* is for more severe operating conditions.

General practice for the **thickness of babbitt lining and shells** is as follows: Fig. 6.4.18, $b = \frac{1}{32}d + \frac{1}{8}$ in, $S = 0.8d$ for bronze or steel $= 0.2d$ for cast iron; Fig. 6.4.18*a*, $t = b/2 + \frac{1}{16}$ in, $W = 1.8t$, $W_1 = 2.2t$.

Solid bronze or steel bushings, when pressed into the bearing housing, must be finished after pressing in. Light press fits and securing by setscrews or keys are preferable to heavy press fits and no keying, since heavy pressure, especially in thin-walled bushings, will set up stresses which will release themselves if bearings should run hot in service and will result in closing in on the journal and scoring when cooling.

Uniform Load Distribution Misalignment between journal and bearing should never be so great as to cause metallic contact. The maximum allowable inclination α of the shaft to the bearing is given by $\tan \alpha = md/l$.

Whenever the deflection angle of the bearing installation is greater than α, either the bearing length should be reduced or, if that is not feasible, the bearing should be mounted on a spherical seat to permit self-alignment.

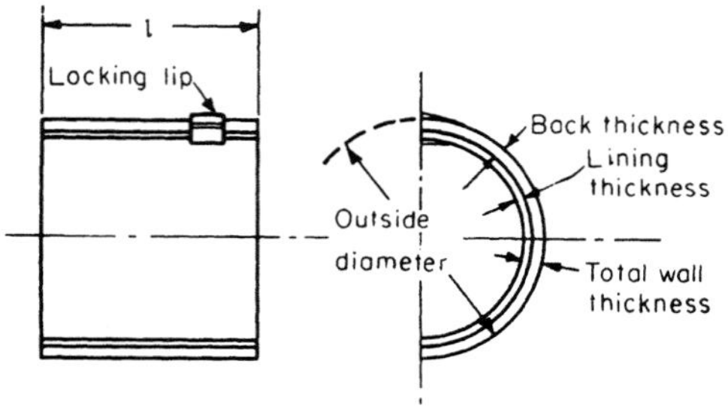

Figure 6.4.16 Bearing insert.

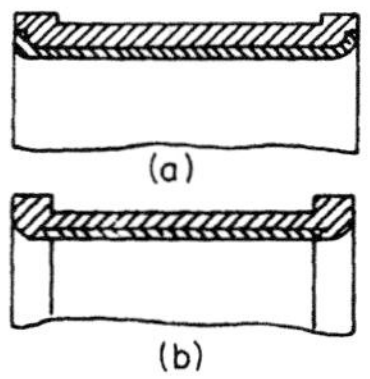

Figure 6.4.17

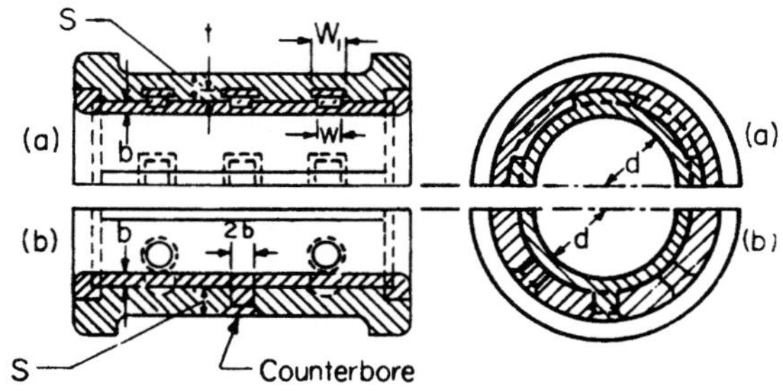

Figure 6.4.18

Oil grooves are of two kinds, axial and circumferential; the former distribute the oil lengthwise in the bearing; the latter distribute it around the shaft at the oil hole, and also collect and return oil which would otherwise be forced out at the ends of the bearing. Grooves have often been put into bearings indiscriminatingly, with the result that they scrape off the oil and interrupt the film.

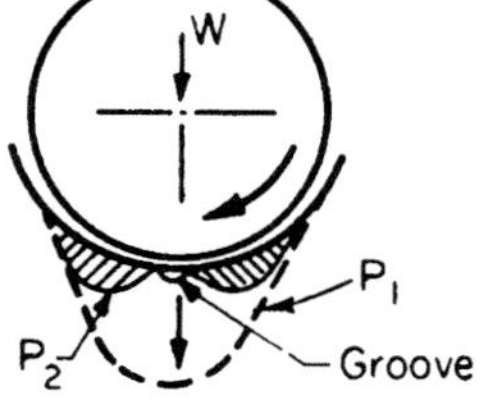

Figure 6.4.19

In Fig. 6.4.19, W is the resultant force or load, pounds, on the bearing or journal. The radial ordinates P_1, to the dotted curve, show the pressures, lb/in², of the journal on the oil film due to the load when there is no axial groove, while the ordinates P_2, to the solid curve, show the pressures with an incorrectly located groove. Since there is no oil pressure near the groove, the permissible load W must be reduced or the film will be ruptured.

Table 6.4.2 Wall Thickness of Bronze Bushings, in

	Diam of journal, in						
	1/4	1/4-1/2	1/2-1	1-1 1/2	1 1/2-2 1/2	2 1/2-4	4-5 1/2
Solid bushing, normal	1/16	3/32	1/8	3/16	1/4	3/8	1/2
Split bushing, normal	3/32	1/8	5/32	7/32	5/16	5/32	5/8
Solid bushing, thin	1/16	3/32	5/32	1/8	3/16	1/4	3/8
Split bushing, thin	1/16	3/32	1/8	3/16	1/4	3/8	1/2

Groove dimensions (Fig. 6.4.20) are given by the following relations: $a = \frac{1}{3}$ wall thickness; $W_o = 2.5a$; $W_d = 3a$; $c = 0.5W_d$; $f = \frac{1}{16}$ in to $0.5W_d$.

In order to maintain the oil film, **the axial distributing groove should be placed in the unloaded sector** of the bearing. The location of grooves in a variety of cases is shown in Figs. 6.4.21 to 6.4.30.

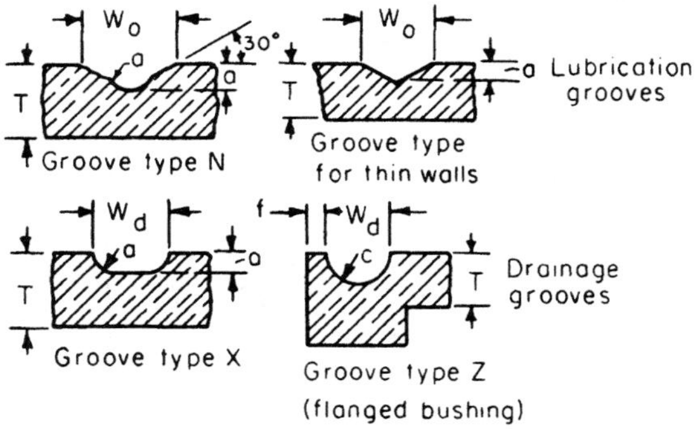

Figure 6.4.20 Lubrication and drainage grooves.

Horizontal Bearings, Rotational Motion

DIRECTION OF LOAD KNOWN AND CONSTANT

Load downward or inside the lower 60° segment as in the case of ring-oiling bearings (Fig. 6.4.21).

Load at an angle more than 45° to the vertical centerline (Fig. 6.4.22).

In force- or drop-feed oiling, the oil inlet may be anywhere within the no-load sector (Fig. 6.4.23).

Oil can be introduced through the center of the revolving shaft (Fig. 6.4.24).

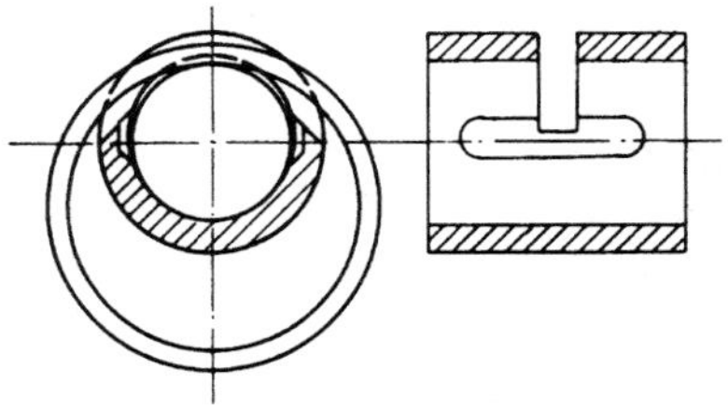

Figure 6.4.21

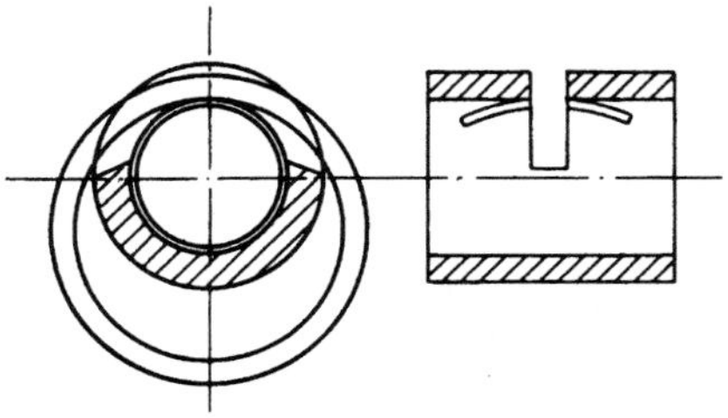

Figure 6.4.22

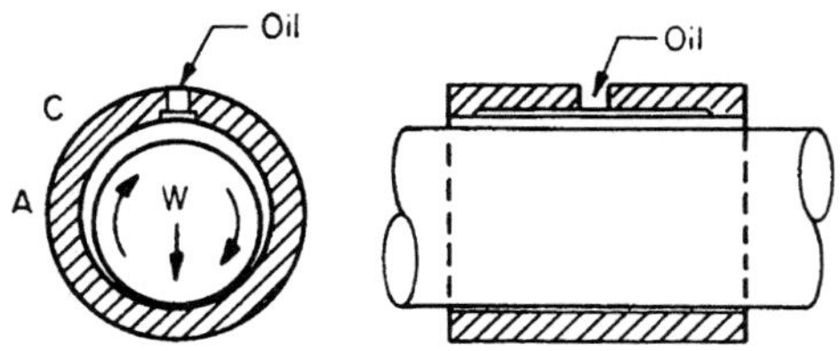

Figure 6.4.23

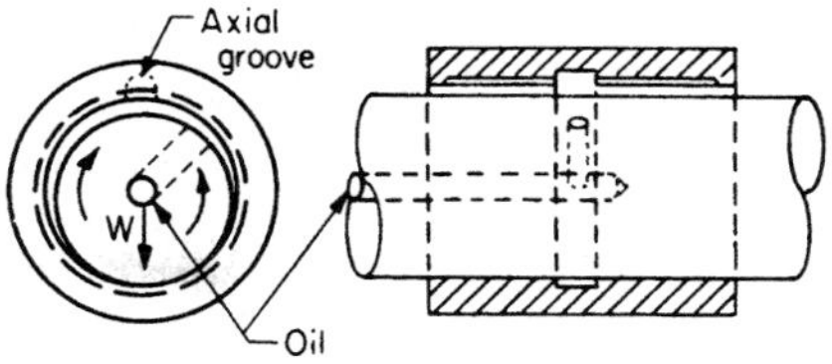

Figure 6.4.24

Where oil-ring electric-motor bearings will be subjected by the purchaser to belt loads varying from vertical downward to horizontal, a continuous type of oil groove developed by General Electric Co. has proved very successful (Fig. 6.4.25). There are no critical spots with this groove because only a small percentage of the babbitt surface is removed along any axial line.

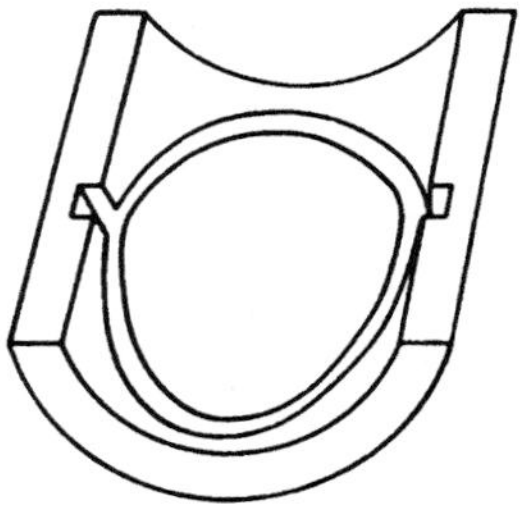

Figure 6.4.25

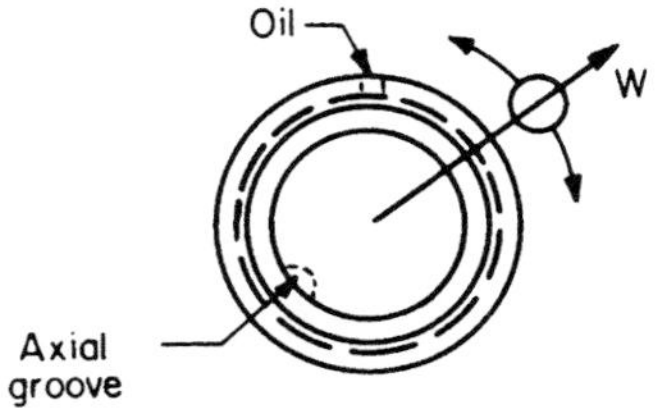

Figure 6.4.26

ROTATING LOAD

For rotating shafts, a circumferential groove at the middle of the bearing and an axial groove on the no-load side (Fig. 6.4.26).

For stationary shafts and rotating bearings, a circumferential groove in the bearing and an axial groove on the no-load side. The oil hole is in the shaft at the midlength of the bearing (Fig. 6.4.27).

LOAD DIRECTION UNCERTAIN

Oil-ring bearings (Figs. 6.4.21 and 6.4.22) may be used, although they have defects under certain load directions. With forced or drop feed, the oil hole enters a circumferential groove at the middle of the bearing and the axial groove is omitted (Fig. 6.4.28). Arrangements for introducing oil through the rotating shaft can be made.

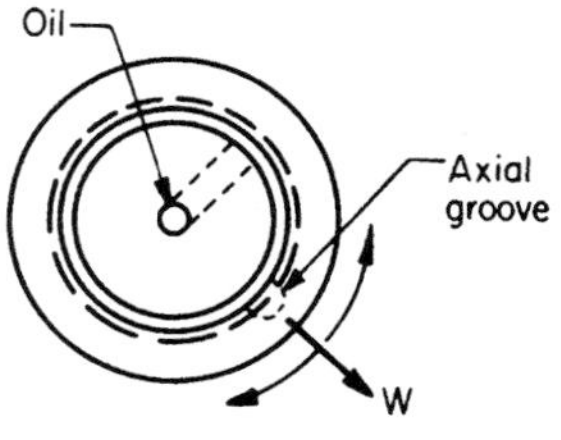

Figure 6.4.27

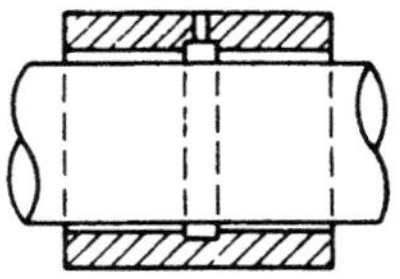

Figure 6.4.28

Bearings with Oscillatory Motion

DIRECTION OF LOAD CONSTANT

No oil film can be built up owing to the small sliding velocity, and boundary lubrication will exist. Axial grooves in the loaded sector distribute the lubricant to all parts of the bearing and avoid dry spots (Fig. 6.4.29).

LOAD DIRECTION REVERSED DURING OSCILLATION

Fluid film lubrication is possible, at least during part of the motion, owing to the vacuum caused by shaft moving back and forth.

Figure 6.4.30 shows grooving which may be modified to suit local conditions. This arrangement is also advisable for bearings under a load which reverses in direction periodically without any rotation of the bearing. The lubrication may then provide an oil cushion to soften shocks.

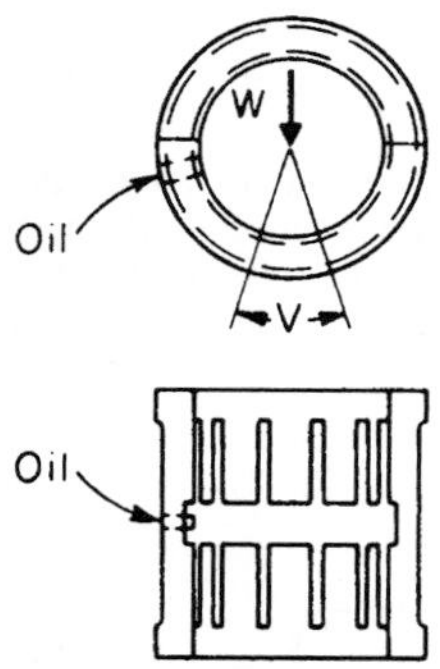

Figure 6.4.29

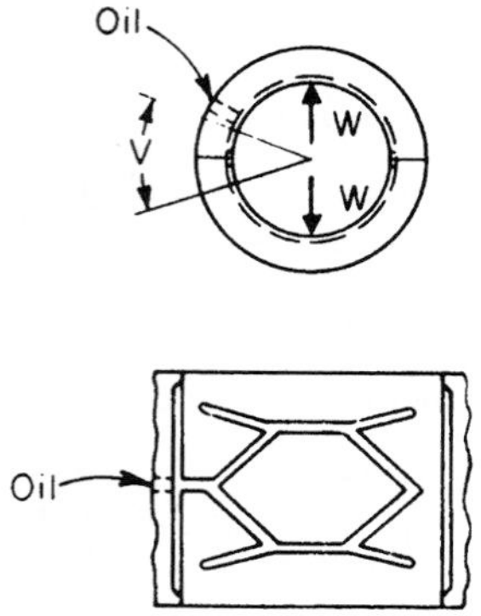

Figure 6.4.30

Bearing seals are used to prevent oil leakage from the bearing housing and to protect the bearing from outside dust, water, vapors, etc. A drainage groove at the end of the bearing is effective to divert the oil passing through the bearing back into the oil well (Fig. 6.4.31*a*). The drain holes at the bottom of the groove must be ample for passage of the oil flow. An oil thrower mounted on the shaft is shown in Fig. 6.4.31*b*. The bearing housing may be provided with a single (Fig. 6.4.31*c*) or double collecting groove, or with brass or aluminum strip scrapers (Fig. 6.4.31*d*), to collect the oil creeping along the shaft.

For protection from dust, etc., felt packing rings are often used (Fig. 6.4.31*e*). The felt ring is soaked in oil to prevent charring by friction heat. In severe cases, additional protection by a labyrinth runner is very effective (Fig. 6.4.31*f*).

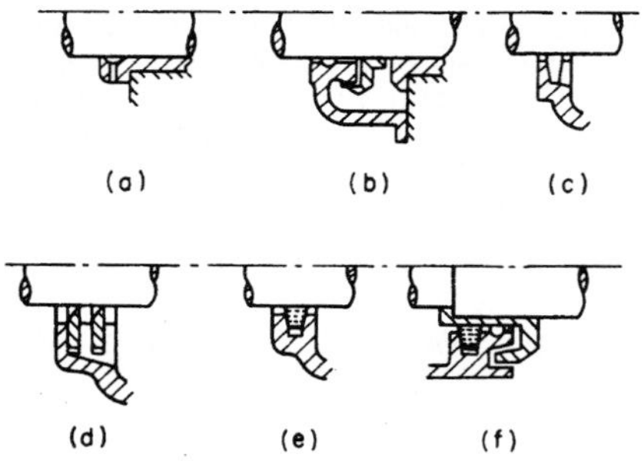

Figure 6.4.31 Sealing end grooves.

Standard seals are available for oil and grease retention as shown in Fig. 6.4.32*a*, *b*, and *c*. The seal material that is pressed against the rotating shaft is typically made of synthetic rubber, which is satisfactory for temperatures as high as about 250°F (121°C). Figure 6.4.32*a* shows the seal material pressed against the shaft by a series of flexible fingers or leaf springs. In Fig. 6.4.32*b* a helical garter spring provides the gripping force. In Fig. 6.4.32*c* the rubber acts as its own spring.

Types of bearings are shown in Figs. 6.4.33 to 6.4.38. They include the principal methods of lubrication and types of construction.

Oiless bearing is the accepted term for self-lubricating bearings containing lubricants in solid or liquid form in their material. Graphite, molybdenum disulfide, and Teflon are used as solid lubricants in one group, and another group consists of porous structures (wood, metal), containing oil, grease, or wax.

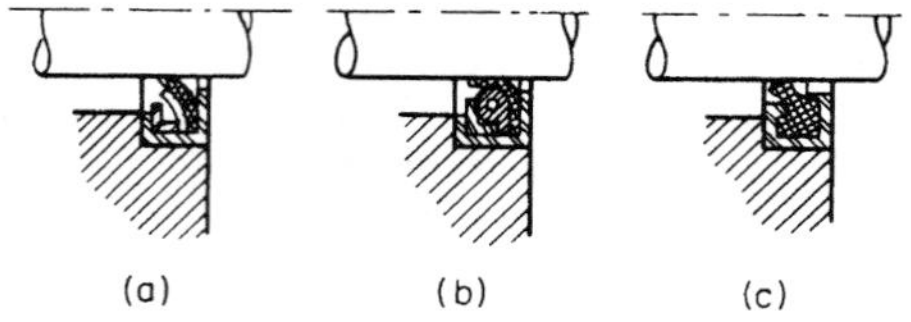

Figure 6.4.32 Seals for oil and grease retention.

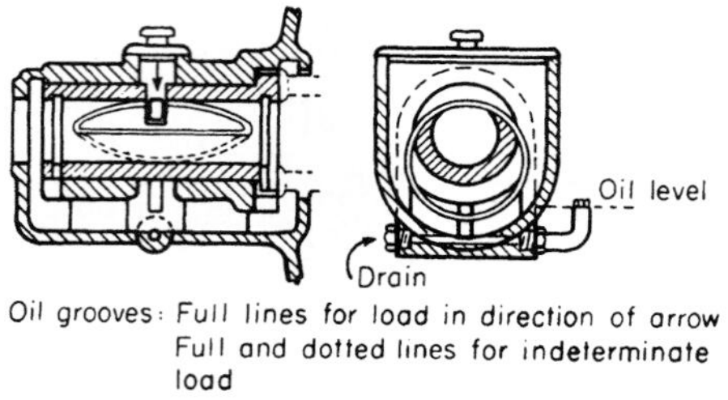

Figure 6.4.33 Ring-oiled bearing solid bushing.

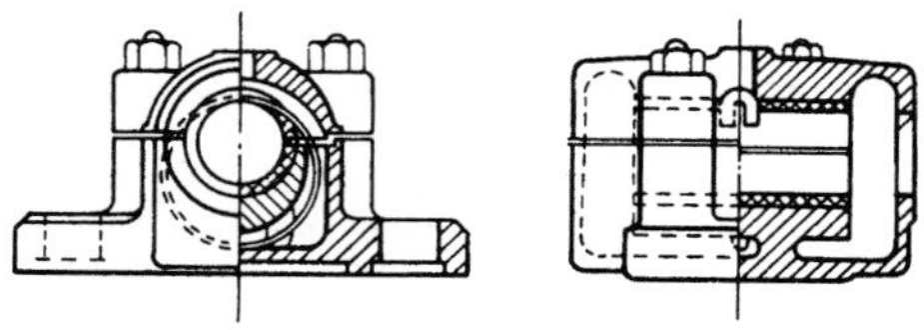

Figure 6.4.34 Rigid ring-oiling pillow block. *(Link Belt Co.)*

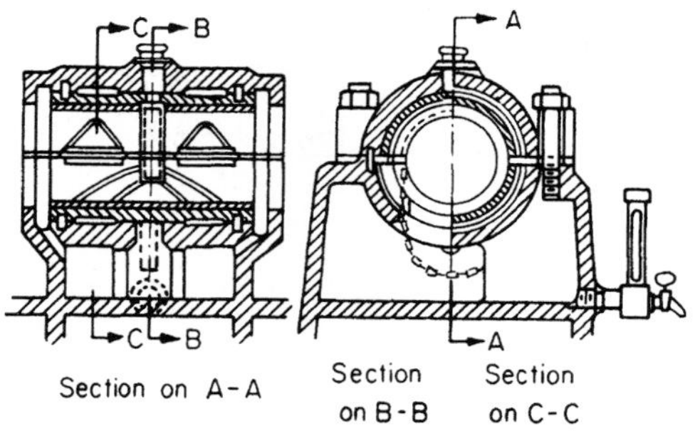

Figure 6.4.35 Split bearing with one chain. Main crankshaft bearing; vertical oil engine.

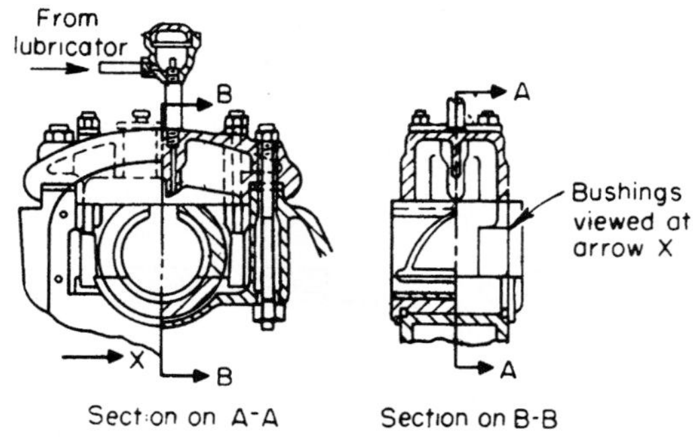

Figure 6.4.36 Crankshaft main bearing. Horizontal engine with drop-feed lubrication.

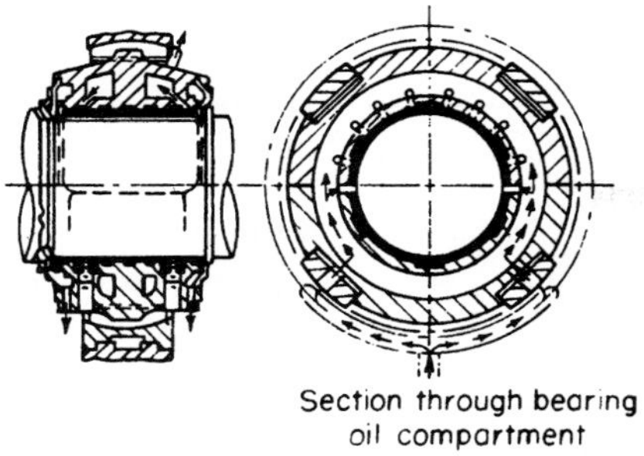

Figure 6.4.37

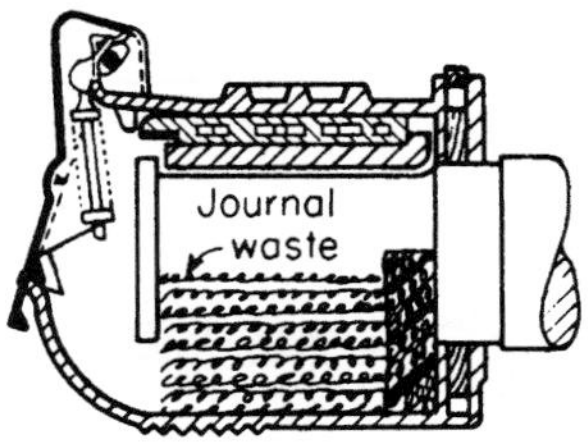

Figure 6.4.38

Graphite-lubricated bearings (bridge bearings, sheaves, trolley wheels, high-temperature applications) consist generally of cast bearing bronze as a supporting structure containing various overlapping designs of grooves which are filled with graphite. The graphite is mixed with a binder, and the plastic mass is pressed into the cavities to the hardness of a lead pencil; 45 percent of the bearing area may be graphite.

Porous-metal bearings, compressed from metal powders and sintered, contain up to 35 percent of liquid lubricant. See ASTM B202-45T for sintered bronze and iron bearings, and also Army and Navy Specification AN-B-7G. The porous metal generally consists of a 90-10 coppertin bronze with 1½ percent graphite. These bearings do not require oil grooves since capillarity distributes the oil and maintains an oil film. If additional lubrication from an oil well should be provided, oil will be absorbed through the porous wall as required. For high temperatures where oil will carburize, a higher percentage of graphite (6 to 15 percent) is used.

Porous-metal bearings are used where plain metal bearings are impractical because of lack of space, cost, or inaccessibility for lubrication, as in automotive generators and motors, hand power tools, vacuum cleaner motors, and the like.

THRUST BEARINGS

At low speeds, shaft shoulders or collars bear against flat bearing rings. The lubrication may be semifluid, and the friction is comparatively high.

For hardened-steel collars on bronze rings, with intermittent service, pressures up to 2,000 lb/in^2 (13,790 kN/m^2) are permissible; for continuous low-speed operation, 1,500 lb/in^2 (10,341 kN/m^2); for steel collars on babbitted rings, 200 lb/in^2 (1,378.8 kN/m^2). In multicollar thrust bearings, the values are reduced considerably because of the difficulty in distributing the load evenly between the several collars.

The performance of the bearing thrust rings is much improved by the introduction of **grooves** with tapered lands as shown in Fig. 6.4.39. The lands extend on either side of the groove. The taper angle of the lands is very slight, so that a pressure oil film is formed between the bearing ring and the collar of the shaft. It is generally known that slightly tapered radial grooves will develop a hydrodynamic load-carrying film, when formed in the manner of Fig. 6.4.39. The taper angle should be on the order of 0.5°. Alternatively, a shallow recessed area that is a couple of film thicknesses deep can be used in place of the taper.

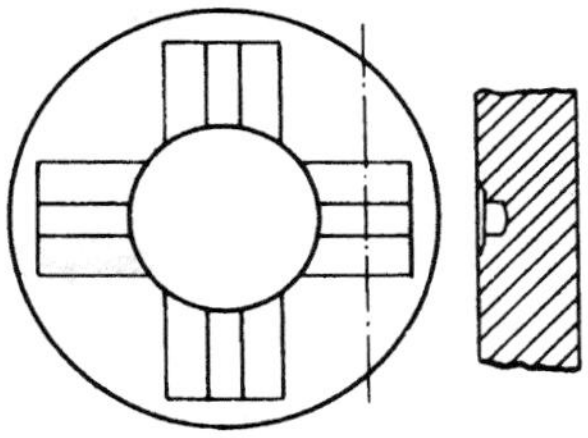

Figure 6.4.39 Thrust collar with grooves fitted with tapered lands.

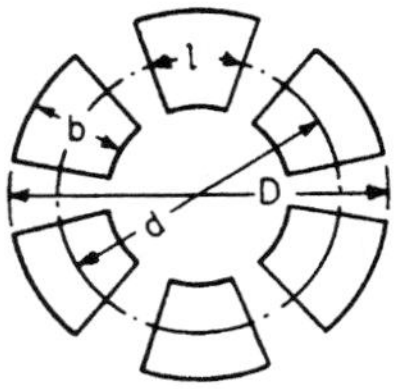

Figure 6.4.40 Kingsbury thrust bearing with six shoes.

For high speeds or where low friction losses and a low wear rate are essential, **pivoted segmental thrust bearings** are used (Kingsbury thrust bearing, or Michell bearing in Europe). The bearing members in this type are tiltable shoes which rest on hard steel buttons mounted on the bearing housing. The shoes are free to form automatically a wedge-shaped oil film between the shoe surface and the collar of the shaft (Figs. 6.4.40 to 6.4.42).

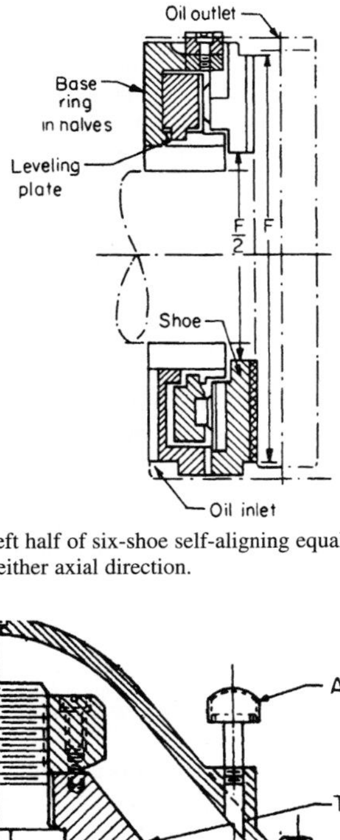

Figure 6.4.41 Left half of six-shoe self-aligning equalizing horizontal thrust bearing for load in either axial direction.

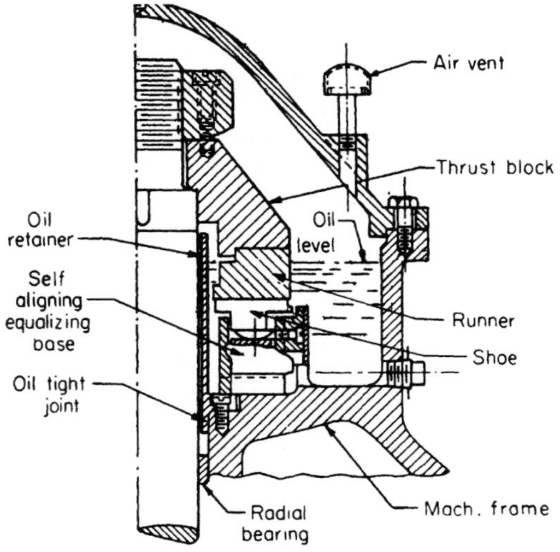

Figure 6.4.42 Half section of mounting for vertical thrust bearing.

Table 6.4.3 Capacities of Six-Shoe Standard-Duty Horizontal and Vertical Thrust Bearings

(Based on viscosity of 150 s Saybolt at operating temperatures. Capacities given may be increased from 10 to 25% if viscosity is increased in same proportion)

Bearing size, in	Area, in²	Speed, r/min						Bearing size, in	Area, in²	Speed, r/min					
		100	200	400	800	1,800	3,600			100	150	200	300	500	700
		Safe load, 10³ lb								Safe load, 10³ lb					
5	12.5	1.44	1.7	2.0	2.4	2.9	3.5	19	180	40.00	44.0	48.0	53.0	60.0	65.0
6	18.0	2.30	2.7	3.2	3.8	4.6	5.5	21	220	51.00	57.0	61.0	68.0	77.0	84.0
7	24.5	3.30	3.9	4.7	5.6	6.8	8.0	23	264	65.00	72.0	77.0	85.0	97.0	105.0
8	32.0	4.60	5.5	6.6	7.8	9.6	11.4	25	312	80.00	88.0	95.0	105.0	119.0	123.0
9	40.5	6.20	7.4	8.8	10.4	13.0	15.0	27	364	97.00	107.0	115.0	127.0	144.0	146.0
10½	55.1	9.20	10.8	13.0	15.4	19.0	22.0	29	420	116.00	128.0	137.0	152.0	168.0	168.0
12	72.0	12.80	15.2	18.0	21.0	26.0	29.0	31	480	137.00	151.0	162.0	180.0	192.0	192.0
13½	91.1	17.20	20.0	24.0	29.0	35.0	36.0	33	544	160.00	177.0	189.0	210.0	220.0	220.0
15	112.5	22.00	26.0	32.0	37.0	45.0		37	684	215.00	235.0	250.0	275.0	275.0	
17	144.5	30.00	36.0	43.0	51.0	58.0		41	840	275.00	305.0	325.0	335.0	335.0	
								45	1,012	345.00	385.0	405.0	405.0		

The **minimum oil-film thickness** h_0, in, between the shoe and the collar, at the trailing edge of the shoe, is approximately

$$h_0 = 0.26\sqrt{\mu u l / P_{avg}}$$

where μ is the absolute viscosity; u is the velocity of the collar, on the mean diam; l is the length of a shoe, at the mean diam of the collar, in the direction of sliding motion; P_{avg} is the average load on the shoes. As indicated in Fig. 6.4.40, $b = l$, approximately. The standard thrust bearings have six shoes. Load-carrying capacities of Kingsbury thrust bearings are given in Table 6.4.3.

The coefficient of friction in Kingsbury thrust bearings, referred to the mean diameter of the shoes, is approximately $f = 11.7h_0/l$, where h_0 is computed as shown above. Figures 6.4.41 and 6.4.42 show typical pivoted segmental thrust bearings. They usually embody a system of rocking levers which are used for alignment and equalization of load on the several shoes (Fig. 6.4.43).

Thrust may be carried on a hydrostatic step bearing as shown schematically in Fig. 6.4.44, where high-pressure oil at P_o is supplied at the center of the bearing from an external pump. The lubricant flows radially outward through the annulus of depth h_0 and escapes at the periphery of the shaft at some pressure P_1 which is usually at atmospheric pressure. An oil film will be present whether the shaft rotates or not. Friction in these bearings can be made to approach zero, depending upon the rotational velocity and the viscosity of the lubricant film. Figure 6.4.45 shows the step bearing of a vertical turbogenerator. The load-carrying capacity is

$$W = \frac{P_o \pi}{2} \frac{R^2 - R_0^2}{\ln(R/R_o)}$$

This equation is valid even when the recess is eliminated, in which case R_o becomes the radius of the inlet oil supply pipe. The volume flow rate of lubricant and the clearance are related thus:

$$Q = P_o \pi h_0^3 [6\mu \ln(R/R_o)]$$

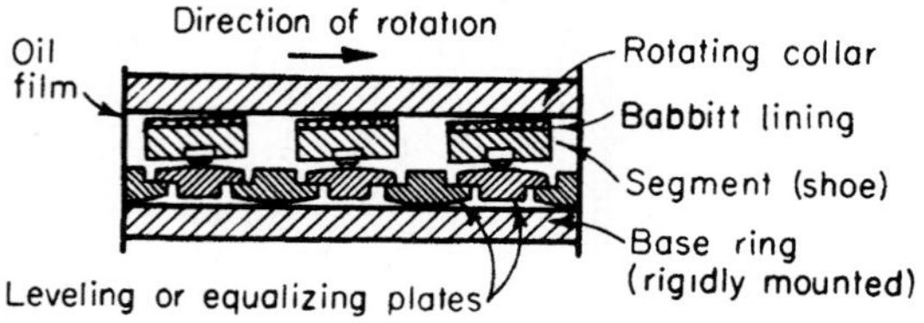

Figure 6.4.43 Kingsbury thrust bearings. (Developed cylindrical sections.)

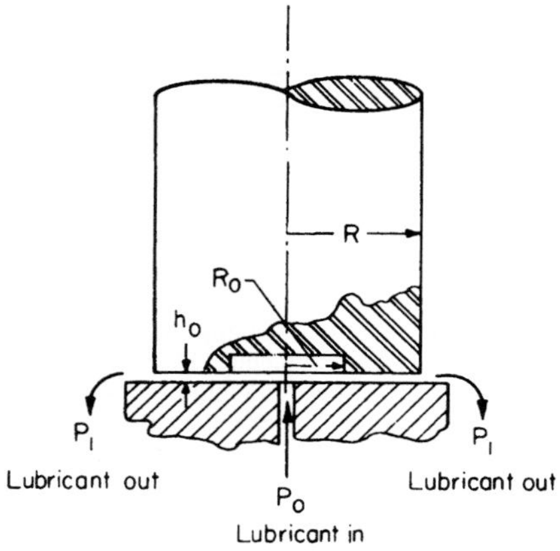

Figure 6.4.44 Hydrostatic step bearing.

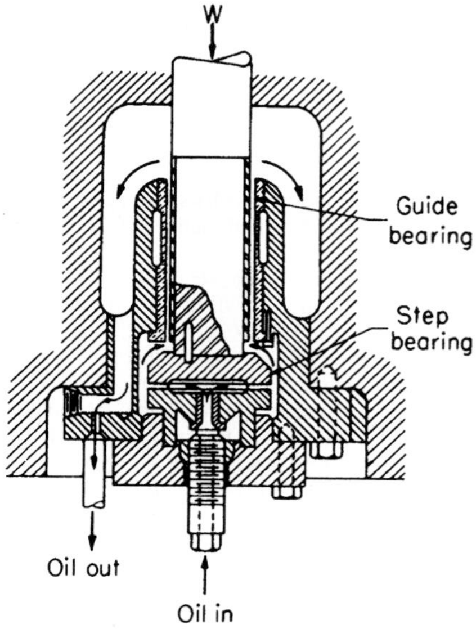

Figure 6.4.45 Step bearing of a vertical turbogenerator.

The friction power loss in the bearing is

$$H_f = \frac{\pi\mu\omega^2(R^4 - R_o^4)}{2h_o}$$

The pumping power loss in forcing the lubricant through the bearing is $H_p = Q(P_o - P_1)/\eta$ where η is the efficiency of the pump.

EXAMPLE 10. A typical 5,000-kW vertical turbogenerator has a thrust load of about 101,000 lb; outside diameter of bearing, 16 in; diam of recess, 10 in; pump efficiency, 0.5; speed, 750 r/min. Substituting these values,

$$101{,}000 = \frac{P_o\pi}{2}\left[\frac{8^2 - 5^2}{\ln(8/5)}\right]$$

or

$$P_o = 774 \text{ lb/in}^2$$

In practice, about 825 lb/in^2 is used on this step bearing to provide some margin of safety. Film thickness in the bearing should be from 0.001 to 0.01 in to protect the surfaces from metal-to-metal contact and allow passage of harmful grit that may find its way into the system. The film thickness determines the oil flow for a given viscosity and pressure. With h_0 = 0.006 in (0.1524 mm) and SAE 20 oil at 130°F (29 centipoises), $Q = 8.25 \times \pi \times (0.006)^3/6 \times 29 \times 1.45 \times 10^{-7} \times 0.470$ = 46.8 in^3/s (766.91 cm^3/s). Flow = 46.8 × 60/231 = 12.15 gal/min (45.99 *L*/min). The horsepower lost due to friction, $H_f = 750^2 \times 29 \times 1.45(8^4 - 5^4)/383{,}000 \times 0.006$ = 3.58 hp (2.669 kW). The horsepower lost due to pumping with pump efficiency of 0.5, H_p = 46.8 × 825/6,600 × 0.5 = 11.7 hp (8.725 kW). The total energy lost = 11.7 + 3.58 = 15.28 hp (11.39 kW).

Evaluation of these equations for other film thicknesses will show that the minimum lost energy will occur between h_0 = 0.004 and h_0 = 0.006 in (0.1016 and 0.1524 mm). The coefficient of friction corresponding to an energy loss of 15.28 hp in the above example is 0.002.

Other configurations for hydrostatically lubricated thrust bearings are shown in Fig. 6.4.46 from Loeb and Rippel. They may be used directly to obtain the value of load-carrying capacity *W* and flow rate *Q*.

LINEAR SLIDING BEARINGS

All sliding bearings (Fig. 6.4.47), to wear true, must have the sliding parts of nearly equal lengths. Bearings made in this way will be found not to wear out of true. Oiling is accomplished in several ways, an acceptable method being that shown in Fig. 6.4.48. Short slides in many machine tools are lubricated by oil pads or direct oil application.

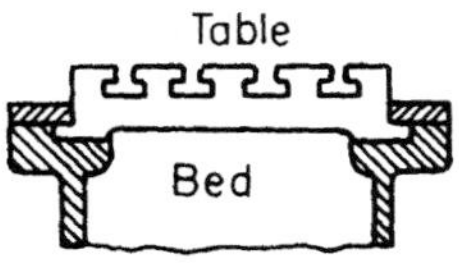

Figure 6.4.47

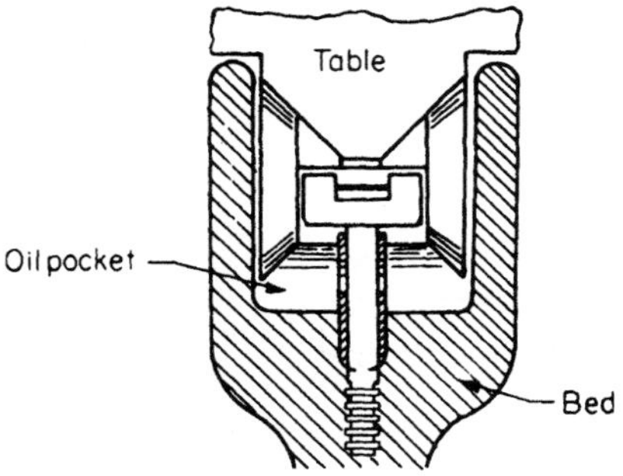

Figure 6.4.48

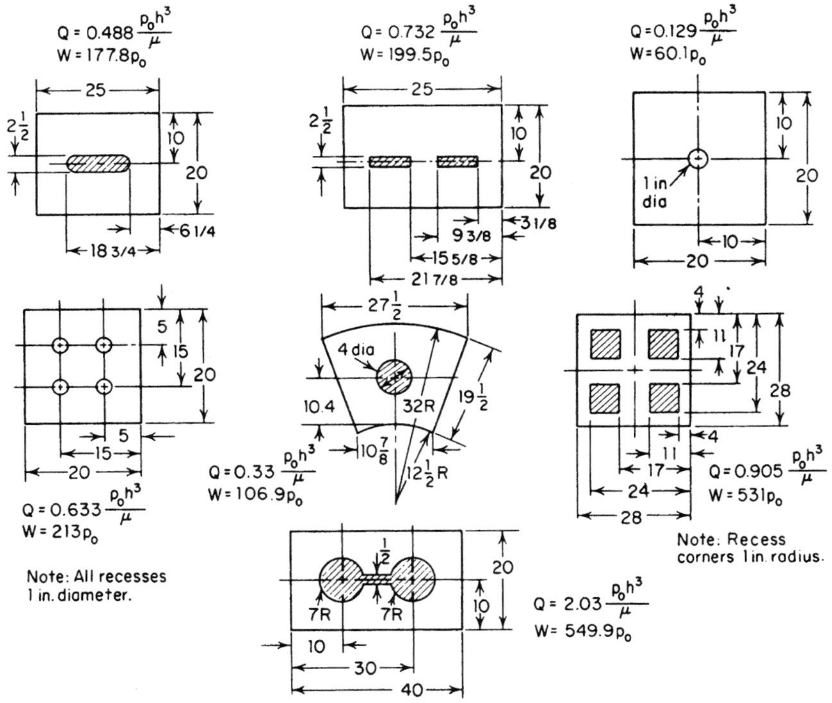

Figure 6.4.46 Load-carrying capacity and flow for several flat thrust bearings *(Loeb)*. Lengths in inches.

The weight of the table and work and thrust of the tool cause wear on the bottom and sides of the guides. To compensate for the wear in both directions, bearings are sometimes made V-shaped, as shown in Fig. 6.4.49.

Simpler sliding bearings in machine tools are made with provision for adjustment (as shown in Fig. 6.4.50) of which there are many modifications. Recent applications involving hydrostatic lubrication on machine-tool ways have been very successful.

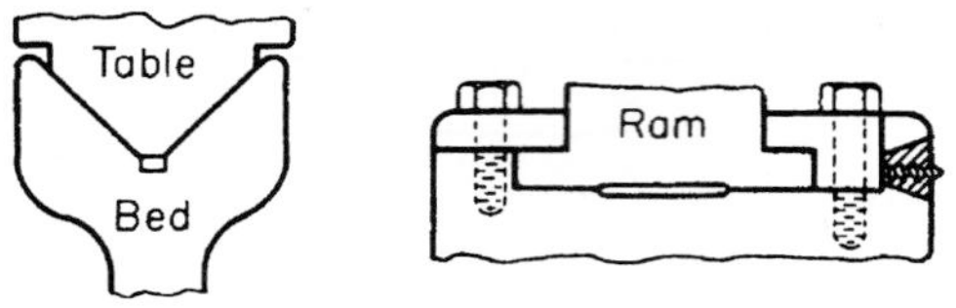

Figure 6.4.49

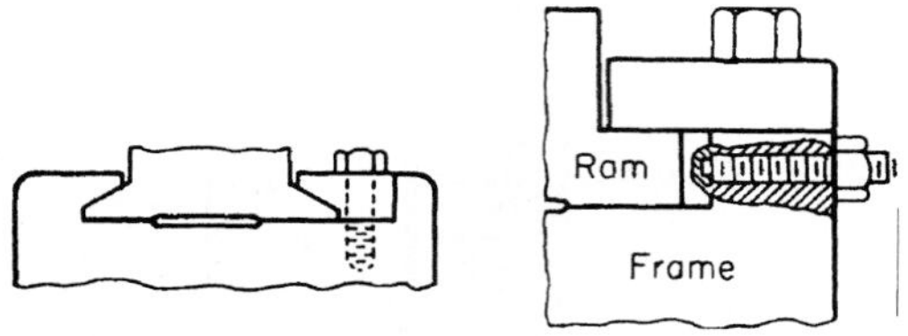

Figure 6.4.50

GAS-LUBRICATED BEARINGS

The fluid-film calculations included in Examples 1 through 10 have assumed that oil (or, in one case, water) was the lubricant. Actually, almost any **process fluid** may be used if proper recognition is given to the viscosity, corrosive action, change in state (where a liquid is close to its boiling point), toxicity, and in the case of a gas, its compressibility. Fluid-film journal and thrust bearings have run successfully, for example, on water, kerosene, gasoline, acid, liquid refrigerants, mercury, molten metals, and a wide variety of gases.

The previous equations for load-carrying capacity, film thickness, friction, and flow may be used for process liquids, but for gases, proper recognition must be made of the compressibility effects.

Because of the great value of gas-lubricated bearings for special applications, and to demonstrate the methods for handling the compressibility action, an introduction to the **design** of **gas-lubricated bearings** follows.

Naturally, if the change in pressure within the bearing clearance is small compared to ambient pressure, the compressibility effect will be likewise small, and lubrication equations based on liquids may be used. A **compressibility parameter** indicates the extent of this action. For hydrodynamic journal bearings it has the form $\Lambda = 6\mu\omega/(P_a m^2)$. For values of Λ less than one, the previous equations of this section for journal bearings may be used. For values of Λ greater than one, compressibility effects are included through the use of Figs. 6.4.51 to 6.4.54. (Data from Elrod and Burgdorfer, Proceedings First International Symposium on Gas-lubricated Bearings, 1959, and Raimondi, *Trans. ASLE*, vol. IV, 1961.)

EXAMPLE 11. Determine the minimum film thickness for a journal bearing 0.5 in (1.27 cm) diameter by 0.5 in long. Ambient pressure 14.7 lb/in² abs (101.34 kN/m² abs). Speed 12,000 r/min. Load 0.4 lb (0.88 kg). Diametral clearance 0.0005 in (0.0127 mm). Lubricant, air at 100°F and 14.7 lb/in² abs (2.68×10^{-9} lb · s/in² from Fig. 6.4.55). $m = 0.0005/0.5 = 0.001$ in/in. $\omega = 12{,}000 \times 2\pi/60 = 1{,}256$ rad/s, $\Lambda = (6 \times 2.68 \times 10^{-9} \times 1{,}256)/14.7 \times 0.001^2 = 1.37$, and $W/(dlP_a) = 0.4/0.5 \times 0.5 \times 14.7 = 0.109$. Then, in Fig. 6.4.53 ($l/d = 1$), we find that $\varepsilon = 0.22$, and the minimum film thickness $h_0 = 0.00025(1 - 0.22) = 0.000195$ in (0.00495 mm).

Gas-lubricated journal bearings should be checked for **whirl stability.** Figure 6.4.56 is applicable with sufficient accuracy to bearings where l/d is equal to or greater than one. It is used in conjunction with Fig. 6.4.51 for $l/d = \infty$. The stability parameter is ω_1^* which, for a bearing having only gravity loading, has the value $\omega_1^* = \omega\sqrt{mr/g}$.

EXAMPLE 12. To determine whether the bearing of Example 11 is stable at the running speed of 12,000 r/min, we compute ω_1^* as $1{,}256\sqrt{0.00025/386} = 1.015$. The value of eccentricity ratio ε_0 for $l/d = \infty$ is computed from Fig. 6.4.51 on the basis of W being the load per inch of bearing length. Thus W(lb/in) = 0.4/0.5 = 0.8 lb/in. For Fig. 6.4.51, $W/(dlP_a) = 0.218$ and in Fig. 6.4.51, we determine $\varepsilon_0 = 0.18$. Then (in Fig. 6.4.56), for $\omega_1^* = 1.015$ and $\Lambda = 1.37$, we find the intersection at about where a curve for $\varepsilon_0 = 0.18$ would be found. The bearing should just be stable. An intersection point on the ε_0 line or to the left should represent a stable condition. An intersection point to the right of the appropriate ε_0 line would predict an unstable condition.

The plain cylindrical journal bearing (360°) is the least stable of possible bearing designs. Control of "half-frequency" whirl has been achieved and the threshold of instability has been raised through modification of the geometry of the plain bearing. The simplest modification is the insertion of axial grooves. Bearings with three or four such grooves have been successful, but lose much stiffness.

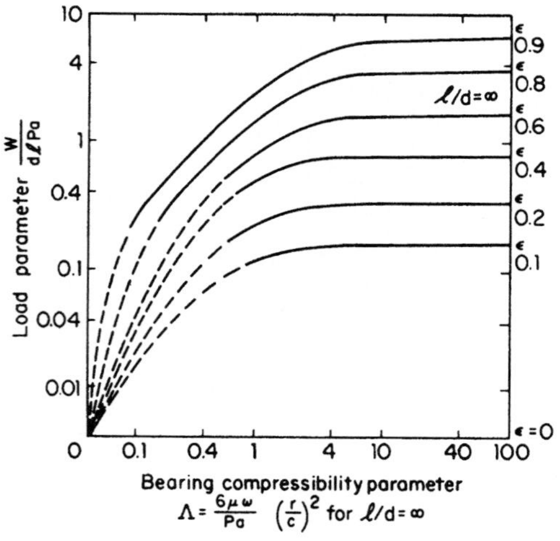

Figure 6.4.51

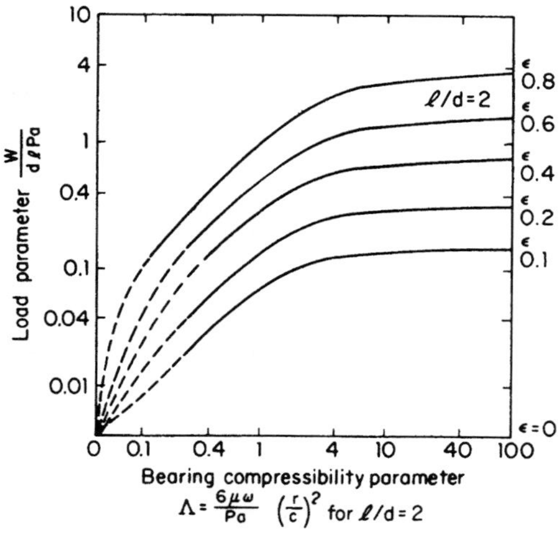

Figure 6.4.52

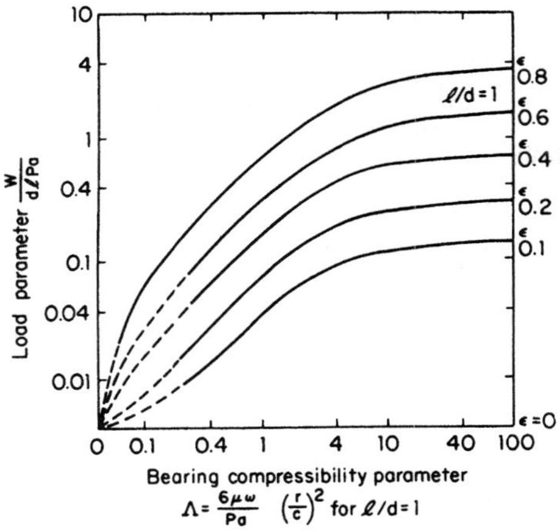

Figure 6.4.53

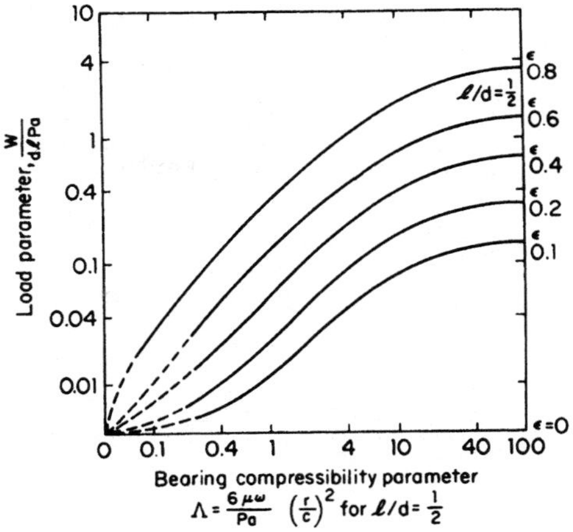

Figure 6.4.54

Figures 6.4.51-6.4.54 Theoretical load-carrying parameter vs. compressibility parameter for a full journal bearing: $l/d = \infty$. (Fig. 6.4.51), $l/d = 2$ (Fig. 6.4.52), $l/d = 1$ (Fig. 6.4.53), and $l/d = 0.5$ (Fig. 6.4.54). *(Elrod and Raimondi.)*

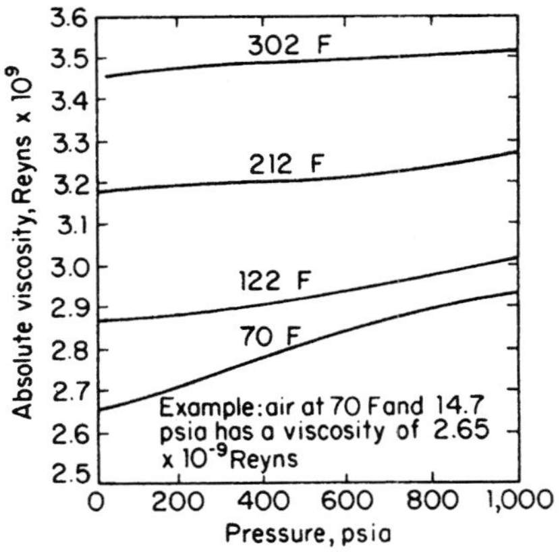

Figure 6.4.55 Absolute viscosity of air. *(Iwaski, Sci. Rpts., Research Inst., Tohuku Univ., Ser. A; Kestin and Pliarczyk, Trans. ASME, 56,* 1954.) Reyns = 1.45×10^{-7} cP.

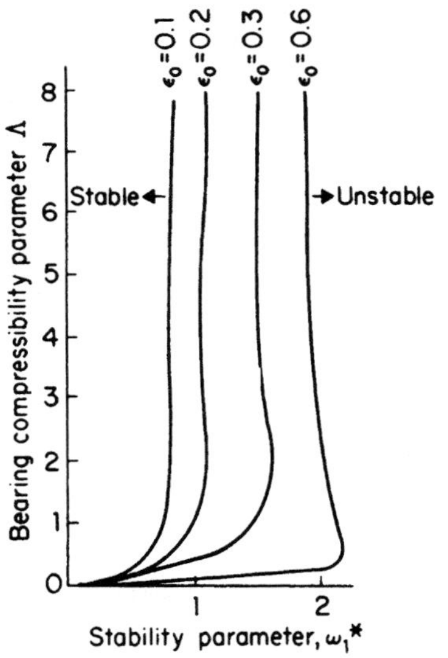

Figure 6.4.56 Half-frequency translatory whirl threshold for infinite length, 360° journal bearing. *(Castelli and Elrod, Solution for the Stability Problem for 360°, Self Acting Gas Lubricated Bearings, Trans. ASME,* **87,** *Mar. 1965.)*

A typical three-groove (three-sector) bearing is shown in Fig. 6.4.57.

Half-frequency whirl indicates a dynamic instability in which the journal orbits at approximately one-half of the shaft rotational speed, which coincides with the average speed of the lubricant in the film. If one looked at the bearing geometry from a coordinate system rotating at this whirl speed, one would see the journal attempting to pump lubricant in one direction and the bearing attempting to do the opposite. The capacity of the film to sustain any load is thus greatly diminished, and failure often occurs. Half-frequency whirl can occur when the dynamic system in which the bearing operates has a natural frequency at approximately one-half the speed of rotation. If the energy dissipation rate is not sufficiently large, instability occurs. In the dynamic system mentioned above, the stiffness and damping characteristics of the bearing, or bearings, play a major role. Damping arises from squeezing the lubricant in and out of the bearing by the action of the journal vibrations against the viscous resistance. When the journal moves, gas can compress and act as a capacitance rather than flow and act as a damper. Therefore, gas bearings are much more prone to instability than are liquid-lubricated ones.

Aside from avoiding resonance conditions, two often successful techniques for whirl prevention are shown in Figs. 6.4.58 and 6.4.59. The first depicts a complete journal bearing with a rotating geometric artifact causing a synchronous disturbance. This artifact can be a variation in the clearance or an asymmetric mass leading to dynamic unbalance. The second depicts a bearing geometry with shallow, inward-pumping spiral grooves. The depth of the grooves is approximately twice the radial clearance, their angle is from 25° to 30° to the axis, the land-to-groove ratio is 0.5, and the axial extent of the grooved area is one-half the length of the bearing. The grooves can be on either the stator or the rotor, depending on manufacturing convenience. Etching is a common method for production of the grooves. Design data for this excellent type of self-acting gas bearing can be found in Sec. 4 of Gross's book. (See References.)

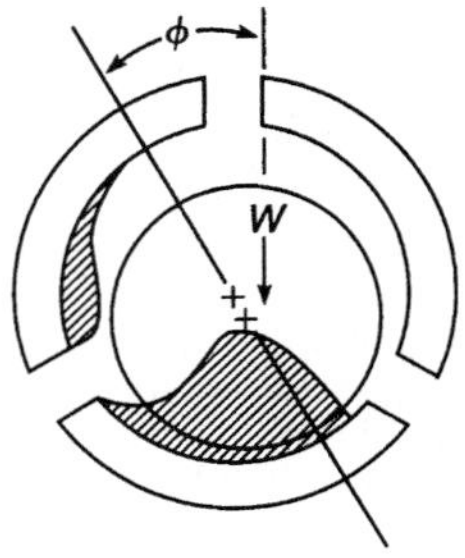

Figure 6.4.57 Sector sleeve bearing.

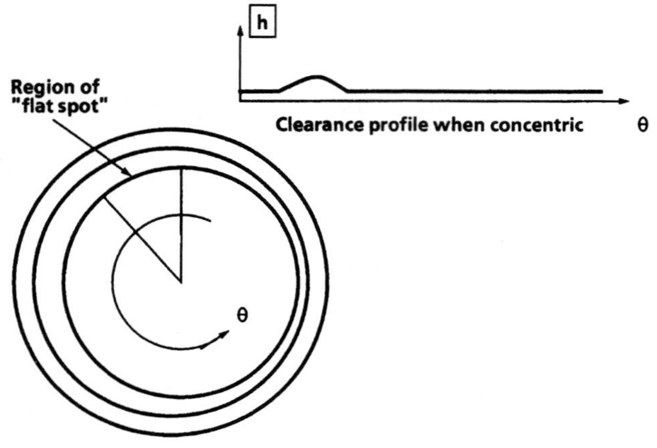

Figure 6.4.58 Stabilizing rotating geometry. (Clearance greatly exaggerated.)

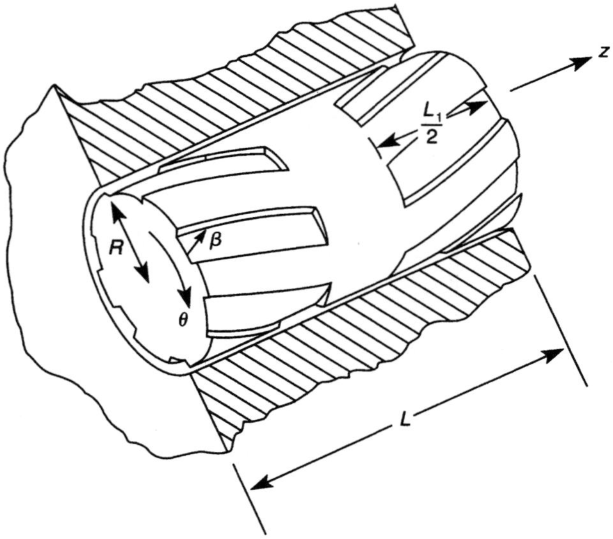

Figure 6.4.59 Spiral groove bearing shown as outward-pumping. (Clearance is greatly exaggerated.)

The tilting-pad journal bearing in Fig. 6.4.60 is considered to be one of the most stable of all possible designs. The shoes, or pads, are supported on rounded pins and are free to pitch and roll through very small angles. Analysis shows that this freedom to move achieves the stability characteristics of these bearings, and also, of course, permits them to be self-aligning. Three-, four-, and five-shoe configurations are often used. Figure 6.4.60 shows an early design used for machine tool grinding spindles.

When applied to gas lubrication, pressurized gas may be supplied through a drilled pivot (hydrostatic lubrication), as an aid in starting and to provide a reserve of load-carrying capacity (Fig. 6.4.61). See Gunter, Hinkle, and Fuller, Design Guide for Gas-Lubricated, Tilting-Pad Journal and Thrust Bearings, NYO-2512-1, U.S. AEC, I-A 2392-3-1, Nov. 1964, Contract AT 30-1-2512.

A bearing design development that is simple, inexpensive, and very stable is the foil bearing. It is also very tolerant of thermal distortions and possible loss of clearance resulting from elevated temperatures. Probably the most widely used of several designs is shown in Fig. 6.4.62. It consists of overlapping metal shims, anchored at the base end like a cantilever beam. The "free" ends deflect and are able to automatically form their own operating clearance. These bearings are widely used in

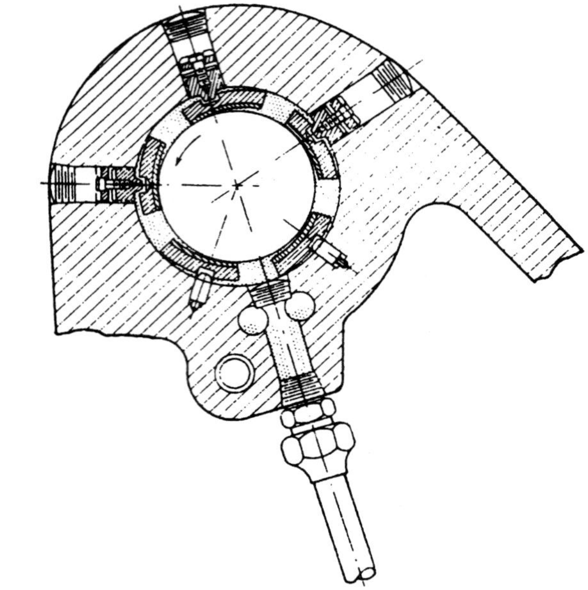

Figure 6.4.60 Filmatic bearing. *(Courtesy Cincinnati Milacron Corp.)*

aircraft cabin cooling and for auxiliary power supply systems (Suriano, Dayton, and Woessner, Test Experience with Turbine-end Foil Bearing Equipped Gas Turbine Engines, ASME Paper 83-GT-73, 1983).

Thrust bearings of the tilting-pad variety are less susceptible to compressibility effects and may be considered as liquid-lubricated for values of Λ (suitable for thrust bearings) less than about 30. $\Lambda = 6u\mu l/(h_0^2 P_a)$ where l is the length of the shoe in the direction of sliding and U is the linear velocity at the mean radius. However, the shoes should not be made flat for gas operation but should have a crowned contour (see Fig. 6.4.63). (Gross, "Gas Film Lubrication," Wiley.) An approximate value for the crown is to make $\delta = \frac{3}{4}h_0$. The tilting-pad bearing design is probably the most common gas bearing presently in existence. Every hard-disk computer memory since the early 1960s has had its read-write heads supported by self-acting tilting-pad sliders. Hundreds of millions of such units, called *flying heads,* have been manufactured to date. Some designs employ the crown geometry while, most commonly, heads with flat multiple sliders with straight ramps in their forward sections are used. The reason for the multiple thin sliders is the achievement of maximum damping possible. The typical minimum film heights have decreased steadily through the years from 1 μm (40 millionths of an inch) 25 years ago to less than 0.2 μm (8 millionths of an inch) currently (1995). This trend is driven by the achievement of the higher and higher recording densities possible at lower flying heights. Design of these devices is done rather precisely from first principles by means of special simulation programs. At these low clearances, allowance must be made for the finiteness of the **molecular mean free path,** which represents the mean distance that a gas molecule must travel between collisions. This effect manifests itself in a lowering of viscosity and wall shear resistance.

Gas-lubricated hydrostatic bearings, unlike liquid-lubricated bearings, cannot be designed on the basis of fixed flow rate. They are designed instead to have a pressure loss produced by an **orifice restrictor** in the supply line. Such throttling enables the bearing to have load-carrying capacity and stiffness. For maximum stiffness the pressure drop in the orifice may be about one-half of the manifold supply pressure. For a circular thrust bearing with a single circular orifice, the load-carrying capacity is given with sufficient accuracy by the equation previously used for liquids (see Fig. 6.4.44). $W = (P_R - P_a/2)[R^2 - R_0^2/\ln(R/R_0)]$ where P_R is the recess pressure, lb/in^2 abs. The flow volume, however, is given by $Q_0 = \pi h_0^3/[6\mu \ln(R/R_0)](P_0^2 - P_1^2)/2P_0$. Q_0 and P_0 refer to recess conditions, and Q_1 and P_1 refer to ambient conditions. Pressures are absolute.

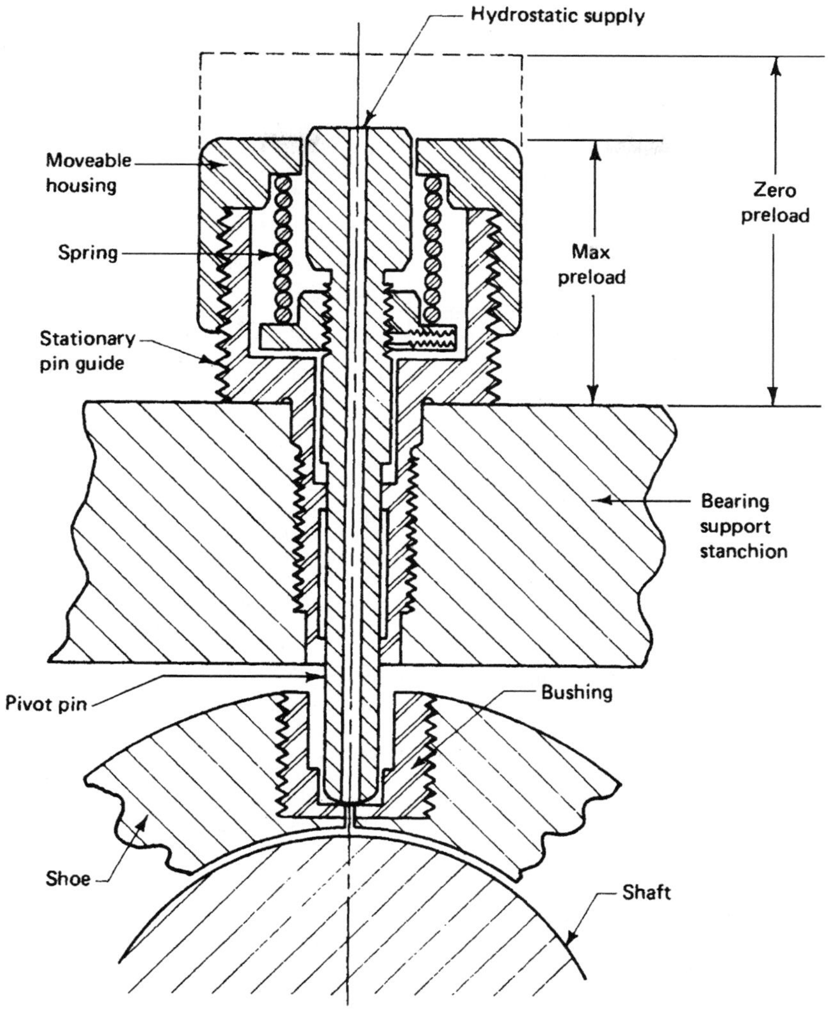

Figure 6.4.61 Cross-sectional view, spring-mounted pivot assembly. *(Courtesy of The Franklin Institute Research Labs.)*

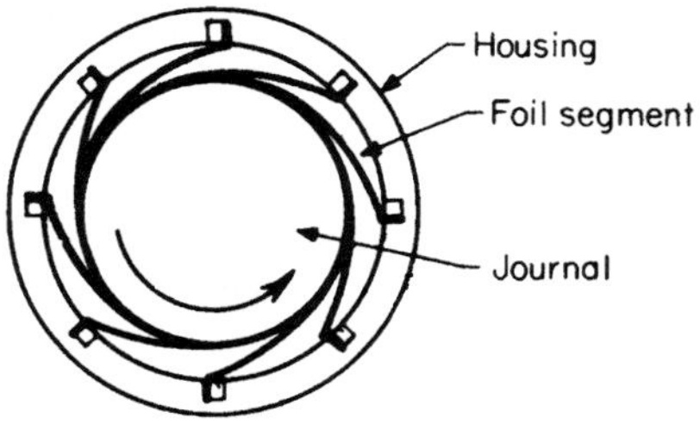

Figure 6.4.62 Bending-dominated segments foil bearing.

EXAMPLE 13. A circular thrust bearing 6 in (15.24 cm) diameter with a recess 2 in (5.08 cm) diameter has a film thickness of $h_0 = 0.0015$ in (0.0381 mm). $P_0 =$ 30 lb/in^2 gage or 44.7 lb/in^2 abs (308.16 kN/m^2). P_1 is room pressure, 14.7 lb/in^2 abs (101.34 kN/m^2 abs). Depth of recess is 0.02 in. Applied load is 375 lb. $Q_0 = (\pi \times 0.0015^3)/(6 \times 2.68 \times 10^{-9}\text{ln}^3)(44.7^2 - 14.7^2)/(2 \times 44.7)$, $Q_0 = 12.3$ in^3/s (201.6 cm^3/s) at recess pressure. Converted to free air, $Q_1 = Q_0(P_0/P_1)$ with isothermal expansion, $Q_1 = 12.3\ (44.7/14.7) = 37.4$ in^3/s (612.87 cm^3/s), or $Q_1 = 37.4 \times 60 =$ 2,224 in^3/min (36.77 L/min). Actual measured flow = 2,440 in^3/min (39.98 L/min).

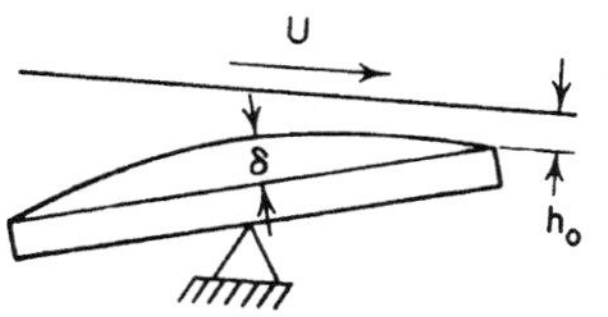

Figure 6.4.63 Schematic of tilting-pad shoe, showing crown height δ.

Externally pressurized gas bearings are not as easily designed as liquid-lubricated ones. Whenever a volume larger than approximately that of the film is present between the restrictor and the film, a phenomenon known as **air hammer** or **pneumatic instability** can take place. Therefore, in practical terms, recesses cannot be used and orifice restrictors must be obtained by the smallest flow cross-section at the very entrance to the film; this area is equal to the perimeter of the inlet holes multiplied by the local height of the film. This technique is called **inherent compensation.** Unfortunately, as one can readily see, the area of the restrictors is smaller where the film is smaller; thus, the stiffness is lower than that obtainable by incompressible lubrication. Design data are available in Sec. 5 of Gross's book (see References).

6.5 BEARINGS WITH ROLLING CONTACT

by Steven V. Boyd, Michael D. Allega, and Sarah A. Meyer

REFERENCES: (ANSI/ABMA 9) – Load and Fatigue Life For Ball Bearings; (ANSI/ABMA 11) – Load and Fatigue Life For Roller Bearings; (ANSI/ABMA 20) – Radial Bearings of Ball, Cylindrical Roller and Spherical Roller Types – Metric Design; ISO 15 – Rolling bearings – Radial bearings – Boundary dimensions, general plan; ISO 355 – Roller bearings – Tapered roller bearings – Boundary dimensions and series designations.

COMPONENTS AND SPECIFICATIONS

Rolling contact bearings are designed to support and locate rotating shafts or parts in machines. They transfer loads between machine components with relative motion and permit rotation with a minimum of friction. They consist of rolling elements between an outer ring and an inner ring. Cages are used to space the rolling elements from each other. Figure 6.5.1 illustrates the common terminology used in describing rolling contact bearings.

Rings The inner and outer rings of a rolling contact bearing are typically manufactured from through hardening or case hardening grades of steel, hardened to Rockwell C 58 to 62. The rolling element raceways are accurately ground or otherwise processed in the rings to a very fine finish.

Rings are available for special purposes in such materials as stainless steel, ceramics, and polymers. These materials are used in applications where extreme conditions of speed, corrosion, or contaminants may exist.

Rolling Elements Typically, the rolling elements are made of the same material and finished like the rings. Other rolling element materials, such as stainless steel, ceramics, and polymers, can be used in conjunction with various ring materials. Bearings that use ceramic rolling elements with steel rings are referred to as hybrid bearings. These harder and lighter rollers offer advantages in high speed and extreme conditions.

Cages Sometimes called separators or retainers, cages are used to space the rolling elements from each other. Cages are furnished in a wide variety of materials and construction. The most common cage designs are pressed steel, filled nylon, and machined. Solid machined cages are used where greater strength or higher speeds are required. They can be manufactured from brass, steel, or phenolic-type materials. Bearings without cages are referred to as full-complement.

A wide variety of rolling contact bearings are normally manufactured to standard boundary dimensions (bore, outside diameter, width) and tolerances that have been standardized by the ABMA. All bearing manufacturers conform to these standards, thereby permitting interchangeability. ANSI has for the most part adopted these and published them jointly as ABMA/ANSI standards as follows:

Title	*Standard*	*Title*	*Standard*
Terminology	1	Ball Standards	10
Gaging Practice	4	Roller Load Ratings	11
Mounting Dimensions	7	Instrument Bearings	12
Mounting Accessories	8.2	Vibration and Noise	13
Ball Load Ratings	9	Basic Boundary Dimensions	20

The International Organization for Standardization (ISO) also has published standards defining these dimensional rules. The boundary dimensions for standard metric bearings are published in ISO 15 for roller bearings, ISO 104 for thrust bearings, and ISO 355 for tapered roller bearings.

The Annular Bearing Engineers Committee (ABEC) of the ABMA has established progressive levels of precision for ball bearings designated as ABEC-1, ABEC-5, ABEC-7, and ABEC-9. Similarly, roller bearings have established precision levels known as RBEC-1 and RBEC-5. These standards specify tolerances for bore, outside diameter, width, and radial run-out.

PRINCIPAL STANDARD BEARING TYPES

The selection of the type of rolling contact bearing depends upon many considerations, as evidenced by the numerous types available. Furthermore, each basic type of bearing is furnished in several standard "series" as illustrated in Fig. 6.5.2. Although the bore is the same, the outside diameter, width, and rolling element size are progressively larger. The result is that a wide range of load-carrying capacity is available for a given size shaft, thus giving designers considerable flexibility in selecting standard-size interchangeable bearings. Figure 5.5.2 shows a sampling of the dimensional series sizes defined by ABMA/ISO.

The basic dimensional rules follow simple nomenclature using a bearing designator prefix followed by a two-digit dimensional series and a two-digit bore size code. A basic part number 22232 indicates a spherical bearing with a 22 dimensional series and a 32 bore code (bore = 32 × 5 mm). An NU2232 would be a cylindrical roller bearing with the same envelope dimensions. These rules apply for ball and roller bearings with the exception of tapered roller bearings, which have slightly different envelope dimensions.

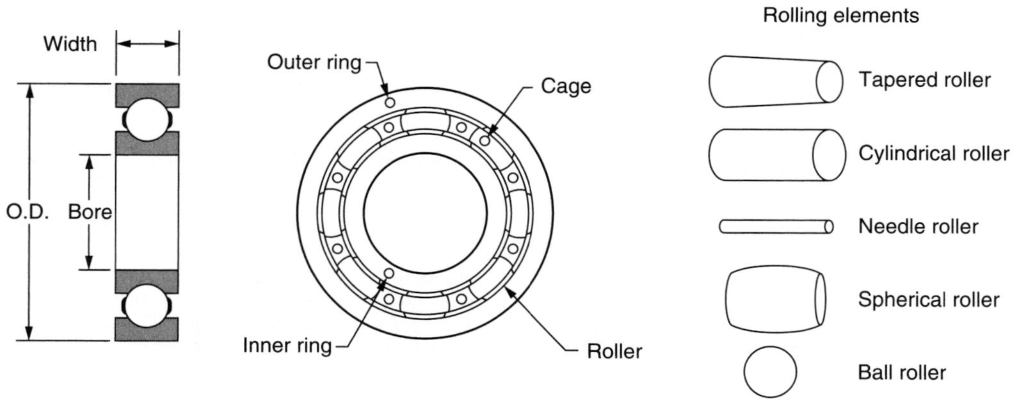

Figure 6.5.1 Rolling contact bearing terminology.

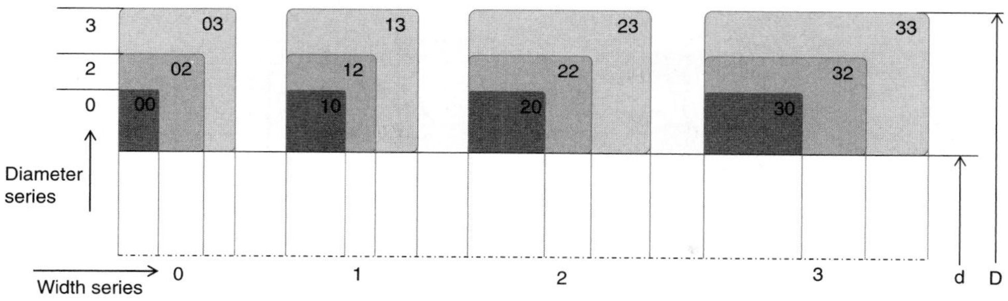

Figure 6.5.2 Bearing standard series.

BALL BEARINGS

Radial Ball Bearing (Fig. 6.5.3) The single row radial ball bearing is often referred to as a deep groove ball bearing or Conrad style ball bearing. These are available in various configurations and sizes and include sealed or shielded bearings. Radial ball bearings are also available in maximum capacity or loading slot designs (Fig. 6.5.4). Conrad style or deep groove ball bearings are suitable for both radial and axial loads in both directions. Maximum capacity designs provide higher radial load capacity, but have limited thrust capacity because of the loading slots.

Double-Row Ball Bearing (Fig. 6.5.5) Radial ball bearings are also available in double-row designs. Double-row bearings are available in both Conrad style and maximum capacity designs. They can be of radial or angular contact design as well. Conrad style designs provide axial load capability in both directions, while maximum capacity designs have limited thrust capability in one direction.

Angular Contact Ball Bearing (Fig. 6.5.6) These bearings are designed to support combined radial and thrust loads or heavy thrust loads depending on the contact angle magnitude. They may be mounted in pairs (Fig. 6.5.7), in which case they are referred to as back-to-back, tandem, or face-to-face duplex bearings. These bearings may be preloaded to minimize axial movement and deflection of the shaft. They are commonly manufactured as hybrid bearings using ceramic balls and steel races.

Internal Self-Aligning Double-Row Ball (Fig. 6.5.8) This bearing design utilizes a spherical curvature on the outer ring raceway that allows for self-alignment of the inner and outer ring, with no detrimental effect, within a specific maximum angle based on design and manufacturer. They are intended primarily to support radial loads with minimal axial load capability. They are often used in mounted units or pillow blocks.

ROLLER BEARINGS

Cylindrical Roller Bearing (Fig. 6.5.9) These bearings utilize cylindrical rollers with approximate length/diameter ratios ranging from 1:1 to 3:1. These bearings are normally used for heavy radial loads. They are especially useful for free axial movement of the shaft. Various design types with single or double flanges on the inner rings or outer rings are available. This allows for limited thrust load capability and system axial positioning. Special configurations of double- and quadruple-row designs are also available. Cylindrical bearings tend to offer lower friction compared to other roller bearings because there is minimal flange contact or sliding.

Needle Roller Bearing (Fig. 6.5.10) These bearings have rollers whose length/diameter ratio is at least 4:1. They are most useful where space is a factor and are available with or without an inner race. If the shaft is used as the inner race, it must be hardened and precisely ground. Caged types are typical in rotational applications, while full-complement types are often used for high loads, oscillating applications, or slow speeds. Needle bearings are not designed to support axial loads.

Spherical Roller Bearing (Fig. 6.5.11) These bearings utilize spherically shaped (barrel shaped) rolling elements, oriented in two rows. These bearings are excellent for heavy radial loads and moderate thrust loads. The design allows for self-alignment of the inner and outer ring, with no detrimental effect, within a specific maximum angle based on design and manufacturer. They are commonly used in mounted units or pillow blocks.

Tapered Roller Bearing (Fig. 6.5.12) These bearings are used for heavy combined radial and thrust loads. The bearing is designed so that all elements in the rolling surface and the raceways intersect at a common point on the axis; thus, true rolling motion is obtained. Bearing designs are available in a multitude of different angle configurations depending on the ratio of radial to axial loads required. The separable races allow for adjustment in bearing clearance. Endplay or preload can be established at the time of assembly or supplied as preset assemblies. Where maximum system rigidity is required, the bearings can be adjusted for a preload. They are commonly available in double row (Fig. 6.5.13) or other multiple-row designs.

Toroidal Roller Bearing (Fig. 6.5.14) These bearings utilize spherically shaped (barrel shaped) rolling elements, oriented in a single row. The raceways have conforming radii. They offer features similar to a spherical bearing and a cylindrical bearing in one design. These features result in a single-row bearing design that can self-align and float axially. The amount of allowable misalignment and axial float are dependent on the specific design series and manufacturer.

THRUST BEARINGS

Cylindrical Roller Thrust Bearing (Fig. 6.5.15) These bearings use cylindrical rolling elements installed between flat races. They can be designed with a single path of rollers, or with multiple shorter roller sets (as seen in Fig. 6.5.16). The shorter rollers can help minimize roller sliding and twisting. They may be used for moderate speeds and loads. Radial loads cannot be supported by this bearing design.

Needle Thrust (Fig. 6.5.17) These bearings utilize needle rollers installed between flat races or mounted directly on other machinery components. They can withstand axial loads in applications where space is very limited. No radial load can be supported by this bearing design.

Thrust Spherical Roller Bearing (Fig. 6.5.18) This design uses an asymmetric spherically shaped rolling element. This design can withstand combined radial and axial loads and allows for self-alignment depending on series, design and manufacturer. Maximum speeds are limited due to raceway sliding and flange/roller end contact.

Thrust Tapered Roller Bearing (Fig. 6.5.19) This thrust bearing design utilizes tapered rollers along with tapered or flat raceways. They can withstand high axial loads. The tapered roller design allows for long rollers and minimal sliding, resulting in a compact bearing with very high thrust capability.

Figure 6.5.3

Figure 6.5.4

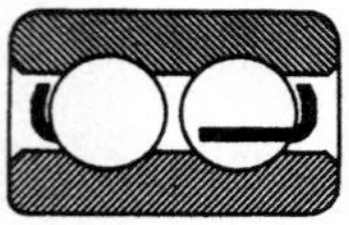
Figure 6.5.5

Figure 6.5.6

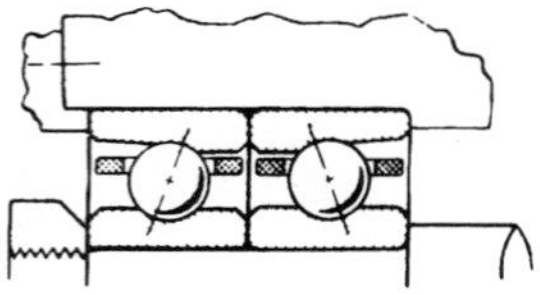
Figure 6.5.7

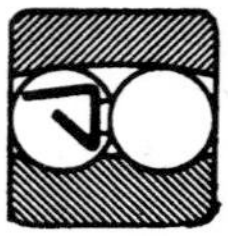
Figure 6.5.8

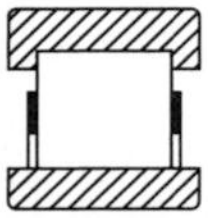
Figure 6.5.9

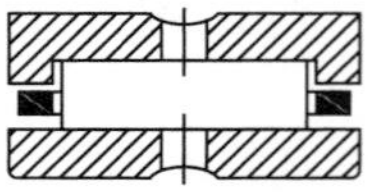
Figure 6.5.10

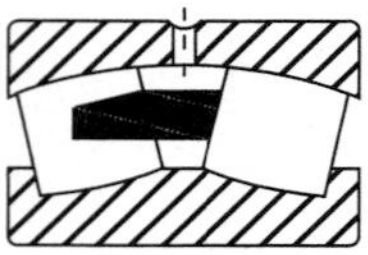
Figure 6.5.11

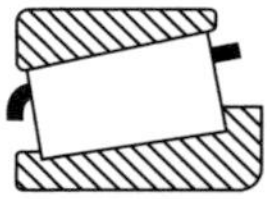
Figure 6.5.12

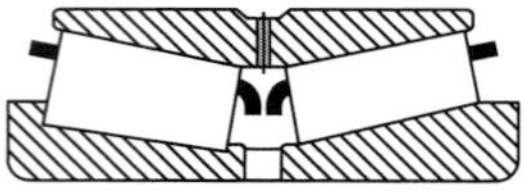
Figure 6.5.13

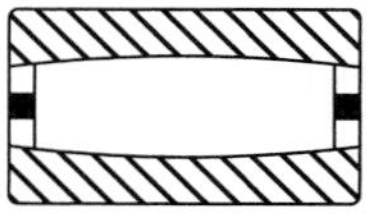
Figure 6.5.14

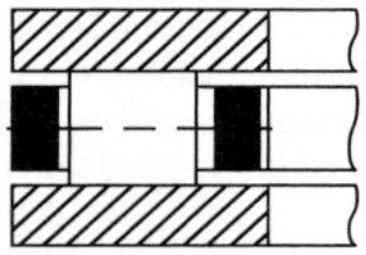
Figure 6.5.15

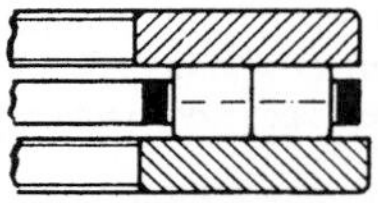
Figure 6.5.16

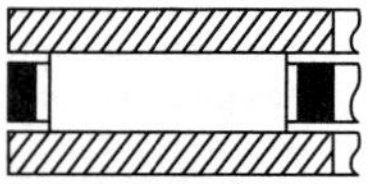
Figure 6.5.17

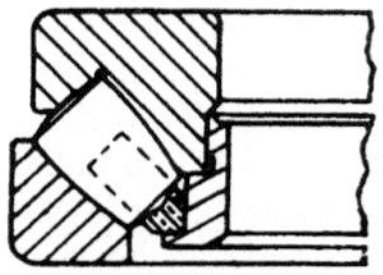
Figure 6.5.18

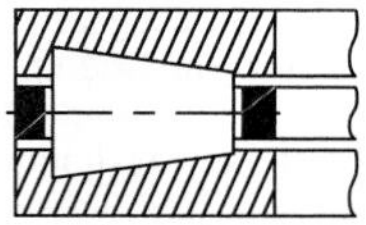
Figure 6.5.19

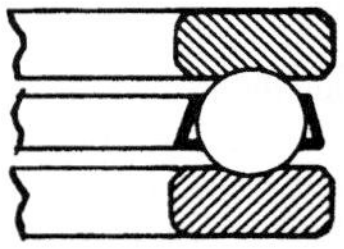
Figure 6.5.20

Thrust Ball Bearing (Fig. 6.5.20) This thrust bearing design utilizes ball rollers. These bearings can withstand moderate thrust loads and a minimal amount of radial load.

SPECIAL BEARINGS

Linear Ball Bearing (Ball Bushing) (Fig. 6.5.21) This type of bearing is used for linear motions on hardened shafts (Rockwell C 58 to 62). Some special types can be used for linear and rotary motion.

Split Type Bearings (Fig. 6.5.22) These special types of bearings have split inner rings, outer rings, and cages. The split halves are assembled by screws over the shaft. Split designs are available in ball bearings, cylindrical bearings, and spherical bearings. This feature can be useful where it is difficult to install or remove a solid bearing.

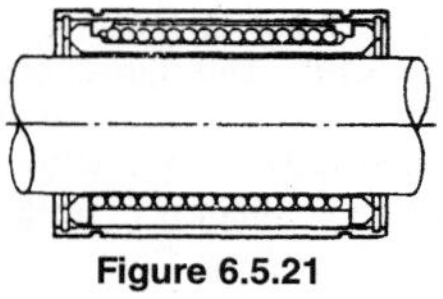
Figure 6.5.21

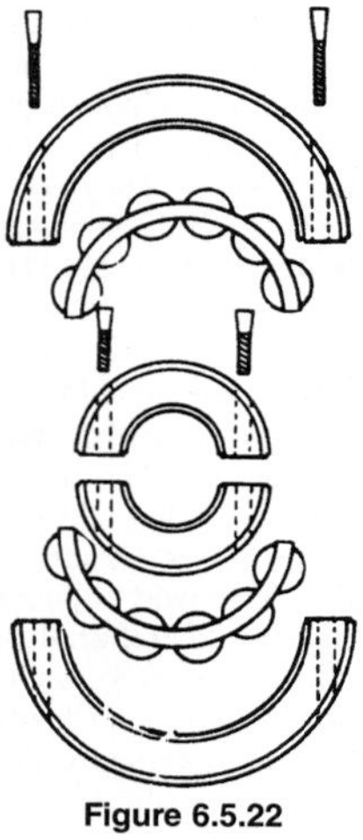
Figure 6.5.22

BEARING TYPE SELECTION AND KEY CHARACTERISTICS

Selection of the type of rolling element bearing is a function of many factors such as load, speed, misalignment sensitivity, space limitations, and the desire for precise shaft positioning. Table 6.5.1 summarizes some key bearing and application characteristics important in bearing selection. There is often more than one bearing solution that could work in an application, and the designer must weigh the factors that are most critical.

Ball bearings function on theoretical point contact and are generally more suitable for higher speeds and lighter loads than roller bearings. Roller bearings operate with line contact, which is generally more suitable for higher loads relative to bearing size. There are roller bearings that allow misalignment, some that float axially, and other types that provide increased stiffness. Table 6.5.1 helps rank these and other key characteristics.

ROLLING CONTACT BEARING LIFE, LOAD, AND SPEED RELATIONSHIPS

An accurate knowledge of the load carrying capacity and expected life is essential in the proper selection of ball and roller bearings. Bearing life is typically defined in terms of material-related fatigue. For this reason, the life of an individual bearing is defined as the total number of revolutions or hours at a given constant speed at which the bearing operates before the first evidence of fatigue develops. Life predictions are based on a statistical evaluation of a large number of bearings operating under the same conditions. The Weibull distribution function is the accepted standard for determining the life of a population of bearings. In practice, damage modes other than material fatigue are very common.

Definitions

Rated Life L_{10} The number of revolutions or hours at a given constant speed that 90 percent of an apparently identical group of bearings will complete or exceed before the first evidence of fatigue develops.

Table 6.5.1 Bearing Type Characteristics

Characteristic	Tapered Roller Bearing	Thrust Tapered Roller Bearing	Cylindrical Roller Bearing	Thrust Cylindrical Roller Bearing	Spherical Roller Bearing	Thrust Spherical Roller Bearing	Radial Ball Bearing	Thrust Ball Bearing	Angular Contact Ball Bearing
Pure Radial Load	◕	○	●	○	◕	○	◕	◔	◑
Pure Axial Load	◕	●	○	◕	◑	●	◑	◑	◕
Combined Load	●	◔	◑	○	◕	◑	◕	◔	●
Moment Load	●	◔	○	○	○	○	◑	◔	◕
Stiffness	●	●	◕	●	◑	◕	◑	◔	◕
Low Friction	◕	◕	●	◔	◑	◑	●	◕	◕
Misalignment	◔	◔	◔	○	●	●	◕	◔	◔
Locating Position (fixed)	●	◑	◑	◕	◕	◕	◕	◕	◕
Non-Locating Position (float)	◕	●	●	○	◕	○	◕	○	◕
Speed	◕	●	●	◔	◑	◑	●	◑	●

Excellent	Good	Fair	Poor	Unsuitable
●	◕	◑	◔	○

In other words, 10 out of 100 bearings will incur material fatigue before rated life is achieved. The terms minimum life and L_{10} life are also used to commonly define rated life.

Basic Dynamic Load Rating *C* The radial load under which a population of bearings will achieve an L_{10} of one million revolutions. Its value depends on bearing type, bearing geometry, accuracy of fabrication, and bearing material. The basic dynamic load rating may also be referred to as the specific dynamic capacity, the basic dynamic capacity, or the dynamic load rating.

Equivalent Radial Load *P* The constant stationary radial load which, if applied to a bearing with a rotating inner ring and stationary outer ring, would give the same life that the bearing will attain under the actual combined loading conditions.

Static Load Rating C_0 The static radial load that produces a maximum contact stress of 580ksi (4,000 MPa) for roller bearings or 609ksi (4,200 MPa) for ball bearings at the center of the most heavily loaded rolling element-raceway contact. This magnitude of stress may leave slight indentation marks on the raceways approximately 0.0001 times the rolling element diameter.

Static Equivalent Load P_0 The static radial load which, if applied, produces a maximum contact stress equal in magnitude to the maximum contact stress in the actual combined loading condition.

Bearing Rated Life

Standard formulas have been developed to predict the statistical rated life of a bearing under any given set of conditions. These formulas are based on an exponential relationship of load to life that has been established from extensive research and testing.

$$L_{10} = \left(\frac{C}{P}\right)^K \times 10^6 \qquad (6.5.1)$$

Where L_{10} = rated life (in revolutions), r; C = basic load rating, lbf; P = equivalent radial load, lbf; K = constant, 3 for ball bearings, 10/3 for roller bearings.

To convert to hours of rated life L_{10}, this formula becomes

$$L_{10} = \frac{10^6}{N \times 60}\left(\frac{C}{P}\right)^K \qquad (6.5.2)$$

Where N = rotational speed, r/min.

Bearing Design Life

The required bearing design life varies significantly based on industry, application, and risk tolerance. Some industries demand specific bearing life targets for machinery. In most cases, the design life is determined by the machine manufacturer or final customer.

Practical examples of bearing design life targets can range from 10,000 hours to 100,000 hours. These targets vary greatly depending on the level of risk associated with the application, or with the detailed understanding of loads and other environmental factors.

Load Rating

The load rating is a function of many parameters, such as number of rolling elements, rolling element diameter, rolling element length, and contact angle. Two load ratings are associated with a rolling contact bearing: basic and static load rating.

Basic Dynamic Load Rating *C* This rating is always used in determining bearing life for all speeds and load conditions, as seen in Eqs. (6.5.1) and (6.5.2).

Basic Static Load Rating C_0 This rating is used only as a check to determine if the maximum allowable stress of the rolling elements will be exceeded. It is never used to calculate bearing life.

Values for C and C_0 are readily attainable in any bearing manufacturer's catalog as a function of size and bearing type. Table 6.5.2 lists some example bearing sizes and load ratings for multiple bearing types. The differences in bearing ratings compared to bore size and type can be seen from this table.

Table 6.5.2 Example Bearing Sizes and Ratings

		Single Row Radial Ball Bearing				Angular Contact Ball Bearing				Double Row Radial Ball Bearing				Cylindrical Roller Bearing				Spherical Roller Bearing				Tapered Roller Bearing			
Bore (mm)	OD (mm)	PN	W (mm)	C (kN)	C0 (kN)	PN	W (mm)	C (kN)	C0 (kN)	PN	W (mm)	C (kN)	C0 (kN)	PN	W (mm)	C (kN)	C0 (kN)	PN	W (mm)	C (kN)	C0 (kN)	PN	W (mm)	C (kN)	C0 (kN)
70	150	**6314**	35	105	70	**7214**	35	75	60	**5314**	63.5	160	165	**NU314**	35	230	225	**21314**	35	290	290	**30314**	38	210	240
70	150	–	–	–	–	–	–	–	–	–	–	–	–	**NU2314**	51	310	330	–	–	–	–	–	–	–	–
100	180	**6220**	34	120	92	**7220**	34	135	125	**5220**	60.32	180	225	**NU220**	34	280	310	**22220**	34	435	505	**30220**	37	250	325
100	180	–	–	–	–	–	–	–	–	–	–	–	–	**NU2220**	46	375	450	–	–	–	–	–	–	–	–
120	180	**6024**	28	85	80	**7024**	28	90	95					**NU1024**	28	150	200	**23024**	46	410	575	**32024**	38	245	410
120	180	–	–	–	–	–	–	–	–	–	–	–	–	–	–	–	–	–	–	–	–	**33024**	48	290	525
120	215	**6224**	40	155	130	–	–	–	–	–	–	–	–	**NU224**	40	380	430	**22224**	58	640	770	–	–	–	–
150	270	**6230**	45	175	166	**7230**	45	220	240	–	–	–	–	**NU230**	45	505	610	–	–	–	–	**30230**	49	470	640
150	270	–	–	–	–	–	–	–	–	–	–	–	–	**NU2230**	73	750	990	**22230**	73	750	1100	**31330**	82	750	1050
150	320	**6330**	65	275	285	–	–	–	–	–	–	–	–	–	–	–	–	–	–	–	–	–	–	–	–

Equivalent Load

In order to calculate bearing L_{10} rated life, the equivalent load must be determined. The combined axial and radial loads are used to determine this equivalent radial load. Under rotation or dynamic conditions, the equivalent radial load P is used in conjunction with C in Eq. (6.5.1) to calculate bearing life. The static equivalent load P_0 is used in comparison with C_0 in applications when a bearing is highly loaded in a static mode.

Equivalent Radial Load *P* All bearing loads are converted to an equivalent radial load. Equation (6.5.3) is the general formula used for both ball and roller bearings.

$$P = XR + YT \tag{6.5.3}$$

Where P = equivalent radial load, lbf; R = radial load, lbf; T = thrust (axial) load, lbf; X and Y = radial and thrust factors (Tables 6.5.3 and 6.5.4).

The empirical X and Y factors in Eq. (6.5.3) depend upon the geometry, loads, and bearing type. Some example X and Y factors can be obtained from Table 6.5.3 for roller bearings and Table 6.5.4 for ball bearings. The values for X and Y vary depending on the ratio of axial to radial load. The tables below show only a sampling of X and Y factors. The full tables of X and Y can be found in the ABMA standards referenced previously.

Table 6.5.3 Radial and Thrust Factors for Roller Bearings

	$\frac{Fa}{Fr} \le e$		$\frac{Fa}{Fr} \ge e$		e
Bearing Type	X	Y	X	Y	
Single-row, $\alpha \ne 0°$	1	0	0.4	0.4 cot α	1.5 tan α
Double-row, $\alpha \ne 0°$	1	0.45 cot α	0.67	0.67 cot α	1.5 tan α

Table 6.5.4 Radial and Thrust Factors for Ball Bearings

	$\frac{Fa}{Fr} \le e$		$\frac{Fa}{Fr} \ge e$		e*
Bearing Type	X	Y	X	Y*	
Single-row	1	0	0.56	1.45	0.3
Double-row radial	1	0	0.56	1.45	0.3
Angular contact, $\alpha = 40°$	1	0	0.35	0.57	1.14

*NOTE: Values for Y depend on the bearing geometry and ratio of axial to radial load. Average values are shown for example only. Refer to the previously mentioned ABMA or ISO standards for further details.

Static Equivalent Load P_0

The static equivalent load may be compared directly to the static load rating C_0. If P_0 is greater than the C_0 rating, permanent deformation of the rolling element is likely to occur. Calculate P_0 as follows:

$$P_0 = X_0 R + Y_0 T \tag{6.5.4}$$

Where; P_0 = static equivalent radial load, lbf; R = radial load, lbf; T = thrust (axial) load, lbf; X_0 and Y_0 = radial and thrust factors (Table 6.5.5).

Oscillating loads, where the motion is such that the rolling element rotates less than half a revolution, approach static load conditions. This type of load is conducive to rapid false brinelling and requires special lubrication techniques.

Table 6.5.5 Static Radial and Thrust Factors for Roller and Ball Bearings

	Single Row		Double Row	
Bearing type	X_0	Y_0	X_0	Y_0
Roller Bearing	0.5	0.22 cot α	1	0.44 cot α
Radial Ball Bearing	0.6	0.5	0.6	0.5
Angular Contact Ball, $\alpha = 40°$	0.5	0.26	1	0.44

Required Capacity

The basic load rating C is very useful in the selection of the type and size of bearing. By calculating the required capacity needed for a bearing in a certain application and comparing this with known capacities, a bearing can be accurately selected. To calculate the required capacity, Eq. (6.5.1) can be solved for C, resulting in the following formula:

$$C = \frac{P(L_{10}N)^{\frac{1}{K}}}{Z} \tag{6.5.5}$$

Where C = required capacity, lbf; L_{10} = rated life, h; P = equivalent radial load, lbf; K = constant, 3 for ball bearings, 10/3 for roller bearings; Z = constant, 25.6 for ball bearings, 18.5 for roller bearings; N = rotational speed, r/min.

LIFE ADJUSTMENT FACTORS

There are many different performance criteria that dictate how a bearing should be selected, such as fatigue life, speed capabilities, and misalignment tolerance. In some cases, the standard reference conditions are appropriate and the basic L_{10} rating life calculation is adequate. To account for the actual environment in which a bearing operates, the traditional life equation is expanded to include adjustments for variables that impact bearing performance. The additional adjustments are made to Eq. (6.5.2) and either have a positive or negative impact on basic rating life. The adjustment factors include:

1. Reliability factors for survival rates greater than 90 percent
2. Life adjustment factor for special material properties or quality
3. Life adjustment factor for environmental operating conditions (including adjustments for lubrication, misalignment, debris or hardness, for example)

Equation (6.5.2) can be rewritten to reflect these influencing factors:

$$L_{na} = a_1 a_2 a_3 \frac{10^6}{N \times 60}\left(\frac{C}{P}\right)^K \tag{6.5.6}$$

Where a_1 = reliability adjustment factor for a chosen survival rate, a_2 = material/quality adjustment factor, and a_3 = environmental operating conditions adjustment factor.

Reliability Factor a_1

The reliability factors listed in Table 6.5.6 represent survival rates from 90 to 99 percent. Using survival rates higher than 90 percent will yield larger bearing selections for equivalent loads.

Material Factor a_2

When a special type and/or quality of material is used, the a_2 factor represents these enhancements. Values other than 1 for this factor should be based on experience, and can usually be obtained from the

Table 6.5.6 Reliability Factor a_1 for Various Survival Rates

Survival rate, %	Bearing life notation	Reliability factor a_1
90	L_{10}	1.00
95	L_5	0.62
96	L_4	0.53
97	L_3	0.44
98	L_2	0.33
99	L_1	0.21

manufacturer. Values greater than 1 may not be applicable when there is a deficiency in the environmental factor (a_3), as this is indicative of increased probability of surface-initiated damage on the raceways and may negate the positive influence of the enhanced material.

Environmental Factor a_3

The factor a_3 is used to modify the basic rating life due to the specific environmental conditions of a particular application. Common variables included in the a_3 factor account for lubrication, mounting condition, and foreign debris in the lubricant. Many other environmental characteristics are included in the a_3 factor according to specific bearing manufacturers.

The calculation of the basic rating life assumes that the lubrication provides a full lubricant film that separates the rolling element and raceway. This factor is used to modify the rated life when this condition is either exceeded or not met. An appropriate value for this factor is based on the bearing geometry, operating speed, operating viscosity of the lubricant, and composite surface roughness of the bearing surfaces. The bearing manufacturers or their catalogs may be consulted to calculate an appropriate factor for a particular application.

The mounting condition impacts the alignment of bearings, and is assumed to be within a specified value, typically 0.0005 radians for roller bearings. Mechanical misalignment from machining practice and system deflections may reduce the performance of the bearing, depending on type selected.

The final mounted clearance of the bearing also has an impact on rating life. The bearing clearance affects how many rolling elements share the load through the bearing. This distribution of rolling element loading and its impact on rating life is often taken into account within the a_3 adjustment factor.

Foreign matter (contamination) within a lubrication system may lead to damage on the surfaces of the bearing raceways and rolling elements, leading to reduced bearing life.

Basic L_{10} rating life equations can be solved with relative ease. The addition of environmental factors adds complexity. In practice, complex modeling software has been put to use to better evaluate bearing performance and life predictions. These tools more accurately evaluate the bearing loading, and also the various impacts due to environmental conditions.

SPEED LIMITS

Thermal Speed Rating

The safe operating speed of a bearing is often limited by its temperature. The thermal speed rating, also known as the thermal reference speed, is the rotational speed at which the heat generated by the friction in the bearing is equal to the heat flow emitted through the bearing seating (shaft and housing) under reference conditions. The thermal reference speed is influenced by ambient temperature, lubrication, and the tolerances of surrounding components. This limit can be exceeded if special precautions are taken. ISO Standard 15312-2003 can be referred to as a method to calculate thermal reference speed.

Limiting Speed

Every bearing has a maximum speed above which the bearing is at an increased risk of seizure or excessive heat generation. The complexity required to maintain a stable temperature at operating conditions (e.g., special lubrication systems) above this speed may be prohibitively high. Many factors combine to determine the limiting speeds of bearings, such as bearing size, inner or outer ring rotation, contacting seals, radial clearance and tolerances, operating loads, cage types and materials, temperature, and type of lubrication. For bearings to function properly at or near the limiting speed, a given minimum load must be applied. Bearing limiting speeds vary by bearing type and manufacturer.

Bearing Torque

One of the main benefits of rolling contact bearings is their low rotational torque. Bearing torque is a complex subject in its own right. Factors that impact bearing torque include the type and geometry of the bearing, speed, load, lubrication, temperature, and the presence of seals.

The major contributors to bearing torque are the rolling element-raceway contact hydrodynamic shear, rolling element-raceway sliding due to curved contacts, and roller end-flange contact when present. The influence of raceway sliding of curved contact surfaces is more prevalent in ball and spherical roller bearings. Lubricant drag losses and elastic hysteresis losses at the rolling element-raceway contact are also considered, although very small. Frictional losses between the cage and rolling elements or the cage and bearing raceways are quite small and typically ignored.

Bearing torque is generally calculated through the use of sophisticated bearing analysis tools created by bearing manufacturers. Some torque models and tools are available in publication or from bearing manufactures for engineers to utilize. A common reference to calculate bearing torque can be found in ISO Standard 15312-2003.

PROCEDURE FOR BEARING SELECTION AND LIFE DETERMINATION

The following basic procedure can be used in determining basic bearing size and calculated life. This procedure does not account for all situations and offers only general guidance.

1. Calculate equivalent radial load P using Eq. (6.5.3).
2. Determine target rating using the equivalent load P and speed using Eq. (6.5.5).
3. Evaluate features of the different bearing types to select bearing type Table (6.5.1).
4. Select bearing size based on required rating (see Table 6.5.2 for some example sizes and ratings).
5. Calculate final bearing L_{10} based on equivalent load P, and bearing rating C using Eqs. (6.5.1) or (6.5.2).
6. Calculate equivalent static load, P_0 using Eq. (6.5.4).
7. Ensure bearing static rating C_0, is greater than equivalent static load P_0.
8. Consider environmental factors like temperature, lubrication, debris, and possible impact on bearing L_{10} life [see Eq. (6.5.6)]. It may be necessary to consult with a bearing manufacturer for extreme environments or complex loading conditions.
9. Finalize mounting and fitting practice.
10. Verify clearance considering fitting practice.

BEARING SEALS AND CLOSURES

Rolling-element bearings are commonly supplied with seals or shields. These are used to keep debris out of the bearings, or to keep lubricant inside the bearings.

Although variations in seal designs are numerous, two of the common basic sealing options are shielded, or sealed (Figs. 6.5.23 and 6.5.24). Bearings with no seals are referred to as open bearings.

Shielded bearings have a small clearance between the stationary, rigid shield and rotating ring. This provides reasonable exclusion of dirt without an increase in friction. Sealed bearings have a flexible lip in contact with the inner ring. Friction is increased, but more effective retention of lubricant and exclusion of contaminants is obtained. Various seal designs are available based on bearing type and manufacturer. Bearings are also available with solid lubricants or oil filled polymers that fill the open spaces and help seal the bearings.

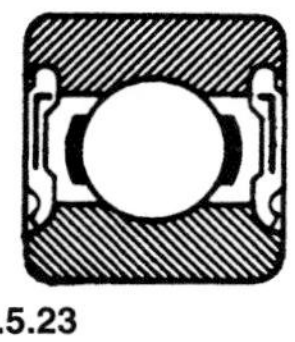

Figure 6.5.23

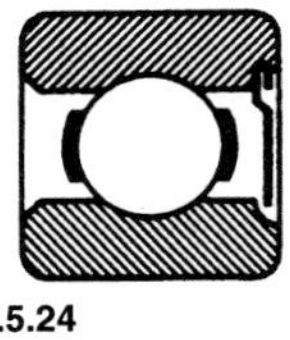

Figure 6.5.24

BEARING MOUNTING

Bearing mounting has three main considerations. The first is to axially locate or fix the bearing into the application in a way that facilitates assembly. The second is the proper selection of shaft and housing tolerances to achieve proper bearing fit. The third consideration is the selection of proper bearing radial internal clearance (RIC) or axial clearance (endplay/preload).

Correct mounting of a rolling contact bearing is essential to obtain its rated life. Many types of mounting methods are available. The selection of the proper method is a function of the accuracy, speed, load, and complexity of the application. A common and acceptable practice for inner ring mounting is a press fit on the shaft against a shaft shoulder. The inner ring can be secured with a locknut or shaft end plate. End plates (end caps) or followers can be used to secure the bearing outer race against the housing shoulder (Fig. 6.5.25). Retaining rings can also be used to retain a bearing to a shaft or housing (Fig. 6.5.26). Some machine designs require axial expansion of the shaft assembly by allowing one of the bearing positions to float. This can be accomplished by allowing the bearing to expand linearly in the housing or by using a cylindrical roller bearing on one end. Care must be exercised when designing a floating installation because it requires a slip fit. An excessively loose fit can cause the bearing to spin on the shaft or in the housing and increases potential for fretting corrosion damage.

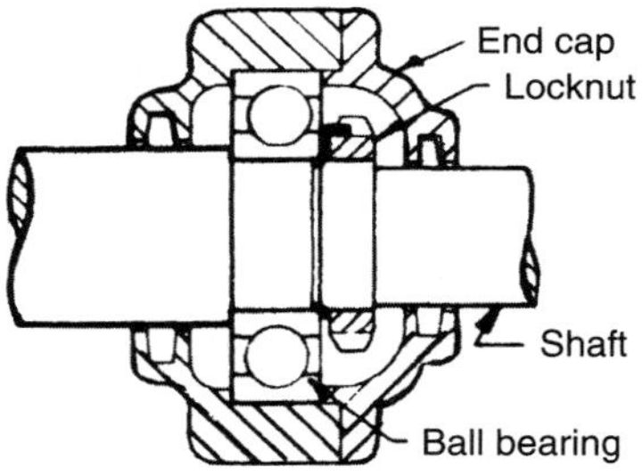

Figure 6.5.25

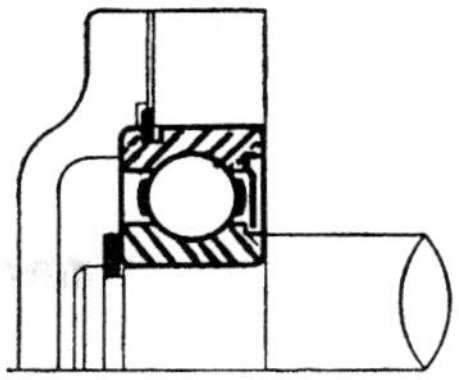

Figure 6.5.26

The selection of shaft and housing tolerances along with their resultant interferences is termed fitting practice. Different levels of interference or clearance are selected based on various factors such as: loads, speeds, temperature, vibration, materials, and mounting considerations. The general rule is that the rotating bearing race has an interference fit with its mating component. Often one of the bearing rings must be loose fit to allow for assembly.

Fitting practice varies by bearing type and application. Bearing manufacturers offer suggested fitting practice based on application; however, some designers and manufacturers have their own guidelines. It is common to use ISO 286 shaft and housing tolerance values to define bearing fit requirements, as shown in Fig. 6.5.27.

The final mounting consideration is the final bearing radial or axial clearance. For tapered roller bearings, clearance is measured axially. The tapered bearing design allows for axial movement of the rings to add or remove clearance at the time of assembly. This way, the amount of final clearance can be adjusted. It is common for tapered roller bearings to be mounted with either axial clearance or axial preload. Other ball and roller bearing designs have RIC that are manufactured into the product, specified in the part number, and measured radially. The resulting shaft and housing fits will impact the final mounted RIC. An interference fit at the bearing bore or OD will cause a reduction in bearing RIC, and must be evaluated. In most applications, it is desirable to have a small amount of radial clearance in the bearing after mounting.

Mounted units, or pillow blocks (Fig. 6.5.28), are available in many different bearing configurations. Figure 6.5.28 shows a ball bearing, spherical bearing, and tapered bearing pillow block. These are used in

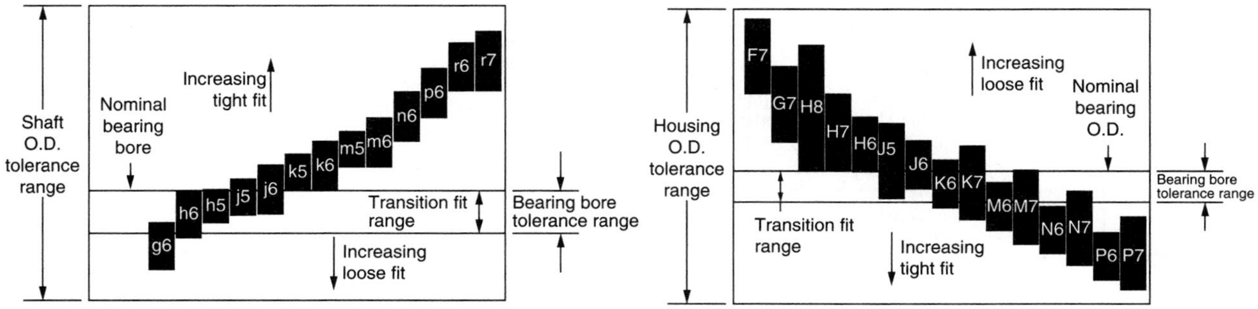

Figure 6.5.27

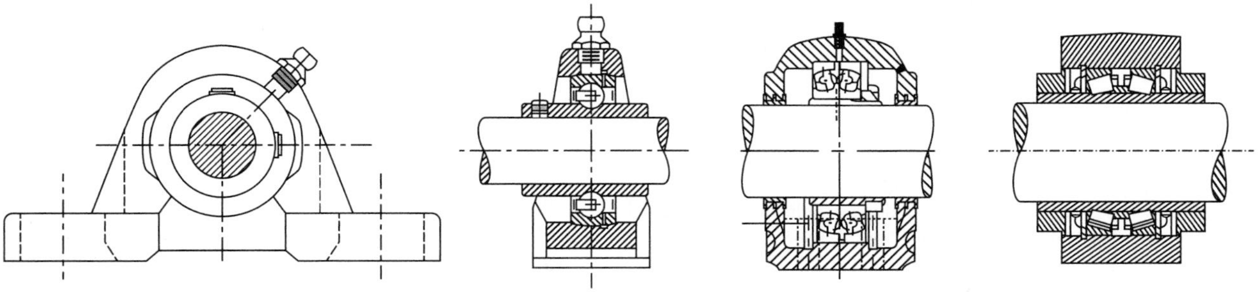

Figure 6.5.28

various industrial applications like fans, conveyors, packaging equipment, and general machinery. While there are others, the three common methods used to attach the bearing to the shaft are setscrew, eccentric locking collar, and taper-sleeve adapter.

Setscrew Figure 6.5.29 illustrates the use of an extended inner-ring bearing held to the shaft with a setscrew. This is the most common locking type for ball bearing pillow blocks, and is suitable in most general industrial applications.

Eccentric Locking Collar Figure 6.5.30 illustrates the use of an extended inner-ring bearing held to the shaft with an eccentric collar. This method tends to keep the shaft centered in the bearing more concentrically than the setscrew method. This design must be installed according to the direction of shaft rotation, and should not be used in reversing applications.

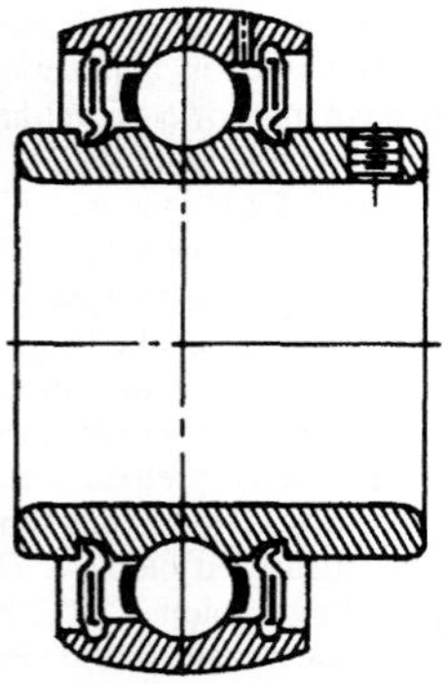

Figure 6.5.29

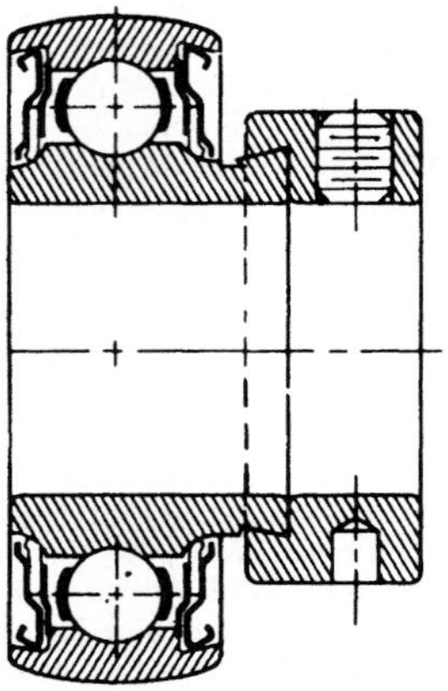

Figure 6.5.30

Taper-Sleeve Adapter Figure 6.5.31 illustrates the use of a taper-sleeve adapter to mount the bearing on the shaft. This method is commonly used in spherical roller bearing pillow block applications. It provides uniform concentric contact between the shaft and bearing bore. However, some care is required to tighten the locking nut enough to keep the sleeve from spinning on the shaft and yet not so tight that the clearance is removed from the bearing. Typically the bearing clearance and reduction of clearance are measured to determine proper mounting.

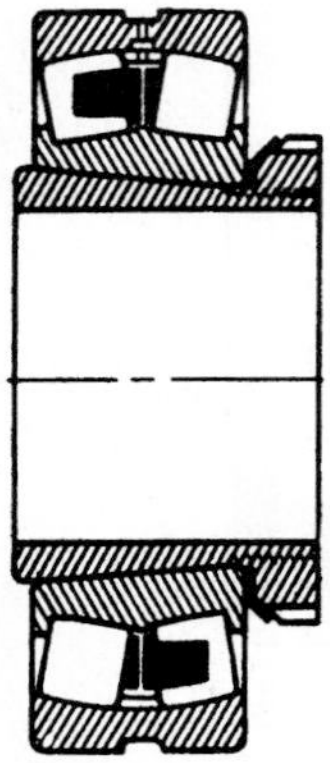

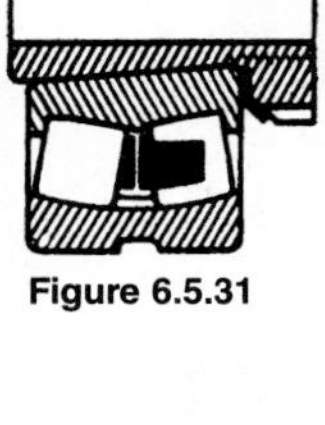

Figure 6.5.31

LUBRICATION

Proper lubrication provides a lubricant film to separate the rolling element-raceway surfaces, and reduces sliding friction present within the bearing. Only a small amount of lubricant within the bearing is required to generate adequate film thickness. Excess lubricant will cause additional heat generation in the bearing and accelerate the deterioration of the lubricant. Optimum lubrication of rolling contact bearings can be predicted by elastohydrodynamic theory (EHD). It has been shown that film thickness is sensitive to the bearing operating speed and lubricant viscosity properties and, moreover, that film thickness is virtually insensitive to load. Grease and oil lubrication are the most common lubrication types, although solid lubricants (graphite, molybdenum disulfide) are used in special circumstances.

Grease is commonly used for lubrication of rolling element bearings because of its convenience and minimum maintenance. A wide variety of lubricating greases are available for rolling element bearings. Common grease thickener types are lithium soap, lithium complex, aluminum complex, polyurea, and calcium sulfonate. Grease with NLGI 2 thickness is quite common, but others ranging from NLGI 1 to NLGI 3 may also be used. The temperature limits of the grease selected and the speed limitations of the bearings must be considered in final grease selection. In applications of higher speed, oil lubrication is often preferred. Equation (6.5.7) can be used as a general guide in selecting the proper base oil viscosity for rolling contact bearing greases.

Oil lubrication not only provides the lubricant film needed for proper rolling contact performance, but also offers the added benefit of removing heat from the bearing system. Various methods of oil lubrication are used in practice, ranging from a simple oil bath/splash system to circulating oil systems, jetted oil systems, and air/oil mist systems. Each has specific advantages and complexities. The selection of base oil viscosity is similar to the selection of base oil for grease-lubricated bearings.

The basic Eq. (6.5.7) can be used to estimate the minimum required kinematic viscosity for the lubrication oil or grease at operating temperature for a given bearing size and speed. Adjustments for operating temperature must be made to determine the proper viscosity specification for a given lubricant.

$$\nu_1 = \begin{cases} 45000 \cdot n^{-0.83} \cdot D_{pw}^{-0.50} & \text{if } n < 1000 \text{ rpm} \\ 4500 \cdot n^{-0.50} \cdot D_{pw}^{-0.50} & \text{if } n \geq 1000 \text{ rpm} \end{cases} \tag{6.5.7}$$

Where ν_1 is the reference kinematic viscosity, cSt; n is rotational speed in, r/min; and D_{pw} is the bearing pitch diameter (mean bearing diameter), mm.

In applications using grease, it is necessary to replenish the lubricant. Relubrication intervals in hours of operation are dependent on temperature, speed, and bearing size. Table 6.5.7 is a general guide that represents the time after which it is advisable to add a small amount of grease in order to safeguard the bearings. The intervals are valid up to 160°F (71°C) and should be divided by 5 for cylindrical or needle roller bearings and by 10 for spherical roller bearings, tapered roller bearings, or thrust roller bearings.

Table 6.5.7 Ball-Bearing Grease Relubrication Intervals (Hours of Operation)

Bearing Bore (mm)	Bearing Speed (RPM) 5,000	3,600	1,750	1,000	200
10	8,700	12,000	25,000	44,000	220,000
20	5,500	8,000	17,000	30,000	150,000
30	4,000	6,000	13,000	24,000	127,000
40	2,800	4,500	11,000	20,000	111,000
50		3,500	9,300	18,000	97,000
60		2,600	8,000	16,000	88,000
70			6,700	14,000	81,000
80			5,700	12,000	75,000
90			4,800	11,000	70,000
100			4,000	10,000	66,000

6.6 PACKINGS, GASKETS, AND SEALS

by Matthew Tones and James Drago

Packings, gaskets, and seals are materials used to control or stop leakage of fluids (liquids and/or gases) or of solid dry products through mechanical clearances when the contained material is under pressure or vacuum. The mechanical clearances to be sealed may be between either fixed or moving components.

Gaskets are compressible materials installed between two sealing surfaces, which normally exist between parallel flanges or concentric cylinders. Sealing of flat flange gaskets is effected by compressive loading achieved through bolting or other mechanical means. The full-face gasket in a flat face flange (Fig. 6.6.1) has gasketing material which extends to the outside edge of the flanges. This configuration is required when one or both of the flanges are made with brittle materials such as cast iron or plastic that would otherwise break if the gasket material was not present at the outside providing support and reducing bending moments. Having the entire full face gasket in compression adds a significant amount of surface area, which in turn reduces the overall stress on the gasket. Therefore, it is common for full face gaskets to be rubber or elastomeric materials that can seal and low stresses, but are unable to handle high stresses. The ring gasket in raised face flanges (Fig. 6.6.2) is very common in higher pressure systems. Contrary to the flat face flange, raised face flanges are typically made of stronger materials, such as forged carbon or stainless steel. In addition, the reduced contact area of the raised face surfaces concentrates the compressive force in a smaller area, thus creating higher gasket stresses. Compressed fiber, PTFE and metallic gaskets are typically much better suited for the stress levels created in raised face flanges. With irregularly shaped flanges, bolt holes may serve to locate the gasket. Bolt holes can be placed in lobes on the outer perimeter or simply cut into a full-face template of the flange. Metal-to-metal fits require a recess whose depth and volume are engineered for the gasket to be used. A gasket, such as an O-ring (Fig. 6.6.13), extends above the groove sufficiently to provide the minimum compression required for initial seating. Depending on the seal configuration, the internal fluid pressure can energize the seal. **Warped, wavy,** or irregular flanges (often resulting from corrosion, wear and tear, welding, other fabrication, or as found in glass-lined equipment) require gaskets that are softer and/or thicker than normal, to compensate for surface imperfections. While thicker gaskets are sometimes necessary to accommodate flange irregularities, thinner gaskets provide improved creep relaxation, bolt load retention, and pressure resistance. Thicker gaskets can compromise the integrity of the flange connection. Tongue-and-groove joints (Fig. 6.6.4) confine the gasket material and can provide added pressure resistance.

The spiral wound gasket (Fig. 6.6.7) is one of the oldest and most widely used gaskets in industry today, as the design works very well in extreme temperature and pressure conditions, and can be produced with various metals and fillers to meet the service needs. Jacketed gaskets (Fig. 6.6.6), while still used, are commonly replaced with either spiral wound or other lower stress metal gaskets, such as corrugated metal gaskets or kammprofile gaskets (Figs. 6.6.5 and 6.6.8) which are not only easier to seal, but can provide improved emissions performance. Not shown in Fig. 6.6.6, but commonly used in the oil and gas industry is the ring type (or RTJ) gasket. RTJ gaskets typically have an oval or octagonal cross section, and seal in a groove like an O-ring. However, since RTJ gaskets are solid metal it is not uncommon for users to have to lap the joint in order to obtain an effective seal. In many cases, if the groove is damaged or worn, it is common to use a special designed spiral wound or kammprofile gasket in place of the RTJ gasket. The envelope gasket (Fig. 6.6.3), usually polytetrafluoroethylene (PTFE) allows the user to insert a variety of core materials as a filler, and even "shim" the filler to make the gasket thicker in certain areas to accommodate flange imperfections or "waviness". The envelope gasket is particularly useful for corrosive, non-contaminating service and average pressure in wavy, glass-lined equipment.

Cylindrical or **concentric** gasketing uses a retaining gland follower and is mechanically loaded, for example, the standard mechanical joint for cast-iron pipe (Fig. 6.6.10) or the condenser tube-sheet ferrule (Fig. 6.6.11). Cup-shaped gaskets are designed to be self-tightening under pressure (Fig. 6.6.12). The O-ring (Fig. 6.6.13) located in an annular groove and pre-compressed as in the grooved flange, is a self-energized gasket. A cylindrical ring with internal single lip or double lips, also automatic in action, is quite common in pipe joints.

A variety of gasketing materials are listed in Table 6.6.1 together with common applications. Beyond rubber are many elastomeric and composite materials generally similar in mechanical behavior but varying as to temperature limits and fluid compatibility (see also Sec. 4).

The **proper design** of a gasketed joint requires flange rigidity to avoid distortion, flange surface finish and texture commensurate with gasket type, adequate compressive stress on the gasket, and adequate bolt loading. The load must seat the gasket, i.e., cause the material to flow into and fill flange irregularities. Once the pressure of the fluid being contained by the joint is applied, the residual compressive stress on the gasket must be sufficient to maintain the seal. These values, known, respectively, as the **seating stress factor** y, expressed in force per unit area, and the **gasket factor** m, which is dimensionless. The factors vary with gasket material, gasket thickness, and tightness required by the joint designer. The ASME Boiler and Pressure Vessel Code, Section VIII, Division I, Appendix 2, gives a brief list of m and y values for various generic gasket materials. However, it is important to note that most if not all of these values were established prior to "emissions compliance requirements". Since the m and y factors are to be used for flange design, not for developing bolt torque values for existing flanges, it is very important to consult the gasket manufacturer obtain accurate tested design factors to ensure the flange design is sufficient. The test media (liquid or gas) and leak rate allowed during tests used to arrive at the m and y values greatly affects the magnitude of these values. Manufacturers should provide that leak rate to designers along with the m and y factors.

The concept of m and y gasket factors was introduced to the ASME Code pertaining to flange design in the 1940s. In the 1980s efforts by the Materials Technology Institute of the Chemical Process Industries (MTI), the Pressure Vessel Research Council (PVRC), and the University of Montreal's Ecole Polytechnique embarked on new methods of characterizing gasket materials. The creation of gasket factors Gb, a, and Gs in conjunction with the tightness parameter Tp gives flange designers an alternative to m and y values. While the new methods of flange design based on the new factors are used, they have not been accepted as part of the ASME Boiler and Pressure Vessel Code.

High bolt loading resulting in high bolt stretch is desirable for tight and enduring gasket joints, but it must not crush the gasket material or permanently deform the bolt material, that is, cause bolt yield. Crush strength values, which vary with thickness and temperature, can be obtained from gasket manufacturers. Consistent with the condition of the flanges, the thinner the gasket, the more efficient the joint. The use of thinner sheet gaskets minimizes leakage and decreases gasket creep, which maximizes gasket performance.

Table 6.6.1 Common Usage of Gasketing Materials

Type	Service principally for
Sheet rubber	Water
Cloth-inserted rubber sheet	Water
Cork composition, cellulose fiber sheet	Oil, low-pressure
Rubberized cloth (Fig. 6.6.9)	Hot water (boiler manholes, etc.)
Compressed fiber sheet	All services up to 750°F (400°C)
Corrugated metal or kammprofile with facing (Fig. 6.6.5)	Steam, oil at high temperatures
Metal jacket over compressible fiber center (Fig. 6.6.6)	Steam, oil at high temperatures
Spirally wound steel strip with flexible graphite filler (Fig. 6.6.7)	Steam, oil at high temperatures

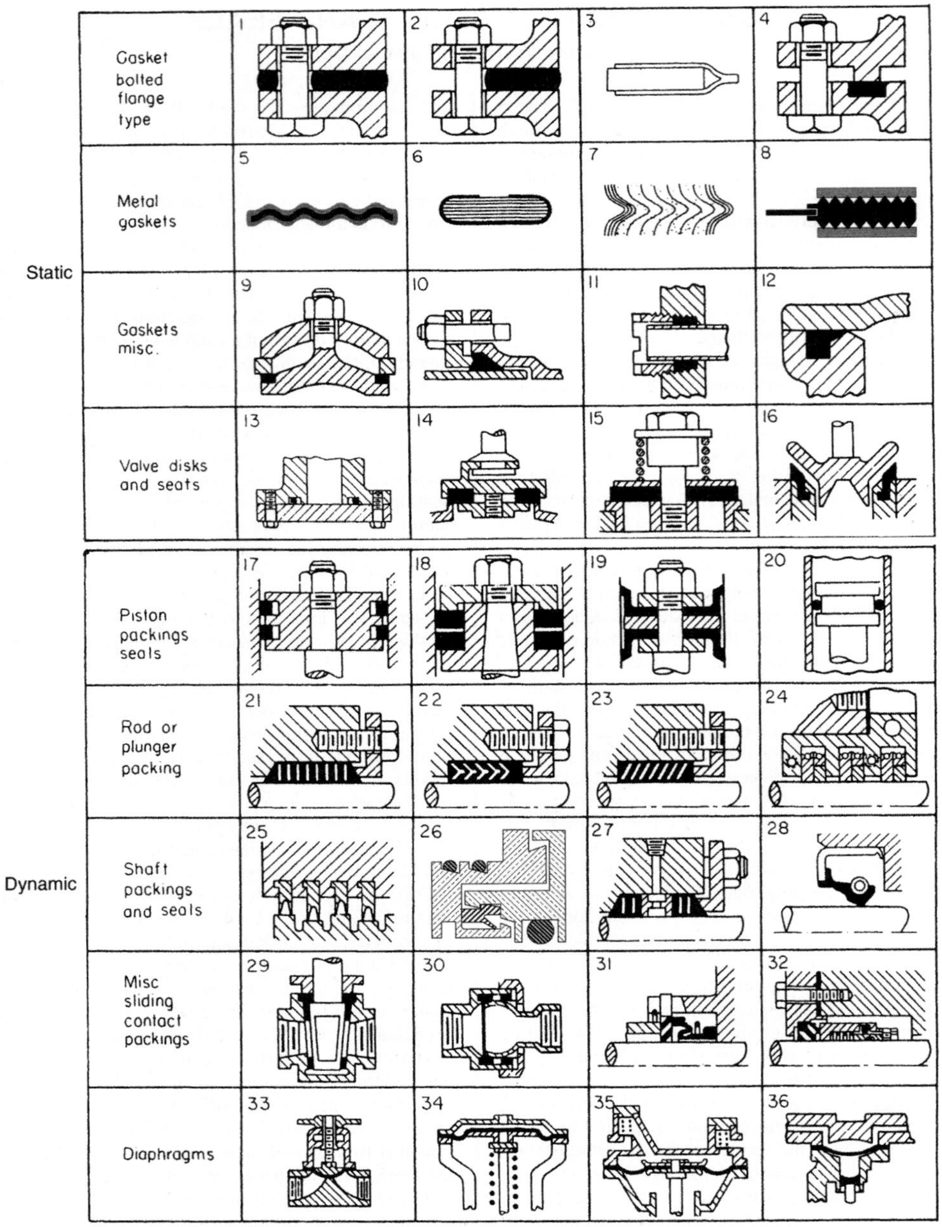

Figure 6.6.1-6.6.36 Packings, gaskets, and diaphragms.

O-rings are used as static seals (Fig. 6.6.13) between flanges and as dynamic or static seals between concentric surfaces (Figs. 6.6.20 and 6.6.32). An O-ring must be either stretched and/or compressed to function. If the O-ring is not compressed or stretched while in service, it is the wrong size. O-ring type seals are available in various profiles (e.g., round, delta, D, T, lobe, cored, hollow ring, and K, to name a few) (see Fig. 6.6.37). They can be molded, extruded, or lathe-cut. Each profile has its application advantages, and users should work with their suppliers to apply the proper geometry and material for each application. Aerospace Materials Specifications (AMS), Air Force/Navy Specifications (AN), Military Specifications (M; MIL; MS) and National Aeronautical Specifications (NAS) are some of the various standards organizations which address O-rings. Following their specifications ensures conformity and uniform quality. O-rings are available in a wide range of elastomeric and thermoplastic compounds as well as metal. Chemical compatibility, temperature limits, size (ID and c/s), durometer, and profile of the selected compound are the main concerns in making O-ring choices. Data on the design of O-ring joints are available from suppliers.

The nominal pressure limit for O-rings based on typical mechanical clearances is 1,500 psig (10 MPa) without backup rings and 3,000 psig (20 MPa) with backup rings. If clearances can be eliminated, as in a flanged joint with close metal-to-metal contact, no limit can be set.

Other self-energizing joints, such as the boiler hand-hole plate (Fig. 6.6.9), need only sufficient load to affect an initial seal.

Valve disks are specialized gaskets designed for joints that are frequently broken and reseated. Disks for globe valves (Fig. 6.6.14) are usually encased in a disk holder with a swivel mounting, which ensures precise reseating without abrasion during the closing and opening

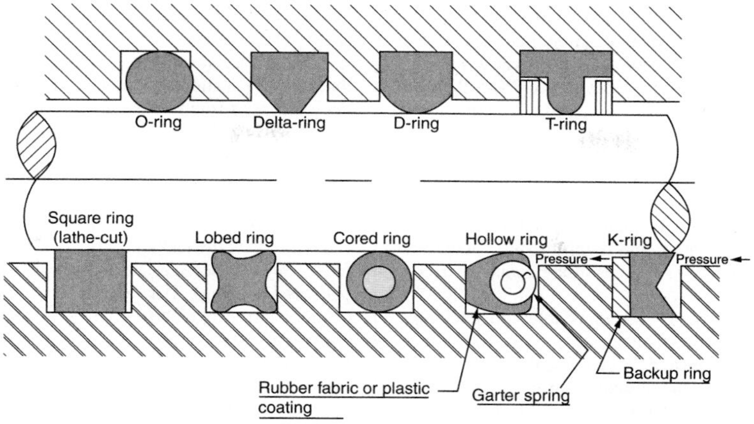

Figure 6.6.37 O-rings in various cross sections.

cycles. They are made of firm rubber for bib washers, hard rubber, and phenolics for more severe service, and plastics such as nylon and PTFE.

Pump valves (Fig. 6.6.15) are described in Sec. 10. **Rubber valve seats** are used with metal valve disks on some pumps, e.g., the rotary drilling pump valve (Fig. 6.6.16). Plastics are also used for seats, notably in ball valves.

Dynamic seals include all seals that operate on moving surfaces. To retain fluid under pressure, they are subjected to the hydraulic load. In the case of an oil seal, (Fig. 6.6.28) when no pressure exists, the seal is mechanically loaded by a spring or by the resiliency of the elastomeric lip. Oil seals operate like bearings, wherein the lubricant serves as both a separating film and a coolant. The fluid film is a hydrodynamic meniscus about 0.00018 in thick and is vital for satisfactory service life. It is important to note that some leakage will occur. Similarly, pumps that use braided packing will also leak, but at a higher level than an oil seal. High-viscosity fluids and high pressures add to leakage problems, as both require thin films to minimize leakage. This causes higher friction and generates heat, which is the single most detrimental factor in packing and oil seal life. Packing sets with many rings (Figs. 6.6.21 to 6.6.23) reduce leakage but increase frictional heat, particularly at high speeds. Five rings of packing are considered optimum. Normally the fluid being sealed serves as the lubricant. Maximum efficiency can be attained when oil is the fluid being sealed; in decreasing order of efficiency are clean water, solvents, and fluids containing solids. These are progressively more unsatisfactory unless supplemental cooling and lubrication in the form of a flush are provided. Lubrication may be introduced by using a lantern ring in the center of the packing set through which fluid is fed to the packings (Fig. 6.6.27). The pressure of the fluid supply should be slightly higher [5 to 10 psig (0.35 to 0.7 barg)] than the system pressure. The process media governs the choice of flushing fluid, as the two need to be compatible. In cases of extreme contamination, the lantern ring is moved to the bottom of the packing set to introduce an outward flow away from the packing set to prevent abrasives from migrating to the dynamic sealing interface. A lantern ring with a barrier or flush fluid can also be used to reduce or prevent air intrusion into the system when equipment is subject to vacuum conditions. Centrifugal pumps equipped in this manner are said to have a **water seal.**

Dynamic packings are classified in three ways:

1. **On the basis of shape of the surfaces:** cylindrical, conical, spherical, or flat. Cylindrical packings are in turn classified according to whether they pack on the outside perimeter, as in piston packings (Figs. 6.6.17 to 6.6.20) or inside perimeter, as on rods or shafts (Figs. 6.6.21 to 6.6.28). Other examples are: conical, where sealing is created by wedging the seal between surfaces (Fig. 6.6.29); spherical, the ball joint (Fig. 6.6.30); and flat, the mechanical seals (Figs. 6.6.31 and 6.6.32).

2. **On the basis of the type of motion:** rotary, oscillating, reciprocating, or helical (as in a rising-stem valve packing).

3. **On the basis of how the seal is energized:** mechanical loading where braided compression packing is compressed and sealed by external means, usually a gland follower; hydraulically loading where a molded shape has geometry that facilitates a seal as the internal service pressure is applied.

The selection of a compression packing type is a matter of performance expectations, service conditions, and economics. In most cases several types are available, some of which, although expensive, yield exceptional performance, whereas a cheaper packing type could yield unsatisfactory performance. Service requirements should always dictate the final choice of a packing material. The packing must be compatible with the fluid being sealed, operating pressure in the system, temperature, type of equipment being sealed, ease of adjustment/maintenance, and ease of replacement.

For **reciprocating elements, the O-ring** (Fig. 6.6.20) is extremely simple. It is a precision part manufactured to close tolerances, as must be the seat into which it is placed. As an elastomeric material completely exposed to the operating fluid, it is subject to chemical degradation. The O-ring material must be carefully chosen to ensure compatibility with the fluid being sealed; the wrong choice will lead to either shrinkage or swelling, with premature failure of the O-ring. It is best suited to medium-pressure service from 1,500 to 3,000 psi (104 to 207 bar) with backup rings and intermittent movement, as in hydraulic cylinder or valve stem service. O-rings are not recommended for rotating services, such as pumps. Backup rings are preferably of block-like cross section in either PTFE or similar material. The split piston ring (Fig. 6.6.17), usually cast iron, is widely used in gas, oil, and steam engines and compressors. Large pistons frequently employ segmental rings similar to floating metal rod packings (Fig. 6.6.24) but facing outward. **Floating metal packing rings** are made of numerous radial or tangential segments, making it possible for them to contract on the shaft; they are assembled in sets of two to break the joints and are held together with garter springs. They are used for steam, gas, or air, in either engines or compressors under the most severe operating conditions and at pressures up to 35,000 lb/in^2 (2,410). Normally oil lubrication is provided; for less severe service, filled PTFE rings perform very well in dry gases without auxiliary lubrication. Step-, scarf-, or butt-cut rings of laminated cotton fabric, bonded with an elastomer or phenolic resin, are employed in water pumps, gasoline pumps, etc. They may float similar to cast iron piston rings, or be retained by a gland follower, as in Fig. 6.6.18. Cups (Fig. 6.6.19) are fully automatic and very tight; cups in their inverted form, with the lip on the ID, are known as **flange packings** and are also fully automatic and very tight. They are used principally for slow-speed applications. **Nested V and conical rings** (Figs. 6.6.22 and 8.6.23) are automatic, though often provided with a gland follower to effect initial fit. They are made of a wide range of materials including homogeneous elastomers and polymers, and reinforced woven fibers (cotton, aramid, or fiberglass) for severe duty. They range in hardness from soft and flexible to semi-rigid. Use of multiple rings allows them to be of the cut or split type for ease of installation and replacement. **Braided, soft,** or **jamb packings** are normally formed in square or rectangular sections with a butt joint staggered from ring to ring at installation. Many materials are employed to make braided packing. One of the main factors considered when selecting materials of construction is "how well will the product perform?" Performance does not end at traditional properties, such as chemical resistance. With the ever growing concern over environmental conservation, "fugitive emissions" is a term that is used synonymously with "performance" when choosing a packing product. Fibers such as synthetic blends, aramid, PTFE graphite, carbon, fiberglass, or flax are saturated with wax, custom lubricating compounds, PTFE suspensoid solutions, or rubber compounds all impact the performance of the finished product. In addition, metal foil and tinsel are sometimes used for high-temperature and high-pressure conditions. Packings containing woven or braided fibers also may contain wire for strength and pressure resistance.

Rotary shafts are generally packed with adjustable soft packings, with the notable exception of the mechanical seals (Figs. 6.6.31 and 6.6.32); where pressures are low, nested V or conical styles may be used. At zero or negligible pressures, the oil seal, a spring-loaded flange packing (Fig. 6.6.28), is very widely used. Where some leakage can be tolerated, the labyrinth (Fig. 6.6.25) and controlled-gap seals are used, particularly on high-speed equipment such as steam and gas turbines. Soft packings are of the same general type as those used for reciprocating service, with the fiber braid lubricated with grease and graphite or with PTFE fibers and suspensoid. Aramid, carbon, and graphite fibers filled with various lubricants and reinforcements are used at higher speeds and fluid pressures. Fiber braid with PTFE suspensoid is widely applied on valve stems operating below 500°F (260°C) and on centrifugal pumps. This material is an insulator, however, and results in high heat buildup on the dynamic surface; a better choice lies in use of a packing with better heat-transfer characteristics, such as one containing carbon or graphite. For continuous rotary service, hydraulically loaded packings (energized by internal pressure) are best restricted to low pressure because their tightness under high pressure can result in overheating. For intermittent service, as on valve stems, they are excellent.

Oil seals (Fig. 6.6.28) are dynamic seals having an elastomer lip generally bonded to a metal cup which is press-fitted into a smooth cylindrical bore. Basically, an oil seal has a flexible lip and a narrow contact area about in (1.6 mm) wide, and perform best in good lubricating media. However, advancements in seal development have resulted in PTFE based lip seal materials that can run without lubrication. This hybrid design is discussed in greater detail in the high performance lip seals section. To accommodate shaft runout up to 0.020 in (0.5 mm) depending on the rotating speed, the lip is spring-loaded with a coil or finger spring. Finger springs are safer as they are molded into the elastomer and are less likely to become dislodged during installation and operation and cause shaft damage. Since the lip is completely exposed to the sealed fluid, particular care should be taken to ensure compatibility between the elastomer and the fluid. Temperature, speed at the surface contacting the seal, equipment type, internal pressure (positive or negative) if any, space available to insert the seal, dynamic runout, and misalignment are all parameters that must be considered when selecting an oil seal. Bearing Isolators (Fig. 6.6.26) are an excellent choice for applications where there are concerns of or evidence of bearing or other internal component damage due to the migration of contaminants such as dirt or water into the system or equipment. Common applications are electric motors, gear boxes and pillow blocks. However, bearing isolators are limited to services that are not flooded, and non-pressurized.

Bearing isolators work on principles similar to those in labyrinth seals, shown in Figs. 6.6.25. An isolator uses a combination of centrifugal force from the shaft rotation and a tortuous path to keep the fluid/lubricant in and the contaminants out. Isolators are very effective in keeping solid contaminants and water spray (used for equipment washdown) out of the bearing area and lubricating fluid.

Mechanical, Rotary, or End Face Seals

The greatest advancements in the design of end face mechanical seals have come about in response to environmental regulations; requirements to minimize energy consumption and operating costs; safety; and concerns over loss of the product which is being sealed. The application of seals to replace packing in rotary equipment has increased dramatically and continues.

All end face mechanical seals (Figs. 6.6.31 and 6.6.32) consist of four parts: a stationary flat face, a rotating flat face, secondary sealing elements (usually elastomeric), and a flexible loading device. The assembled seal is placed and effects proper leak control. The two flat-face seal rings (one stationary, one rotating) rub and create the primary seal. Normally, the flat seal rings have different hardness values, and the soft one is narrower than the hard one. Secondary sealing elements prevent leakage between the rotating shaft and the rotating seal ring, and they block the leakage path around the outside of the stationary seal face. They also serve as gaskets between the assembled parts (i.e., gland plate and housing). The flexible loading device usually consists of one or more springs which press the flat seal rings together. Spring loading ensures a seal when there is little or no hydraulic pressure available to press the faces together and helps maintain constant pressure between the faces as the soft (sacrificial) face wears down. The springs also act as vibration dampers to mitigate against the intrusion of transmitted vibrations, which may affect the efficient operation of the seal assembly.

Types of End Face Mechanical Seals

1. *Inside-mounted.* The seal head is mounted inside the stuffing box (Fig. 6.6.38*a*).
2. *Outside-mounted.* The seal head is mounted outside the stuffing box (Fig. 6.6.38*b*).
3. *Unbalanced seal.* The full hydraulic pressure in the seal chamber is transmitted to the seal faces (Fig. 6.6.39*a*).
4. *Balanced.* The seal elements are designed to reduce the hydraulic forces transmitted to the seal faces (Fig. 6.6.39*b*). A complete balance is not practical.
5. *Rotary seal.* In this design, the springs rotate with the shaft.
6. *Stationary seal.* In this design, the springs do not rotate with the shaft.
7. *Metal bellows.* Welded or formed metal bellows exert a spring load; there is no dynamic secondary seal element (Fig. 6.6.40).
8. *Double seal.* Two mechanical seals are mounted back to back, face to face, or in tandem, between which a barrier fluid (liquid or gas) can be introduced for environmental control (Fig. 6.6.41).

End face mechanical seal materials must satisfy a number of design requirements, including chemical compatibility between the sealed fluid and the seal materials, ability of the seal materials to remain serviceable under the worst operating conditions, and ability to provide a reasonably long life in service at the operating conditions. Mating faces of the seals can be made from ordinary materials like bronze and PTFE for mild service on up to carbon, carbides, stainless steels, and

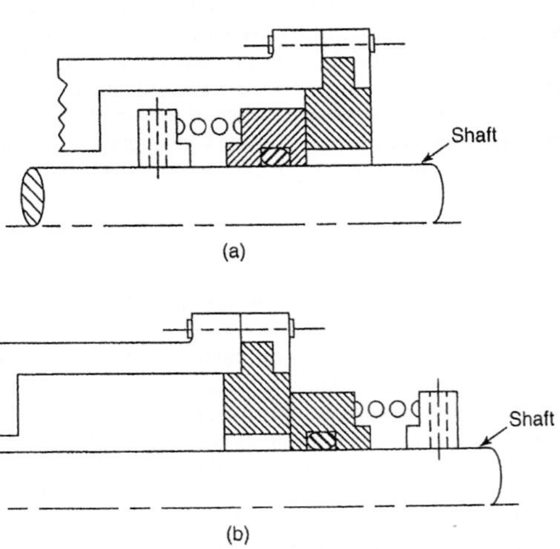

Figure 6.6.38 Rotary end face seal. (*a*) Inside the seal chamber/stuffing box; (*b*) outside the seal chamber/stuffing box.

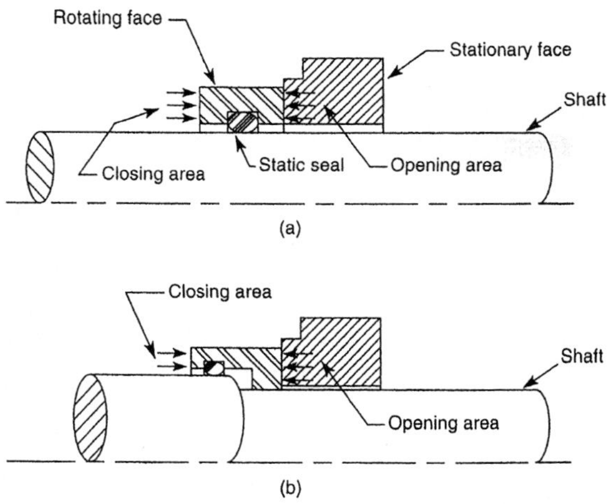

Figure 6.6.39 Rotary end face seals showing (*a*) unbalanced and (*b*) balanced configurations.

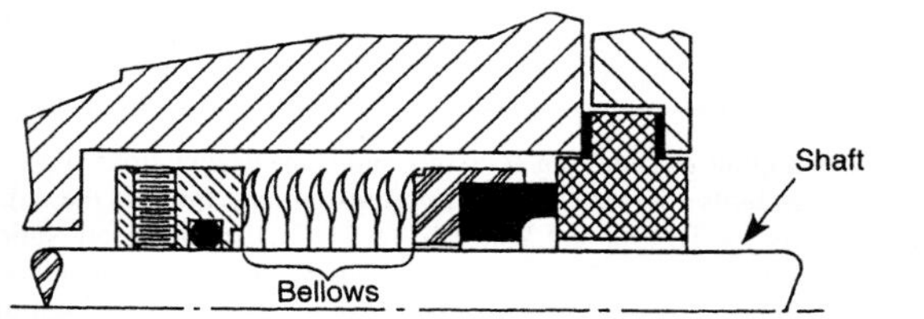

Figure 6.6.40 Rotary end face seal with metal bellows and "t" clamp stationary.

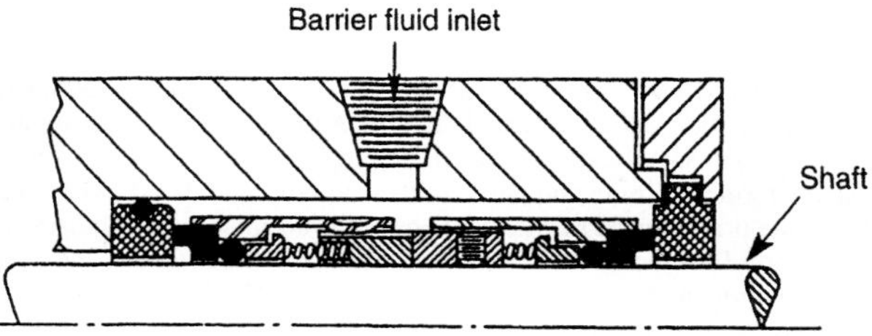

Figure 6.6.41 Back-to-back double seal. Barrier fluid must have an inlet and an outlet.

other exotic alloys as service conditions become more severe. Hard faces can utilize ceramics, tungsten and silicon carbides, and hard coatings over base metals (chromium oxide over stainless steel 316SS, subsequently lapped flat).

Secondary seal materials are usually elastomeric and include these:

Elastomer	*Temperature range*
Nitrile (NBR)	−20°F to 250°F (−29°C to 121°C)
Neoprene® (CR)	−20°F to 250°F (−29°C to 121°C)
Viton® (FKM)	−15°F to 400°F (−26°C to 204°C)
Ethylene propylene (EP)	−40°F to 300°F (−40°C to 149°C)
Kalrez® (FFKM)	32°F to 550°F (0°C to 288°C)
Aflas® (FEPM)	−23 to 400°F (−31 to 205°C)

Environmental controls as applied to seals are special techniques used to control the environment in which the seal operates. Some examples include discharge return recirculation, suction return, flush from an outside source, flush, and quench and drain. The controlling techniques often require supplementary equipment such as heaters, coolers, pumping rings, external circulation devices, various special solenoids and valves compatible with the working environment, and so on. Sealed products that solidify, vaporize, abrade, or carry contaminants are successfully sealed with end face seals augmented by environmental controls.

High Performance Lip Seals The nature of some sealed products is such that end face mechanical seals are not applicable. In many difficult instances of that type, sealing can be achieved with high-performance modified PTFE lip seals (Fig. 6.6.42). GYLON® is such a material which can serve in seals operating over a wide range of pressures, temperatures, and rotating speeds. It is particularly useful to seal against some dry products (air purge may be required), viscous resins, salting solutions, and products which tend to solidify on seal faces. Dry running is possible under some circumstances. Unlike conventional lip seal material, restructured PTFE lip seals in multiples can operate from high vacuums (10^{-3} InHg) up to 10 bar (150 psi), within a temperature range of −130°F to +500°F (−90°C to +260°C), and exhibit excellent compatibility with a wide range of sealed fluids. Manufacturers' literature will provide data showing the effect of temperature and rotating speed on the permissible operating pressure.

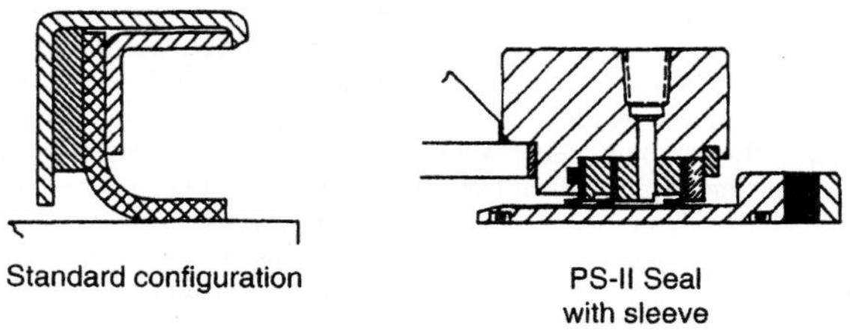

Figure 6.6.42 High-performance lip seal with modified PTFE elastomer. Illustration shows single and staged elements.

For extremely high speeds, where it is desirable to eliminate all rubbing contact, the labyrinth seal (Fig. 6.6.25) is chosen. This seal is not fluid-tight but restricts serious flow by means of a torturous path and induced turbulence. It is widely used on steam turbines (Sec. 7.4). Where no leakage is permissible, a liquid seal based on the U-tube principle (Fig. 6.6.26) may be used. The natural weight of the liquid is amplified by centrifugal force so that under high rotating speed a fair pressure differential can be sealed. Another noncontacting seal is the **controlled gap seal** which is being used on gas turbines where pressure differentials are not excessive and a small amount of leakage can be tolerated. The seal consists of a ring with a shaft clearance in the range of 0.0005 to 0.0015 in (0.013 to 0.038 mm) and is made of exotic heat resisting materials capable of maintaining that clearance at all operating temperatures. Usually one end of the ring is faced to form an axial seal against the inside of its housing.

Diaphragms are a form of dynamic sealing device that includes an outer diameter that acts as a gasket where the diaphragm is gripped or held in position. Diaphragms are designed to be impervious to the system media while being mechanically actuated to transfer the system fluid. By literally rolling one cylinder inside another, considerable increase in travel is possible. This type is often called a bellows, and a simple application is the mechanical seal suspension shown in Fig. 6.6.31. In the diaphragm valve (Fig. 6.6.33) the diaphragm can replace both the conventional stem packing and valve disk in certain designs. **Diaphragms** of fabric such as cotton or nylon covered with an elastomer suitable for the fluids and temperatures involved are used in pumps (fuel pump, Fig. 6.6.35) and in motors (Fig. 6.6.34) to operate valves, switches, and other controls. Correctly designed diaphragms are made with slack to permit a natural rolling action. Flat sheet stock should be used only where limited travel is desired. An example of a hydraulic actuated diaphragm is shown in Fig. 6.6.36, where there are no mechanical components attached to the diaphragm. Actuation is caused by using pressure and vacuum behind the diaphragm which pushes the diaphragm forward (forcing fluid out of the cavity) and pulls fluid and the diaphragm back (pulling fluid into the cavity). While conventional PTFE is used for applications requiring an economical solution in non-critical services, the GYLON® diaphragms are specified and used with chemically aggressive, toxic fluids where long service life, low permeability, and reliability are critical.

6.7 PIPE, PIPE FITTINGS, AND VALVES

Revised by Ali M. Sadegh

REFERENCES: M. L. Nayyar, "Piping Handbook," McGraw-Hill. ANSI Code for Power Piping. ASTM Specifications. Tube Turns Division, Natural Cylinder Gas Co., catalogs. Crane Co., catalogs and bulletins. Grinnell Co., Inc., "Piping Design and Engineering." M. W. Kellogg Co., "Design of Piping Systems," Wiley. United States Steel Co., catalogs and bulletins.

EDITOR'S NOTE: The several piping standards listed in this section are subject to continuing periodic review and/or modification. It is suggested that the reader make inquiry to the issuing organizations as to the currency of a given standard as listed.

PIPING STANDARDS

Codes for various piping services have been developed by nationally recognized engineering societies, standardization bodies, and trade associations. The sound engineering practices incorporated in these codes generally cover minimum safety requirements for the selection of materials, dimensions, design, fabrication, erection, and testing of piping systems. By means of interpretation and revision these codes continually reflect the knowledge gained through experience, testing, and research.

Generally, piping codes form the basis for many state and municipal safety laws. Compliance with a code which has attained this status is mandatory for all systems included within the jurisdiction. Although some of today's piping installations are not within the scope of any mandatory code, it is advisable to comply with the applicable code in the interests of safety and as a basis for contract negotiations. Contracts with various agencies of the federal government are regulated by federal specifications or rules. These often do not have a direct connection with the codes enumerated below.

The reader is cautioned that the **piping standards** are changing more often than in previous years. Although the formulas and other data provided are in accordance with the code rules in effect at the time of publication, it must be recognized that code rules may change, and piping engineering and design work performed in accordance with information contained herein does not provide complete assurance that all extant code requirements have been met. The reader is urged to become familiar with the specific code edition and addenda applicable in a particular project, for they may contain mandatory requirements applicable to the particular project.

The **ASME Boiler and Pressure Vessel Code** is mandatory in many cities, states, and provinces in the United States and Canada. Local application of this code into law is not uniform, making it necessary to investigate the city or state laws which have jurisdiction over the installation in question. Compliance with this code is required in all locations to qualify for insurance approval.

Section I: "Power Boilers" concerns all piping connections to power boilers or superheaters including the first stop valve on single boilers, or including the second stop valve for cross-connected multiple-boiler installations. Section I refers to ASME B31.1 which contains rules for design and construction of "boiler external piping." "Boiler external piping" is under the jurisdiction of Section I and requires inspection and code stamping in accordance with Section I even though the rules for its design and construction are contained in the ASME Code for Pressure Piping, section B31.1.

Section II: "Material Specifications" provides detailed specifications of the materials which are acceptable under this code. (These specifications generally are identical to the corresponding ASTM Standards.)

Section III: "Nuclear Components" includes all nuclear piping. It is the responsibility of the designer to determine whether or not a particular piping system is "nuclear" piping, since Section III makes this determination the responsibility of the designer. In general, piping whose failure could result in the release of radiation which would endanger the public or plant personnel is considered "nuclear" piping.

Section VIII: "Unfired Pressure Vessels" concerns piping only to the extent of the flanged or threaded connections to the pressure vessel, except that the entire section will apply in those special cases where unfired pressure vessels are made from pipe and fittings.

Section IX: "Welding and Brazing Qualifications" establishes the minimum requirements for ASME Code welding.

Section XI: "Rules for Inservice Inspection of Nuclear Power Plant Components" contains rules for the examination and repair of components throughout the life of the plant.

The ASME Code for Pressure Piping B31 is, at present, a nonmandatory code in the United States except where U.S. state legislative bodies and Canadian provinces have adopted this code as a legal requirement. The minimum safety requirements of these codes have been accepted by the industry as a standard for all piping outside the jurisdiction of other codes. The piping systems covered by the separate sections of this code are listed below:

Power Piping	B31.1
Fuel Gas Piping	B31.2
Chemical Plant and Petroleum Refinery Piping	B31.3
Liquid Petroleum Transportation Piping Systems	B31.4
Refrigeration Piping	B31.5
Gas Transmission and Distribution Piping Systems	B31.8
Building Service Piping	B31.9

Several other engineering societies and trade associations have also issued standards covering piping. Foremost among these is the American Society for Testing and Materials (ASTM), the American National Standards Institute (ANSI), the American Water Works Association (AWWA), the American Petroleum Institute (API), and the Manufacturers Standardization Society of the Valve and Fitting Industry (MSS).

Additional piping specifications have been issued by the American Welding Society (AWS), the Pipe Fabrication Institute (PFI), the National Fire Protection Association (NFPA), the Copper Development Association (CDA), the Plastics Pipe Institute (PPI), and several others.

The piping standards issued by the ASTM are most commonly referred to in specifications covering piping for power plants, chemical plants, refineries, pulp and paper mills, and other industrial plants. The large majority of ASTM Standards has also been issued by the ASME in Section II of the ASME Boiler and Pressure Vessel Code. The same specification numbers are applied by the ASME as were originally assigned by the ASTM.

The ANSI formerly prepared the various standards of the B31 Code for Pressure Piping. These standards are now issued by the ASME. The ANSI, however, continues to prepare and issue various standards covering pipe fittings, flanges, and other piping components. Note that ASME B16 prepares and issues standards for fittings, flanges, etc.

The AWWA has issued various standards for waterworks applications. The majority of these involve ductile iron pipe, ductile iron and cast iron pipe fittings, etc.

The MSS has prepared various standards for valves, hangers, and fittings, generally involving the lower range of pressures and temperatures.

PIPING, PIPE, AND TUBING

The term **piping** generally is broadly applied to pipe, fittings, valves, and other components that convey liquids, gases, slurries, etc.

The term **pipe** is applied to tubular products of dimensions and materials commonly used for pipelines and connections, formerly designated as **iron pipe size (IPS).** The outside diameter of all weights and kinds of IPS pipe is of necessity the same for a given pipe size on account of threading. Nevertheless, most pipe is furnished unthreaded with ends prepared for butt welding.

The word **tube** (or **tubing**) is generally applied to tubular products as utilized in boilers, heat exchangers, instrumentation, and in the machine, aircraft, automotive, and related industries.

Pipe and Tube Products—General

Commercial pipe and tube products are grouped into various classifications generally based on the application or use and not on the manufacturing method. Most tubular products fall into one of three very broad classifications: (1) pipe, (2) pressure tubes, and (3) mechanical tubes. Each classification falls into various subgroupings, which may have been defined and standardized differently by the different trade or user groups. The same standard materials specifications may apply to several of the (user) classifications. For example, ASTM A120 or A53 pipe may be used for applications representing refrigeration, pressure, and nipple service.

Cost considerations enter into the selection of specific piping materials. In some sizes, prices of pipe made to different materials specifications may vary, whereas in other sizes, they may be identical.

Within the broad **use classifications** listed above, the **production method classifications** are also recognized. These are primarily (1) seamless wrought pipe, (2) seamless cast pipe, and (3) seam-welded pipe or tubes. The large variety of single and combination pipe- or tube-forming methods can produce different characteristics and properties in essentially identical pipe materials. In addition, the final finishing can result in hot-finished or cold-finished products. Cold-finishing may be accomplished by reducing or by expanding. Heat treatments may also affect the properties of the finished product.

Piping

On the basis of user classification, the more commonly used types of pipe are tabulated in Table 6.7.1. This listing ignores method of manufacture, size range, wall thickness, and finish, for which the different user groups may have developed different standard requirements.

Table 6.7.1 Major Pipe Classification and Examples of Applications

Identification of pipe	Uses
Standard	Mechanical (structural) service pipe, low-pressure service pipe, refrigeration (ice-machine) pipe, ice-rink pipe, dry-kiln pipe
Pressure	Liquid, gas, or vapor service pipe, service for elevated temperature or pressure, or both
Line	Threaded or plain end, gas, oil, and steam pipe
Water well	Reamed and drifted, water-well casing, drive pipe, driven well pipe, pump pipe, turbine-pump pipe
Oil country tubular goods	Casing, well tubing, drill pipe
Other pipe	Conduit, piles, nipple pipe, sprinkler pipe, bedstead tubing

Standard Pipe Mechanical service pipe is produced in three classes of wall thickness—standard weight, extra strong, and double extra strong. It is available as welded or seamless pipe of ordinary finish and dimensional tolerances, produced in sizes up to 12-in nominal OD. This pipe is used for structural and mechanical purposes. Certain applications have other requirements for size, surface finish, or straightness.

Refrigeration Pipe This pipe is also known as ice-machine pipe or ammonia pipe. It may be butt-welded, lap-welded, electric-resistance-welded, or seamless and is intended for use as a conveyor of refrigerants. This pipe is suitable for coiling, bending, and welding. The sizes commonly used range from ¾ to 2 in. The piping is produced in random and double random lengths in standard line pipe sizes and weights. Double random lengths are used as ice-rink pipe. It can be produced with plain ends, with threaded ends only, or with threaded ends and line pipe couplings, as desired.

Dry-Kiln Pipe This pipe is butt-welded, electric-resistance-welded, or seamless pipe for use in the lumber industry. It is produced in standard-weight pipe sizes of ¾, 1, and 1¼ in. Joints are designed to permit subsequent "makeup" after expansion has occurred. Dry-kiln pipe is commonly produced with threaded ends and couplings and in random lengths.

Pressure Pipe Pressure pipe is used for conveying fluids or gases at normal, subzero, or elevated temperatures and/or pressures. It generally is not subjected to external heat application. The range of sizes is ⅛-in nominal size to 36-in actual OD. It is produced in various wall thicknesses. Pressure piping is furnished in random lengths, with threaded or plain ends, as required. Pressure pipe generally is hydrostatically tested at the mill.

Line Pipe Line pipe is seamless or welded pipe produced in sizes from ⅛-in nominal OD to 48-in actual OD. It is used principally for conveying gas, oil, or water. Line pipe is produced with ends which are plain, threaded, beveled, grooved, flanged, or expanded, as required for various types of mechanical couplers, or for welded joints. When threaded ends and couplings are required, recessed couplings are normally supplied.

Water-Well Pipe Water-well pipe is welded or seamless steel pipe used for conveying water for municipal and industrial applications. Pipelines for such purposes involve flow mains, transmission mains, force mains, water mains, or distribution mains. The mains are generally laid underground. Sizes range from ⅛- to 106-in OD in a variety of wall thicknesses. Pipe is produced with ends suitably prepared for mechanical couplers, with plain ends beveled for welding, with ends fitted with butt straps for field welding, or with bell-and-spigot joints with rubber gaskets for field joining. Pipe is produced in double random lengths of about 40 ft, single random lengths of about 20 ft, or in definite cut lengths, as specified. Wall thicknesses vary from 0.068 in for ⅛-in nominal OD to 1.00 in for 106-in actual OD.

When required, water-well pipe is produced with a specified coating or lining or both. For example, cement-mortar lining and coatings are extensively used.

Oil Country Goods Casing is used as a structural retainer for the walls of oil or gas wells. It is also used to exclude undesirable fluids, and to confine and conduct oil or gas from productive subsurface strata to the ground level. Casing is produced in sizes 4½- to 20-in OD. Size designations refer to actual outside diameter and weight per foot. Ends are commonly threaded and furnished with couplings. When required, the ends are prepared to accommodate other types of joints.

Drill Pipe Drill pipe is used to transmit power by rotary motion from ground level to a rotary drilling tool below the surface and also to convey flushing media to the cutting face of the tool. Drill pipe is produced in sizes 2⅜- to 6⅝-in OD. Size designations refer to actual outside diameter and weight per foot. Drill pipe is generally upset, either internally or externally, or both, and is furnished with threaded ends and couplings, threaded only, or prepared to accommodate other types of joints.

Tubing is used within the casing of oil wells to conduct oil to ground level. It is produced in sizes 1.050- to 4.500-in OD in several weights per foot. Ends are threaded and fitted with couplings and may or may not be upset externally.

Other Pipe Classifications **Rigid conduit pipe** is welded or seamless pipe intended especially for the protection of electrical wiring systems. Conduit pipe is not subjected to hydrostatic tests unless so specified. It is furnished in standard-weight pipe sizes from ¼- to 6-in OD in 10-ft lengths, with plain ends or with threaded ends and couplings, as specified. Although some specifications of rigid conduit pipe list lengths to 20 ft, the National Electric Code, 1965, limits lengths to 10 ft.

Piling pipe is welded or seamless pipe for use as piles, where the cylinder section acts as a permanent load-carrying member or where it acts as a shell to form cast-in-place concrete piles. Specifications provide for the choice of three grades by minimum tensile strength, in which the sizes listed are 8⅝- to 24-in OD in a variety of wall thicknesses and in two length ranges. Ends are plain or beveled for welding.

Nipple pipe is standard-weight, extra-strong, or double-extra-strong welded or seamless pipe produced for the manufacture of pipe nipples. Standard-weight pipe with threaded ends is also used in sprinkler systems. Nipple pipe is commonly produced in random lengths with plain ends in nominal sizes ⅛- to 12-in OD. Close OD tolerances, sound welds, good threading properties, and surface cleanliness are essential in this product. It is commonly coated with oil or zinc and well protected in shipment. When reference is made to ASTM Specifications for this application, Specification A120 is generally used for diameters to 5-in OD and A53 for diameters of 5 in and over.

Standard Pipe Sizes Standard pressure, line, and other pipe with plain ends for welding or with threaded ends is standardized in two ranges. Diameters of 12 in and less have a nominal size which represents approximately that of the inside diameter of standard-weight pipe. The nominal outside diameter is standard, regardless of weight. Increase in wall thickness results in a decrease of the inside diameter.

The standardization of pipe sizes over 12 in is based on the actual outside diameter, the wall thickness, and the weight per foot.

Samples of the principal dimensions, weights, and characteristics of commercial piping materials are summarized in Table 6.7.2.

The weights of welding reducers are for one size reduction and are thus only approximately correct for other reductions.

Hot-finished or cold-drawn seamless low-alloy steel tubes generally are process-annealed at temperatures between 1,200 and 1,350°F.

Austenitic stainless-steel tubes are usually annealed at temperatures between 1,800 and 2,100°F, with specific temperatures varying somewhat with each grade. This is generally followed by pickling, unless bright-annealing was done.

Mechanical Tubing

Unlike pipe and pressure tubes, mechanical tubing is generally classified by the method of manufacture and the degree of finish. Examples of classifications are "seamless hot-finished," "cold-drawn welded," "flash-in-grade," etc.

Seamless Tubes Seamless tubes are available as either hot- or cold-finished. They are normally made in sizes from 0.187-in OD to 10.750-in OD.

Dimensions for hot-finished mechanical tubes are provided in Table 6.7.3. Dimensions for cold-finished tubes are listed in Table 6.7.4.

Welded Tubes Welded tubes generally are produced by electric resistance methods. Where required, the welding flash is removed with a cutting tool. Industry practice normally recognizes a number of finish conditions which are summarized in Table 6.7.5.

Flash-in Type Tubing This tubing is generally limited to applications where nothing is inserted in the tube.

Flash-Controlled Tubing This tubing is used where moderate control of the inside diameter is required. Generally, the outside and inside diameters are specified.

For special materials, the equations listed below for weights of tubes and weights of contents of tubes are helpful.

$$\text{Weight of tube, lb/ft} = F \times 10.68 \times T \times D - T$$

where T = wall thickness, in; D = outside diameter, in; F = relative weight factor.

The weight of tube calculation is based on low-carbon steel weighing 0.2833 lb/in^3 and is extended to other materials through the factor F.

Relative weight factor F	
Aluminum	0.35
Brass	1.12
Cast-iron	0.91
Copper	1.14
Ferritic stainless steel	1.02
Steel	1.00
Wrought iron	0.98

$$\text{Weight of contents of tube, lb/ft} = G \times 0.3405 \times (D - 2T)^2$$

where G = specific gravity of contents; T = tube wall thickness, in; D = tube outside diameter, in.

The weight per foot of steel pipe is subject to the tolerances listed in Table 6.7.6.

The designation *sink-draw tubes* is specified where close control over the outer diameter is required with normal tolerance applying to the wall thickness. Smoothness of the inside surface is not controlled, except that the flash is generally controlled to a height of 0.005 or 0.010 in maximum.

Mandrel-drawn tubes usually are normalized after welding by passing the tubes through a continuous atmosphere-controlled furnace. After descaling, the tubes are cold-drawn through a die with a mandrel on the inside of the tube. These tubes provide maximum control over surface finish, outside or inside diameters, and wall thickness. The normalizing heat treatment removes the effects of welding and provides a uniform microstructure around the tube circumference.

The different finish classifications may result in substantial differences in the mechanical properties of the steel material.

Typical examples for low-carbon steel material are given in Table 6.7.7. Differences in carbon content and other chemistry, heat treatment, etc., may significantly change these typical values.

Other Tubing Types Among other tube classifications are sanitary tubing usually made of 18% Cr-8% Ni stainless steel and available as seamless or welded tubing. This tubing is used extensively in the dairy, beverage, and food industries. Sanitary tubing is generally available in sizes from 1- to 4-in OD. It may be furnished either hot- or cold-finished. The tubes are normally annealed at temperatures above 1,900°F.

Some welded tube is also produced by fusion-welding methods utilizing either the inert-gas tungsten-arc-welding or gas-shielded consumable metal-arc-welding process. This tubing is generally more expensive than the resistance-welded types.

The butt-welded cold-finished tubes are made from hot-rolled or cold-rolled strip and fusion-welded. This tubing is usually furnished as sink-drawn or mandrel-drawn.

Butt-welded tubing is made in heavier wall thicknesses than the resistance-welded tube.

Several tubing materials used in the automobile industry are covered by specifications of the Society of Automotive Engineers, "SAE Handbook."

Pressure Tubing

Pressure-tube applications commonly involve external heat applications, as in boilers or superheaters.

Pressure tubing is produced to the actual outside diameter and minimum wall or average wall thickness specified by the purchaser. Pressure tubing may be hot- or cold-finished.

The wall thickness is normally given in decimal parts of an inch rather than as a fraction or gage number. When gage numbers are given without reference to a gage system, Birmingham wire gage (BWG) is implied.

Pressure tubing is usually made from steel produced by the basic oxygen or electric-furnace processes.

Seamless pressure tubing may be either hot-finished or cold-drawn. Cold-drawn steel tubing is frequently process-annealed at temperatures above 1,200°F. To ensure quality, maximum hardness values are frequently specified. For example, in ASME Specification SA192, Specification for Seamless Carbon Steel Boiler Tubes for High-Pressure Service, the following maximum hardness values are given. Specifications for boiler tubes generally reference the ASME Standards such as SA178, SA192, SA210, etc.

Boiler tubes	Brinell hardness no. (tubes 0.200 in and over in wall thickness)	Rockwell hardness no. (tubes less than 0.200 in in wall thickness)
Hot-finished tubes	137	B77
Cold-drawn (normalized) tubes	125	B72

Piping Fabrication Methods Commercial steel pipe is furnished most commonly as seamless pipe. Seamless pipe can be produced by (1) piercing, (2) hollow forging, and (3) forging, turning, and boring. Commercial pipe is also produced by welding involving (1) electric resistance welding, (2) electric fusion welding, or (3) submerged arc welding. Some mills are prepared to extrude small-diameter pipe and tubing in a variety of geometric shapes or to cast large-diameter steel pipe.

Electric-Fusion Welding Flat plate, known as skelp, is prepared in proper width and thickness for the desired pipe inside and outside

Table 6.7.2 Sample of Properties of Commercial Steel Pipe

Nominal pipe size, outside diameter, in	Schedule number* a	b	c	Wall thick-ness, in	Inside diameter, in	Inside area, in^2	Metal area, in^2	Outside surface, ft^2/ft	Inside surface, ft^2/ft	Weight, lb/ft†	Weight of water, lb/ft	Moment of inertia, in^4	Section modulus, in^3	Radius of gyration, in
⅛			10S	0.049	0.307	0.0740	0.0548	0.106	0.0804	0.186	0.0321	0.00088	0.00437	0.1271
0.405	40	Std	40S	0.068	0.269	0.0568	0.0720	0.106	0.0705	0.245	0.0246	0.00106	0.00525	0.1215
	80	XS	80S	0.095	0.215	0.0364	0.0925	0.106	0.0563	0.315	0.0157	0.00122	0.00600	0.1146
¼			10S	0.065	0.410	0.1320	0.0970	0.141	0.1073	0.330	0.0572	0.00279	0.01032	0.1694
0.540	40	Std	40S	0.088	0.364	0.1041	0.1250	0.141	0.0955	0.425	0.0451	0.00331	0.01230	0.1628
	80	XS	80S	0.119	0.302	0.0716	0.1574	0.141	0.0794	0.535	0.0310	0.00378	0.01395	0.1547
⅜			5S	0.065	0.710	0.396	0.1582	0.220	0.1859	0.538	0.1716	0.01197	0.0285	0.2750
0.675			10S	0.065	0.545	0.2333	0.1246	0.177	0.1427	0.423	0.1011	0.00586	0.01737	0.2169
	40	Std	40S	0.091	0.493	0.1910	0.1670	0.177	0.1295	0.568	0.0827	0.00730	0.02160	0.2090
	80	XS	80S	0.126	0.423	0.1405	0.2173	0.177	0.1106	0.739	0.0609	0.00862	0.02554	0.1991
½			5S	0.065	0.710	0.3959	0.1583	0.220	0.1859	0.538	0.171	0.0120	0.0285	0.2750
0.840			10S	0.083	0.674	0.357	0.1974	0.220	0.1765	0.671	0.1547	0.01431	0.0341	0.2692
	40	Std	40S	0.109	0.622	0.304	0.2503	0.220	0.1628	0.851	0.1316	0.01710	0.0407	0.2613
	80	XS	80S	0.147	0.546	0.2340	0.320	0.220	0.1433	1.088	0.1013	0.02010	0.0478	0.2505
	160			0.187	0.466	0.1706	0.383	0.220	0.1220	1.304	0.0740	0.02213	0.0527	0.2402
		XXS		0.294	0.252	0.0499	0.504	0.220	0.0660	1.714	0.0216	0.02425	0.0577	0.2192
¾			5S	0.065	0.920	0.655	0.2011	0.275	0.2409	0.684	0.2882	0.02451	0.0467	0.349
1.050			10S	0.083	0.884	0.614	0.2521	0.275	0.2314	0.857	0.2661	0.02970	0.0566	0.343
	10	Std	40S	0.113	0.824	0.533	0.333	0.275	0.2157	1.131	0.2301	0.0370	0.0706	0.334
	80	XS	80S	0.154	0.742	0.432	0.435	0.275	0.1943	1.474	0.1875	0.0448	0.0853	0.321
	160			0.218	0.614	0.2961	0.570	0.275	0.1607	1.937	0.1284	0.0527	0.1004	0.304
		XXS		0.308	0.434	0.1479	0.718	0.275	0.1137	2.441	0.0641	0.0579	0.1104	0.2840

* Schedule numbers: Standard-weight pipe and schedule 40 are the same in all sizes through 10 in; from 12 in through 24 in, standard weight pipe has a wall thickness of ⅜ in. Extra-strong-weight pipe and schedule 80 are the same in all sizes through 8 in; from 8 in through 24 in, extra-strong-weight pipe has a wall thickness of ½ in. Double-extra-strong-weight pipe has no corresponding schedule number.

[a]ANSI B36.10 steel pipe schedule numbers.

[b]ANSI B36.10 steel pipe nominal wall thickness designation.

[c]ANSI B36.19 stainless steel pipe schedule numbers.

†The ferritic steels may be about 5% less and the austenitic stainless steels about 2% greater than the values shown in this table, which are based on weights for carbon steel.

NOTE: The following formulas are used in the computation of the values shown in the table:

Weight of pipe per foot (pounds) $= 10.6802t(D - t)$

Weight of water per foot (pounds) $= 0.3405d^2$

Square feet outside surface per foot $= 0.2618D$

Square feet inside surface per foot $= 0.2618d$

Inside area (square inches) $= 0.785d^2$

Area of metal (square inches) $= 0.785(D^2 - d^2)$

Moment of inertia (inches^4) $= 0.0491(D^4 - d^4) = A_m R_g^2$

Section modulus (inches^3) $= \dfrac{0.0982(D^4 - d^4)}{D}$

Radius of gyration (inches) $= 0.25\sqrt{D^2 + d^2}$

where A_m = area of metal, in^2; d = inside diameter, in; D = outside diameter, in; R_g = radius of gyration, in; t = pipe wall thickness, in.

Table 6.7.3 Diameter and Wall-Thickness Tolerances for Seamless Hot-finished Mechanical Tubing of Carbon and Alloy Steel (AISI)*

				Wall thicknesses tolerance, %							
		OD tolerance		0.109 in and under		Over 0.109 to 0.172 in, incl.		Over 0.172 to 0.203 in, incl.		Over 0.203 in	
Specified size, OD, in	Ratio of wall thickness to OD	Over	Under	Over	Under	Over	Under	Over	Under	Over	Under
Under 3	All wall thicknesses	0.023	0.023	16.5	16.5	15	15	14	14	12.5	12.5
3-5½, excl.	All wall thicknesses	0.031	0.031	16.5	16.5	15	15	14	14	12.5	12.5
5½-8 excl.	All wall thicknesses	0.047	0.047					14	14	12.5	12.5
8-10¾, incl.	5% and over	0.047	0.047							12.5	12.5
8-10¾, incl.	Under 5%	0.063	0.063							12.5	12.5

* The common range of sizes of hot-finished tubes is 1½ to and including 10¾ in outside diameter with wall thickness not less than 0.095 in (no. 13 BWG) or 3% or more of the outside diameter. For sizes under 1½ or over 10¾ in outside diameter, the tolerances are commonly negotiated between the purchaser and producer.
SOURCE: AISI, "Steel Products Manual."

diameters. It is then charged into an electric furnace and, when the proper welding temperature has been reached, is drawn through a funnel-shaped die so shaped that the plate is gradually formed into the shape of a tube, with the edges of the plate being forced squarely together and fused. The formed pipe then passes through a series of rolls in which it is sized or drawn to final dimensions.

Electric-Resistance Welding For pipes or tubes sized 4-in (10.2-cm) OD and under, strip is fed onto a set of forming rolls which consists of horizontal and vertical rollers so placed as to gradually form the flat strip into a tube. The tube form then passes to the welding electrodes. The electrodes are copper disks connected to the secondary of a revolving transformer assembly. The copper-disk electrodes make contact on each side of the seam of the tube form; a flow of current takes place across the seam, and temperature is raised to the welding point. Outside flash is removed by a cutting tool as the tube leaves the electrodes; inside flash is removed either by an air hammer or by passing a mandrel through the welded tube after the tube has been cooled.

Submerged-Arc Electric Welding This process is used for pipes from 24- to 36-in (61.0- to 9.1.4-cm) OD. Flat plate is first pressed into a U and later into an O shape. The O shape is placed in an automatic welder and backed up on the inside by a water-cooled copper shoe. Two electrodes in close proximity are used. The electrodes are not in actual contact with the pipe. The current passes from one electrode through a granular flux and across the gap in the pipe to the second electrode. The high temperature of the arc heats the edges of the plate; a welding rod placed just over the seam is thereby melted and metal is deposited in the groove. After the outside weld has been made, the pipe is conveyed to an inside welder where a similar operation is carried on, except that no backup shoe is needed.

Seamless Tubing and Pipe A heated billet is brought into contact with tapered revolving rolls in such a way that the billet is pulled into the space allowed between the rolls. A piercing mandrel is placed in this space; the soft center of the billet makes it possible for the rolls to draw the billet over the mandrel, producing a hollow shell. When the billet has entirely passed over the mandrel, it is in the form of a thick-walled seamless tube. The heavy-walled tube is then passed to a rolling mill which reduces the tube to pipe of proper outside diameter and wall thickness.

The method of fabrication described above is limited as to diameter and thickness. For seamless alloy tubes and for heavy-wall carbon-steel tubes or pipe, a process known as **cupping and drawing** is frequently used. A circular flat plate of proper diameter and thickness is heated, placed in a hydraulic press, and pressed by a ram through a die. The cup so formed is reheated and pressed through a smaller die, thus elongating the cup so that it becomes a short cylinder with one closed end. This short cylinder is then placed in a horizontal drawbench, and with reheating as necessary, is pushed by a ram through dies of successively smaller diameters until the desired outer diameter is reached.

Forged, Turned, and Bored Tubing In this process, the ingot is heated and forged to a rough cylindrical shape, oversize in both diameter and length. The forging is then placed in a lathe and the outside turned down to the desired outer diameter. Rough ends are then removed so that the finally desired length is obtained. The cylinder is then placed in a boring mill, and the inside bored out until the desired wall thickness is secured. Because of the relatively high cost, this process is now rarely used.

Hollow-Forged Pipe and Tubing In this process, ingots are cast and their ends cropped; then they are placed in a furnace and heated to a specified temperature. The heated ingot is placed in a press where it is pierced. This hollow cylinder, open at one end, is then descaled and drawn over a mandrel on a horizontal drawbench. The closed end is then burned off, and the hollow forging is chemically descaled. Following this, the forging is straightened, placed in a lathe, and the outer diameter machined to a true dimension. The inside is dressed to remove scale, but no machining is done on the inside.

Carbon-steel piping is most frequently used as manufactured in accordance with ASTM Specifications A106 and A53 (or ASME Specifications SA106 and SA53). The chemical compositions of these two materials are identical except for the deoxidation practice which applies to the A106 pipe. Both are subjected to physical tests, but those for A106 are more rigorous. A53 and A106 are made in grades A and B; grade B has higher strength properties but is less ductile and, for this reason, grade A is permitted only for cold bending or close coiling. When carbon steel is intended for use in welded construction at temperatures in excess of 775°F (413°C), consideration should be given to the possibility of graphite formation.

Chromium-molybdenum steel has been used for temperatures up to 1,100°F (593°C). In the small diameters, the material is usually available in the seamless construction; because of the inability of the seamless mills to fabricate large-diameter and heavy-walled pipe, it may be necessary to resort to the more expensive hollow-forged or forged-and-bored piping for higher pressures and temperatures. The material for a high-temperature piping system should be selected after a careful review of technical and economic considerations; the following is intended only as being indicative of recent and current practice.

For temperatures up to 1,000°F, 1¼% Cr-½% Mo (A335, grade P11) is used. For temperatures from 950 to 1050°F, 2¼% Cr-1% Mo (A335, P22) generally is used. Where there is a combination of high temperatures and erosive action, 5% Cr-½% Mo (A335, grade 5) or other more highly chromium-molybdenum or chromium stainless steels have been used.

Stainless-steel piping is available in a variety of compositions, most popular of which are ASTM A213, grade TP304 (16% Cr-8% Ni), and ASTM A213, grade TP316 (18% Cr-12% Ni and 3% Mo). For high-temperature service, type 34 stainless steels are used (18% Cr-8% Ni and stabilized with columbium). This material may be used up to 1,200°F (649°C); particular care must be given to choice of welding filler metal to avoid brittleness in the welds.

The permissible stress values for a large variety of piping materials at low and elevated temperatures are provided in a basic reference: ASME B31.3.

Table 6.7.4 Diameter and Wall-Thickness Tolerances for Seamless Cold-Worked Mechanical Tubing of Carbon and Alloy Steel (AISI)*

	Unannealed of finish-annealed				Soft annealed or normalized				Quenched and tempered				Wall thickness all conditions, %	
	OD, in		ID, in		OD, in		ID, in		OD, in		ID, in			
Size, OD, in	Over	Under	Over	Under	Over	Under	Over	Under	Over	Under	Over	Under	Over	Under
3⁄16-½, excl.†‡	0.004	0			0.006	0.002			0.010	0.010			15	15
½-1½, excl.†‡§¶	0.005	0	0	0.005	0.008	0.002	0.002	0.008	0.015	0.015	0.015	0.015	10	10
1½-3½, excl.†‡§¶	0.010	0	0	0.010	0.015	0.005	0.005	0.015	0.030	0.030	0.030	0.030	10	10
3½-5½, excl.§¶	0.015	0	0.005	0.015	0.023	0.007	0.015	0.025	0.045	0.045	0.045	0.045	10	10
5½-8, excl.,¶ wall less than 5% OD	0.030	0.030	0.035	0.035	0.060	0.060	0.070	0.070					10	10
5½-8, excl., wall from 5 to 7.5% OD	0.020	0.020	0.025	0.025	0.040	0.040	0.050	0.050					10	10
5½-8, excl.,§ wall over 7.5% OD	0.030	0	0.015	0.030	0.045	0.015	0.037	0.053					10	10
8-10¾, incl.,¶ wall less than 5% OD	0.045	0.045	0.050	0.050									10	10
8-10¾, incl., wall from 5 to 75% OD	0.035	0.035	0.040	0.040									10	10
8-10¾, incl.,§ wall over 7.5% OD	0.045	0	0.015	0.040									10	10

* For tolerances closer than those indicated, availability, and applicable tolerances for tubing less than 3⁄16 in OD or larger than 10¾-in OD, the producer should be consulted.

† For those tubes with inside diameter less than ½ in (or less than ⅝ in when the wall thickness is more than 20% of the outside diameter), which are not commonly drawn over a mandrel. Note § is not applicable. Unless otherwise agreed upon by the purchaser and producer, the wall thickness may vary 15% over and under that specified, and the inside diameter is governed by the outside diameter and wall-thickness tolerances shown.

‡ For tubes with inside diameter less than ½ in (or less than ⅝ in when the wall thickness is more than 20% of the outside diameter), which can be produced by the rod or bar mandrel process, the tolerances are as shown in the above table except that the wall-thickness tolerances are 10% over and under the specified wall thickness.

§ Many tubes with inside diameter less than 50% of outside diameter, or with wall thickness more than 25% of outside diameter, or with wall thickness over 1¼ in, or weighing more than 90 lb/ft are difficult to draw over a mandrel. Unless otherwise agreed upon by the purchaser and producer the inside diameter may vary over or under by an amount equal to 10% of the wall thickness and the wall thickness may vary 12½% over and under that specified.

¶ Tubing having a wall thickness less than 3% of the outside diameter cannot be straightened properly without a certain amount of distortion. Consequently, such tubes, while having an average outside diameter and inside diameter within the tolerances shown in the above table, require an ovality tolerance of 0.5% over and under nominal outside diameter, this being in addition to the tolerances indicated in the above table.

SOURCE: AISI, "Steel Products Manual."

Table 6.7.5 Finish Classifications Normally Recognized as Welded Mechanical Tubing

	Type	Condition
A	Flash-in, hot-rolled	Made by longitudinally forming and welding hot-rolled strip; the interior welding flash remains in place
B	Flash-in, cold-rolled	Made by longitudinally forming and welding cold-rolled strip; the interior flash remains in place
C	Flash controlled–0.010 in max, hot-rolled	Same as A, except that the interior flash is partially removed so that the height remaining does not exceed 0.010 in
D	Flash controlled–0.010 in max, cold-rolled	Same as C, except that cold-rolled strip has been used
E	Flash controlled–0.005 in max, hot-rolled	Same as C, except that flash height on tube inside does not exceed 0.005 in
F	Flash controlled–0.005 in max, cold-rolled	Same as E, except that cold-rolled strip has been used
G	Sink-drawn, hot-rolled	Tubing with flash controlled to 0.010 in max, which has been descaled and cold-drawn through a die to cold-finish the exterior surface
H	Sink-drawn, cold-rolled	Tubing with flash controlled to 0.010 in max, which has been cold-drawn through a die to cold-finish the exterior surface
I	Mandrel-drawn	Cold-rolled tubing with the interior flash removed and drawn through a die and over a mandrel to cold-finish the exterior and interior surfaces

Table 6.7.6 Weight Tolerances of Steel Piping

Specification	Size	Tolerance, %	
ASTM A53	Std wt	+5	−5
	XS wt	+5	−5
ASTM A120	XXS wt	+10	−10
ASTM A106	Sch 10-120	+6.5	−3.5
	Sch 140-160	+10	−3.5
ASTM A335	12 in and under	+6.5	−3.5
	Over 12 in	+10	−5
ASTM A312 ASTM A376	12 in and under	+6.5	−3.5
	Std. wt		
	Reg wt		
	XS wt	+10	−3.5
API 5L	XXS wt		
	Spec. plain end pipe	+10	−5

PIPE FITTINGS

The various major piping materials are also produced in the form of standard fittings. Among the more widely used are wrought-steel fittings, welded-steel fittings, cast-steel fittings, cast-iron fittings, ductile-iron fittings, malleable-iron fittings, brass and copper fittings, aluminum fittings, etc. Other major nonferrous piping materials are also produced in the form of cast and wrought fittings.

Cast-iron, ductile-iron, and malleable-iron fittings are made by conventional founding methods for a variety of joints including bell-and-spigot, flanged, and mechanical (gland-type), or other proprietary joint designs.

Schedule Designations Over 100 years ago piping was designated as standard, extra-strong, and double extra-strong. There was no provision for thin-walled pipe, and no intervening standard thicknesses between the three schedules, which covered too great a spread to be economical without intermediate weights. Piping lists as a function of the schedule number is given, approximately, by the following relationship: Schedule no. = 1,000 $P/(SE)$, where P is operating pressure, lb/in^2 gage, S is allowable stress, lb/in^2, and E is the quality factor which is approximately between 0.75 and 1. For more details please see ASME B31.3.

Commercial sizes of steel pipe are known by their nominal inside diameter (ID) from in (0.3175 cm) to 12 in (30.5 cm). Above 12-in ID, pipe is usually known by its outside diameter (OD). All classes of pipe of a given nominal size have the same OD, the extra thickness for different weights being on the inside.

Thickness of Pipe The following notes, covering power piping systems, have been abstracted from Part 2 of the Code for Power Piping (ASME B31.1).

For inspection purposes, the minimum thickness of pipe wall to be used for piping at different pressures and for temperatures not exceeding those for the various materials listed in ASME B31.3 shall be determined by the formula

$$t_m = \frac{PD}{2(SE + Py)} + A \qquad (6.7.1)$$

where t_m = minimum pipe-wall thickness, in, allowable on inspection; P = maximum internal service pressure, lb/in^2 gage (plus water-hammer allowance in case of cast-iron conveying liquids); D = OD of pipe, in; S = maximum allowable stress in material due to internal pressure, lb/in^2; E = quality factor (given in ASME B31.3 and is approximately between 0.75 and 1), y = a coefficient, values for which are listed in Table 6.7.8; A = allowance for threading, mechanical strength, and corrosion, in, with values of A listed in Table 6.7.9.

The thickness of ductile-iron pipe conveying liquid may be taken from Table 6.7.14, using the pressure class next higher than the maximum internal service pressure in pounds per square inch. Where ductile-iron pipe is used for steam service, the thickness should be calculated by Eq. (6.7.1).

Plain-end pipe includes pipe joined by flared compression couplings, lapped joints, and by welding, that is, by any method that does not reduce the wall thickness of the pipe at the joint.

Physical and Chemical Properties of Pipes, Tubes, etc. The design of piping for operation above 750°F (399°C) presents many problems

Table 6.7.8 Values of *y*
(Interpolate for intermediate values) (ASME B31.1)

	Temp, °F (°C)					
	900 (482) and below	950 (510)	1,000 (538)	1,050 (566)	1,100 (593)	1,150 (621) and above
Ferritic steels	0.4	0.5	0.7	0.7	0.7	0.7
Austenitic steels	0.4	0.4	0.4	0.4	0.5	0.7

Table 6.7.7 Typical Mechanical Properties of Resistance-Welded Mechanical Carbon-Steel Tubing

Type	Yield strength, lb/in^2	Tensile strength, lb/in^2	Elongation, %	Hardness, Rockwell B
Flash-in or flash-controlled, hot-rolled	48,000	58,000	29	68
Flash-in or flash-controlled, cold-rolled	68,000	76,000	17	84
Normalized	34,000	52,000	39	61
Sink-drawn	73,000	76,000	20	84
Mandrel-drawn	80,000	83,000	15	86

Table 6.7.9 Values of *A*
(ASME B31.1)

Type of pipe	*A*, in
Cast-iron pipe, centrifugally cast	0.14
Cast-iron, pit-cast	0.18
Threaded-steel, wrought-iron, or nonferrous pipe:	
⅜ in and smaller	0.05
½ in and larger	Depth of thread
Grooved-steel, wrought-iron or nonferrous pipe	Depth of groove
Plain-end steel, wrought-iron or tube:	
1 in and smaller	0.05
1¼ in and larger	0.065
Plain-end nonferrous pipe or tube	0.000

not encountered at lower temperatures. For the properties of steel applicable to high-temperature service (as well as to ordinary service) for pipes, tubes, fittings, bolting material, etc., see Sec. 4. For a discussion of creep properties, see Sec. 3.

Piping of thickness designed in accordance with Eq. (6.7.1) may be used for any combination of pressure and temperature. The following summarizes piping industry practice.

Steam Pressures above 250 lb/in² (1,724 kPa), and Not above 2,500 lb/in² (17,238 kPa), Temperatures Not above 1,100°F (593°C) For pressures in excess of 100 lb/in² (690 kPa), the pipe may be seamless steel (A106), (A312), (A335), or (A376); or electric-fusion-welded steel (A691); or forged-and-bored steel (A369); or automatic-welded steel (A312). For pressures between 250 and 600 lb/in² (9,224 and 22,137 N/m²) the pipe may be seamless steel (A106) or (A53); electric-fusion-welded steel (A155); electric-resistance-welded steel (A135) or (A53). For pressures of 250 lb/in² (1,724 kPa) and lower and for service up to 750°F (399°C), any of the following may be used: electric-fusion-welded steel (A134) or (A139); electric-resistance-welded steel (A135); seamless or welded steel (A53). Grade A seamless pipe (A106) or (A53); or grade A electric-welded pipe (A53), (A135), or (A139) is used for close coiling or cold bending. Pipe permissible for services specified may be used for temperatures higher than 750°F (399°C), unless otherwise prohibited, if the *S* and *E* values of ASME B31.3 are used when calculating the required wall thickness.

Because of several failures in seam-welded 1¼% Cr-½% Mo or 2¼% Cr-1% Mo piping produced to ASTM Specification A155 operating at temperatures above 950°F, a preference has developed for seamless piping in these applications. Nevertheless, in general, seam-welded piping has provided entirely satisfactory service in high-temperature applications above 950°F.

Valves and fittings must have flange openings or welded ends, and valves must have external stem threads. Valves must be of cast or forged steel or may be fabricated from plate and pipe. Valves of nonferrous materials are generally cast or forged. Forged and cast-steel threaded valves and fittings may be used up to 300 lb/in² and 500°F for 3 (2) [1½] in pipe, and pressure from 250 to 400 (400 to 600) [600 to 2,500] lb/in². Malleable-iron threaded fittings (300 lb/in²) may be used for pressures not greater than 300 lb/in² and temperatures not over 500°F. Valves 8 in and larger should have the bypass of at least ¾ in, commercial size. See Manufacturers Standardization Society SP-45 for recommended size of bypass valves. Welded fittings may be used of the same material and thickness as the pipe to which they are to be connected.

Steam Pressures from 125 to 250 lb/in² (862 to 1,724 kPa), Temperature Not above 450°F (232°C) Pipe may be electric-fusion-welded steel (A134 or A139). Copper and brass may be used if the temperature does not exceed 406°F. Cast iron may also be used. For close coiling or cold bending, grade A seamless steel (A53); or grade A electric-welded steel (A53), (A135), or (A139) is suitable. Pipe permissible for this service may be used for temperatures above 450°F (232°C) if the proper *S* and *E* are used in calculating the pipe-wall thickness.

Valves below 3 in may have inside stem screws. Stop valves 8 in and over must be bypassed. Bodies, bonnets, and yokes are of cast iron, malleable iron, steel, bronze, brass, or Monel. Flanged-steel fittings must conform to the class 300 ANSI Standard B16.5; if of cast iron, to the class 250 ANSI Standard B16.1; or, for threaded fittings, to the ANSI Standard B16.4. Malleable-iron threaded fittings must conform to the class 300 ANSI B16.3 standard, except that the class 150 ANSI Standard B16.3 may be used for pressures not greater than 150 lb/in². Welded fittings may be used.

Steam Pressures from 25 to 125 lb/in² (172 to 802 kPa) Temperatures Not above 450°F Pipe may be of steel, ductile iron, copper, or brass; valve bodies of cast iron, malleable iron, ductile iron, steel, or brass. Fittings are of class 125 or class 150 American Standard cast iron with screwed or flanged ends, or of ductile or malleable iron with screwed ends.

Steam Pressures 25 lb/in² (172 kPa) and Less, Temperature up to 450°F Pipe may be of steel, brass, copper, or cast iron. Flanged fittings conform to the class 25 ANSI Standard B16.1. Screwed fittings are of the class 125 ANSI Standard B16.4 or of the class 150 ANSI Standard B16.3 for malleable iron, or conform to B16.15 for cast bronze. Welded-steel fittings are extensively used.

Pipe coils are made from any of the commercial sizes of iron, steel, brass, and copper pipe and tubing. Limiting center-to-center dimensions, to which pipe coils can be fabricated in sizes ¾ to 2 in, are given in Table 6.7.10. Steel tubing cannot be bent to the absolute limits of brass or copper.

Table 6.7.10 Center-to-Center Dimensions of Pipe Coils

Nominal pipe size, in	Recommended and advisable minimum, in	
	Schedule 40	Schedule 80
¾	3½	2½
1	4	3
1¼	5	4
1½	6	5
2	8	6

Seamless mechanical tubing is obtainable in outside diameters ranging from ¼ to 10¾ in and in wall thickness from 20 gage to 2 in (0.091 to 5.08 cm). Oval, square, rectangular, and other special shapes can be obtained in various sizes and wall thicknesses. Mechanical tubing is available either hot-finished or cold-drawn, but is furnished principally cold-drawn. It is readily adaptable to varied treatment by expansion, cupping, tapering, swaging, flanging, coiling, welding, and similar manipulations. Typical of the many uses are aircraft tubing, automobile axle housings, driveshafts, drive-shaft housings, tie rods, steering columns, piston rods and pins, gear rings, roller-bearing cases and cones, cylinders for various purposes, machine parts, sleeves, bushings, spacers, surgical instruments, and hypodermic needles. Table 6.7.11 lists weights and dimensions of round seamless-steel tubing for sizes that have by common usage become standard. Detailed information on mechanical tubing for any particular applications can be obtained from manufacturers.

Dimensions and weights of **condenser** and **heat-exchanger tubes** are given in Table 6.7.12 and of **boiler** tubes in Table 6.7.13.

Spiral Pipe Spiral pipe is strong lightweight steel pipe with a single continuous welded helical seam from end to end stiffening it throughout. It is listed in sizes 6- to 42-in ID (15.24- to 106.7-cm), in various thicknesses, and in lengths up to 40 ft (12.19 m). It is used for high- and low-pressure water lines, vacuum lines, exhaust-steam lines, low-pressure air lines, sand and gravel slurry conveying and similar services. It is also used extensively by the petroleum industry, for oil and gas lines, for low-pressure steam lines, etc.

Spiral pipe may be asphalt-coated or galvanized. The pipe is designed for special joints, flanges, and lightweight fittings, but the ANSI flanges and fittings can be furnished, if desired.

The **sleeve-type coupling** illustrated in Fig. 6.7.1 is particularly suitable for plain-end pipe and is widely used. A gasket is used to make a tight joint. Advantages of this coupling are low cost, the use of unskilled labor in making the connections, and the fact that small changes in alignment and grade can be made with regular straight

Table 6.7.11 Approx Weight of Round Seamless Cold-Finished Carbon-Steel Mechanical Tubing, lb/ft*
(Carbon 0.25% max. Standard sizes for warehouse stocks random lengths. United States Steel Corporation)

Wall thickness		OD, in																	
in	g or in	3/8	1/2	5/8	3/4	7/8	1	1 1/8	1 1/4	1 3/8	1 1/2	1 5/8	1 3/4	1 7/8	2	2 1/8	2 1/4	2 3/8	2 1/2
0.035	20 g	0.127	0.174	0.221	0.267	0.314	0.361	0.407	0.454	—	0.548								
0.049	18 g	0.171	0.236	0.301	0.367	0.432	0.498	0.563	0.629	0.694	0.759	0.825	0.890	—	1.02				
0.058	17 g	—	0.274	0.351	0.429	—	0.584												
0.065	16 g	0.215	0.302	0.389	0.476	0.562	0.649	0.736	0.823	0.909	0.996	1.08	1.17	1.26	1.34	1.43	1.52	1.60	1.69
0.083	14 g	—	0.370	0.480	0.591	0.702	0.813	0.924	1.03	1.15	1.26	—	—	—	1.70				
0.095	13 g	—	0.411	0.538	0.665	0.791	0.918	1.05	1.17	1.30	1.43	1.55	1.68	1.81	1.93	2.06	2.19	2.31	2.44
0.109	12 g	—	0.455	0.601	0.746	0.892	1.04	1.18	1.33	—	1.62								
0.120	11 g	—	0.487	0.647	0.807	0.968	1.13	1.29	1.45	1.61	1.77	1.93	2.09	2.25	2.41	2.57	2.73	2.89	3.05
0.134	10 g	—	—	0.703	0.882	—	1.24	1.42	1.60	1.78	1.96	2.13	2.31	—	2.67				
0.156	5/32	—	—	0.781	0.990	1.20	1.41	1.61	1.82	2.03	2.24	2.45	2.66	2.86	3.07	3.28	3.49	3.70	3.91
0.188	3/16	—	—	—	1.13	1.38	1.63	1.88	2.13	2.38	2.63	2.89	3.14	3.39	3.64	3.89	4.14	4.39	4.64
0.219	7/32	—	—	—	—	1.53	1.83	2.12	2.41	2.70	3.00	3.29	3.58	3.87	4.17	4.46	4.75	5.04	5.34
0.250	1/4	—	—	—	1.34	1.67	2.00	2.34	2.67	3.00	3.34	3.67	4.01	4.34	4.67	5.01	5.34	5.67	6.01
0.281	9/32	—	—	—	—	—	—	—	2.91	—	3.66	4.03	4.41	4.78	5.16	—	5.91	6.28	6.66
0.313	5/16	—	—	—	—	—	—	—	3.13	3.55	3.97	4.39	4.80	5.22	5.64	6.06	6.48	6.89	7.31
0.375	3/8	—	—	—	—	—	—	—	3.50	4.01	4.51	5.01	5.51	6.01	6.51	7.01	7.51	8.01	8.51
0.438	7/16	—	—	—	—	—	—	—	—	—	4.97	5.53	6.14	6.72	7.31	7.89	8.48	9.06	9.65
0.500	1/2	—	—	—	—	—	—	—	—	—	5.34	6.01	6.68	7.34	8.01	8.68	9.35	10.0	10.7
0.625	5/8	—	—	—	—	—	—	—	—	—	—	—	—	—	9.18	—	10.8	11.7	12.5

To obtain weights in kg/m, multiply tabular values shown by 1.42.

* Other standard sizes, in certain standard wall thicknesses, vary by 1/8-in increments for 2½ to 3½ in; by ¼-in increments from 3½ to 7½ in; ½-in increments from 7½ to 10½ in OD. There are also standard sizes for every in from 3/8 to 1 5/8 in OD.

Table 6.7.12 Steel Condenser and Heat-Exchanger Tubes
(Dimensions and weights. United States Steel Corporation)

		Avg wall			Min wall		
OD, in	Thickness, in	ID, in	Area of metal, in²*	Weight per ft, lb†	ID, in	Area of metal, in²*	Weight per ft, lb†
½	0.035	0.430	0.0511	0.1738	0.423	0.0558	0.1898
	0.050	0.400	0.0707	0.2403	0.390	0.0769	0.2614
	0.065	0.370	0.0888	0.3020	0.357	0.0963	0.3272
5/8	0.035	0.555	0.0649	0.2205	0.548	0.0709	0.2412
	0.050	0.525	0.0903	0.3071	0.515	0.0985	0.3348
	0.065	0.495	0.1144	0.3888	0.482	0.1243	0.4227
	0.085	0.455	0.1442	0.4902	0.438	0.1561	0.5308
¾	0.050	0.650	0.1100	0.3738	0.640	0.1201	0.4082
	0.065	0.620	0.1399	0.4755	0.607	0.1524	0.5181
	0.085	0.580	0.1776	0.6037	0.563	0.1928	0.6556
	0.095	0.560	0.1955	0.6646	0.541	0.2119	0.7204
7/8	0.050	0.775	0.1296	0.4406	0.765	0.1417	0.4817
	0.065	0.745	0.1654	0.5623	0.732	0.1805	0.6136
	0.085	0.705	0.2110	0.7172	0.688	0.2296	0.7804
	0.095	0.685	0.2328	0.7914	0.666	0.2530	0.8599
1	0.050	0.900	0.1492	0.5073	0.890	0.1633	0.5551
	0.065	0.870	0.1909	0.6491	0.857	0.2086	0.7090
	0.085	0.830	0.2443	0.8306	0.813	0.2663	0.9052
	0.095	0.810	0.2701	0.9182	0.791	0.2940	0.9994
1¼	0.050	1.150	0.1885	0.6408	1.140	0.2065	0.7020
	0.065	1.120	0.2420	0.8226	1.107	0.2647	0.8999
	0.085	1.080	0.3111	1.058	1.163	0.3397	1.155
	0.095	1.060	0.3447	1.172	1.041	0.3761	1.278
	0.105	1.040	0.3777	1.284	1.019	0.4117	1.399
1½	0.050	1.400	0.2278	0.7743	1.390	0.2497	0.8488
	0.065	1.370	0.2930	0.9962	1.357	0.3209	1.091
	0.085	1.330	0.3779	1.285	1.313	0.4053	1.378
	0.095	1.310	0.4193	1.426	1.291	0.4581	1.557
	0.105	1.290	0.4602	1.564	1.269	0.5024	1.708
1¾	0.065	1.620	0.3441	1.170	1.606	0.3803	1.293
	0.085	1.580	0.4446	1.512	1.561	0.4907	1.668
	0.095	1.560	0.4939	1.679	1.539	0.5448	1.852
	0.105	1.540	0.5426	1.845	1.517	0.5902	2.007
	0.120	1.510	0.6145	2.089	1.484	0.6766	2.300
2	0.065	1.870	0.3951	1.343	1.856	0.4370	1.486
	0.085	1.830	0.5114	1.738	1.811	0.5649	1.920
	0.095	1.810	0.5685	1.933	1.789	0.6275	2.133
	0.105	1.790	0.6251	2.125	1.767	0.6896	2.344
	0.120	1.760	0.7087	2.409	1.734	0.7812	2.656

* Multiply values shown by 0.0645 to obtain areas in cm².
† Multiply values shown by 1.42 to obtain weights in kg/m.

Table 6.7.13 Seamless-Steel Boiler Tubes

Outside diam, in	Thickness BWG	Thickness in	Mfg.* wt, lb/ft	Outside diam, in	Thickness BWG	Thickness in	Mfg.* wt, lb/ft	Outside diam, in	Thickness BWG	Thickness in	Mfg.* wt, lb/ft
1	13	0.095	1.037	2½	12	0.109	3.171	4½	10	0.134	7.103
	12	0.109	1.168		11	0.120	3.457		9	0.148	7.817
	11	0.120	1.263		10	0.134	3.835		8	0.165	8.702
	10	0.134	1.384		9	0.148	4.207		7	0.180	9.447
1¼	13	0.095	1.323	2¾	12	0.109	3.504	5	9	0.148	8.720
	12	0.109	1.502		11	0.120	3.823		8	0.165	9.711
	11	0.120	1.628		10	0.134	4.244		7	0.180	10.550
	10	0.134	1.793		9	0.148	4.658		6	0.203	11.810
1½	13	0.095	1.619	3	12	0.109	3.838	5½	9	0.148	9.622
	12	0.109	1.836		11	0.120	4.189		8	0.165	10.720
	11	0.120	1.994		10	0.134	4.652		7	0.180	11.650
	10	0.134	2.201		9	0.148	5.110		6	0.203	13.050
1¾	13	0.095	1.910	3¼	11	0.120	4.555	6	7	0.180	12.750
	12	0.109	2.169		10	0.134	5.061		6	0.203	14.290
	11	0.120	2.360		9	0.148	5.061		5	0.220	15.410
	10	0.134	2.610		8	0.165	6.179		4	0.238	16.640
2	13	0.095	2.201	3½	11	0.120	4.921				
	12	0.109	2.503		10	0.134	5.469				
	11	0.120	2.726		9	0.148	6.012				
	10	0.034	3.018		8	0.065	6.683				
2¼	13	0.095	2.492	4	10	0.134	6.286				
	12	0.109	2.837		9	0.148	6.915				
	11	0.120	3.092		8	0.165	7.693				
	10	0.134	3.427		7	0.180	8.347				

* Multiply values shown by 1.42 to obtain weights in kg/m.
SOURCE: United States Steel Corporation.

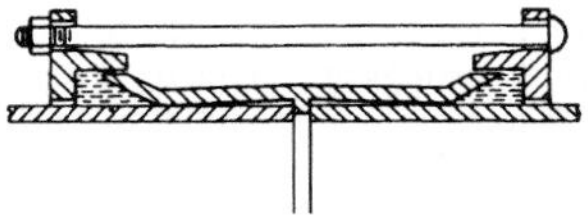

Figure 6.7.1 Sleeve type, plain end coupling.

lengths of pipe by a movement in the coupling. This type of coupling is used extensively in long oil lines.

CAST-IRON AND DUCTILE-IRON PIPE

Cast-iron pipe was extensively produced and used in certain applications until about 1960, but now shares the field with **ductile-iron pipe**. Both provide excellent service when appropriately applied, although—ductile-iron pipe has largely supplanted cast-iron pipe when employed because of its superior strength and ductility, that is, when adapted to underground and submerged service because of its superior corrosion resistance compared to steel pipe. Ductile-iron pipe is now extensively utilized for water, gas, sewage, culverts, drains, etc. It is produced in a wide range of sizes for varying pressures. Nevertheless, steel pipe, when properly coated and wrapped, can also provide adequate resistance to corrosion when placed in certain soils.

Pipe fittings are available as cast-iron pipe fittings and ductile-iron pipe fittings. Both types of fittings are used with ductile-iron pipe.

Ductile-iron pipe may be obtained in various thicknesses and weights with (1) flanges cast on, (2) ends threaded for screwed-on flanges, (3) ends prepared for mechanical joint, (4) ends grooved or shouldered for patented coupling, (5) one end bell, other end spigot, and (6) one end hub, other end spigot. **Bell-and-spigot** ends are most popular for underground work; hub-and-spigot ends are most frequently used for sewage systems in enclosed spaces. Spigot-end joints are prepared by tightly tamping in hemp or jute at the bottom of the recess with a yarning iron and then pouring in molten lead; the lead, when cooled, is caulked in tightly with a caulking iron and makes a gastight joint. Alternatively, an elastomeric **gasket** (usually Neoprene) may be assembled in the bell-and-spigot joint. For exposed piping, flanged ends are used, the joints being made up with gaskets. Flanged pipe has superior strength and tightness of the joint and is used where pipelines can be well supported. The bell-and-spigot joint possesses greater flexibility and provides for expansion and contraction. It is therefore suitable for water pipe and is largely used for that purpose. Figure 6.7.2 shows a typical form of this joint for ordinary pressures. Figure 6.7.3 shows one form of this joint for ordinary pressures. Figure 6.7.3 shows one

Figure 6.7.2 Standard bell-and-spigot joint.

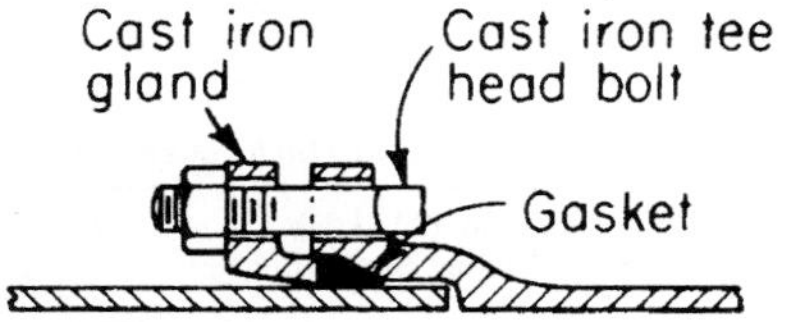

Figure 6.7.3 Mechanical joint.

form of **mechanical joint** suitable for water, gas, or oil. Other forms of joint, plain-end pipe with couplings, and threaded pipe also are manufactured. Cast-iron and ductile-iron pipe, fittings, and valves have been found unsuitable for superheated steam service. The Code for Pressure Piping, B31.1 (Power Piping), states that cast-iron or ductile-iron pipe may be used for steam service not over 250 lb/in^2 or 406°F (1,724 kPa or 208°C) provided that it meets the requirements as dictated by Eq. (6.7.1).

Wall thicknesses for the various conditions which ductile-iron pipe is designed to meet are determined in accordance with the requirements of ANSI 21.50 (AWWA C150).

Hubless pipe (often termed **no-hub**), cast or ductile iron, is rapidly replacing the bell-and-spigot form of cast pipe. The ends of the cast pipe are thickened to accommodate an elastomeric gasket and a mechanical compression coupling; see Fig. 6.7.4*b*. In such nonpressure installations, the outer metallic sleeve of the coupling is configured to allow the gasket to be compressed at assembly and render the joint

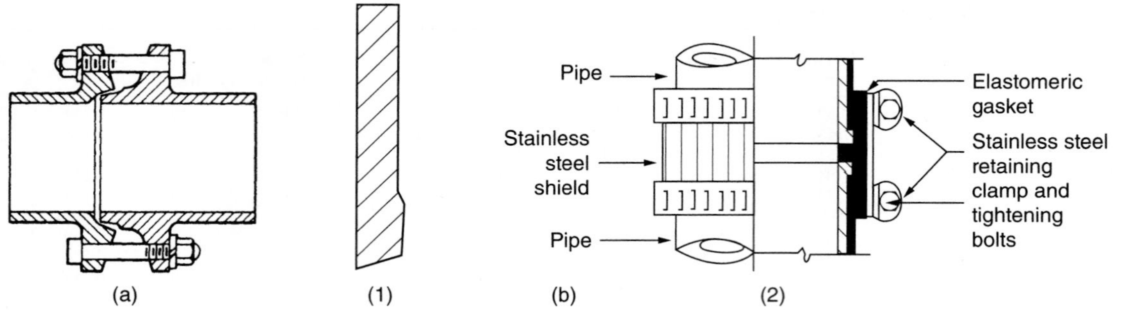

Figure 6.7.4 (*a*) Universal ductile-iron pipe and joint. (*b*) Hubless cast pipe: (1) Cast pipe end; (2) assembled compression fitting.

leak-free. This type of joint is utilized primarily above ground; underground, the metallic sleeve and bolts, albeit of select noncorroding material, ultimately may fail from corrosion in the embedding soil or backfill. A complete line of fittings are available for hubless pipe.

Ductile-iron pipe is made by **centrifugal casting,** in which molten iron is admitted to the interior of a sand-lined or metal-lined mold, the mold being rotated at high speeds so that the molten metal is thrown by centrifugal force against the lining. ANSI specifications have been prepared for the various combinations of fabrication procedure and intended end use.

Table 6.7.14 lists thicknesses and weight data for centrifugally cast ductile-iron pipe intended for use with water or other liquids.

The employment of ductile-iron pipe for gas supply and distribution is second in importance only to its use for carrying water. Bell-and-spigot gas pipe is similar in design to bell-and-spigot water pipe (Fig. 6.7.2). For flanged gas pipe, the class 25 ANSI B16.1 Standard flanges are approved for maximum gas pressures of 25 lb/in^2 (172 kPa). The class 125 ANSI B16.1 Standard flanges are approved for gas pressures of 125 lb/in^2 (862 kPa), up to 4 in nominal pipe size; 100 lb/in^2 (689 kPa), 6 to 12 in; and 80 lb/in^2 (552 kPa), 16 to 48 in. The type of joint shown in Fig. 6.7.2 is also widely used for gas.

Flexible-Joint Pipe The necessity for crossing streams and other waterways and of laying pipelines into them has led to the development of various forms of flexible-joint pipe adapted to laying under water, which are capable of motion through several degrees without leakage. Figure 6.7.5 shows one style of such joint which has an adjustment of about 15° in standard sizes.

In selecting the thickness of a pipe for a submerged line, the internal-service pressure is seldom the determining factor, as ample allowance should be made to minimize the risk of breakage in laying and to withstand external shocks from floating ice or other objects. The dimensions and weights given in Table 6.7.15 are typical of those listed by several manufacturers.

"Universal" pipe (Fig. 6.7.4) is ductile-iron pipe with hub-and-spigot ends, the contact surfaces of which are machined on a taper, giving an iron-to-iron joint. By making the tapers of slightly different pitch, the joint provides for flexibility while remaining tight. Two bolts to the joint are sufficient, except for pressures above 175 lb/in^2 (1,207 kPa). Universal ductile-iron pipe is used largely for carrying gas and water and is suitable for all pressures and services. The pipe is tested with hydrostatic pressure of 300 to 500 lb/in^2 (2,067 to 3,448 kPa). All universal pipe and special castings of a given diameter and of any class are interchangeable with those of a different class. Standard laying lengths are 6 ft (1.83 m). Thicknesses and weights of standard types up to 16 in are given in Table 6.7.16. Information on other types and sizes and on fittings may be obtained from pipe producers.

Fittings for Ductile-Iron Water Pipe Flanged fittings of the dimensions of the ANSI standard for steam are not often used with ductile-iron water pipe. The longer fittings of the AWWA are generally preferred because of low friction loss. The dimensions of the flanged fittings of this class conform very closely to the dimensions of the bell-and-spigot fittings of the AWWA. The flange thicknesses and drillings conform to those of the ANSI standards. These fittings, both flange and bell-and-spigot type, are made in a great variety of forms known as "standard

Table 6.7.14 Standard Weights and Thicknesses of Ductile-Iron Bell-and-Spigot Pipe for Water

	Class 50, 50 lb/in^2 (345 kPa) 115-ft (35.05-m) head			Class 100, 100 lb/in^2 (690 kPa) 231-ft (70.41-m) head			Class 150, 150 lb/in^2 (1,034 kPa) 346-ft (105.5-m) head				
Nominal size, in	Thickness, in	OD, in	Wt, lb per avg ft	Thickness, in	OD, in	Wt, lb per avg ft	Thickness, in	OD, in	Wt, lb per avg ft	Approx lb lead per joint 2 in thick	Approx lb hemp or jute per joint
3	0.32	3.96	12.4	0.32	3.96	12.4	0.32	3.96	12.4	6.2	0.17
4	0.35	4.80	16.5	0.35	4.80	16.5	0.35	4.80	16.5	7.5	0.21
6	0.38	6.90	25.9	0.38	6.90	25.9	0.38	6.90	25.9	10.3	0.31
8	0.41	9.05	37.0	0.41	9.05	37.0	0.41	9.05	37.0	13.3	0.44
10	0.44	11.10	49.1	0.44	11.10	49.1	0.44	11.10	49.1	16.0	0.53
12	0.48	13.20	63.7	0.48	13.20	63.7	0.48	13.20	63.7	19.0	0.61
14	0.48	15.30	74.6	0.51	15.30	78.8	0.51	15.65	80.7	22.0	0.81
16	0.54	17.40	95.2	0.54	17.40	95.2	0.54	17.80	97.5	30.0	0.94
18	0.54	19.50	107.6	0.58	19.50	114.8	0.58	19.92	117.2	33.8	1.00
20	0.57	21.60	125.9	0.62	21.60	135.9	0.62	22.06	138.9	37.0	1.25
24	0.63	25.80	166.0	0.68	25.80	178.1	0.73	26.32	194.0	44.0	1.50
30	0.79	32.00	257.6	0.79	32.00	257.6	0.85	32.00	275.4	54.3	2.06
36	0.87	38.30	340.9	0.87	38.30	340.9	0.94	38.30	365.9	64.8	3.00
42	0.97	44.50	442.0	0.97	44.50	442.0	1.05	44.50	475.3	75.3	3.62
48	1.06	50.80	551.6	1.06	50.80	551.6	1.14	50.80	589.6	85.5	4.37

Pipe weights indicated are approximate and include allowance for bell based on a 16-ft laying length. Calculations are for pipe laid without blocks, on flat-bottom trench, with tamped backfill under 5 ft of cover. Thicknesses given above include allowance for water hammer and factory tolerance.

To obtain weights in kg/m, multiply values shown in lb per avg ft by 1.42.

SOURCE: Condensed from Table 8.2 of Specification AWWA C 108.

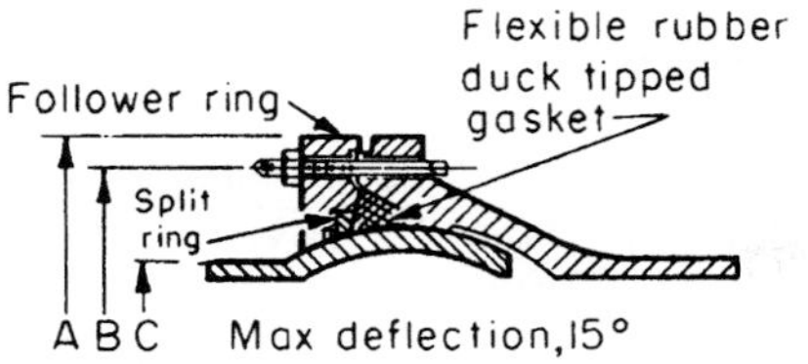

Figure 6.7.5 Flexible joint. (See Table 6.7.15 for dimensions.)

special fittings." For dimensions and weights, see manufacturers' catalogs or standard specifications of the AWWA.

Cast-iron soil pipe and fittings are of the hub-and-spigot form, similar in design to the pipe shown in Fig. 6.7.2. Tapped openings and pipe plugs are threaded in accordance with the taper pipe thread requirements of ANSI B1.20.1.

The ANSI standard, Threaded Cast-Iron Pipe for Drainage, Vent, and Waste Services, ANSI A40.5, covers two types of pipes having threaded joints in nominal pipe sizes 1¼ to 12 in and in lengths 5 to 27 ft. One type has external threads on both ends; the other type has external threads on one end and an internal threaded drainage hub on the other end.

A large volume of piping put in place forms part of the drainage, waste, and vent, or **DWV,** system. The pipe material most commonly includes cast and ductile iron, thin-wall copper, and a wide range of plastic compositions (ABS, PE, PVC, etc.)

PIPES AND TUBES OF NONFERROUS MATERIALS

Brass tubing is commercially available in the form known as yellow brass, an alloy consisting of approximately 65 percent copper and 35 percent zinc, and is used principally for ornamental work and hand railings, It has a density of approximately 0.3 lb/in^3 (8.31 g/cm^3), the exact density being dependent upon the specific chemical composition. **Brass piping** is most frequently furnished as red brass, an alloy consisting of approximately 85 percent copper and 15 percent zinc. This alloy, having a density of about 0.32 lb/in^3 (8.86 g/cm^3), has been found to be structurally superior to the yellow brasses and is used where the fluid being conveyed has corrosive properties.

Copper is available either as pipe or as tubing. In the form of piping, it has the same outer diameter as that of standard steel pipe (Tables 6.7.17 and 6.7.18). As tubing, it is used for a variety of purposes, such as for

Table 6.7.15 Dimensions and Weights of Flexible-Joint Pipe*
(Dimensions refer to Fig. 6.7.5.)

Nominal diam, in	Class	Dimensions, in			Bolts			Average metal thickness, in	Weight† of pipe, incl. bell, lb per 12-ft length
		A	*B*	*C*	No. required	Size, in	Length, in		
4	B	12.13	9.75	5.00	8	0.75	4.50	0.45	290
4	C	12.13	9.75	5.00	8	0.75	4.50	0.48	305
4	D	12.13	9.75	5.00	8	0.75	4.50	0.52	325
6	B	14.25	11.75	7.10	12	0.75	4.50	0.48	440
6	C	14.25	11.75	7.10	12	0.75	4.50	0.51	460
6	D	14.25	11.75	7.10	12	0.75	4.50	0.55	490
8	B	17.25	14.75	9.30	12	0.75	5.25	0.51	635
8	C	17.25	14.75	9.30	12	0.75	5.25	0.56	680
8	D	17.25	14.75	9.30	12	0.75	5.25	0.60	720
10	B	20.56	18.00	11.40	16	0.75	5.25	0.57	905
10	C	20.56	18.00	11.40	16	0.75	5.25	0.62	965
10	D	20.56	18.00	11.40	16	0.75	5.25	0.68	1,035
12	B	23.75	21.00	13.50	16	0.75	6.25	0.62	1,200
12	C	23.75	21.00	13.50	16	0.75	6.25	0.68	1,280
12	D	23.75	21.00	13.50	16	0.75	6.25	0.75	1,375

* United States Pipe and Foundry Co.
† Weights do not include follower rings, bolts, or gaskets. For sizes above 12 in, see manufacturers' catalogs. Multiply weights shown by 38.7 to obtain weight in kg per m or by 0.4536 to obtain weight in kg per 12-ft length.

Table 6.7.16 Standard Weights and Thicknesses of Universal Ductile-Iron Pipe
(Central Foundry Co.)

Nominal ID, in	Class 150, 150 lb/in^2 (1,034 kPa) Approx thickness, in	Class 150 Estimated wt, lb per* ft	Class 150 Estimated wt, lb per* 6-ft length	Class 250, 250 lb/in^2 (1,724 kPa) Approx thickness, in	Class 250 Estimated wt, lb per* ft	Class 250 Estimated wt, lb per* 6-ft length
2	0.25	7	42	0.31	8	48
3	0.30	11¼	67½	0.34	12½	75
4	0.32	16½	99	0.37	18	108
6	0.36	26.6	160	0.43	30	180
8	0.39	38½	231	0.47	44¼	265½
10	0.43	53½	321	0.50	60½	363
12	0.47	69	414	0.53	77½	465
14	0.50	87	522	0.565	98½	591
16	0.53	106	636	0.60	121	726

ID, in	2	3	4	6	8	10	12	14	16
Bolt sizes, in	½ × 4	½ × 4¼	⅝ × 5	¾ × 6	⅞ × 6¾	1 × 7½	1 × 8	1⅛ × 9	1¼×9½

* Multiply wts. by 0.4536 to obtain wt. in kilograms.

Table 6.7.17 Sizes and Weights of SPS Copper and 85 Red Brass Pipe*

	Nominal dimensions, in			Theoretical areas based on nominal dimensions				Allowable internal pressure, lb/in²			
Standard size, in	Outside diameter	Inside diameter	Wall thickness	Cross-sectional area of bore, in²	External surface, ft²/lin. ft	Internal surface ft²/lin. ft	Theoretical weight, lb/ft	100°F, S = 6,000	200°F, S = 4,800	300°F, S = 4,700	400°F, S = 3,000
¼	0.540	0.410	0.065	0.132	0.141	0.107	0.376	1,500	1,200	1,180	740
¾	0.675	0.545	0.065	0.233	0.177	0.143	0.483	1,170	940	910	580
½	0.840	0.710	0.065	0.396	0.220	0.186	0.613	920	740	720	460
¾	1.050	0.920	0.065	0.665	0.275	0.241	0.780	730	580	560	360
1	1.315	1.185	0.065	1.10	0.344	0.310	0.989	580	460	450	290
1¼	1.660	1.530	0.065	1.84	0.435	0.401	1.26	450	360	350	220
1½	1.900	1.770	0.065	2.46	0.497	0.464	1.45	400	310	300	190
2	2.375	2.245	0.065	3.96	0.622	0.588	1.83	300	240	240	140
2½	2.875	2.745	0.065	5.92	0.753	0.719	2.22	250	200	190	120
3	3.500	3.334	0.083	8.73	0.916	0.874	3.45	260	210	210	130
3½	4.000	3.810	0.095	11.4	1.05	0.998	4.52	270	210	210	130
4	4.500	4.286	0.107	14.4	1.18	1.12	5.72	270	210	210	130
5	5.562	5.298	0.132	22.0	1.46	1.39	8.73	270	210	210	130
6	6.625	6.309	0.158	31.3	1.73	1.65	12.4	270	210	210	130
8	8.625	8.215	0.205	53.0	2.26	2.15	21.0	270	210	210	130
10	10.750	10.238	0.256	82.3	2.81	2.68	32.7	270	210	210	130
12	12.750	12.124	0.313	115.	3.34	3.18	47.4	280	220	220	130

* The values in the table above are based on the formula in The Code for Pressure Piping, ANSI B31:

$$t_m = \frac{PD}{2S + 0.8P} + C \qquad \text{or when } C \text{ is 0} \qquad P = \frac{2St_m}{D - 0.8t_m}$$

where t_m = minimum pipe wall thickness, in; P = maximum rated internal working pressure, lb/in²; D = outside diameter of pipe, in; S = allowable stress in material due to internal pressure, at operating temperature, lb/in²; C = allowance for threading, mechanical strength, and/or corrosion, in. The allowable internal pressures apply to the pipe itself after brazing.

SOURCE: Copper Development Association.

compressed-air instrumentation lines, hydraulic control lines around machinery, domestic oil-burner and heating systems, and for general plumbing purposes. **Copper tubing** (Table 6.7.19) is furnished in 12-ft (3.7-m) and 20-ft (6.1-m) straight lengths or in coils of 100-ft (30.5-m) length. **Type K** tubing, in coils, is used for underground work where the minimum number of joints, combined with greater thickness of type K tubing, is of distinct advantage. **Type L** tubing, usually in straight lengths, is used as a principal piping material for plumbing systems in homes and buildings; this is largely due to the economy of installation made possible by the use of soldered fittings. Copper deteriorates rapidly under high temperatures and repeated stresses. At a temperature of 360°F (182°C) its strength is reduced 15 percent, and on this account it should never be used for high steam pressures and temperatures.

Rigid, thin-wall copper tubing (99.9 percent pure copper) is widespread in **DWV** systems. The plain ends are meant to be soldered to a wide range of available fittings. Tubing from 1½- to 4-in dia (≈0.050- to 0.060-in wall) is most common; sizes up to 8-in diam are available for extreme conditions, but the cost then becomes quasi-prohibitive.

Commercial sizes of **aluminum tubing** are listed by the manufacturers in even outside diameters and in wall thicknesses conforming to Stubs gage. To obtain the approximate weight per foot of aluminum pipe or tubing, a weight of 0.098 lb/in³ (2.71 g/cm³) may be used.

Lead pipe is supplied in straight lengths, in coils, or in reels.

Block tin is a term used in the metal trade to refer to products made wholly from strictly pure high-grade tin. While tin pipe has many and varied uses, its most important applications are in types of equipment handling liquids intended for human consumption. Tin pipe does not corrode, and therefore does not contaminate most of the liquids passing through it.

Zirconium tubing has applications primarily in nuclear power.

Titanium tubing, as pure metal or alloyed, is used in the chemical processing industries and in some heat-exchange equipment such as power plant surface condensers. Titanium alloys are stronger, but suffer reduced corrosion resistance and weldability versus the unalloyed metal. A range of sizes, wall thicknesses, and associated fittings (generally cast) are available.

Glass pipe has unique applications in the chemical and food processing industries. Borosilicate glass imparts its chemical resistance. A small number of fittings are available; by and large, however, fittings are custom-made.

Plastic pipes and **tubes** are available in a wide range of diameters and thicknesses, with Table 6.7.20 generally representative. The plastic used is resistant to attack by many chemicals, light in weight, flexible, and available in coiled form so that installation time is low. It is used for a variety of purposes including drainage, irrigation, sewage, and for conveying chemical solutions or waters that would attack metal piping. Plastic pipe used in gas service is listed in Table 6.7.21. Caution should be used in selection of plastic piping insofar as service temperature is concerned: e.g., polyethylene is suitable for a maximum temperature of 120°F (49°C). Table 6.7.22 lists corrosion-resistance data for polyethylene plastic piping.

Rigid **plastic tubing** can serve in DWV systems. Many plastic compositions are available, although ABS, PE, and PVC predominate. Connections are made with solvents compatible with the plastic materials bonded. A wide range of fittings are available.

Pipes with Special Linings For use in lines through which are passed solutions containing more or less free acid or other corrosive agents, standard pipe, valves, and fittings may be **lead-lined, tin-lined, or rubber-lined,** to resist corrosive action. This lining prolongs the life of the pipe and also gives it additional strength. For mine service in coal districts where the drainage water is more or less impregnated with sulfur or free sulfuric acid, **wood-lined pipe** and fittings are sometimes used. For special service, **seamless-copper-lined pipe** is also used. The **cement lining** of ductile-iron and steel pipe for water and other services is advantageous because of its protection against unusual destructive agencies and its ability to prevent tuberculation. Standard **hard-rubber pipe** and fittings have been developed for working pressures of 50 lb/in² (345 kPa) at normal temperatures. Standard sizes run from ¼- to 4-in diam (0.635- to 10.2-cm), in 10-ft (3.05-m) lengths. For temperature above 120°F (49°C), the use of hard-rubber-lined steel

Table 6.7.18 Allowable Internal Pressures, lb/in², for Temperatures up to 400°F for Red Brass and Copper Pipe*

	Pipe with threaded ends								Pipe with plain ends for use with welded, brazed, or soldered fittings							
	Red brass				Copper				Red brass				Copper			
Standard size, in	100°F $S = 8{,}000$	200°F $S = 8{,}000$	300°F $S = 8{,}000$	400°F $S = 5{,}000$	100°F $S = 6{,}000$	200°F $S = 4{,}800$	300°F $S = 4{,}700$	400°F $S = 3{,}000$	100°F $S = 8{,}000$	200°F $S = 8{,}000$	300°F $S = 8{,}000$	400°F $S = 5{,}000$	100°F $S = 6{,}000$	200°F $S = 4{,}800$	300°F $S = 4{,}700$	400°F $S = 3{,}000$
								Regular								
⅛	370	370	370	230	280	210	210	130	2,640	2,640	2,640	1,650	1,980	1,570	1,540	980
¼	870	870	870	550	650	510	510	320	2,610	2,610	2,610	1,640	1,960	1,560	1,530	970
⅜	890	890	890	570	670	530	510	320	2,280	2,280	2,280	1,440	1,710	1,350	1,330	840
½	900	900	900	570	670	530	520	320	2,160	2,160	2,160	1,350	1,620	1,280	1,260	800
¾	810	810	810	520	610	480	470	300	1,800	1,800	1,800	1,130	1,350	1,070	1,050	670
1	630	630	630	400	480	370	370	230	1,580	1,580	1,580	1,000	1,190	940	920	590
1¼	690	690	690	430	520	410	400	250	1,440	1,440	1,440	900	1,080	890	840	530
1½	630	630	630	400	480	370	370	230	1,280	1,280	1,280	800	960	750	740	470
2	540	540	540	350	410	320	310	190	1,050	1,050	1,050	670	790	620	610	380
2½	450	450	450	280	340	260	250	160	1,040	1,040	1,040	650	780	620	610	380
3	510	510	510	320	380	300	290	180	1,000	1,000	1,000	630	750	590	580	370
3½	570	570	570	370	430	330	330	200	1,000	1,000	1,000	630	750	590	580	370
4	510	510	510	320	380	300	290	180	880	880	880	550	660	530	520	320
5	410	410	410	270	310	230	240	140	710	710	710	450	540	420	410	260
6	340	340	340	220	260	190	200	120	600	600	600	380	450	350	340	220
8	360	360	360	236	270	270	210	130	560	560	560	350	420	320	320	200
10	360	360	360	230	270	210	210	130	520	520	520	330	390	300	300	190
12	320	320	320	200	240	190	200	120	450	450	450	280	340	260	250	160
								Extra-Strong								
⅛	1,960	1,960	1,960	1,240	1,470	1,160	1,140	720	4,630	4,630	4,630	2,900	3,470	2,750	2,710	1,730
¼	2,210	2,210	2,210	1,340	1,660	1,310	1,290	820	4,200	4,200	4,200	2,640	3,150	2,500	2,460	1,570
⅜	1,840	1,840	1,840	1,150	1,380	1,090	1,070	680	3,360	3,360	3,360	2,100	2,520	2,000	1,960	1,250
½	1,760	1,760	1,760	1,100	1,320	1,040	1,030	660	3,130	3,130	3,130	1,970	2,350	1,860	1,830	1,160
¾	1,510	1,510	1,510	950	1,130	900	880	560	2,560	2,560	2,560	1,600	1,920	1,520	1,500	960
1	1,340	1,340	1,340	850	1,010	790	780	490	2,360	2,360	2,360	1,490	1,770	1,400	1,370	880
1¼	1,160	1,160	1,160	730	880	730	680	430	1,950	1,950	1,950	1,220	1,460	1,160	1,140	720
1½	1,090	1,090	1,090	680	820	660	640	410	1,770	1,770	1,770	1,120	1,330	1,050	1,030	660
2	1,000	1,000	1,000	630	750	590	570	360	1,520	1,520	1,520	950	1,140	910	890	560
2½	970	970	970	620	730	580	560	360	1,600	1,600	1,600	1,000	1,200	950	930	590
3	910	910	910	570	680	530	530	340	1,420	1,420	1,420	900	1,070	840	830	530
3½	860	860	860	550	650	510	500	310	1,300	1,300	1,300	820	980	790	760	480
4	840	840	840	530	630	490	480	300	1,230	1,230	1,230	770	920	730	710	460
5	770	770	770	480	580	450	440	340	1,080	1,080	1,080	680	810	640	630	400
6	800	800	800	500	600	470	460	290	1,060	1,060	1,060	670	800	620	610	380
8	710	710	710	450	530	430	410	260	910	910	910	570	680	540	540	340
10	550	550	550	350	420	340	330	200	710	710	710	450	540	440	420	240

* The values above are based on the formula in The Code for Pressure Piping, ANSI B31:

$$t_m = \frac{PD}{2S + 0.8P} + C \quad \text{or} \quad P = \frac{2S(t_m - C)}{D - 0.8(t_m - C)} \quad \text{or when } C \text{ is 0} \quad P = \frac{2St_m}{D - 0.8t_m}$$

where t_m = minimum pipe wall thickness, in; P = maximum rated internal working pressure, lb/in²; D = outside diameter of pipe, in; S = allowable stress in annealed material due to internal pressure, at operating temperature, lb/in²; C = allowance for threading, mechanical strength and/or corrosion, in. C equals 0.05 for sizes in and smaller and the depth of the thread for all other sizes, and equals 0 for pipe with plain ends. These allowable internal pressures apply to the pipe itself and do not take into account the limitations which may be imposed by the type of joint and the joining material.

SOURCE: Copper Development Association.

Table 6.7.19 Sizes and Weights of Copper Tubes

Nominal size, in	OD, in, types* K, L	ID, in		Wall thickness, in		Permissible variation of mean OD, in — Types K, L		Weight, † lb/ft	
		Type K	Type L	Type K	Type L	Annealed	Hard-drawn	Type K	Type L
⅜	0.500	0.402	0.430	0.049	0.035	0.0025	0.001	0.269	0.198
½	0.625	0.527	0.545	0.049	0.040	0.0025	0.001	0.344	0.285
⅝	0.750	0.652	0.666	0.049	0.042	0.0025	0.001	0.418	0.362
¾	0.875	0.745	0.785	0.065	0.045	0.003	0.001	0.641	0.455
1	1.125	0.995	1.025	0.065	0.050	0.0035	0.0015	0.839	0.655
1¼	1.375	1.245	1.265	0.065	0.055	0.004	0.0015	1.04	0.884
1½	1.625	1.481	1.505	0.072	0.060	0.0045	0.002	1.36	1.14
2	2.125	1.959	1.985	0.083	0.070	0.005	0.002	2.06	1.75
2½	2.625	2.435	2.465	0.095	0.080	0.005	0.002	2.93	2.48
3	3.125	2.907	2.945	0.109	0.090	0.005	0.002	4.00	3.33
3½	3.625	3.385	3.425	0.120	0.100	0.005	0.002	5.12	4.29
4	4.125	3.857	3.905	0.134	0.110	0.005	0.002	6.51	5.38
5	5.125	4.805	4.875	0.160	0.125	0.005	0.002	9.67	7.61
6	6.125	5.741	5.845	0.192	0.140	0.005	0.002	13.9	10.2
8	8.125	7.583	7.725	0.271	0.200	0.006	0.003	25.9	19.3
10	10.125	9.449	9.625	0.338	0.250	0.008	0.004	40.3	30.1
12	12.125	11.315	11.565	0.405	0.280	0.008	0.004	57.8	40.4

* Type K recommended for underground service and general plumbing. Type L suitable for interior plumbing and other services.
† Multiply these values by 1.48 to obtain weight in kg/m.
SOURCE: American Brass Co.

Table 6.7.20 Commercial Sizes (IPS) and Weights of Polyvinyl Chloride (PVC) Pipe*

Nominal size, in	Schedule	Wall thickness,* in	OD, in	ID, in	Theoretical weight,† lb/ft	Calculated min bursting pressure, lb/in² — Note 1	Note 2
¼	40	0.088	0.540	0.364	0.076	2,490	1,950
	80	0.119	0.540	0.302	0.096	3,620	2,830
½	40	0.109	0.840	0.622	0.153	1,910	1,490
	80	0.147	0.840	0.546	0.185	2,720	2,120
¾	40	0.113	1.050	0.824	0.203	1,540	1,210
	80	0.154	1.050	0.742	0.265	2,200	1,720
1	40	0.133	1.315	1.049	0.305	1,440	1,130
	80	0.179	1.315	0.957	0.385	2,020	1,580
1¼	40	0.140	1.660	1.380	0.409	1,180	920
	80	0.191	1.660	1.278	0.550	1,660	1,300
1½	40	0.145	1.900	1.610	0.489	1,060	830
	80	0.200	1.900	1.500	0.653	1,510	1,180
2	40	0.154	2.375	2.067	0.640	890	690
	80	0.218	2.375	1.939	0.910	1,290	1,010
3	40	0.216	3.500	3.068	1.380	840	660
	80	0.300	3.500	2.900	1.845	1,200	940
4	40	0.237	4.500	4.026	1.965	710	560
	80	0.337	4.500	3.826	2.710	1,040	810

* Thicknesses listed are minimum values. Tolerance is generally – 0 + 10%.
† These representative values are not specified in ASTM D1785.
NOTES:
1. Materials are PVC 1120, 1220, and 4116. A fiber stress of 6,400 lb/in² was used in bursting pressure calculations.
2. Materials are PVC 2112, 2116, and 2120. A fiber stress of 5,000 lb/in² was used in bursting pressure calculations.
SOURCE: Abstracted from ASTM Specification D1785.

pipe is recommended. This pipe is suitable for conveying strong acids and chemicals.

Glass-lined metallic pipe, likewise, finds special applications. The fittings are subject to specialty design in view of the delicate nature of the material. **Rubber-lined pipe** is a specialty item, but is available for discrete end uses. **Plastic-lined pipe** can be obtained with a variety of such linings. Saran and polypropylene are among the more common such plastics used.

VITRIFIED, WOODEN-STAVE, AND CONCRETE PIPE

Vitrified pipe is used extensively for drains and sewerage systems. Burnt-clay tile, being rendered impervious to water by glazing, is by far the best material for sewage purposes as it is not attacked by acids. Dimensions are given in Table 6.7.23 (see also Fig. 6.7.6). For sizes larger than 36 in (91.4 cm) and other data, refer to the publications of the Clay Products Assoc., Chicago (see ASTM Standard C-700).

Table 6.7.21 Dimensions of Plastic Pipe for Gas Service (AGA Requirements)

Nominal size, in	Nominal OD, in	Nominal ID, in	Sleeved weight, lb	Max working pressure at 73°F, lb/in²
½	0.625	0.500	0.054	250
¾	0.875	0.750	0.079	175
1	1.125	1.000	0.103	125
1¼	1.375	1.250	0.127	110
1½	1.625	1.500	0.152	90
1¾	1.875	1.750	0.177	80
2	2.125	2.000	0.200	75
2½	2.660	2.500	0.320	75
3	3.190	3.000	0.457	75
4	4.250	4.000	0.800	75

Wood-stave pipe (Fig. 6.7.7) is used to a large extent for municipal water supply, outfall sewers, mining, irrigation, and various other uses providing for the transportation of water. The water carried may be hot, cold, or acid. It is made either untreated or creosoted by a vacuum and pressure process. This process uses 8 lb of creosote per cubic foot of wood treated. The untreated pipe is most used where the pipe is constantly full of water, and the wood therefore completely saturated, although in many such instances the creosoted wood is used to give assurance of permanence. (See also Sec. 4.)

Wood-stave pipe is made in two types: machine-banded pipe and continuous-stave pipe. **Machine-banded pipe** is banded with wire and is made with wood or metal collars, or with inserted joints. **Continuous-stave pipe** is manufactured in units consisting of staves, bands, and shoes, shipped in knocked-down form, and constructed in the trench. In building this type of pipe, the staves are laid so as to break joints and the completed pipe is without joints. Continuous-stave pipe is banded with individual bands, ranging in size from ⅜ to 1 in (0.95 to 2.54 cm), depending upon the size of the pipe. A factor of safety of 4 is maintained in the band, based on an ultimate strength of 60,000 lb/in² (414 MPa) of cross section. The maximum pressure to which a continuous-stave pipe may be subjected depends upon the size of the pipe. The head for small pipes may run as high as 400 ft (121.9 m) while in the largest sizes the head would be less than 200 ft (61 m).

Machine-banded pipe is made for pressures of 50 to 400 ft (15.2 to 121.9 m). The staves are made from redwood or Douglas-fir lumber, dried and carefully selected. The inside and outside of the staves are dressed to conform to the circumferential lines, and the edges of the staves dressed to conform to the radial lines.

Wooden pipe is largely built in western sections of the United States, close to the natural lumber market. The sizes of machine-banded pipe range from 2 to 24 in (5.08 to 61 cm), and of the continuous-stave pipe from 6- to 20-ft (15.2-cm to 6.1-m) inside diameter.

Pipe made from **plywood** is molded in lengths up to 11 ft (3.35 m) in diameters 3 in (7.62 cm) and up, and in wall thicknesses to specifications. Tubes made from fiber, by a molding process, are obtainable in a variety of sizes and lengths.

Concrete pipe is an important factor in sewer, conduit, railroad, culvert, and water-pipe construction. The pipe, as usually made, is constructed of concrete reinforced longitudinally with bars and transversely with wire mesh or steel bands. It is made in sections of definite length, with the longitudinal reinforcement so disposed as to provide for the interlocking of one section with another, and so formed that when these are locked together and cemented they form a continuous line of pipe free from leakage or seepage. Various forms of joints are used, all capable of taking care of expansion. Figure 6.7.8 shows one type

Table 6.7.22 Corrosion-Resistance Data, Polyethylene Pipe

Reagent	Performance at: 75°F (24°C)	Performance at: 120°F (49°C)	Reagent	Performance at: 75°F (24°C)	Performance at: 120°F (49°C)	Reagent	Performance at: 75°F (24°C)	Performance at: 120°F (49°C)
Acetic acid, glacial*	F	NG	Ferrous sulfate, 15% aq.	E	E	Potassium chloride, saturated	E	E
Acetic acid, 10%*	E	E	Fluorine	E	NG	Potassium dichromate	E	E
Acetone	NG	NG	Fluosilicic acid, concentrated	E	F	Potassium hydroxide	E	E
Ammonia, dry gas	E	E	Formaldehyde, 40%	E	E	Potassium nitrate	E	E
Ammonium hydroxide, 10%	E	E	Formic acid, 50%	E	E	Potassium permanganate	E	E
Ammonium hydroxide, 28%	E	E	Furfuryl alcohol	NG	NG	Silicic acid	E	E
Amyl acetate	NG	NG	Gasoline	NG	NG	Silver nitrate	E	E
Aniline	E	F	Hydrobromic acid	E	E	Sodium benzoate	E	E
Benzene	NG	NG	Hydrochloric acid, 10%	E	E	Sodium bisulfite	E	E
Bromine	NG	NG	Hydrochloric acid, 37%	E	E	Sodium carbonate, concentrated	E	E
Butyraldehyde	E	G	Hydrofluoric acid, 48%	E	E	Sodium chloride, saturated solution	E	E
Calcium chloride, saturated	E	E	Hydrofluoric acid, 75%	E	F	Sodium hydroxide, 10%	E	E
Calcium hydroxide	E	E	Hydrogen peroxide, 30%	E	G	Sodium hydroxide, 50%	E	E
Calcium hypochlorite	E	E	Hydrogen peroxide, 90%	G	NG	Sodium sulfate	E	E
Carbon disulfide	NG	NG	Lactic acid, 90%	E	E	Stannic chloride, saturated	E	E
Carbon tetrachloride	NG	NG	Linseed oil	E	E	Stearic acid, 100%	E	E
Carbonic acid	E	E	Lubricating oil	NG	NG	Sulfuric acid, 10%	E	E
Chlorine, dry gas	F	NG	Magnesium chloride	E	E	Sulfuric acid, 30%	E	E
Chlorine, liquid	NG	NG	Magnesium sulfate	E	E	Sulfuric acid, 60%	E	F
Chlorosulfonic acid	NG	NG	Methyl bromide	NG		Sulfuric acid, 98%	F	NG
Citric acid, saturated	E	E	Methyl isobutyl ketone	F	NG	Tannic acid	E	E
Copper sulfate	E	E	Nitric acid, 10%	E	E	Toluene	NG	NG
Cyclohexanone	NG	NG	Nitric acid, 30-50%	E	E	Trichlorobenzene	F	NF
Diethylene glycol	E	E	Nitric acid, 70%	E	E	Trichloroethylene	NG	NG
Dioxane	E	G	Oleic acid	F	NG	Vinegar	E	E
Ethyl acetate	F	NG	Phosphoric acid, 30%	E	E	Xylene	NG	NG
Ethyl alcohol, 35%	NG	NG	Phosphoric acid, 90%	E	NG	Zinc chloride	E	E
Ethyl butyrate	F	NG	Photographic developer	E	E	Zinc sulfate	E	E
Ethylene dichloride	NG	NG	Potassium borate	E	E			
Ferric chloride	E	E	Potassium carbonate	E	E			

Corrosion-resistance data given in this table based on laboratory tests conducted by the manufacturers of the materials covered, and are indicative only of the conditions under which the tests were made. This information may be considered as a basis for recommendation, but not as a guarantee. Materials should be tested under actual service to determine suitability for a particular purpose.
E = excellent, G = good, F = fair, NG = not good.
* Polyethylene is permeable to acetic acid.

Table 6.7.23 Standard-Strength Vitrified Clay Pipe
(Dimensions refer to Fig. 6.7.6)

Size D, in	Laying length L: Nominal, ft	Laying length L: Limit of minus variation,* in/ft length	Max difference in length of two opposite sides, in	OD of barrel, in: Min	OD of barrel, in: Max	ID of socket in above base D_s, in: Min	ID of socket in above base D_s, in: Max	Depth of socket L_s, in: Nominal	Depth of socket L_s, in: Min	Thickness of barrel T, in: Nominal	Thickness of barrel T, in: Min	Thickness of socket at in from outer end T_s, in: Nominal	Thickness of socket at in from outer end T_s, in: Min
4	2, 2 1/2 3	1/4	5/16	4 7/8	5 1/8	5 3/4	6 1/8	1 3/4	1 1/2	1/2	7/16	7/16	3/8
6	2, 2 1/2, 3	1/4	3/8	7 1/16	7 7/16	8 3/16	8 5/8	2 1/2	2	5/8	9/16	1/2	7/16
8	2, 2 1/2, 3	1/4	7/16	9 1/4	9 3/4	10 1/2	11	2 1/2	2 1/4	3/4	11/16	9/16	1/2
10	2, 2 1/2, 3	1/4	7/16	11 1/2	12	12 3/4	13 1/4	2 5/8	2 3/8	7/8	13/16	5/8	9/16
12	2, 2 1/2, 3	1/4	7/16	13 3/4	14 5/16	15 1/8	15 3/4	2 3/4	2 1/2	1	15/16	3/4	11/16
15	3, 4	1/4	1/2	17 3/16	17 13/16	18 5/8	19 1/4	2 7/8	2 5/8	1 1/4	1 1/8	15/16	7/8
18	3, 4	1/4	1/2	20 5/8	21 7/16	22 1/4	23	3	2 3/4	1 1/2	1 3/8	1 1/8	1 1/16
21	3, 4	1/4	9/16	24 1/8	25	25 7/8	26 3/4	3 1/4	3	1 3/4	1 5/8	1 5/16	1 3/16
24	3, 4	3/4	9/16	27 1/2	28 1/2	29 3/8	30 3/8	3 3/8	3 1/8	2	1 7/8	1 1/2	1 3/8
27	3, 4	3/4	9/16	31	32 1/8	33	34 1/8	3 1/2	3 1/4	2 1/4	2 1/8	1 11/16	1 9/16
30	3, 4	3/4	5/8	34 3/8	35 5/8	36 1/2	37 3/4	3 5/8	3 3/8	2 1/2	2 3/8	1 7/8	1 3/4
33	3, 4	3/4	5/8	37 5/8	38 15/16	39 7/8	41 1/4	3 3/4	3 1/2	2 5/8	2 1/2	2	1 13/16
36	3, 4	3/4	11/16	40 3/4	42 1/4	43 1/4	44 3/4	4	3 3/4	2 3/4	2 5/8	2 1/16	1 7/8

When ordering standard-strength vitrified-clay pipe, give the size of pipe (ID) and the laying strength wanted, and refer to ASTM Specification C-700. Standard lengths of pipe shown meet normal practice in various sections of the country. Manufacturers' stocks include those lengths conforming to local practice.

* There is no limit for plus variation.

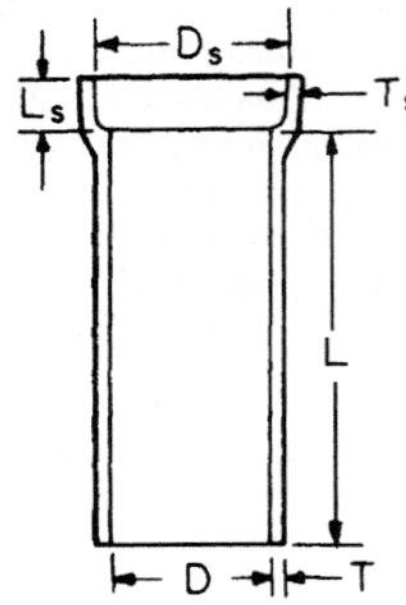

Figure 6.7.6 Vitrified pipe.

of construction. Concrete pipe is manufactured in a great variety of diameters, thicknesses, and lengths to suit almost any requirement arising in practice. (See also Sec. 4.)

Asbestos-cement pipe, known by the trade name **Transite** pipe in this country, was developed initially in Europe. It was widely used in the United States for many years in a large variety of services. It was made of a mixture of portland cement and asbestos fiber, was highly resistant to corrosion, and provided outstanding service in mine drainage systems, waterworks systems, gas lines, and sewerage systems. It was manufactured in diameters from 3 to 36 in (7.6 to 91 cm), in stock lengths of 13 ft (4 m), and in pressure classes of 50, 100, 150, and 200 lb/in² (349, 689, 1,034, and 1,379 kPa). While it is no longer manufactured, it is still in place in some of the above-cited applications. Repairs to existing Transite piping generally devolves into replacing it with pipe made of a material compatible with its service, often with fiberglass-reinforced pipe, plastic pipe, or some form of metal pipe.

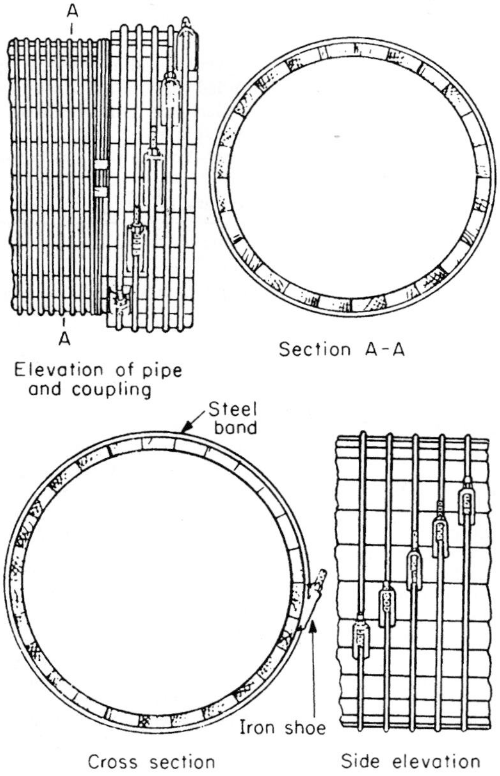

Figure 6.7.7 Wood-stave pipes.

FITTINGS FOR STEEL PIPE

American National Standard Cast-Iron Pipe Flanges and Flanged Fittings for Maximum Working Saturated Steam Pressure of 25, 125, and 250 lb/in² (172, 862, and 1,724 kPa)

INTRODUCTORY NOTES

ANSI B16.1 lists complete dimensional data and drilling template details for these classes of pipe.

Sizes The sizes of the fittings in the following tables are nominal pipe sizes. In the class 25 standard, the nominal pipe size is the same as the port diameter of the fittings, for *all* sizes. In the class 125 and class 250 standards the nominal pipe size is the same as the port diameter of fittings for pipe having inside diameters of 12 in (30.5 cm) and smaller. For pipe 14 in (35.6 cm) and larger, the corresponding outside diameter of the pipe is given, and consequently the fittings will have a smaller port diameter.

Pressure Rating In the class 25 standard the sizes 36 in (91.4 cm) and smaller may also be used for maximum nonshock working hydraulic pressures of 43 lb/in² gage (296 kPa) or a maximum gas pressure of

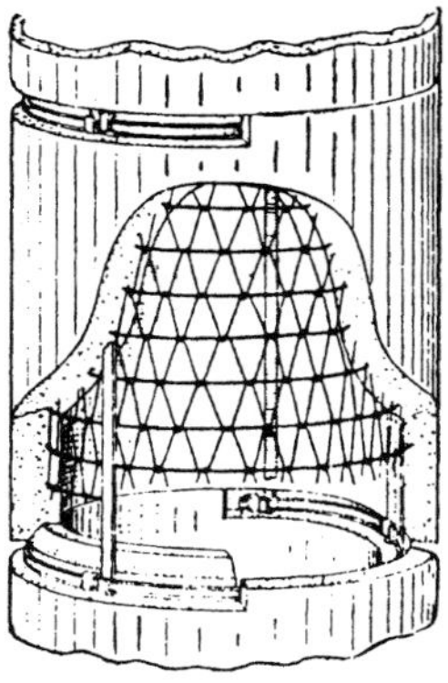

Figure 6.7.8 Reinforced-concrete pipe.

25 lb/in^2 gage (172 kPa), at or near the ordinary range of air temperatures. In the class 125 standard, the sizes 12 in (30.5 cm) and smaller may also be used for maximum nonshock working hydraulic pressure of 175 lb/in^2 gage (1,207 kPa) at or near the ordinary range of air temperatures. In the class 250 standard, the sizes 12 in and smaller may be used for maximum nonshock working hydraulic pressures of 400 lb/in^2 gage (2,758 kPa) at or near the ordinary range of air temperatures.

Facing All class 25 and class 125 cast-iron flanges and flanged fittings are plain faced, that is, without projection or raised face. All class 250 cast-iron flanges and flanged fittings have a raised face (0.16 cm) high. The raised face is included in the minimum flange thickness and center-to-face dimensions.

An **inspection limit** of ± ⅓₂ in (0.08 cm) is allowed on all center-to-contact-surface dimensions for sizes up to and including 10 in (25.4 cm) and ± ⅛ in on sizes larger than 10 in.

Dimensions In the class 25 standard, the flange diameters, bolt circles, and number of bolts are the same as in the class 125 ANSI Standard B16.1 with a reduction in the thickness of flanges and bolt diameters, thereby maintaining interchangeability between the two standards.

The center-to-face and face-to-face dimensions of class 25 standard fittings are the same as for the class 125 standard.

Bolting Drilling templates are in multiples of four, so that fittings may be made to face in any quarter. Bolt holes straddle the centerline. For bolts smaller than 1¾ in (4.45 cm) the bolt holes are drilled ⅛ in larger in diameter than the nominal size of the bolt. Holes for bolts 1¾ in and larger are drilled ¼ in (0.64 cm) larger than nominal diameter of bolts. Bolts of steel are with standard "rough square heads" and the nuts are of steel with standard "rough hexagonal" dimensions; all as given in the American Standard on Wrench Head Bolts and Nuts and Wrench Openings of the National Screw Thread Commission. For bolts, 1¾-in (4.45-cm) diam and larger, bolt studs with a nut on each end are recommended.

Hexagonal nuts for pipe sizes 1 to 48 in (2.54 to 122 cm) in the class 125 standard and 1 to 16 in (40.6 cm) in the class 250 standard can be conveniently pulled up with open wrenches of minimum design of heads. Hexagonal nuts for pipe sizes 48 to 96 in (244 cm) in the class 125 standard and 18 to 48 in (45.7 to 122 cm) in the class 250 standard can be conveniently pulled up with box wrenches.

Spot Facing The bolt-holes of classes 25, 125, and 250 cast-iron flanges and flanged fittings are not spot-faced for ordinary service. When required, the flanges and fittings in sizes 30 in (76.2 cm) and larger may be spot-faced or back-faced to the minimum thickness of flange with a plus tolerance of ⅛ in (0.32 cm).

Reducing Fittings Reducing elbows and side-outlet elbows carry same dimensions center to face as straight-size elbows corresponding to the size of the larger opening.

Tees, side-outlet tees, crosses, and laterals sizes 16 in (40.6 cm) and smaller, reducing on the outlet or branch, have the same dimension center to face and face to face as straight-size fittings corresponding to the size of the larger opening. Sizes 18 in (45.7 cm) and larger, reducing on the outlet or branch, are made in two lengths depending on the size of the outlet as given in the tables of dimensions.

Tees, crosses, and laterals, reducing on the run only, have the same dimensions center to face and face to face as straight-size fittings corresponding to the size of the larger opening.

Reducers and eccentric reducers for all reduction have the same face-to-face dimensions for the larger opening.

Special double-branch elbows whether straight or reducing have the same dimension center to face as straight-size elbows corresponding to the size of the larger opening.

Side-outlet elbows and side-outlet tees have all openings on intersecting centerlines.

Elbows Special degree elbows ranging from 1 to 45° have the same center-to-face dimensions given for 45° elbows, and those over 45° and up to 90° have the same center-to-face dimensions given for 90° elbows. The angle designation of an elbow is its deflection from straight-line flow and is the angle between its flange faces.

Threaded Companion Flanges Threaded companion flanges in the class 25 standard should not be thinner than those in the class 125 standard on sizes 24 in (61 cm) and smaller. Other types of flanges may vary in thickness.

Laterals Laterals (Y branches) both straight and reducing sizes 8 in and larger are reinforced to compensate for the inherent weakness in the casting design.

The American National Standard B16.1 covers also dimensions of base elbows and base tees and anchorage bases for straight tees and reducing tees.

American National Standard cast-iron pipe flanges and flanged fittings are available for maximum nonshock working hydraulic pressure of 800 lb/in^2 gage (5,516 kPa) at ordinary air temperatures.

Assembly of Flanged Joints The optimum degree of tightening occurs when a stress of 30,000 lb/in^2 (207 MPa) is uniformly reached in each flange stud or bolt. For a modulus of elasticity of 30,000,000 lb/in^2 (207 GPa) a stress of 30,000 lb/in^2 occurs when the elongation, determined with a dial indicator or micrometer, is 0.001 in/in of stud length measured between centers of nuts. Uniform tension in flange bolts may also be obtained by use of a torque wrench; bearing surfaces of nuts must have a good machine finish, and threads must be properly lubricated for reliable results with a torque wrench. The following torque values have been found to give 30,000 lb/in^2 stress in studs:

Study diam, in	*Threads per inch*	*Torque, lb · ft**
⅝	11	89
¾	10	107
⅞	9	162
1	8	244
1⅛	8	322
1¼	8	410
1⅜	8	510
1½	8	615

* Multiply by 0.149 to obtain torque in kilogram-metres.

American National Standard Steel Pipe Flanges and Flanged Fittings

INTRODUCTORY NOTES

Pressure Ratings and Tests These standards are known as "American Class 150, 300, 400, 600, 900, 1,500, and 2,500 Steel Flange Standards" (ANSI B16.5).

Sizes The size of the fittings and companion flanges are listed in the standard.

Materials The flanged fittings and flanges should be either steel castings or steel forgings of the grade complying with the ASTM specifications recommended under these standards for the various pressure-temperature ratings for which these standards are designed.

Bolting material including nuts and washers are based on a high-grade product equal to that given in ASTM Standard Specifications for Alloy-Steel Bolting Material for High Temperature Service (Table 6.7.24) and with physical and chemical requirements in accordance with the tables

Table 6.7.24 Mechanical Properties of Alloy Steel Bolting Material for High-Temperature Service

Ferritic steels

Grade	Diameter, in (mm)	Minimum tempering temperature,* °F (°C)	Tensile strength, min ksi (MPa)	Yield strength, min 0.2% offset, ksi (MPa)	Elongation in 2 in (50.8 mm), min %	Reduction of area, min %	Hardness, max
B5, 4 to 6% chromium	Up to 4 (101.6), incl.	1,100 (593)	100 (690)	80 (550)	16	50	
B6, 13% chromium	Up to 4 (101.6), incl.	1,100 (593)	110 (760)	85 (585)	15	50	
B6X, 13% chromium	Up to 4 (101.6), incl.	1,100 (593)	90 (620)	70 (485)	16	50	26 HRC
B7, chromium-molybdenum	2½ (63.5) and under	1,100 (593)	125 (860)	105 (720)	16	50	
	Over 2½ to 4 (63.5 to 101.6)	1,100 (593)	115 (790)	95 (655)	16	50	
	Over 4 to 7 (101.6 to 117.8)	1,100 (593)	100 (690)	75 (515)	18	50	
B7M,† chromium-molybdenum	2½ (63.5) and under	1,150 (620)	100 (690)	80 (550)	18	50	235 HB or 99 HRB
B16, chromium-molybdenum-vanadium	2½ (63.5) and under	1,200 (650)	125 (860)	105 (720)	18	50	
	Over 2½ to 4 (63.5 to 101.6)	1,200 (650)	110 (760)	95 (655)	17	45	
	Over 4 to 7 (101.6 to 177.8)	1,200 (650)	100 (690)	85 (585)	16	45	

Austenitic steels

Class and grade, diameter, in (mm)	Heat treatment‡	Tensile strength, min ksi (MPa)	Yield strength, min 0.2% offset, ksi (MPa)	Elongation in 2 in (50.8 mm), min %	Reduction of area, min %	Hardness, max
Class 1: B8, B8C, B8M, B8P, B8T, B8LN, B8MLN, all diameters	Carbide solution treated	75 (515)	30 (205)	30	50	223 HB or 96 HRB
Class 1A: B8A, B8CA, B8MA, B8PA, B8TA, B8LNA, B8MLNA, B8NA, B8MNA, all diameters	Carbide solution treated in finished condition	75 (515)	30 (205)	30	50	192 HB or 90 HRB
Class 1B: B8N, B8MN, all diameters	Carbide solution treated	80 (550)	35 (240)	30	40	223 HB or 96 HRB
Class 1C: B8R, all diameters	Carbide solution treated	100 (690)	55 (380)	35	55	271 HB or 28 HRC
B8RA, all diameters	Carbide solution treated in finished condition	100 (690)	55 (380)	35	55	271 HB or 28 HRC
B8S, all diameters	Carbide solution treated	95 (655)	50 (345)	35	55	271 HB or 28 HRC
B8SA, all diameters	Carbide solution treated in finished condition	95 (655)	50 (345)	35	55	271 HB or 28 HRC
Class 2: B8, B8C, B8P, B8T, B8N, ¾ (19.05) and under	Carbide solution treated and strain-hardened	125 (860)	100 (690)	12	35	321 HB or 35 HRC
Over ¾ to 1 (19.05 to 25.4) incl.		115 (790)	80 (550)	15	35	321 HB or 35 HRC
Over 1 to 1¼ (25.4 to 31.6) incl.		105 (720)	65 (450)	20	35	321 HB or 35 HRC
Over 1½ to 1¼ (31.6 to 37.9) incl.		100 (690)	50 (345)	28	45	321 HB or 35 HRC
Class 2: B8M, B8MN, ¾ (19.05) and under	Carbide solution treated and strain-hardened	110 (760)	95 (655)	15	45	321 HB or 35 HRC
Over ¾ to 1 (19.05 to 25.4) incl.		100 (690)	80 (550)	20	45	321 HB or 35 HRC
Over 1 to 1¼ (25.4 to 31.6) incl.		95 (655)	65 (450)	25	45	321 HB or 35 HRC
Over 1¼ to 1½ (31.6 to 37.9) incl.		90 (620)	50 (345)	30	45	321 HB or 35 HRC

* For sizes ¾ in. in diameter and smaller, a maximum hardness of 241 HB (100 HRB) is permitted.
† To meet the tensile requirements, the Brinell hardness shall be over 201 HB (94 HRB).
‡ Class 1 is solution treated. Class 1A is solution treated in the finished condition for corrosion resistance: heat treatment is critical due to physical property requirement. Class 2 is solution-treated and strain-hardened. Austenitic steels in the strain-hardened condition may not show uniform properties throughout the section particularly in sizes over ¾ in (19.05 mm) in diameter.

given under ANSI B16.5. **Commercial steel bolts should not be used at steam pressures over 250** lb/in² (1,724 kPa) and temperatures over 450°F (232°C). Nuts should be of carbon or alloy steel. Washers when used under nuts should be of forged or rolled carbon steel.

Bolting Drilling templates are in multiples of four, so that fittings may be made to face in any quarter. Bolt holes straddle the centerlines. Bolt holes are drilled ⅛ in (0.32 cm) larger in diameter than the nominal size of bolt. Bolts or bolt studs threaded at both ends may be used and should be equipped with cold-punched or cold-pressed semifinished nuts of American National Standard rough dimensions, chamfered and trimmed.

All bolts and bolt studs having diameters 1 in and smaller, and the corresponding nuts are threaded with the American National Standard screw thread, coarse thread series, medium fit, class 3, while those bolts and bolt studs whose diameters are 1⅛ in (2.86 cm) and larger have special threads of the American National form whose pitch is ⅛ in (8 threads per inch). It is recommended that these special threads be allowed a pitch-diameter tolerance of –0.006 in and a lead tolerance of ±0.002 in.

Bolt studs with a nut at each end are recommended for high-temperature service.

The allowable working fiber stress, considering internal allowable working pressure only, in bolting material for valve bonnet flanges, cleanout flanges, etc., is not to exceed 9,000 lb/in² (62 MPa) assuming the pressure to act upon an area circumscribed by the periphery of the outside of the contact surface.

Metal Thickness Minimum metal thicknesses specified in the tables are based on an allowable fiber stress of 7,000 lb/in² (48 MPa), using the modified Barlow formula of the ASME Boiler Construction Code for cylindrical sections and adding 50 percent to the thickness thus determined to compensate for the shape of the fittings. The minimum commercial casting thickness is considered to be ¼ in; therefore, the standards do not show thicknesses less than this. The minimum thickness in these standards means the minimum thickness in any part of the finished casting.

The modified Barlow formula is as follows: For pipes having nominal diameters of ¼ to 5 in, $P + 125 = 2S(t - 0.065)/D$. For pipes of nominal diameters over 5 in, $P = 2S(t - 0.1)/D$, where P is the working pressure, lb/in², t is the thickness of wall of pipe, in; D is the actual outside diameter of pipe, in; and S is 7,000 lb/in² (48 MPa).

Ring Joints The dimensions used for ring and groove joint facings were developed by a committee of the API. The corresponding dimensions and ring numbers incorporated in the ANSI Standard are identical with those given in API Standard 5G. The dimension for the depth of groove is added to the basic flange thickness, which makes it necessary to include separate tables of dimensions for fittings having the ring joint facing.

Fitting Dimensions An inspection limit of ±1/32 in is allowed on all center-to-contact surface dimensions for sizes up to and including 10 in, and ±1/16 in on sizes larger than NPS 10. An inspection limit of ±1/16 in is allowed on all contact-surface to contact-surface dimensions for sizes up to and including NPS 10, and ±⅛ in, on sizes larger than NPS 10.

When elbows having longer radii than specified in the standards are required, the use of pipe bends is recommended.

Laterals The 45° laterals of the larger sizes may require additional reinforcement to compensate for the inherent weakness in this shape of casting.

Valve Dimensions Face-to-face and end-to-end dimensions of ferrous valves for the various pressures are in accordance with the requirements of ANSI B16.10 for ferrous flanged valves.

Reducing Fittings Reducing fittings have the same center-to-flange edge dimensions as those of straight-size fittings of the largest opening.

Side-Outlet Fittings All side-outlet fittings have all openings on the intersecting centerlines.

Welding Neck Flanges The materials, facings, spot facings, etc., conform to the requirements given for other flanges, with the additional provision that the carbon content of the steel shall not exceed 0.35 percent.

Templates for drilling and center to contact-surface dimensions of the American Standard class flanges and flanged fittings are listed in the Standard.

Flanged Pipe Joints

The usual form of pipe joint is that made up by bolting together flanges cast or forged integral with the pipe or fitting, threaded flanges, loose flanges on pipes with lapped ends, and flanges arranged for welding. These forms are illustrated in Fig. 6.7.9. The threaded joint is satisfactory for low and medium steam pressures. The lapped joint is permitted in the same sizes and service ratings as for joints with integral flanges. It is extensively used in high-class work. With the ring joint a higher pressure can be maintained with the same total bolt stress than is possible with the flat gasket type of joint. The welded joint eliminates possibility of leakage between flange and pipe. It is very successful in lines subject to high temperatures and pressures and heavy expansion strains. The welding-neck flange is available in the various pipe sizes. Specific requirements covering the application of all the types of joints in common use are outlined in the Code for Power Piping (ASME B31.1 and B31.3).

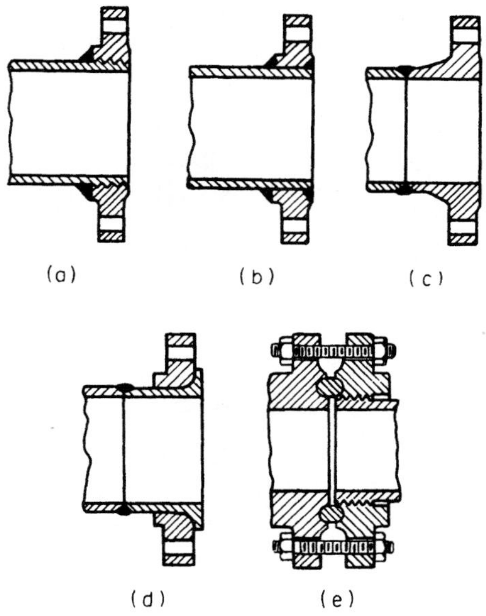

Figure 6.7.9 Welded flange joints and ring joint. (*a*) Forged steel, screwed flange, back-welded and refaced; (*b*) forged steel, slip-on welding flange, welded front and back, refaced; (*c*) forged steel, welding neck flange, butt-welded to pipe; (*d*) lap-welding nipple, butt-welded to pipe; (*e*) ring joint.

Facing of Flanges Various styles of finish are used on the faces of flanges, for the purpose of the retention of the gasket used to make a tight joint. Those in general use are as follows: plain straight face, plain face corrugated or scored, male and female, tongue and groove, and raised face.

The **plain straight face** has the entire face of the flange faced straight across and may be used with either a full face or ring gasket. The **plain face, serrated or V-grooved,** is a plain face upon which concentric grooves have been cut with either a round-nose or V-shaped tool. This finish is sometimes of advantage when the service demands an exceptionally thick, loosely woven fibrous or soft metallic gasket, because the roughening of the faces of the flanges tends to keep the gasket from blowing out. The **male-and-female** facing consists of a recess in one flange and a corresponding raised face or projection on the other mating flange extending from the inside of the pipe nearly to the inside of the bolt holes. In the **tongue-and-groove** facing, the tongue or raised face and the groove or recess are narrow rings located between the bolt holes and the port. The male-and-female and the tongue-and-groove facing have been extensively used, particularly on hydraulic lines. To a more limited extent they have been used also on high-pressure steam lines. Both these types, however, have in common several objectionable features from the standpoint of manufacture, erection, and maintenance. These objections are removed by the use of the **raised-face** facing, which consists of a high narrow raised ring on each of the mating flanges, whose inside diameters is the same as that of the pipe or port. It is particularly recommended for high-pressure steam and hydraulic lines. **Gaskets** used in this type of joint are either soft fibrous material or soft metal and extend from the inside of the pipe to the bolt holes. Only the small portion in contact with the narrow raised face is subjected to the compressive effect of the bolts.

The following advantages are claimed for the raised-face type of facing: all mating of flanges has been eliminated; any valve or fitting may be removed from the line without springing the line apart; the gasket is automatically centered by its outer edge coming in contact with the bolts; the outside edges of the flanges are far enough apart to make it possible to determine whether the joint has been properly made.

Unions may be classified as **screw** and **flange.** Typical designs are shown in Fig. 6.7.10, where at the top left is represented a female screw union of the gasket type, at the top right a female screw union

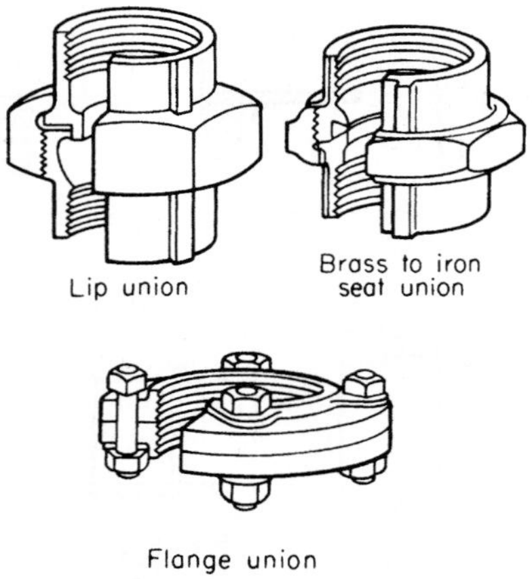

Figure 6.7.10 Types of pipe unions.

having a brass to iron seat that is noncorrosive and a ground joint that eliminates the need for a gasket, and at the bottom a flange union of the gasket type. As in the case of other pipe fittings, unions and union fittings are available in the various pipe sizes and in materials and designs suitable for any service conditions. Very large flange unions can be made by bolting together two screwed companion flanges.

Threaded Fittings

Threaded fittings are made of cast iron, malleable iron, cast steel, forged steel, or brass. Plain standard fittings are generally used for low-pressure gas and water, as in house plumbing and railing work, while the beaded fitting is the standard steam, air, gas, or oil fitting. Screwed fittings are supplied with a large factor of safety. The questions of strength involve much more than the pressure from within the pipe which induces a comparatively low stress in the material. The greater strains come from expansion, contraction, weight of piping, settling, water hammer, etc.

The dimensions of ferrous plugs, bushings, locknuts, and caps with pipe threads are covered by ANSI B16.14.

The normal **amount of thread engagement** necessary to make a tight joint for ANSI Standard pipe thread joints as recommended by Crane Co. is as follows:

Size of pipe, in	1/8	1/4	3/8	1/2	3/4	1	1 1/4	1 1/4	2
Length of thread, in	1/4	3/8	3/8	1/2	9/16	11/16	11/16	11/16	3/4
Size of pipe, in	2 1/2	3	3 1/2	4	5	6	8	10	12
Length of thread, in	15/16	1	1 1/16	1 1/8	1 1/4	1 5/16	1 7/16	1 5/8	1 3/4

The Manufacturers' Standardization Society of Valve and Fitting Industry (MSS) has standardized malleable-iron and brass threaded fittings for several pressures.

Cast-bronze threaded fittings are made in both the class 125 and 250 standards. They are used for any water pipe where bad water makes steel pipe undesirable. Bronze fittings may be had in iron pipe sizes. Forged-steel threaded fittings are made for cold water or oil-working pressures up to 6,000 lb/in^2 (41.4 MPa) hydrostatic. The ANSI has approved standard B16.26 for cast copper alloy fittings for flared copper tubes for maximum cold-water service pressure of 175 lb/in^2 (1,207 kPa).

Railing Fittings Fittings of special construction and of tighter weight than standard steam, gas, and water pipe fittings are widely used for hand railings around areaways, on stairs, for office enclosures with gates, and for permanent ladders. Railing fittings are made in various styles, generally globe-shaped in body, with ends reduced to take thread and recessed to cover all threads. They are furnished in malleable iron, black and galvanized, and in brass.

Special railing-fitting joints are available, such as the slip-and screwed joint, where the post connection is screwed and the rim of the fitting is so made that the rail will slip into the fitting and allow for an angular variation of several degrees, being fastened by pins which are riveted over and filed smooth. The flush-joint stair-rail fitting is another special style of fitting which provides a hand rail with even surfaces at the joints.

Drainage fittings, have no pockets for the lodgment of solids, and the length of the thread chamber is such that when the pipe is threaded to the American National Standard dimensions, the end of the pipe will practically touch the shoulder when screwed in. They are especially adapted to plumbing work and vacuum-cleaning pipe installations. Dimensions conform to ANSI Standard B16.12, Cast-Iron Threaded Drainage Fittings.

The development of standards for **cast-iron long-turn sprinkler fittings** was begun by the National Fire Protection Assoc. in 1914 with a study of the peculiar needs of fittings intended for fire-protection purposes. These fittings (screwed and flanged) are rated at 175 and 250 lb/in^2 (1,207 and 1,724 kPa).

American National Standard Air Gaps and Backflow Preventers in Plumbing Systems, ANSI A40.4 and A40.6, were prepared to establish minimum requirements for plumbing, including water-supply distributing systems, drainage and venting systems, fixtures, apparatus, and devices, and the standardization of plumbing equipment in general.

Ammonia valves and fittings must provide a high margin of safety against accidents. Flanged valves and fittings have tongue-and-groove faces to assure tightness at the joints and against blowing out gaskets. Gaskets are compressed asbestos sheet. Threaded valves and fittings have long threads and are recessed so that the joints may be soldered. These valves and fittings are made of malleable iron, ductile iron, ferrosteel, or forged steel; depending on the size and style. Valves are all iron, with steel stems, and have special lead disk faces or steel disks. Copper or brass must not be used in their construction. Flanged valves are generally interchangeable with flanged fittings. All valves and fittings for ammonia are tested to 300 lb/in^2 (2,069 kPa) air pressure under water. For dimensions of valves, fittings, and specialties for ammonia, refer to manufacturers' catalogs.

Soldered-Joint Fittings The American standard for these fittings (ANSI B16.18) covers certain dimensions of soldered-joint wrought metal and cast brass fittings for copper water tubing including (1) detailed dimensions of the bore, (2) minimum specifications for materials, (3) minimum inside diameter of the fitting, (4) metal thickness for both wrought metal and cast brass fittings, and (5) general dimensions for cast brass fittings including center-to-shoulder dimensions for both straight and reducing cast fittings. Pressure and temperature ratings are also given. Sizes of the fittings are identified by the nominal tubing size as covered by the Specifications for Copper Water Tube (ASTM B88).

Valves

The face-to-face dimensions of ferrous flanged and welding end valves are given in ANSI B16.10. The types covered are:

Wedge-Gate Valves Cast iron, for 125-, 175-, and 250-lb/in² (862, 1,207, and 1,724 kPa) steam service pressure and 800-lb/in² (5,516 kPa) hydraulic pressure, and steel, for 150-, 300-, 400-, 600-, 900-, and 1,500-lb/in² (1,034-, 2,068-, 2,758-, 4,137-, 6,206-, and 10,343-kPa) steam service pressures (see Fig. 6.7.11).

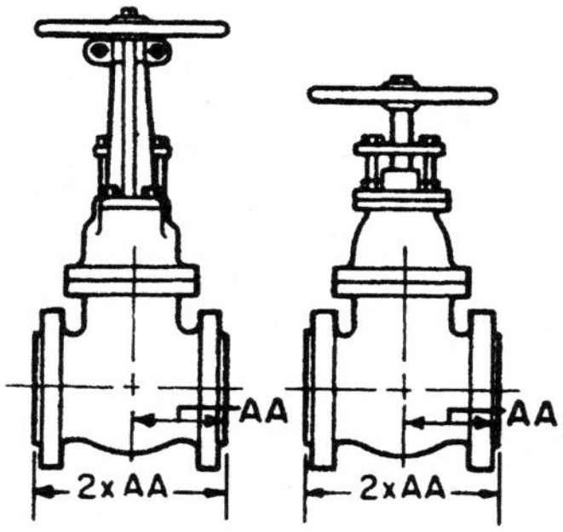

Figure 6.7.11 Wedge gate valves.

Double-Disk Gate Valves Cast iron, for 125- and 250-lb/in² (862- and 1,724-kPa) steam service pressure and 800-lb/in² (5,516-kPa) hydraulic pressure.

Globe and Angle Valves Cast iron, for 125- and 250-lb/in² (862- and 1,724-kPa) steam service pressure, and steel, for 150-, 300-, 400-, 600-, 900-, 1,500-, and 2,500-lb/in² (862-, 2,068-, 2,758-, 4,137-, 6,206-, 10,343-, and 17,238-kPa) steam service pressures (see Fig. 6.7.12).

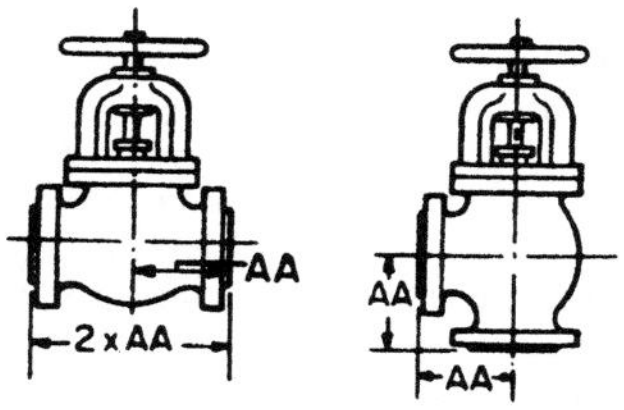

Figure 6.7.12 Globe valve and angle valve.

Swing-Check Valves Cast iron, for 125- and 250-lb/in² (862- and 1,724-kPa) steam service pressure and 800-lb/in² (5,516-kPa) hydraulic pressure, and steel, for 150-, 300-, 400-, and 600-lb/in² (1,034-, 2,068-, 2,758-, and 4,137-kPa) steam service pressures.

Except for ring-joint facings to the face-to-face dimension for flanged valves is the distance between the faces of the connecting end flanges upon which the gaskets are actually compressed, i.e., the "contact surfaces."

All flanges for class 125 cast-iron valves are plain-faced. The facings of the class 250 cast-iron, and the class 150 and 300 steel valves have a 1⁄16-in raised face which is included in the contact-surface to contact surface dimensions. The contact-surface to contact-surface dimensions of steel valves for class 400 and higher pressures and for cast-iron valves for class 800 hydraulic pressure include a ¼-in raised face.

The end-to-end dimensions for **welding-end** valves for sizes NPS 1 to 8 are the same as the contact-surface to contact-surface dimensions given in the tables for steel valves. For details of welding bevel see ANSI B16.10 and Fig. 6.7.15.

A plus or minus **tolerance** of 1⁄16 in is allowed on all face-to-face dimensions of valves NPS 10 and smaller, and a tolerance ⅛ on sizes NPS 12 and larger.

Cocks The ordinary plug cock operated by a handle or wrench is a form of valve in comparatively small sizes suitable for ordinary service only. The ASME Code for Pressure Piping requires that where cocks are used for high-temperature service they shall be so designed as to prevent galling, either by making the plugs of different material from the body of the cock or by treating the plugs to ensure different physical properties. By means of special design features that eliminate the tendency to leak and stick, the plug-cock type of valve has become available in large sizes and for severe service conditions. Sizes are listed as high as 30 in and are gear-operated in the larger sizes. For further details, refer to manufacturers' catalogs.

Expansion and Flexibility

Piping systems must be designed so that they (1) will not fail because of excessive stresses, (2) will not produce excessive thrusts or moments at connected equipment, or (3) will not leak at joints because of expansion of the pipe. Flexibility is provided by changes of direction in the piping through the use of bends or loops, or provision may be made to absorb thermal strains by use of expansion joints. All, or portions, of the pipe may be corrugated to improve flexibility; in many systems, however, sufficient change is provided by the geometry of the layout to make unnecessary the use of either expansion joints or corrugated sections of piping. Proper cold springing is beneficial in assisting the piping system to attain its most favorable condition. Because of plastic flow of the piping material, hot stresses tend to decrease with time while cold stresses tend to increase with time; their sum, called the stress range, remains substantially constant. For this reason no credit is warranted with regard to stresses; for calculation of forces and moments, the effect of cold spring is recognized by use of a cold-spring factor varying from 0 to 1 for cold spring varying from 0 to 100 percent.

The allowable stress range S_A is calculated by

$$S_A = f(1.25S_c + 0.25S_h)$$

where S_c and S_h are the S values for the minimum cold and maximum hot conditions, respectively. The stress-reduction factor f is a function of the number of hot-to-cold-to-hot (full) temperature cycles anticipated over the life of the plant, as follows:

Total no. of full temp cycles over expected life	*Stress-reduction factor*
7,000 and less	1.0
14,000 and less	0.9
22,000 and less	0.8
45,000 and less	0.7
100,000 and less	0.6
Over 100,000	0.5

The bending and torsional stresses calculated (see paragraph 119.6.4 of ANSI B31.1.0-1983) are used to determine the maximum computed expansion stress $S_E = \sqrt{S_b^2 + 4S_t^2}$ where S_b and S_t are bending and torsional stresses, respectively. S_E must not exceed the allowable stress range S_A.

In recent years, many principal high-temperature steam lines have either been analyzed, tested in a model-testing machine, or both. No rigid rule is stipulated for the requirement of analysis or model test; however, the Code for Pressure Piping suggests that when the following criterion is not satisfied, need for an analysis is indicated: $DY/(L - U)^2 \leq 0.03$, where D is the nominal pipe size, in; Y is the resultant of movements to be absorbed by pipeline, in; U is the length of straight line joining the anchor points, ft; and L is the length of the developed line axis, ft. A wide assortment of software is available which makes computer analysis attractive, rapid, and widely applicable.

Expansion Joints for Steam Pipelines In many instances it may be economical to care for thermal expansion by use of expansion joints. For low-pressure steam lines, the use of packed expansion joints may be feasible; experience has indicated that packed joints are difficult to maintain when used on high-pressure lines. Figure 6.7.13 shows a type of joint that has been successfully used for high-pressure, high-temperature service. The bellows is designed to take either axial, lateral, or combined axial and lateral deflections. The internal sleeve guides movement of the joint and also protects the flexible bellows from direct contact with the fluid being handled. Face-to-face dimensions and permissible axial and lateral deflections are available from manufacturers' specifications and catalogs.

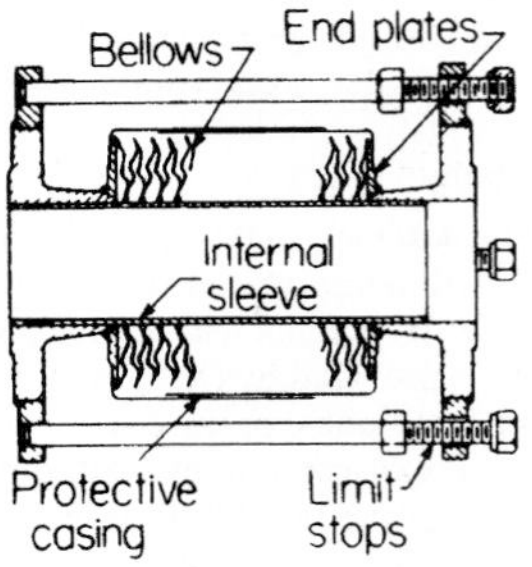

Figure 6.7.13 Expansion joint for steam line. *(Croll-Reynolds, Inc.)*

Where large lateral deflections are to be absorbed, two expansion joints separated by a length of pipe as shown in Fig. 6.7.14 may be used. With such an arrangement, the lateral deflection permissible with one joint only may be increased many times. Tie rods, as shown, should always be installed to protect the joint against overtravel and externally to guide movement of the joint.

Table 6.7.25, extracted from the Code for Pressure Piping, lists thermal-expansion data for both ferrous and nonferrous piping. For expansion at temperatures intermediate between those shown, straight-line interpolation is permitted.

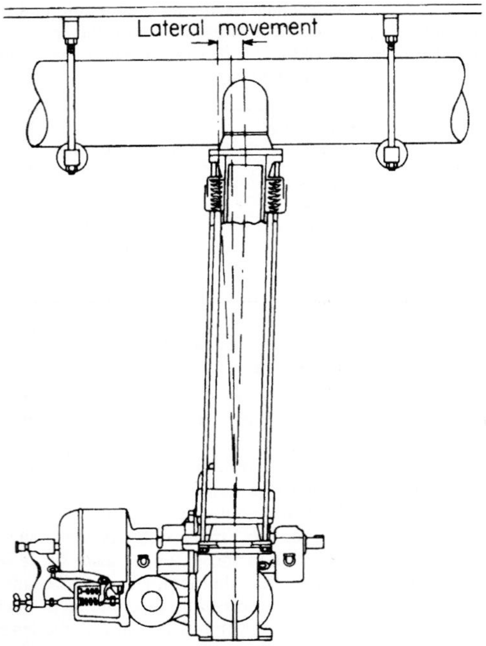

Figure 6.7.14 Arrangement of expansion joints for large lateral deflection. *(Croll-Reynolds, Inc.)*

The **rubber expansion joint** has become an established part of pipeline equipment. Its special field of application is on low-pressure and vacuum lines in condenser applications, etc., and it is recommended for pressures up to 25 lb/in^2 (172 kPa) gage where the maximum temperature does not exceed 250°F. Standard joints for pressure installations are reinforced to withstand working pressures up to 125 lb/in^2 (862 kPa) gage and temperatures up to 200°F. Joints are available in all standard pipe sizes.

Welding in Power-Plant Piping

The majority of main-cycle and service steel piping in modern steam power plants is of welded construction. Steel pipe of NPS 2 and smaller is generally socket-welded; larger-size piping is usually butt-welded. Frequently, depending on location and scheduling, piping larger than NPS 2 is prefabricated; smaller piping is shipped to the construction site in random lengths and is fabricated concurrently with installation. Small-sized chromium-molybdenum piping requiring bending is frequently also shop-fabricated so as to avoid high field preheat, welding, and stress-relieving costs. It is desirable to schedule shipment of hangers so that they will be available at the job site upon arrival of the prefabricated piping; this avoids the expense of providing, installing, and later removing temporary hangers and supports. Aside from the economy of welded construction, it is a virtual necessity in high-pressure, high-temperature work because of danger of leakage if joints are flanged.

Shop welds are frequently made by automatic or semiautomatic submerged-arc or inert-gas shielded-arc processes; field welds are generally of the manual type and are usually done by the shielded metal-arc and/or inert-gas metal-arc processes. Welding in power piping systems, whether in the shop or at the job site, must be done by welders who have qualified under provisions of the Code for Pressure Piping or the ASME Boiler and Pressure Vessel Code.

End Preparation for Butt Welds Figure 6.7.15 shows the end preparation recommended (not required) for piping whose wall thickness is in or less, and Fig. 6.7.16 shows that required for piping with wall thickness above in. During the welding process, to avoid entrance of welding material into the pipe, backing rings may be used as shown in Fig. 6.7.17*a*, *b*, and *c*. Consumable inserts are also available. They are recommended for installation in piping systems which require a smooth, unobstructed interior surface. Note that thick-walled pipes (over ¾ in) are taper-bored on the inside in order that they may receive a tapered, machined backing ring.

Preheating Prior to start of welding, many materials require preheat to a specified temperature: preheat may be done by electrical-resistance or induction heating or by ring-type gas burners placed concentrically with the pipe. The preheat temperature is measured by indicating crayons or by thermocouple pyrometers and must be maintained during the welding operation. Table 1, Appendix D, of the Code for Pressure Piping lists materials used in piping systems and the appropriate temperatures for preheat. In general, the following is indicative of the intent only; for specific instances, the Code must be consulted.

Carbon steel and **wrought iron** should be preheated to a "hand-hot" condition if the ambient temperature at time of field installation is 32°F (0°C) or less: carbon steels which have minimum tensile properties of 70,000 lb/in^2 (483 MPa) or higher should be preheated to 250°F (121°C); under other conditions, preheat is not mandatory, but some purchasers insist that the contractor preheat heavy-walled piping such as boiler feed.

Low-alloy steels with a chromium content not exceeding ¾ percent and low-alloy steels with a total alloy content not exceeding 2 percent are required to be preheated to a minimum temperature of 300°F (149°C).

Alloy steels with a chromium content between ¾ and 2 percent and low-alloy steels with a total alloy content not exceeding 2¾ percent require preheating to 375°F (191°C) minimum. Those with a total alloy content greater than 2¾ percent but not exceeding 10 percent require preheating to a temperature of 450°F (232°C) minimum.

Table 6.7.25 Thermal Expansion Data

Material	Coefficient	Temp range: 70°F (21°C) to: 70 (21)	200 (93)	300 (149)	400 (205)	500 (260)	600 (316)	700 (371)	800 (427)	900 (482)	1,000 (538)	1,100 (593)	1,200 (649)	1,300 (705)	1,400 (760)
Carbon steel: carbon-moly steel low-chrome steels (through 3% Cr)	*A*		6.38	6.60	6.82	7.02	7.23	7.44	7.65	7.84	7.97	8.12	8.19	8.28	8.36
	B	0	0.99	1.82	2.70	3.62	4.60	5.63	6.70	7.81	8.89	10.04	11.10	12.22	13.34
Intermediate alloy steals: 5 Cr Mo-9 Cr Mo	*A*		6.04	6.19	6.34	6.50	6.66	6.80	6.96	7.10	7.22	7.32	7.41	7.49	7.55
	B	0	0.94	1.71	2.50	3.35	4.24	5.14	6.10	7.07	8.06	9.05	10.00	11.06	12.05
Austenitic stainless steels	*A*		9.34	9.47	9.59	9.70	9.82	9.92	10.05	10.16	10.29	10.39	10.48	10.54	10.60
	B	0	1.46	2.61	3.80	5.01	6.24	7.50	8.80	10.12	11.48	12.84	14.20	15.56	16.92
Straight chromium stainless steels: 12 Cr, 17 Cr, and 27 Cr	*A*		5.50	5.66	5.81	5.96	6.13	6.26	6.39	6.52	6.63	6.72	6.78	6.85	6.90
	B	0	0.86	1.56	2.30	3.08	3.90	4.73	5.60	6.49	7.40	8.31	9.20	10.11	11.01
25 Cr-20 Ni	*A*		7.76	7.92	8.08	8.22	8.38	8.52	8.68	8.81	8.02	9.00	9.08	9.12	9.18
	B	0	1.21	2.18	3.20	4.24	5.33	6.44	7.60	8.78	9.95	11.12	12.31	13.46	14.65
Monel 67: Ni-30 Cu	*A*		7.84	8.02	8.20	8.40	8.58	8.78	8.96	9.16	9.34	9.52	9.70	9.88	10.04
	B	0	1.22	2.21	3.25	4.33	5.46	6.64	7.85	9.12	10.42	11.77	13.15	14.58	16.02
Monel 66: Ni-29 CuAl	*A*		7.48	7.68	7.90	8.09	8.30	8.50	8.70	8.90	9.10	9.30	9.50	9.70	9.89
	B	0	1.17	2.12	3.13	4.17	5.28	6.43	7.62	8.86	10.16	11.50	13.00	14.32	15.78
Aluminum	*A*		12.95	13.28	13.60	13.90	14.20								
	B	0	2.00	3.66	5.39	7.17	9.03								
Gray cast iron	*A*		5.75	5.93	6.10	6.28	6.47	6.65	6.83	7.00	7.19				
	B	0	0.90	1.64	2.42	3.24	4.11	5.03	5.98	6.97	8.02				
Bronze	*A*		10.03	10.12	10.23	10.32	10.44	10.52	10.62	10.72	10.80	10.90	11.00		
	B	0	1.56	2.79	4.05	5.33	6.64	7.95	9.30	10.68	12.05	13.47	14.92		
Brass	*A*		9.76	10.00	10.23	10.47	10.69	10.92	11.16	11.40	11.63	11.85	12.09		
	B	0	1.52	2.76	4.05	5.40	6.80	8.26	9.78	11.35	12.98	14.65	16.39		
Wrought iron	*A*		7.32	7.48	7.61	7.73	7.88	8.01	8.13	8.29	8.39				
	B	0	1.14	2.06	3.01	3.99	5.01	6.06	7.12	8.26	9.36				
Copper-nickel (70/30)	*A*		8.54	8.71	8.90										
	B	0	1.33	2.40	3.52										

A = mean coefficient of thermal expansion × 10^6, in/(in · °F) in going from 70°F (21°C) to indicated temperature.
B = linear thermal expansion, in/100 ft in going from 70°F (21°C) to indicate temperature.
Multiply values of *A* shown by 1.8 to obtain coefficient of expansions in cm/(cm · °C).
Multiply values of *B* shown by 8.33 to obtain linear expansion in cm per 100 m.
SOURCE: ANSI B31.1.

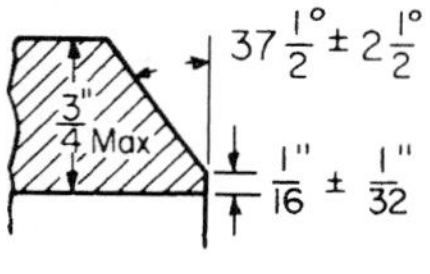

Figure 6.7.15 Recommended end preparation for pipe wall thickness of ¾ in or less.

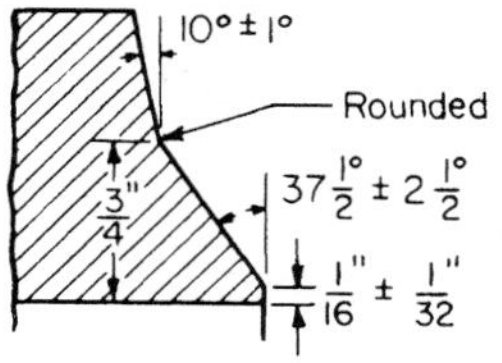

Figure 6.7.16 Recommended end preparation for pipe wall thickness greater than ¾ in.

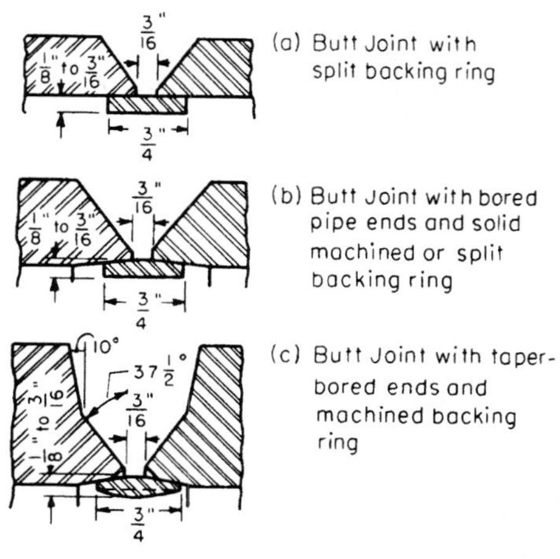

Figure 6.7.17 Recommended backing ring types. (*a*) Butt joint with split backing ring; (*b*) butt joint with bored pipe ends and solid machined or split backing ring; (*c*) butt joint with taper-bore ends and machined backing ring.

High-alloy steels containing the martensitic phase require preheating to 450°F (232°C) minimum; preheating is a matter of agreement between the purchaser and contractor in the case of welding high-alloy ferritic steels (ASTM A240 and A268). The possible advantages of preheat have not been established in the case of welding high-alloy austenitic steels, and for this reason the Code for Pressure Piping states that preheat is optional for these materials.

Welding procedure varies with material and welding process. In general, the pipe ends must be cleaned of oil or grease, and excessive amounts of scale or rust should be removed. The size and type of welding rod must be stated; the number of layers or passes is determined by the thickness of the pieces being joined. All slag or flux remaining on any bead of welding must be removed before laying down the next successive bead; any cracks or blowholes that appear on the surface of any bead must be chipped or ground away before the next bead of weld material is deposited. Throughout the welding process, it is essential that the minimum specified preheat temperature be maintained.

Stress Relieving Welded joints in all carbon-steel material whose thickness is ¾ in (1.91 cm) or greater must be stress-relieved at a temperature of 1,100°F (593°C) or over for a period of time proportioned on the basis of at least 1 h/in of pipe-wall thickness (but in no case less than ½ h) and then allowed to cool slowly (generally under a blanket) and uniformly. No stress relief is required for joints in carbon-steel piping whose wall thickness is less than ¾ in.

Welded joints in alloy steels with a wall thickness of ½ in (1.27 cm) or greater, having a chromium content not exceeding ¾ percent, and low-alloy steels with a total alloy content not exceeding 2 percent require stress-relieving at a temperature of 1,200°F (649°C) or over for a period of time proportioned on the basis of at least 1 h/in (0.4 h/cm) of wall thickness, but in no case less than ½ h.

Welded joints in alloy steels having a chromium content exceeding ¾ percent, or a total alloy content exceeding 2 percent, except high-alloy ferritic (ASTM A240, A268) and austenitic steels, regardless of wall thickness, require stress relief at a temperature of 1,200°F or over for a period of time proportioned on the basis of at least 1 h/in of wall thickness, but in no case less than ½ h. Stress relief of high-alloy ferritic steel (A240, A268) and austenitic steels is not required but may be performed as agreed upon by purchaser and contractor. In welds between austenitic and ferritic materials, stress relieving is optional and, if used, shall be a matter of agreement between the purchaser and contractor. Because of the difference between the coefficients of thermal expansion of the two dissimilar materials, careful consideration should be given to the selection of a heat treatment, if any, that will be beneficial to the welded joint.

Graphitization is precipitation of carbon at the grain boundaries in the heat-affected zone during the welding process. Such a phenomenon occurs when some metals operate at high temperatures for extended periods. It has been observed particularly in carbon-molybdenum steels that operate at 900°F (482°C) or higher.

Graphitization does not generally occur in carbon-molybdenum steels with over 1 percent molybdenum. It also has generally not occurred in the chromium-molybdenum low-alloy steels operated at temperatures between 900°F (482°C) and 1,050°F (566°C). Where graphitization has occurred, the two most commonly used methods for rehabilitation of the pipe are (1) gouging out the heat-affected zone of the weld deposit and rewelding the area with electrodes depositing carbon-molybdenum weld metal, followed by a stabilization heat treatment at 1,300°F (704°C) for 4 h, or (2) solution annealing the weld joints at 1,800°F (982°C), followed by a stabilization heat treatment.

Pipe Supports

The Code for Pressure Piping includes many types of supports and gives directions for their application. A proper pipe support must have a strong rigid base properly supported, and an adjustable roll construction which will maintain the alignment in any direction. It is important to avoid friction caused by the movement of the pipe in the support and to have all parts of sufficient strength to maintain alignment at all times. Wire hangers, band iron hangers, wooden hangers, hangers made from small pipe, and hangers having one vertical pipe support do not maintain alignment.

The direction of expansion in a pipe run can be predetermined by anchoring one end, both ends, or the middle. **Anchors** must be firmly fastened to a rigid and heavy part of the power-plant structure, and must also be securely fastened to the pipe; otherwise the equipment for absorbing expansion is useless, and severe stresses may be thrown on parts of the piping system. Some methods of support are shown in Figs. 6.7.18 and 6.7.19. Welded steel brackets (Fig. 6.7.18*a*) are available in light, medium, and heavy weights. Many types of supports can be mounted on these brackets, such as the anchor chair shown on the bracket at (*a*), pipe roller supports of the type at (*c*), pipe roll stands of various types such as shown in Fig. 6.7.19, pipe seats, etc. Figure 6.7.18*b* illustrates one of the many types of adjustable ring hangers in use. The split ring hanger can be applied after the pipeline is in place. At (*c*) in Fig. 6.7.18 is shown a spring cushion pipe roll hanger recommended for service where constant support is required and compensation must be made for movement of the piping. The support afforded by the hanger of Fig. 6.7.18*c* is constant only in the sense that some degree of support is always present. It might be more appropriately termed a variable-support device. The springs provide an efficient means of absorbing the vibration. Figure 6.7.18*d* shows one of the many types of pipe saddle supports available. Figure 6.7.19 shows a cast-iron pipe roll stand designed for cases where vertical adjustment is not necessary but where provision must be made for expansion and contraction of the pipeline. Several designs of such stands with provision for vertical adjustment and of the same general dimensions are also available. One type of cast-iron roll and plate, illustrated in Fig. 6.7.19, provides for expansion and contraction where vertical adjustment is not necessary. If necessary, the baseplate can be raised or lowered by use of shims. Detailed information and dimensions of a great variety of pipe supports can be found in manufacturers' catalogs.

In supporting a high-temperature piping system, it is necessary to provide for expansion and contraction due to cyclic changes. It is often possible to find a point of zero movement along the run of a long line and to support a considerable portion of the total load by a rigid hanger or support of the type shown in Figs. 6.7.18 and 6.7.19. However, for other portions of the run, some form of spring support is often indicated. For relatively light lines, which are not subjected

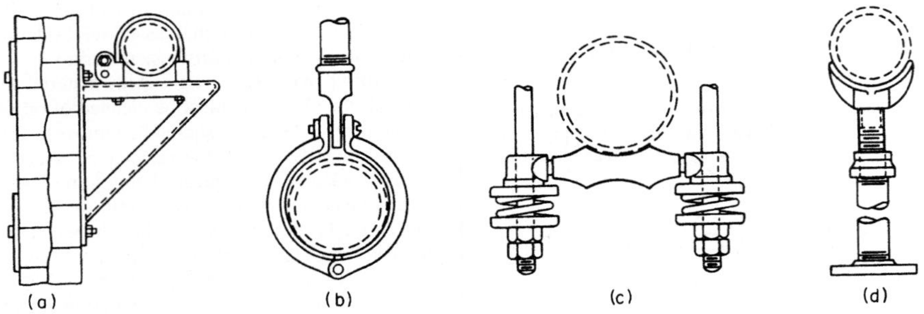

Figure 6.7.18 Methods of supporting pipes.

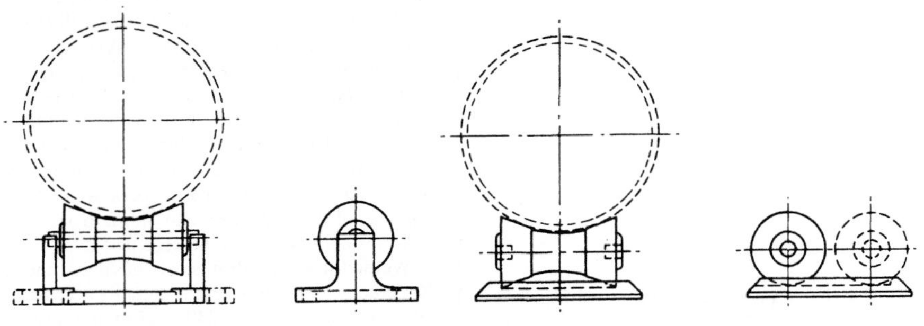

Figure 6.7.19 Pipe supports on cast-iron rolls.

to excessive movements from hot to cold positions, a variable spring hanger will frequently suffice; for heavy lines, or those in which expansion movements are great, it is advisable to use constant support of counterweighted hangers so that transfer of weight to other hangers or equipment connections is prevented. Parts (*a*) and (*b*) of Fig. 6.7.20 indicate, respectively, a horizontal and vertical run of piping supported

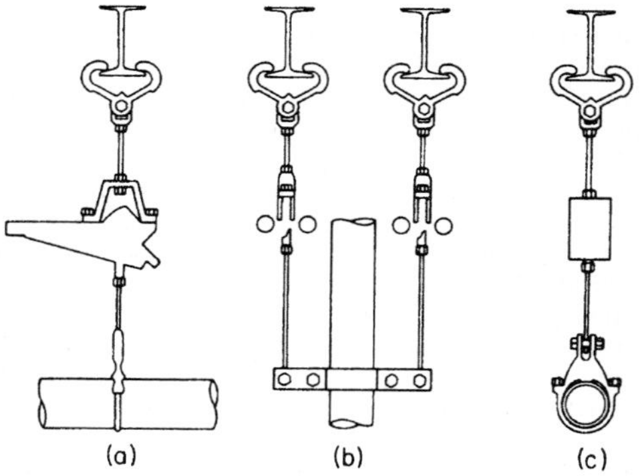

Figure 6.7.20 Constant support and variable-spring hangers.

by a **constant-support hanger.** Figure 6.7.20*c* and 6.7.21*a* indicate horizontal runs supported by **variable-spring hangers.** Figure 6.7.21*b* shows a riser supported by a variable spring beneath a base elbow. Figure 6.7.21*c* indicates a **sway brace** that is used to control vibration and undesirable movement in a piping system.

The principal supports utilized for the support of critical piping involve constant-support hangers, variable-spring hangers, rigid hangers, and restraints.

Constant-Support Hangers This type of hanger provides a constant supporting force for the piping system throughout its full range of vertical pipe movement. This is accomplished through use of a spring coil working in conjunction with a lever in such a way that the spring force times its distance to the lever pivot is always equal to the pipe load times its distance to the lever pivot. This type of support is "thermally invisible," as the supporting force equals the pipe weight throughout its entire expansion or contraction cycles.

These hangers are used on systems or at locations where stresses are considered critical. Pipe weight reactions or transfer of loads are not imposed in the system or connections with this type of device.

As the load is considered constant with the unit in travel, readings from inspections are based on travel. The readings are taken from the position of an indicator and its relation to the numbers on the travel scale. The scale is divided into 10 divisions. H or high on the scale is equal to (0.0), M or midway on the scale is equal to (5.0), and L or low on the scale is equal to 10.

The design settings were obtained from the position of the factory-installed buttons that are placed adjacent to the travel scale.

"Perfect" readings would be if the indicator were to line up with the white button (cold) and the red (hot); however, this is rarely the case. Generally, readings are considered acceptable and not noteworthy as long as they reflect movement consistent with design in both direction and length. A general rule is that when the hot setting is higher than the cold setting, then movement is down from cold to hot. If the cold setting is higher, movement is up.

The following terms are normally applied to these devices:

Actual travel: Anticipated movement of the pipe from design. The hot and cold position stickers are a function of this movement.

Total travel: The maximum movement a support can accept without danger of topping or bottoming out. The scale from H (0.0) to L (10.0) is a function of this.

Topped out: The indicator is above the high point and in contact with the end of the slot. This condition means the support is unloaded or is in the process of unloading.

Bottomed out: The indicator is below the (10.0) point and in contact with the end of the travel slot; the support is overloaded.

Variable-Spring Hangers These devices are installed at locations where stresses are not considered to be critical or where movement and economics permit their use.

The inherent characteristics of a variable are such that the supporting force varies with the spring's deflection. Movement of the pipe causes the spring to extend or compress. This results in a change or variance in the supporting force. Since the weight of the pipe is the same in either condition, hot or cold, the variation results in pipe weight transfer to equipment and adjacent hangers and consequently additional stresses in the piping system. The effects of this variation are usually considered during the original design.

In addition, as it is desirable to support the actual weight of the pipe when the system is hot, when the stresses tend to become most critical, the hot load is the dead weight of the pipe. The cold load is actually under- or oversupporting the pipe, depending on the movement from cold to hot.

The general rule for determining movement is similar to that of constant supports. If the hot load is higher than the cold, then pipe movement is down from cold to hot. If the cold load is higher, then movement is up.

Unlike constant supports, the readings from variables are measured in pounds. The readings are taken by noting the position of the indicator relative to a load scale that is adjacent to the travel slot.

The **distance between supports** will vary with the kind of piping and the number of valves and fittings. Supports should be provided near changes in direction, branch lines, and particularly near valves. The weight of piping must not be carried through valve bodies. In establishing the location of pipe supports, the designer should be guided by two requirements: (1) the horizontal span must not be so long that sag in the pipe will impose an excessive stress in the pipe wall and (2) the pipeline must be pitched downward so that the outlet of each span is lower than maximum sag in the span. Otherwise entrapped water can result in severe water hammer and pipe swings, particularly during plant start-up of steam piping.

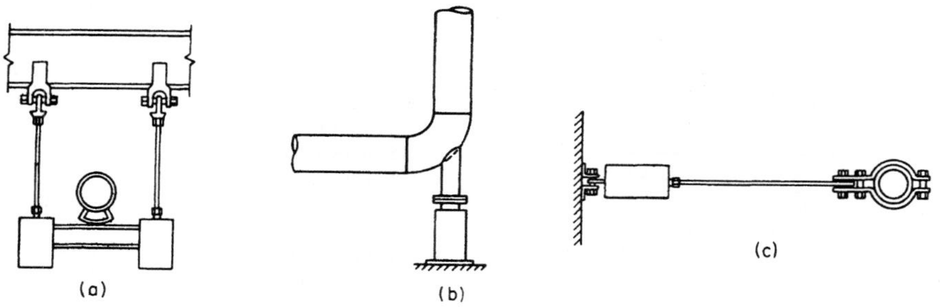

Figure 6.7.21 Spring hangers and sway brace.

Table 6.7.26 Maximum Spacing of Pipe Supports at 750°F (399°C)*

Nominal pipe size, in	1	1½	2	3	4	6	8	10	12	14	16	18	20	24
Maximum span, ft	7	9	10	12	14	17	19	22	23	25	27	28	30	32
Maximum span, m	2.13	2.74	3.05	3.66	4.27	5.18	5.79	6.71	7.01	7.62	8.23	8.53	9.14	9.75

* This tabulation assumes that concentrated loads, such as valves and flanges, are separately supported. Spacing is based on a combined bending and shear stress of 1,500 lb/in² when pipe is filled with water; under this condition, sag in pipeline between supports will be approximately 0.1 in.

Fabrication and installation practices are provided in MSS Standard Practice SP-89. Table 6.7.26 lists spacing for standard-weight pipe supports.

Pipe Insulation

(See Secs. 2 and 4 for heat-transmission data.)

The value of a steam-pipe covering is measured by its ability to reduce heat losses. This might range from 50 percent for small, low-temperature lines to 90 percent for large, high-temperature lines. Many **pipe-insulating materials** are available: 85 percent magnesia, foam glass, calcium silicate, and various forms of diatomaceous earths. Some of these materials are suited for relatively low temperatures only, others are best suited for high temperatures, and still others are suitable over a considerable temperature range.

Pipe insulation is applied in molded sections 3 ft long. For high-temperature work, the insulation is applied in at least two layers with the joints staggered so as to prevent a direct channel for heat loss. Because of its maximum-temperature limitation of about 600°F (316°C), 85 percent magnesia is used as the second layer with a high-temperature-resistant material placed in direct contact with the pipe. The molded insulation is fastened securely in place with copper or galvanized wire and is then given a surface finish; indoor pipes are first sheathed with resin paper and covered with canvas, either pasted or sewed; outdoor pipes may be weather-protected by a coating of asphaltic-type waterproofing compound, they may be sheathed and canvased and then given a weatherproof surface, so they may be encased in metallic (steel or aluminum) jackets.

The heat loss from an insulated pipe appears in three phases: heat passes by *conduction* through the metallic pipe walls and through the insulating material; it then is dissipated from the outdoor surface of the insulation by *convection* and by *radiation*. Extremely accurate calculations must also take into account the temperature drop by convection through the film on the inside surface of the pipe. The task of accurately calculating heat losses is somewhat tedious, since the convection and radiation losses are related to the surface temperature (outside of insulation), which is unknown until conduction losses are balanced against surface losses.

For combined convection and radiation coefficients for bare pipes, and all necessary formulas to permit trial-and-error calculations, see Sec. 2. Insulation manufacturers publish data which give heat losses for wide ranges of pipe size and temperature.

Identification of Piping

The American National Standards Institute has approved a **Scheme for the Identification of Piping Systems** (ANSI A13.1). This scheme is limited to the identification of piping systems in industrial plants, not including pipes buried in the ground, and electric conduits. Fittings, valves, and pipe coverings are included, but not supports, brackets, or other accessories.

Classification by Color All piping systems are classified by the nature of the material carried. Each piping system is placed, by the nature of its contents, in the following classifications:

Class	*Color*
F—Fire-protection equipment	Red
D—Dangerous materials	Yellow (or orange)
S—Safe materials	Green (or the achromatic colors, white, black, gray, or aluminum)
P—Protective materials	Bright blue
V—Extra valuable materials	Deep purple

Method of Identification At conspicuous places throughout a piping system, color bands should be painted on the pipes to designate to which one of the five main classes it belongs. If desired, the entire length of the piping system may be painted the main classification color.

Further, the actual contents of a piping system may be indicated by, preferably, a stenciled legend of standard size giving the name of the contents in full or abbreviated form. These legends should be placed on the color bands. The identification scheme may be extended by the use of colored stripes placed at the edges of the colored bands.

The bands, legends, and stripes should be placed at intervals throughout the piping system, preferably adjacent to valves and fittings to ensure ready recognition during operation, repairs, and at times of emergency.

A recommended classification, under this color scheme of materials carried in pipes, includes, as dangerous, combustible gases and oils, hot water and steam above atmospheric pressure; as safe, compressed air, cold water, and steam under vacuum.

Pressure Hose

Hose with durable rubber lining may be obtained to withstand any needed pressure. If the rubber compound is properly made, the life of a hose will be 7 to 10 years or more.

American National Fire-Hose Coupling Screw Thread (ANSI B1.20-7) This standard is intended to cover the threaded part of fire-hose couplings, hydrant outlets, standpipe connections, and at other special fittings on fire lines, where fittings of the nominal diameters given in Table 6.7.27 are used. It also includes the limiting dimensions of the field inspection gages. The American National Standard form of thread must be used.

Table 6.7.27 Dimensions of Standard Fire-Hose Couplings
(All dimensions in inches. Letters refer to Fig. 6.7.22)

Inside diam, *C*	Diam of thread, *D*	No. of threads per inch	*L*	*I*	*H*	*J*	*T*
2½	3$\frac{1}{16}$	7½	1	¼	$\frac{15}{16}$	$\frac{3}{16}$	$\frac{11}{8}$
3	3⅝	6	1⅛	$\frac{5}{16}$	1$\frac{1}{16}$	¼	$\frac{13}{8}$
3½	4¼	6	1⅛	$\frac{5}{16}$	1$\frac{1}{16}$	¼	$\frac{13}{8}$
4½	5¾	4	1¼	$\frac{7}{16}$	1$\frac{3}{16}$	⅜	$\frac{15}{8}$

SOURCE: ANSI B1.20.7

American National Standard Hose-Coupling Screw Threads (ANSI **B 1.20.7)** These standards apply to the threaded parts of hose couplings, valves, nozzles, and all other fittings used in direct connection with hose intended for fire protection or for domestic, industrial, or general service in nominal sizes given in Table 6.7.28. The American National Standard thread form is used. This coupling is similar in design to the fire-hose couplings illustrated in Fig. 6.7.22.

Flexible metal hose and tubing are available for a wide range of conditions of temperature, pressure, vibration, and corrosion, and are made in two basic constructions, corrugated or interlocked, and in either bronze or steel.

The corrugated type (Fig. 6.7.23) may have either annular or helical corrugated formations, usually covered with metal braid, and is adapted to high-pressure high-temperature leak-proof service. Some typical applications include diesel-engine exhaust hose, reciprocating flexible connections, loading and unloading hose, saturated and superheated steam lines, lubricating lines, gas and oil lines, vibration connections, etc.

Table 6.7.28 Dimensions of Standard Hose Couplings
(All dimensions in inches. Letters refer to Fig. 6.7.22)

Service and nominal size	Inside diam, *C*	Diam of thread, *D*	No. of threads per inch	*L*	*I*	*H*	*T*
Garden:							
½, ⅝, ¾	25/32	1 1/16	11½	9/16	1/8	17/32	3/8
Chemical:							
¾, 1	1 1/32	1 3/8	8	5/8	5/32	19/32	15/32
Fire:							
1½	1 17/32	2	9	5/8	5/32	19/32	15/32
Other connections:							
½	17/32	3/16	14	1/2	1/8	15/32	5/32
¾	25/32	1 1/32	14	9/16	1/8	17/32	3/8
1	1 1/32	1 9/32	11½	9/16	5/32	17/32	3/8
1¼	1 9/32	1 5/8	11½	5/8	5/32	19/32	15/32
1½	1 17/32	1 7/8	11½	5/8	5/32	19/32	15/32
2	2 1/32	2 11/32	11½	3/4	3/16	23/32	19/32

SOURCE: ANSI B1.20.7.

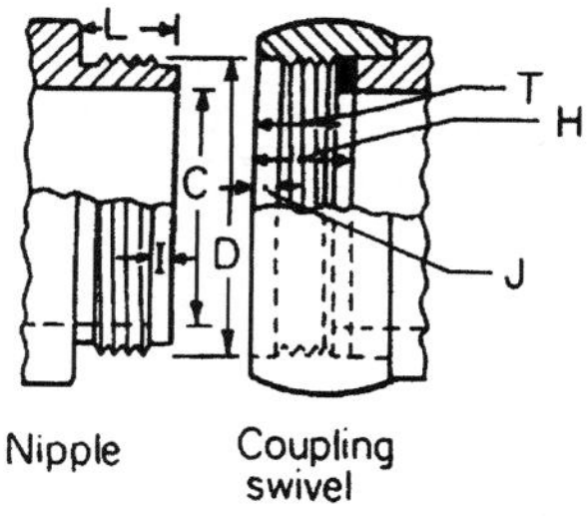

Figure 6.7.22 Typical form of standard coupling.

The interlocked type is made in several ways; the fully interlocked type is illustrated in Fig. 6.7.24. Typical applications include wiring conduit, cable armor, decorative wiring covering, dust-collective tubing, grease and oil connections, flexible spouts, and moderate-pressure oil lines.

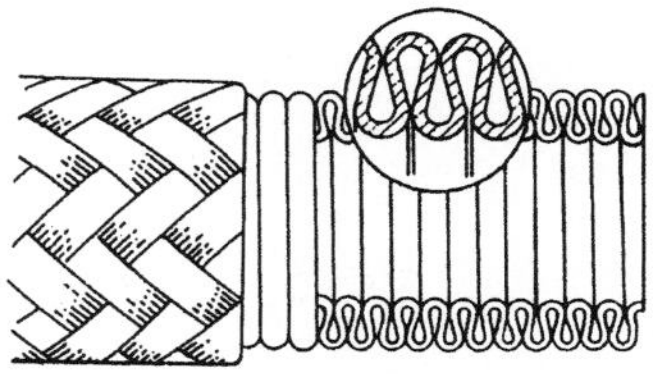

Figure 6.7.23 Flexible metal hose.

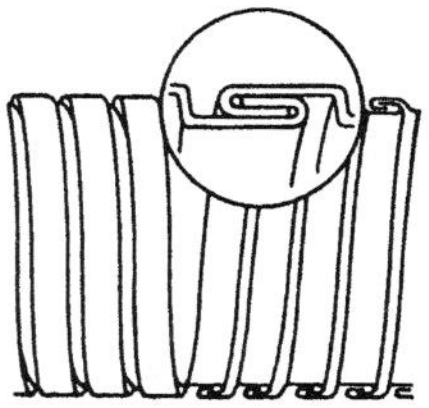

Figure 6.7.24 Interlocked flexible metal hose.

Standard couplings and fittings can be attached to flexible metal hose or tubing by various methods such as brazing or welding. Each type of hose construction has limits of service use and proved application usages. Information and recommendations as to the type and size to use under any given conditions should be obtained from the manufacturers.

Section 7
Power Generation

BY

PATRICK E. PHELAN *Professor of Mechanical and Aerospace Engineering, Arizona State University*
E. A. AVALLONE *Ardsley, NY*
EZRA S. KRENDEL *University of Pennsylvania*
ASFAW BEYENE *Professor of Mechanical Engineering, San Diego State University*
RAMA RAMAKUMAR *PSO/Albrecht Naeter Professor and Director, Engineering Energy Laboratory, Oklahoma State University*
CHARLES P. BUTTERFIELD *Chief Engineer, National Wind Technology Center, National Renewable Energy Laboratory*
EDUARD MULJADI *Senior Engineer, National Wind Technology Center, National Renewable Energy Laboratory*
THOMAS BUTCHER *Brookhaven National Laboratory*
ISRAEL URIELI *Professor Emeritus, Mechanical Engineering, Ohio University*
WILLIAM M. WOREK *Professor of Mechanical Engineering, Texas A&M University—Kingsville*
ERICH A. FARBER *Distinguished Professor Emeritus, Mechanical Engineering, University of Florida*
KENNETH A. PHAIR *Global Power Solutions LLC, Golden, Colorado*
M. ZENOUZI *Professor of Mechanical Engineering and Technology, Wentworth Institute of Technology*
ROY MCALISTER *Founder and President, American Hydrogen Association*
M. ZEBARJADI *Professor of Electrical & Computer Engineering and Materials Science Engineering, University of Virginia*
TRUNG NGUYEN *Professor of Chemical and Petroleum Engineering, University of Kansas*
ANKUR JAIN *Associate Professor of Mechanical and Aerospace Engineering, University of Texas at Arlington*
JOHN B. KITTO, JR. *Consulting Engineer, Retired, The Babcock and Wilcox Company*
IVO A. GARZA *Manager/Consultant, Sargent and Lundy*
FREDERICK G. BAILY *Retired Engineer/Manager, General Electric, Schenectady, NY*
HEINZ P. BLOCK *Process Machinery Consulting, Houston, TX and Westminster, CO*
WILLIAM J. BOW *Late Director, Heat Transfer Products Dept., Foster Wheeler Energy Corp.*
DONALD E. BOLT *Late Engineering Manager, Heat Transfer Products Dept., Foster Wheeler Energy Corp.*
CHARLES F. BOWMAN *Consultant, President, Chuck Bowman Associates, Inc.*
DENNIS N. ASSANIS *President, University of Delaware*
DAVID E. COLE *Chairman Emeritus, Center for Automotive Research (CAR)*
TIMOTHY J. JACOBS *Associate Professor, Department of Mechanical Engineering, Texas A&M University*
SITIRIOS MAMALIS *Assistant Professor, Mechanical Engineering, Stony Brook University*
DONALD J. PATTERSON *Emeritus Professor, Mechanical Engineering, University of Michigan*
JOHN H. LEWIS *Late Engineering Consultant; Formerly, Engineering Staff, Pratt & Whitney Division, United Technologies Corp.*
WILLIAM H. DAY *President, Longview Energy Associates LLC*
ANDREW C. KLEIN *Professor, Nuclear Science and Engineering, Oregon State University*
BRYAN A. ERLER *President, Erler Engineering Ltd.*
JOHN M. (JACK) TUOHY, JR. *Energy Industry Consultant*
ROBERT D. STEELE *Project Manager, Voith Siemens Hydro Power Generation, Inc.*
JULIO C. GUERRERO *Principal, Exponent Inc.*
JUNG-YANG SAN *Professor, Mechanical Engineering Department, National Chung Hsing University, Taiwan, Republic of China*

7.1 SOURCES OF ENERGY
Contributors are shown at the head of each category.

7.2 STEAM BOILERS
by John B. Kitto, Jr.

7.3 EXTERNAL COMBUSTION HEAT ENGINES
Revised by Ivo A. Garza

7.4 STEAM TURBINES
by Frederick G. Baily, Revised by Heinz P. Block

7.5 POWER PLANT HEAT EXCHANGERS
by William J. Bow, Assisted by Donald E. Bolt and Charles F. Bowman

7.6 INTERNAL COMBUSTION ENGINES
by Dennis N. Assanis, David E. Cole, Timothy J. Jacobs, Sitirios Mamalis, and Donald J. Patterson

7.7 GAS TURBINES
by John H. Lewis and William H. Day

7.8 NUCLEAR POWER
by Andrew C. Klein, Bryan A. Erler, and John M. (Jack) Tuohy, Jr.

7.9 HYDRAULIC TURBINES
by Robert D. Steele

7.10 FRACKING
by Julio C. Guerrero

7.11 SECOND-LAW ANALYSIS AND AVAILABLE ENERGY
by Jung-Yang San

7.1 SOURCES OF ENERGY

Contributors are shown at the head of each category.

REFERENCES: Latest available published data from the following: *International Energy Outlook 2016*, US Energy Information Administration. *The Oil & Gas Journal*.

INTRODUCTION

by Patrick E. Phelan

The world is in the midst of significant changes in energy sources, caused in part by increasing concern over global climate change, the emergence of new natural gas resources and production technologies, chiefly the use of hydraulic fracturing ("fracking") to produce natural gas from shale formations, and finally the decreasing cost of renewable energy technologies, chiefly wind and solar photovoltaics. Renewable energy is the fastest-growing energy resource, with projected average growth of 2.6 percent/year between 2012 and 2040 according to the US Energy Information Administration's (EIA's) *2016 International Energy Outlook*. Nuclear energy is the second fastest-growing energy source, at 2.3 percent/year. Meanwhile coal is projected to be the world's slowest growing energy source at only 0.6 percent/year, compared with growth of 1.9 percent/year in natural gas consumption. Electricity remains the fastest-growing energy end use, and the largest increase in electricity generation is forecast to occur in developing countries [i.e., those that do not belong to the Organization for Economic Cooperation and Development (OECD)].

Liquid fuels, mostly from fossil fuels and primarily for transportation and industrial applications, still provide about one third of global energy needs. The emergence of electric vehicles, including plug-in hybrid vehicles, will eventually reduce the predominance of liquid fuels for transportation. But, the IEA forecasts that electricity will consume only 1 percent of the energy used for light-duty vehicles in 2040. Similar to the growth in electricity consumption, the vast majority of the increase in transportation energy is projected to occur in non-OECD countries.

Fossil Fuels

PETROLEUM

Estimates of proved crude oil reserves in all countries of the world are published by *The Oil and Gas Journal*. Table 7.1.1 shows the estimated proved oil reserves for the top 20 countries in the world, as of Jan 1, 2017. "Proved reserves," as defined by the *Society of Petroleum Engineers*, are known reserves (developed or undeveloped) that can be reasonably estimated to be recoverable under existing economic and regulatory conditions. Interestingly, a 2016 analysis by the Norwegian consultancy Rystad Energy estimated that the USA now includes more recoverable oil reserves, including yet undiscovered fields, than any other country, with more than 50 percent of the remaining reserves considered "unconventional shale oil."

NATURAL GAS

The Oil and Gas Journal also provides estimates of natural gas reserves around the world. Table 7.1.2 lists the top 20 countries for proved natural gas reserves. The world total of proved reserves declined slightly from Jan 1, 2016 (6,938,623 billion cubic feet, or bcf) to 2017 (6,895,740 bcf). In the USA, according to the EIA proved natural gas reserves decreased 16.6 percent from 2014 to 2015, including decreases in each of the top-five gas-producing regions (Texas, North Dakota, Gulf of Mexico, California, and Alaska), in large part because of a 42 percent drop in the price of natural gas over the same period.

COAL

Information on worldwide coal reserves is reported in the EIA's *International Energy Outlook*. The world recoverable coal reserves totaled 978 billion short tons in 2012, and considering annual coal production of 8.898 billion short tons yields a reserves-to-production ratio of 110 years. Seventy-two percent of the world's coal production comes from only five regions: the USA (26 percent), Russia (18 percent), China (13 percent), non-OECD Europe and Eurasia outside of Russia (10 percent), and Australia/New Zealand (10 percent). It is likely that improved mining technologies and geological assessments will lead to an increase in estimated recoverable coal reserves.

Table 7.1.1 Top Countries by Proved Oil Reserves (1,000 bbl)

Venezuela	300,878,000
Saudi Arabia	266,455,000
Canada	169,708,702
Iran	158,400,000
Iraq	142,503,000
Kuwait	101,500,000
United Arab Emirates	97,800,000
Russia	80,000,000
Libya	48,363,000
Nigeria	37,062,000
United States	35,271,600
Kazakhstan	30,000,000
China	25,620,000
Qatar	25,244,000
Brazil	12,999,800
Algeria	12,200,000
Angola	8,273,000
Ecuador	8,273,000
Mexico	7,640,000
Azerbaijan	7,000,000

Table 7.1.2 Top Countries by Proved Gas Reserves (bcf)

Russia	1,688,228
Iran	1,183,019
Qatar	858,098
Saudi Arabia	303,284
United States	293,120
Turkmenistan	265,000
United Arab Emirates	215,098
Venezuela	201,343
Nigeria	186,610
China	183,419
Algeria	159,054
Iraq	111,522
Mozambique	100,000
Indonesia	98,000
Kazakhstan	85,000
Egypt	77,200
Canada	77,065
Australia	70,230
Norway	65,543
Uzbekistan	65,000

Nuclear Fuels

by John M. (Jack) Tuohy, Jr.

URANIUM

Uranium reserves are reported by the US Energy Information Administration, the International Atomic Energy Agency (IAEA), the World Nuclear Association, and the International Energy Agency among others. Reserve estimates of uranium represent uranium resources that can be recovered economically from orebodies. Thus estimates are a function of deposits that are currently identified, the cost of recovery and the market price of uranium.

According to the Nuclear Energy Agency (NEA)[a] there is sufficient uranium to meet high case demand requirements out till 2035 while consuming less than 30 percent of 2015 resource estimates. At the beginning of 2015, world uranium production was 55,975 tU compared to world reactor requirements of 56,585 tU.

Natural Uranium, that is, uranium as it currently exists in ore deposits, is principally comprised of ^{238}U which has an abundance of 99.3 percent while ^{235}U is present at 0.7 percent. There are trace amounts of ^{234}U present. ^{238}U is "fissionable," that is, it can fission and release energy as part of the fission process, but it must absorb a high-energy neutron for fission to be energetically possible. ^{235}U on the other hand is not only "fissionable," but also "fissile." By "fissile" it is meant that a neutron with zero kinetic energy can trigger a fission. That is, the binding energy of the absorbed neutron alone is sufficient to split the nucleus.

Thus reactors need fissile fuel to operate. It is possible to operate a reactor using natural uranium with its small ^{235}U content. This has been demonstrated by using heavy water as a moderator and coolant on a commercial operation in Canada. Throughout the balance of the world, uranium enriched in the ^{235}U fraction is used in most practical reactors.

As noted above, ^{235}U is vital to the power industry. Nonetheless, ^{238}U is also quite essential. Within an operating reactor ^{238}U absorbs neutrons and becomes ^{239}U which beta decays twice and becomes ^{239}Pu which is fissile. Thus it is possible to increase the fissile fuel resources of the world by utilizing the more abundant ^{238}U isotope. In such a scenario ^{238}U is referred to as "fertile." A "breeder" reactor is one that generates more fissile fuel from fertile fuel than it consumes generating power for electric generation.

THORIUM

Although ^{235}U is the only naturally occurring fissile isotope, it was demonstrated that ^{238}U is also important because it is fertile.

Such is also the case for ^{232}Th which is more abundant in nature than uranium. Just as we saw for ^{238}U, the ^{232}Th isotope will absorb a neutron within an operating reactor and become ^{233}Th which undergoes two beta decays and becomes ^{233}U a "fissile" isotope. This extends the potential fissile fuel supply enormously.

The EIA estimates (based on 2008 data) uranium resources to be 102 million pounds of U_3O_8 recoverable at costs up to $30 per pound and 1,227 million pounds of U_3O_8 recoverable at costs under $100 per pound. These estimates vary with time as new orebodies are identified and as extraction technology advances. The length of time that the known uranium resources will meet demands for the resource is projected to be at least 135 years, at the 2014 level of uranium requirements.

[a]Uranium 2016: Resources, Production and Demand
A Joint Report by the Nuclear Energy Agency and the International Atomic Energy Agency
© OECD 2016 NEA No. 7301
NUCLEAR ENERGY AGENCY (NEA) ORGANISATION FOR ECONOMIC CO-OPERATION AND DEVELOPMENT (OECD)

Renewable Energy

Renewable energy consists of those resources that are replenishable on time scales short enough to be useful, that is, hours, days, weeks, months, and even years, but not on geological time scales like those required to replenish fossil fuels. Renewable energy resources largely consists of hydroelectric power, solar energy (photovoltaic and thermal), wind energy, biomass, kinetic energy from rivers and tides, energy from thermal and concentration gradients in the ocean and elsewhere, energy from waste streams such as municipal solid waste, and even animal and human muscle power. There are likely numerous other resources that not yet been exploited. Here renewable resources due only to hydroelectric, solar, and wind energy are described; the interested reader is encouraged to consult the *International Energy Outlook*, and corresponding websites at the Energy Information Administration (EIA) and the International Energy Agency (IEA) for further details on these and other renewable resources.

Hydroelectric Energy Hydropower is the predominant renewable energy resource around the world, and in 2012 contributed 77 percent of the world's total renewable energy generation of 4,727 billion kWh. Hydropower is projected to contribute 33 percent of the increase in renewable energy generation between 2012 and 2040, with the greatest increase anticipated for non-OECD Asia (growth of 2.2 percent/year).

Wind Energy Wind energy contributed 11 percent of the world's total renewable energy generation in 2012, and is projected to grow at an annual rate of 5.7 percent from 2012 to 2014. An annual growth rate of 7.7 percent is projected for non-OECD countries, and 4.5 percent for OECD countries. The average annual wind capacity factor—the ratio of generated power to capacity—varies widely among regions due to factors such as differences in operating costs and load patterns, and is approximately 27 percent for the USA.

Solar Energy In 2012, solar energy contributed only 2.2 percent of the total renewable energy generation, but it is the world's fastest-growing source of renewable energy, with average growth of 8.3 percent/year. The majority of solar power is generated by solar photovoltaic modules, with some contribution from thermal concentrating solar power (CSP). The growth rate in non-OECD countries is projected to be 15.7 percent/year from 2012 to 2040, while that in OECD countries is 4.5 percent/year (the same as for wind energy). Ambitious solar targets by India and China contribute significantly to the difference in growth rates between OECD and non-OECD nations.

ALTERNATIVE ENERGY, RENEWABLE ENERGY, AND ENERGY CONVERSION: AN INTRODUCTION

by E. A. Avallone

Many sources of raw energy have been proposed or used for the generation of power. Only a few sources—fossil fuels, nuclear fission, and elevated water—are dominant in practical applications today.

A more complete list of sources would include fossil fuels (coal, petroleum, natural gas); nuclear (fission); wood and vegetation; elevated water supply; solar; winds; tides; waves; geothermal; muscles (human, animal); industrial, agricultural, and domestic wastes; oceanic thermal gradients; oceanic currents. There are others.

Historically, wood, muscles, elevated water, and wind were prominent. These sources were superseded in the industrial era by fossil fuels, with nuclear energy the most recent addition. This dominance rests in the suitability of the thermal sources for practical stationary and transportation power plants. Features of acceptability include reliability, flexibility, portability, maneuverability, size, bulk, weight, efficiency, economy, maintenance, and costs. The plant for transportation service must be self-contained. For stationary service there is wider latitude for choice.

The dominant end product, especially for stationary applications, is electricity, because of its favorable distribution and control features. Reliability and continuity of service consequently dictate the need for reserve, alternate, and interconnection supports. Pumped storage, coal piles, and tanks of liquid and gaseous fuels, for example, offer the necessary continuity, flexibility, and reliability.

Raw energy sources, other than fuels (fossil and nuclear) and elevated water, are particularly deficient in this storage aspect. For example, wind power is best for jobs that can wait for the wind, for example, pumping water or grinding grain. Solar power, to avoid foul weather and the darkness of night, could call for desert locations or extraterrestrial satellites.

The International Energy Agency (IEA) has predicted a five-year growth (2015-2021) for renewables due to strong policy support in key countries and sharp cost reductions. In 2015, renewables have passed coal to be the largest source of installed power capacity in the world. The latest edition of IEA's *Medium Renewable Market Report* predicts renewables growing at 13 percent more from 2015 to 2021 than originally forecasted. This growth is due in large part to strong policies in the United States, China, India, and Mexico. For the period 2015 to 2021, the cost of solar photovoltaic (PV) is expected to drop by 25 percent and onshore wind by 15 percent.

The year 2015, led by wind and solar saw more than half of the new power capacity installed worldwide being renewable, reaching 153 GW. Of this amount, 66 GW was due to wind and 49 GW was due to solar PV.

Despite such limitations an energy-intensive society can expect to see increasing efforts to harness more and new sources of energy. Several of these topics are treated in the following pages to show the factual and technical progress that has been made to adapt sources to practicality.

References

1. International Energy Agency, *Medium-Term Renewable Energy Market Report*, "Strong Future Forecast for Renewable Energy."

2. O. Erdinc and M. Uzunoglu, "A new perspective in optimum sizing of hybrid renewable energy systems: Consideration of component performance degradation issue," *Int. J. of Hydrogen Energy,* 37 (2012) 10,479-10,488.

3. R.-D. Chang, J. Zuo, Z.-Y. Zhao, G. Zillante, X.-L. Gan and V. Soebato, "Evolving theories of sustainability and firms: History, future directions and implications for renewable energy research," *Renewable and Sustainable Energy Reviews*, 72 (2017), 48-56.

4. P. Hernandez and P. Kenny, "From new energy to zero energy buildings: Defining life cycle zero energy buildings (LC-ZEB)," *Energy and Buildings*, 42 (2010) 815-821.

5. H. Djamila, "Indoor thermal comfort predictions: Selected issues and trends," *Renewable and Sustainable Energy Reviews*, 74 (2017) 569-580.

MUSCLE-GENERATED POWER

by Ezra S. Krendel, References updated by Patrick E. Phelan

REFERENCES: Franco et al., 2016, "Human-powered press for producing straw bales for use in construction during post-emergency conditions," Biosystems Engineering **150**, pp. 170-181. Brooks, 2012, "Bioenergentics of exercising humans," Comprehensive Physiology **2**, pp. 537-562. Wilson, "Bicycling Science," 3d ed., MIT Press, 2004. Harrison, Maximizing Human Power Output by Suitable Selection of Motion Cycle and Load, *Human Factors,* **12**, 1970. Krendel, Design Requirements for Man-Generated Power, *Ergonomics,* **3**, 1960. Wilkie, Man as a Source of Mechanical Power, *Ergonomics,* **3**, 1960. Brody, "Bioenergetics and Growth," Reinhold, 1945.

The use of human muscles to generate work will be examined from two points of view. The first is that of measuring the energy expended in gross, long duration physical activities such as marching, forestry work, freight handling, and factory work. The second is that of determining the useful mechanical work which can be performed by specified muscle groups for brief or extended periods of time in well-defined work situations, such as pedaling or cranking.

Labor

Over an 8-h day for a 48-h week, a useful norm for a 35-year-old laborer for total power expenditure, including basal metabolism energy, is 0.49 hp (366 W). Of this total expenditure, approximately 0.1 hp (75 W) is available for useful work. A 20-year-old man can generate about 15 percent more power than this norm, and a 60-year-old man about 20 percent less. The total energy or power expenditure is needed for determining nutritional requirements for classes of labor. A rule of thumb for power developed by European males can be expressed as a function of age and duration of effort in minutes for work lasting from 4 to about 480 min, assuming that 20 percent of the total output is useful power.

Age, years	Useful horsepower (t in min)
20	$hp = 0.40 - 0.10 \log t$
35	$hp = 0.35 - 0.09 \log t$
60	$hp = 0.30 - 0.08 \log t$

For a well-trained man, useful power production by pedaling, hand cranking, or a combination of the two for working durations of from 20 to 120 s may be summarized as follows (t is in seconds):

Arms and legs	$hp = 4.4t^{-0.40}$
Legs only	$hp = 2.8t^{-0.40}$
Arms only	$hp = 1.5t^{-0.40}$

There are examples of well-trained athletes generating between 1.5 and 2 hp for efforts of 5 to 20 s, using both arms and legs to generate power.

For pedaling efforts of from 1 to about 100 min, the useful power generated may be expressed as $hp = 0.53 - 0.13 \log t$ (t is in minutes).

Work scheduling, either as rhythmic work activity or with rest stops for recuperation, the temperature and humidity of the environment, and the detailed nature of the laborer's diet are factors which influence the ability to generate and maintain the above nominal power values. These considerations should be factored in for specific work situations.

Steady State and Transient

When a human and a passive mechanism are working together to generate power, the following conditions obtain: Energy is available both from stores residing in the muscles [a total usefully available energy of about 0.6 hp min (27 kJ), usually applied in transient bursts of activity] and from the oxidation of foods (for producing steady-state power). For an aerobic transient activity, energy production depends on the mass of muscle which can be brought into effective contact with the power transmission mechanism. For example, bicycle pedaling is an effective use of a large muscle mass. For steady-state activity, assuming adequate food for fuel energy, power generated depends on the oxygen supply and the efficiency with which oxygenated blood can be transported to the muscles as well as on the muscle mass.

The **physiological limit,** determined by oxygen-respiration capacity, for steady-state useful mechanical power generation is between 0.4 and 0.54 hp (300 and 400 W), depending on the man's physical condition. Useful power production may be achieved by such methods as rowing, cranking, or pedaling. The **highest values** of human-generated horsepower using robust subjects have been achieved using a rowing assembly which restrained nonuseful motions of the torso and major limbs. Under these conditions up to 2 hp (1,500 W) was generated over intervals of 0.6 s, and averages of about 1 hp (750 W) were generated over 2 min.

In order to approach an optimal **conversion efficiency** (mechanical work/food energy) of 25 percent, a mechanism would be required to store and to transmit energy from the body muscle masses when they were operating at optimal efficiency. This condition occurs when the force exerted by the muscle is about one-half of its maximum and the speed of muscle movement one-quarter of its maximum. Data on both force and speed for a given set of muscles are best measured in situ. Optimal conversion efficiency and maximum output power do not occur together.

Examples of High Output

Data for human-generated power come from measurements of subjects with different kinds of training, skill, body builds, diets, and motivation using a variety of mechanical devices such as bicycles, ergonometers, and variations on rowing machines. For strong, healthy young men, aggregations of such data for power produced in an interval t of 10 to 120 s can be approximated as follows:

$$\text{hp} = 2.5t^{-0.40}$$

For world-class athletes this becomes hp = $0.25 + 2.5\ t^{-40}$. These values can be exceeded for bursts of power of less than 10 s. For long-term efforts of from 2 to about 200 min, the aggregated data for useful power generated by strong, healthy young men can be approximated as follows (t in minutes):

$$\text{hp} = 0.50 - 0.13 \log t$$

For world-class athletes this becomes hp = $0.65 - 0.13 \log t$. The pilot of the Gossamer Albatross, who flew 22 mi from England to France in 2 h 55 min on August 12, 1979, entirely by pedal-generated power, sustained an output of about ⅓ hp (250 W) during the flight.

Maximum power output occurs at a load impedance of 5 to 10 times the size of the human being's source impedance.

Brody has developed detailed nomograms for determining the energetic cost of muscular work by farm animals; these nomograms are useful for precise cost-effectiveness comparisons between animal and mechanical power generation methods. A 1,500- to 1,900-lb horse can work continuously for up to 10 h/day at a rate of 1 hp, or equivalently pull 10 percent of its body weight for a total of 20 mi/day, and retain its vigor to an advanced age. Brody's work allows the following approximations for estimating the useful power output of work animals of varying sizes: The ratio of the power exerted in maximal energy production for a few seconds to the maximum steady-state power maintained for 5 to 30 min to the power produced in sustained heavy work over a 6- to 10-h day is approximately 25:4:1. For any one of these conditions, it has been found that, for healthy, mature specimens,

$$\text{hp}_{\text{animal}} = \text{hp}_{\text{man}}(\text{mass of animal/mass of man})^{0.73}$$

Thus, from the previously given horsepower magnitudes for men, one can compute the power generated by ponies, horses, bullocks, or elephants under the specified working conditions.

WIND POWER

by Asfaw Beyene, Rama Ramakumar, Charles P. Butterfield, and Eduard Muljadi

REFERENCES: NREL technical information at Internet address: http://gopher.nrel.gov.70. AWEA information at Internet address: awea.wind-net@notes.igc.apc.org. Hansen and Butterfield, Aerodynamics of Horizontal-Axis Wind Turbines, Ann. Rev. Fluid Mech., 25, 1993, pp. 115-149. Touryan, Strickland, and Berg, "Electric Power from Vertical-Axis Wind Turbines," J. Propulsion, 3, no. 6, 1987. Betz, "Introduction to the Theory of Flow Machines," Pergamon, New York. Eldridge, "Wind Machines," 2d ed., Van Nostrand Reinhold, New York. Glauert, "Aerodynamic Theory," Durand, ed., 6, div. L, p. 324, Springer, Berlin, 1935. Richardson and McNerney, Wind Energy Systems, Proc. IEEE, 81, no. 3, Mar. 1993, pp. 378-389. Elliott et al., "Wind Energy Resource Atlas," Wilson and Lissaman, Applied Aerodynamics of Wind Power Machines, Oregon State University Report, 1974. Eggleston and Stoddard, Wind Turbine Engineering Design, New York, Van Nostrand Reinhold. Spera, Wind Turbine Technology, ASME Press, New York. Gipe, "Wind Power for Home and Business," Chelsea Green Publishing Company. Ramakumar et al., Economic Aspects of Advanced Energy Technologies, Proc. IEEE, 81, no. 3, Mar. 1993, pp. 318-332, Beyene et al, Adaptive Motion and Their Theoretical Application to Low Speed Turbine Design, American Society of Civil Engineers, *Journal of Energy Engineering*, 135, no 4, Dec. 2009, pp. 112-118.

Wind is considered to be a source of clean energy. Although its use is many centuries old, it has not been a dominant factor in power production in the past because of the abundance of more affordable and easily accessible fossil fuels. Recently, the impact of fossil fuel combustion products on climate as well as the realization that fossil fuels are in limited supply has stimulated interest in wind energy as a clean and renewable resource. There has been a resurgence of wind energy conversion fields all over the world, over the past few years in particular. The state of knowledge is rapidly increasing, and the reader is referred to the current literature and the NREL Internet address cited above for information on the latest technology.

The resurgence of wind turbines over the last years has also contributed to the price drop and its affordability making it one of the lowest on a $/kW basis. In fact, wind energy conversion systems have reached important operating efficiency levels sharing increasingly larger share of power production. A record high amount of new wind capacity, roughly 63,000 MW, was added globally in 2015, yielding a cumulative total of 434,000 MW. Total cumulative installations stand at 369,597 MW by the end of 2014, which also demonstrates a tremendous potential for wind energy recourse.

Wind energy systems, the horizontal axis systems in particular, have a low footprint posing minimum burden on the land. This makes it convenient for the system to coexist on land normally reserved for other activities such as grazing or farming.

For the latest status on worldwide use of wind energy, the reader is referred to the American Wind Energy Association (AWEA) at the Internet address cited above. The fundamental principles of wind power technology do not change and are discussed here.

Use of Wind Energy Conversion Systems Historically, wind energy conversion systems were first used for milling grain and for pumping water. These tasks were ideally suited for wind power sources, since the intermittent nature of the wind did not adversely affect the operation.

The largest impact of wind power on the energy picture in the developed countries of the world is expected to be in the generation of electric power. In most cases, this involves feeding power into the power grid, and requires induction or synchronous generators. These generators require that the rotor turn at a constant speed. Wind turbines operate more efficiently (aerodynamically speaking) if they turn at an optimum ratio of tip speed to wind speed. Thus the use of variable-speed operation, using power electronics to obtain constant-frequency utility-grade ac power, has become attractive. Richardson describes the modern use of variable speed in wind turbines.

Gipe explains that in remote locations, where the power grid is not accessible and the first few units of electric energy may be very valuable, dc generation with storage and/or wind and diesel "village power systems" have been used. These systems are now being optimized to supply stable, constant-frequency ac electric energy.

Power in the Wind Since wind is air in motion, the power in wind can be expressed as

$$P_w = \tfrac{1}{2}\rho V^3 A \tag{7.1.1}$$

where P_w = power, W; A = reference area, m²; V = wind speed, m/s; ρ = air density, kg/m³. Since V appears to the third power, the wind speed is clearly very important. Figure (7.1.1) is a map of the United States showing regions of annual average available wind power.

The wind speed at a location is random; thus, it can be modeled as a continuous random variable in terms of a density function $f(v)$ or a distribution function $F(v)$. The Weibull distribution is commonly used to model wind:

$$F(v) = 1 - \exp[-(v/\alpha)^\beta] \tag{7.1.2}$$

$$f(v) = \beta(v^{\beta-1}/\alpha^\beta)\exp[-(v/\alpha)^\beta] \tag{7.1.3}$$

In Eqs. (7.1.2) and (7.1.3), α and β are two parameters which can be adjusted to fit available data over the study period, typically one month. They can be calculated from the sample mean m_v and the sample variance σ_v^2 using the following equations:

$$m_v = \alpha\Gamma(1 + 1/\beta) \tag{7.1.4}$$

$$(\sigma_v/m_v)^2 = \Gamma(1 + 2/\beta)\Gamma^2(1 + 1/\beta) - 1 \tag{7.1.5}$$

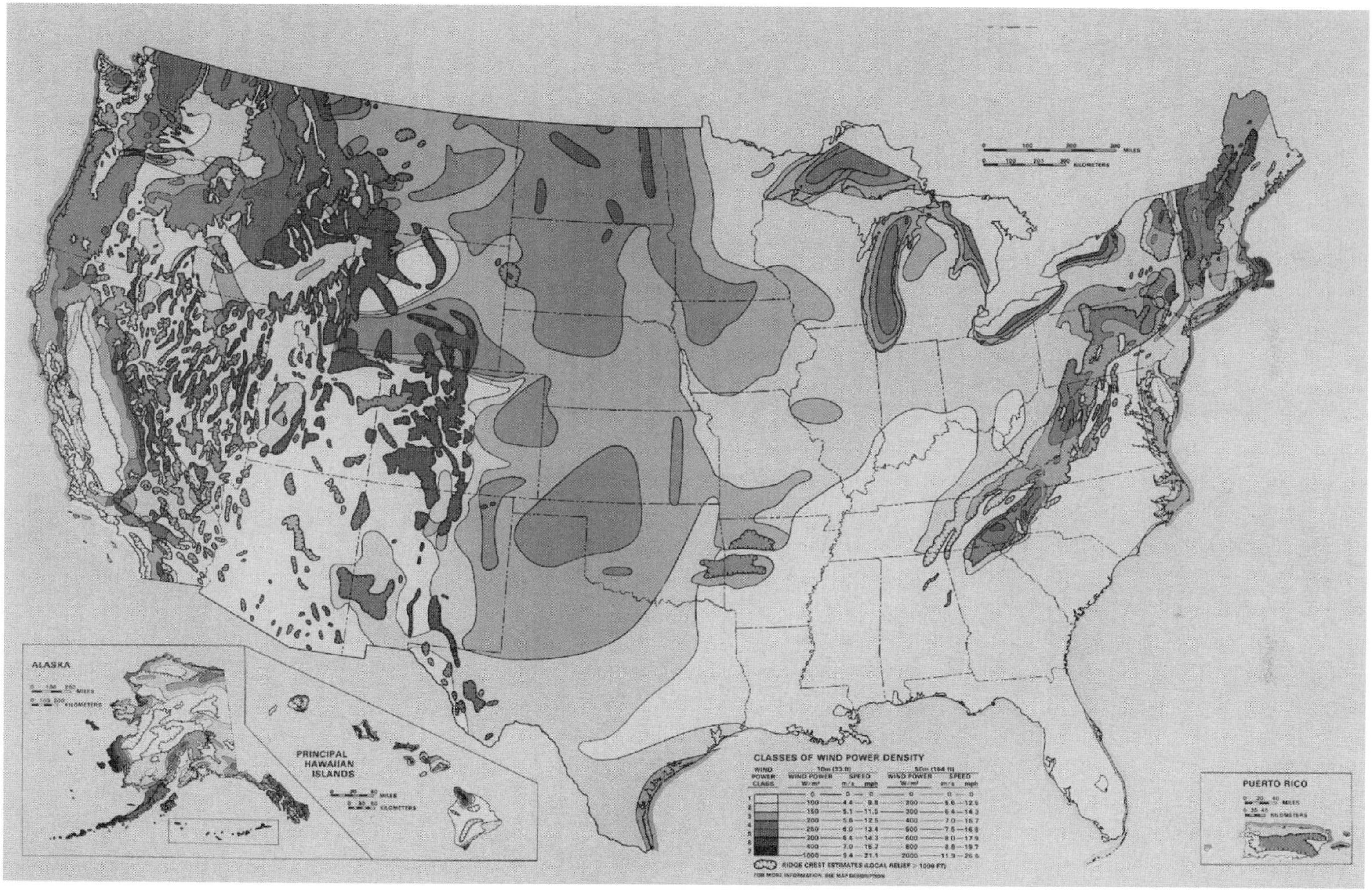

Figure 7.1.1 Gridded map of annual average wind energy resource estimates in the United States and Paerto Rico.

Typically, the sample mean is the only piece of information readily available for many potential sites. In such cases, a knowledge of the variability of the wind speed can be used to select an appropriate value for β, which can be used in (Eq. 7.1.4) to obtain α. A good compromise value for β is about 4 for wind regimes with low variances.

In addition to α and β, several other parameters are used to characterize wind regimes. Some of the important ones are listed below.

$$\text{Mean cubed wind speed} = \langle v^3 \rangle = \int_0^\infty v^3 f(v)dv \quad (7.1.6)$$

$$\text{Cube factor } K_c = (\langle v^3 \rangle)^{1/3}/m_v \quad (7.1.7)$$

$$= \alpha^3 \Gamma(1 + 3/\beta)$$

for Weibull model

$$\text{Average power density} = P_{av} = \tfrac{1}{2}\rho \langle v^3 \rangle \quad \text{W/m}^2 \quad (7.1.8)$$

$$\text{Energy pattern factor} = K_{ep} = \langle v^3 \rangle / m_v^3 = K_c^3 \quad (7.1.9)$$

Values of K_{ep} range from 1.5 to 3 for typical wind regimes.

The annual average available wind power for the continental United States is shown in Fig. 7.1.1). The values shown must be regarded as averages over large areas. The possibility of finding small pockets of sites with excellent wind regimes because of special terrain anywhere in the country should not be overlooked. The variability of the wind can also be shown in terms of a wind rose (Fig. 7.1.2). The "wind rose" illustrates wind speed and direction jointly. The duration for which the wind comes from a given radial sector is represented by the length of the spoke while the speed is shown by the thickness of the spoke. The output of a wind farm is sensitive to the shape of the wind rose.

Wind speed varies with the height above ground level (Fig. 7.1.3). Anemometers are usually located at a height of 10 m above ground level. The long-term average wind speed at height h above ground can be expressed in terms of the average wind speed at 10-m height using a one-seventh power law:

$$(v/v_{10\,\text{m}}) = (h/10)^{1/7} \quad (7.1.10)$$

The power 1/7 in the power law equation above depends on surface roughness and other terrain-related factors and can range from 0.1 to 1.3. The value of 1/7 should be regarded as a compromise value in the absence of other information regarding the terrain. Clearly, it is advantageous to construct an adequately high support tower for a wind energy conversion system.

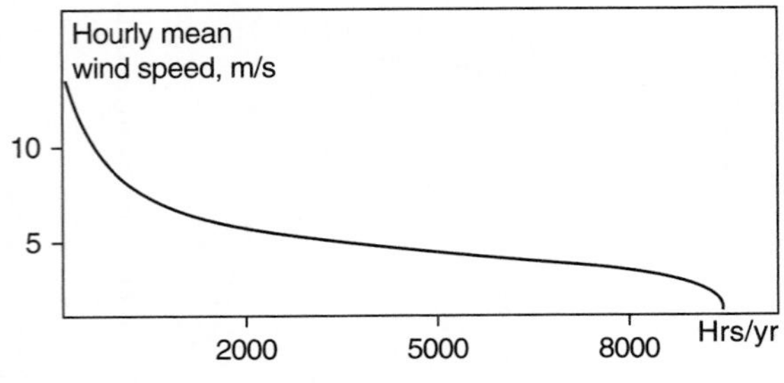

Figure 7.1.2 Typical wind variability.

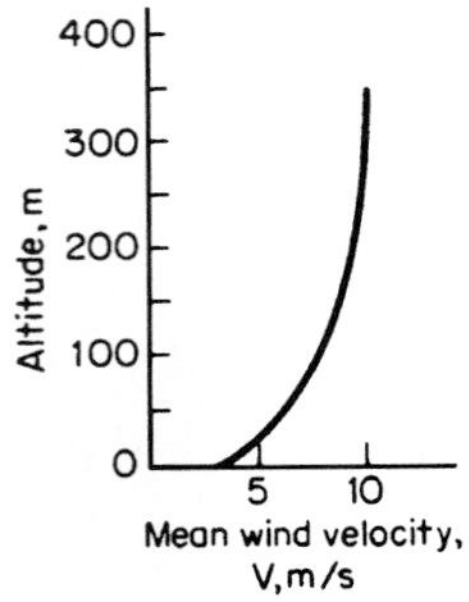

Figure 7.1.3 Typical variation of mean wind velocity with height.

Table 7.1.3 shows average and peak wind velocities at locations within the continental United States.

Classification Wind energy technology is classified in two categories: Horizontal Axis Wind Turbines (HAWTs) and Vertical Axis Wind Turbines (VAWTs).

HAWTs have radial blades mounted from the rotor and the axis is parallel to the wind stream. HAWTs are typically used for large scale electric grids. Currently, HAWTs cover the main wind turbines spectrum and are mainly used in large wind farms because of their high efficiency. On the other hand, HAWTs require high "quality" wind to achieve higher efficiencies, so turbulence, wind speed fluctuations and strong directional variability can drop their performance. For convenience of transportation and installation offshore wind turbines tend to lead in size compared to onshore ones. HAWT sizes and ratings have been steadily increasing. From 15 to 30-m-diam systems rated at 50 to 300 kW in the 1980s, the dawn of the new millennium saw the popular emergence of 1.5-MW units with 66-m rotor diameter. Larger units rated at 3.5 MW (100-m diam) and 5 MW (120-m diam) are in various stages of development. The V164 located in the Danish North Sea is currently the largest wind turbine in serial production today. The blades are about 80 meters in length and weigh about 33 tons each.

HAWT rotors are generally classified according to rotor orientation (upwind or downwind of the tower), blade articulation (rigid or teetering), and number of blades (generally two or three). Downwind turbines were favored initially in the United States, but the trend has been toward greater use of upwind turbines.

Downwind orientation allows blades to deflect away from the tower when thrust loading increases. Coning can also be easily introduced to decrease mean blade loads by balancing aerodynamic loads with centrifugal loads. Figure 7.1.4 shows typical upwind and downwind configurations along with definitions for blade coning and yaw orientation.

Free yaw, or passive orientation with the wind direction, is also possible with downwind configurations, but yaw must be actively controlled with upwind configurations. Free-yaw systems rely on rotor thrust loads and blade moments to orient the turbine. Net yaw moments for rigid rotors are sensitive to inflow asymmetry caused by turbulence, wind shear, and vertical wind. These are in addition to the moments caused by changes in wind direction which are commonly, though often incorrectly, considered the dominant cause of yaw loads.

Some early downwind turbine designs developed a reputation for generating sub-audible noise as the blades passed through the tower shadow (tower wake). Most downwind turbines operating today have greater tower clearances and lower tip speeds, which result in negligible infrasound emissions (Kelley and McKenna, 1985).

Wind turbine blade and system design is undergoing systematic improvement, and there are several concepts currently under development for HAWTs and VAWTs (Beyene et al.). These developments will allow increased conversion efficiency as well as higher rate capacity factors.

Blade Articulation Several different rotor blade articulations have been tested. Only two have survived—the three-blade, rigid rotor and the two-blade, teetered rotor. The rigid, three-blade rotor attaches the

Table 7.1.3 Average and peak wind velocities at locations within the continental United States

Albany, N.Y.	9.0	S	71	Louisville, Ky.	8.7	S	68
Albuquerque, N.M.	8.8	SE	90	Memphis, Tenn.	9.9	S	57
Atlanta, Ga.	9.8	NW	70	Miami, Fla.	12.6	—	132
Boise, Idaho	9.6	SE	61	Minneapolis, Minn.	11.2	SE	92
Boston, Mass.	11.8	SW	87	Mt. Washington, N.H.	36.9	W	150
Bismarck, N. Dak.	10.8	NW	72	New Orleans, La.	7.7	—	98
Buffalo, N.Y.	14.6	SW	91	New York, N.Y.	14.6	NW	113
Burlington, Vt.	10.1	S	72	Oklahoma City, Okla.	14.6	SSE	87
Chattanooga, Tenn.	6.7	—	82	Omaha, Neb.	9.5	SSE	109
Cheyenne, Wyo.	11.5	W	75	Pensacola, Fla.	10.1	NE	114
Chicago, Ill.	10.7	SSW	87	Philadelphia, Pa.	10.1	NW	88
Cincinnati, Ohio	7.5	SW	49	Pittsburgh, Pa.	10.4	WSW	73
Cleveland, Ohio	12.7	S	78	Portland, Maine	8.4	N	76
Denver, Colo.	7.5	S	65	Portland, Ore.	6.8	NW	57
Des Moines, Iowa	10.1	NW	76	Rochester, N.Y.	9.1	SW	73
Detroit, Mich.	10.6	NW	95	St. Louis, Mo.	11.0	S	91
Duluth, Minn.	12.4	NW	75	Salt Lake City, Utah	8.8	SE	71
El Paso, Tex.	9.3	N	70	San Diego, Calif.	6.4	WNW	53
Galveston, Tex.	10.8	—	91	San Francisco, Calif.	10.5	WNW	62
Helena, Mont.	7.9	W	73	Savannah, Ga.	9.0	NNE	90
Kansas City, Mo.	10.0	SSW	72	Spokane, Wash.	6.7	SSW	56
Knoxville, Tenn.	6.7	NE	71	Washington, D.C.	7.1	NW	62

U.S. Weather Bureau records of the average wind velocity, and peak velocity, at selected stations. The period of record ranges from 6 to 84 years, ending 1954. No correction for height of station above ground.

See also ASCE Standard (ASCE/SEI 7-05, with Supplement No. 1) "Minimum Design Loads for Buildings and Other Structures," for updated basic wind velocities (ASCE, 2005).

blade to a hub by using a stiff cantilevered joint. The first bending natural frequency of such a blade is typically greater than twice the rotor rotation speed 2*p*. Cyclic loads on rigid blades are generally higher than on teetering blades of the same diameter. Richardson and McNerney describe a 33-m, 300-kW turbine currently under development which reflects a mature version of this configuration.

Teetered, two-blade rotors use relatively stiff blades rigidly connected to a hub, but the hub is attached to the main drive shaft through a teeter hinge. This type of rotor is commonly used in tail rotors and some main rotors on helicopters. Two-blade rotors usually require teeter hinges or flexible root connections to reduce dynamic loading resulting from non-axisymmetric mass moments of inertia. In normal operation, the cyclic loads on the teetering rotor are low, but there is risk of teeterstop bumping ("mast bumping" in helicopter terminology) that can greatly increase dynamic loads in unusual situations.

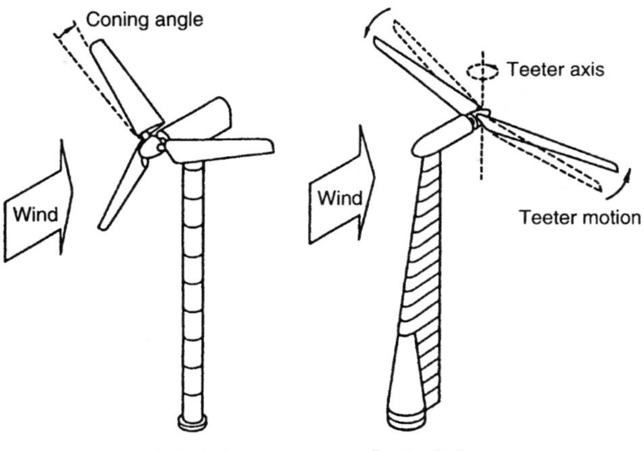

Figure 7.1.4 HAWT configurations. *(Courtesy of Atlantic Orient Corp.)*

Number of Blades Most two-blade rotors operating today use teetering hinges, but all three-blade rotors use rigid root connections. For small turbines (smaller than 50-ft diameter) rigid, three-blade rotors are inexpensive and simple and have the lowest system cost. As the turbines become larger, blade weight (and hence cost) increases in proportion to the third power of the rotor diameter, whereas power output increases only as the square of the diameter. This makes it cost-effective to reduce the number of blades to two and to add the complexity of a teeter hinge or flex beams to reduce blade loads. In the midscale rotor size (15 to 30 m), it is difficult to determine whether three rigidly mounted blades or two teetered blades are more cost-effective. In many cases, the choice between two- and three-blade rotors has been driven by designers' lack of experience and the potential risk of high development cost rather than by technical and economic merit. Currently 10 percent of the turbines installed are two-bladed, yet approximately 60 percent of all new designs being considered in the United States are two-blade, teetered rotors.

Design Problems A key design consideration is survival in severe storms. Various systems for furling the rotor, feathering the blades, or braking the shaft have been employed; failure of these systems in a high wind has been known to cause severe damage to the turbine.

A different, but related, consideration is the control of the turbine after a loss of electrical load, which also could cause severe overspeeding and catastrophic failure.

The other major cause of mechanical failure is the high level of vibration and alternating stresses. Loosening of inappropriately chosen fasteners is common. Fatigue considerations must be taken into account, especially at the rotor blade root.

Resonant oscillations are also possible if exciting frequencies and structural frequencies coincide. The dominant exciting frequency tends to be the blade passage frequency, which is equal to the number of blades times the revolutions per second. An important structural frequency in HAWTs is the natural frequency of the tower. One design approach is to make the tower so stiff that the exciting frequency is always below the lowest natural frequency of the tower. Another is to permit the tower to be more flexible, but manage the speed of the turbine such that the exciting frequency is never at a structural frequency for any significant length of time.

VAWTs have advantages in harsh operating conditions like urban environments and hilly terrains. However, they produce a variable torque over a rotational period, and the efficiency is typically lower than the HAWTs. Thus, generally VAWT are used in small-scale power production when cost and reliability are more important than efficiency. Despite their shortcomings in efficiency, VAWT technologies are promising because of their feasibility in distributed small-scale systems. One such promising design for distributed systems is the Savonius rotor which claims many technological advantages including easy ground-based access to the generator, affordability, high torque, impartiality to wind direction, and low noise. VAWTs are not as common as their counterpart, and have not grown in size. Installed capacities remain below a few dozen kW.

The Darrieus rotor looks somewhat like an eggbeater (Fig. 7.1.5). The blades are high-performance symmetric airfoils formed into a gentle curve to minimize the bending stresses in the blades. There are usually two or three blades in a turbine, and as shown in Fig. 7.1.6, the turbines operate efficiently at high speed. Wilson shows that VAWT performance analysis also takes advantage of the same momentum principles as the horizontal axis wind turbines. However, the blade element momentum analysis becomes much more complicated (see Touryan et al.).

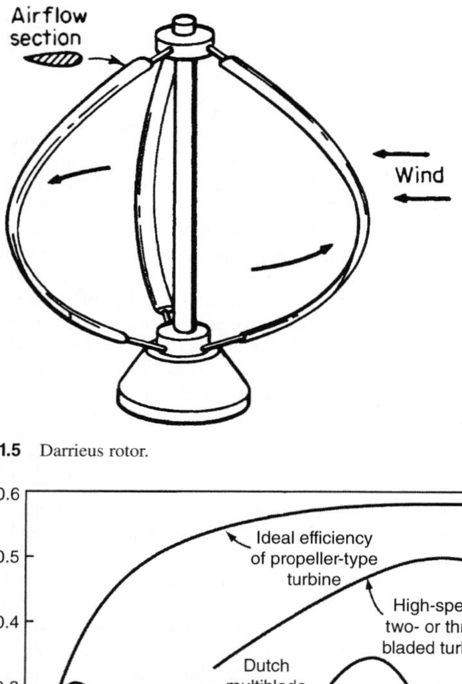

Figure 7.1.5 Darrieus rotor.

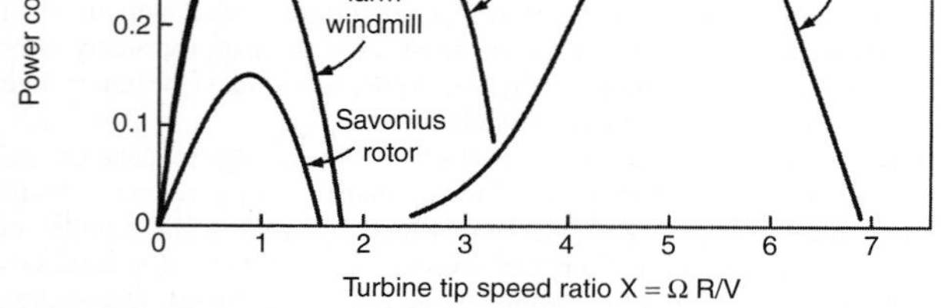

Figure 7.1.6 Turbine tip speed ratio versus power coefficient.

Care must be taken not to overemphasize the aerodynamic efficiency of wind turbine configurations. The most important criterion in evaluating WECSs is the power produced on a per-unit-cost basis.

Drag Devices Rotors utilizing drag rather than lift have been constructed since antiquity, even though they are bulky and limited to low coefficients of performance. The Savonius rotor is a modern variation of these devices; in practice it is limited to small sizes. Eldridge describes the history and theory of this type of windmill.

Augmentation Occasionally, the use of structures designed to concentrate and equalize the wind at the turbine is proposed. For its size, the most effective of these has been a short diffuser (hollow cone) placed around and downwind of a wind turbine. The disadvantage of such augmentation devices is the cost of the bulky static structures required.

Rotor Configuration Trends Hansen and Butterfield describe some trends in turbine configurations which have developed from 1975 to 1995. Although no single configuration has emerged which is clearly superior, HAWTs have been more widely used than VAWTs. Despite a growing research attention they have enjoyed in recent years, there are no known utility scale VAWTs operating today.

Design and Modeling One essential component of wind energy conversion (WEC) systems in both HAWTs and VAWTs is the turbine, traditionally called the windmill. In the performance analysis of wind turbines, the propeller devices were studied first, and their analysis set the current conventions for the evaluation of all turbines. Blade design follows strict mathematical processes governed by physics. The two dominating theories are blade element theory and momentum theory, used to calculate the optimum blade shape. **The blade element momentum theory** combines blade element theory with momentum theory to calculate the local forces on a propeller or wind-turbine blade, thereby alleviating some of the difficulties faced in calculating the induced velocities using each theory separately.

General Momentum Theory for Horizontal-Axis Turbines Conventional analysis of horizontal-axis turbines begins with an axial momentum balance originated by Rankine using the control volume depicted in Fig. 7.1.7. The analysis assumes that the rotor is enclosed in a control volume with one inlet and outlet. The surface of the control volume is a stream tube. The turbine is represented by a porous disk of area A which converts the kinetic energy of air thereby reducing its pressure. On the upstream side the pressure is raised above atmospheric at the cost of reduced air speed. On the downstream side pressure is lower, and the pressure rises while the velocity decreases. Momentum analysis predicts the axial thrust on the turbine of radius R to be

$$T = 2\pi R^2 \rho V^2 a(1-a) \tag{7.1.11}$$

where air density, ρ, equals 1.221 kg/m^3 at sea-level standard-atmosphere conditions. V is original wind speed, decelerated to $V(1-a)$ at the turbine disk, and to $V(1-2a)$ in the wake of the turbine (a is called the interference factor).

Application of the mechanical energy equation to the control volume depicted in Fig. 7.1.7 yields the prediction of power to the turbine of

$$P = 2\pi R^2 \rho V^3 a(1-a)^2 \tag{7.1.12}$$

This power can be nondimensionalized with the energy flux E in the upstream wind covering an area equal to the rotor disk, that is,

$$E = \tfrac{1}{2}\rho V^3 \pi R^2 \tag{7.1.13}$$

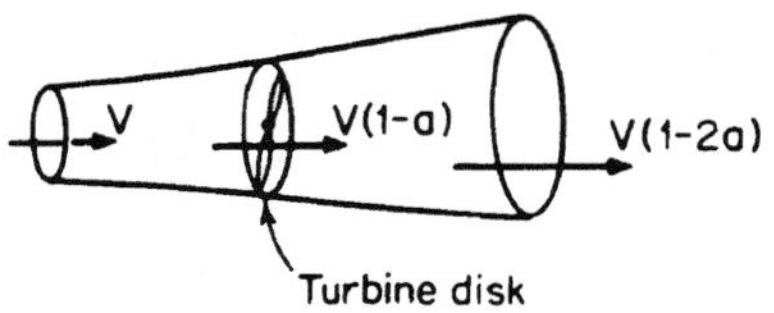

Figure 7.1.7 Control volume.

The resulting power coefficient is

$$C_p = \frac{P}{E} = 4a(1-a)^2 \tag{7.1.14}$$

This power coefficient has a theoretical maximum at $a = 1/3$ of C_p = 0.593. This result was first predicted by Betz and shows that the load placed on a wind turbine must be optimized to obtain the best power output; if the load is too small (small a), too much of the power is carried off with the wake; if the load is too large (large a), the flow is excessively obstructed and most of the approaching wind passes around the turbine.

This above power coefficient determination includes some important assumptions which limit its accuracy and applicability. In particular, the portion of the kinetic energy in the swirl component of the wake is neglected. Partial accounting for the rotation in the wake has been included in the analysis of Glauert with the resulting prediction of ideal power coefficient as a function of turbine tip speed ratio $X = DR/V$ (where D is the angular velocity of the turbine) shown in Fig. 7.1.6. Clearly, the swirl is made up of wasted kinetic energy and is largest for a high-torque, low-speed turbine. Actual farm, multiblade, and two- or three-bladed turbines show somewhat lower than ideal performance because of drag effects neglected in ideal flow analysis, but the high-speed two- or three-bladed turbines do tend to yield higher efficiency than low-speed multiblade windmills.

Blade Element Theory for Horizontal-Axis Turbines Wilson and Eggleston describe blade element theory as a mechanism for analyzing the relationship between the individual airfoil properties and the interference factor a, the power produced P, and the axial thrust T of the turbine. Rather than the stream tube of Fig. 7.1.7, the control volume consists of the annular ring bounded by streamlines depicted in Fig. 7.1.8. It is assumed that the flow in each annular ring is independent of the flow in all other rings.

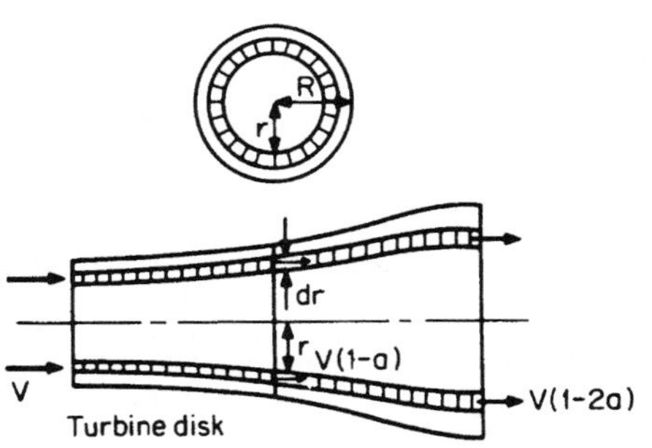

Figure 7.1.8 Annular ring control volume.

A schematic of the velocity and force vector diagrams is given in Fig. 7.1.9. The turbine is defined by the number B of its blades, by the variation of chord c and blade angle θ, and by the shape of blade sections $a' = \omega/(2\Omega)$, where ω is the angular velocity of the air just behind the turbine and Ω is the turbine angular velocity. Also, W is the velocity of the wind relative to the airfoil. Note that the angle will be different for each blade element since the velocity of the blade is a function of the radius. In order to keep the local flow angle of attack $\alpha = \theta - \phi$ at a suitable value, it will generally be necessary to construct twisted blades, varying θ with the radius. The sectional lift and drag coefficients C_L and C_D are obtained from empirical airfoil data and are unique functions of the local flow angle of attack $\alpha = \theta - \phi$ and the local Reynolds number of the flow. The entire calculation requires trial-and-error procedures to obtain the axial interference factor a and the angular velocity fraction a'. It can, however, be reduced to programs for small computers.

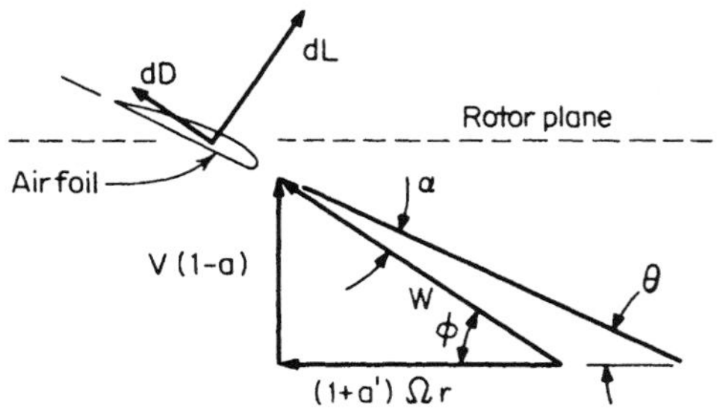

Figure 7.1.9 Velocity and force vector diagrams.

A typical solution for steady-state operation of a two- or three-bladed wind axis turbine is shown in Fig. 7.1.6. When optimized, these turbines run at high tip speed ratios. The curve shown in Fig. 7.1.6 for the two- or three-bladed wind turbine is for constant blade pitch angle. These turbines typically have pitch change mechanisms which are used to feather the blades in extreme wind conditions. In some instances the blade pitch is continuously controlled to assist the turbine to maintain constant speed and appropriate output. Turbines with continuous pitch control typically have flatter, more desirable operating curves than the one depicted in Fig. 7.1.6.

The traditional U.S. wind farm has a large number of blades with a high solidity ratio σ. (σ is the ratio of area of the blades to swept area of the turbine πR^2.) It operates at slower speed with a lower power coefficient than high-speed turbines and is primarily designed for good starting torque.

The curves depicted in Fig. 7.1.6 representing the performance of high- and low-speed wind axis turbines are theoretically predicted performance curves which have been experimentally confirmed.

BEM theory uses both axial and angular momentum balances to determine the flow and the energy balance at the blade. BEM theory represents a simple model for HAWTs, and it is based on the assumption that the flow at a given annulus does not affect the flow at adjacent annuli. The rotor blade is then analyzed in separate sections; the calculated forces of each section are then summed up to determine the overall forces of the rotor.

The thrust and torque induced by the blade will induce a secondary flow in the approaching wind that in turn impacts the flow behavior near the blade. This flow behavior affects the force distribution on the blade. This interaction between the momentum and the local blade forces and their conservation principles dictate computers simulate the resulting momentum and airfoil equations simultaneously. The simplest and most widely used BEM equations ignore the pressure drop from wake rotation and can be presented as:

$$a = \frac{1}{\frac{4}{c_n 6}\sin^2\phi + 1} \tag{7.1.15}$$

$$a' = \frac{1}{\frac{4}{c_t 6}\sin\phi\,\cos\phi - 1} \tag{7.1.16}$$

where a is the axial component of the induced flow, a' is the tangential component of the induced flow, 6 is the solidity of the rotor, ϕ is the local inflow angle, C_n the coefficient of normal force and C_t is the coefficient of tangential force.

Wind Energy to Electric Power Conversion The electric power output of a wind-to-electric conversion system can be expressed as

$$P_e = \tfrac{1}{2}\rho\eta_g\eta_m\eta_p A C_p V^3 \tag{7.1.17}$$

where P_e = electric power output, W; η_g and η_m = efficiencies of the electric generator and the mechanical interface, respectively; η_p = efficiency of the power conditioning equipment (if employed). The product of these efficiencies and the coefficient of performance (Fig. 7.1.6) usually will be in the range of 20 to 40 percent.

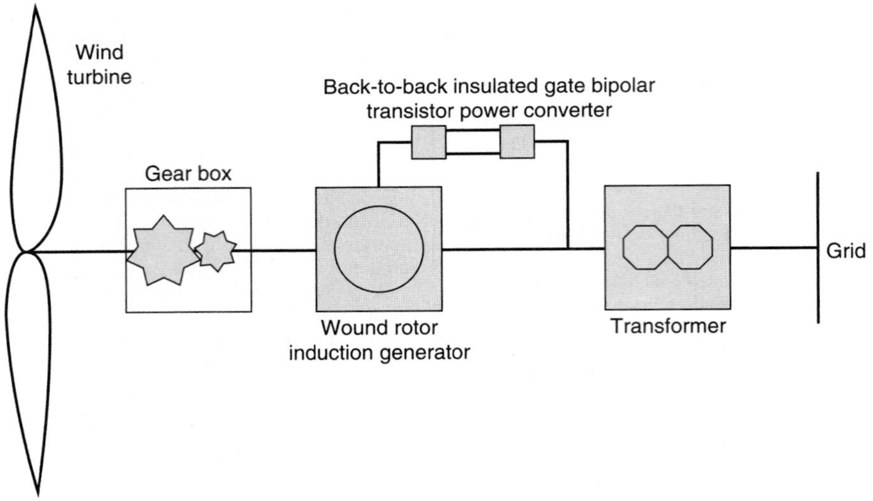

Figure 7.1.10 Doubly fed induction generator.

The electrical equipment needed for wind-to-electric conversion depends, above all, on whether the aeroturbine is operated in the constant-speed, nearly constant-speed, or variable-speed mode. With constant-speed and nearly constant-speed operation, the power coefficient C_p in Eq. (7.1.17) becomes a function of wind speed. If variable-speed mode is used, it is possible to operate the turbine at a constant optimum C_p over a range of wind speeds, thus extracting a larger fraction of the energy in the wind.

Most of the small wind turbines use permanent magnet alternators and are operated in isolation for battery charging, water pumping, ice making, etc. Sometimes they are connected to the grid or operated in conjunction with a local diesel generator. Utility-scale systems are equipped with different types of generators, often with power electronic subsystems, and are always connected to inject power into a conventional grid.

Generator Types Wind turbine generator begins producing power in winds of about 3 m/s, peaking at 10 to 13 m/s. The capacity of these generators has doubled every about five years, and continues to rise for offshore wind installations. Typical commercial generators can reach 8 MW. Power from wind fields is collected and transmitted to the grid through an alternating current (AC) or direct current (DC) lines, after voltage step-up at the substation. There are four types of wind turbine generators: fixed speed induction generator, active pitch controlled generator, doubly fed induction generator, and full-power conversion generators.

The fixed speed induction generators can operate with nearly fixed speed, and as a result their power output fluctuates as wind speed varies. There is a double-speed version of this generator type, with some control options such as an active stall design which allows the blades to pitch toward stall by a control system. Furthermore since induction generators absorb a lot of reactive power when generating active power, the fixed speed induction generator is generally equipped with reactive power compensators. Active pitch controlled generators are equipped with a wound rotor induction generator. The magnitude of the rotor current is controlled, which allows a speed variation of plus or minus 10 percent. This improves the power quality and reduces the mechanical loading of the turbine components compared to the fixed speed control. Currently, the doubly fed induction generator (Fig. 7.1.10) is the most popular wind type. Up to about 40 percent of the power load to the grid can pass through the converter while the rest is loaded directly to the grid. This allows about plus or minus 40 percent speed variation. In full power conversion all the power output goes to the grid through the converter. These types of generators are completely decoupled from the grid. The output current can be modulated to zero which limits the short-circuit current input to the grid. It also allows full speed variation as well as reactive power and voltage control capability.

Control (Fig. 7.1.11) shows a simplified block diagram of control systems implemented in a typical variable-speed wind turbine.

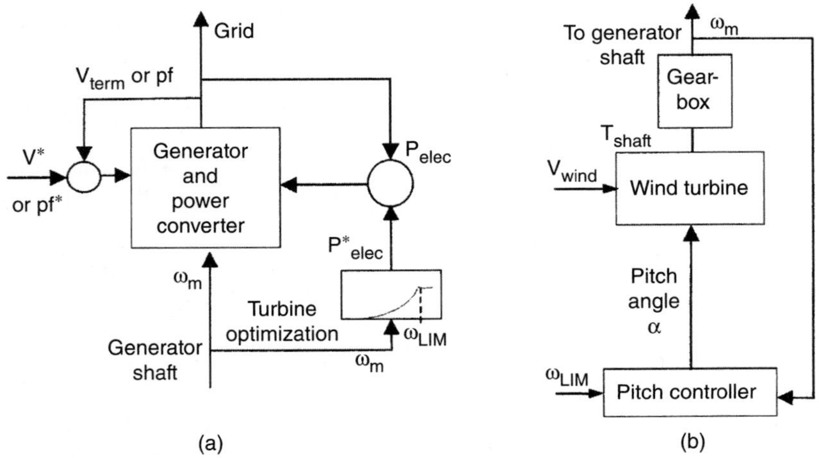

Figure 7.1.11 Control schemes for wind electric conversion systems. (a) Generator control (b) Turbine control.

Variables with an asterisk indicate reference (or command) quantities, and the others indicate measured quantities. There are two main subsystems, one controlling the generator and the other, the turbine.

The turbine control is accomplished by pitching the blade. At low and medium wind speeds, the pitch angle is controlled to operate at maximum energy capture. At high wind speeds, after the rotor reaches its rated speed, the pitch angle is controlled to shave the aerodynamic power and limit the rotor speed of the wind turbine. Employing separate generator and turbine controls works very harmoniously. Having the turbine control capability is a big plus because the ratings of the generator and power converter can be minimized. Thus the generator and power converter do not have to be used to control turbine aerodynamics. The USAairnet gives current wind speeds in the USA http://www.usairnet.com/weather/maps/current/wind-speed/

System Integration Typically, wind turbines generate electrical energy at low voltages (480 to 690 V). Pad-mounted transformers at the base of the towers step up the voltage to distribution voltage level (11 to 66 kV). Outputs of pad-mounted transformers are connected to a substation at the point of interconnection where the output power is stepped up to transmission line level (115 to 345 kV) to be transmitted over long distances.

The total rating of wind farms, consisting of several wind turbines, may approach 30 to 300 MW. As the level of penetration into conventional power systems increases, the impact of wind farms on the overall power system in terms of stability, reliability, availability, security, power quality, economics, and protection against faults should be studied thoroughly, and the results incorporated in the planning process.

Economics Costs of wind energy systems are often divided into two categories: annualized fixed costs and operation and maintenance (O&M) costs. Annualized fixed costs are comprised primarily of the cost of capital required to purchase and install the turbines. In addition, they include certain fixed costs such as taxes and insurance. O&M costs include scheduled and unscheduled maintenance and the levelized cost for major equipment overhauls.

Average wind turbine prices hit a low of $750/kW from 2000 to 2002, increasing to more than $1,500/kW by the end of 2008. By 2014, this average price dropped to $850-$1,250/kW range while exhibiting substantial increase in several design features including turbine efficiency, hub heights and rotor diameters. The decrease in turbine prices have also helped reduce project costs with a weighted average cost of $1,710/kW (2014). Larger capacity installations have benefitted from economies of scale. Generally, operations and maintenance costs also seem to decline over time.

In addition to capital and O&M costs, an economic assessment of wind energy systems must account for system performance. A commonly used parameter that describes the production of useful energy by wind and other energy systems is the capacity factor C, also called the plant factor or load factor. It is the ratio of the annual energy produced (AEP) to the energy that would be produced if the turbine operated at full-rated output throughout the year:

$$C_f = \frac{\text{AEP}}{8{,}760 P_R} \qquad (7.1.18)$$

where AEP is in kWh, 8,760 is the number of hours in 1 year, and P_R is the unit's nameplate rating in kW.

In order of decreasing importance, C_f is affected by the average power available in the wind, speed versus duration curve of the wind regime, efficiency of the turbine, and reliability of the turbine. Variable-speed turbines which tend to have low cut-in speeds and high efficiency in low winds exhibit better capacity factors than constant-speed turbines. Modern utility-grade turbines at good sites can achieve capacity factors in the range of 30 to 50 percent.

The combination of cost and performance can be used to calculate the cost of energy (COE) as follows:

$$\text{COE} = \frac{\text{FCR} \times \text{ICC}}{8{,}760 C_f} + (\text{O\&M}) \qquad (7.1.19)$$

Figure 7.1.12 A typical modern large on-shore wind farm consisting of 1.5-MW units.

where FCR is the fixed charge rate for the cost of capital and for other fixed charges such as taxes and insurance, ICC is the installed capital cost of the turbine and balance of plant in dollars per kW, and (O&M) is the prorated operation and maintenance cost per kWh of electric energy generated. This method is useful to estimate the cost of energy for different technologies or sites. However, for investment decisions, more detailed analyzes that include the effects of various investment strategies, tax incentives, and environmental factors should be performed. Ramakumar et al. discuss the economic aspects of advanced energy technologies, including wind energy systems.

Levelized cost of wind-generated electricity has come down from U.S. $0.38/kWh in 1980 to about U.S. $0.024/kWh in 2013.

A photograph of a typical, modern large on-shore wind farm consisting of 1.5-MW units is shown in Fig. 7.1.12. A view of the world's largest commercial offshore wind turbine, over 8 MW capacity, is shown in Fig. 7.1.13.

Figure 7.1.13 The Vestas V164, the largest wind turbine currently installed (2014) with a rated capacity of 8 MW, overall height of 220 m (722 ft), a diameter of 164 m (538 ft), permission NOT secured.

POWER FROM BIOMASS

by E. A. Avallone, Revised by Thomas Butcher

Vegetation offers, by photosynthesis, a natural process for the storage of solar energy. The efficiency of the photosynthetic process for the conversion of the sun's rays into a usable fuel form is low (less than 2 percent is probably realistic). Biomass is currently the largest U.S. renewable energy source. It is estimated that energy from biomass has the potential to displace 30 percent of the U.S. 2005 petroleum consumption (https://energy.gov/eere/bioenergy/bioenergy-tchnologies-office accessed Jan 27, 2017). This is based on an annual resource of 1 billion tons (U.S. Department of Energy, 2016 Billion Ton Report—Advancing Domestic Resources for a Thriving Bioeconomy, July 2016). A wide range of direct combustion technologies are in use (Nussbaumer, Thomas, Biomass Combustion in Europe—Overview on Technologies and Regulations, Final Report, 2008, www.nyserda.ny.gov, accessed 1/27/17).

Wood, wood waste, sawdust, hogged fuel, bagasse, straw, and tanbark have heating values ranging to 10,000 Btu/lb (see Sec. 5.1). They may be incinerated for disposal as waste material or burned directly for the subsequent production of steam or hot water, most often used in the processing activities of the plant, for example, hot water soak of logs for plywood peeling and steam for drying in paper mills. In food processing, fruit pits and nut shells have been used to generate a portion of the in-house requirements for steam. Most power from biomass is produced by direct combustion.

The alternative to direct burning of the so-called biofuels lies in their possible conversion to gaseous fuel by gasification at high temperature in the presence of air. Pyrolytic treatment can render biofuels to fractions of liquids and gases that have thermal value. With upgrading, these liquid fuels can be used as direct replacements for petroleum-derived fuels (Ofei D. Mante, Thomas A. Butcher, George, Wei, Rebecca Trojanowski, and Vicente Sanchez, Evaluation of biomass-derived distillate fuel as renewable heating oil, Energy & Fuels, 2015, 29(10) pp. 6,536-6,543). The solid residue remaining also has some thermal value which can be utilized in normal combustion.

Tree farming, with controlled growth and cutting, proposes to balance harvesting plans to load demands; for example, Szego and Kemp (Chem. Tech., May 1973) project a 400-mi^2 "energy plantation" to serve a 400-MW steam electric plant. Such proposals would utilize proven steam power plant cycles and equipment for novel breeding, growing, harvesting, preparation, and combustion of vegetation. (See also Sec. 5.1.) The photosynthesis process is basic to all agricultural practice. The human animal has long known how to convert grain to alcohol. It can be said that as long as we can grow green stuff we should be able to harness some of the sun's energy. The prohibition era in the United States saw many efforts to use the alcohol production capacity of the nation to offer alcohol as a substitute or supplementary fuel for internal combustion engines. Ethanol (C_2H_5OH) and methanol (CH_3OH) have properties that are basically attractive for internal combustion engines, to wit smokeless combustion, high volatility, high octane ratings, high compression ratios (Rv10). Heating values are 9,600 Btu/lb for methanol and 12,800 Btu/lb for ethanol. On a volume basis these translate, respectively, to 63,000 and 85,000 Btu/gal for methanol and ethanol. Gasoline, by comparison, has 126,000 Btu/gal (20,700 Btu/lb). (See Sec. 5 for values.) The blending of ethanol and methanol with gasolines (9 gasoline to 1 alcohol) has been used particularly in Europe since the 1930s as a suitable internal combustion engine fuel. The miscibility of the lighter alcohols with water and gasoline introduces corrosion problems for engine parts and lowers the octane number. Higher-carbon alcohols (e.g., butyl) which are immiscible with water are possible blending substitutes, but their availability and cost are not presently attractive. Properties such as flash point would introduce further problems. While these constitute some of the unsolved technical problems, the basic principle of harnessing the sun's energy through vegetation will continue as a provocative challenge not only in the field of power generation but also as a solution for the perennial problem of waste from farms and wood product manufacturers.

STIRLING ENGINES

Information from Israel Urieli, Edited by William M. Worek

Background

Robert Stirling published his famous patent in 1816. In his treatise, *Reflections on the Motive Power of Fire* in 1824,[1-5] Carnot stated that a heat engine can only attain the ideal maximum efficiency if the heat is transferred isothermally with the source and sink. The ingenious aspect of Stirling's patent is the regenerator, which allows the non-isothermal heat transfer to be done internally, enabling the ideal Stirling engine cycle to attain the ideal maximum efficiency. One of the main reasons for the lack of development of Stirling engines was that the importance of the regenerator was not understood for about 100 years after Stirling's original patent.

From the 1950s through the 1980s there was an ambitious effort to develop automobile Stirling engines in the USA and Europe [Ford Motor Company[6] and the Mechanical Technology Inc. (MTI)[7]]. There are a number of reasons why this program was ultimately unsuccessful: a significantly high pressure of Hydrogen gas is required in order to obtain an acceptable specific power output. This leads ultimately to sealing problems for the output crankshaft, hydrogen embrittlement of the casing, and a complex valve system to increase or decrease pressure for acceleration.

The current focus is on fully sealed Stirling cycle machines for electrical power production or refrigeration using Helium or Nitrogen gas. Since the Stirling engine is the only conceivable heat engine that can operate on a low temperature difference, enabling CHP (combined heat and power), low temperature (flat plate) solar, or waste heat recovery applications. Stirling cycle refrigerator systems using Helium gas have higher COP (coefficient of performance) values than regular vapor-compression refrigerators, and can operate at cryogenic temperature levels.

The ideal Stirling cycle machine is easily analyzed using basic thermodynamics; however, the analysis of actual Stirling cycle machines is extremely complex, mainly because of heat transfer between the external heat source/sink and the working gas, the regenerator, and the nonsteady reversing flow of the working gas, requiring sophisticated computer analysis. A unique alternative approach was presented by David Berchowitz at the 17th International Conference Stirling Engine Conference and Exhibition in the paper entitled: "A Phasor Description of the Stirling Cycle." This approach uses a phasor description includes both thermodynamic processes and mechanical dynamics, resulting in a useful guide to the understanding of these machines.

Different Types of Stirling Engines

The mechanical configurations of Stirling engines are generally divided into three groups known as the Alpha, Beta, and Gamma arrangements.

Alpha Stirling Engine

Alpha engines have two pistons in separate cylinders which are connected in series by a heater, regenerator, and cooler as shown in Fig. 7.1.14. Both Beta and Gamma engines use displacer-piston

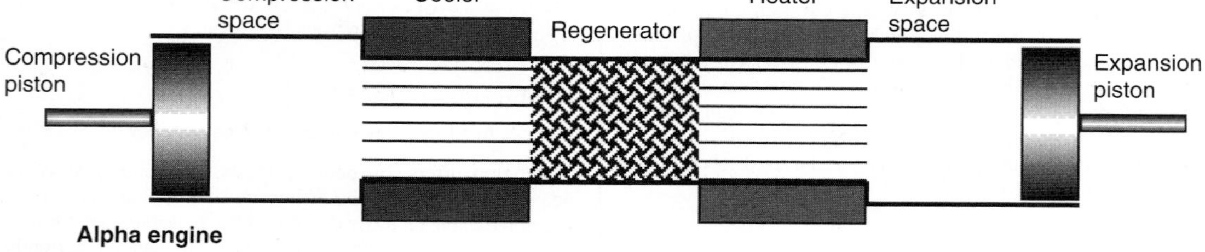

Figure 7.1.14 Representation of an Alpha Stirling engine.

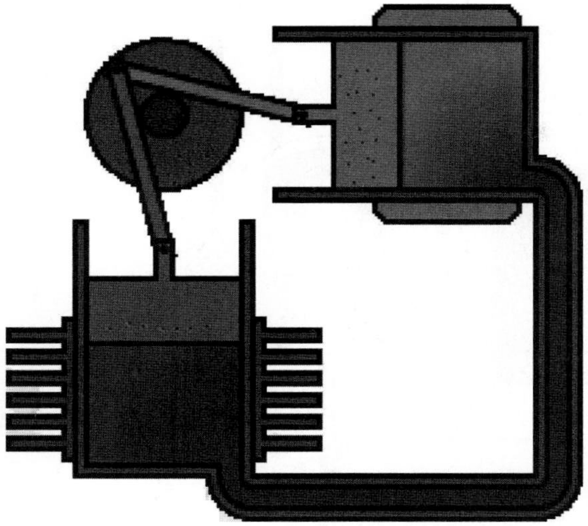

Figure 7.1.15 Representation of a Gamma Stirling engine.

arrangements, the Beta engine having both the displacer and the piston in an in-line cylinder system, whilst the Gamma engine uses separate cylinders.

The Alpha engine is conceptually the simplest Stirling engine configuration, however, suffers from the disadvantage that both the hot and cold pistons need to have seals to contain the working gas. There are a number of mechanical mechanisms which enable this type of engine to operate correctly with the correct phasing of the two pistons. An excellent animation of the V-type Alpha engine developed by Richard Wheeler, see Fig. 7.1.15.

Andy Ross of Columbus, Ohio, has been designing and building small air engines since the 1970s, including extremely innovative Alpha designs. Ross is the inventor of the classical Ross Yoke drive engine as well as a balanced "Rocker-V" mechanism, both shown in Fig. 7.1.16.

Multiple Cylinder Alpha Stirling Engines The Alpha engine can also be compounded into a compact multiple cylinder configuration, enabling an extremely high specific power output. A schematic diagram of this configuration is shown in Fig. 7.1.17. Notice that the four cylinders are interconnected, so that the expansion space of one cylinder is connected to the compression space of the adjacent cylinder via a series connected heater, regenerator, and cooler. The pistons are typically driven by a swashplate, resulting in a pure sinusoidal reciprocating motion having a 90-degree phase difference between the adjacent pistons.

One example of the swashplate four-cylinder Alpha engine is shown below. This engine was originally developed by Stirling Thermal Motors and is currently being continued by Stirling Power of Ann Arbor, Michigan. See Fig. 7.1.18.

During the 1970s, N V Philips of Holland and the Ford Motor Company developed an experimental automotive engine—a four-cylinder swashplate-drive engine as shown in Fig. 7.1.19.

Beta Type Stirling Engines

The Beta configuration is the classic Stirling engine configuration and has enjoyed popularity from its inception until today. Stirling's original engine from his patent drawing of 1816 shows a Beta arrangement. From Fig. 7.1.19, unlike the Alpha machine, the Beta engine has a single power piston and a displacer, whose ideal purpose is to "displace" the working gas at constant volume, and shuttle it between the expansion and the compression spaces through the series arrangement cooler, regenerator, and heater. In actual engines, the linkage driving the piston and displacer will move them such that the gas will compress while it is mainly in the cool compression space and expand while in the hot expansion space. This is clearly illustrated in Fig. 7.1.20.

Free Piston Beta Stirling Engines Probably the most ingenious Stirling engines yet devised are the free-piston engines invented and developed by William Beale at Ohio University in the late 1960s. The main aspect of the free piston machine is that the output power can be obtained through a linear alternator, allowing the entire system to

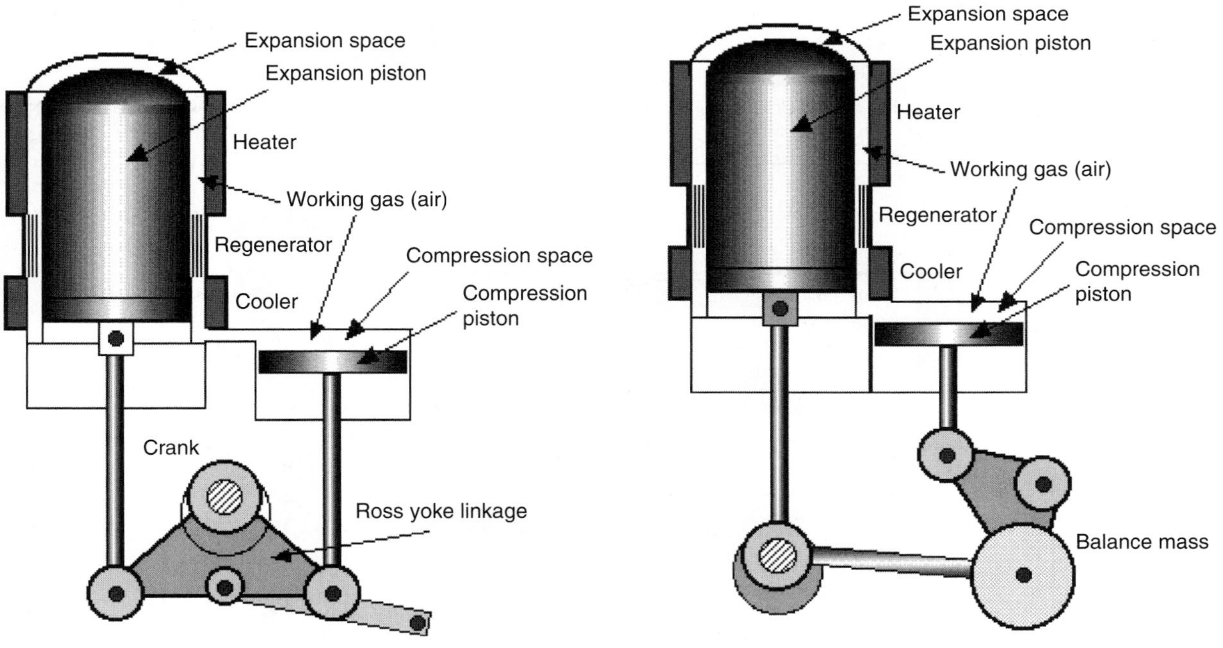

Figure 7.1.16 Representation of the Ross Yoke drive engine (left) and the balanced "Rocker-V" mechanism (right).

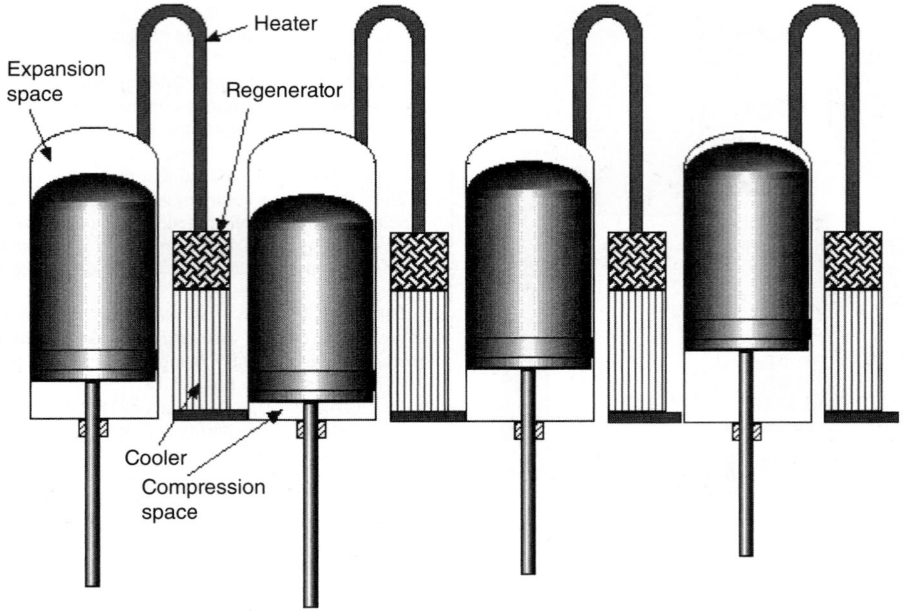

Figure 7.1.17 Representation of a multiple-cylinder Alpha Stirling engine.

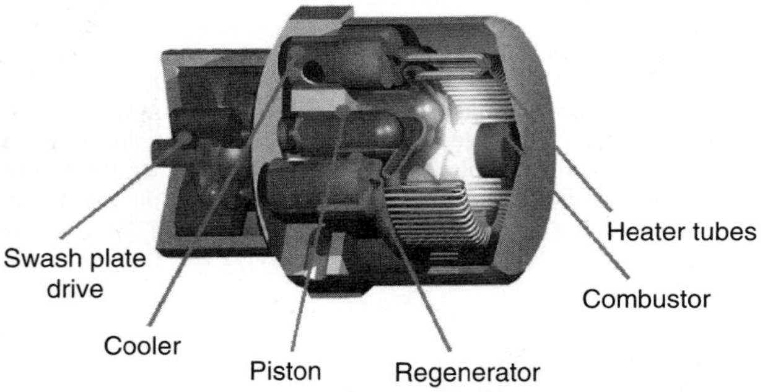

Figure 7.1.18 Representation of a swashplate 4-cylinder Alpha Stirling engine.

Figure 7.1.19 N V Philips of Holland and the Ford Motor Company experimental automotive engine—a four-cylinder swashplate-drive engine.

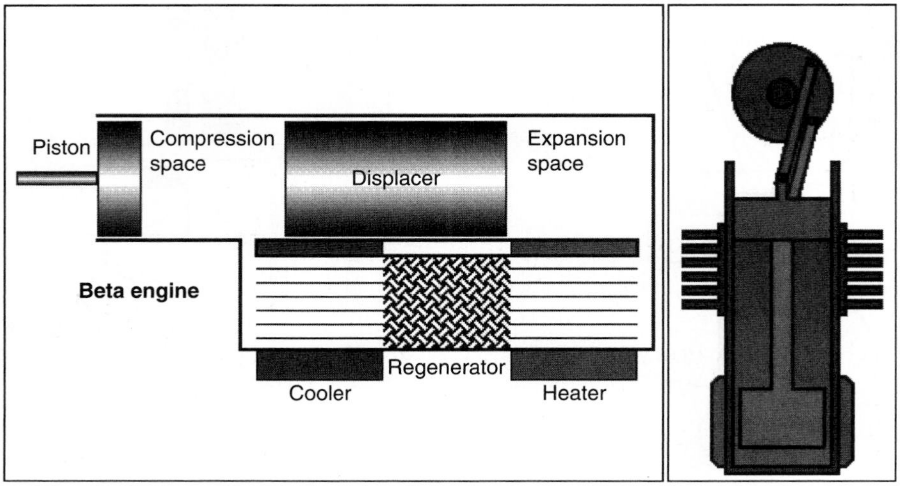

Figure 7.1.20 Representation of how an actual Beta Stirling engine is realized.

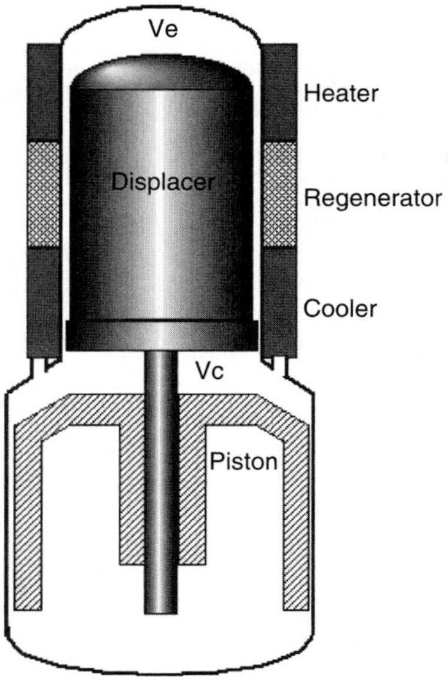

Figure 7.1.21 Representation of a "free-piston" Stirling engine.

be hermetically sealed. In fact, this is the only Stirling configuration to reach commercialization in any numbers. This is mainly because it avoids the fatal flaws of the crank, proven again and again over the years to be near-insurmountable—sealing and lubrication. This engine is shown in Fig. 7.1.21.

Gamma Type Stirling Engines

Gamma type engines have a displacer and power piston, similar to Beta machines, however, in different cylinders. This allows a convenient complete separation between the heat exchangers associated with the displacer cylinder and the compression and expansion work space associated with the piston. Therefore, they tend to have somewhat larger dead (or unswept) volumes than either the Alpha or Beta engines. See Fig. 7.1.22.

Furthermore, during the expansion process some of the expansion must take place in the compression space leading to a reduction of specific power. Gamma engines are therefore used when the advantages of having separate cylinders outweigh the specific power disadvantage. Because of the structural convenience of two cylinders, this enables completely independent sizing and construction of the displacer and piston assemblies.

References

1. Carnot, Sadi (1824). *Réflexions sur la puissance motrice du feu et sur les machines propres à développer cette puissance* (in French). Paris: Bachelier.

2. Carnot, Sadi; Thurston, Henry (editor and translator) (1890). *Reflections on the Motive Power of Heat.* John Wiley & Sons, New York. OL 14037447M. (full text of 1897 ed.) (Full text of 1897 edition on Wikisource.)

3. Carnot, Sadi; Magie, William Francis (editor and translator) (1899). "Reflections on the Motive Power of Heat and on Engines Suitable for Developing this Power." *The Second Law of Thermodynamics: Memoirs by Carnot, Clausius, and Thomson.* Harper & Brothers. OL 7072574M.

4. Carnot, Sadi; Clapeyron, E.; Clausius, R. (1960). Mendoza, Eric, ed. *Reflections on the Motive Power of Fire – and other Papers on the Second Law of Thermodynamics.* Dover Publications, New York. ISBN 0-486-44641-7.

5. Carnot, Sadi; Fox, Robert (1986), *Reflections on the Motive Power of Fire: a Critical Edition with the Surviving Scientific Manuscripts*, Manchester University Press; Lilian Barber Press, New York, ISBN 978-0-936508-16-0.

6. Ford Motor Company, CONS/4396-1 (NASA-CR-135331) AUTOMOTIVE STIRLING ENGINE DEVELOPMENT PEOGIAM Quarterly Technical Progress Report, Oct. 1977 - Dec. 1977 (Ford Motor Co.) 98 p HC A05/F A01 CSCL 13F Unclassified, G3/85 15,653.

7. DOE/NASA/0032-28, NASA CR-175106, MT186ASE58SRI, "Automotive Stirling Engine Mod II Design Report," Noel P. Nightingale, Mechanical Technology Incorporated, Latham, New York 12110, October 1986, Prepared for National Aeronautics and Space Administration, Lewis Research Center, Cleveland, Ohio 44135, Under Contract DEN 3-32, for U.S. DEPARTMENT OF ENERGY, Conservation and Renewable Energy, Office of Vehicle and Engine R&D, Washington, D.C. 20545, Under Interagency Agreement DE-AI01 -85CE50112.

8. The 17th International Conference Stirling Engine Conference and Exhibition, August 24-26, 2016, Northumbria University, Newcastle, UK.

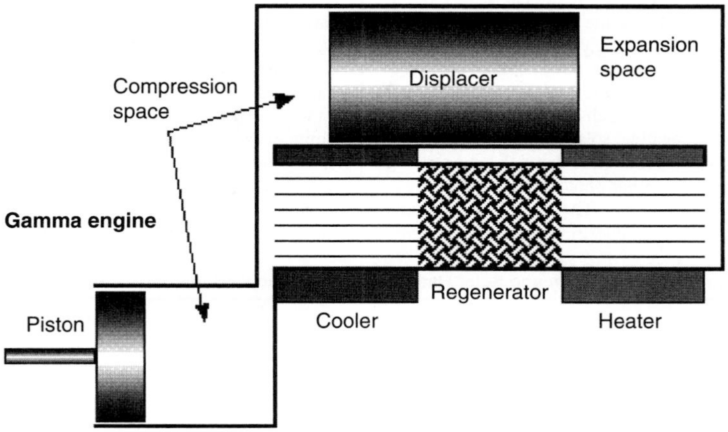

Figure 7.1.22 Representation of a Gamma Stirling engine.

SOLAR THERMAL ENERGY

by Erich A. Farber, Revised by Patrick E. Phelan

REFERENCES: Philbert, *Solar Energy Perspectives*, 2011, IEA. Sukhatme and Nayak, *Solar Energy: Principles of Thermal Collection and Storage*, 3rd ed., 2009, Tata McGraw-Hill. Duffie and Beckman, *Solar Engineering of Thermal Processes*, 4th ed., 2013, Wiley. Walker, *Solar Energy: Technologies and Project Delivery for Buildings*, 2013, Wiley. Kaligorou, *Solar Energy Engineering: Processes and Systems*, 2nd ed., 2014, Elsevier. Chwieduk, *Solar Energy in Buildings: Thermal Balance for Efficient Heating and Cooling*, 2014, Elsevier. Goswami, Kreith, and Kreider, *Principles of Solar Energy*, 3rd ed., 2015, CRC Press. Sørensen, *Solar Energy Storage*, 2015, Elsevier. Tiwari, Tiwari, and Shyam, *Handbook of Solar Energy: Theory, Analysis and Applications*, 2016, Springer.

Notation

A, R, T = subscripts denoting absorbed, reflected, and transmitted solar radiation
$\quad C$ = concentration ratio
$\quad c$ = subscript denoting collector cover
$\quad c_p$ = specific heat of fluid, J kg^{-1} °C^{-1}
$\quad I_{DN}$ = direct normal solar intensity, W/m²
$\quad I_d$ = diffuse radiation, W/m²
$\quad I_0$ = radiation intensity beyond earth's atmosphere, W/m²
$\quad I_r$ = reflected solar radiation, W/m²
$\quad I_{SC}$ = solar constant; normal incidence intensity at average earth-sun distance, W/m²
$\quad I_t$ = total solar radiation, W/m²
$\quad \phi$ = latitude, deg
$\quad m$ = air mass
$\quad o, i$ = subscripts denoting outgoing and incoming fluid conditions
$\quad q$ = heat flux, W/m²
$\quad q_1$ = heat flow through insulation, W/m²
$\quad T_p$ = temperature of absorbing surface, K
$\quad U$ = overall coefficient of heat transfer, W m^{-2} °C^{-1}
$\quad w_f$ = flow rate of collecting fluid, kg s^{-1} m^{-2}
$\quad \gamma_s$ = solar azimuth, deg from south
α, ρ, τ = absorptance, reflectance, and transmittance for solar radiation
$\quad \alpha_s$ = solar altitude, deg
$\quad \delta$ = solar declination, deg
$\quad \varepsilon$ = emittance for long-wave radiation
$\quad \gamma$ = surface azimuth, deg
$\quad \lambda$ = unit of wavelength, μm
$\quad \beta$ = angle of tilt from horizontal, deg
$\quad \theta$ = incident angle, deg, from perpendicular to surface

Introduction and Scope

The sun exerts forces upon the earth and radiates solar energy produced within the sun by nuclear fusion. A small fraction of that energy is intercepted by the earth and is converted by nature to heat, winds, ocean currents, waves, tides; makes plants grow, some of which over millions of years produced fossil fuels (oil, coal, and gas); and creates biomass which can be burned to generate heat and/or power. Solar energy is implicit in many subject areas treated elsewhere in the Handbook; only the more direct uses of solar thermal energy such as water heating, space heating and cooling, swimming pool heating, solar distillation, solar drying and cooking, solar furnaces, solar engines, and thermal concentrating solar power (CSP), will be treated here. Solar photovoltaic conversion is discussed under "Direct Energy Conversion." The total field is widely termed *alternative or renewable sources of energy and their conversion.*

Figure 7.1.23   Spectral distribution of solar radiation and radiation emitted by a blackbody at 35°C (95°F).

Solar Energy Utilization   Solar energy reaches the earth's surface as shortwave electromagnetic radiation in the wavelength band between 0.3 and 3.0 μm; its peak spectral sensitivity occurs at 0.48 μm (Fig. 7.1.23). Total solar radiation intensity on a horizontal surface at sea level varies from zero at sunrise and sunset to a noon maximum which can reach 1,070 W/m² (340 Btu h^{-1} ft^{-2}) on clear summer days. This inexhaustible source of energy, despite its variability in magnitude and direction, can be used in three major processes: (1) **Thermal,** in which the sun's radiation is absorbed and converted into heat which can then be used for many purposes, such as evaporating seawater to produce salt or distilling it into potable water; heating domestic hot water supplies; house heating by warm air or hot water; cooling by absorption refrigeration; cooking; generating electricity by vapor cycles and thermoelectric processes; attaining temperatures as high as 3,600°C (6,500°F) in solar furnaces. (2) **Thermochemical,** in which the shorter wavelengths can cause chemical reactions, sustain the growth of plants and animals, convert biomass to solar fuels, convert carbon dioxide to oxygen by photosynthesis, cause degradation and fading of fabrics, plastics, and paint, detoxify waste, and increase the rate of chemical reactions. (3) **Electrical,** in which part of the energy between 0.33 and 1.3 μm can be converted directly to electricity by photovoltaic cells, as discussed in the section "Direct Energy Conversion."

Solar Radiation Intensity   In space at the average earth-sun distance, 150 million km (92.957 million mi), solar radiation intensity on a surface normal to the sun's rays is 1,370 ± 3 W/m² (434.6 ± 1 Btu h^{-1} ft^{-2}). This quantity, called the solar constant I_{SC}, undergoes small (±1 percent) periodic variations which affect primarily the shortwave portion of the spectrum. Since the earth-sun distance varies throughout the year, the intensity beyond the earth's atmosphere I_0 also varies by ±3.3 percent (Table 7.1.4). The great seasonal variations in terrestrial solar radiation intensity are due to the earth's tilted axis, which causes the solar declination δ (the angle between the earth's equatorial plane and earth-sun line) to change from 0° on Mar. 21 and Sept. 21 to −23.5° on Dec. 21 and +23.5° on June 21.

In passing through the earth's atmosphere, the sun's radiation is partially and selectively absorbed, scattered, and reflected by water vapor and ozone, air molecules, natural dust, clouds, and artificial pollutants. Some of the scattered and reflected energy reaches the earth as diffuse or sky radiation I_d.

The intensity of the direct normal radiation I_{DN} depends upon the clarity and the amount of precipitable moisture in the atmosphere and the length of the atmospheric path, which is determined by the solar

Table 7.1.4   Annual Variation in Solar Declination and Solar Radiation Intensity beyond the Earth's Atmosphere

Date	Jan. 1	Feb. 1 Nov. 10	Mar. 1 Oct. 13	Apr. 1 Sept. 12	May 1 Aug. 12	June 1 July 12	July 1
Declination, deg	−23.0	−17.1	−7.7	+4.4	+15.0	+22.0	+23.1
Ratio, I_0/I_{sc}	1.033	1.029	1.017	1.000	0.983	0.971	0.967
Intensity I_0,	449	447	442	435	427	422	420
(Btu h^{-1} ft^{-2}) (W/m²)	1,416	1,409	1,393	1,370	1,347	1,331	1,324

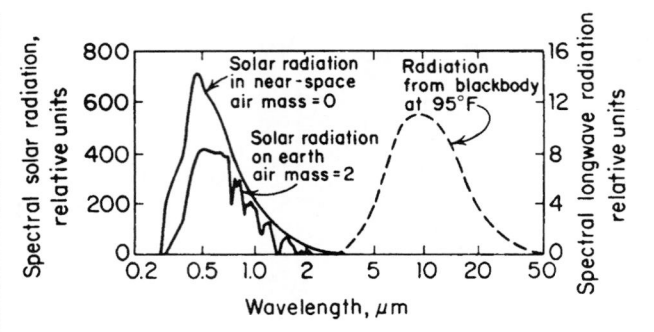

Table 7.1.5 Solar Altitude and Azimuth, Direct Normal Radiation, and Total Solar Radiation on Horizontal and Vertical South-Facing Surfaces, June 21 and Dec. 21, for Phoenix, Arizona, USA (33.45° North Latitude)

			June 21, declination = +23.45°				
Local Civil Time	6 AM	8 AM	10 AM	12 PM	2 PM	4 PM	6 PM
Solar altitude, deg	9.5	31.7	54.5	72.3	64.8	43.1	20.3
Solar azimuth, deg	–112.7	–95.3	–73.6	–22.9	55.8	85.8	104.1
Direct normal irradiation, W/m²	55	296	603	609	712	730	535
Total irradiation, W/m²							
On horizontal surface	24	285	749	965	993	761	343
On vertical south surface	0.4	89.9	312	458	435	223	0
			Dec. 21, declination = –23.45°				
Local Civil Time	6 AM	8 AM	10 AM	12 PM	2 PM	4 PM	6 PM
Solar altitude, deg			18.0	26.3	22.9	9.4	
Solar azimuth, deg			–35.0	–6.7	23.4	48.3	
Direct normal irradiation, W/m²	0	0	723	808	765	667	0
Total irradiation, W/m²							
On horizontal surface	0	0	300	523	524	296	0
On vertical south surface	0	0	665	954	978	917	0

Solar data from Typical Meteorological Year 3 (TMY3), http://rredc.nrel.gov/solar/old_data/nsrdb/1991-2005/tmy3/by_state_and_city.html; assumes ground reflectance of 0.2.

zenith θ_z and expressed in terms of the air mass m, which is the ratio of the existing path length to the path length when the sun is at the zenith. Except at very low solar altitudes, $m = 1.0/\cos(\theta_z)$.

Figure 7.1.23 shows relative values of the spectral intensity of solar radiation in space for $m = 0$ and at sea level for a solar altitude of 30° ($m = 2.0$). Table 7.1.5 shows the variation for Phoenix, Arizona, USA (33.45°N latitude) at 40° north latitude for summer (June 21) and winter (Dec. 21) days of solar altitude and azimuth (measured from the south), direct normal radiation, and total solar irradiation of horizontal and vertical south-facing surfaces. Solar radiation data are taken from Typical Meteorological Year 3 (TMY3) data available from the National Renewable Energy Laboratory for locations in the USA, and a ground reflectance of $\rho_g = 0.2$ is assumed to compute the total irradiation on a vertical surface using the isotropic sky model:

$$I_t = I_b R_b + I_d\left(\frac{1+\cos(\beta)}{2}\right) + I\rho_g\left(\frac{1-\cos(\beta)}{2}\right)$$

where R_b is the ratio of beam radiation on a surface tilted at angle β to that on a horizontal surface, I_b the beam radiation on a horizontal surface, I_d the diffuse radiation on a horizontal surface, and $I = I_b + I_d$.

Direct beam solar radiation intensity is measured by **pyroheliometers** with collimating tubes to exclude all but the direct rays from their sensors, which may use calorimetric, thermoelectric, or photovoltaic means to produce a response proportional to the irradiation rate. Similar but uncollimated instruments called **pyranometers** are used to measure the total radiation from sun and sky; when their sensors are shaded from the sun's direct rays, they also can measure the diffuse component.

Incident Angle Determination The incident angle θ affects both the direct solar intensity and the solar optical properties of the irradiated surface. Standard equations relate θ to the angle of tilt of the surface, β, and other angles that describe the position of the sun in the sky. The local solar time, LST, must also be determined. LST depends on the Local Civil Time (i.e., the usual time), the day of the year, and the longitude.

Solar Optical Properties of Transparent Materials When solar radiation with total intensity I_t falls on a transparent material, part of the energy is reflected, part is absorbed, and the remainder is transmitted. At any instant,

$$I_t = q_\tau + q_A + q_R = I_t(\tau + \alpha + \rho)$$

The sum of the solar optical properties τ, α, and ρ must equal unity, but the individual values depend upon the incident angle and wavelength of the radiation, the composition of the material, and the nature of any coatings which may be applied to the surfaces.

For uncoated single-strength (3/32-in or 2.4-mm) clear window glass (Fig. 7.1.24), solar transmittance at normal incidence ($\theta = 0°$) is approximately 0.90, but the transmittance for longwave thermal radiation (5 μm) is virtually zero. Thus glass acts as a "heat trap" by admitting solar radiation readily but retaining most of the heat produced by the absorbed sunshine. This "greenhouse effect," which is also exhibited but to a lesser degree by some plastic films, is the basis for most solar thermal processes. Heat absorbing glass [1/4 in (6.3 mm) thick (Fig. 7.1.24)],

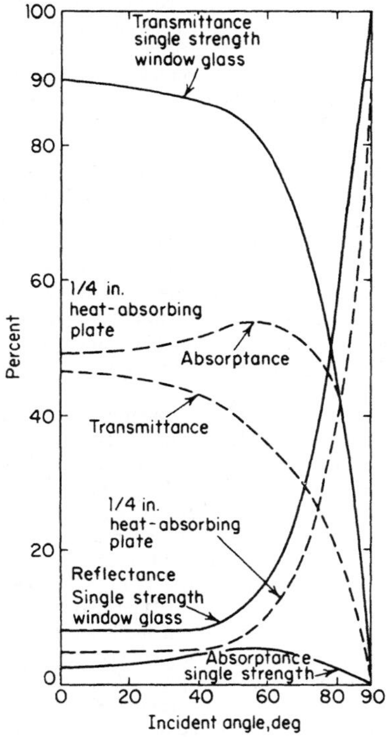

Figure 7.1.24 Variation with incident angle of solar optical properties of 3/32-in (2.4-mm) clear glass and 1/4-in (6.3-mm) heat-absorbing glass.

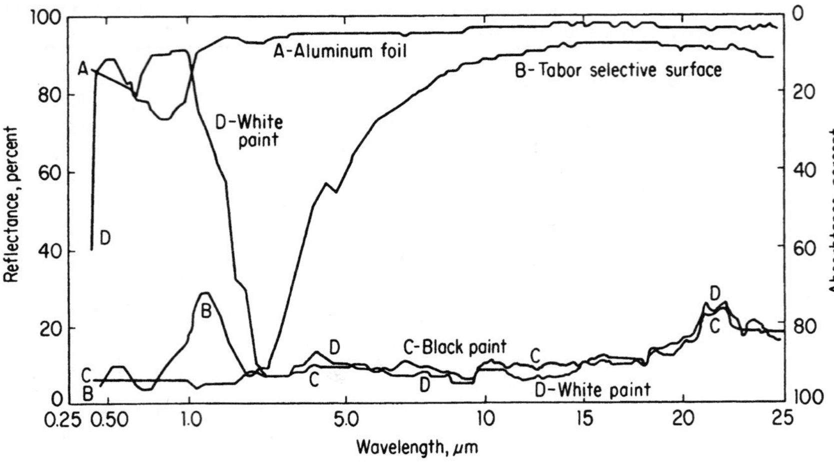

Figure 7.1.25 Variation with wavelength of reflectance and absorptance for opaque surfaces.

which absorbs more than 50 percent of the incident solar radiation, is widely used by architects to reduce the heat and glare admitted through unshaded windows. Low-emittance ("low-e") coatings have been developed to serve similar purposes.

For all types of glass, transmittance falls and reflectance rises as θ exceeds about 30°. Absorptance increases somewhat owing to the increased path length and then drops off sharply toward zero as θ exceeds 60°.

Absorptance and Emittance of Opaque Surfaces Opaque materials absorb or reflect all the incident sunshine. The absorptance α for solar radiation and the emittance ε for longwave radiation at the temperature of the receiving surface are particularly important. For a true blackbody, the absorptance and emittance are equal and do not change with wavelength. Most real surfaces have reflectances and absorptances which vary with wavelength (Fig. 7.1.25). Aluminum foil has a consistently low absorptance and high reflectance over the entire spectrum from 0.25 to 25 μm, while black paint has a high absorptance and low reflectance. White paint, however, has low shortwave (solar) absorptance, but beyond 3 μm its absorptance and reflectance are virtually the same as for black paint.

Solar collectors require a high α/ε ratio, while surfaces which should remain cool, such as rooftops or space vehicles, should have low ratios since their objective usually is to absorb as little solar radiation and emit as much longwave radiation as possible. Special surface treatments have been developed for which the ratio α/ε is above 7.0, making them suitable for solar collectors; others with ratios as low as 0.15 are useful as heat rejectors for space applications (see Table 7.1.6). In addition, the absorptance can be changed and controlled by paint composition (grain material and size, binder, thickness, etc.) and surface configuration, both large and small.

Equilibrium Temperatures for Concentrating Collectors When a surface is irradiated, its temperature rises until the rate of solar radiation absorption equals the rate at which heat is removed from the surface. If no heat is intentionally removed, the maximum temperature which can be attained by a blackbody ($\alpha = \varepsilon$) is found from $I_{DN}C\alpha = 5.67\varepsilon(T_p/100)^4$, where C is the concentration ratio. Figure 7.1.26 shows the variation of blackbody equilibrium temperatures for earth and near space where $I_{DN} = 1{,}000$ and $I_0 = 1{,}370$ W/m^2 (320 and 435 Btu h^{-1} ft^{-2}).

For flat-plate collectors, $C = 1.0$; so their maximum attainable temperatures are below 100°C (212°F) unless a selective surface is used with $\alpha/\varepsilon > 1.0$, or both radiation and convection loss are suppressed by the use of multiple glass cover plates. Only the direct component of the total solar radiation can be concentrated, and concentrating collectors must follow the sun's apparent motion across the sky or use heliostats which serve the same function. Diffuse radiation cannot be concentrated effectively.

Table 7.1.6 Solar Absorptance, Longwave Emittance, and Radiation Ratio for Typical Surfaces

Surface or material	Shortwave (solar) absorptance α	Longwave emittance ε	Radiation ratio, α/ε
Flat, oil-based paints:			
Black	0.90	0.90	1.00
Red	0.74	0.90	0.82
Green	0.50	0.90	0.55
Aluminum	0.45	0.90	0.50
White	0.25	0.90	0.28
Whitewash on galvanized iron	0.22	0.90	0.25
Building materials:			
Asbestos slate	0.81	0.96	0.84
Tar paper, black	0.93	0.93	1.00
Brick, red	0.55	0.92	0.59
Concrete	0.60	0.88	0.68
Sand, dry	0.82	0.90	0.92
Glass	0.04–0.70	0.84	
Metals:			
Copper, polished	0.18	0.04	4.50
Copper, oxidized	0.64	0.60–0.90	1.03–0.71
Aluminum, polished	0.30	0.05	6.00
Selective surfaces:			
Tabor, electrolytic	0.90	0.12	7.50
Silicon cell, uncoated	0.94	0.30	3.13
Black cupric oxide on copper	0.91	0.16	5.67

Flat-Plate Collectors Direct, diffuse, and reflected solar radiation can be collected and converted into heat by flat-plate collectors (Fig. 7.1.27). These generally use blackened metal plates which are finned, tubed, or otherwise provided with passages through which water, air, or other fluids may flow and be heated to temperatures as much as 55 to 86°C (100 to 150°F) above the ambient air. The actual temperature rise may be estimated from the heat balance for a unit area of collector surface:

$$q_A = I_t\tau_c\alpha_p = w_f c_p(t_o - T_i) \div q_l + U(t_p - t_a)$$

The loss from the back of the collector plate q_l can be minimized by the use of adequate insulation. The radiation component of the loss from the upper surface can be reduced by using selective reflectance coatings with high α/ε ratios and by using covers which are transparent to solar radiation but opaque to longwave emissions (see Table 7.1.7). Both convection and radiation can be reduced by the use of honeycomb structures in the airspace between the cover and the collector plate. For the transparent covers, glass is best since it lets through the shortwave

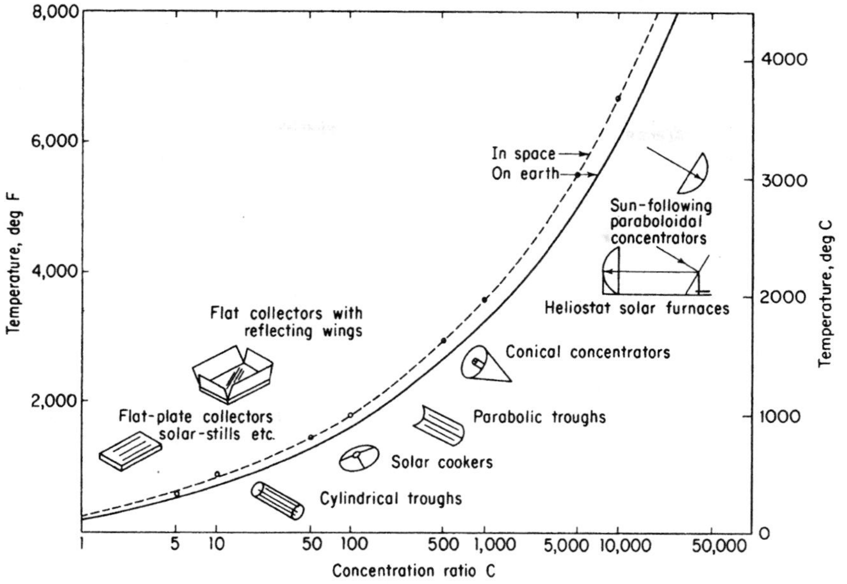

Figure 7.1.26 Variation with concentration ratio of equilibrium temperatures for earth and space.

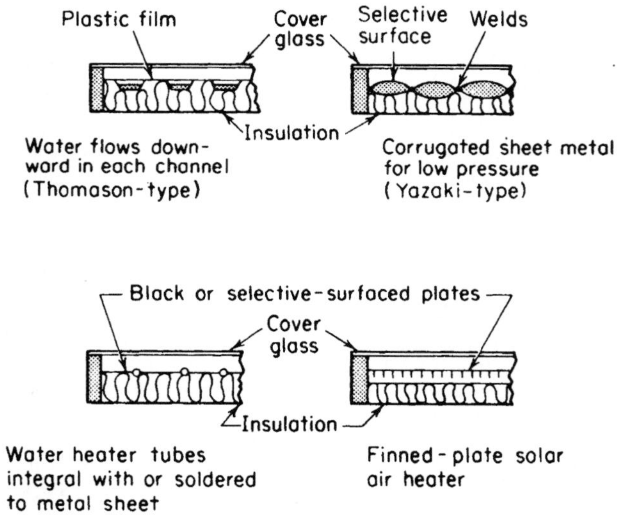

Figure 7.1.27 Typical flat-plate solar radiation collectors.

Table 7.1.7 Transmittance and Overall Heat-Transfer Coefficients for Collectors with Glass and Plastic Covers

Type and number of covers	None	One glass	One plastic	Two glass	Two plastic
Solar transmittance	1.00	0.90	0.92	0.81	0.85
Overall coefficient U	3.90	1.12	1.30	0.71	0.87

solar radiation but stops the longwave radiation given off by the collector plate. Plastics do not have this trapping characteristic, do have a shorter life, may lose their transparency, and outgas when heated. The fumes may condense on other surfaces, forming a film to reduce the collector performance.

Evacuated-Tube Collectors Another type of nonconcentrating solar thermal collector is the evacuated-tube collector, in which an evacuated glass tube contains a heat pipe (or thermosyphon). The incident sunlight heats an absorber within the evacuated tube, which transfers heat to the evaporator of the heat pipe. The heat is efficiently transported to the other end of the heat pipe (the condenser), where it can be transferred via a header to a circulating heat transfer fluid and thus utilized for other purposes. Evacuated-tube collectors can reach temperatures higher than flat-plate collectors, and exceed the boiling point of water. Thus, evacuated-tube collectors can be used not only for domestic hot water, but also potentially to drive solar cooling systems based on absorption or adsorption cycles.

Compound Parabolic Concentrators This type of collector (CPC) is a nonimaging type of concentrator that does not require tracking, which is required for all other types of concentrators to date. Two parabolic mirrors face each other and focus light on a receiver located near their bottom junction. Concentration ratios are typically limited to less than 10, making CPCs potentially useful for relatively lower temperature applications but probably not for electric power generation.

Applications of Solar Thermal Energy

Solar Drying Probably the largest use of solar energy over the centuries—the drying of agricultural crops, evaporation of ocean and salt lake ponds for salt production, etc.—has led more recently to more efficient dryers. The newer dryers prevent rain and dew from rewetting the materials. Simple inexpensive solar dryers—essentially transparent covers—sometimes are supplemented by air heaters providing hot air to dry crops, fish, etc., especially in tropical regions where daily rains prevent efficient natural drying. They are used for curing wood and to remove moisture from mining ores (especially if they have been washed) to reduce shipping costs. Some uneconomical mining operations have been made profitable by the use of solar drying.

Swimming Pool Heating Probably the widest U.S. commercial application of solar energy today is in swimming pool heating. To extend the swimming season, a transparent cover floating on the surface of the pool, with as few air bubbles underneath as possible, will raise the temperature of the water by up to 11°C (20°F). Flat-plate type of heaters on roofs, often made of rubber or plastics, can be used instead of or in addition to the pool cover. Approximately 3.1 kWh day^{-1} m^{-2} (1,000 Btu day^{-1} ft^{-2}) can be expected on average from a reasonably good collector. Since in many places a fence is required around a pool, the fence can incorporate collectors. They are not as efficient because of less favorable orientation, but they serve a dual purpose. Some developing countries are interested in solar swimming pool heating since they lack fossil fuels and the currency to buy them, but want to attract tourists with modern conveniences.

Solar Ponds If water in ponds or reservoirs contains salts in solution, the warmer layers will have higher concentrations and, being heavier, will sink to the bottom. The hot water on the bottom is insulated against heat losses by the cooler layers above. Heat can be extracted from these ponds for power generation. Large solar ponds have been used in the Mideast along the Dead and Red seas, along the Salton Sea; a number of artificial ponds have been established elsewhere. The inexpensive extraction of heat at over 94°C (200°F) still requires improvement. Solar ponds are, effectively, large inexpensive solar collectors. Their heat can be used to power vapor engines and turbines; these, in turn, drive electric generators.

Solar Stills Covering swimming pools, ponds, or basins with airtight covers (glass or plastic) will condense water vapor on the underside of the covers. The condensate produced by the solar energy can be collected in troughs as distilled water. Deep-basin stills have a water depth of several feet (between approximately 0.5 and 1.5 m) and require renewal only every few months. Shallow-basin stills have a water depth of about 1 to 5 cm (0.5 to 2.0 in) and have to be fed and flushed out frequently. Solar stills can be designed to also collect rainwater. If the supply water is not contaminated and not too high in solids content, it can be mixed with the distilled water to increase the actual output. This is often done for farm animal water and water used for irrigation.

The glass-covered roof-type solar still (Fig. 7.1.28*a*) is in wide use in arid areas for the production of drinking water from salty or brackish sources. The sun's rays enter through the cover glasses, warm the water, and thus produce vapor which condenses on the inner surface of the cover. The water droplets coalesce and flow downward into the discharge troughs, while the remaining brine is periodically replaced with a new supply of nonpotable water. Daily yield ranges from 2 kg/m^2 (0.4 lb/ft^2) of water surface in winter to 5 kg/m^2 (1.0 lb/ft^2) in summer.

Inflated plastic films (Fig. 7.1.28*b*) have also been used to cover solar stills, but their greatest success has been achieved in controlled-environment greenhouses where the vapor which transpires from plant leaves is condensed and reused at the plant roots. Stills made of inflatable plastic also are equipment in survival kits, on lifeboats, etc. Wicks in some of the solar stills can improve their performance. Most plastics have to be surface-treated for this application to produce film condensation (for good solar transmission) rather than dropwise condensation, which reflects a considerable fraction of the impinging solar radiation.

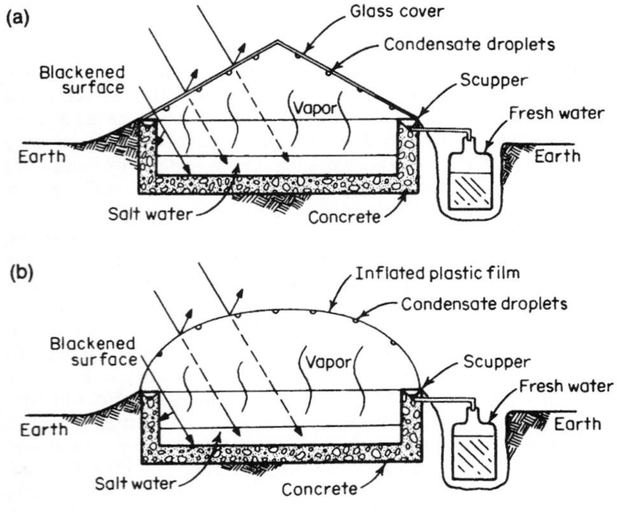

Figure 7.1.28 Shallow-basin horizontal-surface solar stills. (*a*) Glass-covered roof type; (*b*) inflated plastic type.

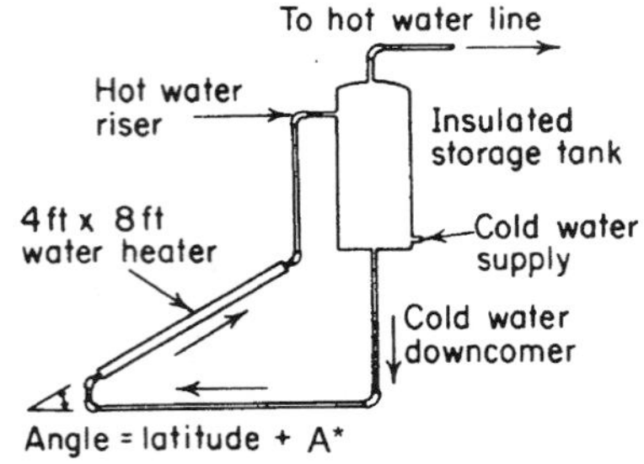

Figure 7.1.29 Thermosiphon type of water heater. (*The collector angle *A* depends on the latitude. More favorable orientation toward the winter sun when days are shorter can produce the same amount of hot water all year round. In Florida, *A* is 10°.)

Solar Water Heaters These can be the pan (batch) type—a tank or basin with transparent cover—or tube collector type, described previously. The simple thermosiphon solar water heater (Fig. 7.1.29) with a glass-covered flat-plate collector is used in thousands of homes all over the world, but mainly in Australia, Japan, Israel, North Africa, and Central and South America. Under favorable climatic conditions (abundant sunshine and moderate winter temperatures) they can produce 110 to 190 L (30 to 50 gal) of water at temperatures up to 70°C (160°F) in summer and 50°C (120°F) in winter. Auxiliary electric heaters are often used to produce higher temperatures during unfavorable winter weather. In the United States and Europe, solar collectors are usually placed on the roof, with the hot water storage tank lower. This requires a small circulating pump. The pump is controlled by a timer, a temperature sensor in the collector, or a differential temperature controller. Ideally, the pump runs when heat can be added to the water in the tank.

To protect the system from freezing, the collectors are drained, manually or automatically. The system can also be designed with a primary circuit containing antifreeze and effecting heat transfer with a heat exchanger in the tank, or by use of a double-walled tank. Another method for protection is a dual system, in which the water drains from the collectors when the pump stops. This is the preferred method for large systems.

Auxiliary heaters, usually electric, are used often in solar water heaters to handle overloads and unusually bad weather conditions.

Solar House Heating and Cooling House heating can be accomplished in temperate climates by collecting solar radiation with flat-plate devices (Fig. 7.1.27) mounted on south-facing roofs or walls (in the northern hemisphere). Water or air, warmed by solar radiation, can be used in conventional heating systems, with small auxiliary fuel-burning apparatus available for use during protracted cloudy periods. Excess heat collected during the day can be stored for use at night in insulated tanks of hot water or beds of heated gravel. Heat-of-fusion storage systems, which use salts that melt and freeze at moderate temperatures, may also be used to improve heat storage capacity per unit volume.

Solar air conditioning and refrigeration can be done with LiBr absorption systems supplied with moderately high-temperature (93°C or 200°F) working fluids from high-performance good flat-plate collectors. Ammonia-water absorption systems require higher temperatures from either evacuated-tube collectors or concentrating collectors. Adsorption cooling systems, in which the refrigerant vapor is adsorbed on the surface of a solid rather than absorbed into a liquid, are also under investigation. Desiccant-based systems that consume water for evaporative cooling have also been developed. The economics of solar energy utilization for domestic purposes become much more favorable when the same collection and storage apparatus can be used for both summer cooling and winter heating.

Figure 7.1.30 is a schematic of a typical active solar house heating, cooling, and hot water system. Flat-plate collectors, roughly 93 m^2 (1,000 ft^2) provide both 3 tons of air conditioning and hot water, properly oriented on the roof (they can actually be the roof), supply

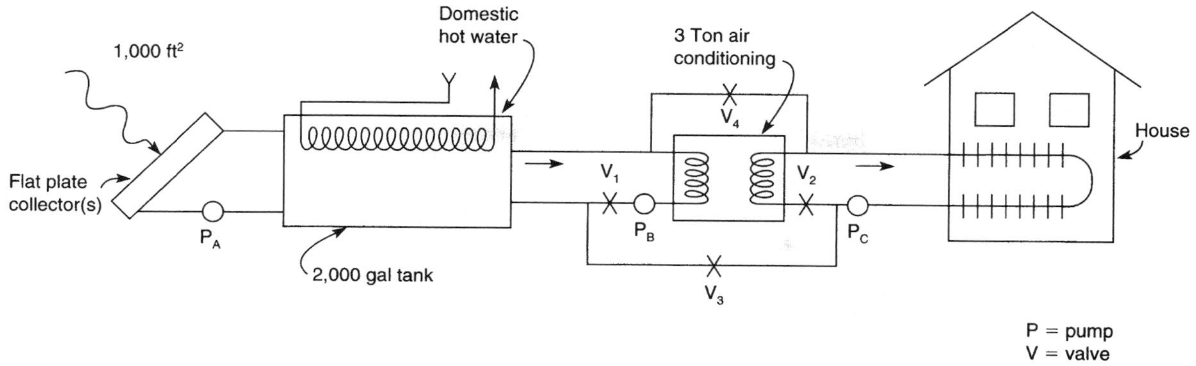

Figure 7.1.30 Schematic of typical solar house heating, cooling, and hot water system.

hot water to a storage tank (usually buried) of about 7,570-L (2,000 gal) capacity. A heat-exchanger coil, submerged in and near the top of the tank, provides domestic hot water. When heat is needed, the thermostat orders the following: Valves V_1 and V_2 close, V_3 and V_4 open, and pump P_C circulates hot water through the house as long as required. When cooling is required, the thermostat orders the following: Valves V_1 and V_2 open, V_3 and V_4 close, and pumps P_B and P_C feed hot water to the air conditioner, which, in turn, produces chilled water for circulation through the house as long as needed. This system can be used over a wide range of latitudes, and only the heating and cooling duty cycles will change (i.e., more heating in the north and more cooling in the south). The proper orientation of the collectors will depend upon the duty cycles. In the northern hemisphere, the collectors will face south. When used mostly for heating, they are inclined to the horizontal by latitude plus up to 20°. When they are used mostly for cooling, the inclination will be latitude minus up to 10°. The basic idea is to orient the collectors more favorably toward the winter sun when mostly heating and more favorably toward the summer sun when mostly cooling. The collectors can be made adjustable at extra cost. Air can be used instead of water with rock bin storage and blowers instead of pumps. Blowers require more energy to run.

The air conditioner can be a conventional type, driven by solar electricity, or a thermally driven system (absorption, adsorption, desiccant, etc.).

Solar cooking utilizes (1) a sun-following broiler-type device with a metallized parabolic reflector and a grid in the focal area where cooking pots can be placed; (2) an oven-type cooker comprising an insulated box with glass covers over an open end which is pointed toward the sun. When reflecting wings are used to increase the solar input, temperatures as high as 204°C (400°F) are reached at midday. Large solar cookers for community cooking in third-world country villages can be floated on water and thus easily adjusted to point at the sun. For cooking when the sun does not shine, oils or other fluids can be heated to a high temperature, 425°C (800°F), with a solar concentrator when the sun shines, and then stored. A range similar to an electric range but with the hot oil flowing through the coils, at adjustable rates, is then used to cook with solar energy 24 h/day.

Solar Furnaces Precise paraboloid concentrators can focus the sun's rays upon small areas, and if suitable receivers are used, temperatures up to 3,600°C (6,500°F) can be attained. The concentrator must be able to follow the sun, either through movement of the paraboloidal reflector itself (Fig. 7.1.26) or by the use of a heliostat which tracks the sun and reflects the rays along a horizontal or vertical axis into the concentrator. This pure, noncontaminating heat can be used to produce highly purified materials through zone refining, in a vacuum or controlled atmosphere. This allows us to grow crystals of high-temperature materials, crystals not existing in nature, or to do simple things such as determining the melting points of exotic materials. Other methods of heating contaminate these materials before they melt. With solar energy the materials can be sealed in a glass or plastic bulb, and the solar energy can be concentrated through the glass or plastic onto the target. The glass or plastic is not heated appreciably since the energy is not highly concentrated when it passes through it.

Solar furnaces are used in high-temperature research, can simulate the effects of nuclear blasts on materials, and at the largest solar furnace in the world (the Odeillo furnace in France) produce considerable quantities of highly purified materials for industry.

Thermal Concentrating Solar Power (CSP) Concentrating Solar Power (CSP) utilizes highly concentrated sunlight to produce high-temperature heat, which can then drive a steam Rankine cycle, an air (or CO_2) Brayton cycle, a Stirling cycle, or other approaches to generate electricity. The chief potential advantage of CSP over utility-scale photovoltaic power plants is the relative ease at which solar energy can be stored in the form of heat. Recent years have seen a renaissance of sorts in CSP, in that development has been spurred by increasing concern over greenhouse gas emissions (GHG) from fossil-fuel power plants. This has led to a number of commercially operating CSP plants in the USA and elsewhere. The type of CSP plant with the longest history (and thus the most operating experience) is that based on parabolic troughs, which typically heat an oil to near 400°C (750°F), which in turn exchanges heat to boil water to drive a steam Rankine power block. One example of such a plant is the Solana Generating Station near Gila Bend, Arizona, USA, which was completed in 2013 and has a gross capacity of 280 MW_e.

Another type of CSP plant is the "power tower" concept, in which an array of heliostats (i.e., mirrors) focuses light on a receiver located at the top of a central tower. The Ivanpah power plant in the Mojave Desert of California became operational in 2013, and has a gross output of 392 MW_e. The heat is supplied directly to water in the receiver to produce superheated steam, which then generates electricity through a conventional Rankine cycle.

Other types of CSP plants have been proposed and developed to some extent, including Stirling engines driven by dish concentrators, huge solar chimneys in which the updraft flow of air through a central chimney spins turbines and hence generates power, and high-temperature Brayton cycles using either air or CO_2 as working fluids. As of this writing, these other approaches have not yet been commercially developed.

Solar Fuels Turning sunlight into gaseous or especially liquid fuels is particularly desirable, since fuels are a highly concentrated source of energy that can be used for transportation or other needs. The production of solar fuels offers an alternative path for storing energy from the sun, as opposed to storing it thermally (in the form of heat) or electrochemically (in batteries). One approach utilizes biomass, such as algae, through solar pyrolysis or gasification. The gasification process involves heating biomass to temperatures above 700°C in the presence of steam, which converts the biomass to a syngas that can then be transformed to a more conventional hydrocarbon fuel. The production of solar fuels from algae can also be done by extracting oil from the algae, which requires CO_2 for its growth and thus provides a "sink" for CO_2. Alternative approaches are thermochemical in nature, and involve either solar thermolysis of water to produce hydrogen and oxygen, or reversible reduction-oxidation reactions with solid metals or ceramics. Thermolysis of water requires temperatures in excess of 2,200°C (4,000°F), and thus poses substantial technical challenges. The alternative reduction-oxidation reactions are two-step processes that enable the solar energy to be stored in the form of a reduced solid material. With appropriate addition of water in the process, hydrogen gas can also be produced.

GEOTHERMAL POWER

by Kenneth A. Phair, Revised by Patrick E. Phelan

REFERENCES: Matek, "2016 Annual U.S. and Global Geothermal Power Production Report," Geothermal Energy Association (also see later editions). *International Energy Outlook 2016*, IEA (and later editions). Interested readers are also encouraged to consult the USA Department of Energy's Geothermal Technologies Office website: https://energy.gov/eere/geothermal/geothermal-energy-us-department-energy.

Geothermal energy is a naturally occurring, renewable source of thermal energy. Thermal energy within the earth approaches the surface in many different geologic formations. Volcanic eruptions, geysers, fumaroles, hot springs, and mud pots are visual indications of geothermal energy.

Significant **geothermal reserves** exist in many parts of the world. Electric power was first generated from geothermal energy in 1904. Active worldwide development of geothermal resources began in earnest in 1960 and continues. Table 7.1.8 presents the top ten countries in terms of operating geothermal capacity in 2015, and shows that the USA leads with 3,567 MWe. The total worldwide operating capacity was about 13.3 GWe as of January 2016, distributed across 24 countries. Table 7.1.8 also shows planned capacity additions for each of these ten countries, as well as estimated potential capacity. There is apparently considerable scope to increase geothermal power production, assuming technical and economic issues can be overcome.

Geothermal power plants are typically found in areas with "recently" active volcanoes, crustal stretching, and continuing seismic activity. Since the mid-1990s, significant additional geothermal power generation facilities were installed in Indonesia, Italy, Mexico, and the Philippines. Many countries not included in Table 7.1.8 also have significant geothermal resources that are not yet developed.

Although geothermal energy is a renewable resource, economic development of geothermal resources usually extracts energy from the reservoir at a much higher rate than natural recharge can replenish it. Therefore, facilities that use geothermal energy should be designed for high efficiency to obtain maximum benefit from the resource.

Geothermal Resources Geothermal resources may be described as hydrothermal, hot dry rock, geopressured, or magma.

Table 7.1.8 Top 10 Countries for Geothermal Operating Capacity (Adapted from 2016 Annual U.S. and Global Geothermal Power Production Report)

Country	Potential Capacity (MWe)	Operating Capacity as of 2015 (MWe)	Planned Capacity Additions as of 2015 (MWe)
USA	9,057	3,567	1,272
Philippines	3,000 to 4,000	1,930	587
Indonesia	29,000	1,375	4,013
Mexico	10,500	1,069	481
New Zealand	3,600	973	285
Italy	2,000	944	unknown
Iceland	4,300 to 7,000	665	575
Turkey	1,500	637	1,153
Kenya	10,000	607	1,091
Japan	23,000	533	57

Hydrothermal resources contain hot water, steam, or a mixture of water and steam. These fluids transport thermal energy from the reservoir to the surface. Reservoir pressures are usually sufficient to deliver the fluids to the surface at useful pressures, although some liquid-dominated resources require downhole pumps for fluid production. Hydrothermal resources may be geologically closed or open systems. In a closed system, the reservoir fluids are contained within an essentially impermeable boundary. Communication and fluid transport within the reservoir occur through fractures in the reservoir rock. There is little, if any, natural replenishment of fluids from outside the reservoir boundary. An open system allows influx of cold subsurface fluids into the reservoir as the reservoir pressure decreases. Hydrothermal reservoirs have been found at depths ranging from 122 m (400 ft) to over 3,050 m (10,000 ft).

Hot dry rock resources are geologic formations that have high heat content but do not contain meteoric or magmatic waters to transport thermal energy. Thus water must be injected to carry the energy to the surface. The difficulty in recovering a sufficient percentage of the injected water and the limited thermal conductivity of rock have hindered development of hot dry rock resources. Because of the vast amount of energy in these resources, additional research and development is justified to evaluate whether it is technically exploitable. Research and development of hot dry rock resources is proceeding in Australia, France, Japan, and the United States.

Geopressured resources are liquid-dominated resources at unusually high pressure. They occur between 1,500 and 6,100 m (5,000 and 20,000 ft), contain water that varies widely in salinity and dissolved minerals, and usually contain a significant amount of dissolved gas. Pressures in such reservoirs vary from about 21 to 96 MPa (3,000 to 14,000 lb/in^2 gage) with temperatures between 60°C (140°F) and 182°C (360°F). The largest geopressured zones in the United States exist beneath the continental shelf in the Gulf of Mexico, near the Texas, Louisiana, and Mississippi coasts. Other zones of lesser extent are scattered throughout the United States.

Magma resources occur as formations of molten rock that have temperatures as high as 700°C (1,300°F). In most regions in the continental United States, such resources occur at depths of 30,500 m (100,000 ft) or more.

Exploration Technology Geothermal sites historically have been identified from obvious surface manifestations such as hot springs, fumaroles, and geysers. Some discoveries have been made accidentally during exploring or drilling for other natural resources. This approach has been replaced by more scientific prospecting methods that appraise the extent, as well as the physical and thermodynamic properties, of the reservoir. Modern methods include geological studies involving aerial, surface, and subsurface investigations (including remote infrared sensing) and geochemical analyses which provide a guide for selecting specific drilling sites. Geophysical methods include drilling, measuring the temperature gradient in the drill hole, and measuring the thermal conductivity of rock samples taken at various depths.

Resource Development Extraction of fluids from a geothermal resource entails drilling large-diameter production wells into the reservoir formation. Bottom hole temperatures in hydrothermal wells and hot dry rock formations can exceed 260°C (500°F). Geopressured resources have lower temperatures but offer energy in the form of fluids at unusually high pressures that frequently contain significant amounts of dissolved combustible gases. Although research and development projects continue to seek ways to efficiently extract and use the energy contained in hot dry rock, geopressured, and magma resources, virtually all current geothermal power plants operate on hydrothermal resources. Advances in technology are required to extract the thermal energy in many known geothermal resources, such as the Enhanced Geothermal Systems under development by the US Department of Energy's Geothermal Technologies Office.

Production Facilities For most projects, a number of wells drilled into different regions of the reservoir are connected to an aboveground piping system. This system delivers the geothermal fluid to the power plant. As with any fluid flow system, the geothermal reservoir, wells, and production facilities operate with a specific flow versus pressure relationship. Resource permeability; the number, depth, and size of the wells; and the surface equipment and piping arrangement all contribute to make the deliverability curve different for each power plant. Production of geothermal fluids over time results in declining deliverability. For one-half or more of the operating life of a reservoir, the deliverability can usually be held constant by drilling additional production wells into other regions of the resource. As the resource matures, this technique ceases to provide additional production. The deliverability curve begins to change shape and slope as deliverability declines. The power plant design must be matched to the deliverability curve if maximum generation from the resource is to be achieved.

Geothermal Power Plants A steam-cycle geothermal power plant is very much like a conventional fossil-fueled power plant, but without a boiler. There are, however, significant differences. The turbines, condensers, noncondensable gas removal systems, and materials used to fabricate the equipment are designed for the specific geothermal application. With geothermal steam delivered to the power plant at approximately 689 kPa (100 lb/in^2 gage), only the low-pressure sections of a conventional turbine generator are used. Additionally, the geothermal turbine must operate with steam that is far from pure. Chemicals and compounds in solid, liquid, and gaseous phases are transported with the steam to the power plant. At the power plant, the steam passes through a separator that removes water droplets and particulates before it is delivered to the turbine. Geothermal turbines are of conventional design with special materials and design enhancements to improve reliability in geothermal service. Turbine rotors, blades, and diaphragms operate in a wet, corrosive, and erosive environment. High-alloy steels, stainless steels, and titanium provide improved durability and reliability. Still, frequent overhauls are necessary to maintain reliability and performance. The high moisture content and the corrosive nature of the condensed steam require effective moisture removal techniques in the later (low-pressure) stages of the turbine. Scale formation on rotating and stationary parts of the turbine occurs frequently. Water washing of the turbine at low-load operation is sometimes used between major over-hauls to remove scale.

Most geothermal power plants use direct-contact condensers. Only when control of hydrogen sulfide emissions has been required or anticipated have surface condensers been used. Surface condensers in geothermal service are subject to fouling on both sides of the tubes. Power plants in The Geysers in northern California use conservative cleanliness factors to account for the expected tube-side and shell-side fouling. Some plants have installed on-line tube-cleaning systems to combat tube-side fouling on a continuous basis, whereas other plants mechanically clean the condenser tubes to restore lost performance.

Noncondensable gas is transported with the steam from the geothermal resource. The gas is primarily carbon dioxide but contains lesser amounts of hydrogen sulfide, ammonia, methane, nitrogen, and other gases. Noncondensable gas content can range from 0.1 percent to more than 5 percent of the steam. The makeup and quantity of noncondensable gas vary not only from resource to resource but also from well to well within a resource. The noncondensable gas removal system for a geothermal power plant is substantially larger than the same system for a conventional power plant. The equipment that removes and compresses the noncondensable gas from the condenser is one of the largest consumers of auxiliary power in the facility, requiring up to 15 percent of the thermal energy delivered to the power plant. A typical system uses two stages of compression. The first stage is a steam jet ejector. The second stage may be another steam jet ejector, a liquid ring vacuum pump, or a centrifugal compressor. The choice of equipment selected for the second stage is influenced by project economics and the amount of gas to be compressed.

The chemicals and compounds in geothermal fluids are highly corrosive to the materials normally used for power plant equipment and facilities. The chemical content of geothermal fluids is unique to each resource; therefore, each resource must be evaluated separately to determine suitable materials for system components. Carbon steel usually will degrade at alarmingly high rates when exposed to geothermal fluids. Corrosion-resistant materials such as stainless steel may perform satisfactorily, but may experience rapid, unpredictable local failures depending upon the composition of the geothermal fluid. Based on experience with a number of geothermal resources, carbon steel with a corrosion allowance is usually suitable for transporting dry geothermal steam. Because noncondensable gas is also corrosive, special materials are usually required. Copper is extremely vulnerable to attack from the atmosphere surrounding a geothermal power plant. Therefore, copper wire and electrical components should be protected with tin plating and isolated from the corrosive atmosphere. Within the context of these generalities, the fluids at each resource must be evaluated before construction materials are chosen.

The steam Rankine cycle used in fossil-fueled power plants is also used in geothermal power plants. In addition, a number of plants operate with binary cycles. Combined cycles also find application in geothermal power plants. The basic cycles are shown in Fig. 7.1.31.

The direct steam cycle shown in Fig. 7.1.31*a* is typical of power plants at The Geysers in northern California, the world's largest geothermal field. Steam from geothermal production wells is delivered to power plants through steam-gathering pipelines. The wells are up to 1.6 km (1 mi) or more from the power plant. The number of wells required to supply steam to the power plant varies with the geothermal resource as well as the size of the power plant. Each 55-MWe power plant in The Geysers receives steam from between 8 and 23 production wells.

A flash steam cycle for a liquid-dominated resource is shown in Fig. 7.1.31*b*. Geothermal brine or a mixture of brine and steam is delivered to a flash vessel at the power plant by either natural circulation or pumps in the production wells. At the entrance to the flash vessel, the pressure is reduced to produce flash steam, which then is delivered to the turbine. This cycle has been used at power plants in California, Nevada, Utah, and many other locations around the world. Increased thermal efficiency is available from the use of a second, lower-pressure flash to extract more energy from the geothermal fluid. However, this technique must be approached carefully because dissolved solids in the geothermal fluids will concentrate and may precipitate as more steam is flashed from the fluid. The solids also tend to form scale at lower temperatures, resulting in clogged turbine nozzles and rapid buildup in equipment and piping to unacceptable levels.

A binary cycle is the economic choice for hydrothermal resources with temperatures below approximately 166°C (330°F). A binary cycle uses a secondary heat-transfer fluid instead of steam in the power generation equipment. A typical binary cycle is shown in Fig. 7.1.31*c*. Binary cycles can be used to generate electric power from resources with temperatures as low as 121°C (250°F). The binary cycle shown in Fig. 7.1.31*c* uses isobutane as the heat-transfer fluid. It is representative of units of about 10-MWe capacity. Many small modular units of 1- or 2-MWe capacity use pentane as the binary fluid. Heat from geothermal brine vaporizes the binary fluid in the brine heat exchanger. The binary fluid vapor drives a turbine generator. The turbine exhaust vapor is delivered to an air-cooled condenser where the vapor is condensed. Liquid binary fluid drains to an accumulator vessel before being pumped back to the brine heat exchangers to repeat the cycle. Binary-cycle geothermal plants are in operation in several countries. In the United States, they are located in California, Nevada, Utah, and Hawaii.

A geothermal combined cycle is shown in Fig. 7.1.31*d*. Just as combustion-based power plants have achieved improved efficiencies by using combined cycles, geothermal combined cycles also show improved efficiencies. Some power plants in the Philippines use a combination of steam and binary cycles to extract more useful energy from the geothermal resource. Existing steam-cycle plants can be modified with a binary bottoming cycle to improve efficiency.

Cycle optimization is critically important to maximize the power generation potential of a geothermal resource. Selecting optimum cycle design parameters for a geothermal power plant does not follow the practices used for fossil-fueled power plants. While a higher turbine inlet pressure will improve the efficiency of the power plant, a lower turbine inlet pressure may result in increased generation over the life of the resource. The resource deliverability curve is used with turbine and cycle performance predictions to determine the flow and turbine inlet

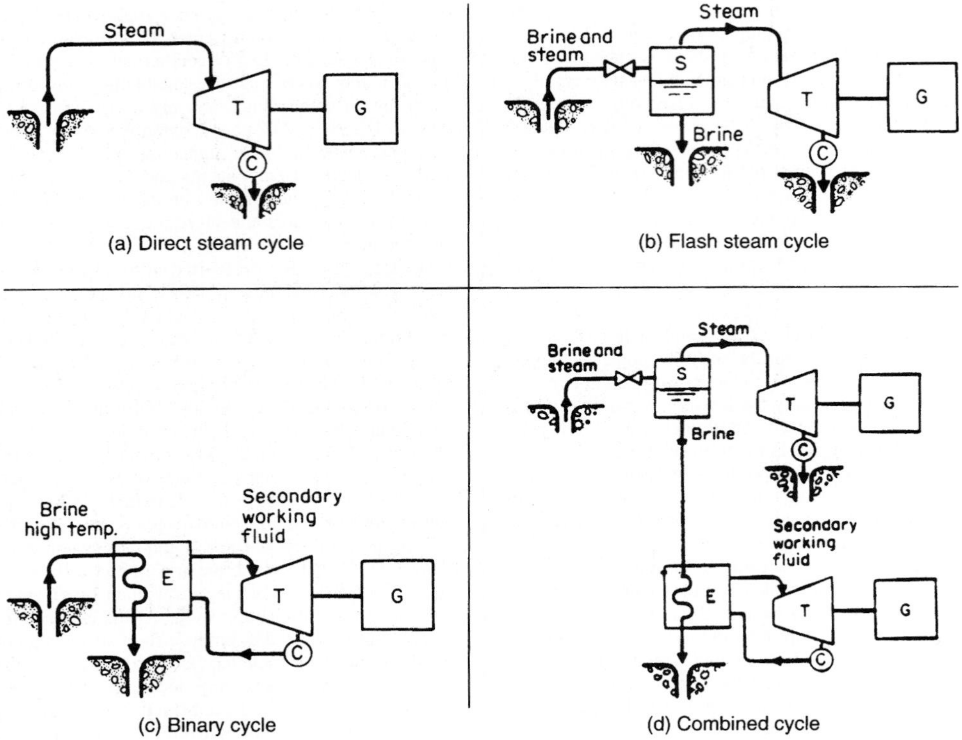

Figure 7.1.31 Geothermal power cycles. T = turbine, G = generator, C = condenser, S = separator, E = heat exchanger.

pressure that will yield maximum generation. The technical optimum must then be subjected to an economic analysis to identify the best parameters for the power plant design. Because the shape and slope of the deliverability curve vary from resource to resource, the optimum turbine inlet conditions will likewise vary.

Direct Use There are substantial geothermal resources with temperatures less than 121°C (250°F). While these resources cannot currently be used to generate electric power economically, they can be used for various low-temperature direct uses. Services such as district heating, industrial process heating, greenhouse heating, food processing, and aquaculture farming have been provided by geothermal fluids. For these applications, corrosion and fouling of surface equipment must be addressed in the system design.

The **geothermal heat pump (GHP)** (also called a ground-source heat pump, GSHP) is another direct use of the earth's thermal energy. The GHP, however, does not require a high-temperature geothermal reservoir. The GHP uses essentially constant-temperature groundwater as a heat source or heat sink in a conventional, reversible, water-to-air heat pump cycle for building heating or cooling. The ground, groundwater, and local climatic conditions must be included in the design of a GHP for a specific location. Systems are currently available for residential (single- and multifamily) dwellings, offices, small industrial buildings, and campuses.

Environmental Considerations Geothermal fluids contain many chemicals and compounds in solid, liquid, and gaseous phases. For both environmental protection and resource conservation, spent geothermal liquids are returned to the reservoir in injection wells. This limits the release of compounds to the environment to a small amount of liquid lost as drift from the cooling tower and noncondensable gases. Problems with arsenic, boron, and mercury contamination have been encountered in the immediate vicinity of cooling towers at geothermal power plants. The noncondensable gases, composed primarily of carbon dioxide, usually also contain hydrogen sulfide. Along with its noxious odor, hydrogen sulfide is hazardous to human and animal life. Although many geothermal power plants do not currently control the release of hydrogen sulfide, others use process systems to oxidize the hydrogen sulfide to less toxic compounds. A number of the process systems produce 99.9 percent pure sulfur that can be sold as a by-product. Using geothermal energy for power generation and other direct applications provides environmental benefits. Carbon dioxide released from a geothermal power plant is approximately 90 percent less than the amount released from a combustion-based power plant of equal size, and they create little, if any, liquid or solid waste.

TIDAL, WAVES, AND OCEAN THERMAL AND SALINITY GRADIENT ENERGY

by E. A. Avallone, Revised by M. Zenouzi

References: R.A. Dunlap, "Sustainable Energy," Cengage Learning, 2015. J.W. Tester, E.M. Drake, M.J. Driscoll, W. Golay, W.A. Peters, "Sustainable Energy—Choosing Among Options," The MIT Press, 2005.

Energy can be harvested from the potential or kinetic energy associated with the movement of water in the ocean because of tides and waves. Ocean thermal energy conversion (OTEC), which is basically a heat engine, is used to convert a thermal energy gradient to mechanical energy in the tropical regions since there is at least 20°C temperature difference between warm surface water (25°C, 77°F) and cold water at a depth of more than 1,000 m [5°C (41°F)].

Mixing river freshwater and salty water (seawater or saline brine) in a controlled fashion would produce an energy known as salinity gradient energy (SGE). The harnessing of this energy for conversion into power can be accomplished by means of Pressure Retarded Osmosis (PRO) and Reverse Electrodialysis (RED), which are two different membrane-based technologies.

All these methods will be discussed in the following sections and the most promising devices for harnessing this renewable energy will be mentioned.

Tidal Power

REFERENCES: The Rance Estuary Tidal Power Project, *Pub. Util. Ftly.*, Dec. 3, 1964. *Mech. Eng.*, Apr. 1984. ASCE Symposium, 1987: Tidal Power. Gray and Gashus, "Tidal Power," Department of Commerce, NOAA, Water for Energy, *Proc. 3d Intl. Symp.*, 1986.

The tides are a renewable source of energy originating in the gravitational pull of the moon and sun, coupled with the rotation of the earth. The consequent portion of the earth's rotation is a mean ocean tide of 2 ± ft (0.6 m). The seashore periodic variation of the tides averages 12 h 25 min or 706 tidal cycles per year.

Tidal power is derivable from the large periodic variations in tidal flows and water levels in certain oceanic coastal basins. Suitable configurations of the continental shelves and of the coastal profiles result in reflection and resonance that amplify normally small bulges to ranges as high as 50 ± ft (15 ± m). Tidal power can be harnessed either through a barrage or through submerged tidal turbines in straits.

Principal tidal-power sites include the North Sea [12 ft (3.6 m) average tidal range]; the Irish Sea [22 ft (6.7 m)]; the west coast of India [23 ft (7 m)]; the Kimberly coast of western Australia [40 ft (12 m)]; San Jose Bay on the east coast of the Argentine [23 ft (7 m)]; the Kislaya Guba (Kisgalobskaia Bay) near the White Sea (no data); St. Michel (including the Rance estuary) on the Brittany coast of France [26 ft (8 m)]; the Bristol Channel (Severn) in England [32 ft (9.8 m)]; the Bay of Fundy (including the Chignecto Bay between New Brunswick and Nova Scotia and the Minas Basin in Nova Scotia) [40 ft (12 m)]; Passamaquoddy Bay between Maine and New Brunswick [18 ft (5.5 m)].

The harnessing of the tides reaches back into ancient history. Tidal mills, typically with undershot water wheels, were used in New England raceway estuaries, with reversible features for ebb and flood conditions. These power applications were suitable for purposes such as grinding grain, but their number and size were small. The world's largest tidal power system, the Sihwa Lake Tidal Power Station located in South Korea, was opened in 2011. The tidal barrage dam has a power of 254 MW, is 7.9 miles long, and uses ten 26 MW turbines positioned on the ocean floor. The area of the basin is 17 square miles (43 km^2). There are only a few other tidal power developments in actual service (1995)—the Rance estuary in France (240 MW), the Kislaya Guba in Russia (400 kW), the Bay of Fundy in Canada (20 MW), and a small pilot plant in Kiangshia, China. The amount of energy available from a tidal development is proportional to the basin area, *A*, and to the square of the tidal range, *h*. Head variations are large in tidal projects during generating cycles and on a daily, monthly, and annual basis owing to various cosmic factors. Annual average generation of a single-basin one-way operation can be estimated by [Tester et al., 2005]:

$$\mathrm{E} = 0.986 \times 10^6 \times COP \times h^2 \times A,\ kWh,$$

where *h* is in meters, *A* is in square kilometers, and *COP* is an overall coefficient of performance, on the order of 0.2-0.35, to account for a variety of factors. This equation predicts annual generation of 517 GWh/yr for the La Rance station in France, using $h = 8.5$ m, $A = 22$ km^2, and COP = 0.33, which is about what is actually achieved.

Intermittent power, as from all single-basin plans and from two-basin plans with low-capacity factor, implies that the output can best be utilized as peaking capacity. Because of low heads, particularly toward the end of any generating cycle when pools have been drawn down, the cost of adding generating units only to tidal projects is well over the total installation cost of alternative peaking capacity. To be economically competitive with alternative capacity, the tidal projects must produce enough energy to pay the power plant costs and also to pay for the dams and other costs, such as general site development, transmission, operation, and replacements. The **risks and uncertainties** involved in designing, pricing, building, and operating capital-intensive tidal works, and the technological developments in alternative types of generating capacity, have tended to defeat tidal developments. Civil works are too extensive; transmission distances to load centers are too great; the required scale of development is too large for existing loads; the coordination of system demands and tidal generation requires interconnections for economic loading; the ultimate capacity of all the world's tidal potentials is practically insignificant to meet the world's demands for electricity. In addition, the matter of interrupting tidal action and storage of tidewaters in large basins for extended periods of time raises the possibility of saltwater infiltration of adjacent underground freshwater supplies which, in many cases, are the source of drinking water for the contiguous areas.

Tidal Current the aforementioned environmental and economic factors make the use of underwater turbines, which could be anchored to the bottom of a river, or streams that empty into the ocean, more attractive than tidal range. They make use of the kinetic energy associated with tidal oscillating currents and rely more on the current's speed and volume than actual tidal range, and thus can be more attractive. The tidal stream average power generated by the turbine can be calculated in a way similar to wind power as:

$$P_e = \left(\frac{1}{4}\right)\eta\ \rho U^3_{\max}$$

where ρ is the density of water, *A* is the area of water intercepted by the device, η is the efficiency of tidal stream power to electricity, and $U_{\max}$ is the maximum speed of the periodic current.

A number of tidal current power systems have been installed and are in operation; SeaGen, on the east coast of Northern Ireland, is the largest tidal current systems. Verdant Power was the first company in the USA to be granted a license for a commercial tidal energy project. They designed and installed several underwater turbines, each with three fixed 5 m (16 ft) blades rotating at approximately 40 rpm in the current. Each turbine/generator produces 35 kW of electricity. The project is scheduled to install up to 300 turbines in New York's East River.

Gorlov's Helical Turbine is a cross flow tidal energy turbine, consisting of two or three long helical blades that run along a cylindrical surface like a screw thread. It is a very efficient turbine, having an airfoil profile most suitable for tidal energy conversion due to its ability to develop unidirectional rotation, even in reversible tidal current.

Wave Energy and Power

REFERENCES: D. Buchla, T. Kissell, T. Floyd, "Renewable Energy Systems," Pearson Education Inc., 2015.

As the wind blows over the surface of the ocean, it transfers some of its energy to the water forming waves that store this energy as potential energy and kinetic energy. The wave energy tends to be consistently available, even more so than tides.

The total potential and kinetic energy of a series of **trochoidal deep-sea waves** may be expressed as follows (Tester, 2005):

$$P = 1.96\ H^2T,\ \mathrm{kW/m}$$

where *H* is the crest-to-trough wave height in meters and *T* is the wave period-time for successive crests to pass a fixed point in seconds. For $H = 4$ m and $T = 6$ s, $P = 188$ kW/m. The lack of enthusiasm for wave power devices has largely been the result of their vulnerability to damage in storms, when linear power densities in excess of 200 kW/m may have to be withstood. Not much more than a quarter of the total energy of such waves would probably be available after reaching shallow water, and apparatus rugged enough for this purpose would doubtless be able to utilize more than a third of this amount.

The main devices for harnessing wave power are: **Oscillating Wave Columns** which are primarily onshore devices that are typically permanently attached to the shoreline or anchored close to the shore, while floats, pitching devices, and wave-focusing devices are typically located offshore.

The oscillating column uses wave motion to force air through a turbine to drive a generator to produce electricity. A wave-actuated air turbine and electric generator have been operating on the Norwegian coast to study feasibility and to gather operating data. Another similar unit has been emplaced recently off the Scottish coast, and while serving to provide operational data, it also feeds about 2 MW of power into the local grid.

Floating wave energy converters are in a fixed position; they move up and down from wave motion. The motion with respect to a fixed reference is captured and the energy is converted to electricity using electromechanical or hydraulic energy converters.

Attenuator devices such as the **Pelamis wave energy system,** are composed of large semi-submerged cylindrical sections joined together at hinge joints and placed perpendicular to the wave front and are anchored to the seafloor. A passing wave causes each of the tubes to rise and fall independently. The movement tugs at the universal joints linking the tubes. The joints use a hydraulic cylinder to act as a pumping system and push high pressure oil through a series of hydraulic motors that drive an electrical generator.

The first prototype of Pelamis wave energy was installed at a site off the Portuguese Atlantic coast and fed electric power of up to 1.5 MW onto the Portuguese power grid. The latest version consists of five semi-submerged cylindrical sections; each section is 180 m long and 4 m in diameter, and weighs 1,350 tons.

Ocean Thermal Energy Conversion (OTEC)

REFERENCES: Claude and Boucherot, *Compt. trend,* 183, 1926, pp. 929-933. *The Engineer,* 1926, p. 584. Anderson and Anderson, *Mech. Eng.*, Apr. 1966. Othmer and Roels, Power, Fresh Water, and Food from Cold, Deep Sea Water, *Science,* Oct. 12, 1973. Roe and Othmer, *Mech. Eng.*, May 1971. Veziroglu, "Alternate Energy Sources: An International Compendium." Department of Commerce. NOAA, Water for Energy, *Proc. 3d Intl. Symp.*, 1986. Current *Proc. of Third World Water Forum.*

Deep seawater, for example, at 1-mile (1.6-km) depth in some tropical regions, may be as much as 50°F (28°C) colder than the surface water. This difference in temperature is a fundamental challenge to the power engineer, as it offers a potential for the conversion of heat into work. The Carnot cycle specifies the limits of conversion efficiency. Typically, with a heat source surface temperature $T_1 = 85°F$ (545°R, 29.4°C), and a heat sink temperature T_2, 50° lower, or $T_2 = 35°F$ (495°R, 1.7°C), the ideal Carnot cycle thermal efficiency = $(T_1 - T_2)/T_1 = [(545 - 495)/545]\ 100 = 9.2$ percent. Some units in experimental or pilot operation have demonstrated actual thermal efficiency in the range of 2 to 3 percent. These efficiencies, both ideal and actual, are far lower than those obtained with fossil or nuclear fuel-burning plants. The fundamental attraction of **ocean thermal energy conversion (OTEC)** is the vast quantity of seawater exhibiting sufficient difference in temperature between shallow and deep layers. In reality, the sea essentially represents a limitless store of solar energy which manifests itself in the warmth of seawater, especially in the top, shallow layers. Water temperatures fluctuate very little over time; thus the thermodynamic properties are relatively constant. In addition, the thermal energy is available on a 24-h basis, can be harnessed to serve a plant on land or offshore, provides "free" fuel, and results in a nonpolluting recycled effluent.

Considering the extent of the ocean between latitudes of 30°S and 30°N, and ascribing a temperature difference between shallow and very deep waters of 20°F, the theoretical energy content comes to about 8×10^{21} Btu. Of this enormous amount of raw energy, the actual amount posited for eventual recovery in the best of circumstances via an OTEC system is a miniscule percentage of that total. The challenge is to develop practical machinery to harness thermal energy of the sea in a competitive way.

The concept was put forward first in the nineteenth century, and it has been reduced to practice in several experimental or pilot plants in the past decades. There are three basic conversion schemes: closed Rankine cycle, open Rankine cycle, and mist cycle. In the closed cycle, warm surface water is pumped through a heat exchanger (boiler) which transfers heat to a low-temperature, high-vapor-pressure working fluid (e.g., ammonia). The working fluid vaporizes and expands, drives a turbine, and is subsequently cooled by cold deep water in another heat exchanger (condenser). The heat exchangers are large; the turbine, likewise, is large, by virtue of the low pressure of the working fluid flowing through and enormous quantities of water are pumped through the system. In the open cycle, seawater itself is the working fluid. Steam is generated by flash evaporation of warm surface water in an evacuated chamber (boiler), flows through a turbine, and is cooled by pumped cold deep water in a direct-contact condenser. The reduction of heat-transfer barriers between working fluid and seawater increases the overall system efficiency and requires smaller volumes of seawater than those used in the closed cycle. Introduction of a closed-cycle heat exchanger into the open-cycle scheme results in a slightly reduced thermal efficiency, but provides a valuable by-product in the form of freshwater, which is suitable for human and animal consumption and may be used to irrigate vegetation. A plant of this last type operates in Hawaii and generates 210 kW of electric power, with a 50-kW net surplus power available after supplying all pumping and other house power. This plant continues to provide operational and feasibility data for further development. An experimental installation is proposed offshore in the island nation of Palau, in the western Pacific. The project is dual-purpose: it will provide electric power as well as potable drinking water, which results from the processing of cold seawater through low-pressure distilling equipment. Favorable results may be extended to other island enclaves that depend on rainwater as their sole source of drinking water. In that case potable water could be the main product while the generation of power the secondary.

The mist cycle mimics the natural cycle that converts evaporated seawater to rain, which is collected and impounded and ultimately flows through a hydraulic turbine to generate power. Problems encountered in the implementation of any OTEC system revolve on material selection, corrosion, maintenance, and significant fouling of equipment and heat-transfer surfaces by marine flora and fauna. Although development of OTEC systems will continue, with a view to their application in fairly restricted locales, there is little prospect that the systems will make any significant impact on power generation in the foreseeable future.

Osmotic Energy

Osmotic energy is the entropy energy extraction technique from mixing of fresh (river) water and salty water (seawater or saline brine) in a controlled fashion. The energy produced from water salinity is a renewable energy source that is free of CO_2 emissions, and is known as Blue Energy. The membrane-based conversion of salinity gradient energy (SGE) is the most promising technology for large-scale applications. Two main membrane-based conversion technologies of SGE are PRO and RED.

In PRO, semipermeable membranes placed between two streams of solutions allow the passage of water from low-pressure diluted solution—referred to as the "feed solution" to high-pressure concentrated solution—referred to as the "draw solution" and rejecting solute molecules or ions. Osmotic pressure is the pressure that would cease the passage of feed solution across the semipermeable membrane if applied to the draw solution.

The osmotic pressure (π) of any solution is approximated by:

$$\pi = iRcT$$

where i is the number of osmotically active particles in the solute (i = 2.0 for NaCl), R is the universal gas constant (8.31441 $J \cdot mol^{-1} \cdot K^{-1}$), c is the molar concentration ($mol \cdot L^{-1}$), and T is the absolute temperature (K). For seawater at 10°C and a molar concentration of 0.51 to 0.68 mol/L (~30 – 40 g L^{-1}), the osmotic pressure is $2.4 - 3.2 \times 10^6\ N/m^2$ or (24 – 32) bar. The osmotic pressure for the Great Salt Lake is about 200 bar, for the Dead Sea about 290 bar and for freshwater is close to zero.

The volume flow rate and pressure of the saltier water increases as it draws the less concentrated water through the semipermeable membrane, so a hydro turbine could be used to produce power.

The amount of water transfer from the low salinity to the high salinity side depends on the osmotic pressure difference ($\Delta\pi$), the hydraulic pressure difference (ΔP), the water permeability of the membrane,

Table 7.1.9 Bulk and Calorific Power of Selected Fuels (Approximate and Comparative)

Fuel	State	Sp. wt., lb/ft³	Btu/lb	Btu/ft³	Btu/gal
Hydrogen gas	Gas (NTP)	0.0052	61,013	320	N/A
Natural gas	Gas (NTP)	0.042	22,453	1,000	N/A
Gasoline (Reg.90 Oct.)	Liquid (NTP)	44.09	20,500	950,000	122,000
Ethanol (99 Oct.)	Liquid (NTP)	49.3	12,800	620,000	84,530
Methanol (98 Oct.)	Liquid (NTP)	49.4	9,600	480,000	65,200
Hydrogen (130 Oct.)	Liquid (20 K, 1 Atm.)	4.4	56,000	240,000	36,020
Coal	Piled	50	12,000	600,000	80,000

and the effective membrane area. This can be given by the following equation:

$$\dot{m}_w = \rho \times \mathcal{A}_w(\Delta\pi - \Delta P) = \rho \times \Delta Q$$

where $\mathcal{A}_w$ (m^3/s Pa) is the water permeability of the membrane multiplied by the effective membrane area and ρ (kg/m^3) is the density of the water.

The power that can be generated is equal to the product of the water flow rate and the hydraulic pressure differential across the membrane:

$$\dot{W}_{PRO}(W) = \Delta P(Pa) \times \Delta Q\left(\frac{m^3}{s}\right)$$

It can be shown that the power reaches a maximum when $\Delta P = \Delta\pi/2$.

In RED, a migration of ions through two different semipermeable membranes generates a DC voltage that can be accessed directly at the electrodes located at each end of the RED system. RED systems are composed of many stacked "cells" which consist of alternating channels of freshwater and saltwater separated by alternating anion exchange membranes (AEM) and cation exchange membranes (CEM). The concentration gradient between the channels forces the diffusion of ions through the membranes to the freshwater channel, while the semipermeable membranes do not allow the flow of H_2O. The power generated by a stack of N cells is given by:

$$\dot{W}_{RED}(W) = I \times (N \times \Delta V)$$

where I is the electrical current or transported charge (A) and $N \times \Delta V$ is the potential difference over the applied external load (V). The potential difference over each membrane (membrane potential) is 80 mV for seawater and freshwater.

The lifetime and power density of the semipermeable membranes are two main factors affecting deployment of PRO and RED. Semipermeable membranes with lifetimes greater than 10 years and power densities higher than 5 W/m^2 would lead to faster development and commercial energy production of these conversion technologies.

POWER FROM HYDROGEN

Updated by Roy McAlister

Hydrogen is an energy carrier for providing energy storage along with many attractive properties as a fuel. Hydrogen is a "clean" fuel because it combusts without producing CO, CO_2 or particulates, and if burned with oxygen, it produces condensable water vapor as the sole end product.

If hydrogen is burned with air at sufficiently high combustion temperature, some of the nitrogen may combine with oxygen to form oxides of nitrogen such as N_2O, a problematic air contaminant and greenhouse gas. If carbon is a fuel constituent, or if it is present from a source such as a lubricant, the carbon source can introduce further contaminant potentials, for example, carbon monoxide, benzene, and greenhouse gas constituents including carbon dioxide and particulates.

The potential cleanliness of hydrogen combustion and fuel cell applications is supported by other properties that make hydrogen an attractive fuel choice, including widespread availability as a constituent of water, fossil hydrocarbons, and renewable organic materials along with high energy-value/mass and a wide flammability range. Hydrogen can be safer than gasoline and other hydrocarbons because it escapes rapidly if leaked as it is lighter than air, has a higher ignition temperature, extremely high escape diffusivity to rapidly dilute below the flammability limit with air, low flame emissivity, and if burned it forms water vapor instead of poisonous substances such as carbon monoxide.

Hydrogen can be directly injected into almost all designs of internal combustion engines without throttling of intake air to achieve better than Diesel engine thermal efficiency and actually clean the air during combustion. Formation of oxides of nitrogen (NO_x) can be prevented by adaptive control of the rate of hydrogen injection to suitably limit the peak combustion temperature.[1,6,10]

Hydrogen offers a uniquely high calorific value of 140,000 kJ/kg (61,000 Btu/lb). However, it has a relatively low energy/volume as a gas with the specific volume of 12 m^3/kg (190 ft^3/lb) or 12,000 kJ/m^3 (319 Btu/ft^3) at normal temperature and pressure (NTP), that is, at 0°C, 1 bar (32°F, 14.7 lb/in²). Moreover this relatively low energy/volume introduces many practical challenges because hydrogen has a critical point of 33 K (–400°F) at 12.8 atm. Thus hydrogen is a gas at all temperatures above the cryogenic temperature of 33 K (–400°F) at 12.8 atm., or about 20 K at 1 bar atmospheric pressure. Properties of other traditional and alternative fuels are compared in Table 7.1.9 in quantities typical for dispensing gasoline and distributing natural gas in the USA.

Table 7.1.9 comparisons demonstrate the relative volumetric shortcoming of gaseous hydrogen and to a lesser extent that of cryogenic liquid hydrogen. High-pressure storage (300 to 700 atm) is a proven approach for replacement of gasoline in automobiles. Liquefaction calls for cryogenic storage and boil-off arrangements. Chemical compounds, ammonia, metallic hydrides, hydrazine, and alcohols[4] are potential alternates, but past approaches have not provided cost savings sufficient to replace traditional gasoline, diesel, and jet fuels, thus carbon constituents continue to be burned.

Both compressed gas and cryogenic liquid hydrogen are being used as fuel for a growing number of vehicles ranging from motor bikes to busses. Its use as a rocket fuel is well documented; but in that application, cost is of little concern. Applications of hydrogen in automotive and heavy commercial vehicles with altered internal combustion engines to improve performance and efficiency have been well proven. Hydrogen as a source of combustion or fuel cell energy for electric power generation has advanced to establish its feasibility, particularly in combined heat and power applications. Since tanks for compressed gas storage of hydrogen are more expensive than petroleum fuel tanks and cryogenic liquid hydrogen fuel must be carried in special pressurized tanks that are highly insulated and provided with boil-off relief systems, the design of vehicles with hydrogen-fueled engines or fuel cells have presented difficult, but not intractable problems. At the very least, the whole new infrastructure for dispensing compressed gas or cryogenic liquid hydrogen fuel at "filling stations" would require massive changes.

Emerging manifestations of hydrogen-powered vehicles anticipate hydrogen fuel cells.[10] Aircraft jet engines have been successfully flight tested on cryogenic hydrogen and it can be expected to be among the first successful commercial applications of hydrogen as a lower weight, clean fuel for jet aircraft to help overcome atmospheric pollution and global climate change damages.

In spite of the cleanliness and demonstrated thermodynamic advantages inherent in hydrogen as an energy carrier, in the current competitive economic market for fuels, highly pressurized gaseous and cryogenic liquid hydrogen still face daunting problems because of the higher preparation costs and infrastructure challenges related to efficient storage and transportation.

These challenges can be solved by economical production of net-hydrogen fuel that can be stored and transported as a liquid at ambient temperature and pressure. In fuel cells and combustion applications such "net-hydrogen liquid fuel" can produce the same environmental impact as pure hydrogen, which is condensable water vapor.[7]

Hydrogen can be combined with nitrogen and carbon dioxide preemptively taken from more concentrated sources such as the exhaust stacks of power plants, wastewater digesters, landfills, breweries, bakeries, ethanol plants, mineral calciners, and decaying permafrost (or the ambient air) to produce a blend of net-hydrogen liquid constituents with suitable vapor pressure, viscosity, and lubricity to replace fuel oil, gasoline, diesel and jet fuels. Such net-hydrogen liquid fuel can provide about the same volumetric and mass energy density as gasoline; and can be transported by existing liquid fuel pipelines, marine tankers, rail cars, and tank delivery trucks to local stations for filling gasoline and diesel fuel tanks.[7]

Both hydrogen and such net-hydrogen liquid fuel can be less expensive than fuel oil, gasoline, diesel and jet fuels as a result of converting substances that ordinarily rot or burn (including renewable and fossil fuels) into high-value carbon products and low-cost hydrogen. The carbon can be utilized to produce durable goods particularly including carbon-enhanced equipment for annually converting solar, wind, moving water, or geothermal energy into hundreds of times more electricity, hydrogen and heat compared to the raw heat released one time by wasteful burning of such carbon.

In transportation applications carbon-reinforced components that are lighter than aluminum and stronger than steel reduce maintenance and fuel cost per delivered ton of payload. Various automobile manufacturers and aerospace companies produce carbon-enhanced composite structures for reducing curb weight and inertia including applications in military and commercial aircraft that provide higher performance and are more economical to operate because of lower costs for maintenance and fuel.

Utilizing such carbon-enhanced equipment to produce much more hydrogen energy every year (for 20 years or more) compared to burning such carbon one time dispels the shortcoming that hydrogen is not economically viable. The energy value of hydrogen produced by steam reformation of hydrocarbon fuels or by electrolysis of water using electricity from a Rankine cycle power plant is less than the hydrocarbon fuel selection or the electricity for electrolysis. Table 7.1.10 shows the minimum energy requirement for economically advantageous co-production of carbon and hydrogen by anaerobic endothermic hydrocarbon dissociation in comparison with steam reformation with the carbon source, along with water dissociation, and water electrolysis.

Important advantages for co-production of carbon and hydrogen from substances that rot or burn are shown in Table 7.1.10 including the collection of valuable carbon and doing so along with economical collection of hydrogen by requiring less energy than required for steam reformation and electrolysis systems both of which require production of carbon dioxide.

Illustratively, steam reformation of CH_4 to produce 2 kg of H_2 causes the release of at least 11 kg of CO_2; and steam reformation of carbon from coal to produce 2 kg of H_2 releases at least 22 kg of CO_2. In contrast, less endothermic energy is required for anaerobic dissociation of CH_4 to produce 4 kg of H_2 and 12 kg of valuable carbon, which prevents production of CO_2.

Energy for such dissociation of hydrocarbons can be provided from a small fraction of the energy produced by carbon enhanced equipment for sustainably converting solar, wind, moving water or geothermal energy into electricity, hydrogen, and heat.[1-3,6,9]

In addition to energy conversion equipment, carbon-reinforced building products, bridges, and transportation components can be stronger than steel, and lighter than aluminum. It can be far more profitable to require less energy to produce such carbon-enhanced equipment and low-cost hydrogen compared to wastefully producing carbon dioxide by steam reformation.

At sufficient scale of production such carbon-reinforced materials require less energy to produce and are less expensive than steel, titanium, and aluminum components. Many more highly desirable jobs will be provided for profit-motivated production of infrastructure, energy-conversion, and transportation equipment; thus to drive sustainable economic development.[5,7-9]

Hydrogen or net-hydrogen liquid fuel is considerably better than a zero emissions fuel because it can be produced by a process that profitably prevents the carbon in renewable fuels and fossil hydrocarbons from wastefully forming carbon dioxide. Illustratively renewable or fossil methane can be anaerobically heated to dissociate into carbon and hydrogen. The carbon can be made into products that are far more valuable than burning the carbon including graphene, nanotubes, diamond-like coatings, carbon fiber, and carbides.

Carbon allotropes have been shown that have greater electrical and thermal conductivity than copper.[9] Such graphene and carbon fiber allotropes can be lighter and operate at higher temperatures than aluminum and copper conductors. These exciting developments are progressing to answer the question of sufficiency regarding the availability of copper to meet the growing demand for electrical apparatus and equipment particularly including electric vehicles and equipment.

Subsequent utilization of the co-produced carbon to reinforce equipment that can annually convert renewable energy into far more electricity, hydrogen, and heat provides sustainable production of hydrogen or net-hydrogen liquid that can be stored in gasoline, diesel or jet fuel tanks. Net-hydrogen liquid fuel can enable 1.3 billion engines (expected soon to be 2 billion engines) in transportation, electricity-generation, farming and mining applications to actually clean the air to overcome local pollution and thus global climate change damages.

References

1. Ogden, Joan. "Hydrogen as an Energy Carrier: Outlook for 2010, 2030 and 2050," Institute of Transportation Studies, University of California, Davis. From workshop proceedings 2001, "The 10-50 Solution: Technologies and Policies for a Low-Carbon Future." The Pew Center on Global Climate Change and the National Commission on Energy Policy.

Table 7.1.10 Minimum CO_2 and Energy Requirement Per Mole of Hydrogen

Resource	Process	Reaction	CO_2/Mole H_2	Thermal Energy Requirement
Methane	Dissociation[a]	$CH_4 >>> C + 2H_2$	Prevents 0.50	37.5 kJ/Mole (H_2) 800-1,000°C
Methane	Steam Reformation	$CH_4 + 2H_2O >>> CO_2 + 4H_2$	Adds 0.25	41.25 kJ/Mole (H_2) 800-1,000°C
Coal	Steam Reformation	$C + 2H_2O >>> CO_2 + 2H_2$	Adds 0.50	45 kJ/Mole (H_2) 1,000-1,500°C
Water	Dissociation[b]	$H_2O >>> H_2 +.5O_2$	0.00[b]	241 kJ/Mole (H_2) 2,800-3,000°C
Water	Electrolysis[c]	$H_2O >>>H_2 +.5O_2$	Adds 1.50[c]	285 kJ/Mole (H_2) @ Power Plant

[a] Anaerobic thermal dissociation of hydrocarbons can efficiently co-produce carbon and hydrogen.
[b] Assume point focus parabolic solar tracking concentrator heat source (future development possibility).
[c] Requires about 3 times more power plant coal combustion energy to make electricity required for electrolysis.

2. Ogden, Joan and Lorraine Anderson. "Sustainable Transportation Energy Pathways: A Research Summary for Decision Makers," Institute of Transportation Studies, University of California, Davis, 2011.

3. Dahl, J., K. Bachler, R. Finley, T. Stanislaus, A. Weimer, A. Lewandowski, C. Bingham, A. Sheets, and A. Schneider. Rapid Solar-thermal Dissociation of Natural Gas in an Aerosol Flow Reactor. Department of Chemical Engineering, University of Colorado, Boulder, CO 80309-0424; National Renewable Energy Laboratory, Golden, CO 80,401-3,393.

4. Bromberg, L. and W.K. Cheng. "Methanol as an alternative transportation fuel in the US: Options for sustainable and/or energy-secure transportation"; Prepared by the Sloan Automotive Laboratory, Massachusetts Institute of Technology, Cambridge, MA 02139, September 27, 2010 (Finalized November 2, 2010; Revised November 28, 2010).

5. McAlister, Roy, US Patent 8,318,131 "Chemical Process Reactors" (2012).

6. McAlister, Roy, US Patent 8,733,331 "Adaptive Control System of Fuel Injection and Ignition" (2014).

7. McAlister, Roy, US Patent 8,623,925 "Preparation of Liquid Fuels" (2014).

8. McAlister, Roy, US Patent 9,416,457 "Renewable Resources Production" (2016).

9. McAlister, Roy, US Patent 9,284,191 "Carbon Fiber and Graphene Materials" (2016).

10. McAlister, Roy, US Patent 9,371,787 "Adaptive Control System to Prevent Oxides of Nitrogen" (2016).

DIRECT ENERGY CONVERSION

by Erich A. Farber, Introduction updated by Patrick E. Phelan

REFERENCES: Decher, *Direct Energy Conversion: Fundamentals of Electric Power Production*, 1996, Oxford University Press. Goldsmid, "Introduction to Thermoelectricity" (Springer Series in Materials Science). Springer, 2009. Poudel et al., "High-thermoelectric performance of nanostructured bismuth antimony telluride bulk alloys," *Science*, Vol. 320, no. 5876, pp. 634-638, 2008. Biswas et al. "High-performance bulk thermoelectrics with all-scale hierarchical architectures," *Nature*, Vol. 489, no. 7416, pp. 414-418, 2012. *Handbook of Fuel Cells: Fundamentals, Technology, Applications*, 4 Volume Set, Wolf Vielstich, Arnold Lamm, Hubert A. Gasteiger (Editors), Wiley, 2003. *Fuel Cell Engines*, Matthew M. Mench, Wiley, 2008. *PEM Fuel Cells: Theory and Practice*, Frano Barbir, Academic Press, 2013. Fu et al., "U.S. Solar Photovoltaic System Cost Benchmark: Q1 2016," NREL/TP-6A20-66532, 2016 (and later reports). Green et al., "Solar Cell Efficiency Tables (Version 49)," *Progress in Photovoltaics: Research and Applications*, 2017, Vol. 25, pp. 3-13 (and later versions).

In contrast to the conventional thermal cycle for the conversion of heat into electricity are several more direct methods of converting thermal and chemical energy into electric power. The methods which seem to have the greatest potential possibilities are solar photovoltaic, fuel cells, thermoelectric, and thermionic devices. The principles of operation of these processes have long been known, but technological and economic obstacles have limited their use. New applications, materials, and technology now provide increased impetus to the development of these processes.

Thermoelectric Generation

Revised by M. Zebarjadi

REFERENCES: Goldsmid, "Introduction to Thermoelectricity" (Springer Series in Materials Science). Springer, 2009. Poudel et al., "High-thermoelectric performance of nanostructured bismuth antimony telluride bulk alloys." Science, vol. 320, no. 5876, pp. 634-638, 2008. Biswas et al., "High-performance bulk thermoelectrics with all-scale hierarchical architectures." Nature, vol. 489, no. 7416, pp. 414-418, 2012.

Thermoelectric generation is based on the phenomenon, discovered by **Seebeck** in 1821, that current is produced in a closed circuit of two dissimilar metals if the two junctions are maintained at different temperatures, as in thermocouples for measuring temperature. A thermoelectric generator is a low-voltage, dc device. To obtain higher voltages, the elements must be stacked. Typical thermocouples produce potentials on the order of 50 to 70 μV/°C and power at efficiencies on the order of 1 percent.

Certain semiconductors have thermoelectric properties superior to conductor materials, with resultant improved efficiency. The criterion for evaluating material characteristics for thermoelectric generation is the **figure of merit Z,** measured in $(°C)^{-1}$ and defined as $Z = S^2/(\rho\kappa)$, where S = Seebeck coefficient, V/°C; ρ = electrical resistivity, Ω cm; κ = thermal conductivity, W/(°C cm). ZT is also commonly used for thermoelectric evaluation, which is Z times temperature, T, and is dimensionless. An ideal thermoelectric material would have a high Seebeck coefficient, low electrical resistivity, and low thermal conductivity. Unfortunately, these three properties are interconnected. For example, materials with low electrical resistivity have a high electronic thermal conductivity since both properties are dependent, to some extent, on the number of free electrons in the material. Another example is the interplay between electrical resistivity and the Seebeck coefficient. It is in general difficult to have high Seebeck coefficient and high electrical conductivity at the same time. A high Seebeck coefficient represent accumulation of charge on one side of the sample. A large electrical conductivity serves as an internal leak to diffuse the carriers back and to avoid extreme charge accumulations. In semiconductor materials, a general observed trend is that as the doping levels increase, the electrical conductivity increases, and the Seebeck coefficient decreases. The maximum conversion efficiency of a thermoelectric generator is proportional to the Carnot efficiency and is an increasing function of the average figure of merit times temperature.

In some types of thermoelectric materials, the voltage difference between the hot and cold junctions results from the flow of negatively charged electrons (n type, hot junction positive), whereas in other types, the voltage difference between the cold and hot junctions results from the flow of positively charged voids vacated by electrons (p type, cold junction positive). Since the voltage output of a typical semiconductor thermoelectric couple is low (about 100 to 300 μV/°C temperature difference between the hot and cold junctions), it is advantageous to use both p- and n-type materials in constructing a thermoelectric generator. The two types of materials make it possible to connect the thermojunctions in series electrically and in parallel thermally (Fig. 7.1.32).

Typical semiconductor thermoelectric materials are compounds and alloys of lead, selenium, tellurium, antimony, bismuth, germanium, tin, manganese, cobalt, and silicon. To these materials, minute quantities of "dopants" such as boron, phosphorus, sodium, and iodine are sometimes added to improve properties. Typical ZT values for the commercially used thermoelectric materials are in the range of 0.5 to 1. Many new ideas to improve ZT have emerged in the last decade and some of them have been demonstrated to be very effective. For example, bismuth telluride alloys owned the highest ZT values for about 50 years. Their ZT was around one at temperatures close to room temperature. In 2008, Chen and Ren groups reported improvement in ZT of bulk bismuth telluride to 1.4 using nanostructuring and creating many interfaces to

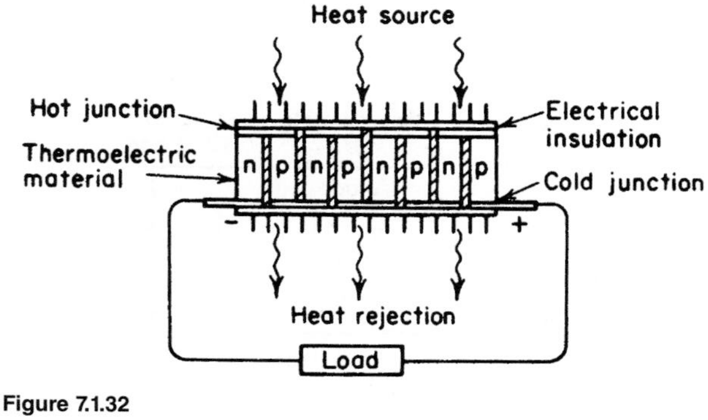

Figure 7.1.32

suppress lattice thermal conductivity with minimally affecting electrical conductivity. Following their pioneer work, ZT values of many bulk materials were improved. Examples include skutterudites, half-heuslers, PbTe, PbSe, and SiGe based alloys. Today, ZT values as high as 2 are reported in research laboratories.

The onset of deleterious thermochemical effects at elevated temperatures, such as sublimation or reaction, limit the materials' application. Bismuth telluride alloys, which have the highest Z values at room temperature, cannot be used beyond a hot-side temperature of about 300°C without encountering undue degradation. Silicon-germanium alloys have high-temperature capability up to 1,000°C that can take advantage of higher Carnot efficiencies. These alloys possess low Z values at room temperatures. Their peak ZT values can reach to 1 at 800°C. Optimized designs of segmented thermoelectric legs are therefore needed to cover a wide temperature range and to prepare mechanically stable legs with large average ZT values over large temperature difference. Optimized designs of thermoelectric junctions using semiconductor materials have resulted in experimental conversion efficiencies as high as 13 percent; however, the efficiency of practical thermoelectric generators is lower, for example, 4 to 9 percent. Materials which have higher figures of merit (2 or 3 × 10^{-3}) and which are capable of operating at higher temperatures (800 to 1,000°C) are required for an appreciable improvement in efficiency.

Thermoelectric-generation technology has matured considerably through its application to nuclear power systems for space vehicles where modules as large as 500 W have been used. It is also used in terrestrial applications such as gas pipeline cathodic protection and power for microwave repeater stations. Development work continues, but the use of this technology is expected to be limited to special cases where power source selection criteria other than efficiency and first cost will dominate.

Thermoelectric Cooling

The **Peltier effect,** discovered in 1834, is the inverse of the Seebeck effect. Peltier coefficient is defined as the ratio of the thermal current to the electrical current under uniform temperature. It is connected to the Seebeck coefficient via Kelvin (Onsager) relation. Peltier heating\ cooling involves the heating or cooling of the junction of two dissimilar thermoelectric materials by passing current through the junction. The effectiveness of the thermojunction as a cooling device has been greatly increased by the application of semiconductor thermoelectric materials. Typical applications of thermoelectric coolers include electronic circuit cooling, small-capacity ice makers, and dew-point hygrometers, small refrigerators, freezers, portable coolers or heaters, etc. These devices make it possible to preserve vaccines, medicines, etc., in remote areas and in third world countries during disasters or military conflicts.

Thermionic Generation

REFERENCES: Mahan, "Thermionic refrigeration," J. Appl. Phys., vol. 76, pp. 4362- 4366, 1994. Shakouri and Bowers, "Heterostructure integrated thermionic coolers," Appl. Phys. Lett., vol. 71, no. 9, p. 1234, 1997. Schwede et al., "Photon-enhanced thermionic emission for solar concentrator systems." Nat Mater., vol. 9, no. 9, pp. 762-767, 2010.

Thermionic generation, proposed by **Schlicter** in 1915, uses a thermionic converter (Fig. 7.1.33), which is a vacuum or gas-filled device

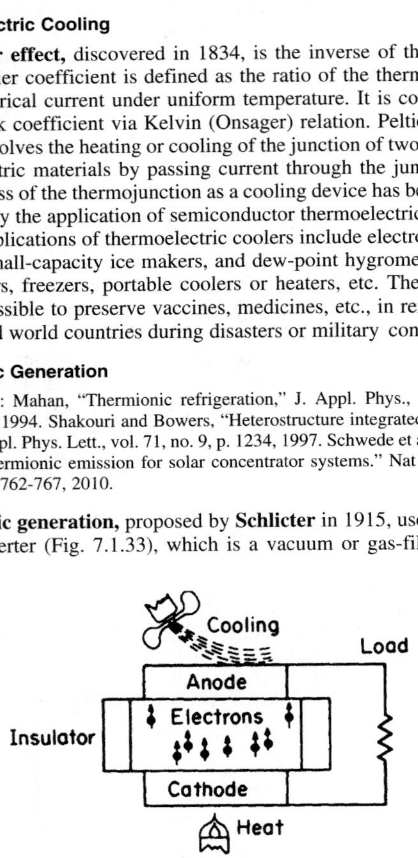

Figure 7.1.33 Representation of a thermionic generator.

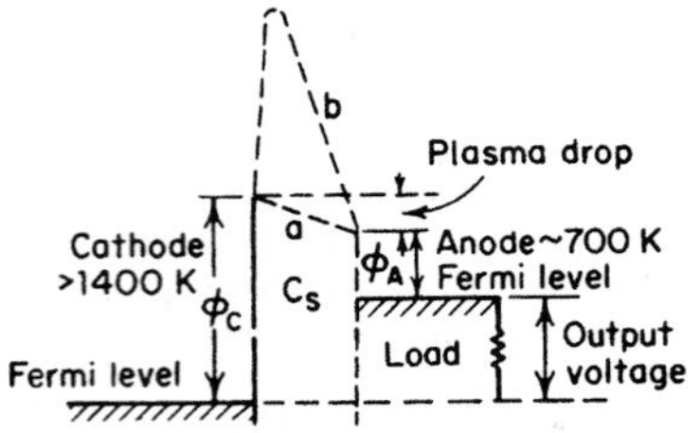

Figure 7.1.34 Electron energy levels.

with a hot electron "emitter" (cathode) and a cold electron "collector" (anode) in or as part of a suitable gastight enclosure, with electrical connections to the anode and cathode, and with means for heating the cathode and cooling the anode. A thermionic generator is a low-voltage dc device.

Figure 7.1.34 is a plot of the electron energy at various places in the converter. The abscissa is cathode-anode spacing, and the ordinate is electron energy. The base line corresponds to the energy of the electrons in the cathode before heating. Heating the cathode imparts sufficient energy to some of the electrons to lift them over the **work function barrier** (retaining force) at the surface of the cathode into the interelectrode space. (The lower the work function, the easier it is for an electron to escape from the surface of the cathode.) If it is assumed that the electrons can follow path *a* to the anode with only a small loss of energy, they will "drop down" the work function barrier as they join the electrons in the anode still retaining some of their potential energy **(Fermi level),** which is available to cause an electric current to flow in the external circuit. The work function of the anode should be as small as possible. The anode should be maintained at a lower temperature to prevent anode emission or back current. This pattern presumes that the electrons could follow path *a* from the cathode to the anode with little interference. Since, however, electrons are charged particles, those in the space between the cathode and anode form a space charge barrier, as shown by *b*. This space charge barrier limits the electrons emitted from the cathode. Space charge formation can be reduced by close spacing of the cathode and anode surfaces or by the introduction of a suitable gas atmosphere that can be ionized by heating and thus neutralize the space charge. In vacuum-type thermionic converters, the spacing between cathode and anode must be less than 0.02 mm to get as many as 10 percent of the electrons over to the cathode and to achieve an efficiency of 4 to 5 percent. In gas-filled converters, the negative electron space charge is neutralized by positive ions. Cesium vapor is used for this purpose. At low pressure, it will also lower the work function of the anode, and at high pressure, it can, in addition, be used to adjust the work function of the cathode. Efficiencies as high as 17 percent have been obtained with gas-filled converters operating at a cathode temperature of 1,900°C (2,173 K). The output voltage is 1 to 2 V, so the units must be connected in series for reasonable utilization voltages.

Photon-enhanced thermionic emission (PETE), proposed by Melosh group in 2010. Electrons are excited by photons to higher energy states and then emitted into the vacuum to overcome the barrier and extend operation to lower temperatures using currently available materials. The idea is proposed originally as a topping cycle for concentration solar power plants. The cathode is replaced by a wide bandgap semiconductor, in which electrons are first excited to the conduction band by absorbing a photon and then thermally excited to the vacuum similar to a traditional thermionic generators.

Although theoretical efficiencies as high as 50 percent are predicted for PETE devices, experimentally measured values are substantially lower. Solid-state thermionic generators/refrigerators are proposed by Mahan and Shakouri in 1994. A solid material fills the vacuum space of traditional thermionic generators. The advantage of solid state thermionic devices is their tunable transport barrier. The barrier could be significantly lowered to extend the operational temperature to low temperatures. When the thickness of the barrier layer increases to values much more than the electron mean free-path, the operational principle of solid-state thermionic devices, is the same as the thermoelectric legs discussed before.

Thermionic development results have been encouraging, but major technical challenges remain to be resolved before reliable, long-life converters become available. Effort has been focused on problem areas such as the limited life of emitter materials, leak tightness of the converter, and dimensional stability of the converter gap. Studies of thermionic converters incorporated into nuclear reactors and as a topping cycle for fossil fuel fired steam generators as well as for space and solar applications have received the greatest attention.

FUEL CELLS

Revised by Trung Nguyen

A **fuel cell** is an electrochemical energy conversion device in which the chemical energy from a chemical reaction is converted directly to electrical energy, and if the fuel cell is reversible, it can also convert electrical energy back to chemical energy by reversing the chemical reaction. The fuel cell is unique in that it converts chemical energy to electric energy without an intermediate conversion to heat energy; its efficiency is therefore independent of the thermodynamic limitation of the Carnot cycle.

Analogous to a combustion engine, a fuel cell does this without altering the basic components (electrodes and electrolyte) of the cell itself and can run continuously if the reactants are fed and products are removed. The fact that electrode and electrolyte are invariant distinguishes the fuel cell from the primary and secondary cell or battery. It is a low-voltage, dc device. To obtain higher voltages, the cells are stacked in series. The fuel cell dates to 1839, when **Grove** demonstrated that the electrolysis of water could be reversed using platinum electrodes.

Figure 7.1.35*a* is a schematic of a hydrogen fuel cell with air or oxygen as the oxidant operating in either an acid or alkaline electrolyte. The fuel is supplied to the anode, where it is oxidized generating electrons, which flow through the external circuit and perform electrical work, and hydrogen ions, which pass through the electrolyte to the cathode, which is supplied with oxygen. The oxygen is reduced by electrons flowing into the cathode from the external circuit, and the reduced oxygen reacts with hydrogen ions from the electrolyte to form water.

Electrodes for this type of cell are usually porous, to achieve high active surface area per unit volume, and impregnated with a catalyst. In a simple cell of this type, chemical and catalytic action takes place only at the solid surface that is accessible to the liquid electrolyte and

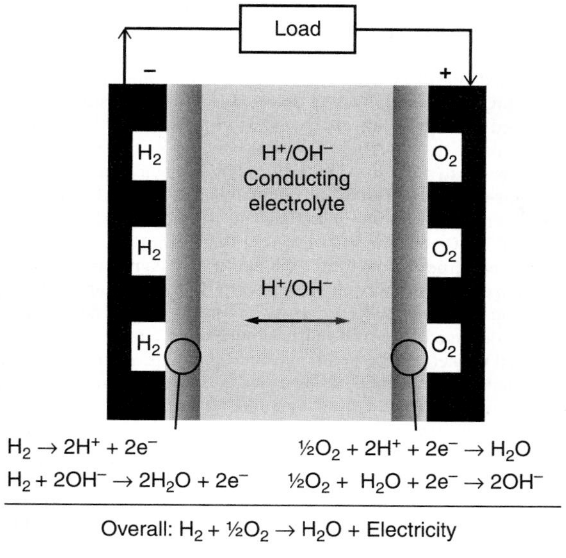

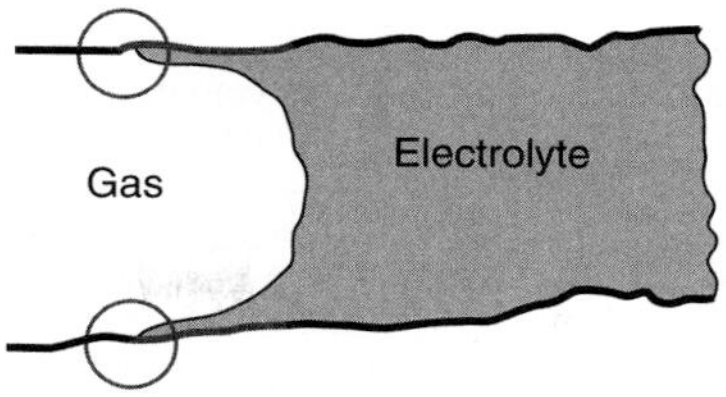

Figure 7.1.35*b*   A schematic of triple-phase reaction interface (circles) in a three-phase solid-gas-solid electrode and two-phase reaction interface (grey line) in a three-phase solid-gas-liquid electrode.

reactants. This surface can change depending on the solubility of the gaseous reactants in the electrolyte and the transport rate of the soluble reactants in the electrolyte relative to the consumption rate at the active surface. The only case where this reaction interface exists as a line of triple-phase boundary is when the electrolyte is also a solid and the gaseous reactants are insoluble in the electrolyte, as in the case of the solid oxide fuel cell system described later below. This phenomenon is illustrated in Fig. 7.1.35*b* where the circles highlight the triple-phase interface line associated with a solid electrolyte and the grey lines highlight the active surface with a liquid electrolyte.

One of the objectives in designing a high-performance fuel cell is to increase the active surface area described above. This has been accomplished in several ways, but usually by the creation of porous electrodes within which, in the case of gas diffusion electrodes, the fuel and oxidant in gaseous state can come in contact with the electrolyte at many sites. If the electrolyte is a liquid, a delicate balance must be achieved in which surface tension and density of the liquid must be considered and gas pressure and electrode pore size must be chosen to hold their interface inside the electrode. If the gas pressure is too high, the electrolyte is excluded from the electrode, gas leaks into the electrolyte, and ion access to the reactive sites stops; if the gas pressure is too low, drowning of the electrode by the liquid electrolyte occurs and access of gaseous reactants is limited to the region near the outer surface of the electrode next to the gas chamber.

Fuel cells may be classified broadly by operating temperature level, type of electrolyte, and type of fuel. Low-temperature (less than 150°C) fuel cells are characterized by the need for expensive noble metal catalysts, such as platinum and relatively simple fuel molecules, such as hydrogen and methanol. High temperatures (500 to 1,000°C) offer the potential for use of hydrocarbon fuels and lower-cost catalysts. Electrolytes may be either acidic or alkaline in liquid, solid, or solid-liquid composite form. In one type of fuel cell, the electrolyte is a solid polymer.

Low-temperature fuel cells of the hydrogen-oxygen type, one of a solid-polymer- electrolyte type, and the other using free KOH as an electrolyte, have been successfully applied in generating systems for U.S. space vehicles. High-temperature fuel cell development has been primarily in molten carbonate cells (500 to 700°C) and solid-electrolyte (zirconia) cells (1,100°C), but no significant practical applications have resulted.

Considerable study and development work has been done toward the application of fuel cell generating systems to bulk utility power systems. Low-temperature cells of the phosphoric acid matrix and solid-polymer-electrolyte types using petroleum fuels and air have been considered. Cell efficiencies of about 50 percent have been achieved; but with losses in the fuel reformers and electrical inverters, the overall system efficiency becomes of the order of 35-40 percent. In this application fuel cells have environmental advantages, such as low noise, low atmospheric emissions, and low heat rejection requirements. Additional development work is necessary to overcome the disadvantage of high catalyst costs and requirements for expensive fuels. More than 50 phosphoric acid fuel cell units, having a capacity of 200 kW, are in use. Companies in Canada, Germany, Japan, Korea, and the United States have demonstrated solid-polymer-electrolyte fuel cells in passenger vehicle propulsion systems. Limited quantities have been deployed in Japan and the USA.

$$H_2 \rightarrow 2H^+ + 2e^-$$
$$H_2 + 2OH^- \rightarrow 2H_2O + 2e^-$$

$$\tfrac{1}{2}O_2 + 2H^+ + 2e^- \rightarrow H_2O$$
$$\tfrac{1}{2}O_2 + H_2O + 2e^- \rightarrow 2OH^-$$

Overall: $H_2 + \tfrac{1}{2}O_2 \rightarrow H_2O$ + Electricity

Figure 7.1.35*a*   A schematic of a hydrogen-oxygen fuel cell operating in an acid or alkaline electrolyte.

Summary

Fuel cells require further development before they will be practical for widespread use, particularly for automotive applications. Certainly, fuel cells are in the forefront in the arena of public discussion, and their great advantages (high conversion efficiency, low— or no—pollution) are extolled. Most fuel cells which have been thoroughly studied and developed use very expensive fuels, fuels that, in many cases, are derived from or made from and with fossil fuels. An example is the hydrogen fuel cell, utilizing derived or produced hydrogen. At this time, hydrogen is produced by the consumption of other fossil fuel, via processes operating at moderate efficiency and which give rise to considerable pollution products. When one considers the overall process, from source to hydrogen to fuel cell to release of energy, the net result varies depending on the hydrogen source. If a hydrocarbon fuel is used, it may be less attractive than if renewable and intermittent energy sources such as wind and solar that may otherwise go unused are used.

References

1. Wolf Vielstich, Arnold Lamm, Hubert A. Gasteiger (Editor). Handbook of Fuel Cells: Fundamentals, Technology, Applications, 4 Volume Set, Wiley, ISBN: 978-0-471-49926-8, 3826 pages, May 2003.

2. Matthew M. Mench. Fuel Cell Engines, Wiley, ISBN: 978-0-471-68958-4, 528 pages, March 2008.

3. Frano Barbir. PEM Fuel Cells: Theory and Practice, Academic Press, 518 pages, 2013.

SOLAR PHOTOVOLTAIC GENERATION

Revised by Patrick E. Phelan

REFERENCES: See previous references for solar thermal energy. In addition, see Fu et al., "U.S. Solar Photovoltaic System Cost Benchmark: Q1 2016," NREL/TP-6A20-66532, 2016 (and later reports). Green et al., "Solar Cell Efficiency Tables (Version 49)," *Progress in Photovoltaics: Research and Applications*, 2017, Vol. 25, pp. 3-13 (and later versions). Interested readers can also find up-to-date information at the *USA Department of Energy SunShot Program* (https://energy.gov/eere/sunshot/sunshot-initiative) and the *BP Statistical Review of World Energy* (http://www.bp.com/en/global/corporate/energy-economics/statistical-review-of-world-energy.html).

Photovoltaic generation utilizes the direct conversion of light energy to electric energy and stems from the discovery by **Becquerel** in 1839 that a voltage is generated when light is directed on one of the electrodes in an electrolyte solution. Subsequent work using selenium led to the development of the photoelectric cell and the exposure meter. It results in low-voltage direct current. To obtain higher voltages, the elements must be stacked.

The photovoltaic effect is the generation of electric potential by the ionization by light energy (photons) of the area at or near the *p-n* junction of a semiconductor. The *p-n* junction constitutes a one-way potential barrier which permits the passage of photon-generated (–) electrons from the *p* to the *n* material and (+) "holes" from the *n* to the *p* material. The resulting excess of (–) electrons in the *n* material and (+) holes in the *p* material produces a voltage at the terminals comparable to the junction potential.

Solar cells have been a useful source of electricity since about 1960. They have proved particularly well suited for use in spacecraft; in fact, much of the early effort in photovoltaic R&D was funded by the space program. Solar cells have been applied also to remote weather monitoring and recording stations (some of them equipped with transmitters to send the data to collecting stations), traffic control devices, buoys, channel markers, navigational beacons, etc. They have also been used for applications such as battery chargers for cars, boats, flashlights, tools, watches, calculators, and emergency radios.

Single-crystal or polycrystalline silicon (Si) is the material for approximately 90 percent of the solar PV cells in use today. The second most common semiconductor material for solar PV is cadmium telluride (CdTe), which can be deposited onto a substrate and thus made into a thin-film configuration that offers convenient manufacturing. Another choice for thin-film manufacturing is copper indium gallium diselenide (CIGS), but the presence of more constituent materials increases manufacturing complexity and cost. Organic solar PV cells are under development, and may potentially lead to lower costs than solar PV based on semiconductor materials. Another configuration is concentrating solar PV (CPV), in which a concentrator focuses sunlight onto a single high-efficiency PC cell, typically one with multiple *p-n* junctions. As of 2016, the highest reported laboratory-measured efficiency of a single-crystal Si PV solar cell was 26.3 percent, for CdTe 21.0 percent, for CIGS also 21.0 percent, for amorphous Si 10.2 percent, and for an organic cell 11.2 percent. An efficiency of 38.8 percent has been reported for a 5-junction cell that could potentially find application in CPV. Of course, the efficiencies of commercially available solar PV modules are lower than these laboratory values.

The USA Department of Energy's SunShot Initiative (https://energy.gov/eere/sunshot/sunshot-initiative) has led the development in the USA of major development efforts in solar photovoltaic (PV) research for broader applications in residential and commercial buildings, and at the utility scale. This has resulted in much wider deployment of solar PV across the USA, and other countries. More than 50 GWe of solar generating capacity was added worldwide in 2015, a 28 percent increase over 2014. This represents a trebling of solar PV capacity during the previous four-year period. In 2015, China had the largest installed capacity at 43.5 GWe, followed by Germany with 39.7 GWe, Japan with 35.4 GWe, the USA with 25.6 GWe, and Italy with 18.9 GWe. The DOE's SunShot stated 2030 goals are to reduce the levelized cost of energy (LCOE) of solar PV to US\$0.05/kWh for residential customers, US\$0.04/kWh for commercial customers, and to US\$0.03/kWh for utility-scale generation. These costs are relative to 2010 costs of US\$0.42/kWh, US\$0.34/kWh, and US\$0.27/kWh, respectively. Clearly, solar PV generation is a much more important factor in the overall electricity supply mix than ever before, and will grow more so in the future.

OTHER ENERGY CONVERTERS

Revised by Patrick E. Phelan

Each of the converters described in the following section has characteristics which makes it suitable for specific tasks. Some produce low-voltage direct current and must be stacked for higher voltage output. Other converters produce high-voltage direct or alternating current depending upon their design. High voltages can be used to drive Klystron or X-ray tubes, or similar equipment, and can be transmitted without step-up transformers. Some medium- or low-temperature converters can be used in low-grade energy (heat) applications; in medical practice, for example, body heat can drive heart pacemakers, artificial heart pumps, internal medicine dispensing devices, and organ monitoring equipment.

Electrohydrodynamic (EHD) Converters When positive ions are transported by neutral hot gases against an electric field, high potential differences result. The charges produced by the ions do work when allowed to flow through a load. These devices are also called **electrogas-dynamic (EGD)** converters. If the gases are allowed to condense, producing small liquid droplets, the devices are often referred to as **aerosol EGD** converters. A number of different basic designs exist.

Magnetohydrodynamic (MHD) Converters Magnetohydrodynamic generation utilizes the movement of electrically conducting gas through a magnetic field. Normally, it results in a high-voltage, dc output, but it can be designed to provide alternating current. In the simple open-cycle MHD generator, hot, partially ionized, compressed gas, which is the product of combustion, is expanded in a duct and forced through a strong magnetic field. Closed-cycle MHD generators are also under study for bulk power generation. They are of two types: first, one in which the working fluid is an inert gas such as argon seeded with cesium; and second, the liquid metal type in which the working fluid is a helium-sodium mixture. Closed-cycle MHD generation offers the potential for high efficiency with considerably lower peak cycle temperatures, lower pressure ratios, and lower average magnetic-flux density.

Van der Graaff Converter This device operates on the same principle as the EHD converter, except that a belt is used instead of hot gases to transport the ions against an electric field. Very high potential differences are produced, which may be utilized in high-energy particle accelerators, atom smashers, artificial lightning generators, and the like.

Ferroelectric Converters Certain materials exhibit a rapid change in their dielectric constant k around their Curie temperature. The voltage produced by a charge Q is the charge divided by the capacitance, or $V = Q/C$. The variation between capacitance and dielectric constant is expressed by $C_h = (k_h/k_c)\,C_c$, where c = cold and h = hot. A capacitor containing ferroelectric material is charged when the capacitance is high. When the temperature is changed to lower the dielectric constant, the capacitance is lowered, with an accompanying increase in voltage. The charge is dissipated through a load when the voltage is high, resulting in the performance of work. Barium titanate, for example, can produce a fivefold voltage swing between temperatures of 100 and 120°C. Thermocycling could be produced by the sun on a spinning satellite.

Ferromagnetic Converters Ferromagnetic material is used to complete the magnetic circuit of a permanent magnet. The ferromagnetic material is heat-cycled through its Curie point, producing flux changes in a coil wrapped around the magnet. The operating conditions can be selected by the Curie temperature of the ferromagnetic material. Gadolinium, for example, has a Curie point near room temperature.

Piezoelectric Converters When axisymmetric crystals are compressed parallel to their polar axes, they become polarized, that is, positive charges are generated on one side, and negative charges are generated on the other side of the crystal. The induced compressive stresses can be produced mechanically or by heating the crystal. The resulting potential difference will do work when allowed to flow through a load. In a reverse process, imposing a potential difference between the ends of the crystal will result in a compressive stress within the crystal. The imposition of alternating current will result in controlled oscillations useful in sonar, ultrasound equipment, and the like. Piezoelectric converters are also being examined in applications for energy harvesting from people, roadways, etc.

Pyroelectric Converters Some materials become electrically polarized when heated, and the conversion of heat to electricity can be utilized as it is in piezoelectric converters.

Bioenergetic Converters Energy requirements for medical devices used to monitor or control the performance of human organs (heart, brain, etc.) range from a few μW to a few W. In many cases, body heat is sufficient to operate an energy converter which, in turn, will power the device.

Nernst Effect Converters When heat flows through certain semiconductors exposed to a magnetic field perpendicular to the direction of heat flow, an electric potential difference will be induced along a third orthogonal axis. This conversion of heat to electricity can be utilized to do work. In a reverse fashion, crossing a magnetic and an electric field will produce a temperature difference (Ettingshausen effect, the reverse of the Nernst effect). This reverse conversion may be useful in electric heating, cooling, and refrigeration.

Thermophotovoltaic Converters A radiant heat source surrounded by photovoltaic cells will result in the radiant energy being converted to electricity. Source radiation and photovoltaic cell characteristics can be controlled to operate anywhere in the spectrum. Ongoing work targets challenges in developing spectral emitters, reflectors, and filters. An efficiency of about 18 percent has been experimentally demonstrated.

Photoelectromagnetic Converters When certain semiconductors (Cu_2O, for example) are placed in a tangential magnetic field and illuminated by visible light, there will result an electric potential difference along an axis orthogonal to the other two axes. The resulting flow of electric current can be used to do work.

Magnetothermoelectric Converters A magnetic field applied to certain thermoelectric semiconductors produces electric potential differences, useful in power generation.

Superconducting Converters The phase transition in a superconductor can be utilized similarly to a ferroelectric converter. Thermal cycling of the superconductor material will produce alternating current in the coil surrounding it. An idealized analysis for niobium at 8 K yields a conversion efficiency of about 44 percent.

Magnetostrictive Converters Changes in dimensions of materials in a magnetic field produce electric potential differences, thus converting mechanical energy to electricity. The effects can be reversed by combining an electric field with a magnetic field to produce dimensional changes.

Electron Convection Converters When a liquid is heated (sometimes to the boiling point), electrons and neutral atoms are emitted from the liquid surface. The flow of vapor transports the electrons upward, where they are collected on screens. The vapor condenses and recycles into the liquid pool. High electric potential differences can be produced in this manner between the screens and the liquid pool, allowing the subsequent flow of current to do work. The process is similar to that for EGD converters.

Electrokinetic Converters Certain fluids flowing through capillary tubes due to pressure gradients produce an electric potential difference between the ends of the capillaries, converting flow (kinetic) energy to electricity.

Particle-Collecting Converters When an alpha, beta, or gamma particle emitter is surrounded by a collector surface, an electric potential difference is produced between the emitter and the collector. Biased screen grids can improve the performance.

EHD Water Drop Converter Two separate streams of water coming from the same reservoir, while falling, are allowed to break up into droplets. At the breakup points, each stream is surrounded by a short metal cylinder. Each cylinder is connected electrically to a screen at the bottom of the opposite stream. High potential differences are produced between the two metal cylinders.

Photogalvanic Converters Photochemical reactions often produced by solar radiation (especially at the shorter wavelengths) can be used to generate electricity. Concentration of solar energy can increase the power of the converters considerably. The actual processes are similar to those in fuel cells.

The field of instrumentation provides other techniques which could become useful as energy conversion devices. While many of the methods cited and described above are not economically competitive with conventional conversion methods in current use, some are adapted to unique situations where the matter of cost becomes inconsequential. Certainly, it is expected that as progress is made in the field of energy conversion, certain techniques will be refined to greater practicality, and others will be developed.

ELECTROCHEMICAL BATTERIES

by Ankur Jain

Electrochemical cells and batteries are commonly used for energy storage and conversion. Despite the commonplace use of the terms "cell" and "battery," it is important to note that the fundamental electrochemical device that converts electrical energy to electrochemical energy and vice versa is a cell, whereas a battery is a combination of appropriately connected cells. Even though the terms cell and battery may often be used interchangeably, the distinction between the two as outlined above is important to recognize.

Electrochemical batteries are ubiquitous, and find applications in consumer electronics, vehicles, defense systems, aircraft, etc. Even though batteries have much lower energy density compared to chemical fuels, the flexibility offered by batteries has resulted in their usage in an extremely wide variety of technological applications. For example, Li-ion batteries are commonly used to power electronic devices such as smartphones, laptops, etc. and are now being investigated for use in electric vehicles. Lead-acid based batteries are used ubiquitously in automobiles to provide startup power. Nickel Cadmium batteries are used as standard batteries for a number of applications, including toys, power tools, etc. Large packs of batteries are used for storing and supplying backup electrical power.

Batteries have been employed extensively for several applications where the portability of energy is important. Batteries also enable standardization of devices. Most consumer goods are designed based on power sourced from one of the several standard batteries, which allows these devices to be used anywhere, without the need of a direct source of energy. The capability of storing electrical energy is also complementary with renewable energy sources such as wind and solar power where excess electrical energy can be stored away into a battery and be drawn upon even when the renewable energy source is absent, such as in the night, in case of solar power. By enabling efficient storage of excess energy, batteries may play a key role in ensuring the success of renewable energy sources. As renewable energy sources become more commonplace in the future, the nature of the electrical grid will become more and more complicated, requiring decentralization and bidirectional flow of power. It is expected that energy storage capability will play a key role in enabling future smart grids. Among the several technologies for energy storage, such as thermal energy storage, mechanical energy storage, chemical energy storage, electrical energy storage, etc., electrochemical energy storage offers several advantages, but also presents shortcomings that must be recognized and addressed.

Fundamentally speaking, an electrochemical cell is a device that is capable of storing energy in electrochemical form when connected to an external source of electrical potential difference through electrochemical reactions that occur in two halves of the cell. Conversely, once charged, the cell is capable of supplying electrical energy when connected to an external load. As shown in Fig. 7.1.36, during the charging process, one of the halves consumes electrons while the other half supplies electrons. During the charging process, an external source of potential difference drives the flow of electrons and positively charged ions move across the cell from one half of the cell to the other. When the cell is connected to an external electrical load, these processes are reversed, and the cell itself acts as a source of electrical potential difference, driving electrons from one half of the cell to the other through the external load. The potential difference generated by the cell is a function of the specific electrochemical reactions that occur in the two halves of the cell.

It is interesting to note that because the conversion of electrical energy to electrochemical energy and vice versa does not involve conversion to or from heat, the efficiency of this process is not limited by the second law of thermodynamics. While this may appear to be a significant advantage over other thermal-based energy storage mechanisms, it is important to note that several sources of losses and inefficiency do exist in an electrochemical cell, such as Joule heating due to electron flow through the internal resistance of the cell. As outlined later, batteries also present several hazards that must be considered in determining their appropriateness for specific applications.

Electrochemical energy conversion has a long history, with the first electrochemical cell being demonstrated by Alessandro Volta in the 1800s. His cell, often referred to as a voltaic pile, was a stack of alternating Copper and Zinc plates separated by a solid saturated with salt water. Since then, a wide variety of cells, differentiated by a number of factors have been developed. Specifically, the development of dry cells that did not require the use of a liquid electrolyte catalyzed the near-universal adoption of batteries as a power source. Today, materials development continues to expand the state-of-the-art in batteries that offer unprecedented capabilities in terms of energy storage density, power output, and reliability.

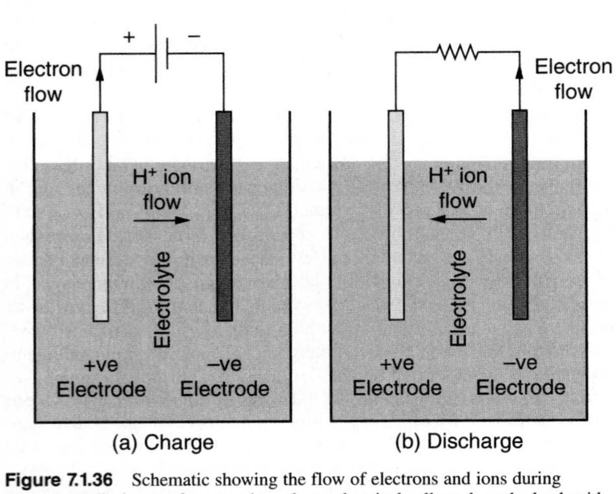

Figure 7.1.36 Schematic showing the flow of electrons and ions during charge and discharge of a secondary electrochemical cell, such as the lead-acid cell.

Battery Fundamentals

The three primary components of a cell are the positive and negative electrodes and an electrolyte. During the charging of the cell, an oxidation reaction at the positive electrode produces excess electrons and ions. The ions move through the electrolyte in which the electrodes are dipped toward the negative electrode. An externally connected power source drives the electrons along the external circuit to the negative electrode. A reduction reaction occurs at the negative electrode that consumes the ions and electrons. These reactions are reversed during the discharge process when an electrical load is connected in the external circuit across the electrodes. During discharge, oxidation and reduction at the negative and positive electrodes respectively result in a potential difference between the external load, thereby converting the stored electrochemical energy into electrical energy.

This is illustrated in Fig. 7.1.36. For the specific case of lead-acid battery, the relevant electrochemical reactions are

Charge:
$PbSO_4 + 2H_2O \rightarrow PbO_2 + H_2SO_4 + 2H^+ + 2e^-$ (positive electrode)
$PbSO_4 + 2H^+ + 2e^- \rightarrow Pb + H_2SO_4$ (negative electrode)

Discharge:
$PbO_2 + H_2SO_4 + 2H^+ + 2e^- \rightarrow PbSO_4 + 2H_2O$ (positive electrode)
$Pb + H_2SO_4 \rightarrow PbSO_4 + 2H^+ + 2e^-$ (negative electrode)

The difference between the electrode potentials of the two electrodes results in the overall potential difference provided by the cell. In the case of a lead-acid battery, this is found to be 2.04V. Note that in this case, charging produces H_2SO_4, which stays in the electrolyte and is used up during the discharge process.

The basic principle described above is utilized with variants of the specific electrochemical reactions in a wide variety of cells. Other commonly used electrochemical cells include NiCd, NiMH (Ni metal hydride) and Li-ion cells that vary in terms of the underlying chemistry and resulting electrochemical performance. Mg-ion, Li-S, Li-air, and Na-ion are some of the battery chemistries that are currently being developed for potential use in future batteries.

In addition to the electrodes and electrolyte, a number of other components are also used in practical electrochemical cells. While early electrochemical cells used a liquid electrolyte, dry electrolyte has also been commonly used, such as in standard dry cells, due to the convenience offered. In some cases, a porous material soaked in electrolyte, commonly known as a separator is used instead of liquid electrolyte to separate the electrodes. In addition, multiple electrodes may exist within a single cell. While Fig. 7.1.36 shows discrete electrodes, in several cells, the electrodes are in the form of thin films coated on flexible metal foils known as current collectors, which are then rolled together, along with a thin polymer separator. In this case, the separator plays a key role of preventing physical contact between electrodes, and permitting passage of ions while not allowing passage of electrons. The stack of electrodes and separator is usually wound tightly in a roll to produce cylindrical cells, or assembled in multiple folds to produce prismatic cells. These cells are either housed in a hollow metal casing, wherein the cap and body serve as connections to the two electrodes, or in a flexible polymer casing, in which protruding metal tabs are provided for electrical connections. In both cases, the cell is hermetically sealed in order to prevent the escape of the electrolyte, which is often volatile and combustible.

An electrochemical battery is a truly multiscale, multiphysics system, analysis and optimization of which calls for an understanding of multiple disciplines, as well as the nature of interactions between these disciplines that occur at multiple physical scales. Relevant physical processes include electrochemical reactions, ion transport, electron transport, heat generation, heat conduction, etc. At the atomistic scale, the nature of fundamental electrochemical reactions and their dependence on materials and material morphologies is important. At the scale of electrodes, it is important to understand how electrode materials can be integrated in an electrode stack. At the scale of cells, the dependence of cell performance on its shape and size is important to understand. When cells are integrated into batteries, modules and packs, the system-level challenges include real-time performance optimization and balancing, thermal management, etc.

Battery Classification

Electrochemical batteries can be classified in a different number of ways.

Primary batteries are those in which the electrochemical reactions responsible for energy conversion may occur only once. Once the battery is discharged, it cannot be charged again and must be discarded. A good example is the commonly used Zinc-Carbon based AA or AAA cell. On the other hand, secondary batteries can be charged and discharged multiple times. The number of such cycles possible in a secondary battery depends on several factors related to the conditions in which the battery is operated, and is an important metric of performance of the secondary battery. The most commonly used cells such as AA or AAA cells were originally primary cells. However, secondary cells in the same configuration are also now commonly available. Several more recent technological applications, such as electric vehicles and battery packs for renewable energy storage, necessarily require secondary batteries.

Cells may also be classified on the basis of the underlying chemistry, which also provides the basis for the voltage supplied by the cell. Some commonly used cell chemistries include lead-acid (used to provide startup power in automobiles), Zinc-Carbon (traditional, primary AA, AAA and similar cells), Nickel Cadmium and Nickel-Metal Hydride (secondary AA, AAA and similar cells), Li-ion (laptops, smartphones, etc.). A number of different variants of these chemistries, as well as entirely new chemistries continue to be investigated for improved performance characteristics.

Cells are also classified based on their geometry. Cylindrical shapes are very common, within which AA and AAA are some of the most commonly used and easily recognizable cell types. Button or coin cells are also cylindrical in nature, but very thin. Cells are also manufactured in a flat shape, which are often referred to as a prismatic cell. Pouch cells are similar to prismatic cells, except that the outer body is made of a flexible plastic material, and electrical interconnection to the electrodes is provided in the form of metal tabs instead of using the body of the cell.

Performance Parameters of a Battery

Several parameters related to the output of the battery characterize its performance and help make a distinction between the large number of battery types available. These performance parameters must be closely matched with the expectations of the application for electrochemical energy conversion and storage.

The capacity of a battery is the amount of charge that the battery can store within itself and supply to an external load at its rated voltage. The battery capacity is often reported in the units of Ampere-Hours (A-hr). The capacity of a battery is a function of operating conditions, such as ambient temperature, and in the case of secondary batteries, the life of the battery, including the number of cycles to which it has already been subjected. Typically, the capacity of a battery reduces as it ages, until it is no longer capable of holding significant charge.

Specific energy and energy density, measured in units of J/kg and J/liter respectively, characterize the amount of energy that the battery can store in a given mass or volume. While an electrochemical cell has an inherent specific energy and energy density, these numbers reduce, often significantly, when the cell is combined with other cells to form a battery, and when multiple batteries are combined to form a battery pack, due to the presence of peripheral components that add weight and volume without contributing to energy storage. Examples include battery management systems, thermal management systems, etc. Specific energy and energy density are important characteristics in space-constrained applications such as aeronautical and space applications, and where weight is an important consideration, such as in an automobile.

The rate at which a battery can supply energy is also an important performance characteristic. This is often reported in the form of the C-rate. C-rate refers to the reciprocal of the number of hours in which the battery will deliver its entire capacity, and therefore go from fully charged to fully discharged. For example, C-rates of 1C, C/10, and 10C mean that the battery will undergo a complete discharge process in 1 hour, 10 hours, and 6 minutes, respectively. If the capacity of this battery is 5A-hr, these processes will produce 5A, 0.5A, and 50A respectively at the rated voltage. The higher the C-rate, the larger is the discharge current, and therefore more aggressive is the discharge process, resulting in greater Joule heating losses and side reactions that result in reduced lifetime of the battery.

The lifetime of a secondary battery is often measured in terms of its cycle life, which refers to the number of charge-discharge cycles that the battery can complete before its output reduces to below a threshold value. Cycle life of a battery often depends on the nature of electrical load to which it supplies power, C-rate, ambient temperature, etc. Self-discharge is also an important phenomenon in a battery, wherein the battery slowly loses charge when simply stored in an open circuit configuration, and ultimately reaches a state of complete discharge. This can be an important characteristic in applications where the battery may be required to remain on standby for a long duration before being required to supply energy.

Finally, the response of the battery to the operating temperature is also an important performance characteristic that must be considered in matching a battery to a specific application. For example, automotive batteries are expected to operate well even at very low temperatures that may be encountered in some regions. Depending on the underlying chemistry, different batteries have different temperature characteristics.

Battery Hazards

Batteries are associated with a number of operational, storage and transportation hazards, some of which have manifested themselves in recent incidents of fire on aircraft, cars, etc. as well as expensive product recalls due to fire and explosion in batteries. Batteries often contain harmful, toxic chemicals that present various hazards upon exposure. Although a well-designed battery is expected to be pressure-tight, an anomalous event may lead to leakage of chemicals such as the electrolyte. For this reason, batteries must be handled with care, and must be operated on a load that is compatible with the characteristics of the battery. Further, disposal of spent batteries is an important challenge, and most communities in the developed world have well-designed processes to collect batteries separately from other garbage and recycling products, in order to prevent the battery chemicals from leaking into the water or soil.

A key operational and transportation hazard in a battery is associated with short circuiting of the battery, which leads to a very large current and rate of energy release from the battery. This usually happens when the battery is mishandled or subjected to abuse such as penetration with a sharp object, external short circuit, or a failure of the separator inside the battery that is designed to prevent cross-over of electrons between the two electrodes. Joule heating caused by such a large discharge current often leads to high temperature rise, which may in turn cause the activation of other exothermic processes inside the battery, such as decomposition of the electrolyte and electrodes. The electrolyte in various battery chemistries is volatile and flammable, which may result in increased pressure inside the battery, followed by venting and explosion. This series of heat-generating processes that feed into each other is often referred to as a thermal runaway, which is a particularly critical problem for specific battery chemistries. Thermal runaway is believed to be responsible for several fire accidents in batteries, which has led to several restrictions on transportation of

Li-ion batteries aboard passenger and cargo aircraft. Various concerns related to these hazards will continue to play a role in the development of future battery chemistries. For example, non-flammable electrolytes are very desirable for ensuring safety of next-generation batteries.

A number of engineering steps are taken to mitigate some of these hazards. While a typical battery is hermetically sealed, a pressure relief vent is provided to minimize damage due to catastrophic failure. The separator inside the battery is often designed such that one of its layers melts when it reaches a certain temperature, thereby blocking the transport of ions through the separator and thus shutting down the battery. External mechanical protective features and current protection circuits are often present to safeguard against short circuiting and excessively large current. In large battery packs, cooling is provided using air or liquid in order to minimize temperature rise of the battery during operation. Battery management systems also monitor the state of each battery in a pack, and optimize the load on each battery, keeping in mind long-term reliability as well as the safety of each battery. A number of battery safety standards exist. These standards cover nearly all aspects of battery manufacturing, testing, qualification, operation, storage, transportation, handling, disposal, etc. Adhering to relevant safety standards is no doubt critical in the design and operation of an engineering system that requires the use of a battery.

References

1. D. Linden, T.B. Reddy, "Handbook of Batteries," 4th Ed., McGraw-Hill, New York, 2010. ISBN: 9780071624213.

2. H.A. Kiehne, "Battery Technology Handbook," 2nd Ed., CRC Press, New York, 2003. ISBN: 9780203911853.

7.2 STEAM BOILERS

by John B. Kitto, Jr.

REFERENCES: The Babcock & Wilcox Company, "Steam—Its Generation and Use." Combustion Engineering, Inc. (now Alstom), "Combustion—Fossil Power Systems." "Standard Handbook of Powerplant Engineering," McGraw-Hill. Cohen, "The ASME Handbook on Water Technology in Thermal Power Systems," American Society of Mechanical Engineers. Gunn, "Industrial Boilers," Longman Scientific. Clapp, "Modern Power Station Practice: Boilers and Ancillary Plant," Pergamon Press. "Boiler and Pressure Vessel Code," "Performance Test Code for Fired Steam Generators," American Society of Mechanical Engineers. Also, *Proc. ASME* and the Joint Power Generation Conference.

Steam boilers or steam generators will convert the thermal energy from the combustion of one or more solid, liquid, or gaseous fuels into steam for a variety of applications. Primary among these are electric power generation and industrial process heating. Steam flow rates and operating conditions vary dramatically from one boiler to another: 1,000 lb/h (0.1 kg/s) to more than 10 million lb/h (1,260 kg/s); 14.7 psi (0.10 MPa) and 212°F (100°C) to 4,500 psi (31.0 MPa); and 1,100°F (593°C) or more in advanced power cycles. Since the combustion of fuels generates air pollutants, a key element in boiler and power system design is the control and reduction of regulated emissions.

Steam boilers are fabricated from pressure parts (tubes, tube sheets, and cylindrical vessels) which separate the products of combustion (or the **flue gas**) from the water while permitting heat transfer to take place and generate steam. Two basic boiler design configurations are used: the **fire-tube** boiler with the **flue gas** passing through the tubes and **water-tube** boilers with the water and steam passing through the tubes. Cost, size, technical, and safety considerations dictate that water-tube boilers, the subject of this section, supply the vast majority of steam used today, especially in thermal power stations.

Water-tube boilers generally consist of a large open **furnace** volume where the fuel combustion takes place, and heat is absorbed primarily by thermal radiation from the flue gas to the water-cooled furnace enclosure walls and surfaces. The furnace is followed by a **convection section or pass** where heat transfer from the flue gas takes place primarily by convective heat transfer as the flue gas passes over tube bundles containing steam and/or water.

A specialized steam generator is used in nuclear power applications where pressurized hot water from a nuclear reactor core is used to generate steam in a lower-pressure water-steam loop for power production. This specialized steam generator is discussed in detail in Sec. 7.8.

In all cases, the function of boiler design is to produce the desired quantity of steam at the specified temperature and pressure conditions while providing the most economical design that is safe and meets environmental control standards.

FUELS AVAILABLE FOR STEAM GENERATION

(See also Secs. 5, 7.1, and 7.8.)

A large variety of materials and heat sources can be used for steam generation. In the absence of other considerations, boilers are designed to use the most economical fuel or combination of fuels available. These include natural gas, residual oil, and solid fuels. Some typical solid fossil fuels are: anthracite coal, bituminous and subbituminous coal, lignite, coke breeze, fluid petroleum coke (4 to 5 percent volatile), and petroleum coke (9 to 14 percent volatile; with auxiliary fuel). Solid fuels which are generally classified as renewable resources in whole or in part include: wood, bark, bagasse, other agricultural wastes and municipal solid waste (MSW). Other less common energy sources, many suitable for steam generation, are described in Secs. 5.4 and 7.1.

In the past fuel sources nearest the plant generally supplied the boiler, but now complex economics of fuel cost, storage, transportation, emissions control, and operating requirements determine fuel use. In many cases, boilers must be designed to efficiently burn a variety of fuels which can change over time. Many industries use their by-products for steam generation. Paper mills burn the black liquor from the cooking of wood pulp. Steel mill boilers may be fired with blast furnace and coke oven gases. Oil refineries burn lean CO gas, a waste product, in combination with a richer gas. The increased cost of basic fuels has caused many other industries to consider the use of their own combustible or heat-bearing by-products. The use of municipal wastes for steam generation is also widespread (Sec. 5.4).

EFFECT OF FUEL ON BOILER DESIGN

Fuel is the governing factor in boiler design. Clean natural gas leads to the simplest design. If it is the only fuel, the boiler can be relatively small and compact. When solid or liquid fuels are to be used, the boilers will be larger because of the need to provide the required furnace volume for combustion and to accommodate slag and ash. Combustion also releases air pollutants such as nitrogen oxides (NO_x) sulfur dioxide (SO_2) particulate, and selected other emissions which must be reduced to regulated levels before the combustion products are released. NO_x is formed from fuel-bound nitrogen and nitrogen in the combustion air. NO_x formation can be limited through effective combustion system and furnace design, but emissions must frequently be further controlled through the use of postcombustion equipment. Where significant amounts of sulfur are present in the fuel, postcombustion control equipment is usually required to control SO_2 release. In some applications, fluidized-bed combustion technology where the fuel is burned in an air fluidized-bed of limestone can be effective for SO_2 emission control. Where the fuel contains a significant amount of inert solids, fly ash collectors, electrostatic precipitators, and fabric filters are required to meet government regulations. These emission control technologies are discussed later.

Slag and Ash

Virtually all solid fuels generate an ash residue which is collected in the bottom of the boiler furnace (**bottom ash**) or is entrained with the products of combustion and carried out of the boiler furnace as fine **fly ash**. Some of the coarser fly ash particles are collected in hoppers arranged under the economizer and air heater. Finer fly ash particles remain in suspension and are carried through the unit and collected in particulate control equipment downstream of the boiler. The relatively small portion of the ash collecting on the furnace walls and other heat-transfer surfaces is removed by effective sootblower and other cleaning equipment operation.

All modern pulverized coal-fired boilers are designed for **dry-ash** or **dry-bottom** furnace operation. Most of the ash, typically 70 to 80 percent, is entrained in the flue gas and carried out of the furnace as **fly ash**. The remaining 20 to 30 percent of the ash (**bottom ash**) is collected in the hopper formed by the front and rear wall tube panels at the bottom of the furnace. A limited number of existing boilers have been designed with **slag-tap** or **wet-bottom** furnaces which burn coals having low ash fusion temperatures and which are designed to have high temperatures near the furnace floor so that the ash is kept molten for tapping.

Ash passing through the boiler is subject to various chemical reactions and physical forces which lead to deposition on heat-transfer surfaces. Such ash deposition is defined as slagging or fouling due primarily to the difference in deposition mechanisms. **Slagging** is the formation of molten, partially fused or resolidified deposits on furnace surfaces exposed to radiant heat from combustion. Slagging can also extend into the convection surfaces if the flue gas temperatures are not reduced sufficiently. **Fouling** is defined as the formation of high-temperature bonded deposits on convection surfaces, such as superheaters and reheaters, which are not exposed to radiant heat. In general, fouling is caused by the vaporization of volatile inorganic elements in the fuel during combustion. As heat is absorbed from the flue gas and the temperatures fall in the convection section of the boiler, components formed by these volatile elements condense on ash particles and heating surfaces, forming a "glue" which drives the deposition process. Both slagging and fouling can reduce the effectiveness of the heat-absorbing surfaces and in the case of extreme conditions can result in uncontrolled ash deposits, which can increase pressure drop and ultimately block portions of the flow passages, requiring the unit to be shut down for manual deposit removal.

Slagging and fouling characteristics are influenced by the chemical composition and physical characteristics of the ash. The amount of ash, as well as its chemical and physical characteristics, varies over a wide range for all solid fuels. In the case of coal, variations exist not only from coal mine to coal mine, but also from different parts of the same mine. Thus, in the design of new boilers or selection of new fuel sources for existing units, it is essential to have a thorough knowledge of the fuel and compare this to the considerable operating data that have been accumulated over time.

Slag and Ash from Coal

Coal ash is classified into two categories based upon chemical composition. **Bituminous** ash is defined as having more Fe_2O_3 than CaO plus MgO. **Lignitic** ash is defined as having more CaO plus MgO than Fe_2O_3. Bituminous ash is generally a characteristic of higher-rank coals (Sec. 5.1 definition) found in the eastern United States. Lower-rank western U.S. coals tend to have lignitic ash. Because the quantitative and qualitative evaluations of mineral forms are extremely difficult, the relatively simple **elemental ash analyses** are performed on coal ash samples prepared in accordance with the ASTM International Standard D-3174 ashing procedure (**ASTM ash**). Elements in the ash are quantitatively measured by using a combination of emission spectroscopy and flame photometry, and they are reported as weight percentages of their oxides. Coal ash is consistently found to be composed mainly of silicon, aluminum, iron, and calcium and smaller quantities of magnesium, titanium, sodium, and potassium. The elemental analysis also identifies phosphorous as P_2O_5 and sulfur as sulfur trioxide, SO_3.

Unfortunately, elemental ash analyses alone do not directly identify the compounds that cause deposition, or directly identify the mechanisms of deposit formation. As a result, coal ash research has developed methods to correlate elemental analysis data and other characteristics (such as ash fusibility, sintering strength, and viscosity) of ASTM ash with the slagging and fouling behavior observed in full-scale boilers and in test facilities. The measurement of ash fusibility temperatures is the most widely used method for predicting ash behavior at elevated temperatures. The preferred procedure in the United States is ASTM D-1857 which defines and provides procedures to determine **initial deformation temperature (IT** or **IDT)**, **softening temperature (ST)**, **hemispherical temperature (HT)**, and **fluid temperature (FT)**. These ash fusion temperatures are determined under both oxidizing and reducing furnace conditions, although reducing conditions are assumed if not specified. In practical terms for dry-bottom furnaces, fusion temperatures provide an indication of the temperature range over which portions of the ash will be a molten fluid or semimolten plastic state. High fusion temperatures indicate that the ash released will cool quickly to a nonsticky state, resulting in minimum potential for slagging. Conversely, low fusion temperatures indicate that ash will remain in a molten state longer, exposing more of the furnace surface or convection surface to potential deposition.

Iron has a dominating influence on the behavior of bituminous-type ash and the ash fusion temperatures, as shown in Fig. 7.2.1. In the completely oxidized form, iron tends to raise fusion temperatures while in the lesser oxidized form it tends to lower them. The iron content and its degree of oxidation also have a great influence on the ash viscosity, which increases with ferric percentage. Based upon the weight

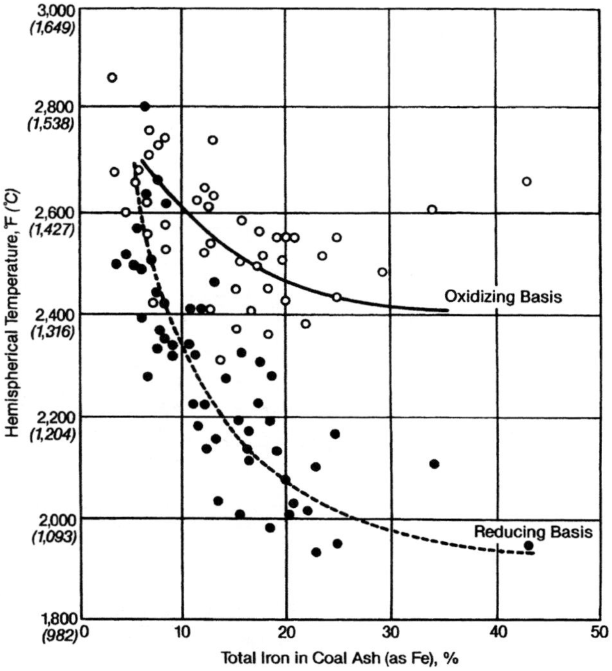

Figure 7.2.1 Influence of iron on coal ash hemispherical fusion temperature.

percentage elemental ash analysis, some of the parameters used to evaluate slagging and fouling potential include

$$\text{Ferric percentage} = \frac{Fe_2O_3 \times 100}{Fe_2O_3 + 1.11FeO + 1.43Fe}$$

$$\text{Base/acid ratio} = B/A = \frac{Fe_2O_3 + CaO + MgO + Na_2O + K_2O}{SiO_2 + Al_2O_3 + TiO_2}$$

$$\text{Dolomite percentage} = \frac{(CaO + MgO) \times 100}{Fe_2O_3 + CaO + MgO + Na_2O + K_2O}$$

The most accurate indicators of coal ash slagging potential, such as ash viscosity, and fouling potential, such as sintering strength, can be complex, time-consuming, and costly to acquire. As a result, indices based upon the ash elemental analysis and ash fusion temperatures have been developed to provide criteria for various aspects of boiler design with slagging indices establishing furnace design criteria and fouling indices establishing criteria for convection pass design. These indices are generally used to establish the **slagging classification** and **fouling classification** as low, medium, high, or severe. Because of their significantly different characteristics, bituminous ash and lignitic ash are treated separately:

Bituminous Ash Indices

$$\text{Slagging index: } R_s = (B/A) \times S$$

where S = wt percent sulfur, on dry coal basis
B/A = base/acid ratio defined above

Slagging classification: low for R_s less than 0.6; medium for 0.6 to 2.0; high for 2.0 to 2.6; severe for greater than 2.6.

$$\text{Fouling index: } R_f = (B/A) \times Na_2O$$

where Na_2O = wt percent from coal ash analysis.

Fouling classification: low for R_f less than 0.2; medium for 0.2 to 0.5; high for 0.5 to 1.0; severe for greater than 1.0.

Lignitic Ash Indices

$$\text{Slagging index: } R_s^* = [(\text{Max } HT) + 4(\text{Min } IT)]/5$$

where Max *HT* equals the higher of the reducing or oxidizing hemispherical softening temperatures and Min *IT* equals the lower of the reducing or oxidizing initial deformation temperature.

Slagging classification: low for R_s^* greater than 2,450°F; medium for 2,250 to 2,450°F; high for 2,100 to 2,250°F; severe for less than 2,100°F

Fouling index: Na_2O percentage in the coal ash

For ($CaO + MgO + Fe_2O_3$) > 20 percent by weight of coal ash, fouling classification: low to medium for $Na_2O < 3$; high for Na_2O of 3 to 6; severe for Na_2O greater than 6.

For ($CaO + MgO + Fe_2O_3$) < 20 percent by weight of coal ash, fouling classification: low to medium for $Na_2O < 1.2$; high for Na_2O of 1.2 to 3; severe for Na_2O greater than 3.

Coal Ash Reflectivity Ashes from certain coals produce furnace deposits that reduce heat transfer by high reflectivity as well as by thermal insulation. This is particularly true of low-sulfur, low-sodium coals found in the western United States, Powder River Basin in Wyoming and Montana. Reflective deposits can significantly reduce furnace heat absorption and increase the furnace exit gas temperature, even when only a very thin deposit is present. This can result in excessive slagging and fouling in the upper furnace and convection pass, respectively. These deposits can be difficult to remove and require special attention in the selection of ash cleaning equipment and cleaning media. Field experience and laboratory tests are used to evaluate the potential impact and boiler design factors.

Effect on Boiler Design The boiler furnace must have sufficient volume to completely burn the fuel, provide sufficient surface to cool the flue gas and ash particles to a temperature suitable for admission to the convection pass, and minimize the formation of NO_x emissions. In general, it is the second criterion that determines the minimum furnace size. The **slagging classification** establishes the upper limit on the furnace exit gas temperature (FEGT) required to minimize the potential for slagging in the superheater and closely spaced convection surfaces. This, in turn, establishes the total furnace volume and heat-transfer surface requirements. In addition, the furnace must be correctly proportioned with respect to width, depth, and height to minimize slagging with ample clearance provided between the burners and the furnace walls and furnace hoppers. The slagging classification also determines the locations, quantity, and spacing of furnace wall sootblowers and long, retractable sootblowers. Fig. 7.2.2 shows the effect of different slagging

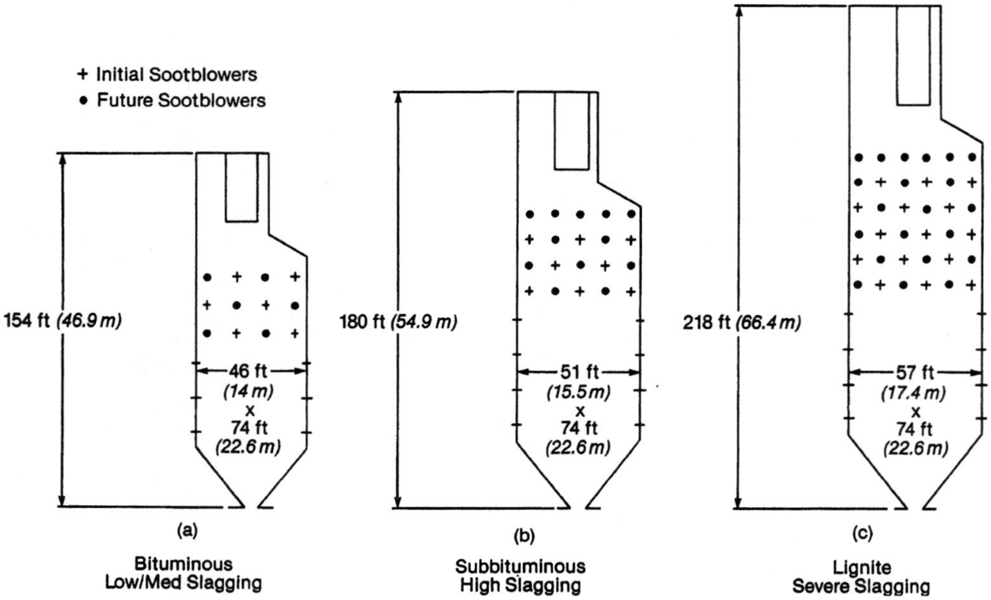

Figure 7.2.2 Influence of slagging potential on furnace size.

characteristics on a 660-MW large utility coal-fired boiler. The boilers are assumed to have the same width for the purposes of illustration with the boiler setting height and furnace depth varied to accommodate the slagging classification and characteristics of the different fuels.

In the convection pass design, the fouling classification determines the specific side spacing (perpendicular to flow) dimension at the temperature entering the tube bank with severe fouling coals requiring the widest spacing. The fouling classification and temperature entering the bank also establish the tube bank depth (parallel to flue gas flow). Cavities are provided between the tube banks for long retractable sootblowers.

Sootblower Systems

While slagging and fouling in boilers can be minimized by proper design and operation, auxiliary equipment for online cleaning of furnace walls and removing deposits from convection surfaces must be supplied to maintain capacity and efficiency. Superheated steam is the most common cleaning medium, but saturated steam, compressed air, and water are also used. The cleaning media jets from the sootblower nozzles dislodge the dry or sintered ash and slag, which then fall into hoppers or travel along with the flue gas to the particulate removal equipment.

Types of sootblowers vary with their location in the boiler unit, the severity of ash or slag conditions, and the arrangement of the heat-absorbing surfaces.

Furnace walls are generally cleaned with **wall blowers** (Fig. 7.2.3) which project a nozzle assembly into the furnace for blowing and then retract it behind the wall tubes for protection after operation.

A water lance sootblower may be used for dense slag layers which are resistant to mechanical impact pressure alone. The rapid quenching and mechanical impact energy create cracks and fissures, dislodging the slag in pieces. A more recent development is the use of a water jet through a sidewall opening to clean surfaces on the opposite side of the furnace.

Tube banks in high-gas-temperature zones, such as slag screens, superheaters, and reheaters, where slag or sintered ash may accumulate, are generally cleaned by **long-lance retracting-type blowers** (Fig. 7.2.3). The lance, which rotates or oscillates as it advances into the boiler, is fitted with nozzles to supply a powerful cleaning action and is retracted from the boiler for protection when it is not operating.

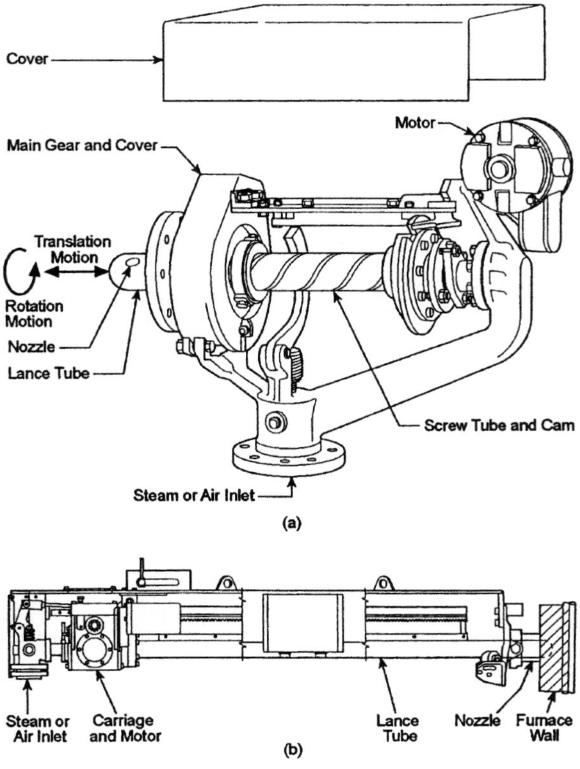

Figure 7.2.3 Retracting sootblowers. (*a*) Furnace wall blower; (*b*) long-lance blower.

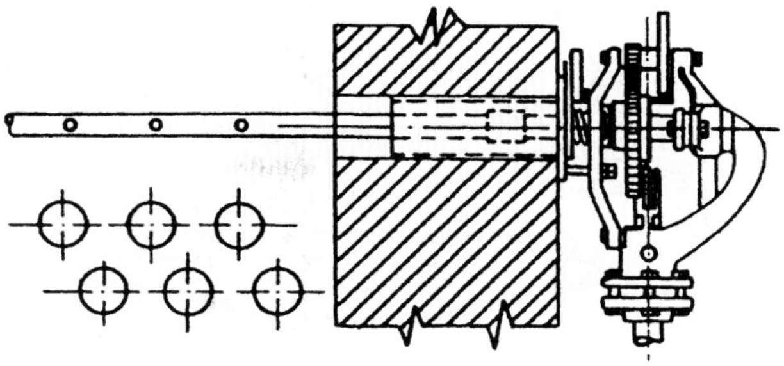

Figure 7.2.4 Rotating sootblower with multiple nozzles.

Tube banks located in low-gas-temperature zones, including the economizer and boiler sections, where uncooled metals have satisfactory life and ash removal is easier, usually can be cleaned by **multiple-nozzle rotating-type sootblowers** (Fig. 7.2.4). However, long-lance retracting-type blowers may be necessary for very wide boilers, for extended cleaning ranges, or where the ash tends to accummulate.

Sootblowers for air heaters generally are arranged to blow through the plate or tube assemblies with single- or multiple-nozzle elements moving in an arc or straight-line motion. These blowers may also supply water for washing the air-heater surface.

Additional technology has evolved to improve efficiency or meet specialized cleaning requirements. Sonic horns use acoustic energy to propagate sound waves (20-200 hertz: 75 hertz typical) within the boiler or emissions control equipment to dislodge dry, loosely-adhering ash deposits. These are used in combination with or in place of sootblowers in the back-end regions of the boiler (such as economizer) as well as back-end electrostatic precipitators, fabric filters, and selective catalytic reduction (SCR) equipment. In waste-to-energy applications, mechanical rapping systems are used to only partially clean superheater surfaces to avoid exposing fresh metal to the very corrosive products of combustion. Rapper anvils strike pins which impart acceleration forces (energy) to superheater assemblies to dislodge most of the ash.

Advancements have occurred in automated sootblower control systems. Automatic control systems have historically been arranged to operate the blowers in a prescribed sequence at time intervals adjusted to blower-cleaning requirements or to receive signals from the blower unit's instruments and controls so as to operate the sootblowers selectively in the various heat-absorbing sections to maintain the required cleanliness and heat absorption. Modern boiler cleaning systems can incorporate on-line diagnostic sensors and advanced control algorithms to increase cleaning effectiveness and boiler efficiency while reducing the consumption of sootblower media and reducing potential damage to the boiler. In addition, flue gas temperature can be used with heat balance and heat-transfer calculations for a particular tube bank to determine a real-time **fouling factor** which can be used to detect deposit accumulation over time. Advanced "intelligent" sootblower control systems then select the individual sootblowers and cleaning cycle to address the specific deposition area.

COMBUSTION AND FUEL PREPARATION SYSTEMS

Stoker Firing Systems

Mechanical stoker/grate system designs are available to burn a wide range of fuels including all forms of coal, sludge, wood, wood waste, other biomass and agricultural wastes, as well as residential, commercial, and industrial refuse. Modern mechanical stoker firing systems consist of

1. A **stoker** or fuel feeding system
2. A stationary or moving **grate** assembly to support the burning mass of fuel and admit air from under the grate

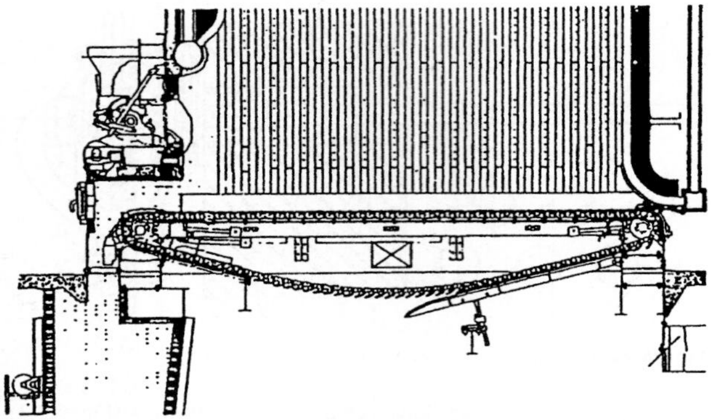

Figure 7.2.5 Spreader stoker with traveling grate.

3. An **overfire air** (OFA) system above the fuel bed to complete the combustion and limit air pollutant emissions
4. An ash discharge system

The term *stoker* often refers to not only the fuel feeder but also to the full feed/grate/air system.

There are two general types of systems: **underfeed,** in which fuel and air are both supplied from below the grate, and **overfeed,** in which fuel is supplied from above and air from below. Overfeed systems are further divided into two types: **mass feed,** in which the fuel is continuously fed to one end of the grate and travels across the grate as it burns, and **spreader,** in which the fuel is thrown or spread across the grate area. Underfeed and small mass feed stokers are uncommon today.

Spreader stokers, in combination with either a traveling or vibrating grate, are the most commonly used systems for a steaming capacity range from 75,000 to 900,000 lb/h (9.5 to 113 kg/s). They respond quickly to changes in steam demand, have good **turndown capability** (maximum to minimum load), and can be used with a variety of fuels. Figure 7.2.5 shows a typical spreader stoker with continuous ash discharge traveling grate. The rotating mechanical feeder on the left side spreads the fuel evenly across the grate area. Fine particles burn in suspension with the coarser particles landing on the grate to complete the combustion. The traveling grate moves the fuel bed back toward the feeder, below which is the ash discharge. The traveling grate is an endless moving belt which consists of a series of chains attached to grate bars with air metering holes. To achieve high combustion efficiency, fine particles carried over from the combustion may be captured in a mechanical dust collector, collected in hoppers, and then returned to the furnace by a **carbon reinjection** system. The split of the air supplied under the grate and injected above the fuel bed is carefully controlled to maximize burnout of the fuel and to minimize the formation of NO_x.

Water- and air-cooled vibrating grates are also frequently used in spreader stoker firing systems. As shown in Fig. 7.2.6, the vibrating grate consists of water-cooled grate panels that create a furnace floor which is mounted on leaf-spring supports. As the fuel bed burns, a vibrating unit creates a back-and-forth motion which moves the fuel bed toward the ash discharge. Air-swept spouts (fuel spreaders) are combined with vibrating grates in many wood- and bark-fired boilers.

Pulverizers

Pulverized coal (PC) is the dominant method of firing coal in modern utility boilers. Pulverizers are used to grind coal into a fine powder with 70 percent of the ground coal passing through a 200-mesh (74-μm) screen for typical PC applications (sample collected using ASTM procedure D-197). Utility systems in operation today are **direct-fired** with the coal ground in the pulverizer directly transported to the burner with a portion of the combustion air (**primary air**). Pulverizer models range in total capacity from 1.5 to 115 tons/h of coal (0.4 to 29 kg/s) with a typical pulverizer shown in Fig. 7.2.7. The primary air entering the pulverizer is preheated to temperatures as high as 700°F (371°C) to dry

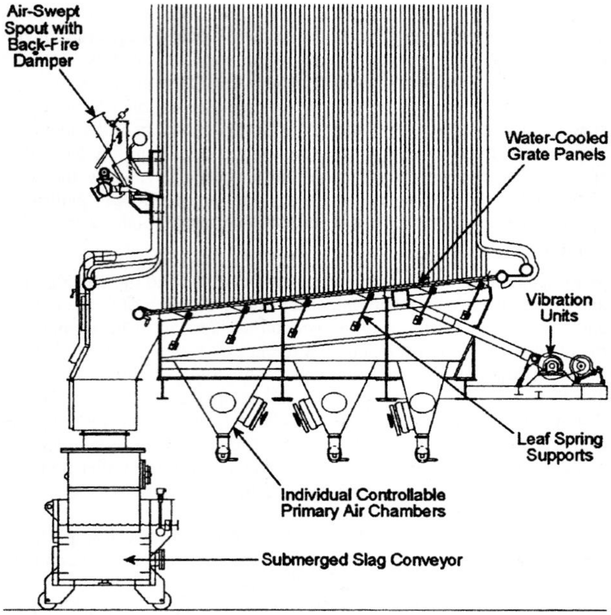

Figure 7.2.6 Water-cooled vibrating grate stoker.

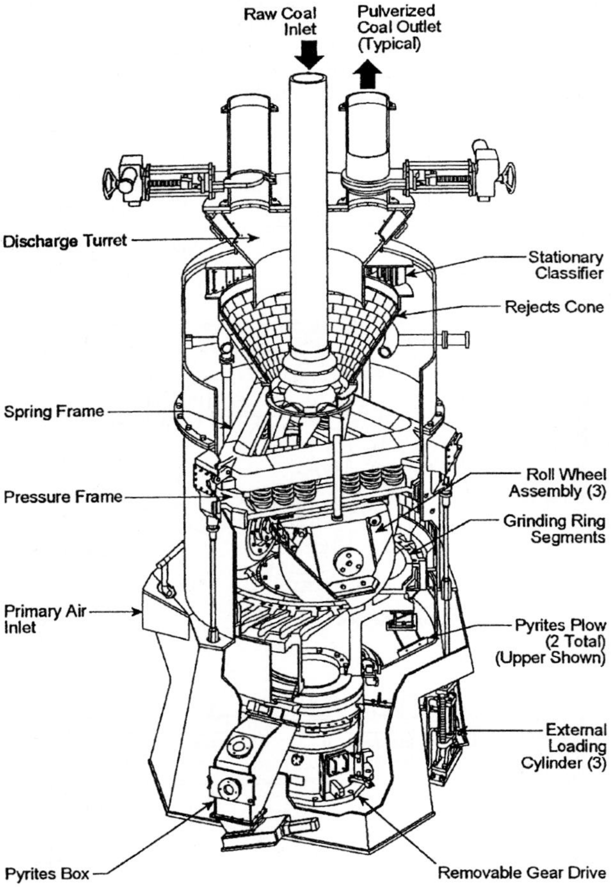

Figure 7.2.7 Slow-speed pulverizer, roller-and-race type. (*Courtesy The Babcock & Wilcox Company.*)

Table 7.2.1 Typical Pulverizer Outlet Temperatures

Fuel type	Volatile content, percent*	Exit temperature, °F (°C)†
Lignite	All values	120-140 (49-60)
Subbituminous	All values	130-150 (54-66)
High-volatile bituminous	Above 31	140-175 (60-79)
Medium- and low-volatile bituminous	14-31	160-200 (71-93)
Anthracite and coal waste	0-14	200-210 (93-99)
Petroleum coke	0-8	200-250 (93-121)

* Volatile content is on a dry, mineral-matter-free basis.
† The capacity of the pulverizer is adversely affected for exit temperatures below 125°F (52°C) when grinding high-moisture lignites.

the moisture in the coal and provide stable, efficient combustion. The inlet temperature is controlled to keep the pulverizer outlet temperature to a specified range which is largely determined by the volatile matter content in the coal. Coal with low volatile matter content may require higher air-coal temperatures to ensure stable combustion. The typical pulverizer outlet temperature is 150°F (66°C). Table 7.2.1 provides outlet temperature ranges for a variety of coal types.

The three principal types of pulverizers may be classified by their speed: low (< 75 r/min), medium (75 to 250 r/min), and high (> 250 r/min). Figure 7.2.7 shows a slow-speed, **roller-and-race** type of pulverizer. Medium-speed pulverizers are generally either the **ball-and-race** type (Fig. 7.2.8) or the **bowl-and-roller** type (Fig. 7.2.9). All three of these pulverizers are referred to as vertical spindle pulverizers. High-speed pulverizers are usually of the attrition type, shown in Fig. 7.2.10. Some older boilers are equipped with **ball-and-tube** coal pulverization equipment, in which the coal is ground in a horizontal rotating cylinder that is partially filled with small-diameter tumbling balls. In all pulverizers, provision is made to remove and dispose of **pyrite** (iron sulfide, FeS_2) and other mineral matter from the coal which is hard to grind and which has no material heating value.

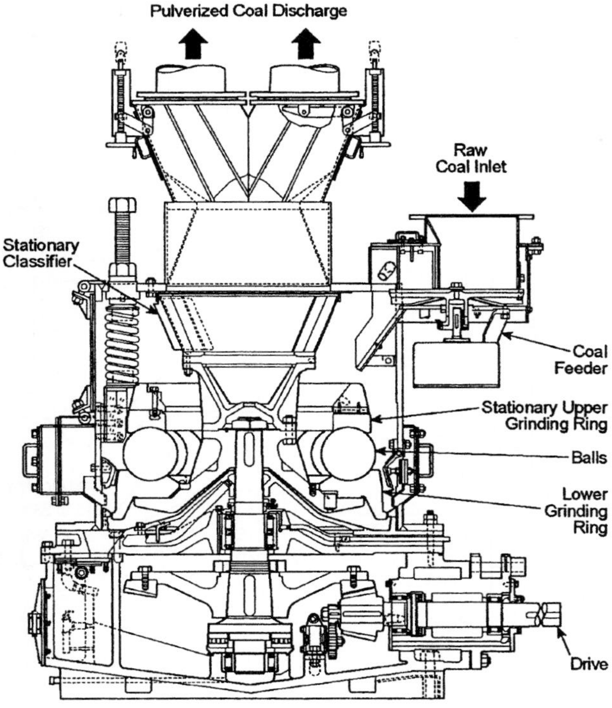

Figure 7.2.8 Medium-speed pulverizer, ball-and-race type. (*Courtesy The Babcock & Wilcox Company.*)

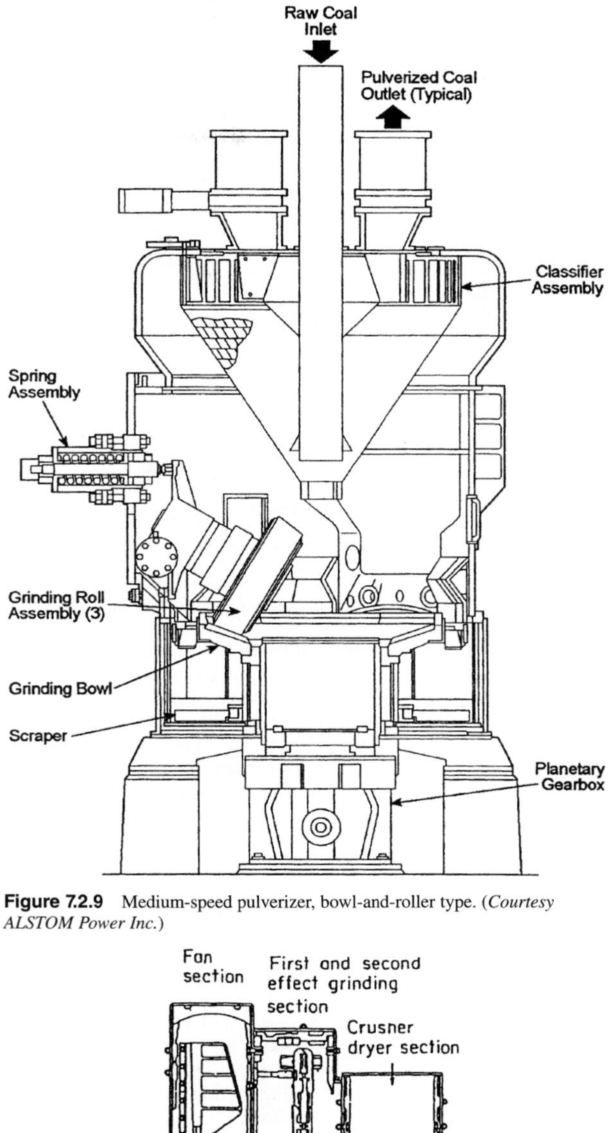

Figure 7.2.9 Medium-speed pulverizer, bowl-and-roller type. (*Courtesy ALSTOM Power Inc.*)

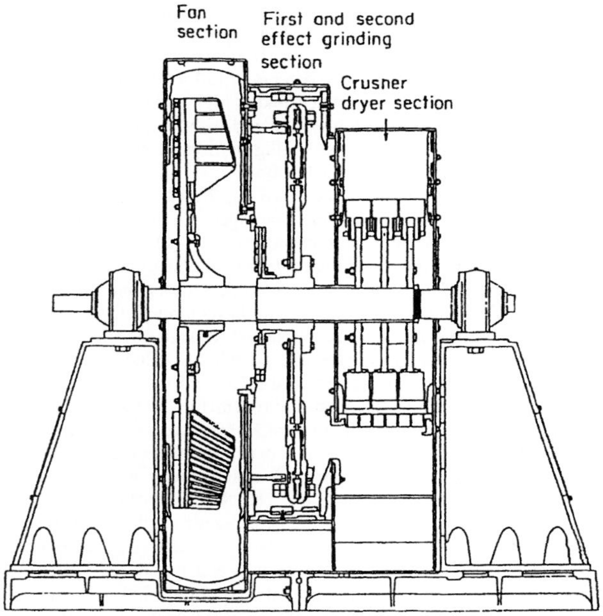

Figure 7.2.10 High-speed pulverizer, attrition type.

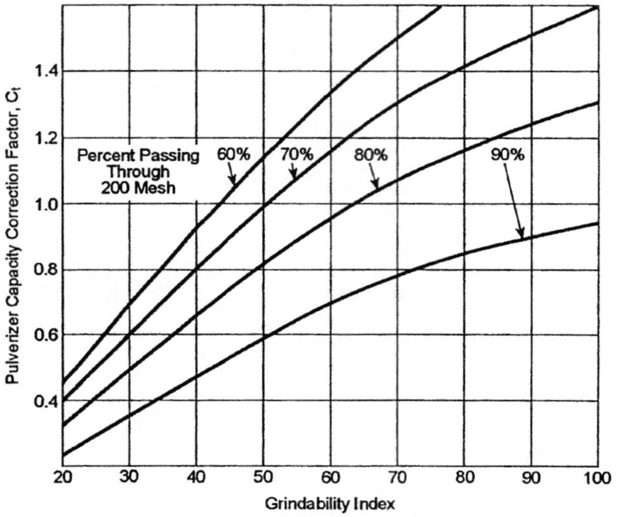

Figure 7.2.11 Pulverizer capacity factors for varying fineness and grindability: roller-and-race and ball-and-race types.

The capacity of a specific pulverizer is a function of the grindability of the coal, the desired coal fineness (percent passing through 200-mesh), and the moisture. **Grindability** defines the ease with which a specific coal can be ground and is quantified by the Hardgrove Grindability Index (HGI), as described in ASTM D-409. Higher HGI indicates easier to grind coals. The required fineness is established based upon the coal and firing system. Figure 7.2.11 correlates the HGI and the fineness with a pulverizer **capacity factor** for roller-and-race and ball-and-race types of pulverizers. The expected pulverizer capacity is then the capacity factor times the pulverizer **base capacity** for the specific size and model, plus a margin. An additional factor must be included where high moisture is present (moisture determined according to ASTM D-3302).

Burners and Combustion Systems

Advanced burners mix air and fuel to ensure rapid ignition and effective combustion (> 99 percent fuel utilization) while minimizing the formation of nitrogen oxides (NO_x). In pulverized coal firing, a portion of the total air flow (typically 15 to 25 percent at full load, but as low as 11 percent for ball-and-tube pulverizers and as high as 33 percent for some lignite fuels), called **primary air**, is mixed with the pulverized coal to convey it to the burners. The remaining portion, or **secondary air,** is introduced at the burner from a **wind box**.

Multiple low-NO_x circular burners (see Fig. 7.2.12) located in the walls of the lower boiler furnace are the centerpiece of the overall combustion system which includes the fuel preparation (including pulverizers), airflow distribution, igniters, furnace design, and integrated combustion control system including monitors, diagnostic sensors, and flame safety equipment. Coal and primary air are introduced through the central coal pipe while the secondary air is introduced through several concentric zones with controlled flow rates and swirl rates induced by the spin vanes. Approximately 75 percent of the NO_x formed from pulverized coal combustion is from fuel-bound nitrogen (**fuel NO_x**) while the balance is from nitrogen in the combustion air (referred to as **thermal NO_x** because the NO_x formation is strongly dependent upon high combustion temperatures). By controlling the local stoichiometry within the flame and the secondary air-fuel mixing rates, NO_x formation can be minimized. NO_x emissions can be further reduced by using **air staging** in the furnace (Fig. 7.2.13)—a portion of the secondary air (referred to as **overfire air,** or **OFA**) is removed from the burners and introduced later in the combustion process through air injection **NO_x ports**. Modern pulverized-coal-fired boilers typically operate with 15 to 20 percent **excess air** above the theoretical amount needed at full load and with higher amounts as

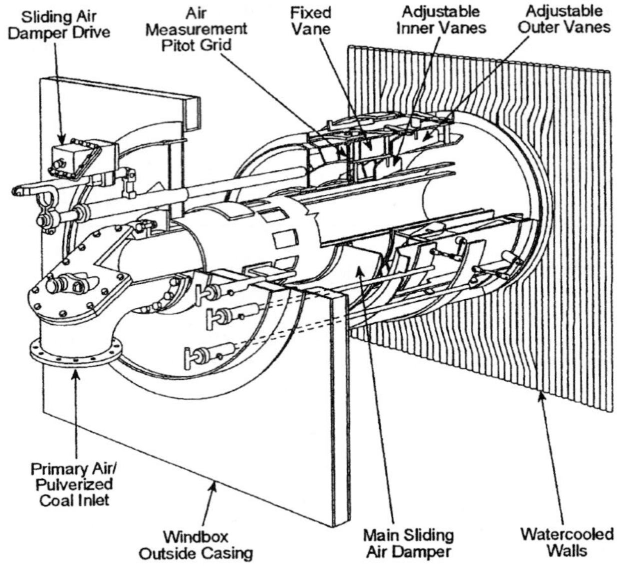

Figure 7.2.12 Low-NO_x circular burner for pulverized coal firing.

boiler load is reduced. Individual pulverized coal circular burners have maximum heat input rates of 50 to 300 million Btu/h (15 to 88 MW). For modern applications with high volatile bituminous or subbituminous coal, combustion should be stable without the use of igniters from about 30 to 100 percent load. The actual minimum is dependent on the coal, burner design, and pulverizer load. Operation below this load is performed with igniters in service. Specialized burners are available for difficult to combust coals (typically low-volatile fuels and high-moisture and/or ash fuels).

Oil and natural gas are also usually burned in specially designed low-NO_x circular burners where all the combustion air is supplied from

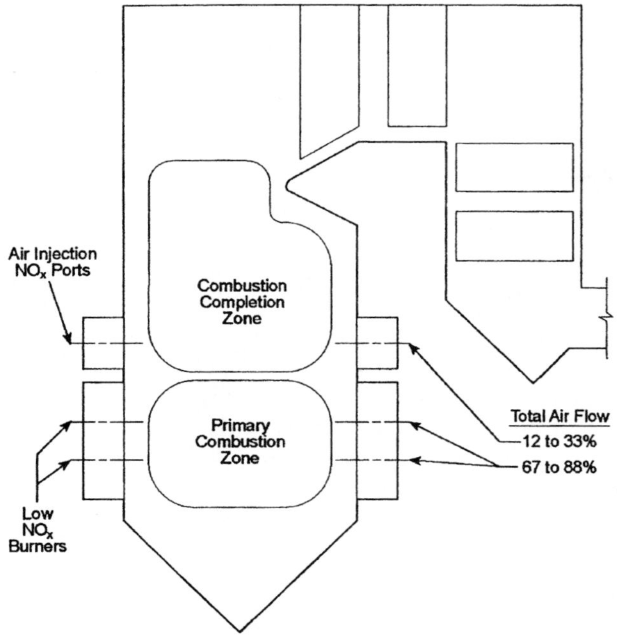

Figure 7.2.13 Air staging for NO_x reduction in pulverized coal wall-fired boilers.

the wind box and specialized combustion components are used. Fuel oil is atomized into a fine mist which is dispersed into the airstream for combustion, using an individual atomizer located in the centerline of the burner. Atomizers are classified as one of two types and are selected based upon economics and the individual application. **Dual fluid atomizers** use steam or compressed air to supply most of the energy to create the fine fuel-oil mist. Steam atomization is generally preferred for its safety characteristics and the creation of a finer mist. Several designs are available with up to 300 million Btu/h (88-MW) heat input. Depending upon the design, steam pressures as high as 200 psig (1.4 MPa gage) are used. Dry steam with 20 to 40°F (11 to 22°C) superheat is recommended. **Mechanical atomizers** use relatively high fuel pressure to create the droplets, using one of several designs. Depending upon the design, oil pressure requirements range from 300 to 1,000 psig (2.1 to 6.9 MPa gage) and maximum firing rates range from 70 to 200 million Btu/h (21 to 59 MW). A typical turndown ratio for oil firing is on the order of 6:1 depending upon the fuel characteristics, flexibility of the delivery system, and atomization technique. The primary source of NO_x with heavy fuel oil firing is fuel NO_x, while thermal and fuel NO_x are important for distillate oils.

Natural gas and other commercially high-heating-value gaseous fuels can be burned with circular burners by admitting the gas through a series of radial spuds or through a centrally located high-capacity gas element, which is sited at the burner centerline. The latter is usually employed where multifuel capacity including the use of coal is needed. The maximum practical limit for gas input per burner is 250 million Btu/h (73.3 MW). Gas pressure at the burner is generally in the range of 8 to 12 psig (55 to 83 kPa gage) for the multiradial spud design and as high as 20 psig (138 kPa gage) for the high-capacity gas element. The design of the gas spuds is critical to minimizing the formation of NO_x from natural-gas-fired burners. A turndown ratio of 10:1 is not uncommon for natural gas firing.

Thermal NO_x is the primary source of NO_x emissions from natural gas and distillate fuel oil firing. Fuel NO_x is the primary source when firing heavy fuel oil. A variety of techniques have been used to reduce NO_x on boilers using older oil and gas firing equipment. Low excess air (LEA) and burners out of service (BOOS) have been used as has air staging. An additional technique referred to as **flue gas recirculation (FGR)** has been successfully used where fuel NO_x is a relatively small contributor to total NO_x emissions. Flue gas from the economizer is returned to the combustion airstream where it helps reduce the peak flame temperatures and thereby the thermal NO_x formed.

When **corner-fired burners** (Fig. 7.2.14) are used, the mixing of the fuel and combustion air takes place in the furnace (Fig. 7.2.14*a*). Coal and air are introduced into the furnace at distinct elevations in the corners of the furnace. Nozzles direct the coal and air streams at slight tangents to an imaginary firing circle in the vertical center of the furnace, to create a rotating fireball that fills the furnace volume. For steam temperature control at lower loads, the nozzles can be tilted down (Fig. 7.2.14*b*) or up (Fig. 7.2.14*c*) in unison to lower or raise the fireball and lower or raise the furnace exit gas temperatures. Air staging can also be used in this design to reduce NO_x emissions.

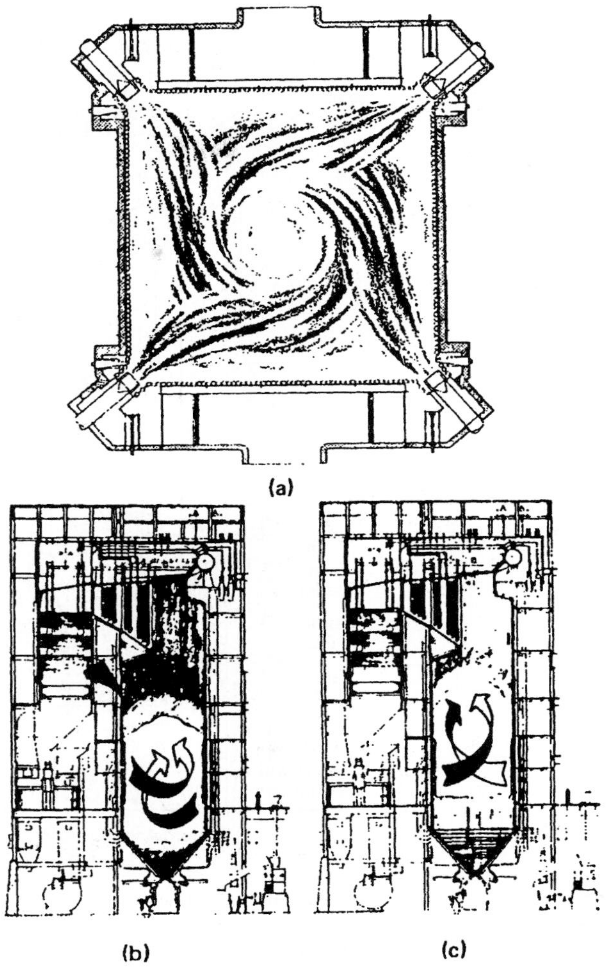

Figure 7.2.14 Corner-fired tangential tilting burners for pulverized coal, oil, or gas. (*a*) Plan section; (*b*) burners tilted down; (*c*) burners tilted up.

Fluidized-Bed Combustion

Fluidized-bed boilers can be designed to burn a wide variety of fuels, including all coals, in an air-suspended mass (bed) of solid particles while simultaneously reducing air pollutant formation and emission. **Fluidized-bed combustion (FBC)** is particularly attractive for difficult to burn fuels with high ash or moisture. Where the fuel contains sulfur, the bed material feed can be limestone **(reagent)** which ultimately absorbs the sulfur dioxide (SO_2) during the combustion process. Where no sulfur is present, an inert material such as sand and/or the ash from the fuel can form the bed of solids. Nitrogen oxides (NO_x) formation and emissions are controlled by utilizing relatively low combustion temperatures and by staging the air addition to the bed. By carefully controlling the bed temperature, emission of nitrogen oxides (NO_x) can be minimized while still optimizing the SO_2 capture. Depending upon the reagent bed stoichiometry, the limestone reactivity, and the fluidized-bed type (see below), SO_2 reductions of 90 to 95 percent are typically achieved. Where tighter NO_x emission limits are required, NO_x can be further reduced through selective noncatalytic reduction (SNCR) of NO_x by injecting ammonia into the furnace at the proper locations.

Fluidized-bed boilers are generally classified into two types: **circulating fluidized-bed (CFB)**. or **bubbling fluidized-bed (BFB)**. Figure 7.2.15 shows the main features of a large CFB boiler which is the dominant technology used for utility applications. Air is supplied through a wind box and air distributor in the furnace floor, creating an initial bubbling fluidized-bed in the bottom of the furnace. The addition of the overfire air then provides sufficient gas velocity to carry the bed solids vertically upward through the full furnace shaft before the bulk of the solids are collected in the primary particle collectors at the furnace exit and returned to the furnace. Secondary solids collection is provided by multicyclone dust collectors downstream of the superheat and reheat convection surfaces with the solids also being returned to the furnace to maintain total solids inventory. The furnace temperature is controlled by heat absorption to the furnace enclosure water walls which in turn is primarily controlled by the furnace local solids density. CFBs are operated to maintain bed temperatures typically in the range of 1,500 to 1,600°F (816 to 871°C) for applications burning coal and other sulfur-containing fuels to maximize SO_2 absorption and minimize NO_x emissions.

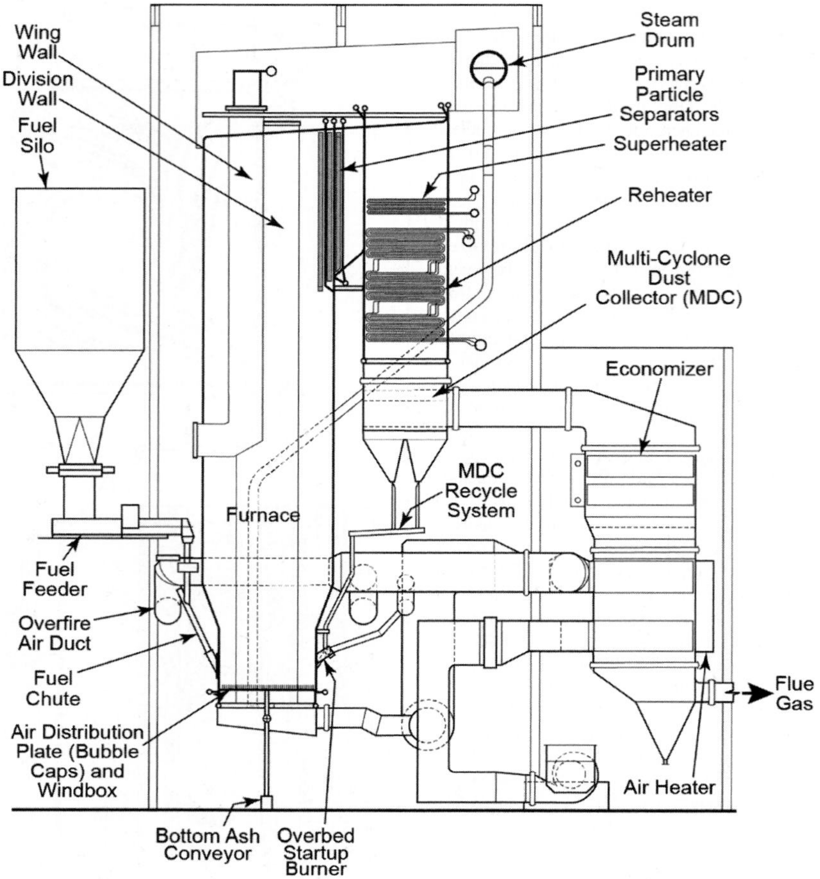

Figure 7.2.15 150 MW utility reheat internal recirculation, circulating fluidized bed (CFB) boiler; coal-fired. (*Courtesy The Babcock & Wilcox Company.*)

BFB boilers have been found to be particularly attractive and economical in retrofit applications and in cases burning biomass and waste fuels containing a high noncombustible fraction. BFBs operate at lower superficial flue gas velocities than CFBs. This creates a distinct stationary bubbling or churning bed of solids in the lower furnace where the combustion takes place. BFB boilers operate with bed temperatures typically in the range of 1,500 to 1,600°F (816° to 871°C) depending upon the fuel moisture, ash content, and alkali content in the ash. Bed temperature is maintained for biomass and other high-volatility fuels by adjusting the split between primary and overfire air and sometimes by the use for flue gas recirculation. For coal, bed temperature is maintained by the use of in-bed heat-transfer surface and compartmentalized wind box designs.

Unburned Combustible Loss

The unburned combustible loss in the fly ash from pulverized-coal firing varies with the furnace heat liberation, type of furnace cooling, volatility and fineness of the coal, excess air, and type of burner. Although the hopper refuse from dry-ash furnaces usually is low in combustible, the combustible may be appreciable in some cases especially where very low NO_x emissions are achieved.

Fly ash combustible from spreader stokers varies widely with the rating, fuel sizing and type of fuel burned. The combustion carryover at high rates of firing can be excessive but reinjection of the fly ash can be used and the loss of efficiency reduced. Historically for coal-fired steam boilers, reinjection reduces combustion loss to 1 to 2 percent. However, today's advance overfire air systems can achieve the same level of performance without reinjection. For wood, bark, and other biomass firing with spreader stokers, unburned carbon loss can be controlled to 1 to 3 percent with furnaces sized for a maximum heat liberation rate of 18,000 Btu/h-ft^3 (18 kW/m^3). Further reductions can be achieved with fly ash reinjection using mechanical dust collectors.

The combustible loss in the fly ash from solid fuels is determined by withdrawing a representative sample of fly ash and flue gas from the boiler outlet flue, or stack, at the same velocity in the sampling tip as the gas velocity in the flue (see ASME Test Code). The rates of flue-gas flow and fly ash collection are measured, and the dust loading in pounds per 1,000 lb of flue gas is then calculated. The combustible in fly ash also can be measured. Combustion efficiency loss from unburned combustible in ash can be calculated by using the following simplified relationship:

Combustion efficiency loss

$$= \frac{14{,}500 \times (\%\text{ carbon in residue})/[100 - (\%\text{ carbon in residue})]}{(\text{coal HHV})/(\%\text{ ash})}$$

where coal HHV = higher heating value of coal as fired, Btu/lb; percent ash = percent of inerts in coal as-fired; percent carbon in residue = weighted average of measured percent carbon in fly ash leaving the steam generator, economizer hopper ash, and boiler bottom ash. For a typical dry-bottom opposed-wall fired boiler:

Electrostatic precipitator = 75 percent

Economizer = 5 percent

Boiler bottom = 20 percent

This assumes that the pyrite loss through the pulverizer rejects is small and/or the percent carbon loss in the pyrite is about the same as the weighted average. The equation accounts for efficiency losses from total combustibles which are primarily in the form of carbon.

BOILER TYPES

Water-tube boilers are available in a wide range of steaming capacities—from as low as 1,000 lb/h (0.1 kg/s) to as high as 10,000,000 lb/h (1,260 kg/s). By integrating the components—pressure parts, furnaces, fuel burners, fans, and controls—boiler manufacturers have produced a broad series of standardized and economical steam generating units. For capacities up to 230,000 lb/h (29 kg/s) most units can be shipped by rail or truck as a completely shop-assembled package (Fig. 7.2.16). For larger units, barges or ocean vessels may be used if the site is suitably located.

For higher steam capacity needs typical of industrial and small utility applications, larger field-erected boilers are used, although the erection of large shop-assembled modules facilitates field erection and reduces cost. Such field-erected units can be designed to burn most solid, liquid, and gaseous fuels alone or in combination. Figure 7.2.17 shows a boiler capable of multifuel firing using biomass (bark), pulverized coal, and natural gas. Such two-drum boilers are typically designed for pressures up to 1,800 psi (12.4 MPa), although pressures as high as 2,200 psi (15.2 MPa) have been used. For pulverized coal, oil, and natural gas firing, steam capacities range up to 1,200,000 lb/h (151 kg/s). For stoker coal firing, steam capacities range up to 400,000 lb/h (50 kg/s), and for stoker wood and biomass, steam capacities can reach 600,000 lb/h (76 kg/s.) The **boiler bank** consisting of a large number of tubes connecting the lower water drum to the steam drum (shown in Fig. 7.2.17) is typically used in lower-pressure boilers to aid in steam generation where the water-cooled furnace enclosure provides insufficient boiling surface.

Designs of high-pressure, high-temperature, utility-type boilers ranging from steam capacities of 700,000 to 10,000,000 lb/h (88 to 1,260 kg/s) are classed as radiant-type boilers. In radiant boilers, virtually all the steam is generated in the tubes forming the furnace enclosure walls from thermal radiation to these tubes from the hot combustion gases. Figure 7.2.18 illustrates a modern large, pulverized coal wall-fired steam-water natural circulation radiant boiler while Fig. 7.2.19 illustrates a smaller natural circulation unit which has been designed for oil and natural gas wall firing. A typical boiler with assisted circulation utilizing tangential firing techniques is shown in Fig. 7.2.20. Such drum-type natural or assisted circulation boilers are restricted to operating pressures below the critical pressure of water [3,200 psia (22.1 MPa)] because of circulation and steam separation characteristics, and generally have superheater outlet pressures below 2,685 psia (18.51 MPa). Main steam and reheat steam superheat temperatures of 1,005°F (541°C) are common.

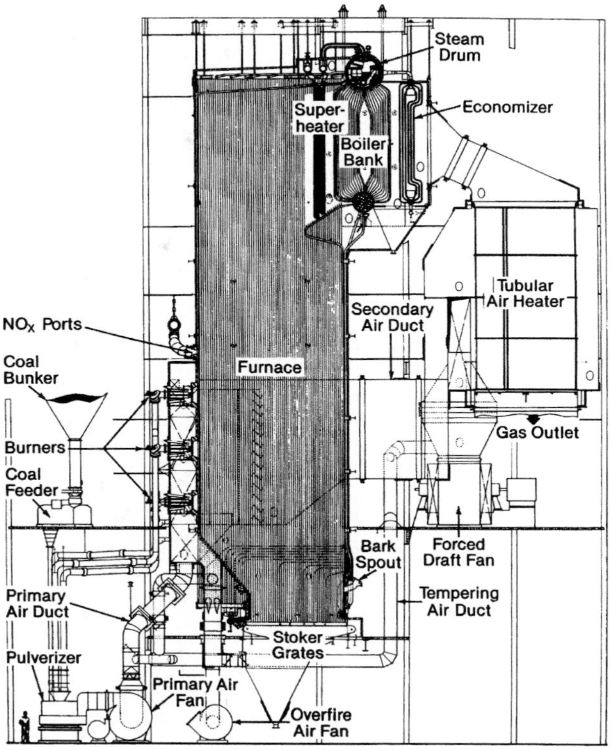

Figure 7.2.17 Two-drum type of boiler designed for multifuel firing. (*Courtesy The Babcock & Wilcox Company.*)

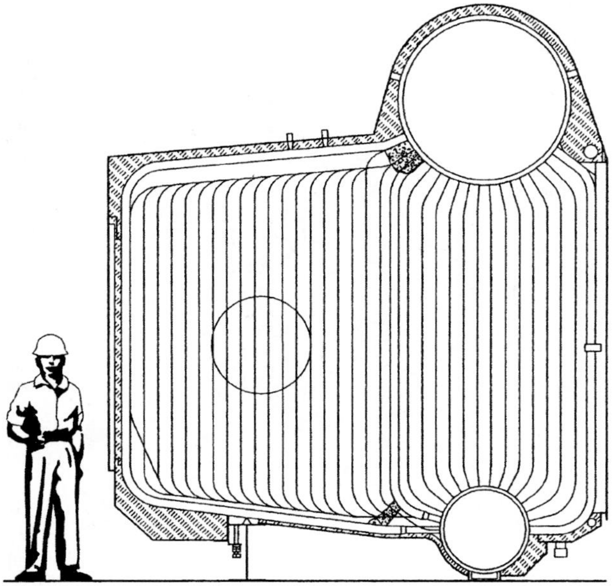

Figure 7.2.16 Water-tube package boiler for natural gas and light oil firing; 60,000 lb/h (7.6 kg/s) steam flow.

To increase steam cycle and power plant efficiency, higher steam pressures and temperatures are often used with once-through forced circulation boilers where the steam-water circuitry is designed so that water is continuously converted to steam without the use of a large steam drum. Such units can be designed for constant-pressure operation at all loads or variable-pressure operation where the operating pressure can be reduced as the load is reduced. Figure 7.2.21 shows a modern universal pressure radiant boiler which is designed for full once-through forced circulation and a main superheat outlet pressure of 3,793 psig (26.2 MPa) at full load. The unit is designed with advanced (**ultra-supercritical**) steam conditions with a main superheat outlet temperature of 1,121°F (605°C) and reheat superheat temperature of 1,117°F (603°C). This boiler employs a spiral circuitry furnace waterwall construction where the tubes spiral around the furnace perimeter (vs. vertical tube orientation in the other boilers discussed so far), which permits the unit to operate in variable-pressure mode with subcritical pressure pump-assisted circulation at low loads.

More recently, ultrasupercritical, variable pressure once-through boilers have been introduced which utilize vertical straight (ribbed or rifled) tubes in place of the spiral wound furnace. This technology can reduce costs and improve furnace steam-water circulation at variable furnace surface heat input rates.

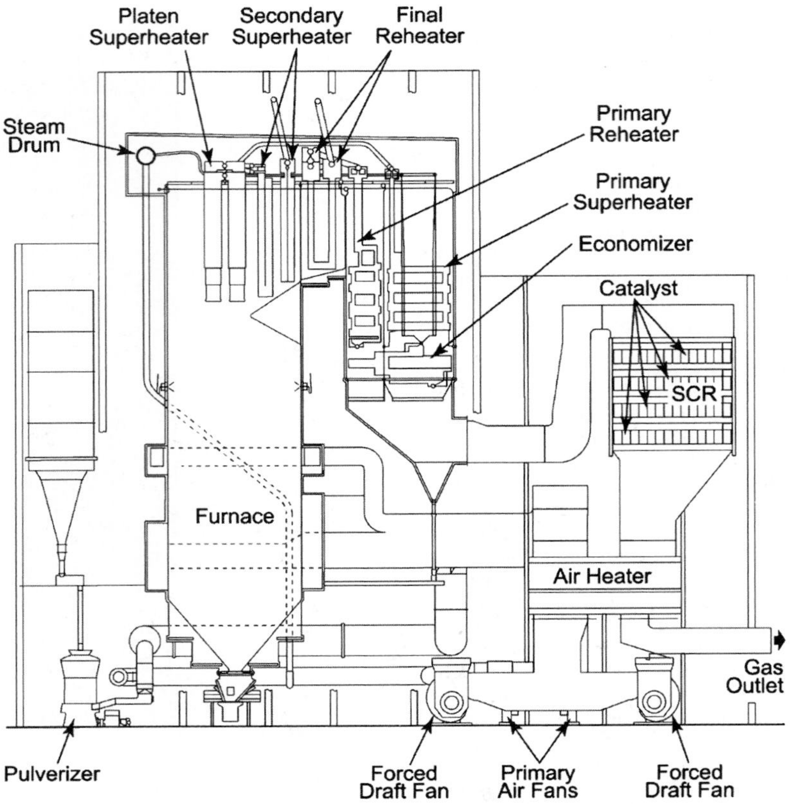

Figure 7.2.18 422 MW natural circulation radiant boiler; coal fired; 2,887,000 lb/h (364 kg/s) steam, 2,483 psig (17.1 MPa) pressure, 1,055°F (568°C) steam temperature, 1,055°F (568°C) reheat steam temperature. (*Courtesy The Babcock & Wilcox Company.*)

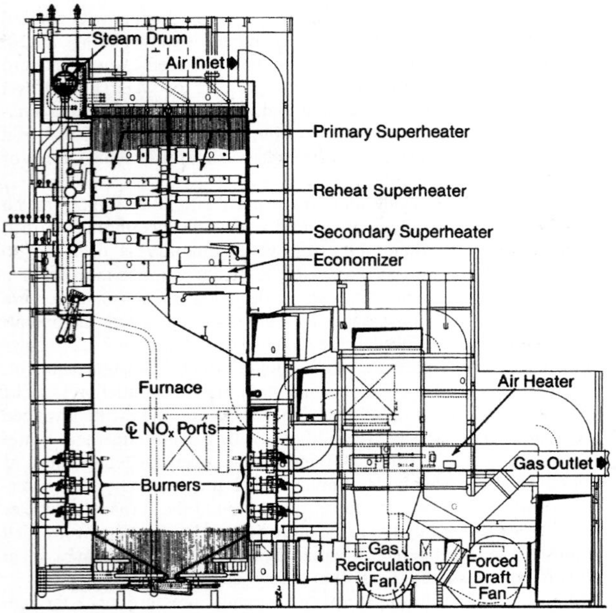

Figure 7.2.19 A 320-MW natural-circulation radiant boiler, oil- and gas-fired; 2,275,000 lb/h (287 kg/s) steam 2,610-psi (18.0-MPa) pressure; 1,006°F (541°C) steam temperature; 1,006°F (541°C) reheat steam temperature. (*Courtesy The Babcock & Wilcox Company.*)

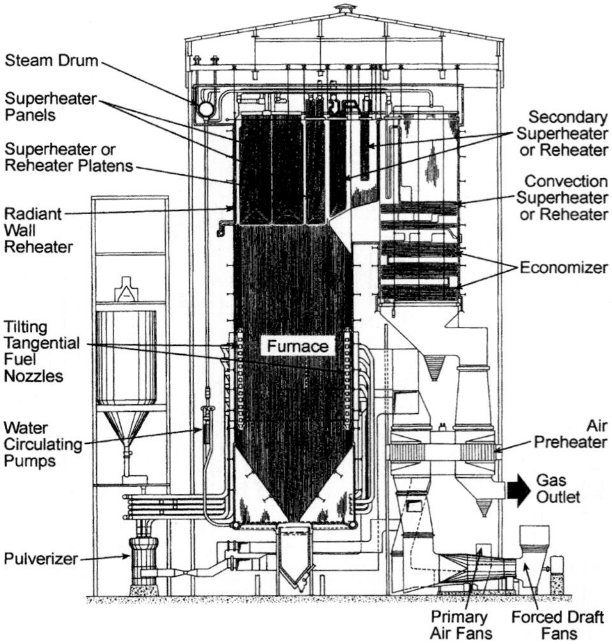

Figure 7.2.20 A 500-MW controlled or assisted-circulation radiant boiler with tangential coal firing. (*Courtesy ALSTOM Power Inc.*)

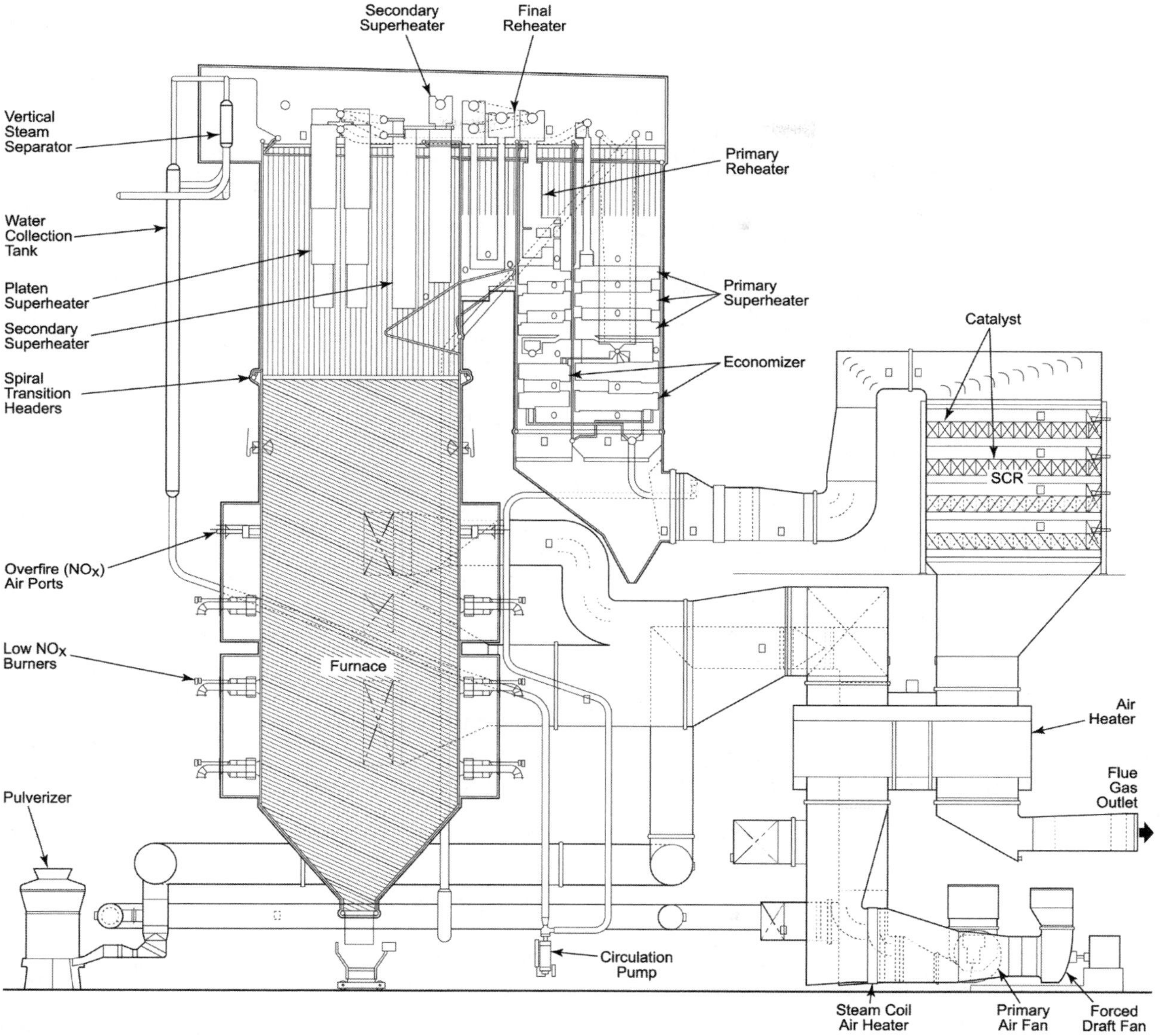

Figure 7.2.21 1,000 MW universal pressure radiant boiler; coal-fired; full sliding pressure with supercritical pressure once-through flow at full load; 6,834,000 lb/h (861 kg/s) steam, 3,793 psig (26.2 MPa) pressure, 1,121°F (605°C) steam temperature, 1,117°F (603°C) reheat steam temperature. (*Courtesy The Babcock & Wilcox Company.*)

A variety of other boiler designs are used to meet the specialized needs for steam generation from waste energy or process by-product fuels. Figure 7.2.22 shows a specialized waste heat recovery boiler [or **heat recovery steam generator (HRSG)**] which is used to recover heat from the exhaust of a gas turbine in a combined cycle power system. The turbine exhaust gas enters the unit at moderate temperatures [typically less than 1,200°F (649°C)], and heat is recovered as the gas passes cross-flow over a series of modular tube bundles where water is heated to saturation temperature, steam is generated, and steam is superheated for power cycle application. The designs have become quite integrated with multiple pressure loops to maximize the energy recovered at a minimum equipment cost. Steam from the HRSG supplies a steam-turbine power cycle which adds to the power produced by the gas turbine cycle.

The exhaust gas from oil refinery fluid catalytic-cracking units is used as a fuel in specialized **carbon monoxide (CO) fired boilers**. Historically these have used refractory-lined circular boiler enclosures, but newer designs are replacing these original units with rectangular water-wall furnace enclosures. Supplemental fuel firing is required to raise the temperature of the CO to the ignition point and to ensure complete burning of the combustibles in the CO gas stream.

Process recovery (PR) boilers used in the pulp and paper industry are specially designed for the recovery of chemicals in the spent cooking liquors from Kraft, sulfite, soda, and other papermaking processes. The unique recovery boiler furnace design permits the combustion of the organic elements in the spent liquor while reducing the oxidized inorganic material in a pile or bed supported on the furnace floor. In the Kraft process, the molten inorganic chemicals or smelt in the bed is

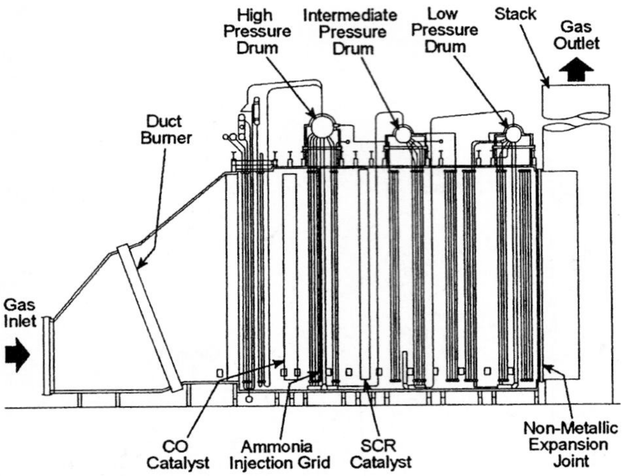

Figure 7.2.22 Triple pressure heat recovery steam generator (HRSG).

discharged to a tank for further processing and ultimate reuse in the papermaking process.

Bubbling fluidized bed boilers (Fig. 7.2.23) have shown the flexibility to burn a wide range of higher moisture and volatile biomass fuels such as bark, wood, and agricultural wastes.

Waste-to-energy (WTE) boilers firing municipal solid waste provide unique challenges to handling the very nonuniform fuel, extremely corrosive products of combustion and stringent emission control requirements. These boilers incorporate specialized stoker grate designs, corrosion-resistant furnace material such as Inconel® weld overlay, corrosion-resistant superheater materials, as well as minimum furnace time-temperature relationships [2 seconds at greater than 1800°F (982°C)]. **Mass burn** technology (see Fig. 7.2.24) uses MSW basically as-received while **refuse derived fuel** (RDF) technology uses materials

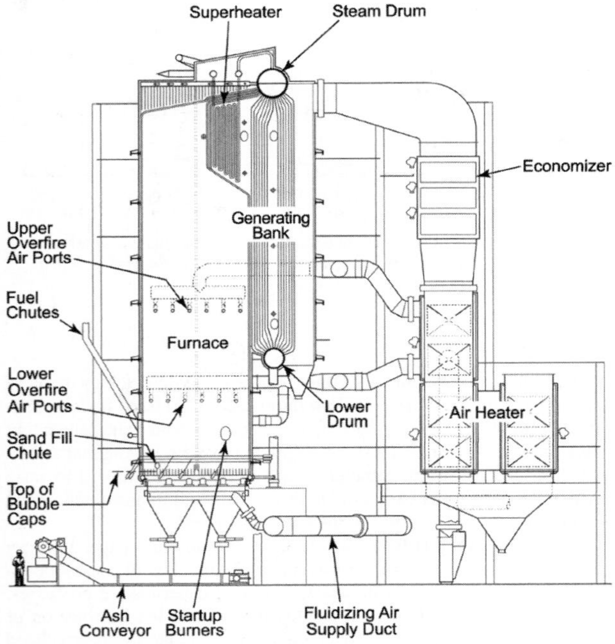

Figure 7.2.23 Bottom-supported bubbling fluidized-bed boiler for biomass firing. (*Courtesy The Babcock & Wilcox Company.*)

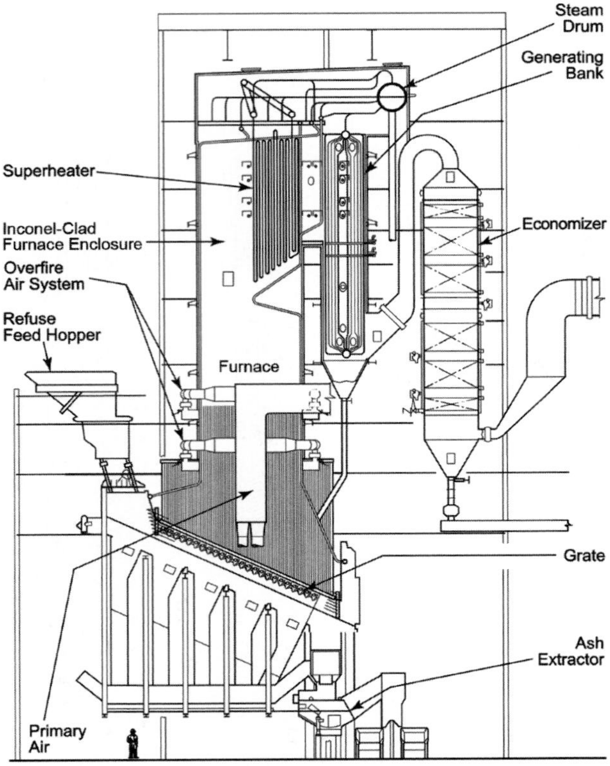

Figure 7.2.24 Single-pass mass burn boiler firing 1,000 ton per day (907 t_m/d) municipal solid waste; 284,400 lb/h (35.8 kg/s) steam, 900 psi (6.2 MPa) pressure, 830°F (443°C) superheat temperature. (*Courtesy The Babcock & Wilcox Company.*)

processing technology to provide a more uniform, higher heating value fuel (see also Sec. 5.4).

BOILER COMPONENTS

Furnaces

A furnace is an enclosure provided for the combustion of fuel. The enclosure confines the products of combustion and is capable of withstanding the high temperatures developed and the pressures involved with operation. Its dimension and geometry are adapted to the rate of heat release, the type of fuel, the method of firing and air pollution control requirements so as to promote complete burning of combustible and provide suitable disposal of the ash.

Water-cooled furnaces are used with most boiler units and for all types of fuel and methods of firing. Water cooling of the furnace walls reduces the transfer of heat to the structural members and, consequently, their temperature can be limited to that which will meet the requirements of strength. Water-cooled tube construction facilitates large furnace dimensions and optimum arrangements of roofs, hoppers, arches, and burners; and the use of tubular screens, platens, or division walls to increase the amount of heat-absorbing surface in the combustion zone. The use of water-cooled furnaces also reduces the external heat losses.

Heat-absorbing surfaces in the furnace receive heat from the products of combustion and contribute directly to steam generation while lowering the (FEGT). In the boiler furnace, all of the principal heat-transfer mechanisms take place simultaneously. These include inter-solid radiation between suspended solid particles, tubes, and refractory materials; nonluminous gas radiation from the products of combustion; convection heat transfer from the gases to the furnace walls; and heat conduction through deposits and tube metals. (See also Sec. 2.) The absorption effectiveness of the furnace surfaces is influenced by the deposits of ash or slag.

Furnaces vary in shape and size, in the location and spacing of burners, in the disposition of heat-absorbing surfaces, and in the general overall arrangement of arches and hoppers. Flame shape and length affect the distribution of radiation and heat absorption by water-cooled surfaces.

Analytical solution of the transfer of heat in the furnaces of steam-generating units is extremely complex, and it is not possible to calculate furnace outlet gas temperature by theoretical methods alone. Nevertheless, the furnace outlet gas temperature must be accurately predicted, since this temperature determines the design of the remainder of the boiler unit, particularly that of the superheater and reheater if used. The design calculations are based upon test results supplemented by data accumulated from operating experience and application of the principles of heat transfer and the characteristics of fuels and slags. Today this is supplemented by numerical computational methods.

The curve in Fig. 7.2.25 shows the gas temperatures at the furnace outlet of typical boiler units when firing coal, and by reference oil and gas. The furnace exit gas temperatures vary considerably with coal firing because of the insulating effect of ash and slag deposits on the heat-absorbing surfaces. The amount of surface is the major factor in overall furnace heat absorption, and the heat released and available for absorption per unit time per unit area of effective heat-absorbing surface is therefore used as the basis for correlation. The heat released and available for absorption is the sum of the heat content of the fuel fired and the sensible heat of the combustion air, less the sum of the heat unavailable owing to the unburned portion of the fuel and latent heat of the water vapor formed from moisture in the fuel and the combustion of hydrogen.

Empirical furnace design methods are being supplemented with numerical computational methods as the level of detail increases and

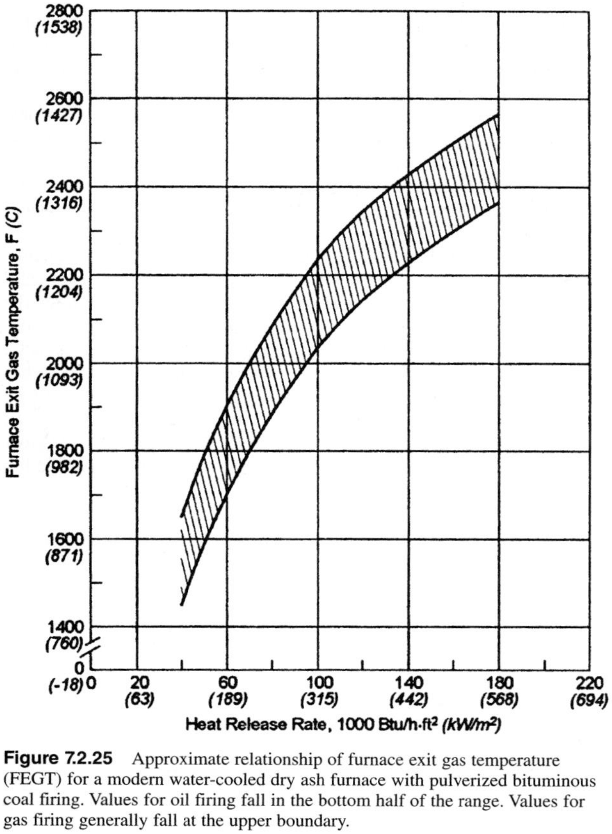

Figure 7.2.25 Approximate relationship of furnace exit gas temperature (FEGT) for a modern water-cooled dry ash furnace with pulverized bituminous coal firing. Values for oil firing fall in the bottom half of the range. Values for gas firing generally fall at the upper boundary.

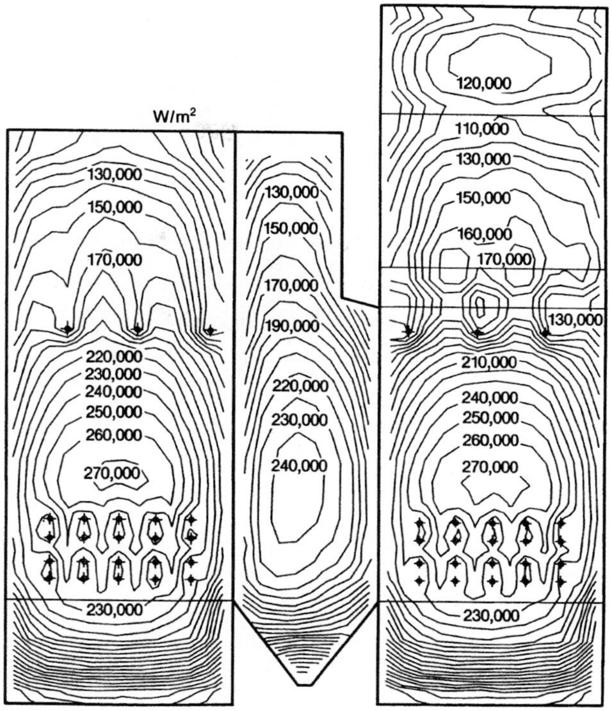

Figure 7.2.26 Numerical model-predicted furnace wall, flat projected heat flux distribution for a modern low-NO_x pulverized bituminous coal-fired dry-bottom furnace [1W/m² = 0.317 Btu (h · ft²)].

confidence improves. Fundamental relationships for thermal radiation, combustion, energy, and turbulent flow are combined with thermophysical properties and other experimentally determined factors to provide useful engineering information. Detailed results include three-dimensional distributions of gas temperatures and combustion species such as air pollutants as well as heat flux on the furnace walls (see Fig. 7.2.26).

The common forms of furnace water-cooled wall construction found in operation today are shown in Fig. 7.2.27. However, most boilers now use **membrane tube walls** in which a steel bar, or membrane, is welded between adjacent tubes. This construction facilitates the fabrication of water-cooled walls in large shop-assembled tube panels and ash removal.

Additional furnace cooling in the form of tubular platens, division walls, or wide-spaced tubular screens may be used. In high heat input zones, the tubes can be protected by refractory coverings anchored to the tubes by studs. Peak heat-absorption rates of furnace-wall tubes in the combustion zone may be well in excess of 130,000 Btu/(h · ft²) (0.4 MW/m²) of projected surface, but the average heat absorption rate for the furnace is considerably lower (Fig. 7.2.28).

Furnace walls must be adequately supported with provision for thermal expansion and with reinforcing buckstays (horizontal steel beams) to withstand the lateral forces caused by the difference between the furnace pressure and the surrounding atmosphere. The furnace enclosure must prevent air infiltration when the furnace is operated under negative pressure and gas leakage when the furnace is operated at positive pressure.

Superheaters and Reheaters

The addition of heat to steam after evaporation increases the temperature and the enthalpy of the fluid. The heat is added to the steam in boiler components called superheaters and reheaters, which are comprised of tubular elements exposed to the high-temperature gaseous products of combustion.

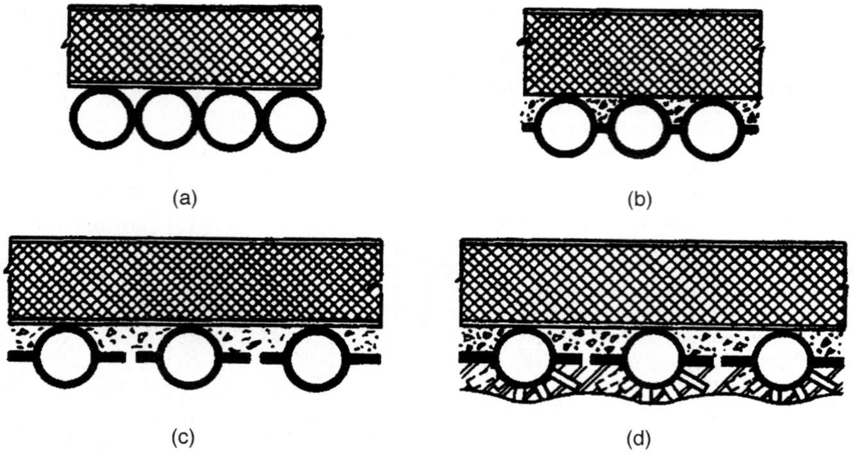

Figure 7.2.27 Water-cooled furnace construction types. (*a*) Tangent tube wall; (*b*) welded membrane tube wall; (*c*) flat studs welded to sides of tubes; and (*d*) full stud tube wall, refractory covered.

The advantages of superheat and reheat in power generation result from thermodynamic gain in the Rankine cycle (see Sec. 2.1) and from the reduction of heat losses due to moisture in the low-pressure stages of the turbine. With high steam pressures and temperatures, more useful energy is available, but the advances to high steam temperature often are restricted by the strength and the oxidation resistance of the steel and the metallic alloys currently available and economically practical for use in boiler pressure-part and turbine-blade construction.

The term superheating is applied to the higher-pressure steam and the term reheating to the lower-pressure steam which has given up some of its energy during expansion in the high-pressure turbine. With high initial steam pressure, one or more stages of reheating may be employed to improve the cycle thermal efficiency.

Separately fired superheaters have been used, but superheaters usually are installed as an integral part of the steam-generating unit and broadly classified as **radiant** or **convection** types, depending upon the predominant method of heat transfer to the heat-absorbing surfaces.

The quantity of heat absorbed and the amount of superheat attained are dependent upon the size, location, and arrangement of the heat-absorbing surfaces; the temperature differentials between the gas, the tube metal, and the steam; and the heat-transfer coefficients. Steam-temperature characteristics of radiant- and convection-type superheaters are shown in Fig. 7.2.29, as well as the effect of using a combination of these types to produce a more uniform steam temperature over a wide operating range.

Superheaters of the predominantly **radiant type** usually are arranged for direct exposure to the very high-temperature furnace gases and, in some designs, form a part of the furnace enclosure. In other designs, the surface is arranged in the form of tubular loops or platens, on wide lateral spacing, extending into the furnace (see Fig. 7.2.21). Such surface is exposed to high-temperature furnace gases traveling at relatively low velocities, and the transfer of heat is principally by thermal radiation.

Convection-type superheaters are installed beyond the furnace exit where the gas temperatures are lower than those in the zones where radiant-type superheaters are used. The tubes are usually arranged in the form of parallel elements on close lateral spacing and in tube banks extending partially or completely across the width of the gas stream, with the gas flowing through the relatively narrow spaces between the tubes. High gas flow rates and, thus, high convection heat-transfer rates are obtained at the expense of gas-pressure drop through the tube bank.

Superheaters, shielded from the furnace combustion zone by arches or wide-spaced screens of steam-generating tubes, which receive heat by thermal radiation from the high-temperature gases in cavities or intertube spaces and also by convection due to the relatively high velocity gas flow through the tube banks, have both radiant and convection characteristics.

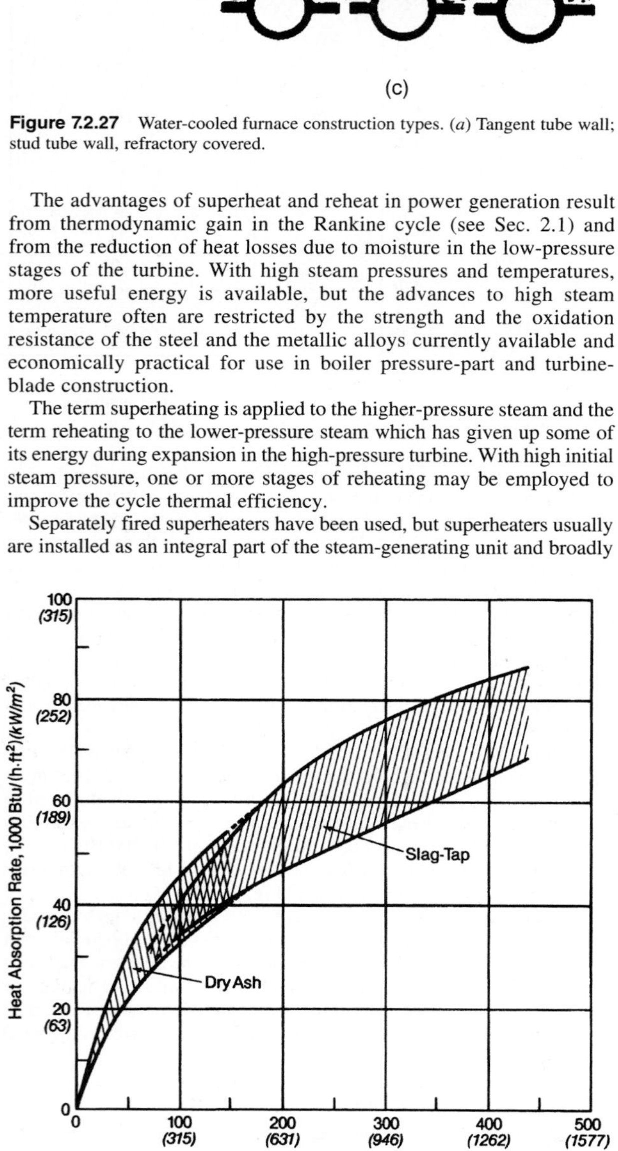

Figure 7.2.28 General range of average heat absorption rates in water-cooled coal-fired furnaces.

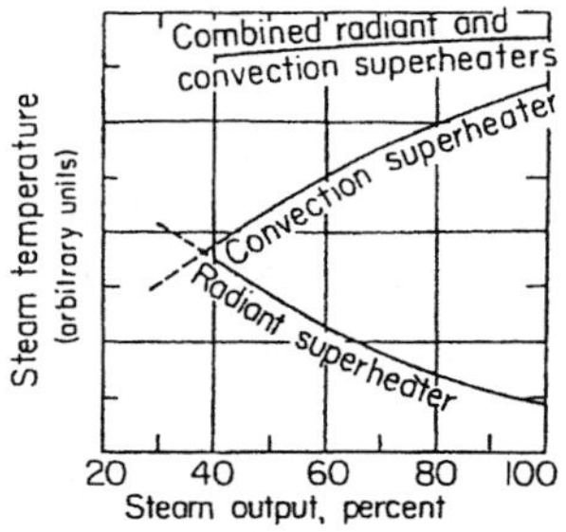

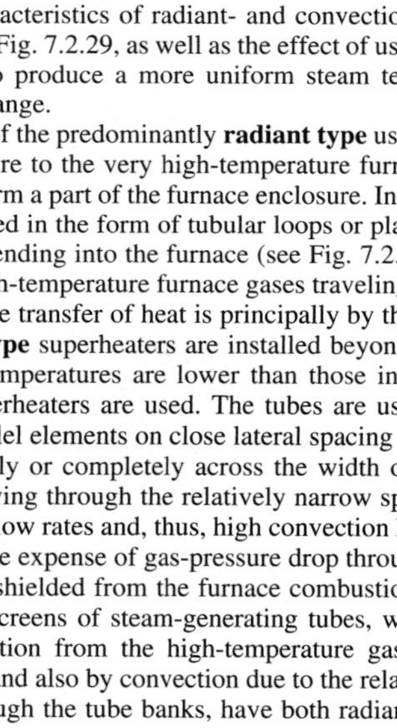

Figure 7.2.29 Comparative radiant and convection superheater characteristics.

Table 7.2.2 Typical Superheater and Reheater Tube Materials—Maximum Allowable Design Stress,1,000 lb/in^2
SI conversions: $C = (F - 32) \times 5/9$; kPa $= 6.895 \times$ lb/in^2

Material	Allowable stresses, 1,000 lb/in² — Temperature, °F														
	700	750	800	850	900	950	1,000	1,050	1,100	1,150	1,200	1,250	1,300	1,350	1,400
SA-210C	18.3	14.8	12.0	9.3	6.7	4.0	2.5								
SA-209T1a	16.8	16.4	15.9	15.4	13.7	8.2	4.8								
SA-213T2	15.7	15.4	14.9	14.5	13.9	9.2	5.9								
SA-213-T12	15.8	15.5	15.3	14.9	14.5	11.3	7.2	4.5	2.8	1.8	1.1				
SA-213-T11	15.1	14.8	14.4	14.0	13.6	9.3	6.3	4.2	2.8	1.9	1.2				
SA-213-T22	16.6	16.6	16.6	16.6	13.6	10.8	8.0	5.7	3.8	2.4	1.4				
SA-213-T91	22.9	22.2	21.3	20.3	19.1	17.8	16.3	14.0	10.3	7.0	4.3				
SA-213 304H	15.8	15.5	15.2	14.9	14.6	14.3	14.0	12.4	9.8	7.7	6.1	4.7	3.7	2.9	2.3
SA-213 347H	16.8	16.8	16.8	16.8	16.7	16.6	16.4	16.2	14.1	10.5	7.9	5.9	4.4	3.2	2.5
SA-213-TP310 HCBN	23.8	23.7	23.6	23.4	23.1	22.8	22.4	22.0	18.4	13.6	10.1	7.6	5.7	4.3	
SA213-S30432	20.4	20.1	19.8	19.5	19.2	18.9	18.6	18.3	18.1	14.7	11.4	8.7	6.5	4.7	3.3

SOURCE: Table 1A, Section IID, "Properties of the ASME Boiler and Pressure Vessel Code," 2015 edition.

Superheaters may utilize tubes arranged in the form of hairpin loops connected in parallel to inlet and outlet headers; or they may be of the continuous-tube type, where each element consists of a number of tube loops in series between the inlet and outlet headers. The latter arrangement permits the use of large tube banks, thus increasing the amount of heat-absorbing surface that can be installed and providing economy of space and reduction of cost. Either type may be designed for the drainage of the condensate which forms within the tubes during outages of the unit, or they may be used in pendant arrangements which are not drainable but, usually, have simpler and better supports. Nondrainable superheaters require additional care during start-up to remove the condensate by evaporation. Both types require that every tube have sufficient steam flow to prevent overheating during operation. In start-up, when flow is insufficient, gas temperatures entering the superheater must be controlled to limit tube metal temperatures to safe values for the material used.

The heat transferred from high-temperature gases by thermal radiation and convection is conducted through the metal tube wall and imparted by convection to the high-velocity steam in the tubes. The removal of heat by the steam is necessary to keep the tube metals within a safe temperature range consistent with the temperature limits of oxidation and the creep or rupture strength of the materials (see Sec. 3). Allowable design stresses for various steels and alloys, as established by the ASME Code, are listed in Table 7.2.2 (see also Sec. 4). The practical use limit for each material is also indicated. Current editions of the code should be checked for latest revisions. For economic reasons, it is customary to use low-carbon steel in the steam inlet sections of the superheater and, progressively, more costly alloys as the metal temperatures increase.

The rate of steam flow through superheater tubes must be sufficiently high to keep the metal temperatures within a safe operating range and to ensure good distribution of flow through all the elements connected in parallel circuits. This can be accomplished by arrangements which provide for multiple passes of the steam flow through the superheater tube banks. Excessive steam-flow rates, while providing lower tube-metal temperatures, should be avoided, since they result in high pressure drop with consequent loss of thermodynamic cycle efficiency.

The spacing of the tubular elements in the tube bank and, consequently, the gas flow rate and convection heat transfer are governed primarily by the types of fuel fired, draft-loss considerations, and the fouling and erosive characteristics of fuel ash carried in the gas stream. Gas flow velocity limits for coal-fired units are typically set to control fly ash erosion in the convection surfaces. The velocity limits are determined based upon the ash quantity (mass per unit heat input basis) and the relative proportion of abrasive constituents in the ash (primarily Al_2O_3 and SiO_2). Typical limits range from 65 ft/s (20 m/s) for relatively nonabrasive low-ash coals to 45 ft/s (14 m/s) or less for coals with high ash quantities and/or abrasive ash. In the higher-gas-temperature zones, the adherence and the accumulation of ash deposits can reduce the gas-flow area and, in some cases, may completely bridge the space between tubes. Thus, as gas temperatures increase, it is customary to increase the tube spacings in the tube banks to avoid excessive draft loss and to facilitate ash removal.

For biomass firing, velocities in the convection pass and boiler bank are kept low, typically below 60 ft/s (18 m/s), because the fuels frequently contain sand or other mineral matter. In municipal solid waste (MSW) fired units, design velocities are much lower and typically range from 10 to 12 ft/s (3.0 to 3.7 m/s) depending upon the gas temperature because of the highly corrosive/erosive nature of the products of combustion. For boiler banks and economizers, MSW velocity limits can be somewhat higher at 10 to 20 ft/s (3.0 to 6.1 m/s).

Economizers

Economizers remove heat from the moderately low-temperature combustion gases after the gases leave the steam-generating and superheating/reheating sections of the boiler. Economizers are, in effect, feedwater heaters which receive water from the boiler feed pumps and deliver it at a higher temperature to the steam generator. Economizers are used instead of additional steam-generating surface, since the feedwater and, consequently, the heat-receiving surface are at temperatures below the saturated-steam temperature. Thus, the gases can be cooled to lower temperature levels for greater heat recovery and economy.

Economizers are forced-flow, once-through, convection heat-transfer devices, usually consisting of steel tubes, to which feedwater is supplied at a pressure above that in the steam-generating section and at a rate corresponding to the steam output of the boiler unit plus any blowdown to remove boiler water contaminants. They are classed as horizontal- or vertical-tube types, according to geometrical arrangement; as longitudinal or crossflow, depending upon the direction of gas flow with respect to the tubes; as parallel or counterflow, with respect to the relative direction of gas and water flow; as steaming or nonsteaming, depending on the thermal performance; and as bare-tube or extended-surface, according to the type of external heat-absorbing surface. Staggered or in-line tube arrangements may be used. The arrangement of tubes affects the gas flow through the tube bank, the draft loss, the heat-transfer characteristics, and the ease of cleaning.

The size of an economizer is governed by economic considerations involving the cost of fuel, the comparative cost and thermal performance of alternate steam-generating or air-heater surface, the feedwater temperature, and the desired exit gas temperature. In many cases, it is more economical to use both an economizer and an air heater.

The temperatures of the economizer tube metals generally approximate those of the water flowing within the tubes, and thus with low feedwater temperatures, condensation and external corrosion are encountered in those locations where the tube-metal temperature is below that of the acid or water dew point of the gas (see Fig. 7.2.30). Internal corrosion and pitting also may be experienced if the feedwater contains significant levels of dissolved oxygen. Therefore, it is

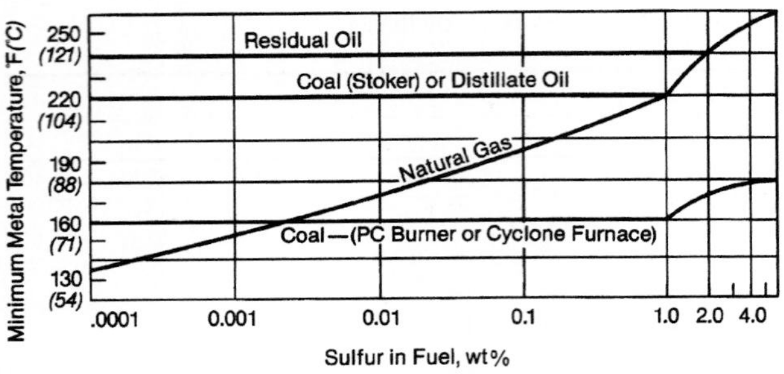

Figure 7.2.30 Limiting metal temperatures to minimize external corrosion on economizers or air heaters when fuel containing sulfur is burned.

imperative to maintain feedwater temperatures above the dew-point temperature of the gas and to provide suitable deaeration of the feedwater for the removal of oxygen.

Air Heaters

Air heaters, like economizers, remove heat from the relatively low-temperature combustion gases. The temperature of the inlet air is less than that of the water to the economizer, and thus it is possible to reduce the temperature of the gaseous products of combustion further before they are discharged to the stack.

The heat recovered from the combustion gases is recycled to the furnace with the combustion air and, when added to the thermal energy released from the fuel, is available for absorption by the steam-generating unit, with a resultant gain in overall thermal efficiency. The use of preheated combustion air accelerates ignition and promotes rapid and complete burning of the fuel.

Air heaters are usually classed as **recuperative** or **regenerative**. Both types utilize the convection transfer of heat from the gas stream to a metal or other solid surface and the convection transfer of heat from this surface to the air. In recuperative air heaters, exemplified by the tubular or plate types (Fig. 7.2.31), the stationary metal parts form a separating boundary between the heating and cooling fluids, and the heat passes by conduction through the metal wall. There are two commonly used types of regenerative air heaters (Fig. 7.2.32). In Fig. 7.2.32*a*, the heat transfer members are moved alternately through the gas and air streams undergoing successive heating and cooling cycles and transferring heat by the thermal-storage capacity of the members. The other type of regenerative air heater (Fig. 7.2.32*b*) has stationary elements, and the alternate flow of gas and air is controlled by rotating the inlet and outlet connections.

Recuperative and regenerative air heaters may be arranged either vertically or horizontally and for either parallel or counterflow of the gas and air. The gases are usually passed through the tubes of tubular

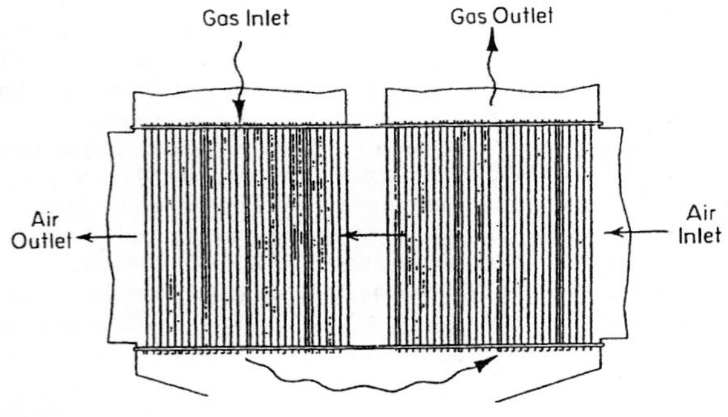

Figure 7.2.31 Recuperative-type tubular air heaters, two gas passes, single counterflow air pass.

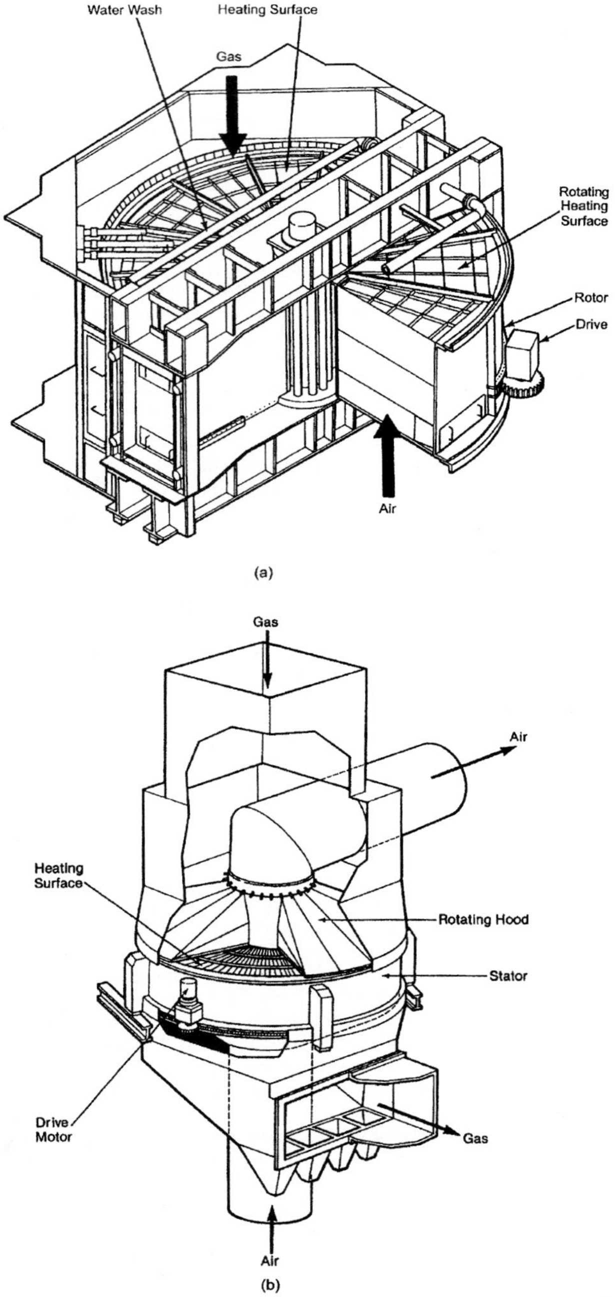

Figure 7.2.32 Two rotary regenerative air heater designs. (*a*) Ljungstrom (rotating heating surface); (*b*) Rothemuhle (rotating hoods).

air heaters to facilitate cleaning, although in some designs, particularly for marine installations, the air flows through the tubes. For most fuel applications, either recuperative or regenerative type air heaters may be used as needed. For biomass firing, recuperative type units (usually tubular) are generally used due to the ash, sand, and char present in the products of combustion.

Improved heat transfer and better utilization of the heat-absorbing surfaces are obtained with a counterflow of the gases and the use of small flow channels. Regenerative type air heaters readily lend themselves to these two principles and thus offer high capability in minimum space. However, regenerative air heaters have the disadvantages of air leakage into the gas stream and the transport of fly ash into the combustion air system.

The products of combustion from most fuels contain a high percentage of water vapor, and thus condensation will be experienced in air heaters if the exposed metal surfaces are cooled below the dew point of the gas. Minute concentrations of sulfur trioxide in the gases, originating from the combustion of sulfur and varying with the sulfur content of the fuel and the method of firing, combine with the water vapor in the combustion gases to form sulfuric acid, which may condense on the metal surfaces at acid-dew-point temperatures as high as 250 to 300°F (121 to 149°C), well above the water dew point. Such condensation leads to corrosion and/or the fouling of the gas-flow area. This is most likely to occur during the winter when the entering-air temperature is low, at low operating loads or in localized sections at the cold-air inlet if there is poor distribution of the air or the gas flowing through the air heater. Corrosion and fouling can be prevented by the use of auxiliary steam-heated air heaters located ahead of the air inlet, by recirculating heated air from the outlet duct, or by bypassing a portion of the cold air to reduce the airflow through the air heater. Both recuperative and regenerative air heaters often are designed with separate corrosion sections arranged to facilitate the replacement of the vulnerable cold-end portions.

Steam Temperature, Adjustment, and Control

The control of steam temperatures is vital to the life of high-temperature equipment and to the economy of power generation. Actual, or operating, steam temperatures below the design temperature reduce the thermodynamic efficiency and increase fuel cost, and temperatures above the design temperature reduce the margins of reserve in the strength of tubes, headers, piping, valves, and turbine elements. Further, sudden or extreme temperature variations may cause destructive stresses, particularly in rotating equipment.

It is sometimes necessary, because of the complexities involved in the design evaluation of heat-transfer rates and variability of fuel characteristics, to modify installed equipment to obtain the required steam temperatures. Such changes might involve the installation of baffles for the distribution of gas through the superheater and the removal, or addition, of tubular elements in the superheater or in those components preceding the superheater which affect the temperature of the gas to the superheater. For routine operation, it is essential to provide a suitable means of controlling steam temperature to compensate for the likely variations in fuel, heat transfer, and surface cleanliness conditions. These may include (1) damper control of the gases to the superheater and/or reheater; (2) recirculation of the low-temperature gaseous products of combustion to the furnace to change the relative amounts of heat absorbed in the furnace, in the superheater, and/or in the reheater; (3) selective use of burners at different elevations in the furnace or the use of tilting burners to change the location of the combustion zone with respect to the furnace heat-absorbing surface; (4) attemperation or controlled cooling of the steam at superheater inlet, at superheater outlet, or between the primary and secondary stages of the superheater; (5) control of the firing rate in divided furnaces; and (6) control of the firing rate relative to the pumping rate of water in forced-flow once-through boilers.

The speed of response differs for the various methods, and the control of steam temperature by gas bypass or flame position is slower than that by spray-water attemperation. The operating controls for these methods can be arranged for manual, automatic, or combination adjustment, and the use of more than one method often facilitates the maintaining of constant steam temperature over a wider range of boiler load.

The attemperation of superheated steam by direct-contact water spray (Fig. 7.2.33) results in an equivalent increase in high-pressure steam generation without thermal loss. Spray attemperation requires the use of essentially pure water, such as condensate, to avoid impurities in the steam.

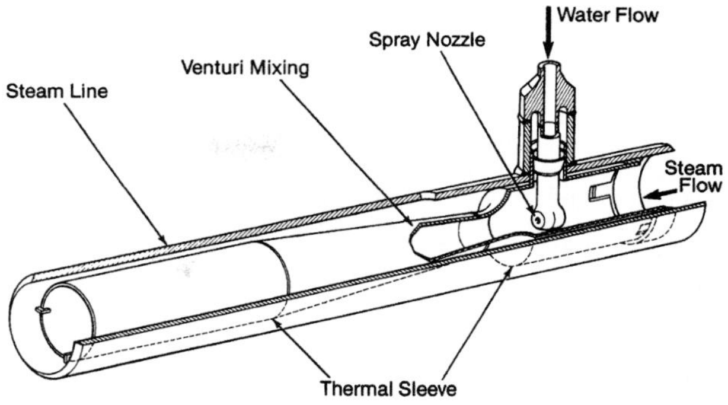

Figure 7.2.33 Spray-type direct-contact attemperator.

Ash and slag deposits on superheater and reheater surfaces reduce heat transfer and lower steam temperatures. Similar deposits on the furnace walls and steam-generating surface ahead of the superheater and/or reheater also reduce heat transfer to those surfaces, resulting in higher-temperature gas to the superheater and reheater and, consequently, increased steam temperatures. Thus the control of surface cleanliness is an important factor in the control of steam temperature.

Increased excess air results in higher steam temperatures because of reduction in furnace radiant-heat absorption, the greater amount of gas, and the increased convection heat transfer in the superheater and/or reheater. Operation with feedwater temperatures below that anticipated also results in increased steam temperatures because of the greater firing rate required to maintain steam generation.

OPERATING CONTROLS

(See also Sec. 12.)

The need for operating instruments and manual or automatic controls varies with the size and type of equipment, method of firing, and proficiency of operating personnel.

Safe operation and efficient performance require information relative to the (1) water level in the boiler drum; (2) burner performance; (3) steam and feedwater pressures; (4) superheated and reheated steam temperatures; (5) pressures of the gas and air entering and leaving principal components; (6) feedwater and boiler-water chemical conditions and particle carryover; (7) operation of feed pumps, fans, and fuel-burning and fuel-preparation equipment; (8) relationship of the actual combustion air passing through the furnace to that theoretically required for the fuel fired; (9) temperatures of the fuel, water, gas, and air entering and leaving the principal components of the boiler unit; and (10) fuel, feedwater, steam, and air flows so as to monitor operating conditions continuously and to make such adjustments as might be necessary.

Control of the various functions to maintain the desired operating conditions may be accomplished on small-capacity boilers by the manual adjustment of valves, dampers, and motor speeds. Most oil- and gas-fired package boilers are equipped with automatic controls to purge the furnace, to start and stop the burners, and to maintain the required steam pressure and water level. The operating requirements of utility and large industrial boilers dictate the use of automatic controls for the major variables, such as feedwater flow, firing rate, and steam temperature. The type of boiler and its components generally establish the basic mode of control. Analog controls of either the pneumatic or the electric type are still in use but are largely being replaced with advanced integrated digital control systems which manage all the normal boiler functions. These systems are designed with the maximum amount of monitoring and control at the lowest possible control system level.

Sequence controls often are applied in the start-up of utility boilers to program the furnace purge, burner light-off, and burner control. Interlocks are essential to ensure the proper starting and firing sequence and to alarm or automatically shut down the unit in the event of the failure of essential auxiliaries.

BOILER PERFORMANCE

Boiler Circulation

The water-steam pressure part components are arranged for the most economical system to provide the continuous supply of steam at the specified operating conditions. As an example, the full circulation system (excluding reheater) for a natural-circulation subcritical pressure drum type coal-fired steam generator is shown in Fig. 7.2.34.

Adequate circulation in the steam-generating section of a boiler (Circuit D.E.F.G.H in Fig. 7.2.34) is required to prevent overheating of the heat-absorbing surfaces, and it may be provided naturally by gravitational forces, mechanically by pumps, or by a combination of both methods. Figure 7.2.35 shows four common boiler circulation systems.

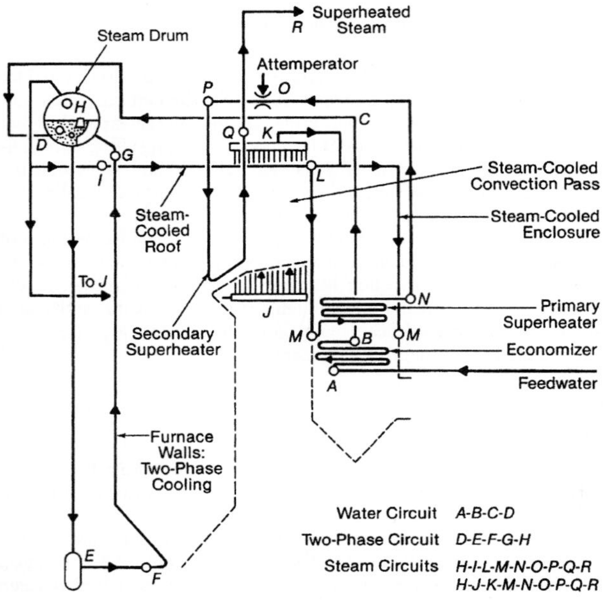

Figure 7.2.34 Simplified diagram of a coal-fired stream-water natural-circulation boiler system.

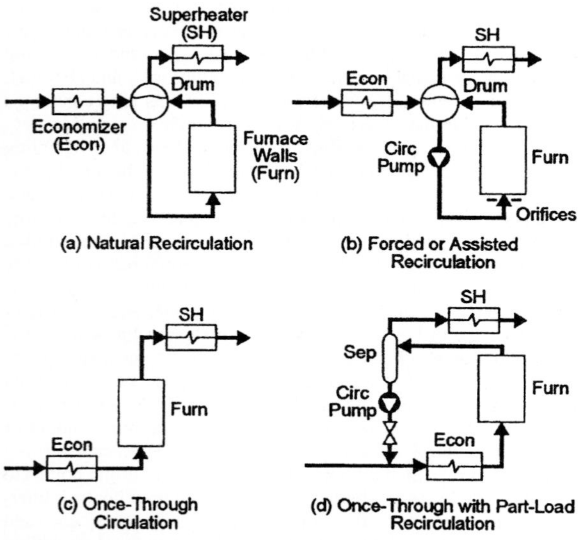

Figure 7.2.35 Common boiler circulation systems.

Natural circulation is produced by the difference in the densities of the water in the unheated downcomers and the steam-water mixture in the heated steam generating tubes. This density differential provides a large circulating force. The downcomers and the heated circuits are so designed that the resistance to flow from friction, local component pressure losses, and acceleration of the two-phase steam water flow through the system balances the circulating force at the desired total circulating flow.

The subcritical pressure forced-recirculation or assisted-circulation type of boiler (Fig. 7.2.35*b*) uses a steam drum similar to that used with natural-circulation boilers. The supply of water to the furnace walls and the boiler surfaces flows from this drum to a circulating pump, which supplies the pressure necessary to force the water through the water-steam mixture circuits and then back to the drum, where the steam and water are separated. The recirculating pump power required is equivalent to about 0.5 percent of the heat input to the boiler. Resistor orifices are required at the entrance to each tube, or circuit, to control flow distribution.

In assisted-circulation designs, the velocities or flows are independent of boiler rating, thus facilitating the use of smaller connecting piping and, sometimes, smaller-diameter furnace-wall tubing than that used with natural-circulation units. Both drum-type natural- and assisted-circulation boilers operate with essentially saturated steam temperatures in all parts of the steam-evaporating sections, and they are used for drum operating pressures up to approximately 3,000 psi (20.7 MPa).

In natural- and assisted-circulation boilers, it is essential to wet inside surfaces continually with water of the two-phase water-steam mixture to prevent dryout and overheating heat-absorbing surfaces. This dryout phenomenon is commonly referred to as departure from nucleate boiling (DNB) or critical heat flux (CHF). Satisfactory cooling of the heat-absorbing steam-generating surfaces is dependent upon the pressure, heat flux, water-steam mass flux, percent steam by weight (quality), tube diameter, and the tube's internal geometry. The heat flux is one of the most predominant of these parameters, and the rate of furnace heat absorption at the maximum firing rate generally dictates design considerations. Satisfactory circulation design also includes the evaluation of two-phase flow stability/instability limits, steam separator and drum flow limits, minimum tube flow velocity limits, and sensitivity (flow increases for heat flux increases at all expected operating conditions).

In forced-flow once-through boilers (Fig. 7.2.35*c*), the water from the feed supply is pumped to the inlet of the heat-absorbing circuits. Evaporation, or change of state, takes place along the length of the circuit, and when evaporation is completed the steam is superheated. These units do not require steam or water drums and, in most cases, use relatively small-diameter tubes. The water flow to the unit is the same as the steam output, and fluid velocities greater than those needed for natural- or assisted-circulation units must be used at full load so as to maintain adequate velocities at the low loads and satisfactory tube-metal temperatures at all loads.

The transition from a liquid to a vapor at, or above, the critical steam pressure of 3,200 psi (221 MPa) is dependent upon temperature and takes place without a change in density. Thus, separation of steam and water is impossible and forced-flow once-through boilers must be used.

Traditional forced-flow once-through boilers must be operated above a specified minimum flow in order to maintain adequate water velocities in the furnace-wall tubes. However, the turbogenerator can be operated at any load by the use of a bypass system that diverts the excess flow to a flash tank for heat recovery. The bypass system also can be used as a pressure relieving system, as the source of low-pressure steam to the turbine during start-up, and as a means of controlling steam temperature to the turbine during hot restarts.

Combined circulation units utilize forced once-through flow with flow recirculation in the furnace walls to provide satisfactory water velocities during start-up and low-load operations. In this design, some of the water at the exit of the furnace circuits is mixed with the incoming feedwater, flows to and through a circulating pump, and then passes to the furnace-wall inlet headers. The use of combined circulation increases the water velocities in the furnace tubes at low loads, and since recirculation is not used at the higher loads, there is no increase in velocity or in the resistance to flow at the higher loads.

Boilers are also designed for variable-pressure operation with operating pressure varying as the boiler load changes. Near full load, supercritical pressure once-through operation occurs. At reduced loads, operating pressure is reduced below the critical pressure, and a part-load recirculation system (Fig. 7.2.35*d*) is used to separate the steam-water flow and return the water with the feedwater to the economizer inlet.

Steam Separation

In drum-type boilers, using either natural or assisted circulation, a mixture of steam and water is delivered to the upper or steam drum where separation of the steam and water takes place and a water level is maintained. The water is recycled through downcomers to the heat-absorbing circuits, and the steam is discharged from the top of the drum for use as saturated steam or the supply to the superheater.

Separation of the steam and water by buoyant or gravitational force requires a relatively large cross-sectional area and, consequently, a low fluid velocity within the drum as well as an effective difference in the fluid densities, which decreases as pressure is increased. Steam entrainment in the recycled water impedes circulation, and water entrainment in the outlet steam transports dissolved or suspended particulate matter into the superheater, steam piping, and turbine, where particle deposition can cause overheating of the tubes or flow obstructions in the turbine blading with subsequent loss of capacity, efficiency, and dynamic balance.

Gravity separation of steam and water may be satisfactory in low-pressure, low-duty boilers. This type of separation can be augmented by the use of baffles which utilize a change in direction to throw out the water droplets, or by dry pipes.

In high-pressure boiler units, particularly those employing high evaporative ratings, a part of the circulating head can be utilized to provide a separating force, many times greater than that of gravity, in centrifugal separating devices such as the cyclone steam separators shown in Fig. 7.2.36. These separators deliver steam-free water to the drum and downcomers, and discharge steam with a minimum of water entrainment. Secondary steam-and-water separation is accomplished by passing the steam at a low velocity through sinuously shaped passages between closely spaced corrugated plates, which provide a large surface area for intercepting entrained boiler-water particles. In modern high-capacity boilers, the steam leaving the steam drum contains less than 0.1 ppm of total solids. This purity is generally acceptable for commercial practice. However, where further refinement in steam purity is required to remove boiler contaminants such as silica, the steam can be washed with condensate or feedwater of sufficient purity.

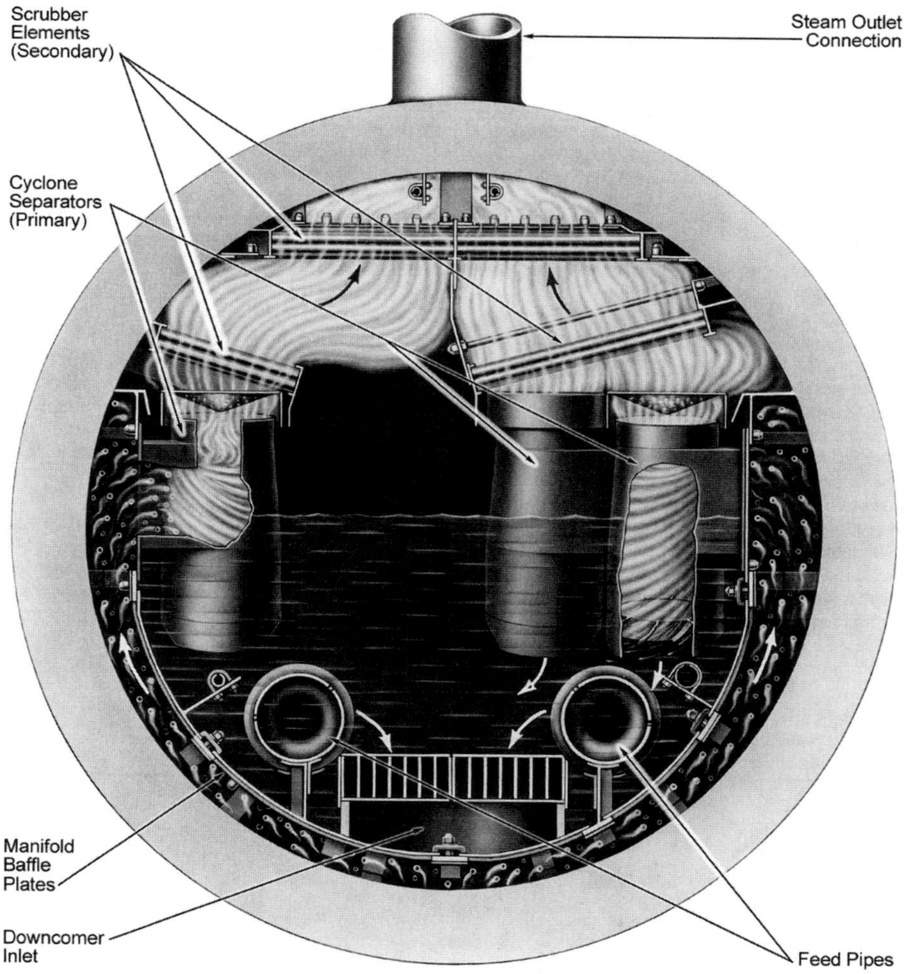

Figure 7.2.36 Steam drum internal arrangement with cyclone steam separators and scrubber elements.

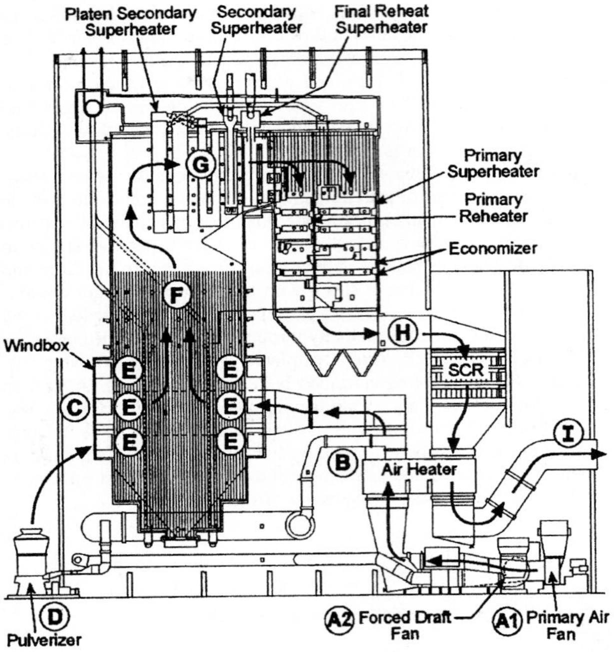

Figure 7.2.37 Simplified boiler air-gas flow path diagram: primary air (A1-B-D-E); secondary air (A2-B-C-E); and products of combustion (E-F-G-H-I).

Flow of Gas Through Boiler Unit

During the combustion of fuel and the transfer of heat to the heat-absorbing surfaces, it is necessary to maintain sufficient pressure to overcome the resistance to flow imposed by the burning equipment, tube banks, directional turns, and the flues and dampers in the system. The resistance to air and gas flows depends upon the arrangement of the equipment and varies with the rates of flow and the temperatures of the air and gas. Figure 7.2.37 provides a simplified air-gas flow schematic for a coal-fired boiler.

The term draft denotes the difference between atmospheric pressure and some lower pressure existing in the furnace or gas passages of a steam-generating unit. Draft loss is defined as the difference in static pressure of a gas between two points in a system and is the result of the resistance to flow. These terms originated with the use of the so-called natural-draft units in which the pressure differentials are obtained from a chimney or stack which produces static pressures throughout the boiler setting that are below that of the atmosphere. The terms are rather loosely applied to modern boilers using induced draft and/or forced draft, mechanically produced by fans, in which the pressures throughout the boiler unit may be well above atmospheric pressure.

Forced-draft fans, handling cold, clean air, provide the most economical source of energy to produce flow through high-capacity units (see Sec. 10.5). Induced-draft fans, handling hot flue gases, require more power and are subject to fly ash erosion. However, they facilitate operation by providing a draft in the boiler setting and thus prevent the outward leakage of gas through joints or crevices in the boiler enclosure. Boilers built for positive-pressure operation are generally referred to as pressure-fired units, while those using induced-draft fans are classed as suction units. When both forced- and induced-draft fans are used, the boilers are designated as balanced-draft units, the most common design today.

The pressure drop throughout the unit is caused by the fluid friction in the gas stream and the shock losses at the turns or the contractions and enlargements of sections. It can be calculated as a function of the fluid mass flow and fluid properties in accordance with the principles of fluid flow (see Sec. 1.3). It is essential in the design of a boiler to determine the sum of all component resistances in the flow system at the maximum load in order to establish the fan requirements. It is customary to specify test-block static head, temperature, and capacity requirements of the fan in excess of those calculated so as to allow for departure from ideal flow conditions and to provide a satisfactory margin of reserve.

Stack effect is caused by the difference in densities resulting from the difference in the temperatures of two vertical columns of gas. In a chimney, or stack, the stack effect is due to the difference between the confined hot gas and the cooler surrounding air and the equal static pressure at the top or free outlet of the stack. The stack effect, which varies with the height and the mean temperature of the columns, can be calculated from the data in Table 7.2.3. The effect is the static draft produced by a stack, at sea level, with no gas flow. When flow occurs, a portion of the stack effect is used to establish gas velocity and the remainder is used to overcome the resistance of the connected system, including the dampers and the stack. The limit of natural-draft capacity is reached when these forces are in balance with the dampers in a wide-open position. Stack performance may be favorably or adversely affected by external factors such as the wind and the atmospheric conditions. The available draft varies directly with the barometric pressure for altitudes above sea level.

Stack effects also exist within the boiler setting and are most pronounced in tall units with vertical gas passes. The individual gas columns within the setting may aid the head produced by the fan or chimney if the flow is upward or may reduce it if the flow is downward. The net stack effect, and its overall influence on the performance of the fan, may be calculated from the data in Table 7.2.3, taking into account the positive or negative effects. The relationship between local static pressures and the atmospheric pressure is important, since gas may blow into the building through an open inspection door at the top of a furnace, even though a strong draft, or negative pressure, exists at some other elevation.

Performance

Specific guarantees relating to the boiler or steam generator scope include the maximum continuous rating (MCR) capacity (steam flows, temperatures, and pressures), boiler efficiency, auxiliary power consumption, water- and steam-side pressure drop, superheater and reheater steam temperature control range, spray water quantities, and emissions of NO_x, CO, and volatile organic compounds (VOCs). If the spray water source is downstream of the feedwater heaters, then the

Table 7.2.3 Stack Effect of Pressure Difference, in of Water for 1 ft of Vertical Height (mm of Water for each 1 m of Vertical Height)
Barometer = 29.92 inHg (759.97 mmHg)

Avg temp in flue, °F (°C)	Air temp outside flue, °F (°C) 40°F (5°C)	60°F (15°C)	80°F (25°C)	100°F (35°C)
250 (125)	0.0041 inH_2O/ft (0.346 mmH_2O/m)	0.0035 (0.303)	0.0030 (0.262)	0.0025 (0.224)
500 (250)	0.0070 (0.563)	0.0064 (0.520)	0.0059 (0.480)	0.0054 (0.442)
1,000 (500)	0.0098 (0.788)	0.0092 (0.774)	0.0087 (0.708)	0.0082 (0.665)
1,500 (750)	0.0112 (0.902)	0.0106 (0.858)	0.0100 (0.818)	0.0095 (0.780)
2,000 (1,000)	0.0120 (0.972)	0.0114 (0.925)	0.0108 (0.887)	0.0103 (0.850)
2,500 (1,250)	0.0125 (1.018)	0.0119 (0.975)	0.0114 (0.934)	0.0109 (0.896)

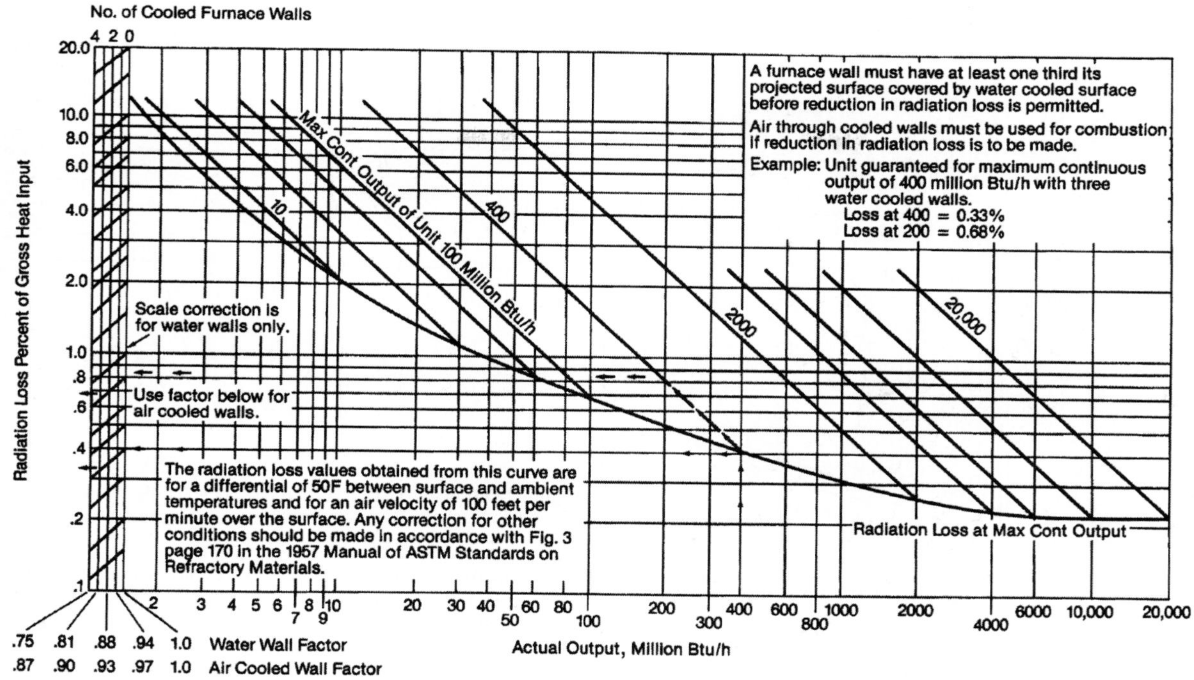

Figure 7.2.38 External heat loss from boiler setting based upon gross heat input (ABMA).

superheater spray is not considered. For subcritical-pressure, drum-type boilers, steam purity is commonly included.

Anticipated-performance data for several rates of operation may be given to the purchaser in addition to the guaranteed-performance data. Guarantees may be demonstrated by acceptance tests, conducted in accordance with established codes, as agreed upon by the parties in the contract. However, acceptance tests are more difficult to perform as unit size and capacity increase, and overall performance usually is determined from the operating data. Warranties of materials and the quality of manufacture and erection are usually considered separately from those pertaining to operating performance.

Heat balances account for the thermal energy entering the system in terms of its ultimate useful heat absorption or thermal loss. Methods of measuring and calculating the quantities involved in heat balances for most fuels are presented in the ASME Performance Test Code PTC4 for Fired Steam Generators. Because of the complexities of municipal solid waste firing, ASME Performance Test Code PTC34 for Waste Combustors with Energy Recovery has been developed for units firing this fuel.

Both the heat input and the heat absorption may be very large quantities. Therefore, unless elaborate precautions are taken in the sampling and the measurement of fuel and steam quantities, it is difficult to obtain test data having the degree of accuracy required to determine the actual efficiency of the boiler unit. For this reason, boiler efficiency usually is established from the heat losses, since each of the thermal losses is a relatively small percentage of the heat entering the system and reasonable errors in measurement will not appreciably affect the final result.

The principal thermal losses are those due to the sensible heat in the gases leaving the unit, latent-heat losses associated with the evaporation of fuel moisture and formation of water vapor resulting from burning of hydrogen in the fuel, unburned-combustible loss, loss from the boiler setting or enclosure due to external convection and radiation, and wet ash pit loss. The first two losses can be derived from the fuel analysis, the exit gas temperature, and the analysis of the flue gas (see Sec. 2). The unburned-combustible loss can be established by a qualitative and quantitative sampling of bottom ash (residue) and fly ash. The radiation from the boiler setting can be estimated in detail, but it also can be approximated from Fig. 7.2.38. The heat loss to the ash pit of large units can be determined by measuring the quantity of quenching water evaporated from the furnace ash hopper or the slag tank; for small units the ashpit loss is included in the heat loss from the boiler setting. The sum of these heat losses, expressed as a percentage of the total heat input, is the total measurable loss. The anticipated or guaranteed performance data include a tolerance for the so-called manufacturer's margin (unaccountable losses) in the order of 0.5 percent, depending upon the type of fuel fired. The thermal efficiency of the unit is established by subtracting the sum of all these losses from 100 percent.

ENVIRONMENTAL CONTROL SYSTEMS

Environmental protection and the control of gaseous effluents or emissions are key elements in the design of all utility and industrial steam boiler systems. Stringent local and federal government rules and requirements are constantly changing. At present, the most significant of these are for fine particulate, nitrogen oxides (NO_x), sulfur oxides (SO_x: primarily SO_2), mercury, and other specially classified hazardous air pollutants (HAPs). The strategies for control are formulated by considering the range of specific fuels to be burned and their sources, specific kind and extent of emission levels mandated, and economic factors such as the specific boiler design, location, new or existing equipment, plant age, and remaining life. In most cases, overall control of individual pollutants for a specific fuel involves some combination of precombustion, combustion, and postcombustion technologies. This section provides a brief overview of the major postcombustion technologies. The control of NO_x emissions by combustion modifications is covered under "Burners and Combustion Systems" and "Stokers." The control of SO_2 during combustion is covered under "Fluidized-Bed Combustion."

Control of Particulate

Particulate emissions from boilers arise from noncombustible, ash-forming mineral matter in the fuel that is released during the combustion

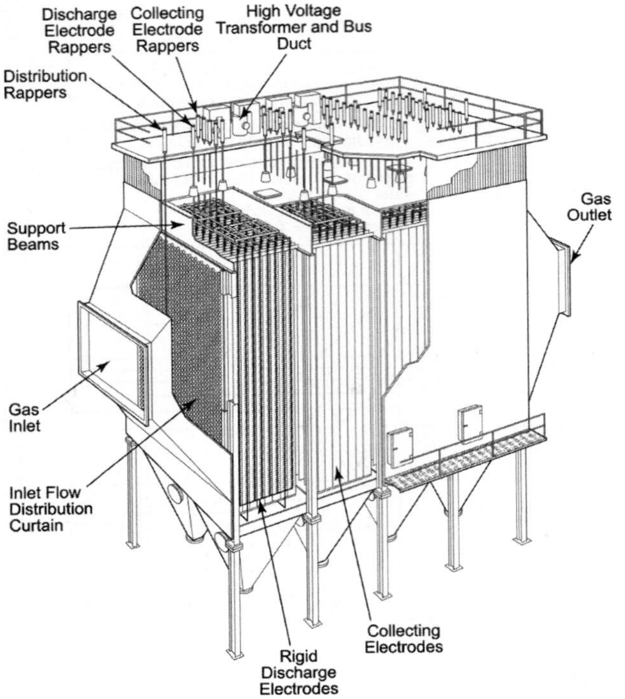

Figure 7.2.39 Top-rapped electroststic precipitator for a large industrial boiler application.

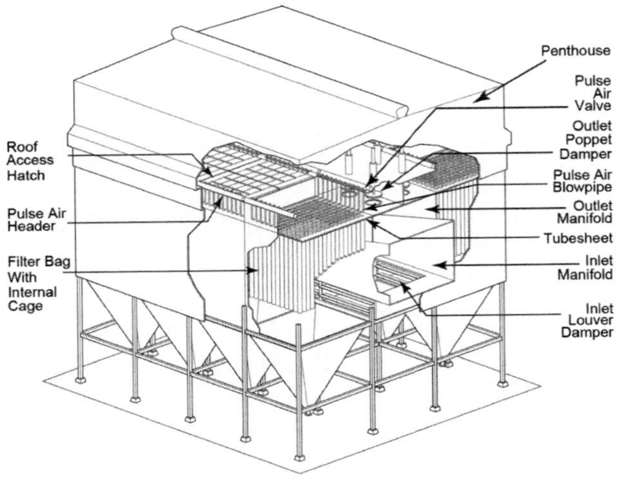

Figure 7.2.40 Pulse jet fabric filter (or baghouse) for fly ash particulate removal.

process and carried with the products of combustion. Three general types of equipment are used to reduce particulate emissions.

Mechanical collectors are generally cyclone collectors and have been widely used on small boilers with less stringent emission limits. They are low-cost, simple, compact, and rugged, but conventional cyclones are limited in collection efficiencies of about 90 percent and are poor at collecting the finest particulate.

Electrostatic precipitators (ESPs) electrically charge the ash particles in the products of combustion to collect and remove them. As shown in Fig. 7.2.39, an ESP normally consists of a series of parallel, vertical metallic plates (collecting electrodes) forming lanes through which the products of combustion flow. Centered between the plates are discharge electrodes which provide the particle charging and electric field. ESPs are available for most utility and industrial applications, and collection efficiencies can be expected to exceed 99.9 percent of the inlet ash loading. Top or bottom rapping systems periodically dislodge the fly ash for collection in the bottom hoppers. ESPs are considered to be less sensitive to plant upsets, and they have low pressure drop. ESP performance is sensitive to fly ash loading, ash resistivity (ability to acquire and retain a charge), and sulfur content. Lower-sulfur fuels result in lower ESP performance or large ESP sizes. ESPs have found application in wood, biomass and other fuel firing system where high temperature (sometimes "glowing") particles may carry over from the boiler.

Fabric filters (or baghouses) collect the ash particles as the products of combustion pass through a filter material. As shown in Fig. 7.2.40, a large utility fabric filter is comprised of a series of individual compartments, with each compartment containing up to several thousand long, vertically supported, small-diameter fabric bags. The ash cake which builds up on the filter bag surface is dislodged and removed by one of three techniques: reversing the gas flow temporarily, shaking the bags, or using a pulse jet of air down the bag. Collection efficiencies can be expected to exceed 99.9 percent of the inlet ash loading. When combined with dry SO_2 scrubbing, they can enhance SO_2 removal as the gas passes through the filter cake on the bag surface. Fabric filters are more sensitive to plant temperature excursions and upsets because of the filter material, and they are a higher pressure-drop device by nature. Fabric filters are more capable of collecting the finest fraction of the ash and are not sensitive to fuel sulfur content.

Control of NO_x

Oxides of nitrogen (NO_x) are formed during the combustion process, as discussed under "Burners and Combustion Systems," and are initially limited by the combustion system design. If further NO_x reduction is necessary, **selective catalytic reduction (SCR)** NO_x control systems can be installed, most frequently between the exit of the boiler economizer and the air heater (see Figs. 7.2.18 and 7.2.21). Such designs are referred to as **high-dust** applications because all the fly ash passes through them on the way to the particulate control system. **Low-dust** systems can be installed after a particulate control system but can involve higher capital costs. However, low dust systems are necessary in some utility boiler retrofits where space is limited as well as in boiler applications firing biomass or MSW. A reagent, typically ammonia, is uniformly injected and mixed into the products of combustion (see Fig. 7.2.21) upstream of the SCR reactor which contains the blocks of high-surface-area catalyst. As the products of combustion pass through the catalyst, the ammonia and NO_x react to produce water vapor and gaseous nitrogen. The most common catalyst types are coated metal **plate** catalyst, homogeneous ceramic **honeycomb** catalyst, and coated **corrugated** glass fiber substrate catalyst.

The catalyst performance is temperature-dependent with an optimal range of 700 to 750°F (371 to 399°C) although specialized catalyst can perform somewhere in the wider temperature window of 450 to 800°F (232 to 427°C) depending upon the gas composition and modified catalyst formulation. Some catalysts are also specially formulated to be resistant to certain constituents of the products of combustion which can poison and deactivate the catalyst surface. SCR units can effectively reduce NO_x levels leaving the boiler by 90 percent or greater depending upon the site- and boiler-specific conditions. Commercial projects are usually designed for 70 to 90 percent reduction. Ammonia is supplied in anhydrous or aqueous forms and from systems which convert urea to gaseous ammonia.

Control of SO_2

Sulfur dioxide (SO_2) comes from oxidation of sulfur in the fuel. The dominant method of SO_2 reduction for utility units is through the installation of a **flue gas desulfurization (FGD)** system downstream of the boiler. There are three basic types of FGD systems in wide use today: wet FGD scrubbers, spray dryer absorption (SDA) systems, and

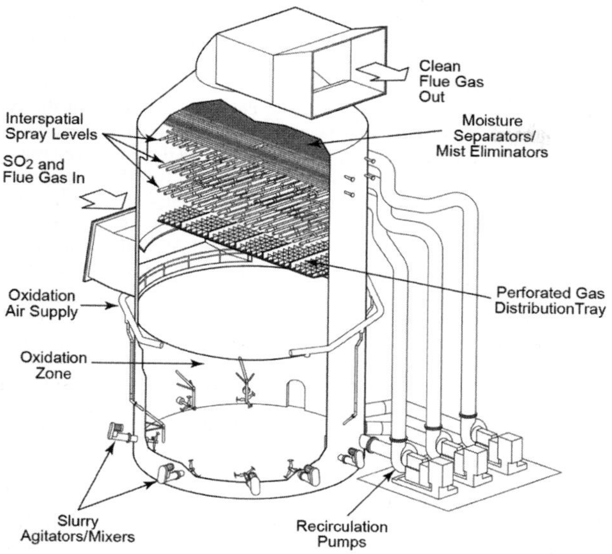

Figure 7.2.41 Wet flue gas desulfurization system (absorber module). (*Courtesy The Babcock & Wilcox Company.*)

circulating dry scrubbers (CDS). The selection of the most appropriate system depends upon the site- and fuel-specific economics.

Wet scrubbing (see Fig. 7.2.41) is centered on a large-diameter absorber module in which a sorbent slurry consisting of water mixed with a reagent (usually limestone, lime, magnesium-enhanced lime, or sodium carbonate) is contacted with the products of combustion. The SO_2 is first absorbed by the aqueous solution and then reacted with the reagent in the slurry. Limestone ($CaCO_3$) based systems with forced oxidation of the residual slurry by-product to produce wallboard-quality gypsum ($CaSO_4 \cdot 2H_2O$) are the dominant systems being installed. Wet scrubbing is highly efficient with upto 99 percent removal of the SO_2 at calcium-to-sulfur molar ratios of close to 1.0. The full scrubber system consists of reagent receiving and preparation equipment, the absorber module with recirculation pumps, and the spent slurry dewatering system. The absorbers are designed with corrosion-resistant linings or materials to prevent chemical attack. One or more absorber modules (more commonly one on new systems) are located between the forced-draft fan and the stack. Performance-enhancing additives can be used to reduce power consumption or increase removal efficiency.

Dry scrubbing (see Fig. 7.2.42), which is more correctly referred to as **spray dryer absorption (SDA)**, involves the spraying of an aqueous sorbent slurry into a reaction vessel under conditions which permit the slurry droplets to dry as they contact the hot gas [~300°F(~149°C)]. The SO_2 reaction occurs during the drying process and results in a dry particulate containing reaction products and unreacted sorbent or reagent entrained in the products of combustion flow along with the fly ash. These materials are then captured in the downstream particulate control equipment. The reagent is usually lime. Scrubber modules are located between the air heater and particulate control equipment, which often is a fabric filter to take advantage of the increased removal possible, since the unreacted lime collected in the bag filter cake reacts with the remaining SO_2. Such SDA plus fabric filter systems can be designed to remove 90 to 96 percent or more of the inlet SO_2. The SDA reactor is made of carbon steel. The full SDA system consists of reagent receiving and preparation equipment, the SDA, a fabric filter, and spent reagent/ash collection and disposal. A key attribute of the system is the dry waste residue.

Circulating dry scrubbing (CDS) is centered on an absorber (see Fig. 7.2.43) in which a vertical circulating bed of dry alkaline reagent contacts the products of combustion which have first been humidified and cooled. In the co-current upward flow, SO_2 and other acid gases

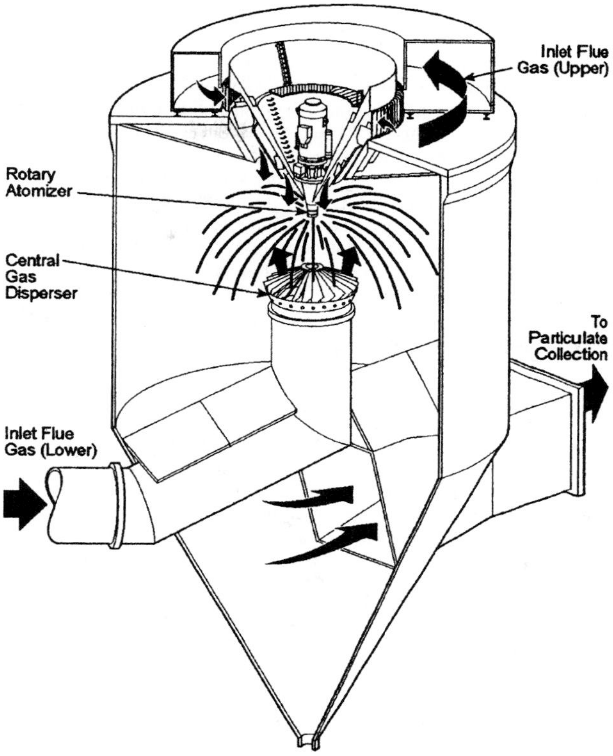

Figure 7.2.42 Dry flue gas desulfurization system (spray dryer module).

(SO_3, HF, HCl) are absorbed. Like SDA systems, the CDS absorber is located upstream of the fabric filter which then collects the spent and unused reagent as well as fly ash. Most of the solids are recirculated to the CDS absorber inlet to create the high solids circulating bed. SO_2 recovery is 97 percent or higher.

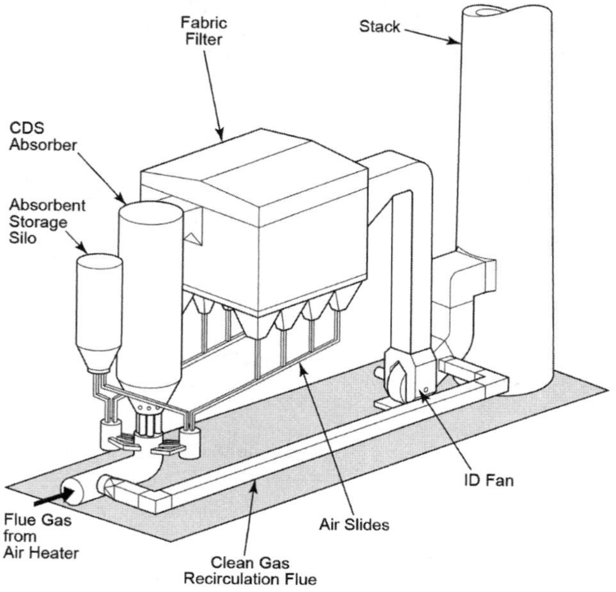

Figure 7.2.43 Circulating dry scrubber (CDS) system.

Control of Mercury and Other Hazardous Air Pollutants

Hazardous air pollutants (HAPs) or air toxics (in particular mercury, Hg) control involves a wide range of technologies and products which typically work with other major post-combustion emissions control equipment. These technologies and products include: (1) powdered activated carbon (PAC) injection usually upstream of an SDA for mercury and dioxin/furan control, (2) dry sorbent injection of calcium hydroxide or trona to remove SO_3/H_2SO_4 prior to the other pollutant control processes, (3) proprietary additives for wet FGDs to enhance mercury removal, (4) halide injection to enhance mercury oxidation within SCR systems, and (5) wet ESPs to remove very fine particulate and aerosols, among others. The key is to optimize control though system selection and integration with other equipment in order to meet site- and fuel-specific needs at a minimum cost.

WATER TREATMENT

The objectives of water treatment in boiler systems include (1) minimizing deposits of scale or sludge on internal tube surfaces, (2) reducing and managing corrosion of the internal boiler surfaces, and (3) aiding in the delivery of steam of required purity to the steam turbine or application. Scale and sludge deposits tend to collect preferentially in the high-heat-flux zones of the boiler where they retard heat flow, raise metal temperatures, and potentially lead to overheating and tube failure. Sludge and solid particles can settle out in some areas and cause flow restrictions. Corrosion (see also Sec. 4.5) due to chemical imbalance, excessive concentration, or dissolved gases can cause general metal loss and ultimately lead to tube failures. The presence of internal deposits can accelerate underdeposit corrosion because of chemical concentration and elevated temperatures. Certain chemical reactions produce intergranular attack of metals and may lead to embrittlement and ultimate fracture.

A comprehensive discussion of water and treatment practices for power boilers is provided in "The ASME Handbook on Water Technology in Thermal Power Systems," American Society of Mechanical Engineers, 1989. A brief overview discussion is provided here.

Raw Water

Raw water treatment is necessary for all natural water sources to remove impurities and prepare water for steam generation. Concentrations are typically expressed as parts weight fraction of the impurity per million parts water or **parts per million (ppm)**. For very low concentration **parts per billion (ppb)** is used. Treatment of raw water for use as boiler feedwater and makeup water involves one or more of the following processes:

1. *Removal of suspended solids*
2. *Chemical treatment for the removal of hardness.* Calcium and magnesium are the principal hardness impurities in water which have retrograde solubility with increasing temperature and form scale deposits on heated surfaces. They must be removed to very low levels.
3. *Cation exchange for the removal of hardness*
4. *Ion-exchange demineralization for complete removal of dissolved solids.* Several types of inorganic resins have the ability to selectively remove undesirable cations and anions of impurities from water by exchange with hydrogen or hydroxyl ions. The resins are later regenerated. **Mixed bed ion-exchange demineralizers** use a mixture of the anion and cation resins.
5. *Evaporation*
6. *Reverse osmosis*

Demineralized water nearly free of all dissolved solids is generally recommended for high pressure boilers, especially for operating pressures > 1,000 psia (6.9 MPa).

Feedwater and Makeup Water

Boiler **feedwater** is generally a mixture of returned steam condensate and fresh makeup water from the treated raw water supply. For utility boilers, most of the steam is usually returned as condensate, and only 1 to 2 percent makeup is necessary. However, for some industrial cycles there is little or no return condensate, so up to 100 percent makeup may be necessary. Overall feedwater parameters which must be controlled include dissolved solids, pH, dissolved oxygen, hardness, suspended solids, total organic carbon (TOC), oil, chlorides, sulfides, alkalinity, and acid- or base-forming tendencies. At a minimum, boiler feedwater must be softened water for low-pressure boilers and demineralized water for high-pressure boilers. Removal of oxygen and other dissolved gases adequate for boiler feedwater applications is generally accomplished by thermal deaeration of the water in open-type feedwater heaters as well as tray or spray-type deaerating heaters. Thermal deaeration can reduce feedwater oxygen concentrations to less than 7 ppb. It also essentially eliminates dissolved carbon dioxide, nitrogen, and argon.

Chemical agents are generally used to scavenge residual oxygen. Traditional oxygen scavengers have been sodium sulfite for low-pressure boilers [< 900 psig (6.2 MPa gage)] and hydrazine for high-pressure boilers. Hydrazine has been identified as a carcinogen, and thus the use of alternate high-temperature oxygen scavengers (erythorbic acid, diethylhydroxylamine, hydroquinone, and carbohydrazide) has increased. Scavengers are generally fed at the exit of the condensate polishing system and/or the boiler feed pump suction.

When the boiler feedwater pH falls below a minimum value, ammonium hydroxide or an alternate alkalizer is added. Common alternative pH control agents include neutralizing amines, such as cyclohexylamine and morpholine. For high-pressure boilers with superheaters, the more complex amines are thermally unstable and decomposition products can be problematic.

For many boilers, a large fraction of the feedwater is returned as condensate. Condensate has been purified by prior evaporation, and in many cases makeup water can be directly mixed with the condensate to supply feedwater to the boiler. However, steam condensate can be contaminated by corrosion products or by in-leakage of condenser cooling water. Where the returning condensate is contaminated to such an extent that it does not meet the feedwater quality requirements, mixed bed ion-exchange demineralization purifiers are used to remove the dissolved impurities and filter out the suspended solids. This is referred to as **condensate polishing** and is essential for satisfactory operation of once-through utility boilers where the feedwater requirements are especially stringent. Condensate polishing is also recommended for all boilers operating with all-volatile treatment (AVT) (see below).

Boiler Water

In boilers, water is converted to steam, and the steam leaves the boiler in a relatively pure state. In drum-type recirculating boilers, impurities other than gases are retained and concentrate in the boiler water. High concentrations of foam-producing solids in the boiler water contribute to particle and water carryover and contamination of steam. To limit the dissolved solids to specified levels, a portion of the boiler water is intermittently or continuously removed (referred to as **blowdown**), disposed of, and replaced with fresh feedwater. Blowdown is normally expressed as a percent relative to the total steam flow. In once-through boilers, basically all the material entering with the feedwater is carried out with the steam or deposited on the internal boiler surfaces. Therefore, more stringent feedwater quality limits are required, and any boiler water treatment must take this into account to ensure that the quality requirements are met. The ASME Guidelines for Water Quality in Modern Industrial Water Tube Boilers are shown in Table 7.2.4. High pressure utility boilers require extremely close control of water quality and steam purity.

Customized chemistry limits and treatment practices must be established for each boiler. These limits depend upon the steam purity requirements, feedwater chemistry, and boiler design. They also depend upon boiler owner/operator preferences in the development of an overall plant water treatment program and system. Because corrosion in boilers is usually minimized by maintaining alkaline boiler water, the pH or total alkalinity (a measure of all reactives that have the ability to neutralize acids, expressed in ppm) is closely monitored and controlled.

Table 7.2.4 ASME Guidelines for Water Quality in Modern Industrial Water Tube Boilers for Reliable Continuous Operation
SI conversion: kPa = 6.895 × lb/in²

	Boiler feedwater			Boiler Water		
Drum pressure (lb/in²)	Iron (ppm Fe)	Copper (ppm Cu)	Total hardness (ppm $CaCO_3$)	Silica (ppm SiO_2)	Total alkalinity (ppm $CaCO_3$)	Specific conductance (micro-ohms/cm) (unneutralized)
0-300	≤ 0.100	≤ 0.050	≤ 0.300	≤ 150	≤ 700	1100-5400
301-450	≤ 0.050	≤ 0.025	≤ 0.300	≤ 90	≤ 600	900-4600
451-600	≤ 0.030	≤ 0.020	≤ 0.200	≤ 40	≤ 500	800-3800
601-750	≤ 0.025	≤ 0.020	≤ 0.200	≤ 30	≤ 200	300-1500
751-900	≤ 0020	≤ 0.015	≤ 0.100	≤ 20	≤ 150	200-1200
901-1000	≤ 0.020	≤ 0.015	≤ 0.050	≤ 8	≤ 100	200-1000
1,001-1,500	≤ 0.010	≤ 0.010	ND	≤ 2	ND	≤ 150
1,501-2,000	≤ 0.010	≤ 0.010	ND	≤ 1	ND	≤ 80

ND = Not Detectable.
SOURCE: "Consensus on Operating Practices for the Control of Feedwater and Boiler Water Chemistry in Modern Industrial Boilers," Publication CRTD-Vol. 34, American Society of Mechanical Engineers, 1994.

Direct boiler water treatment **(internal treatment)** practices commonly used to control boiler water chemistry include:

1. **Traditional all-volatile treatment:** No solid chemicals are added, and boiler water chemistry is controlled by feedwater treatment alone. Feedwater pH is controlled with ammonia and hydrazine, or a suitable alternative is added to scavenge residual oxygen.
2. **Oxygen treatment:** For boilers with copper-free feedwater trains and in the effective absence of impurities [cation conductivity < 0.15 μS/cm at 77°F (25°C)], a controlled low level of oxygen is maintained in the boiler which helps establish and maintain an especially protective Fe^{3+} iron oxide coating. Usually used in high-pressure once-through boilers, it has also been used successfully in drum boilers. The target oxygen concentration is 0.030 to 0.150 ppm for once-through boilers and 0.030 to 0.050 ppm for drum-type boilers.
3. **Coordinated phosphate treatment:** This treatment controls the boiler water alkalinity with mixtures of disodium and trisodium phosphate added to the drum through a chemical feed pipe. The objective is to maintain the pH of the boiler water and underdeposit boiler water concentration within an acceptable range.
4. **High-alkalinity phosphate treatment (low-pressure boilers only):** Used in industrial boilers less than 1,000 psig (6.9 MPa gage) where full demineralized makeup water is not cost-effective, trisodium phosphate and sodium hydroxide are added to the drum to maintain the pH at 10.8 to 11.4 and phosphate at acceptable levels.
5. **High-alkalinity dispersant, polymer, and chelant treatment (low-pressure boilers only):** In industrial boilers with substantial hardness (e.g., 0.1 ppm as $CaCO_3$) and pressures below 1,000 psig (6.9 MPa gage), the additives promote the formation of stable calcium and magnesium complexes which remain in solution or suspension and can be removed by blowdown.

CARE OF BOILERS

The care of power boilers is delineated in Section VII of the ASME Boiler and Pressure Vessel Code. Principal considerations include the initial preparation of new equipment for service; normal operation, including routine start-up and shutdown; emergency operations; inspection and maintenance; and idle storage. In all these phases, the handling of equipment is the responsibility of the operator, but recommendations and operating instructions supplied by the manufacturer should be thoroughly understood and followed.

The initial preparation of new boiler units for service, or of old equipment after completing major alterations or repairs, involves the removal of construction or foreign material from the setting and from the interior of pressure parts, hydrostatic testing and inspection for leaks, and the boiling out of the unit with an alkaline solution for the removal of grease and other deposits in the steam-generating pressure parts. The boiler unit is fired at a low rate during boilout. This procedure also facilitates the desired slow drying of any refractories used in the setting. After boilout and flushing, corrosion products remain in the feedwater system and boiler in the form of iron oxide and mill scale. Chemical cleaning to remove these should be delayed until after full-load operation in most cases, so that the scale and oxides are carried into the boiler from the feedwater system. Exceptions include boilers with full-flow condensate polishing systems and boilers in which the preboiler system has been chemically cleaned. These units can be chemically cleaned immediately.

After the boiler is filled with high-purity water and brought back into operation, the pressure is raised to test and set the safety valves to code requirements; the superheater and the steam piping are blown out to remove any foreign material; and the boiler is placed on the line for a period of low-load operation, during which the auxiliary equipment, controls, and interlocks are test-operated. After these operations, it is advisable to shut down, cool, and drain the boiler unit prior to a thorough internal and external inspection and any adjustments or modifications required to the equipment.

Normal operation involves the orderly start-up and shut-down of equipment and operation, under controlled conditions, to meet plant requirements. Statistics show that about 80 percent of the recorded furnace explosions occur during start-up and low-load operation, and particular care must be taken during such operations to prevent such explosions. The National Fire Protection Association "NFPA 85: Boiler and Combustion Systems Hazard Code" delineates the preferred sequence of starting fuel-burning equipment, the recommended minimum flame-monitoring equipment and safety interlocks, the recommended fuel-transport piping systems, the recommended purging procedures, and the procedures to be taken in the event of burner or furnace flame-out.

Normal operation also entails the maintenance of specified feedwater and boiler water conditioning, design steam and metal temperatures, clean gas passages and heat-absorbing surfaces, and ash removal.

In most cases the rate of firing during start-up is limited by the allowable metal temperatures in the superheater and reheater until steam flow through the turbine is established. In some cases, the time for both start-up and shutdown may be determined by the time required to limit the thermal stresses in the drum and headers. Then, after the steam flow is established, the temperatures and the temperature differentials in the various parts of the boiler and turbine control the permissible rate of firing.

Emergency operations are usually the direct result of abnormal conditions such as the failure of the feedwater supply, the rupture of a pressure part, the interruption of the fuel supply, the loss of air, or a burner flame-out. Automatic safety interlocks usually are installed which trip the fuel supply and shut down the unit if these or other hazardous conditions are experienced. Abnormal operating conditions, which might become hazardous if allowed to persist, such as low (fuel-rich) or high (air-rich) air-fuel ratios or the failure of essential auxiliaries, require appropriate action to correct operating conditions. An operator who cannot correct an abnormal condition must determine whether operation can continue and, if not, must shut down the unit in the proper manner or activate the emergency trip through the automatic interlock system.

Inspection and maintenance should be performed during regularly scheduled outages. A list of the known items requiring repair or maintenance should be prepared before the outage and should be supplemented by any additional items noted in thorough inspection of the boiler and auxiliary equipment during the outage. A major item on the work list should be the maintenance of internal and external cleanliness. External cleaning is usually accomplished by water washing or air lancing. The internal surfaces of large-capacity units are generally chemically cleaned under competent supervision.

Boilers removed from service for more than 7 days should be laid up using one of several accepted practices. Alternatives for fluid-side layup include nitrogen blanketing, wet layup with treated demineralized water, and dry dehumidified layup. For extended outages it is important that the preboiler feedwater train, balance of the water-steam cycle, auxiliary equipment, and fire side of the boiler be adequately preserved. For idle periods longer than 6 months, dry dehumidified storage is usually recommended. For boilers with nondrainable components, water removal issues must be weighed against the advantages of dry layup. Reheaters are generally stored dry because they cannot easily be isolated from the turbine. The use of nitrogen blanketing to maintain a positive pressure in the system for either dry or wet storage can be very effective in preventing air in-leakage and the associated corrosion. However, the boiler must be well sealed to prevent excessive leakage.

CODES

The ASME Boiler and Pressure Vessel Code, initiated in 1914 and supplemented by continuing revisions, contains the basic rules for the safe design, the construction, and the materials for steam generating units. Its legal status depends upon its adoption by state or municipal authority. The Code is administered by the National Board of Boiler and Pressure Vessel Inspectors. This organization also has established the "National Board Inspection Code," or NBIC, which provides rules and guidelines for in-service inspection, repair, and alterations of pressure-retaining items including steam generators. The adoption of both codes has been widespread, and they form the basis for the pertinent legal requirements in all but a few localities throughout the United States and Canada.

7.3 EXTERNAL COMBUSTION HEAT ENGINES

Revised by Ivo A. Garza

REFERENCES: Ewing, "The Steam Engine and Other Engines," Cambridge. Ripper, "Steam Engine Theory and Practice," Longmans. Heck, "The Steam Engine and Turbine," Van Nostrand. Allen, "Uniflow, Back Pressure, and Steam Extraction Engines," Pitman. Peabody, "Valve Gears for Steam Engines," Wiley. Zeuner, "Treatise on Valve Gears," Spon. Spangler, "Valve Gears," Wiley. Dalby, "Valves and Valve Gear Mechanisms," Longmans. Urieli and Berchowitz, "Sirling Cycle Machine Analysis," Adam Hilger, 1984. Carslaw, H.S. and Jaeger, J.C., "Conduction of Heat in Solids," 2nd ed., Clarendon Press, London, 1959. Martini, W.R, "Stirling Engine Design Manual," NASA CR-168088, 1983. Organ, A.J., Sirling Cycle Engines, "Inner Workings and Design," Wiley, 2014. Dyson, R.W., Wilson, S.D., Tew, R.C., "Review of Computational Stirling Analysis Methods," NASA/TM-2004-213300, 2004.

This section considers heat engines with external heat sources utilizing a gas expanding within cylinders/pistons. It includes steam and gas cycles. Figure 7.3.1 shows the idealized PV (indicator) diagram of a steam engine. Figure 7.3.9 shows the idealized PV diagram for three idealized gas cycles.

The reciprocating steam engine is a major, if not the major, hallmark in the pantheon of machinery most intimately associated with the profession and practice of mechanical engineering, and it was at the heart of the early industrial era. It dominated power generation for stationary and transportation service for more than a century until the development of the steam turbine and the internal-combustion engine. The mechanisms were numerous (see Patent Office listings), but practicality essentially standardized the positive-displacement, double-acting, piston and cylinder design in a vertical or horizontal configuration. These engines were heavy, cast-iron structures, for example 50 to 100 lb/hp; they had low piston speed (600 to 1,200 ft/min); long stroke (up to 6 ft); low turning speeds (50 to 500 r/min); steam conditions less than 300 lb/in^2 (dry saturated, or 100 to 200°F superheat); noncondensing or condensing (25 ± in Hg vacuum); sized from children's toys to 25,000 hp. Diversity of valve gear was an inventor's paradise with many suitable for reversible operation, as in locomotive, rolling-mill, and ship applications. Most of the machine elements known today, such as cylinders, piston rods, crossheads, connecting rods, crankshafts, flywheels, and governors, were developed in steam engines.

These engines utilize the expansive power of steam. Theoretically, the more the steam can be expanded in the engine cylinder, the better will be the economy. Practical losses, which occur in every steam engine, limit the expansion ratio and result in a minimum steam rate for a definite degree of expansion. Cylinder dimensions preclude the practical utilization of high-vacuum conditions because of the high specific volume (e.g., 339 ft^3/lb dry and saturated at 2 in Hg abs). The steam turbine can effectively utilize maximum vacuum.

WORK AND DIMENSIONS OF THE STEAM ENGINE

Thermodynamic principles define the limits of the conversion efficiency of heat into work (see Rankine and Carnot cycles, Sec. 2). In fact, the historical development of thermodynamic principles was primarily aimed in the nineteenth century at defining the thermal performance of steam engines and steam power plants. It can be said that the steam engine did more for thermodynamics than thermodynamics did for the steam engine. Steam tables and steam charts (e.g., temperature-entropy; enthalpy-entropy; and pressure-volume) are essential to evaluate theoretical and actual equipment performance.

The pressure-volume diagram, or indicator card, is of primary significance in both the design and operation of the reciprocating piston and cylinder mechanism—not only steam engines, but also internal-combustion engines, air compressors, and pumps. The utility of the *p-v* diagram is often lost in attempts to improve fluid-dynamic reciprocating mechanisms. The work and power are the consequence of the difference in pressure on the two sides of the piston, expressed as mean effective pressure (mep or p_m). This is applied to the cylinder dimensions and rotating speed in the **"plan" equation**

$$\text{hp} = \frac{p_m Lan}{33{,}000} \tag{7.3.1}$$

where p_m = mep, lb/in^2; L = stroke, ft; a = effective piston area, in^2; n = number of cycles completed per min; 33,000 = mechanical equivalent of horsepower, ft · lb/min, by definition.

A typical steam-engine indicator card is shown in Fig. 7.3.1, identifying the established nomenclature for the events of the cycle. Superimposed is a theoretical diagram, with expansion, but without clearance or compression. The terminal pressure controls the steam, or water, rate. The term "back pressure" is used for pressures above the

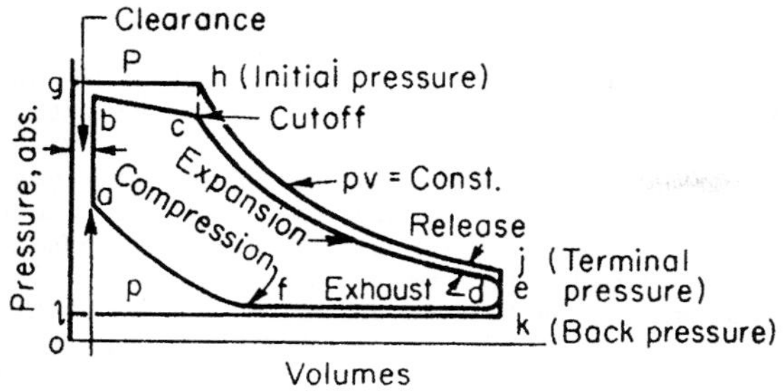

Figure 7.3.1 Typical steam engine indicator diagram.

atmosphere, while "condenser pressure" is used when engines operate with negative exhaust pressure. The volume ratio $v_j/v_h = R$ is called the **ratio of expansion.**

The average ordinate, or **mean effective pressure** p_m [applicable in Eq. (7.3.1)] for the ideal cycle diagram *ghjkl* is

$$p_m = p[(1 + \log_e R)/R] - p \qquad (7.3.2)$$

The expansion phase *h* to *j* of the cycle is *logarithmic* where pv = constant. It is not isothermal but reasonably approximates expansion (and compression) in actual engines where condensation and evaporation on cylinder walls, heads, pistons, and valves are cyclically recurrent.

The value of R is commonly around 4 in simple engines. It should increase as P increases and as p decreases, usually between 3 and 5. It should be higher in jacketed than in unjacketed engines. The efficiency of the engine depends largely on the value chosen for R. This is dictated by service requirements, for example, load variation, load fluctuation, engine governing type (cutoff vs. throttle), overload capacity, steam cost, and economics. Efficiency, however, is not the sole requirement for all installations. The high starting torque of a steam locomotive dictates a valve gear that allows full-stroke admission of steam with zero expansion, a rectangular indicator card with consequent maximum p_m.

Figure 7.3.1 shows that the theoretical card has a larger area than the actual card. The value of p_m obtained from Eq. (7.3.2) must be multiplied by the **diagram factor** to obtain the actual p_m under the assumed conditions. This factor may have a value between 0.7 and 0.95. The actual p_m is obtained from the card drawn on the indicator drum under running conditions of the engine. The card area, graphically measured by a planimeter (see Sec. 12), is divided by the card length to get the average equivalent height of the card. The spring scale of the instrument, applied to this average height, gives the mean effective pressure actually obtaining within the engine cylinder. This value, when introduced in Eq. (7.3.1), gives the indicated horsepower (ihp) of the engine.

Losses in steam-engine cylinders are (1) incomplete expansion; (2) initial condensation; (3) throttling, affected by valve and port area; and (4) radiation, which can be considered constant. The point of best steam economy occurs with the p_m at which the total of all losses is a minimum.

Mechanical Efficiency and Shaft Output

$$\text{Brake horsepower (bhp)} = \text{ihp} \times \text{mechanical efficiency} \qquad (7.3.3)$$

$$\text{Friction horsepower} = \text{ihp} - \text{bhp} \qquad (7.3.4)$$

bhp	25	50	75	100
Friction hp	6	6	6	6
hp	31	56	81	106
Mechanical efficiency, percent	81	89.5	92.5	94

Figure 7.3.2 shows the effect of mechanical efficiency and generator efficiency on the steam rate of an engine-generator set, both as to relative magnitude and as to location of the minimum values.

Engine economy may be improved by a number of methods: (1) **separation of inlet and outlet ports;** (2) **steam jackets,** applied to cylinders and heads to keep surfaces hot and dry; (3) **multiple expansion.** Condensation losses are related to the temperature difference existing in the cylinder. Cylinders in series reduce this temperature difference and allow more complete overall expansion of the steam. Two, three, and four cylinders in series, as in **compound-, triple-,** and **quadruple-expansion** engines were common constructions, but the successive improvement is smaller for each additional stage. (4) **Superheating** gives the vapor the properties of a gas, reduces cylinder condensation, and necessitates decisive changes in engine design. The overall improvement in performance and water rate is so substantial that superheat is prevalent in practice. (5) The **uniflow** arrangement was the last great improvement in design (Figs. 7.3.3 and 7.3.4). Its high economy results from the high temperature of the residual steam at the end of compression. This temperature, aided by jackets, is higher than the live steam temperature, so that initial condensation is reduced to a negligible amount. For **condensing operation** the engines are built with steam valves only (two-valve type), Fig. 7.3.3. Clearance pockets are provided, with either hand-operated or automatic valves to permit operation with atmospheric exhaust or against back pressure.

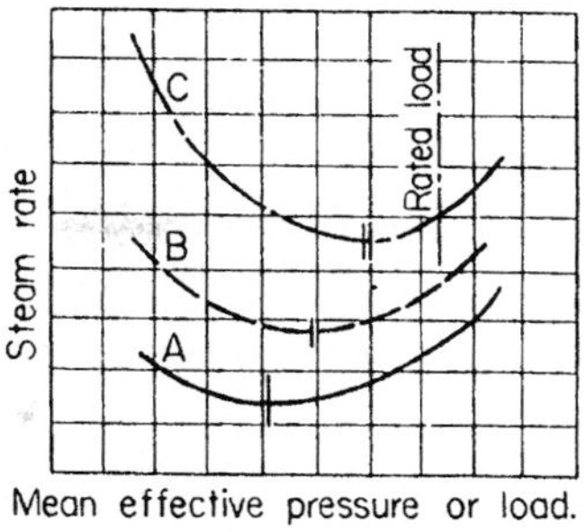

Figure 7.3.2 Engine steam rate curves (a) at the steam cylinder, lb/(ihp · h); (b) at the engine shaft, lb/(bhp · h); (c) at the generator, lb/kWh.

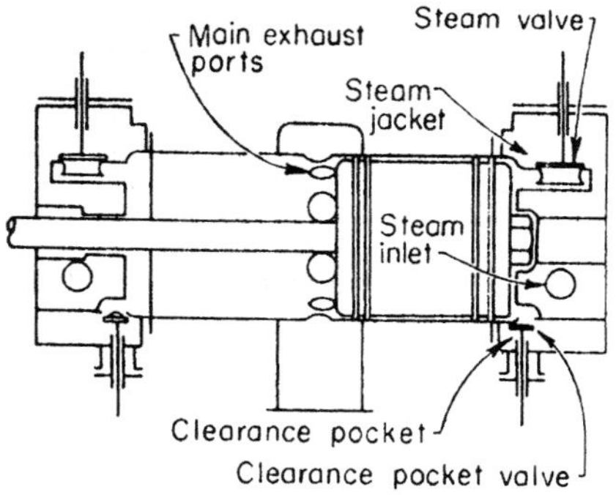

Figure 7.3.3 Condensing uniflow engine cylinder, with clearance pockets.

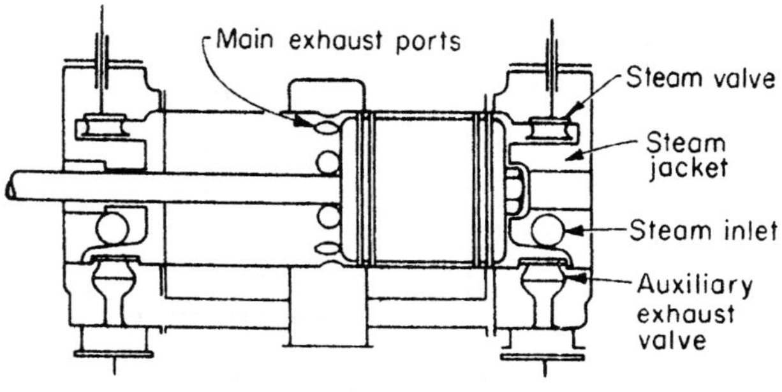

Figure 7.3.4 Noncondensing uniflow engine cylinder, with auxiliary exhaust valves.

Noncondensing uniflow engines have, in addition, auxiliary exhaust valves to reduce the otherwise large clearance required (four-valve type). If back pressure is variable, exhaust-valve *gears* are designed to change the length of compression while the engine is in operation (Fig. 7.3.4).

Engine Steam (Water) Rates The efficiency of steam engines has generally been expressed in terms of pounds of steam per horsepower-hour. The term water rate was frequently used because of the convenient accuracy in weighing liquid water in the condensing plant. Figure 7.3.5 shows, on a percentage basis, two types of performance curves, one in lb/hph and the other in lb/h. The latter is identified as the "total steam" or **Willans line** and is straight for an engine with fixed cutoff and variable initial pressure. Figure 7.3.6 reflects the impact of steam pressure and superheat on the steam consumption of a condensing uniflow engine.

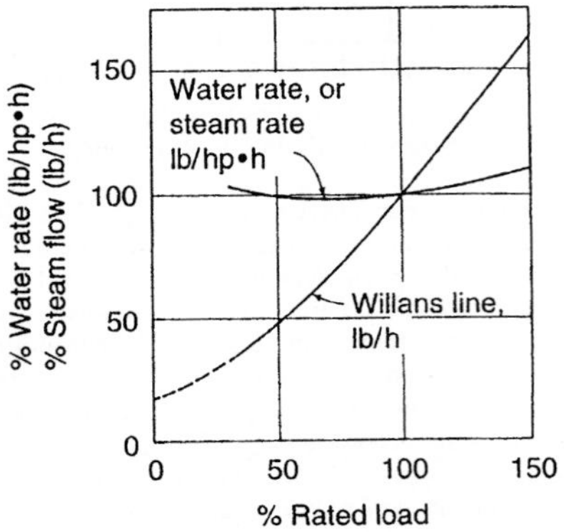

Figure 7.3.5 Actual steam (water) rate, lb/(hp · h), and Willans line, lb/h, for a noncondensing uniflow engine, percentage basis.

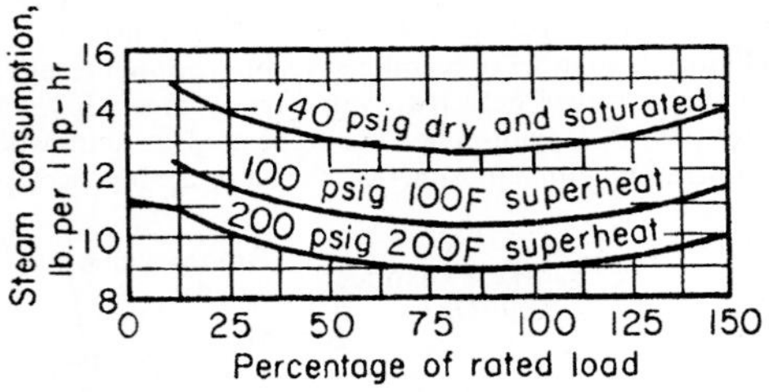

Figure 7.3.6 Steam consumption of uniflow engine with 27.5-in vacuum.

The **Rankine cycle** (see Sec. 2) is the accepted thermodynamic standard for comparing the performance of steam prime movers (engines and turbines). Figure 7.3.7 shows the PV diagram for a steam engine operating on the Rankine cycle with steam between 498°F (550 K) and 212°F (373 K). The steam pressure would be 886 psig (60.3 atm). This cycle includes condensing from the exhaust point (2) to the point (3) where pumping power is input to pressurize the water. It enters the boiler and undergoes isobaric boiling and expansion between points (4) and (1). It is transported to the cylinders where produces work through adiabatic expansion from points (1) to (2). The steam that exits this expansion is superheated and may be passed through a feedwater heater prior to entering the condenser to recover the heat that would be otherwise wasted. The cycle shown in Figure 7.3.7 was analyzed with a digital steam table and an electronic spread sheet to facilitate the numerical integration of the PV work of expansion. Since the specific volume of the pumped water is essentially constant the pumping work per mass of pumped water is estimated as

$$W_p = f \times \Delta P \times V \tag{7.3.5}$$

W_p is the work per unit mass in ft-lb$_f$/lb$_m$ (kJ/kg); ΔP is the differential pressure in psi (atm); V is the specific volume in ft^3/lb$_m$ (m^3/kg); and f is a unit conversion factor of 144 (101.3). This of course is the pumping power to the water, the pump and driver efficiency would need to be considered. The ideal net power from this example is 203,000 ft-lbf/lbm (607 kJ/kg) and the heat input is 704,000 ft-lbf/lbm (2,105 kJ/kg). The resulting efficiency is 28.8 percent. The water rate for this cycle may also be estimated by the use of the Mollier chart (Sec. 2) where …

$$\text{Rankine steam rate} = 2{,}545/(h_1 - h_2) \quad \text{lb/(hp} \cdot \text{h)} \tag{7.3.6}$$

and h_1 and h_2 are the initial and exhaust enthalpies, Btu/lb (constant entropy). The lowest point of the actual steam-rate curve (Fig. 7.3.6) is usually referred to as the Rankine rate, and the ratio of Rankine rate to actual rate is the **Rankine efficiency ratio (RER).** This ratio may vary from 0.5 to 0.9, depending on the type of engine.

When clearance, compression, and incomplete expansion are introduced, the methods of Sec. 2 should be used to evaluate steam rate. This involves steam tables and charts to find the net work of the indicator

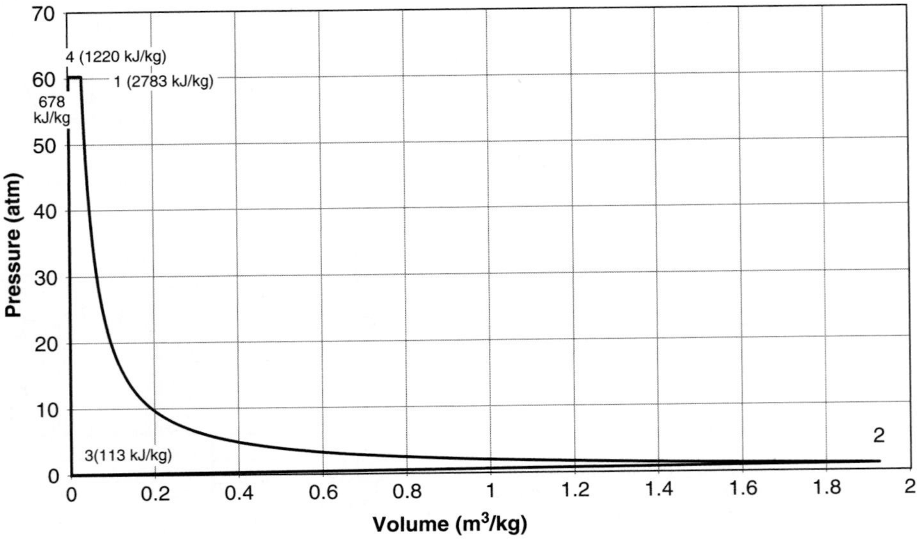

Figure 7.3.7 Rankine cycle, *p-v* diagram.

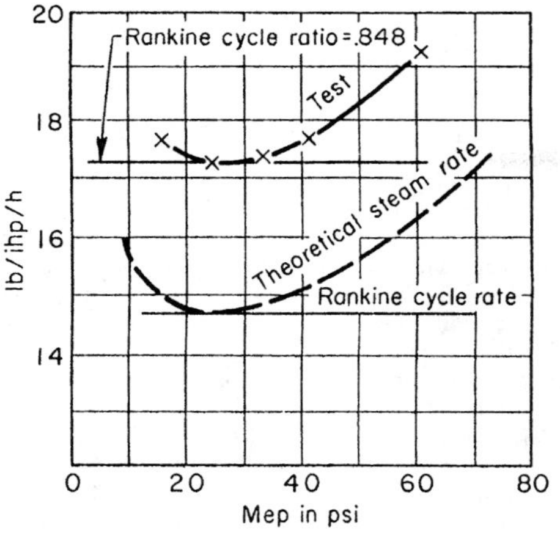

Figure 7.3.8 Theoretical and test steam rate curves. Uniflow engine 20 × 24 at 200 r/min; throttle conditions 150 lb/in² saturated steam; exhaust to atmosphere.

card from the positive and negative areas of the several component phases. Figure 7.3.8 shows a theoretical steam-rate curve, computed by such methods, together with the actual test curve for the engine.

Engine Details Valves and valve-gear types range from the simplest D slide to gridiron, double-slide (Meyer or Rider), rocking, piston, releasing and nonreleasing Corliss, and poppet.

Volumetric clearance should be made as small as possible. Slide and piston valves bring clearance volumes to 12 to 15 percent, Corliss valves 6 to 10 percent, and poppet valves 4 to 8 percent.

Valve and Port Sizes Flow area of valves and cross section of ports are usually determined by port area = AS/C in², where A is net piston area, in², S is mean piston speed, ft/min, and C is a constant, ft/min. Values of C are approximately 9,000 to 15,000 ft/min for inlet and 6,000 to 7,000 ft/min for outlet. Small valves and ports represent lost work.

Superheated Steam With high-temperature steam, the cylinder and parts design must allow for free expansion. Poppet and piston-valve cylinders can easily meet these requirements. The orthodox type of Corliss cylinder, however, is not suitable for highly superheated steam.

With the advent of automated calculations, the equation of state can be calculated along the steam cycle to provide more accurate results. In conjunction with the automated equations of state, second order models using simplified geometry can be used to be estimate losses in the engine. These models would include estimates of heat transfer to and from the cylinder walls during expansion based on empiric heat transfer formula. The cylinder pressure and steam volume would be corrected for the heat transfer. Back pressure from the cylinder to the exhaust would be estimated based on the exhaust valve and exhaust port geometry, and on the required steam flow. Condenser surface area would be estimated from empiric heat transfer equations. Friction losses in the piping and boiler would be estimated based on empiric equations and included in the required pumping power. The pump efficiencies, generally from pump vendor literature would be used to determine the required driver power. Losses from the boiler through the transfer piping, cylinder ports, and valves would be estimated with empiric equations and subtracted from the boiler pressure to determine the actual cylinder pressures. Computational fluid dynamics allows further refinement by more accurately modeling the geometry of the engine along with micro calculation of local heat transfer and mass transport.

EXTERNAL COMBUSTION GAS ENGINES

Three idealized thermal cycles are considered in this section; the Carnot, the Ericsson, and the Stirling Cycles. Figure 7.3.9 shows these three cycles operating between 498°F (550 K) and 48°F (300 K), and 64 ft³/lb-mole (4 l/gm-mole) and 320 ft³/lb-mole (20 l/gm-mole). The

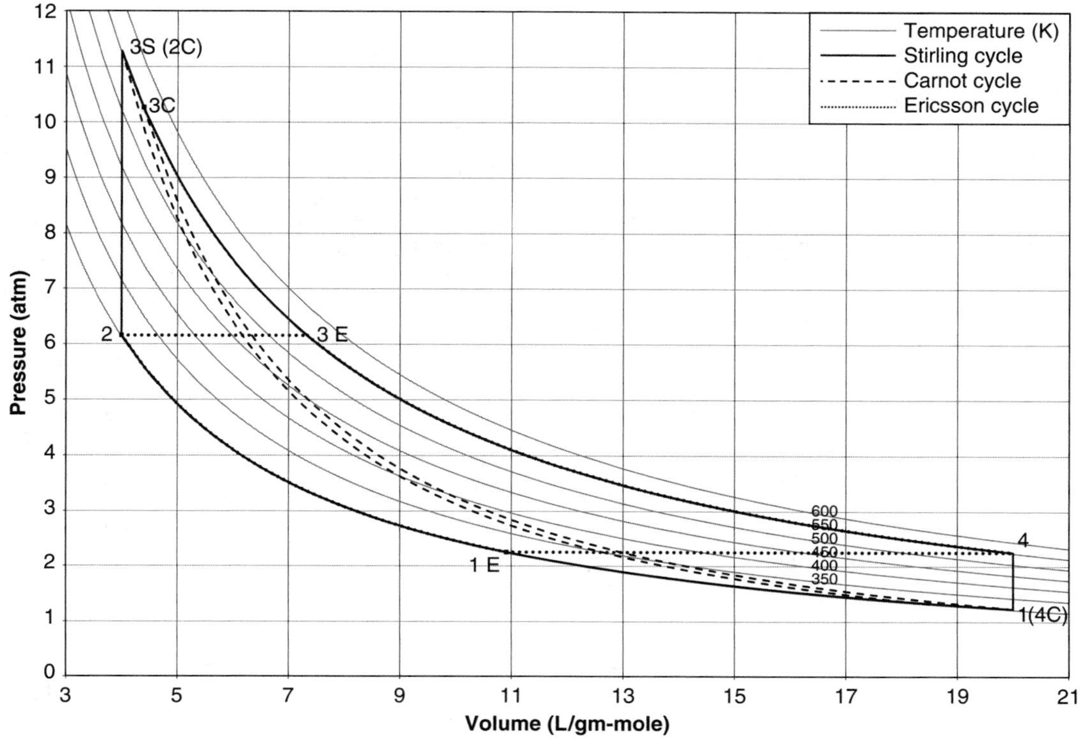

Figure 7.3.9

Carnot cycle consists of adiabatic compression between points 1C and 2C, isothermal heating between points 2C and 3C, adiabatic expansion between points 3C and 4C, and then isothermal cooling between points 4C and 1C. The Ericsson cycle consists of isothermal compression between points 1E and 2, isobaric heating between points 2 and 3E, isothermal expansion between points 3E and 4, and then isobaric cooling between points 4 and 1E. The Stirling cycle consists of isothermal compression between points 1 and 2, isochoric heating between points 2 and 3S, isothermal expansion between points 5S and 4, and then isochoric cooling between points 4 and 1.

If we assume an ideal gas we can derive the equations for net work and efficiency for these cycles. The following ideal gas equations are used to analyze these cycles:

$$PV = nRT, \text{ equation of state} \quad (7.3.7)$$

$$Q = C_v\Delta T, \text{ isochoric heating} \quad (7.3.8)$$

$$Q = C_p\Delta T, \text{ isobaric heating} \quad (7.3.9)$$

$$\gamma = C_p/C_v \quad (7.3.10)$$

$$PV^{\gamma} = \text{constant, adiabatic expansion} \quad (7.3.11)$$

$$TV^{(\gamma-1)} = \text{constant, adiabatic expansion} \quad (7.3.12)$$

C_p – Constant pressure specific heat
C_v – Constant volume specific heat
n – number of molecules of gas
P – Absolute pressure
Q – heat
R – Universal gas constant
T – Absolute temperature
V – Volume

The results in this section are presented on a molar basis.

Utilizing the above equations we can derive the following equations for work and heat.

Isothermal work

$$W = nRT\ln(V_f/V_i) \quad (7.3.13)$$

Adiabatic work

$$W = [nRT_i/(1-\gamma)] \times [(V_f/V_i)^{(1-\gamma)} - 1] \quad (7.3.14)$$

These equations are for the work done by or on the gas. Note that this work is opposed by both the atmospheric pressure on the other side of the piston as well as the work that is output to the shaft. The work of expanding against the atmosphere is regained during compression in our specific example.

Heat added during isothermal expansion

$$Q = nRT\ln(V_f/V_i) \quad (7.3.15)$$

f – The final state
i – The initial state

The equations for indicated work per cycle and efficiency are derived from these relations with the following results:

Carnot

$$W = nR(T_H - T_C)\ln[r_v(T_C/T_H)^{1/(\gamma-1)}] \quad (7.3.16)$$

$$Q = (T_H - T_C)/T_H \quad (7.3.17)$$

Ericsson

$$W = nR(T_H - T_C)\ln[r_v(T_C/T_H)] \quad (7.3.18)$$

$$Q = nR(T_H - T_C)\ln[r_v(T_C/T_H)] / [C_p(T_H - T_C) + nR(T_H - T_C)\ln[r_v(T_C/T_H)] \quad (7.3.19)$$

Stirling

$$W = nR(T_H - T_C)\ln(r_v) \quad (7.3.20)$$

$$Q = nR(T_H - T_C)\ln(r_v) / [C_v(1-\eta_e)(T_H - T_C) + nR(T_H)\ln(r_v)] \quad (7.3.21)$$

In the Stirling engine there is an economizer that recovers some of the heat that is removed from the gas when it is cooled. The efficiency of that heat exchanger is defined as η_e. The economizer will operate with some temperature gradient across it but will have losses due to the required temperature difference between the economizer and the air and an efficiency of 50 percent is assumed in this example. The ratio between the maximum and minimum volumes in the cycle is the volumetric compression ratio and is denoted by r_v. The larger the volumetric compression ratio, the more closely the Ericsson and Stirling cycles approach the efficiency of the Carnot cycle. Independent of the compression ratio, the more efficient the Stirling economizer is, the more closely the Stirling cycle approaches the efficiency of the Carnot cycle.

We can compare the net work per cycle and efficiency of the three cycles indicated in Fig. 7.3.9 and of the steam engine indicated in Fig. 7.3.7. For a diatomic gas, such as hydrogen, nitrogen, or oxygen, C_v is approximately $5/2R$ and C_p is approximately $7/2R$. R is 154.5 ft-lbf/lb-mole °R (8.314 J/gm 9m-mole °K). Applying these values to equations 7.3.16 through 7.3.21 the following table is produced:

Cycle	Work per cycle	Efficiency (percent)
Carnot	65,000 ft-lb/lb-mol (200 J/gm-mole)	45.5
Ericsson	700,000 ft-lb/lb-mole (2,100 J/gm-mole)	17.6
Stirling	1,100,000 ft-lb/lb-mole (3,300 J/gm-mole)	33.6
Steam Engine	2,800,000 ft-lb/lb-mole (8,400 J/gm-mole)	21.8

While it is efficient, the Carnot cycle produces much less work for the same size of engine when compared to the Ericsson or Stirling cycle. The economizer in the Stirling cycle results in more efficient operation than the Ericsson cycle for the same operating conditions and volumetric compression ratio. The Stirling cycle will be the basis for the remainder of this section. For the same size engine a steam cycle will provide significantly more work per stroke than these three gas cycles. It should be noted that at the temperature in the example the steam engine would be operating at a pressure that is not typical of steam engines that have been built. The steam engine also has to operate at temperatures above the boiling point of water so that combustion is typically used to heat a steam engine. The Stirling cycle, while operating less efficiently, can operate at temperatures below 212°F (373 K) and can use a wider variety of heat sources. This makes it well suited for waste heat recovery, solar heat usage, and low tech heat source usage. The Stirling cycle has been the basis for several practical engines. The Stirling engine has five essential parts; the hot section for isochoric heating, the cool section for isochoric cooling, the volume control (power) piston, a means to transfer gas between the sections and an economizer to capture heat from the hot gas as it passes to the cool section and return of the heat back when the gas returns to the hot section. Stirling engines are typically built without the use of valves. There are several configurations of Stirling engines, a few are shown in Fig. 7.3.10.

The way the Stirling engine works is that when the power piston is just before its lowest volume, the gas is transferred to the hot end, where the gas is heated as the power piston passes through the top dead center (TDC). The gas is heated at a relatively constant volume. The piston passes over the top and is driven outward as a result of the expanding gas. The gas is still in the hot side so heat is added to approximate isothermal expansion. Just before the piston reaches the bottom the gas is transferred to the cold section, where it is cooled with a relatively constant volume. The gas is routed to pass through a heat exchanger that captures some of the heat from the hot gas as the gas passes to the cold section. The piston passes the bottom dead center and starts to compress the gas. The gas is in the cold section so that the heat of compression is removed by the walls of the cold section so the

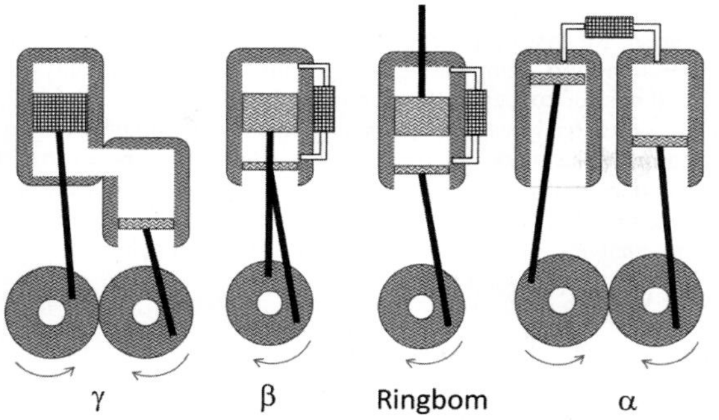

Figure 7.3.10

compression is approximately isothermal. As the piston nears the top of the stroke the gas is transferred back to the hot side through the economizer where it picks up a fraction of the heat that was captured when the gas was transferred from the hot section. The cycle then repeats.

Among the design challenges of a Stirling engine are the efficient transfer of heat into the hot section and out of the cold section, insulating the hot section from the cold section, the transfer mechanism and timing, designing an economizer that maximizes heat transfer and minimizes pressure loss, and minimizing the dead volumes in the engine so that the volumetric compression ratio can be maintained.

Timing is accomplished in several ways. In the case of the β and γ configurations the displacer that transfers the gas is controlled by the angle between power and displacer crank arms. The displacer in the beta configuration lags the power piston so that the displacer is transferring the gas as the volume control piston is in the ends of the cylinder. In the α configuration the hot end piston lags the cold end piston to provide the proper gas transfer timing. The α configuration can also include the use of more than two cylinders and the use of double acting cylinders to provide more power for the size and to provide additional parameters for optimizing the gas transfer timing.

Assuming an infinitely long connecting rod, the length that a piston is from the top of the cylinder can be expressed by

$$R(1 - \cos(\beta - \alpha)) + O \tag{7.3.22}$$

R – crank arm offset from the center of rotation
β – angle of the power piston or cold piston from top dead center
α – angular offset of the hot or displacer cylinder from the power or cold piston. This would be 0 for the power or cold piston.
O – applicable to the power piston of a beta configuration. It is the distance from the top of the power piston at TDC to the bottom of the displacer piston at TDC required to prevent contact between the two pistons. It can be found by subtracting distance between the power piston and the top of the cylinder (without the offset) from the distance between the displacer and the top of the cylinder at each crank angle. The offset is the maximum of these differences.

The length of the crank arms can also be varied to aid in optimizing the timing. Figures 7.3.11 and 7.3.12 show the clearance in the hot and cold ends as well as the total clearance in the cylinder. Since the cylinder is a constant area, this corresponds to volume. In Fig. 7.3.11 the displacer piston travels half as much as the power piston and lags the power piston by 160 degrees in the beta configuration. In Fig. 7.3.12, the hot side piston travels one-fourth as much as the cold piston and lags the cold piston by 135 degrees in the alpha example. Varying the lag angle changes the timing of the gas transfer compared to the occurrence of the maximum and minimum volume in the piston so that the transfer occurs just before the extremes so that isochoric heat transfer is approximated. The relative length of the crank arms has the greatest affect on the volumetric compression ratio.

In the Ringbom configuration the transfer piston is actuated based on the differential pressure between the cylinder and a fixed pressure source, usually atmospheric pressure. When pressure in the cylinder is high the transfer piston moves up and the gas is transferred to the cold end. After expansion the pressure decreases and the piston moves down transferring gas to the hot end.

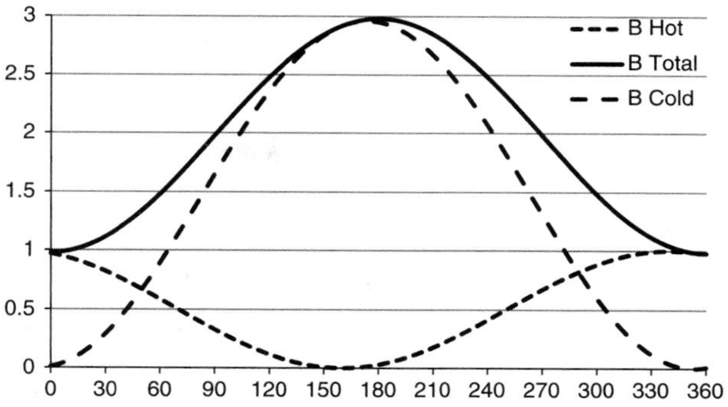

Figure 7.3.11

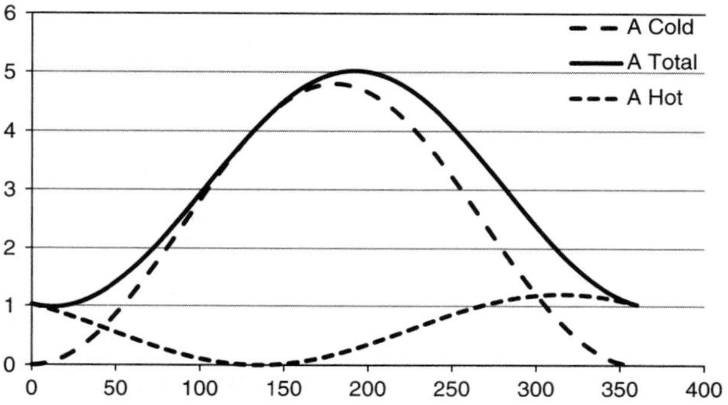

Figure 7.3.12

Much of the economizer design is independent of the engine configuration. In order to optimize the economizer efficiency the heat needs to be transferred from and back to the gas without allowing heat transfer along the direction of gas flow. This allows the hot side of the economizer to remain above the average temperature and the cold side to remain below average temperature. Ideally a series of masses along the flow path which are insulated from each other, but which have a large heat transfer surface with the gas would perform this function. This design needs to be balanced with the need to keep friction drop low and dead volume to a minimum. A series of screens is often used as a compromise in the economizer. For maximum efficiency, the mass of the screen must be enough to absorb the heat required to cool the gas from the hot to the cold temperature. Due to a need for differential temperature to drive the heat transfer this whole amount of heat will not be transferred and an economizer sized with this method will require several cycles to come up to a quasi-steady state temperature. The wire used in the economizer must be thin enough so that the whole wire can approach the local gas temperature. Using the methods of Carslaw and Jaeger the temperature at the center of the wire can be calculated by the following equation:

$$\Theta(0,\tau) = 2\Sigma_i e^{-\mu_i^2 \tau}/\mu_i J_1(\mu_i) \tag{7.3.23}$$

$$\Theta = (T - T_g)/(T_i - T_g) \tag{7.3.24}$$

$$\tau = \alpha\, t/r^2 \tag{7.3.25}$$

μ_i – the roots of the J_0 function
T – the temperature of the center of the wire
T_g – the surface temperature of the wire due to heat transfer from the gas
T_i – the initial temperature of the whole wire
α – the thermal diffusivity of the wire
r – the radius of the wire

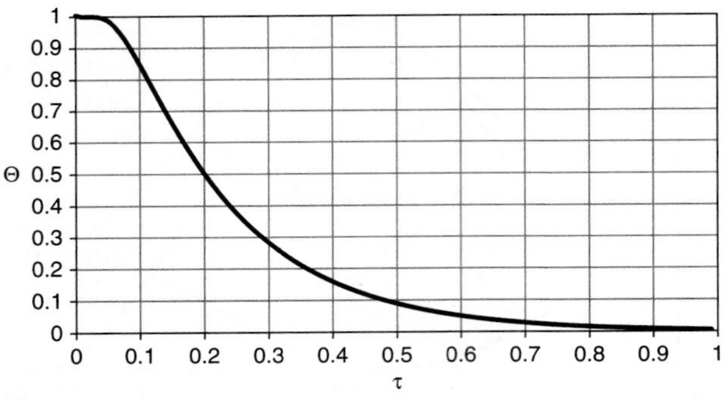

Figure 7.3.13

This equation is graphed in Fig. 7.3.13. In order to reach 90 percent of the quasi-steady state temperature within the half cycle of gas transfer the value of τ must be greater than 0.5. Based on the operating speed of the engine and Eq. 7.3.25 the wire size for maximum efficiency can be found. Mechanical strength considerations may require a thicker wire.

Heat transfer from the cylinder wall and heads needs to be maximized while keeping heat transfer between the hold and cold sections to a minimum. This becomes more critical in β and Ringbom configurations that have the hot and cold sections in the same cylinder. For a given displaced volume, long cylinders decrease the cross sectional area of the cylinders between the hot and cold end but decrease the head area for heat transfer. Insulating sections between the hot and cold sections, particularly in low pressure engines, can provide more design freedom. Gamma and alpha configurations allow more freedom in the cylinder design. For production machines the cylinder walls are bored smooth and only the head and piston top are available for enhanced heat transfer area. Heaters and coolers can be placed between the economizer and the cylinders for increased heat transfer. There is a disadvantage when gas is transferred in the other direction and the gas passes through the heater or cooler prior to entering the economizer.

The analysis of Stirling engines can be classified by the complexity and accuracy in modeling. The work and efficiency of the ideal cycle (Eqs. 7.3.20 and 7.3.21) would be a 0 order analysis. It is based on an ideal cycle and does not consider losses in the engines. The operating differential temperature might be reduced by 40°F (22 K), the economizer efficiency might only be 25 percent, mechanical efficiency may only be 80 percent and the actual size of the PV during expansion and compression would be rounded off due to the use of piston timing for gas transfer. A real engine might only produce half of the power predicted by the ideal cycle. Another 0 order method is the Beale method. It is equivalent to the PLAN formula for steam engines. The work provided by an engine is proportional to the PLAN (Eq. 7.3.1) of the engine. If the actual work produced by a prototype is divided by the PLAN, the result is an efficiency, termed the Beale Number. For a particular design of a prototype Stirling engine the Beale Number is assumed to remain constant for variations in the size of the engine. This is helpful in determining what size engine is needed if you have a power requirement and the results of prototype testing. The Beale Number was originally proposed for sizing engines from scratch based on engines that had been produced at the time, but is really not appropriate for that purpose. It does not account for variances in the compression ratio, or operating temperatures. Zero order analyses are fast methods for conceptual evaluations.

First order analyses use an actual sinusoidal compression cycle to determine the work out of a Stirling engine in closed form. The analysis does not include mechanical, temperature, or pressure losses in the engine. Berchowitz and Urieli provide a modern derivation of the Schmidt analysis. The resulting formulae are shown below.

$$W = W_e + W_c$$

$$W_c = \pi V_{\text{swc}} P_{\text{mean}} \sin(\beta)[\sqrt{(1 - b^2)} - 1]/b$$

$$W_e = \pi V_{\text{swe}} P_{\text{mean}} \sin(\beta - \alpha)[\sqrt{(1 - b^2)} - 1]/b$$

$$b = c/s$$

$$c = 0.5[(V_{\text{swe}}/T_h)^2 + 2(V_{\text{swe}}/T_h)(V_{\text{swc}}/T_c)\cos(\alpha)+(V_{\text{swc}}/T_c)]^{0.5}$$

$$s = [V_{\text{swc}}/2T_c + V_{\text{clc}}/T_c + V_{\text{cool}}/T_c + V_r/T_{\text{lm}} + V_{\text{heat}}/T_h + V_{\text{swe}}/2T_h + V_{\text{cle}}/T_h]$$

$$\beta = \tan^{-1}[(V_{\text{swe}}\sin(\alpha)/T_h)/(V_{\text{swe}}\cos(\alpha)/T_h + V_{\text{swc}}/T_c)]$$

$$P_{\text{mean}} = nR/[s\sqrt{(1 - b^2)}]$$

Where

α – angular offset between pistons
sw – subscript prefix for swept
cl – subscript prefix for clearance
e – subscript for expansion (hot side)
c – subscript used for volume to indicate compression or cold side
cool – subscript for cooler
heat – subscript for heater
r – subscript for regenerator (economizer)

These equations would be solved with the proposed design and then sensitivity charts prepared to allow for optimization of the engine geometry.

Second order analysis includes temperature differences into the global analysis. Third order analysis beaks the engine into homogeneous nodes and connects them in one dimension. The nodes would include the cylinder gas volumes, the heater and cooler volumes, the economizer gas and solid volumes, the interconnecting piping. Simplified geometry along with gas equations of state (typically ideal gas), empirical equations for heat transfer and pressure losses in the individual components of the Stirling engine are used in simultaneous mass and energy balances to determine the pressures, volumes, and temperatures in the nodes. The solution can either be a periodic solution for overall work and efficiency or an explicit time step solution. The references at the front of this section provide pointers to works describing second and third order analyses. These methods are appropriate for optimizing physical prototypes. Fourth order analysis would include transient analysis of the full engine geometry with multi-physics computation fluid dynamics (CFD) software. CFD can also be used on the individual components to define the empirical equations used in third order analysis. This level of analysis allows for prototyping in analysis space, which results in less trial and error in the physical prototype.

7.4 STEAM TURBINES

by Frederick G. Baily, Revised by Heinz P. Block

REFERENCES: Stodola, "Steam and Gas Turbines," trans. by L. C. Loewenstein, McGraw-Hill. Bartlett, "Steam Turbine Performance and Economics," McGraw-Hill. Warren, Development of Steam Turbines for Main Propulsion of High-Powered Combatant Ships, *Trans. Soc. Naval Architects Marine Engrs.,* 1946. Newman, Modern Extraction Turbines, *Power Plant Eng.,* Jan.-Apr., 1945. Salisbury, "Steam Turbines and Their Cycles," Wiley. Campbell and Heckman, Tangential Vibration of Steam Turbine Buckets, *Trans. ASME,* **47**, 1925, pp. 643-671. Campbell, Protection of Steam Turbine Disk Wheels from Axial Vibration, *Trans. ASME,* **46**, 1924, pp. 31-139. Deak and Baird, A Procedure for Calculating the Packet Frequencies of Steam Turbine Exhaust Blades, *Trans. ASME,* **85**, series A, Oct. 1963. Bloch, Heinz P., "Petrochemical Machinery Insights," Elsevier Publishing, 2017. Bloch, Heinz P. and Murari P. Singh, "Steam Turbines: Design, Applications and Re-Rating," 2nd Ed., McGraw-Hill, 2008.

Steam turbines have established a wide usefulness as prime movers, and are manufactured in many different forms and arrangements. They are used to drive many different types of apparatus, e.g., electric generators, pumps, compressors, and for driving ship propellers, through suitable gears. When designed for variable-speed operation, a turbine may be operated over a considerable speed range, which may be of advantage in many applications. Steam turbines range in output capacity from a few horsepower to over 1,500 MW. The largest ones are used for generator drive in central power stations.

Turbines are classified descriptively in various ways.

1. **By steam supply and exhaust conditions,** e.g., condensing, noncondensing, automatic extraction, mixed pressure (in which steam is supplied from more than one source at more than one pressure), regenerative extractions, reheat.
2. **By casing or shaft arrangement,** e.g., single casing, tandem compound (two or more casings with the shaft coupled together in line), cross-compound (two or more shafts not in line, often at different speeds).
3. **By number of exhaust stages in parallel,** as regards steam flow, e.g., two-flow, four-flow, six-flow.
4. **By details of stage design,** e.g., **impulse** or **reaction**.
5. **By direction of steam flow,** in the turbine, e.g., axial flow, radial flow, tangential flow. In this country (the United States of America), radial flow steam turbines have not been used; there are quite a few such machines abroad. Axial-flow units predominate; some small turbines in this country operate on the tangential-flow principle.
6. **Whether single-stage or multistage.** Small turbines, or those designed for small energy drop, may have only one stage; larger units are always multistage.
7. **By type of driven apparatus,** e.g., generator, mechanical, or ship drive.
8. **By nature of steam supply,** e.g., fossil-fuel-fired boiler, or light-water nuclear reactor.

Any particular turbine unit may be described under one or more of these classifications, e.g., a single-casing, condensing, regenerative extraction fossil unit, or a tandem-compound, three-casing, four-flow steam-reheat nuclear unit.

Turbine-Stage Design A turbine stage consists of a **stationary set of blades (S),** often called **nozzles,** and a moving set adjacent thereto, called **buckets,** or **rotor blades (M)**. These stationary and rotating blades act together to allow the steam flow to do work on the rotor, which can be transmitted to the **load** through the **shaft** on which the rotor assembly is carried. Classical turbine-stage design recognized two distinct designs of turbine stage, "impulse" and "reaction" (see classification 4 above). In the **impulse** stage, the total pressure drop for the stage is taken across the nozzles or stationary element, the flow through the buckets or rotor blades then being substantially at constant static pressure. This may be extended to include flow through an additional set of stationary "intermediate" blades and another row of buckets, or rotor blades (Curtis or two row stages). See velocity diagrams, Figs. 7.4.1 and 7.4.2.

In the **reaction** stage, the total pressure drop assigned to the stage is divided equally between the stationary blades and the rotor blades, giving rise to a velocity diagram, as shown in Fig. 7.4.3*a*. As can be

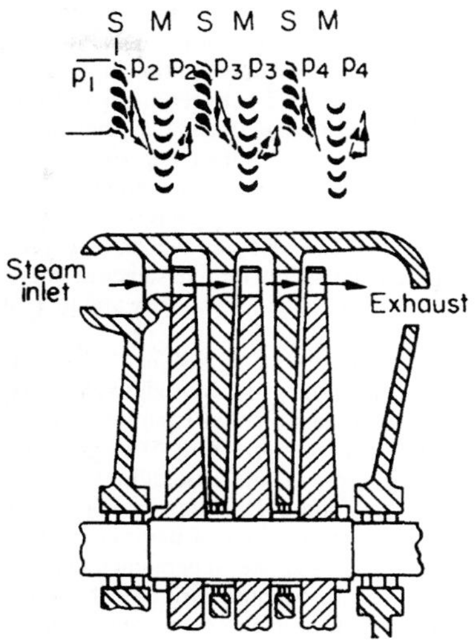

Figure 7.4.1 Impulse turbine with single-velocity stages.

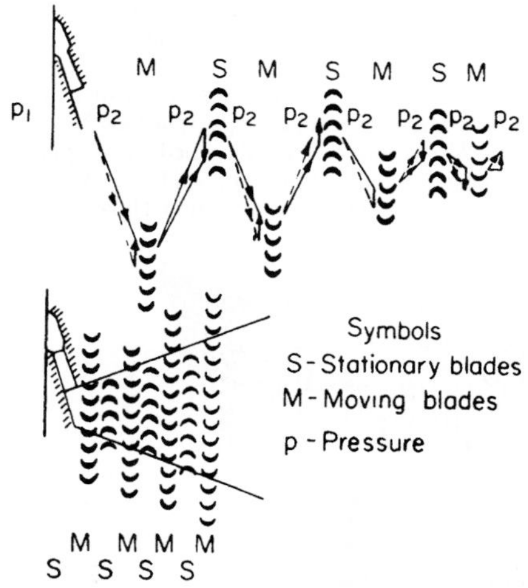

Figure 7.4.2 Impulse turbine with multivelocity stages.

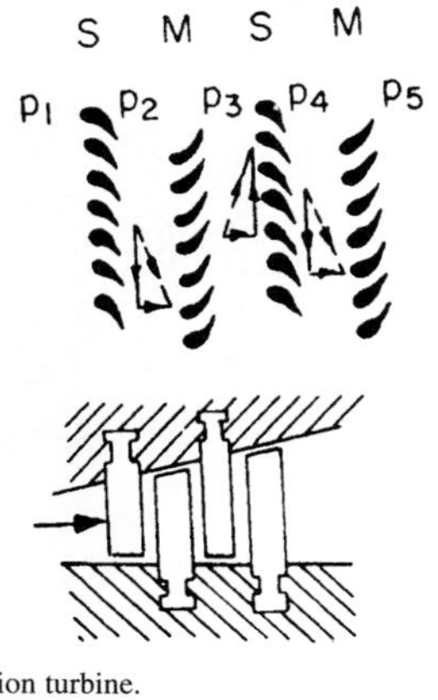

Figure 7.4.3*a* Reaction turbine.

seen, there arises a marked difference in the shapes of the rotor blades in the two classical designs; the impulse buckets do much more turning of the stream; the reaction-bucket shape is more nearly the same as the nozzle-blade shape.

Fluid flow theory recognizes that only in rare cases can an axial-flow turbine stage be either pure impulse or pure reaction. The annulus following the nozzle exit is filled with steam flowing with a high tangential velocity, i.e., a vortex, confined between inner and outer boundaries, and for equilibrium to exist, there must be a gradient in static pressure from a lower-than-average value at the inner boundary to a higher-than-average value at the outer boundary. The amount of this depends upon the boundary **radius ratio** R_{outer}/R_{inner}. Thus only for a radius ratio near 1.0 (small-height blades) can it be said that any one pressure condition exists for the stage. All axial-flow turbine stages of larger radius tend to be more nearly impulse at the inner diameter and more nearly reaction at the outer diameter.

The designers of multistage steam turbines have continued to refine the efficiency of their steam paths over the years. For example, the 1950s saw the broad introduction of **twisted free-vortex staging**. Vane profiles have been improved since the 1950s by drawing on developments made in their aerodynamic laboratories and those of others. Complex computational fluid-dynamics computer codes based on true three-dimensional formulations of the inviscid Euler and viscous Navier-Stokes equations, and the supercomputers on which to run them, are now common. Manufacturing technology advancement permits the creation of steam-path components of almost any desired three-dimensional configuration. The result has been the introduction of **controlled-vortex staging** in the 1990s, in which the flow is biased toward the more efficient midsection of the blade, and secondary flow and profile losses are minimized by refinement in nozzle and bucket profiles, as illustrated in Fig. 7.4.3*b*.

Differences in basic mechanical construction of axial-flow turbine stages exist. Generally, the reaction type of turbine has continued as in the past with a "drum rotor" (often designed for weight optimization) and stationary blades fixed in the casing, while the impulse type continues as a "diaphragm-and-wheel" construction. However, the mechanical construction bears no fixed relation to the degree of impulse or reaction adopted in the blading design. Designers employ the mechanical construction which they deem suitable for best reliability and efficiency.

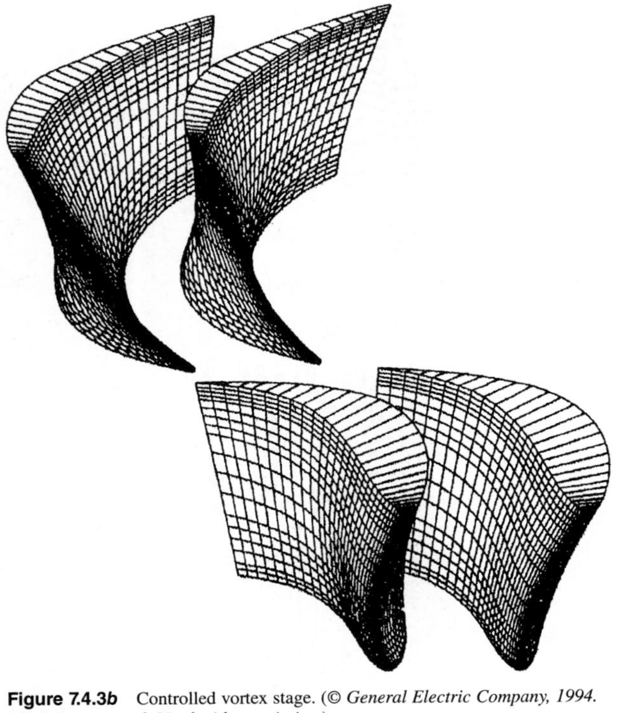

Figure 7.4.3*b* Controlled vortex stage. (© *General Electric Company, 1994. All rights reserved. Used with permission.*)

An impulse stage, when employed for the first expansion, permits nozzle group control, i.e., steam admission to each group, opening and closing, successively, in response to load changes. This improves the efficiency at low loads through the reduction of the loss due to throttling.

A greater enthalpy drop can be employed per stage with the impulse element, particularly in the case of multivelocity elements, thus reducing the number of stages in a turbine. This is of special importance in the first expansion if the nozzle chamber is not integral with the turbine casing, so that the casing is not subjected to the initial steam conditions. If turbine elements of equal blade speed could have equal efficiency, one 2 rows impulse element would equal four 1 row impulse elements or 16 rows (8 pairs) of reaction elements.

General Advantages of Steam Turbines Compared with reciprocating engines steam turbines require less floor space, lighter foundations, and less attendance; have a lower lubricating-oil consumption, with no internal lubrication, the exhaust steam being free from oil; have no reciprocating masses with their resulting vibrations; have uniform torque; have no rubbing parts excepting the bearings; have great over-load capacity, great reliability, low maintenance cost, and excellent regulation; are capable of operating with higher steam temperature and of expanding to lower exhaust pressure than the reciprocating steam engine. Their efficiencies may be as good as steam engines for small powers, and much better at large capacities. Single units can be built of greater capacity than can any other type of prime mover. Small turbines cost about the same as reciprocating engines; larger turbines cost much less than corresponding sizes of reciprocating engines, and they can be built in capacities never reached by reciprocating engines. Combustion-gas turbines possess many of the advantages of steam turbines but their applicability must be determined on a case-by-case basis. They are now available in ratings exceeding 400 MW for constant-speed 60- and 50-Hz service.

STEAM FLOW THROUGH NOZZLES AND BUCKETS IN IMPULSE TURBINES

Nozzles For the general treatment of the flow of steam and for maximum weight of flow of saturated steam, see Sec. 2.

The **theoretical work** obtainable from the expansion of 1 lb of steam is equal to the enthalpy drop in isentropic expansion $h_1 - hs_2$, in Btu/lb, and the spouting velocity in 223.8 $\sqrt{h - h_{s2}}$ *ft/s* (m/s = 44.7 $\sqrt{h_1' - h_{s2}'}$, where h' is in kJ/kg). The actual expansion is not isentropic but follows a path such as h_1h_2 on the enthalpy-entropy diagram (Fig. 7.4.4), and the available work becomes $h_1 - h_2$.

The **nozzle efficiency** is $(h_1 - h_2)/(h_1 - h_{s2})$.

The required **throat area** of the nozzle is $A_t = Wv_t/V_t$, and the **mouth area** is $A_m = Wv_m/V_m$, where v is specific volume, V is velocity, and the subscripts relate to throat and mouth, respectively.

In the case of a subcritical pressure ratio, throat and mouth conditions will coincide; with a supercritical pressure ratio, the mouth area will be larger than the throat area. For **nozzle velocity coefficients** based on

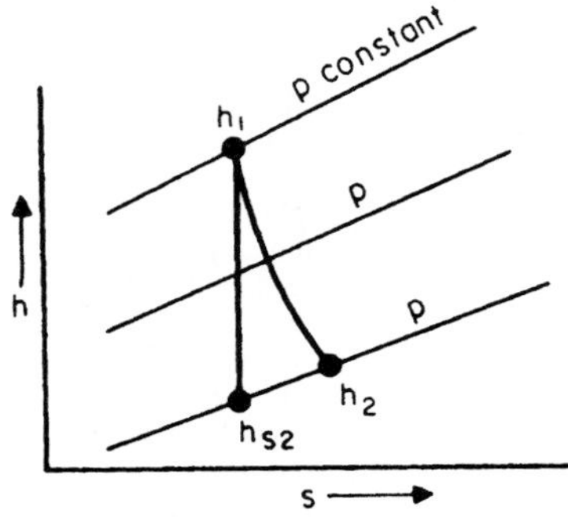

Figure 7.4.4 Enthalpy-entropy diagram.

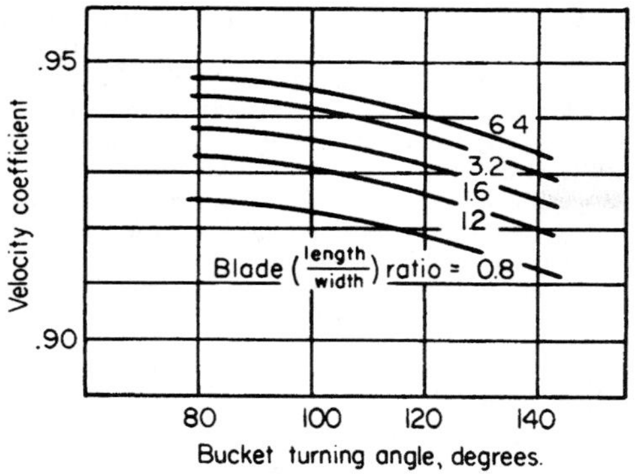

Figure 7.4.5 Impulse-bucket velocity coefficients.

tests, see Keenan and Kraft, *Trans. ASME,* **71,** 1949, pp. 773-787. The **bucket-velocity coefficient** is the ratio of the average exit velocity from the bucket divided by the velocity equivalent of the total energy available to the bucket, i.e., the sum of inlet-velocity energy and pressure-drop energy. Typical values of tests on impulse buckets are shown in Fig. 7.4.5, when the bucket inlet angle is 3 to 5° larger than the exit angle. These are for subsonic flow. Reaction buckets can have the same coefficients as nozzles. For intermediate cases between pure impulse and pure reaction, interpolate between the values for these limiting cases.

If, in the velocity diagram for a usual turbine stage, V_1 = actual nozzle exit velocity, V_2 = bucket relative entrance velocity, V_3 = bucket relative exit velocity, V_4 = absolute leaving velocity, V_0 = velocity corresponding to total stage available energy, then the **diagram efficiency** is $(V_1^2 - V_2^2 + V_3^2 - V_4^2)/V_0^2$.

Diagram efficiencies calculated with nozzle and bucket coefficients given above will be higher than the efficiencies derived by turbine tests, because of losses not existing in stationary tests of nozzles and buckets.

For supersonic bucket velocities, the impulse bucket-velocity coefficient is lower by the factors in Table 7.4.1.

Turbine Stage Efficiency Single-row stages of short blade length have relatively lower efficiency, owing to inner and outer sidewall losses; stages with longer blades are therefore higher in efficiency. Figure 7.4.6 shows typical values of stage efficiency for single-row stages, plotted against the wheel-speed steam-speed ratio, with pitch diameter, inches/nozzle area, square inches, as a parameter. These curves reflect the net total of losses: (1) friction losses in nozzles (stationary blades); (2) friction losses in buckets (rotor blades); (3) rotation loss of rotor; (4) leakage loss between inner circumference of stationary element and rotor; (5) leakage loss between tip of rotor blades and casing; (6) moisture and supersaturation losses, if steam is wet (not included in curves of Fig. 7.4.6).

Table 7.4.1 Impulse Bucket-Velocity Coefficient Factors

Mach no.	< 1.0	1.2	1.3	1.5	1.75	2.0
Factor	1.0	0.997	0.995	0.978	0.928	0.816

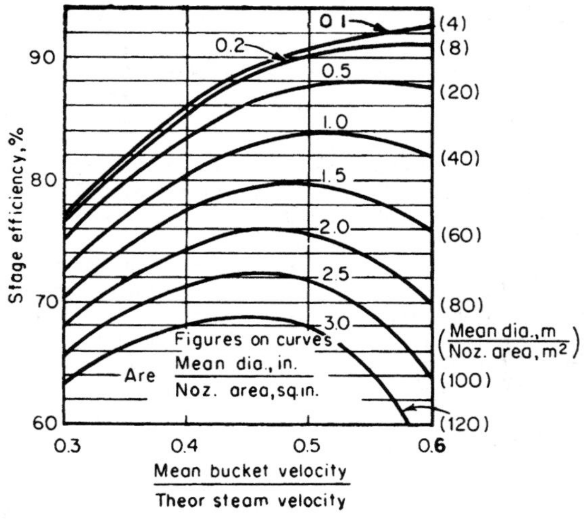

Figure 7.4.6 Turbine single-row stage efficiency.

Nozzles and bucket friction losses are minimized by good aerodynamic design and by increasing **aspect ratio** (blade length/steam passage width). Rotation losses depend upon disk or rotor dimensions and surrounding stationary parts. Exact values for rotation loss depend on several factors; the following formulas may be relied upon for all usual purposes:

$$L_d = 0.042D^2w(U/100)^{2.9} = (1.4 \times 10^{-12})(D')^2w'(U')^{2.9}$$
$$L_b = 0.187Dwh^{1.25}(U/100)^{2.9}$$
$$= 3.34 \times 10^{-11}D'w'(h')^{1.25}(U')^{2.9}$$

where L_d = rotation loss of disk carrying buckets, kW; L_b = rotation loss of one row of buckets, kW; U=wheel speed at pitch diameter, ft/s; U', m/s; D=pitch diameter (at center line of nozzle), ft; D', mm; w= density of steam, lb/ft³; w', kg/m³; h=mean bucket height, in; h', mm. L_b must be figured for each row of buckets. L_d plus the sum of the values of L_b gives the total rotation loss in kilowatts for dry saturated steam. The formula for L_b is approximate.

Leakage loss of steam between inner circumference of stationary element and rotor is minimized by maintaining minimum practical clearance and by use of labyrinth packings (see Figs. 7.4.14 and 7.4.15 and accompanying text). Leakage loss of steam between tip of rotor blades and casing is similar to that through labyrinths between shaft and stationary parts; the magnitude depends upon the clearance area and the amount of reaction; other things being equal, the larger the percentage of reaction, the larger the leakage (see Fig. 7.4.9). Thus designers often employ considerable amounts of reaction in stages with long blades where such losses are small; this improves the net efficiency. In stages with short blades, the best net efficiency obtains with near impulse design. The curves in Fig. 7.4.6 reflect the effects of these practices.

The presence of moisture in the steam causes extra losses. These are probably mainly due to three factors:

1. Effect of supersaturation; i.e., the steam in expanding rapidly does not remain in equilibrium but tends to be more or less supercooled; thus less than theoretical equilibrium energy is available.

2. The presence of water drops increases friction losses in the steam itself.

3. Water drops tend to move more slowly than the vapor; they strike the rotor blades at unfavorable velocities and exert a braking effect.

Figure 7.4.7 gives correction factors which may be applied to the values from Fig. 7.4.6 to arrive at stage efficiencies in the "wet" region. Curves are identified by initial superheat, °F, or by initial quality, percent.

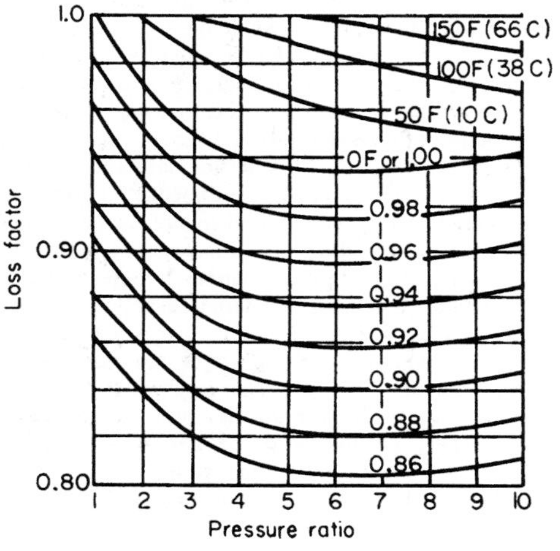

Figure 7.4.7 Supersaturation and moisture loss.

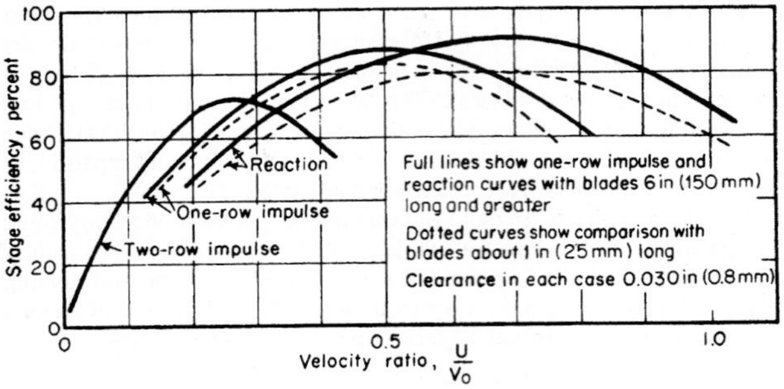

Figure 7.4.8 Variation of the efficiency of turbine elements with velocity ratio.

Two-row stages, with one set of nozzles, have lower basic efficiency than single-row stages; they are useful for the first, or governing, stage in small to medium units. They are no longer employed in large central station designs. The approximate relative efficiency level of these stages is shown in Fig. 7.4.8.

The losses occurring in a turbine stage are partially recoverable in succeeding stages in a multistage turbine because the energy available to succeeding stages is increased above that resulting from isentropic expansion. The amount of this factor depends upon the temperature at which the loss takes place, and the turbine exhaust temperature. Values for the reheat or energy regain factor are shown in Fig. 7.4.9.

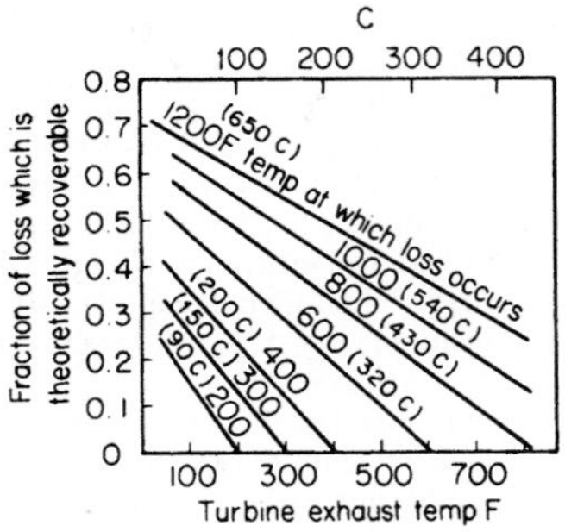

Figure 7.4.9 Energy regain chart.

Turbine Steam-Flow Requirements The steam flow required by a turbine is related to its power output and steam conditions by the following expressions:

$$\text{Flow, lb/h} = (\text{TSR/efficiency}) \times \text{power output, hp or kW}$$

where TSR = theoretical steam rate, lb/(hp·h) = 2,545/available energy, Btu/lb, or TSR, lb/kWh = 3,412.14/available energy, Btu/lb.

Values of TSR are given in tables or may be calculated. In case of mixed-pressure or extraction turbines, the various sections of the turbine where flows are added or subtracted must be treated separately.

Turbine Steam-Path Design The basic quantities required for this are the steam conditions (i.e., inlet pressure and temperature and exhaust pressure), the required flow (see above), and the turbine speed. The latter is often fixed by the requirements of the driven machine; if not, the choice of speed by the designer is based upon experience, and factors such as space or weight limitations, efficiency requirements, stress limits, or required exhaust area. Often several preliminary layouts are needed to arrive at the best design. If it appears that a single stage will suffice, the problem is simple, since the entire available energy is allotted to the one stage. If a multistage machine is required, the total available energy must be divided properly between the various stages.

For a given wheel-speed/steam-speed ratio, the energy to be allotted to the stage is directly proportional to the square of the product (r/min × D), D being the rotor mean diameter. If available energy AE is in Btu/lb, D, in, and velocity ratio, 0.50, then

$$AE = (\text{r/min} \times D)^2/(6.57 \times 10^8)$$

The sum of the AE values per stage for all the stages must equal the total energy on the turbine, which is greater than the isentropic available energy by the amount of the reheat factor.

Having determined the energy to be assigned to each stage, the steam pressure in each stage is fixed, and this, together with the enthalpy determined from the efficiency of the machine from inlet, determines the steam specific volume. The velocity ratio and the degree of reaction decided upon fix the velocity diagram, from which the velocities through the stationary and rotating blades are determined. With the flow Q, lb/h, the velocity V, ft/s, and the volume v, ft^3/lb, determined, theoretical areas required are given by

$$A = Qv/(25V) \quad \text{in}^2$$

The actual area required will be larger than theoretical because of friction losses; a value of 0.98 for nozzle flow coefficient is reasonable.

There is no fixed criterion for the number of stages to be used in a steam turbine design, and experienced designers differ in the number they will choose for any particular design. The stage velocity ratio, the mean diameter, and often the r/min are subject to judgment, bearing in mind the general relationships shown in Figs. 7.4.6, 7.4.7, and 7.4.9. Cross-compound arrangements are possible, allowing higher speed for the high-pressure section, where steam volume flow is smaller, and a lower speed for the low-pressure section, where the final exhaust area desired may require long blades on a large diameter.

A detailed calculation of the efficiencies and energy outputs of each stage can be summed up to the total "internal used energy" of the turbine, which, when divided by the isentropic available energy, results in an **"internal efficiency."** Then, account must be taken of other losses to arrive at the turbine overall efficiency. These losses are

1. Exhaust loss, i.e., the kinetic energy corresponding to the absolute velocity of the steam leaving the last stage, plus the pressure drop through the exhaust connection to the turbine outlet flange, where exhaust pressure is by custom measured as a static pressure
2. Pressure drops through interconnecting piping if turbine has more than one casing
3. Shaft end-packing leakages
4. Valve-stem leakages, if any
5. Inlet-valve and intermediate-valve pressure-drop losses
6. Bearing, oil-pump, and coupling power losses
7. If the turbine drives a generator or gear, the losses of these elements

LOW-PRESSURE ELEMENTS OF TURBINES

In condensing turbines expanding to high vacuum, the steam increases in specific volume as it passes through the stages, exhausting at about 1,000 times the inlet volume. The increase in specific volume from stage to stage is greatest in the latter few stages, which are commonly designed for pressure ratios of about 2:1. Considerations of efficiency and economics dictate providing reasonably low interstage steam velocity in these stages, and reasonable leaving loss from the last-stage blades. These requirements are satisfied by providing a steam path whose cross-sectional area increases in proportion to the specific volume. The area increase may be achieved by increasing blading length, by increasing the mean diameter of the steam path, by arranging the last expansions in multiple flow, or by some combination of two or more. The relatively small single casing unit of Fig. 7.4.10 is provided with increasing area by the first two techniques.

The ratio of the last-stage blade height to the mean diameter of the steam path may be as high as 0.35 in the last stage of large central station units. With such a ratio, there is a variation in blade velocity between the inner and the outer radii of the steam path that cannot be properly satisfied by any one steam velocity. Losses incidental to this may be minimized or eliminated by providing blades with warped surfaces, the sections at the inner radius partaking of impulse form with relatively small inlet angles, and those at the outer radius of reaction form with large inlet angles. The longest last-stage blades in service today have tips whose speed exceeds sonic velocity. Many are of transonic design employing subsonic profiles toward the inner radius, blending into supersonic profiles at the tip.

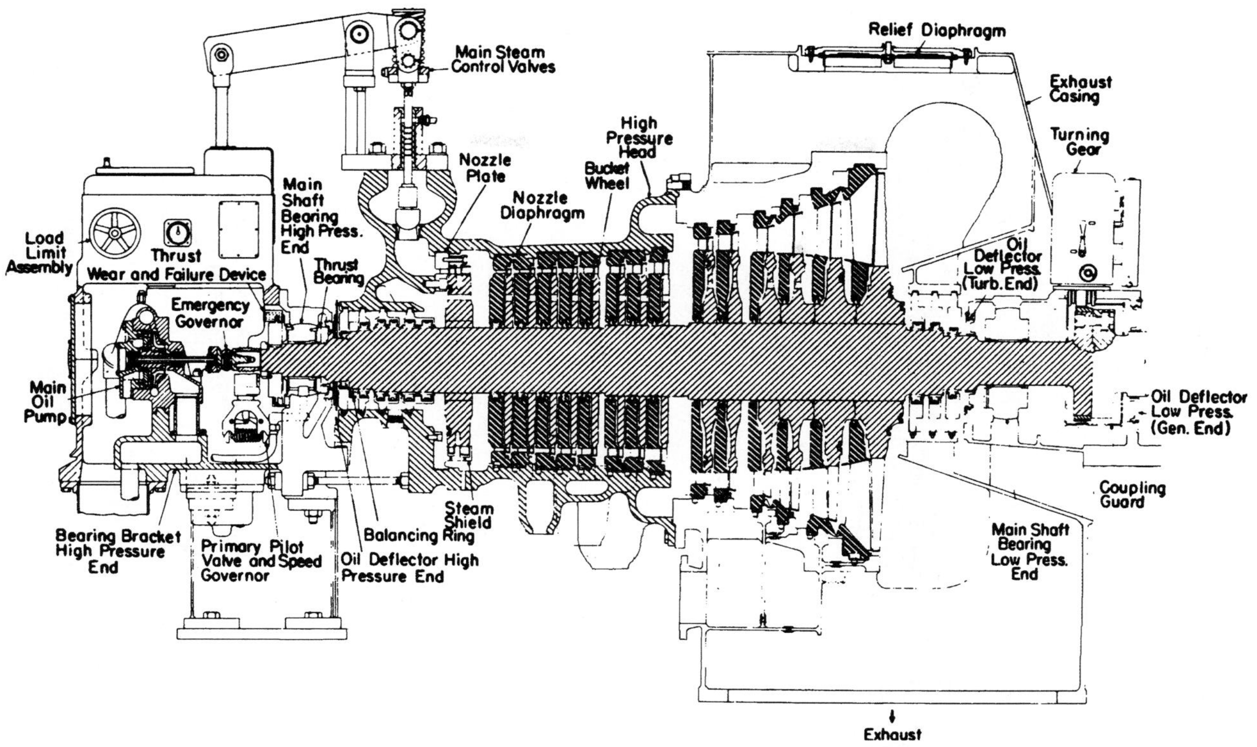

Figure 7.4.10 Cross section of a multistage impulse condensing turbine rated at 30,000 kW. (*General Electric Co.*)

Among the methods of obtaining increased low-pressure blade areas through multiple flowing are:

1. A single-casing turbine with the last elements arranged fordouble flow.

2. The steam expansion divided between two casings, coupled in tandem, and driving a single main generator, the low-pressure casing being arranged for double flow. This arrangement is commonly extended to provide tandem-compound turbines with four and six exhaust ends. Figure 7.4.11 illustrates the application of a double-flow low-pressure casing to a unit of medium rating.

3. Cross-compound turbines, in which the steam expansion is divided between two or more separate casings driving separate generators, electrically synchronized. The low-pressure casings are usually double-flow and can be arranged so as to provide two, four, or six exhaust ends. This system permits the turbine elements to be operated at different speeds, selected as appropriate to the respective steam volumes. It lends itself to geared applications, such as marine propulsion machinery, when two or more pinions of different diameters and speeds drive a single-output gear wheel.

4. Divided-flow turbines in which steam expands in a series of elements in a single flow to a point where the flow is divided, with, perhaps, one-third continuing expansion within the same casing to condenser pressure, the remaining two-thirds expanding to condenser pressure in a separate double-flow casing. This construction has been used to provide triple-flow exhaust ends.

The **leaving loss** at the exit from the last row of blades is $V_{c2}^2/50{,}100$ Btu/lb of steam, where V_{c2} is the absolute terminal velocity in feet per second [or $(V'_{c2})^2/2{,}000$ kJ/kg, where V'_{c2} is in m/s]. The presence of moisture in the steam causes **moisture loss.** The acceleration of moisture particles is less as the density of the steam decreases; hence the difference between the velocities of the particles and the steam increases. As indicated in Fig. 7.4.12, with steam velocity, V_{s1} and moisture velocity V_m leaving the stationary blades and with bucket velocity V_b, the velocities relative to the moving blades of the steam are V_{c1} and of the moisture V_{m1}. The component V_{m2} of the moisture relative to the moving blades is opposite to the direction of their motion and is proportional to the force acting on the back of the blades, needed to accelerate the water to blade speed, and results in negative work. This negative work can be calculated when the weight of moisture per pound of steam and V_m are known. The results of many tests indicate that the *efficiency of a stage is reduced about 1 percent for each 1 percent moisture present in the steam.*

The presence of moisture particles will result in erosion of the blades along their inlet edges if V_{m1} is too large—unless the moisture content is very small. The rate of erosion is reduced by using materials of inherently high erosion resistance, or protective shielding of hardened materials along the inlet edge. Attached Stellite shields and thermally hardened edges are commonly employed. Of course, many different protective materials and processes are available. All are carefully chosen to minimize the possibility of corrosion.

Experience with the usual 12-chrome alloy bucket steels indicates that a threshold velocity exists at about V_b = 900 ft/s (270 m/s), below which impact erosion does not normally occur. With alloys specifically selected for their erosion resistance and with Stellite shields, satisfactory service has been obtained with values of V_b in excess of 1,900 ft/s (580 m/s). Centrifugal stress limits the application of steel blades to a tip velocity of about 2,000 ft/s (610 m/s). Titanium alloys provide high strength, low density (and hence low centrifugal stress), combined with excellent inherent moisture erosion resistance. Titanium bucket designs are available with a tip speed of 2,200 ft/s (670 m/s), without the need for separate erosion protection shielding. The severity of erosion penetration is dependent upon the thermodynamic properties of the stage and the effectiveness of reducing moisture content by means of interstage collection and drainage as well.

Rotative Speed The speed selected greatly influences weight and cost. With two geometrically similar turbines, one having twice the linear dimensions of the other, the steam path areas of the larger, and hence its capacity, would be 4 times and its weight 8 times as great as those of the

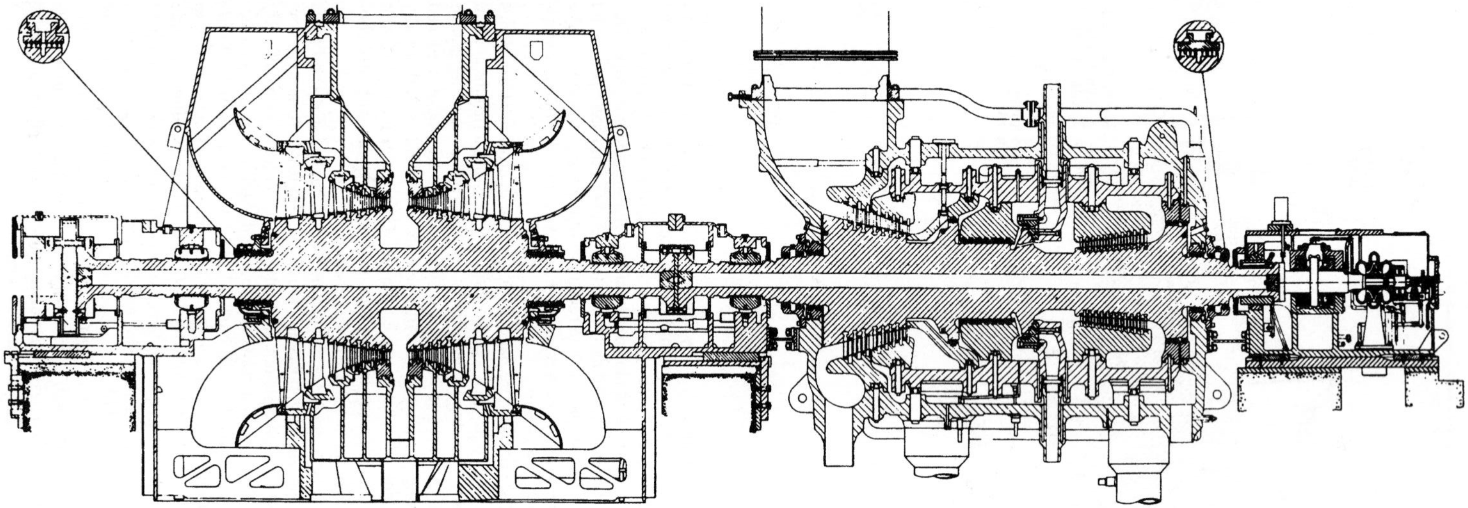

Figure 7.4.11 Cross section of a 160,000-kW tandem-compound, double-flow reheat turbine. (*Westinghouse Electric Corp.*)

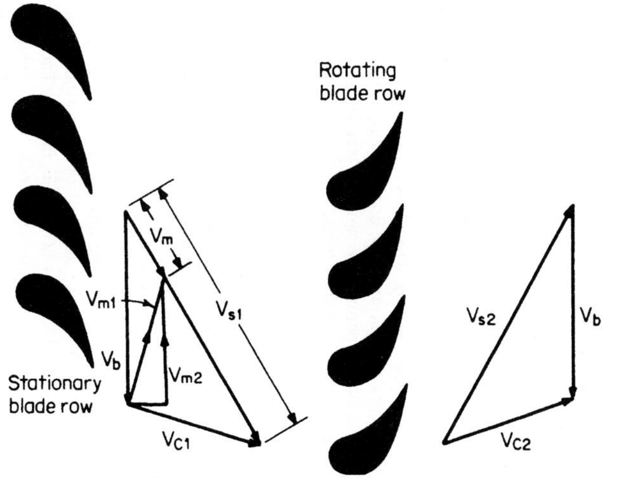

Figure 7.4.12 Velocity of steam and of the moisture in steam.

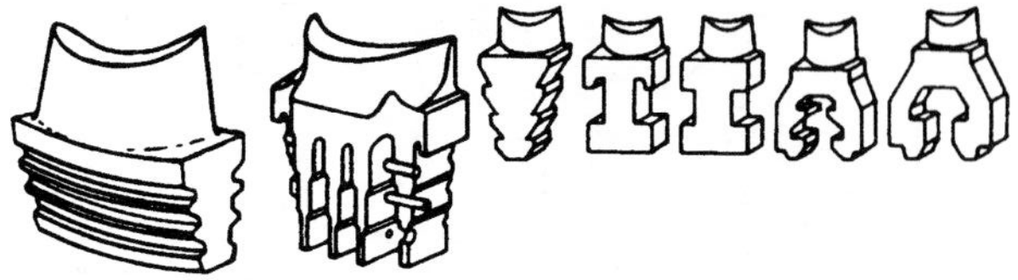

Figure 7.4.13 Steam-turbine blade fastenings, commonly called blade roots.

smaller. The weight per unit of capacity with similar machines increases inversely as the speed with strict geometrical similarity. For this reason, the highest possible low-pressure blade speeds and r/min are selected. Machines of different speeds are not usually made strictly geometrically similar, and the reduction of specific weight is not as rapid as the above rule would indicate. With large-capacity turbines, blade speeds, at the outer radius, can reach 2,200 ft/s (670 m/s). With high speeds and small dimensions, the turbine can operate with higher steam temperature and greater temperature fluctuations because of lighter casing walls and less mass of rotor; the amount of distortion is less with more uniform heating; the turbine can be heated and put in service more quickly; space requirements are less; and dynamic loadings on foundations are reduced.

Balancing (See also Secs. 1 and 3). With a rotor, or a component of a rotor, of relatively short axial length (such as a disk), static balancing may suffice. Single bodies of more than half the diameter in axial length are usually dynamically balanced by the use of balancing machines. The balancing may have to be done at less than maximum running speed since a bladed rotor cannot be rotated in air at a speed approaching the running speed. Balancing is often done at high speed in an evacuated spin facility, or with a combination of both. Balance at full speed may not be satisfactory unless the balance weights are applied at points diametrically opposite the errors in balance, so that balance corrections must frequently be provided along the length of the rotor. Five balance planes are commonly used for the rotors of large central-station turbines, of which three are in the rotor body between bearings and two are in the overhung couplings.

Unsatisfactory operation of turbine rotors in service, resembling a simple imbalance, may be caused by non-uniform material density or non-uniform heating of the rotor. The latter may be caused by permitting a rotor to remain stationary in a hot casing, or by packing rubbing which may apply frictional heating to the "high" side of the rotor, leading to further bowing into the rub. Care must be taken to see that non-uniformity of material which could cause rotor distortion with heat is avoided, and to avoid rubbing of the packings on the rotor. Turning gears are generally used to keep the rotor turning at low speed to maintain uniform temperature when the turbine is shut down and cooling, and for a prolonged period before starting.

TURBINE BUCKETS, BLADING, AND PARTS

Figure 7.4.13 shows various blade fastenings ("roots"). Blades are subject to vibration and possible fatigue fracture if their natural frequency is resonant with some applied vibration force. Forced vibrations may arise from the following causes (see also Sec. 3):

1. Variations in steam forces. The blade frequencies should not be even multiples of the running speed, nor should they be resonant with the frequency of passing nozzle partitions or exhaust-hood struts.
2. Shock, the result of blades being subjected to discontinuous steam flow, such as may be caused by incomplete peripheral steam admission or extraction.
3. Torsional vibrations of the rotor.

High-speed, low-pressure blades of condensing turbines are usually of tapering section and have a warped surface in order to provide appropriate blade angles throughout their length. Long blades of this type frequently have their natural frequency between 3 and 4 times the running speed or even lower. Such blades should always be specially tuned to have a margin in frequency away from running-speed stimuli.

Margins from running speed to assure freedom from fatigue due to resonant vibration and transverse to the plane of the wheel are as follows:

Frequency, cycles per revolution	2	3	4	5
Margin between critical and running speeds, percent:				
Within wheel plane (tangential)	15	10	5	5
Transverse to wheel plane (axial)	20	10	10	5

Higher-frequency buckets whose frequencies cannot assuredly be made nonresonant should be designed with adequate strength to resist such stimuli as may occur under service conditions.

Blade Materials The material in most general use is a low-carbon stainless steel of the following composition: Cr, 12 to 14 percent; C, 0.10 to 0.12; Mn, 0.08 max; P, 0.03 max; S, 0.05 max; Si, 0.25 max. Its physical characteristics in the heat-treated condition at room temperature may be tensile strength, 100,000 lb/in^2 (690 MPa); yield point, 80,000 lb/in^2 (550 MPa); elongation, 21 percent; reduction of area, 60 percent. For the higher-temperature blades, particularly on large machines, it is practice to use alloyed chrome steel to achieve the required strength and oxidation-erosion resistance (see also Sec. 4).

Rotor Materials Since steam turbines operate at high speeds, rotor materials must be of very high integrity and of basically high strength. In addition, the material should be "tough" at the temperatures at which it is to be highly stressed. A measure of this toughness may be obtained by running Charpy notch impact tests at various temperatures (see Sec. 3). In large modern machines the rotor forgings are almost exclusively made of steel melted in basic electric furnaces and vacuum poured to achieve freedom from internal defects. Turbine rotor forgings are usually made with small amounts of alloying elements such as Ni, Cr, V, or Mo.

Casing and Bolting Materials High-temperature and -pressure casings are almost always made of castings in order to achieve the complicated shapes required by these components. The alloy compositions used are selected so as to provide good weldability and castability as well as good physical properties. Low-temperature and -pressure casings are usually fabricated from steel plate. Bolts are made of forged or rolled materials.

The practical use of higher steam pressures and temperatures is limited by the strength and cost of available materials.

Leakage Metallic labyrinth packings are employed to (1) reduce internal steam leakage from stage to stage, (2) prevent steam from escaping the turbine from elevated-pressure ends, and (3) prevent air from leaking into the turbine at subatmospheric-pressure shaft ends. Interstage packings usually employ single rings with multiple teeth. End packings are arranged in multiple rings. At the high-pressure end, the leakage of steam past the first few rings may be carried to a lower-pressure stage of the turbine so that the outer rings need only prevent the leakage of low-pressure steam to atmosphere. The annulus above the last ring, or group of rings, is connected to a packing exhauster which maintains a pressure slightly below atmospheric. In consequence, no steam leaks out along the shaft, while a small amount of air is drawn past the final packing to the exhauster. Vacuum packings are provided with an inner annulus which is supplied with steam above atmospheric pressure. Steam flows inward from that annulus to supply the leakage toward the vacuum end, and outward toward a packing-exhauster connection. Thus steam is prevented from escaping along the shaft, while air is prevented from being drawn into the turbine.

Some turbines use carbon end packings or water seals. Carbon packings consist of one or more rings of pure carbon made in sections of 90 or 120° and held toward the shaft with small clearances by means of springs. The springs should have an axial component of force to hold the rings against the side of the box.

Labyrinths dependent upon radial clearances are shown in Fig. 7.4.14. In the design with "high and low" teeth, heavy teeth are cut on the turbine shaft, and thin teeth are part of a renewable packing ring which is made in segments backed and held inward by flat springs. The key indicated in the figure prevents turning of the segments. These types require that the rotor remain sensibly concentric with the stator but do not require a close axial adjustment.

Labyrinths dependent upon axial clearances are shown in Fig. 7.4.15*a*. These require the maintenance of a close axial adjustment of the rotor.

The flow through a labyrinth may be approximately determined by the formula

$$W = 25KA\sqrt{\frac{(P_1/V_1)[1-(P_2/P_1)^2]}{N-\ln(P_2/P_1)}}$$

where W = mass flow of steam, lb/h; K = experimentally determined coefficient; A = area through packing clearance space, in^2; P_1 = initial pressure, lb/in^2 abs; V_1 = initial specific volume of steam, ft^3/lb; P_2 = final pressure, lb/in^2 abs; N = number of throttlings.

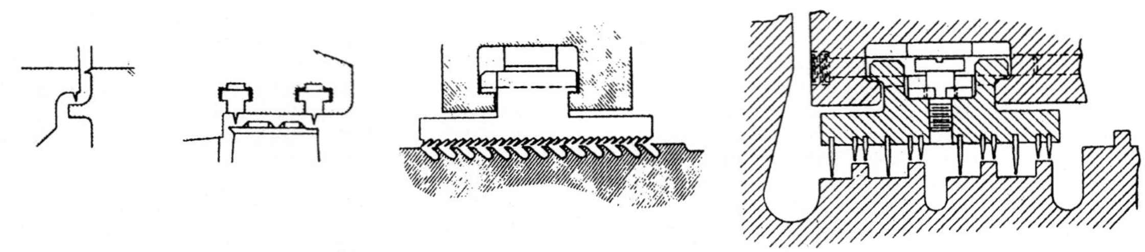

Figure 7.4.14 Labyrinths with radial clearance.

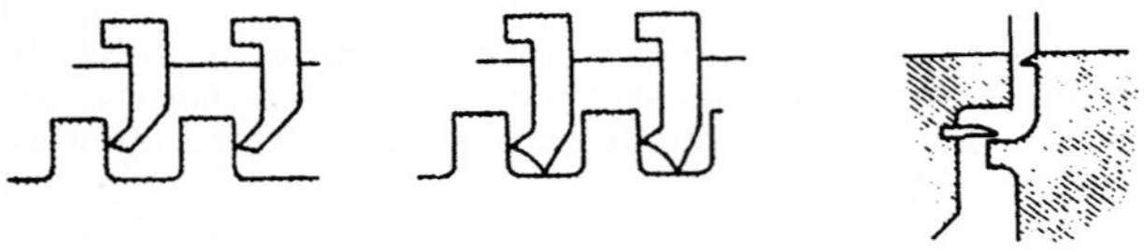

Figure 7.4.15*a* Labyrinths with axial clearance.

The value of K for interlocking labyrinths where the flow velocity is effectively destroyed between throttlings is approximately 50 and is independent of clearance for usual clearance values.

For non-interlocking labyrinths, i.e., stationary teeth against a straight cylindrical shaft, the value of K varies with the ratio of tooth spacing to radial clearance, being about 120 for tooth spacing of five times the radial clearance, reducing to approximately 50 for a tooth spacing fifty times the clearance.

Brush Seals The 1990s and 2000s have seen the application of brush seals to steam turbines. They consist of a multiplicity of fine wires (bristles) inclined at an angle of approximately 45° to the surface of the shaft or to the bucket covers. The use of resilient bristles permits closer clearances than do the rigid teeth of a labyrinth seal, and the resulting leakage flow is typically one-third that of the labyrinth. Figure 7.4.15*b* shows brush seals incorporated into the bodies of two labyrinth packing rings in series, as is frequently retrofitted to existing high-pressure units. Flow and pressure drop from left to right. The bristles are shown supported by a backing plate on the downstream side and protected by a forward plate on the upstream side. Both plates and the bristles are welded into segments for the same arc length as the labyrinth segments.

Turning Gears Large turbines are equipped with turning gears to rotate the rotors slowly during warming up, cooling off, and particularly during shutdown periods of several days when it may be necessary to start the turbine again on short notice. The objective is to maintain the shaft or rotor at an approximately uniform circumferential temperature, so as to maintain straightness and preserve the balance. Turning gears permit an appreciable reduction in starting time, particularly following a relatively brief shutdown.

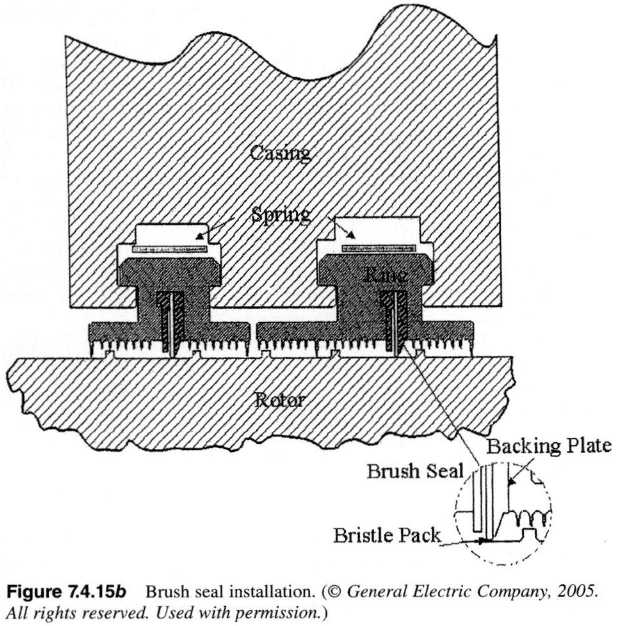

Figure 7.4.15*b* Brush seal installation. (© *General Electric Company, 2005. All rights reserved. Used with permission.*)

It is seldom necessary to use high-pressure oil to lift the journals off their bearings when using a turning gear. A low-pressure motor-driven oil pump is used which floods the bearings with about half their usual flow of oil. The turning gear is made powerful enough to start the rotor and rotate it at low speed.

High-Temperature Bolting The bolting of high-pressure, high-temperature joints, particularly turbine-shell or valve-bonnet joints is very exacting. It is worthwhile to taper the threads of either the male or the female element so that the engagement of the threads throughout the length of the engaged thread portion will give approximately uniform loading throughout. The reliability of taper-threaded bolts is superior to that of parallel-threaded bolts.

Thrust Bearings They must be designed to carry axial rotor thrust in either direction, with sufficient margin to take care of unusual operating conditions. Thrust runners may be machined solid on the shaft or can be a separate piece shrunk on and secured from endwise motion. The stationary bearing surfaces may be of the pivoted-shoe type, or made in solid plates with babbitt or other bearing-material facing, with grooves for oil supply and "lands" to carry the thrust load.

Axial thrust on the turbine rotor is caused by pressure and velocity differences across the rotor blades, pressure differences from one side to the other on wheels or rotor bodies, and pressure differences across shaft labyrinths which have steps in diameter. The net thrust is the sum of all these effects, some of which may be in one direction and some in the opposite direction. Rotor-blade and wheel or body thrust are usually in the direction of steam flow. It is customary to balance this thrust either partially or completely by proper choice of shaft packing diameters and pressure differences so that the net thrust is not too large. The thrust bearing must be made large enough so that it is not overloaded by the net thrust. In this respect it is necessary to foresee all operating conditions, including moderate wear, which may influence the net thrust. Thrust bearing selection must allow for these; in addition some margin must be allowed for abnormal or unforeseen circumstances which may occur in service. (See also Sec. 6.)

Controls Steam turbines are nearly always equipped with speed-control governors, and with separate overspeed governors. The only exceptions are special cases where it is judged that the possibility of overspeed due to loss of load is exceedingly remote. The speed-control governor may be arranged for a wide range of speed setting in the case of a variable-speed turbine. The steam flow-control valve or valves are operated by this governor, usually through a hydraulic relay mechanism. The overspeed governor is usually of an overisochronous type, arranged to trip at 10 percent over normal full speed (on some small turbines, 15 percent), actuating quick-closing stop valves to shut off the steam supply to the turbine. Speed-control governing systems are usually designed so that the overspeed-governing system is not brought into action on sudden loss of full load.

On automatic extraction machines, the speed governor must be correlated with the extraction-pressure controlling-valve system.

On fossil-reheat turbines, because of the large stored steam volume in the reheater and piping, and on nuclear turbines, because of the moisture separator/steam reheater and piping volumes, it is necessary to protect the turbine from overspeed on sudden loss of load by shutting off this stored steam ahead of the lower-pressure stages. This is done by intermediate intercept valves, actuated by a governor set slightly higher than the speed-control governor. An intermediate stop valve actuated by the overspeed trip is usually added in series for additional protection.

Speed-governing systems can be supplied in various sensitivities and speed ranges, selected to suit the requirements of the driven apparatus. Both mechanical-hydraulic and electrohydraulic systems are employed. Analog electrohydraulic systems were introduced during the 1960s when electronic components of sufficient reliability for turbine service became available. Digital systems were introduced in the 1980s and have become the standard control technology. Modern systems control transient thermal stress and the expenditure of low-cycle fatigue life of high-temperature components, in addition to controlling speed and load. (See "Steam Temperature—Starting and Loading.")

Steam turbines are usually provided with a supervisory instrumentation system. For a large central station unit such systems may process several hundred channels of information, providing alarms and/or trips for abnormal parameters. Data may be stored on paper charts, on magnetic media, or in computer memory depending upon the design of the system. Displays frequently employ multicolor screens.

INDUSTRIAL AND AUXILIARY TURBINES

Low-capacity turbines are employed for services such as engine-room auxiliaries and small generating sets. Usually they comprise a single turbine element. Their efficiency may be less than that of a corresponding reciprocating engine, but they are employed because of their compactness and because they require no internal lubrication. The exhaust steam is free from oil and grease and is available for heating purposes. These turbines are frequently coupled to the driven machine by means of speed-reducing gears. Turbines of this type are usually of the axial-flow type, but the **tangential helical-flow** turbine is also used. Here, the steam is directed tangentially and radially inward by nozzles against buckets milled in the wheel rim and made to flow in a helical path reentering the buckets one or more times. Such machines have generally been limited to small single-stage designs. They are usually very simple and rugged; some models incorporate speed-up or speed reduction gears to accommodate the needs of the driven machine.

In **back-pressure turbines,** the exhaust steam is employed for some heating process and the turbine work may be a by-product. If all the exhaust steam is condensed in heat-absorbing apparatus and returned to the system, the thermal efficiency of the system may be over 90 percent. One application is the **superposition** of a high-pressure system on lower-pressure power units, with the exhaust from the high-pressure turbine power units going to the low-pressure steam mains. By this device, an old power station can be rehabilitated and its capacity increased. Two methods of operation are in use: (1) with constant intermediate pressure as when the lower-pressure power units operate also with steam from existing lower-pressure boilers and (2) with variable intermediate pressure as when the low-pressure units receive steam only from the back-pressure turbine.

Boiler-feed-pump-drive turbines have been used extensively as part of the power-plant system, especially for large, high-pressure plants where the required feed-pump power may amount to 4 percent of the gross plant output, and for large nuclear units. The turbine and pump can be matched as to rotative speed. These turbines are variously integrated into the main cycle. The most common present practice is to use condensing, nonextracting turbines supplied with steam in the range of 150 to 200 lb/in^2 abs (1,000 to 1,400 kPa) taken from the exhaust of the intermediate sections of fossil turbines, or from the inlet of the low-pressure sections of nuclear units. These turbines normally have a connection to the main steam supply for starting and low-load operation. Similar auxiliary turbines are frequently used to drive the forced- or induced-draft fans of large fossil-fuel-fired boilers.

With **extraction turbines,** partly expanded steam is extracted for external process use at one or more points. The turbines may be either condensing or noncondensing. Extraction turbines are usually designed to sustain full rated output, with or without extraction, and are provided with automatic regulating mechanisms to deliver steam from the extraction points at constant pressure, as long as there is sufficient power load to permit the necessary flow. The use of such extraction turbines, particularly with high initial pressures in connection with many industrial processes requiring moderate- or low-pressure steam, results frequently in a high efficiency of power production. In some cases the only heat required in such a plant over and above that to provide the required process steam is the heat equivalent of the power generated by the steam before extraction. This means that such power can be produced at nearly 100 percent thermal efficiency.

Figure 7.4.16 illustrates a typical double-automatic, condensing extraction turbine, providing two controlled extraction pressures. In this case, the unit is equipped with internal spool valves at both extraction points. Grid and poppet-type valves have been used for this purpose. The extraction-stage valves are under the control of an extraction-pressure-actuated governor; they determine the flow to the subsequent stages of the turbine and maintain the pressure in the extraction stage. The operation of the valves is by means of a pilot valve controlling the admission of high-pressure fluid to an actuating cylinder which, in turn, opens or closes the valves to the nozzle ports to the succeeding stage. Extractions of this kind are called pressure-controlled extractions, and the pressure is maintained practically constant over a wide range if the load is sufficient to permit the required steam flow.

Mechanical-drive turbines are commonly applied where moderate to high power and/or precise speed control of the driven machine are needed. Typical applications include the powering of papermaking machines and the driving of process gas compressors in petrochemical plants. So many sizes and types are available from the manufacturers, and they have been adapted to so many applications, that it is impossible to give here more than a general description. These turbines are commonly built in sizes from a few hundred to well over 100,000 hp. If the speed of the driven machine is low, a reduction gear may be used in order to reduce the size and cost of the driving turbine and to improve its efficiency. Mechanical-drive turbines have a wide range of application, being adaptable for any steam conditions and a very wide range of speeds. They can be equipped with speed governors suited to the requirements, i.e., of very good stability and accuracy, if this is desirable, and arranged for infinitely variable speed settings over a relatively wide range of process-dependent operating speeds.

Main Propulsion Marine Turbines (See also Sec. 8.3.) Steam turbines were commonly employed for the propulsion of naval and merchant ships into the 1970s. The availability of aero-derivative gas turbines of light weight, high capacity, and acceptable efficiency has led to their use for the propulsion of oil-fired naval vessels. The rapid rise in the price of fuel oil in the 1970s led merchant ship operators away from steam propulsion to the use of the more efficient diesel engine. In later years the situation again changed with fossil fuel price fluctuations and also with legislation relating to exhaust emissions. Steam turbines continue to be used for the propulsion of all nuclear-powered vessels. Marine turbines are basically the same as central-station or industrial turbines except that usually the turbine is divided into a high-pressure and a low-pressure element, each geared through a common low-speed gear to the propeller shaft. The advantages of this compound arrangement are that two high-speed pinions divide the load on a common low-speed gear, thus reducing gear weight when compared with a single turbine. The high-pressure turbine can be made higher-speed than the low-pressure turbine and so be better adapted to the low volume flow. Subject to coupling deflection limitations, each turbine can have a relatively short rugged shaft, and either turbine can be used to propel the ship in an emergency.

Geared marine turbines require a reversing element for operating the vessel astern. This is typically a two-stage impulse turbine with two 2-row, or one 2-row and one 1 row, velocity stages arranged in the exhaust space of the low-pressure ahead turbine, so as to operate under vacuum under normal ahead conditions. The rotation loss of such an astern turbine is about ½ percent under normal ahead conditions.

Being directly geared to the propeller shaft, marine turbines must work at variable speeds. Overspeed governors are not required but are sometimes applied as a precautionary measure. Control is effected in most cases by sequentially operated nozzle valves. A typical marine turbine rated 16,000 hp is shown in Fig. 7.4.17. Ratings of 70,000 hp have been built, and larger sizes are realistic.

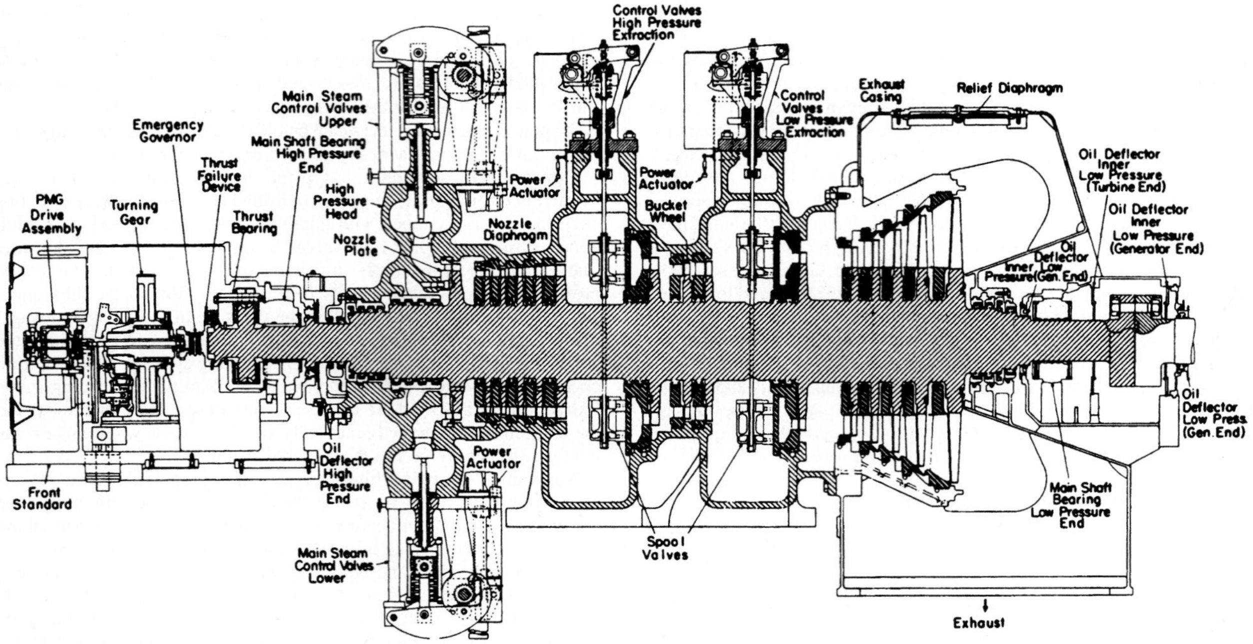

Figure 7.4.16 Double automatic condensing extracting turbine, 25,000 kW. (*General Electric Co.*)

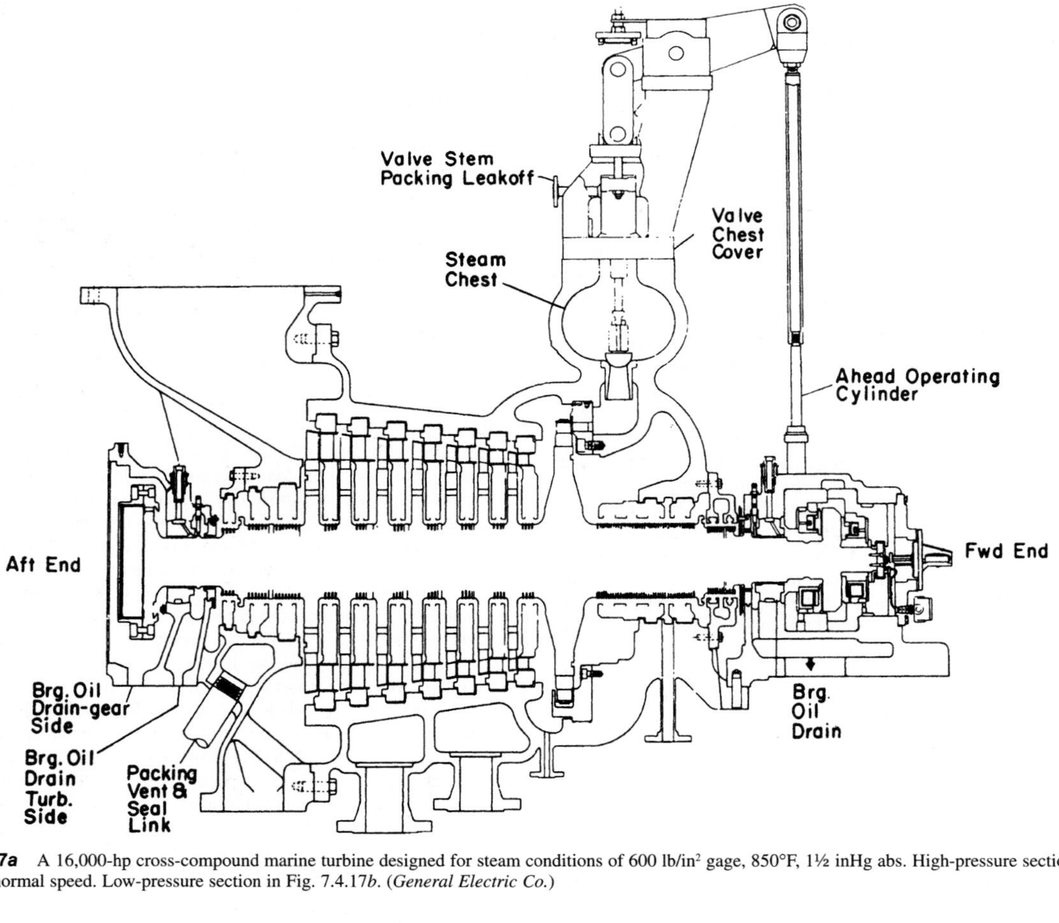

Figure 7.4.17*a* A 16,000-hp cross-compound marine turbine designed for steam conditions of 600 lb/in^2 gage, 850°F, 1½ inHg abs. High-pressure section for 6,550-r/min normal speed. Low-pressure section in Fig. 7.4.17*b*. (*General Electric Co.*)

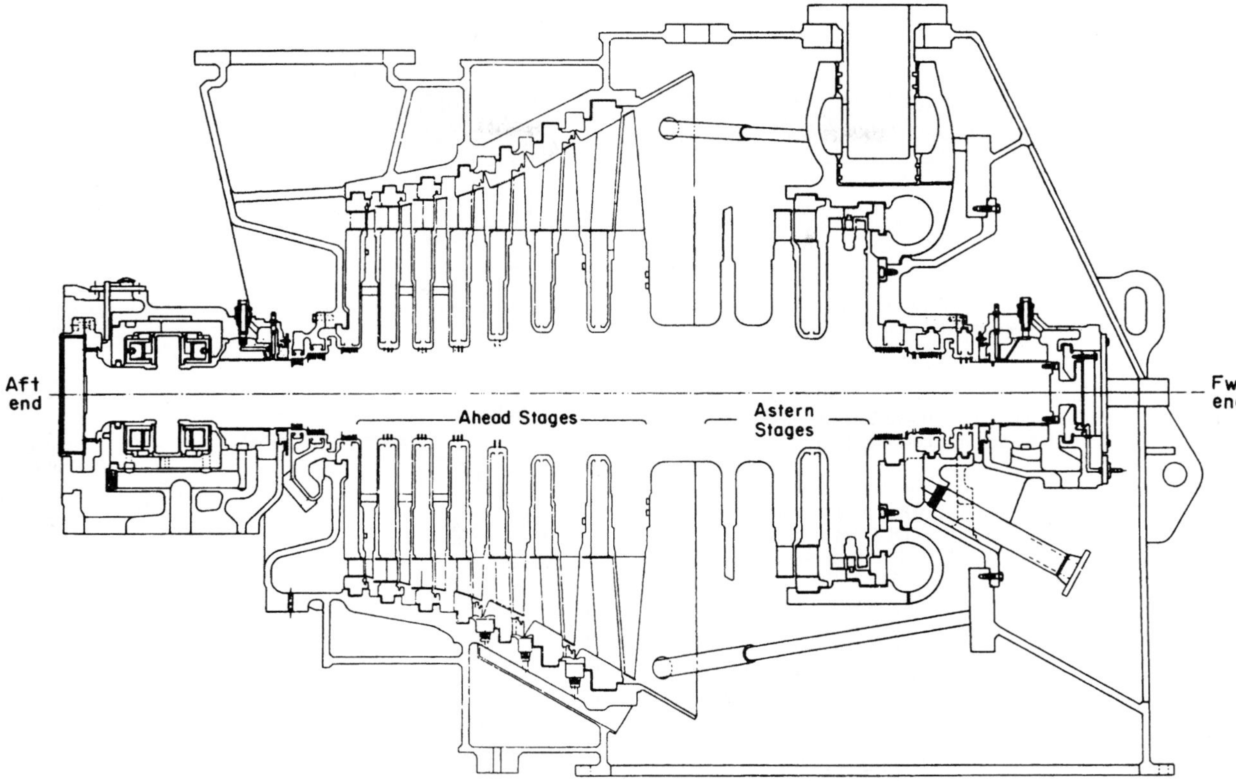

Figure 7.4.17*b* Same marine turbine as in Fig. 7.4.17*a*, but showing low-pressure section for 3,750-r/min normal speed. Low-pressure section contains two-stage reversing element. (*General Electric Co.*)

LARGE CENTRAL-STATION TURBINES

Figures 7.4.18 and 7.4.19 show examples of large central-station turbines as built by two manufacturers.

Figure 7.4.18 illustrates a 3,600-r/min tandem-compound four-flow unit rated 500 MW. It is a single-reheat fossil unit for nominal inlet-steam conditions of 2,400 lb/in^2 gage (16,650 kPa) 1,000°F (538°C), reheat to 1,000°F (538°C). The right-hand casing is a combined high-pressure and reheat section. Steam flows from the left center to the right through the impulse-type governing stage, then reverses, flowing to the left through the nine reaction-type high-pressure stages, and exhausts from the casing to the reheat section of the boiler. Reheated steam reenters that casing at its right center, flowing to the right through five reheat stages, turning once more to flow to the left between the inner and outer casings, finally exhausting up and to the left to the two double-flow low-pressure casings on the left end of the unit. The inlet stop and control valves, the reheat stop and interceptor valves, and the generator are not shown.

Four-flow units of this general type employ last-stage rotor blades from 25 to 40 in. (600 to 900 mm) long and in ratings up to about 700 MW. Significantly higher ratings require dividing the functions of the combined casing into a separate single-flow high-pressure casing and a separate two-flow reheat casing, for a total of four casings. The use of the long titanium last-stage buckets makes this configuration suitable for ratings up to 1,200 MW. In cases requiring additional exhaust area, a third double-flow low-pressure element can be employed for a total of five casings. Tandem-compound 3,600- and 3,000-r/min units are

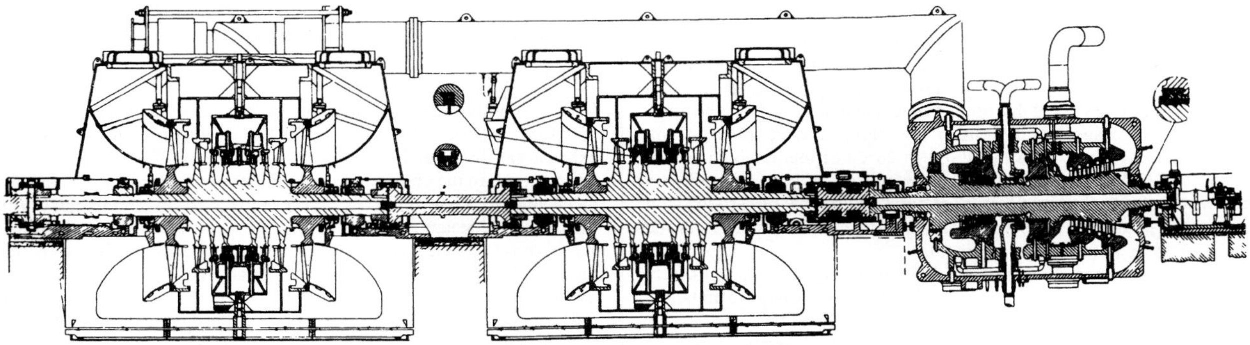

Figure 7.4.18 Cross section of a 500-MW tandem-compound, quadruple-flow 3,600-r/min reheat turbine. (*Westinghouse Electric Corp.*)

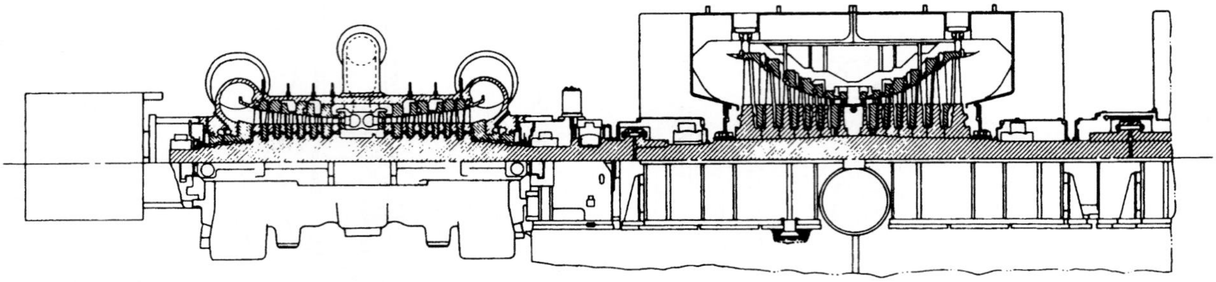

Figure 7.4.19 Cross section of a 1,300-MW class tandem-compound, six-flow 1,800-r/min turbine for steam from light-water nuclear reactors, showing one of the three low-pressure sections. (*General Electric Co.*)

commonly offered for ratings up to about 1,200 MW. Somewhat larger ratings can be accommodated by 3,600/3,600- or 3,600/1,800-r/min cross-compound units, but economic considerations makes their application infrequent.

Figure 7.4.19 illustrates an 1,800-r/min tandem-compound six-flow turbine designed for steam from light-water nuclear reactors. Reactors of both the boiling and pressurized-water types raise steam at 1,000 lb/in^2 abs (7,000 kPa) approximately with little or no initial superheat, so that the initial temperature is about 545°F (285°C), with a fraction of 1 percent of moisture frequently present. The poorer steam conditions result in higher steam rates than seen by fossil turbines. The lower initial pressure causes larger initial specific volume. In consequence, a typical nuclear turbine must accommodate 2.5 to 4 times the initial volume flow, and about 1½ times the exhaust volume flow of a fossil unit of the same rating. These considerations and the fact that the low temperature of the steam results in high moisture content in the expansion lead to the choice of 1,800 r/min. In halving the speed, diameters are less than doubled, balancing the advantages of larger steam-path area to accommodate large flow, while reducing velocities to minimize the occurrence of impact moisture erosion. The shortened energy range due to the lower initial conditions requires only two kinds of casings, high pressure and low pressure, compared with the three needed by fossil-reheat units.

Referring to Fig. 7.4.19, the steam enters the double-flow nozzle boxes of the high-pressure section, to the left, through stop and control valves which are not shown. It flows in both directions through the impulse-type stages, exhausting through four connections on each end of the shell. At the exhaust, the pressure is reduced to 200 lb/in^2 abs (1,400 kPa) approximately, and the moisture content is increased to 12 percent. That moisture poses an erosion risk and a performance loss to the low-pressure section following. It is current practice to dry the steam in an external moisture separator, frequently combined with one or two stages of steam-to-steam reheating, before admission to the low-pressure casings. Figure 7.4.20 is a cross section through a combination moisture separator and two-stage steam reheater. The exhaust from the high-pressure turbine enters the shell at the bottom, flowing upward through the inclined corrugated-plate moisture-separating elements, which remove essentially all the entrained water. It continues upward, flowing over the tubes of the first-stage bundle, which are supplied with steam extracted from the high-pressure turbine, approaching to within 20 to 50°F (11 to 28°C) of that temperature. Next, it flows over the tubes of the second stage bundle, which are supplied with initial steam, approaching to within 20 to 50°F (11 to 28°C) of that temperature, or to 495 to 525°F (257 to 274°C). The steam leaves the vessel at the top and is admitted to the low-pressure sections of Fig. 7.4.19, through stop and intercept valves, not shown. The last-stage blade length is in the range of 38 to 52 in (960 to 1,320 mm).

Units of the type described are built to ratings of approximately 1,300 MW with larger sizes available. Similar four-flow units are employed for ratings up to approximately 1,000 MW.

Other types of nuclear reactors, such as the high-temperature gas-cooled reactor and the liquid-metal-cooled fast-breeder reactor, produce steam conditions at temperature and pressure levels comparable with fossil-fuel-fired boilers, leading to the use of 3,600-r/min units similar to fossil practice.

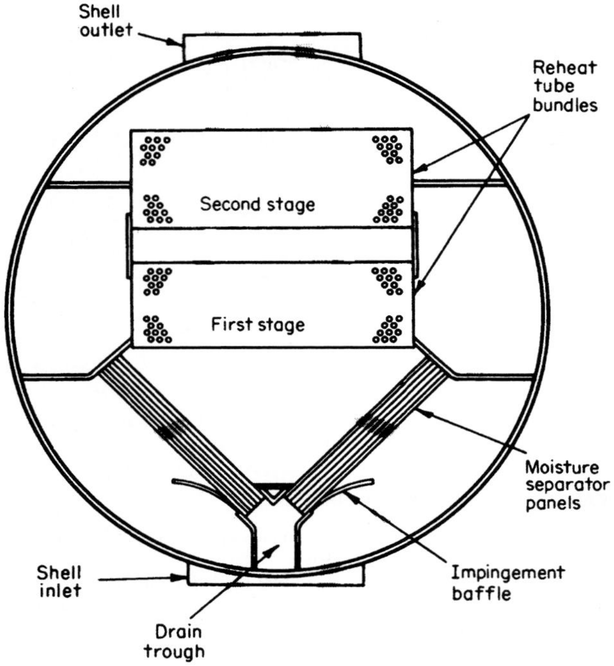

Figure 7.4.20 Cross section of a combination moisture separator and two-stage steam reheater for use with nuclear reactor turbines.

STEAM TURBINES FOR COMBINED CYCLES

(See also Sec. 7.7.)

Modern combustion **gas turbines (GTs)** have ratings approaching 250 MW. Considerable sensible heat is available in the gas-turbine exhaust-gas flow which ranges from 1,000 to 1,100°F (530 to 600°C) in temperature. In the **combined cycle,** the gas-turbine exhaust is directed to an unfired steam boiler called a **heat-recovery steam generator (HRSG).** The steam generated in the HRSG is admitted to a steam turbine, whose electrical output is approximately one-half that of the gas turbine. The resulting **combined thermal efficiency** can range from 50 to approaching 60 percent, better than that of any other available power cycle. High efficiency combined with the clean exhaust of gas turbines burning natural gas has created a broad field of application for combined cycles and the associated steam turbines.

Combined cycles can be arranged in several ways. In the *single-shaft* configuration, the gas and steam turbines are coupled together, driving a single generator. Units rated 413 MW have been manufactured for 50-Hz power systems, in which the gas turbine contributes 263 MW while the steam turbine generates 150 MW. The *multiple-shaft* arrangement uses a separate generator for each steam turbine and gas turbine. Each gas turbine has its own HRSG. The steam output from two or more GT/HRSG pairs can be manifolded to a single steam turbine. For example, a typical application for 60-Hz power systems combines two 181 MW gas turbines with one 201-MW steam turbine for a combined output of 563 MW. Typical steam conditions are 1,800 lb/in^2 gage (12,510 kPa), 1,050°F (566°C) with reheat in the HRSGs back to 1,050°F (566°C). Some cycles use inlet pressure as high as 2,200 lb/in^2 gage (15,270 kPa).

Steam turbines in combined-cycle service operate following the gas turbine(s), with their admission valves wide open accepting the full instantaneous output of the HRSG(s). The valves are located away from the turbine casings and are used only for speed control upon starting and for overspeed protection in the event of loss of load. The result is a very simple symmetric casing arrangement which reduces transient thermal stress and helps the steam turbine follow the potentially rapid changes in gas-turbine output. These features can be seen in Fig. 7.4.21. A two-casing reheat unit with single-flow exhaust is illustrated.

STEAM-TURBINE PERFORMANCE

The ideal **steam rate** (steam consumption, lb/kWh) of a simple turbine cycle is $3{,}412.14/(h_1 - h_{s2})$, where h is in Btu/lb [or kg/kWs = $1/(h_1' - h_{s2}')$, h' in kJ/kg]. (See Fig. 7.4.4.)

The actual steam rate is $3{,}412.14/[\eta_t(h_1 - h_{s2})]$, where η_t is the **engine efficiency** of the turbine only, inclusive of mechanical losses {or kg/kWs = $1/[\eta_t(h_1' - h_{s2}')]$}.

The actual steam rate of turbine and generator is $3{,}412.14/[\eta_e(h_1 - h_{s2})]$, where η_e is the engine efficiency including all mechanical and electrical losses {or kg/kWs = $1/[\eta_e(h_1' - h_{s2}')]$}.

The enthalpy of the steam leaving the turbine elements h_2 is

$$h_2 = h_1 - \eta_s(h_1 - h_{s2}) = (Wh_1 - 3{,}412.14P_g)/W$$

where h_s is the engine efficiency of the steam path, inclusive of leakages and losses but exclusive of mechanical and electrical losses; W is the steam flow, lb/h; and the gross output P_g is the net output plus mechanical and electrical losses, kW [or $h_2' = h_1' - \eta_s(h_1' - h_{s2}') = (W'h_1' - P_g)/W'$, where W' is flow in kg/s].

Table 7.4.2 gives steam rates for the ideal simple turbine cycle through a wide range of operating conditions. The performance of a turbine is usually expressed as a steam rate in the case of machines having no extraction or admission of steam between inlet and exhaust, which is generally true of small units and most noncondensing turbines.

Turbines having automatic pressure-controlled extractions or admissions of steam between inlet and exhaust usually have their performance expressed by a chart showing required throttle flow vs. load forvarying amounts of steam extracted or admitted at specified conditions.

Turbines working on regenerative and/or reheat cycles, condensing, usually have performance expressed as a heat rate, based upon a carefully specified heat cycle arrangement. This is usually illustrated by a diagram that defines all the surrounding conditions. See, for example, Fig. 7.4.26, actual cycle.

Table 7.4.2 Theoretical Steam Rates for Typical Steam Conditions, lb/kWh*

	Initial pressure, lb/in^2 gage															
	150	250	400	600	600	850	850	900	900	1,200	1,250	1,250	1,450	1,450	1,800	2,400
	Initial temp, °F															
	365.9	500	650	750	825	825	900	825	900	825	900	950	825	950	1,000	1,000
	Initial Superheat, °F															
	0	94.0	201.9	261.2	336.2	297.8	372.8	291.1	366.1	256.3	326.1	376.1	232.0	357.0	377.9	337.0
Exhaust pressure	Initial enthalpy, Btu/lb															
	1,195.5	1,261.8	1,334.9	1,379.6	1,421.4	1,410.6	1,453.5	1,408.4	1,451.6	1,394.7	1,438.4	1,468.1	1,382.7	1,461.2	1,480.1	1,460.4
inHg abs																
2.0	10.52	9.070	7.831	7.083	6.761	6.580	6.282	6.555	6.256	6.451	6.133	5.944	6.408	5.900	5.668	5.633
2.5	10.88	9.343	8.037	7.251	6.916	6.723	6.415	6.696	6.388	6.584	6.256	6.061	6.536	6.014	5.773	5.733
3.0	11.20	9.582	8.217	7.396	7.052	6.847	6.530	6.819	6.502	6.699	6.362	6.162	6.648	6.112	5.862	5.819
4.0	11.76	9.996	8.524	7.644	7.282	7.058	6.726	7.026	6.694	6.894	6.541	6.332	6.835	6.277	6.013	5.963
lb/in^2 gage																
5	21.69	16.57	13.01	11.05	10.42	9.838	9.288	9.755	9.209	9.397	8.820	8.491	9.218	8.351	7.874	7.713
10	23.97	17.90	13.83	11.64	10.95	10.30	9.705	10.202	9.617	9.797	9.180	8.830	9.593	8.673	8.158	7.975
20	28.63	20.44	15.33	12.68	11.90	11.10	10.43	10.982	10.327	10.490	9.801	9.415	10.240	9.227	8.642	8.421
30	33.69	22.95	16.73	13.63	12.75	11.80	11.08	11.67	10.952	11.095	10.341	9.922	10.801	9.704	9.057	8.799
40	39.39	25.52	18.08	14.51	13.54	12.46	11.66	12.304	11.52	11.646	10.831	10.380	11.309	10.134	9.427	9.136
50	46.00	28.21	19.42	15.36	14.30	13.07	12.22	12.90	12.06	12.16	11.284	10.804	11.779	10.531	9.767	9.442
60	53.90	31.07	20.76	16.18	15.05	13.66	12.74	13.47	12.57	12.64	11.71	11.20	12.22	10.90	10.08	9.727
75	69.4	35.77	22.81	17.40	16.16	14.50	13.51	14.28	13.30	13.34	12.32	11.77	12.85	11.43	10.53	10.12
80	75.9	37.47	23.51	17.80	16.54	14.78	13.77	14.55	13.55	13.56	12.52	11.95	13.05	11.60	10.67	10.25
100		45.21	26.46	19.43	18.05	15.86	14.77	15.59	14.50	14.42	13.27	12.65	13.83	12.24	11.21	10.73
125		57.88	30.59	21.56	20.03	17.22	16.04	16.87	15.70	15.46	14.17	13.51	14.76	13.01	11.84	11.28
150		76.5	35.40	23.83	22.14	18.61	17.33	18.18	16.91	16.47	15.06	14.35	15.65	13.75	12.44	11.80
160		86.8	37.57	24.79	23.03	19.17	17.85	18.71	17.41	16.88	15.41	14.69	16.00	14.05	12.68	12.00
175			41.16	26.29	24.43	20.04	18.66	19.52	18.16	17.48	15.94	15.20	16.52	14.49	13.03	12.29
200			48.24	29.00	26.95	21.53	20.05	20.91	19.45	18.48	16.84	16.05	17.39	15.23	13.62	12.77
250			69.1	35.40	32.89	24.78	23.08	23.90	22.24	20.57	18.68	17.81	19.11	16.73	14.78	13.69
300				43.72	40.62	28.50	26.53	27.27	25.37	22.79	20.62	19.66	20.89	18.28	15.95	14.59
400				72.2	67.0	38.05	35.43	35.71	33.22	27.82	24.99	23.82	24.74	21.64	18.39	16.41
425				84.2	78.3	41.08	38.26	38.33	35.65	29.24	26.21	24.98	25.78	22.55	19.03	16.87
600						78.5	73.1	68.11	63.4	42.10	37.03	35.30	34.50	30.16	24.06	20.29

* From Theoretical Steam Rate Tables—compatible with the 1967 ASME Steam Tables, ASME 1969.

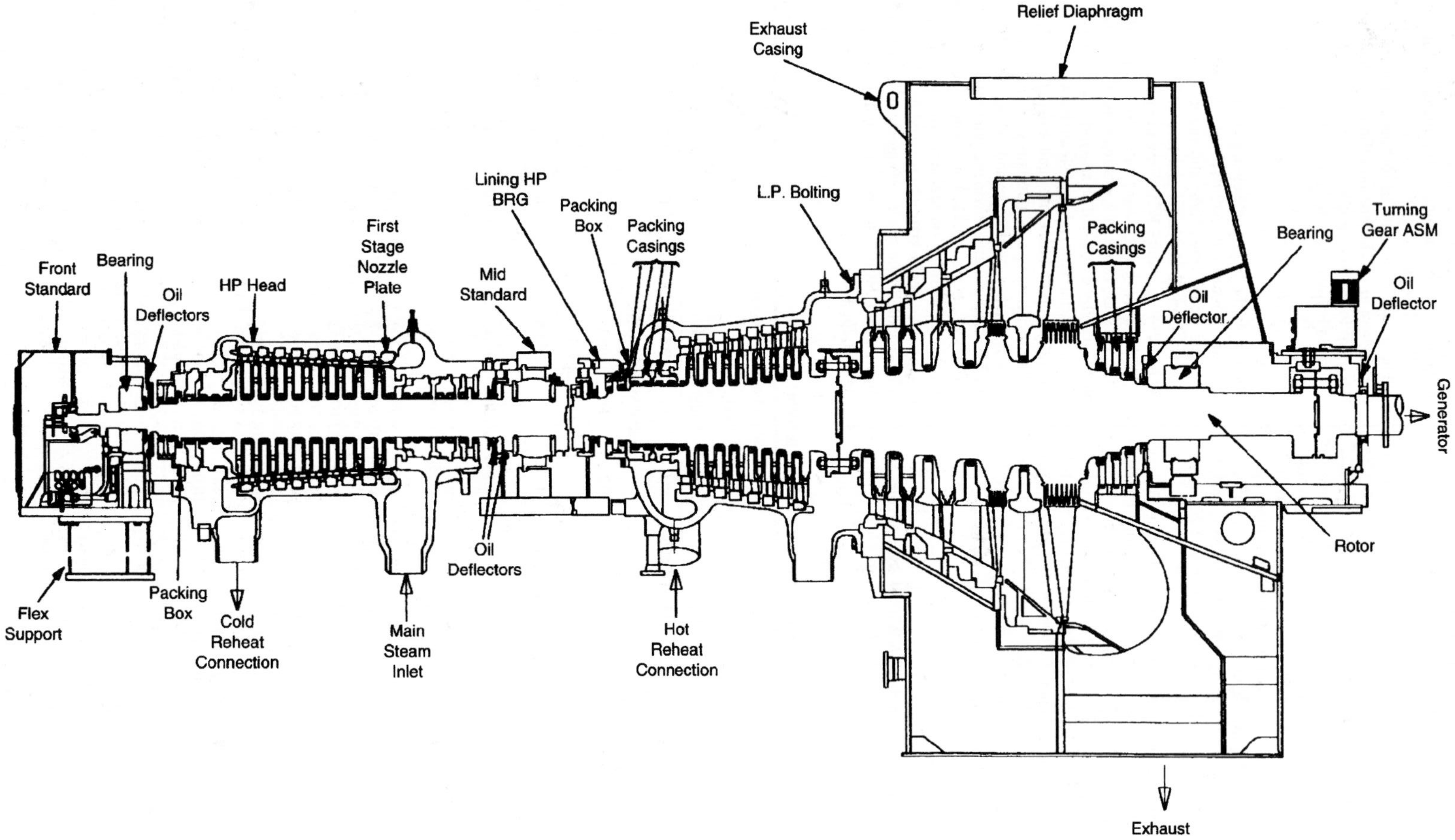

Figure 7.4.21 Two-casting reheat combined-cycle turbine with single-flow down exhaust. (*General Electric Co.*)

Table 7.4.3 Basic Efficiency for Steam Turbines, Straight Condensing at Rated Load*

kW capacity	Equivalent mechanical drive, hp	Initial steam conditions (gage pressure and temp.) 250 lb/in² 500°F	400 lb/in² 650°F	600 lb/in² 750°F	800 lb/in² 825°F	1,250 lb/in² 900°F
875	1,200	63	63	62		
1,875	2,600	76	67	66		
2,500	3,500	69	69	68		
5,000	6,900		74	73	73	
7,500	10,300		76	75	75	
12,500	17,200		78	78	78	77
15,625	21,500		79	79	79	77
20,000	27,100		79	80	79	79

* Efficiency correction factors, mechanical drive and auto extraction-condensing turbines—multiply basic efficiencies by:

	At 80 percent rating	At 100 percent rating
Single autoextraction-condensing	0.92	0.96
Double autoextraction-condensing	0.88	0.92

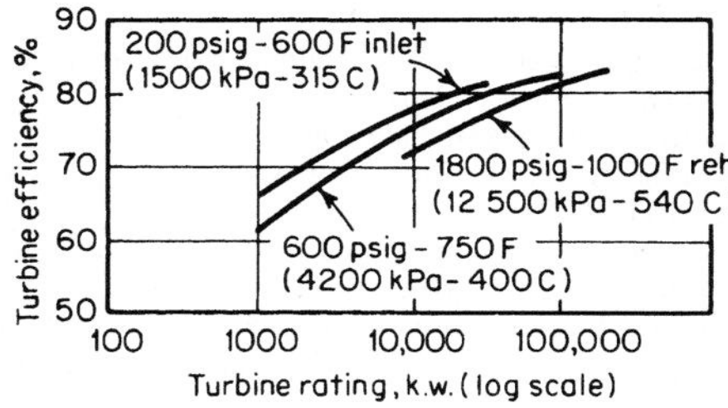

Figure 7.4.22 Turbine efficiencies vs. capacity.

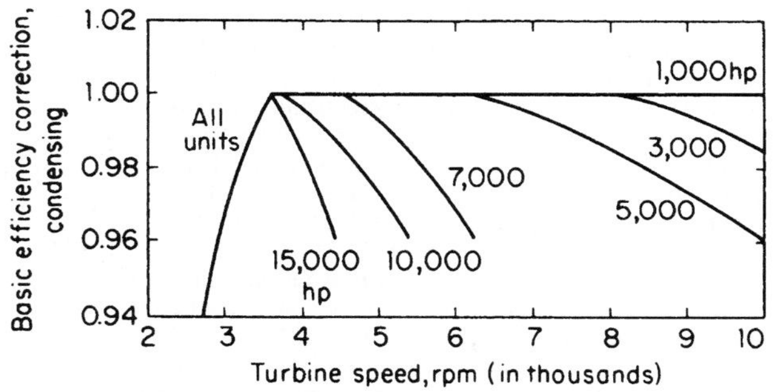

Figure 7.4.23 Correction factor for condensing mechanical-drive turbines.

The above methods for expressing turbine performance are more satisfactory for most application than the use of turbine "engine efficiencies." However, it is useful to know the general range of turbine efficiency realized in practice. The engine efficiency of a turbine depends mainly upon the flow areas and diameter of stages, the average velocity ratio, as can be deduced from Figs. 7.4.6, 7.4.7, and 7.4.8, the number of turbine stages, and the steam conditions. With so many variables, it is not possible to do more than show a general picture of efficiency as a function of rating, as in Fig. 7.4.22, for multistage condensing turbines. Noncondensing turbines will usually have similar efficiency levels; automatic extraction turbines will generally be slightly lower because of extra losses in the control-stage sections.

Approximate steam rates for turbines operating without auxiliary admissions or extractions of steam between inlet and exhaust may be estimated for any turbine rating by dividing theoretical steam rate, corresponding to inlet steam pressure and temperature and exhaust pressure, by the appropriate turbine efficiency from Fig. 7.4.22.

A short method for calculating extraction-turbine performance is illustrated by the following example:

EXAMPLE. Assume a 12,500-kW automatic-extraction-condensing unit operating at 10,000 kW, with 175,000 lb/h extraction for process at 150 psig with no extraction for feedwater heating, and throttle steam conditions of 850 lb/in², 825°F, exhaust at 2 inHg abs.

PROCEDURE. Find theoretical steam rates (TSR) from Table 7.4.2 or steam charts; TSR_1 for 850 lb/in², 825°F, 2 inHg is 6.58 lb/kWh; TSR_2 for 850 lb/in², 825°F to 150 lb/in² is 18.61 lb/kWh.

Turbine-generator efficiency from Table 7.4.3, single autoextraction at 80 percent rating (10,000 kW on a 12,500-kW unit), is 78 percent. Efficiency correction for autoextraction (see Table 7.4.3) is 0.92. Then actual steam rate (ASR) is TSR/(efficiency × correction); ASR_1 = 6.58/78 percent × 0.92 = 9.17 lb/kWh; ASR_2 = 18.61/78 percent × 0.92 = 25.9 lb/kWh; kW generation from extraction flow = extraction flow/ASR_2 = 175,000/25.9 = 6,760 kW.

kW to be generated by condenser flow = 10,000 − 6,760 = 3,240.

Condenser steam flow required is 3,240 × 9.17 = 29,700 lb/h, or (say) 30,000 lb/h. Total steam flow to throttle then is 175,000 + 30,000 = 205,000 lb/h.

Figure 7.4.23 gives correction factors for speeds which differ from 3,600 r/min and is representative of units designed for about 4 inHg abs exhaust pressure.

Mechanical-Drive Turbines Table 7.4.3 and Figs. 7.4.22 and 7.4.23 provide efficiency-estimating data for typical condensing turbines, primarily for 3,600-r/min generator drive. Figure 7.4.24 shows approximate values for turbine efficiency to be expected for noncondensing mechanical-drive units designed for a broad range of horsepower rating, speed, and inlet steam conditions.

LARGE CENTRAL-STATION TURBINE-GENERATORS

REFERENCES: Spencer, Cotton, and Cannon, A Method for Predicting the Performance of Steam Turbine-Generators · · · 16,500 kW and Larger, *Trans. ASME,* ser. A, Oct. 1963. Baily, Cotton, and Spencer, Predicting the Performance of Large Steam Turbine-Generators Operating with Saturated and Low Superheat Steam Conditions, *Proc. Amer. Power Conf.,* 1967; discussion of foregoing, *Combustion,* Sept. 1967. Spencer and Booth, Heat Rate Performance of Nuclear Steam Turbine-Generators, *Proc. Amer. Power Conf.,* 1968. Baily, Booth, Cotton, and Miller, "Predicting the Performance of 1,800-rpm Large Steam Turbine-Generators Operating with Light Water-Cooled Reactors," General Electric publication GET-6020, 1973. "Heat Rates for Fossil Reheat Cycles Using General Electric Steam Turbine-Generators 150,000 kW and Larger," General Electric publication GET-2050C, 1974.

The performance of central-station turbine-generators is generally expressed as **heat rate,** Btu/kWh, the ratio of the heat added to the cycle in Btu/h, to generation, in kW. Heat rate may be converted to **thermal efficiency** using the relationship, Efficiency = 3,412.14/heat rate (or 1/heat rate expressed in kJ/kWs). Heat rates are calculated by the preparation of a **heat balance,** which considers steam conditions, steam flow, turbine-expansion efficiency, packing leaking losses, exhaust loss

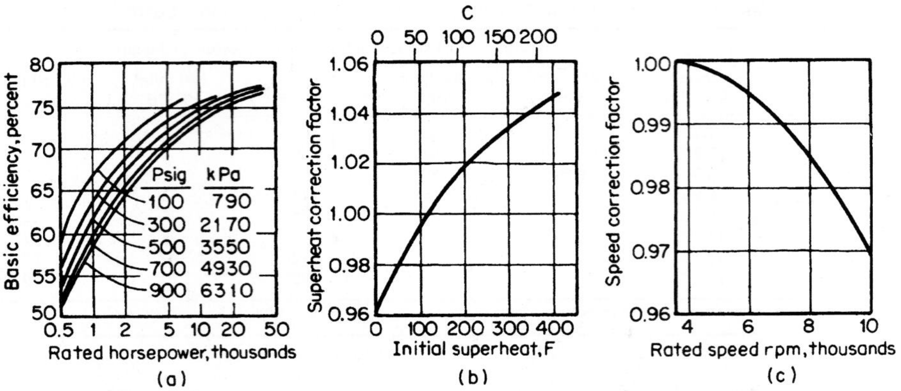

Figure 7.4.24 Mechanical-drive turbine efficiencies. (*a*) Basic efficiency, 3,600 r/min. Figures on curve are inlet steam pressure in lb/in^2 gage. (*b*) Superheat correction factor. (*c*) Rated-load speed-correction factor.

at the end of the low-pressure expansion (perhaps other casings as well), mechanical losses, electrical losses associated with the generator, moisture separation and reheat if present, and extraction for feedwater heating. **Gross heat rate** is calculated without consideration of the power consumed by the boiler feed pump. **Net heat rate** does consider pump power and is higher (poorer) than gross by a factor related to the initial steam pressure. If the pump is driven by an auxiliary turbine, as is the present usual practice, net heat rate is the natural result of the heat-balance calculation, and gross heat rate has little meaning. **Net station heat rate** considers the auxiliary power required by the rest of the power-plant equipment, and the boiler efficiency of fossil plants. It is generally about 3 percent higher than net heat rate in nuclear plants (3 percent auxiliary power, 100 percent "boiler" efficiency), and about 16 percent higher than net heat in the case of a coal-fired plant (4 percent auxiliary power, 90 percent boiler efficiency).

A typical current value of net station heat rate for the fossil steam conditions is 9,000 Btu/kWh (2.64 kJ/kWs), equivalent to a thermal efficiency of 38 percent. A typical net station heat rate for a large light-water nuclear-reactor plant is about 10,100 Btu/kWh (2.96 kJ/ kWs), or about 34 percent thermal efficiency.

Table 7.4.4 lists representative net heat rates for large fossil turbines of various types and steam conditions. Steam pressures in excess of 3,500 lb/in^2 gage (24,200 kPa) and initial and reheat temperatures in excess of 1,000°F (538°C) were frequently employed in the past. However, a number of operating and economic considerations have led to near standardization on the single reheat cycle with initial pressure of 2,400 or 3,500 lb/in^2 gage (16,650 or 24,240 kPa), with initial and reheat temperature of 1,000°F (538°C), although there is some renewed interest in "ultra-supercritical" units in the Far East.

Table 7.4.5 lists some representative net heat rates for large nuclear turbines for service with steam from boiling-water reactors (BWR), at 950 lb/in^2 gage (6,650 kPa), $\frac{1}{2}$ percent initial moisture. Values for other light-water reactors may be approximated by reducing heat rate by 1 percent for each 100 lb/in^2 (690 kPa) pressure increase, reducing heat rate by 0.15 percent for reducing initial moisture to 0 percent, reducing heat rate by 0.3 percent for each 10°F (6°C) of initial superheat provided.

Reheating with Regenerative Cycle

References: Reynolds, Reheating in Steam Turbines, *Trans. ASME*, **71**, 1949, p. 701. Harris and White, Development in Resuperheating in Steam Power Plants, *Trans. ASME*, **71**, 1949, p. 685.

Reheating is commonly used on large fossil central-station turbines. It is accomplished by taking the steam from the turbine after partial expansion, reheating it in a separate section of the boiler, and returning it to the next lower-pressure section of the turbine. Reheating results in lowering of the turbine heat rate by approximately 5 percent; the exact improvement is dependent on several factors. Roughly speaking, 40 percent of the improvement comes from having added heat to the cycle at a higher-than-average temperature (thermodynamic gain), and the remaining 60 percent comes from improvement in turbine efficiency due to reduced moisture loss and increased reheat factor.

Reheating can theoretically be done any number of times, but because of extra cost of apparatus and piping, and the steam pressure drops

Table 7.4.4 Representative Net Heat Rates for Large Fossil Central-Station Turbine-Generators

Nominal rating MW, at 1.5 inHg abs	Steam conditions			Tandem compound, 3,600 r/min. last-stage buckets				Boiler feed-pump drive	Net heat rate, Btu/kWh, at rated load and steam conditions, and at exhaust pressure, inHg abs				
	Throttle pressure lb/in2 gage	Temp, °F	Reheat temp, °F	No. of rows	Length in	Exhaust area, ft^2	Approx kW/ft^2		1.5	2	3	4	5
150	1,800	1,000	1,000	2	26	82	1,820	Motor	8,010	8,060	8,230	8,440	8,630
235	1,800	1,000	1,000	2	26	82	2,860	Motor	8,240	8,240	8,290	8,380	8,500
250	1,800	1,000	1,000	2	30	111	2,250	Motor	8,080	8,100	8,220	8,400	8,620
250	1,800	1,000	1,000	2	30	111	2,250	Turbine	8,030	8,060	8,200	8,390	8,610
250	2,400	1,000	1,000	2	30	111	2,250	Turbine	7,850	7,890	8,030	8,240	8,450
500	2,400	1,000	1,000	4	30	222	2,250	Turbine	7,790	7,830	7,970	8,170	8,370
700	2,400	1,000	1,000	4	33.5	264	2,650	Turbine	7,860	7,870	7,970	8,130	8,320
1,000	2,400	1,000	1,000	6	30	334	3,000	Turbine	7,920	7,930	8,000	8,100	8,250
500	3,500	1,000	1,000	4	30	222	2,250	Turbine	7,620	7,660	7,820	8,030	8,220
700	3,500	1,000	1,000	4	33.5	264	2,650	Turbine	7,670	7,690	7,810	7,980	8,170
1,000	3,500	1,000	1,000	6	30	334	3,000	Turbine	7,710	7,730	7,810	7,940	8,090
1,100	3,500	1,000	1,000	6	33.5	397	2,770	Turbine	7,680	7,700	7,310	7,960	8,140

Table 7.4.5 Representative Net Heat Rates for Large Nuclear Central-Station Turbine-Generators

Warranted reactor thermal power, MWt	Nominal turbine rating, MWe at 2 inHg abs	Tandem compound, 1,800 r/min, last stage buckets				Net heat rate, Btu/k/Wh, at warranted reactor thermal power, at rated steam-conditions, and at exhaust pressure, inHg abs				
		No. of rows	Length, in	Exhaust, area, ft²	Approx kW/ft²	1.5	2	3	4	5
2,440	840	4	38	423	1,980	9,950	9,950	10,090	10,190	10,410
2,440	850	4	43	495	1,720	9,810	9,820	9,950	10,170	10,440
2,890	1,010	6	38	634	1,590	9,750	9,780	9,950	10,200	10,480
2,890	990	4	43	495	2,000	9,980	9,980	10,050	10,200	10,410
3,580	1,230	6	38	634	1,540	9,910	9,920	10,000	10,170	10,380
3,580	1,250	6	43	743	1,680	9,780	9,790	9,930	10,160	10,430
3,830	1,310	6	38	634	2,070	9,990	9,990	10,050	10,190	10,390
3,830	1,330	6	43	743	1,790	9,840	9,850	9,960	10,170	10,420

All units boiling-water reactor steam conditions of 965 lb/in² abs, 1,190.8 Btu/lb, and two-stage steam reheat with 25°F approach to reheating steam temperature.

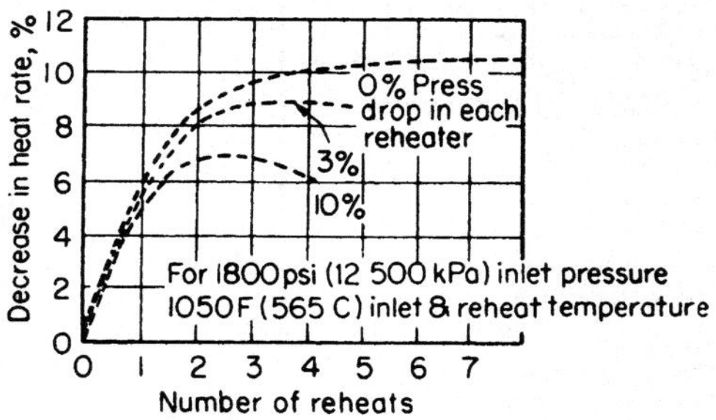

Figure 7.4.25 Approximate gains due to reheating.

required in practice (8 to 10 percent of the reheat pressure), the economic gains diminish rapidly with more than one reheating (see Fig. 7.4.25). In a few cases, two reheatings are employed. (See also Sec. 2.)

The throttle and condenser steam-flow rates for a given turbine output are reduced approximately 17 and 13 percent, respectively, by reheating once to the initial temperature, as compared with no reheat with the same initial steam conditions.

The maximum gain in heat rate from one reheating with a fixed-percentage pressure drop through the reheating system occurs when the reheat pressure is about 0.15 of the initial pressure. In practice, however, the reheat pressure is higher, 0.20 to 0.30 times initial pressure, because of the extra cost of larger piping, valves, etc., required for lower reheater pressures owing to the larger steam volume.

Regenerative Feedwater Heating

(See also Sec. 2.)

The heat consumption of a turbine may be reduced by heating the condensate (feedwater) in stages by the condensation of steam extracted at various points from the turbine. This is shown diagrammatically for an ideal cycle and a more practical cycle in Fig. 7.4.26. The difference between the two is in the use of mixing heaters in the ideal cycle with each discharge pumped back while the practical cycle has closed heaters with cascaded drains in the upper and pumped drains in the lowest heater, together with some pressure drop between turbine and heaters and a terminal temperature difference between saturated-steam temperature in the heater and feedwater temperature coming out. Usually the difference between such an ideal and a practical cycle is about 1½ percent. A deaerating type of contact heater with no terminal difference may be substituted for one of the closed heaters as is shown in Fig. 7.4.26. Other variations are the use of (1) all open-contact heaters, or (2) drain coolers to reduce the loss due to cascading the drips, or (3) a desuperheating section on the top heater to get a higher final feed temperature, thereby approaching most closely to the ideal cycle.

Figures 7.4.27 and 7.4.28 and Table 7.4.6 supply data on the results of regenerative heating based on the ideal cycle of Fig. 7.4.26. Figure 7.4.27 shows the reduction in heat rate for various initial pressures and temperatures at 1 inHg abs (3.4 kPa) exhaust pressure, for various feedwater temperatures and number of heaters. The increase in throttle flow necessary to maintain the same power output when extracting steam for feedwater heating is shown in Fig. 7.4.28.

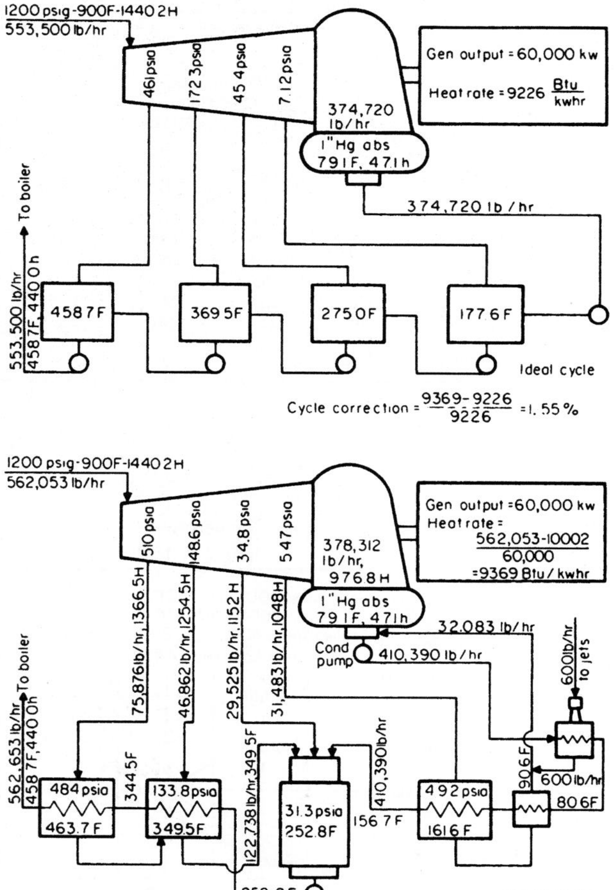

Figure 7.4.26 Comparison of ideal and actual cycles for regenerative feedwater heating.

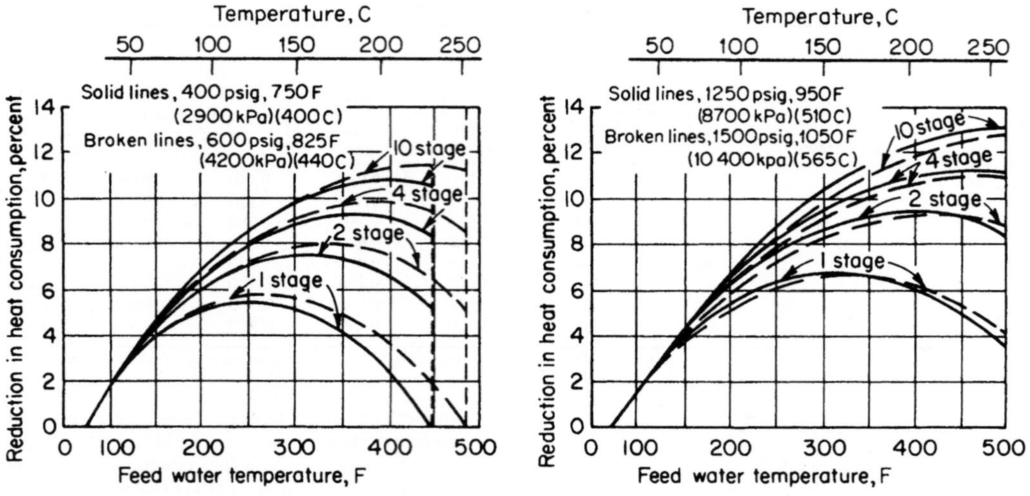

Figure 7.4.27 Reduction in heat rate by use of ideal regenerative cycle, with 1-inHg back pressure.

Table 7.4.6 Total Steam Bled, Percent of Throttle Flow

Final feed temp		Steam pressure and temperature: 400 lb/in² gage (2,900 kPa)		600 lb/in² gage 825°F (4,200 kPa, 440°C)		1,250 lb/in² gage 950°F (8,700 kPa, 510°C)		1,500 lb/in² gage 1,050°F (10,400 kPa, 565°C)	
		Stages of feedwater heating							
°F	°C	2	10	2	10	2	10	2	10
150	65	7.0	7.1	6.9	7.0				
200	93	11.4	11.8	11.3	11.6	11.2	11.5	10.7	11.0
250	121	15.6	16.2	15.5	16.0	15.5	15.9	12.6	13.1
300	149	19.6	20.6	19.5	20.4	19.5	20.3	18.8	19.5
350	177	23.6	24.8	23.5	24.6	23.6	24.6	20.7	21.6
400	204	27.1	29.0	27.1	28.8	27.4	28.9	26.4	27.8
450	232			30.2	32.5	31.1	33.2	30.0	32.0
500	260					35.0	37.4	33.9	36.1

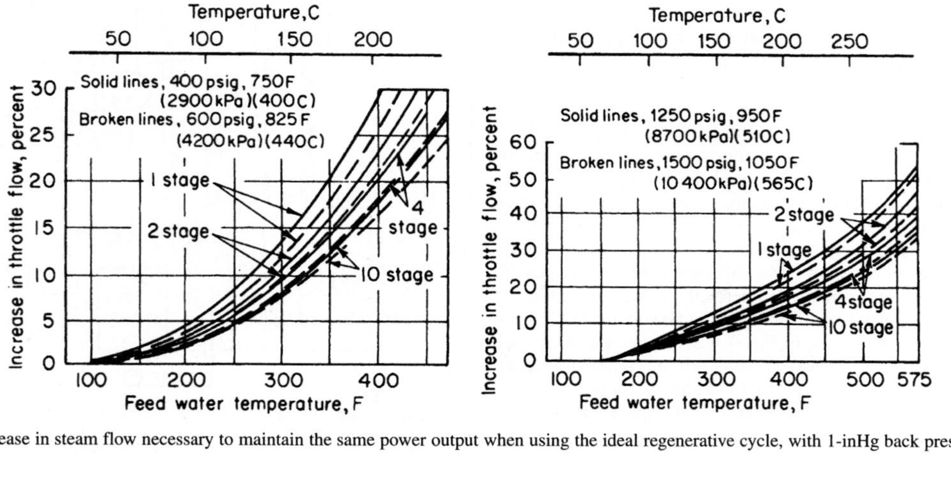

Figure 7.4.28 Increase in steam flow necessary to maintain the same power output when using the ideal regenerative cycle, with 1-inHg back pressure.

INSTALLATION, OPERATION, AND MAINTENANCE CONSIDERATIONS

Steam turbines are capable of long life and high reliability with relatively little maintenance, if proper attention is paid to their installation, operation, and preventive maintenance. This section considers four areas proved to be of particular importance by operating experience.

Steam Temperature—Starting and Loading

REFERENCES: Mora et al., "Design and Operation of Large Fossil-Fueled Steam Turbines Engaged in Cyclic Duty," ASME/IEEE Joint Power Generation Conference, Oct. 1979. Spencer and Timo, Starting and Loading of Large Steam Turbines, *Proc. Amer. Power Conf.,* 1974. Ipsen and Timo, The Design of Turbines for Frequent Starting, *Proc. Amer. Power Conf.,* 1969. Timo and Sarney, "The Operation of Large Steam Turbines to Limit Cyclic Shell Cracking," ASME Paper 67-WA/PWR-4, 1967.

Changing steam temperature at constant load or changing load at constant temperature subjects rotors and shells to thermal transients. Whereas temperature and load may be changed in seconds, it may take heavy metal sections hours to reach equilibrium with the new temperatures imposed on their surfaces. Parts are subjected to transient thermal stresses which may deplete their low-cycle thermal fatigue life. Repeated thermal cycles may lead to full expenditure of the available fatigue life of the part, followed by surface cracking. Further cycles tend to drive the cracks deeper into the affected part, leading to steam leakage through shells or vibration problems with rotors. Cracks tend to be driven deeper by downward steam-temperature changes, since the surface chills faster than the underlying material and is stressed in tension.

Steam-turbine manufacturers publish specific starting and loading instructions for their units. Data are provided such that the operator may select loading rates that stay within an acceptable expenditure of total low-cycle fatigue life per starting or loading cycle. For example, if a unit is expected to be started and loaded daily for a 30-year life, total cycles will be about 10,000, and it would be desirable to avoid exceeding 0.01 percent life expenditure per daily cycle.

Water-Induction Damage

REFERENCES: Recommended Practices for the Prevention of Water Damage to Steam Turbines Used for Electric Power Generation, ANSI/ASME TDP-1-1985 Fossil Fueled Plants. ASME Standard No. TWPDS-1, Part 2—Nuclear Fueled Plants, Apr. 1973.

Any connection to a steam turbine is a potential source of water either by induction from external equipment or by accumulation of condensed steam. The sources include the following along with their piping and drains: main and reheat steam systems; reheat attemperating system (fossil units); bypass systems, crossaround piping, moisture separator/reheater system (nuclear units); extraction system and feedwater heaters; steam-seal system; turbine-drain system. Water induction may lead to steam-path damage such as broken buckets, thrust-bearing failure, rotor bowing, and shell distortion, which may be indicated by abnormal vibration or differential expansion, or inability to turn the rotor on turning gear.

Water induction may be prevented by proper system installation, provision of protective and indicating devices, and periodic testing, inspection, and maintenance. Detailed recommendations are given in the ANSI/ASME and ASME standards and in manufacturers' instructions.

Lubricating-Oil and Hydraulic-Fluid Purity

Most steam turbines are provided with a lubricating-oil system consisting of reservoir, pumps, coolers, and piping to provide the thrust and journal bearings with a generous supply of oil at the proper temperature and viscosity. Some units also use the lube-oil system as a source of fluid power for control devices and steam-valve actuation. It is most important to assure the cleanliness and purity of the lube oil at all times to avoid bearing and journal damage, or control-system malfunction. Bearings have failed because of oil starvation caused by clogged lines. Many overspeed failures have resulted from the silting of control devices with rust caused by the entry of water into the oil system. Dewatering (oil purification) technology should be used.

Units employing an electrohydraulic control system frequently use a synthetic fire-resistant fluid for the high-pressure control hydraulics, separate from the petroleum oil used in the bearing lubrication system. High-pressure hydraulic systems employing synthetic fluid offer advantages in size reduction, speed of response, and fire safety. However, because of the need for very close clearances between small parts at high pressures, and because of the poorer rust-preventing properties of the fluid, cleanliness is of even greater importance than with oil-based systems.

Turbine manufacturers provide equipment such as oil conditioners to maintain bulk lubricating oil purity and full-flow filters to remove contaminants before they enter bearings. Further, they provide instructions for cleaning entire systems by oil flushing before commissioning (first operation) and for maintaining the required oil and hydraulic-fluid purity. It is important that these be followed carefully.

Steam Purity

REFERENCES: Lindinger and Curren, "Corrosion Experience in Large Steam Turbines," ASME/IEEE Joint Power Generation Conference, Oct. 1981. Bussert et al., "The Effect of Water Chemistry on the Reliability of Modern Large Steam Turbines," ASME/IEEE Joint Power Generation Conference, Sept. 1978. McCord et al., "Stress Corrosion Cracking of Steam Turbine Materials," National Association of Corrosion Engineers, Apr. 1975.

Boiler feedwater treatments have traditionally been designed to remove solids that might clog steam passages in the boiler, remove salts that could cause scaling of tube surfaces and interfere with heat transfer, prevent corrosion of tube surfaces by reducing oxygen content and by maintaining pH at a high level, and provide "clean" steam to the turbine.

At one time the main concern with steam quality in the turbine was the level of silica present, since that chemical tends to deposit in the steam passages and causes reduction in capacity and efficiency. In general, the monitoring and control of silica has been well developed, and as steam turbines have increased in rating, the passages have increased in area, so that the net effect has been a reduction in the extent of losses in efficiency and capacity from deposits.

On the other hand, the growth in unit ratings has been accomplished without a proportional growth in physical size, and has resulted in greater power densities per casing, per stage, and per pound of material. Such increased duty has required the use of higher-strength alloys operating at higher stresses. As a result, modern turbine components are more susceptible to stress-corrosion cracking than those in older, smaller units, and therefore require better control of contaminants in the steam.

The **feedwater treatment** in most fossil fuel stations is designed to provide a sufficiently low level of contaminants so that stress-corrosion cracking of turbine components should not be a problem. Both the "zero solids" and the "coordinated phosphate" treatments can be controlled to provide steam of acceptable chemistry. Unfortunately, there are a number of situations in which undesirable chemicals can be introduced in the steam in spite of well-intentioned "normal" water-control practices. For example, the composition of coatings put on turbine components for corrosion protection during shipment, storage, and installation must be controlled. The chemistry of solutions used for the removal of coatings during installation, and the methods used, must be carefully regulated. The turbine must be protected during the chemical cleaning of related components such as the boiler, condenser, and feedwater heaters. Critical components have been damaged by fumes from cleaning. The feedwater system must be designed so that only water of high purity is used for boiler desuperheater sprays, and for the turbine exhaust-hood sprays used to limit temperature during light-load operation. Condensate demineralizers must be operated and regenerated so as to ensure that they do not introduce the harmful chemical they are intended to remove. In the event of a leak into the condenser of impure cooling water, it is important to avoid changing the feedwater treatment in such a way that the turbine is subjected to harmful contaminants introduced to protect other station components.

In each of these undesirable instances, the average concentration of chemicals in the steam can be quite low, but high local concentrations can be developed through several mechanisms. For example, dilute solutions can enter crevices not washed by flowing steam; as water evaporates on heating, the concentration of the solution wetting the surfaces tends to increase. In the case of expansion-joint bellows, chemicals contained in steam condensing on shutdown or cold start-up tend to be trapped and concentrated in the bottom of the bellows convolutions.

Succeeding cycles can lead to dry residue or to concentrated solutions during operating conditions which provide moisture. In the case of the steam path, as the expansion crosses into the moisture region, the first droplets of water condensed from the steam will tend to contain most of the contaminants. Concentration-enhancement factors of 100 to 1,000 can be achieved. In modern reheat turbines the early-moisture region occurs in one of the later few stages of the low-pressure section, and at light loads can occur on the most highly stressed last-stage buckets.

Extreme care must be exercised in protecting turbines from chemical contamination during installation, operation, and maintenance. Any deviation from sound feedwater-treatment practice during condenser leaks should be done with the full realization that damage to the turbine may result.

STEAM TURBINE RELIABILITY IMPROVEMENT

by Heinz P. Bloch

With saving money the primary objective and reliability also among the desired attributes for steam turbines, purchasers often face conflicts. Whoever had final say over budgets may have opted for lowest initial cost and earliest possible startup dates. Even the selection of EPCs (Engineering-Procurement-Construction) contractors is largely influenced by cost and schedule. Ever since the expression "lean and mean" was coined in the late 1990s, these terms have been misapplied or taken out of context. While they originally implied cutting out wasteful and unnecessary steps, the two words are often invoked to the detriment of long-term reliability. Steam turbines are machines that suffer when reliability thinking ranks low among the purchaser's priorities. The proper long-term reliability approach will include paying close attention to steam turbine auxiliaries such as control elements, methods to maintain turbine performance (efficiency), and the prevention of lube oil contamination.

Selecting Steam Turbines in a "Lean" Environment

Engineers involved in power plant designs are often asked to clear up the intent of turbine technical specifications. Combined-cycle power plant projects are but one of many instances where guidance may be sought.

Suppose we were dealing with an 86 MW condensing-type steam turbine with one reheat entry. The HP (high-pressure) inlet steam is at 110 bar/540°C and the reheat is being designed for 24 bars. In a competitive bidding situation the EPC (Engineering/Procurement/Construction) organization is likely to communicate with several steam turbine bidders or manufacturers and some of them will probably propose a "standard" machine. A "standard" turbine will often comprise a single casing and a single rotor direct-coupled to the generator. But there will probably also be some manufacturers proposing a "Cross-Compound Type" machine. Such machines consist of a turbine with two casings and two rotors. Conceivably, the HP rotor will be coupled to the generator by a gearbox and the IP/LP casing will be direct-coupled to the generator.

Making a good choice requires a number of steps:

(a) looking at the guaranteed efficiencies of the two different offers and keeping in mind the overall steam balance of the facility,

(b) making a decision as to how well trained the operators will be, and

(c) closely examining the respective field and service experience histories of the two different steam turbine offers.

Complying with the basic requirements of (a), (b), and (c) requires considerable diligence, time, and effort. The reviewer would include a thorough check of the gearbox design and should accept the fact that time is needed to draw up a comprehensive comparison between the two offers. It would even be appropriate to ask if the original inquiry or bid invitation went to the right bidders. It is always prudent to solicit bids from manufacturers that have ample experience with both direct-drive generator turbines and the more complex compound/reheat multicasing machines.

Time permitting, the engineer making the selection might consider adding a few bidders willing to comment on the very advisability of double-shell machines. A double-shell construction machine prevents inlet steam coming into direct contact with the outer casing joint. Double-shell machines are (probably) less in need of careful operator attention but, during the maintenance cycle, might require rather competent maintenance skills.

"Cross-Compound" machines are more likely found on shipboard, but predominantly at inlet pressures slightly lower than 110 bars. Again, there is much inquiring to be done before a well-researched decision can be made for a land-based installation. Ships have downtime while in a harbor, whereas land-based utilities desire long uninterrupted run lengths. For the latter, downtime caused by maintenance is very expensive. As regards items to be reviewed one might investigate, among other things, the lubrication system. In a cross-compound machine the input and output shafts are at different levels and the lubrication system serves not only the turbine and generator bearings, but also the gearbox. Who makes the gearbox and how are the gears lubricated? Bearings often require thin oils and gears require thicker oils. How is a mutually satisfactory compromise reached? Will an expensive synthetic satisfy both?

Initial equipment cost, operating cost (efficiency), and long-term reliability are of great interest and must be considered as life cycle cost factors. All are of equal concern and, without making a final judgment one way or the other, many different options should be explored before reaching a conclusion. Although one should make good use of vendor input and carefully weigh a vendor's demonstrated experience, expect double-shell machines to cost more money and cross-compound machines to require more than the average maintenance commitment. And the "simple" machine would also stay in the running until all the relevant data have been obtained and ready for comparison.

Consider Sweeping Upgrades to Small Steam Turbines

In or around 2009, forward-looking mechanical engineering consultants took tangible steps to infuse modern reliability thinking in old general purpose (GP) steam turbines. Modern machines now incorporate pneumatic actuators at the trip valve and speed indicator cogs (take-offs) on the outboard bearing. These take-offs will be used in conjunction with magnetic or passive sped sensors. Best Practices plants many improvements were applied after 2009 by encouraging manufacturers of small steam turbines to comply with a competent user's GP turbine standard. In general, these modern in-plant GP standards now require many changes to the manufacturers' old standard designs.

At Best Practices plants, knowledgeable corporate consulting engineers often work with responsive OEMs (Original Equipment Manufacturers) to achieve a reliable heavy duty design which eliminates the conventional spring-loaded trip valve. The new design applies, instead, a pneumatic trip actuator. Its fail-safe spring return also eliminates all mechanical trip devices normally found on older steam turbine shafts. The first of the resulting "hybrid" GP machines was commissioned in 2015 at a major refinery on the Red Sea where it drives a deaerator pump. Note again that there no longer are mechanical trip linkages and no more unsafe overspeed events after incorporating these low-cost upgrades. The redesign has taken World War II-vintage steam turbine designs into the twenty-first century.

In the decades since 1990, the often simplified interpretations and applications of "think lean and mean" have led to costly shortcuts, oversights and omissions. No time or budget is allocated to understand what happens when steam turbine speeds are reset for operation away from the original governor adjustment range. The result has been a much higher probability of steam turbine blade failures due to the incidence of resonant blade vibration occurring at speeds that were not anticipated in the original design.

Steam Turbine Overspeed Protection Retrofits

Not too long ago, the overwhelming majority of small turbine overspeed trip devices were mechanical; they incorporated a spring-activated trip bolt which moves out and contacts a trip lever whenever a preset overspeed is reached. A number of factors make it necessary to check, and possibly recalibrate, these built-in devices on a yearly basis. Needless to say, process plants are not always performing this preventive maintenance activity in timely and procedurally correct fashion. Unscheduled downtime events and serious injury, even deaths, have resulted from such oversights.

With electronics and even robotics having made huge gains in the recent past, well-engineered and highly reliable electronic Over-Speed Prevention (OSP) systems are available for new and retrofit applications. These systems offer end users the security and dependability of a two-out-of-three voting system with the flexibility and ease of do-it-your-self retrofit installation and configuration.

A typical modern OSP system incorporates a two-out-of-three voting feature that monitors turbine speed and will initiate a trip command to prevent overspeed events. One such system includes three identical speed modules which individually measure a frequency input signal from the passive or active magnetic pickup sensor mentioned earlier. A supervisory module continually monitors the three speed modules for proper operation, which helps to eliminate unnecessary downtime and increases system availability.

Multilevel Password and Desirable Self-diagnostics

A modern OSP system provides users with a multilevel password function for added security. Each level typically allows access to higher system functions, including Peak Speed Reset, Over-speed Test Mode, Configuration Mode, and Set New Passwords. The three speed modules provide a greater level of dependability. Open-pickup detection on "passive" sensors is provided as well as "dynamic" sensor failure detection via the supervisory module, which knows when a speed module should be receiving a valid frequency input. Active sensor power is provided as an output as well. Each speed module can display its current speed and current set point. Additionally, each speed module includes four dedicated LEDs for indicating overspeed test, trip, MPU failure, and fault conditions. A good overspeed protection system also features complete continuous self-diagnostics. The supervisory module monitors proper operation of the speed modules and power supplies. A suitable alphanumeric display is usually integrated into the operator interface. One or more test modes facilitate testing of each speed module, including the verification of the voting relays' operation. This module typically includes a dedicated LED for indicating an alarm condition.

Power supply modules usually provide redundant power to the relay board, speed modules, and supervisor. In modern devices a failed power supply module can be replaced without interrupting the operation of the OSP system. Users should be able to select de-energize-to-trip or energize-to-trip logic. By meeting a broad range of installation requirements, the better OSP systems can be installed with any turbomachinery train for virtually any application. A good OPS package opens up safety, reliability, and maintenance enhancement possibilities that merit serious consideration.

Consider Bearing Protection for Small Steam Turbines

In most small steam turbine bearing housings and their associated lubrication auxiliaries there is a need to limit both contaminant ingress and oil leakage. As of 2010 inexpensive lip seals are used very infrequently for sealing at the bearing housing because lip seals typically last only about 2,000 operating hours—three months. Moreover, when lip seals are too tight they cause shaft wear and, in some cases, lubricant discoloration known as "black oil." Once lip seals have worn and no longer seal tightly, oil is lost through leakage. This fact is recognized by certain API (American Petroleum Institute) standards which disallow lip seals and call for either rotating labyrinth-style or contacting face seals. Modern small steam turbines incorporate rotating bearing housing protector seals; these are a subject of interest here.

Leakage Events

Small steam turbines often suffer from steam leakage at both drive and governor-end sealing glands. Each bearing housing (Fig. 7.4.29) is located adjacent to one of these two glands, which contain carbon

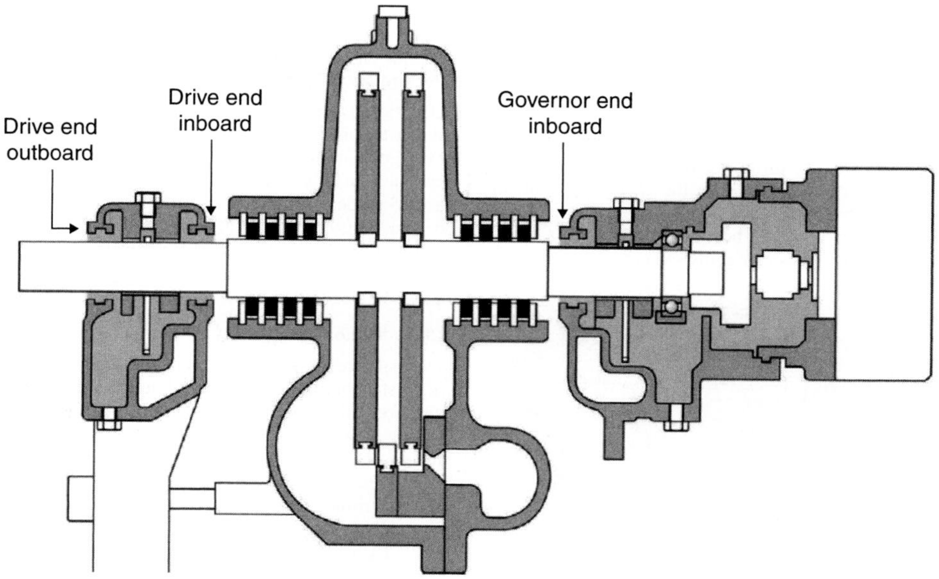

Figure 7.4.29 Generic sketch of a small steam turbine. The drive end is on the left. A mechanical or electronic governor can be accommodated on the thrust end at right.

Figure 7.4.30 Steam leakage will often contaminate the adjacent bearing housing.

rings or high temperature mechanical seals. It is a well-known fact that, as soon as the internally split carbon rings start to wear, high-pressure and high-velocity leakage steam (Fig. 7.4.30) finds its way into the bearing housings. Traditional labyrinth seals have proven ineffective in many such cases and only solidly engineered bearing protector seals will serve to block leakage steam passage for many years of operation.

The bearing housing protector seal in Fig. 7.4.31 was designed for steam turbines. It incorporates a small and a large diameter dynamic O-ring. This bearing housing protector seal, generally found in thousands of modern process pumps, is highly stable and not likely to wobble on the shaft; it is also field-repairable. With sufficient shaft rotational speed, one of the rotating ("dynamic") O-rings is flung outward and away from the larger O-ring. The larger cross-section O-ring is then free to move axially and a micro-gap opens up.

When the turbine is stopped, the outer of the two dynamic O-rings will move back to its stand-still position. At stand-still, the outer O-ring contracts and touches the larger cross-section O-ring. In this highly purposeful design, the larger cross-section O-ring touches a relatively large contoured area. Because Contact Pressure = Force/Area, a good designs aims for low pressure. Good designs differ greatly from technologically outdated configurations wherein contact with the sharp edges of an O-ring groove will cause O-ring damage.

Retrofitting or upgrading to advanced bearing protector seals is not only possible but can be affected very rapidly. Many such protector

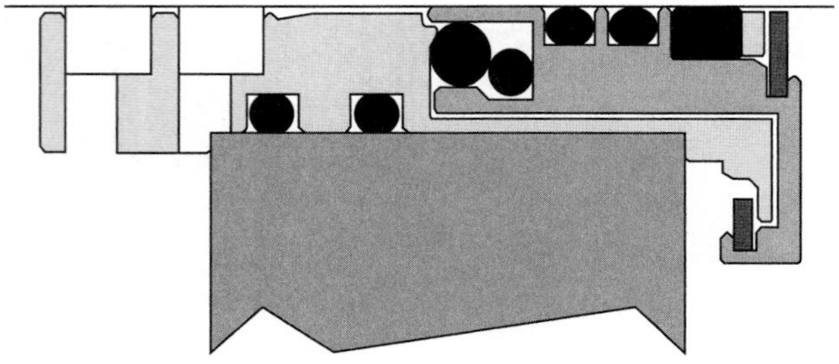

Figure 7.4.31 Modern steam turbine bearing protector seal made to fit into each of the three bearing housing end caps of most small steam turbines. (SOURCE: AESSEAL Inc., Rotherham, UK and Rockford, TN.)

seals have been in highly successful operation for years; they represent essential components for systematic failure reduction and equipment reliability improvement.

Our point is that highly cost-effective equipment upgrades are possible at hundreds of refineries and petrochemical plants. Nevertheless, superior bearing protector products for use in steam turbines must be purposefully developed. The type described here has important advantages compared with standard products:

- It is suitable for high temperatures.
- It incorporates Aflas® fluoroelastomer O-rings as the optimum secondary sealing material.
- Extra axial clearance is provided to accommodate thermal expansion.
- High-temperature graphite gaskets are incorporated in this design.

"Water Washing" for Steam Turbine Performance Restoration

Occasions may arise when deposits form on the internal parts of steam turbines. The accumulation of these deposits may be indicated by a gradual increase in stage pressures over a period of time with no evidence of vibration, rubbing, or other distress. Such deposits have a marked detrimental effect on turbine efficiency and capacity. When deposits cause extensive plugging, thrust bearing failure, wheel rubbing, and other serious problems can result. Fortunately, the deposits can be washed off with the steam turbine staying in operation and with great care being exercised while this water washing procedure is in progress.

Classification of Deposits

Deposits are classified as water insoluble and water soluble. The characteristics of the deposits should be determined by analysis of samples, and corrective measures should be taken to eliminate these deposits during future operation. When it has been determined that deposits have formed on the internal parts of the turbine, three methods may be employed to remove the deposits:

1. Turbine shut down, casing opened, and deposits removed manually.
2. Turbine shut down and allowed to cool, the deposits cracking off because of temperature changes.
3. On-line water washing (while running); essentially removal of deposits when deposits are water soluble.

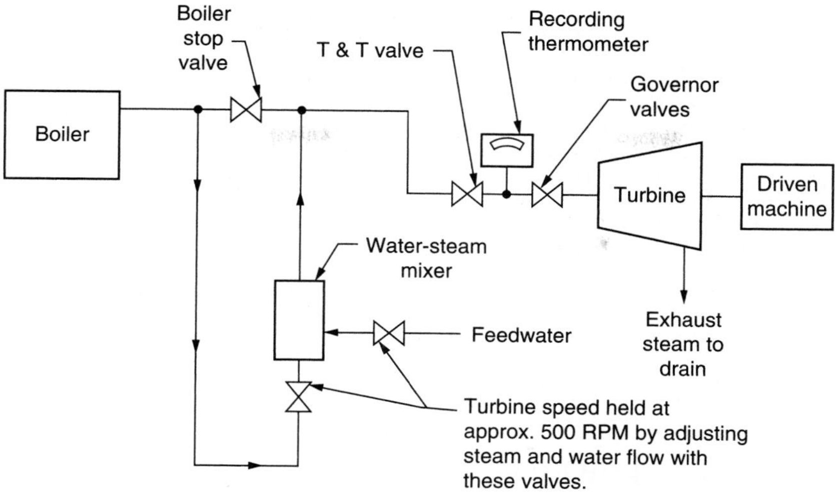

Figure 7.4.32 Schematic of low-speed water wash system.

Since deposits tend to accumulate to a greater extent during steady high-load conditions (e.g., base load generator drive, ethylene process drive), the application and plant operating conditions will dictate which method will best serve to restore the turbine to optimum performance. In a generator-drive service, shutting down the unit or water washing at low speed and reduced load may create minimum plant upset. The size of process drivers and plant-operating conditions present different circumstances.

Methods of Water Washing

Turbine washing at full speed (on-stream cleaning) can be and has been successfully accomplished on many mechanical drive steam turbines. Considerable hazard attends any method of water washing and full-speed washing is more hazardous than washing at reduced speed. But either method can be accomplished, provided great care and judgment are exercised. While we know of no steam turbine manufacturer who would guarantee the safety of turbines for any washing cycle, capable manufacturers recognize that deposits do occur. They will, therefore, help operators as much as possible in dealing with the problems until effective prevention is established.

Saturated steam washing by water injection is the conventional and well-tried method of removing water soluble deposits from turbines. The amount and rate of superheat to be removed and the amount of steam flow required for operation determine the water injection rate. It is the injection of large quantities of liquid (such as may be required on process drivers) that creates potential problems.

The nature of a typical impulse turbine lends itself to full-speed water washing. Axial clearances between first-stage buckets and nozzles and between moving buckets and diaphragms will range from 0.050 to 0.090 in. The typical NiResist labyrinth packing radial clearance when the unit is cold will be approximately 0.007 in. The labyrinth will seal on the shaft only; the moving blades will not require seals. With impulse turbines, these liberal clearances help minimize the hazards associated with water washing. Nevertheless, numerous reaction turbines have also been successfully water washed.

Water injection is accomplished by a piping arrangement for the atomization and injection of water into the steam supply to ensure a gradual and uniform reduction in the temperature of the turbine inlet steam until it reaches 10 to 15°F superheat. It is probably a safe rule that the temperature should not be reduced faster than 25°F in 15 minutes or 100°F/h. Figures 7.4.32 and 7.4.33 show suggested piping arrangements for the admission of water and steam and a simple assembly of fabricated pipes to form a desuperheater is found in the narrative below. Failure of water injection pumps presents a great hazard, especially at maximum injection rates. To guard against pump failure, untreated boiler feedwater is used since these pumps are usually the most reliable in a plant.

If plant operating conditions allow, the vacuum on a condensing turbine should be reduced to 5 to 10 mmHg; for non-condensing turbines, the exhaust pressure should be reduced to atmospheric pressure. Note that on any non-condensing unit requiring full-speed washing, the manufacturer should be consulted about minimum allowable exhaust

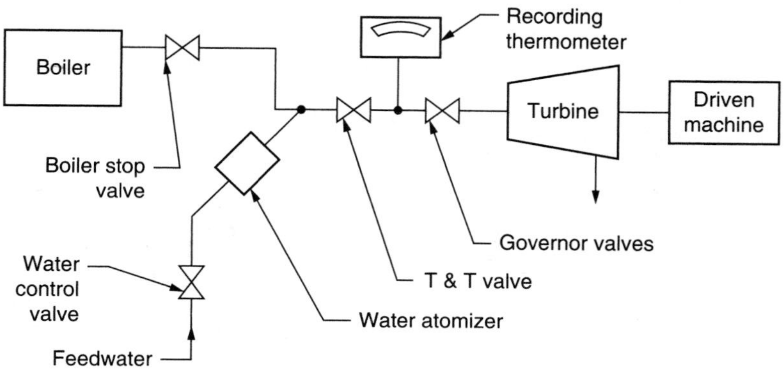

Figure 7.4.33 Full-speed water wash arrangement.

pressures. Extraction turbines should be run with the extraction line shut off.

A steam gauge and thermometer should be installed between the trip-throttle valve and the governor-controlled valves. The thermometer should preferably be a recording type and should be very responsive to small changes in temperature. Low speed wash, as illustrated in Fig. 7.4.32, represents a well-understood method of deposit removal.

To start the washing procedure, it is normally recommended to operate the turbine on trip-throttle valve control at one-fifth to one-fourth normal speed with no load. The live-steam valve to the mixer would now be opened and the boiler stop valve closed, after which the trip-throttle valve and the governor valves may be opened wide and the speed controlled by the small live-steam valve to the mixer. Water is then supplied to the mixing chamber in quantities sufficient to reduce the steam temperature at the recommended rate until 10 to 15°F (6 to 9°C) superheat at turbine inlet is reached. During the washing cycle the exhaust steam should be discharged to the sewer.

Full speed water washing (Fig. 7.4.33) uses a water atomizer instead of the water-steam mixer arrangement. Both Figs. 7.4.32 and 7.4.33 illustrate operating principles and the required minimum instrumentation. Advanced controls usually go beyond the instruments shown here, but will adhere to the same principles.

Summary

There are large financial incentives for upgrading steam turbine controls, for incorporating water wash features, and for precluding water intrusion into the bearing housings of small steam turbines. The initial incremental cost outlays for the reliability improvements described above are relatively low. Reliability-focused industrial facilities have documented rapid payback for each of these improvement or upgrade measures.

7.5 POWER PLANT HEAT EXCHANGERS

by William J. Bow, Assisted by Donald E. Bolt and Charles F. Bowman

NOTE: Standards for this industry retain USCS units except as indicated in the text.

SURFACE CONDENSERS

REFERENCE: Heat Exchange Institute Standards for Steam Surface Condensers.

The power plant surface condenser is attached to the low-pressure exhaust of a steam turbine (see Figs. 7.5.1 and 7.5.2). Its purposes are (1) to produce a vacuum or desired back pressure at the turbine exhaust for the improvement of plant heat rate, (2) to condense turbine exhaust steam for reuse in the closed cycle, (3) to deaerate the condensate, and (4) to accept heater drains, makeup water, steam drains, and start-up and emergency drains.

An economical turbine back pressure is from 1.0 to 3.5 inHg abs. The factors involved in establishing this pressure are involved and will not be discussed here.

An equipment diagram of a closed power plant cycle is shown in Fig. 7.5.1.

For a condenser to deaerate the condensate, it must remove oxygen and other noncondensable gases to an acceptable level compatible with material selection and/or chemical treatment of the feedwater (condensate). Depending on materials and treatment, the dissolved O_2 level must normally be kept below 0.005 cm^3/L for turbine units operating with high-pressure and -temperature steam.

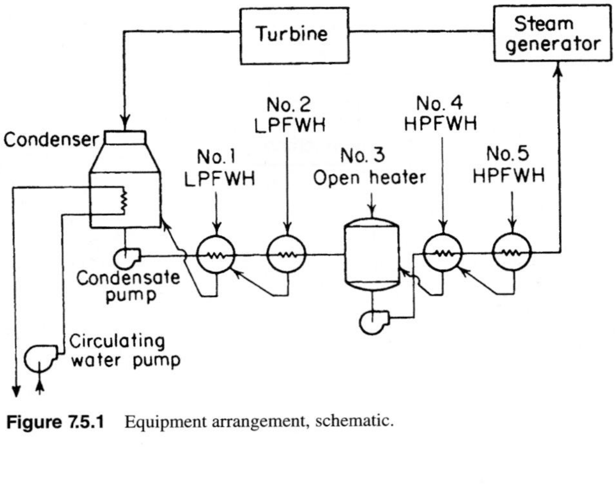

Figure 7.5.1 Equipment arrangement, schematic.

Deaeration in a condenser is accomplished by applying Henry's law, which states that the concentration of the dissolved gas in a solution is directly proportional to the partial pressure of that gas in the free space above the condensate level in the hot well, with the exception of those gases (e.g., $CO_2 + NH_3$) which unite chemically with the solvent. In a condenser droplets of condensate are continually scrubbed with steam, liberating the O_2 and permitting it to flow to the low-pressure air-removal section, where it is discharged to the atmosphere by the air-removal equipment.

To remove the last traces of O_2 from the condensate, an ammonia compound such as hydrazine is normally added. Free ammonia is liberated in the cycle and is either removed with the noncondensables as a gas or is condensed and retained in the condensate, depending on the detailed design of the condenser air-removal section. If the ammonia is concentrated as a liquid, it can be very corrosive to certain copper-base materials.

Most condenser manufacturers have **tube-bundle configurations** unique to their design philosophy. Basically, pressure losses from turbine exhaust to the air offtake are kept to a minimum and tubes are arranged to promote good heat-transfer rates. Small condensers are usually cylindrical, whereas large ones are rectangular for better utilization of space. Most turbines exhaust downward into the condenser, but condensers are also built to accommodate side as well as axial exhaust turbines.

Because of the inherent strength of cylindrical shapes as opposed to flat plates, condenser **water boxes** are generally made with curved surfaces. This has come about because of the increased pressure resulting from cooling towers, which in turn, are the result of environmental influences. With a cooling tower, pressures are in the 60 to 80 lb/in^2 range, whereas with water from lakes, rivers, etc., where a siphon system can be employed, water-box design pressures are in the 20 to 30 lb/in^2 range. As a general rule, **tube selection** is based on economics; 18 BWG admiralty metal has been satisfactory for freshwater service and 90-10 copper-nickel material likewise for seawater. The current trend is to use 22 BWG titanium or one of the new specially formulated stainless-steel tube materials for this application. Titanium has found favor as condenser tubing material. Its long life offsets the higher initial cost. Titanium in alloyed form suffers from reduced corrosion resistance and weldability but generally exhibits higher tensile strength when compared to the unalloyed form. Material prices fluctuate greatly, and selection can be influenced by first cost. Lost revenue due to downtime caused by tube leaks or other causes, particularly in larger units, can usually justify the use of more exotic and expensive materials.

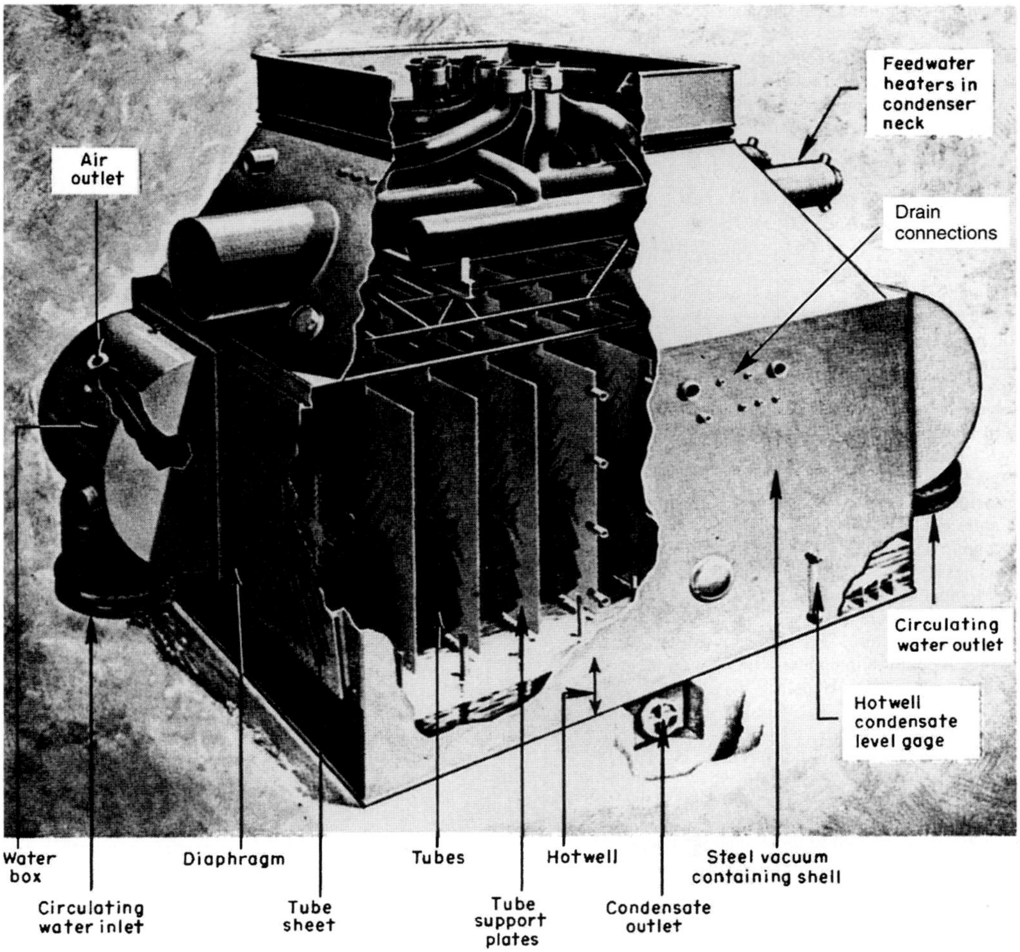

Figure 7.5.2 One-pass rectangular surface condenser.

To minimize leaks between the tubes and tube sheets, copper-based alloyed tubes are normally rolled into groves in the tube sheet, whereas titanium and stainless steel alloyed tubes are normally welded to a tube sheet with a weld overlay.

Low-pressure feedwater heaters are frequently located in the steam-inlet neck of a condenser. This is done to minimize pressure drop in the extraction steam piping and to utilize floor space surrounding the condenser better.

A sufficient number of **tube supports** must be provided within the condenser so that the tubes will not vibrate excessively, which will cause tubes to rub or crack circumferentially. During low-water-temperature operation, the steam entering the condenser will often reach sonic velocities, causing severe **flow-induced vibration** and ultimate tube failures if the tube support system is inadaquate.

Where once-through boiler or nuclear steam generators are used, it is imperative to dispose of large quantities of steam during starting and stopping of a turbine unit. The condenser, because of its large volume, has been used as a convenient dumping place for this steam. Means must be provided within the condenser to accommodate the high-energy steam without damage to condenser tubing, structural members, or the low-pressure end of the turbine.

A major factor in condenser performance is **tube cleanliness.** Tubes can be plugged with leaves, marine life, and other debris deposited on the face of the tube sheets. By valving, various arrangements for back washing, or flushing, are effected to remove these materials.

The inside surfaces of the tubes can also become fouled by scaling, corrosion products, silt, products of biofouling such as slimes, or a combination of these. To clean the inside surfaces of the tubes, it is necessary to force brushes or plugs through them individually by the use of water or air jets. To do this, the affected portion of the condenser must be shut down. As an alternative, recirculating sponge rubber ball or brush systems are available that can be used during normal operation. To remove scale formations, sponge rubber balls with abrasive coatings are effective, but their continued use will shorten tube life.

Movement due to **temperature differences** between turbine and condenser is usually accommodated by a stainless-steel bellow or rubber-belt-type **expansion joint.** For small units the condenser may be supported on springs and rigidly connected to the turbine. To accommodate differential expansion between condenser shell and tubes, a flexible diaphragm or other expansion element may be installed.

The **tube-to-tube sheet joint** is usually rolled but has been welded in certain installations. Proper material selection must be made regardless of whether the joint is rolled or welded. Currently, considerable emphasis is placed on inward leakage of circulating water into the condenser steam space. Double tube sheets have been supplied for some large central-station condensers for nuclear units. They are standard in the Navy's nuclear-powered ships and submarines to minimize this source of contamination.

Condensers for some large turbine units having two or three low-pressure ends are designed for **multipressure application.** Usually there will be a gain, in the form of either heat-rate improvement or a reduction in capital investment, with a multipressure condenser. A rather complex analysis must be made for each application before a decision can be made.

The current version of "Standards for Steam Surface Condensers" is published by the Heat Exchange Institute in Cleveland, OH. These are available to the public and provide guidance in solving most structural and sizing problems related to surface condensers.

Performance Calculations and Sizing

Notaion

S = condenser tube surface area, ft^2
C_c = cleanliness factor
C_1 = heat-transfer-rate constant
C_m = material and gage factor
C_t = temperature correction factor
c_p = specific heat, Btu/lb, °F
G = circulating-water quantity, gal/min
h = enthalpy, Btu/lb
h_r = heat rejected by steam, Btu/lb
L = length of water travel, ft
NPSH = net positive suction head, ft
OD = tube outside diameter, in
Q = heat transferred, Btu/h
R = temperature rise ($t_o - t_i$), °F
TDH = total dynamic head, ft
t = tube thickness, in
TTD = terminal temperature difference = $t_s - t_o$
t_i = inlet-water temperature, °F
t_o = outlet-water temperature, °F
t_s = saturation temperature in condenser, °F
U_o = overall heat-transfer rate, Btu/(ft^2 h °F)
V = water velocity, ft/s
W_s = steam to be condensed, lb/h
Δt_m = logarithmic mean temperature difference, °F

In **sizing** a condenser, the steam flow and heat rejected to the condenser are obtained from the turbine heat balance. Table 7.5.1 gives representative steam flows; heat rejected to the condenser is approximately 950 Btu/lb of steam for nonreheat turbines and 975 Btu/lb for reheat machines. Figure 7.5.3 shows approximate average water temperatures for the United States; local water temperature should be used, when known. The number of passes is usually dictated by the plant arrangement, with total length of water travel and tube diameter dictated by economic considerations. Normally, small-diameter tubes, single-pass condensers are used where water is plentiful, and larger-diameter tubes, two-pass condensers when water is scarce. The vacuum, or back pressure, is determined by economic evaluation, but Table 7.5.2

Table 7.5.1 Steam Flow to Condensers*

Turbine throttle conditions	Weight flow, lb/kWh
600 lb/in^2, 825°F	7.08
850 lb/in^2, 900°F	6.45
1,250 lb/in^2, 950°F	5.80
1,450 lb/in^2, 1,000°F	5.44
1,450 lb/in^2, 1,000°F, 1,000°F	4.70

* Approximate values for unit sizes to 100,000 kW and exhaust at 1.0 inHg abs.

NOTE: For exhaust pressures other than 1.0 inHg abs, multiply tabular values by 1.02 (1.04)[1.06] for 1.5 (2.0)[2.5] in.

Table 7.5.2 Normal Condenser Pressures and Circulating-Water Temperatures

Water inlet temp t_i, °F	Normal back pressure, inHg abs
55	1.0
70	1.5
80	2.0
85	2.5
90	3.0
95	3.5

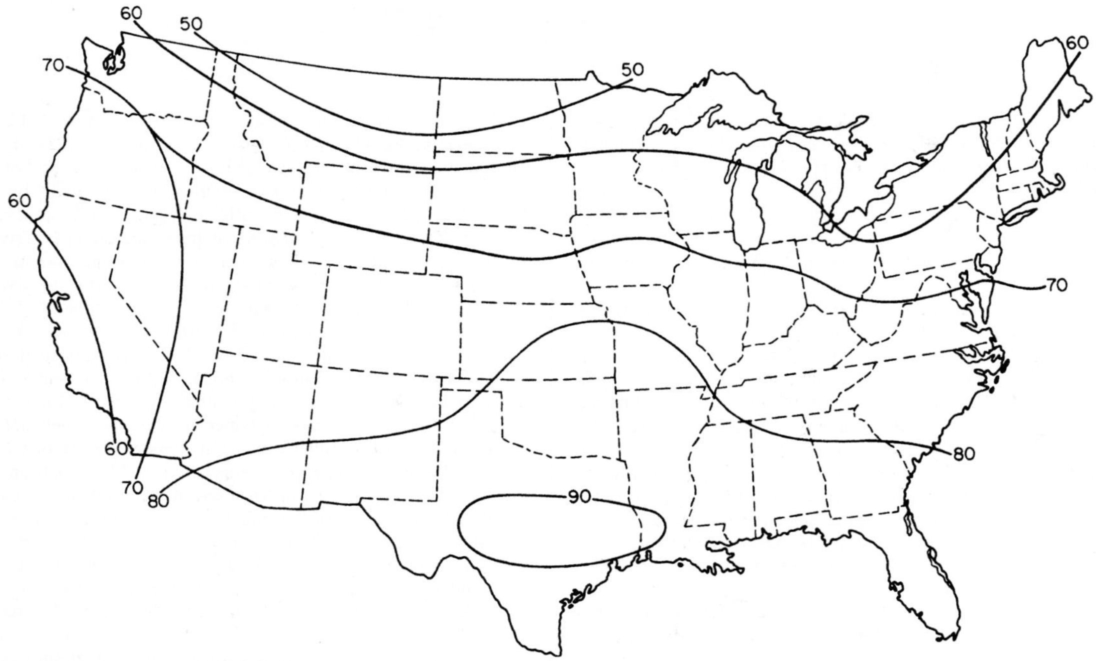

Figure 7.5.3 Average inlet temperature of circulating water, °F, United States.

Table 7.5.3 Pressure-Temperature Conversion Table for Water*

Abs press, inHg	Sat temp, °F	Abs press, inHg	Sat temp, °F	Abs press, inHg	Sat temp, °F	Abs press, inHg	Sat temp, °F
0.20	34.57	1.35	88.36	2.50	108.71	8.00	152.24
0.25	40.23	1.40	89.51	2.55	109.38	9.00	157.09
0.30	44.96	1.45	90.64	2.60	110.06	10.00	161.49
0.35	49.06	1.50	91.72	2.65	110.72	11.00	165.54
0.40	52.64	1.55	92.77	2.70	111.37	12.00	169.28
0.45	55.89	1.60	93.81	2.75	112.01	13.00	172.78
0.50	58.80	1.65	94.80	2.80	112.63	14.00	176.05
0.55	61.48	1.70	95.78	2.85	113.25	15.00	179.14
0.60	63.96	1.75	96.73	2.90	113.86	16.00	182.05
0.65	66.26	1.80	97.65	2.95	114.45	17.00	184.82
0.70	68.41	1.85	98.56	3.00	115.06	18.00	187.45
0.75	70.43	1.90	99.43	3.20	117.35	19.00	189.96
0.80	72.32	1.95	100.30	3.40	119.51	20.00	192.37
0.85	74.13	2.00	101.14	3.60	121.57	21.00	194.68
0.90	75.84	2.05	101.96	3.80	123.53	22.00	196.90
0.95	77.48	2.10	102.77	4.00	125.43	23.00	199.03
1.00	79.03	2.15	103.56	4.20	127.21	24.00	201.09
1.05	80.53	2.20	104.33	4.40	128.94	25.00	203.08
1.10	81.96	2.25	105.09	4.60	130.61	26.00	205.00
1.15	83.33	2.30	105.85	4.80	132.20	27.00	206.87
1.20	84.64	2.35	106.58	5.00	133.76	28.00	208.67
1.25	85.93	2.40	107.30	6.00	140.78	29.00	210.43
1.30	87.17	2.45	108.01	7.00	146.86	29.921	212.00

* See also Sec. 2 for further data.

gives normal recommended values for average water temperatures. Table 7.5.3 is a pressure-temperature conversion table.

A **cleanliness factor** is applied to the heat-transfer rate of new, clean tubes to allow for gradual decrease by fouling. A standardized cleanliness factor of 85 percent is frequently used, but this can often be misleading and even erroneous. The fouling is attributable to (1) sedimentation, (1) scaling, (3) steam-side deposits, (4) corrosion, and (5) biological growth. The fouling is correctly determined by use; in some cases the cleanliness factor may be 90 percent especially if an online tube cleaning system is employed, whereas in other cases it may never rise above 75 percent.

Velocities normally used are: for clean water, 7.0 ft/s; for very clean water with cooling towers, 8.0 ft/s; and for seawater, with entrained sand, as low as 6.0 ft/s to minimize erosion. Prevalent velocities are 6.5 ft/s with aluminum-brass tubes, 7.0 ft/s with admiralty metal, and 8 + ft/s with stainless steel or titanium. **Water-temperature rise** is about 10°F for single-pass condensers and 15°F for two-pass condensers, with a minimum 5° terminal temperature difference (TTD).

Approximate general rules for condensers serving turbines rated up to 100 MW. The **surface area,** ft^2, is equal to the steam flow, lb/h, divided by 10 for single-pass condensers and by 7.5 for two-pass condensers. **Circulating-water quantity,** gal/min, is equal to the area, ft^2, for a two-pass condenser and is twice the area for a single-pass condenser. Condenser **proportions** are given in Table 7.5.4. Empty **weight** of an installed condenser is 5 to 6 lb/ft^2 of surface.

Table 7.5.4 Typical Small Condenser Proportions

Surface area, ft^2	Effective tube lengths, ft		
	¾-in OD	⅞-in OD	1-in OD
1,000–1,750	8, 10, 12, 14		
2,000–2,500	10, 12, 14, 16		
2,750–4,750	12, 14, 16, 18	12, 14, 16, 18	
5,000–7,000	14, 16, 18, 20	14, 16, 18, 20	14, 16, 18, 20
7,250–14,000	16, 18, 20, 22	16, 18, 20, 22	16, 18, 20, 22
15,000–19,000		18, 20, 22, 24	18, 20, 22, 24
20,000–27,500		20, 22, 24, 26	20, 22, 24, 26
30,000–47,500		22, 24, 26, 28	22, 24, 26, 28
50,000–75,000		24, 26, 28, 30	24, 26, 28, 30

On large power plants that employ cooling towers, economics dictate a relatively low circulating water flow rate and higher rise compensated for by a larger condenser. To avoid a condenser design having a terminal temperature difference of less than 5°F a multi-pressure condenser is sometimes employed in which the circulating water passes through multiple condenser shells in series.

Condenser Calculations

The basic diagram is shown in Fig. 7.5.4 and the applicable equations are as follows:

P is defined as

$$P = \frac{t_o - t_i}{t_s - t_i} \tag{7.5.1}$$

Therefore,

$$t_s = \frac{t_{out} - t_{in}(1 - P)}{P} \tag{7.5.2}$$

Where for a condenser

$$P = 1 - e^{-NTU} \tag{7.5.3}$$

and

$$NTU = \frac{U\,S}{500G} \tag{7.5.4}$$

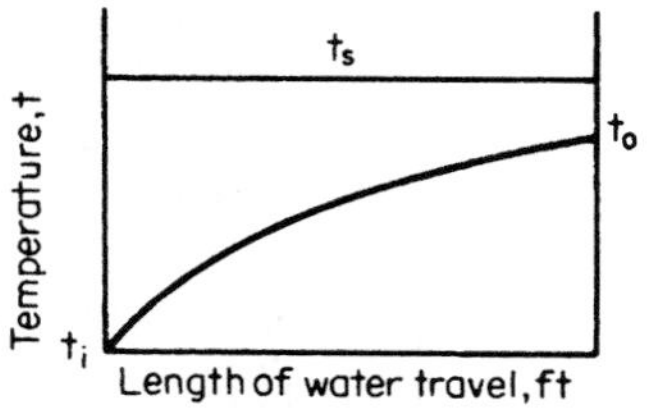

Figure 7.5.4 Temperature vs. water travel, surface condenser.

and

$$U = C\sqrt{V}\,C_t\,C_m\,C_c \qquad (7.5.5)$$

Note that the specific heat c_p is normally neglected in condenser work, it being considered insignificant.

EXAMPLE. Calculate the surface and circulating-water requirements to condense 445,000 lb of steam per h to an absolute pressure of 2.00 inHg abs. The heat rejected is 980 Btu/lb of steam and is absorbed by circulating seawater at an average inlet temperature of 75°F. The condenser is to be two-pass with ⅞-in 18-gage, 26-ft-active-length aluminum-brass tubes; cleanliness factor = 85 percent.

SOLUTION. The heat transfer rate U_o is a function of the tube diameter, material, gage, water velocity, and cleanliness factors within the velocity limits of 3 and 10 ft/s. For ⅝- and ¾-in tubes $C_1 = 267$; for 7/8- and 1-in tubes $C_1 = 263$, and for 1⅛- and 1¼-in tubes $C = 259$. The inlet-water-temperature correction factor is obtained from Table 7.5.7. The material and gage correction factor C_m is obtained from Table 7.5.6, and the cleanliness factor C_c is given in the conditions of the example.

Assume a surface of 59,000 ft² based on the typical steam loading for such a unit and a water velocity of 6.5 ft/s. For the selected surface and water velocity by using the information in Table 7.5.5, the number of gallons per minute required by this selection is determined. It is now possible to determine the steam temperature that this selection would produce, by using Eq. (7.5.1). The results show a steam temperature of 99°F, indicating that a smaller surface is required to attain a 2-inHg abs condenser pressure. Try a smaller unit until a surface with no more than three significant figures will yield a steam temperature of no more than 101.14°F. The surface arrived at is 54,400 ft² with 43,900 gal/min of circulating water.

Table 7.5.5 Tube Characteristics

OD and ext. surface, ft²/ft	BWG no.	ID, in	Water, gal/min @ 1 ft/s velocity
	16	0.495	0.600
⅝-in OD	18	0.527	0.680
0.1636	20	0.555	0.754
	22	0.569	0.793
	16	0.620	0.941
¾-in OD	18	0.652	1.041
0.1963	20	0.680	1.132
	22	0.694	1.179
	16	0.745	1.359
⅞-in OD	18	0.777	1.478
0.2291	20	0.805	1.586
	22	0.819	1.642
	16	0.870	1.853
1-in OD	18	0.902	1.992
0.2618	20	0.930	2.117
	22	0.944	2.182

Table 7.5.6 Material and Gage Factor

	C_m						
	BWG no.						
Tube material	24	22	20	18	16	14	12
Admiralty metal	1.03	1.02	1.01	1.00	0.98	0.96	0.93
Arsenical copper	1.04	1.03	1.03	1.02	1.01	1.00	0.98
Copper iron 194	1.04	1.04	1.03	1.03	1.02	1.01	1.00
Aluminum brass	1.02	1.02	1.01	0.99	0.97	0.95	0.92
Aluminum bronze	1.02	1.01	1.00	0.98	0.96	0.93	0.89
90-10 Cu-Ni	0.99	0.98	0.96	0.93	0.89	0.85	0.80
70-30 Cu-Ni	0.97	0.95	0.92	0.88	0.83	0.78	0.71
Cold-rolled carbon steel	1.00	0.98	0.97	0.93	0.89	0.85	0.80
Stainless steel type 304/316	0.90	0.86	0.82	0.75	0.69	0.62	0.54
Titanium	0.94	0.91	0.88	0.82	0.77	0.71	0.63

Table 7.5.7 Inlet-Water-Temperature Correction Factors

Inlet temp t_i, °F	Correction factor C_t
40	0.744
50	0.830
60	0.920
70	1.000
80	1.047
90	1.078
100	1.102

If this condenser is single-shell with one main turbine exhaust inlet and has no auxiliary turbine exhausts entering it (such as from a turbine-driven boiler feed pump), then the air-removal equipment must have a minimum capacity of 10 scfm of air at 1-inHg abs suction pressure. (See Table 7.5.8.)

Performance curves (Fig. 7.5.5) are drawn for a condenser to show the back pressure for various condenser steam loads and inlet-water temperatures maintaining a minimum TTD of 5°. The zero-load back pressure and the cutoff pressure are shown in Fig. 7.5.6, where the cutoff limitation is set by the air-removal equipment. A combined turbine condenser performance curve is sometimes drawn in which heat rejected versus back pressure for a given turbine load is superimposed on the condenser performance.

Table 7.5.8 Venting-Equipment Capacities*

	One condenser shell		
Effective steam flow each main exhaust opening, lb/h		Total number of exhaust openings	
		1	2
Up to 25,000	scfm†	3.0	4.0
	Dry air, lb/h	13.5	18.0
	Water vapor, lb/h	29.7	39.6
	Total mixture, lb/h	43.2	57.6
25,001-50,000	scfm†	4.0	5.0
	Dry air, lb/h	18.0	22.5
	Water vapor, lb/h	39.6	49.5
	Total mixture, lb/h	57.6	72.0
50,001-100,000	scfm†	5.0	7.5
	Dry air, lb/h	22.5	33.8
	Water vapor, lb/h	49.5	74.4
	Total mixture, lb/h	72.0	108.2
100,001-250,000	scfm†	7.5	12.5
	Dry air, lb/h	33.8	56.2
	Water vapor, lb/h	74.4	123.6
	Total mixture, lb/h	108.2	179.8
250,001-500,000	scfm†	10.0	15.0
	Dry air, lb/h	45.0	67.5
	Water vapor, lb/h	99.0	148.5
	Total mixture, lb/h	144.0	216.0
500,001-1,000,000	scfm†	12.5	20.0
	Dry air, lb/h	56.2	90.0
	Water vapor, lb/h	123.6	198.0
	Total mixture, lb/h	179.8	288.0
1,000,001-2,000,000	scfm†	15.0	25.0
	Dry air, lb/h	67.5	112.5
	Water vapor, lb/h	148.5	247.5
	Total mixture, lb/h	216.0	360.0

* Heat Exchange Institute.
† 14.7 lb/in² abs at 70°F.
NOTE: These tables are based on air leakage only and the air-vapor mixture at 1 inHg abs and 71.5°F.

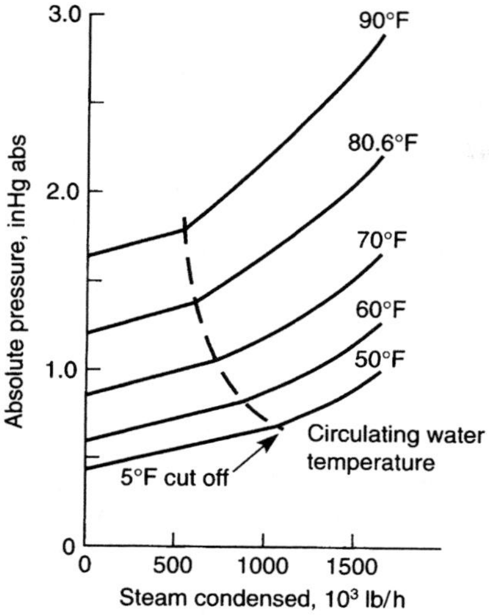

Figure 7.5.5 Representative performance curves for a 157,500-ft^2 one-pass surface condenser, 1-in, 18-gage, 37.8-ft-active-length aluminum-brass tubes. Performance is based on 85 percent clean tubes, 221,400 gal/min cooling water at a velocity of 7 ft/s, and 968 Btu/lb rejected to the cooling water.

Circulating-water pumps are usually half capacity, with 3 or 4 percent additional allowance on each pump for miscellaneous heat exchangers. The total dynamic head (TDH) is the sum of the condenser and water- box friction (Figs. 7.5.7 to 7.5.11), the pipe loss, and any unrecovered static head. Normal heads are about 20 ft; for cooling-tower installations, 60 ft. (See also Sec. 10 and Fig. 7.5.12.)

Condensate pumps are usually full capacity, determined by condenser flow plus any heater drains dumped into the condenser and 5 percent additional margin. Vertical-pit-type designs are commonly used because of low net-positive-suction-head (NPSH) requirements with water at the boiling point. TDH ranges from a normal value of 100 ft on small plants to 800 ft on larger plants. (See Sec. 10.)

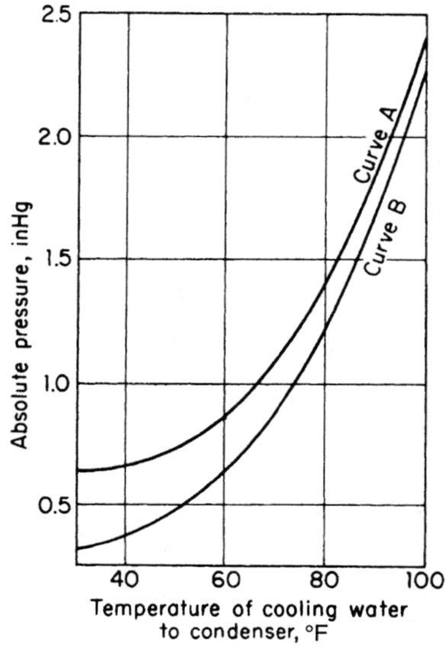

Figure 7.5.6 Cutoff and zero load vacuums. Curve *A*, cutoff (except where the absolute pressure is limited to 5°F TTD); curve *B*, zero load.

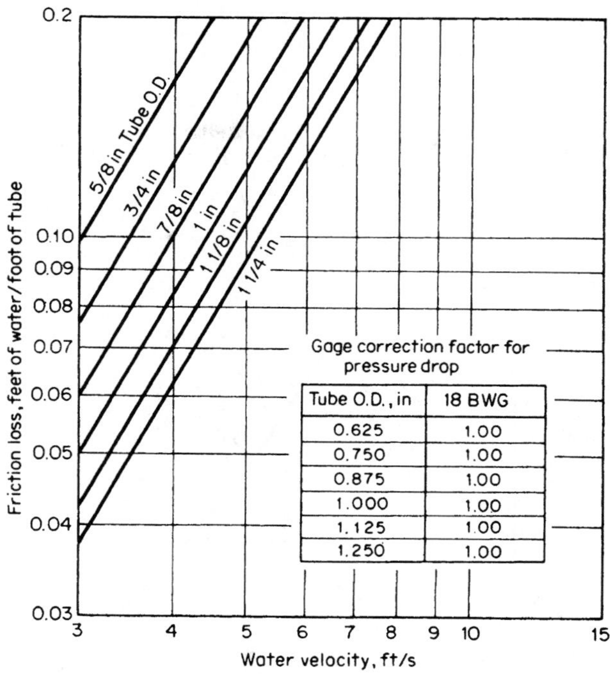

Figure 7.5.7 Friction loss for water flowing in 18 BWG tubes, low velocity. *(Heat Exchange Institute.)*

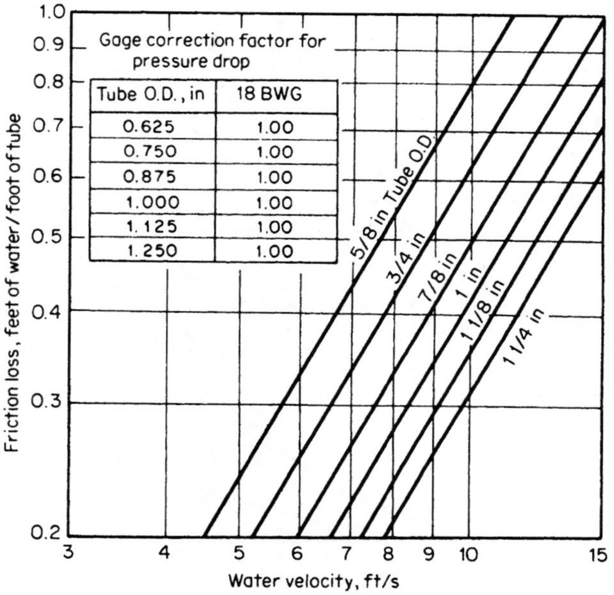

Figure 7.5.8 Friction loss for water flowing in 18 BWG tubes, high velocity. *(Heat Exchange Institute.)*

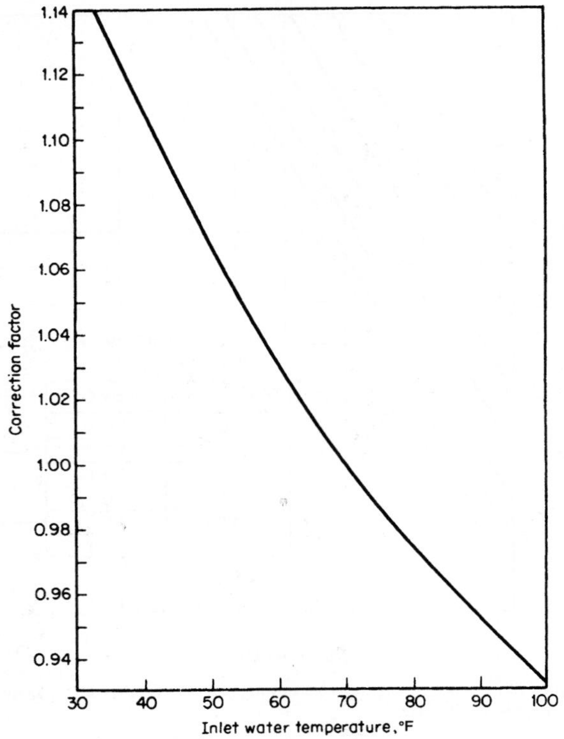

Figure 7.5.9 Temperature correction for friction losses in tubes (see Figs. 7.5.7 and 7.5.8). *(Heat Exchange Institute.)*

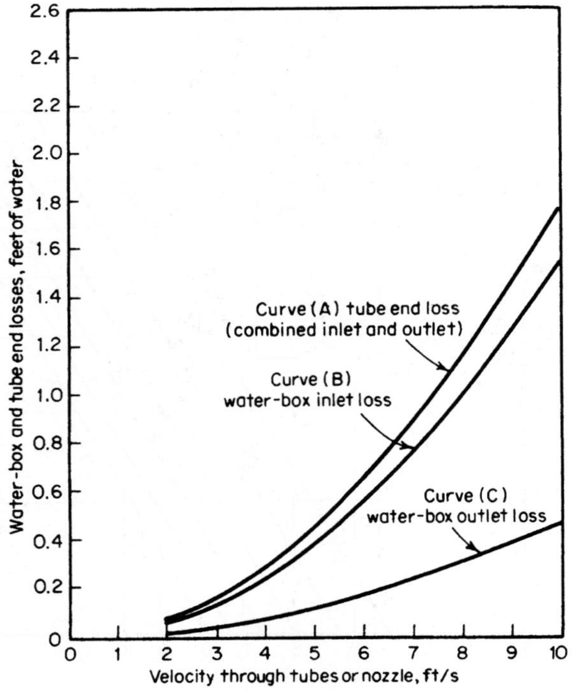

Figure 7.5.10 Water-box and tube end losses, single-pass condensers. *(Heat Exchange Institute.)*

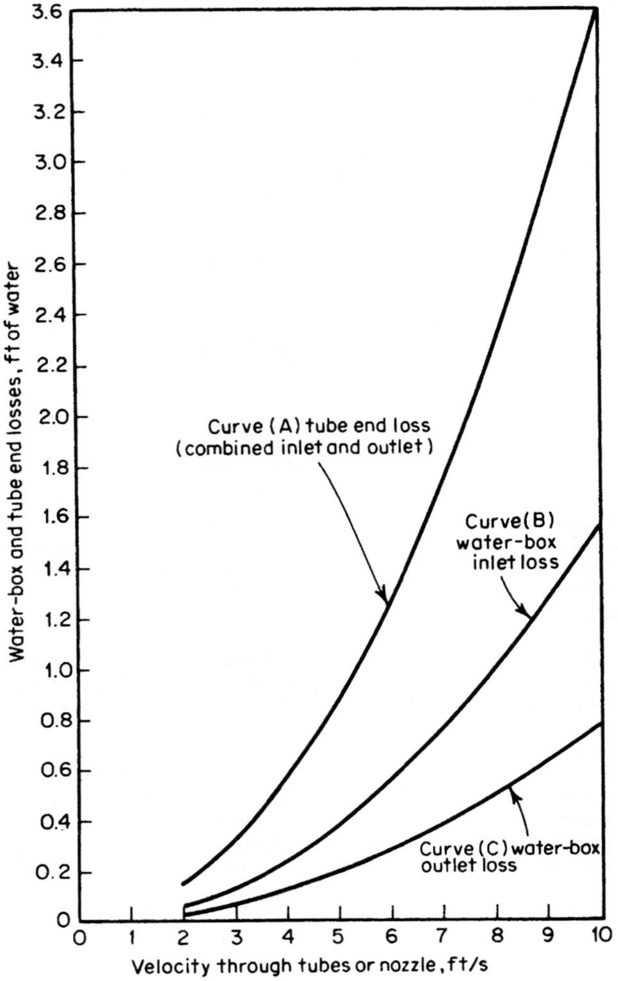

Figure 7.5.11 Water-box and tube end losses, two-pass condensers. *(Heat Exchange Institute.)*

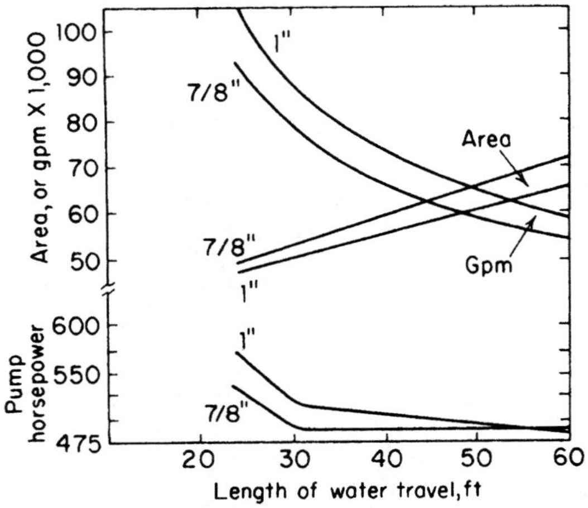

Figure 7.5.12 Condenser configuration selection for a given heat load, W_s = 477,355 lb/h; h = 975 Btu/lb; t_i = 70.0°F; 1.5 inHg abs; 18-gage admiralty metal tubes; 7.0-ft/s water velocity; 85 percent cleanliness factor.

AIR-COOLED CONDENSERS

The air-cooled condenser is used where adequate water is not available in sufficient quantities to permit the use of conventional cooling towers. The water lost by evaporation and cooling tower blow-down is approximately equal to the steam condensed from the turbine exhaust.

In an air-cooled condenser, the steam is condensed inside tubes while cooling air flows over fins on the external surfaces. These tubes are arranged in the area where packing or fill would be arranged in a conventional cooling tower, and propeller fans (see Sec. 10) supply air across the tube surfaces.

Aluminum is generally the material of choice for tubing because of its excellent heat transfer properties and availability. Items to be considered in design are the overall heat-transfer coefficient, steam-side pressure drop, uniformity of steam distribution, noncondensible gas concentration for air removal, and, of course, provision for freeze protection. Turbine design back pressures are normally in the range of 5 inHg abs when served by air-cooled condensers.

DIRECT-CONTACT CONDENSERS

REFERENCES: Heat Exchange Institute Standards for Direct Contact and Low Level Condensers.

When (1) low investment is desired and (2) condensate recovery is not a factor, direct-contact condensers are effective. They are relatively simple to build and operate, are limited to sizes less than 250,000 lb of steam per h, and are built in three types: (1) barometric, (2) low-level, and (3) jet.

Figure 7.5.13 shows a self-supported counterflow **barometric condenser** with tail pipe, hot well, and air ejector. Steam and cooling water flow in opposite directions, with the coldest water for final condensation and cooling of noncondensables. The air pump must handle that part of the air disengaged from the cooling water as well as air leakage. The head required to pump water is pipe friction plus static head, minus 75 percent approx of the design vacuum. The barometric condenser is usually placed outdoors, and the water leg must be more than 34 ft high. The **low-level condenser** substitutes a pump for the water leg of the barometric condenser to remove the liquid from the vacuum space. The **jet condenser** utilizes the aspirating effect of a jet for the entrainment of noncondensables and the consequent elimination of a separate air pump.

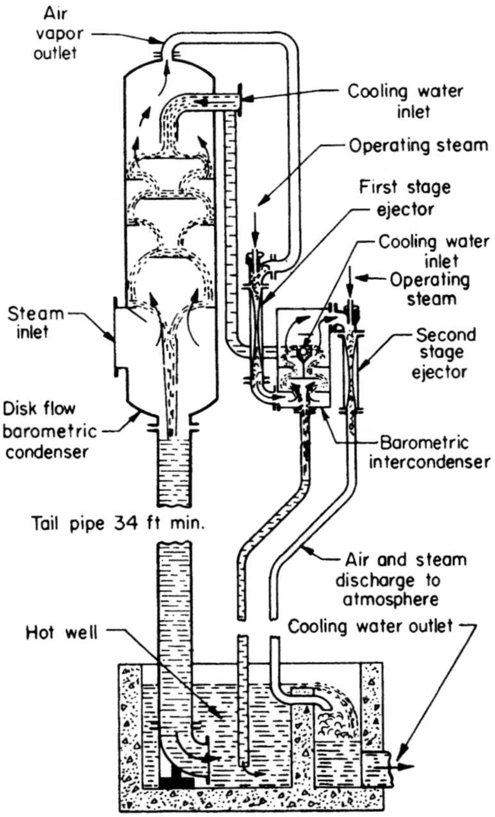

Figure 7.5.13 Counterflow barometric condenser with two-stage air ejector. (*Ingersoll-Rand Co.*)

Table 7.5.9 Air in Cooling Water

Water temp, °F	Air, ft³/min per 1,000 gal/min
35	4.0
40	3.78
50	3.35
60	2.97
70	2.68
80	2.41
90	2.21
100	2.0

In the usual direct-contact condenser, where steam and raw circulating water are mixed, the recovery of pure condensate is precluded; greater feedwater makeup is necessary, and poorer vacuums are attained than with surface condensers. Direct-contact-condenser installations are not found in large plants, but there is some recent interest in their use with dry cooling towers.

Table 7.5.9 gives values of expected air content of cooling waters; Fig. 7.5.14 gives typical performance curves; Fig. 7.5.15 defines flows W, temperatures t, and enthalpies h needed for the basic heat-balance equation $W_s(h_s - h_2) = W_w(t_2 - t_1)$. The latent heat $(h_s - h_2)$ is frequently taken as 950 Btu/lb of steam.

Shell materials are usually steel plate; with dirty or corrosive water, bronze, stainless steel, or linings of ceramic, plastic, or rubber may be used.

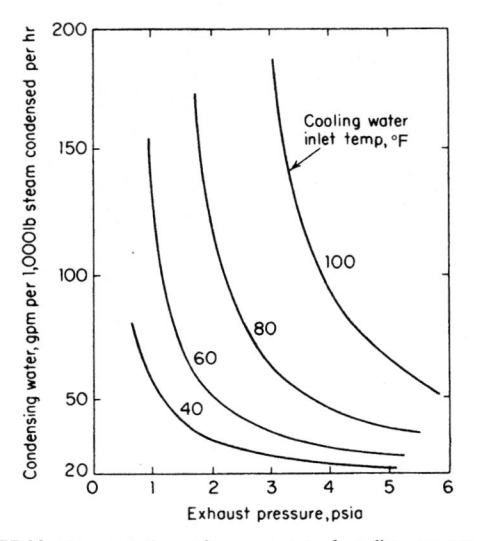

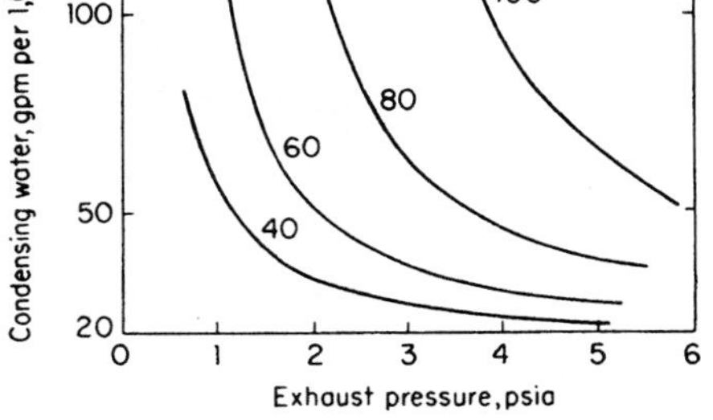

Figure 7.5.14 Representative performance curves for a direct-contact condenser.

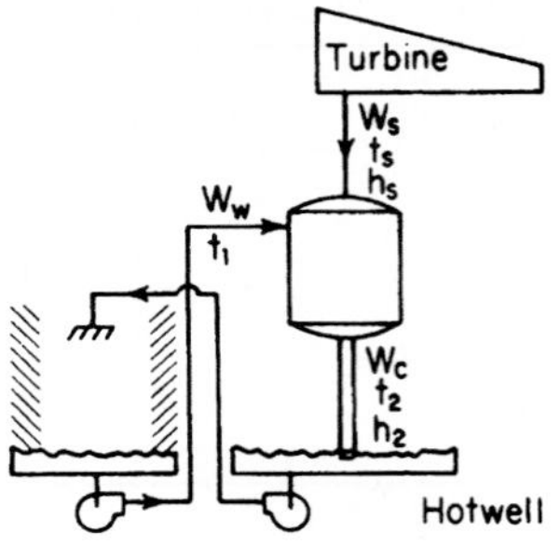

Figure 7.5.15 Direct-contact condenser performance terms.

AIR EJECTORS

REFERENCE: Heat Exchange Institute, Standards for Steam Jet Ejectors.

The steam-jet ejector is used to remove noncondensable air and gases from condensers. It consists of a suction chamber, diffuser, and steam nozzle. The high-velocity jet of steam issuing from the nozzle entrains the gas, and the kinetic energy of the mixture is converted to pressure energy in the diffuser. Ratings are in lb/h at a given suction pressure. Ejectors are operated in series or are staged for absolute suction pressures of 1 ± inHg abs. They will handle wet or dry mixtures, they are easy to operate, installation costs are low, and there are no moving parts—with consequent long life, high sustained efficiency, and low maintenance.

Two-stage ejectors, with surface-type inter- and aftercondensers (Fig. 7.5.16), are common in steam-surface-condenser application. The main condensate serves as a coolant, returning heat regeneratively to the boiler feed system. Two-stage ejectors have a shutoff pressure of 1.5 inHg abs approx. Two elements are usually provided, one serving as a spare.

Computation of **ejector performance** requires application of Dalton's law of mixtures (see Sec. 2), where basically the total pressure is the sum of the condensable vapor (steam) pressure and the noncondensable gas (air) pressure. In power plant condensers, the mixture is saturated with steam so that the temperature of the mixture fixes the partial pressure of the vapor.

Air ejectors are usually rated with suction conditions of 1 inHg abs and 7.5°F subcooling. Since the saturation temperature of steam at 1 inHg abs is 79°F, the mixture entering the ejector is at 71.5°F. The partial pressure of the vapor at 71.5°F is 0.78 inHg abs, and the partial pressure of the air is 1 – 0.78 = 0.22 inHg abs. Therefore, an ejector with a standard rating (see Table 7.5.8) of 12.5 ft^3/min of free dry air at 30 inHg abs handles, by the gas laws, 12.5(30/0.22) = 1,704 ft^3/min at the ejector suction.

Motive steam to the ejector is usually supplied from turbine throttle conditions, for example, 600 lb/in^2, 750°F; 1,500 lb/in2, 1,000°F; 2,500 lb/in^2, 1,050°F, through a reducing valve to a pressure not higher than 600 lb/in^2. The **inter- and aftercondenser,** one- or two-pass, is a small heat exchanger which condenses the motive steam and allows the noncondensable gases, such as O_2 and CO_2 to be expelled to the atmosphere.

However, the highly soluble ammonia gas may be returned to the feedwater system. The water-box pressure of tube-side pressure must be designed for the condensate-pump shutoff head. The pressure drop through the inter- and aftercondensers varies from 1 to 10 ft of water, with 3 ft as a reasonable average.

Normal **materials** used are: nozzle, stainless steel type 303; steam chest, steel; inlet-air chamber, cast steel; diffuser, aluminum bronze; inter- and aftercondenser shell, steel; inter- and aftercondenser tubes, stainless steel type 304; inter- and aftercondenser tube sheet, carbon steel.

Hogging and Priming Ejectors Single-stage steam-jet ejectors (Fig. 7.5.17) are used for evacuating, or **hogging,** the air from the steam side of a surface condenser, in 15 or 20 min time, so that steam flow can be established and the condenser brought on the line. These ejectors are usually noncondensing and exhaust to the atmosphere. Ratings (Table 7.5.10) are typically in lb/h of free dry air at 70°F, suction conditions of 15 inHg abs, shutoff pressure about 2 inHg abs, motive steam pressures and temperatures as with multistage ejectors.

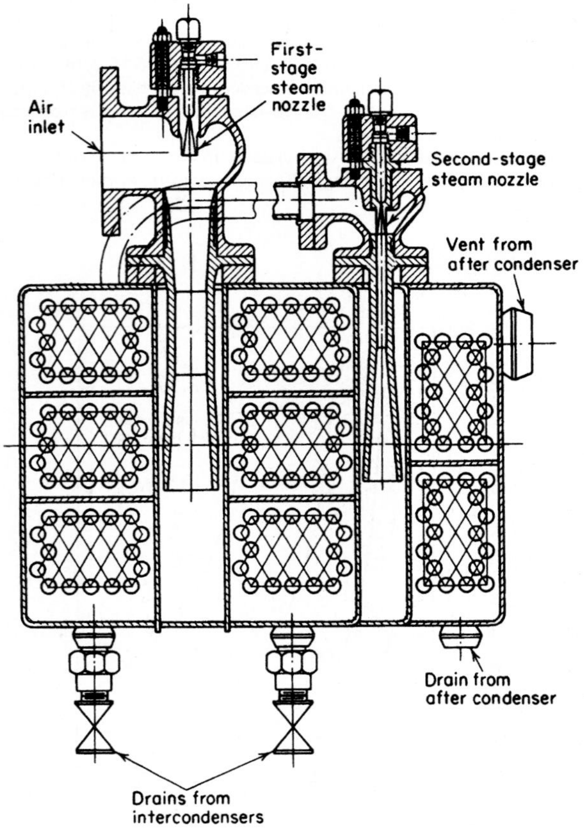

Figure 7.5.16 Two-stage air ejector with surface-type inter- and aftercondensers. (*Westinghouse Electric Corp.*)

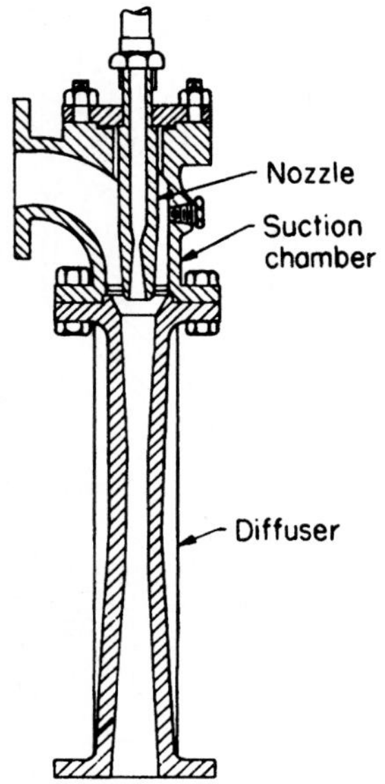

Figure 7.5.17 Single-stage air ejector.

Table 7.5.10 Hogger Capacities*

Total steam condensed, lb/h	scfm†—dry air at 10 inHg abs design suction pressure
Up to 100,000	50
100,001–250,000	100
250,001–500,000	200
500,001–1,000,000	350
1,000,001–2,000,000	700
2,000,001–3,000,000	1,050
3,000,001–4,000,000	1,400
4,000,001–5,000,000	1,750
5,000,001–6,000,000	2,100
6,000,001–7,000,000	2,450
7,000,001–8,000,000	2,800
8,000,001–9,000,000	3,150
9,000,001–10,000,000	3,500

* HEAT Exchange Institute.
† 14.7 lb/in² abs at 70°F.
NOTE: In the range of 500,000 lb/h steam condensed and above, the table provides evacuation of the air in the condenser and low-pressure turbine from atmospheric pressure to 10 inHg abs in about 30 min if the volume of condenser and low-pressure turbine is assumed to be 26 ft³/(1,000 lb/h) of steam condensed.

Priming ejectors are also single-stage noncondensing units and are used to withdraw air from the water side and to fill the condenser with circulating water during start-up periods. The hogging ejector may also serve this priming function by a suitable system of piping and valves.

Materials normally employed are: for the nozzle, stainless steel, 18-8 chrome nickel; steam chest, steel; air-inlet chamber and diffuser, cast steel.

VACUUM PUMPS

REFERENCE: Woodman, Rotary or Steam-Jet Air Pumps, *Power,* Aug. 1948.

Mechanical vacuum pumps, used for hogging, holding, and priming, are of several types for power-plant service: (1) reciprocating—piston, diaphragm; (2) rotary—sliding-vane, oval-water-seal, eccentric-rotor. (See Sec. 10 for details.) Multistage pumps usually need intercooling. A silencer is generally provided on the air discharge. Normally, two half-capacity pumps are installed, both operating during hogging and one adequate for holding service. The relative capabilities of a vacuum pump and hogging and holding ejectors are illustrated in Fig. 7.5.18 (note the considerable hogging capacity of the vacuum pump on start-up).

Advantages of vacuum pumps compared to steam-jet air ejectors include (1) system independent of a steam supply; (2) start-up when steam is not available, as on a once-through boiler system; (3) system capable of completely automatic operation; (4) high-pressure steam piping eliminated; (5) recycling of noncondensables and ammonia eliminated; (6) quieter operation. Disadvantages include (1) damage by

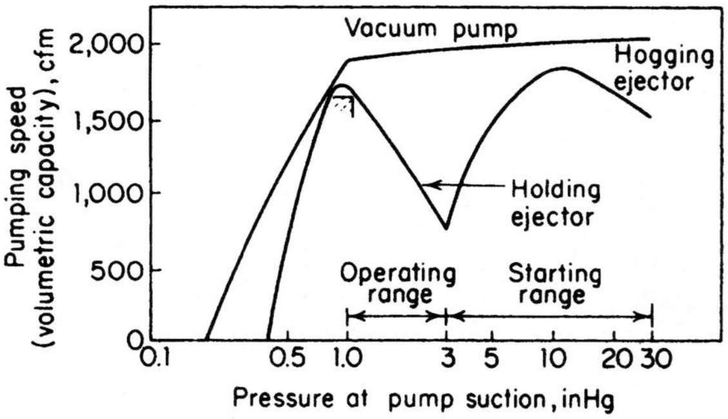

Figure 7.5.18 Relative capabilities of air-removal pumps.

water entering the intake (except when pumps having a liquid rotating seal); (2) higher initial cost; (3) higher maintenance cost. Operating costs are approximately equal. Various combination arrangements are found in practice, for example, an air-operated air-ejector first stage followed by a vacuum-pump second stage, with consequent savings in motor-horse-power and space requirements.

COOLING TOWERS

REFERENCES: "Evaluated Weather Data for Cooling Equipment Design," Fluor Products Co. Cooling Towers, *Power,* Mar. 1963. Baker and Shryock, A Comprehensive Approach to the Analysis of Cooling Tower Performance, *Trans. ASME,* 1961.

The prevalent type of cooling tower (see Figs. 7.5.19 and 7.5.22) dissipates heat by the evaporation of some of the water sprayed into the air circulated through the tower. It is used where water is in limited supply, where temperature pollution of natural water bodies is to be avoided, where water conservation is to be effected, or where otherwise polluted sources must be avoided. Figure 7.5.20 illustrates the functional cycle and the significant basic terms (see also Sec. 2).

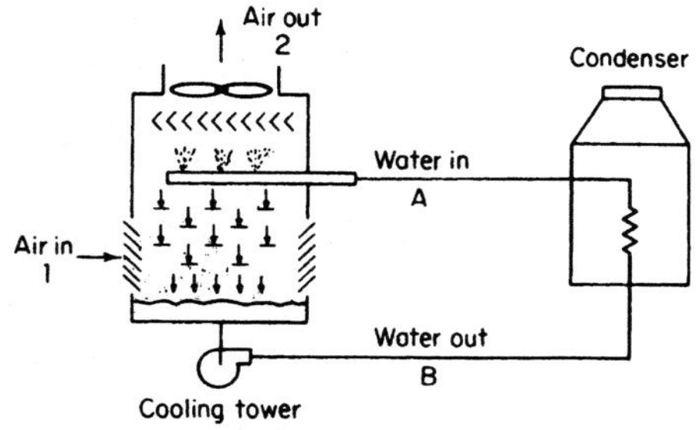

Figure 7.5.19 Schematic for induced-draft, counterflow, cooling tower installation.

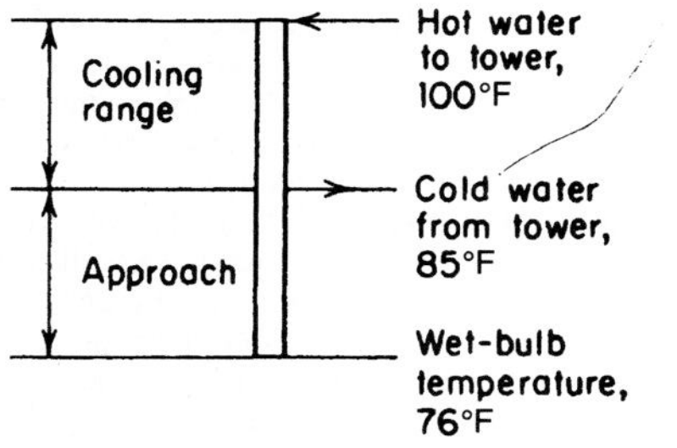

Figure 7.5.20 Cooling tower terms.

Wet-bulb temperature, for design purposes, should not exceed the maximum expected value more than 5 percent of the time during summer (June to September). (See Fig. 7.5.21.)

Approach is the difference in temperature between the cold water leaving the tower and the ambient wet bulb.

Cooling range is the difference in temperature between the hot water entering and the cold water leaving the tower.

Drift is the water lost as mist or droplets entrained by the circulating air and discharged to the atmosphere. It is in addition to the evaporative loss and is minimized by good design.

Makeup is the water required to replace total losses by evaporation, drift, blowdown, and small leaks.

Blowdown is the water that must be discharged from the system to avoid unduely concentrating dissolved solids such as calcium in the system that could be deposited on the condenser tubes, inhibiting heat transfer.

The early, simple atmospheric towers have, because of reliance on natural air circulation, high pumping heads, excessive spray losses, and

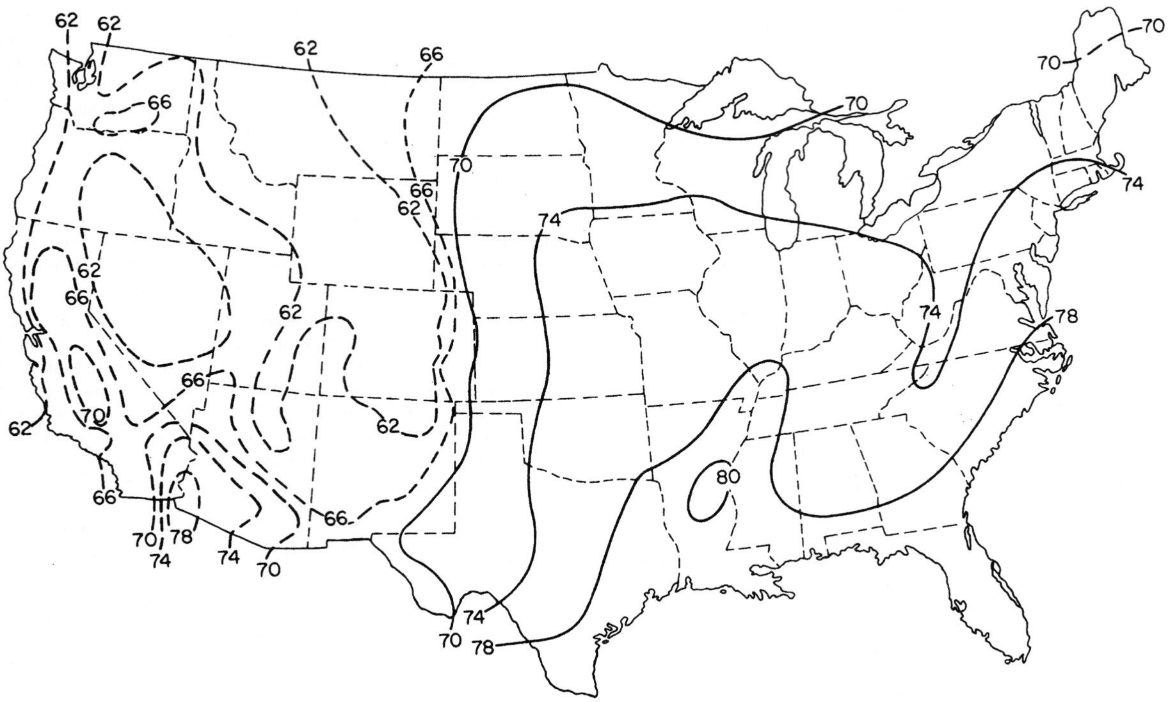

Figure 7.5.21 Wet-bulb temperature, °F, isoclines at the 5 percent level, United States. *(Adapted from Fluor Products Co., by permission.)*

makeup, been largely superseded by three important types: (1) forced-draft, (2) induced-draft, and (3) hyperbolic (Fig. 7.5.22).

The **mechanical forced-draft tower** (Fig. 7.5.22*a*) gives (1) a controllable air supply from fans conveniently located for inspection and maintenance at ground level, and (2) reduced water-pumping head. Nonuniform distribution of air over the ground area of the tower cell, recirculation of vapor from the tower discharge to the tower inlet, with its deleterious effects and fan icing in cold weather, and limitations on the physical diameter of fans are all problems with forced-draft towers.

The **induced-draft tower** (Fig. 7.5.22*b*) is prevalent in U.S. practice. The fan is mounted on the top (discharge) of the cell, with consequent improved air distribution within the cell; drift eliminators reduce makeup requirements; spray nozzles, downspouts, splash plates, and splash bars ensure ample evaporative surface for the water, with maximum volumetric heat-transfer rates. In the **counterflow** design, air is introduced beneath the cell fill, but in the **crossflow** design (Fig. 7.5.22*b*), air is introduced at the sides of the fill.

The **hyperbolic tower** (Fig. 7.5.22*c*) utilizes the chimney effect (height, 500 ± ft) for natural circulation. It has been favored in large European installations where the prevalent lower dry-bulb ambient temperature gives a greater difference in density between entrance to and exit from the tower. The wide approach keeps the unit size within practical bounds, and the savings in fan power support the higher investment. Hyperbolic cooling towers may be of either the counter-flow design as shown in Figure 7.5.22 (c) or of the cross-flow design as shown in Figure 7.5.22 (b) but with a chimney (or veil) replacing the fan. Both types of hyperbolic cooling towers have gained wide acceptance in the United States and are employed mainly at large fossil and nuclear plants. However, hyperbolic cooling towers of the cross-flow design have proven to be particularly susceptible to damage due to freezing.

Precautions are necessary to avoid freezing troubles in cold weather, fire hazards with intermittent operations, corrosion, scale, and microbiological growth problems. **Location** must avoid recirculation: for tower lengths to 250 ft, the long-axis orientation should be parallel

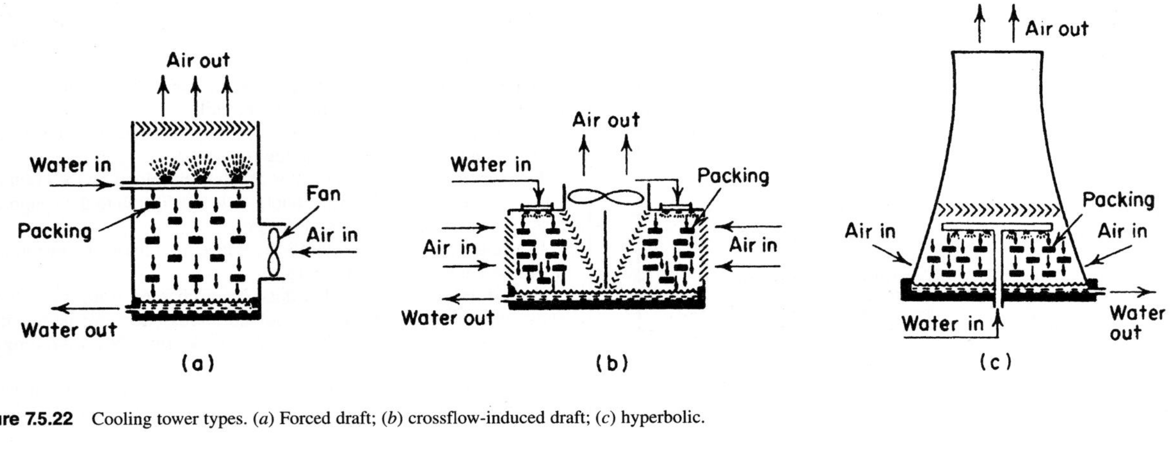

Figure 7.5.22 Cooling tower types. (*a*) Forced draft; (*b*) crossflow-induced draft; (*c*) hyperbolic.

to the prevailing wind; towers longer than 250 ft should be arranged broadside to the prevailing wind. The tower should be isolated as much as possible; adjacent heat sources compel the specification of higher design wet-bulb temperatures.

Performance Calculations

(See Fig. 7.5.19.)

Applicable equations are

$$W_{a1}h_{a1} + W_{v1}h_{v1} + W_{wA}h_{fA} = W_{a2}h_{a2} + W_{v2}h_{v2} + W_{wB}h_{fB} \tag{7.5.6}$$

$$W_{wB} = W_{wA} - (W_{v2} - W_{v1}) \qquad W_{a1} - W_{a2} \tag{7.5.7}$$

$$W_{wA}(h_{fA} - h_{fB}) = W_{a2}h_{a2} + W_{v2}h_{v2} - (W_{a1}h_{a1} + W_{v1}h_{v1}) - (W_{v2} - W_{v1})h_{fB} \tag{7.5.8}$$

$$h_{fA} - h_{fB} = t_{wA} - t_{wB} \tag{7.5.9}$$

and

$$W_a h_a + W_v h_v = \text{total heat, from psychrometric chart (See Sec. 2.)} \tag{7.5.10}$$

$$W_{wA}(t_{wA} - t_{wB}) = \text{total heat at 2} - \text{total heat at 1} - (W_{v2} - W_{v1})h_{fB} \tag{7.5.11}$$

where W = flow, lb/h; a = air; w = water; n = vapor; h = enthalpy, Btu/lb; f = fluid; t = temperature, °F; 1, 2, A, B = locations.

EXAMPLE. Given: flow = 80,000; wet bulb = 76°F; dry bulb = 86°F; range (also condenser rise) = 100 – 85 = 15°F; approach = 9°F. Find: economical tower size, circulating-pump power, fan capacity and power, and makeup water.

SOLUTION. (1) *Economical tower size:* A study must be made to tie in with condenser. Heat to be dissipated is constant. Vary rise (range), approach, and then compare capital costs of cooling tower, circulating pump, and condenser with operating costs for fan and pump power. (2) *Circulating-water-pump power;* bhp = gal/min × TDH/3,960 × eff. TDH = condenser friction + pipe friction + static head. (3) *Fan capacity and power:* If air leaves tower saturated at 95°F, then from psychrometric charts (Sec. 2),

$$\text{Total heat} = 1 \text{ lb dry air } C_p(t_a - 0°)\ 1\ W_v h_g \quad \text{at } t_a \tag{7.5.12}$$

For air and vapor at 1, vapor = 120 grains/lb dry air. By Eqs. (7.5.10) and (7.5.12), total heat = 1 × 0.24(86 – 0) + (120/7,000)1,099.2 = 39.5 Btu/lb dry air. For air and vapor at 2, vapor = 256 grains/lb dry air. Total heat = 63.1 per lb dry air. Sp vol (air and vapor) = 14.80 ft^3/lb dry air. By Eq. (7.5.11), W_{wA}(100 – 85) = 63.1 – 39.5 – (256 – 120) $\frac{63}{7{,}000}$ = 1.50 lb water/lb dry air. 80,000 gal/min/(1.50 × 500) = 26.6 × 10^6 lb dry air/h, or 26.6 × 10^6 × 14.80/60 = 6.58 × 10^6 ft^3/min. With test data on static-pressure requirements for the tower and with the fan characteristics, the horsepower to drive the fan can be calculated. (4) *Makeup:* $W_{v2} - W_{v1}$ = (256 – 120) 26.6 × 10^6/7,000 = 515,000 lb/h.

Normally, a cooling tower is purchased for only one guarantee point. It is well, however, to have **performance** curves (Fig. 7.5.23) showing operation for various wet-bulb temperatures and cooling ranges. The

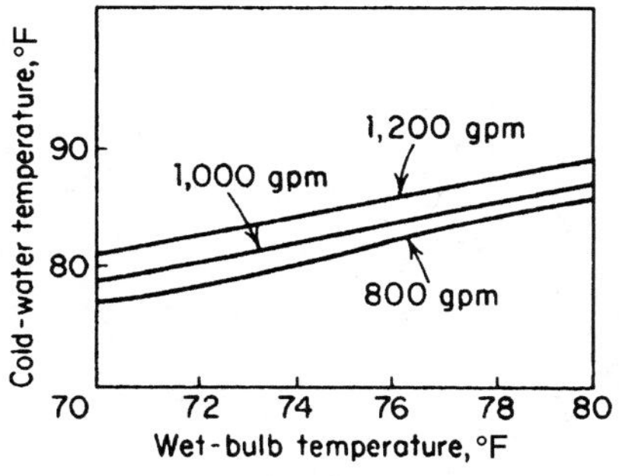

Figure 7.5.23 Cooling tower performance. Design for 85°F cold water, 95°F hot water, 78°F wet-bulb temperature, 10° range; 1,000-gal/min cell.

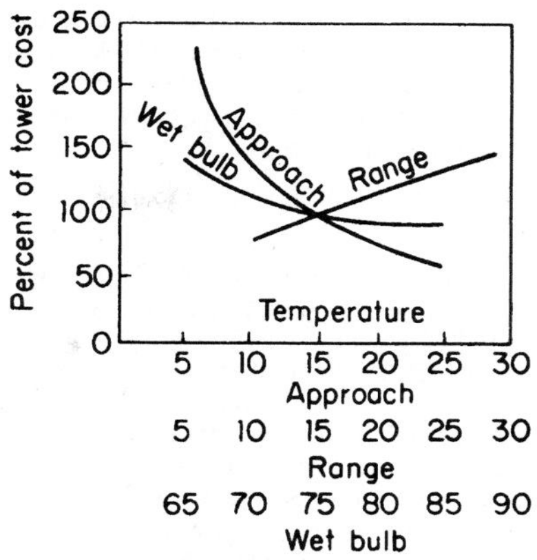

Figure 7.5.24 Comparative cooling tower costs. *(Foster Wheeler Corp.)*

investment for a cooling tower is essentially a matter of water flow and is influenced by approach, range, and wet bulb (Fig. 7.5.24). The **cost evaluation** should include consideration of tower frame and fill, fans, motors, basin and pump pit, pump head, fan horsepower, freight, labor, and erection. In the choice of a tower for a power plant, there should be coordinated study and evaluation of the turbine and condenser for best overall economy.

The **height** of a field-erected induced-draft tower, from basin curb to fan deck, ranges from 8 to 50 ft; widths vary from 6 to 60 ft; lengths from 8 to 500 ft; fan-stack height between 2 and 15 ft.

Materials used are: frame—redwood (treated or untreated), Douglas fir (treated), steel (galvanized), concrete; fill—red wood, plastics (polyethylene, polypropylene), cement asbestos* board (extruded); casing—cement asbestos* board (corrugated, slat), redwood, fiber glass, aluminum, concrete; fan blades—aluminum, glass-reinforced polyester, stainless steel, monel; fan hubs—cast iron, galvanized steel, stainless steel; fan stack—redwood, steel, masonite, drift eliminators—redwood, cement asbestos* board; spray nozzles— ceramic, bakelite; louvers—redwood, cement asbestos* board.

DRY COOLING TOWERS, WITH DIRECT-CONTACT CONDENSERS

REFERENCE: Ritchings and Lotz, Economics of Closed vs. Open Cooling Water Cycles, *Power Eng.,* May and June 1963.

The thermodynamic requirement for a heat sink can generally best be met with a natural body of water such as a river, lake, or ocean. When water supplies are limited, a **"wet" cooling tower** (water losses less than 2 or 3 percent) may be used for reclamation. A **"dry" cooling tower** will further reduce this loss. Figure 7.5.25 shows an application with a direct-contact condenser. The finned surfaces of the dry cooling tower are substituted for the tubes of a surface condenser. Heat-transfer rates (see Sec. 2) are appreciably lower, but the extended surface,

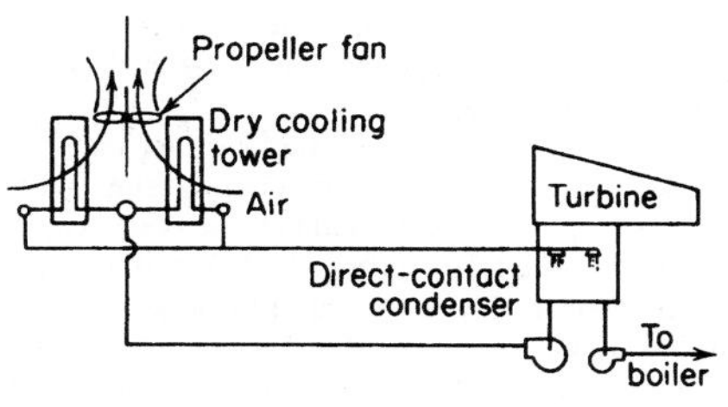

Figure 7.5.25 Dry cooling tower system.

*Existing plants only.

relatively low in cost, may be installed in an induced-draft tower or in a hyperbolic tower, as in England for a 125,000-kW unit. Problems of recirculation, freezing, and air leakage must be solved. If a minor evaporative loss is permissible, peaking capacity is obtained by wetting the heat-exchange surface in the tower.

SPRAY PONDS

If land area is available, a spray pond may serve as an alternative reclamation system. Pipelines are typically arranged on 50-ft centers, with five nozzles (50 ± gal/min each) every 12 ft of pipeline. Cooling is limited by the relatively short period of contact between spray and air. Drift, especially with adverse wind conditions, can make losses higher than on towers. Generally, if cooling efficiencies above 50 percent are required, a tower is favored, where efficiency = $(t_h - t_c) \times 100/(t_h - t_{wb})$, where the temperature t subscripts are h = hot water, c = cold water, wb = wet bulb of the ambient air. (See Wright and Kirsopp, Pond Surface Cooling of a Chemical Plant Cooling Water, *Proc. Am. Power Conf.*, 1960.) A common mistake is to apply the efficiency that may be achieved by a single spray nozzle to a large field of spray nozzles. With large spray ponds, ambient air flowing to the nozzles located away from the outer regions of the spray field is inhibited by the spray. For the air that reaches the interior of the spray field, the local wet-bulb temperature is higher than the ambient. One way to overcome this problem is to place the spray nozzles on vertical spray trees arranged in a circular pattern and directed toward the central axis of the circle. However, this arrangement typically requires the spray nozzles to operate with a greater pressure.

CLOSED FEEDWATER HEATERS

REFERENCES: Heat Exchange Institute Standards for Closed Feedwater Heaters. ASME Boiler and Pressure Vessel Code, Sec. VIII.

Feedwater heaters are used (1) to effect the thermodynamic gains of the regenerative steam cycle (see Secs. 2 and 7), and (2) to raise water temperatures to a sufficient value for the avoidance of thermal shock to the boiler metal. The **number and types of heaters** usually employed are: (1) plant sizes of up to 70 mW, two closed low-pressure heaters, one open heater, one closed high-pressure heater; (2) plant sizes of 75 to 300 mW, two or three closed low-pressure heaters, one open heater, two closed high-pressure heaters; (3) fossil-fueled plant sizes above 300 mW, three or four closed low-pressure heaters, one open heater, two or three high-pressure heaters. Nuclear units require very large feedwater flows, necessitating double or triple parallel strings of heaters. There are generally five or six low-pressure and one high-pressure heaters (no open heaters) in each string.

Minimum heater cost prevails with minimum restrictive specifications, for example, horizontal, two-pass, high water velocity (10 ft/s at 60°F), no length limits. Overall heater length is limited by maximum available tube lengths of 100 ft for copper alloys, admiralty metal, and copper-nickel and 85 ft for Monel. With U-tube construction, this results in heater lengths of about 48 and 40 ft, respectively. A general rule, to ensure good steam distribution, is that the length in feet shall not exceed the shell diameter in inches plus 2; that is, with a 30-in diam shell, the length should not exceed 32 ft. **Pressure drop** through the tubes must be economically evaluated as it varies approximately with the square of the water velocity.

If a **length restriction** is imposed, the designer may have to substitute a four-pass arrangement for the two-pass design, with consequent large-diameter shell and water chamber, heavier walls, more tube holes to be drilled, more tubes to be installed, and a cost increase. If a **pressure-drop restriction** is imposed, a lower water velocity results, with more tubes, larger shell and chamber diameter, and more surface because of the lower heat-transfer rate. Vertical heaters, with appropriate construction details, are also higher in cost.

The **condensing heater** (Fig. 7.5.26), like a surface condenser, performs at constant saturation temperature outside the tubes with no condensate subcooling. Feedwater is generally heated to within 5°F of the saturation temperature. The addition of a **drain-cooling section** (Fig. 7.5.27), with a cover or enveloping baffle around the tubes of the inlet pass, can reduce the drip temperature to within 10 or 15°F of the incoming water. When the steam is at sufficient pressure (above 125 lb/in^2 abs) and contains enough superheat (more than 250°), a **desuperheating section** (Fig. 7.5.27), using an enveloping baffle around the tubes at feedwater outlet, can raise the water temperature above the saturation temperature of the entering steam. Drain coolers and desuperheating sections improve overall plant heat rate by reduction of energy degradation. Subcooling of drains reduces flashing, erosion, vibration, and noise when drains are dropped to a lower-pressure heater. Special attention should be given to drains flashed to the main condenser, as this is a direct thermodynamic loss. As an alternate, these drains can be pumped forward into the feedwater line.

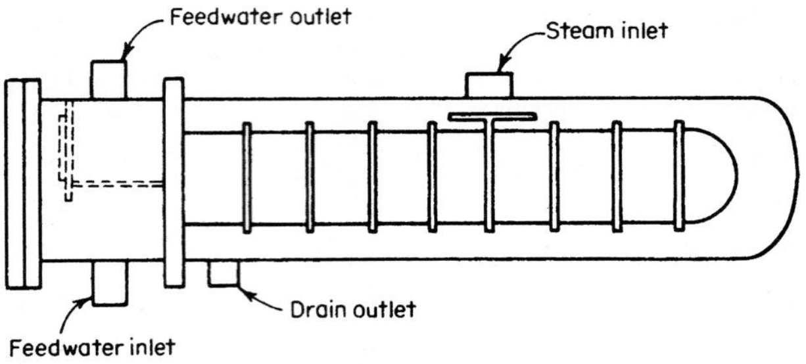

Figure 7.5.26 Basic heater section for condensing heater.

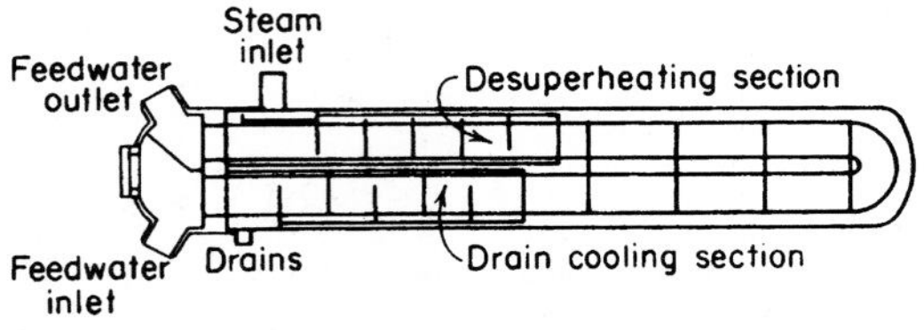

Figure 7.5.27 Basic heater section with drain cooling and desuperheater sections added.

Heat-transfer rates for a condensing feedwater heater range, overall, between 500 and 900 Btu/(h·ft^2 °F). (See Figs. 7.5.28 and 7.5.29.) Heat-transfer rates for drain-cooling and desuperheating sections must be separately evaluated, with overall ranges from 300 to 500 on the former and 80 to 140 on the latter. (See Figs. 7.5.30 and 7.5.31.) Normally, the three sections are included in the same shell (Fig. 7.5.27); when the cooler becomes too large, as with the lowest-pressure heater, a separate shell-and-tube exchanger may be justified.

Low-pressure heaters (upstream of the boiler feed pump) are usually designed for tube-side pressure of less than 90 lb/in^2 gage. This is based on the shutoff head of the condensate pump plus 10 percent. For pressures up to 300 to 400 lb/in^2 gage, a bolted cover with flange and

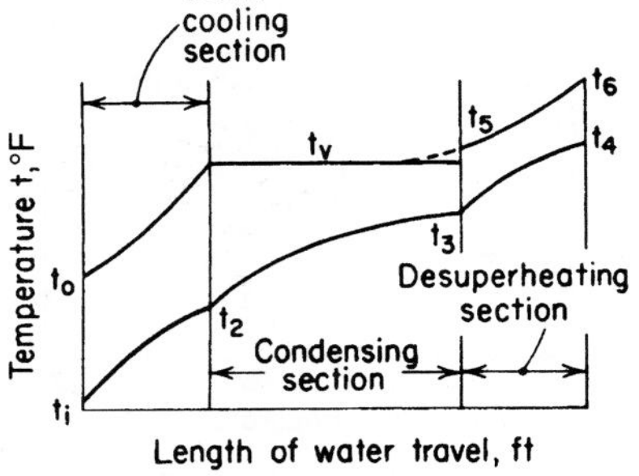

Figure 7.5.28 Temperature vs. water travel, closed-feed heater.

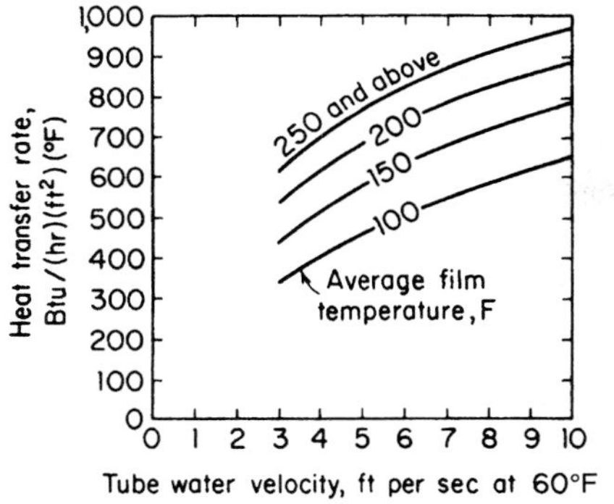

Figure 7.5.29 Approximate condenser heat-transfer rates with 18 BWG admiralty metal tubes. Average film temperature = $t_s - 0.8llt_m$.

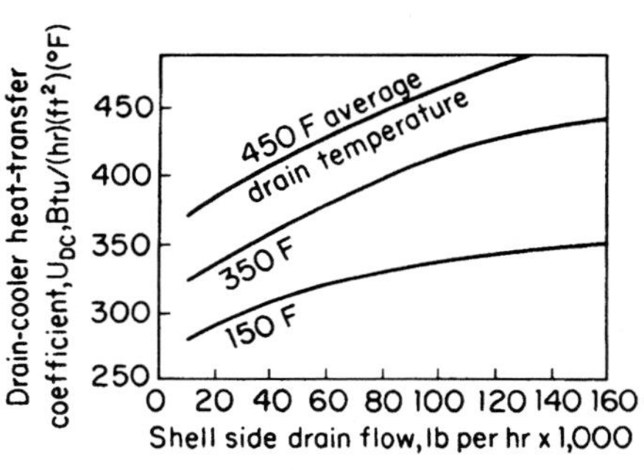

Figure 7.5.30 Approximate drain-cooler heat-transfer rates.

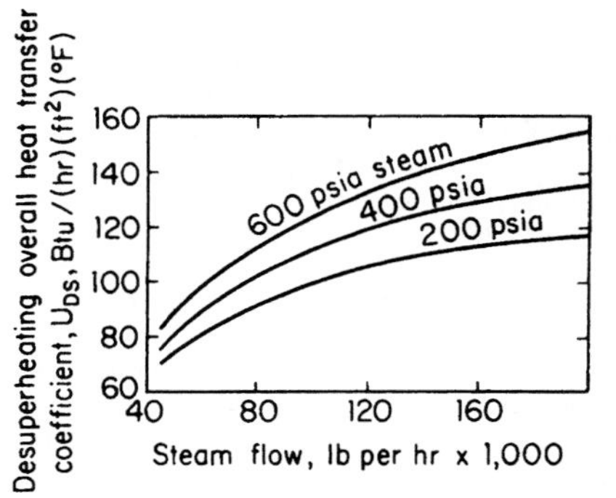

Figure 7.5.31 Approximate desuperheater section heat-transfer rates.

gasket (Fig. 7.5.32*a*) is used, while higher pressures usually justify a welded joint to eliminate the gasket (Fig. 7.5.32*b* and *c*). A manway is utilized for access to the tube sheet. Tubes are generally rolled into the tube sheet.

High-pressure heaters (downstream of the boiler feed pump) are designed for shutoff head of the pump plus 10 percent, resulting in about 1,500 lb/in² gage for a nuclear plant and up to 5,000 lb/in² gage for fossil-fuel plants. Hemispherical head construction (Fig. 7.5.32*c*) and welded tube-to-tube sheet joints are almost exclusively used. Since the cost of the tubes represents the largest portion of the heater cost, material selection is important (see Table 7.5.11). Depending on the desired water-chemistry requirements, low-pressure heaters usually use admiralty metal, 90-10 Cu-Ni, stainless or carbon steel. High-pressure heaters require the use of the higher-strength alloys of 70-30 Cu-Ni, monel, stainless, or carbon steel. The thinnest wall possible [$t = Pd/(2s + 0.8p)$] or the industry minimum standard of 18 BWG (20 BWG for stainless steel) is utilized.

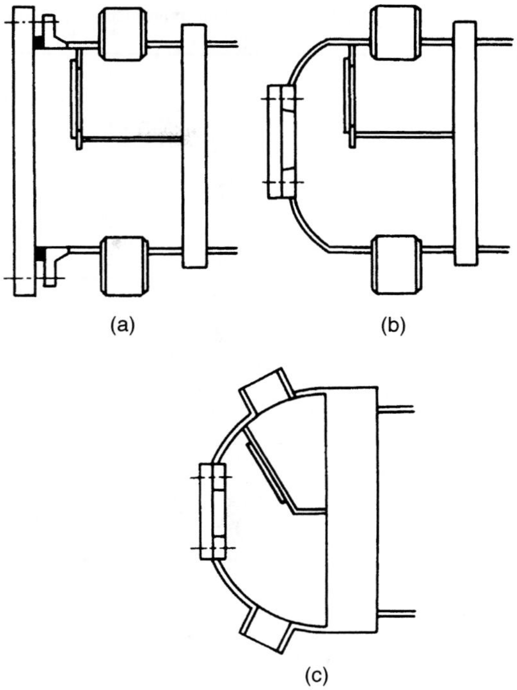

Figure 7.5.32 Feedwater heater channels. (*a*) Bolted; (*b*) dished; (*c*) hemispherical.

Performance Calculations

Notation

A = heat-transfer surface, ft²
C_m = material and gage correction factor, Table 7.5.11
D = tube ID, in, Table 7.5.12
d = OD of tube, in (⅝ and ¾ in are most common)
F_1 = friction factor, Table 7.5.13
F_2 = water-temperature correction factor, Table 7.5.14
h = enthalpy, Btu/lb
k = tube diameter and gage factor, Table 7.5.15
L = active length of tubes per pass, ft
N = number of passes
n = number of tubes
P = design pressure, lb/in²
p = pressure, lb/in²
Q = total heat transferred, Btu/h
S = allowable design stress at tube design temperature, lb/in², from ASME Code
t = wall thickness, in (see Table 7.5.16); temperature, °F
U_o = overall heat-transfer rate, Btu/(h·ft² °F) (the material correction factor, Table 7.5.11, must be applied to the overall heat-transfer rate)
V = water velocity, ft/s
W = steam flow, lb/h
W_w = feedwater flow, lb/h
Δp = pressure drop, lb/in²
Δt_m = logarithmic mean temperature difference, °F (Fig. 7.5.28)

Heat-balance data will provide the major parameters of steam pressure, feedwater flows, and temperatures. The choice of tube diameter and material must be made first. Tube thickness is calculated by

$$t = Pd/(2S + 0.8P)$$

Test pressure is 1.5 times design pressure.

Table 7.5.11 Material and Gage Correction Factor C_m
(For 5/8- to 1-in tubes)

BWG no.	Tube material							
	Ars-Cu	Admiralty	90-10 Cu-Ni	80-20 Cu-Ni	70-30 Cu-Ni	Monel	18-8 SS	Carbon steel
18	1.00	1.00	0.97	0.95	0.92	0.89	0.85	0.96
17	1.00	1.00	0.94	0.91	0.87	0.85	0.80	0.92
16	1.00	1.00	0.91	0.88	0.84	0.82	0.77	0.89
15	1.00	0.99	0.89	0.86	0.82	0.79	0.74	0.87
14	1.00	0.96	0.85	0.82	0.77	0.75	0.70	0.83
13	0.98	0.93	0.81	0.78	0.73	0.70	0.65	0.79

Table 7.5.12 Feedwater-Heater Tube Constants

Tube BWG	5/8-in OD		3/4-in OD		7/8-in OD	
	D	$D^{1.24}$	D	$D^{1.24}$	D	$D^{1.24}$
18	0.527	0.452	0.652	0.589	0.777	0.731
17	0.509	0.433	0.634	0.567	0.759	0.710
16	0.495	0.418	0.620	0.554	0.745	0.695
15	0.481	0.404	0.606	0.538	0.731	0.678
14	0.458	0.380	0.584	0.514	0.709	0.652
13	—	—	0.560	0.488	0.685	0.625

Table 7.5.13 Friction Loss, lb/in² per ft of Straight Travel

Water velocity (at 60°F), ft/s	F_1
3.0	0.018
4.0	0.031
5.0	0.047
6.0	0.065
7.0	0.085
8.0	0.108
9.0	0.134
10.0	0.162

Table 7.5.14 Water-Temperature Correction Factor*

Average water temp, °F	F_2
100	0.905
150	0.840
200	0.795
250	0.770
300	0.755
350	0.750
400	0.755
450	0.757
500	0.795
550	0.830
600	0.885

* Average water temperature = $t_s - \Delta t_m$.

Table 7.5.15 Tube Diameter and Gage Factor k

Diam, in	BWG no.				
	20	18	17	16	15
5/8	0.217	0.2407	0.2580	0.2728	0.289
3/4	0.1732	0.1887	0.1995	0.2089	0.218
7/8	0.1445	0.1550	0.1624	0.1686	0.1748
1	0.1238	0.1314	0.1369	0.1413	0.1458

Table 7.5.16 Tube-Wall Thickness

Gage, BWG	Thickness, in
15	0.072
16	0.065
17	0.058
18	0.049
19	0.042
20	0.035
22	0.028

Condensing-heater-size equations are

$$Q = UA\,\Delta\,t_m = W_w(h_3 - h_2) = W_s(\Delta h_s)$$

(See Figs. 7.5.28 and 7.5.29.) The minimum terminal temperature difference without a superheating section is 2°F.

The active length of tubes is calculated from $L = 500AV/(NW_w k)$, where maximum velocity is 10 ft/s at 60° F; the number of tubes, from $n = 3.82A/(Ld)$; the pressure drop, from $\Delta p = (L + 5.5D)F_1F_2N/D^{1.24}$.

The shell design pressure is 20 percent greater than the maximum operating pressure, rounded to the higher 25 lb/in² as a minimum.

A **drain-cooling section,** when added, is treated as a separate heat exchanger where the quantities t_v, t_o, t_i, W_s, and W_w are known; t_2 is the only unknown and is to be calculated from a heat balance on the section or $Q = W_s(h_v - h_o) = W_w(h_2 - h_1)$. Compute Δt_m; find the area of the drain-cooler section A_{dc} from $Q = U_o A_{dc} \Delta t_m$, with overall U_o from Fig. 7.5.30.

A **desuperheating section** is similarly calculated with t_v, t_6, t_4, p_6, h_6, h_4, W_s known. Allow approximately 1 percent pressure drop through the section to get p_v, and 80 to 90 percent desuperheating (30°F superheat entering condensing section) to obtain t_5 and h_5. By a heat balance on section, calculate h_3 from $Q = W_s(h_6 - h_5) = W_w(h_4 - h_3)$. With equivalent t_3, calculate Δt_m and substitute in $Q = U_o A\,\Delta t_m$ for area of desuperheating section (overall U_o from Fig. 7.5.31).

Materials

Heaters are designed in accordance with the Heat Exchange Institute Standards for Closed Feedwater Heaters and the ASME Boiler and Pressure Vessel Code, Sec. VIII, using the following materials:

Tubes:

Material		Max temp, °F Rolled joint	Max temp, °F Welded joint
Admiralty metal	ASME SB-395	350	
90-10 Cu-Ni	ASME SB-395	400	450
80-20 Cu-Ni	ASME SB-395	400	475
70-30 Cu-Ni	ASME SB-395	400	525
70-30 Cu-Ni, stress-relieved	ASME SB-395	400	525
Monel, ASME SB-163		400	600
Carbon steel, ASME SA-556, grades A, B, and C		350	650
Stainless steel, ASME SA-249; TP 304 (welded)		350	650

Shell, nozzles: carbon-steel pipe and plant, ASME SA-515 grade 70; SA-106 grade B.

Channels, or heads, channel covers: carbon-steel plate, forgings, and pipe, ASME SA-515 grade 70; SA-105 grade II; SA-106 grade B.

Baffles and tube supports: steel plate, ASME SA-283 grade C.

Tube sheets: carbon-steel plate and forgings, ASME SA-515 grade 70; SA-105 grade II.

Carbon steel is used for temperatures up to 800°F. Chromium-molybdenum steel (ASME SA-387 grade C) is used for higher temperatures.

OPEN, DEAERATING, AND DIRECT-CONTACT HEATERS

REFERENCE: Heat Exchange Institute Standards for Deaerators and Deaerating Heaters.

Deaerators or deaerating heaters serve (1) to degasify feedwater and thus reduce equipment corrosion (see Sec. 4), (2) to heat feedwater regeneratively and improve thermodynamic efficiency (see Sec. 7), and (3) to provide storage, positive submergence, and surge protection on the boiler feed-pump suction (see Sec. 10).

Removal of oxygen and carbon dioxide from boiler feedwater and process water at elevated temperature is essential for adequate conditioning. In some power-plant applications, a well-designed surface condenser gives adequate deaeration and the accompanying exclusive use of closed feedwater heaters.

A modern **deaerator** will, by mechanical action, reduce O_2 content of effluent to less than 0.005 cm^3/L and CO_2 content to a negligible amount. Water must (1) be heated to and kept at saturation temperature, as the gas solubility is zero at the boiling point of the liquid, and (2) be mechanically agitated by spraying or cascading over trays for effective scrubbing, release, and removal of gases. Gases must be swept away by an adequate supply of steam. Since the water is heated to saturation conditions, the terminal temperature difference is zero with maximum improvement in associated turbine heat rate. Extremely low partial gas pressures, dictated by Henry's law, call for large volumes of scrubbing steam. **Vent condensers** of the shell-and-tube or direct-contact types, cooled by incoming feed, serve to recover heat and water before release of gases to the atmosphere.

The **tray-type deaerator** (Fig. 7.5.33) is prevalent. While it has some tendency to scale, it will operate at wide load conditions and is practically independent of water-inlet temperature. Trays can be loaded to some 10,000 lb/(ft^2 h), and the deaerator seldom exceeds 8 ft in height.

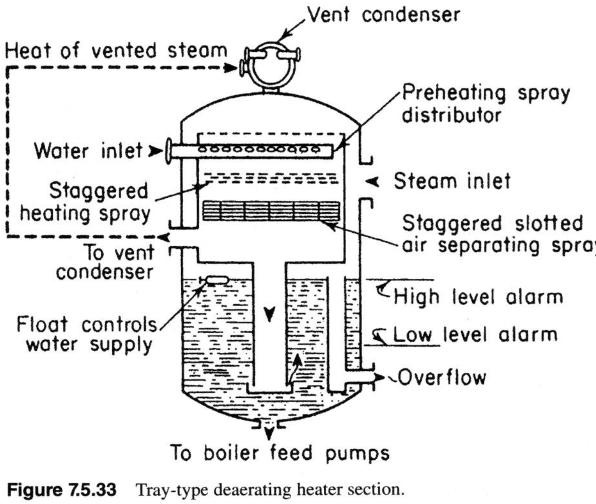

Figure 7.5.33 Tray-type deaerating heater section.

The **spray type** uses a high-velocity steam jet to atomize and scrub the preheated water. Applications are (1) marine service, where it is unaffected by ship roll and pitch, and (2) industrial plants where operating pressures are stable. It requires a temperature gradient, for example, 50°F minimum, to produce the fine sprays and vacuums with the cold water required.

Materials

Open-feed heaters are designed to the ASME Boiler and Pressure Vessel Code. Materials used are: shell—steel, ASME SA-285 grade C and SA-515 grade 70; trays—stainless steel, type 304; baffles in vent condenser—stainless steel, type 304; spray valves—stainless steel, type 304.

EVAPORATORS

For general treatment of evaporators used in the preparation of makeup water, see Sec. 4.

Plate Heat Exchangers

REFERENCE: Heat Transfer – Professional Version, Lindon C Thomas, Capstone Publishing Corp.

The plate heat exchangers are built of thin corrugated plates that are clamped together to form passages. The plates are fitted with gaskets which are used to seal the unit and to direct the hot and cold fluids through alternate narrow passages between plates consisting of herringbone grooves that are inclined at an angle to the direction of flow. The chevrons on adjacent plates are oriented in opposite directions such that the grooves actually cross. The arrangement produces multiple contact points between facing plates, resulting in highly effective flowpatterns and enhanced convection heat transfer. Plate heat exchangers are easily dismantled for cleaning or inspection and are relatively compact. However, because of limitations in gasket material, they are generally restricted to pressures below 25 atmospheres and temperatures below 250°C.

Notation

L_W = Plate width, ft.
L_c = Compressed length of plate, ft.
N_p = Number of plates
N_{cp} = Number of channel passes
ΔX = Thickness of plate, ft.
A = Total effective area, ft^2
r_w = Wall resistance, hr-ft^2- °F/BTU
r_f = Fouling resistance, hr-ft^2- °F/BTU
b = Channel plate spacing, ft.
D_e = Hydraulic diameter, ft.
G_c = Cold stream mass velocity, Lbm/Ft^2-hr
G_h = Hot stream mass velocity, Lbm/Ft^2-hr
k = Thermal conductivity, BTU/hr-ft °F
T = Hot side temperature, °F
t = Cold side temperature, °F

Analysis

$$b = \left(L_c / N_p \right) - \Delta X$$

$$D_e = \frac{4bL_W}{2(b + L_W)}$$

$$U = \frac{1}{1/h_c + 1/h_h + r_w + r_f}$$

From the Colburn Analogy for turbulent heat transfer through flat plates

$$h_c = C\,\mathrm{Re}_c^{3/4}\,\mathrm{Pr}_c^{1/3}\left(\frac{k_h}{D_e}\right)$$

$$h_h = C\,\mathrm{Re}_h^{3/4}\,\mathrm{Pr}_h^{1/3}\left(\frac{k_h}{D_e}\right)$$

$$\mathrm{Re} = \frac{D_e G}{\mu}$$

$$\mathrm{Pr} = \frac{\mu c_P}{k}$$

The value of the coefficient, C, is normally provided by the heat exchanger vendor but may be proprietary. However, it may be back-calculated from guaranteed performance data or from a heat transfer test in the clean condition. (See ASME PTC 12.5-2000, Single Phase Heat Exchangers.)

Solving for U and referencing the cold side of the heat exchanger

$$R = \frac{G_c}{G_h}$$

$$NTU = \frac{U\,A}{G_c}$$

$$P = \frac{1 - e^{[-NTU(1-R)]}}{1 - R e^{[-NTU(1-R)]}}$$

$$Q = G_c P (T_i - t_i)$$

Example

$L_W = 3.118$ ft.
$L_c = 6.320$ ft.
$N_p = 391$
$N_{cp} = 1$
$\Delta X = 0.00164$ ft.
$A = 7{,}704.4$, ft^2
$r_w = 1.34 \times 10^{-4}$ hr-ft^2- °F/BTU
$r_f = 2.23 \times 10^{-4}$ hr-ft^2- °F/BTU
$G_c = 326{,}357$ Lbm/Ft^2-hr
$G_h = 163{,}023$ Lbm/Ft^2-hr
$K_c = 0.358$ BTU/hr-ft °F
$K_h = 0.361$ BTU/hr-ft °F
$T_i = 100.51$ °F
$t_i = 83.51$ °F

$$b = \left(\frac{L_c}{N_p}\right) - \Delta X = \left(\frac{6.320}{391}\right) - 0.00164 = 0.1069\ ft.$$

$$D_e = \frac{4bL_W}{2(b + L_W)} = \frac{4(0.01069)(3.118)}{2(0.01069 + 3.118)} = 0.02131\ \text{ft.}$$

$$\mathrm{Re}_c = \frac{D_e G_c}{\mu_c} = \frac{(0.02131)(326{,}357)}{1.929} = 3{,}605$$

$$\mathrm{Re}_h = \frac{D_e G_h}{\mu_h} = \frac{(0.02131)(163{,}023)}{1.810} = 1{,}919$$

$$\mathrm{Pr}_c = \frac{\mu_c c_{p,c}}{k_c} = \frac{(1.929)(1.0)}{0.358} = 5.39$$

$$\mathrm{Pr}_h = \frac{\mu_h c_{p,h}}{k_h} = \frac{(1.810)(1.0)}{0.361} = 5.01$$

$$h_c = C\,\mathrm{Re}_c^{3/4}\,\mathrm{Pr}_c^{1/3}\left(\frac{k_c}{D_e}\right) = 0.127\,(3{,}605)^{3/4}\,(5.39)^{1/3}\left(\frac{0.358}{0.02131}\right)$$

$$= 1739.3\ \text{BTU/hr} - \text{ft}^2 - °\text{F}$$

$$h_h = C\,\mathrm{Re}_h^{3/4}\,\mathrm{Pr}_h^{1/3}\left(\frac{k_h}{D_e}\right) = 0.127\,(1{,}919)^{3/4}\,(5.01)^{1/3}\left(\frac{0.361}{0.02131}\right)$$

$$= 1067.2\ \text{BTU/hr} - \text{ft}^2 - °\text{F}$$

$$U = \frac{1}{\frac{1}{h_c} + \frac{1}{h_h} + r_w + r_f}$$

$$= \frac{1}{\left(\frac{1}{1739.3}\right) + \left(\frac{1}{1067.2}\right) + 0.000134 + 0.000223}$$

$$= 535.1 \frac{\text{BTU}}{\text{hr} - \text{ft}^2 - °\text{F}}$$

$$R = \frac{G_c}{G_h} = \frac{326{,}357}{163{,}023} = 2.00$$

$$NTU = \frac{U\,A}{G_c} = \frac{(535.1)(7{,}704.4)}{326{,}357} = 12.63$$

Since almost all plate heat exchangers are designed for counter-flow

$$P = \frac{1 - e^{[-NTU(1-R)]}}{1 - R e^{[-NTU(1-R)]}} = \frac{1 - e^{[-12.63(1-2.0)]}}{1 - R e^{[-12.63(1-2.9)]}} = 0.50$$

$$Q = G_c\,P\,(T_i - t_i) = (326{,}357)(0.50)(100.51 - 83.51)$$

$$= 2.77 x 10^6\ \text{BTU/hr}$$

7.6 INTERNAL COMBUSTION ENGINES

by Dennis N. Assanis, David E. Cole, Timothy J. Jacobs, Sitirios Mamalis, and Donald J. Patterson

REFERENCES: Lichty "Internal-Combustion Engines," McGraw-Hill. Obert, "Internal-Combustion Engines and Air Pollution," Harper & Row. Taylor and Taylor, "The Internal-Combustion Engine," International Textbook. Vincent, "Supercharging the Internal-Combustion Engine," McGraw-Hill. Crouse, "Automotive Engine Design," McGraw-Hill. Patterson and Henein, "Emissions from Combustion Engines," Ann Arbor Science. Starkman, "Combustion Generated Air Pollution," Plenum. *Trans. ASME Trans. SAE Proc. I Mech E,* Automobile Division, *Automotive Inds.* Ferguson, "Internal Combustion Engines," Wiley. Heywood, "Internal Combustion Engine Fundamentals," McGraw-Hill. Blair, "The Basic Design of Two-Stroke Engines," SAE. Watson and Janota, "Turbo-charging the Internal Combustion Engine," Wiley. Benson and Whitehouse, "Internal Combustion Engines," Pergamon. Eastwood, "Critical Topics in Exhaust Gas Aftertreatment," Research Studies Press.

GENERAL FEATURES

Cycles

(See also Sec. 2.1)

In reciprocating **internal combustion engines,** the combustion process can be approximated by the constant volume cycle (Fig. 7.6.1), the constant pressure cycle (Fig. 7.6.2), or by a combination of these

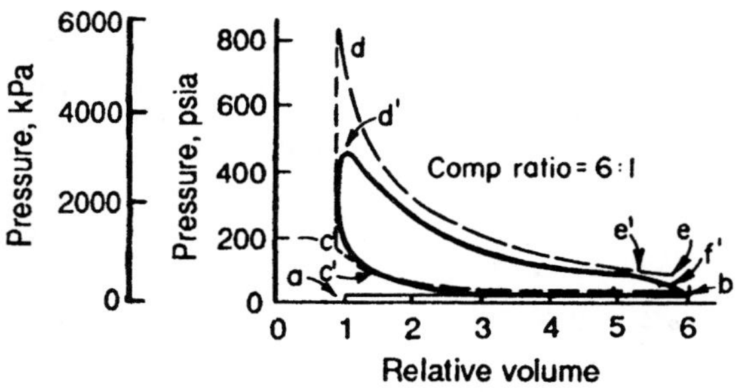

Figure 7.6.1 Otto cycle (constant-volume combustion, no excess air). Process of the ideal cycle: *a-b,* suction stroke, admission of the charge; *b-c,* compression stroke; at *c,* ignition of the compressed charge; *c-d,* combustion; *d-e,* expansion; at *e,* exhaust valve opens; *b-a,* exhaust stroke.

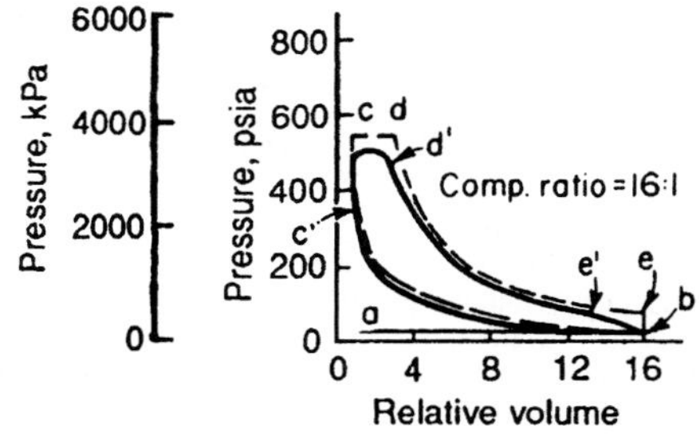

Figure 7.6.2 Diesel cycle (constant-pressure combustion, 50 percent excess air). Processes of the ideal cycle: *a-b,* suction stroke, admission of air; *b-c,* compression of air; *c-d,* injection and combustion of the fuel; *d-e,* expansion; at *e,* exhaust valve opens; *b-a,* exhaust stroke.

(Fig. 7.6.3). The constant-volume or Otto cycle is often used to approximate **spark-ignition** engines; the constant-pressure cycle is used to approximate **compression-ignition** or **diesel engines;** when both processes are combined, the cycle is called a **mixed, combination,** or **limited pressure cycle.** The cylinder pressure trace obtained from a high-speed, mixed-cycle, compression ignition engine may be similar

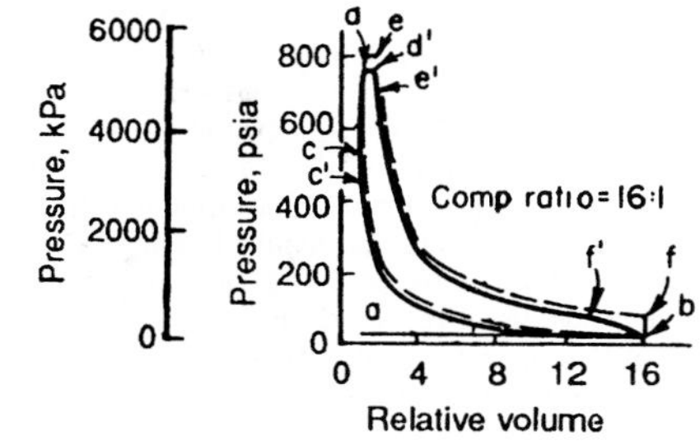

Figure 7.6.3 Mixed cycle (constant-volume and constant-pressure combustion, 50 percent excess air). Processes of the ideal cycle: *a-b,* intake stroke, admission of air; *b-c,* compression of air; *c-d,* ignition and constant-volume combustion; *e-f,* expansion; at *f,* exhaust begins; *b-a,* exhaust stroke. Broken lines represent ideal cycle; solid lines show characteristics of actual cycle. Primes attached to letters show actual locations of events.

to that obtained from a spark ignition engine. The fundamental differences between spark-ignition and compression-ignition engines are the mixture preparation of fuel and air (premixed or direct fuel injection in spark-ignition engines, and direct injection in diesel engines) and the methods of ignition.

The nominal **compression ratio** is defined as the displacement plus clearance volume divided by the clearance volume. The actual (or dynamic, effective) compression ratio is often less than the nominal value because of late intake valve or port closing.

The compression **pressure** may be estimated from the relation $p = p_m * r_a^{1.33}$, where p_m is the intake manifold pressure and r_a is the actual compression ratio.

The actual compression pressure can be determined with a compression pressure gage that traps the gases by means of a check valve, thus indicating the maximum pressure under motoring conditions.

Spark ignition engines use volatile liquids or gases as fuel; have compression ratios between 9:1 and 14:1 (limited by end-gas knock) and compression pressures up to 493 psi (3,400 kPa); and use port fuel injection or direct injection systems. Gasoline is the fuel commonly used in airplane, automobile, small marine, small stationary, and tractor engines. Commercial gases such as blast furnace gas, coal gas, coke oven gas, carbureted water gas, producer gas, and natural gas are used in large stationary engines. The fuel-air mixtures used are stoichiometric. Combustion pressures can be up to 5 times the compression pressures. Load and speed are controlled by either throttling the intake air flow at port fuel injected engines or by controlling the fuel flow at direct injected engines; piston speeds above 3,000 ft/min (15 m/s) are permissible.

Advantages are low cost, low specific weight, low cranking effort required, wide operating range in speed and load, high mechanical efficiency, and fairly low specific fuel consumption at high compression ratios and loads.

Compression ignition engines use liquid fuels of low volatility varying from fuel oil and distillates to crude oil (see Sec. 3.1), have compression ratios between 11.5:1 and 22:1 and compression pressures up to 1,088 psi (7,500 kPa). Generally no ignition devices are used, although the majority of compression ignition engines use glow plugs for cold start. Load and speed are controlled by varying the fuel quantity injected directly into the cylinder.

Dual-fuel engines are often based on diesel engines with added port fuel injectors. Premixed high volatility fuel-air charge enters the cylinder, while a pilot (small) injection of liquid fuel of low volatility is used to initiate the combustion process.

Advantages of compression ignition engines are low specific fuel consumption, high thermal efficiency at part loads, no preignition, low CO and moderate hydrocarbon emissions at low and moderate loads without exhaust aftertreatment, suitability for two-stroke operation, and excellent durability (by design). The lower-compression engines are of simpler construction (usually valveless two-stroke cycle), are more lightweight, and have lower first cost, lower operating expense, and higher mechanical efficiency than the higher-compression engines.

The **four-stroke** engine requires four piston strokes or two crankshaft revolutions per cycle (Figs. 7.6.1 and 7.6.2). This engine cycle is used in automobiles, trucks, tractors, and some marine engines with the exception of larger outboard and handheld engines. The pumping loss, indicated by the area between the exhaust and the intake curves (Fig. 7.6.12), depends on valve events and engine speed and amounts to 3-7 percent of the engine indicated power. Throttling increases the pumping loss and reduces indicated power. Fuel is introduced by direct injection into the cylinder.

Advantages compared with two-stroke crankcase compression spark ignition engines include wider variation in speed and load, cooler pistons, common crankcase in multicylinder engines, good lubrication without oil addition to the fuel, no fuel loss during exhaust stroke, lower specific fuel consumption, lower pumping losses, less exhaust dilution, lower hydrocarbon emissions, positive-displacement inlet and exhaust processes, and easier power regulation.

The **two-stroke** engine requires only one crankshaft revolution for each cycle. Exhaust ports in the cylinder wall are uncovered by the piston, or exhaust valves in the cylinder head are opened near the end of the expansion stroke, permitting the escape of exhaust gases and reducing the pressure in the cylinder (Fig. 7.6.4). The charge of air or

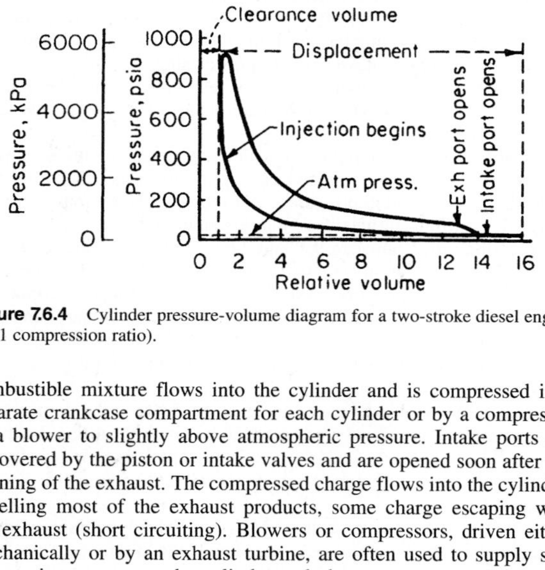

Figure 7.6.4 Cylinder pressure-volume diagram for a two-stroke diesel engine (16:1 compression ratio).

combustible mixture flows into the cylinder and is compressed in a separate crankcase compartment for each cylinder or by a compressor or a blower to slightly above atmospheric pressure. Intake ports are uncovered by the piston or intake valves and are opened soon after the opening of the exhaust. The compressed charge flows into the cylinder, expelling most of the exhaust products, some charge escaping with the exhaust (short circuiting). Blowers or compressors, driven either mechanically or by an exhaust turbine, are often used to supply sufficient air to scavenge the cylinder and clearance space and, in some cases, to supercharge the engine. Although small engines use only ports and crankcase compression (Fig. 7.6.5), large engines usually have separate compressors or blowers and either ports or both valves and ports. With crankcase compression, low scavenging of the cylinder occurs, and the volumetric efficiency is low (30 to 70 percent). The crankcase compression work amounts to 7-12 percent of the total work, but scavenging blowers with displacements 20-80 percent greater than the piston displacement may require as high as 30 percent of the indicated work in high-speed engines. To eliminate hydrocarbon emissions from scavenging, fuel is added by direct cylinder injection or timed intake port injection (gasoline).

Advantages compared with four-stroke engines include the 50 to 80 percent greater power output per crank revolution, smoother operation, low cost for valveless designs employing crankcase compression, low NO_x emissions (spark ignited), and light weight.

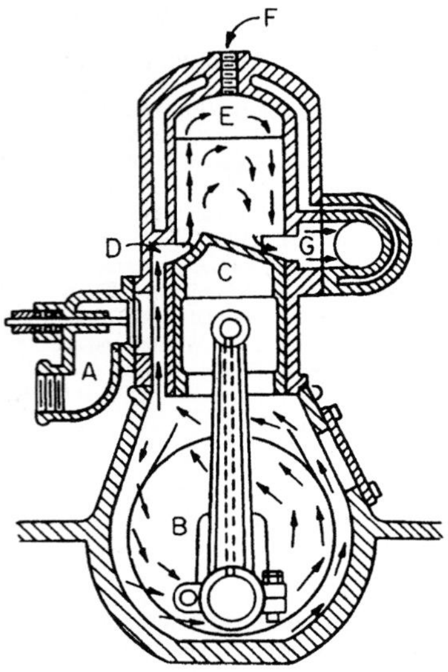

Figure 7.6.5 Small two-cycle gasoline engine with crankcase compression.

ANALYSIS OF ENGINE PROCESS

Cycle Analysis The theoretical thermal efficiency η_a of the **air standard Otto cycle** depends only on the volumetric compression ratio r_c and the ratio of specific heats of the working fluid γ, and is given by

$$\eta_\alpha = 1 - \left(\frac{1}{r_c}\right)^{\gamma-1} \tag{7.6.1}$$

For the **air standard diesel cycle** (constant-pressure combustion),

$$\eta_\alpha = 1 - \frac{r_d^\gamma - 1}{\gamma r_d^{\gamma-1}(r_d - 1)} \tag{7.6.2}$$

where r_d is the volumetric expansion ratio of the constant-pressure combustion process.

However, the diesel engine is more closely approximated by the limited pressure combustion (dual) cycle for which the thermal efficiency is given by

$$\eta_\alpha = 1 - \frac{r_p r_d^\gamma - 1}{r_c^{\gamma-1}[r_p - 1 + \gamma r_p(r_d - 1)]} \tag{7.6.3}$$

where r_p is the pressure ratio during the constant-volume heat addition.

The efficiencies indicated by the air standard analysis are much higher than can be attained in practice, principally because of variable specific heats, dissociation, and heat losses. The improved fuel-air cycle analysis takes into additional consideration variable fuel/air mixture properties and dissociation of combustion products and results in lower, more realistic values for efficiencies (Figs. 7.6.6 and 7.6.7). It more closely approximates a fast-burn, low-heat-rejection engine. Real cycle analyses also include combustion, heat losses, and blow-by losses, resulting in thermal efficiencies approximately 20 percent lower than predicted by the fuel-air cycle analyses, but requiring complex and time-intensive computer computations.

The **indicated mean effective pressure (IMEP)** is equal to net indicated work divided by volumetric displacement, and is proportional to torque. Indicated power can be calculated from

$$P_{i,n} = IMEP \cdot S \cdot A \cdot N \cdot \frac{1}{K} \tag{7.6.4}$$

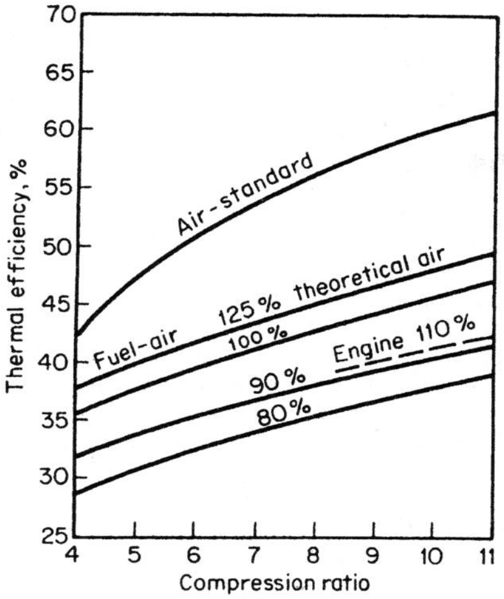

Figure 7.6.6 Indicated thermal efficiencies of various Otto cycle analyses and engine test results.

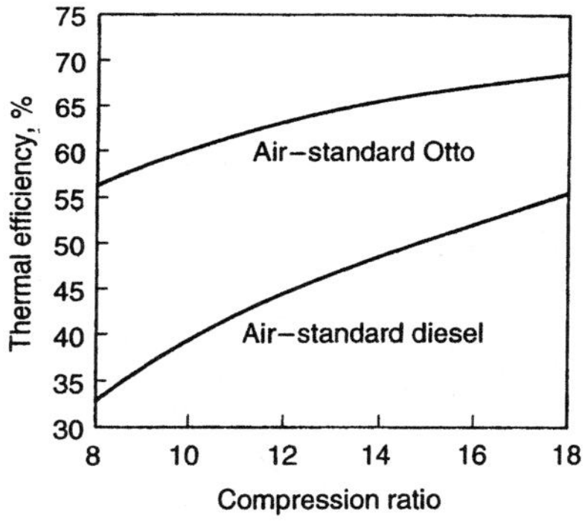

Figure 7.6.7 Indicated thermal efficiency of the air standard diesel cycle with constant-pressure combustion.

Where *IMEP* = Indicated Mean Effective Pressure, lb/in² (kPa); *S* = stroke, ft (m); *A* = total piston area, in², (m²); *N* = engine speed, rpm; *K* = 33,000 (0.4566 for SI units).

Increasing the compression ratio and the air-fuel ratio increases the thermal efficiency of both spark-ignition (Fig. 7.6.6) and diesel cycles (Fig. 7.6.7). Mean effective pressures depend on charge properties and thermal efficiency, both of which depend on compression ratio, air-fuel ratio (Figs. 7.6.8 and 7.6.9), and intake pressure.

Deviations from Ideal Processes

Changes in atmospheric pressure and temperature change power output and affect thermal efficiency.

Pumping work (area EFGAE, Fig. 7.6.10) for the constant-volume cycle is required to overcome the resistance to flow fresh charge into and exhaust products out of the cylinder. It varies from almost 0 for an engine running at low speed with wide-open throttle to about 3 lb/in² (21 kPa) mep at high speed, and to 10 lb/in² (69 kPa) with idle throttle position. The negative loop (area AXFGA, Fig. 7.6.10) in a normally aspirated engine represents about 60 percent of the pumping work.

Volumetric efficiency is the ratio of the actual mass of air entering the cylinder to the maximum air mass that can be trapped in the cylinder, referenced at atmospheric pressure and temperature. High manifold and cylinder temperatures and resistance to flow reduce volumetric efficiency. High-output engines have maximum volumetric efficiencies of 85-90 percent at rated speeds. Super charged engines can

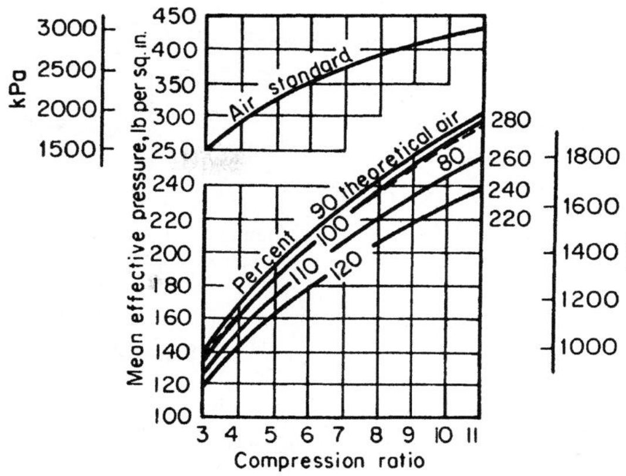

Figure 7.6.8 MEP for Otto cycle engine [air and liquid iso-octane supplied at 60°F (16°C)].

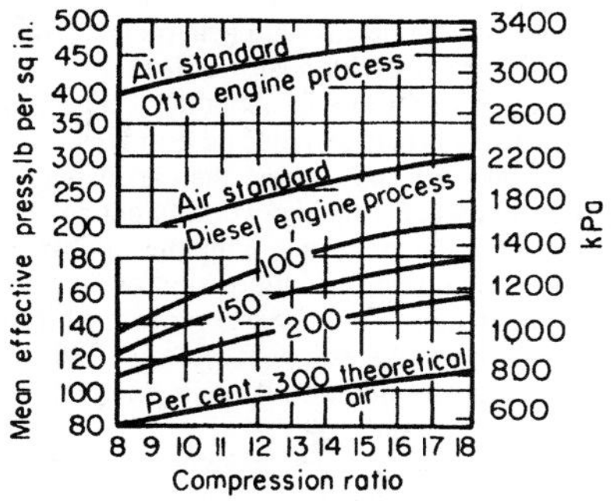

Figure 7.6.9 MEP for diesel cycle engine [air and liquid dodecane supplied at 60°F (16°C)].

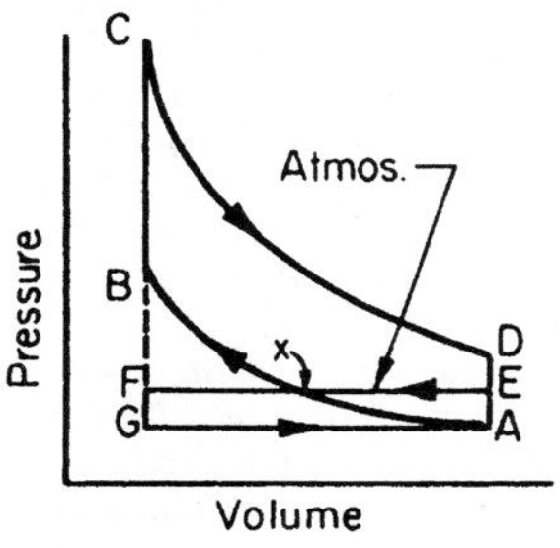

Figure 7.6.10 Throttled Otto cycle.

have volumetric efficiencies above 100 percent or can be designed to maintain high volumetric efficiencies with an increase in speed.

With fixed valve timing, volumetric efficiencies of automobile engines (Fig. 7.6.11) reduce as piston speeds exceed 1,200 to 1,600 ft/min (6 to 8.1 m/s), decreasing to approximately 60 percent at top speeds of 2,500 to 3,200 ft/min (13 to 16 m/s). Volumetric efficiency varies appreciably with valve timing and the number of valves, especially intake valves. Variable valve timing (camshaft phasing) is used to increase high-and low-speed volumetric efficiency.

Two-stroke crankcase compression engines have volumetric efficiencies of about 60 percent at low speeds and 50 percent or below at high speeds.

Fuels with high latent heats of vaporization and low boiling temperatures reduce the charge temperature and increase volumetric efficiencies (Table 7.6.1).

Air filters, governors, and carburetors increase the resistance to flow and decrease the volumetric efficiency.

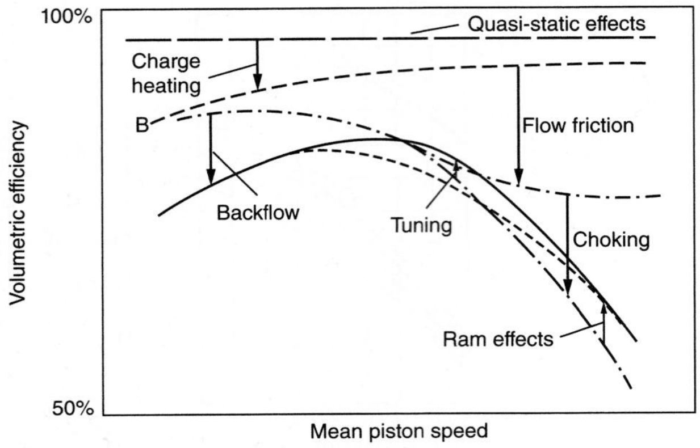

Figure 7.6.11 The effect of engine factors on volumetric efficiency as a function of mean piston speed at a fixed valve timing. Quasi-static effects include exhaust residual and fuel evaporation factors.

The **compression stroke** begins with heat transfer from the cylinder walls to the charge but ends with heat transfer from the charge to the walls. The net result is a heat loss of about 0.5-1.5 percent of the heating value of the charge. Heat loss during compression increases the required compression work, but lowers the expansion work a greater amount, thus slightly reducing the net work output and efficiency.

The mean value of the exponent n in the approximate equation pv^n = constant for the polytropic compression of a correct mixture of octane and air is about 1.33. For the diesel cycle in which the gas under compression is mainly air, the exponent is about 1.38 for a 15:1 compression ratio. Heat transfer into the charge increases the exponent, but blow by losses lowers its value.

The **combustion process** in the spark-ignition engine begins before the end of the compression stroke and ends after the beginning of the expansion stroke. Rassweiler, Withrow, and Cornelius (*Trans. SAE,* **34,** 1939) found a maximum variation of 30 percent in combustion duration for a single-cylinder engine while variations of both maximum cylinder pressure and rate of pressure rise of about 30 percent were found by Patterson (SAE paper 660129) in a V-8 engine. These effects are thought to be due primarily to cyclic variability in swirl and turbulence at the spark plug, but can also result from variations in air/fuel ratio and homogeneity of the mixture. Poor distribution in multicylinder engines also causes variations between cylinders. It has been shown that the loss due to the actual finite combustion duration is 3-6 percent of the ideal efficiency of constant-volume combustion. Any energy released before or after top dead center has an availability corresponding to the expansion ratio for the piston position at which the energy is released. Heat loss during the combustion process amounts to 5-30 percent of the energy content of the charge at rated outputs and speeds for automotive engines. Combustion in the diesel engine begins almost simultaneously at several locations depending on cylinder pressure, temperature, and mixture composition.

Flame propagation in spark ignition engines begins at the spark plug and travels in all directions through the mixture. At the end of combustion, the first part of the charge to burn has reached a higher temperature than the last portion to burn. Rassweiler and Withrow (Trans. *SAE,* **30,** 1935, p. 125) observed temperature differences of over 400°F (200°C) (non-knocking) and over 600°F (315°C) (knocking) in a nonturbulent combustion chamber.

The combustion of part of the charge results in expanding gases that compress the remainder unburned mixture. Due to temperature difference, the volume fraction occupied by the burned gases is higher than the mass fraction burned. This can increase the temperature of the last charge fraction to burn (end gas) above its autoignition temperature under certain operating conditions. This may result in rapid combustion of the end gas, which could produce the abnormal combustion condition called **end gas knock.**

The mean adiabatic exponent for the **expansion process** is about 1.22, varying from about 1.20 at the beginning to 1.25 at the end of expansion. Heat transfer and blowby losses reduce the exponent. The heat transfer during the expansion stroke amounts to 5-30 percent of the energy content of the charge.

The exhaust valve opens and blowdown begins at 70-90 percent of the expansion stroke. High exhaust gas velocities through the valves [1,200-1,500 ft/s (360-460 m/s)] are attained, and heat transfer, including that through exhaust port walls, amounts to 10-20 percent of the energy content of the charge, but represents practically no loss of work or availability.

The **exhaust stroke continues with** displacement of exhaust gases by the piston. The exhaust pressure is usually above atmospheric (Fig. 7.6.12) but may drop below atmospheric for part of the stroke.

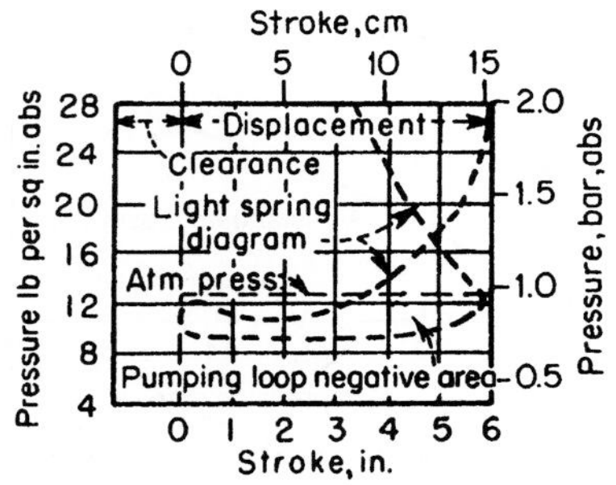

Figure 7.6.12 Pumping loop of Otto cycle engine.

Heat loss during the exhaust amounts to 3-5 percent of the energy content of the charge but does not represent any loss of work or availability. At full load about 80 percent of the gases in the cylinder escape during blowdown, and about 15 percent are pushed out of the cylinder during exhaust, the remainder being left in the clearance volume at the end of the exhaust stroke. At part load, the blowdown pulse is less vigorous and the residual gas fraction higher. Variable valve timing and/or lift can change the residual fraction considerably.

Exhaust gas temperatures vary with engine speed and load (Figs. 7.6.13 and 7.6.14), with high loads and speeds resulting in the highest temperatures.

Table 7.6.1 Relative Ideal Volumetric Efficiencies with Liquid Fuels
(Air and liquid fuel at 60°F)

Fuel	Stoichiometric air-fuel ratio	Heat required per lb of fuel for complete evaporation, Btu/lb (J/g)	Percent evap with 200 Btu/lb fuel fuel	Suction temperature, °F (°C): Complete evaporation	Suction temperature, °F (°C): 200 Btu/lb fuel	Relative volumetric efficiency: Complete evaporation	Relative volumetric efficiency: 200 Btu/lb fuel
Gasoline	14.7	150 (349)	81	106 (41)	81 (27)	100	100
Benzene	13.26	169 (392)	100+	50 (10)	68 (20)	111	103
Ethyl alcohol	8.99	425 (989)	55	71 (22)	55 (13)	107	105
Methyl alcohol	6.46	519 (1,207)	49	68 (20)	45 (7)	108	107

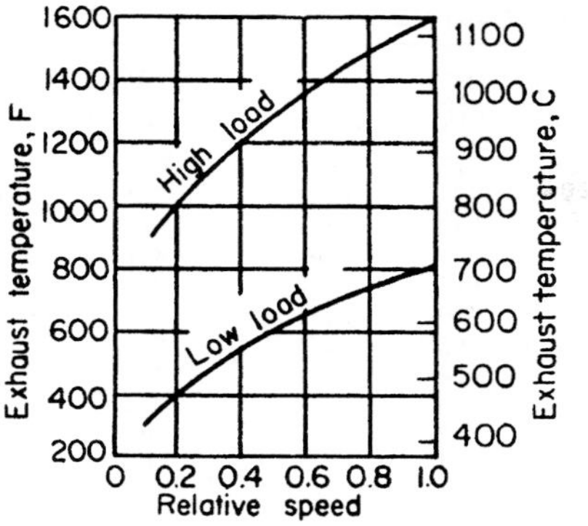

Figure 7.6.13 Exhaust temperatures for Otto cycle engines.

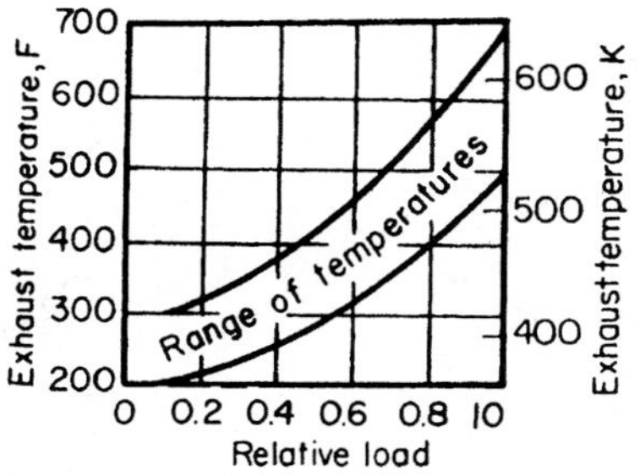

Figure 7.6.14 Exhaust temperatures for diesel cycle engines.

Exhaust Gas Analysis The composition of exhaust varies depending upon the equivalence ratio and the hydrogen/carbon ratio of the fuel. Figure 7.6.15 shows the calculated composition including water (wet basis) for complete combustion. Charts of theoretical exhaust composition for various fuels have been prepared by D'Alleva (*General Motors Res. Rept.*, **372,** 1960) and Eltinge (*SAE Trans.*, 1968).

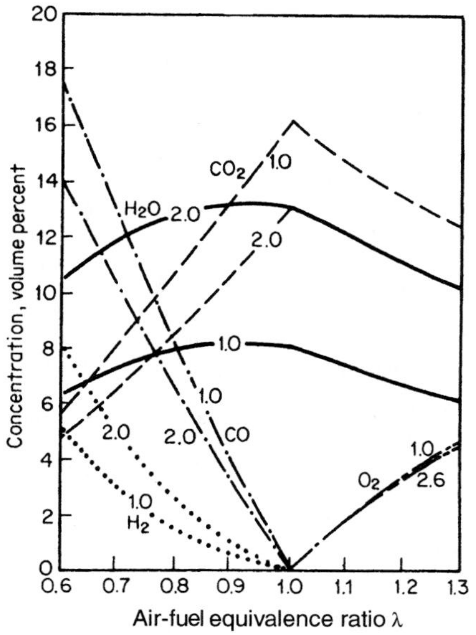

Figure 7.6.15 Theoretical mole percent of principal combustion products of hydrocarbon fuels for fuel hydrogen/carbon ratios. Air/fuel equivalence ratio, i.e., the inverse of the fuel equivalence ratio *f*, is the actual mixture air/fuel ratio divided by the chemically correct air/fuel ratio.

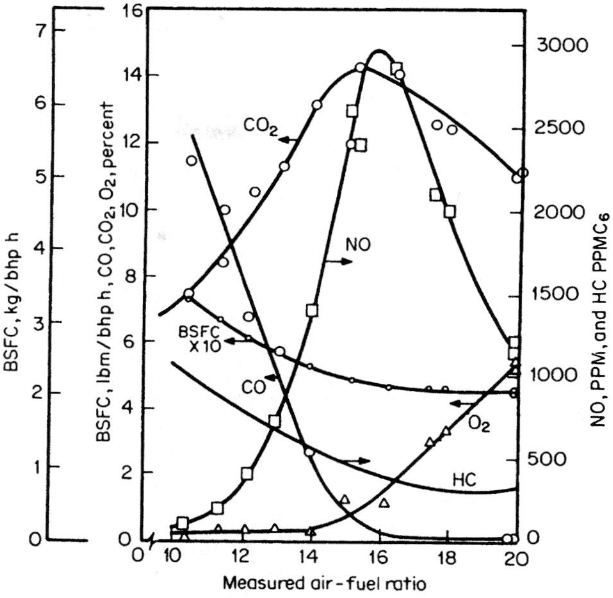

Figure 7.6.16 Gasoline engine basic specific fuel consumption and emissions vs. air/fuel ratio without exhaust after treatment.

In practice, non-equilibrium products, namely, unburned hydrocarbons and nitrogen oxide, appear in trace quantities. Figure 7.6.16 shows measured exhaust composition on a dry basis for a fuel with a hydrogen/carbon ratio of 1.85, typical of gasoline used for emissions certification.

Equations for calculating the air/fuel ratio from exhaust CO, CO_2, HC, and O_2 have been developed by Spindt (*SAE Trans.*, 1966) and others. Such determinations and those for calculation of remaining but unmeasured exhaust species depend upon carbon, hydrogen, oxygen, and nitrogen balances employing measured values for some exhaust constituents (Obert, "Internal Combustion Engines").

Table 7.6.2 summarizes the principal exhaust gas constituents and combustion efficiency of an older gasoline engine without exhaust after treatment. HC and NO were not measured. Modern engines, both gasoline and diesel, have combustion efficiencies approaching 99 percent and, with exhaust aftertreatment and the excellent fuel distribution made possible with fuel injection systems, have exhaust constituents that approach the theoretical levels of Fig. 7.6.15. Consequently, there is little fuel economy gain with better combustion efficiency per se.

U.S. AUTOMOBILE ENGINES

The size (displacement) of U.S. light-duty automobile engines generally varies from 61 in^3 (1.0 L) to 488 in^3 (8.0 L). Rated horsepower and displacement vary greatly depending on application, with horsepower per unit of displacement generally increasing during the period from 1994 to 2005 (Fig. 7.6.17). Most U.S. automobiles are spark-ignited and are fueled with gasoline, or a blend with ethanol (minor). Compression-ignited engines have shown low market penetration.

Emissions control and the need for better fuel economy resulted in introduction of (1) variable valve timing and lift, (2) cylinder deactivation, (3) direct fuel injection, (4) controlled inlet air temperatures, (5) electronically controlled fuel management (most with closed-loop feedback control to maintain a stoichiometric mixture), (6) exhaust gas recirculation (EGR), and (7) more stable ignition timing. Current advertised horsepower is based on engine brake power, which is measured with a fully equipped engine, including all parts needed to perform its intended function unaided, such as fuel pump, oil pump, intake air system, and exhaust system.

Table 7.6.2 Exhaust Gas Constituents and Combustion Efficiency of Older Gasoline Engines*

Air/Fuel	Percent by volume						H_2O/CO_2	H_2O/Fuel	Combustion† eff., percent
	CO_2	O_2	CO	H_2	N_2	H_2O			
11	8.76	0.15	9.14	4.66	77.08	13.78	1.57	0.972	66.7
12	10.18	0.44	6.65	3.39	79.13	13.93	1.37	1.043	73.8
13	11.60	0.59	4.31	2.20	81.09	14.16	1.22	1.122	81.5
14	13.02	0.63	2.09	1.07	82.99	14.46	1.11	1.205	89.6
15	13.23	1.35	0.99	0.50	83.72	14.09	1.06	1.247	93.8
16	12.62	2.49	0.68	0.35	83.65	13.30	1.05	1.256	94.8
17	12.00	3.55	0.48	0.25	83.51	12.54	1.05	1.261	95.5
18	11.45	4.49	0.30	0.16	83.39	11.88	1.04	1.267	96.2
19	10.90	5.36	0.20	0.10	83.23	11.25	1.03	1.269	96.5
20	10.40	6.15	0.11	0.06	83.07	10.68	1.03	1.272	96.9
21	9.92	6.86	0.08	0.04	82.90	10.16	1.03	1.271	96.9
22	9.44	7.55	0.06	0.03	82.71	9.65	1.02	1.268	96.8
23	9.00	8.18	0.05	0.03	82.53	9.19	1.02	1.266	96.7
24	8.60	8.74	0.06	0.03	82.37	8.78	1.02	1.264	96.6

* Gerrish and Voss, *NACA Rept.* 616, 1937.
† Low-heat-value basis.

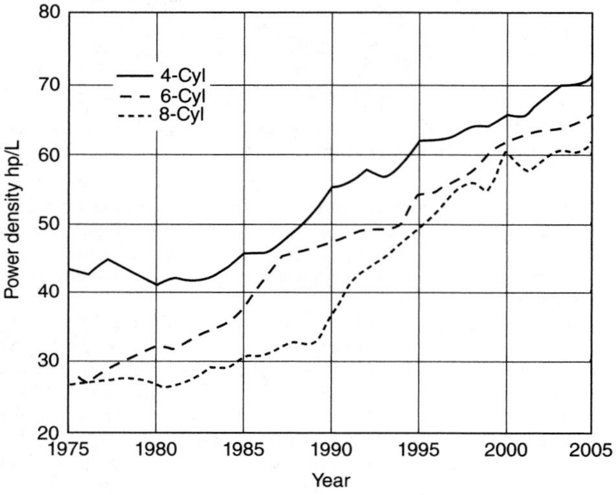

Figure 7.6.17 Trend of average U.S. automobile power density. *(Source: Center for Automotive Research, Ann Arbor, MI.)*

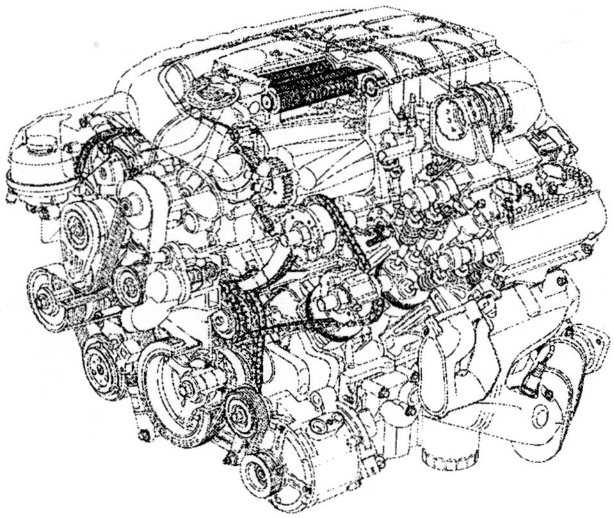

Figure 7.6.18 GM Northstar V8, engine (2006). *(Courtesy: General Motors.)*

With a fixed valve timing system, the closure of intake valves (approximately 50° crank angle after bottom dead center) is selected to give a rising volumetric efficiency and torque as the rotation speed is increased. At high speeds, restriction in the intake system, developed primarily by the intake valves, begins to throttle the engine. Automotive engines typically develop maximum torque at approximately 3,000 rev/min. Despite reduced torque at higher speeds, the power peaks at 5,000 to 6,000 rev/min. Higher power can be obtained with electronic fuel injection, variable valve timing, and other technologies, such as intake manifold tuning. Hybrid vehicles combine smaller conventional gasoline or diesel engines, augmented by battery-powered electric motors, and are in the marketplace in increasing numbers.

Engines have been produced with various numbers of cylinders over the years (from 2 to 16), but the most common engines in modern automobiles are in-line four-cylinder designs for smaller cars and V-6, V-8, or V-12 designs (Fig. 7.6.18) for larger cars. V-8 engines often have an angle of 90° between the two banks of four cylinders, whereas the V-6 engines often have an angle of either 60° or 90° between two banks of three cylinders. Smaller engines, in terms of both cylinder number and displacement, provide improved fuel economy, which is particularly important because of rigid federal corporate average fuel economy (CAFE) standards. In-line six-cylinder, V-8, and V-12 engines have characteristically smooth operation because of effective dynamic balancing of torque fluctuations from different cylinders. With the rapid growth in the use of transverse engines with front drive, and smaller and lighter vehicle structures with sloping aerodynamic front hoods, packaging constraints on automobile engines are severe. For this reason, push-rod engines remain popular in some applications (e.g., Chevrolet Corvette V-8), as they require less physical vehicle space.

All modern engines employ compact, rigid structures. Cylinder blocks and heads are made of cast iron or aluminum. Aluminum use is increasing in order to reduce weight. Other materials, such as aluminum-magnesium alloys have been used for engine blocks. Four-cylinder engines use three or five crankshaft bearings; the V-6 uses four, and the V-8 uses five. Cylinder bore can be larger than, smaller than, or equal to the piston stroke.

Liquid cooling passages are provided around the bore of each cylinder as well as at the hot regions of the combustion chambers and exhaust ports in the head. Modern light-duty engine pistons are almost always made of aluminum. Crankshafts and connecting rods are usually made of cast iron or forged steel. Four-cylinder engines and some six-cylinder engines use a separate crank throw (eccentric portion of the

Table 7.6.3 Selected U.S. Automobile Engines

Make and model	Type and no.*	Bore	*B/S* ratio	Displacement, in³ (*L*)	Valves per cyl.	Comp. ratio	bhp	r/min	bhp/L	Vehicle weight, lb
General Motors										
Saturn L-Series	DOHC I-4	86.00 mm	0.91	134 (2.2)	4	10	140	6,500	63.6	3,237
Cadillac CTS	DOHC V-6	89 mm	1.02	170 (2.8)	4	10	210	7,000(auto) 7,200(man)	75	3,960
Buick Century	OHV V-6	89.00 mm	1.06	191 (3.1)	2	9.6	175	6,000	54.5	3,342
Chevrolet Impala	OHV V-6	92.00 mm	1.1	204 (3.4)	2	9.5	180	6,000	52.9	3,389
Pontiac Grand Prix	OHV V-6	96.52 mm	1.12	231 (3.8)	2	9.4	200	6,000	56.6	3,583
Chevrolet Silverado	OHV V-6	101.6 mm	1.15	272 (4.3)	2	9.2	195	5,600	45.3	4,360
Buick Rainier	OHV V-8	96.01 mm	1.04	325 (5.3)	2	9.95	300	5,750	56.6	4,417
Ford										
Ford Focus	I-4	3.44 in	1.05	121 (2.0)	4	10	136		68	
Ford Mustang GT	OHV V-8	3.6 in	1	281 (4.6)	3	9.8	300		65.2	
Ford Taurus	V-6	3.50 in	1.13	181 (3.0)	2	10	201		67	
Ford Five Hundred	V-6	3.5 in	1.12	183 (3.0)	4	10	203		67.7	
Ford GT	DOHC V-8 supercharged	90.2 mm	0.85	330 (5.4)	4	8.4	550		101.9	
Fiat Chrysler										
Dodge Neon	SOHC I-4	87.5 mm	1.05	2.0 L	4	9.3	132	5,600	66	2,581
Dodge Charger	SOHC V-6	96 mm	1.19	5.7 L	4	10	250	6,400	53.2	3,800
Chrysler 300	DOHC V-6	86 mm	1.1	2.7 L	4	9.7	190	6,400	70.4	3,721
Chrysler Sebring	DOHC I-4	87.4 mm	0.87	2.4 L	4	9.4	150	5,500	62.5	3,172
Chrysler Pacifica	SOHC V-6	96 mm	1.19	3.5 L	4	10	250	6,400	71.4	4,383
Jeep Grand Cherokee	SOHC V-6	93 mm	1.02	3.7 L	2	9.6	210	5,200	56.76	4,195
Jeep Wrangler	DOHC I-4	88 mm	0.87	2.4 L	4	9.5	147	5,200	61.25	3,235

* I = in-line, V = vee cylinder configuration.
SOURCE: 2016 Manufacturers' websites.

crankshaft) for each connecting rod, whereas in V-8s, connecting rods for corresponding cylinders on each side of the V are side by side on the same crank throw.

Detachable cylinder heads contain the inlet and exhaust valves (overhead valves) and ports. Most modern engines use four valves per cylinder (Table 7.6.3) actuated by belt or chain-driven overhead camshafts. Exhaust gas is directed through short passages to the exhaust collector or manifold, thereby minimizing heat loss and improving catalytic converter performance.

Modern spark-ignition engines have port fuel injection or direct injection. In the port fuel injected engines, the intake air is delivered through a compact inlet manifold and fuel is injected into the port when the intake valves are closed. Fuel management is provided by an electronic fuel injection system with one injector for each inlet port and injection timing for each cylinder. Feedback from an exhaust gas oxygen sensor is used to provide precise control of air/fuel ratio and optimize performance of the emissions control system. The control system maintains the mixture at stoichiometry as is controlled by the throttle position. In direct injection engines, the fuel is injected directly into the cylinder. The throttle is often eliminated and the control system governs the amount of fuel that is direct injected. Direct injection engines can operate at stoichiometry with homogeneous charge or overall lean with stratified charge. Electronic ignition virtually eliminates the need for periodic ignition system maintenance. Much attention has been directed to reducing friction losses in pistons, rings, valve train, bearings, belts, and engine accessories. Lower friction improves engine brake efficiency and part-load fuel economy.

Quiet operation is essential in automobile engines and requires close manufacturing control of clearances. Most engines have hydraulic valve lifters and means of rotating valves to prolong their life. Exhaust valve seats are hardened or incorporate inserts to prevent recession.

It is standard practice to provide pistons with three rings above the wrist pin—two narrow compression rings above one oil scraper ring, with several drain holes from the back of the ring groove to return oil to the crankcase. Compression rings are usually narrow, are made of cast iron or steel from 0.063 to 0.0787 in (0.160 to 0.200 cm) wide, and have a wear-resistant coating, such as chromium (particularly for the top ring) and tin, or compounds (iron oxide, phosphates), which minimizes the running-in period without danger of scuffing or scoring the cylinders. Oil scraper rings are made of steel or cast iron, generally are less than 3⁄16 in (0.476 cm) wide, and are provided with internal expander springs to increase pressure against the cylinder bore. Many oil scraper rings, especially those of steel, also use wear-resistant coatings. Low friction bore or piston skirt coatings are often used.

Liquid coolant under a pressure of 15 lb/in² (103 kPa) is standard practice. Coolant jacket temperatures are controlled by thermostats which open at approximately 195°F (90.6°C), permitting coolant from the cylinder heads to flow to the radiator. All engines are designed with coolant all around each cylinder or are "siamesed" with jackets having the full length of the piston travel. The water jacket cools the oil also.

FOREIGN AUTOMOBILE ENGINES

Foreign automobile engines sold in the United States are similar to engines produced in the United States (see Table 7.6.4.). These similarities include size and basic design features. They are primarily four-stroke, spark ignited engines and secondarily diesels. All engines are liquid-cooled. Engines produced for areas of the world other than North America exhibit a broader range of characteristics. Very small automobiles with engines of less than 40 in³ (0.66 L) and 35 hp are found in some areas of the world. Engines produced for the U.S. market are larger on average, in terms of piston displacement, than those used elsewhere. Foreign engines often have very high specific output in terms of power per unit displacement, because of taxation based on engine displacement, higher vehicle speed limits, and emissions regulations. Many countries tax larger engines at a higher rate than smaller engines.

Most modern foreign spark-ignition engines are turbocharged although some are naturally aspirated. All modern foreign diesel engines are turbocharged. Spark-ignition engines are often sized about 0.5 to 1.5 L/cyl. Vehicle size and power requirements along with smoothness, efficiency, cost, size, and weight limitations determine the number of cylinders.

The maximum power output of foreign automobile engines ranges from approximately 100 to over 800 hp, equivalent to the range for U.S. engines. During the last few decades power output of foreign engines has increased in response to consumer demand. Foreign automobiles were the first to introduce hybrid powertrains to the North American market.

Table 7.6.4 Selected Foreign Automobile Engines

Make and model	Type and no.	Bore	*B/S* ratio	Displacement, L	Valves per cyl.	Comp. ratio	bhp	Rated rpm	bhp/L	Curb weight, lb
VW Golf	SOHC I-4	82.55 mm	0.89	2.0	2	10	115	5,200	57.5	2,972
VW Jetta 1.8t	DOHC I-4 Turbo	86.4 mm	0.94	1.8	5	9.5	180	5,500	100	3,197
VW Jetta Diesel	SOHC I-4 Turbo	79.5 mm	0.83	1.9	2	18.5	100	4,000	52.6	3,197
BMW 325i	DOHC I-6	3.35 in	0.97	3.0	4	10.7	215	6,250	71.7	3,285
BMW 750i	DOHC V-8	3.66 in	1.05	4.8	4	10.5	360	6,300	75	4,486
Toyota Camry	DOHC I-4	88.4 mm	0.92	2.4	4	9.6	160	5,700	66.7	3,109
Toyota Corolla	DOHC I-4	79.0 mm	0.86	1.8	4	10	130	6,000	72.2	2,502
Toyota Prius	DOHC I-4 hybrid Electric	75.0 mm	0.89	1.5	4	13	76 (+67 electric)†	5,000	70.7 (95.3)†	2,890
Subaru Outback	SOHC Flat-4	99.5 mm	1.26	2.5	4	10	168	5,600	67.2	3,309
Mercedes CL-Class	SOHC V-8	97 mm	1.15	5.0	3	10	302	5,600	60.4	4,085
Mercedes E-Class	DOHC V-6	92.9 mm	1.08	3.5	4	10.7	268	6,000	76.6	3,691
Lexus GS300	DOHC I-6	87.4 mm	1.06	3.0	4	11.5	245	6,200	81.7	3,536
Nissan Maxima	DOHC V-6	95.5 mm	1.17	3.5	4	10.3	265	5,800	2.7	3,432
Lotus Elise	DOHC I-4	80 mm	0.9	1.8	4	11.5	190	7,800	105.6	1,777
Acura TL	SOHC V-6	89 mm	1.03	3.2	4	11	270	6,200	84.4	3,482
Audi A6	DOHC V-6	84.5 mm	0.91	3.1	5	12.5	255	6,500	82.3	3,957
Honda Civic	SOHC I-4	75.0 mm	0.79	1.7	4	9.5	115	6,100	67.65	2,449

* I = in-line, V = vee type.
† Numbers in parentheses are with the addition of an electric motor.
SOURCE: 2005 Manufacturers' websites.

TRUCK AND BUS ENGINES

Truck and bus engines (Figs. 7.6.19 and 7.6.20) are similar to automobile engines but, in general, are larger, having 300- to 1,000-in^3 (4.6- to 16-L) displacement, are more robust, and run at lower speeds. Both spark-ignition and diesel combustion are used, with the diesel predominant in the large and medium sizes. All modern heavy-duty engines are four-stroke and water cooling is universally used. The few heavy-duty gasoline engines are naturally aspirated and have compression ratios lower than automobile engines, with similar valvetrains and similar performance curves. They can attain maximum power (bhp) (55 to above 300) at a mean piston speed averaging about 2,600 ft/min (13 m/s). Maximum torque is obtained at about 1,500 rev/min, appreciably lower than for automobile engines. Modern heavy-duty diesel engines are turbocharged and have mean piston speeds ranging from 1,500 to 2,100 ft/min (7.6 to 11 m/s) and operate at somewhat lower speeds than gasoline engines. Truck and bus gasoline engines can operate at speeds slightly higher than those at which maximum power is attained. Some diesel engines can be operated near the maximum power speed. Most are operated below (Fig. 7.6.21).

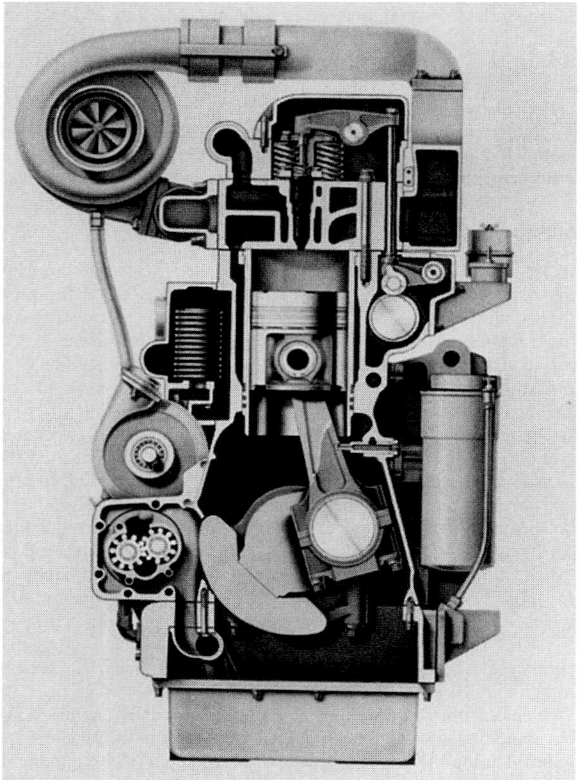

Figure 7.6.19 Cummins model KT 450-hp turbocharged, four-stroke diesel engine.

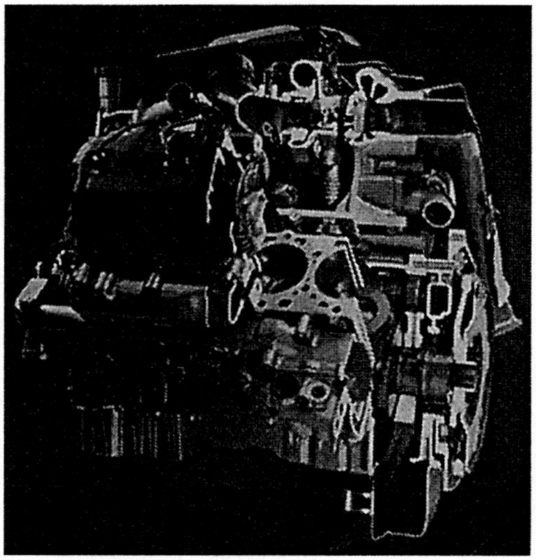

Figure 7.6.20 GM Duramax medium-duty four-stroke diesel engine with bowl-in-piston combustion chamber, direct fuel injection, and turbocharging. (*Courtesy of GM.*)

Truck and bus engines are usually built with four or six cylinders in line or with eight cylinders in V-type construction. The bore/stroke

Table 7.6.5 Selected Diesel Engines for Truck, Bus, Tractor, Marine, and Industrial Uses

Make and model	Designed for*	Cylinders: No.	Bore, in	*B/S* ratio	Displacement, in^3	Compression ratio	Max continuous bhp @ r/min	Max torque: lb · ft	(N · m)	At r/min
Case 445/M2	Tr	4	4.09	0.79	272		76 @ 2,200	236	(320)	1,400
Caterpillar C7	B, T	6	4.33	0.87	441	16.5	190–210 @ 2,500	520	(705)	1,440
Deere 4024T	Ind, Tr	4	3.4	0.83	149	18.2	41 @ 2,800	129	(175)	1,500
Detroit Diesel Series 60	T	6	5.12	0.81	778	17.25	390 @ 1,800	1,350		1,200
Deutz BF4M1013	Ind, T	4	4.25	0.83	290	17.6	125 @ 2,300		(464)	1,400
International DT 570	Ind, T	6	4.59	0.8	570	17.5	310 @ 2,200	950		2,200
Mack AI-350	T	6	4.875	0.75	728	16	350 @ 1,950	1,260	(1,708)	1,300
Perkins 1104C-44T	Ind	4	4.09	0.83	269	18.25	90 @ 2,400	274	(371)	1,400
Allis-Chalmers 2800	Ind, T	6	3.88	0.91	301	16.25	68 @ 2,200	225	(305)	1,400
Cummins B-785-C	Ind	8	5.50	1.33	785	15.9	220 @ 2,300	560	(759)	1,600
MAN DO826	B, T	6	4.25	0.86	419	16	270 @ 2,400	710	(960)	1,500

* B = bus; T = truck; Ind = industrial; Tr = tractor.

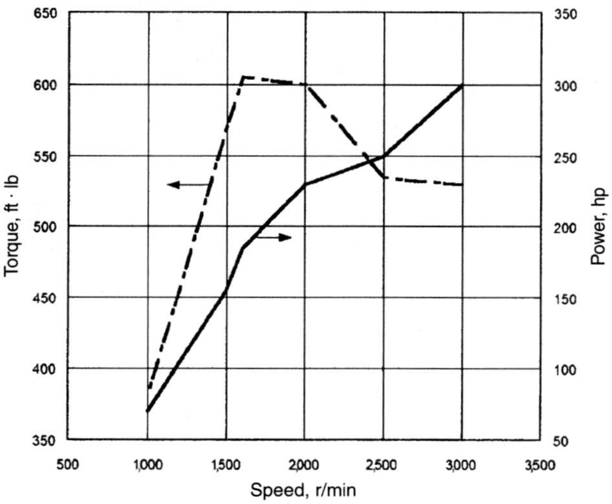

Figure 7.6.21 Torque and power profiles for 2006 GM 6600 6.6-L V-8 Turbo Duramax medium-duty four-stroke diesel engine. (See Fig. 7.6.20.)

ratio ranges from about 0.75 to 1.4, the latter engines having the larger values (see Tables 7.6.5 and 7.6.6).

TRACTOR ENGINES

Both spark ignition (gasoline or liquefied petroleum gas) and compression ignition (diesel) engines are used in tractors. Almost all engines above 100 hp are diesels. Most engines are four-stroke, water-cooled, with valve-in-head arrangements. Integral bore, dry-sleeve, or wet-sleeve cylinder liner arrangements are utilized. One-, two-, three-, four-, five-, six-, and eight-cylinder engines are used. All engines except the eight-cylinder are usually in-line in order to minimize the width of the hood in the interest of driver visibility. Tractor applications have a high load factor, are relatively low-speed, and deliver high low-speed torque (Fig. 7.6.22). Off-highway engine equipment now must meet emission standards in the United States.

STATIONARY ENGINES

Stationary engines may be either spark-ignited or compression-ignited, use either liquid or gaseous fuel, and use the two- or four-stroke cycle. The power output ranges up to about 48,000 bhp per engine. They commonly use pistons and connecting rods, which permit the crankcase

Table 7.6.6 Selected Gasoline Engines for Truck, Bus, Tractor, Marine, and Industrial Uses

Make and model	Designed for*	Cylinders: No.	Bore, in (cm)	*B/S* ratio	Displacement, in^3 (L)	Compression ratio	Rating† bhp	at r/min	Max torque, lb · ft (N · m)	at r/min
Allis-Chalmers G-153	Ind	4	3.44 (8.74)	0.83	153 (2.5)	8	52	2,400	132 (179)	1,460
Jeep Cherokee	T, Tr, Gp, Ind	I-6	3.88 (9.53)	1.13	242 (4.0)	8.8	190	4,750	215 (292)	1,400
Case 201 G	Tr, Gp, Ind	4	4.0 (10.16)	—	251 (4.1)	7.5	87	2,400	221 (300)	1,400
Chevrolet 350	Gp	V-8	4.0 (10.16)	1.15	350 (5.7)	9.1	210	4,000	—	
Chevrolet 454	T	8	4.25 (10.79)	1.06	454 (7.4)	8.3	250	4,000	365 (495)	2,800
Chrysler H-225	M, Ind	6	3.41 (8.66)	0.83	225 (3.7)	8.4	132	4,000	203 (275)	2,000
Continental R 821-46	Tr, Ind	4	3.03 (7.70)	0.92	95.5 (1.6)	8.6	67.2	5,000	84 (114)	2,400
Dodge CT 900	T	8	4.19 (10.64)	1.12	413 (6.8)	7.5	238	3,600	355 (481)	2,000
Ford 158-G	Tr	3	4.20 (10.67)	1.10	158 (2.6)	7.8	45.7	2,100	120 (163)	1,350
Ford Aerostar	Gp	V-6	3.5 (8.89)	1.11	182 (3.0)	9.2	135	4,600	—	
Ford F150	T	I-6	4.0 (10.16)	1.00	300 (4.9)	8.8	145	3,400	—	
Hercules G-3400	Tr, Ind	6	4.0 (10.16)	0.89	339 (5.6)	7.5	143	2,800	311 (422)	1,400
International V5-401	T	8	4.13 (10.49)	1.10	400 (6.6)	7.7	226	3,600	355 (481)	2,000
Minneapolis-Moline HD 425	Tr	6	4.25 (10.79)	0.85	425 (7.0)	8.1	133	2,000	378 (512)	800
Oliver 1655	Tr	6	3.75 (9.53)	0.94	265 (4.4)	8.5	84	2,200	220 (298)	1,300
Volkswagen 124A	Ind	4	3.38 (8.59)	1.24	96.7 (1.6)	7.7	53	3,600	75.5 (102)	2,500

* B = bus; T = truck; Tr = tractor; M = marine; Gp = general purpose; Ind = industrial.
† hp without accessories.

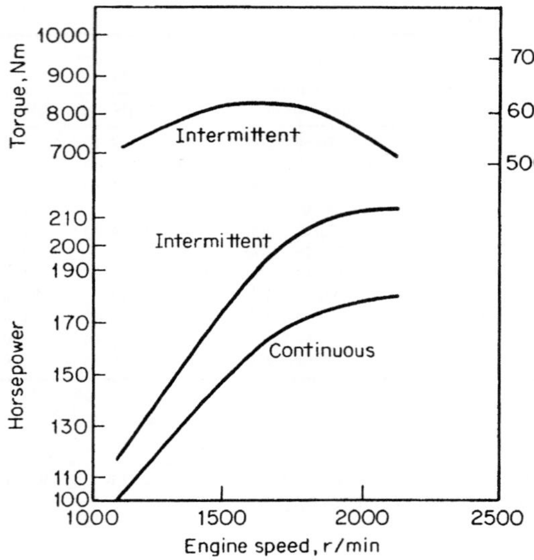

Figure 7.6.22 Typical performance of a tractor diesel engine.

to be isolated from blowby gases and the bottom face of the piston to function as an air pump. In the larger sizes, cylinders are built up of several parts, cylinder liners are used, and long pistons for single-acting diesel engines may have up to five piston rings. The valves are usually massive and in the larger sizes may be cooled. The exhaust pipe is sometimes water-cooled, and the pistons for large-bore engines are always liquid-cooled with oil or water. In general, there is much less integral construction (Fig. 7.6.23) in the large stationary engine than in the automobile type.

An interesting medium-speed, four-stroke diesel engine, V-type design with rotating pistons, is illustrated in Fig. 7.6.23. The symmetric piston is cooled with lubricating oil, and besides the reciprocating motion, it performs a slow rotating motion around its longitudinal axis. At each stroke, a new oil-wetted portion of the piston is always in contact with the pressure side of the liner, thus minimizing wear, oil consumption, and local overheating of the liner resulting from blowby.

Stationary engines usually operate at constant speed and are controlled by throttling the intake air flow in spark ignition engine and by varying the amount of fuel injected into the diesel engines. Compression ratios as low as 6:1 are used with spark ignition engines and as low as 12:1 with compression ignition engines. Stationary engines are usually built with up to 12 cylinders in-line, or with a V-type arrangement having up to 16 or more cylinders.

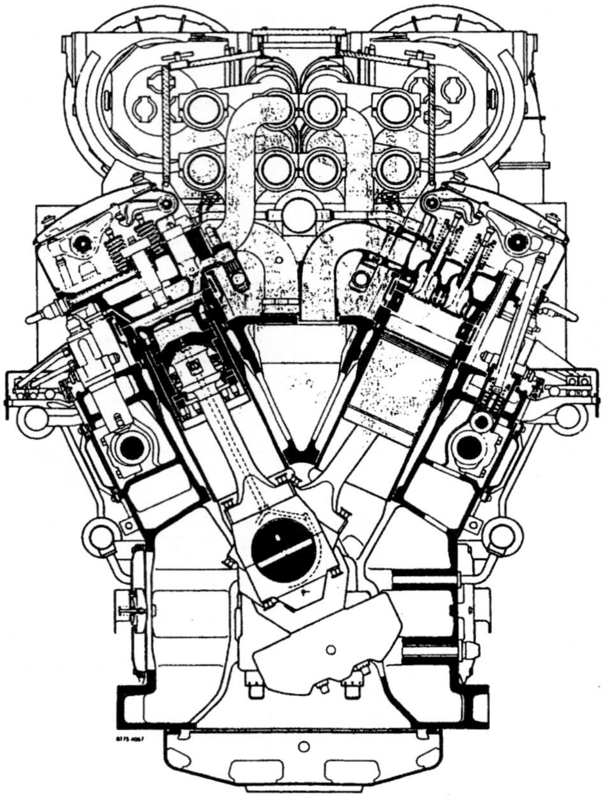

Figure 7.6.23 Cross section of Sulzer V-type ZV-40 engine. Bore = 400 mm; stroke = 480 mm.

MARINE ENGINES

(See also Sec. 8.3)

Marine engines may be either spark ignited or compression ignited and may be normally aspirated or boosted. Outboard engines range in power from less than 1 to more than 200 bhp and are generally of the two-stroke gasoline type. Modified automotive gasoline engines are also used in boats. Diesel engines range in power from below 20 to 48,000 bhp and may be either two- or four-stroke. Both automotive and stationary types are used in marine work; consequently, wide variations in specific weights are found. (See Tables 7.6.5 to 7.6.7).

Table 7.6.7 Selected Outboard Engines

	Cylinders						
Make and model	Type and no.*	Bore, in (cm)	*B/S* ratio	Displacement, in^3 (cm^3)	Rating, bhp at r/min		Weight, lb (kg)
Mercury							
Verado 135	I-4	3.23 (8.2)	1	105.7 (1,732)	135	5,200	510 (231)
OptiMax 150	V-6	3.50 (8.9)	1.32	153 (2,507)	153	5,250	431 (195)
OptiMax 90	I-3	3.63 (9.2)	1.21	92.96 (1,526)	90	5,000	375 (170)
40 EFI	I-3	2.56 (6.5)	0.9	45.6 (747)	40	5,500	216 (98)
Johnson							
J25PL4	I-3	2.56 (6.5)	1.08	36.4 (597)	25	5,000	212 (96)
J115PL	V-4	3.6 (9.1)	1.4	105.4 (1,727)	115	4,500	335 (152)
J175PL	V-6	3.6 (9.1)	1.4	158 (2,589)	175	4,500	383 (174)
Evinrude							
E60DPL	I-2	3.6 (9.1)	1.4	52.7 (863)	60	5,750	240 (109)
E200DPL	V-6	3.6 (9.1)	1.4	158 (2,589)	200	5,000	419 (190)

* I = in-line, V = vee type.
SOURCE: 2005 Manufacturers' websites.

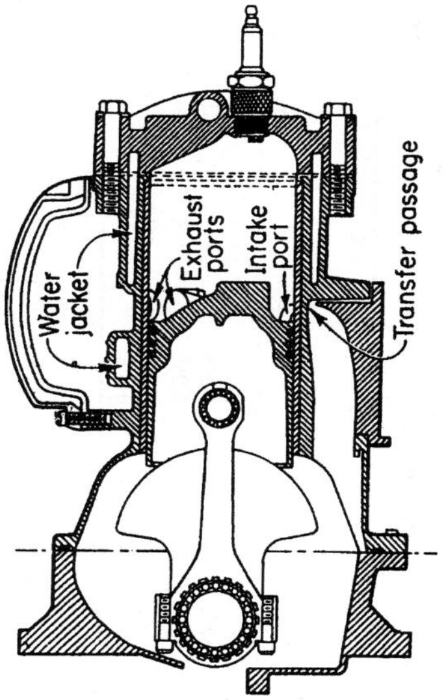

Figure 7.6.24 Johnson two-cycle, 30-hp outboard motor.

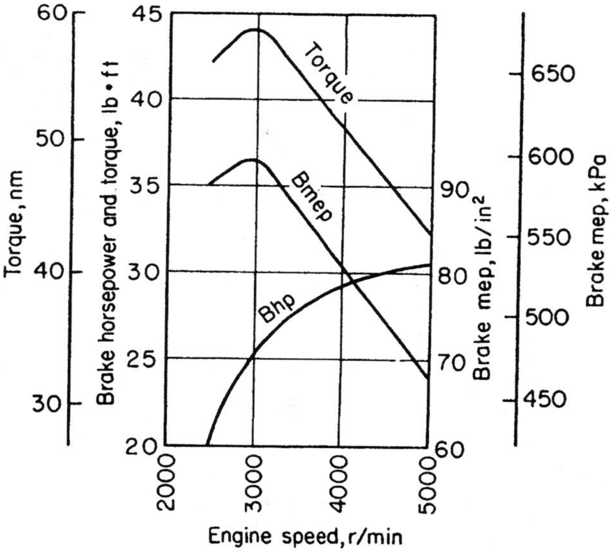

Figure 7.6.25 Performance of Johnson outboard motor.

The outboard two-stroke engines (see Table 7.6.7) have crankcase compression (Fig. 7.6.24) and may have two ports, three ports, or two ports with a rotary or reed crankcase inlet valve for the highest outputs (Fig. 7.6.25). Spark-ignition four-stroke engines have either L-head or valve-in-head construction, and diesel four-cycle engines are invariably of valve-in-head construction.

Regular gasoline is required for spark ignition engines, the fuel and lubricating oil being mixed for outboard two-stroke engines. A comparison of a number of fuel consumption curves for various types of engines shows the large low-speed marine diesel engine as having the lowest specific fuel consumption or highest efficiency (Fig. 7.6.26).

Low-speed engines can be directly connected to the propellers, and high-speed engines are connected through reduction gearing. Outboard engines have maximum speeds up to 6,000 rev/min with the piston speeds ranging from 1,000 to about 2,500 ft/min (5.0 to 12.7 m/s).

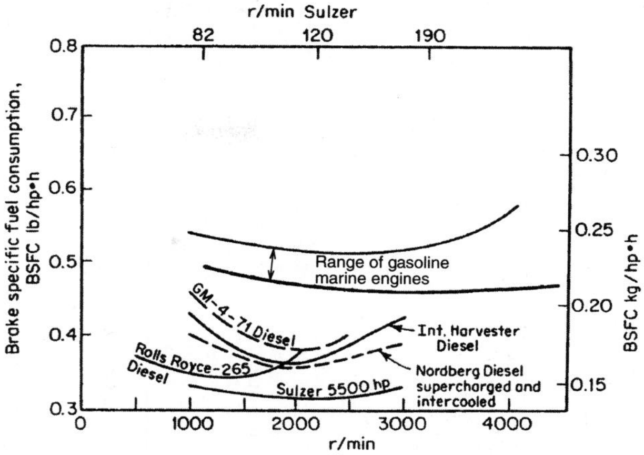

Figure 7.6.26 Specific fuel consumption curves.

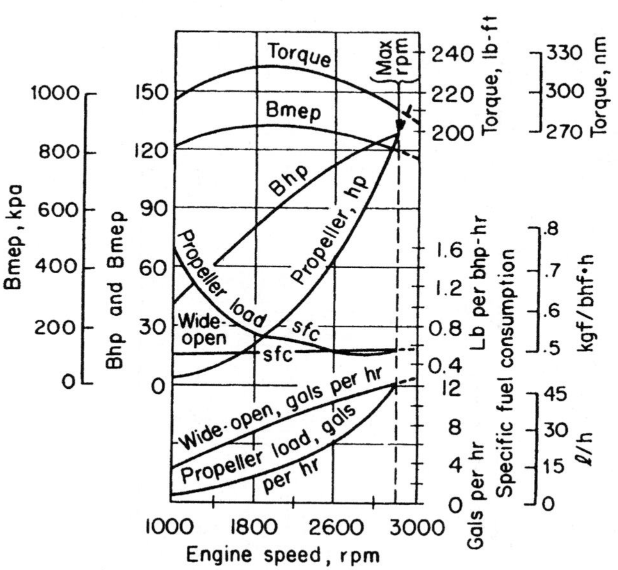

Figure 7.6.27 Performance of six-cylinder $3^7/_{16} \times 4^3/_4$ in (8.73 × 12.07 cm) Chrysler marine engine with 7:1 compression ratio.

Maximum engine speed is obtained at the intersection of the propeller/horsepower and the maximum horsepower curves (Fig. 7.6.27). This may occur well below the speed for maximum horsepower or beyond this speed. The effect of throttling to reduce speed is to increase the specific fuel consumption considerably.

SMALL INDUSTRIAL, UTILITY, AND RECREATIONAL VEHICLE GASOLINE ENGINES

The spark ignition engine with two or fewer cylinders is predominant in this class (Table 7.6.8). Fuel economy is generally not a major concern, whereas simplicity and performance are important. For reciprocating engines, the L-head design is predominant. Both high- and low-speed engines are used, depending on the application. Engines of this type generally employ air cooling of the cylinder and charge cooling of the interior. Water or liquid cooling is used on a few high-output small engines. Higher load air-cooled engines may use forced fan cooling. Two-stroke engines are often used in applications where minimum weight and cost are important.

Table 7.6.8 Selected Small Gasoline Engines

Make and model	Cycles	Type	Cylinders				Compression ratio	Valve location*	Continuous rating, bhp at r/min		Max torque, lb · ft (N · m), at r/min		Weight, lb (kg)
			No.	Bore, in (cm)	*B/S* ratio	Displacement, in³ (cm³)							
Briggs & Stratton 100900	4	V	1	2.5 (6.50)	1.17	10.43 (171)	6.2	L	3.40	3,600	5.9 (8.0)	3,100	30.5 (13.9)
Briggs & Stratton 80400	4	H	1	2.38 (6.04)	1.36	7.75 (127)	6.2	L	2.55	3,600	4.6 (6.2)	3,100	25.25 (11.9)
Chrysler 500	2	H, V	1	2.00 (6.45)	1.26	5.00 (82)	6.5		3.33	7,000	3 (4.1)	5,000	8.75 (3.9)
Homelite XL-123	2	H	1	1.81 (4.60)	1.31	3.60 (59)	8.3						13.0 (5.9)
Kohler MV18	4	H, V	2	3.12 (7.92)	1.13	42.2 (692)	6	L	18	3,600	26 (35)	2,600	130 (59)
O & R 13A	2	V	1	1.25 (3.18)	1.15	1.34 (22)	9		1.0	6,300	0.88 (1.2)	5,200	3.75 (1.70)
Onan AI	4	V	1	2.75 (6.99)	1.10	14.90 (244)	6.25	L	3.86	3,600	8 (10.8)	2,100	150 (68.2)
Rockwell Zf-400	2	V	2	2.56 (6.50)	1.00	24.29 (398)	7.5		24.0	6,250	28 (38)	6,250	62 (28.2)
Tecumseh LAV 35-1A	4	V	1	2.50 (6.35)	1.36	9.06 (149)	—	L	2.27	2,400	5.25 (7.12)	3,100	21.5 (9.77)
Tecumseh HH160	4	H	1	3.5 (8.89)	1.22	27.66 (453)	8.7	I	16.0	3,600	25 (33.9)	2,400	84 (38.2)
WE400 Wisconsin TRA-10D	4	V	1	3.3 (8.4)	1.20	23.7 (388)	6.5	L	10.0	3,400	16.5 (22.4)	2,800	80 (36.4)
Fichtel & Sachs KM 914	4	H	1	—	—	18.5 (303)	8	RC	20	5,000	—	—	56 (25.5)

* I = valve in head; L = L head; RC = rotary combustion.
SOURCE: *Auto. Ind.*, **48**, no. 7, Apr. 1, 1973, pp. 104-106.

Table 7.6.9 Selected Medium-Size Diesel Engines*

Make and model	Cylinder arrangement†	Cycles	Cylinders				Compression ratio	Max rating, bhp, at r/min		Weight, lb (kg)
			No.	Bore, in (cm)	*B/S* ratio	Displacement, in³ (L)				
Alco 251	V	4	12	9.0 (22.86)	0.86	8,016 (131.4)	11.5	3,150	1,000	35,581 (16,152)
DeLaval GVB-16	V	4	16	13.0 (33.0)	0.87	31,856 (522.0)	—	2,900	1,000	141,000 (64,005)
Electro-Motive 12-645	V	2	12	9.125 (22.0)	0.92	7,740 (128.4)	16	2,950	950	28,600 (11,350)
Fairbanks-Morse 12-38D 1/8	OP	2	12	8.13 (20.65)	0.81	12,443 (203.9)	16.1	3,840	900	45,000 (20,430)
White-Superior 40-VX-12	V	4	12	10.0 (25.4)	0.95	9,896 (162.2)	—	1,800	1,000	—

* "Diesel and Gas Turbine Catalog," 1994. All engines are turbocharged.
† V = V type; OP = opposed.

Fuel and ignition systems and other auxiliary equipment are usually less sophisticated than those on larger engines. A given engine is often used in a wide range of applications.

Engines used in motorcycles and in all-terrain vehicles are generally very sophisticated and constructed with great precision. Features such as overhead valves and overhead cams are common. The specific output of these engines is often high as measured by power per unit of piston displacement.

LOCOMOTIVE ENGINES

Diesel engines are used for locomotive service because of high thermal efficiency (overall diesel brake thermal efficiency of up to 50 percent with the net locomotive traction efficiency of about 32 percent for a diesel locomotive, compared with 6 percent for a steam locomotive), high availability for service, elimination of destructive pounding of rails, and elimination of shutdown fuel losses. The locomotive diesels are usually lighter, have lower displacement per cylinder, but higher outputs per unit displacement than the corresponding stationary diesel engines. These engines are built with up to 20 cylinders. Usually one engine is used per locomotive, and one or more locomotives are used per train. On an experimental basis, locomotive diesel engines have used coal-water slurries, conventional diesel, and natural gas as fuels. (See Table 7.6.9.)

AIRCRAFT ENGINES

Small aircraft piston engines are air-cooled and invariably four-stroke. All types are naturally aspirated or boosted. These engines have the lowest specific weight of piston engines (Table 7.6.10). Aircraft engines always have valve-in-head construction; they also have low specific fuel consumption, and the guaranteed minimum may be as low as 0.40 lb (0.18 kg)/(bhp · h). Maximum cruising bmep ranges from 100 to 150 lb/in^2 (690 to 1,030 kPa), while takeoff bmep ranges from 120 to 200 lb/in^2 (830 to 1,380 kPa), depending principally on the compression ratio, fuel, and boost level.

Aircraft engines are valved to obtain high mean effective pressure at rated speed. High-output engines usually operate with compression ratios and boosting levels that prohibit full-throttle operation at or near sea level with the specified fuel. Such engines can be operated at full throttle at or above a critical altitude, beyond which the power decreases with an increase in altitude. Below the critical altitude, the power decreases with a decrease in altitude, at constant intake-manifold pressure, because of the increase in exhaust back pressure and ambient temperature. Turbocharged engines, however, maintain essentially constant-power with constant manifold pressure from sea level to a critical altitude, which can be as high as 20,000 ft (6,100 m).

The performance of an aircraft engine varies with speed, manifold pressure, and altitude. At any given speed and with a fixed supercharger or blower ratio (supercharger rev/min to engine rev/min) and wide-open throttle, the engine power varies almost linearly with the density of air. This characteristic makes it possible to estimate the power curves at altitude if the sea-level output is known. The power at 20,000-ft (6,100-m) altitude is about 50 percent of the sea-level wide-open-throttle output for any given speed.

Low specific fuel consumption is obtained by operating at reduced speed for the desired power with a lean mixture setting, the optimum output at this point being about one-half to two-thirds of the normal rated output of the engine.

Aircraft engine cylinders are limited to a maximum diameter of about 6 in (0.15 m) by piston cooling and knocking difficulties that arise with large cylinders, high compression ratios, and high intake manifold pressures. The bore/stroke ratio is commonly above 1.0 (oversquare design). The larger piston engine sizes are now replaced by jet engines.

Piston aircraft engines are mostly of the horizontally opposed design (desirable profile) with an even number of cylinders, usually four, six, or eight. They are equipped with a dual ignition system, resulting in more reliable combustion.

WANKEL (ROTARY) ENGINES

The Wankel (rotary) SI engine has been used in some Mazda models. A triangular rotor rotates on an eccentric shaft inside an epitrochoidal housing. The rotor tips are in constant contact with the housing (liner) and form three combustion chambers. The operating cycle (Fig. 7.6.28)

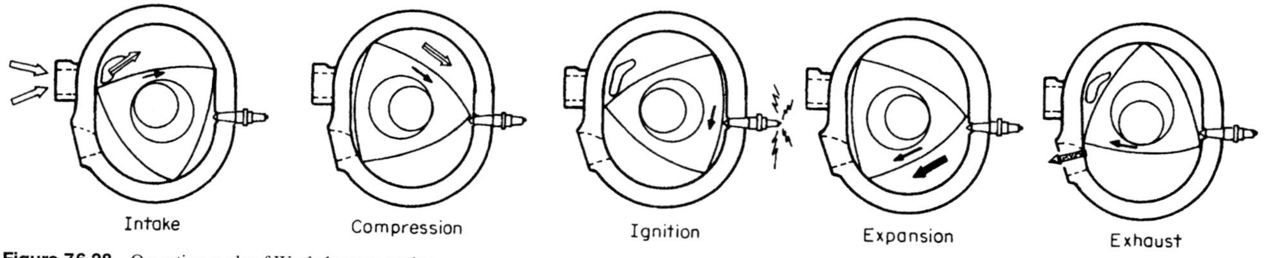

Figure 7.6.28 Operating cycle of Wankel rotary engine.

Table 7.6.10 Selected Piston-Type Aircraft Engines

Make and model	Cylinders No.	Arrangements*	Bore, in (cm)	Stroke, in (cm)	Displacement, in^3 (cm^3)	Ratings (METO)† Comp. ratio	bhp‡	r/min	At sea level (SL) or altitude, ft	Weight§ bmep at max bhp, lb/in^2 (kPa)	Engine, dry, lb (kg)	Fuel octane no. required
Avco Lycoming												
0-320-E	4	H	5.13 (13.03)	3.88 (9.86)	319.8 (5,240.6)	7.0	150	2,700	SL	138 (952)	244 (110.4)	80/87
I0-360-A	4	H	5.13 (13.03)	4.38 (9.86)	361.0 (5,915.7)	8.7	200	2,700	SL	163 (1,120)	293 (132.6)	100/130
TIGO-541-E	6	H	5.13 (13.03)	4.38 (9.86)	541.5 (8,873.6)	7.3	425	3,200	15,000	195 (1,340)	700 (316.8)	100/130
Teledyne Continental												
0-200	4	H	4.06 (10.31)	3.88 (9.86)	201 (3,293.8)	7.1	100	2,750	SL	143 (936)	190 (86.0)	80/87
10-470	6	H	5.0 (12.7)	4.0 (10.16)	471 (7,718.3)	8.6	260	2,600	SL	166 (1,140)	465 (210.5)	100/130
GTSIO-520-F	6	H	5.25 (13.34)	4.0 (10.16)	520 (8,521.3)	7.5	435	3,400	19,000	195 (1,340)	647 (292.8)	100/130

* H = horizontal (production of radial and vertical cylinder arrangements has been discontinued).
† METO = maximum except during takeoff.
‡ Takeoff power is usually higher than the recommended maximum continuous power.
§ Weights are for dry engines without hub or starter.

Table 7.6.11 Data on Fuel Properties

Fuel	Formula (phase)	Molecular weight	Specific gravity (density* kg/m^3)	Heat of vaporization, kJ/kg†	Specific heat: Liquid, kJ/(kg · K)	Specific heat: Vapor c_p, kJ/(kg · K)	Higher heating value, MJ/kg	Lower heating value, MJ/kg	LHV of stoich. mixture, MJ/kg	$(A/F)_s$	$(F/A)_s$	Fuel octane rating: RON	Fuel octane rating: MON
Practical fuels													
Gasoline	$C_nH_{1.87n}(l)$	~110	0.72-0.78	305	2.4	~1.7	47.3	44.0	2.83	14.6	0.0685	92-98	80-90
Light diesel	$C_nH_{1.8n}(l)$	~170	0.84-0.88	270	2.2	~1.7	44.8	42.5	2.74	14.5	0.0690	—	—
Heavy diesel	$C_nH_{1.7n}(l)$	~200	0.82-0.95	230	1.9	~1.7	43.8	41.4	2.76	14.4	0.0697	—	—
Natural gas‡	$C_nH_{3.8n}N_{0.1n}(g)$	~18	(~0.79†)	—	—	~2	50	45	2.9	14.5	0.069	—	—
Pure hydrocarbons													
Methane	$CH_4(g)$	16.04	(0.72*)	509	0.63	2.2	55.5	50.0	2.72	17.23	0.0580	120	120
Propane	$C_3H_8(g)$	44.10	0.51 (2.0*)	426	2.5	1.6	50.4	46.4	2.75	15.67	0.0638	112	97
Isooctane	$C_8H_{18}(l)$	114.23	0.692	308	2.1	1.63	47.8	44.3	2.75	15.13	0.0661	100	100
Cetane	$C_{16}H_{34}(l)$	226.44	0.773	358		1.6	47.3	44.0	2.78	14.82	0.0675	—	—
Benzene	$C_6H_6(l)$	78.11	0.879	433	1.72	1.1	41.9	40.2	2.82	13.27	0.0753		115
Toluene	$C_7H_8(l)$	92.14	0.867	412	1.68	1.1	42.5	40.6	2.79	13.50	0.0741	120	109
Alcohols													
Methanol	$CH_4O(l)$	32.04	0.792	1,103	2.6	1.72	22.7	20.0	2.68	6.47	0.155	106	92
Ethanol	$C_2H_6(l)$	46.07	0.785	840	2.5	1.93	29.7	26.9	2.69	9.00	0.111	107	89
Other fuels													
Carbon	$C(s)$	12.01	~2‡	—	—	—	33.8	33.8	2.70	11.51	0.0869		
Carbon monoxide	$CO(g)$	28.01	(1.25*)	—	—	1.05	10.1	10.1	2.91	2.467	0.405		
Hydrogen	$H_2(g)$	2.015	(0.090*)	—	—	1,44	142.0	120.0	3.40	34.3	0.0292		

(l) = liquid phase; (g) = gaseous phase; (s) = solid phase; RON = research octane number; MON = motor octane number.
* Density in kg/m^3 at 0°C and 1 atm.
† At 1 atm and 25°C for liquid fuels; at 1 atm and boiling temperature for gaseous fuels.
‡ Typical values.
SOURCES: E. M. Goodger, "Hydrocarbon Fuels; Production, Properties and Performance of Liquids and Gases," Macmillan, London, 1975. E. F. Obert, "Internal Combustion Engines and Air Pollution," Intext Educational Publishers, 1973. C. F. Taylor, "The Internal Combustion Engine in Theory and Practice," vol. 1, MIT Press, 1966. J. W. Rose and J. R. Cooper (eds.), "Technical Data on Fuel," 7th ed., British National Committee, World Energy Conference, London, 1977.

follows the four-stroke cycle principle, although the output shaft rotates at 3 times the rotor speed, providing one power stroke per crankshaft revolution (as in the two-stroke engine). A fixed gear meshes with an internal gear in the rotor to maintain the proper relationship between the rotor and housing. The housing is cooled by either water or air and the rotor is cooled by oil or fuel/air charge. Auxiliary equipment (fuel system, ignition system, etc.) is similar to that used on piston engines. A major problem is the sealing grid (90° intersection of seals and single-line seal at apexes of rotor), and this tends to produce high hydrocarbon emissions.

Advantages of the Wankel engine include small size, low weight, simple construction, less components compared with piston-type four-stroke engines, no valves (rotor and its seals serve as valves), lower friction, and no reciprocating unbalance.

FUELS

Fuels for internal combustion engines are predominantly petroleum products. Natural and manufactured gases are also used in limited quantities. Data on properties of fuels which are commonly used for internal combustion engines are summarized in Table 7.6.11. Additional details on fuels are given in Sec. 5.1.

Gasoline Fuels Gasoline consists of various amounts of many hydrocarbons, each having its vapor pressure and temperature characteristics. Also included are small amounts of additives such as deposit modifiers, antioxidants, metal deactivators, antirust agents, anti-icing agents, detergents, upper cylinder lubricants, and dyes. Different grades of gasoline are marketed on the basis of octane number. The overall volatility of a gasoline sample is specified in U.S. industry by the ASTM Distillation Test (D86) and the ASTM Vapor Pressure Test (D323, Reid Method). The distillation test records the increasing temperature of the vapor in the neck of a flask vs. percent evaporated as 100 mL of fuel is gradually heated and distilled. The vapor pressure test gives the pressure in a split-chamber bomb which contains the fuel sample and room air in thermal equilibrium with a 100°F (311 K) temperature bath. The pressure, corrected to standard initial bomb air conditions, gives the **Reid vapor pressure (RVP)** of the gasoline. Typical results for summer- and winter-grade gasolines are given in Table 7.6.12.

In general, volatility is specified by giving the initial boiling point (IBP); the 10, 50, and 90 percent evaporation points; and the endpoint (EP) temperatures as well as the Reid vapor pressure. These data indicate the effects of fuel volatility on engine performance, e.g., ease of starting, warm-up, and fuel economy (Figs. 7.6.29 and 7.6.30.)

Reformulated Gasoline (RFG) The composition of gasoline is known to affect the exhaust and evaporative emissions. The C_4 and C_5 paraffins and olefins (photochemically active) are the main contributors to evaporative losses. Reformulated gasoline fuels have much reduced amounts and thus lower vapor pressures (Reid vapor pressure maximum of 48 kPa) and a maximum of 6 percent by volume olefins. Benzene, a toxic compound, is limited to a maximum of 0.85 percent by volume, while aromatics (photochemically active, deposits) as a class are limited to 25 percent by volume. Oxygenated fuel components (methanol, ethanol, etc.) are sometimes added at a level of 1.8 to 2.2 percent by weight. This reduces CO somewhat, especially from older vehicles without feedback fuel controls. For the federal RFG standards, sulfur is limited to 130 ppm; this lowers acid emissions and lengthens engine life. Further, the ASTM 90 percent distillation temperature is limited to 165°C. This minimizes deposits and leads to less enrichment required upon cold starting. The California Air Resources Board (CARB) applies even more stringent limits, with sulfur held to 30 ppm and the 90 percent distillation temperature limited to 143°C.

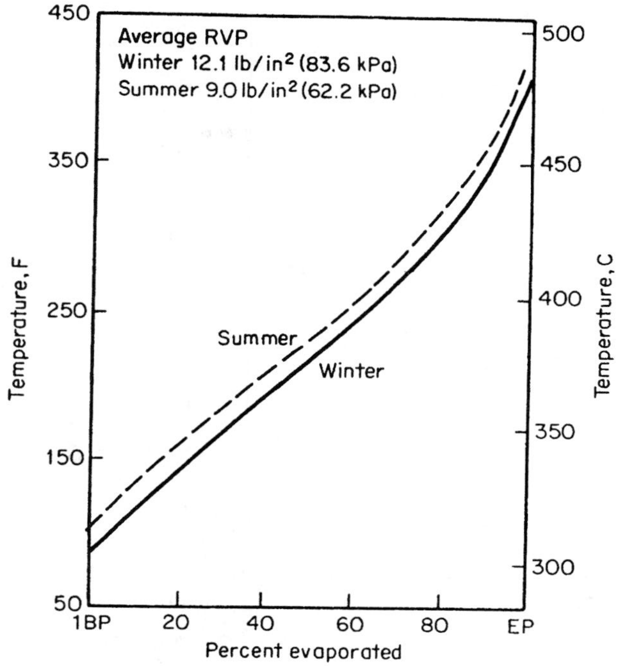

Figure 7.6.29 Distillation curves for typical winter and summer gasolines. *(Ethyl Corp.)*

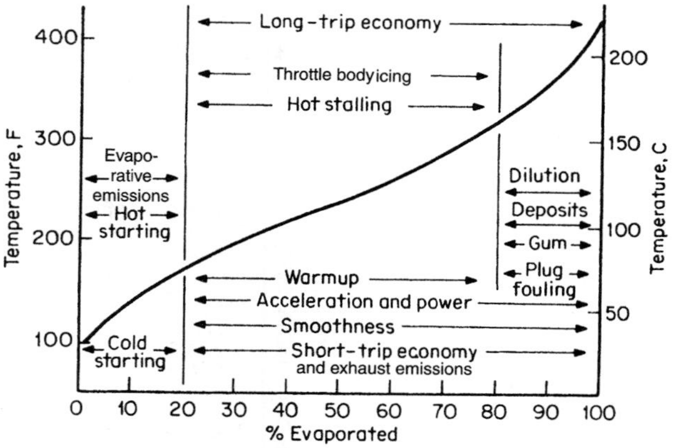

Figure 7.6.30 Effects of volatility characteristics on engine performance.

Table 7.6.12 Average Volatility Characteristics of U.S. Motor Gasolines

	Summer		Winter	
	Premium	Regular	Premium	Regular
IBP temp, °F (°C)	93 (34)	94 (34)	82 (28)	82 (28)
10 percent temp, °F (°C)	126 (52)	125 (52)	107 (42)	107 (42)
50 percent temp, °F (°C)	216 (102)	207 (97)	207 (97)	198 (92)
90 percent temp, °F (°C)	321 (160)	338 (170)	316 (157)	333 (167)
Endpoint temp, °F (°C)	402 (205)	414 (212)	397 (203)	407 (209)
RVP, lb/in² (kPa)	8.9 (61.3)	8.8 (60.6)	12.3 (84.8)	12.1 (83.4)

Table 7.6.13 Detailed Requirements for Diesel Fuel Oils[a,f,g] (ASTM D975)

Grade of diesel fuel oil	Flash point, °C (°F)	Cloud point °C (°F)	Water and sediment, vol percent	Carbon residue on, 10 percent residuum, percent	Ash, wt percent	Distillation temperatures, °C (°F), at 90 percent point	Distillation temperatures, °C (°F), at 90 percent point	Viscosity kinematic, cSt (mm²/s), at 40°C	Viscosity kinematic, cSt (mm²/s), at 40°C	Viscosity Saybolt, SUS at 100°F	Viscosity Saybolt, SUS at 100°F	Sulfur,[d] wt percent	Copper strip corrosion	Cetane no.
	Min	Max	Max	Max	Max	Min	Max	Min	Max	Min	Max	Max	Max	Min
No. 1-D—a volatile distillate fuel oil for engines in service requiring frequent speed and load changes	38 (100)	[b]	0.05	0.15	0.01	—	288 (550)	1.3	2.4	—	34.4	0.0015	No. 3	40[e]
No. 2-D—a distillate fuel oil of lower volatility for engines in industrial and heavy mobile service	52 (125)	[b]	0.05	0.35	0.01	282[c] (540)	338 (640)	1.9	4.1	32.6	40.1	0.0015	No. 3	40[e]
No. 4-D—a fuel oil for low- and medium-speed engines	55 (130)	[b]	0.05	—	0.10	—	—	5.5	24.0	45.0	125.0	2.0	—	30[e]

[a] To meet special operating conditions, modifications of individual limiting requirements may be agreed upon between purchaser, seller, and manufacturer.

[b] It is unrealistic to specify low-temperature properties that will ensure satisfactory operation on a broad basis. Satisfactory operation should be achieved in most cases if the cloud point (or wax appearance point) is specified at 6°C (10°F) above the tenth percentile minimum ambient temperature for the area in which the fuel will be used. The tenth percentile minimum ambient temperatures for the months of October, November, December, January, and March are given in ASTM D 975 in maps for the 48 contiguous states and Alaska, This guidance is of a general nature; some equipment designs, use of flow improver additives, fuel properties, and/or operations may allow higher or require lower cloud point fuels. Appropriate low-temperature operability properties should be agreed on between the fuel supplier and purchaser for the intended use and expected ambient temperatures.

[c] When cloud point less than −12°C (10°F) is specified, the minimum viscosity shall be 1.7 cSt and the 90 percent point shall be waived.

[d] In countries outside the U.S., other sulfur limits may apply.

[e] Low atmospheric temperature as well as engine operation at high altitudes may require system use of fuels with higher cetane ratings.

[f] The values in SI units are to be regarded as the standard. The values in U.S. Customary system units are for information only.

[g] Nothing in this specification shall preclude observance of federal, state, or local regulations which may be more restrictive.

Diesel Fuels Automotive and railroad diesel fuels are derived from petroleum refinery products which are referred to as middle distillates. They have a higher distillation temperature range than gasoline. The properties of commercial distillate diesel fuels depend on the refinery practices employed and the nature of the crude oil from which they are derived, and thus vary substantially. Such fuels generally boil over a range between 163 and 371°C (325 and 700°F). Their makeup can represent various combinations of volatility, ignition quality, viscosity, sulfur level, density, and other characteristics. Additives may be used to enhance specific properties. The SAE recommended practices are described in SAE J313 (July 2004). Permissible limits of significant diesel fuel properties are summarized in ASTM's classification chart D975, reproduced as Table 7.6.13.

Alternative Fuels Other than gasoline and diesel fuels, alternative fuels used in various parts of the world include propane, frequently referred to as liquid petroleum gas (LPG); hydrous ethyl alcohol (ethanol), particularly in Brazil; gasohol, a blend of 10 percent by volume of gasoline and 90 percent by volume of denatured anhydrous ethanol. Other alternative fuels include compressed or liquefied natural gas (CNG or LNG)—primarily methane, other alcohols—particularly methanol, which is also used in racing, and biomass fuels. Details can be found in SAE J1297.

GAS EXCHANGE PROCESSES

Internal combustion engines exchange gases with the environment during the open part of their cycle. Fresh air is distributed to the cylinders through the intake manifold and the intake valves; exhaust gas is expelled from the cylinders through the exhaust valves and collected in the exhaust manifold.

Intake Manifolds The intake manifold distributes the air (diesel) or the air and fuel (SI) to the engine cylinders. In spark-ignited engines with port fuel injection, the issue of manifold wetting is virtually eliminated. Runners may be long for tuning at low engine speeds or short to minimize flow losses (and improve tuning at high engine speeds). In diesel engines the runners are normally short, especially in medium-speed engines.

The overlapping of the intake valve periods produces complicated disturbances in the intake manifold. This results in dynamic air charging, which may lead to cylinder-to-cylinder variation in trapped mass and effects on the combustion process.

Port fuel-injected engines have reduced volumetric efficiency due to the presence of vaporized fuel in the intake manifold (see Fig. 7.6.11). However, vaporizing the fuel in the air partially offsets the effect on volumetric efficiency, as fuel vaporization lowers the mixture temperature, thereby increasing its density. The dew points or condensation temperatures for fuel in an air/fuel mixture depend on the fuel, air/fuel ratio, and total pressure (Table 7.6.14), and they may be determined from equilibrium air distillation data. Direct injection engines have greater volumetric efficiency than port fuel injected engines.

Exhaust Manifolds The most important considerations in exhaust manifold design are adequate flow area and expansion of the walls. Small multicylinder engines often employ a one-piece exhaust manifold. This is anchored at the center of the engine, and elongated holes are provided at other points for expansion. Large engines employ multipiece manifolds with expansion joints. They are usually water-cooled in marine applications to prevent strain for safety reasons. Exhaust manifolds can be insulated to conserve energy for turbocharging and/or exhaust aftertreatment.

Manifold flow area should be large enough to avoid local back pressure buildup during exhaust and in connection with the total exhaust system should not raise back pressure excessively at maximum flow. A flow area of 0.7 times the intake manifold area may be used for four-stroke engines. For two-stroke engines, the exhaust flow area usually exceeds the intake flow area.

Mufflers In order to be efficient as sound silencers, exhaust mufflers must decrease the exhaust gas velocity and either absorb the sound waves or cancel them by interference with other waves from the same source. Mufflers usually have volumes 6 to 8 times the piston displacement and may contain baffles with or without holes. Mufflers that cancel sound waves by interference usually break the waves into two parts which follow different paths and meet again, out of phase, before leaving the muffler (Davis et al., NACA Rept. 1192).

Exhaust muffle pits are used with large engines, are usually made of concrete, have a volume about 20 times the piston displacement, and may be open or provided with baffling or partly filled with loose stone through which the gases must pass. A stack is usually provided for the escape of the gases, and a drain for the discharge of the condensed water from the exhaust products.

Exhaust back pressure should be kept to a minimum since an increase of 1 lb/in^2 (6.9 kPa) in back pressure may decrease the maximum power Output by up to 2 percent, due to increased pumping work and reduced volumetric efficiency.

Supercharging

Supercharging increases the amount of charge per cycle (above that of the normally aspirated engine) by increasing its pressure and hence its density. Supercharging permits more fuel to be burned and is a practical means to increase engine power, but it also increases mechanical and thermal stresses in the engine. For aircraft piston engines, it is a means to offer high-power output for takeoff and to offset the rare atmosphere at high altitude. For diesel engines supercharging enables the design of smaller and lighter power plants. It is especially practical for diesel engines because supercharging does not add to the fuel quality required, and because increased cylinder pressure aids fuel spray development and the combustion process.

Three alternative methods can be used to achieve supercharging. In mechanical supercharging, a positive-displacement type of Roots blower or compressor is geared directly to the engine shaft and thus is driven by engine power. This type is desirable for variable-speed engines, where high torque is required at various speeds, since the pressure and capacity characteristics of this type do not decrease with speed. In turbocharging, exhaust energy drives a turbine, which, in turn, provides the necessary work to drive the turbocharger's compressor, mounted on the same shaft as the turbine (Fig. 7.6.31). The centrifugal compressor is particularly adaptable to aircraft engines because of high capacity, small size, and low weight. It is also used with truck diesel engines to improve the volumetric efficiency and power at high engine speeds. Centrifugal blowers may have one or two stages with cooling between stages or after the single or last stage. The rotors run at maximum speeds of 15,000 to 70,000 rev/min. Finally, in pressure wave super-charging, wave action and resonant tuning in the intake

Table 7.6.14 Condensation Temperatures of Liquid Fuels, °F (°C)

						Condensation temp.			
						Atmospheric pressure		Pressure, ²⁄₃ atmospheric	
		ASTM distillation				Air/fuel ratio		Air/fuel ratio	
Fuel	Deg API	Initial point	50 percent	90 percent	Endpoint	12:1	15:1	12:1	15:1
Gasoline	60	100 (38)	249 (121)	354 (179)	400 (204)	118 (48)	114 (46)	111 (44)	107 (42)
Kerosene	43	350 (177)	450 (232)	510 (266)	550 (288)	204 (96)	200 (93)	201 (94)	197 (92)

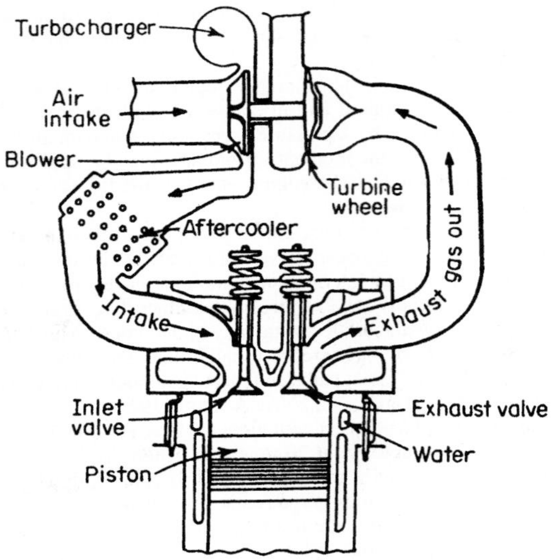

Figure 7.6.31 Schematic of exhaust turbosupercharger installation on diesel engine cylinder.

and exhaust systems are used to compress the charge and thereby boost volumetric efficiency.

Supercharging spark ignition engines may lead to end gas knock because it makes the end gas in the cylinder more reactive and may necessitate lowering the compression ratio, enriching the mixture, or using a higher octane number fuel. Supercharging a diesel engine increases the peak cylinder pressure and permits lowering the compression ratio or changing the fuel injection timing. The use of variable engine compression ratio with supercharging is very desirable but difficult and costly.

Turbocharging is used in most applications of diesel engines including truck, locomotive, and marine engines. Turbocharging enables higher cylinder trapped mass without the mechanical work penalty

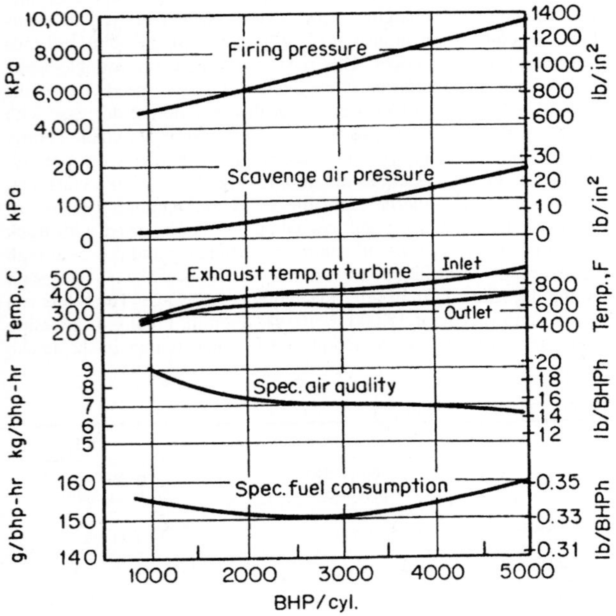

Figure 7.6.32 Performance curves of a turbocharged diesel engine. *(Sulzer.)*

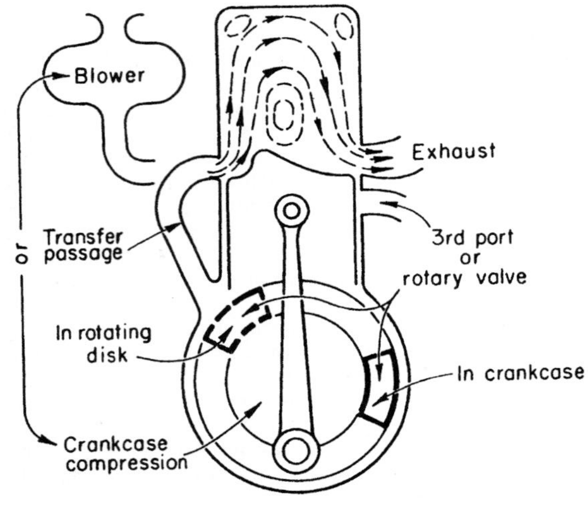

Figure 7.6.33 Common scavenging systems.

of supercharging and thus aids thermal efficiency. Higher output also increases the engine mechanical efficiency (Buchi, *ASME Trans.*, **59**, pp. 85-96). The fuel economy and other performance characteristics of a supercharged Sulzer large marine diesel engine of 30-in (760-mm) cylinder diameter are shown in Fig. 7.6.32.

Scavenging Two-Stroke Cycle Engines

Crankcase compression for each cylinder or an engine driven blower may be used for scavenging and charging the cylinders of a two-stroke cycle engine (Fig. 7.6.33). With crankcase compression, used on some gasoline outboard marine engines, a third port uncovered by the piston near the top of the stroke, a rotary valve, or an automatic poppet valve may be used. The top of the piston is shaped to deflect the gases entering the cylinder and to prevent short-circuiting to the exhaust port. A "dead" spot of gases may remain in the lower center of the cylinder or in the upper corners.

The intake may be inclined and tangential, giving the entering charge an upward helical motion (Fig. 7.6.34). This motion keeps the entering charge near the cylinder walls, forcing the exhaust gases to the center of the cylinder and down to the exhaust port. Some engines employ loop scavenging (Fig. 7.6.35).

Some large diesel engines, such as the Detroit Diesel engine speci-

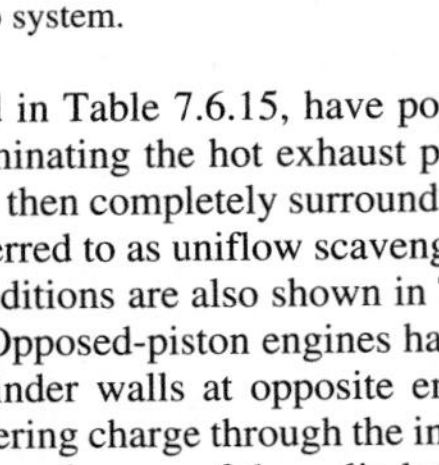

Figure 7.6.34 Hesselman helical-loop system.

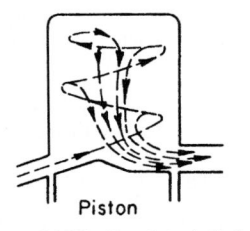

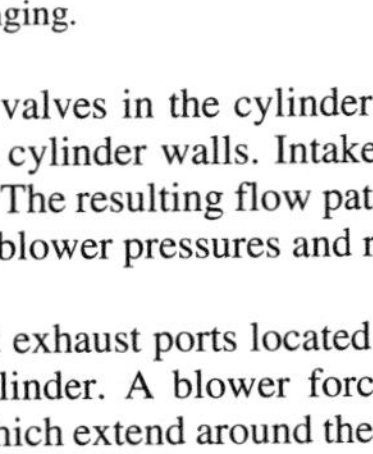

Figure 7.6.35 M-A-N loop scavenging.

fied in Table 7.6.15, have poppet exhaust valves in the cylinder head, eliminating the hot exhaust ports from the cylinder walls. Intake ports can then completely surround the cylinder. The resulting flow pattern is referred to as uniflow scavenging. Typical blower pressures and related conditions are also shown in Table 7.6.15.

Opposed-piston engines have intake and exhaust ports located in the cylinder walls at opposite ends of the cylinder. A blower forces the entering charge through the intake ports, which extend around the entire circumference of the cylinder.

Table 7.6.15 Detroit Diesel Two-Stroke Cycle Diesel Engine

Engine type	Air-box pressure at 500 to 2,300 r/min		Max bmep*		Max imep*	
	inHg	cmHg	lb/in²	kPa	lb/in²	kPa
Naturally aspirated	1-9	2.54-22.86	101	697	129	890
Turbocharged	1.5-40	3.81-101.62	136	938	163	1,125

* At 2,300 r/min.

The two-stroke cycle Sulzer diesel engine (Fig. 7.6.36) makes use of a crosshead and piston rod design. The air in the space between the piston and the diaphragm carrying the piston rod stuffing box is displaced by the descending piston and is forced into the cylinder through the intake ports as they are uncovered by the piston, which aids the scavenging of the cylinder. Boosting is provided by exhaust gas turbine-driven blowers. Air from these blowers passes through intercoolers and is admitted through check valves to the scavenging chambers. Electric motor-driven air blowers are sometimes used on large engines to assist scavenging at idle and low power operating points.

The crosshead design, with its stuffing box, also prevents combustion products from contaminating the crankcase oil system and is commonly used on large engines. This RL design (Fig. 7.6.36) is used for cylinder

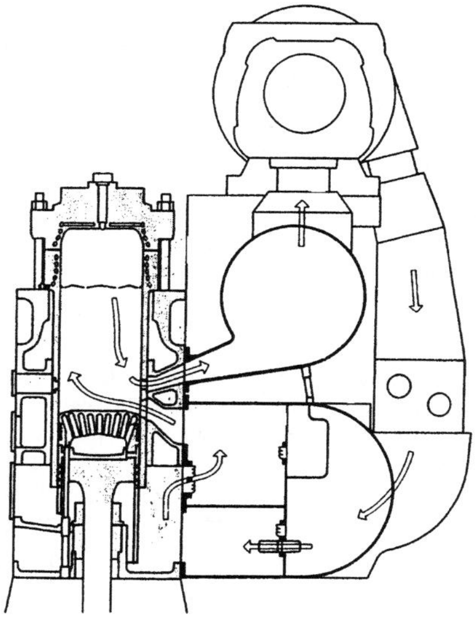

Figure 7.6.36 A Sulzer series RL two-stroke crosshead diesel engine.

diameters of 560, 660, 760, and 1,900 mm. The largest cylinder, with a stroke of 1,900 mm, delivers a normal output of 4,000 bhp at 102 rev/min corresponding to a BMEP of 1,430 kPa (207 lb/in²). Figure 7.6.32 shows the performance curves of this type up to an overload of 5,000 bhp per cylinder. Smaller cylinders produce correspondingly lower outputs with roughly the same BMEP and piston speed.

FUEL-AIR MIXTURE PREPARATION

Perfect distribution of fuel-air mixture consists of equal distribution of both air and fuel to the various cylinders of a multicylinder engine for each cycle. The compression pressure of each cylinder is an indication of the air distribution if all cylinders have identical valve timing and compression ratios and no leakage. Cylinders receiving lean mixtures develop less power than those receiving correspondingly rich mixtures. Poor distribution can be masked by enriching the mixture, and high-power output attained at the cost of high fuel consumption rates and HC and CO emission increases. The CO, CO_2, and O_2 analysis of the exhaust gases from the individual cylinders and spark plug temperatures (thermocouple in the center electrode) may be used to determine the mixture distribution.

Distribution of load (and fuel) and variation in injection timing between cylinders of a diesel engine are similarly indicated by the exhaust gas composition and temperature of the individual cylinders, the mixtures in the various cylinders being lean under all conditions.

Gasoline and diesel fuel lines should be designed for maximum velocities under 1 ft/s (0.3 m/s) for gravity feed. Gasoline lines should be in a cool location, because heating above 100°F (38°C) may vaporize the lighter fractions and cause vapor lock. Sediment and water traps as well as strainers or filters should be located in the fuel line ahead of the fuel injector or carburetor, fuel pump, or injection pump.

Gas lines should be designed for gas velocities of 30 to 60 ft/s (9 to 18 m/s). A pressure drop of 0.07 to 0.12 in (1.8 to 3 mm) of water should be allowed per 100 ft (30 m) of pipe. Close calculation of the size of gas pipe for industrial gas installations is undesirable, because of the reduction of the normal pipe cross section by tar and dust deposits. A large-diameter storage chamber should be located in the pipe near the engine. The gas line should be free of traps and as free as possible from changes in direction of flow, and it should be pitched slightly from both ends toward a low point in the center, where a sealed drain should be located. All water seals in the gas line should be as cool as possible, to minimize evaporation and the breaking of the seal.

Mixture Preparation in Spark Ignition Engines

Spark ignition engines use stoichiometric mixtures at idle and part-load operation, but transition to slightly rich mixtures near and at full power. A stoichiometric mixture, relative to slightly rich mixtures, provides good fuel economy and low engine-out emissions of CO and HC. Exhaust gas is recirculated (EGR) into the fresh mixture to control nitrogen oxides. The dilution reduces the need for throttling and thus improves part-load fuel economy. In smaller engines, a slightly rich mixture is required for idling and small throttle openings, transitioning to near stoichiometric at medium loads, and may be stoichiometric or rich at full load to improve power but is limited by emissions considerations (Fig. 7.6.37).

In the past, carburetion has been used to meter, atomize (if liquid), and mix the fuel with the air flowing to the engine. This method is

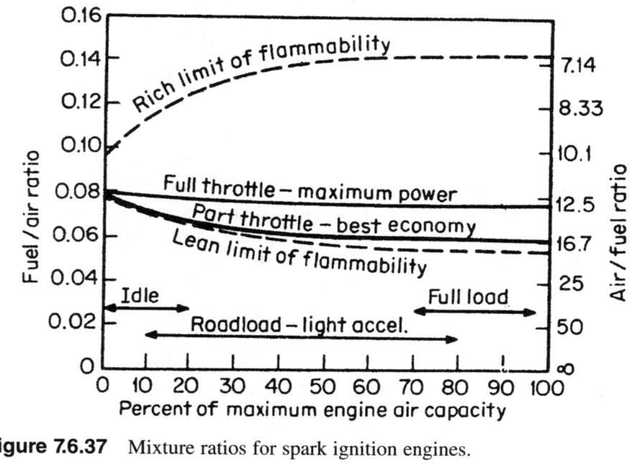

Figure 7.6.37 Mixture ratios for spark ignition engines.

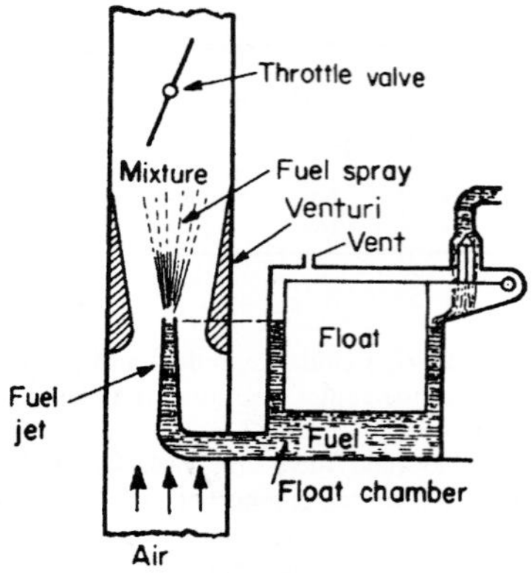

Figure 7.6.38 Principle of carburetion.

still in use for mixing gaseous fuels with air, for small engines such as those used in household and off-road applications, for piston-type air-cooled engines, and for older passenger cars. However, carburetion has been replaced by port fuel injection and direct injection for automotive applications. The two methods of fuel-air mixture preparation are described below.

Carburetion

The basic principle of carburetion is illustrated in Fig. 7.6.38. A float maintains a constant fuel level in a float chamber which is vented to the atmosphere or air entrance of the carburetor. The float chamber fuel level is slightly below the outlet of the fuel jet. The engine air flows through a venturi tube which has a throat diameter of 0.75 to 0.85 of the diameter of the carburetor bore. The reduction in pressure at the venturi throat causes fuel to flow from the float chamber through the main fuel jet into the airstream. Fuel atomization is accomplished by the velocity difference between the air and fuel. Vaporization of the fuel continues while flowing to the engine cylinder but is usually not completed in the intake manifold at wide-open throttle.

The metering characteristic of a carburetor is the weight ratio of fuel to air supplied over the operating range of airflow. These carburetor flow characteristics are commonly shown on a graph, similar to Fig. 7.6.37, which gives the fuel/air ratios at idle, along the "road-load curve" for an automobile engine, and at wide-open throttle. Although computations can be made to establish the fuel/air ratio under specified conditions, estimates are approximate because air is usually bled into the fuel channels to help control the fuel flow. Also, airflow through the venturi is pulsating in nature and makes both airflow and fuel flow somewhat unpredictable. Actual carburetor characteristics approach the desirable characteristics indicated in Fig. 7.6.37.

When the throttle is opened quickly, there is a momentary lag of the fuel flow in relation to the airflow due to the formation of a liquid film on the intake manifold walls. This necessitates the use of some device (accelerating pump or well) to enrich the mixture during sudden accelerations and thereby avoid engine backfire. At idle, the depression of the throat of the venturi is insufficient to overcome the viscosity of the fuel, thus requiring special idle circuits to overcome this problem and to enrich the otherwise very dilute, high-residual cylinder contents. Provisions are also made for mixture enrichment at high loads. In the 1980s, with the enforcement of more stringent emissions standards and the introduction of closed-loop control of the fuel/air ratio to ensure operation in a narrow window about the stoichiometric ratio, the electronic carburetor was introduced (Fig. 7.6.39). In this device, the flow area for fuel metering is computer-controlled by a pulse-width-modulated solenoid.

Carburetors for piston-type air-cooled aircraft engines provide fuel/air ratios that vary from 0.11 at idle to 0.085 for cruising rich (cruising lean being about 0.072). As the full-power condition is approached, the mixture richness is increased to aid cylinder cooling, becoming about 0.108 at maximum output. Aircraft carburetors must compensate

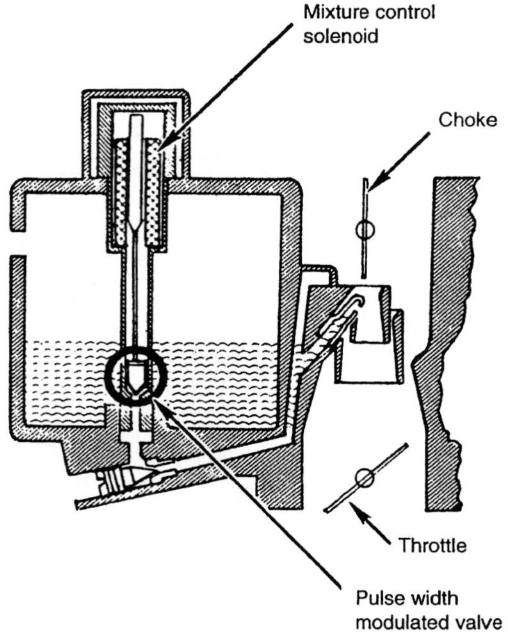

Figure 7.6.39 Electronic carburetor. *(Amann, "The Automotive Spark-Ignition Engine—An Historical Perspective," ASME-ICE vol. 8, 1989, pp. 33-45.)*

for the effect of altitude on the air density and the resulting change in the mixture ratio. Compensation for the enrichment that occurs with increasing altitude is usually provided by reducing the area of the main fuel-metering orifice. This is accomplished by moving a tapered needle in the orifice, which may be done manually or automatically by means of an air-pressure-sensitive bellows. A carburetor providing a 0.067 fuel/air ratio at sea level would provide a fuel/air ratio of about 0.091 at an altitude of 20,000 ft (6,100 m).

Fuel Injection in Spark Ignition Engines Modern automobile and truck engines and some small marine and utility engines employ fuel injection. Fuel is injected either upstream of the throttle body, into the port or directly into the cylinder. Fuel injection equipment consists of an in-tank pump, common-rail recirculating supply, and injection nozzles. With single-point, central fuel injection above the throttle, as typified by the throttle body injector of Fig. 7.6.40, one or two injectors with injection pressure of about 70 kPa are used for multicylinder engines. This system provides an electronically controlled injector at the intake manifold entrance where the carburetor was formerly

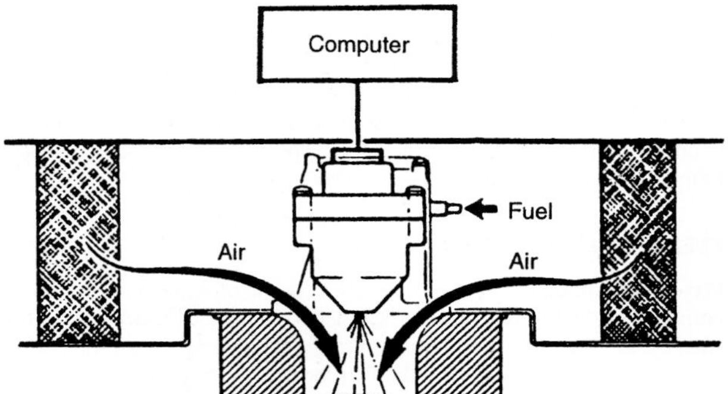

Figure 7.6.40 Throttle body injector. *(Amann, "The Automotive Spark-Ignition Engine—An Historical Perspective," ASME-ICE vol. 8, 1989, pp. 33-45.)*

positioned. With multipoint fuel injection, an injector is positioned in each port, and pressures from 400 to 700 kPa are used to avoid vapor formation in the tip located in the hot spit-back region just upstream of the intake valve. In certain cases, an additional injector is used to supplement the fuel flow during starting.

The advantages of fuel injection over carburetion are the more uniform distribution of fuel between the engine cylinders, the reduction of combustion knock caused by nonuniform fuel distribution, the elimination of heat transfer to the intake manifold to obtain the desirable fuel vaporization (thus obtaining higher-power outputs because of higher charge density), better throttle response, and the use of less volatile fuel. With an injection system, it is easier to cut off the fuel flow during deceleration (coasting) than is the case with the carburetor with its fuel-wetted intake manifold. Note that throttle body injection encounters problems associated with the slower transport of fuel than air during transients, as with carburetors.

Fuel injection systems may have mechanical or electronic control of fuel flow, with the electronic type being predominant. In mechanical systems, such as the Bosch K-Jetronic, fuel is continuously injected into the port where fuel quantity is mechanically determined by movement of an air valve. In electronic systems, an electromagnetic injector is pulsed on and off by the computer to supply the desired fuel quantity in proportion to the air. The injector pulse width is controlled by an exhaust sensor signal to achieve the desired air-fuel ratio. The airflow must be sensed to establish the correct air-fuel ratio for various engine conditions. This can be done by sensing the airflow with a venturi or equivalent, a hot-film, hot-wire, or other air anemometer (such as in the L-Jetronic system), or by sensing the principal engine variables which control engine air consumption. The latter is referred to as a speed-density control system. A system of this type is shown schematically in Fig. 7.6.41. Engine speed and manifold pressure and air temperature are used to calculate the engine airflow from prior dynamometer data, a method which has proved less accurate in practice.

For central fuel injection designs, the injector opening angle may be independent of engine cylinder events (asynchronous). For port injection designs, the opening is normally synchronized with cylinder events to open before the intake valve opens. This minimizes axial mixture stratification which occurs when the injector supplies fuel when the intake valve is open. Such axial stratification may be used

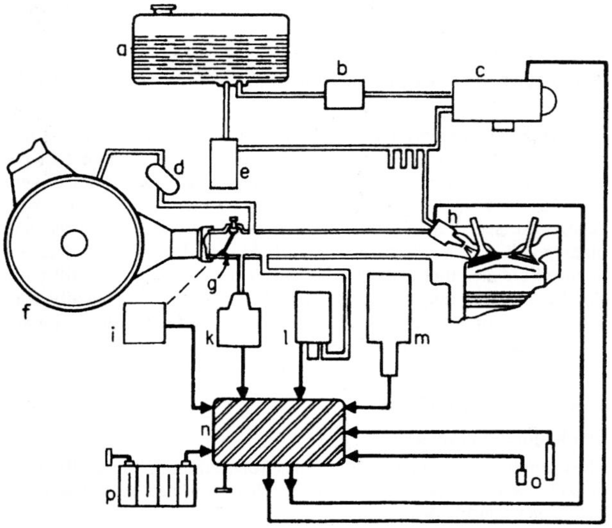

Figure 7.6.41 Schematic of electronic gasoline injection system for an automobile. (*a*) Gasoline tank; (*b*) fuel filter; (*c*) fuel supply pump; (*d*) auxiliary idling air valve; (*e*) fuel pressure regulator; (*f*) air filter, (*g*) throttle plate; (*h*) fuel injection valve; (*i*) throttle position switch; (*k*) full load pressure switch; (*l*) manifold pressure sensor; (*m*) ignition distributor with trigger contacts; (*n*) electronic control unit; (*o*) temperature sensor; (*p*) battery. (*Robert Bosch.*)

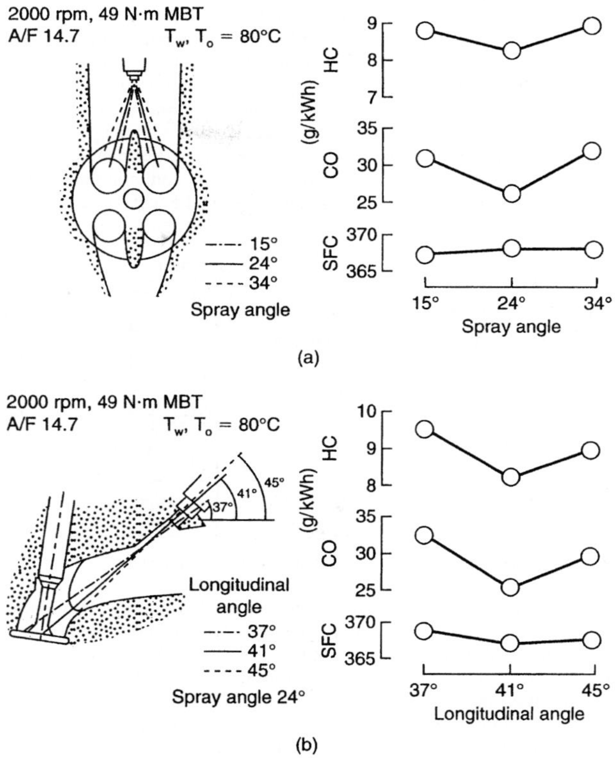

Figure 7.6.42 Effects of spray angle and injection targeting on engine emissions. (*a*) Effect of dual stream separation angle; (*b*) effect of injection direction. (*Iwata et al, SAE paper 870125, 1987.*)

advantageously to create a lean-burn engine. Injection of fuel into the intake port usually occurs early in the intake stroke, to provide the maximum time for fuel evaporation and mixing to occur. Injection into the intake port provides good mixing as the fuel and air are drawn at high velocity through the intake valve passages, and it isolates the injector from combustion gases at high pressure and temperature. Direct injection into the cylinder occurs during intake or early compression, and it usually results in charge stratification.

Spray targeting in port fuel injection is especially important for fuel evaporation, wall wetting, and engine emissions. Typically when the spray is directed at the center of the intake valve, toward its valve stem base, hydrocarbon and CO emissions are minimized (Fig. 7.6.42). Injector design, spray angle, and droplet size distribution are important. Quader (1989) also showed that targeting the fuel spray at the intake valve centerline toward the valve head resulted in the lowest smoke and HC emissions under high-speed and heavy-load conditions. Some designs are less affected by deposits. Air-assisted fuel sprays provide smaller droplets and may be an advantage in both port and direct cylinder injection. Air-assisted direct cylinder injection has shown promise for improved economy and lower emissions in two-stroke, spark ignition, orbital combustion process (OCP) engines.

Fuel Injection in Diesel Engines The fuel injection system for diesel engines usually consists of a pump, fuel line, and nozzle. This system provides control of the start, rate and duration of injection, and desired fuel quantity per injection, and atomizes and partially distributes the fuel in the combustion chamber. Methods of fuel injection include pump or unit injectors and common-rail systems. In the pump injection system (Fig. 7.6.43), a pump forces a desired quantity of fuel through a high-pressure fuel line and an atomizing nozzle into the combustion chamber. Unit injectors combine the plunger pump and spray nozzle, thus eliminating the high-pressure fuel line. Injection pressures usually range from 1,500 to 7,000 lb/in^2 (10,300 to 48,000 kPa) but may run above 30,000 lb/in^2 (206,000 kPa) at high injection rates. Fuel quantity control with this type of injection is usually affected by controlling a bypass valve in the pump or the pump discharge line or by varying the time of closure or opening of the fuel-pump inlet port. A separate fuel-pump is commonly required for each engine cylinder, although rotating plunger pumps serving from two to eight cylinders per plunger are in common use, especially on small engines. In the common-rail system (Fig. 7.6.44), a constant pressure is maintained in the fuel discharge line,

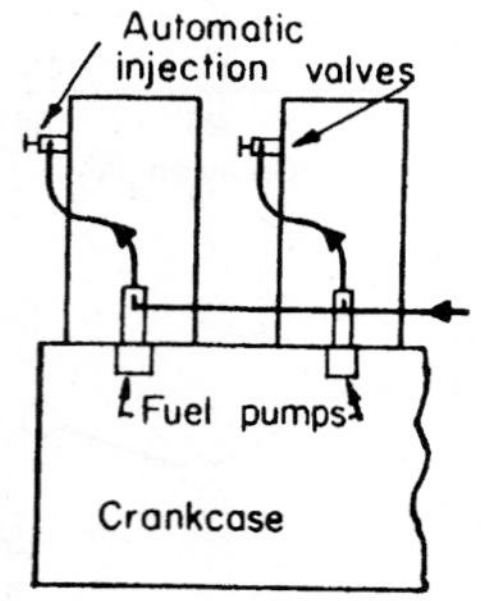

Figure 7.6.43 Pump injection.

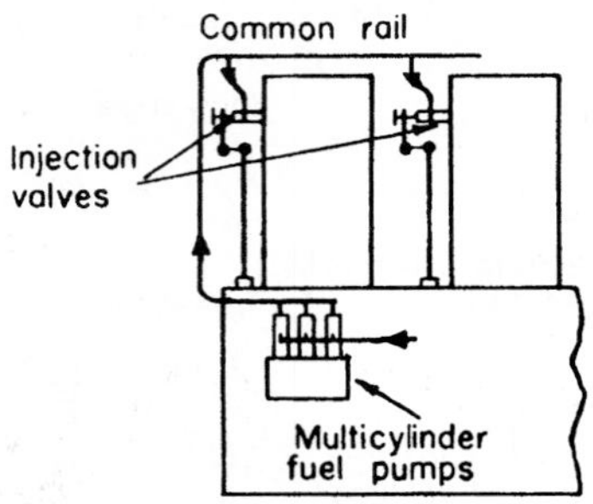

Figure 7.6.44 Common rail.

and the discharge of fuel to the engine cylinder is controlled by the lift, timing, and duration of opening of the fuel valve. Considerable capacity of the discharge line is necessary in this case, a triplex pump being considered desirable to supply fuel at a constant rate. A different injection system, used in Caterpillar and International engines, is the hydraulic electronic unit injector (HEUI) system. Fuel is pressurized up to 3,000 to 3,500 psi (21,000 to 24,000 kPa). The pressurized oil acts on a stepped piston which pressurizes fuel up to 25,000 psi (170,000 kPa). Injection control via a small shuttle piston is done electronically.

Pumps Fuel pumps for pump injection systems require rugged construction, are usually actuated by a cam, and have constant mechanical stroke. The pump plunger and barrel (Fig. 7.6.45) are a very close lapped fit to minimize leakage. The inlet port is uncovered near the bottom of the plunger stroke, the low vapor pressure in the pump cylinder causing fuel flow into the barrel. On the upward plunger stroke, fuel flows past the discharge valve and out of the spring-loaded nozzle. Fuel spray from the nozzle continues until the pump bypass port is uncovered. The pressure then subsides, the nozzle closes, and the pump delivery valve closes to maintain a desired residual line pressure until the next injection. Fuel quantity control is obtained by turning the pump plunger in the barrel, which varies the time of uncovering the bypass

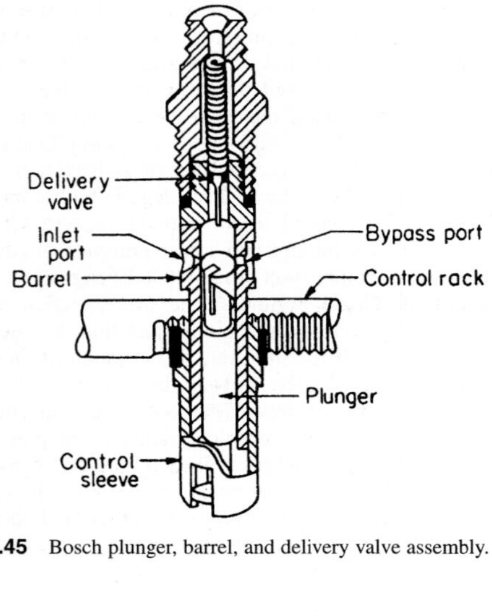

Figure 7.6.45 Bosch plunger, barrel, and delivery valve assembly.

port by the plunger helix. This in turn varies the duration of the effective plunger stroke and the amount of fuel injected. The high pressures which occur in these injection systems, together with the compressibility of the fuel, give rise to transient pressure wave phenomena in the connecting lines which change injection characteristics.

Fuel Lines The tubing connecting the injection pump with the spray nozzle must be thick-walled, of smooth and uniform bore, and made of ductile material to permit bending. Tubes range in size from 1/4-in (0.64-cm) OD and 1/16-in (0.16-cm) ID to 1/2-in (1.3-cm) OD and 3/16-in (0.48-cm) ID. Fuel lines should be the same length for each cylinder, to obtain the same transient hydraulic characteristics. Pressure waves due to plunger motion and nozzle efflux characteristics are developed, travel back and forth, and create disturbances during the injection process (De Juhasz, *Trans. ASME, Oil and Gas Power,* **60**, p. 2; Bolt et al., *SAE Trans.,* 1971, paper 710569), making short lines desirable.

Fuel Injection Nozzles Injection nozzles are usually hydraulically operated, having a differential-diameter valve (Fig. 7.4.46), held on its seat by a spring, which opens when the fuel line pressure is increased by the injection pump. The valve closing pressure is always lower than the opening pressure, depending on the relative effective valve areas exposed to fuel pressure. The valves may be either the pintle or multipleorifice type (Fig. 7.6.46). The former has a single orifice through which the end of the valve protrudes, and it was used in the past with precombustion chambers. A pintle may be designed to restrict the fuel flow as it opens; this results in a throttling nozzle. The pintle nozzle provides a hollow conical spray with an included angle which varies from 4° to 60° with various pintle sizes. The multiple-orifice type of nozzle (Fig. 7.6.46*c*) is used principally with the open combustion chamber. The resulting fuel spray consists of a core and surrounding envelope of atomized fuel.

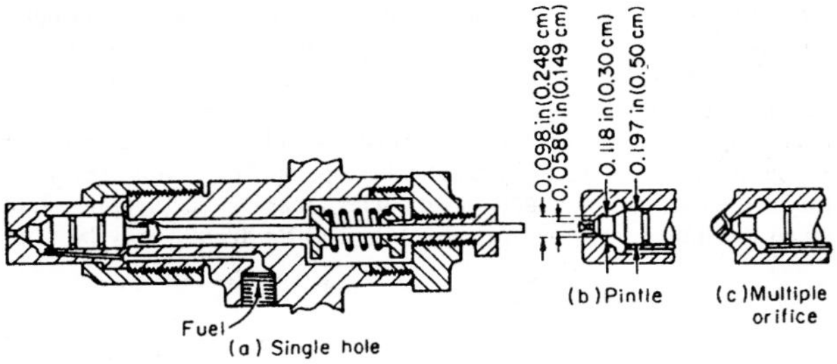

Figure 7.6.46 Solid injection closed nozzle.

Diesel Fuel Injection System Characteristics At any given setting, most pump and nozzle fuel injection systems deliver more fuel per cycle with a decrease in speed, except under the idling condition of very low speed and small fuel delivery. The rising fuel delivery curve with decrease in speed at full load contributes to the good torque rise and lugging ability of the high-speed diesel. At the idle condition, decrease in speed reduces fuel delivery and causes the engine to run still more slowly, etc. The injection quantity and timing of electronically controlled injection systems are determined by a variety of driver inputs, sensors, and software algorithms.

COMBUSTION CHAMBERS

Spark Ignition Engines

Most combustion chambers for four-stroke engines, especially automotive, have an open design and employ overhead valves actuated by either pushrods or overhead cams. This design provides fast burning of the charge because of the concentrated combustion volume and minimizes surface area and thus heat losses and wall quenching. Figure 7.6.47 shows classical, two-valve chamber designs. In each case, the spark plug location and chamber shape yield a relation between flame front area and flame travel similar to that in Fig. 7.6.48. For a given engine, the maximum rate of pressure rise is proportional to the

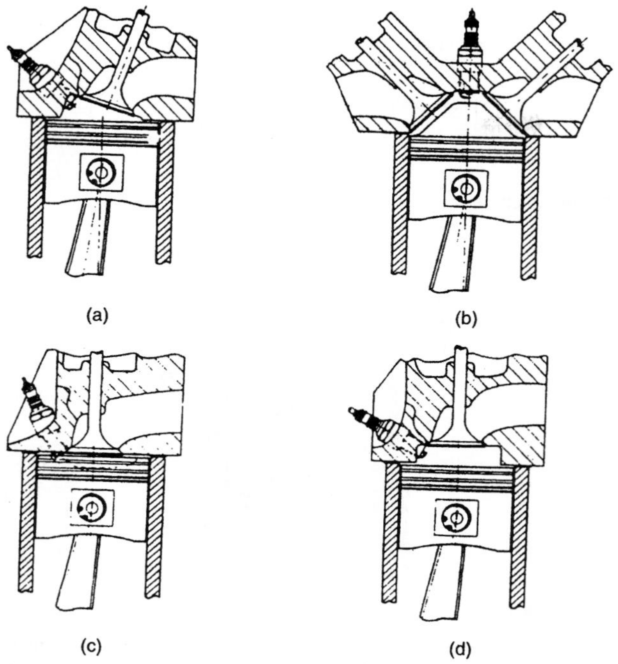

Figure 7.6.47 Classical combustion chambers for spark-ignition engines; (*a*) Wedge; (*b*) hemispherical head; (*c*) bowl in piston; (*d*) bathtub head. *(Courtesy: R. Stone, SAE, 1995.)*

peak flame-front area viewed at TDC. For the chambers of Fig. 7.6.47*a*, *b*, and *d*, virtually all the volume is in the head. This design allows large valves and minimizes piston inertia. The hemispherical or domed head of Fig. 7.6.41*b* allows very large valves and with a domed piston provides a combustion chamber of normal compression ratio. Aircraft engines usually have this chamber shape. The chamber of Fig. 7.6.47*c*, used by Jaguar in its V-12 design, is in the piston bowl. The bathtub chamber of Fig. 7.6.47*d* is very compact for good fuel economy. To suppress knock, each chamber has the spark close to the exhaust valve and a well-cooled, quench region in the end gas far from the spark plug. While all these chambers have a squish area to generate turbulence, compression-induced turbulence is low and mixture motion depends on intake-induced swirl, which is generally quite considerable.

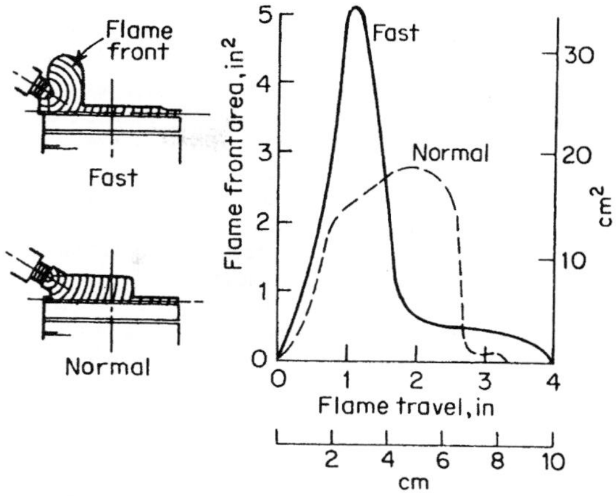

Figure 7.6.48 Flame-front area curves for fast- and normal-burn combustion chambers.

More recently, pent-roof heads with four valves per cylinder and a centrally located spark plug have become very popular (Fig. 7.6.47). This arrangement gives minimum flame travel distance and hence fast burn, as well as large valve flow area and thus high volumetric efficiency. In such designs, turbulence is generated through the complex interaction of the two inlet flows which produce a tumbling motion. Other fast-burn chambers have been designed by Nissan and Ricardo. In the Nissan NAPS-Z combustion system, there are twin spark plugs and an induction system producing a high level of axial swirl (Fig. 7.6.49). In Ricardo's high-ratio compact chamber (HRCC), the combustion chamber is centered on the exhaust valve. Chamber compactness, high turbulence, and lean burning allow this design to operate at compression ratios exceeding 11 without knock.

Figure 7.6.50 L-head engine.

Where cost or engine height is a principal concern, a design such as the L-head (Fig. 7.6.50), in which the valves are actuated from below directly by the cam, is commonly used. Spark plug location can compensate to some extent for the slow burn rate of the elongated chamber. The L-head has a relatively high surface area, and thus greater heat loss and wall quenching, and a small space for valves.

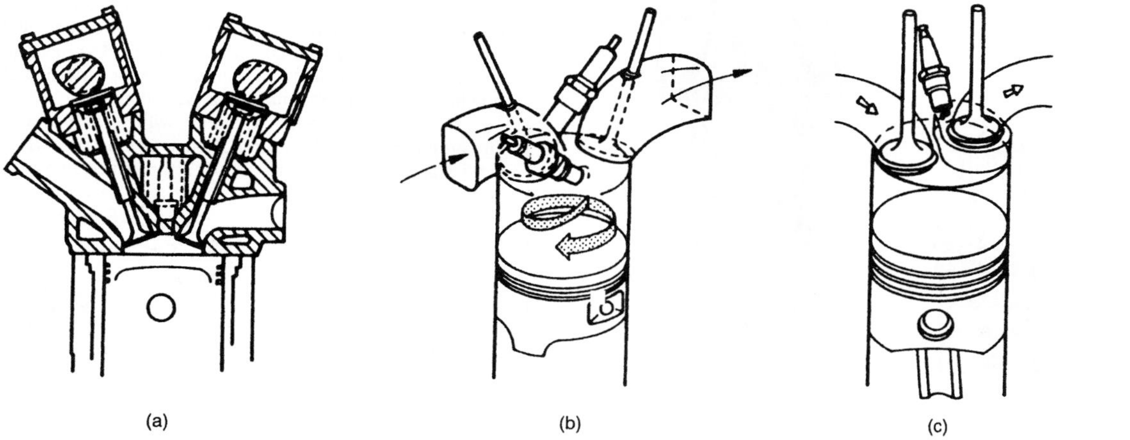

Figure 7.6.49 Modern fast-burn combustion chambers. (*a*) Four-valve pent-roof; (*b*) NAPS-Z with twin spark plugs and high axial swirl; (*c*) high-ratio compact chamber (HRCC). *(Collins and Stokes, SAE, 1983.)*

Combustion Roughness The high rate of pressure rise during combustion at relatively low engine speeds (1,000 rev/min) and the maximum combustion pressure near peak indicated torque engine speeds subject the engine mechanism to severe stresses which may result in combustion roughness noise unless the pressure rate is controlled. Andon and Marks (*SAE Trans.* **72**, 1964, p. 636) found that spark plug location, combustion chamber shape, and degree of turbulence are the principal factors that can be varied to control combustion roughness and the resulting noise. In a given engine, the maximum combustion rate or pressure rise can be reduced by retarding the spark timing or decreasing the volumetric efficiency.

A rate of pressure rise of 30 lb/in^2 (207 kPa) per crank angle degree results in near-maximum efficiency, and higher rates of pressure rise usually result in combustion roughness noise. Increasing the rigidity of the crankshaft increases the natural frequency of vibration and reduces the deflection of the structure caused by combustion and inertial forces and thereby reduces roughness noise.

Die-cast aluminum transmits roughness noise more readily than cast iron. Structurally stiff compact engines such as the V-block are better than the in-line engines in this regard.

Compression Ignition Engines

Compression ignition engines have valve-in-head construction if operating on four strokes. Two types of combustion chamber arrangements are used: a single chamber, referred to as the *open* or *direct* injection type, and a two-chamber, referred to as a *prechamber* or *divided-chamber* type. In open combustion chambers (direct injection), fuel is injected under high pressure, usually through a multiple-orifice nozzle, directly into the clearance space or chamber between the piston and the cylinder head. The piston head is usually conformed (Fig. 7.6.51) to fit the fuel spray, and swirl moves the air into the fuel spray. Air swirl is accomplished by intake port design, by shrouding the intake valve, or, in two-stroke engines, by using tangential intake ports. High turbulence is accomplished by having the piston closely approach part of the cylinder head. This forces the gases out of the small clearances and agitates the mixture. Open chambers are used for small automotive as well as heavy-duty diesel engines.

Precombustion chambers are divided into two parts, the major volume being between the piston and the cylinder head and connected by a small passageway to the minor volume located in the cylinder head (Fig. 7.6.52). Fuel is injected into only the smaller chamber, and except under light loads, partial combustion occurs and discharges the burning mixture into the large chamber in which combustion is completed. This type of combustion chamber produces smooth combustion but has fairly high fluid friction and heat transfer losses.

Turbulence or divided combustion chambers are modifications of the precombustion chamber, having the major chamber in the cylinder head

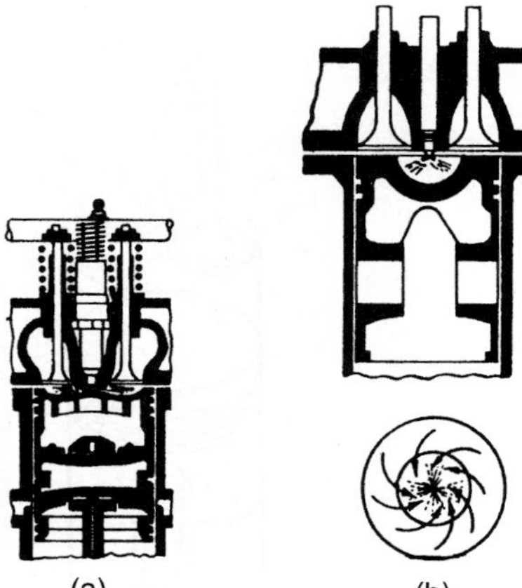

Figure 7.6.51 (*a*) Open combustion chamber with swirl produced by intake ports (left-hand image); (*b*) open combustion chamber with high turbulence and some swirl (right-hand image).

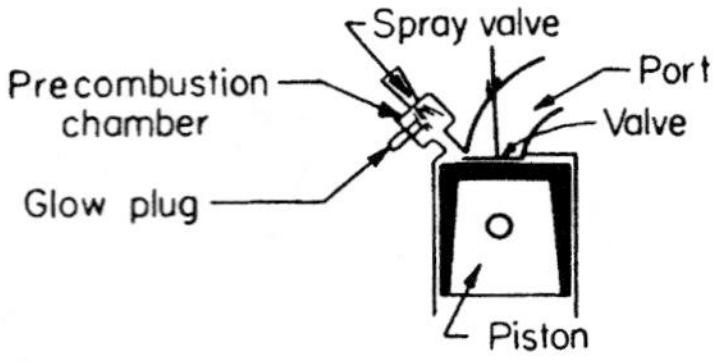

Figure 7.6.52 Precombustion chamber.

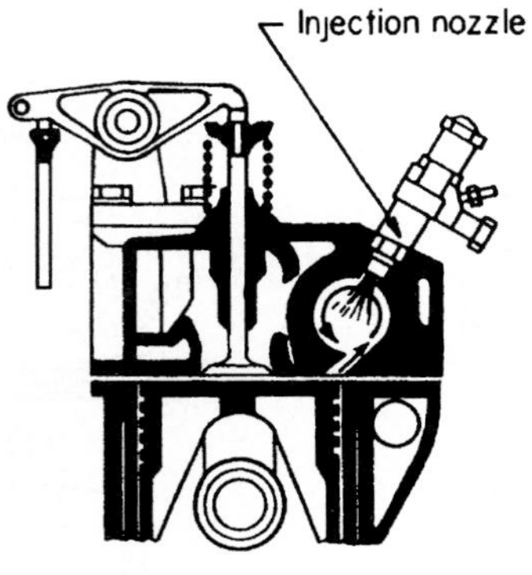

Figure 7.6.53 Ricardo divided combustion chamber.

and usually only a small clearance space between piston and cylinder head (Fig. 7.6.53). The passage between the two chambers is considerably larger than that in the precombustion chamber. Close piston clearance produces high turbulence in the prechamber and promotes rapid combustion. Part of the prechamber containing the transfer passage commonly is thermally insulated. An electrically heated hot-wire glow plug is used to assist cold starting, together with compression ratios up to 24:1.

Stratified-Charge Engines

A stratified-charge spark-ignited engine has potential for burning lean mixtures for improved fuel economy and reduced emissions. In theory, an overall lean mixture is formed in the combustion chamber, while controlled to be slightly richer than stoichiometric in the vicinity of the spark plug at the time of ignition. The richer portion is thus easily ignited, and this in turn ignites the remaining lean mixture. Some of the benefits claimed are (1) lower flame temperature, (2) improved cycle efficiency, (3) less heat loss, (4) less dissociation, and (5) less pumping loss. In addition, less NO_x would be expected because of the lower combustion temperatures; likewise, less unburned hydrocarbon and CO emissions would result because of the overall excess of oxygen.

For effective stratification, very careful development has proved to be necessary to obtain complete combustion of fuel under the wide range of speed and load condition required of an automotive type of engine. Distinct means for accomplishing the stratified charge condition have been explored, including the following.

1. A single combustion chamber with a well-controlled rotating air motion is used. This arrangement is illustrated (Fig. 7.6.54) by the Texaco combustion process (TCP), patented in 1949. Ford Motor

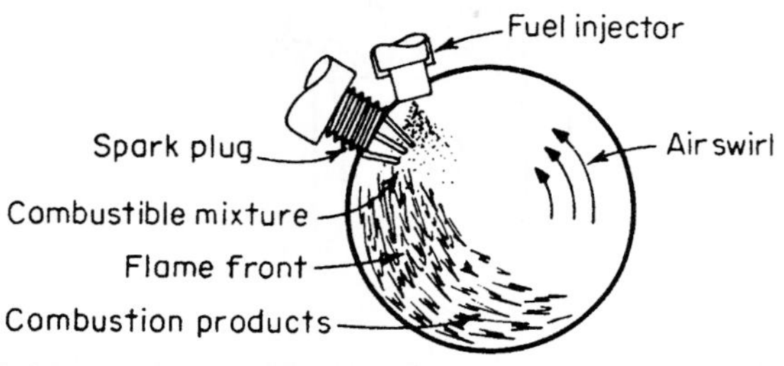

Figure 7.6.54 Texaco combustion process (TCP).

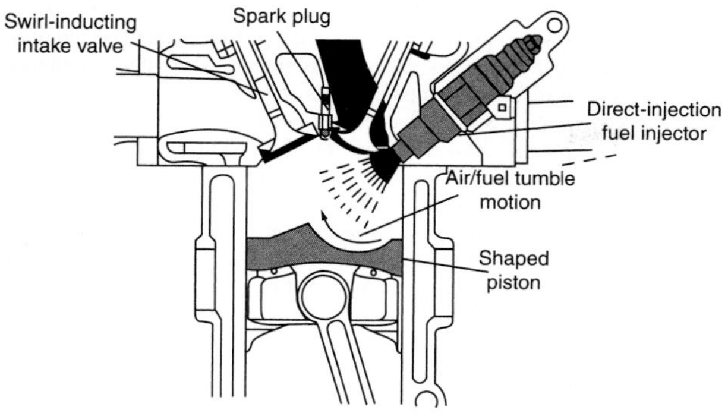

Figure 7.6.55 Mitsubishi gasoline direct injection (GDI) engine concept.

Company's programmed combustion (PROCO) is a similar, single-chamber concept, but has only been produced in very limited quantities.

2. In the wall-guided concept (Figs. 7.6.55 and 7.6.56), the preparation of the stratified fuel-air charge relies on a wall-guided system. The fuel spray is injected from the side and directed toward the piston surface, so as to impinge on the spherical piston cavity and be reflected toward the spark plug by the tumble.

3. In the air-guided concept (Fig. 7.6.56), the spray does not impinge on the piston. The injector is mounted half way between the side and the center. Preparation of the stratified charge is accomplished via a combination of swirl and tumble in the charge.

4. In the spray/jet guided system, the injector is mounted at the center and is the primary means for providing the sprayed fuel prominent swirl motion to enable fuel-air mixing. Note that the spark plug is off to the side, but placed nearer the injector than the other two concepts.

5. There is axially stratified charge. By turning on the port fuel injector toward the end of the filling process, a rich mixture is produced near the spark plug, while the overall mixture may be lean. This method is known as the *Toyota lean-burn system.*

Some production vehicles employing stratified-charge concepts have appeared in other global markets such as Japan, but are not sold in North America because of current emission regulations.

Homogeneous Charge Compression Ignition

Recent developments in engine control technology have enabled the prospects of a fairly novel form of engine combustion: homogeneous charge compression ignition (HCCI). HCCI operates by inducting a homogeneous mixture of fuel and air (similar to conventional spark-ignited engines), but using compression ignition to ignite the fuel (similar to conventional diesel engines). Major challenges, such as precise control over the start of combustion, continue to keep HCCI under rigorous development without any use in production vehicles yet.

One practical implementation of an HCCI-type combustion mode is low-temperature premixed compression ignition (PCI) combustion. Used with direct-injection configurations, PCI operates on the principle of utilizing an extremely long ignition delay (the time between start of injection and start of combustion) via strategic injection timings and large rates of EGR to create a combustion process that burns premixed fuel and air at ultralow temperatures. The major benefit of PCI combustion is dramatically reduced emissions of nitrogen oxides and particular matter (Jacobs et al., *SAE Trans. Jour. Eng.*, **114,** 2005). Nissan has demonstrated its PCI combustion concept in production use.

SPARK IGNITION COMBUSTION

Ignition System The ignition system usually consists of a 12-V storage battery, induction coil, and high-tension distributor, or a high-tension magneto with a distributor. Recent automotive practice employs electronic means to interrupt primary current in place of a contact breaker, and modern engines use individual coils for each spark plug. For aircraft engines, the low-tension distribution system to transformer spark plugs, or other apparatus for obtaining a high-tension spark at the plug, appears to eliminate some challenges of the conventional ignition system in aircraft engines at high altitude. For details on ignition systems, see Sec. 11.1.

The usual firing order for in-line engines is 1, 3, 4, 2 or 1, 2, 4, 3 for four cylinders; 1, 5, 3, 6, 2, 4 or 1, 4, 2, 6, 3, 5 for six cylinders in line; and 1, 6, 5, 4, 3, 2 for V-6 engines. A common procedure for V-8 engines is to number the cylinders from front to rear, with the odd numbers in the left bank, as viewed from the driver's seat. For this arrangement a typical firing order is 1, 8, 4, 3, 6, 5, 7, 2. Radial engines have firing orders that skip every other cylinder. For a nine-cylinder, single-row engine the order is 1, 3, 5, 7, 9, 2, 4, 6, 8.

Spark plug gaps may vary from 0.020 to 0.080 in (0.5 to 2.0 mm). The smaller gaps require 4,000 to 8,000 V while the larger gaps require 10,000 to 34,000 V. In wedge-shaped, two-valve combustion chambers, spark plugs were typically located near the exhaust valve (hottest spot) so that the flame propagates toward the cooler part of the combustion chamber; this arrangement permitting higher compression ratios without end-gas knock. In modern, four-valve, pent-roof combustion chambers, a central spark plug location results in lower burn duration and good tolerance for dilution. Dual ignition (two spark plugs located opposite to each other) also reduces burn duration and may result in small thermal efficiency gains compared with single ignition with optimum spark advance. Rotary (Wankel) engines employ one or two spark plugs per rotor. Single spark plugs are centrally located either above or below the trochoid waist. In modern, direct injection, stratified-charge, lean-burn engines, optimum placement of the spark plug is crucial to provide a combustible mixture near its vicinity.

Flame Speed Flame speeds in spark ignition engines are low immediately after ignition (laminar flame), attain maximum values when about half of the combustion chamber has been inflamed (turbulent flame), and decrease toward the end of the piston. Mean flame speeds peak for mixtures 10-20 percent richer than stoichiometric (φ = 1.1-1.2), and they vary with fuel, engine speed, and turbulence.

At 900-ft/min (4.5-m/s) mean piston speed, mean flame speeds with maximum-power gasoline-air mixtures and optimum spark advance

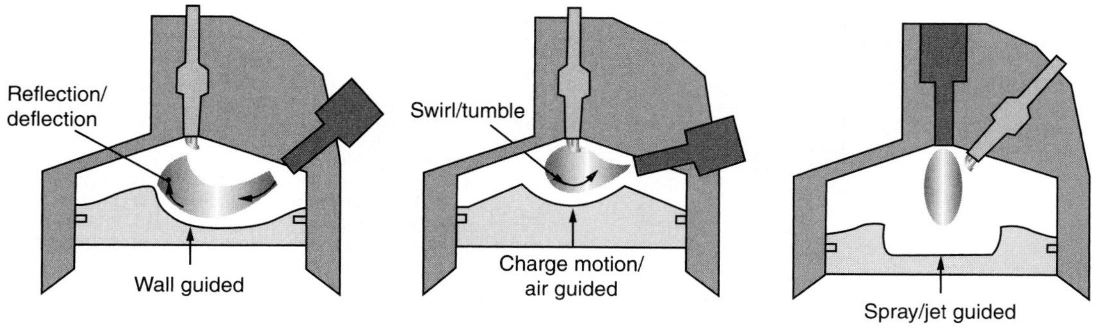

Figure 7.6.56 Various designs implementing the stratified-charge combustion concept.

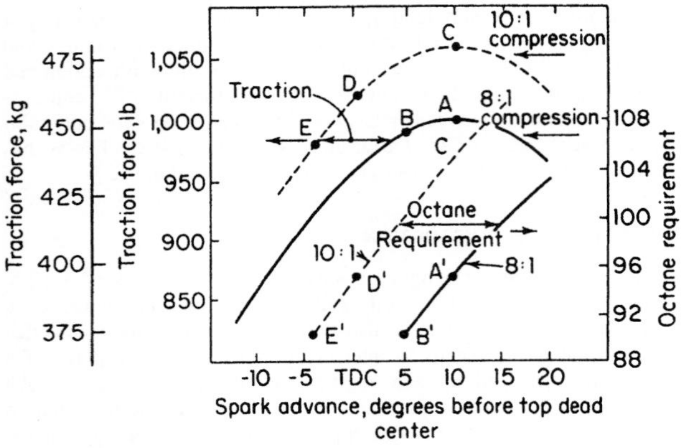

Figure 7.6.57 Effect of spark advance on traction force and octane requirement.

range from about 50 to 100 ft/s (15 to 30 m/s). Mean flame speeds under the same mixture, spark, and speed conditions in a CFR Waukesha engine are estimated at 55, 70, 80 ft/s (17, 21, 24 m/s) for a 4.6:1, 6.5:1, and 8:1 compression ratio, while with ethyl alcohol–air mixtures the corresponding mean flame speeds are 50, 65, 75 ft/s (15, 20, 23 m/s).

The mean flame speed in an automotive engine under the foregoing conditions is estimated from the optimum spark advance to be 90 ft/s (27 m/s) for a 6.5:1 compression ratio. In the same engine, mean flame speeds with optimum conditions vary from about 60 ft/s (18 m/s) at 500 ft/min (2.5 m/s) mean piston speed to about 170 ft/s (0.9 m/s) at 3,000 ft/min (15 m/s) mean piston speed.

Spark Advance Since combustion requires finite time, it is necessary to ignite the mixture in the cylinder before the piston reaches the end of its compression stroke (top dead center). This should distribute the combustion process before and after top dead center in order to obtain maximum torque and efficiency. Spark advance is measured in crank angle degrees between the time of the spark and top dead center. Optimum spark advance is the minimum advance which develops the maximum brake torque (MBT). Usually this causes the maximum pressure to peak at about 15-20° after top dead center and 50 percent of the charge to be burned at 5-10° after top dead center (Heywood, 1988). The torque lost by a spark timing earlier or later than optimum is at first only slight but increases rapidly as the timing further deviates from optimum (see Fig. 7.6.57).

Optimum spark advance depends principally on air-fuel mixture, amount of residual gas, emission control requirements, combustion chamber design, turbulence, engine speed, number of spark plugs, and spark plug location. The optimum spark advance is also affected by knock considerations, as described later. Maximum power air/fuel ratios require the minimum spark advance. Low-speed engines require 10° to 20° spark advance, and high-speed automotive engines require 30° to 40° spark advance. Full load requires less spark advance than part throttle. Spark advance in contemporary automotive engines is set by the engine control unit and is a function of speed, load, and other engine operating variables.

Autoignition The autoignition temperature of an air-fuel mixture is the lowest temperature at which chemical reactions proceed at a rate sufficient to result in ignition. This temperature depends principally on the air-fuel ratio of the mixture, fuel properties, and mixture pressure. Subjecting a mixture to a temperature higher than the autoignition temperature will result in bulk ignition after a time lag (Table 7.6.16). An increase in pressure decreases the ignition temperature of an air-fuel mixture (Table 7.6.17).

Table 7.6.16 Ignition Temperatures of Air-Gas Mixtures at Atmospheric Pressure

		With ignition lag*		Instantaneous† ignition	
Gas	Chemical symbol	percent gas in mixture	Avg temp, °F (K)	percent gas in mixture	°F (K)
Hydrogen	H_2	8-24	1,130 (893)	10	1,377 (1,020)
Carbon monoxide	CO	13-47	1,204 (924)	50	1,708 (1,204)
Methane	CH_4	5-38	1,202‡ (923)	25	>1,832 (1,273)
Ethane	C_2H_6	§	968‡ (793)		
Propane	C_3H_8	§	914‡ (763)		
Acetylene	C_2H_2	4-22	804 (702)		
Ethylene	C_2H_4	6-19	1,004 (813)	10	1,832 (1,273)
Benzene	C_6H_6			50	1,943 (1,335)
Ether	$C_4H_{10}O$			50	1,892 (1,306)
Gasoline:					
92 octane					734 (663¶)
100 octane					804 (1,702¶)

* Dixon and Coward, *Trans. Chem. Soc.*, **95,** 1909, p. 514.
† David, *Trans. Chem. Soc.*, **111,** 1917, p. 1003.
‡ Minimum value, since spread in values was appreciable.
§ Composition not reported.
¶ Jones, *U.S. Bur. Mines Bull,* 1946.

Table 7.6.17 Ignition Temperatures at Various Air Pressure, °F (K)

	Air pressure, atm								
Gas	1	3	5	7	10	15	20	25	30
Hydrogen*	1,134 (885)	1,132 (884)	1,112 (873)	1,100 (866)					
Methane*	1,343 (1,001)	1,269 (960)	1,215 (930)	1,175 (908)					
Gasoline†			590 (583)		480 (522)	420 (489)			
Kerosine†			670 (628)		490 (528)	430 (494)	400 (478)	385 (469)	
Gas oil†			580 (578)		500 (533)	450 (505)	435 (497)	415 (486)	
Machine oil†			710 (650)		610 (594)	550 (561)	520 (544)		
Benzene†			685 (636)		585 (580)	545 (558)	530 (550)	515 (541)	500 (533)
Cylinder oil†			825 (714)		710 (650)	645 (614)	620 (600)	595 (586)	572 (573)
Benzol†					1,095 (864)	970 (794)	930 (772)	905 (758)	880 (744)

* Bone and Townsend, "Flame and Combustion in Gases," p. 69.
† Tausz and Schulte, *Zeit. V.D.I.*, 1924, p. 574.

Compression ignition engines rely on autoignition for starting the combustion process. However, in spark ignition engines, autoignition of the gas ahead of the flame front (end gas) is undesirable as it leads to combustion knock. This abnormal combustion phenomenon is described in greater detail in the following section.

COMBUSTION KNOCK

Knock in internal combustion engines is produced by the spontaneous autoignition of an appreciable portion of the charge (Rassweiler and Withrow, *Trans, SAE,* **31,** 1936, p. 297), which results in rapid local pressure rise and sharp metallic noise. In spark ignition engines, knock occurs when the last fraction of the charge to burn (end gas) is compressed above its autoignition temperature by the fraction that has burned. The occurrence of knock depends on the reactivity of the end gas and its induction time in the combustion chamber. End gas reactivity is affected by fuel properties, the air-fuel ratio of the mixture, and mixture pressure and temperature. Induction time is dependent on flame speed and combustion chamber geometry. The use of knock-suppressing blending agents or additives is common in modern fuels. They reduce end gas reactivity, which allows the flame from the spark plug to travel through the unburned fraction before it can autoignite.

The tendency of a spark ignition engine to knock increases almost directly with spark advance (Fig. 7.6.57). Prior to the introduction of exhaust emission control requirements, it was general practice in automobile engine design to use a spark advance a few degrees less than optimum at wide-open throttle. This materially reduces the tendency to knock with only a negligible loss of torque. At part-throttle operation, where combustion knock may not be a problem, the spark is advanced closer to optimum. However, a compromise may be made between spark advance for good torque and spark retard for knock control. Knock suppression in spark ignition engines requires fuels with high octane number (low reactivity). Modern automotive engines employ knock sensors, which provide feedback to the engine control unit to alter spark advance.

In compression ignition engines, combustion knock occurs at the beginning of the combustion process. The fuel spray formed in the combustion chamber breaks down to fuel droplets, which evaporate and mix with the surrounding air. The physical processes of atomization, evaporation, and mixing result in ignition delay between fuel injection timing and autoignition of the first part of the mixture. The fuel-air mixture fraction that ignites first is the one closer to stoichiometry. Upon autoignition, this fraction of the mixture burns at high rates resulting in knock and metallic noise. Knock suppression in diesel engines requires fuels with high cetane number (high reactivity).

Knock Rating of Fuels

For detailed information regarding methods, apparatus, reference fuels, and suppliers of apparatus and reference fuels, see ASTM Manual of Engine Test Methods for Rating Fuels.

Octane Number

The octane number (ON) of a fuel is found by comparing the knock tendency of the fuel with that of the primary reference fuels isooctane (2,2,4 trimethylpentane) and n-heptane. Isooctane has an octane number of 100 while n-heptane has an octane number of 0. Fuels that exhibit less knocking tendency than isooctane may have octane numbers higher than 100. A fuel of high octane will have a low cetane number and vice versa (Fig. 7.6.58). An octane number of X indicates that a fuel will have the same knocking tendency as a blend of X percent isooctane and (100-X) percent n-heptane by volume. The octane number measurement is done using a CFR spark ignition engine (Waukesha Motor Co.) at prescribed inlet pressure and temperature, humidity, coolant temperature, engine speed, spark advance, and air-fuel ratio. Two methods are commonly used, the research method and the motor method, which differ in their prescribed operating conditions, and yield the RON and MON metrics respectively (ASTM D908 and 357). Modern fuels are marketed by reporting their antiknock index (AKI), which

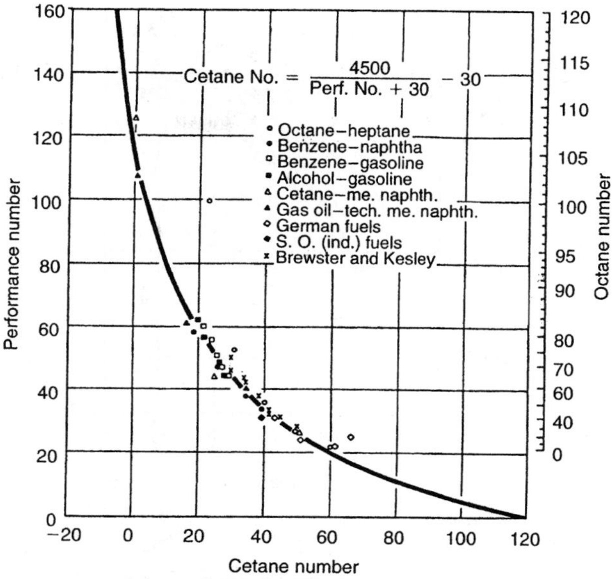

Figure 7.6.58 Relationship between antiknock scales (octane number and performance number) and cetane number. *(Brewster and Kesley, SAE, 1963.)*

is the average of RON and MON. Commercial gasoline fuels available in the North American market have AKI between 87 and 93. The fuel sensitivity S is defined as RON-MON.

The performance number scale for aviation gasoline fuels (Fig. 7.6.58) is designed to relate fuel rating to average knock-limited performance (imep). A gasoline of the 100/130 grade indicates that in the aviation test (lean mixture, normally aspirated) the fuel is equivalent to isooctane in imep while in the boosted test (rich mixture) it permits an imep of 130 percent of that of iso-octane. Although most gasoline fuels rate lower by the motor method than by the research method, refinery component streams that are taken directly from crude oil without further processing may have about the same rating by both methods. For modern fuels, the antiknock index provides the best correlation with road ratings. This value is posted on the fuel station pump, with the vehicle's recommended fuel published in the owner's manual.

The use of the aviation and boosted methods for aviation gasoline fuels results in two performance numbers (such as 100/130). This grading indicates that the fuel is less sensitive than iso-octane to the difference in engine severity in the two test methods.

Road ratings of gasoline are determined by means of borderline knock curves for the test fuel and various blends of the reference fuels at various spark advances. The car is accelerated with wide-open throttle, and the speed at which steady knock ceases is recorded as the knock die-out speed (CRC procedure E-2). A plot of these data on a spark advance vs. knock die-out speed chart (Fig. 7.6.59) makes possible the determination of the octane number rating of the test fuel over the desired speed range.

Mechanical octane number is the resulting decrease in fuel octane requirements resulting from mechanical changes in engine design or control. Design variables include ignition control, combustion chamber design (including spark plug location and cooling), valve timing, fuel system, and engine-transmission combinations. (See Table 7.6.18.)

Knock Suppression The antiknock characteristics of a fuel can be improved by blending in either another fuel of better antiknock characteristics or a knock-suppressing fuel additive (Fig. 7.6.60). In the past, tetraethyl lead (TEL) and tetramethyl lead (TML) were used as effective knock suppressors (Table 7.6.19) in small amounts (up to 3 mL/gal of automotive fuels) for gasoline fuels. Their addition has been eliminated because of adverse effects on human health and exhaust emissions. Other compounds can raise antiknock quality, but

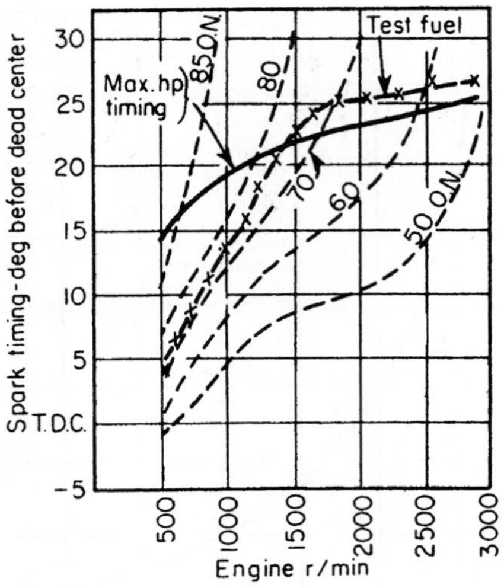

Figure 7.6.59 Road octane rating.

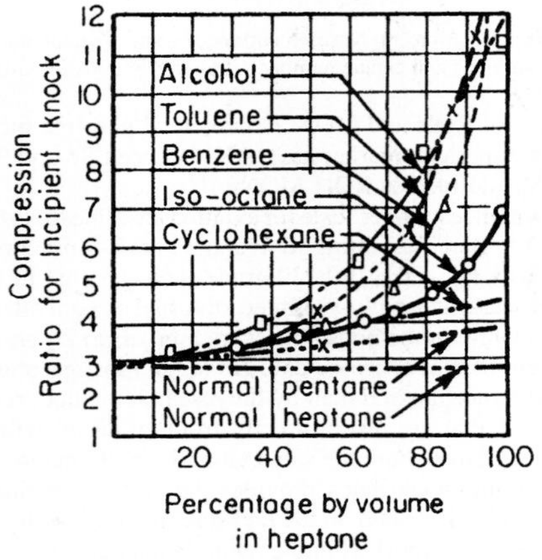

Figure 7.6.60 Effect of fuel composition on knock-limited compression ratio.

the much larger amount needed typically changes volatility and stoichiometry significantly and may affect emissions formation, lubrication, and engine life.

Cetane Number The cetane number (CN) of a diesel fuel is determined by matching its ignition quality with that of a blend of two reference fuels, normal cetane (CN = 100) and heptamethylnonane (CN = 15). The comparison is made in a CFR diesel engine (Waukesha Motor Co.) using a specified procedure (ASTM D 613 or CRC-F-S). Cetane number for the reference blend is calculated by

$$CN = (\text{vol percent } n\text{-cetane}) + 0.15(\text{vol percent heptamethylnonane})$$

The ignition quality of a fuel for a compression ignition engine is indicated by the time delay between the start of injection and the start of rapid pressure rise caused by combustion. The compression ratio in the test engine is adjusted to result in a 13° crank angle delay for the unknown fuel when injection begins at 13° before top dead center. The mixture of reference fuels which has the same lag under the same conditions indicates the cetane number of the fuel.

The factors that tend to aggravate knocking in a spark ignition gasoline engine tend to suppress it in a diesel engine. Thus, isooctane has a low, while n-heptane has a high, cetane number. High CN fuels are more reactive and can be readily used for engine cold start.

Table 7.6.18 Antiknock Indices (ASTM D4814)

Unleaded fuel* (for vehicles that can or must use unleaded fuel)	
Antiknock index (RON + MON)/2	Application
87	Designed to meet antiknock requirements of most 1971 and later model vehicles
89	Satisfies vehicles with somewhat higher antiknock requirements
91+	Satisfies vehicles with high antiknock requirements

* Unleaded fuel having an antiknock index of at least 87 should also have a minimum motor octane number of 82 in order to adequately protect those vehicles that are sensitive to motor octane quality.

SOURCE: Abstracted from ASTM D4814, with permission.

Table 7.6.19 Relative Effect of Antiknock Compounds

Compound	Chemical symbol	Weight for a given effect, g
Tetraethyllead (TEL)	$Pb(C_2H_6)_4$	0.0295
Aniline	$C_6H_5NH_2$	1.0000
Ethyl iodide	C_2H_5I	1.55
Ethyl alcohol	C_2H_5OH	4.75
Xylene	$C_6H_4(CH_3)_2$	8.0
Toluene	$C_6H_5CH_3$	8.8
Benzene	C_6H_6	9.8

SOURCE: "International Critical Tables," vol. 2, pp. 162-163.

Cetane number has been correlated with the mid-boiling point and the API gravity of the fuel. The relation for computing the approximate cetane number is (*Jour. Inst. Petroleum,* **30,** 1944, p. 193)

$$CN = 175.4(\log \text{mid-boiling pt, °F}) + 1.98(\text{API gr}) - 496$$

The deviation from engine test CN for 579 fuels was less than ±5 CN for 94 percent of the fuels (see also Sec. 5.1).

Ignition accelerators increase the cetane number of diesel fuels, but the susceptibility of the fuel to the accelerator decreases with increasing concentration of the accelerator (Fig. 7.6.61). CN improves that have

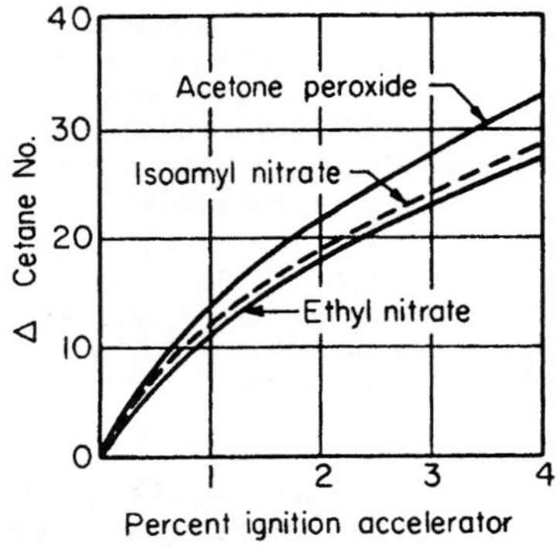

Figure 7.6.61 Improvement in CN with ignition accelerators. *(Bogen and Wilson.)*

been used in the past are commercial amyl nitrate and pure isoamyl nitrate (Bogen and Wilson, *Petroleum Refiner,* July 1944).

Permissible Compression Ratios Data obtained on single-cylinder research engines and multi-cylinder automotive engines show wide variations in the compression ratios producing incipient knocking of fuels of various octane numbers. High compression ratios in engines using fuels of a given octane number do not necessarily indicate good design and high performance, since valve timing, manifold and port design may act as throttling devices, or spark timing may be retarded. Compression ratios vary considerably depending on the engine design and fuel used. An increase in cylinder size may necessitate a lower

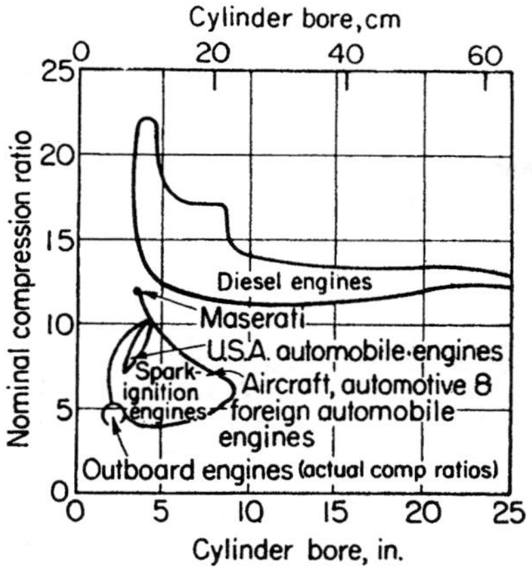

Figure 7.6.62 Range of compression ratios as a function of cylinder bore. (replace?)

compression ratio to control knock and pressure rise during combustion (Fig. 7.6.62). Compression ratios in diesel engines are higher than spark ignition engines due to higher cylinder pressures required to achieve mixture autoignition.

OUTPUT CONTROL

Speed and load control in four-stroke gasoline engines are achieved by throttling the intake air. The air/fuel ratio may vary with load, depending on the engine requirements and emissions control system. Intake throttling reduces the pressure in the cylinder at the end of the intake stroke and thereby reduces the mass of charge trapped per cycle. While throttling is the simplest means for regulation and provides a stable idling condition, it increases the pumping work and consequently decreases the efficiency of the engine. Presently, there is a trend toward using variable valve timing and variable valve lift as a means of controlling load without increasing pumping losses. In direct injection, stratified-charge engines, load control may be accomplished through control of the amount of fuel injected per cylinder.

Power control of spark-ignited, two-stroke engines is difficult because intake throttling reduces or prevents cylinder scavenging. Qualitative governing is used in some cases with constant-speed gas engines. The fuel is throttled while the air is unrestricted. Qualitative governing is limited by the lean limit of flammability. In addition, lean mixtures burn more slowly, causing a decrease in fuel conversion efficiency. Advancing the ignition timing, as the mixture is made leaner, reduces this loss of efficiency.

Governing by variation of ignition timing, as used on some small two-stroke gasoline engines, is undesirable since this is achieved by variation of thermal efficiency and may result in overheating the engine. Variation of ignition timing is required for each change in speed and load to obtain optimum results.

Output control of diesel engines is obtained by varying the amount of fuel injected per cycle, which provides control of the air-fuel ratio as the engine load varies.

Diesel engines require a speed governor to provide a stable idle. Mechanical, hydraulic, and electronic speed-control governors are used on diesel engines. These control the fuel pump injection quantity and provide speed regulations at all operating conditions.

COOLING SYSTEMS

Engine cylinders must be cooled to maintain a lubricant film on the cylinder walls and other sliding surfaces. The cylinder heads, pistons, and exhaust valves are cooled to prevent combustion knock or destruction of these parts due to overheating. The lubricant must be cooled to maintain the desirable viscosity under operating conditions. Liquid or air cooling systems are used. Pistons, exhaust valves, and lubricants in comparatively small or low-duty engines are sufficiently cooled by contact with other engine parts or the lubricant between them and do not require separate cooling systems.

Liquid Cooling The heat removed by the cooling liquid from the cylinders and cylinder heads can be 15-20 percent of the energy input for large diesel engines, 20-35 percent for automotive engines, and may run as high as 40 percent for automotive engines at part load. An increase in speed decreases the heat loss to the coolant. Heat loss can be 40-50 percent of the brake output for large diesel engines and 100-150 percent of the brake output for automotive engines.

Pistons in small engines are cooled by heat transfer to the cylinder walls and lubricant. In some cases, an appreciable quantity of oil is directed against the piston head to maintain desirable head temperatures.

Some large engines have even used liquid-cooled pistons. Liquid is usually fed to the piston by swing joints or telescopic connections. The coolant jacket surrounding the high-temperature areas of the engine is maintained at a pressure considerably above atmospheric—10 to 15 lb/in^2 (70 to 100 kPa). This permits operation at higher coolant temperatures and/or decreases the tendency for local boiling in the region of hot spots, which could lead to component failures. The average percentage distribution of the coolant to the various jackets is approximately 60 percent to the cylinder jacket and 40 percent to the cylinder-head jacket with uncooled piston; 50 percent to the cylinder jacket, 25 percent to the cylinder-head jacket, and 25 percent to the piston and piston rod with liquid-cooled piston.

The cooling system may be noncirculating, in which case cold water is supplied, flows through the water jacket, and is wasted. With a temperature rise in large engines of 90°F (32°C), from 2 to 4 gal/(bhp · h) [7.6 to 15 L/(bhp · h)] of coolant must be supplied. In some engines, coolant outlet temperatures as low as 120°F (49°C) must be maintained, which, with high inlet temperatures, may increase the required quantity of coolant to as much as 10 gal/(bhp · h) (38 L)/(bhp · h). Coolant is circulated at a rate of 25 to 50 gal/(bhp · h) [95 to 190 L/(bhp · h)] with a temperature rise of 10 to 20°F (5.6 to 11°C) while flowing through the water jacket.

Natural circulation (thermosiphon), with low velocities, requiring large connections, may be used if the coolant forms a complete circuit. In this case, hot coolant rises in the engine jacket, flows to the radiator, is cooled, descends, and flows back to the engine jacket.

Small stationary engines are often hopper-cooled. This is an evaporative cooling system requiring about 4 to 6 lb/(bhp · h) [1.8 to 2.7 kg/(bhp · h)] of makeup water. The ASTM CFR knock-testing engines are cooled by evaporation, the vapor being condensed by a reflex condenser (copper coil with cold water flowing through) and returned to the cylinder jacket.

In high-output engines, the coolant is directed at the hottest spot, usually the exhaust valve seat, with either an external or internal inlet manifold from a common cylinder block jacket; otherwise vapor bubbles will form, cling to the surface, and cause overheating. In vertical engines, the coolant usually flows upward, around the cylinder barrel into the cylinder head jacket, and to the outlet. Higher output and reduced knocking tendency have resulted from directing the entering coolant at the hot spots in the cylinder head and then letting the coolant flow downward around the cylinders.

The size of piping required for the inlet to the jacket may be calculated for 3-ft/s (1-m/s) coolant velocity, and for the discharge 2 ft/s (0.6 m/s) for large engines. In engines with recirculation, these values are increased up to 10 to 15 ft/s (3 or 4.5 m/s), and the size of the inlet line to the circulating pump is usually larger than the outlet from the coolant jacket. It is desirable to have an open or visible outlet from each separately cooled part, showing the flow of coolant. To ensure totally filled systems which are as air-free as possible, overflow bottles may be used. Water is an outstanding liquid coolant. However, where freeze protection is necessary or where higher jacket temperatures are desired, other cooling media are generally used. For example, light-duty engines typically use a mixture of water and ethylene glycol. (See Sec. 4.7)

Air Cooling Many small industrial engines and most motorcycle and aircraft engines are air-cooled. However, higher specific output engines are increasingly liquid-cooled. Air cooling eliminates the necessity for water or other liquid cooling media, coolant jackets, pumps, radiators, and coolant connections, but necessitates single-cylinder construction, finning, baffles, and, in some cases, blowers. Air-cooled engines also tend to be more noisy since the combustion chamber walls radiate sound and the engine structure is less rigid. The permissible compression ratio and output of air-cooled aircraft engines depend on the efficiency of the cooling of the cylinder head, exhaust valves, and seats. Long fins [1 to 2 in (25 to 50 mm) depending on cylinder size], closely spaced [0.10 to 0.20 in (2.5. to 5 mm)], with baffles directing the air at high velocity at the hot spots, have made high-output aircraft engines possible at the expense of considerable air resistance or drag. Properly designed cooling ducts may make use of the heat added to the cooling air to obtain a jet effect, thereby reducing drag.

Oil Cooling Shearing of various oil films by moving parts and oil contact with hot parts of an engine cause a rise in oil temperature until the energy input to the oil equals the energy transferred by the oil to the cooler parts that it contacts. Oil coolers are required for high-duty engines to maintain oil temperatures of 200°F (94°C) or under. Temperatures of 250°F (120°C) are not uncommon in automobile engines on hot days: 300°F (150°C) is considered too high, particularly for oils which decompose rapidly under conditions causing the high oil temperatures. The desired oil temperature is maintained by circulating the oil through a radiator or cooler to an oil sump and then through the engine. Transmission oil cooling is also accomplished by placing plate coolers in the end tanks of radiators or with separate radiators.

LUBRICATION

(See also Sec. 6.4)

General A suitable liquid lubricant is vital to the satisfactory operation of an internal combustion engine. It prevents excessive wear and deposit accumulation and removes heat from the areas of relatively high temperature within the engine.

Most engine oils are composed of base oils and additives. The base oils are usually mineral oils, but some are synthetic oils. Chemical additives are incorporated in engine oil formulations to impart desirable performance characteristics which are not provided by base oils alone.

Physical Properties Both the physical and chemical properties of an engine oil affect its performance. The principal physical property involved is viscosity. It must be low enough at low temperatures to permit cranking and starting and high enough at high temperatures to provide an adequate oil film between rubbing surfaces. Sufficiently high viscosity is also required to help prevent excessive oil consumption at high temperatures. Another physical property affecting consumption is volatility. Consumption can be undesirably high with an oil of excessive volatility. Both viscosity and volatility are influenced by base oil selection. Viscosity can also be modified by means of additives.

Chemical additives perform a number of other functions in an engine oil. For example, oxidation inhibitors reduce oxidative and thermal degradation, which can lead to varnish and sludge deposition and to excessive thickening. Detergent-dispersant additives suspend insoluble products formed during engine operation, so they can be removed from the engine when the used oil is drained. Antiwear agents help protect rubbing surfaces, especially those subject to boundary lubrication. Antifoam agents prevent foam and air entrainment and their adverse effects on oil pressure and heat transfer. Rust inhibitors eliminate the formation of corrosion products on ferrous surfaces and reduce corrosive wear. Friction-reducing additives lower boundary friction.

SAE Classifications Engine oils are classified by the Society of Automotive Engineers in two general groups, viscosity (SAE J300 standard) and performance (SAE J183 standard). Viscosity is measured at both 0°F (-18°C) and 210°F (98.9°C). SAE 5W, 10W, 15W, 20W, and 25W viscosity numbers are related to 0°F (−18°C) measurements; SAE 20, 30, 40, 50, and 60 to 210°F (98.9°C) measurements. Multigrade oils, such as SAE 5W-20, SAE 10W-30, and SAE20W-40, are those which have viscosity characteristics meeting requirements at both temperatures. Multigrade oils can be used throughout the year, reduce friction and wear, and provide better lubrication and fuel economy during a cold start; for example, the viscosity of an SAE 20W-40 oil will be less than that of an SAE 40 oil at ambient conditions.

The SAE performance classification includes engine oils ranging in performance quality from that of a straight mineral oil plus a small amount of pour-point and/or foam depressants, to those required in severe passenger car service and heavy-duty truck and off-highway operation. Letter designations such as SG apply to active test techniques for oils for gasoline engines existing since 1988; designations CD and CE apply to oils for naturally aspirated and turbocharged diesel engine service; and designation CD-II applies to oils for two-stroke cycle diesel engines. Performance of these oils is evaluated in both single- and multicylinder engine tests. Factors measured include rust, wear, deposit formation, oil consumption, oil thickening, and fuel economy.

References: Most of the foregoing information on engine lubrication is covered in greater detail, and references are provided in the "SAE Handbook." See also the list of references in Sec. 4.8.

AIR POLLUTION

General Relationship

Automobile engine exhaust was first detected to contribute to air pollution in 1942, where smog created serious visibility and health problems in Los Angeles, CA (Haagen-Smit, *Sci. Amer.,* **210,** no. 1, 1964, p. 25). Smog itself is a combination of smoke (particulates) and ground-level ozone. Ground-level ozone, unlike the beneficial stratospheric ozone, arises from sunlight-induced photochemical reactions between nitrogen dioxide and hydrocarbons (Caplan, SAE Pt. 12, p. 20) and can cause severe eye, lung, and plant life irritation. Engines are a contributor of both nitrogen dioxide and hydrocarbons to the atmosphere.

In addition to nitrogen dioxide and hydrocarbons, engines emit the toxic gas carbon monoxide, particulates (unburned carbon particles), and sulfur dioxide. Table 7.6.20 identifies the historical trend of mobile engine emissions and the relative contribution they make to total atmospheric emissions. In spite of increased driven mileage from 1 trillion mi in 1970 to a projected 4 trillion mi in 2000, nearly all mobile engine source emissions decreased over the span of the three decades. This trend has resulted from improved engine design, improved fuel technology, and the universal use of catalytic converters in new automobiles.

Table 7.6.20 Annual U.S. Emissions from Mobile Engine Sources, Million Tons and Percent of Total Emissions*

	1970		1980		1990		2000	
	(Million tons)	percent	(Million tons)	percent	(Million tons)	percent	(Million tons)	percent
PM10	0.644	5.3	0.689	11.2	0.715	22.2	0.552	17.8
CO	174.6	88.5	160.5	90.3	131.7	91.7	92.2	90
NO_x	15.3	56.8	14.85	54.8	13.4	53.2	12.6	56.2
SO_2	0.551	1.8	0.717	2.8	0.874	3.8	0.697	4.3
VOC	18.5	54.9	16.1	53.4	12.1	52.1	8.0	47.2

* Stated values are calculated.
Source: EPA.

Table 7.6.21 Carter Scale of Reactivity

Nonmethane organic gas	g ozone/g NMOG (approx. range)
Methane	0
C_2-C_5 paraffins	0-1.5
Acetylene	0
Benzene	0.42
C_6 + C_8 paraffins	2-0.2
Toluene (and higher aromatics)	2-10
Alcohols	0.5-2.5
I-alkenes	9-1
Diolefins	11-9
Aldehydes	3-7
Internally double-bonded olefins	10-3

SOURCE: Carter and Atkinson, *Emu. Sc. Tech.* **23,** 1989, p. 864.

Furthermore, the trend is expected to continue as more stringent government regulations become fully enforceable in 2007.

Locally, human-made pollutants can exceed those from natural sources and can become dominant. However, emitted mass does not reflect pollution severity. For example, 1 unit mass of oxides of sulfur may equal 30 units of carbon monoxide or more in terms of health effects.

Emissions from internal combustion engines include those from blowby, evaporation, and exhaust. These can vary considerably in amount and composition depending on engine type, design, and condition; fuel-system type; fuel volatility; and engine operating point. For an automobile without emission control it is estimated that of the organic emissions, 20 to 25 percent arise from blowby, 60 percent from the exhaust, and the balance from evaporative losses primarily from the fuel tank and to a lesser extent from the carburetor (if used). All other non-hydrocarbon emissions emanate from the exhaust.

At least 200 hydrocarbon (HC) and other organic compounds have been identified in exhaust. Some of those compounds, such as the olefins, are referred to as **reactive** since they react rapidly and to a great extent in the atmosphere to form ozone. Others such as the paraffins are virtually unreactive. Methane is deemed unreactive. Of several reactivity scales, the Carter scale (Table 7.6.21) has become most widely used. It ranks nonmethane organic gases (NMOGs) on a scale from 0 to 11 g ozone/g NMOG. The reactivity of a mixture is computed as the sum of the grams of ozone from individual species, with lower levels typical of the larger molecules. A value of 3.42 g NMOG/mi is representative of non-catalyst-equipped cars.

Atmospheric oxides of nitrogen (commonly termed NO_x) are a mixture of nitrogen oxide (NO) and nitrogen dioxide (NO_2). Oxides of sulfur (SO_Z) arising from fuel sulfur are a mixture of SO_2, SO_3, and SO_4. Sulfuric acid may be a product in some oxidizing catalytic exhaust converters.

Emission quantities are most often reported as mole percent for the principal exhaust constituents (CO, H_2O, CO_2, H_2, O_2, N_2) and as moles per million moles (ppm) for trace constituents (HC, NO). Hydrocarbons are reported commonly as either ppm equivalent hexane (C_6H_{14}) or as ppm carbon (C_1). One ppm C_6 is equivalent to 6 ppm C. Other reporting parameters are (1) specific emission rate, g pollutant/kWh or g pollutant/vehicle mi (km), and (2) emission index, g pollutant/1,000 g fuel.

Emission Sources

Homogenous Combustion and Spark Ignition Engines ORGANIC GAS EMISSIONS. Emissions of unburned hydrocarbon and other organic gases can be generated from several sources, as shown in Fig. 7.6.63. The largest source is thought to be flow during compression of unburned fuel-air mixture into the crevice volume between the piston and liner above the top ring; the crevice mixture returns to the combustion during expansion at a time when it is too cool to burn. A similar mechanism is caused by the absorption of fuel vapor into oil layers and combustion chamber deposits during compression, and subsequent desorption during expansion and exhaust.

Emissions also are generated from flame extinction or wall quenching due to the inability of the flame to propagate totally to a surface or into a crevice. Daniel ("Sixth Symposium on Combustion," Reinhold, 1957, p. 886) described the result of wall quenching to be a visibly dark layer of unburned and partially burned fuel-air mixture left adjacent to all engine combustion chamber surfaces. Subsequent studies showed that the layer adjacent to single walls is largely burned prior to being exhausted. Thus the organic emission from the engine is mainly due to crevice quenching, of which the top piston land clearance is the major source.

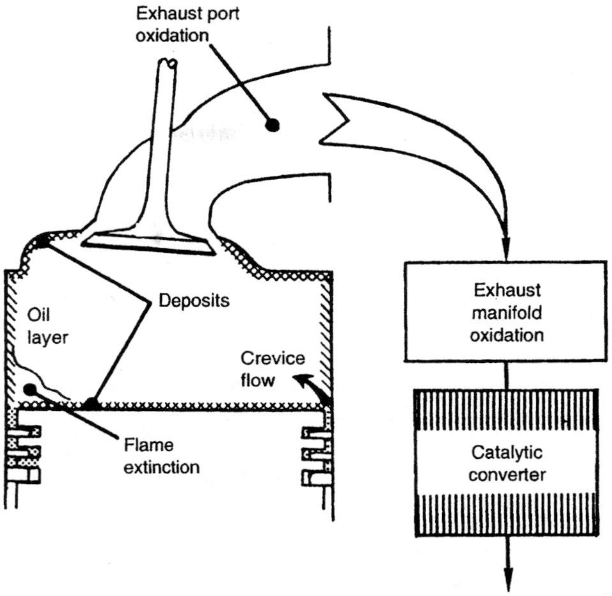

Figure 7.6.63 Emission sources of unburned hydrocarbon and other organic gases. *(Amann, "The Automotive Spark-Ignition Engine—An Historical Perspective," ASME-ICE vol. 8, 1989, pp. 33-45.)*

Under light-load engine operating conditions where cylinder exhaust residual is high, incomplete flame propagation may leave a portion of the combustion chamber contents unburned. This can produce organic emissions of several thousand ppm C_6. When residual exceeds 15 percent by weight, some incomplete combustion may occur. In some stratified-charge, direct-injection designs, bulk quenching as the flame comes in contact with cooler or leaner gases may also be present.

Wankel engine organic emissions arise from the above sources, but significant additional quantities may arise from mixture leakage through the leading rotor apex seal directly into the exhaust.

Two-stroke engines take in a mixture of fuel and air, and where intake pressure exceeds exhaust during the scavenging or overlap period, raw-mixture loss to the exhaust creates very high organic emissions. Emissions from such two-cycle engines may be 10 times higher than those from four-cycle engines. Note also that most two-stroke designs utilize once-through lubrication; for example, the oil is mixed with the fuel, typically in a fuel/oil ratio of 50:1. Hence, increased deposits and oil films generate organic emissions by physical and chemical storage and release of fuel components. Fuel injection with proper timing can reduce engine-out organic emissions to near four-stroke engine levels.

Any unburned hydrocarbons and other organic gases escaping the cylinder during exhaust can still oxidize in the exhaust temperature and manifold if the exhaust temperature is high enough for the available residence time and if there is sufficient oxygen present. In addition, where combustion is close to stoichiometric, oxidation of unburned gases is accomplished in catalysts (see later).

CARBON MONOXIDE EMISSIONS. The principal source of exhaust carbon monoxide (CO) is rich-mixture combustion. Figure 7.6.16 shows exhaust composition as mixture strength is varied. Cycle-to-cycle and cylinder-to-cylinder maldistribution of the fuel-air mixture, especially during transients, increases CO. Additional CO may arise from incomplete combustion of organics in an exhaust treatment device.

OXIDES OF NITROGEN EMISSIONS. The principal oxide of nitrogen produced inside the engine cylinder is nitrogen oxide (NO). However, since the residence time of gases within the reaction zone is very short, NO forms primarily in postflame combustion gases and does not attain chemical equilibrium. The formation depends primarily on peak combustion gas temperature and to a lesser extent on oxygen content. Once formed, NO tends to be frozen and persists in high, nonequilibrium concentrations during expansion and exhaust. Nitrogen oxide variation with mixture strength is shown in Fig. 7.6.16. Other important variables that affect NO emissions include the amount of exhaust gas recirculation and spark timing. When mixtures are more than 30 percent lean or where air is injected into the exhaust, about 5 percent of NO may be oxidized to NO_2 prior to leaving the exhaust pipe. Typically NO and NO_2 emissions are measured together with a chemiluminescent analyzer, and their sum is reported as NO_x emissions.

ALDEHYDE EMISSIONS. Aldehydes are incomplete combustion products of the oxidation of organics. They arise from the same wall quenching and incomplete combustion mechanisms as the unburned organic emissions themselves, but in addition may be formed in low-temperature exhaust system oxidation reactions. Typically gasoline engine exhaust contains 100 ppm C aldehyde, less for rich and more for lean mixtures. Methanol fuel tends to produce high emission levels of formaldehyde.

PARTICULATE EMISSIONS. Particle emissions from engines consist primarily of carbon, liquid hydrocarbons, sulfates, metals, and metal oxides. Metals arise from fuel and lubricant additives such as lead, phosphorus, zinc, and barium, as well as engine wear and corrosion. Sulfuric acid particles may be emitted as a result of reactions involving SO_2 over platinum metal catalysts. SO_2 is formed during combustion from fuel sulfur. Liquid aerosols of heavy hydrocarbons are emitted not only during starting, but also during running. Of these, polynuclear aromatics (PNAs) are health concerns. Catalytic converters virtually eliminate most hydrocarbons, especially those with long molecular chains which tend to be the most reactive.

Heterogeneous Combustion and Diesel Engines The hydrocarbon, carbon monoxide, and nitrogen oxide emission formation in heterogeneous combustion processes in diesel engines derives from the same basic mechanisms as in the spark ignition engine. However, the relative importance of each is different. In the diesel engine, the emission formation in combustion can be considered to arise from a broad distribution of individual fuel-air elements, some of which burn rich and others lean. Patterson and Henein (Emissions from Combustion Engines and Their Control, *Ann Arbor Sci.,* 1972, p. 260) have suggested the various emission sources in the combustion of a fuel spray injected into swirling air, as shown in Fig. 7.6.64. In addition, fuel impingement on chamber surfaces contributes to unburned hydrocarbons and carbon smoke particles. Hydrocarbon emissions from diesels contain many heavy fuel types. Normally, heated sampling lines are required to avoid HC condensation during sample analysis.

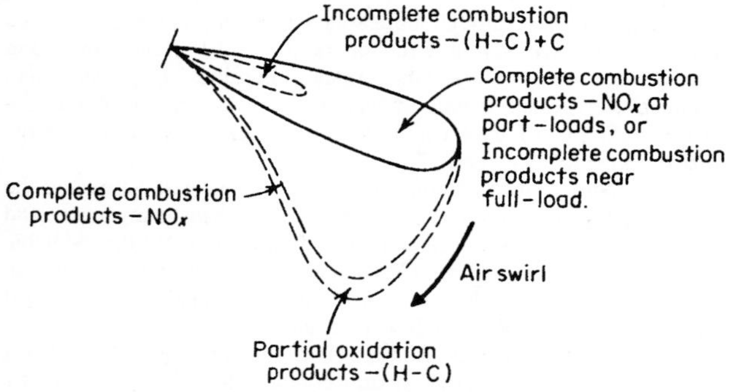

Figure 7.6.64 Emission formation in a fuel spray injected in swirling air.

The amount of carbon, carbon monoxide, hydrocarbon, and nitrogen oxide emissions from diesel engines varies considerably with engine and fuel system design as well as with the operating point. Moreover, because of the high air mass flow rate at partial load, emissions from diesel and gasoline engines cannot be compared directly except on a mass rate emission basis. Diesel emissions as mole percent or ppm may be corrected for dilution to an equivalence ratio of unity for comparison with spark ignition engine data. On a corrected basis, diesel engines emit in the range of 50 to 500 ppm C_6 and NO_x up to 3,000 ppm. Diesel engines, since they run lean, tend to show higher NO_2/NO ratios (between 10 percent and 30 percent), especially at light loads when cooler regions quench any NO_2 formed in the reaction zone (Hiliard and Wheeler, SAE 790691, *SAE Trans.,* **88,** 1979). Carbon monoxide emission from the diesel engine is very low, typically under 1,000 ppm (0.1 percent).

Diesel engines emit considerable amounts of carbon smoke and odor constituents. Most carbon arises from pyrolysis in rich combustion regions. Spindt, Barnes, and Somers (SAE paper 710605, 1971) have identified low concentrations of many potential oxidation products and certain fuel fractions as being the odor-producing compounds. Human-panel techniques are commonly used to measure odor.

Strategies for Controlling Spark Ignition Engine Emissions: Effects of Engine Design Variables

Combustion Chamber Shape The shape of the combustion chamber affects organic emissions through a change in surface area and thus heat loss. Surface area relative to the entire clearance volume is an indicator of the volume fraction of organic emissions. This may be stated as

$$\text{Organic, ppm} \propto \frac{A_s}{V_c}$$

where A_s is the chamber surface area and V_c the clearance volume at TDC. According to Sheffler (SAE, Pt. 12, p. 60), the following change the surface/volume ratio, thereby changing hydrocarbon emissions: chamber shape, bore/stroke ratio, compression ratio, and cylinder displacement. In automotive reciprocating engines, surface/volume ratios range from 5 to 8 in^2/in^3 (2 to 3 cm^2/cm^3). Wankel rotary engines have surface/volume ratios about 50 percent higher. In well-designed engines, combustion chamber shape has little, if any, effect on CO. A fast-burning chamber increases NO because of an increase in peak temperature.

Ignition System Under most running conditions, improvements in spark energy, rise time, and spark duration produce little improvement in emissions. However, a high-energy, long-duration spark in conjunction with extended electrodes and wider spark gap has been found to minimize the increase in organic emissions due to misfire, as mixtures are leaned well beyond what is chemically correct. (See Quader, *SAE Trans.,* **85,** 1976.)

Exhaust back pressure and valve overlap affect organics and NO through changes in the exhaust residual. Increasing overlap or back pressure lowers NO_x and may lower organic emissions. At very light loads, excessive residual leads to incomplete combustion, thereby increasing organic emissions.

Induction Systems and Fuel-Air Mixture Preparation Optimum mixture preparation and distribution are a key to minimum emissions. With carburetors, high-velocity manifolds, increased manifold heating, and staged throttles have shown emission reductions (see Bartholomew, SAE Pt. 12, 1971). Port fuel injection should be well-targeted on the intake valve to minimize hydrocarbon and soot emission. (See Kosowski, SAE paper 850294, 1985.)

Effects of Operating Variables

Mixture ratio and ignition timing are the most important operating variables affecting emissions. Figure 7.6.16 shows the effect of the mixture ratio on major emission amounts. Retarding ignition timing from the best-efficiency setting reduces HC and NO_x emissions significantly. Excessive retard increases HC and CO emissions. Increasing engine speed reduces HC emissions as ppm. NO_x emissions increase as load increases. Increasing coolant temperature normally reduces HC emissions and increases NO_x emissions.

Limiting fuel enrichment during warm-up is critical to low emissions, especially without exhaust treatment devices. With exhaust treatment, emissions may be reduced to virtually zero once the system is warmed up. Thus the only significant emissions from such systems are those generated during starting and warm-up.

Exhaust Treatment Devices

Catalytic or noncatalytic exhaust treatment devices may be used to clean up remaining exhaust emissions. Thermal reactors are noncatalytic devices which rely on homogeneous bulk gas reactions to oxidize CO and HC. Commonly they appear to be enlarged exhaust manifolds, but they may have internal baffling. In thermal reactors, NO_x is unaffected. Reactions are enhanced by increased exhaust temperature (reduced compression ratio, retarded timing) or by increasing exhaust combustibles (rich mixtures). Typically, temperatures of 1,500°F (800°C) or more are required for peak efficiency. Usually the engine is run rich to give 1 percent CO, and air is injected into the exhaust. A typical thermal reactor residence time is 100 ms. Required reactor volume is commonly 1.5 times engine displacement. Maximum allowable reactor temperature is governed by materials and is typically 1,750°F (950°C). Thermal reactors are seldom used because the required engine calibration produces low fuel economy.

Catalytic systems are capable of reducing NO_x as well as oxidizing CO and HC. However, a reducing environment for NO_x treatment is required which necessitates a richer than chemically correct engine mixture ratio. A two-bed converter may be used in which air is injected into the second stage to oxidize CO and HC. It is highly efficient but results in lower fuel economy. Single-stage, three-way catalysts (TWCs) are widely used, but they require extremely precise fuel control to be effective. As illustrated in Fig. 7.6.65, only in the vicinity of stoichiometric ratio is the efficiency high for all three pollutants. Such TWC systems employ either a zirconia or a titanium oxide exhaust oxygen sensor (EGO, HEGO, or UEGO) and a feedback fuel system to maintain the required mixture strength near stoichiometric. Figure 7.6.66 identifies the typical placement of various emission control technologies on an automotive engine. Figure 7.6.18 also shows the close-coupled three-way catalyst with oxygen sensor. Conversion efficiencies for typical automotive three-way catalyst emissions (HC, CO, and NO_x) range between 70 percent and 99 percent over the U.S. transient emissions drive cycle for a hot, stabilized engine/catalyst. Conversion efficiencies range between 40 percent and 99 percent for vehicles operating in high-speed accelerations.

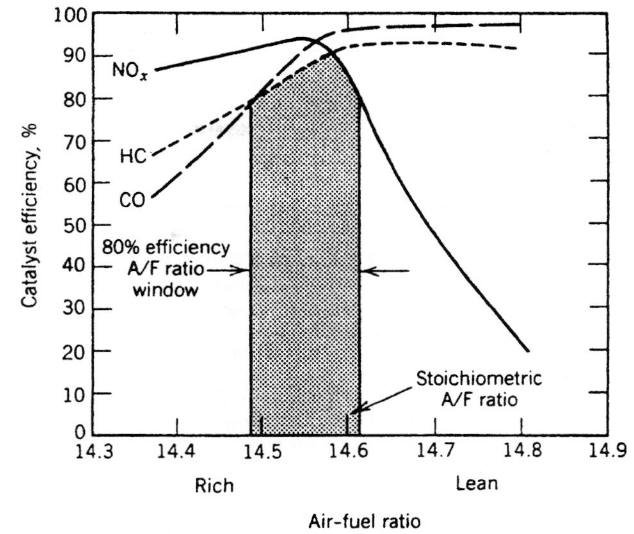

Figure 7.6.65 Conversion efficiency of three-way catalyst (TWC) as a function of air/fuel ratio. *(Kummer, Prog. Energy Combust. Sci., **6**, 1981, pp. 177-199. Reprinted with permission.)*

Catalyst support beds may be of the pellet or honeycomb (monolith) type. Materials for reducing catalysts include rhodium, Monel, and ruthenium. Materials for oxidizing catalysts include platinum and palladium. The amount of active catalyst material required may be as low as 0.05 troy oz (1.4 g) per vehicle system. Catalyst volume is about 50 to 100 percent of the engine displacement. The weight of the catalyst bed typically is about 2.2 to 5 lb (1 to 2.3 kg) excluding the metal container. Many aged catalysts can oxidize at 85 percent or higher efficiency at temperatures as low as 800 to 1,000°F (430 to 540°C). Above 1,500°F (820°C), they deteriorate rapidly. Compared with thermal reactors, catalytic converters warm up slowly because of the larger mass of material. Catalytic converters may be located farther from the engine than thermal reactors, since converters operate efficiently at

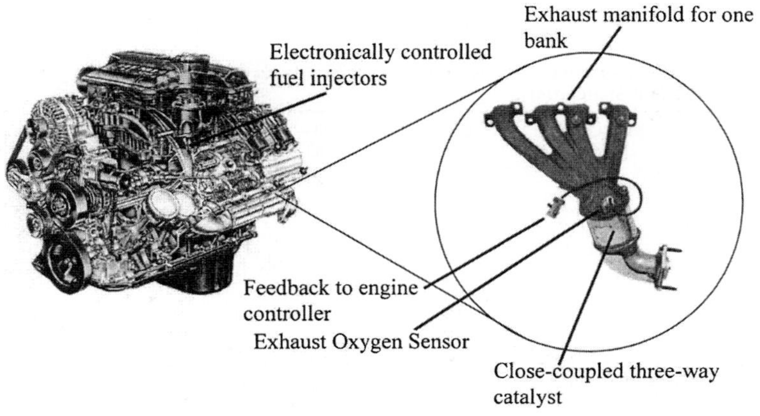

Figure 7.6.66 Typical placement of engine emission control technology.

lower temperatures. However, this slows warm-up. Monolithic converters, which can normally be operated hotter, can be smaller and can be placed closer to the engine. Catalysts with metal substrates (monoliths) warm up even faster and are more resistant to high-temperature deterioration. A small cold-start catalyst located near the engine, especially one with supplementary heating, is very effective. (See Laing, *Auto. Eng.*, Apr. 1994, p. 51.)

Strategies for Controlling Diesel Emissions: Engine Design and Operating Variables

Horrocks (*Proc. Inst. Mech. Engr.*, **208,** 1994, pp. 289-297) has summarized the strategies for controlling particulate and HC emissions as follows: (1) optimization of fuel injection and air motion; (2) good atomization of the fuel spray, particularly at idle, at partial load, and during transients; (3) precise control of fuel injection timing, particularly during transients; (4) minimization of parasitic losses in the combustion chamber; (5) low sac volume or valve cover orifice nozzles for direct injection; (6) minimization of contribution of lubricating oil; (7) central injector and combustion chamber for high-speed, direct injection engines; (8) modulated EGR; (9) electronic controls; (10) rapid warm-up of engine.

Control of NO_x emissions from diesels presents additional challenges since suppression of NO formation tends to cause an increase in particulates. Nevertheless, the following strategies are frequently used (Horrocks, op. cit.): (1) injection retard; (2) injection rate shaping; (3) intercooling of turbocharged engines; (4) centralized combustion for high-speed, direct injection; (5) modulated EGR; (6) electronic controls.

Exhaust Treatment Devices

Since diesel engines burn lean overall, the exhaust will always contain excess oxygen. Hence the reduction of NO_x with conventional three-way catalysts is not feasible. Currently, NO_x is removed from the diesel exhaust by either selective catalytic reduction (e.g., Feeley et al., SAE-paper 950747) or catalyzed thermal decomposition of NO into O_2 and N_2. In addition, lean NO_x catalysts are under development.

Diesel particulate traps have been developed which employ ceramic or metal filters (see Wade et al., SAE 830083, 1983). Thermal and catalytic regeneration can burn out the material stored. Particulate standards of 0.2 g/mi may necessitate such traps. Both fuel sulfur and aromatic content contribute strongly to particulates. Catalysts have been developed for diesels which are very effective in oxidizing the organic portion of the particulate (see Ball and Stack, SAE paper 902110, 1990).

Emission Regulations

Passenger Vehicles (Cars, Trucks, and SUVs) Federal regulations on automotive engine emissions have been enforced in the United States since the implementation of the Clean Air Act in 1970. The state of California has had its own regulations in place since before 1970 and has often applied more stringent standards on automobiles driven there. In 1990, the Clean Air Act Amendments gave authority to the EPA to enforce further restrictions on the allowable emissions from engine-drive automobiles. Additionally, standards were stiffened for heavy-duty on highway truck engines, off-road equipment engines, handheld engines, and marine engines.

The most recent U.S. EPA standards are collectively known as Tier II, and they have been fully implemented since 2009. Tier I standards were the first set of reductions enforced under the 1990 Clean Air Act Amendments. Under Tier I and older standards, allowable vehicle emissions were a function of both vehicle weight and engine (or fuel) type. Thus heavier vehicles were allowed to emit higher per-mile pollution due to the larger engine. Similarly, emission control technology was required to function for at least 50,000 mi. Under Tier II regulations, vehicles are no longer differentiated based on vehicle weight or engine type. Instead, a vehicle is certified to one of eight emission categories, or bins (see Table 7.6.22). Any vehicle is allowed to be certified in any one of the eight bins; however, a manufacturer's passenger vehicle fleet average must meet bin 5 standards. Additionally, Tier II requires emission control technology to function for 120,000 mi. Table 7.6.22 lists the standards associated with each bin after full implementation in 2009 that a vehicle must meet for 120,000 mi. Tighter standards are required for most emissions within the vehicle's first 50,000 mi. During the phase-in period from 2004 to 2009, additional transition bins (bins 9 and 10) as well as relaxed standards for certain vehicle classes in other bins were in place to smooth the implementation.

A vehicle is certified by measuring the emissions, with diluted bag samples, produced as it is driven through a predetermined drive cycle, and then calculating the average emission on a gram per mile basis. Traditionally, passenger vehicles only underwent the **Federal Test Procedure (FTP)** 75 drive cycle, which involves a 20-min test of accelerations, decelerations, and steady cruises on a chassis dynamometer. The vehicle is started after a 12-h soak in a 60 to 86°F (16 to 30°C) ambient temperature room. Top cycle speed is 56.7 mi/h (91.3 km/h). To reflect the assumed typical behavior of a U.S. driver, vehicles are now required to also pass **supplemental FTP (SFTP)** drive cycles, one (SC03) which includes more aggressive accelerations and decelerations to higher speeds and another (US06) which includes high-speed steady cruises reaching greater than 75 mi/h (120.7 km/h).

Heavy-Duty Engines Similar to passenger vehicles, medium and heavy-duty trucks must meet stringent emission standards. One major difference between passenger vehicle engines and heavy-duty truck engines is the burden of certification; passenger vehicles are certified by the vehicle manufacturer (on a gram per mile basis) whereas heavy-duty engines are certified by the engine manufacturer (on a gram per horsepower per hour basis). Similar to passenger vehicles, truck engines are "driven" through an FTP cycle on an engine dynamometer. Diluted emissions are recorded and normalized by the engine cycle's average

Table 7.6.22 U.S. Passenger Vehicle Emission Standards for Cars and Light Trucks (EPA Tier II Standards)*

Bin no.	NOMG,† g/mi	CO, g/mi	NO_x, g/mi	PM, g/mi	HCHO,‡ g/mi
8	0.125	4.2	0.20	0.02	0.018
7	0.090	4.2	0.15	0.02	0.018
6	0.090	4.2	0.10	0.01	0.018
5	0.090	4.2	0.07	0.01	0.018
4	0.070	2.1	0.04	0.01	0.011
3	0.055	2.1	0.03	0.01	0.011
2	0.010	2.1	0.02	0.01	0.004
1	0	0	0	0	0

*Standards apply to all vehicle classes, regardless of weight or engine type. Values shown are for 2009 full implementation and apply to the vehicle for 120,000 mi.

† NMOG = nonmethane organic gases, which include unburned hydrocarbons.

‡ HCHO = formaldehyde.

SOURCE: EPA (www.epa.gov).

horsepower. With the advent of 2004 regulations, truck engines must also certify to a steady-state emissions cycle. The engine is operated at several predetermined steady-state points, each having a different weighting factor. The weighted average of emissions from the several steady-state points must not exceed the same standard applied to the FTP cycle. Furthermore, the engine cannot emit more than 1.5 times the standard, a "not-to-exceed" specification, at any operating point.

Heavy-duty truck engines were first regulated in 1980, and the standards have since undergone several reductions. The current phase-in of emission reductions began in 2007, with full fleet compliance by 2010. Table 7.6.23 summarizes the standards that truck engines had to meet by 2010. The engines must comply with the listed standards for the full useful "life" as 435,000 mi.

Other Regulations In addition to the standards applied to road vehicles, such as passenger vehicles and on-highway trucks, emission regulations are now applied to off-road vehicles (such as heavy construction equipment), utility and handheld engines, outboard and personal watercraft engines, and large marine engines. Similarly, engine regulations are enforced in other countries or political alignments such as Japan, Argentina, Australia, Brazil, Canada, China, Hong Kong, India, Korea, Mexico, Russia, and the European Union. The standards vary considerably relative to the regulations imposed by the United States, which has the most stringent regulations.

Table 7.6.23 U.S. Heavy-Duty Truck Engine Regulations*

NO, g/(hp · h)	PM, g/(hp · h)	NMHC, g/(hp · h)	CO, g/(hp · h)
0.2	0.01	0.14	15.5

*Full compliance required by 2010. Full useful life of heavy-duty truck engines is 435,000 mi.
SOURCE: EPA (www.epa.gov) and DieselNet (www.dieselnet.com).

7.7 GAS TURBINES

by John H. Lewis and William H. Day

REFERENCES: Kerrebrock, "Aircraft Engines and Gas Turbines," MIT Press. Koff, Spanning the Globe with Jet Propulsion, 1991, AIAA paper 2987. "Turbine Engine Directory," Flight Int., June 8-14, 1994. Hill and Peterson, "Mechanics and Thermodynamics of Propulsion," Addison-Wesley. Bejan, "Advanced Engineering Thermodynamics," Wiley. Horlock, "Combined Cycle Powerplants," Pergamon. "Gas Turbine World Handbook," published annually, Pequot Publishing, Inc.

INTRODUCTION

The **gas turbine** is a turbine-type engine which operates with gas (usually air) as distinguished from steam or water as the working substance. In its most basic configuration, the gas-turbine engine consists of a compressor to draw in and compress the gas, a combustor or heat source to add energy to the compressed gas, and a turbine to extract power from the heated gas flow. Practical applications of gas turbines first occurred from 1939 to 1941. In 1939, the firm of Brown Boveri of Switzerland used a gas turbine to generate electricity. Also in 1939, the first flight of an aircraft powered by a gas turbine developed by Hans von Ohain took place in Germany. Another aircraft gas turbine was developed by Frank Whittle, who powered an aircraft in 1941 in England. From these early applications, the gas turbine has been developed to the point where today it is the most important aircraft power plant in use. In addition to its aviation applications, the gas turbine is a significant power plant for electric power generation, pipeline pumping power, and marine propulsion. It is a key component of combined-cycle power plants and cogeneration plants (see Fig. 7.7.18 and Sec. 7.4).

Depending upon its development roots and applications, the gas turbine is referred to by many names. In aviation, it is called a jet engine, a turbojet, a jet turbine, a turboprop, a turbofan, and a fanjet. In land and marine applications it is called a gas turbine, a turboshaft, a combustion turbine, and an industrial turbine. In the industrial world, engines generally fall into two categories. **Heavy-duty** or **frame machines** refer to gas turbines designed and built strictly for ground applications. **Aeroderivative machines**, as the name implies, are gas turbines originally designed for aircraft use and adapted for industrial applications. A typical aeroderivative is shown in Fig. 7.7.1. Future trends to marry heavy industrial and aerospace technologies will blur this distinction between the two types.

Gas turbines offer significant advantages compared with other types of engines because they are compact, lightweight, reliable, and efficient. They are capable of rapid start-up, follow transient loading well, and can be operated remotely or left unattended. They have a long service life, long service intervals, and low maintenance costs.

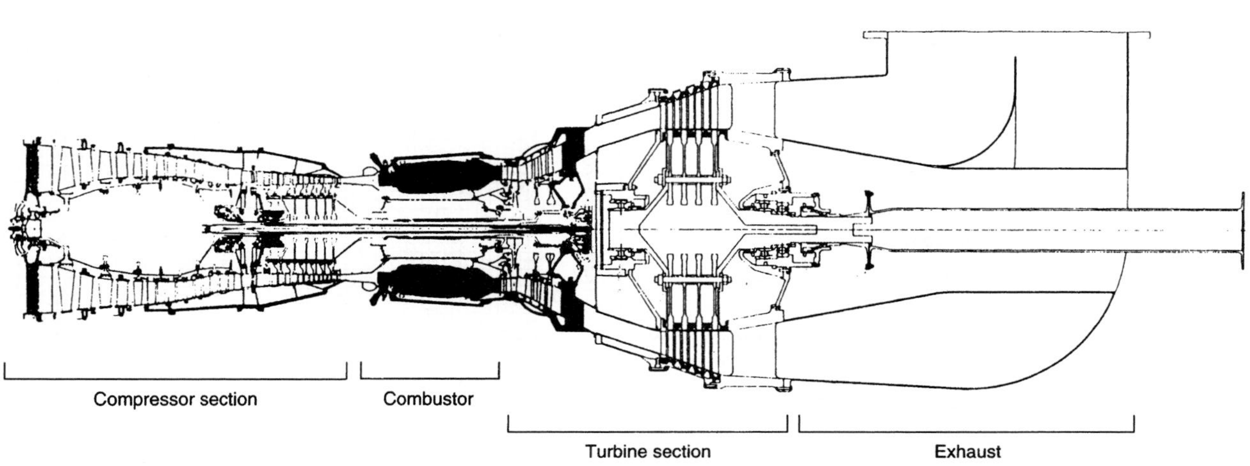

Figure 7.7.1 Cross section of Pratt and Whitney FT8 gas turbine derived from the JT8D commercial aircraft engine.

Cooling fluids are usually not required. Their compact size high power-to-weight ratio, and high reliability make gas turbines the ideal power plant for aircraft and for portable power plants. In industrial applications, the engine's small footprint allows relatively small foundations and enclosure buildings.

FUELS

Gas turbines use a wide variety of **gaseous and liquid fuels**. For ground and marine applications, natural gas and light distillate oils are most commonly used, but gasified coal and heavy residual oils can also be used. When fuels that contain significant amounts of sulfur, salt, vanadium, and other metals are used, the hot-section components such as the combustor and turbine must be designed to withstand the corrosive effects of these impurities. For aircraft applications, specific kerosine-type fuels have been developed that control important qualities of the fuel, to obtain maximum performance.

THERMODYNAMIC CYCLE BASIS

The gas turbine operates on the principle of the **Brayton** (or **Joule**) **cycle**. The working substance, usually ambient air, is compressed adiabatically to high pressure where heat is added at constant pressure to elevate the temperature, at which point it is expanded, also adiabatically, back to the original pressure. Because of the expansion from a higher temperature, the work extracted in the expansion process exceeds the work required for compression by an amount that equals the net output work of the cycle. The cycle is completed at the original pressure with heat rejection in the exhaust. See Fig. 7.7.2. At the end of the expansion process, the working substance temperature is higher than its initial temperature by an amount that corresponds to the waste energy to be dissipated at constant pressure to the surroundings. Unique to the Bray-ton cycle are the heat addition and heat rejection processes at (ideally) constant pressure. The constant-pressure process lends itself to continuous flow of the working substance. When coupled with compressors and turbines which operate on the rotating turbomachinery

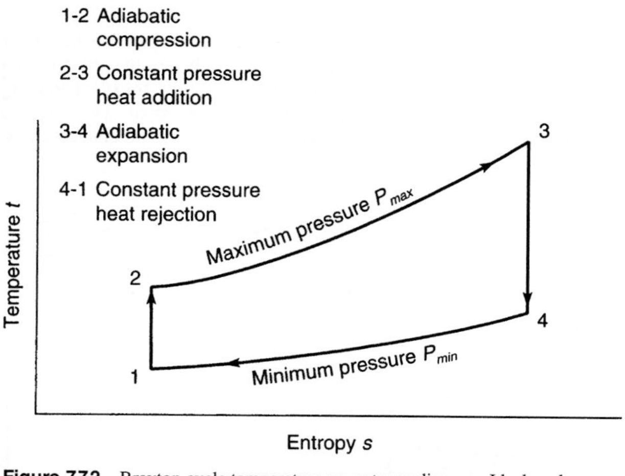

Figure 7.7.2 Brayton cycle temperature vs. entropy diagram. Ideal cycle.

principle (see Sec. 10), the entire gas turbine acts as a **continuous flow engine**. The continuously flowing working substance (air) results in very high power density. The gas turbine's intrinsic compactness favors its use over other engine types in many power generation applications.

Temperature-entropy diagrams describe the Brayton cycle. Figure 7.7.2 shows the ideal case where all component efficiencies are 100 percent. Figure 7.7.3 shows the real cycle with nonideal components,

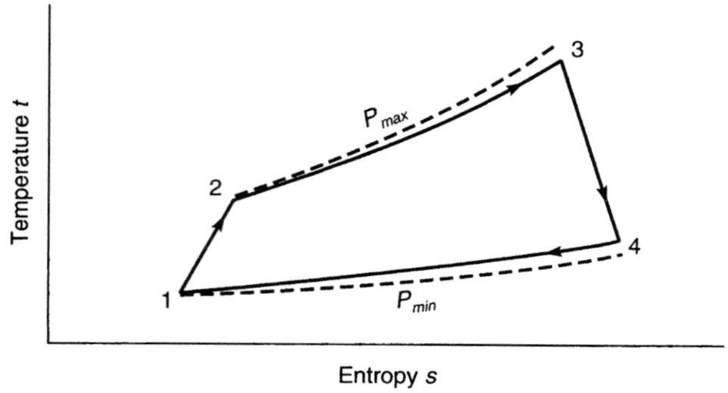

Figure 7.7.3 Brayton cycle temperature vs. entropy diagram. Real cycle reflects component inefficiencies and pressure losses.

including friction losses in the ducting between the compressor and turbine.

The **ideal Brayton cycle efficiency** η is related to the cycle pressure ratio by the following expression, where P_{max}/P_{min} = pressure ratio, R = gas constant, and c_p (the latter being specific heat at constant pressure properties of the working substance):

$$\eta \text{ Brayton, ideal} = 1 - 1/(P_{max}/P_{min})^{R/C_p} \tag{7.7.1}$$

Thus with ideal components, the gas-turbine efficiency will be increasingly high as its characteristics maximum/minimum pressure ratio continues to increase approaching 100 percent when the pressure ratio becomes infinitely large. With real components, however, when the pressure ratio increases, the component inefficiencies become larger contributors to exhaust waste heat. The net result is that efficiency reaches a peak at some level and then falls off. The point where this peaking occurs is directly related to the quantity of cycle heat addition, which, in turn, is related to the maximum cycle temperature. As illustrated in Fig. 7.7.4, the peak shifts to a higher efficiency level at high pressure ratio as the turbine temperature increases. At the same time, the specific power output has a similar trend of increasing up to a peak and then falling off as the pressure ratio increases, while the amount of specific output power increases continuously with heat addition as determined by the maximum cycle temperature. These trends are illustrated in Fig. 7.7.5.

Among manufacturers and users, the **maximum cycle temperature** is variously referred to as turbine inlet temperature, combustor exit

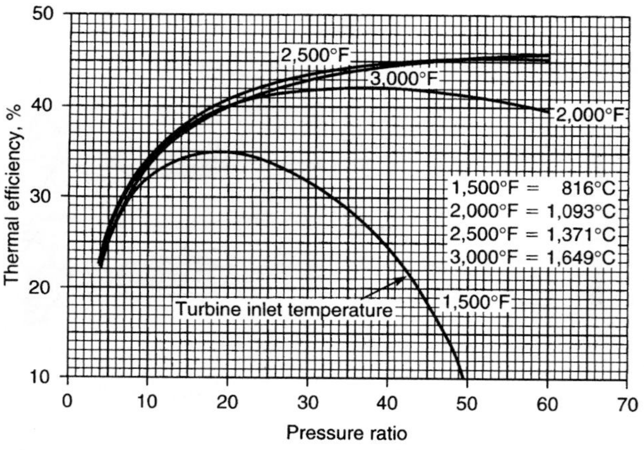

Figure 7.7.4 Representative Brayton cycle variation of thermal efficiency with pressure ratio at different turbine inlet temperatures.

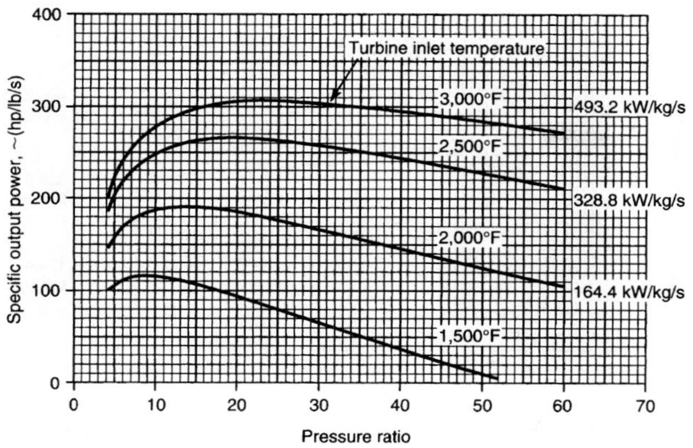

1,500°F = 816°C
2,000°F = 1,093°C
2,500°F = 1,371°C
3,000°F = 1,649°C

Figure 7.7.5 Representative Brayton cycle variation of specific output power with different pressure ratios and turbine inlet temperatures.

temperature, turbine rotor inlet temperature, or firing temperature. Precise definition depends on the design details for a specific engine.

Specific power refers to output per unit mass flow of the working substance, expressed in hp/(lb · s) or its electrical equivalent kW/(kg · s). As shown in Fig. 7.7.5, a given cycle fixes specific power. When specific power is constant, absolute power then varies directly with the mass flow rate of the working substance. The mass flow quantity, in turn, sets the overall size of the machine

$$\begin{aligned} \text{Power} &= \text{specific power} \times \text{mass flow rate} \\ &= \text{hp/(lb·s)} \times \text{(lb/s)} \\ &= \text{kW/(kg·s)} \times \text{(kg/s)} \end{aligned}$$

Generally, the gas turbine's working substance is atmospheric air which is continuously drawn into the front of the compressor through inlet ducting and is expelled back to atmosphere through exhaust ducting. In this case, the minimum cycle pressure is identical to the surrounding ambient pressure while the initial working substance temperature is the ambient temperature of the surrounding atmosphere. The schematic in Fig. 7.7.6 shows the typical component arrangement of a continuous flow simple-cycle gas turbine. In this typical arrangement, the heat addition section consists of an internal combustor where fuel is added directly into the through-flowing air and is continuously burned. The rotating compressor is attached to the turbine by a shaft which transmits the power needed for compression. The excess power available from expansion through the turbine at high temperature is fed by a rotating power output shaft to an external load.

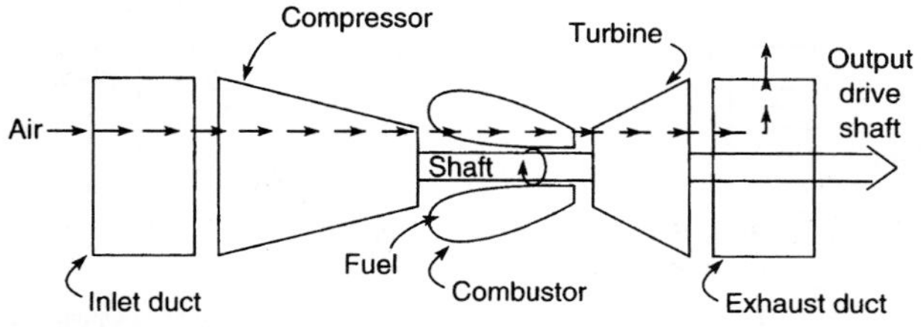

Figure 7.7.6 Simple Brayton cycle gas-turbine component arrangement.

BRAYTON CYCLE VARIATIONS

The following paragraphs describe variations to the Brayton cycle. In some cases these variations are represented by practical machines that have been or are now seeing commercial use. The combined cycle (Fig. 7.7.18) has come to be most widely used in the electric power generation market. There are three main reasons why combined-cycle (CC) plants have become so popular: (1) The gas-turbine manufacturer need not make any special modifications to the basic simple cycle to adapt it to the CC application; (2) combined cycles are by far the most efficient of all the various possible variants; and (3) as the temperature of the gas-turbine exhaust gas reduces, while it passes through and gives up heat to the heat recovery steam generator, a **selective catalytic reactor (SCR)** can be placed at an appropriate location in the boiler system. The SCR system is designed to reduce NO_x emissions to extremely low levels. As a consequence, combined-cycle plants can gain local permits to operate in environmentally sensitive areas.

Regenerative Cycle (Fig. 7.7.7). In this cycle, air exiting the compressor is directed through a heat exchanger or **regenerator** before being introduced to the combustion section. Then the heated air exhausting from the turbine passes through the other side of the regenerator. By this means, some of the exhaust heat is recovered and used as an additional source of energy to heat the high-pressure air,

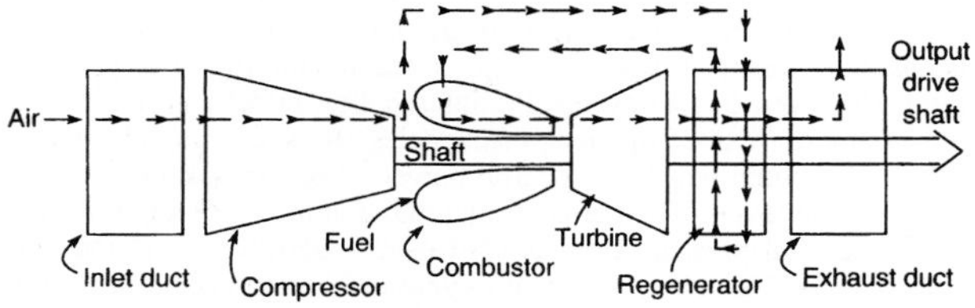

Figure 7.7.7 Regenerative Brayton cycle gas-turbine component arrangement.

instead of being dissipated as waste heat in the low-pressure exhaust. The effect of regeneration is to increase cycle efficiency, but it is only effective at low levels of pressure ratio. This is because as the pressure ratio increases, the compressor exit air temperature increases and the turbine exhaust temperature is lowered, thereby reducing the regenerator temperature difference necessary for heat transfer. At some point, depending on the maximum cycle temperature, the compressor exit and exhaust temperatures become equal and there can be no further regenerative heat recovery. Figure 7.7.8 compares regenerative cycle trends with those of the simple cycle.

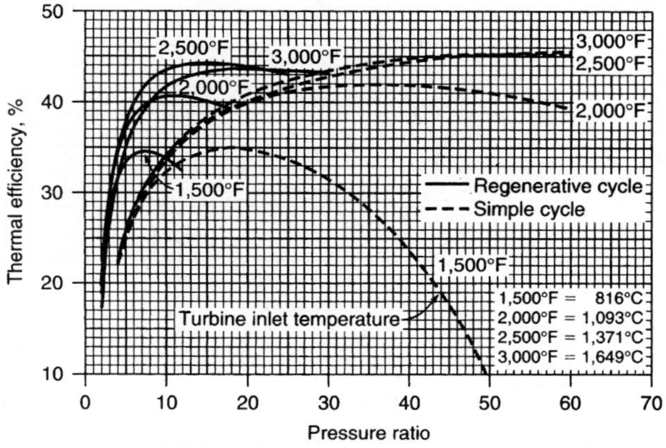

Figure 7.7.8 Regenerative Brayton cycle variation of thermal efficiency with pressure ratio at different turbine inlet temperatures.

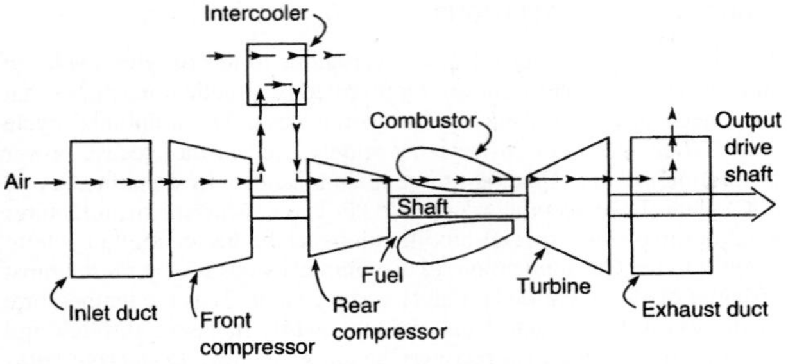

Figure 7.7.9 Intercooled Brayton cycle gas-turbine component arrangement.

Intercooled Cycle (Fig. 7.7.9) In this cycle, the compression process is split between two compressors. An **intercooling heat exchanger** is placed between the front and rear compressors to reduce the temperature of the working substance. This cycle takes advantage of the thermodynamic principle that the work of compression is directly proportional to the level of the entering temperature. With a lowered inlet temperature, the work input needed to drive the rear compressor is reduced for the same pressure ratio. The reduced compressor work is directly available as increased net cycle output. This is true for both ideal and real cycles. In the ideal cycle, because the intercooler is a source of added heat rejection, the intercooling cycle efficiency is always lower than that of a simple cycle having the same pressure ratio. However, in real cycles, the intercooling effect can help offset the effect of real component efficiencies. In some circumstances, intercooling improves both specific power output and efficiency over simple cycle levels. Theoretically, the maximum power output occurs when the total pressure ratio is split evenly between front and rear compressors. However, maximum cycle efficiency occurs at a split which depends on the turbine inlet temperature as well as the efficiency levels of compressors and turbines. Typically, peak efficiency occurs when intercooling takes place well in front of the midpoint of the compression process. Figures 7.7.10 and 7.7.11 compare intercooled thermal efficiency and specific power output trends for pressure-ratio splits optimized for power output and for pressure-ratio splits optimized for efficiency.

Reheat Cycle (Fig. 7.7.12) In this cycle, the turbine expansion process is interrupted at some intermediate point before it reaches

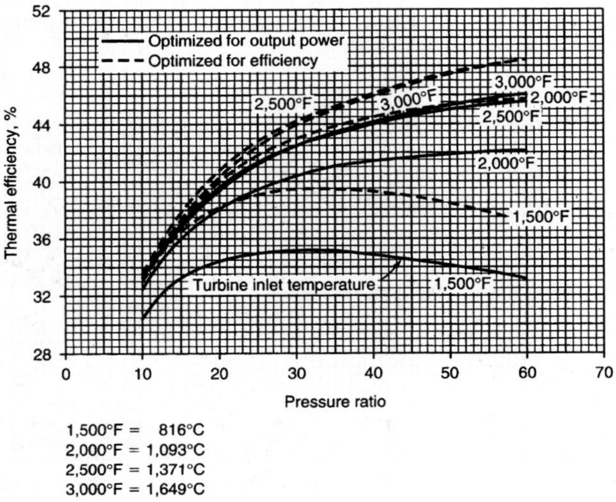

Figure 7.7.10 Intercooled Brayton cycle variation of thermal efficiency with pressure ratios at different turbine inlet temperatures, optimized for output power and for efficiency.

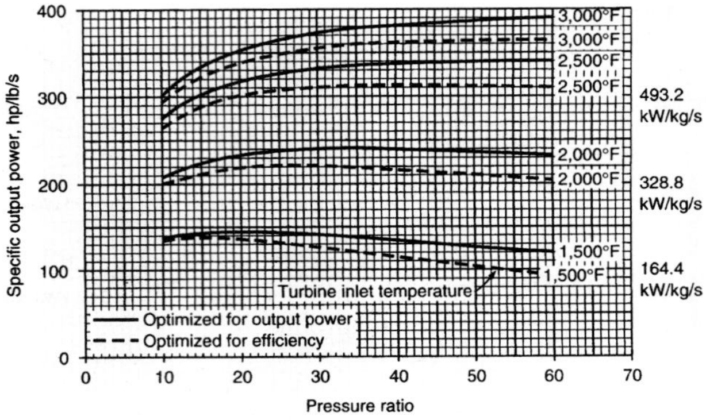

Figure 7.7.11 Intercooled Brayton cycle variation of specific output power with pressure ratio at different turbine inlet temperatures, optimized for output power and for efficiency.

minimum (or ambient) pressure in the exhaust. At this point, additional heat is added to the working substance by combustion at constant pressure in a second combustor. This takes advantage of the basic thermodynamic principle that the work of any expansion process increases in direct proportion to the temperature entering the turbine. (This is the converse of the principle that applies to compression, as cited above for the intercooled process.) The elevated temperature entering the second turbine increases its work output; this additional work appears directly as greater power on the output shaft. Since the extra work is more than

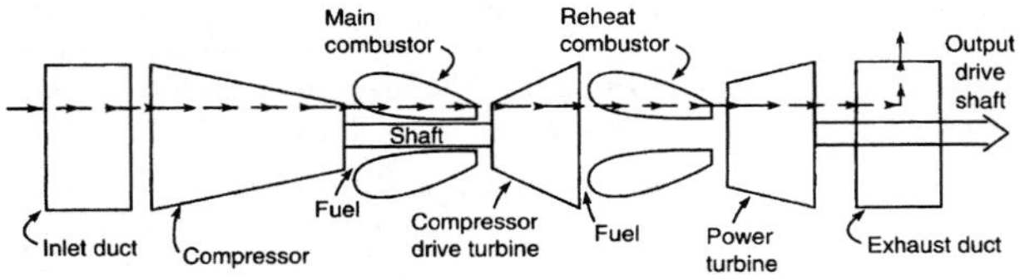

Figure 7.7.12 Reheat Brayton cycle gas-turbine component arrangement.

offset by the added heat energy of the second combustor, the result is also reduced cycle efficiency.

Associated with the **reheat cycle** is an increase in exhaust temperature which ordinarily represents higher exhaust waste heat. However, in the case of a **combined cycle**, the higher exhaust temperature is a source of waste heat recovery that can be used to generate steam. The higher exhaust temperature can improve cycle efficiency of the steam system to the extent that the **overall system efficiency** can be improved. Reheat thus lends itself to use with exhaust waste heat recovery systems.

Intercooled and Regenerative Cycle This cycle (Fig. 7.7.13) has both an intercooler in the middle of the compression process and a

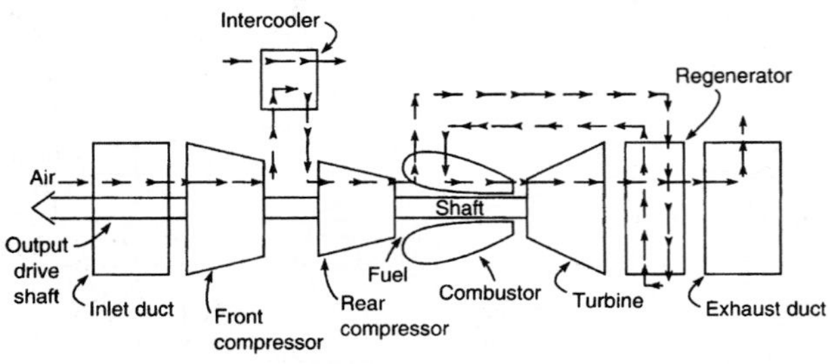

Figure 7.7.13 Intercooled regenerative Brayton cycle gas-turbine component arrangement.

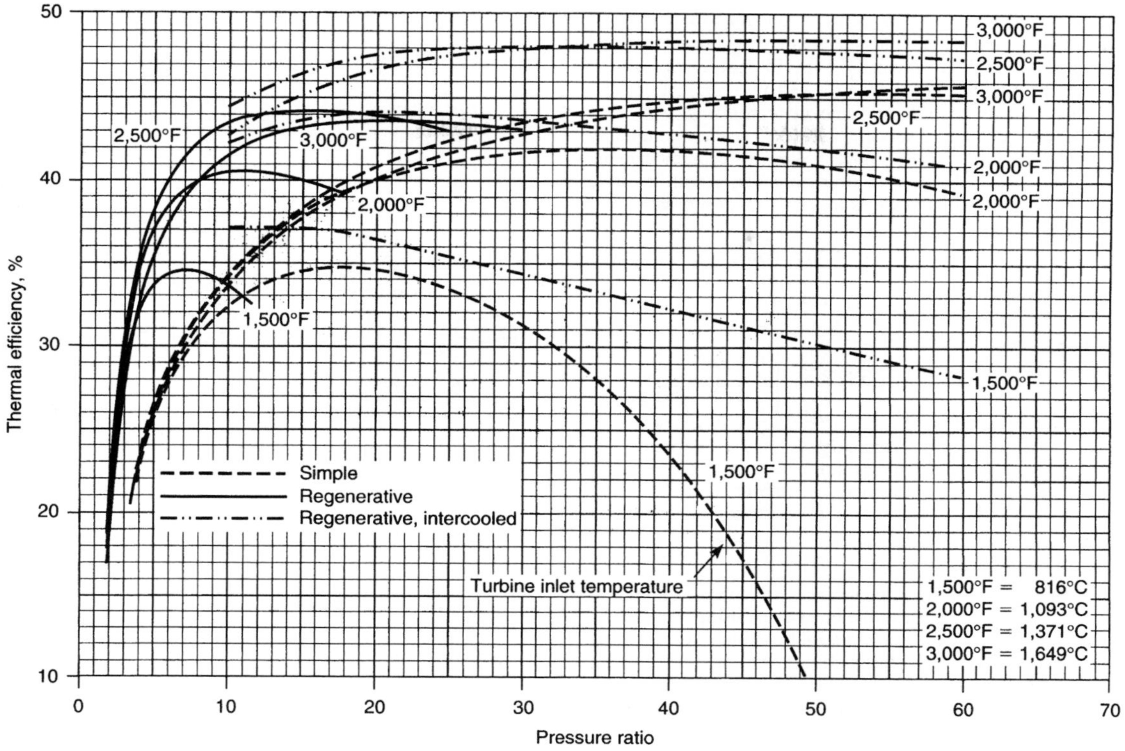

Figure 7.7.14 Variation of Brayton cycle thermal efficiency with pressure ratio and at different turbine inlet temperatures, for regenerative cycle and for simple cycle.

regenerator to capture exhaust waste heat. The intercooling results in a lower temperature leaving the rear compressor. This circumvents the temperature difference limitation reached in regeneration. That is, at increasingly high pressure-ratio levels, the turbine exhaust temperature remains sufficiently higher than the compressor exit temperature to sustain heat exchange and thus waste heat recovery. Overall, the cycle efficiency of the intercooled regenerative cycle is superior to that of the simple Brayton cycle and all its common variants over a wide range of cycle pressure-ratio levels. Figure 7.7.14 shows variation of efficiency for an intercooled regenerative cycle compared to simple and pure regenerative cycles.

In actual practice, the choice of a simple cycle or one of its variants depends on overall economics applied to a specific use. The extra cycle benefits must trade favorably against the additional cost and complexity of the added heat exchangers or combustors and the penalties of the added ducting in between.

CONFIGURATION VARIATIONS

Since the gas turbine is made up of rotating compressors and turbines connected by driveshafts, there are various ways to arrange these elements. The simplest, Fig. 7.7.15, is the single shaft or "spool" arrangement. In this case, the compressor, turbine, and output driveshaft all turn at the same rotational speed.

Free Turbine (Fig. 7.7.16) To provide more flexibility by allowing the load to turn independently at a speed different from that of the compressor and its drive turbine, a **free turbine** configuration connects the compressor and its turbine on one driveshaft, while hot air continues to expand through a separate turbine which delivers net output power. The free turbine finds use either in electric generator power applications, where the generator must operate at a fixed speed (either 3,600 r/min or 3,000 r/min for 60- or 50-Hz installations, respectively), or in pumping applications,

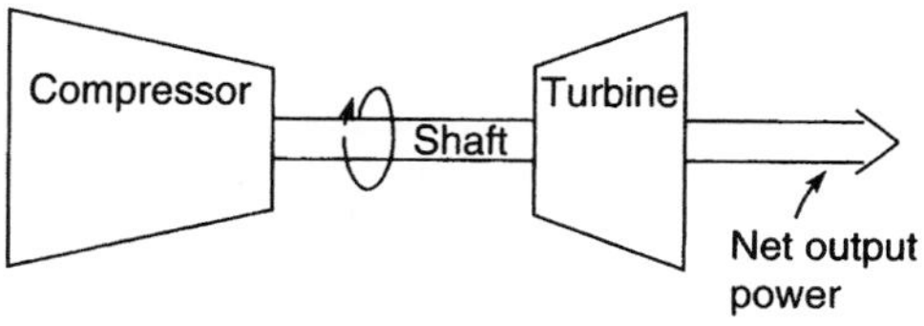

Figure 7.7.15 Single-spool gas-turbine component arrangement.

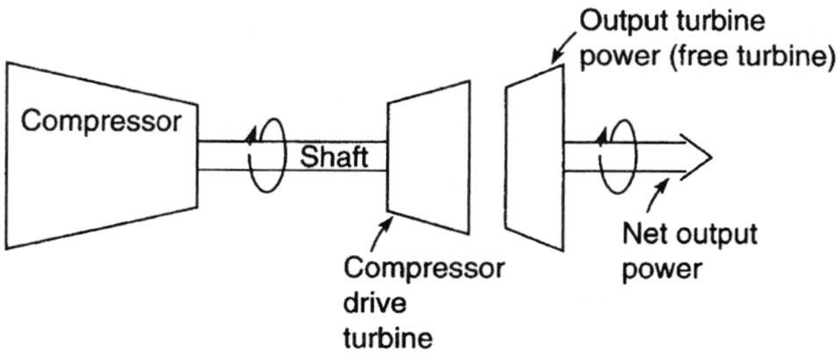

Figure 7.7.16 Free turbine gas-turbine component arrangement.

where speed can vary. In either case, the compressor at partial load can operate at its most favorable speed independent of the driveshaft load.

Dual Spool (Fig. 7.7.17) At very high compressor pressure ratios, the part load operation of a compressor on a single spool tends to become unstable. When a compressor designed for high pressure ratio is forced to operate at low pressure ratio (as load is reduced), the front and rear portions of the single-spool compressor become aerodynamically unbalanced. This is alleviated by dividing the compression process into two or more spools on separate shafts driven by separate turbines. With this arrangement, the front compressor can slow down more rapidly than the one behind, thus maintaining a more stable overall aerodynamic balance. Conventionally, to avoid unnecessary turning and ducting of the air, the spools are arranged in a straight-through alignment. This is made possible by **concentric shafting**. The shaft connecting the second compressor and its drive turbine (commonly known as the *high-pressure spool*) is made hollow to accommodate the driveshaft which connects the first or front compressor with its drive turbine at the rear (commonly known as the *low-pressure spool*).

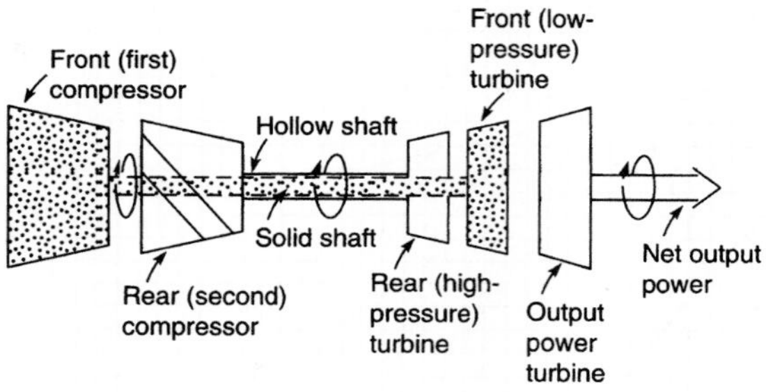

Figure 7.7.17 Dual-spool gas-turbine component arrangement.

Closed vs. Open System In circumstances where heat addition by internal combustion is not practical, the open cycle using atmospheric air as the working substance may be replaced by another working substance continuously flowing in a **closed circuit**. An example of such a power plant would be the helium-cooled nuclear reactor. As the name implies, helium under high pressure replaces air. With the working substance completely contained, the minimum cycle pressure can be made much higher than ambient. Even such a low-molecular-weight gas as helium can be compressed into a low volume and thus keep the overall power plant relatively compact. Helium lends itself to the nuclear reactor application by virtue of its being almost completely chemically inert, which prevents radioactive transport.

External Combustion The gas turbine can maintain its continuous flow feature without internal combustion; such is the case with the direct use of coal as a fuel. Early attempts to use coal directly in internally fired gas turbines always resulted in severe and rapid damage to the turbine materials. Associated with the coal burning are not only the erosive particulates but also highly corrosive impurities such as sulfur. As an alternative, gas-turbine power plants using indirect firing of coal were also considered, and at least one was built in Germany. It, too, was a failure due to the corrosive attack on the heat exchanger, largely by the alkali metals in the coal. There is now no significant development taking place on external combustion with coal. The alternative for use of coal in gas turbines now receiving considerable attention is coal gasification. See the description in the "Applications" section of this chapter and in Sec. 5.2.

WASTE HEAT RECOVERY SYSTEMS

A characteristic of gas-turbine engines is the incentive to operate at as high a "firing" (turbine inlet) temperature as the prevailing technology will allow. This incentive comes from the direct benefit to both specific output power and cycle efficiency, as illustrated in Figs. 7.7.4 and 7.7.5. Associated with the high maximum temperature is a high exhaust temperature which, if not utilized, represents waste heat dissipated to the atmosphere. Schemes to capture this high-temperature waste heat are prevalent in industrial applications of the gas turbine.

Combined Cycle This term commonly refers to the scheme that places a boiler and superheater directly behind the gas turbine to generate steam and produce additional power in a steam turbine (Rankine cycle). A simplified version of a combined cycle is shown in Fig. 7.7.18. Figure 7.7.19 illustrates the combined-cycle process on a temperature vs. entropy diagram. The steam portion is sometimes called the **bottoming** cycle. Early combined cycles had **fired** as well as **unfired** versions. The supplementary firing was required because of

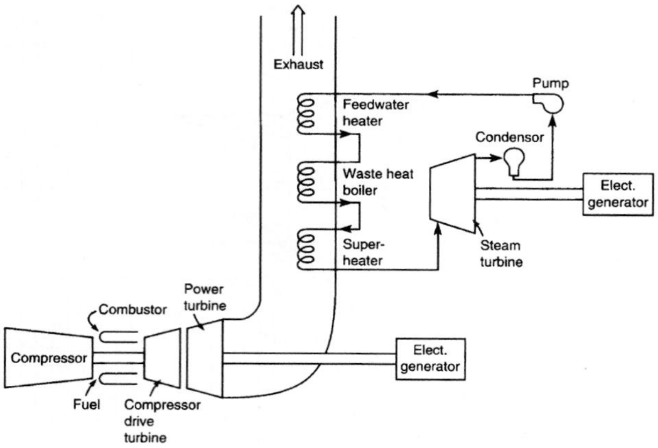

Figure 7.7.18 Schematic of combined-cycle heat recovery system.

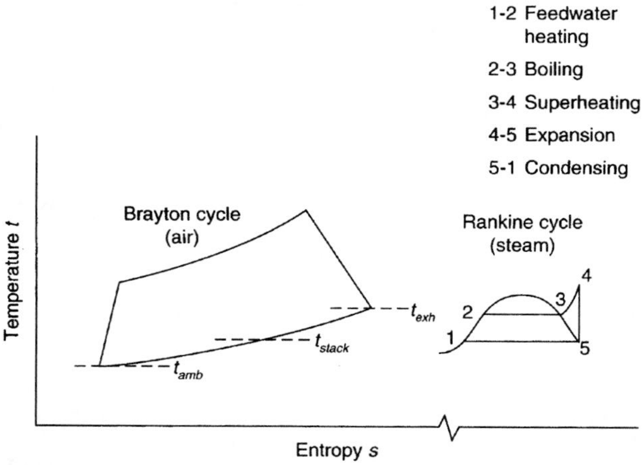

Figure 7.7.19 Combined-cycle temperature vs. entropy diagram.

the high sensitivity of the Rankine cycle to maximum superheat temperature, as shown in Fig. 7.7.20. When the gas-turbine exhaust temperature is too low to sustain a good steam cycle efficiency, the extra combustion in the gas-turbine exhaust could result in an overall system efficiency improvement. Modern gas turbines, with their increasingly high turbine inlet and exhaust temperatures, can approach or exceed 60 percent in **overall combined-cycle plant efficiency** without supplementary firing.

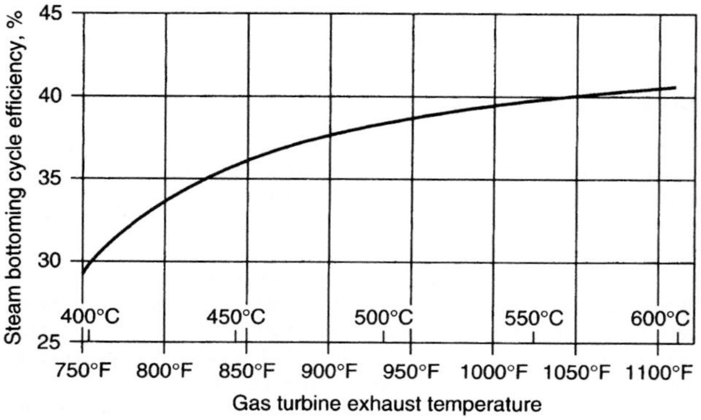

Figure 7.7.20 Variation of typical steam system heat recovery with turbine exhaust temperature.

If η_{gt} is the gas-turbine (Brayton) cycle efficiency and η_{st} is the steam bottoming cycle (Rankine) efficiency, then the overall combined-cycle plant efficiency η_{cc} is:

$$\eta_{cc} = \eta_{gt} + (1 - \eta_{gt})\eta_{st}\left(\frac{t_{exh} - t_{stack}}{t_{exh} - t_{amb}}\right) \tag{7.7.2}$$

where t_{stack} is the minimum practical exhaust temperature to be released into the atmosphere after heat recovery, t_{exh} is the temperature of the gas turbine exhaust gases leaving the turbine, and t_{amb} is the outside ambient temperature. (Typically, stack exhaust temperatures are limited to 200 to 300°F (93 to 150°C) to avoid condensation of the potentially corrosive combustion products.)

Steam Injection (Fig. 7.7.21) As a variation of the conventional combined cycle described above, some gas-turbine installations generate steam by exhaust waste heat recovery and, instead of using that steam in a separate steam turbine, inject the steam directly back into the turbine section of the gas turbine. In this system, the turbine acts not just on air (plus combustion products) as the working substance, but instead on a mixture of air products and steam. Such a scheme eliminates the need for additional equipment and is thus less expensive than the normal combined cycle. This system has a drawback in that direct steam injection consumes water at the plant site and thus raises environmental concerns.

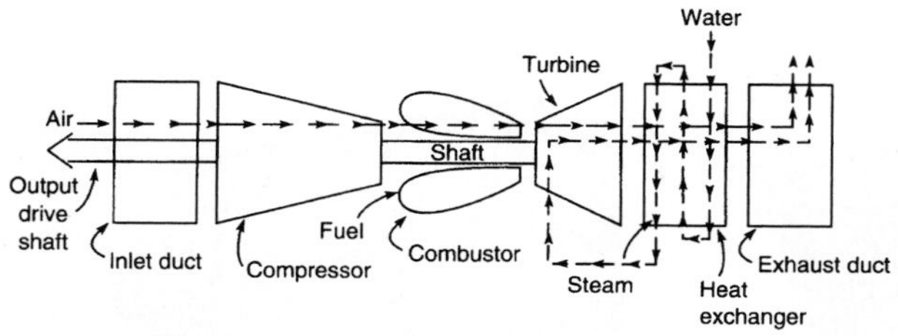

Figure 7.7.21 Arrangement of components for gas turbine with steam injection.

Cogeneration There are many power plants used to provide the total energy needs of industries which use process heat. In these instances the exhaust heat of the gas turbine provides a source of high energy that can be used directly in the plant process rather than for producing more power. This is particularly true in the chemical industry, where many operational **total energy plants** can be found. In addition, some of or all the exhaust waste heat can be used for **district heating** where, especially in colder climates, steam or warm water generated from the gas-turbine exhaust waste heat can be circulated around the community to heat homes and buildings. Use of gas-turbine exhaust for heating is especially suited to provide the total energy needs of moderate-size to large institutions such as schools, hospitals, or commercial complexes like shopping malls.

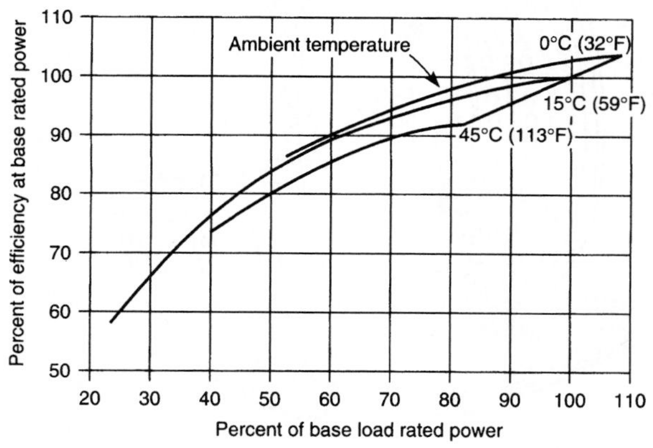

Figure 7.7.22 Typical partial-load performance characteristics for fixed output speed.

OPERATING CHARACTERISTICS

Partial Load Gas turbines respond rapidly to throttle variations. They are capable of accelerating from idle to full power in minutes or, in the case of the lightweight flight engine designs, seconds. They can be started in a matter of minutes and, via automatic controls, can operate for long periods of time unattended. Partial-load fuel efficiency remains high. Typical partial-load performance characteristics are shown in Fig. 7.7.22 for fixed output rotating speed (r/min) and in Fig. 7.7.23 for a variable-speed (free-turbine) drive.

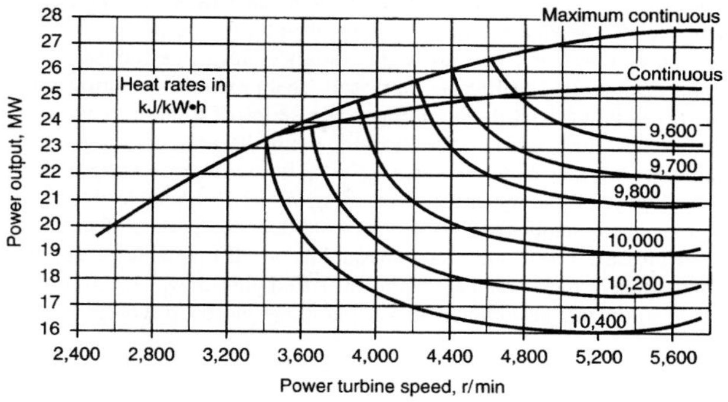

Figure 7.7.23 Typical partial-load performance characteristics for variable output speed in a free-turbine drive.

Ambient Conditions Gas-turbine operation is sensitive to ambient conditions. Being directly dependent on mass flow for a fixed cycle, power output varies directly with air density and thus ambient pressure. Variations in ambient temperature affect the thermodynamic cycle and engine operating characteristics. Normally, any particular gas-turbine model will have an associated **power rating curve** which varies as shown in Fig. 7.7.24.

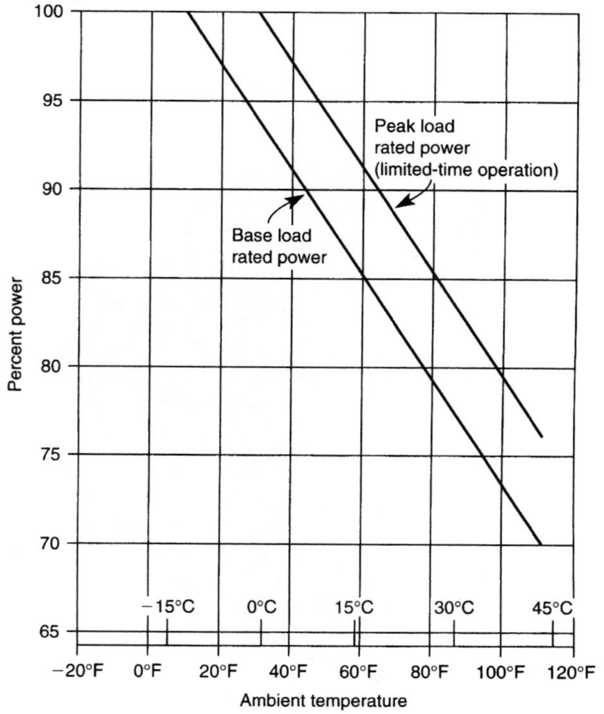

Figure 7.7.24 Typical variation of output power with ambient temperature.

GAS-TURBINE COMPONENTS

The primary components of a gas turbine consist of a **compressor**, a **combustor** and a **turbine.** Other elements of a gas turbine, depending on the installation, are the electronic control, inlet to the compressor, exhaust nozzle, power turbine, regenerators, and intercoolers. The component that poses the greatest aerodynamic and thermodynamic design challenge, as well as having the greatest influence on engine performance, is the compressor.

Compressors Compressors can be centrifugal flow, axial flow, or a combination centrifugal and axial flow. In a **centrifugal flow compressor**, the airflow enters at the hub and is then turned from axial to radial by the compressor rotor. Early gas turbines and some modern small engines employ centrifugal compressors. In an **axial flow compressor**, air flows in an axial direction through a series of rotating blades and stationary vanes which are concentric with the axis of rotation. All large gas turbines employ multistage axial flow compressors because of their higher efficiency and capacity. Figure 7.7.25 illustrates the rotor, stators, and assembly of a multistage axial flow compressor. To achieve high pressure ratios and maintain stability at all operating conditions, **dual-spool** or three-spool arrangements are employed, which permit each spool to operate at optimum speed. Interstage bleeds, discharge bleeds, and variable-angle vanes are also employed to avoid surge. Compression ratios in any one spool can exceed 20:1 with spools of 10 or more stages, and polytropic efficiency can be as high as 92 percent.

Combustors A number of arrangements are used for the **combustion** chamber including side-by-side can annular combustors or a single annular combustion chamber (Fig. 7.7.26). Fuel is introduced through a manifold and an arrangement of multiple fuel nozzles. The **combustor** must maintain high combustion efficiency, and low levels of pollutant emissions and smoke, have a minimum pressure loss, maintain stable combustion over a wide range of operating conditions, and produce a uniform temperature in the hot gases distributed to the turbine. The combustor must be designed to utilize some of the airflow to atomize the fuel to achieve efficient combustion, generate turbulent mixing needed to produce a uniform exit temperature, and cool the metal parts of the combustor for durability at high operating temperature. Burner efficiency usually exceeds 99 percent except at low power in the idle region.

Turbines **The turbine** consists of one or more stages of blades and interstage guide vanes located immediately to the rear of the combustor section. Figure 7.7.27 illustrates the rotors or wheels, sets of stationary vanes or nozzles, and their assembly. The turbine extracts kinetic energy from the heated gas stream and converts it to shaft power to drive the

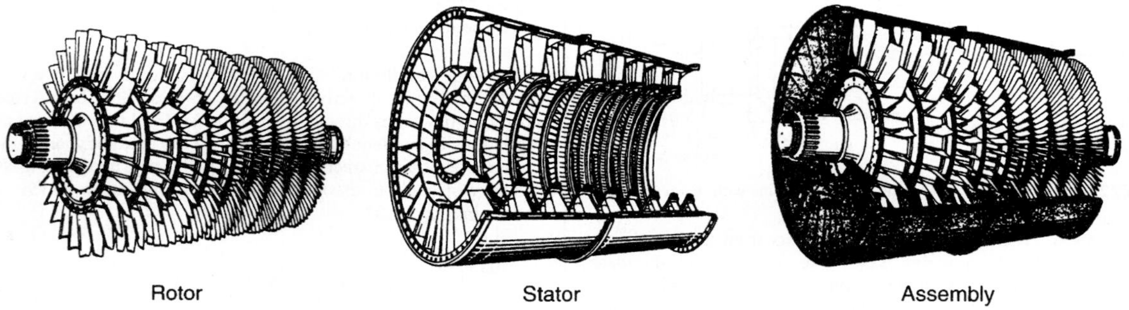

Figure 7.7.25 Components and assembly of axial flow compressor.

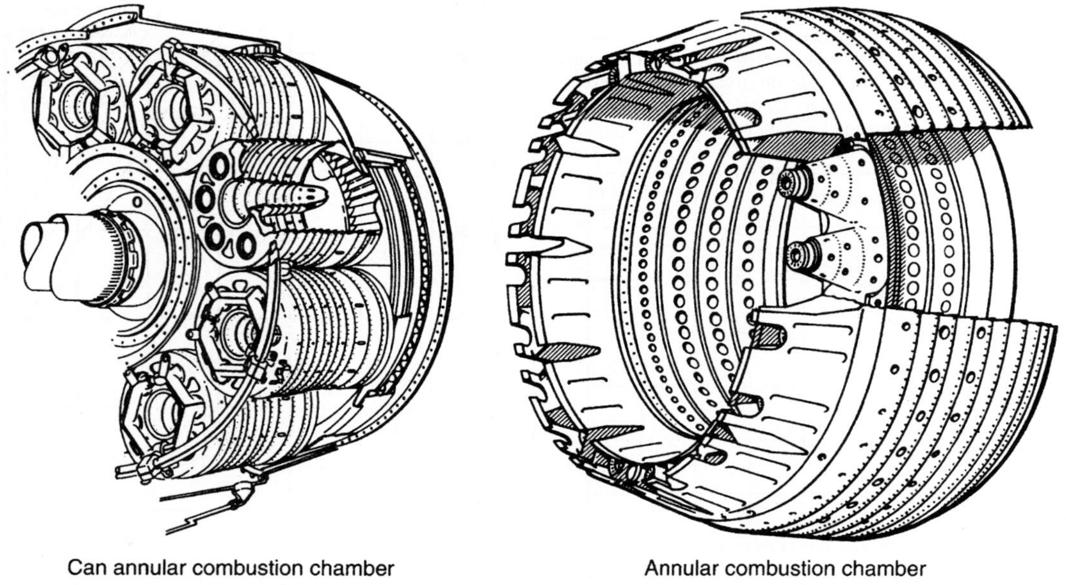

Figure 7.7.26 Combustion chamber (combustor) configurations.

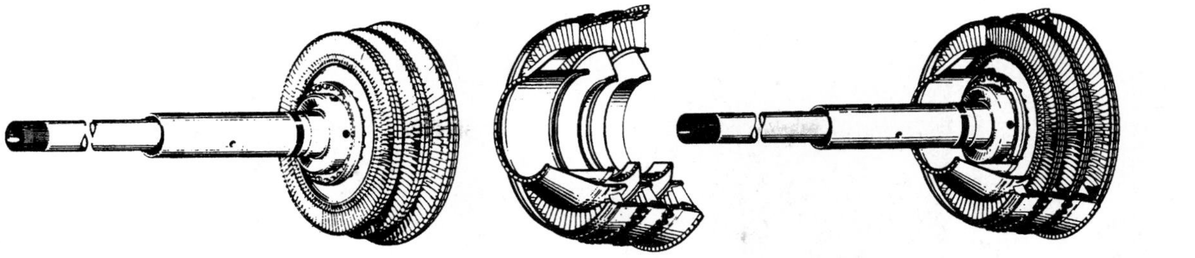

Figure 7.7.27 Turbine elements and assembly.

compressors and accessories. In the case of a turboprop, turbofan, or turboshaft, power is also extracted to drive a propeller, a fan, or an output shaft (see Sec. 8.5). The arrangement of the turbine is similar to that of the compressor with two major differences. First, since the expansion process is aerodynamically easier than the compression process, turbines have fewer stages than compressors. Second, the gases are much hotter and, therefore, require materials and cooling schemes to maintain the life of the turbine parts. High-temperature alloys, internal cooling passages (for air usually supplied from the compressor), and ceramic thermal barrier coatings are used in the turbine. The normal range of turbine efficiency is 90 to 95 percent.

APPLICATIONS

A worldwide directory of currently available **gas-turbine aircraft engines** published in *Flight International* (June 8-14, 1994) lists 37 manufacturers and 166 models. A much larger number of companies are manufacturers or suppliers of parts or components to the original equipment manufacturers. More than 700 model designations are now or have been in service.

Military Aviation Military aircrafts, including fixed-wing combat and transport as well as helicopters, are almost exclusively gas-turbine powered. Fighter aircraft are powered with afterburning turbofan engines. Modern heavy-lift cargo and troop transports are powered by high-bypass-ratio turbofans often derived from commercial engines. Many older cargo, troop, tanker, and utility aircraft still in service are powered by turboprop engines. Most helicopters are powered by turboshaft engines. New developments in military engine technology aim at increasing overall performance by increasing the power of the core engine components through use of higher-temperature materials and turbine technology. The military strives to achieve a balanced tradeoff of performance, weight, and stealth against reliability, maintainability, and cost.

Commercial Aviation **Commercial aircraft** are primarily powered by turbofan and turboprop engines. Design of large commercial wide-body transports tends toward long-range twin-engine turbofan power, but three- and four-engine aircraft are still being manufactured. With twin engine aircraft approaching the size and range of four-engine aircraft, thrust requirements are ever increasing. Turbofan engines in the 60,000- to 90,000-lb (267- to 400-kN) thrust range have been certified, and versions to 100,000 lb (444.8 kN) are planned. Medium-size commercial transports are also turbofan-powered, with engines in the 20,000- to 40,000-lb (89- to 178-kN) thrust range. Small regional or commuter aircraft are primarily powered by turboprop engines from 500 to 3,000 hp (372 to 2,237 kW). Commercial engine technology is aimed at higher bypass ratios for increased thrust and reduced emissions and noise; higher component efficiency for better fuel consumption; and increased reliability. Future developments include advanced ducted propulsors (ADPs) which couple a gear-driven prop fan to the gas turbine, geared fan engines, and engines for the second generation of supersonic aircraft.

For **industrial gas turbines** above 500 kW, the "2014-2015 Gas Turbine World Handbook" lists 85 gas-turbine models on the market, from 18 original equipment manufacturers (OEMs). In addition there are a number of companies that manufacture or package gas turbines under license from the OEMs. Frame-type gas turbines account for more than three-fourths of the gas turbines shipped annually, measured in megawatt capacity of the shipments. Aeroderivatives, from four OEMs, are limited in power output to a maximum of about 60 MW due to the size of the aircraft engines from which they are derived, while frame-type gas turbines are available from less than 1 MW to over 300 MW. This limits the applicability of the aeroderivatives from small to medium size power plants for electric power generation, while much of the power generation market is in large combined-cycle power plants. Figures 7.7.28 and 7.7.29 show cutaway views of an aeroderivative engine (see also Figure 7.7.1) and a large, heavy-duty industrial design respectively.

Electric Power Of the three types of industrial gas-turbine applications—electric power generation, mechanical drive, and marine propulsion—electric power generation typically amounts to about 90 percent of the annual shipments of new capacity, as measured in megawatts. Gas turbines, including combined cycles, are now the fastest-growing type of power plant in the world. According to the Energy Information Agency, part of the U.S. Department of Energy, in 2012 natural gas fuel (virtually all used in gas turbines), accounted for 22 percent of the world's electric power generation while coal had 40 percent and renewables had 22 percent. The forecast for 2040 is natural gas at 28 percent, coal at 29 percent, and renewables at 29 percent. Among fossil fuels natural gas is taking considerable market share from coal. The reasons are that gas turbines and combined cycles, compared to other types of power plants, are low in capital cost, easy to install, and highly efficient, with combined-cycle efficiencies up to over 61 percent compared to at best about 40 percent for coal-fired steam-turbine power plants. Combined cycles produce less than half the CO_2 per kWh compared to coal fueled plants and far lower "criteria pollutants," for example, NO_x, SO_2, heavy metals, particulates. Some gas-turbine installations use oil fuel, but natural gas has become more widely available worldwide and is usually the preferred fuel where it is available, because it results in lower maintenance cost for the gas turbine and produces lower emissions. When gas turbines began to be used in the late 1950s to early 1960s, they were mostly simple cycle in "peaking" service, that is, operated only at times of high demand for electricity. But by the 1990s very large, efficient, and low-capital-cost combined cycles became available, and these products now dominate the market for new central station power plants.

Integrated Gasification Combined Cycles (IGCCs) The advent of large natural-gas-fired combined cycles, coupled with the worldwide desire to reduce emissions from power generation plants, gave new life to a relatively old concept—coal gasification—in the early 2000s. The concept of gasifying coal started in the 1940s but was not considered seriously as a power plant option until the late 1990s to early 2000s, when the price for natural gas escalated substantially at the same time

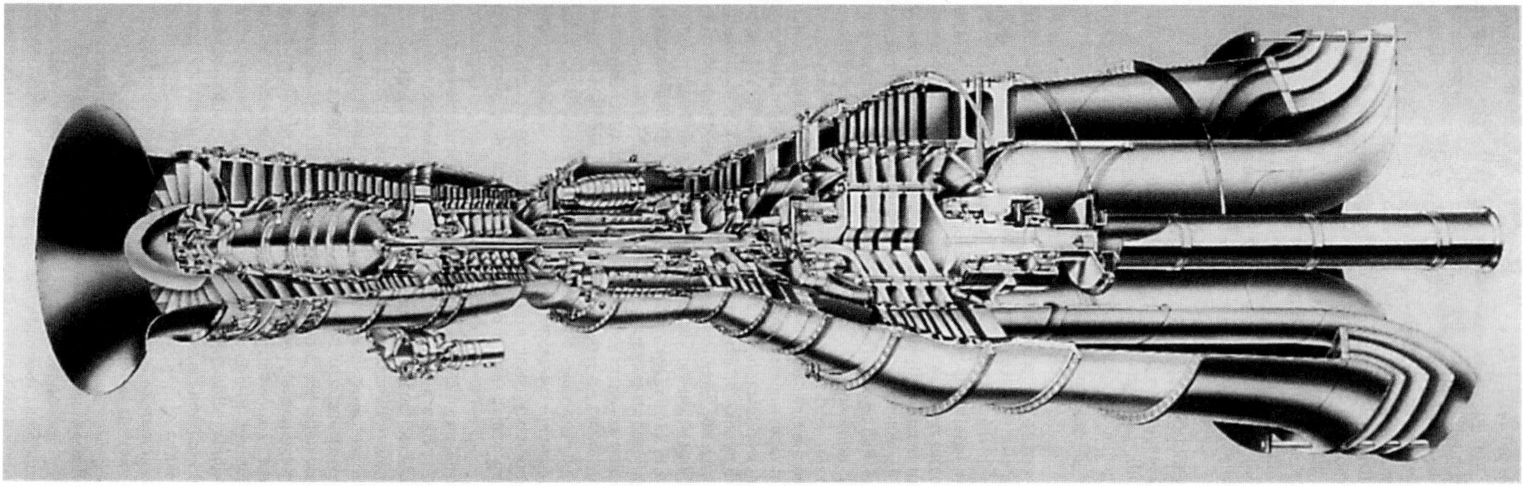

Figure 7.7.28 Cutaway view of FT8 aeroderivative industrial gas turbine. (*Courtesy Pratt & Whitney Aircraft.*)

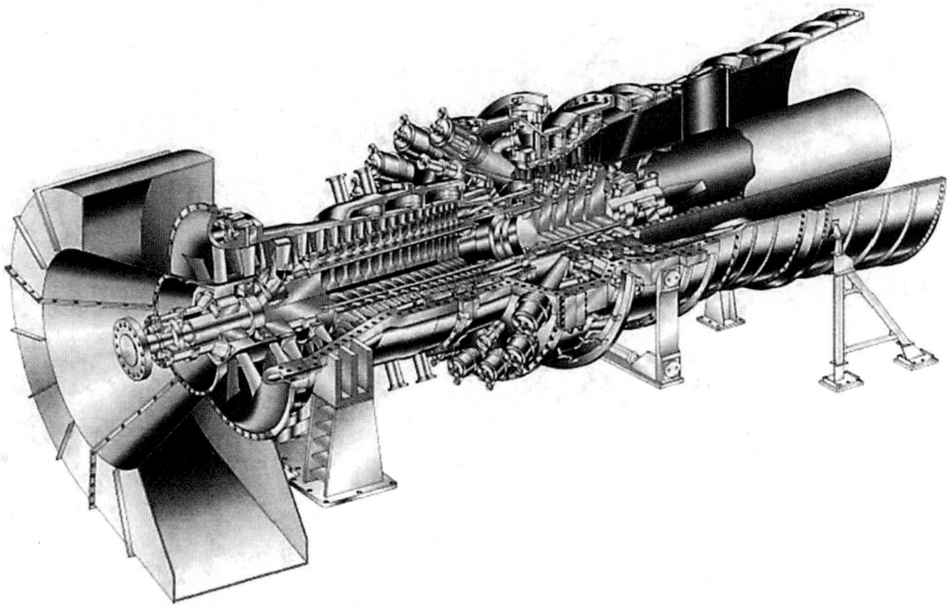

Figure 7.7.29 Cutaway view of M701 large heavy-duty industrial design. (*Courtesy Mitsubishi Heavy Industries.*)

as the Kyoto Treaty focused attention on reducing emissions of CO_2. The concept is shown in Fig. 7.7.30.

A major incentive for IGCCs is that by gasifying the coal it is possible to remove, before combustion, "criteria pollutants" than is possible in a conventional coal-fired plant, where the contaminants are substantially diluted in the exhaust gas.

Development of IGCCs started in the 1970s following the energy crisis in 1973 and was accelerated by the second energy crisis in 1979. The resulting incentive in the United States and in Europe to find ways to produce electric power from coal with increased efficiency and reduced emissions prompted government funding for IGCC plants. Two coal-based IGCCs were built in the United States, in Florida (Tampa Electric) and Indiana (Cinergy), and in Europe, Nuon in the Netherlands and Elcogas in Spain. All the plants were relatively small, 250- to 320-MW plant output, but large enough for one large gas-turbine combined cycle. None of them would have been competitive economically without government subsidies, but the plants provided valuable demonstration and experience that the technology held promise. Also, IGCCs fed with petroleum coke or with tars or asphalt have been built without government subsidies in the United States, Europe, and Asia. Additionally, in the United States there was a DOE-funded coal gasification plant in Kingsport, TN, which produced product gases and final chemical products clean enough for human consumption in pharmaceuticals, although at high capital cost. The plant demonstrated 98 percent reliability and helped convince many utilities that IGCC is a viable technology on a technical basis, the question being whether it could be competitive economically versus conventional coal-fired steam plants. As of the early 2000s the capital cost per kilowatt of an IGCC was estimated to be about 10 to 15 percent higher than that of a conventional coal-fired plant.

In 2005, the U.S. Department of Energy (DOE) launched a solicitation for an R&D program for the next level of technology that could be included in IGCCs to be built in about 2010 with the intent to enable IGCCs to be fully competitive with conventional coal-fired plants and produce far lower emissions, including the possibility of stripping off some of the CO_2 and sequestering it (carbon capture and storage, or CCS; see the cycle diagram in Fig. 7.7.30).

As of 2016, the status of IGGC includes the following in the United States: (1) The boom in shale gas production and the resulting significant drop in natural gas prices has caused interest in IGCCs to come to a screeching halt; that is, IGCCs can't compete versus natural gas-fueled combined cycles so no one is building them. (2) However, prior to the shale gas boom two large IGCC plants were built. A plant in Edwardsport, Indiana, is running successfully and includes a small slipstream of CCS, and a plant in Kemper County, Mississippi, which includes 67 percent CCS, is nearing completion. (3) DOE is continuing to fund university research projects on CCS. One promising finding is that by including water with the right characteristics with the CO_2 being injected, the CO_2 seems to be mineralized (turned into limestone) in 5 years instead of 100 years that would occur otherwise. This would reduce the risk of the CO_2 escaping from storage.

As of 2016, outside the United States there is activity in Japan and Korea on IGCC, since their natural gas prices are far higher than those in the United States, and in China many gasification plants are being installed.

Marine gas turbines are almost exclusively aeroderivatives because of their light weight and compact size. They range in size up to 69,000 hp, as listed in the "2014-2015 Gas Turbine World Handbook." They provide power for most of the surface warships (e.g., cruisers, destroyers, frigates, corvettes) recently added to the world's fleets. Two new market opportunities are emerging. The first is the growing market for

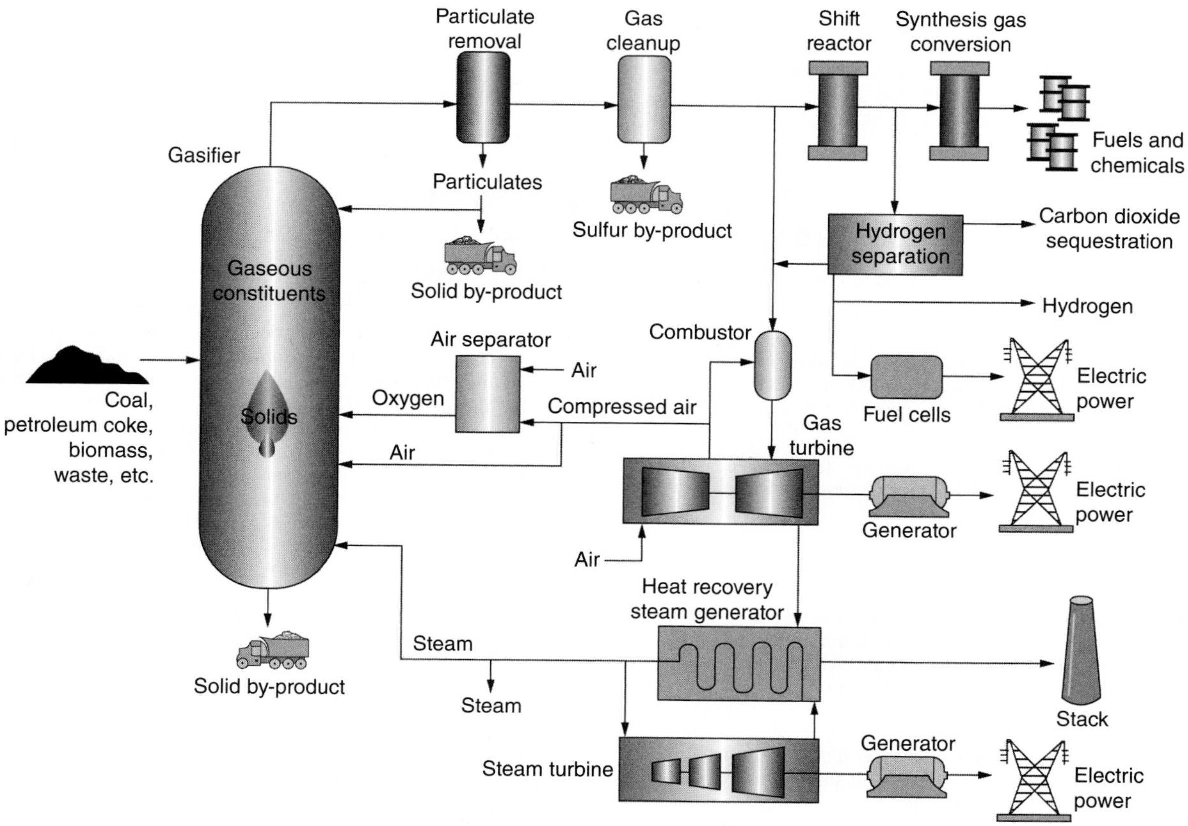

Figure 7.7.30 Integrated gasification combined cycle (IGCC) system diagram. (*U.S. Department of Energy.*)

commercial, lightweight, very fast (40+ kn), novel ferries and cargo ships—catamarans, slim monohulls, etc.

The second is being created by the increasingly severe emissions standards for ships. Most of the world's cargo and cruise ships are diesel-powered due to their ability to use poor-quality, less expensive fuels. However, as emissions regulations are tightened for SO_x and NO_x, the fuels must be lower in sulfur, and the power plants must reduce NO_x. Gas turbines are more capable than diesels of reducing NO_x levels, so they are gaining in market share. As of 2016 a roll-on – roll-off ferry has converted its diesel engines to burn methanol, which sharply reduces emissions.

Mechanical Drive Gas Turbines The increased popularity of natural gas as a preferred fuel for power generation and home heating has stimulated the market for gas turbines in mechanical drive applications, largely for natural gas pipeline pumping but also for driving large compressors in process plants such as oil refineries and liquefied natural gas (LNG) installations. The "2014-2015 Gas Turbine World Handbook" lists mechanical drive gas turbines from 1,500 to 175,000 hp, in a mix of aeroderivative and frame-type units. Most applications are simple cycle, although a number that are in process plants are combined-cycle configurations. The aeroderivatives have higher efficiency, but frame-type units are also popular for remote of critical applications due to their durability and low maintenance requirements. A characteristic of these applications is sometimes unstaffed stations and often the high economic cost consequences of failure. Consequently many owners are willing to sacrifice some efficiency for lower risk of downtime.

Microturbines Gas turbines in smaller sizes than 1 MW are used in several applications. The gas turbines usually have centrifugal compressors and, in some cases, radial inflow turbines. There are three manufacturers who specialize in this type of gas turbine, although some of the bigger companies are beginning to develop microturbines to add to their existing product lines. The most common applications are for **distributed generation**, often for small commercial or industrial companies that have a need for the exhaust heat for a process, for heating or for cooling via absorption refrigeration. It is generally not economical for a user to buy a microturbine for peaking power, although it can be a good economic choice for backup power to protect against the loss of power from the grid. Due to the economy-of-scale effects, the cost per kilowatt is high and the efficiency is low for this type of gas turbine; but if the user's requirement for exhaust heat is a good match for the gas turbine, and if the user's process runs 24/7, a microturbine can be a worthwhile investment. Some users have a varying electric or heat load and buy a microturbine plus exhaust heat recovery system strictly for baseload operation. A current issue is the **interconnect requirement**—the requirements imposed by the local utility on the way that users can connect to the grid, and the price for backup power in case in microturbine has downtime when the power need is present. This is as much a regulatory and political issue as a technical one.

Small Engines for Aerospace Applications Gas turbines in smaller sizes [less than 1,000 hp (746 kW)] are used in several fields of application. Such engines usually feature centrifugal compressors and are all manufactured by specialized companies. In the ground transportation field, automobile and truck manufacturers have experimented since 1950 with gas turbines, primarily of the low-pressure-ratio (less than 10:1) regenerative type. While several prototypes have been tried over the years, displacement of the widespread, well entrenched automobile piston engine has not taken place, with one exception: The U.S. Army has found the gas turbine to be well suited as a tank engine. As with the automobile engine, gas turbines have been experimentally tried in railroad trains, but have yet to make inroads to replace the diesel engine.

Small gas turbines are used as a compact, simple, and reliable power source on board aircraft for start-ups and auxiliary power. They are known as auxiliary power units or APUs. Similar engines find use as quick-start standby units to drive generators for emergency power. Finally, very small jet engines provide propulsion for remotely controlled aircraft with surveillance or cruise missile military missions.

7.8 NUCLEAR POWER

by Andrew C. Klein, Bryan A. Erler, and John M. (Jack) Tuohy, Jr.

REFERENCES: Cacuci, "Handbook of Nuclear Engineering," Springer Science, 2010. International Atomic Energy Agency, "Thermophysical Properties of Materials for Nuclear Engineering," IAEA, 2008. Ellis, "Nuclear Technology for Engineers," Literary Licensing, 2012. Glasstone, "Source Book on Atomic Engineering," Krieger, 1979 Annual reports of the International Atomic Energy Agency, Vienna. American Nuclear Society, "Nuclear News," LaGrange Park, IL. Glasstone and Sesonske, "Nuclear Reactor Engineering," Springer US, 1994. Shultis and Faw, "Fundamentals of Nuclear Science and Engineering," CRC Press, 2008. Lamarsh and Baratta, "Introduction to Nuclear Engineering," Prentice Hall, 2001. Lewis, "Fundamentals of Nuclear Reactor Physics," Academic Press, 2008. Bodansky, "Nuclear Energy: Principles, Practices, and Prospects," Springer, 2004.

FISSION AND FUSION ENERGY

Nuclear power is the energy derived from the fission (or splitting) of the nuclei of heavy elements, such as uranium or thorium, or from the fusion (or combining) of the nuclei of light elements, such as deuterium or tritium. Particles set in motion by these processes yield heat energy virtually instantaneously. The amount of energy released per atom exceeds by a factor of several million the amount of energy obtainable per atom in a chemical reaction, such as the burning of fossil fuels. While control of the fusion process is still surrounded by tremendous technical problems, considerable success has been achieved in utilizing heat energy produced by fission for power generation, propulsion, industrial production, and scientific experiments.

NUCLEAR PHYSICS

A nuclear reaction occurs when the particles making up the nucleus of an atom are rearranged—as distinct from a chemical reaction, which is caused by changes in the electron structure surrounding the nucleus.

Fundamental (elementary) particles are, theoretically, the irreducible constituents of the material world. New experimental techniques lead to discovery of more particles; advances in theory indicate that some are compounds of particles. Except for the electron and proton, all the fundamental particles are unstable.

An **electron** carries one unit of negatively charged electricity [defined as 1.6×10^{-19} coulomb (C)] and has a mass at rest equal to 1/1,836 times the mass of the hydrogen atom. Electrons emitted by radioactive atoms are called **beta (β) rays** and have energies up to several MeV (million electron volts).

A **proton** carries one unit of positively charged electricity equal in magnitude to that of the electron and has a mass, at rest, of 1.0072764 atomic mass units [u when 1 u (1.66057×10^{-27} kg) is defined as 1/12 the mass of a neutral atom of the most abundant isotope of carbon]. It is identical to the nucleus of the hydrogen atom. Protons are not emitted spontaneously from atomic nuclei.

A **neutron** has a mass approximately equal to that of the proton but lacks an electric charge. It cannot be detected by subjecting it to electric or magnetic fields, nor can its presence be shown by its passage through cloud chambers. The presence of neutrons can be shown only by their interaction with other particles.

An **atom**, which is the basic unit of any chemical element, consists of a central nucleus surrounded by planetary electrons. The number of its electrons determines its chemical characteristics and in electrically neutral atoms equals the positive charge on its nucleus. The nuclei of all atoms except the hydrogen atom are composed of protons and neutrons. The nucleus of the hydrogen atom consists of a single proton, which gives it an atomic number of 1. This number is designated by the letter *Z*. The number of neutrons in the nucleus is represented by the letter *N*. From this it follows that the mass number *A*, which represents the atomic weight, is equal to $N + Z$. All with atomic numbers above 92 (uranium) must be produced artificially and generally have a short half-life.

Isotopes are elements which have identical chemical characteristics but different atomic weights, for example, certain atoms of the same element have different numbers of neutrons in their nuclei. There are considerably more than a thousand isotopes of the known elements. Only 320 of these exist in nature, of which approximately 40 are unstable and radioactive.

For example, hydrogen has three isotopes. One is ordinary hydrogen with a proton nucleus. Another is deuterium, whose nucleus consists of a proton and a neutron. The third is tritium, which consists of one proton and two neutrons. At the heavy end of the natural periodic table is uranium, with 92 protons. When it has 146 neutrons, it is $_{92}U^{238}$, which comprises 99.28 percent of natural uranium. The remainder of natural uranium consists of 0.006 percent U^{234} and 0.71 percent U^{235}. The latter isotope, with 92 protons and 143 neutrons, is the only one of the three which is readily fissionable and is the one utilized in most of the nuclear power reactors operating in the United States. It is separated from the natural element in the enrichment desired by gaseous diffusion or centrifuge separation processes in government facilities.

Energy and mass are used interchangeably in nuclear physics, and the total energy of a nucleus can be determined by measurement of its exact mass *m*, which is related to its energy *E* by the Einstein equation $E = mc^2$, where *c* is the velocity of light, 3×10^{10} cm/s. This law of equivalence of mass and energy unifies two long-accepted laws: the conservation of energy and the conservation of matter. Singly, the two older laws must be regarded as high-order approximations, adequate in all engineering fields except atomic energy. The exact mass *m* of a nucleus differs by only a few hundredths of a percent from an integral number *A* of proton masses, but this small deviation is of significance in nuclear theory, since it measures the difference in energy between the nucleus and its separate components, that is, its binding energy. The energies of nuclides are recorded in the tables in terms of their masses.

Physical atomic weights do not quite equal chemical atomic weights because the oxygen used in the chemical determinations is a mixture of isotopes, whereas in mass spectrographic measurements the single isotope C^{12} is used for reference. The relationship is: Physical atomic weight = 1.000280 × chemical atomic weight.

Planck's Constant Since radiation has the properties of both waves and particles, it is theorized that in the emission or absorption of energy by atoms or molecules, the process takes place by steps, each step being the emission or absorption of an **individual quantity of electromagnetic energy**, which is considered the elementary quantum of action. It is known as **Planck's constant** and has the value $h = 6.626 \times 10^{-34}$ J · s. It is used in virtually all quantum relationships, including Schrödinger's equation and Heisenberg's uncertainty principle, and in the formula for the energy levels of atomic hydrogen. The standard notion is $U = h/2\pi$, where U signifies the angular momentum of an orbiting electron.

Relativistic Mass When moving particles approach the speed of light, the approximation of kinetic energy, $E = \frac{1}{2}m_0v^2$, fails; it then becomes necessary to use the equation $E = mc^2$ and to use the relativistic mass, a function that increases with velocity according to the law $m - m_0/\sqrt{1-(v/c)^2}$, where m_0 is the rest mass (i.e., the mass of newtonian mechanics) and *v* is the coordinate velocity of the particle.

Binding energy is the work required to disintegrate an atom completely into *Z* protons and $A - Z$ neutrons. It is a measure of the total kinetic and potential energy of the nucleons in the nucleus and may be calculated from the equations

$$B = c^2\Delta \quad \text{and} \quad \Delta = Zm_H + (A - Z)m_n - m$$

where *m* is the mass of the nuclide, m_H the mass of the hydrogen atom, and m_n the mass of the neutron, all on the physical atomic-mass scale (u). The quantity of Δ is known as the **mass defect**. The energy involved is usually stated in MeV.

Fusion Figure 7.8.1 shows that if two deuterium (H^2) nuclei react to produce one He^3 nucleus plus one neutron, the total binding energy of the four particles (two neutrons plus two protons) involved in the reaction is increased from approximately 4.4 to 7.6 MeV, releasing 3.2 MeV for the reaction. Such a process is called **fusion**. The deuterium-deuterium fusion reaction may also produce an H^3 plus an H^1 nucleus, with a release of 4.0 MeV.

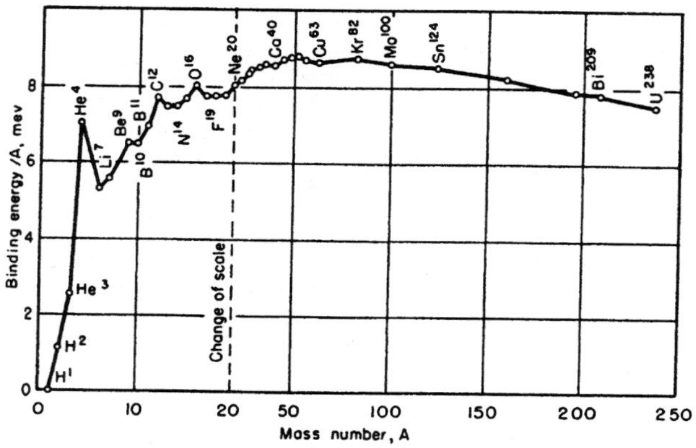

Figure 7.8.1 Binding energy per nuclear particle for stable nuclei.

For fusion to occur, the two nuclei must approach each other with exceptionally high kinetic energy in order to overcome their electrostatic repulsion; and since kinetic energy is proportional to absolute temperature, effective fusion can occur only at extremely high temperatures. For fusion to continue as a chain reaction, it is further necessary that nuclear collisions be sufficiently frequent to maintain the high temperature in spite of heat radiation. The condition necessary for frequent collisions is high pressure.

A light-element chain reaction is the principal source of heat in the sun, hydrogen being converted to helium as the net result of a multistage process. The temperature and pressure at which this reaction is sustained at the center of the sun are 10 million °F and 1.0×10^{10} atm respectively.

A deuterium-deuterium reaction can be sustained at temperatures and pressures which are considerably lower than those in the solar reactions but still fantastically high by terrestrial standards. Achievement of a controlled light-element chain reaction for power generation by fusion depends on producing high particle velocity and high density locally by electromagnetic means, without development of correspondingly high temperatures and pressures at the container walls.

Fission is the division of an atomic nucleus into parts of comparable mass, either naturally (spontaneously) or under bombardment (induced) with neutrons, α particles, gamma rays, deuterons, or protons. In elements with $Z > 90$, fission can be produced by neutrons of low or moderate energy.

Fission is important because in the process neutrons are emitted that may, under certain conditions, be utilized to produce further fissions, thus leading to a self-sustaining, or "chain," reaction. It is thus possible to "burn" uranium to obtain an energy release per atom approximately 10^8 greater than from a chemical fuel. The energy released in the fission of U^{235} is

	MeV
Kinetic energy of fission products	168
Kinetic energy of fission neutrons	5
Energy of *gamma* rays	14
Energy of *beta* rays	8
Energy of neutrinos	12
	207

The total of 199 MeV is equivalent to 3.2×10^{-11} W · s, which indicates that it requires the fissioning of 1.12×10^{17} atoms to release a kilowatt-hour of energy. This number of atoms forms only a speck of matter six thousands of an inch in diameter.

Radioactivity occurs when an unstable nucleus undergoes atomic disintegration by emitting α, β, or $\beta+$ particles, γ or X-ray electromagnetic radiation.

The **alpha** (α) **particle** is identical with the nucleus of a helium atom. The emission of an α particle creates a new nucleus, with Z reduced by 2 and A decreased by 4 mass units.

Beta (β) **decay** involves emission of an electron from the nucleus of an atom, thereby increasing Z by unity.

The **gamma** (γ) **ray** is electromagnetic radiation originating in the nucleus—as differentiated from X-rays, which usually are less energetic and arise from energy adjustments as electrons move between orbital shells outside the nucleus. Gamma rays are emitted instantaneously when a neutron is captured in a nucleus and are frequently emitted following ejection of a particle from a radioactive nucleus.

The **positron** ($\beta+$) is a particle whose mass is the same as that of an electron and whose charge is equal in magnitude but opposite in sign. Positron emission decreases Z by unity.

Electron capture (EC) results when an unstable atom decays by capturing an orbital electron in the nucleus, resulting also in a decrease of Z by unity. This capture produces a vacancy in the orbital shell which is filled by an electron moving from an outer shell, giving the rise to X-ray characteristics.

Half-life is the time required for half the original nuclei in a sample of an isotope to decay; it is given as $0.693/\lambda$ s. (where λ is the decay constant for the particular radioisotope). Half-lives of the radioactive nuclides vary from 10^{-7} s to 10^{10} years.

UTILIZATION OF FISSION ENERGY

Nonpower Types

Nuclear reactors having a purpose other than the generation of usable power may be classified into two general types: research and test reactors and production reactors.

The latter category of **production reactors** includes both light- and heavy-water and gas-cooled reactors of high thermal power (comparable in thermal output to large power reactors). Past reactors in the production reactor category used varied fuel cycles adjusted to produce plutonium and tritium of weapons quality for nuclear weapons programs. More recently medical-isotope production reactors have been developed to produce radioisotopes for use in medical procedures such as ^{99}Mo/^{99m}Tc and others used in diagnosis and clinical treatments. Most production reactors operate with outlet temperatures near or below the boiling point of water, so the thermal heat is wasted to the environment. A few **"dual-purpose"** production reactors operate at temperatures which allow the coupling of useful electric power output facilities.

Teaching reactors have the characteristics of very little excess reactivity, very low power capability (a few to a few thousand watts of heat energy), and relatively easy access to the core. Teaching reactors are used mainly for the purpose of instruction in the behavior of a nuclear reactor. With a teaching reactor novice reactor operators may experience the response of reactor instrumentation as reactivity is inserted and criticality is approached, "just critical" behavior, and power increases or decreases.

Teaching reactors are generally owned and operated by universities. In addition to being used for operator training, they are used in a variety of nuclear-engineering instructional experiments. Typical experiments include measurements of neutron-flux distribution; measurement of neutronic characteristics such as generation time, leakage fraction, thermal-utilization factor, resonance escape probability, and other characteristics affecting criticality and kinetics; and measurement of parameters affecting control such as control-rod worth and reactivity feedback coefficients. Although the neutron flux available from a teaching reactor is relatively small, it is nevertheless significant and can be used for limited production of radioactive isotopes.

Research reactors generally have a power capability of more than 1.0 MW. They are for the purpose of providing an intense radiation source, both neutrons and gamma rays. A research reactor is usually designed to have a high leakage flux in order to facilitate its use as a radiation source.

The neutrons from a research reactor have a variety of uses including neutron activation analysis, neutron radiography, neutron diffractometry, and production of radioactive isotopes. The gamma rays are applied in materials testing and in studies using the interaction of gamma rays with the nuclei of various elements.

In **neutron activation analysis**, an unknown sample is irradiated in a calibrated neutron flux, and the constituents of the sample become radioactive through neutron absorption. The radiation from the sample is then analyzed. Since the radiation from a specific radioactive isotope is unique to that isotope, a qualitative and quantitative determination of the constituents of the sample is possible. Extremely small quantities of radioactivity can be detected; therefore, neutron activation analysis is a very sensitive analytical tool.

Neutron radiography is a technique similar to X-radiography. The perturbation of a collimated beam of neutrons that has passed through the radiographed item is determined by film exposure or other means. Unlike x-rays that serve to outline dense, highly absorptive materials, neutron radiography serves to outline light materials that preferentially scatter neutrons from the incident beam.

Neutron diffractometry is a technique used to determine the molecular and crystalline structure of materials. The analysis is accomplished through a precise determination of the diffraction pattern from a collimated monoenergetic beam of neutrons that has interacted with the sample.

Certain radioactive isotopes are useful and effective in the **treatment of some diseases**. Isotopes that do not occur naturally can be produced in large quantities by exposure of the precursor element to the neutron fluxes available from research or test reactors.

Test reactors generally have a power capability ranging from 25 to hundreds of thermal MW. The Idaho National Laboratory Advanced Test Reactor can operate at power levels up to 250 MW. Reactors such as these are used largely for the purpose of providing an experimental radiation environment similar to that which exists in the core of a power reactor. They are also used for large scale radioactive isotope production.

Power Cycles

Although the energy of fission appears as kinetic energy (85 percent) and photon energy of *gamma* rays, there is currently no practical method of converting this energy directly into useful work on a large scale. Therefore, a nuclear reactor must be treated as a heat source which differs from a chemical heat source in that no oxygen is required and the heat does not have to be removed from gaseous combustion products which possess poor heat-transfer properties.

In the large power capacities associated with nuclear power plants, the **Rankine vapor cycle** is the one in use. The Rankine cycle appears to be best suited for nuclear power because (1) the maximum practicable operating temperature for a reactor corresponds to or is lower than the temperature used in the conventional Rankine vapor cycle; (2) the problem of containment of radioactivity is better solved by this cycle than by other cycles; (3) reactors tend to be most economical in large sizes, as are Rankine engines.

Three variations of the Rankine cycle (Fig. 7.8.2) are being used. The **direct cycle** is the least complicated and thermodynamically the most desirable. Its disadvantages are the facts that the boiling process does not permit the high power densities which are attainable with liquid cooling and that radioactive steam is carried into the turbine and condenser. The **indirect cycle** gives greater power density and eliminates radioactivity in the turbine, although it too produces steam at lower pressures and temperatures than those considered most efficient for modern turbines. Both concepts are highly developed and commercially available. Plants utilizing sodium or other liquid metals as the heat-exchange medium use an intermediate-link system to isolate the water system from radioactive sodium, but the ability to operate

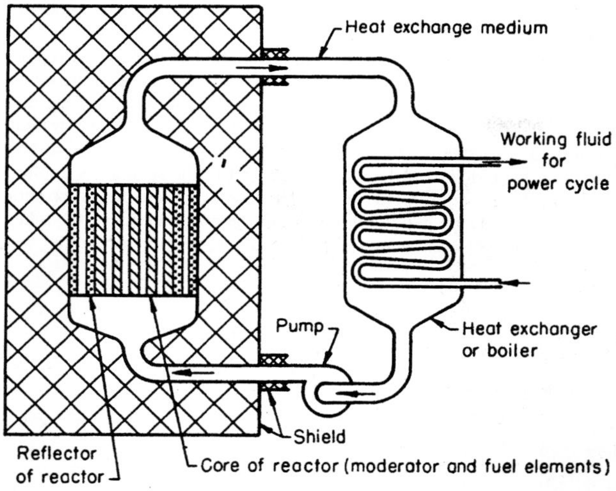

Figure 7.8.2 Plant cycles—coupling with a working fluid. *(Amorosi, Selection of Reactors, "Nuclear Engineering Handbook," McGraw-Hill, 1957.)*

at high temperature with liquid-metal coolants permits the generation of high-temperature steam to utilize the present state of steam turbine technology.

The only other cycle which has been given consideration is the **Brayton cycle**. This cycle is being examined with renewed interest because of the rapid advances in gas-cooled reactor technology and the availability of large gas turbines suitable for reliable utility industry service. The use of a closed Brayton cycle with the latest high-temperature gas-cooled reactors (HTGRs) has potential advantages in that the gas temperatures can match the latest gas-turbine designs and a very compact plant can be built.

The ability of a nuclear reactor to produce high temperatures not limited by chemical reaction equilibrium has generated interest in advanced concepts such as magnetohydrodynamics and thermionic generators. These systems are limited by the availability of suitable materials of construction and appear destined to remain in the research phase until such materials are developed.

Power Reactor Types

During the early years in the development of power reactors, a large number of possible reactor designs were investigated. In the present state of the industry (2017), there are many new concepts and designs under consideration ranging from small modular reactors (SMRs), which operate at less than approximately 300 MWe, to large 1,300 to even 1,700 MWe designs. The currently operating fleet of reactors, utilize, generally boiling or pressurized water for their primary coolant and neutron moderator, while some reactors utilize heavy water or graphite for neutron moderation and a few reactors use gas or liquid metals for their coolant. Most of these reactors also use solid fuel in the form of uranium dioxide. Recent research and development activities have looked at developing nuclear systems that utilized an increasing variety of fuel types (solid, molten, particulate and composite), coolants (including molten salts, liquid sodium or lead, helium and super-critical water) and may be focused on fast neutron spectrum (no moderator), thermal neutron spectrum (including a moderator) and some intermediate energy neutron spectrum. Light-water (moderated and cooled) reactors (LWRs), comprised of the boiling water reactor (BWR) and pressurized water reactor (PWR), currently represent all of the operating reactors in the United States. A high-temperature gas-cooled reactor (HTGR) is also a feasible concept in which there is significant interest. The liquid-metal-cooled fast reactor (LMFR) concept is employed in operating test reactors and has been successfully implemented in full-scale operating plants in France and Russia. Molten salt fueled and/or cooled concepts are also receiving significant interest world-wide. For the complete utilization of the energy represented by the world's uranium reserves, a successful breeder reactor is desirable since the present generation of reactors are all consumers of only the naturally occurring fissionable isotope of uranium, U^{235}.

For utilization of natural uranium as a fuel, D_2O is substituted for H_2O as the coolant moderator. The reactor must be refueled more frequently than enriched reactors to maintain an adequate operating reactivity margin. However, D_2O cooled and moderated reactors such as the Canadian designs (CANDU) utilize on-line refueling to minimize shutdown time.

Boiling water reactors have a simple design and can utilize relatively thin-walled vessels and pipes because they operate at moderate pressures compared with pressurized water reactors. Fuel cladding temperatures are only slightly higher than steam temperatures, and there is an inherent safety factor because the steam-void volume increases on a transient power increase. Some of the problems caused by radioactivity carryover, such as maintenance on the turbine and condenser and the prevention of radioactive leakage, are offset by the cost of the steam generator (boiler) in pressurized water reactors. However, boiling water reactor vessels, despite their lower design pressure, are larger and heavier than pressurized water vessels of similar rating and extensive water purification equipment is required to remove corrosion products and other impurities from the feedwater before introduction into the reactor.

Limitations on the **power density** imposed by exit voids (i.e., the vapor volume of the exiting steam-water mixture from the reactor core) are a disadvantage in that low-power density contributes to high fuel inventory charges. Load changes have been accomplished by steam bypass control or control rod movements. Adjustment of the turbine throttle and consequent reduction in steam flow will cause an increase in pressure in the reactor, which in turn will collapse the steam bubbles and increase reactivity. The advanced boiling water reactor (ABWR) uses nine recirculation pumps located internal to the reactor vessel to control power output. This approach is very effective and stable and allows the BWR to operate in load following mode. The net decomposition of the water into O and H is much greater than in a pressurized water reactor.

Pressurized Water Reactors Properties of water and steam which led to their predominance as general-purpose heat-transfer media have also caused their widespread application as a reactor coolant. A major disadvantage of water results from its relatively high vapor pressure. However, this can be partially overcome by allowing boiling in the reactor. Thermal efficiencies up to 36 percent are possible.

The use of H_2O as a coolant and moderator is based on well-developed technology which indicates that the ultimate size (or capacity) will be dictated primarily by heat transfer requirements.

The **light-water** (H_2O) **reactors** require slightly enriched fuel (3 to 5 percent U^{235}). These enrichment levels have resulted in refueling outages being extended from annual occurrences to 18 month and 24 month intervals. Recent (2017) developments in the merchant fleet markets have caused some utilities to reexamine the economics of longer intervals between refueling outages.

Control rods, burnable neutron absorbers, and soluble boron in the coolant are used to control excess reactivity, achieve high fuel burnup, and maintain power distribution within the design limits.

Nuclear Steam Generators Pressurized water reactors (PWRs) provide a specialized application for steam generators, or boilers. Each PWR incorporates two, three, or four steam generators to isolate the primary coolant system from the secondary side. The primary loop operates at a higher pressure than the secondary side and is comprised of four major equipment items namely the reactor vessel, reactor coolant pumps, a pressurizer and steam generators. The secondary side of the steam generator operates at a lower pressure resulting in steam generation that feeds the steam turbines. The combination of primary and secondary loops comprises the nuclear steam supply system (NSSS).

Approximately two-thirds of the power reactors in the United States are PWRs. The balance are boiling water reactors (BWRs). BWRs operate at about 1,000 psig. They directly generate the steam that feeds the steam turbines and thus there are no large steam generators associated with the BWR. Canada has a reactor design that uses heavy water (D_2O) as a moderator. Their reactor system design utilizes steam generators as with the conventional PWR.

Steam generators are designed to transfer reactor generated heat from the primary loop to the secondary loop to deliver steam at a designated flow rate and quality. The steam conditions must meet the specifications of the steam turbines and other elements of the balance of plant (BOP) including feedwater heaters. Thermal efficiency is typically around 33 percent.

The steam generator is most often a U-tube design with recirculating secondary coolant. In this design primary coolant enters the lower plenum through the nozzle designated primary inlet in Fig. 7.8.3 and exits the nozzle designated primary outlet. primary coolant travels up through the steam generator within the tubes looping around and traveling back down while transferring heat to the secondary coolant surrounding the tubes. This occurs in what is called the evaporator section of the steam generator. The steam generated outside the tubes travels up through the vessel entering the steam drum where the wet

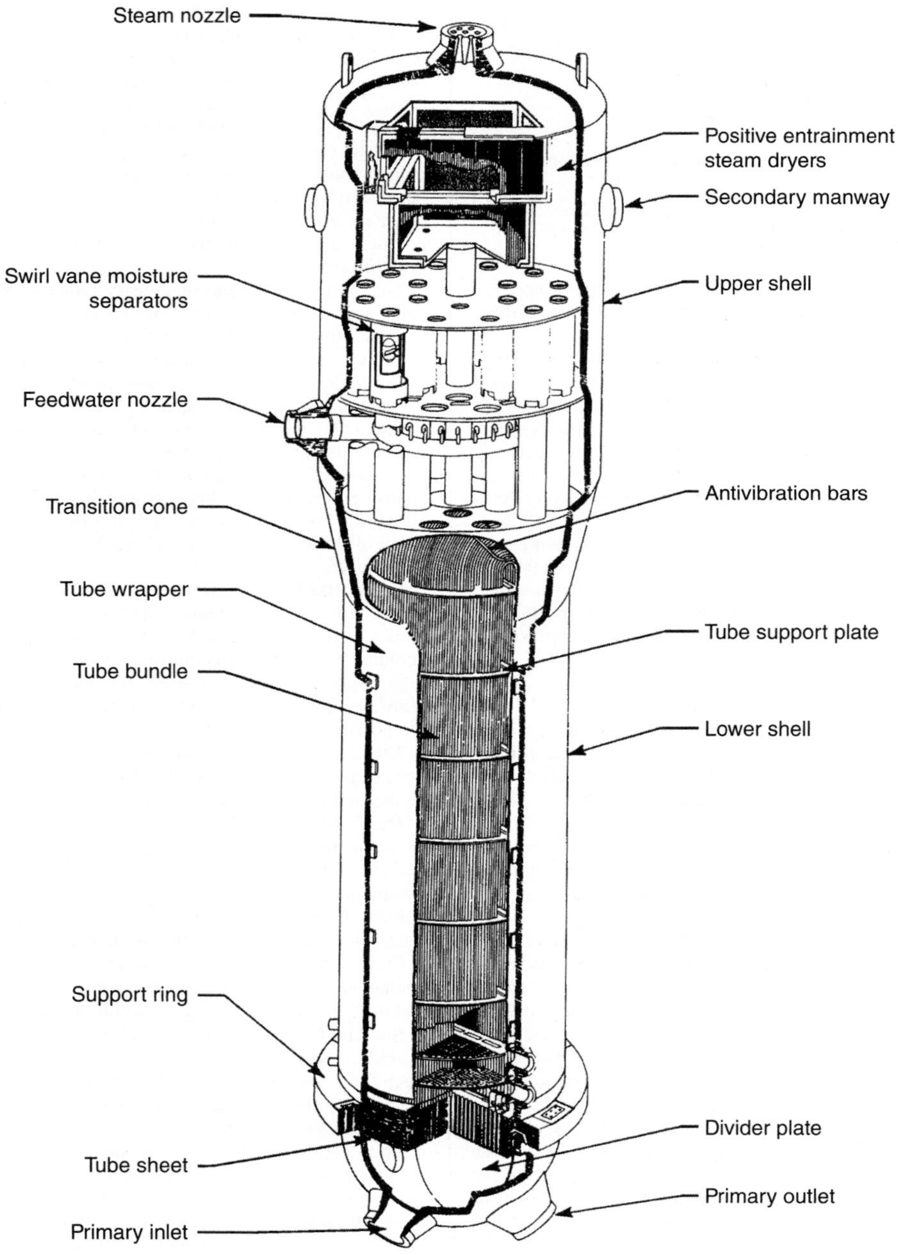

Figure 7.8.3

steam (10 to 40 percent steam by weight) is passed through moisture separators and dryers with the balance of the secondary coolant traveling back down to the base of the evaporator section (lower shell) between the inner shroud and the vessel wall. The U-tubes are sealed to the tube sheet at the bottom of the bundle which separates the primary and secondary flow loops. The effectively moisture free steam (generally <0.25 percent water and as low as <0.10 percent water) is sent to the power turbine while the recirculated water is returned to the bottom tube bundle through the downcomer annulus described above. Feedwater is most often injected into the downcomer near the top of the annulus to be mixed with the recirculating flow from the separators.

Horizontal tube supports are provided at various elevations along the vertical straight section of the tube bundle to provide lateral support. Additional tube supports or restraining systems are required in the upper U-shaped part of the bundle to position the tubes and prevent excessive flow induced vibration.

The primary operating pressure is typically 2,235 psig (15.4 MPa gage) with the average temperature around 580°F (304°C). The secondary operating pressure is generally between 740 and 1070 psig (5.1 and 7.4 MPa gage), depending on the system supplier.

The steam generator is designed in accordance with Section III of the ASME Boiler and Pressure Vessel Code.

A competing steam generator design is the once-through steam generator (OTSG) offered by B&W in the early years of the industry. The secondary side coolant inventory in this design is less than that for the recirculating design resulting in a reduced thermal inertia for the system (i.e., smaller heat sink).

Gas-cooled reactors offer low fuel and operating costs because of their ability to utilize natural uranium as fuel. This advantage is offset by higher capital costs caused by the large system size, in turn resulting from the lower power density usually used. More advanced designs such as the high-temperature gas-cooled reactor and its variants can use low enriched uranium and are able to use higher temperatures and the required structural material for such temperatures. These newer gas-cooled designs are much more compact than the natural-uranium reactors which are no longer in service.

Carbon dioxide has been the coolant most often used in natural uranium gas reactors, as it is nontoxic, only mildly radioactive due to activation, nonflammable, non-contaminating in case of leakage, and relatively inexpensive. Helium makes an attractive coolant, and its superior performance at high temperatures can be utilized. Helium, being inert, does not chemically react with graphite and the reactor structure at high temperatures as does carbon dioxide. Thus helium is the preferred choice of gas cooled reactor designers at the present time (2017).

The latest gas-cooled reactor designs are modular designs which use direct Brayton cycles to achieve high efficiency. These advanced modular designs use either solid blocks of graphite-fuel mixtures with coolant channels, or small balls (~8-cm diam) of fuel-graphite to achieve nuclear criticality. Modifications of these high-temperature reactors can be used for process heat applications which include thermochemical reactions to split water, thus providing a means for producing hydrogen for proposed transportation fuel needs and also the Fischer–Tropsch process which is employed to produce liquid fuel from coal.

Breeder reactors operate at extremely high power densities (as they are unmoderated) and are generally liquid-metal-cooled. Their attractiveness stems from an ability to "breed," that is, produce more fuel than they consume.

Aside from savings in fuel cost, the ability to "breed" fissile material will increase by orders of magnitude the energy available from the fission process.

The selection of a coolant for a breeder reactor is restricted by the requirement that it not act as a "moderator" reducing the average energy of the neutrons initiating fission within the reactor core. This requirement rules out the use of water, although the attainment of some breeding in a water-cooled reactor has been investigated. The choice of coolant is generally helium or liquid metal.

Choice of a liquid-metal coolant is based mainly on nuclear properties and on the engineering difficulties associated with melting temperatures and corrosion effects. Those of prime interest are sodium, sodium-potassium alloy, bismuth, lead, and lead-bismuth alloy. Sodium has been selected as the preferred coolant because of its availability, good nuclear properties, and the short half-life of its induced radioactivity. However, in the presence of oxygen, it burns and can react violently with water.

Advantages of operating in the fast neutron spectrum include (1) structural material for the reactor core can be selected without consideration of neutron absorption; (2) very little excess reactivity is needed to compensate for fission product buildup; and (3) reprocessed fuel in which fission product removal may not be complete can be used. The disadvantages are the increased damage to structural materials by the high fast flux and the control problems resulting from the short neutron lifetime.

Currently (2017) efforts are underway to develop next-generation (so-called Generation IV) reactor designs for future deployment around 2030. Currently second- and third-generation nuclear reactor power plants are used effectively around the world. Generation-III+ reactors have been designed and are beginning to enter service in the US, China, France, Finland, and other countries. Many countries are in the process of developing and expanding their nuclear power programs, aimed at generating reasonably priced electric power, clean water, and other useful by-products, all the while reducing discharge of greenhouse gases.

Molten salt cooled and/or fueled reactors are under development to build upon the early Molten Salt Reactor Experiment concepts from the 1960s. Some of these designs utilize molten fluoride salts containing uranium as both fuel and coolant, while others use a solid fuel to generate the heat and use the fluoride molten salt as the primary coolant. Significant advances on both of these concepts are expected in the years to come.

Small modular reactors (SMRs) are a special class of reactors that has recently been developed by the nuclear industry. They can be any type of reactor that is designed to operate at low power, less than about 350 MWe. SMR concepts are being developed for different markets including electricity generation, process heat applications and areas and countries with small electrical grids. Of special note are the integrated pressurized water reactor designs that are moving rapidly toward design certification. These designs incorporate the entire primary loop, including steam generator and pressurizer, within an integral reactor vessel. Some of these integral PWRs are designed to operate in full natural circulation mode and others have integrated primary pumps into the integral reactor vessel. SMRs are also under development that use gas-cooled reactor concepts as well as liquid sodium and molten salt coolants. Many of these concepts also plan to utilize factory construction to a large degree that would allow barge, rail and truck shipment to the operational site.

Fourth-generation reactors are under design to meet exacting requirements related to low capital cost construction, short construction times, easier fuel cycle management, minimization of radioactive waste, enhanced safety, and better control against proliferation of radioactive waste. These concepts hold great promise for even safer and more reliable fission nuclear power plant deployment in the twenty-first century.

Fuel and Waste Cycle

The steps and processes in a typical fuel cycle for light-water reactors are described below. Figure 7.8.4 illustrates the relationship of these steps and processes.

Natural Uranium Uranium occurs naturally in ores in very low concentrations, less than ½ percent by weight. In certain very unusual cases, the uranium content in the ore, called ore grade, may be as high as 10 percent, but these occurrences are extremely rare.

Uranium ores are mined by conventional techniques and then crushed and ground in a mill, which has special equipment for the recovery and purification of the ore into uranium concentrates. The product is in the form of a powder called yellow cake, which has the chemical form of either ammonium or sodium diuranate, both of which are yellow, hence the name.

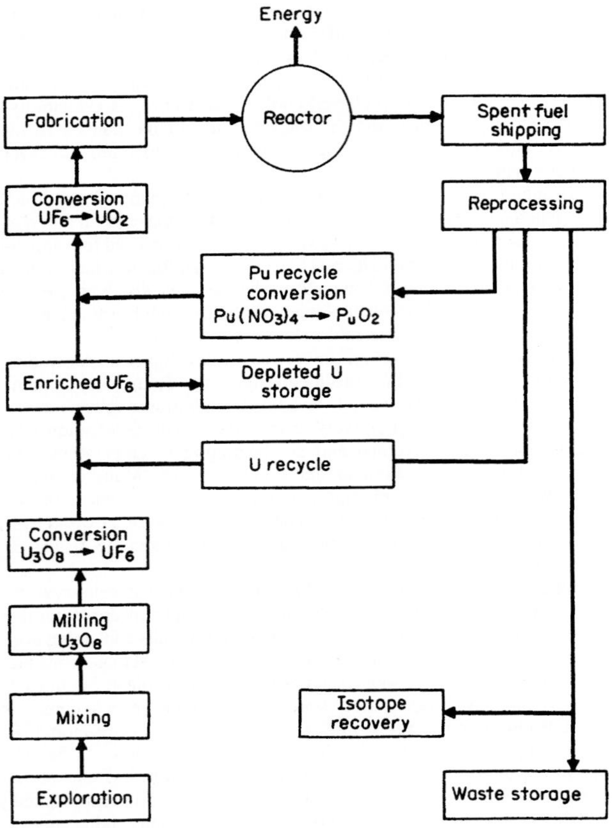

Figure 7.8.4 Nuclear fuel cycle.

Conversion of U_3O_8 to UF_6 The Gaseous Diffusion Process of enriching the U^{235} isotope (from 0.711 percent to the 3 to 5 percent used by water reactors) requires that the uranium be converted to a gaseous form. The gas UF_6 is used; so the natural U_3O_8 must be converted to natural UF_6.

Enrichment alters the relative amounts of U^{235} and U^{238} in uranium, raising the U^{235} concentration from 0.711 percent to the percentage required for use in a reactor. By so doing, the input (feed) stream of natural UF_6 is broken into two output (product plus tails) streams, one enriched and one depleted.

The process of enrichment makes use of the mass difference between U^{235} and U^{238} isotopes. In a gaseous form, the average energies of all UF_6 molecules are equal whether the U atom is U^{235} or U^{238}. Since the energies are equal, the average velocity of the lighter $U^{235}F_6$ molecule is greater than that of the $U^{238}F_6$. Because of this greater velocity and, hence, a greater momentum, the $U^{235}F_6$ molecules can diffuse through a barrier more easily. **Gas diffusion** plants work on this principle.

In the **gas centrifuge** process, gaseous UF_6 is fed into centrifuges which rotate at very high speed. The heavier U^{238}-bearing molecules drift to the outside wall, while those containing the U^{235} atoms accumulate close to the axis of the machine. The separation process is enhanced by longitudinal gas circulation along the wall and near the axis, in currents counter to each other. This also allows end scoops to collect the enriched and depleted streams. Centrifuges are connected in cascade to obtain specific enrichment assay levels and the cascades are operated in parallel to obtain adequate amounts of enriched product. A plant will contain several thousand stages. Electric power demands for the centrifuge enrichment process are typically 5 to 10 percent those of the diffusion process for similar enrichment and amount. The term **separative work unit (SWU)** is used to express the quantity. Centrifuge plants are in use in European countries. The US shutdown the only remaining American owned enrichment plant in 2013. It utilized the gaseous diffusion process. Centrus is currently working with the US DOE to develop and eventually deploy a new plant based on advanced gas centrifuge technology. In addition URENCO began operation of its gas centrifuge facility in New Mexico in 2010.

There are a number of other possible separation processes for uranium (or for that matter, other isotopes of any material). These include various thermal separation processes and plasma processes, activated either by electronic or laser energy, which work on atomic or molecular ions with separation by magnetic field. The lighter isotope will have a slightly smaller radius of curvature and can be caught in a pocket separate from the heavier. Several of these processes have been proven at the experimental level, and some have been used to produce significant amounts of material in the past.

Fabrication The gaseous enriched UF_6 that comes out of the enrichment plant is delivered in standard cylindrical gas bottles. The UF_6 is then converted to UO_2, the fuel form in which it is used. The UO_2 is in the form of a powder which is then milled to provide suitable sintering properties. The powder is formed into pellets by cold pressing, sintered, and ground to final dimensions. The pellets are loaded into a zirconium alloy tube (cladding) which is backfilled with helium and sealed at each end with a welded-end plug. The fuel rods are assembled into complete fuel assemblies. The number of fuel rods in an assembly varies from 50 to 200 or more, depending on the reactor type and specific design.

Spent Fuel Storage and Transportation Several options are available for managing spent fuel after its discharge from the reactor. It can be stored at the reactor site for a minimum time of 3 months and then shipped in a shielded cask to a reprocessing plant for recovery and recycle of Uranium and Plutonium. This procedure is used for most power reactors outside the United States. Alternatively, it may be stored in pools of water (used fuel pools) or in dry storage at reactor sites, or in an away-from-reactor storage facility for an extended period of time, and then shipped to an interim dry storage site, possibly after fuel mechanical compaction, or sent to a geological repository for long-term disposal. This latter procedure is mandated by current U.S. law.

Reprocessing and Reconversion Spent nuclear fuel has substantial residual value in its uranium and plutonium, and possibly in other fission products or transuranic elements. The plutonium and other transuranics are the longer-lived radioactive components. The spent fuel is mechanically chopped and dissolved in acid, and the solution processed through subsequent steps to purify the plutonium, uranium, and any other desired product. The remaining fission wastes remain in the high-level waste streams. The plutonium product is usually in the form of a nitrate, and the uranium, either directly or in a subsequent process as UF_6.

Uranium and Plutonium Recycle As described above, the components of value in the spent fuel are the remaining uranium, now of lower enrichment, and the plutonium generated through neutron capture in U^{238}. The uranium can either be re-enriched or blended with new enriched uranium and used to make new fuel elements. The plutonium can also be used either alone or mixed with uranium in making new fuel elements. A breeder reactor, by definition, will reuse the plutonium as fuel. Water reactors are nearly all designed to reuse fuel, but this is currently not general practice in the United States. By law the United States has mandated the use of a once-through cycle, without reprocessing the spent fuel. The economics of the two approaches depend on the relative cost of newly mined uranium.

Waste Disposal Disposal of **low-level radioactive wastes**, from power plants or other radioactive material uses, such as medical is usually done by dry burial in a dedicated site.

High-level waste in the form of solid spent fuel containing transuranics as well as high-level fission products, or reprocessing plant high-level waste converted to a solid form as a glass or ceramic, can be disposed of in a mined geologic repository to ensure long-term isolation of the waste from the environment. Current U.S. policy calls for acceptance of commercial high-level waste by the Department of Energy for disposal in mined repositories. At this time (2017) it seems unlikely that the U.S. government will be able to use those repositories for long-time storage of spent nuclear fuel. The electric power producing utilities have paid a fee on the basis of kilowatt-hours generated to the US federal government to fund the development of a disposal facility. Isolation will be by the use of both engineered (waste package and container) and geologic (geologic media, such as salt, basalt, or granite) barriers. Other countries, after separating the uranium and plutonium, generally plan to store the waste for extended periods of time before final disposal. Key technical areas in implementation include the development of highly stable materials for encapsulating the waste (e.g., glass, ceramic, corrosion-resistant metals), methodologies for predicting repository performance over extended time periods, and repository closure techniques. Demonstration of the ability to safely dispose of waste has become critical to public acceptance of nuclear power in many countries.

PROPERTIES OF MATERIALS

(See also Table 7.8.1.)

Materials used in nuclear reactors can be divided into six basic categories: fuel, moderator, control, coolant, structural, and shielding. While the nature of the materials may be different, they must be physically and chemically stable and compatible with their environment for a sufficient period to enable the reactor to operate predictably, stably, and economically under the particular set of parameters imposed by the design.

In addition to the environmental conditions to which materials may ordinarily be exposed, reactor materials must also sustain the extraordinary bombardment of X-rays (photons) and other atomic particles which are products of nuclear fission. In their passage through matter, these photons and atomic particles may inflict severe damage, since the energy of these particles is spent mainly in ejecting or pulling electrons from their orbits, for example, producing **ionization**. Gamma photons and neutrons are highly penetrating, whereas fission fragments have a path under a few μm in solids and liquids, but cause intense ionization over that path.

The **behavior of materials** following breakage of a bond by ionization may differ markedly. The fragments of simple, highly stable molecules such as water (H_2O) or carbon dioxide (CO_2) will most probably recombine. However, it is highly probable that the parts of complex **organic compounds** such as lubricants and electrical insulation will suffer permanent fragmentation and/or recombination in a different manner to produce new molecules.

Metals, on the other hand, are essentially unaffected by ionization since the probability of displacement of excited atoms in a metal lattice is small and the atoms will merely revert to their original status on ridding themselves of their excess energy. The atoms of **nonconducting crystals** and glasses are similarly immobile, and most of their properties are insensitive to the ionization. However, since these crystals are non-conducting, some of the electrons displaced by ionization become trapped in lattice vacancies, producing changes of optical properties (such as darkening of glass by gamma irradiation) and change of electrical resistivity.

Table 7.8.1 General Thermophysical Properties of Reactor Materials before Irradiation*,†

Materials	Density, g/cm³	Melting point, °C	Heat capacity, J/(kg · K)	Thermal conductivity,† W/(m · K)	Linear expansion coefficient, 10^{-6} (1/K)	Electrical resistivity, 10^{-8} Ω · m
			Fuels and fertile materials			
Uranium, metal	19.3	1,132	155 (673 K)	31.2 (673 K)	13.9 (300 K)	35
Plutonium, metal	19.84	640	135 (300 K)	6.5 (508 K)	33.9 (470 K)	150
Thorium, metal	11.72	1,750	118 (298 K)	54 (298 K)	11.2 (298 K)	13
Uranium oxide (UO_2)	10.96	2,850	328 (1,523 K)	2.6 (1,523 K)	9.8 (300 K)	7.32
Plutonium oxide (PuO_2)	11.44	2,390	344 (1,523 K)	2.2 (1,500 K)	6.7 (300 K)	
Thorium oxide (ThO_2)	10.60	3,650	266 (1,500 K)	3.2 (1,500 K)	8.9 (300 K)	
Uranium nitride(UN)	14.42	2,850	238 (1,000 K)	20.9 (1,000 K)	7.5 (300 K)	146
Uranium carbide (UC)	13.63	2,520	240 (700 K)	23.0 (700 K)	10.5 (300 K)	250
			Cladding materials			
Zirconium	6.52	1,855	278	24	5.7	40
Zircalloy 2	6.55	1,845	290	17	5.8	74
Zircalloy 4	6.58	1,845	293	14.1	5.8	74
Aluminum	1.74	660	897	220	23.3	2.5
Magnesium	6.24	650	1,040	156	24.8	3.5
			Solid moderators			
Graphite	1.71	4,260t	710	103	9.6	16.5
$ZrH_{1.85}$	5.62	650	410	32		54.7
Beryllium	1.84	1,287	1,825	157	34	0.15
BeO†	2.86	2,550	1,229	93	24	
			Structural materials			
Stainless steel, martensitic	7.75	1,427	480	42	11.2	57.0
Stainless steel, austenitic	7.95	1,395-1,425	500	16	16.0	72.0
Molybdenum	10.2	2,622 ± 10	249	147	5.2	4.9
Nickel	8.9	1,690	440	88	13.5	6.2
Niobium	8.57	2,415	265	54	7.3	12.5
			Control materials			
Boron Carbide (B_4C)	2.5	2,450	960	92	4.5	
Hafnium	13.09	2,220	363	22.3	5.9	35.1
Alloy AgInCd	10.17	800	230	60	22.5	
Eu_2O_3	7.34	2,050	413	2.2	1.01	

* All properties change with irradiation at room temperature except when specified.
† Thermal conductivity, strength, and the modulus of elasticity vary directly with the density; however, the coefficient of thermal expansion appears unaffected.
SOURCE: "Thermophysical Properties of Materials for Nuclear Engineering," IAEA, 2008.

On direct collision, **fast neutrons** may displace atoms from a chemical compound without affording opportunity for immediate recombination. Similarly, **energetic neutrons** may displace atoms from their normal lattice positions in a metal or other crystalline material and push them into abnormal interstitial positions in the lattice, generally producing an effect on metals similar to work hardening. The hardness and tensile strength of the metal will increase, but the metal becomes more brittle. **Thermal neutrons** affect the properties of metals only in those cases where the probability of absorption is high enough to cause significant transmutation to other elements.

Most severe damage occurs when the atoms of a fuel material fission and produce two or more new atoms from each of the original fissioned atoms. Volumetric changes occur by the substitution of these additional atoms in the lattice for the original atom, and the lattice is further damaged by the intrusion of the high-velocity fission fragments.

Fuel Materials

Fuel materials are generally divided into two categories: fissionable and fertile. The **fissile materials** are those isotopes of uranium and plutonium which fission upon interaction with thermal neutrons. These are the isotopes U^{233} U^{235}, Pu^{239}, and Pu^{241}, which have odd atomic weights. The **fertile materials** are those isotopes of uranium and plutonium which have even atomic weights. These absorb neutrons under proper conditions and undergo a series of nuclear reactions, with the eventual formation of fissile isotopes. Thus U^{238} forms Pu^{239}, and Pu^{240} forms Pu^{241}. In addition, thorium 232 forms the fissile isotope U^{233}.

Both fissile and fertile materials fall short of possessing the **properties considered ideal** for a nuclear fuel, namely, corrosion resistance, good thermal conductivity, strength, and ductility. To overcome these problems as well as to prevent fission products from entering the coolant, various techniques are used, including plating, cladding, and alloying.

Materials for this use must be corrosion-resistant, be compatible with the fuel materials, be physically stable under irradiation, have good heat transfer characteristics, have good nuclear properties, and be reasonably fabricable. From a consideration of cross section alone, only aluminum, beryllium, magnesium, and zirconium are attractive for use in thermal reactors. Magnesium usually is ruled out on the basis of strength and corrosion resistance but has been widely used in some gas-cooled systems. In fast reactors, a number of other materials become attractive, the most important being stainless steel, which also is used under special conditions in thermal reactors. Consideration also is being given to ceramic materials, BeO, Be_2C, SiC, and $MoSi_2$, which are very attractive in all phases except resistance to thermal shock.

Moderating material must be capable of reducing neutron energy very rapidly and should have good strength at high temperatures, corrosion resistance, thermal and radiation stability, and reasonable cost. Such properties are not available in a single material. Graphite is the most widely used solid moderator. Attention also is being given to beryllium and its compounds, deuterium, oxygen, and hydrogen.

The nuclear and physical properties considered desirable for **fuel diluents and cladding** are, in general, desirable for structural materials. For the latter, however, more emphasis is placed on strength and corrosion resistance than on nuclear properties. Stainless steel, aluminum, zirconium, molybdenum, titanium, niobium, and their alloys are used. As operating temperatures continue to rise, ceramics and cermets will have to be developed for this purpose.

Materials used for **control purposes** must have a high cross section for absorption of neutrons, adequate strength, low mass to permit movement, good corrosion resistance, chemical and dimensional stability under heat and irradiation, and low cost. Boron, cadmium, and hafnium are given the most consideration in various metallic and ceramic forms and as compounds dissolved in the coolant.

Coolant Properties

The choice of a coolant also is a compromise, since no known material possesses all the **desirable characteristics**. These are good heat transfer coefficient and good heat capacity on a volume basis, low absorption cross section, low vapor pressure at the operating temperature, minimum of chemical reaction with the material contacted by the coolant, good resistance to forming γ emitters with long half-lives, stability under irradiation and high temperatures, and workable melting points.

Coolants are divided into three broad **classes:** aqueous, liquid metals, and gases. **Water** has found the widest application, largely because it is readily available and economical, has fairly good heat transfer properties, and offers no corrosion problems which cannot be handled. Its most serious drawback is its high vapor pressure. **Heavy water** has superior qualifications except for cost, which is high. Diphenyl and other **organic liquids** have demonstrated an ability to stand up under irradiation but are handicapped by a tendency to deposit decomposition products on the fuel element plates. See Table 7.8.2.

Liquid metals are attractive because they provide high heat-transfer rates and do not require pressurizing to operate at high temperatures. Sodium is most attractive because of its high boiling point and high heat conductivity, but it reacts violently with water and has poor lubricating qualities. Lithium 7 better meets qualifications but is harder to obtain and is costly.

Gas coolants offset their advantage of being inert at elevated temperatures with poor heat transfer qualities. They must be operated at high pressures to be useful in power reactors.

FISSION REACTOR DESIGN

Neutron Balance The main objective of reactor design is to achieve neutron balance during reactor operation. At steady state, the number of neutrons produced during fission is equal to the neutrons lost by absorption and leakage processes. The net rate of change of neutron density n (neutrons per cubic centimeter) is given by the neutron conservation equation:

$$\frac{\partial n}{\partial t} = \text{production} - \text{absorption} - \text{leakage}$$

If $\partial n/\partial t$ is positive, the reactor is said to be supercritical; if it is zero, the reactor is at steady state and is said to be critical; if it is negative, the reactor is subcritical.

Cross Section The microscopic cross section σ for a particular nuclear process is the effective target area of the nucleus with which a neutron must interact to produce the given reaction. The unit of σ barn (1 barn = 10^{-24} cm^2). Neutron cross sections (absorption, scattering, and fission) constitute the basic nuclear data for reactor design. Absorption cross sections for thermal neutrons range from 0.00046 barn for deuterium to 3.3×10^6 barns for $xenon^{135}$. Moderating atoms for the slowing down of fast neutrons have low atomic mass, small absorption, and large scattering cross sections. Examples are hydrogen, deuterium, beryllium, and carbon. Table 7.8.3 gives thermal neutron cross sections in barns for selected elements (based upon the naturally occurring combinations of isotopes).

Multiplication Factor The number of neutrons produced in any one generation in a reactor for each neutron produced in a previous generation is called the multiplication factor k. If $k = 1$, the reactor is critical; if $k < 1$, it is subcritical; and if $k > 1$, it is supercritical. For an infinite thermal reactor, multiplication factor $k_\infty = \eta f p \varepsilon$ where η is the number of fission neutrons produced per neutron absorption in fuel, f is the thermal utilization factor (neutron absorption rate in fuel/total absorption rate), ε is the fast-fission factor, and p is the resonance capture escape probability. The leakage of neutrons from the reactor core is accounted for by multiplying k_∞ by the factor $1/(1 + M^2B^2)$, where M^2 is the migration area (sum of the squares of the diffusion length L and the slowing-down length τ) and B^2 is the geometric buckling ($B^2 = 0$ for an infinite reactor).

Breeding Ratio If the ratio of a number of fissile nuclides produced by the capture of neutrons by the fertile nuclides to the number of fissile

Table 7.8.2 Properties of Nuclear Coolants*

	Aqueous			
	Water	Heavy water		
Density, kg/mt³	42.4 (600°F)	46.2 (600°F)		
Viscosity @ 450°C, m²/s	0.20 (600°F)	0.220 (600°F)		
Melting point, °C	32	39		
Boiling point, °C	212	214		
Heat capacity, kJ/(kg-K)	1.54 (600°F)	1.68 (600°F)		
Thermal conductivity, W/(m-K)	0.294 (600°F)	0.294 (600°F)		
Capture cross section σ_a, 0.025 eV (barns)	0.66	0.0011		
	Gases			
	Helium	Helium	CO_2	Air
Density, kg/m³	0.00375	0.00375	0.041	0.0275
Viscosity @ 450°C, m²/s		0.094	0.081	0.089
Heat capacity, kJ/(kg-K)		1.245	0.283	0.263
Thermal conductivity, W/(m-K)		0.159	0.0328	0.0337
Prandtl no. @ 450°C		0.735	0.6988	0.6898
	Liquid metals			
	Lithium	Sodium	Na (56 percent) K (44 percent)	Pb (44.5 percent) Bi (55.5 percent)
Density, kg/m³	29.9	51.2	48.8	625.5
Viscosity @ 450°C, m²/s	1.188	0.5040	0.435	2.88
Melting point, °F	354	208	66.2	257
Boiling point, °F	2,403	1,621	1,518	3,038
Heat capacity, kJ/(kg-K)	1.0	0.3005	0.2484	0.035
Thermal conductivity, W/(m-K)	18.0	37.7	16.35	8.05
Capture cross section σ_a, 0.025 eV (barns)	71.0†	0.505		0.094

* Adapted from "Thermophysical Properties of Materials for Nuclear Engineering, IAEA, 2008, which contains a detailed bibliography.

Table 7.8.3 Absorption (σ_a) and Scattering (σ_s) Cross Section for Thermal Neutrons, in barns

Element	σ_α	σ_s	Element	σ_α	σ_s	Element	σ_α	σ_s
Aluminum	0.23	1.4	Helium	0.007	0.8	Potassium	2.1	1.5
Beryllium	0.01	6.1	Hydrogen	0.332		Plutonium^{239}	1011.3	7.7
Bismuth	0.033	0	Iron	2.55	10.9	Sodium	0.53	3.2
Boron	759	3.6	Lead	0.17	11.4	Thorium	7.40	12.67
Cadmium	2,450	5.6	Magnesium	0.063	3.42	Tin	0.63	
Carbon	0.0034	4.75	Molybdenum	2.65	5.8	Titanium	6.1	4
Chromium	3.1	3.8	Nickel	4.6	17.5	Uranium	7.59	8.9
Copper	3.79	7.9	Niobium	1.15		Zirconium	0.185	6.4
Hafnium	102	8	Oxygen	<0.0002	3.76			

NOTE: Individual isotopes have markedly different cross section. For example, U^{235} has total σ_α = 687 barns and fission cross section σ_f = 587 barns.

nuclides consumed is greater than unity, it is called the **breeding ratio;** if it is less than unity, it is called the conversion ratio. For breeding, the value of η (defined above) must be greater than 2, as one neutron is required to maintain the chain reaction by fission, another neutron maintains breeding by its capture by the fertile nuclide, and the remaining to compensate for the loss of neutrons due to absorption and leakage. Neutron economy favors the breeding of Pu^{239} fissile nuclides from fast neutron capture by U^{238} fertile nuclides. This breeding process is the basis of the LMFBR (liquid-metal fast breeder reactor). On the other hand, thermal breeders are based upon the breeding of uranium 233 fissile nuclides from thorium 232 fertile nuclides by the capture of a thermal neutron.

Diffusion Theory Multienergy group diffusion theory is presently the basis of reactor design. For a critical reactor the neutron balance equation reduces to

$$D_g \nabla^2 \phi_g - \sum R_g \phi_g + S_g = 0$$

where ϕ_g and S_g are the gth energy group flux and the slowing-down neutron source; D_g and R_g are the diffusion coefficient and removal cross section. The latter are called group constants, which are obtained by the analysis of the $\Sigma R_g \phi_g$ basic fuel lattice. The solution of the above equations provides the reactor multiplication factor, reactor critical size, neutron fluxes, power distributions, isotopics of elements, and fuel-cycle length.

Control Extra fuel is required for depletion to compensate fuel lost in producing energy, to account for the loss of reactivity due to the buildup of fission products, and to account for the change in reactivity due to changes in fuel and moderator temperatures. The control of excess reactivity is undertaken by the control rods of strongly absorbing materials having large absorption cross sections (such as boron, hafnium, or silver-indium-cadmium) and the burnable poison rods containing boron or gadolinium. In PWR, soluble boron is also employed. Control rods are inserted to shut down the reactor and are withdrawn to make the reactor critical.

A fraction of fission neutrons (0.0065) are delayed neutrons, that is, they are emitted as the fission fragments decay not at the moment of fission (prompt neutrons). The presence of the delayed neutrons makes control of the reactor easier.

Thermal

Sources of Heat The energy produced by fissioning appears in several forms, each of which eventually degrades to heat. To provide for removal of this heat, the core, or active portion of the reactor, usually consists of uranium bearing fuel elements with passages around these elements for flow of coolant. The rate of coolant flow and the temperature of the coolant must be such as to permit removal of all the heat generated in the elements without exceeding allowable materials properties.

Hot Spot and Hot Channel Factors The problem of heat removal is complicated by the fact that heat is not generated uniformly throughout the core. The geometric configuration of most power reactors is approximately represented by a finite cylinder of height H and radius R. Diffusion theory predicts the following thermal neutron flux pattern for an unreflected, unrodded, and homogenized cylindrical reactor as a function of axial Z and radial r variables:

$$\phi(z,r) = \phi_0\left(\cos\frac{\pi Z}{H}\right)J_0\frac{2.405r}{R}$$

where ϕ_0 is peak flux and J_0 is the Bessel function.

In addition, the power generation is influenced by mechanical factors such as variables in fuel element dimensions, fuel concentrations, and flow. These nuclear and mechanical factors are generally called hot spot factors. Thus the peak power density of heat flux in the core is greater than the average by an amount indicated by the hot spot factor. In addition, hot channel factors are identified which indicate how much greater the temperature rise of coolant is in the hot channel than in the average channel.

Design Criteria The maximum fuel element surface temperature T_{sm} is of interest because it may be limited by corrosion effects or by a desire to avoid film boiling, a condition which lowers the heat transfer from the fuel to the coolant. However, other limiting conditions also exist. In addition the internal fuel element temperature during all anticipated reactor transients is generally limited to a maximum value (generally below the melting point of the fuel) or to a maximum average value.

Coolant temperature rise generally is limited indirectly by fuel temperature limits. Occasionally, however, thermal stresses within the fuel element may impose a direct limit. In pressurized water reactors, the coolant temperature is limited not to exceed the saturation temperature of the fluid; thus it is more significant to speak of the maximum enthalpy rise, which is related directly to the steam quality of the fluid leaving the hot channel.

Density changes within the coolant may be limited to provide control stability in reactors that use boiling heat transfer.

Coolant velocity usually is limited to a maximum value by erosion or by vibration resonances, and sometimes is required to exceed a minimum value because of crud deposition. The burnout condition is determined in a given fuel element geometry for a particular combination of fuel element heat flux, primary coolant flow, and primary coolant enthalpy. Steady-state, local, or bulk boiling may have to be limited, even when burnout is not a danger, because of pitting of fuel element surfaces or crud deposition on the fuel surfaces. The burnout problem is most acute during reactor transients; hence steady state burnout generally is not limiting.

Mechanical

The mechanical design of nuclear reactors is influenced by three factors not encountered with other apparatus: the need to control criticality, the effects of irradiation, and the high power density produced in the core.

Control of Criticality The following requirements are introduced by the need to control criticality:

1. Not only must components perform reliably, but also if they fail to function as designed, the reactor will shut down. Small parts located elsewhere in the core should be so designed that if they fail they do not interfere with the action of control rods in shutting down the reactor.
2. Emergency shutdown mechanisms must be fast-acting devices to reduce the time delay in initiating shutdown.
3. During normal operation, components of the core must be rigid enough so that deflections due to mechanical or thermal stresses do not introduce unpredicted criticality variations.

Effects of Irradiation Neutron and gamma irradiations cause the following effects, which must be considered in the design of reactor systems and reactor components:

1. Radiation effects on properties of materials—particularly on the brittle fracture characteristics of the reactor vessel, on the relaxation characteristics of bolts, and on crevice corrosion.
2. Heat generation brought about by attenuation of radiation with in components materials—probably insofar as such heat influences heat removal requirements, thermal stresses, and thermal distortions of parts.
3. Induced radioactivity in corrosion and wear products that could be carried by the coolant and deposited in such regions as pumps and valves.

High Power Density A nuclear reactor requires a large amount of heat transfer surface for removal of the high power per unit volume which can be produced. Thus the reactor consists of many small, closely packed fuel elements. Mechanical tolerances are often appreciable fractions of the dimensions and spacing of these elements and therefore influence thermal performance. The influence of such tolerances on performance must be evaluated. Parts also must be rigid and vibration-free in the face of the high fluid flow rates required for power removal. The high power density may also lead to large thermal stresses in fuel elements.

Furthermore, moving parts may have to operate without benefit of organic lubricants because such lubricants tend to sludge under irradiation, and if deposited on heat surfaces could interfere with removal of heat from the core.

Residual radioactivity also influences programs and facilities for refueling of the core and for periodic maintenance of components in the reactor system. In addition, personnel exposure by the above radiation requires adequate shielding.

Shielding To protect reactor operating personnel against damaging biological effects of neutrons and gamma rays, shielding is required around a nuclear reactor. Neutron and gamma ray fluxes in the range of 10^{13} to 10^{14} must be attenuated to 10^3 particles/(cm$^2 \cdot$ s) to meet the tolerance radiation level.

To attenuate gamma rays, which interact primarily with the orbital electrons of atoms, a material with a high atomic number containing a high density of these electrons is required. Examples are lead, tungsten, depleted uranium, or concrete containing high-Z elements in the form of scrap or heavy ore.

To attenuate neutrons, they must be slowed down and then absorbed. Hydrogenous materials such as water, concrete, or polyethylene are excellent moderators. The slowed neutrons must be absorbed without producing high-energy-capture gamma rays by using boron 10. Therefore, some gamma shielding outside the neutron shield is generally required.

Electrical

Neutron level instruments, located adjacent to but outside the core or core vessel, are used for the basic control of nuclear reactors. Thermocouples, flowmeters, pressure taps, and fission chambers are used inside the core for power calibration of nuclear instruments and for determining the distribution of power and/or temperature within the core; in turn, the power distribution is used to compute heat fluxes and fuel temperatures for comparison with limiting values.

Neutron level instruments utilize the fact that neutrons can produce ionizing particles to indicate their presence. The two most common processes used for this purpose are the (n, α) reaction with boron and the fission reaction. The resulting ion pair is used to produce ionization of a gas in a tube containing a central anode and outer electrode. The electric discharge produced by the ionization is used for measurement purposes. The detailed design features vary widely with the range of the instrument and the type of reactor in which it is to be used.

Different neutron level instruments are employed during the reactor operation at various power levels, because of the wide range of sensitivity required to monitor neutron fluxes from start-up to full-power condition. In power reactors at least three instrument ranges are used with the overlap between the adjacent ranges. These are (1) source range, (2) intermediate range, and (3) power range. During the reactor start-up (source range) the BF_3 proportional counters are used to measure neutron flux in the range of 10^{-1} to 5×10^4 neutrons/(cm$^2 \cdot$ s). The compensated ionization chambers are employed to monitor neutron flux in the intermediate power range of 2.5×10^2 to 2.5×10^{10} neutrons/(cm$^2 \cdot$ s). In the power range, where neutron flux ranges from 2.5×10^7 to 2.5×10^{10} neutrons/(cm$^2 \cdot$ s), uncompensated ionization chambers are employed.

Compensated ionization chambers are designed to count neutrons against a background of gamma radiation. Uncompensated chambers are useful for providing axial power distribution.

All the above instruments are located outside the core and monitor neutron flux, hence reactor power. The latter is also obtained from the measure of ΔT (change in temperature) across the reactor core from the temperature measuring instruments located in the primary loop of the reactor. However, these instruments do not provide local flux peaking and local anomalous coolant temperature situations. In-core instrumentation must be used for this purpose.

In-core instrumentation is employed to provide information on neutron flux and fuel-assembly-coolant outlet temperature at a few selected locations in the reactor core. Thermocouples measure the coolant outlet temperature and thus provide the rise in local coolant temperature, a measure of the increase in local reactor power. Miniature, fixed, and movable fission chambers containing U_3O_8 which is 90 to 95 percent enriched U^{235} constitute neutron flux detectors. The experimental data obtained during the reactor operation from the in-core instrumentation provide the check against the reactor design parameters such as the assembly distributions and hot channel factors.

In addition to thermocouples and fission chambers, other in-core instruments such as pressure, displacement, and strain transducers are also desirable. None of these instruments except thermocouples have operated satisfactorily over a long period. The installation of in-core instrumentation is vital to the safe and economic operation of the reactor.

NUCLEAR POWER PLANT ECONOMICS

The decision to deploy a nuclear generating station is complex. It is not a decision that weighs merely the economics of each alternative. It is influenced by National Energy Policy as well as local policies that can incentivize development. These policies have economic consequences. Nonetheless, the decision to introduce any energy source has social and political impact and these considerations need to be addressed by citizens and their elected representatives.

Generally speaking each country and each Regional Transmission Organization (RTO) which can be transnational needs to have a strategy to ensure that the current and future energy needs of the citizens are met. This strategy needs to incorporate disparate factors such as diversity of energy supply, rate stabilization, balance between base load generation and load following generation, availability of suitable sites, potential impact on climate, sophistication of the society and grid size where deployment is proposed (i.e., developed vs. developing nation), availability of trained workforce to construct and operate the facility.

Diversity

Nuclear generating stations rely on a unique fuel supply. Commercial units operate up to two years between refueling. Thus they do not need continuous access to a transportation infrastructure to provide a constant flow of fuel to maintain plant output as is the case with coal and other fossil fuels. Nuclear Plants are refueled on a cyclical basis of typically 12, 18, or 24 months depending on considerations that weigh the economics of carrying higher cost fuel versus lost revenue during "down time" to refuel at more frequent intervals. This characteristic has proved to be important when barge traffic carrying coal on inland waterways has been interrupted during extensive and deep cold spells.

Stability of Electric Rates

Another characteristic of nuclear plants is that their Operations and Maintenance costs represent a small and stable component in the overall cost of nuclear generation (~\$13/MWh [c]). Consequently the cost of generation at an operating nuclear plant is not volatile and contributes to stable electric rates.

Baseload vs. Load Following vs. Displacement

Nuclear generating stations are expensive to build. Consequently, to be viable they must be base loaded to generate the revenues needed to service the debt and return on equity. Other competing technologies such as Combined Cycle Gas Turbine (CCGT) Technology can accommodate periods of reduced generation (hence revenue) due to their lower capital cost and consequently, lower fixed costs. CCGT fuel costs are a larger component of the overall cost of operating the facility. Thus reduced load results in significantly reduced operating costs allowing them to be profitable while load following.

Renewables such as Wind and Solar fall into a third category that "Displaces" active generation on the grid when the wind blows or the sun shines. For reasons stated above, displacing nuclear generation is very damaging to the economics of nuclear plants. In addition in some countries such as the US, Renewables receive a Production Tax Credit. The inflation adjusted credit in the US in 2016 is \$23/MWH[d] for wind generated electricity. The duration of this tax credit is 10 years.[d] Thus Wind generation can bid to provide electricity to the grid at no cost to the consumer and still derive income in the form of credits. This skews capital markets and is a deterrent to building any capital intensive generation.

Siting

Siting a nuclear plant may not be an option in some areas. It requires approval from the NRC as noted under the Licensing section. Siting weighs:

- the nature and likelihood of external events occurring in the region, both natural and human generated
- Population density and ability to carry out emergency evacuations
- the hydrology, seismology, geology, meteorology, and other characteristics that influence safe performance and migration of radionuclides offsite
- Access to a water supply and a heat sink (e.g., body of water such as an ocean or river, cooling towers).

A typical nuclear plant requires on average 1.3 square miles to accommodate a 1,000 MWe unit. A comparable wind farm would occupy 260 to 360 square miles depending on availability factor. Similarly, Solar Photovoltaic (PV) technology would require 45 to 75 square miles.[e]

Climate Considerations

If the external social costs of electric generation are reflected in the price, nuclear generation becomes more competitive when compared to coal and natural gas generation. Nuclear plants have essentially no carbon emissions. The EPA estimates that a carbon tax of \$50/tC will reduce US CO_2 emissions by about 1 billion metric tons per year.[b]

Developed vs. Emerging Nations

Nuclear generating stations are being deployed in many places throughout the world for the first time. This includes emerging nations in Asia, the Middle East, and Africa. Although these countries may have vibrant economies, they do not necessarily have the infrastructure, technical expertise, and organizations to support the deployment and regulation of nuclear plants. Thus states such as these are investing to establish the necessary prerequisites to successfully deploy and regulate a complex industry. Coupled with this effort is necessarily an educational initiative to bring their populations along in their understanding of the technology, its benefits and its risks.

Workforce

A trained workforce is needed to support deployment of nuclear plants. This means not only the technical capability to design, maintain, and operate the plant but also craft labor with the breadth of skills and associate qualifications that are essential to construct the plant. In most countries nuclear plants have not been under the continuous deployment that would allow the technical workforce and craft labor to function as a knowledgeable, coordinated unit as they moved from one project to the next. This is a significant element responsible for the delays in plant construction and the associated escalation in plant cost.

Economic Benefits to a Region

A nuclear plant needs 450 to 700 people to run an operational plant depending on the plant design and whether it is an add-on unit or a standalone unit. In addition as many as 2,400 onsite construction jobs are created during peak periods of construction. Consequently, there is a boost to the local and regional economy due to the influx of people both during and following construction. Tax revenues to the local community are transformational.

Regulated Markets

Virtually all the nuclear plants operating today where procured in a regulated market. Plants were constructed by state owned or investor owned, vertically integrated, regulated, utility monopolies. Thus the construction risk was borne by the consumer. For the most part, electric rates for regulated utilities were set by Public Utility Commissions and were based on rate cases submitted by the utility. Rates would be set to allow the utility to cover its costs of service and make a reasonable profit for its shareholders, many of whom were retired individuals seeking a safe return on their savings. As the cost of nuclear construction escalated over the decades, Public Utility Commissioners put increasing pressure on utilities to demonstrate that they were "prudent" in their management decisions and actions. Thus it became evident over time that shareholder dividends were not necessarily certain.

In the 1980s "Prudence Hearings" justifying costs on nuclear construction became contentious. This was followed by a long period where no nuclear plants were ordered and the industry consolidated. An example of this consolidation is the sale of the Seabrook Nuclear Station in New Hampshire to Florida Power and Light. The nominal output of the plant in 2002 when the deal was completed was 1,024 MWe. The sale price was $749.1 million ($732/KWe). This turned out to be a very profitable transaction in a deregulated market. The EIA estimates that the Total Overnight costs in 2013 dollars for a plant coming online in 2022 is $ 5,366/KWe.[j]

"New" construction began to appear in 2002 when TVA decided to invest in a unit that had been shutdown since 1985 (Browns Ferry Unit 1). The Browns Ferry Unit was restored to service in 2007 after an investment of $1.8 billion. It is rated at 1,155 MWe. In 2007 TVA's Board approved the expenditure to complete a previously cancelled, partially constructed unit (Watts Bar Unit 2). It became the utility's 7th operating unit when it was issued a full power operating license on October 22, 2015, and represented the first electricity produced by a newly completed nuclear unit in 20 years.

On April 8, 2008, Georgia Power entered into an agreement to build two Westinghouse AP1000 units at its Vogtle site in Georgia. These units are designated Plant Vogtle 3 & 4. [The Georgia Public Service Commission (PSC) rules required market bids to be compared with Georgia Power's self-build proposal, but no market bids were received.] An important element in the financing of these units was the passage of a state law which allowed Georgia Power to recover financing costs during the construction of nuclear units. This policy decision runs contrary to past practice which reasons that current customers should not pay for tomorrow's power generating capacity. In February 2010, Georgia Power was awarded DOE loan guarantees which are projected to save consumers millions in interest payments annually.

Plant Vogtle units 3 & 4 will be the first in the industry to use the Westinghouse AP1000 advanced pressurized water reactor technology. This advanced technology allows nuclear cores to be cooled even in the absence of operator interventions or mechanical assistance. The units are projected to go into operation in 2019 and 2020 respectively.

There are two additional Westinghouse units under construction at the Virgil C. Summer Nuclear Generating Station in South Carolina. These units are also within a regulated market

Deregulated Markets

Deregulation was an attempt to substitute market forces for regulated generation rates. Price deregulation requires open markets.

Exchange markets take the form of bid-offer auctions where sellers can bid against each other and market clearing prices are known by all parties.

These markets place a heavy burden on anyone contemplating the construction of new generation. They will be required to compete in competitive wholesale markets where the risk formerly borne by consumers is now shifted to suppliers. These suppliers will use a risk adjusted cost comparison in evaluating prospects for new generating capacity. The cost of capital will be greater than in a regulated environment since the risk of regulatory delays, construction cost escalation, low capacity factors, and competition from gas and coal are no longer borne by consumers. This leads investors to prefer less capital intensive projects.

Levelized Cost of Electricity (LCOE)

The real levelized cost of electricity is the constant dollar price of electricity required over the plant life needed to cover interest and principal payments on debt, O&M costs, taxes and an attractive rate of return for equity investors. Appendix 5 in reference f notes the need to use "real" levelized costs where "revenues and expenses are projected over the life of the project and discounted at rates sufficient to satisfy interest and principal repayment obligations to debt investors and the minimum hurdle rate (cost of equity capital) required by equity investors." This approach is used in deregulated markets as a measure of the economic competitiveness of different electric generation alternatives with different capital investment profiles. It is expressed as the per kilowatt-hour cost of building and operating a plant over its financial life and duty cycle.

Levelized Avoided Cost of Electricity (LACE)

The specific conditions under which new generation will interact within a region where it is postulated can vary dramatically. Consequently a comparison of the LCOE for competing technologies is a good academic tool for broadly comparing technology options, but not a representative way to assess the relative value of the technology when operating in the environment in which it is deployed. A better assessment of economic competitiveness is to predict the value of what it would cost the grid to generate the energy that would be displaced by the new generation project. Avoided cost is a measure of the economic value of a proposed project over the life of the plant and is divided by the average annual output of the plant to establish its levelized avoided cost of electricity (LACE).

The LACE when compared to the LCOE provides an indication of whether the value of the project exceeds its cost or not. This approach can be used for candidate technologies to determine which one provides the best economic incentive for development.

The U.S. Energy Information Agency (EIA) in its August 2016 report has calculated the LCOE (weighed average of regional values) for Nuclear Generation entering service in 2022 to be \$99.7/MWh.[g] The breakdown is as follows:

Levelized Capital Cost	\$75/MWh
Fixed O&M	\$12.4/MWh
Variable O&M (including fuel)	\$11.3/MWh
Transmission Investment	\$1.0/MWh
Total System LCOE	\$99.7/MWh

Similarly, the EIA has calculated the capacity weighted average LACE for Advanced Nuclear capacity entering service in 2022 as \$61.4/MWh. One gains an indication of the economic viability of a technology by comparing the difference between capacity-weighted levelized avoided costs of electricity (LACE) and capacity-weighted levelized costs of electricity (LCOE). For nuclear plants entering service in 2022 this difference is determined to be \$99.7/MWh - \$61.4/MWh (i.e., negative \$38.3/MWh). This number must be positive for a project to be viable.

Given the challenges represented by the economic hurdle identified above, countries with market based generation are taking steps to incentivize nuclear construction to meet energy policy objectives as delineated in the example below for the UK namely:

- stable and secure supply of electricity
- dispatchable technology with high availability
- low carbon
- cost competitive
- grid stability

A press release [h] dated October 23, 2013, from the UK Department of Energy and Climate Change noted that the UK Government and EDF had reached agreement on the commercial terms of providing new nuclear capacity without upfront money form the UK taxpayer. The deal represents agreement for EDF to provide an investment of £16 billion to build twin nuclear units each with a capacity of 1.6 GWe. These two units can generate 7 percent of the UK's electric demand when running at full power. The key to the deal is agreement on a "Strike Price" of £92.50/MWh. Government reforms to the electricity markets have created an investment contract or "Contract for Difference (CfD). Under a CfD the generator is guaranteed to be compensated at a wholesale electric rate designated as the "Strike Price." The project developer (in this example EDF) would receive the difference between the Strike Price and a reference market price when market prices are lower than the Strike Price. Conversely, when market rates rise above the Strike Price, the developer is required to refund the difference to the consumer. The CfD transaction does not involve the taxpayer but only the consumer.

Advances in Construction

In countries such as Japan there have been decades of continuous nuclear plant construction contributing to the continuity and skills of the workforce. This factor along with innovation in construction methods have led to a remarkable record of performance in adhering to construction schedules and addressing the largest risk factor associated with nuclear construction. Unfortunately, the 2011 Tōhoku earthquake and tsunami has interrupted new nuclear construction in Japan.

In addition to the continuity that comes with a thriving industry Japan had incorporated new construction methods beginning with the open-top method where construction is sequenced to allow components to be placed into position from above. The effectiveness of this approach was enhanced with the introduction of crawler cranes with large capacities and long reach. This allowed large (>500 ton) components to be efficiently set in position by the crane. The long boom and mobility of the crawler gave the construction crew access to virtually every location within the facility. The next advancement was the introduction of parallel construction whereby multiple large modules were constructed in parallel on site and sequenced to be set in place by the high capacity crane. These large modules constructed on site were complemented by smaller, more complex modules constructed at an offsite module factory where the environment and quality could be controlled and monitored. Modules were transported from the factory to the site over land or by barge. More recently RFID technology is being used for inventory control as well as for monitoring individuals and their qualifications for various tasks. These innovations have resulted in predictable schedules. A prime example is the record of ABWRs constructed in Japan beginning with the first unit that went into commercial operation in November 1996 through the forth unit that went into commercial operation in March 2006. The time from First Concrete (FC) through Fuel Load (FL) for these four ABWRs were 37 months, 37 months, 43.5 months and 43 months, respectively. [k] To replicate this performance outside Japan is a challenge being undertaken on the Horizon Project ABWR units at Wylfa in the UK.

(a) http://www.nei.org/CorporateSite/media/filefolder/Policy/Papers/Land_Use_Carbon_Free_Technologies.pdf?ext=.pdf

(b) Summary and Analysis of McCain-Leiberman 'Climate Stewardship Act of 2003', William Pizer and Raymond Kopp, Resources for the Future, January 28, 2003.

(c) Updated Capital Cost Estimates for Utility Scale Electricity Generating Plants US Energy Information Agency Report, April 2013.

(d) http://energy.gov/savings/renewable-electricity-production-tax-credit-ptc

(e) "Land Needs for Wind, Solar Dwarf Nuclear Plant's Footprint," Nuclear Energy Institute.

(f) The Future of Nuclear Power, an Interdisciplinary MIT Study 2003.

(g) US EIA

(h) Press Release dated October 23, 2013 from Department of Energy & Climate Change, "Initial Agreement reached on new nuclear power station at Hinkley."

(i) Policy paper: Nuclear Industrial Strategy - The UK's Nuclear Future from Department for Business, Innovation & Skills and Department of Energy & Climate Change first published March 2013.

(j) http://www.eia.gov/outlooks/aeo/assumptions/pdf/electricity.pdf EIA Electricity Market Module.

(k) Lessons Learned from Deployment of Four Operating ABWRs, EUCI conference October 2006, Jack Tuohy, Director Hitachi America, Ltd.

NUCLEAR POWER PLANT SAFETY

Through the end of 2015, over 4,000 reactor-years of operation had been accumulated by commercial nuclear power plants in the United States, and about 15,000 worldwide, not including military or shipboard reactors (International Atomic Energy Agency, "Nuclear Power Reactors of the World"). This operational history has been accompanied by an impressive safety record with respect to health and safety impacts on the public, yet accidents have occurred. The most disastrous occurred in 1986 at the Chernobyl nuclear power plant complex located in the then U.S.S.R., resulting in loss of life and long-term radioactive contamination over a large area in the vicinity of the plant as well as transport of short-lived radioactive contamination dictated by the prevailing local weather patterns. Ultimate disposition of the destroyed reactor, now encased in an added containment structure ("sarcophagus"), remains a concern. In the United States, two of the worst commercial accidents occurred at the Brown's Ferry facility in 1975 and the Three Mile Island (TMI) facility in 1979. In the Brown's Ferry accident, a fire burned through hundreds of electrical power and control cables, threatening the plant's capability to be shut down and maintained in a safe condition. There was ultimately no damage to the reactor and no off-site radioactive releases occurred. At Three Mile Island significant reactor damage occurred through a sequence of events that began with a stuck-open primary system relief valve. Radioactivity was released to the environment, mostly in the form of noble gas fission products. However, estimated exposures to members of the public were found by a Federal interagency working group to be exceedingly small, even for the maximally exposed individual. The TMI reactor vessel did not fail despite relocation of approximately 19 tons of molten core materials. Although these accidents had minimal public impact, safety systems and plant operators were severely challenged during the events. The most recent serious commercial nuclear power accidents occurred in March 2011 at the Fukushima Daiichi power plant in Japan. A massive earthquake off the northeast coast of Japan generated a sizable tsunami which overwhelmed the sea wall defenses designed to protect the plant from this occurrence. As a result of the loss of offsite power and the inability to provide long-term cooling of the reactors, the fuel in three of the reactors melted at the Fukushima Daiichi site causing containments of these reactors to fail resulting in the release of radioactivity into the surrounding environment. Ongoing efforts at the Fukushima Daiichi site include efforts to stem the release of additional radioactivity, particularly from the contaminated water in the three failed units.

Commercial reactors have inherent safety features that prevent them from exploding like bombs. There are two fundamental inherent safety features that contribute to the controllability of commercial power reactors. The first of these is the **Doppler effect**. As the power level of a reactor increases rapidly, the temperature of the fuel increases very rapidly. This rise in temperature causes the low-energy resonant peaks in the microscopic cross section to broaden, effectively creating a larger probability that neutrons will suffer nonfission captures by the U^{238} in the fuel. The second inherent safety feature is caused by the physical densities of materials, particularly the moderator, that decrease as the temperature increases. This results in higher probabilities that neutrons will escape from the reactor rather than cause fission in the fuel. Both of these inherent safety features provide a negative feed back to the chain reaction in a way to reduce the power level of the reactor, thus automatically shutting the reactor down. This is called a **negative temperature coefficient**. Steam voids in a water moderator have a similar effect.

In spite of the inherent safety features, a reactor still provides a potential hazard to plant workers and members of the public in the vicinity of the plant. This is because of the generation of highly radioactive fission products within the fuel. The essential safety challenge is to assure that these fission products are kept confined at all times.

Nuclear power plant safety is based on a defense-in-depth philosophy. The defense-in-depth relies not only upon the inherent safety features, but also on high quality standards that provide a high degree of assurance that accidents are not likely to occur, and that if they do occur, the radioactivity will be contained locally.

High quality begins with the selection of highly qualified personnel to design, construct, and operate a plant. Each applicant for a license to build a nuclear plant must institute a quality assurance program to be applied to design, fabrication, construction, and testing of the structures, systems, and components of the facility. An applicant for a license to operate a plant must also provide for appropriate managerial and administrative controls to be used to assure safe operation. The function of the quality assurance program is to provide a high degree of confidence that the structures, systems, and components in the plant will perform as intended during the service life of the plant.

Nevertheless, accidents can occur, and other safety features are designed to keep fission products confined in the event of a broad range of accidents that could conceivably occur. Two categories of anticipated accidents are undercooling events and positive reactivity addition events. An undercooling event can be caused by such things as ruptures in the reactor cooling system, loss of flow through the reactor core, or loss of feedwater flow. Positive reactivity addition events are the result of such things as inadvertent control-rod withdrawals or dissolved boron dilution, main steam line breaks (pressurized water reactor), or overfeeding of a steam generator.

There are various types of engineered safety features found in today's commercial reactors. They can be grouped into four major areas: (1) reactor protection system, (2) emergency core-cooling systems, (3) containment, and (4) emergency power supplies. The following discussion of these features is generally applicable to light-water-cooled reactors, pressurized water and boiling water reactors, but the same general features are applicable to other types.

The purpose of the **reactor protection system** is to rapidly insert the control rods into the reactor upon receipt of a signal indicative of a malfunction that potentially threatens the integrity of the core. For example, a low reactor coolant system pressure is indicative of a loss-of-coolant accident. High temperatures in the reactor coolant system or the core indicate the possibility of a loss of coolant or loss of flow through the core. The reactor protection system processes the various input signals and compares them to predetermined set points. If a set point is reached or exceeded, the system is designed to automatically insert the control rods into the reactor. The control rods absorb neutrons and quickly reduce the fission rate in the core. This rapidly drops the power level in the reactor and aids in prevention of damage to the fuel.

In the event of a loss-of-coolant accident, it is not sufficient to shut the reactor down in order to prevent damage to the fuel. Decay heat produced from the radioactive decay of fission products (initially some 4 percent of the previous operating power, but decreasing rapidly with time after stopping the chain reaction) would cause damage to the fuel if the coolant inventory losses were not made up by an injection system. **Emergency core-cooling systems** are designed to permit injection of coolant at high reactor coolant system pressures that exist during small ruptures as well as at low reactor coolant system pressures resulting from large breaks. Redundancy is provided to assure that equipment failures do not result in the loss of capability to keep the core covered with water. Large sources of water are stored to meet the long-term cooling requirements to remove the decay heat. In the event that a pipe rupture results in the depletion of the stored injection water, provisions are made to allow recirculation of water from the emergency sumps in the containment building back through the injection pumps. In time the rate of heat generation drops to the point where forced-circulation cooling is no longer required. After the Three Mile Island core damage event the system was cooling by natural convection to the surroundings at atmospheric pressure after about 3 years.

Containment features provide an important function essential to the defense-in-depth philosophy. There are actually three levels in the containment scheme. The first level is the fuel cladding, which not only mechanically holds the fuel pellets in the core, but also provides a boundary to prevent migration of fission products to the coolant. The second level is the reactor cooling system pressure boundary. This consists of the reactor vessel system piping, and other components designed to withstand pressures in the order of hundreds of atmospheres without failure. Parts of these systems that are outside containment in normal operation, as in a boiling water reactor, are isolated at the containment building boundary in case of an accident. Thus even if the fuel cladding should fail, the pressure boundary is likely to keep the fission products contained. In the event that the accident would cause both of these boundaries to fail, a third level is provided in the form of a containment building. The typical containment building consists of a steel shell surrounded by a concrete structure subjected to precompression by steel cables. Thick steel only, reinforced concrete, and combinations have all been used in containment schemes, most of which are designed to withstand 5- to 10-atm pressure, based on containing the total energy available during an accident. The destroyed Chernobyl reactor did not have a containment building; it had only confinement provisions for small local pipe breaks.

Emergency power supplies also exhibit redundancy and defense-in-depth features. The power requirements for auxiliary systems within a power plant are normally supplied from the reactor output itself through the main electrical generator and auxiliary station transformer. In the event of an accident, the reactor would not be available to produce useful power; thus the preferred power source for supplying emergency electrical loads is from the transmission lines from other power plants. At least two off-site power supplies are required. In case of loss of these sources, redundant on-site sources are also required, usually powered by quick-start diesel generators. Battery systems are also needed to provide certain direct-current loads and some alternating-current loads through inverters. Control, computer, and instrument supplies are typically so served.

Protection is provided not only for events originating in the plant but also for external events. Plants are designed to withstand safely the worst anticipated earthquakes, floods, and winds. Containment and other safety features of nuclear power plants generally offer significant protection against consequences of accidents exceeding the severity of design-basis accidents. In recent years, experimental work and probabilistic studies have been carried out to determine the extent of such margins and discover residual vulnerabilities that would then become the basis for possible further safety improvements, generally on a plant-specific basis.

As the operating years of plants have accumulated, there has been increasing attention to detecting and counteracting age-related degradation of plant materials and equipment. A notable example is the varying degrees of radiation-induced embrittlement of reactor vessel materials, which if unchecked should increase vulnerability to overcooling accidents (pressurized thermal shock).

In the aftermath of the events at the Fukushima Daiichi plants in Japan in 2011, the US nuclear industry instituted a flexible strategy to implement the Nuclear Regulatory Commission's Fukushima recommendations. This FLEX strategy addressed the main safety challenges at Fukushima, notably the loss of cooling capacity and electrical power resulting from a severe natural event. The FLEX strategy is site-specific to account for the differences in the design and most likely risks at each US nuclear facility and provides reliable backup electrical power and cooling capability in the case where an extreme natural event disables the multiple power and cooling systems at each reactor.

Another development in recent years has been the increasing use of probabilistic safety assessments as a partial guide to plant safety actions as well as safety regulation. In such assessments, **fault trees** are used to trace malfunctions through all antecedent events to the initiating causes and **event trees** to trace undesired events through all their consequences to ultimate outcomes. Known or estimated probabilities are assigned at each step in the chain of causation, to derive probabilities of sequences of safety (or risk) interest.

The various inherent and engineered safety features, along with the high quality standards which are required in the design, construction, and operation of a nuclear power plant, make accidental releases of radioactivity to the environment extremely unlikely. Nonetheless, additional defense-in-depth is provided in the form of trained operators and procedural adherence. As a final feature of the defense-in-depth philosophy, emergency planning is required that establishes the protective measures that would be taken to protect workers and members of the public against possible radioactive releases. Such emergency measures are required to be practiced at least annually in the United States.

An additional recent development is the incorporation of risk-informed, performance-based regulation approaches by the Nuclear Regulatory Commission. As described by the NRC, this "is an approach in which risk insights, engineering analysis and judgment, and performance history are used to (1) focus attention on the most important activities, (2) establish objective criteria for evaluating performance, (3) develop measurable or calculable parameters for monitoring system and licensee performance, (4) provide flexibility to determine how to meet the established performance criteria in a way that will encourage and reward improved outcomes, and (5) focus on the results as the primary basis for regulatory decision-making."

NUCLEAR POWER PLANT LICENSING

Dozens of licenses and permits are needed to secure approval to construct and operate a nuclear generating station. In some localities the number may exceed 100. This section addresses the requirements developed to address the unique challenges of nuclear technology thereby ensuring adequate protection of public health and safety.

Licensing of nuclear generating stations is of particular importance due to the large radioisotope inventory present within the reactor core. International consensus on minimum standards is important since accidents at nuclear generating plants can have far-reaching consequences. Thus the safe operation of nuclear facilities is an international issue.

The Multinational Design Evaluation Program (MDEP) was established in 2006 to address this issue. It is a multinational initiative that brings the regulators together to pool their knowledge and experience in an effort to ensure that the regulations within each country will result in a level of safe operation that is consistent with best practices in the global community. Fifteen countries participate in MDEP. The Nuclear Energy Agency (NEA) oversees the program and provides technical secretariat services. Collaboration on safety standards facilitates commerce in that it helps to ensure that a particular technology can be licensed outside the borders of the country where the technology was originally developed.

The regulatory process varies greatly from country to country. Nonetheless the goal is the same, safe operation. The balance of this section goes into some detail in describing the regulatory system within the United States. It is a prescriptive approach to licensing and is used as a model by other countries particularly those just introducing nuclear generation. It is not uncommon for a country to stipulate that a technology be licensed within the US before considering its deployment within their own.

Licensing of commercial nuclear facilities in the United States is conducted in accordance with the Atomic Energy Act of 1954, as amended. The applicable regulations appear in Title 10 of the Code of Federal Regulations (10 CFR) which codifies Federal laws and regulations that are in effect as of the date of the publication pertaining to energy.

In the United States a license issued by the Nuclear Regulatory Commission (NRC) is required to operate a nuclear plant. The process to secure this license is governed by either Part 50 or Part 52 of Title 10 of the Code of Federal Regulations. Part 50 describes a process requiring two steps, namely the issuance of a Construction Permit followed by the issuance of an Operating License. Over 120 Operating Licenses have been granted for Nuclear Electric Generating Stations in the US under the two step process. In 1989 the NRC established an alternate process that combined the Construction Permit and Operating License into a single license, a Combined Construction and Operating (COL). In addition to the COL other licensing alternatives were introduced, namely the opportunity to secure an Early Site Permit (ESP) which allowed a utility to gain approval of a reactor site for future use and also to secure certification of standard reactor designs that can be referenced.

- Part 50 DOMESTIC LICENSING OF PRODUCTION AND UTILIZATION FACILITIES (10CFR50) is a two-step process. The first step is to secure a construction permit thereby giving the utility permission to build the nuclear facility. After completing the construction phase the utility (licensee) would need to undertake a second hearing process before being granted an operating license. During both the construction and operating permit application process opportunities are provided for public involvement which introduced uncertainties in the construction and startup schedule. A delay in plant startup for any reason is a very costly proposition especially if it occurs after a large investment in the project. This is due to the rapid accrual of Interest during construction (IDC) charges.
- Part 52 LICENSES, CERTIFICATIONS, AND APPROVALS FOR NUCLEAR POWER PLANTS (10CFR52) was adopted in 1989. It provides for the issuance of a Combined Construction and Operating Licensed (COL) that grants permission to construct and operate the plant provided that it can be demonstrated that the plant was constructed in accordance with the license. This process is intended to engage the regulator and the public before construction begins, thus aiming at reducing the impact of cost escalation if delays are encountered. However, the complete execution of this process leading to the operation of new nuclear power plants has not, as yet (2017), been fully exercised.

Since Part 52 was not an option up until 1989, the vast majority of nuclear generating stations were licensed under the Part 50 two-step process involving the issuance of a construction permit (CP) followed by an operating license (OL). This process begins with the filing by a utility of an application for the construction permit. If the NRC judges that the filing contains the required information, a Notice of Receipt is published in the Federal Register.

The NRC then conducts a review of the safety, environmental, security, and antitrust aspects of the proposed facility. The safety review is primarily based on information contained in a Safety Analysis Report (SAR). Section 50.34 of 10 CFR Part 50 of the Commission's regulations requires that each application for a construction permit for a nuclear facility shall include a Preliminary Safety Analysis Report (PSAR) and that each application for a license to operate such a facility shall include a Final Safety Analysis Report (FSAR). The SAR is the principal document upon which the NRC relies to assess whether the plant can be built and operated without undue risk to the health and safety of the public.

NUREG-0800 "Standard Review Plan for the Review of Safety Analysis Reports for Nuclear Power Plants" (SRP) provides guidance for NRC staff reviewers in the Office of Nuclear Reactor Regulation. It helps to assure uniformity in performing safety reviews by staff. Once a review is completed the U.S. Nuclear Regulatory Commission (NRC) staff issues a Safety Evaluation Report (SER) which documents their technical review of the site safety analysis reports (SAR) at both steps in the licensing process.

Advisory Committee on Reactor Safeguards (ACRS)

The ACRS is mandated by the Atomic Energy Act of 1954 as amended. The committee has four purposes one of which is to review and report on safety studies and reactor facility license applications. The ACRS is comprised of 15 members who represent diverse technical areas. They meet on average about 10 times per year. The ACRS is independent of the NRC staff and report directly to the Commission (i.e., the five Commissioners), which appoints its members.

Following completion of the SERs, the ACRS independently completes its reviews. These ACRS reviews typically include meetings with the NRC staff and applicant. At the conclusion of the ACRS review, a letter is sent to the chairman of the NRC providing the ACRS recommendation as to whether the CP or OL should be issued. Supplements to the SERs are issued by the staff which incorporate appropriate changes as a result of the ACRS recommendations.

Finally a mandatory public hearing is held by a three-member Atomic Safety and Licensing Board (ASLB), which reports to the NRC. If all issues are resolved satisfactorily, the staff is authorized by the NRC Commissioners to issue the construction permit. The applicant may then begin construction, provided other local, state, and federal requirements are met.

Several years before the plant construction is completed the applicant files an application to the NRC for an operating license. After completing the construction phase the licensee goes through a second hearing process before being granted an operating license.

Part 52 licensing

The licensing process under Part 52 is represented in the Fig. 7.8.5.

It was developed to make the licensing process more predictable and efficient. In addition to combining the construction and operating licenses it also incorporates the utilization of Early Site Permits and Certified Designs.

Early Site Permit (ESP)[a]

Potential licensees can seek the issuance of an ESP by the US Nuclear Regulatory Commission (NRC). The ESP grants the licensee permission to locate a nuclear power facility at that site. The design of the facility is unspecified, that is, the ESP is independent of an application for a construction permit or combined license. The ESP addresses generic issues pertaining to the site where the plant will be located. Once granted the ESP is valid for 10 to 20 years from the date of issuance and can be renewed an additional 10 to 20 years.[a] Some utilities have undertaken the process of securing an ESP for several sites and have "banked" the sites for future use. The ESP application addresses

- site safety issues
- environmental protection issues
- emergency plans

Stakeholders generally participate in this process.

Design Certification Applications for New Reactors[b]

The NRC can approve a nuclear plant design independent of an application to construct or operate a plant by issuing a Design Certification. A design certification is valid for 15 years from the date of issuance but can be renewed.

The application to certify a design addresses the safety issues associated with the proposed nuclear plant design independent of the specific site. Standardized designs are expected to result in a more streamlined licensing process, interchangeability of spare parts, operator training

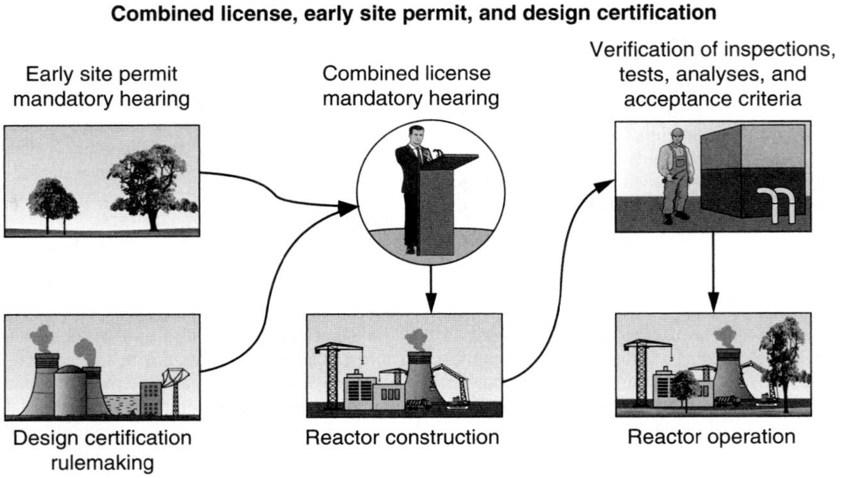

Figure 7.8.5 General process for licensing a new nuclear reactor (NUREG/ BR 0468).

programs shared among multiple units, reduced costs attributed to production quantities, improved maintenance and operation by sharing lessons learned among a fleet with significant commonality.

Although the evaluation of the standard design is independent of the site upon which it is ultimately constructed, it is necessary to establish bounding site conditions intended to represent a range of potential sites. Thus the applicant must demonstrate that the actual site falls within the bounding site conditions.

The NRC has certified several standard nuclear power plant designs, and it strongly encourages future applicants for licenses to select a standard design and reference it in their license application. This dramatically reduces the resources assigned to the license application.

Environmental Review (ER) This review is held at both the CP and OL steps pursuant to the National Environmental Policy Act of 1969. The staff review is based on the Safety Analysis Report (SAR) prepared by the applicant. Upon completion of the staff s analysis, a Draft Environmental Impact Statement (DEIS) is prepared and distributed for comment to the various federal, state, and local agencies and to members of the public. Comments are taken into account in the preparation of a Final Environmental Impact Statement (FEIS).

Antitrust Reviews This review is conducted by the NRC and the Attorney General (Department of Justice) to ensure that operation of the facility will not be in violation of the antitrust laws.

Public Hearings Separate hearings may be held on safety and environmental matters, or the issues may be consolidated into a single hearing. If an antitrust hearing is required, it is always held separately. Hearings are normally held close to the reactor site to allow local public participation.

License Issuance and Commission Reviews Since 1981 it has been NRC practice for the staff to initially issue an operating license restricting operations of the plant to power levels not exceeding 5 percent of the full power rating. This low power license is issued after the completion of the safety, environmental, and antitrust reviews, and after completion of the public hearings, and resolution of any appeals. The exception to this is that public hearings on off-site emergency planning need not be completed prior to the issuance of a low-power license. The low-power license permits plant heat-up and reactor system testing, which typically takes 2 or 3 months. After the remaining licensing issues are resolved and the staff has confidence that the applicant can operate the plant in compliance with the law, the staff brings the matter before the five-member Commission (NRC) for review. Following this review and a favorable finding by the Commission, the staff is authorized to issue a license amendment permitting full-power operation of the plant. Federal courts may review the Commission decision on appeal from interested parties, with or without a stay of execution of the license as the court may decide. As the initial 40-year license is expiring, the licensee can apply to the NRC to extend the period of the license for an additional 20 years. License extensions leading to plant operation of up to 80 years are currently being contemplated.

The NRC has issued a safety goal policy statement that established goals that broadly define an acceptable level of radiological risk that might be imposed on the public as a result of nuclear power plant operation. To implement the goals, numerical objectives on core damage frequency and containment performance have been established. The safety goals are among the considerations underlying NRC's generic regulatory actions, but do not supplant NRC regulations and are not applied to individual plants.

Continuing Review A licensed reactor is subject to continual oversight by the NRC throughout its operating life to assure that regulations and license conditions are observed.

The NRC assigns "onsite resident inspectors" to each operating nuclear generating plant. They are responsible for executing a continuous inspection program. These resident inspectors can be augmented by NRC inspectors with specialized expertise. These inspectors are based in the NRC regional offices and NRC headquarters in Rockville, MD. Off normal events are investigated by the licensee to determine their causes and to implement corrective actions. Resident inspectors will monitor such activities in addition to their normal oversight role.

The NRC has significant authority over the licensee including the authority to order a shutdown of the plant.

Licensees are required to implement a formal program to monitor the performance and condition of specified structures, systems, and components. This provides a level of assurance that systems important to safety are able to perform their intended safety functions.

In addition to site specific oversight the NRC maintains an active program for evaluating and resolving generic safety issues, that is, issues relevant to all plants or to a class of plants. Resolution of Generic Safety Issues may be in the form of physical plant modifications or procedural changes. A safety vs. cost analysis is necessarily part of the process to address generic safety issues.

Reactor licensees are also required to perform probabilistic risk assessments of their facilities. These are generally called **individual plant examinations**. These risk assessments are a result of the NRC policy regarding assessment of beyond-design-basis accidents (severe accidents) and are intended to identify plant vulnerabilities.

Operator Licenses

In addition to issuing operating licenses to the owners of nuclear power plants, the NRC also licenses persons who are permitted to manipulate the controls of the reactor. There are two levels of responsibility associated with reactor operation: reactor operator and senior reactor operator. The licensing process requires training, experience and passing both written and practical examinations. "Simulators" of the plant control room are commonly employed for initial training of licensed reactor operators as well as part of a continuing training program to maintain operator skills at their highest level. As part of the training, operators are required to respond to off normal conditions that are triggered by an instructor. Modern computational capability allows a very realistic simulated response to the triggering event as well as the actions of the operator reacting to it.

Decommissioning

Permanent removal of an operating plant from service and terminating its license is a process designated as decommissioning the plant. The NRC has regulations and guidance designed to ensure that this process and outcome are safe and environmentally sound. Decommissioning involves removal and disposal of plant components and structures. Generally there is a process to remove radioactive contamination from the structures and components to the extent it is practical. In this way the volume of radioactive waste is reduced along with the associated cost. The licensee (owner) remains accountable to the NRC until the plant is decommissioned and the license is formally terminated.

The licensee is required to set aside funds to finance the decommissioning effort while the plant is operating and generating revenue. Five years before the scheduled termination of plant operations the licensee is required to provide a cost estimate to decommission the plant to the NRC.

The two currently employed approaches to decommissioning are DECON and SAFSTOR. SAFSTOR is characterized by removing fuel from the reactor and storing it in the spent fuel pool while placing fuel that has been decaying for 5 years or more in dry cask storage. The SAFSTOR approach allows the facility to sit in an inoperative safe state for up to 60 years all the while benefiting from the reduction in radiation levels due to the natural decay of residual radioactivity. This option also allows the funds set aside in the decommissioning trust fund to grow. The DECON option begins with a more aggressive approach of removing radioactivity associated with contaminated components and decontaminating surfaces that are accessible. This can take several years. Following this phase the utility may opt to complete the decommissioning effort by dismantling all the structures and components and moving all the fuel to dry cask storage or to opt for the SAFSTOR option.

Radioactive Waste

Aside from normal industrial waste, employing nuclear technology generates low level radioactive waste (LLW). Such applications include medical diagnosis and treatment, smoke detectors, instrumentation, electric generation, food preservation and insect eradication. Nuclear power plants produce about half the volume of LLW generated in the US and most of the radioactivity.[c]

LLW waste is further classified into three subcategories. Class A waste contains the lowest level of activity of the three and also contains radionuclides with relatively short half-lives. Class B waste contains higher concentrations of radionuclides with generally longer half-lives. Class B waste accounts for less than 5 percent of the volume of LLW. Class C waste quantities are smaller still but their radionuclide concentrations are higher. Disposal of all three Classes of LLW is the responsibility of the generator. Class B waste must meet certain waste form criteria to ensure stability after disposal. Class C waste must meet not only this same waste form criteria but also must be dispositioned with engineered features to inhibit intrusion into the disposal area after active monitoring of the site is terminated.[d]

Some of the radioactive waste can also contain chemical constituents that are designated as hazardous under Title 40 of the Code of Federal Regulations. In such instances the LLW is classified as Mixed Waste and must be disposed in a facility licensed to receive the waste.

Used Fuel

Fuel assemblies are removed from the reactor when the concentration of fissile material is diminished but not exhausted. These assemblies are designated as Used Fuel and/or Spent Fuel. Used Fuel is highly radioactive when first removed from a reactor and is stored in a pool of water that is about 40 feet deep. The fuel assemblies are about 14 feet high and are situated in a rack at the bottom of the pool. That leaves about 20 feet of water above the top of the fuel assemblies to shield personnel on the deck around the pool. The water not only serves as shielding for personnel but also cools the fuel pins comprising each assembly. After about 5 years the radioisotopes within the used fuel have decayed to a level that allow them to be removed from the pool and stored in dry casks at an independent spent fuel storage installation or ISFISI as they are generally called. If the NRC approves the license for the ISFISI, it can be valid for up to 40 years.[e]

(a) http://www.nrc.gov/reactors/new-reactors/esp.html (2017).
(b) http://www.nrc.gov/reactors/new-reactors/design-cert.html (2017).
(c) NEI Fact Sheet: Disposal of Commercial Low-Level Radioactive Waste April 2014.
(d) 10 CFR 61.65 Waste Classification (2017).
(e) http://www.nrc.gov/waste/spent-fuel-storage/licensing.html (2017).

NATIONAL STANDARDS FOR DESIGN AND CONSTRUCTION

The design and construction of a nuclear power plant is a complex process requiring sophisticated analysis and great attention to detail layout. There are in excess of 400 plants worldwide with 99 in the US. As you can see from the discussion in the earlier parts of this chapter there are a variety of designs with various advantages of each of them.

The system design is fundamentally the same for the various types of light water reactors and heavy water reactors. The purpose of this section is to provide an overview of the design process and give some guidance as to the national codes and standards that are used in order to assure the appropriate level of safety.

A nuclear power plant is made up of systems, structures and components that are linked together linked together by pipes, cables and structures resulting in a machine that generates steam to drive a steam turbine which turns a generator to produce the electricity. This subsection provides information that will assist you in keeping up with the processes and standards that will assure the knowledge necessary for the design of a safe nuclear power plant.

System Design

The system design for a nuclear power plant is essentially identical to that of a coal/oil or gas fired power plant. The discussions of the power cycle components of a power plant including boilers, steam turbines/generators and heat exchangers are covered in Secs 7.2, 7.4, and 7.5. The only difference is the source of energy for producing the steam. In the case of nuclear power uranium fission reaction is used to heat the water into steam, which drives the turbine/generator to produce electricity. Therefore this section will discuss the nuclear system codes and standards required for nuclear power.

The reactor provides the energy to turn the water into steam, which drives the turbine/generator. In earlier sections of this chapter the different type of reactors are covered. Boiling water reactors turn the water directly into steam in the reactor while pressurized water reactors have a heat exchanger call the steam generator to produce the steam.

The nuclear power and safety systems are those required in order to assure the safe control of the fission chain reaction in the reactor. These systems include the primary coolant loop, the main steam and feed water systems, which continually provide cooling water to the reactor. The numerous safety system provide backup cooling to cool the core incase of a design basis accident. The containment system encloses the reactor to prevent release of radiation to the environment. The ultimate heat sink assures a source of cooling water to the safety systems.

The American Nuclear Society, ANS, has extensive standards defining detailed requirements for the safe operation of power reactors.

Standard ANS 58.14 "Safety and Pressure Integrity Classification for Light Water Reactors" defines the classification for the various components, which make up the safety systems. There are also over 160 standards that cover numerous safety areas requirements such cooling, radiation protection, seismic design, tsunami and flooding protection, and many more.

ANS and ASME have developed a standard to evaluate the safety risk against internal operating events, external events, and fire. Clear understandings of risks are obtained by developing a Probability Risk Analysis, PRA, for each plant. The standard ASME/ANS RA-S provides the rules to develop an appropriate value of risks and consequences for events.

The significant portion of the concept of defense in depth is the containment system, which surrounds the nuclear reactor and its primary coolant components. The containment system is there to assure that if there is a loss of coolant accident that there would be no release to the public of fission products. The design requirements for the containment system are defined by standard ANS 56.8 "Containment Systems Leakage Testing Requirements" while the requirements for structural components of the containment are defined in ASME standards.

Pressure Components

ASME Boiler and Pressure Vessel Code Section III cover the design of the pressure components that make up the reactor safety systems. The code defines the design strength requirements for the various classes of pressure components. ANS 58.14 defines the requirements for the safety classification the reactor and its components. The primary reactor cooling system components are Class 1. The safety systems supporting the requirements to assured a safe shutdown of a reactor are Class 2 and the other support safety systems are Class 3.

ASME B&PV Section III has design and fabrication rules in Subsections NB, NC and ND for Class 1, 2, and 3 components respectively. In each of those subsections there are also rules for the piping that ties the components together.

The containment that encloses the reactor systems is usually made of steel or concrete. Steel containments design and fabricated in accordance with ASME B&PV Section III Division 1 Subsection NE. Reinforced or post-tension concrete containments design and construction requirements are defined in ASME B&PV Section III Division 2 Subsection CC. This standard was developed jointly with the American Concrete Institute, ACI, and ASME.

Each of these subsections for the various classes of components has different levels of requirements depending on the importance of their function to the safety of the reactor.

Electrical Power and Control Devices

The performance of the safety systems are controlled and operated by the electrical components such as motors, transformers, switchgear, instrumentation, batteries, control panels, and computers. These are electrical components are tied together by cables, wiring, and tubing. The safety system instrumentation continuously monitors the status of the reactor systems. Including pressure, temperature, flow, voltage, current, criticality, etc. This provides information to the operators and computers and the controls to assure that the appropriate action necessary to keep the reactor safely producing power.

There're also a significant amount of monitoring of the incore condition of the fuel elements this is to make sure that the performance of the nuclear core will operate as necessary to provide the energy necessary to turn the water into steam to drive the turbine. The condition of the uranium and its cladding is important to ensure the safe operation of the reactor.

The control of the chain reaction in the reactor is performed by control rods, which penetrate the reactor core by going between the fuel elements to absorb neutrons and stop the fission.

All of this operation is provided by electrical and mechanical components. The Institute of Electrical and Electronic Engineers, IEEE, has developed and maintains over 261 standards for nuclear power plants. The electrical components are classified dependent on their safety function. IEEE 308 "Standard Criteria Class 1E Power Systems for Nuclear Power Stations" provides the criteria for the most important electrical components. IEEE has many standards that drill down into the details of the design, qualification, testing and reliability requirements necessary to ensure the safe operation of the electrical and controls necessary for the operation of the safe operation of a nuclear power plant.

Structures

The equipment of a nuclear power plant is enclosed in steel or concrete structures. These are large buildings that provide support for the equipment, piping, cables which make up a power plant. In order to assure the equipment can function and is stable even when subject to number of environmental challenges these buildings are design to numerous extreme conditions. Including high earthquake and wind conditions such the highest earthquake possible at the site, tornado and hurricane wind levels, tsunami, and other extreme flooding. The structure is there to support and protect the components.

The foundations of these structures are typically a mat foundation supported directly on rock or compact soil. These mat foundations are made of reinforced concrete and which are designed constructed in accordance with the rules of American Concrete Institute ACI 349. The standard provides the requirements the loadings and load combinations that the structure must withstand with the appropriate safety margin against the ultimate strength requirement. Concrete structures are also required to provide shielding requirements from radiation and therefore the thickness can be defined by either strength or shielding. The shielding requirements are defined in ANS-6.4 "Nuclear Analysis and Design of Concrete Radiation Shielding for Nuclear Power Plants."

Buildings that do not have a shielding requirement provide the support function by a structural steel frame. They need to be designed to the same extreme loadings that the concrete buildings are designed to. The rules for the structural steel safety margin for these extreme conditions are defined in American Institute of Steel Construction, AISC N690 "Specifications for Safety Related Steel Structures for Nuclear Facilities."

The containment structure, which was discussed in previous sections, is designed in accordance with ASME B&PV Section III Division 1 NE for steel containments or Division 2 CC for concrete containments.

Fabrication and Construction

The quality of fabrication and construction of a power plant are as important as the design. The quality of all work is controlled by the ASME quality assurance standard NQA-1 "Nuclear Quality Assurance." Due to the size of the pressure components, steel and concrete structures the quality requirements are unique. They require the fabricating shops and constructors with high quality production capabilities to produce the quality of work required.

The welding requirements for the types of steel and thickness are extremely sophisticated and are controlled by the ASME standards for pressure components and American Welding Society for structural components. ASME Section IX "Welding and Brazing Calculations" provides the rules for ASME pressure components. ASME has a very strong conformity assessment program to qualify and continually monitor fabricators to make sure they have their processes, machines and personnel to perform the work in accordance with the ASME rules.

AISC N690 "Specifications for Safety Related Steel Structures for Nuclear Facilities" references AWS for their welding rules for structural steel construction. The steel fabricators and erectors must have a quality assurance program meeting 10CFR50 Appendix B requirements.

Because of the significant strength requirements and the thickness needed for radiation shielding, the concrete structures are very large and heavily reinforced. Therefore require special concrete mix design and placing controls to assure the strength of the structures are adequate. The reinforced concrete structures construction requirements are in ACI 349 "Code Requirements for Nuclear Safety Related Concrete Structures and Commentary" and ASME B&PV Section III Division 2 CC.

OTHER POWER APPLICATIONS

Nuclear Ships The earliest and most successful of the efforts to develop nuclear power for ship propulsion were those by the U.S. AEC and U.S. Navy. These initial efforts toward this end predated and contributed to the development of large electric power generators. The U.S. Navy launched the *USS Nautilus* in 1954 and followed it with over 200 operational vessels including nuclear submarines and surface vessels. Additional submarines and surface vessels have been committed by the U.S. Navy. Nuclear submarines and ships are also being operated by the British, Russian, and French and Chinese navies.

Although a ship can be propelled by diesel engines, steam engines, gas turbines, and other devices, the steam turbine has emerged as the principal modern marine propulsion unit. When a nuclear reactor of a type developed for generating stations is used to produce steam, the combustion process is eliminated. Such a propulsion plant is especially advantageous for a submarine, since it is then completely independent of the earth's atmosphere. The U.S. Navy's nuclear ships and submarines all employ the pressurized water type of nuclear reactor.

Other Vehicles The use of nuclear power in airplanes, automobiles, and railroad trains has been deterred by the requirements for heavy shielding to protect against radioactivity and collisions.

Remote Power Sources Small nuclear power plants have been built and operated by the U.S. Army Corps of Engineers in such remote regions as the Greenland ice cap and the Antarctic ice sheet. A barge-mounted plant was also built and operated in the Panama Canal Zone, but was scrapped.

Isotopic Power Generators Isotopic power generators are direct conversion devices that convert the energy of decaying radioisotopes to electricity, using direct thermoelectric or thermionic devices, or thermal conversion systems. (See Sec. 7.1.) The first isotopic power generator, completed in 1959, produced 2.5 W of power and used polonium 210 as fuel. The first commercial isotopic power generator, fueled by strontium 90, became available in 1966, and produced 25 W of power. There are numerous radioisotopes available in this area (see Table 7.8.4).

Plutonium 238 was used in the long-range space probes Voyager 1 and 2, on the lunar surface during the Apollo program, as well as on the more recent Mars surface rovers and the Galileo, Cassini and New Horizons planetary science missions.

Space Power Sources Most satellites and space vehicles use solar and battery power sources. However for very high powers, into the multikilowatt region, and for very deep space probes where solar energy

Table 7.8.4 Radioisotopic Materials

Radioisotope	Half-life, yr	Typical power density,* W/cm^3	Radiation present†	Shielding required
Strontium^{90}	28	1.4	Beta and bremsstrahlung	Heavy
Cesium^{137}	30	0.21	Beta and gamma	Heavy
Cerium^{144}	0.78	24.5	Beta and gamma	Heavy
Promethium^{147}	2.7	1.8	Beta	Minor
Polonium^{210}	0.38	1,210	Alpha	Minor
Plutonium^{238}	89	3.5	Alpha	Minor
Americium^{241}	432.2	0.115	Alpha and gamma	Minor
Curium^{242}	0.45	1,150	Alpha	Minor
Curium^{244}	18	26.4	Alpha and neutron	Moderate

* Reduced below theoretical maxima in order to account for expected isotopic impurities.
† Of dominant importance in heat generation and for consideration of radiation shielding.
SOURCE: AEC TID-20079.

is not available, nuclear sources (either radioisotopes or reactor heat systems) are required. Several dozen such systems have been deployed since 1970. Most are isotopic, but the United States orbited one reactor system in 1965. The Soviet Union has put several reactor-powered systems into orbit. On at least two occasions such U.S.S.R. systems have decayed back into the earth's atmosphere, and in one instance radioactive parts landed in Canada.

NUCLEAR FUSION

Development of nuclear fusion reactors for the production of useful energy will require significant advances from the present state of the art in both plasma physics and engineering. A large fusion effort, especially in physics research, has been carried on since 1956 in many industrial nations, notably the Soviet Union, Japan, the United States, and several in western Europe; considerable progress has been made.

Table 7.8.5 Possible Fusion Fuel Cycles

Reactor equation	Approx threshold plasma temp*	Approx avg energy gain per fusion
$D + T \rightarrow {}^4He\ (3.5\ MeV) + n(14.1\ MeV)$	10 keV	1,800
$D + D \rightarrow {}^3He\ (0.82\ MeV) + n(2.45\ MeV)$ or $\rightarrow T\ (1.01\ MeV) + p(3.02\ MeV)$	50 keV	70
$D + {}^3He \rightarrow {}^4He\ (3.6\ MeV) + p(14.7\ MeV)$	100 keV	180

* 1 keV = 11,000,000 K approximately.

There are a number of possible **fuel cycles** involving the fusion of light ions into heavier atoms, the most promising of which are shown in Table 7.8.5. The most accessible of these is the deuterium-tritium (DT) reaction, which involves fusion of two hydrogen isotopes. Even this lowest-temperature reaction of the possible fusion cycles has a threshold temperature of about 50 million K, and a practical reactor would require operating temperatures of about 100 million K. At these temperatures the ionized gas or plasma cannot be contained in vessels of conventional materials.

Fortunately, ionized plasmas can be contained by magnetic or inertial forces on earth. (The level of gravitation forces which produce fusion in the sun and stars cannot be achieved on earth.) Research seeking how this can be done at high enough combinations of temperatures, pressures, and densities to obtain practical net energy output is the main effort of fusion programs worldwide. This research is concentrated in three major approaches:

1. **Toroidal systems**, especially of the tokamak variety, in which very high currents are circulated through an endless plasma confined by toroidal (doughnut-shaped) magnetic systems of various physical cross sections and magnetic shaped fields.
2. **Magnetic mirror systems**, in which the ends of a solenoidal (linear) magnetic field are "plugged" by a combination of electrostatic and magnetic forces to make the plasma reflect back and forth.
3. **Inertial systems**, in which small pellets of fusionable material are rapidly compressed by beams of laser-produced light, electrons, heavy ions, or other compressive forces. This is the principle employed in nuclear fusion weapons.

For practical energy production, a DT plasma should have a temperature of about 100 million K. This is usually expressed as equivalent ion temperature in electron volts (1 eV = 11,000 K approx.). Coincidentally with this temperature the product of density (in ions per cubic centimeter) and confinement time (in seconds) should reach 10^{14}. These conditions have been achieved experimentally.

Three large tokamak installations (TFTR in the United States, JET in Europe, and JT-60 in Japan) have operated. Each of these machines created plasma conditions approximating those required in practical power producing reactors, although they are not in themselves net power producing devices. At the Lawrence Livermore National Laboratory in the United States a very large laser device, the National Ignition Facility, was developed to attempt to establish similar conditions by inertial means, using very tiny pellets filled with DT mixtures. The International Thermonuclear Experimental Reactor (ITER) is currently (2017) under construction in France. ITER is intended to be the first fusion device capable of producing more fusion energy than the energy required to maintain the machine's heating systems, a concept known as creating net energy. ITER also should be the first fusion device to maintain fusion energy production for long periods of time. It also is envisioned to be the first fusion device to test the integrated technologies, materials, and physics processes that will be required for the future commercial production of fusion-based electricity. All of these devices are experimental in nature.

After practical reactor conditions have been achieved in a confined plasma, significant engineering developments associated with removing the energy and converting it into useful form will be required.

Developmental work on fusion engineering continues, but formidable obstacles lie ahead, especially in the requirements for materials capable of surviving the environment of burning plasmas for long periods of time.

Although developing fusion as a practical energy source is likely feasible, the very considerable engineering hurdles yet unresolved put the operation of a demonstration fusion power reactor well into the twenty-first century or beyond. Since uranium resources appear adequate for many centuries, there is time to develop this essentially unlimited energy resource (fusion).

7.9 HYDRAULIC TURBINES

by Robert D. Steele

REFERENCES: "Water Power Engineering," McGraw-Hill. Creager and Justin, "Hydroelectric Handbook," Wiley. Daugherty and Ingersoll, "Fluid Mechanics," McGraw-Hill. Hammitt, "Cavitation and Multiphase Flow Phenomena," McGraw-Hill. Kovalev, "Hydroturbines, Design and Construction," Russian translated by U.S. Department of the Interior and National Science Foundation. Mosonyi, "Water Power Development," Hungarian Academy of Science.

GENERAL

Notation

B = width of distributor or height of wicket gates, in (mm)
D = diameter of runner, in (mm)
D_l = diameter of runner, at centerline of guide case, in (mm)
D_{th} = throat diameter of runner, in (mm)
D_r = relative diameter of runner, in (mm) (= entrance diameter for low-speed Francis turbines; throat diameter for high-speed Francis turbines; bore of throat ring at centerline of blade for propeller turbines)
D_d = diameter top of draft tube, in (mm)
D_p = pitch diameter of impulse-turbine runner, in (mm)
d = jet diameter of impulse turbine, in (mm)
e = overall efficiency of turbine = $e_h e_m$
e_h = hydraulic efficiency (including draft tube)
e_m = mechanical efficiency
g = acceleration of gravity, ft/s² (m/s²)
H = net effective head, ft (m)
b = head change due to load change, ft (m)
n = r/min
n_l = r/min at 1 ft (m) head = $n/\sqrt{H}$
n_s = specific speed = $n\sqrt{P}/H^{5/4}$
P = horsepower = kW/0.746
P_1 = horsepower (kW) at 1 ft (m) head = $P/H^{3/2}$
Q = discharge, ft³/s (m³/s)
Q_1 = discharge at 1 ft (m) head = $Q/\sqrt{H}$, ft³/s (m³/s)
t = thickness, in (mm), or time, s
u = circumferential velocity of a point on runner, ft/s (m/s)
V = absolute velocity of water, ft/s (m/s)
V_u = tangential components of absolute velocity = $V \cos \alpha$
V_r = component of V in radial plane
v = velocity of water relative to runner, ft/s (m/s)
w = weight of water lb/ft³ (kg/m³)
z = number of runner buckets (blades for propeller turbines)
α = angle between V and u (measured between positive directions)
β = angle between v and u (measured between positive directions)
θ = angle of water deflection relative to bucket
ϕ = peripheral coefficient $[\pi Dn/(720\sqrt{2gH})][\pi Dn/(60{,}000\sqrt{2gH})]$
σ = cavitation coefficient

Fundamental Formulas The actual power of a hydraulic turbine is the theoretical power P_t multiplied by the turbine efficiency e.

Power in U.S. Customary System (USCS) Units

$$P = eP_t = \frac{eHQw}{550}$$

In the power equation above, **power** is calculated in horsepower. In the following text, when power P is used in stated equations, it is in horsepower.

Power in Metric Units

$$P = eP_t = \frac{eHQw \times .746}{76.04} = \frac{eHQw}{101.93}$$

The metric equation stated above for power calculates power in kilowatts. In the following text, where power P is used in metric equations (shown in parentheses), the power stated in the equations is in kilowatts. [The customary unit of horsepower in the United States should not be confused with the term *metric horsepower* (see the conversion and equivalency tables of this handbook).]

The laws of proportionality for homologous turbines are:

For constant runner diam	For constant head	For variable diam and head
$p \propto H^{3/2}$	$P \propto D^2$	$P \propto D^2H^{3/2}$
$n \propto H^{1/2}$	$n \propto 1/D$	$n \propto H^{1/2}/D$
$Q \propto H^{1/2}$	$Q \propto D^2$	$Q \propto D^2H^{1/2}$

Nomenclature The nomenclature used throughout this section is based on NEMA's standards publication HT1-1957, "Hydraulic Turbines, Governors and Accessory Equipment," which also contains definitions.

Types of Turbines Three characteristic types of hydraulic turbines are now in general use: the **Francis reaction** type (Figs. 7.9.1 and 7.9.2); the **propeller reaction** type (Fig. 7.9.3) and the **impulse** type (Fig. 7.9.12). The propeller type may be further divided into fixed and adjustable blade types. All three types have in common a stationary guide case (or nozzle in the case of the impulse type) in which the static head is transformed partly, or wholly, into velocity, and a revolving part, the runner.

In the guide case of the impulse turbine, the static head is completely transformed into velocity, so that air surrounds both the jet issuing from the nozzle and the runner.

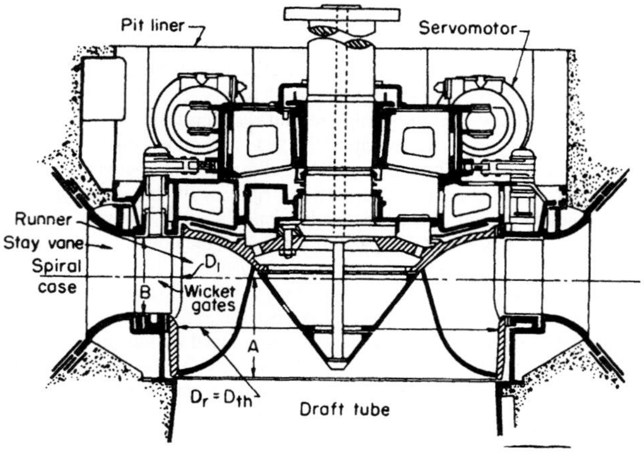

Figure 7.9.1 Medium-head Francis turbine.

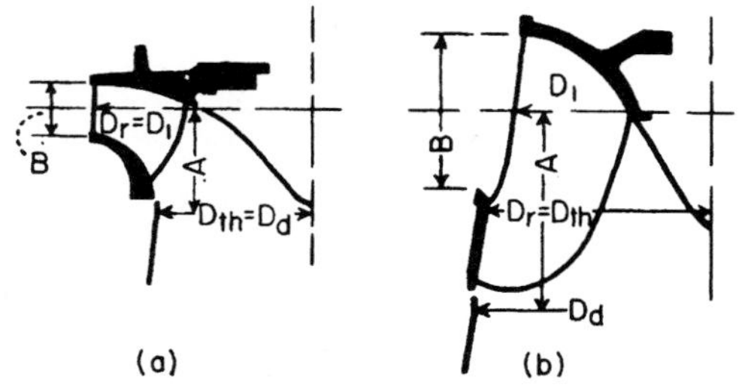

Figure 7.9.2 Typical profile of Francis runners. (*a*) Low specific speed; (*b*) high specific speed.

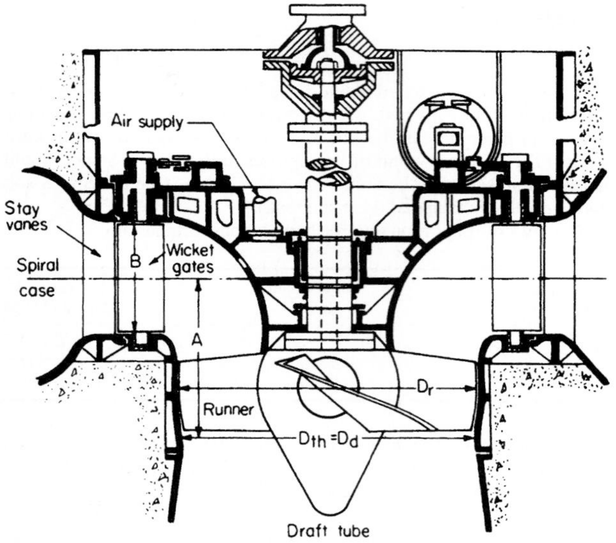

Figure 7.9.3 Adjustable-blade Kaplan turbine.

In the guide case of the reaction types, the static head is only partly transformed into velocity, leaving an overpressure between the guide case and the runner. This overpressure causes an acceleration of the relative velocity of the water passing through the runner, the discharge area of which is smaller than the entrance area. Except when operating vented at low loads, the water passages are completely filled with water from the intake to the end of the draft tube.

Selection of Turbine Type and Casing Impulse turbines receive their water supply directly from a pipeline. Francis and propeller reaction types are set in a case of concrete or metal. While the limits of head to which the impulse and reaction types are adopted may be fairly well defined in practice, as roughly outlined in Table 7.9.1, there is no definite line where the application of one type ends and the other begins. The choice between impulse and reaction types depends upon the size of the unit as well as upon the head and other considerations.

Specific Speed The common basis of comparison between turbine runners of different types and between runners of the same type but different design and characteristics is termed *specific speed* n_s. This is the relationship between the speed of a runner at the point of highest efficiency and the maximum power output at this speed regardless of size. However, since both power and speed vary with head, specific speed is defined as the relationship between the speed n_1 and power P_1 at 1 ft (m) head. Subscript 1 denotes that the value is reduced by the similarity law to a 1 ft (m) head basis. Since $n \propto 1/D$ and $P \propto D^2$, the product $n_1\sqrt{P_1}$ remains a constant regardless of the size of the runner and is designated the specific speed of the runner. The term *specific speed* for this relationship stems from the fact that $n_1\sqrt{P_1}$ also is the value of the speed in r/min at best efficiency which the runner would have if operated under 1 ft (m) head, the runner being of such size as to develop 1 hp (kW) ($P_1 = 1$).

Since $n \propto \sqrt{H}$ and $P \propto H^{3/2}$, the specific speed of any runner operating under head H will be $n_s = n\sqrt{P}/H^{5/4}$ where n is the best efficiency speed and P the maximum power output at this speed, all at head H.

Selecting the Speed Hydraulic turbines are usually connected to ac generators. The turbine speed must agree with one of the synchronous speeds required for the system frequency. The prevailing frequency for most systems in the United States is 60 Hz. Synchronous speeds are determined by the formula $n = 120 \times$ frequency/number of poles in generator. The number of poles must be even.

The speed should be as high as practicable, as the higher the speed the less expensive will be the turbine and generator and the more efficient will be the generator.

A convenient way of determining the highest practicable speed is by the relation of the specific speed to the head. For Francis turbines this may be taken as $n_s = 900/\sqrt{H}\,(1{,}900/\sqrt{H})$; for propeller-type turbines as $n_s = 1{,}000/\sqrt{H}\,(2{,}100/\sqrt{H})$. The adoption of these values of specific speed is predicated on the use of a reasonable setting of the unit with reference to tailwater level, i.e., selection of proper cavitation coefficient.

Number of Units From the standpoint of reducing the number of auxiliaries and the amount of associated equipment and also reducing initial and maintenance costs for the entire plant, the number of units should be kept to a minimum. Also the larger the unit, the higher the efficiency and generally the lower the cost per unit power output. However, other considerations, such as flexibility of operation, higher efficiency operation during low-load demands, and minimum loss of capacity during shutdown for repair or maintenance, might dictate the use of multiple units where one unit would be feasible in terms of physical size. For some projects, the physical size of the unit has been limited to the maximum size runner that could be shipped in one piece; this is largely due to the extra manufacturing costs involved in furnishing split runners. However, since split runners present no serious mechanical difficulties, the tendency in recent years has been to disregard this limitation.

REACTION TURBINES

Francis Figure 7.9.1 shows a Francis-type reaction turbine for medium head. The runner consists of a relatively large number of shrouded buckets. Movable wicket gates with axes parallel to the turbine shaft control the flow. This type of turbine is normally used for heads ranging from 75 to 1,600 ft (25 to 500 m). Specific speeds vary from 15 to 100 (50 to 400).

Propeller Turbines, Fixed and Adjustable Blade The propeller turbine has a runner which is normally provided with from 3 to 10 unshrouded blades, either fixed or adjustable. This type of turbine usually is used for heads from 5 ft (1.5 m) up to 120 ft (40 m), although in a few cases they have been used for heads up to 200 ft (60 m). The higher the head, the greater the number of blades used. Specific speeds

Table 7.9.1 General Arrangements of Turbine Installations and Usual Head Limits Employed

Type	Setting	Construction	No. of runners	Usual head limits for direct-connected units, ft	m
Reaction turbines, 5 to 1,000 ft (1.5 to 300 m) head	Axial flow	Vertical or horizontal or slanted	1	5-60	1.5-20
	Encased	Concrete vertical	1	15-130	5-40
		Cast or welded plate steel. Stainless steel. Vertical or horizontal	1 or 2	30-1,600	10-500
Impulse turbines, 500 to 6,000 ft (150 to 1,800 m) head	Encased	Horizontal (1-2 nozzles)	2	500-6,000	150-1,800
		Vertical (1-6 nozzles)	1	500-6,000	150-1,800

vary from 80 to 250 (300 to 1,000). Propellers have very steep efficiency vs. power curves (4 and 5, Fig. 7.9.5). Adjustable-blade propeller runners are used to produce a flat efficiency curve over a wide range of power (6, Fig. 7.9.5) and to produce considerably more power beyond the maximum efficiency point than can be obtained with a fixed-blade runner of equal diameter. For fixed-blade runners, the blade angle is usually set between 16 and 28°, where maximum efficiency occurs. For adjustable blade runners, the blade angle may vary from −10° min to +45° max. Normally, automatic oil-pressure-operated blades are used in a **Kaplan turbine** (Fig. 7.9.3). The blades are adjusted by means of an oil-operated piston located either within the main shaft or in the runner hub. The oil is admitted to and discharged from the piston by means of a distributor head either on top of the generator shaft or surrounding the main shaft. The oil pressure is supplied from the governor oil pressure system. The controls are so arranged that the blade tilt varies automatically with the wicket gate opening to produce a maximum efficiency envelope curve (Fig. 7.9.9). A large operating-cylinder capacity is required. The capacity S, ft · lb (J) may be approximated from the formula $S = 20Pn_s^{1/4}/\sqrt{H}\,(14n_s^{1/4}/\sqrt{H})$. If antifriction bearings are used in the hub instead of bronze bearings for the blade trunnions, this formula becomes $S = 10Pn_s^{1/4}/\sqrt{H}\,(7n_s^{1/4}/\sqrt{H})$. However, since the hydraulic unbalance on the blades varies with speed, gate opening, and the location of the blade pivot axis, the oil-operated cylinder size should be calculated from model test data.

Axial-Flow Turbines Axial-flow turbines use the propeller type runner with either fixed or adjustable blades. Their characteristic feature is the straight-through, or nearly straight-through, water passageway from intake to discharge. The shaft is, therefore, either horizontal, vertical, or inclined (Fig. 7.9.4). The spiral or semispiral case and elbow draft tube, which require substantial widths and depth of excavation, are eliminated. Therefore, with a reduction in height and area of the powerhouse and the turbine's suitability for location directly within the dam, an overall construction-cost savings may be obtained for the power plant part of the project compared with conventional vertical units. This may make it possible to build or redevelop power plants for low heads or in small sizes that have previously been considered uneconomical.

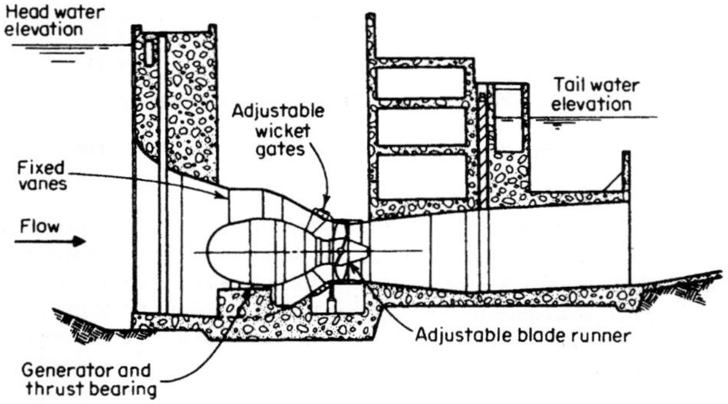

Figure 7.9.4 Axial-flow bulb turbine.

In recent years, axial-flow turbines have been installed for use with tidal power since they can be arranged to operate with water flowing in either direction and, when required, to pump with water flowing in either direction.

There are four general types of axial-flow turbines. The first type has the generator rotor mounted around the periphery of the turbine runner. This type of turbine and generator arrangement has been called a **rim** turbine. The second type is the **bulb** turbine, where the generator is directly connected to the turbine at a submerged elevation. The bulb turbine has the generator enclosed in a streamlined, watertight housing located in the water passageway on either the upstream or downstream side of the runner (Fig. 7.9.4). The third type of axial-flow turbine is the **pit** turbine. In the pit turbine arrangement, a high-speed generator is driven from the high-speed output shaft of a speed increaser, and the low-speed input torque to the speed increaser is supplied from the lowspeed shaft of the turbine. The generator of an axial-flow pit turbine can be mounted in line with the turbine, or in smaller units the generator may be mounted inclined or vertical.

The fourth type of axial-flow turbine is the **tube** type, which has the generator located outside the water passages. The construction characteristic of this turbine is the slight bend of the water passageway that permits the extension of the turbine shaft to the outside of the water passageway. The tube turbine is typically arranged so that the generator is on the downstream side of the turbine, but it could be constructed to have the generator on the upstream side. The shaft from the turbine to the generator of a tube turbine can be inclined to reduce excavation, thereby raising the generator higher above tailwater elevations. In some cases, a gear-type speed increaser is used to reduce the combined equipment cost and generator size and weight.

Axial-flow turbines are suitable for heads up to at least 90 ft (30 m) with basically the same limitations that apply to the conventional Kaplan or other propeller-type turbines. Maximum unit capacity is limited, however, by maximum practical speed increaser torques and maximum practical horizontal generator capacities. This appears to be approximately 100 MW at present. Either fixed-position or movable wicket gates can be used. Power and efficiency performance is comparable to conventional vertical shaft propeller turbines.

Design The water entering the runner is given a whirl or tangential component by the guide vanes. This whirl is taken out by the runner so that, at the design point (point of best efficiency), the water leaves the runner without appreciable whirl. According to Euler's theorem, the power of a turbine in steady motion equals the angular velocity multiplied by the change in angular momentum experienced by the mass of water flowing in a unit time in its passage through the turbine. This is expressed in the following equations:

$$\begin{aligned} P &= 62.4Q(u_1V_{u1} - u_2V_{u2})/(550g) \\ &= QHe_n/8.81 \end{aligned} \tag{7.9.1}$$

and

$$u_1V_{u1} - u_2V_{u2} = e_h gH$$

$$\begin{aligned} P &= 1{,}000Q(u_1v_{ul} - u_2v_{u2})/(101.93g) \\ &= QHe_h w/101.93 \end{aligned} \tag{7.9.2}$$

(See notation at beginning of this section.) Subscript 1 refers to the inlet edge of the bucket and 2 to the discharge edge.

In practice, e_h may be taken as 0.92 to 0.94, and the term u_2V_{u2} may be omitted as it becomes zero for discharge in a radial plane, $\alpha_2 = 90°$.

The right hand member of Eq. (7.9.2) may be regarded as a fixed value for a given head. Consequently, in order to have a high-speed-type runner (u_1, large) the whirl V_{u1} placed in the water by the guide case must be small. This means that for the best load, low-speed low-capacity (small n_s) runners will have relatively small gate openings and short wicket gates, whereas high-speed high-capacity (large n_s) runners will have relatively large gate openings and long wicket gates.

Equations (7.9.1) and (7.9.2) are used in designing the runner and guide case. The P and Q used in Eq. (7.9.1) are for the point of best efficiency and not for the rated values. For specific speeds up to about 60 (250) the power at best efficiency may be taken as 85 percent of the rated power. Above that specific speed the point of best efficiency occurs at higher loads as indicated in Fig. 7.9.5. Adjustable-blade propeller turbine-runner blades are laid out for an intermediate load of 75 to 80 percent of the rated load. The actual design of a runner should be worked out by an experienced designer on the basis of coordinated test data.

The general proportions of a normal line of turbine runners may be determined as follows for any combination of net head H and rated power P. (Rated power is usually considered to be 95 percent of full gate power.)

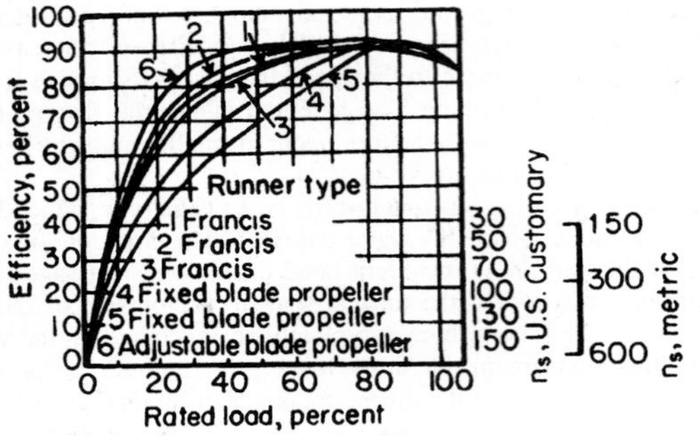

Figure 7.9.5 Variation of efficiency vs. load for reaction turbines.

First select the specific speed appropriate to the head (see "Specific Speed" above). Next calculate n from the formula $n_s = n\sqrt{P}/H^{5/4}$. If $n/H^{5/4}$ is not a synchronous speed, select the nearest synchronous speed and calculate the corresponding n_s. This value of n_s may be used in Fig. 7.9.6 to determine the peripheral coefficient ϕ for D_r and D_1. Then the diameter may be determined from the formula

$$\phi = \pi Dn/720\sqrt{2gH}\,[= \pi Dn/(60{,}000\sqrt{2gH})]$$

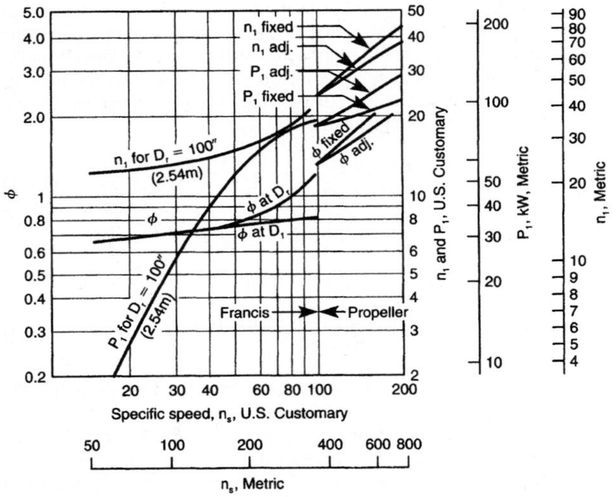

Figure 7.9.6 Turbine characteristics and specific speed.

Other proportions of the runner may be determined from Fig. 7.9.7, which shows the ratio to D_r of width of distributor B; diameter at top of draft tube D_d; and centerline of distributor to top of draft tube A. See Fig. 7.9.1 for D_r, B, and A.

The values of unit power P_1 and unit speed n_1 for D_r = 100 in. (2,540 mm), which correspond to the specific speed of the runner may be determined from Fig. 7.9.6. The following formulas may then be used: $P = P_1H^{3/2}(D_r/100)^2\,[= P_1H^{3/2}(D_r/(2{,}540)^2]$ and $n = n_1\sqrt{H}(100/D_r)[n_1\sqrt{H}(2{,}540/D_r)]$. These formulas may be used to verify the diameter calculated from the peripheral coefficient, and will prove useful in selecting runner sizes for units which must meet power requirements other than rated power at design head. In such cases the values of n_1 and P_1 from Fig. 7.9.6 should be used in conjunction with Fig. 7.9.8.

For preliminary powerhouse layouts these ratios of diameter to D_r may be used for all specific speeds: gate circle, 1.20; stay ring, 1.65; pit liner, 1.50; unit spacing, 3.5 to 4.5.

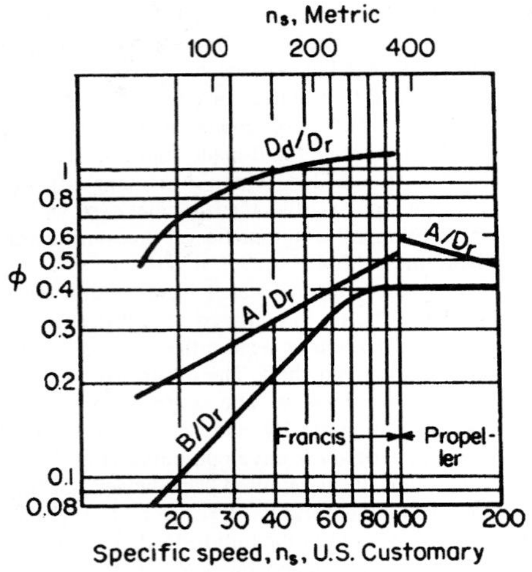

Figure 7.9.7 Turbine characteristics and specific speed.

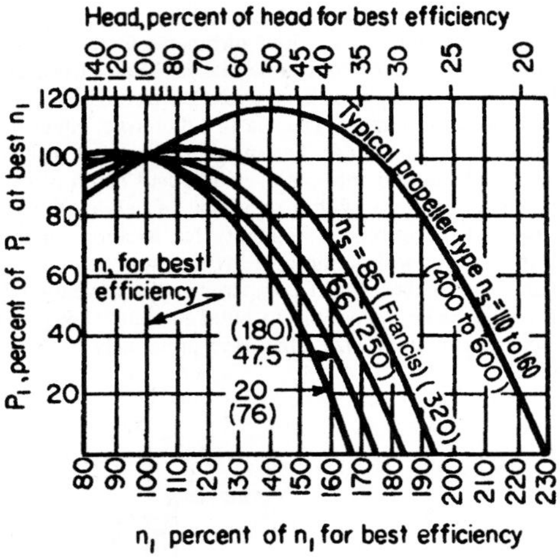

Figure 7.9.8 Variation of unit power with unit speed and specific speed.

The clearance space between the discharge tips of the wicket gates and the entrance edge of the runner buckets is treated as a free vortex space in which the tangential component of the absolute velocity of discharge from the gates (V_u) increases in inverse proportion to the radius while the component in a radial plane (V_r) increases in inverse proportion to the area.

To obtain the best efficiency, the water must enter the runner without shock and leave it with as little velocity as possible; i.e., the entrance angles of the buckets must be approximately in line with the relative direction of the entering flow and the direction of the absolute discharge velocity should be approximately in a radial plane (axial in the case of propeller runners). This condition prevails only at the most efficient load. Above that load the water enters the draft tube with reverse whirl and below that load with a forward whirl.

The shape of the buckets should be kept as simple as possible, consistent with meeting the proper angles. The curvature of the bucket along a flow line is made greatest near the entrance and reduces as the discharge edge is approached, somewhat the shape of a portion of a parabola.

Turbine Characteristics Speed, power, and efficiency characteristics vary widely with specific speed as shown in Figs. 7.9.5 and 7.9.6. The variation of unit power P_1 with the unit speed (as a percent of normal basis) is shown in Fig. 7.9.8 for typical specific speeds ranging from 20 to 160 (70 to 600). For low specific speed Francis-type turbines, P_1

decreases as n_1 increases, as when operating at heads below normal. This is due to the centrifugal effect, which tends to decrease the flow as n_1 increases. This characteristic of low specific speed Francis runners makes them less suitable for operation under extreme variations in head. With the higher specific speeds and particularly with the propeller type, P_1 increases as n_1 increases above normal. This makes the higher specific speeds particularly suitable for operation at widely varying heads.

Figure 7.9.9 shows the advantage of blade adjustment for propeller-type turbines. By angular adjustment of the blades as the load changes, the peaked power-efficiency curve of the fixed blade propeller is transformed into a flat power-efficiency curve, and the maximum power output is considerably increased.

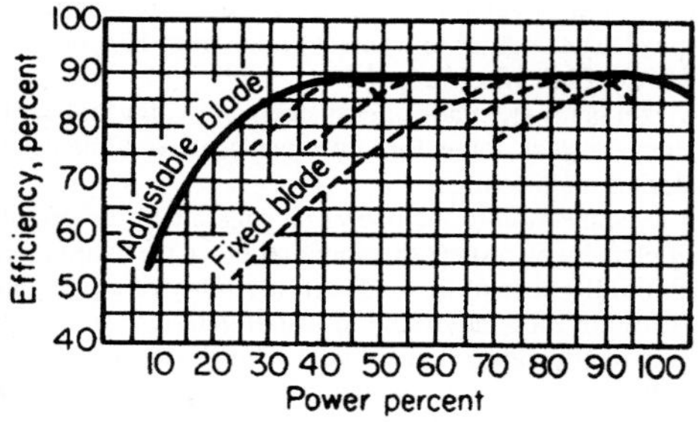

Figure 7.9.9 Variation of efficiency vs. power for fixed- and adjustable-blade propeller turbines.

Runaway Speed If a turbine runner is allowed to revolve freely without load and with the wicket gates wide open, it will overspeed to a value called the *runaway speed.* The runaway speed, at normal head, varies with the specific speed and for Francis turbines ranges from 170 percent (normal speed = 100 percent) at low specific speed [n_s = 20 (76)] to some 195 percent at high specific speed [n_s = 100 (383)]. For propeller turbines, the runaway speed varies with blade angle—the steeper the blade angle, the lower the runaway speed. For fixed-blade propellers with the blades set from 16 to 28°, where maximum efficiency is usually obtained, the runaway speed ranges approx from 255 to 180 percent, respectively. For adjustable blade turbines, where the minimum blade angle is sometimes as low as 6 to 16° in order to obtain high efficiency at part load, the maximum possible runaway speed will be about 290 to 270 percent, respectively. However, with adjustable-blade propeller turbines, there is from the standpoint of efficiency an optimum relationship between runner-blade angle and wicket gate opening, usually controlled by a cam in the operating mechanism; the higher the gate opening, the steeper the blade angle. Thus the combination of wide-open gate and minimum design blade angle can only occur in the so-called "off-cam" position, which is an extremely rare possibility. In some units, this maximum possible off-cam runaway speed is reduced by limiting the minimum design blade angle to 16°. Another method is the use of a runaway speed limiter. There are several designs in use. One of the most frequently used is a valve designed to open by centrifugal force at speeds of 120 percent (approximately). The open valve bypasses the runner blade servomotor piston, thus equalizing the oil pressure on both sides of this piston. If the runner blades are designed to open because of hydraulic unbalance at overspeeds, they will then go to a higher blade angle under off-cam overspeed conditions. The off-cam runaway speed thus can be limited to some 185 percent if the runner is designed for a maximum blade angle of 28 to 40°.

For all turbines, if the maximum head is higher than the normal head, the runaway speed will be increased in proportion to the square root of the head. Therefore, runaway speeds should be based on the maximum operating head rather than on the normal head. Any runaway speed above 180 percent can increase appreciably the cost of the generator.

Turbine Thrust The water passing through the turbine creates a thrust load, that is carried by a thrust bearing, typically supplied with the generator. For vertical units, the thrust T consists of the hydraulic thrust T_h plus the weight of the turbine runner and shaft T_w, lb (kg). Units have been constructed with the thrust bearing supporting the total weight of rotating components of the turbine and generator, plus the hydraulic thrust, located on the turbine head cover.

For horizontal or inclined units, the thrust T consists of the hydraulic thrust T_h and a component of the weight T_w, of the rotating components, if the unit is inclined. The thrust bearing can be located in the generator, in a speed increaser, in the hydraulic turbine, or in a separate thrust bearing housing.

The hydraulic thrust may be obtained by using a **thrust coefficient K_t** multiplied by the weight, lb (kg), of a circular column of water whose diameter is the diameter of the runner D_r, in (mm), and whose height is equal to the maximum head H_m, ft (m), under which the turbine must operate, or

$$T_h = \frac{K_t D_r^2 H_m}{2.94} \quad \text{lb} \qquad \left(T_h = \frac{K_t D_r^2 H_m}{1{,}273} \quad kg\right)$$

The thrust coefficient for Francis turbines varies with the specific speed and may be taken as approximately $K_t = n_s/250$ ($K_t = n_s/954$). In using this value of K_t, it is assumed that the runner seals are properly arranged and that the crown of the runner inside the seals is properly drained; otherwise, the thrust may be much higher.

For propeller turbines, with either fixed or adjustable blades, the thrust coefficient K_t may be taken as 0.90 for all specific speeds.

Reaction Turbine Elements

Runner and Wearing Rings The number of buckets z for Francis runners ranges from about 21 for low specific speed to 12 for high specific speed. The approximate number may be taken as $z = 55/n_s^{1/3}$ ($86/n_s^{1/3}$).

Materials that can readily be repaired by welding with either carbon-steel or stainless-steel electrodes are selected for the construction of runners. Most runners are made of cast carbon steel, cast stainless steel, formed carbon steel plate, formed stainless-steel plates, or a combination of carbon steel and stainless steel. The majority of new equipment is supplied manufactured of stainless steel, using a combination of martensitic and austenitic stainless-steel casting, forgings, and plates. The use of high-strength stainless steel for runners is increasing for high heads and for conditions where cavitation pitting may be a design consideration.

The number of blades for propeller-type turbines ranges from 10 for low specific speeds to 3 for high specific speeds. Propeller runner blades are usually made of cast carbon steel, formed carbon steel, formed stainless steel, or cast stainless steel. If they are made of cast steel or formed carbon plate steel, the surfaces over which pitting may be expected are overlay-welded with stainless steel before finishing to final contour. In place of overlay welding, solid stainless-steel inserts are sometimes welded into propeller blades, and stainless trim is added to the periphery of the blades.

The functions of the runner seals for Francis turbines are to prevent excessive leakage loss and thus improve the efficiency and control the hydraulic thrust.

Rolled, forged, and cast steels make excellent wearing rings for low- and medium-head units. To prevent seizure in case of contact the rotating ring material should differ from that of the stationary rings. Stainless steel, bronze, or steel with bronze inserts make excellent wearing rings. For extremely high heads, stainless steel should be used to prevent undue wear and erosion.

Seal clearances are made as small as practicable to reduce leakage, particularly on high-head units. Larger clearances are required with water-lubricated main bearings, subject to considerable wear before being readjusted.

Main Shaft and Bearings The main shaft, for a typical vertical hydraulic turbine unit, must be rigid and should be made of a medium-grade carbon-steel forging. The size of the shaft is determined by limiting the torsional stress from 3,000 to 6,000 lb/in^2 (20 to 41 N/mm^2).

The shafting system of a horizontal or an inclined hydraulic turbine unit is subject to the torsional stress produced by the torque transmitted and the bending stress produced by the weight of the runner, the shaft, and the components that are attached to the shafting system. The design of a shaft for a horizontal unit must be subjected to review by experienced mechanical engineering designers to ensure acceptable stress levels, to account for points of stress concentration, and to determine the overall safety margin of a design. The allowable stress levels in a horizontal shaft must be evaluated after the fatigue properties of the shafting material and the points of stress concentration are taken into consideration. Should the bearing be designed to be water-lubricated, a stainless steel sleeve is installed on the shaft to provide a bearing surface for the guide bearing. A stainless-steel sleeve is typically placed on the shaft in the area of a packing box or stuffing box.

Vertical hydraulic turbines are usually provided with one main bearing located in the head cover, as near the runner as practicable. The main turbine bearing is generally a babbitted oil bearing, supplied in halves, or as independently adjustable segmented babbitted shoes. The oil is supplied to the bearing by an independent low-pressure oiling system. Self-lubricated babbitted shell or shoe bearings, containing an oil reservoir, and pumping grooves in the babbitt are sometimes used.

Water-lubricated bearings of wood, lignum vitae, rubber, or special composition are sometimes preferred, particularly for small and medium-size propeller-type turbines where the bearing is located at the bottom of the head cover cone, where the packing box and the oil-lubricated bearing would be inaccessible, and where it would be difficult to avoid water contamination. With the water-lubricated bearing, the packing box is placed above the bearing.

Spiral Case The spiral case must be proportioned so as to cause relatively low friction losses, as well as to prevent eddying which would travel into the runner and affect its efficiency. The cases generally are made of (1) metal: cast steel, cast iron, or steel plate, and (2) concrete. Metal cases are made as complete spirals, customarily with uniform or slightly increasing velocity from the throat to the small end. When a valve is used at the case entrance, the net area at the centerline of the valve should be at least equal to the case entrance area.

Concrete cases may be complete spirals or semispirals, rectangular or oval in cross section. There should be no piers close to the turbine.

Rated load velocities for general practice for cases of various types are shown in Fig. 7.9.10. Higher velocities are sometimes used with the larger units to reduce size and cost.

Stay Ring That part of the guide apparatus between the spiral case and the wicket gates, containing stationary stay vanes, is called the stay ring. The water is accelerated within this space as it approaches the gates. The number of stay vanes employed is usually made equal to or one-half the number of gates. They are placed at the angle which will cause the least obstruction to the flow.

The stay ring is cast integral with cast cases and is made separately of welded or cast steel for concrete or steel plate spiral cases. It should be a continuous ring to facilitate erection, and very rigid, because it serves as a foundation for the rest of the turbine and generator.

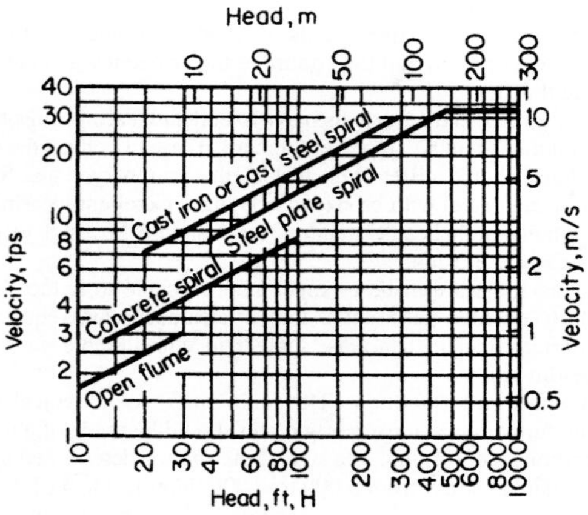

Figure 7.9.10 Case velocities.

Wicket Gates and Operating Mechanism Wicket gates control the power and speed of the turbine. The number of gates m ranges from 12 for small units to 28 for large units. The overall dimensions of the turbine decrease as the number of gates increases.

To prevent interference between the gates and the runner buckets, which may cause noise and vibration, the discharge tips of the fully open gates should be kept well away from the inlet edges of the runner buckets of Francis turbines, the radial clearance being large enough to prevent the gate tips from overhanging the curved part of the discharge ring. Gate tip diameter = 1.1 (maximum inlet diameter D_r).

The height and angular movement of the gates increase with the specific speed. The angular movement varies from 15° for low specific speed to 80° for high specific speed.

Most wicket gates are made of cast steel; a few, for higher heads, are made of stainless steel; some, for lower heads, are weldments built from rolled materials and castings.

Each gate connection to the operating ring should be provided with a breaking element to protect the gate and other mechanism in case of an obstruction. Because of the inherent instability of a free gate and its tendency to flutter, it is advisable to install gate-restraining devices to hold the gate if a breaking element fails. This device also avoids potential resonant flutter with the natural frequencies in the system. Each gate should also be provided with stops to prevent it from striking the runner or reversing after the breaking element fails.

The capacity G, ft · lb (J), of the oil pressure cylinder or cylinders (sometimes known as governor capacity) may be computed from the formula

$$G = 18Pn_s^{1/4}/\sqrt{H} \quad (13Pn_s^{1/4}/\sqrt{H})$$

Some units now provide individual servomotors on each wicket gate. This synchronized system allows the hydraulic control system to dampen a tendency to gate flutter.

One or more vacuum breakers or air valves are installed in the head cover to admit air to the runner or draft tube, to improve efficiency at low gate openings or to alleviate draft tube vortex cavitation. An air valve is also necessary on propeller-type units to break the vacuum under the head cover and to help prevent the runner from "screwing up" in the water when the gates are suddenly closed. The air valve is piped to the outside of the powerhouse above the floodwater level.

Draft Tubes The draft tube may serve the double purpose of (1) allowing the turbine to be set above tailwater level, without loss of head, to facilitate inspection and maintenance, and (2) regaining, by diffuser action, the major portion of the kinetic energy delivered to it from the runner.

At rated load the velocity at the upstream end of the tube for modern units ranges from 24 to 40 ft/s (7 to 12 m/s), representing from 9 to 25 ft (3 to 8 m) head. As the specific speed is increased and the head reduced, it becomes increasingly more important to have an efficient draft tube. Good practice limits the velocity at the discharge end of the tube to 5 to 10 ft/s (2 to 3 m/s), representing less than 1 ft (0.3 m) velocity head loss.

The elbow type of tube is now used with most vertical turbine installations. With this type the vertical portion begins with a conical section which gradually flattens in the elbow section and then discharges horizontally through substantially rectangular sections to the tailrace. Most of the regain of energy takes place in the vertical portion, very little in the elbow section which is shaped to deliver the water to the horizontal portion so that the regain may be efficiently completed. Figure 7.9.11 shows proportions of a good elbow tube. One or two vertical piers are placed in the horizontal portion of the tube for structural reasons.

Most tubes are made of concrete with a steel plate lining extending from the upper end to a point where the velocity has been sufficiently reduced [say 20 ft/s (7 m/s)] to prevent erosion of the concrete. Pier noses are also lined where necessary to prevent erosion and for structural reasons.

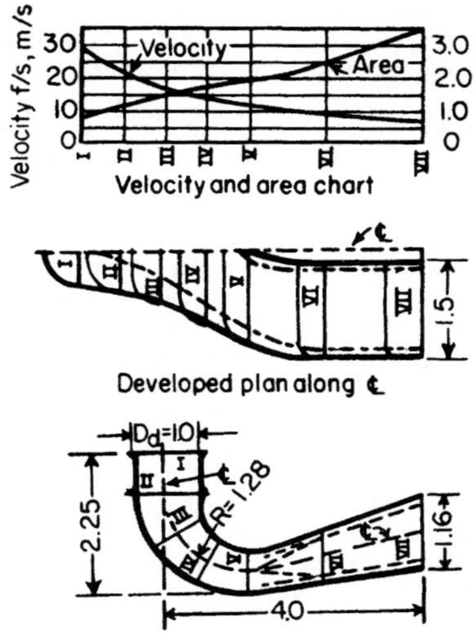

Figure 7.9.11 Elbow draft tube layout.

IMPULSE TURBINES

Impulse turbines are utilized when the head is too high for the practical use of Francis turbines, which is normally a head exceeding 1,600 ft (500 m). Impulse turbines are also sometimes used for heads below 1,600 ft (500 m) where excessive erosion due to foreign materials in the water presents a problem. The main disadvantage of impulse turbines, especially for low heads, is their low specific speed. In the past, this was overcome on conventional horizontal shaft units by the use of two runners or two jets per runner. In recent years, the vertical shaft multijet impulse turbine (Fig. 7.9.12) has become popular.

The efficiency obtainable from a horizontal shaft impulse turbine is about 90 percent. Field tests on multijet vertical units have shown efficiencies as high as 91.5 percent. The use of multiple jets on the vertical units reduces the percentage loss due to windage of the runner. The unit can be operated with a reduced number of jets at part load, thus increasing part load efficiency. Six jets are about the maximum number that can be used on one runner without jet interference.

Selection of Speed The first consideration is the selection of suitable n_s. As with reaction turbines, there is a relation which determines approximately the limiting value of specific speed n_s (per jet) for any head. For heads around 1,000 ft (300 m), $n_s = 5.0$ to 5.5 (19 to 21) gives high efficiency. For heads around 2,000 ft (600 m), maximum efficiency is attained near $n_s = 4.0$ to 5.0 (15 to 19). Care has to be exercised in selecting an n_s (per jet), especially for the higher heads, so that the proper number of buckets can be used on the wheel disk. The higher the n_s, the fewer the number of buckets required (Fig. 7.9.13) but the smaller the pitch circle. The latter can create stress problems in the attachment of the buckets to the disk. If the resulting speed n is quite low for the power to be developed, the n_s of the unit and consequently the speed n can be increased by increasing the number of runners or the number of jets per runner. The n_s of the unit will then be n_s (per jet) times the square root of the number of jets.

Basic Dimensions The pitch diameter D_p of the runner is twice the distance from the center of the runner to the axial centerline of the jet and is determined by the value of the coefficient ϕ selected, $D_p = 1{,}840\phi\sqrt{H}/n\,(84{,}578\phi\sqrt{H}/n)$. The variation of ϕ with n_s is established by experience and model tests (Fig. 7.9.13).

The quantity of water Q discharged per jet is $550P/(wHe)$ $[101.93P/(wHe)]$, where P is the horsepower (kilowatts) developed by each jet.

The diameter of the jet is $4.85\sqrt{Q}/H^{1/4}$ in ($0.55\sqrt{Q}/H^{1/4}$ m), using a velocity coefficient of 0.97 for the jet.

The **velocity** in the inlet should not exceed $0.12\sqrt{2gH}$ (approximately), and it is considered good practice not to exceed an absolute value of 40 ft/s (12 m/s).

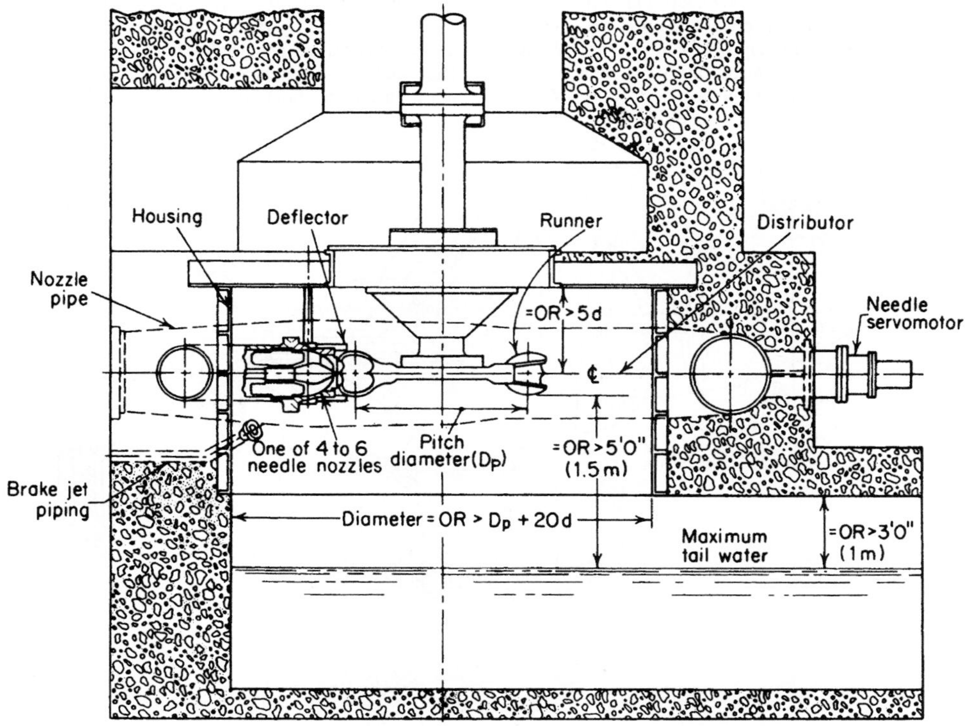

Figure 7.9.12 Vertical-shaft multijet impulse turbine installation.

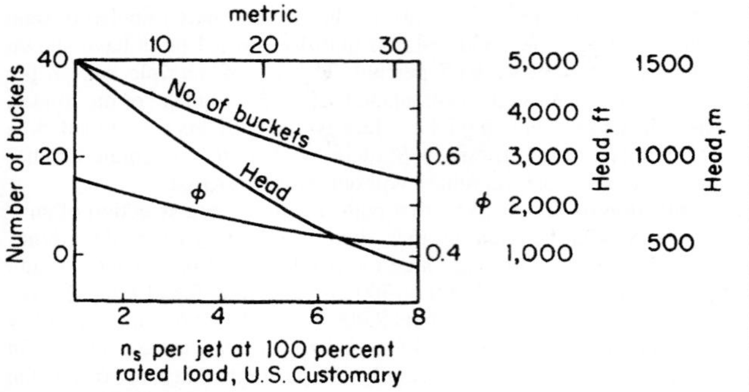

Figure 7.9.13 Impulse turbine characteristics and proportions.

The **valve** employed in the inlet pipe should be of a type which, when open, permits a smooth, uninterrupted flow, free of eddies or obstructions. Modern preference is for rotary sphere valves for this purpose.

The **housing** serves primarily to carry off the discharged water to the tail pit below. On horizontal shaft units, at the place where the runner receives the discharge from the jet, the housing should be about 10 to 12 times the jet diameter. At the place where the runner has been cleared of the discharge, the housing should be as narrow as possible to decrease windage. Suitable baffling should be used to carry the discharge away from the buckets. The housing should be vented adequately near the center of the runner.

For vertical-shaft units, the housing should be of ample size to prevent discharge water from interfering with the buckets. The housing should also be vented adequately near the center of the runner.

The **setting** of the horizontal-shaft unit should be such that the lower edges of the buckets are at least 3 ft (1 m) above maximum tailwater elevation. For vertical shaft units, this distance should be at least 5 ft (1.5 m). Impulse turbine discharge entrains large amounts of air, and unless this is completely replaced, a vacuum will be produced in the housing which will draw the tailwater up and drown out the runner. Thus the roof of the discharge tunnel or passageway, for both horizontal- and vertical-shaft units, should be at least 3 ft (1 m) above the maximum operating tailwater elevation to permit the free circulation of air to the runner (Fig. 7.9.12).

In cases where extremely high tailwater occurs for a short period, compressed air can be used to depress the tailwater.

Experience has shown that the ft^3/s (m^3/s) of air (at the discharge pressure of the compressor) required to maintain the depressed tailwater is about 15 percent of the ft/3/s (m^3/s) of water (Q) discharged by the turbine.

Runner The force F exerted by a stream upon a stationary bucket (Fig. 7.9.14) is

$$F = MV(1-\cos\theta) = wQV(1-\cos\theta)/g$$

where M = mass per s and θ = angle in deg through which the water is turned relative to the bucket (see section *Y-Y,* Fig. 7.9.14). If the bucket is moving in the direction of the stream with a velocity u, $F = wQ(V-u)(1-\cos\theta)/g$. The work done by the stream $= Fu = wQn(V-u)(1-\cos\theta)/g$. When θ is greater than 90°, the cosine becomes negative. Thus the closer the angle θ is to 180°, the greater the amount of work done with the same Q and V. However, in order to prevent the water from striking the succeeding bucket, θ should never be 180° (see section *Y-Y,* Fig. 7.9.14). Also, since the water discharged from the bucket must have some velocity, the value of u is made somewhat less than $V/2$.

The angle β_1 at the centerline of the bucket should be calculated from the velocity triangle (Fig. 7.9.14). The value of β_1 is greatest as the entrance lip E enters the jet. The angle of the underside of the bucket should be greater than β_1 to avoid pitting.

In terms of the jet diameter which will give maximum guaranteed or rated capacity, the approximate proportions of the buckets are $B = 3d$, $L = 2.6d$, $D = 0.85d$ (Fig. 7.9.14). The contour of the entrance lip E is important, since the manner in which this lip comes in contact with the jet affects the path of the water passing to the preceding bucket. The size of the bucket with relation to the jet diameter also determines the percentage of full load at which maximum efficiency occurs. The larger the bucket, the higher this percentage of load.

The number of buckets on a runner should be such that no water can pass through the buckets without being deflected by the buckets. Figure 7.9.13 indicated approximately the most efficient number.

The **needle nozzle** should be placed as close to the buckets as possible, as the jet tends to lose its compactness of form shortly after emerging from the nozzle. The needle and the seat of the nozzle should be designed for easy replacement and should be of a material highly resistant to erosion. The needle should terminate in a cone of 30 to 45°. The nozzle tip should have a subtended angle of 45 to 60° and a diameter at discharge about 20 percent greater than the calculated diameter

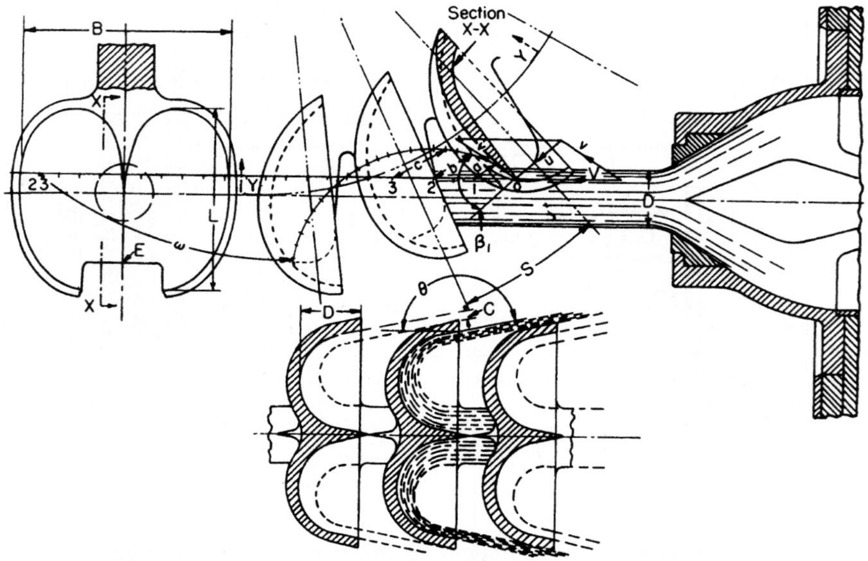

Figure 7.9.14 Impulse turbine bucket design.

of the jet. The maximum diameter of the needle should be about 15 percent greater than the discharge diameter of the nozzle tip. The diameter of the upstream portion of the nozzle should be such that the velocity does not exceed $0.10\sqrt{2gH}$. (See Lowy, Efficiency Analysis of Pelton Wheels, *Trans. ASME,* Aug. 1944, for additional information on the design of impulse turbines.)

Regulation The inertia of the water flowing through the long penstocks usually employed with impulse turbines prohibits rapid reduction in velocity because of the pressure rise which would occur. Therefore, to minimize the speed rise following a sudden load rejection, it is necessary to reduce the hydraulic power delivered to the runner without changing the flow in the penstock too rapidly. This is usually accomplished by placing a governor controlled jet deflector between the needle nozzle and the runner. The governor moves this deflector rapidly into the jet, cutting off the load. It is not unusual for the deflector to cut off the entire jet in 1½ s. Since the deflector acts on the jet after it leaves the nozzle, there is no change of flow in the penstock; hence, there is no pressure rise. The governor then moves the needle at a permissible rate (in terms of pressure rise), with simultaneous automatic withdrawal of the deflector. The jet is finally reduced the necessary amount to correspond to the reduced load. The needle must also move slowly in the opening direction for oncoming loads to avoid penstock collapse because of large pressure drops.

Runaway Speed The runaway speed for impulse turbines ranges from 160 to 190 percent of normal speed, depending upon the specific speed of the runner; the higher the specific speed, the higher the runaway speed.

REVERSIBLE PUMP/TURBINES

In this type of project, surplus low-value energy from either hydro or thermal plants is used to pump water during off-peak periods to an elevated reservoir, where it becomes available for generating high-value peaking energy. While separate pumps and hydraulic turbines of conventional design can and have been used for this purpose, the development of single reversible pump/turbines has made many pumped storage projects economically feasible.

Figure 7.9.15 is a cross section of a reversible pump/turbine, showing that, with the exception of the runner, it is essentially a conventional turbine. The wicket gates must be designed for flow in both directions. A few units have been built without movable wicket gates. The runner is essentially a pump runner modified for optimum performance while generating power. (See Sec. 10 for the design of centrifugal pumps and their characteristics.) Conventional turbine runners, because of their short blades, are not suited for the pumping cavitation requirements.

The reversible pump/turbines have certain fundamental performance characteristics which are inherent in the design. The relationship between pumping and generating performance for a given specific speed is more or less fixed and can be modified only to a minor degree by alterations in the design. Figure 7.9.16 shows the expected generating performance, and Fig. 7.9.17 shows the expected pumping performance based on model tests of a reversible pump/turbine with a specific speed $n_s = 47.3$ (180) generating and $n_s = 2{,}600$ (50) pumping (see "Centrifugal and Axial Pumps" for definition of pump specific speed).

The best efficiency, with reversible pump/turbines, occurs at a lower speed when generating than when pumping. This can be compensated for by using a generator/motor capable of operating at two speeds with a constant frequency. Several units of this type have been built, but since two-speed generator/motors cost considerably more than those built for single speed, a careful study should be made of the advantages to be obtained in operating at two speeds.

Single reversible pump/turbines can be built for any heads up to 2,000 ft (600 m). Beyond this, either multiple-stage reversible units or

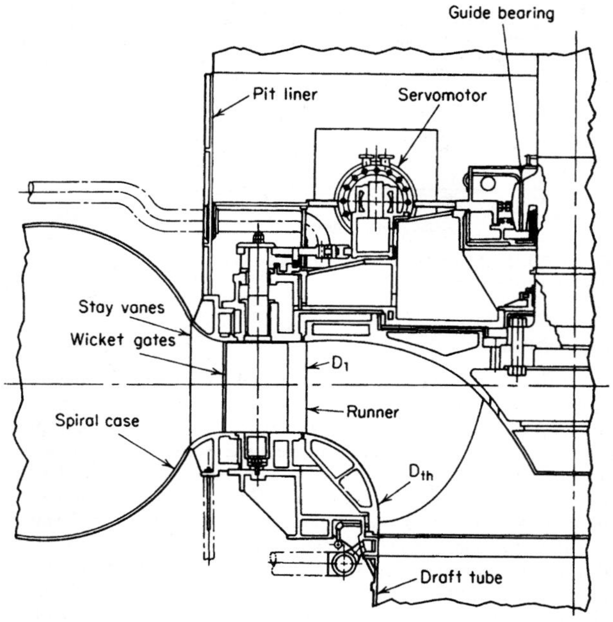

Figure 7.9.15 Reversible pump turbine.

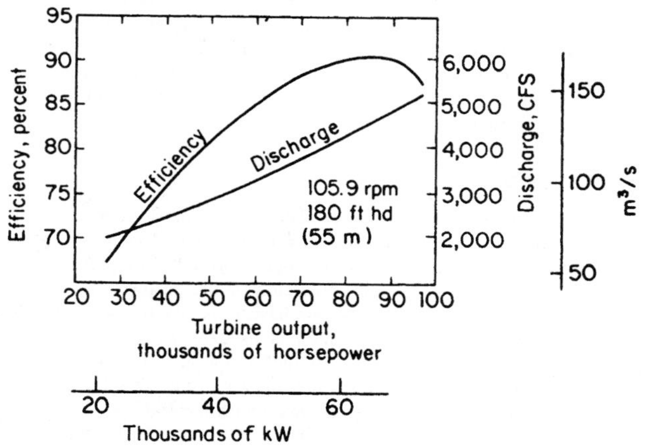

Figure 7.9.16 Generating.

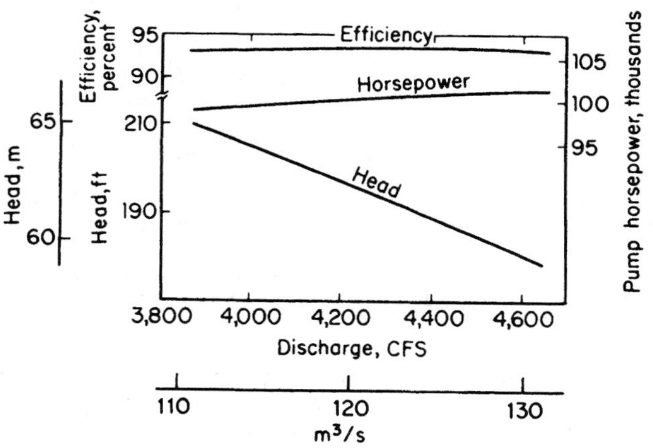

Figure 7.9.17 Pumping.

Figures 7.9.16 and 7.9.17 Typical performance curves for a pump turbine.

Table 7.9.2 Reversible Pump/Turbine Installations

	Generating				Pumping				
	Power		Head		Capacity		Head		Year in operation
Plant	1,000 hp	MW	ft	m	ft³/s	m³/s	ft	m	
Pedreira "A"	5.2	3.9	49	15	690	19.5	49	15	1939
Hiwassee	62	46	190	58	3,900	111	205	62.5	1956
Taum Sauk	220	165	790	241	2,650	75.0	765	233	1963
Cruachan	100	74.6	1,130	343	1,010	28.6	1,175	358	1966
Cabin Creek	166	125	1,190	363	1,375	39.0	1,060	323	1967
Villarino	135	100	1,250	382	1,020	29	1,325	404	1969
Kisenyami	240	180	755	220	3,880	110	645	197	1969
Drakensburg	360	269	1,380	422	1,685	42	1,443	440	1977
Dinorwic	405	302	1,682	513	1,647	46.7	1,771	540	1975
Bajina Basta	422	385	1,968	600	1,520	43	2,132	650	1976
Raccoon Mt.	465	350	940	287	3,850	110	1,000	305	1974
Bath Co.	510	380	1,080	329	4,100	116	1,100	335	1983
Palmiet	335	250	988	301	2,012	57	1,000	305	1988
La Muela	284	212	1,725	525	1,165	33	1,637	499	1990
BadCreek	482	360	1,200	367	3,531	100	1,240	378	1991
Mingtan	368	275	1,322	403	2,896	82	1,345	410	1992
Shisanling	273	204	1,411	430	1,335	37.8	1,588	484	1996
Guangzhou II	410	306	1,673	509	2,030	57.5	1,742	531	2000
Goldisthal	360	269	990	302	2,825	80	1,004	306	2003
Taian	335	250	1,344	410	3,397	96.2	820	250	2005

separate pumps and turbines should be used. Some of the more notable reversible pump/turbine installations are shown in Table 7.9.2.

Runaway Speed The runaway speed for pump/turbines is considerably lower than that for conventional turbines. It ranges from 150 percent of normal for low specific speed runners to 175 percent for high specific speed runners.

MODEL TESTS

Model tests serve several purposes. They are primarily used to check turbine runner, wicket gate, draft tube, casing, and (sometimes) inlet works designs for optimum performance. Correctly interpreted, they may also be used as a reliable indication of the performance of the units in the field. In many cases, purchasers specify the performance of a homologous model test, which is used as an acceptance test of the unit in lieu of field tests. In such cases, the field conditions, particularly the casing and draft tube, must be reproduced faithfully. Model tests should be run in accordance with the International Code for Model Acceptance Tests of Hydraulic Turbines, Publication 193; International Code for Model Acceptance Tests of Storage Pumps, Publication 497 of the International Electrotechnical Commission (IEC); and International Standard Publication, IEC 995, that supplements IEC 193 and 497.

Figure 7.9.18 shows typical model test results of a reaction-type turbine, in which the power P at 1 ft (0.3 m) head and the efficiency are each plotted against the speed n at 1 ft (0.3 m) head, for each of several gate openings.

Laws of Proportionality for Homologous Turbines The laws of proportionality shown previously are used to calculate from model tests the power, speed, and discharges of a homologous turbine of different diameter under a different head. The field unit will have a somewhat higher efficiency and power owing to proportionally smaller frictional and bearing losses. Expected field efficiency is customarily computed from the model efficiency by the Moody formula

$$c' = 1 - \frac{2}{3}\left[(1-e)\left(\frac{D}{D'}\right)^{1/5}\right]$$

in which the prime letters refer to the field installation. The step-up efficiency is usually computed for the point of best efficiency only, and the corresponding differential is applied as a constant value from, say, half load to full load.

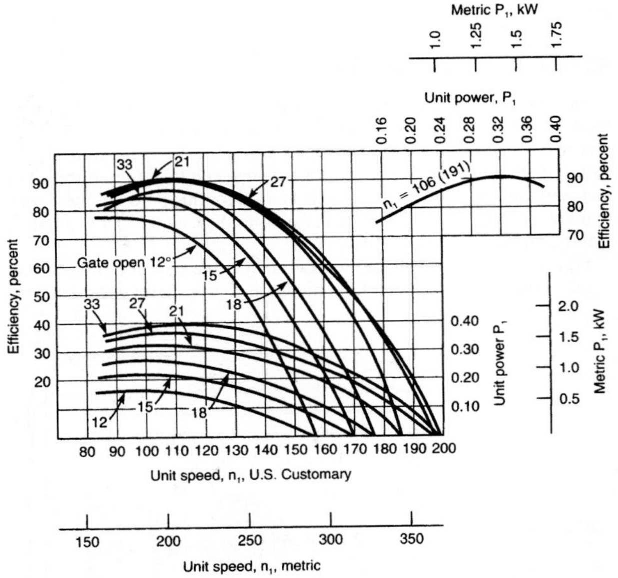

Figure 7.9.18 Model test curves for a Francis runner [$D_r = D_{th}$ = 16.2 in (0.41 m)] and n_s = about 65 (250).

CAVITATION

Cavitation occurs when the pressure at any point in the flowing water drops below the vapor pressure of the water.

The relationship which produces cavitation is between vapor pressure, barometric pressure, setting of the runner with respect to tailwater, and net effective head on the turbine and is expressed by the **Thoma cavitation coefficient** $\sigma = (H_b - H_v - H_S)/H$, where H_b = barometric head, ft (m) of water; H_v = vapor pressure of water, abs; H_s = elevation, ft (m) of the runner above tailwater, measured at the centerline of the distributor of a Francis runner and at the centerline of the blades of a propeller runner (if the runner is submerged, H_s becomes negative); and H = total or net effective head, ft (m), on the turbine. The setting of the runner depends upon the value of σ, which varies with specific speed

n_s of the runner and the individual characteristics of a particular runner design. In practice, the model of the proposed runner is first tested with relatively high back pressure (H_s small or negative). Then the back pressure is reduced in increments until the breaking point, as indicated by a drop in power, efficiency, and discharge, is reached. This breaking point is designated as the critical σ and will vary with gate opening and speed and, on propeller turbines, with blade angle. Consequently, σ must be determined for a range of limiting conditions.

In the absence of cavitation tests, the value of σ should not be lower than $\sigma = n_s^{3/2}/2{,}000\,(n_s^{3/2}/15{,}000)$ for Francis and propeller runners and $\sigma = n_s^2/25{,}000\,(n_s^2/350{,}000)$ for adjustable-blade propeller runners.

The value of σ at which a plant operates, depending largely upon the setting of the runner with respect to tailwater, is called the **plant** σ. To avoid excessive cavitation, the plant σ should exceed the critical σ. The greater this margin, the less possibility of cavitation during operation. For a general discussion of cavitation phenomena, see Knapp, Recent Investigations of the Mechanics of Cavitation and Cavitation Damage, *Trans. ASME,* Oct. 1955. Laboratory tests and experience have shown that materials having a high resistance to cavitation erosion (pitting) and suitable for use in hydraulic turbines are the stainless steels and aluminum bronzes, especially when used as welding overlays. (See Rheingans, "Resistance of Various Materials to Cavitation Damage," ASME Report of 1956 Cavitation Symposium.)

SPEED REGULATION

(See also Sec. 12.)

Regulation is accomplished by changing the flow of water to the turbine. The flow is controlled by the wicket gates of reaction turbines and by the needle valve or jet deflector of impulse turbines. The governor moves the gates or needle in response to speed changes resulting from load or head changes.

A schematic diagram of a classic governor is shown in Fig. 7.9.19. The parts consist of a speed-responsive device, a power element that changes the gates or needle position, and a follow-up or compensating device that prevents hunting.

The **speed-responsive device** originally was a pair of spring-loaded flyballs mounted directly on the turbine shaft, or driven from the shaft by belt or gears, or driven by an electric motor that receives its power either from the bus line or from an independent generator driven from the main turbine generator shaft.

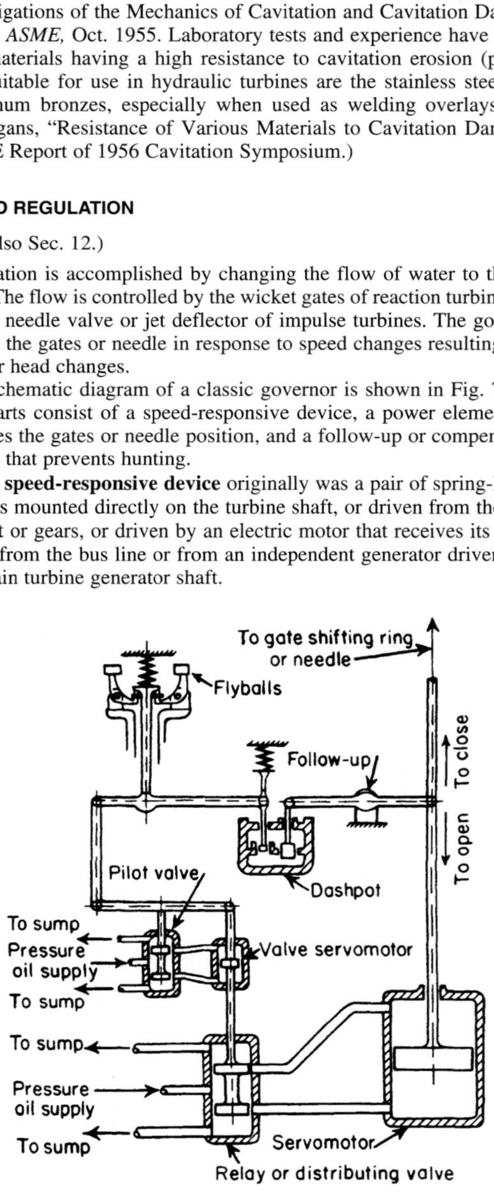

Figure 7.9.19 Schematic diagram of governor.

The **power element** consists of oil-operated power cylinders or servomotors which operate the turbine gates or needle. Oil pumps and a pressure tank or accumulator maintain a supply of oil. A valve operated by the speed sensor controls the flow of oil to the servomotors or acts as a pilot valve controlling a larger relay valve, which in turn controls the oil to the servomotors. The pump capacity is usually 3 servomotor volumes per minute. The capacity of the pressure tank is generally made 20 times the servomotor volume, allowing for 8 volumes of oil and 12 volumes of air. The velocity of oil in the pipelines is kept below 15 ft/s (5 m/s). When using individual servomotors on each gate, these values are usually increased 50 percent to balance the usual damping effect of the common massive two-servomotor system.

The **follow-up** or **compensating device** connects the power piston of the servomotor to the control valve, usually through a dashpot, and causes the motion of gates or needle to stop when they have moved sufficiently to compensate for the load change.

The time for a full stroke or traverse of the governor is controlled by the rate of flow of oil to the servomotors; most governors have provisions for varying this time. The gate opening changes at a uniform rate over the major portion of the stroke and at a somewhat slower rate at the ends of the stroke. The governor *dead time,* or the elapsed time from the initial speed change to the first movement of the gates, is usually less than 0.2 s.

Electric Governor In recent years, the functional requirements placed upon the hydraulic-turbine governor have increased to the point where electrical control of the hydraulic turbine is attractive in view of the simplicity with which electrical signals can be manipulated. The basic elements of an electric-hydraulic governing system are (1) permanent magnet generator (PMG) or equivalent for measurement of turbine speed and the means for transmittal of such speed signals to the electrical portion of the governor; (2) an electric circuit sensitive to speed variations about some adjustable reference point; (3) amplifying circuits to convert speed reference changes, speed error signals, and auxiliary signals into a useful electric current; (4) an electrohydraulic transducer to transform the electric current into a hydraulic output signal; (5) hydraulic amplifying equipment to deliver suitable power and the desired signal to the gate servomotors as a function of the output of the electro-hydraulic transducer; (6) power supplies for the electric and hydraulic portions of the control. For further particulars of the electric governor, see Leum, Electric Governors for Hydro Turbines, ASME Paper 62-WA165.

Electronic Governors With the advent of minicomputers, microprocessors, and programmable controllers, many of the traditional mechanical functions in the hydraulic turbine governor have been replaced electronically. While the basic elements of the governor remain the same, the use of electronics in the feedback sensing and control circuits has greatly reduced the size and mechanical complexity of these units. The basic elements of the modern electronic governing system are: (1) Inductive pickups or permanent magnetic generators or equivalent to measure the turbine speed and to transmit the information to the processor. (2) Electronic circuitry to establish the speed-control function of the governor (integrated circuits or digital programming). Basically the circuitry consists of a proportional, an integrator, and a derivative function. (3) A servo valve or equivalent to translate the electronic signal from the processor to a hydraulic signal. (4) Where required, a means to hydraulically amplify the output from the servo valve so that a suitable power level can be delivered to the gate and blade servomotors. (5) Auxiliary equipment as required in the form of power supplies, etc. to support the system.

Speed Regulation Requirements Usually, a sufficient measure of the regulation provided is the maximum speed rise resulting from sudden rejection of full load, as from the breaker tripping. A maximum speed rise of 30 percent of normal speed for this condition is a common limitation.

Speed Rise Following Load Reduction For sudden load reductions, the approximate speed rise is

$$n_t/n = [1 + 1{,}620{,}000 T_t P_t (1 + h/H)^{3/2}/WR^2 n^2]^{1/2} \quad \text{(USCS)}$$
$$= [1 + 365{,}000 T_t P_t (1 + h/H)^{3/2}/GD^2 n^2]^{1/2} \quad \text{(SI)}$$

where n_t is the speed, rpm, at the end of time T_t; n is the speed, rpm, before the load decrease; T_t is the time interval, seconds, for the governor to adjust the flow to the new load; P_t is the reduction in load, hp (kW); h is the head rise caused by the retardation of the flow, ft (m); H is the net effective head before the load change, ft (m); WR^2 is the product of the revolving parts weight, lb, and the square of their radius of gyration, ft; GD^2 is the product of the revolving parts weight, kg, and the square of their diameter of gyration, m. For approximate values of h, see below. Very rapid gate closure produces a reduction of pressure in the draft tube and the possibility of breaking the water column, with subsequent violent resurge which may damage the turbines.

Speed Drop Following Load Increase For sudden load increases, the approximate speed drop is

$$n_t/n = [1 - 1{,}620{,}000 T_t P_t / WR^2 n^2 (1 - h/H)^{3/2}]^{1/2} \quad \text{(USCS)}$$
$$= [1 - 365{,}000 T_t P_t / GD^2 n^2 (1 - h/H)^{3/2}]^{1/2} \quad \text{(SI)}$$

where P_t is the actual load increase and h is the head drop caused by the increase of the flow. If the speed drop is to be determined for a given increase in gate opening, the governor time T_t for making this increase and the normal change in load for the change in gate opening, under constant head H, can be used in the following formula:

$$n_t/n = [1 - 1{,}620{,}000 T_t P_t (1 - h/H)^{3/2} / WR^2 n^2]^{1/2} \quad \text{(USES)}$$
$$= [1 - 365{,}000 T_t P_t (1 - h/H)^{3/2} / GD^2 n^2]^{1/2} \quad \text{(SI)}$$

The actual change in load, however, will be $P_t(1 - h/H)^{3/2}$.

For derivation of the above speed variation formulas and for a more accurate determination, see Strowger and Kerr, Speed Changes of Hydraulic Turbines for Sudden Changes of Load, *Trans. ASME,* 1926; and Rich, "Hydraulic Transients," McGraw-Hill.

Water Hammer in Penstocks If a gate movement is considered as a series of instantaneous movements with a very small interval between each movement, the pressure variation in the penstock following the gate movement will be the effect of a series of pressure waves, each caused by one of the instantaneous small gate movements. For a steel penstock, the velocity of the pressure wave $a = 4{,}660/\sqrt{1 + d/100t}$ $(1{,}420/\sqrt{1 + d/100t})$, where d is the penstock diameter, in (m), and t is the penstock wall thickness, in (m). The pressure change at any point along the penstock at any time after the start of the gate movement may be calculated by summing up the effect of the individual pressure waves. See "Symposium on Water Hammer," ASME, 1933; and Parmakian, "Water Hammer Analysis," Dover.

Approximate formulas (De Sparre) for the increase in pressure h, ft (m), following gate closure, are given below. They are quite accurate for pressure rises not exceeding 50 percent of the initial pressure, which includes most practical cases.

$$h = aV/g \qquad \text{for } K < 1, N < 1$$
$$h = aV/\{g[N + K(N - 1)]\} \qquad \text{for } K < 1, N > 1$$
$$h = aV/[g(2N - K)] \qquad \text{for } K > 1, N > 1$$

where $K = aV/(2gH)$; $N = aT/(2L)$. V and H are the penstock velocity, ft/s (m/s), and head, ft (m), prior to closure; L is the penstock length, ft (m), and T is the time of gate closure. For full load rejection, T may be taken as 85 percent of the total gate traversing time to allow for nonuniform gate motion.

For pressure drop following a complete gate opening, the following formula (S. Logan Kerr) may be used with T not less than $2L/a$:

$$h = \frac{aV}{g}\left(\frac{-K + \sqrt{K^2 + N^2}}{N^2}\right) = \text{pressure, drop, ft(m)}$$

Pressure variations exceeding 40 percent rise and about 25 percent drop should be avoided. When the control, directly by the governor, causes undesirable pressure variations, a surge tank, a pressure regulator, or a jet deflector may be used. A **surge tank** is a standpipe with an atmospheric tank, attached to the penstock as close as possible to the casing inlet. The tank provides a reservoir and expansion chamber for the water demand or the water rejection following sudden gate movements, so that sudden accelerations or decelerations of the flow in the penstock are avoided.

Pressure regulators may be of either the water-wasting or water-saving type. The **water-wasting type** is a synchronous bypass, generally attached to the turbine casing. It is operated directly from the governor, or the gate mechanism of the turbine, and wastes such an amount as to keep the total water discharge equal at all times to the full-load discharge of the turbine. The bypass is a needle nozzle or a mushroom-shaped disk valve or a cone valve which opens and is partly balanced hydraulically by a piston under pipeline pressure. The **water-saving type** permits the regulator to open upon rapid closure of the turbine gates, and then close slowly, so that the total water discharge is gradually reduced and finally limited to that through the turbine, adjusted for the new load.

AUXILIARIES

Valves and head gates are provided for shutting off the water to each turbine for safety, for ease of maintenance, and to reduce water leakage losses. Motor-operated steel head gates are generally used for low- and medium-head plants with concrete scroll cases, although a few still use stop logs. The butterfly type of valve placed close to the turbine casing is suitable for medium- and high-head units with metal casings of circular inlet diameter. Butterfly valves of 8-ft (2.5 m) diameter for 1,000-ft (305 m) head and 27-ft (8.2 m) diameter for 100-ft (30 m) head have been built. The new flow through valve or biplane valve is becoming more popular for this application. In recent years, rotary sphere valves have replaced gate valves for high heads and where the loss through the butterfly valve is excessive because of the obstruction to flow by the valve disk.

COMPUTER-AIDED DESIGN

The continued development, refinement, and use of proven high-technology computer-aided design and analysis tools help create new generations of designs. Design objectives today are to produce equipment which has characteristics of reliability, easy servicing, operational flexibility, and high performance. The success of this approach is made possible by integrating design history and field experience into hydraulic, mechanical, and operational computer design and analysis systems.

TURBINE TESTS

Field testing of hydraulic turbines to determine the absolute efficiency and output involves careful and accurate measurement of the power available in the water supplied to the turbine [water hp (kW)] and the turbine output [developed hp (kW)] e = developed hp (kW)/water hp (kW) = $550P/(wQH)$ [$101.93P/(QHw)$]. The tests should be conducted in accordance with the International Code for Field Acceptance Tests of Hydraulic Turbines and/or Storage Pumps, IEC Publications 41 and 198.

Because of the difficulties and costs involved in making accurate measurements of horsepower, net head, and discharge in the field, there has been a trend in recent years to dispense with the field test, especially where a laboratory test on a homologous model turbine is available. Instead, an index test is made on the unit in the field, which measures the turbine output and relative discharge under various conditions. Index tests should be conducted in accordance with the International Code for Field Acceptance Tests of Hydraulic Turbines, IEC Publication 41.

7.10 FRACKING

by Julio C. Guerrero

Fracking, which is also called hydraulic fracking, fractures the underground formation apart with a fracking fluid at very high pressure and flow rate. This process stimulates the formation, so it allows producing more hydrocarbons from the underground. It stimulates the formation by increasing its porosity and permeability—interconnectedness of the pore space of the formation, so that the hydrocarbon—gas—can flow through the rock mass into the production well. The fracking fluid must be pumped into the formation through a cased drilled well that has perforations facing the formation. Once the formation is fracked, the fracking fluid is kept inside the fracked formation for about 1 hour, so the fracking fluid can migrate into the formation as far as possible. After that time, the pressure exerted by the fracking fluid is released, the formation exerts pressure back on the fracking fluid, and makes the fluid come back to the surface. Once the fracking is completed, particles that were carried in the fracking fluid are left inside the formation. These particles are called proppant and they contribute to the change in permeability and porosity that stimulates the formation to render more hydrocarbons. Proppant is solid material in particulate form that is strong enough to keep fractures propped open in hydrocarbon formations, during or following an induced hydraulic fracturing treatment. There are four types of proppant: (1) Sand, which has irregular size and shape, dependent on source; (2) resin-coated sand, which has irregular size and shape, and is smooth and rounded; (3) ceramic, which has uniform size and shape, round; and (4) resin-coated ceramic.

Fracking is performed on shale gas formations. In USA, there are 12 main shale gas formations: (1) Antrim Shale; (2) Barnett Shale; (3) Caney Shale; (4) Conesauga Shale; (5) Fayetteville Shale; (6) Floyd Shale; (7) Gothic Shale; (8) Haynesville Shale; (9) New Albany Shale; (10) Pearsall Shale; (11) Eagle Ford Shale; and (12) Marcellus Shale, which is the largest one. Shale is a clastic sedimentary rock, which is composed of mud, clay minerals flakes, and particles of other minerals such as quartz and calcite. Shale varies depending on the ratio of clay to other minerals. One of the most important characteristics of shale is that it breaks along thin laminae or parallel layering or bedding less than one centimeter in thickness, called fissility. Other formations have similar composition to Shale, but they do not have fissility, Mudstones are an example of them. The United States, Canada, China, and Argentina are the countries with commercial shale gas production. However, technological improvements will help the development of shale resources in other countries, primarily in Mexico and Algeria. These six countries are projected to account for 70 percent of global shale production by 2040.[5]

There are two main types of formations that contain natural gas: shale natural gas and tight natural gas. Large-scale natural gas production from shale began around 2000, when Mitchell Energy experimented with alternative methods of hydraulically fracturing in the Barnett Shale. By 2000, the company had developed a hydraulic fracturing technique that produced commercial volumes of shale gas. By 2005, the Barnett Shale was producing almost half a trillion cubic feet (Tcf) of natural gas per year. As natural gas producers were able to profitably produce natural gas in the Barnett Shale, they started developing other shale formations, including the Haynesville in eastern Texas and north Louisiana, the Woodford in Oklahoma, the Eagle Ford in southern Texas, and the Marcellus and Utica shales in northern Appalachia. Tight natural gas resources have been in production since the early 1980s, primarily from low-permeability sandstones and carbonate formations, with a small production volume coming from eastern Devonian shale. Tight natural gas generically still pertains to natural gas produced from low-permeability sandstone and carbonate reservoirs. Most of the natural gas and oil wells are on land; however, some wells are drilled into the ocean floor.

On one had, in the conventional gas reservoirs, natural gas migrate to the earth surface from an organic-rich source formation to highly permeable reservoir rock. It is trapped because it finds an overlying layer of impermeable rock on top. On the other hand, the shale gas forms within the organic-rich shale source rock. The low permeability of the shale prevents the migration of the gas to more permeable reservoir rocks. Horizontal drilling and hydraulic fracturing, shale gas are the two technologies that enable economically natural gas production. Without these two technologies there would not be flow from the formation at high enough rates to justify the cost of drilling.

The depths at which one wants to frack determine in great part the necessary pressure at which the fracking fluid must be pumped. The formation presents a resistance to fracking due to two main components: (1) the matrix pressure and (2) the pore pressure. As a rule of thumb one can estimate that for each foot of depth, the pressure presented by the underground is at least 1 psi. Therefore, the pressure at 10,000 ft. of depth is at least 10,000 psi. If one wants to frack the formation at that depth, it will be necessary to exert at least 10,000 psi of hydraulic pressure with pumps on the surface. The depths of most shale gas containing formations are between 7,000 and 10,000 ft., many oil wells are deeper than 10,000 ft.—water aquifers exist at an average depth of 500 ft. The fracking fluid contains in general 99 percent water and 1 percent of other chemical substances. To have a mental picture, one can imagine the Empire State building, which has a height of 1,250 ft. Fracking usually takes place at a depth between 7 and 10 Empire State buildings. The depths of the aquifers are at about half the Empire State building—fracking fluid at a depth of about eight Empire State buildings and under almost 10,000 psi does not migrate by itself to the aquifers that are close to the surface. The largest chance of pollution of an aquifer by fracking comes from the surface operations.

Hydraulic fracturing requires large amounts of freshwater. About four to six million gallons of water are used to frack wells. It is estimated that, at its peak production, the Marcellus Shale activity will require five million gallons of water per well, would be about 650 million barrels per year.

The quality of the water quality is one of the most important factors for fracturing to increase the efficiency of the additives used in the process. Most of the time, hydraulic fracturing uses water from lakes, rivers, aquifers, and municipal supplies. Most of the time, the water is transported from long distances to the wells that are being fracked. In some situations, under groundwater is used to supplement water from lakes, rivers, etc.

The amount of water that Fracking uses is small compared to the water needs of farming or power plant cooling, but in areas that have little water available, Fracking can strain water supplies. In USA, almost 38 percent of all water withdrawn is for power plants. Natural gas power plants use less water for cooling than nuclear plants per megawatt-hour of electricity generated. Since 2005, total electricity generated from natural gas in the U.S. has risen dramatically, with larger changes in some states than others. Natural gas requires the least water to produce energy.[6]

Flowback water is the one that comes back to the surface with the initially produced hydrocarbons after the fracking is completed. The flowback water can come back in amounts as large as 50 percent of the one placed into the well. The fracking companies have to separate from the formation the flowback and produced waters, both part of the production stream. Most of the time, flowback and produced water are disposed into an injection well, put in evaporation ponds, or treated and disposed of according regulations. Water management is a significant component of the fracking cost as well as the environmental impact of oil and gas production. Engineering innovations can play a very important role in water management to provide significant economic and environmental advantages.

Some of the treatment technologies recycle water recovered from hydraulic fracturing. Some of the methods for recycling fracking water: clarification, filtration, electrocoagulation, blending (directly diluting wastewater with freshwater), evaporation, and anaerobic and aerobic biologic treatment.

To pump the fracking fluid into the formation, the fracking companies use pumps connected to tanks containing it. The whole pumping

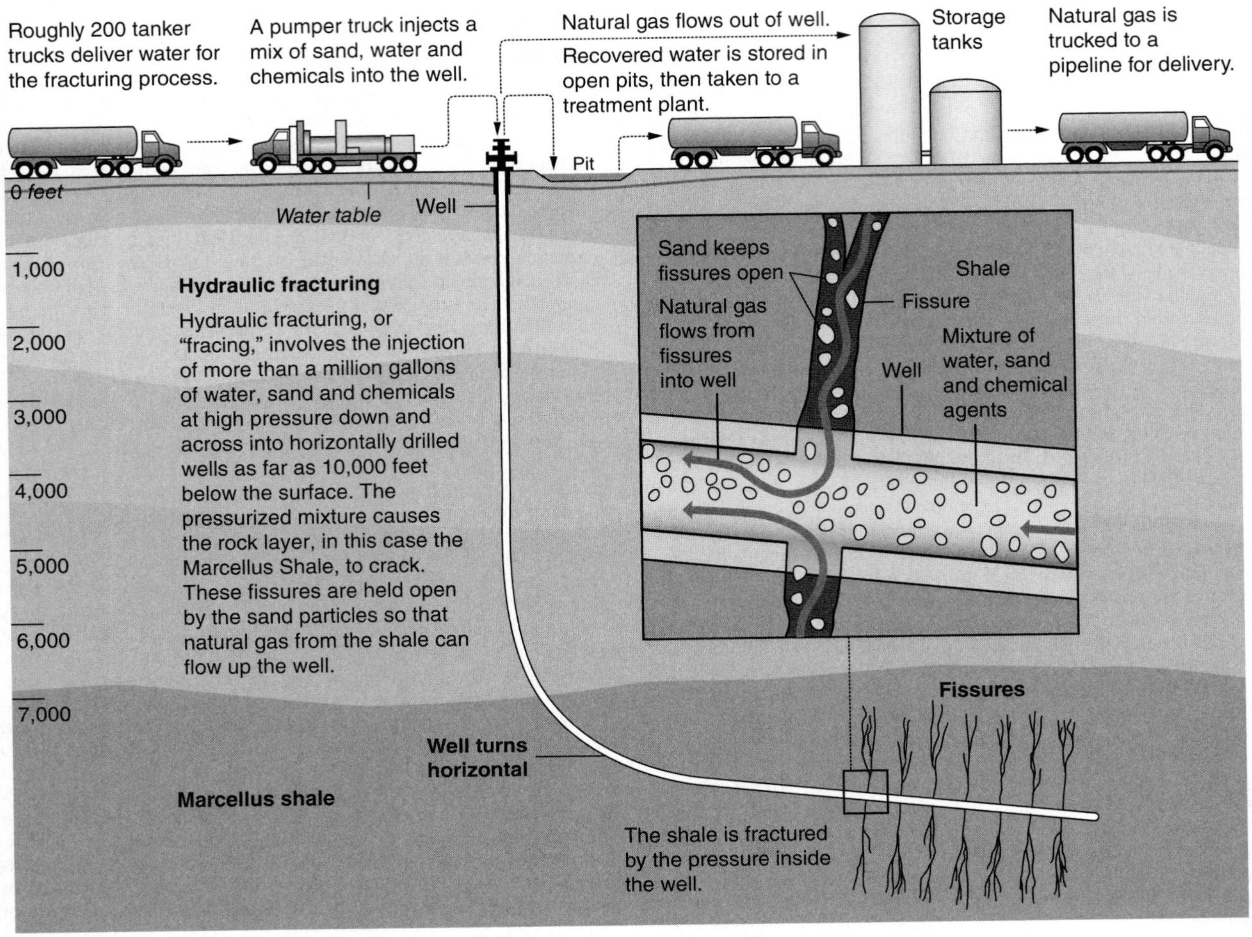

Roughly 200 tanker trucks deliver water for the fracturing process.
A pumper truck injects a mix of sand, water and chemicals into the well.
Natural gas flows out of well.
Recovered water is stored in open pits, then taken to a treatment plant.
Storage tanks
Natural gas is trucked to a pipeline for delivery.
Pit
0 feet
Water table
Well
1,000
2,000
3,000
4,000
5,000
6,000
7,000
Hydraulic fracturing
Hydraulic fracturing, or "fracing," involves the injection of more than a million gallons of water, sand and chemicals at high pressure down and across into horizontally drilled wells as far as 10,000 feet below the surface. The pressurized mixture causes the rock layer, in this case the Marcellus Shale, to crack. These fissures are held open by the sand particles so that natural gas from the shale can flow up the well.
Sand keeps fissures open
Natural gas flows from fissures into well
Shale
Fissure
Mixture of water, sand and chemical agents
Well
Fissures
Well turns horizontal
Marcellus shale
The shale is fractured by the pressure inside the well.

Figure 7.10.1 Fracking process.

equipment lies on the surface next to the entrance to the well through which the fracking fluid is pumped, Fig. 7.10.1 shows and example of the surface and underground Fracking system. Hydraulic fracking has been happening for at least 60 years in USA. However, its significant growth has happened since 2005. Almost 137,000 fracking jobs have been performed since 2005 in USA. Fracking became more cost effective in the last 10 years due to new enabling technologies: (1) Robotics that allows directional drilling and logging while drilling LWD at competitive costs; (2) Perforating processes that are more efficient than the ones that existed 10 years ago; (3) Materials Sciences and technology that have produced better and more cost effective proppant, and perforating plugs materials; (4) Mechanical systems and design; and (5) An innovation driven market in USA that stimulated the creation of almost 2000 companies in USA that accelerated the technology development.

Robotics has allowed directional drilling by integrating mud-pulse telemetry, drilling mud motors and their controlled actuation, and sophisticated guidance and navigation. Highly deviated and horizontal wells are drilled now mainly due to Robotics driven directional drilling. This type of wells allows crossing several different types of geological layers until the hydrocarbon producing section in the underground formation is reached.

Once the wells are drilled and cased, information collected from sophisticated logging processes—LWD is one of them—allows determining the specific section where the formation stimulation needs to take place. Once that section is determined, the companies implement zonal isolation in that section. The zonal isolation permits to produce hydrocarbons from a specific section. Zonal isolation is implemented with any of a variety of available methods; one of them uses elastomeric packers. Once the zonal isolation is completed, perforating guns are conveyed to the desired section, they are fired, and perforations are produced in the casing. The perforations are performed while the correct balance of pressure is maintained between the formation and the fluids inside the well. The differential hydraulic pressure is most of the time maintained at about 200 psi. It is considered positive when the pressure in the well is higher than the formation pressure. Once the pressure of the well is reduced allowing the formation pressure to be higher than the well's, the formation's fluids start flowing into the well. It is important to remark that the differential pressure is 200 psi with respect to 10,000 psi.

Currently there are several different sequences used to perforate the wells and synchronize it with the fracking process. One of the key drivers in this synchronization is the minimization of time, which translates into cost saving, safety is another driver. Offshore Oil and Gas platforms may cost the operators between $10,000 and $20,000 per hour, for having access only to the infrastructure, so this synchronization optimization is paramount. Currently, the development of perforating-plugs is playing a key role in the competition between fracking companies. One of the diverse perforating-plug technologies allows conveying it with downhole fluid as it is pumped into the well. After the fracking-plug has completed its function, it is dissolved automatically in a predetermined time—in a few hours—or pumping fluids that provoke it dissolves it. Another option drills through the perforating plug after it has completed its function—also within a few hours. Mechanical Systems and Design and Material Sciences are the key technologies that are driving the development of perforating plugs. Mechanical Engineering is and will continue to be the key discipline providing the technology enablers for fracking. The most relevant specialties inside Mechanical Engineering for fracking: (1) materials that can withstand the downhole environment pressure and temperature while providing the necessary structural integrity; (2) materials that can be used to produced mechanical devices that can be dissolved inside the well after performing their function, and at a desire time and rate; (3) mechanisms that can be deployed inside the well and expanded or contracted as needed to perform different tasks, deployable structures and compliant mechanisms play and will play a key role; (4) passive sensors that can be inserted in the formation to monitor pressure or temperature, but do not carry energy sources—i.e., batteries— to power them, and instead are energized by substances or EM fields conveyed into the formation when they need to be interrogated; (5) the most important technical grand challenge in Fracking is to determine the 3D configuration of the fractured formation, hence the need for passive sensors that are cost competitive and can withstand the downhole and formation conditions.

On the Economics side, the impact of Fracking: (1) has increased the supply and diversification of energy worldwide; (2) has impacted the Gas markets; and (3) has disrupted the global price of oil by more than 50 percent price reduction in the last 3 years since 2014.

Four of the engineering and technology enablers necessary to make Fracking more efficient include: (1) mechanical systems and design that will make the perforating plugs better, more efficient and more cost effective; (2) technologies that allow implementing electronics and electrical devices capable of withstanding high temperatures, close to 500°F; (3) telemetry that allows communicating with and commanding the Bottom Hole Assemble (BHA), its computer, actuators, and guidance and navigation to have faster, and more accurate and efficient directional drilling; (4) better downhole conveyance equipment that allows to convey the perforating guns to the desired place in the wells, one of the three most used conveyance equipment is called wireline Downhole Tractors, another one is Coil-Tubing, and a third one is the drilling pipe itself. Mechanical Engineering has been the most impact in Fracking, and will continue to be so. Figure 7.10.2 shows how technology enablers will favor the growth of hydrocarbons produced with Fracking.

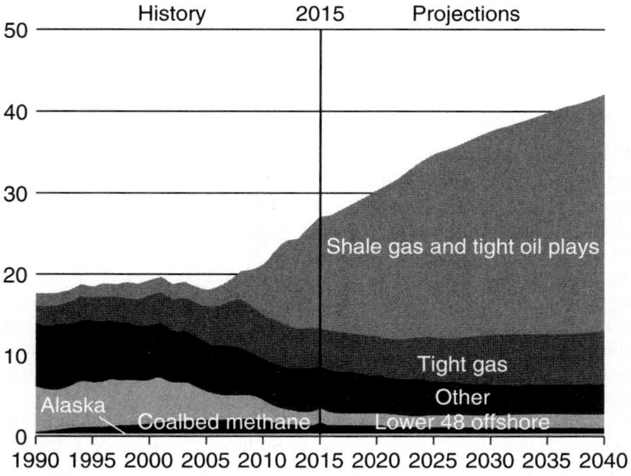

Figure 7.10.2 Technology improvements have allowed, and are likely to continue to allow, the expansion of tight and shale gas production, as indicated in this figure. (trillion cubic feet).[5]

Figure 7.10.3 Surface Fracking system: Valves.

The fracking fluid used for each Fracking operation requires large amounts of water. It is so large that there is a market for fracking water, its distribution and treatment. In many Fracking operations, the water is transported to the site from other places, and mixed with the other chemicals and proppant on site. As previously mentioned, the largest risk to pollute during Fracking jobs is due to the potential leaking of contaminants on the surface where the Fracking trucks and tanks operate. Among all the systems and parts used in those operations, the valves and the moment in which they are connected to the pipes as well as the return of the Fracking fluid from the formation are the ones where more engineering, codes and standards can reduce contamination risks the most, Fig. 7.10.3 shows the challenges presented by valves used by Fracking surface equipment.

There is a large range of variations among wells that use Fracking. The amount of water used in each depends of several factors: (1) the rock formation; (2) the operator; (3) if the well is vertical, highly deviated, or horizontal; (4) the number of stages fractured in the well; and (5) whether the water is recycled from fluids produced by the well or if most of the water is brought to the well site. The amount of water used can vary between 2.7 and 15 million gallons per job. Examples of average reported water usage per well include: (1) Marcellus Shale, Pennsylvania, 4.5 million gallons;[1] (2) Barnett Shale, Texas, 2.8 million gallons;[2] and (3) Horn River Shale, British Columbia, Canada, 15.8 million gallons.[3] Once the Fracking is completed, and the hydrocarbons production starts, large amount of water are not necessary. However, it is relevant to mention that water is again used when steam is sometimes to stimulate formations from wells drilled parallel and close to the producing wells.

Fracking causes extremely small earthquakes, but they are almost always too small to be a safety concern. In addition to natural gas, fracking fluids and saltwater are trapped in the same formation as the gas is returned to the surface. These wastewaters are frequently disposed of by injection into deep wells. The injection of wastewater and saltwater into the subsurface can cause earthquakes that are large enough to be felt and may cause damage. A few important points about Fracking and earthquakes:[4] (1) Fracking is not causing most of the induced earthquakes; (2) wastewater disposal is the primary cause of the recent increase in earthquakes in the central United States; (3) not all wastewater injection wells induce earthquakes; (4) wastewater is produced at all oil wells, not just hydraulic fracturing sites; (5) the content of the wastewater injected in disposal wells is highly variable; (6) induced seismicity can occur at significant distances from injection wells and at different depths; and (7) wells not requiring surface pressure to inject wastewater can still induce earthquakes. Finally, but not least, earthquakes with magnitudes between 4.5 and 5.0 have been induced by fluid injection in Arkansas, Colorado, Kansas, Oklahoma, and Texas. The largest earthquake induced by fluid injection that has been documented in the scientific literature was the November 6, 2011, earthquake in central Oklahoma. It had a magnitude of 5.6. Earlier that year, a magnitude 5.3 earthquake was induced by fluid injection in the Raton Basin, Colorado.

CONCLUSIONS

The newest Fracking technology has increased significantly the production of hydrocarbons, especially in Shale formations, in the last 10 years more than the previous 60 years, and will continue to do so in great part due to innovation brought up by Mechanical Engineering and related sciences. Fracking as any other technology brings risks, but if managed properly, it can continue to be one of the components in the portfolio of energy sources for the world.

REFERENCES

1. Marcellus Shale, Pennsylvania, 4.5 million gallons (Risser, 2012, USGS Public Lecture, "Shale gas, Hydraulic Fracturing, and Induced Earthquakes").

2. Barnett Shale, Texas, 2.8 million gallons (Nicot and Scanlon, 2012, Environmental Science and Technology).

3. Horn River Shale, British Columbia, Canada, 15.8 million gallons (Horn River Basin Producers Group, 2010).

4. U.S. Geological Survey (U.S.G.S) Link: https://www.usgs.gov. Last accessed February 7, 2017.

5. U.S. Energy Information Administration I Annual Energy Outlook 2016, Link http://www.eia.gov/outlooks/aeo/pdf/0383(2016).pdf.

6. Jones, W., "How Much Water Does It Take to Make Electricity?" IEEE Spectrum, April 1st, 2008.

7.11 SECOND-LAW ANALYSIS AND AVAILABLE ENERGY

by Jung-Yang San

A thermodynamic system is composed of several processes. In the processes, energy is never destroyed. Through an energy-balanced analysis (first-law analysis), the thermodynamic states of the processes and the overall efficiency of the system can be obtained. Nevertheless, in real processes, the capability of the working substance to convert other forms of energy into work is continuously decreased. The result of the energy-balanced analysis is unable to provide an evaluation of the inefficiencies in the processes. For accurately locating the major inefficiencies in the system and identifying the causes of the inefficiencies, in addition to the energy-balanced analysis, a second-law analysis for the system is needed. The second-law analysis is based on the concept of available energy which is the potential for achieving the maximum possible work in a thermodynamic process.

AVAILABLE ENERGY OF AN AMOUNT OF HEAT

Figure 7.11.1 shows a reversible heat engine (Carnot engine). The heat engine receives an amount of heat Q_H from a reservoir at temperature T_H and dumps an amount of heat Q_o into the environment at temperature T_o (i.e., $T_H > T_o$). From the first law and second law of thermodynamic, it is known that the work produced by the reversible heat engine (W_{rev}) is:

$$W_{\text{rev}} = W_{\max} = Q_H\left(1-\frac{T_o}{T_H}\right)$$

The work produced by the reversible heat engine is the maximum work ($W_{\max}$) that can be extracted from the Q_H. It is also called the available energy or availability of the amount of heat delivered from the reservoir at temperature T_H.

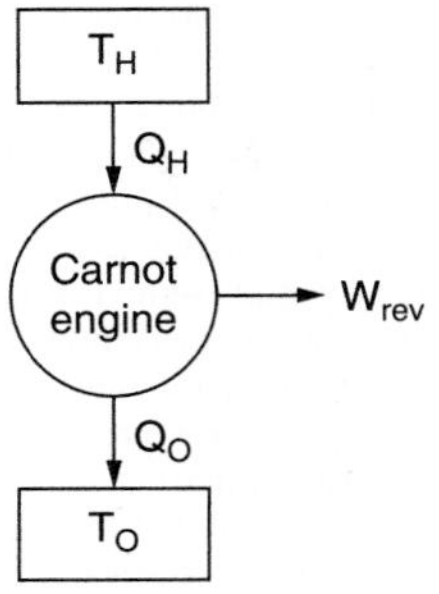

Figure 7.11.1 A Carnot cycle receiving an amount of heat Q_H from a reservoir at T_H.

Figure 7.11.2 shows the T-S diagram of the four processes in the cycle of the reversible heat engine. The dotted area (area 1) represents the part of energy in the Q_H that can be converted into work; while the blank area below area 1 represents the rest in the Q_H that is unavailable for producing work. For fixed Q_H and T_o, the work produced by the reversible heat engine increases with the T_H.

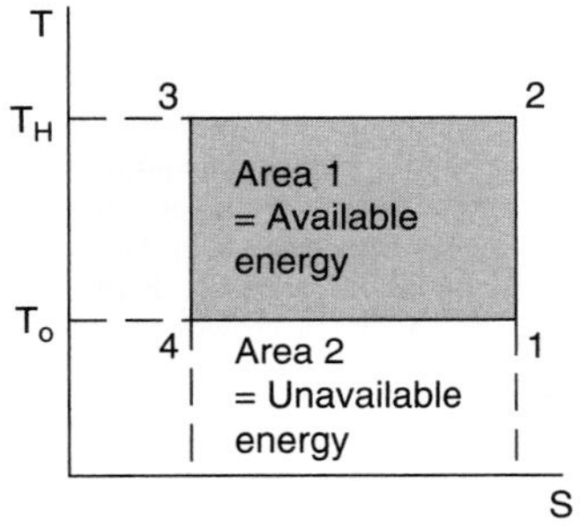

Figure 7.11.2 Available energy and unavailable energy of an amount of heat Q_H.

AVAILABLE ENERGY OF A FLOW

Consider a steady flow that undergoes an internally and externally reversible process (Fig. 7.11.3). At the inlet of the process, the specific enthalpy and specific entropy of the flow are designated as h and s, respectively; the velocity is v and the height is z (relative to the height of the flow at the exit). At the exit of the process, the flow reaches the state of the environment at temperature T_o and pressure P_o. The state of the environment (state "o") is also called the dead state. At the dead state, the velocity and the height of the flow are zero ($v = z = 0$), and the working substance no longer possesses the capability to extract work through an energy conversion. The specific enthalpy and specific entropy at the dead state are denoted as h_o and s_o, respectively.

Within a time interval (Δt) in the process, "m" kg of the working substance passes through the control volume of the process and an amount of heat $(Q_{\text{cv}})_{\text{rev}}$ is transferred from the flow to a reversible heat engine. The reversible heat engine converts a part of the heat into work (W_{rev}) and dumps the rest (Q_o) into the environment. From the first law of thermodynamics, the work produced in the reversible process is the

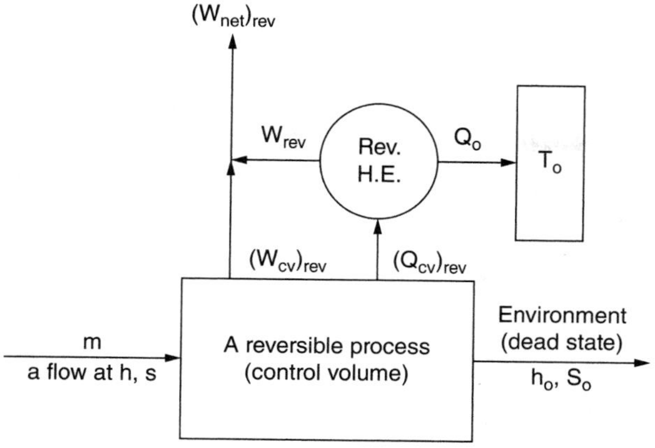

Figure 7.11.3 Available energy of a flow.

sum of the work produced by the reversible heat engine and the work done by the control volume.

$$(W_{\text{net}})_{\text{rev}} = W_{\text{rev}} + (W_{\text{cv}})_{\text{rev}}$$

$$= m\left(h + \frac{v^2}{2} + gz\right) - mh_o - Q_o$$

From the second law of thermodynamics, the decrease in entropy of the flow in the reversible process must equal Q_o/T_o. That means

$$\frac{Q_o}{T_o} = \frac{(Q_{\text{cv}})_{\text{rev}}}{T_{\text{cv}}} = ms - ms_o$$

Combining the two equations shown above, it yield

$$(W_{\text{net}})_{\text{rev}} = m\left[\left(h + \frac{v^2}{2} + gz\right) - h_o\right] - mT_o(s - s_o)$$

The work extracted from the flow (W_{rev}) is the maximum work that can be produced in the process. It is also the part of energy in the flow available for producing work. The maximum possible work per unit mass of the flow generally is called the specific exergy of the flow and it is assigned by the symbol ψ as

$$\psi \equiv \frac{(W_{\text{net}})_{\text{rev}}}{m} = \left[\left(h + \frac{v^2}{2} + gz\right) - h_o\right] - T_o(s - s_o)$$

The units of the specific exergy is J/kg.

As the kinetic energy and potential energy of the flow are small, neglecting the two factors won't cause a significant change of the maximum possible work. Under this circumstance, the specific exergy of the flow (ψ) is simplified into the following form:

$$\psi \equiv (h - h_o) - T_o(s - s_o)$$

As shown in the equation, the simplified form of the specific exergy is a combination of several thermodynamic properties. Hence, it is also a thermodynamic property.

In thermodynamic systems, gas and liquid are two common types of working substance. By assuming the gas to be an ideal gas and the liquid to be an incompressible fluid, the simplified specific exergy of the flow can be converted into a function of temperature and pressure. This form of specific exergy facilitates the second-law analysis of the processes in the systems and the acquired data are still reliable. For ideal gas flow and incompressible flow, the specific exergy can be separately expressed as:

For ideal gas:

$$\psi = c_p T_o\left[\left(\frac{T - T_o}{T_o}\right) - \ln\left(\frac{T}{T_o}\right)\right] + c_p T_o\left(1 - \frac{1}{k}\right)\left[\ln\left(\frac{P}{P_o}\right)\right]$$

where k = ratio of specific heats = c_p/c_v

For incompressible flow:

$$\psi = c_p T_o\left[\left(\frac{T - T_o}{T_o}\right) - \ln\left(\frac{T}{T_o}\right)\right] + c_p T_o\left[\frac{P - P_o}{\rho c T_o}\right]$$

where $c = c_p = c_v$

In the equations, the specific exergy are divided into two parts. One is related to thermal energy and the other is related to mechanical energy. The former is a function of temperature, while the latter is a function of pressure. In this arrangement, the magnitudes of the two forms of exergy in the flow can be clearly known.

EXERGY LOSS IN A UNIFORM-STATE AND UNIFORM-FLOW PROCESS

Figure 7.11.4 shows a control volume undergoing a uniform-state and uniform-flow process. The process is an irreversible process. Within a time interval, the control volume receives an amount of heat Q_H from a high-temperature reservoir at temperature T_H and dumps an amount of heat Q_{leak} to the environment at temperature T_o. In addition, the working substance in the control volume varies from state 1 to state 2. During the period, the work produced in the process is W_{real}. From the first law of thermodynamics, the W_{real} is expressed as:

$$W_{\text{real}} = \left[\Sigma m_i\left(h_i + \frac{v_i^2}{2} + gz_i\right) - \Sigma m_e\left(h_e + \frac{v_e^2}{2} + gz_e\right)\right] + \left[m_1\left(u_1 + \frac{v_1^2}{2}\right) - m_2\left(u_2 + \frac{v_2^2}{2}\right)\right] + Q_H - Q_{\text{leak}}$$

Figure 7.11.5 shows an ideal process which remains all the states the same as those shown in Fig. 7.11.4. However, the process is reversible and two Carnot heat engines are arranged to avoid external irreversibility due to heat transfer through a finite temperature difference. Based on the first law and second law of thermodynamics, the work produced in the reversible process $(W_{\text{net}})_{\text{rev}}$ is expressed as:

$$(W_{\text{net}})_{\text{rev}} = (W_{\text{cv}})_{\text{rev}} + W_{\text{rev},1} + W_{\text{rev},2}$$

$$= \left[\Sigma m_i\left(h_i + \frac{v_i^2}{2} + gz_i\right) - \Sigma m_e\left(h_e + \frac{v_e^2}{2} + gz_e\right)\right] + \left[m_1\left(u_1 + \frac{v_1^2}{2}\right) - m_2\left(u_2 + \frac{v_2^2}{2}\right)\right] + Q_H\left(1 - \frac{T_o}{T_H}\right)$$

$$+ T_o[(m_2 s_2 - m_1 s_1) + (\Sigma m_e s_e - \Sigma m_i s_i)]$$

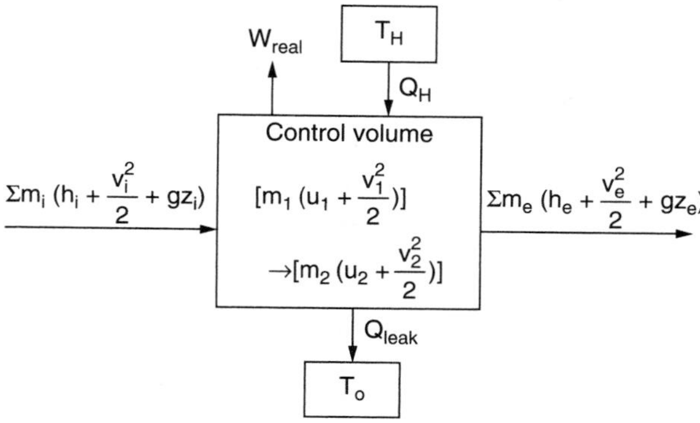

Figure 7.11.4 A uniform-state and uniform-flow process.

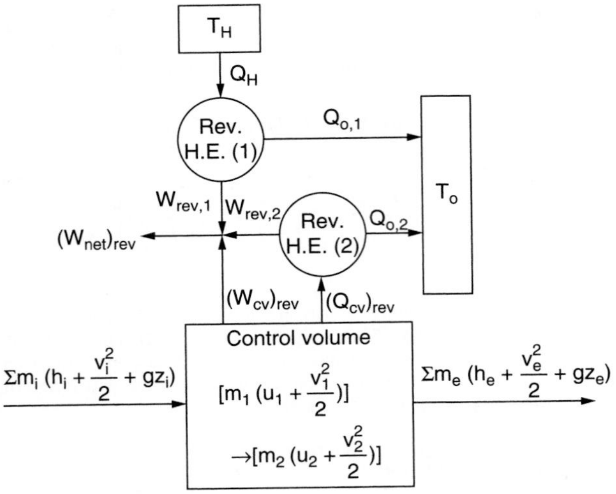

Figure 7.11.5 A reversible process corresponding to Fig. 7.11.4.

The work produced in the reversible process is the maximum work that can be done under the given conditions. The deviation between the work produced in the irreversible process (W_{real}) and the work produced in the reversible process [$(W_{net})_{rev}$] is the lost work of the irreversible process. The lost work is commonly called the exergy loss or exergy destruction ($\Delta\Psi$). It mathematical form is expressed as:

$$\Delta\Psi = T_o\left[(m_2s_2 - m_1s_1) + \left(\Sigma m_e s_e - \Sigma m_i s_i\right)\right] - Q_H\left(\frac{T_o}{T_H}\right) + Q_{\text{leak}}$$

The equation can be used to analyze exergy losses in most of engineering processes. The units of the exergy loss is Joules.

For steady-state and steady-flow processes, the exergy loss rate ($\Delta\dot{\Psi}$) is deduced from the equation shown above and it is expressed as:

$$\Delta\dot{\Psi} = T_o\left(\Sigma \dot{m}_e s_e - \Sigma \dot{m}_i s_i\right) - \dot{Q}_H\left(\frac{T_o}{T_H}\right) + \dot{Q}_{\text{leak}}$$

The units of the exergy loss rate is Watts.

EXERGY LOSS IN A STEADY-FLOW PROCESS AND SECOND-LAW EFFICIENCY

For a steady flow undergoing a process without receiving heat from a high temperature reservoir (Fig. 7.11.6), the exergy loss rate in the process is expressed as:

$$\Delta\dot{\Psi} = \dot{m}[(h_i - h_e) - T_o(s_i - s_e)] + \left[\left(\frac{v_i^2}{2} + gz_i\right) - \left(\frac{v_e^2}{2} + gz_e\right)\right] - W_{\text{real}}$$

The equation is applicable to analyze the exergy loss in mechanical devices such as pump, compressor, fan, turbine, etc. These devices are either used to convert electricity into mechanical work or vice versa.

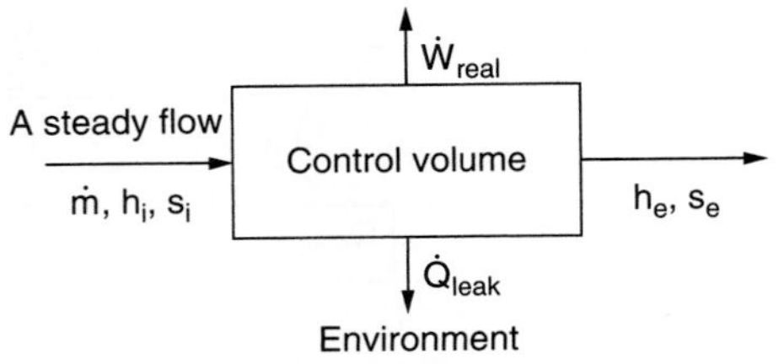

Figure 7.11.6 Exergy loss in a steady flow.

The exergy loss obtained by using the equation only counts the input/output mechanical work (exergy) in the process of a mechanical device, instead of the generated/consumed electricity. However, the electricity can be calculated from the mechanical work, provided that the conversion factor between the mechanical work and the electricity is known.

The second-law efficiency (η_{II}) of a power producing mechanical device, such as turbine, is defined as the actual work output of the device divided by the decrease in exergy from the same inlet state to the same exit state. For the process shown in Fig. 7.11.6, the second-law efficiency is expressed as:

$$\eta_{\text{II}} = \frac{\dot{W}_{\text{real}}}{\dot{m}(\psi_i - \psi_e)}$$

Similarly, the second-law efficiency of a power consuming mechanical device, such as compressor, pump, etc., is defined as the increase in exergy of the flow divided by the actual work input of the device. It is expressed as:

$$\eta_{\text{II}} = \frac{\dot{m}(\psi_e - \psi_i)}{\dot{W}_{\text{real}}}$$

The second-law efficiency provides a measure of the real process in terms of the actual changes of the state, which is different from the definition of the isentropic efficiency commonly used for indicating the performance of mechanical devices.

EXAMPLE. Consider a steam turbine that is used for producing power (Fig. 7.11.7). The pressure and temperature of the steam at the inlet of the turbine are 30 bar and 400°C; at the outlet, the steam is 100°C and saturated vapor. In the analysis, the heat leakage and the changes of kinetic energy and potential energy are neglected.

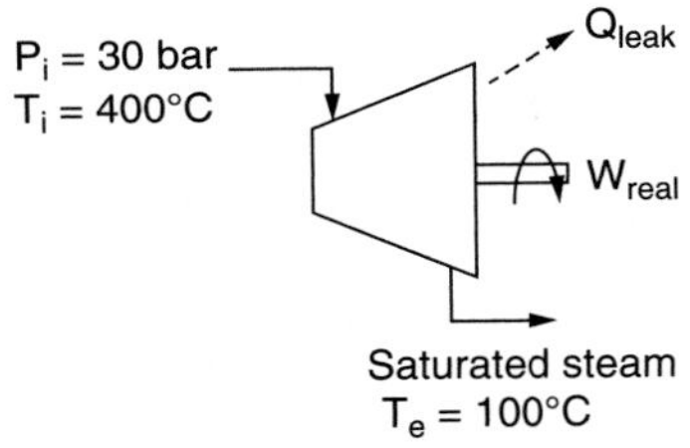

Figure 7.11.7 Inlet and exit states of a steam turbine.

From the steam table, the specific enthalpy and specific entropy of the steam at the inlet and exit are found to be:

$$h_i = 3{,}230.9 \text{ kJ/kg},\ s_i = 6.9212 \text{ kJ/kg} - \text{K}$$

$$h_e = 2{,}676.1 \text{ kJ/kg},\ s_i = 7.3549 \text{ kJ/kg} - \text{K}$$

From the first law of thermodynamics, the real work output of the turbine is:

$$W_{\text{real}} = h_i - h_e = 554.8 \text{ kJ/kg}$$

The decrease in exergy of the flow is:

$$\psi_i - \psi_e = (h_1 - h_2) - T_o(s_i - s_e)$$
$$= 684.1 \text{ kJ/kg}$$

The decrease in exergy is the maximum work that can be extracted from the flow. Subtracting the real work output from the maximum possible work, it yields an exergy loss of 129.3 kJ/kg. Then, the second-law efficiency of the turbine is calculated to be:

$$\eta_{\text{II}} = \frac{W_{\text{real}}}{(\psi_i - \psi_e)} = 81.8\%$$

EXERGY LOSS IN A HEAT EXCHANGE DEVICE

Heat exchanger is one of the most common and important devices used in thermodynamic systems. The device involves heat exchange between two or more flows passing through it. For the flows in the heat exchange process, the changes of kinetic energy and potential energy usually are small, thus they can be neglected. Under the circumstance, the exergy loss rate in the process can be expressed as:

$$\Delta\dot{\Psi} = \Sigma\,\dot{m}[(h_i - h_e) - T_o(s_i - s_e)]$$

The equation can be applied to heat exchangers with heat leakage. With the heat leakage, the specific enthalpy at the exit (h_e) would be different from that without the heat leakage. This is why the heat leakage term does not appear in the equation.

In industrial thermodynamic systems, a poor insulation of the pipes or ducts using for connecting mechanical devices could cause a severe heat leakage to the environment. It eventually results in an exergy loss of the flow. Without the summation term, the equation shown above can also be used to calculate the exergy loss due to the heat leakage through the external surface of the pipes or ducts.

RELATION BETWEEN EXERGY LOSS AND ENTROPY GENERATION

The equation used for calculating the exergy loss rate in steady-state and steady-flow processes can also be arranged into the following form:

$$\Delta\dot{\Psi} = T_o\left[\left(\Sigma\,\dot{m}_e s_e - \Sigma\,\dot{m}_i s_i\right) - \frac{\dot{Q}_H}{T_H} + \frac{\dot{Q}_{\text{leak}}}{T_o}\right]$$

$$= T_o(\dot{S}_{\text{gen}})$$

In the equation, the sum of the four terms in the parenthesis is the rate of irreversible increase of entropy in the process. The rate of irreversible increase of entropy is also called the entropy generation rate ($\dot{S}_{\text{gen}}$). The exergy loss rate in the process is proportional to the entropy generation rate and the proportional constant is the dead state temperature. The entropy generation causes the process irreversible, consequently results in the exergy loss. In the literature, for optimizing thermodynamic processes, much work on minimizing the entropy generation has been done. For irreversible processes, the energy generation is always a positive value; for reversible processes, it is zero.

CONCEPT OF EXERGY TRANSPORT

The exergy loss rate can be calculated by using the concept of exergy transport. For the steady-flow process shown in Fig. 7.11.8, the exergy loss rate in the process is equal to the net exergy inflow rate. By applying the first law of thermodynamics to the control volume of the process, the exergy loss rate in the process can be expressed in terms of different forms of exergy flows (Fig. 7.11.9) as:

$$\Delta\dot{\Psi} = (\Sigma\,\dot{m}_i\psi_i - \Sigma\,\dot{m}_e\psi_e) + \dot{Q}_H\left(1 - \frac{T_o}{T_H}\right) - \dot{W}_{\text{real}}$$

On the right-hand side of the equation, the first term represents the net exergy input rate due to the flows; the second term is the exergy input rate through the heat transfer across the boundary of the control volume; the last term denotes the output mechanical power. For

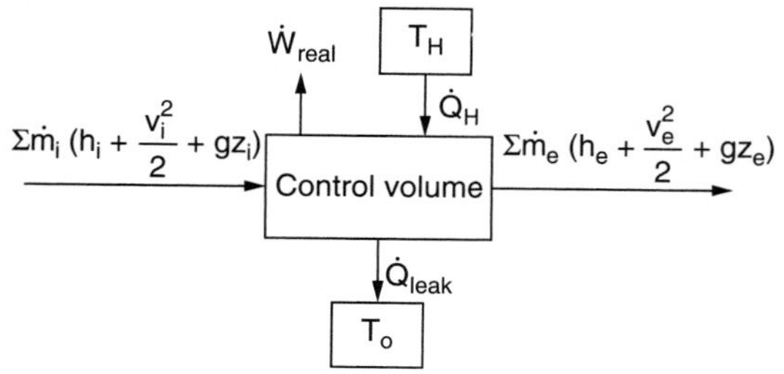

Figure 7.11.8 Energy balance for a steady-flow process.

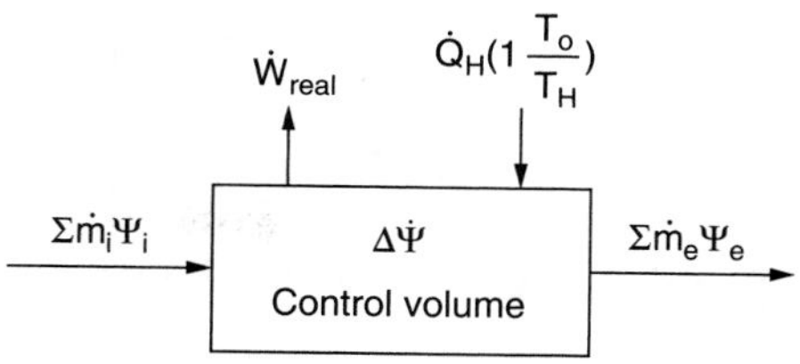

Figure 7.11.9 Exergy balance for a steady-flow process.

irreversible processes, the exergy loss rate is always a positive value and it equals the sum of the three forms of exergy flow.

EXERGY LOSS RATE FOR IDEAL GAS FLOW AND INCOMPRESSIBLE FLOW

By neglecting the changes of kinetic energy and potential energy, the general expressions of the exergy loss rate for an ideal gas flow and an incompressible flow can be expressed as:

$$\Delta\dot{\Psi} = (\dot{m}c_p)T_o\left[\left(\frac{T_i - T_e}{T_o}\right) - \ln\left(\frac{T_i}{T_e}\right)\right] + (\dot{m}c_p)T_o F_{\Delta P}$$

$$= \Delta(\dot{\Psi})_{\text{thermal loss}} + \Delta(\dot{\Psi})_{\text{mech. loss}}$$

(i) for ideal gas flow: $F_{\Delta P} = \left(1 - \frac{1}{k}\right)\left[\ln\left(1 + \frac{\Delta P}{P_e}\right)\right]$

where $k = c_p/c_v = 1.4$, $\Delta P = P_i - P_e$

(ii) for incompressible flow: $F_{\Delta P} = \dfrac{\Delta P}{\rho c_p T_o}$

The equation shown above separates the effect of temperature and the effect of pressure drop on the exergy loss in a flow. This expression facilitates identifying the cause of exergy loss. In addition, the form of the equation is also quite easy to be used with the aid of a computer for analyzing the exergy losses of the processes in thermodynamic systems. In the equation, the first term on the right-hand side represents the thermal exergy loss rate due to heat transfer through a finite temperature difference, while the second term represents the mechanical exergy loss rate resulting from pressure drop in the flow. The symbol, $F_{\Delta P}$, is defined as the pressure-drop factor of the flow, which dominates the magnitude of the mechanical exergy loss rate. The pressure-drop factor of an ideal gas flow usually is much larger than that of an incompressible flow. At the same pressure drop (ΔP), the former could be in the order 10^4 of the latter.

EXAMPLE. Consider a heat exchanger that is used for exchanging heat between a hot flow "1" and a cold flow "2" (Fig. 7.11.10). The inlet and exit temperatures of the two flows are $T_{1,i} = 580°C$, $T_{1,e} = 315.6°C$, $T_{2,i} = 100°C$ and $T_{2,e} = 392.8°C$, respectively; the inlet and exit pressures are $P_{1,i} = 101{,}825$ Pa, $P_{1,e} = P_o = 101{,}325$ Pa, $P_{2,i} = 102{,}075$ Pa, $P_{2,e} = 101{,}695$ Pa, respectively; the mass flow rates of the two flows are $\dot{m}_1 = 1.57$ kg/s and $\dot{m}_2 = 1.555$ kg/s, respectively. In addition, the environmental temperature is $T_o = 300\ K$ and the constant-pressure specific heats of the

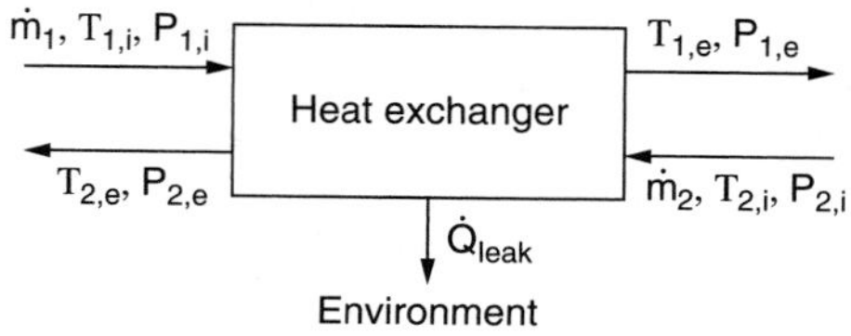

Figure 7.11.10 Thermodynamic states of two flows in a heat exchanger.

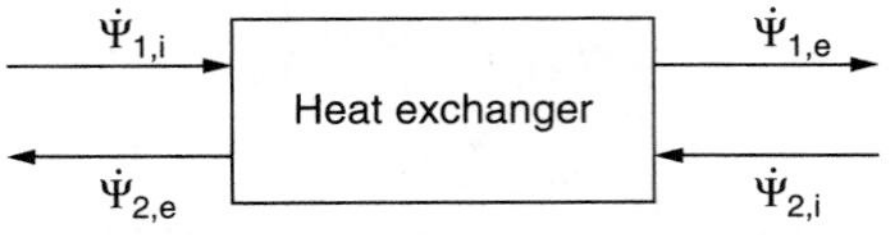

Figure 7.11.11 Change of exergy rates of two flows in a heat exchanger.

two flows are $c_{p,1} = 1.11$ kJ/kg-K and $c_{p,2} = 1.012$ kJ/kg-K, respectively. Applying the first law of thermodynamics to the control volume shown in Fig. 7.11.10 and neglecting the heat leakage to the environment, the heat exchange rate between the two flows is calculated to be 460.8 kW. The exergy loss rate in the heat exchange process ($\Delta\dot{\Psi}$) is the sum of the exergy loss rates of the two flows (Fig. 7.11.11). By assuming the two flows to be ideal gas flows, the exergy loss rate in the process can be calculated using the following equation:

$$\Delta\dot{\Psi} = (\Delta\dot{\Psi})_1 + (\Delta\dot{\Psi})_2 = (\dot{\Psi}_{1,i} - \dot{\Psi}_{1,e}) + (\dot{\Psi}_{2,i} - \dot{\Psi}_{2,e})$$

$$= (\dot{m}_1 c_{p,1})T_o\left[\left(\frac{T_{1,i} - T_{1,e}}{T_o}\right) - \ln\left(\frac{T_{1,i}}{T_{1,e}}\right) + \left(\frac{R}{c_{p,1}}\right)\ln\left(1 + \frac{(\Delta P)_1}{P_{1,e}}\right)\right]$$

$$+ (\dot{m}_2 c_{p,2})T_o\left[\left(\frac{T_{2,i} - T_{2,e}}{T_o}\right) - \ln\left(\frac{T_{2,i}}{T_{2,e}}\right) + \left(\frac{R}{c_{p,2}}\right)\ln\left(1 + \frac{(\Delta P)_2}{P_{2,e}}\right)\right]$$

In the equation, the gas constant (R) is 0.287 kJ/kg-K. The pressure drops of the two flows are calculated to be $(\Delta P)_1 = 360$ Pa and $(\Delta P)_2 = 250$ Pa, respectively. Using the data, the exergy loss rate in the process ($\Delta\dot{\Psi}$) is found to be 80.68 kW and the rate of exergy gained in the cold flow ($= \dot{\Psi}_{2,e} - \dot{\Psi}_{2,i}$) is 186.83 kW. In the process, due to a large temperature difference between the two flows, the exergy loss due to heat transfer dominates the total exergy loss. The exergy loss rate due to pressure drop is only 1.15 kW which is a small fraction (~1.4 percent) of the total exergy loss rate.

It is worth to mention that, for heat exchange processes with small temperature variations in the flows, the exergy loss due to pressure drop could be at the same level as the increase of exergy in the cold flow. In this situation, a check of the magnitude of the two terms is crucial. In waste heat recovery processes, the exergy loss due to pressure drop is the energy expended for acquiring the increase of exergy in the cold flow. From the economic viewpoint, the former needs to be smaller than the latter.

The exergy loss in the heat exchanger is due to two factors. One is the heat transfer through a finite temperature difference and the other is the fluid friction loss in the two flows. From the heat exchanger theory, it is known that, the effectiveness of the heat exchanger increases with the heat transfer area. At balanced flow operation ($\dot{m}_1 c_{p,1} = \dot{m}_2 c_{p,2}$), as the heat transfer area increases, the exit temperature of the hot flow will approach the inlet temperature of the cold flow ($T_{1,e} = T_{2,i}$), as well as the exit temperature of the cold flow will approach the inlet temperature of the hot flow ($T_{2,e} = T_{1,i}$). This would make the thermal exergy loss decrease. However, an increase of the heat transfer area frequently causes a large increase in pressure drop of the flows and consequently results in a significant increase of the mechanical exergy loss. The opposite trend between the thermal exergy loss and the mechanical exergy loss ensures the existence of an optimum size of the heat exchanger. In the literature, much work relating to the trade-off between the two factors for minimizing the total exergy loss of various types of heat exchanger can be found.

EXERGY LOSS IN A DUMPING PROCESS

In many industrial thermodynamic systems, exhaust gas at moderate temperature is directly dumped into the environment. This causes a significant waste of energy. The waste of energy in the dumping process can also be measured by using an analysis based on exergy destruction.

In the example shown in the last section, after passing through the heat exchanger, flow "1" is dumped into the environment (Fig. 7.11.12).

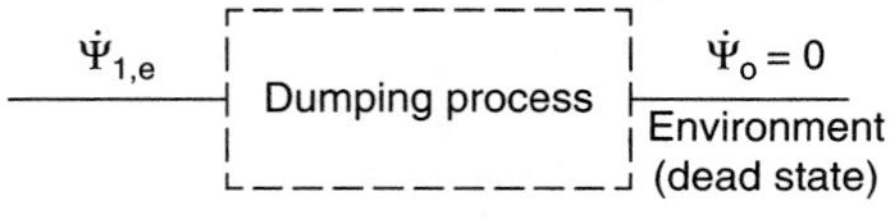

Figure 7.11.12 Exergy loss in a dumping process.

From the second-law viewpoint, the dumping process is an irreversible process which results an exergy loss. The exergy loss rate due to the dumping process $(\Delta\dot{\Psi})_{\text{dumping}}$ is the rate of exergy of the flow at the exit of the heat exchanger ($\dot{\Psi}_{1,e}$), which can be expressed as:

$$(\Delta\dot{\Psi})_{\text{dumping}} = \dot{\Psi}_{1,e}$$

$$= (\dot{m}_1 c_{p,1})T_o\left[\left(\frac{T_{1,e} - T_o}{T_o}\right) - \ln\left(\frac{T_{1,e}}{T_o}\right)\right] + (\dot{m}_1 c_{p,1})T_o\left(\frac{R}{c_{p,1}}\right)\left[\ln\left(\frac{P_{1,e}}{P_o}\right)\right]$$

Using the equation, the exergy loss rate due to the dumping process is calculated to be 150.7 kW. Comparing to the rate of exergy gained in the cold flow (186.83 kW), it reveals that the dumping exergy loss is quite large. This implies that more effort on recovering the exergy in the exhaust gas is needed.

Heat exchangers are frequently used for recovering waste heat from flue gases. From published data in the literature, it is known that, at a near balanced-flow operation $[(\dot{m}c_p)_1/(\dot{m}c_p)_2 \approx 1]$, the exergy recovered by the heat exchangers reaches a maximum. In a waste heat recovery process, usually the flow capacity rate of the flue gas $[(\dot{m}c_p)_1]$ is fixed. For design of a heat exchanger used for recovering the waste heat and for selection of a device used for converting the recovered thermal energy into work, the provided optimum ratio of flow capacity rates is a very useful information.

EXERGY OF A BINARY MIXTURE

A mixture is composed of several different species. Theoretically, the species in the mixture can be reversibly separated by using an ideal semi-permeable membrane. The exergy of a mixture is the sum of the maximum work that can be extracted from each specie in the mixture. For a binary mixture, the exergy can be expressed as:

$$\text{maximum work} = \sum_{i=1}^{2}(n_i\mu_i - n_i\mu_{o,i})$$

where n_i and μ_i represent the number of moles and the chemical potential of specie "i"; subscript "i,o" denotes the dead state or reference state of specie "i".

In air-conditioning and refrigeration systems, it often involves condensation and evaporation of water, so as the moisture content of the air flows in the systems changes. By assuming moist air to be an ideal gas, its specific exergy based on per unit mass of dry air can be expressed as a function of temperature (T), humidity ratio (w) and total pressure of the mixture (P_{total}) as:

$$\psi = c_{p,m}\left[(T - T_o) - T_o\ln\left(\frac{T}{T_o}\right)\right] + T_o R_a(1 + 1.608w)$$

$$\times\left[\ln\left(\frac{w_o + 0.622}{w + 0.622}\right) + \left(\frac{w}{w + 0.622}\right)\ln\left(\frac{w}{w_o}\right) + \ln\left(\frac{P_{\text{total}}}{P_{\text{total,o}}}\right)\right]$$

In the equation, $c_{p,m} \equiv c_{p,a} + wc_{p,v}$ where $c_{p,a}$ and $c_{p,v}$ denote the constant-pressure specific heats of dry air (1.0 kJ/kg-K) and water vapor (1.88 kJ/kg-K), respectively; R_a is the gas constant of dry air (0.287 kJ/kg-K); T_o, w_o, and $P_{\text{total,o}}$ are the dead-state temperature, humidity ratio and total pressure for the moist air.

EXAMPLE. Consider a finned-tube evaporator that is used for cooling and dehumidification of an air flow (Fig. 7.11.13) with mass flowrate (based on dry air) of 0.4 kg/s. The thermodynamic properties of the moist air, condensate and HFC-32 refrigerant at the inlets and exits of the evaporator are specified as:

Moist air:

$w_1 = 0.013$ kg H_2O/kg dry air, $T_1 = 26°C$, $P_1 = 101.525$ kPa,

$w_2 = 0.009$ kg H_2O/kg dry air, $T_2 = 14°C$, $P_2 = 101.325$ kPa,

$\dot{m}_{ad} = 0.4$ kg/s

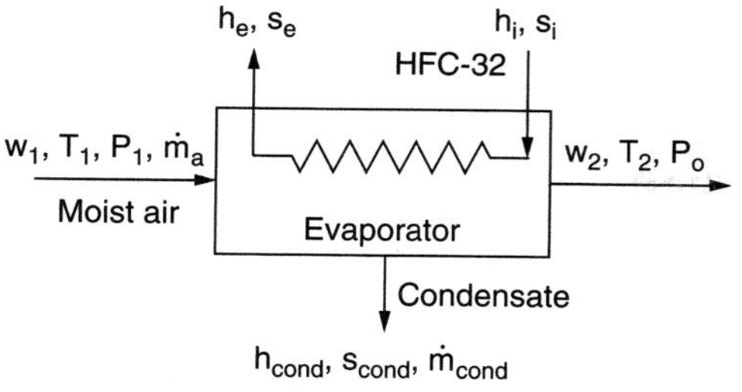

Figure 7.11.13 Exergy loss in a finned-tube evaporator.

Condensate:

$$T_{\text{cond}} = 5°\text{C}, \; h_{\text{cond}} = 21.01 \text{ kJ/kg}, \; s_{\text{cond}} = 0.0762 \text{ kJ/kg-K}$$

$$\dot{m}_{\text{cond}} = 0.0016 \text{ kg/s}$$

Refrigerant (HFC-32):

$$h_i = 300 \text{ kJ/kg}, \quad s_i = 1.359 \text{ kJ/kg-K},$$

$$h_e = 516.1 \text{ kJ/kg}, \quad s_e = 2.136 \text{ kJ/kg-K},$$

$$\dot{m}_{\text{refri}} = 0.0406 \text{ kg/s}, \quad \Delta P \sim 0$$

Dead state for moist air:

$$w_o = 0.0142 \text{ kg H}_2\text{O/kg dry air}, \quad T_o = 35°\text{C}, \quad P_o = 101.325 \text{ kPa},$$

Dead state for condensate (state of water vapor in the environment):

$$T_{v,o} = T_o = 35°\text{C}, \; P_{v,o} = 2.262 \text{ kPa},$$

$$h_{v,o} \approx h_{\text{sat,o}} = 2{,}565.4 \text{ kJ/kg},$$

$$s_{v,o} \approx s_{\text{sat},o} + R\ln(P_{\text{sat},o}/P_{v,o})$$
$$= 8.3544 + 0.462\ln(5.629/2.262) = 8.776 \text{ kJ/kg}$$

From the energy balance of the refrigerant, the cooling power in the process is found to be 8.77 kW. Based on the principle of exergy flow, the exergy loss rate in the evaporator is calculated by using the following equation:

$$\Delta\dot{\Psi} = \dot{m}_a(\psi_1 - \psi_2) + \dot{m}_{\text{refri}}(\psi_i - \psi_e) - \dot{m}_{\text{cond}}\psi_{\text{cond}}$$

In the equation, the first term on the right-hand side represents the rate of exergy input into the evaporator from the air flow; the second term is the rate of exergy input into the evaporator from the refrigerant flow; the last term is the output exergy rate due to the condensate leaving the evaporator.

The values of specific exergy of the moist air at the inlet and exit of the evaporator are calculated and the result is shown as:

$$\psi_1 = 0.322 \text{ kJ/kg dry air}, \quad \psi_2 = 0.916 \text{ kJ/kg dry air}$$

The dead state for the condensate is the dead state for the vapor in the moist air. That means the dead-state specific enthalpy and specific entropy of the condensate are the specific enthalpy and specific entropy of the vapor at temperature of 35°C and pressure of 2.262 kPa, respectively. Based on the concept, the exergy per unit mass of the condensate is:

$$\psi_{\text{cond}} = (h_{\text{cond}} - h_{v,o}) - T_o(s_{\text{cond}} - s_{v,o})$$
$$= 136.5 \text{ kJ/kg condensate}$$

The exergy input into the evaporator is through the refrigerant flow. The input exergy per unit mass of the refrigerant is:

$$\psi_i - \psi_o = (h_i - h_e) - T_o(s_i - s_e)$$
$$= 23.333 \text{ kJ/kg refri.}$$

Using the obtained exergy values, the exergy loss rate in the evaporator is found to be:

$$\Delta\dot{\Psi} = \dot{m}_a(\psi_1 - \psi_2) + \dot{m}_{\text{refri}}(\psi_i - \psi_o) - \dot{m}_{\text{cond}}\psi_{\text{cond}}$$
$$= (0.4)(0.322 - 0.916) + (0.0406)(23.333) - (0.0016)(136.5)$$
$$= 0.491 \text{ kW}$$

The exergy input into the evaporator through the refrigerant flow is 0.947 kW which is obtained from the second term of the equation shown above. Comparing this value to the exergy loss rate in the evaporator (0.491 kW), it reveals that a large portion of the input exergy has been destructed due to the heat transfer through a large temperature difference between the air flow and the refrigerant flow.

An alternate approach can also be used to calculate the exergy loss in the evaporator by dividing the process into an air flow process and a refrigerant flow process. In the air flow process, heat is transferred from the air flow to a reservoir at the evaporating temperature T_L (Fig. 7.11.14). The heat transfer rate is 8.774 kW which is the cooling power previously obtained. In the refrigerant flow process, the same amount of heat is transferred from the reservoir to the refrigerant flow. The exergy loss in the evaporator is the sum of the exergy losses in the two processes.

In the refrigerant flow process, the phase change of the refrigerant is reversible, thus the exergy loss is zero. This makes the exergy loss in the air flow process equal the exergy loss in the evaporator. In the air flow process, the heat transfer between the control volume and the low temperature reservoir ($T_L < T_o$) has a potential to produce work (Fig. 7.11.15). The work can be treated as an exergy flow. Expressing the process in terms of exergy flows as shown in Fig. 7.11.16, the exergy loss rate in the evaporator is calculated by using the following equation:

$$\Delta\dot{\Psi} = \dot{m}_a(\psi_1 - \psi_2) + \dot{Q}_L\left(\frac{T_o}{T_L} - 1\right) - \dot{m}_{\text{cond}}\psi_{\text{cond}}$$

The result obtained from the equation is still 0.491 kW.

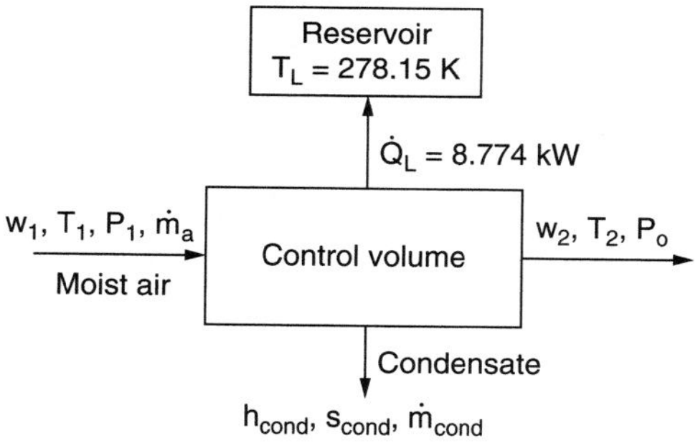

Figure 7.11.14 Control volume for air flow process.

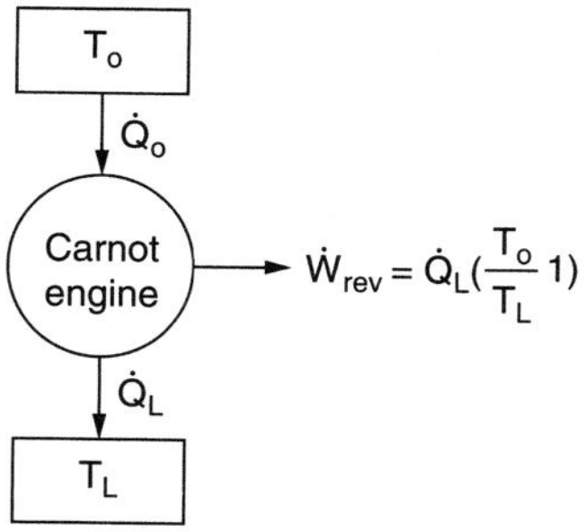

Figure 7.11.15 Maximum possible work produced by $\dot{Q}_L$.

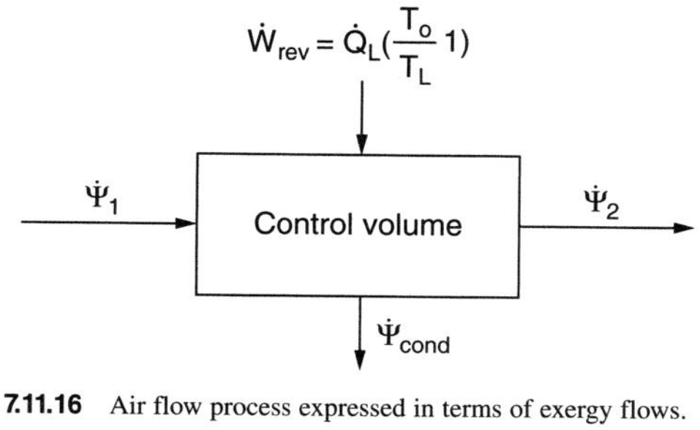

Figure 7.11.16 Air flow process expressed in terms of exergy flows.

AVAILABLE ENERGY IN A CHEMICAL REACTION PROCESS

Many thermodynamic systems involve chemical reactions, such as combustion of hydrocarbon fuels with air in power generation systems and electrochemical processes in fuel cells. The combustion process can be assumed to be a steady-state process. In the absence of changes of kinetic and potential energy, if the reactants and products are in temperature equilibrium with the environment, the maximum possible work produced in an isothermal combustion process (Fig. 7.11.17) can be expressed as:

$$W_{\text{rev,iso}} = \sum_{R} n_i \bar{g}_i - \sum_{P} n_e \bar{g}_e$$

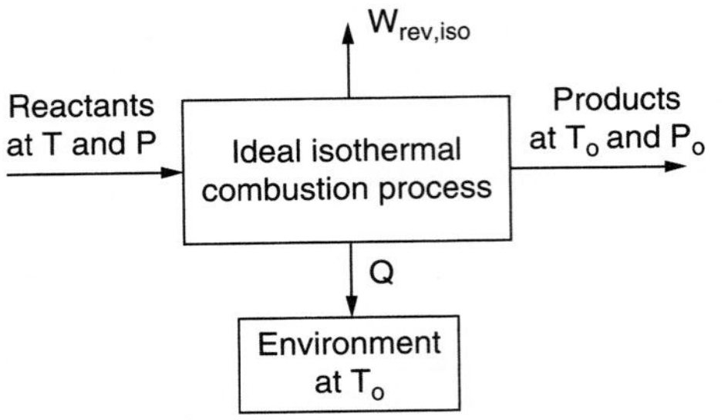

Figure 7.11.17 Maximum work produced in an isothermal combustion process.

The maximum possible work is the available energy (or exergy) in the combustion process. In the equation shown above, n_i and n_e represent the number of moles of the reactants and the products, respectively; $\bar{g}_i$ and $\bar{g}_e$ are the Gibbs functions of the reactants and products in a molar basis, respectively. The Gibbs function is related to enthalpy and entropy in the following manner: $\bar{g} = \bar{h} - T\bar{s}$.

The isothermal combustion process is an ideal chemical reaction process. In a real combustion process, the reactants rapidly turn into the products and the temperature spontaneously rises. From the description, it is more reasonable to consider the process to be adiabatic, instead of isothermal. For an adiabatic combustion process, the maximum possible work extracted from the reactants is the maximum possible work extracted from the products at the adiabatic flame temperature (Fig. 7.11.18). The maximum possible work produced in the adiabatic combustion process ($W_{\text{rev,adi}}$) is much less than that produced in the isothermal combustion process ($W_{\text{rev,iso}}$). The inherent imperfection of the chemical reaction in the real combustion process results in a loss of the available energy of the reactants.

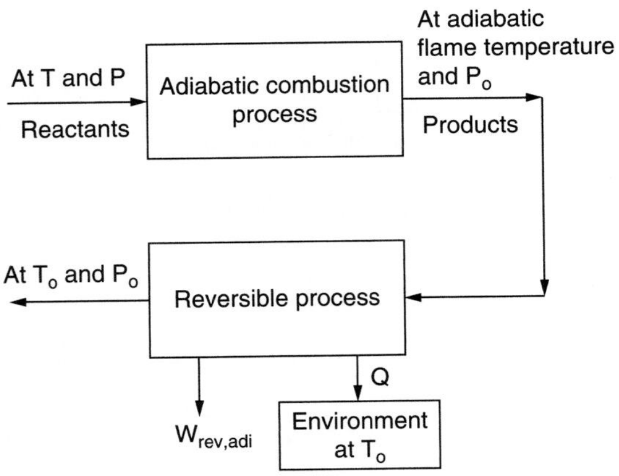

Figure 7.11.18 Maximum work produced in an adiabatic combustion process.

In the second law analysis of thermodynamic systems involving a combustion process, the available energy (or exergy) in the combustion process can be treated as the maximum possible work extracted from the products at the adiabatic flame temperature (T_{adi}). In the absence of changes of kinetic and potential energy, the exergy per mole of the fuel ($\bar{\psi}_{\text{adi}}$) can be expressed as

$$\bar{\psi}_{\text{adi}} = \sum_{P} n_j[(\bar{h}_e - \bar{h}_o) - T_o(\bar{s}_e - \bar{s}_o)]$$

In the equation, $\bar{h}_e$ and $\bar{s}_e$ are the molar enthalpy and molar entropy of product "j" at adiabatic flame temperature and pressure $P_{e,j}$, respectively; $\bar{h}_e$ and $\bar{s}_e$ are the molar enthalpy and molar entropy of product "j" at environmental temperature T_o and pressure $P_{o,j}$, respectively.

A simplified approach is introduced by assuming the products to be an ideal gas. From the result previously obtained, the specific exergy of the products in the adiabatic combustion process can be calculated by using the following equation:

$$\psi = c_{p,av} T_o \left[\left(\frac{T_{\text{adi}} - T_o}{T_o} \right) - \ln\left(\frac{T_{\text{adi}}}{T_o} \right) \right] + c_{p,av} T_o \left(1 - \frac{1}{k} \right) \ln\left(\frac{P}{P_o} \right)$$

$$\text{where } c_{p,av} = \frac{\sum n_j M_j c_{p,j}}{\sum n_j M_j}, \quad k = \frac{c_{p,av}}{c_{v,av}} = \frac{c_{p,av}}{c_{p,av} - R_{av}},$$

$$R = \frac{R_u}{M_{av}}, \quad M_{av} = \sum \left(\frac{n_j}{N_t} \right) M_j,$$

$$N_t = \sum n_j, \quad R_u = 8.314 \text{ kJ/kmol-K}$$

In the equation, the $c_{p,av}$ is the average constant-pressure specific heat of the products; n_j and M_j are the number of moles and molecular weight of specie "j" in the products; $c_{p,j}$ is the constant-pressure specific heat of specie "j," which is evaluated at $T_{av} = (T_{\text{adi}} + T_o)/2$; R_u is the universal gas constant; M_{av} is the average molecular weight of the products; N_t is the total number of moles of the products; R is the average gas constant of the products.

EXAMPLE: Gaseous-phase propane at environmental temperature T_o and environmental pressure P_o (25°C and 0.1 MPa) is burn with 200 percent theoretical air. The products are assumed to be an ideal gas at the adiabatic flame temperature T_{adi} and P_o. In the combustion process, the chemical reaction equation is:

$$C_3H_8 + (5)(2)(O_2 + 3.76N_2) \rightarrow 3CO_2 + 4H_2O + 5O_2 + 37.6N_2$$

For the adiabatic combustion process, the enthalpy of the reactants H_R is equal to the enthalpy of the products H_P. By setting the products at various temperatures, the $(H_P - H_R)$ values are calculated and the obtained values are used to interpolate the adiabatic flame temperature. As shown in Fig. 7.11.19, the adiabatic flame temperature is found to be 1510 K.

The average constant-pressure specific heat $c_{p,av}$ is evaluated at 904 K which is the arithmetic mean value between the T_{adi} and the T_o. At 904 K, the c_p values of CO_2, H_2O, O_2, and N_2 are 1.205, 2.22, 1.075, and 1.137 kJ/kg-K, respectively. Using the data, the $c_{p,av}$ value is found to be 1.19 kJ/kg-K and the R value is 0.291 kJ/kg-K. Then, the specific exergy of the products at the adiabatic flame temperature ψ_{adi} is calculated using the following equation:

$$\psi_{\text{adi}} = c_{p,av} T_o \left[\left(\frac{T_{\text{adi}} - T_o}{T_o} \right) - \ln\left(\frac{T_{\text{adi}}}{T_o} \right) \right]$$

$$= 867 \text{ kJ/kg of products}$$

The specific exergy of the products is the maximum work that can be extracted from per kg of the products at the adiabatic flame temperature.

The specific exergy of the products can also be calculated by using tabulated values of enthalpy and entropy of the products. The result is:

$$\psi_{\text{adi}} = \bar{\psi}_{\text{adi}} / \Sigma(n_j M_j)$$

$$= \sum_{P} n_j[(\bar{h}_e - \bar{h}_o) - T_o(\bar{s}_e - \bar{s}_o)] / \Sigma(n_j M_j)$$

$$= 884 \text{ kJ/kg of products}$$

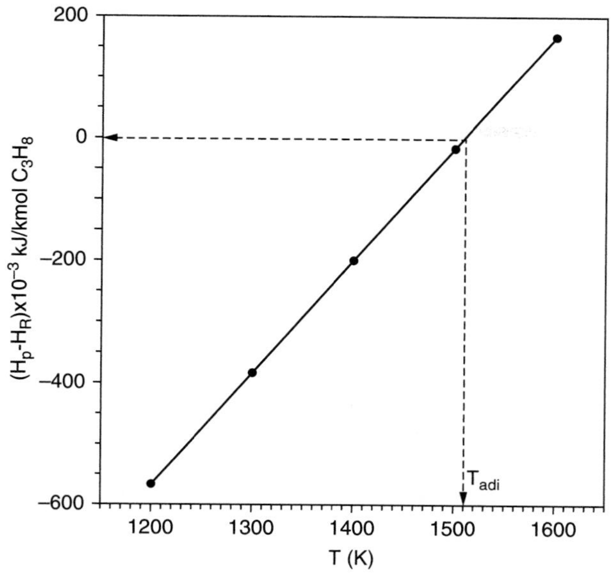

Figure 7.11.19 Interpolation of adiabatic flame temperature.

Comparing the ψ_{adi} value to the result previously obtained by assuming the products to be an ideal gas, it shows that the latter is 1.9 percent smaller than the former. In the equation shown above, $\bar{h}_e$ and $\bar{s}_e$ are the molar enthalpy and molar entropy of product "j" at adiabatic flame temperature T_{adi} and pressure P_j [i.e., $P_j = (n_j/\Sigma n_j)P_o$]; $\bar{h}_e$ and $\bar{s}_e$ are the molar enthalpy and molar entropy of product "j" at environmental temperature T_o and pressure P_j. Tabulated entropy data of a combustion product are usually given for a pressure of 0.1 MPa. In this example, the tabulated entropy data needs to be modified by using the equation $\bar{s}_{P_j} - \bar{s}_{P_o} \approx -R\ln(P_j/P_o) = -R_u(n_j/\Sigma n_j)$.

By assuming the products to be an ideal gas, the heat released in the process is calculated to be:

$$Q = c_{p,avg}(T_{adi} - T_o) = 1442 \text{ kJ/kg of products}$$

Divided the Q value by the ψ_{adi} value, it reveals that 61.3 percent of the heat released in the process is available for converting into work. This exergy conversion efficiency is equivalent to the conversion efficiency of a Carnot engine operated between a high-temperature reservoir at 770 K and a low-temperature reservoir at T_o.

The heat released in the process is also calculated by using tabulated values of enthalpy for propane and the products. The result is 1442 kJ/kg products which is the same as the Q value obtained by assuming the products to be an ideal gas.

SECOND-LAW ANALYSIS OF THERMODYNAMIC CYCLES

A thermodynamic cycle is composed of several thermodynamic processes. In the processes, exergy is either input into the working substance from an energy source, or converted into mechanical work, or transferred from one flow to another flow. In real situation, inevitably, a part of the exergy is destructed during the exergy conversion or exergy transfer. The summation of the exergy losses in all the processes is the total exergy loss of the cycle.

The second law analysis of a thermodynamic cycle includes an exergy analysis of all the processes in the cycle and a comparison of the performance factors between the real cycle and an ideal cycle without any exergy loss. The result of the analysis not only shows how far the performance of the cycle from its limit, but also provides a measure of all the processes from the viewpoint of exergy loss. From the information, the major source of the exergy loss in the cycle can be identified. Then the most effective strategy to upgrade the performance of the cycle can be made.

Example. Consider a simple power generation cycle that exchanges heats $\dot{Q}_H$ and $\dot{Q}_o$ with two reservoirs at temperatures T_H and T_o, respectively (Fig. 7.11.20). The symbol, T_o, is designated to be the environmental temperature.

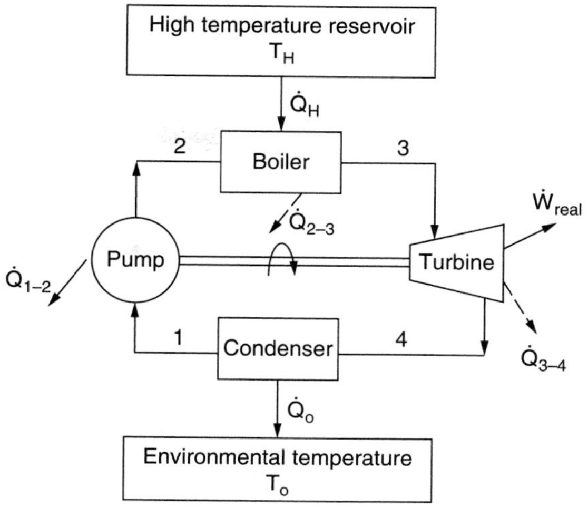

Figure 7.11.20 A simple power generation cycle.

In the cycle, the required input power of the pump is supplied by the output power of the turbine. From the first law of thermodynamics, the net output power produced by the turbine in the cycle is:

$$\dot{W}_{real} = \dot{W}_{turbine} - \dot{W}_{pump}$$
$$= \dot{Q}_H - \dot{Q}_o - \dot{Q}_{1\text{-}2} - \dot{Q}_{2\text{-}3} - \dot{Q}_{3\text{-}4}$$

where $\dot{Q}_{1-2}$, $\dot{Q}_{2-3}$ and $\dot{Q}_{3-4}$ denote the rates of heat leakage from the pump, boiler and turbine to the environment, respectively. Neglecting the changes of kinetic and potential energy, the exergy loss rates in the four processes of the cycle are shown as:

$$\Delta(\dot{\Psi})_{1\text{-}2} = \dot{m}T_o(s_2 - s_1) + \dot{Q}_{1\text{-}2} \qquad \Delta(\dot{\Psi})_{2\text{-}3} = \dot{m}T_o(s_3 - s_2) - \dot{Q}_H\left(\frac{T_o}{T_H}\right) + \dot{Q}_{2\text{-}3}$$
$$\Delta(\dot{\Psi})_{3\text{-}4} = \dot{m}T_o(s_4 - s_3) + \dot{Q}_{3\text{-}4}$$
$$\Delta(\dot{\Psi})_{4\text{-}1} = \dot{m}T_o(s_1 - s_4) + \dot{Q}_o$$

In the equations shown above, $\dot{m}$ denotes the mass flow rate of the working substance. The total exergy loss rate in the cycle is the cumulative exergy loss rate of the four processes. It is shown as:

$$\Delta(\dot{\Psi})_{cycle} = \Sigma(\Delta\dot{\Psi})$$
$$= \dot{Q}_o - \dot{Q}_H\left(\frac{T_o}{T_H}\right) + \dot{Q}_{1\text{-}2} + \dot{Q}_{2\text{-}3} + \dot{Q}_{3\text{-}4}$$

The ideal power generation cycle is a Carnot cycle (Fig. 7.11.21) which operates between the same two reservoirs as those in the real cycle. The power produced by the Carnot cycle is expressed as:

$$\dot{W}_{rev} = \dot{Q}_H - \dot{Q}_{o'} = \dot{Q}_H\left(1 - \frac{T_o}{T_H}\right)$$

Figure 7.11.21 Output power of an ideal Carnot cycle.

From the result shown above, it is easily verified that:

$$\dot{W}_{\text{real}} = \dot{W}_{\text{rev}} - \Delta(\dot{\Psi})_{\text{cycle}}$$

The equation shows that a reduction of the total exergy loss in the four processes increases the power output of the cycle. Hence, for effectively improving the performance of the power generation cycle, the exergy loss in each process needs to be known. From the result, the major source of the exergy loss in the cycle can be clearly identified.

Many expressions are used to define the efficiencies of thermodynamic cycles. As the efficiencies are based on energy, they are classified as the first-law efficiency (η_{I}). In general, the first-law efficiency is defined as the ratio of the achieved energy to the input energy and it reflects the fraction of energy being converted from one form to the other. For the simple power generation cycle, the first-law efficiency or thermal efficiency (η_{th}) is defined as:

$$\eta_{\text{I}} = \eta_{\text{th}} = \frac{\dot{W}_{\text{real}}}{\dot{Q}_H}$$

The definitions of efficiencies can also be based on a measure of performance of thermodynamic cycles by taking into account the limitations imposed by the second law of thermodynamics. This type of efficiency is classified as the second-law efficiency (η_{II}). For the simple power generation cycle, the second-law efficiency is defined as the output power of the real cycle ($\dot{W}_{\text{real}}$) divided by the output power of the Carnot cycle ($\dot{W}_{\text{rev}}$). The second-law efficiency is related to the first-law efficiency and total exergy loss of the cycle as:

$$\eta_{\text{II}} = \frac{\dot{W}_{\text{real}}}{\dot{W}_{\text{rev}}} = \frac{\eta_{\text{I}}}{\eta_{\text{Carnot}}} = 1 - \frac{\Delta(\dot{\Psi})_{\text{cycle}}}{\dot{W}_{\text{rev}}}$$

The numerical value of the second-law efficiency falls into the range of 0 to 1. Comparing to the first-law efficiency (thermal efficiency), the second-law efficiency gives a better scale to indicate the performance of the cycle.

Commercial power generation cycles are much more complicated than the simple power generation cycle. In addition, the heat input into the boiler usually is transferred from a hot flow, instead of being provided by a reservoir at T_H. However, the method of the second-law analysis for the commercial power generation cycles basically is the same as that illustrated in this example. The exergy input into the boiler can also be calculated using the equations provided in the previous sections. In the second-law analysis of a complicated thermodynamic system, after the exergy loss in each process is obtained, it is urged to check the exergy balance of the entire system. The exergy input into the system must be equal to or close to the total exergy loss plus the net output work. The exergy balance is an easy way to verify the result of the analysis.

EXAMPLE. Figure 7.11.22 shows a simple refrigeration cycle which is composed of a compressor, a condenser, an expansion device, and an evaporator. The exergy loss rates in the four processes of the cycle are expressed as:

$$\Delta(\dot{\Psi})_{1\text{-}2} = \dot{m}T_H(s_2 - s_1) + \dot{Q}_{1\text{-}2} \qquad \Delta(\dot{\Psi})_{2\text{-}3} = \dot{m}T_o(s_3 - s_2) + \dot{Q}_H$$

$$\Delta(\dot{\Psi})_{3\text{-}4} = \dot{m}T_o(s_4 - s_3) - \dot{Q}_{3\text{-}4}$$

$$\Delta(\dot{\Psi})_{4\text{-}1} = \dot{m}T_o(s_1 - s_4) - \dot{Q}_L\left(\frac{T_H}{T_L}\right)$$

where T_H and T_L are the temperatures of the environment and cooling space, respectively. $\dot{Q}_{1\text{-}2}$ and $\dot{Q}_{3\text{-}4}$ are the rates of heat leakage in the compressor and expansion device, respectively. Usually, the heat leakages in the two components are small, thus they can be neglected. The total exergy loss in the cycle is the sum of the exergy losses in the four processes and it can be expressed as:

$$\Delta(\dot{\Psi})_{\text{cycle}} = \dot{Q}_H - \dot{Q}_L(T_H/T_L) + \dot{Q}_{1\text{-}2} - \dot{Q}_{3\text{-}4}$$

The first-law efficiency of the refrigeration cycle is also known as the coefficient of performance (COP) which is defined as the ratio of the cooling power to the input power. It is expressed as:

$$\eta_{\text{I}} = \text{COP} = \frac{\dot{Q}_L}{|\dot{W}_{\text{real}}|}$$

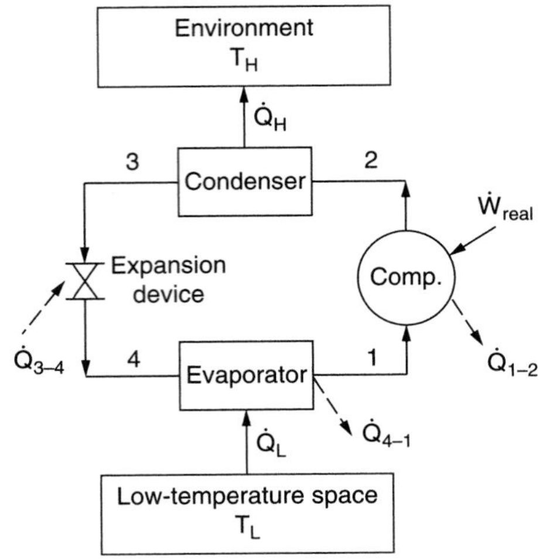

Figure 7.11.22 A simple refrigeration cycle operated between T_H and T_L.

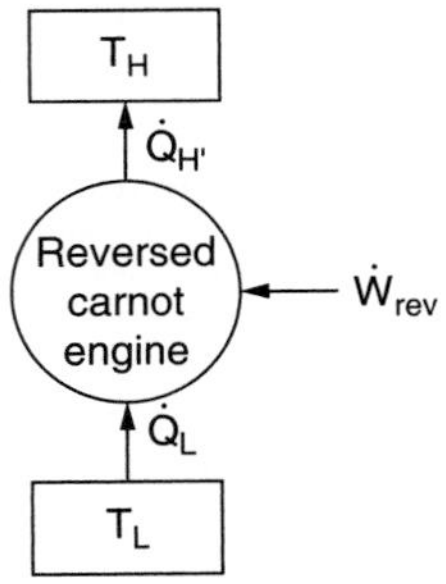

Figure 7.11.23 A reversed Carnot cycle operated between T_H and T_L.

Figure 7.11.23 shows an ideal reversed Carnot cycle operated between the same high temperature reservoir (T_H) and the low temperature reservoir (T_L), and receiving the same amount heat transfer rate ($\dot{Q}_L$) from T_L. Similar to the simple power generation cycle previously illustrated, the second-law efficiency of the refrigeration cycle is defined as the input power in the reversed Carnot cycle ($\dot{W}_{\text{rev}}$) divided by the input power in the real cycle ($\dot{W}_{\text{real}}$). The second-law efficiency of the refrigeration cycle can also be related to the first-law efficiency and total exergy loss of the cycle as:

$$\eta_{\text{II}} = \frac{|\dot{W}_{\text{rev}}|}{|\dot{W}_{\text{real}}|} = \frac{\eta_{\text{I}}}{\eta_{\text{Carnot}}} = 1 - \frac{\Delta(\dot{\psi})_{\text{cycle}}}{|\dot{W}_{\text{real}}|}$$

REFERENCES

1. Sonntag, R.E., Borgnakke, C., Van Wylen, G.J., Fundamentals of Thermodynamics, 6th ed., John Wiley & Sons, Inc., 2003.

2. Kestin, J., A course in Thermodynamics, Vol. 1, Chapter 13, Mc Graw Hill, 1979.

3. Bejan, A., Advanced Engineering Thermodynamics, John Wiley & Sons, Inc., 1988.

4. San, J.Y., Second-law Performance of Heat Exchangers for Waste Heat Recovery, Energy, Vol. 25, 2010, pp. 1936-1945.

Section 8
Transportation

BY

GENE LIAO *Professor, Wayne State University, Detroit, Michigan*
V. TERREY HAWTHORNE *Late Senior Engineer, LTK Engineering Services*
KEITH L. HAWTHORNE *Late Vice President—Technology, Transportation Technology Center, Inc.*
MEHDI AHMADIAN *Professor and Director, Railway Technologies Laboratory, Virginia Tech*
RICHARD J. BURKE *ABS Professor of Naval Architecture & Marine Engineering, State University of New York Maritime College*
M. W. M. JENKINS *Professor, Aerospace Design, Georgia Institute of Technology*
JOAQUIM R. R. A. MARTINS *Professor of Aerospace Engineering, University of Michigan*
SANFORD FLEETER *Late Professor Emeritus, Purdue University*
GLENN J. MILLER *National Aeronautics and Space Administration*
MIKE YOON *Fellow, American Society of Mechanical Engineers*

8.1 AUTOMOTIVE ENGINEERING
by Gene Liao

8.2 ROADMAP OF VEHICLE ELECTRIFICATION AND HYBRIDIZATION
by Gene Liao

8.3 RAILWAY ENGINEERING
by V. Terrey Hawthorne, Keith L. Hawthorne, and Mehdi Ahmadian (in collaboration with E. Thomas Harley, Charles M. Smith, Robert B. Watson, and C. A. Woodbury)

8.4 MARINE ENGINEERING
by Richard J. Burke

8.5 AERONAUTICS
by M. W. M. Jenkins and Joaquim R. R. A. Martins

8.6 JET PROPULSION AND AIRCRAFT PROPELLERS
by Sanford Fleeter

8.7 ASTRONAUTICS
by Glenn J. Miller and Joaquim R. R. A. Martins

8.8 PIPELINE TRANSMISSION
by Mike Yoon

8.1 AUTOMOTIVE ENGINEERING

by Gene Liao

REFERENCES: Sovran and Blazer, "A Contribution to Understanding Automotive Fuel Economy and Its Limits," SAE Paper 2003–01–2070. Toboldt, Johnson, and Gauthier, "Automotive Encyclopedia," Goodheart-Willcox Co., Inc. "Bosch Automotive Handbook," Society of Automotive Engineers. Gillespie, "Fundamentals of Vehicle Dynamics," Society of Automotive Engineers. Hellman and Heavenrich, "Light-Duty Automotive Technology Trends: 1975 Through 2003," Environmental Protection Agency. Davis et al., "Transportation Energy Book," U.S. Department of Energy. Various publications of the Society of Automotive Engineers, Inc. (SAE International).

AUTOMOTIVE VEHICLES

The dominant mode of personal transportation in the United States is the light-duty automotive vehicle. This category is composed not only of passenger cars, but of light-duty trucks as well. Light-duty trucks include passenger vans, SUVs (sport utility vehicles), and pickup trucks. Among light-duty trucks, the distinction between personal and commercial vehicles is blurred because they may be used for either purpose. Beyond the light-duty truck, medium- and heavy-duty trucks are normally engaged solely in commercial activities.

According to the U.S. Bureau of Transportation Statistics for 2014, there were around 260 million registered vehicles. Of these, 183 million were classified as "Light duty vehicle, short wheel base," while another 50.5 million were listed as "Light duty vehicle, long wheel base." Another 8 million were classified as vehicles with two axles and six or more tires and 2.4 million were classified as "Truck, combination." There were 8.4 million motorcycles also listed along with 0.76 million buses. The U.S. Department of Transportation's definition of a passenger vehicle, to mean a car or truck, used for passengers, excluding buses and trains.

Data released by the U.S. Department of Transportation's Federal Highway Administration show that U.S. driving reached 3.148 trillion miles in 2015. Of this mileage, 85 percent was accumulated by passenger vehicles, and 13.6 present by heavy trucks and buses.

In 2013, 33.9 percent of U.S. households operated one vehicle, 37.3 percent had two, and 19.7 percent had three or more, leaving only 9.1 percent of households without a vehicle. The 2009 National Household Travel Survey found that the most common use of privately owned vehicles was in trips to or from work, which accounted for 26.2 percent of vehicle miles. Other major uses were for shopping (22.8 percent), other family or personal business (21.9 percent), visiting friends and relatives (5.7 percent), and other social or recreational outlets (14.9 percent).

The U.S. freight transportation system moved nearly 20.1 billion tons of goods valued at $18.0 trillion in 2013, according to estimates derived from the Freight Analysis Framework. Freight transportation is carried by a variety of networks. The largest percentage of U.S. freight is carried by trucks (60%), followed by pipelines (18%), rail (10%), ship (8%), and air (0.01%). Much of the balance was transported multimodally, with truck transport often being one of those modes.

Trucks are often grouped by **gross vehicle weight (GVW)** rating into light duty (0 to 14,000 lb), medium-duty (14,001 to 33,000 lb), and heavy-duty (over 33,000 lb) categories. They are further subdivided into classes 1 through 8, as outlined in Table 8.1.1. Examples of Classes 6 through 8 are illustrated by the silhouettes of Fig. 8.1.1.

Table 8.1.1 Gross Vehicle Weight Classification

GVW group	Class	Weight kg (max)
6,000 lb or less	1	2,722
6,001-10,000 lb	2	4,536
10,001-14,000 lb	3	6,350
14,001-16,000 lb	4	7,258
16,001-19,500 lb	5	8,845
19,501-26,000 lb	6	11,794
26,001-33,000 lb	7	14,469
33,001 lb and over	8	14,970

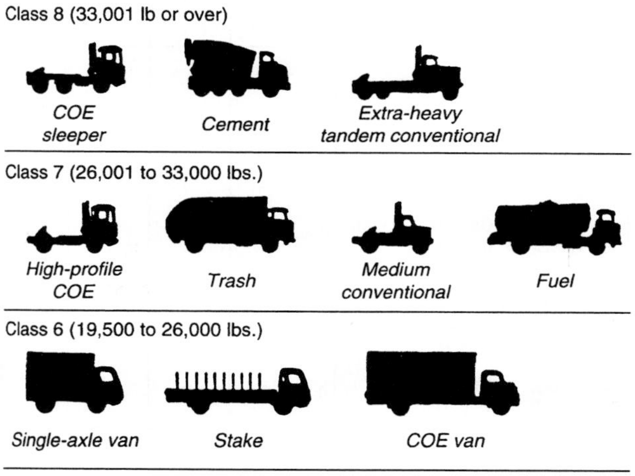

Figure 8.1.1 Medium and heavy-duty truck classification by gross weight and vehicle usage. (*Reprinted from SAE, SP-868, © 1991, Society of Automotive Engineers, Inc.*)

TRACTIVE FORCE

The **tractive force** (F_{TR}) required to propel an automobile, measured at the tire-road interface, is composed of four components—the **rolling resistance** (F_R), the **aerodynamic drag** (F_D), the **inertia force** (F_I), and the **grade requirement** (F_G). That is, $F_{TR} = F_R + F_D + F_I + F_G$.

The force of rolling resistance in a flat road is $C_R W$, where C_R is the rolling resistance coefficient of the tires and W is vehicle weight. For a given tire, C_R is normally considered constant with speed, although it does rise slowly at very high speeds. In the days of bias-ply tires, it had a typical value of 0.015, but with modern radial tires, coefficients as low as 0.006 have been reported. The rolling resistance of cold tires is greater than these warmed-up values, however. On a 70°F day, it may take 20 minutes of driving for a tire to reach its equilibrium running temperature. Under-inflated tires also have higher rolling resistance. Figure 8.1.2 illustrates how the rolling resistance of a particular tire falls off with increasing inflation pressure.

Aerodynamic drag force is given by $C_D A \rho V^2/2$, where C_D is the aerodynamic drag coefficient, A is the projected vehicle frontal area, ρ is the ambient air density, V is the vehicle velocity, and g is the gravitational constant. The frontal area of most cars falls within the range of 20 to 30 ft^2 (1.9 to 2.8 m^2). The frontal areas of light trucks generally range from 27 to 38 ft^2 (2.5 to 3.5 m^2). For heavy-duty trucks and truck-trailer combinations, frontal area can range from 50 to 100 ft^2 (4.6 to 9.3 m^2). The drag coefficient of a car is highly dependent on its shape, as illustrated in Fig. 8.1.3. The power required to overcome aerodynamic drag is the product of the aerodynamic component of tractive force and vehicle velocity. That power is listed at various speeds in Fig. 8.1.3 for a car with a frontal area of 2 m^2. For most contemporary cars, C_D falls between 0.30 and 0.35, although values around 0.20 have been reported for certain advanced concepts. For light trucks, C_D

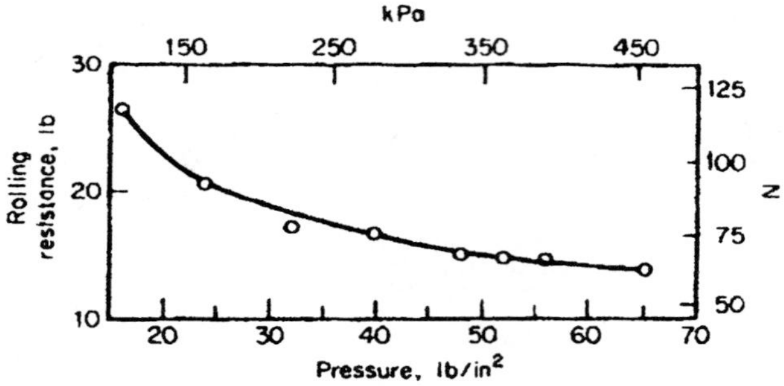

Figure 8.1.2 Dependence of rolling resistance on inflation pressure for FR78-14 tire, 1280-lb load and 60-mi/h speed. (*"Tire Rolling Losses," SAE Proceedings P-74.*)

	Drag coefficient	Drag power in kW, average values for $A = 2m^2$ at various speeds		
		40 km/h	120 km/h	160 km/h
Open convertible	0.5 ... 0.7	1	27	63
Station wagon (2-box)	0.5 ... 0.6	0.91	24	58
Conventional form (3-box)	0.4 ... 0.55	0.78	21	50
Wedge shape, headlights & bumpers integrated in body	0.3 ... 0.4	0.58	16	37
Optimum streamlining	0.15 ... 0.20	0.29	7.8	18

Figure 8.1.3 Drag coefficient and aerodynamic power requirements for various body shapes. (*Bosch, "Automotive Handbook," SAE.*)

generally falls between 0.40 and 0.50. Drag coefficients for heavy-duty trucks and truck-trailers range from 0.6 to over 1.0, with 0.7 being a typical value for a Class 8 tractor-trailer.

The tractive force at the tire patch that is associated with changes in vehicle speed is given by $F_I = (W/g)\,a$, where a is vehicle acceleration. If the vehicle is decelerating, then a (hence F_I) is negative.

If the vehicle is operated on a grade, an additional tractive force must be supplied. Within the limits of grades normally encountered, that force is $F_G = W \sin \theta$, where θ is the angle of the grade, measured from the horizontal.

For constant-speed driving over a horizontal roadbed, F_I and F_G are both zero, and the tractive force is simply $F_{TR} = F_R + F_D$. This is termed the **road load** force. In Fig. 8.1.4, curve T shows this road-load force requirement for a car with a test weight of 4,000 lb (1,814 kg). Curves A and R represent the aerodynamic and rolling resistance components of the total resistance, respectively, on a level road with no wind. Curves T′ paralleling curve T show the displacement of curve T for operation on 5, 10, and 15 percent grades. Curve E shows the tractive force delivered from the engine at full throttle. The differences between the E curve and the T curves represent the tractive force available for acceleration. The intersections of the E curve with the T curves indicate maximum speed capabilities of 98, 87, and 70 mi/h on grades of 0, 5, and 10 percent. Operation on a 15 percent grade is seen to exceed the capability of the engine in this transmission gear ratio. Therefore, operation on a 15 percent grade requires a transmission downshift. Tractive forces

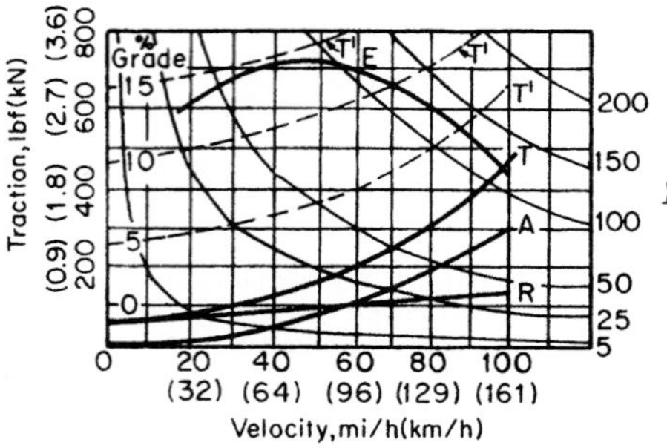

Figure 8.1.4 Traction available and traction required for a typical large automobile.

on the ordinate of Fig. 8.1.4 can be converted to power by multiplying by the corresponding vehicle velocities on the abscissa.

PERFORMANCE

In the United States, a commonly used performance metric is acceleration time from a standstill to 60 mi/h. A zero to 60 mi/h acceleration at wide-open throttle is seldom executed. However, a vehicle with inferior acceleration to 60 mi/h is likely to do poorly in more important performance characteristics, such as the ability to accelerate uphill at highway speeds, to carry a heavy load or pull a trailer, to pass a slow-moving truck, or to merge into freeway traffic.

The most important determinant of zero-to-60 mi/h acceleration time is vehicle weight-to-power ratio. To calculate this acceleration time rigorously, the appropriate power is the power available from the engine at each instant, diminished for the inefficiencies of the transmission and driveline. Furthermore, the actual weight of the vehicle is replaced by its effective weight (W_{eff}), which includes the rotational inertia of the wheels, driveline, transmission, and engine. In the highest (drive) gear, W_{eff} may exceed W by about 10 percent. In lower gears (higher input/output speed ratios), the engine and parts of the transmission rotate at higher speeds for a given vehicle velocity. Because rotational energy varies as the square of angular velocity, operation in lower gears increases the effective mass of the vehicle more significantly.

The maximum acceleration capability calculated in this manner may be limited by loss of traction between the driving wheels and the road surface. This is dependent upon the coefficient of friction between tire and road. For a new tire, this coefficient can range from 0.85 on dry concrete to 0.5 if water is puddled on the road. Under the same conditions, the coefficient for a moderately worn tire can range from 1.0 down to 0.25. Using special rubber compounds, coefficients as high as 1.8 have been attained with special racing tires.

To avoid all these variables when estimating the acceleration performance of a typical light-duty vehicle under normal driving conditions on a level road, various empirical correlations have been developed by fitting test data from large fleets of production cars. Such correlations for the time from zero to 60 mi/h (t_{60}) have generally taken the form, $C(P/W)^n$. In one simple correlation, t_{60} is approximated in seconds when P is the rated engine horsepower, W is the test weight in pounds, $C = 0.07$ and $n = 1.0$. According to the U.S. Environmental Protection Agency (EPA), the fleet-average t_{60} for new passenger cars has decreased rather consistently from 14.2 s in 1975 to 10.3 s in 2001, and to 9.3 s in 2011. Over that same time span, it has decreased from 13.6 to 10.6 s for light-duty trucks.

FUEL CONSUMPTION

Automotive fuels are refined from petroleum. In 1950, less than 10 percent of U.S. petroleum consumed was imported, but that fraction has grown to 40 percent in 2012. Much of this imported petroleum comes from politically unstable parts of the world. Therefore, it is in the national interest to reduce automotive fuel consumption, or conversely, to increase automotive fuel economy.

In the United States, passenger-car fuel-economy standards were promulgated by Congress in 1975 and took effect in 1978. Standards for light-duty trucks took effect in 1982. The fuel economies of new vehicles are monitored by the U.S. EPA. Each vehicle model is evaluated by running that model over prescribed transient driving schedules of velocity versus time on a chassis dynamometer. Those driving schedules are plotted in Figs. 8.1.5 and 8.1.6. The **urban driving schedule** (Fig. 8.1.5) simulates 7.5 mi of driving at an average speed of 19.5 mi/h. It is the same schedule used in measuring

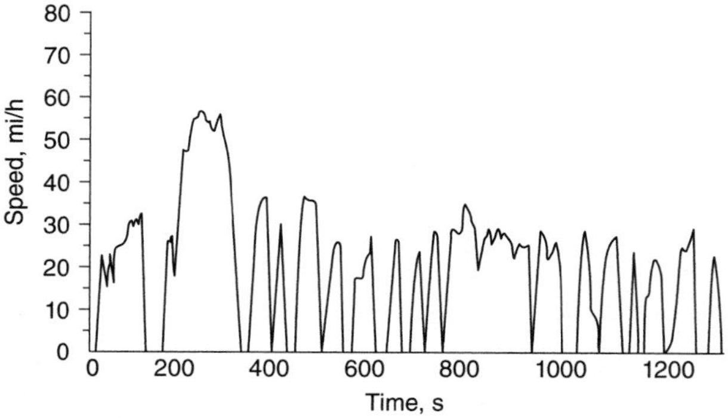

Figure 8.1.5 Urban driving schedule.

emissions compliance. It requires a 12-h soak of the vehicle at room temperature, with driving initiated 20 s after the cold start. The 18 start-stop cycles of the schedule are then followed by engine shutdown and a 10-min hot soak, after which the vehicle is restarted and the first five cycles repeated. The **highway driving schedule** (Fig. 8.1.6) is free of intermediate stops, covering 10.2 mi at an average speed of 48.2 mi/h.

The **combined fuel economy** of each vehicle model is calculated by assuming that the 55 percent of the distance driven is accumulated in the urban setting and 45 percent on the highway. Consequently, the combined value is sometimes known as the **55/45 fuel economy.** It is derived from $MPG_{55/45} = 1/[(0.55\ MPG_U + 0.45\ MPG_H)]$, where

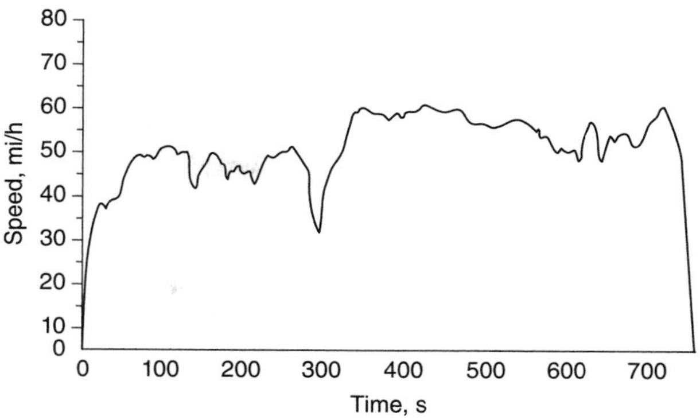

Figure 8.1.6 Highway driving schedule.

MPG represents fuel economy in mi/gal, and subscripts U and H signify the urban and highway schedules, respectively.

Each manufacturer's **CAFE (corporate average fuel economy)** is the sales-weighted average of the 55/45 fuel economies for its vehicle sales in a given year. CAFE standards are set by the U.S. Congress. From 1990-2004, the CAFE standard for passenger cars was 27.5 mi/gal. For light-duty trucks, it was 20.7 mi/gal from 1976-2004. In model years 2017 through 2025, EPA's standards apply to passenger cars, light-duty trucks, and medium-duty passenger vehicles. The final standards are projected to result in an average industry fleet wide level of 54.5 mi/gal. Manufacturers failing to meet the CAFE standard in any given year are fined for their shortfall.

The 55/45 fuel economy of a vehicle depends on its tractive-force requirements, the power consumption of its accessories, the losses in its transmission and driveline, the vehicle-to-engine speed ratio, and the fuel-consumption characteristics of the engine itself. For a given level of technology, of all these variables, the most significant factor is the EPA test weight. This is the curb weight of the vehicle plus 300 lb. In a given model year, test weight correlates with vehicle size, which is loosely related for passenger cars to interior volume. The various EPA classification of vehicles are listed in Table 8.1.2, along

Table 8.1.2 Adjusted Fuel Economy and Production Share by Vehicle Classification and Type for MY 2014

	Cars		Car SUVs		All Cars		Truck SUVs		Pickups		Minivans/Vans		All Trucks	
Manufacturer	Adj FE (MPG)	Prod Share	Adj FE (MPG)	Prod Share	Adj FE (MPG)	Prod Share	Adj FE (MPG)	Prod Share	Adj FE (MPG)	Prod Share	Adj FE (MPG)	Prod Share	Adj FE (MPG)	Prod Share
GM	26.9	43.8%	24.4	13.4%	26.3	57.2%	19.8	16.6%	19.2	25.4%	15.4	0.8%	19.3	42.8%
Ford	28.4	41.3%	24.6	12.6%	27.4	53.9%	20.9	21.3%	17.5	23.1%	22.7	1.7%	19.1	46.1%
Toyota	31.4	59.4%	25.5	5.4%	30.8	64.8%	21.2	16.9%	17.3	12.6%	21.1	5.8%	19.6	35.2%
Fiat-Chrysler	23.5	22.2%	24.8	8.9%	23.8	31.0%	20.3	38.7%	18.1	15.8%	20.9	14.4%	19.6	69.0%
Honda	31.5	48.7%	26.3	11.3%	30.4	60.0%	24.1	29.4%	17.3	1.0%	23.2	9.6%	23.7	40.0%
Nissan	31.1	62.9%	25.4	7.6%	30.4	70.6%	22.3	22.1%	17.8	5.8%	23.4	1.4%	21.3	29.4%
Hyundai	29.8	73.8%	23.2	19.2%	28.1	93.0%	21.5	7.0%	—	—	—	—	21.5	7.0%
Kia	27.9	70.0%	22.3	24.6%	26.2	94.6%	22.6	2.8%	—	—	20.2	2.6%	21.4	5.4%
Subaru	28.2	23.4%	—	—	28.2	23.4%	27.5	76.6%	—	—	—	—	27.5	76.6%
BMW	27.5	78.4%	—	—	27.5	78.4%	22.9	21.6%	—	—	—	—	22.9	21.6%
Mercedes	24.9	71.0%	22.1	4.1%	24.8	75.1%	19.3	24.9%	—	—	—	—	19.3	24.9%
Mazda	32.9	56.5%	28.6	16.9%	31.8	73.4%	24.4	22.8%	—	—	24.8	3.8%	24.5	26.6%
Other	31.7	42.2%	26.5	10.0%	30.5	52.2%	20.9	47.8%	—	—	—	—	20.9	47.8%
All	28.7	49.2%	24.6	10.1%	27.9	59.3%	21.7	23.9%	18.0	12.4%	21.3	4.3%	20.4	40.7%

SOURCE: "Light-Duty Automotive Technology, Carbon Dioxide Emissions, and Fuel Economy Trends: 1975 through 2015," U.S. EPA, December 2015.

with their fleet-average test weights and fuel economies, for the 2014 model year.

After the CAFE regulation had been put in place, the driving public complained of its inability to achieve the fuel economy on the road that had been measured in the laboratory for regulatory purposes. Over-the-road fuel consumption is affected by ambient temperature, wind, hills, and driver behavior. Jack-rabbit starts, prolonged engine idling, short trips, excessive highway speed, carrying extra loads, not anticipating stops at traffic lights and stop signs, tire underinflation and inadequate maintenance all contribute to higher fuel consumption. To compensate for the difference between the EPA 55/45 fuel economy and the over-the-road fuel economy achieved by the average driver, EPA tested a fleet of cars in typical driving. It determined that on average, the public should expect to achieve 90 percent of MPG_U in city driving and 78 percent of MPG_H on the highway. Applying these factors to the urban and highway economies measured in the laboratory results in **adjusted fuel economy** for each of the two driving schedules. These can be combined to derive an adjusted 55/45 fuel economy, as listed for 2014 in Table 8.1.2. Typically, the adjusted 55/45 fuel economy is about 15 percent less than the measured laboratory fuel economy on which CAFE regulation is based. Figure 8.1.7 shows the production-weighted distribution of adjusted fuel economy by model year, for gasoline (including conventional hybrids) and diesel vehicles. It is important to note that the methodology used for calculating adjusted fuel economy values has changed over time. The fuel economy difference between the least efficient and most efficient car increased from about 20 mpg in MY 1975 to nearly 40 mpg in MY 2014. The overall fuel economy distribution for trucks is narrower than that for cars. Like cars, half of the trucks built each year have always been within a few mpg of each year's average fuel economy value.

TRANSMISSIONS

The torque developed by the engine is fed into a **transmission.** One function of the transmission is to control the ratio of engine speed to vehicle velocity, enabling improved fuel economy and/or performance. This may be done either with a **manual transmission,** which requires the driver to select that ratio with a gear-shift lever, or with **automatic transmission,** in which the transmission controls select that ratio unless overridden by the driver. Another function of the transmission is to provide a means for operating the vehicle in reverse.

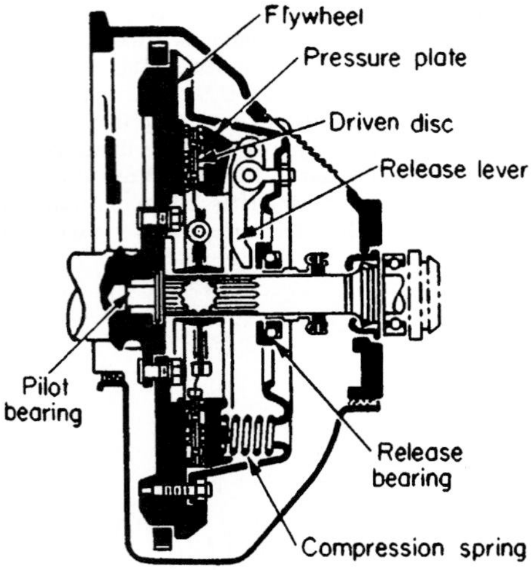

Figure 8.1.8 Single-plate dry-disk friction clutch.

Because an internal-combustion engine cannot sustain operation at zero rotational speed, a means must be provided that allows the engine to run at its idle speed while the driving wheels of the vehicle are stationary. With the manual transmission, this is accomplished with a **dry friction clutch** (Fig. 8.1.8). During normal driving, the pressure plate is held tightly against the engine flywheel by compression springs, forcing the flywheel and plate to rotate as one. At idle, depression of the clutch pedal by the driver counters the force of the springs, freeing the engine to idle while the pressure plate and driving wheels remain stationary. The transmission gearbox on the output side of the clutch changes its output/input speed ratio, typically in from three to six discrete gear steps. As the driver shifts the transmission from one gear ratio to the next, he disengages the clutch and then slips it back into engagement as the shift is completed.

With an automatic transmission, such a dry clutch is unnecessary at idle because a fluid element between the engine and the gearbox accommodates the required speed disparity between the engine and the driving wheels. However, the **wet clutch,** cooled by transmission fluid, is used in the gearbox to ease the transition from one gear set to another during shifting. Typically, the hydraulically operated **multiple-disk**

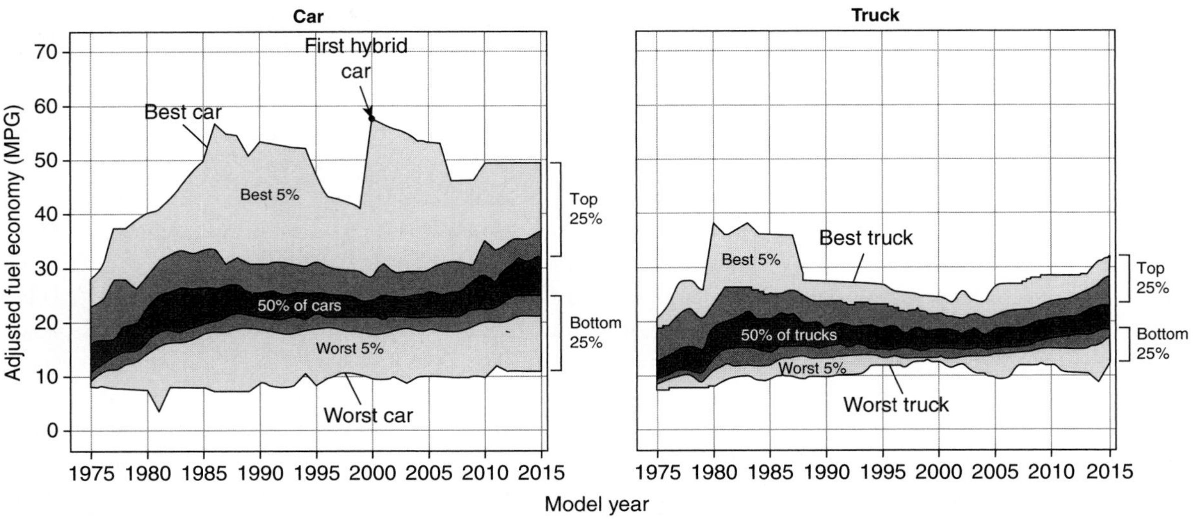

Figure 8.1.7 Adjusted fuel economy distribution by model year. (*U.S. EPA, December 2015.*)

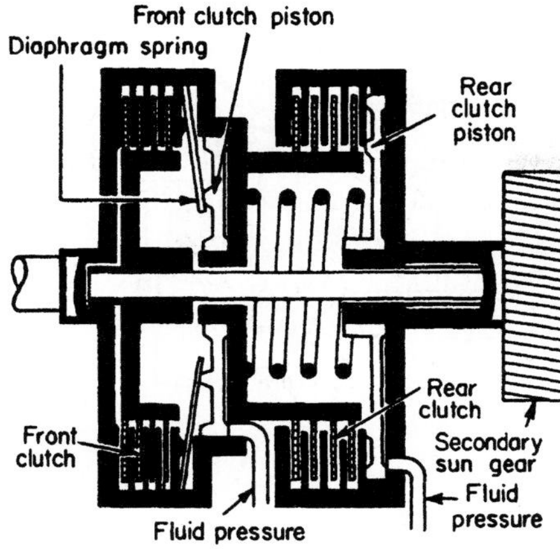

Figure 8.1.9 Schematic of two hydraulically operated multiple-disk clutches in an automatic transmission. (*Ford Motor Co.*)

clutch is used (Fig. 8.1.9). The **band clutch** is also used in automatic transmissions.

The fluid element used between the engine and gearbox in early automatic transmissions was the **fluid coupling** (Fig. 8.1.10). It assumes the shape of a split torus, filled with transmission fluid. Each semitorus is lined with radial blades. As the engine rotates the driving member, fluid is forced outward by centrifugal force. The angular momentum of

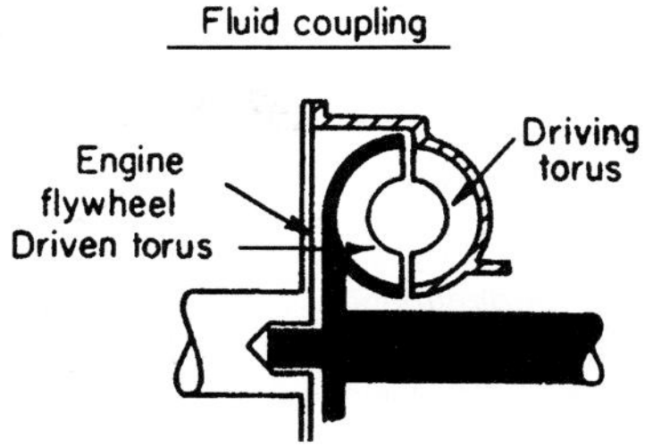

Figure 8.1.10 Fluid coupling.

this swirling flow leaving the periphery to the driving member is then removed by the slower rotating blades of the driven member as the fluid flows inward to reenter the driving member at its inner diameter. This transmits the torque on the driving shaft to the driven shaft with an input/output torque ratio of unity, independent of input/output speed ratio. That gives the fluid coupling the torque delivery characteristic of a slipping clutch. Transmitted torque being proportional to the square of input speed, very little output torque is developed when the engine is idling. This minimizes vehicle creep at stoplights. Because torque is transmitted from the fastest to the slowest member in either direction, the engine may be used as a brake during vehicle declaration. This characteristic also permits starting the engine by pushing the car.

In modern automatic transmissions, the usual fluid element is the three-component **torque converter,** equipped with a one-way clutch (Fig. 8.1.11*a*). As with the fluid coupling, transmission fluid is centrifuged outward by the engine-driven pump and returned inward through a turbine connected to the output shaft delivering torque to the gearbox. However, a stator is interposed between the turbine exit and the pump inlet. The shaft supporting the stator is equipped with **sprag clutch.** The blading in the torque converter is nonradial, being oriented to provide torque multiplication. When the output shaft is stalled, the sprag locks the stator in place, multiplying the output torque to from 2.0 to 2.7 times the input torque (Fig. 8.1.11*b*). As the vehicle accelerates and the output/input speed ratio rises, the torque ratio falls

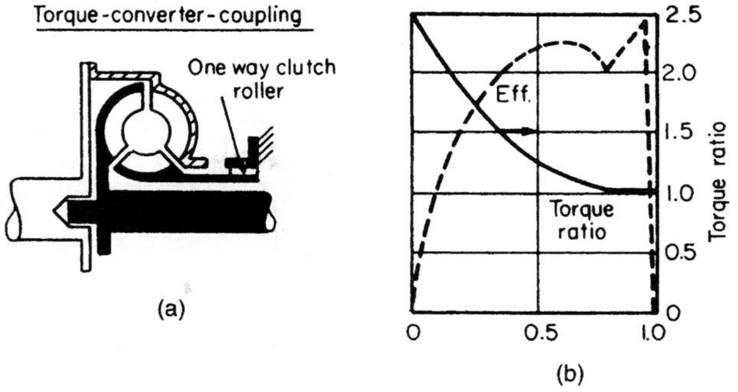

Figure 8.1.11 Torque converter coupling: (*a*) section; (*b*) characteristics.

toward 1.0. Beyond that speed ratio, known as the **coupling point,** the direction of the force imposed on the stator by the fluid reverses, the sprag allows the stator to free wheel, and the torque converter assumes the characteristic of a fluid coupling.

It is clear from Fig. 8.1.11*b* that multiplication of torque by the torque converter comes at the expense of efficiency. Therefore, most modern torque-converter transmissions incorporate a hydraulically operated **lockup clutch** capable of eliminating the speed differential between pump and turbine above a preset cruising speed, say 40 mi/h in high gear. In the lower gears, lockup is less likely to be used. When a significant increase in engine power is requested, the clutch is also disengaged to capitalize on the torque multiplication of the torque converter. During its engagement, the efficiency-enhancing elimination of torque-converter slip comes with a loss of the torsional damping quality of the fluid connection. Sometimes the loss of damping is countered by allowing modest slippage in the lockup clutch.

A cross section through the gearbox of a three-speed manual transmission is shown in Fig. 8.1.12. The input and output shafts share a common axis. Below, a parallel countershaft carries several gears of differing diameter, one of which is permanently meshed with a gear on the input shaft. Additional gears of differing diameter are splined to the output shaft. Driver operation of the gear-shift lever positions these splined gears along the output shaft, determining which pair of gears on the output shaft and the countershaft is engaged. In direct drive (high gear), the output shaft is coupled directly to the input shaft. Reverse is secured by interposing an idler gear between gears on the output shaft and the countershaft. Helical gears are favored to minimize noise production. During gear changes, a **synchromesh** device, acting as a friction clutch, brings the gears to be meshed approximately to the correct speed just before their engagement. Typically, transmission gear ratios follow an approximately geometric pattern. For example, in a four-speed gearbox, the input/output speed ratios might be 2.67 in first, 1.93 in second, 1.45 in third, and 1.0 in fourth, or direct, drive.

The gearbox for a typical automatic transmission uses planetary gear sets, any one set of which can supply one or two gear-reduction ratios, and reverse, by actuating multiple-disk or band clutches that lock various elements of the planetary system in Fig. 8.1.13. These clutches are actuated hydraulically according to a built-in control schedule that utilizes such input signals as vehicle velocity and engine throttle position. A schematic of such an automatic transmission, combining a torque converter and a compound planetary gearbox, is shown in Fig. 8.1.14.

Increasingly, **electronic transmission control** is taking over from totally hydraulic control. Although hydraulic actuation is retained, electronic modules control gear selection and modulate hydraulic pressure in accordance with torque flow, providing smoother shifts.

The influence of the transmission on the interaction between a vehicle and its roadbed is illustrated by Fig. 8.1.15, where the tractive force required to propel the vehicle is plotted against its road speed. The hyperbolas of constant horsepower superimposed on these coordinates are independent of engine characteristics. The family of curves rising

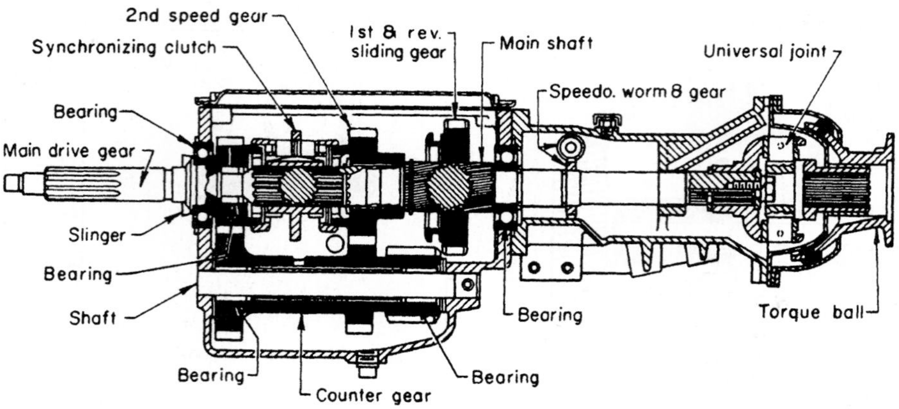

Figure 8.1.12 Three-speed synchromesh transmission. (*Buick.*)

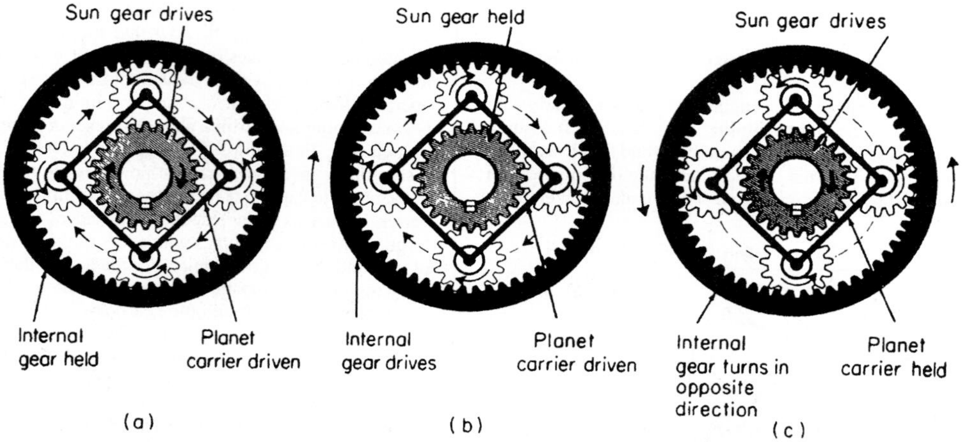

Figure 8.1.13 Planetary gear action: (*a*) Large speed reduction: ratio = 1 + (internal gear diam.)/(sun gear diam.) = 3.33 for example shown. (*b*) Small speed reduction: ratio = 1 + (sun gear diam.)/(internal gear diam.) = 1.428. (*c*) Reverse gear ratio = (internal gear diam.)/(sun gear diam.) = −2.33.

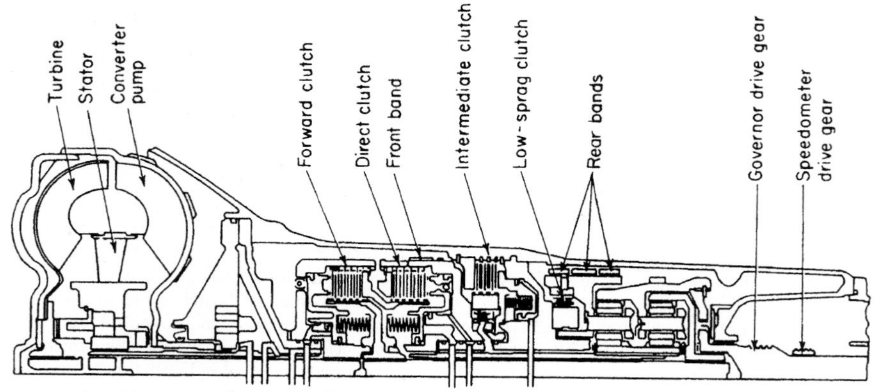

Figure 8.1.14 Three-element torque converter and planetary gear. (*General Motors Corp.*)

from the left axis at increasing rates traces the traction requirements of the vehicles as it travels at constant speed on grades of increasing steepness. Two curves representing the engine operating at full throttle in high gear are overlaid on this basic representation of vehicle characteristics. One is for a manual transmission followed by a differential with an input/output speed ratio of 3.6. The other is for the same engine with a torque-converter transmission using a differential with a speed ratio of 3.2. With either transmission, partial closing of the engine throttle allows operation below these two curves, but operation above them is not possible. The difference in tractive force between the engine

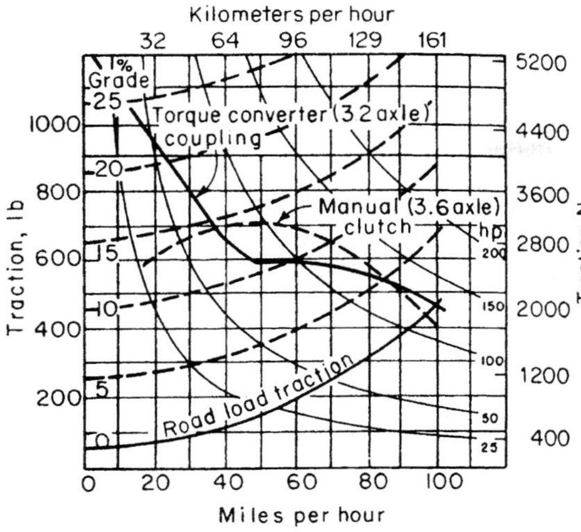

Figure 8.1.15 Comparative traction available in the performance of a fluid torque-converter coupling and a friction clutch.

curve and the road-load curve indicates the performance potential—the excess power available for acceleration or for hill climbing on a level road. The high-speed intersection of these two curves indicates that with either transmission, the vehicle can sustain a speed of 87 mi/h on a grade of nearly 5 percent. From 87 mi/h down to 37 mi/h, the manual transmission shows slightly greater performance potential. However, at 49 mi/h the torque converter reaches its coupling point, torque multiplication begins to occur, and below 37 mi/h its performance potential in high gear exceeds that of the manual transmission by an increasing margin. This performance disadvantage of the manual transmission may be overcome by downshifting. This transmission comparison illustrates why the driver is usually satisfied with an automatic transmission that has one fewer forward gear ratio than an equivalent manual transmission. Combining the benefits of both AT and MT, an Automated Manual Transmission (AMT) is actually an MT with an add-on control unit that automates the clutch and shift operations. An AMT provides the same user interface as an AT, that is, accelerator and brake pedals but no clutch. A vehicle equipped with AMT can be operated in the same way as a vehicle with AT, but the fuel economy will be improved compared with the AT. This fuel economy improvement is due to the gear shift schedule. An AMT is able to shift to a higher gear earlier while the AT has to stick with a mandatory gear schedule. At highway cruise the AMT will be no better than the MT. It may be slightly worse if parasitic losses are present to operate a hydraulic pump continuously.

An evolving transmission development is the **continuously variable transmission** (CVT), which changes the engine-to-vehicle speed ratio without employing finite steps. This greater flexibility promises improved fuel economy. Its most successful implementation in passenger cars to date varies transmission input/output speed ratio by using a steel V-belt composed of individual links held together by steel bands. This belt runs between variable-geometry sheaves mounted on parallel input and output shafts. Each of these two sheaves is split in the plane of rotation such that moving its two halves apart decreased its effective diameter, and vice versa. The control system uses hydraulic pressure to move the driving sheaves apart while concurrently moving the driven sheaves together in such a way that the constant length of the steel belt is accommodated. Thus the effective input/output diameter ratio, hence speed ratio, of the sheaves can be varied over a range of about 6:1. To accommodate the infinite speed ratio required when the vehicle is stopped and the engine is idling, and to manage vehicle acceleration from a standstill, this type of CVT typically includes a torque converter or a wet multidisk clutch.

DRIVELINE

The **driveline** delivers power from the transmission output to the driving wheels of the vehicle, whether they be the rear wheels or the front. An important element of the driveline is the **differential,** which in its basic form delivers equal torque to both left and right driving wheels while allowing one wheel to rotate faster than the other, as is essential during turning. Figure 8.1.16 is a cross section through the differential of a rear-wheel driven vehicle as viewed in the longitudinal plane. It

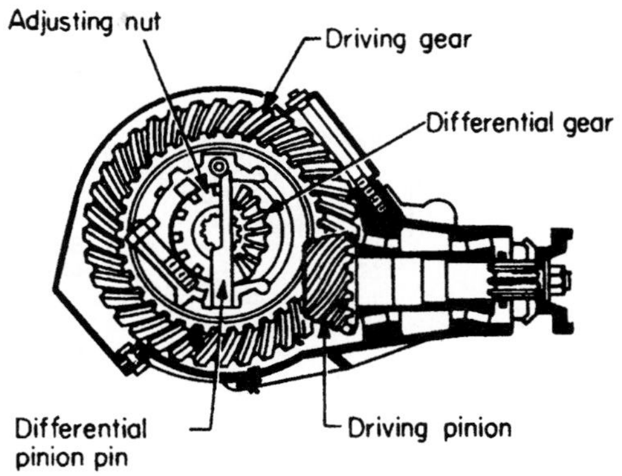

Figure 8.1.16 Rear-axle hypoid gearing.

employs hypoid gearing, which allows the axis of the driving pinion to lie below the centerline of the differential gear. This facilitates lowering the intruding hump in the floor of a passenger car, which is needed to accommodate the longitudinal drive shaft. Figure 8.1.17 is a cross-sectional view of the differential from above, showing its input connection to the drive shaft, and its output connection to one of the driving wheels. The driving pinion delivers transmission-output torque to a ring gear, which is integral with a differential carrier and rotates around a transverse axis. A pair of beveled pinion gears is mounted on the pinion pin, which serves as a pinion shaft and is fixed to the carrier. The pinion gears mesh with a pair of beveled side gears splined to the two rear axles. In straight-ahead driving, the pinion gears do not rotate on the pinion gear shaft. In a turn, they rotate in opposite directions, thus accommodating the difference in speeds between the two rear wheels.

If one of the driving wheels slips on ice, the torque transmitted to that wheel falls essentially to zero. In the basic differential, the torque on the opposite wheel then falls to the same level and the tractive force is insufficient to move the vehicle. Various **limited-slip differentials** have been devised to counter this problem. In one type, disk clutches are introduced into the differential. When one wheel begins to lose traction, energizing the clutches transmits power to the wheel with the best traction.

Figure 8.1.17 illustrates a **semifloating rear axle.** This axle is supported in bearings at each of its ends. The vehicle weight supported by the rear wheel is imposed on the axle shaft and transmitted to the axle housing through the outboard wheel bearing between the shaft and that housing. In contrast, most commercial vehicles use a **full-floating rear axle.** With this arrangement the load on the rear wheel is transmitted directly to the axle housing through a bearing between it and the wheel itself. Thus freed of vehicle support, the full-floating axle can be withdrawn from the axle housing without disturbing the wheel. In both types of rear axle, of course, the axle shaft transmits the driving torque to the wheels.

With **rear-wheel drive,** a **driveshaft** (or **propeller shaft**) connects the transmission output to the differential input. Misalignment between the latter two shafts is accommodated through a pair of **universal joints** (U joints) of the Cardan (or Hooke) type, one located at either end of the driveshaft. With this type of universal joint, the driving and the driven shaft each terminate in a slingshot-like fork.

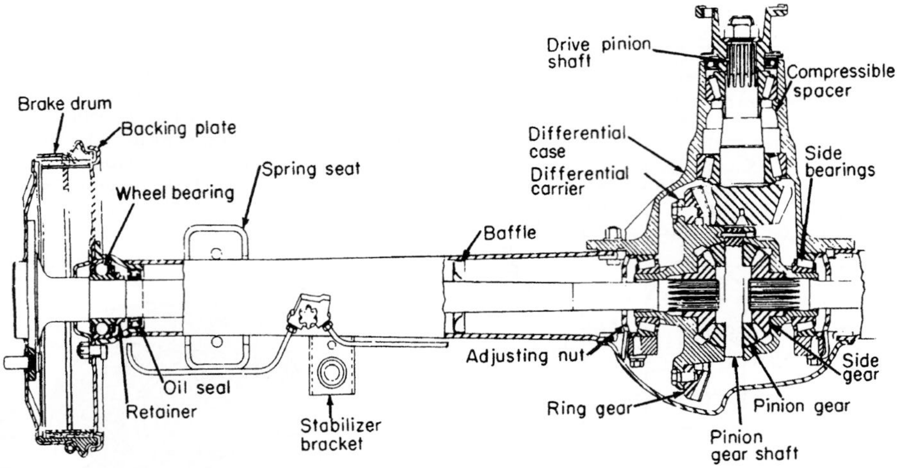

Figure 8.1.17 Rear axle. (*Oldsmobile.*)

The planes of the forks are oriented perpendicular to one another. Each fork is pinned to the perpendicular arms of a cruciform connector in such a way that rotary motion can be transmitted from the driving to the driven shaft, even when they are misaligned. Vehicles with a long wheel base may use a two-piece driveshaft. Then a third U joint is required, and an additional bearing is placed near that third U joint to support the driveshaft.

With **front-wheel drive,** the traditional driveshaft becomes unnecessary as the engine is coupled directly to a **transaxle,** which combines the transmission and the differential in a single unit that is typically housed in a single die-cast aluminum housing. A cutaway view of a transaxle with an automatic transmission appears in Fig. 8.1.18. Output from a transversely mounted engine is delivered to the torque converter, which shares a common centerline with the gearbox and its various gears and clutches. The gearbox output is delivered to the transfer-shaft gear, which rotates about a parallel axis. Its shaft contains the parking sprag. The gear at the opposite end of this shaft feeds torque into the differential, which operates around a third rotational axis, as seen at the bottom of Fig. 8.1.19. The axle shafts driving

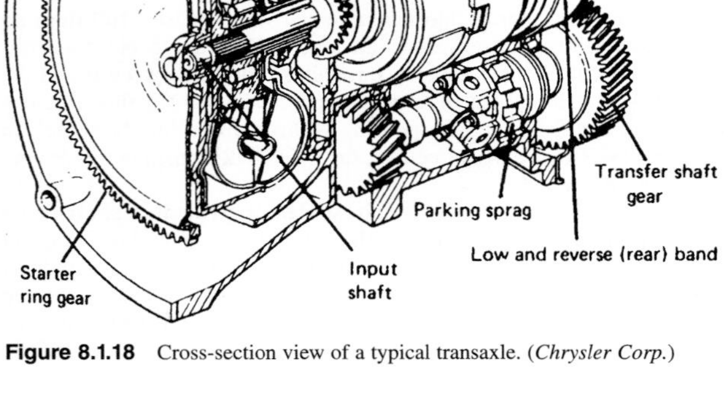

Figure 8.1.18 Cross-section view of a typical transaxle. (*Chrysler Corp.*)

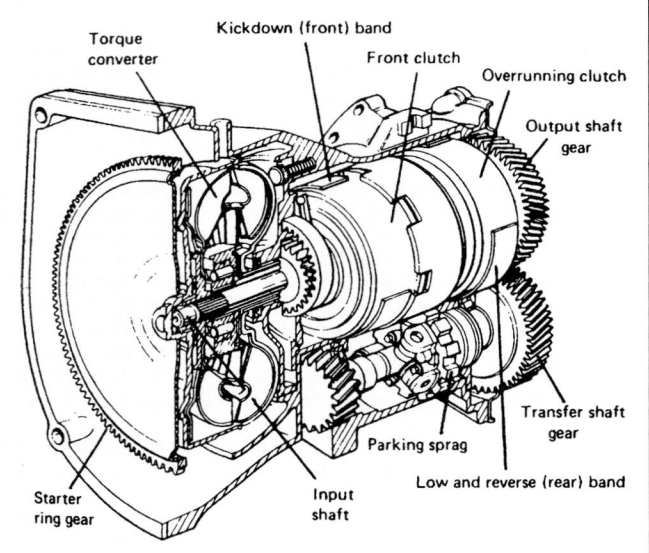

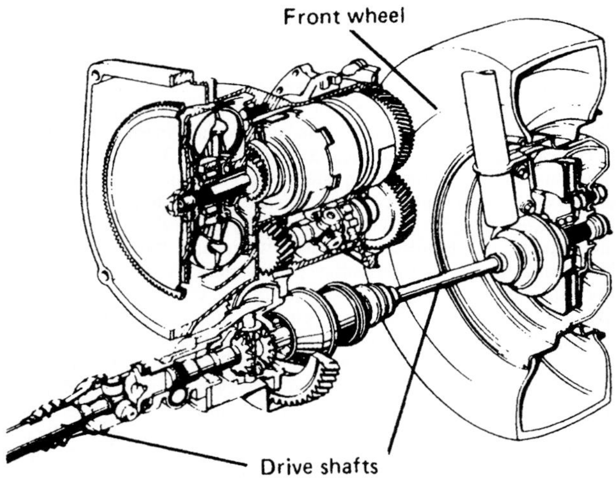

Figure 8.1.19 Automatic transaxle used on a front-wheel-drive car. (*Chrysler Corp.*)

the front wheels may be of unequal length. A **constant-velocity (CV) joint** is located at each end of each front-drive axle. During driving, the front axles may be required to flex through angles up to 20° while transmitting substantial torque—a duty too severe for the type of U joint commonly used on the driveshaft of a rear-wheel-drive vehicle. Two types of CV joints are in common use. In the **Rzeppa joint,** the input and output shafts are connected through a balls-in-grooves arrangement. In the **tripod joint,** the connection is through a rollers-in-grooves arrangement.

To enhance operation on roads with a low-friction-coefficient surface, various means for delivering torque to all four wheels have been devised. In the **four-wheel drive** system, two-wheel drive is converted to four-wheel drive by making manual adjustments to the driveline, normally with the vehicle at rest. In **all-wheel drive** systems, torque is always delivered to all four wheels, by a system capable of distributing power to the front and rear wheels at varying optimal ratios according to driving conditions and the critical limits of vehicle dynamics. In vehicles capable of driving through all four wheels, a **transfer case** receives the transmission output torque and distributes it through drive shafts and differentials to the front and rear axles.

SUSPENSIONS

The suspension system improves ride quality and handling in driving on irregular roadbeds and during maneuvers like turning and braking. The vehicle body is supported on the front and rear **springs,** which are important elements in the suspension system. Three types used are the **coil spring,** the **leaf spring,** and the **torsion bar.**

Also important are the **shock absorbers.** These piston-in-tube devices dampen spring oscillation, road shocks, and vibration. As the hydraulic shock-absorber strut telescopes, it forces fluid through a restrictive orifice, thus damping its response to deflections imposed by the springs.

Especially in the front suspension, most vehicles incorporate an **antiroll bar** (sway bar, stabilizer). This is a U-shaped bar mounted transversely across the vehicle, each end attached to an axle. During high-speed turns, torsion developed in the bar reduces the tendency of the vehicle to lean to one side, thus improving handling.

A wide variety of suspension systems have seen service.

Front Suspension All passenger cars use independent front-wheel suspension, which allows one wheel to move vertically without disturbing the opposite wheel. Rear-wheel drive cars typically use the **short-and-long-arm (SLA) system** illustrated in Fig. 8.1.20. The **steering knuckle** is held between the outboard ends of the upper and lower **control arms** (wishbones) by spherical joints. The upper control arm is always shorter than the lower. The load on these control arms is generally taken by coil springs acting on the lower arm, or by torsion-bar springs mounted longitudinally.

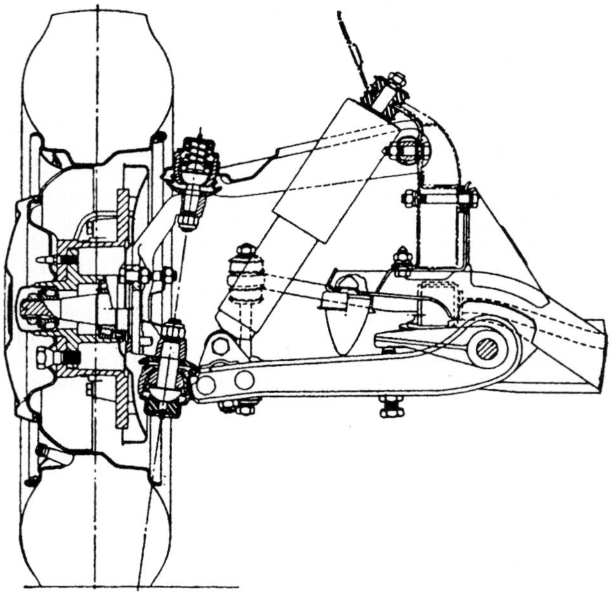

Figure 8.1.20 Front-wheel suspension for rear-wheel-drive car.

Figure 8.1.21 shows a representative application of the **spring strut (MacPherson) system** for the front suspension of front-wheel drive cars. The lower end of the shock absorber (strut) is mounted within the coil spring and connected to the steering knuckle. The upper end is anchored to the body structure.

Rear Suspensions The trailing-arm rear suspension illustrated in Fig. 8.1.22 is used on many passenger cars. Each longitudinal trailing arm pivots at its forward end. Motion of its opposite end is constrained by a coil spring. One or more links may be added to enhance stability.

Using leaf springs in place of coil springs is common practice on trucks. The Hotchkiss drive, used with a solid rear axle, employs longitudinal leaf springs. They are connected to the chassis at their front and rear ends, with the rear axle attached near their midpoints.

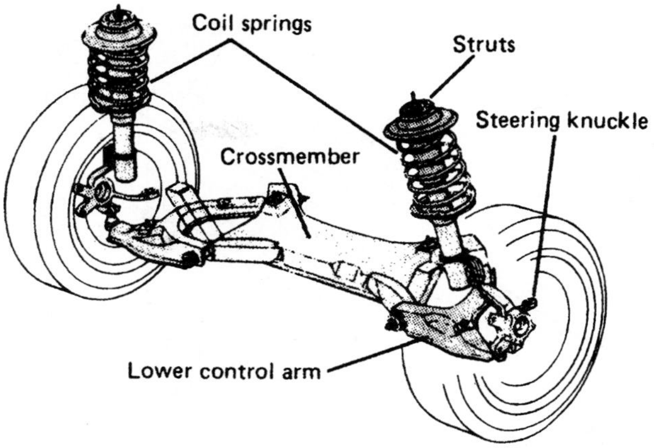

Figure 8.1.21 Typical MacPherson strut, coil spring front suspension for front-wheel-drive car. (*Chrysler Corp.*)

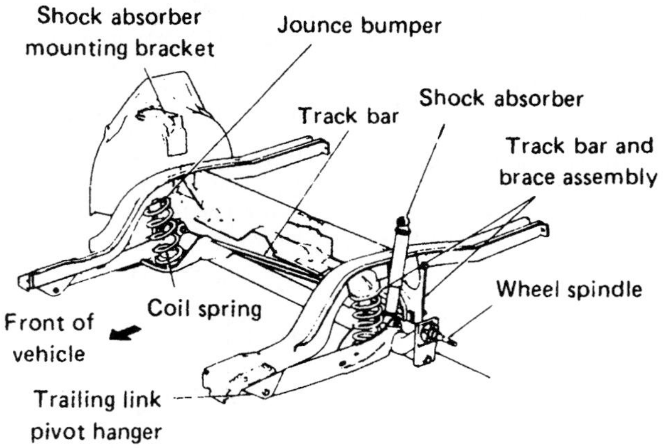

Figure 8.1.22 Trailing-arm type rear suspension with coil springs, used on some front-wheel-drive cars. (*Chrysler Corp.*)

Active Suspension A variety of improvements to the traditional passive suspension system have been introduced. Generally they require some form of input energy. The air suspension widely used in heavy-duty trucks, for example, requires an air compressor. To an increasing degree, advanced systems employ sensors and electronic controls. Some of the improvements available with active suspension are improved ride quality (adaptation to road roughness), control of vehicle height (low for reduced aerodynamic drag on highways; high for off-road driving), and vehicle leveling (to correct change in pitch when the trunk is heavily loaded; avoid pitching during braking and heavy accelerations).

STEERING

When turning a corner, each wheel traces an arc of different radius, as illustrated in Fig. 8.1.23. To avoid tire scrubbing when turning a corner of radius R, the point of intersection M for the projected front-wheel axes must fall in a vertical plane through the center of the rear axle. This requires each front wheel to turn through a different angle. For small turning angles of the vehicle, α and β can be shown to be simple functions of W, L, and R. A steering geometry conforming to Fig. 8.1.23 is known as **Ackerman geometry,** and L/R defines the **Ackerman angle.** The difference in front-wheel angles, $\beta - \alpha$, varies with turning radius R. Having the α and β corresponding to Ackerman geometry improves handling in low-speed driving, as in parking lots.

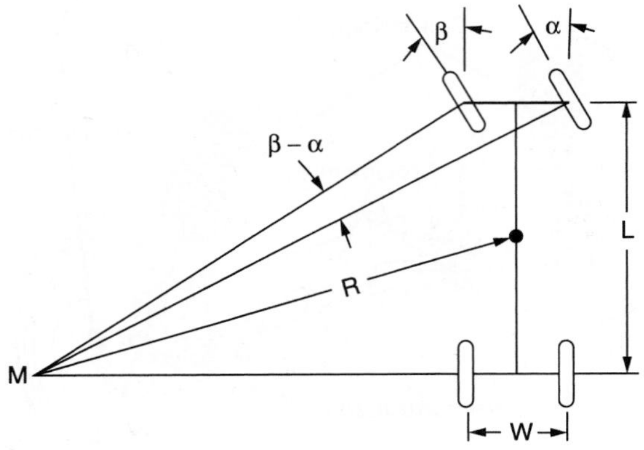

Figure 8.1.23 Ackerman steering geometry.

At higher speeds, dynamic effects become significant. In a turn, the vehicle may encounter either **understeer** or **oversteer.** With understeer, the vehicle fails to turn as sharply as the position of the steering wheel suggests it should. With oversteer, the vehicle turns more sharply than intended, and in extreme cases may spin around. These tendencies are amplified by increased speed and are affected by many variables, for example, tire pressure, weight distribution, design of the suspension system, and the application of brakes.

The physical implementation of a front-wheel steering system is shown schematically from above in Fig. 8.1.24. The front wheels are steered around the axes of their respective **kingpins.** For improved handling, the axes of the two kingpins are not vertical. In the transverse plane they lean inward, and in the longitudinal plane they lean rearward. The **steering arms** are the lever arms for turning the front wheels. Each **tie rod** is pinned at one end to a steering arm and at its opposite end to some type of motion transformer, the purpose of which is to convert the rotational input of the **steering column** into translational motion at the inner ends of the tie rods.

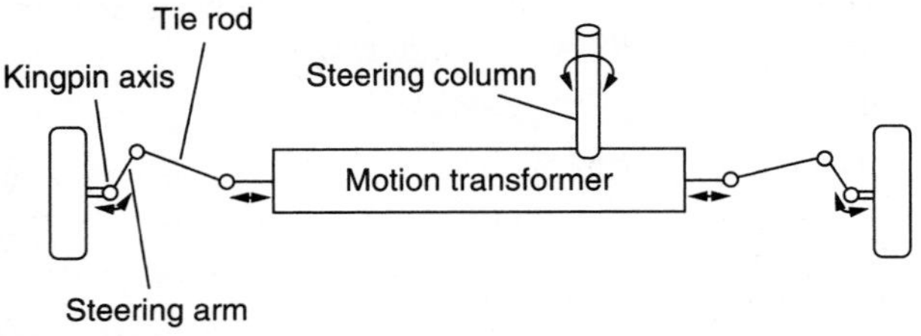

Figure 8.1.24 Schematic of front-wheel steering system.

A common motion transformer is the **rack and pinion,** in which a **pinion gear** attached to the steering column engages a geared bar, or **rack,** causing it to slide to the left or right as the **steering wheel** is turned. The ends of the rack are attached to the inner ends of the tie rods. In the variation of the rack-and-pinion illustrated in Fig. 8.1.25*a*, the rotation of the pinion on the end of the steering shaft causes a rotatable **worm gear** to control the tie-rod positions as it moves left and right like the rack.

With the **rotatable ball** system of Fig. 8.1.25*b*, a worm is mounted on the end of the steering column. Its spiral grooves contain ball bearings. The outer surfaces of the balls rotate in spiral grooves lining the inside of a **ball nut.** The outer surface of the ball nut contains gear teeth that mesh with a **gear sector** on the **pitman arm.** The end of pitman arm is connected to the tie rods through a series of links.

With **variable-ratio steering,** the response of the front wheels to steering-wheel rotation is increased as the steering wheel approaches the extremes of its travel. This is especially helpful in curbside parallel parking. Variable-ratio steering can be achieved by nonuniformly spacing the worm grooves in Fig. 8.1.25*a*, or by varying the lengths of the teeth on the sector of the pitman arm in Fig. 8.1.25*b*.

Most modern U.S. cars have **power-assisted steering.** A pressurized hydraulic fluid is used to augment the steering effort input by the driver, but if the engine is off or if there is a failure in the power-steering system, the driver retains steering control, albeit with greater effort.

The power-steering circuit contains an engine-driven hydraulic **pump,** a fluid reservoir, a hose assembly, return lines, and a **control valve.** One of a variety of pumps that has been used is the rotary-vane pump of Fig. 8.1.26. As the rotor is turned by a belt drive from the engine, the radially slidable vanes are held against the inner surface of the eccentric case by centrifugal force, springs, and/or hydraulic pressure. Not shown are the inlet and exit ports, and a pressure-relief valve.

A balanced-spool control valve is represented in Fig. 8.1.27. In straight-ahead driving, centering springs hold the spool in the neutral position. In the position illustrated, however, the driver has turned the steering wheel to position the spool valve at the right extremity of its travel. That directs all of the pump discharge flow to the left side of a piston, moving the piston rod to the right. The piston illustrated can, for example, be

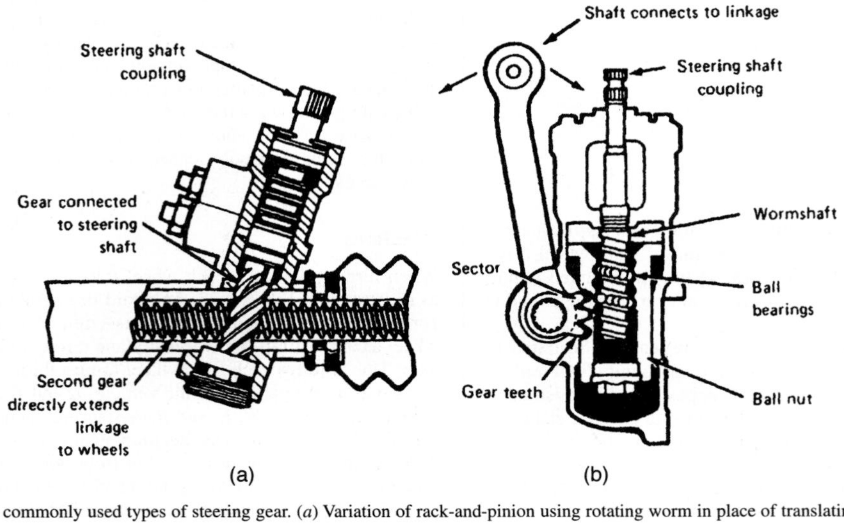

Figure 8.1.25 Two commonly used types of steering gear. (*a*) Variation of rack-and-pinion using rotating worm in place of translating rack; (*b*) recirculating ball.

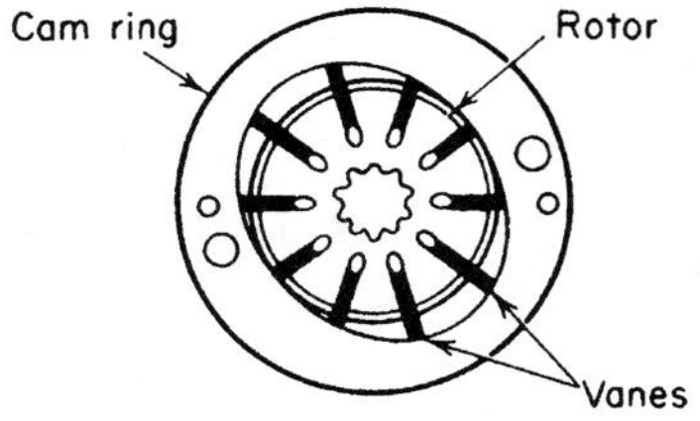

Figure 8.1.26 Rotary power-steering pump. (*General Motors, Saginaw.*)

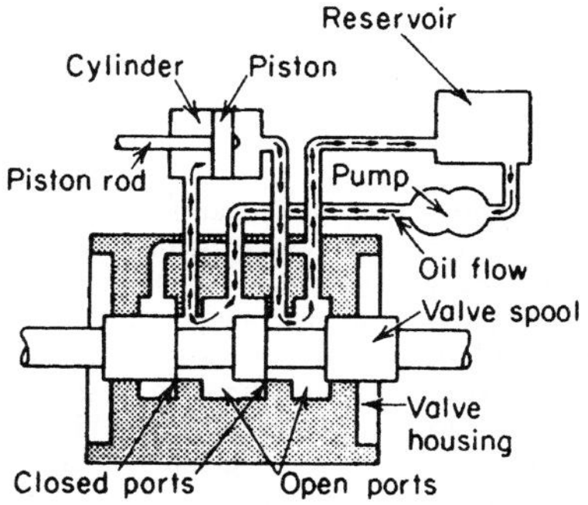

Figure 8.1.27 Control valve positioned for full-turn power steering assistance using maximum pump pressure.

incorporated into the rack of a rack-and-pinion steering system to add to the turning effort put into the steering wheel by the driver.

Some cars now use electric power steering (EPS). Both the need for hydraulic fluid and the equipment pictured in Fig. 8.1.27 are eliminated. They are replaced by a variable-speed electric motor, and electronic sensors that match the appropriate degree of assistance to driver demand and road resistance. Compared to hydraulic power steering, electric power steering effects a small improvement in fuel economy by reducing the parasitic load on the engine.

BRAKING SYSTEMS

Brakes The moving automobile is stopped with hydraulic **service brakes** on all four wheels. The service brakes are engaged by depressing and holding down a self-returning foot pedal. The mechanical **parking brake**, used to prevent the vehicle from rolling when parked, acts on the rear wheels. It is engaged with an auxiliary foot pedal or a hand pull and is held by a ratchet until released. **Drum brakes**, once used extensively on all four wheels, are now used primarily on rear wheels. Typically, **disk brakes** are used on the front wheels, but sometimes on the rear wheels.

A **Bendix dual-servo** drum brake is illustrated in Fig. 8.1.28*c*. It is **self-energizing** because the rotating drum tries to drag the brake lining and shoe counter-clockwise, along with it. Because the anchor pin prevents this for the secondary shoe, and the adjusting screw prevents it for the primary shoe, both are called **leading shoes**. This causes both shoes to push more strongly against the drum, increasing the friction force. This self-energization is lost when the wheel turns in the reverse direction. With this brake configuration, typically a linkage turns a notched wheel on the adjusting screw when the brake is applied with the car moving in reverse. This automatically adjusts for lining wear.

In the arrangement of Fig. 8.1.28*a*, the anchor pin has been moved 180°. The leading shoe continues to be self-energized, but that quality is lost on the **trailing** (following) **shoe**. In this arrangement the leading shoe provides most of the braking torque and experiences most of the wear when traveling forward. In reverse, however, the trailing shoe in forward travel becomes the leading shoe and provides most of the braking torque. This arrangement is used on the rear wheels of some front-wheel drive cars.

The **Caliper disk brake** is illustrated in Fig. 8.1.29. The disk rotates with the wheel. The fixed caliper straddles the disk. Within the caliper, the pistons in opposing hydraulic cylinders press the attached pads against opposite sides of the disk when actuated.

As friction heats the brake lining and the surface on which it rubs during stopping, the coefficient of friction often falls. This undesirable characteristic causes what is known as **brake fade**. Because brakes are normally applied only intermittently, the time available for cool down between successive brake applications keeps brake fade at a tolerable level. However, closely spaced panic stops from high speeds can cause objectionable fade. The "disk" illustrated in Fig. 8.1.29 often consists of two parallel disks that are joined together by radial vanes. The ability of air to access the nonfriction side of each of these disks, and the air pumped between the disks by the vanes when the wheel rotates, help to avoid brake overheating. Brake fade is more probable with drum brakes because the more enclosed nature of their construction makes brake cooling more difficult.

Many modern disk brakes use a single-cylinder caliper. One cylinder and its actuating piston and attached pad are retained on the inboard side. However, the outboard cylinder is eliminated, the pad on that side instead being fastened directly to the inner surface of the caliper. Instead of being fixed, the caliper is allowed to slide inboard

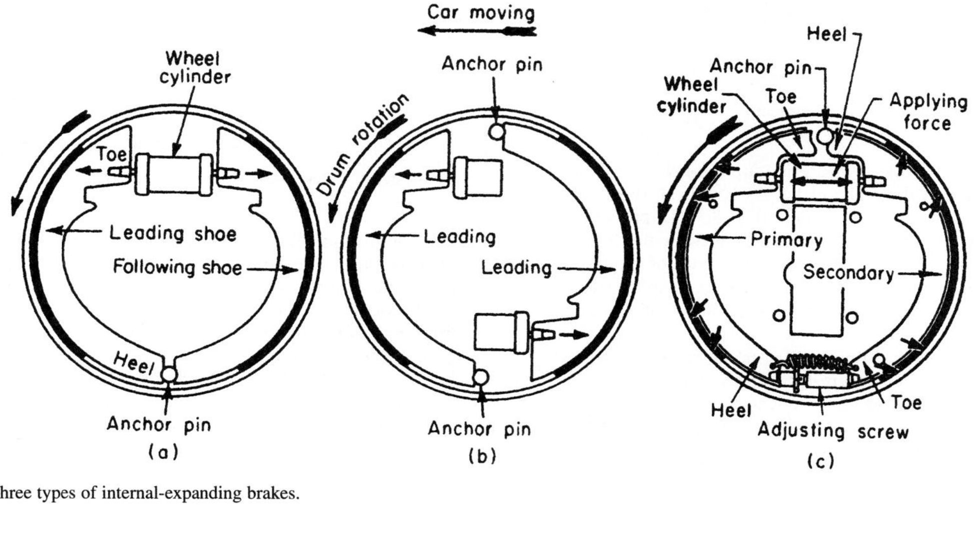

Figure 8.1.28 Three types of internal-expanding brakes.

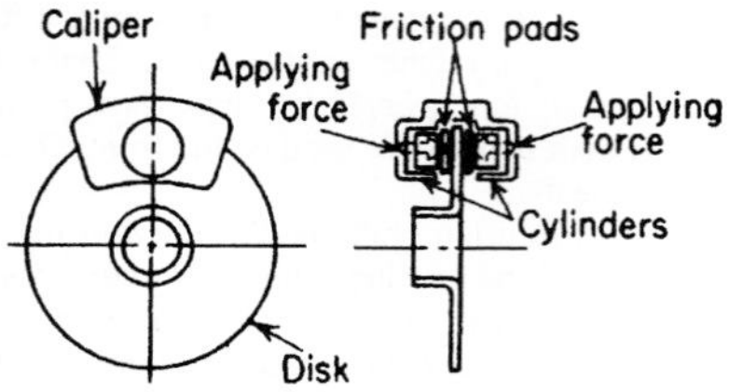

Figure 8.1.29 Schematic of caliper disk brake.

on mounting bolts when the single piston is actuated. Transverse movement of the caliper on its mounting bolts then causes each pad to rub on its adjacent disk surface.

Because disk brakes are not self-energizing, and the rubbing area of the pads is less than the rubbing area of the shoes on an equivalent drum brake, a higher hydraulic pressure is required with disk brakes than with drums. If disk brakes are used on the front wheels and drums on the rear, then provisions are made in the upstream hydraulic system to provide that difference.

Cast iron is the preferred material for the rubbing surface on both rotors and drums. The **brake lining** on pad or shoe may contain high-temperature resins, metal powder, solid lubricant, abrasives, and fibers of metal, plastic, and ceramic. Friction coefficients from 0.35 to 0.50 are typical with these combinations. The asbestos once common in brake linings has been abandoned because of the adverse health effect of asbestos wear particles.

Hydraulics When the brake pedal is depressed, it moves pistons in the master cylinder, as illustrated in Fig. 8.1.30. This forces pressurized brake fluid through lines to the wheel cylinders for brake application.

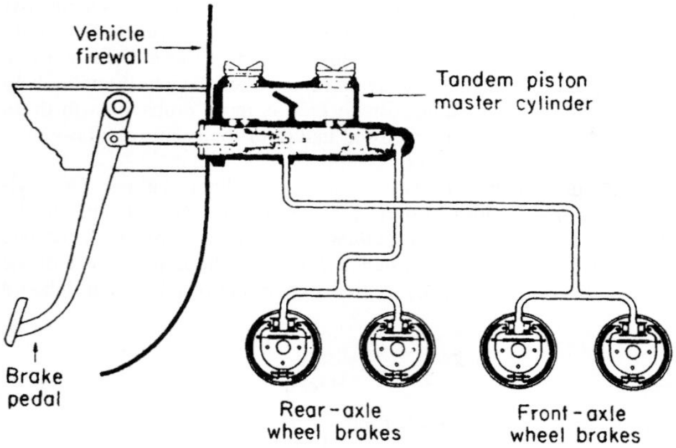

Figure 8.1.30 "Split" hydraulic brake system.

The magnitude of the fluid pressure is proportional to the magnitude of the force applied to the brake pedal. As is typical, this dual master cylinder contains two pistons in tandem (Fig. 8.1.31), providing a split system. One operates the rear brakes and the other the front brakes, providing the ability to brake the vehicle even if one of the two systems should fail. Late model cars often use a diagonal-split arrangement, where each piston in the dual master cylinder serves diagonally opposite wheels.

When the brake pedal is released, a spring in the master cylinder returns the pistons to their original positions. This uncovers ports in the master cylinder, allowing communication with the brake-fluid reservoirs on the master cylinder and relieving the pressure in the wheel cylinders. A check valve maintains a minimum pressure in the system to prevent the entrance of air. Air in the system makes the brakes feel spongy, and, if it gains entrance, must be bled from the system.

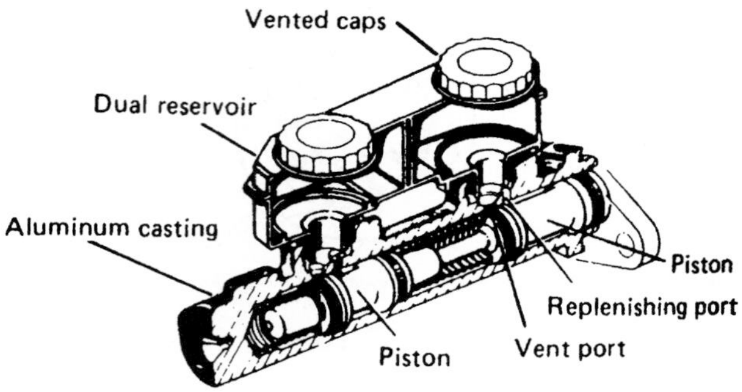

Figure 8.1.31 Typical design of a dual master cylinder for a split brake system.

A **combination valve** is usually included in vehicles using disks in the front and drums in the rear. Among its functions, the **proportioning valve** regulates pressure individually to the front and rear wheels to compensate for the different pressure requirements of disks and drums, and the **pressure differential valve** checks to ensure that both sides of a diagonally split system experience the same pressures.

The conventional cast-iron master cylinder with integral reservoir has been largely replaced by the aluminum master cylinder with translucent plastic reservoirs for visual inspection of brake-fluid level. The specially formulated brake fluid must be free of petroleum-based oil, which would cause swelling of rubber parts in the braking system. Double-walled steel lines are used to distribute the brake fluid. Near the wheels, these rigid lines are replaced by flexible hydraulic hoses to accommodate wheel movement.

Power Brakes Power brakes relieve the driver of much physical effort in retarding or stopping a car. Several types have been used, but on passenger cars, the **vacuum-suspended type** illustrated in Fig. 8.1.32 is the most common. The power section contains a large diaphragm. In

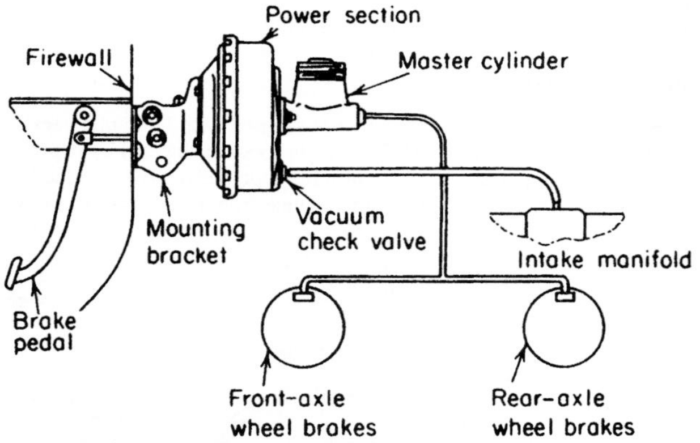

Figure 8.1.32 Power-assisted brake installation.

the unapplied condition, the chambers on each side of the diaphragm are exposed to intake-manifold vacuum. When the brake pedal is depressed, first the vacuum passage is closed and then, with further depression, air is admitted to one side of the diaphragm. The pressure difference across the diaphragm, acting over its area, creates an axial force that is applied to the piston in the master cylinder. This augments the force provided directly by the driver's foot. Without requiring an additional reservoir, the vacuum captured allows at least one stop after the engine operation has terminated.

Another type of power brake used a **hydraulic booster**. The accumulator, a reservoir of pressurized fluid provided by the power-steering pump, provides the force that moves the piston in the master cylinder.

Heavy-duty tractor-trailer combinations require braking forces far in excess of what can be supplied from a simple foot pedal. **Air brakes** are common in such applications. Air supplied from an engine-driven air compressor is stored in a high-pressure tank, from which the air is withdrawn for brake augmentation as needed.

Retarders Using traditional wheel brakes to control the road speed of a heavily laden tractor-trailer rig on long downhill runs promotes brake fade and accelerates brake-lining wear. Several types of retarders are available that decelerate the vehicle independent of wheel brakes.

With the **compression-release braking system** (often called the **Jake brake**), fuel injection in the diesel engine is terminated and engine valve timing is temporarily changed so that the exhaust valve opens near the end of the compression stroke. This releases the air compressed during compression directly into the exhaust system and imposes the negative work of compression on the crankshaft. A disadvantage of this system is that the sudden intermittent release of high-pressure air from the cylinder creates considerable noise.

The **hydrodynamic retarder** consists essentially of a centrifugal fluid pump that is run by the driveshaft. The high tangential velocity of the discharge fluids is removed by stator vanes as the fluid is returned to the pump inlet. Because such a pump consumes little torque at low rotational speeds, a booster retarder that is geared to rotate at a higher speed may be added in parallel. Since the braking energy generated appears as heat in the recirculating fluid, adequate cooling must be provided for it.

The **electrodynamic retarder** resembles an eddy-current dynamometer. Voltage applied from the battery or the alternator to the field coils controls braking torque. The electrodynamic retarder offers greater braking torque at low speeds than the hydrodynamic retarder. However, its braking ability falls off significantly as it is heated by the energy absorbed, and heat rejection is problematic.

Antilock Brakes At a steady forward speed, vehicle velocity equals product of the rolling radius of the tire and its angular velocity. When the brakes are applied, the vehicle velocity exceeds that product by a difference known as **slip**. As slip in any wheel increases, its braking force rises to a peak, then slowly decreases until the wheel locks up and the tire merely slides over the road surface. At lockup, the brakes have not only lost much of their effectiveness, but steering control is also sacrificed. The **antilock braking system** (ABS) incorporates wheel-speed sensors, an electronic control unit (ECU), and solenoids to release and reapply pressure to each brake. When the ECU determines from a wheel-speed signal that lockup is imminent, it signals the corresponding solenoid to release hydraulic pressure in that brake line. Once that wheel has accelerated again, the solenoid reapplies its brake-line pressure. This sequence is repeated in each wheel at a high frequency, thus avoiding lockup and generally shortening stopping distance during severe braking or on slippery roads.

Traction Control The **traction control system** (TCS) helps maintain traction and directional stability, especially on slippery roads. It uses many of the same components as ABS. If one driven wheel spins, the ECU applies the brake on that wheel until traction is regained. If both driven wheels spin, the ECU temporarily overrides the command signal from the accelerator pedal to reduce engine output torque.

Electronic Stability Control The step beyond ABS and TCS is electronic stability control, which helps the driver to maintain steering during severe vehicle maneuvers. By adding sensor measurements of steering-wheel angle and vehicle yaw rate to the ECU input, it is possible to reduce the incidence of inadvertent departure of the vehicle from the road, or of vehicle rollover. This is accomplished by controlling the wheel brakes as with TCS, and in addition, by temporarily reducing engine torque.

Stopping Distance The service brakes are the most important element in arresting a moving vehicle, but they derive some help from other sources. Aerodynamic drag provides a retarding force. Because it varies with the square of speed, it is most influential at highway speeds. The engine, when it is driven by the inertia of the vehicle during braking, and transmission and driveline losses, all contribute resistance to forward motion. On the other hand, during forward motion the inertia of these components, as well as that of the wheels, all oppose the action of the brakes. The braking force from the motoring engine can be especially significant if it is joined to the driven wheels by a manual transmission. The force can be amplified by downshifting that transmission to increase the speed of the crankshaft. With the conventional automatic transmission, however, engine braking is minimal because the torque converter is relatively ineffective in transmitting torque in the reverse direction. A small retarding force is also derived from the rolling resistance of the tires.

In braking, the sequence of events involves, successively, (1) the driver reaction time t_r, during which the driver responds to an external signal by engaging the brake pedal, (2) a response time from pedal engagement to full pedal depression, (3) a buildup time, during which brake line pressure increases to develop a substantial braking force, and (4) a prolonged active braking time. In estimating minimum stopping distance from initial vehicle velocity V, the vehicle may be assumed to move at its original speed during the reaction time, response time, and half of the buildup time. The sum of the latter two is the delay time t_d. During the second half of the buildup time and the active braking time, the braking force depends on μ, the average coefficient of friction between tire and roadbed. Then the stopping distance is $S = V(t_r + t_d) + V^2/2\mu g$, where g is the gravitational constant. The reaction time typically consumes 1 ± 0.7 s. For brakes in good condition, the delay time may range from 0.35 s for passenger cars to 0.55 s for large vehicles. The static friction coefficient depends on speed and the condition of tires and road. For new tires on dry paved roads, it is in the range of 0.75 to 0.85, increasing about 20 percent with wear. On moderately wet pavement the coefficient for new tires may fall to the 0.55 to 0.65 range. On wet pavement, the coefficient falls even more, especially at high speeds. The sliding friction coefficient, applicable when the wheels are locked and skid over the roadbed, is less than the static coefficient.

WHEELS AND TIRES

The total weight of a vehicle is the sum of the **sprung weight** supported by the tires and suspension system, and the **unsprung weight**. Wheels and tires are primary contributors to unsprung weight. For a given vehicle, the lower the unsprung weight, the better the acceleration performance, and the ride and handling qualities on rough roads. Because of cost considerations, most automotive wheels are fabricated from sheet steel. Special lower weight wheels have been forged or cast in aluminum, however, and even made of magnesium.

Maintenance of proper front-wheel alignment is important for optimum vehicle performance and handling. Normally, proper **camber, caster**, and **toe-in** are specified by the manufacturer.

Camber is the tilt angle of the wheel plane with respect to the longitudinal plane of the vehicle. When the plane of the wheel is perpendicular to the ground, the camber angle is 0°. During operation, that angle is affected by the changing geometry of the suspension system as the vehicle is loaded or traverses a bump in the road, or as the body rolls during cornering. Taking such factors into account, the manufacturer specifies the proper camber angle for each vehicle design when the vehicle is at rest. With positive camber, the distance between wheels is greater at their tops than between their bottoms, where they contact the roadbed. Normally, a zero to slightly negative camber is specified.

The **caster** angle is the angle between the kingpin axis (axis around which the wheel is steered) and a radial line from the center of the wheel to the ground, when both are projected onto a longitudinal plane. When the caster is positive, that radial line intersects the ground behind the intersection of the kingpin axis with the ground. Increasing the caster angle increases the tendency of a turned wheel to restore itself to the straight-ahead position as the vehicle moves forward. This self-restoring tendency can also be influenced somewhat by the inclination of the kingpin in the transverse plane.

When viewing an unloaded vehicle at rest from above, the front wheels may be set either to **toe in** or **toe out** slightly to compensate for the effects of driving forces and kingpin inclination. This adjustment influences the tendency of the wheel to shimmy. Toe-in is typical for rear-wheel-drive cars, while the front wheels of front-wheel-drive cars may be set to toe out.

The automotive pneumatic **tire** performs four **main functions**: supporting a moving load; generating steering forces; transmitting driving and braking forces; and providing isolation from road irregularities by acting as a spring in the total suspension system. A tire made up of two basic parts: the tread, or road-contacting part, which must provide traction and resist wear and abrasion; and the body, consisting of rubberized fabric that gives the tire strength and flexibility.

The three **principal types** of automobile and truck tires are the cross-ply or bias-ply, the radial-ply, and the bias-belted (Fig. 8.1.33). In the **bias-ply** design, the cords in each layer of fabric run at an angle from

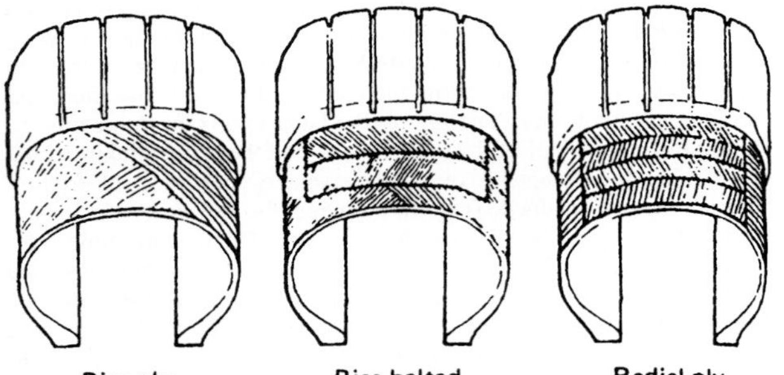

Figure 8.1.33 Three types of tire construction.

one bead (rim edge) to the opposite bead. To balance the tire strength symmetrically across the tread center, another layer is added at an opposing angle of 30 to 38°, producing a two-ply tire. The addition of two criss-crossed plies makes a four-ply tire.

In the **radial tire** (used on most automobiles), the cords run straight across from bead to bead. The second layer of cords runs the same way; the plies do not cross over one another. Reinforcing belts of two or more layers are laid circumferentially around the tire between the radial body piles and the tread. The **bias-belted** tire construction is similar to that of the conventional bias-ply tire, but it has circumferential belts like those in the radial tire.

Of the three types, the radial-ply offers the longest tread life, the best traction, the coolest running, the highest gasoline mileage, and the greatest resistance to road hazards. The bias-ply provides a softer, quieter ride and is least expensive. The bias-belted design is intermediate between the good quality bias-ply and the radial tire. It has a longer tread life, and it gives a smoother low-speed ride than the radial tire.

Tire specifications are molded into the sidewall. Consider, for example, the specification "P225/60TR16." The initial P designates "passenger car" (LT = light truck, ST = trailer). Next, 225 is the tire width in mm. Following the slash, 60 is the aspect ratio, that is, the ratio of sidewall height, from bead to tread, to tire width, expressed as a percentage. T indicates a maximum speed rating of 190 km/h. Speed designators progress from S = 180 km/h through T (190 km/h), U (200 km/h), H (210 km/h), V (240 km/h) to Z (>240 km/h). On some tires the speed rating is omitted. R indicates radial construction (D = bias ply; B = bias belted). Finally, 16 indicates a wheel with a 16-in. rim diameter. Alternatively, such a tire could be labeled "P225/60R 89T," where the speed rating comes last and is preceded by a numerical load index indicating carrying capacity.

For passenger cars, the recommended **tire pressure** is posted on the rear edge of the driver's door. **Underinflation** causes excessive tire flexing and heat generation in the tire, more rapid wear, poor handling, and poor fuel economy. **Overinflation** causes a rough ride, unusual wear, poor handling, and increased susceptibility to tire damage from road hazards.

As a safety measure, a dashboard signal of tire under-inflation can be provided for the driver. Early versions of such a system made use of the wheel-speed signal necessary for traction control. Any wheel that consistently rotates at a higher speed than the rest suggests a reduced running radius, as would occur with an underinflated tire. Later versions measure the pressure of individual tires with wheel-mounted sensors, electronically transmitting their signals to an on-board receiver for dashboard display.

An out-of-balance wheel causes uneven tire wear and adversely affects ride quality. Wheel-tire assemblies are balanced individually by attaching balance weights to the rim of the wheel. For **static balance**, a proper weight on the outside rim of the wheel suffices. For the **dynamic balance** required at high speeds, weights are necessary on both the outside and inside rims. Dynamic balancing necessitates spinning the wheel to determine correct placement of the balance weights.

The trunk space occupied by a fifth wheel on which a **spare tire** is mounted to replace a flat tire has long bothered consumers and manufacturers alike. To regain part of that space, some cars have used a collapsible tire that must be inflated when pressed into service, using either a compressed-air storage reservoir or an onboard air compressor. Alternatively, some cars store a smaller-than-standard compact spare tire that has been preinflated to a high pressure. The run-flat tire is another option. It eliminates the need to carry a jack. It may rely upon an extra stiff sidewall to carry the load if the tire is punctured, or upon a plastic donut built into the tire. Such run-flat tires eliminate not only the storage-space requirement for carrying any spare, but also the need to carry a jack. When used, all of these alternatives to carrying a fifth wheel/tire combination entail some sacrifice in ride quality, and a limitation on allowable travel distance and speed.

HEATING, VENTILATING, AND AIR CONDITIONING (HVAC)

The **heating, ventilation, and air conditioning** (HVAC) system is installed to provide a comfortable environment within the passenger compartment. Fresh air for HVAC usually enters through an opening in front of the windshield and into a plenum that separates rain from the ambient air. The airflow rate is augmented by a variable-speed electrically driven blower.

In cold weather, heater air is passed through a finned heat-exchanger core, where it is heated by the engine coolant. Compartment temperature is controlled by (1) mixing ambient air with heated air, (2) mixing heated air with recirculated air, or (3) changing blower speed. Provision is made to direct heated air against the interior surface of the windshield to prevent the formation of ice or fog. Figure 8.1.34 schematically illustrates a system in which a multi-speed blower drives fresh air through (1) a radiator core or (2) a bypass. The flow rate and direction of the air entering the passenger compartment are controlled with manually operated doors and vanes in the discharge ducts.

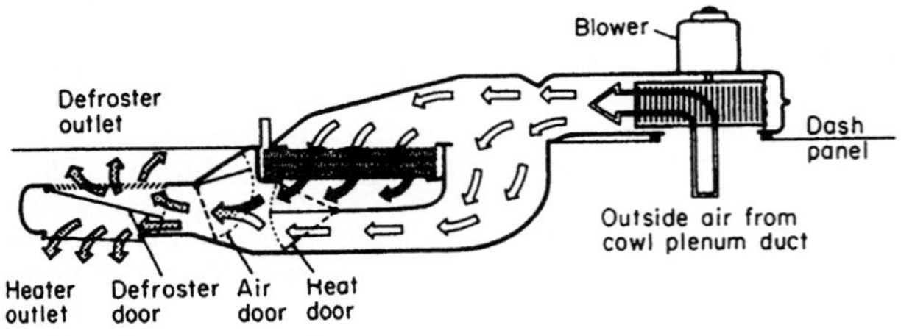

Figure 8.1.34 Heater airflow diagram.

For cooling in hot weather, an electric clutch is engaged to operate the air-conditioning compressor, which is belt-driven from the engine (Fig. 8.1.35). The pressurized, hot gaseous refrigerant exiting the compressor enters the condenser, a heat exchanger typically located in front of the engine radiator. Ambient air passed through this heat exchanger by a fan cools the refrigerant, liquefying it. The high-pressure liquid then passes through a liquid reservoir and on to an expansion valve that drops the refrigerant pressure as it sprays the refrigerant into the

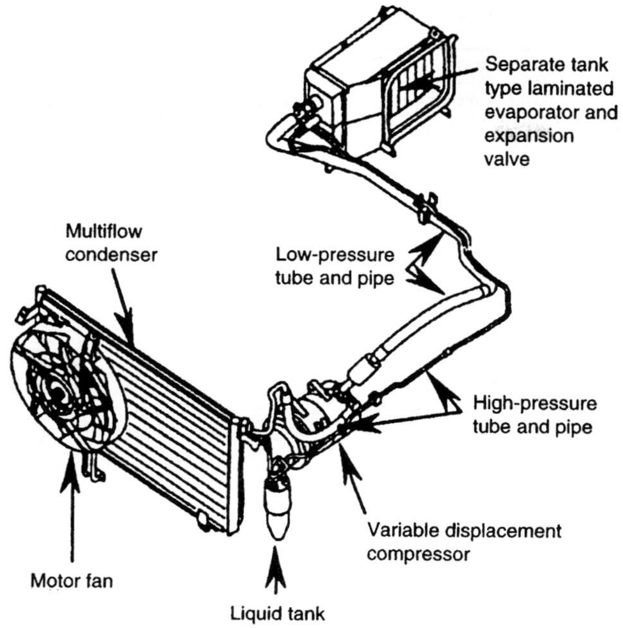

Figure 8.1.35 Schematic of air conditioning system using HFC-134a refrigerant. *(Reprinted with permission from SAE, SP-916, © 1992, Society of Automotive Engineers, Inc.)*

evaporator. The evaporator is a heat exchanger normally mounted on the firewall. Compartment air circulated through the evaporator rejects its heat to the colder refrigerant, thus cooling as it vaporizes the refrigerant in preparation for reentry of the refrigerant into the compressor. Moisture condensed from the ambient air during cooling is leaked onto the roadbed.

Figure 8.1.36 shows schematically a **combined air-heating and air-cooling system**. Various dampers control the proportions of fresh and recirculated air to the heater and evaporator cores. Compartment-air temperature is controlled electronically by a thermostat that modulates

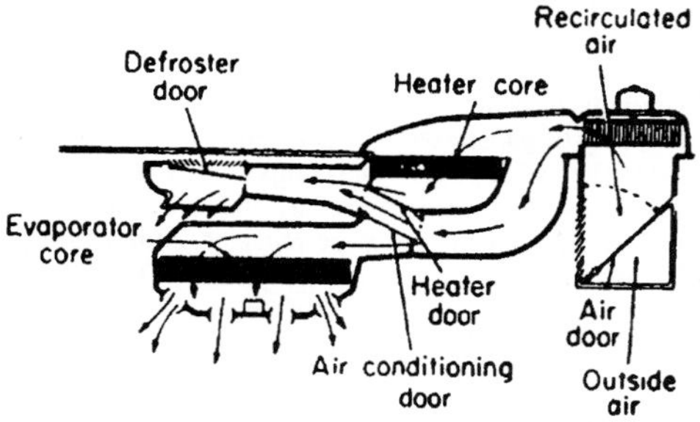

Figure 8.1.36 Combined heater and air conditioner. *(Chevrolet.)*

the mixing of cooled and heated air, and cycles the compressor on and off through a magnetic clutch. During defrosting, mixing cooled air in with heated air helps to remove humidity from the air directed to the interior of the windshield.

For many years the refrigerant in automotive air conditioners was CFC-12, a chlorinated fluorocarbon also known as F-12. About 1990 the depletion of stratospheric ozone, which shields humankind from harmful solar ultraviolet radiation, became an environmental concern. The chlorine in CFC-12 having been identified as a contributor to the problem, the use of CFC-12 was banned worldwide. About the same time, concern mounted about the threat of global warming from the anthropogenic production of greenhouse gases. Carbon dioxide from the combustion of fossil fuels is a major contributor. It was estimated that on a mass basis, CFC-12 was about 10,000 times as serious a greenhouse gas as carbon dioxide. In the United States, an immediate response to the ozone depletion concern was to require mechanics to capture spent CFC-12 for recycling, rather than venting it to the atmosphere, every time an air conditioner was serviced. Beyond that, within a few years, CFC-12 was replaced by chlorine-free HFC-134a, which has a global warming effect about one-eighth that of CFC-12. Because HFC-134a is still 1,300 times as serious a greenhouse gas as carbon dioxide, a replacement for HFC-134a is being sought. The EPA has listed HFC-134a as unacceptable for newly manufactured light duty vehicles in 2021 model year. As of 2026 HFC-134a will be banned entirely for new vehicles. The majority opinion is that HFO refrigerants are the future, and for automobile applications 1,234YF will be the new go to. In Europe 1,234YF is being used across the continent.

AUTOMOBILE CONSTRUCTION

Light-duty vehicles include passenger cars and light-duty trucks. Light-duty trucks include passenger vans, SUVs, and pickup trucks. The **layout of light-duty vehicles** generally falls into one of four categories: (1) front engine, rear-wheel drive—used mostly in large cars and light-duty trucks, (2) front engine, front-wheel drive—used mostly in small cars and passenger vans, (3) rear or midengine, rear-wheel drive—sometimes used in sports cars, and (4) four-wheel drive—used mostly in light-duty trucks and premium large cars.

The two most common basic body constructions, illustrated in Figs. 8.1.37 and 8.1.38, are **body-and-frame** and **unit construction**. The body-and-frame design uses a body that is bolted to a separate frame. Most of the suspension, bumper, and brake loads are transmitted to the

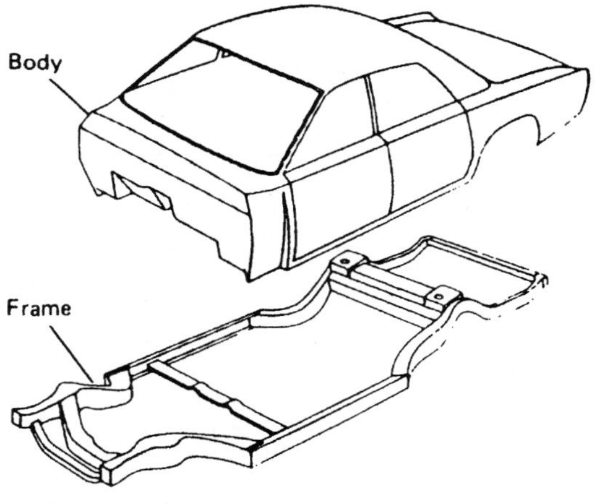

Figure 8.1.37 Separate body and frame design.

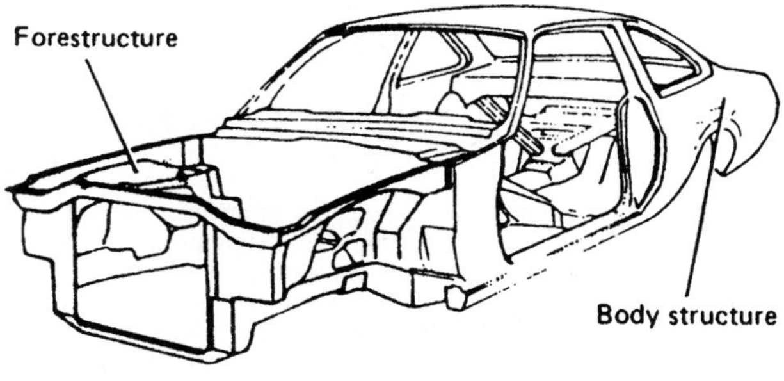

Figure 8.1.38 Unit-construction design for automobiles.

Table 8.1.3 Average Material Consumption for a Domestic Light Vehicle Model Years 1995, 2000, and 2013

	1995		2000		2013	
Material	Pounds	Percentage	Pounds	Percentage	Pounds	Percentage
Regular steel	1,630.0	44.1%	1,655.0	42.4%	1,361.0	34.1%
High and medium strength steel	324.0	8.8%	408.0	10.5%	649.0	16.3%
Stainless steel	51.0	1.4%	62.0	1.6%	74.0	1.9%
Other steels	46.0	1.2%	26.0	0.7%	32.0	0.8%
Iron castings	466.0	12.6%	432.0	11.1%	283.0	7.1%
Aluminum	231.0	6.3%	268.0	6.9%	379.0	9.5%
Magnesium castings	4.0	0.1%	8.0	0.2%	11.0	0.3%
Copper and brass	50.0	1.4%	52.0	1.3%	73.0	1.8%
Lead	33.0	0.9%	36.0	0.9%	36.0	0.9%
Zinc castings	19.0	0.5%	13.0	0.3%	8.0	0.2%
Powder metal parts	29.0	0.8%	36.0	0.9%	45.0	1.1%
Other metals	4.0	0.1%	4.0	0.1%	5.0	0.1%
Plastics and plastic composites	240.0	6.5%	286.0	7.3%	336.0	8.4%
Rubber	149.0	4.0%	166.0	4.3%	203.0	5.1%
Coatings	23.0	0.6%	25.0	0.6%	32.0	0.8%
Textiles	42.0	1.1%	44.0	1.1%	50.0	1.3%
Fluids and lubricants	192.0	5.2%	207.0	5.3%	222.0	5.6%
Glass	97.0	2.6%	103.0	2.6%	96.0	2.4%
Other materials	64.0	1.7%	71.0	1.8%	93.0	2.3%
Total	3,694.0	100.0%	3,902.0	100.0%	3,988.0	100.0%

SOURCE: "Ward's Communications, Ward's Motor Vehicle Facts and Figures 2014," Detroit, MI, 2014.

Table 8.1.4 Materials Disposition from Recycled Vehicles

Recycled materials ←		→	
Ferrous metals	Nonferrous metals	Fluids	Shredder residue
• Steel • Iron	• Aluminum • Copper • Lead • Zinc	• Engine oil • Coolant • Refrigerant	• Plastic • Rubber • Glass • Fluids • Other

frame. Body-and-frame construction dominates in pickups and SUVs. With unit construction, most common in passenger cars, the frame is eliminated. Its function is assumed by a framework consisting of stamped sheet-metal sections joined together by welding. Outer panels are then fastened to this framework, primarily by bolting.

A third construction option that has been borrowed from aircraft practice and used to a limited extent is the **space frame** and **monocoque** combination. The space frame is a skeleton to which the body panels are attached, but the body assembly is designed in a manner that causes it to contribute significantly to the strength of the vehicle.

An important aspect of auto design is passenger protection. Crumple zones are designed into the structure, front and rear, to absorb collision impacts. Beams may be built into doors to protect against side impact.

The 2000s drive for improved passenger-car fuel economy, hence smaller and lighter cars, and for improved corrosion resistance has led to a gradual shift in material usage. This trend is evident in Table 8.1.3. The weight percentage of ferrous content has declined consistently, but remains over half the total weight of the average car. The shift from conventional to high-strength steel is evident. In areas likely to experience rust, the steel may be galvanized. The percentage use of lighter aluminum has increased. The use of plastics and composites, while still less than 10 percent of the car weight, has increased by 40 percent on a mass basis.

Automobiles are among the most recycled commodities. Indeed, automobiles maintain a recycling rate of nearly 100 percent. Approximately 75 percent of vehicle content was recycled. Table 8.1.4 indicates the disposition of materials from recycled vehicles.

8.2 ROADMAP OF VEHICLE ELECTRIFICATION AND HYBRIDIZATION

by Gene Liao

The automotive industries, in response to economic, political, and environmental pressures, are expanding their development and production of more fuel efficient and environmentally friendly automobiles through vehicle electrification or hybridization. The hybrid vehicles are vehicles powered by two or more sources of energy. The potential benefits of hybrid vehicles include fuel economy improvement, exhaust emissions reduction, power performance enhancement, and combination of all. "Hybrid" can mean Hybrid Electric Vehicle (HEV), mechanical hybrid vehicles, or Hydraulic Hybrid Vehicle (HHV). The first HEV reached the mass market in 1999 in the United States.[1,2] The HEVs are different from traditional automobiles in that the HEVs utilize Internal Combustion Engines (ICEs) and electric machines for propulsion systems and involve high electrical voltage. Electric motors deliver maximum torque at low speed so they are an ideal complement to ICEs, which generate best torque at higher speed. The level of electrification (or hybridization) is defined as the percentage of the electric power over the entire vehicle powertrain power. Hence, the electrification level is 100 percent for the Electric Vehicle (EV), Battery Electric Vehicle (BEV), or Fuel Cell Electric Vehicle (FCEV). Depending on the level of electrification, a HEV could have operating features of engine idle stop-start, regenerative brake, power boost by electric motor assist, engine-off driving, and battery charging from the electric grid. Specifically, vehicle electrification or hybridization can attempt to address the following points:

- Operate the engine nearer its best efficiency region.
- Downsize the engine (some hybrid powertrain architectures) and still meet the maximum power requirements.
- Use regeneration to restore some of the vehicle kinematic energy during deceleration instead of dissipating it as heat in the brakes.
- Eliminate or mitigate the idling losses by turning the ICE on and off.

Depending on the level of electrification, HEVs can be classified into stop-start (or micro hybrid), mild hybrid, full (or strong) hybrid, Plug-in HEV (PHEV), and EV. With increasing the level of electrification, a HEV will have more operating features as shown in Fig. 8.2.1. All types of HEVs have the engine idle stop-start feature. This feature is

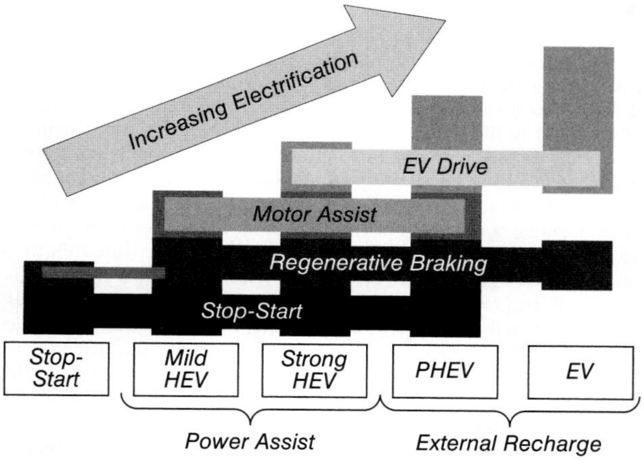

Figure 8.2.1 Electrification level and operating features of HEVs.

not needed in an EV as it has no engine. The regenerative braking feature is available almost in all types of HEVs. A stop-start (or micro) HEV might have minor regenerative braking depending on the types of onboard energy storage systems. The feature of power boost by electric motor assist is available in mild HEV, full HEV, and PHEV. Engine-off driving (or electric vehicle mode) is only available in full HEV and PHEV that has higher power electric machine and larger capacity of energy storage system.

Energy economy and emissions are and will continue to be primary concerns in the foreseeable future. The only long-term solution is to utilizing sustainable mobility and renewable energy sources. In sustainable mobility, hybrid vehicles present a medium-term solution due to their complication of powertrain architecture for dual energy sources, complexity of control algorithms, high cost of engineering, components and implementations, as well as high vehicle maintenance cost. Single energy source (fossil fuel, natural gas, biofuel, electricity, etc.) is still the best choice for automotive by means of simplicity and reliability. Hybrid vehicles have been called a "bridge" toward pure electric vehicles. The EV, using single electrical energy source, is currently believed to be the long-term solution for ground transportation while the low-cost and highly reliable onboard energy storage system for electricity is still under development. The onboard electrical energy source for EVs could be hydrogen fuel cells or variety types of batteries.

OPERATING FEATURES OF HYBRID AND ELECTRIC VEHICLES

This section gives more details of hybrid and electric vehicle operating features including engine idle stop-start, regenerative brake, engine-off driving, power boost by electric motor assist, optimal engine operation, and battery charging from the electric grid.

Engine Idle Stop-Start

The engine idle stop-start is the simplest, lowest-cost hybrid, and an important energy-saving building block used in all types of HEVs. This function conserves energy by automatically shut off the engine whenever the vehicle comes to a stop, for example, at traffic lights. Then it automatically restarts the engine as soon as the brake is released. Several sensors are needed for battery State-of-Charge (SOC), ambient temperature, clutch, neutral gear, and brake pressure. When starting a conventional ICE, a starter motor turns the crankshaft at about 250 rpm. On a HEV, the motor/generator turns the crankshaft at 500-1,000 rpm, faster than normal. This feature allows the engine to stop running when it would normally be idling as the vehicle is at rest or decelerating. The engine can seamlessly restart without the driver or passengers being aware. Automotive makers currently use three types of techniques to perform engine idle stop-start.

1. Using electric machines equipped in mild, full, and PHEVs. The types of HEVs and their powertrain architectures are later elaborated upon in other section. Dependent upon the hybrid powertrain architectures, the power of an electric machine in the HEV ranges from 10 kW to 150 kW. The electric machine could directly connect to the engine crankshaft with or without a clutch as example shown in Fig. 8.2.2*a*, or though other mechanisms (such as planetary gear set) as example shown in Fig. 8.2.2*b*. When the vehicle master controller calls for the

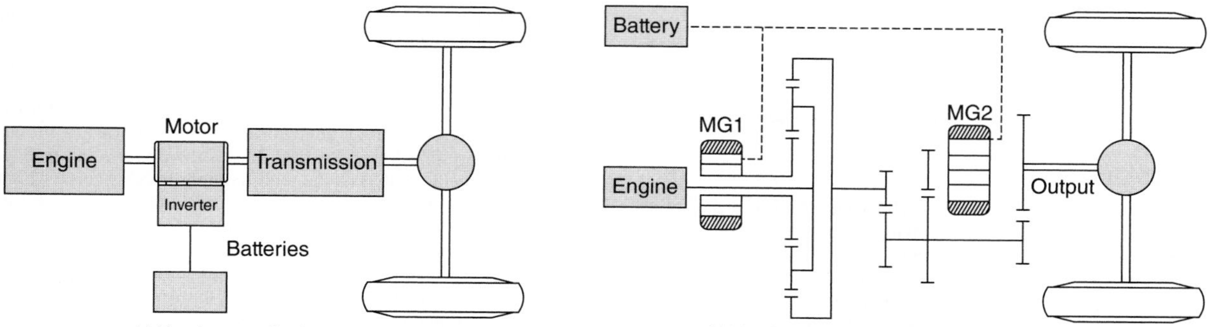

(a) Motor/generator directly connected to engine

(b) Motor/generator connected to engine through planetary gear set

Figure 8.2.2 Using electric machine to restart the engine.

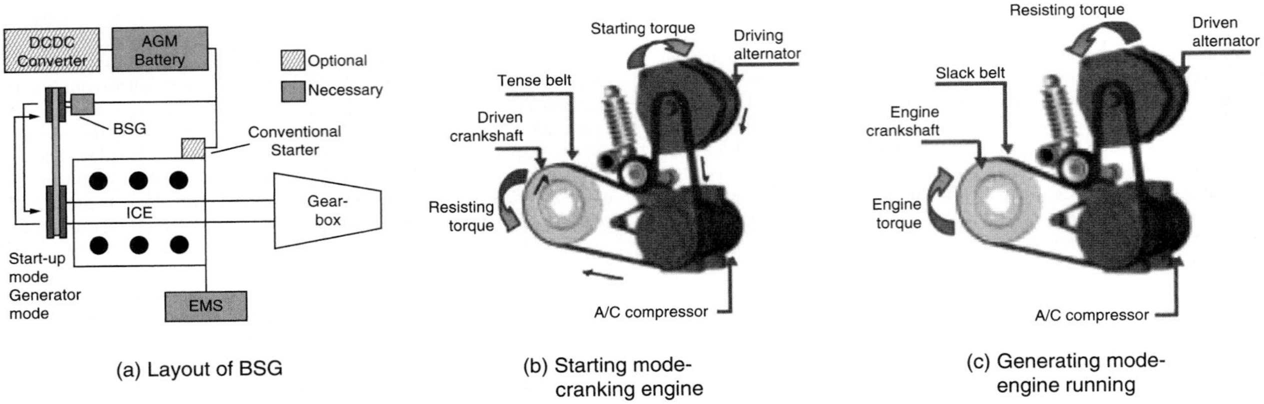

Figure 8.2.3 BSG for engine idle stop-start.[3,4]

engine to be restarted, the electric machine provides a powerful motoring torque to start the engine.

2. Replacing the regular alternator by a small electric machine (motor/generator dual function). The electric machine acts as a starter (motoring mode) which starts the engine through engine accessory belt. When the engine is running, the electric machine is then in generating mode to act as an alternator. The layout of this Belt-Starter-Generator (BSG) for engine stop-start is shown in Fig. 8.2.3*a*. The Valeo StARS system,[3-5] as illustrated in Figs. 8.2.3*b* and *c*, is an example. The pros are easy to integrate and smooth restart while the cons include that belt drive system has to be changed, cold start is difficult with bigger engines, and flywheel has to be modified.

3. Using an enhanced starter motor with modified parts to allow re-cranking of the engine. Figure 8.2.4 is the basic layout of this enhanced starter motor for engine stop-start. The pro is low system cost while the con is not very comfortable restart. The major automotive suppliers for this type of stop-start system are Bosch stop-start system,[4,6] Denso Advanced Engagement starter, Tandem Solenoid starter, and Permanent Engaged starter.[7]

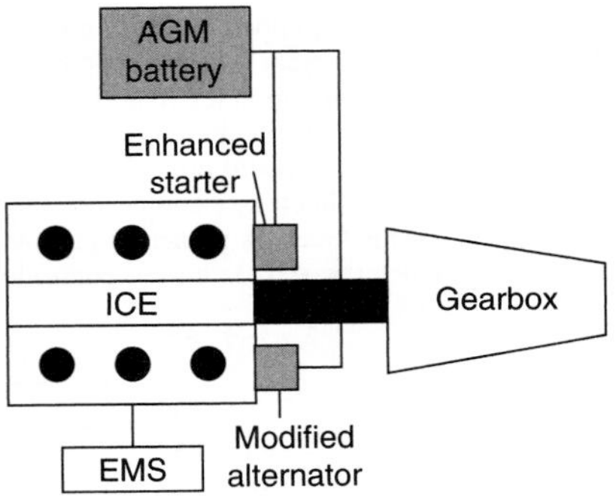

Figure 8.2.4 Layout of enhanced starter engine idle stop-start.[4]

Regenerative Brake

Studies show that, in urban driving, about one third to a half of the powertrain energy is consumed in braking. Recovery of vehicle kinetic energy becomes possible through regenerative braking in hybrid and electric vehicles. During acceleration, the powertrain must supply the kinetic energy to vehicle plus the energy lost to dissipative forces and through the overall inefficiencies of the powertrain from fuel/electricity to wheel. During deceleration, this kinetic energy (minus the energy lost to dissipative forces) must be either dissipated as heat in the brakes or converted back to useful energy (mechanical, electrical, etc.) through "regeneration" system. The design and control target is not only to achieve a good braking performance, but also recover braking energy as much as possible. A portion of the vehicle kinetic energy must be removed during the decelerations of wheel braking. With regenerative braking, however, the retarding force is the reaction to that used to drive an energy converter (such as electric machine). Consequently, some of this braking energy can be converted, stored, and made available for later use. A traction motor can be controlled to function as a generator, in which the kinetic or potential energy of the vehicle can be converted into electrical energy and stored into electric energy storage (usually chemical batteries). The recovered electric energy can be used for later propulsion. Thus, the vehicle energy efficiency can be greatly enhanced. Figure 8.2.5*a* shows the Toyota Prius braking system that consists of conventional frictional brake (hydraulic system in yellow shaded area) and regenerative brake (electric system in blue shaded area).[8] The MG2 (motor/generator 2) connecting to the front axle acts as traction motor during vehicle acceleration and becomes a generator, which is turned by wheels, as the vehicle in deceleration. Figure 8.2.5*b* is the control block diagram. Based on the signals received from the sensors, the communication maintains among the vehicle master ECU, brake ECU, conventional brake control, ABS control, and regenerative brake control.

In HEV/PHEV/EV, the electric motor becomes a generator when the brake pedal is applied and the vehicle kinetic energy is used to generate electricity that is then used to recharge the batteries. Conventional frictional (hydraulic) brake must also be used to ensure that the vehicle slows down or stops as driver's demand. To meet these two requirements (regenerative brake and driver's demand), most of the automakers implement the technique of blending brake which generates the required braking torque by mixing the regenerative braking with frictional brake. The brake ECU performs an arbitration, which results in brake blending between friction brake (pressure apply) and regenerative braking. This allows for a completely variable regulation of friction brake torque, while the sum of regen brake torque and friction brake torque conforms to driver's demand. The basic interaction of blending brake is shown in Fig. 8.2.6. In Fig. 8.2.6*a*,[8] the dash line is the driver's input to slow and stop the vehicle, that is, the total required braking force which is the combination of hydraulic/frictional force and regenerative force. The portions of hydraulic/frictional and regenerative forces are varied based on the vehicle speed. An example shown in Fig. 8.2.6*b*, the vehicle speed starts from 100 kilometer per hour (kph) and slows down to a fully stop in 8.3 s as black line shown. The blue line is the total deceleration demand that is the blend of frictional (green curve) and regenerative (red curve) braking. The curve of regenerative brake is determined by the brake calibration, or regenerative brake control

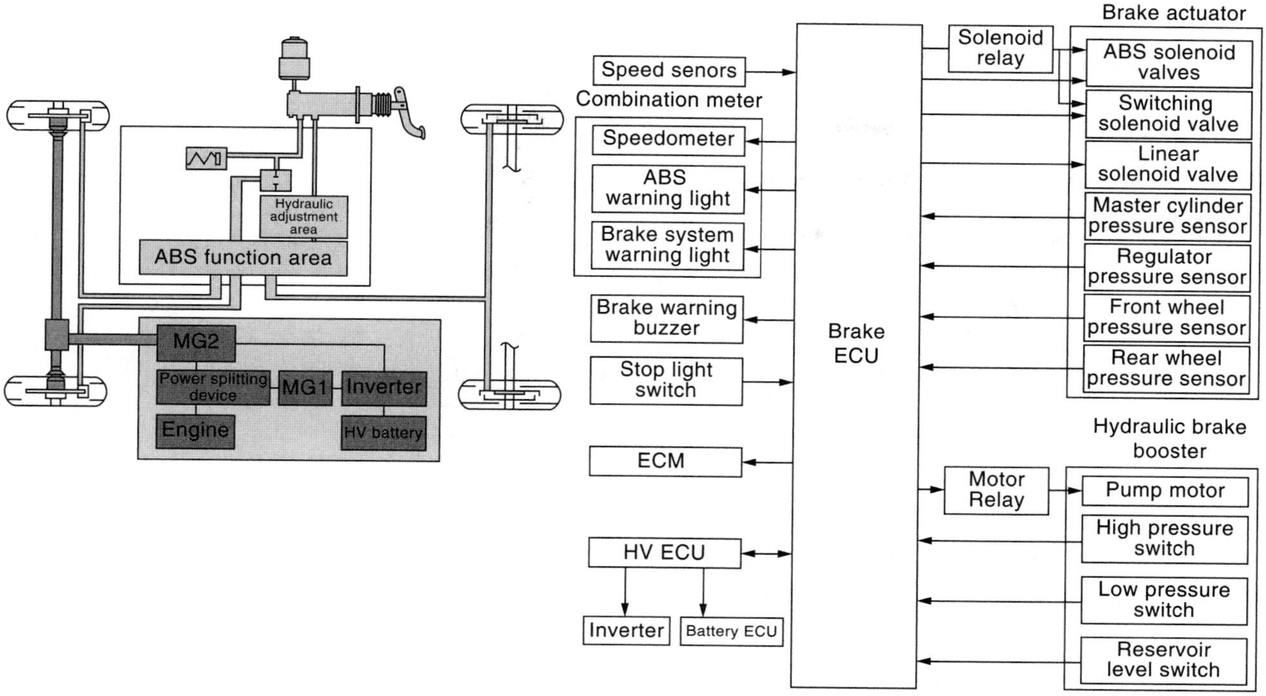

(a) Toyota Prius brake system

(b) Control block diagram of hydraulic and regenerative brakes

Figure 8.2.5 Layout of hydraulic and regenerative brakes.[8]

algorithm, which needs to consider the driverability and not to create driver panic.

Depending on the brake calibration, vehicle speed, and battery pack capacity, the regenerative brake is not always occurred as the driver presses the brake pedal. When applying brake at high vehicle speed, the electric current generated by the electric machine is too large that could not charge to battery pack because the battery has specified acceptance capacity. On the other hand, applying brake at too low vehicle speed, the generated electric current is too small to charge to the battery pack. In fact, it is difficult to regenerate at low speeds because of the low motor electromotive force (voltage) generated at low motor rotational speeds. The regenerative brake could dominate large portion of braking force only in certain vehicle speed ranges. Not all of the vehicle kinetic energy can be used for the batteries because some of it is "lost" to waste heat. Some energy is also lost to resistance/friction as the energy travels from the wheel/axle, through the drivetrain and electric motor, and into the battery. Additionally, the front-wheel-drive hybrid and electric vehicles capture more vehicle kinetic energy (than rear-wheel-drive vehicles) because the vehicle weight shifts to the front axle during braking. When calculating the efficiency of regenerative braking, it needs to consider the Wheel-to-Wheel (W2W) efficiency—a closed loop that the kinetic (mechanical) energy is captured from vehicle wheels, regenerative electric energy is stored in the battery, and then electric energy is discharged and converted back to mechanical energy to the vehicle wheels for traction.[9] The regenerative efficiency depends upon regenerative brake calibration, driving cycles, battery charging/discharging efficiency, electric machine efficiency, and power electronics efficiency. With a NiMH battery pack, the maximum overall efficiency of reuse is about 46.2 percent. In practice, Toyota Prius could reach 36 percent regenerative efficiency in UDDS.[10] The average efficiency of regenerative brake is about 20 percent. For the lithium-based battery, the regenerative efficiency is higher.

Engine-Off Driving and Power Boost by Electric Motor Assist

The feature of engine-off driving is only available in full (strong) HEV and PHEV. As the electric motors deliver maximum torque at low speed, so they are used to propel vehicle in low driving speed. During low vehicle speed or low road load, the full HEVs and PHEVs operate in the electric vehicle mode, that is, only the electric propulsion/traction motor moves the vehicle. The engine-off driving normally effects during vehicle in launching (vehicle starting from standstill to certain speed) and low cruising speed such as in parking lots driving. The vehicle speed for engine-off driving might be up to 40 kph (25 mile per hour—mph) which is constrained by the vehicle acceleration, road load, and battery capacity. The feature of electric motor assist covers a broader range of speed and vehicle operations such as hill climbing and driving in snow. Vehicle launch is part of motor assist, but applies to very low speed. The power boost by electric motor is occurred when the vehicle needs more acceleration such as in highway merging (driving from ramp to highway), passing other vehicles in highway (passing speed about from 80 to 100 kph), top speed driving, and maximum acceleration with speed from zero to 100 kph. When the HEVs in high acceleration driving, the engine provides the major power while the electric motor offers additional power to boost the vehicle driving.

The application of traction motors on hybrid and electric vehicles requires the motors to operate in multi-quadrants, including forward motoring, forward braking (regenerative mode), backward motoring, and backward braking (regenerative mode), as shown in Fig. 8.2.7. For vehicles with reverse mechanical gears, two-quadrant operation (forward motoring and forward braking, or quadrant-I and IV) is required. However, for the vehicles without reverse mechanical gears, three-quadrant operation (forward motoring, forward braking, and backward motoring, or quadrant-I, III, and IV) is needed. The backward braking (quadrant-II) normally is not occurred for typical hybrid and electric

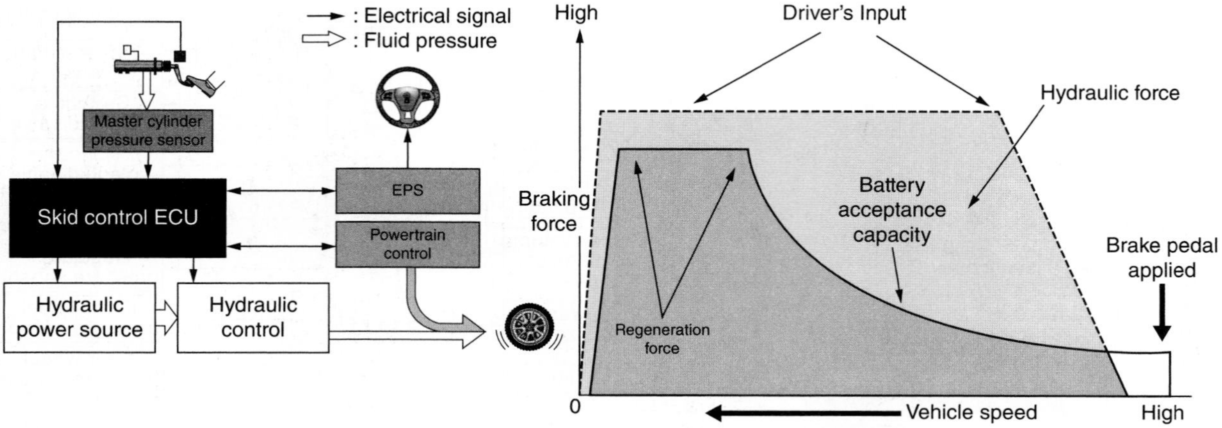

(a) Blending of hydraulic/frictional force and regenerative force[8]

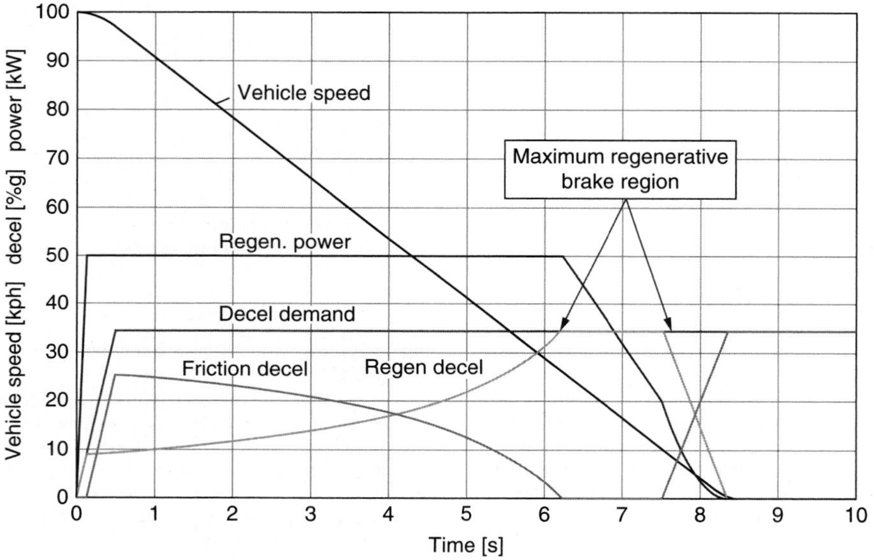

(b) Deceleration demand consisting of frictional and regenerative

Figure 8.2.6 Examples of blending brakes.

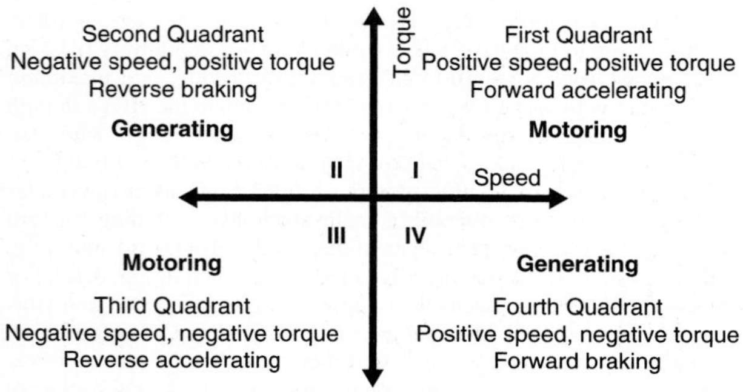

Figure 8.2.7 Four-quadrant operation of traction motor.

vehicles because vehicles backward usually in low speed which does not energize regeneration. Multi-quadrant operation of a motor is implemented by controlling the input current frequency through power electronics.

Optimal Engine Operation

The fuel economy characteristic of an engine is evaluated by the amount of fuel per kWh energy output, which is referred to as the specific fuel consumption (g/kWh). The typical fuel economy characteristic of a gasoline engine varies widely with engine speed and load as shown in Fig. 8.2.8. The fuel consumption is quite different from one operating point to another. The choices of the engine operating point to meet the demand at the wheel are discrete (only a few gear ratios available to match engine speed along a power curve).

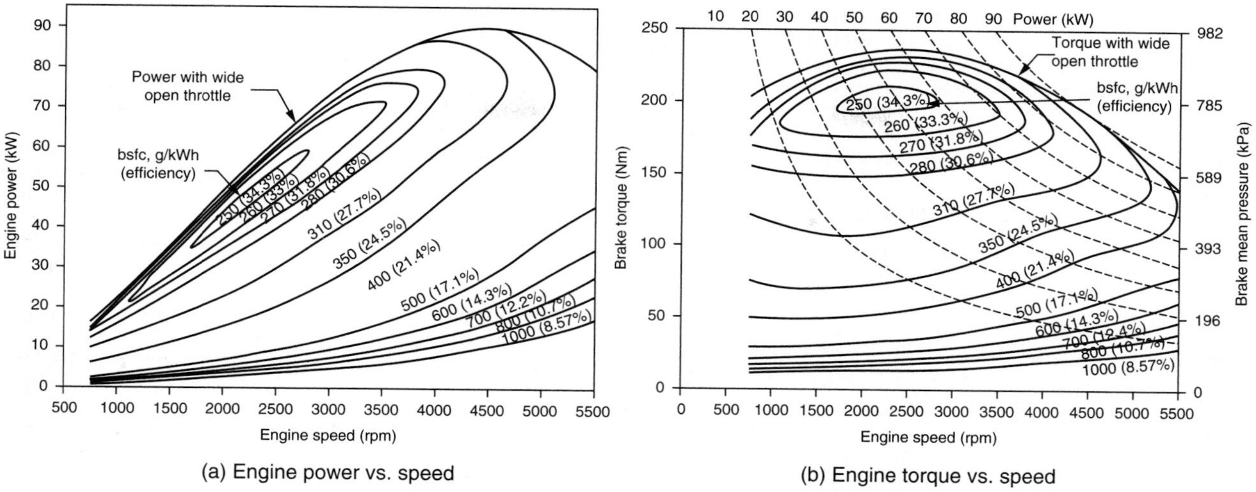

Figure 8.2.8 A typical gasoline engine map.[11]

An engine generally has its optimal operating region in which the fuel consumption and the losses in the induction/exhaust strokes are minimized, that is, the percentage of losses to the total indicated power is small. This region usually is located in the middle of the speed range and close to full load operation (wide-open throttle), corresponding to the maximum torque. For a gasoline engine map shown in Fig. 8.2.8, the best possible efficiency is about 34 percent for an intermediate speed and for an output torque close to the maximum.[11] In vehicle design, the engine operating points should be close to its optimal region in order to achieve high operating fuel economy. Utilizing the battery pack as an energy buffer, the HEV maintains operating its engine at optimal region for better efficiency while delivering the required traction power to the wheels. If the demand traction power is less than the best power level (optimal operating region) of the engine, the excess power from the engine can be used to charge the battery pack. This situation occurs at the lower-left portion shown in Fig. 8.2.8. For the case of demand traction power more than the engine can deliver at its optimal operating region, the electric motor complements the traction power. This situation is at the upper-right area shown in Fig. 8.2.8.

Conventional vehicles typically operate in the lower left area of the engine operating map to have plenty of "reserve" power for fast accelerations, high speed, etc. In other words, engines are typically oversized compared to the average power required. The engine size is determined in terms of the maximum power requirements (at vehicle top speed, maximum acceleration, and towing requirements). This forces the engine to operate at relatively low efficiencies on the average, significantly lower than the peak efficiency. An engine in the HEV is often smaller than what would normally be required to run a conventional vehicle of similar weight. Participating with an electric motor, a reasonable amount of power is available to accelerate the vehicle. Such as climbing a hill where extra power is needed, the electric motor assists the small engine's power.

Fuel Economy Improvement

The operating features of a HEV can be shown in a simple drive cycle like Fig. 8.2.9 where the shaded areas represent the electric propulsion systems (electric machine and battery pack) participating the vehicle movements. Vehicles in an urban driving cycle idle typically 20 percent

Zone number	Operation features
1	Engine off driving in vehicle launch
2	Power boost by electric motor assist
3	Regenerative braking
4	Engine idle stop-start

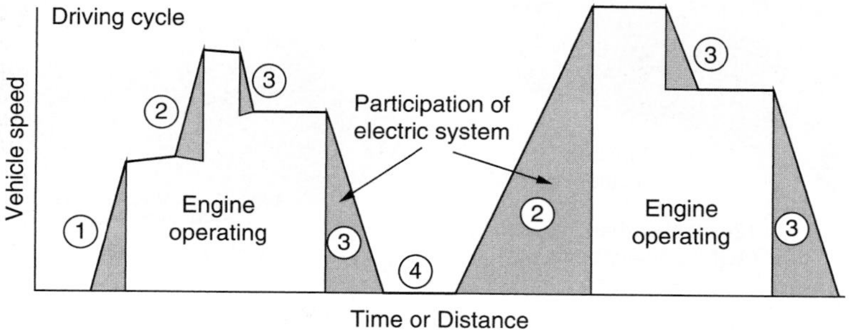

Figure 8.2.9 HEV operation features in a simple driving cycle.

of the time. By engine idle stop-start, the fuel economy can improve 5-8 percent.[12] Regenerative braking may yield 10-15 percent fuel savings. The optimized engine operation may improve 8-10 percent fuel economy. However, HEVs are heavier than conventional vehicles because the HEVs require additional components including electric machine, battery pack, power electronics, and electric cables. Weight is a major impediment to vehicle fuel economy and performance.

TYPES OF HYBRID AND ELECTRIC VEHICLES

According to the level of electrification, HEVs can be classified into stop-start (or micro hybrid), mild hybrid, full (or strong) hybrid, Plug-in HEV (PHEV), and EV. This section provides more detail descriptions and examples of each type of HEV.

Micro Hybrid or Stop-Start

Some consider a micro hybrid barely even a hybrid. A micro hybrid vehicle automatically shuts the engine off when the vehicle comes to a stop in traffic or at a light. Either an enhanced starter or a small electric machine (motor/generator dual function) is equipped to cope with the frequent starting demands. The engine stop-start function varies slightly based on the type of transmission. For a Manual Transmission (MT), the system works if the vehicle speed goes below 8 kph (5 mph), and the driver brakes after shifting the gear lever to neutral. This situation is known as coast-down and the engine will automatically shut down. The engine will start up again as soon as the driver engages the clutch. For an Automatic Transmission (AT), then the engine does not turn off automatically until the vehicle is at a complete standstill. The engine restarts the instant the driver released the brake pedal. The micro hybrid normally has smallest battery size as it only supports engine idle stop-start function. The battery is usually a 12 V Absorbed Glass Mat (AGM) battery, a Valve Regulated Sealed Lead-Acid Battery (VRSLAB), to operate engine restart function and other electric accessories such as radio, heater, fans, and interior/exterior lights. The AGM battery has abilities of deep cycle for 80 percent Depth-of-Discharge (DOD) and delivery of high current on demand. It can be mounted in any position and function without leaking electrolyte. Some micro hybrids use two 12 V AGM batteries to operate the engine restart and auxiliary function separately, as an example of the Chevrolet Malibu with stop-start shown in Fig. 8.2.10*a*.[13] For a micro hybrid, the air conditioning is not functional when the engine stops because the air condition compressor is driven by engine accessory belt. The driver normally can disable the engine stop-start function in a micro hybrid and then operate it as a conventional vehicle, that is, the engine is still running to provide air conditioning during vehicle standstill. Figure 8.2.10*b* shows a control button for driver to turn on/off the engine stop-start function in a Mercedes-Benz ECO Start/Stop vehicle.[14]

The European automakers initiated micro hybrids that have become popular in Europe, such as the Mercedes-Benz ECO Start/Stop. Vehicles in Europe typically equipped with MTs or Continuously-Variable-Transmissions (CVTs), therefore the engine stop-start function is cheap and easy for manufacturers to implement. For a vehicle equipped with AT, the transmission oil pump (mechanical type) is

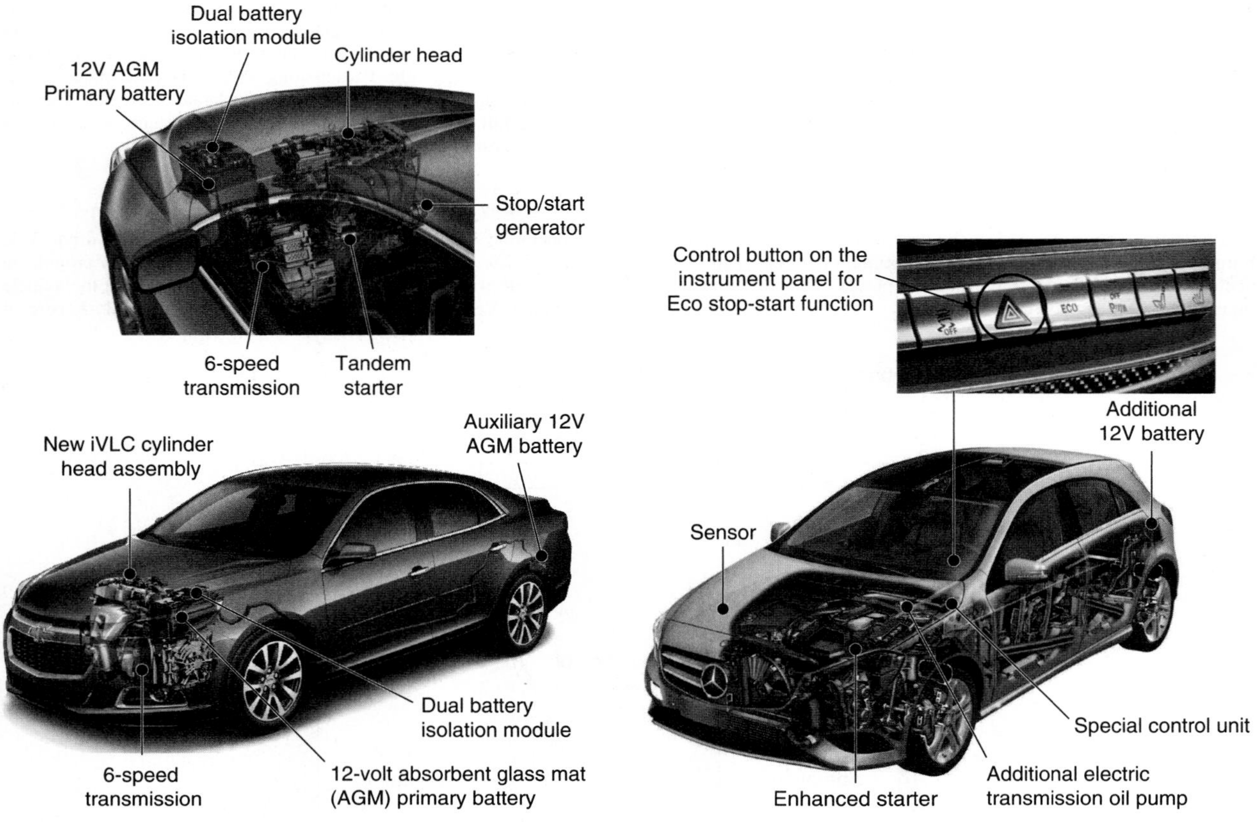

(a) The Chevrolet malibu stop-start[13]

(b) The Mercedes-Benz ECO start/stop[14]

Figure 8.2.10 Examples of micro hybrid or stop-start.

mounted behind the torque converter and driven by the engine. The AT can function properly only the transmission oil is pumped to reach certain pressure level. The transmission oil pressure drops when the engine stops running. For a vehicle with AT to have engine stop-start feature, a small electric motor/pump unit (mounted to the transmission housing) is needed. Figure 8.2.11 lays out the architecture of electric oil pump, enhanced starter, oil pressure sensor, and AGM battery in

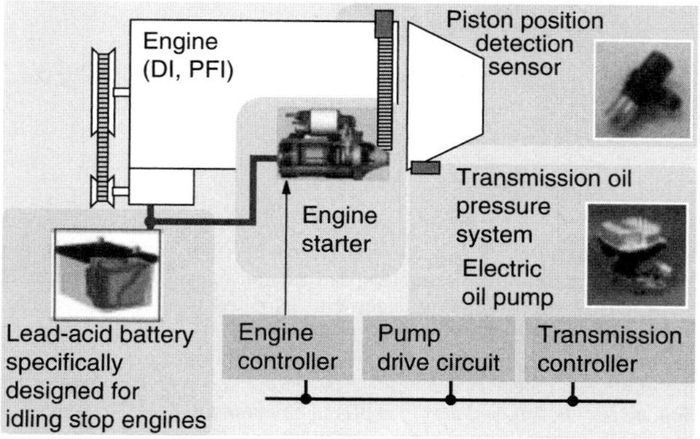

Figure 8.2.11 Electric pump for transmission oil in micro hybrid with AT.

a micro hybrid with AT. The electric motor/pump needs to function before restarting the engine so the vehicle can move forward smoothly. As soon as the engine runs, the electric motor/pump stops and conventional mechanical pump, driven by the engine, starts to operating. The additional requirements of electric motor/pump, sensors, and control algorithm increase the complication and cost to implement the engine idle stop-start feature in the vehicles with ATs. While micro hybrids are relative cheap and less complex to implement, they provide a modest improvement in fuel economy from 3-10 percent. The micro hybrids (engine stop-start) are best for urban delivery vehicles in heavy traffic. A micro hybrid might have minor regenerative braking dependent upon types of onboard energy storage systems such as adding supercapacitor (ultracapacitor) pack.[15,16]

Mild HEV

The mild HEV is the simplest form of hybridization and has features of engine idle stop-start, regenerative brake, and power boost by electric motor assist. The feature of engine-off driving is not available in the mild HEVs. A mild HEV has battery pack and electric machine but these are not powerful enough to propel the vehicle without operating the engine. The engine always runs and provides most of the power when the vehicle is moving. Most of the mild HEVs use Nickel–Metal-Hydride (NiMH) battery packs with 1 kWh capacity, 20 kW continuous power, and voltage ranging from 36 V to 140 V. Some of them recently start to use lithium-based battery packs, such as the Honda 2015 Civic hybrid has a 0.68 kWh, 144 V, 20 kW lithium-ion battery pack, and the Mercedes-Bens 2010 S400 hybrid has a 0.8 kWh, 126 V, 19 kW lithium-ion battery pack. The electric machine can be driven by a rubber belt, similar to convention's alternators, or it can be mounted to the crankshaft between the engine and the transmission. The mild HEVs normally can be classified as Belt-Integrated-Starter-Generator (B-ISG)[17-19] and Crankshaft-ISG (C-ISG).[20-22] The ISG, or Integrated-Starter-Alternator (ISA), is an electric machine with the combined functions of motoring (starter) and generating (generator or alternator). The B-ISG, also called Belt-Alternator-Starter (BAS), has an ISG mounted in the front of engine through accessory drive belt. This ISG directly replaces alternator and requires minor modifications to the belt and tensioner. Although the powertrain modification is minimal, the B-ISG has limited power between 1.5 kW and 15 kW and torque limited by the belt. The Toyota 2002 Crown, General Motors 2009 Saturn Vue Hybrid, and 2016 Chevrolet Malibu BAS+ are the examples of mild HEVs using B-ISG technique. Figure 8.2.12 shows configuration examples of the B-ISG mild HEVs.

The C-ISG, also called Flywheel-Alternator-Starter (FAS), replaces the conventional starter motor and alternator with a larger electric machine located between the engine flywheel and transmission. A C-ISG system typically utilizes conventional transmissions with significant packaging changes to accommodate the increased envelope

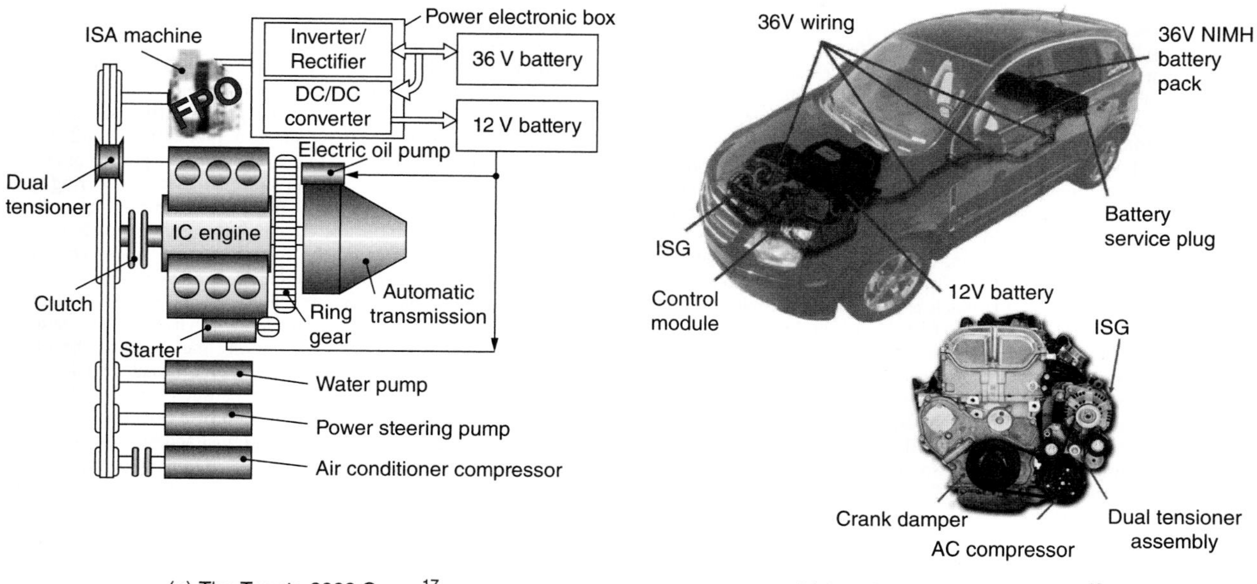

(a) The Toyota 2002 Crown[17]

(b) The Saturn 2009 Vue hybrid[19]

Figure 8.2.12 Examples of B-ISG mild HEVs.

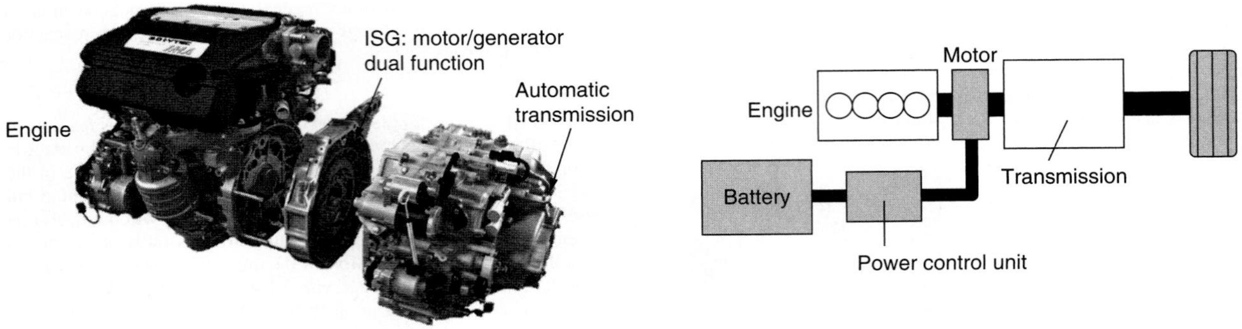

Figure 8.2.13 C-ISG powertrain architecture of the Honda mild HEVs.[22]

of the electric machine. The electric machine in the C-ISG system may be packaged around torque converter (using a ring type motor) such as,[21] or between the engine and the transmission where the torque converter is removed, such as Honda Integrated-Motor-Assist (IMA) system.[22] The C-ISG can reach high power level between 10 kW and 30+ kW. It provides engine idle stop-start and two additional functions. During braking or deceleration, the electric machine becomes a generator to charge the battery pack. During acceleration, the electric motor can pitch in to supplement the engine's torque. That yields improved performance and better fuel economy. The Honda's IMA system in Insight, Accord hybrid, Civic hybrid, CR-Z, and the Mercedes-Bens 2010 S400 hybrid are examples of C-ISG mild HEVs. Figure 8.2.13 shows the C-ISG powertrain architecture of the Honda mild HEVs. Like conventional vehicles, mild HEVs usually get better fuel economy on the highway than they do in city. The mild HEV typically can get about a 10-15 percent fuel economy improvement over a conventional vehicle.

Full HEV

The main difference of a full (or strong) HEV from mild HEV is its engine-off driving (EV driving). This type of HEV has larger battery pack and electric machine to drive the vehicle in pure EV mode for some time that also increases the fuel efficiency. That is why the full HEV provides better fuel economy. Unlike mild HEVs and conventional vehicles, the full HEVs usually get better fuel economy on the urban driving than they do in highway. Other features like engine idle stop-start, regenerative brake, and power boost by electric motor assist are also available. Most of the production full HEVs initially use NiMH battery packs with 2 kWh capacity, 40 kW continuous power, and voltage ranging from 200 V to 300 V. Such as the Toyota 2017 Prius V has a 1.3 kWh, 201 V, 27 kW NiMH battery pack. Some of them recently start to use lithium-ion battery packs like the Honda 2017 Accord hybrid, Toyota 2017 Prius three, BMW and Mercedes-Bens hybrids. The way a Prius, Accord hybrid, or Ford C-Max hybrid works is that the vehicle operates among engine, electric machine, and combination. This ensures its battery SOC is maintained within predetermined band (charge sustaining mode), which is part of the multiple tasks of its sophisticated control algorithm. The vehicle controller will not let the vehicle run out of electric power, and will switch on the engine as needed. Combining the operation of the vehicle's two power sources in a control manner, the vehicle operates as efficiently as possible. According to the hybrid powertrain architectures, the current production full HEVs can be classified into two types: Electric-Variable-Transmission (EVT) or Electronically-Controlled-Continuously Variable-Transmission (e-CVT),[23-32] and strong C-ISG.[33-42] The full hybrid C-ISG architecture is changed from mild C-ISG to strong hybridization by adding starting clutch and increasing the power rating of the electric machine and the corresponding battery capacity.

Through mechanical connections, the hybrid powertrain controller converts the engine mechanical power to three forms: constant rotational speed with increasing torque, constant torque with increasing rotational speed at the wheels, and electricity (thru generator). In the EVT, the continuously variable speed of transmission output shaft is achieved by controlling the electricity supplied to electric machine. The EVTs exploit the kinematic relationships afforded by planetary gear sets that play an important role in distributing the power among the engine and electric machines. The torque and speed inputs can be applied to sun, ring and carrier shafts in any combination. This multiple input and multiple output system provides a large number of combinations by appropriately driving one of the shafts with the engine and another by an electric machine. Owing to the existence of (at least) one electric motor as part of the drivetrain in the HEVs, these EVT configurations are quite attractive. Based on the numbers of planetary gear sets and their connections, the EVTs can be classified into one-mode and multiple-mode function. The "mode" means the continuously variable gear ratios, so a one-mode EVT indicates this EVT has continuously variable gear ratios within "one" specific speed range. A one-mode EVT is typically implemented with a single planetary gear set. A multiple-mode EVT contains multiple planetary gear sets and clutches. Both types of systems are capable of full hybrid functionality. The types of EVTs also are classified into input-, output-, and compound-split depending on the location of the power split device.[23,24]

The simplest configuration of the EVT is one-mode input-split type that has only one simple planetary gear set and two electric machines, which are all packaged within the transmission housing connected to the engine. A classic example of the one-mode input-split EVT is typically implemented in the first-generation (1997-2003) of Toyota Prius or any of Toyota's Hybrid Synergy Drive vehicles.[25,26] Figure 8.2.14*a* illustrates the connections of propulsion components with one planetary gear set where engine connected to the pinion carrier shaft, motor/generator 1 (MG1) connected to the sun gear, motor/generator 2 (MG2) and vehicle drive shaft connected to the ring gear. The portion of engine torque transmitted and split into the sun gear (MG1) and ring gear (MG2) is determined by the ratio of the sun gear to ring gear, which is fixed. Thus, the fraction of the engine power transmitted to the sun gear and ring gear depends on the speeds of these gears. The MG2 (traction motor) is used to control the required torque at vehicle wheels and MG1 (generator) is utilized to regulate the speed. The planetary unit relates the rotational speed of the vehicle wheel to the rotational speeds of the engine and MG1. By varying the speed of the MG1, the engine speed can maintain constant over a broad vehicle speed range. The MG1 absorbs the difference between wheel and engine speed. Toyota released an improved version of its one-mode EVT in 2004. The main difference between the original and the new one is the electric machine sizing rather than the powertrain architecture.[27] In the first and second-generation of Toyota's one-mode EVT, the drawbacks are the maximum speed and transmission efficiency in high speed driving. The vehicle maximum speed is mainly limited by the speed of the MG1 (usually functioning as a generator). The efficiency of the transmission heavily depends on the amount of power being transmitted over the electrical path. Especially in higher speed regimes (>120 kph or 70 mph), the efficiency (of the transmission alone) therefore drops

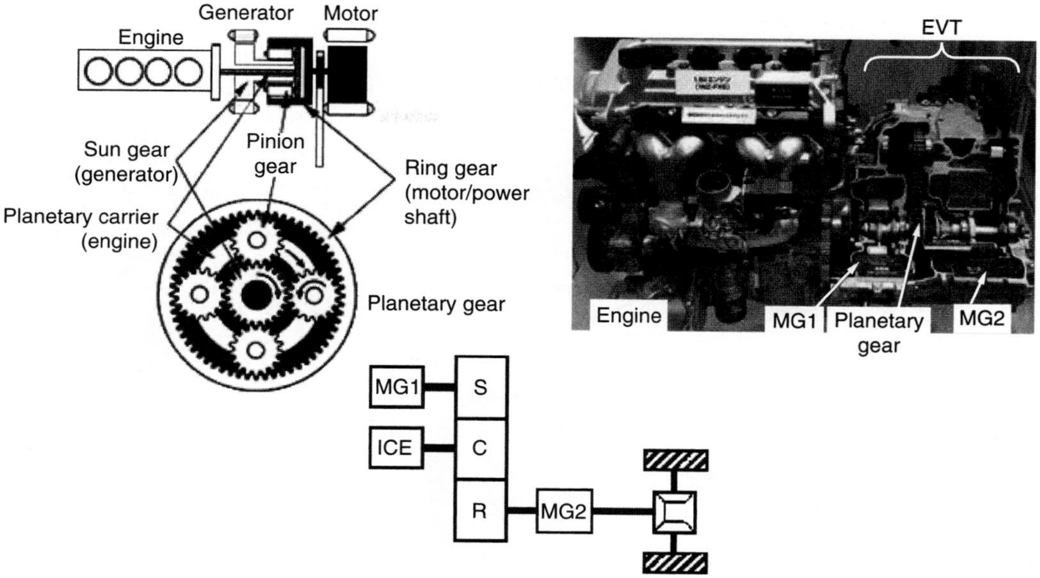

(a) The first-generation of Toyota Prius[25, 26]

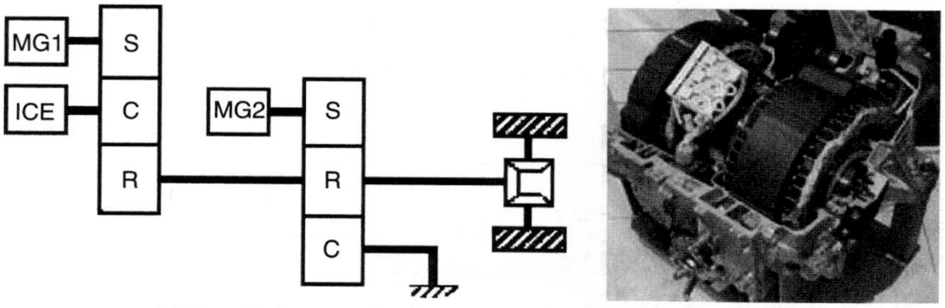

(b) The third-generation of Toyota Prius and Camry hybrid[28]

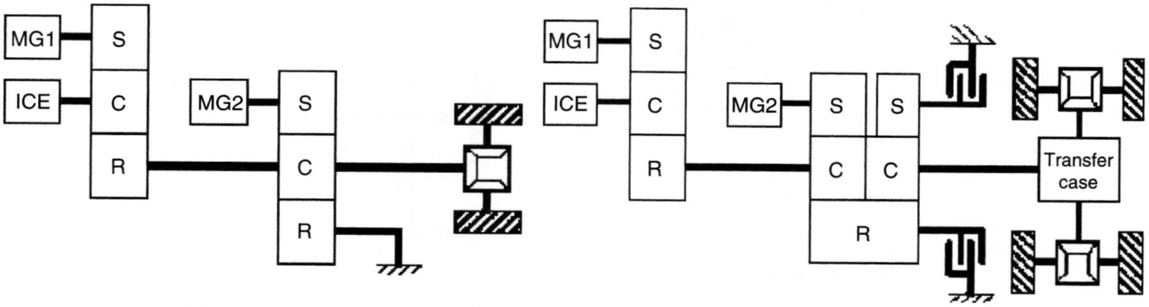

(c) Lexus hybrid RX400h (Highlander hybrid), GS450h/LS600h (four-wheel drive)[28]

Figure 8.2.14 Examples of one-mode input-split EVTs.

below that of a generic AT. This is due to two stages of energy conversions to meet the larger power demand in high speed driving, converting engine mechanical power to electricity by MG1 and then converting electricity to mechanical power by MG2. In 2010, Toyota released the third-generation of its EVT that both the component sizing and the powertrain architecture (with two planetary gear sets) were changed, and the performance has been improved.[28] Figure 8.2.14*b* shows the third-generation of Toyota's EVT. The structure of this EVT has the engine driving the carrier 1; the MG1 is on the sun gear 1; the planetary output ring gear 1 is on a common shaft with another planetary gear whose carrier (carrier 2) connects to the housing; and the MG2 is on the sun gear 2. As the MG2 directly couples to the output axle, there is a linear relation between the MG2 speed and vehicle speed. However, the requirement of the MG2 speed range is quite high which induces poor performance and poor efficiency at high vehicle speed. For a desired engine speed, the MG1 in the third-generation EVT could have a lower absolute speed than that of the first-generation. Equation (1) and (2) respectively is the speed equation of the first- and third-generation

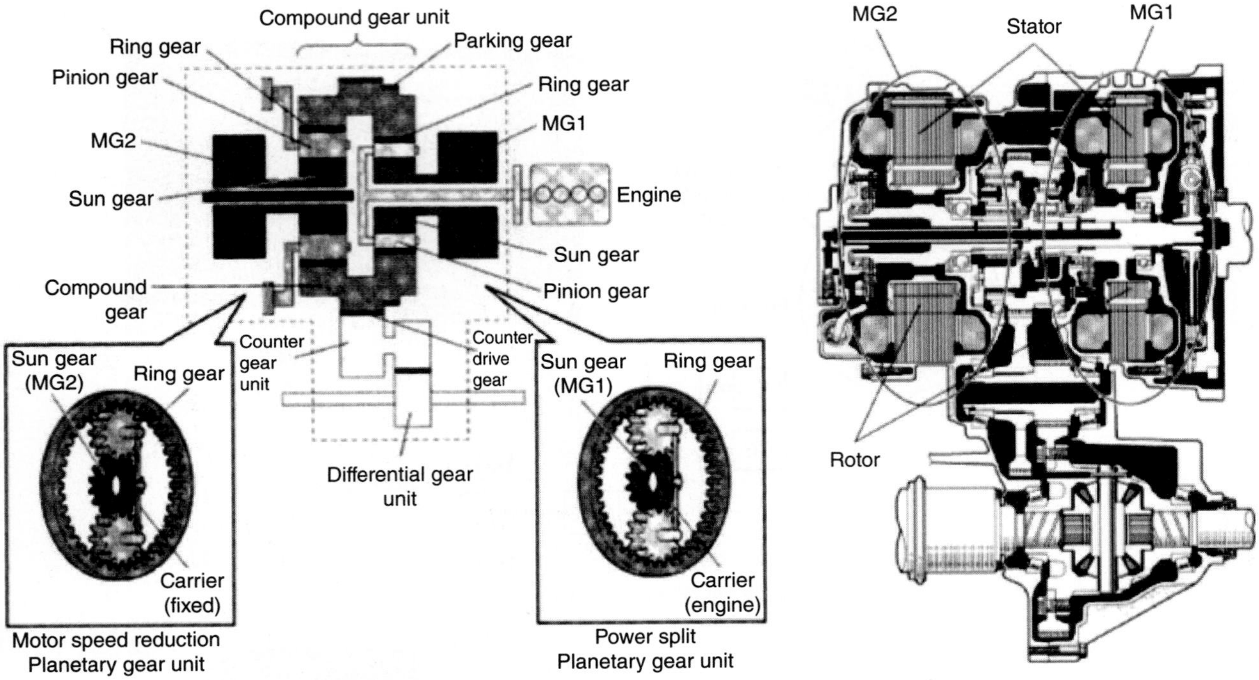

(d) The Nissan Altima hybrid

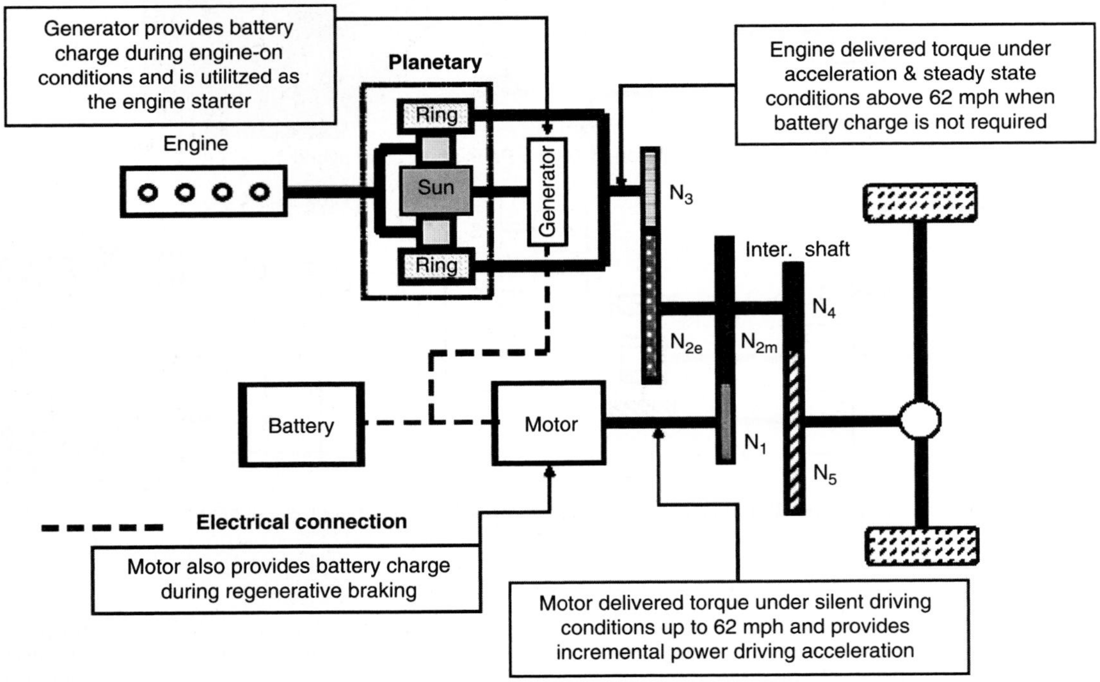

(e) The Ford C-Max hybrid

Figure 8.2.14 (*Continued*)

of one-mode input-split EVT. Other full HEVs use one-mode EVTs include Ford Escape hybrid, C-Max hybrid, Fusion hybrid, Lexus hybrid, and Nissan Altima hybrid.

$$w_{Engine} = \frac{N_r}{N_s + N_r} w_{MG2} + \frac{N_s}{N_s + N_r} w_{MG1} \quad (8.2.1)$$

$$w_{Engine} = \frac{N_{s2}}{N_{s1} + N_{s2}} w_{output} + \frac{N_{r2}}{N_{s1} + N_{s2}} w_{MG2} + \frac{N_{r1} + N_{s1} + N_{s2}}{N_{s1} + N_{s2}} w_{MG1} \quad (8.2.2)$$

In the Eqs (8.2.1 and 8.2.2), where N and w represents the teeth number and angular velocity of gear respectively, and subscripts s and r is respectively for sun and ring gear.

To overcome two shortcomings of one-mode input-split EVTs, various types of multiple-mode EVTs have been developed,[29,30] such as Chevrolet Tahoe hybrid, GMC Yukon hybrid, GMC Sierra 1,500 hybrid, Dodge Durango hybrid, Mercedes-Benz ML450 hybrid, etc. With additional planetary gear sets and clutches, the multiple-mode EVTs are able to switch the percentage of mechanically or electrically transmitted power. Figure 8.2.15 shows a General Motors two-mode EVT consisting of three planetary gear sets and three clutches.[30] In planetary gear set 1, the carrier shaft is connected to the engine output shaft through a clutch (C1). This EVT may perform in a hybrid or purely electrical manner depending on the on/off status of the C1. The ring gear shaft (in planetary gear set 1) and the sun gear shaft (in planetary gear set 3) are connected to the transmission output shaft through clutch 2 (C2). Similarly, the carrier shaft in planetary gear set 2 is connected to the output shaft through clutch 3 (C3). When C1 and C3 are engaged (with C2 is off), the EVT runs in the first-mode such that the engine drives the output shaft through planetary gear sets 1 and 2. Electric machine B, serving as a traction motor, also gives additional power to the output shaft through planetary gear set 2 and C3. Simultaneously, through the ring gear in planetary gear set 3, part of the engine power drives electric machine A which acts as a generator. Electric machine A converts the mechanical energy into electrical energy that is stored in the battery pack. In the second-mode, the engine drives the output shaft through planetary gear sets 1 and 3. Electric machine A, serving as a motor, also gives additional power to the output shaft through planetary gear set 3 and C2. Simultaneously, through the ring gear in planetary gear set 2, part of the engine power drives electric machine B that acts as a generator, or the power of unit B will drive the output shaft as B serves as a motor. The output shaft speed in the first- and second-mode is respectively derived in Eqs. (8.2.3) and (8.2.4).

$$w_{Output} = \frac{N_{r1} + N_{r2}}{N_{r1} + N_{r2} + N_{s2}} * w_B + \frac{N_{s2}}{N_{r1} + N_{r2} + N_{s2}} * w_{Engine} \quad (8.2.3)$$

$$w_{Output} = \frac{(N_{r1} + N_{r2} + N_{s2} + N_{s1})}{N_{r1} + N_{r2} + N_{s2}} * w_{Engine} - \frac{N_{s1}}{N_{r1} + N_{r2} + N_{s2}} * w_B \quad (8.2.4)$$

To package a two-mode EVT in the front-drive vehicle, the housing size of the EVT needs to be scaled down using two or one planetary gear set owing to the constrained space of engine compartment. Figure 8.2.16 shows a two-mode EVT consisting of one planetary gear

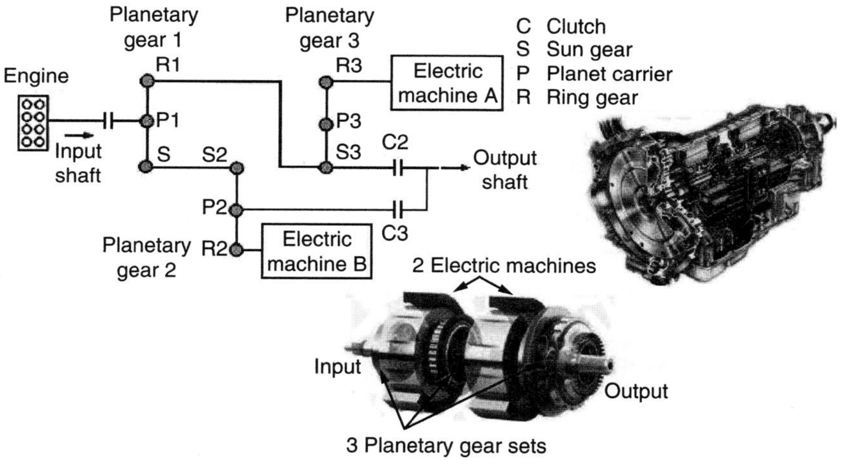

Figure 8.2.15 Schematic of a two-mode EVT (from US Patent 5,558,588)[30]

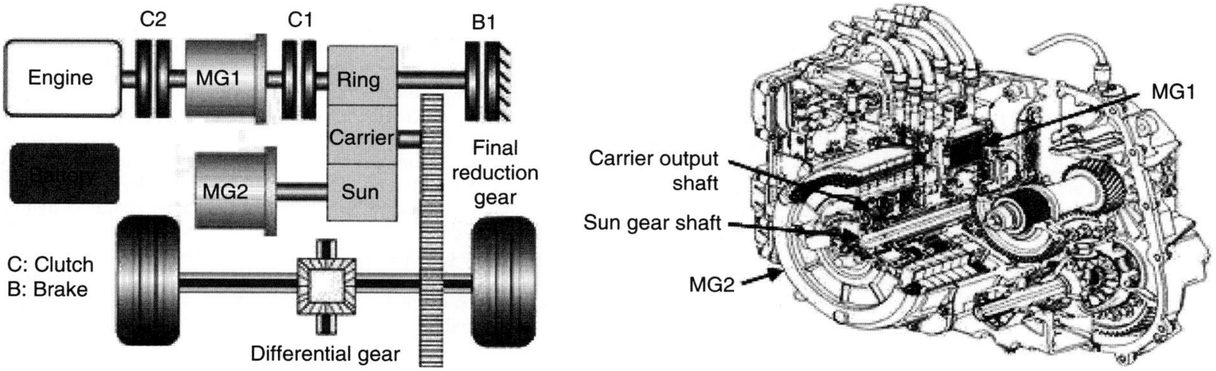

Figure 8.2.16 Schematic and operation conditions of the Chevrolet Volt two-mode EVT.[31]

set, two clutches, and one brake. This output-split EVT, with engine power split at the output, is implemented in Chevrolet Volt.[31] The engine and the electric machine MG1 are connected through clutch C2, and MG1 is connected to the ring gear through clutch C1. The ring gear is also connected to brake B1. The MG2 is connected to the sun gear, and the carrier is connected to the vehicle's final reduction gear. This configuration allows both the traction motor (MG2) and the generator (MG1) to provide tractive effort while simultaneously reducing electric motor speeds and the total associated electric motor losses. By setting the engagement and disengagement of the clutches and brake as shown Fig. 8.2.16*a*, four driving conditions are provided: (1) the first-mode, pure electric vehicle driven by MG2 at low road load, (2) the second-mode, pure electric vehicle driven by MG1 and MG2 at high road load, (3) series hybrid driven by MG2 with MG1 as a generator powered by engine, and (4) output-split or parallel hybrid driven by both engine and MG2. The two continuously variable modes are charge-depleting operation, the series and parallel are charge-sustaining operation.[32] In the charge-depleting operation, the battery is the only power source for vehicle propulsion. In the charge-sustaining operation, the engine serves as the main power source while also sustaining the battery SOC. The speed equations of these four driving conditions are illustrated as:

$$w_{Output} = \frac{N_s + N_r}{N_s} * w_{MG2} \quad \text{The first mode and series hybrid} \tag{8.2.5}$$

$$w_{Output} = \frac{N_s}{N_s + N_r} * w_{MG2} + \frac{N_r}{N_s + N_r} * w_{MG1} \quad \text{The second mode} \tag{8.2.6}$$

$$w_{Output} = \frac{N_s}{N_s + N_r} * w_{MG2} + \frac{N_r}{N_s + N_r} * (w_{MG1} + w_{Engine}) \quad \text{Parallel hybrid or output-split} \tag{8.2.7}$$

Operations	Clutch C1	Clutch C2	Brake B1
First mode	Off	Off	On
Second mode	On	Off	Off
Series	Off	On	On
Parallel/output-split	On	On	Off

Other configuration of full HEV is similar to the C-ISG, where an electric machine is mounted between the engine and transmission, with adding starting clutch and increasing the capacities of the electric machine and battery pack.[33-42] Some call this "parallel HEV" because engine or electric motor (or both of them simultaneously) can provide torque through transmission to propel the vehicle. The strong C-ISG configuration is attractive because one electric machine can be used to propel the vehicle, to start the engine, and to be function as generator. The transmission in this full HEV can be AT, Automatic-Manual-Transmission (AMT), or CVT. Figure 8.2.17*a* illustrates a full hybrid powertrain with an electric machine mounted between the engine and transmission. An automatic clutch usually is implemented between the engine and the electric machine.[34] Figure 8.2.17*b* shows the operations. The electric machine performs motoring or generating according to five conditions: (1) motoring for engine-off driving, (2) motoring to restart the engine during vehicle standstill, (3) generating during deceleration and braking, (4) motoring for power boost, and (5) generating during vehicle cruising. Therefore, the control strategy for electric machine drive (or power electronics) is critical and complicated for this full HEV. Due to the simple architecture of fewer components, this type of full HEVs have been deployed by many automakers, such as examples shown in Fig. 8.2.18 for BMW ActiveHybrid,[37] Hyundai Sonata/Kia Optima hybrid,[38] Infiniti Direct Response hybrid,[39] Mercedes-Benz P2-20 OM651 hybrid powertrain,[40] Nissan Tino,[41] and Volkswagen/Audi hybrid.[42]

PHEV

The PHEV is designed to be used both as an EV for city driving and as a HEV when the battery pack is depleted or for highway driving. The PHEVs have been on the market since December 2010. The PHEV differs from a full HEV only by the fact that it allows one to plug in a cable to an external source of electric power, usually a regular wall socket at home or elsewhere to charge the vehicle's battery pack. A PHEV is operated exclusively (except during maximum acceleration) on electric power alone until its battery SOC is depleted to a predetermined level, at which time the engine will engages. The vehicle continues to travel on the engine and battery/electric motor as a full HEV. The battery pack can then be recharged by the engine whenever the vehicle is at cruising speed or parked, depending on the hybrid powertrain architectures. Since plug-in allows a fair amount of external utility system energy to charge the vehicle battery pack, it is helpful to use a larger battery than in a full HEV. The larger size of battery pack and electric machine in the PHEV increases the fuel economy and initial electric range of the vehicle when the battery pack is fully charged. The PHEV also has features of engine stop-start, regenerative brake, electric motor power boost, and engine-off driving.

The PHEV uses battery pack to power the electric motor for the initial "All Electric Range (AER)" of the vehicle. This is one step further along the "bridge" toward a pure electric vehicle. The capacity of the battery pack, vehicle weight, and driving pattern determines how many kilometers (or miles) a PHEV can travel purely on electric power before the engine must be used to power further travel of the vehicle. A full HEV has the engine-off drive feature but its typical AER is only about 2.5-3.2 kms (1.6-2 miles). The number of kilometers (or miles) of the AER typically defines a PHEV. For an example, a PHEV20 has 20 kms (or miles) AER. Cost and weight of the battery pack versus AER needed by the customers determines the AER offered in the various PHEVs made by automakers. Examples as Toyota 2017 Prime PHEV has 35-km (22-mile) AER, 2016 Chevrolet Volt has 85-km (53-mile) AER, and 2016 Ford C-Max has 32-km (20-mile) AER. The dual requirements for an extended AER, 32-80 kms (20-50 miles), as well as maintaining high power availability at low SOC (e.g., 30 percent SOC), impose very stressful conditions on the battery. All the current production PHEVs use lithium-based battery packs with about 20 kWh capacity, 100 kW continuous power, and 300 V. Examples like that 2012 Prius PHEV is 4.4 kWh, and Chevrolet 2016 Volt is 18.4 kWh with 136 kW peak power.

EV

The EV is currently considered to be a long-term solution for ground transportation because of its single onboard electrical energy source, high energy efficiency, and environmental friendly. EVs convert about 59-62 percent of the electrical energy from the grid to power at the wheels while conventional gasoline vehicles only convert about 17-21 percent of the energy stored in gasoline to power at the wheels.[43] EVs emit no tailpipe pollutants although the power plant producing the electricity may emit them. Electric motors also provide quiet, smooth operation and larger acceleration and require less maintenance than ICEs. The drawbacks of the EVs involve limited driving range, high cost, and recharging time. The driving range is typically limited to 190 kms (120 miles) on a full charge, although a few models can go 320-640 kms (200-400 miles). The large lithium-based battery packs are expensive and may need to be replaced one or more times. The cost of battery pack dominates one third to a half of the entire cost of a typical BEV. Fully recharging the battery pack can take 4-8 hours. Even a fast charge to 80 percent capacity can take 30 min. All the current production BEVs use lithium-based battery packs with about 20-60 kWh capacity, 100 kW continuous power, and over 300 V.

The onboard electrical energy source for EVs could be hydrogen fuel cells or variety types of battery packs. The concept of the BEV is essentially simple as shown in Fig. 8.2.19. The vehicle consists of an electrochemical battery pack for energy storage, electric machine(s), power electronics, and multiple controllers. The battery pack is normally recharged from mains electricity via a plug similar to the PHEV. The controllers manage the power supplied to the motor to meet vehicle speed in forward and reverse. Referring to Fig. 8.2.7, this is normally known as a two-quadrant control in forward (quadrant-I) and backward

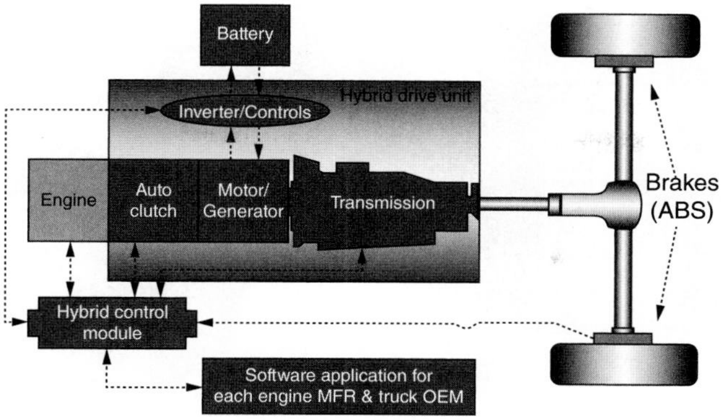

(a) Layout of strongC-ISG with clutch for a full HEV[34]

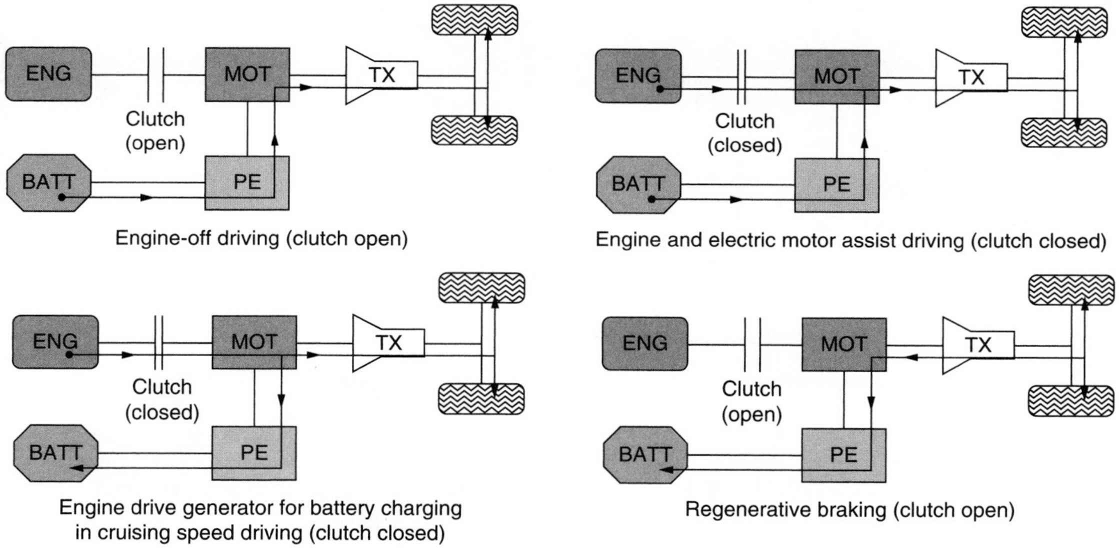

(b) Operations of strong C-ISG with clutch for a full HEV

Figure 8.2.17 Layout and operations of strong C-ISG with clutch for a full HEV.

(quadrant-III). As the regenerative braking applied, it is known as the four-quadrant control. BEVs obviously do not have a fossil-fuel tank, so they have limited driving range that is mainly dependent upon the types of battery chemistries and size of battery pack. Additionally, driving condition and ambient temperature (e.g., of high-speed driving and cold weather conditions) will reduce the range. In addition, the total amount of battery capacity quoted by automakers does not mean all those kWhs are usable. Automakers place extra battery capacity in the EV because the battery life will last longer if it is never fully charged or discharged. The Nissan Leaf BEV with its 24 kWh battery pack (16 kWh available to 67 percent DOD) can go around 100 miles before the batteries must be recharged. Tesla's Model S has a range of up to about 300 miles on a fully charged battery. A BEV typically requires large capacity battery pack to achieve reasonable range. Traction batteries are usually optimized for high capacity in the case of BEVs and for high power in the case of HEVs. The BEV battery operates down to a deep DOD for long range whereas the HEV operates at a shallow DOD for long life. A typical BEV uses about 150-250 Wh per mile depending on the terrain and the driving pattern. The battery must be capable of regular deep discharge (80 percent DOD) operation. It is designed to maximize energy content and deliver full power even with deep discharge to ensure long range. A range is required to satisfy the needs of different size vehicles and usage patterns.

Series versus Parallel HEV

According to the operating features, HEVs can be classified into stop-start (or micro hybrid), mild hybrid, full (or strong) hybrid, PHEV, and EV. The HEVs also simply group into series, parallel, and series/parallel hybrid based on the path of traction power flow. In a series HEV, the engine is only used to drive the generator for electricity generation. The electrical system provides a variable speed and the electric motor provides the full driving power. In other words, the engine never mechanically turns the drive wheels and only electric motor propels the vehicle. The battery pack requirement is similar to EV but lower capacity needed since the charge is kept topped up by the engine. The series HEV needs electric summation (like electrical coupling) to gather

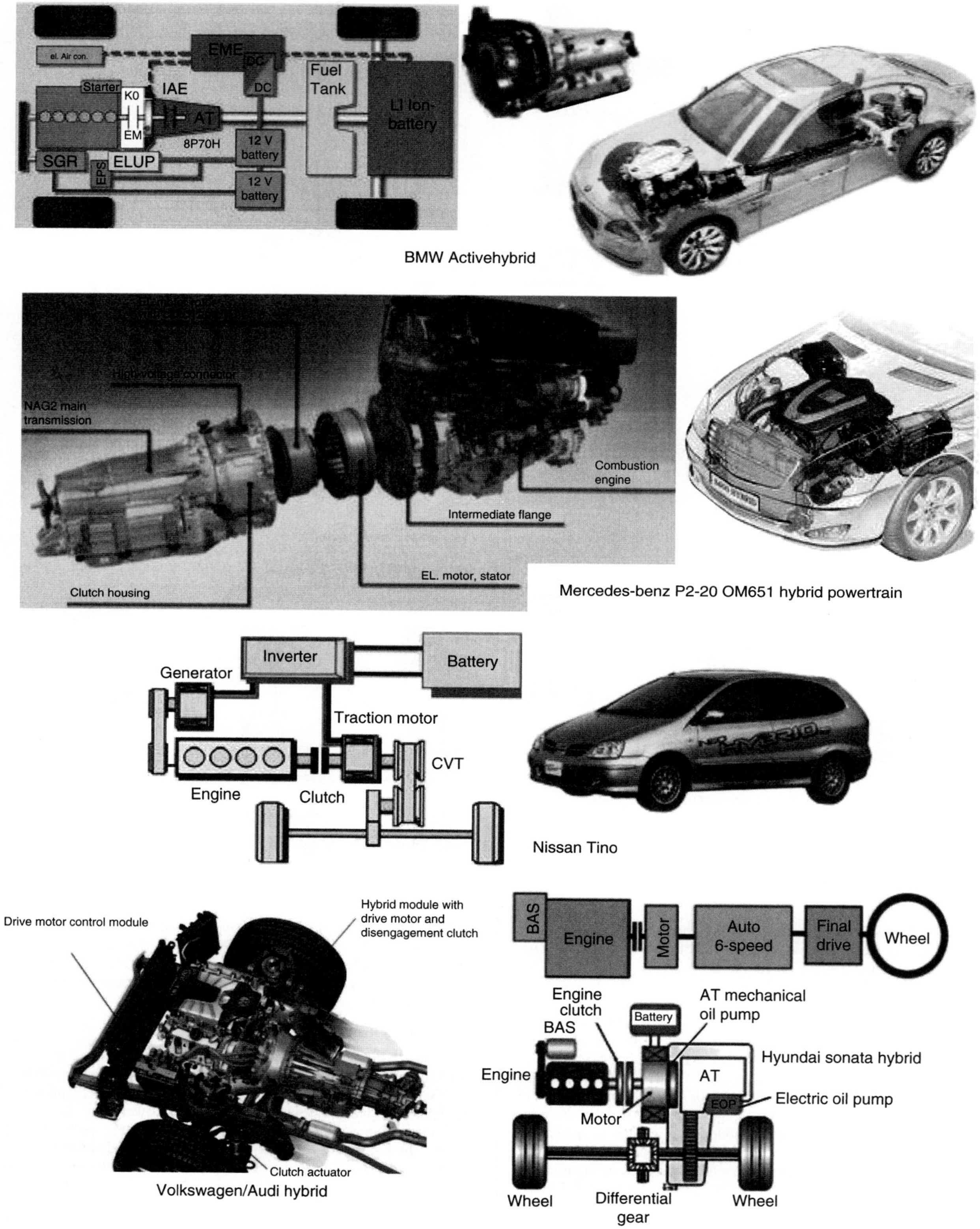

Figure 8.2.18 Examples of strong C-ISG with clutch for full HEVs.[37-42]

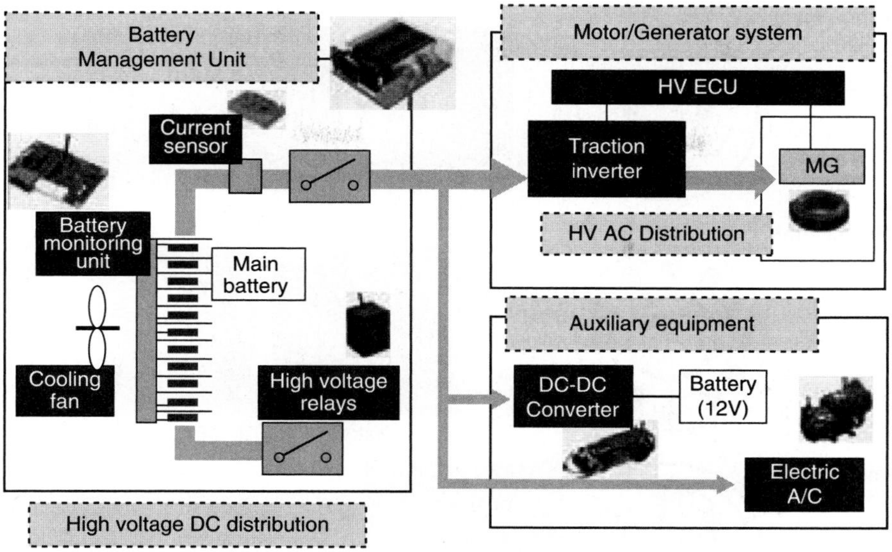

Figure 8.2.19 Example of EV propulsion architecture.

electricity from battery pack and engine/generator set (or fuel cell stacks). A FCEV and diesel locomotive is actually a series hybrid. For a parallel HEV, both the engine and the electric motor provide power to the wheels. The parallel HEV needs mechanical summation (like gear train) to collect the mechanical torque from engine and electric motor. The share of the load taken by the electric motor can range from zero to 100 percent depending on the operating conditions and the design goals. The battery capacity may be as low as 2 KWh but it must deliver short duration power boosts requiring very high currents for acceleration and hill climbing. Most of the full HEVs and PHEVs have series, parallel, and combined series/parallel power flow path.

COMPONENTS OF HYBRID AND ELECTRIC VEHICLES

The major components of hybrid and electric vehicles are categorized in two groups: drivetrain and onboard energy storage. The drivetrain is an assembly of components permitting motion and mechanical power transmission. It comprises the energy conversion devices (ICE, electric machine, or fuel cell), coupling mechanism, mechanical transmission, differential gear, and wheels. The onboard energy storage system accumulates and releases energy in mechanical, electrochemical, or chemical form. Since a topic of ICE is presented in other section, the transmissions, electric machines, and onboard energy storage systems are briefly discussed here.

Transmission

The mechanical transmissions are needed in conventional vehicles and most of the HEVs. The purposes of the transmission are to modify the engine torque and speed to match the requirements of the vehicle, allow the engine to be started without being connected to the drivetrain by means of a neutral gear position, and allow the vehicle to pull away from rest. The engine torque is relatively low at low speeds, so the transmission provides a torque multiplication that allows the engine to accelerate the vehicle from rest. A HEV's transmission interacts with the electric machines (motor/generator), engine, and controllers to provide traction power to the vehicle. Various types of transmissions including AT, AMT, CVT, EVT (e-CVT), and MT are available for HEVs based on the hybrid powertrain architectures. The series HEV and EV might only need simple gearbox to change vehicle speed and torque.

Electric Machine

The electric propulsion components in HEVs/PHEVs/EVs include electric machine (motor/generator), onboard energy storage, power electronics, safety devices, instrumentation, and auxiliaries. An electric machine in motoring mode converts electrical energy to mechanical energy to propel the vehicle wheels or start the engine. The electric machine also can act as a generator to create electricity from mechanical torque that it receives from engine during vehicle cruising or from the wheels during regenerative braking. The electric machines are broadly organized into two main categories: AC (alternating current) and DC (direct current) electric machines. The DC synchronous and AC induction electric machines all find potential uses in vehicles as starters, alternators, and motors for auxiliaries (such as windshield wiper, power steering, power seat, power window, etc.). In the case of traction electric machines, AC induction, Permanent Magnet (PM), and Switched Reluctance (SR) are the relevant category, as shown in Fig. 8.2.20.

The Permanent Magnet Brushless (PMBL) electric machine essentially is a permanent magnet AC motor that is integrated with an electronic controller which runs on DC, and basically makes the AC (required by the motor) by alternately switching on and off different windings on the motor. The permanent magnets, mounted on the rotor surface, provide excitation resulting in high power factor, high torque density, and cooler rotor. The PMBL electric machines are widely used in current production HEVs and PHEVs. Because a HEV or PHEV has engine installed in the engine compartment, the space for electric machine is constrained. The packaging space is particularly limited in the EVTs. The PMBL machine has the best total system performance, packaging advantages, lower inverter size, and high efficiency because of the absence of rotor copper losses (5-10 percent better than typical AC induction motors and 8-12 percent better than brush DC motors). The drawbacks of PMBL machine includes expensive rare-earth permanent magnets, difficult to handle in manufacturing, higher complexity of electronic system, and the possibility of demagnetization of the permanent magnets. The AC induction machine operates by the electric currents induced from the stator field in the rotor that is in a squirrel cage or a wound structure. It has advantages of low cost per horsepower, low torque ripple, and available in wide range of power ratings. The cons include inefficient at light loads, complicated for speed control (varying stator input current frequency required), difficult for position control (field orientation required), and larger size than PMBL machine (with the same rated power). Most current production

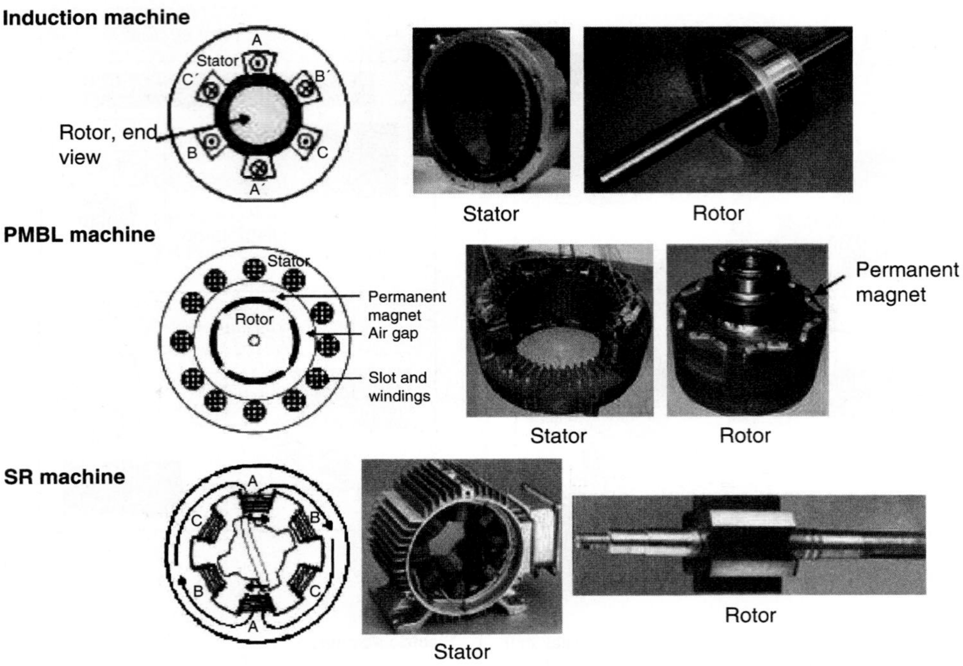

Figure 8.2.20 Electric machines for traction in hybrid and electric vehicles.

EVs use AC induction machines because of their lower cost. An EV has no engine and thus more rooms for electric machines in the front compartment. Although the SR machines are not found in current production hybrid and electric vehicles, they have highly possible applications in the future. The SR machine is a simple construction with the rotor formed in a solid steel shaft. The operating heat is generated mainly in the stator therefore the heat can be easily transferred away. The SR machine has benefits of low cost on motor structure, high reliability, high starting and overload torque, and wide speed range. The main drawbacks are acoustically noisy, difficult to control, complex and expensive power electronics for motor drive, high vibration, and torque ripple.

Onboard Energy Storage System

The electrochemical batteries, ultracapacitors (supercapacitors), electromechanical flywheels, mechanical flywheels, and fluid accumulators are the candidates of onboard energy storage systems in hybrid vehicles. The mechanical flywheels and fluid accumulators are respectively used in mechanical hybrids and hydraulic hybrids. For the hybrid and electric vehicles, the energy storage system is typically a battery pack, ultracapacitors, and/or electromechanical flywheel. The onboard energy storage system stores electricity from the generator, which helps power the electric motor and other auxiliary electrical devices in the hybrid and electric vehicle. The ultracapacitor and electromechanical flywheel have high specific power, which means that they can take in (recharge) and give out (discharge) energy very quickly. Therefore, they are suitable for vehicle electric launch, power boost, and regenerative braking. However, the amount of energy they can store is currently rather small. The poor specific energy of the ultracapacitor and electromechanical flywheel precludes their applications in the EVs, PHEVs, or full HEVs having large AERs. The electrochemical battery has relative higher specific energy and energy density. The lead-acid (PB-A) batteries are used for conventional vehicle starters, micro HEV or engine idle stop-start function. The NiMH batteries are found in many current production HEVs. All the current production EVs and PHEVs use Lithium-ion batteries to meet their designed AER.

REFERENCES

1. Walters J., Husted H., and Rajashekara K. Comparative study of hybrid powertrain strategies, Society of Automotive Engineers (SAE) paper 2001-01-2501, 2001.
2. Berman B. History of hybrid vehicles, http://www.hybridcars.com/history-of-hybrid-vehicles/, 2011 (accessed July 30 2016).
3. Valeo i-StARS and ReStart, Valeo. http://www.valeo.com/en/our-activities/powertrain-systems/technologies/systems-stop-start-48.html/, 2016 (accessed July 30 2016).
4. Kremer M. and Hulshorst T. In-market application of start-stop systems in European market, FEV Inc. 2011.
5. Richard D. and Dubel Y. Valeo StARS technology: a competitive solution for hybridization. In: *IEEE Power Conversion Conference*, Nagoya, Japan. 2-5 April 2007.
6. Bosch stop-start system, Robert Bosch GmbH. http://de.bosch-automotive.com/en/parts_and_accessories/motor_and_sytems/start_stop_system/overview_start_stop_sytem_1/ (accessed July 30 2016).
7. Tandem solenoid starter for stop/start systems, Denso. http://www.globaldenso.com/en/newsreleases/events/otherexhibitions/2012/fisita12/files/fisita12_stop_start_system.pdf/ (accessed July 30 2016).
8. Brake ECU, Toyota technical training. http://techno-fandom.org/~hobbit/cars/cours-section6.pdf/andhttp://priuschat.com/attachments/brakeecu-pdf.9308/, 2016 (accessed July 30 2016).
9. Gray T. and Shirk M. 2010 Toyota Prius VIN0462 hybrid electric vehicle battery test results, The Idaho National Laboratory, 2013.
10. Sharer P., Leydier R., and Rousseau A. Impact of drive cycle aggressiveness and speed on HEVs fuel consumption sensitivity, SAE paper 2007-01-0281, 2007.
11. Ehsani M., Gao Y., Gay S., and Emadi A. Modern Electric, Hybrid Electric, and Fuel Cell Vehicles: Fundamentals, Theory, and Design, 2nd ed, CRC Press, ISBN 978-1-4200-5398-2, 2010.
12. Dhand A., Cho B., Walker A., Kok D., Burgess M., and Semar B. Optimization potential of the vehicle launch performance for start-stop micro-hybrid vehicles. *Proc IMechE Part D: J. Automobile Engineering*. 2010; 224(2): 1059-1070.
13. Simger D. Chevrolet 2014 Malibu stop/start technology, http://gmauthority.com/blog/2015/04/gm-eyeing-issue-with-outgoing-chevrolet-malibu-stopstart-system-is-a-recall-next/, 2015 (accessed July 30 2016).
14. Mercedes-Benz ECO start/stop function, https://techcenter.mercedes-benz.com/en/eco_start_stop/detail.html/ (accessed July 30 2016).

15. Rizoug N., Feld G., Barbedette B., and Sadoun R. Association of batteries and supercapacitors to supply a micro-hybrid vehicle. In: *2011 IEEE Vehicle Power and Propulsion Conference*, Chicago, USA. 6-9 Sept. 2011.

16. Burke A. and Zhao J. Supercapacitors in micro- and mild hybrids with lithium titanate oxide batteries: vehicle simulations and laboratory tests. In: *European Battery, Hybrid and Fuel Cell Electric Vehicle Congress*, Brussels, Belgium. 2-4 Dec. 2015.

17. Shida Y., Kanda M., Ohta K., Furuta S., and Ishii J. Development of inverter and power capacitors for mild hybrid vehicle (MHV)—Toyota "Grown". *Int. J. Automotive Technology.* 2003; 4(1): 41-45.

18. Walters J., Krefta R., Gallegos-Lopez G., and Fattic G. Technology considerations for belt alternator starter systems, SAE paper 2004-01-0566, 2004.

19. Tamai G., Jeffers M., Lo C., Thurston C., Tarnowsky S., and Poulos S. Development of the hybrid system for the Saturn Vue hybrid, SAE paper 2006-01-1502, 2006.

20. Ao G., Qiang J., Zhong H., Yang L., et al. "Exploring the fuel economy potential of ISG hybrid electric vehicles through dynamic programming." *Int. J. Auto. Technol.* 2007; 8(6):781-790.

21. Evans D. and Van Maanen K. Electric machine powertrain integration for GM's hybrid full-size pickup truck, SAE paper 2003-01-0084, 2003.

22. Hanada K., Kaizuka M., Ishikawa S., Imai T., Matsuoka H., and Adachi, H. Development of a hybrid system for the V6 midsize sedan, SAE paper 2005-01-0274, 2005.

23. Hong S., Choi W., Ahn S., Kim Y., and Kim H. Mode shift control for a dual-mode power-split-type hybrid electric vehicle. *Proc IMechE Part D: J. Automobile Engineering.* 2014; 228(10): 1217-1231.

24. Zhu F., Chen L., and Yin C. Design and analysis of a novel multimode transmission for a HEV using a single electric machine. *IEEE Trans Vehicular Technology.* 2013; 62(3): 1097-1110.

25. Toyota hybrid synergy drive, http://www.toyota-global.com/innovation/environmental_technology/hybrid/ (accessed July 30 2016).

26. Ozeki T. and Umeyama M. Development of Toyota's transaxle for mini-van hybrid vehicles. SAE paper 2002-01-0931, 2002.

27. Olszewski M. Evaluation of 2004 Toyota Prius hybrid electric drive system. Oak Ridge National, Laboratory U.S. Department of Energy, FreedomCAR and Vehicle Technologies, EE-2G, Washington, D.C. May 2005, http://www.osti.gov/scitech/servlets/purl/890029/ (2016, accessed January 7 2016).

28. Kim N., Rousseau A., and Rask E. Vehicle-level control analysis of 2010 Toyota Prius based on test data. *Proc IMechE Part D: J. Automobile Engineering.* 2012; 226(11): 1483-1494.

29. Tsai W., Schultz G., and Higuchi N. A novel parallel hybrid transmission, *Trans. ASME, J. Mech. Des.* 2001; 123: 161-168.

30. Zhang D., Chen J., Hsieh T., Rancourt J., and Schmidt MR. Dynamic modelling and simulation of two-mode electric variable transmission, *Proc IMechE Part D: J. Automobile Engineering.* 2001; 215: 1217-1223.

31. Zhang X., Li T., Kum D., and Peng H. Prius$^+$ and Volt$^-$: configuration analysis of power-split hybrid vehicles with a single planetary gear. *IEEE Trans. Vehicular Technology.* 2012; 61(8): 3544-3552.

32. Kim N., Kwon J., and Rousseau A. Comparison of powertrain configuration options for plug-in HEVs from a fuel economy perspective. SAE paper 2012-01-1027, 2012.

33. Lefief P. P2 hybrid electrification system cost reduction potential constructed on original cost assessment, Analysis report BAV 10-683-001_4, FEV Inc. 2014.

34. Eaton electric hybrid, Eaton. http://www.eaton.com/Eaton/ProductsServices/HybridPower/ElectricHybrid/index.htm/ (accessed July 30 2016).

35. Liao Y., O'Malley M., and Quail, A. "Experimental assessments of parallel hybrid medium-duty truck," *SAE Int J Commer Veh.* 2014; 7(1): 305-314.

36. Lin C., Peng H., Grizzle, J., Liu, J., and Busdiecker M. Control system development for an advanced-technology medium-duty hybrid truck, SAE paper 2003-01-3369, 2003.

37. Griebel CO., Rabenstein F., Kluting M., Kessler F., Kretschmer J., and Hockgeiger E. The full-hybrid powertrain of the new BMW ActiveHybrid 5. In: *20th Acahen Colloquium Automobile and Engine Technology*, Aachen, Germany. 10-12 October 2011.

38. Hyundai Sonata Hybrid 2011-2012, AESwave.com, http://resources.aeswave.com/ACDC-2012-Sample-Pages-149-152.pdf/ (accessed July 30 2016).

39. Q50 Direct Response Hybrid, Infiniti, http://www.infinitiusa.com/now/technology/direct-response-hybrid/ (accessed July 30 2016).

40. Keller U., Godecke T., Weiss M., Enderle C., and Henning G. Diesel hybrid—the next generation of hybrid powertrains by Mercedes-Benz. In: *33rd Int. Vienna Motor Symposium*, Vienna, Germany. 26-27 April 2012.

41. Matsuo I., Miyamoto T., and Maeda H. The Nissan hybrid vehicle, SAE paper 2000-01-1568, 2000.

42. Bahns A. and Gruszczynski B. Additional powertrain components on Volkswagen Touareg hybrid. In: *SAE 2011 Electronic Systems for Vehicle Propulsion Symposium*, Troy, Michigan, USA. 8-9 November 2011.

43. All-Electric Vehicles, Energy Efficiency & Renewable Energy, U.S. Department of Energy. https://www.fueleconomy.gov/feg/evtech.shtml/ (accessed August 1 2016).

8.3 RAILWAY ENGINEERING

by V. Terrey Hawthorne, Keith L. Hawthorne, and Mehdi Ahmadian

(in collaboration with E. Thomas Harley, Charles M. Smith, Robert B. Watson, and C. A. Woodbury)

REFERENCES: *Proceedings,* Association of American Railroads (AAR), Mechanical Division. *Proceedings,* American Railway Engineering Association (AREA). *Proceedings,* American Society of Mechanical Engineers (ASME). "The Car and Locomotive Cyclopedia," Simmons-Boardman. J. K. Tuthill, "High Speed Freight Train Resistance," *University of Illinois Engr. Bull.,* **376,** 1948. A. I. Totten, "Resistance of Light Weight Passenger Trains," *Railway Age,* **103,** Jul. 17, 1937. J. H. Armstrong, "The Railroad—What It Is, What It Does," Simmons-Boardman, 1978. Publications of the International Government-Industry Research Program on Track Train Dynamics, Chicago. Publications of AAR Research and Test Department, Chicago. Max Ephraim, Jr., "The AEM-7—A New High Speed, Light Weight Electric Passenger Locomotive," ASME RT Division, 82-RT-7, 1982. W. J. Davis, Jr., *General Electric Review,* October 1926. R. A. Allen, Conference on the Economics and Performance of Freight Car Trucks, October 1983. "Engineering and Design of Railway Brake Systems," The Air Brake Association, Sept. 1975. Code of Federal Regulations, Title 40, Protection of Environment. Code of Federal Regulations, Title 49, Transportation. *Standards and Recommended Practices and Proceedings,* Association of American Railroads (AAR) Mechanical Division, www.aar.com. "Standard for the Design and Construction of Passenger Railroad Rolling Stock," American Public Transportation Association (APTA), www.apta.com. "Manual for Railway Engineering," American Railway Engineering and Maintenance-of-Way Association (AREMA), www.arema.com. *Proceedings*, American Society of Mechanical Engineers (ASME), www.asme.org. "The Car and Locomotive Cyclopedia of American Practices," Simmons-Boardman Books, Inc., www.transalert.com. Publications of AAR Research and Test Department, www.aar.com. Code of Federal Regulations, Title 40, Protection of Environment and Title 49, Transportation, published by the Office of the Federal Register, National Archives and Records Administration. "The Official Railway Equipment Register—Freight Connections and Freight Cars Operated by Railroads and Private Car Companies of North America," R.E.R. Publishing Corp., East Windsor, NJ, www.cbizmedia.com.

DIESEL-ELECTRIC LOCOMOTIVES

Diesel-electric locomotives and electric locomotives are classified by wheel arrangement; letters represent the number of adjacent driving axles in a rigid truck (A for one axle, B for two axles, C for three axles, etc.). Idler axles between drivers are designated by numerals. A plus sign indicates articulated trucks or motive power units. A minus sign indicates separate nonarticulated trucks. This nomenclature is fully explained in RP-5523, issued by the Association of American Railroads (AAR). Virtually all modern locomotives are of either B-B or C-C configuration.

The high efficiency of the diesel engine is an important factor in its selection as a prime mover for traction. This efficiency at full or partial load makes it ideally suited to the variable service requirements of routine railroad operations. The diesel engine is a constant-torque machine that cannot be started under load and hence requires a variably coupled transmission arrangement. The electric transmission system allows it to

Table 8.3.1 Locomotives in Service in North America

Locomotive*					Tractive effort			
Builder	Model	Service	Arrangement	Weight min./max. (1,000 lb)	Starting† for min./max. weight (1,000 lb)	At continuous‡ speed (mi/h)	Number of cylinders	Horsepower rating (r/min)
Bombardier	HR-412	General purpose	B-B	240/280	60/70	60,400 (10.5)	12	2,700 (1,050)
Bombardier	HR-616	General purpose	C-C	380/420	95/105	90,600 (10.0)	16	3,450 (1,000)
Bombardier	LRC	Passenger	B-B	252 nominal	63	19,200 (42.5)	16	3,725 (1,050)
EMD	SW-1001	Switching	B-B	230/240	58/60	41,700 (6.7)	8	1,000 (900)
EMD	MP15	Multipurpose	B-B	248/278	62/69	46,822 (9.3)	12	1,500 (900)
EMD	GP40-2	General purpose	B-B	256/278	64/69	54,700 (11.1)	16	3,000 (900)
EMD	SD40-2	General purpose	C-C	368/420	92/105	82,100 (11.0)	16	3,000 (900)
EMD	GP50	General purpose	B-B	260/278	65/69	64,200 (9.8)	16	3,500 (950)
EMD	SD50	General purpose	C-C	368/420	92/105	96,300 (9.8)	16	3,500 (950)
EMD	F40PH-2	Passenger	B-B	260 nominal	65	38,240 (16.1)	16	3,000 (900)
GE	B18-7	General purpose	B-B	231/268	58/67	61,000 (8.4)	8	1,800 (1,050)
GE	B30-7A	General purpose	B-B	253/280	63/70	64,600 (12.0)	12	3,000 (1,050)
GE	C30-7A	General purpose	C-C	359/420	90/105	96,900 (8.8)	12	3,000 (1,050)
GE	B36-7	General purpose	B-B	260/280	65/70	64,600 (12.0)	16	3,600 (1,050)
GE	C36-7	General purpose	C-C	367/420	92/105	96,900 (11.0)	16	3,600 (1,050)
EMD	AEM-7	Passenger	B-B	201 nominal	50	33,500 (10.0)	NA¶	7,000 §
EMD	GM6C	Freight	C-C	365 nominal	91	88,000 (11.0)	NA	6,000 §
EMD	GM10B	Freight	B-B-B	390 nominal	97	100,000 (5.0)	NA	10,000 §
GE	E60C	General purpose	C-C	364 nominal	91	82,000 (22.0)	NA	6,000 §
GE	E60CP	Passenger	C-C	366 nominal	91	34,000 (55.0)	NA	6,000 §
GE	E25B	Freight	B-B	280 nominal	70	55,000 (15.0)	NA	2,500 §

* Engines: Bombardier—model 251, 4-cycle, V-type, 9 × 10½ in cylinders.
EMD—model 645E, 2-cycle, V-type, 9$\frac{1}{16}$ × 10 in cylinders.
GE—model 7FDL, 4-cycle, V-type, 9 × 10½ in cylinders.
† Starting tractive effort at 25% adhesion.
‡ Continuous tractive effort for smallest pinion (maximum).
§ Electric locomotive horsepower expressed as diesel-electric equivalent (input to generator).
¶ Not applicable.

make use of its full rated power output at low track speeds for starting as well as for efficient hauling of heavy trains at all speeds. Examples of the most common diesel-electric locomotive types in service are shown in Table 8.3.1. A typical diesel-electric locomotive is shown in Fig. 8.3.1.

Diesel-electric locomotives have a DC generator or rectified alternator coupled directly to the diesel engine crankshaft. The generator/alternator is electrically connected to DC series traction motors having nose suspension mountings. Many recent locomotives utilize gate turn-off inverters and AC traction motors to obtain the benefits of increased adhesion and higher tractive effort. The gear ratio for the axle-mounted bull gears to the motor pinions that they engage is determined by the locomotive speed range, which is related to the type of service. A high ratio is used for freight service where high tractive effort and low speeds are common, whereas high-speed passenger locomotives have a lower ratio.

Diesel Engines Most new diesel-electric locomotives are equipped with either V-type, two strokes per cycle, or V-type, four strokes per cycle, engines. Engines range from 8 to 20 cylinders each. Output power ranges from 1,000 hp to over 6,000 hp for a single engine application. These medium-speed diesel engines have displacements from 560 to over 1,000 in^3 per cylinder.

Two-cycle engines are aspirated by either a gear-driven blower or a turbocharger. Because these engines lack an intake stroke for natural aspiration, the turbocharger is gear-driven at low engine speeds. At higher engine speeds, when the exhaust gases contain enough energy to drive the turbocharger, an overriding clutch disengages the gear train. Free-running turbochargers are used on four-cycle engines, as at lower speeds the engines are aspirated by the intake stroke.

The engine control governor is an electrohydraulic or electronic device used to regulate the speed and power of the diesel engine.

Electrohydraulic governors are self-contained units mounted on the engine and driven from one of the engine camshafts. They have integral oil supplies and pressure pumps. They utilize four solenoids that are actuated individually or in combination from the 74-V auxiliary generator/battery supply by a series of switches actuated by the engineer's throttle. There are eight power positions of the throttle, each corresponding to a specific value of engine speed and horsepower. The governor maintains the predetermined engine speed through a mechanical linkage to the engine fuel racks, which control the amount of fuel metered to the cylinders.

Computer engine control systems utilize electronic sensors to monitor the engine's vital functions, providing both electronic engine speed control and fuel management. The locomotive throttle is interfaced with an on-board computer that sends corresponding commands to the engine speed control system. These commands are compared to input from timing and engine speed sensors. The pulse width and timing of the engine's fuel injectors are adjusted to attain the desired engine speed and to optimize engine performance.

One or more centrifugal pumps, gear driven from the engine crankshaft, force water through passages in the cylinder heads and liners to provide cooling for the engine. The water temperature is automatically controlled by regulating shutter and fan operation, which in turn controls the passage of air through the cooling radiators. The fans are driven by electric motors.

The lubricating oil system supplies clean oil at the proper temperature and pressure to the various bearing surfaces of the engine, such as the crankshaft, camshaft, wrist pins, and cylinder walls. It also provides oil internally to the heads of the pistons to remove excess heat. One or more gear-type pumps driven from the crankshaft are used to move the oil from the crankcase through filters and strainers to the bearings and the piston cooling passages, after which the oil flows by gravity back to the crankcase. A heat exchanger is part of the system as all of the oil passes through it at some time during a complete cycle. The oil is cooled by engine cooling water on the other side of the exchanger. Paper-element cartridge filters in series with the oil flow remove fine impurities in the oil before it enters the engine.

Electric Transmission Equipment A DC main generator or three-phase AC alternator is directly coupled to the diesel engine crankshaft. Alternator output is rectified through a full-wave bridge to keep ripple

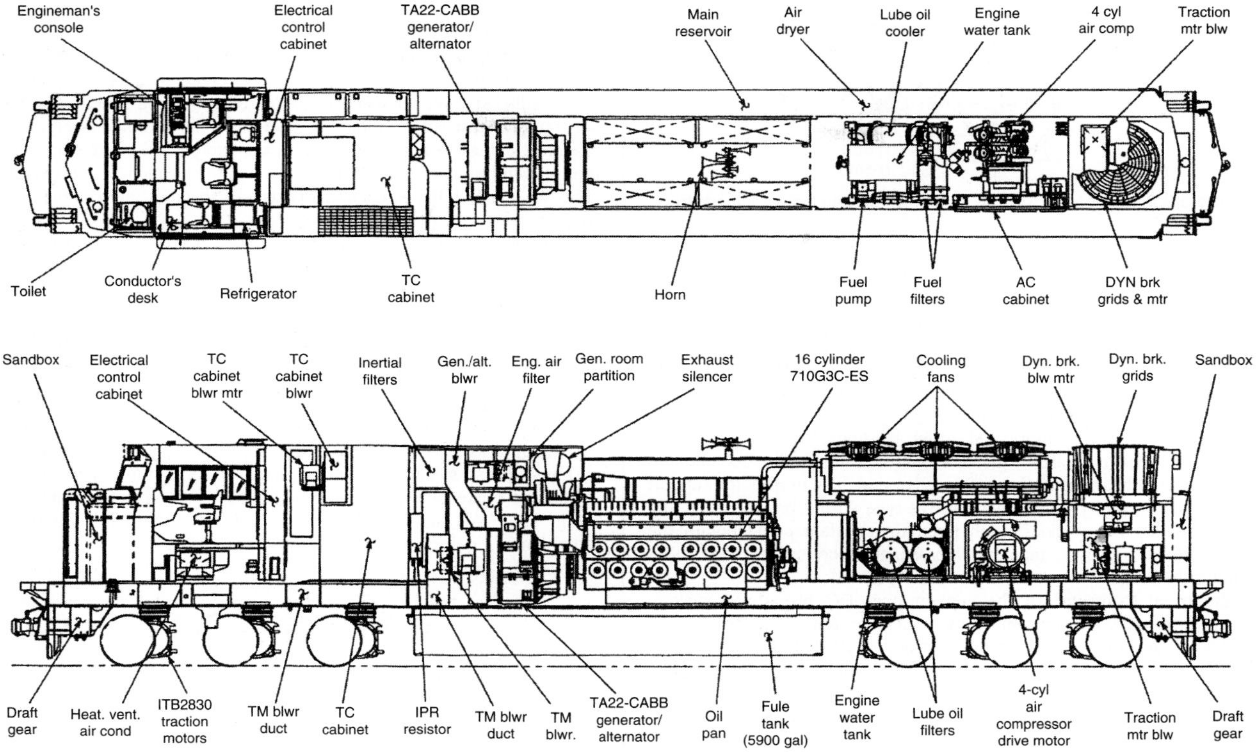

Figure 8.3.1 A typical diesel-electric locomotive. (*Electro-Motive Division, General Motors Corp.*)

to a level acceptable for operation of series field DC motors or for input to a solid-state inverter system for AC motors.

Generator power output is controlled by (1) varying engine speed through movement of the engineer's controller and (2) controlling the flow of current in its battery field or in the field of a separate exciter generator. Shunt and differential fields (if used) are designed to maintain constant generator power output for a given engine speed as the load and voltage vary. The fields do not completely accomplish this, thus the battery field or separately excited field must be controlled by a load regulator to provide the final adjustment in excitation to load the engine properly. This field can also be automatically deenergized to reduce or remove the load to prevent damage to the power plant or other traction equipment when certain undesirable conditions occur.

The engine main generator or alternator power plant functions at any throttle setting as a constant-horsepower source of energy. Therefore, the main generator voltage must be controlled to provide constant power output for each specific throttle position under the varying conditions of train speed, train resistance, atmospheric pressure, and quality of fuel. The load regulator, which is an integral part of the governor, accomplishes this within the maximum safe values of main generator voltage and current. For example, when the locomotive experiences an increase in track gradient with a consequent reduction in speed, traction motor counter emf decreases, causing a change in traction motor and main generator current. Because this alters the load demand on the engine, the speed of the engine tends to change to compensate. As the speed changes, the governor begins to reposition the fuel racks, but at the same time a pilot valve in the governor directs hydraulic pressure into a load regulator vane motor, which changes the resistance value of the load regulator rheostat in series with the main generator excitation circuit. This alters the main generator excitation current and consequently, main generator voltage and returns the value of main generator power output to normal. Engine fuel racks return to normal, consistent with constant values of engine speed.

The load regulator is effective within maximum and minimum limit values of its rheostat. Beyond these limits the power output of the engine is reduced. However, protective devices in the main generator excitation circuit limit its voltage to ensure that values of current and voltage in the traction motor circuits are within safe limits.

Auxiliary Generating Apparatus DC power for battery charging, lighting, control, and cab heaters is provided by a separate generator, geared to the main engine. Voltage output is regulated within 1 percent of 74 V over the full range of engine speeds. Auxiliary alternators with full-wave bridge rectifiers are also utilized for this application. Modern locomotives with on-board computer systems also have power supplies to provide "clean" DC for sensors and computer systems. These are typically 24-V systems.

Traction motor blowers are mounted above the locomotive underframe. Air from the centrifugal blower housings is carried through the underframe and into the motor housings through flexible ducts. Other designs have been developed to vary the air output with cooling requirements to conserve parasitic energy demands. The main generator/alternators are cooled in a similar manner.

Traction motors are nose-suspended from the truck frame and bearing-suspended from the axle (Fig. 8.3.2). The traction motors employ series exciting (main) and commutating field poles. The current in the series field is reversed to change locomotive direction and may be partially shunted through resistors to reduce counter emf as locomotive speed is increased. Newer locomotives dispense with field shunting to improve commutation. Early locomotive designs also required motor connection changes (series, series/parallel, parallel) referred to as transition, to maintain motor current as speed increased.

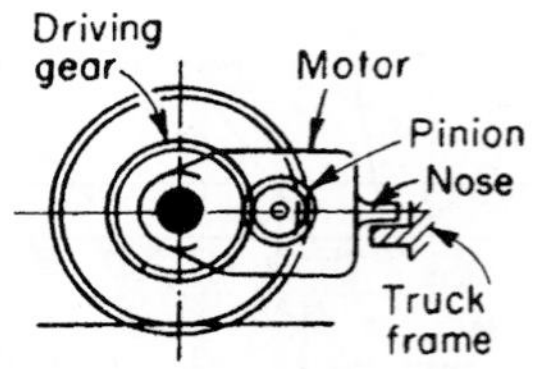

Figure 8.3.2 Axle-hung traction motors for diesel-electric and electric locomotives.

AC traction motors employ a variable-frequency supply derived from a computer controlled solid-state inverter system, fed from the rectified alternator output. Locomotive direction is controlled by reversing the sequence of the three-phase supply.

The development of the modern traction alternator with its high output current has resulted in a trend toward DC traction motors that are permanently connected in parallel. DC motor armature shafts are equipped with grease-lubricated roller bearings, while AC traction motor shafts have grease-lubricated roller bearings at the free end and oil-lubricated bearings at the pinion drive end. Traction motor support bearings are usually of the plain sleeve type with lubricant wells and spring-loaded felt wicks, which maintain constant contact with the axle surface; however, many new passenger locomotives have roller support bearings.

Electrical Controls In the conventional DC propulsion system, electropneumatic or electromagnetic contactors are employed to make and break the circuits between the traction motors and the main generator. They are equipped with interlocks for various control circuit functions. Similar contactors are used for other power and excitation circuits of lower power (current). An electropneumatic or electric motor-operated cam switch, consisting of a two-position drum with copper segments moving between spring loaded fingers, is generally used to reverse traction motor field current ("reverser") or to set up the circuits for dynamic braking. This switch is not designed to operate under load. On some DC locomotives these functions have been accomplished with a system of contractors. In the AC propulsion systems, the power control contractors are totally eliminated since their function is performed by the solid-state switching devices in the inverters.

Locomotives with AC traction motors utilize inverters to provide phase-controlled electrical power to their traction motors. Two basic arrangements of inverter control for AC traction motors have emerged. One arrangement uses an individual inverter for each axle. The other uses one inverter per truck. Inverters are typically GTO (gate turn-off) or IGBT (insulated-gate bipolar transistor) devices.

Cabs In order to promote uniformity and safety, the AAR has issued standards for many locomotive cab features, RP-5104. Locomotive cab noise standards are prescribed by 49 CFR § 229.121.

Propulsion control circuits transmit the engineer's movements of the throttle lever, reverse lever, and transition or dynamic brake control lever in the controlling unit to the power-producing equipment of each unit operating in multiple in the locomotive consist. Before power is applied, all reversers must move to provide the proper motor connections for the direction of movement desired. Power contractors complete the circuits between generators and traction motors. For DC propulsion systems, excitation circuits then function to provide the proper main generator field current while the engine speed increases to correspond to the engineer's throttle position. In AC propulsion systems, all power circuits are controlled by computerized switching of the inverter.

To provide for multiple unit operation, the control circuits of each locomotive unit are connected by jumper cables. The AAR has issued Standard S-512 covering standard dimensions and contact identification for 27-point control jumpers used between diesel-electric locomotive units.

Wheel slip is detected by sensing equipment connected either electrically to the motor circuits or mechanically to the axles. When slipping occurs on some units, relays automatically reduce main generator excitation until slipping ceases, whereupon power is gradually reapplied. On newer units, an electronic system senses small changes in motor current and reduces motor current before a slip occurs. An advanced system recently introduced adjusts wheel creep to maximize wheel-to-rail adhesion. Wheel speed is compared to ground speed, which is accurately measured by radar. A warning light and/or buzzer in the operating cab alerts the engineer, who must notch back on the throttle if the slip condition persists. With AC traction the frequency selected prevents wheel slip if the wheel tries to exceed a speed dictated by the input frequency.

Batteries Lead-acid storage batteries of 280 or 420 ampere-hour capacity are usually used for starting the diesel engine. Thirty-two cells on each locomotive unit are used to provide 64 V to the system. (See also Sec. 11.) The batteries are charged from the 74-V power supply.

Air Brake System The "independent" brake valve handle at the engineer's position controls air pressure supplied from the locomotive reservoirs to the brake cylinders on only the locomotive itself. The "automatic" brake valve handle controls the air pressure in the brake pipe to the train (Fig. 8.3.3). On more recent locomotives, purely pneumatic braking systems have been supplanted by electropneumatic systems. These systems allow for more uniform brake applications, enhancing brake pipe pressure control and enabling better train handling. The AAR has issued Standard S-5,529, "Multiple Unit Pneumatic Brake Equipment for Locomotives."

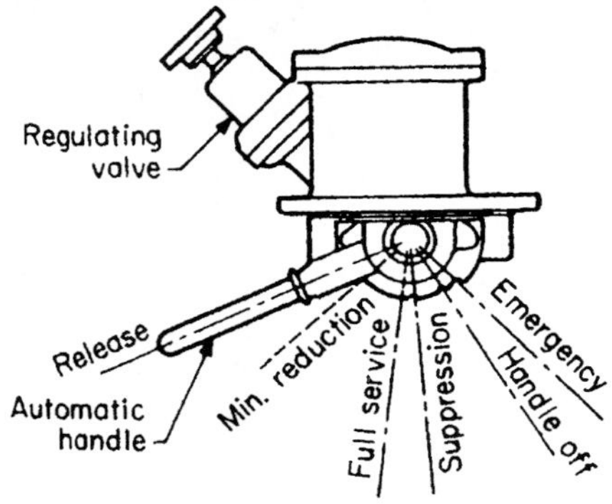

Figure 8.3.3 Automatic-brake valve-handle positions for 26-L brake equipment.

Compressed air for braking and for various pneumatic controls on the locomotive is usually supplied by a two-stage, three-cylinder compressor connected directly or through a clutch to the engine crankshaft, or with electric motor drive. A compressor control switch is activated to maintain a pressure of approximately 130 to 140 lb/in^2 (896 to 965 kPa) in the main reservoirs.

Dynamic Braking On most locomotives, dynamic brakes supplement the air brake system. The traction motors are used as generators to convert the kinetic energy of the locomotive and train into electrical energy, which is dissipated through resistance grids located near the locomotive roof. Motor-driven blowers, designed to utilize some of this braking energy, force cooler outside air over the grids and out through roof hatches. By directing a generous and evenly distributed air stream over the grids, their physical size is reduced, in keeping with the relatively small space available in the locomotive. On some locomotives, resistor grid cooling is accomplished by an engine-driven radiator/braking fan, but energy conservation is causing this arrangement to be replaced by motor-driven fans that can be energized in response to need, using the parasitic power generated by dynamic braking itself.

By means of a cam-switch reverser, the traction motors are connected to the resistance grids. The motor fields are usually connected in series across the main generator to supply the necessary high excitation current. The magnitude of the braking force is set by controlling the traction motor excitation and the resistance of the grids. Conventional dynamic braking is not usually effective below 10 mi/h (16 km/h), but it is very useful at 20 to 30 mi/h (32 to 48 km/h). Some locomotives are equipped with "extended range" dynamic braking which enables dynamic braking to be used at speeds as low as 3 mi/h (5 km/h) by shunting out grid resistance (both conventional and extended range are shown on Fig. 8.3.4). Dynamic braking is now controlled according to the "tapered" system, although the "flat" system has been used in

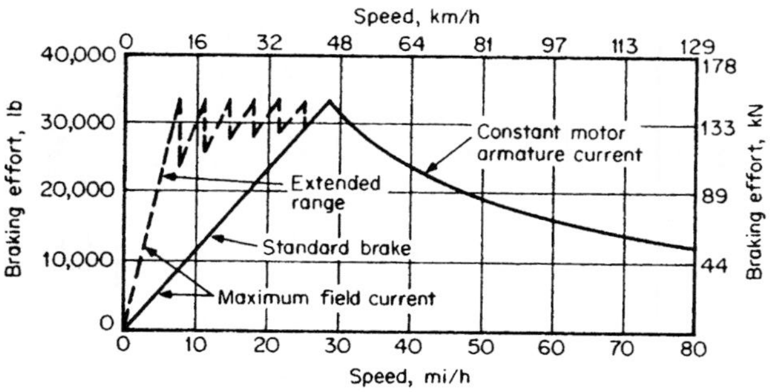

Figure 8.3.4 Dynamic-braking effort versus speed. (*Electro-Motive Division, General Motors Corp.*)

the past. Dynamic braking control requirements are specified by AAR Standard S-5,018. Dynamic braking is especially advantageous on long grades where accelerated brake shoe wear and the potential for thermal damage to wheels could otherwise be problems. The other advantages are smoother control of train speed and less concern for keeping the pneumatic trainline charged. Dynamic brake grids can also be used for a self-contained load-test feature, which permits a standing locomotive to be tested for power output. On locomotives equipped with AC traction motors, a constant dynamic braking (flat-top) force can be achieved from the horsepower limit down to 2 mi/h (3 km/h).

Performance

Engine Indicated Horsepower The power delivered at the diesel locomotive drawbar is the end result of a series of subtractions from the original indicated horsepower of the engine, which takes into account the efficiency of transmission equipment and the losses due to the power requirements of various auxiliaries. The formula for the engine's indicated horsepower is:

$$\text{ihp} = \frac{PLAN}{33{,}000} \tag{8.3.1}$$

where P = mean effective pressure in the cylinder, lb/in^2; L = length of piston stroke, ft; A = piston area, in^2; and N = total number of cycles completed per min. Factor P is governed by the overall condition of the engine, quality of fuel, rate of fuel injection, completeness of combustion, compression ratio, etc. Factors L and A are fixed with design of engine. Factor N is a function of engine speed, number of working chambers, and strokes needed to complete a cycle.

Engine Brake Horsepower In order to calculate the horsepower delivered by the crankshaft coupling to the main-generator, frictional losses in bearings and gears must be subtracted from the **indicated horsepower** (ihp). Some power is also used to drive lubricating oil pumps, governor, water pumps, scavenging blower, and other auxiliary devices. The resultant horsepower at the coupling is **brake horsepower** (bhp).

Rail Horsepower A portion of the engine bhp is transmitted mechanically via couplings or gears to operate the traction motor blowers, air compressor, auxiliary generator, and radiator cooling fan generator or alternator. Part of the auxiliary generator electrical output is used to run some of the auxiliaries. The remainder of the engine bhp transmitted to the main generator or main alternator for traction purposes must be multiplied by generator efficiency (usually about 91 percent), and the result again multiplied by the efficiency of the traction motors (including power circuits) and gearing to develop rail horsepower. Power output of the main generator for traction may be expressed as

$$\text{Watts}_{\text{traction}} = E_g \times I_m \tag{8.3.2}$$

where E_g is the main-generator voltage and I_m is the traction motor current in amperes, multiplied by the number of parallel paths or the DC link current in the case of an AC traction system.

Rail horsepower may be expressed as

$$\text{hp}_{\text{rail}} = V \times \text{TE}/375 \tag{8.3.3}$$

where V is the velocity, mi/h, and TE is tractive effort at the rail, lb.

Thermal Efficiency The thermal efficiency of the diesel engine at the crankshaft, or the ratio of bhp output to the rate at which energy of the fuel is delivered to the engine, is about 33 percent. Thermal efficiency at the rail is about 26 percent.

Drawbar Horsepower The drawbar horsepower represents power available at the rear of the locomotive to move the cars and may be expressed as

$$\text{hp}_{\text{drawbar}} = \text{hp}_{\text{rail}} - \text{locomotive running resistance} \times \frac{V}{375} \tag{8.3.4}$$

where V is the speed in mi/h. Train resistance calculations are discussed later under "Vehicle/Track Interaction." Theoretically, therefore, drawbar horsepower available is the power output of the diesel engine less the parasitic loads and losses described above.

Speed-Tractive Effort At full throttle the losses vary somewhat at different values of speed and tractive effort, but a curve of tractive effort plotted against speed is nearly hyperbolic. Figure 8.3.5 is a typical speed-tractive effort curve for a 3,500-hp (2,600-kW) freight locomotive. The diesel-electric locomotive has full horsepower available over the entire speed range (within the limits of adhesion described below). The reduction in power as continuous speed is approached is known as *power*

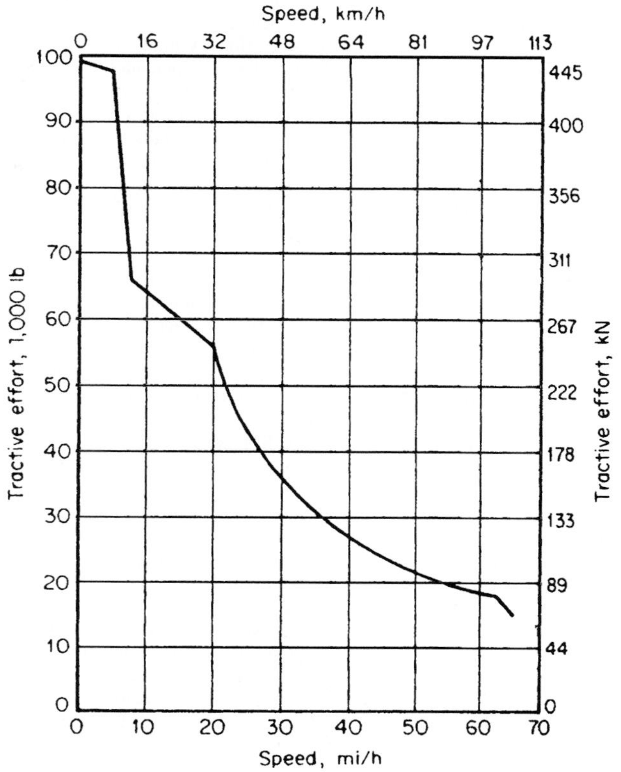

Figure 8.3.5 Tractive effort versus speed.

matching. This allows multiple operation of locomotives of different ratings at the same continuous speed.

Adhesion In Fig. 8.3.6 the maximum value of tractive effort represents the level usually achievable just before the wheels slip under average rail conditions. Adhesion is usually expressed as a percentage of vehicle weight on drivers, with the nominal level being 25 percent. This means that a force equal to 25 percent of the total locomotive weight on drivers is available as tractive effort. Actually, adhesion will vary widely with rail conditions, from as low as 5 percent to as high as 35 percent or more. Adhesion is severely reduced by lubricants which spread as thin films in the presence of moisture on running surfaces. Adhesion can be increased with sand applied to the rails from the locomotive sanding system. More recent wheel slip systems permit wheel creep (very slow controlled slip) to achieve greater levels of tractive effort. Even higher adhesion levels are available from AC traction motors; for example, 45 percent at start-up and low speed, with a nominal value of 35 percent.

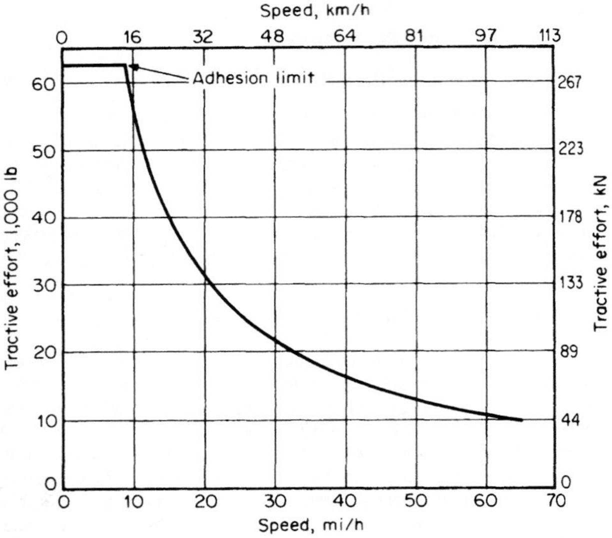

Figure 8.3.6 Typical tractive effort versus speed characteristics. (*Electro-Motive Division, General Motors Corp.*)

Traction Motor Characteristics Motor torque is a function of armature current and field flux (which is a function of field current). Since the traction motors are series-connected, armature and field current are the same (except when field shunting circuits are introduced), and therefore tractive effort is solely a function of motor current. Figure 8.3.7 presents a group of traction motor characteristic curves with tractive effort, speed and efficiency plotted against motor current for full field (FF) and at 35 (FS1) and 55 (FS2) percent field shunting. Wheel diameter and gear ratio must be specified when plotting torque in terms of tractive effort. (See also Sec. 11.)

Traction motors are usually rated in terms of their maximum continuous current. This represents the current at which the heating due to electrical losses in the armature and field windings is sufficient to raise the temperature of the motor to its maximum safe limit when cooling air at maximum expected ambient temperature is forced through it at the prescribed rate by the blowers. Continuous operation at this current level ideally allows the motor to operate at its maximum safe power level, with waste heat generated equal to heat dissipated. The tractive effort corresponding to this current is usually somewhat lower than that allowed by adhesion at very low speeds. Higher current values may be permitted for short periods of time (as when starting). These ratings are specified in intervals of time (minutes) and are posted on or near the load meter (ammeter) in the cab.

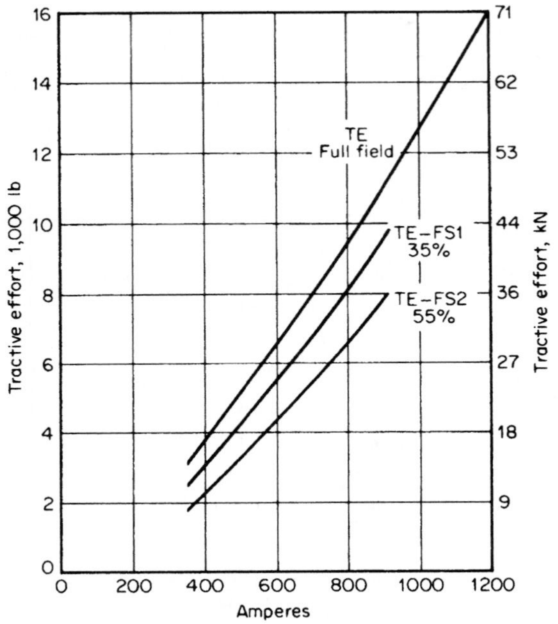

Figure 8.3.7 Traction motor characteristics. (*Electro-Motive Division, General Motors Corp.*)

Maximum Speed Traction motors are also rated in terms of their maximum safe speed in r/min, which in turn limits locomotive speed. The gear ratio and wheel diameter are directly related to speed as well as the maximum tractive effort and the minimum speed at which full horsepower can be developed at the continuous rating of the motors. Maximum locomotive speed may be expressed as follows:

$$(\text{mi/h})_{\text{max}} = \frac{\text{wheel diameter (in)} \times \text{maximum motor r/min}}{\text{gear ratio} \times 336} \tag{8.3.5}$$

where the gear ratio is the number of teeth on the gear mounted on the axle divided by the number of teeth on the pinion mounted on the armature shaft.

Locomotive Compatibility The AAR has developed two standards in an effort to improve compatibility between locomotives of different model, manufacture, and ownership: a standard 27-point control system (Standard S-512, Table 8.3.2) and a standard control stand (RP-5,132). The control stand has been supplanted by a control console in many road locomotives (Fig. 8.3.8).

Energy Conservation Efforts to improve efficiency and fuel economy have resulted in major changes in the prime movers, including more efficient turbocharging, fuel injection, and combustion. The auxiliary (parasitic) power demands have also been reduced by improvements, which include fans and blowers that move only the air required by the immediate demand, air compressors that declutch when unloaded, and a selective low-speed engine idle. Fuel-saver switches permit dropping trailing locomotive units off the line when less than maximum power is required, while allowing the remaining units to operate at maximum efficiency.

Emissions The U.S. Environmental Protection Agency has promulgated regulations aimed at reducing diesel locomotive emissions, especially oxides of nitrogen (NO_x). These standards also include emissions reductions for hydrocarbons (HC), carbon monoxide (CO), particulate matter (PM), and smoke. These standards are executed in three "tiers." Tier 0 going into effect in 2000, Tier 1 in 2002, and Tier 3 in 2005. EPA locomotive emissions standards are published under 40 CFR § 85, § 89, and § 92.

Table 8.3.2 Standard Dimensions and Contact Identification of 27-Point Control Plug and Receptacle for Diesel-Electric Locomotives

Receptacle point	Function	Code	Wire size, AWG
1	Power reduction setup, if used	(PRS)	14
2	Alarm signal	SG	14
3	Engine speed	DV	14
4*	Negative	N	14 or 10
5	Emergency sanding	ES	14
6	Generator field	GF	12
7	Engine speed	CV	14
8	Forward	FO	12
9	Reverse	RE	12
10	Wheel slip	WS	14
11	Spare		14
12	Engine speed	BV	14
13	Positive control	PC	12
14	Spare		14
15	Engine speed	AV	14
16	Engine run	ER	14
17	Dynamic brake	B	14
18	Unit selector circuit	US	12
19	2d negative, if used	(NN)	12
20	Brake warning light	BW	14
21	Dynamic brake	BG	14
22	Compressor	CC	14
23	Sanding	SA	14
24	Brake control/power reduction control	BC/PRC	14
25	Headlight	HL	12
26	Separator blowdown/ remote reset	SV/RR	14
27	Boiler shutdown	BS	14

*Receptacle point 4—AWG wire size 12 is "standard" and AWG wire size 10 is "Alternate standard" at customer's request. A dab of white paint in the cover latch cavity must be added for ready identification of a no. 10 wire present in a no. 4 cavity.

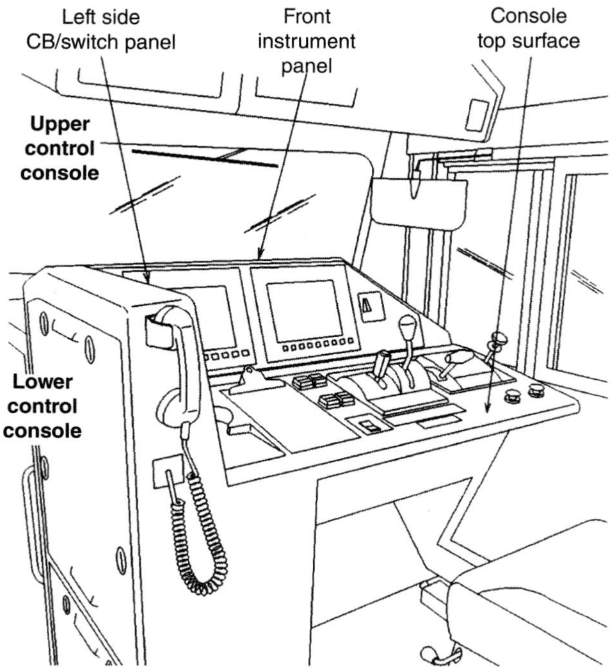

Figure 8.3.8 Locomotive control console. (*Electro-Motive Division, General Motors Corp.*)

ELECTRIC LOCOMOTIVES

Electric locomotives are presently in very limited use in North America. Freight locomotives are in dedicated service primarily for hauling coal or mineral products. Electric passenger locomotives are used in high-density service in the northeastern United States.

Electric locomotives draw power from overhead catenary or third-rail systems. While earlier systems used either direct current up to 3,000 V or single-phase alternating current at 11,000 V, 25 Hz, the newer systems in North America use 25,000 or 50,000 V at 60 Hz. The higher voltage levels can be used only where clearances permit. A three-phase power supply was tried briefly in this country and overseas many years ago, but was abandoned because of the complexity of the required double catenary.

While the older DC locomotives used resistance control, AC locomotives have used a variety of systems including Scott-connected transformers; series AC motors; motor generators and DC motors; ignitrons and DC motors; silicon thyristors and DC motors; and, more recently, chopper control and inverter with AC motors. Examples of the various electric locomotives in service are shown in Table 8.3.1.

An electric locomotive used in high-speed passenger service is shown in Fig. 8.3.9. High short-time ratings (Figs. 8.3.10 and 8.3.11) render electric locomotives suitable for passenger service where high acceleration rates and high speeds are combined to meet demanding schedules.

The modern electric locomotive in Fig. 8.3.9 obtains power for the main circuit and motor control from the catenary through a pantograph. A motor-operated switch provides for the transformer change from series to parallel connection of the primary windings to give a constant secondary voltage with either 25-kV or 12.5-kV primary supply. The converters for armature current consist of two asymmetric-type bridges for each motor. The traction-motor fields are each separately fed from a one-way connected field converter.

Identical control modules separate the control of motors on each truck. Motor sets are therefore connected to the same transformer winding. Wheel slip correction is also modularized, utilizing one module for each two-motor truck set. This correction is made with a complementary wheel-slip detection and correction system. A magnetic pickup speed signal is used for the basic wheel slip. Correction is enhanced by a magnetoelastic transducer used to measure force swings in the traction motor reaction rods. This system provides the final limit correction.

To optimize the utilization of available adhesion, the wheel-slip control modules operate independently to allow the motor modules to receive different current references, depending on their respective adhesion conditions.

All auxiliary machines—air compressor, traction motor blower, cooling fans, etc.—are driven by three-phase, 400-V, 60-Hz induction motors powered by a static inverter that has a rating of 175 kVA at a 0.8 power factor. When cooling requirements are reduced, the control system automatically reduces the voltage and frequency supplied to the blower motors to the required level. As a backup, the system can be powered by the static converter used for the head-end power requirements of the passenger cars. This converter has a 500-kW, 480-V, three-phase, 60-Hz output capacity and has a built-in overload capacity of 10 percent for half of any 1-hour period.

Dynamic brake resistors are roof-mounted and are cooled by ambient airflow induced by locomotive motion. The dynamic brake capacity relative to train speed is shown in Fig. 8.3.12. Regenerative braking can be utilized with electric locomotives by returning braking energy to the distribution system.

Many electric locomotives utilize traction motors identical to those used on diesel-electric locomotives. Some, however, use frame-suspended motors with quill-drive systems (Fig. 8.3.13). In these systems, torque is transmitted from the traction motor by a splined coupling to a quill shaft and rubber coupling to the gear unit. This transmission allows for greater relative movement between the traction motor (truck frame) and gear (wheel axle) and reduces unsprung weight.

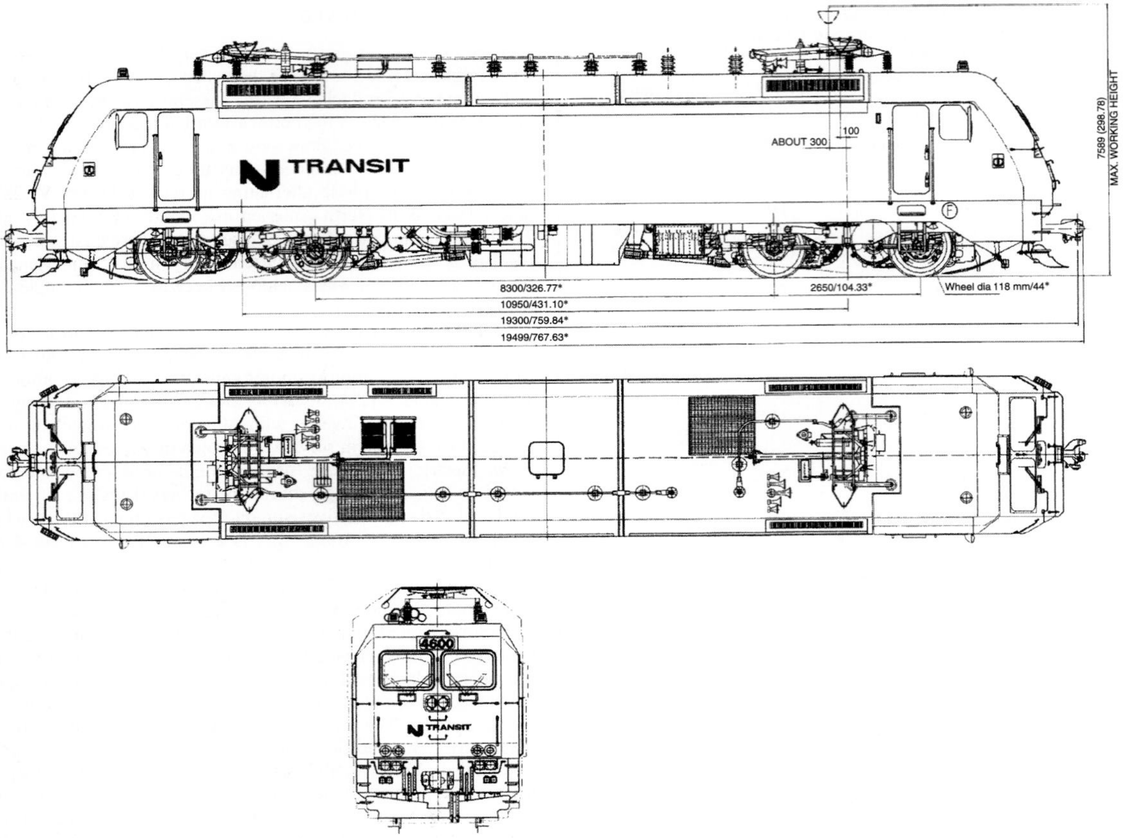

Figure 8.3.9 Modern electric high-speed locomotive. (*Courtesy of Bombardier Transportation.*)

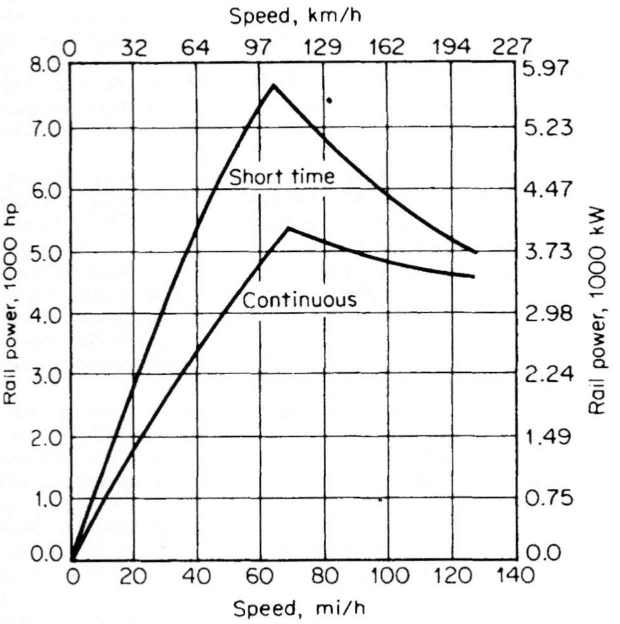

Figure 8.3.10 Speed-horsepower characteristics (half-worn wheels). [Gear ratio = 85/36; wheel diameter = 50 in (1270 mm); ambient temperature = 60°F (15.5°C).] (*Courtesy of M. Ephraim, ASME, Rail Transportation Division.*)

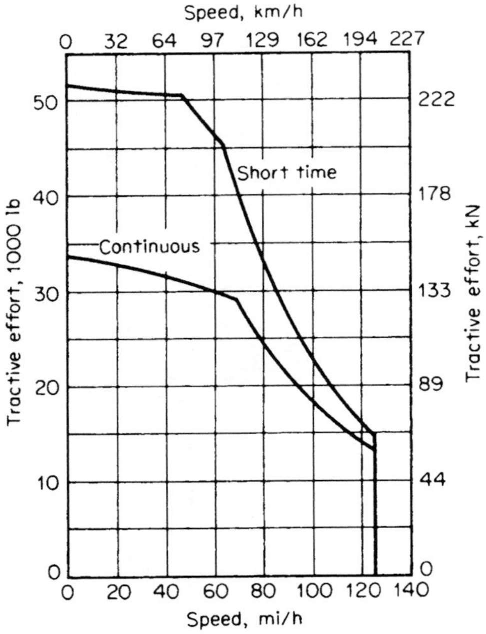

Figure 8.3.11 Tractive effort versus speed (half-worn wheels). [Voltage = 11.0 kV at 25 Hz; gear ratio = 85/36; wheel diameter = 50 in (1270 mm).] (*Courtesy of M. Ephraim, ASME, Rail Transportation Division.*)

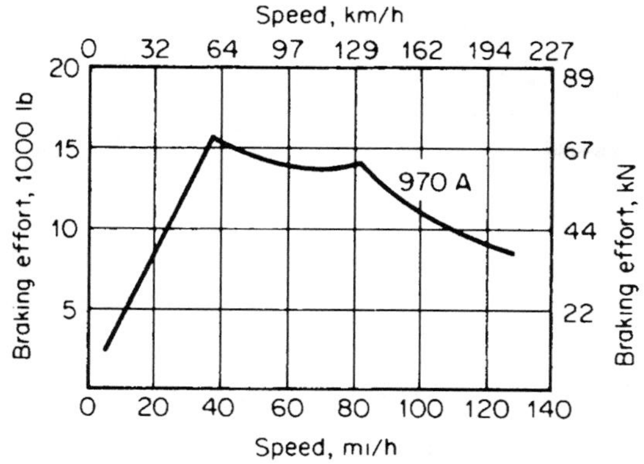

Figure 8.3.12 Dynamic-brake performance. (*Courtesy of M. Ephraim, ASME, Rail Transportation Division.*)

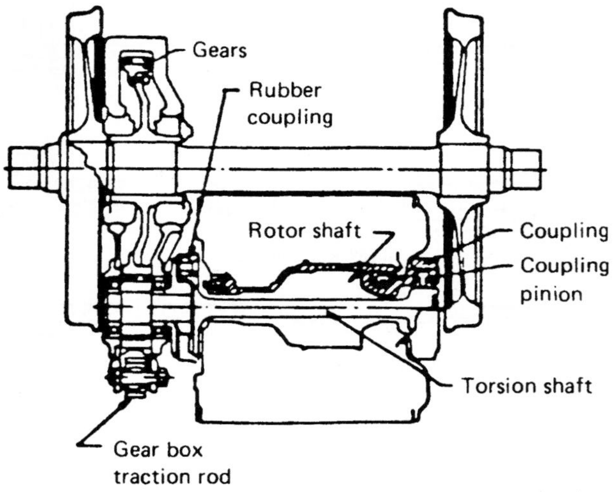

Figure 8.3.13 Frame-suspended locomotive transmission. (*Courtesy of M. Ephraim, ASME, Rail Transportation Division.*)

FREIGHT CARS

Freight Car Types

Freight cars are designed and constructed for either general service or specific ladings. The Association of American Railroads (AAR) has established design and maintenance criteria to assure the interchangeability of freight cars on all North American railroads. Freight cars that do not conform to AAR standards have been placed in service by special agreement with the railroads over which the cars operate. The AAR "Manual of Standards and Recommended Practices" specifies dimensional limits, weights, and other design criteria for cars, which may be freely interchanged between North American railroads. This manual is revised annually by the AAR. Many of the standards are reproduced in the "Car and Locomotive Cyclopedia," which is revised periodically. Safety appliances (such as end ladders, sill steps, and uncoupling levers), braking equipment, and certain car design features and maintenance practices must comply with the Safety Appliances Act, the Power Brake Law, and the Freight Car Safety Standards covered by the Federal Railroad Administration's (FRA) Code of Federal Regulations (Title 49 CFR). Maintenance practices are set forth in the "Field Manual of the AAR Interchange Rules" and repair pricing information is provided in the companion "AAR Office Manual."

The AAR identifies most cars by nominal capacity and type. In some cases there are restrictions as to type of load (i.e., food, automobile parts, coil steel, etc.). Most modern cars have nominal capacities of either 70 or 100 tons.* A 100-ton car is limited to a maximum gross rail load of 263,000 lb (119.3 t*) with 6½ by 12 in (165 by 305 mm) axle journals, four pairs of 36-in (915-mm) wheels and a 70-in (1,778-mm) rigid truck wheel base. Some 100-ton cars are rated at 286,000 lb and are handled by individual railroads on specific routes or by mutual agreement between the handing railroads. A 70-ton car is limited to a maximum gross rail load of 220,000 lb (99.8 t) with 6- by 11-in (152- by 280-mm) axle journals, four pairs of 33-in (838-mm) wheels with trucks having a 66-in (1,676-mm) rigid wheel base.

On some special cars where height limitations are critical, 28-in (711-mm) wheels are used with 70-ton axles and bearings. In these cases wheel loads are restricted to 22,400 lb (10.2 t). Some special cars are equipped with two 125-ton two-axle trucks with 38-in (965-mm) wheels in four-wheel trucks having 7 by 14 in (178 by 356 mm) axle journals and 72-in (1,829-mm) truck wheel base. This application is prevalent in articulated double-stack (two-high) container cars. Interchange of these very heavy cars is by mutual agreement between the operating railroads involved.

The following are the most common car types in service in North America. Dimensions given are for typical cars; actual measurements may vary.

Box cars (Fig. 8.3.14*a*) come in six popular types:

1. Standard box cars may have either sliding or plug doors. Plug doors provide a tight seal from weather and a smooth interior. Unequipped box cars are usually of 70-ton capacity. These cars have tongue-and-groove or plywood lining on the interior sides and ends with nailable floors (either wood or steel with special grooves for locking nails). The cars carry typical general merchandise lading: packaged, canned, or bottled foodstuffs; finished lumber; bagged or boxed bulk commodities; or, in the past when equipped with temporary door fillers, bulk commodities such as grain [70-ton: $L = 50$ ft 6 in (15.4 m), $H = 11$ ft 0 in (3.4 m), $W = 9$ ft 6 in (2.9 m), truck centers = 40 ft 10 in (12.4 m)].

2. Specially equipped box cars usually have the same dimensions as standard cars but include special interior devices to protect lading from impacts and over-the-road vibrations. Specially equipped cars may have hydraulic cushion units to dampen longitudinal shock at the couplers.

3. Insulated box cars have plug doors and special insulation. These cars can carry foodstuffs such as unpasteurized beer, produce, and dairy products. These cars may be precooled by the shipper and maintain a heat loss rate equivalent to 1°F (0.55°C) per day. They also can protect loads from freezing when operating at low ambient temperatures.

4. Refrigerated box cars are used where transit times are longer. These cars are equipped with diesel-powered refrigeration units and are primarily used to carry fresh produce and meat. They are often 100-ton cars [$L = 52$ ft 6 in (16.0 m), $H = 10$ ft 6 in (3.2 m), truck centers = 42 ft 11 in (13.1 m)].

5. "All door" box cars have doors which open the full length of the car for loading package lumber products such as plywood and gypsum board.

6. High cubic capacity box cars with an inside volume of 10,000 ft^3 (283 m^3) have been designed for light density lading, such as some automobile parts and low-density paper products.

Box car door widths vary from 6 to 10 ft for single-door cars and 16 to 20 ft (4.9 to 6.1 m) for double-door cars. "All door" cars have clear doorway openings in excess of 25 ft (7.6 m). The floor height above rail for an empty uninsulated box car is approximately 44 in (1,120 mm) and for an empty insulated box car approximately 48 in (1,220 mm). The floor height of a loaded car can be as much as 3 in (76 mm) lower than the empty car.

Covered hopper cars (Fig. 8.3.14*b*) are used to haul bulk commodities that must be protected from the environment. Modern covered hopper cars are typically 100-ton cars with roof hatches for loading and from two to six bottom outlets for discharge. Cars used

*1 ton = 1 short ton = 2,000 lb; 1 t = 1 metric ton = 1,000 kg = 2,205 lb.

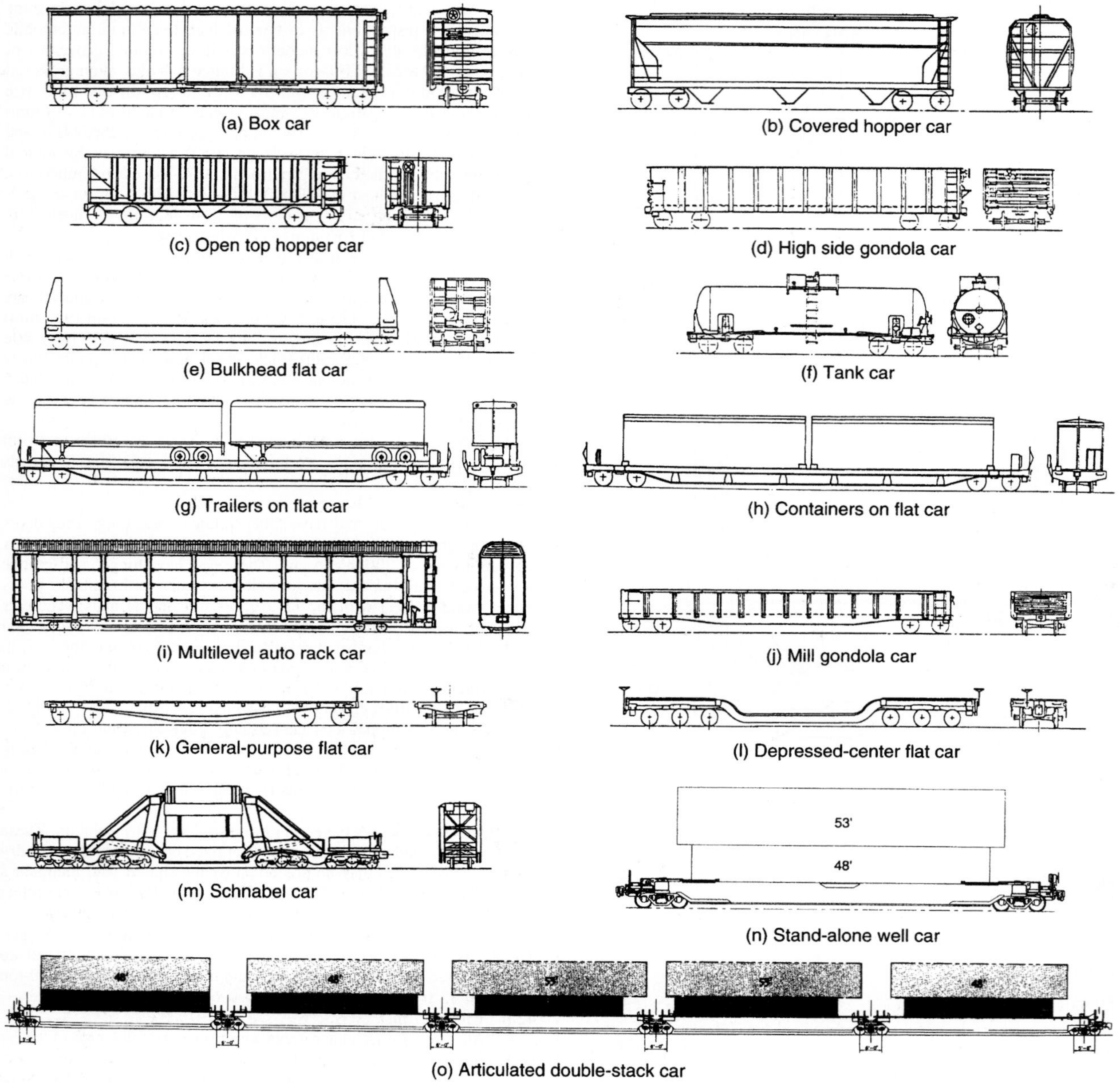

Figure 8.3.14 Typical freight cars.

for dense commodities, such as fertilizer or cement, have two bottom outlets, round roof hatches, and volumes of 3,000 to 4,000 ft^3 (84.9 to 113.2 m^3) [100-ton: L = 39 ft 3 in (12.0 m). H = 14 ft 10 in (4.5 m), truck centers = 26 ft 2 in (8.0 m)]. Cars used for grain service (corn, wheat, rye, etc.) have three or four bottom outlets, longitudinal trough roof hatches, and volumes of from 4,000 to 5,000 ft^3 (113 to 142 m^3). Cars used for hauling plastic pellets have four to six bottom outlets (for pneumatic unloading with a vacuum system), round roof hatches, and volumes of from 5,000 to 6,000 ft^3 (142 to 170 m^3) [100-ton: L = 65 ft 7 in (20.0 m), H = 15 ft 5 in (4.7 m), truck centers = 54 ft 0 in (16.5 m)].

Open-top hopper cars (Fig. 8.3.14*c*) are used for hauling bulk commodities such as coal, ore, or wood chips. A typical 100-ton coal hopper car will vary in volume depending on the light weight of the car and density of the coal to be hauled. Volumes range from 3,900 to 4,800 ft^3 (110 to 136 m^3). Cars may have three or four manually operated hopper doors or a group of automatically operated doors. Some cars are equipped with rotating couplers on one end of allow rotary dumping without uncoupling [100-ton: L = 53 ft ½ in (16.2 m), H = 12 ft 8½ in (3.9 m), truck centers = 40 ft 6 in (12.3 m)]. Cars intended for aggregate or ore service have smaller volumes for the more dense commodity. These cars typically have two manual bottom outlets [100-ton: L = 40 ft

8 in (12.4 m), $H = 11$ ft 10 in (3.6 m), truck centers = 29 ft 9 in (9.1 m)]. Hopper cars used for wood chip service are configured for low-density loads. Volumes range from 6,500 to 7,500 ft³ (184 to 212 m³).

High-side gondola cars (Fig. 8.3.14*d*) are open-top cars typically used to haul coal or wood chips. These cars are similar to open-top hopper cars in volume but require a rotary coupler on one end for rotary dumping to discharge lading since they do not have bottom outlets. Rotary-dump coal gondolas are usually used in dedicated, unit-train service between a coal mine and an electric power plant. The length over coupler pulling faces is approximately 53 ft 1 in (16.2 m) to suit the standard coal dumper [100-ton: $L = 50$ ft 5½ in (15.4 m), $H = 11$ ft 9 in (3.6 m), $W = 10$ ft 5 in (3.2 m), truck centers = 50 ft 4 in (15.3 m)]. Wood chip cars are used to haul chips from sawmills to paper mills or particle board manufacturers. These high-volume cars are either rotary-dumped or, when equipped with end doors, end-dumped [100-ton: $L =$ 62 ft 3 in (19.0 m), $H = 12$ ft 7 in (3.8 m), $W = 10$ ft 5 in (3.2 m), truck centers = 50 ft 4 in (15.3 m)]. Rotary dump aggregate or ore cars, called "ore jennies," have smaller volumes for the high-density load.

Bulkhead flat cars (Fig. 8.3.14*e*) are used for hauling such commodities as packaged finished lumber, pipe, or, with special inward canted floors, pulpwood. Both 70- and 100-ton bulkhead flats are used. Typical deck heights are approximately 45 to 50 in (1,143 to 1,270 mm) [100-ton: $L = 66$ ft 2 in (20.2 m), $W = 10$ ft 5 in (3,175 m), $H = 14$ ft 2⁹⁄₁₆ in (4.3 m), truck centers = 48 ft 2 in (14.7 m)]. Special "center beam" bulkhead flat cars designed for pulpwood and lumber service have a full-height, longitudinal beam from bulkhead to bulkhead.

Tank cars (Fig 8.3.14*f*) are used for liquids, compressed gases, and other ladings, such as sulfur, which can be loaded and unloaded in a molten state. Nonhazardous liquids such as corn syrup, crude oil, and mineral spring water are carried in nonpressure cars. Cars used to haul hazardous substances such as liquefied petroleum gas (LPG), vinyl chloride, and anhydrous ammonia are regulated by the U.S. Department of Transportation. Newer and earlier-built retrofitted cars equipped for hazardous commodities have safety features including safety valves; especially designed top and bottom "shelf" couplers, which increase the interlocking effect between couplers and decrease the danger of disengagement due to derailment; head shields on the ends of the tank to prevent puncturing; bottom outlet protection if bottom outlets are used; and thermal insulation and jackets to reduce the risk of rupturing in a fire. These features resulted from industry-government studies in the RPI-AAR Tank Car Safety Research and Test Program.

Cars for asphalt, sulfur, and other viscous liquid service have heating coils on the shell so that steam may be used to liquefy the lading for discharge [pressure cars, 100-ton: volume = 20,000 gal (75.7 m³), $L = 59$ ft 11¾ in (18.3 m), truck centers = 49 ft ¼ in (14.9 m)] [nonpressure, 100-ton: volume = 21,000 gal (79.5 m³), $L = 51$ ft 3¼ in (15.6 m), truck centers = 38 ft 11¼ in (11.9 m)].

Conventional intermodal cars are 89 ft 4 in (27.2-m) intermodal flat cars equipped to haul one 45-ft (13.7-m) and one 40-ft (12.2-m) trailer with or without end-mounted refrigeration units, two 45-ft (13.7-m) trailers (dry vans), or combinations of containers from 20 to 40 ft (6.1 to 12.2 m) in length. Hitches to support trailer fifth wheels may be fixed for trailer-only cars or retractable for conversion to haul containers or to facilitate driving trailers onto the cars in the rare event where "circus" loading is still required. In some cars a drawbar connects two units so three 57-ft trailers can be loaded, with one trailer over the drawbar. Trailer hauling service (Fig. 8.3.14*g*) is called TOFC (trailer on flat car); container service (Fig. 8.3.14*h*) is called COFC (container on flat car) [70-ton: $L = 89$ ft 4 in (27.1 m), $W = 10$ ft 3 in (3.1 m), truck centers = 64 ft 0 in (19.5 m)].

Introduction of larger trailers in highway service has led to the development of alternative TOFC and COFC cars. Other techniques involve articulation of well-type car bodies or skeletonized spine car bodies, assembled into multiunit cars for lift-on loading and lift-off unloading (stand-alone well car, Fig. 8.3.14*n* and articulated well car, Fig. 8.3.14*o*). These cars are typically composed of from three to ten units. Well cars consist of a center well for double-stacked containers [10-unit spine car; $L = 46$ ft 6 in (14.2 m) per end unit, $L = 465$ ft 3½ in (141.8 m)].

In the 1980s another approach was to haul larger trailers on two axle spine cars. These cars, used either singly or in multiple combinations, can haul a single trailer from 40 to 48 ft long (12.2 to 14.6 m) with end-mounted refrigeration unit and 36 or 42-in (914 to 1,070 mm) spacing between the kingpin and the front of the trailer.

Bilevel and trilevel auto rack cars (Fig. 8.3.14*i*) are used to haul finished automobiles and other vehicles. Most recent designs of these cars feature fully enclosed racks to provide security against theft and vandalism [70-ton: $L = 89$ ft 4 in (27.2 m), $H = 18$ ft 11 in (5.8 m), $W = 10$ ft 7 in (3.2 m), truck centers = 64 ft 0 in (19.5 m)]. Because of increased clearances provided by some railroads for double-stack cars, auto rack cars with a height of 20 ft 2 in (6.1 m) are now being produced.

Mill gondolas (Fig. 8.3.14*j*) are 70- or 100-ton open-top cars principally used to haul pipe, structural steel, scrap metal, and when specially equipped, coils of aluminum or tinplate and other steel materials [100-ton: $L = 52$ ft 6 in (16.0 m), $W = 9$ ft 6 in (2.9 m), $H = 4$ ft 6 in (1.4 m), truck centers = 43 ft 6 in (13.3 m)].

General-purpose or **machinery flat cars** (Fig. 8.3.14*k*) are 70- or 100-ton cars used to haul machinery such as farm equipment and highway tractors. These cars usually have wood decks for nailing lading restraint dunnage. Some heavy-duty six-axle cars are used for hauling off highway vehicles such as army tanks and mining machinery [100-ton, four-axle: $L = 60$ ft 0 in (18.3 m), $H = 3$ ft 9 in (1.1 m), truck centers = 42 ft 6 in (13.0 m)] [200-ton, 8-axle: $L = 44$ ft 4 in (13.5 m), $H = 4$ ft 0 in (1.2 m), truck centers = 33 ft 9 in (10.3 m)].

Depressed-center flat cars (Fig. 8.3.14*l*) are used for hauling transformers and other heavy, large materials that require special clearance considerations. Depressed-center flat cars may have four- or six-axle or dual four-axle, trucks with span bolsters, depending on weight requirements.

Schnabel cars (Fig. 8.3.14*m*) are special cars for transformers and power plant components. With these cars the load itself provides the center section of the car structure during shipment. Some Schnabel cars are equipped with hydraulic controls to lower the load for height restrictions and shift the load laterally for wayside restrictions [472-ton: $L = 22$ ft 10 in to 37 ft 10 in (7.0 to 11.5 m), truck centers = 55 ft 6 in to 70 ft 6 in (16.9 to 21.5 m)]. Schnabel cars must be operated in special trains.

Freight Car Design

The AAR provides specifications to cover minimum requirements for design and construction of new freight cars. Experience has demonstrated that the AAR specifications alone do not ensure an adequate design for all service conditions. The designer must be familiar with the specific service and increase the design criteria for the particular car above the minimum criteria provided by the AAR. The AAR requirements include stress calculations for the load-carrying members of the car and physical tests that may be required at the option of the AAR Equipment Engineering Committee (EEC) approving the car design. In the application for approval of a new or untired type of car, the EEC may require either additional calculations or tests to assess the design's ability to meet the AAR minimum requirements. These tests might consist of a static compression test of 1,000,000 lb (4.4 MN), a static vertical test applied at the coupler, and impact tests simulating yard impact conditions.

In some cases, it is advisable to operate an instrumented prototype car in service to detect problems which might result from unexpected track or train-handling input forces. The car design must comply with width and height restrictions shown in AAR clearance plates furnished in the specifications, Fig. 8.3.15. In addition, there are limitations on the height of the center of gravity of the loaded car and on the vertical and horizontal curving capability allowed by the clearance provided at the coupler. The AAR provides a method of calculating the minimum radius curve that the car design can negotiate when coupled to another car of the same type or to a standard AAR "base" car. In the case of horizontal curves, the requirements are based on the length of the car over the pulling faces of the couplers.

Freight cars are designed to withstand single-end impact or coupling loads on the basis of the type of energy absorption, cushioning, provided

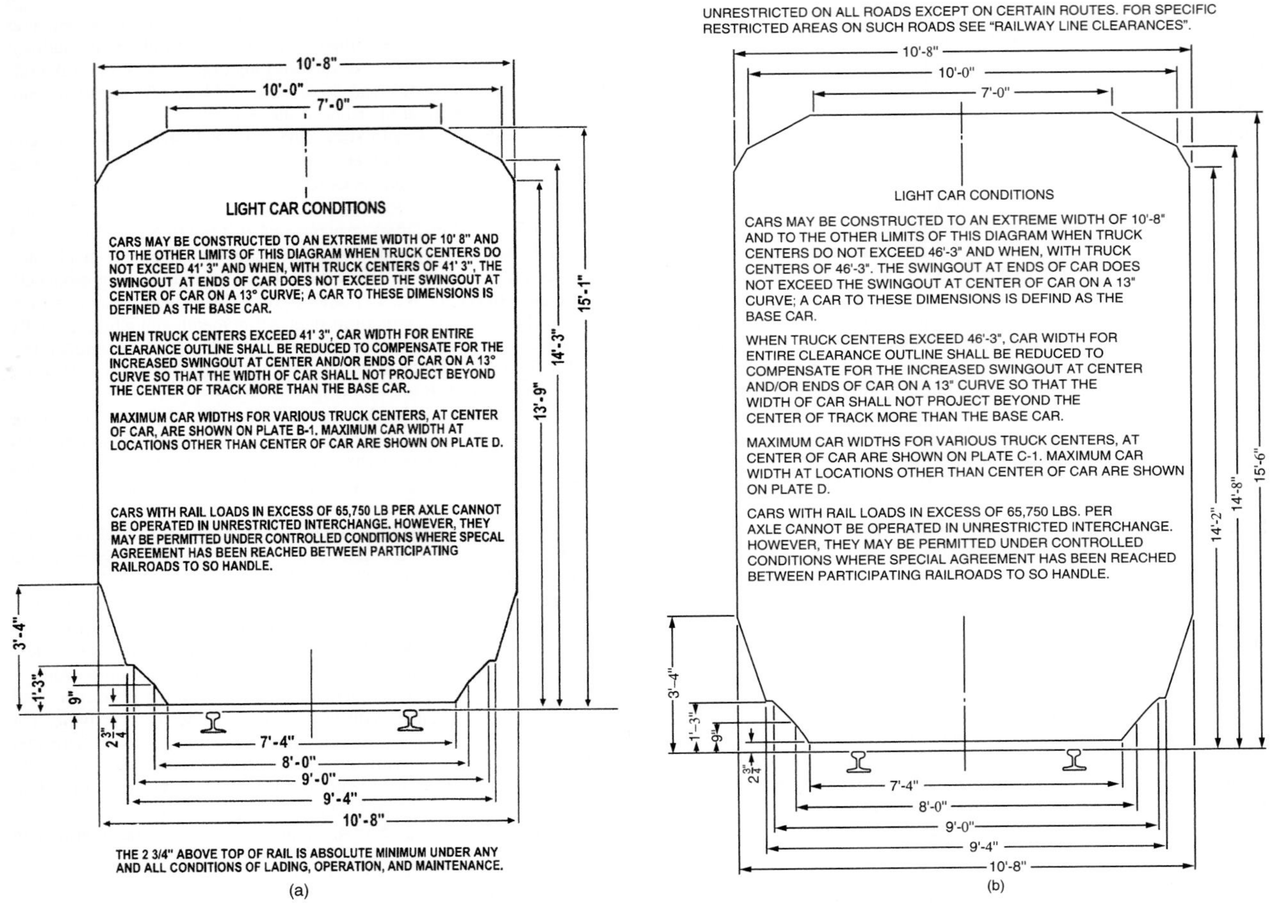
10'-8"
10'-0"
7'-0"
LIGHT CAR CONDITIONS
CARS MAY BE CONSTRUCTED TO AN EXTREME WIDTH OF 10' 8" AND TO THE OTHER LIMITS OF THIS DIAGRAM WHEN TRUCK CENTERS DO NOT EXCEED 41' 3" AND WHEN, WITH TRUCK CENTERS OF 41' 3", THE SWINGOUT AT ENDS OF CAR DOES NOT EXCEED THE SWINGOUT AT CENTER OF CAR ON A 13° CURVE; A CAR TO THESE DIMENSIONS IS DEFINED AS THE BASE CAR.
WHEN TRUCK CENTERS EXCEED 41' 3", CAR WIDTH FOR ENTIRE CLEARANCE OUTLINE SHALL BE REDUCED TO COMPENSATE FOR THE INCREASED SWINGOUT AT CENTER AND/OR ENDS OF CAR ON A 13° CURVE SO THAT THE WIDTH OF CAR SHALL NOT PROJECT BEYOND THE CENTER OF TRACK MORE THAN THE BASE CAR.
MAXIMUM CAR WIDTHS FOR VARIOUS TRUCK CENTERS, AT CENTER OF CAR, ARE SHOWN ON PLATE B-1. MAXIMUM CAR WIDTH AT LOCATIONS OTHER THAN CENTER OF CAR ARE SHOWN ON PLATE D.
CARS WITH RAIL LOADS IN EXCESS OF 65,750 LB PER AXLE CANNOT BE OPERATED IN UNRESTRICTED INTERCHANGE. HOWEVER, THEY MAY BE PERMITTED UNDER CONTROLLED CONDITIONS WHERE SPECAL AGREEMENT HAS BEEN REACHED BETWEEN PARTICIPATING RAILROADS TO SO HANDLE.
15'-1"
14'-3"
13'-9"
3'-4"
1'-3"
9"
2 3/4"
7'-4"
8'-0"
9'-0"
9'-4"
10'-8"
THE 2 3/4" ABOVE TOP OF RAIL IS ABSOLUTE MINIMUM UNDER ANY AND ALL CONDITIONS OF LADING, OPERATION, AND MAINTENANCE.
(a)
UNRESTRICTED ON ALL ROADS EXCEPT ON CERTAIN ROUTES. FOR SPECIFIC RESTRICTED AREAS ON SUCH ROADS SEE "RAILWAY LINE CLEARANCES".
10'-8"
10'-0"
7'-0"
LIGHT CAR CONDITIONS
CARS MAY BE CONSTRUCTED TO AN EXTREME WIDTH OF 10'-8" AND TO THE OTHER LIMITS OF THIS DIAGRAM WHEN TRUCK CENTERS DO NOT EXCEED 46'-3" AND WHEN, WITH TRUCK CENTERS OF 46'-3". THE SWINGOUT AT ENDS OF CAR DOES NOT EXCEED THE SWINGOUT AT CENTER OF CAR ON A 13" CURVE; A CAR TO THESE DIMENSIONS IS DEFIND AS THE BASE CAR.
WHEN TRUCK CENTERS EXCEED 46'-3", CAR WIDTH FOR ENTIRE CLEARANCE OUTLINE SHALL BE REDUCED TO COMPENSATE FOR THE INCREASED SWINGOUT AT CENTER AND/OR ENDS OF CAR ON A 13" CURVE SO THAT THE WIDTH OF CAR SHALL NOT PROJECT BEYOND THE CENTER OF TRACK MORE THAN THE BASE CAR.
MAXIMUM CAR WIDTHS FOR VARIOUS TRUCK CENTERS, AT CENTER OF CAR ARE SHOWN ON PLATE C-1. MAXIMUM CAR WIDTH AT LOCATIONS OTHER THAN CENTER OF CAR ARE SHOWN ON PLATE D.
CARS WITH RAIL LOADS IN EXCESS OF 65,750 LBS. PER AXLE CANNOT BE OPERATED IN UNRESTRICTED INTERCHANGE. HOWEVER, THEY MAY BE PERMITTED UNDER CONTROLLED CONDITIONS WHERE SPECIAL AGREEMENT HAS BEEN REACHED BETWEEN PARTICIPATING RAILROADS TO SO HANDLE.
15'-6"
14'-8"
14'-2"
3'-4"
1'-3"
9"
2 3/4"
7'-4"
8'-0"
9'-0"
9'-4"
10'-8"
(b)

Figure 8.3.15 AAR Equipment diagrams. (*a*) Plate B unrestricted interchange service; (*b*) Plate C limited interchange service; (*c*) Plate H controlled interchange service.

in the car design. Conventional friction, elastomer, or combination draft gears or short-travel hydraulic cushion units that provide less than 6 in (152 mm) of travel require a structure capable of withstanding a 1,250,000-lb (5.56-MN) impact load. For cars with hydraulic units that provide greater than 14 in (356 mm) of travel, the required design impact load is 600,000 lb (2.7 MN). In all cases, the structural connections to the car must be capable of withstanding a static compressive (squeeze) end load of 1,000,000 lb (4.44 MN) or a dynamic (impact) compressive load of 1,250,000 lb (5.56 MN).

The AAR has adopted requirements for unit trains of high-utilization cars to be designed for 3,000,000 mi (4.8 Gm) of service based upon fatigue life estimates. General-interchange cars that accumulate less mileage in their life should be designed for 1,000,000 mi (1.6 Gm) of service. Road environment spectra for various locations within the car are being developed for different car designs for use in this analysis. The fatigue strengths of various welded connections are provided in the AAR "Manual of Standards and Recommended Practices," Section C, Part II.

Many of the design equations and procedures are available from the AAR. Important information on car design and approval testing is contained in the AAR "Manual of Standards and Recommended Practice," Section C-II M-1001, Chapter XI.

Freight-Car Suspension

Most freight cars are equipped with standard three-piece trucks (Fig. 8.3.16) consisting of two side-frame castings and one bolster casting. Side-frame and bolster designs are subjected to both static and fatigue test requirements specified by the AAR. The bolster casting is equipped with a female centerplate bowl upon which the car body rests and with side bearings generally located 25 in (635 mm) on each side of the centerline. In most cases, the side bearings have clearance to the car body and are equipped with either flat sliding plates or rollers. In some cases, constant-contact side bearings provide a resilient material between the car body and the truck bolster.

The centerplate arrangement may have a variety of styles of wear plates or friction materials and a vertical loose or locked pin between the truck centerplate and the car body.

Truck springs nested into the bottom of the side-frame opening support the end of the truck bolster. Requirements for spring designs and the grouping of springs are generally specified by the AAR. Historically, the damping provided within the spring group has utilized a combination of springs and friction wedges. In addition to friction wedges, in more recent years, some cars have been equipped with hydraulic damping devices that parallel the spring group.

A few trucks have a "steering" feature that includes an interconnection between axles to increase the lateral interaxle stiffness and decrease the interaxle yaw stiffness. Increased lateral stiffness improves the lateral stability and decreased yaw stiffness improves the curving characteristics.

Freight-Car Wheel-Set Design

A freight-car wheel set consists of two wheels, one axle, and two bearings. Cast- and wrought-steel wheels are used on freight cars in North

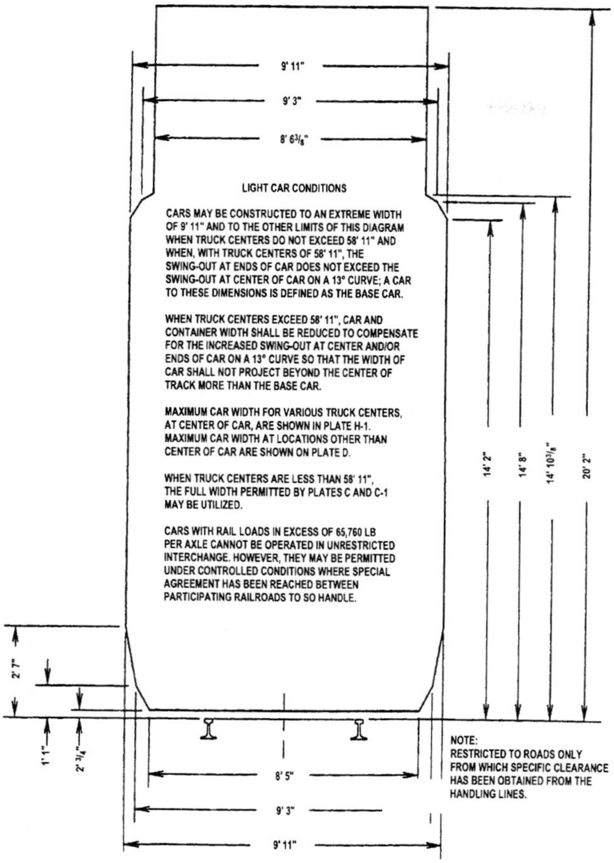
9' 11"
9' 3"
8' 6 3/8"
LIGHT CAR CONDITIONS
CARS MAY BE CONSTRUCTED TO AN EXTREME WIDTH OF 9' 11" AND TO THE OTHER LIMITS OF THIS DIAGRAM WHEN TRUCK CENTERS DO NOT EXCEED 58' 11" AND WHEN, WITH TRUCK CENTERS OF 58' 11", THE SWING-OUT AT ENDS OF CAR DOES NOT EXCEED THE SWING-OUT AT CENTER OF CAR ON A 13° CURVE; A CAR TO THESE DIMENSIONS IS DEFINED AS THE BASE CAR.
WHEN TRUCK CENTERS EXCEED 58' 11", CAR AND CONTAINER WIDTH SHALL BE REDUCED TO COMPENSATE FOR THE INCREASED SWING-OUT AT CENTER AND/OR ENDS OF CAR ON A 13° CURVE SO THAT THE WIDTH OF CAR SHALL NOT PROJECT BEYOND THE CENTER OF TRACK MORE THAN THE BASE CAR.
MAXIMUM CAR WIDTH FOR VARIOUS TRUCK CENTERS, AT CENTER OF CAR, ARE SHOWN IN PLATE H-1. MAXIMUM CAR WIDTH AT LOCATIONS OTHER THAN CENTER OF CAR ARE SHOWN ON PLATE D.
WHEN TRUCK CENTERS ARE LESS THAN 58' 11", THE FULL WIDTH PERMITTED BY PLATES C AND C-1 MAY BE UTILIZED.
CARS WITH RAIL LOADS IN EXCESS OF 65,760 LB PER AXLE CANNOT BE OPERATED IN UNRESTRICTED INTERCHANGE. HOWEVER, THEY MAY BE PERMITTED UNDER CONTROLLED CONDITIONS WHERE SPECIAL AGREEMENT HAS BEEN REACHED BETWEEN PARTICIPATING RAILROADS TO SO HANDLE.
14' 2"
14' 8"
14' 10 3/8"
20' 2"
2' 7"
1' 1"
2' 3/4"
NOTE:
RESTRICTED TO ROADS ONLY FROM WHICH SPECIFIC CLEARANCE HAS BEEN OBTAINED FROM THE HANDLING LINES.
8' 5"
9' 3"
9' 11"

(c)

Figure 8.3.15 *(Continued)*

America (AAR "Manual of Standards and Recommended Practice," Sec. G). Freight-car wheels are subjected to thermal loads from braking, as well as mechanical loads at the wheel-rail interface. Experience with thermal damage to wheels has led to the requirement for "low stress" or curved-plate wheels (Fig. 8.3.17). These wheels are less susceptible to the development of circumferential residual tensile stresses which render the wheel vulnerable to sudden failure if a flange or rim crack occurs. New wheel designs introduced for interchange service must be evaluated by using a finite-element technique employing both thermal and mechanical loads (AAR S-660).

Freight-car wheels range in diameter from 28 to 38 in (711 to 965 mm), depending on car weight (Table 8.3.3). The old AAR standard tread profile (Fig. 8.3.18*a*) has been replaced with the AAR-1B (Fig. 8.3.18*c*) profile which represents a "worn" profile to minimize early tread loss due to wear and provide a more stable profile over the life of the tread. Several variant tread profiles, including the AAR-1B, were developed from the basic Heumann design, Fig. 8.3.18*b*. One of these, for application in Canada, provided increasing conicity into the throat of the flange, similar to the Heumann profile. This reduces curving resistance and extends wheel life.

Wheels are also specified by chemistry and heat treatment. Low-stress wheel designs of Classes B and C are required for freight cars. Class B wheels have a carbon content of 0.57 to 0.67 percent and are rim-quenched. Class C wheels have a carbon content of 0.67 to 0.77 percent and are also rim-quenched. Rim quenching provides a hardened running surface for a long wear life. Lower carbon levels than

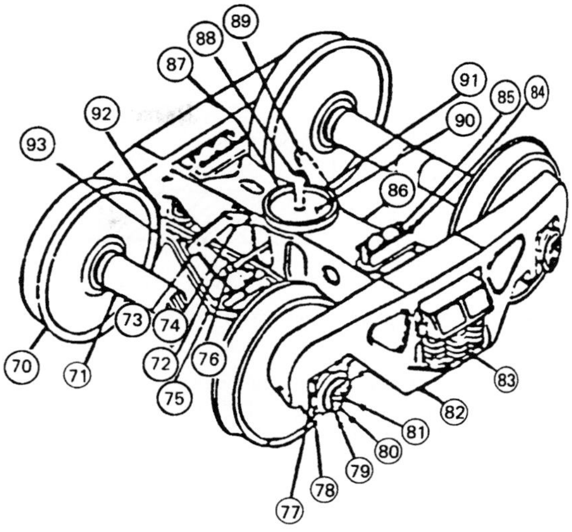
70
71
72
73
74
75
76
77
78
79
80
81
82
83
84
85
86
87
88
89
90
91
92
93

Unit beam roller bearing truck components

70. Wheel	82. Truck side frame
71. Axle	83. Truck springs
72. Truck dead lever	84. Truck side bearing
73. Dead lever fulcrum	85. Side bearing roller
74. Dead lever fulcrum bracket	86. Truck bolster
75. Brake beam	87. Truck center plate cast
76. Bottom rod	88. Truck live lever
77. Roller bearing adapter	89. Center pin
78. Roller bearing assembly	90 Horizontal wear plate
79. End cap	91. Vertical wear plate
80. End cap retaining bolt	92. Brake shoe key
81. Locking plate	93. Brake shoe

Figure 8.3.16 Three-piece freight car truck. (*Courtesy AAR Research and Test Department.*)

(a) (b) (c)

Figure 8.3.17 Wheel-plate designs. (*a*) Flat plate; (*b*) parabolic plate; (*c*) S-plate curved wheel.

Table 8.3.3 Wheel and Journal Sizes of Eight-Wheel Cars

Nominal car capacity, ton	Maximum gross weight, lb	Journal (bearing) size, in	Bearing AAR class	Wheel diameter, in
50	177,000	5½ × 10		33
*	179,200	6 × 11	E	28
70	220,000	6 × 11	E	33
100	263,000	6½ × 12	F	36
110†	286,000	6½ × 12	F	36
110	286,000	6½ × 9	K	36
125†	315,000	7 × 12	G	38

* Load limited by wheel rating.
† Not approved for free interchange.

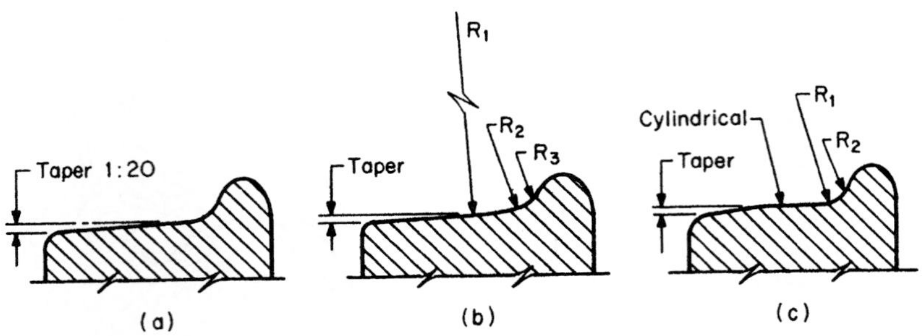

Figure 8.3.18 Wheel-tread designs. (*a*) Obsolete standard AAR; (*b*) Heumann; (*c*) new AAR-1B.

those in Class B may be used where thermal cracking is experienced, but freight-car equipment generally does not require their use.

Axles used in interchange service are solid steel forgings with raised wheel seats. Axles are specified by journal size for different car capacities (Table 8.3.3).

Most freight car journal bearings are grease-lubricated, tapered-roller bearings (see Sec. 6). Current bearing designs eliminate the need for periodic field lubrication.

Wheels are mounted and secured on axles with an interference fit. Bearings are mounted with an interference fit and retained by an end cap bolted to the end of the axle. Wheels and bearings for cars in interchange service must be mounted by an AAR-inspected and -approved facility.

Special Features

Many components are available to enhance the usefulness of freight cars. In most cases, the design or performance of the component is specified by the AAR.

Coupler Cushioning Switching of cars in a classification yard can result in relatively high coupler forces at the time of the impact between the moving and standing cars. Nominal coupling speeds of 4 mi/h (6.4 km/h) or less are sometimes exceeded, with lading damage a possible result. Conventional cars are equipped with an AAR-approved draft gear, usually a friction-spring energy-absorbing device, mounted between the coupler and the car body. The rated capacity of draft gears ranges between 20,000 ft-lb (27.1 kJ) for earlier units to over 65,000 ft · lb (88.1 kJ) for later designs. Impact forces of 1,250,000 lb (5.56 MN) can be expected when a moving 100-ton car strikes standing cars at 8 to 10 mi/h (12.8 to 16 km/h). Hydraulic cushioning devices are available to reduce the impact force to 500,000 lb (2.22 MN) at coupling speeds of 12 to 14 mi/h (19 to 22 km/h). These devices may be mounted either at each end of the car (end-of-car devices) or in a long beam which extends from coupler to coupler (sliding centersill devices).

Lading Restraint Many forms of lading restraint are available, from tie-down chains for automobiles on rack cars to movable bulkheads for boxcars. Most load-restraining devices are specified by the AAR "Manual of Standards and Recommended Practices" and approved car loading arrangements are specified in AAR "Loading Rules," a multi-volume publication for enclosed, open-top, and TOFC and COFC cars.

Covered Hopper Car Discharge Gates The majority of covered hopper cars are equipped with rack-and-pinion–operated sliding gates, which allow the lading to discharge by gravity between the rails. These gates can be operated manually, with a simple bar or a torque-multiplying wrench, or mechanically with an impact or hydraulic wrench. Many special covered hopper cars have discharge gates with nozzles and metering devices for vacuum or pneumatic unloading.

Coupling Systems The majority of freight cars are connected with AAR standard couplers. A specification has been developed to permit the use of alternative coupling systems such as articulated connectors, drawbars, and rotary-dump couplers; see AAR M-215.

Freight-Train Braking

The retarding forces acting on a railway train are rolling and mechanical resistance, aerodynamic drag, curvature, and grade, plus that force resulting from friction of brake shoes rubbing the wheel treads. On locomotives so equipped, dynamic or rheostatic brake using the traction motors as generators can provide all or a portion of the retarding force to control train speed.

Quick-action automatic air brakes of the type specified by the AAR are the common standard in North America. With the automatic air brake system, the brake pipe extends through every vehicle in the train, connected by hoses between each locomotive unit and car. The front and rear end brake pipe angle cocks are closed.

Air pressure is provided by compressors on the locomotive units to the main reservoirs, usually at 130 to 150 lb/in^2 (900 to 965 kPa). (Pressure values are gage pressures.) The engineer's automatic brake valve, in "release" position, provides air to the brake pipe on freight trains at reduced pressure, usually at 75, 80, 85, or 90 lb/in^2 (520, 550, 585, or 620 kPa), depending on the type of service, train weight, grades, and speeds at which a train will operate. In passenger service, brake pipe pressure is usually 90 or 110 lb/in^2 (620 to 836 kPa).

When brake pipe pressure is increased, the control valve allows the reservoir capacity on each car and locomotive to be charged and at the same time connects the brake cylinders to exhaust. Brake pipe pressure is reduced when the engineer's brake valve is placed in a "service" position, and the control valve cuts off the charging function and allows the reservoir air on each car to flow into the brake cylinder. This moves the piston and, through a system of levers and rods, pushes the brake shoes against the wheel treads.

When the engineer's automatic brake valve is placed in the emergency position, the brake pipe pressure (BP) is reduced very rapidly. The control valves on each car move to the emergency-application position and rapidly open a large vent valve, exhausting brake pipe pressure to atmosphere. This will serially propagate the emergency application through the train at from 900 to 950 ft/s (280 to 290 m/s.) With the control valve in the emergency position both auxiliary- and emergency-reservoir volumes (pressures) equalize with the brake cylinder, and higher brake cylinder pressure (BCP) results, building up at a faster rate than in service applications.

The foregoing briefly describes the functions of the fundamental automatic air brake based on the functions of the control valve. AAR-approved brake equipment is required on all freight cars used in interchange service. The functions of the control valve have been refined to permit the handling of longer trains by more uniform brake performance. Important improvements in this design have been (1) reduction of the time required to apply the brakes on the last car of a train, (2) more uniform and faster release of the brakes, and (3) availability of emergency application with brake pipe pressure greater than 40 lb/in^2 (275 kPa). Faster brake response has been achieved by a number of railroads using "distributed power," where radio-controlled locomotives are spaced within the train or placed at both ends of the train.

The braking ratio of a car is defined as the ratio of brake shoe (normal) force to the car's rated gross weight. Two types of brake shoes, high-friction composition and high-phosphorus cast iron, are used in interchange service. As these shoes have very different friction characteristics, different braking ratios are required to assure uniform train braking performance (Table 8.3.4). Actual or net shoe forces are measured with

Table 8.3.4 Braking Ratios, AAR Standard S-401

Type of brake rigging and shoes	With 50 lb/in² brake cylinder pressure: Percent of gross rail load, Min.	Max.	Maximum percent of light weight	Hand brake*: Minimum percent of gross rail load
Conventional body-mounted brake rigging or truck-mounted brake rigging using levers to transmit brake cylinder force to the brake shoes				
Cars equipped with cast iron brake shoes	13	20	53	13
Cars equipped with high-friction composition brake shoes	6.5	10	30	11
Direct-acting brake cylinders not using levers to transmit brake cylinder force to the brake shoes				
Cars equipped with cast iron brake shoes				
Cars equipped with high-friction composition brake shoes	6.5	10	33	11
Cabooses†				
Cabooses equipped with cast iron brake shoes			35-45	
Cabooses equipped with high-friction composition brake shoes			18-23	

* Hand brake force applied at the horizontal hand brake chain with AAR certified or AAR approved hand brake.
† Effective for cabooses ordered new after July 1, 1982, hand brake ratios for cabooses to the same as lightweight ratios for cabooses.
NOTE: Above braking ratios also apply to cars equipped with empty and load brake equipment.

calibrated devices. The calculated braking ratio R (nominal) is determined from the equation

$$R = PLANE \times 100/W \tag{8.3.6}$$

where P = brake cylinder pressure, 50 lb/in² gage; L = mechanical ratio of brake levers; A = brake cylinder area, in²; N = number of brake cylinders; E = brake rigging efficiency $= E_r \times E_b \times E_c$; and W = car weight, lb.

To estimate rigging efficiency consider each pinned joint and horizontal sliding joint as a 0.01 loss of efficiency; that is, in a system with 20 pinned and horizontal sliding joints, $E_r = 0.80$. For unit-type (hangerless) brake beams $E_b = 0.90$ and for the brake cylinder $E_c = 0.95$, giving an overall efficiency of 0.684 or 68.4 percent.

The total retarding fore in pounds per ton may be taken as:

$$F = (PLef/W) = F_g G \tag{8.3.7}$$

where P = total brake-cylinder piston force, lbf; L = multiplying ratio of the leverage between cylinder pistons and wheel treads; ef = product of the coefficient of brake shoe friction and brake rigging efficiency; W = loaded weight of vehicle, tons; F_g = force of gravity, 20 lb/ton/percent grade; G = ascending grade, percent.

Stopping distance can be found by adding the distance covered during the time the brakes are fully applied to the distance covered during the equivalent instantaneous application time.

$$S = \frac{0.0334V_2^2}{\left[\frac{W_n B_n (p_a/p_n) ef}{W_a}\right] + \left(\frac{R}{2000}\right) \pm (G)} + 1.467\, t_1 \left[V_1 - \left(\frac{R + 2000G}{91.1}\right)\frac{t_1}{2}\right] \tag{8.3.8}$$

where S = stopping distance, ft; V_1 = initial speed when brake applied, mi/h; V_2 = speed, mi/h, at time t_1; W_n = weight on which braking ratio B_n is based, lb (see Table 8.3.5 for values of W_n for freight cars; for passenger cars and locomotives: W_n is based on empty or ready-to-run weight); B_n = braking ratio (total brake shoe force at stated brake cylinder, lb/in², divided by W_n); P_n = brake cylinder pressure on which B_n is based, usually 50 lb/in²; P_a = full brake cylinder pressure; e = overall rigging and cylinder efficiency, decimal; f = typical friction of brake shoes, see below; R = total resistance, mechanical plus aerodynamic and curve resistance, lb/ton; G = grade in decimal, + upgrade, – downgrade; t_1 = equivalent instantaneous application time, s.

Table 8.3.5 Capacity versus Load Limit for Railroad Freight Cars

Capacity, ton	W_n, 1,000 lb
50	177
70	220
100	263
110	286
125	315

Equivalent instantaneous application time is that time on a curve of average brake cylinder buildup versus time for a train or car where the area above the buildup curve is equal to the area below the curve. A straight-line buildup curve starting at zero time would have a t_1 of half the total buildup time.

The friction coefficient f varies with the speed; it is usually lower at high speed. To a lesser extent, it varies with brake shoe force and with the material of the wheel and shoe. For stops below 60 mi/h (97 km/h), a conservative figure for a high-friction composition brake shoe on steel wheels is approximately

$$ef = 0.30 \tag{8.3.9}$$

In the case of high-phosphorus iron shoes, this figure must be reduced by approximately 50 percent.

P_n is based on 50 lb/in² (345 kPa) air pressure in the cylinder; 80 lb/in² is a typical value for the brake pipe pressure of a fully charged freight train. This will give a 50 lb/in² (345 kPa) brake cylinder pressure during a full service application on AB equipment, and a 60 lb/in² (415 kPa) brake cylinder pressure with an emergency application.

To prevent wheel sliding, $F_R \leq \phi W$, where F_R = retarding force at wheel rims resisting rotation of any pair of connected wheels, lb; ϕ = coefficient of wheel-rail adhesion or friction (a decimal); and W = weight upon a pair of wheels, lb. Actual or adhesive weight on wheels when the vehicle is in motion is affected by weight transfer (force transmitted to the trucks and axles by the inertia of the car body through the truck center plates), center of gravity, and vertical oscillation of body weight upon truck springs. The value of ϕ varies with speed as shown in Fig. 8.3.19.

The relationship between the required coefficient ϕ of wheel-rail adhesion to prevent wheel sliding and rate of retardation A is miles per hour per second may be expressed by $A = 21.95\phi_1$.

There has been some encouraging work to develop an electric brake that may eventually obsolete the present pneumatic brake systems.

Test Devices Special devices have been developed testing brake components and cars on a repair track.

End-of-Train Devices To eliminate the requirement for a caboose crew car at the end of the train, special electronic devices have been developed to transmit the end-of-train brake-pipe pressure to the locomotive operator by telemetry.

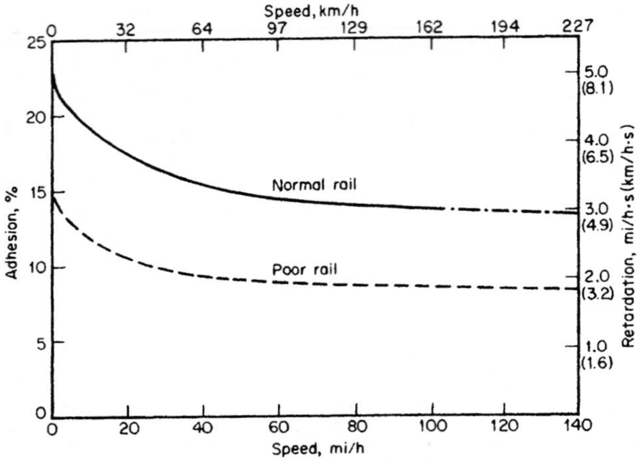

Figure 8.3.19 Typical wheel-rail adhesion. Track has jointed rails. (*Air Brake Association.*)

PASSENGER EQUIPMENT

During the past two decades most mainline or long-haul passenger service in North America has become a function of government agencies, that is, Amtrak in the United States and Via Rail in Canada. Equipment for intraurban service is divided into three major categories: commuter rail, heavy rail rapid transit, and light rail transit, depending upon the characteristics of the service. Commuter rail equipment operates on conventional railroad rights-of-way, usually intermixed with other long-haul passenger and freight traffic. Heavy-rail rapid transit (HRT), often referred to as "metro," operates on a dedicated right-of-way, which is commonly in subways or on elevated structures. Light-rail transit (LRT) utilizing light rail vehicles (LRV) has evolved from the older trolley or streetcar concepts and may operate in any combination of surface, subway, and elevated dedicated rights-of-way, semireserved surface rights-of-way with grade crossing, and shared rights-of-way with other traffic on surface streets. In a few cases LRT shares the trackage with freight operations, which are time-separated to comply with FRA regulations.

Since mainline and commuter rail equipment operates over conventional railroad rights-of-way, the structural design is heavy to provide the FRA-required "crashworthiness" of vehicles in the event of collisions with other trains or at grade crossings with automotive vehicles. Although HRT vehicles are designed to stringent structural criteria, the requirements are somewhat different, since the operation is separate from freight equipment and there are usually no highway grade crossings. Minimum weight is particularly important for transit vehicles so that demanding schedules over lines with close station spacing can be met with minimum energy consumption.

Main-Line Passenger Equipment There are four primary passenger train sets in operation across the world: diesel locomotive hauled, diesel multiple unit (DMU), electric locomotive hauled, and electric MU (EMU). In recent years the design of mainline passenger equipment has been controlled by specifications provided by APTA and the operating authority. Most of the newer cars provided for Amtrak have had stainless steel structural components. These cars have been designed to be locomotive-hauled and to use a separate 480-V three-phase power supply for heating, ventilation, air conditioning, food car services, and other control and auxiliary power requirements. Figure 8.3.20*a* shows an Amtrak coach car of the Superliner class; Fig. 8.3.20*b* shows an intercity high-speed coach.

Trucks for passenger equipment are designed to provide a superior ride to freight-car trucks. As a result, passenger trucks include a form of "primary" suspension to isolate the wheel set from the frame of the truck. A softer "secondary" suspension is provided to isolate the truck from the car body. In most cases, the primary suspension uses either coil springs, elliptical springs, or elastomeric components. The secondary suspension generally utilizes either large coil ring springs or pneumatic springs with special leveling valves to control the height of the car body. Hydraulic dampers are also applied to improve the vertical and lateral ride quality.

Commuter Rail Passenger Equipment Commuter rail equipment can be either locomotive-hauled or self-propelled (Fig. 8.3.21). Some locomotive-hauled equipment is arranged for "push-pull" service. This configuration permits the train to be operated with the locomotive either pushing or pulling the train. For push-pull service, some of the passenger cars must be equipped with cabs to allow the engineer to operate the train from the end opposite the locomotive during the push operation. All cars must have control trainlines to connect the lead (cab) car to the trailing locomotive. Locomotive-hauled commuter rail cars use AAR-type H couplers. Most self-propelled (single- or multiple-unit) cars use other automatic coupler designs, which can be mechanically coupled to an AAR coupler with an adaptor.

Heavy Rail Rapid Transit Equipment This equipment is used on traditional subway/elevated properties in such cities as Boston, New York, Philadelphia (Fig. 8.3.22), and Chicago in a semiautomatic mode of operation over dedicated rights-of-way which are constrained by limiting civil features. State-of-the-art subway-elevated properties include such cities as Washington, Atlanta, Miami, and San Francisco, where the equipment provides highly automated modes of operation on rights-of-way with generous civil alignments.

The cars can operate bidirectionally in multiple with as many as 12 or more cars controlled from the leading cab. They are electrically propelled, usually from a DC third rail, which makes contact with a shoe insulated from and supported by the frame of the truck. Occasionally, roof-mounted pantographs are used. Voltages range from 600 to 1,500 V DC.

The cars range from 48 to 75 ft (14.6 to 22.9 m) over the anticlimbers, the longer cars being used on the newer properties. Passenger seating varies from 40 to 80 seats per car, depending upon length and the local policy (or preference) regarding seated to standee ratio. Older properties require negotiation of curves as sharp as 50 ft (15.2 m) minimum radius with speeds up to only 50 mi/h (80 km/h) on tangent track, while newer properties usually have no less than 125 ft (38.1 m) minimum radius curves with speeds up to 75 mi/h (120 km/h) on tangent track.

All North American properties operate on standard-gage track with the exception of the 5 ft 6 in (1.7 m) San Francisco Bay Area Rapid Transit (BART), the 4 ft 10 in (1.5 m) Toronto Transit Subways, and the 5 ft 2½ in (1.6 m) Philadelphia Southeastern Pennsylvania Transportation Authority (SEPTA) Market-Frankford line. Grades seldom exceed 3 percent, and 1.5 to 2.0 percent is the desired maximum.

Typically, newer properties require maximum acceleration rates of between 2.5 and 3.0 mi/(h · s) [4.0 to 4.8 km/(h · s)] as nearly independent of passenger loads as possible from 0 to approximately 20 mi/h (32 km/h). Depending upon the selection of motors and gearing, this rate falls off as speed is increased, but rates can generally be controlled at a variety of levels between zero and maximum.

Deceleration is typically accomplished by a blended dynamic and electropneumatic friction tread or disk brake, although a few properties use an electrically controlled hydraulic friction brake. In some cases energy is regenerated and sent back into the power system. All of these systems usually provide a maximum braking rate of between 3.0 and 3.5 mi/(h · s) [4.8 and 5.6 km/(h · s)] and are made as independent of passenger loads as possible by a load-weighing system that adjusts braking effort to suit passenger loads. Some employ regenerative braking to supplement the other brake systems.

Dynamic braking is generally used as the primary stopping mode and is effective from maximum speed down to 10 mi/h (16 km/h) with friction braking supplementation as the characteristic dynamic fade occurs. The friction brakes provide the final stopping forces. Emergency braking rates depend upon line constraints, car subsystems, and other factors, but generally rely on the maximum retardation force that can be provided by the dynamic and friction brakes within the limits of available wheel-to-rail adhesion.

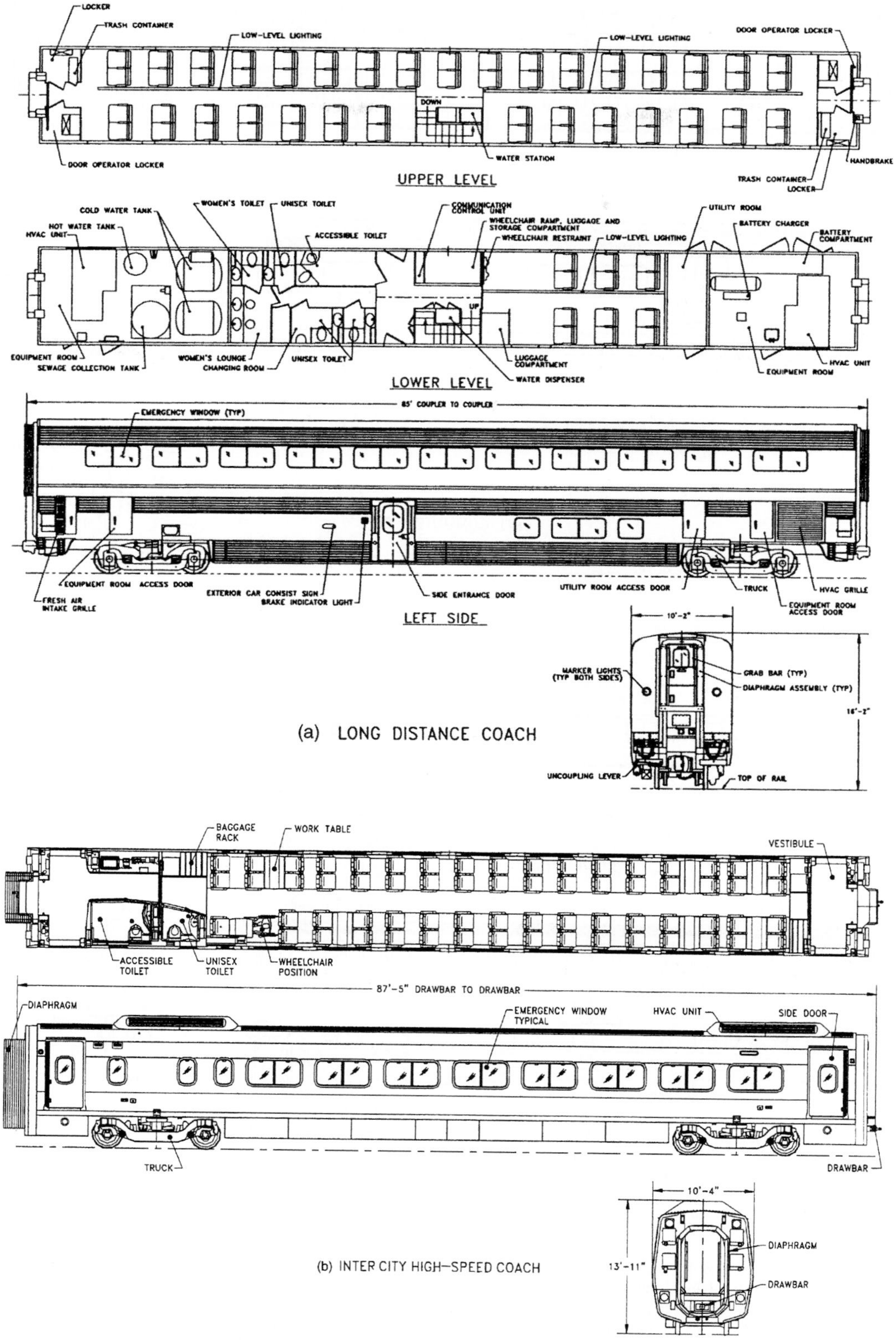

Figure 8.3.20 Mainline passenger cars. (*a*) Double-deck coach; (*b*) Single-deck coach. (*Courtesy of Bombardier Transportation.*)

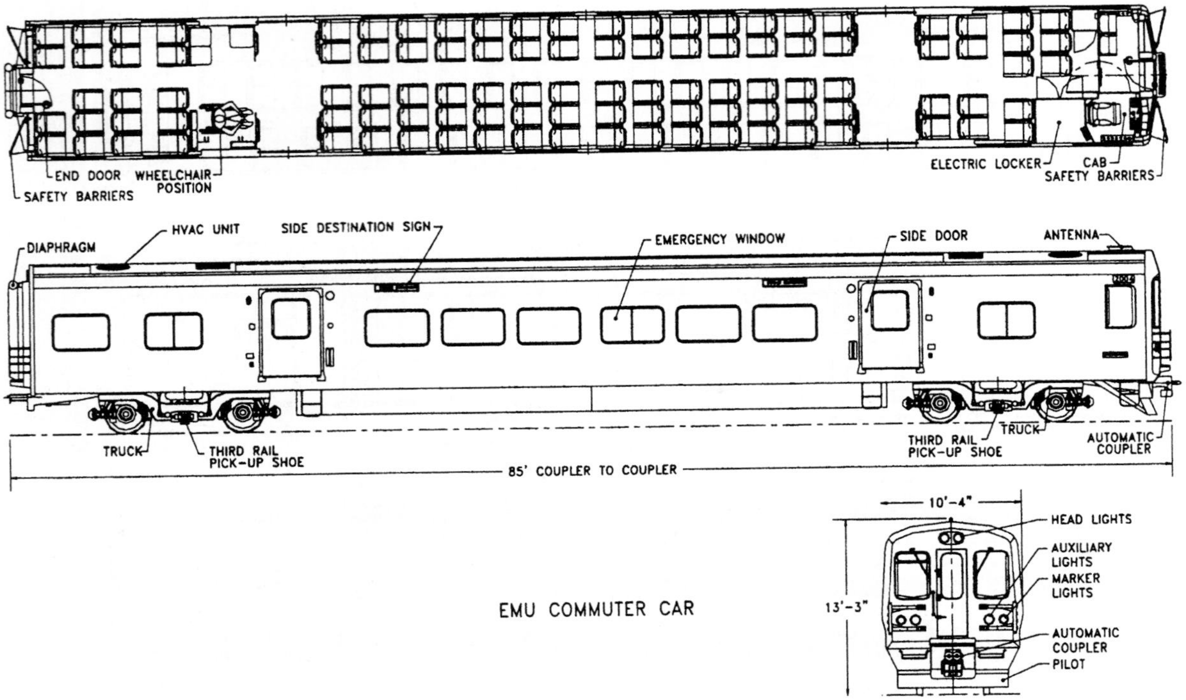

Figure 8.3.21 Commuter rail car. (*Courtesy of Bombardier Transportation.*)

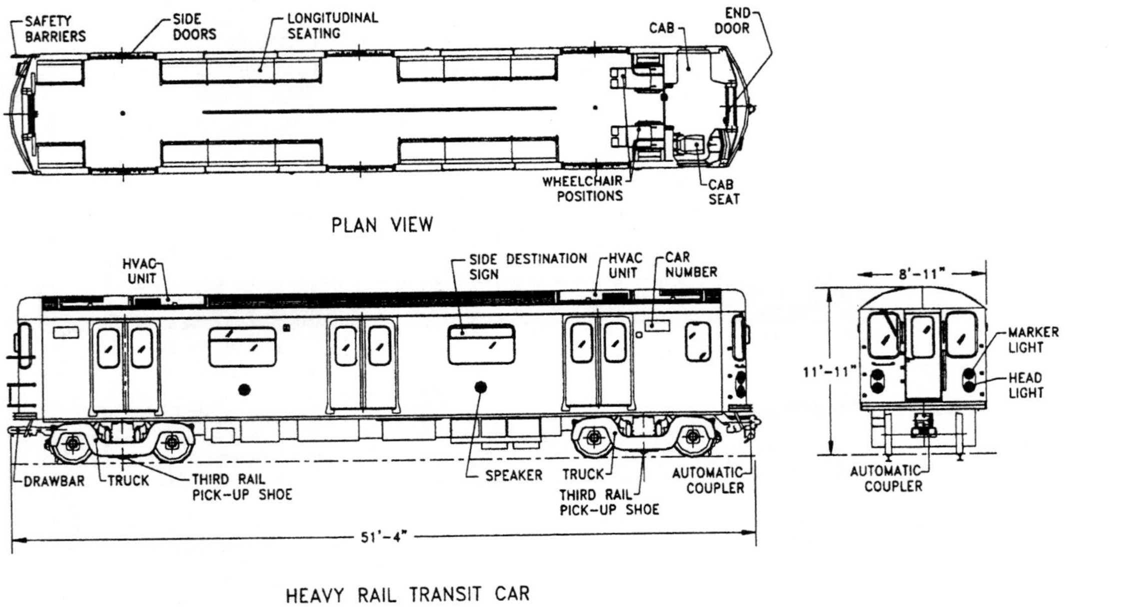

Figure 8.3.22 Heavy rail transit car. (*Courtesy of Bombardier Transportation.*)

Acceleration and braking on modern properties are usually controlled by a single master controller handle that has power positions in one direction and a coasting or neutral center position and braking positions in the opposite direction. A few properties use foot-pedal control with a "deadman" pedal operated by the left foot, a brake pedal operated by the right foot, and an accelerator (power) pedal also operated by the right foot. In either case, the number of positions depends upon property policy and control subsystems on the car. Control elements include motor-current sensors and some or all of the following: speed sensors, rate sensors, and load-weighing sensors. Signals from these sensors are processed by an electronic control unit (ECU), which provides control functions to the propulsion and braking systems. The propulsion systems currently include pilot-motor-operated cams which actuate switches or electronically controlled unit switches to control resistance steps, or chopper or inverter systems that electronically provide the desired voltages and currents at the motors.

In some applications, dynamic braking utilizes the traction motors as generators to dissipate energy through the onboard resistors also used in acceleration. It is expected that regenerative braking will become more common since energy can be returned to the line for use by other cars. Theoretically, 35 to 50 percent of the energy can be returned, but the present state-of-the-art is limited to a practical 20 percent on properties with large numbers of cars on close headways.

Car bodies are made of welded stainless steel or low-alloy high-tensile (LAHT) steel of a design that carries structural loads. Earlier problems with aluminum, primarily electrolytic action among dissimilar metals and welding techniques, have been resolved and aluminum is also used on a significant number of new cars.

Trucks may be cast or fabricated steel with frames and journal bearings either inside or outside the wheels. Axles are carried in roller-bearing journals connected to the frames so as to be able to move vertically against a variety of types of primary spring restraint. Metal springs or air bags are used as the secondary suspension between the trucks and the car body. Most wheels are solid cast or wrought steel. Resilient wheels have been tested but are not in general service on heavy rail transit equipment.

All heavy rail rapid transit systems use high-level loading platforms to speed passenger flow. This adds to the civil construction costs, but is necessary to achieve the required level of service.

Light Rail Transit Equipment The cars are called light rail vehicles (Fig. 8.3.23 shows typical **LRVs**) and are used on a few remaining streetcar systems such as those in Boston, Philadelphia, and Toronto on city streets and on state-of-the-art subway, surface, and elevated systems such as those in Calgary, Edmonton, Los Angeles, San Diego, and Portland (Oregon) and in semiautomated modes over partially or wholly reserved rights-of-way.

As a practical matter, the LRV's track, signal systems, and power systems utilize the same subsystems as heavy rail rapid transit and are not "lighter" in a physical sense.

Most LRVs are designed to operate bidirectionally in multiple, with up to four cars controlled from the leading cab. The LRVs are electrically propelled from an overhead contact wire, often of catenary design, that makes contact with a pantograph or pole on the roof. For reasons of wayside safety, third-rail pickup is not used. Voltages range from 550 to 750 V DC.

The cars range from 60 to 65 ft (18.3 to 19.8 m) over the anticlimbers for single cars and from 70 to 90 ft (21.3 to 27.4 m) for articulated cars. The choice is determined by passenger volumes and civil constraints.

Articulated cars have been used in railroad and transit applications since the 1920s, but have found favor in light rail applications because of their ability to increase passenger loads in a longer car that can negotiate relatively tight-radius curves. These cars require the additional mechanical complexity of the articulated connection and a third truck between two car-body sections.

Passenger seating varies from 50 to 80 or more seats per car, depending upon length, placement of seats in the articulation, and the policy regarding seated/standee ratio. Older systems require negotiation of curves down to 30 ft (9.1 m) minimum radius with speeds up to only 40 mi/h (65 km/h) on tangent track, while newer systems usually have no less than 75 ft (22.9 m) minimum radius curves with speeds up to 65 mi/h (104 km/h) on tangent track.

The newer properties all use standard-gage track; however, existing older systems include 4 ft 10 in (1,495 mm) and 5 ft 2½ in (1,587 mm) gages. Grades have reached 12 percent but 6 percent is now considered the maximum and 5 percent is preferred.

Typically, newer properties require maximum acceleration rates of between 3.0 to 3.5 mi/(h·s) [4.8 to 5.6 km/(h·s)] as nearly independent of passenger load as possible from 0 to approximately 20 mi/h (32 km/h). Depending upon the selection of motors and gearing, this rate falls off as speed is increased, but the rates can generally be controlled at a variety of levels.

Unlike most heavy rail rapid transit cars, LRVs incorporate three braking modes: dynamic, friction, and track brake, which typically provide maximum service braking at between 3.0 and 3.5 mi/(h·s) [4.8 to 5.6 km/(h·s)] and 6.0 mi/(h·s) [9.6 km/(h·s)] maximum emergency braking rates. The dynamic and friction brakes are usually blended, but a variety of techniques exist. The track brake is intended to be used primarily for emergency conditions and may or may not be controlled with the other braking systems.

The friction brakes are almost exclusively disk brakes, since LRVs use resilient wheels that can be damaged by tread-brake heat buildup. No single consistent pattern exists for the actuation mechanism. All-electric, all-pneumatic, electropneumatic, electrohydraulic, and electropneumatic over hydraulic are in common use.

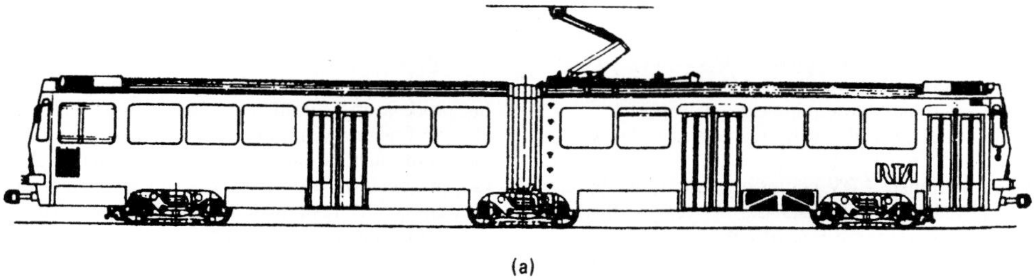

(a)

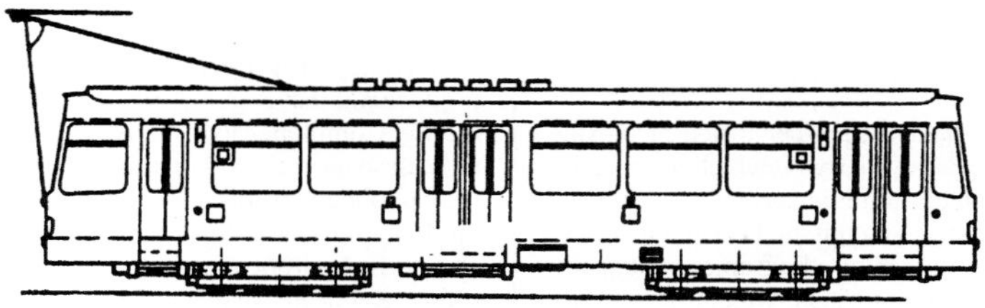

(b)

Figure 8.3.23 Light-rail vehicle (LRV). (*a*) Articulated; (*b*) nonarticulated

Dynamic braking is generally used as the primary braking mode and is effective from maximum speed down to about 5 mi/h (8 km/h) with friction braking supplementation as the characteristic dynamic fade occurs.

As with heavy rail rapid transit, the emergency braking rates depend upon line constraints, car-control subsystems selected, and other factors, but the use of track brakes means that higher braking rates can be achieved because the wheel-to-rail adhesion is not the limiting factor.

Acceleration and braking on modern properties are usually controlled by a single master controller handle that has power positions in one direction and a coasting or neutral center position and braking positions in the opposite direction. A few properties use foot-pedal control with a "deadman" pedal operated by the left foot, a brake pedal operated by the right foot, and an accelerator (power) pedal also operated by the right foot. In either case, the number of positions depends upon property policy and control subsystems on the car.

Control elements include motor-current sensors and some or all of the following: speed sensors, rate sensors, and load-weighing sensors. Signals from these sensors are processed by an electronic control unit (ECU), which provides control functions to the propulsion and braking systems.

The propulsion systems currently include pilot-motor-operated cams that actuate switches or electronically controlled unit switches to control resistance steps. Chopper and inverter systems that electronically provide the desired voltages and currents at the motors are also used.

Most modern LRVs are equipped with two powered trucks. In two-section articulated designs (Fig. 8.3.23*a*), the third (center) truck may be left unpowered but usually has friction and track brake capability. Some European designs use three powered trucks but the additional cost and complexity have not been found necessary in North America.

Unlike heavy rail rapid transit, there are three major DC-motor configurations in use: the traditional series-wound motors used in bimotor trucks, the European-derived monomotor, and a hybrid monomotor with a separately excited field—the last in chopper-control version only.

The bimotor designs are rated between 100 and 125 shaft hp per motor at between 300 and 750 V DC, depending upon line voltage and series or series-parallel control schemes (electronic or electromechanical control). The monomotor designs are rated between 225 and 250 shaft hp per motor at between 300 and 750 V DC (electronic or electromechanical control).

The motors, gear units (right angle or parallel), and axles are joined variously through flexible couplings. In the case of the monomotor, it is supported in the center of the truck, and right-angle gearboxes are mounted on either end of the motor. Commonly, the axle goes through the gearbox and connection is made with a flexible coupling arrangement. Electronic inverter control drives with AC motors have been applied in recent conversions and new equipment.

Dynamic braking is achieved in the same manner as in heavy rail rapid transit.

Unlike heavy rail rapid transit, LRV bodies are usually made only of welded LAHT steel and are of a load-bearing design. Because of the semireserved right-of-way, the risk of collision damage with automotive vehicles is greater than with heavy rail rapid transit and the LAHT steel has been found to be easier to repair than stainless steel or aluminum. Although LAHT steel requires painting, this can be an asset since the painting can be performed in a highly decorative manner pleasing to the public and appropriate to themes desired by the cities.

Trucks may be cast or fabricated steel with either inside or outside frames. Axles are carried in roller-bearing journals, which are usually resiliently coupled to the frames with elastomeric springs as a primary suspension. Both vertical and a limited amount of horizontal movement occur. Since tight curve radii are common, the frames are usually connected to concentric circular ball-bearing rings, which, in turn, are connected to the car body. Air bags, solid elastomeric springs, or metal springs are used as a secondary suspension. Resilient wheels are used on virtually all LRVs.

Newer LRVs have low-level loading doors and steps, which minimize station platform costs.

TRACK

Gage The gage of railway track is the distance between the inner sides of the rail heads measured 0.626 in (15.9 mm) below the top running surface of the rail. North American railways are laid with a nominal gage of 4 ft 8½ in (1,435 mm), which is known as *standard gage*. Rail wear causes an increase in gage. On sharp curves [over 8°, see Eq. (8.3.10)], it has been the practice to widen the gage to reduce rail wear.

Track Structure In basic track structure is composed of six major elements: rail, tie plates, fasteners, cross ties, ballast, and subgrade (Fig. 8.3.24).

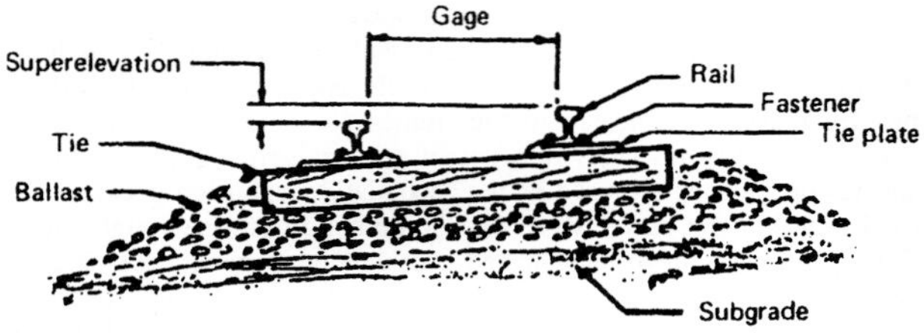

Figure 8.3.24 Track structure.

Rail The American Railway Engineering and Maintenance of Way Association had defined standards for the cross section of rails of various weights, which are identified by their weight (in pounds) per yard. The AREMA standards provide for rail weights varying from 90 to 141 lb/yd. However, recommendations for new rail purchases are limited to weights of 115 lb/yd and above. Many railroads historically used special sections of their own design; however, the freight industry is converging on the use of the 115-, 136-, and 141-lb/yd sections. The standard length in use is 80 ft (31.5 m), although some mills are now considering making longer lengths available.

The prevailing weight of rails used in mainline track is from 112 to 141 lb/yd. Secondary and branch lines are generally laid with 90- to 115-lb/yd or heavier rail that has been previously used and partly worn in mainline service. In a very limited application, rail weights as low as 60 lb/yd exist in some tracks having very light usage. In the past, rail sections were connected by using bolted joint bars. Current practice is the use of continuous welded rail (CWR). In general, rails are first welded into lengths averaging 1,440 feet (440 m) or more. After these strings, or "ribbons," are installed, they are later field welded to eliminate conventional joints. At one time the gas-fusion welding process was common, but today almost all rail is welded by the electric-flash or aluminothermic processes. CWR requires the use of more rail anchors than jointed rail in order to clamp the base of the rail and position it on the cross ties to prevent rail movement. This is necessary to restrain longitudinal forces that result from thermal expansion and contraction with ambient temperature changes. Failure to adequately restrain the rail can result in track buckling in extremely hot weather or in rail "pull aparts" in extremely cold weather. Care must be exercised to install the rail at temperatures that do not approach high or low extremes.

A variety of heat-treating and alloying processes have been used to produce rail that is more resistant to wear and less susceptible to fatigue failure. Comprehensive studies indicate that both gage face and top-of-rail lubrication can dramatically reduce the wear rate of rail in curves (see publications of AAR Research and Test Department). Friction modification using top-of-rail lubrication must be carefully managed so that it does not interfere with traction required to move and brake trains. Reduced wear and higher permissible axle loads have led to fatigue becoming the dominant cause for rail replacement at extended life. Grinding the running surface of rail has been greatly increased to reduce the damage caused by fatigue.

Tie Plates Rail is supported on tie plates that restrain the rail laterally, transmit the vehicle loads to the cross tie, and position (cant) the rail for optimum vehicle performance. Tie plates cant the rail toward the gage side. Cants vary from 1:14 to 1:40, depending on service conditions and operating speeds, with 1:40 being most common.

Fasteners Rail and tie plates are fixed to the cross ties by fasteners. Conventional construction with wood cross ties utilizes steel spikes driven into the tie through holes in the tie plate. In some cases, screw-type fasteners are used to provide more resistance to vertical forces. Elastic, clip-type fasteners are used on concrete and, at times, on wooden cross ties to provide a uniform, resilient attachment and longitudinal restraint in lieu of anchors. This type of fastening system also reduces the likelihood of rail overturns.

Cross Ties To retain gage and provide a further distribution of vehicle loads to the ballast, lateral cross ties are used. The majority of cross ties used in North America are wood. However, concrete cross ties are being used more frequently in areas where track surface and lateral stability are difficult to maintain and in high-speed passenger corridors. Slab track designs are now being given scrutiny for high-utilization areas.

Ballast Ballast serves to further distribute vehicle loads to the subgrade, restrain vertical and lateral displacement of the track structure, and provide for drainage. Typical mainline ballast materials are limestone, trap rock, granite, or slag. Lightly built secondary lines have used gravel, cinders, or sand in the past.

Subgrade The subgrade serves as the interface between the ballast and the native soil. Subgrade material is typically compacted or stabilized soil. Instability of the subgrade due to groundwater permeation is a concern in some areas. Subgrade instability must be analyzed and treated on a case-by-case basis.

Track Geometry Railroad maintenance practices, state standards, AREMA standards, and, more recently, FRA safety standards (49 CFR Title § 113) specify limitations on geometric deviations for track structure. Among the characteristics considered are:

Gage
Alignment (the lateral position of the rails)
Profile (the elevation of the rail)
Curvature
Easement (the rate of change from tangent to curve)
Superelevation (the elevation of the outer or "high" rail in a curve compared to the inner or "low" rail)
Runoff (the rate of change in superelevation)
Cross level (the relative height of opposite rails)
Twist (the change in cross level over a specified distance)

FRA standards specify maximum operating speeds by class of track ranging from Class 1 (lowest acceptable) to Class 9. Track geometry is a significant consideration in vehicle suspension design. For operating speeds greater than 110 mi/h, FRA has promulgated regulations (track Classes 6 through 9) that require qualification of locomotives and cars through a test program over each operating route. These tests call for comprehensive measurement of wheel/rail forces and vehicle accelerations (49 CFT § 213).

Curvature The curvature of track is designated in terms of "degrees of curvature." The degree of curve is the number of degrees of central angle subtended by a chord of 100-ft length (measured on the track centerline). Equation (8.3.10) gives the approximate radius R in feet for ordinary railway curves.

$$R = \frac{5{,}730}{\text{degrees per 100-ft chord}} \tag{8.3.10}$$

On important main lines where trains are operated at relatively high speed, the curves are ordinarily not sharper than 6 to 8°. In mountainous territory, and in other rare instances, curves as sharp as 18° occur. Most diesel and electric locomotives are designed to traverse curves up to 21°. Most uncoupled cars will pass considerably sharper curves, the limiting factor in this case being the clearance of the truck components to car body structural members or equipment or the flexibility of the brake connections to the trucks (AAR "Manual of Standards and Recommended Practices," Sec. C).

Clearances For new construction, the AREMA standard clearance diagrams provide for a clear height of 23 ft (7.0 m) above the tops of the railheads and for a width of 18 ft (5.5 m). Where conditions require, as in tunnels and bridges, these dimensions are reduced at the top and bottom of the diagram. For tracks entering buildings, an opening 16 ft (4.9 m) wide and 18 ft (5.5 m) high is recommended and will ordinarily suffice to pass the largest locomotives and most freight cars. A standard car clearance diagram for railway bridges (Fig. 8.3.25) represents the AAR recommendation for new construction. Minimum clearances are dictated by state authorities, and are summarized in Chapter 28, Table 3.3, of the AREMA Manual. Individual railroads have their own standards for existing lines and new construction, which are generally more stringent than the legal requirements. The actual clearance limitations on a particular railroad can be found in the official publication "Railway Line Clearances," Railway Equipment and Publishing Co.

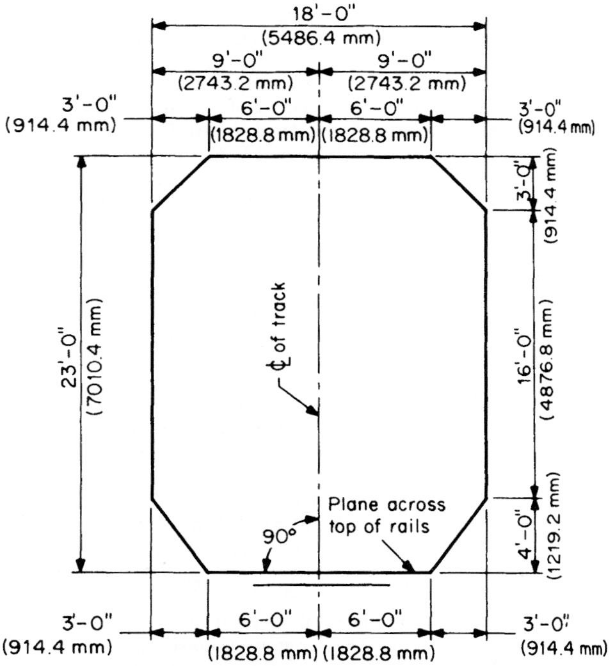

Figure 8.3.25 Wayside clearance diagram. (*AREA "Manual for Railway Engineering—Fixed Properties."*)

Track Spacing The distance between centers of mainline track commonly varies between 12 and 15 ft (3.6 and 4.6 m). Twelve-foot spacing exists, however, only on older construction. Most states require a minimum of 14 ft 0 in (4.3 m) between main tracks; however, many railroads have adopted 15 ft 0 in (3.6 m) as their standard. In a few cases, where space permits, 20 ft (6.1 m) is being used.

Turnouts Railway turnouts are defined by frog number. The frog number is the cotangent of the frog angle, which is the ratio of the lateral displacement of the diverging track per unit of longitudinal distance along the nondiverging track. Typical turnouts in yard and industry trackage range from number 7 to number 10. For mainline service, typical turnouts range from number 10 to number 20. The other important part of a turnout is the switch. The switch has movable points that direct the wheel down the straight or diverging side of the turnout. The higher the turnout number, the longer the switch, and the higher the allowable speed for trains traversing the diverging track. Allowable speeds are governed by the curvature of the diverging tack, and the angle of the switch. As a general rule, the allowable speed (in mi/h) is slightly less than twice the turnout number. In recent track construction for high-speed service special turnouts with movable point frogs have been applied.

Noise Allowable noise levels from railroad operation are prescribed by 40 CFR § 201.

VEHICLE-TRACK INTERACTION

Train Resistance The resistance to a train in motion along the track is of prime interest, as it is reflected directly in locomotive energy requirement. This resistance is expressed in terms of pounds per ton of train weight. **Gross train resistance** is that force that must be overcome by the locomotives at the driving-wheel–rail interface. **Trailing train resistance** must be overcome at the rear drawbar of the locomotive.

There are two classes of resistance that must be overcome: inherent and incidental. **Inherent resistance** includes the rolling resistance of bearings and wheels and aerodynamic resistance due to motion through still air. It may be considered equal to the force necessary to maintain motion at constant speed on level tangent track in still air. **Incidental resistance** includes resistance due to grade, curvature, wind, and vehicle dynamics.

Inherent Resistance Of the elements of inherent resistance, at low speeds rolling resistance is dominant, but at high speeds aerodynamic resistance is the predominant factor. Attempts to differentiate and evaluate the various elements through the speed range are a continuing part of industry research programs to reduce train resistance. At very high speeds, the effect of air resistance can be approximated as an aid to studies in its reduction by means of cowling and fairing. The resistance of a car moving in still air on straight, level track increases parabolically with speed. Because the aerodynamic resistance is independent of car weight, the resistance in pounds per ton decreases as the weight of the car increases. The total resistance in pounds per ton of a 100-ton car is much less than twice as great as that of a 50-ton car under similar conditions. With known conditions of speed and car weight, inherent resistance can be predicted with reasonable accuracy. Knowledge of track conditions will permit further refining of the estimate, but for very rough track or extremely cold ambient temperatures, generous allowances must be made. Under such conditions normal resistance may be doubled. A formula proposed by Davis (General Electric Review, Oct. 1926) and revised by Tuthill ("High Speed Freight Train Resistance." *Univ. of Illinois Engr. Bull.*, **376**, 1948) has been used extensively for inherent freight-train resistances at speeds up to 40 mi/h:

$$R = 1.3W + 29n + 0.045WV + 0.0005\,AV^2 \tag{8.3.11}$$

where R = train resistance, lb/car; W = weight per car, tons; V = speed, mi/h; n = total number of axles; A = cross-sectional area, ft².

With freight train speeds of 50 to 70 mi/h (80 to 112 km/h), it has been found that actual resistance values fall considerably below calculations based on the above formula. Several modifications of the Davis equation have been developed for more specific applications. All of these equations apply to cars trailing locomotives.

1. Davis equation as modified by I. K. Tuthill (*University of Illinois Engr. Bull*, **376**, 1948):

$$R = 1.3W + 29n + 0.045WV + 0.045V^2 \tag{8.3.12}$$

Note: In the Totten modification, the equation is augmented by a matrix of coefficients when the velocity exceeds 40 mi/h.

2. Davis equation as modified by the Canadian National Railway:

$$R = 0.6W + 20n + 0.01WV + 0.07V^2 \tag{8.3.13}$$

3. Davis equation as modified by the Canadian National Railway and former Erie-Lackawanna Railroad for trailers and containers on flat cars:

$$R = 0.6W + 20n + 0.01WV + 0.2V^2 \tag{8.3.14}$$

Other modifications of the Davis equation have been developed for passenger cars by Totten ("Resistance of Light Weight Passenger Trains," *Railway Age*, **103**, July 17, 1937). These formulas are for passenger cars pulled by a locomotive and do not include head-end air resistance.

1. Davis equations as modified by Totten for streamlined passenger cars:

$$R = 1.3W + 29n + 0.045WV + [0.00005 + 0.060725(L/100)^{0.88}]V^2 \tag{8.3.15}$$

2. Davis equations as modified by Totten for nonstreamlined passenger cars

$$R = 1.3W + 29n + 0.045WV + [0.00005 + 0.1085(L/100)^{0.7}]V^2 \tag{8.3.16}$$

where L = car length in feet.

Aerodynamic and Wind Resistance Wind-tunnel testing has indicated a significant effect on freight train resistance resulting from vehicle spacing, open tops of hopper and gondola cars, open boxcar doors, vertical side reinforcements on railway cars and intermodal trailers, and protruding appurtenances on cars. These effects can cause significant increases in train resistance at higher speeds. For example, the spacing of intermodal trailers or containers greater than approximately 6 ft can result in a new frontal area to be considered in determining train resistance. Frontal or cornering ambient wind conditions can also have an adverse effect on train resistance, which is increased with discontinuities along the length of the train.

Curve Resistance Train resistance due to track curvature varies with speed and degree of curvature. The behavior of rail vehicles in curve negotiation is the subject of several ongoing AAR studies. Lubrication of the rail gage face or wheel flanges has become common practice for reducing friction and the resulting wheel and rail wear. Recent studies indicate that flange and/or gage face lubrication can significantly reduce train resistance on tangent track as well (Allen, Conference on the Economics and Performance of Freight Car Trucks, October 1983). In addition, a variety of special trucks (wheel assemblies) that reduce curve resistance by allowing axles to steer toward a radial position in curves have been developed. For general estimates of car resistance and locomotive hauling capacity on dry (unlubricated) rail with conventional trucks, speed and gage relief may be ignored and a figure of 0.8 lb/ton per degree of curvature used.

Grade resistance depends only on the angle of ascent or descent and relates only to the gravitational forces acting on the vehicle. It equates to 20 lb/ton for each percent of grade, or 0.379 lb/ton for each foot per mile rise.

Acceleration The force (tractive effort) required to accelerate the train is the sum of the forces required for linear acceleration and that required for rotational acceleration of the wheels about their axle centers. A linear acceleration of 1 mi/h · s (1.6 km/h · s) is produced by a force of 91.1 lb/ton. The rotary acceleration requirement adds 6 to 12 percent, so that the total is nearly 100 lb/ton (the figure commonly used) for each mile per hour per second. If greater accuracy is required, the following expression is used:

$$R_a = A(91.05W + 36.36n) \tag{8.3.17}$$

where R_a = the total accelerating force, lb; A = acceleration, mi/h · s; W = weight of train, tons; n = number of axles.

Acceleration and Distance If a distance of S ft the speed of a car or train changes from V_1 to V_2 mi/h, the force required to produce acceleration (or deceleration if the speed is reduced) is

$$R_a = 74(V_2^2 - V_1^2)/S \tag{8.3.18}$$

The coefficient, 74, corresponds to the use of 100 lb/ton. This formula is useful in the calculation of the energy required to climb a grade with the assistance of stored energy. In any train-resistance calculation or analysis, assumptions with regard to acceleration will generally submerge all other variables; for example, an acceleration of 0.1 mi/h · s (0.16 km/h · s) requires more tractive force than that required to overcome inherent resistance for any car at moderate speeds.

Starting Resistance Most railway cars are equipped with roller bearings requiring a starting force of 5 or 6 lb/ton.

Vehicle Suspension Design The primary consideration in the design of the vehicle suspension system is to isolate track input forces from the vehicle car body and lading. In addition, there are a few specific areas of instability that railway suspension systems must address; see AAR "Manual of Standards and Recommended Practice," Sec. C-II-M-1001, Chap. XI.

Harmonic roll is the tendency of a freight car with a high center of gravity to rotate about its longitudinal axis (parallel to the track). This instability is excited by passing over staggered low rail joints at a speed that causes the frequency of the input for each joint to match the natural roll frequency of the car. Unfortunately, in many car designs, this occurs for loaded cars at 12 to 18 mi/h (19.2 to 28.8 km/h), a common speed for trains moving in yards or on branch lines where tracks are not well maintained. Many freight operations avoid continuous operation in this speed range.

This adverse behavior is more noticeable in cars with truck centers approximately the same as the rail length. The effect of harmonic roll can be mitigated by improved track surface and by damping in the truck suspension.

Pitch and bounce are the tendencies of the vehicle to either translate vertically up and down (bounce), or rotate (pitch) about a horizontal axis perpendicular to the centerline of track. This response is also excited by low track joints and can be relieved by increased truck damping.

Yaw is the tendency of the car to rotate about its axis vertical to the centerline of track. Yaw responses are usually related to truck hunting.

Truck hunting is an instability inherent in the design of the truck and dependent on the stiffness parameters of the truck and on wheel conicity (tread profile). The instability is observed as "parallelogramming" of the truck components at a frequency of 2 to 3 Hz, causing the car body to yaw or translate laterally. This response is excited by the effect of the natural frequency of the gravitational stiffness of the wheel set when the speed of the vehicle approaches the kinematic velocity of the wheel set. This problem is discussed in analytic work available from the AAR Research and Test Department.

Superelevation As a train passes around a curve, there is a tendency for the cars to roll toward the outside of the curve in response to centrifugal force acting on the center of gravity of the car body (Fig. 8.3.26*a*). To compensate for this effect, the outside rail is superelevated, or raised, relative to the inside rail (Fig. 8.3.26*b*). The amount of superelevation for a particular curve is based upon the radius of the curve and the operating speed of the train. The "balance" or equilibrium speed for a given curve is that speed at which the weight of the vehicle is equally distributed on each rail. The FRA allows a railroad to operate with 3 in of unbalance, or at the speed at which equilibrium would exist if the superelevation were 3 in greater than that which exists. The maximum superelevation is usually 6 in but maybe lower if freight operation is used exclusively.

Longitudinal Train Action Longitudinal train (slack) action is associated with the dynamic action between individual cars in a train.

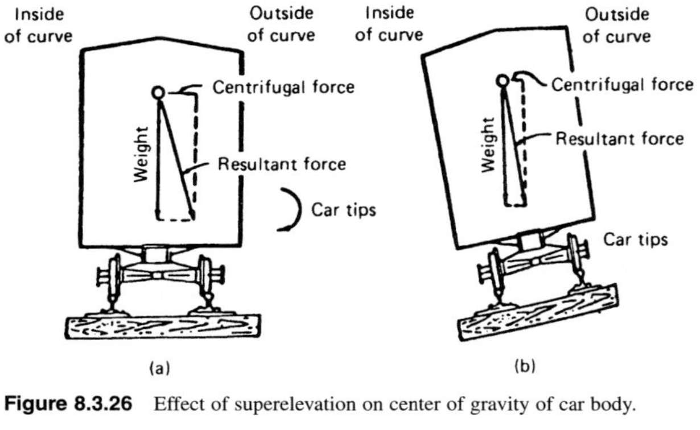

Figure 8.3.26 Effect of superelevation on center of gravity of car body.

An example would be the effect of starting a long train in which the couplers between each car had been compressed (i.e., bunched up). As the locomotive begins to pull the train, the slack between the locomotive and the first car must be traversed before the first car begins to move. Next the slack between the first and second car must traversed before the second car beings to move, and so on. Before the last car in a long train begins to move, the locomotive and the moving cars may be traveling at a rate of several miles per hour. This effect can result in coupler forces sufficient to cause the train to break in two.

Longitudinal train (slack) action is also induced by serial braking, undulating grades, or by braking on varying grades. The AAR Track Train Dynamics Program has published guidelines titled "Track Train Dynamics to Improve Freight Train Performance" that explain the causes of undesirable train action and how to minimize its effects. Analysis of the forces developed by longitudinal train action requires the application of the Davis equation to represent the resistance of each vehicle based upon its velocity and location on a grade or curve. Also, the longitudinal stiffness of each car and the tractive effort of the locomotive must be considered in equations that model the kinematic response of each vehicle in the train. Computer programs are available from the AAR to assist in the analysis of longitudinal train action.

POSITIVE TRAIN CONTROL (PTC)

What Is PTC?

Positive train control (PTC) is a collision avoidance system that integrates command, control, communication, and information system for improved train movement, safety, and efficiency. PTC employs communication-based/processor-based train control technology to prevent train-to-train collisions and derailments caused by over speeding. PTC is designed to detect erroneous train movement through switches as well as prevent train-to-human accidents caused by train incursion into work areas or individuals trespassing into railroad properties. PTC systems can widely vary in complexity and sophistication based on their level of automation and functionality. The system architecture depends on the degree of train control and wayside system complexity, for elements such as non-signaled, block signal, cab signal.

The PTC system is able to determine the precise location, direction, and speed of trains. It is designed to warn the train engineer and conductor of any potential risk of collision. If needed, the system takes immediate action to stop the train, if the train operator fails to take corrective action in response to the system warnings within 15 seconds. The warnings can be generated by traveling too fast for the condition of the rail (such as in speed-restricted areas or in curves), or not stopping the train before a stop signal.

The Rail Safety Improvement Act of 2008 (RSIA) mandates that the U.S., as shown in the timeline in Fig. 8.3.27. It is important to note that the implementation deadline of December 31, 2015 has been postponed to the end of 2020. Class I railroads (the 7 largest railroads in the U.S. and Canada) and passenger railroads such as Amtrak are required to fully install an operational PTC on their mainlines, or on lines that carry hazardous materials (such as chemicals and crude oil), or where intercity passenger commuter trains operate.

It is important to note that PTC will not prevent accidents at grade crossing caused by improper vehicle movement, trespassing on railroad property, track or rolling stock equipment failure, and operator errors that go beyond the system's operating capability.

As shown in Fig. 8.3.28, the four main elements of PTC are:

On-board or Locomotive System These include GPS-based equipment that monitor the train position and speed, and automatic braking systems that can activate the brakes, in order to reduce train speed to enforce speed restrictions or stop the train when an unauthorized train movement is detected.

The onboard computer that is most often located in the locomotive receives the movement authorities from the Back Office Server (BOS) and communicates them with the train engineer. The engineer uses notifications such as changes in speed limit or location of work crew

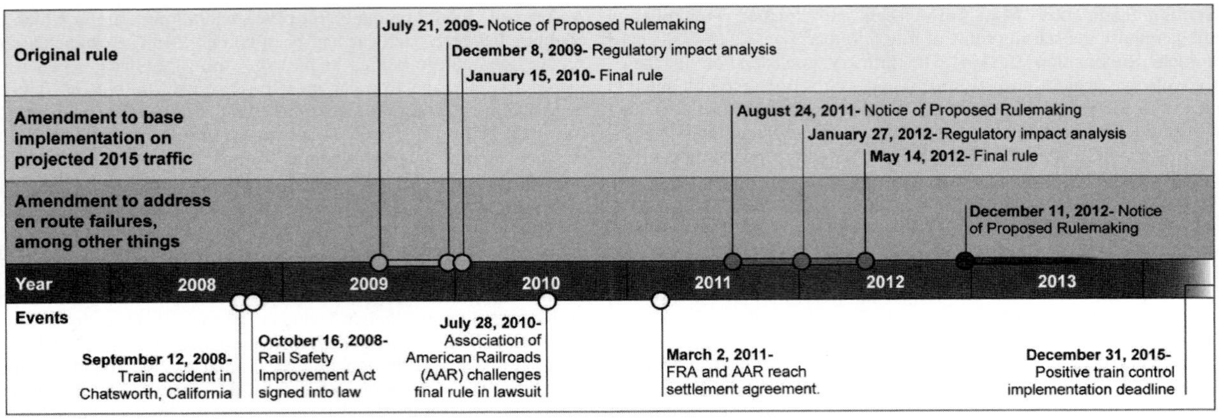

Figure 8.3.27 Positive train control's regulatory timeline, 2008-2015.[1]

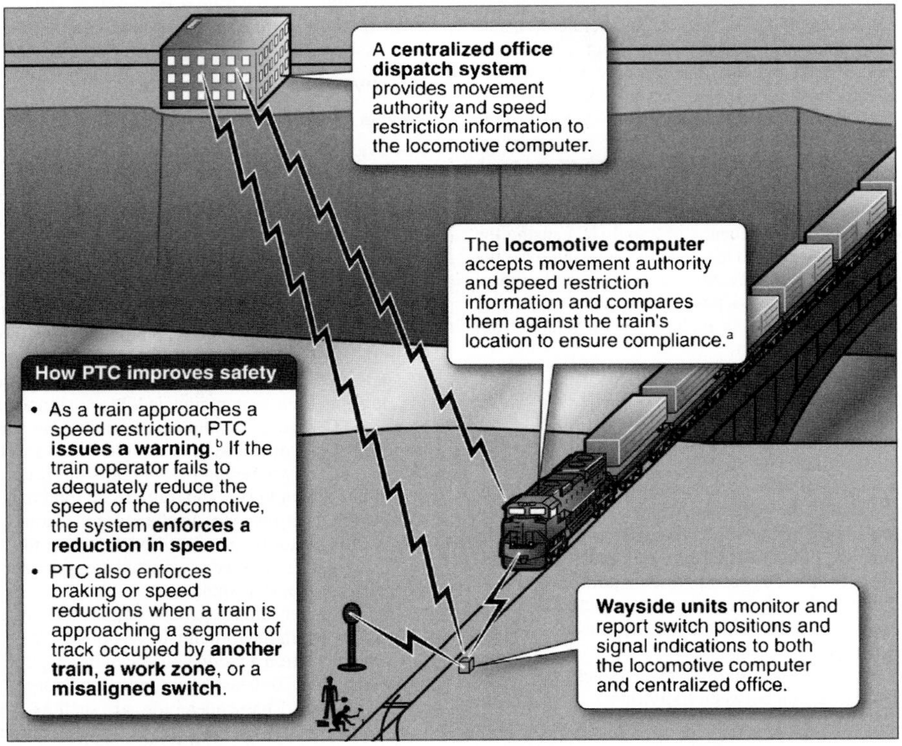

[a]Train location information is determined through various methods depending on the specific PTC system, including through satellite-based positioning systems and sensors installed along the track.
[b]Although the law does not require PTC systems to issue such warnings, the PTC systems that most railroads are implementing will do so.

Figure 8.3.28 Basic operation of a Positive Train Control (PTC) System.
SOURCE: GAO Report 13-720.

to safely operate the train. Similar to the existing signaling systems, the engineer relies on wayside devices to take the appropriate action in terms of speed, approaching a signal block, etc. If the onboard system loses communication with BOS, then the train is stopped prior to entering the next signal block.

[1] GAO Report, POSITIVE TRAIN CONTROL: Additional Authorities Could Benefit Implementation, GAO-13-720 (Washington, D.C.: August 16, 2013).

Wayside System The wayside system includes the signaling devices on and around the track. These include signal equipment, switches, and track circuits to authorize train movement, similar to what is done with conventional signal equipment in which the presence of a train shunts the train circuit, resulting in changing the track signal from green (or yellow) to red and alerting other trains not to approach the territory. In the current signaling system, which dates back to the nineteenth century, a central dispatch serves as the communication node among the trains. In PTC, the communication among the trains will be facilitated automatically.

The wayside system connects to a wayside-messaging server through the use of a Wayside Interface Unit (WIU), in order to send any critical information about the trackside equipment to the BOS. The information is also broadcast over a radio interface so that the train operator can receive the information directly.

Back Office Server This is used for two main purposes. First, to store the information related to the train and track, such as train composition, length, speed, and movement and rail network information such as speed limits and general track composition. Second, it provides authorization for each train to move from one rail segment into next. Much as is the case with the existing signal systems, PTC "segmentizes" the track network and prevents train movement into a segments that is already occupied by another train. The system prevents another train to approach an occupied segment, through alerting the train engineer and conductor as well as automatically applying the brakes if needed.

Communication There is bi-directional communication between the back office, locomotive, and wayside devices. This is achieved through four different interfaces, including Ethernet, Wi-Fi, 220 MHz PTC Radio, and 3G/4G Cellular in order to ensure the needed redundancy and resiliency.

The wayside equipment communicates with the BOS through one of the four communication modes. The onboard equipment uses either the 220 MHz radio or 3G/4G cellular interface, in order to provide the ability to roam horizontally from one base station to another. The standardization to 220 MHz radio frequency is due to its long range capability, reaching as far as 20-30 miles. The onboard system are also able to roam vertically among the four modes of communications, commonly using one mode as back up to another.

How Does PTC Work?

U.S. freight and passenger trains often share track, dispatchers, and signals that control train movement. Some railroads also use additional technologies to improve efficiency and achieve business benefits. Currently, dispatchers in centralized offices issue train movement authorities that allow trains to enter specific track segments, or blocks. These authorities are communicated to train operators through signals alongside the track, or in non-signaled territory through track warrants generally issued by verbal radio communication (see Fig. 8.3.29).

PTC can be implemented either as an overlay over existing track equipment or as a completely standalone system. The overlay implementation allows PTC to work in conjunction with the existing signal system and the trains' method of operation. In contrast, a standalone system eliminates the need for the existing signal system through putting onboard the locomotive the information currently communicated through the signal system. RSIA requires that systems have the ability to communicate with one another, allowing trains seamlessly move across tracks with different PTC systems. The interoperability is a vital part of the effective implementation of PTC, considering that there are 37 freight, intercity passenger, and commuter railroads that

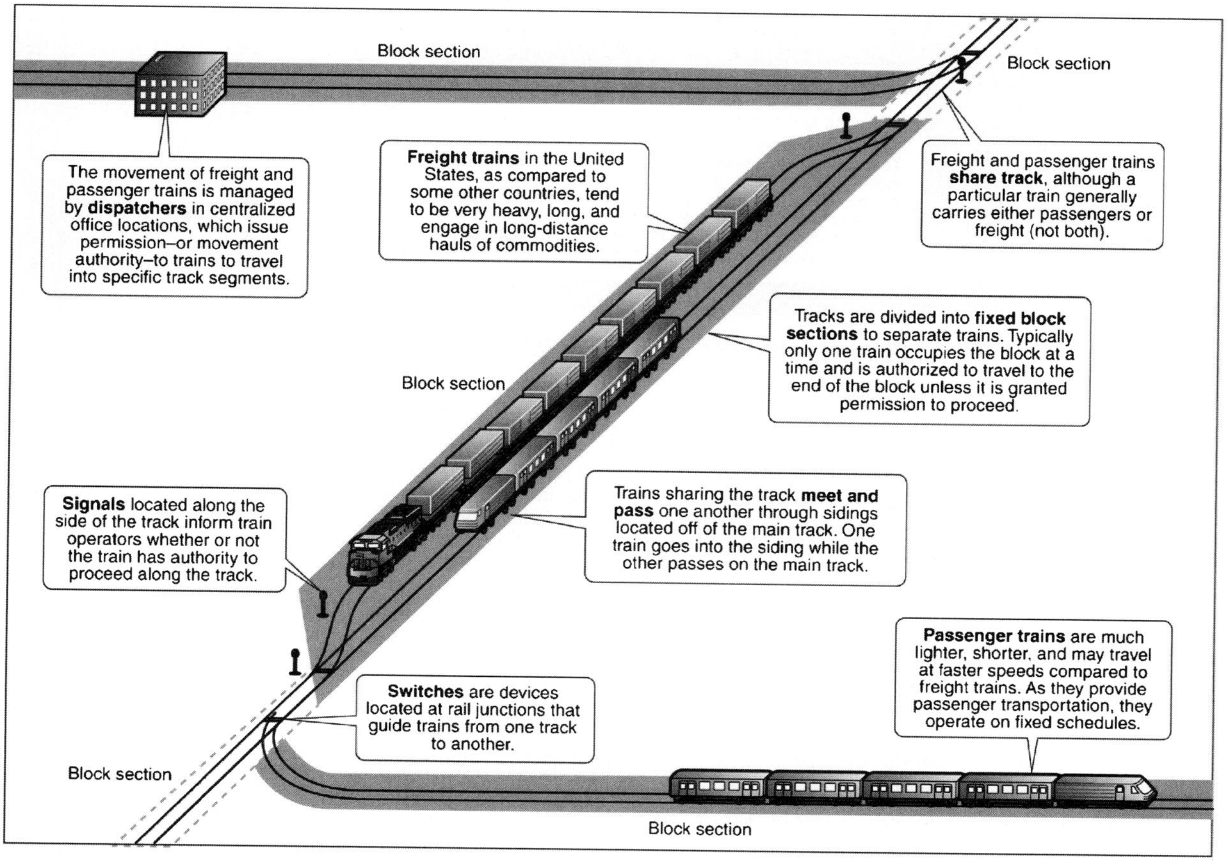

Figure 8.3.29 Key components of current U.S. railroad operations.
SOURCE: GAO Report 13-720.

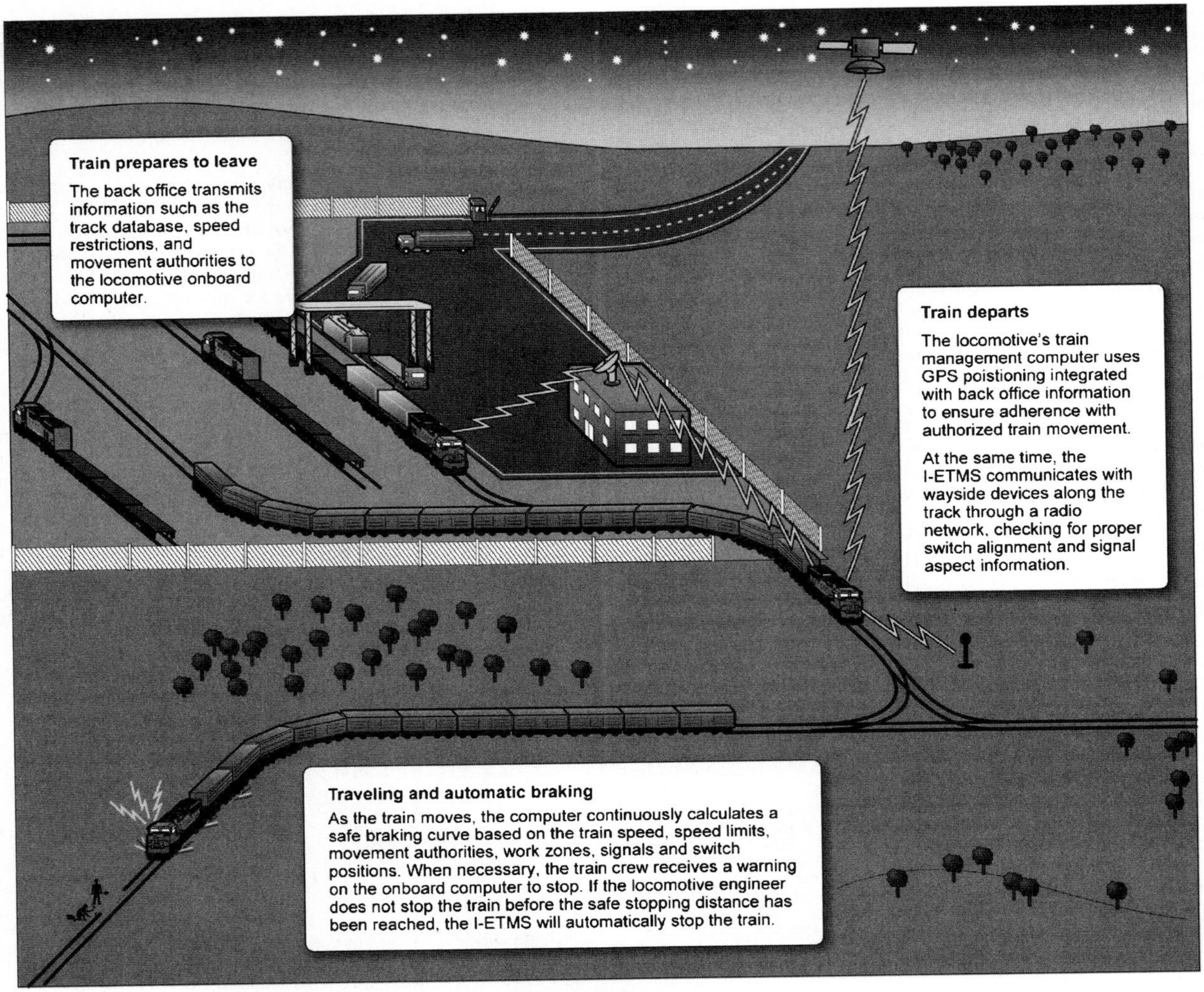
Train prepares to leave
The back office transmits information such as the track database, speed restrictions, and movement authorities to the locomotive onboard computer.
Train departs
The locomotive's train management computer uses GPS poistioning integrated with back office information to ensure adherence with authorized train movement.
At the same time, the I-ETMS communicates with wayside devices along the track through a radio network, checking for proper switch alignment and signal aspect information.
Traveling and automatic braking
As the train moves, the computer continuously calculates a safe braking curve based on the train speed, speed limits, movement authorities, work zones, signals and switch positions. When necessary, the train crew receives a warning on the onboard computer to stop. If the locomotive engineer does not stop the train before the safe stopping distance has been reached, the I-ETMS will automatically stop the train.

Figure 8.3.30 Basic operation of the Interoperable Electronic Train Management System (I-ETMS).
SOURCE: *GAO Report 13-720.*

are required to implement PTC, according to the Federal Railroad Administration.

All seven major freight railroads in the United States plan to implement Interoperable Electronic Train Management System (I-ETMS), shown in Fig. 8.3.30, on most of the approximately 60,000 miles of track that need to have PTC. Amtrak is implementing Advanced Civil Speed Enforcement System (ACSES) on the Northeast Corridor. Although the two systems are functionally similar, they differ technologically. ACSES relies on track-embedded transponders for determining the train location, while I-ETMS uses Global Positioning System (GPS) information.

8.4 MARINE ENGINEERING

by Richard J. Burke

REFERENCES: Harrington, "Marine Engineering," SNAME, 1992. Lewis, "Principles of Naval Architecture," SNAME, 1988/89. Myers, "Handbook of Ocean and Underwater Engineering," McGraw-Hill, 1969. "Rules for Building and Classing Steel Vessels," American Bureau of Shipping. Rawson and Tupper, "Basic Ship Theory," Oxford: Butterworth-Heinemann, 2001. Gillmer and Johnson, "Introduction to Naval Architecture," Naval Institute, 1982. Barnaby, "Basic Naval Architecture," Hutchinson, London, 1967. *Jour. Inst. Environ, Sci.* Taggert, "Ship Design and Construction," SNAME, 1980. *Trans. Soc. Naval Architects and Marine Engrs.*, SNAME. *Naval Engrs. Jour.*, Am. Soc. of Naval Engrs., ASNE. *Trans. Royal Inst. of Naval Architects*, RINA. Figures and examples herein credited to SNAME have been included by permission of The Society of Naval Architects and Marine Engineers.

Marine engineering is an ambiguous designation of multiple engineering disciplines. Most generally, it refers to all engineering efforts carried out in the marine environment, a designation somewhat shared with *ocean engineering*. More specifically, it is concerned with the design of propulsion and auxiliary systems for ships, boats, and other marine vehicles and platforms, while the design of the vehicle or platform structure is a task assigned to members of the *naval architecture* discipline (sometimes referred to as *naval engineering*). Applicable degree programs are often entitled *Naval Architecture & Marine Engineering*, although naval architecture and marine engineering may be taught separately. Another use of the marine engineering designation refers to the qualification, training, and employment of licensed officers who, as members of the crew, operate and maintain the propulsion and auxiliary systems of ships, other marine vehicles, and ocean platforms.

In all cases, *marine engineering* requires integration of many areas of engineering science directed toward applications in transportation, warfare, and the exploration and retrieval of natural resources, and having common elements with respect to operation in or on water. Marine engineers not only are responsible for design of engineering systems for propulsion, generally referred to as engines, but also are responsible for the design of systems for the provision of electrical and other modes of auxiliary power, provision of heating, ventilation, air conditioning, and other environmental necessities for crew and passengers, provision of systems for handling and protection of cargo and stores, provision of systems for vehicle maneuvering, including anchoring and mooring, and other related requirements. They usually have joint responsibility with naval architects in areas of propulsor design, hull vibration excited by the propeller or main propulsion plant, noise reduction, shock hardening, and general dynamic responses of machinery and associated structure.

Marine engineering is a distinct and a continuously advancing engineering discipline having historical roots in the early industrial revolution. Practitioners must be broadly well versed in the following areas: (1) rigid body mechanics including vibrations, (2) structural mechanics, (3) fluid mechanics, (4) thermodynamics, (5) heat transfer, (6) electrical engineering, (7) environmental engineering including HVAC, (8) materials science, (9) naval architecture, and (10) engineering economy.

THE PHYSICAL MARINE ENVIRONMENT

The marine environment is notoriously unforgiving and difficult to predict. Also, when difficulties develop on a vehicle or platform (hereafter a *vessel*) at sea, the crew are often isolated and far from assistance. As a consequence, marine engineers are committed to three priorities:

1. Safety
2. Reliability
3. Economy

Winds and waves at sea are closely linked, although wave conditions may be challenging when the winds are calm because of storms at other locations. The **irregular sea** can be described by statistical and mathematical techniques based upon the superposition of a large number of regular waves of different frequencies, directions, and amplitudes; thus, the understanding of realistic irregular seas begins with the characteristics of idealized regular waves. Actual sea states consisting of a combination of different size waves running in different directions may momentarily superimpose themselves into exceptionally large waves which can be modeled probabilistically.

The effects of the marine environment also vary with water depth, sea temperature, and air temperature. Shallow water not only affects the nature of the prevailing sea state, but it also may alter the vessel's response to the sea. Water and air temperatures have significant effects with respect to the operability of the vessel's equipment, and the efficiency of operation. The likely presence of sea ice imposes significant constraints on design and operation. Additionally, silt, sea life, and bottom growth may affect seawater systems. Water depth also affects navigation, maneuvering, and mooring.

Lastly, salt water is especially corrosive to most materials, and has important effects on ferrous materials with respect to fatigue and fracture.

THE REGULATORY MARINE ENVIRONMENT

Just as important as the physical environment, marine engineers must be aware of a regulatory environment where the body of national and international regulation is growing rapidly in size and complexity. Not surprisingly, protection of the physical environment has been a critical driver of the regulatory environment.

International agreements concerning safety and lifesaving date back to the aftermath of the loss of the *Titanic* in 1912. Since approximately 1972, regulations have (a) made it illegal to release oil of even minor quantities into the ocean, (b) imposed a regimen of training and qualification for seagoing marine engineers, (c) forbidden the dumping of plastics and other types of trash at sea, (d) imposed requirements of the energy efficiency of new ship designs, (e) imposed air quality regulations on ships operating at sea as well as in port, and (f) regulated the treatment of ballast water discharges due to the prevalence of invasive species. New regulatory requirements are forthcoming regularly.

Regulations are generally developed, imposed, and enforced by:

- *International Maritime Organization* (hereinafter *IMO*), part of the United Nations based in London, responsible for developing and imposing international agreements on maritime affairs.
- *National Authorities*, a term for national governmental bodies that develop, impose, and enforce laws and treaties as national regulations. In the United States, the *U.S. Coast Guard* is referred to most frequently as the national authority, although other governmental agencies have limited jurisdictions.
- *Classification Societies*, a group of Non-Governmental Bodies that act to create and enforce safety requirements for the design, operation, and maintenance of vessels, and under various circumstances may act on behalf of national authorities, vessel owners, marine underwriters, or other interested parties. Classification societies belong to the *International Association of Classification Societies* (hereinafter IACS); the member of IACS based in the United States is the *American Bureau of Shipping* (ABS).

MARINE VEHICLES

The taxonomy of ships, boats, other marine vehicles, and ocean platforms can be considered as follows:

- *Buoyantly supported*, also called displacement hull forms, meaning that the vessel is fully supported in accordance with *Archimedes Principle* in that the total weight of the vessel is counteracted by a buoyant force equal to the weight of the displaced volume of liquid. Ships, boats, buoys, and barges are typical examples.
- *Dynamically supported*, meaning that for at least a portion of its operational life the vessel is supported by the action of a lifting surface or other dynamically-maintained force. Examples are hydrofoils, submarines, planning boats, and air cushion vehicles, which all return to buoyant support after dynamic operation ceases.
- *Bottom supported*, meaning that the vessel rests on the bottom for at least partial static support, either temporarily or permanently, but was buoyantly supported prior to installation.

Displacement Hull Forms

Displacement hull forms are the familiar monohull and the catamaran or other multihull form. The displacement **monohull** form provides the best possible combination of payload-carrying ability, economical powering characteristics, and good sea keeping qualities. A more slender hull form achieves a significant reduction in hydrodynamic resistance to achieve increased speed; however, it is limited in its ability to carry topside weight because of the low transverse stability of its narrow beam. **Multihull ships** provide a solution to the problem of low transverse stability. These types are increasingly popular in sailing yachts, high-speed passenger ferries, and research and small support ships. Sailing catamarans, with their superior transverse stability permitting large sail-plane area, gain a speed advantage over monohull craft of comparable size. A powered catamaran has the advantage of increased deck space and relatively low roll angles over a monohull ship.

The **Small-Waterplane-Twin-Hull** vessel, that is, SWATH, is a high performance displacement hull form that gains the advantages of the catamaran's twin displacement hulls with the further advantage of minimized

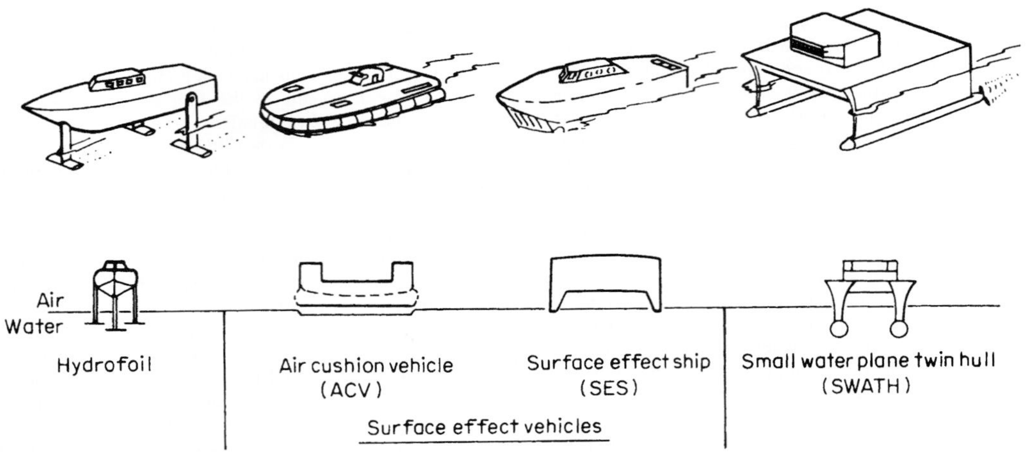

Figure 8.4.1 Family of high-performance ships.

wave-making resistance and wave-induced motions achieved by submarine-shaped hulls beneath the surface with small water-plane area supporting struts penetrating the air-water interface. This obviates the power requirements required for dynamic support which characterizes most high performance designs (see Fig. 8.4.1).

A more generalized, lower performance version of this concept is the **semi-submersible drilling rig**, a large and complex vessel used for offshore oil exploration. Such vessels may be self-propelled or not; if self-propelled, they also may be dynamically positioned, although most dynamically-positioned exploratory drilling is conducted by **drill ships.**

Floating production platforms or vessels are designed to produce oil and place it into storage until it can be moved to shore. If undersea pipelines are not available, a **floating storage vessel** is used in conjunction with a **single-point mooring buoy.** For deep water drilling, a vessel type known as a **floating production storage and offloading vessel** (abbreviated **FPSO**) has been developed.

Dynamically Supported Vessels

Submarines The submarine is buoyantly supported while operating on the surface, and surrenders that buoyancy to submerge by flooding ballast tanks to achieve nearly neutral buoyancy. Once submerged, the submarine is generally dynamically supported by the action of lifting surfaces called *diving planes* or by the effect of vectorable thrusters. A submerged submarine enjoys the benefit of stealth, and does not generate surface waves of significant size if the vessel is proceeding slowly or at sufficient depth. In submerged operation the submarine shares many characteristics with the airplane. The return to the surface the submarine uses compressed air stored in bottles to displace water from the flooded ballast tanks, however, in practice the vessel is usually flown close to the surface prior to blowing ballast tanks.

Planing Hull Forms The planing hull form, although most commonly used for recreational and racing craft, is used increasingly in small, fast commercial craft and in coastal patrol craft. The weight of the planing hull ship is partially borne by the dynamic lift of the water streaming against a relatively flat or V-shaped bottom at high speed. The effective displacement is hence reduced below that of a ship supported statically, with significant reduction in wave-making resistance at higher speeds.

Hydrofoils and Air-Cushion Vehicles In a search for higher speeds in rougher seas, several advanced dynamic support concepts have been developed to minimize wave-making resistance. The resulting vessel types are the **hydrofoil craft** and the **air-cushion vehicle** (also called the surface-effect vehicle or hovercraft). The hydrofoil craft has a planing hull that is raised clear of the water through dynamic lift generated by an underwater foil system. Air cushion vehicles ride on a cushion of compressed air generated and maintained in the space between the vehicle and the surface over which it hovers or moves. The most practical vehicles employ a peripheral-jet principle, with flexible skirts for obstacle or wave clearance. A rigid sidewall craft, achieving some lift from hydrodynamic effects, is more adaptable to marine construction techniques.

Bottom Supported Vessels

Bottom supported vessels are commonplace among the offshore oil industry, and are used both for exploratory drilling and for oil production and storage.

The **submersible drilling rig** was developed for shallow water exploratory drilling. Designed generally in a catamaran fashion, the submersible rig would be located at a drilling site in sufficiently shallow water that the vessel could be ballasted down to rest on the bottom to provide a static drilling platform above sea level. When the drilling operation was complete, the vessel could be de-ballasted and towed to another site. This type of drilling rig is used rarely today.

A more common type is the **jack-up drilling rig**. This type consists of a floating hull and a number of legs that could be jacked down to the bottom, and that would provide sufficient support to allow the hull to be jacked above the water surface. Jack-up rigs provide a similar configuration on site that a submersible rig does, but it has advantages in that it can drill in deeper water and the legs can accommodate irregularities in bottom topography. The disadvantage is that the legs extend upward from the rig when floating, reducing stability under tow, especially in high winds and seas. More jack-up rigs have been built for exploratory drilling service than any other design; such vessels are also used for offshore construction.

Both submersible and jack-up drilling rigs vessels of a type known as **mobile offshore drilling rigs** (abbreviated as **MODU**). Other types of MODUs include semi-submersible rigs and drilling ships, both buoyantly supported.

Bottom supported platforms used for offshore oil production include **compliant platforms, jacket structures,** and **gravity-based structure.** A compliant platform is attached to the ocean bottom, but receives most of its support from the buoyancy of a long, flexible structure that will move and bend considerable with lateral sea and wind loads. A jacket structure is essential a large truss constructed of welded tubular sections. When uprighted onto the bottom and pinned into place by piles, the jacket structure extends above the water surface to a safe height to support the production deck, which is subsequently installed. A gravity-based structure is floated into place and then permanently ballasted into position to combine the roles of oil production and storage. Gravity-based vessels are usually constructed of steel and concrete, and once in place, will probably never be refloated.

NAVAL ARCHITECTURE

Seaworthiness is the quality of a marine vehicle being fit to accomplish its intended mission. In meeting the responsibilities to produce seaworthy vehicles, marine engineers must have a basic understanding of the effects of the marine environment with regard to the vehicle's (1) structure, (2) stability, (3) sea keeping, and (4) resistance and powering requirements.

Units and Definitions

Throughout the world, metric units are used in marine engineering, including naval architecture. At this writing, the United States remains the last place where Imperial units are routinely employed. However, even where metric units are used, the *Systeme Internationale* (**SI units**) is not rigorously applied.

The most common measure of weight or displacement is the metric ton, or *tonne,* equivalent to the weight of a mass of 1,000 kg. Lengths are typically given in meters.

In Imperial units, weight or displacement is measured in *long tons*, a ton of 2,240 lb. Lengths would be given in decimal feet, or in feet-inches-eighths.

Of all units, the *knot*, or nautical mile per hour, is still routinely used worldwide, although for technical calculations this in converted to meters per second (metric) or feet per second (Imperial).

The **displacement** Δ is the weight of the water displaced by the immersed part of the vehicle. It is equal (1) to the buoyant force exerted on the vehicle and (2) to the weight of the vehicle (in equilibrium) and everything on board. Displacement is expressed in long tons (1.01605 metric tons). Since the specific weight of seawater averages about 64 lb/ft^3, displacement in saltwater is measured by the displaced volume ∇ in ft^3 divided by 35. In freshwater, divide by 35.9.

The standard maritime industry metric weight unit is the metric ton (tonne) of 1,000 kg. Under the SI system, the mass displacement of a ship is commonly represented in metric tons, which under gravity, leads to a force displacement represented in metric tonneforce or newtons. Two measurements of a merchant ship's earning capacity that are of significant importance to its design and operation are deadweight and tonnage. The **deadweight** of a ship is the weight of cargo, stores, fuel, water, personnel, and effects that the ship can carry when loaded to a specific load draft. Deadweight is the difference between the load displacement, at the minimum permitted freeboard, and the light displacement, which comprises hull weight and machinery. Deadweight is expressed in long tons (2,240 lb each) or MN. The volume of a ship is expressed in tons of 100 ft^3 (2.83 m^3) each and is referred to as its **tonnage.** Charges for berthing, docking, passage through canals and locks, and for many other facilities are based on tonnage. **Gross tonnage** was based on cubic capacity below the tonnage deck, generally the uppermost complete deck, plus allowances for certain compartments above, which are used for cargo, passengers, crew, and navigating equipment. Deduction of spaces for propulsion machinery, crew quarters, and other prescribed volumes from the gross tonnage leaves the **net tonnage**. The dimensions of a ship may refer to the molded body (or form defined by the outside of the frames), to general outside or overall dimensions, or to dimensions on which the determination of tonnage or of classification is based. There are thus (1) molded dimensions, (2) overall dimensions, (3) tonnage dimensions, and (4) classification dimensions. The published rules and regulations of the classification societies and the U.S. Coast Guard should be consulted for detailed information.

The **designed load waterline** (DWL) is the waterline at which a ship would float freely, at rest in still water, in its normally loaded or designed condition. The keel line of most ships is parallel to DWL. Some keel lines are designed to slope downward toward the stern, termed **designed drag**.

A vertical line through the intersection of DWL and the foreside of the stem is called the **forward perpendicular, FP**. A vertical line through the intersection of DWL with the afterside of the straight portion of the rudder post, with the afterside of the stern contour, or with the centerline of the rudder stock (depending upon stern configuration), is called the **after perpendicular, AP**.

The **length** on the designed load waterline, L_{WL}, is the length measured at the DWL, which, because of the stern configuration, may be equal to the **length between perpendiculars,** L_{PP}. This is generally the same as classification-society-defined length, with certain stipulations. The extreme length of the ship is the **length overall,** L_{OA}.

The **molded beam** B is the extreme breadth of the molded form. The extreme or maximum breadth is generally used, referring to the extreme transverse dimension taken to the outside of the plating.

The **draft** T (molded) is the distance from the top of the keel plate or bar keel to the load waterline. It may refer to draft amidships, forward, or aft.

Trim is the longitudinal inclination of the ship usually expressed as the difference between the draft forward, T_F, and the draft aft, T_A.

Coefficients of Form Assume the following notation: L = length on waterline; B = beam; T = draft; ∇ = volume of displacement; A_{WP} = area of water plane; A_M = area of midship section, up to draft T; v = speed in ft/s (m/s); and V = speed in knots.

Block coefficient, $C_B = \nabla/LBT$, may vary from about 0.38, for high-powered yachts and destroyers, to greater than 0.90 for slow-speed seagoing cargo ships and is a measure of the fullness of the underwater body.

Midship section coefficient, $C_M = A_M/BT$, varies from about 0.75 for tugs or trawlers to about 0.99 for cargo ships and is a measure of the fullness of the maximum section.

Prismatic coefficient, $C_P = \nabla/L\,A_M = C_B/C_M$, is a measure of the fullness of the ends of the hull, and is an important parameter in powering estimates.

Water-plane coefficient, $C_{WP} = A_{WP}/LB$, ranges from about 0.67 to 0.95, is a measure of the fullness of the water plane, and may be estimated by $C_{WP} \approx \frac{2}{3}C_B + 1\frac{1}{3}$.

Displacement/length ratio, $\Delta = \nabla/(L/100)^3$, is a measure of the slenderness of the hull, and is used in calculating the power of ships and in recording the resistance data of models.

A similar coefficient is the **volumetric coefficient,** $C_V = \nabla/(L/10)^3$, which is commonly used as this measure.

Table 8.4.1 presents typical values of the coefficients with representative values for Froude number ($v/\sqrt{gL}$), discussed under "Resistance and Powering."

Structure

The structure of a ship is a complex assembly of small pieces of material. Common hull structural materials for small boats are wood,

Table 8.4.1 Coefficients of Form

Type of vessel	$\frac{v}{\sqrt{gL}}$	C_B	C_M	C_P	C_{WP}	$\frac{\Delta}{(L/100)^3}$
Great Lakes ore ships	0.116-0.128	0.85-0.87	0.99-0.995	0.86-0.88	0.89-0.92	70-95
Slow ocean freighters	0.134-0.149	0.77-0.82	0.99-0.995	0.78-0.83	0.85-0.88	180-200
Moderate-speed freighters	0.164-0.223	0.67-0.76	0.98-0.99	0.68-0.78	0.78-0.84	165-195
Fast passenger liners	0.208-0.313	0.56-0.65	0.94-0.985	0.59-0.67	0.71-0.76	75-105
Battleships	0.173-0.380	0.59-0.62	0.99-0.996	0.60-0.62	0.69-0.71	86-144
Destroyer, cruisers	0.536-0.744	0.44-0.53	0.72-0.83	0.62-0.71	0.67-0.73	40-65
Tugs	0.268-0.357	0.45-0.53	0.71-0.83	0.61-0.66	0.71-0.77	200-420

aluminum, and fiberglass-reinforced plastic. Large ships are nearly always constructed of steel, with common application of alloyed aluminum in superstructures of some ship types, particularly for KG-critical designs.

The methods for design and analysis of a ship's structure have undergone significant improvement over the years, with research from the ocean environment, shipboard instrumentation, model testing, computer programming and probabilistic approaches. Early methods consisted of standard strength calculations involving still water and quasi-static analysis along with conventions and experience from proven designs. These methods were extremely useful and remain the basis, as well as a starting tool, for comparative analysis with the newer applications and rules in use today. For detailed discussion of these methods, see the SNAME publications referenced earlier.

Weight, buoyancy, and load curves (Figs. 8.4.2 and 8.4.3) are developed for the ship for the determination of shear force and bending moment. Several extreme conditions of loading may be analyzed. The weight curves include the weights of the hull, superstructure, rudder, and castings, forgings, masts, booms, all machinery and accessories, solid ballast, anchors, chains, cargo, fuel, supplies, passengers, and baggage. Each individual weight is distributed over a length equal to the distance between frames at the location of that particular weight.

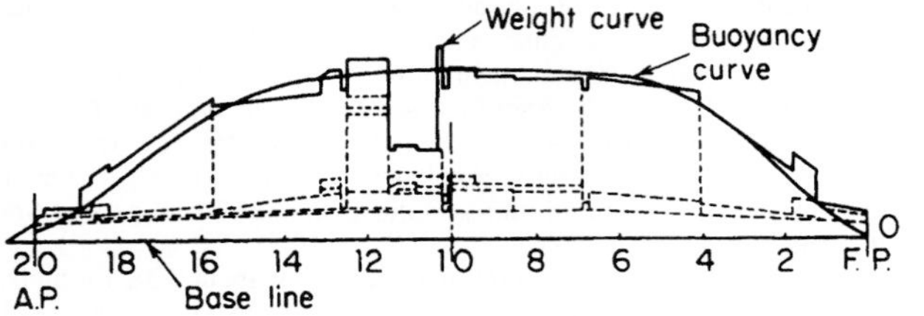

Figure 8.4.2 Representative buoyancy and weight curves for a ship in still water.

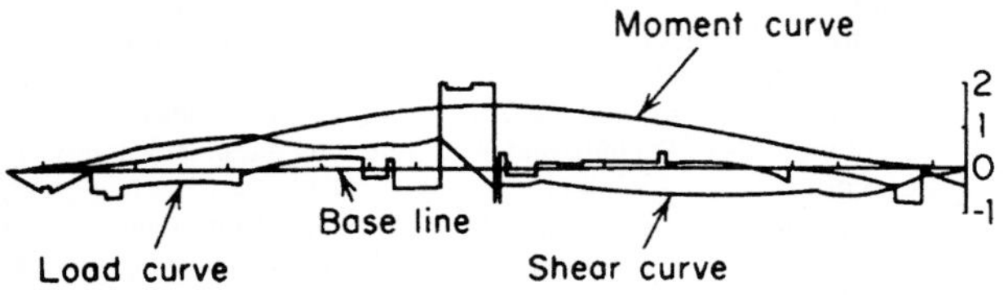

Figure 8.4.3 Representative moment and net-load curves for a ship in still water.

The traditional method of calculating the **maximum design bending moment** involved the determination of weight and buoyancy distributions with the ship poised on a wave of trochoidal form whose length L_w is equal to the length of the ship, L_{pp}, and whose height $H_w = L_w/20$. To more accurately model waves whose lengths exceed 300 ft, $H_w = 1.1\sqrt{L}$ has been widely used. With the wave crest amidships, the ship is in a **hogging** condition (Fig. 8.4.4*a*); the deck is in tension and the bottom shell in compression. With the trough amidships, the ship is in a **sagging** condition (Fig. 8.4.4*b*); the deck is in compression and the bottom in tension. For cargo ships with the machinery amidships, the hogging condition produces the highest bending moment, whereas for tankers and ore carriers with the machinery aft, the sagging condition gives the highest bending moment.

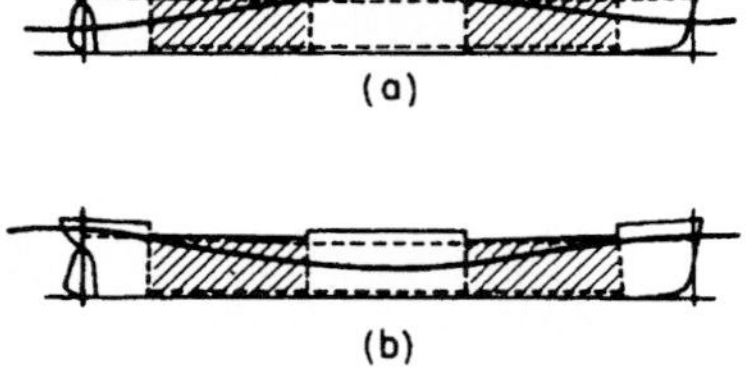

Figure 8.4.4 Hogging (*a*) and sagging (*b*) conditions of a vessel. (*From "Principles of Naval Architecture," SNAME, 1967.*)

The limitations of the above approach should be apparent; nevertheless, it has been an extremely useful one for many decades, and a great deal of ship data have been accumulated upon which to base refinements. Many advances have been made in the theoretical prediction of the actual loads a ship is likely to experience in a realistic, confused seaway over its life span. Today it is possible to predict reliably the bending moments and shear forces a ship will experience over a short term in irregular seas. Long-term prediction techniques use a probabilistic approach which associates load severity and expected periodicity to establish the design load.

The maximum longitudinal bending stress normally occurs in the deck or bottom plating in vicinity of the midship section. In determining the **design section modulus** Z which the continuous longitudinal material in the midship section must meet, an **allowable bending stress** u_{all} must be introduced into the bending stress equation. Based on past experience, an appropriate choice of such an allowable stress is $1.19\,L^{1/3}$ tons/in^2 ($27.31\,L^{1/3}$ MN/m^2) with L in ft (m).

For ships under 195 ft (90 m) in length, strength requirements are dictated more on the basis of locally induced stresses than longitudinal bending stresses. This may be accounted for by reducing the allowable stress based on the above equation or by reducing the K values indicated by the trend in Table 8.4.2 for the calculation of the ship's bending moment. For aluminum construction, Z must be twice that obtained for steel construction. Minimum statutory values of section modulus are published by the U.S. Coast Guard in "Load Line Regulations." Shipbuilding classification societies such as the American Bureau of Shipping have section modulus standards somewhat greater. Rules for section modulus have a permissible (allowable) bending stress based on the hull material.

Table 8.4.2 Constants for Bending Moment Approximations

Type	Δ tons (MN)	L ft (m)	K*
Tankers	35,000-150,000 (349-1,495)	600-900 (183-274)	35-41
High-speed cargo passenger	20,000-40,000 (199-399)	500-700 (152-213)	29-36
Great Lakes ore carrier	28,000-32,000 (279-319)	500-600 (152-183)	54-67
Trawlers	180-1,600 (1.79-16)	100-200 (30-61)	12-18
Crew boats	65-275 (0.65-2.74)	80-130 (24-40)	10-16

*Based on allowable stress of with $\sigma_{all.} = 1.19\sqrt[3]{L}$ tons/in^2 ($27.31\sqrt[3]{L}$ MN/m^2) L in ft (m).

The **maximum bending stress** at each section can be computed by the equation

$$\sigma = M/Z$$

where σ = maximum bending stress, lb/in^2 (N/m^2) or tons/in^2 (MN/ m^2); M = maximum bending moment, ft · lb (N · m) or ft · tons; Z = section modulus, in^2·ft (m^3); $Z = I/y$, where I = minimum vertical moment of inertia of section, in^2 · ft^2 (m^4), and y = maximum distances from neutral axis to bottom and strength deck, ft (m).

The **maximum shear stress** can be computed by the equation

$$\tau = VAc/It$$

where τ = horizontal shear stress, lb/in^2 (N/m^2); V = shear force, lb (N); AC = moment of area above shear plane under consideration taken about neutral axis, ft^3 (m^3); I = vertical moment of inertia of section, in^2·ft^2 (m^4); t = thickness of material at shear plane, ft (m).

For an approximation, the **bending moment** of a ship may be computed by the equation

$$M = \Delta L/K \quad \text{ft} \cdot \text{tons (MN} \cdot \text{m)}$$

where Δ = displacement, tons (MN); L = length, ft (m); and K is as listed in Table 8.4.2.

The above formulas have been successfully applied in the past as a convenient tool for the initial assessment of strength requirements for various designs. At present, computer programs are routinely used in ship structural design. Empirical formulas for the individual calculation of both still water and wave-induced bending moments have been derived by the classification societies. They are based on ship length, breadth, block coefficient, and effective wave height (for the wave-induced bending moment). Similarly, permissible bending stresses are calculated from formulas as a function of ship length and its service environment. Finite element techniques in the form of commercially available computer software packages combined with statistical reliability methods are being applied with increasing frequency.

For ships, the maximum longitudinal bending stresses occur in the vicinity of the midship section at the deck edge and in the bottom plating. Maximum shear stresses occur in the shell plating in the vicinity of the quarter points at the neutral axis. For long, slender girders, such as ships, the maximum shear stress is small compared with the maximum bending stress.

The structure of a ship consists of a grillage of stiffened plating supported by longitudinals, longitudinal girders, transverse beams, transverse frames, and web frames. Since the primary stress system in the hull arises from longitudinal bending, it follows that the longitudinally continuous structural elements are the most effective in carrying and distributing this stress system. The strength deck, particularly at the side, and the keel and turn of bilge are highly stressed regions. The shell plating, particularly deck and bottom plating, carry the major part of the stress. Other **key longitudinal elements,** as shown in Fig. 8.4.5, are longitudinal deck girders, main deck stringer plate, gunwale angle, sheer strake, bilge strake, inner bottom margin plate, garboard strake, flat plate keel, center vertical keel, and rider plate.

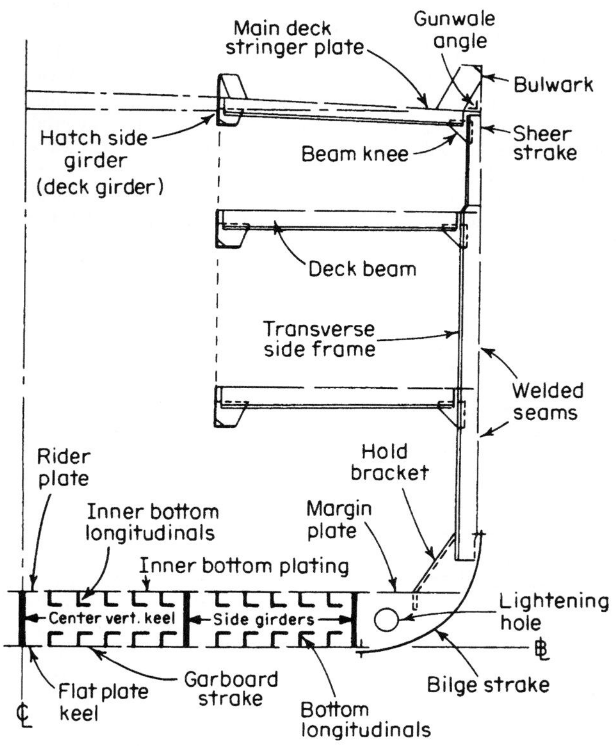

Figure 8.4.5 Key structural elements.

Transverse elements include deck beams and transverse frames and web frames which serve to support and transmit vertical and transverse loads and resist hydrostatic pressure.

Good structural design minimizes the structural weight while providing adequate strength, minimizes interference with ship function, includes access for maintenance and inspection, provides for effective continuity of the structure, facilitates stress flow around deck openings and other stress obstacles, and avoids square corner discontinuities in the plating and other stress concentration "hot spots."

A longitudinally framed ship is one which has closely spaced longitudinal structural elements and widely spaced transverse elements. A transversely framed ship has closely spaced transverse elements and widely spaced longitudinal elements. Longitudinal framing systems are generally more efficient structurally, but because of the deep web frames supporting the longitudinals, it is less efficient in the use of internal space than the transverse framing system. Where interruptions of open internal spaces are unimportant, as in tankers and bulk carriers, longitudinal framing is universally used. However, modern practice tends increasingly toward longitudinal framing in other types of ships also.

Stability

A ship afloat is in vertical equilibrium when the force of **gravity,** acting at the ship's **center of gravity** G, is equal, opposite, and in the same vertical line as the force of **buoyancy,** acting upward at the **center of buoyancy** B. Figure 8.4.6 shows that an upsetting, transverse couple acting on the ship causes it to rotate about a longitudinal axis, taking a list ϕ. G does not change; however, B moves to B_1, the centroid of the new underwater volume. The resulting couple, created by the transverse separation of the two forces ($\overline{GZ}$) opposes the upsetting couple, thereby righting the ship. A static stability curve (Fig. 8.4.7), consisting of values for the righting arm $\overline{GZ}$ plotted against angles of inclination (f), gives a graphic representation of the static stability of the ship. The

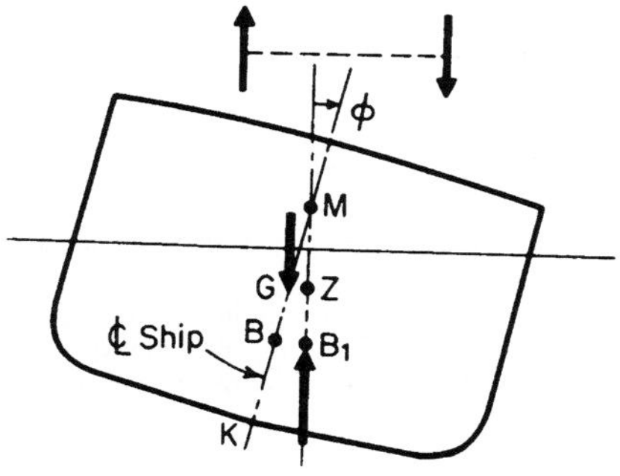

Figure 8.4.6 Ship stability.

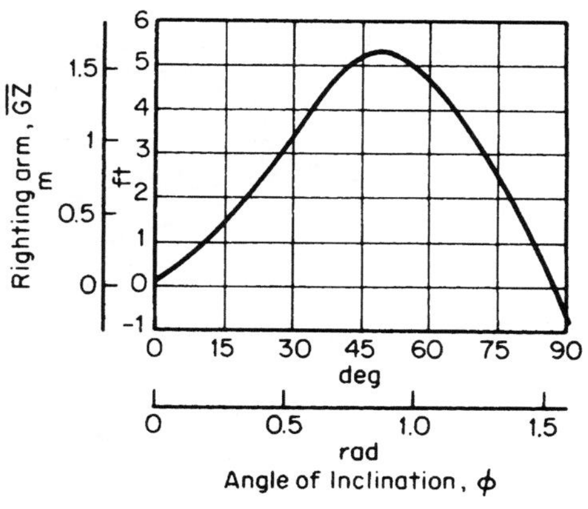

Figure 8.4.7 Static stability curve.

primary indicator for the safety of a ship is the maximum righting arm it develops and the angle heel at which this righting arm occurs.

For small angles of inclination ($\phi < 10°$), centers of buoyancy follow a locus whose instantaneous center of curvature M is known as the transverse **metacenter**. When M lies above G ($\overline{GM} > 0$), the resulting gravity-buoyancy couple will right the ship; the ship has positive stability. When G and M are coincident, the ship has neutral stability. When G is above $M(\overline{GM} < 0)$, negative stability results. Hence $\overline{GM}$, known as the **transverse metacentric height,** is an indication of **initial stability** of a ship.

The transverse **metacentric radius** $\overline{BM}$ and the vertical location of the center of buoyancy are determined by the design of the ship and can be calculated. Once the vertical location of the ship's center of gravity is known, then $\overline{GM}$ can be found. The vertical center of gravity of practically all ships varies with the condition of loading and must be determined either by a careful calculation or by an inclining experiment.

Minimum values of $\overline{GM}$ ranging from 1.5 to 3.5 ft (0.46 to 1.07 m), corresponding to small and large seagoing ships, respectively, have been accepted in the past. The $\overline{GM}$ of passenger ships should be under 6 percent of the beam to ensure a reasonable (comfortable) rolling period. This relatively low value is offset by the generally large range of stability due to the high freeboard of passenger ships.

The question of **longitudinal stability** affects the trim of a ship. As in the transverse case, a longitudinal metacenter exists, and a longitudinal metacentric height $\overline{GM}_L$ can be determined. The moment to alter trim 1 in, $\overline{MT1}$, is computed by $\overline{MT1} = \Delta\overline{GM}_L/12L$, ft·tons/in and moment to alter trim 1 m is $\overline{MT1} = 10^6\Delta\overline{GM}_L/L$, N·m/m. Displacement Δ is in tons and MN, respectively.

The **location of a ship's center of gravity** changes as small weights are shifted within the system. The vertical, transverse, or longitudinal component of movement of the center of gravity is computed by $\overline{GG_1} = w \times d/\Delta$, where w is the small weight, d is the distance the weight is shifted in a component direction, and Δ is the displacement of the ship, which includes w.

Vertical changes in the location of G caused by **weight addition** or **removal** can be calculated by

$$KG_1 = \frac{\Delta\overline{KG} \pm w\overline{K}g}{\Delta_1}$$

where $\Delta_1 = \Delta \pm w$ and $\overline{K}g$ is the height of the center of gravity of w above the keel.

The free surface of the liquid in fuel oil, lubricating oil, and water storage and service tanks is deleterious to ship stability. The weight shift of the liquid as the ship heels can be represented as a virtual rise in G, hence a reduction in $\overline{GM}$ and the ship's initial stability. This virtual rise, called the **free surface effect,** is calculated by the expression

$$GG_v = \gamma_i i/\gamma_w \nabla$$

where γ_i, γ_w = specific gravities of liquid in tank and sea, respectively; i = moment of inertia of free surface area about its longitudinal centroidal axis; and ∇ = volume of displacement of ship. The ship designer can minimize this effect by designing long, narrow, deep tanks or by using baffles.

Seakeeping

Seakeeping is how a ship moves in response to the ocean's wave motion and determines how successfully the ship will function in a seaway. These motions in turn directly affect the practices of marine engineers. The motion of a floating object has six degrees of freedom. Figure 8.4.8 shows the conventional ship coordinate axes and ship motions. Oscillatory movement along the x axis is called **surge**; along the y axis, **sway**; and along the z axis, **heave**. Rotation about x is called **roll**; about y, **pitch**; and about z, **yaw**.

Rolling has a major effect on crew comfort and on the structural and bearing requirements for machinery and its foundations. The natural period of roll of a ship is $T = 2\pi K/\sqrt{g\overline{GM}}$, where K is the radius of gyration of virtual ship mass about a longitudinal axis through G. K varies from 0.4 to 0.5 of the beam, depending on the ship depth and transverse distribution of weights.

Angular acceleration of roll, if large, has a very distressing effect on crew, passengers, machinery, and structure. This can be minimized by increasing the period or by decreasing the roll amplitude. Maximum angular acceleration of roll is

$$d^2\phi/dt^2 = -4\pi^2(\phi_{max}/T^2)$$

where ϕ_{max} is the maximum roll amplitude. The period of the roll can be increased effectively by decreasing $\overline{GM}$; hence the lowest value of $\overline{GM}$ compatible with all stability criteria should be sought.

Pitching is in many respects analogous to rolling. The natural period of pitching, bow up to bow down, can be found by using the same expression used for the rolling period, with the longitudinal radius of gyration K_L substituted for K. A good approximation is $K_L = L/4$, where L is the length of the ship. The natural period of pitching is usually between one-third and one-half the natural period of roll. A by-product of pitching is **slamming,** the reentry of bow sections into the sea with heavy force.

Heaving and **yawing** are the other two principal rigid-body motions caused by the sea. The amplitude of heave associated with head seas,

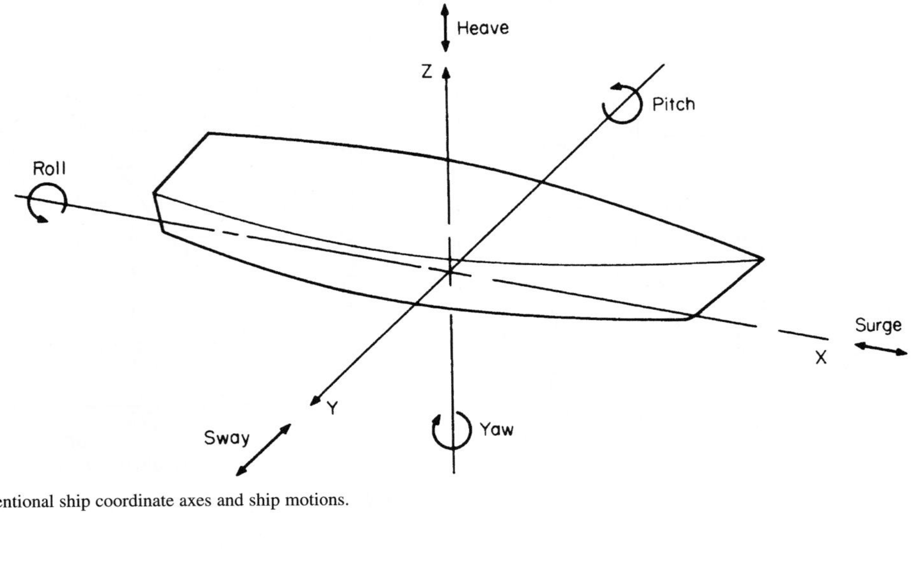

Figure 8.4.8 Conventional ship coordinate axes and ship motions.

which generate pitching, may be as much as 15 ft (4.57 m); that arising from beam seas, which induce rolling, can be even greater. Yawing is started by unequal forces acting on a ship as it quarters into a sea. Once a ship begins to yaw, it behaves as it would at the start of a rudder-actuated turn. As the ship travels in a direction oblique to its plane of symmetry, forces are generated which force it into heel angles independent of sea-induced rolling.

Analysis of ship movements in a seaway is performed by using probabilistic and statistical techniques. The seaway is defined by a mathematically modeled wave energy density spectrum based on data gathered by oceanographers. This wave spectrum is statistically calculated for various sea routes and weather conditions. One way of determining a ship response amplitude operator or transfer function is by linearly superimposing a ship's responses to varying regular waves, both experimentally and theoretically. The wave spectrum multiplied by the ship response transfer function yields the ship response spectrum. The ship response spectrum is an energy density spectrum from which the statistical character of ship motions can be predicted. (See Edward V. Lewis, Motions in Waves, "Principles of Naval Architecture," SNAME, 1989.)

Resistance and Powering

Resistance to ship motion through the water is the aggregate of (1) wave-making, (2) frictional, (3) pressure or form, and (4) air resistances.

Wave-making resistance is primarily a function of Froude number, $\mathrm{Fr} = v/\sqrt{gL}$, where v = ship speed, ft/s (m/s); g = acceleration of gravity, ft/s² (m/s²); and L = ship length, ft (m). In many instances, the dimensional speed-length ratio $V/\sqrt{L}$ is used for convenience, where V is given in knots. A ship makes at least two distinct wave patterns, one from the bow and the other at the stern. There also may be other patterns caused by abrupt changes in section. These patterns combine to form the total wave system for the ship. At various speeds there is mutual cancellation and reinforcement of these patterns. Thus, a plot of total resistance of the ship versus Fr or $V/\sqrt{L}$ is not smooth but shows humps and hollows corresponding to the wave cancellation or reinforcement. Normal procedure is to design the operating speed of a ship to fall at one of the low points in the resistance curve.

Frictional resistance is a function of Reynolds number (see Sec. 1). Because of the size of a ship, the Reynolds number is large and the flow is always turbulent.

Pressure or **form resistance** is a viscosity effect but is different from frictional resistance. The principal observed effects are boundary-layer separation and eddying near the stern.

It is usual practice to combine the wave-making and pressure resistances into one term, called the **residuary resistance,** assumed to be a function of Froude number. The combination, although not strictly legitimate, is practical because the pressure resistance is usually only 2 to 3 percent of the total resistance. The frictional resistance is then the only term which is considered to be a function of Reynolds numberand can be calculated. Based on an analysis of the water resistance of flat, smooth planes, Schoenherr gives the formula

$$R_f = 0.5\rho S v^2 C_f$$

where R_f = frictional resistance, lb (N); ρ = mass density, lb/s² · ft⁴ (kg/m³); S = wetted surface area, ft² (m²); v = velocity, ft/s (m/s); and C_f is the frictional coefficient computed from the ITTC formula

$$C_f = \frac{0.075}{(\log_{10}\mathrm{Re} - 2)^2}$$

and Re = Reynolds number = vL/ν.

Through towing tests of ship models at a series of speeds for which Froude numbers are equal between the model and the ship, **total model resistance** (R_{tm}) is determined. Residuary resistance (R_{rm}) for the model is obtained by subtracting the frictional resistance (R_{fm}). By **Froude's law of comparison,** the residuary resistance of the ship (R_{rs}) is equal to R_{rm} multiplied by the ratio of ship displacement to model displacement. Total ship resistance (R_{ts}) then is equal to the sum of R_{rs}, the calculated ship frictional resistance (R_{fs}), and a correlation allowance that allows for the roughness of ship's hull opposed to the smooth hull of a model.

$$R_{tm} = \text{measured resistance}$$
$$R_{rm} = R_{rm} - R_{fm}$$
$$R_{rs} = R_{tm}(\Delta s/\Delta m)$$
$$R_{ts} = R_{rs} + R_{fs} + R_a$$

A nominal value of 0.0004 is generally used as the correlation allowance coefficient C_a. $R_a = 0.5\rho S v^2 C_a$.

From R_{ts}, the **total effective power** P_E required to propel the ship can be determined:

$$P_E = \frac{R_{ts}v}{550} \quad \text{ehp} \quad \left(P_E = \frac{R_{ts}v}{1{,}000} \quad \mathrm{kW_E}\right)$$

where R_{ts} = total ship resistance, lb (N); v = velocity, ft/s (m/s); and $P_E = P_s \times$ P.C., where P_s is the shaft power (see Propulsion Systems) and P.C. is the propulsive coefficient, a factor which takes into consideration mechanical losses, propeller efficiency, and the flow interaction with the hull. *P.C.*= 0.45 to 0.53 for high-speed craft; 0.50 to 0.60, for tugs and trawlers; 0.55 to 0.65, for destroyers; and 0.63 to 0.72, for merchant ships.

Figure 8.4.9 illustrates the specific effective power for various displacement hull forms and planing craft over their appropriate speed regimes. Figure 8.4.10 shows the general trend of specific resistance versus Froude number and may be used for coarse powering estimates.

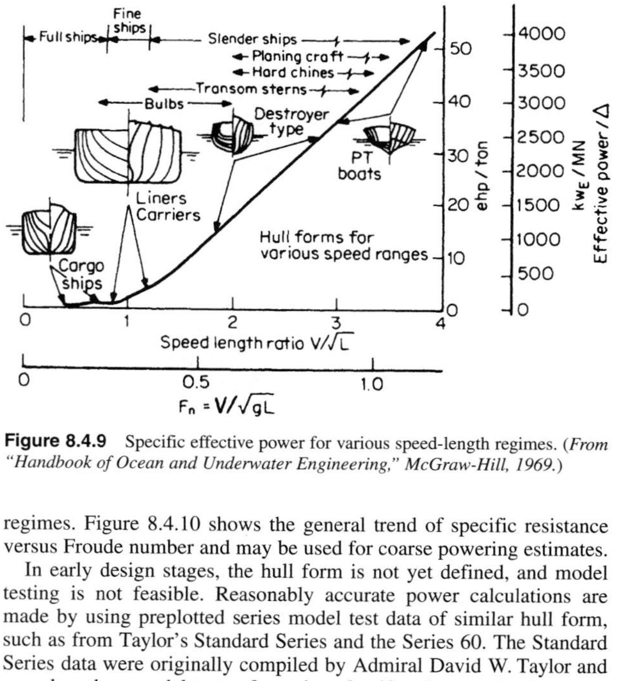

Figure 8.4.9 Specific effective power for various speed-length regimes. (*From "Handbook of Ocean and Underwater Engineering," McGraw-Hill, 1969.*)

In early design stages, the hull form is not yet defined, and model testing is not feasible. Reasonably accurate power calculations are made by using preplotted series model test data of similar hull form, such as from Taylor's Standard Series and the Series 60. The Standard Series data were originally compiled by Admiral David W. Taylor and were based on model tests of a series of uniformly varied twin screw, cruiser hull forms of similar geometry. Revised Taylor's Series data are available in "A Reanalysis of the Original Test Data for the Taylor Standard Series," Taylor Model Basin, Rept. no. 806.

Air Resistance The air resistance of ships in a windless sea is only a few percent of the water resistance. However, the effect of wind can have a significant impact on drag. The wind resistance parallel to the ship's axis is roughly 30 percent greater when the wind direction is about 30° ($\pi/6$ rad) off the bow vice dead ahead, since the projected above-water area is greater. Wind resistance is approximated by $R_A = 0.002B^2V_R^2\,\mathrm{lb}\,(0.36B^2V_R^2\,N)$, where B = ship's beam, ft (m) and V_R = ship speed relative to the wind, in knots (m/s).

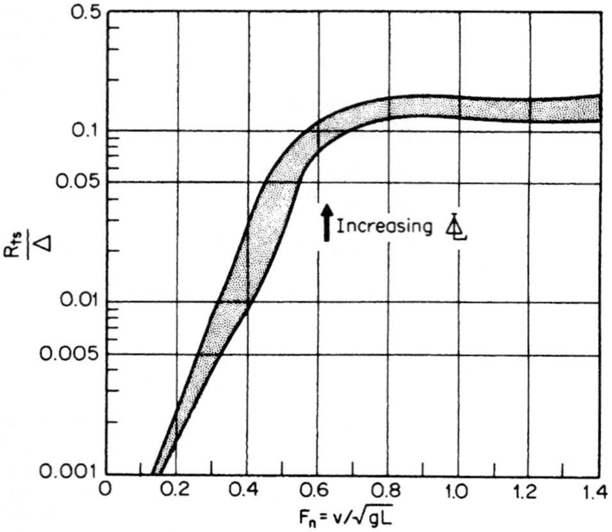

Figure 8.4.10 General trends of specific resistance versus Froude number.

Powering of Small Craft The American Boat and Yacht Council, Inc. (ABYC) provides guidance for determining the maximum power for propulsion of outboard boats, to evaluate the suitability of power installed in inboard boats, and to determine maneuvering speed. Figures 8.4.11 and 8.4.12 demonstrate such data. The latest standard review dates are available at www.abycinc.org.

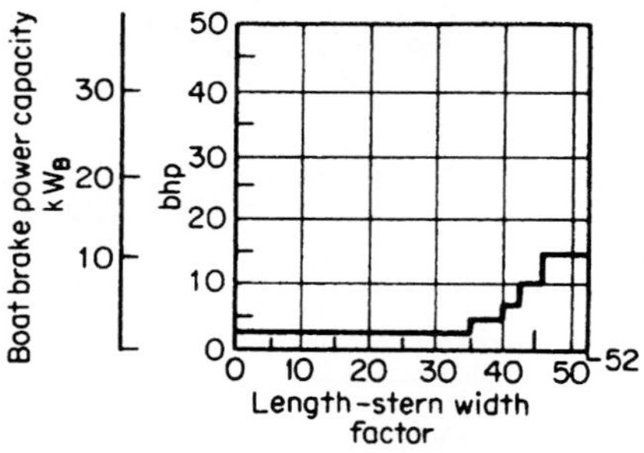

Figure 8.4.11 Boat brake power capacity for length-width factor under 52. *(From "Safety Standards for Small Craft," ABYC, 1974.)*

The development of high-power lightweight engines facilitated the evolution of planing hulls, which develop dynamic lift.

Small high-speed craft primarily use gasoline engines for powering because of weight and cost considerations. However, small high-speed diesel engines are commonly used in pleasure boats.

ENGINEERING CONSTRAINTS

The constraints affecting marine engineering design are too numerous, and some too obvious, to include in this section. Three significant categories, however, are discussed. The geometry of the hull forms immediately suggests **physical constraints.** The interaction of the vehicle with the marine environment suggests **dynamic constraints,** particularly vibration, noise, and shock. The broad topic of **environmental protection** is one of the foremost engineering constraints of today, having a very pronounced effect on the operating systems of a marine vehicle.

Physical Constraints

Formerly, tonnage laws in effect made it economically desirable for the propulsion machinery spaces of a merchant ship to exceed 13 percent

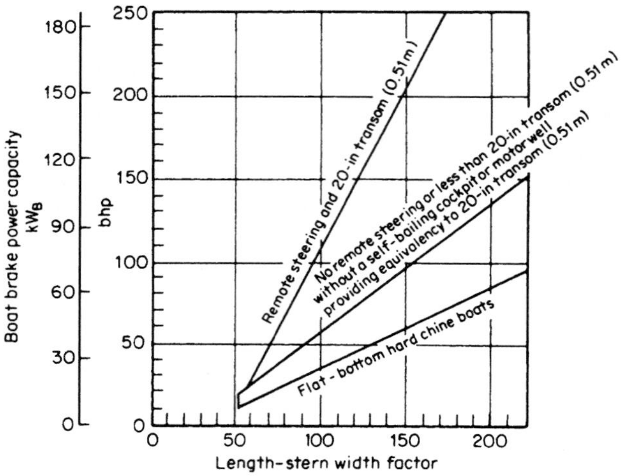

Figure 8.4.12 Boat brake power capacity for length-width factor over 52. *(From "Safety Standards for Small Craft," ABYC, 1974.)*

of the gross tonnage of the ship so that 32 percent of the gross tonnage could be deducted in computing net tonnage. In most design configurations, however, a great effort is made to minimize the space required for the propulsion plant in order to maximize that available to the mission or the money-making aspects of the ship.

Specifically, **space** is of extreme importance as each component of support equipment is selected. Each component must fit into the master compact arrangement scheme to provide the most efficient operation and maintenance by engineering personnel.

Weight constraints for a main propulsion plant vary with the application. In a tanker where cargo capacity is limited by draft restrictions, the weight of machinery represents lost cargo. Cargo ships, on the other hand, rarely operate at full load draft. Additionally, the low weight of propulsion machinery somewhat improves inherent cargo ship stability deficiencies. The weight of each component of equipment is constrained by structural support and shock resistance considerations. Naval shipboard equipment, in general, is carefully analyzed to effect weight reduction.

Dynamic Constraints

Dynamic effects, principally mechanical vibration, noise, shock, and ship motions, are considered in determining the dynamic characteristics of a ship and the dynamic requirements for equipment.

Vibration (See Sec. 7.) Vibration analyses are especially important in the design of the propulsion shafting system and its relation to the excitation forces resulting from the propeller. The main propulsion **shafting** can vibrate in longitudinal, torsional, and lateral modes. Modes of **hull vibration** may be vertical, horizontal, torsional, or longitudinal; may occur separately or coupled; may be excited by synchronization with periodic harmonics of the propeller forces acting either through the shafting, by the propeller force field interacting with the hull afterbody, or both; and may also be set up by unbalanced harmonic forces from the main machinery, or, in some cases, by impact excitation from slamming or periodic-wave encounter.

It is most important to reduce the excitation forces at the source. Very objectionable and serious vibrations may occur when the frequency of the exciting force coincides with one of the hull or shafting-system natural frequencies.

Vibratory forces generated by the propeller are (1) alternating pressure forces on the hull due to the alternating hydrodynamic pressure fields of the propeller blades; (2) alternating propeller-shaft bearing forces, primarily caused by wake irregularities; and (3) alternating forces transmitted throughout the shafting system, also primarily caused by wake irregularities.

The most effective means to ensure a satisfactory level of vibration is to provide adequate clearance between the propeller and the hull surface and to select the propeller revolutions or number of blades to avoid synchronism. Replacement of a four-blade propeller, for instance, by a three- or five-blade, or vice versa, may bring about a reduction in vibration. Singing propellers are due to the vibration of the propeller blade edge about the blade body, producing a disagreeable noise or hum. Chamfering the edge is sometimes helpful.

Vibration due to variations in **engine torque reaction** is hard to overcome. In ships with large diesel engines, the torque reaction tends to produce an athwartship motion of the upper ends of the engines. The motion is increased if the engine foundation strength is inadequate.

Foundations must be designed to take both the thrust and torque of the shaft and be sufficiently rigid to maintain alignment when the ship's hull is working in heavy seas. Engines designed for foundations with three or four points of support are relatively insensitive to minor working of foundations. Considerations should be given to using flexible couplings in cases when alignment cannot be assured.

Torsional vibration frequency of the prime mover–shafting-propeller system should be carefully computed, and necessary steps taken to ensure that its natural frequency is clear of the frequency of the main- unit or propeller heavy-torque variations; serious failures have occurred. Problems due to resonance of engine torsional vibration (unbalanced forces) and foundations or hull structure are usually found after ship trials, and correction consists of local stiffening of the hull structure. If possible, engines should be located at the nodes of hull vibrations.

Although more rare, a **longitudinal vibration** of the propulsion shafting has occurred when the natural frequency of the shafting agreed with that of a pulsating axial force, such as propeller thrust variation.

Other forces inducing vibrations may be vertical inertia forces due to the acceleration of reciprocating parts of an unbalanced reciprocating engine or pump; longitudinal inertia forces due to reciprocating parts or an unbalanced crankshaft creating unbalanced rocking moments; and horizontal and vertical components of centrifugal forces developed by unbalanced rotating parts. Rotating parts can be balanced.

Noise The noise characteristics of shipboard systems are increasingly important, particularly in naval combatant submarines where remaining undetected is essential, and also from a human-factors point of view on all ships. Achieving significant reduction in machinery noise level can be costly. Therefore desired noise levels should be analyzed. Each operating system and each piece of rotating or reciprocating machinery installed aboard a submarine are subjected to intensive airborne (noise) and structure borne (mechanical) vibration analyses and tests. Depending on the noise attenuation required, similar tests and analyses are also conducted for all surface ships, military and merchant.

Shock In naval combatant ships, shock loading due to noncontact underwater explosions is a major design parameter. Methods of qualifying equipment as "shock-resistant" might include "static" shock analysis, "dynamic" shock analysis, physical shock tests, or a combination.

Motions Marine lubricating systems are specifically distinguished by the necessity of including list, trim, roll, and pitch as design criteria. The American Bureau of Shipping requires satisfactory functioning of lubricating systems when the vessel is permanently inclined to certain angles athwartship and fore and aft. In addition, electric generator bearings must not spill oil under momentary roll specifications. Military specifications are similar, but generally add greater requirements for roll and pitch.

Bearing loads are also affected by ship roll and pitch accelerations, and thus rotating equipment often must be arranged longitudinally.

PROPULSION SYSTEMS

(See Sec. 7 for component details.)

The basic operating requirement for the main propulsion system is to safely and reliably propel the vehicle at the required sustained speed for the range or endurance required and to provide suitable maneuvering capabilities. In meeting this basic requirement, the marine propulsion system integrates the power generator/prime mover (generally called the ship's engines or engines), the transmission system, the propulsor, and other shipboard systems with the ship's hull or vehicle platform.

Steam propulsion has virtually disappeared from the seas, other than for nuclear powered naval vessels. Diesel engines provide the most economical source of marine power, and are almost universally adopted for non-naval use. Vessels requiring very high power, such as small naval vessels, are likely to use gas turbine propulsion, often in conjunction with diesel engines for use in cruising or patrolling mode.

Unlike steam plants, which are custom-designed for a given ship design, diesel engines and gas turbines are selected from commercial sources at discrete powers. The selection of such engines also includes a choice of fuel types, including the increasing use of **liquefied natural gas (LNG)** as a marine fuel in diesels and gas turbines.

A current development is the all-electric ship concept. Advances in power electronics have enabled infinite speed control of large AC electric motors, and electric power is in increasing demand for auxiliary and hotel loads aboard ships. Economical production of electrical power by a large number of medium speed diesel-driven generators can service all uses; additionally, the continuing development of electrically-driven vectorable thrusters provides opportunities for both underway propulsion and maneuvering without tug assistance while providing significant operational economies. This concept is being implemented for large cruise ship newbuildings.

Figure 8.4.13 shows propulsion system alternatives with the most popular drives for fixed-pitch and controllable-pitch propellers.

Definitions for Propulsion Systems

Brake power P_B, bhp (kW_B), is the power delivered by the output coupling of a prime mover before passing through speed-reducing and transmission devices and with all continuously operating engine-driven auxiliaries in use.

Shaft power P_s, shp (kW_s), is the net power supplied to the propeller shafting after passing through all reduction gears or other transmission devices and after power for all attached auxiliaries has been taken off. Shaft power is measured in the shafting within the ship by a torsionmeter as close to the propeller or stern tube as possible.

Delivered power P_D, dhp (kW_D), is the power actually delivered to the propeller, somewhat less than P_s because of the power losses in the stern tube and other bearings between the point of measurement and the propeller. Sometimes called **propeller power,** it is also equal to the effective power P_E, ehp (kW_E), plus the power losses in the propeller and the losses in the interaction between the propeller and the ship.

Normal shaft power or **normal power** is the power at which a marine vehicle is designed to run most of its service life.

Maximum shaft power is the highest power for which propulsion machinery is designed to operate continuously.

Service speed is the actual speed maintained by a vehicle at load draft on its normal route and under average weather and sea conditions typical of that route and with average fouling of bottom.

Designed service speed is the speed expected on trials in fair weather at load draft, with clean bottom, when machinery is developing a specified fraction of maximum shaft power. This fraction may vary but is of order of 0.8.

MAIN PROPULSION PLANTS

Diesel Engines

(See Sec. 7.)

Diesel engines are classified by:

- Speed, generally referring to average piston speed. Note that for a given stroke length this implies engine speed in r/min. Large low-speed engines with long strokes will operate at rotational speeds suitable for direct connection to the propeller shaft, while medium speed and high

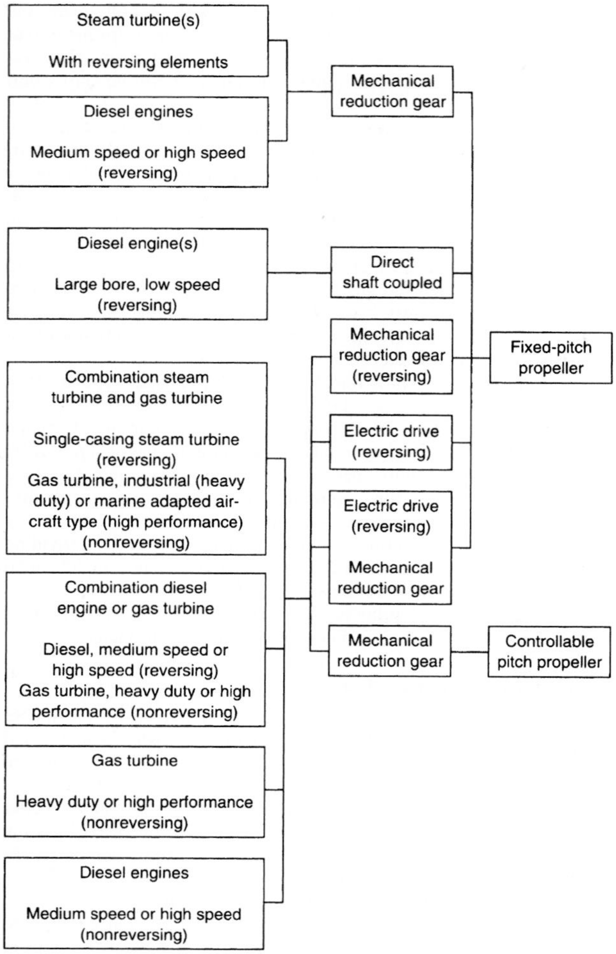

Figure 8.4.13 Alternatives in the selection of a main propulsion plant. (*From "Marine Engineering" SNAME, 1992.*)

speed engines may require reduction gears to mechanically drive a propeller.

- Configuration, either in-line or v-type.
- Two- or four-stroke cycle.
- Reversing or non-reversing. Engines that are not reversing by develop stern power through a reversing gear, or by means of a controllable pitch propeller.

High powered low-speed diesel engines are well-suited to marine propulsion, and such engines have inherently better combustion performance with lower-grade diesel fuels; however, these engines are large, heavy, and expensive in terms of capital cost. Additionally, air quality regulations have eliminated the operational cost advantages of such engines with respect to using less expensive heavy fuel. Medium speed engines generally have lower capital and operating costs. Low speed engine parts take longer to wear to the same percentage of their original dimension than higher speed engine parts. Medium speed engines also offer advantages in terms of reliability because of redundancies, and reduced ship weight and volume requirements for machinery spaces.

Engine manufacturers can offer a variety of stock engines to customers by altering a basic design in (1) cylinder bore, thus varying the power available per cylinder, and in (2) the number of cylinders, thus varying the power available per engine. Engine volume and weight can be altered by offering both in-line and v-type configurations. Marine engines can be found from 4 to 20 cylinders, available at ratings up from a few horsepower to an excess of 109,000 hp (81.3 MW).

Diesel Propulsion Plant Design

Diesel engines may install either singly or in multiple to drive propeller shafts, but usually only low-speed diesel engines are directly coupled to the propeller shaft. For all but high-speed boats, the rpm of the modern diesel is too fast to drive the propeller directly with efficiency, and some means of speed reduction, either mechanical or electrical, is necessary.

If a single engine of the power required for a given application is available, then a decision must be made whether it or several smaller engines should be used. The decision may be dictated by the available space. The diesel power plant is flexible in adapting to specific space requirements. When more than one engine is geared to the propeller shaft, the gear serves as both a speed reducer and combining gear. The same series of engines could be used in an electric-drive propulsion system, with even greater flexibility. Each engine drives its own generator and may be located independently of other engines and the propeller shaft.

A rigid hull foundation, with a high resistance to vertical, athwartship, and fore-and-aft deflections, is required for a low-speed diesel engine. Such engines are also several decks high; the engine room must be designed with sufficient overhaul space and with access for large and heavy replacement parts to be lifted by cranes.

Every diesel propulsion plant requires certain auxiliary equipment. Since the engines are started by air, several large air compressors and banks of high-pressure starting air bottles must be provided. Since fuel injectors on diesel engines are subject to damage by contaminants in liquid fuel, several centrifuges must be provided for fuel treatment. Redundancy of centrifuges is important, not only because different levels of treatment may be needed, but also because since they also serve as the fuel oil transfer pumps for the vessel. Lastly, seawater circulating pumps are essential for engine cooling.

Electric power may be produced by a generator mounted directly on the line shafting, although separate ship's service diesel generators must be provided for times when the main engine is not operating. Diesel plants may also be provided with exhaust gas and auxiliary boilers. The steam from the boilers can be used for cargo, fuel, or space heating, and may be used to increase plant efficiency by powering a ship's service turbo generator.

Because of the low-frequency noise generated by low speedengines, an operating platform could be located at the engine itself; however, separate control rooms are often preferred for noise level reduction and a convenient grouping of plant indicators and controls. Operation of the entire plant may be automatic and remotely controlled from the bridge. The engine room is often completely unattended for 16 h a day.

Gas Turbines

(See Sec. 7.)

The gas turbine has developed since World War II to join the steam turbine and the diesel engine as alternative prime movers for various shipboard applications.

In gas turbines, the efficiency of the components is extremely important since the compressor power is very high compared with its counterpart in competitive thermodynamic cycles. For example, a typical marine propulsion gas turbine rated at 20,000 bhp (14,914 kW_B) might require a 30,000 hp (22,371 kW) compressor and, therefore, 50,000 hp (37,285 kW) in turbine power to balance the cycle.

The basic **advantages** of the gas turbine for marine applications are its simplicity as well as small size and light weight for the level of power delivered. As an internal-combustion engine, it is a self-contained power plant in one package with a minimum number of supporting auxiliaries and the ability to start and go on line very quickly. A further feature of the gas turbine is its low personnel requirement and ready adaptability to automation.

The relative simplicity of the gas turbine has enabled it to attain outstanding records of reliability and maintainability when used for aircraft propulsion and in industrial service. The same level of reliability and maintainability is being achieved in marine service if the unit is properly applied and installed.

Marine units derived from aircraft engines usually have the gas generator section, comprising the compressor and its turbine, arranged to be removed and replaced as a unit. Maintenance on the power turbine, which usually has the smallest part of the total maintenance requirements, is performed aboard ship. Because of their light weight, small gas turbines used for auxiliary power or the propulsion of small boats can also be readily removed for maintenance.

Units designed specifically for marine use and those derived from industrial gas turbines are usually maintained and overhauled in place. Since they are somewhat larger and heavier than the aviation-type units, removal and replacement are not as readily accomplished. For this reason, they usually have split casings and other provisions for easy access and maintenance. The work can be performed by the usual ship repair forces.

Both **single-shaft** and **split-shaft** gas turbines can be used in marine service. Single-shaft units are most commonly used for generator drives. When used for main propulsion, where the propeller must operate over a very wide speed range, the single-shaft unit must have a controllable-pitch propeller or some equivalent variable-speed transmission, such as an electric drive, because of its limited speed range and poor acceleration characteristics. A multishaft unit is normally used for main propulsion, with the usual arrangement being a split-shaft unit with an independent variable-speed power turbine; the power turbine and propeller can be stopped, if necessary, and the gas producer kept in operation for rapid load pickup. The use of variable-area nozzles on the power turbine increases flexibility by enabling the compressor to be maintained at or near full speed and airflow while the power turbine is operating at low power. Nearly full power is available by adding fuel, without waiting for the compressor to accelerate and increase the airflow. Where low-load economy is of importance, the controls can be arranged to reduce the compressor speed at low loads and maintain the maximum turbine inlet and/or exhaust temperature for best efficiency. Since a gas turbine inherently has a poor part-load fuel rate performance, this variable-area nozzle feature can be very advantageous.

The **physical arrangement** of the various components, that is, compressors, combustion systems, and turbines, that make up the gas turbine is influenced by the thermodynamic factors (i.e., the turbine connected to a compressor must develop enough power to drive it), by mechanical considerations (i.e., shafts must have adequate bearings, seals, etc.), and also by the necessity to conduct the very high air and gas flows to and from the various components with minimum pressure losses.

In marine applications, the gas turbines usually **cannot be mounted rigidly** to the ship's structure. Normal movement and distortions of the hull when under way would cause distortions and misalignment in the turbine and cause internal rubs or bearing and/or structural failure. The turbine components can be mounted on a subbase built up of structural sections of sufficient rigidity to maintain the gas-turbine alignment when properly supported by the ship's hull.

Since the gas turbine is a high-speed machine with output shaft speeds ranging from about 3,600 r/min for large machines up to 100,000 r/min for very small machines (approximately 25,000 r/min is an upper limit for units suitable for the propulsion of small boats), a **reduction gear is necessary** to reduce the speed to the range suitable for a propeller. Smaller units suitable for boats or driving auxiliary units, such as generators in larger vessels, frequently have a reduction gear integral with the unit. Larger units normally require a separate reduction gear, usually of the double- or triple-reduction type.

A gas turbine, in common with all turbine machinery, is **not inherently reversible** and must be reversed by external means. Electric drives offer ready reversing but are usually ruled out on the basis of weight, cost, and to some extent efficiency. From a practical standpoint there are two alternatives, a reversing gear or a controllable reversible-pitch (CRP) propeller. Both have been used successfully in gas-turbine-driven ships, with the CRP favored.

Steam Plants

The basic steam propulsion plant contains main boilers, steam turbines, a condensate system, a feedwater system, and numerous auxiliary components necessary for the plant to function, as shown in Fig. 8.4.14. A heat balance calculation, the basic tool for determining the effect of various configurations on plant thermal efficiency, is demonstrated for the basic steam cycle. Both fossil-fuel and nuclear energy sources are successfully employed for marine applications.

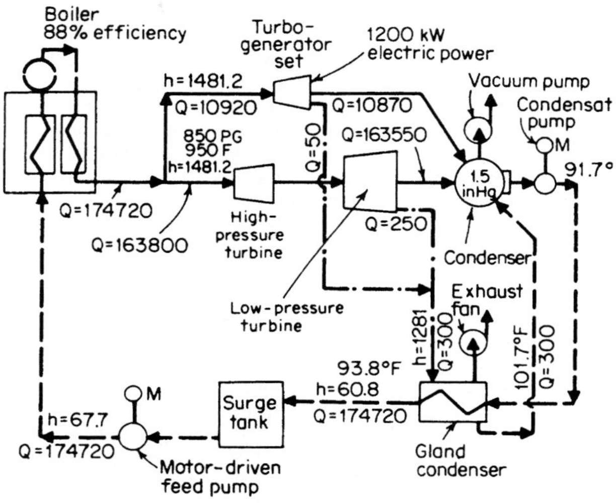

Figure 8.4.14 Simple steam cycle. (*From "Marine Engineering," SNAME, 1992.*)

Main Boilers (See Sec. 7.) The pressures and temperatures achieved in steam-generating equipment have increased steadily over recent years, permitting either a higher-power installation for a given space or a reduction in the size and weight of a given propulsion plant. The **trend in steam pressures** and **temperatures** has been an increase from 600 psig (4.14 MN/m^2) and 850°F (454°C) during World War II to 1,200 psig (8.27 MN/m^2) and 950°F (510°C) for naval combatants in and since the postwar era. For merchant ships, the progression has been from 400 psig (2.76 MN/m^2) and 750°F (399°C) gradually up to 850 psig (5.86 MN/m^2) and 850°F (454°C) in the 1960s, with some boilers at 1,500 psig (10.34 MN/m^2) and 1,000°F (538°C) in the 1970s.

The **quantity of steam** produced by a marine boiler ranges from approximately 1,500 lb/h (680 kg/h) in small auxiliary boilers to over 400,000 lb/h (181,500 kg/h) in large main propulsion boilers. Outputs of 750,000 lb/h (340,200 kg/h) or more per boiler are practical for high-power installations.

Most marine boilers are **oil-fired.** Compared with other fuels, oil is easily loaded aboard ship, stored, and introduced into the furnace, and does not require the ash-handling facilities required for coal firing.

Gas-fired boilers are used primarily on power or drill barges which are fixed in location and can be supplied from shore (normally classed in the Ocean Engineering category). At sea, tankers designed to carry liquefied natural gas (LNG) may use the natural boil-off from their cargo gas tanks as a supplemental fuel **(dual fuel system).** The cargo gas boil-off is collected and pumped to the boilers where it is burned in conjunction with oil. The quantity of boil-off available is a function of the ambient sea and air temperatures, the ship's motion, and the cargo loading; thus, it may vary from day to day.

For economy of space, weight, and cost and for ease of operation, the trend in **boiler installations** is for fewer boilers of high capacity rather than a large number of boilers of lower capacity. The minimum installation is usually two boilers, to ensure propulsion if one boiler is lost; one boiler per shaft for twin-screw ships. Some large ocean-going ships operate on single boilers, requiring exceptional reliability in boiler design and operation.

Combustion systems include forced-draft fans or blowers, the fuel oil service system, burners, and combustion controls. Operation and maintenance of the combustion system are extremely important to the efficiency and reliability of the plant. The best combustion with the least possible excess air should be attained.

Main Turbines (See Sec. 7.) Single-expansion (i.e., single casing) marine steam turbines are fairly common at lower powers, usually not exceeding 4,000 to 6,000 shp (2,983 to 4,474 kW_S). Above that power range, the turbines are usually double-expansion (cross-compound) with high- and low-pressure turbines each driving the main reduction gear through its own pinion. The low-pressure turbine normally contains the reversing turbine in its exhaust end. The main condenser is either underslung and supported from the turbine, or the condenser is carried on the foundations, with the low-pressure turbine supported on the condenser.

The inherent **advantages** of the steam turbine have favored its use over the reciprocating steam engine for all large, modern marine steam propulsion plants. Turbines are not size-limited, and their high temperatures and pressures are accommodated safely. Rotary motion is simpler than reciprocating motion; hence unbalanced forces can be eliminated in the turbine. The turbine can efficiently utilize a low exhaust pressure; it is lightweight and requires minimum space and low maintenance.

The **reheat cycle** is the best and most economical means available to improve turbine efficiencies and fuel rates in marine steam propulsion plants. In the reheat cycle, steam is withdrawn from the turbine after partial expansion and is passed through a heat exchanger where its temperature is raised; it is then readmitted to the turbine for expansion to condenser pressure. Marine reheat plants have more modest steam conditions than land-based applications because of lower power ratings and a greater reliability requirement for ship safety.

The reheat cycle is applied mostly in high-powered units above 25,000 shp (18,642 kW_s) and offers the maximum economical thermal efficiency that can be provided by a steam plant. Reheat cycles are not used in naval vessels because the improvement in efficiency does not warrant the additional complexity; hence a trade-off is made.

Turbine foundations must have adequate rigidity to avoid vibration conditions. This is particularly important with respect to periodic variations in propeller thrust which may excite longitudinal vibrations in the propulsion system.

Condensate System (See Sec. 7.) The condensate system provides the means by which feedwater for the boilers is recovered and returned to the feedwater system. The major components of the condensate system of a marine propulsion plant are the **main condenser,** the **condensate pumps,** and the **deaerating feed tank** or **heater.** Both singlepass and two-pass condenser designs are used; the single-pass design, however, allows somewhat simpler construction and lower water velocities. The single-pass condenser is also adaptable to scoop circulation, as opposed to pump circulation, which is practical for higher-speed ships. The deaerating feedwater heater is supplied from the condensate pumps, which take suction from the condenser hot well, together with condensate drains from steam piping and various heaters.

The deaerating feed heater is normally maintained at about 35 psig (0.241 MN/m^2) and 280°F (138°C) by auxiliary exhaust and turbine extraction steam. The condensate is sprayed into the steam atmosphere at the top of the heater, and the heated feedwater is pumped from the bottom by the feed booster or main feed pumps. In addition to removing oxygen or air, the heater also acts as a surge tank to meet varying demands during maneuvering conditions. Since the feed pumps take suction where the water is almost saturated, the heater must be located 30 to 50 ft (9.14 to 15.24 m) above the pumps in order to provide enough positive suction head to prevent flashing from pressure fluctuations during sudden plant load change.

Feedwater System The feedwater system comprises the pumps, piping, and controls necessary to transport feedwater to the boiler or steam generator, to raise water pressure above boiler pressure, and to control flow of feedwater to the boiler. **Main feed pumps** are so vital that they are usually installed in duplicate, providing a standby pump capable of feeding the boilers at full load. Auxiliary steam-turbine-driven centrifugal pumps are usually selected. A typical naval installation consists of three main feed booster pumps and three main feed pumps for each propulsion plant. Two of the booster pumps are turbine-driven and one is electric. Additionally, an electric-motor-driven emergency booster pump is usually provided. The total capacity of the main feed pumps must be 150 percent of the boiler requirement at full power plus the required recirculation capacity. Reliable feedwater regulators are important.

Steam Plant—Nuclear (See Sec. 7.) The compact nature of the energy source and not requiring air for combustion are the most significant characteristics of nuclear power for marine applications. The fission of one gram of uranium per day produces about one megawatt of power. (One pound produces about 600,000 horsepower.) In other terms, the fission of 1 lb of uranium is equivalent to the combustion of about 86 tons (87,380 kg) of 18,500 Btu/lb (43.03 MJ/kg) fuel oil. This characteristic permits utilization of large power plants on board ship without the necessity for frequent refueling or large bunker storage. Economic studies, however, show that nuclear power, as presently developed, is best suited for military purposes, where the advantages of high power and endurance override the pure economic considerations. As the physical size of nuclear propulsion plants is reduced, their economic attractiveness for commercial marine application will increase.

Major differences between the nuclear power and fossil-fuel plants are: (1) The **safety aspect** of the nuclear reactor system—operating personnel must be shielded from fission product radiation, hence the size and weight of the reactor are increased and maintenance and reliability become more complicated; (2) the steam produced by a pressurized-water reactor plant is saturated and, because of the **high moisture content** in the turbine steam path, the turbine design requires more careful attention; (3) the steam pressure provided by a pressurized-water reactor plant varies with output, the maximum pressure occurring at no load and decreasing approximately linearly with load to a minimum at full power; hence, **blade stresses** in a nuclear turbine increase more rapidly with a decrease in power than in a conventional turbine. Special attention must be given to the design of the control stage of a nuclear turbine. The *N.S. Savannah* was designed for 700 psig (4.83 MN/m^2) at no load and 400 psig (2.76 MN/m^2) at full power.

Combined Propulsion Plants

In some shipboard applications, diesel engines, gas turbines, and steam turbines can be employed effectively in various combinations. The prime movers may be combined either mechanically, thermodynamically, or both. The leveling out of specific fuel consumption over the operating speed range is the goal of most combined systems.

The gas turbine is a very flexible power plant and consequently figures in most possible combinations which include combined diesel and gas-turbine plants **(CODAG);** combined gas- and steam-turbine plants **(COGAS);** and combined gas-turbine and gas-turbine plants **(COGAG).** In these cycles, gas turbines and other engines or gas turbines of two different sizes or types are combined in one plant to give optimum performance over a very wide range of power and speed. In addition, combinations of diesel or gas-turbine plants **(CODOG),** or gas-turbine or gas-turbine plants **(COGOG),** where one plant is a diesel or small gas turbine for use at low or cruising powers and the other a large gas turbine which operates alone at high powers, are also possibilities. Even the combination of a small nuclear plant and a gas turbine plant **(CONAG)** has undergone feasibility studies.

COGAS Gas and steam turbines are connected to a common reduction gear, but thermodynamically are either independent or combined. Both gas- and steam-turbine drives are required to develop full power. In a thermodynamically independent plant, the boost power is furnished by a lightweight gas turbine. This combination produces a significant reduction in machinery weight.

In the thermodynamically combined cycle, commonly called STAG (steam and gas turbine) or COGES (used with electric-drive systems), energy is recovered from the exhaust of the gas turbine and is used to augment the main propulsion system through a steam turbine. The principal advantage gained by the thermodynamic interconnection is the potential for improved overall efficiency and resulting fuel savings. In this arrangement, the gas turbine discharges to a heat-recovery boiler where a large quantity of heat in the exhaust gases is used to generate steam. The boiler supplies the steam turbine that is geared to the propeller. The steam turbine may be coupled to provide part of the power for the gas-turbine compressor. The gas-turbine may be used to provide additional propulsion power, or its exhaust may be recovered to supply heat for various ship's services (Fig. 8.4.15).

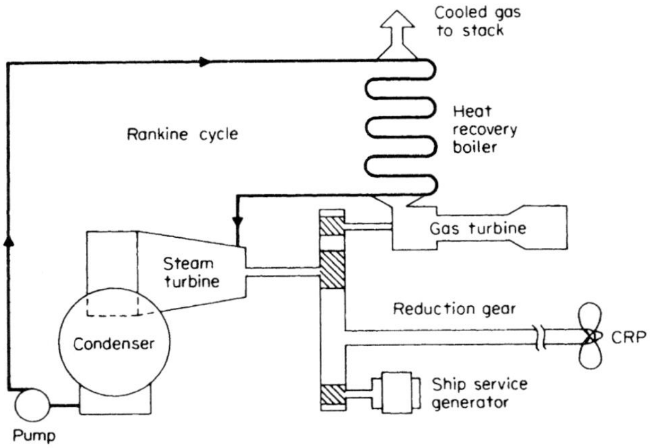

Figure 8.4.15 COGAS system schematic.

Combined propulsion plants have been used in several applications in the past. **CODOG** plants have serviced some Navy patrol gunboats and the Coast Guard's Hamilton class cutters. Both **COGOG** and **COGAS** plants have been used in foreign navies, primarily for destroyer-type vessels.

PROPULSORS

The force to propel a marine vehicle arises from the rate of change of momentum induced in either the water or air. Since the force produced is directly proportional to the mass density of the fluid used, the reasonable choice is to induce the momentum change in water. If air were used, either the cross-sectional area of the jet must be large or the velocity must be high. A variety of propulsor types are used to generate this stream of water aft relative to the vehicle: screw propellers, controllable-pitch propellers, water jets, vectorable thrusters, vertical-axis thrusters, and other thrust devices. Figure 8.4.16 presents a comparison of performance for different types of propulsors.

Screw Propellers

The screw propeller, the basis of most types of propulsors, may be regarded as part of a helicoidal surface which, as it rotates, appears to "screw" its way through the water, driving water aft and the vehicle forward. A propeller is termed **"right-handed"** if it turns clockwise (viewed from aft) when producing ahead thrust; if counterclockwise, **"left-handed."** In a twin-screw installation, the starboard propeller is normally right-handed and the port propeller left-handed. The surface of the propeller blade facing aft, which experiences the increase in pressure, producing thrust, is the **face** of the blade; the forward surface is the **back.** The face is commonly constructed as a true helical surface of constant pitch; the back is not a helical surface. A **true helical surface** is generated by a line rotated about an axis normal to itself and advancing in the direction of this axis at constant speed. The distance the line advances in one revolution is the **pitch.** For simple propellers, the pitch is constant on the face; but in practice, it is common for large propellers to have a reduced pitch toward the hub and less usually toward the tip. The pitch at 0.7 times the maximum radius is usually a representative

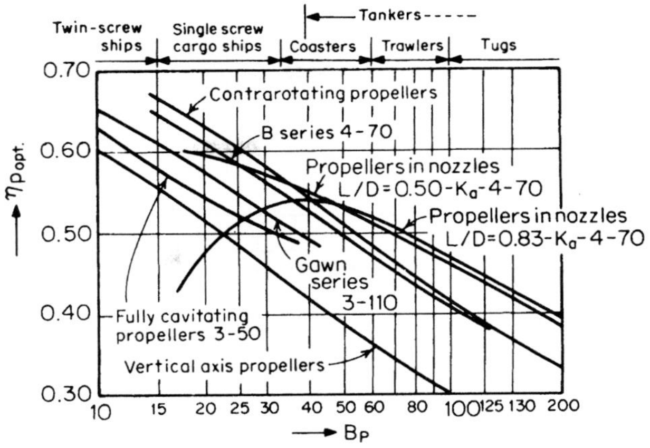

Figure 8.4.16 Comparison of optimum efficiency values for different types of propulsors. (*From "Marine Engineering," SNAME, 1992.*)

mean pitch; maximum lift is generated at that approximate point. Pitch may be expressed as a dimensionless ratio, P/D.

The **shapes** of blade outlines and sections vary greatly according to the type of ship for which the propeller is intended and to the designer's ideas. Figure 8.4.17 shows a typical design and defines many of the terms in common use. The **projected area** is the area of the projection of the propeller contour on a plane normal to the shaft, and the **developed area** is the total face area of all the blades. If the variation of helical cord length is known, then the true blade area, called **expanded area,** can be obtained graphically or analytically by integration.

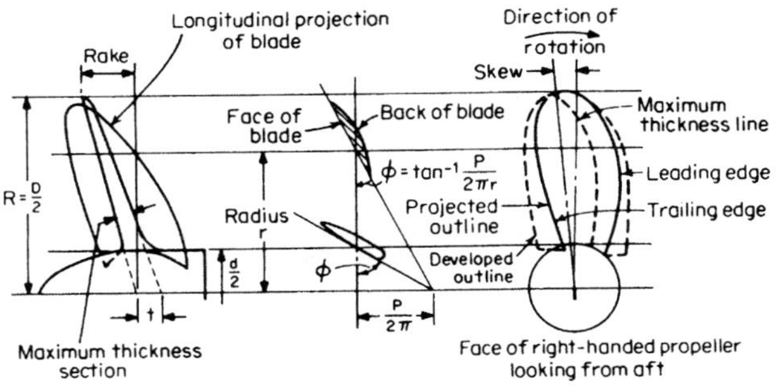

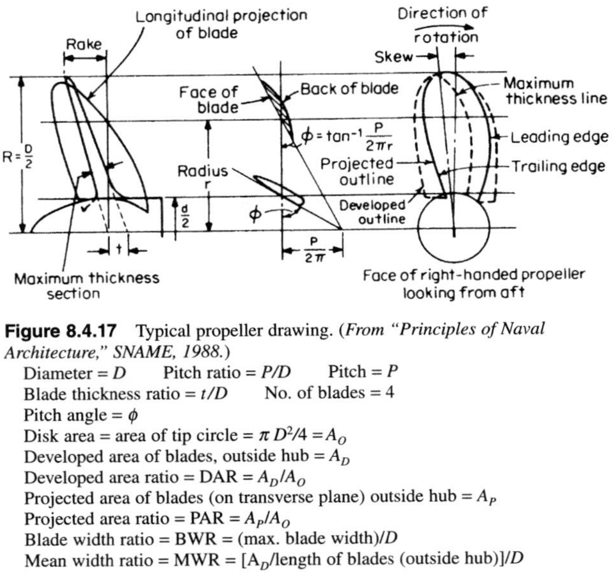

Figure 8.4.17 Typical propeller drawing. (*From "Principles of Naval Architecture," SNAME, 1988.*)

Diameter = D Pitch ratio = P/D Pitch = P
Blade thickness ratio = t/D No. of blades = 4
Pitch angle = ϕ
Disk area = area of tip circle = $\pi D^2/4 = A_O$
Developed area of blades, outside hub = A_D
Developed area ratio = DAR = A_D/A_O
Projected area of blades (on transverse plane) outside hub = A_P
Projected area ratio = PAR = A_P/A_O
Blade width ratio = BWR = (max. blade width)/D
Mean width ratio = MWR = [A_D/length of blades (outside hub)]/D

Consider a section of the propeller blade at a radius r with a pitch angle ϕ and pitch P working in an unyielding medium; in one revolution of the propeller it will advance a distance P. Turning n revolutions in unit time it will advance a distance $P \times n$ in that time. In a real fluid, there will be a certain amount of yielding when the propeller is developing thrust and the screw will not advance $P \times n$, but some smaller distance. The difference between $P \times n$ and that smaller distance is called the **slip velocity. Real slip ratio** is defined in Fig. 8.4.18.

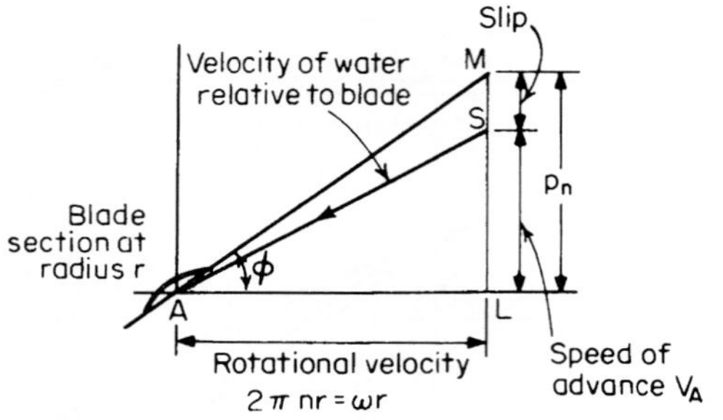

Figure 8.4.18 Definition of slip. (*From "Principles of Naval Architecture," SNAME, 1988.*)

$$\tan\varphi = Pn/2\pi nr = P/2\pi r$$

$$\text{Real slip ratio} = S_R = MS/ML = (Pn - V_A)/Pn = 1 - V_A/Pn$$

A wake or a frictional belt of water accompanies every hull in motion; its velocity varies as the ship's speed, hull shape, the distance from the ship's side and from the bow, and the condition of the hull surface. For ordinary propeller design the **wake velocity** is a fraction w of the ship's speed. Wake velocity = wV. The **wake fraction** w for a ship may be obtained from Fig. 8.4.19. The velocity of the ship relative to the ship's wake at the stern is $V_A = (1 - w)V$.

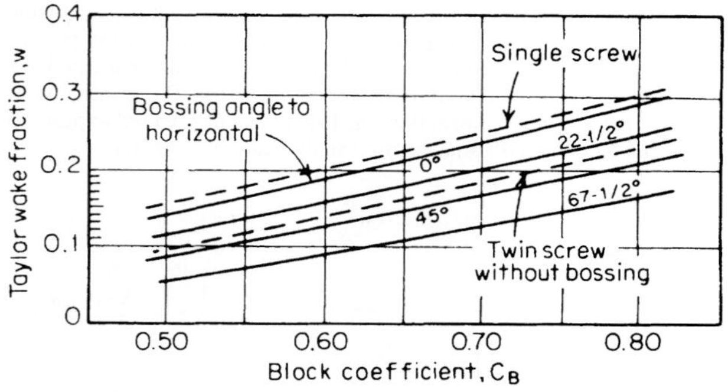

Figure 8.4.19 Wake fractions for single- and twin-screw models. (*From "Principles of Naval Architecture," SNAME, 1988.*)

The **apparent slip ratio** S_A is given by

$$S_A = (Pn - V)/Pn = 1 - V/Pn$$

Although real slip ratio, which requires knowledge of the wake fraction, is a real guide to ship performance, the apparent slip ratio requires only ship speed, revolutions, and pitch to calculate and is therefore often recorded in ships' logs.

For P in ft (m), n in r/min, and V in knots, $S_A = (Pn - 101.3\ V)/Pn$; [$S_A = (Pn - 30.9V)/Pn$].

Propeller Design The design of a marine propeller is usually carried out by one of two methods. In the first, the design is based upon charts giving the results of open-water tests on a series of model propellers. These cover variations in a number of the design parameters such as pitch ratio, blade area, number of blades, and section shapes. A propeller which conforms with the characteristics of any particular series can be rapidly designed and drawn to suit the required ship conditions.

The second method is used in cases where a propeller is heavily loaded and liable to cavitation or has to work in a very uneven wake pattern; it is based on purely theoretical calculations. Basically this involves finding the chord width, section shape, pitch, and efficiency at a number of radii to suit the average circumferential wake values and give optimum efficiency and protection from cavitation. By integration of the resulting thrust and torque-loading curves over the blades, the thrust, torque, and efficiency for the whole propeller can be found.

Using the first method and **Taylor's propeller** and **advance coefficients** B_P and δ, a convenient practical design and an initial estimate of propeller size can be obtained

$$B_P = \frac{n(P_D)^{0.5}}{(V_A)^{2.5}} \quad \left[\frac{1.158n(P_D)^{0.5}}{(V_A)^{2.5}}\right]$$

$$\delta = \frac{nD}{V_A} \quad \left(\frac{3.281nD}{V_A}\right)$$

$$\eta_O = \frac{TV_A}{325.7P_D\eta_R} \quad \left(\frac{TV_A}{1{,}942.5P_D\eta_R}\right)$$

where B_p = Taylor's propeller coefficient; δ = Taylor's advance coefficient; n = r/min; P_D = delivered power, dhp (kW$_D$); V_A = speed of advance, knots; D = propeller diameter, ft (m); T = thrust, lb (N); η_O = open-water propeller efficiency; η_R = relative rotative efficiency ($0.95 < \eta_R < 1.0$, twin-screw; $1.0 < \eta_R < 1.05$, single-screw), a factor which corrects η_O to the efficiency in the actual flow conditions behind the ship.

Assume a reasonable value for n and, using known values for P_D and V_A (for a useful approximation, $P_D = 0.98P_S$), calculate B_p. Then enter Fig. 8.4.20, or an appropriate series of Taylor or Troost propeller charts, to determine δ, η_O, and P/D for optimum efficiency. The charts and parameters should be varied, and the results plotted to recognize the most suitable propeller.

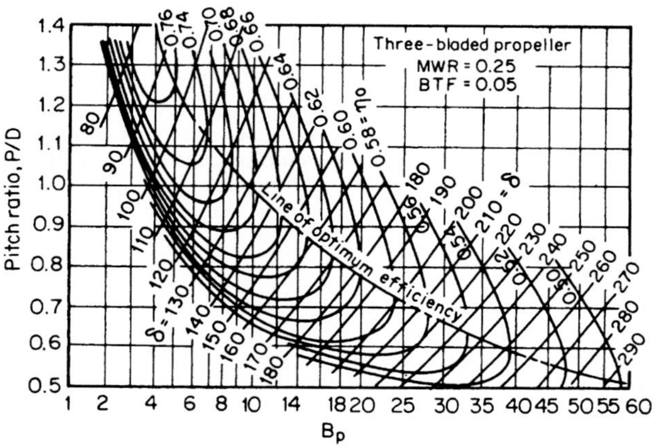

Figure 8.4.20 Typical Taylor propeller characteristic curves. (*From "Handbook of Ocean and Underwater Engineering," McGraw-Hill, 1969.*)

Propeller cavitation, when severe, may result in marked increase in rpm, slip, and shaft power with little increase in ship speed or effective power. As cavitation develops, noise, vibration, and erosion of the propeller blades, struts, and rudders are experienced. It may occur either on the face or on the back of the propeller. Although cavitation of the face has little effect on thrust and torque, extensive cavitation of the back can materially affect thrust and, in general, can be avoided by either increasing blade area or reducing propeller rpm. The erosion of the backs is caused by the collapse of cavitation bubbles as they move into higher pressure regions toward the trailing edge.

Avoidance of cavitation is an important requirement in propeller design and selection. The Maritime Research Institute Netherlands (MARIN) suggests the following **criterion** for the minimum blade area required **to avoid cavitation:**

$$A_P^2 = \frac{T^2}{1{,}360(p_0 - p_v)^{1.5}V_A} \quad \text{or} \quad \left[\frac{T^2}{5.44(p_0 - p_v)^{1.5}V_A}\right]$$

where A_p = projected area of blades, ft^2 (m^2); T = thrust, lb (N); p_0 = pressure at screw centerline, due to water head plus atmosphere, lb/in^2 (N/m^2);

p_v = water-vapor pressure, lb/in^2 (N/m^2); and V_A = speed of advance, knots (1 knot = 0.515 m/s); and

$$p_0 - p_v = 14.45 + 0.45h \quad \text{lb/in}^2$$

or

$$(p_0 - p_v = 99{,}629 + 10{,}179h \quad \text{N/m}^2)$$

where h = head of water at screw centerline, ft (m).

A four-blade propeller of 0.97 times the diameter of the three-blade, the same pitch ratio, and four-thirds the area will absorb the **same power** at the same rpm as the three-blade propeller. Similarly, a two-blade propeller of 5 percent greater diameter is approximately equivalent to a three-blade unit. Figure 8.4.20 shows that propeller efficiency increases with decrease in value of the propeller coefficient B_p. A low value of B_p in a slow-speed ship calls for a large-diameter (optimum) propeller. It is frequently necessary to limit the diameter of the propeller and accept the accompanying loss in efficiency.

The number of propeller blades is usually three or four. Four blades are commonly used with single-screw merchant ships. Recently five, and even seven, blades have been used to reduce vibration. Alternatively, highly skewed blades reduce vibration by creating a smoother interaction between the propeller and the ship's wake. Highly loaded propellers of fast ships and naval vessels call for large blade area, preferably three to five blades, to reduce blade interference and vibration.

Propeller fore-and-aft clearance from the stern frame of large single-screw ships at 70 percent of propeller radius should be greater than 18 in. The stern frame should be streamlined. The clearance from the propeller tips to the shell plating of twin-screw vessels ranges from about 20 in for low-powered ships to 4 or 5 ft for high-powered ships. U.S. Navy practice generally calls for a propeller tip clearance of 0.25D. Many high-speed motorboats have a tip clearance of only several inches. The immersion of the propeller tips should be sufficient to prevent the drawing in of air.

EXAMPLE. Consider a three-blade propeller for a single-screw installation with P_S = 2,950 shp (2,200 kW$_S$), V = 17 knots, and C_B = 0.52. Find the propeller diameter D and efficiency η_O for n = 160 r/min.

From Fig. 8.4.19, the wake fraction w = 0.165. $V_A = V(1 - 0.165)$ = 14.19 knots. $P_D = 0.98P_S$ = 2,891 dhp (2,156 kW$_D$). At 160 r/min, $B_P = 160(2{,}891)^{0.5}/(14.19)^{2.5}$ = 11.34. [$1.158 \times 160\,(2{,}156)^{0.5}/(14.19)^{2.5}$ = 11.34.] From Fig. 8.4.20, the following is obtained or calculated:

r/min	B_p	P/d	η_O	δ	D, ft (m)	P, ft (m)
160	11.34	0.95	0.69	139	12.33 (3.76)	11.71 (3.57)

For a screw centerline submergence of 9 ft (2.74 m) and a selected projected area ratio PAR = 0.3, investigate the cavitation criterion. Assume η_R = 1.0.

Minimum projected area:

$$T = 325.7\,\eta_O \eta_R P_D / V_A = 325.7 \times 0.69 \times 1.0 \times 2{,}891/14.19$$
$$= 45{,}786 \text{ lb}$$
$$(T = 1{,}942.5 \times 0.69 \times 2{,}156/14.19 = 203{,}646 \text{ N})$$
$$p_0 - p_v = 14.45 + 0.45(9) = 18.5 \text{ lb/in}^2$$
$$[p_0 - p_v = 99{,}629 + 10{,}179(2.74 = 127{,}519 \text{ N/m}^2)$$
$$A_p^2 = (45{,}786)^2/1{,}360(18.5)^{1.5}(14.19) = 1{,}365, A_p$$
$$= 36.95 \text{ ft}^2 \text{ minimum}$$
$$A_p^2 = (203{,}646)^2/5.44(127{,}519)^{1.5}(14.19) = 11.79, A_p$$
$$= 3.43 \text{ m}^2 \text{ minimum]}$$

Actual projected area: PAR = A_p/A_0 = 0.3, $A_p = 0.3A_0 = 0.37\pi\,(12.33)^2/4$ = 35.82 ft^2 (3.33 m^2).

Since the selected PAR does not meet the minimum cavitation criterion, either the blade width can be increased or a four-blade propeller can be adopted. To absorb the same power at the same rpm, a four-blade propeller of about 0.97 (12.33) = 11.96 ft (3.65 m) diameter with the same blade shape would have an $A_p = 35.82 \times 4/3 \times (11.96/12.33)^2$ = 44.94 ft^2 (4.17 m^2), or about 25 percent increase. The pitch = 0.95 (11.96) = 11.36 ft (3.46 m). The real and apparent slip ratios for the four-blade propeller are:

$$S_R = 1 - (101.3)(14.19)/(11.36)(160) = 0.209$$
$$S_A = 1 - (30.9)(17.0)/(3.46)(160) = 0.051$$

that is, 20.9 and 5.1 percent (S_A calculated using SI values).

Since the projected area in the three-blade propeller of this example is only slightly under the minimum, a three-blade propeller with increased blade width would be the more appropriate choice.

Controllable-Pitch Propellers

Controllable-pitch propellers are screw propellers in which the blades are separately mounted on the hub, each on an axis, and in which the pitch of the blades can be changed, and even reversed, while the propeller is turning in the same direction.

The pitch is changed by means of an internal mechanism consisting essentially of hydraulic pistons in the hub acting on crossheads. Controllable-pitch propellers are most suitable for vehicles which must meet varying operating conditions, such as tugs, trawlers, ferries, minesweepers, and landing craft. As the propeller pitch is changed, the engine can still run at its most efficient speed. Maneuvering is more rapid since the pitch can be changed more rapidly than could the shaft revolutions. By use of controllable-pitch propellers, neither reversing mechanisms are necessary in reciprocating engines, nor astern turbines in turbine-powered vehicles, especially important in gas-turbine installations. Except for the larger hub needed to house the pitch-changing mechanism, the controllable-pitch propeller can be made almost as efficient as the solid, fixed-blade propeller. Gas turbines and controllable reversible-pitch propellers (CRPs) are generally used in all recent U.S. Navy surface ship designs (except carriers).

Vectorable Thrusters

A relatively recent development in the application of propellers is the development of marine vectorable thrusters, sometimes called **azimuth thrusters, pods** or **pod drives.** The propellers may be fixed pitch or controllable, may be ducted with nozzles or unducted, and may be retractable or not. Although thrusters with mechanical transmissions, generally involving gearing to change shaft direction, have been designed and installed for some time, the current trend is toward electrically powered thrusters. By gimballing the thrusters to change direction, the need for a rudder may be obviated.

Recent developments include vessels powered entirely by vectorable thrusters, with power levels of individual units exceeding 28,500 hp (21.3 MW). Various designs are sold by major power plant manufacturers under a variety of trade names.

Water Jet

This method usually consists of an impeller or pump inside the hull, which draws water from outside, accelerates it, and discharges it astern as a jet at a higher velocity. It is a reaction device like the propeller but in which the moving parts are contained inside the hull, desirable for shallow-water operating conditions and maneuverability. The overall efficiency is lower than that of the screw propeller of diameter equal to the jet orifice diameter, principally because of inlet and ducting losses. Other disadvantages include the loss of volume to the ducting and impeller and the danger of fouling of the impeller. Water jets have been used in several of the U.S. Navy's hydrofoil and surface-effect vehicles.

Vertical-Axis Propellers

There are two types of vertical-axis propulsor systems consisting of one or two vertical-axis rotors located underwater at the stern. Rotor disks are flush with the shell plating and have five to eight streamline, spade-like, vertical impeller blades fitted near the periphery of the disks. The blades feather during rotation of the disk to produce a maximum thrust effect in any direction desired. In the **Kirsten-Boeing** system the blades are interlocked by gears so that each blade makes a half revolution about its axis for each revolution of the disk. The blades of the **Voith-Schneider** system make a complete revolution about their own axis for each revolution of the disk. A bevel gear must be used to transmit power from the conventional horizontal drive shaft to the horizontal disk; therefore, limitations exist on the maximum power that can be transmitted. Although the propulsor is 30 to 40 percent less efficient than the screw propeller, it has obvious maneuverability advantages. Propulsors of this type have also been used at the bow to assist in maneuvering.

Other Thrust Devices

Pump Jet In a pump-jet arrangement, the rotating impeller is external to the hull with fixed guide vanes either ahead and/or astern of it; the whole unit is enclosed in a duct or long shroud ring. The duct diameter increases from the entrance to the impeller so that the velocity falls and the pressure increases. Thus the impeller diameter is larger, thrust loading less, and the efficiency higher; the incidence of cavitation and noise is delayed. A penalty is paid, however, for the resistance of the duct.

Kort Nozzles In the Kort nozzle system, the screw propeller operates in a ring or nozzle attached to the hull at the top. The longitudinal sections are of airfoil shape, and the length of the nozzle is generally about one-half its diameter. Unlike the pump-jet shroud ring, the Kort nozzle entrance is much bigger than the propeller, drawing in more water than the open propeller and achieving greater thrust. Because of the acceleration of the water into the nozzle, the pressure inside is less; hence a forward thrust is exerted on the nozzle and the hull. The greatest advantage is in a tug, pulling at rest. The free-running speed is usually less with the nozzle than without. In some tugs and rivercraft, the whole nozzle is pivoted and becomes a very efficient steering mechanism.

Tandem and Contrarotating Propellers Two or more propellers arranged on the same shaft are used to divide the increased loading factor when the diameter of a propeller is restricted. Propellers turning in the same direction are termed **tandem,** and in opposite directions, **contrarotating.** In tandem, the rotational energy in the race from the forward propeller is augmented by the after one. Contrarotating propellers work on coaxial, contrary-turning shafts so that the after propeller may regain the rotational energy from the forward one. The after propeller is of smaller diameter to fit the contracting race and has a pitch designed for proper power absorption. Such propellers have been used for years on torpedoes to prevent the torpedo body from rotating. Hydrodynamically, the advantages of contrarotating propellers are increased propulsive efficiency, improved vibration characteristics, and higher blade frequency. Disadvantages are the complicated gearing, coaxial shafting, and sealing problems; therefore, such installations are normally restricted to special applications.

Supercavitating Propellers When cavitation covers the entire back of a propeller blade, an increase of rpm cannot further reduce the pressure at the back but the pressure on the face continues to increase as does the total thrust, though at a slower rate than before cavitation began. Advantages of such fully cavitating propellers are an absence of back erosion and less vibration. Although the characteristics of such propellers were determined by trial and error, they have long been used on high-speed racing motor boats. The blade section design must ensure clean separation of flow at the leading and trailing edges and provide good lift-drag ratios for high efficiency. Introducing air to the back of the blades (**ventilated propeller**) will ensure full cavitation and also enables use at lower speeds.

Partially Submerged Propellers The appendage drag presented by a propeller supported below high-speed craft, such as planing boats, hydrofoil craft, and surface-effect ships, led to the development of partially submerged propellers. Although vibration and strength problems, arising from the cyclic loading and unloading of the blades as they enter and emerge from the air-water interface, remain to be solved, it has been demonstrated that efficiencies in partially submerged operation, comparable to fully submerged noncavitating operation, can be achieved. The performance of these propellers must be considered over a wide range of submergences.

Outboard Gasoline Engine Outboard gasoline engines of 1 to 300 bhp (0.75 to 225 kW_B), combining steering and propulsion, are popular for small pleasure craft. Maximum speeds of 3,000 to 6,000 r/min are typical, as driven by the propeller horsepower characteristics (see Sec. 7).

PROPULSION TRANSMISSION

In modern ships, only large-bore, slow-speed diesel engines are directly connected to the propeller shaft. Transmission devices such as mechanical speed-reducing gears or electric drives (generator/motor transmissions) are required to convert the relatively high rpm of a compact, economical prime mover to the relatively low propeller rpm necessary for a high propulsive efficiency. In the case of steam turbines, medium and high-speed diesel engines, and gas turbines, speed reduction gears are used. Gear ratios vary from relatively low values for medium-speed diesels up to approximately 50:1 for a compact turbine design. Where the prime mover is unidirectional, the drive mechanism must also include a reversing mechanism, unless a CRP is used.

Reduction Gears

Speed reduction is usually obtained with reduction gears. The simplest arrangement of a marine reduction gear is the **single-reduction,** single-input type, that is, one pinion meshing with a gear (Fig. 8.4.21). This arrangement is used for connecting a propeller to a diesel engine or to an electric motor but is not used for propelling equipment with a turbine drive.

Figure 8.4.21 Single-reduction, single-input gear. *(From "Marine Engineering," SNAME, 1992.)*

The usual arrangement for turbine-driven ships is the **double-reduction,** double-input, articulated type of reduction gear (Fig. 8.4.22). The two input pinions are driven by the two elements of a cross-compound turbine. The term *articulated* applies because a flexible coupling is generally provided between the first reduction or primary gear wheel and the second reduction or secondary pinion.

The **locked-train** type of double-reduction gear has become standard for high-powered naval ships and, because it minimizes the total weight and size of the assembly, is gaining increased popularity for higher-powered merchant ships (Fig. 8.4.23).

Figure 8.4.22 Double-reduction, double-input, articulated gear. *(From "Marine Engineering," SNAME, 1992.)*

Figure 8.4.23 Double-reduction, double-input, locked-train gear. *(From "Marine Engineering," SNAME, 1992.)*

Electric Drive

Electric propulsion drive is the principal alternative to direct- or geared-drive systems. The prime mover drives a generator or alternator which in turn drives a propulsion motor which is either coupled directly to the propeller or drives the propeller through a low-ratio reduction gear.

Among its **advantages** are the ease and convenience by which propeller speed and direction are controllable; the freedom of installation arrangement offered by the electrical connection between the generator and the propulsion motor; the flexibility of power use when not used for propulsion; the convenience of coupling several prime movers to the propeller without mechanical clutches or couplings; the relative simplicity of controls required to provide reverse propeller rotation when the prime mover is unidirectional; and the speed reduction that can be provided between the generator and the motor, hence between the prime mover and the propeller, without mechanical speed-reducing means.

The disadvantages are the inherently higher first cost, increased weight and space, and the higher transmission losses of the system. In general, electric propulsion drives are employed in marine vehicles requiring a high degree of maneuverability such as ferries, icebreakers, and tugs; in those requiring large amounts of special-purpose power such as fireboats, and large tankers; in those utilizing nonreversing, high-speed, and multiple prime movers; and in deep-submergence vehicles. Diesel-electric drive is ideally adapted to bridge control.

Reversing

Reversing may be accomplished by stopping and reversing a reversible engine, as in the case of many reciprocating engines, or by adding reversing elements in the prime mover, as in the case of steam turbines. It is impracticable to provide reversing elements in gas turbines; hence a reversing capability must be provided in the transmission system or in the propulsor itself. Reversing reduction gears for such transmissions are available up to quite substantial powers, and controllable-reversible propellers are also used with diesel or gas-turbine drives. Electrical drives provide reversing by dynamic braking and energizing the electric motor in the reverse direction. Waterjets reverse inserting reversing thrust buckets into the discharge stream.

Marine Shafting

A main propulsion shafting system consists of the equipment necessary to transmit the rotative power from the main propulsion engines to the propulsor; support the propulsor; transmit the thrust developed by the propulsor to the ship's hull; safely withstand transient operating loads (e.g., high-speed maneuvers, quick reversals); be free of deleterious modes of vibration; and provide reliable operation.

Figure 8.4.24 is a shafting arrangement typical of multishaft ships and single-shaft ships with transom sterns. The shafting must extend outboard a sufficient distance for adequate clearance between the propeller and the hull. Figure 8.4.25 is typical of single-screw merchant ships.

The shafting located inside the ship is **line shafting.** The outboard section to which the propeller is secured is the **propeller shaft** or **tail shaft.** The section passing through the stern tube is the **stern tube shaft** unless the propeller is supported by it. If there is a section of shafting between the propeller and stern tube shafts, it is an **intermediate shaft.**

The principal dimensions, kind of material, and material tests are specified by the classification societies for marine propulsion shafting. The American Bureau of Shipping Rules for Steel Vessels specifies the minimum diameter of propulsion shafting, tail shaft bearing lengths, bearing liner and flange coupling thicknesses, and minimum fitted bolt diameters.

Tail or propeller shafts are of larger diameter than the line or tunnel shafting because of the propeller-weight-induced bending moment; also, inspection during service is possible only when the vessel is drydocked.

Propellers are fitted to shafts over 6 in (0.1524 m) diameter by a taper fit. Aft of the propeller, the shaft is reduced to 0.58 to 0.68 of its specified diameter under the liner and is fitted with a relatively fine thread, nut, and keeper. Propeller torque is taken by a long flat key, the width and total depth of which are, respectively, about 0.21 and

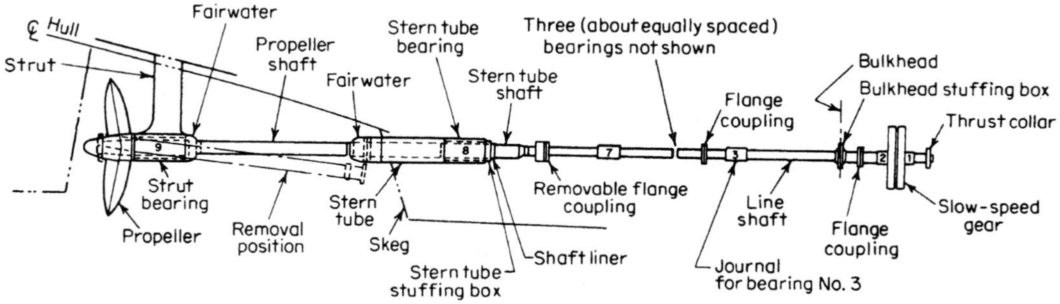

Figure 8.4.24 Shafting arrangement with strut bearings. *(From "Marine Engineering," SNAME, 1992.)*

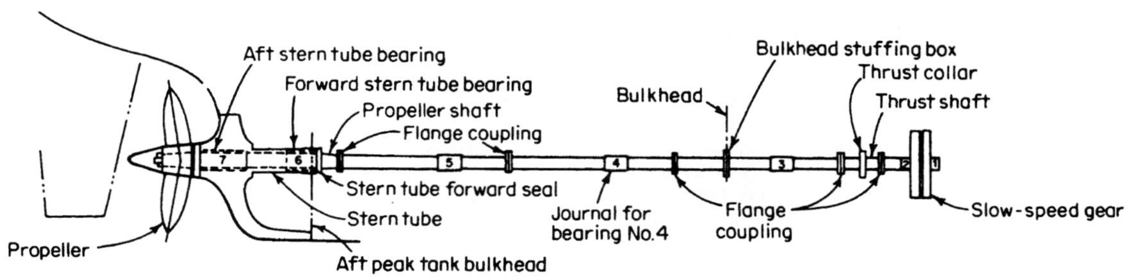

Figure 8.4.25 Shafting arrangement without strut bearings. *(From "Marine Engineering," SNAME, 1992.)*

0.11 times the shaft diameter under the liner. The forward end of the key slot in the shaft should be tapered off in depth and terminate well aft of the large end of the taper to avoid a high concentrated stress at the end of the keyway. Every effort is made to keep saltwater out of contact with the steel tail shaft so as to prevent failure from corrosion fatigue. Modern design practice has moved to keyless shafts to which propellers are hydraulically force-fit on a taper.

A cast-iron, cast-steel, or wrought-steel-weldment **stern tube** is rigidly fastened to the hull. In single-screw ships it supports the propeller shaft and is normally provided with a packing gland at the forward end as seawater lubricates the sterntube bearings. The aft stern tube bearing is made four diameters in length; the forward stern tube bearing is much shorter. A bronze liner, ¾ to 1 in (0.0191 to 0.0254 m) thick, is shrunk on the propeller shaft to protect it from corrosion and to give a good journal surface. Lignum-vitae inserts, epoxy applications, or rubber linings have been used for stern tube bearings; the average bearing pressure is about 30 lb/in^2 (206,842 N/m^2). Rubber and phenolic compound bearings, having longitudinal grooves, are also extensively used. White-metal oil-lubricated bearings and, in Germany, roller-type stern bearings have been fitted in a few installations.

The line or tunnel shafting is laid out, in long equal lengths, so that, in the case of a single-screw ship, withdrawal of the tail shaft inboard for inspection every several years will require only the removal of the next inboard length of shafting. For large outboard or water-exposed shafts of twin-screw ships, a bearing spacing up to 30 shaft diameters has been used. A greater ratio prevails for steel shafts of power boats. Bearings inside the hull are spaced closer—normally under 15 diameters if the shaft is more than 6 in (0.1524 m) diameter. Usually, only the bottom half of the bearing is completely white-metaled; oil-wick lubrication is common, but oil-ring lubrication is used in high-class installations.

Bearings are used to support the shafting in essentially a straight line between the main propulsion engine and the desired location of the propeller. Bearings inside the ship are most popularly known as line shaft bearings, steady bearings, or spring bearings. Bearings which support outboard sections of shafting are stern tube bearings if they are located in the stern tube and strut bearings when located in struts.

The propeller thrust is transmitted to the hull by means of a main thrust bearing. The main thrust bearing may be located either forward or aft of the slow-speed gear.

HIGH-PERFORMANCE SHIP SYSTEMS

High-performance ships have spawned new hull forms and propulsion systems which were relatively unknown just a generation ago. Because of their unique characteristics and requirements, conventional propulsors are rarely adequate. The prime thrust devices range from those using water, to water-air mixtures, to large air propellers.

The total propulsion power required by these vehicles is presented in general terms in the Gabrielli and Von Karman plot of Fig. 8.4.26. Basically, there are those vehicles obtaining lift by displacement and those obtaining lift from a wing or foil moving through the fluid. In order for marine vehicles to maintain a reasonable lift-to-drag ratio over a moderate range of speed, the propulsion system must provide increased thrust at lower speeds to overcome low-speed wave drag. The weight of any vehicle is converted into total drag and, therefore, total required propulsion and lift power.

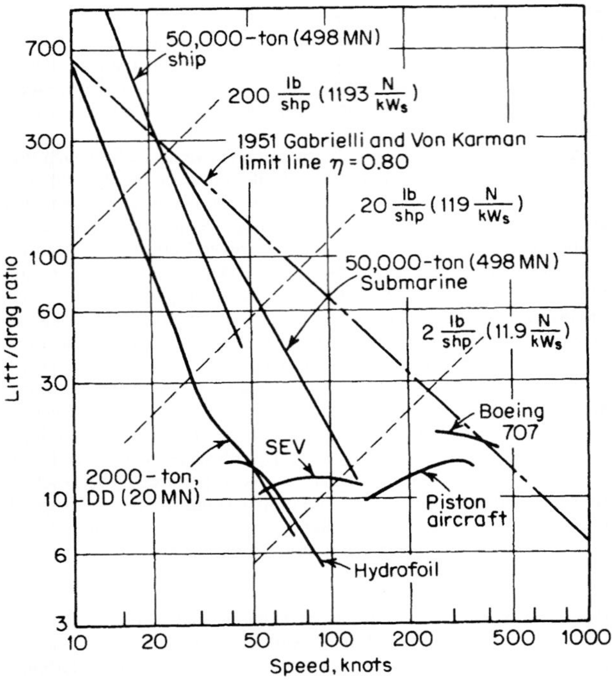

Figure 8.4.26 Lift-drag ratio versus calm weather speed for various vehicles.

Lightweight, compact, and efficient propulsion systems to satisfy extreme weight sensitivity are achieved over a wide range of power by the marinized aircraft gas-turbine engine. Marinization involves addition of a power turbine unit, incorporation of air inlet filtration to provide salt-free air, addition of silencing equipment on both inlet and exhaust ducts, and modification of some components for resistance to salt corrosion. The shaft power delivered is then distributed to the propulsion and lift devices by lightweight shaft and gear power transmission systems.

A comparison of high-performance ship propulsion system characteristics is given in Table 8.4.3.

Hydrofoil Craft

Under foilborne conditions, the support of the hydrofoil craft is derived from the dynamic lift generated by the foil system, and the craft hull is lifted clear of the water. Special problems relating to operation and

Table 8.4.3 U.S. Navy High-Performance Ship Propulsion System Characteristics

Ship	Engine	Propulsor/lift
Hydrofoil, PHM, 215 tons (2.1 MN), 50 knots	Foilborne: one LM-2,500 GT, 22,000 bhp (15,405 kW$_B$), 3,600 r/min	Foilborne: one 2-stage axial-flow waterjet, twin aft strut inlets
Surface-effect ship, SES-200, 195 tons (1.94 MN), 30 + knots	Propulsion: two 16V-149TI diesel, 1,600 bhp (1,193 kW$_B$) Lift: two 8V-92TI diesel, 800 bhp (597 kW$_B$)	Two waterjet units Two double-inlet centrifugal lift fans
Air-cushion vehicle, LCAC, 150 tons (1.49 MN), 50 knots	Propulsion and lift: four TF-40B GT, 3,350 bhp (2,498 kW$_B$), 15,400 r/min	Two ducted-air propellers, 12 ft (3.66 m) dia., 1,250 r/min; four double-entry centrifugal lift fans, 1,900 r/min
SWATH, SSP, 190 tons (1.89 MN), 25 knots	Propulsion: two T-64 GT, 3,200 bhp (2,386 kW$_B$), 1,000 r/min at output of attached gearbox	Two subcavitating controllable-pitch propellers, 450 r/min

directly related to the conditions presented by waves, high fluid density, surface piercing of foils and struts, stability, control, cavitation, and ventilation of foils have given rise to foil system configurations defined in Fig. 8.4.27.

The hydrofoil craft has three modes of operation: **hullborne, takeoff,** and **foilborne.** The thrust and drag forces involved in hydrofoil operation are shown in Fig. 8.4.28. Design must be a compromise between the varied requirements of these operating conditions. This may be difficult, as with propeller design, where maximum thrust is often required at takeoff speed (normally about one-half the operating speed).

A propeller designed for maximum efficiency at takeoff will have a smaller efficiency at design speeds, and vice versa. This is complicated by the need for supercavitating propellers at operating speeds over 50 knots. Two sets of propellers may be needed in such a case, one for hull-borne operation and the other for foilborne operation. The selection of a surface-piercing or a fully submerged foil design is determined from specified operating conditions and the permissible degree of complexity. A **surface-piercing system** is inherently stable, as a deviation from the equilibrium position of the craft results in corrective lift forces. **Fully submerged foil** designs, on the other hand, provide the best lift-drag ratio but are not stable by themselves and require a depth-control system for satisfactory operation. Such a control system, however, permits operation of the submerged-foil design in higher waves.

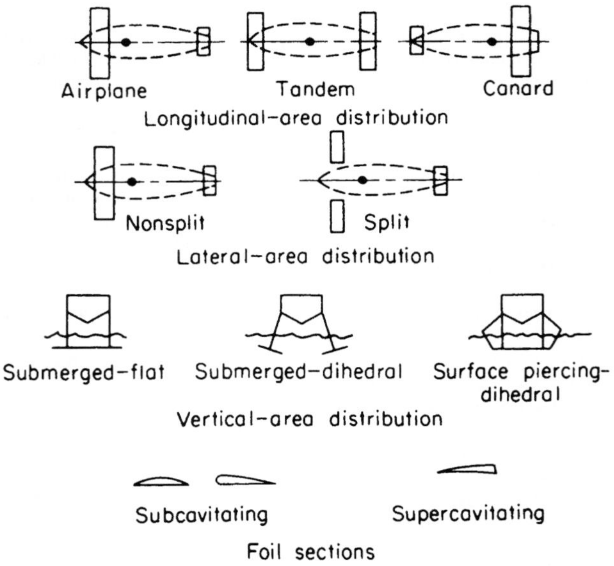

Figure 8.4.27 Some foil arrangements and sections.

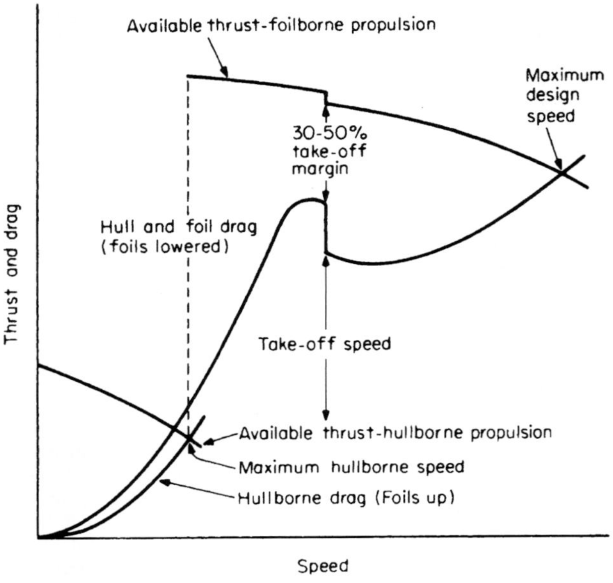

Figure 8.4.28 Thrust and drag forces for a typical hydrofoil.

Surface-Effect Vehicles (SEV)

The air-cushion-supported **surface-effect ship (SES)** and **air-cushion vehicle (ACV)** show great potential for high-speed and amphibious operation. Basically, the SEV rides on a cushion of air generated and maintained in the space between the vehicle and the surface over which it moves or hovers. In the SES, the cushion is captured by rigid sidewalls and flexible skirts fore and aft. The ACV has a flexible seal or skirt all around. Figure 8.4.29 illustrates two configurations of the surface-effect principle.

Since the ACV must rise above either land or water, the propulsors use only air for thrust, requiring a greater portion of the total power allocated to lift than in the SES design as shown in Fig. 8.4.30.

While SES technology has greatly improved the capability of amphibious landing vehicles, it is rapidly being developed for high-speed large cargo vessels.

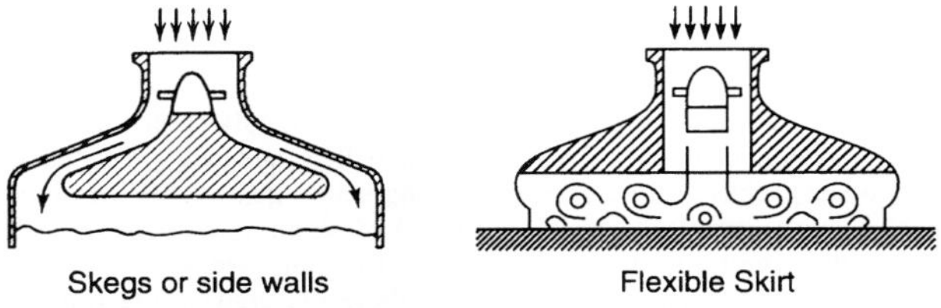

Figure 8.4.29 Two configurations of the surface-effect principle.

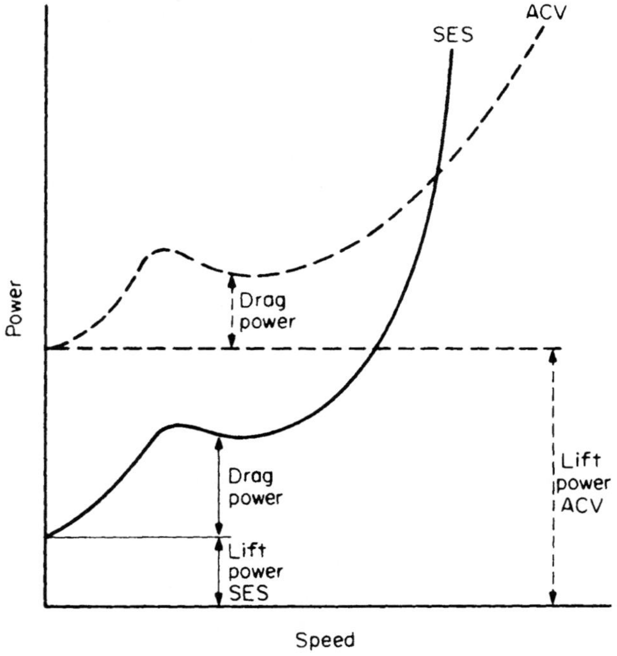

Figure 8.4.30 Comparison of power requirements: SES versus ACV.

Small Water-Plane Area Twin Hull

The **SWATH,** configured to be relatively insensitive to seaway motions, gains its buoyant lift from two parallel submerged, submarinelike hulls which support the main hull platform through relatively small vertical struts. The small hull area at the waterline lowers the response of the hull to waves. The same small waterplane area makes the SWATH hull form sensitive to changes in weight and load distribution. In this configuration, a propulsion arrangement choice must be with regard to the location of the main engines. In the U.S. Navy's SSP, power from two independently operated gas turbines is transmitted down each aft strut by a group of three wide-belt chain drives with speed reduction to the propellers. By contrast, the T-AGOS 19 series SWATH vessels employ diesel-electric drive, with both motors and propulsors in the lower hull.

CARGO SHIP SYSTEMS

Cargo vessels, generally referred to as merchants ships to distinguish them from naval vessels, fall into four generally categories:

1. *Passenger* ship include passenger liners, which operate on regular transoceanic runs, *cruise* ships which operate as a means of entertainment and relaxation for the passengers, and ferries which operate to transport passengers and perhaps vehicles between ports when land transport is difficult or impossible. Specialized systems aboard such ships may include large HVAC plants with specialized controls for passenger comfort; high capacity freshwater distilling plants to assure a sufficient supply of freshwater for bathing and cooking; extensive berthing spaces; entertainment facilities, including auditoriums, swimming pools, gambling casinos, and sports and fitness facilities, and extensive restaurant, bar, and cooking facilities.

2. *General cargo vessels,* which include break-bulk ships that load, discharge, and carry crates, barrels, sacks, pallets, and cargo in virtually any other package into cargo holds using ship's gear; container ships which carry cargo prestowed into standard shipping containers each resembling the trailer body of an articulated trailer truck, and usually loaded and discharged by specialized shoreside cranes; and roll-on/roll-off ships (also called **RO-ROs**) that carry commercial trucks, trailers, and vehicles that are loaded and discharged by driving aboard for stowage and securing. Specialized systems aboard such vessels may include, on break bulk ships, cargo booms or cranes, specialized cargo tanks for carrying edible oils, and refrigerated and insulated cargo holds for refrigerated cargo (called **reefer cargo**); on container ships, cell guides within cargo holds to maintain stacks of containers in place, specially strengthened hatch covers to permit deck loading of stacks of containers with specialized cargo securing devices, and electrical outlets on deck for providing electric power to refrigerated containers; and on roll-on/roll-off ships, watertight doors and hatches to provide sufficiently large openings to drive vehicles aboard, internal ramps systems to permit internal vehicles movement, and high capacity ventilation systems to remove engines exhaust fumes.

3. *Dry bulk carriers*, some times simply called **bulk carriers**, which carry ore, grain, coal, chemicals, and other non-liquid bulk commodities in large open holds. Such vessel may be loaded and discharged by shoreside equipment, or may be designed to discharge cargo with ships' equipment as **self-unloaders**. Other than self-unloaders, which are provided with a cargo boom, conveyors, and lifting screws, bulk carriers are remarkably devoid of specialized cargo equipment.

4. *Tankers,* vessels designed to carry liquid bulk cargoes, most often petroleum and chemical cargoes, in tanks integral to the ship's hull. Tankers are normally loaded by shoreside pumping, but equipped with pumps and piping systems to discharge cargoes. Vessels designed to cargo crude oil, after cargo is discharged but before the cargo tanks are cleaned, have a potentially explosive gaseous atmosphere in the cargo tanks, so inert gas systems are used to reduce the tank atmospheric conditions to a safe condition until the tanks are cleaned of residually cargo using a crude oil washing system. These systems include inert gas generators, cargo heaters, pumps, spray nozzles, and associated piping. Cargo pumps may either be centralized, or individual deep well pumps may be provided to each tank. For chemical tankers and specialized refined product tankers, deep well pumps are normally installed together with segregated piping systems designed to prevent cargo contamination or environmental discharge, and to facilitate tank cleaning procedures. Specialized gas tankers carry liquid petroleum gas (**LPG**) under pressure or liquefied natural gas (**LNG**) under cryogenic conditions. These ships have extensive and sophisticated cargo containment systems that may include gas re-liquefaction plants, pressure relief and venting systems, provision for burning vented cargo as ship's fuel, and extensive cryogenic insulation.

One aspect of merchant ship profitability is to combine the carriage of different cargoes to assure profitability over a larger portion of the total voyage. This has led to the combination of ship types. For instance, although a crude oil tanker will be in ballast for approximately half of each voyage, and **oil-bulk-ore carrier** (also called an **OBO**) may have revenue producing legs over as much as 70 percent of a voyage. Such combination of cargoes, however, not only challenges the naval architect with respect to the design of such a ship, but requires duplications or synergies in terms of specialized cargo systems. Typical combinations include **RO-Pax ferries**, a combination of RO-RO general cargo with passenger ferry service, **Con-RO vessels** that combine the carriage of containerized general cargo with RO-RO cargo, **O-O carriers** designed specifically to combine services in the carriage of heavy ores with the carriage of relatively lighter petroleum liquids on different voyage legs, and vessel designed for the combination of containerized and break bulk cargoes on the same voyage legs.

8.5 AERONAUTICS

by M. W. M. Jenkins and Joaquim R. R. A. Martins

REFERENCES: National Advisory Committee for Aeronautics, *Technical Reports* (designated *NACA-TR* with number), *Technical Notes* (*NACA-TN* with number), and *Technical Memoranda* (*NACA-TM* with number). British Aeronautical Research Committee *Reports and Memoranda* (designated *Br. ARC-R & M* with number). *Ergebnisse der Aerodynamischen Versuchsanstalt zu Göttingen.* Diehl, "Engineering Aerodynamics," Ronald. Reid, "Applied Wing Theory," McGraw-Hill. Durand, "Aerodynamic Theory," Springer. Prandtl-Tietjens, "Fundamentals of Hydro- and Aeromechanics," and "Applied Hydro- and Aeromechanics," McGraw-Hill. Goldstein, "Modern Developments in Fluid Dynamics," Oxford. Millikan, "Aerodynamics of the Airplane," Wiley. Von Mises, "Theory of Flight," McGraw-Hill. Hoerner, "Aerodynamic Drag," Hoerner. Glauert, "The Elements of Airfoil and Airscrew Theory," Cambridge. Milne-Thompson, "Theoretical Hydrodynamics," Macmillan. Munk, "Fluid Dynamics for Aircraft Designers," and "Principles of Aerodynamics," Ronald. Abraham, "Structural Design of Missiles and Spacecraft," McGraw-Hill. "U.S. Standard Atmosphere, 1976," U.S. Government Printing Office.

DEFINITIONS

Aeronautics is the science and art of flight and includes the operation of heavier-than-air aircraft. An **aircraft** is any weight-carrying device designed to be supported by the air, either by buoyancy or by dynamic action. An **airplane** is a mechanically driven fixed-wing aircraft, heavier than air, which is supported by the dynamic reaction of the air against its wings. An aircraft supported by buoyancy is called an **airship,** which can be either a balloon, a hot-air type, or of semirigid or rigid construction. These airships can be free floating or, if of semirigid or rigid design, propelled by fixed or vectoring propeller or jet propulsion. Ascent or descent is achieved primarily through helium gas or hot air, venting, and by the use of disposable ballast. A **helicopter** is a kind of aircraft lifted and moved by a large propeller mounted horizontally above the fuselage. It differs from the **autogiro** in that this propeller is turned by motor power and there is no auxiliary propeller for forward motion. A **ground-effect machine (GEM)** is a heavier-than-air surface vehicle that operates in close proximity to the earth's surface (over land or water), never touching except at rest, being separated from the surface by a cushion or film of air, however thin, and depending entirely upon aerodynamic forces for propulsion and control.

Aerodynamics is the branch of aeronautics that is concerned with the motion of air and other gaseous fluids, and the forces acting on solids in motion relative to such fluids. The branches of aerodynamics are divided according to velocity ranges, depending upon whether the velocity is below or above the local speed of sound in the fluid. The velocity range below the local speed of sound is called the **subsonic** regime. Where the velocity is above the local speed of sound, the flow is said to be **supersonic.** The term **transonic** refers to flows in which both subsonic and supersonic regions are present. The **hypersonic** regime is a high supersonic speed range usually above five times the speed of sound.

STANDARD ATMOSPHERE

The U.S. Standard Atmosphere is a model that describes how the atmospheric properties (temperature, pressure, density, and speed of sound) vary with altitude. It is an idealized, steady-state representation of the atmosphere that extends to 621.4 miles (1,000 km[1]) above the surface and is assumed to rotate with the earth. Table 8.5.1*a* lists the atmospheric properties according to the latest version of the U.S. Standard Atmosphere (1976) as a function of altitude. This standard is largely consistent with International Standard Atmosphere, which differs in the assumed temperature profile at the higher altitudes.

Figure 8.5.1 shows the variation of base temperature *K* with altitude Z up to approximately 76 miles (122 km). Above 281,952 feet, four distinct altitude zones are identified (Table 8.5.1*b*) in which temperature altitude relationships are defined relative to the base altitude of each zone. These four zones constitute the current standard atmosphere in terms of base temperature *K* from 53.4 miles to 621.4 miles (1,000 km). For further definition of the standard atmosphere in this four-zone region the reader is referred to U.S. Department of Commerce document "U.S. Standard Atmosphere, 1976," ADA035728.

Table 8.5.1*a* U.S. Standard Atmosphere, 1976

Altitude h, ft*	Temp T, °F†	Pressure ratio, p/p_0	Density ratio, ρ/ρ_0	$(\rho_0/\rho)^{0.5}$	Speed of sound V_s, ft/s‡
0	59.00	1.0000	1.0000	1.000	1,116
5,000	41.17	0.8320	0.8617	1.077	1,097
10,000	23.34	0.6877	0.7385	1.164	1,077
15,000	5.51	0.5643	0.6292	1.261	1,057
20,000	−12.62	0.4595	0.5328	1.370	1,036
25,000	−30.15	0.3711	0.4481	1.494	1,015
30,000	−47.99	0.2970	0.3741	1.635	995
35,000	−65.82	0.2353	0.3099	1.796	973
36,089	−69.70	0.2234	0.2971	1.835	968
40,000	−69.70	0.1851	0.2462	2.016	968
45,000	−69.70	0.1455	0.1936	2.273	968
50,000	−69.70	0.1145	0.1522	2.563	968
55,000	−69.70	0.09001	0.1197	2.890	968
60,000	−69.70	0.07078	0.09414	3.259	968
65,000	−69.70	0.05566	0.07403	3.675	968
65,800	−69.70	0.05356	0.07123	3.747	968
70,000	−67.30	0.04380	0.05789	4.156	971
75,000	−64.55	0.03452	0.04532	4.697	974
80,000	−61.81	0.02725	0.03553	5.305	977
85,000	−59.07	0.02155	0.02790	5.986	981
90,000	−56.32	0.01707	0.02195	6.970	984
95,000	−53.58	0.01354	0.01730	7.600	988
100,000	−50.84	0.01076	0.01365	8.559	991

* × 0.3048 = meters.
† × (°F − 32)/1.8 = °C.
‡ × 0.3048 = m/s.

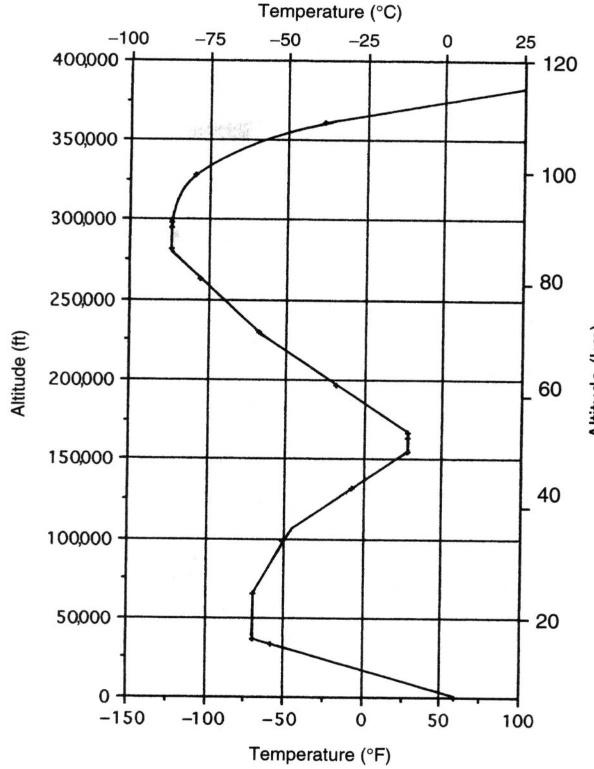

Figure 8.5.1 Temperature as a function of altitude. *(COESA-1976.)*

High-altitude atmospheric data have been obtained directly from balloons, sounding rockets, and satellites and indirectly from observations of meteors, aurora, radio waves, light absorption, and sound effects. At relatively low altitudes, the earth's atmosphere is, for aerodynamic purposes, a uniform gas. Above 250,000 ft, day and night standards differ because of dissociation of oxygen by solar radiation. This difference in density is as high as 35 percent, but it is usually negligible above 250,000 ft in terms of the aerodynamic forces, since these forces become less than 0.05 percent of their sea-level value for the same velocity. The temperature profile of the standard atmosphere is shown in Fig. 8.5.1. Other properties of the atmosphere can be calculated from temperature data.

Speed in aeronautics may be given in knots (now standard for U.S. Armed Forces and FAA), miles per hour, feet per second, or meters per second. The international nautical mile = 6,076.1155 ft (1,852 m). The basic relations are: 1 knot = 0.5144 m/s = 1.6877 ft/s = 1.1508 mi/h. Higher speeds are often given as **Mach number,** or the ratio of the particular speed to the speed of sound in the surrounding air. For additional data see "Supersonic and Hypersonic Aerodynamics" below.

SUBSONIC AERODYNAMIC FORCES

Axes The forces and moments acting on an airplane (and the resultant velocity components) are referred to a set of three mutually perpendicular axes having the origin at the airplane center of gravity (cg) and moving with the airplane. Three sets of axes can be used, depending on which direction is considered to be the longitudinal one:

Table 8.5.1b Reference Levels and Function Designations for Each of Four Zones of the Upper Atmosphere

Base altitude (Z), km	Base altitude, ft	Base temperature K	Base temperature °F	Form of function relating K to Z (four zones)	Base values ρ/ρ_0	Base values p/p_0
86	281,952	186.9	−123.3	Linear-constant	5.680×10^{-6}	3.685×10^{-6}
91	298,320	186.9	−123.3	Elliptical	2.335×10^{-6}	1.518×10^{-6}
110	361,152	240.0	−27.7	Linear-increasing	0.0793×10^{-6}	0.0701×10^{-6}
120	393,888	360.0	188.3	Exponential	0.0181×10^{-6}	0.0251×10^{-6}
1,000	3,280,992	1000.0	1340.3	—	2.907×10^{-15}	7.416×10^{-14}

Wind axes: The x axis lies in the direction of the relative wind. This is the system most commonly used, and the one used in this section. It is shown on Fig. 8.5.2.

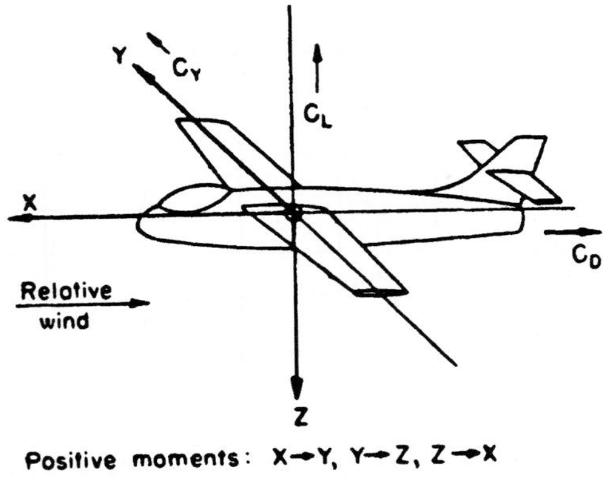

Figure 8.5.2 Wind axes.

Body axes: The x axis is fixed in the body, usually parallel to the thrust line.

Stability axes: The x axis points in the initial direction of flight. This results in a simplification of the equations of motion.

Absolute Coefficients Aerodynamic force and moment data are usually presented in the form of absolute coefficients. The main force coefficients are: lift $C_L = L/qS$; drag $C_D = D/qS$; side force $C_Y = Y/qS$; where q is the dynamic pressure $\frac{1}{2}\rho V^2$, ρ = air mass-density, and V = air speed. The main moment coefficients are $C_m = M/qSc$; roll $C_l = L/qSb$; and yaw $C_n = N/qSb$; where c = mean wing chord and b = wing span.

Section Coefficients Basic test data on wing sections are usually given in the form of section lift coefficient C_l and section drag coefficient C_d. These apply directly to an infinite aspect ratio or to two-dimensional flow, but aspect-ratio and lift-distribution corrections are necessary in applying to a finite wing. For a wing having an elliptical lift distribution, $C_L = (\pi/4)C_l$, where C_l is the centerline value of the local lift coefficient.

When an airfoil moves through the air, the motion produces a pressure at every point of the airfoil that acts normal to its surface. In addition, a frictional force tangential to the surface opposes the motion. The sum of these pressure and frictional forces gives the **resultant force** R acting on the body. The point at which the resultant force acts is defined as the **center of pressure,** c.p. The resultant force R is, in general, inclined to the airfoil and the relative wind velocity V, and can be resolved (Fig. 8.5.3) either along wind axes into

L = **lift** = component normal to V
D = **drag** or resistance = component along V, or along body axes into

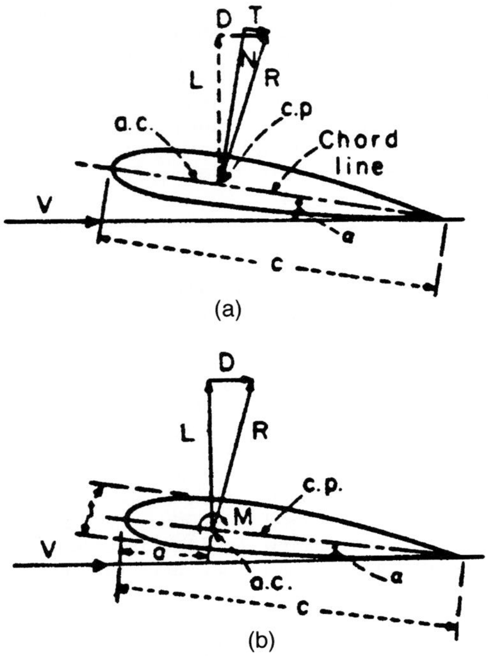

Figure 8.5.3 Forces acting on an airfoil. (*a*) Actual forces; (*b*) equivalent forces through the aerodynamic center plus moment.

N = **normal force** = component perpendicular to airfoil chord
T = **tangential force** = component along airfoil chord

Instead calculating the moment of the aerodynamic forces about the center of pressure, it is more convenient to calculate the moment of the forces about the *aerodynamic center,* a.c. This point lies at a distance a (about a quarter-chord length) behind the leading edge of the airfoil, and is defined as the point about which the moment of the air forces remains constant when the angle of attack α is changed. The force R acting at the c.p. is equivalent to the same force acting at the a.c. plus a moment equal to that force times the distance between the c.p. and the a.c. (see Fig. 8.5.3). The location of the aerodynamic center, in terms of the chord c and the section thickness t, is given for NACA 4- and 5-digit series airfoils approximately by

$$a/c = 0.25 - 0.40(t/c)^2$$

The distance $x_{c.p.}$ from the leading edge to the center of pressure expressed as a fraction of the chord is, where $M_{c/4}$ is the moment about the quarter chord point,

$$M_{c/4} = (\tfrac{1}{4} - x_{c.p.}) \cdot cN$$

$$x_{c.p.} = \frac{1}{4} - \frac{M_{c/4}}{cN}$$

Based on dimensional analysis the **air force on a body** of length dimension l moving with velocity V through air of density ρ can be expressed as

$$F = \varphi \cdot \rho V^2 l^2$$

where φ is a coefficient that depends upon all the dimensionless factors of the problem. In the case of a wing these are:

1. **Angle of attack** α, the inclination between the chord line and the velocity V
2. **Aspect ratio** $A = b/c$, where b is the span and c the mean chord of the wing
3. **Reynolds number** $\text{Re} = \rho V l/\mu$, where μ is the coefficient of dynamic viscosity of air
4. **Mach number** V/V_s, where V_s is the speed of sound
5. Relative surface roughness
6. Relative turbulence

The dependence of the force coefficient φ upon α and A can be theoretically determined; the variation of φ with the other parameters must be established experimentally, or through numerical simulation.

AIRFOILS AND WINGS

Bernoulli's equation can applied along any streamline in the flow around a body to relate the air pressure and speed between the freestream (i.e., the undisturbed air) and any point around the body as follows,

$$p + \tfrac{1}{2}\rho V^2 = p_1 + \tfrac{1}{2}\rho V_1^2$$

where p is the freestream atmospheric pressure, V is the speed of the body relative to the air, and p_1 and V_1 are the static pressure and speed of the flow at any point around the body (Fig. 8.5.4).

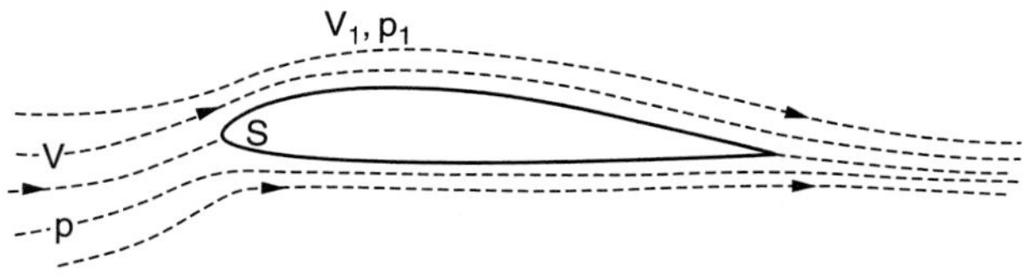

Figure 8.5.4 Fluid flow around an airfoil.

For a 2D body, the maximum pressure occurs at a point s on the body at which the velocity is zero, which is called the **stagnation point.** The maximum pressure increase occurs at this point and is

$$p_s - p = \tfrac{1}{2}\rho V^2$$

which is called the **dynamic pressure,** and is denoted by q. It is customary to express all aerodynamic forces in terms of $\tfrac{1}{2}\rho V^2$, hence

$$F = \tfrac{1}{2}\varphi\rho V^2 l^2 = q\varphi l^2$$

For the case of a wing, the forces and moments are expressed as

$$\text{Lift} = L = C_L \tfrac{1}{2}\rho V^2 S = C_L q S$$
$$\text{Drag} = D = C_D \tfrac{1}{2}\rho V^2 S = C_D q S$$
$$\text{Moment} = M = C_M \tfrac{1}{2}\rho V^2 S c = C_M q S c$$

where S = wing area and c = wing chord.

The lift can be determined from the intensity of **circulation** Γ by the relation $L' = \rho V \Gamma$, where L' is the lift per unit width of wing. In a **wing of infinite aspect ratio,** the flow is two-dimensional and the lift reaction is at right angles to the line of relative motion. The **lift coefficient** is a function of the angle of attack α and by mathematical analysis $C_L = 2\pi \sin \alpha$, which for small angles becomes $C_L = 2\pi\alpha$. Experiments show that $C_L = 2\pi\eta(\alpha - \alpha_0)$, where α_0 is the angle of attack corresponding to zero lift. $\eta \approx 1 - 0.64(t/c)$, where t is airfoil thickness.

In **a wing of finite aspect ratio,** the circulation flow around the wing creates a strong vortex trailing downstream from each wing tip (Fig. 8.5.5). The effect of this is that the direction of the resultant velocity at the wing is tilted downward by an **induced angle of attack** $\alpha_i = w_i/V$ (Fig. 8.5.6). If friction is neglected, the resultant force R is now tilted to the rear by this same angle α_i. The lift force L is approximately the same as R. In addition, there is also a drag component D_i, called the **induced drag,** given by

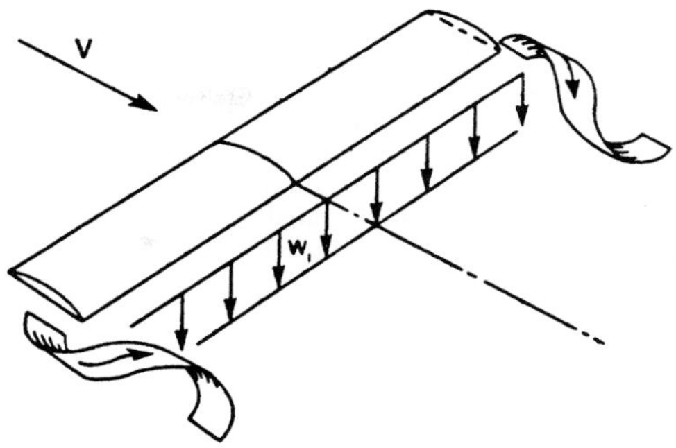

Figure 8.5.5 Vortex formation at wing tips.

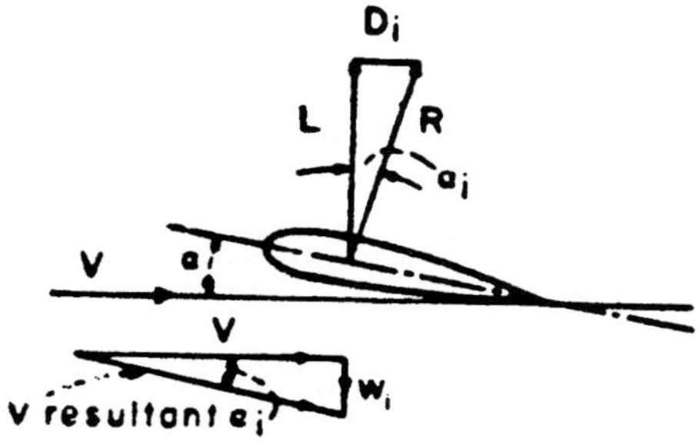

Figure 8.5.6 Induced angle of attack.

$$D_i = L \tan \alpha_i = L w_i / V$$

For a given geometrical angle of attack a the **effective angle of attack** is reduced by α_i, and thus

$$C_L = 2\pi\eta(\alpha - \alpha_0 - \alpha_i)$$

According to the Lanchester-Prandtl theory, for **wings having an elliptical lift distribution**

$$\alpha_i = w_i/V = C_L/\pi A \quad \text{rad}$$
$$D_i = L^2/\pi b^2 q \quad \text{and} \quad C_{Di} = D_i/qS = C_L^2/\pi A$$

where b = span of wing and $A = b^2/S$ = aspect ratio of wing. For **square tips,** correction factors are required:

$$\left.\begin{aligned} \alpha_i &= \frac{C_L}{\pi A}(1+\tau) \quad \text{where } \tau = 0.05 + 0.02A \\ C_{Di} &= \frac{C_L^2}{\pi A}(1+\sigma) \quad \text{where } \sigma = 0.01A - 0.01 \end{aligned}\right\} \quad \text{for } A < 12$$

These formulas are the basis for transforming the characteristics of rectangular wings from an **aspect ratio** A_1 to an aspect ratio A_2:

$$\alpha_2 = \alpha_1 + \frac{57.3 C_L}{\pi}\left(\frac{1+\tau_2}{A_2} - \frac{1+\tau_1}{A_1}\right) \quad \text{deg}$$
$$C_{D2} = C_{D1} + \frac{C_L^2}{\pi}\left(\frac{1+\sigma_2}{A_2} - \frac{1+\sigma_1}{A_1}\right)$$

For elliptical wings the values of τ and σ are zero. Most wing-section data are given in terms of aspect ratios 6 and ∞. For other values, the preceding formulas must be used.

Characteristics of Wings Wings have **airfoil sections** that can be constant from wing root to tip, or sections that vary across the wing to tailor the aerodynamic properties of the wing. Wing aerodynamic properties are expressed in terms of the dimensionless coefficients C_L, C_D, and C_M. **Airfoil** coefficients, termed section characteristics, are usually denoted as C_l, C_d, and C_m, and refer to wings of infinite aspect ratio. Figure 8.5.10 presents tunnel test results from early NACA studies on 6 percent and 12 percent thick two-dimensional sections.

The lift coefficient C_L is a linear function of the angle of attack up to a critical angle called the **stalling angle** (Fig. 8.5.7). The **maximum lift coefficient** $C_{L\max}$ that can be reached is one of the important characteristics of a wing because it determines the landing speed of the airplane.

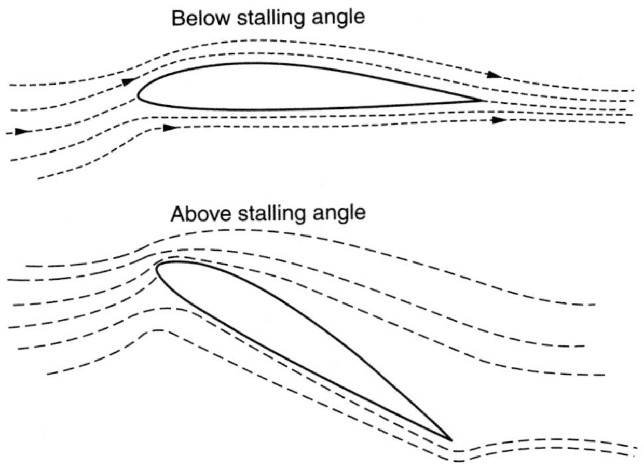

Figure 8.5.7 Stalling angle of an airfoil.

The **drag of a wing** is made up of three components: the **profile drag** D_0, the **induced drag** D_i, and the **wave drag** D_W, which becomes significant as the wing approaches the speed of sound. (Not true: in addition to skin friction, a significant proportion of profile drag is viscous pressure drag). For aspect ratio ∞ or at zero lift the induced drag is zero, and thus the entire drag is profile drag. In Fig. 8.5.8 the coefficient of induced drag $C_{Di} = C_L^2/\pi A$ is also plotted for a wing. The difference $C_D - C_{Di} = C_{D0}$, the profile-drag coefficient. A small value of the minimum profile-drag coefficient and a large value of C_L/C_D (L/D) are two desirable wing performance characteristics.

The moment characteristics of a wing are obtainable from the curve of the center of pressure as a function of α or by the **moment coefficient taken about the aerodynamic center** $C_{M\text{a.c.}}$ as a function of C_L. A forward motion of the c.p. as α is increased corresponds to an unstable wing section. This instability is undesirable because it requires a large downforce on the tail to counteract it.

The characteristics of a wing section (Fig. 8.5.9) are determined principally by its **mean camber line,** that is, the curvature of the median line of the profile, and secondly by the thickness distribution along the chord. In the NACA system of designation, when a four-digit number is used such as 2,412, the significance is always first digit = maximum camber in percent of chord, second digit = location of position of maximum camber in tenths of chord measured from the leading edge (i.e., 4 stands for 40 percent), and the last two figures indicate the maximum thickness in percent of chord. NACA also developed a 5-digit and 6-digit series of airfoils, the latter referred to as *laminar flow sections.* More recently, NASA work has centered on developing low-speed sections for general aviation aircraft. Today most major aerospace companies no longer use the early NACA section data and prefer to develop their own geometric and aerodynamic section properties, through either computational or experimental means. Extensive reshaping of airfoil leading edge geometries has yielded significant increases in maximum lift coefficients, and high-speed sections, termed "supercritical" sections have permitted cruise Mach numbers to increase significantly. These

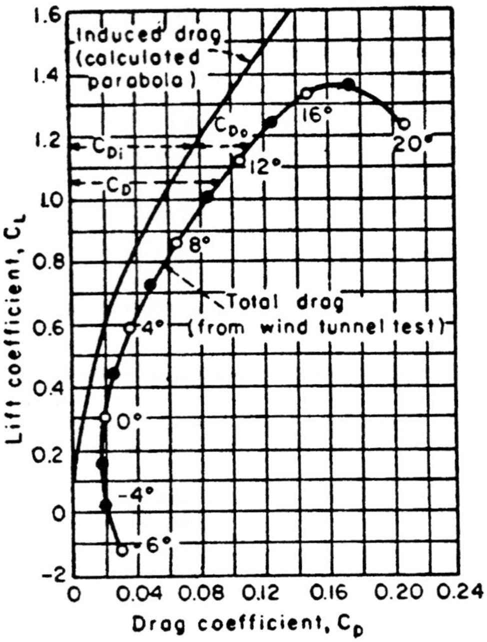

Figure 8.5.8 Polar-diagram plot of airfoil data.

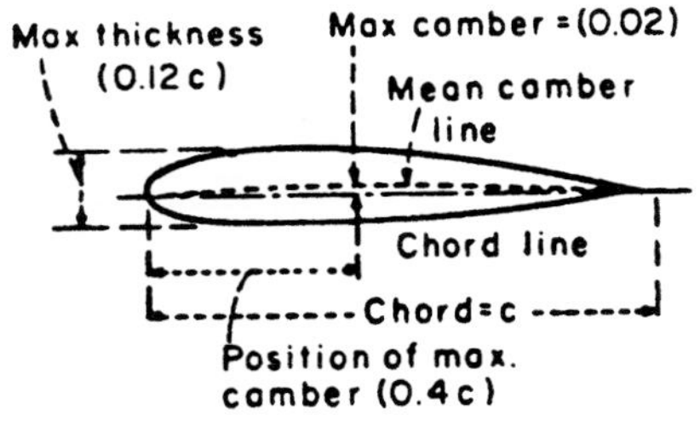

Figure 8.5.9 Characteristics of a wing section.

high-speed sections exhibit a flatter top surface, which reduces the maximum local flow speed for a given lift coefficient and decreases the strength of the inevitable shocks, reducing the wave drag.

Selection of Wing Section

In selecting a wing section for a particular airplane the following factors are generally considered:

1. Maximum lift coefficient $C_{L\max}$
2. Minimum drag coefficient $C_{D\min}$
3. Moment coefficient at zero lift C_{m0}
4. Maximum value of the ratio C_L/C_D

For certain special cases it is necessary to consider one or more factors from the following group:

5. Value of C_L for maximum C_L/C_D
6. Value of C_L for minimum profile drag
7. Maximum value of $C_L^{3/2}/C_D$
8. Maximum value of $C_L^{1/2}/C_D$
9. Type of lift curve peak (stall characteristics)
10. Drag divergence Mach number M_{DD}

Characteristics of airfoil sections are given in *NACA-TR* 586, 647, 669, 708, and 824. Figure 8.5.10 gives data on two typical sections: 0006 is often used for tail surfaces, and *NACA*$_I$ 65-212 is especially suitable for the wings of subsonic airplanes. (Not necessarily. Up to 15 percent t/c can be uses in modern transonic jet transports).

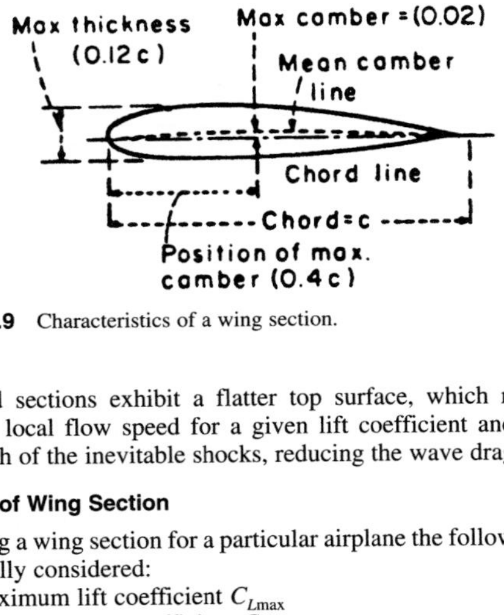

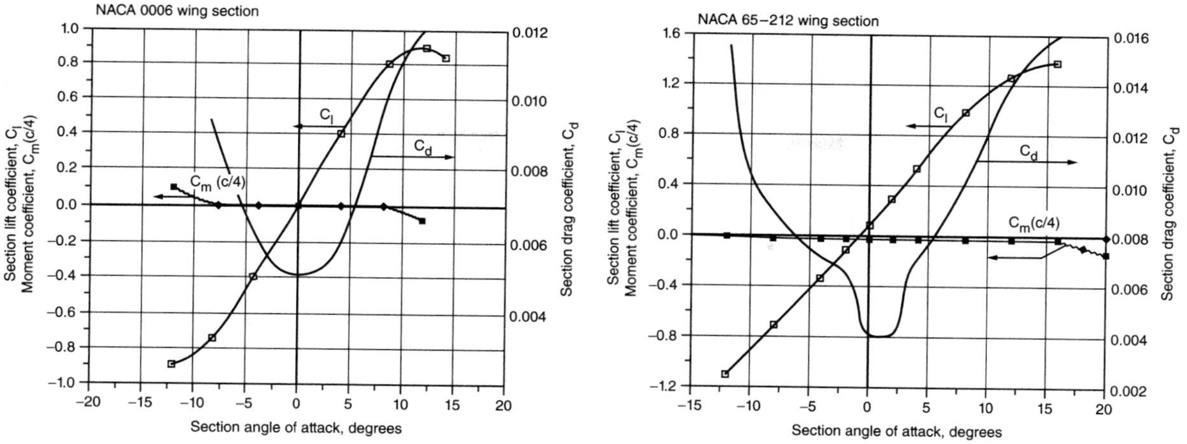

x, % c	0	1.25	2.5	5	7.5	10	15	20	25
y, % c(+/–)	0	0.947	1.307	1.777	2.1	2.341	2.673	2.869	2.971
x, % c	30	40	50	60	70	80	90	95	100
y, % c(+/–)	3.001	2.902	2.647	2.282	1.832	1.312	0.724	0.403	–0.063
L.E. Radius: 0.40c									

Figure 8.5.10 Properties of two airfoil sections. (*Courtesy of NACA.*)

The dimensionless coefficients C_L and C_D are functions of the **Reynolds number** $Re = \rho Vl/\mu$. For wings, the characteristic length l is taken to be the chord. In standard air at sea level

$$Re = 6{,}378 V_{ft/s} \cdot l_{ft} = 9{,}354_{mi/h} \cdot l_{ft}$$

For other heights, multiply by the coefficient $K = (\rho/\mu)/(\rho_0/\mu_0)$. Values of K are as follows:

Altitude, ft*	0	5,000	10,000	15,000	20,000	25,000	30,000	40,000	50,000	60,000
K	1.000	0.884	0.779	0.683	0.595	0.515	0.443	0.300	0.186	0.116

* × 0.305 = m.

The variation of C_L and C_D with the Reynolds number is known as **scale effect.** These are shown in Figs. 8.5.11 and 8.5.12 for some typical airfoils. Figure 8.5.12 shows in dashed lines also the theoretical variation of the drag coefficient for a smooth flat plate for laminar and turbulent flow, respectively, and also for the transition region.

In addition to the Reynolds number, airfoil characteristics also depend upon the **Mach number** (see "Supersonic and Hypersonic Aerodynamics" below).

Laminar-flow Wings NACA developed a series of **low-drag wings** in which the distribution of thickness along the chord is so selected as to maintain **laminar flow** over as much of the wing surface as possible.

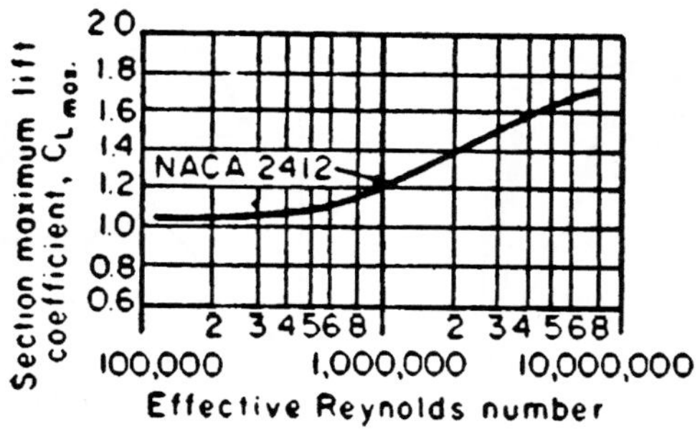

Figure 8.5.11 Scale effect on section maximum lift coefficient.

The low-drag wing under controlled conditions of surface smoothness may have drag coefficients about 30 percent lower than those obtained on normal conventional wings. Low-drag airfoils are so sensitive to roughness in any form that the full advantage of laminar flow is challenging in practice.

Transonic Airfoils At airplane Mach numbers of about M = 0.75, normal airfoil sections begin to show a greater drag, which increases

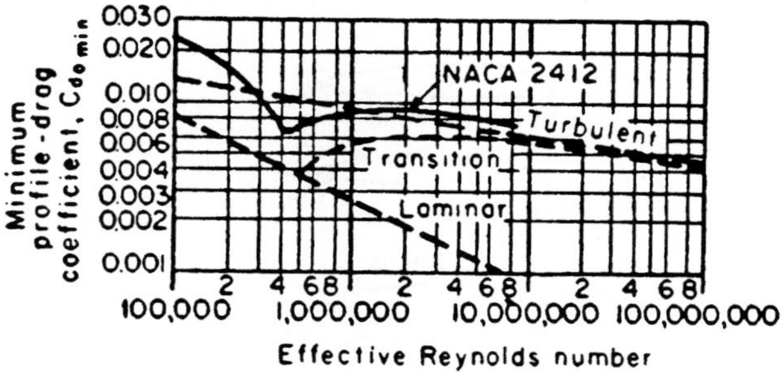

Figure 8.5.12 Scale effect on minimum profile-drag coefficient.

sharply as the speed of sound is approached. Airfoil sections known as **transonic airfoils** have been developed which significantly delay this **drag rise** to Mach numbers of 0.9 or greater (Fig. 8.5.13). Advantage may also be taken of these airfoil sections by increasing the thickness of the airfoil at a given Mach number without suffering an increase in drag.

Wing sweepback (Fig. 8.5.14) or sweepforward also delays the onset of the drag rise to higher Mach numbers. The flow component normal to the wing leading edge, M cos Γ determines the effective Mach number seen by the wing. The parallel component, M sin Γ, does not significantly influence the drag rise. Therefore for a flight Mach

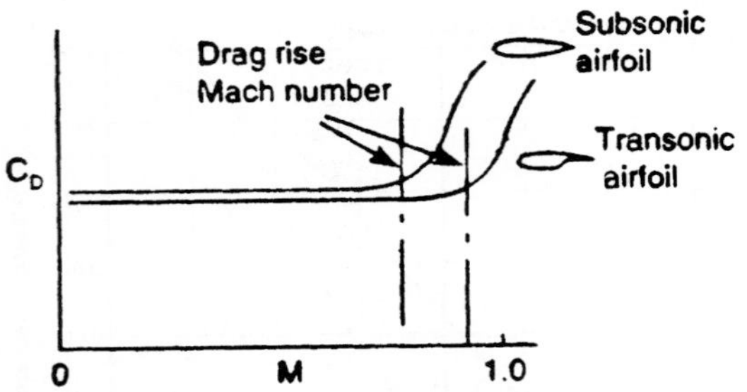

Figure 8.5.13 Drag rise delay due to transonic section.

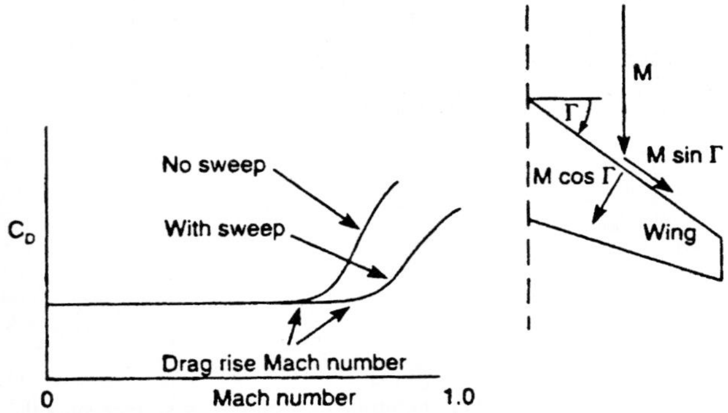

Figure 8.5.14 Influence of wing sweepback on drag rise Mach number.

number M, the wing airfoil is operating at a lower Mach number given by M cos Γ.

Flaps and Slats

The maximum lift of a wing can be increased by the use of slats on the leading edge, or **flaps** on the trailing edge. The trailing-edge flap is used to give increased lift at moderate angles of attack and to increase $C_{L\max}$. Theoretical analyses of the effects are given in *NACA-TR* 938 and *Br. ARCR&M* 1,095. Flight experience indicates that the four types shown in Fig. 8.5.15 have special advantages over other known types. Values

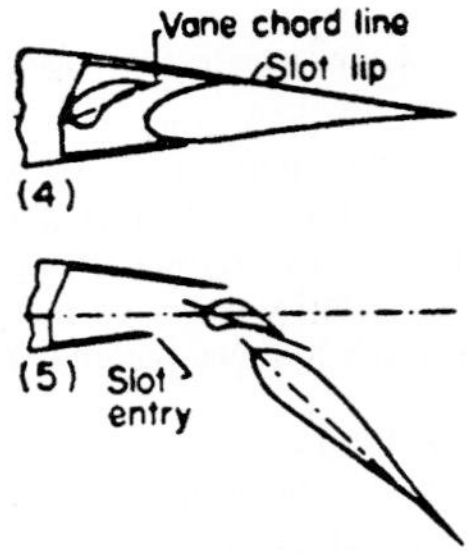

Figure 8.5.15 Trailing-edge flaps: (1) plain flap; (2) split flap; (3) single-slotted flap; (4) double-slotted flap (retracted); (5) double-slotted flap (extended).

Table 8.5.2 Flaps*

Type of flap	Section $C_{L\max}$ (approx)	Finite wing $C_{L\max}$ (approx)	Remarks
Plain	2.4	1.9	$C_{f/c}$ = 0.20. Sensitive to leakage between flap and wing.
Split	2.6	2.0	$C_{f/c}$ = 0.20. Simplest type of flap.
Single slotted	2.8 3.0	2.2 2.4	$C_{f/c}$ = 0.25. Shape of slot and location of deflected flap are critical.
Double slotted	3.0 3.4	2.4 2.7	$C_{f/c}$ = 0.25. Shape of slot and location of deflected flap are critical.

* The data is for wing thickness t/c = 0.12 or 0.15. Lift increment is dependent on the leading-edge radius of the wing.

of $C_{L\max}$ for these types are shown in Table 8.5.2. All flaps cause a diving moment ($-C_m$), which must be trimmed out by a downforce on the tail, reducing $C_{L\max}$. The correction is $\Delta C_{L\max}$ (trim) = $\Delta C_{mo}(l/c)$, where l/c is the tail length in mean chords.

Boundary-Layer Control (BLC) This includes numerous schemes for (1) maintaining laminar flow in the boundary layer in the flow over a wing or (2) preventing flow separation. Schemes in the first category try to obtain the lower frictional drag of laminar flow either by providing favorable pressure gradients, as in the NACA "low-drag" wings, or by removing part of the boundary layer. The boundary-layer thickness can be partially controlled by the use of suction applied either to spanwise slots or to porous areas. The flow so obtained approximates the laminar skin-friction drag coefficients (see Fig. 8.5.12) (drag reduction covered in the previous few sentences is out of context, which is high lift devices). Schemes in the second category try to delay or improve the stall. Examples are the leading-edge slot, slotted flaps, and various forms of suction or blowing applied through transverse slots. The slotted flap is a highly effective form of boundary-layer control that delays flow separation on the flap.

Spanwise Blowing Another technique for increasing the lift from an aerodynamic surface is spanwise blowing. A high-velocity jet of air is blown out along the wing. The location of the jet may be parallel to and near the leading edge of the wing to prevent or delay leading-edge flow separation, or it may be slightly behind the juncture of the trailing-edge flap to increase the lift of the flap. Power for spanwise blowing is sensitive to vehicle configuration and works most efficiently on swept wings. Other applications of this technique may include local blowing on ailerons or empennage surfaces (Fig. 8.5.16).

The **pressure distributions** on a typical airfoil are shown in Fig. 8.5.17. In this figure the ratio p/q is given as a function of distance along the chord. In Fig. 8.5.18 the effect of addition of an external airfoil flap is shown.

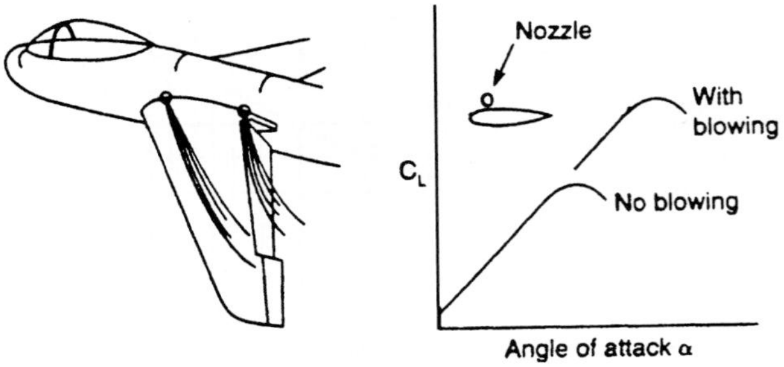

Figure 8.5.16 Influence of spanwise blowing.

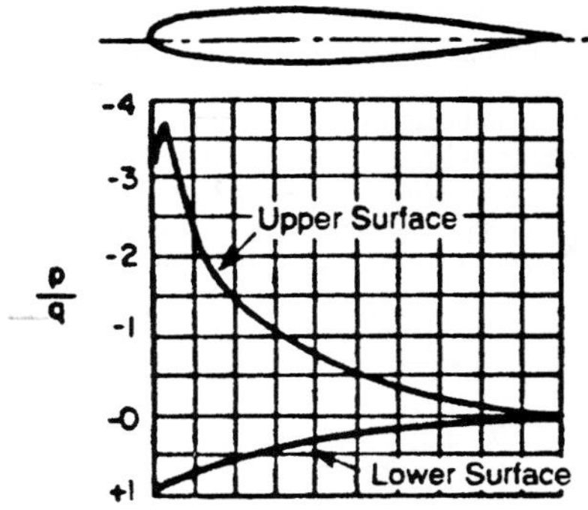

Figure 8.5.17 Pressure distribution over a typical airfoil.

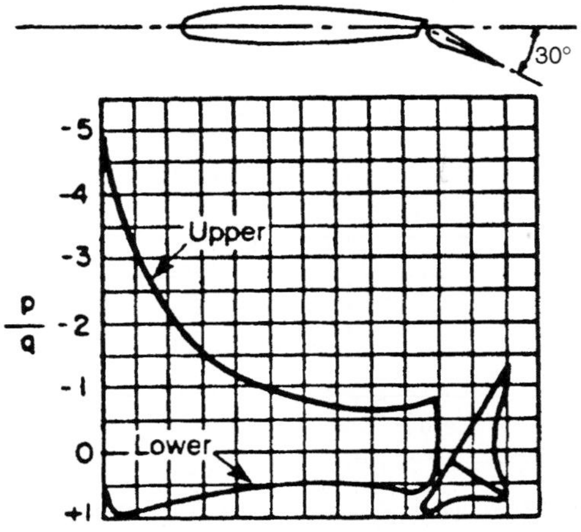

Figure 8.5.18 Pressure distribution on an airfoil-flap combination.

Airplane Performance

In **horizontal flight** the lift of the wings must be equal to the weight of the airplane, or $L = nW$, where W is the weight (lb or kg) of the airplane and n is 1.00. When the airplane is in a turn or is performing a pull-up, $n > 1.00$. The following equation determines the minimum speed (the stall speed) of the airplane at the appropriate value of n:

$$nW = C_{L\max}\frac{1}{2}\rho V_{\min}^2 S \quad \text{or} \quad V_{\min} = \sqrt{\frac{2nW}{C_{L\max}\rho S}} \tag{8.5.1}$$

This stall *speed* should be multiplied by a suitable safety factor, such as 1.2, to determine the **minimum safe flight speed.**

The interplay of airplane **drag** D and **thrust** T dictates the airplane performance. This interplay can be shown graphically by superimposing the airplane **drag polar** with the available thrust, as shown in Fig. 8.5.19. In level flight at constant speed, $T = D$, where T can be generated either by propeller, rocket, or jet propulsion. For propeller-driven airplanes, T becomes TV and D becomes DV in performance analyses. Our discussion will center on **jet propulsion** with a summary showing significant propeller-induced equation changes.

Figure 8.5.19 shows that at maximum speed B and minimum speed A, the required thrust equals the available thrust. The minimum thrust-dictated speed is usually lower than the stall speed; the latter then dictates the lower performance boundary speed. Between the maximum and minimum speeds $T > D$, and this excess in T available can be used to climb, or to accelerate in level flight. Similar curves can be constructed for a range of altitudes (a 30,000-ft set is shown), resulting in

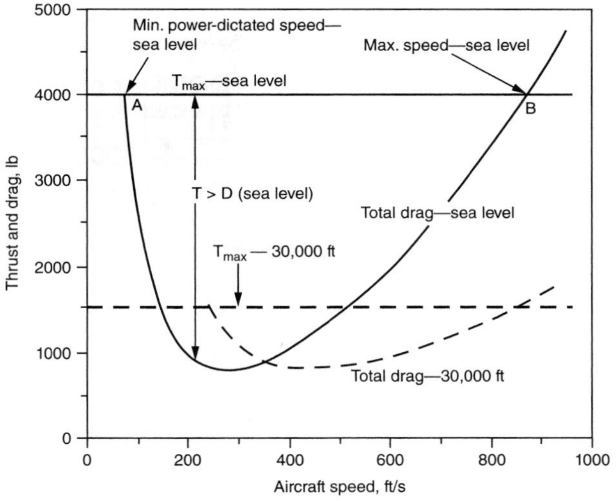

Figure 8.5.19 Thrust-drag relationships for airplane performance at sea level and 30,000 ft.

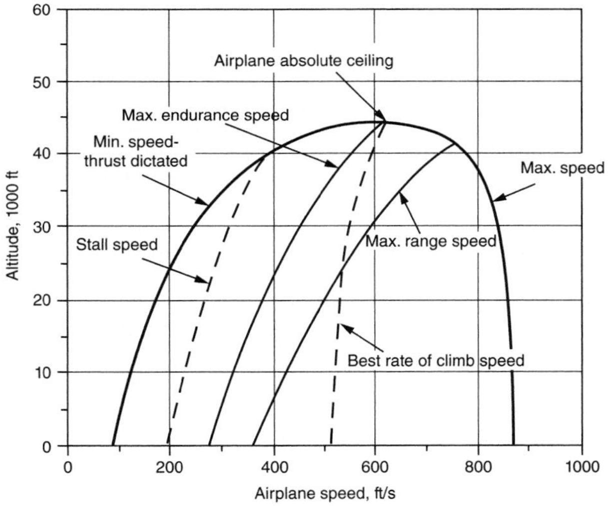

Figure 8.5.20 Summary of airplane speed variations with altitude.

the performance boundaries shown in Fig. 8.5.20. This figure shows minimum and maximum speeds, best climb speed, stall speed, and best cruise speed. Maximum and minimum speeds are equal at the absolute ceiling where the airplane can no longer climb.

The Rate of climb, or airplane **vertical velocity,** is given by

$$r = \frac{dh}{dt} = V\left(\frac{T-D}{W}\right) \quad \text{ft/s} \tag{8.5.2}$$

where V is in ft/s.

The **Climb angle** θ is given by

$$\theta = \sin^{-1}\left(\frac{T-D}{W}\right) \quad \text{deg} \tag{8.5.3}$$

Jet engine thrust reduces with increase in altitude as density decreases. At some altitude $(T - D)$ becomes zero (and therefore

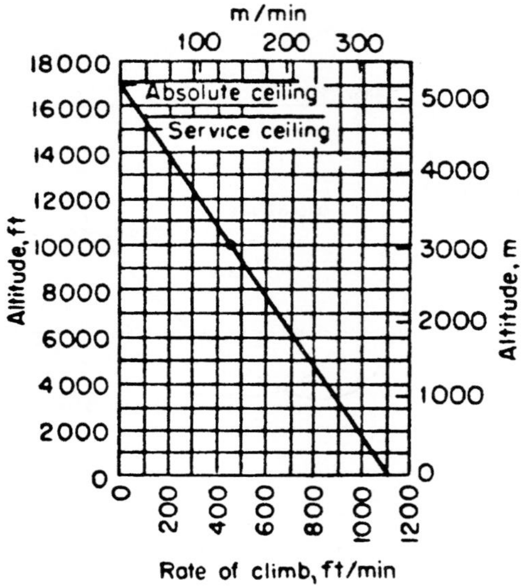

Figure 8.5.21 Characteristic rate of climb curves.

$dh/dt = 0$) and the **absolute ceiling** of the airplane is reached, beyond which the aircraft cannot fly in steady flight. The Climb angle becomes zero also and the airplane can fly at one speed only. A practical maximum altitude, termed **service ceiling,** is assumed to occur when $dh/dt = 100$ ft/min. The absolute ceiling can be determined from the curve of maximum rate of climb as a function of altitude when this is approximated by a straight line, usually accurate at the higher altitudes (see Fig. 8.5.21). For a linear decrease in the rate of climb the following approximations hold:

$$\text{Absolute ceiling } H = (r_0\, h)/(r_0 - r) \quad \text{ft} \tag{8.5.4}$$

where r is the rate of climb at altitude h and r_0 is the rate of climb at seal level.

$$\text{Service ceiling } h_s = H\,(r_0 - 100)/r_0 \quad \text{ft} \tag{8.5.5}$$

The Altitude climbed in *t* minutes is

$$h = H\,(1 - e^k) \quad \text{ft} \tag{8.5.6}$$

where $k = (-r_0\, t)/H$.
The **Time to climb to altitude** h is

$$t = 2.303\left(\frac{H}{r_0}\right)\log\left(\frac{H}{H-h}\right) \quad \text{min} \tag{8.5.7}$$

Power off **minimum glide angle** γ is given by

$$\gamma = \tan^{-1}\left(\frac{1}{(L/D)_{\max}}\right) \quad \text{deg} \tag{8.5.8}$$

The minimum glide angle corresponds to maximum straight line **glide distance measured along the ground,** s_g, given by

$$s_g = h_g(L/D)_{\max} \quad \text{ft} \tag{8.5.9}$$

where h_s is the start of glide altitude in feet.

For performance purposes, we require the relationship between lift and drag for the airplane, which is given by the **drag polar.** This polar varies with Mach and Reynolds number, and also changes when flaps and landing gears are extended. The drag polar can be approximated by the quadratic,

$$C_D qS = (C_{D0} + C_L^2/Ae\pi)qS \tag{8.5.10}$$

where C_{D0} is the profile drag coefficient, which includes the drag of the following components: clean aircraft, and landing gear. Here, e is Oswald's span efficiency factor and S is the reference area.

A typical clean aircraft drag variation with V, complete with the corresponding L/D is shown in Fig. 8.5.22 which shows that a maximum value of L/D occurs at a given velocity called the best **endurance speed.** The **best range speed** occurs at 0.866 $(L/D)_{\max}$ as shown. Table 8.5.3 shows typical values of C_{D0} for various types of jet aircraft.

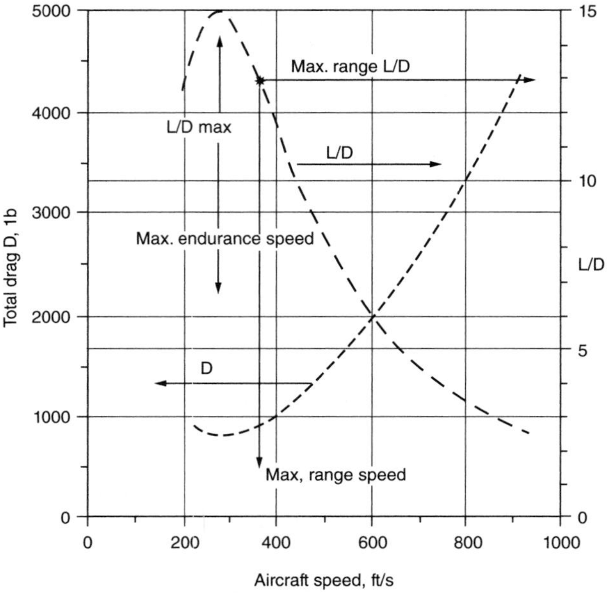

Figure 8.5.22 Aircraft aerodynamic properties: drag polar and *L/D* ratio.

Table 8.5.3 C_{D0} Values for Typical Aircraft Types (Clean Aircraft)

Aircraft type	Subsonic C_{D0}
Supersonic twin-jet fighter	0.0237
Narrow-body twin-jet transport	0.0166
Wide-body four-jet transport	0.0170

The ratio L/D, (C_L/C_D), is sometimes called the **aerodynamic efficiency** of the airplane. Mission requirements dictate the magnitude for competitive values of this ratio which is primarily controlled by the value of the **wing aspect ratio** and C_{D0}. Airliners usually have high aspect ratio values and fighter aircraft low values (see Fig. 8.5.23). As seen in the following equations, maximizing L/D will maximize range and endurance. The **maximum range** is the maximum **distance** flown on a given quantity of fuel. **Maximum endurance** is the maximum **time** spent in the air for a given quantity of fuel.

The maximum range at constant altitude is

$$R = 2\sqrt{\frac{2}{\rho S}}\,\frac{1}{c}\,\frac{C_L^{1/2}}{C_D}(W_0^{1/2} - W_1^{1/2}) \quad \text{ft} \tag{8.5.11}$$

where ρ = air density in slugs/ft^3, c = specific fuel consumption in lb of fuel/lb thrust · s, V = aircraft velocity in ft/s. Subscript 0 refers to the segment weight at start of mission and subscript 1 to segment weight at end of mission.

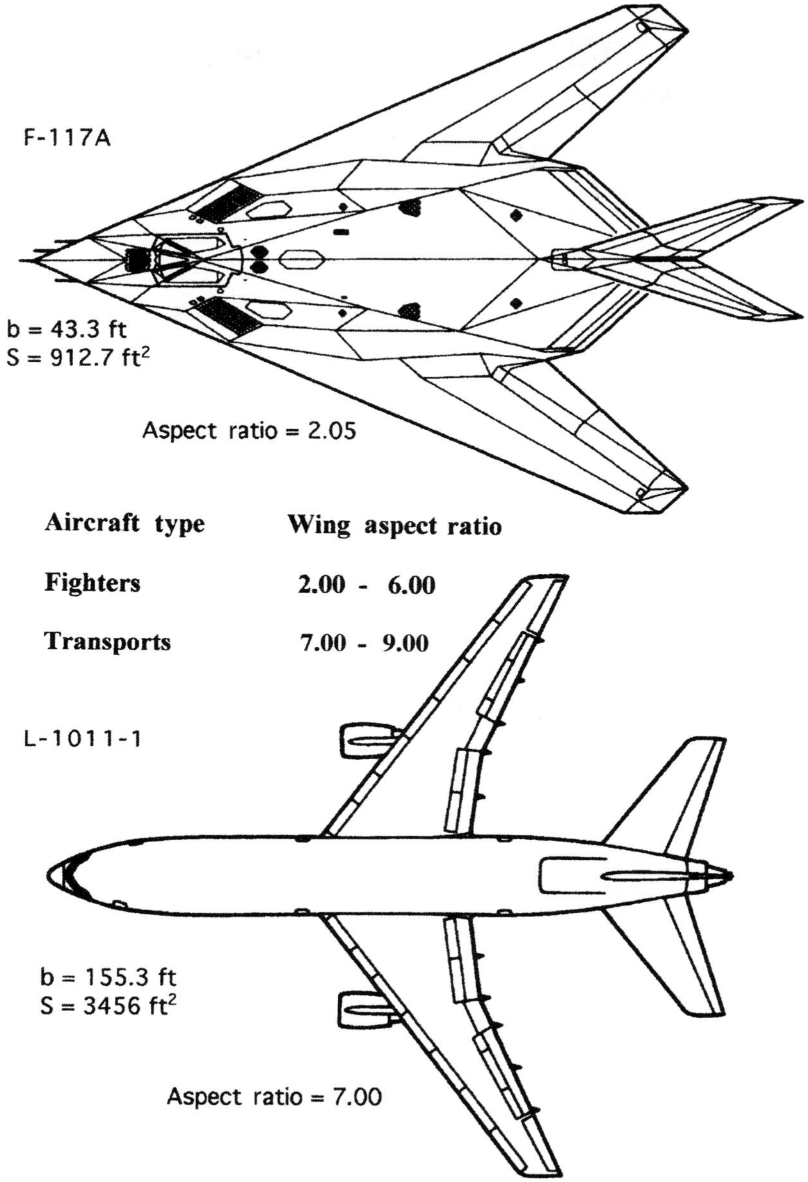

Figure 8.5.23 Typical aircraft configurations. (*Drawings courtesy of Lockeed-Martin.*)

Endurance E, in hours, is given by

$$E = \frac{1}{c}\left(\frac{C_L}{C_D}\right)\ln\left(\frac{W_0}{W_1}\right) \quad \text{h} \qquad (8.5.12)$$

In each instance, c and (L/D) correspond to the appropriate altitude and speed, and $(W_0 - W_1)$ represents fuel used.

For **propeller-powered airplanes,** the above equations, starting at Eq. (8.5.2) become

$$\frac{dh}{dt} = \left(\frac{TV - DV}{W}\right) \quad \text{ft/s} \qquad (8.5.13)$$

where V is in ft/s. If thrust horsepower available $P_{Ta} = \eta \cdot P_a$, where P_a = **available engine horsepower,** η is the propeller efficiency, and P_r = horsepower required, then

$$\frac{dh}{dt} = 550\left(\frac{P_{Ta} - P_r}{W}\right) \quad \text{ft/s} \qquad (8.5.14)$$

$$\text{Range } R = \frac{\eta}{c}\left(\frac{C_L}{C_D}\right)\ln\left(\frac{W_0}{W_1}\right) \quad \text{ft} \qquad (8.5.15)$$

where the units of c are (lb fuel)/ft · lb/s) · (s).

$$\text{Endurance } E = \left(\frac{\eta}{c}\right)\frac{C_L^{3/2}}{C_D}(2\rho S)^{1/2}(W_1^{-1/2} - W_0^{-1/2})\left(\frac{1}{3600}\right) \quad \text{h} \qquad (8.5.16)$$

See Table 8.5.4 for dimensions and performance of selected airplanes.

Table 8.5.4 Principal Dimensions and Performance of Typical Air Vehicles

	General Dynamics	Lockheed Martin	Raytheon Aircraft	Bell Helicopter Textron	Boeing	Boeing	Airbus	Airbus	Northrop-Grumman	Agusta-Westland
Designation model no.	Gulfstream G 500	F-35C	Beech King Air C-90B	427	747-400	737-800	A380-800	A310-200	B-2A	Super Lynx 300
Type	Executive transport	Fighter	General aviation	Utility helicopter	Wide-body turbofan	Narrow-body turbofan	Wide-body turbofan	Wide-body turbofan	Bomber	Naval helicopter
No passengers	14-19	Pilot only	12	6	416	162	555	220-280	2 crew	9
Cargo capacity, lb	N.A.	15,000	N.A.	N.A.	6,025 ft^3	1,555 ft^3	N.A.	N.A.	40,000 payload	N.A.
Span, ft	93.5	35	50.3	37*	211.4	112.6	261.8	144	172	42*
Overall length, ft	96.4	51.4	35.5	42.9	231.8	129.5	239.5	153.1	69	50
Overall height, ft	25.8	15	14.3	11.4	63.7	41.2	79.1	51.1	17	12
Wing area, ft^2	1136.5	620	294	N.A.	5,650	1,341	9,095	2,360	5,140	N.A.
Weight empty, lb	47,800	30,000	6,810	3,485	398,800	91,660	611,000	174,700	N.A.	N.A.
Weight gross, lb	85,500	60,000	10,100	6,000	875,000	174,200	1,234,600	313,100	336,500	11,750
Power plant	2 × RR Tay–BR710	1 × PWF 135–PW-400	2 × PWC–PT6A-21	2 × PWC PW–206D	4 × PW4056	2 × CFM56-7B	4 × RR Trent 970	2 × GE CF6	4 × GE-F118	2 × LHTEC CTS800-4N
High speed, mi/h (Mach)	(0.89)	(1.6+)	283	160	(0.92)	(0.82)	(0.89)	(0.84)	N.A.	173
Cruise speed, mi/h (Mach)	(0.85)	N.A.	283	N.A.	(0.85)	(0.788)	(0.85)	N.A.	N.A.	N.A.
Range, mi	7,200	N.A.	1,640	411	8,356	3,522	9,200	N.A.	N.A.	550

NOTES: *rotor diameter.
N.A. = not available.
SOURCE: *Aviation Week, 2003 Source Book,* Copyright 2003 by The McGraw-Hill Companies.

Power Available The maximum efficiency η_m and the diameter of a propeller to absorb a given power at a given speed and rpm are found from a propeller-performance curve. The thrust horsepower at maximum speed P_{Tm} is found from $P_{Tm} = \eta_m P_m$. The **thrust horsepower** at **any speed** can be approximately determined from the ratios given in the following table:

V/V_{max}		0.20	0.30	0.40	0.50	0.60
P_T/P_{TM}	Fixed-pitch propeller	0.29	0.44	0.57	0.68	0.77
	constant-rpm propeller	0.47	0.62	0.74	0.82	0.88
V/V_{max}		0.70	0.80	0.90	1.00	1.10
P_T/P_{Tm}	Fixed-pitch propeller	0.84	0.90	0.96	1.00	1.03
	constant-rpm propeller	0.93	0.97	0.99	1.00	1.00

The **brake horsepower** of an engine decreases with increase of altitude. The variation of P_T with altitude depends on engine and propeller characteristics with average values as follows:

h, ft	0	5,000	10,000	15,000	20,000	25,000
P_T/P_{T0} (fixed pitch)	1.00	0.82	0.66	0.52	0.41	0.30
P_T/P_{T0} (controllable pitch)	1.00	0.85	0.71	0.59	0.48	0.38

Parasite Drag

The drag of the nonlifting parts of an airplane is called the **parasite drag.** It consists of two components: the frictional and the eddy-making drag.

Frictional drag, or **skin friction,** is due to the viscosity of the fluid. It is the force produced by the viscous shear in the layers of fluid immediately adjacent to the body. It is always proportional to the wetted area, that is, the total surface exposed to the air.

Eddy-making drag, sometimes called **form drag,** is due to the disturbance or wake created by the body. It is a function of the shape of the body.

The total drag of a body may be composed of the two components in any proportion, varying from almost pure skin friction for a plate edgewise or a good streamline form to 100 percent form drag for a flat plate normal to the wind. A **streamlined form** is a shape having very low form drag. Such a form creates little disturbance in moving through a fluid.

When air flows past a surface, the layer immediately adjacent to the surface adheres to it, or the tangential velocity at the surface is zero. In the transition region near the surface, which is called the **boundary layer,** the velocity increases from zero to the velocity of the stream. When the flow in the boundary layer proceeds as if it were made up of laminae sliding smoothly over each other, it is called a **laminar boundary layer.** If there are also irregular motions in the layers normal to the surface, it is a **turbulent boundary layer.** Under normal conditions the flow is laminar at low Reynolds numbers and turbulent at high Re with a transition range of values of Re extending between 5×10^5 and 5×10^7. The profile-drag coefficients corresponding to these conditions are shown in Fig. 8.5.12.

For laminar flow the friction-drag coefficient is practically independent of surface roughness and for a flat plate is given by the Blasius equation,

$$C_{DF} = 2.656/\sqrt{\text{Re}}$$

The turbulent boundary layer is thicker and produces a greater frictional drag. For a smooth flat plate with a turbulent boundary layer

$$C_{DF} = 0.91/\log \text{Re}^{2.58}$$

The drag coefficient increases with surface roughness. The drag coefficients as given above are based on *projected area* of a double-surfaced plane. If *wetted* area is used, the coefficients must be divided by 2. The frictional drag is $D_F = C_{DF}\, qS$, where S is the *projected area,* or $D_F = \frac{1}{2} C_{DF}\, qA$, where A is the wetted area.

Values of C_{DF} for double-surfaced planes may be estimated from the following tabulation:

Laminar flow (Blasius equation):

Re	10	10^2	10^3	10^4	10^5	10^6
C_{DF}	0.838	0.265	0.0838	0.0265	0.0084	0.0265

Turbulent flow:

Re	10^5	10^6	10^7	10^8	10^9	10^{10}
C_{DF}	0.0148	0.0089	0.0060	0.0043	0.00315	0.0024

These are double-surface values, which facilitate direct comparison with wing-drag coefficients based on projected area. For calculations involving wetted area, use one-half of double-surface coefficients.

Drag Coefficients of Various Bodies

For **bodies with sharp edges,** the drag coefficients are almost independent of the Reynolds number, because most of the resistance is due to p the difference pressure on the front and rear surfaces. Table 8.5.5 gives $C_D = D/qS$, where S is the maximum cross section perpendicular to the wind.

For **rounded bodies** such as **spheres, cylinders,** and **ellipsoids** the drag coefficient depends markedly upon the Reynolds number, the surface roughness, and the degree of turbulence in the airstream. A sphere and a cylinder, for instance, experience a sudden reduction in C_D as the Reynolds number exceeds a certain critical value. The reason is that at low speeds (small Re) the flow in the boundary layer adjacent to the body is laminar and the flow separates at about 83° from the front (Fig. 8.5.24). A wide wake thus gives a large drag. At higher speeds (large Re) the boundary layer becomes turbulent, gets additional energy from the outside flow, and does not separate on the front side of the sphere. The drag coefficient is reduced from about 0.47 to about 0.08 at a critical Reynolds number of about 400,000 in free air. Turbulence in the airstream reduces the value of the critical Reynolds number (Fig. 8.5.25). The Reynolds number at which the sphere drag $C_D = 0.3$ is taken as a criterion of the amount of turbulence in the airstream of wind tunnels.

Cylinders The drag coefficient of a cylinder with its axis normal to the wind is given as a function of Reynolds number in Fig. 8.5.26. Cylinder drag is sensitive to both Reynolds and Mach number. Figure 8.5.26 gives the Reynolds number effect for M = 0.35. The increase in C_D because of Mach number is approximately:

M	0.35	0.4	0.6	0.8	1.0
C_D increase, percent	0	2	20	50	70

(See *NACA-TN* 2960.)

Streamlined Forms The drag of a streamline body of revolution depends to a large extent on the Reynolds number. The difference between extreme types at a given Reynolds number is of the same order as the change in C_D for a given form for values of Re from 10^6 to 10^7.

Tests reported in *NACA-TN* 614 indicate that the shape for minimum drag should have a fairly sharp leading and trailing edges. At Re = 6.6×10^6 the best forms for a fineness ratio (ratio of length to diameter) 5 have a drag of

$$D = 0.040qA = 0.0175qV^{2/3}$$

where A = max cross-sectional area and V = volume. This value is equivalent to about 1.0 lb/ft² at 100 mi/h. For Re > 5×10^6, the drag coefficients vary approximately as $\text{Re}^{0.15}$.

Minimum drag on the basis of cross-sectional or frontal area is obtained with a fineness ratio of the order of 2 to 3. Minimum drag on the basis of contained volume is obtained with a fineness ratio of the order of 4 to 6 (see *NACA-TR* 291).

Table 8.5.5 Drag Coefficients

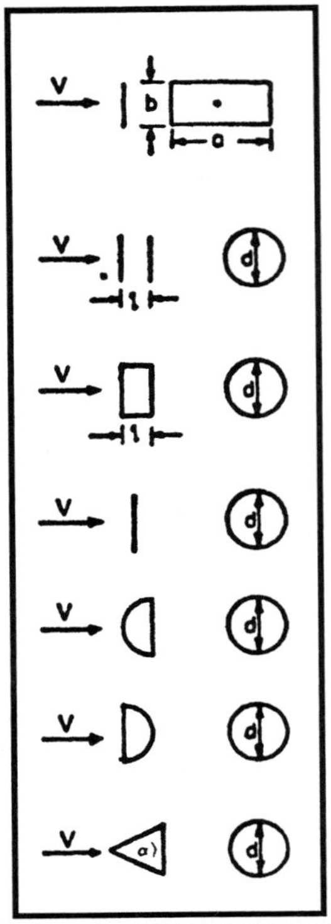

Object	Proportions	Attitude	C_D
Rectangular plate, sides a and b	$\frac{a}{b}$ = 1		1.16
	4		1.17
	8		1.23
	12.5		1.34
	25		1.57
	50		1.76
	∞		2.00
Two disks, spaced a distance l apart	$\frac{l}{d}$ = 1		0.93
	1.5		0.78
	2		1.04
	3		1.52
Cylinder	$\frac{l}{d}$ = 1		0.91
	2		0.85
	4		0.87
	7		0.99
Circular disk			1.11
Hemispherical cup, open back			0.41
Hemispherical cup, open front, parachute			1.35
Cone, closed base			$\alpha = 60°$, 0.51
			$\alpha = 30°$, 0.34

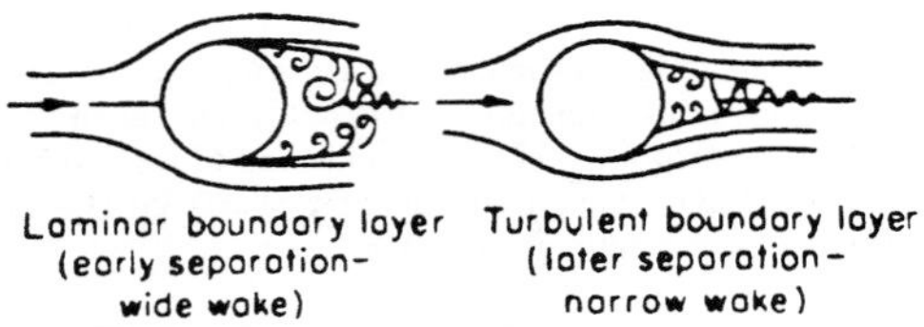

Figure 8.5.24 Boundary layer of a sphere.

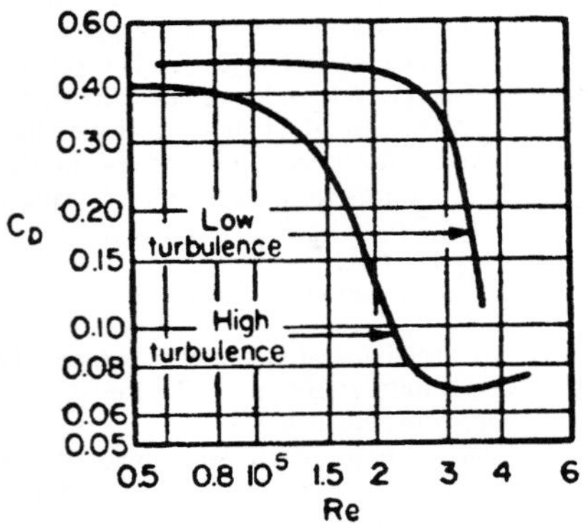

Figure 8.5.25 Drag coefficients of a sphere as a function of Reynolds number and of turbulence.

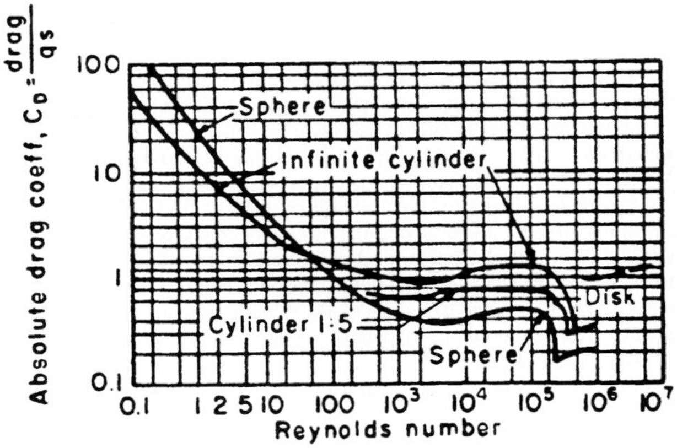

Figure 8.5.26 Drag coefficients of cylinders and spheres.

The following table gives the ordinates for good streamline shapes: the **Navy strut,** a two-dimensional shape; and the Class C **airships,** a three-dimensional shape. Streamline shapes for high Mach numbers have a high fineness ratio.

Percent length		1.25	2.50	4.00	7.50	10.0	12.50
Percent of max ordinate	Navy strut	26	37.1	52.50	63.00	72.0	78.50
	Class C airship	20	33.5	52.60	65.80	75.8	83.50

		20	40	60	80	90
Percent length		20	40	60	80	90
Percent of max ordinate	Navy strut	91.1	99.5	86.1	56.2	33.8
	Class C airship	94.7	99.0	88.5	60.5	49.3

Test data on the *RM*-10 shape are given in *NACA-TR* 1,160. This shape is a parabolic-arc type for which the coordinates are given by the equation $r_x = x/7.5(1.0 - x/L)$.

Nonintrusive flow field measurements about airfoils can be made by laser Doppler velocimeter (LDV) techniques. A typical LDV measurement system setup is shown in Fig. 8.5.27*a*. A seeded airflow generates the necessary reflective particles to permit two-dimensional flow directions and velocities to be accurately recorded.

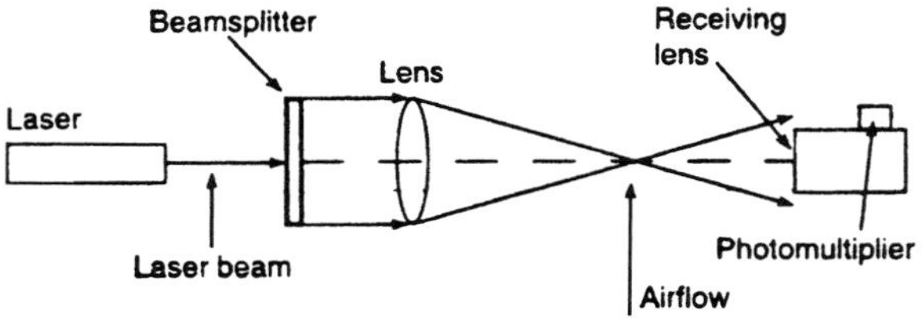

Figure 8.5.27*a* Typical laser Doppler velocimeter measurement system.

CONTROL, STABILITY, AND FLYING QUALITIES

Control An airplane is controlled in flight by imposing yawing, pitching, and rolling moments by use of rudders, elevators, and ailerons. Special airplane configurations can use mixtures or combinations of these control methods. It is possible to control aircraft through the use of two controls by suitable blending of aileron and rudder inputs.

Stability An airplane is **statically stable** if the airframe moments, due to a disturbance, tend to return the airplane to its original attitude. It is **dynamically** stable if the oscillations produced by the static stability are rapidly damped out. A **statically unstable** aircraft produces upsetting moments instead of restoring moments. A **dynamically unstable** aircraft oscillations increasing in amplitude with time. A statically unstable or a dynamically unstable aircraft can be made stable by the use of suitable automatic flight control systems.

Longitudinal static and dynamic stability, and longitudinal trim, are commonly achieved through the use of a suitably sized horizontal tail with a large enough moment arm relative to the airplane's center of gravity. Under these conditions the pitching moment equation, representing forces and moments on the airframe, is conditioned as follows: $C_m = 0$ for trim, and $(\partial C_m/\partial \alpha) < 0$ for stability, where C_m is the overall pitching moment coefficient about the aircraft center of gravity and α is the aircraft angle of attack. When a definitive value is chosen for $(\partial C_m/\partial \alpha)$ in the design process (such as –0.03), this is known as the design **static margin.** In a similar manner, the vertical tail size is determined from lateral/directional stability and lateral trim requirements. The trim requirement is usually overriding in fixing vertical tail size for multiengine aircraft when critical engine-out design requirements are evaluated.

It can be shown that the forces and moments, inertial and aerodynamic, acting on an aircraft can be represented by six simultaneous, linear, small perturbation differential equations. These are called *the equations of motion of an aircraft,* and they represent the three forces and three moments assumed to act on the aircraft center of gravity (Fig. 8.5.27*b*). For conventional aircraft configurations, it is common practice to separate these into two sets of three each: one set representing longitudinal motions and one set the lateral-directional motions of the aircraft.

In this separation, the assumption is that these motions do not normally influence each other, and therefore they can be solved separately to establish the separate dynamic behaviors. The longitudinal set is characterized by two distinct oscillations called *short period* and *long period.* Short period motions usually influence angle of attack and pitch angle θ, whereas the long period, called "phugoid," primarily influences the height excursions and small speed changes. The

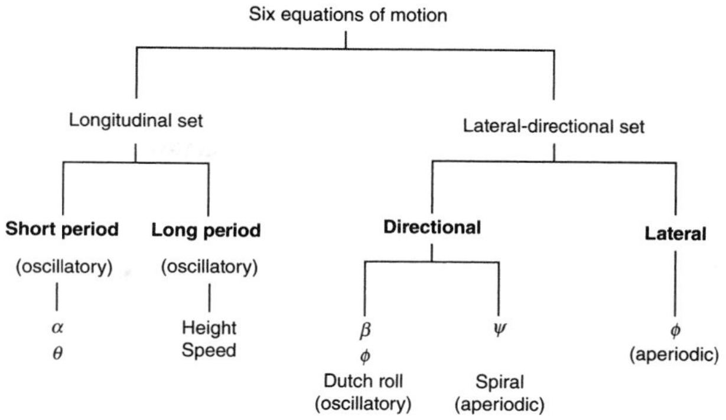

Figure 8.5.27*b* Six equations of motion.

lateral-directional set characterizes the motions in spiral ψ the so-called Dutch roll mode, a mixture of bank angle ϕ and sideslip β motions, and in the bank angle response. All response behavior is characterized by the oscillatory parameters, relative damping ratio ξ and undamped natural frequency ω_n as well as time to one-half amplitude ($t_{1/2}$), or time to double amplitude (t_2). It is the magnitudes of these parameters that establish the **flying** or **handling qualities** of an aircraft.

Flying Qualities Requirements These apply to both military aircraft and civil aircraft through the Federal Aviation Requirements (FARs) and Military Standards (Mil. STD), respectively. These requirements imply certain levels for the response parameters to ensure that the aircraft will achieve satisfactory pilot acceptability in the operation of the aircraft. The response parameter requirements vary with mission segments. Typical longitudinal response values for two aircraft, in cruise, are:

Large Jet Transport, *W* = 630,000 lb, *V* = 460 knots, altitude = 40,000 ft

	$t_{1/2}$, s	ω_n, rad/s	ξ
Short period	1.86	1.00	0.39
Phugoid	211.00	0.07	0.05

Small Civil Propeller Aircraft, *W* = 2,800 lb, *V* = 104 knots, altitude = sea level

	$t_{1/2}$, s	ω_n, rad/s	ξ
Short period	0.28	3.6	0.69
Phugoid	40.3	0.22	0.08

Rolling (or **banking**) does not produce any lateral shift in the center of the lift, so that there is no restoring moment as in pitch. However, when banked, the airplane sideslips toward the low wing. A fin placed above the center of gravity generates a lateral restoring moment that can correct the roll and stop the slip. The same effect can be obtained by **dihedral,** that is, by raising the wing tips to give a transverse *V*. An effective dihedral of 1 to 3° on each side is generally required to obtain stability in roll. In low-wing monoplanes, 2° of effective dihedral may require 8° or more of geometrical dihedral owing to interference between the wing and fuselage.

In a **yaw** or **slip,** the line of action of the lateral force depends on the size of the effective vertical fin area. Insufficient fin surface aft allows the skid or slip to increase. Too much fin surface aft swings the nose of the plane around into a tight spiral. Sound design demands sufficient vertical tail surface for adequate directional control and then enough dihedral to provide lateral stability.

When moderate positive effective dihedral is present, the airplane possesses **static lateral stability,** and a low wing comes up automatically with very little yaw. If the dihedral is too great, the airplane may roll considerably in gusts, but there is little danger of the amplitude ever becoming excessive. **Dynamic lateral stability** is not assured by static stability in roll and yaw but requires that these be properly proportioned to the damping in roll and yaw.

Spiral instability is the result of too much fin surface and insufficient dihedral.

HELICOPTERS

REFERENCES: *Br. ARC-R & M* 1,111, 1,127, 1,132, 1,157, 1,730, and 1,859. *NACA-TM* 827, 836, 858. *NACA-TN* 626, 835, 1,192, 3,323, 3,236. *NACA-TR* 434, 515, 905, 1,078. *NACA Wartime Reports* L-97, L-101, L-110. NACA, "Conference on Helicopters," May, 1954. Gessow and Myers, "Aerodynamics of the Helicopter," Macmillan.

Helicopters derive lift, propulsion force, and control effect from adjustments in the blade angles and rotational speed of the rotor system. At least two rotors are required, and these may be arranged in any form that permits control over the reaction torque. The common arrangements are main lift rotor; auxiliary torque-control or tail rotor at 90° relative to the main rotor; two main rotors side by side; two main rotors fore and aft; and two main rotors, coaxial and oppositely rotating.

The helicopter rotor is a momentum device that follows the same general laws as a propeller. In calculating the rotor performance, the major variables\are the diameter D, ft (or radius R); the tip speed, $Vt = \Omega R$, ft/s; the angular velocity of rotor $\Omega = 2\pi n$, rad/s; and the rotor solidity q (= ratio of blade area/disk area). The rotor performance is stated in terms of coefficients similar to propeller coefficients. That is,

$$\text{Thrust coefficient } C_T = T/\rho(\Omega R)^2 \pi R^2$$

$$\text{Torque coefficient } C_Q = Q/\rho(\Omega R)^2 \pi R^3$$

$$\text{Torque Q} = 5{,}250 \text{ bhp/rpm}$$

Hovering The **hovering flight** condition may be calculated from basic rotor data given in Fig. 8.5.28. These data are taken from

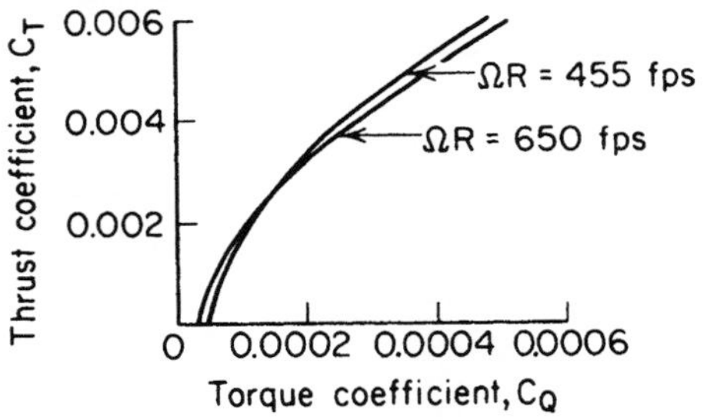

Figure 8.5.28 Static thrust performance of NACA 8-H-12 blades.

full-scale rotor tests. Surface-contour accuracy can reduce the total torque coefficient by 6 to 7 percent. The power required for hovering flight is greatly reduced near the ground. Observed flight-test data from various sources are plotted in Fig. 8.5.29. The ordinates are heights above the ground measured in rotor diameters.

Effect of Gross Weight on Rate of Climb Figure 8.5.30 from Talkin (*NACA-TN* 1,192) shows the rate of climb that can be obtained by reducing the load on a helicopter that will just hover. Conversely, given the rate of climb with a given load, the curves determine the increase in load that will reduce the rate of climb to zero; that is, they determine the maximum load for which hovering is possible.

Performance with Forward Speed The performance of a typical single-main-rotor-type helicopter may be shown by a curve of C_{TR}/C_{QR} plotted against $\mu = v/\pi nD$, where v is the speed of the helicopter, as in Fig. 8.5.31. This curve includes the parasite-drag effects, which are appreciable at the higher values of μ. It is only a rough approximation to the experimentally determined values. C_{TR} may be calculated to determine C_{QR}, from which Q is obtained.

The performance of rotors at forward speeds involves a number of variables. For more complete treatment, see *NACA-TN* 1,192 and *NACA Wartime Report* L-110.

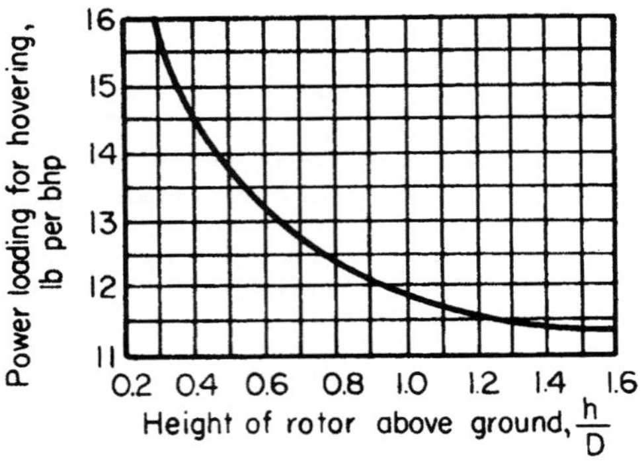

Figure 8.5.29 Observed ground effect on Sikorsky-type helicopters.

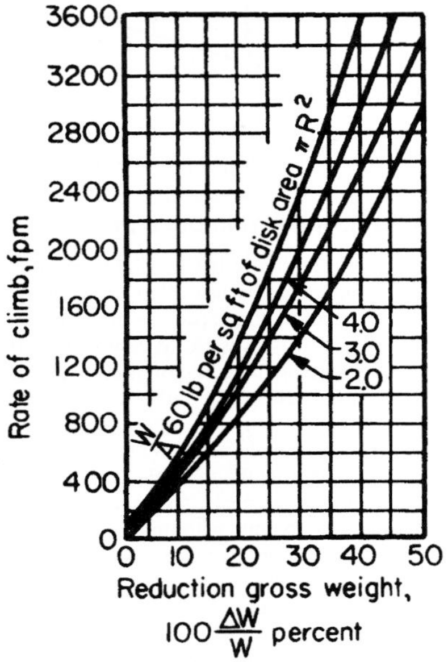

Figure 8.5.30 Effect of gross weight on the rate of climb of a helicopter.

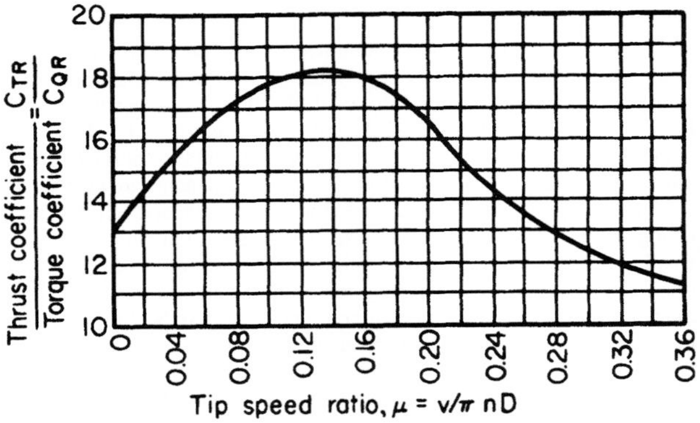

Figure 8.5.31 Performance curve for a single-main-rotor-type helicopter.

GROUND-EFFECT MACHINES (GEM)

For data on air-cushion vehicles and hydrofoil craft, see Sec. 8.4 "Marine Engineering."

SUPERSONIC AND HYPERSONIC AERODYNAMICS

REFERENCES: Liepmann and Roshko, "Elements of Gas Dynamics," Wiley. "High Speed Aerodynamics and Jet Propulsion" 12 vols., Princeton. Shapiro, "The Dynamics and Thermodynamics of Compressible Fluid Flow," Vols. I, II, Ronald. Howarth (ed.), "Modern Developments in Fluid Mechanics—High Speed Flow," Vols. I, II, Oxford. Kuethe and Schetzer, "Foundations of Aerodynamics," Wiley. Ferri, "Elements of Aerodynamics of Supersonic Flows," Macmillan. Bonney, "Engineering Supersonic Aerodynamics," McGraw-Hill.

The effect of the compressibility of a fluid upon its motion is determined primarily by the **Mach number M.**

$$\mathrm{M} = V/Vs \tag{8.5.17}$$

where V = speed of the fluid and V_s = speed of sound in the fluid which for air is $49.02\sqrt{T}$ where T is in °R and V_s in ft/s. The Mach number varies with position in the fluid, and the compressibility effect likewise varies from point to point.

If a *body moves through the atmosphere,* the overall compressibility effects are a function of the Mach number $\overline{\mathrm{M}}$ of the body, defined as

$$\overline{\mathrm{M}} = \text{velocity of body/speed of sound in the atmosphere}$$

Table 8.5.6 lists the useful **gas dynamics relations** between velocity, Mach number, and various fluid properties for isentropic flow. (See also Sec. 2, "Heat.")

$$V^2 = 2(h_0 - h) = 2C_p(T_0 - T)$$

$$\frac{V^2}{V_{s0}^2} = \frac{\mathrm{M}^2}{1 + \dfrac{\gamma-1}{2}\mathrm{M}^2}$$

$$\left(\frac{p}{p_0}\right)^{(\gamma-1)/\gamma} = \frac{T}{T_0} = \left(\frac{\rho}{\rho_0}\right)^{\gamma-1} = \left(\frac{V_s}{V_{s0}}\right)^2 = \frac{1}{1 + \dfrac{\gamma-1}{2}\mathrm{M}^2}$$

$$\frac{\tfrac{1}{2}\rho V^2}{p_0} = \frac{(\gamma/2)\mathrm{M}^2}{\left(1 + \dfrac{\gamma-1}{2}\mathrm{M}^2\right)^{\gamma/(\gamma-1)}} \tag{8.5.18}$$

$$\left(\frac{A}{A^*}\right)^2 = \frac{1}{\mathrm{M}^2}\left[\frac{2}{\gamma+1}\left(1 + \frac{\gamma-1}{2}\mathrm{M}^2\right)\right]^{(\gamma+1)/(\gamma-1)}$$

where h = enthalpy of the fluid, A = stream-tube cross section normal to the velocity, and the subscript $_0$ denotes the isentropic stagnation condition reached by the stream when stopped frictionlessly and adiabatically (hence isentropically). V_{s0} denotes the speed of sound in that medium.

The superscript * denotes the conditions occurring when the speed of the fluid equals the speed of sound in the fluid. For M = 1, Eqs. (8.5.18) become

$$\frac{T^*}{T_0} = \left(\frac{V_s^*}{V_{s0}}\right)^2 = \frac{2}{\gamma+1}$$

$$\frac{p^*}{p_0} = \left(\frac{2}{\gamma+1}\right)^{\gamma/(\gamma-1)} \tag{8.5.19}$$

$$\frac{\rho^*}{\rho_0} = \left(\frac{2}{\gamma+1}\right)^{1/(\gamma-1)}$$

For air $p^* = 0.52828p_0$; $\rho^* = 0.63394\rho_0$; $T^* = 0.83333T_0$.

For subsonic regions flow behaves similarly to the familiar hydraulics or incompressible aerodynamics: In particular, an increase of velocity is associated with a decrease of stream-tube area, and friction causes a pressure drop in a tube. There are no regions where the local flow velocity exceeds sonic speed. For **supersonic regions,** an increase of velocity is associated with an increase of stream-tube area, and friction causes a pressure rise in a tube. M = 1 is the dividing line between these two regions. For **transonic flows,** there are regions where the local flow exceeds sonic velocity, and this mixed-flow region requires special analysis. For **supersonic flows,** the entire flow field, with the exception of the regime near stagnation areas, has a velocity higher than the speed of sound.

The hypersonic regime is that range of very high supersonic speeds (usually taken as M > 5) where even a very streamlined body causes disturbance velocities comparable to the speed of sound, and stagnation temperatures can become so high that the gas molecules dissociate and become ionized.

Shock waves may occur at locally supersonic speeds. At low velocities the fluctuations that occur in the motion of a body (at the start of the motion or during flight) propagate away from the body at essentially the speed of sound. At higher subsonic speeds, fluctuations still propagate at the speed of sound in the fluid away from the body in all directions, but now the waves cannot get so far ahead; therefore, the force coefficients increase with an increase in the Mach number. When the Mach number of a body exceeds unity, the fluctuations instead of traveling away from the body in all directions are actually left behind by the body. These cases are illustrated in Fig. 8.5.32. As the supersonic speed is increased, the fluctuations are left farther behind and the force coefficients again decrease.

Where, for any reason, waves form an envelope, as at M > 1 in Fig. 8.5.32, a wave of finite pressure jump may result.

The speed of sound is higher in a higher-temperature region. For a compression wave, disturbances in the compressed (hence high-temperature) fluid will propagate faster than and overtake disturbances in the lower-temperature region. In this way shock waves are formed.

For a **stationary normal shock wave** the following properties exist, where subscripts 1 and 2 refer to conditions in front of and behind the shock, respectively,

$$\left(\frac{V_{S2}}{V_{S1}}\right) = \sqrt{\left(\frac{T_2}{T_1}\right)} = \left(\frac{a_2}{a_1}\right) \tag{8.5.20a}$$

$$\left(\frac{V_2}{V_1}\right) = \left(\frac{\mathrm{M}_2}{\mathrm{M}_1}\right)\cdot\left(\frac{a_2}{a_1}\right) \tag{8.5.20b}$$

In the following equations, also for a normal shock, the second form shown is for air with $\gamma = 1.4$.

$$\left(\frac{p_2}{p_1}\right) = \frac{2\gamma\mathrm{M}_1^2 - (\gamma-1)}{\gamma+1} = \left(\frac{7\mathrm{M}_1^2 - 1}{6}\right) \tag{8.5.20c}$$

$$\left(\frac{\rho_2}{\rho_1}\right) = \frac{(\gamma+1)\mathrm{M}_1^2}{(\gamma-1)\mathrm{M}_1^2 + 2} = \left(\frac{6\mathrm{M}_1^2}{\mathrm{M}_1^2 + 5}\right) \tag{8.5.20d}$$

$$\mathrm{M}_2 = \left[\frac{(\gamma-1)\mathrm{M}_1^2 + 2}{2\gamma\mathrm{M}_1^2 - (\gamma-1)}\right]^{1/2} = \left(\frac{\mathrm{M}_1^2 + 5}{7\mathrm{M}_1^2 - 5}\right)^{1/2} \tag{8.5.20e}$$

$$\left(\frac{p_{20}}{p_{10}}\right) = \left[\frac{(\gamma+1)\mathrm{M}_1^2}{(\gamma-1)\mathrm{M}_1^2 + 2}\right]^{\gamma/(\gamma-1)} \cdot \left(\frac{\gamma+1}{2\gamma\mathrm{M}_1^2 - (\gamma-1)}\right)^{1/(\gamma-1)} = \left(\frac{6\mathrm{M}_1^2}{\mathrm{M}_1^2 + 5}\right)^{3.5} \cdot \left(\frac{6}{7\mathrm{M}_1^2 - 1}\right)^{2.5} \tag{8.5.20f}$$

$$\left(\frac{T_2}{T_1}\right) = \frac{1 + \left(\dfrac{\gamma-1}{2}\right)\mathrm{M}_1^2}{1 + \left(\dfrac{\gamma-1}{2}\right)\mathrm{M}_1^2} = \left(\frac{1 + 0.2\mathrm{M}_1^2}{1 + 0.2\mathrm{M}_2^2}\right) \tag{8.5.20g}$$

The subscript 20 refers to the stagnation condition after the shock, and the subscript 10 refers to the stagnation condition ahead of the shock.

In a normal shock M, V, and p_0 decrease; p, ρ, T, and s increase, whereas T_0 remains unchanged. These relations are given numerically in Table 8.5.7.

If the shock is moving, these same relations apply *relative to the shock.* In particular, if the shock advances into stationary fluid, it does so at the speed V_1 which is always greater than the speed of sound in the stationary fluid by an amount dependent upon the shock strength.

When an air vehicle is traveling at high subsonic, transonic, or supersonic speeds, shock waves will form on it. A shock wave is a

Table 8.5.6 Isentropic Gas Dynamics Relations

M	$\frac{p}{p_0}$	$\frac{V}{V_{s0}}$	$\frac{A}{A^*}$	$\frac{\frac{1}{2}\rho V^2}{p_0}$	$\frac{\rho V}{\rho_0 V_{s0}}$	$\frac{\rho}{\rho_0}$	$\frac{T}{T_0}$	$\frac{V_s}{V_{s0}}$
0.00	1.000	0.000	∞	0.000	0.000	1.000	1.000	1.000
0.05	0.998	0.050	11.59	0.002	0.050	0.999	0.999	1.000
0.10	0.993	0.100	5.82	0.007	0.099	0.995	0.998	0.999
0.15	0.984	0.150	3.91	0.016	0.148	0.989	0.996	0.998
0.20	0.972	0.199	2.964	0.027	0.195	0.980	0.992	0.996
0.25	0.957	0.248	2.403	0.042	0.241	0.969	0.988	0.994
0.30	0.939	0.297	2.035	0.059	0.284	0.956	0.982	0.991
0.35	0.919	0.346	1.778	0.079	0.325	0.941	0.976	0.988
0.40	0.896	0.394	1.590	0.100	0.364	0.924	0.969	0.984
0.45	0.870	0.441	1.449	0.123	0.399	0.906	0.961	0.980
0.50	0.843	0.488	1.340	0.148	0.432	0.885	0.952	0.976
0.55	0.814	0.534	1.255	0.172	0.461	0.863	0.943	0.971
0.60	0.784	0.580	1.188	0.198	0.487	0.840	0.933	0.966
0.65	0.753	0.624	1.136	0.223	0.510	0.816	0.922	0.960
0.70	0.721	0.668	1.094	0.247	0.529	0.792	0.911	0.954
0.75	0.689	0.711	1.062	0.271	0.545	0.766	0.899	0.948
0.80	0.656	0.753	1.038	0.294	0.557	0.740	0.887	0.942
0.85	0.624	0.795	1.021	0.315	0.567	0.714	0.874	0.935
0.90	0.591	0.835	1.009	0.335	0.574	0.687	0.861	0.928
0.95	0.559	0.874	1.002	0.353	0.577	0.660	0.847	0.920
1.00	0.528	0.913	1.000	0.370	0.579	0.634	0.833	0.913
1.05	0.498	0.950	1.002	0.384	0.578	0.608	0.819	0.905
1.10	0.468	0.987	1.008	0.397	0.574	0.582	0.805	0.897
1.15	0.440	1.023	1.018	0.407	0.569	0.556	0.791	0.889
1.20	0.412	1.057	1.030	0.416	0.562	0.531	0.776	0.881
1.25	0.386	1.091	1.047	0.422	0.553	0.507	0.762	0.873
1.30	0.361	1.124	1.066	0.427	0.543	0.483	0.747	0.865
1.35	0.337	1.156	1.089	0.430	0.531	0.460	0.733	0.856
1.40	0.314	1.187	1.115	0.431	0.519	0.437	0.718	0.848
1.45	0.293	1.217	1.144	0.431	0.506	0.416	0.704	0.839
1.50	0.272	1.246	1.176	0.429	0.492	0.395	0.690	0.830
1.55	0.253	1.274	1.212	0.426	0.478	0.375	0.675	0.822
1.60	0.235	1.301	1.250	0.422	0.463	0.356	0.661	0.813
1.65	0.218	1.328	1.292	0.416	0.443	0.337	0.647	0.805
1.70	0.203	1.353	1.338	0.410	0.433	0.320	0.634	0.796
1.75	0.188	1.378	1.387	0.403	0.417	0.303	0.620	0.788
1.80	0.174	1.402	1.439	0.395	0.402	0.287	0.607	0.779
1.85	1.161	1.425	1.495	0.386	0.387	0.272	0.594	0.770
1.90	0.149	1.448	1.555	0.377	0.372	0.257	0.581	0.762
1.95	0.138	1.470	1.619	0.368	0.357	0.243	0.568	0.754
2.00	0.128	1.491	1.688	0.358	0.343	0.230	0.556	0.745
2.50	0.059	1.667	2.637	0.256	0.219	0.132	0.444	0.667
3.00	0.027	1.793	4.235	0.172	0.137	0.076	0.357	0.598
3.50	0.013	1.884	6.790	0.112	0.085	0.045	0.290	0.538
4.00	0.007	1.952	10.72	0.074	0.054	0.028	0.238	0.488
4.50	0.003	2.003	16.56	0.049	0.035	0.017	0.198	0.445
5.00	0.002	2.041	25.00	0.033	0.023	0.011	0.167	0.408
10.00	0.00002	2.182	536.00	0.002	0.001	0.005	0.048	0.218

SOURCE: Emmons, "Gas Dynamics Tables for Air," Dover Publications, Inc., 1947.

thin region of air through which air physical properties change. Shock waves can be normal or oblique to the air vehicle direction of flight. If oblique, it behaves exactly like a normal shock to the normal component of the stream. The tangential component is left unchanged. Thus the resultant velocity not only drops abruptly in magnitude but also changes discontinuously in direction. Figure 8.5.33 gives the relations between M_1, M_2, θ_W, and δ.

In **supersonic flow past a two-dimensional wedge** with semiangle δ one **strong** and one **weak** oblique shock wave will theoretically form attached to the leading edge. See Fig. 8.5.34*a*. In practice only the weak one will be present and downstream of this oblique shock the resultant flow will be parallel to the wedge surface. It is possible to use Eqs. (8.5.20) to generate the flow characteristics about wedges with various wedge angles or more easily with the help of Fig. 8.5.33.

The relationship between δ, M_1, and θ_W is shown in Fig. 8.5.34*b*. These are related through the equation

$$\tan\delta = \left[\frac{2\cot\theta_W (M_1^2 \sin^2\theta_W - 1}{M_1^2(\gamma + \cos 2\theta_W) + 2}\right]$$

The results of a numerical evaluation of the above equation for $1.5 < M_1 < 20$, $\gamma = 1.4$ (for air) are shown graphically in Fig. 8.5.35. The locus of the maximum wedge angle line differentiates between weak and strong shocks. For small wedge angles, the shock angle differs only slightly from $\sin^{-1}$ (1/M_1), the **Mach angle,** and the velocity component

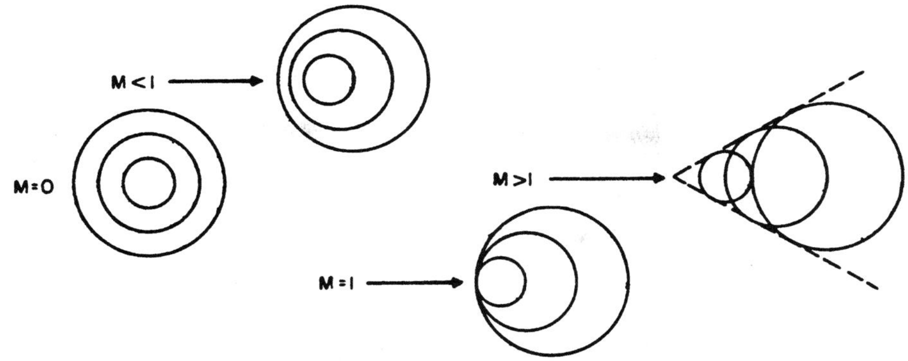

Figure 8.5.32 Propagation of sound waves in moving streams.

Table 8.5.7 Normal Shock Relations

M	p_2/p_1	p_{20}/p_1	p_{20}/p_{10}	M_2	V_{s2}/V_{s1}	V_2/V_1	T_2/T_1	ρ_2/ρ_1
1.00	1.000	1.893	1.000	1.000	1.000	1.000	1.000	1.000
1.05	1.120	2.008	1.000	0.953	1.016	0.923	1.033	1.084
1.10	1.245	2.133	0.999	0.912	1.032	0.855	1.065	1.169
1.15	1.376	2.266	0.997	0.875	1.047	0.797	1.097	1.255
1.20	1.513	2.408	0.993	0.842	1.062	0.745	1.128	1.342
1.25	1.656	2.557	0.987	0.813	1.077	0.700	1.159	1.429
1.30	1.805	2.714	0.979	0.786	1.091	0.660	1.191	1.516
1.35	1.960	2.878	0.970	0.762	1.106	0.624	1.223	1.603
1.40	2.120	3.049	0.958	0.740	1.120	0.592	1.255	1.690
1.45	2.286	3.228	0.945	0.720	1.135	0.563	1.287	1.776
1.50	2.458	3.413	0.930	0.701	1.149	0.537	1.320	1.862
1.55	2.636	3.607	0.913	0.684	1.164	0.514	1.354	1.947
1.60	2.820	3.805	0.895	0.668	1.178	0.492	1.388	2.032
1.65	3.010	4.011	0.876	0.654	1.193	0.473	1.423	2.115
1.70	3.205	4.224	0.856	0.641	1.208	0.455	1.458	2.198
1.75	3.406	4.443	0.835	0.628	1.223	0.439	1.495	2.279
1.80	3.613	4.670	0.813	0.617	1.238	0.424	1.532	2.359
1.85	3.826	4.902	0.790	0.606	1.253	0.410	1.573	2.438
1.90	4.045	5.142	0.767	0.596	1.268	0.398	1.608	2.516
1.95	4.270	5.389	0.744	0.586	1.284	0.386	1.647	2.592
2.00	4.500	5.640	0.721	0.577	1.299	0.375	1.688	2.667
2.50	7.125	8.526	0.499	0.513	1.462	0.300	2.138	3.333
3.00	10.333	12.061	0.328	0.475	1.637	0.259	2.679	3.857
3.50	14.125	16.242	0.213	0.451	1.821	0.235	3.315	4.261
4.00	18.500	21.068	0.139	0.435	2.012	0.219	4.047	4.571
4.50	23.458	26.539	0.092	0.424	2.208	0.208	4.875	4.812
5.00	29.000	32.653	0.062	0.415	2.408	0.200	5.800	5.000
10.00	116.500	129.220	0.003	0.388	4.515	0.175	20.388	5.714

SOURCE: Emmons, "Gas Dynamics Tables for Air," Dover Publications, Inc., 1947.

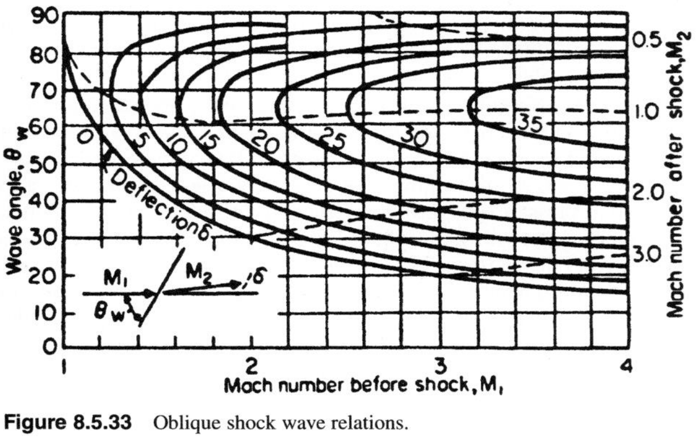

Figure 8.5.33 Oblique shock wave relations.

normal to this wave is the speed of sound. The pressure jump is small and is given approximately by

$$p - p_1 = \left[\frac{\gamma p_1 \overline{M}_1^2}{\sqrt{\overline{M}_1^2 - 1}}\right]\delta$$

where δ is the wedge semiangle.

For wedge semiangles that exceed the corresponding Mach number lines, a detached shock will form at the leading edge (Fig. 8.5.36). For example, if $\delta = 40°$ and $M_1 = 3.00$, then a detached bow wave will form in front of the wedge leading edge (see Fig. 8.5.35). Behind this detached shock there will be a region of subsonic flow reaching the wedge leading edge.

Exact solutions also exist for **supersonic flow past a cone.** Above a certain supersonic Mach number, a conical shock wave is attached

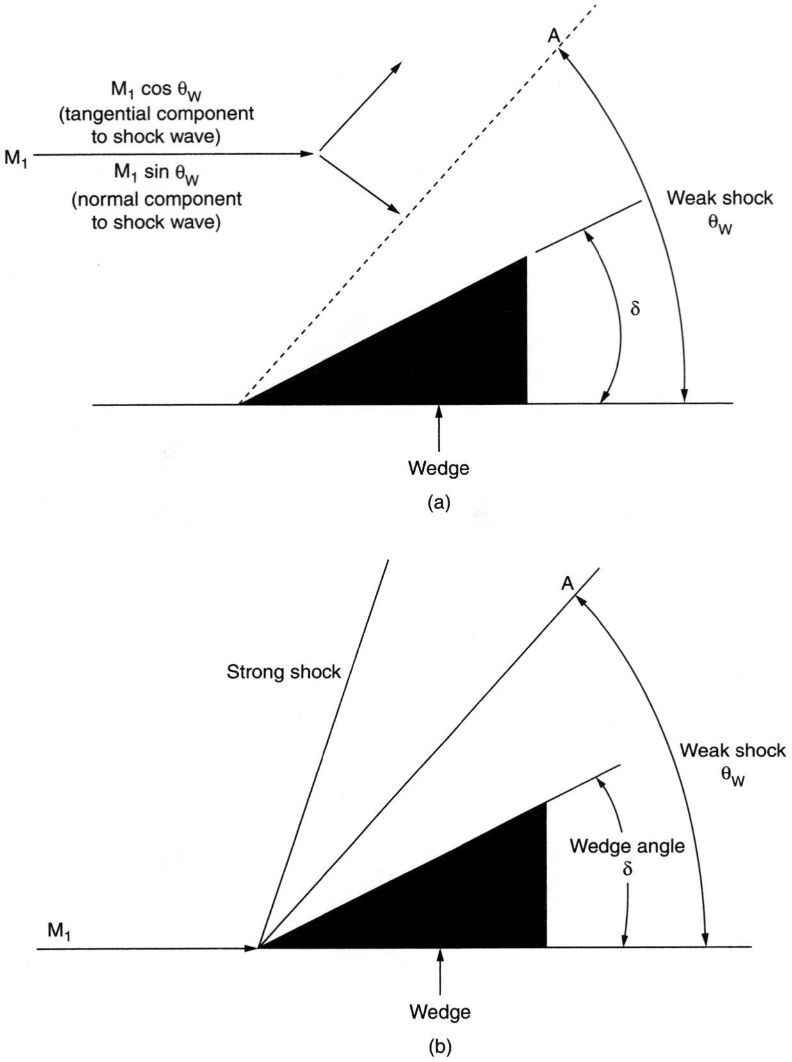

Figure 8.5.34 Shock waves on a wedge.

to the apex of the cone. Figures 8.5.37 and 8.5.38 show these exact relations; they are so accurate that they are often used to determine the Mach number of a stream by measuring the shock-wave angle on a cone of known angle. For small cone angles the shock differs only slightly from the **Mach cone,** that is, a cone whose semiapex angle is the Mach angle; the pressure on the cone is then given approximately by

$$p - p_1 = \gamma p_1 \overline{M}_1^2 \delta^2 (\ln 2/\delta\sqrt{\overline{M}_1^2 - 1}) \qquad (8.5.21)$$

where δ is the cone semiangle.

A **nozzle** consisting of a single contraction will produce at its exit a jet of any velocity from M = 0 to M = 1 by a proper adjustment of the pressure ratio. For use as a subsonic wind-tunnel nozzle where a uniform parallel gas stream is desired, it is only necessary to connect the supply section to the parallel-walled or open-jet test section by a smooth gently curving wall. If the radius of curvature of the wall is nowhere less than the largest test-section cross-sectional dimension, no flow separation will occur and a good test gas stream will result.

When a **converging nozzle** connects two chambers with the pressure drop beyond the critical $[(p/p_0) < (p^*/p_0)]$, the Mach number at the exit of the nozzle will be 1; the pressure ratio from the supply section to the nozzle exit will be critical; and all additional expansion will take place outside the nozzle.

A nozzle designed to supply a supersonic jet at its exit must converge to a minimum section and diverge again. The area ratio from the minimum section to the exit is given in the column headed A/A^* in Table 8.5.6. A converging-diverging nozzle with a pressure ratio p/p_0 = (gas pressure)/(stagnation pressure), Table 8.5.6, gradually falling from unity to zero will produce shock-free flow for all exit Mach numbers from zero to the subsonic Mach number corresponding to its area ratio. From this Mach number to the supersonic Mach number corresponding to the given area ratio, there will be shock waves in the nozzle. For all smaller pressure ratios, the Mach number at the nozzle exit will not change, but additional expansion to higher velocities will occur outside of the nozzle.

To obtain a uniform parallel shock-free supersonic stream, the converging section of the nozzle can be designed as for a simple converging

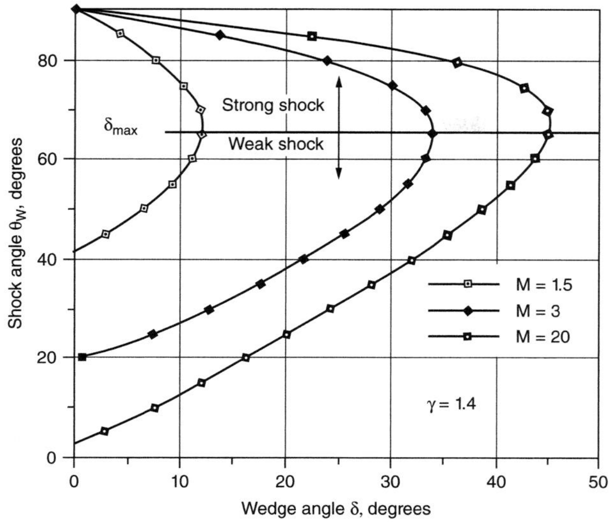

Figure 8.5.35 Oblique shock properties for air: strong and weak shock variables.

nozzle. The diverging or supersonic portion must be designed to produce and then cancel the expansion waves. These nozzles would perform as designed if it were not for the growth of the boundary layer. Experience indicates that these nozzles give a good first approximation to a uniform parallel supersonic stream but at a somewhat lower Mach number.

For **rocket nozzles** and other thrust devices the gain in thrust obtained by making the jet uniform and parallel at complete expansion must be balanced against the loss of thrust caused by the friction on the wall of the greater length of nozzle required. A simple conical diverging section, cut off experimentally for maximum thrust, is generally used (see also Sec. 8.6 "Jet Propulsion and Aircraft Propellers").

For transonic and supersonic flow, **diffusers** are used for the recovery of kinetic energy. They follow the test sections of supersonic wind tunnels and are used as inlets on high-speed planes and missiles for ram recovery. For the first use, the diffuser is fed a nonuniform stream from the test section (the nonuniformities depending on the particular body under test) and should yield the maximum possible pressure-rise ratio. In missile use the inlet diffuser is fed by a uniform (but perhaps slightly yawed) airstream. The maximum possible pressure-rise ratio is important but must provide a sufficiently uniform flow at the exit to assure good performance of the compressor or combustion chamber that follows.

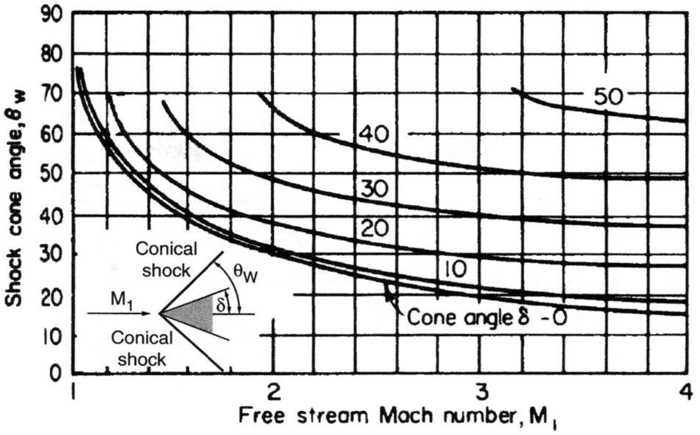

Figure 8.5.37 Wave angles for supersonic flow around cones.

In simplest form, a **subsonic diffuser** is a diverging channel, a nozzle in reverse. Since boundary layers grow rapidly with a pressure rise, subsonic diffusers must be diverged slowly, 6 to 8° equivalent cone angle, that is, the apex angle of a cone with the same length and area ratio. Similarly, a **supersonic diffuser** in its simplest form is a supersonic nozzle in reverse. Both the convergent and divergent portions must change cross section gradually. In principle it is possible to design a shock-free diffuser. In practice shock-free flow is not attained, and the design is based upon minimizing the shock losses. Oblique shocks should be produced at the inlet and reflected a sufficient number of times to get compression nearly to M = 1. A short parallel section and a divergent section can now be added with the expectation that a weak normal shock will be formed near the throat of the diffuser.

For **a supersonic wind tunnel,** the best way to attain the maximum pressure recovery at a wide range of operating conditions is to make the diffuser throat variable. The ratio of diffuser-exit static pressure to the diffuser-inlet (test-section outlet) total pressure is given in Table 8.5.8. These pressure recoveries are attained by the proper adjustment of the throat section of a variable diffuser on a supersonic wind-tunnel nozzle.

A **supersonic wind tunnel** consists of a compressor or compressor system including precoolers or aftercoolers, a supply section, a supersonic nozzle, a test section with balance and other measuring equipment, a diffuser, and sufficient ducting to connect the parts. The **minimum pressure ratio** required from supply section to diffuser exit is given in Table 8.5.8. Any pressure ratio greater than this is satisfactory. The extra pressure ratio is automatically wasted by additional shock waves that appear in the diffuser. The compression ratio required of the compressor system must be greater than that of

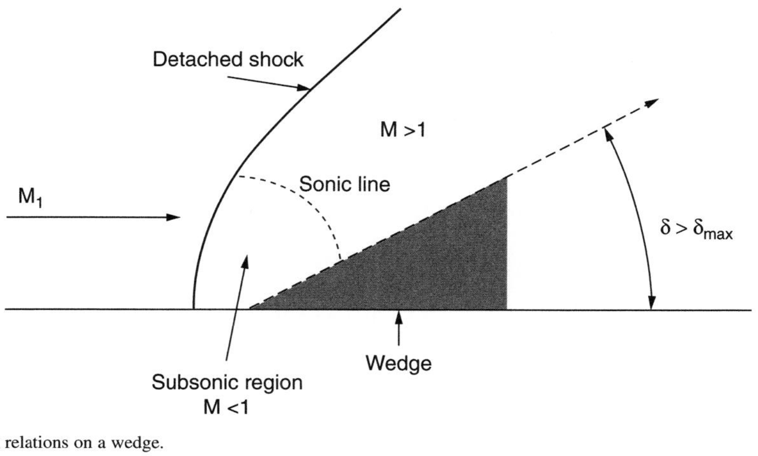

Figure 8.5.36 Detached shock relations on a wedge.

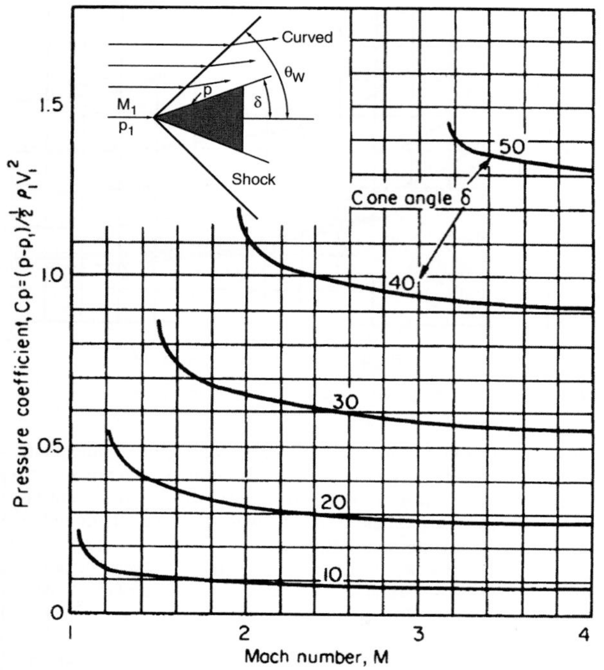

Figure 8.5.38 Pressure calculations for supersonic flow around cones.

Table 8.5.8

M	1	1.5	2	3	4
P_e/P_0	0.83	0.69	0.50	0.23	0.10

Table 8.5.8 by at least an amount sufficient to take care of the pressure drop in the ducting and valves. The latter losses are estimated by the usual hydraulic formulas.

After selecting a compressor system capable of supplying the required maximum pressure ratio, the test section area is computed from

$$A = 1.73 \frac{Q}{V_{s0}} \frac{A}{A^*} \frac{p_e}{p_0} \quad \text{ft}^2 \tag{8.5.22}$$

where Q is the inlet volume capacity of the compressors, ft³/s; V_{s0} is the speed of sound in the supply section, ft/s; A/A^* is the area ratio given in Table 8.5.6 as a function of M; p_e/p_0 is the pressure ratio given in Table 8.5.8 as a function of M.

The nozzle itself is designed for uniform parallel airflow in the test section. Since for each Mach number a different nozzle is required, the nozzle must be flexible or the tunnel so arranged that fixed nozzles can be readily interchanged.

The Mach number of a test is set by the nozzle selection, and the Reynolds number is set by the inlet conditions and size of model. The Reynolds number is computed from

$$\text{Re} = \text{Re}_0 D(p_0/14.7)(540/T_0)^{1.268} \tag{8.5.23}$$

where Re_0 is the Reynolds number per inch of model size for atmospheric temperature and pressure, as given by Fig. 8.5.39, D is model diam, in; p_0 is the stagnation pressure, lb/in²; T_0 is the stagnation temperature, °R.

For a closed-circuit tunnel, the Reynolds number can be varied independently of the Mach number by adjusting the mass of air in the system, thus changing p_0.

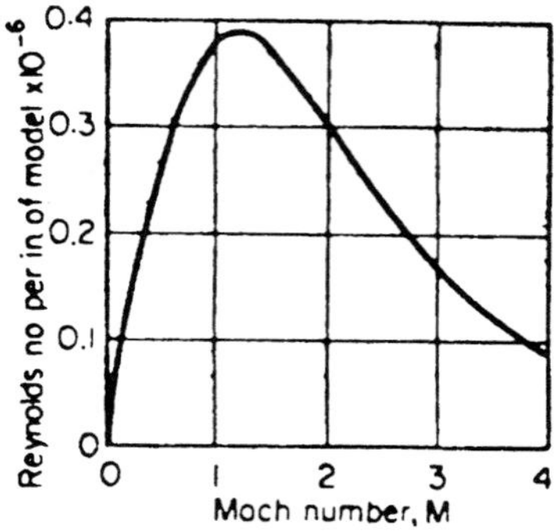

Figure 8.5.39 Reynolds number–Mach number relation (for a fixed model size and stagnation condition).

Intermittent-wind tunnels for testing at high speeds do not require the large and expensive compressors associated with continuous-flow tunnels. They use either a large vacuum tank or a large pressure tank (often in the form of a sphere) to produce a pressure differential across the test section. Such tunnels may have steady flow for only a few seconds, but by careful instrumentation, sufficient data may be obtained in this time. A **shock tube** may be used as an intermittent-wind tunnel as well as to study shock waves and their interactions; it is essentially a long tube of constant or varying cross section separated into two parts by a frangible diaphragm. High pressure exists on one side; by rupturing the diaphragm, a shock wave moves into the gas with the lower pressure. After the shock wave a region of steady flow exists for a few milliseconds. Very high stagnation temperatures can be created in a shock tube, which is not the case in a wind tunnel, so that it is useful for studying hypersonic flow phenomena.

Wind-tunnel force measurements are subject to errors caused by the model support strut. Wall interference is small at high Mach numbers for which the reflected model head wave returns well behind the model. Near M = 1, the wall interference becomes very large. In fact, the tunnel chokes at Mach numbers given in Table 8.5.6, at

$$\frac{A}{A^*} = \frac{1}{1 - \dfrac{\text{area of model projected on test-section cross section}}{\text{area of test-section cross section}}} \tag{8.5.24}$$

There are two choking points: one subsonic and one supersonic. Between these two Mach numbers, it is impossible to test in the tunnel. As these Mach numbers are approached, the tunnel wall interference becomes very large.

For tests in this range of Mach number, specially constructed **transonic wind tunnels** with perforated or slotted walls have been built. The object here is to produce a mean flow velocity through the walls that comes close to that which would have existed there had the body been moving at that speed in the open. Often auxiliary blowers are needed to produce the necessary suction on the walls and to reinject this air into the tunnel circuit in or after the diffuser section.

Drag is difficult to predict precisely from wind-tunnel measurements, especially in the transonic regime. **Flight tests** of rocket-boosted models which coast through the range of Mach numbers of interest are used to obtain better drag estimates; from telemetered data and radar or optical sighting the deceleration can be determined, which in turn yields the drag.

LINEARIZED SMALL-DISTURBANCE THEORY

As the speed of a body is increased from a low subsonic value (Fig. 8.5.40*a*), the local Mach number becomes unity somewhere in the fluid along the surface of the body, that is, the lower critical Mach number is reached. Above this there is a small range of transonic speed in which a supersonic region exists (Fig. 8.5.40*b*). Shock waves appear

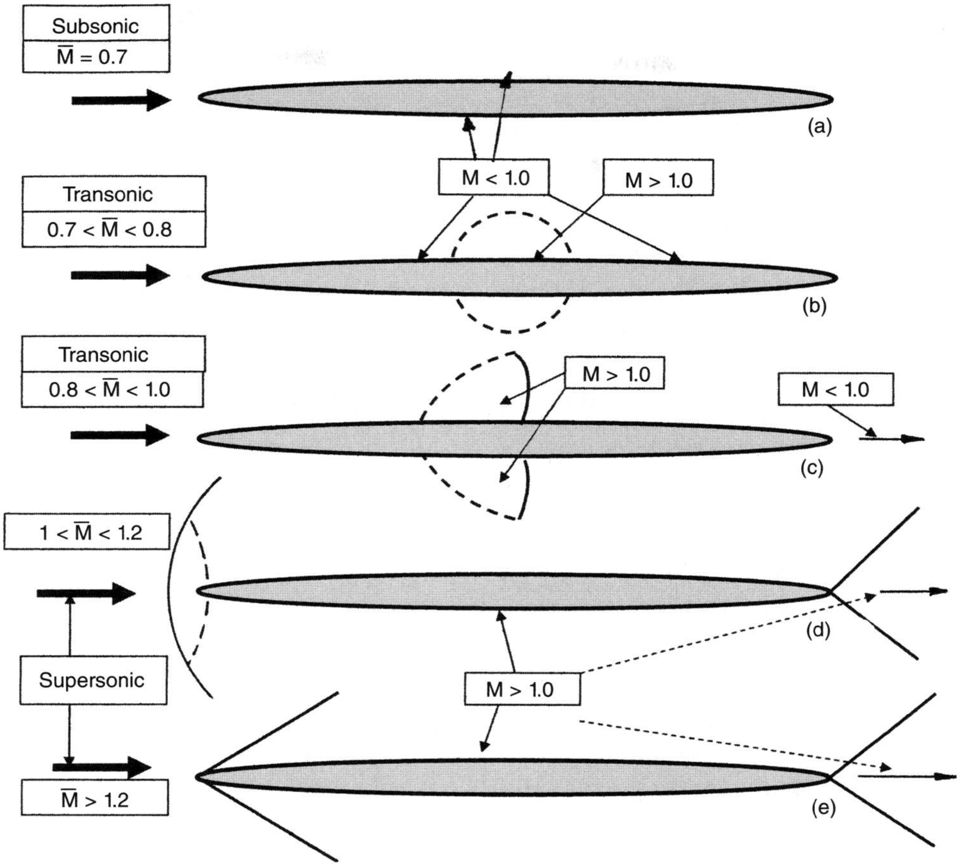

Figure 8.5.40 Regions of flow about an airfoil (Mach numbers are approximate).

in this region attached to the sides of the body (Fig. 8.5.40*c*) and grow with increasing speed. At still higher transonic speeds a detached shock wave appears ahead of the body, and the earlier side shocks either disappear or move to the rear (Fig. 8.5.40*d*). Finally, for a sharp-nosed body, the head wave moves back and becomes attached (Fig. 8.5.40*e*). The flow is now generally supersonic everywhere, and the transonic regime is replaced by the supersonic regime. With the appearance of shock waves there occurs a considerable alteration of the pressure distribution, and the center of pressure on airfoil sections moves from the one-fourth chord point back toward the one-half chord point. There is an associated increase of drag and, often, flow separation at the base of the shock.

The redistribution of pressures and the motion of shock waves over the wing surfaces through the transonic regime demands special consideration in the design of control surfaces so that they do not become ineffective by separation or inoperative by excessive loading.

If a body is slender (i.e., planes tangent to its surface at any point make small angles with the flight direction), the disturbance velocities caused by this body will be small compared with the flight speed and, excluding the hypersonic regime, small compared with the speed of sound. This permits the use of linearized small-disturbance theory to predict the approximate flow past the body. This theory relates the steady flow past the body at subsonic speeds to the incompressible flow past a distorted version of this body (the "generalized Prandtl-Glauert rule"). For **thin airfoil sections** at subsonic speeds, this theory predicts

$$C_L = 2\pi\alpha/\sqrt{1-\mathrm{M}^2} \quad \alpha \text{ in radians} \tag{8.5.25}$$

where $L = C_L q S$ and $q = (\gamma/2)p_0\mathrm{M}^2$.

An approximation to the **subsonic** lift curve slope for wings is given by:

$$C_{L\alpha} = \left(\frac{2\pi A}{\left[2+\left\{\left(\frac{A^2\beta^2}{0.90}\right)\left(1+\frac{\tan^2\Lambda_{c/2}}{\beta^2}\right)+4\right\}^{1/2}\right]}\right) \quad \text{per radian}$$

where A = aspect ratio, $\beta = (1-\mathrm{M}^2)^{0.5}$, $\Lambda_{c/2}$ = semichord sweep angle. This can be applied to wings of all aspect ratios, taper ratios, and for quarter chord sweep angles of up to 60°.

The corresponding **induced** drag, or **drag due to lift,** is given by

$$C_{Di} = \frac{C_L^2}{\pi A e}$$

where e is the span efficiency factor, which is usually between 0.85 and 0.95.

At supersonic speeds, the Prandtl-Glauert rule is also applicable if we replace β by $\lambda = \sqrt{\mathrm{M}^2-1}$. For **thin airflow sections** (see Fig. 8.5.41) this theory predicts

$$C_L = 4\alpha/\sqrt{\mathrm{M}^2-1} \quad \alpha \text{ in radians} \tag{8.5.26}$$

$$C_D = \frac{2}{\lambda}\int_0^1\left[\left(\frac{dy_u}{dx}\right)^2+\left(\frac{dy_l}{dx}\right)^2\right]d\left(\frac{x}{c}\right)+\frac{4a^2}{\lambda}+C_{DF} \tag{8.5.27}$$

where dy_u/dx and dy_l/dx are the slopes of the upper and lower surfaces of the airfoil, respectively, and C_{DF} is the skin-friction drag coefficient.

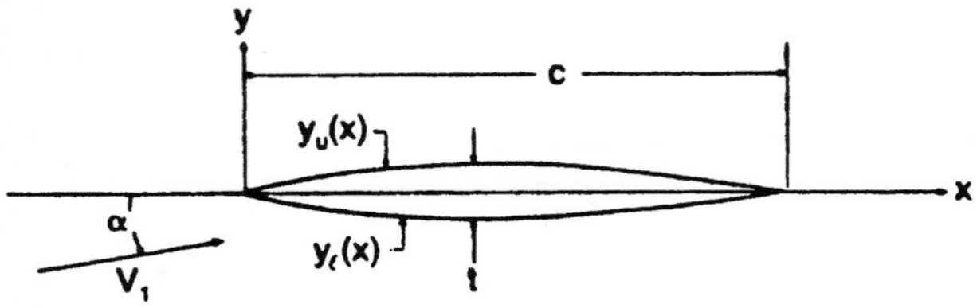

Figure 8.5.41 Supersonic airfoil section.

Note that there is a drag due to thickness and a drag due to lift at supersonic speeds for an airfoil section where there is none at subsonic speeds. For symmetric double-wedge airfoil sections, the thickness drag coefficient becomes $(4/\lambda)(t/c)^2$, and for symmetric biconvex airfoil sections it is $(16/3\lambda)(t/c)^2$, where t is the maximum thickness.

Within the small disturbance approximation, a disturbance at a point in supersonic flow can affect only the points in the downstream **Mach cone**. Thus a rectangular wing with constant airfoil section has two-dimensional flow on all parts of the wing except the points within the Mach cones. At Mach numbers near unity these tip Mach cones cover nearly the entire wing, whereas at high Mach numbers they cover only a small part.

Research at NACA (NACA TR 839, NACA TR 900) established that for **supersonic** wings, swept and unswept, the relative location of the Mach lines and the wing planform (Fig. 8.5.42) is important in the determination of the magnitudes of the wing lift curve slopes and the drag due to lift. For swept wings, the wing possesses a **subsonic** leading edge when $\mu > \varepsilon$ (Fig. 8.5.43), and a **supersonic** leading edge when $\varepsilon > \mu$. When $\varepsilon > \mu$ the leading edge is termed **sonic.**

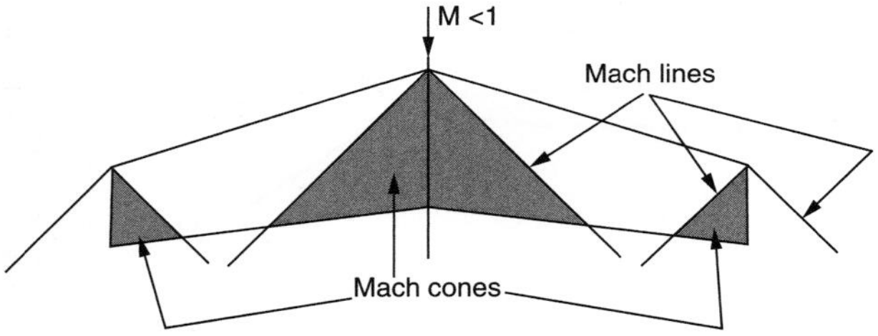

Figure 8.5.42 Mach cones and Mach lines for wing with supersonic leading edge.

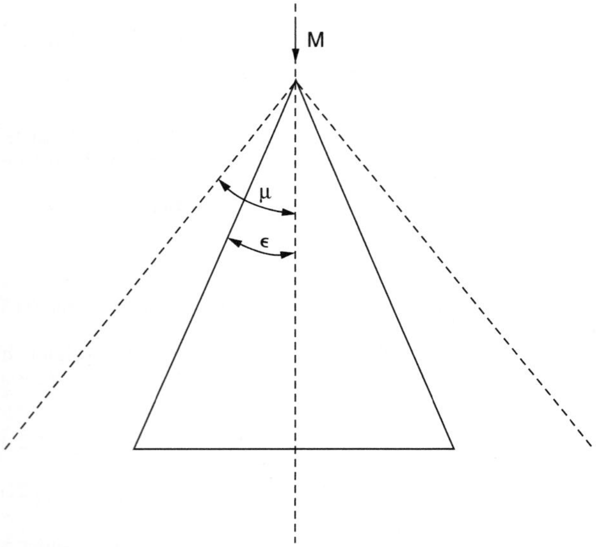

Figure 8.5.43 Wing sweep/Mach line relationship.

For **thin airfoil section supersonic delta wings** with **subsonic leading edges,** lift can be approximated by

$$\beta\left(\frac{C_L}{\alpha}\right) = \left(\frac{2\pi^2}{\pi+\lambda}\right)\left(\frac{\tan\varepsilon}{\tan\mu}\right)$$

where λ is given in Fig. 8.5.44 and $\beta = (\mathrm{M}^2 - 1)^{0.5}$, and the corresponding drag due to lift approximated by

$$C_{Di} = \left(\frac{C_L^2}{\pi A}\right)\left[2\left(1+\frac{\lambda}{\pi}\right) - \sqrt{1-\beta^2\tan^2\varepsilon}\right]$$

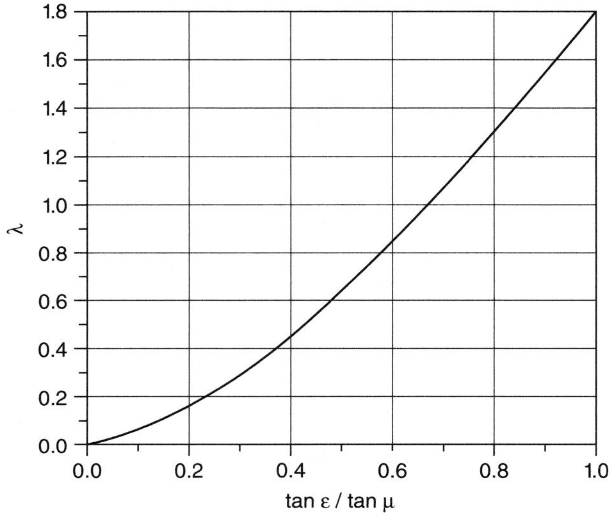

Figure 8.5.44 Lambda (λ) variation with sweep and Mach number.

For **delta wings with supersonic leading edges,** the corresponding first approximations are

$$\left(\frac{C_L}{\alpha}\right) = \frac{4}{\sqrt{\mathrm{M}^2-1}} \text{ per radian} \quad \text{and} \quad C_{Di} = \frac{4\alpha^2}{\sqrt{\mathrm{M}^2-1}}$$

For delta wings with sharp leading edges, research at NASA (NASA TN-D 4,739) established the presence of a **vortex lift** component at low speed. Figure 8.5.45 illustrates the formation of these vortices which generates a total lift coefficient given by

$$C_L = K_p \sin\alpha\cos^2\alpha + K_v \cos\alpha\sin^2\alpha$$

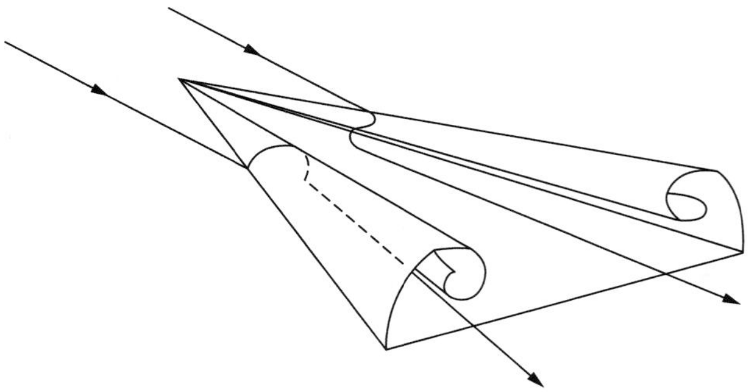

Figure 8.5.45 Vortex-induced lift.

Drag due to this lift is given by

$$\Delta C_{Di} = K_p \sin^2 \alpha \cos \alpha + K \sin^3 \alpha$$

where K_p and K_v are given in Fig. 8.5.46.

Rectangular wings in supersonic flight are not amenable to simple closed form solutions for either lift or drag due to lift and the reader is referred to NACA TN 2,114 for lift and drag estimations.

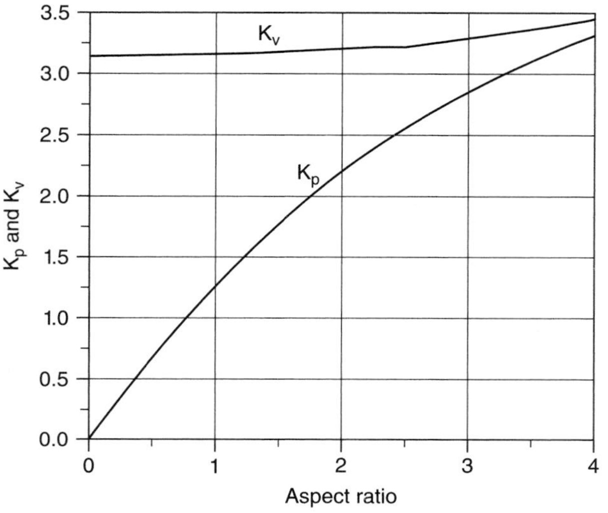

Figure 8.5.46 Variations of K_p and K_v with wing aspect ratio.

The lift of a slender **axially symmetric body** at small angle of attack is nearly independent of Mach number and is given approximately by

$$C_L = 2\alpha \tag{8.5.28}$$

where the lift coefficient is based on the cross-sectional area of the base. The drag for $\alpha = 0$ at subsonic speeds is made up of skin friction and **base drag;** a dead-air region exists just behind the blunt base, and the pressure here is below ambient, causing a rearward suction which is the base drag. At supersonic speeds a wave drag is added, which represents the energy dissipated in shock waves from the nose. Figure 8.5.47 shows a typical drag-coefficient curve for a body of revolution where the **fineness ratio** (ratio of length to diameter) is 12.2. The skin-friction drag coefficient based on wetted area is the same as that of a flat plate, within experimental error.

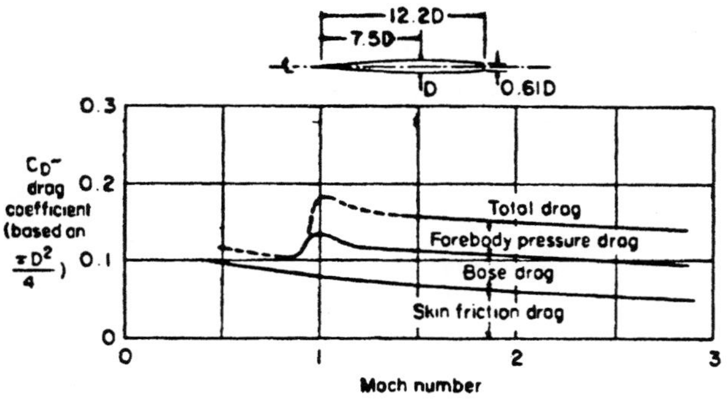

Figure 8.5.47 Drag coefficients for parabolic-arc (*NACA-RM* 10) body (calculations from experimental data for a 30,000-ft altitude, $D = 12$ in, from *NACA-TR* 1,160 and 1,161, 1954).

At transonic and moderate supersonic speeds, the wave drag of a **wing body combination** can be effectively reduced by making the cross-sectional area distribution (including wings) a smooth curve when plotted versus fuselage station. This is the **area rule;** its application results in a decided indentation in the fuselage contour at the wing juncture. **Lift interference** effects also occur, especially when the body diameter is not small compared with the wing span. If the wing is attached to a cylindrical body, and if the wing-alone lift coefficient would have been C_L, then the lift carried on the body is $K_B C_L$, and the lift carried on the wing is $K_W C_L$. Figure 8.5.48 shows the variation of K_B and K_W with body diameter to wing-span ratio.

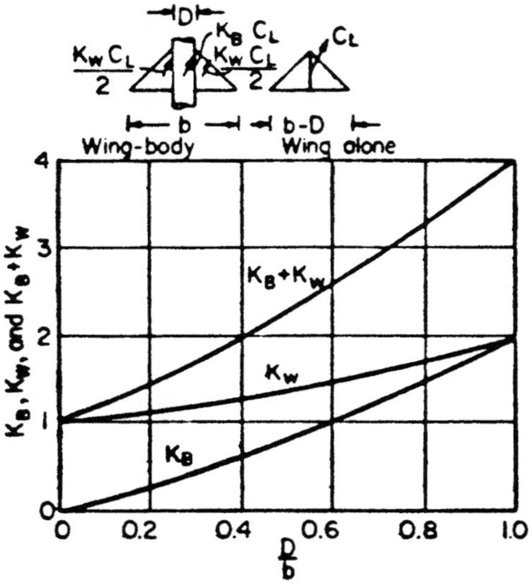

Figure 8.5.48 Wing-body lift-interface factors.

The **effect of compressibility on skin-friction drag** is slight at subsonic speeds, but at supersonic speeds, a significant reduction in skin-friction coefficient occurs. Figure 8.5.49 shows the turbulent-boundary-layer mean skin-friction coefficient for a cone as a function of wall temperature and Mach number. An important effect at high speed is **aerodynamic heating.**

Figure 8.5.50 illustrates the velocity profile behind the shock wave of a body traveling at hypersonic speed. The shock-wave front represents an area of high-temperature gas which radiates energy to the body, but boundary-layer convective heating is usually the major contributor. Behind this front is shown the velocity gradient in the boundary layer. The decrease in velocity in the boundary layer is brought about by the forces of interaction between fluid particles and the body (viscosity).

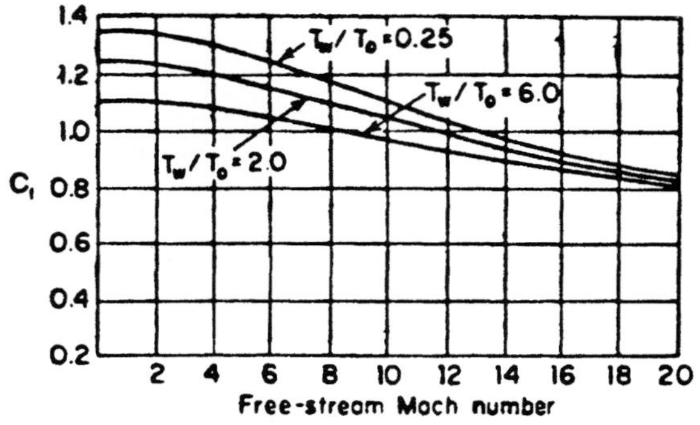

Figure 8.5.49 Laminar-skin-friction coefficient for a cone.

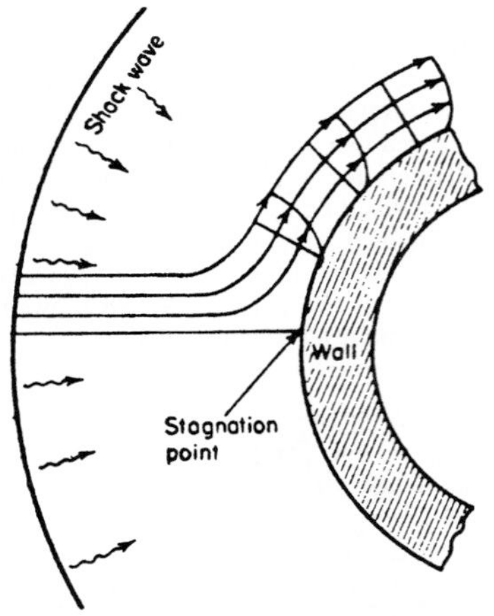

Figure 8.5.50 Velocity profile behind a shock wave.

This change in velocity is accompanied by a change in temperature and is dependent on the characteristics of the boundary layer. For example, heat transfer from a turbulent boundary layer may be of an order of magnitude greater than for laminar flow.

If the gas is brought to rest instantaneously, the total energy in the flow is converted to heat and the temperature of the air will rise. The resulting temperature is known as the **stagnation temperature** (see stagnation point, Fig. 8.5.50).

$$T_S = T_\infty\left(1 + \frac{\gamma - 1}{2}\mathrm{M}^2\right) \tag{8.5.29}$$

T_∞ is the ambient temperature of the gas at infinity, and γ is the specific-heat ratio of the gas. For undissociated air, $\gamma = 1.4$.

In general, this simple, one-dimensional relationship between velocity and temperature does not hold for temperature in the boundary layer. The laminae of the boundary layer are not insulated from each other, and there is cross conduction. This is associated with the **Prandtl number** Pr, which is defined as

$$\mathrm{Pr} = C_p\mu/k_r \tag{8.5.30}$$

where k_r = thermal conductivity of fluid, C_p = specific heat of fluid at constant pressure, and μ = absolute viscosity of fluid.

Defining the **recovery factor** r as the ratio of the rise in the idealistic wall and stagnation temperature over the free-stream temperature,

$$r = (T_{aw} - T_\infty)/(T_s - T_\infty) = (\mathrm{Pr})^n \tag{8.5.31}$$

For laminar flow, $r = (\mathrm{Pr})^{1/2}$; and for turbulent flow, $r = (\mathrm{Pr})^{1/3}$.

Prandtl number Pr greatly complicates the thermal computations, but since it varies only over a small range of values, a recovery factor of 0.85 is generally used for laminar flow and 0.90 for turbulent flow.

For a thermally thin wall, the rate of change of the surface temperature is a function of the rate of total heat input and the surface's ability to absorb the heat.

$$dT_w/dt = q_T/wcb \tag{8.5.32}$$

where t = time, q_T = forced convective heating + radiation heating − heat radiation from the skin, w = density of skin material, c = specific heat of skin material, b = skin thickness, and T_w = skin temperature.

The heat balance may then be written

$$wcb\left(\frac{dt_w}{dt}\right) = k_c\left[T_0\left(1 + r\frac{\gamma_B - 1}{2}\mathrm{M}_0^2\right) - T_w\right] + \alpha G_s A_p - \epsilon\sigma T_w^4 \tag{8.5.33}$$

where A_p = correction factor to account for area normal to radiation source, γ_B = specific heat ratio of boundary layer, ϵ = radiative emissivity of surface, σ = Stefan-Boltzmann constant [17.3×10^{-10} Btu/(h · ft^2 · °R^4), α = surface absorptivity, G_s = solar irradiation Btu/ (h·ft^2)].

For small time increments Δt,

$$\frac{dT_w}{dt} = \frac{T_{w2} - T_{w1}}{\Delta t}$$

Then

$$T_{w2} = T_{w1} + \frac{\Delta t}{wcb}\left\{h_c\left[T_0\left(1 + \frac{\gamma_B - 1}{2}\mathrm{M}_0^2\right) - T_w\right] + \alpha G_s A_p - T_w^4\right\} \tag{8.5.34}$$

The local heat-transfer coefficient h_c is defined as

$$h_c = \frac{k_r}{r^4/3}C_p C_f \rho_0 V_g \tag{8.5.35}$$

where C_p = specific heat of air; C_f = local skin-friction coefficient; ρ_0 = density outside of boundary layer; V = velocity outside of boundary layer; g = acceleration of gravity.

For a cone, k_r = 1,800 (laminar Re < 2 × 10^6) and $k_r = 1{,}800r^{4/3}$ (turbulent).

For a flat-plate transition, Reynolds number is 1 × 10^6.

Figure 8.5.49 gives laminar-skin-friction coefficient for a cone. For a flat plate, multiply C_f by $\sqrt{3/2}$. For turbulent skin-friction coefficient for a cone,

$$\frac{0.242}{\sqrt{A^2 C_f T_w/T_0}}(\sin^{-1}\psi + \sin^{-1}\theta) = 0.41\log\frac{\mathrm{Re}}{2}C_f - 1.26\log\frac{T_w}{T_0}$$

where

$$\psi = \frac{2A^2 - B}{\sqrt{B^2 + 4A^2}} \qquad A^2 = \frac{(\gamma_0 - ½)\mathrm{M}_0^2}{T_w/T_0}$$

$$\theta = \frac{B}{\sqrt{B^2 + 4A^2}} \qquad B = \frac{1 + (\gamma_0 - ½)\mathrm{M}_0^2}{T_w/T_0} - 1$$

For a flat plate, use Re instead of Re/2 in the above equation.

Figure 8.5.51 shows data on stagnation and adiabatic wall temperatures.

The primary **measurements in aerodynamics** are pressure measurements. A well-aligned pitot tube with an impact-pressure hole at its nose and static-pressure holes 10 diameters or more back from the nose will accurately measure the **impact pressure** and the **static pressure** of a uniform gas stream. Up to sonic speed the impact pressure is

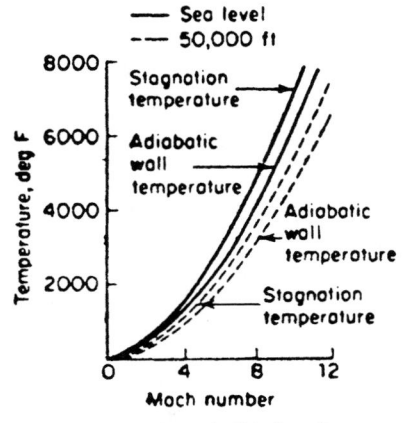

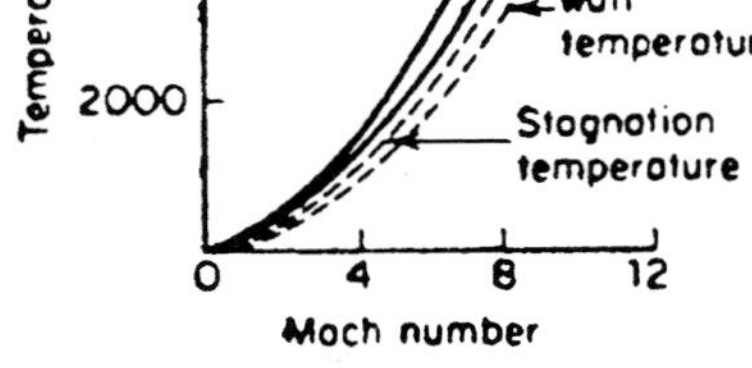

Figure 8.5.51 Variation of stagnation and adiabatic wall temperature with Mach number.

identical with stagnation pressure, but at supersonic speeds, a detached shock wave forms ahead of the probe, through which there is drop in stagnation pressure; the portion of the shock wave just ahead of the probe is normal, so that the equation for water flow in layer form over horizontal tubes (see "Transmission of Heat by Conduction and Convection" in Sec. 2) gives the relation of the measured impact pressure to the isentropic stagnation pressure.

Force measurements of total lift, drag, side force, pitching moment, yawing moment, and rolling moment on models are made on wind-tunnel balances just as at low speed. Small internal-strain-gage balances are often used to minimize strut interference.

An open thermocouple is unreliable for determining **stagnation temperature** if it is in a stream of high velocity. Figure 8.5.52 shows a simple temperature probe in which the fluid is decelerated adiabatically

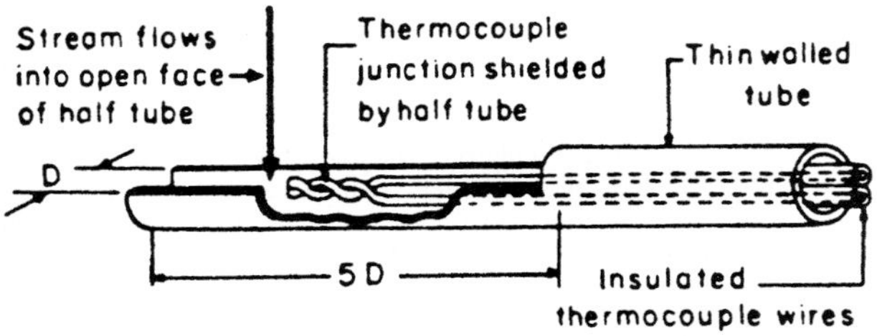

Figure 8.5.52 Stagnation-temperature probe (recovery factor = 0.98).

before its temperature is measured. The recovery factor used in equations, for this probe, accurately aligned to the stream, is 0.98.

Optical measurements in high-speed flow depend on the variation of index of refraction with gas density. This variation is given by

$$n = 1 + k\rho \tag{8.5.36}$$

where $k = 0.116$ ft^3/slug for air. A Mach-Zehnder **interferometer** (Fig. 8.5.53*a*), is capable of giving accurate density information for two-dimensional and axially symmetric flows. The **Schlieren** optical system (Fig. 8.5.53*b*) is sensitive to density gradients and is the most commonly used system to determine location of shock waves and regions of compression or expansion. The **shadowgraph** optical system is the simplest system (Fig. 8.5.53*c*) and is sensitive to the second space derivative of the density.

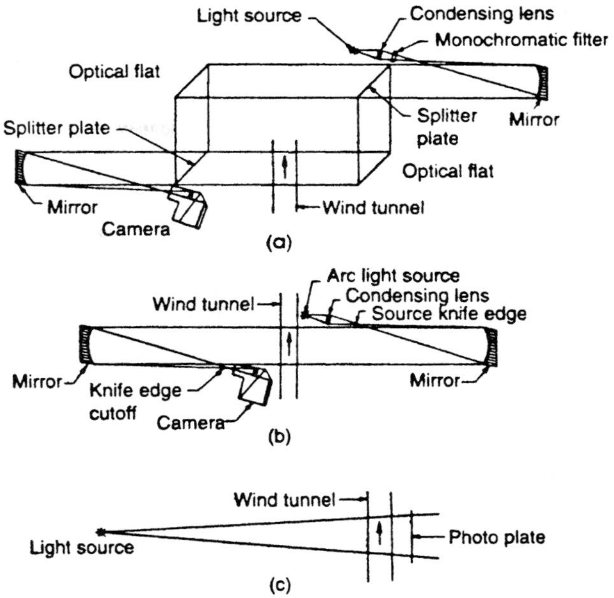

Figure 8.5.53 Optical systems for observing high-speed flow phenomena. (*a*) Interferometer; (*b*) Schlieren (two-mirror) system; (*c*) shadowgraph.

Further Reading: "Aerospace Source Book 2003," Aviation Week & Space Technology, McGraw-Hill, New York. "Air International," 1989-2001, Vol. 36-60, Key Publishing Ltd, Stamford, Lincolnshire, U.K. Anderson, J. D., Jr., 1999, "Aircraft Performance and Design," WCB/McGraw-Hill, Boston. Jenkinson, D., Simpson, P., and Rhodes, D. "Civil Jet Aircraft Design," AIAA Education Series, Washington, DC. Loftin, L. K., Jr., 1980. "Subsonic Aircraft: Evolution and the Matching of Size to Performance," NASA, Reference Publication 1060. Nicolai, L. M., 1984, "Fundamentals of Aircraft Design," 1st ed., METS, San Jose, CA. Raymer, D. P., 1989, "Aircraft Design—A Conceptual Approach," AIAA Educational Series, Washington, DC. Rolls-Royce, plc, 1986, "The Jet Engine," Rolls-Royce, plc, Derby, U.K. Roskam, J., 1980, "Aircraft Design, Parts I through VIII," DAR Corporation, Lawrence, KS. Gessow, A., and Myers, G. C., Jr., 1952, "Aerodynamics of the Helicopter," Macmillan, New York. Huenecke, K., 1987, "Modern Combat Aircraft Design," Naval Institute Press, Annapolis, MD. Khoury, G. A., and Gilbert, J. D., 1999, "Airship Technology," Cambridge Aerospace Series 10, Cambridge University Press, Cambridge, U.K. Mair, W. A., and Birdshall, D. J., 1992, "Aircraft Performance," Cambridge Aerospace Series 5, Cambridge University Press, Cambridge, U.K. McCormick, B. W., Jr., 1967, "Aerodynamics of V/STOL Flight," Academic Press, New York. Schaufele, R., 2000, "The Elements of Aircraft Preliminary Design," Aires, Santa Ana, CA. Shapiro, J., 1955, "Principles of Helicopter Engineering," McGraw-Hill, New York. Smetana, F. O., 2001, "Flight Vehicle Performance and Aerodynamic Control," AIAA Educational Series, Washington, DC. Stinton, D., 1988, "The Anatomy of the Airplane," 2nd ed., Blackwell Science, Oxford, and AIAA, Washington, DC. Stinton, D., 1983, "The Design of the Airplane," Van Nostrand, New York, Stinton, D., 1996, "Flying Qualities and Flight Testing of the Airplane," AIAA Educational Series, Washington, DC.

8.6 JET PROPULSION AND AIRCRAFT PROPELLERS

by Sanford Fleeter

References: Zucrow, "Aircraft and Missile Propulsion," Vol. 2, Wiley. Hill and Peterson, "Mechanics and Thermodynamics of Propulsion," Addison-Wesley. Sutton, "Rocket Propulsion Elements," Wiley. Shorr and Zaehringer, "Solid Propellant Technology," Wiley. Forestor and Kuskevics, "Ion Propulsion," AIAA Selected Reprints. Theodorsen, "Theory of Propellers," McGraw-Hill. Smith, "Propellers for High Speed Flight," Princeton. Zucrow, "Principles of Jet Propulsion," Wiley. "Aircraft Propeller Handbook," Depts. of the Air Force, Navy, and Commerce, U.S. Government Printing Office, 1956. Kerrebrock, "Aircraft Engines and Gas Turbines," MIT Press. Oates, ed., "The Aerothermodynamics of Aircraft Gas Turbine Engines," AFAPL-TR-78–52, 1978. Fleeter, "Aeroelasticity for Turbomachine Applications," *AIAA J. of Aircraft,* **16,** no. 5, 1979. Bathie, "Fundamentals of Gas Turbines," Wiley. "Aeronautical Technologies for the Twenty-First Century," National Academy Press. Mattingly, Heiser, and Daley, "Aircraft Engine Design," AIAA.

Newton's reaction principle, based on the second and third laws of motion, is the theoretical basis for all methods of propelling a body either in (or on) a fluid medium or in space. Thus, the aircraft propeller, the ships screw, and the jet propulsion of aircraft, missiles, and boats are all examples of the application of the reaction principle to the propulsion of vehicles. Furthermore, all jet engines belong to that class of power plants known as **reaction engines.**

Newton's second law states that a change in motion is proportional to the force applied; that is, a force proportional to the rate of change of the velocity results whenever a mass is accelerated: $F = ma$. This can be expressed in a more convenient form as $F = d(mv)/dt$, which makes it clear that the application of the reaction principle involves the time rate of increase of the momentum (mv) of the body. It should be noted that the same force results from either providing a small acceleration to a large mass or a large acceleration to a small mass.

Newton's third law defines the fundamental principle underlying all means of propulsion. It states that for every force acting on a body there is an equal and opposite reaction force.

The application of this reaction principle for propulsion involves the indirect effect of increasing the momentum of a mass of fluid in one direction and then utilizing the reaction force for propulsion of the vehicle in the opposite direction. Thus, the reaction to the time rate of increase of the momentum of that fluid, called the **propulsive fluid,** creates a force, termed the **thrust,** which acts in the direction of motion desired for the propelled vehicle. The known devices for achieving the propulsion of bodies differ only in the methods and mechanisms for achieving the time rate of increase in the momentum of the propulsive fluid or matter.

The function of a propeller is to convert the power output of an engine into useful thrust. To do this it accelerates a mass of air in the direction opposite to the direction of flight, generating thrust via Newton's reaction principle. Figure 8.6.1 illustrates schematically, in the relative coordinate system for steady flow, the operating principle of the **ideal aircraft propeller.** Power is supplied to the propeller, assumed to be equivalent to an **actuator disk,** which imparts only an axial acceleration to the air flowing through it. The rotation of the actuator disk produces a **slipstream** composed of the entire mass of air flowing in the axial direction through the area of the actuator disk, that is, the circle swept by the propeller. Atmospheric air enters the slipstream with the flight speed V_0 and mass flow rate m_a. It leaves the slipstream with the **wake velocity** w. The thrust is given in Eq. (8.6.5). Propulsion systems utilizing propellers will be termed **propeller propulsion**. (For detailed discussions of the airplane propeller, see the section "Aircraft Propellers.")

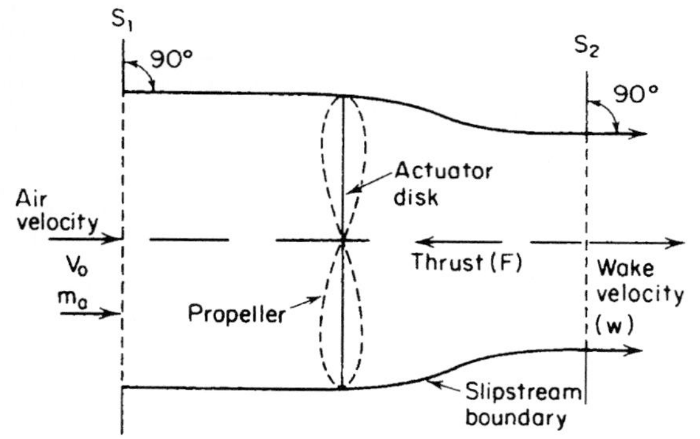

Figure 8.6.1 Ideal propeller in the relative coordinate system.

Jet propulsion differs from propeller propulsion in that the propulsive fluid (or matter), instead of being caused to flow around the propelled body, is ejected from within the propelled body in the form of one or more high-speed jets of fluid or particles. In fact, the jet propulsion engine is basically designed to accelerate a large stream of fluid (or matter) and to expel it at a high velocity. There are a number of ways of accomplishing this, but in all instances the resultant reaction or thrust exerted on the engine is proportional to the time rate of increase of the momentum of the fluid. Thus, jet engines produce thrust in an analogous manner to the propeller and engine combination. While the propeller gives a small acceleration to a large mass of air, the jet engine gives a larger acceleration to a smaller mass of air.

Theoretically, there are no restrictions on either the type of matter, called the **propellant,** for forming a high-velocity jet or the means for producing the **propulsive jet**. The selection of the most suitable propellant and the most appropriate jet propulsion engine is dictated by the specific mission for the propelled vehicle. For example, in the case of the jet propulsion of a boat, termed **hydraulic jet propulsion,** the propulsive jet is formed from the water on which the boat moves. For the practical propulsion of bodies through either the atmosphere or in space, however, only two types of propulsive jets are suitable:

1. For propulsion within the atmosphere, there is the jet formed by expanding a highly heated, compressed gas containing atmospheric air as either a major or a sole constituent. Such an engine is called either an **airbreathing** or a **thermal-jet engine**. If the heating is accomplished by burning a fuel in the air, the engine is a **chemical thermal-jet engine**. If the air is heated by direct or indirect heat exchange with a nuclear-energy source, the engine is termed a **nuclear thermal-jet engine.**

2. For propulsion both within and beyond the atmosphere of earth, there is the exhaust jet containing no atmospheric air. Such an exhaust jet is termed a **rocket jet,** and any matter used for creating the jet is called a **propellant**. A rocket may consist of a stream of gases, solids, liquids, ions, electrons, or a plasma. The assembly of all the equipment required for producing the rocket jet constitutes a **rocket engine**.

Modern airbreathing engines may be segregated into two principal types: (1) **ramjet engines,** and (2) **turbojet engines**. The turbojet engines are of two types: (1) the **simple turbojet engine,** and (2) the **turbofan,** or **bypass, engine**.

Rocket engines can be classified by the form of energy used for achieving the desired jet velocity. The three principal types of rocket engines are (1) **chemical rocket engines,** (2) **nuclear heat-transfer rocket engines,** and (3) **electric rocket engines**.

The propulsive element of a jet-propulsion engine, irrespective of type, is the **exhaust nozzle** or orifice. If the exhaust jet is gaseous, the assembly comprising all the other components of the jet-propulsion engine constitutes a gas generator for supplying highly heated, high-pressure gases to the exhaust nozzle.

These classifications of jet-propulsion engines apply to the basic types of engines. It is possible to have combinations of the different types of thermal-jet engines and also combinations of thermal-jet engines with rocket engines; only the principal types are discussed here.

ESSENTIAL FEATURES OF AIRBREATHING OR THERMAL-JET ENGINES

In the subsequent discussions a *relative coordinate system* is employed wherein the atmospheric air flows toward the propulsion system with the flight speed V_0, and the gases leave the propulsion system with the velocity w, *relative to the walls of the propulsion system.* Furthermore, steady-state operating conditions are assumed.

Ramjet Engine Figure 8.6.2 illustrates schematically the essential features of the simple ramjet engine. It comprises three major

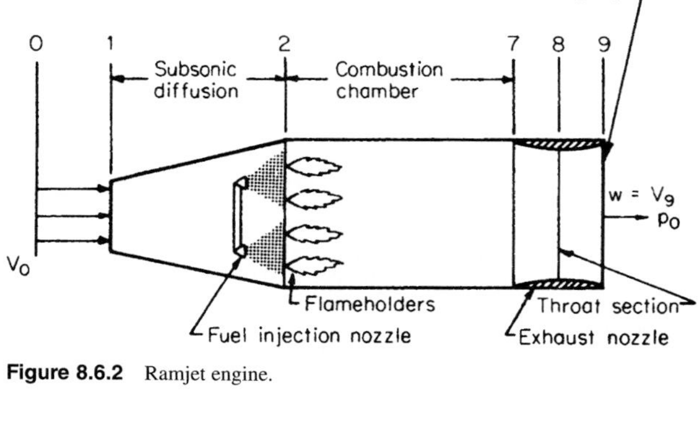

Figure 8.6.2 Ramjet engine.

components: a diffusion or inlet system, (0-2); a combustion chamber (2-7); and an exhaust nozzle (7-9). In this simple ramjet engine, apart from the necessary control devices, there are no moving parts. However, for accelerating the vehicle or for operating at different Mach numbers, a variable geometry diffuser and exhaust may be required.

The operating principle of the ramjet engine is as follows. The freestream air flowing toward the engine with a velocity V_0 and Mach number M_0 is decelerated by the inlet diffuser system so that it arrives at the entrance to the combustion chamber with a low Mach number, on the order of $M_2 = 0.2$. When the inlet-flow Mach number is supersonic, this diffuser system consists of a supersonic diffuser followed by a subsonic diffuser. The supersonic diffuser decelerates the inlet flow to approximately unity Mach number at the entrance to the subsonic diffuser; this deceleration is accompanied by the formation of shock waves and by an increase in the pressure of the air (diffusion). In the subsonic diffuser the air is further diffused so that it arrives at the entrance to the combustion chamber with the required low Mach number. If P_0 is the total pressure of the free-stream air having the Mach number M_0 and P_2 that for the air entering the combustion system with the Mach number M_2, then it is desirable that P_2/P_0 be a maximum.

In the combustion chamber a fuel is burned in the air, thereby raising the total temperature of the gases entering the exhaust nozzle (see Sec. 5) to approximately $T_7 = 4{,}260°R$ (2,366 K). Most generally a liquid hydrocarbon fuel is used, but experiments have been conducted with solid fuels, liquid hydrogen, liquid methane, and "slurries" of metallic fuels in a liquid fuel. The combustion process is not quite isobaric because of the pressure drops in the combustion chamber, because of the increase in the momentum of the working fluid due to heat addition, and because of friction. The hot gases are discharged to the atmosphere, after expanding in the exhaust nozzle, with the relative velocity $w = V_9$.

Since the ramjet engine can function only if there is a ram pressure rise at the entrance to the combustion chamber, it is not self-operating at zero flight speed. It must, therefore, be accelerated to a flight speed which permits the engine to develop sufficient thrust for accelerating the vehicle it propels to the design flight Mach number. Consequently, a ramjet-propelled missile, for example, must either be launched by dropping it from an airplane or be **boosted** to the required flight speed by means of **launching,** or **bootster,** rockets. It appears that the most appropriate flight regime for the ramjet is at low to moderate supersonic Mach numbers, between approximately $M_0 = 2$ and 5. The lower limit is associated with the necessary ram pressure rise while the upper limit is set by the problem of either cooling or protecting the outer skin of the engine body.

Scramjet Low-cost access to space is one of the major reasons for the interest in **hypersonic propulsion** (speeds above Mach 5). A vehicle to provide access to space must be capable of reaching very high Mach numbers, since orbital velocity (17,000 mi/h) is about Mach 24. The ramjet engine, which must slow the entering air to a low subsonic velocity, cannot be used much above Mach 5 or 6. The solution is to decelerate the flow in the engine to a supersonic Mach number lower than the flight Mach number but greater than the local speed of sound. This propulsion device is the **supersonic combustion ramjet,** or **scramjet,** which is capable, in principle, of operation to Mach 24. As shown schematically in Fig. 8.6.3, in contrast to the ramjet, the scramjet engine requires that fuel be added to the air, mixed, and burned, all at supersonic velocities, which is a significant technical challenge.

The advantage derived from supersonic combustion of the liquid fuel (usually liquid hydrogen) is that the diffuser of the scramjet engine is required to decelerate the air entering the engine from M_0 to only approximately $M_2 = 0.35M_0$ instead of the $M_2 = 0.2$, which is essential with subsonic combustion. This elimination of the subsonic diffusion increases the diffuser efficiency, reduces the static pressure in the combustor (and, therefore, the engine weight and heat transfer rate), and increases the velocity in the combustor (thereby decreasing engine frontal area). It is the combination of the above effects that makes ramjet propulsion at high flight Mach numbers (>7) feasible.

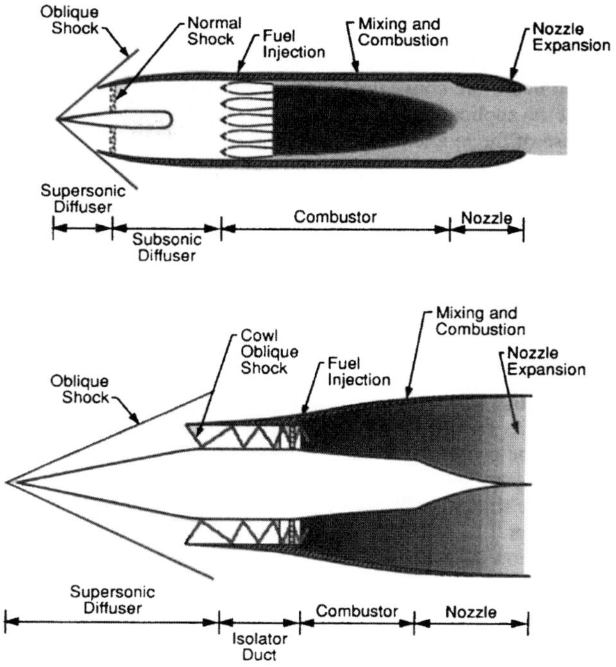

Figure 8.6.3 Ramjet and scramjet flow configurations.

Pulsejets may be started and operated at considerably lower speeds than ramjets. A pulsejet is a ramjet with an air inlet which is provided with a set of shutters loaded to remain in the closed position. After the pulsejet engine is launched, ram air pressure forces the shutters to open, fuel is injected into the combustion chamber, and is burned. As soon as the pressure in the combustion chamber equals the ram air pressure, the shutters close. The gases produced by combustion are forced out the jet nozzle by the pressure that has built up within the chamber. When the pressure in the combustion chamber falls off, the shutters open again, admitting more air, and the cycle repeats at a high rate.

Simple Turbojet Engine Figure 8.6.4 illustrates schematically the principal features of a simple turbojet engine, which is basically a gasturbine engine equipped with a propulsive nozzle and diffuser. Atmospheric air enters the engine and is partially compressed in the diffusion system, and further compressed to a much higher pressure by the air compressor, which may be of either the axial-flow or centrifugal type. The highly compressed air then flows to a combustion chamber wherein sufficient fuel is burned to raise the total temperature of the gases entering the turbine to approximately $T_4 = 2{,}160°R$ (1,200 K) for an uncooled turbine. The maximum allowable value for T_4 is limited by metallurgical and stress considerations; it is desirable, however, that T_4 be as high as possible. The combustion process is approximately

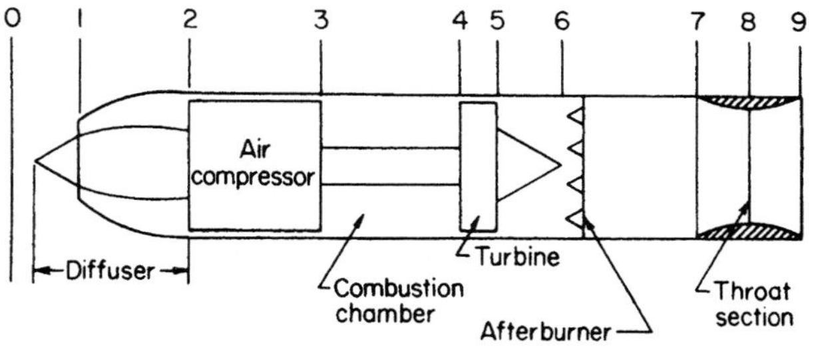

Figure 8.6.4 Simple turbojet engine.

isobaric. The highly heated air, containing approximately 25 percent of combustion products, expands in the turbine, which is directly connected to the air compressor, and in so doing furnishes the power for driving the air compressor. From the turbine the gases pass through a tailpipe which may be equipped with an **afterburner**. The gases are expanded in a suitably shaped exhaust nozzle and ejected to the atmosphere in the form of a high-speed jet.

Like the ramjet engine, the turbojet engine is a continuous-flow engine. It has an advantage over the ramjet engine in that its functioning does not depend upon the ram pressure of the entering air, although the amount of ram pressure recovered does affect its overall economy and performance. The turbojet is the only airbreathing jet engine that has been utilized as the sole propulsion means for piloted aircraft. It appears to be eminently suited for propelling aircraft at speeds above 500 mi/h (805 km/h). As the design flight speed is increased, the ram pressure increases rapidly, and the characteristics of the turbojet engine tend to change over to those of the ramjet engine. Consequently, its top speed appears to be limited to that flight speed where it becomes more advantageous to employ the ramjet engine. It appears that for speeds above approximately 2,000 mi/h (3,219 km/h) it is advantageous to use some form of a ramjet engine.

The thrust of the simple turbojet engine increases rapidly with T_4, because increasing T_4 increases the jet velocity V_j. Actually, V_j increases faster than the corresponding increase in T_4. It is also an inherent characteristic of the gas-turbine engine which produces shaft power, called a **turboshaft** engine, that its useful power increases proportionally faster than a corresponding increase in its turbine inlet temperature T_4. Because of the decrease in strength of turbine materials with increase in temperature, the turbine blades, stators, and disks require cooling at $T_4 > 2{,}160°\text{R}$ (1,200 K) approximately. A typical advanced turbine environment has **gas temperatures** of 3,600°R (2,000 K), with cooling provided to maintain a blade **metal temperature** of 2,250°R (1,250 K). Note that gas temperatures are being pushed toward the 4,600°R (2,556 K) stoichioemetric limit of JP4 fuel. The cooling air is bled from the compressor at the appropriate stage (or stages) and used to cool the stator blades or rotor blades by convective, film, or transpiration heat transfer. Up to 10 percent of the compressor air may be bled for turbine cooling, and this air is "lost" for turbine work for that blade row where it is used for cooling. Consequently "trade" studies must be made to "weigh" the increased complexity of the engine and the turbine work loss due to air bleed against the increased engine performance associated with increased T_4.

As in the case of any gas-turbine power plant, the efficiencies of the components of the turbojet engine have an influence on its performance characteristics, but its performance is not nearly as sensitive to changes in the efficiency of its component machines as is a gas turbine which delivers shaft power. (See Sec. 7.)

As noted previously, two types of compressors are currently employed, the axial-flow compressor and the centrifugal compressor. Irrespective of the type, the objectives are similar. The compressor must be reliable, compact, easy to manufacture, and have a small frontal area. Because of the limited air induction capacity of the centrifugal compressor, also called the radial compressor, engines for developing thrusts above 7,000 lb (31 kN) at static sea level employ axial-flow compressors (see Sec. 10).

The turbojet engine exhibits a rather flat thrust-versus-speed curve. Because the ratio of the takeoff thrust to thrust in flight is small, certain operational problems exist at takeoff. Since the exhaust gases from the turbine contain considerable excess air, the jet velocity, and consequently the thrust, can be increased by burning additional fuel in the tailpipe upstream from the exhaust nozzle. By employing **"tailpipe burning,"** or **"afterburning,"** as it is called, the thrust can be increased by 35 percent and at 500 mi/h, in a tactical emergency, by approximately 60 percent. With afterburning, the temperature of the gases entering the nozzle T_7, is of the order of 3,800°R (2,110 K).

Turbofan, or Bypass, Engine For a fixed turbine inlet temperature, the jet velocity from a simple turbojet engine propelling an airplane at subsonic speed is relatively constant. The propulsive efficiency depends on the ratio of the flight speed to the jet velocity and increases as the ratio increases. On the other hand, the thrust depends on the difference between the jet velocity and the flight speed; the larger the difference, the larger the thrust per unit mass of air induced into the engine. By reducing the jet velocity and simultaneously increasing the mass rate of airflow through the engine, the **propulsive efficiency** can be increased without decreasing the thrust. To accomplish this, a **turbofan or bypass engine** is required.

Figure 8.6.5 illustrates schematically the components of a turbofan engine. There are two turbines, a low-pressure turbine (LPT) and a high-pressure turbine (HPT); one drives the air compressor of the hot-gas generator and the other drives the fan. Air enters the fan at the rate of $\dot{m}_{aF}$ and is ejected through the nozzle of area A_{7F} with the jet velocity $V_{jF} < V_j$, where V_j is the jet velocity attained by the air flowing through the hot-gas generator with the mass flow rate $\dot{m}_{a1}$; the hot-gas generator is basically a turbojet engine.

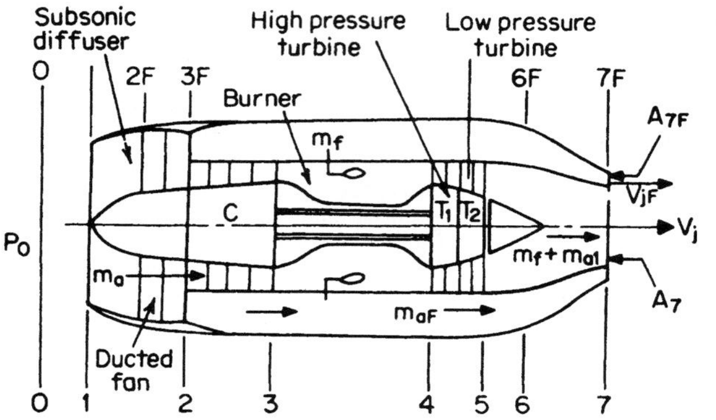

Figure 8.6.5 Schematic arrangement of components of turbofan engines (subsonic flight).

A key parameter is the **bypass ratio,** defined as the ratio of the air mass flow rate through the fan bypass duct to that through the hot-gas generator $(\dot{m}_{aF}/\dot{m}_{a1})$. Bypass ratios of 5 to 10 are in use, with values of about 10 utilized in modern large commercial transport engines, for example, the Trent 800, the GE-90, and the PW-4,000. The hot-gas generator is basically a turbojet engine. The fuel is added to $\dot{m}_{a1}$ at the rate $\dot{m}_f$. The hot-gas stream $(\dot{m}_{a1} + \dot{m}_f)$ is discharged to the atmosphere through A_7 with the velocity V_j. Turbofan engines produce an "overall" jet velocity V_{jTF} which is smaller than that for a turbojet engine operating with the same P_3/P_2 and T_4. The arrangement shown in Fig. 8.6.5 is for subsonic propulsion. Fuel can, of course, be burned in the fan air m_{aF} for increasing the thrusts of the engines. Practically all the newer commercial passenger aircraft, including regional jets, are propelled by turbofan engines. The advantage of the lower effective jet velocity V_{jTF} is twofold: (1) It increases the propulsive efficiency η_p by reducing $v = V_{jTF}/V_0$ and, consequently, raises the value of η_0. (2) The reduced jet velocity reduces the jet noise; the latter increases with approximately the eighth power of the jet speed.

The next generation of large transport engines will be conventional high-bypass engines, but will have a larger thrust capability and will offer further improvements in specific fuel consumption, achieved by using even higher overall pressure ratios (over 50:1) and turbine inlet temperature (over 1,850 K or 2,870°F) combined with even more efficient components (Fig. 8.6.6).

A highly loaded, multiblade swept variable-pitch propeller, the **propfan** or **advanced turboprop,** could be combined with the latest turbine engine technology. The resulting advanced turboprop would offer a potential fuel saving of 50 percent over an equivalent-technology turbofan engine operating at competitive speeds and altitudes because of the turboprop's much higher installed efficiency. To eliminate the gearbox development for a 20,000 shaft horsepower engine, the propfans of the **unducted fan (UDF)** engine were

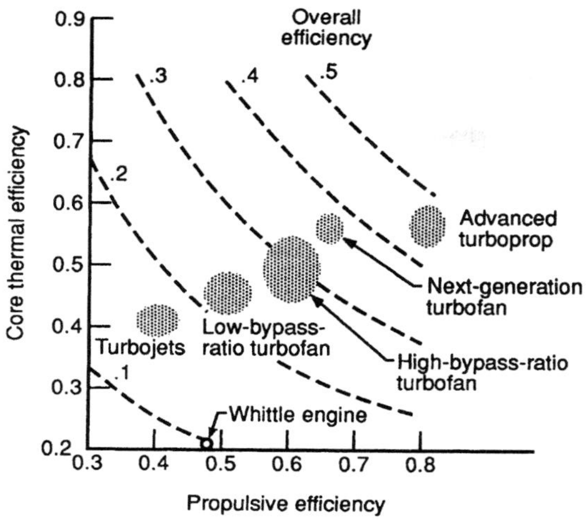

Figure 8.6.6 Efficiency trends.

directly driven with counterrotating turbine stages, a unique concept (Fig. 8.6.7). The projected specific fuel consumption at cruise for the resulting gearless, counterrotating UDF engine was 30 percent lower than that of the most modern turbofan engines, and about 50 percent lower than that of engines presently in use on 150-passenger airplanes. The fuel-saving potential of the turboprop has already been demonstrated. If economic conditions change, this concept could replace conventional high-bypass-ratio turbofans on small aircraft and military transports.

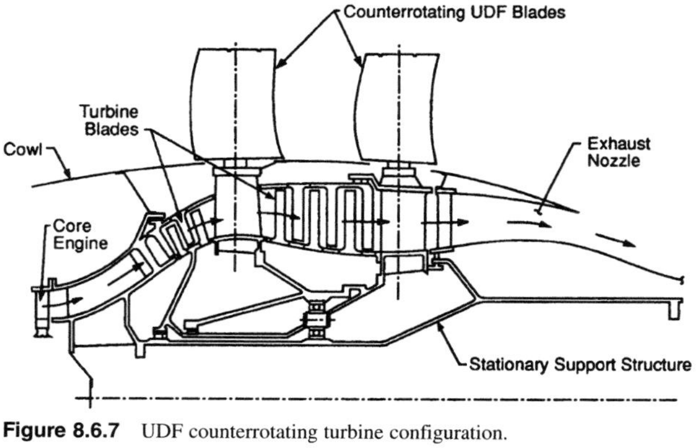

Figure 8.6.7 UDF counterrotating turbine configuration.

The work of traditional gas turbines is based on the constant-pressure **Brayton cycle.** However, the **Humphrey cycle**'s constant-volume detonation combustion, or **pulse detonation,** has potential of increased efficiency and fuel consumption as compared to constant-pressure-combustion turbofan and turbojet engines, while having fewer moving parts and being able to operate at both subsonic and supersonic speeds. The **pulse detonation engine** (PDE) is, thus, a new engine technology currently being developed.

PDEs are essentially tubes with fuel and air admitted into one end, with **detonation combustion** occurring rather than the deflagration combustion utilized in current gas turbines. Note that there are two types of combustion: deflagration and detonation. Consider a tube containing a fuel-air mixture. The fuel is ignited by a spark at one end and the combustion reaction then propagates down the tube. With deflagration, the burning is relatively slow. Detonation combustion, in contrast, results in a high-pressure detonation or shock wave that travels down the tube at supersonic speeds, resulting in a nearly constant-volume heat addition process that produces a high pressure in the combustor and provides the thrust. This wave compresses and ignites the fuel and air in a narrow high-pressure zone, thereby resulting in burning at thousands of meters per second. Exhaust gases are expelled at the open end of the tube and, as the exhaust gases clear out, a new cycle begins. A schematic of PDE operation is shown in Fig. 8.6.8. The operation of multitube configurations at high frequencies (100+ Hz) produces near-constant thrust.

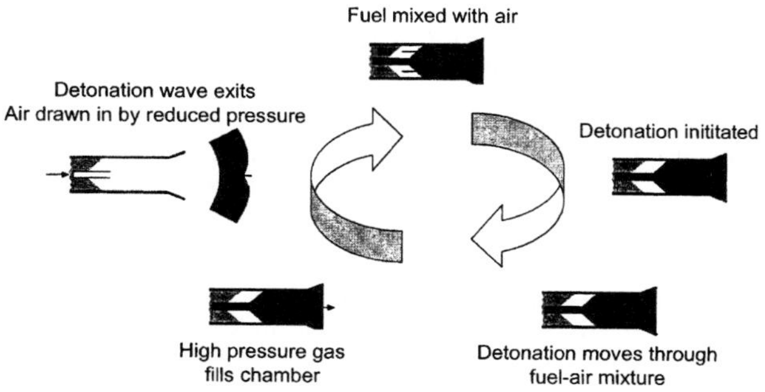

Figure 8.6.8 Pulse detonation engine schematic.

To generate the critical high-pressure combustion zone, the fuel, air, and the ignition spark must be precisely coordinated, thereby generating the **deflagration-to-detonation** transition **(DDT)**. The DDT process, whereby an ordinary flame suddenly accelerates into a detonation, has four phases. In the first phase, deflagration is initiated by an electric spark. The second phase is characterized by flame acceleration and shock wave formation. This is followed by the formation and amplification of explosion centers from pockets of reactants that create small blast waves. Finally, in the last phase, a strong detonation wave is formed by coupling between the amplified blast waves and the reaction zone.

There are three categories of pulse-detonation-based propulsion systems being considered. **Pure PDEs** are the simplest and comprise an array of detonation tubes, an inlet, and a nozzle. The upper speed limit of a pure PDE is about Mach 4 with a hydrocarbon fuel, but could be higher with hydrogen. **Combined-cycle PDEs** consist of a pure PDE combined with a ramjet/scramjet flow path or other propulsion cycle, where each cycle operates in a different speed range to optimize overall engine performance. **Hybrid PDEs** involve pure PDEs in combination with turbomachinery. For example, a hybrid turbofan-PDE would combine a turbofan with a PDE: the turbofan core engine would still power the large fan, but the bypass air would flow into a ring of PDEs, thereby producing significantly more thrust.

ESSENTIAL FEATURES OF ROCKET ENGINES

Figure 8.6.9 is a block-type diagram illustrating the essential features of a rocket engine, which comprises three main components: (A) a supply of propellant material contained in the rocket-propelled vehicle, (B) a

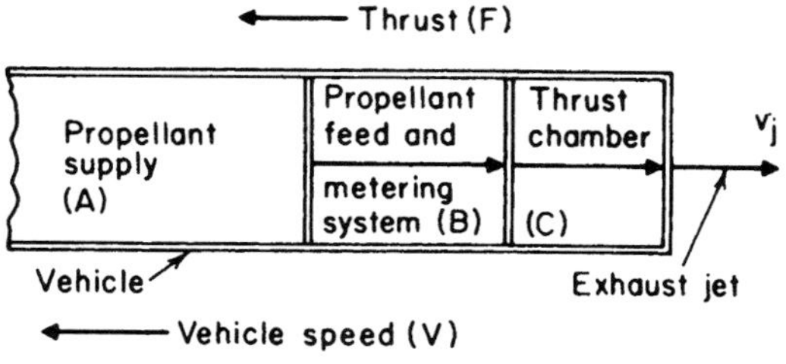

Figure 8.6.9 Essential features of a rocket engine.

propellant feed and metering system, and (C) a **thrust chamber,** also called a **rocket motor, thrustor,** or **accelerator.** In any rocket engine, energy must be added to the propellant as it flows through the thrustor. In Fig. 8.6.9 propellant material from the supply is metered and fed to the thrust chamber, where energy is added to it. As a consequence of the energy addition, the propellant is discharged from the thrustor with the jet velocity V_j. The thrustor F acts in the direction opposite to that for the jet velocity V_j.

Chemical Rocket Engines

All chemical rocket engines have two common characteristics: (1) They utilize chemical reactions in a thrust chamber to produce a high-pressure, high-temperature gas at the entrance to a converging-diverging exhaust nozzle; (2) the hot propellant gas expands in flowing through the exhaust nozzle, and the expansion process converts a portion of the thermal energy, released by the chemical reaction, into the kinetic energy associated with a high-velocity gaseous-exhaust jet. Chemical rocket engines may be grouped into (1) liquid-bipropellant rocket engines, (2) liquid-monopropellant rocket engines, and (3) solid-propellant rocket engines.

Liquid-Bipropellant Rocket Engine Figure 8.6.10 illustrates the essential features of a liquid-bipropellant rocket engine employing **turbopumps** for feeding two propellants, an **oxidizer** and a **fuel,** to a rocket motor. The motor comprises (1) an injector, (2) a combustion chamber, and (3) a converging-diverging exhaust nozzle. The liquid propellants are fed under pressure, through the injector, into the combustion chamber, where they react chemically to produce large volumes of high-temperature, high-pressure gases. For a given propellant combination, the **combustion temperature** T_c depends primarily on the oxidizer-fuel ratio (by weight), termed the **mixture ratio,** and to a lesser extent upon the combustion static pressure p_c. When the mass rate of flow of the liquid propellants equals that of the exhaust gases, the combustion pressure remains constant—the mode of operation that is usually desired.

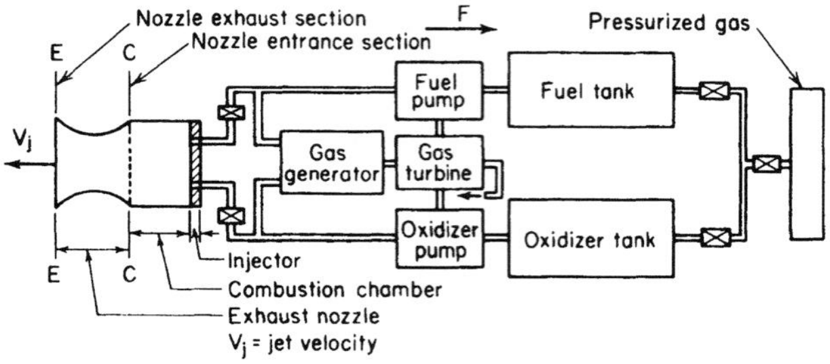

Figure 8.6.10 Essential features of a liquid-bipropellant rocket engine.

If the bipropellants react chemically when their liquid streams come in contact with each other, they are said to be **hypergolic.** Propellants which are not hypergolic are said to be **diergolic,** and some form of ignition system is required to initiate combustion.

Except in those cases where the operating duration of the rocket motor is very short or where the combustion temperature is low, means must be provided for protecting the interior walls of the motor. The two most common methods are ablative cooling and regenerative cooling. In **ablative cooling** the inner surfaces of the thrust chamber are covered with an ablative material which vaporizes, thereby providing some cooling. Ordinarily, the material leaves a "char" which acts as a high-temperature insulating material. For high-performance engines which are to operate for relatively long periods, regenerative cooling is employed. In **regenerative cooling,** one of the propellants is circulated around the walls before injection into the motor. In some engines the regenerative cooling is combined with local liquid film cooling at critical areas of the thrust chamber. See Table 8.6.1 for bipropellant combinations.

Refer to Figure 8.6.10. By removing the oxidizer tank, the oxidizer pump, and the plumbing associated with the liquid oxidizer, one obtains

Table 8.6.1 Calculated Values of Specific Impulse for Liquid Bipropellent Systems[a]
[P_c = 1,000 lbf/in² (6,896 × 10³ N/m²) to P_e = 1 atm]
Shifting equilibrium; isentropic expansion; adiabatic combustion; one-dimensional flow

Oxidizer	Fuel	$\dot{m}_o/\dot{m}_f$	ρ	Temperature T_c °R	Temperature T_c K	c^*	I (s)
Chlorine trifluoride (CTF) (ClF_3)	Hydrazine	2.80	1.50	6,553	3,640	5,961	293.1
	MMH[b]	2.70	1.41	5,858	3,254	5,670	286.0
	Pentaborane (B_5H_9)	7.05	1.47	7,466	4,148	5,724	289.0
Fluorine (F_2)	Ammonia	3.30	1.12	7,797	4,332	7,183	359.5
	Hydrazine	2.30	1.31	8,004	4,447	7,257	364.0
	Hydrogen	7.70	0.45	6,902	3,834	8,380	411.1
	Methane	4.32	1.02	7,000	3,889	6,652	343.8
	RP-1	2.62	1.21	6,839	3,799	6,153	318.0
Hydrogen peroxide (100% H_2O_2)	Hydrazine	2.00	1.26	4,814	2,674	5,765	287.4
	MMH	3.44	1.26	4,928	2,738	5,665	284.8
IRFNA[c]	Hydine[d]	3.17	1.26	5,198	2,888	5,367	270.3
	MMH	2.57	1.24	5,192	2,885	5,749	275.5
Nitrogen tetroxide (N_2O_4)	Hydrazine	1.30	1.22	5,406	3,002	5,871	292.2
	Aerozine-50[e]	2.00	1.21	5,610	3,117	5,740	289.2
	MMH	2.15	1.20	5,653	3,141	5,730	288.7
Oxygen (LOX) (O_2)	Hydrazine	0.91	1.07	5,667	3,148	6,208	312.8
	Hydrogen	4.00	0.28	4,910	2,728	7,892	291.2
	Methane	3.35	0.82	6,002	3,333	6,080	310.8
	Pentaborane	2.21	0.91	7,136	3,964	6,194	318.1
	RP-1	2.60	1.02	6,164	3,424	5,895	300.0

[a] Based on "Theoretical Performance of Rocket Propellant Combinations," Rocketdyne Corporation, Canoga Park, CA.
[b] MMH: Monomethyl hydrazine ($N_2H_3CH_3$).
[c] IRFNA: Inhibited red fuming nitric acid; 84.4% (HNO_3), 14% (N_2O_4), 1% (H_2O), 0.6% (HF).
[d] Hydine: 60% UDMH[f], 40% Deta[g].
[e] Aerozine-50: 50% UDMH[f], 50% hydrazine (N_2H_4).
[f] UDMII: Unsymmetrical dimethylhydrazine ($(CH_3)_2N_2H_4$).
[g] DETA: Diethylenetriamine ($C_2H_{13}N_3$).

the essential elements of a liquid-monopropellant rocket engine. Such an engine does not require a liquid oxidizer to cause the monopropellant to decompose and release its thermochemical energy.

There are basically three groups of monopropellants: (1) liquids which contain the fuel and oxidizer in the same molecule, for example, hydrogen peroxide (H_2O_2) or nitromethane (CH_3NO_2); (2) liquids which contain either the oxidizer or the fuel constituent in an unstable molecular arrangement, for example, hydrazine (N_2H_4); and (3) synthetic mixtures of liquid fuels and oxidizers. The most important liquid monopropellants are hydrazine and hydrogen peroxide (up to 98% H_2O_2). Hydrazine can be decomposed by a suitable metal catalyst, which is ordinarily packed in a portion of the thrust chamber at the injector end. The decomposition of hydrazine yields gases at a temperature of 2,260°R (1,265 K); the gases are a mixture of hydrogen and ammonia. Hydrogen peroxide can be readily decomposed either thermally, chemically, or catalytically. The most favored method is catalytically; a series of silver screens coated with samarium oxide is tightly packed in a decomposition chamber located at the injector end of the thrust chamber. At P_c = 300 psia, the decomposition of hydrogen peroxide (98%) yields gases having a temperature of 2,260°R (1,256 K).

The specific impulse obtainable for a liquid monopropellant is considerably smaller than that obtainable from liquid-bipropellant systems. Consequently, monopropellants are used for such auxiliary purposes as thrust vernier control, attitude reaction controls for space vehicles and missiles, and for gas generation.

Solid-propellant Rocket Engine Figure 8.6.11 illustrates schematically a solid-propellant rocket engine employing an **internal-burning case-bonded grain;** the latter burns radially outward at a substantially constant rate. A solid propellant contains both its fuel and the requisite oxidizer. If the fuel and oxidizer are contained in the molecules forming the solid propellant, the propellant is termed a **double-base propellant.** Those solid propellants wherein a solid fuel and a solid oxidizer form an intimate mechanical mixture are termed either **composite or heterogeneous** solid propellants. The chemical reaction of a solid propellant is initiated by an igniter. In general, the configuration of a propellant grain can be designed so that the area of the burning surface of the propellant varies to give a prescribed thrust-versus-time curve.

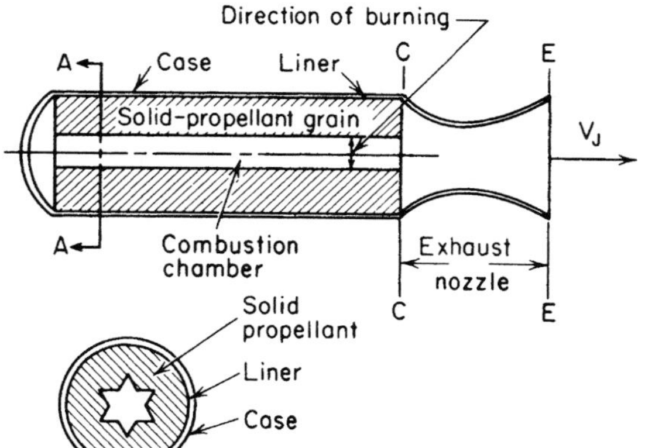

Figure 8.6.11 Essential features of an internal-burning case-bonded solid-propellant rocket engine.

Nuclear Heat-transfer Rocket Engine

Figure 8.6.12 illustrates schematically a nuclear heat-transfer rocket engine employing a **solid-core reactor**. The heat generated by the fissions of the uranium nucleus is utilized for heating a gaseous propellant, such as hydrogen, to a high temperature of 4,000°R (2,200 K) approx at the entrance section of the exhaust nozzle. The hot gas is ejected to the surroundings after expansion in a converging-diverging exhaust nozzle. The basic difference between the operating principles of a nuclear heat-transfer and a chemical rocket engine is the substitution of nuclear fission for chemical reaction as the source of heat for the propellant gas. Since the exhaust nozzle is the propulsive element, the remaining components of the nuclear heat-transfer rocket engine constitute the hot-gas generator.

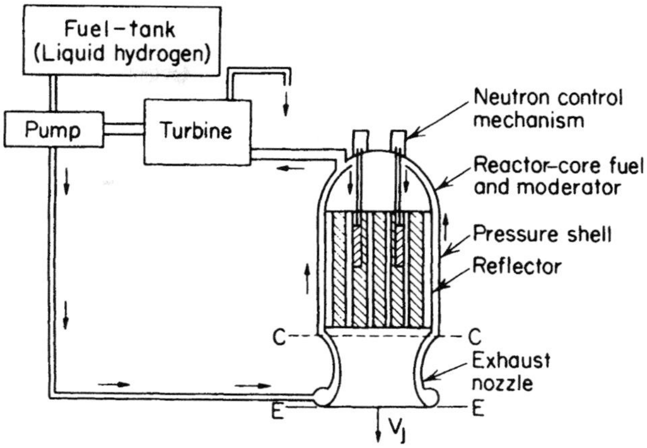

Figure 8.6.12 Schematic arrangement of a nuclear heat-transfer rocket engine.

The propellant in a nuclear heat-transfer rocket engine functions basically as a fluid for cooling the solid-core nuclear reactor. Its reaction is not based on energy considerations alone, but on such properties as specific heat, latent heat, molecular weight, and liquid density and on certain practical considerations.

Electric Rocket Engines

Several different types of electric rocket engines have been conceived; the principal types are described below. All of them require some form of power plant for generating electricity. Thus the power plant may be nuclear, solar-cell batteries, thermoelectric, or other. The choice of the type of power plant depends upon the characteristics of both the electric rocket engine and the space-flight mission. Figure 8.6.13 illustrates diagrammatically an electric rocket engine, comprising (1) a **nuclear power source,** (2) an energy-conversion unit for obtaining the desired form of electric energy, (3) a propellant feed and metering system, (4) an electrically operated thrustor, and (5) the requisite control devices. Electric rocket engines may be classified as (1) electrothermal, (2) electromagnetic, and (3) electrostatic.

Electrothermal Rocket Engine Figure 8.6.14*a* illustrates the type of engine which uses electric power for heating gaseous propellant to a high temperature before ejecting it through a converging-diverging exhaust nozzle. If an electric arc is employed for heating the propellant, the engine is called a **thermal arc-jet rocket engine.**

Electromagnetic Rocket Engine Figure 8.6.14*b* illustrates an electromagnetic, or plasma, rocket engine. There is a wide variety of such

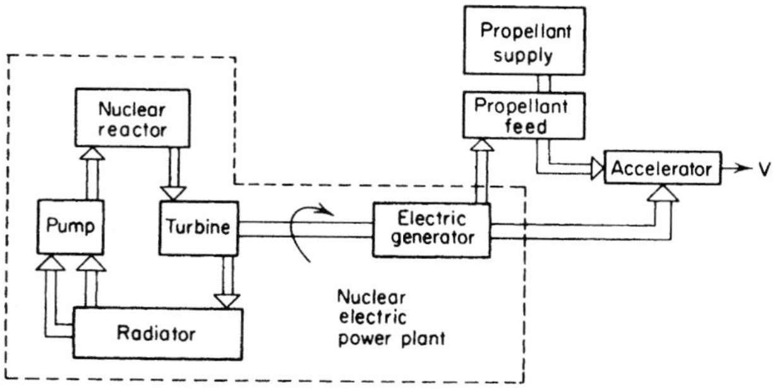

Figure 8.6.13 Essential features of an electric rocket engine.

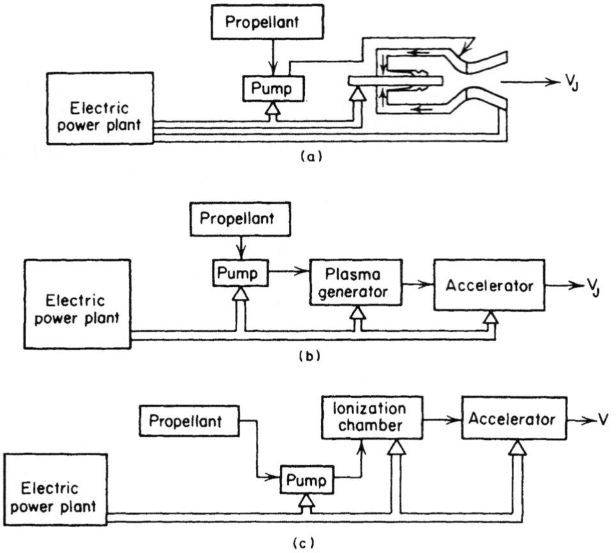

Figure 8.6.14 Electric rocket engine. (*a*) Electrothermal rocket engine; (*b*) electromagnetic rocket engine; (*c*) electrostatic (ion) engine.

engines, but all of them utilize the same operating principle. A plasma (a neutral ionized conducting gas) is accelerated by means of its interaction with either a stationary or a varying magnetic field. Basically, a plasma engine differs from a conventional electric motor by the substitution of a conducting plasma for a moving armature.

Electrostatic Rocket Engine Figure 8.6.14*c* illustrates schematically the electrostatic, or ion, rocket engine, comprising (1) an electric power plant; (2) a propellant supply; (3) ionization apparatus; (4) an ion accelerator; and (5) an electron emitter for **neutralizing** the ion beam ejected from the accelerator. Its operating principle is based on utilizing electrostatic fields for accelerating and ejecting electrically charged particles with extremely large velocities. The overall objective is to transform thermal energy into the kinetic energy associated with an extremely high-velocity stream of electrically neutral particles ejected from one or more thrustors. **Mercury ion engines** are currently being tested. The engine provides thrust by a rapid, controlled discharge of ions created by the ionization of atoms of the mercury fuel supply. Solar cells convert sunlight into electricity, which is then used to bombard the mercury with free electrons. This causes one electron to be stripped from each mercury atom, resulting in ionized mercury atoms. The ions mix with other electrons to form a plasma. A magnetic field set up between two metal screens at the aft end of the engine draws the positively charged ions from the plasma and accelerates them at very high exit speeds out the back of the engine. This creates a high-velocity exit beam producing low-level but extremely long-term thrust.

All electric rocket engines are low-thrust devices. The interest in such engines stems from the fact that chemical-rocket technology, for example, the space shuttle, is now so advanced that it has become feasible to place heavy *payloads,* such as a satellite equipped with an electric rocket engine, into an earth orbit. An electric rocket engine which is virtually inoperable terrestrially can be positioned in space and can then operate effectively because of the absence of aerodynamic drag and strong gravitational fields. Consequently, if a vehicle equipped with an electric rocket engine is placed in an earth orbit, the low-thrust electric rocket engine, under the conditions in space, will accelerate it to a large vehicle velocity if the operating time for the engine is sufficiently long.

In a chemical rocket engine, the energy for propulsion, as well as the mass ejected through the exhaust nozzle, is provided by the propellants, but the energy which can be added to a unit mass of the propellant gas is a fixed quantity, limited by the nature of the chemical bonds of the reacting materials. For that reason, chemical rocket engines are said to be **energy-limited rocket engines**. Although a nuclear heat-transfer rocket engine is also energy-limited, the limitation is imposed by the amount of energy that can be added per unit mass of propellant without exceeding the maximum allowable temperature for the materials employed for the solid-core reactor.

In an electric rocket engine, on the other hand, the energy added to the propellant is furnished by an electric power plant. The power available for heating the propellant is limited by the maximum power output of the electric power plant, accordingly, electric rocket engines are **power-limited**.

NOTATION

a = acoustic speed
a_0 = acoustic speed in free-stream air
a_2 = burning rate constant for solid propellant
A_e = cross-sectional area of exit section of exhaust nozzle
A_p = area of burning surface for solid propellant
A_t = cross-sectional area of throat of exhaust nozzle
c_p = specific heat at constant pressure
c_v = specific heat at constant volume
$c^* = P_c A_t / \dot{m}$ = characteristic velocity for rocket motor
C_d = discharge coefficient for exhaust nozzle
C_F = thrust coefficient
C_{Fg} = gross-thrust coefficient for ramjet engine
C_{Fn} = net thrust coefficient
D or $\mathcal{D}$ = drag
$D_i = \dot{m}_a V_0$ = ram drag for airbreathing or thermal-jet engine
d = diameter
d_p = diameter of propeller
E_{in} = rate at which energy is supplied to propulsion system
E_f = calorific value of fuel
E_p = calorific value of rocket propellants
$f = \dot{m}_f / \dot{m}_a$ = fuel-air ratio
$f' = f/\eta_B$ = fuel-air ratio for ideal combustion chamber
F = force or thrust
$F_i = m_e w - m_i V_0 + (p_e - p_0)Ae$ = thrust due to internal flow
$F_j = m_e V_j$ = jet thrust
$F_p = (p_e - p_0)A_e$ = pressure thrust
$F_g = F_i$ = gross thrust for ramjet engine
$F_g SFC$ = gross-thrust specific-fuel consumption for ramjet engine
g_c = correction factor defined by Newton's second law of motion
h = static specific enthalpy
Δh_n = enthalpy change for exhaust nozzle
H = total (stagnation) specific enthalpy
ΔH_c = lower heating value of fuel
I = specific impulse
$I_a = F/gm_a$ = air specific impulse
J = mechanical equivalent of heat; advance ratio, propeller
$k = c_p/c_v$ = specific heat ratio
L or $\mathcal{L}$ = lift
$\mathfrak{M}$ = mass
m = mass rate of flow
$\dot{m}_a$ = mass rate of air consumption for thermal-jet engine
$\dot{m}_e$ = mass rate of flow of gas leaving propulsion system
$\dot{m}_f$ = mass rate of fuel consumption
$\dot{m}_i$ = mass rate of flow of gas into propulsion system
$\dot{m}_o$ = mass rate of oxidizer consumption for rocket engine
$\dot{m}_p = \dot{m}_o + \dot{m}_f$ = mass rate of propellant consumption for rocket engine
$\bar{m}$ = molecular weight
M_e = momentum of gases leaving propulsion system in unit time
M_i = momentum of gases entering propulsion system in unit time
$M_0 = V_0/a_0$ = Mach number of free-stream air (flight speed)

n = revolutions per unit time
p = absolute static pressure; pitch of propeller blade
p_c = absolute static pressure of gases in combustion chamber
p_e = absolute static pressure in exit section of exhaust nozzle
p_0 = absolute static pressure of free-stream ambient air
P = absolute total (stagnation) pressure
P_c = absolute total pressure at entrance to exhaust nozzle
$P_0 = p_0\left(1+\frac{k-1}{2}M_0^2\right)^{k/(k-1)}$ = absolute total pressure of free-stream air
$\mathcal{P}$ = propulsion power
$\mathcal{P}_L$ = leaving loss
$\mathcal{P}_T$ = thrust power
$q = \rho V^2/2$ = dynamic pressure
$q_0 = \rho_0 V_0^2/2 = k_0 p_0 M_0^2/2$ = dynamic pressure of free-stream air
Q = torque
Q_i = heat supplied to actual combustor
Q_i' = heat supplied to ideal combustor (no losses)
r_0 = linear burning rate for solid propellant
$R = R_u/\bar{m}$ = gas constant
R_u = universal gas constant $\bar{m}R$
t = absolute static temperature
t_p = temperature of solid propellant prior to ignition
T = absolute total (stagnation) temperature
T_c = absolute total temperature of gas entering exhaust nozzle of rocket motor
T_2 = absolute total temperature at entrance to air compressor (see Fig. 8.6.4)
T_3 = absolute total temperature at exit section of air compressor (see Fig. 8.6.4)
T_3' = absolute total temperature at exit section of ideal compressor (see Fig. 8.6.4) operating between same pressure limits as actual compressor
T_4 = absolute total temperature at entrance to turbine (see Fig. 8.6.4)
T_5 = absolute total temperature at exit from turbine (see Fig. 8.6.4)
T_5' = absolute total temperature at exit from ideal turbine
$u = \pi n d_p$ = propeller tip speed
V = velocity
V_F = forward speed of propeller
V_j = effective jet velocity
V_0 = velocity of free-stream air (flight speed)
w = velocity of exit gases relative to walls of exhaust nozzle or velocity of air in ultimate wake of propeller
$\dot{W}$ = weight rate of flow
$\dot{W}_o$ = weight rate of flow of oxidizer
$\dot{W}_p$ = weight rate of flow of propellants

Greek

$\alpha = T_4/t_0 = \alpha_d\alpha_1$ = cycle temperature ratio; angle of attack
$\alpha_d = T_2/t_0$ = diffusion temperature ratio
$\alpha_1 = T_4/T_2$
β = bypass ratio for turbofans; helix angle for propeller blade
$\delta = P/P_{std}$ corrected pressure
η = efficiency
$\eta_B = f'/f$ = efficiency of combustion for thermal-jet engine
$\eta_c = (T_3' - T_2)/(T_3 - T_2)$ = isentropic efficiency of compressor
η_d = isentropic efficiency of diffuser
$\eta_n = \varphi^2$ = isentropic efficiency of exhaust nozzle
$\eta_o = \mathcal{P}_T/E_{in}$ = overall efficiency of propulsion system
$\eta_p = \mathcal{P}_T/(\mathcal{P}_T + \mathcal{P}_L)$ = ideal propulsive efficiency
$\eta_t = (T_4 - T_5)/(T_4 - T_5')$ = isentropic efficiency of turbine
$\eta_{th} = \mathcal{P}_T/E_{in}$ = thermal efficiency of propulsion engine
$\lambda = gI_a/\sqrt{2gJc_p t_0}$ = thrust parameter for turbojet engine; $½ + ½ \cos\phi$ = divergence coefficient for exhaust nozzle for rocket engines
$v = V_0/w$ = speed ratio
ρ = density
$\bar{\rho}$ = mean density
$\Omega = \sqrt{k}\left(\frac{2}{k+1}\right)^{(k-1)(2k-1)}$
ω = angular velocity of propeller shaft
ω' = rate of rotation of slipstream at propeller
ω'' = rate of rotation of slipstream in ultimate slipstream
Φ = fan velocity coefficient = V_F/u
ϕ = semiangle of exhaust-nozzle divergence; effective helix angle
$\varphi = \sqrt{\eta_n}$ = velocity coefficient for exhaust nozzle
ρ = density
σ = solidity of propeller
$\Theta = (P_3/p_0)^{(k-1)/k} = \Theta_d\Theta_c$ = cycle pressure ratio parameter
$\Theta_c = (P_3/P_2)^{(k-1)/k}$ = compressor pressure-ratio parameter
$\Theta_d = (P_2/P_0)^{(k-1)/k}$ = diffuser pressure-ratio parameter
$\Theta_n = (P_7/p_0)^{(k-1)/k} = (P_7/p_9)^{(k-1)/k}$ = nozzle pressure-ratio parameter
$\Theta_t = (P_4/P_5)^{(k-1)/k}$ = turbine pressure-ratio parameter
$\theta = T/T_{std}$ = corrected temperature

Subscripts

(a) numbered
0 = free stream
1 = entrance to subsonic diffuser
2 = exit from subsonic diffuser
3 = entrance to combustion chamber
4 = entrance to turbine of turbojet engine
5 = exit from turbine of turbojet engine
6 = tail-pipe entrance
7 = entrance to exhaust nozzle
8 = throat section of exhaust nozzle
9 = exit section of exhaust nozzle

(b) lettered
a = air
B = burner or combustion chamber
b = blade
c = compressor
d = diffuser
e = exit section; effective
F = fan
f = fuel
h = hydraulic
n = nozzle
o = overall or oxidizer
p = propellant
P = propulsive
std = standard
t = turbine or throat, as specified in text

Statement on Units The dynamic equations employed in the following sections are written for *consistent sets of units,* that is, for sets in which 1 *unit of force* = 1 *unit of mass* × 1 *unit of acceleration.* Consequently, the gravitational correction factor g_c in Newton's equation $F = (1/g_c)ma$ (see notation) has the numerical value unity and is omitted from the dynamic equations. When the equations are used for calculation purposes, g_c should be included and its appropriate value employed.

THRUST EQUATIONS FOR JET-PROPULSION ENGINES

Refer to Fig. 8.6.15, which illustrates schematically a rotationally symmetrical arbitrary propulsion engine immersed in a uniform flow field. Because of the reactions between the fluid flowing through the engine, called the **internal flow,** and the interior surfaces wetted by the internal flow a resultant **axial force** is produced, that is, a force collinear with the longitudinal axis of the engine. If an axial force acts

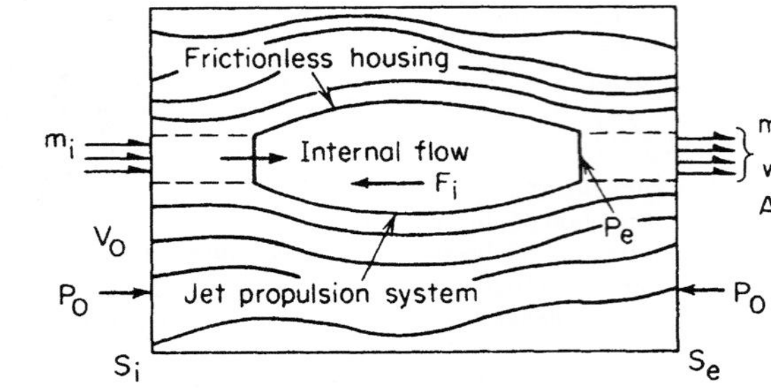

Figure 8.6.15 Generalized jet-propulsion system.

in the **forward direction,** employing a relative coordinate system, it is called a **thrust;** if it acts in the **backward direction** it is called a **drag**. Similarly, the resultant axial force due to the **external flow,** the flow passing over the external surfaces of the propulsion system, is a thrust or drag depending upon whether it acts in the forward direction or the backward direction.

Application of the momentum equation of fluid mechanics to the generalized propulsion system illustrated in Fig. 8.6.15 gives the following equation for the thrust F_i due to the **internal flow**. Thus, if it is assumed that the external flow is frictionless, there is no change in the rate of momentum for the external flow between S_i and S_e. Hence, the thrust $F = F_i$ is due entirely to the internal flow, and

$$F = F_i = m_e w - \dot{m}_i V_0 + (p_e - p_0)A_e \tag{8.6.1}$$

In Eq. (8.6.1), $\dot{m}_e w = F_j =$ **jet thrust,** $\dot{m}_i V_0 = D_i =$ **ram drag,** and $(p_e - p_0)A_e = F_p =$ **pressure thrust**. Hence,

$$F_i = F_j - D_i + F_p \tag{8.6.2}$$

It is convenient to introduce a fictitious **effective jet velocity** V_j, which is defined by

$$F = \dot{m}_e V_j - \dot{m}_i V_0 = \dot{m}_e w - \dot{m}_i V_0 + (p_e - p_0)A_e \tag{8.6.3}$$

If the gases are expanded completely, in the exhaust nozzle, then $w = V_j$ and $F_p = 0$. No appreciable error is introduced, in general, if it is assumed that $F_p = 0$. For **thermal jet engines,** $\dot{m}_i = \dot{m}_a$ (see notation), and $\dot{m}_e = \dot{m}_a + \dot{m}_f$. Let $f = \dot{m}_f / \dot{m}_a$, $v = V_0/V_j$; then

$$F_i = \left(\frac{1+f}{v} - 1\right)\dot{m}_a V_0 + (p_e - p_0)A_e \tag{8.6.4}$$

In the case of uncooled turbojet engines, $\dot{m}_f$ is not significantly different from the fraction of m_a utilized for cooling the bearings and turbine disk. Consequently, no significant error is introduced by assuming that $\dot{m}_e = \dot{m}_a + \dot{m}_f \approx \dot{m}_a = \dot{m}$. Hence, for a simple **turbojet engine,** one may write

$$F_e = \dot{m}_a(V_j - V_0) = \dot{m}V_0(1/v - 1) \tag{8.6.5}$$

Equation (8.6.5) is also the thrust equation for an **ideal propeller;** in that case V_j is the **wake velocity** for the air leaving its slipstream. The thrust equation for a hydraulic jet propulsion system has the same form as Eq. (8.6.5). Let V_0 denote the speed of a boat, and $\dot{m}_w$ the mass rate of flow of the water entering the hydraulic pump and discharged by the exit nozzle with the velocity $w = V_j$ relative to the boat. Hence, for hydraulic jet propulsion

$$F = \dot{m}_s(V_j - V_0) = \dot{m}_w V_0(1/v - 1) \tag{8.6.6}$$

where $v = V_0/V_j$.

In the case of **rocket engines,** since they do not consume atmospheric air, $\dot{m}_a = 0$, and the flow of gas out of the rocket motor, under steady-state conditions, is equal to $\dot{m}_p - \dot{m}_o + \dot{m}_f$ (see notation).

The effective jet velocity V_j is larger than $V_e =$ **exit velocity** if $p_e > p_0$; that is, if the gases are **underexpanded**. The effective jet velocity is a useful criterion because it can be determined accurately from the measured values of F_i and $\dot{m}_p$ obtained from a static firing test of the rocket motor.

The ratio $F_i/\dot{m}_p$ is denoted by I and called either the **specific thrust** or the **specific impulse**. Hence

$$I = F/\dot{m}_p = V_j/g_c \tag{8.6.7}$$

Although the dimensions of specific impulse are force/(mass)(s), it is conventional to state its units as *seconds.*

Equation (8.6.5) for the thrust of a **turbojet engine,** when expressed in terms of the effective jet velocity V_j, becomes

$$F = \dot{m}_a(V_j - V_0) \tag{8.6.8}$$

POWER AND EFFICIENCY RELATIONSHIPS

In a jet-propulsion engine the **propulsion element** is the **exhaust nozzle,** and the rate at which energy is supplied to it is called the **propulsion power,** denoted by $\mathcal{P}$. The rate at which the propulsion system does useful work is termed the **thrust power** $\mathcal{P}_T$ and is given by

$$\mathcal{P}_T = FV_0 \tag{8.6.9}$$

Assume that $p_e = p_0$, and that the only energy loss in a propulsion system is the **leaving loss** $\mathcal{P}_L = \dot{m}(V_j - V_0)^2/2$ that is, the kinetic energy associated with the jet gases discharged from the system; then the propulsive power is given by

$$\mathcal{P} = \mathcal{P}_L + \mathcal{P}_T \tag{8.6.10}$$

The **ideal propulsive efficiency** is defined, in general, by

$$\eta_P = \mathcal{P}_T/(\mathcal{P}_T + \mathcal{P}_L) = \text{thrust power/propulsion power} \tag{8.6.11}$$

For a turbojet engine and hydraulic jet propulsion,

$$\eta_P = 2v/(1+v) \tag{8.6.12}$$

For a chemical rocket engine,

$$\eta_P = 2v/(2+v^2) \tag{8.6.13}$$

The propulsive efficiency η_P is of more or less academic interest. Of more importance is the overall efficiency η_o, which for airbreathing and rocket engines is defined by

$$\eta_o = \eta_{\text{th}}\eta_P \tag{8.6.14}$$

where

$$\eta_{\text{th}} = \mathcal{P}/E_{in} = \text{thermal efficiency of system} \tag{8.6.15}$$

and E_{in} is the rate at which energy is supplied to the propulsion system.

The overall efficiency of a hydraulic jet-propulsion system is given by

$$\eta_o = \eta_{\text{th}}\eta_h\eta_p = \eta_h\eta_{\text{th}}[2v/(1+v)] \tag{8.6.16}$$

where η_{th} is the thermal efficiency of the power plant which drives the water pump, and η_h is the hydraulic efficiency of the water pump. To achieve a reasonable fuel consumption rate, η_h must have a larger value. For an airbreathing jet engine, $E_{in} = \dot{m}_f(E_f + V_0^2/2)$, and

$$\eta_0 = \frac{2v}{1+v}\frac{\mathcal{P}}{\dot{m}_f(E_f + V_0^2/2)} \tag{8.6.17}$$

The ratio $\dot{m}_f/F$ is called the **thrust specific-fuel consumption** (TSFC) and is measured in mass of fuel per hour per unit of thrust. Hence,

$$\text{TSFC} = \dot{m}_f/F = f\dot{m}_a/F \tag{8.6.18}$$

For a rocket engine, $E_{in} = \dot{m}_p(E_p + V_0^2/2)$, so that

$$\eta_0 = \frac{2v}{1+v^2}\frac{\mathcal{P}}{\dot{m}_p(E_p + V_0^2/2)} \tag{8.6.19}$$

PERFORMANCE CHARACTERISTICS OF AIRBREATHING JET ENGINES

The Ramjet Engine In ramjet technology the thrust due to the internal flow F_i is called the **gross thrust** and denoted by F_g. Refer to Fig. 8.6.2 and assume $\dot{m}_0 \approx \dot{m}_9 = \dot{m}_a$. Then

$$F_g = F_i \approx \dot{m}_a (V_j - V_0) \tag{8.6.20}$$

In level unaccelerated flight F_g is equal to the external drag of the propelled vehicle and the ramjet body.

It is customary to express the thrust capabilities of the engine in terms of the **gross-thrust coefficient** C_{Fg}. *If* A_m is the maximum cross-sectional area of the ramjet engine, $q_0 = \rho_0 V_0^2/2 = k_0 p_0 M_0^2/2$ = the **dynamic pressure** of the free-stream air, then

$$C_{Fg} = \frac{F_g}{q_0 A_m} = \frac{2F_g}{A_m k_0 p_0 M_0^2} \tag{8.6.21}$$

In terms of the Mach numbers M_0 and M_9,

$$C_{Fg} = \frac{2A_9/A_m}{k_0 M_0^2}\left(\frac{P_9}{p_0}\frac{1+k_9 M_9^2}{P_9/p_9} - 1\right) - 2\frac{A_0}{A_m} \tag{8.6.22}$$

In a fixed-geometry engine, M_9 depends upon the total temperature T_7, the total pressure P_7, the fuel-air ratio f, the nozzle area ratio A_e/A_t, and the efficiency of the nozzle η_n. For estimating purposes, $k_0 = 1.4$ and $k_9 = 1.28$, when the engine burns a liquid-hydrocarbon fuel. The manner in which C_{Fg} varies with altitude M_0 and fuel-air ratio f is shown schematically in Fig. 8.6.16.

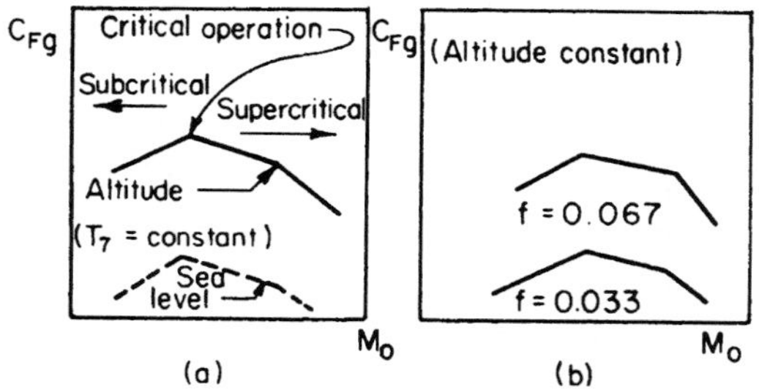

Figure 8.6.16 Effect of altitude and fuel-air ratio on the gross-thrust coefficient of a fixed-geometry ramjet engine. (*a*) Effect of altitude; (*b*) effect of fuel-air ratio.

If $\alpha = T_7/t_0$ = cycle temperature ratio, and k_B is the mean value of k for the combustion gases, the rate at which heat is supplied to the engine is

$$Q_i = Q_i'/\eta_B = \dot{m}_a c_{pB} t_0 (\alpha - T_2)/\eta_B \tag{8.6.23}$$

It is readily shown that

$$Q_i = \frac{A_0 p_0 V_0}{\eta_B J}\frac{k_B}{k_B - 1}\left[\alpha - \left(1 + \frac{k_0 - 1}{2}M_0^2\right)\right] \tag{8.6.24}$$

The **gross-thrust specific-fuel consumption** (F_gSFC) is, by definition,

$$F_g\text{SFC} = \dot{m}_f / F_g \tag{8.6.25}$$

The principal sources of loss are aerothermodynamic in nature and cause a decrease in total pressure between stations 0 and 7. For estimating purposes, assuming $M_0 = 2.0$ and $T_6 = 3{,}800$°R (2,111 K), the total pressures across different sections of the engine may be assumed to be approximately those tabulated below.

Part of engine	Total pressure ratio
Supersonic diffuser (0-1)	$P_1/P_0 = 0.92$
Subsonic diffuser (1-2)	$P_2/P_1 = 0.90$
Flameholders (2-7)	$P_7/P_2 = 0.97$
Combustion chamber (2-7)	$P_7/P_2 = 0.92$
Exhaust nozzle (7-9)	$P_9/P_7 = 0.97$

The Simple Turbojet Engine A good insight into the design performance characteristics of the turbojet engine is obtained conveniently by making the following assumptions: (1) The mass rate of flow of working fluid is identical at all stations in the engine; (2) the thermodynamic properties of the working fluid are those for air; (3) the air is a perfect gas and its specific heats are constants; (4) there are no pressure losses due to friction or heat addition; (5) the exhaust nozzle expands the working fluid completely so that $p_9 = p_0$; and (6) the auxiliary power requirements can be neglected. Refer to Fig. 8.6.4 and let

$$\Theta = \left(\frac{P_3}{p_0}\right)^{(k-1)/k} \quad \Theta_d = \left(\frac{P_2}{p_0}\right)^{(k-1)/k} \quad \Theta_c = \left(\frac{P_3}{P_2}\right)^{(k-1)/k}$$

$$\Theta_t = \left(\frac{P_4}{P_5}\right)^{(k-1)/k} \quad \Theta_n = \left(\frac{P_7}{p_0}\right)^{(k-1)/k} = \left(\frac{P_7}{p_9}\right)^{(k-1)/k}$$

$$\alpha = T_4/t_0 \qquad \alpha_d = T_2/t_0 \qquad \alpha_1 = T_4/t_0$$

In view of the assumptions,

$$\Theta = \Theta_d \Theta_c = \Theta_t \Theta_n \tag{8.6.26}$$

and

$$\alpha = \alpha_a \alpha_1 \tag{8.6.27}$$

The diffuser pressure-ratio parameter Θ_d is given by

$$\Theta_d = 1 + \eta_d[(k-1)/2]M_0^2 \tag{8.6.28}$$

where η_d = isentropic efficiency of diffuser (0.75 to 0.90 for well-designed systems).

The turbine pressure-ratio parameter Θ_t is given by

$$1/\Theta_t = 1 - [(\Theta_c - 1)/\alpha_1 \eta_c \eta_t] \tag{8.6.29}$$

where η_t = the isentropic efficiency of the turbine (0.90 to 0.95).

Heat supplied, per unit mass of air, is

$$Q_1 = (c_p/\eta_B)(T_4 - T_3)$$

or

$$Q_i = \frac{c_p t_0}{\eta_B \eta_c}\alpha_d\left(\frac{\alpha}{\alpha_d}\eta_c - \eta_c - \Theta_c + 1\right) \tag{8.6.30}$$

where η_c is the isentropic efficiency of the air compressor (0.85 to 0.90), and η_B is the efficiency of the burner (0.95 to 0.99).

The enthalpy change for the exhaust nozzle Δh_n is given by

$$\frac{\Delta h_n}{c_p t_0} = \frac{\eta_n}{\eta_c}[\alpha\eta_c - \alpha_d(\Theta_c - 1)] \times \left\{1 - \frac{\alpha_1 \eta_c \eta_t}{\Theta_c[1 + \eta_d(\alpha_d - 1)][\alpha_1 \eta_c \eta_t - (\Theta_c - 1)]}\right\} \tag{8.6.31}$$

The **specific thrust,** also called the **air specific impulse,** is

$$I_a = F_i/\dot{m}_a g = (\sqrt{2\Delta h_n} - V_0)/\dot{m}_a \tag{8.6.32}$$

The overall efficiency η_o is given by

$$\eta_o = \eta_{\text{th}}\eta_p = \frac{I_a V_0}{JQ_i} = 2\eta_B\eta_c \frac{\lambda M_0\left(\frac{k-1}{2}\right)^{1/2}}{\alpha\eta_c - \left(1 + \frac{k-1}{2}M_0^2\right)(\eta_c + \Theta_c - 1)} \tag{8.6.33}$$

where η_B ranges from 0.95 to 0.99 and λ is the dimensionless turbojet specific thrust parameter.

The TSFC for a turbojet engine is given by

$$\text{TSFC} = \frac{\dot{m}_f}{F} \tag{8.6.34}$$

It can be shown by dimensional analysis, if the effects of Reynolds numbers are neglected, that the variables entering into the performance of a given turbojet engine may be grouped as indicated in Table 8.6.2.

Table 8.6.2 Performance of a Given Turbojet Engine

Nondimensional group	Uncorrected form	Corrected form
Flight speed	$V_0/\sqrt{t_0}$	$V_0/\sqrt{\theta}$
Rotational speed	$N/\sqrt{t}$	$N/\sqrt{\theta}$
Air flow rate	$\dot{W}_a\sqrt{T}/D^2P$	$\dot{W}_a\sqrt{\theta}/\delta$
Thrust	F/D^2P	F/δ
Fuel flow rate	$\dot{W}_f J\Delta H_c/D^2P\sqrt{T}$	$\dot{W}_f/\delta\sqrt{\theta}$

$\theta = T/T_{std} = T/519$ ($T/288$) = corrected temperature (exact value for T_{std} is 518.699°R).
$\delta = P/p_{std} = P/14.7$ ($P/1.013 \times 10^5$) = corrected pressure.

Figure 8.6.17 is a design-point chart presenting λ and η_o as functions of α, P_3/P_2, and Θ_c for the subsonic performance of simple turbojet engines. The curves apply to propulsion at 30,000 ft (9.14 km) altitude for engines having characteristic data indicated in the figure. Curve A illustrates the effect of pressure ratio P_3/P_2 or Θ_c for engines operating with $\alpha = 5.0$. It is seen that increasing the compressor pressure ratio increases η_o but it approaches a maximum value at P_3/P_2 approximately equal to 19. Curve B applies to the design point of engines having a fixed compressor pressure ratio of 10:1 but having different values of α, that is, turbine inlet temperature. It shows that increasing α much above $\alpha = 5$ reduces the overall efficiency and consequently TSFC. Curve C presents the design-point characteristics for engines operating with $\alpha = 5$ but having different compressor pressure ratios. The maximum value of λ is obtained at a relatively low pressure ratio: approximately $P_3/P_2 = 5$. Curve D presents the design-point characteristics of engines having a fixed compressor pressure ratio of 10:1 but different cycle temperature ratios. Increasing α gives significant increases in λ. Hence, large values of turbine inlet temperature T_4 offer the potential for large thrusts per unit of frontal area for such turbojet engines.

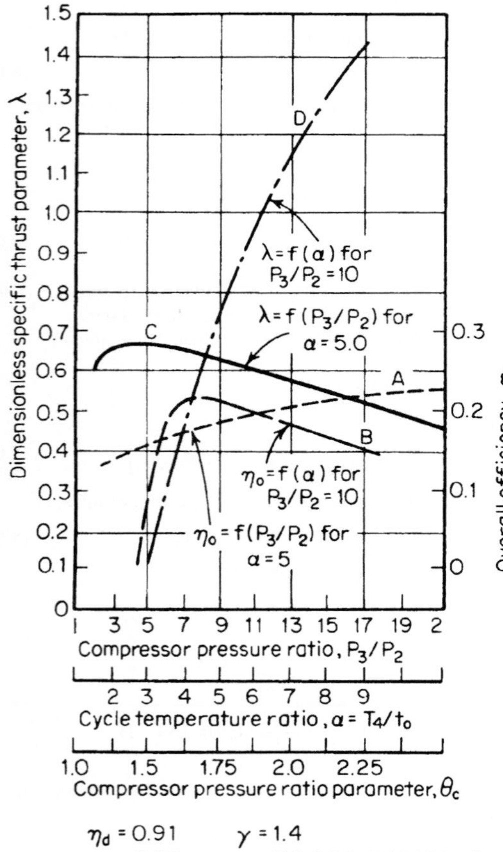

Figure 8.6.17 Dimensionless thrust parameter λ and overall efficiency η_0 as functions of the compressor pressure ratio (simple turbojet engine).

Figure 8.6.18 presents the design-point characteristics for engines having $\eta_c = 0.85$, $\eta_t = 0.90$, $\eta_r = 0.93$, and $\eta_B = 0.99$. Different values of diffuser efficiency are presented for flight Mach numbers $M_0 = 0.8$ ($\eta_d = 0.91$), $M_0 = 1.60$ ($\eta_d = 0.90$), and $M_0 = 2.4$ ($\eta_d = 0.88$). Two different classes of engines are considered, those with $P_3/P_2 = 4.0$ and those with $P_3/P_2 = 12.0$. The dimensionless thrust λ is plotted as a function of α with M_0 as a parameter. It is evident that increasing λ, that is, the turbine inlet temperature T_4, offers the advantage of a large increase in thrust per unit area of the engine under all the conditions considered in Fig. 8.6.18.

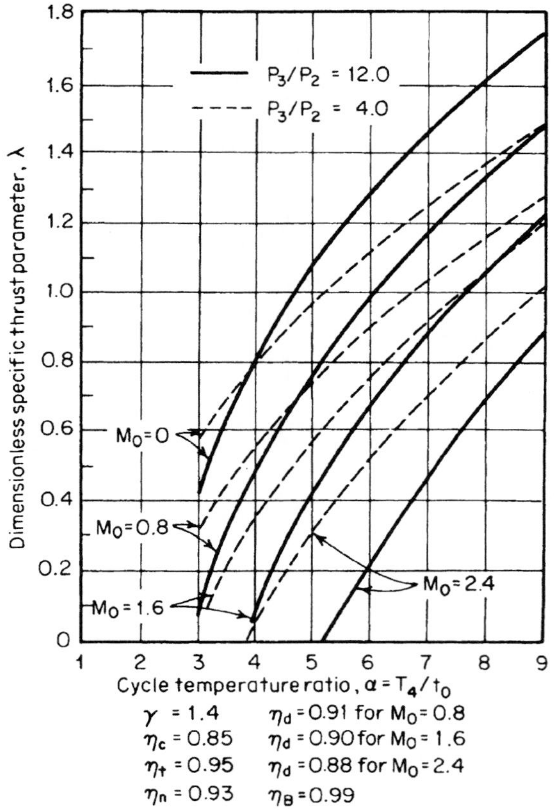

Figure 8.6.18 Dimensionless thrust parameter λ for a simple turbojet engine and its overall efficiency η_0 as functions of the cycle temperature ratio α, with the flight Mach number as a parameter.

Figure 8.6.19 presents the effect of flight speed on the design-point performance, at 30,000 ft (9.14 km) altitude, for engines having $\eta_c = 0.85$, $\eta_t = 0.90$, $\eta_B = 0.95$, $T_4 = 2{,}000$°R (1,111 K) and burning a fuel having a heating value $\Delta H_c = 18{,}700$ Btu/lbm (10,380 kcal/kgm). The magnitude of the compressor pressure ratio P_3/P_2 approaches unity when V_0 is approximately 1,500 mi/h, indicating that at higher speeds the compressor and turbine are superfluous; that is, the propulsion requirements would be met more adequately by a ramjet engine. Raising T_4 tends to delay the aforementioned condition to a high value of V_0.

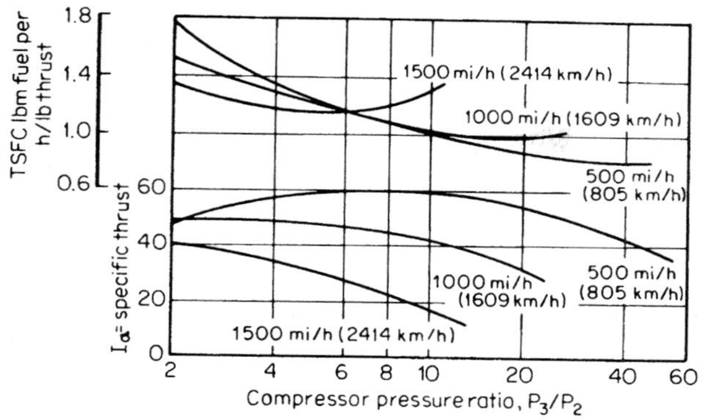

η_c = 0.85 $\quad \gamma$ = 1.40 for compressions
η_t = 0.90 $\quad \gamma$ = 1.34 for expansions
η_B = 0.95 $\quad z$ = 30,000 ft (9.14 km)
T_4 = 2000°R ΔH_c = 18,700 Btu/lbm (8.93 × 10^6 j/kgm)
= 1111 K

Figure 8.6.19 *TSFC* and specific thrust I_a functions of the compressor pressure ratio, with flight speed as a parameter (simple turbojet engine).

It is a characteristic of turbojet engines, since $\eta_o = \eta_{th}\eta_p$, that for a given M_0 increasing η_{th} causes Δh_n to increase, and hence the jet velocity V_j. As a consequence, η_p is decreased. It is this characteristic which causes $\eta_o = f(\alpha)$, curve B of Fig. 8.6.17, to be rather flat for a wide range of values of $\alpha = T_4/t_0$.

The curves of Figs. 8.6.17-8.6.19 are design-point performance curves; each point on a curve is the design point for a different turbojet engine. The performance curves for a specific engine are either computed from experimental data pertaining to its components operating over a wide range of conditions or obtained from testing the complete turbojet engine. It is customary to present the data in standardized form.

The Turbofan or the Ducted-Fan Engine Refer to Fig. 8.6.5*b*. Assume that the hot-gas flow $(\dot{m}_{a1} + \dot{m}_f)$ is ejected through the converging nozzle, area A_t, with the effective jet velocity V_j, and that the airflow $\dot{m}_{aF}$ is ejected through the converging annular nozzle, area A_{7F}, with the jet velocity V_{jF}. Let $\beta = \dot{m}_{aF}/\dot{m}_{a1}$ denote the bypass ratio, approximately 5 to 10 for modern transport engines.

The thrust equation for the turbofan engine is

$$F = \dot{m}_{a1}(1+f)V_j + \beta\dot{m}_{a1}V_{jF} - \dot{m}_{a1}(1+\beta)V_0 \tag{8.6.35}$$

where $V_{jF} < V_j$. The thrust per unit mass of airflow, or air specific impulse, is

$$I_a = F/\dot{m}_{a1}(1+\beta) \tag{8.6.36}$$

The TSFC is given by

$$\text{TSFC} = \dot{m}_f/F \tag{8.6.37}$$

One of the most important applications of turbofans is for **high-subsonic transport aircraft** where the fuel consumption in cruise is a major consideration. In general, as the bypass ratio increases, the TSFC decreases but the engine weight and external drag increase. The **optimum bypass ratio** for any particular application is determined by a trade-off between fuel weight, which is reduced by increasing the bypass ratio, and engine weight and external drag, which increase with bypass ratio for a given thrust. In general, the optimum value of β is that which makes the difference $(F - D_e)$ a maximum, where D_e is the external drag of the engine. Up to a certain point, the more sophisticated and efficient the engine, the larger the bypass ratio.

To achieve the maximum potential for the turbofan engine, high values of turbine inlet temperature T_4 are essential; that is, cooled turbine stator and rotor blades must be developed. There is an upper limit to T_4; above it the TSFC begins to increase. As T_4 is increased, however, the optimum value of β must also be increased to avoid excessive leaving losses in the propulsive jet.

Mechanical Design

Blades New technologies being developed to achieve higher thrust-to-weight ratios in advanced gas turbine engines also cause higher vibratory blade row responses and stresses that delay engine development and may adversely affect engine reliability. In particular, the design trend toward higher stage loadings and higher specific flow are being attained through increased tip speeds, lower radius ratios, lower aspect ratios, and fewer stages. The resultant axial-flow compressor blade designs utilize thin, low-aspect-ratio blading with corresponding high steady-state stresses. Also, the mechanical damping is considerably reduced in newer rotor designs, particularly those with integral blade-disk configurations **(blisks)** and in those without shrouds. These are similar trends in turbine blading, but they are further complicated by increased inlet and cooling temperatures that have resulted in thin-walled, complex cooling passage designs. As a result, advanced axial-flow blade designs, both compressors and turbines, feature low-aspect-ratio blading which affect the **structural integrity** of the blading. These problems can be classified into two general cateories: (1) **flutter** and (2) **forced response**. A fundamental parameter common to both categories is the **reduced frequency,** $k = \omega b/V$, where b is the blade semichord, ω the frequency of vibration, and V the relative free-stream velocity. For engine applications, the frequency corresponds generally to one of the lower-blade or coupled-blade disk **natural frequencies**.

Under certain conditions, a blade row operating in a completely uniform flow can enter into a self-excited vibrational **instability** termed **flutter.** The vibration is sustained by the extraction of energy from the uniform flow during each vibratory cycle. The outstanding feature of flutter is that **high stresses** exist in the blading, leading to short-term, high-cycle, **fatigue failures.** Its importance is evidenced by the fact that if a portion of a single blade fails due to flutter, the result may be instantaneous and total loss of engine power.

Figure 8.6.20 is a typical compressor map showing schematically the more common types of flutter. At subsonic Mach numbers, up to 0.8 to 0.9, either positive or negative **stall flutter** in the **bending** or **torsion** mode may occur. It is caused by operating beyond some critical airfoil angle of attack. A **critical reduced frequency** value of 0.8 has been cited for stall flutter; that is, if $k < 0.8$ stall flutter is possible. **Choking flutter** usually occurs at negative incidence angles at a part-speed condition with the blade operating either subsonically or transonically. The physical mechanism of choke flutter is not fully understood; both high negative incidence angles and choked flow are viable candidates. **Supersonic unstalled flutter** is largely associated with thin **fan blades**

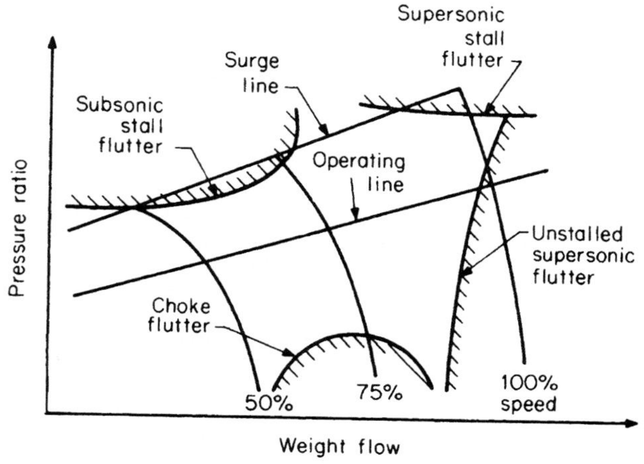

Figure 8.6.20 Common types of compressor flutter.

operating at supersonic relative flow velocities at the tip section. It is caused by a lagging of the unsteady aerodynamic forces relative to the blade motion and imposes a limit on the **high-speed operation** of the engine. The stress encountered during this type of flutter can be catastrophically large, decreasing rapidly as a constant speed line is traversed toward higher pressure ratios. High-speed operation near the surge line can lead to **supersonic stall flutter.** Because this flutter is associated with **high pressure ratios,** there can be strong shocks present within the blade passages, and in some circumstances, ahead of the blade row.

Destructive aerodynamic **forced responses** have been noted in all engine blading. These failure-level vibratory responses occur when a periodic forcing function, with frequency equal to a natural blade **resonant frequency,** acts on a blade row. Responses of sufficient magnitude to fail blades have been generated by a variety of sources including upstream and downstream blade rows, flow distortion, bleeds, and mechanical sources.

The importance of forced response lies in the **multiple sources** of excitation for any rotating system. The rotor speeds at which forced responses occur are predicted with **"Campbell"** or **speed-frequency** diagrams. These display the **natural frequency** of each blade mode versus rotor speed and, at the same time, the forcing function frequency (or **engine order** E lines) versus rotor speed, as indicated schematically in Fig. 8.6.21. These E lines represent the loci of available **excitation energy** at any rotational speed for 1, 2, 3, etc. excitations per revolution. Wherever these curves intersect, a potential source of destructive forced

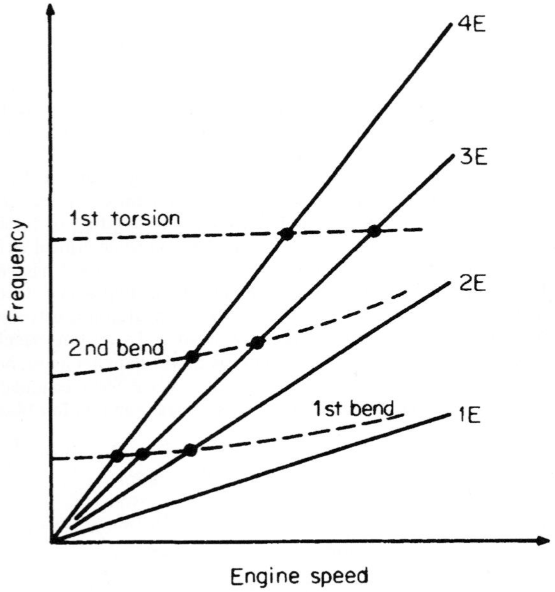

Figure 8.6.21 Schematic speed-frequency diagram.

response exists. Not all intersections can be avoided. Design practice is to eliminate the lower-order excitations from the operating range whenever possible. This is because the sources of most forced response energy usually result from these lower-order resonances, particularly the 2E line.

High-cycle fatigue (HCF) of turbomachine blading resulting from flow-induced vibrations is a significant problem throughout the gas turbine industry. To address this problem, various approaches have been developed. In these, the response of a tuned airfoil row, that is, a rotor with all airfoils having the same structural properties and thus identical natural frequencies, is analyzed. In fact, there are small airfoil-to-airfoil structural property variations that result, for example, from the manufacturing process or as a consequence of in-service wear. These are collectively referred to as **mistuning** and are known to lead to significant increases in airfoil resonant response amplitudes as compared to that of the tuned airfoil row, with mistuning thus often cited as a source of HCF in gas turbine engines. Hence, the key metric that characterizes the resonant response of mistuned bladed disks is the **amplification factor,** defined as the ratio of a blade's response amplitude on a mistuned bladed disk to its response amplitude on a tuned bladed disk.

Turbomachinery rotors typically have been bladed disks, with individual airfoils inserted into a slotted disk and retained by means of a dovetail or fir-tree attachment. However, advances in manufacturing techniques have resulted in bladed-disk assemblies with increased uniformity; that is, the blade-to-blade natural frequency variation, termed mistuning, is small. In addition, new design and manufacturing techniques have enabled the development and implementation of **integrally bladed rotors** (IBRs) wherein the blades and disk are machined from a single piece of material. IBRs have even less blade-to-blade mistuning than do bladed disks. Unfortunately, smaller mistuning does not translate into smaller amplification factors. On the contrary, as mistuning decreases, the maximum amplification factor on a disk tends to increase until some threshold is reached, at which point it decreases rapidly toward 1. Figure 8.6.22 shows this trend; it is based on studies wherein many hundreds of bladed disks

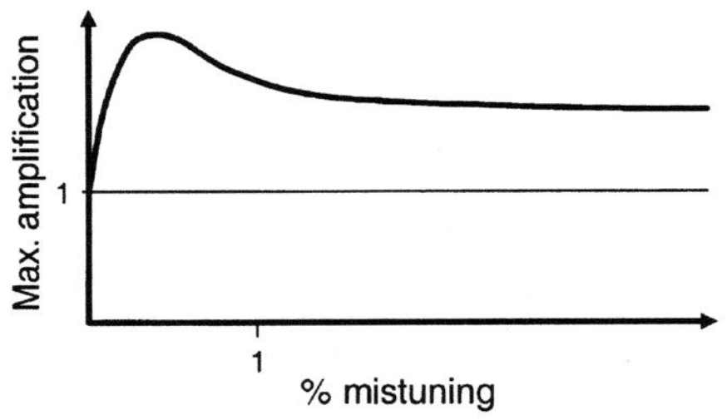

Figure 8.6.22 Amplification factor versus mistuning level.

are simulated for each percent mistuning, with the maximum resulting amplification factor taken. The decrease in mistuning due to new manufacturing methods is on the negative slope portion of Fig. 8.6.22, thus rendering the new manufacturing methods a strong impetus for improving models which predict mistuned bladed disk response. In contrast to the maximum amplification factor of Fig. 8.6.22, there is a **theoretical maximum amplification factor** that depends only on the number of blades N; specifically, amplification factor = $0.5(1+\sqrt{N})$. Taking these theoretical maxima into account during design is an important first step in building blade rows resistant to HCF failure.

Environmental Issues

The increasing worldwide concern for the environment has led to two technical issues becoming significant: (1) airport noise and (2) the Earth's ozone layer.

Aeroacoustics is an increasingly important issue in the design of advanced gas turbine engines for commercial aircraft. This is because engine certification requires meeting prevailing **International Civil Aviation Organization (ICAO) Chapter 3 noise regulations** established in 1978. Aircraft are certified to a standard that is actually a combination of three measures: the noise generated as the plane approaches the runway radiating to the community under its flight path; the noise it generates during takeoff radiating to the areas surrounding the runway; and the noise under the flight path further out in the community during takeoff (Fig. 8.6.23).

The ICAO Chapter 3 regulations specify approach, sideline, and takeoff perceived noise levels (EPNL) as a function of the aircraft takeoff gross weight (Fig. 8.6.24). The total or cumulative margin for a given aircraft defined as the sum of the margin at all three conditions. Note that all current-technology aircraft meet ICAO Chapter 3 regulations.

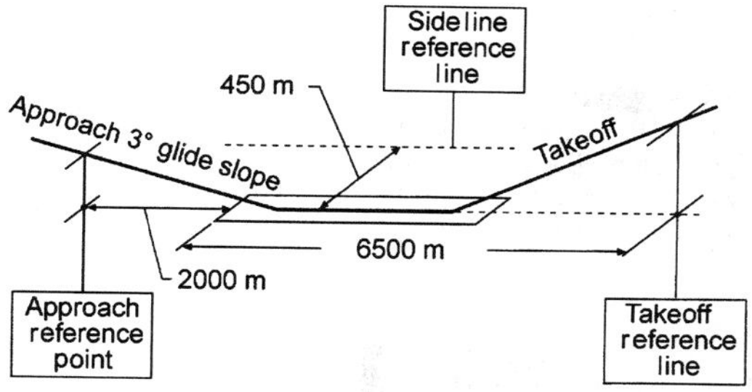

Figure 8.6.23 Aircraft certification noise measurement locations.

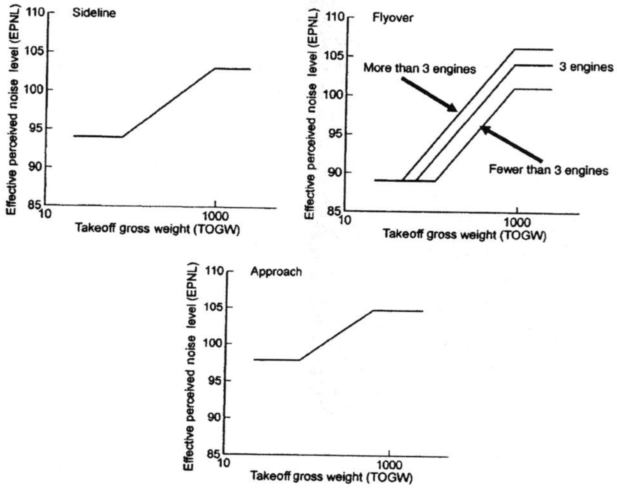

Figure 8.6.24 Chapter 3 regulations specify EPNL at approach, sideline, and takeoff.

Air travel has increased dramatically since 1978. To meet the demands of community groups for stricter aircraft noise requirements, ICAO has established a new set of limits which are to become effective in 2006—the **Chapter 4 rule**. These new limits require aircraft to be 10 dB (cumulative) quieter than current Chapter 3 levels. The use of trades to meet the new rules was also eliminated. With Chapter 3, an aircraft could exceed the limits on one measure of noise, takeoff, as long as it was lower than the limits at the other conditions by an offsetting amount. With Chapter 4, an aircraft must operate below the Chapter 3 curve at all three conditions: approach, sideline, and takeoff.

The high-bypass-ratio turbofans introduced in the late 1960s had a major effect on engine noise reduction. ICAO recognized this when it established the Chapter 3 regulations in 1978. Turbofans direct more air through the bypass than through the core. As a result, the maximum jet exhaust velocity is substantially reduced and the **turbomachinery noise** is more important. The fan, the low-pressure or boost compressor, and the low-pressure turbine are the primary noise sources for a high-bypass turbofan. Their noise signatures include a broadband noise level with large spikes or tones at multiples of blade passing frequencies (Fig. 8.6.25).

Current noise control and reduction methods for high-bypass turbofans are a combination of turbomachinery noise source control and suppression. Source control is accomplished by increasing the axial spacing between adjacent blade rows and by selecting blade and vane number combinations to produce *cutoff*, whereby the highest-energy acoustic modes decay exponentially with distance along the ducting. Source suppression is achieved with acoustic liners in the fan inlet and exhaust ducts and the core exhaust duct. As a result, future advanced turbofans will have much lower levels of turbomachinery noise. Thus, the dominant engine sources in these future engines are the fan at takeoff and approach and the turbulent mixing in the jet exhaust stream at takeoff.

The **emissions** generated by the combustion of aviation fuel include carbon dioxide (CO_2), carbon monoxide (CO), oxides of nitrogen (NO_x), and particulates such as soot and oxides of sulfur (SO_x). The effect that each has on the atmosphere depends on the altitude at which it is emitted and its concentration. Most commercial aircraft cruise in the **troposphere**, which ranges from an altitude of 11 mi (58,000 ft) at the equator to 6 mi (33,000 ft) at the poles. In this region, the temperature decreases with increasing altitude, resulting in large air motions and mixing. Increased NO_x concentration results in increased ozone concentration through complex photochemical reactions with CO and hydroxyls (HO_x). Ozone also acts as a global warming gas in the troposphere. Methane is another global warming gas, although NO_x reduces the concentration of methane.

It has been projected that, by 2050, aircraft emissions will increase by a factor of 4 because of an increase in air traffic and that the total contribution to **global warming** by aircraft will be a factor of 2 greater than from CO_2 alone. In addition, aircraft exhaust plumes contribute to global warming. Contrails are formed by water in the exhaust freezing in cold atmospheric conditions, with their formation aided by the particulates present in exhaust plumes. These contrails may turn into cirrus clouds, which lead to a net increase in warming.

In summary, pollutants generated by aircraft engines contribute to global warming because of the altitude at which they operate and the projected growth of the aviation industry. Hence, there is a distinct need to develop ultralow-emission combustor designs for aircraft engines. The two primary approaches to ultralow-emission combustors are rich and lean front-end designs. Each has demonstrated the potential to achieve low emissions.

The International Civil Aviation Organization (ICAO) standards limit aircraft engine emissions in the vicinity of an airport, up to 3,000 ft altitude. A standardized **landing-takeoff cycle (LTO)** is specified, which includes time at each of four engine power settings. Typical levels of engine emissions as a function of engine thrust are presented in Fig. 8.6.26. Emissions of carbon monoxide (CO) and unburned hydrocarbons (UHC) are dominant at low power conditions such as idle, and rapidly decrease to negligible levels as engine power increases. Most modern combustor designs result in very low levels of CO and UHC at all cruise power settings through efficient combustion that minimizes fuel burn. Nitrogen oxides (NO_x) and soot particle emissions or smoke both exhibit the opposite trend, increasing with engine power.

Aircraft spend much more time at takeoff, climb, and cruise settings than at idle. Thus, NO_x is the dominant pollutant by mass in a flight. In fact, the **NO_x produced** at cruise is significantly higher than that produced during the LTO cycle. NO_x emissions are a result of the high temperatures in the combustor, with the rate of NO_x production highly nonlinear with temperature. This is shown in Fig. 8.6.27, with the NO_x production rate presented as a function of the combustion reaction **equivalence ratio,** which is the ratio of fuel to air in the actual mixture divided by the fuel-to-air ratio for stoichiometric conditions. The peak is in the region of stoichiometric combustion, that is, an equivalence ratio of 1.0.

Rich-front-end combustors are found in most current engines. Fuel is introduced at the front and mixed with some of the total combustor airflow to create a stable rich reaction zone (Fig. 8.6.28), with the remainder of the airflow introduced in a mixing zone. The combination of rich oxygen-depleted products from the front end quicky react with the mixing air and are then diluted to exit conditions. This mixing zone can be highly nonuniform, resulting in local stoichiometric combustion with high NO_x production rates.

The **lean-front-end combustor** focuses on mixing the air and fuel in the front end of the combustor exactly to create a combustion zone that is, on average, below the peak stoichiometric equivalence ratio of 1.0. The lower flame temperature at these lean conditions significantly reduces the NO_x production rate. Lean-front-end combustors require multiple fuel stages to achieve the operability requirements for aircraft application. Typical modern engine designs require the combustor to

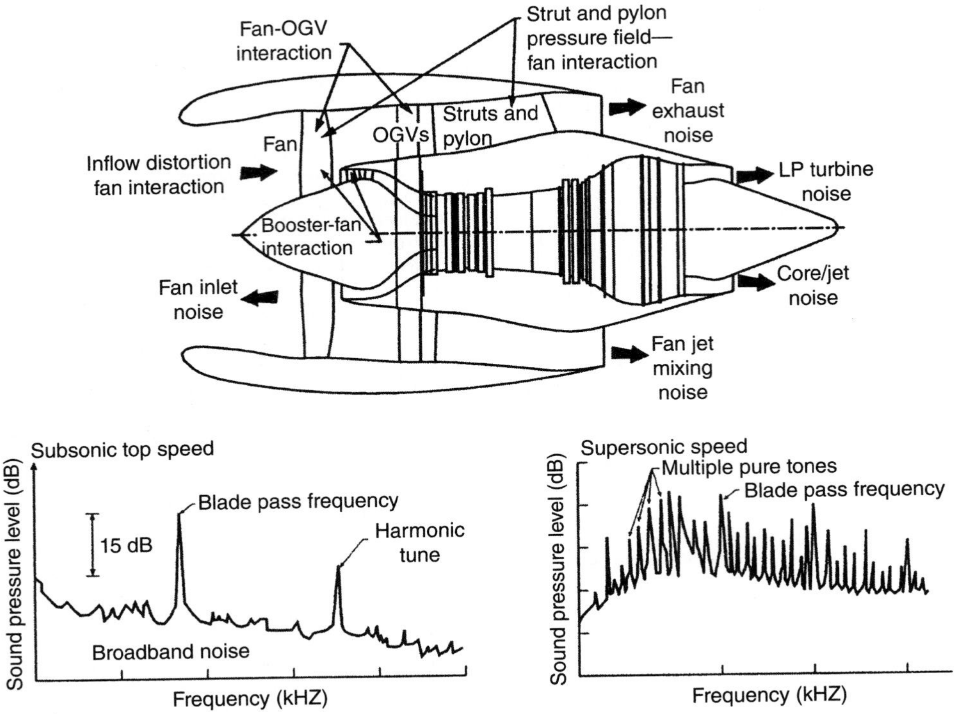

Figure 8.6.25 High-bypass-ratio turbofan noise sources and noise spectrum.

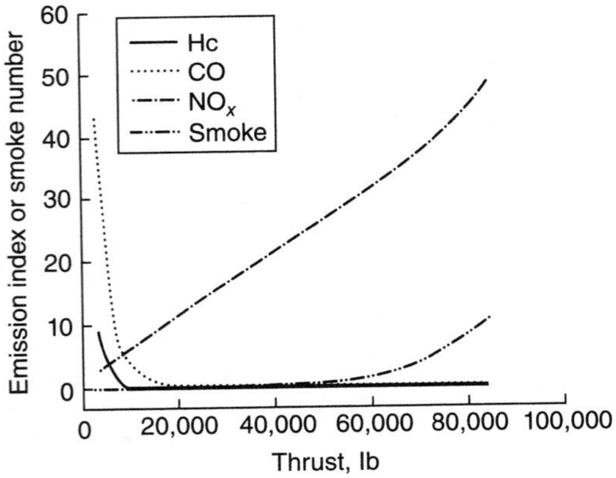

Figure 8.6.26 Typical aircraft combustor emission characteristics.

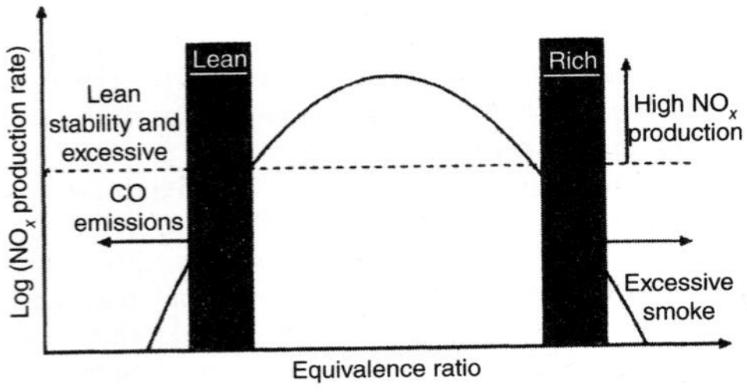

Figure 8.6.27 NO_x production rate versus equivalence ratio.

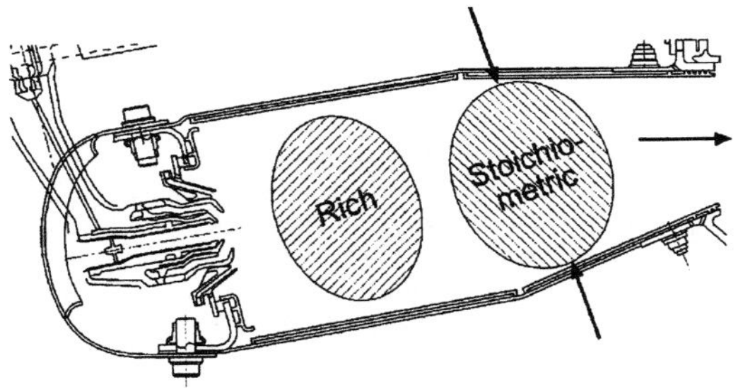

Figure 8.6.28 Schematic of rich-front-end combustor.

operate with a factor of 3 higher exit fuel-air ratio at maximum power than at minimum power. A lean combustion system must be optimized to a much narrower range of fuel-air ratios to avoid high CO production, if too lean, and high NO_x production, if too rich. This issue has been addressed by combustors with many small air and fuel injectors and sequentially fueling a subset of these as engine power increases.

CRITERIA OF ROCKET-MOTOR PERFORMANCE

In most applications of rocket motors the objective is to produce a large thrust for a limited period. The criteria of rocket-motor performance are, therefore, related to thrust-producing capabilities and operating duration. If $\dot{W}_p$ denotes the rate at which the rocket motor consumes propellants, P_c the total pressure at the entrance to the exhaust nozzle, and C_w the **weight flow coefficient,** then

$$\dot{W}_p = C_w P_c A_t \tag{8.6.38}$$

Ordinarily, the values of C_w as a function of P_c and **mixture ratio** $\dot{W}_o/\dot{W}_f$ are determined experimentally. If no experimental data are available, the values of C_w can be calculated with good accuracy from the calculated thermodynamic properties of the combustion gases. If T_c is the total temperature, and P_c the total pressure of the combustion gases at the entrance section of the nozzle, then, for steady operating conditions, $\dot{W}_p = \dot{W}_g$.

$$\dot{W}_g = C_d A_t P_c \Omega \sqrt{\frac{g}{RT_c}} \tag{8.6.39}$$

where

$$\Omega = \left(\frac{2}{k+1}\right)^{(k-1)/[2(k-1)]} \sqrt{k} \tag{8.6.40}$$

An equation similar to Eq. (8.6.38) can be written for the thrust developed by the rocket motor. Thus

$$F = C_F P_c A_t \tag{8.6.41}$$

The values of the **thrust coefficient** C_F for specific propellant combinations are determined experimentally as functions of P_c and the ratio $\dot{W}_o/\dot{W}_f$. When no experimental values are available, C_F can be calculated from the following equation. Therefore

$$C_F = C_d \lambda \phi \Omega \sqrt{\frac{2k}{k-1} Z_t} + \left(\frac{p_e}{P_c} - \frac{p_0}{P_c}\right)\frac{A_e}{A_t} \tag{8.6.42}$$

where

$$Z_t = 1 - \left(\frac{p_e}{P_c}\right)^{(k-1)/k}$$

Equation (8.6.42) shows that C_F is independent of the combustion temperature T_c and the molecular weight of the combustion gases. The nozzle area ratio A_e/A_t is given by the formula

$$\frac{A_e}{A_t} = \frac{(P_c/p_e)^{1/k}}{\left(\frac{k+1}{2}\right)^{1/(k-1)} \sqrt{[(k+1)/(k-1)]Z_t}} \tag{8.6.43}$$

Figure 8.6.29 presents C_F calculated by means of the preceding equations, as a function of the area ratio A_e/A_t, with P_c/p_0 as a parameter, for gases having $k = 1.28$; the calculations assumed $\lambda = 0.983$ and $C_d = \varphi = 1.0$. The curves do not take into account the separation phenomena which occur when A_e/A_t is larger than that required to expand the gases to p_0.

The specific impulse I is defined by Eq. (8.6.8) and can be shown to be given by

$$I = F/\dot{m}_p = C_F/C_w = V_j/g_c \tag{8.6.44}$$

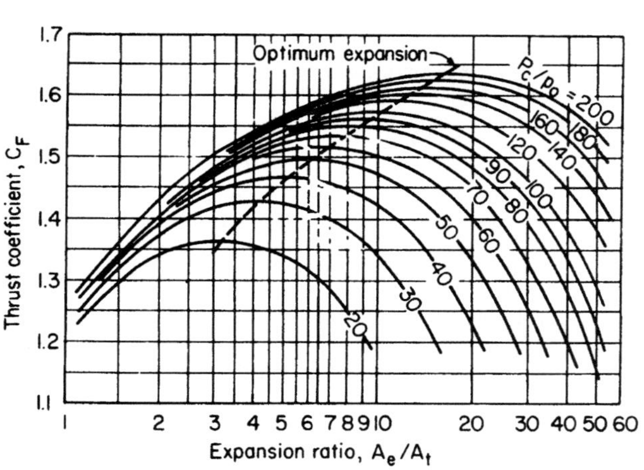

Figure 8.6.29 Calculated thrust coefficient versus expansion ratio for different pressure ratios (based on $k = 1.28$, $\lambda = 0.9830$, $C_d = \varphi = 1.0$).

A criterion which is employed for stating the performance of rocket motor is the characteristic velocity c^*. By definition

$$c^* = V_j/C_F = P_c A_t/\dot{m} = I/C_F \tag{8.6.45}$$

The values of c^* for a given propellant combination are determined experimentally. It should be noted that the value of c^* is independent of the thrust. Basically, c^* measures the effectiveness with which the thermochemical energy of the propellants is converted into effective jet kinetic energy. When no experimental data are available the value of c^* can be estimated quite closely from

$$c^* = \sqrt{RT_c}/\Omega \tag{8.6.46}$$

Table 8.6.1 presents values of specific impulse for some possible rocket-propellant combinations.

Solid-propellant rocket motors may be segregated into (1) **double-base powders** and (2) **composite, or heterogeneous, propellants.** Double-base powders are gelatinized colloidal mixtures of nitroglycerin and cellulose to which certain stabilizers have been added. Heterogeneous, or composite, propellants are physical mixtures of a solid oxidizer in powder form and some form of solid fuel, such as a plastic or rubber-like material.

Solid-propellant rocket motor technology has made tremendous strides. It is possible to produce solid propellants with a wide range of linear burning rates and with good physical properties. It has been demonstrated experimentally that large thrust rocket motors with chamber diameters of approximately 22 ft (6.7 m) are feasible. As a result, the solid-propellant rocket motor has displaced the liquid rocket engine from a large number of applications, which heretofore were considered the province of the liquid rocket engine. In military applications, there is a strong tendency to supplant liquid rocket engines with solid rocket engines. Only in such areas as the large thrust launching rockets for boosting manned vehicles into space have the liquid rocket engines reigned supreme; the space shuttle is equipped with liquid-bipropellant rocket engines, but it is boosted with auxiliary solid-propellant rocket motors.

The following considerations are what make solid-propellant rocket motors attractive: (1) They are simpler in construction than liquid rocket engines; (2) they are easier to handle in the field; (3) they have instant readiness for use; (4) they have very good storage properties; (5) they have a larger average density than most liquid-propellant combinations; (6) they have considerably fewer parts; and (7) their reliability in practice has been very good. Furthermore, once the performance characteristics and physical properties of a solid propellant have been established, it is a relatively straightforward engineering problem to design and develop solid-propellant rocket motors of widely different thrust levels and burning times t_R.

Rocket motors which burn double-base propellants find the greatest use in weaponry, for example, artillery rockets, antitank weapons, etc.

Rocket motors burning heterogeneous, also called **composite,** solid propellants are used for propelling all kinds of military missiles and for sounding rockets. Modern composite propellants have three basic types of ingredients: (1) a fuel which is an organic polymer, called the **binder;** (2) a finely powdered oxidizer, ordinarily ammonium perchlorate (NH_4NO_3); and (3) various additives, for catalyzing the combustion process, for increasing the propellant density, for increasing the specific impulse (e.g., powdered aluminum), for improving the physical properties of the propellant, and the like. After the ingredients have been thoroughly mixed, the resulting viscous fluid is poured, usually under vacuum to prevent the formation of voids, directly into the chamber of the rocket motor which contains a suitable mandril for producing the desired configuration for the propellant grain. The propellant is cured by polymerization to form an elastomeric material. Some of the more common binders are butadiene copolymers and polyurethane.

Modern composite solid propellants have densities ranging from 1.60 to 1.70, depending upon the formulation, and specific impulse values of approximately 245 s, based on a combustion pressure of 1,000 psia and expansion to 14.7 psia.

In the case of solid-propellant rockets the rate of propellant consumption is related to the **linear burning rate** for the propellant r_0, that is, the rate at which the burning surface of the propellant recedes normal to itself as it burns. For practical purposes it can be assumed that the linear burning rate r_0 is given by

$$r_0 = a_2 p_c^n \tag{8.6.47}$$

where a_2 and n are constants which are determined by experiment.

If ρ_p denotes the density of the solid propellant and A_p the area of the burning surface, then the propellant consumption rate is given by

$$\dot{m}_p = A_p \rho_p r_0 = a_2 A_p \rho_p p_c^n \tag{8.6.48}$$

Figure 8.6.30 presents the linear burning rate as a function of the combustion pressure for several propellants.

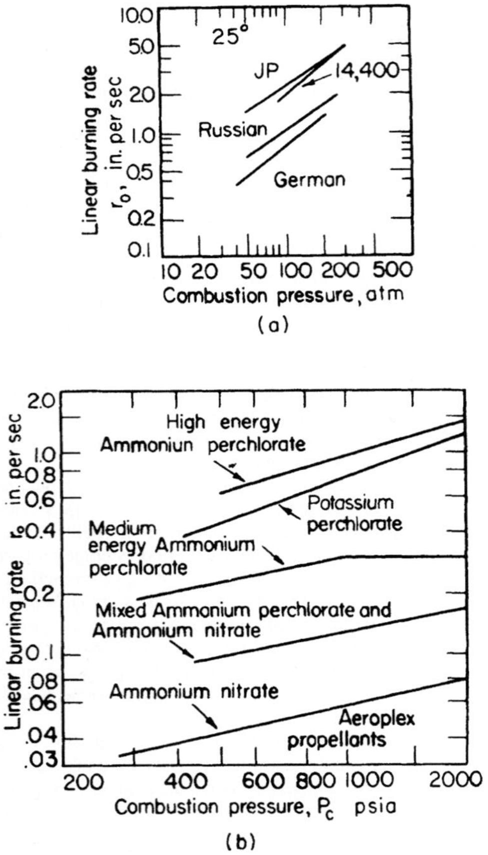

Figure 8.6.30 (*a*) Linear burning rate as a function of combustion pressure for double-base propellants; (*b*) burning-rate characteristics of several heterogeneous propellants at 60°F (15.5°C). (*Aerojet-General Corp.*)

The linear burning rate r_0 is influenced by the temperature of the solid propellant t_p prior to its ignition. Low values of t_p reduce the burning rate, and vice versa. Consequently, the temperature of the propellant must be given in presenting data on linear burning rates. Figure 8.6.31 presents r_0 as a function of p_c for $t_p = -40$, 60, and 140°F for a composite propellant manufactured by the Aerojet-General Corp., Azusa, Calif.

AIRCRAFT PROPELLERS

The function of the aircraft propeller (or airscrew) is to convert the torque delivered to it by an engine into the thrust for propelling an airplane. If the airplane is in steady, level flight, the propeller thrust and

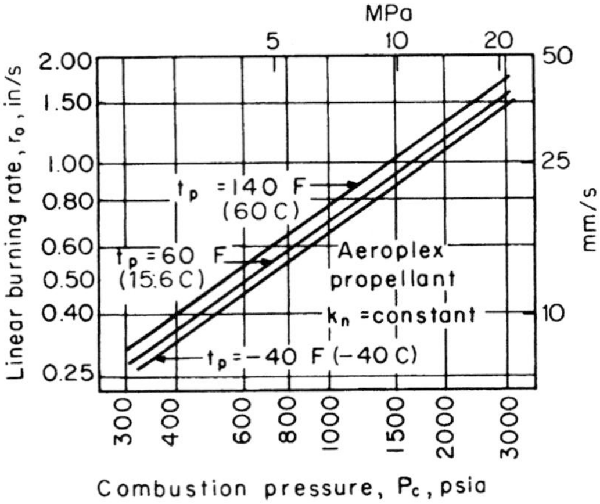

Figure 8.6.31 Effect of propellant temperature upon the linear burning rate of a heterogeneous propellant. (*Aerojet-General Corp.*)

the airplane drag equal each other. If the airplane is climbing, the thrust must overcome the drag plus the weight component of the airplane.

The propeller may be the sole thrust-producing element, but it can also be employed in conjunction with another thrust-producing device, such as the exhaust jet of a **turboprop** or **gas-turbine engine.** It may also be employed as a **windmilling** device for producing power.

The conventional propeller consists of two or more equally spaced radial blades, which are rotated at a substantially uniform angular velocity. At any arbitrary radius, the section of a blade has the shape of an airfoil, but as the hub is approached, the sections become more nearly circular because of considerations of strength rather than aerodynamic performance. The portions of the blades in the vicinity of the hub contribute, at best, only a small portion of the thrust developed by the propeller blades.

Each section of the propeller blade experiences the aerodynamic reactions of an airfoil of like shape moving through the air in a similar manner. Furthermore, at each radius, the section of one blade forms with the corresponding sections on the other blades a series of similar airfoils, which follow one another as the propeller rotates.

The torque of the rotating propeller imparts a rotational motion to the air flowing through it and, furthermore, causes the pressure immediately behind the propeller to increase while that in front of it is reduced; that is, the air is sucked toward the front of the propeller and pushed away behind it. A slipstream is created as illustrated in Fig. 8.6.1. The ratio of the cross-sectional area of the slipstream at any station to that of the actuator disk is termed the **slipstream contraction.**

The thrust produced by the propeller follows the relationships developed from the momentum theorem of fluid mechanics for jet engines as presented in Fig. 8.6.15 and Eq. (8.6.5). The thrust equation for the airbreathing engine (simplified) and the propeller is

$$F = \dot{m}(w - V_0) = \rho A V_0 (w - V_0) \tag{8.6.49}$$

In the case of a propeller the mass flow m is large and the velocity rise across the propeller is small, as compared with a jet engine of the same thrust. However, the thrust of propellers falls off rapidly as air compressibility effects on the propeller become large (flight speeds of 300 to 500 mi/h).

Propeller efficiency is defined as the useful work divided by the energy supplied to the propeller:

$$\eta_p = \frac{P_T}{P} = \frac{FV_0}{P} = \frac{m(v - V_0)V_0}{m/2(w^2 - V_0^2)} = \frac{2v}{1+v} \tag{8.6.50}$$

where $v = V_0/w$.

Since w is only slightly larger than V_0, the ideal efficiency is of the order of 0.95, and actual efficiencies are in the range of 0.5 to 0.9.

Propeller Theory

Axial-Momentum Theory The axial-momentum theory of the propeller is basically a special case of the generalized jet-propulsion system presented in Fig. 8.6.15. It is only a first approximation of the action of the propeller and cannot be employed for design purposes. It neglects such factors as the drag of the blades, energy losses due to slipstream rotation, blade interference, and air compressibility effects. Because of those losses, an actual propeller requires power to rotate at zero thrust, and at zero thrust its propulsive efficiency is, of course, zero.

Blade-Element Theory This theory, called **strip theory**, takes into account the profile losses of the blade sections. It is the theory most commonly employed in designing a propeller blade and for assessing the off-design performance characteristics of the propeller.

Each section of the propeller is considered to be a rotating airfoil. It is assumed that the radial flow of air may be neglected, so that the flow over a blade section is two-dimensional. Figure 8.6.32 illustrates

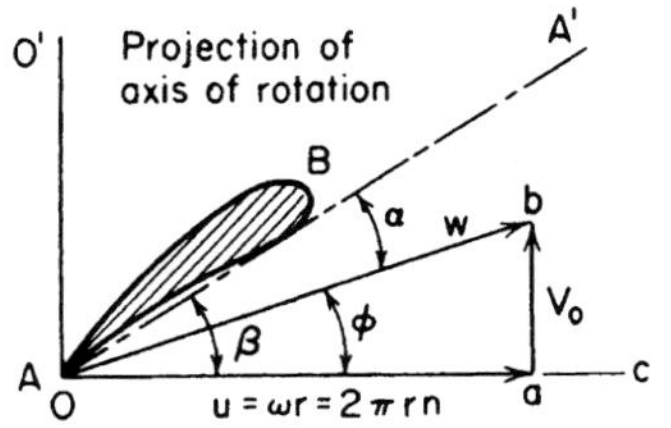

Figure 8.6.32 Vector diagram for a blade element of a propeller.

diagrammatically the velocity vectors pertinent to a blade section of length dr, located at an arbitrary radius r. The projection of the axis of rotation is OO′, the plane of rotation is OC, the blade angle is β, the angle of attack for the air flowing toward the blade (in the relative coordinate system) with the velocity V_0 is α, the tangential velocity of the blade element is $u = 2\pi nr$, and the **pitch** or **advance angle** is ϕ. From Fig. 8.6.32

$$\alpha = \beta - \phi \tag{8.6.51}$$

and

$$\phi = \tan^{-1}(V_0/\pi nd) \tag{8.6.52}$$

Figure 8.6.33 illustrates the forces acting on the blade element. If b denotes the width of the blade under consideration and C_L and C_D are the lift and drag coefficients of the blade (airfoil) section, then the thrust force dF acting on the blade element is given by

$$dF = \tfrac{1}{2}\rho V_0^2 b\, dr C_L \frac{\cos(\phi+\gamma)}{\cos\gamma \sin^2\phi} \tag{8.6.53}$$

The corresponding torque force dQ is given by

$$dQ = \tfrac{1}{2}\rho V_0^2 b\, dr C_L \frac{\sin(\phi+\gamma)}{\cos\gamma \sin^2\phi} \tag{8.6.54}$$

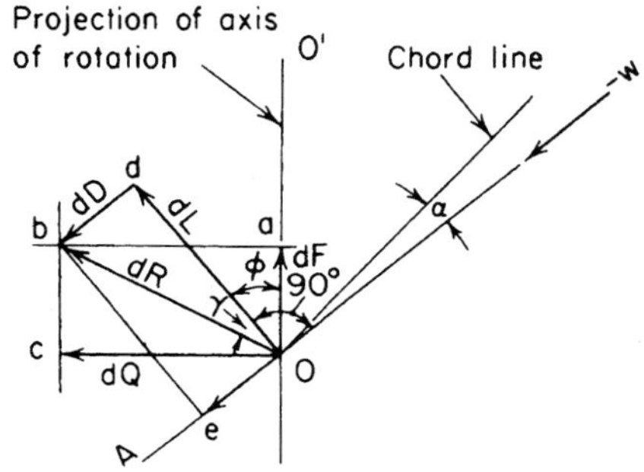

Figure 8.6.33 Aerodynamic forces acting on a blade element.

where

$$\tan\gamma = d\mathcal{D}/d\mathcal{L} = C_D/C_L = \mathcal{D}/\mathcal{L} \tag{8.6.55}$$

where $\mathcal{D}$ and $\mathcal{L}$ denote drag and lift, respectively.

The propulsion efficiency of the blade element, termed the **blading efficiency,** is defined by

$$\eta_b = \frac{V_0 dF}{u\, dQ} = \frac{\tan\phi}{\tan(\phi+\gamma)} = \frac{\mathcal{L}/\mathcal{D} - \tan\phi}{\mathcal{L}/\mathcal{D} + \cot\phi} \tag{8.6.56}$$

The value of ϕ which makes η_b a maximum is termed the **optimum advance angle** ϕ_{opt}. Thus

$$\phi_{\text{opt}} = \frac{\pi}{4} - \frac{\gamma}{2} = 45° - \frac{57.3}{2(\mathcal{L}/\mathcal{D})} \tag{8.6.57}$$

The maximum blade efficiency is accordingly

$$(\eta_b)_{\max} = \frac{2\gamma - 1}{2\gamma + 1} = \frac{2(\mathcal{L}/\mathcal{D}) - 1}{2(\mathcal{L}/\mathcal{D}) + 1} \tag{8.6.58}$$

A force diagram similar to Fig. 8.6.33 can be constructed for each element of the propeller blade, taking into account the variation in ϕ with the radius. The resultant forces acting on the propeller are obtained by summing (integrating) those forces acting on each blade element.

The integrated lift coefficient is

$$C_{Li} = \int_{r_i/R}^{r/R=1.0} C_{Lr}\left(\frac{r}{R}\right)^3 d\left(\frac{r}{R}\right) \tag{8.6.59}$$

where C_{Lr} = propeller-blade-section lift coefficient and r/R = fraction of propeller-tip radius.

The integrated lift coefficient is in the range 0.15 to 0.70. There is a dilemma in the design and selection of blade sections since high static thrust requires high C_{Li} which gives relatively low η_p at cruise. Cruise conditions require low C_{Li} for good η_p. Variable camber propellers permit optimization of C_{Li}, but at the expense of weight and complexity.

Vortex Theory The simple two-dimensional blade-element theory can be further improved by taking account of the actual velocity distribution in the slipstream and thus determining the actual loading on the blading by the so-called *vortex theory.* The blades are represented by bound vortex filaments sweeping out vortex sheets to infinity. The vortex sheets are distorted both by the slipstream contraction and by the profile drag of the blades. If such distortions are neglected, the induced velocity in the wake can be calculated by assuming the vortex sheets to be tubular. Thus, in Fig. 8.6.32, the induced axial velocity becomes the arithmetic mean of the axial velocities at far distances upstream and downstream of the rotor disk; and the induced rotational velocity at the disk becomes $\omega - 0.5\omega'$, where ω' is the rotation of the slipstream in the plane of the propeller. Hence, the angle of attack α for a rotor element is given by

$$\alpha = \beta - \tan^{-1}[V_F/r(\omega - 0.5\omega')] \tag{8.6.60}$$

where V_F is the induced axial velocity in the plane of the propeller.

The addition of the vortex effects improves the representation of the propeller by the blade sections. There are, however, three-dimensional flow effects which are not included, and the following coefficients have been suggested to include this effect (which reduces the lift coefficient).

Lift Curve Slope Correction Factors for Three-Dimensional Airfoils

Radial station (ratio r/R)	Correction factor
(0.2)(0.3)(0.45)(0.6)(0.7)	0.85
0.8	0.80
0.9	0.70
0.95	0.65

SOURCE: Departments of the Air Force, Navy, and Commerce, "Propeller Handbook."

Performance Characteristics

The pitch and the angle ϕ (Fig. 8.6.32) have different values at different radii along a propeller blade. It is customary, therefore, to refer to all parameters determining the overall characteristics of a propeller to their values at either $0.7r$ or $0.75r$.

For a given blade angle β, the angle of attack α at any blade section is a function of the **velocity coefficient** V_0/u, where $u = \pi n d_p$; it is customary, however, to replace V_0/u by its equivalent, the **advance ratio** $J = V_0/nd_p$.

For given values of β and n, the angle of attack α attains its maximum value when $V_0 = 0$, with $\phi = 0$. In other words, the propeller thrust F is maximum at takeoff. As V_0 increases, α decreases and so does the thrust. Finally, a forward speed is reached for which the thrust is zero.

Increasing $J = V_0/nd_p$, for a constant blade angle, causes the propeller to operate successively as a fan, propeller, brake, and windmill. Most of the operation, however, is conducted in the propeller state. At takeoff, the propeller is in the fan state. Windmilling must be avoided because it may overspeed the engine and damage it.

The lift coefficient C_L is a linear function of α up to the stalling angle, and C_D is a quadratic function of α. Furthermore, the lift $\mathcal{L}$ is zero at a negative value for α. Figure 8.6.34 presents the $\mathcal{L}/\mathcal{D}$ ratios of typical propeller sections. Figure 8.6.35 presents the blade (section) efficiency as a function of the pitch angle ϕ for different values of $\mathcal{L}/\mathcal{D}$. Propellers normally operate at $\mathcal{L}/\mathcal{D}$ ratios of 20 or more. Supersonic propellers operate with $\mathcal{L}/\mathcal{D}$ ratios of approximately 10, and from Fig. 8.6.34 it is apparent that they must operate close to the optimum pitch angle ϕ to obtain usable blade efficiencies.

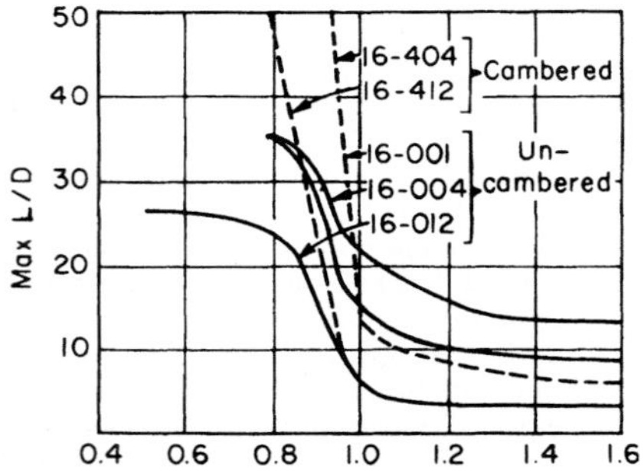

Figure 8.6.34 Maximum lift-drag ratios for representative sections.

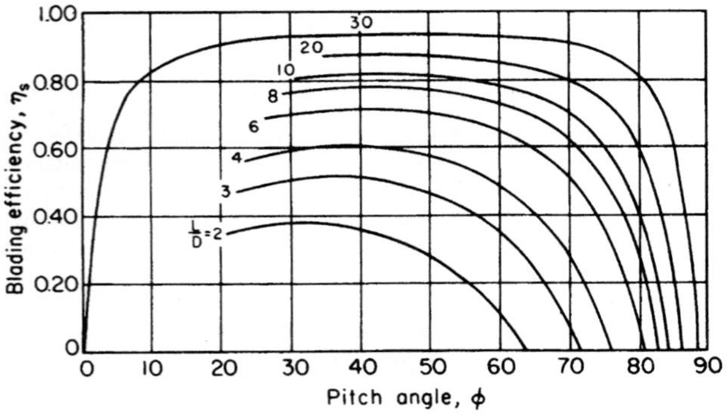

Figure 8.6.35 Blade-element efficiency versus pitch angle.

Propeller Coefficients It can be shown, neglecting the compressibility of the air, that

$$f(V_0, n, d_p, \rho, F) = 0 \tag{8.6.61}$$

By dimensional analysis, the following coefficients are obtained for expressing the performances of propellers having the same geometry:

$$F = \rho n^2 d_p^4 C_F \qquad Q = \rho n^2 d_p^5 C_Q \qquad \mathcal{P} = \rho n^3 d_p^5 C_p \tag{8.6.62}$$

C_F, C_Q, and C_P are termed the **thrust, torque,** and **power coefficients,** respectively. They are independent of the size of the propeller. Consequently, tests of small-scale models can be employed for obtaining the values of F, Q, and $\mathcal{P}$ for geometrically similar full-scale propellers. Hence, the ideal propulsive efficiency of a propeller is given by

$$\eta_p = \frac{FV_0}{\mathcal{P}} = \frac{C_F}{C_p}\frac{V_0}{d_p n} = J\frac{C_F}{C_p} \tag{8.6.63}$$

Other useful coefficients are the **speed-power coefficient** C_S and the **torque-speed coefficient** C_{QS}. Thus,

$$C_S = V_0 5\sqrt{\rho/\mathcal{P}n^2} \qquad C_{QS} = V_0\sqrt{\rho d_p^3/Q} \tag{8.6.64}$$

A commonly employed factor related to the geometry of a propeller is the **solidity** σ. By definition,

$$\sigma = c/p \tag{8.6.65}$$

The solidity varies from radius to radius; at a given radius, σ is proportional to the power-absorption capacity of the annulus of blade elements.

The **activity factor** of a blade AF is an arbitrary measure of the blade solidity and hence of its ability to absorb power. It takes the form

$$AF = \frac{100{,}000}{16}\int_{r_h/R}^{r/R=1.0} \frac{c}{d_p}\left(\frac{r}{R}\right)^3 d\left(\frac{r}{R}\right) \tag{8.6.66}$$

where $R = d_p/2$, r = radius at any section, and r_h = spinner radius.

Performance

Static Thrust From the simple axial-momentum theory, the static thrust F_0 of an actuator disk of diameter d_p is given by

$$F_0 = (\pi\rho/2)^{1/3}(\mathcal{P}d_p)^{2/3} = 10.4\text{bhp} \times d_p^{2/3} \quad \text{(at sea level)} \tag{8.6.67}$$

The **thrust horsepower** (thp) is accordingly

$$\text{thp} = \frac{F_0}{\text{bhp}} = (38{,}400/Nd_p)\,(1/C_p^{1/3}) \tag{8.6.68}$$

Equations (8.6.67) and (8.6.68) assume uniform axial velocity, no rotation of the air, and no profile losses. Consequently, they may be in error by as much as 50 percent. Hesse suggests the use of the equations

$$\eta_{p\,\text{static}} = \sqrt{\frac{2}{\pi}}\frac{C_F^{3/2}}{C_P} = 0.798\frac{C_F^{3/2}}{C_p} \tag{8.6.69}$$

and

$$F = 10.42[d_p(\eta_{p\,\text{static}})\text{shp}]^{2/3}(\rho/\rho_0)^{1/3} \tag{8.6.70}$$

together with data computed for a specific propeller, as illustrated in Fig. 8.6.36. The static thrust for a given power input increases with the diameter and the number of blades. Controllable pitch propellers selected for good overall performance will develop 3 to 4 lb of thrust per shp.

At flight conditions the performance may be obtained from generalized charts presenting C_P as a function of the advance ratio J, with the propeller efficiency and blade pitch angle at 75 percent radius, $\theta_{3/4}$, as parameters. Figure 8.6.37 is such a chart.

Reverse Thrust By operating the propeller blades at large negative angles of attack, reversed thrust can be developed. In that condition, the blades are stalled, and as in the case of static thrust, the forces acting on the blades cannot be calculated. In practice, the reverse-pitch stop is set by trial and error so that the propeller can absorb the rated power at zero speed, that is, at $V_0 = 0$. With fixed-pitch reversing, the power absorbed when there is forward speed will be smaller than the rated value, but the reversed thrust is usually substantially larger than the static forward thrust corresponding to the rated power. The landing

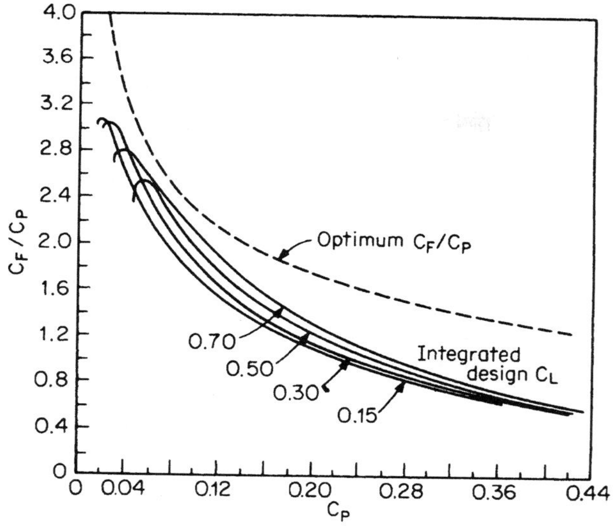

Figure 8.6.36 Static-thrust performance of a propeller; four-bladed, 100 activity-factor propellers of various integrated design C_L's. (*From Hamilton Standard Division, United Aircraft Corporation.*)

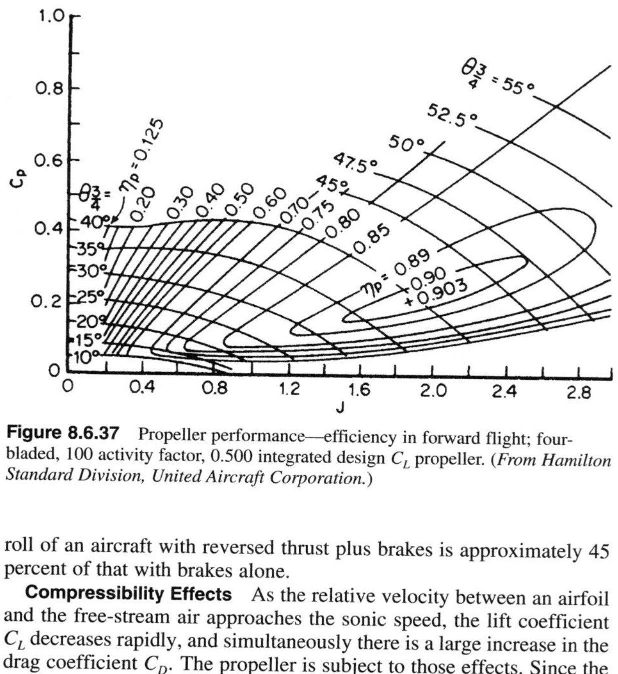

Figure 8.6.37 Propeller performance—efficiency in forward flight; four-bladed, 100 activity factor, 0.500 integrated design C_L propeller. (*From Hamilton Standard Division, United Aircraft Corporation.*)

roll of an aircraft with reversed thrust plus brakes is approximately 45 percent of that with brakes alone.

Compressibility Effects As the relative velocity between an airfoil and the free-stream air approaches the sonic speed, the lift coefficient C_L decreases rapidly, and simultaneously there is a large increase in the drag coefficient C_D. The propeller is subject to those effects. Since the acoustic speed decreases from sea level to the isothermal altitude, the effects appear at lower tip speeds at high altitude. Furthermore, the effect is more pronounced at higher tip speeds. Shock waves are formed at the leading edge of a propeller blade when its pitch-line velocity approaches the local sonic velocity of the air. As a consequence, the coefficient C_F is reduced and C_P is increased, thereby adversely affecting the propulsive efficiency of the propeller.

Because of the high propulsive efficiency of propellers, the application of advanced technologies may provide **turboprop**-powered transports with cruise speeds and other desirable characteristics comparable to advanced turbofan-powered transports, but with a 15 to 20 percent reduction in fuel consumption.

Advanced turboprop propeller blades operate at high power loading, are extremely thin, and have swept leading edges to minimize compressibility losses. Shaping the spinner and nacelle minimizes choking and compressibility losses. Efficiencies of about 70 percent at Mach 0.8 and design power loadings have been achieved in research propellers. Such propellers may be relatively noisy because the tips may be slightly supersonic at Mach 0.8 cruise. Attenuating the noise within the cabin wall will likely add airframe weight and reduce the fuel savings.

Mechanical Design

Blades The principal blade loadings are (1) steady tensile, due to centrifugal forces; (2) steady bending, due to the aerodynamic thrust and torque forces; and (3) vibratory bending, due to cyclic variations in airloads and other excitations originating in the engine. The most serious and limiting stresses usually result from the vibratory loadings. The principal vibratory loading results from the cyclic variation in angle of attack of the blades when the axis of rotation is pitched or yawed relative to airstream. When the axis is pitched up, such as when the aircraft is in a high-angle-of-attack climb, the blades are at a higher angle of attack on the downstroke than when on the upstroke. This results in a once per revolution, $1E$, variation in aerodynamic loading on the blades, usually referred to as $1E$ **aerodynamic excitation,** and a steady **side force** and **yawing moment,** which are transmitted through the shaft to the aircraft.

The degree of pitch or yaw of the propeller axis is usually measured in terms of the **excitation factor,** defined as

$$\text{Excitation factor} = \psi(V_i/400)^2 \tag{8.6.71}$$

where ψ = angle of pitch or yaw, deg; V_i = indicated flight velocity, mi/h. Another factor sometimes used is Aq, where $A = \psi°$, $q = 12\rho V_i^2$, Aq = excitation factor × 410.

Because of the restoring moment of the **centrifugal loads** on the blade elements when the blade deflects in bending, centrifugal forces have the effect of apparently stiffening the blade in **bending.** In Fig. 8.6.38, the bending natural frequency increase with rpm. At some rpm, the bending natural frequency will come into resonance with the $1E$ excitation, at which point small loadings will be greatly magnified (limited only by the damping present in the blade).

Propellers are designed so that the $1E$ resonant speed is above any expected operating speed. However, there will always be some magnification of the vibratory loads (due to proximity to resonant speed), resulting in disproportionately high vibratory stresses. As indicated in Fig. 8.6.38, blades normally pass through a $2E$ **resonance** in coming up to speed (usually not critical because it is a transient condition) and may pass through or be close to resonance with other modes.

The **natural frequencies** of the rotating propeller blade can be computed with good accuracy, but since the excitation varies from aircraft to aircraft and is usually not well defined, vibration surveys must be made of all new installations. These often result in the definition of ranges at which continuous operation should be avoided. **Allowable vibratory** stresses are limited by the fatigue strength of the material. Unlike the steady-stress limits, **fatigue strength** is not a unique

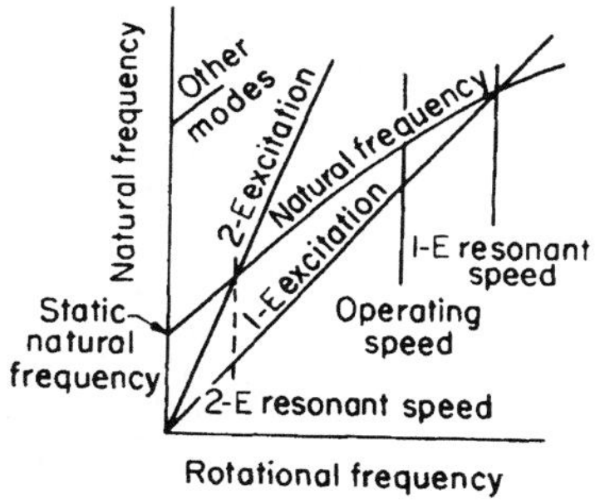

Figure 8.6.38 Relation between blade natural frequency and major excitations.

function of the material but is a function of surface finish as well. A sharp notch in the surface reduces the fatigue strength to a very low value. If propellers were designed to hold vibratory stresses below the lowest possible fatigue strength, the blades would be unacceptably heavy. Thus propellers are designed on the maintenance of a reasonably smooth surface finish. Since the blade surfaces are constantly subject to **store damage,** blades must be regularly inspected to make sure that the finish does not deteriorate below design standards.

Propeller blades are stalled when operating at low forward speeds and when reversing (see Performance, above). Under these conditions torsion-mode **stall flutter** can occur.

In general, $1E$ **considerations** control the design of the **inboard** portions of a blade, and **stall flutter** the **outboard** sections.

There are five types of **blade construction** in common use today: **solid aluminum; fabricated steel; one-piece steel; monocoque, fiberglass construction,** supported on a steel shank; and a construction with a **hollow steel central spar and a lightweight fiberglass cover** that forms the aerodynamic contour of the blade. As in any beam designed on the basis of bending loads, it is desirable to concentrate the structural material at the maximum radii, as in the flanges of a simple I beam, and omit it from the center, where it carries no load. This is accomplished, with steel blades, by making them hollow. The structural material is concentrated in the outer fibers, or surface. In general, steel blades weigh about 80 to 85 percent of the equivalent solid aluminum blade. The advantage of steel increases with size. The blades normally constitute from one-third to one-half of the total propeller weight.

The **hub** retains the blades and contains the pitch-change motor. A **pitch-change mechanism** is provided so that the blades can be positioned at the proper angle of attack α to absorb the desired power at the desired rpm, regardless of flight speed. If the pitch is unchanged, as in Fig. 8.6.39, the angle of attack is excessive at low speed. This prevents the engine from reaching rated speed and delivering rated power. The centrifugal loads on the blade elements tend to rotate the blades toward flat pitch (see Fig. 8.6.40). The pitch-change motor acts against this moment by applying a pitch-increasing moment. The addition of **counterweights,** shown dotted in Fig. 8.6.40, reduces the moment and the size of the pitch-change motor. Counterweights, however, add to the radial centrifugal load of the blades, necessitating a heavier retention, and consequently do not necessarily reduce the net weight of the propeller.

Most pitch-change mechanisms employ a **hydraulic piston,** mounted on the hub, with feed through the propeller shaft and with rotation of the blades by means of gears or links. **Electric motors** and **direct mechanical drives** have been used successfully. In the past, the weight of all systems, when fully developed, has turned out to be about the same.

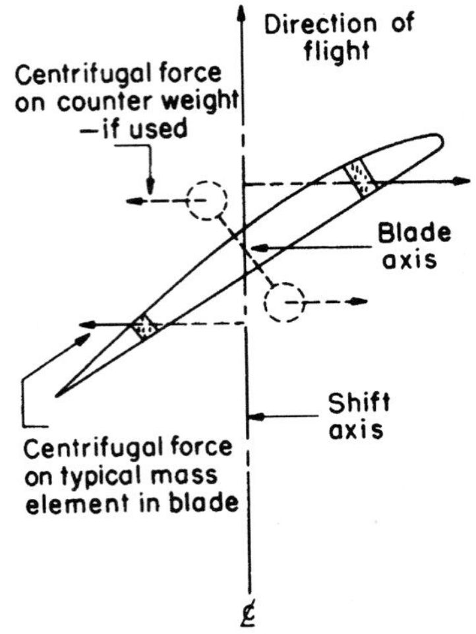

Figure 8.6.40 Twisting effect of centrifugal loads on blades.

Action of the pitch-change motor is limited by **low-** and **high-pitch stops**. These prevent the blades from assuming negative angles with reverse thrust. On reciprocating engines, the low-pitch stop is set to allow the propeller to absorb approximately rated power at zero forward speed. A lower setting results in higher windmilling drags in gliding flight, which could be dangerous if carried too far. In reversing propellers the low-pitch stop is automatically removed when the controls are set for reverse pitch. The high-pitch stop prevents the propeller from reversing through the positive range. In feathering propellers it is set at the angle which gives zero aerodynamic moment about the shaft.

Propeller feathering is provided in order to reduce the drag of a dead engine. To accomplish it, the blade-angle range is extended to 90°, approximately. Auxiliary pitch-change power is provided to complete the feathering as the engine slows down and to unfeather when the engine is stationary.

Reversing is provided in order to give propeller braking on the landing roll. It is accomplished by removing the flight low-pitch stop and driving the blades to a high negative **reverse-pitch angle**. The principal mechanical problem is to provide a mechanism which has the least possible chance of going into reverse inadvertently.

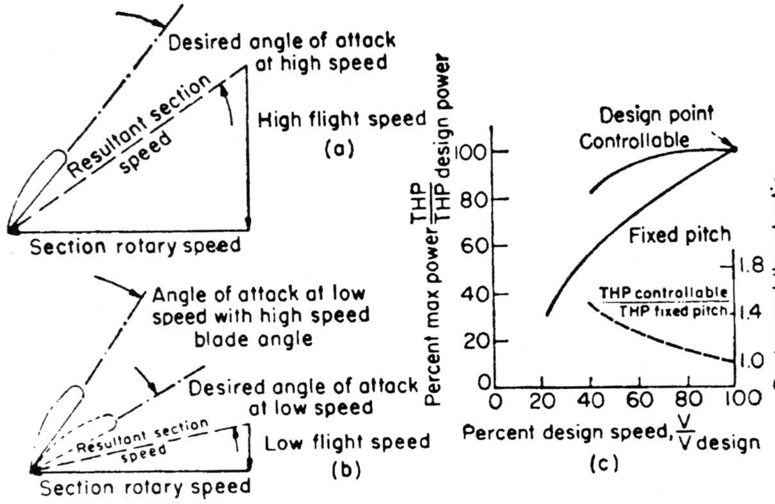

Figure 8.6.39 Effect of flight speed on the desired blade angle.

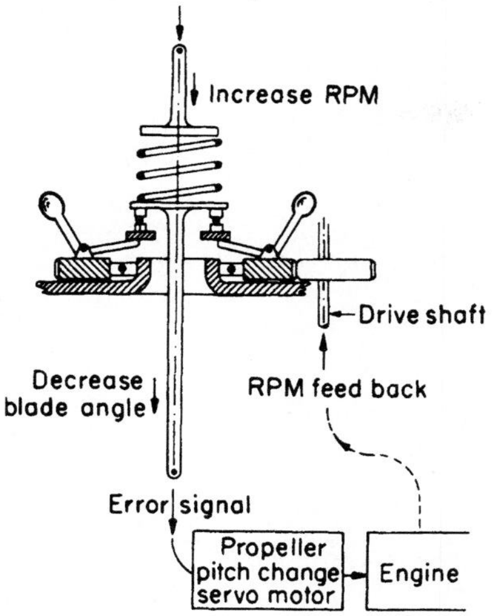

Figure 8.6.41 Propeller-speed governor.

The **propeller control** is that portion of the system which regulates the blade angle. In light aircraft, this may consist merely of a mechanism which sets a given blade angle corresponding to a given position of the cockpit control. In such cases the propeller acts as a fixed-pitch propeller except when activated by the pilot. In most cases, the propeller control regulates blade angle so as to maintain a preset rotary rpm, irrespective of flight speed or shaft power. The **basic elements** of this system normally consist of a spring-balanced, engine-drive flyweight delivering an error signal which directs the servo, or pitch-change motor, to increase or decrease blade angle (Fig. 8.6.41). The control loop is closed by rpm feedback through the main engine. This basic system may be embellished with anticipating and delay devices to maintain speed in rapid throttle movements, synchronizing to coordinate two or more engines, and with overriding features to activate the reversing and feathering sequences. (See also Sec. 12.)

Propeller **controls** are fundamentally the same, with several added secondary features. First, because the engine is capable of a very rapid increase or decrease in torque and because of the high gear ratio between the propeller and engine shaft, anticipating features are necessary to prevent overspeeds in the propeller from producing large overspeeds and possible failure in the turbine. Second, because of the high idle rpm needed to maintain the pressure ratio in the compressor, the low-pitch stop of the turboprop engine propeller must be set substantially lower than for a piston engine. This is not true of a free-turbine drive and is only partially true of a split-compressor engine. In the event of an **engine failure,** the propeller, sensing the loss in rpm, will govern down to the low-pitch stop. This may lead to large, often uncontrollable, negative thrust. To avoid this possibility, an engine-shaft torque signal is put into the control, a negative torque automatically activating the feathering system. This portion of the system is known as the **ENT (emergency negative-thrust)** system. Third, because of the high idle rpm characteristic of the turboprop engine, and the resulting low blade angles at low power in the ground-taxi regime, provision must be made to cut out the governing system and substitute direct blade-angle regulation (known as **beta control**). This is required because the propeller will not govern under these conditions as the torque-to-blade-angle relationship goes to zero. The propeller control of a turboprop engine is normally mechanically coordinated directly to the engine throttle, leaving only a single control, with so-called power lever, in the aircraft.

The shaft on which the propeller is mounted is subjected to the steady-thrust load, vibratory loads due to the propeller side, and **gyroscopic forces** (the 1*E* moment, acting on the blades, is reactionless in propellers with three or more blades). In large aircraft, the shaft size is usually determined by the propeller side force. In aircraft used for aerobatics, gyroscopic loads may become limiting. The **gyroscopic moment** on the propeller shaft is given by $2\omega_1 \cdot \omega_2(I_1 - I_2)$, where ω_1 = angular velocity of maneuver, ω_2 = angular velocity of shaft, I_1 = polar moment of inertia of propeller about shaft, I_2 = moment of inertia of propeller about diameter in plane of rotation. (See also Sec. 1.)

8.7 ASTRONAUTICS

by Glenn J. Miller and Joaquim R. R. A. Martins

REFERENCES: "Progress in Astronautics and Aeronautics," AIAA series, Academic Press. "Advances in Astronautical Sciences," American Astronautical Society series. "Handbook of the British Astronautical Association." Koelle, "Handbook of Astronautical Engineering," McGraw-Hill. Purser, Faget, and Smith, "Manned Spacecraft: Engineering Design and Operation," Fairchild Publications. Herrick, "Astrodynamics," Vol. 1, Van Nostrand Reinhold. Cohen, "Space Shuttle," Yearbook of Science and Technology, 1979, McGraw-Hill.

METEOROID/ORBITAL DEBRIS SHIELDING

by Eric L. Christiansen

NASA

REFERENCES: Moorhead et al., NASA Meteoroid Engineering Model Release 2.0, NASA TM-2015-218214, 2015. Stansbery et al., NASA Orbital Debris Engineering Model ORDEM 3.0 – Verification and Validation, NASA TP-2015-218592, 2015. Christiansen, Meteoroid/Debris Shielding, NASA TP-2003-210788, 2003. Christiansen et al., Handbook for Designing MMOD Protection, NASA TM-2009-214785, 2009.

Notation

d = projectile diameter (cm)
ρ = density (g/cm^3)
m = mass per unit area (g/cm^2)
M = projectile mass (g)
S = shield standoff (cm)
σ = rear wall yield strength (ksi = 1,000 psi)
t = thickness (cm)
V = impact velocity (km/s)
θ = impact angle measured from target normal (i.e., $\theta = 0$ is impact directly into target)

Subscripts

b = bumper (first wall)
p = projectile
w = rear wall (second wall)

Spacecraft shielding is designed to prevent a majority of the meteoroid/orbital debris (M/OD) particles that can impact the spacecraft during its lifetime from causing serious damage that would endanger the continuing operation of the spacecraft and/or endanger crew survivability. Because of spacecraft size and mass constraints, it is not possible with current shielding technology to eliminate completely the risk from M/OD impact. Current practice calls for designing shields to meet or exceed protection requirements, which are generally expressed in terms of a minimum acceptable reliability level or success criteria (i.e., a probability of *not* being struck by M/OD that will completely penetrate through the spacecraft shielding) or in some cases, the maximum allowable risk of M/OD failure. While shielding is an important component of the overall strategy used to reduce the risk from M/OD impact, other means to reduce risk are also employed, such as collision warning and avoidance to reduce the risk from orbital debris impact, or inspection/repair of a returned spacecraft's thermal protection system (TPS).

Spacecraft M/OD Environment

NASA standard meteoroid and debris environment models are used in the risk assessments to determine the cumulative flux of particles with a diameter that exceeds the ballistic limits. The environment models indicate that there are many smaller particles than larger particles. Thus, raising the shielding performance in terms of the M/OD particles sizes the shielding can "stop" decreases the flux of penetrating particles and improves spacecraft reliability.

Meteoroid Environment Model Meteoroids are the natural particles in orbit about the sun, which have quite high impact velocities relative to spacecraft in orbit about earth. Meteoroid velocities range from 11 km/s to 72 km/s, with an average for earth orbiting spacecraft of 19 km/s. A majority of meteoroids impacting a spacecraft are thought to originate from comets, with relatively low particle densities, ranging from 2 g/cm^3 (for particles 1 *mg* and less) to 0.5 g/cm^3 (particle mass 0.01 g and greater). This is contrasted with the meteorites that survive atmospheric entry and are found on earth's surface, which are higher density and thought to originate mainly from asteroids. The meteoroid environment model to be used for shielding design is defined in the NASA documentation (TM-2015-218214).

Orbital Debris Environment Model Orbital debris is human-created pieces of spacecraft that are in orbit about earth (Fig. 8.7.1). Orbital debris impact velocities are lower than meteoroids, generally impacting spacecraft in low earth orbit at relative speeds of less than 1 km/s to just over 15 km/s, with an average velocity of about 9 km/s for a 400-km altitude orbit. The debris environment threat is composed of metallic fragments, paint, aluminum oxide, and other components of spacecraft and solid-rocket motor exhaust. For debris risk assessments, orbital debris particle density is assumed to be 2.8 g/cm^3, corresponding to aluminum metal, and 7.8 g/cm^3, corresponding to steel (the fraction of the debris flux in each density bin is a function of particle size as given in the NASA reference). Because the orbital debris environment is tied to human activities in space, it is much more dynamic then the meteoroid environment, and the orbital debris environment definitions are subject to periodic updates and revisions as more data is collected on the amount and evolution of orbital debris in earth orbit. The current debris environment model for purposes of shielding design is defined by NASA TP-2015-218592 and referred to as ORDEM 3.0.

Figure 8.7.1 Orbital debris "snapshot."

Design Requirements

Vehicles from the early years of space exploration have used the probabilistic approach to design meteoroid shielding. Table 8.7.1 provides a listing of historical M/OD protection design requirements. Generally, each program defines "critical" penetrations as those that would endanger the survivability of the vehicle and/or crew, although requirements for mission success and functionality have been applied as well.

Space Station Design Requirements The Space Station has meteoroid/debris protection requirements consistent with past programs. As such, it carries by far the most capable M/OD shields ever flown. This is because ISS is larger and exposed longer than other space vehicles. It operates at higher altitudes in general than other spacecraft and its operation will extend into the future. These factors increase the expected number of M/OD impacts. To meet comparable protection requirements, ISS shielding must be more effective. For instance, most ISS critical hardware exposed to the M/OD flux in the velocity vector (front) or port/starboard (side) directions are protected by shields effective at stopping 1- to 1.3-cm-diameter aluminum debris particle at typical impact

Table 8.7.1 Historical Meteoroid/Debris Shielding Requirements

Vehicle	Environments	Required probability of no penetration (PNP)
Apollo command module	Meteoroid	0.996 per 8.3-day mission
Skylab module	Meteoroid	0.995 for 8-month mission
Shuttle orbiter	Meteoroid	0.95 for 500 missions (design requirement)
	Meteoroid and debris	0.995 per mission (for critical penetration)
	Meteoroid and debris	0.984 per mission (for radiator leak—mission abort)
Spacelab module	Meteoroid	0.999 for 7-day mission
Hubble space telescope	Meteoroid and debris	0.95 for 2 years
International Space Station	Meteoroid and debris	0.98 to 0.998 per critical element over 10 years

velocity and angle (9 km/s, 45 deg). In comparison, the Mir space station was able to stop 0.3-cm particles, the space shuttle orbiter was capable of stopping 0.2- to 0.5-cm particles, and Apollo and Skylab were able to stop 0.15 to 0.2-cm particles under similar impact conditions. ISS will also have the ability to maneuver from ground-trackable debris particles (typically >10 cm diameter). By virtue of the large internal volume of ISS, its crews have time to locate and isolate leaks, if they were to occur, by closing hatches. Hole repair kits are available and crews are trained to repair a leak in a module if it occurs. Crew escape vehicles are docked to ISS in the event of a major event requiring evacuation.

Risk Assessment Process

The approach to evaluating and designing protection systems to meet requirements is illustrated in Fig. 8.7.2. By using this methodology, the analyst will be able to accurately evaluate spacecraft risks from M/OD impact, identify zones or areas of the spacecraft that are the "risk drivers" that control the M/OD risk, and evaluate options to reduce risk. Each major step in the risk assessment process is described in the following sections.

Space Geometry Spacecraft size is directly proportional to the number of impacts as given in Eq. (8.7.1). Risk of M/OD impact failure increases as the area and time exposed to the M/OD flux increases.

$$N = \sum_{i=1}^{n} N_i = \sum_{i=1}^{n} (FAt)_i \tag{8.7.1}$$

where N is the number of impacts causing failure and is equal to the sum of the impact failures in each region (N_i) over all regions ($i = 1$ to n) of the spacecraft. N_i is found from the product of flux (F, number/m^2-year) of meteoroid and debris impacts that exceed the failure ballistic limits of a region, exposed area (A, m^2), and time (t, year). It is important to break down the spacecraft geometry into different regions, each with a different exposed area, materials of construction, shielding configurations, and thickness. For instance, there are over 1,000 different shields protecting International Space Station (ISS) habitable modules and external critical items (pressure vessels, control moment gyros) which have been optimized according to their location and exposure to the M/OD threat to provide the necessary M/OD protection.

Impact Directionality The front and sides of a spacecraft are more exposed to debris impact, while the front, sides, and top (zenith) are more exposed to meteoroid impact. The Long Duration Exposure Facility (LDEF) had 20 times more craters observed on the forward face compared to the aft, and 200 times more craters on the forward than the earth-facing side (Fig. 8.7.3). Heavier shielding is typically placed on front and sides of spacecraft with fixed orientation relative to the velocity direction to protect from higher impact rates on these surfaces.

Figure 8.7.3 Long duration exposure facility.

Spacecraft Failure Criteria A key step in the risk assessment process is to precisely define what is meant by "failure," that is, to define potential damage modes from M/OD impact and, for each, to quantify the minimum amount of damage that can lead to failure of spacecraft hardware. This definition establishes the failure criterion or criteria that are assigned to each region of the spacecraft geometry model. For instance, M/OD damage modes and failure criteria for elements of the Space Station are given in Table 8.7.2.

Hypervelocity Impact Tests Hypervelocity impact (HVI) tests are an integral part of the analyzes conducted to ensure adequate design of spacecraft M/OD shielding. Two-stage light-gas guns (LGGs) are used to accelerate projectiles up to 7 km/s (Fig. 8.7.4). LGGs are capable of launching a variety of different and well-controlled projectile shapes. A disadvantage is that LGGs are capable of velocities that cover only a fraction of the orbital debris threat. Since average orbital debris velocity in low-earth orbit is on the order of 9 km/s, LGG can directly simulate only 40 percent of the orbital debris threat.

Other techniques exist to launch projectiles at speeds over 10 km/s. For instance, an inhibited shaped-charge launcher (ISCL) at Southwest Research Institute (SwRI) is capable of launching aluminum projectiles, 0.25 g to 2 g mass, in the shape of a cylinder to 11.5 km/s (Fig. 8.7.5). Another high-speed launcher that has provided useful information on shield capabilities in excess of 10 km/s is the three-stage hypervelocity launcher (HVL) developed at Sandia National Laboratories. This launches thin disks (L/d = 0.1 to 0.2, mass = 0.2 to 1 g) of aluminum and titanium from 10 to 15 km/s with some bowing and tilting of the projectile. A three-stage light-gas launcher at the University of Dayton Research Institute (UDRI) launches 1 to 3 mm diameter spherical projectiles to 10 km/s. Hydrocode simulations are used to assess projectile shape effects on shield response.

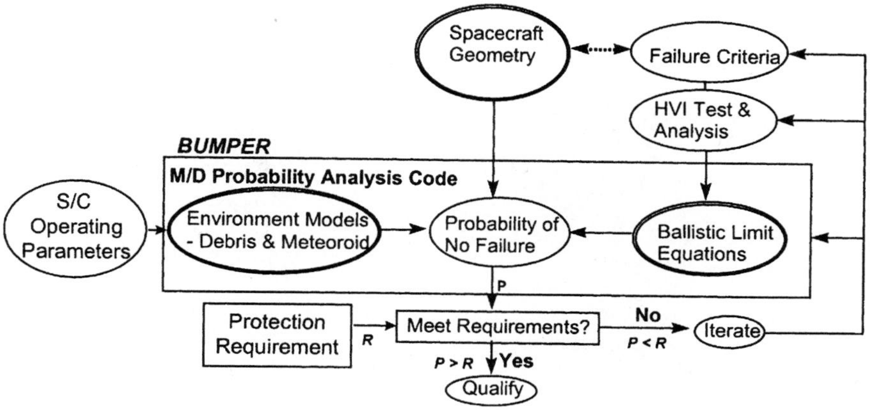

Figure 8.7.2 Meteoroid/orbital debris risk assessment methodology.

Table 8.7.2 Damage Modes and Failure Criteria for Space Station Elements
(Note: Critical failure criteria marked with*.)

Critical elements	Thermal protection system (TPS) damage	Spall/perforation of shield exterior of pressure shell	Perforation of pressure shell	Uncontrolled decompression	Catastrophic rupture	Detonation
Crew modules			X*	X	X	
Pressure vessels (liquids and gases)		X*1	X*1		X	X
Control moment gyros (rotating equipment)		X*			X	
Crew return vehicles	X		X*	X	X	

[1] Failure criterion for Russian PVs is perforation of pressure shell. Failure criterion of other designs is perforation or detached spall of shielding surrounding PV.

Functional elements	Surface degradation	Leak	Short or open circuit
Radiator panels	X	X	
Radiator flex/hard lines		X	
Thermal loop lines and exchangers		X	
Power cables			X
Batteries		X	X
Solar arrays	X		X
Data lines/cables			X
Window outer panes	X		

Figure 8.7.4 NASA JSC-WSTF two-stage light-gas gun.

Ballistic Limit Equations

Ballistic limit equations (BLEs) or "penetration" equations are based on the HVI test results, numerical simulations, and analytical assessments (Christiansen, 2003). BLEs describe the particle sizes that are on the failure threshold of a particular spacecraft component. The ability of a particular shield to withstand hypervelocity impact is predicted by a BLE. BLEs are a function of the target "failure criteria"; target configuration, materials, and thickness; and projectile parameters such as velocity, angle, shape, and density. M/OD risk assessments require a BLE defined for each surface of a spacecraft, for each failure criterion assessed. Shield design equations have also been developed that provide a means to estimate shielding parameters needed to defend against specific projectile threats.

Because of the complexity of spacecraft and their HVI failure modes, many different ballistic limit equations exist for a wide variety of shielding types, spacecraft components and hardware, and differing damage/failure modes. A description of ballistic limit equations for one particular type of shield, the basic Whipple shield, is provided in the following section. This example illustrates the considerations involved in developing ballistic limit equations.

Whipple Shield Ballistic Limit Equations

Fred Whipple proposed in the 1940s a meteoroid shield for spacecraft consisting of a thin "sacrificial" bumper at a distance from a rear wall. The Whipple shield is shown in Fig. 8.7.6. The function of the first sheet, or bumper, is to break up the projectile into a cloud of material containing both projectile and bumper debris. This cloud expands while moving across the standoff, resulting in the impactor momentum being distributed over a wide area of the rear wall (Figs. 8.7.7 to 8.7.9). The back sheet must be thick enough to withstand the blast loading from the debris cloud and any solid fragments that remain. For most conditions, a Whipple shield results in a significant weight reduction over a single

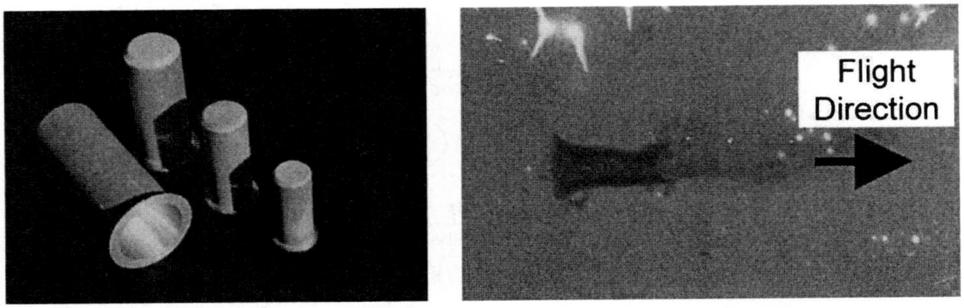

Figure 8.7.5 SwRI's inhibited shaped-charge launcher. Various size charges are available that are capable of launching 0.25 to 2 g aluminum projectiles up to 11.5 km/s (left view). Projectiles are typically in the shape of a hollow cylinder (right view).

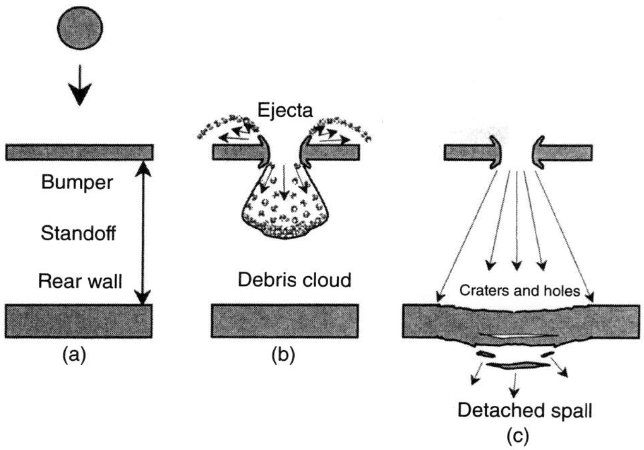

Figure 8.7.6 Whipple shield. (*a*) Whipple shields consist of a bumper, standoff (gap or spacing), and real wall. (*b*) Hypervelocity impacts will generate a cloud of bumper and projectile debris that can contain solid fragments, liquid, and vapor particles. (*c*) The rear wall must survive the fragments and debris cloud impulsive loading. It could fail by perforation from solid fragments, spall, or tear and petal from the impulsive loading.

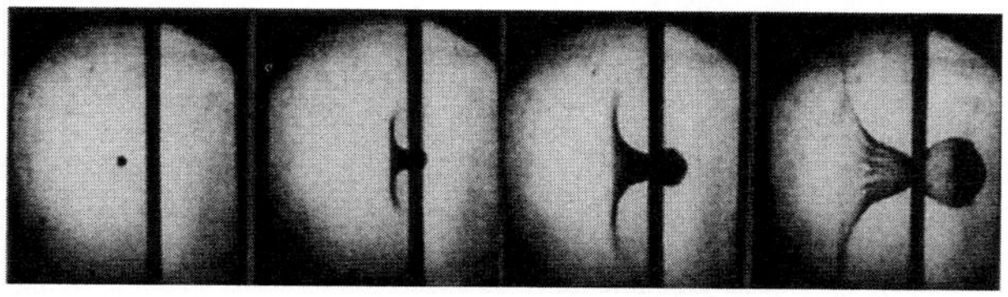

Figure 8.7.7 A 0.32-cm projectile impacts bumper at 6.8 km/s, left to right, with 1 μs between frames. [High-speed camera film *(photograph by Cordin)*.]

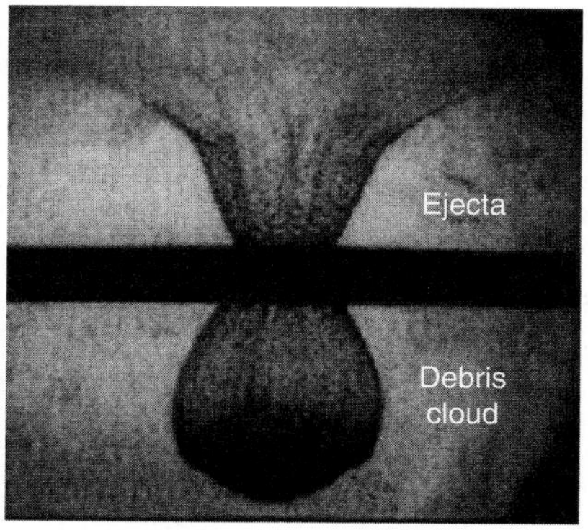

Figure 8.7.8 Ejecta and debris cloud.

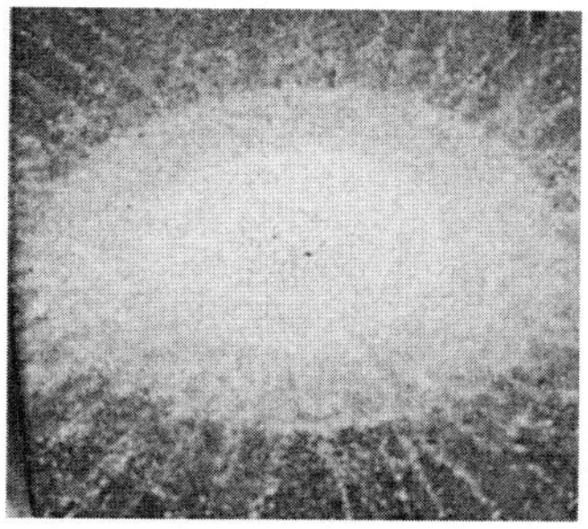

Figure 8.7.9 Rear wall after impact.

plate, which must contend with deposition of the projectile kinetic energy in a very localized area.

Whipple shields were used to provide protection to the Apollo command module and lunar lander. Apollo shields were capable of stopping 0.16-cm-diameter aluminum particles at 7 km/s, normal impact angle (0°).

Whipple Shield Design Equations For design purposes, Eq. (8.7.2) to (8.7.4) provide bumper and rear wall thickness to defeat a given threat particle (assuming $V_n \geq 7$ km/s).

$$t_b = c_b m_p/\rho_b = c_b d\rho_p/\rho_b \tag{8.7.2}$$

$$t_w = c_w d^{0.5} \mathrm{M}^{1/3} (\rho_p \rho_b)^{1/6} \rho_w^{-1} (V\cos\theta) S^{-3/4} \sigma_h'^{-1/2} \tag{8.7.3}$$

where c_b = 0.25 when S/d < 30, c_b = 0.20 when $S/d \geq 30$, and c_w = 0.79 cm$^{-3/4}$·s·g$^{1/3}$ · km^{-1}. Normalized yield stress (unitless) $\sigma_h' = (\sigma/70\ \text{ksi})$.

The ballistic limit criterion is no perforation or spall of the rear wall. Bumper fragments become the primary source of rear wall damage at impact angles greater than 65°. Therefore, for oblique angles over 65°, the calculated rear wall thickness should be constrained to 65°. Coefficient c_w should be adjusted by the factor $[(S/d)/(S/d)_0]$ where $(S/d)_0 = 15$, when $S/d < 15$, as follows.

$$c_w = 0.79\ [(S/d)/(S/d)_0]^{-0.185} \tag{8.7.4}$$

Whipple Shield Ballistic Limits for All Impact Velocities Typical ballistic limits for a Whipple shield at all impact velocities are illustrated in Fig. 8.7.10 for normal impact by an aluminum sphere. The Whipple shield ballistic limit is compared to a monolithic, single aluminum plate of same mass as the combined Whipple shield's bumper and rear wall. Protection capabilities are defined for three penetration regimes based on normal component velocity (V_n) for an aluminum impact on a Whipple shield: ballistic regime (V_n < 3 km/s), fragmentation regime (3 km/s < V_n < 7 km/s), and melt/vaporization regime (V_n > 7 km/s). Important shield physical and material properties that influence ballistic limits of the shield vary as a function of impact velocity. Equations used to quantify ballistic limits for various types of shields are provided in the literature (Christiansen et al., 2009).

A key factor governing the performance of spaced shields is the "state" of the debris cloud projected from the bumper toward the rear wall. The debris cloud may contain solid, liquid, or vaporized projectile and bumper materials, or a combination of the three states, depending on the initial impact pressure. Solid fragments in the debris cloud are generally more penetrating when they contact the rear wall than liquid or vapor particles. Blast loading in the rear wall is a function of shield standoff, which becomes of increasing importance as impact velocity increases above about 5 km/s.

Bumper Code

The BUMPER code has been the standard in use by NASA and contractors to perform meteoroid/debris risk assessments since the early 1960s. Over that period of time, it has undergone extensive revisions and updates. NASA has applied BUMPER to risk assessments for space station, commercial crew/cargo vehicles, shuttle, Mir, extravehicular mobility units (i.e., "space suits"), and other satellites and spacecraft. Significant effort has been expended to validate BUMPER and "benchmark" it to other M/OD risk assessment codes used by some ISS international partners. Figure 8.7.2 illustrates where BUMPER fits in the risk assessment process. The BLEs and M/OD environment models are included in BUMPER. A finite-element model (FEM) that describes the spacecraft geometry is created in separate software and input into BUMPER. BUMPER calculates the number of failures by determining the number of M/OD particles that exceed the ballistic limits for each element of the FEM, and calculates the total number of failures by summing the individual elements. BUMPER results are used to determine the risk drivers, that is, the areas on the spacecraft that control the risk. Emphasis on risk reduction is placed on lowering the risk for the drivers.

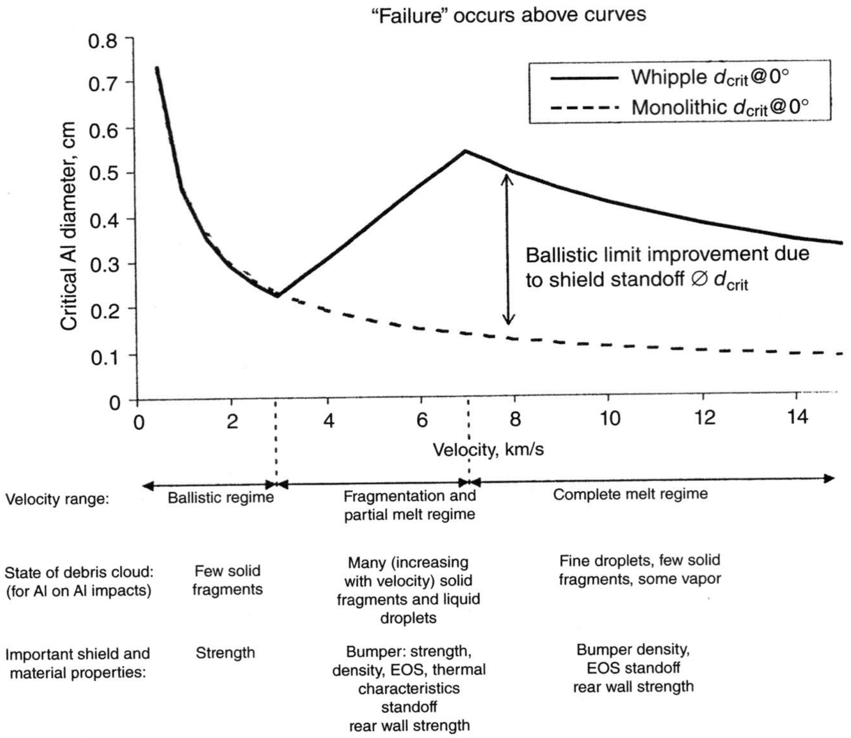

Figure 8.7.10 Ballistic limits for equal mass monolithic target and Whipple shield. Failure criterion is threshold perforation or detached spall from rear wall. Monolithic target is 0.44-cm-thick Al 6061T6. Whipple shield consists of 0.12-cm-thick A1 6061T6 bumper followed at 10 cm by 0.32-cm-thick A1 6061T6 rear wall.

Probability of No Penetration (PNP)

Probability of no penetration (PNP) failure is assessed on the basis of Poisson statistics [Eq. (8.7.5)], where N is defined by Eq. (8.7.1). This is NASA's standard approach used to assess meteoroid shielding since Apollo. The same probabilistic approach has been extended to designing spacecraft protection systems for the combined meteoroid and debris environments, assessing risks of functional failure, or calculating risks of impact damage exceeding a user-defined size (diameter or depth). The same statistical approach is used to assess risks of any kind of damage extent that can be expressed in terms of different ballistic limit equations.

$$\text{PNP} = \exp(-N) \tag{8.7.5}$$

The risk of penetration is:

$$\text{Risk} = 1 - \text{PNP} \tag{8.7.6}$$

Analysis Iteration and Design Qualification

Assessed PNP is compared to requirements for M/OD protection. The shielding design effort is successful when the assessed PNP is greater than the required PNP. As illustrated in Fig. 8.7.2, iteration of the risk assessment and risk reduction process is often necessary to meet protection requirements and optimize the design, that is, meet the requirements with less weight, lower volume (less standoff), less cost, etc. The last step in the process is to conduct final shield verification by analysis (supported by HVI test as appropriate), prior to final design certification and acceptance. During and after the mission has flown, the vehicle should be inspected if possible for M/OD damage to compare to predictions and determine if the protection design is performing as intended.

Risk Reduction Practices

Engineering practices to decrease risk, improve shielding performance, and decrease shielding mass are summarized below.

Find M/OD Risk Drivers Perform detailed assessment of penetration risks for the overall vehicle to determine the zones that control the risks. Selectively improve the protection capability in the areas identified as risk drivers.

Shielding Enhancement

1. Optimize the shielding weight distribution to account for the directional M/OD distribution. Shielding on each critical item is tailored for the environment and its location on station. Because M/OD impact rate is highest on forward and side surfaces, more capable shielding (heavier or with greater standoff) is applied to these surfaces and less on the earth-facing surface. Shielding is also reduced in areas where shadowing from neighboring structures will reduce impacts. The goal of this effort is to equalize the ratio of risk to area for the various vehicle zones. The risk/area ratio for each zone is assessed and optimized by repeated BUMPER code runs.
2. Incorporate enhancements such as increasing the spacing between bumper and rear wall, or using higher-performance alloys and materials for the rear wall.
3. Implement more-fundamental changes such as incorporating more efficient, multibumper shielding concepts.
4. Incorporate toughening materials, such as ceramic fabric and high-strength fabric into the multilayer insulation (MLI) thermal blanket that is often integral to the M/OD shielding.

Impact Reduction

1. *Maximize shadowing.* Take advantage of shielding/shadowing from neighboring items. Locate external critical equipment to trailing

or earth-facing surfaces to reduce M/OD impact rates, or put them in areas highly shadowed by other hardware.

2. *Select low frontal area orientations.* It has been shown that the best orientation for a relatively short cylinder (length/diameter of 1.8) in terms of lowest meteoroid/debris impact risk is with cylindrical length axis oriented perpendicular to the orbital plane. A vertical (gravity-gradient) orientation (with length axis parallel to earth radial) has a 30 percent higher M/OD impact risk. An orientation with length axis parallel to velocity vector is in the middle.

3. *Flight attitude.* Selecting the best flight attitude by pointing the most vulnerable surfaces aft or toward earth is standard procedure for the shuttle. Other spacecraft can take a similar approach.

Inert Stored Energy Equipment After use, stored energy equipment should be made inert if possible. For instance, completely depressurize any storage tank after it is emptied. The risk of catastrophic rupture is eliminated when stress levels in the pressure wall are made negligible by depressurizing to a small value. This would require design modifications to implement for propellant tanks and other fluid storage tanks. Another example is to keep spare flywheels, gyros, and other momentum storage devices in an inactive state until required. A corollary idea for gaseous pressure vessels is to use them in series (not parallel) and deplete first the pressure vessels in the locations that are most exposed to M/OD impact (those in the forward or side positions), followed by less exposed positions.

Reduce Hazards if Shield Penetration Occurs Design and operational options may exist to reduce hazards if a penetration occurs. For instance, some hatches to unoccupied modules can be kept closed to prevent a depressurization of an entire station if a penetration occurred to the module. A perforation into an unoccupied module with hatch closed would not result in loss of crew from the fragments/shrapnel, light flash, acoustic overpressure, or depressurization. Vent lines between modules could be left open to allow for some air circulation and to keep pressures equalized to facilitate hatch opening during normal operations.

Critical Damage Detection, Repair, and Replacement (R&R) Inspection and repair of impact damage to critical areas of reentry vehicles (such as the crew return vehicle attached to ISS) can be used as a supplement to M/OD shielding for maintaining flight worthiness. Some impact damage to thermal protection systems (TPS) on earth-return vehicles is not a hazard while on orbit (and may therefore be undetected) but could become hazardous later during reentry aerodynamic heating phases. Properly placed instrumentation with correct sensitivity to detect critical TPS damage can help support the inspection/detection process.

Enhanced Shielding

The primary purpose of developing improved shielding is to provide higher levels of spacecraft meteoroid/debris protection with less weight.

To illustrate the issue, consider the shielding required to stop 1-cm-diameter aluminum projectiles at 7 km/s, 08 impact angle. Four shield concepts to meet the requirement are given in Fig. 8.7.11: a conventional aluminum Whipple shield with a 10.2-cm standoff, a Nextel/Kevlar stuffed Whipple shield with the same standoff, a Whipple shield with a 30-cm standoff, and a Nextel multishock shield concept. The stuffed Whipple shield incorporates a blanket between the outer aluminum bumper and inner pressure wall that combines two materials: Nextel™ ceramic fabric and Kevlar™ high-strength fabric. The shielding mass estimates are made assuming the shielding encloses a cylinder, with 4.2-m inside diameter by 8.5 m long. Stuffed Whipple and multishock shields are described in more detail later in this section. But it is clear for this example that there are significant mass savings by using advanced shielding concepts (i.e., up to 50 percent reduction).

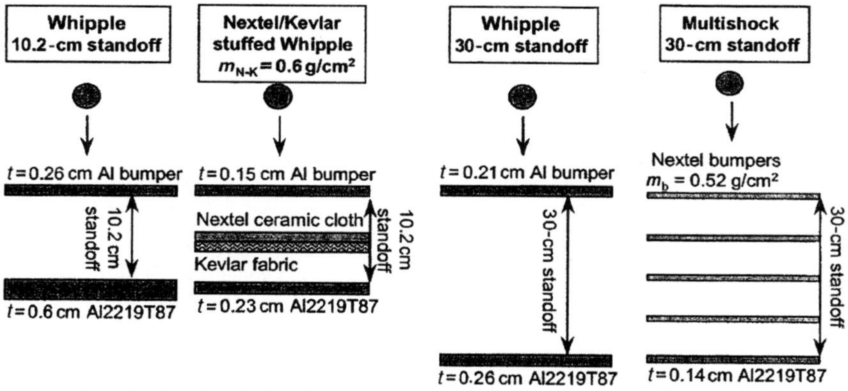

	Whipple S = 10 cm	Stuffed Whipple S = 10 cm	Whipple S = 30 cm	Multishock S = 30 cm
Areal density (kg/m²)				
Bumper:	7.0	10.6	5.6	5.2
Rear wall:	17.2	6.6	7.5	3.8
Total:	24.2	17.3	13.1	9.0
Surface area (m²)				
Bumper:	152	152	175	175
Rear wall:	141	141	141	141
Mass (kg) including 30% of bumper for support mass				
Bumper:	1060	1620	980	910
Support:	320	490	300	270
Rear wall:	2420	940	1060	540
Total: (include support)	3800	3050	2340	1720

Figure 8.7.11 Shielding concepts comparison. (Projectile in all cases 1-cm-diameter aluminum, 1.5 g, 7 km/s, normal impact.)

Also, it is possible to trade weight for protection capability (i.e., capability and shield PNP are related), so it can be shown that lower weight and more effective protection in terms of higher PNP are possible by using Nextel/Kevlar stuffed Whipple and multishock shields, compared to conventional Whipple shields.

SPACE FLIGHT

by Aaron Cohen, Revised by Glenn J. Miller

The science of astronautics deals with the design, construction, and operation of craft capable of flight through interplanetary or interstellar space. In addition to orbits around, and trajectories between, such bodies as stars, planets, and planetoids, the ascents from and descents to the surface of such bodies are considered to be part of the field.

The purposes of such craft are numerous and include the transportation of people and cargo, the transmission or relaying of signals, and the carrying of instruments for the measurement of the characteristics of (1) the environment through which the craft flies, (2) the surface of the celestial body over which the craft flies, or (3) astronomical objects or phenomena. The commercial applications of space are now being realized. Many communication, weather, and earth-sensing satellites have been deployed. With the completion of the International Space Station (ISS) scientists have access to a continuously manned research laboratory in Low Earth Orbit. The ISS consists of multiple modules joined together to form a pressurized volume of 930 cubic meters. Although the last Space Shuttle flight was in 2011, transportation of crew and cargo to ISS continues with spacecraft launched on Expendable Launch Vehicles provided by International and Commercial Partners.

Processing of materials in a microgravity environment is proving to be very reliable. In microgravity the effects of thermal convection and gravity are eliminated, hence purer materials can be manufactured. Containerless processing can also be accomplished which will lead to additional applications in microgravity materials technology. Using expendable launch vehicles satellites can be placed in various orbits. A polar orbit is one whose orbital plane includes the earth's axis and permits line-of-sight contact between the spacecraft and every point on earth on a periodic basis. Synchronous orbits have angular rates equal to the rotational rate of the earth about its axis, providing a spacecraft locus of positions at nearly constant longitude. Orbital-parameter variation provides a wide variety of other orbits, permitting the selection of characteristics most suitable for the specific flight objective, such as communication relays, meteorological observations, earth surface monitoring, or interplanetary trajectories.

SHUTTLE THERMAL PROTECTION SYSTEM TILES

by William C. Schneider

Texas A & M University

Revised by Glenn Miller

REFERENCES: "National Space Transportation System Reference Document," NASA, June 1988. Moser and Schneider, "Strength Integrity of the Space Shuttle Orbiter Tiles," AIAA First Flight Testing Conference, Nov. 11-12, 1981, Las Vegas, paper AIAA-81-2469. Schneider and G. J. Miller, "The Challenging Scales of the Bird (Shuttle Tile Structural Integrity)," Space Shuttle Technical Conference, June 28-30, Houston. O'Hearn, "Application of Probabilistic Analysis to Aerospace Structural Design," AIAA 81-0231.

The shuttle orbiter was the first reusable spacecraft. Prior to the first shuttle orbiter launch in 1981, humans were transported to space in single use capsules. For launch the orbiter was attached to an external tank and two solid rocket boosters. The combined stack was launched in an upright position. During ascent both solid rocket boosters and the external tank separated from the shuttle orbiter and did not continue into orbit. After deorbit and entry, the unpowered orbiter glided back to Earth and landed on a runway like an airplane. The primary structure of the orbiter was designed using typical aircraft construction techniques. However, the unique aspect of the orbiter design was the addition of high temperature materials to the outer structural skin to protect the metallic and composite airframe from the extreme temperatures of entry. These high temperature materials comprise the orbiter thermal protection system (TPS). During entry, the TPS materials maintain the structural skin of the orbiter to temperatures at or below 350°F. In addition, they are reusable for 100 missions, with refurbishment and maintenance. These materials perform in temperature ranges from −250°F in the cold soak of space to entry temperatures that reach nearly 3,000°F. The TPS also withstands the forces induced by deflections of the orbiter airframe as it responds to the various external environments. Because the thermal protection system is installed on the outside of the orbiter skin, it establishes the aerodynamics over the vehicle. The orbiter interior temperatures also are controlled by internal insulation, heaters, and purging techniques in the various phases of the mission.

The TPS is a passive thermal insulation system consisting of materials selected for stability at high temperatures and weight efficiency. Additionally, the TPS has to be reusable flight after flight. These highly constraining requirements drove the design to the highly reliable, passive, lightweight, reusable system that, flight after flight, protects the Shuttle Orbiter. However, the TPS was never designed to withstand significant impacts of any kind any more than an automobile windshield was designed to withstand significant impacts. The region of the orbiter vehicle where each material is used is depicted in Fig. 8.7.12. These materials are as follows:

1. **Reinforced carbon-carbon** (RCC) protects areas where the surface temperatures exceed 2,300°F during entry, such as the wing leading edges and the nose cap.

2. Black **high-temperature reusable surface insulation** (HRSI) tiles protect areas where the surface temperatures are below 2,300°F but above 1,200°F. These tiles have a black surface coating, necessary for entry emittance. HRSI tiles are used in areas on the upper forward fuselage, including around the forward fuselage windows; the entire underside of the vehicle where RCC is not used; portions of the orbital maneuvering system and reaction control system pods; the leading and trailing edges of the vertical stabilizer; wing glove areas; elevon trailing edges; adjacent to the RCC on the upper wing surface; the base heat shield; the interface with wing leading edge RCC; and the upper body flap surface.

3. Black tiles, called **fibrous refractory composite insulation** (FRCI), with higher strength and weight were developed later in the thermal protection system program. FRCI tiles replace some of the HRSI tiles in selected areas of the orbiter.

4. **Low-temperature reusable surface insulation** (LRSI) tiles protect areas where surface temperatures are below 1,200°F but above 700°F and have a white surface coating to provide better thermal characteristics on orbit. These white tiles are used on the upper wing and vertical tail and in selected areas of the forward, mid-, and aft fuselages, etc.

5. After the initial delivery of Columbia to the assembly facility, an **advanced flexible reusable surface insulation** (AFRSI) was developed. This material consists of composite quilted fabric insulation batting sewn between two layers of white fabric. AFRSI was used on Discovery and Atlantis to replace the vast majority of the LRSI tiles. Following its seventh flight, Columbia also was modified to replace most of the LRSI tiles with portions on the upper wing. The AFRSI blankets provide improved producibility and durability, reduced fabrication and installation time and costs, and a weight reduction over that of the LRSI tiles.

6. White blankets made of coated Nomex **felt reusable surface insulation** (FRSI) protect areas where surface temperatures are below 700°F and are used on the upper payload bay doors, portions of the mid-fuselage and aft fuselage sides.

7. Additional materials are used in other special areas such as the thermal panes for the windows; metal for the forward reaction control system fairings and elevon seal panels on the upper wing-to-elevon interface; and a combination of white- and black-pigmented silica cloth for thermal barriers and gap fillers around operable penetrations, such as main and nose landing gear doors, egress and ingress flight crew side hatch, umbilical doors, elevon cove, forward RCS, RCS thrusters, mid-fuselage vent doors, payload bay doors, rudder/speed brake, and gaps between TPS tiles in high differential pressure areas.

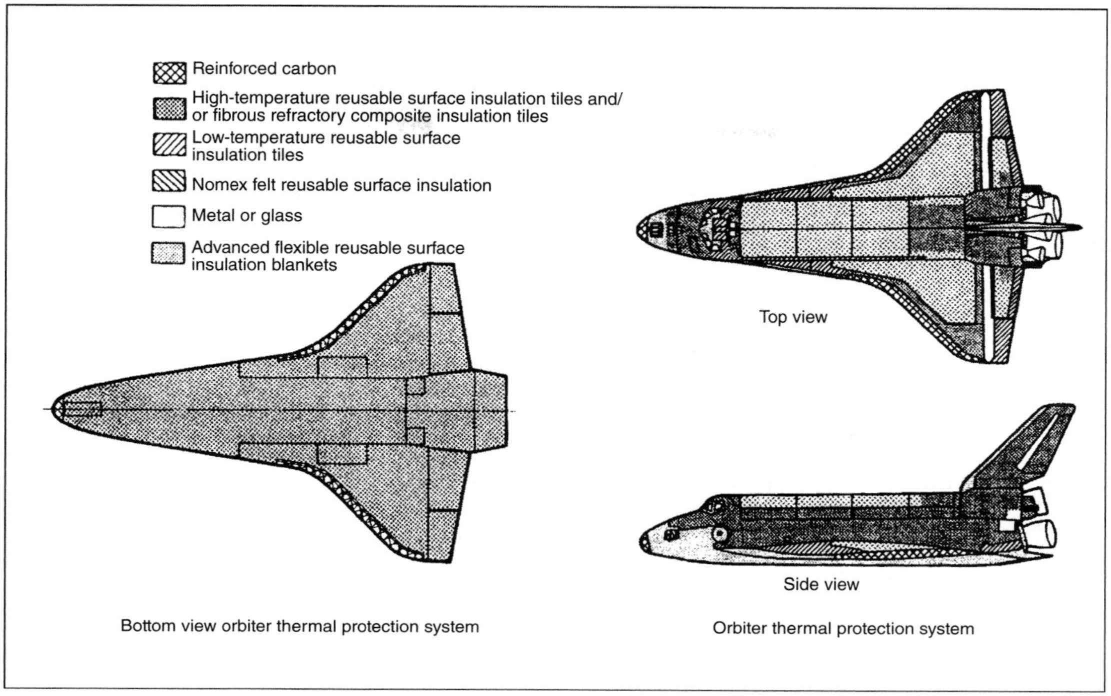

Figure 8.7.12 Thermal protection system regions.

The RSI tiles, shown conceptually in Fig. 8.7.13, are designed to radiate 80 to 90 percent of the reentry heat back to space before it can be conducted to the Orbiter's structure. Since the structure deforms in-plane during a mission, continuous covering for large areas would be prohibitive because of the weak RSI material; therefore the nominal 6-in-square tile with a 0.05-in gap between was used. To accommodate the flight-induced out-of-plane structural deformation, the individual tiles are individually mounted on a compliant strain isolation pad (SIP) made of Nomex felt, the same material as FRSI. This design concept of using individual tiles on SIP becomes fairly obvious with a limited amount of stress analysis using the RSI physical properties shown in Table 8.7.3.

As the LI-900 RSI 13-psi ultimate flatwise tension strength allowable was being established, the strength properties of the tile/SIP system were also being established. The effective joint allowable of this interface was approximately 50 percent that of the LI-900 material. This reduction in strength was caused by stress concentrations in the RSI because of "stiff spots" introduced into the SIP by the needling felting process. The challenge of accommodating these stiff spots for the more highly loaded tiles, as opposed to using the heavier, higher-strength LI-2200, was met by locally densifying (strengthening) the underside of the tile. A solution of colloidal silica particles is applied to the noncoated tile underside and baked in an oven at 3,500°F for 3 hours. The densified layer produced

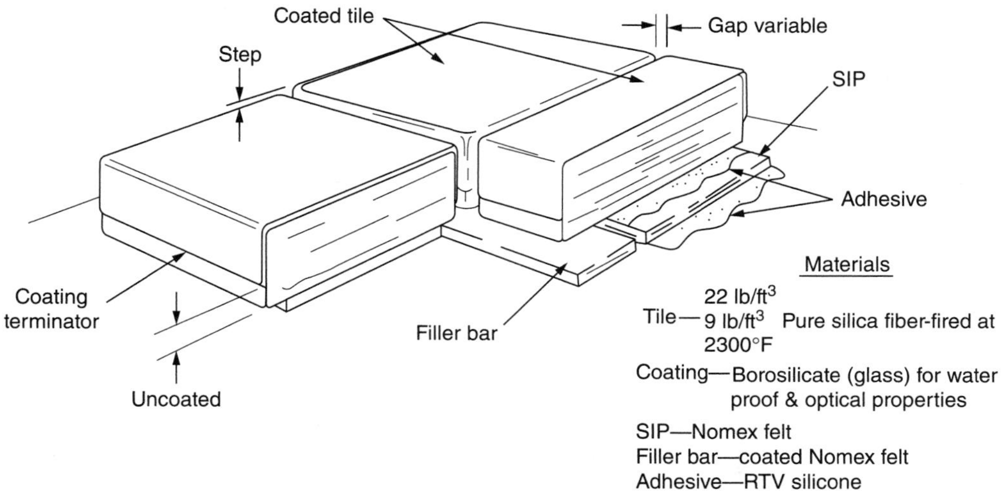

Figure 8.7.13 Tile design. (*From Moser et al.; reprinted by permission of the American Institute of Aeronautics and Astronautics, Inc.*)

Table 8.7.3 Properties of Reusable Surface Insulation

	Low density	High density
Material	High-purity silica fiber, LI-900	High-purity silica fiber, LI-2200
Density	9 lb/ft^3	22 lb/ft^3
Strength		
Flatwise tension	13 psi	35 psi
In-plane tension	41 psi	100 psi
Modulus of elasticity	6000 psi	36,000 psi
Thermal coefficient of expansion	2.7×10^{-7} in/in/°F	2.7×10^{-7} in/in/°F

SOURCE: AIAA Paper 81–2469. From Moser et al., reprinted by permission of the American Institute of Aeronautics and Astronautics, Inc.

is 0.1 in thick and increases the weight of a typical 6- by 6-in tile by only 0.06 lb. For load distribution, the densified layer serves as a structural plate that distributes the concentrated SIP loads evenly into the weaker unmodified RSI material. As shown in Fig. 8.7.14, this forces the failure surface to occur above the densified layer. This enhancement in itself offers somewhat of a fail-safe feature for thermal protection, but the major enhancement is in the improved tile/SIP system strength shown in Fig. 8.7.15.

There were many tiles that had standard 6- by 6-in square planforms but the numerous special tiles created a greater structural design challenge. It was necessary to design thousands of unique tiles that had compound curves, which interfaced with thermal barriers and hatches, which had penetrations for instrumentation and structural access, etc. The overriding challenge then was to assure the strength integrity of the tiles with a confidence that there was no greater chance than 1/10,000 of losing any one of the more than 10,000 more critical tiles (i.e., a probability of tile failure of no greater than $1/10^8$). The statistical treatment of the tile strength integrity is addressed in the first reference.

To accomplish this magnitude of system reliability and still minimize the weight it was necessary to define the detail loads and environments on each tile. Each tile experienced stresses induced by the following combined sources:

- Substrate or structure out-of-plane displacement
- Aerodynamic loads on the tile
- Tile accelerations due to vibration and acoustics
- Mismatch between tile and structure at installation
- Thermal gradients in the tile
- Residual stress due to tile manufacture
- Substrate in-plane displacement

Literally years of complex structural analysis, hundreds of tile specimens on high-speed-aircraft flow tests, tens of multitile wind tunnel tests, approximately 40 full-scale structural vibroacoustic tests (containing hundreds of tiles each), and approximately 10,000 on-the-vehicle structural tile proof tests served in concert to design and verify the integrity of the TPS tile design. The Space Shuttle Transportation System is no longer an active flight program. However, the techniques developed to attach and analyze the attachment of TPS material to primary structure have been incorporated into some of the capsules that currently resupply the International Space Station.

DYNAMIC ENVIRONMENTS

by Michael B. Duke
NASA

REFERENCES: Harris and Crede, "Shock and Vibration Handbook," McGraw-Hill. Beranek, "Noise Reduction," McGraw-Hill. Crandall, "Random Vibration," Technology Press, MIT. Neugebauer, The Space Environment, *Tech. Rep. 34-229,* Jet Propulsion Lab., Pasadena, 1960. Hart, Effects of Outer Space Environment Important to Simulation of Space Vehicles, *AST Tech Rep. 61-201.* Cornell Aeronautical Lab. Barret, Techniques for Predicting Localized Vibratory Environments of Rocket Vehicles, *NASA TN D-1836.* Wilhold, Guest, and Jones, A Technique for Predicting Far Field Acoustic Environments Due to a Moving Rocket Sound Source, *NASA TN D-1832.* Eldred, Roberts, and White, Structural Vibration in Space Vehicles, *WADD Tech. Rep. 61-62.* Bolt, Beranek, and Newman, Exterior Sound and Vibration Fields of a Saturn Vehicle during Static Firing and during Launching, *U.S.A. Ord. Rep. 764.*

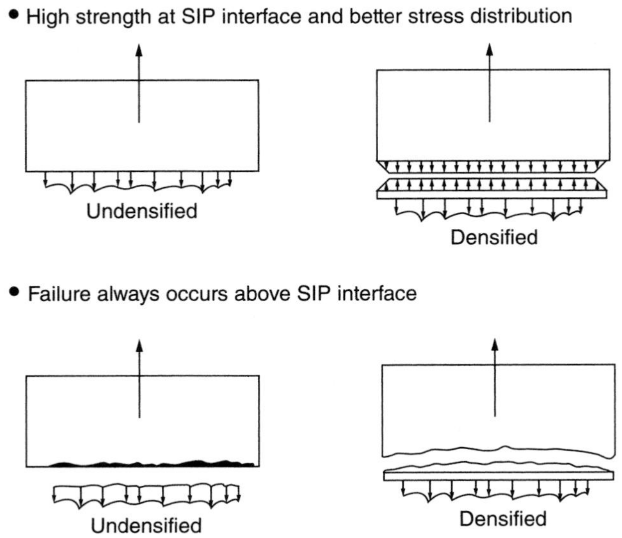

Figure 8.7.14 Salient features of densified tiles. (*From Moser et al.; reprinted by permission of the American Institute of Aeronautics and Astronautics, Inc.*)

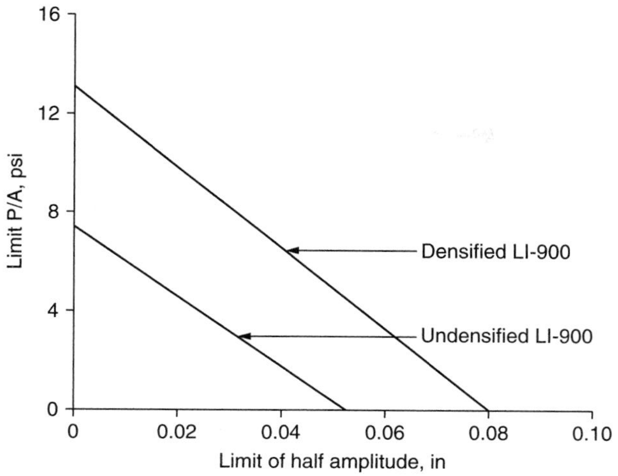

Figure 8.7.15 Allowable flatwise tension in presence of substrate deflection. *(From Moser et al.; reprinted by permission of the American Institute of Aeronautics and Astronautics, Inc.)*

The forces required to propel a space vehicle from the launch pad are tremendous. The dynamic pressures generated in the atmosphere by large rocket engines are exceeded only by nuclear blasts. Slow initial ascent from the launch pad is followed by rapid acceleration and high *g* loading, and acoustic and aerodynamic forces drive every point of the vehicle surface. The space-vehicle velocity quickly becomes supersonic and continues to hypersonic velocities. Gradually, as the space vehicle leaves the atmosphere, the aerodynamic forces recede. Then suddenly the rocket thrust decays and is followed by the ignition of another rocket, which quickly develops thrust and continues to accelerate the space vehicle. Finally, the space vehicle becomes weightless. In completing a mission, a space vehicle is subjected to the additional environmental extremes of planetary landings, escape, and earth reentry.

These environments represent basic criteria for space-vehicle design. For operational reliability, the most important are the shock, vibration, and acoustic environments which are present to varying magnitudes in all operational phases and which constitute major engineering problems.

It is not possible to have original environmental data prior to fabrication and testing of a particular structure. Vibration, shock, and acoustic noise adversely affect structural integrity and vehicle reliability. These environments must be considered prior to design and fabrication and then again for design verification through ground testing. Precise vibration and acoustic environmental predictions are complicated because of structural nonlinearities, random forcing functions, and multiple degree-of-freedom systems. Current environmental-prediction techniques consist mainly of extrapolating measured data obtained from existing vehicles. Acoustic environments produced during ground test are important factors in evaluating ground-facility locations and design, personnel safety, and public relations.

A source of sound common to all jet and rocket-engine propulsion systems is the turbulent exhaust flow. This high-velocity flow produces pressure fluctuations, referred to as noise, and has adverse effects on the vehicle and its operations. Also, far-field uncontrolled areas may be subjected to intense acoustic-sound-pressure levels which may require personnel protection or sound-suppression devices. The sound-generation mechanism is presented in empirical and analytical form, utilizing measured sound pressures and the known engine operational parameters. This combination of empirical terms and analytical reasoning provides a generalized relationship to predict sound-pressure levels for the vehicle structure and also for far-field areas. It has been shown by Lighthill (*Proc. Royal Soc.*, 1961) that acoustic-power generation of a subsonic jet is proportional to $\rho V^8 D^2/a_0^3$, where ρ is the density of the ambient medium, a_0 is the speed of sound in the ambient medium, V is the jet or rocket exit exhaust velocity, and D is a characteristic dimension of the engine. Attempts have been made to extend this relationship to allow supersonic-jet acoustic-power predictions to be made, but no one scheme is accepted for broad use.

From acoustic-data-gathering programs utilizing rocket engines, various trends have been noted which facilitate acoustic predictions. For rocket engines of up to 5×10^9 W mechanical power, the acoustic efficiency shows a nonlinear rate of increase. For greater mechanical power, the acoustic efficiency approaches a value slightly higher than 0.5 percent (Fig. 8.7.16).

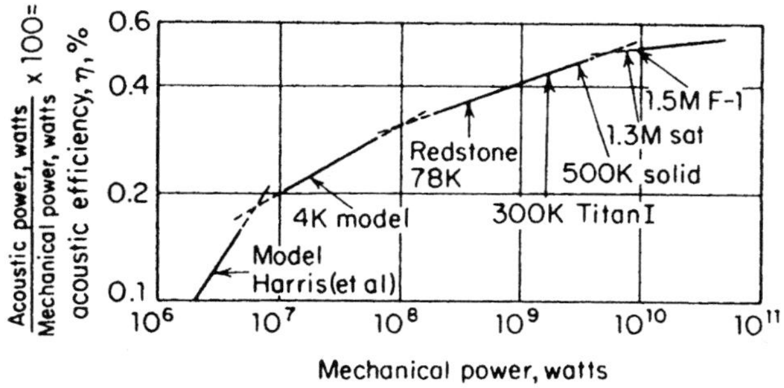

Figure 8.7.16 Acoustic-efficiency trends.

Turbulent-exhaust-flow sound-source mechanisms also exhibit directional characteristics. Figure 8.7.17 shows data from a Saturn rocket vehicle launch with the microphone flush-mounted on the vehicle skin at approximately 100 ft above the nozzle plane.

During the period between on-pad firing and post-lifting-off or main-stage flight, the exhaust-flow direction changes. The directivity has also shifted 90° to retain its inherent relationship to the exhaust stream. The sound-pressure-level variation during a small time interval—between on-pad condition and shortly after liftoff—depends on the changes which have occurred in the directivity and the jet-exhaust-velocity vector (Fig. 8.7.17). The change in sound-pressure level is 20 dB approximately in this case and is indicative of the significance of a sound source's directional properties, which depend on velocity and temperature of the jet.

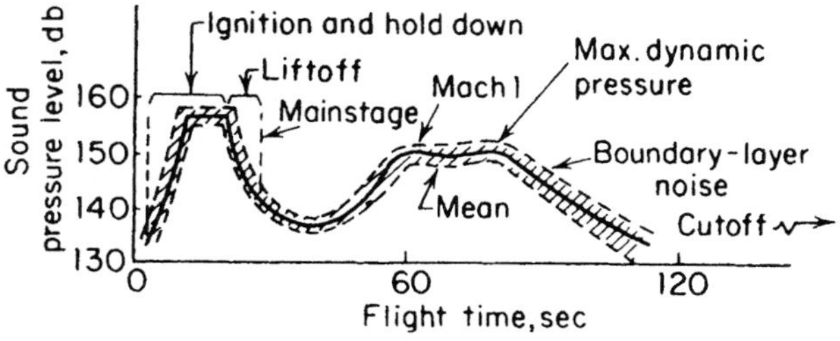

Figure 8.7.17 Vehicle acoustic environment.

The acoustic environments produced by a given conventional engine can be empirically described by that engine's flow parameters and geometry. It is difficult to predict the acoustic environments for non-conventional engines such as the plug-nozzle and expansion-deflection engines.

The turbulent-boundary-layer problem of high-velocity vehicle flights is difficult to evaluate in all but the simplest cases. Only for attached, homogeneous, subsonic flows can the boundary-layer noise be estimated with confidence.

The vibratory environment on rocket vehicles consists of total-vehicle bending vibrations, in which the vehicle is considered a nonuniform beam, with low-frequency responses (0 to 20 Hz) and localized vibrations with frequencies up to thousands of hertz. (See also Sec. 3.)

The response of a simple structure to a dynamic forcing function may be described by the universal equation of motion given as $M\ddot{x} + c\dot{x} + Kx = F(t)$ or $F(t)x(t) = (M - K\omega^2) - j(c/\omega)$ and is a complex function containing real and imaginary quantities. In the case of rocket-vehicle vibrations, $F(t)$ is a random forcing phenomenon. Thus, the motion responses must also be random. Random vibration may be described as vibration whose instantaneous magnitudes can be specified only by probability-distribution functions giving the probable fraction of the total time that the magnitude, or some sequence of magnitudes, lies within a specified range. In this type of vibration, many frequencies are present simultaneously in the waveform (Fig. 8.7.18). Power

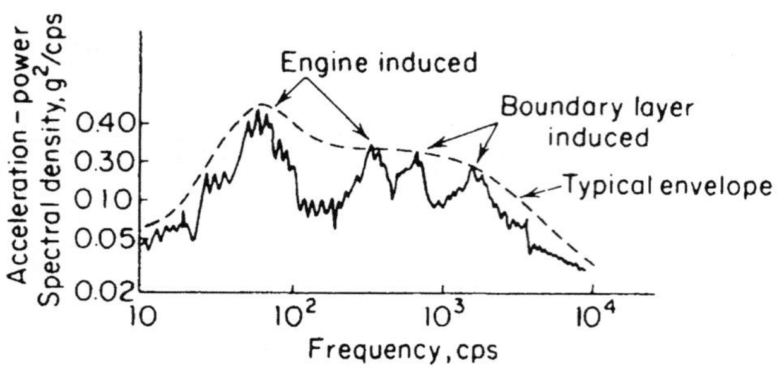

Figure 8.7.18 Random-vibration spectrum of a space-vehicle structure.

spectral density (PSD; in units of g^2/Hz) is defined as the limiting mean-square value of a random variable, in this case acceleration per unit bandwidth; that is,

$$\mathrm{PSD} = \lim_{\Delta f \to 0} \bar{g}^2/\Delta f = d\bar{g}^2/df$$

The sources of vibration excitation are (1) acoustic pressures generated by rocket-engine operation; (2) aerodynamic pressures created by boundary-layer fluctuations; (3) mechanically induced vibration from rocket-engine pulsation, which is transmitted throughout the vehicle structure; (4) vibration transmitted from adjacent structure; and (5) localized machinery. Since the mean square acceleration is proportional to power, the total vibrational power at any point on the vehicle may be expressed as

$$Pv = Pv(1) + Pv(2) + Pv(3) + Pv(4) + Pv(5)$$

This is illustrated in Fig. 8.7.19 for an arbitrary vehicle. In many instances, the structure is principally excited by only one of these sources and the remaining can be considered negligible (Fig. 8.7.20). The vibration characteristics of structure B, taken from a skin panel, closely correspond to the trends of the acoustic pressures during flight. However, structure A, taken from a rocket-engine component, exhibits stationary trends, indicating no susceptibility to the impinging acoustic field. Thus, for this structure, it can be assumed that only $Pv(3)$ is significant, whereas for structure B, only $Pv(1)$ is the principal exciting source during the on-pad phase and only $Pv(2)$ is significant during the maximum dynamic-pressure phase. This also holds true for captive firings of the vehicle, in which only $Pv(1)$ is significant for structure B and $Pv(3)$ for structure A.

SPACE-VEHICLE TRAJECTORIES, FLIGHT MECHANICS, AND PERFORMANCE

by O. Elnan, W. R. Perry, J. W. Russell, A. G. Kromis, and D. W. Fellenz
University of Cincinnati

REFERENCES: Russell, Dugan, and Stewart, "Astronomy," Ginn. Sutton, "Rocket Propulsion Elements," Wiley. White, "Flight Performance Handbook for Powered Flight Operations," Wiley. "U.S. Standard Atmosphere, 1962," NASA, USAF, U.S. Weather Bureau. Space Flight Handbooks, *NASA SP*-33, *SP*-34, and *SP*-35. Gazley, Deceleration and Heating of a Body Entering a Planetary Atmosphere from Space, "Visits in Astronautics," Vol. I, Pergamon. Chapman, An Approximate Analytical Method for Studying Entry into Planetary Atmospheres, *NACA-TN* 4276, May 1958. Chapman, An Analysis of the Corridor and Guidance Requirements for Super-circular Entry into Planetary Atmospheres, *NASA-TN D*-136, Sept 1959. Loh, "Dynamics and Thermodynamics of Planetary Entry," Prentice-Hall. Grant, Importance of the Variation of Drag with Lift in Minimization of Satellite Entry Acceleration, *NASA-TN D*-120, Oct. 1959. Lees, Hartwig, and Cohen, Use of Aerodynamic Lift during Entry into the Earth's Atmosphere, *Jour. ARS*, **29**, Sept 1959. Gervais and Johnson, Abort during Manned Ascent into Space, *AAS*, Jan. 1962.

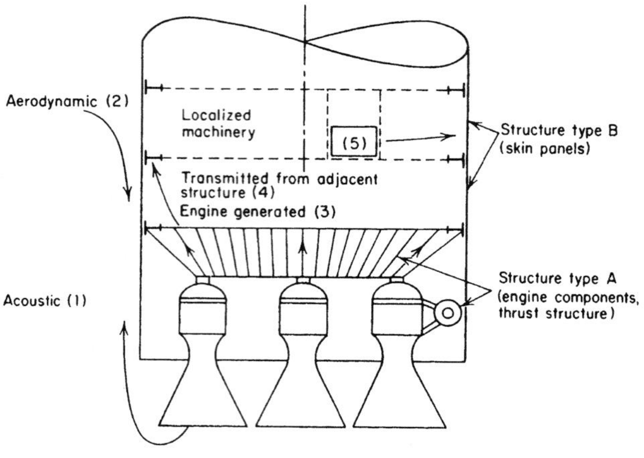

Figure 8.7.19 Sources of vibration excitement.

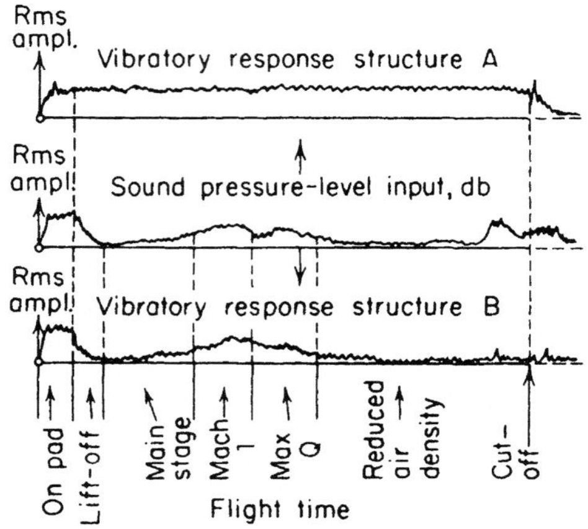

Figure 8.7.20 Typical structural responses to a random-vibration environment.

Notation

A = reference area
A_z = launch azimuth
B = ballistic coefficient
C_L = lift coefficient
e = eccentricity
f = stage-mass fraction (WP/W_A)
H = total energy
H_∞ = hyperbolic excess energy
I = specific impulse
Δ_i = plane-change angle
L = lift force
m = mass
n = mean motion $(2\pi/p_{\mathrm{sid}})$
p = ambient atmospheric pressure

q = dynamic pressure
R = earth's equatorial radius
S = wing area
T = transfer time, time, absolute temperature
t_{pp} = time after perigee passage
V_c = injection velocity
V = velocity
V_g = velocity loss to gravity
W = weight
W_L = payload weight
W_P = propellant weight
Y = distance
α_Ω = right ascension of the ascending node
$\gamma = \theta - 90°$ = flight-path angle used in reentry analysis
v = true anomaly
ω = earth's rotational velocity
ψ = central angle
$\sigma = \rho/\rho_0$ = relative atmospheric density
Υ = vernal equinox
A_e = nozzle exit area
a = semimajor axis
C_D = drag coefficient
D = drag force
E = eccentric anomaly
F = thrust
g = acceleration due to gravity
H_{esc} = escape energy
h = altitude
i = inclination
J_2 = oblateness coefficient of the earth's potential (second-zonal harmonic)
M = molecular weight, mean anomaly
nm = nautical miles
P = period of revolution
P_e = nozzle exit pressure
q_s = stagnation-point heat-transfer rate
r = radius
s = range
t = time
V_∞ = hyperbolic excess velocity
V_{id} = ideal velocity
ΔV = impulsive velocity
W_A = stage weight ($W_O - W_L$)
W_O = gross weight
X = distance
α = angle of attack, right ascension
β = exponent of density-altitude function
λ = longitude
τ = hour angle
ω = argument of perigee
ϕ' = latitude
ρ = atmospheric density
θ = flight-path angle

Subscripts

a = apogee
e = entry
f = final
o = sea level conditions at 45° latitude
s = space fixed
vac = vacuum
circ = circular orbital condition
esc = escape
i = initial
p = perigee
M = mean
SL = sea level
sid = sidereal

ORBITAL MECHANICS

by O. Elnan and W. R. Perry
University of Cincinnati

The motion of the planets about the sun, as well as that of a satellite in its orbit about a planet, is governed by the inverse-square force law for attracting bodies. In those cases where the mass of the orbiting body is small relative to the central attracting body, it can be neglected. This simplifies the analysis of the orbital motion. The shape of the orbit is always a conic section, that is, ellipse, parabola, or hyperbola, with the central attracting body at one of the foci. Parabolic and hyperbolic orbits are open, terminate at infinity, and represent cases where the orbiting body escapes the central force field of the attracting body.

The **laws of Kepler** for satellite orbits are (1) the radius vector to the satellite from the central body sweeps over equal areas in equal times; (2) the orbit is an ellipse, with the central attracting body at one of its foci; and (3) the square of the period of the satellite is proportional to the cube of the semimajor axis of the orbit.

Six **orbital elements** are required to describe the orbit in the orbit plane and the orientation of the plane in inertial space. The three elements that define the orbit are the semimajor axis a, eccentricity e, and period of revolution P. The orientation of the orbit (Fig. 8.7.21) is defined by the right ascension of the ascending node α_Ω, inclination of the orbit plane to the earth's equatorial plane i, and the argument of perigee v, which is the central angle measured in the orbit plane from ascending node to perigee. The following equations compute these orbital elements and other orbital parameters.

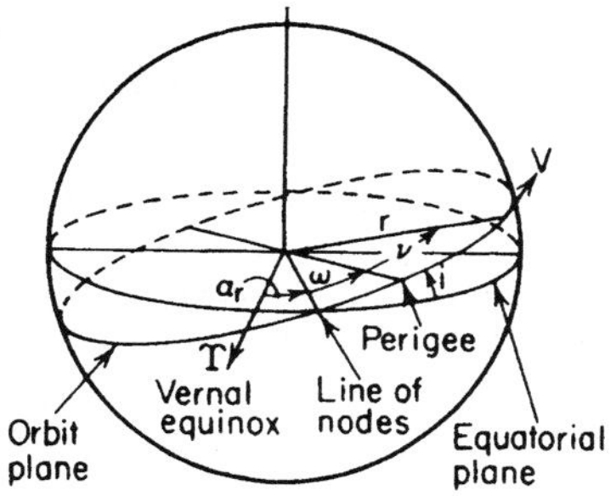

Figure 8.7.21 Elements of satellite orbit around a planet.

$$a = (r_a + r_p)/2 = \mu/V_aV_p = r_a/(1+e) = r_p/(1-e)$$

$$e = (r_a/a) - 1 = (V_p - V_a)/(V_p + V_a) = (r_pV_p^2/\mu) - 1$$

$$r = a(1-e^2)/(1+e\cos v)$$

$$r_a = a(1+e) = r_p(1+e)/(1-e)$$

$$r_p = a(1-e) = r_a(1-e)/(1+e)$$

$$V = [\mu(2/r - 1/a)]^{0.5}$$

$$P = 2\pi(a^3\mu)^{0.5} \quad \text{(for circular orbits, } r = a\text{)}$$

$$V_a = V_p(1-e)/(1+e) = [\mu(1-e)/a(1+e)]^{0.5}$$

$$V_p = V_a(1+e)/(1-e) = [\mu(1+e)/a(1-e)]^{0.5}$$

$$v = \cos^{-1}\{[2r_ar_p - r(r_a + r_p)]/r(r_a - r_p)\}$$

$$t_{pp} = (a^3/\mu)^{0.5}\{\cos^{-1}[(a-r)/ae] - e[1-(a-r)^2/(ae)^2]^{0.5}\}$$

$$M = nt_{pp} = E - e\sin E$$

$$r = a(1-e\cos E)$$

$$V_{esc} = (2\mu/r)^{0.5}$$

The elements may be determined from known injection conditions. For example,

$$1/a = V_i^2\mu - 2/r_i$$

$$e^2 = \cos^2 V_i(r_iV_i^2/\mu - 1)^2 + \sin^2 V_i$$

General precession in longitude of vernal equinox:

$$x = 50''.2575 + 0''.0222T$$

where T is in Julian centuries of 36,525 mean solar days reckoned from 1900 Jan. 0.5 UT.

Measurement of Time

The concept of time is based on the position of celestial bodies with respect to the observer's meridian as measured along the geocentric celestial equator. The mean sun and the vernal equinox (♈) are used to define mean solar time and sidereal time. The mean sun, rather than the apparent (true) sun, is used since the latter's irregular motion from month to month gives a variation in the length of a day. The mean sun moves with uniform speed along the equator at a rate equal to the average of the true sun's angular motion during the year.

One sidereal day is the interval between two successive transits of the vernal equinox across the observer's meridian. A mean solar day is the interval between two successive transits of the mean sun. Local civil time (LCT) is defined to be the hour angle of the mean sun plus 12 h. Local sidereal time is the hour angle of the vernal equinox plus 12 h. For an observer at the Greenwich meridian, the local civil time is Greenwich mean time (GMT) or universal time (UT). The earth is divided into 24 time zones, each 15° of longitude wide. Each zone keeps the time of the standard meridian in the middle of the zone. (See Fig. 8.7.22.)

Time $T = \tau + 12$ h
Local sidereal time LST $= \tau_{\Upsilon} + 12$ h
Greenwich sidereal time GST = LST + $\lambda(w)$
Local civil time LCT $= \tau_M + 12$ h = mean solar time
Equation of time ET $= \tau - \tau_M = \alpha_M - \alpha$
Greenwich mean time GMT = UT = τ_M(Greenwich) + 12 h = universal time

$$\text{LCT} = \text{UT} - \lambda(w)$$

$$\text{GMT} = \text{ZT} \pm \lambda°/15$$

The "American Ephemeris and Nautical Almanac" gives the equation of time for every day of the year and conversions of mean solar time to and from sidereal time.

Hohmann Transfer This is a maneuver between two coplanar orbits where the elliptical transfer orbit is tangent at its perigee to the lower orbit and tangent at its apogee to the higher orbit. To transfer from a low circular orbit to a higher one, the impulsive velocity required is the difference between the velocities in the circular orbits and the velocities in the corresponding points on the transfer orbit. Using the general velocity equation above, the transfer velocities are $\Delta V_i = \{\mu[2/r_i - 2/(r_i + r_f)]\}^{0.5} - (\mu/r_i)^{0.5}$ and $\Delta V_f = (\mu/r_f)^{0.5} - \{\mu[2/r_f - 2/(r_i + r_f)]\}^{0.5}$, where V_i is the orbiting velocity in low orbit and V_f is the orbiting velocity in the high orbit.

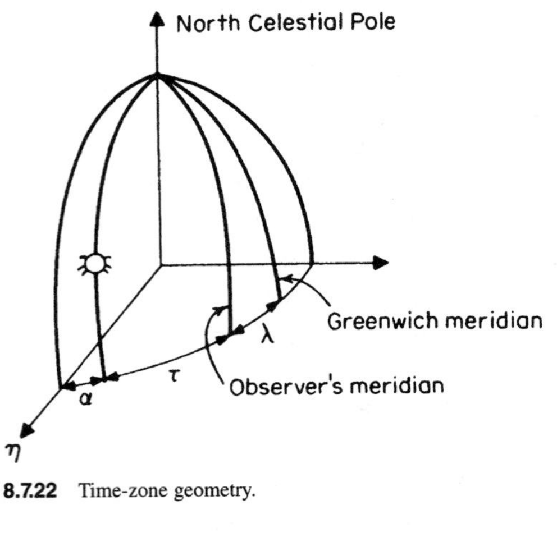

Figure 8.7.22 Time-zone geometry.

Orbital Lifetime The satellite orbit will gradually decay to lower altitudes because of the drag effects of the atmosphere. This drag force has the form $D = \frac{1}{2} C_D A_\rho V^2$, where values of the drag coefficient C_D may range from 2.0 to 2.5. The atmospheric density, as a function of altitude, can be used for computing an estimated orbital lifetime of a satellite (Fig. 8.7.23). (See NASA, "U.S. Standard Atmosphere" and "Space Flight Handbooks.")

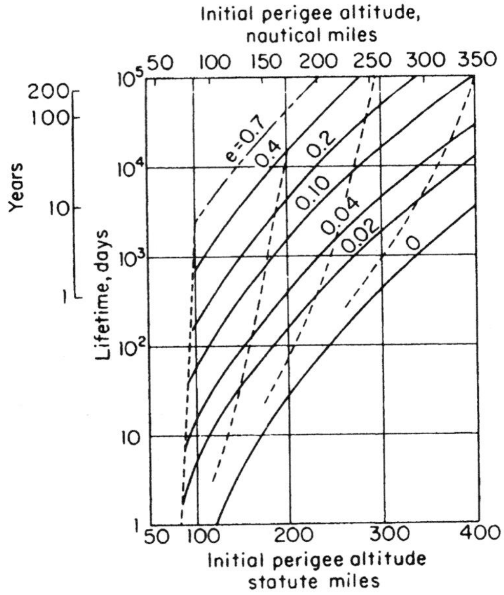

Figure 8.7.23 Satellite lifetimes in elliptic orbits. ρ = ARDC, 1959; $B = C_DA/2m = 1.0$ ft²/slug; e = initial eccentricity; — = decay histories.

Perturbations of Satellite Orbits There are many secular and periodic perturbations of satellite orbits due to the effects of the sun, the moon, and some planets. The most significant perturbation on close-earth satellite orbits is caused by the oblateness of the earth. Its greatest effects are the precession of the orbit along the equator (nodal regression) and the rotation of the orbit in the orbit plane (advance of perigee). The nodal-regression rate is given to first-order approximation by

$$\dot{\alpha}_\Omega = -\tfrac{3}{2}(\mu/a^3)^{0.5} J_2[R/a(1-e^2)]^2 \cos i$$

The mean motion of the perigee is

$$\dot{\omega} = -J_2 n(2 - (\tfrac{5}{2})\sin^2 i)/a^2(1-e^2)^2$$

LUNAR- AND INTERPLANETARY-FLIGHT MECHANICS

by J. W. Russell
University of Cincinnati

The extension of flight mechanics to the areas of lunar and interplanetary flight must include the effects of the sun, moon, and planets on the transfer trajectories. For most preliminary performance calculations, the **"sphere-of-influence,"** or **"patched-conic,"** method—whereby the multibody force field is treated as a series of central force fields—provides sufficient accuracy. By this method, the trajectories in the central force field are calculated to the "sphere of influence" of each body as standard Keplerian mechanics, and the velocity and position of the extremals are then matched to give a continuous trajectory. After the missions have been finalized, the precision trajectories are obtained by numerically integrating the equations of motion which include all the perturbative elements. It is necessary to know the exact positions of the sun, moon, and planets in their respective orbits.

Lunar-Flight Mechanics The moon moves about the earth in an orbit having an eccentricity of 0.055 and an inclination to the ecliptic of about 5.145°. The sun causes a precession of the lunar orbit about the ecliptic, making the inclination of the lunar plane to the earth's equatorial plane oscillate between 18.5 and 28.5° over a period of 18.5 years approx. To compute precision earth-moon trajectories, it is necessary to know the precise launch time to be able to include the perturbative effects of the sun, moon, and planets. The data presented are based on the moon being at its mean distance from earth and neglect the perturbation of the sun and planets; they are therefore only to be considered as representative data. The injection velocity (see Gazley) at 100 nmi for earth-moon trajectories and the impulsive velocity for braking into a 100-nmi orbit about the moon are shown in Fig. 8.7.24 as a function of transfer time. Additional energy is required to offset losses due to gravity and aerodynamic drag as well as to provide plane-change and launch-window capabilities.

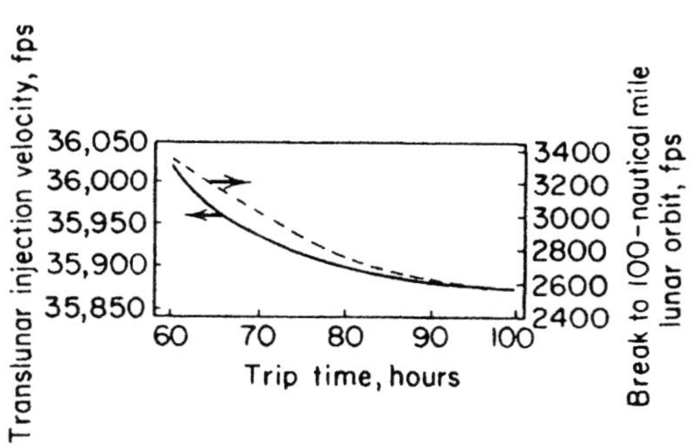

Figure 8.7.24 Lunar-mission velocity requirements.

Interplanetary-Flight Mechanics The fact that planetary orbits about the sun are not coplanar greatly restricts the number of feasible interplanetary trajectories. The plane of the interplanetary trajectory must include the position of the departure planet at departure, the sun, and the position of the target planet at arrival. The necessary plane change can be prohibitive even though the relative inclinations of the planetary orbits are small. The impulsive velocity required to effect a plane change for a spacecraft departing earth can be approximated by $\Delta V = 195{,}350 \sin (\Delta_i/2)$, where Δ_i is the amount of plane change required and ΔV is in ft/s.

For interplanetary flight, the "ideal" total energy that must be imparted to the spacecraft is the energy required to escape the gravitational field of the departure planet plus the energy required to change path about the sun so as to arrive at the target planet at the desired position and time. The energy required to escape the gravitational attraction of a planet can be determined by Keplerian mechanics to be $H_{esc} = \mu/r$, where μ is the gaussian gravitational constant and r is the distance from the center of the planet. After escaping the departure planet, the velocity must be altered in both magnitude and direction—in order to arrive at the target planet at the chosen time—by supplying additional energy $H_\infty = \frac{1}{2}V_\infty^2$.

For determination of vehicle size necessary to inject the spacecraft into the interplanetary trajectory, it is convenient to express the total energy $H = H_{esc} + H_\infty$ in terms of the required injection velocity

$$V_c = \sqrt{2(H_{esc} + H_\infty)} \quad \text{or} \quad V_c = \sqrt{(2\mu/r) + V_\infty^2}$$

Hyperbolic excess velocities V_∞ for earth-to-Mars missions are shown in Table 8.7.4 as a function of date. These velocities are near optimum for the trip time and year given but do not represent absolute-optimum trajectories. Mars has a cyclic period with respect to earth of 17 years approx; hence the energy requirements for Mars missions from earth also follow approximate 17-year cycles.

Table 8.7.4 Hyperbolic-Excess-Velocity Requirements (ft/s) for Typical Mars Missions

Trip time, days	Opposition year 1982	1984	1986	1988	1990
100	27,241	23,012	18,070	15,569	19,955
150	15,940	13,088	11,437	10,520	13,792
200	12,268	10,021	11,486	9,308	12,463
250	10,510	12,688	13,410	10,480	15,784

SOURCE: Gammal, "Space Flight Handbooks."

Figure 8.7.25 shows how the hyperbolic excess velocities of Table 8.7.4 can be converted to stage characteristic velocities for a vehicle leaving a 100-nmi earth parking orbit. These are simple cases.

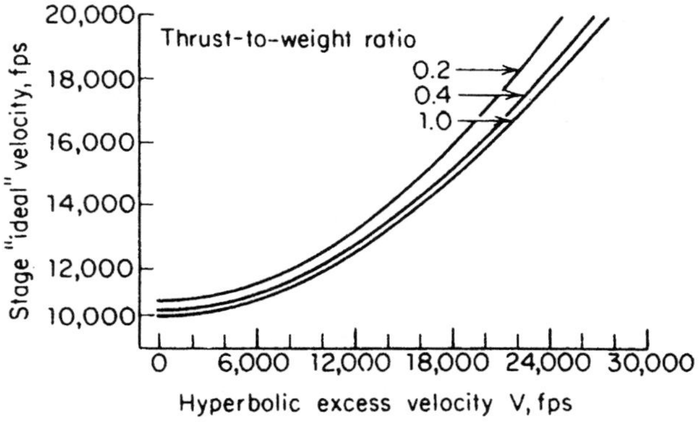

Figure 8.7.25 Stage-velocity for earth departure from a 100-nmi parking orbit (specific impulse = 450 s).

In many recent planetary missions, use has been made of planetary flybys to gain velocity and aim the space probe. These "gravity assists" significantly increase the capability of interplanetary flight.

ATMOSPHERIC ENTRY

by D. W. Fellenz

University of Cincinnati

A vehicle approaching a planet possesses a considerable amount of energy. The entry vehicle must be designed to dissipate this energy without exceeding its limits with respect to maximum decelerations or heating.

The trajectory parameters of an entering vehicle are determined largely by its initial trajectory conditions (suborbital, orbital, superorbital), by the ratio of gas-dynamic forces acting upon it, and by its mass (ballistic factor, lift-drag ratio) and the type of atmosphere it is entering.

Planetary atmospheres (Table 8.7.5) can be assumed, as a first approximation, to have exponential density-altitude distributions: $\sigma = \rho/\rho_{SL} = e^{-\beta h}$, where

$$\beta = -(1/\rho)(d\rho/dh) = Mg/RT$$

For more exacting analyzes, empirical atmospheric characteristics (e.g., U.S. Standard Atmosphere) have to be used.

The trajectory of the vehicle in flight-path fixed notation with the assumption of a nonrotating atmosphere is described as $dV/dt = g \sin \gamma - (C_D A/m)(\rho/2)V^2$ and $(V/\cos \gamma)(d\gamma/dt) = g - (V^2/r) - (C_L A/m)(\rho/2)(V^2/\cos \gamma)$. Solutions exist for direct ($\gamma > 5°$) ballistic entry and for equilibrium glide-lifting entry for $L/D > 1$. General solutions of the

Table 8.7.5 Planetary Atmospheres

Planet	V_{esc}, ft/s	Gases	M, gmol^{-1}	T, K	β^{-1}, ft	βr	ρ_{SL}, lb/ft^3
Venus	34,300	2% N_2 98% CO_2	40	250-350	2×10^4	1,006	1.0
Earth	36,800	78% N_2 21% O_2	29	240	2.35×10^4	880	0.0765
Mars	16,900	95% N_2 3% CO_2	28	130-300	6×10^4	132	0.0062
Jupiter	195,000	89% H_2 11% He	2.2	100-200	6×10^4	3,600	

equations of motion have been obtained for shallow entry of both ballistic and lifting bodies. For a survey of analytical methods available, see Chapman, Loh, Grant, Lees, and Gervais.

The **energy of an earth satellite** at 200 mi altitude is about 13,000 Btu/lb, and a vehicle entering at escape velocity possesses twice this energy. This energy is transformed, through the mechanism of gas-dynamic drag, into thermal energy in the air around the vehicle, of which only a fraction enters the vehicle surface as heat. This fraction depends on the characteristics of the boundary-layer flow, which is determined by shape, surface condition, and Reynolds number. Figure 8.7.26 illustrates the **energy conversion** where it is necessary to manage a given amount of energy in a way that minimizes structural and heat-protection weights, operational restraints, and cost. It would be most desirable if this energy could be dissipated at a constant rate, but with constant vehicle parameters, decelerations vary proportionally to ρV^2 and heating rates proportionally to ρV^3. The selection of a heat-protection system depends on the type of entry flown. Lifting entry from orbit results in relatively long flight times (in the order of ½ h) as compared with 10 min in the case of a steep ballistic entry. **Radiative-heat-transfer systems** favor low heating rates over long time periods. **Ablative systems** favor short heat pulses. In fact, longer soaking periods may melt the ablation coating without getting the benefit of heat absorption through multiple-phase changes. A more uniform dissipation of energy can be achieved by modulation of vehicle parameters.

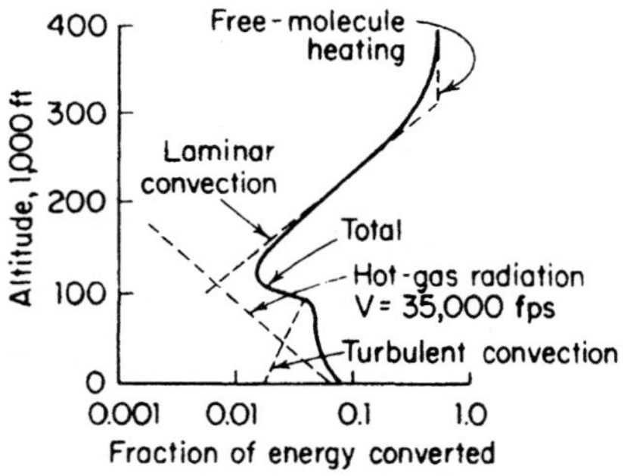

Figure 8.7.26 Energy conversion during reentry. (*After Gazley in Knoelle, "Handbook of Astronautical Engineering."*)

Determination of an **entry-vehicle configuration** is a process of iteration. The entry-flight profile and the entry and recovery procedures are mainly determined by whether experiments or passengers are carried. The external shape is determined by requirements for hypersonic glide capability and subsonic handling and landing characteristics and also by the relations between aerodynamic shape, heat input, and structural-materials characteristics. Intermediate results are fed back into the evaluation of performance and operational effectiveness of the total transportation system.

If reentry capability from any point of the ascent trajectory is desired for a manned vehicle, vehicle constraints and trajectory requirements must be compatible. **Performance-optimized trajectories** encompass combinations of relatively small velocities and large flight-path angles, which would result in considerable decelerations. In such cases, either reshaping the ascent trajectory or adding velocity at the time of abort, resulting in lower entry-flight-path angles, is effective (see Fig. 8.7.27). Lower flight-path angles during ascent depress the trajectory and increase drag losses. **Atmospheric entry is initiated** by changing the vehicle-velocity vector so that the virtual perigee of the descent ellipse comes to lie inside the atmosphere. The retrovelocity requirement for entry from low orbit is between 250 and 500 ft/s, depending on the range desired from deorbit to landing. For nonlifting entry, maximum deceleration and heat input into the vehicle are largely a function of entry angle, with the ballistic factor W/C_DA determining the altitude at which the maximum decelerations and heating rates occur. Deceleration can be readily determined through the approximate relation $-(1/g)(dv/dt) \approx q/(W/C_D A)$. Decelerations and temperatures are drastically reduced with increasing lift-drag ratio (Fig. 8.7.28) and decreasing $W/C_D A$ (Figs. 8.7.29 and 8.7.30). This effect is particularly beneficial at steeper entry angles. The combination of longer flight times with lower heating rates, however, may actually result in a larger total heat input into the vehicle.

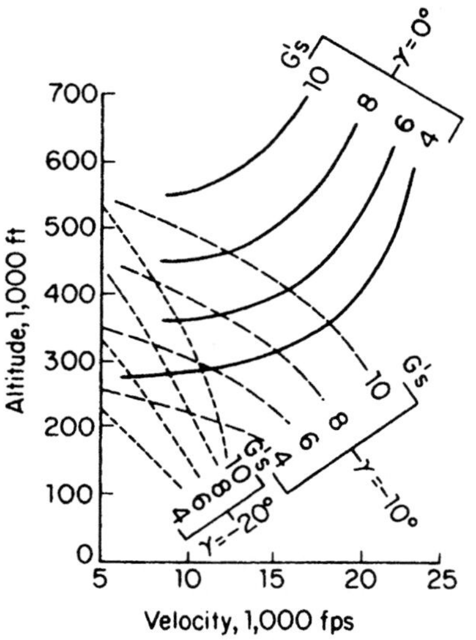

Figure 8.7.27 Suborbital entry of a lifting vehicle. W/C_DA = 28 lb/ft; L/D = 0.7.

The **influence of lift** on the reduction of decelerations is greatest for the step up to $L/D \approx 1$. The influence of W/C_DA on decelerations disappears beyond $L/D \approx 0.6$. Higher, hypersonic L/D ratios serve to improve maneuverability. For entry from low orbit, the landing area selection (footprint) can be increased (see Table 8.7.6).

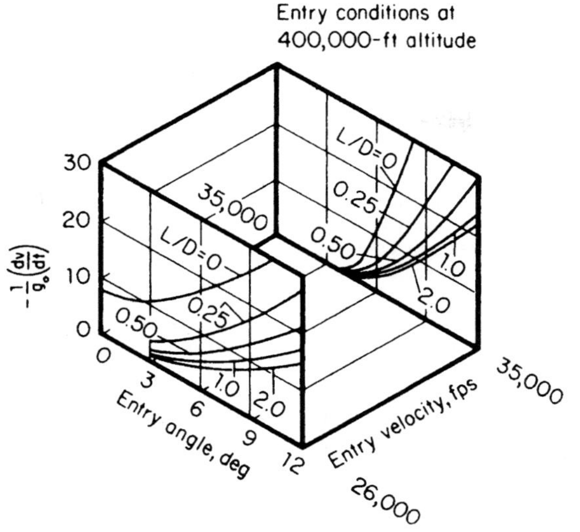

Figure 8.7.28 Peak decelerations for entry at constant *L/D*.

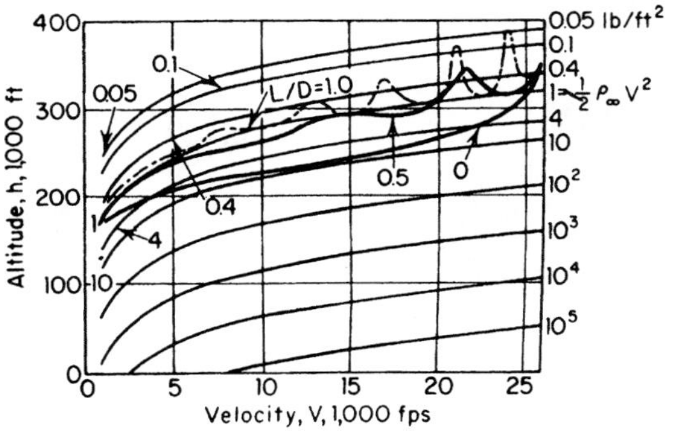

Figure 8.7.29 Glide reentry, dynamic pressure. W/C_DA = 1 lb/ft²; γ= 2°; $½\rho_\infty V = (W/C_DA)(½\rho_\infty V^2)$; $h = h_1 - 23{,}500 \ln (W/C_DA)$; $V = V_1$.

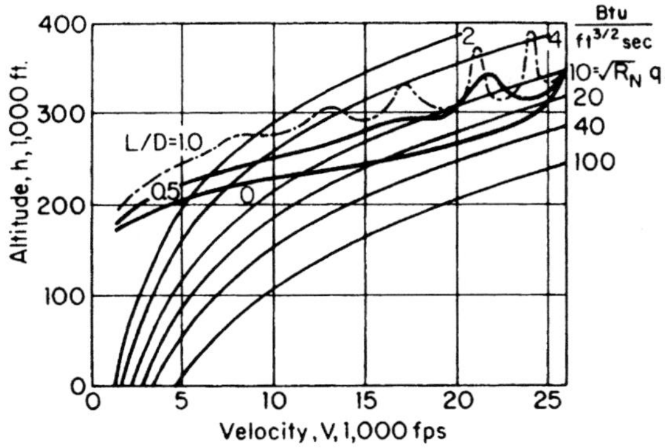

Figure 8.7.30 Glide reentry, stagnation-heating rate. W/C_DA = 1 lb/ft²; γ= 2°; $q_s = q_{s1} W/C_DA$; $h = h_1 - 23{,}500 \ln (W/C_DA)$; $V = V_1$.

For entry at parabolic or hyperbolic speeds, it is necessary to dissipate sufficient energy so that the planet can capture the vehicle. Vehicle mass and aerodynamic characteristics determine the minimum allowable entry angle, the skip limit. The vehicle must be steered between the skip limit and the angle for maximum tolerable deceleration and heating. (See Figs. 8.7.31 and 8.7.32.)

Table 8.7.6 Landing Area Selection Footprint

L/D	Lateral, nmi	Longitudinal, nmi
0.5	≈ 200	≈ 2,000
1	≈ 600	≈ 4,000
1.5	≈ 1,200	≈ 6,000
2	≈ 2,000	≈ 9,000

* Assumptions: γ_e = 1°; b_e = 400,000 ft; V_e = 26,000 ft/s.

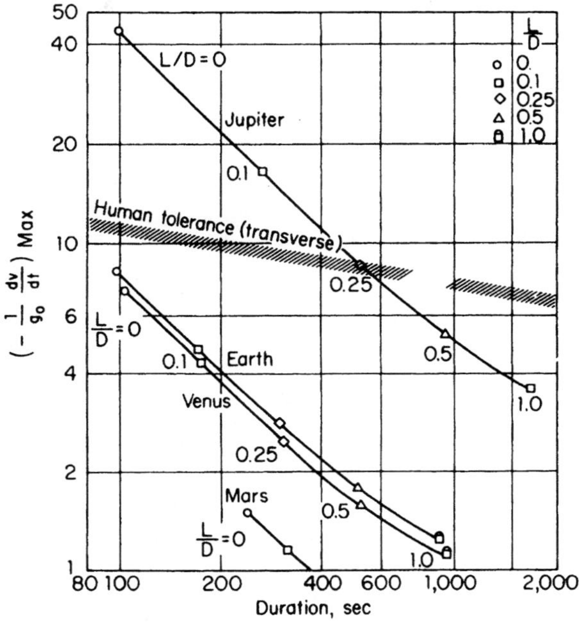

Figure 8.7.31 Comparison of decelerations and duration for entry into various atmospheres from decaying orbits. (*After Chapman.*)

ATTITUDE DYNAMICS, STABILIZATION, AND CONTROL OF SPACECRAFT

by M. R. M. Crespo da Silva
University of Cincinnati

REFERENCES: Crespo da Silva, Attitude Stability of a Gravity-stabilized Gyrostat Satellite, *Celestial Mech.,* 2, 1970; Non-linear Resonant Attitude Motions in Gravity-stabilized Gyrostat Satellites, *Int. Jour. Non-linear Mech.,* **7,** 1972; On the Equivalence Between Two Types of Vehicles with Rotos, *Jour. Brit. Int. Soc.,* **25,** 1972. Thomson, Attitude Dynamics of Satellites, in Huang and Johnson (eds.), "Developments in Mechanics," Vol. 3, 1965. Kane, Attitude Stability of Earth Pointing Satellites, *AIAA Jour.,* **3,** 1965. Kane and Mingori, Effect of a Rotor on the Attitude Stability of a Satellite in a Circular Orbit, *AIAA Jour.,* **3,** 1965. Lange, The Drag-free Satellite, *AIAA Jour.,* **2,** 1964. Fleming and DeBra, Stability of Gravity-stabilized Drag-free Satellites, *AIAA Jour.,* **9,** 1971. Kendrick (ed.), TWR Space Book, 3d ed., 1967. Greensite, "Analysis and Design of Space Vehicle Flight Control Systems," Spartan Books, 1970.

A spacecraft is required to maintain a certain angular orientation, or attitude, in space in order to perform its mission adequately. For example, within a certain tolerance, it may be required to point a face toward the earth for communications and observation purposes, as well as a solar panel toward the sun for power generation. A vehicle in space is subject to several external forces which produce a moment about its center of mass tending to change its attitude. The environmental moments that can act on a spacecraft can be due to solar radiation pressure, aerodynamic forces, magnetic forces, and the gravity-gradient effect. The relative importance of each of the above moments depends on the spacecraft design and on how close it is to a central attracting body. Most often, the effect on the vehicle's attitude of the first three

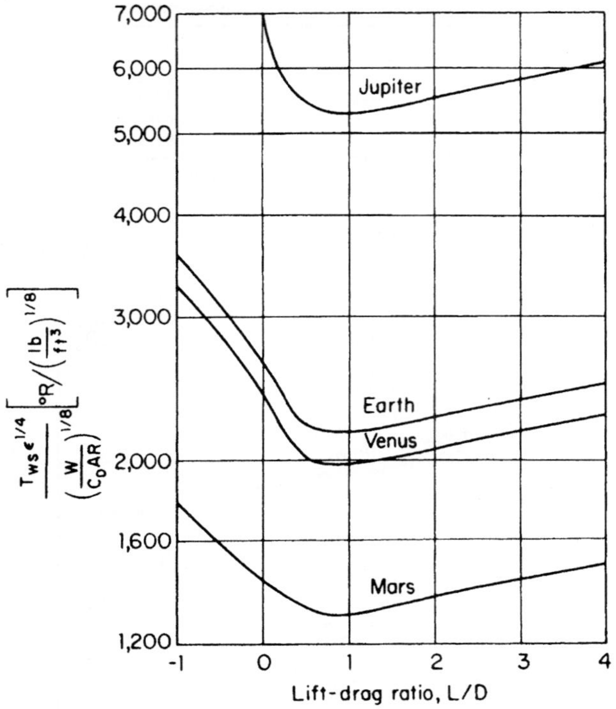

Figure 8.7.32 Maximum surface temperature for entry into various planets from decaying orbits. *(After Chapman.)*

moments mentioned above is undesirable, although they have been used occasionally for attitude control (e.g., Mariner IV, Tiros, OAO). Micrometeorite impacts can also produce a deleterious effect on the vehicle's attitude.

The most common ways to stabilize a vehicle's attitude in space are the gravity-gradient and the spin-stabilization methods. When a spacecraft is subject to the gravitational force of a central attracting body E, its mass element near the center of the gravity field will be subject to a greater force than that acting on the mass elements farther from E. This creates a moment about the center of mass, C, of the spacecraft, which depends on the inertia properties of the vehicle and also on the distance r from E to C. This gravity-gradient moment is given by

$$\mathbf{M} = (3GM_e/r^3)\hat{\mathbf{r}} \times (\mathbf{I} \cdot \hat{\mathbf{r}})$$

where $\mathbf{I}$ is the inertia dyadic of the spacecraft, referred to its mass center, and $\hat{\mathbf{r}}$ is the unit vector in the direction from E to C. As a specific example, consider the elongated orbiting spacecraft shown in Fig. 8.7.33. The body-fixed x axis (yaw) departs an angle θ_3 from the local vertical, and the body-fixed z axis (pitch) remains normal to the orbital plane.

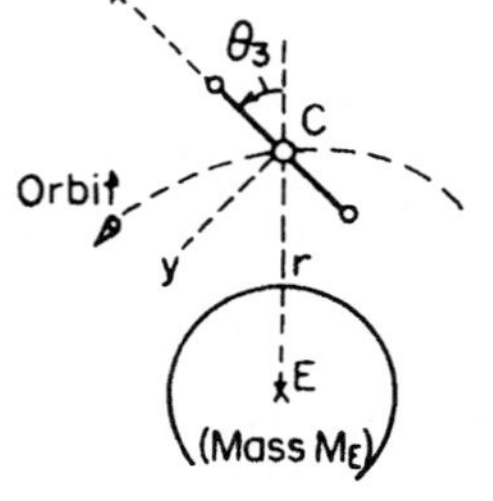

Figure 8.7.33 Dumbell satellite.

The gravitational moment about C is readily found to be as given in the equation below (it is assumed that the maximum dimension of the vehicle is much smaller than the distance r).

$$\mathbf{M} = (3GM_e/2r^3)(I_x - I_y)(\sin 2\theta_3)\hat{\mathbf{z}}$$

(I_x and I_y are the spacecraft's moments of inertia about the x and y axes, respectively.) It is seen that unless the spacecraft's x axis is vertical or horizontal, this moment is nonzero, and for $I_x < I_y$ it tends to force the x axis of the vehicle to oscillate about the local vertical. Thus, the gravity-gradient moment provides a passive, and therefore very reliable, means for stabilizing the vehicle's attitude.

The orientation of a rigid body with respect to a reference frame is described by a set of three Euler angles. Figures 8.7.34 and 8.7.35 show a rigid body in orbit around a central attracting point E and the coordinate systems used to describe its attitude. The unit vectors $\hat{\mathbf{a}}_1$, $\hat{\mathbf{a}}_2$, and $\hat{\mathbf{a}}_3$, with origin at the spacecraft's center of mass, define the orbital reference frame. The vector $\hat{\mathbf{a}}_1$ points in the direction of the vector from E to C; $\hat{\mathbf{a}}_3$ in the direction of the orbital angular velocity (whose magnitude is denoted by n); and $\hat{\mathbf{a}}_2$ is such that $\hat{\mathbf{a}}_2 = \hat{\mathbf{a}}_3 \times \hat{\mathbf{a}}_1$. The unit vectors $\hat{\mathbf{x}}$, $\hat{\mathbf{y}}$, and $\hat{\mathbf{z}}$ in Figs. 8.7.34 and 8.7.35 are directed along the three principal axes of inertia of the spacecraft. The attitude of the vehicle with respect to the orbiting frame is defined by the three successive rotations θ_1 (yaw), θ_2 (roll), and θ_3 (pitch) as shown in Fig. 8.7.35.

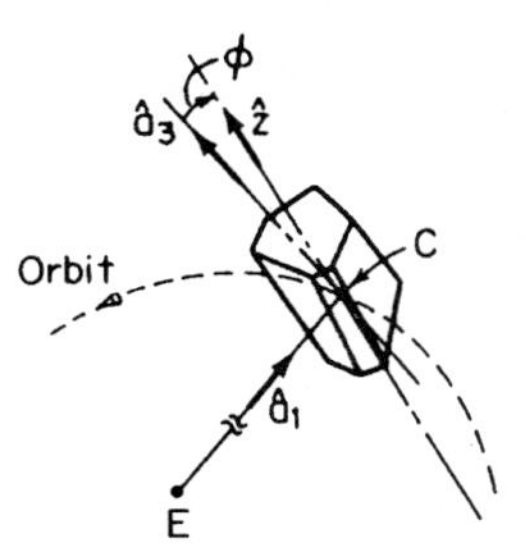

Figure 8.7.34 Orbiting satellite.

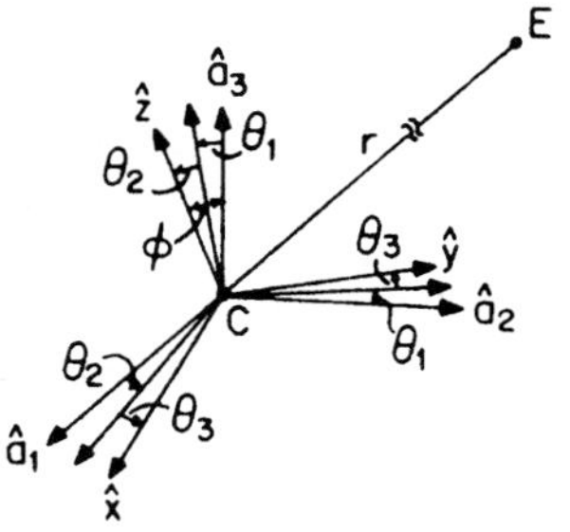

Figure 8.7.35 Coordinate frames and Euler angles.

Using the above Euler angles, the gravity-gradient moment (about C) acting on the vehicle is given as

$$\mathbf{M} = \tfrac{3}{2}n^2[\hat{\mathbf{x}}(I_y - I_z)\sin 2\theta_2 \sin\theta_3 + \hat{\mathbf{y}}(I_x - I_z)\sin 2\theta_2 \cos\theta_3 + \hat{\mathbf{z}}(I_x - I_y)\cos^2\theta_2 \sin 2\theta_3]$$

If the spacecraft is subject to an external (other than the gravitational attraction of E) torque $\mathbf{T} = T_x\hat{\mathbf{x}} + T_y\hat{\mathbf{y}} + T_z\hat{\mathbf{z}}$ about its center of mass C, and if its attitude deviations θ_1, θ_2, and θ_3 remain small, its linearized equations of motion are obtained as given in the following equations.

$$I_x\ddot{\theta}_1 + n\dot{\theta}_2(I_z - I_y - I_x) + n^2\theta_1(I_z - I_y) = T_x$$
$$I_y\ddot{\theta}_2 + n\dot{\theta}_1(I_x - I_z - I_y) - 4n^2\theta_2(I_x - I_z) = T_y$$
$$I_z\ddot{\theta}_3 + 3n^2(I_2 - I_x)\theta_3 = T_z$$

Care must be taken in the design of a gravity-gradient stabilized satellite in order to guarantee that $I_x/I_z < I_y/I_z < 1$ for the attitude motions to be stable in the presence of the least amount of damping. For augmenting the gravity-gradient moments, booms are added to the satellite in order to make its inertia ellipsoid thinly shaped. It is interesting to note that theoretically stable motions can also be obtained when the axis of minimum moment of inertia is normal to the orbital plane. However, this orientation is unstable if damping is present, and this is what occurs in practice since all actual vehicles have parts that can move relative to each other, such as an antenna.

It is seen from the equations above that for infinitesimal oscillations the pitch motion of the satellite is uncoupled from its coupled roll-yaw motions. The natural frequency ω_p of the infinitesimal pitch oscillations is given by

$$\omega_p = n\sqrt{3(I_y - I_x)/I_z}$$

and those of the coupled roll-yaw oscillations are given by the roots of the polynomial $\omega^4 - a_2 n^2 \omega^2 - a_0 n^4 = 0$, where

$$a_2 = 1 - [3 + (I_z - I_y)/I_x](I_x - I_z)/I_y$$
$$a_0 = 4(I_z - I_y)(I_x - I_z)/(I_x I_y)$$

For non-infinitesimal oscillations the pitch motion is coupled to the roll-yaw motion through nonlinear terms in the equations of motion. This coupling may give rise to an internal undesirable energy interchange between the modes of the oscillation, causing the roll-yaw motion, if uncontrolled, to oscillate slowly between two bounds. The upper bound of this nonlinear resonant roll-yaw motion may be much greater than (and independent of) the roll-yaw initial condition. For given values of I_x, I_y, and I_z, it can be decreased only by reducing the initial conditions of the pitch motion. This phenomenon can be excited if $\omega_p \approx 2\omega_1$ or $\omega_p \approx 2\omega_2$, where ω_1 and ω_2 are the two natural frequencies of the roll-yaw oscillations. The same phenomenon can also be excited if $\omega_p \approx \omega_1$ (or $\omega_p \approx \omega_2$). However, the observation of this latter resonance requires a much finer "tuning" between the modes of oscillation, and therefore it is of lesser importance.

As the altitude of the orbit increases, the gravity-gradient moment decreases rapidly. For precise pointing systems, spin stabilization is used. Historically, spin stabilization was the first method used in space and is still the most commonly employed today. Simplicity and reliability are the advantages of this method when only two-axis stabilization is needed. By increasing the spin rate, the vehicle can be made very stiff in resisting disturbance moments.

A rigid body spins in a stable manner about either its axis of maximum or of minimum moment of inertia. However, in the presence of energy dissipation, even if it is infinitesimal, a spin about the axis of minimum moment of inertia leads to an unstable roll-yaw motion. Spinning satellites are built in such a way as to achieve symmetry about the spin axis which, in practice, is the axis of maximum moment of inertia. A spin of a rigid body about its intermediate axis of inertia leads to an unstable roll-yaw motion. Since damping can destroy roll-yaw stability, an individual analysis of the equations of motion must be performed to guarantee the stability of the motions when the spacecraft houses a specific damper.

For planet-orbiting spinning satellites (for which the gravitational moment can now be viewed as a disturbance) the stability of the roll-yaw motion depends not only on the magnitude of the spin rate but also on the direction of its angular velocity (relative to the orbital reference frame) due to spin. Denoting the satellite's angular velocity due to spin by $\mathbf{s} = s\hat{\mathbf{z}}$, the following inequalities must be satisfied for the roll-yaw motions of a spinning symmetric ($I_x = I_y$) satellite in a circular orbit to be stable in the presence of an infinitesimal amount of damping:

$$\left(1 + \frac{s}{n}\right)\frac{I_z}{I_x} - 1 > 0$$
$$\left(4 + \frac{s}{n}\right)\frac{I_z}{I_x} - 4 > 0$$

Very often, a spinning pitch wheel is added to a satellite to provide additional stiffness for resisting motions out of the orbital plane. Also, the inclusion of such a wheel provides more freedom to the spacecraft designer when specifying the range of the inertia ratios to guarantee that the attitude motions of the vehicle are stable.

Let us assume that a pitch wheel with axial moment of inertia I_w is connected to the spacecraft shown in Fig. 8.7.34. It is assumed that the wheel is driven by a motor at a constant angular velocity $\mathbf{s} = s\hat{\mathbf{z}}$ relative to the main body of the spacecraft. If the attitude deviations of the main body with respect to the orbiting reference frame remain small, the linearized equations for the attitude motion of the vehicle are now given as

$$I_x\ddot{\theta}_1 + n\dot{\theta}_2(I_z - I_y - I_x + \beta I_z) + n^2\theta_1(I_z - I_y + \beta I_z) = T_x$$
$$I_y\ddot{\theta}_2 + n\dot{\theta}_1(I_x - I_z + I_y - \beta I_z) - n^2\theta_2[4(I_x - I_z) - \beta I_z] = T_y$$
$$I_z\ddot{\theta}_3 + 3n^2(I_y - I_x)\theta_3 = T_z$$

In these equations, the parameter β is defined as $\beta = I_w s/I_z n$, and I_x, I_y, I_z refer to the principal moments of inertia of the entire vehicle (rotor included).

Figure 8.7.36 shows a "stability chart" for this spacecraft when its center of mass, C, is describing a circular orbit around the central attracting point E. In this figure, a cut along the plane $\beta = 0$ produces the stability conditions for a gravity-gradient satellite, whereas a cut along the plane $K_1 + K_2 = 0$ (i.e., $I_x = I_y$) gives rise to the inequalities for a spin-stabilized vehicle. Note that a rotor's spin rate with opposite sign to the orbital angular speed ($s < 0$) may destabilize the roll-yaw motions. Nonlinear resonances, as previously discussed, can also be excited in this vehicle. However, since the natural frequencies of the roll-yaw motions now also depend on the magnitude and sign of the internal angular momentum due to the rotor, the resonance lines are displaced in the region shown in Fig. 8.7.36 when the spin rate s of the rotor is changed to a different constant. Therefore, this phenomenon, when excited, can be avoided simply by changing the parameter β.

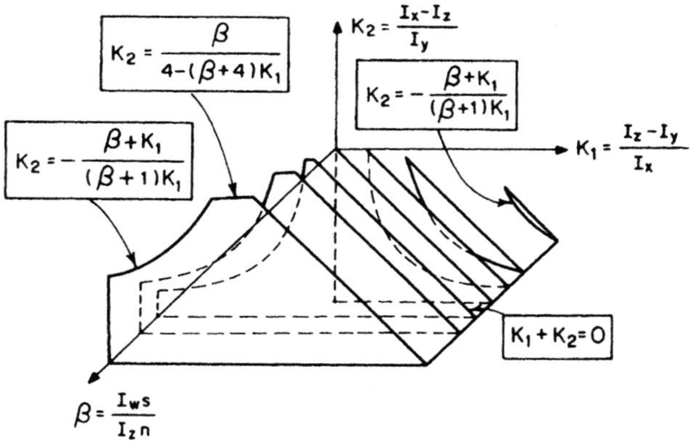

Figure 8.7.36 Stable region of the vehicle parameter space.

For a spacecraft to perform its mission adequately during a long period of time, an attitude control system to maintain the vehicle's attitude within specified limits is necessary. Typically, this is done by means of a set of mass expulsion jets mounted on the vehicle, which are actuated by the output of a controller that receives information from the attitude sensors.

Drag-Free Satellites For scientific applications that require the use of a satellite that follows a purely gravitational orbit, a translational control system is incorporated into the vehicle. Typical applications include satellite geodesy and navigation where accurate ephemeris prediction is needed. A drag-free satellite consists of a vehicle that contains a spherical cavity which houses an unsupported spherical proof mass. Sensors in the satellite detect the relative position of the proof mass inside the cavity and activate a set of thrustors placed in the vehicle, forcing it to follow the proof mass without touching it. Since the proof mass is shielded by the satellite from nongravitational forces, it follows a purely gravitational orbit.

METALLIC MATERIALS FOR AEROSPACE APPLICATIONS

by Stephen D. Antolovich, Revised by Robert L. Johnston
NASA

REFERENCES: MIL-HDBK-5C, "Metallic Materials and Elements for Aerospace Vehicle Structures," Sept. 1976. Wolf and Brown (eds.), "Aerospace Structural Metals Handbook," 1978, Mechanical Properties Data Center, Traverse City, Michigan. (This handbook gives information on mechanical, physical, and chemical properties; fabrication; availability; etc. in five published volumes encompassing ferrous, nonferrous, light metal alloys, and nonferrous heat-resistant alloys.) "Damage Tolerant Design Handbook," MCIC-HB-01, Metals and Ceramics Information Center, Battelle Columbus Laboratories, 1975, parts 1 and 2. (These two documents provide the most up-to-date and complete compilation of fracture-mechanics data for alloy steels, stainless steels, aluminum, and titanium alloys.) "Metals Handbook," 8th ed., 11 vols., American Society for Metals: Vol. I, Properties and Selection of Metals, 1961; Vol. II, Heat Treating, Cleaning and Finishing, 1964; Vol. III, Machinery, 1967; Vol. IV, Forming, 1969; Vol. V, Forging and Casting, 1970; Vol. VI, Welding and Brazing, 1971; Vol. VII, Atlas of Microstructures, 1972, Vol. VIII, Metallography, Structure and Phase Diagrams, 1973; Vol. IX, Fractography and Atlas of Fractographs, 1974; Vol. X, Failure Analysis and Prevention, 1975; Vol. XI, Nondestructive Testing and Quality Control, 1976. "Metals Handbook," 9th ed., 5 vols., American Society for Metals: Vol. I, Properties and Selection: Nonferrous Alloys and Pure Metals, 1979; Vol. III, Properties and Selection: Stainless Steels, Tool Materials and Special Purpose Metals, 1980; Vol. IV, Heat Treating; Vol. V, Surface Cleaning, Finishing and Coating. "Structural Alloys Handbook," Mechanical Properties Data Center, Belfour Stulen, Inc., Traverse City, Michigan, 1978. "Titanium Alloys Handbook," MCIC-HB-05, Metals and Ceramics Information Center, Battelle Columbus Laboratories, 1972.

A wide variety of materials are utilized in aerospace applications where performance requirements place extreme demands on the materials. Some of the more important engineering properties to be considered in the selection of materials are (1) strength-to-weight ratio, (2) density, (3) modulus of elasticity, (4) strength and toughness at operating temperature, (5) resistance to fatigue damage, and (6) environmental effects on strength, toughness, and fatigue properties. Factors such as weldability, formability, castability, quality control, and cost must be considered.

Although it is desirable to maximize the strength-to-weight ratio to achieve design goals, materials with high strength-to-weight ratios generally are sensitive to the presence of small cracks and may exhibit catastrophic brittle failure at stresses below the nominal design stress. The problem of premature brittle fracture is accentuated as the temperature decreases and the yield and ultimate strength increase. For these reasons it is imperative that the tendency toward brittle failure be given major consideration for low-temperature applications. For example, liquid oxygen, a common oxidizer, boils at −297°F (−183°C) whereas liquid hydrogen, an important fuel, boils at −423°F (−253°C), and the problem of brittle failure is predominant. The fracture-mechanics approach has gained wide acceptance, and nondestructive inspection (NDI) characterization of flaws allows safe operating stresses to be calculated. A considerable amount of fracture-mechanics data has been collected and evaluated ("Damage Tolerant Design Handbook"). Stress corrosion cracking and corrosion fatigue become important for repetitive-use components such as aircraft airframes, landing gear, fan shafts, fan and compressor blades, and disks, in addition to brittle fracture problems such as fatigue crack formation and propagation.

The presence of flaws must be assumed (they may occur either during processing or fabrication) and design must be based on **fatigue crack propagation** (FCP) properties (Sec. 3). FCP of many materials follows an equation of the form

$$da/dN = R(\Delta K)^n \tag{8.7.7}$$

where a refers to crack length, N is the number of cycles, K is the fluctuation in the stress-intensity parameter, and R and n are material constants. Equation (8.7.7) can be integrated to determine the cyclic life of a component provided that the stress-intensity parameter is known (Sec. 3). Although fatigue-crack initiation is generally retarded by having smooth surfaces, shot-peening, and other surface treatments, FCP of commercial aerospace alloys is relatively insensitive to metallurgical treatment and, to a reasonable approximation, whole classes of materials (i.e., aluminum alloys, high-strength steels, etc.) can be represented within the same scatter band.

It should be noted that increases in the plane-strain fracture toughness K_{IC} result in increased stress, through its linear relationship to the stress-intensity parameter, which in turn causes an exponential increase in the FCP rate and may introduce a fatigue problem where previously there was none.

The effect of environment on materials used for aerospace problems is usually to aggravate the FCP rate. It should be emphasized that fatigue loads have a synergistic effect on the environmental component of cracking and that the crack growth rate during corrosion fatigue is rarely the sum of the FCP and stress-corrosion cracking rates. In weldments of castings or wrought material, tests are needed to simulate anticipated use conditions and to prove weld soundness.

Materials in the turbine section of jet engines are subjected to deleterious gases such as O_2 and SO_2, high temperatures and stresses, and cyclic loads. Consequently the most significant metallurgical factors to consider are microstructural stability, resistance to oxidation and sulfidization, creep resistance, and resistance to low cycle fatigue. Nickel-base superalloys are generally used in the turbine section of the engine for both disks and blades. (See "Aerospace Structural Metals Handbook," Vols. II and IIA, for specific properties.) Although the Nibase superalloys generally exhibit excellent strength and oxidation resistance at temperatures up to 1,800°F (982.22°C), the interactive effects of sustained and cyclic loads are difficult to handle because, at high temperature, plastic deformation of these materials is time-dependent. An approach based on strain range partitioning is described in ASTM STP 520, 1973, pp. 744-782. The utility of this approach is that frequency and temperature dependence are incorporated into the analysis by the way in which the strain is partitioned and reasonable envelopes of the expected life can be calculated.

STRUCTURAL COMPOSITES

by Terry S. Creasy
Texas A & M University

REFERENCES: MIL-HDBK-17, "Plastics for Aerospace Vehicles," Sept. 1973. Advanced Composites Design Guide, Vols. I–V, 3d ed., Sept. 1976. Herakovich, "Mechanics of Fibrous Composites," 1998.

High-performance composite materials are fiber-filled matrix materials with specific strength (strength/mass) and specific stiffness (modulus/mass) greater than that found in monolithic materials such as high-performance alloys or polymers (Fig. 8.7.37). The fibers contribute high strength and stiffness to the composite. The fibers and matrix must be chemically compatible and structurally sound at the application temperature.

A first-order estimate of the composite's effective Young's modulus in the fiber (primary) direction can be found using the rule of mixtures relation:

$$E_1 = (E_f - E_m)V_f + E_m$$

where the Es indicate the Young's modulus of the composite (1), the fiber (f), and the matrix (m), and the volume fraction of the fiber in the composite is V_f. The fiber volume fraction of fiber composites can vary between 0.5 and 0.7, that is, 50 to 70 percent. Inspection of this relation shows that the effective Young's modulus E_1 is less than the fiber's modulus but greater than the modulus of the matrix material. Fiber composites are anisotropic—their properties are strong functions of the material orientation. The composite's stiffness perpendicular to the fiber direction (transverse direction) comes from this equation:

$$E_2 = \frac{E_m}{V_f\left(\dfrac{E_m}{E_f} - 1\right) + 1}$$

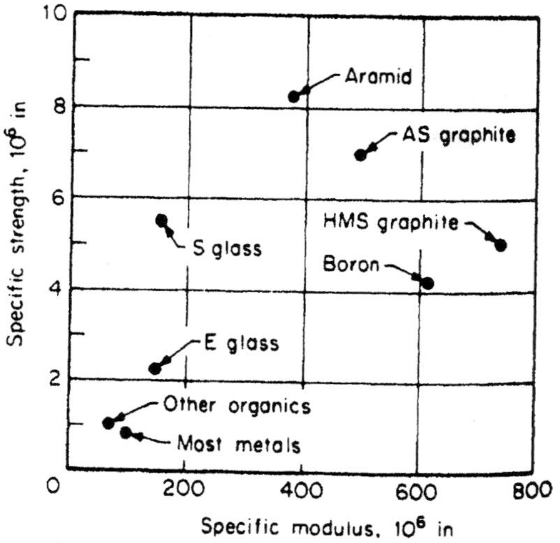

Figure 8.7.37 Specific properties of nonmetallic reinforcement materials compared with most metals. (*Hercules Aerospace Co.*)

In the range of typical fiber volume fraction, the transverse modulus E_2 is only moderately larger than that of the matrix alone. Combined layers of composite produce a laminated structural material. As the primary and secondary modulus equations suggest, the properties will vary with the orientation of the material. The fiber placement must address the state of stress in the design.

Material selection must address two issues: (1) the fibers must provide the specific strength and stiffness needed for the design and (2) the matrix must be compatible with the fibers and they must survive the maximum use temperature for the application.

Tables 8.7.7 and 8.7.8 list typical properties of fiber and matrix materials.

The fatigue behavior of several fiber/matrix systems appears in Fig. 8.7.38. The response of these composites can be compared with 2024-T3 aluminum alloy.

Finally, composite materials provide two unique capabilities for astronautic structures. Some carbon fibers have a negative coefficient of thermal expansion of approximately -0.2×10^{-6} in/in. With careful design a composite structure can compensate for expansion and contraction of other components. In addition, carbon fibers provide high thermal conductivity in the fiber direction. Thermal management of astronautic systems can take advantage of this capability.

Table 8.7.8 Properties of Matrix Materials Used in Composites

Material	Maximum use temperature, °F
Polyester	250
Polysulfone	250
Epoxy	350
Polyarylsulfone	400
Phenolic	450
Silicone	450
Polyimide	650
Aluminum	700
Magnesium	700
Glass	2,500
Carbon* (W/C)	3,200

* With oxidation resistance.

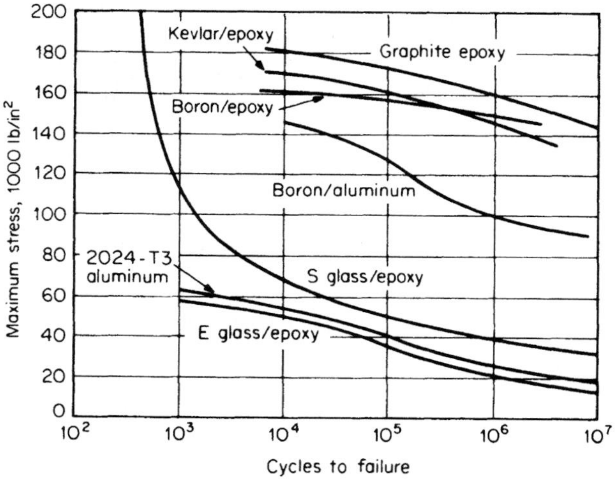

Figure 8.7.38 Fatigue behavior of unidirectional composites and aluminum. (*Hercules Aerospace Co.*)

SPACE-VEHICLE STRUCTURES

by Thomas L. Moser and Orvis E. Pigg

NASA

REFERENCES: MIL-HDBK-5D, "Metallic Materials and Elements for Aerospace Vehicle Structures." Advanced Composites Design Guide, Sept. 1976, Library Accession no. AD916679 through AD916683. Ashton, Halpin, Pertit, "Primer on Composite Materials Analysis," Technomic Pur. Co. Aeronautical Structures Manual (Vols. I–III), NASA TMX-73305. Bruhn, "Analysis and Design of Flight Vehicle Structures," Tri-State Offset Co., Cincinnati. Barton, "Fundamentals of Aircraft Structures," Prentice-Hall, 1948. Roark, "Formulas for Stress and Strain," McGraw-Hill. "Nastran Theoretical Manual," Computer Software Management and Information Control (COSMIC), University of Georgia. Perry, "Aircraft Structures," McGraw-Hill. Gathwood, "Thermal Stress," McGraw-Hill. Zienkiewicz, "The Finite Filement," McGraw-Hill. MIL-STD-810D, "Environmental Test Methods and Engineering Guidelines." Chapman, "Heat Transfer," Macmillan.

Space vehicles include a large variety of vehicle types, like their predecessors, earth vehicles. Airplanes were the first generation of space

Table 8.7.7 Properties of Fibers Used in Composites

Fiber	Modulus, 10^6 lb/in^2	Tensile strength, ksi	Tensile failure strain, %	Specific gravity	Density, lb/in^3
Boron (4 mil)	60	460	0.8	2.60	0.094
Carbon (Thornel 300)	34	470	1.1	1.74	0.063
Carbon (IM8)	45	750	1.9	1.79	0.065
Carbon (pitch)	120	350	0.4	2.18	0.079
Kevlar 49	19	400	1.8	1.45	0.052
Kevlar 29	12	400	3.8	1.45	0.052
Glass (E)	10.5	250-300	2.4	2.54	0.092
Glass (S)	12.5	450	3.6	2.48	0.090

vehicles, and our definition of space vehicle has changed as the airplane approached the limits of the earth's atmosphere. Space, in this discussion, is the region beyond the earth's atmosphere.

Many space vehicles are hybrids since they are required to operate in the earth's atmosphere and beyond. The space shuttle orbiter is such a hybrid space vehicle. The Apollo command and service module is also a hybrid, but the Apollo lunar module, which primarily functioned outside the earth's atmosphere, was a space vehicle.

The design procedure for any vehicle structure is basically the same:

1. Establish the design requirements and criteria
2. Define the loads and environments
3. Perform the design and analysis
4. Iterate the previous two steps (as required) because of the interrelation between configuration and loads
5. Verify the design

There are conditions and requirements which are unique for hybrid and space-vehicle structures.

Minimum weight and high reliability are significant but opposing requirements for space-vehicle structures which must be balanced to achieve an efficient, reliable structure design. Depending on the mission of the particular space vehicle, the ratio between the weight of the earth launch vehicle and the payload (e.g., interplanetary probe) can be as high as 400 to 1. The large weight leverages require that the design be as efficient as possible. Structural materials, therefore, must be stressed (or worked) as close to the capabilities as possible. Like aircraft structures, space-vehicle structures should be designed for a zero margin of safety (MS):

$$\text{MS} = \frac{\text{material ultimate allowable}}{\text{FS (limit stress)}} - 1$$

1. *Material ultimate allowable.* The stress at which the material will fail.
2. *FS (factor of safety).* This factor usually ranges between 1.25 and 1.50, for space structures, depending on the maturity of the design, the behavior of the material, the confidence in the loads, etc.
3. *Limit stress.* The maximum (reasonable) stress that the structural component would be expected to encounter.

Failure can be based on rupture, collapse, or yield as dictated by the functional requirements of the component. High reliability is necessary for most space vehicle structures since a failure is usually catastrophic or nonrepairable. Total structural reliability is composed of many of the design elements—loads, materials, allowables, load paths, factor of safety, analysis techniques, etc. If the reliability is specified and the statistical parameters are known for each design element, then the process is straightforward. A total structural reliability is seldom specified and good engineering practice is used for the solution of the reliability of each design element. Commonly used to assure structural reliability are (1) material allowables equivalent to A values from MIL-Handbook-5A for safe-life structure and B values for fail-safe structures; (2) limit loads which correspond to events and conditions which would not be exceeded more often than 3 times in 1,000 occurrences; (3) factor of safety equal to 1.4 for crew-carrying space vehicles and 1.25 for those without crews; and (4) verification of the structural design by test demonstration. A good structural design is one which has the correct balance between minimum weight and high reliability.

Space vehicles are usually manufactured and assembled on earth and transported to space. The launch environments of acceleration, vibration, aerodynamic pressure, rapid pressure change, and aerodynamic heating coupled with the space environments of extremely low pressure, wide ranges in temperature, meteroids, and radiation dictate that the design engineers accurately define the magnitudes of these environments and the correct time-compatible combinations of these environments if realistic loads are to be determined. Many natural environments have been defined and are documented in the references. The induced environments, which are a function of the system and/or structural characteristics (e.g., dynamic pressure, vibration, etc.), must be determined by analysis, test, or extrapolation from similar space vehicles. Quantifying the induced loads for the structural designers is one of the major consumers of time and money in any aerospace program and one of the most important. Load determination can be a very complex task. For further guidance the reader is directed to the references.

For a space vehicle which is not exposed to the launch loads, the load determination is greatly simplified. In the absence of an appreciable atmosphere and gravitational accelerations the predominant external force on a space-vehicle structure is that of the propulsion and attitude control systems and the resulting vibration loads. Even though not induced by an external force, thermal stresses and/or deflections can be a major design driver. Space vehicles commonly experience larger thermal gradients because of the time- or attitude-related exposure to solar radiation.

The thermal effects of stress and deflections of the structure require that temperature distributions be defined consistent with the strength and deflection requirements. From a strength integrity standpoint, temperature distribution on a space vehicle can be as important as the pressure distribution on an aircraft. The significance and sensitivity of thermal effects are greatly reduced by the use of composite materials for the structure. The materials can be layered to achieve, within limits, the desired thermal coefficient of expansion and thereby reduce the thermal effects.

Space vehicles which return to the earth must withstand the extremely high temperatures of atmospheric reentry. This produces dynamic temperature distributions which must be combined with the aerodynamic pressures and vehicle dynamic loads.

Innovative, clever, and efficient structural designs minimize the effects of loads associated with launch for a vehicle which is to remain in space. This is accomplished by attaching the space vehicle to the booster rocket to minimize scar weight and to reduce the aerodynamic loads by the use of shrouds, fairings, or controlling the attitude during atmospheric flight.

Efficiencies of weight-critical structures have been greatly improved by more accurately quantifying the loads and stress distributions in complex structural configurations. **Finite element methods** (FEM) of analysis and large capacity, high-speed digital computers have enabled the increased fidelity of analyzes. In this process, the structure is modified with finite elements (beams, bars, plates, or solids) by breaking the structure into a group of nodal points and connecting the nodal points with the finite elements which best represent the actual structure. Loads are applied at the nodal points or on the elements which then distribute the loads to the nodal points. Constraints are also applied to the nodal points to support or fix the structure being analyzed. The finite element program then assembles all the data into a number of linear simultaneous equations and solves for the unknown displacements and then for internal loads and stresses, as required. Many finite element programs have been developed to perform a variety of structural analyzes including static, buckling, vibration, and response.

The loads in a structure depend on the characteristics of the structure and the characteristics of the structure are largely dictated by the loads. This functional dependency between loads and structures requires an iterative design process.

Classically, when structures are worked to stresses near the capability of the materials, for example, airplane structures, the design is verified by test demonstration. This method of verification was common before high-fidelity structural mathematical models were employed and is still used when practical to apply the design loads and environments to a dedicated test article to verify the design. Design deficiencies are then corrected depending on the importance and the manufacturing schedule.

Space-vehicle structures often do not lend themselves to the classical test approach because of the multitude of loads and environments which must be applied, because testing in earth's gravity precludes realistic loadings, and because of the size and configuration. Verification by analyzes with complementary component tests and tests of selected load conditions are common.

INFLATABLE SPACE STRUCTURES FOR HUMAN HABITATION

by William C. Schneider
Texas A & M University

Inflatable space structures have been considered for many years and even some have been put into practice for applications such as large reflecting satellites, shock attenuation, and reentry vehicles. Their clear advantage over rigid structures from the standpoint of weight, compactness, and large volume potential have been recognized for years, but only recently, as a result of the development of very high strength fabrics, has serious development of inflatable structures for human application taken place. NASA, as well as some private aerospace companies, recognizes the increasing demand for human-inhabited inflatable space structures in astronautics. As newer, even higher strength, fabrics are developed, the full utilization of inflatable structures for human space will be realized. In the following subsections the TransHab and the portable hyperbaric chamber are discussed, and they represent only two examples of the emerging inflatable space structure technology for human habitation.

Inflatable Flight Demonstrators

by Glenn J. Miller

The TransHab module (Fig. 8.7.40) was developed in the late 1990s and demonstrated the viability of inflatable structures as potential manned habitats in the environment of space. Ground testing and analysis provided confidence that similar designs would survive the ride to orbit and inflate once in the near zero gravity of space. In 2006, the Genesis inflatable habitat module was placed in low earth orbit by Bigelow Aerospace using a Dnepr launch vehicle. Manufactured by Bigelow Aerospace, the Genesis I flight article used a multilayer softgood design similar to TransHab. For launch the Genesis softgoods were packed around a central structural core with an outer diameter of 1.6 meters and an overall length of 4.4 meters. Once inflated on orbit the diameter increased to 2.5 meters and provided 11.5 cubic meters of usable pressurized volume. A second Genesis spacecraft was launched in 2007 and also successfully inflated in low earth orbit. Although not manned, the Genesis vehicles moved the technology readiness of inflatables ahead by demonstrating:

(a) The central structural core could meet the stiffness and strength requirements of a launch vehicle. The central core structure also served as the backbone of the inflated vehicle housing avionics, guidance, and payloads.

(b) The performance of the pressure seals located at the softgoods to metallic core interfaces.

(c) The expandable softgoods would survive the vibrational and acoustic environments experienced during launch without damage to the individual layers.

(d) Proprietary folding, packing and deployment techniques. These packaging details allowed air located within the softgoods to vent during ascent, thus preventing a premature expansion of the habitat.

(e) A successful expansion and pressurization in microgravity.

(f) The fully inflated module stiffness did not impede remote control capabilities.

(g) That the environment within the inflated spacecraft was capable of sustaining a human presence (verified by the collection of temperature and pressure data).

In 2011, NASA and Bigelow Aerospace entered into a partnership to send an inflatable module to the International Space Station (ISS). The objective of the Bigelow Expandable Aerospace Module (BEAM) was to serve as a pathfinder for the "man-rating" of inflatable structures. Although ground testing and the success of the Genesis spacecraft had demonstrated inflatable structures were valid, the BEAM project would verify that the technology had matured to the point that man-rating was possible. Man-rating establishes the reliability of the structure from manufacturing through operational life and insures that the pressurized environment after inflation is not dangerous to astronauts. In meeting requirements for certification, the BEAM inflatable structure underwent:

(1) Multiple ground tests where the softgoods were pressurized to greater than 4 times the nominal ISS environment to demonstrate strength and repeatability of the manufacturing process.

(2) Failure tolerance testing where members of the structural restraint layer were cut while at ambient pressure to demonstrate the ability of the design to redistribute load around a potential damage site without catastrophic consequences.

(3) Testing of the bladder layers to insure that the outgassing of the material would not be a threat to the health of the ISS crew.

(4) Long-term creep testing of the restraint straps to verify that exposure to sustained pressure loading would not cause a detrimental reduction in strength and potential failure of structure during the module mission life.

While the details of the creep testing performed on the BEAM restraint material are proprietary, the hypothetical plot in Fig. 8.7.39 can be used to illustrate how the structural margin for a generic restraint is determined.

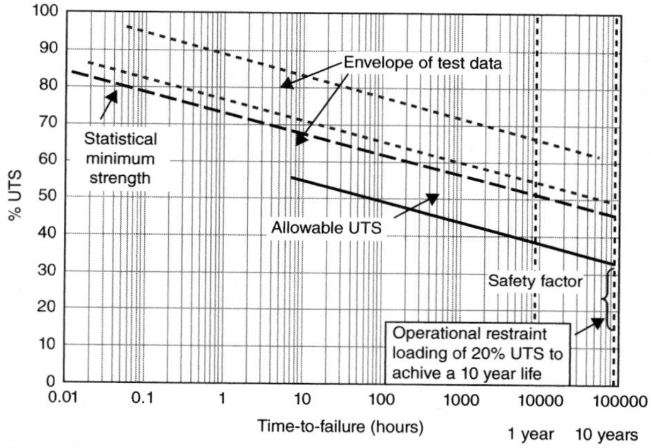

Figure 8.7.39 Hypothetical creep to failure plot.

The semi-log plot of ultimate tensile strength (UTS) verses time under load would be generated from test results of multiple straps held under a constant tensile load until failure. The small dashed lines would represent an envelope of the single strap test data. The larger dashed line would be a statistically derived set of minimum single strap strengths. However when multiple straps are integrated into a restraint additional reduction in strength can occur due to design details (such as stitching or weaving). A further reduction in test derived strap strength maybe required depending on the operational restraint environment (temperature extremes, atomic oxygen, etc.). Therefore the allowable UTS value for a specific restraint design operating in a space environment may be represented by the solid line. Finally a safety factor should be included to account for uncertainties in the test data and strength analysis. In this example the calculated restraint load under operational pressures, should not exceed 20 percent of the restraint strap UTS in order to sustain a 10-year pressurized mission.

The BEAM module was launched in June of 2016, attached to ISS and successfully expanded from a packed volume of 4 cubic meters to an inflated volume of approximately 16 cubic meters. After verifying a habitable environment existed within the module, ISS astronauts entered the module to install and activate instrumentation designed to gather data in radiation, temperature, module dynamics, and MMOD impacts. Periodically over the life of BEAM, astronauts will collect samples from the interior surface of the habitat to characterize the growth of microbes (standard practice for all ISS modules). The success of the BEAM program established inflatable habitats on a comparative footing with the more classic metallic pressurized volumes as manratable structures.

TransHab

by Horacio de la Fuente, NASA

One example of the significant advantages derived from the use of inflatable technology for human space applications was demonstrated by the TransHab inflatable habitation module. Figure 8.7.40 shows a unit developed at the NASA Johnson Space Center. The TransHab is an inflatable human habitation module (30 ft long × 25 ft in diameter) originally conceived as an interplanetary module for use in spaceflight missions beyond low earth orbit. The long transit times required for these types of missions create several significant design challenges. Because multiple crewmembers will be living in a confined space for long periods of time, psychological health required the spacecraft to have a large-volume living space yet be lightweight. The ability of inflatable structures to be folded and packaged during the launch environment is a key feature that allows this requirement to be met. The advantages of inflatable modules, however, go beyond the obvious volumetric packing efficiency. The inflatable TransHab is folded and packaged tightly during the launch phase of the mission and is inflated only after it is in the quiescent space environment. Weight savings are realized when compared to a large rigid shell structure, since the rigid shell must be designed to withstand the severe steady-state, transient, and vibration launch loads as a large, rigid shell. There is, however, no requirement for the inflated TransHab to react to the launch loads. It must only withstand the differential pressure experienced in space.

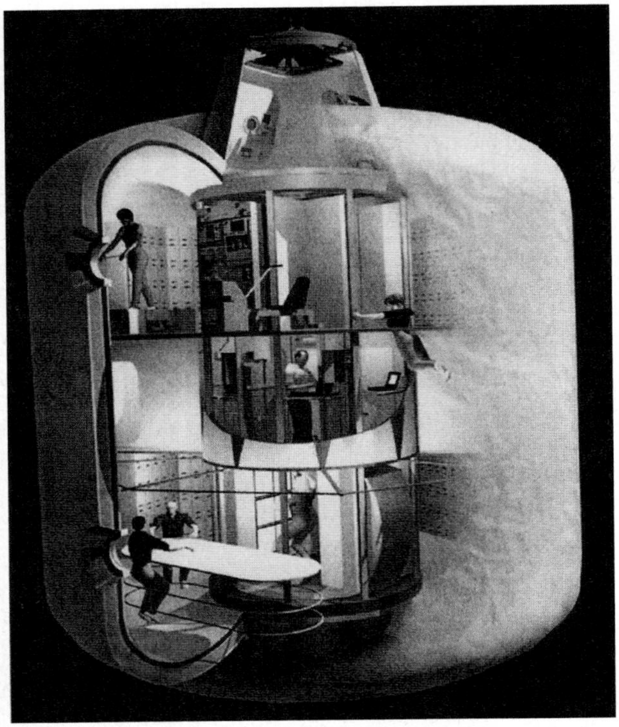

Figure 8.7.40 TransHab inflatable habitation module.

The TransHab module design was later adapted to serve as a habitation module for the International Space Station. This change primarily resulted in modifications to some of the inflatable shell layers. As described later, the shell layers most affected by this change were the orbital debris protection layer, the atomic oxygen protection layer, and the restraint layer, which was strengthened to be compatible with the 14.7-lb/in^2 internal pressure of the International Space Station.

Central Structure Core Turning the large volume into a functional living environment creates additional challenges. Once inflated, the space habitation module requires several internal integrated subsystems such as environmental control and life support, power distribution, communication, and data handling, as well as crew mobility and translations aids. In addition, to maintain psychological health during the long stay times, dedicated living spaces for exercise, personal hygiene, food preparation and consumption, and sleeping are required. All of the above require internal structure as well as a significant amount of preintegration. For this reason the TransHab is designed as a preintegrated system. The integrated systems approach for TransHab used a hard central structural core, shown in Fig. 8.7.41, to provide the necessary backbone structure to allow for the preintegration and checkout of all the subsystems before launch, to support the subsystems during launch, and to provide the necessary internal structure for the module after inflation.

The inflatable shell of the TransHab is folded and wrapped around the central structural core during launch, then released and inflated on orbit, leaving a central structural core running down the central axis of

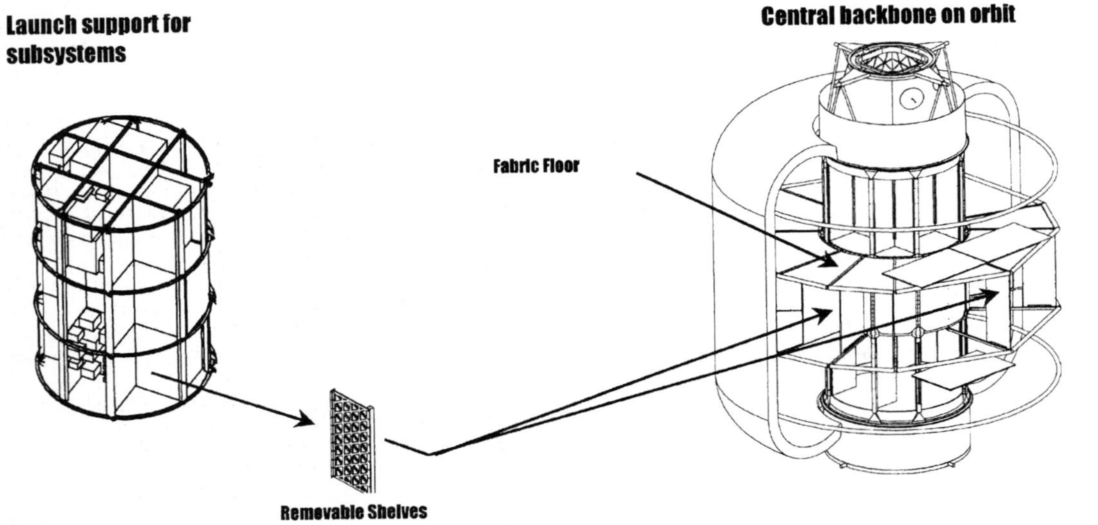

Figure 8.7.41 Central structural core.

the inflated module. All of the necessary structural elements needed to support the internal distributed systems, crew translation aids, and dedicated living spaces are obtained from the central structural core, which is already inside the module. Many of the structural elements of the central structural core are repositioned after inflation to provide a new function within the module. A good example is the use of removable shelf elements in the core. These shelves, which contain preintegrated subsystem components, are used to strengthen the core during launch and are then repositioned adjacent to the core after inflation. Another example is a stretched fabric flooring system that partitions the internal volume into three separate "floors" after inflation. These floor system elements were also used to provide a protective barrier between the central structural core and the inflatable shell during launch.

The approach of reusing elements for multiple purposes is a common feature used throughout the TransHab design.

Inflatable Shell The design of the inflatable shell is driven by several requirements. These requirements include the need for flexure resistance, high tensile strength, low air permeability, internal puncture resistance, protection from meteoroid and orbital debris, thermal insulation, and protection against atomic oxygen. Finding a single material construction that meets all these requirements is difficult. As shown in Fig. 8.7.42, the TransHab design uses a multilayer construction to satisfy all of these requirements. This allows each layer to be optimized for its specific function. In addition, the use of separate layers loosely connected to each other allows the layers to slip relative to each other during the folding process. The different layers used in the TransHab design are an inner protective liner, a redundant bladder system, a strong restraint layer, a multilayer meteoroid and orbital debris layers, a multilayer insulation system, and an atomic oxygen protection layer. These layers are described briefly below.

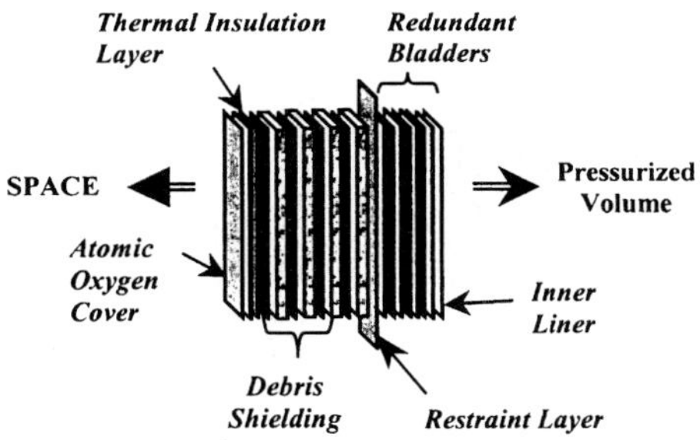

Figure 8.7.42 Multilayer inflatable shell.

Inner protective liner. The primary function of the inner liner is to protect the bladder layers from internal puncture and from flammability hazards. It uses a Nomex fabric backed by Kevlar felt. This layer is directly attached to the innermost bladder layer.

Bladder system. The primary function of the bladder system is to contain the internal gas, in this case air, from escaping. Because this layer is so critical, the TransHab design uses more than one bladder for redundancy in the event of a puncture. The bladder layers are made from a polyurethane-based laminated film, which provides extremely low permeability as well as excellent flexure characteristics at low temperature. The bladder system is oversized so that it transfers the entire pressure load to the restraint layer.

Restraint layer. The primary function of the restraint layer is to react the shell loads created by internal pressure. The restraint layer is composed of interwoven Kevlar straps running in a circumferential and a longitudinal direction.

Orbital debris protection layer. The primary function of the orbital debris protection layer is to act as a shield to protect the inboard layers from space debris impacts. To shield against hypervelocity impacts, the debris protection layer consists of multiple layers of Nextel and Kevlar™ cloth separated by foam spacers. While the spacing of the debris protection layers is needed for the shield to function, it also makes the inflatable shell difficult to fold and package. The use of open cell foam inside individual bags allows the foam spacers to be vacuum packed and collapsed for easy folding. After deployment of the inflatable shell, the lack of external atmospheric pressure allows the foam to expand back to its original state. As a result, the debris protection layer contributes approximately 16 inches to the overall thickness of the inflatable shell after deployment.

Thermal insulation layer. The primary function of the thermal insulation layer is to protect the inboard layers from the extreme on-orbit temperatures. The thermal insulation layer for TransHab uses standard multilayer insulation blankets with some modifications to account for having the shell constrained during launch and then released on orbit.

Atomic oxygen layer. The primary function of the atomic oxygen layer is to protect the inboard layers from atomic oxygen degradation in low earth orbit. The TransHab module uses a standard Beta-cloth layer for this purpose.

Integrity Tests Two of the key TransHab features, the ability to protect against space debris and the structural integrity of the TransHab restraint layer design, were verified by test. Sample cross sections of the debris protection layer were shot with a hypervelocity gun. The debris protection layer successfully stopped an approximately 2-cm-diameter aluminum projectile traveling at 7 km/s from penetrating into the bladder layers. A 1-g particle traveling at 11.5 km/s was also tested successfully. The structural integrity of the restraint layer was test-verified through hydrostatic pressure testing. A full-scale-diameter, yet shorter length, test article, shown in Fig. 8.7.43, was fabricated and hydrostatically pressurized. While space habitation modules are

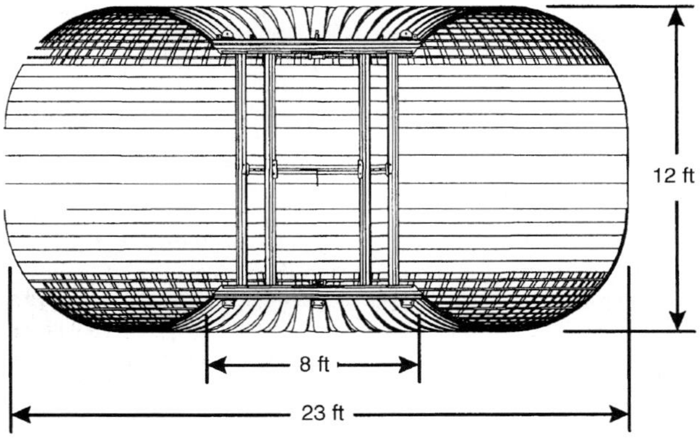

Figure 8.7.43 Structural integrity test article.

designed to withstand a structural factor of safety equal to 2 (2X the operating pressure) and are typically 14 ft in diameter, the TransHab test article was 23 ft in diameter and was tested to a factor of safety equal to 4 (4X the operating pressure). Because pressure vessel stresses increase linearly with diameter, the test showed that the test article had much greater structural capability than its metallic counterpart. Since the structural weight of the TransHab module was comparable to that of existing metallic habitation modules, the test showed the significant advantages of inflatable habitation modules.

Portable Hyperbaric Chamber

by Christopher P. Hansen and James P. Locke, NASA

The assembly and maintenance of the International Space Station requires that astronauts perform many extravehicular activities (EVAs), or spacewalks. The interior habitable compartment contains a nitrogen/ oxygen atmosphere maintained at sea level atmospheric pressure (14.7 lb/in^2 or 760 mm Hg). Current spacesuit designs must be operated at significantly lower pressures, so that they are flexible enough to

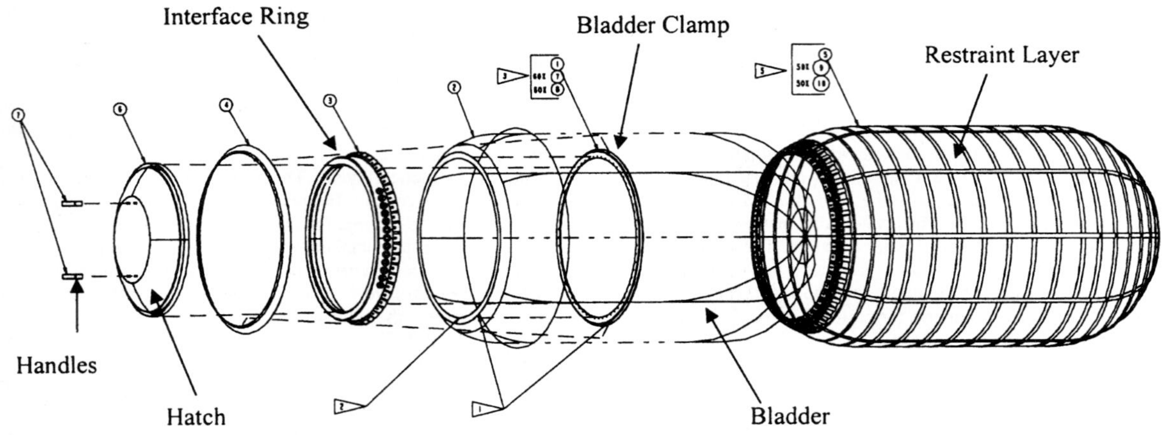

Figure 8.7.44 Portable hyperbaric chamber structure (shown in an exploded view).

allow the performance of meaningful work. The NASA extravehicular mobility unit (EMU) spacesuit operates in space at a pressure of 4.3 lb/in^2 (220 mm Hg), or an earth altitude equivalent of approximately 30,000 ft. Depressurizing from sea level to 30,000 ft can cause the nitrogen gas dissolved in body tissues to leave solution and form gas bubbles, which can lead to decompression sickness (DSC), or "the bends." The risk of such altitude DCS can be mitigated by breathing pure oxygen prior to depressurization. This safely decreases tissue nitrogen content and decreases bubble formation. However, even with these procedures, the risk of serious, or even fatal, DCS remains. DCS can be effectively treated by using a hyperbaric chamber, a room containing air at high pressure (typically equivalent to a depth of 60 ft of seawater). Breathing pure oxygen under such high atmospheric pressure forces the nitrogen bubbles back into solution. In general, patients can be easily cured of even severe DCS if the hyperbaric chamber treatment is initiated quickly. Since it takes more than 24 hours to leave the space station, reenter the earth's atmosphere, land, and reach a terrestrial hyperbaric chamber, placing such a chamber on the station would be highly desirable. Unfortunately, hyperbaric chambers are typically very large and heavy, making their use on the space station impractical.

However, with the development of lightweight, high-strength inflatable construction techniques developed for the TransHab project, it was possible to design and build a portable hyperbaric chamber that was light enough to put on the space station, and could be stowed away in a relatively small space. In the late 1990s, a small NASA group was funded to develop a prototype for a portable hyperbaric chamber (PHC) that could be tested and eventually used on the space station. In the first phase of the project, the team concentrated on the structural components of the chamber, since it was considered to be the most difficult technical challenge. While TransHab had revolutionized inflatable space structures, the PHC had to be greatly scaled down from TransHab while still utilizing, as much as possible, the materials research and construction techniques previously developed. The PHC had to be just big enough to accommodate two astronauts: one (a patient) with DCS and a second to tend to the possibly incapacitated patient.

The structure of the PHC is shown in Fig. 8.7.44. It is composed of an inner, airtight polyurethane bladder surrounded by a woven Kevlar restraint layer. The restraint layer is actually composed of two sets of orthogonal interwoven Kevlar straps. One end of the chamber has a high-strength 7,050 aluminum interface ring that acts as a doorway to the chamber and provides the sealing surface for the hatch and feedthroughs for air, pure oxygen, communication equipment, etc. The hatch, also high-strength aluminum, is machined in an oval shape, allowing the hatch to be internally sealing, yet allowing it to be removed by rotating it sideways and sliding it through the oval doorway. By having an internally sealing hatch, the hatch structure is greatly reduced in size and the need for numerous heavy latches is eliminated. The interface ring has series of steel brackets with Delrin rollers that connect the restraint layer to the interface ring. The Delrin rollers reduce the stress concentration in the straps at the interface ring enough to allow the full pressure of the chamber to be developed without failure of the Kevlar straps. The seal between the hatch and the interface ring is formed with an O ring. The O-ring interface and the Delrin roller are shown in Fig. 8.7.45.

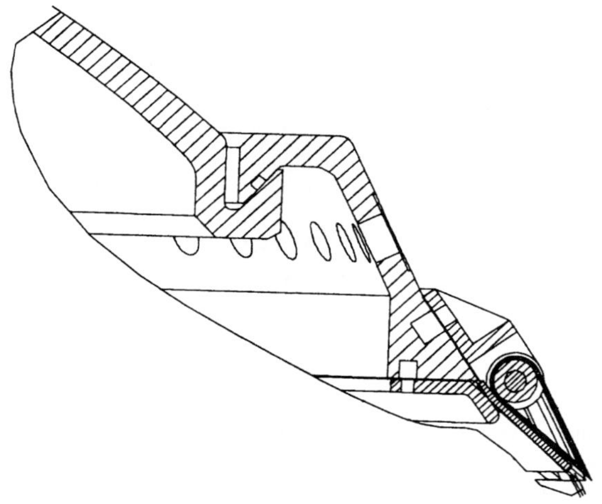

Figure 8.7.45 Cross section of the hatch, interface ring, and restraint layer interface.

One of the most challenging parts of the structural design was developing a method of manufacturing the restraint layer. Made from 1-in Kevlar straps, the restraint layer had to be woven into the cylindrical shape of the chamber. Unlike TransHab, the restraint layer shape could not be formed into a flat pattern because it did not have an interface ring at each end. The second interface ring was eliminated to save weight, although it complicated the restraint layer design. However, this decision led to an estimated 30 percent reduction in weight.

After completion of the structural prototype, a pressurized hydrostatic test was performed. The working pressure of the chamber is 1.8 atm, and the chamber is designed to withstand 4.0 atm without failure. The final prototype is shown in Fig. 8.7.46 during a low-pressure leak test. All testing of the PHC was successful and demonstrated the flexibility of the inflatable structure technology developed during the TransHab project.

Figure 8.7.46 Portable hyperbaric chamber structural prototype during low-pressure leak test.

VIBRATION OF STRUCTURES

by Lawrence H. Sobel
University of Cincinnati

REFERENCES: Crandall and McCalley, Numerical Methods of Analysis, chap. 28, in Harris and Crede (eds.), "Shock and Vibration Handbook," Vol. 2, McGraw-Hill. Hurty and Rubinstein, "Dynamics of Structures," Prentice-Hall. Meirovitch, "Analytical Methods in Vibrations," Macmillan. Przemieniecki, "Theory of Matrix Structural Analysis," McGraw-Hill. Wilkinson, "The Algebraic Eigenvalue Problem," Oxford.

As a result of the vast improvements in large-scale, high-speed digital computers, numerical methods are being used at an exponentially increasing rate to analyze complex structural vibration problems. The numerical methods most widely used in structural mechanics are the finite element method and the finite difference method. In the finite element method, the actual structure is represented by a finite collection, or assemblage, of structural components whereas, in the finite difference method, spatial derivatives in the differential equations governing the motion of the structure are approximated by finite difference quotients. With either method, the continuous structure with an infinite number of degrees of freedom is, in effect, approximated by a discrete system with a finite number of degrees of freedom (unknowns). The linear, undamped free and forced vibration analysis of such a discrete system is discussed herein. The analysis is most conveniently carried out with the aid of matrix algebra, which is well suited for theoretical and computational purposes. The following discussion will be brief, and the reader is referred to the cited references for more comprehensive treatments and examples.

Equation of Motion

Consider a conservative discrete system undergoing small motion about a state of equilibrium (neutral or stable). *Let* $q_1(t), \ldots, q_n(t)$, where t is time, be the minimum number of independent coordinates (linear or rotational) that completely define the general dynamical configuration of the system and that are compatible with any geometrical constraints imposed on the system. Then the n coordinates $q_1, \ldots, q_n$ are called generalized coordinates, and the system is said to have n degrees of freedom. Let $\{\mathbf{Q}(t)\}$ be the vector of generalized forces, and let $\{\mathbf{q}(t)\}$ be the vector of generalized displacements, $\{\mathbf{q}(t)\} = \{\mathbf{q}_1(t), \ldots, \mathbf{q}_n(t)\}$. Corresponding elements of the generalized force and displacement vectors are to be conjugate in the energy sense, which simply means that if one of the elements of $\{\mathbf{q}\}$ is a rotation (for instance), then the corresponding element of $\{\mathbf{Q}\}$ must be a moment, so that their product represents work or energy. Methods of computing $\{\mathbf{Q}\}$ are discussed in the references (e.g., Meirovitch). The matrix equation governing the small undamped motion of the system is given by

$$[M]\{\ddot{q}\}+[K]\{q\}=\{Q\} \tag{8.7.8}$$

In this equation a dot denotes single differentiation in time, and $[M]$ and $[K]$ are the mass and stiffness matrices, respectively. $[M]$ and $[K]$ are real and constant matrices, which are assumed to be symmetric. Such symmetry will arise whenever an energy approach (e.g., Lagrange's equations) is employed to derive the equation of motion. The mass matrix is assumed always to be positive definite (and hence nonsingular, or can be made to be positive definite; see Przemieniecki).

Free-Vibration Analysis

The following equation governing the free vibration motion of the conservative system is obtained from Eq. (8.7.2) with $\{\mathbf{Q}\} = \{\mathbf{0}\}$:$[M]\{\ddot{\mathbf{q}}\} + [K]\{\mathbf{q}\} = \{\mathbf{0}\}$. To obtain a solution of this equation, we note that, for a conservative system, periodic motion may be possible. In particular, let us assume a harmonic solution for $\{\mathbf{q}\}$ in the form $\{\mathbf{q}\} = \{\mathbf{A}\}\cos(\omega t - \alpha)$. That is, we inquire whether it is possible for all coordinates to vibrate harmonically with the same angular frequency ν and the same phase angle α. This means that all points of the system will reach their extreme positions at precisely the same instant of time and pass through the equilibrium position at the same time. Substitution of this trial solution into the free vibration equation yields the eigenvector equation $[K]\{\mathbf{A}\} = \omega^2[M]\{\mathbf{A}\}$. In this equation $\{\mathbf{A}\}$ is called the eigenvector and ω^2 is the eigenvalue (the square of a natural frequency). This homogeneous equation always admits the trivial solution $\{\mathbf{A}\} = \{\mathbf{0}\}$.

Nontrivial solutions are possible provided that ω^2 takes on certain discrete values, which are obtained from the requirement that the determinant of the coefficient matrix of $\{\mathbf{A}\}$ vanishes. This yields the eigen-equation (frequency equation) $\Delta(\omega^2) = |[K] - \omega^2[M]| = 0$. It is an nth-degree polynomial in ω^2, and its n roots are denoted by ω_r^2, $r = 1,\ldots,n$. The eigenvector $\{\mathbf{A}\}_r$, $r = 1,\ldots, n$, corresponding to each eigenvalue is determined from $[K]\{\mathbf{A}\}_r = \omega_r^2[M]\{\mathbf{A}\}_r$. However, the homogeneous nature of this equation precludes the possibility of obtaining the explicit values of all the elements of $\{\mathbf{A}\}_r$. It is possible to determine only their relative values or ratios in terms of one of the components. Thus, we see that for each eigenvalue ω_r^2 there is a corresponding eigenvector $\{\mathbf{A}\}_r$, which has a unique shape (based on the relative values of its components) but which is determined only to within a scalar multiplicative factor that may be regarded as being the amplitude of the shape. This unique shape is called the mode shape, and it can be thought of as being the position of the system when it is in its extreme position, and hence momentarily stationary, as in a static problem. Each frequency along with its corresponding mode shape is said to define a so-called natural mode of vibration. Methods of determining these natural modes are presented in the references (e.g., Wilkinson). If the initial conditions are prescribed in just the right way, it is possible to excite just one of the natural modes. However, for arbitrary initial conditions all modes of vibration are excited. Thus, the general solution of the free-vibration problem is given by a linear combination of the natural modes, that is,

$$\{\mathbf{q}\} = \sum_{r=1}^{n} C_r\{\mathbf{A}\}_r \cos(\omega_r t - \alpha_r)$$

The $2n$ arbitrary constants C_r and α_r are determined through specification of the $2n$ initial conditions $\{\mathbf{q}(0)\}$, and $\{\dot{\mathbf{q}}(0)\}$. An explicit representation for $\{\mathbf{q}\}$ in terms of $\{\mathbf{q}(0)\}$ and $\{\dot{\mathbf{q}}(0)\}$. is given later.

Some of the properties of the natural modes of vibration will now be listed. Unless stated otherwise, the following properties are all consequences of the assumptions that $[M]$ and $[K]$ are real and symmetric and that $[M]$ is positive definite: (1) the n eigenvalues and eigenvectors are real. (2) If $[K]$ is positive definite (corresponding to vibration about a stable state of equilibrium), all eigenvalues are positive. (3) If $[K]$ is positive semidefinite (corresponding to vibration about a neutral state of equilibrium), there is at least one zero eigenvalue. All zero eigenvalues correspond to rigid-body modes. (4) Eigenvectors $\{\mathbf{A}\}_r$, $\{\mathbf{A}\}_s$ corresponding to different eigenvalues $\omega_r^2, \omega_s^2 (\omega_r^2 \neq \omega_s^2; r \neq s; r, s = 1,\ldots,n)$ are orthogonal by pairs with respect to the mass and stiffness matrices (weighting matrices), that is, $\{\mathbf{A}\}_r^T[M]\{\mathbf{A}\}_s = 0$ and $\{\mathbf{A}\}_r^T[K]\{\mathbf{A}\}_s = 0$ for $r \neq s; r, s = 1,\ldots,n$. Since the eigenvectors are orthogonal, and hence independent, they form a basis of the n-dimensional vector space. Thus, any vector in the space, such as the solution vector $\{\mathbf{q}\}$, can be represented as a linear combination of the eigenvectors (base vectors). This theorem (expansion theorem) is of vital importance for the response problem to be considered next.

Response Analysis

The response of the system due to arbitrary deterministic excitations in the form of initial displacements $\{\mathbf{q}(0)\}$, initial velocities (impulses) $\{\dot{\mathbf{q}}(0)\}$, or forcing functions $\{\mathbf{Q}(t)\}$ may be obtained by means of the above expansion theorem. That is, the displacement solution of the response problem is expanded with respect to the eigenvectors of the corresponding free-vibration problem. The time-dependent coefficients in this expansion are the so-called normal coordinates of the system. The equations of motion when expressed in terms of the n normal coordinates are uncoupled, and each one of the equations is mathematically identical in form to the equation governing the motion of a simple one-degree-of-freedom spring-mass system. Hence solutions of these equations are readily obtained for the normal coordinates. The normal coordinates are then inserted into the eigenvector expansion to obtain the following explicit solution for displacement response $\{\mathbf{q}(t)\}$ (Przemieniecki) for undamped constrained or unconstrained (free) structures:

$$
\begin{aligned}
\{\mathbf{q}(t)\} = &\sum_{r=1}^{m} \frac{\{\mathbf{A}\}_{ro}\{\mathbf{A}\}_{ro}^T}{\{\mathbf{A}\}_{ro}^T[M]\{\mathbf{A}\}_{ro}}\Bigg([M](\{\mathbf{q}(0)\} + t\{\mathbf{q}(0)\}) \\
&+ \int_{\tau 2=0}^{\tau 2=t}\int_{\tau 1=0}^{\tau 1=\tau 2}\{\mathbf{Q}(\tau_1)\}\,d\tau_1\,d\tau_2\Bigg) \\
&+ \sum_{r=m+1}^{n} \frac{\{\mathbf{A}\}_{re}\{\mathbf{A}\}_{re}^T}{\{\mathbf{A}\}_{re}^T[M]\{\mathbf{A}\}_{re}}\Bigg([M]\Big(\{\mathbf{q}(0)\}\cos\omega_r t + \frac{1}{\omega_r}\{\dot{\mathbf{q}}(0)\}\sin\omega_r t\Big) \\
&+ \frac{1}{\omega_r}\int_{\tau=1}^{\tau=t}\{\mathbf{Q}(\tau)\}\sin\omega_r(t-\tau)\,d\tau\Bigg)
\end{aligned}
\qquad (8.7.9)
$$

From this equation, it is seen that the displacement response is expressed directly in terms of the arbitrary excitations and the free-vibration frequencies and eigenvectors. The eigenvectors have been separated into two types denoted by $\{\mathbf{A}\}_{ro}$ and $\{\mathbf{A}\}_{re}$. The "rigid-body" eigenvectors $\{\mathbf{A}\}_{ro}$ correspond to the m rigid-body modes (if there are any) for which $\omega_r = 0$, $r = 1, \ldots, m$. The "elastic" eigenvectors $\{\mathbf{A}\}_{re}$ correspond to the remaining $n - m$ eigenvectors, for which $\omega_r \neq 0$, $r = m + 1, \ldots, n$. The general solution given by Eq. (8.7.9) is very useful provided that the second integral (Duhamel's integral) can be evaluated. (The first integral is easier to evaluate.) Closed-form solutions of this integral for some of the simpler types of elements of $\{\mathbf{Q}(t)\}$, such as step functions, ramps, etc., are available (see Przemieniecki). Solutions of Duhamel's integral for more complicated forcing functions can be obtained through superposition of solutions for the simpler cases or, of course, from direct numerical integration.

The displacement response obtained from Eq. (8.7.9) may be employed to obtain other response variables of interest, such as velocities, accelerations, stresses, strains, etc. Velocities and accelerations are obtained from successive time differentiations of Eq. (8.7.9). The method used to evaluate the stresses depends on the type of spatial discretization employed to approximate the structural continuum. For example, difference quotients are used in the finite difference approach. Note that the rigid-body components do not contribute to the stress or strain response and hence the first summation in Eq. (8.7.9) can be omitted for computation of these responses.

SPACE PROPULSION

by Henry O. Pohl

NASA

The propulsion system provides the maneuverability, speed, and range of a space vehicle. Modern propulsion systems have diverse forms, however, they all convert an energy source into a controlled high velocity stream of particles to produce force. Among the many possible energy sources, five are considered useful for space-vehicle propulsion:

1. Chemical reaction
2. Solar energy
3. Nuclear energy
4. Electrical energy
5. Stored energy (cold gas)

Accordingly, the various propulsion devices are categorized into chemical energy propulsion, nuclear energy propulsion, stored energy propulsion, solar energy propulsion and electrical energy propulsion.

Chemical propulsion can be further categorized into liquid propulsion or solid propulsion systems. Many combinations of fuels and oxidizers are available. Likewise, electrical energy propulsion can be subdivided into systems using resistive heating, arc heating, plasma, or ion propulsion. Nuclear energy and solar energy are external energy sources used to heat a gas (usually hydrogen).

All space propulsion systems, with the exception of ion propulsion, use nozzle expansion and acceleration of a heated gas as the mechanism for imparting momentum to a vehicle. The fundamental relationships of space propulsion systems are commonly expressed by a number of parameters as follows: The basic performance equation can be derived from the classical newtonian equation $F = MA$. It is defined as

$$F = \left(\frac{\dot{W}}{g}\right)V_e + A_e(P_e - P_a)$$

where F = thrust, lbf; $\dot{W}$ = propellant weight flow rate; g = gravitational acceleration; V_e = exhaust velocity; A_e = nozzle exit area; P_e = pressure at nozzle exit; and P_a = ambient external pressure.

The specific impulse is one of the more important parameters used in rocket design. It is defined as $I_s = F/W$, where I_s = specific impulse in seconds, F = thrust, and W = propellant weight flow rate per second of operation. I_s increases in engines using gas expansion in a nozzle as the pressure and temperature increase and/or the molecular weight of the gas decreases. The specific impulse values for representative space propulsion systems are shown in Table 8.7.9.

The total impulse $I_t = \int F(t)\,dt$ is useful to determine candidate rocket firing duration and thrust combinations required to satisfy a given change in momentum, where F = thrust and t = time.

Mass fraction is the ratio of propellant to total weight of the vehicle and is defined as $\gamma = (W_i - W_f)/W_i$, where γ = mass fraction; W_i = total weight of the vehicle including the propellant, structure, engine, etc., at the start of operation; and W_f = total weight of the vehicle less the weight of propellant consumed.

In the simplified case of a vehicle operating free of atmospheric drag and gravitational attractions, the velocity increment obtained from a propulsion system is defined as $\Delta V = I_s g \ln(W_i/W_f)$, where ΔV = velocity increment obtained; W_i = total weight of vehicle including propellant, structure, engine, etc., at start of operation; and W_f = total weight of vehicle less weight of propellant consumed during operations. This

Table 8.7.9 Specific Impulse for Representative Space Propulsion Systems

Engine type	Working fluid	Specific impulse
Chemical (liquid)		
Monopropellant	Hydrogen peroxide-	110-140
	hydrazine	220-245
Bipropellant	O_2-H_2	440-480
	O_2-hydrocarbon	340-380
	N_2O_4-Monomethyl hydrazine	300-340
Chemical (solid)	Fuel and oxidizer	260-300
Nuclear	H_2	600-1000
Solar heating	H_2	400-800
Arc jet	H_2	400-2000
Cold gas	N_2	50-60
Ion	Cesium	5,000-25,000

equation brings together the mission parameters, vehicle parameters, and rocket-engine parameters.

Figure 8.7.47 graphically shows the effect specific impulse and mass fraction have on the ability of a propulsion system to change the velocity of a payload in space.

Chemical Propulsion Only chemical energy has found wide acceptance for use in space propulsion systems. Modern materials and current technologies have made it possible to operate rockets at relatively high pressures. When these rockets are fitted with very large expansion-ratio nozzles (nozzle exits 40 to 200 times the area of the throat) the force produced per unit of propellant consumed is greatly increased when operating in the vacuum of space.

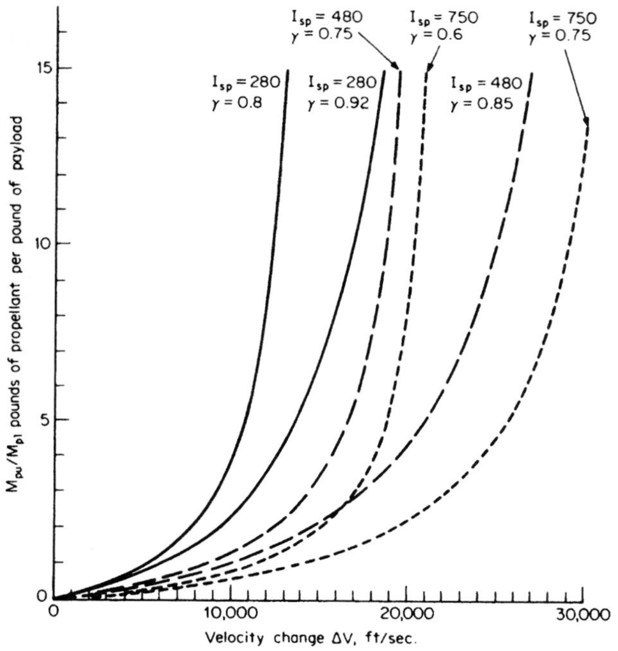

Figure 8.7.47 Performance curves for a range of specific impulse and mass fractions.

Solid-propellant rocket motors are used for moderate total impulse requirements, in spite of their low specific impulse, provided that the total impulse requirements can be precisely set before the rocket is built. The low specific impulse for these applications is largely offset by the high mass fraction (ratio of propellant weight to total propulsion system weight including propellant weight). They are simple to start, have good storage characteristics and multiple velocity requirements can be accommodated by staging, that is, using a separate motor for each burn.

Liquid-propellant systems are generally used for moderate to high total impulse requirements, multiple burns, velocity-controlled thrust terminations, variable thrust requirements or for applications where the total impulse requirement is likely to change after the rocket is built or launched.

Bipropellant systems are the most commonly desired systems to satisfy moderate to high total impulse requirements. A fuel and oxidizer are injected into a combustion chamber where they burn, producing large volumes of high-temperature gases. Oxygen and hydrogen produce the highest specific impulse of any in-service chemical rocket (440 to 460 s). These systems also have good mass fractions (0.82 to 0.85). Oxygen and hydrogen, as stored in the rocket propellant tanks, are cryogenic, with boiling points of –296.7°F and –442°F, respectively. Storing these propellants for long periods of time requires thermally insulated propellant containers and specially designed thermodynamic vents or refrigeration systems to control boil-off losses. The low density of hydrogen (4.4 lb/ft^3) requires the use of bulky containers. Because of these characteristics, oxygen/hydrogen rocket systems can best satisfy moderate to high total-impulse requirements. Where designs are constrained by volume, very long durations of time between firings, many short burns, or precise impulse requirements, nitrogen tetroxide and monomethylhydrazine are often preferred, if the total impulse requirements are moderate.

Monopropellant systems generally use a single propellant that decomposes in the presence of a catalyst. These are usually low total-impulse systems used to provide attitude control of a spacecraft or vernier velocity corrections. For attitude control, small thrustors are arranged in clusters around the perimeter of the vehicle. These small thrustors may, in the course of a single mission, produce over 1,000,000 impulses, to keep the vehicle oriented or pointed in the desired direction.

Solar-heating propulsion would use a solar collector to heat the working fluid which is exhausted through a conventional nozzle. The concentrated solar energy heats the rocket propellant directly through the heat exchanger. The heated propellant is then exhausted through the nozzle to produce thrust. Many design problems make the practical development of the solar-heating rocket quite difficult. For example, the solar collector must be pointed toward the sun at all times and the available solar energy varies inversely with the square of the distance from the sun. Using hydrogen as a working fluid, this type system would have a specific impulse of 400 to 800 s and could satisfy very high total impulse requirements at low thrust levels.

Nuclear propulsion is similar to solar-heating propulsion except that the energy source is replaced by a nuclear reactor. A propellant is injected into the reactor heat exchanger, where the propellant is heated under pressure and expanded through the nozzle. Liquid hydrogen is the best choice for the nuclear rocket's working fluid because its low molecular weight produces the highest exhaust velocity for a given nozzle-inlet temperature. For example, for a given entrance nozzle pressure of 43 atm and temperatures of 1,650, 3,300, and 4,950 K the following specific impulses are obtainable: 625, 890, and 1,216 s, respectively.

The nuclear rocket engine is started by adjusting the reactor neutron-control drums to increase the neutron population. Propellant flow is initiated at a low reactor-power level and is increased in proportion to the increasing neutron population until the design steady-state reactor power output is obtained.

For shutdown of the engine, control drums are adjusted to poison the core and decrease the neutron population. Steady-state reactor thermal power, in megawatts, is determined by the relation $P1 = Kw\,(h_{out} - h_{in})$, where w and h are core flow rate and enthalpy, respectively, and K is the appropriate conversion factor to megawatts.

The reactor is a high-power-density, self-energizing heat exchanger which elevates the temperature of the hydrogen propellant to the limit of component materials.

SPACECRAFT LIFE SUPPORT AND THERMAL MANAGEMENT

by Walter W. Guy

NASA

REFERENCES: *NASA SP-3006,* "Bioastronautics Data Book." Chapman, "Heat Transfer," Macmillan. Purser, Faget, and Smith, "Manned Spacecraft: Engineering Design and Operation," Fairchild.

Life Support—General Considerations

To sustain human life in a spacecraft two functions must be provided. The first is to replenish those substances which humans consume and the second is to control the conditions of the environment at levels consistent with human existence. Humans consume oxygen, water, and food in the metabolism process. As a by-product of this metabolism, they generate CO_2, respired and perspired water, urine, and feces (Table 8.7.10). The provisioning of man's necessities (and the elimination of his wastes) is a primary function of life support. However, an equally important function is the maintenance of the environment in which he lives.

Table 8.7.10 Metabolic Balance for Humans

Consumed		Produced	
Oxygen	1.84	Carbon dioxide	2.20
Drinking water	5.18	Perspiration	2.00
Food (⅔ H_2O)	3.98	Respiration	2.00
		Urine	4.53
		Feces	0.27
	11.00		11.00

NOTE: Metabolic rate = 11,200 Btu/human-day and RQ = 0.87; RQ (respiratory quotient) = vol CO_2 produced/vol O_2 consumed. All values are in units of lb/human-day.

There are three aspects of environmental control—ambient pressure, atmospheric composition, and thermal condition of the environs. In a typical spacecraft design the ambient pressure is maintained generally in the range between ⅓ atm and sea-level pressure, with the oxygen partial pressure varying in accordance with Fig. 8.7.48 in order to provide the appropriate partial pressure of oxygen in the lungs for breathing.

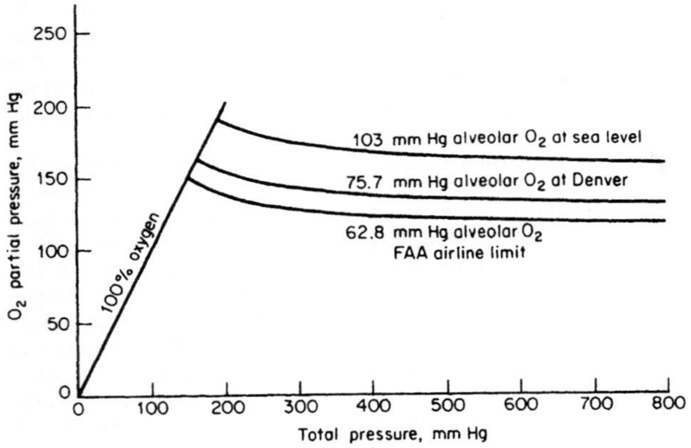

Figure 8.7.48 Atmospheric oxygen pressure requirements for humans.

The other constituents of concern in the atmosphere are the diluent, CO_2, water vapor, and trace amounts of noxious and toxic gases. The percent of the diluent in the atmosphere (or indeed whether a diluent is present at all) is selectable after proper consideration of the design parameters. For instance, short-duration spacecraft can take advantage of human adaptability and use atmospheric pressures and compositions significantly different from those at sea level, with the attendant advantage of system design simplicity. The Mercury, Gemini, and Apollo spacecrafts utilized 5-lb/in^2 pure oxygen as the basic cabin atmosphere. Skylab used an oxygen-nitrogen mixture although at a reduced pressure, while the shuttle is designed for an essentially sea-level equivalent atmosphere. When selecting the pressure and composition, it is important to remember that although a two-gas system adds complexity, it can eliminate the concerns of (1) flammability in an oxygen-rich environment, (2) oxygen toxicity of the crew, and (3) decreased atmospheric cooling potential for electronic equipment in a reduced-pressure cabin. In the selection of a diluent, nitrogen should be considered first, since its properties and effects on humans and equipment are well understood. However, other diluents, such as helium, which have characteristics attractive for a special application, are available.

The other atmospheric constituents—CO_2, water vapor, and trace gases—must be controlled within acceptable ranges for comfort and well-being. Although the partial pressure of CO_2 is less than half a millimeter of mercury in an ambient earth environment, humans are generally insensitive to CO_2 levels up to 1 percent of the atmosphere (or 7.6 mm Hg) unless mission durations increase to more than a month. As spacecraft missions are further extended, the partial pressure of CO_2 should approach the earth ambient level to ensure physiological acceptance. Atmospheric water vapor (or humidity) should be maintained high enough to prevent drying of the skin, eyes, and mucous membranes, but low enough to facilitate evaporation of body perspiration, maximizing crew comfort. The generally accepted range for design is 40 to 70 percent relative humidity. In the closed environment of the spacecraft, cabin noxious and toxic gases are of particular concern. It should be noted that 8-h/day industrial exposure limits are not appropriate for the continuous exposure which results from a sealed spacecraft cabin. Therefore, materials which can offgas should be screened and limited in the crew compartment. The small quantities which cannot be completely eliminated, as well as certain undesirable products of metabolism, must be actively removed by the life-support system to maintain an acceptable atmosphere for the crew.

The third of the environmental control aspects—atmospheric thermal condition—is subject to several design variables, all of which must be taken into account. Thermal comfort is a function of the temperature and relative humidity of the surrounding atmosphere, the velocity of any ventilation present, and the temperature and radiant properties of the cabin walls and equipment. The interrelationship of these parameters is shown in Fig. 8.7.49 for a reasonable range of flight clothing ensembles. Because of the effective insulations available to the vehicle designers, the problem of thermal control in the spacecraft cabin is generally one of heat removal. Utilizing increased ventilation rates to enhance local film coefficients is one means of improving thermal comfort within the cabin. However, the higher power penalties associated

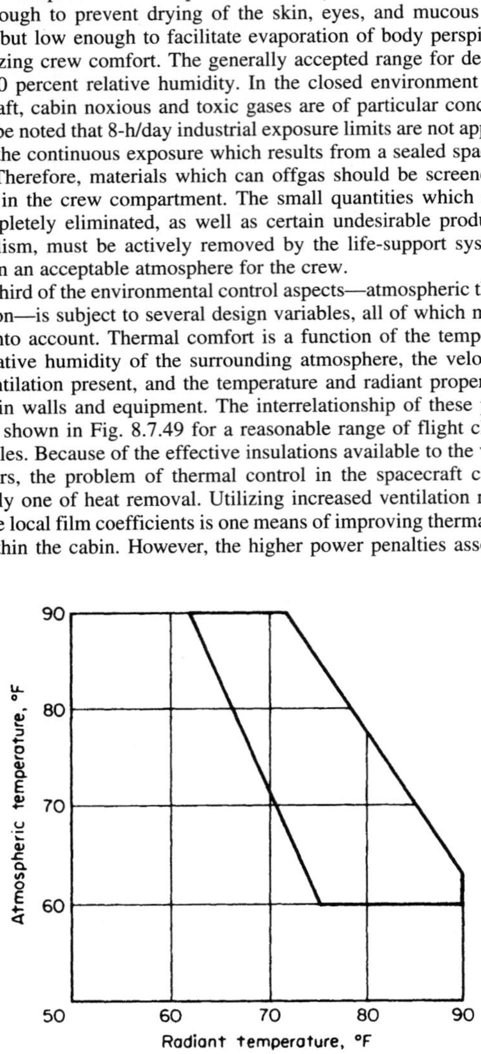

Figure 8.7.49 Thermal comfort envelope for humans. (Note: metabolic rate 300-600 Btu/h; air velocity 15-45 ft/min; pressure 14.7 lb/in^2.)

with gas circulation blowers, as compared to liquid pumps, should suggest using the spacecraft thermal control liquid circuit in reducing the temperature of the radiant environment as potentially a lower-power alternative. However, if the walls or equipment are chilled below the dew point, condensation will occur.

As an additional benefit, maintaining an acceptable thermal environment through radiative means reduces the psychological annoyance of a drafty, noisy environment. Ventilation rates between 15 and 45 ft/min are generally considered comfortable, and exceeding this range should be avoided.

Life Support—Subsystems Selection

In designing the life-support system for a spacecraft, subsystems concepts must be selected to accomplish each of the required functions. This process of selection involves many design criteria. Size, weight, power consumption, reliability of operation, process efficiency, service and maintenance requirements, logistics dependency, and environmental sensitivity may each be very important or inconsequential depending on the spacecraft application. In general, for life support subsystems, the dominant selection criterion is mission duration. Various subsystem choices for accomplishing the life support functions are shown in Fig. 8.7.50.

To illustrate the relationship between mission duration and subsystem selection, the various concepts for removal of CO_2 from the atmosphere shown in Fig. 8.7.50*b* will be discussed. The simplest method of controlling the level of CO_2 in the spacecraft cabin is merely to purge a small amount of the cabin atmosphere overboard. This concept is simple and reliable and requires almost no subsystem hardware, although it does incur a fairly significant penalty for atmospheric makeup. For missions such as Alan Sheppard's first Mercury flight, this concept is probably optimum. However, even a minor increase in mission duration will result in a sizable weight penalty to the spacecraft. By adding a simple chemical absorption bed to the spacecraft ventilation loop (with a chemical such as lithium hydroxide as the absorbent), the expendable penalty for CO_2 removal can be reduced to the weight of the chemical canisters consumed in the absorption process. This concept was used on the Apollo spacecraft. The penalty for this type of concept also becomes prohibitive if mission durations are further extended. For missions such as Skylab, the expendable chemical absorbent can be replaced by a regenerable adsorbent. This change eliminates the lithium hydroxide

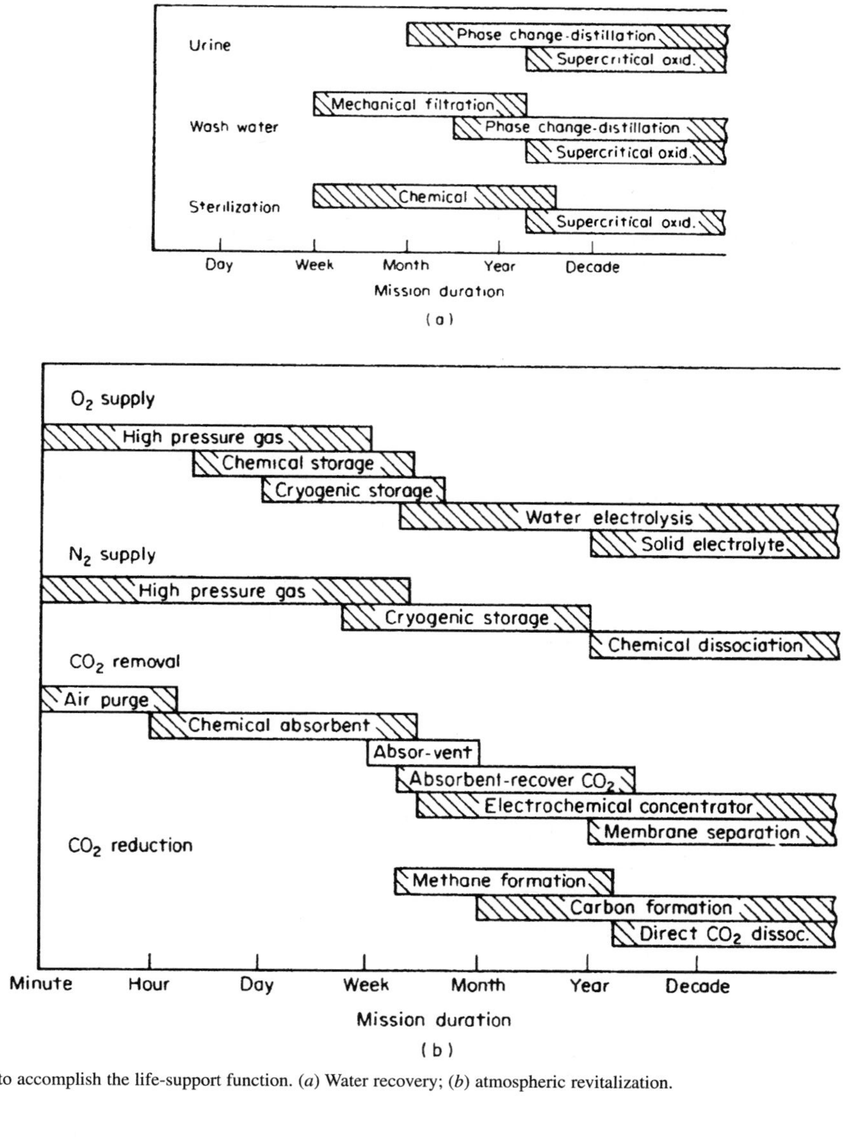

Figure 8.7.50 Concepts to accomplish the life-support function. (*a*) Water recovery; (*b*) atmospheric revitalization.

weight penalty but incurs a small atmospheric ullage loss during the vacuum regeneration of the adsorbent. In addition, the regenerable system is more complex, requiring redundant beds (one for adsorption while the alternate bed desorbs), and mechanical-timing circuits and switching valves. This result is typical in that reducing expendable requirements is generally at the expense of system complexity.

The three CO_2 removal concepts discussed above all prevent the subsequent use of the collected CO_2 after removal from the atmosphere. Should the CO_2 be required for reclaiming the metabolic oxygen (such as will be the case for a space station design), the regenerable concepts can be adapted to accommodate retrieval of the CO_2 during desorption. However, this enhancement further complicates the subsystem design. There is an alternative CO_2 removal concept available which utilizes a fuel-cell-type reaction to concentrate the CO_2 in the atmosphere through electrochemical means. This process is mechanically simple (as compared to the adsorption-desorption system) but does require more electrical power. However, it has an integration advantage in that it provides the CO_2 collected from the atmosphere mixed with hydrogen, which makes its effluent a natural feedstock for either one of several CO_2 reduction subsystems (which will be required for oxygen recovery on a space station). Although Fig. 8.7.50 identifies the various life-support subsystem choices for candidate spacecraft designs of different mission durations, this information should be used only as a guide. The crew size as well as other design parameters can skew the range of applicability for each subsystem (e.g., a large crew can render expendable dependent subsystems inappropriate even for relatively short missions).

There is another major aspect of life support system design for spacecraft (in addition to subsystem selection) and that is process integration and mass balance. For example, a space-station-class life-support system design must accommodate wastewater with a wide range of contamination levels—from humidity condensate which includes only minor airborne contamination to various wash and rinse waters (including personal hygiene, shower, clothes washing, and dish washing) and urine. The degree of contamination generally influences the reclamation process selection; therefore, multiple contaminated sources tend to result in multiple subsystem processes. For example, highly contaminated water, such as urine, is most reliably reclaimed through a distillation process since evaporation and condensation can be more effectively utilized to separate dissolved and particulate material from the water, while less contaminated wash waters can be effectively reclaimed through osmotic and mechanical filtration techniques at significantly less power penalties in watts/lb H_2O recovered. Regardless of the techniques selected, water must be made available for drinking, food preparation, and cleaning the crew and the equipment they use, as well as for generating the oxygen they breathe. In fact, the space station life-support design process can be looked upon, in one respect, as a multifaceted water balance since the dissociation of CO_2 and generation of O_2 both involve water in the respective processes. Figure 8.7.51 is a typical space station schematic depicting the interrelationship of the subsystems and the resulting mass balance.

Thermal Management—General Considerations

Although there is an aspect of life support associated with thermal comfort, the area of spacecraft thermal management encompasses a much broader scope. In the context of this section, thermal management includes the acquisition and transport of waste heat from the various sources on the spacecraft, and the disbursing of this thermal energy at vehicle locations requiring it, as well as the final rejection of the net waste heat to space. Although there are thermal implications to the design of most spacecraft hardware, the Life Support and Thermal Management section only deals with the integration of the residual thermal requirements of the various vehicle subsystems to provide an energy-efficient spacecraft design (i.e., one which utilizes waste heat to the maximum and therefore minimizes requirements for the vehicle heat rejection system).

Thermal management systems are typically divided into two parts. The first, associated with the spacecraft cabin, has design requirements associated with maintaining the crew environment and equipment located in the cabin at acceptable temperatures, and in addition has special requirements imposed on its design because of the location of the system in the pressurized compartment and its proximity to humans. These additional requirements are usually dominant. Any cooling medium which is used in the crew compartment must have acceptable toxicological and flammability characteristics in order to not impose a hazard to the crew in the event of malfunction or leakage. Although there are thermal control fluids which have low toxicity potential,

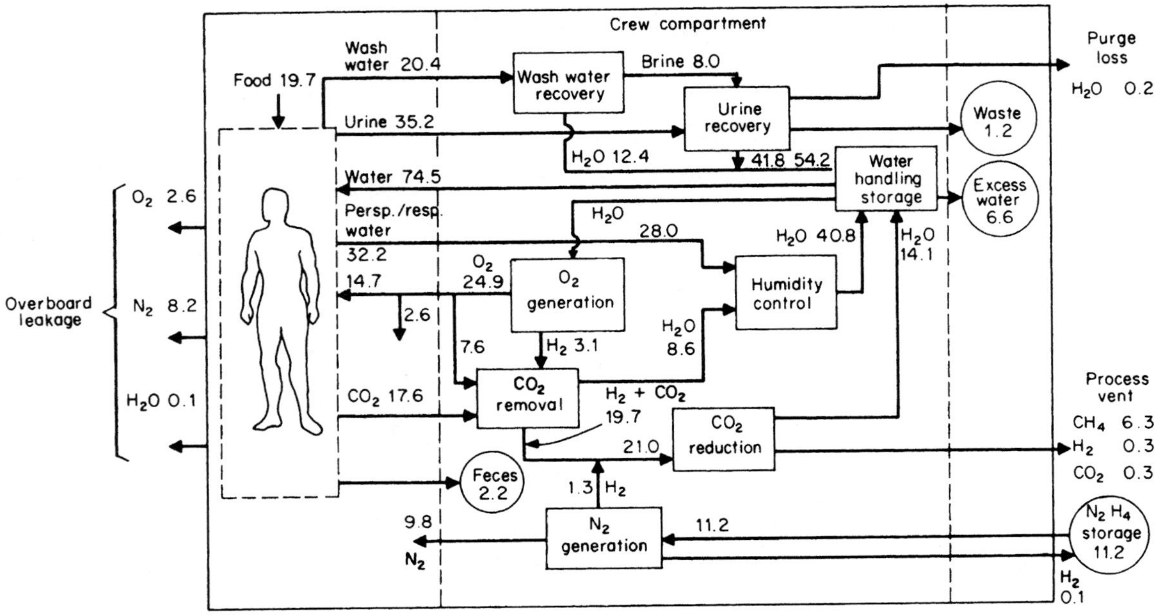

Figure 8.7.51 Life-support systems mass balance. (All values are in pounds per day for an eight-person crew.)

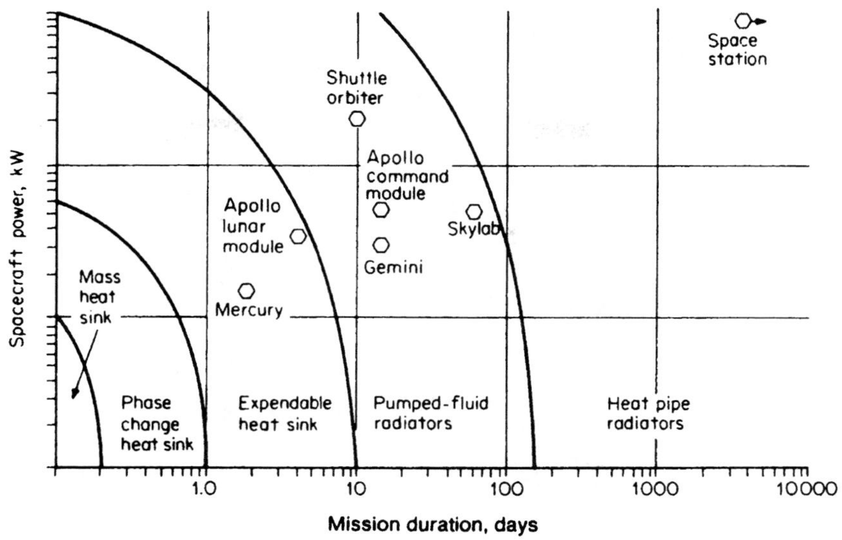

Figure 8.7.52 Range of application of various heat-rejection methods.

reasonable flammability characteristics and adequate thermal transport properties, none are better than water.

The second aspect of the thermal management system involves those spacecraft systems located outside the cabin. The dominant design criteria in thermal control fluid selection for these systems are thermal transport efficiency and low-temperature characteristics. The thermal transport efficiency is important since the size, weight, and power requirements of the pumping and distribution system depend on this parameter. However, in the environs of deep space the potential exists for very low temperatures. Therefore, for maximum system operating flexibility, the characteristics of the thermal transport fluid must be such that operation of the system can continue even in a cold environment; thus, low-temperature properties become a fluid selection constraint. Although the range of acceptable fluids is broader here, several fluids in the Freon family meet these requirements. For single-phase, pumped-fluid systems, the temperature levels of the heat sources are important in determining the optimum sequence for locating equipment in the thermal circuit. The equipment with the lowest-temperature heat sink requirements are accommodated by placing them immediately after the heat rejection devices, while equipment which can accept a higher-temperature heat sink are located downstream. Also, those subsystems which require heating must be located at a point in the circuit which can provide the thermal energy at the required temperature level.

Typically spacecraft heat sources include cabin-atmosphere heat exchangers, the avionics system as well as other electronic black boxes, power generation and storage equipment, etc., while heat sinks include propulsion system components, various mechanical equipment requiring intermittent operation, dormant fluid systems which have freezing potential, etc. The ultimate elimination of the net thermal energy (i.e., sources minus sinks) is by rejection to space. This is accomplished through expendables or radiation.

Recent developments in thermal management technology have resulted in the consideration of a two-phase acquisition, transport, and rejection system. This concept uses the heat of evaporation and condensation of the transport fluid to acquire and reject thermal energy, reducing by several orders of magnitude the amount of fluid required to be pumped to the various heat sources and sinks within the spacecraft. This concept has the added advantage of operating nearly isothermally, permitting the idea of a thermal "bus" to be included in future spacecraft design, with indiscriminate sequencing of the heat sources and sinks within the vehicle.

Thermal Management Subsystem Selection

The subsystem selection process associated with the thermal management area depends largely on the magnitude of thermal energy being managed and the duration of the mission. Figure 8.7.52 graphically depicts the areas of applicability for various types of heat-rejection devices as a function of these parameters.

Small heat-rejection requirements can sometimes be met by direct heat sinking to the structure (or a fluid reservoir). Often, passive heat sinks need to be supplemented with fusible heat sinks (to take advantage of the latent heat of fusion) or evaporative heat sinks (to take advantage of the latent heat of evaporation). Of the three types of heat sinks, the evaporative heat sink is the only one which has been used as the primary vehicle heat-rejection device in the manned spaceflight program to date. Both the Mercury spacecraft and the lunar module utilized a water evaporative heat sink in this manner. However, the Apollo spacecraft and the shuttle have also used evaporative heat sinks and Skylab used a fusible heat sink as supplemental heat rejection devices.

Since space is the ultimate heat sink for any nonexpendable heat rejection, devices which allow the thermal energy which has been collected in spacecraft coolant circuits to be radiated to space become the mainstay of the thermal management system. These "radiators" take many forms. The simplest design involves routing the cooling circuits to the external skin of the spacecraft so that the thermal energy can be radiated to deep space. Because of spacecraft configuration restrictions, only partial use can be made of the external skin of the spacecraft, and the amount of thermal energy which can be rejected by this technique is thereby limited. Although the Gemini and Apollo spacecraft successfully used integral skin radiators, the unique characteristics of the outer surface of the orbiter required a different technique be adopted for its radiator design. Accordingly, the radiators were designed to be deployed from inside of the payload bay doors. This not only provides good exposure to space, but also, in the case of the forward panels, allows two-sided exposure of the panels, thereby doubling the active radiating area.

Heat rejected Q_{rej}, according to the Stefan-Boltzmann law, is given by the following equation:

$$Q_{rej} = \varepsilon \gamma A T^4$$

where γ = constant, 0.174×10^{-8} Btu/(h·ft²·°R⁴); ε = emittance property of surface (termed emissivity), A = area available for radiation;

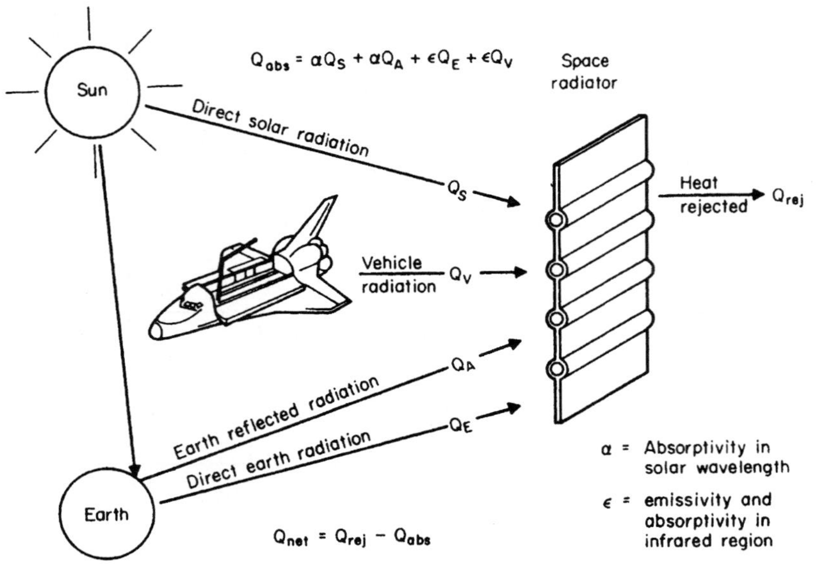

Figure 8.7.53 Space radiator energy balance.

and T = radiating temperature. From this equation it is obvious that a radiator with both sides exposed will reject twice as much heat to space as one of the same dimensions which is integral with the skin of the vehicle. A space radiator can absorb as well as reject thermal energy. The net heat rejected Q_{net} from a radiator is given by the following equation:

$$Q_{net} = Q_{rej} - \sum_{N}^{A=1} Q_{abs}$$

The energy absorbed by the radiator, Q_{abs}, can come from various sources (see Fig. 8.7.53).

The most common heat sources for an earth-orbiting spacecraft are direct solar impingement, reflected solar energy from the earth (i.e., albedo), direct radiation from the earth, radiation from other parts of the spacecraft, and radiation from other spacecraft in the immediate vicinity. These energy sources are absorbed on the radiator surface in direct proportion to the absorptance characteristic of the surface, termed absorptivity. Emissivity and absorptivity are unique to each surface material but are variables directly linked to the temperature of the source of the radiation (i.e., wavelength of the radiation). Therefore, since absorptivity and emissivity are characteristics of the surface, the careful selection of a surface coating can significantly enhance the effectiveness of the radiator. The coating should have a high emissivity in the wavelength of the rejected energy from the radiator (for good radiation efficiency) and a low absorptivity in the wavelength of the impinging radiation (for minimal absorbed energy). Since generally the earth and the surrounding spacecraft are producing thermal energy of essentially the same wavelength as the radiator, it is necessary to restrict the view of the radiating surface to these objects since absorptivity and emissivity of a surface are equal for the same wavelength energy. Thus, radiators which can be located so that they are not obstructed from viewing deep space by elements of the spacecraft or other spacecraft in the vicinity will be much more effective in rejecting waste heat. The sun, however, emits thermal energy of a dramatically different wavelength and, therefore, coatings can be selected which absorb less than 10 percent of the solar energy while radiating nearly 90 percent of the energy at the emitting temperature of the radiator.

For applications which require large radiating surface areas as compared to the vehicle's size, integral skin radiators offer insufficient area and thus space radiators which use deployment schemes must be considered. However, as radiators become larger the mechanisms become very complicated. To circumvent the mechanical problems of deployment, a new radiator concept has been developed which allows in-orbit construction. This concept utilizes a single-tube heat pipe as the primary element (see Fig. 8.7.54).

There are two distinct advantages of heat pipe radiators. The first is that assembly of large surface areas in-orbit can be affected without breaking into the thermal control circuit. This is practical since the heat pipe radiator element is completely self-contained. The second advantage is that in-orbit damage from accidental collision or meteoroid puncture affects only that portion of the radiator actually damaged and does not drain the entire radiating surface of fluid as it would the radiators which have been used on spacecraft to date. Because of this advantage, future missions which require either of the three types of radiators (i.e., integral skin, deployable, or space constructible) will utilize the heat pipe concept to transport the thermal energy to the exposed radiating surface.

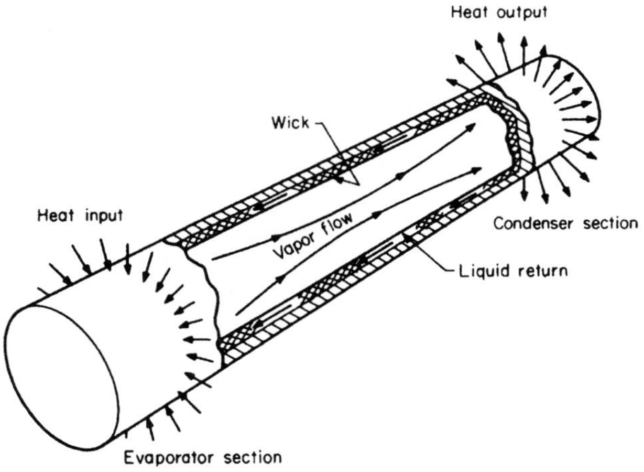

Figure 8.7.54 Basic heat-pipe operation.

The design of space radiators can take many forms, but the basic concept is an exposed surface made of a conducting material with regularly spaced fluid tubes to distribute the heat over the surface (see Fig. 8.7.55).

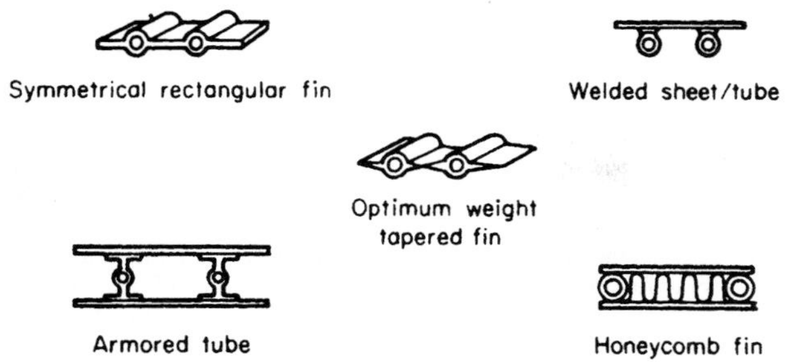

Figure 8.7.55 Typical radiator tube/fin configurations.

The selection of the material and configuration of the radiating surface (i.e., honeycomb, box structure, thin sheet, etc.) determine its thermal conductance. This conductance, when coupled with tube spacing and the internal heat transfer characteristics within the tube and fluid system, determines the temperature gradient on the radiating surface between adjacent tubes. This so-called "fin effectiveness" represents a loss in radiating potential for any particular radiator design; therefore, these parameters should be selected carefully to optimize the radiator design within the constraints of each application. Figure 8.7.56 demonstrates the effect of the configuration parameters of the radiator on panel weight.

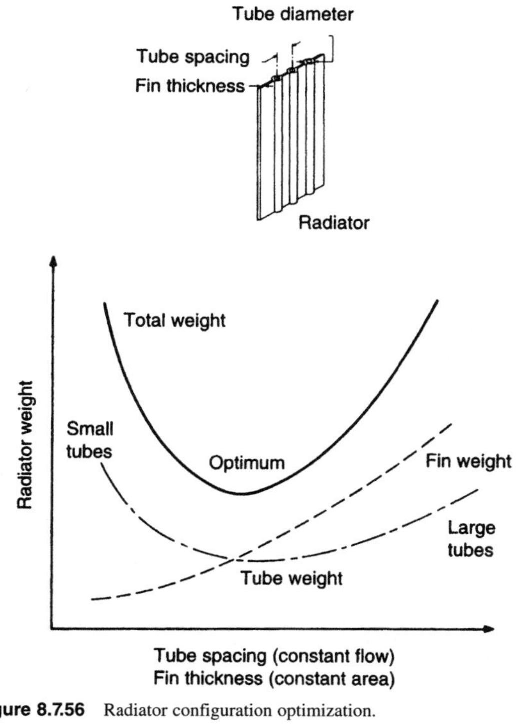

Figure 8.7.56 Radiator configuration optimization.

DOCKING OF TWO FREE-FLYING SPACECRAFT

by Siamak Ghofranian and Matthew S. Schmidt

Rockwell Aerospace

REFERENCES: Ghofranian, Schmidt, Briscoe, and Shliesing, "Space Shuttle Docking to MIR, Mission-1"; Ghofranian, Schmidt, Briscoe, and Shliesing, "Simulation of Shuttle/MIR Docking"; Ghofranian, Schmidt, Lin, Khalessi, and Razi, "Probabilistic Analysis of Docking Mechanism Induced Loads for MIR/Shuttle Mission"; *35th Structures, Structural Dynamics, and Materials Conf.* AIAA/ASME/ASCE/AHS/ASC. Haug, "Computer Aided Kinematics and Dynamics of Mechanical Systems," vol 1: "Basic Methods," Allyn and Bacon. Greenwood, "Principles of Dynamics," Prentice-Hall.

The information in this section covers the mechanics of docking two free-flying unconstrained vehicles in space. Different methods of mating are addressed with the emphasis applied to docking. The content does not encompass all facets of the procedure, but it does bring forth the physics of the operation which make the process an engineering challenge.

Types of Mating

The understanding of mating is a precursor to preliminary design. Different techniques of mating free-flying bodies are often discussed when planning the rendezvous of two vehicles in space. The mating of two structures can be performed in three ways: docking, berthing, or a combination of both. **Docking** relies on the transfer of vehicle momentum to provide the means to force the docking ports into alignment. If a significant amount of momentum is required for capture, jet firings can be used to introduce the impulse into the system as needed. **Berthing** relies on a secondary system, such as a robotic arm to enforce the docking ports into alignment. The active vehicle grapples the other with a robot manipulator and guides the structural interfaces together. If manipulator usage is attempted, specific design factors need to be considered: arm configuration, gearbox stiffness, linkage flexibility, and controllability.

Mating of two structures in space is by every means considered a multibody dynamic event. Docking involves a significant level of momentum transfer, while berthing involves relatively little. The dynamics result from the deliberate impact of the two vehicles. As docking occurs the system responds with rigid-body motion and mechanical motion of the docking device. Depending on the relative states between the vehicles and the mode in which the maneuver is attempted, the response of the system and the physics vary considerably.

For discussion purposes the two docking vehicles can be considered as a target vehicle (passive) and a docking vehicle (active). The passive vehicle maintains attitude control as the active vehicle is navigated along an approach corridor. The active vehicle holds an attitude control mode with translational control authority. As the corridor narrows, sighting aids, range, and range-rate feedback on both vehicles are used to guide the docking ports into contact.

As the vehicles approach, a closing velocity exists. From a design standpoint the operation of a docking system involves various mechanism operational phases: mechanism deployment, interface contact and capture, attenuation of relative motion, mechanism retraction, and structural lockup. The **deployment phase** is where the mechanism is driven from its stowed position to its ready-to-dock position. Vehicle interface contact and capture are initiated when the docking ports of each vehicle are maneuvered into contact. Ideally, a positive closing velocity should be maintained from initial contact through capture. Once an opening velocity develops, the probability of a failed capture increases. As the vehicles interact the mechanism interfaces are indexed into alignment as the system responds to the impact loads. The compliance of the system occurs in two forms: mechanism movement and rigid body vehicle motion.

When the docking interfaces interact, the system responds in six degrees of freedom and the docking system is stroked as the mechanism absorbs the relative misalignments. Once the interfaces are fully aligned, capture latches lock the opposing interfaces together, and the vehicles are "soft-docked." After capture, residual energy in the system allows relative motion between the structures to continue. Depending on the location of the docking system relative to the vehicle center of masses, the motion may dissipate due to the inherent hysteresis in the docking system; otherwise, separate energy-dissipation devices may be necessary. After attenuation the vehicles are likely to be misaligned due to the new relative equilibrium position of the spacecraft. Separate procedures or devices can be used to bring the structural interfaces into alignment.

The **retraction process** is where the active mechanism is driven back to its stowed position. Once they are fully retracted, the structural interfaces are latched at hard interfaces, and the vehicles are "harddocked."

The mass of each vehicle is an essential part of the docking system design. The momentum of the docking vehicle is the medium which provides the force required by the docking mechanism to comply with the relative misalignments between the vehicles at the docking interface. The operation can be complicated if a vehicle center of mass is offset from the longitudinal axis of the docking system. As a result, the vehicle effective mass at the docking interface is reduced. Furthermore, the center-of-mass offset induces relative rotation after contact which complicates the capture and attenuation process. Ideally two vehicles should dock such that the velocity vector and longitudinal axis of the docking system have minimum offset from the center of mass. The relative rotation of the vehicles after contact is known as *jackknifing*.

The docking mechanism center-of-mass offset and the vehicle size significantly influence the design of the docking interfaces, attenuation components, docking aids, piloting procedures, and capture-assist methods. The relative vehicle misalignments and rates at the moment the docking interfaces first come into contact (initial conditions) should be known in order to size the mechanism attenuation components, to establish piloting procedures and training criteria, and to study the mechanics of the system before, during, and after docking.

The initial conditions result from vehicle as-built tolerances, thermal and dynamic distortions, variation in piloting performance, and control system tolerances. The misalignments are composed of lateral misalignments in the plane of the docking interface and angular misalignments about three axes. The rates are composed of translational velocities along three axes and angular velocities about three axes.

The mass and motion of the system are the basic ingredients for providing the energy necessary to join the interfaces. Ideally the interfaces should capture before an opening velocity develops between the docking interfaces. Adequate momentum at the docking port is required to actuate the mechanism. The impulse introduced to the system during the impact must be less than that required to reduce the relative interface velocity to zero. If the relative interface velocity goes to zero before capture, the vehicles will separate. For low closing velocities, a small axial impulse due to initial contact can reduce the relative interface velocity to zero and cause a "no capture" condition.

A drive system along the longitudinal axis of the docking mechanism or reaction control jets increases the probability of capture. It extends the docking system toward the target vehicle once initial contact is made and thus improves capture performance for a docking vehicle with a low closing velocity. However, a drawback of a drive system is that the reaction has the tendency to push the vehicles apart. The action of jet thrusting has the opposite effect; it tends to push the vehicles together. If the momentum required to provide a low possibility of no capture becomes operationally prohibitive, thrusting can be used as a substitute and applied as necessary.

As energy introduced into the system increases, whether in the form of vehicle momentum or jets pulsing, a tradeoff exists between acceptable internal vehicle loads, allowable component stroke limits, vehicle jackknifing, and piloting capability.

Vehicle Mass and Inertia Effects

Concentrated spherical masses colliding in two dimensions can be used to describe the dynamic effects of an impact. The same class of examples are consistently used in beginning physics textbooks to describe the macroscopic concepts of impulse, momentum, and energy. The same theory should be applied to understand the requirements of mating two vehicles and perform conceptual and preliminary design studies. A single-degree-of-freedom mechanical equivalent exhibiting simple harmonic motion is a useful example for understanding the mechanics.

Consider two spacecraft in a two-dimensional plane on a planned course of docking without any misalignments and with some closing velocity. If only the docking-axis degree of freedom is considered, the effective mass of each body, M_1 and M_2, can be computed from

$$M_1(\text{or } M_2) = M - (ML)^2/(I + ML^2)$$

where M is the vehicle mass, I is the vehicle mass moment of inertia about the center of mass, and L is the mass center offset from the docking axis. In this case the process of capture and attenuation can be thought of as a perfectly plastic impact of two bodies. The masses dock and stick together. After impact the two bodies continue their trajectory with some final velocity. The combined effective mass of the two bodies represented as a single degree of freedom can be computed from $M_{\text{eff}} - M_1M_2/(M_1 + M_2)$. Since momentum is conserved during a perfectly plastic impact, kinetic energy loss during docking can be computed from $\frac{1}{2}M_{\text{eff}}v^2$, where v is the capture velocity. Since at capture we know the effective mass of the system and the capture velocity, the unknown parameter becomes the work required to bring the system to rest.

Design Parameters

The energy loss during docking is equal to the work done, $F \times d$, by the mechanism-attenuation components. If a hysteresis brake is used to bring the system to rest, then a trade can be performed on the force-distance relationship by using

$$\tfrac{1}{2}M_{\text{eff}}\, v^2 = Fd$$

where F represents the force necessary to bring the system to rest over distance d. Using the hysteresis brake as a starting point provides the average about which other devices would oscillate. For an underdamped spring/damper attenuation system, the relative motion and maximum force can be computed from

$$x = v/\omega_d \times e^{-\zeta\omega} \sin \omega_d t \qquad \text{and} \qquad F = Kx + C\, dx/dt$$

respectively. In this representation,

$$\omega_d = \omega(1 - \zeta^2)^{1/2} \qquad \text{and} \qquad \omega^2 = K/M_{\text{eff}}$$

where ζ represents the damping factor and K and C represent the attenuation component characteristics. As can be seen, minimal knowledge of the basic system characteristics can be used to design the type and arrangement of the attenuation components. During the design of the attenuation components the following components may prove useful: preloaded centering springs, staged springs, ON/OFF dampers, and variable-level hysteresis brakes. The material in this section is only a first look at understanding the design of a docking system. As more detail is developed, the effects of relative vehicle motion, interface contact dynamics, mechanism performance, and vehicle flexibility need to be simulated, which goes beyond the scope of this section. For a comprehensive background, the reader should refer to the cited references.

"ACTIVE" DOCKING MECHANISM DESIGN

by James Lewis

The previous discussion of the docking of two free-flying spacecraft has described the interaction between two docking spacecraft that use a heritage docking attenuation design. These heritage designs require the use of momentum to force indexed docking interfaces into alignment. These mechanisms are designed to comply during momentum transfers to account for spacecraft relative misalignments, provide spacecraft-to-spacecraft motion dissipation, achieve capture, and a final relative position. An example of a heritage design is shown in Fig. 8.7.57.

Newer docking system designs are incorporating additional components to change the "passive" operational nature of the capture mechanism for all the phases and processes to become more "active." These newer components allow for the attenuation mechanism's systems to have modifiable performance characteristics during different phases of the docking processes. This "active" or "smart" docking mechanism design has components driven by (position, force, velocity, acceleration) feedback/input, either directly measured or derived from various sensing components or a camera/vision system, using simple (or complex) control electronics and software and can make adjustments in

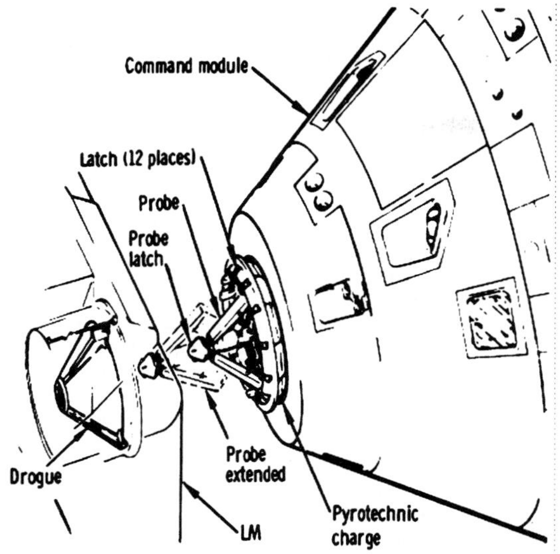

Figure 8.7.57 Heritage docking mechanism.

voltage for controlling electric motors real-time in various attenuation components.

"Smart" docking mechanisms can reduce or eliminate the need for momentum transfer to enable docking operations and offer flexibility for the various initial condition design factors (e.g., closing velocity, impulse momentum, and coupled body dynamics) which impact the alignments, dissipation, and capture processes; and component selection, sizing, performance, and power usage. Furthermore, mitigating momentum transfer requirements shortens the length of critical operations and can simplify piloting and training.

Given the limited adjustability of passive heritage designs, "smart" docking mechanism enable easier reconfiguration of component or system responses to better match a wider range of spacecraft center-of-mass offsets and vehicle sizes by decoupling the individual control/response for all six degrees of freedom (lateral misalignments in the three axes and angular velocities along and about three axes of the plane of the docking interface. "Smart" components can be localized to key components or distributed across the entire attenuation system depending upon the specific design implementation; desired amount of decoupled system control and response; and ability to integrate the additional electronics, software, and their interfaces into the spacecraft.

For a spacecraft, the application of "smart" docking can support the need for deliberate impact of two docking spacecraft should their navigation control authority still require it or obviate the need for impact by offering a "lunge" or reach out capability if no closing velocity is provided or desired as is the case for Berthing mating. An example of a "smart" docking mechanism is shown in Fig. 8.7.58.

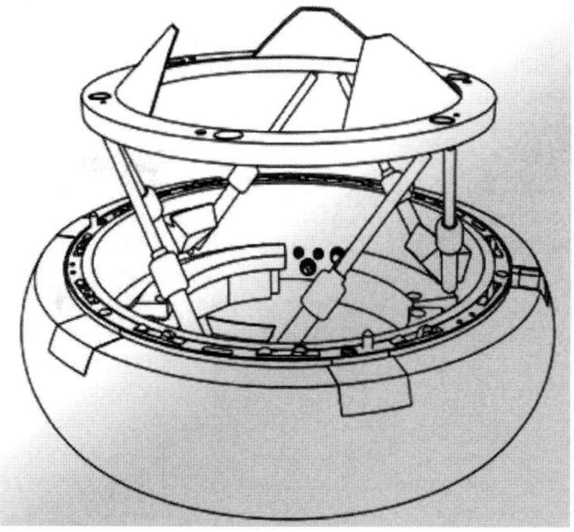

Figure 8.7.58 "Smart" docking mechanism.

International Docking System Standard

The International Docking System Standard—IDSS is one of the first standards developed to promote and enhance future multipartner or multinational space missions where "space" vehicles, modules, segments, or elements are developed and provided by separate space mission partners which need to be docked in space. A several year, joint effort by the International Space Station Partners from the U.S., Europe, Russia, Japan, and the Canadian Space Agency resulted in a baseline definition of the key physical interface geometry and dimensions, as well as key performance parameters which would allow participating members to build and/or utilize a standard docking interface with a higher probability of docking mated interoperability and compatibility when compliant with the docking standard.

The first standard compliant docking interface, the International Docking Adapter—IDA, was installed on the International Space Station in 2016 for use by space station International and Commercial visiting vehicles. The IDSS Interface Definition Document—IDSS IDD is assessable online and each Partner Nation has technical representatives.

The development and use of this Standard is still fairly recent and it is expected that as more compliant applications and uses occur that the standard definition will evolve as missions and members dictate; and that other mechanical interface standards may be develop as needed in the future.

8.8 PIPELINE TRANSMISSION

by Mike Yoon

REFERENCES: US DOT PHMSA, "49 CFR Part 192 - TRANSPORTATION OF NATURAL AND OTHER GAS BY PIPELINE: MINIMUM FEDERAL SAFETY STANDARDS," including amendment table of section revisions, is available at: http://www.phmsa.dot.gov/staticfiles/PHMSA/DownloadableFiles/Files/Pipeline/192%20Master%20_192-114_%20highlight.pdf; "49 CFR Part 195 - TRANSPORTATION OF HAZARDOUS LIQUIDS BY PIPELINE," available at: http://www.phmsa.dot.gov/staticfiles/PHMSA/DownloadableFiles/Files/195_Master%20_195_86c_%20Clean.pdf; US EPA, "40 CFR Parts 72 Permit Regulations," available at: http://www.ecfr.gov/cgi-bin/retrieveECFR?gp=&SID=5584e589aef37add39257c1f0c1617e4&r=PART&n=40y17.0.1.1.1; US DOE FERC "FERC Order 636 Restructuring of Pipeline Services," and "FERC Order 637 Regulation of Short-Term Natural Gas Transportation Services and Regulation of Interstate Natural Gas Transportation Services," are available at https://www.ferc.gov/legal/maj-ord-reg.asp; ASME B31.4 "Pipeline Transportation Systems for Liquids and Slurries"; ASME B31.8 "Gas Transmission and Distribution Piping"; API Recommended Practice (RP) 1130, "Computational Pipeline Monitoring for Liquids"; API RP 1175, "Pipeline Leak Detection Management Program"; Yoon, "Pipeline System Automation and Control," ASME Press; Mohitpour, "Hydrocarbon Liquid Transmission Pipeline and Storage Systems," ASME Press.

OVERVIEW

Typical transmission pipeline systems are composed of pipes, joint and fittings, valves, instrumentation and control, and facilities. Transmission pipelines operate at pressures, so high strength carbon steel pipes are used in their construction. Pipe sizes for transmission pipelines in the United States range from 6" up to 48" in diameter and grades of steel with specified minimum yield strengths (SMYSs) range from 35,000 up to 120,000 psi (241.3 MPa to 827.4 MPa). Although the trend has been toward the use of larger pipe diameters and pipe steel of greater tensile strength, the economics of pipe production and coating and the ease of handling, including transportation as well as installation, has led to greater utilization of 36-in-diameter (914-mm) pipe with a SMYS of 80,000 psi (551.6 MPa). Pipes are usually coated to prevent corrosion, and welded or joined with joint and fittings such as flanges.

The types of valves used are block valves, check valves, and control valves. Flow and pressure control valves, as well as overpressure protection, may also be required. If a large temperature drop is anticipated across any of the pressure-reducing valves, then a radiant heater or indirect-fired heater may be required. In most new construction, mainline valves are usually ball or gate valves and are of the full-opening variety to allow for the passage of cleaning pigs and inspection tools. Ball and gate valves are also used in addition to check and plug valves on piping other than mainline piping. In addition, several pig launching and receiving facilities are installed to clean or inspect the pipeline.

Main facilities may include meter stations to measure and monitor the flow of petroleum products in pipelines, compressor or pump stations to boost pressure, valve stations to control flow direction, and gas or liquid storages for peak shaving or other purposes. Because of the significant differences in product properties between natural gas and petroleum liquids, the standards applicable to their respective pipeline systems are usually different, and so are pressure boosting and storage facilities. Natural gas storage is mainly held in-line in the form of line pack and underground caverns. The role of storage facilities is to balance the seasonal and intra-seasonal demand swings of gas end-users. On the other hands, petroleum liquids storage includes various types of above-ground storage tanks and underground caverns for storing large volumes of liquids.

Meter stations to measure the volume or flow rate are installed at product receiving and delivery points, where custody transfer takes place. Measurements are required for producers, customers/shippers and transportation companies for custody transfer (change in ownership or responsibility) as well as pipeline system monitoring and control. Also, measurements may be required if petroleum products move across national boundaries. Pipeline companies charge their shippers for transportation services based on the measured quantities of the products they have transported, assuming that other transportation requirements such as the product quality are met. Therefore, a custody transfer quality measurement system should generate accurate measured values. The quantities typically measured for custody transfer and monitoring or controlling facilities are volume or mass flow rate, pressure, temperature, density, and quality at the custody transfer or designated control points along the pipeline.

The measurement systems used to establish custody transfer are dependent on the fluids being measured and/or the different regulations or contractual conditions applicable. For certain products such as ethylene, custody transfer is based on mass, but for the majority of petroleum liquids, custody transfer is based on volume measurement. Most pipeline systems are monitored and controlled for flow and pressure, while liquid density is required for batch control.

The flow of natural gas is measured by flow meters at meter stations. These stations usually consist of pressure control valves and volume or flow rate measurement devices, which are based on the following standards reported by the American Gas Association (AGA):

- Orifice meters—AGA Report 3 Gas Flow Equation for Orifice Plates,
- Turbine meters—AGA Report 7 Turbine Gas Meter,
- Ultrasonic meters—AGA Report 9 Measurement of Gas by Multipath Ultrasonic Meters,
- Coriolis meters—AGA Report No. 11: Measurement of Natural Gas by Coriolis Meter.

Since natural gas is highly compressible, the above flow meters require AGA Report Number 8, Compressibility Factors for Natural Gas and Other Hydrocarbon Gases. The first three meters convert uncorrected flow to corrected flow at base conditions, while the fourth meter may need to convert mass flow to volume at base conditions because a Coriolis meter directly measures mass flow.

Positive-displacement and turbine meters are in general use on products systems. Totalizers for multimeter installations and remote transmission of readings are an industry standard. Greater application of meters with crude-oil gathering facilities is being made as automatic-custody units are installed. The petroleum liquid measurement standards, a compendium of petroleum liquid and also gas measurement specifications, are described in Manual of Petroleum Measurement Standards (API MPMS). The standards for the following meter types are given in the manual:

- 5.2 Measurement of Liquid Petroleum by Displacement Meters
- 5.3 Measurement of Liquid Petroleum by Turbine Meters
- 5.6 Measurement of Liquid Petroleum by Coriolis Meters
- 5.8 Measurement of Liquid Petroleum by Ultrasonic Flow Meters

The above AGA reports are applicable as measurement standards for light hydrocarbon liquids such as natural gas liquids or liquid petroleum gas.

Much progress has been made in automated operations of all types of pipeline facilities. Valves, measuring and regulating stations, and other facilities are operated remotely or automatically. Various types of instruments measure flow rate, pressure, temperature, and densities or gas compositions using a gas chromatograph. A gas chromatograph is normally installed at gas receipt points, while a densitometer is installed for liquid transmission pipelines. Factory-packaged compressor or pump station assemblies, most of which are highly automated, are now commonplace. Whole stations with thousands of horsepower are operated unattended by coded dispatching systems handled by high-speed communications. The station measurements and statuses are transmitted to a centralized control center, where the entire pipeline system is controlled by means of a Supervisory Control and Data Acquisition (SCADA) system. SCADA systems have become commonplace for automatic supervision and operation of pipelines in today's climate of microprocessor-based remote terminal units (RTUs) and Programmable Logic Controllers (PLCs). They provide the capability to scan a system of several hundred locations and report measurements and statuses in a few seconds. These ultrahigh-speed data-gathering systems can be used to gather operational data in real-time, do calculations, provide process control, and undertake other auxiliary functions.

Nowadays, pipeline system control is automated to provide the capabilities of operating pipeline systems reliably, efficiently and thus economically. Automated system control enables monitoring and control of pipelines, pump stations, metering stations, and other facilities through a SCADA system. Pipeline system control is accomplished by means of a set point mechanism. In other words, the dispatcher sets pressure, flow or temperature at the desired level and the control system responds to reach the set point. It is now a generally accepted practice that stations are automated and operated under remote control from a central SCADA control center. This is possible because a centralized system allows the pipeline operator to monitor the complete pipeline system and control the entire pipeline system in a safe and efficient manner. Only under abnormal conditions or during some maintenance tasks will the station be under local control. Some stations may be completely unmanned whereas others will have maintenance staff on site. The centralized control functions can be extended to other applications such as storage management, energy optimization, electronic flow measurement (EFM) and volume accounting, leak detection, etc.

EFM has become commonplace, meeting the requirements of current regulations. An EFM system is defined as a flow measurement and related system that electronically collects data and performs flow calculations. It is often part of the SCADA system, remotely monitoring, correcting, and controlling flow in real time using the latest calculation methods of the AGA or American Petroleum Institute (API) reports. The standard for EFM is API MPMS Chapter 21 Flow Measurement Using Electronic Metering Systems—Section 1: Electronic Gas Measurement, and Section 2: Electronic Liquid Volume Measurement. These sections describe the minimum specifications for electronic gas and liquid measurement systems used in the measurement and recording of flow parameters of natural gas and petroleum liquids.

To accommodate a rapidly changing business condition or environment, corporate-wide information access has become critical to the efficient operation and management of a pipeline system. Not only is it important to provide accurate information to operation and management staff, but also timely access to this information is of vital importance. Companies that are able to acquire, process, and analyze information more efficiently than their competitors have a distinct market advantage. Information such as real-time measurement, business and decision support are integrated and used for both operational and economic aspects of the business.

The operational nerve center of today's pipelines is the pipeline control center, where the integrated information is made available in a timely manner. It is from this central location that a geographically diverse pipeline is monitored and operated. It is also the center for gathering information in real time that is used for real-time operation, for making business decisions and for operational planning. The integrated information may include normal operation data such as pressure and flow as well as pipeline hydraulic modeling, leak detection, batch tracking, historical data gathering, and other required data. A control center houses most of the equipment used by the operators on a daily basis. The equipment required includes the SCADA system computers and terminals, printers, communication devices, and network equipment used to implement LANs (Local Area Networks) and/or WANs (Wide Area Networks). In addition, pipeline system maps and schematics may be displayed, and operator manuals and other information required for performing dispatching functions can be made available. A backup control center is required in order to operate the pipeline system continuously in the event that the main control center is severely disrupted. This backup is normally in a physically separate location from the main control room.

The pipe and fittings for natural gas and petroleum liquids transmission lines are manufactured in accordance with specifications of the API and are fabricated, installed, and operated in the United States in conformance with Federal Regulations, DOT, Part 192 for natural gas and Part 195 for liquids pipelines. Many factors influence the working pressure of an oil line, and the ANSI ASME B31.4 and B31.8 codes for pressure piping is used as a guide in these matters. Studies of the hydraulic gradient, land profile, and static-head conditions are made in conjunction with the selection of the mainline pipe.

Pipe bends are usually made on site by the construction contractor. However, in case of short distance pipelines, or where a large degree of bend is required, shop-fabricated bends are produced by induction heating.

Repairs to a pipeline segment, according to current regulations of the Department of Transportation (DOT), are to either replace the segment, or use alternative methods that have been shown by reliable testing and analysis to restore pipeline serviceability. The installation of steel repair sleeves over localized pipeline defects has been used extensively to affect permanent repairs. Repair sleeves may be installed by welding or bolting in place. A newer repair method being used by many pipeline operators is a steel or polymer composite reinforcement warp (pipe sleeve, Clock Spring® or equivalent) which is wrapped around the repair area. For this type of repair, welding is not required.

NATURAL GAS

General

Natural gas is a mixture of naturally occurring light hydrocarbons and impurities, but there is no one mixture that can be referred to as natural gas. Natural gas is composed mostly of methane gas with the addition of ethane, propane, butane, and small amounts of heavier hydrocarbons, and impurities such as carbon dioxide, nitrogen, hydrogen sulfide, and water vapor. Natural gas compositions can vary with the gas reservoirs, or different wells from the same reservoir where the natural gas stream is produced.

Until recently, natural gas has been produced from:

- gas reservoirs which contain predominantly methane and light hydrocarbons,
- condensate reservoirs, or
- oil reservoirs in which gas is dissolved in the oil or above the oil in the reservoir as a free gas cap, or
- coalbeds.

However, with the recent development of drilling and production technologies, a significant amount of natural gas is produced from tight shale formation, called shale gas. There are several other unconventional sources of natural gas such as coalbeds.

The naturally occurring natural gas must be processed to remove impurities and heavier hydrocarbons to meet the pipeline quality natural gas specification, which is set forth in 40 CFR Parts 72 "Permit Regulations." The definition of "Pipeline Natural Gas" in Sec. 72.2 states, "Pipeline natural gas contains 0.5 grains or less of total sulfur per 100 standard cubic feet. Additionally, pipeline natural gas must either be composed of at least 70 percent methane by volume or have a gross calorific value between 950 and 1,100 Btu per standard cubic foot." Only the natural gas that satisfies this specification can be transported to consumers through transmission and distribution pipelines.

U.S. Energy Information Administration (EIA) reported that proved reserves of the U.S. total natural gas increased 34.8 trillion cubic feet (Tcf) to 388.8 Tcf in 2014. This increase boosts the national total of proved natural gas reserves in the United States for the second consecutive year (Source: U.S. EIA Form EIA-23L Annual Survey of Domestic Oil and Gas Reserves, http://www.eia.gov/naturalgas/crudeoilreserves/pdf/table_1.pdf). Table 8.8.1 shows the U.S. natural gas production and consumption volumes from 2011 to 2015, during which period the production volume has increased significantly, and the consumption volume has steadily increased. Shale gas produced mainly from Texas, Pennsylvania and Louisiana is attributed to the increase.

As shown in Table 8.8.1, huge volumes of natural gas produced from gas wells are transported to consumers mainly through pipeline networks in the United States. Three types of pipelines are used to gather natural gas from the production wells and to deliver it to the end users:

Table 8.8.1 Yearly Natural Gas Marketed Production, Supply and Consumption (Billion Cubic Feet)

Year	Production (BCF)	Dry Production (BCF) (1)	Supply (BCF) (2)	Consumption (BCF)
2011	24,036	22,902	29,519	24,477
2012	25,283	24,033	30,015	25,538
2013	25,562	24,206	30,966	26,155
2014	27,498	25,890	32,046	26,593
2015	28,753	27,060	32,792	27,306

(1) Natural gas plant liquids (NGPL) are removed as part of gas processing to meet the pipeline natural gas specification.
(2) Supply volumes include imports, withdrawal from storage, and other supply sources. These volumes are the same as the total disposition volume that includes the total consumption, exports, and others.
(SOURCE: https://www.eia.gov/naturalgas/annual/pdf/table_001.pdf).

- gathering pipelines,
- transmission pipelines, and
- local distribution pipelines.

The gathering pipelines consolidate raw natural gas from production wells and transport it to gas processing plants, where the gathered raw gas is processed to pipeline quality gas. The pipeline quality gas is injected into large transmission pipelines, which are generally large and buried underground. They transport natural gas from the receiving points over long distances to storage or local distribution points. There, local distribution pipeline networks receive natural gas at a reduced pressure, and distribute to homes and businesses.

The need to transport natural gas from the producing regions to consuming areas has generated a burgeoning network of transmission pipelines throughout the continent. The U.S. natural gas pipeline network is highly integrated to transport natural gas almost anywhere in the lower 48 states and between the United States and Canada. The U.S. pipeline network comprises 305,000 miles of interstate and intrastate transmission pipelines with more than 1,400 compressor stations to boost and maintain the pipeline pressure. Throughout the United States, the pipeline network also includes:

- more than 11,000 delivery points, 5,000 receipt points, 1,400 interconnection points,
- 24 hubs or market centers that provide additional interconnections,
- 400 underground natural gas storage facilities,
- 49 locations where natural gas can be imported/exported via pipelines, and
- 8 LNG (liquefied natural gas) import facilities and 100 LNG peaking facilities.

The U.S. EIA website provides detailed information on the U.S. natural gas pipelines (Source: www.eia.gov/pub/oil_gas/natural_gas/analysis_publications/ngpipeline/index.html).

Gas Pipe Flow

A mathematical model for pipeline systems is often used to determine the relationship between flow, density, pressure, and temperature with respect to the pipeline length and time. The mathematical models are based on equations derived from the fundamental principles of fluid flow and thermodynamics. Flow through a pipeline is described by the momentum, mass and energy conservation equations. These three equations can be solved, by use of numerical methods and the aid of computers, for pressure, temperature, flow rate and density together with the equations of state appropriate to the fluids, either natural gas or liquids, in the pipeline. The solutions for these variables are expressed in terms of location within the pipeline and time.

The solutions obtained from the fundamental principles are suitable to analyze operational changes when the flow rate varies with time and position (transient or unsteady-state flow). Gas flowing in a pipeline undergoes a number of transient, or unsteady-state, flow conditions caused by such things as prime movers starting and stopping, valves opening and closing, and flow rate changes either into or out of the pipeline. Simulation software allows an engineer to model simple as well as complex pipeline situations. These programs can be used to perform capacity studies, to evaluate the effects of the transients mentioned above, and to test proposed system expansions such as loops or additional compressor stations.

Various forms of steady state gas pipeline flow equations were developed by applying momentum equation and gas laws, and by making some assumptions and simplification. In most gas pipeline design, isothermal steady state flow equations are satisfactory to predict the hydraulic behaviors of long transmission pipelines, because the gas temperature reaches constant values in long transmission lines. These steady state flow equations express the pipeline flow rate in terms of gas properties including compressibility factor and gravity, pipe size and length, pressure difference along the pipe length, etc. One of these steady state flow equation is the general flow equation given in Monograph 9 (1956) developed by U.S. Bureau of Mines.

The general flow equation for constant-volume (steady-state), isothermal, horizontal flow of natural gas is given as

$$Q = 38.7744\frac{T_b}{P_b}\left[\frac{(P_1^2 - P_2^2)D^5}{fZGTL}\right]^{0.5} \qquad (8.8.1)$$

where Q = rate of flow, ft^3/day; T_b = temperature base, °R; P_b = pressure base, psia; P_1 = inlet pressure, psia; P_2 = outlet pressure, psia; D = internal pipe diameter, in; f = friction factor, dimensionless; Z = average gas compressibility, dimensionless; G = gas gravity (for air, G = 1.0); T = average flowing temperature, °R; and L = length of pipe, mi.

Monograph 9 also gives a good discussion of friction factors and elevation corrections. An AGA paper, Steady Flow in Gas Pipelines (1965), gives a good general review of constant-volume gas flow equations, both with and without elevation corrections. Equations and test data are included for both the partially turbulent (Reynolds number–dependent) flow regime and the fully turbulent flow regime using effective internal pipe roughness. In this reference, the drag factor is introduced to account for pressure losses for fittings such as bends and valves and is applied in the partially turbulent flow regime. However, the friction factor can be calculated with the Colebrook-White equation using a computer. This equation is expressed in terms of Reynolds number and pipe relative roughness, and can be applied to both gas and liquid pipe flow for turbulent flow regimes.

Other flow equations may be derived from the general equation by substitution of an empirical expression for the friction factor. The Panhandle A equation has the form

$$Q = 435.87\left(\frac{T_b}{P_b}\right)^{1.0788}\left[\frac{(P_1^2 - P_2^2)}{ZG^{0.8539}TL}\right]^{0.5394} D^{2.6182}E \qquad (8.8.2)$$

where E = pipeline flow efficiency, a dimensionless decimal fraction (design values of 0.88 to 0.96 are common). In the Panhandle A equation, the compressibility is included in the pipeline efficiency, and the other symbols are as previously defined. The Panhandle A flow equation is an empirical relationship, and requires neither friction factor nor effective pipe roughness. This equation is suitable for small to medium diameter pipelines (14<NPS<24) with moderate flow rate, operating under medium to high pressures.

Panhandle B equation is another empirical flow equation, which has the form

$$Q = 737\left(\frac{T_b}{P_b}\right)^{1.02}\left[\frac{(P_1^2 - P_2^2)}{ZG^{0.961}TL}\right]^{0.51} D^{2.53}E \qquad (8.8.3)$$

This flow equation requires neither friction factor, and is suitable for large diameter pipes (larger than or equal to NPS 24), high flow and high pressure systems. The pipeline efficiency, E (0.92-0.96), is multiplied to this equation to match the measured value. The degree of accuracy depends on how precisely the efficiency is measured.

Weymouth equation is another empirical flow equation that has been used in the natural gas pipeline industry. It has the following form:

$$Q = 433.5\frac{T_b}{P_b}\left[\frac{(P_1^2 - P_2^2)}{ZGTL}\right]^{0.5} D^{2.667}E \qquad (8.8.4)$$

These three flow equations calculate the flow rate for a given pressure drop without an explicit friction factor. As a result, their application ranges can be limited. The general flow equation is valid and tends to be accurate for any flow rate, pipe size, and pressure range. However, it is suggested that more than one equation be run and the results be compared for validation.

Compression

Compression is required to keep the natural gas flowing, and compressors are used to provide sufficient pressure to allow a certain volume of natural gas to move through a pipeline. Due to pressure losses within the pipeline over long distances as well as elevation changes, the pressure within the pipeline will drop across distance and thus compressor stations are needed to boost the pressure. Compressor stations are usually placed every 40 to 100 mile intervals along gas pipelines. The locations and pressures of compressor stations are determined by maximum allowable pipe pressures together with environmental and geotechnical factors.

Two types of compressor (reciprocating and centrifugal) are used in natural gas transmission pipelines. Lately, more centrifugal compressors are installed and operated due to lower capital and maintenance costs than reciprocating compressors. Major components of a typical compressor station include:

- compressors to boost the gas pressure,
- prime movers to drive the compressors,

- a flow meter to measure the flow of gas going through the compressor station, and
- fuel lines to supply fuel to the drivers.

In addition,

- suction and discharge headers with piping and valves,
- scrubbers to clean the natural gas by removing liquids,
- air coolers for cooling the gas to protect internal coatings of pipes and maintain the pipeline transmission efficiency, and
- blowdown and station relief valves to vent gas to protect the station from overpressure.

Centrifugal compressor stations are equipped with a recycling line to protect the compressors from surge.

The theoretical horsepower requirements for a station compressing natural gas may be calculated by polytropic or adiabatic compression formulas. The change of state that takes place in almost all reciprocating compressors is close to polytropic. Compression by centrifugal compressors is even closer to polytropic than that by reciprocating compressors due to heat of compression taken away by the jacket cooling and by a small amount of heat added by piston-ring friction.

For general design, the horsepower requirements for a station compressing natural gas with reciprocating compressors can be calculated as follows:

$$\text{Reciprocating station horsepower} = \frac{h_p Z_s T_s}{520} \tag{8.8.5}$$

The value of h_p can be taken from Fig. 8.8.1, typical manufacturer's curve; T_s is suction temperature, °R; and Z_s can be taken from Fig. 8.8.2 or calculated with the procedure as described in AGA Report 8, Compressibility Factor of Natural Gas and Related Hydrocarbon Gases (1994).

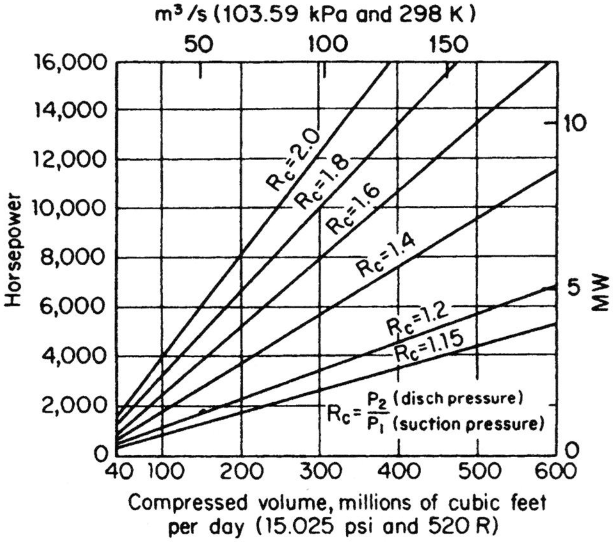

Figure 8.8.1 Reciprocating-compressor horsepower (MW) graph. $R_c = P_2/P_1$ (discharge pressure, psia/suction pressure, psia).

Similarly, the horsepower requirements for a station compressing natural gas with centrifugal compressors can be calculated as follows:

$$\text{Centrifugal station horsepower} = \frac{h_p Z_s}{E_c}\frac{T_s}{520} \tag{8.8.6}$$

The value for h_p can be taken from Fig. 8.8.3, typical centrifugal horsepower curve; E_c is the centrifugal-compressor shaft efficiency. (Design shaft efficiencies of 80 to 85 percent are common.)

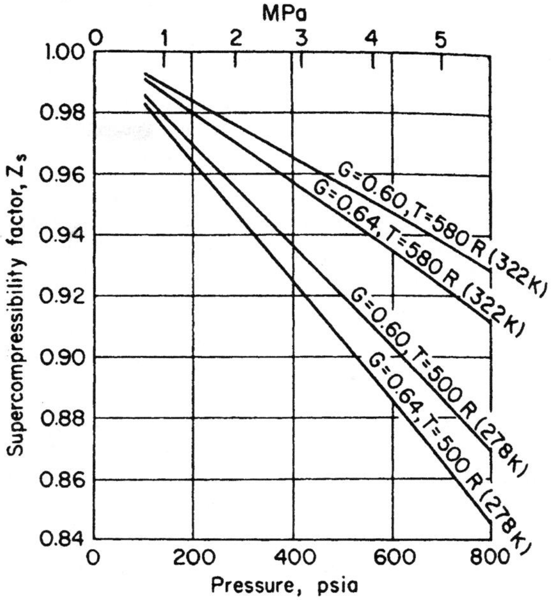

Figure 8.8.2 Supercompressibility factor Z_s for natural gas. (Natural gas, 1 percent N_2.)

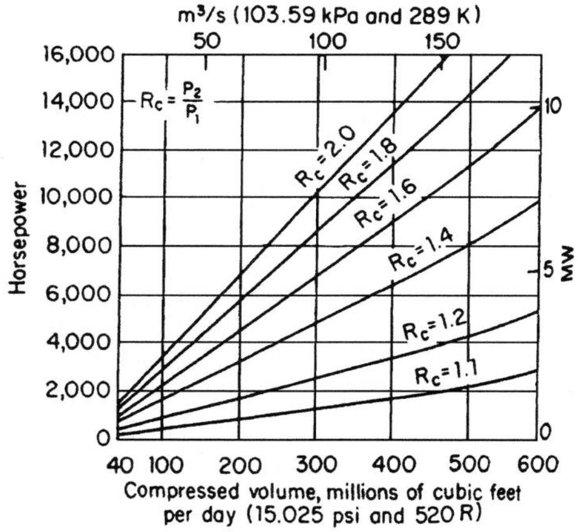

Figure 8.8.3 Centrifugal-compressor horsepower (MW) graph. $R_c c = P_2/P_1$ (discharge pressure, psia/suction pressure, psia).

Compression ratios across a single machine normally do not exceed 3.0 because of outlet temperature limitations. Practically, an acceptable compression ratio for the natural gas pipeline application ranges from 1.2 to 1.6, with an optimum ratio of 1.4. For greater ratios, compressors can be arranged in series operation with an interstage gas cooler placed between the individual units. Multiple units can also be placed in parallel operation in order to increase flexibility and provide for routine maintenance.

Pipeline System Design

The design of a pipeline system is concerned with the determination of an optimum pipe size which allows the pipeline system to operate most economically for the given flow profile over its life span. Complex engineering and economic studies are necessary to decide on the optimum pipe size that results in its maximum yield strength and allowable working pressure, compression requirements for the initiating and intermediate compressor stations, and locations of compressor stations and other facilities. The design must satisfy all the engineering and legal requirements, as a minimum, complying with the applicable codes and regulations.

The required data for the design includes the gas supply and demand locations and flow profiles, design conditions such as natural gas properties and their variations, route and topography with ground and ambient conditions, and pipe properties, pressure and temperature limitations, and pipe material and construction cost. In addition, information on environmental impact, economics, operating plan and equipment, and pipe protection is also required. If a unique condition, such as extreme weather in the Arctic, is present, additional information on equipment and facilities must be included in the design.

The pipe and fittings for gas transmission lines are manufactured in accordance with the specifications of the American Petroleum Institute (API), American Society of Mechanical Engineers (ASME), American National Standards Institute (ANSI), American Society for Testing and Materials (ASTM), and Manufacturers Standardization Society (MSS). Interstate pipelines are installed and operated in conformance with regulations of the U.S. Department of Transportation (DOT). The ANSI ASME B31.8 Code and the Gas Measurement Committee Report 3 (Orifice), Report 7 (Turbine Meter), Report 9 (Ultrasonic Meter), and Report 11 (Coriolis Meter) of the AGA provide supplemental design guidelines. Pipeline design pressure is dictated by the DOT in 49 Code of Federal Regulations (CFR) Part 192 for natural gas pipeline and Part 195 for liquid pipeline. The equation given below with the safety factors relates the internal design pressure that a pipe can safely withstand to the strength of its material.

$$P = \frac{2St}{D} FET \tag{8.8.7}$$

where P = design pressure, psig; S = specified minimum yield strength, psi; t = nominal wall thickness, in; D = nominal outside diameter, in; F = design factor based on population density in the area through which the line passes, dimensionless; E = longitudinal joint factor, dimensionless; and T = temperature derating factor (applies above 250°F only), dimensionless. Note that the temperature derating factor is not included in the internal design pressure equation for liquid pipelines (Refer to section 195.106 of 49 CFR Part 195).

The selection of the diameter, steel strength, and wall thickness of a line and the determination of the optimum station spacing and sizing are based on transportation economics. Figures 8.8.4 through 8.8.6

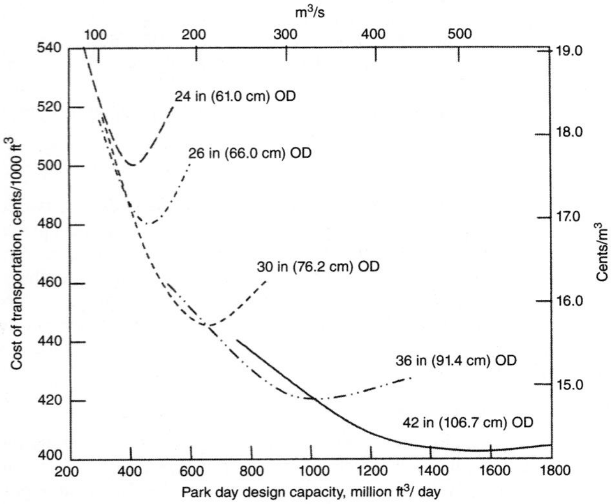

Figure 8.8.4 Schematic relationship of cost of transportation versus design capacity for a given length of line with various pipeline diameters at equal design pressures.

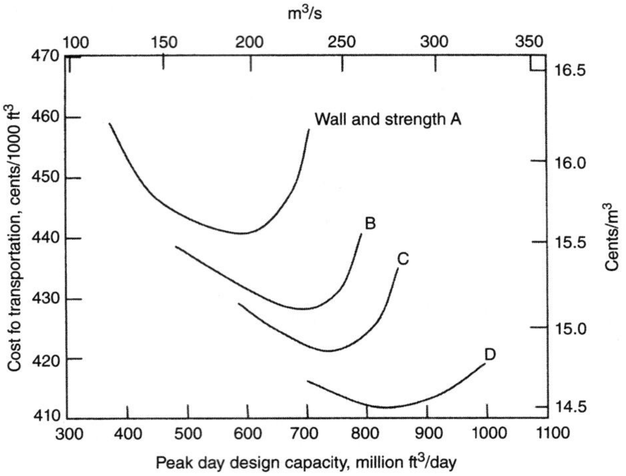

Figure 8.8.5 Schematic relationship of cost of transportation versus pipeline steel strength and wall thickness.

A = 36-in dia., 0.385-in wall thickness, 52,000-psi min. yield (800-psi operating pressure).

B = 36-in dia., 0.385-in wall thickness, 60,000-psi min. yield (924-psi operating pressure).

C = 36-in dia., 0.404-in wall thickness, 65,000-psi min. yield (1,050-psi operating pressure).

D = 36-in dia., 0.438-in wall thickness, 70,000-psi min. yield (1,220-psi operating pressure).

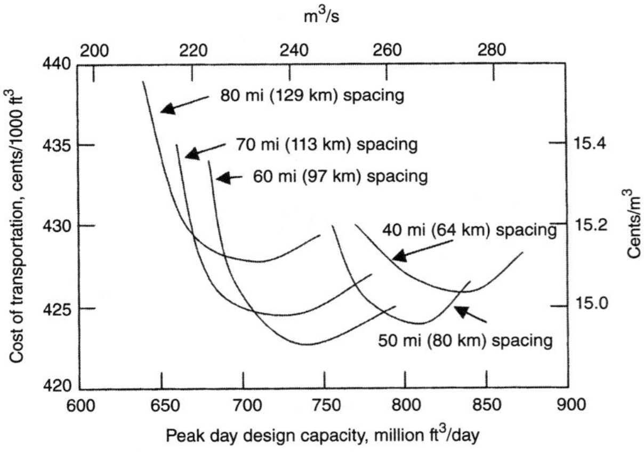

Figure 8.8.6 Schematic relationship of cost of transportation versus design capacity for a given pipeline diameter and length, with several alternatives of compressor station spacing.

schematically show methods of analysis commonly used to relate design factors to the cost of transportation.

As an example, a perspective of various elements of the cost of transportation for a long-distance, large-diameter, fully developed pipeline system is presented in Fig. 8.8.7. The cost of installed pipelines varies greatly with the location, the type of terrain, the design pressure, the total length to be constructed, and many other factors. In general, figures from \$32,000 to \$90,000 per in (per 2.54 cm) of OD represent the range of approximate costs per mile (per 1.61 km) of installed pipeline. Offshore pipeline installed costs vary with such items as length, water depth, bottom conditions, pipeline crossings, depth of bury, and many other factors. These costs range from \$53,000 to \$76,000 per inch (per

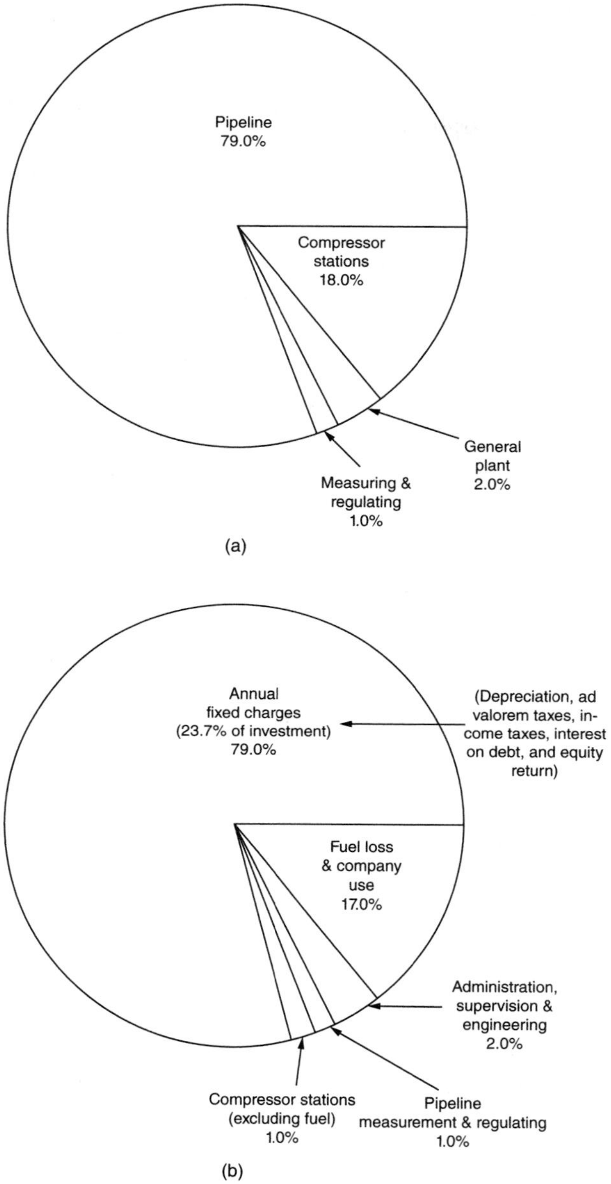

Figure 8.8.7 Analysis of (*a*) investment and (*b*) cost of transportation.

2.54 cm) of OD per mile (per 1.61 km) for a typical offshore pipeline. Compressor-station installed costs vary widely with the amount and type of horsepower to be installed and with the location, the type of construction, and weather conditions. The installed costs of gas turbine stations range from \$800 to \$3,300 per installed hp (\$1,070 to \$4,425 per kW); gas engine stations, from \$1,700 to \$7,800 per hp (\$2,280 to \$10,455 per kW); and electric motor stations, from \$1,300 to \$5,000 per hp (\$1,740 to \$6,700 per kW).

After a pipeline system is constructed and fully utilized, expansion of delivery capacity can be accomplished by adding pipeline loops (connecting segments of line parallel to the original line), by adding compressor stations between the existing stations, or by a combination of these two methods.

Pipeline System Operation

Implementation of FERC Order 636, issued in 1992, has had a significant impact on natural gas transmission pipelines. Order 636 mandated interstate natural gas pipeline companies to transform themselves from natural gas marketers to strictly transporters of natural gas. Pipeline companies that built and operated gathering systems to support contractual obligations are now restricted from including costs for these activities with the costs of buying, transporting, and selling gas, because of the mandate that gathering be provided as a separate service. Open access marketing has replaced bundled sales services to meet the instantaneous needs of buyers at competitive prices. The ultimate objective of the FERC Order 636 was to provide consumers with access to an adequate and reliable supply of natural gas at a reasonable price. As a means of achieving this objective, the order mandates that transmission pipeline companies open their transmission services to all shippers regardless of the ownership of the gas or its quantity.

Increasing competition brought about as a consequence of open-access marketing has been accompanied by major shifts in pipeline ownership and changes in the way pipeline companies do business. In addition to a large number of mergers and acquisitions, many pipeline companies created affiliated marketing arms, or subsidiaries, and transferred merchant functions to these new divisions to comply with FERC Order 636. Many of these marketing entities engage in marketing other types of energy as well, and some are much larger now than the pipeline divisions from which they were originally developed. Corporate expansion strategies have helped pipeline companies to maintain market share and reduce overhead costs as they search to find more efficient ways to operate.

There are two types of transportation services: firm and interruptible. Pipeline companies provide guaranteed transportation service with a firm service contract, but can interrupt transportation service with an interruptible service contract. They typically offer interruptible service supported by system line pack and storage facilities in both gathering and market areas to provide the flexibility required to operate efficiently in an open access market. Market centers and hubs have evolved to provide gas shippers the physical and administrative capabilities they require. The two primary functions of these hubs are to provide interconnection with other pipelines and to provide short-term receipt/delivery balancing needs. Many of the hubs also developed Internet-based access to gas trading platforms and capacity release programs that have improved the efficiency of the entire gas transportation and marketing processes.

The swing in market demand and compliance with FERC Order 637 has magnified the need for pipeline companies to provide "parking" and "lending" type services. FERC now requires all natural gas pipeline companies to have an Internet site providing information on its rates (tariffs), operational capabilities, and availability and nomination/scheduling information. To address business and operation issues effectively, the North American Energy Standards Board (NAESR) standards have been implemented. They have helped the natural gas industry to change a paper-driven business process to an internet based one. They provide many benefits for pipeline industry and shippers, including: increased profitability, information transparency, and fast business processing with short response time. North American gas transmission companies and shippers have adopted the NAESR Standards.

The NAESR provides business practice standards for contracts including:

- short-term sale and purchase of natural gas,
- nominations and capacity release,
- invoicing,
- data interchange such as data syntax,
- time synchronization with time-stamp,
- security, and
- other functions for effective operation.

The daily flow starts with nominations, the process whereby a shipper requests transportation or other services of the pipeline company. Each nomination may include information on the services to be performed: gas volume to be transported or stored, locations of receipts and deliveries, lists of customers, etc. Based on the nominations received from all shippers, the pipeline company performs several tasks internally. These include: nomination confirmation, gas scheduling, and the receipt and delivery operations required to fulfill the nomination. After all the nominations including receipt and delivery quantities have been confirmed and scheduled, the pipeline operator physically transports the confirmed volumes from/to the designated points.

The revenue accounting phase includes volume accounting and allocation, volume balancing, billing and invoicing, and account settlement. After nominated volumes are physically received and delivered, the physical volumes are allocated to confirmed nominations based on the contracted allocation method such as proration or ranking. Allocation is a process designed for balancing and billing purposes. Once allocation is completed, the cumulative volume difference (called imbalance) between nominated and allocated volumes and the difference between receipts and deliveries are calculated through a balancing process. Billing and invoicing for the services take place after volume balancing is completed.

One of the primary responsibilities of a gas pipeline company is to monitor and ensure the quality of the gas being transported. Gas quality management includes all aspects of determining gas quality from the measurement device through to the proper application of industry standards. Having gas quality data allows operators to identify potential contamination, which could cause measurement and operating problems. Such data can also be used to calculate corrected volume and heating values accurately at remote terminals and/or SCADA.

The natural gas quality is also measured by one of several means. For small-volume stations, a continuous gas sampler accumulates a small volume for laboratory testing. Medium- and large-volume stations will use a gas chromatograph. Gas composition is expressed in terms of mole percentage or fraction of methane, ethane, propane, N-butane, I-butane, and trace amount of heavier hydrocarbons. Other important characteristics of gas quality include specific gravity, heating value, and hydrogen sulfide content.

LIQUID PETROLEUM AND PRODUCTS

General

Petroleum products are made up of a range of substances, mostly mixtures of hydrocarbons, or molecular compounds of hydrogen and carbon. They range from light gases through liquids to crude oil. Crude oil is a complex mixture of hydrocarbons of varying density and viscosity, and ranges in color from almost clear to amber, green, brown, or black. Crude oil can be sweet or sour, depending on the sulfur content; sweet has less than 0.5 percent by weight, intermediate has greater than 0.5 percent but less than 1.0 percent, and sour has more than 1.0 percent. Crude oil is classified as light crude, medium crude, heavy crude, and bitumen.

Crude oil produced from well heads is composed of natural gas, oil and water. Before being transported, it needs to be processed to remove Basic Sediment & Water (BS&W), H_2S, and other impurities to an acceptable level. Certain specifications and physical properties need to be met before the pipeline companies take the product from the producer. This includes the separation of the different contents of crude oil and natural gas. Crude oil is classified in terms of API gravity or density as discussed in Table 8.8.2.

Table 8.8.2 Crude Oil Classifications

Classification	Gravity (°API)	Density (Kg/m^3)
Light crude	> 31.1°	< 870
Synthetic crude	31.0°-32.0°	865.4-870.8
Intermediate crude	22.3°-31.0°	870.8-920.0
Heavy crude	10.0°-22.2°	920.6-1,000.0
Bitumen	9.9°-6.0°	1,000.7-1,029.1

Petroleum liquids are classified as follows:

- Natural gas liquids (NGL)—mixture mainly of ethane, propane, N-butane, I-butane, etc., separated from the gas state in the form of liquids at gas processing plants.
- Liquefied Petroleum Gas (LPG) —propane, butane, or mixture of these gases, separated from natural gas at processing or oil refining plants.
- Condensate—mixture of mainly pentanes and heavier hydrocarbons recovered as a liquid from field separators or other gathering facilities at crude oil and natural gas wells.
- Distillate—fraction of crude oil; a term generally used for naphtha, kerosene, diesel, and fuel oils.
- Alkene—ethylene, propylene, butylene, etc.
- Refined products—gasoline, jet fuel, diesel.

Table 8.8.3 shows the U.S. average daily crude oil production and supply from 2011 to 2015. The total supply has slightly increased during the same period, while the total domestic production has increased significantly during the five year period. For example, the net growth from 2012 to 2013 was about 1 million barrels a day, and from 2013 to 2014 another 1.3 million barrels. Increasing production of light crude oil in low-permeability or tight resource formations in regions like the Bakken, Permian Basin, and Eagle Ford account for nearly all the net increase in the U.S. crude oil production.

Table 8.8.3 Average Daily Crude Oil Production and Supply (Thousand Barrels)

Year	Production (1,000 Barrels) (1)	Supply (1,000 Barrels) (2)
2011	5,646	19,144
2012	6,487	18,798
2013	7,468	19,222
2014	8,764	19,371
2015	9,415	19,802

(1) Source: https://www.eia.gov/dnav/pet/pet_crd_crpdn_adc_mbblpd_a.htm. Production shows combined output of crude oil and condensate.

(2) Source: https://www.eia.gov/dnav/pet/hist/LeafHandler.ashx?n=PET&s=MTTUPUS1&f=M, Supply volumes include imports, withdrawal from storage, and other supply sources.

The major markets in which petroleum products are consumed are remote from the proven reserves. In 2014 and 2015, about 55 and 52 percent, respectively, of all petroleum supplied to the United States was imported and had to be moved from ports or points of entry to the centers of consumption. API reported that in 2014, the total length of liquid transmission pipelines was 199,243 miles (320,582 km) in the United States, with 66,649 miles (107,238 km) transporting crude oil, 61,681 miles (99,245 km) transporting refined petroleum products (gasoline, diesel, jet fuel, etc.), 65,595 miles (105,542 km) delivering natural gas liquids (propane, ethane, butane, etc.), and 5,318 miles (8,557 km) delivering other liquids such as carbon dioxide and liquefied natural gas (LNG). (Source: http://www.api.org/~/media/Files/Oil-and-Natural-Gas/pipeline/Whats-New/2015-AOPL-API-Pipeline-Usage-Mileage-Report.pdf.)

Consumption of refined petroleum products in the United States has been divided into five Petroleum Administration for Defense Districts, or PADDs, based on historical, demographic, geological, and topographical characteristics, by the Energy Information Administration. The East Coast (PADD I) is the most heavily populated and has the highest petroleum consumption. With virtually no local natural petroleum resources and limited refineries, the Northeast area must cover its deficit in refined product supply through pipeline deliveries from the Gulf Coast and imports of refined products by tanker. The Midwest (PADD II) has a better balance of supply and demand to meet the needs of its less heavily populated area. Logistics hubs as well as refining hubs in the region, such as the Chicago hub, depend on deliveries from pipelines to provide their services.

The Gulf Coast (PADD III) is the nation's primary oil supply region. The design and sophistication of refineries in this region are unrivaled in the world. Production from this region alone is greater than all but two members of the Organization of Petroleum Exporting Countries (OPEC). Low regional demand enables it to provide for the needs of the other regions. The Rocky Mountain States (PADD IV) are thinly populated and major portion of the consumption within the area is low-sulfur diesel fuel for highway consumption. Pipelines serving this region distribute products from refineries in Denver, CO, Billings, MT, and Casper, WY.

The West Coast (PADD V) oil market is separated from the rest of the country geographically as well as politically. For many years, the Rockies have been a challenge to the transport of oil in and out of the region. More recently, California's stringent oil product specifications have exacerbated its isolation from oil supplies. There is high demand in the region, due to the high population in California.

Liquids Pipe Flow

Assuming that the liquid density does not change with pressure, the pressure-flow equation under an isothermal condition can be obtained from the steady-state momentum conservation equation:

$$P_x = P_0 - \rho g(h_x - h_0) - f\rho V^2(X - X_0)/2D \qquad (8.8.8)$$

where P_x is the pressure at the downstream point, X, P_0 is the pressure at the upstream point, X_0, ρ is the liquid density, g is the gravitational constant, h_x and h_0 are the elevation at the downstream point and the upstream point, f is the friction factor, V is the liquid velocity, and D is the inside pipe diameter. The second term on the right hand side of this equation represents the static pressure difference, and the third term is the Darcy-Weisbach equation for the friction pressure drop between two points.

The isothermal flow assumption can be justified for fluid which is transported near ground temperature. It is especially valid for a long transmission pipeline with multiple pump stations, because the temperature approaches close to the ground temperature within the first section and the temperature increases at the subsequent pump stations are in the order of a few degrees. However, large changes in liquid temperature can affect liquid density and/or viscosity, which will subsequently affect pressure drop. Therefore, the following hydraulic problems should be treated as temperature dependent flow:

- Heavy hydrocarbon liquids or waxy crudes whose viscosity changes significantly with temperature.
- Light hydrocarbon liquids whose density changes significantly with temperature.
- Injection temperature is significantly higher than the soil temperature.
- Pipelines with a large pipe size running in a hot ambient temperature condition.

The liquid temperature rises or falls along a pipeline and rises through a pump station. Temperature profile along the pipeline is influenced by external factors such as ground temperature and soil conductivity as well as heat generated by friction. Fluid temperature rises through a pump station mainly because of the inefficiency of the pump and the small temperature drop through station piping. The temperature change along a liquid pipeline occurs mainly due to heat conduction with the surrounding ground and ambient temperatures.

When an operation change takes place, the flow rate and pressure change significantly in liquid pipelines, and consequently the change will have an impact on the liquid pipeline system. The steady-state assumption is not valid for analyzing short-term operation problems, designing control systems, and locating facilities, because these operation scenarios are strongly dependent upon time. Therefore, the transient or surge effects should be included in the design and operation evaluation.

A transient state occurs when the flow is disturbed or changed in the pipeline system; these are often caused by changes in hydraulic facilities such as flow or pressure control equipment. A large transient, often called a pressure surge, is a change in pressure in the pipeline that occurs abruptly during a change from a normal steady state or another transient state flow. All surges travel at acoustic speed through the flowing liquid in the pipeline. When a pump is tripped or a valve is suddenly closed, for example, the suction pressure increases sharply and the pressure wave propagates upstream and downstream in the pipeline. The initial suction pressure increases by up to potential surge pressure, which is given by the following equation:

$$\Delta P = \rho a v \quad \text{or} \quad \Delta h = \frac{av}{g} \qquad (8.8.9)$$

where ΔP = potential surge in pressure, psf, ρ = density of liquid, lb/ft^3, a = acoustic speed of the liquid in the pipe, ft/sec, v = liquid velocity before the closure or pump trip, ft/sec, Δh = surge in head, ft, and g = gravitational constant, 32.2 ft/sec^2. The suction pressure keeps increasing until the upstream flow stops.

Excessively large pressure increases due to operation change can cause damage to pipe and/or other equipment, resulting in potential pipeline system failure. On the other hand, when an upstream flow in a pipe is suddenly stopped, the fluid downstream will attempt to continue flowing, creating a vacuum that may cause the pipe to collapse. Therefore, liquid pipelines are installed with surge relief devices to control surge pressure: surge relief valve, surge tank, or rupture disk. In addition, pump stations are equipped with a proportional-integral-derivative (PID) controller, and valves are closed or opened slowly to minimize the pressure surges.

Pumping

A pump transforms energy to increase pressure of a liquid and is used extensively to transport liquid through a pipeline system. The pressure of a liquid has to be increased either to overcome frictional losses or to raise the liquid from one elevation to a higher elevation. As the flow rate increases, more pumps are required to produce the required pressure along the pipeline. Intermediate pump station locations on a pipeline system are determined by the hydraulic design of the transmission pipeline and the frictional pressure losses that occur with the product proposed for transportation.

Intermediate pump stations are generally simpler than compressor stations, because pump stations do not need components such as scrubbers and air coolers. The main components of pump stations include two or more pump units, station piping connecting various components, station isolation valves, bypass valves, pressure control valve, and check valves. In addition, most intermediate pump stations are installed with a pig receiver and launcher set. Pump stations pumping heavy crudes may be equipped with heaters.

If a station has more than one pump unit, it is operated in one of two modes: parallel operation or series operation. Multiple pumps in a pump station can be arranged in parallel, series, or combination of parallel and series arrangement. Some combination of these arrangements is needed to increase both flow and pressure. The primary purpose of operating units in parallel is to allow a wider range of flow than would be possible with a single unit for systems with widely varying flows. Generally, pumps are arranged to operate in parallel to increase system flow rates by maintaining discharge head through a much greater pipeline throughput. The main reason for operating units in series is to increase the pumping head from what would be possible with a single unit. Pumps operating in series add head and capacity to the system output.

Both centrifugal pumps and positive displacement pumps are used in liquid pipelines, but centrifugal pumps have been used extensively for most liquid pipeline applications. Generally, these units are installed in series or parallel, depending on the application. Positive displacement pumps can be classified as reciprocating or rotary pumps. Reciprocating pumps are installed in parallel; that is, a common suction and a common discharge header are utilized by all pumps. There are instances where rotary pumps, including gear and screw-type pumps, have been employed to fulfill specific requirements. For example, screw pumps are often employed for highly viscous petroleum fluid such as bitumen.

A mainline centrifugal pump can be driven by an electric motor, gas turbine or a diesel engine. Liquid hydrocarbon transmission pipelines are typically driven by electric motors where electrical power is available. This is primarily due to their lower initial capital cost, their compactness, ease of control, and inherent reliability. As well, electric motor drivers have benefits over gas turbine and diesel drivers due to ever more stringent emission level limits. Most electrically driven centrifugal pumps are coupled with fixed speed drivers, but lately more variable speed pumps are used as the price of variable frequency drive (VFD) motors drops. Variable speed pumps control the flow and pressures by varying the speed of the drivers. Variable speed motors are more energy efficient, because pump capacity can be controlled without the disadvantage of pressure loss incurred by the throttling through a discharge control valve. For a pump station that contains both fixed-speed and variable speed motors, the control strategy is to run the fixed speed units at a base load with minimal throttling and use the variable speed units to adjust for the required station set point.

The brake horsepower for pumping oil is

$$hp = \frac{RSh}{3{,}960E} \tag{8.8.10a}$$

or

$$hp = \frac{BP}{2{,}450E} \tag{8.8.10b}$$

where R = flow rate, gal/min; S = specific gravity of the oil; h = pump head, ft; E = pump efficiency, decimal fraction; B = flow rate, bbl/h; and P = pump differential pressure, psi.

Pipeline System Design

Determining the pipe size and station spacing to transport a given oil or oil product at a specified rate of flow and to provide the lowest cost of transportation is complex matter. Usually the basic approach is to prepare:

1) a series of pipeline cost estimates covering a range of pipe diameters and wall thicknesses, and

2) a series of station cost estimates for evaluation of the effect of station spacing.

By applying capital charges to the system cost estimates and estimating the system operating costs, a series of transportation-cost curves similar to those in Fig. 8.8.8 can be drawn. Estimates of $30,000 to $120,000 per in (per 2.54 cm) of OD represent the cost per mile (1.61 km) of installed trunk pipeline. The tabulation of pump-station investment costs for automated electric-motor, centrifugal-type installations presented in Table 8.8.4 can be used for preliminary evaluations.

Careful design and good operating practice must be followed to minimize contamination due to interface mixing in oil-product pipelines.

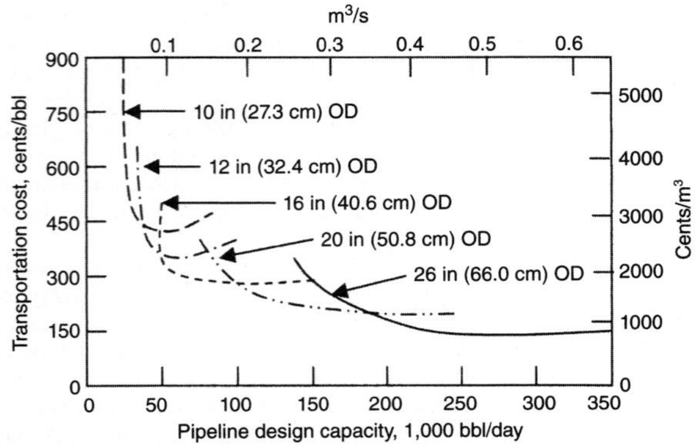

Figure 8.8.8 Schematic relationship of pipeline capacity to cost of transportation.

Table 8.8.4 Pump-Station Investment Costs

Pump Station Installed *hp*, total	Cost per Installed *hp* (1 *hp* = 0.746 kW)
250-500	$800-1,400
500-750	$600-1,050
750-1,000	$500-850
1,000-2,000	$350-600
Above 2,000	$300

When the throughput capacity of an oil line becomes fully utilized, additional capacity can be obtained by installing a parallel pipeline or a partial-loop line. A partial loop is a parallel line that runs for only part of the distance between stations. In the case of product lines, careful attention must be paid to the design of the facilities where the loop line is tied into the original pipeline to prevent commingling of products.

Pipeline System Operation

Petroleum products, including crude oil and refined products, are gathered at central points. The gathered products are then scheduled and dispatched to various destinations based on requirements estimation. The scheduling activity begins with the preparation of a product pumping schedule based on pipeline capacity and shippers' requirements, while the dispatching activity involves organizing various operations of the pipeline facilities to conform to the schedule. In scheduling and dispatching, the pipeline system needs to be operated in a safe and efficient manner.

The control of pipeline pressure is crucial to ensure safe, reliable, and economical operations because all pipelines are subject to minimum and maximum operating limits. Maintaining these limits is essential to preserve the integrity of the pipeline and pumping equipment, and at the same time to operate the system economically. The critical operating limits may include the maximum allowable operating pressure (MAOP), vapor pressures of the petroleum products, and maximum discharge pressure at pump stations.

Broadly, operation of liquid pipeline systems is concerned with pipeline system or facility start-up and shut-down, flow rate changes, product receipt and delivery including batch movements, equipment failure, emergency shutdown, etc. Pipeline operations involve transporting products from the lifting points to the delivery points; lifting products into the pipeline and delivering the received products to the designated delivery points. The receiving and delivery points may be tank farms, refineries, or another pipeline. For transporting products, the pipeline operator monitors product movements and pipeline states to ensure that products are adequately supplied from the lifting points and delivered to the designated points, while controlling flow and pressure at pump and regulator stations for safe and reliable transportation.

Oil and petroleum product pipelines lift their products from and deliver to tank farms. A tank farm refers to a collection of tanks located at a refinery, shipping terminal or pipeline terminal. Crude oil and petroleum products, including light hydrocarbons, are often stored in tanks in various locations such as producing areas, refineries, petrochemical plants, and/or distribution centers. Storage allows for flexible pipeline transportation and efficient transportation management through the existing pipeline system and minimizes supply/delivery disruptions. A tank farm at a refining operation is used to store various products produced by the refinery and to hold them until they are scheduled for injection into a pipeline for transportation. Similarly, tank farms at shipping terminals hold products until a shipping route is scheduled. Tank farm operation covers tank control, volume measurement, and inventory.

Nomination All transportation service begins with transportation requests in the form of nominations. Nowadays, nominations are entered via browsers on the Internet or through the on-line shipper information system, and are communicated directly to pipeline operations. The delivery of nominated volumes is scheduled and monitored using a system which performs the following functions: consolidate system-wide nominations information, calculate capacity proration, divide monthly nominations into a batch schedule, generate an operational movement plan, calculate line fill with respect to any time in the schedule "window," track confirmed nomination status, and provide reports for operations and accounting functions. The information such as confirmed nominations and batch schedule/tracking is made available to the shippers through the on-line shipper information system.

Batch Operation Petroleum liquid pipelines typically carry multiple products or grades of product in the same pipeline. When multiple products are transported in a batch mode, nominated products are pumped within the pipeline capacity during a fixed period. This period is called a batching cycle. This is accomplished by introducing the different products sequentially at the origin of the pipeline so that a given "batch" of the product occupies a section of the pipeline during the transport. A 25,000-barrel (4,000-m^3) product batch will occupy nearly 50 miles (80 km) of 10-in-diameter (273-mm) pipeline. Product batches can be transported with a batch separation pig or without any special means or devices to separate them. As multiple products move along the pipeline without a separation pig, interfacial mixing takes place at the interface boundaries between two adjacent batches. The size of interface mix depends on the flowing condition, pipeline system operation, pipe size and traveled distance, and product properties of the two interfacing batches. The batching sequence is arranged in such a way that it is likely to result in the least batch interface mixing. Turbulent flow minimizes interface mixing, whereas flows that start and stop will generate greater interface contamination. In most pipelines, interface contamination is relatively minor. It is typical for 150 barrels (24 m^3) of mixed material to be generated in a 10-in (273-mm) pipeline over a shipment of 100 miles (161 km). The interface mix may be cut into one or the other product, or divided between the two adjacent products at the mid-gravity point, depending on the product quality specifications. The mixed material may not be lost or wasted, as it reflects the two materials from which it is generated. The interface mix between batches of different grades of the same material may be blended back into the lower-grade material. The interface mix between batches of different products which cannot be blended is segregated and reprocessed in a full-scale refinery or special facility built for that purpose. Figure 8.8.9 shows a typical batching sequence with interface mixes between batches.

Batches can be injected or delivered anywhere along the pipeline as long as injection or delivery facilities are available. If a batch is injected at an intermediate location along the main pipeline, the injection can be full stream or side stream injection. For full stream injection, the upstream section of the injection location shuts down, producing zero upstream flow and the downstream flow rate is the same as the full stream injection rate, while the downstream flow for a side stream injection is the sum of the upstream flow and side stream injection rate. Similarly, either full stream or strip (side stream) delivery can be made at delivery points along the pipeline. For full stream delivery, the upstream flow of the delivery location is the same as the delivery rate and the downstream flow is zero. Similarly, the upstream flow for the strip delivery is the sum of the downstream flow and the delivery rate.

Terminals receiving products typically distinguish between batches by monitoring the specific gravity of the product. The pipeline operator "swings" batches from one pipeline or storage facility to another based on this signal from the interface location. This method of separating batches of low-sulfur diesel fuel may become inadequate; however, as ultra-low-sulfur diesel (ULSD) fuel specifications are mandated. The ULSD may be contaminated by the sulfur in the adjacent product by the time the shift in specific gravity is discernable. This is one of the challenges facing products pipelines as product quality specifications become more critical.

Pipelines deliver batches of product based on contract deliveries. These delivery contracts fall into two modes of operation: batched service and fungible service. Under batched service, a specific volume of refined petroleum products tracked throughout the transportation process. The product delivered is the same material designated for the contract at the origin of the shipment. Under fungible service, the carrier does not deliver the same material that is designated for the contract at the origin of the shipment, but instead delivers material that has the same product specifications. Operating and economic efficiency are the bases for which mode of operation is used. In general, the fungible mode is most efficient since it tends to minimize the generation of interface contamination and permits split-stream deliveries to multiple customers from the same batch. Trunk lines in fungible service provide greater flexibility than delivering pipelines operating in batch mode.

Batch tracking monitors the volume in each batch, its origin, its destination, its current location, and its estimated time of arrival to delivery locations. The launch information includes batch launch time and batch ID. The batch ID identifies the product of the lifted batch, batch number, size or quantity of the batch, and shipper or owner of the batch. The launch is detected based on a change in density and/or valve status or on a batch schedule. Batch volumes are updated based on injection and delivery volumes obtained from metering locations. Batch tracking and interface detection is needed to deliver to correct locations at accurate times. Batch interfaces can be detected by a densitometer if batches have different densities or by a dye detector if the batch densities are similar. If batches are separated by a batch separation pig, the pig has to be inserted at the time of batch launch and removed when it arrives at the delivery point. Upon completion of a batch delivery, the batch is removed in the batch tracking list and an over/short volume is calculated, reflecting the difference between metered injections and deliveries.

Petroleum Products Quality All petroleum products to be transported through pipelines should satisfy product quality requirements. The product quality can be specified in terms of basic sediment and water (BS&W) contents, product density or gravity, viscosity, and sulfur contents. In addition, vapor pressure and vapor content, size of batch interface mix, and/or injection temperature are specified as a quality measure. For batch pipelines, sequencing and interface cutting procedures are used to maintain the product quality.

Drag Reducing Agent A drag reducing agent (DRA) is used primarily to increase the capacity of petroleum liquid pipelines. It is reported that the Trans-Alaska Pipe Line System (TAPS) started to use a commercially available DRA in 1979, after its effectiveness in increasing throughput was proven. Nowadays, many North American pipeline companies use DRA to increase throughput well beyond their

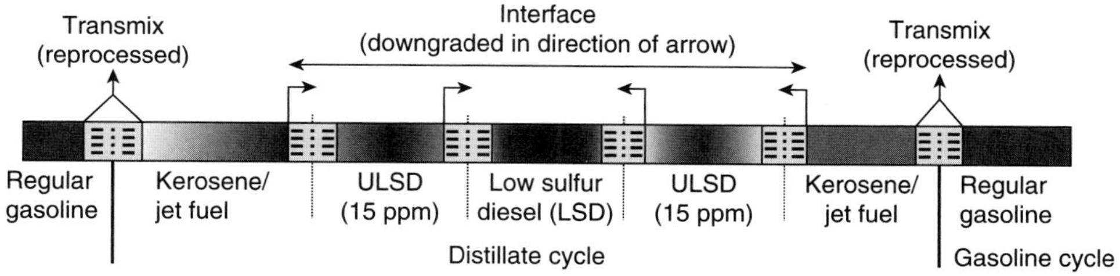

Figure 8.8.9 Batch sequencing with interfaces. (It may be noted that low-sulfur and high-sulfur diesels (LSD/HSD) are no longer produced in North American refineries.)

design capacities. Most DRA products consist of a long chain and high molecular weight polymer which is injected into the petroleum liquid in small amount (in the range of 10 to 30 ppm of the petroleum liquid) for reducing the frictional pressure drop. Since a DRA can significantly reduce pressure drop, it can save power costs and also relieve the pipeline company of capacity restriction.

The key benefits of using a DRA are operating cost saving and flexibility in dealing with throughput with relatively small investment in facilities. A traditional approach to increase the throughput is to install new intermediate pump stations along the entire pipeline. The costs of constructing and operating pump stations are high. However, significant increase in pipeline throughput can be handled with relatively small investment in the construction of DRA facilities, while the DRA facility construction time is shorter and the cost is significantly lower compared to the time and cost for the construction of pump stations.

A DRA can be degraded due to shearing while flowing through the pipeline and facilities such as pump and valve stations. The degradation through pipes is roughly proportional to travelling distance. It is indicated that high wax or water contents in the fluid reduces the effectiveness of the DRA, and at present a DRA is not allowed to be injected into jet fuel because of potential safety issues.

Leak Detection Pipeline leak detection is only one aspect of a pipeline leak management program; it encompasses leak prevention, detection, and mitigation procedures. To minimize the consequences of a leak, pipeline companies require a comprehensive leak management program. A leak detection system by itself does not improve on a pipeline's integrity nor reduce potential failures of a pipeline system. However, such a program will not only help prevent and monitor the degradation of a pipeline that may eventually lead to failure, but will also minimize the consequences of pipeline leaks if they occur.

Broadly, there are three different types of leak detection methods: Inspection Methods, Sensing Devices, and Computational Pipeline Monitoring Methods. Inspection techniques include visual inspection, magnetic flux technique, ultrasonic technique, hydrostatic test, and others. In general, inspection methods provide very accurate, sensitive and reliable results. Particularly, ultrasonic and magnetic inspection techniques can detect both actual and incipient leaks by determining the pipe wall thickness. However, internal inspection methods are very expensive requiring specialized tools and expertise, and a pipeline cannot be inspected continuously. Due to the nature of intermittent operation, only leaks that occurred prior to the inspection will be detected and any occurring after will remain undetected until the next inspection.

Sensing Devices can be used to continuously sense particular characteristics of leaks such as sudden pressure drop, noise, electrical impedance, or other signals caused by a leak or interference around a pipe. Some sensing devices can detect not only leaks but also third party interference around the pipeline system. However, some of these techniques have not been fully proven reliable yet, while some are impractical and expensive for existing pipelines as the pipeline has to be retrofitted with the sensing devices.

No single leak detection system could detect leaks quickly and accurately or provide reliable leak detection continuously and cost effectively. Therefore, more systematic approaches to leak detection are required, such as a combination of line patrol, sensing devices and/or SCADA-based systems with automated leak detection capability. Since SCADA systems have become an integral part of pipeline operations, a particular consideration has to be given to leak detection methods that can be easily implemented on the SCADA system. API RP 1130 addresses various Computational Pipeline Monitoring (CPM) methodologies, integrated with a host SCADA system, and API RP 1175 the pipeline leak management issues.

The CPM methods are based on mathematical or statistical computations of certain quantities using commonly available measured values such as flows and pressures obtained through the host SCADA system. A CPM system receives an updated set of SCADA data continuously in real-time, and sends a set of the modeled results back to SCADA through the SCADA interface software. In general, the cost is relatively reasonable but the sensitivity is lower than other methods. The key advantage of the CPM methods is that they seldom need additional instruments and equipment to those that already exist for normal pipeline operations. As a result, the implementation and operating costs are typically lower than the costs for inspection and sensor methods.

The CPM and sensing methods of leak detection take advantage of real-time capability and the effectiveness of the SCADA system as a monitoring and controlling tool. As the historical data indicates, the current CPM technologies are far from satisfactory in their performance. They need further improvement in their reliability and leak detection sensitivity. In 2015, the API issued Recommended Practice 1175, Pipeline Leak Detection Management Program to assist the pipeline companies in managing their leak detection system effectively in terms of leak detection culture and strategy, control center procedures, performance target and evaluation, alarm management, training, and other subjects. An effective leak detection system and management should help pipeline companies operate their pipeline systems safely by reducing the consequences related to a leak.

Section 9

Fans, Pumps, and Compressors

BY

TERRY L. HENSHAW *Consulting Engineer, Magnolia, TX*
HEIMIR FANNAR *Chief Design Engineer, Ariel Corp.*
PIOTR CYKLIS *Professor, Mechanical Engineering, Cracow University of Technology*
MARSBED HABLANIAN *Retired Manager of Engineering and R&D, Varian Vacuum Technologies*
ROBERT JORGENSEN *Engineering Consultant*

9.1 DISPLACEMENT PUMPS
by Terry L. Henshaw

9.2 CENTRIFUGAL PUMPS
by Terry L. Henshaw

9.3 COMPRESSORS
by Heimir Fannar and Piotr Cyklis

9.4 HIGH-VACUUM PUMPS
by Marsbed Hablanian

9.5 FANS
by Robert Jorgensen

9.1 DISPLACEMENT PUMPS

by Terry L. Henshaw

REFERENCES: Buse, "Positive-Displacement Pumps," Marks' Handbook, 9th ed., McGraw-Hill. Henshaw, "Reciprocating Pumps," Van Nostrand Reinhold. "Hydraulic Institute Standards," 14 ed., Hydraulic Institute. Chesney, Water Injection—Pump Development, paper 68-Pet-11, ASME. Lees, Designing for High Pressures, paper 93-DE-9, ASME. "Positive Displacement Pumps—Reciprocating," API Standard 674, American Petroleum Institute. Brennan, "Multiple Screw Pumps," Pumps & Systems Magazine. Henshaw, "Power Pump Valve Dynamics," 25th International Pump Users Symposium.

NOTE: Much of the text and some of the illustrations are from the author's book: Henshaw, "Reciprocating Pumps," Van Nostrand Reinhold, 1987.

GENERAL

A **displacement pump** (also called **positive-displacement,** or just **p-d**) is a pump which imparts energy to the pumpage (the material pumped) by trapping a fixed volume at suction (inlet) conditions, compressing it to discharge pressure, then pushing it into the discharge (outlet) line. A displacement pump does not rely on velocity to achieve pumping action, as does a centrifugal pump or ejector.

Displacement pumps fall into two major classes: reciprocating and rotary, as illustrated in Fig. 9.1.1.

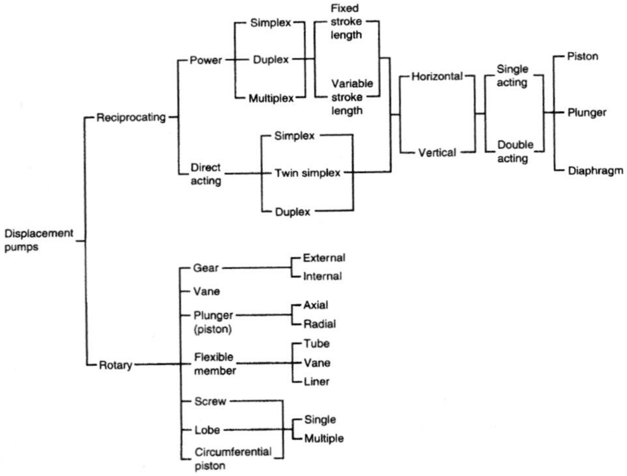

Figure 9.1.1 Classification diagram of displacement pumps.

Uses and Applications

Displacement pumps serve primarily in applications of low capacity and high pressure, those mostly beyond the capabilities of centrifugal pumps. Some of these services could be performed by centrifugals, but not without an increase in power requirements and/or maintenance. Because displacement pumps achieve high pressures with low pumpage velocities, they are well-suited for abrasive-slurry and high-viscosity services. A reciprocating pump must have special fittings to be suitable for most slurries.

Net Positive Suction Head

Net positive suction head (NPSH), also called **net positive inlet pressure (NPIP)** and **net inlet pressure (NIP),** is the difference between suction pressure and vapor pressure, at the pump suction nozzle, when the pump is running. In a reciprocating pump, NPSH is required to push the suction valve from its seat and to overcome the friction losses and acceleration head within the pump liquid end. In a rotary pump, NPSH is required to push the pumpage into the cavities created by the pumping elements.

If sufficient NPSH is not provided by the system, the pumpage will begin to flash (boil) as it flows into the pump. The vapor will cause a deterioration of pump performance. As the vapor flows into regions of higher pressure in the pump, it condenses back to a liquid. This collapse of the vapor bubbles occurs with significant impact, impinging upon metal surfaces with enough energy to break out small pieces of the metal. This is called **cavitation** damage. The shock created by the bubble collapse may be severe enough to crack a fluid cylinder or break a crankshaft.

All pumps require the system to provide some NPSH. The NPSH provided by the system is called NPSHA (the A is for available). The NPSH required by the pump is called NPSHR. For displacement pumps, NPSHR is normally expressed in pressure units (lb/in^2 or kPa).

Since water usually contains dissolved air, the vapor pressure of the solution is higher than for deaerated water, but this is often overlooked when NPSH calculations are performed. The Hydraulic Institute recommends an NPSH margin of 3 lb/in^2 (20 kPa) for power pumps in systems where the pumpage has been exposed to a gas other than the liquid's own vapor. A liquid (such as propane) at its bubble point in the suction vessel requires no such margin.

RECIPROCATING PUMPS

A **reciprocating pump** is a displacement pump which reciprocates the pumping element (piston, plunger, or diaphragm). The capacity of a reciprocating pump is proportional to its speed, and is relatively independent of discharge pressure.

A **power pump** is one that reciprocates the pumping element with a crankshaft or camshaft (see Figs. 9.1.2 and 9.1.3). It requires a driver which has a rotating shaft, such as a motor, engine, or turbine. A constant-speed power pump will deliver essentially the same capacity at any pressure within the capability of the driver and the strength of the pump.

A **direct-acting** pump is a reciprocating pump driven by a fluid which has a differential pressure (see Fig. 9.1.4). The motive fluid

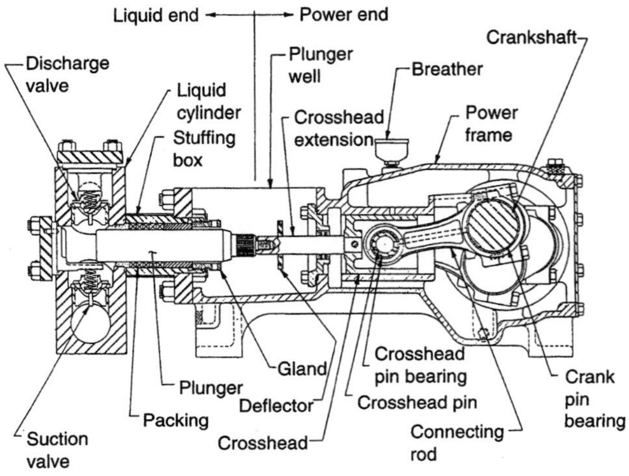

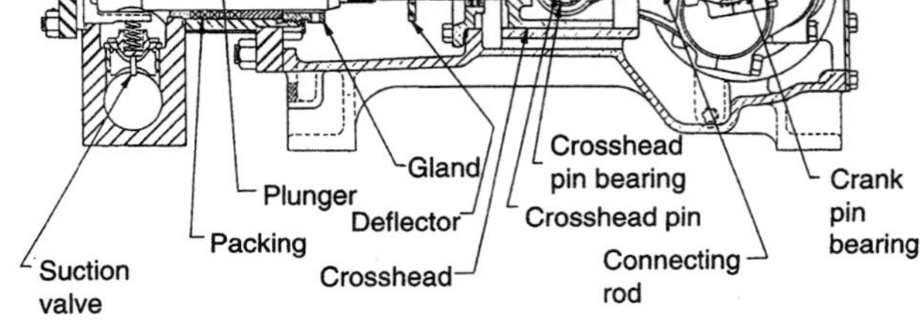

Figure 9.1.2 Horizontal power pump.

pushes on a piston (or diaphragm) which pushes the pumping element through a rod (or directly on the pumpage).

Reciprocating pumps are classified by the following features:

Drive end (power or direct-acting)

Orientation of centerline of the pumping element (horizontal or vertical)

Number of discharge strokes per cycle of each drive rod (single-acting or double-acting)

Pumping element (piston, plunger, or diaphragm)

Number of drive rods (simplex, duplex, triplex, quintuplex, etc.)

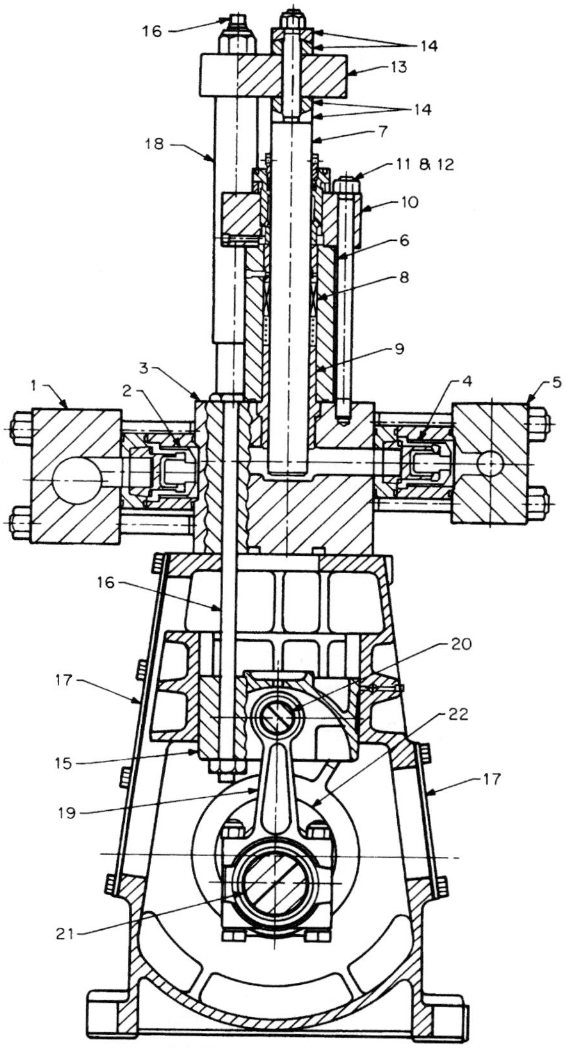

Figure 9.1.3 Vertical power pump. *(Ingersoll-Rand Company.)*

Uses and Applications for Reciprocating Pumps

The justification for selecting a reciprocating pump instead of a centrifugal or rotary is cost, including costs of power and maintenance.

Reciprocating pumps are best suited for high-pressure/low-capacity services. Such services include high-pressure water-jet cleaning and cutting; glycol injection and charge, as well as amine and lean oil charge, in gas processing; ammonia and carbamate charging for fertilizer production; nuclear-reactor charging and standby control; oil-field saltwater injection and disposal; oil well blowout preventers; pipelines (slurries and crude oil, and injection of ammonia and light hydrocarbons); steel and aluminum mill hydraulic systems; knockout drums in process plants; hydrostatic testing; process slurries; metering; food and

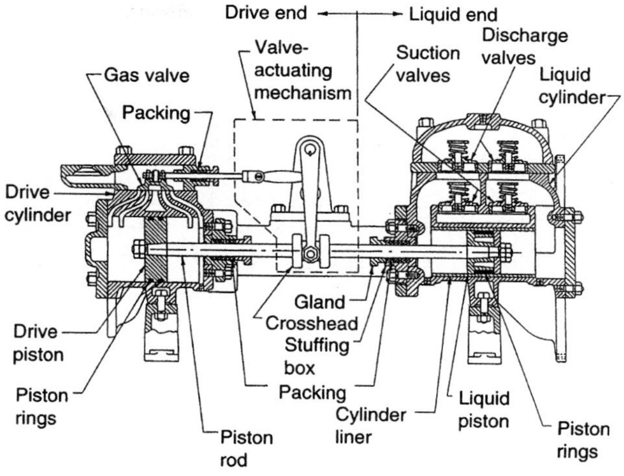

Figure 9.1.4 Horizontal direct-acting gas-driven pump.

chemical homogenizing; well-drilling mud; and car washes.

Applications that practically mandate reciprocating units are abrasive and/or viscous slurries above about 500 lb/in^2 (3,500 kPa). Examples of these services range from powdered coal to peanut butter.

High-Pressure Applications

Hand-powered pumps, for pressures to 40,000 lb/in^2 (300 MPa), are used for small jacks, deadweight testers, and hydrostatic testing of small components. Small, air-driven direct-acting pumps are used for low-flow, high-pressure systems such as hydrostatic testing. The liquid pressure is approximately the air pressure multiplied ("intensified") by the ratio of the air piston area to the liquid plunger area. They are very simple and will stall and hold a fixed pressure without using power; but when pumping, they consume relatively large amounts of air. Available up to 10 hp (7 kW), the most popular sizes are up to 2 hp (1.5 kW). Pressures range to 100,000 lb/in^2 (700 MPa).

Hydraulically driven **intensifiers** operate on the same principle, but are more efficient, and because of the higher pressures commonly available from hydraulic systems, can produce high pressures with lower intensification ratios. Intensifiers are available to 300,000 lb/in^2 (2,100 MPa) for laboratory-scale hydrostatic applications. Intensifier systems for water jetting are available from 30,000 to 60,000 lb/in^2 (210 MPa to 410 MPa) up to 200 hp (150 kW). They are powered by electric motors or diesel engines. Figure 9.1.5 shows the pumping end of an intensifier.

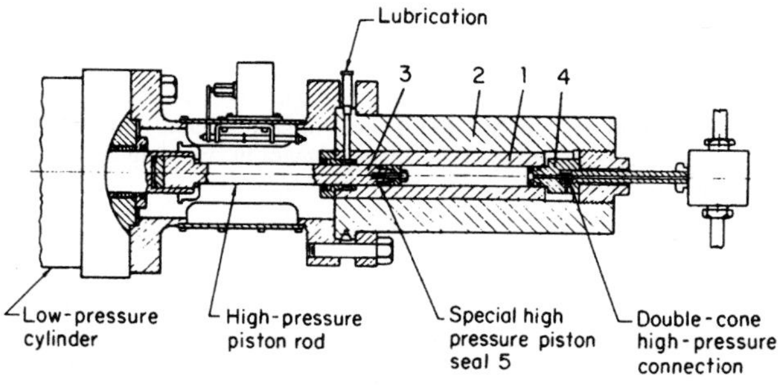

Figure 9.1.5 Pumping end of a direct-acting, liquid-driven, "intensifier" pump. The check valves are in the tubing.

Power pumps (usually triplex or quintuplex) are available from about 5 to 150 hp (5 to 100 kW) with pressures to 40,000 lb/in² (280 MPa), and up to 2,000 hp (1,500 kW) at 20,000 lb/in² (140 MPa). Power pumps are efficient and mechanically simple. Figure 9.1.6 shows a power pump liquid end for a 36,000 lb/in² (250 MPa) water-jetting pump.

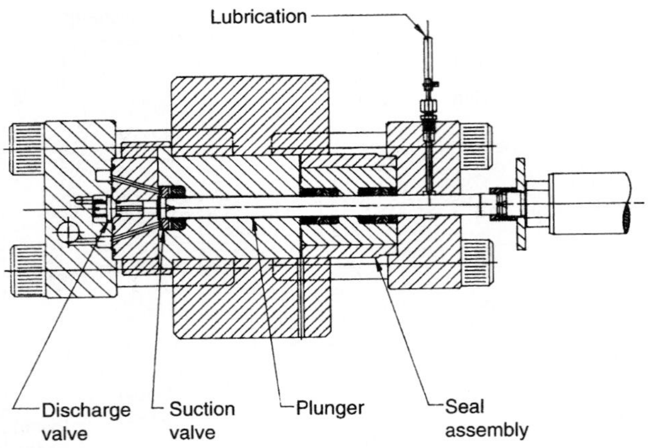

Figure 9.1.6 Power pump liquid end for 36,000 lb/in² (250 MPa) water jetting. *(NLB Corporation.)*

Slurry Applications

The standard reciprocating pump is not designed to handle slurries. Modifications to standard designs, and in some cases special designs, are required to achieve satisfactory operation and component life.

To achieve satisfactory packing and plunger life, abrasive slurry must be prevented from entering the packing. Methods include a wiper ring between the pumpage and packing, a long throat bushing, injection of a clean liquid into the throat area, insertion of a diaphragm or floating piston between the plunger and the pumpage, and complete removal of the valve assembly from the stuffing box area. This last arrangement requires a liquid column between the valve assembly and the stuffing box, which increases the clearance volume and acceleration head within the pump.

Special pump valves are usually required for slurries. Depending on the nature of the solids, they can be ball, bell, bevel seat with elastomer insert, wing-guided with reduced seating area, or disk with special seating surfaces. Special construction can prevent the slurry from contacting the packing, but the valves cannot avoid contact with the slurry.

Problems and How to Avoid Them

Reciprocating pumps have some disadvantages, the most common being pulsating flow. Because of the pulsation, special consideration must be given to system design. Guidelines are provided later in this section.

In most applications, the initial and maintenance costs for a reciprocating pump will be greater than for a centrifugal or rotary pump. The packing in a typical power pump lasts about 2,500 hours, less than a mechanical seal on a rotating shaft.

Most problems with reciprocating pumps can be minimized by selecting pumps to operate at conservative speeds, by carefully designing the pumping system, by careful operating procedures, and by maintenance practices which preserve the alignment of the plunger (or rod) with the stuffing box.

Liquid-End Components

All reciprocating pumps contain one or more pumping elements (pistons, plungers, or diaphragms) that reciprocate into and out of the pumping chambers. (In reciprocating pump terminology, a **piston** is a cylindrical disk which mounts on a rod. A **plunger** is just a smooth rod.)

Each pumping chamber contains at least one suction and one discharge valve.

The liquid cylinder is the major pressure-retaining part of the liquid end, and forms the major portion of the pumping chamber.

A piston pump is normally equipped with a replaceable liner (sleeve) that absorbs the wear from the piston rings. Because a plunger contacts only stuffing-box components, plunger pumps do not require liners.

Sealing between the pumping chamber and atmosphere is accomplished by a **stuffing box** (Fig. 9.1.7). The stuffing box contains rings of packing that conform to and seal against the stuffing box bore and the rod (or plunger).

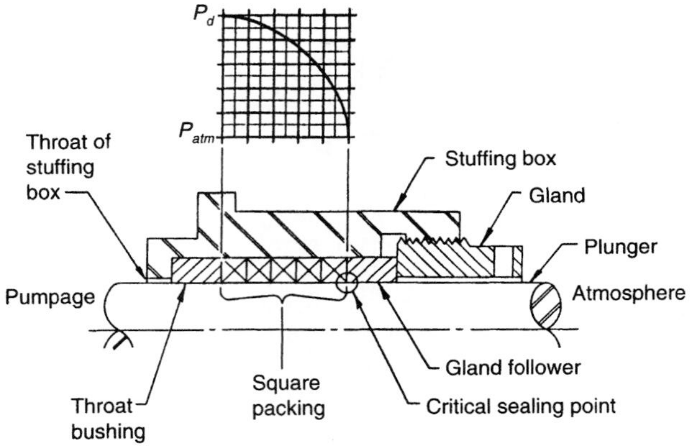

Figure 9.1.7 Reciprocating pump stuffing box (nonlubricated) showing pressure gradient during discharge stroke.

The valves in a reciprocating pump are simply check valves which are opened by the liquid differential pressure. Most valves are spring-loaded. They have a variety of shapes, including spheres, hemispheres, and disks. Disks may be center-guided, outside-diameter-guided, or wing-guided. Figures 9.1.8 through 9.1.13 illustrate some of the valves used in reciprocating pumps.

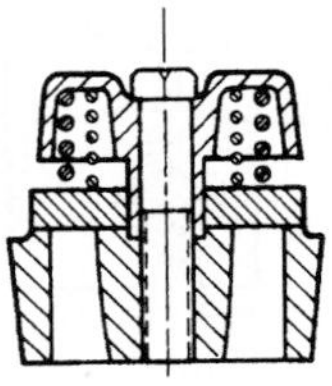

Figure 9.1.8 Reciprocating pump disk valve. *(Ingersoll-Rand Company.)*

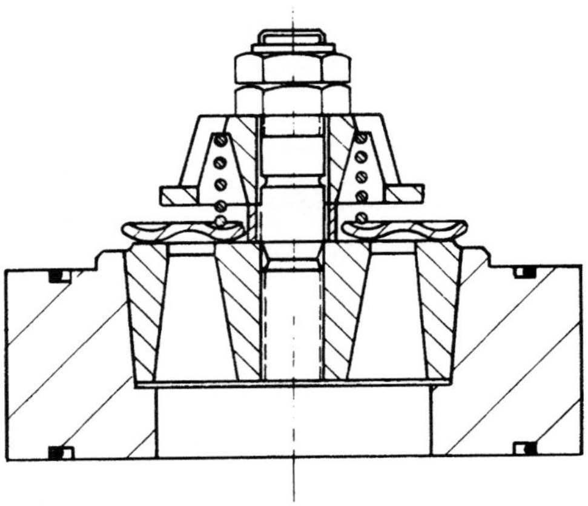

Figure 9.1.9 Reciprocating pump disk valve. *(Durabla Manufacturing Company.)*

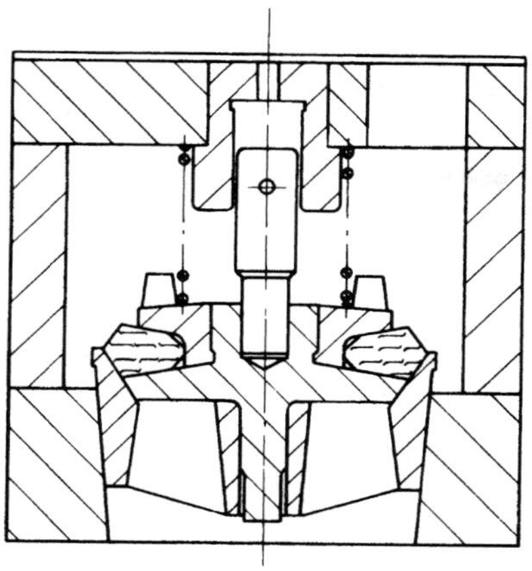

Figure 9.1.10 Reciprocating pump elastomeric-insert valve. *(TRW-Mission Manufacturing Co.)*

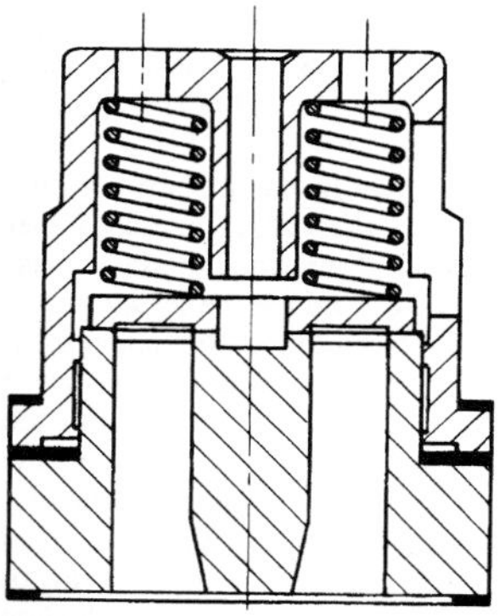

Figure 9.1.11 Reciprocating pump double-ported disk valve.

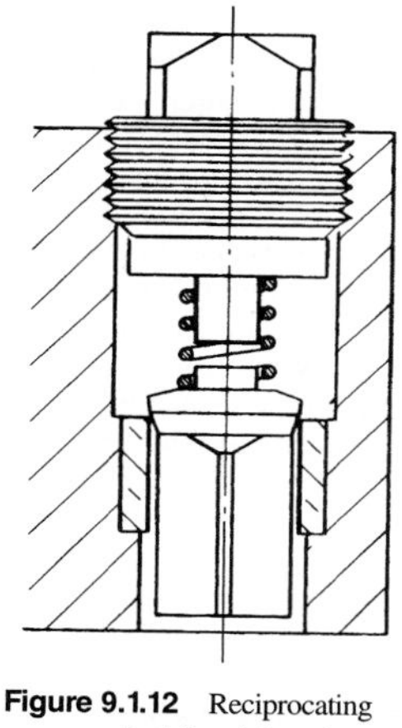

Figure 9.1.12 Reciprocating pump conical-faced (bevel-seat) wing-guided valve.

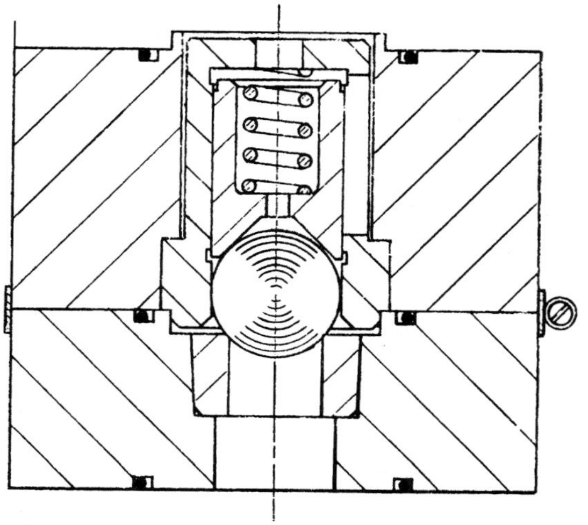

Figure 9.1.13 Reciprocating pump ball valve. *(Ingersoll-Rand Company.)*

The Pumping Cycle

As the pumping element reverses from the discharge stroke to the suction stroke, liquid within the pumping chambers expands, and the pressure drops. Since most liquids are relatively incompressible, little movement of the element is required to reduce the pressure. When the pressure drops sufficiently below suction pressure, the differential pressure (suction pressure minus chamber pressure) pushes the suction valve open. This occurs when the element is moving slowly, so that the valve opens slowly and smoothly as the velocity of the element increases. Liquid then flows through the valve and follows the element on its suction stroke. As the element decelerates near the end of the stroke, the suction valve returns to its seat. When the element stops, the suction valve closes.

The pumping element then reverses and starts its discharge stroke. The liquid trapped in the pumping chamber is compressed until chamber pressure exceeds discharge pressure by an amount sufficient to push the discharge valve from its seat. The action of the discharge valve, during the discharge stroke, is then the same as that of the suction valve during the suction stroke.

Power Pumps

The drive end of a power pump is called a **power end** (see Fig. 9.1.2). Its function is to convert rotary motion from a driver into reciprocating motion for the liquid end. The main component of the power end is the **power frame,** which supports all other power end parts and, usually, the liquid end. The second major component of the power end is the **crankshaft** (sometimes a camshaft). The crankshaft transmits power from the driver to the connecting rods.

The **connecting rod** is driven by a throw (journal) of the crankshaft on one end, and drives a crosshead on the other. The crankshaft moves in pure rotating motion, the crosshead in pure reciprocating motion. The connecting rod is the link between the two. **Main bearings** support the shaft in the power frame.

The **crosshead** transmits the force from the connecting rod into the pumping element. It is fastened directly to a plunger or piston rod, or to a rod called an **extension, stub,** or **pony rod.** The other end of this rod is fastened to the pumping element.

Power pump overall efficiencies normally range from 85 to 94 percent, higher than any other type of pump.

Selection Graphs Different manufacturers produce these graphs in different configurations. All graphs consist of straight lines since displacement is directly proportional to speed. Figure 9.1.14 shows one such graph for a triplex power pump. Displacement is plotted against crankshaft speed for the different plunger diameters available in the

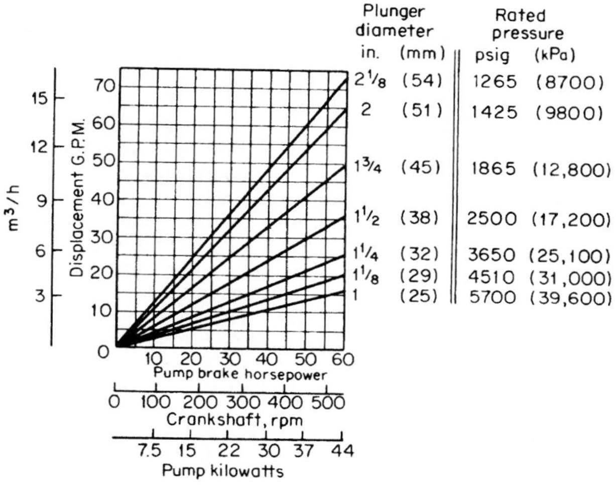

Figure 9.1.14 Power pump selection graph. *(Ingersoll-Rand Company.)*

pump. The pressures listed are those at which each plunger size would operate to load the power end to its rated value (4,500 lbf or 20,000 N).

NPSH Curves Figure 9.1.15 provides insight into power pump NPSHR characteristics, valve action, and the effect of valve springs on pump performance. These curves are for a 3-in (76-mm) stroke, horizontal, triplex plunger power pump with $2\frac{1}{4}$-in-diameter (57-mm) plungers, and with suction valves that operate vertically. The valves are wing-guided, and have a diameter approximately equal to the plunger diameter.

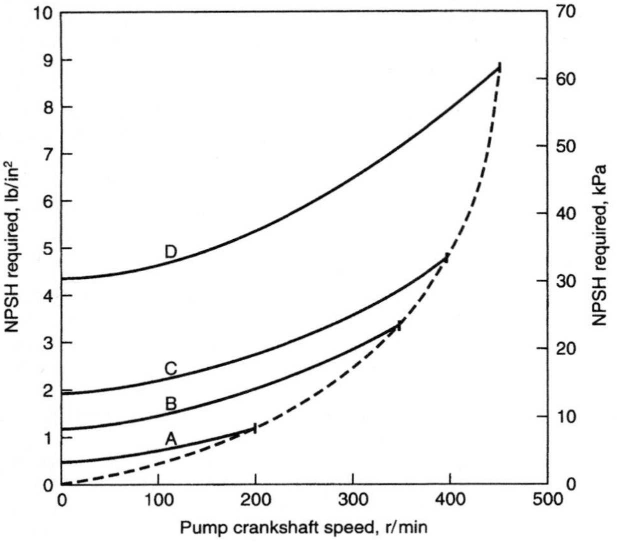

A = no suction valve springs (std. disch.)
B = weak suction valve springs (std. disch.)
C = standard suction valve springs (std. disch.)
D = heavy valve springs—suction and discharge

Figure 9.1.15 NPSH curves for a 3-in- (75-mm-) stroke triplex power pump.

Because the axis of the suction valve is vertical, the valve can operate without a spring if the pump speed is kept low. Curve A represents the NPSH requirements with no springs on the suction valves. NPSHR at 100 r/min is only 0.7 lb/in² (5 kPa) (1.6 ft or 0.5 m of water), less than for most centrifugals.

The speed of the pump in this configuration is limited by the ability of the suction valve to keep up with the plunger. Since there is no spring to push the valve back onto its seat, gravity is the only force tending to close the valve against the entering fluid. If the pump is run too fast, the valve will still be off its seat when the plunger reverses to the discharge stroke. The liquid will then momentarily flow backward through the seat, and the valve will be slammed onto the seat, sending a shock wave into the suction manifold and piping. At that moment, the plunger is moving at a finite velocity, but the discharge valve is still closed. The pressure in the pumping chamber will quickly exceed discharge pressure, and the discharge valve will be driven from its seat. A shock wave will be transmitted from the pumping chamber through the discharge manifold, and into the discharge line. The vertical line at the end of Curve A indicates the maximum speed for proper suction valve operation.

Curve B is for weak springs added to the suction valves. Because both spring force and valve weight must now be overcome to open the suction valve, NPSHR has increased about 100 percent over Curve A. These springs get the suction valves onto their seats quicker, so that operation is smooth at higher speeds.

If speeds beyond the end of Curve B are desired, stronger suction valve springs are required. Stronger springs allow operation to 400 r/min. NPSH requirements are about 3 times those of Curve A, ranging from about 2 to 5 lb/in² (14 to 35 kPa).

Curves A through C represent a pump equipped with the same (standard) discharge valve spring. Only the suction spring has been changed.

If operation is required at speeds exceeding the limits of Curve C, extra-strong springs are required on both suction and discharge valves. As shown by Curve D, NPSHR is about double that required for the standard springs, ranging from about 4.5 to 9 lb/in² (31 to 62 kPa).

The dashed line passing through the ends of the NPSH curves illustrates the dramatic increase of NPSHR as pump speed is increased. When it is necessary to change suction springs, NPSHR increases to the fourth power of speed, for example, if the pump speed is doubled, to retain smooth valve action, the spring force must be increased by a factor of sixteen.

To minimize the problem of dissolved air, the NPSH tests that produced these curves were performed with water at, or near, its boiling point in the suction vessel.

Reciprocating pumps, under the correct conditions, can operate with a suction pressure below atmospheric. Such a situation, though, can lead to air being drawn through the stuffing box packing and into the pumping chamber on the suction stroke. This air will cause as many problems as air entrained in the pumpage. Capacity will drop, the pump may operate noisily, the system may vibrate, and damage may occur to pump and system components.

This inward air leakage can be reduced by an external sealing liquid, such as lubricating oil, being directed onto the plunger surface or into the packing.

Test Criteria for NPSH Power pump NPSH tests are performed by holding the pump speed and discharge pressure constant, and varying the NPSH available (NPSHA) in the system. Capacity remains constant for all NPSHA values above a certain point. Below this NPSHA value, capacity beings to fall. NPSHR is defined as the NPSH available when the capacity has dropped 3 percent.

Acceleration Head Because the velocities in the suction and discharge piping are not constant, the pumpage must accelerate a number of times for each revolution of the crankshaft. Since the liquid has mass and, therefore, inertia, energy is required to produce the acceleration. This energy is returned to the system upon deceleration. Sufficient pressure, however, must be provided to accelerate the liquid on the suction side of the pump to prevent cavitation in the suction pipe and/or the pumping chambers. The drop in pressure, below the average, on the suction side, caused by this acceleration, is called **acceleration head**.

An approximation of acceleration head for power pumps is given by the equation $h_a = LvNC/gK$, where h_a = acceleration head, ft (m) of liquid being pumped; L = actual length of suction line (not equivalent length), ft (m); v = average liquid velocity in suction line, ft/s (m/s); N = speed of pump crankshaft, r/min; C = constant, depending on type of pump; g = gravitational constant, ft/s² (m/s²); and K = fluid compressibility correction factor. Values for the constants C and K are:

Pump type	C
Simplex, single-acting	0.400
Duplex, single-acting	0.200
Triplex	0.066
Quintuplex	0.040
Septuplex	0.028
Nonuplex	0.022

Fluid	K
Slightly compressible liquids such as deaerated water	1.4
Most liquids	1.5
Highly compressible liquids (such as ethane)	2.5

Note that increasing the pump speed, without changing the suction line, increases h_a as the square of speed, because both v and N increase proportionately with speed.

Because this equation does not adequately compensate for pumpage and system elasticity, it is recommended only for short, rigid suction lines.

To reduce acceleration head in a suction line, a bottle or suction stabilizer may be installed in the pipe, adjacent to the pump.

Figure 9.1.16 shows the theoretical flow rates for various types of power pumps during one revolution of the crankshaft. Pumps with an odd number of cranks have the same flow variations for both single-acting and double-acting liquid ends (except the simplex). Flow curves for septuplex and nonuplex pumps are similar to the quintuplex curve, with more but smaller variations, as indicated in the tabulation. The values tabulated will vary slightly with variations in the ratio of connecting rod length to stroke length. These curves indicate flow-rate variations, not **pressure pulsations.** Pressure pulsations are a product of the acceleration created by the pump and the mass of the pumpage in the system.

Actual flow rates may vary considerably from these curves, particularly with high-speed pumps, where valve closing and opening may lag the crankshaft rotation as much as 15°, and at high pressures, where liquid compression may cause a further lag of 5° to 10° (or more for a large clearance volume).

	Number of plungers
A = Simplex-double acting Max. rate of flow above mean 59.96% Min. rate of flow below mean 100.0 %	1
B = Duplex-double acting-cranks 90° apart Max. rate of flow above mean 26.72% Min. rate of flow below mean 21.56%	2
C = Triplex-single acting-cranks 120° apart Max. rate of flow above mean 6.64% Min. rate of flow below mean 18.42%	3
D = Quintuplex-single acting-cranks 72° apart Max. rate of flow above mean 1.88% Min. rate of flow below mean 5.63%	5
Septuplex-single acting-cranks 51 4/7° apart Max. rate of flow above mean 1.2% Min. rate of flow below mean 2.8%	7
Nonuplex-single acting-cranks 40° apart Max. rate of flow above mean 0.7% Min. rate of flow below mean 1.5%	9

Figure 9.1.16 Flow-rate variations in reciprocating power pumps.

Power Pump Speeds Probably the most controversial factor in the selection of power pumps is the maximum allowable speed. Some manufacturers, for example, offer 3-in- (76-mm-) stroke triplex pumps for operation at 500 r/min. One builder of portable water-jetting units routinely operates this size pump at 600 r/min. Most users of this size pump, in continuous-duty applications, prefer to operate it no faster than about 400 r/min.

Top speeds quoted by one manufacturer for the 3-in-stroke triplex increased over a period of years from 150 to 520 r/min. Although improvements in power end and liquid end construction made the pump capable of operating at higher speeds, albeit with shorter packing life, a major obstacle to higher speeds is system design. Initially, some thought that a smaller pump, running faster to achieve the same capacity, would produce less pulsation in the suction and discharge pipes. Unfortunately, the opposite is true.

The velocity variation in the piping for a triplex pump is 25 percent, regardless of the pump size or speed (see Fig. 9.1.16). For the same capacity, a smaller pump, running faster, will produce the same maximum and minimum flow rates and more pulses per second. Since the acceleration head is proportional to frequency, doubling the pump speed doubles the acceleration head, thereby reducing the NPSH available from the system. Also, doubling the speed requires a much stiffer valve spring, thereby significantly increasing the NPSHR of the pump. If the NPSHA drops below the NPSHR, cavitation and pounding will occur.

The best solution for excessive acceleration is an effective dampener in the suction line; but such devices often are less effective at higher frequencies, due to the inertia of their moving parts and inertia of the liquid inside them, which must oscillate for them to function.

Through field observations, the author developed a set of **maximum recommended speeds** for plunger-type power pumps in continuous-duty services. These speeds, as adopted by API 674, are shown in Fig. 9.1.17, along with the Hydraulic Institute basic speeds. Intermittent and cyclic operation is often satisfactory above these speeds. Lower speeds may be dictated by factors such as pump construction, system design, low NPSHA, entrained gas, high temperature, entrained solids, high viscosity, and requirements for a low sound level.

The **minimum speed** of a power pump is determined by its ability to provide sufficient lubrication to all bearing surfaces in the power end. Some units can run satisfactorily at 20 r/min. Others must be maintained at 100 r/min or more.

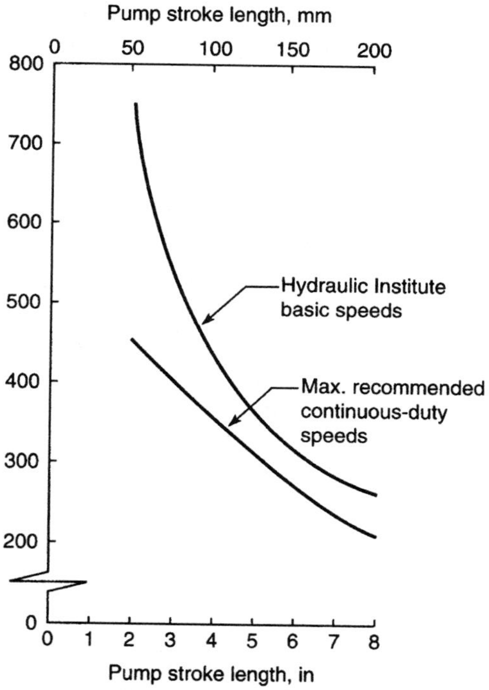

Figure 9.1.17 Maximum recommended speeds for plunger-type power pumps in continuous-duty services.

Plunger load (also called **frame load** and **rod load**) is the force transmitted to the power end by one plunger. For a single-acting pump, the discharge plunger load is calculated by multiplying discharge pressure by the cross-sectional area of one plunger. Suction plunger load is the suction pressure multiplied by the plunger area.

A power pump is rated by the maximum discharge plunger load that the power end is capable of absorbing when the suction pressure is zero. Some units are rated for continuous operation, some for intermittent operation, and some are rated both ways.

Torque Characteristics For fixed suction and discharge pressures, a power pump requires an **average** input torque that is independent of speed (except for increases at very low and very high speeds). The torque actually has a variation that mirrors the discharge flow-velocity curve. A power pump will require the same torque at one-half or one-quarter rated speed and, therefore, will require one-half or one-quarter rated power, respectively. Figure 9.1.18 illustrates the variation of average torque with speed for a typical triplex power pump.

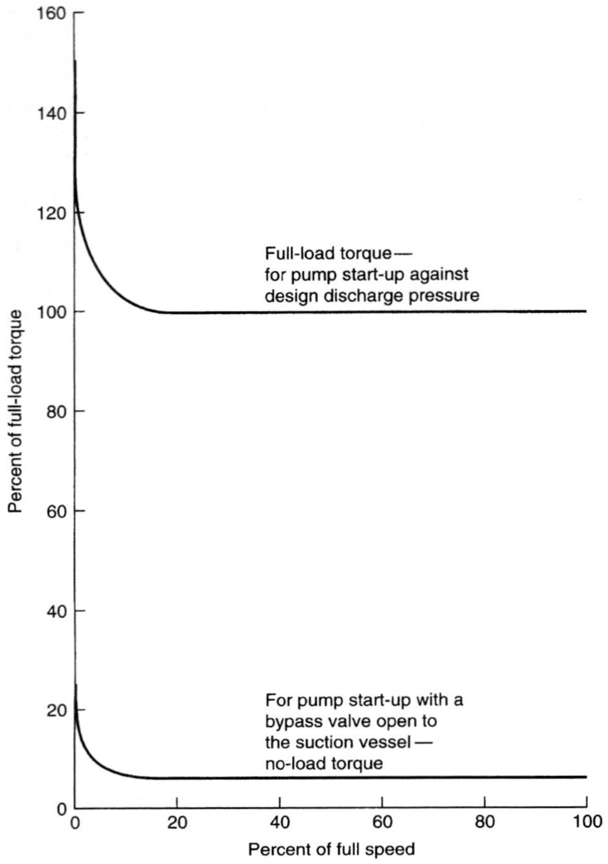

Figure 9.1.18 Torque-speed curves for a typical triplex power pump.

The curve for full-load torque in Fig. 9.1.18 is for starting against full discharge pressure. Breakaway torque is about 150 percent of average full-load running torque. As speed is increased, and proper lubrication is established in the power end and packing, torque drops to the full-load, full-speed value, and is thereafter constant up to full speed.

For a start-up that is easier on equipment, pump discharge is piped back to the suction vessel, making the discharge pressure near suction pressure. The no-load curve in Fig. 9.1.18 illustrates the resulting torque requirements imposed by the pump on the drive train. Breakaway torque is approximately 25 percent of full-load torque. This will vary, depending on the type of packing and bearings in the pump and the length of time the pump has been idle. As speed increases, the torque drops to less than 10 percent of the full-load torque.

Packing The component normally requiring the most maintenance on reciprocating pumps is packing. Although the life of packing in a power pump is typically about 2,500 h (3 months), some installations, with special stuffing box arrangements, have experienced lives of more than 18,000 h (2 years), at discharge pressures to 4,000 lb/in^2 (28 MPa).

Short packing life can result from any of the following conditions: (1) misalignment of plunger (or rod) with stuffing box; (2) worn plunger, rod, stuffing-box bushings, or stuffing-box bore; (3) improper packing for the application; (4) insufficient or excessive lubrication; (5) packing gland too tight or too loose; (6) excessive speed or pressure; (7) high or low temperature of pumpage; (8) excessive friction (too much packing); (9) packing running dry (pumping chamber gas-bound); (10) shock conditions caused by entrained gas or cavitation, broken or weak valve springs, or system problems; (11) solids from the pumpage, environment, or lubricant; (12) improper packing installation or break-in; and (13) icing caused by volatile liquids that refrigerate and form ice crystals on leakage to atmosphere, or by pumpage at temperatures below 32°F (0°C). As is evident from this list, short packing life can indicate problems elsewhere in the pump or system.

To achieve a low leakage rate, the clearance between the plunger (or rod) and packing must be essentially zero. This requires that the sealing rings be relatively soft and pliant. Because the packing is pliant, it tends to flow into the stuffing box clearances, especially between the plunger and follower bushing. If this bushing does not provide an effective barrier, the packing will extrude, and leakage will increase.

A set of square or V-type packing rings will experience a pressure gradient, during operation, as indicated in Fig. 9.1.7. The last ring of packing, adjacent to the gland-follower bushing, will experience the largest axial loading, resulting in greater deformation, tighter sealing and, therefore, the largest pressure drop. The gap between the plunger and the follower must be small enough to prevent packing extrusion. Most packing failures originate at this critical sealing point.

Because this last ring of packing is the most critical, does the most sealing, and generates the most friction, it requires more lubrication than the others. In a nonlubricated arrangement (Fig. 9.1.7), this ring must rely on the plunger to drag some of the pumpage back to it in order to provide cooling and lubrication. Therefore, to maximize packing life, the overall stack height of the packing should not exceed the stroke length of the pump.

Because the last ring of packing requires more lubrication than do the others, lubrication of the packing from the atmospheric side is more effective than injecting oil into a lantern ring located in the center of the packing. Care must be exercised to get the lubricant onto the plunger surface and close enough to the last ring, so that the stroke of the plunger will carry the lubricant under the ring.

During the first few hours of pump operation with new packing, each stuffing box should be monitored for temperature. It is normal for some boxes to run warmer than others—as much as 50°F (30°C) above the pumping temperature. Only if this exceeds the maximum temperature rating of the packing are steps required to reduce box temperature.

The best lubricant for packing in most services has been found to be steam cylinder oil.

The concepts that a higher discharge pressure requires more rings of packing and that a larger number of rings lasts longer are true for long-stroke, low-speed machines, but have been disproven in a number of moderate- to high-speed power-pump applications. Unless they are profusely lubricated, the larger number of rings create additional frictional heat and wipe lubricant from the plunger surface, thus depriving some rings of lubrication. On numerous saltwater injection pumps operating at pressures above 4,000 lb/in^2 (28 MPa), Chesney reported that packing life was only two weeks with twelve rings of packing in each stuffing box. With three rings in each box, packing life was approximately 6 months.

Plungers Next to packing, the plunger is the component of a power pump that typically requires most frequent replacement. The high speed of the plunger and the friction load of the packing tend to wear the plunger surface. For longer life, plungers are sometimes hardened, although it is more common to apply a hard coating. Such coatings are of chrome, various ceramics, nickel-based alloys, or cobalt-based alloys. Desired characteristics of the coatings include hardness, smoothness, high bond strength, low porosity, corrosion resistance, and low cost. No one coating optimizes all characteristics.

Ceramic coatings are harder than the metals, but are brittle, porous, and sometimes lower in bond strength. Porosity contributes to shorter packing life. Mixing of hard particles, such as tungsten carbide, into the less-hard nickel or cobalt alloys has resulted in longer plunger life at the expense of shorter packing life.

Stuffing Boxes Various stuffing box designs (various lubrication and bleed-off arrangements to minimize leakage and extend packing life) are shown in Fig. 9.1.19.

The most significant advance in packing arrangements in recent years has been spring loading. Spring loading is applied most to V-ring (chevron) packing but also works well with square packing. The spring must be located on the pressure side of the packing. Springs of various types

Ⓐ Standard Lubricated Stuffing Box

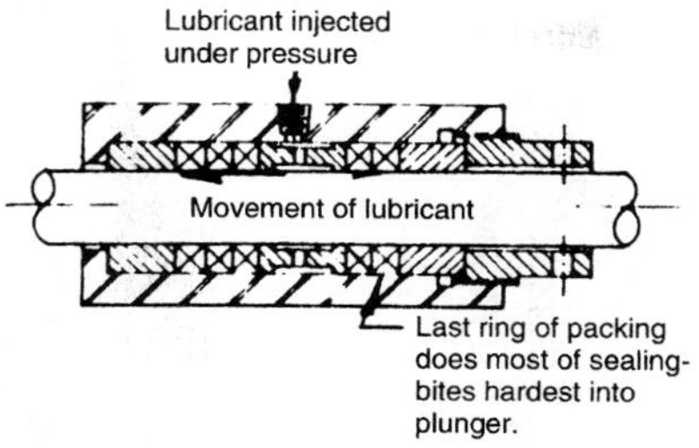

1. Good design.
2. Most of lubricant migrates into pumpage.
3. Packing may be square, chevron, or nonadjustable.

Ⓑ Alternate Lubricated Stuffing Box

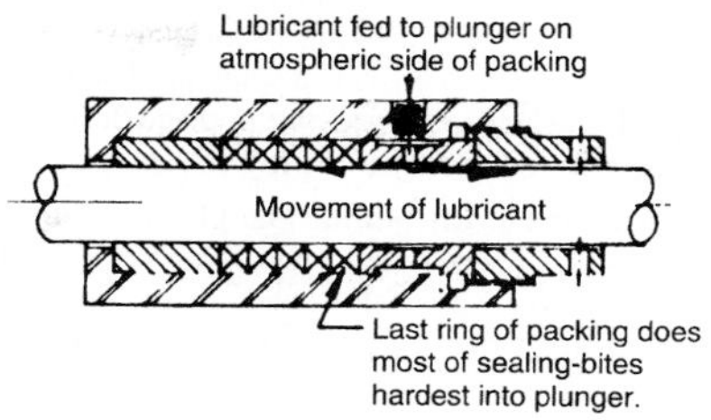

1. Good design—performs better than standard arrangement in most cases.
2. Puts lubricant under last ring—where it's needed most.
3. Allows use of low-pressure and drip-type lubricators.
4. Very little lubricant migrates into pumpage.
5. Packing may be square, chevron, or nonadjustable.

Ⓒ Spring-Loaded Packing

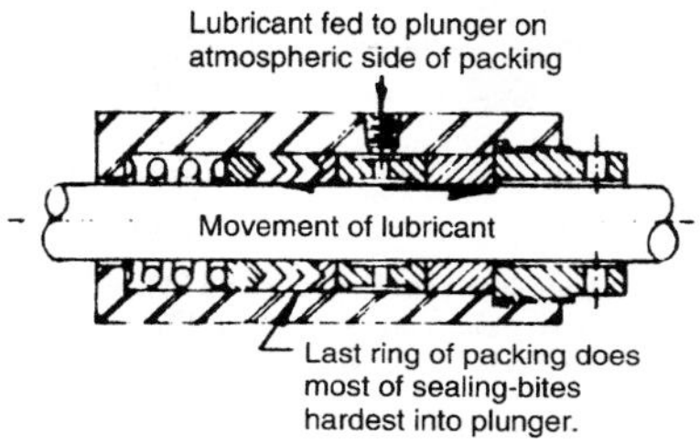

1. Good design—long life—minimal leakage.
2. Puts lubricant under last ring—where it's needed most.
3. Allows use of low-pressure and drip-type lubricators.
4. No adjustment required (self adjusting).

Ⓓ Standard Box Used to Bleed Pumpage to Low-Pressure Point—Bad Application

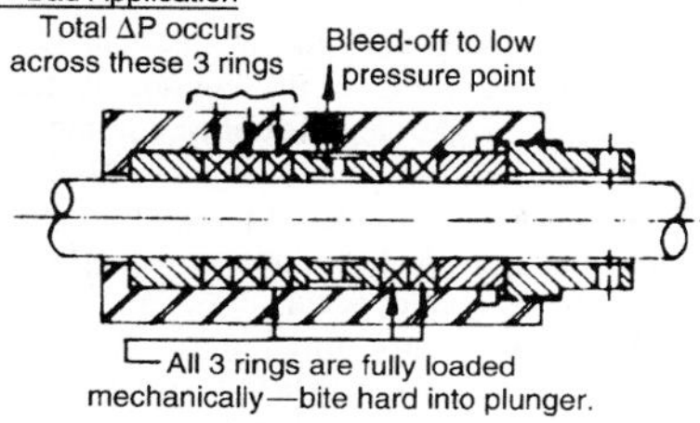

1. High friction causes excess heat.
2. Short life of packing and plungers.
3. Improper use of standard box—poor application.

Ⓔ Modified Gland Follower to Allow Bleed-Off to Low Pressure Point

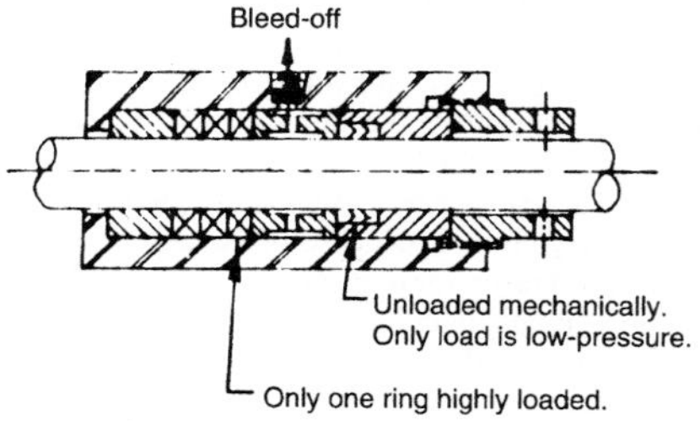

1. Less friction and lower temp. than "D".
2. Longer life of packing and plungers.
3. Big improvement over "D", but secondary packing not adjustable to compensate for wear.

Ⓕ Two-Gland Stuffing Box—Lubricated

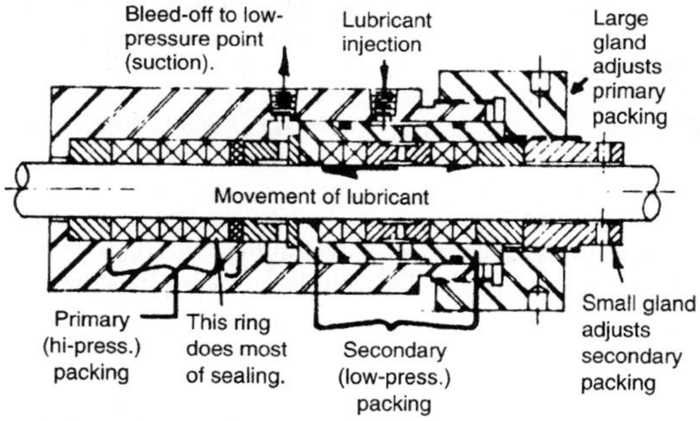

1. The old standard for high-pressure, critical services.
2. Provides independent adjustment of primary and secondary packing. (Proper adjustment requires skilled mechanic.)
3. Full-size secondary packing.
4. Positive packing lubrication.
5. Negligible leakage to atmosphere.
6. Long packing and plunger life.
7. Excellent for volatile fluids.

Ⓖ Double Tandem Spring-Loaded Packing

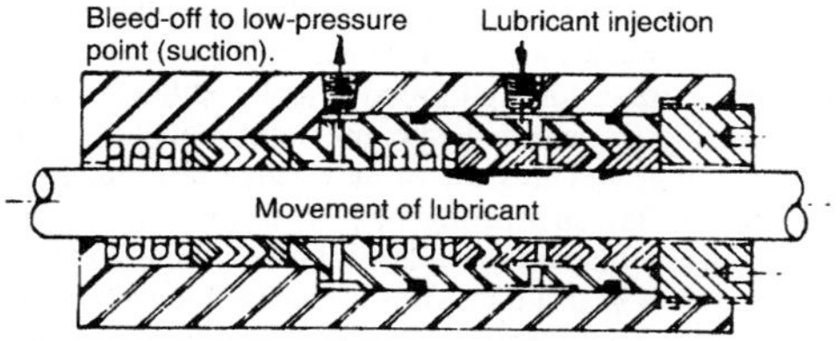

1. The best design available for most high-pressure, critical services.
2. Combines best features of two-gland box and spring-loaded packing—negligible leakage, long life, and the packing is self-adjusting.

Figure 9.1.19 Various stuffing box designs for reciprocating pumps.

can be used, including single-coil, multiple coil, wave-washer, belleville, and thick-rubber-washer. The force provided by the spring is small compared to the force imposed on the packing by discharge pressure. The major functions of the spring are to provide a small preload to help set the packing, and to hold all bushings and packing in place during operation.

Spring loading of packing has the following advantages:

Requires no adjustment of the gland. The gland is tightened until it bottoms, then is locked. This removes one of the biggest variables in packing life: personnel skill.

Alloys expansion. If the packing expands during the initial break-in, the spring allows for the expansion.

Adjusts for wear. As the packing wears, adjustment automatically occurs from inside the box. The problem of transmitting gland movement through the top packing ring is eliminated.

Provides a cavity. The spring cavity provides an annular space for the injection of a clean liquid for slurry applications.

Eliminates the need for a gland, if allowed by pump design. The stuffing box assembly (if a separate component) can be disassembled and reassembled on a workbench.

Disadvantages of spring-loading packing are associated with the cavity created by the spring. Since this cavity communicates directly with the pumping chamber, the additional clearance volume will cause a reduction in volumetric efficiency if the pumpage is sufficiently compressible. This cavity also provides a place for vapors to accumulate. If the pump design does not provide for venting this space, a further reduction in volumetric efficiency may occur.

Spring-loaded packing is the reciprocating pump's equivalent to the mechanical seal for rotating shafts. Leakage is low, life is extended, and adjustments are eliminated. Packing sets can be stacked in tandem (they must be independently supported) for a stepped pressure reduction, or to capture leakage from the primary packing that should not escape to the environment.

Controlled-Volume Pumps A **controlled-volume** pump (also called a **metering, proportioning,** or **chemical-injection** pump) is a power pump with an adjustable stroke length (or adjustable *effective* stroke length). It is used to provide an accurate, and adjustable, capacity. The capacity can be changed manually, as shown in Fig. 9.1.20, or with an electric, pneumatic, or hydraulic controller.

Such pumps can develop pressures to 30,000 lb/in² (200 MPa) with capacity controlled with ± 1 percent. The piston or plunger may come in direct contact with the pumpage, or may be separated from the pumpage by a diaphragm. The diaphragm may be flat, tubular, or conical. It can be actuated mechanically, as shown in Fig. 9.1.20, or hydraulically, as shown in Fig. 9.1.21. Mechanically actuated diaphragms are

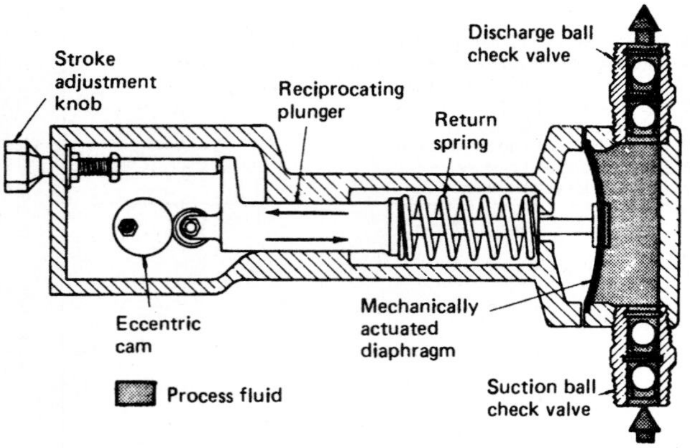

Figure 9.1.20 Controlled-volume pump with mechanically actuated diaphragm and manual-adjustment stroke length. *(Reprinted from James P. Poynton, "Metering Pumps: Selection and Application," Marcel Dekker, Inc., Copyright 1983. Used by permission.)*

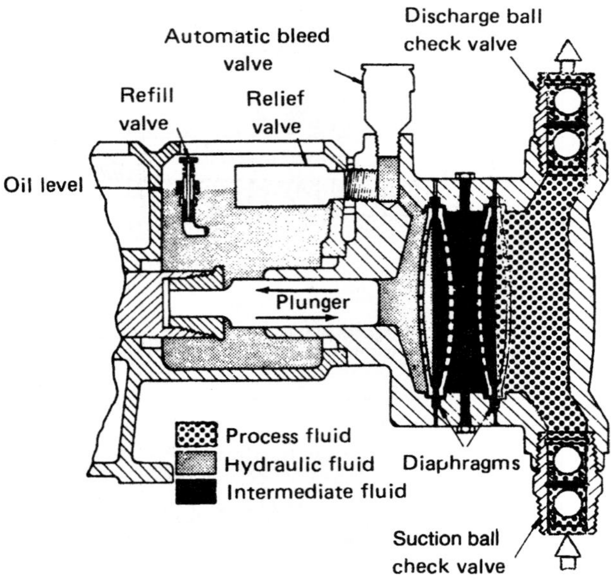

Figure 9.1.21 Controlled-volume pump with hydraulically actuated double diaphragms. *(Reprinted from James P. Poynton, "Metering Pumps: Selection and Application," Marcel Dekker, Inc., Copyright 1983. Used by permission.)*

generally limited to capacities of 25 gal/h (100 L/h) and pressures of 250 lb/in² (1,700 kPa).

Direct-Acting Pumps

The direct-acting pump (see Fig. 9.1.4) has some of the same advantages as the power pump, plus others. These units are also well suited for high-pressure, low-flow applications. Discharge pressures normally range from 100 to 5,000 lb/in² (0.7 to 35 MPa), but may exceed 10,000 lb/in² (70 MPa). Capacity is proportional to speed from stall to maximum speed, affected little by discharge pressure. Speed is controlled by controlling the motive fluid. The unit is normally self-priming, particularly the low-clearance-volume type.

Direct-acting pumps are negligibly affected by hostile environments, such as corrosive fumes, because of the absence of a bearing housing, crankcase, or oil reservoir (except for units requiring lubricators). Some direct-acting pumps, inundated by floodwater, have continued to operate without adverse effects. The direct-acting pump is quiet, simple to maintain, and its low speed often results in a long life.

Since the direct-acting pump was originally designed to be driven by steam (initially, most pumps were), it was known as a **steam pump**—not for the pumped fluid, but for the motive fluid. Other fluids are now also used to drive direct-acting pumps. Fuel gas, which would otherwise be throttled through a pressure regulator for plant use, is often piped through a direct-acting pump to provide "free" pumping. Compressed air is frequently used to drive small pumps for services such as hydrostatic testing and chemical metering. Hydraulic oil is used to drive **intensifiers** for high-pressure water-jetting.

Direct-acting air-driven diaphragm pumps handle a variety of liquids at pressures up to the pressure of the driving air. This pump, as shown in Fig. 9.1.22, has no stuffing box or packing, and therefore has no leakage, unless a diaphragm fails. Failure of a diaphragm may result in pumpage in the exhaust air and/or air in the pumpage.

Direct-acting pumps operate at speeds that normally range from 0 to 50 cycles/min, depending on the stroke length. The ability of these units to operate at speeds down to zero (i.e., stall conditions) is desirable for some applications.

The direct-acting pump has a low thermal efficiency when driven by a gas (such as steam or air). The mechanical efficiency (output force divided by input force) is high; but because the unit has no device to

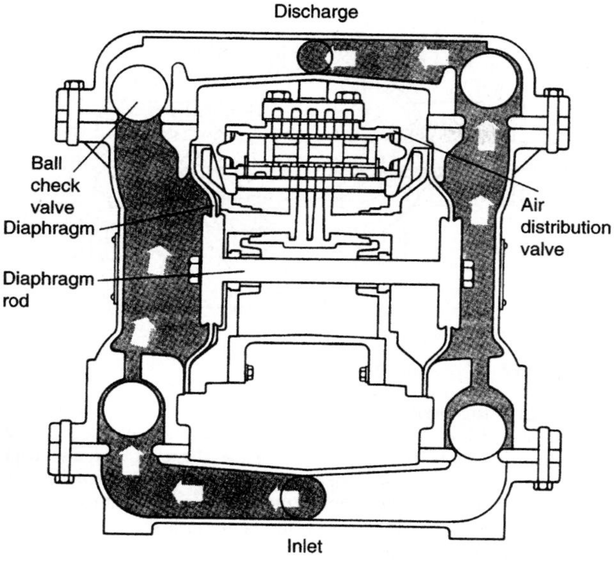

Figure 9.1.22 Direct-acting, air-driven diaphragm pump. *(Used by permission of Warren Rupp, Inc., a unit of IDEX Corp., Mansfield, OH.)*

store energy (such as a flywheel), the motive gas must remain at full inlet pressure in the drive cylinder through the entire stroke. At the end of the stroke, the gas expands to exhaust pressure, but no work is performed during this expansion. Hence, the thermal energy of the gas is lost to friction. Steam pumps consume about 100 lb/h of steam for each hydraulic horsepower (hhp) (60 kg/h · kW) developed on the liquid end. When natural gas or air is the motive fluid, consumption is about 3,500 std. ft^3/h · hhp (130 m^3/h · kW).

The **drive end** (or steam end, or gas end) of a direct-acting pump converts the differential pressure of the motive fluid to reciprocating motion for the liquid end. It is similar in construction to the liquid end, containing a piston (or diaphragm) and valving. The major difference is that the valve is mechanically actuated by a control system which senses the location of the drive piston, causing the valve to reverse the flow of the motive fluid when the drive piston reaches the end of its stroke.

The main component of the drive end is the drive cylinder. This cylinder forms the major portion of the pressure boundary, and supports the other drive end parts.

Mechanical Efficiency—Direct-Acting Pumps For a direct-acting pump, mechanical efficiency is the ratio of the force transmitted to the liquid by the piston (or plunger or diaphragm) to the force transmitted to the drive piston (or diaphragm) by the motive fluid. For double-acting pumps, the differential pressures are used at both ends of the pump, and since fluid friction losses in valves and ports must be charged to the pump, the pressures are as measured at the inlet and outlet ports. In equation form:

$$\eta_m = \frac{A_L \Delta p_L}{A_{dr} \Delta p_{dr}} \qquad (9.1.1)$$

where η_m = mechanical efficiency of pump, A_L = area of liquid piston or plunger, Δp_L = differential pressure across liquid end, A_{dr} = area of drive piston, and Δp_{dr} = differential pressure across drive end.

The area of the piston rod is usually small relative to the piston area, and is often ignored. However, it must be considered when the rod area becomes a significant portion of the piston area.

The mechanical efficiency of a typical direct-acting pump is shown in Fig. 9.1.23. Note that η_m rises as speed is reduced. It is this characteristic that allows the unit to be controlled by throttling the motive fluid. The reduction in the available differential driving pressure forces the pump to operate more efficiently, for example, at a lower speed.

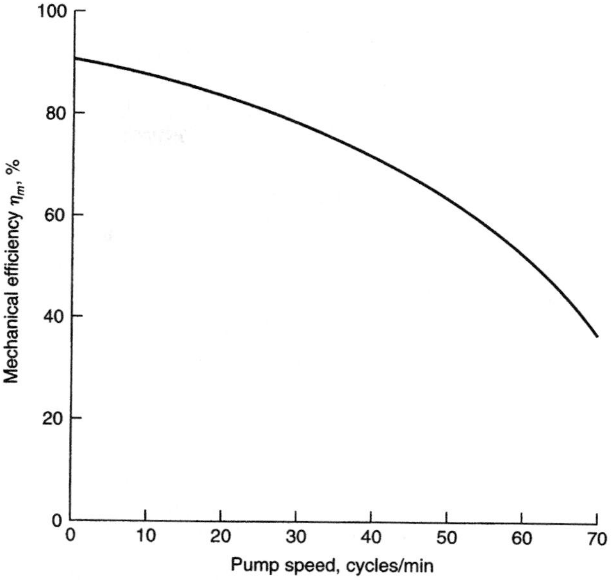

Figure 9.1.23 Mechanical efficiency of a direct-acting pump.

ROTARY PUMPS

Rotary pumps are displacement pumps which have rotating pumping elements, such as gears, lobes, screws, vanes, or rollers. They do not contain inlet and outlet check valves, as do reciprocating pumps. They are built for capacities from a fraction of a gallon per minute (cubic metre per hour) to about 18,000 gal/min (4,000 m^3/h). Though used for pressures up to 20,000 lb/in^2 (140 MPa), their particular field is for pressures of 25 to 500 lb/in^2 (170 to 3,500 kPa).

Most rotary pumps rely on close running clearances to prevent the pumpage from leaking from the discharge side back to the suction side of the pump. Because of the close clearances, the pumpage must be clean. Most rotary pumps are noted for their ability to handle viscous liquids, and many actually *require* a viscous liquid to achieve peak performance. Viscosity affects the mechanical efficiency, volumetric efficiency, and NPSHR of a rotary pump.

The **rotor** is the pumping element of the rotary pump, and is usually the feature by which the pump is classified.

Gears The teeth of the gear trap and displace the pumpage. If the teeth are on the outside of the gear, as shown in Fig. 9.1.24, it is called an **external gear**. As the teeth disengage, suction pressure pushes liquid into the cavities between the teeth. The teeth then carry the liquid around to the discharge side of the pump. As the teeth engage, the fluid is pushed into the discharge line. In the most common type of gear pump, one of the pumping gears drives the other pumping gear. This direct contact between the teeth requires the pumpage to be a good lubricant. In some units, though, the two pumping gears are driven by gears (called **timing gears**) mounted on the two shafts external to the

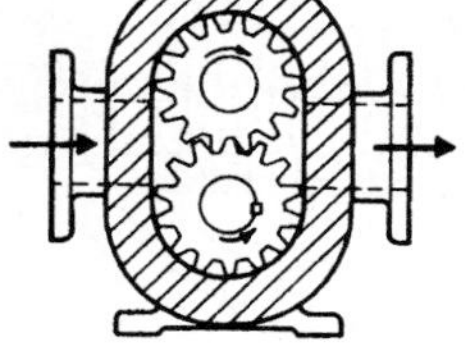

Figure 9.1.24 External-gear rotary pump. *(Reprinted from "Hydraulic Institute Standards for Centrifugal, Rotary and Reciprocating Pumps," 14th ed., copyright 1983, with the permission of Hydraulic Institute.)*

pumpage. With this arrangement, the two pumping gears do not contact each other, and the unit is more suitable for liquids with low lubricity. This arrangement does, however, require a larger, more complex pump.

If the teeth are on the inside of a ring, as shown in Fig. 9.1.25, it is called an **internal gear**. In this unit, the larger internal-tooth gear drives the smaller external-tooth gear. Both gears transfer liquid from the suction to the discharge.

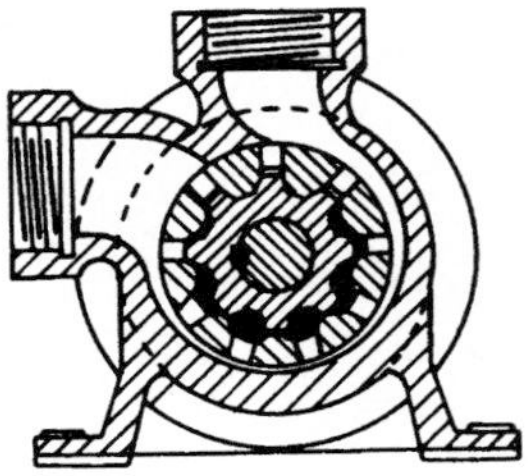

Figure 9.1.25 Internal-gear rotary pump.

Both external- and internal-gear rotary pumps are used in lubrication systems of engines, compressors, and larger pumps.

Lobe pumps are similar in construction and pumping action to external gear pumps, but one lobe does not drive the other. They do not even touch each other, and therefore must have their shafts independently driven by external timing gears. The lobes are often made of elastomers, and operate at low speeds. These units are used to transfer delicate items such as cherries and other foods—and even live fish. Figure 9.1.26 shows a rotary pump with three-lobe rotors.

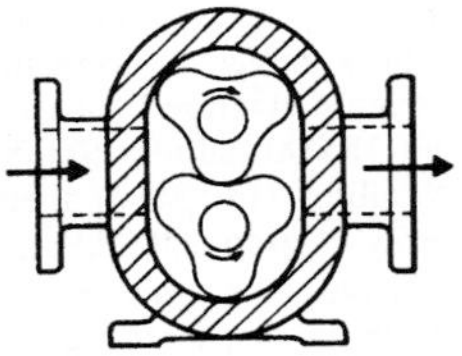

Figure 9.1.26 Three-lobe rotary pump. *(Reprinted from "Hydraulic Institute Standards for Centrifugal Rotary and Reciprocating Pumps," 14th ed., copyright 1983, with the permission of Hydraulic Institute.)*

Screw pumps are constructed with one, two, or three screws. The single-screw pump is more commonly called a progressing-cavity pump, and is discussed separately below. The two-screw pump, as illustrated by Fig. 9.1.27, is used to handle liquids with viscosities to greater than 1,000,000 SSU (200,000 mm²/s), with capacities to about 18,000 gal/min (4,000 m³/h), pressures to 3,500 lb/in² (25 MPa), and temperatures to 600°F (300°C). It is particularly suited for high-viscosity liquids.

The three-screw rotary pump, as illustrated in Fig. 9.1.28, is available, in the smaller sizes, for operation up to 8,000 r/min for use in lubrication systems on turbines, compressors, and centrifugal pumps. Larger units, as found in such services as crude oil pipelines are available with capacities to 3,300 gal/min (750 m³/h) and pressures to 4,500 lb/in² (31 MPa).

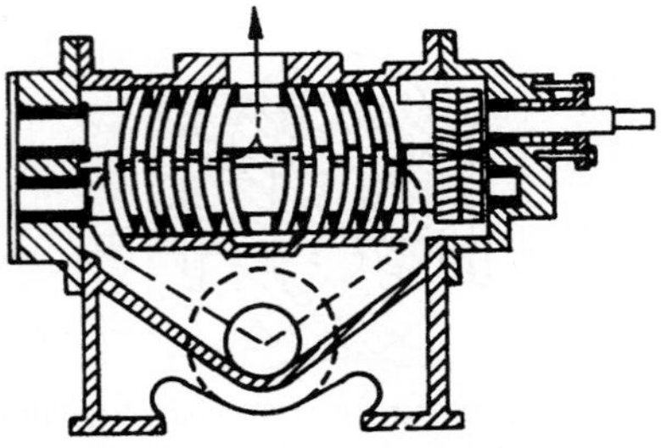

Figure 9.1.27 Two-screw rotary pump.

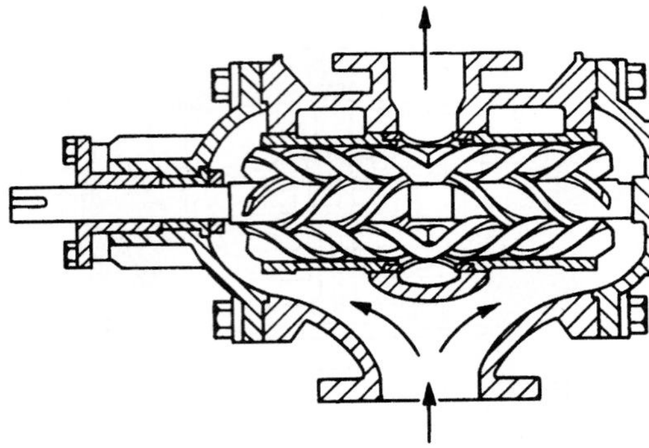

Figure 9.1.28 Three-screw rotary pump.

A **progressing-cavity** pump is illustrated in Fig. 9.1.29. The rotor is made of polished steel, and in the shape of a single-lead screw. The stator, usually made of elastomer, is in the shape of a double-lead screw. As the rotor turns, its axis prescribes a circle. This characteristic requires the rotor to be driven by an internal universal joint (or equivalent). The pumpage moves through the pump in a spiral. The rubber stator makes this unit well-suited for abrasive services.

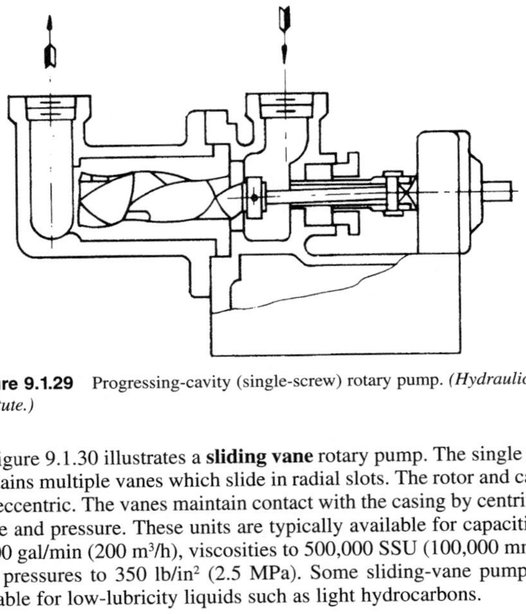

Figure 9.1.29 Progressing-cavity (single-screw) rotary pump. *(Hydraulic Institute.)*

Figure 9.1.30 illustrates a **sliding vane** rotary pump. The single rotor contains multiple vanes which slide in radial slots. The rotor and casing are eccentric. The vanes maintain contact with the casing by centrifugal force and pressure. These units are typically available for capacities to 1,000 gal/min (200 m³/h), viscosities to 500,000 SSU (100,000 mm²/s), and pressures to 350 lb/in² (2.5 MPa). Some sliding-vane pumps are suitable for low-lubricity liquids such as light hydrocarbons.

Figure 9.1.30 Sliding-vane rotary pump. *(Reprinted from "Hydraulic Institute Standards for Centrifugal, Rotary and Reciprocating Pumps," 14th ed., copyright 1983, with the permission of Hydraulic Institute.)*

Flexible Member Some rotary pumps are built with flexible vanes, liners, and tubes. The flexible tube pump, also called a **peristaltic** pump, is shown in Fig. 9.1.31.

A **radial-plunger** rotary pump is shown in Fig. 9.1.32. An **axial-plunger** rotary pump (also called an **axial-piston** or **swash-plate** pump) is shown in Fig. 9.1.33. In these pumps, the inner body rotates,

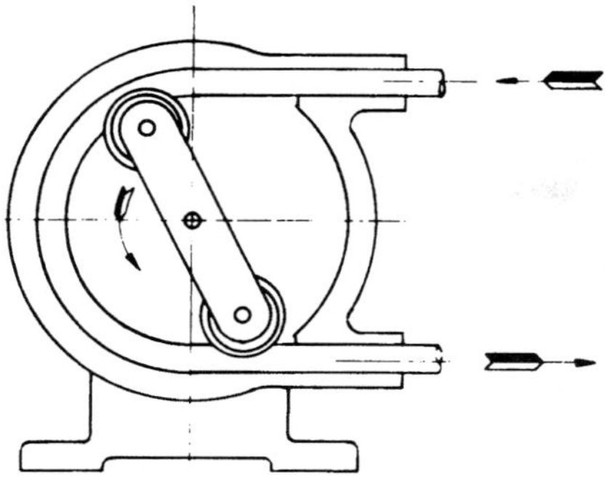

Figure 9.1.31 Flexible-tube (peristaltic) rotary pump. *(Hydraulic Institute.)*

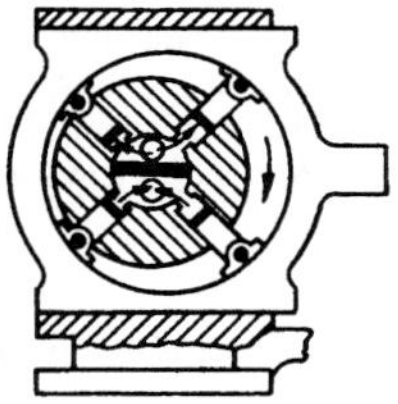

Figure 9.1.32 Radial-plunger rotary pump.

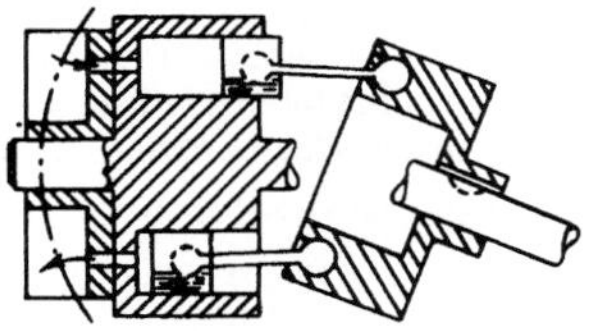

Figure 9.1.33 Axial-plunger (swash-plate) rotary pump.

causing each plunger to alternately accept pumpage from the inlet and deliver it to the discharge port. In the radial pump, the length of stroke of each plunger is established by the eccentricity of the rotor. In the axial pump, the stroke length is established by the angle of the plate or shaft. On some pumps, the eccentricity (or angle) is adjustable, providing variable displacement. Rotary plunger pumps are used in hydraulic systems to provide power to hydraulic motors and cylinders.

Circumferential-Piston Pump Although sometimes considered a lobe pump, this unit (shown in Fig. 9.1.34) differs from the lobe unit in that there is no close clearance between the two rotors. Close clearance does exist, though, between each rotor and adjacent stationary parts.

The **casing** (also called a **body** or **housing**) is the main pressure-retaining component of a rotary pump. It supports the rotor, contains the suction and discharge nozzles, and usually forms a portion of the boundaries of the pumping chambers.

The **bearings** of a rotary pump may be internal or external. If internal, they are lubricated by the pumpage. If the pumpage is not suitable for bearing lubrication, the bearings must be external. External bearings require two pumpage seals on each shaft. Some rotary pumps contain four seals.

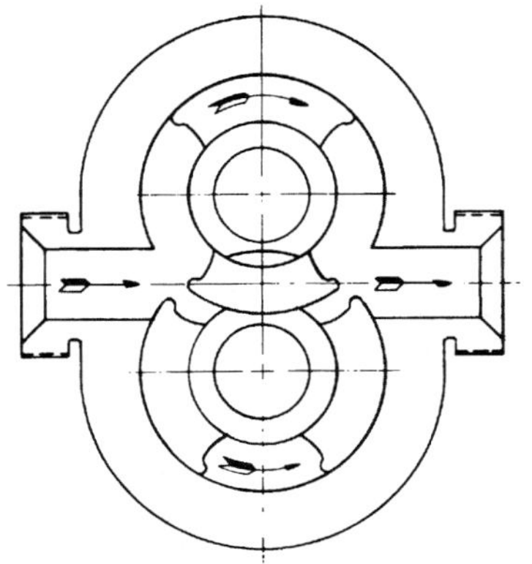

Figure 9.1.34 Circumferential-piston rotary pump. *(Hydraulic Institute.)*

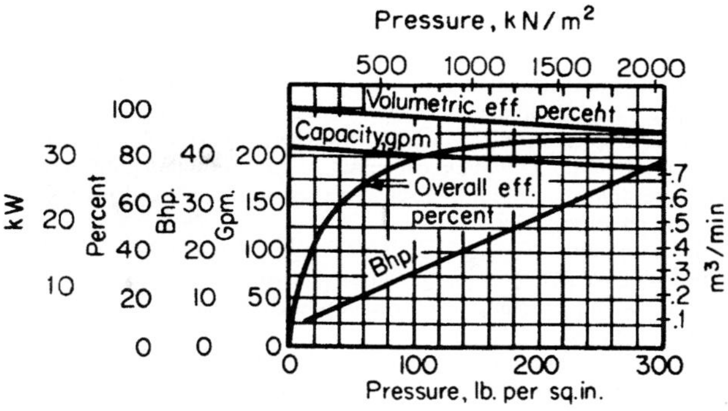

Figure 9.1.35 Rotary pump performance curves for a herringbone gear pump at 600 r/min with a 400-SSU (90-mm²/s) liquid.

Figure 9.1.35 shows typical **rotary pump performance curves.** Capacity, mechanical efficiency, and power are plotted against pressure.

The difference between displacement and capacity is called **slip.** Power input is the sum of power output and power losses (due to friction and slip).

As defined by the Hydraulic Institute, the NPSH required by a rotary pump is that value of NPSHA when cavitation noise, a sharp drop in capacity, or a 5 percent reduction in capacity occurs, whichever occurs first.

Increased Running Clearances An increase in the clearances between the rotating parts, and/or between the rotating and stationary parts, of a rotary pump will cause the slip to increase. An increase in slip causes the volumetric efficiency to drop, the mechanical efficiency to drop, and the NPSHR to rise. If the unit is running at a constant speed, the lower volumetric efficiency (lower capacity) will result in a lower discharge pressure if the system head varies with capacity.

VOLUMETRIC EFFICIENCY

Volumetric efficiency η_v of a displacement pump is the ratio of capacity Q to displacement D: $\eta_v = Q/D$. Capacity is the volume flow rate in the suction pipe. Displacement is the volume swept by the pumping elements per unit time. For a single-acting triplex plunger pump, displacement may be calculated as follows: $D = 3(\pi/4)d^2LN$, where D = displacement; d = diameter of plunger; L = stroke length; and N = crankshaft speed, r/min.

To predict the capacity of a reciprocating pump, in the selection process, it is necessary to predict the volumetric efficiency. Volumetric efficiency may be calculated with the following equation:

$$\eta_v = 1 - \frac{S}{D'} - \frac{C}{D'}\left(\frac{v_s}{v_d} - 1\right) \tag{9.1.2}$$

where η_v = volumetric efficiency; S = slip through both suction and discharge valves, typically about 3 percent of D'; D' = displacement of one pumping element during one stroke; C = clearance volume (dead space) in the pumping chamber (when the pumping element is at the end of the discharge stroke); v_s = specific volume of pumpage at suction conditions; and v_d = specific volume of the pumpage at discharge conditions.

Although volumetric efficiency has an effect on mechanical efficiency, in reciprocating pumps the two do not necessarily move in unison. It is possible to have a high η_m with a low η_v (due to liquid compressibility), or a low η_m with a high η_v (such as would occur in a high-suction-pressure, low-differential-pressure application).

The same equation can be applied to rotary pumps, but usually $v_s = v_d$, so that Eq. (9.1.2) reduces to $\eta_v = 1 - S/D'$. For a rotary pump, S = leakage of pumpage from the discharge side of the pump to the suction side. Because rotary pumps have no check valves, they must rely on close clearances to restrict slip. Slip is a function of the differential pressure, the viscosity of the pumpage, and the clearance between the parts. Slip in a rotary pump is usually a higher fraction of displacement than in a reciprocating pump.

POWER OUTPUT AND INPUT

The **power output** (also called **hydraulic power, hydraulic horsepower,** and **water horsepower**) of any pump is the usable power it imparts to the pumpage. It is the product of capacity and differential pressure.

In U.S. units, $P_o = Q \times \Delta p/1{,}714 = Q(p_d - p_s)/1{,}714$, where P_o = power output, hp; Q = pump capacity, gal/min; Δp = pump differential pressure, lb/in^2; p_d = pump discharge pressure, lb/in^2; and p_s = pump suction pressure, lb/in^2.

In metric units, $P_o = Q \times \Delta p/3{,}600 = Q(p_d - p_s)/3{,}600$, where P_o = power output, kW; Q = pump capacity, m^3/h; Δp = pump differential pressure kPa; p_d = pump discharge pressure, kPa; and p_s = pump suction pressure, kPa.

The **power input** (sometimes called **brake horsepower**) is the power required to drive power and rotary pumps. Normally, the power input includes the losses in an integral gear unit, but not the losses in a separate gear unit or variable-speed drive. V-belt losses may be included. If the mechanical efficiency is known, power input may be calculated by dividing the power output by the mechanical efficiency.

MECHANICAL EFFICIENCY—POWER AND ROTARY PUMPS

Mechanical efficiency (also called **pump efficiency** and **overall efficiency**) of a power-driven pump is the ratio of output power P_o to input power P_i: $\eta_m = P_o/P_i$.

As shown in Fig. 9.1.36, mechanical efficiency drops as the frame load is reduced, because the power output falls faster than friction losses, and becomes a smaller part of the power input. Mechanical efficiency is zero when the differential pressure is zero.

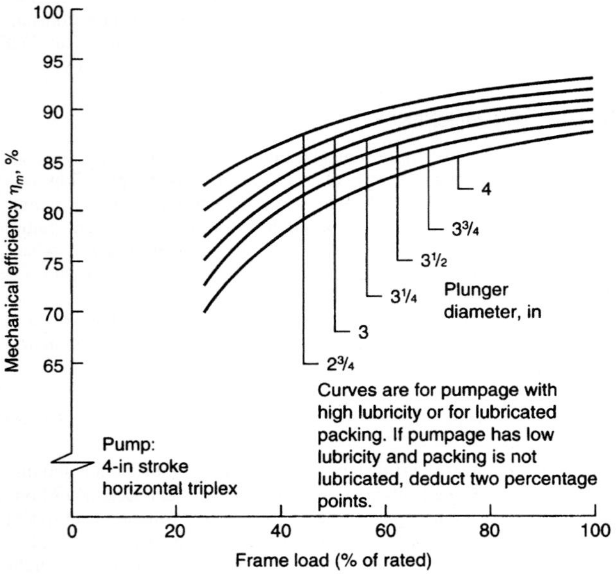

Figure 9.1.36 Mechanical (overall) efficiency of a triplex power pump.

PULSATION DAMPENERS

A **pulsation dampener** is a device which reduces the pressure pulsations created, in a suction or discharge pipe, by the interaction of a pump and a system. The dampener performs this function by reducing the velocity variations of the pumpage in the pipe.

Most reciprocating pumps, and many rotary pumps, create a variation in the velocity of the pumpage in the piping. Unless the suction and discharge vessels are very close to the pump, these velocity variations will be converted by the system into pressure variations. To reduce these pulses, a small vessel can be installed in the suction and/or discharge line. Such a vessel mounted in the discharge line is called a **pulsation dampener.** If the suction pressure is high enough, this same type of device can be used in the suction line, but a special dampener, called a **suction stabilizer** will perform the dual function of reducing pulses and separating free gas from the pumpage.

For optimum effectiveness, each dampener must be installed close to the pipe and adjacent to the pump. There must be minimal restriction between the dampener and pipe. For a low suction pressure, a simple air-filled standpipe is often adequate. For the discharge side, and for high suction pressures, the dampener often contains a diaphragm or bladder that is precharged with nitrogen to about 70 percent of system pressure.

SYSTEM DESIGN

A properly designed system is necessary for a satisfactory installation. Improper design will result in a system that vibrates and is noisy. Pulsations may be severe enough to damage pump components and instrumentation.

Field experience and information from the Hydraulic Institute standards are condensed below into a list of general design guidelines for the (1) suction vessel, (2) suction piping, and (3) discharge piping. The asterisks in the following lists show features that *do not* apply to rotary pumps that produce a uniform flow rate (constant velocity).

The suction vessel should:

Be large enough to provide sufficient retention time to allow free gas to rise to the liquid surface.

Have the feed and return lines enter below the minimum liquid level.

Include a vortex breaker over the outlet (pump suction) line and/or provide sufficient submergence to preclude vortex formation.

Contain a weir plate to force gas bubbles toward the surface. The top of the weir must be sufficiently below the minimum tank level to avoid disturbance.

The suction piping should:

Be as short and direct as possible.

Be one or two pipe sizes larger than the pump suction connection.

Have an average liquid velocity less than the values in Fig. 9.1.37.

Contain a minimum number of turns, which should be long-radius elbows or laterals.

Be designed to preclude the collection of vapor in the piping. (There should be no high points, unless vented. The reducer at the pump should be the eccentric type, installed with the straight side up.)

Be designed so that NPSHA, allowing for acceleration head, exceeds NPSHR.

Include a suitable suction stabilizer/gas separator located in the suction pipe adjacent to the pump liquid end if the acceleration head is excessive or the pumpage contains free gas.

Contain a full-opening block valve so that flow to the pump is not restricted.

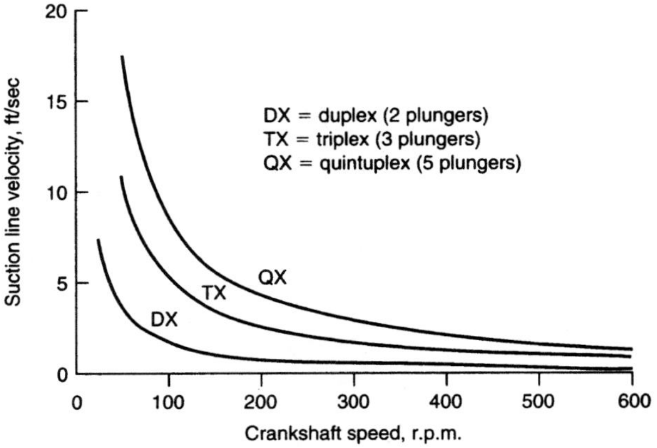

Figure 9.1.37 Suction line velocities for constant acceleration head, for reciprocating power pumps.

Not include a strainer or filter unless regular maintenance is assured. (The starved condition resulting from a plugged strainer can cause more damage to a pump than solids.)

The discharge piping should:

Be one or two pipe sizes larger than the pump discharge connection.

Have an average velocity less than 3 times the maximum suction line velocity.

Contain a minimum number of turns, using long-radius elbows and laterals where practical.

Include a suitable pulsation dampener (or provisions for adding one) adjacent to the pump liquid end.*

Contain a relief valve, sized to pass full pump capacity at a pressure that does not exceed 110 percent of cracking pressure (the opening pressure of the relief valve). The discharge from the relief valve should be piped back to the suction vessel so that gases liberated through the valve are not fed back into the pump.

Contain a bypass line and valve so that the pump may be started against negligible discharge pressure.

Contain a check valve to prevent the imposition of system pressure on the pump during start-up.

The features of a good system are illustrated in Fig. 9.1.38.

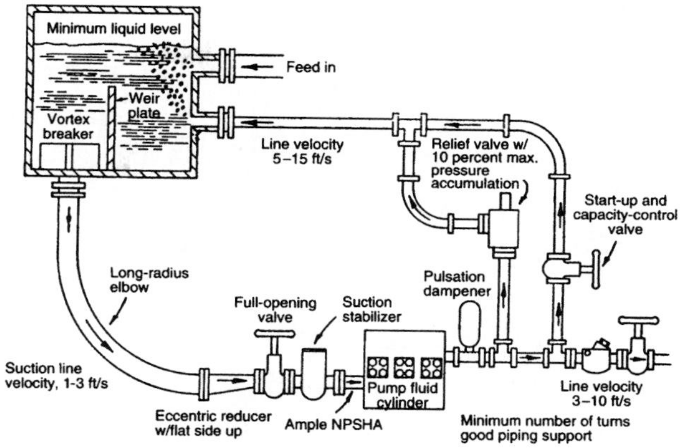

Figure 9.1.38 A good system for a reciprocating pump.

Remedies for Low NPSHA/High NPSHR

In designing the suction system for a displacement pump, if NPSHA is found to be less than NPSHR, a remedy may be found in the following list:

Increase the diameter of the suction line.

Reduce the length of the suction line by providing a more direct route, or by moving the pump closer to the suction vessel.

Install a suction bottle or stabilizer adjacent to the pump liquid end. A fabricated bottle has often been successfully used at pressures below about 50 lb/in^2 (300 kPa), but maintenance of a liquid level is required.

A section of rubber hose in the suction line, adjacent to the pump, will often reduce the acceleration head.

Elevate the suction vessel or the level of liquid in the suction vessel.

Reduce temperature of pumpage. (If the pumpage is at bubble point in the vessel, it must be cooled after it leaves the vessel.)

Reduce speed of the pump, or select a larger pump running slower. At a lower speed, it may be possible to operate a reciprocating pump with light suction valve springs, or none at all.

If the above steps are insufficient, impractical, or impossible, a booster pump must be provided. A booster for a displacement pump is normally a centrifugal, although direct-acting reciprocating and rotary pumps are sometimes used. The NPSHR of the booster must be less than that available from the system. The head of the booster should exceed the main pump NPSHR plus suction-line losses by about 20 percent. The booster should be installed near the suction vessel, and a pulsation bottle or stabilizer should be installed in the suction line, adjacent to the main pump, to protect the booster from pulsating flow.

9.2 CENTRIFUGAL PUMPS

by Terry L. Henshaw

REFERENCES: Stepanoff, "Centrifugal and Axial Flow Pumps," Wiley. Wislicenus, "Fluid Mechanics of Turbomachinery," McGraw-Hill, and Marks' Handbook, 6th ed., McGraw-Hill. Pfleiderer, "Die Kreiselpumpen," Springer. Spannhake, "Centrifugal Pumps, Turbines and Propellers," Technology Press. Hicks, "Pump Selection and Application" and "Pump Operation and Maintenance," McGraw-Hill. Standards of the Hydraulic Institute. Karassik, "Centrifugal Pump Clinic," Marcel Dekker.

NOMENCLATURE AND MECHANICAL DESIGN

The term "centrifugal pumps" as used here includes radial-flow, mixed-flow, and axial-flow (propeller) pumps. Centrifugal pumps are sometimes called **impeller pumps.**

A centrifugal pump consists primarily of a vane-carrying wheel, called the **impeller,** and a **casing** (Fig. 9.2.1). The impeller imparts velocity to the **pumpage,** with most of that velocity being converted to pressure inside the impeller and casing. The casing guides the pumpage to and from the impeller.

If a pump contains a single impeller, it is called a **single-stage** pump (Figs. 9.2.1 and 9.2.2). When two or more impellers are arranged in series (for higher pressure), it is called a **multistage** pump (Figs. 9.2.9 and 9.2.12).

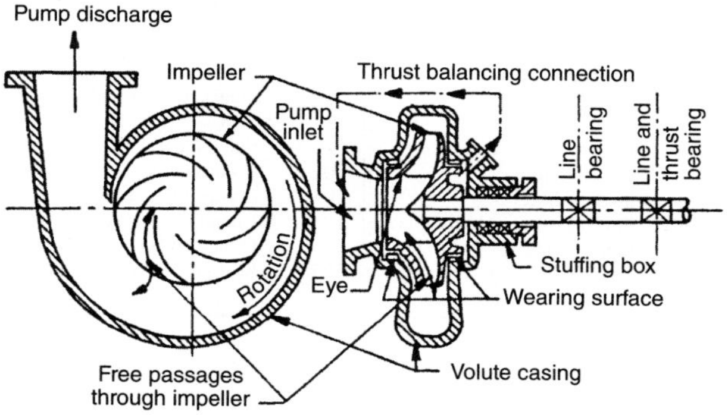

Figure 9.2.1 Single-stage single-suction overhung pump with volute casing.

Size designations of centrifugal pumps usually contain the nominal sizes of the discharge and suction nozzles, and sometimes the approximate maximum diameter of the impeller—or the number of stages. For example, in U.S. units, a 2 × 3 × 8 pump would have a 2-in discharge

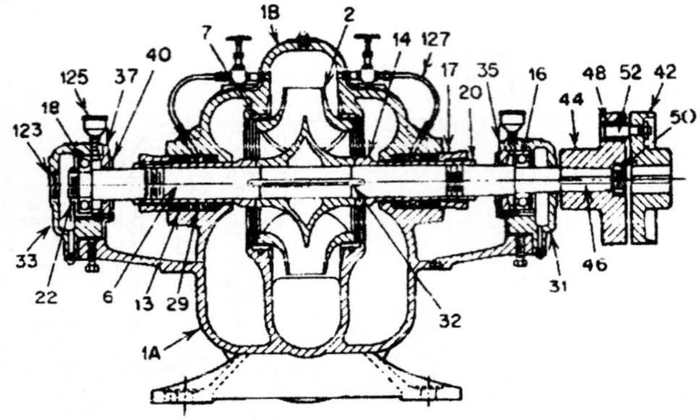

Figure 9.2.2 Horizontal single-stage double-suction between-bearings pump with a volute casing. (Numbers refer to parts listed in Table 9.2.1.)

nozzle, a 3-in suction nozzle, and an 8-in (approximately) maximum-diameter impeller.

If the impeller is mounted near the middle of the shaft, with a bearing near each end of the shaft, the pump is said to be of the **between-bearings** design (Fig. 9.2.2). If the impeller is mounted at the end of the shaft, with both bearings between the impeller and the coupling, the pump is called **overhung** or **cantilevered** (Fig. 9.2.1).

If the pump shaft is horizontal, it is a **horizontal pump.** If the shaft is vertical, it's a **vertical pump.**

Table 9.2.1 is a list of names used to describe centrifugal pump parts that are discussed below.

Table 9.2.1 Recommended Names of Centrifugal-Pump Parts
(These parts are called out in Figs. 9.2.2, 9.2.4, and 9.2.7)

Item no.	Name of part	Item no.	Name of part
1	Casing	33	Bearing housing (outboard)
1A	Casing (lower half)	35	Bearing cover (inboard)
1B	Casing (upper half)	36	Propeller key
2	Impeller	37	Bearing cover (outboard)
4	Propeller	39	Bearing bushing
6	Pump shaft	40	Deflector
7	Casing ring	42	Coupling (driver half)
8	Impeller ring	44	Coupling (pump half)
9	Suction cover	46	Coupling key
11	Stuffing-box cover	48	Coupling bushing
13	Packing	50	Coupling lock nut
14	Shaft sleeve	52	Coupling pin
15	Discharge bowl	59	Handhole cover
16	Bearing (inboard)	68	Shaft collar
17	Gland	72	Thrust collar
18	Bearing (outboard)	78	Bearing spacer
19	Frame	85	Shaft-enclosing tube
20	Shaft-sleeve nut	89	Seal
22	Bearing lock nut	91	Suction bowl
24	Impeller nut	101	Column pipe
25	Suction-head ring	103	Connector bearing
27	Stuffing-box-cover ring	123	Bearing end cover
29	Lantern ring	125	Grease (oil) cup
31	Bearing housing (inboard)	127	Seal piping (tubing)
32	Impeller key		

Impellers

An impeller is either **single-suction** (Fig. 9.2.1) or **double-suction** (Fig. 9.2.2). A single-suction impeller can have a **radial-flow** (Fig. 9.2.1), a **mixed-flow** (Fig. 9.2.3), or an **axial-flow** (Fig. 9.2.4) profile. A **closed impeller** has full **shrouds** (sidewalls) enclosing the vanes (Figs. 9.2.1 and 9.2.2), an **open impeller** has no shrouds, and a **semi-open impeller** has partial shrouds (Figs. 9.2.8 and 9.2.11). A semi-open impeller is often called an open impeller.

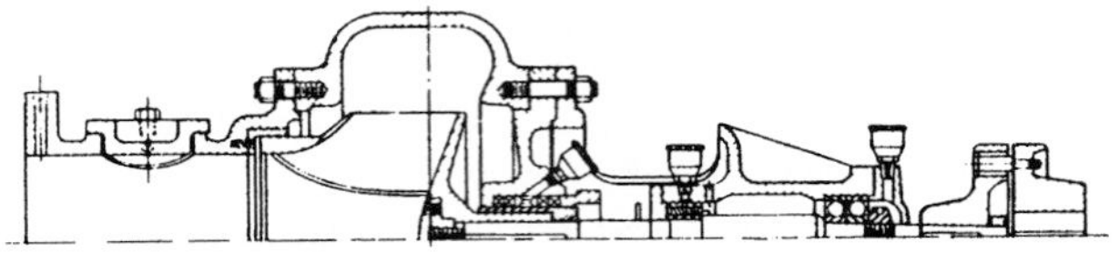

Figure 9.2.3 End-suction pump with a mixed-flow impeller and removable suction and stuffing-box heads.

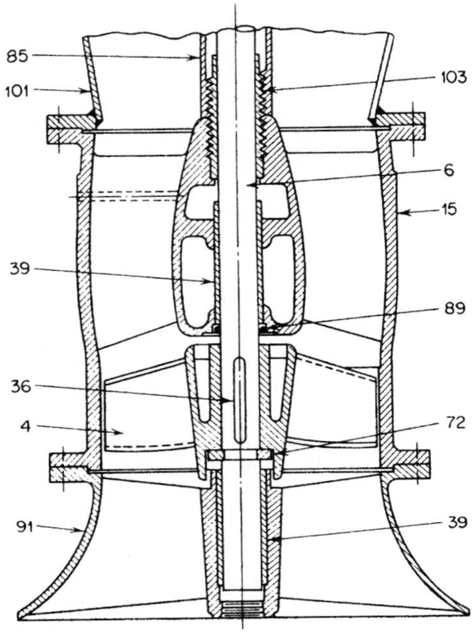

Figure 9.2.4 Vertical wet-pit diffuser pump bowl assembly with an axial-flow impeller (Numbers refer to parts listed in Table 9.2.1.)

Some impellers are designed for specific applications. Impellers for sewage have blunt edges and large passages to minimize clogging from rags and other fibrous matter. Paper stock impellers are open and may have screw conveyor vanes that extend into the suction nozzle. Impellers for abrasive slurries have thick vanes for extended life.

Casings

The casing usually is either of the **volute** (Fig. 9.2.1) or **diffuser** (guide vane) (Fig. 9.2.5) design, although some casings are circular. Diffuser pumps were once commonly called turbine pumps, but this term has recently been more selectively applied to vertical diffuser pumps, now called **vertical turbine pumps** (Fig. 9.2.9).

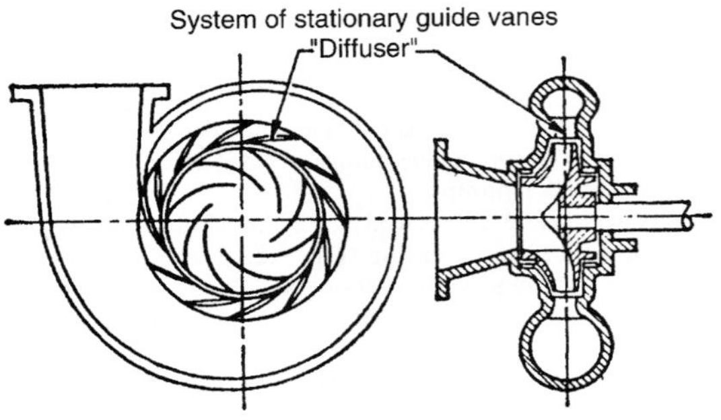

Figure 9.2.5 Single-stage single-suction pump with a multivane diffuser in a volute casing.

If the casing separates along a plane that is parallel to the axis of the shaft, the casing is said to be **axially split** (Fig. 9.2.2). If it parts in a plane that is at a right angle to the axis, it is called **radially split** (Figs. 9.2.6 and 9.2.7).

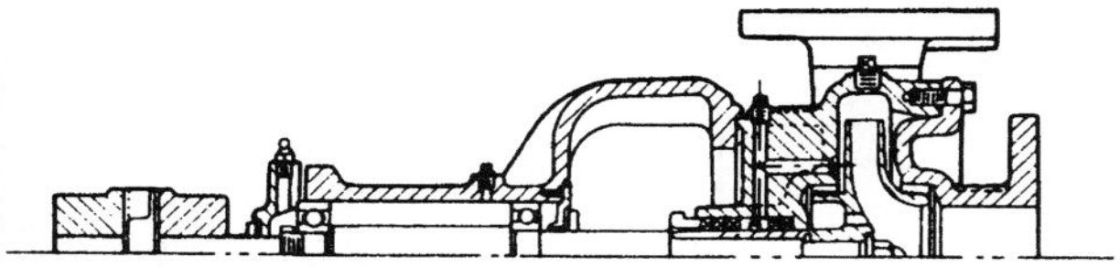

Figure 9.2.6 Single-stage end-suction pump with a radially split casing.

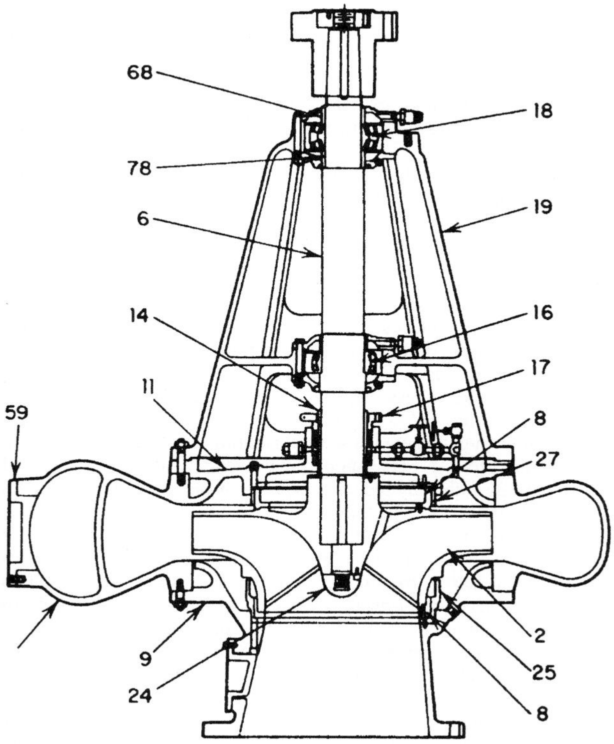

Figure 9.2.7 Vertical end-suction pump with a double-volute radially split casing. (Numbers refer to parts listed in Table 9.2.1.)

Multistage pumps also are made with both axially split casings (Figs. 9.2.12 and 9.2.13) and radially split casings (Fig. 9.2.14). The axially split units are limited to approximately 2,000 lb/in^2 (15 MPa). Higher-pressure pumps consist of a radially split outer casing **(barrel)** that contains a radially split (Fig. 9.2.14) or axially split inner casing assembly.

The location of the suction (inlet) nozzle determines whether the pump is classified as **end suction, side suction, bottom suction,** or **top suction.**

End-suction pumps usually have radially split casings. A cover (head) is bolted to the casing on the suction side and/or the back side. A **suction cover (casing suction head)** contains the suction nozzle. A back cover contains the seal or packing, and is called a **stuffing-box cover, casing cover,** or **stuffing-box head.** Both designs are illustrated in Figs. 9.2.3 and 9.2.7.

The discharge nozzle of end-suction single-stage horizontal pumps is usually in a top-vertical position (Fig. 9.2.6), although with some units the casing can be rotated so the discharge nozzle is top horizontal, bottom horizontal, or bottom vertical. Most double-suction, axially split pumps (Fig. 9.2.2) have a side discharge nozzle and either a side

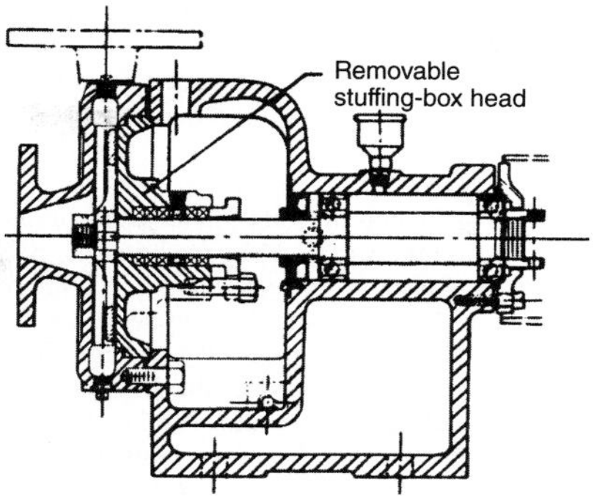

Figure 9.2.8 End-suction frame-mounted pump with a semi-open impeller and a removable stuffing-box head.

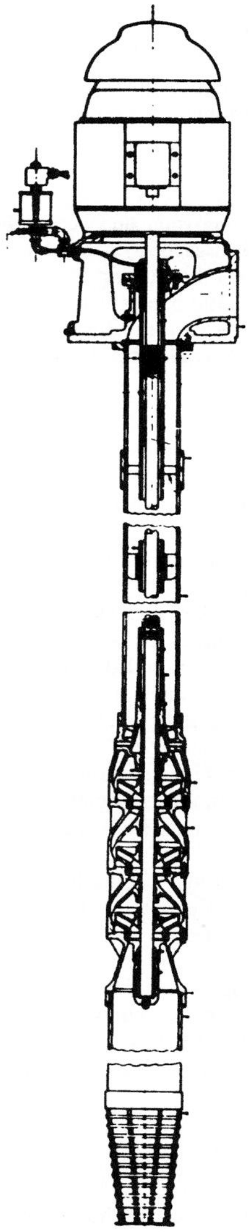

Figure 9.2.9 Vertical turbine multistage pump with closed impellers and oil-lubricated enclosed line shaft.

or bottom suction nozzle. The bottom-suction units are rarely made in sizes below a 10-in (25-mm) discharge nozzle.

If the casing is supported by a foot directly below the casing, it is said to be **foot mounted** (Fig. 9.2.2). If the support is under the bearing housing, it is called **frame mounted** (Fig. 9.2.8). If the casing is supported at the centerline of the shaft, it is a **centerline-mounted** pump.

Centerline support is provided for high-temperature services, so that coupling alignment is retained as the casing expands at the higher temperature.

With most pumps the liquid flows to the unit through piping, but some vertical pumps are submerged in their suction supply. Vertical pumps are therefore called either **dry-pit** or **wetpit** type (Fig. 9.2.9). The pumpage from a wet-pit pump may discharge above or below its support surface, creating designations of **above-grade discharge** (Fig. 9.2.9) and **below-grade discharge**.

Wearing Rings

Wearing rings (wear rings) (Figs. 9.2.2 and 9.2.7) are used with closed impellers to restrict the leakage of the pumpage from the discharge side to the suction side of the impeller. These rings may be integral with the casing and/or impeller, or may be separate parts. The stationary ring is called a **casing ring** or **cover ring.** The rotating ring, on the impeller, is called the **impeller ring.** Normally only small pumps have integral casing rings. Integral impeller rings are common on small pumps and on high-speed, high-energy pumps. High-energy, and some lower-energy multistage, pumps have experienced loosening and breakage of separate impeller rings, resulting in pump failures. Therefore, these units are often made with integral impeller rings, with hard surfaces to extend the life.

Although available in a variety of configurations, the most common ring configurations are the flat type (Fig. 9.2.6) and the L type (Fig. 9.2.3).

The simple semi-open-impeller pump of Fig. 9.2.8 minimizes leakage from the discharge side to the suction side with a small axial clearance between the impeller and casing. **Pump-out vanes** on the back of the impeller reduce axial thrust and pressure on the seal.

Shafts

The shaft in a centrifugal pump supports and drives the impeller and other rotating parts, and is, in turn, supported by the bearings. Because rubbing contact can lead to accelerated wear, or seizure, of wearing rings, the shaft may be larger than needed to transmit the torque, so as to reduce radial deflection of rotating parts. The diameter may be large enough to prevent contact between wearing rings and between the shaft and internal bushings.

The "dry" first **critical speed** of a rotor is related to its static deflection in the horizontal position. A shaft for a multistage pump, designed for a static deflection of, for instance, 0.006 in. (0.15 mm), will have a dry first critical speed of about 2,400 r/min. If operated within 20 percent of 2,400 r/min, excessive vibration and damage can occur. Such a pump, operating at 3,600 r/min, or faster, will be running above this dry critical speed. Smooth wearing rings and internal bushings can increase the actual ("wet") critical speed to a value above running speed. Subsequent wear of these surfaces (increased radial clearances), or grooves in the surfaces, will cause the wet critical speed to drop. If the wet critical speed is within 20 percent of running speed, excessive vibration and damage can result.

Shaft Sleeves

Pump shafts are often protected from erosion, corrosion, and wear at the sealing device, at leakage joints, and at liquid passages by **shaft sleeves** (Fig. 9.2.2). The most common application is at the stuffing box. Shaft sleeves serving other purposes are given specific names that indicate their purposes. For example, a sleeve between two impellers, which, in conjunction with a bushing, limits leakage between the impellers, is called an **interstage** (or distance) **sleeve.**

Bearings

(See also Sec. 6.)

Centrifugal pumps are provided with **rolling-contact** (ball and roller) **bearings** (Figs. 9.2.2, 9.2.6, and 9.2.7), and **fluid-film** (hydrodynamic) sliding **bearings** (Fig. 9.2.14). Their functions are to position the rotor and to absorb residual radial and axial forces created by the rotor.

The typical horizontal pump contains ball bearings (Fig. 9.2.2), lubricated by either grease or oil. If the rating of the ball bearings is exceeded (or by user preference), the pump is provided with fluid-film bearings. Some pumps are available with a combination of sleeve radial- and ball axial-thrust bearings, or sleeve radial and tilting-pad axial-thrust bearings (Fig. 9.2.14).

A **tilting-pad** (axial) **thrust bearing** requires a pressure-lubrication system, with its attendant components of pumps, filters, coolers, gages, and switches, which significantly increases the complexity and cost of the installation.

Some pumps contain internal fluid-film bearings that are lubricated by the pumpage. Examples include vertical turbine pumps (Fig. 9.2.9) and sealless pumps. These bearings can be damaged by solids or gas in the pumpage.

For a between-bearings pump, the bearing nearer the coupling is called the **inboard bearing.** The one farther from the coupling is called the **outboard bearing.** For an overhung pump, the bearing nearer the impeller is the inboard bearing; the one farther from the impeller is the outboard bearing.

Couplings

(See also Sec. 6.)

Some centrifugal pumps are constructed with the impeller mounted directly on the end of the driver shaft (Fig. 9.2.10). These units are called **close-coupled,** and require no coupling.

If the pump shaft and driver shaft are separate pieces, the two shafts must be connected by a mechanical coupling. If the pump and driver each contains its own bearing system, the coupling must be flexible. **Flexible couplings** are available in designs such as gear or grid, which require lubrication, and disk or diaphragm, which require no lubrication.

If the pump shaft relies on the driver shaft for support, the coupling must be a **rigid coupling.**

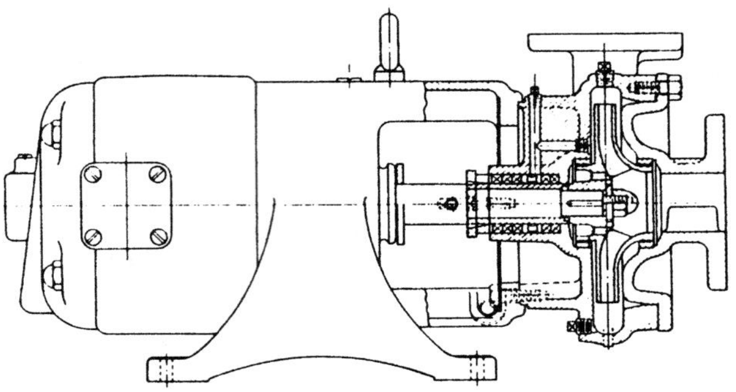

Figure 9.2.10 Close-coupled (motor-mounted) pump.

Sealing Devices

(See also Sec. 6.)

Packing is just fancy square rope, available in a variety of fibers, or foil, cut and formed into rings that fit snugly into the bore of the **stuffing box** and onto the surface of the shaft or sleeve (Figs. 9.2.2 and 9.2.6). Packing is held in the stuffing box by a **gland** (Figs. 9.2.2 and 9.2.6), which, in turn is retained and adjusted by gland bolts or studs and nuts.

Packed stuffing boxes have the primary function of minimizing leakage at the point where the shaft passes through the casing, although packing also provides radial support to the shaft, reducing vibration and radial deflection thereby reducing bending of the shaft (which reduces the probability of a fatigue failure of the shaft).

Packing on a rotating shaft (or sleeve) must be allowed to leak, or it will overheat and fail. Leakage must be a tiny stream, or at least a fast drip. If the pressure at the "throat" of the stuffing box (normally suction pressure) is below atmospheric pressure, the packing must prevent air from leaking into the pump. The rings of packing must therefore be separated by a **lantern ring (seal cage)** (Figs. 9.2.2 and 9.2.6), and clean, cool liquid must be provided to the lantern ring at a positive pressure. This "sealing liquid" flows along the shaft (or sleeve) both into the pump and to the atmosphere. If the pumpage is clean, cool water, the sealing liquid can be fed from the discharge of the pump into the lantern ring; either through external piping (Fig. 9.2.2), or through an internal passage (Fig. 9.2.6).

If the pumpage is hot, a slurry, toxic, flammable, or otherwise it should not be allowed to flow into the packing or escape to the atmosphere. A clean, cool sealing liquid must be provided to a lantern ring, at a pressure higher than throat pressure.

Mechanical seals are used in centrifugal pumps when the leakage rate from packing is unacceptable, the requirement for manual adjustment is undesirable, or a longer-life sealing device is desired. A mechanical seal provides no radial support for the pump rotor, so should be used only when the shaft and bearings provide adequate rigidity and support.

All mechanical seals leak. The leakage may be so small as to be visibly undetectable. In fact, a seal *must* leak, or it will overheat and fail prematurely. The sealing faces rely on a thin film of liquid between the faces for cooling and lubrication. Some of this liquid leaks past the faces.

For decades, a mechanical seal was required to fit inside the bore of a stuffing box that was designed for packing. This required the seal cross section to be small, and provided an inadequate environment for the seal, both contributing to shorter seal life. Newer pump designs provide larger cavities for seals, allowing more robust cross sections and a better environment, resulting in longer seal life.

A centrifugal pump seal chamber normally contains only one mechanical seal, but safety and environmental concerns have led to seal chambers containing two seals. These dual arrangements can be provided in the **double seal** arrangement, in which the seals are mounted back to back (Fig. 9.2.11) or face to face, or in the **tandem seal** arrangement, in which both seals face the same direction. The back-to-back double seal requires a liquid between the seals at a pressure above the throat pressure. The barrier liquid between tandem seals is normally at atmospheric pressure.

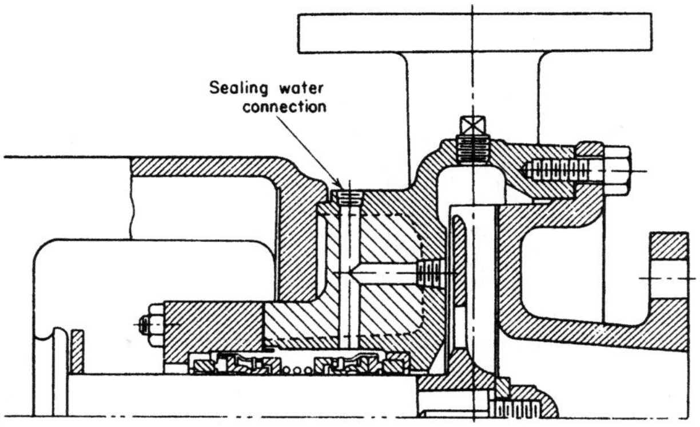

Figure 9.2.11 Single-stage end-suction pump with semi-open impeller and double mechanical seal.

Power plant boiler feed pumps are often equipped with a long labyrinth-type throttle bushing for a seal. Cool condensate is injected near the center of the bushing, flowing both into the pump and to the atmosphere. The atmospheric leakage is collected and returned to the system. Such a seal has high leakage, but a longer life than packing or a mechanical seal.

Sealless Pumps

Sealless centrifugal pumps can be provided for services where no leakage to the atmosphere can be tolerated. These pumps are available in both **canned-motor** and **magnet-drive** designs. In the canned-motor design, the motor rotor is inside the pump, surrounded by the pumpage, separated from the motor stator by a thin "can." The motor rotor is connected directly to the pump impeller. In the magnet-drive design, there are two sets of magnets, separated by a thin can. The inner set is connected to the pump impeller. The outer set is coupled to the pump driver. As the outer set rotates, lines of flux pass through the can and rotate the inner set. Both designs utilize fluid-film bearings that are lubricated by the pumpage.

HYDRAULIC THRUST

Radial thrust on an impeller is created by a non-uniform pressure around the periphery of the impeller. In a **single-volute** casing (Fig. 9.2.1), when the pump is operating at or near its capacity correspondent to peak efficiency **(best efficiency point, bep),** both velocity and pressure are uniform around the impeller, resulting in zero hydraulic radial thrust. At capacities both above and below bep, a non-uniform pressure distribution creates a radial force (thrust), which will increase shaft deflection and bearing loads. To reduce radial thrust, at off-bep capacities, casings may be designed with a **double volute** (Fig. 9.2.7) or a **diffuser** (Fig. 9.2.5).

Axial thrust is created by pressures acting axially on impeller shrouds and vanes. A double-suction impeller creates very little axial thrust. A single-suction impeller can create a significant axial thrust. This thrust is created by near-discharge pressure acting on a back shroud that is larger than the front shroud, and by vanes that twist and extend into the impeller eye. The thrust can be balanced by a back wearing ring and **balance holes** (Figs. 9.2.6 and 9.2.10) drilled through the back shroud, allowing the pumpage to flow from the back balance chamber into the impeller eye. For suction pressures near atmospheric, the back wearing ring diameter is normally equal to or larger than the front (eye) ring. For higher suction pressures on overhung pumps, the back ring may be reduced in diameter to compensate for the thrust created by the higher suction pressure pushing on the end of the shaft.

Most multistage pumps are built with single-suction impellers. To balance the axial thrust of these impellers, two arrangements are used:

(1) The impellers all face in the same direction and are mounted in the ascending order of the stages. The axial thrust is balanced by a hydraulic balancing device (Figs. 9.2.12 and 9.2.14). (2) An even number of single-suction impellers is used, one half of these facing in a direction opposite to the second half (Fig. 9.2.13). This mounting of single-suction impellers back to back is frequently called **opposed impellers.**

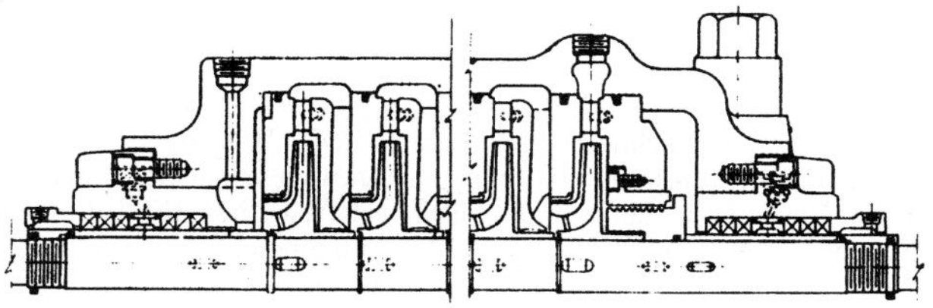

Figure 9.2.12 Multistage pump with axially split casing, single-suction impellers facing one direction, and hydraulic axial balancing device.

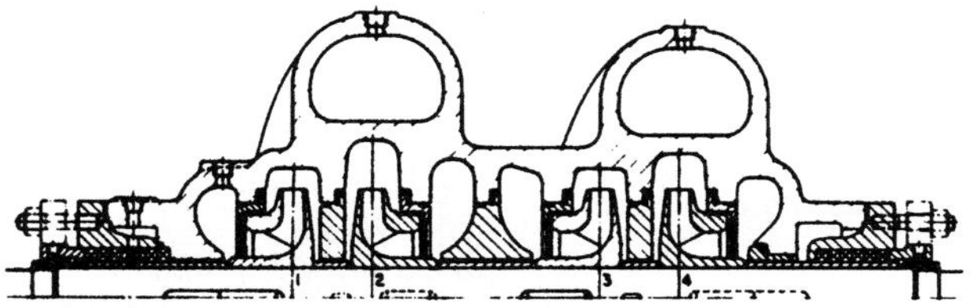

Figure 9.2.13 Four-stage pump with opposed impellers.

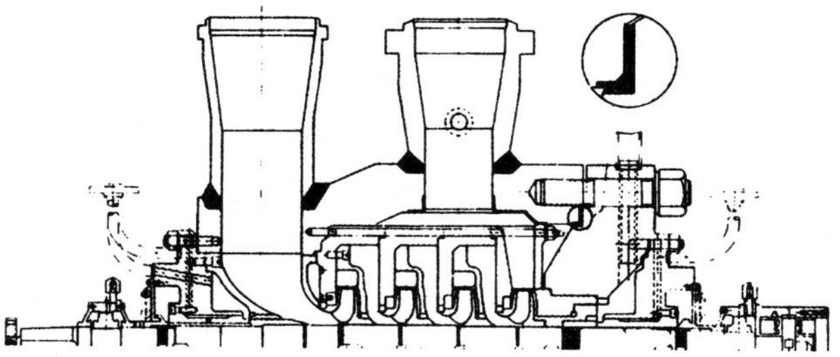

Figure 9.2.14 Double-casing multistage pump with radially split inner and outer casing.

Hydraulic balancing devices may take the form of a **balancing drum,** a **balancing disk,** or a combination of the two. A **balancing drum** is illustrated in Fig. 9.2.15. The balancing chamber is separated from the last-stage impeller by a drum mounted on the shaft. The drum has a close radial clearance with a stationary bushing.

The balancing chamber is connected either to the pump suction or to the vessel from which the pump takes its suction. The force acting on area B is pump discharge pressure. The force acting on area C is suction pressure. The difference between these two forces balances the axial thrust created by the impellers. The drum diameter can be selected to balance the axial thrust completely or to balance 90 to 95 percent of the thrust, depending on whether a slight thrust load on the thrust bearing is desired.

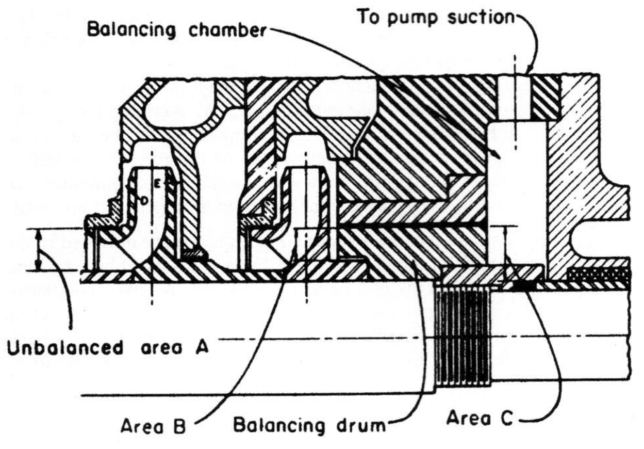

Figure 9.2.15 Balancing drum.

The simple **balancing disk** is illustrated in Fig. 9.2.16. The rotating disk is separated from the balancing-disk head by a small axial clearance. The leakage through this clearance flows into the balancing chamber, through an orifice, then either to the pump suction or to the suction vessel. The back of the balancing disk is subject to the balancing-chamber back pressure which varies as the axial clearance varies. If the rotor moves to the right, the axial clearance increases, causing the bck pressure to increase, which pushes the rotor back to the left.

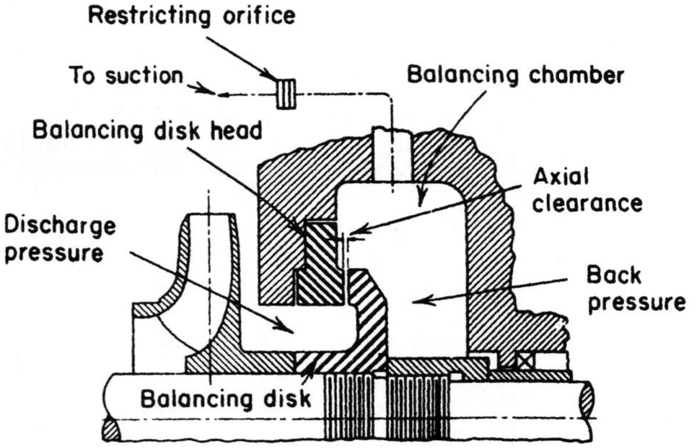

Figure 9.2.16 Simple balancing disk.

The **combination disk and drum** (Fig. 9.2.14) incorporates an orifice, eliminating the need for an external orifice. The radial clearance remains constant regardless of any axial displacement of the rotor. Such displacement, however, changes the axial clearance within the balancing device which changes the leakage, which in turn changes the pressure drop across the radial clearances, thus increasing or decreasing the pressure acting on the disk face. These changes in the intermediate pressure on the disk face act to move the balancing device in whichever direction is required to restore equilibrium and axial balance.

HYDRAULIC PERFORMANCE

The performance of a centrifugal pump is generally described in terms of the following **characteristics:** (1) rate of flow, or **capacity** Q, expressed in units of volume per unit of time, most frequently ft^3/s, gal/min, or m^3/h ($1\ ft^3/s = 449$ gal/min; $1\ m^3/h = 4.403$ gal/min); (2) increase of energy content in the fluid pumped, or **head** H, expressed in units of energy per unit of mass, usually ft · lb/lb or, more simply, ft, or m; (3) input **power** P, expressed in units of work per unit of time, horsepower or kW; (4) **efficiency** η, the ratio of useful work performed to power input; (5) **rotative speed** N, in r/min.

Since all other characteristics are a function of capacity, it is customary to represent the performance of a centrifugal pump by means of **characteristic curves** similar to that shown in Fig. 9.2.17. While it is possible, within certain limits, for the pump designer to regulate the shape of these curves to suit the needs of a particular application, the shapes are essentially a function of design capacity, head, and speed (specific speed).

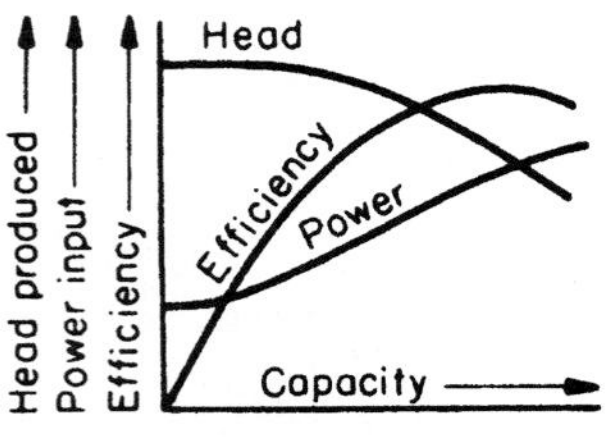

Figure 9.2.17 Typical characteristic curves at constant speed.

For any given capacity, the relationship between performance characteristics is expressed by the equation $\eta = \gamma QH/3{,}960P$, where η is efficiency expressed as a decimal, γ is the specific gravity of the fluid pumped, Q is in gal/min, H in ft, and P in horsepower. In metric units, $\eta = \gamma QH/367P$, where Q is in m^3/h and H in meters and P in kW.

Inasmuch as the actual performance of centrifugal pumps is determined by testing, it is highly desirable to be able to use the results of past tests as a basis for predicting the performance of future designs. To this end, an interesting and widely used characteristic number known as **specific speed,** $N_s = N\sqrt{Q}/H^{3/4}$, has been developed. In this expression, values of N, Q, and H are all for the point of best efficiency. While not dimensionless, N_s can be made dimensionless by the appropriate selection of units. Specific speed is of interest to both the pump designer and the pump user essentially in two ways: (1) All geometrically similar pumps, regardless of their size, will have identical specific speeds (but all pumps of the same specific speed will not necessarily be geometrically similar). (2) Within reasonable limits, pump geometry and performance can be predicted as a function of N_s and Q. For N in r/min, Q in gal/min, and H in ft, the practical range of N_s is approximately 500 to 15,000. In metric units Q is in m^3/s and H is in metres, so that this range becomes approximately 10 to 300 ($N_{sm} = N_s/51.66$) The shape of the profiles for impellers varies over the range of specific speeds, as shown in Fig. 9.2.18. The commercially attainable

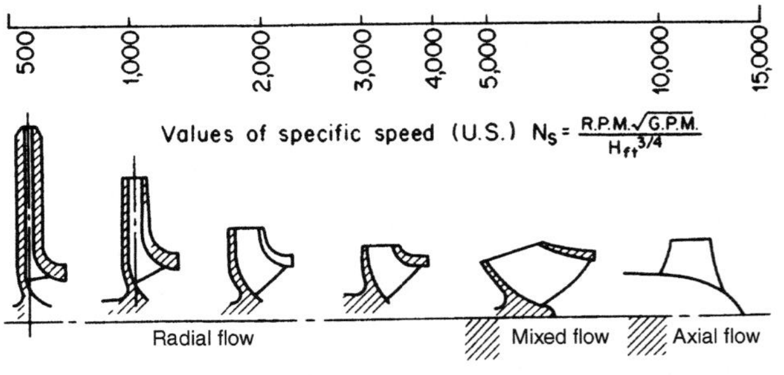

Figure 9.2.18 Approximate impeller shapes versus specific speed.

efficiencies also vary with the specific speed (Fig. 9.2.19). It must be remembered that these efficiencies are attainable at the capacity at best efficiency. The actual capacity may differ from this best-efficiency flow. In addition, the general shape of the characteristic curves will vary widely from one end of this range to the other, as illustrated by Figs. 9.2.24 and 9.2.25.

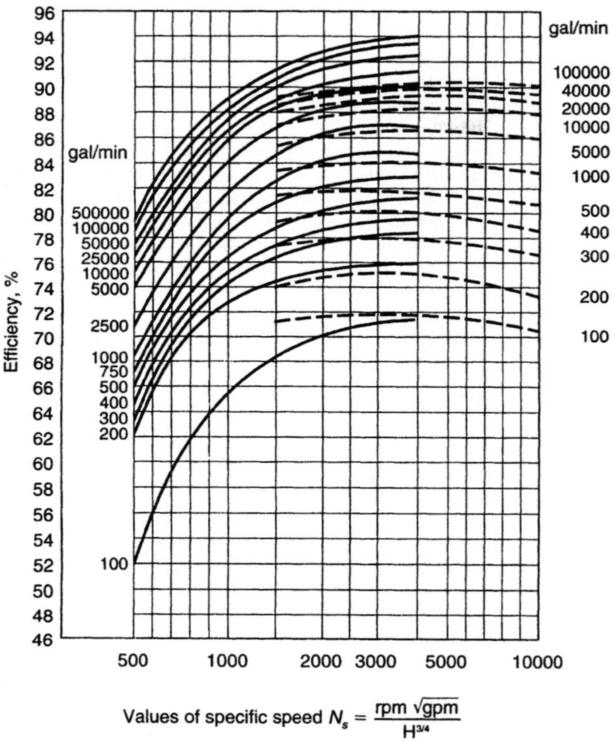

Figure 9.2.19 Efficiencies of single-stage end-suction and double-suction centrifugal pumps (solid lines) and wet pit pumps (dashed lines).

Hydraulic Design

The purpose of a centrifugal pump in any fluid-handling system is to add energy to the fluid, and since it is a dynamic machine, the pump depends entirely on changes in velocity to provide the energy. While the measurable evidence of energy addition is in most cases largely in the form of static pressure, this is partially the result of velocity reductions and constraints occurring in the diffuser and to this extent represents a conversion from the velocity energy produced by the impeller. Thus, any discussion of centrifugal-pump theory generally becomes a discussion of velocities occurring at various points within the pump.

The true velocity relationships existing within a pump are extremely complex; but for practical purposes, a one-dimensional analysis serves to illustrate the basic concepts and, indeed, has served as the basis of design for virtually all centrifugal pumps ever built.

For radial and mixed-flow impellers ($500 \le N_s \le 7{,}500$ or $10 \le N_{sm} \le 150$) the velocities at the inlet and outlet of the impeller are shown by the vector diagrams in Fig. 9.2.20. The head produced by such an impeller is represented by

$$H = \eta_H (U_2 V_{u2} - U_1 V_{u1})/g \qquad (9.2.1)$$

where H = head, ft (m); U = circumferential velocity of impeller at radius being considered, ft/s (m/s), V_u = average value of the circumferential component of absolute fluid velocity, ft/s (m/s), and g = 32.174 ft/s^2 (9.80 m/s^2). Subscript 1 refers to the impeller inlet section, and subscript 2 to the impeller discharge section. The coefficient η_H is the hydraulic efficiency of the rotating-vane system and, for the range of specific speeds indicated above, will generally fall between 0.85 and 0.95. This hydraulic efficiency is considerably higher than pump efficiency η since it does not include mechanical losses due to bearing or packing friction, volumetric losses due to internal wearing-ring clearances, impeller-disk friction, or fluid-friction losses due to velocity conversion or boundary-layer considerations ahead of or following the impeller. For pumps in this specific-speed range, the hydraulic losses $(1 - \eta_H)$ will generally be between one-quarter and three-quarters of total pump losses $(1 - \eta)$.

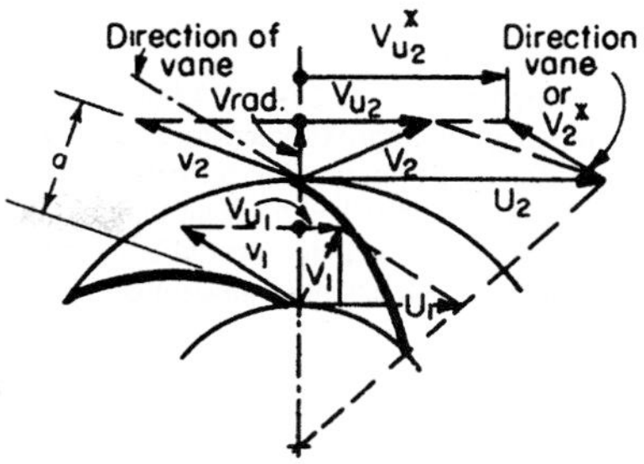

Figure 9.2.20 Velocity diagram of a radial-flow impeller.

For pumps arranged with an axial inlet to the impeller (such as that shown in Fig. 9.2.7), it is generally assumed that the entering flow will have no rotational component, and V_{u1} is therefore zero. Equation (9.2.1) can thus be reduced to

$$H = \eta_H U_2 V_{u2}/g \qquad (9.2.2)$$

In practice, Eq. (9.2.2) will provide a close approximation to total head for any pump up to a specific speed of 2,000 ($N_{sm} \approx 40$) since the term $U_1 V_{u1}$ in Eq. (9.2.1) is very small compared with $U_2 V_{u2}$.

It should also be noted in Fig. 9.2.20 that the relative velocity υ_2 does not coincide in direction with the vane angle at the impeller discharge. The angular difference between υ_2 and the direction of the vane is due to the irrotational nature of the flow between the vanes. The effect of this difference can be taken into account as the vector difference $V_{u2}^* - V_{u2} = Na/229$, where a is the shortest distance in inches taken in the radial plane between the discharge tip of any vane and the upper surface of the following vane. ($V_{u2}^* - V_{u2} = Na/19{,}100$, where a is in mm and velocities are in m/s.)

The foregoing relationships provide the basis for determining impeller diameters and vane angles required to produce a given total head requirement at a specified rotative speed. It is equally necessary, of course, to provide in the design for handling a specified flow volume, and this is readily accomplished by providing the necessary cross-sectional area A between vanes to pass the required flow at velocities previously determined. A useful relationship for this purpose, in light of the units commonly employed in pump design, is $Q = AV/0.321$, where Q is in gal/min, V in ft/s, and A in in^2 ($Q = AV/278$, where Q is in m^3/h, V in m/s, and A in mm^2).

Upon leaving the impeller, the liquid pumped enters either (1) a system of diffusing vanes surrounded by an outer casing or (2) directly into a casing designed to contain the fluid and control its velocities. Where diffusing vanes are used, they are designed on the basis of velocity relationships very similar to those employed in impeller design but with the objective being to slow the fluid down to convert velocity energy to pressure energy and, further, to reduce frictional losses in the discharge system following the diffuser. Where the impeller discharges directly into the casing, this component of the machine is most frequently designed in the form of a volute to provide constant velocity all around the impeller periphery up to the point of entry into the discharge nozzle. From this point, commonly called the **casing throat,** to the discharge flange or to the inlet of a succeeding stage, the velocity is gradually reduced. Special circumstances related to pump design or application often result in modifications to the constant-velocity design of such a casing, and variations may be found covering the entire range from constant velocity to constant area. In addition, many casings are now designed with one or more spiral vanes placed in such a way as to approximate a condition of geometric similarity in relation to the impeller, which is advantageous when a pump is operated at capacities other than those for which it is designed. A casing of this nature represents an effort by the designer to obtain an optimum balance between the desirable geometric similarity of the diffuser discharge and the manufacturing simplicity and generally high efficiency of the volute-type casing.

The most common form of such a casing is the double-volute type discussed under Hydraulic Thrust.

For **axial-flow impellers** ($7{,}500 \le N_s \le 15{,}000$ or $150 \le N_{sm} \le 300$) velocity relationships can be approximated in a manner similar to that used for lower-specific-speed pumps, but refinement of these approximations is approached in a somewhat different manner, largely because of the considerable body of knowledge available in the form of airfoil data which can be applied. Velocity diagrams for pumps of this type are shown in Fig. 9.2.21.

Considering a cylindrical stream tube intersecting the vanes of an axial-flow impeller, we can rewrite Eq. (9.2.1) in the form

$$H = \eta_H U \Delta V_u / g \qquad (9.2.3)$$

where ΔV_u represents the increase in the tangential component of the absolute velocity as the fluid passes through the impeller. For pumps in this specific-speed range, η_H will generally fall between 0.80 and 0.90 and $1 - \eta_H$ will generally be between one-half and three-quarters of $1 - \eta$.

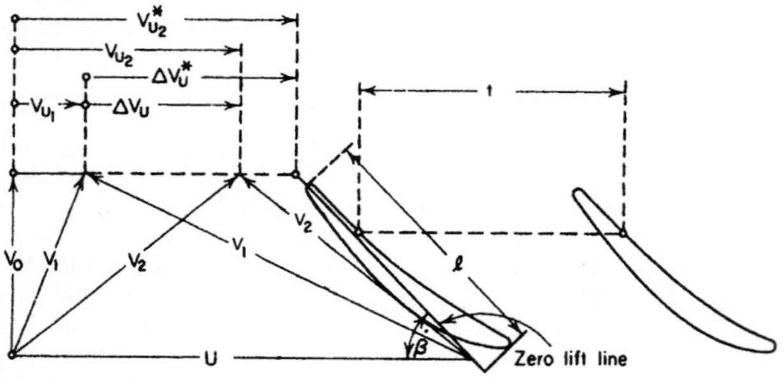

Figure 9.2.21 Velocity diagram of an axial-flow vane system.

As in the case of radial-flow impellers, the relative velocity v_2 at the impeller exit does not coincide with the vane angle. In this case, the necessary correction can be applied by means of the expression

$$\Delta V_u = 2\Delta V_u^* / [(t/l)(2/\pi K)(1/\sin\beta) + 1] \qquad (9.2.4)$$

where t is vane spacing, l is vane length, β is the discharge vane angle, and K is the coefficient determined from Fig. 9.2.22, which provides for the fact that the impeller blades are, in effect, arranged in a continuous lattice.

In the design of axial-flow pumps, it is generally assumed that the head developed by the blade elements within all the cylindrical stream tubes between the impeller hub and its outer diameter will be the same. For this condition to be achieved, it will be evident from Eq. (9.2.3)

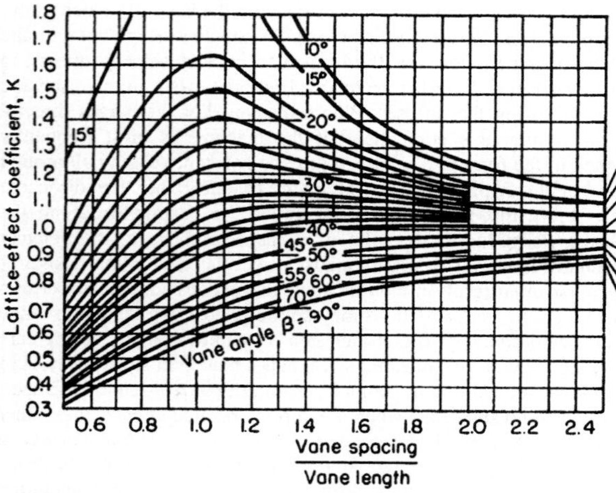

Figure 9.2.22 Lattice-effect coefficient. *(Weinig.)*

that since U will vary directly with the radius, V_u must vary inversely with the radius. Thus vane camber (or curvature) will be greater near the hub than at the periphery. It is further generally assumed that the axial velocity is constant throughout the impeller, and to satisfy this condition, the blade angles will be greater at the hub than at the periphery, giving rise to the twist of the vane.

By designing the vane element at the hub, the periphery, and any reasonable number of radial stations between, it is possible to define the vane over its entire surface to produce the total head desired. The design capacity can be simply established from the constant axial velocity and the annular area between the hub and the periphery.

Axial-flow impellers will almost invariably discharge into a vaned diffuser designed primarily to convert into pressure the tangential component of the absolute velocity leaving the impeller, thus producing a uniform, nonrotating velocity profile at the pump discharge or at the entrance to any succeeding stage.

Pump Application

In applying any centrifugal pump to a fluid-handling system, the engineer generally has two variables in the pump which may be used to effect a match between it and the system, namely, speed of operation and impeller diameter. (In axial-flow pumps, impeller diameter cannot conveniently be changed, but similar results can be accomplished within a limited range by reducing vane length.) To take advantage of these variables, it is necessary to understand the similarity relationships which govern pump behavior.

The effect of change in speed can be most readily explained by referring back to Eq. (9.2.1). For a fixed impeller diameter, it will immediately be evident that any increase in rotative speed will result in a directly proportional increase in the peripheral velocities of the impeller at both inlet and outlet, U_1 and U_2, respectively. Furthermore, since the directions of both the relative and absolute velocities v and V, respectively, are controlled by the vane angles, it follows that both the inlet and outlet velocity diagrams remain geometrically similar, in other words, as pump speed changes, all fluid velocities change in direct proportion. Thus the effect of a speed change on pump capacity can be represented for two speeds, N_1 and N_2 by

$$Q_1/Q_2 = N_1/N_2$$

Referring again to Eq. (9.2.1), it will be evident that the change in head will be proportional to the square of the change in speed, resulting in

$$H_1/H_2 = (N_1/N_2)^2$$

Also, since P is proportional to QH, then

$$P_1/P_2 = Q_1 H_1 / Q_2 H_2 = (N_1/N_2)^3$$

The effect of a reduction in impeller diameter is for practical purposes, identical to that of a reduction in speed, and the above equations can all be rewritten in the same form, substituting D_1 and D_2 for N_1 and N_2. In this case, however, there is somewhat less latitude for change, the practical limit for cut-down being approximately 25 to 30 percent in the case of low-specific-speed impellers and decreasing from this value as specific speed increases.

From the foregoing it should be noted that running a pump at speeds far below its normal rated speed would be uneconomical since the pump would obviously be overdesigned for the service. Conversely, the same pump could not be run at speeds far above its rating since it would soon exceed its power capability. Thus the normal approach to pump application calls for selection of a unit which will meet the system requirements at or near the pump's maximum rated speed and preferably as close to maximum impeller diameter as possible.

A further aspect of the similarity laws, of interest primarily to pump designers, is the matter of geometrically similar pumps, or as they are commonly called, **factors of each other.** Assuming two such pumps with a size ratio f, then **at the same rotative speed** $H_1/H_2 = f^2$, which is the same as for a change in speed of the same ratio. In the case of capacity, however, we have, in addition to the linear increase due to velocity change, a further increase due to the change in the cross-sectional area of all fluid channels which is proportional to the square of the size factor, resulting in the relationship $Q_1/Q_2 = f^3$. The ratio of power requirements therefore becomes $P_1/P_2 = f^5$.

To simplify selection of pumps, it is customary to plot, for any given speed, performance curves to show pump characteristics over the available range of impeller diameters rather than at the single diameter which would be implicit in a curve of the type shown in Fig. 9.2.17. A typical rating curve of this nature is shown in Fig. 9.2.23. Such rat-

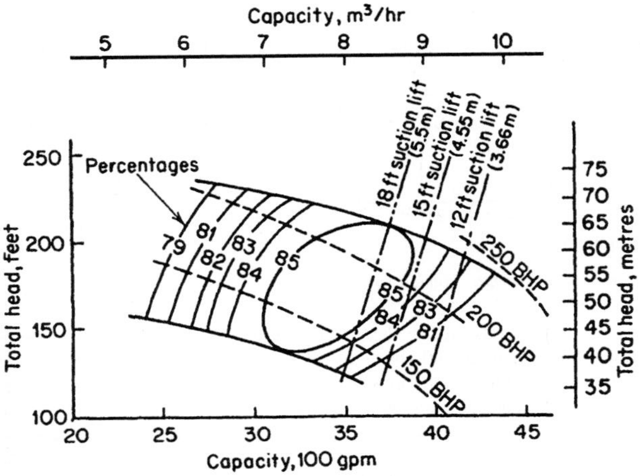

Figure 9.2.23 Rating curve of 10-in double-suction single-stage pump.

ing curves will vary widely in their nature with changes in N_s, and an understanding of these variations is essential to the selection of properly sized pump drivers and, in many cases, proper design of discharge piping and/or the determination of limiting ranges of pump operation. Some idea of the diversity of characteristics which is available can be obtained by a comparison between Figs. 9.2.24 and 9.2.25, both of which are plotted entirely in percentages of design values occurring at the point of best efficiency.

In the case of the lower-specific-speed pump shown in Fig. 9.2.24, a driving motor selected for the rated condition would be more than adequate at lower capacities, and discharge piping would be subjected to only modest overpressures. On the other hand, if the pump shown in Fig. 9.2.25 is to be operated at reduced capacities, then both motor

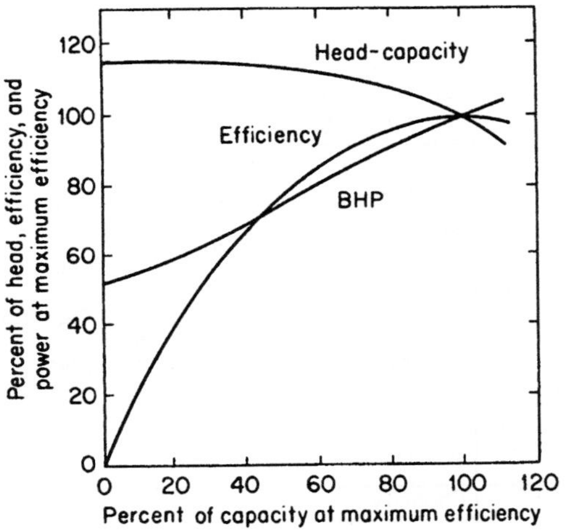

Figure 9.2.24 Typical characteristics for $N_s = 1{,}550$ ($N_{sm} = 30$).

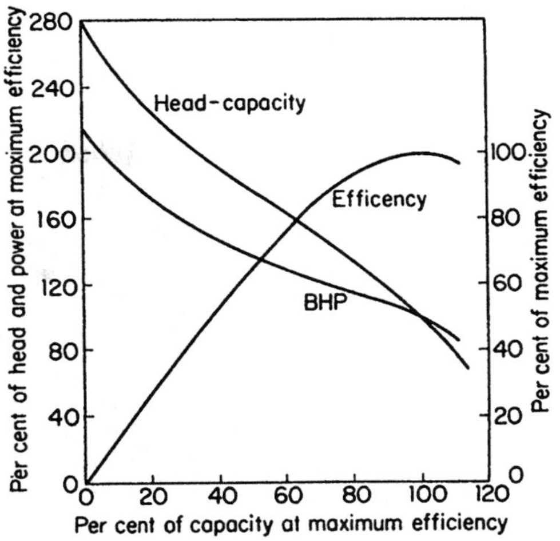

Figure 9.2.25 Typical characteristics for $N_s = 10{,}000$ ($N_{sm} = 194$).

size and discharge piping must be designed for the minimum flows to be encountered.

NPSH

In order for a pump to deliver its rated output, it is obviously necessary that it be supplied with fluid at its inlet at the same rate. It is further necessary that the absolute pressure (including velocity head, $V^2/2g$) of the fluid at the inlet exceed the vapor pressure by an amount sufficient to overcome (1) any entrance or frictional losses between the point of entry into the pump and the impeller, and (2) the shock losses occurring at the impeller inlet. This gives rise to the definition of **net positive suction head (NPSH)** which is the absolute pressure at the pump inlet expressed in feet (meters) of liquid, plus velocity head, minus the vapor pressure of the fluid at pumping temperature, and corrected to the elevation of the pump centerline in the case of horizontal pumps or to the entrance to the first-stage impeller for vertical pumps.

NPSH required is determined by the pump manufacturer and is a function of both pump speed and pump capacity. **NPSH available** represents the energy level of the fluid over the vapor pressure at the pump inlet and is determined entirely by the system preceding the pump. Unless NPSH available at least equals NPSH required at any condition of operation, some of the fluid will vaporize in the pump inlet and bubbles of vapor will be carried into the impeller. These bubbles will collapse violently at some point downstream of the pump inlet (usually at some point within the impeller) and produce very sharp, crackling noises, frequently accompanied by physical damage of adjacent metal surfaces. This phenomenon is known as **cavitation** and is generally highly undesirable.

A term similar to that used for pump specific speed has been developed for pump NPSHR characteristics and has been identified as **suction specific speed,** $S = N\sqrt{Q}/H_{sv}^{3/4}$, where H_{sv} is NPSH. For double-suction impellers, $S = N\sqrt{Q/2}/H_{sv}^{3/4}$. The value S, like N_s, is expressed simply as a number and, for general-purpose pumps, does not generally exceed 10,000, where N is in r/min, Q is in gal/min, and H_{sv} is in ft. When Q is in m³/s and H_{sv} is in meters, the equivalent value S_m is 194.

Within the limits of NPSH capabilities, it is desirable to select a pump of the highest specific speed since this will produce the highest pump speed and consequently use the smallest pump. As a convenience in accomplishing this, the Hydraulic Institute has published two charts indicating the recommended NPSH as a function of pump capacity and operating speed and, conversely, the maximum recommended operating speed for different capacities if the NPSH value is known or first selected. These charts are based on a recommended value of suction specific speed of 8,500 ($S_m = 165$). The chart for double-suction pumps is reproduced in Fig. 9.2.26.

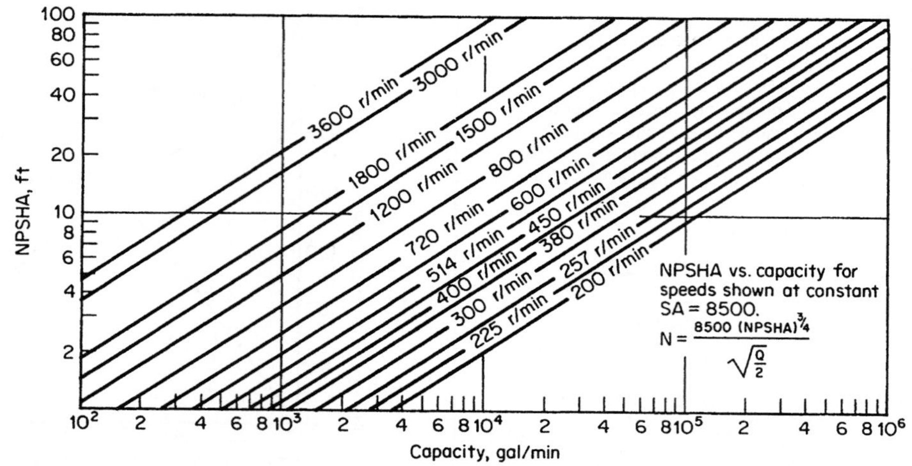

Figure 9.2.26 Recommended maximum operating speeds for double-suction pumps. *(Hydraulic Institute Standards.)*

Recirculation A better understanding has been recently reached of the flow pattern in a pump impeller when the pump is operated at capacities below that at its best efficiency. All pumps exhibit the phenomenon of recirculation which is a flow reversal at the inlet or at the discharge tips of the impeller. The capacities at which flow recirculation occurs at the suction or discharge are not necessarily coincidental. It is now possible to predict the flow patterns that must exist to produce this flow reversal. Depending on the size and speed of the pump, the effects of recirculation can be very damaging not only to the operation but also to the life of the impeller and casing. The symptoms associated with recirculation and a method for calculating the capacity at which it takes place at both the suction and discharge of the impeller are given in Fraser, "Recirculation in Centrifugal Pumps," ASME Winter Annual Meeting, Nov. 16, 1981, Washington, D.C.

System Head Curve

Just as the head characteristic of a pump is represented by a curve indicating reducing head developed with increasing capacity, the head requirements of the system served by the pump can be depicted by a curve showing increased head requirements for increased flow, that is, a **system-head curve** (Fig. 9.2.27). The capacity at which a pump will operate in this system is that at which the pump head curve intersects the system-head curve.

Every system-head curve consists of (1) a total static head (which may in some cases be equal to zero) plus (2) friction losses. The total static head $H_{stat} = H_d - H_s + P_d - P_s$, where H_d and H_s are the elevations at the system's termination and origin, respectively, and P_d and P_s are the gage pressures at the same points.

The **friction losses** are the sum of the line losses and fitting losses.

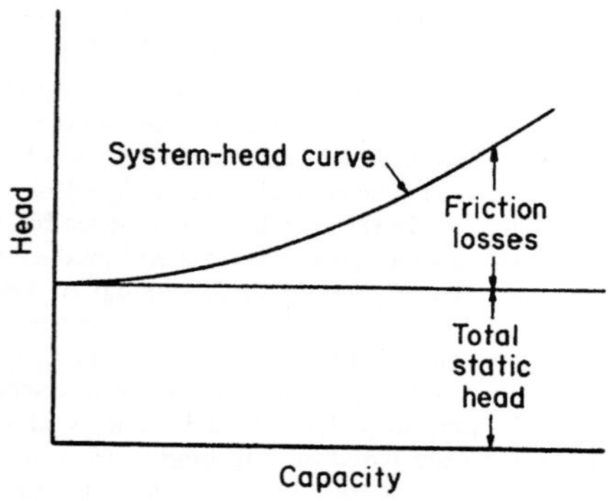

Figure 9.2.27 System-head curve.

In complex pumping systems, such as municipal water-supply operations, both the total static-head and the friction-loss components may be variable, for example, the former by changes in reservoir levels and the latter by the particular combination of lines being served at any given moment. This results in upper and lower limits of system-head requirements, and the intersection of these limiting curves with the pump head curve will define the capacity range over which the pump will be required to operate.

Since the capacity at which a pump will operate corresponds to the intersection of the system-head curve with the pump head-curve, any changes in pump capacity can only be obtained by varying one or the other of these two curves (Fig. 9.2.28). Thus the capacity of a centrifugal pump operating in a system can be regulated by (1) changing the pump speed or (2) throttling in the discharge piping. The former method is preferable whenever the driver permits it, because throttling always involves an appreciable waste of power.

Pump capacity should never be permitted to be reduced to zero because the fluid within the pump would have to absorb the entire power input and therefore would heat up rapidly with injurious effects on the pump. A bypass line should be provided from the pump discharge line to permit a predetermined amount of liquid to flow through the pump if the discharge valve is closed entirely. This bypass may be operated manually or automatically.

Centrifugal pumps may sometimes be operated in **parallel** or in **series.** To construct the head-capacity curve of two pumps in parallel, it is merely necessary to add the capacities of the individual pumps for various total heads. The head-capacity curve of two pumps in series is

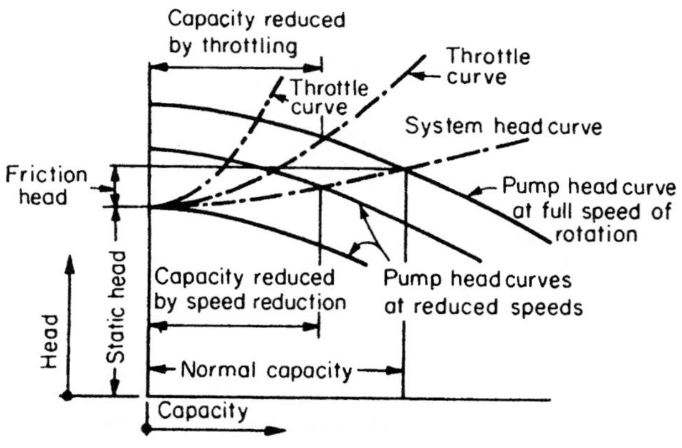

Figure 9.2.28 Pump operation in a system.

constructed by adding the individual pump heads for various capacities. Figure 9.2.29 shows series and parallel operation of two centrifugal pumps with both flat and steep system-head curves.

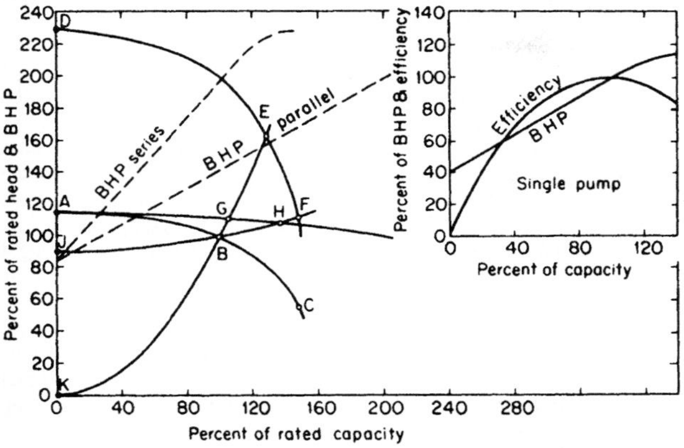

Figure 9.2.29 Series and parallel operation of centrifugal pumps.

Performance on Other Than Water

Selection of drivers, as well as pumps, is also influenced by **properties of the fluid** handled. The required pump power varies directly with the specific gravity of the liquid and is influenced in a more complex manner by viscosity. The effect of viscosity is illustrated in Fig. 9.2.30. For the example shown in the chart, the pump in question has a best-efficiency-point (bep) of 750 gal/min (170 m³/h) at 100-ft (30-m) head. When handling a fluid having a viscosity of 1,000 SSU (220 cS), its capacity is reduced to 95 percent of that of water for all capacities. Its head is reduced to 96, 94, 92, and 89 percent of its head on cold water at 60, 80, 100, and 120 percent of the bep capacity, and its efficiency is reduced to 63.5 percent of its efficiency on water for all capacities. Thus, if the pump had a bep efficiency on cold water of 70 percent, it would now deliver 712.5 gal/min (161.8 m³/h) at 92-ft (28.0-m) head with an efficiency of 44.5 percent at the bep.

MATERIALS OF CONSTRUCTION

Centrifugal pumps are available in many common metals and alloys, as well as porcelain, glass, and synthetics. A listing of materials recommended for various services can be found in the Standards of the Hydraulic Institute. Table 9.2.2 has some of the materials commonly used for pump parts.

INSTALLATION, OPERATION, MAINTENANCE

Proper installation, operation, and maintenance of centrifugal pumps will vary widely over the complete range of services to which the pumps may be applied, and satisfactory results in these areas can be fully achieved only by following the manufacturer's instructions for the size and type of unit involved. There are, however, certain general considerations that should be observed and will seldom need to be modified under any circumstances.

In general, the location selected for installation should be as close to the source of the fluid as possible, consistent with the requirement that adequate space be made available to provide accessibility for operation, inspection, and maintenance.

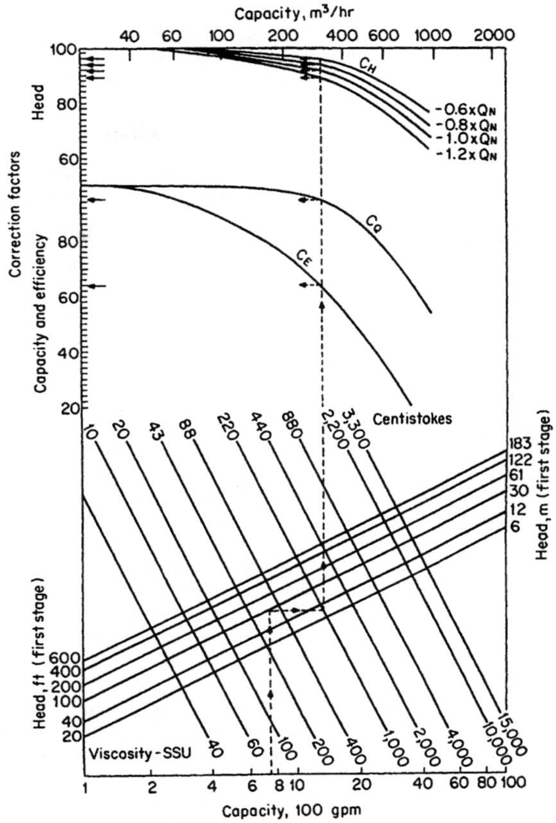

Figure 9.2.30 Performance-correction chart for effect of viscosity. *(Hydraulic Institute.)*

Shaft Alignment

Alignment of the driver shaft with the pump shaft is crucial for component longevity. Misaligned shafts will result in vibration and shorter life of bearings and seals. For a horizontal **separately coupled pump** (one in which the pump and driver bolt independently to a baseplate), the coupling can be precisely aligned when the unit is idle, but become misaligned when the pump runs. Running misalignment can be caused

Table 9.2.2 Materials for Various Fittings
(Materials for bearing housings, bearings, and other parts are not usually affected by the liquid handled)

Part	Standard fitting	All-iron fitting	All-bronze fitting
Casing	Cast iron	Cast iron	Bronze
Suction head	Cast iron	Cast iron	Bronze
Impeller	Bronze	Cast iron	Bronze
Impeller ring	Bronze	Cast iron or steel	Bronze
Casing ring	Bronze	Cast iron	Bronze
Diffuser	Cast iron or bronze	Cast iron	Bronze
Stage piece	Cast iron or bronze	Cast iron	Bronze
Shaft, with sleeve	Steel	Steel	Steel, bronze, or Monel
Without sleeve	Stainless steel or steel	Stainless steel or steel	Bronze or Monel
Shaft sleeve	Bronze	Steel or stainless steel	Bronze
Gland	Bronze	Cast iron	Bronze

by temperature changes in the driver, pump, and connected piping; by hydraulic forces produced by the piping; and by the torque between the driver and the pump. Piping should be supported so as to minimize piping loads on the pump, and any expansion joint should be installed so as not to transmit any force to the pump.

Supporting the Pump

To retain coupling alignment, a separately coupled unit must be supported by a rigid **baseplate** (and pedestals) and by a **foundation** of concrete (or an equivalent support).

A single-stage in-line pump is typically designed to be supported solely by the piping, like a valve. The driver mounts directly on the pump, and moves with the pump as the pump moves with the piping, thereby retaining shaft alignment. Such a unit does not require a baseplate or foundation, although the piping support must be adequate for the additional weight of the pump and driver.

Priming

A centrifugal pump is primed when the casing, all impellers, and the seal chambers are filled with the liquid to be pumped. Failure to evacuate all air, or other gases, from the pump can result in wear or seizure of wear rings and bushings, and accelerated wear of mechanical seals. If the suction supply is above atmospheric pressure, priming is accomplished by venting the pump at all high points in the casing and seal chambers. If the suction supply is below atmospheric pressure, priming must be accomplished by providing a vacuum at all the high points; by providing a foot valve, below the suction liquid level, and filling the pump and suction line with liquid; or by providing a priming chamber in the suction line.

Starting

The normal starting sequence is as follows: (1) open valves in all auxiliary sealing, cooling, flushing, and bypass lines; (2) open suction valve; (3) close discharge valve for low-specific-speed pumps where no check valve is installed after the pump, or open discharge valve for high-specific-speed pumps or wherever a discharge check valve has been provided; (4) prime or vent as necessary; (5) energize the driver; and (6) open discharge valve if it was previously closed in step 3.

Following start-up and until stable operation has been adequately established, it is desirable to monitor bearing temperature, stuffing-box leakage, and other outward symptoms of the unit's behavior. Securing of the pump is accomplished by a reversal of the start-up sequence, encompassing steps 6, 5, 3, and 1, in that order.

Minimum Flow

Until the early 1970s there were only four factors to consider when setting an acceptable minimum flow for centrifugal pumps:

Higher radial thrust developed by single-volute pumps at reduced flows.

Temperature rise in the liquid pumped as capacity is reduced.

Desire to avoid overload of drivers on high-specific-speed pumps, which have a rising power requirement as capacity is reduced.

For pumps handling liquids with significant amounts of dissolved or entrained air or gas, there is the need to maintain sufficiently high velocities in the pump casing to wash out this air or gas with the liquid.

Since then, the phenomenon of internal recirculation in the impeller has been discovered. The recirculation occurs at flows less than the capacity at best efficiency. Problems include pressure pulsations at the suction and discharge and rapid deterioration of the pump and casing. The unfavorable effects of this internal recirculation have brought a fifth factor into the picture when setting minimum flows. This is discussed more fully above under "Recirculation."

Each of the effects mentioned above may dictate a different minimum operating capacity. Obviously, the final decision must be based on the greatest of the individual minimums. The internal recirculation usually sets the recommended minimum. Assuming that the pump is properly furnished with the necessary instrumentation, such as flowmeters, pressure gages with sufficient sensitivity to show pulsations, and vibration- and noise-monitoring equipment, an experienced test engineer should be able to pinpoint the onset of internal recirculation.

If the flow demand of the service in which the pump is installed is expected to fall below the minimum permissible flow, some means must be provided to accommodate the difference between the minimum permissible flow and that required by the service. This is accomplished by installing a bypass in the discharge line from the pump, located on the pump side of the check and gate valves and leading to some lower pressure point in the installation where excess heat absorbed through operation at low flows may be dissipated. A valve is located in this bypass line. For small installations, the bypass valve is operated as fully open or fully closed. For large installations, a modulating valve is usually used to bypass the difference between the required flow and the minimum permissible.

Unnecessary Maintenance

A generally accepted cardinal rule is that as long as operation continues normal, the unit should be left alone. Thus, except in special circumstances, periodic overhauls are not recommended. The amount and degree of maintenance likely to be required are influenced primarily by the nature of the service to which the pump is applied, and maintenance practices must therefore be determined largely by the user as a result of personal experience.

9.3 COMPRESSORS

by Heimir Fannar and Piotr Cyklis

REFERENCES: Chlumsky, "Reciprocating and Rotary Compressors," E&FN Spons Ltd., Loomis, "Compressed Air and Gas Data," 3rd ed., Ingersoll-Rand. Rollins, "Compressed Air and Gas Handbook," 5th ed., Compressed Air and Gas Institute. Sheppard, "Principles of Turbomachinery," Macmillan. Stepanoff, "Turboblowers," Wiley. Koppers Research and Engineering Staff, "Engineers Handbook of Piston Rings, Seal Rings, Mechanical Shaft Seals," 8th ed., Koppers Company, Inc. Lundberg and Glanvall, "A Comparison of SRM and Globoid Type Screw Compressors at Full Load," Proceedings of the 1978 Purdue Compressor Technology Conference, Purdue University, W. Lafayette, IN. Von Nimitz, "Pulsation and Vibration Control Requirements in the Design of Reciprocating Compressor and Pump Installations," Proceedings of the 1982 Purdue Compressor Technology Conference, Purdue University, W. Lafayette, IN. Mechanics, Simulation and Design of Compressor Valves, Gas Passages and Pulsation Mufflers, short course, July 12-14, 1992, Purdue University, West Lafayette, IN.

Notation

a = speed of sound
A = piston area
A_v = valve flow area, cell area
c_p = constant pressure specific heat
c_v = constant volume specific heat
d = depth of ring section, diameter
D = diameter
e = eccentricity
E = modulus of elasticity
F = thrust
f = valve resistance, in velocity heads
g = ring free gap less end clearance

g_c = gravitational constant
ghp = gas horsepower
h = enthalpy
H_p = polytropic head (ft · lb/lbm or kJ/kg)
k = ratio of specific heats
K = area constant (decimal)
L = length, leakage coefficient, connecting rod length
L_{eq} = equivalent length of pipe
m = mass
MW = molecular weight
n = polytropic coefficient
n_f = number of first stage cylinders
N_m = number of lobes on main rotor
n_s = number of stages
n_v = number of vanes
N = rotational speed
N_s = specific speed
P = pressure
Q = heat flow, volume flow at inlet conditions
Q_{sa} = actual volumetric flow rate (inlet conditions)
Q_{st} = theoretical volumetric flow rate (inlet conditions)
r = external compression ratio (P_d/P_s)
r_i = compression ratio per i^{th} stage
r_t = pressure ratio across entire compressor
r_v = volumetric internal compression ratio (V_{min}/V_{max})
r_δ = ratio of cylinder pressure at bottom dead center to inlet pressure
R = gas constant
RC = crank radius
s = entropy, stress
sg = specific gravity relative to air
T = absolute temperature
T'_d = isentropic temperature
δ_v = vane thickness
U = piston velocity, peripheral velocity
V = volume
V_{max} = maximum volume of the volumetric compressor chamber
V_{min} = minimum volume of the volumetric compressor chamber (clearance volume)
$\bar{U}$ = velocity
W = work, mass flow
Z = compressibility factor

Greek

α = ratio of piston area over valve flow area
β = blade angle, degrees
λ = wave length
Δ = pressure loss, differential pressure (lb/in²)
η_c = adiabatic efficiency
η_p = polytropic efficiency
η_v = volumetric efficiency
μ = pressure coefficient (gH_p/U^2)
ϕ = flow coefficient (Q/ND^3)
ω = angular speed (radians/second)

Subscripts

s = supply conditions
d = discharge conditions
A = axial
R = radial

Compressed-Air and Gas Usage

Compressed-air is used for machine and tool operation, drilling, painting, soot blowing, pneumatic conveying, food processing, instrument operations, and in situ operations (e.g., underground combustion). Pressures range from 25 psig (172 kPa) to 60,000 psig (413,790 kPa). The largest usage is at 90 to 110 psig (612 kPa to 748 kPa) which is the normal plant air-pressure range.

Gas compressors are used for refrigeration, air conditioning, heating, pipeline conveying, natural gas gathering, catalytic cracking, polymerization, and in other chemical processes.

Standard Units and Conditions

In the ISO system, the standard unit of pressure for compressors is the kilopascal (kPa). In some countries this is the only unit which can appear on the compressor pressure gages by law. In Europe, the European Committee of Manufacturers of Compressors, Vacuum Pumps and Pneumatic Tools (PNEUROP) and, in the United States, the Compressed Air and Gas Institute (CAGI) prefer the bar as the standard unit of pressure. PNEUROP and CAGI have selected as standard conditions 1 bar (14.5 lb/in²) (100 kPa), 20°C (68°F), and 0 percent relative humidity. The unit of flow in the ISO system is m³/s. Other units still in common usage are m³/h, m³/min, and l/s. In the United States the most commonly used units are ft³/min (cfm) and ft³/h (cfh). Power is normally expressed in kilowatts (PNEUROP) and horsepower (CAGI).

Thermodynamics of Compression

Most compressors are analyzed using the ideal-gas law and assuming constant specific heat. Real-gas deviations are handled by applying a compressibility factor (also called *super compressibility*). The ideal-gas law will give satisfactory results for nonhydrocarbon gases for pressures to approximately 1,000 psig (6,900 kPa) at normal temperatures. Most hydrocarbon gases and refrigerants deviate strongly from the ideal-gas laws even at moderate pressures. For these cases, thermodynamics property tables, Mollier Charts, or compressibility charts should be used. Unfortunately these are not available for many gases of industrial importance. In this case a generalized compressibility chart is used. Compressibility charts for various gases can be found in Loomis, "Compressed Air and Gas Data," 3d ed., Ingersoll-Rand, and Rollins, "Compressed Air and Gas Handbook," 5th ed., Compressed Air and Gas Institute.

Adiabatic Analysis

The equation of state for an ideal gas is:

$$PV = mRT \tag{9.3.1}$$

The first law of thermodynamics for a steady-flow process is (per unit mass):

$$Q = (h_d - h_s) + \frac{\bar{U}_d^2 - \bar{U}_s^2}{2g_c} + W \tag{9.3.2}$$

Neglecting the kinetic energy of the gas and assuming constant specific heat, for a reversible adiabatic process, Eq. (9.3.2) becomes:

$$W = c_p(T_s - T'_d) \tag{9.3.3}$$

Defining $r = P_d/P_s$ and substituting $r^{(k-1)/k}$ for T'_d which is the isentropic temperature and $k/(k-1)$ for Cp/R. Eq. (9.3.3) becomes:

$$W = P_s V_s \frac{k}{1-k}[r^{(k-1)/k} - 1] \tag{9.3.4}$$

The same results can be obtained by evaluating $W = \int V\,dp$ on an idealized P-V diagram (Fig. 9.3.1).

The isentropic efficiency (ratio of isentropic work to actual work) is given by:

$$\eta_c = T_s \frac{r^{(k-1)/k} - 1}{T_d - T_s} \tag{9.3.5}$$

Polytropic Process

As an alternative to the adiabatic analysis, a polytropic process can be defined as

$$PV^n = \text{const} \tag{9.3.6}$$

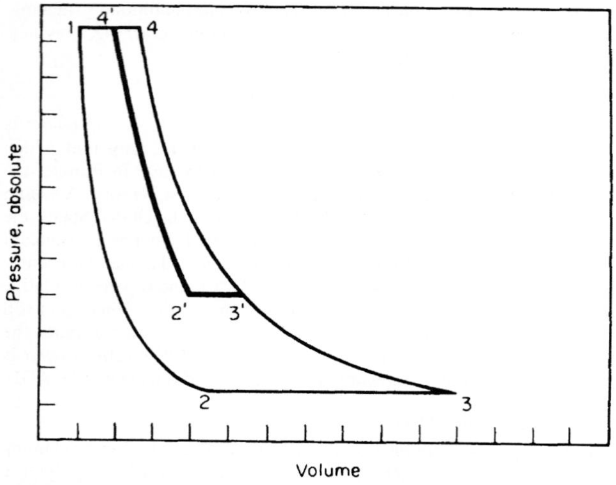

Figure 9.3.1 Idealized indicator diagram for a positive-displacement compressor.
State Points:
1. End of discharge = discharge valve closing, beginning of re-expansion
2. Beginning of suction = suction valve opening
3. End of suction = suction-valve closing, beginning of compression
4. End of compression = discharge valve opening,

3′-2′ - ideal interstage cooling
2′-4′ - compression on the second stage after interstage cooling

All the adiabatic equations apply if $(n-1)/n$ is substituted for $(k-1)/k$ where

$$\frac{n-1}{n}=\frac{k-1}{k}\frac{1}{\eta_p}$$

If the inlet and discharge temperatures and pressures are known, n can be calculated from:

$$n=\frac{1}{1-\ln(T_d/T_s)/\ln r} \tag{9.3.7}$$

where n is directly related to the heat transferred during the process by

$$n=\frac{C_p(T_d-T_s)-Q}{C_v(T_d-T_s)-Q} \tag{9.3.8}$$

The reversible work done along the polytropic path is called the **polytropic work.**

The polytropic efficiency η_p (ratio of polytropic work to actual work) is given by:

$$\eta_p=\frac{\frac{k-1}{k}\ln r}{\ln(T_d/T_s)} \tag{9.3.9}$$

The polytropic efficiency and the isentropic efficiency can be related by:

$$\eta_c=\frac{r^{(k-1)/k}-1}{r^{(k-1)(k\eta_p)}-1} \tag{9.3.10}$$

Equation (9.3.10) is strictly valid only for ideal gases with constant specific heats. However, because the real-gas errors occur in both the numerator and denominator, the equation gives reasonably accurate results for real gases which deviate moderately from ideal gases.

Positive-displacement compressors are almost universally analyzed using the adiabatic model. Dynamic compressors are analyzed using both the adiabatic and polytropic models with the trend toward the polytropic model. The advantage of the polytropic model is that the discharge temperature is immediately available and η_p is essentially constant for different gases. The polytropic efficiency is independent of the thermodynamic state of the gas. Also, the total polytropic head is the sum of the polytropic heads for the stages. This is not the case for isentropic heads.

The advantage of the isentropic model is that the isentropic work can be read immediately from thermodynamic tables or charts. Comparing adiabatic efficiencies at different pressure ratios is not a valid procedure.

Real-Gas Effects

Deviations from the ideal-gas law are accounted for by introducing a compressibility factor

$$Z=\frac{PV}{RT} \tag{9.3.11}$$

The isentropic work of compression for a real gas becomes:

$$W=P_sV_s\frac{k}{1-k}[r^{(k-1)/k}-1]\frac{Z_s+Z_d}{2Z_s} \tag{9.3.12}$$

where Z_1 and Z_2 are the compressibilities at conditions 1 and 2. Compressibility will also affect the volume flow of the compressor because of the reexpansion of the clearance volume gas.

Multistaging and Intercooling

Temperature rise and mechanical stresses limit the maximum pressure differential across a single stage of any compressor type. The pressure rise across a dynamic compressor stage is further limited by the available polytropic head which the stage can develop. The volumetric efficiency of a reciprocating compressor decreases with increasing pressure ratio placing a practical limit on the maximum ratio per stage.

Multistaging is used to overcome these limitations and to save power. With perfect intercooling and no pressure losses between stages, theoretically the minimum power is obtained when

$$r_i=\sqrt[n_s]{r_t} \tag{9.3.13}$$

where n_s = the number of stages; r_i = the pressure ratio per stage; and r_t = the ratio across the compressor.

For every 10°F (5.6°C) of imperfect intercooling, the horsepower increases approximately 1 percent. Perfect intercooling is achieved when the temperature of the gas leaving the intercooler equals the temperature of the gas at the compressor inlet. The maximum theoretical work that can be saved by perfect intercooling is:

$$W=\left(\frac{k}{1-k}P_sV_s\right)\left[\left(\frac{P_d}{P_s}\right)^{(k-1)/k}-n_s\left(\frac{P_d}{P_s}\right)^{(k-1)/n_sk}+n_s-1\right] \tag{9.3.14}$$

On the ideal indicator diagram (Fig. 9.3.1), this is area 3′4 4′2′.

In practice, perfect intercooling is seldom obtained. Normal practice is to cool the interstage gas to within 15°F to 20°F (8.3°C to 11.1°C) of the inlet gas. For water-cooled machines, the coldest water available should be used for intercooling.

Positive-Displacement Compressors versus Dynamic Compressors

Positive-displacement compressors collect a fixed volume of gas within a chamber and compress it by reducing the chamber volume. The ideal work for such a process is given by Eq. (9.3.4). The work could also be obtained from the area enclosed by an ideal indicator diagram (Fig. 9.3.1). The actual horsepower, taking into account real-gas deviations, is

$$\text{hp}=\frac{144P_sV_sk}{33{,}000(1-k)}\left[\left(\frac{P_d}{P_s}\right)^{(k-1)/k}-1\right]\frac{1}{\eta_c}\left(\frac{Z_s+Z_d}{2Z_s}\right) \tag{9.3.15}$$

with pressures in psig and volumes in cfm. With the volume rate V in m³/s and the pressures in pascals (dropping the 144/33,000), the power will be in watts.

For a given volume flow, the ideal power is affected by inlet pressure, the k value of the gas, and the compression ratio. The ideal power is not affected by the inlet temperature of the gas or its molecular weight.

The actual power is increased due to losses through the intake and discharge valves or ports as can be seen on the indicator diagram Fig. 9.3.2.

Additional losses are caused by turbulence within the compression chamber, preheating of the inlet gas, and leakage from the compression chamber. The isentropic compression efficiency (ratio of isentropic

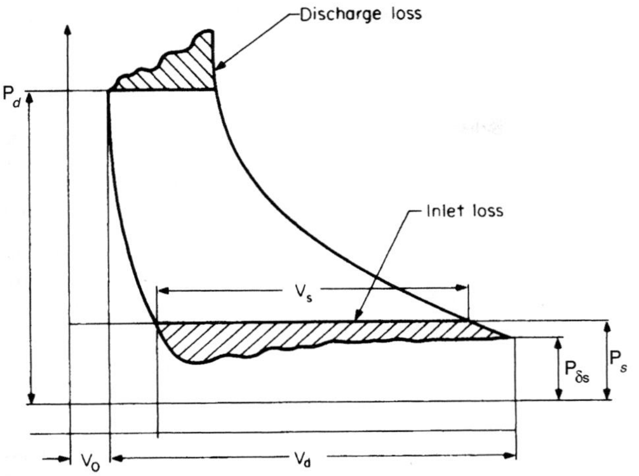

Figure 9.3.2 Typical indicator diagram showing intake and discharge losses.

work to actual compression work) can be determined from the indicator diagram or calculated from Eq. (9.3.5) provided that there is no liquid injection. In addition to the thermodynamic losses, the mechanical losses must be added to the power requirement.

Theoretical compressor volumetric capacity is given as:

$$Q_{st}(V_{\max}) \times 2\pi\omega \times n_f$$

In case of double acting compressors Q_{st} shall be calculated separately for crank and head side of the cylinders considering the piston rod dimensions. Volumetric efficiency for reciprocating compressor is defined as:

$$\eta_v = \frac{\text{actual delivery at intake conditions}}{\text{theoretical volumetric capacity}} = \frac{Q_{sa}}{Q_{st}}$$

For multistaging compressors, only the first-stage displacement is used.

Positive-displacement compressors are essentially constant-volume, variable-pressure machines as shown in Fig. 9.3.3. For fixed inlet and discharge pressures and varying speed, positive-displacement compressors are essentially constant-torque machines.

Dynamic compressors operate by transferring momentum to the gas via a high-speed rotor. The process is steady flow with the work given by Eq. (9.3.4). Head is defined as the energy per unit mass of the fluid. Head is imparted to the fluid through the change in magnitude and radius of the velocity components of the fluid as it passes through the rotor. The head developed by the rotor is

$$H = (U_s U_{Us} - U_d U_{Ud})/g_c \tag{9.3.16a}$$

where U_1 and U_2 are the linear velocity of the rotor at radii 1 and 2. U_{Us} and U_{Ud} are the fluid tangential velocity at radii 1 and 2.

The polytropic head required for a given compression ratio is

$$H_p = \frac{Z_{\text{ave}} R T_s [r^{(n-1)/n} - 1]}{(n-1)/n} \tag{9.3.16b}$$

The gas horsepower on a CFM basis is

$$\text{ghp} = Q_s P_s [r^{(n-1)/n} - 1] \frac{n}{229\eta_p (n-1)} \left(\frac{Z_s + Z_d}{2Z_s} \right) \tag{9.3.17}$$

and on a mass flow basis:

$$\text{ghp} = \frac{W H_p}{33{,}000\eta_p} \tag{9.3.18}$$

With H_p in J/kg and W in kg/s, the power in kilowatts is

$$\text{kW} = H_p \left(\frac{W}{\eta_p} \right) \tag{9.3.19}$$

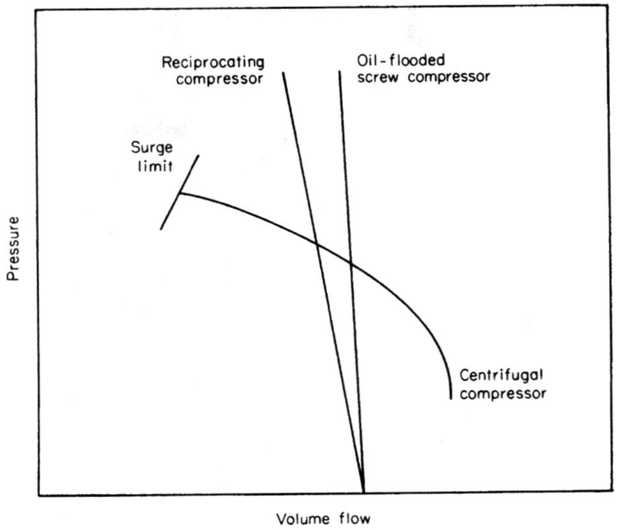

Figure 9.3.3 Pressure capacity curves for different compressor types.

Mechanical losses must be added to the gas power to obtain the shaft power. Volumetric efficiency is not defined for dynamic compressors.

For a fixed impeller design and a fixed speed, the energy that is transferred to a unit mass of fluid as it passes through the impeller (i.e., the head) is constant. Pressure rise and power vary directly with inlet gas density independent of the cause of the density change. An increase in k will cause the pressure rise to decrease but will not affect the power.

A centrifugal compressor is essentially a constant-pressure, variable-capacity machine (see Fig. 9.3.3). An axial compressor is a constant-capacity, variable-pressure machine over a significant discharge pressure range. Dynamic compressors at least qualitatively follow the fan laws.

$$\frac{Q_1}{Q_2} = \frac{N_1}{N_2} \qquad \frac{H_{p1}}{H_{p2}} = \left(\frac{N_1}{N_2} \right)^2 \qquad \frac{\text{ghp}_1}{\text{ghp}_2} = \left(\frac{N_1}{N_2} \right)^3$$

The geometry of a dynamic compressor is dictated by the required flow, polytropic head, and speed. The specific speed is a dimensionless parameter used to select the proper geometry.

$$N_s = \frac{N Q^{(1/2)}}{H_p^{(3/4)}} \tag{9.3.20}$$

For specific speeds from 400 to 900, radial impellers are used. Between 800 and 1,400, mixed-flow impellers are preferred. For specific speeds above 1,400, axial impellers are normally used.

Surging

All dynamic compressors have a minimum flow point called the **surge limit** below which the operation of the machine is unstable. The surge limit is a function of the compressor type, design pressure ratio, gas properties, inlet temperature, blade angle, and speed. Operation at or below the surge limit must be avoided. The existence of the surge limit makes it essential to know the demand curve for the particular installation involved.

Reciprocating Compressors

Reciprocating compressors as shown in Figs. 9.3.4 and 9.3.5 range in size from fractional cfm to 15,000 cfm (25,485 m³/h) with discharge pressures as high as 60,000 psig (413,790 kPa). The majority of applications fall in the pressure range of 10 to 300 psig (690 to 2,069 kPa) and capacities less than 2,500 cfm (4,250 m³/h). Single-acting compressors (which compress gas on one side of the piston only) have their widest application below 50 hp (37 kW). Larger compressors are usually double-acting (both sides of the piston are used to compress gas). Most plant air systems operate at 100 to 150 psig. Oil-fl

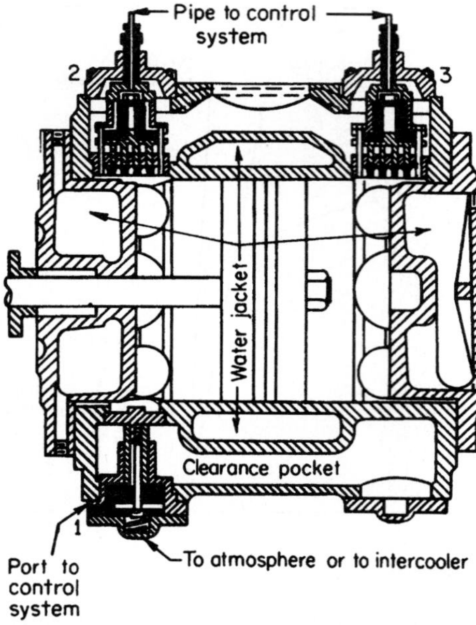

Figure 9.3.4 Cross section of gas-compressor cylinder, showing water jackets, clearance pocket with pneumatic control, and suction-valve lift devices.

screw compressors dominate this market. Above 200 psig, reciprocating compressors dominate.

Single-stage compressors of 100 hp (75 kW) operating at 125 psig were once common. Modern practice is to use multistaging with intercooling for pressures above 80 psig and sizes as low as 10 hp (7.5 kW). The intercooling saves power, and the reduced discharge temperatures improve safety and prolong compressor life.

The capacity of a reciprocating compressor is determined by the first-stage displacement multiplied by the volumetric efficiency.

$$\eta_v = (1+r_v)(r_\delta)^{(1/k)} - \frac{r_v(r)^{(1/k)}}{(Z_d/Z_s)} - L \tag{9.3.21}$$

where $r_\delta = P_{\delta s}/P_s$; $r = P_d/P_s$; and L = a leakage coefficient = 0.05 for lubricated compressors and 0.10 for nonlubricated. P_s = the pressure that exists in the cylinder when the piston is at bottom dead center. r_v volumetric internal compressor ratio as a decimal fraction and includes valve passage volumes up to the sealing elements (point P_dV_c in Fig. 9.3.2). Clearance volume for normal service is 5 to 15 percent. High-pressure machines have as small a clearance volume as practical whereas pipeline compressors may have 100 percent clearance. The indicator volumetric efficiency from Fig. 9.3.2 is V_s/V_d. In general, this will be greater than the value obtained by Eq. (9.3.21).

The compression efficiency varies between 0.85 and 0.95. The mechanical efficiency falls in the range 0.88 to 0.95. The overall efficiency is the product of the compression and mechanical efficiencies.

The discharge temperature can be calculated from Eq. (9.3.22):

$$T_d = T_s\left[1+\frac{r^{(k-1)/k}-1}{\eta_c}\right] \tag{9.3.22}$$

If the inlet pressure is varied while holding the discharge pressure constant, the horsepower of the compressor will pass through a maximum value. This can be seen by substituting displacement times η_v for V_1 in the horsepower equation (9.3.12). For correct sizing of the driver, it is necessary to know the range of inlet pressures as well as discharge pressure. Interstage pressures are selected approximately by Eq. (9.3.13) with empirical adjustments made for interstage piping, intercooler, and valve losses. The loss of volume due to condensation in the intercoolers should be accounted for.

Valve losses can be approximated by:

$$\Delta = 1.26\times10^{-6}\alpha f U_{\text{ave}}^2 \text{sg}\left(\frac{P}{Z}\right)\text{psi} \tag{9.3.23}$$

where α = ratio of piston area to valve flow area; f = valve resistance in velocity heads (≅4); sg = specific gravity relative to air; and U_{ave} = average piston speed, ft/s.

The piston velocity as a function of time is:

$$U = R\omega\left(\sin\omega t + \frac{R}{2L}\sin 2\omega t\right) \tag{9.3.24}$$

where R = crank radius; L = connecting rod length; and ω = crank speed (rad/s).

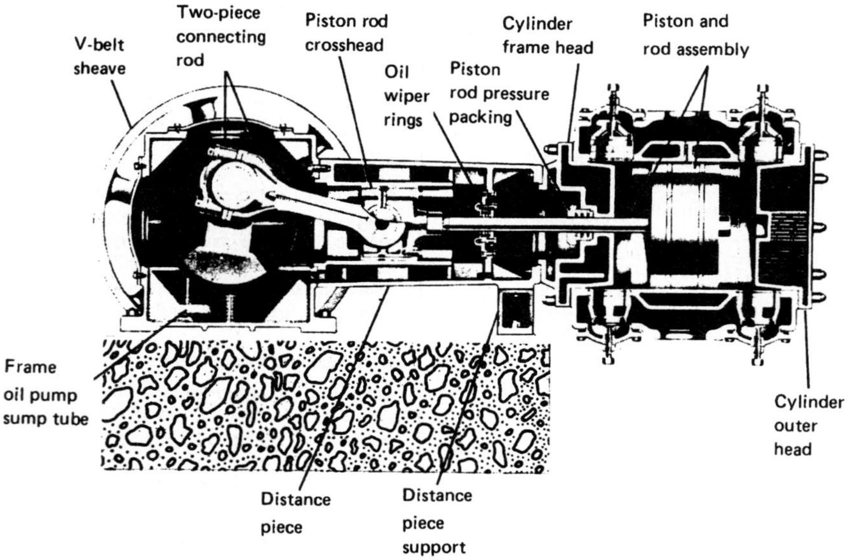

Figure 9.3.5 Single-stage double-acting reciprocating compressor.

If there were no sealing elements in the valves, the velocity through the valves would be:

$$U_v = \left(\frac{A}{A_v}\right)U$$

This pulsating velocity is amplified by the valve sealing elements so that pulsating flow is inherent in reciprocating compressors. The full wavelength for a double-acting cylinder is $\lambda = 60\ a/2N$, where a is the speed of sound. For a single-acting cylinder, it is two times this value. Pipes with total equivalent lengths equal to multiples of $\lambda/4$ should be avoided.

For a double-acting cylinder, the major problems occur at $\lambda/4$ and $3\lambda/4$ although resonance can occur at the other multiples of $\lambda/4$. Equivalent length L_{eq} is given by

$$L_{eq} = \frac{\lambda}{360}\tan^{-1}\left(\frac{2\pi V}{S\lambda}\right) \tag{9.3.25}$$

where V = equivalent cylinder and passage volume; and S = cross-sectional flow area of the pipe.

If it is not feasible to avoid the designated lengths, a volume bottle or an orifice can be installed in the piping system. If an orifice is used, it should be ½ the pipe diameter and located at a node (one node occurs at the open end of the pipe). A volume bottle should be located as close to the cylinder as possible. Von Nimitz ("Pulsation and Vibration Control Requirements in the Design of Reciprocating Compressor and Pump Installations," Proceedings of the 1982 Purdue Compressor Technology Conference, Purdue University) recommends the following equations for sizing:

$$V_{sb} = 4\text{V(\# of cylinders)}\left(\frac{kT_s}{\text{MW}}\right)^{1/4} \qquad V_{db} = \frac{V_{sb}}{r^{1/k}} \tag{9.3.26}$$

V(# of cylinders) is the total double-acting displacement volume of all cylinders to be manifolded to the surge bottle (ft³). Subscript *sb* refers to suction, *db* to discharge. *r* is the compression ratio. Temperatures are in degrees Rankine.

Compressor Valves

The valve lift area is defined as the product of the sealing element lift and the sum of the valve seat peripheries. The flow coefficient can be determined by measuring the pressure drop through the valve in a wind tunnel. Once the flow coefficient is determined, the "equivalent," or effective, flow area can be derived. The flow coefficient equals the square of the lift area divided by the square of the equivalent area. Poppet valves (Fig. 9.3.6) are ideal for low-speed compressors used in low compression ratios, such as pipeline applications. Concentric ring valves (Fig. 9.3.7) are used in process, natural gas, and high-pressure applications, and can better tolerate certain types of contamination than some of the other valve designs. Plate valves (Fig. 9.3.8) are used in a wide variety of applications including gas gathering, compressed natural gas, pipeline, gas storage, process, and fuel booster applications.

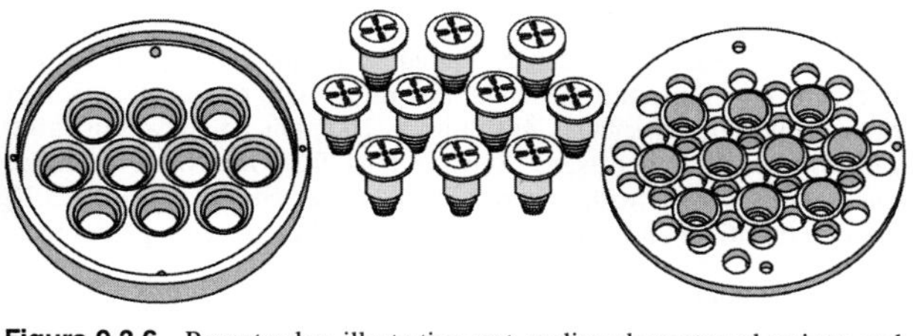

Figure 9.3.6 Poppet valve, illustrating seat, sealing elements and springs, and guard. *(Hoerbiger Corporation.)*

Figure 9.3.7 Concentric ring valve, illustrating concentric rings and seat. *(Hoerbiger Corporation.)*

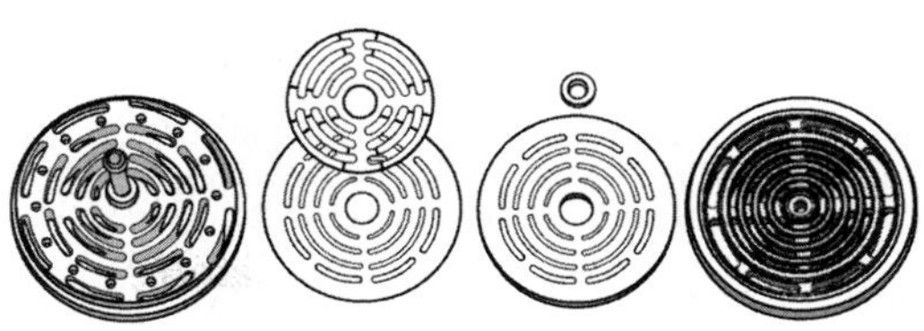

Figure 9.3.8 Plate valve, illustrating guard, cushion plate and wafer spring, sealing element, and seat. *(Hoerbiger Corporation.)*

Valve lift ranges from approximately 0.039 in (1.0 mm) to 0.157 in (4.0 mm) in plate valve and concentric ring valves, and approximately 0.098 in (2.5 mm) to 0.25 in (6.3 mm) in poppet valves.

Springs are used to close the sealing element against the valve seat in the last few degrees of crank angle, before the piston momentarily stops and reverses direction. This prevents backflow from slamming the valve closed, which can damage or destroy the sealing element. A single sealing element design, such as a plate valve, is the simplest to control.

Most compressor valve designs now use sealing elements made of nylon, peek, or other nonmetallic materials. Some designs, however, still use steel sealing elements. These are less common.

Adequate valve lift and spring load may be calculated on the basis of pressure, specific gravity of the gas, speed of the compressor, and mean velocity of the gas through the valve.

Piston Rings

Reciprocating compressors employ compression rings, oil rings, and rider rings. Compression rings are always present, but whether oil rings or rider rings are used will depend on the type of compressor and its service.

The standard compression ring is the one-piece square-cut or angle-cut ring although two-piece and segmental rings are sometimes used. The one-piece ring usually has a rectangular cross section, but occasionally the cylinder contact face is crowned to prevent edge loading and to improve lubrication. Cast-iron rings are still widely used but are being displaced in many applications by filled TFE rings. Carbon-filled TFE rings are preferred for process compressors. For air compressors, carbon, glass, and bronze fillers are used. TFE rings can be used from −450 to 500°F (−268 to 260°C). Bronze-filled TFE is good for higher-temperature operations because of its higher thermal conductivity. Carbon-filled TFE is preferred for oil-free service. Many compressors that handle gases with corrosive components (e.g., H_2S) use piston rings made of cloth-reinforced thermoset laminates with molybdenum disulfide added to improve the lubricity. Polyamide, polyimide, and poly(amide-imide) rings work well with marginal or no lubrication.

A one-piece ring must be stretched over the piston to install it in the piston groove. The maximum stress occurs 180° from the joint and is given by Koppers Research and Engineering Staff ("Engineers Handbook of Piston Rings, Seal Rings, Mechanical Shaft Seals," 8th ed., Koppers Company, Inc.) as:

$$S_o = \frac{0.424E(8d - g)/d}{(D/d - 1)^2} \quad \text{psi} \tag{9.3.27}$$

When closed to the cylinder diameter, the ring is stressed to supply the initial seating pressure. The maximum stress (180° from the joint) is given by the Koppers "Engineers Handbook" as:

$$S_c = \frac{0.482E(g/d)}{(D/d - 1)^2} \quad \text{psi} \tag{9.3.28}$$

Oil rings are used in lubricated compressors to scrape oil from the cylinder wall and to meter a small quantity of oil to the compression rings. The oil rings may be above or below the wrist pin. Operating conditions for oil rings are less severe than for compression rings.

Nonlubricant compressors use rider rings to prevent piston cylinder contact. To prevent pressure buildup in front of the rider rings, longitudinal grooves are cut into the rings. Usually two or more rider rings are required.

Below 300 psig (2,068 kPa) differential, two compression rings and one or two oil rings are normally used. From 300 to 600 lb/in^2 differential, three to four compression rings are used. From 600 to 1,500 psig (4,137 to 10,342 kPa) differential, four to five rings are used. Above 1,500 psig (10,342 kPa), six or more rings may be required.

Piston-Rod Packing

Double-acting compressors are driven through a crosshead (Fig. 9.3.5) and require a rod packing to seal the piston rod connecting the piston to the crosshead. The operating conditions for rod packings are more severe than those for piston rings. In a double-acting compressor, the piston rings seal against the pressure differential across the stage whereas the rod packing seals the differential between discharge pressure and atmospheric pressure.

Rod packings use the same materials as piston rings. Metallic packing can tolerate rod wear of 0.15 percent of the rod diameter, with less tolerance above 2,000 lb/in^2 (13,790 kPa). The rod should be hardened to Rockwell 40 C and ground to a 10-rms or less finish.

Nonlubricated Cylinders

NL-1 class does not permit any direct lubrication and tolerates the minute contamination introduced along the piston rod. This can be eliminated by an additional spacer and rod wiper, qualifying the cylinder as a class NL-2. The pistons for this service use various Teflon rings and riders. The life of carbon rings is reduced by low humidities. The cylinder should have a ground 10- to 16-rms finish and a minimum hardness of Rockwell 20 C.

Cylinder Lubrication

Lubricants serve to (1) prevent wear by providing a supporting film between the rubbing surfaces, (2) seal close clearances, (3) protect against corrosion, and (4) transmit heat of friction and minute wear particles away from points of contact. Industrial cylinders and packings require force-feed lubricators. Viscosity is the best index of suitability: general service at pressures below 500 lb/in^2 (3,448 kPa) requires 400 SSU (86 centistokes) at 100°F (38°C) (SAE 40); 700 SSU (151 centistokes) for 2,000 lb/in^2 (13,793 kPa); and 1,000 SSU (216 centistokes) (SAE 50) for 8,000 lb/in^2 (55,000 kPa) operations. These oils have very high viscosity at 40°F (4.5°C), which will result in poor lubrication. Good winterizing practice provides continuous circulation of warm jacket water and an immersed electric heater in the oil reservoir. Crankcase oils should include a foam inhibitor, sludge dispersant, and rust inhibitor. Phosphorous extreme-pressure agents are added to high-pressure lubricants to avoid wear and scuff damage. Castor-bean and rapeseed oils are additives resistant to solvent action of condensing hydrocarbons.

Cellulube, Houghto-safe, Pydraul AC, Anderol, and Fluorolubes are used to resist the hazard of exothermic reaction in air compression. The crankcase lubricant consumption for 10 plants over a score of years averaged 20,000 bhp/(h)(gal). Table 9.3.1 shows lubricant requirements for general-process and natural-gas operations.

General Lubricant Issues

The choice of lubricant is important, in particular in those compressor designs (screw, scroll, rolling piston, rotary vane, etc.) where the lubricant comes into direct contact with the working fluid. Some gases, such as heavy hydrocarbons and chlorinated refrigerants (HCFCs) are highly soluble in oil, and can significantly affect operational viscosity. They can also affect compressor capacity, as the gas comes out of solution in the lubricant during cyclic pressure fluctuations. Synthetic lubricants such as polyalkylene glycol (PAG), polyalfaolefins (PAO), and polyolesters (POE) are commonly used. Care must be taken to select oil such that its viscosity remains within limits after full saturation of gas in the oil.

Table 9.3.1 Lubrication Required for Compressor Cylinders

Range of cylinder diam, in	Film thickness, μm	10^3 hp/(h) per gal	Pints per cylinder per day per 100 hp of load	Oil drops/ min
24-36	0.40	13	1.5	13
15-23	0.45	19	1.0	9
10-14	0.55	24	0.8	7
7-9	0.60	32	0.6	5
4-7	1.20	24	0.8	7
3-5	1.40	27	0.7	6

NOTE: Based on 12,500 drops per pint of SAE 40 oil at 75°F with vacuum-type lubricator. Reduced drop count to roughly one-half for pressure-type lubricator. Example: 500 bhp, 20-in cylinder, requires 5 pint/day and feed of 45 drops/min. Each installation will vary on lubrication requirements. Manufacturers' recommendations on oil type and quantity should be followed.

Because of the adverse environmental effects of chlorinated refrigerants (HCFCs), chlorine-free hydrofluorocarbon (HCF) refrigerants (R410A etc.) have been made in order to reduce global warming and ozone depletion. The removal of the chlorine causes the refrigerant/lubricant mix to lose some lubricity, and consequently an increase in component wear can occur.

The elastomer material in sealing elements (O rings, lip seals, etc.) is commonly affected by both the operating gas and the lubricant. The results include shrinkage, hardening, or chemical breakdown, and can cause failure to seal if compatibility is not thoroughly checked.

Chemical compounds (additives) are added to lubricants in order to enhance their properties. These include anticorrosion agents, EP (extreme pressure) agents, oxidation inhibitors, stabilizers, and antiwear additives. Care should be taken to ensure that these chemicals do not attack materials such as bearings in the compressor.

The choices of lubricant, operating temperatures, viscosity range, and replacement intervals are usually clearly specified by the compressor manufacturer.

Compressor Accessories

Most state laws and safe practice require a relief valve ahead of the first stop valve in every positive-displacement compressor. It is set to release at 1.25 times normal discharge or at the maximum working pressure of the cylinder, whichever is lower. The relief-valve piping system sometimes includes a manual vent valve and/or a bypass valve to the suction to facilitate start-up and shutdown operations. Quick line-sizing equations are (1) line connection, *d*/1.75; (2) bypass, *d*/4.5; (3) vent, *d*/6.3; (4) relief-valve port, *d*/9.

Volume flow is controlled by variable-speed drivers; steam engines can operate at 20 percent of rated speed and gas engines at 60 percent, and electric-motor speed can be varied by means of eddy-current and hydraulic couplings and special wound rotors with rheostats, which are both costly and inefficient. Unloading can be applied by means of valve-lift and clearance pockets, as shown in Figs. 9.3.4 and 9.3.9. Utility air plants are throttled with suction unloaders, as shown in Fig. 9.3.10. Suction throttling and bypass controls are used in process operations.

Cylinder Cooling

Single-acting compressors, especially smaller ones, are usually air cooled by fins cast into the cylinder. Cooling air is drawn across the fins by a fan. Double-acting compressors usually have cylinders that are water jacketed.

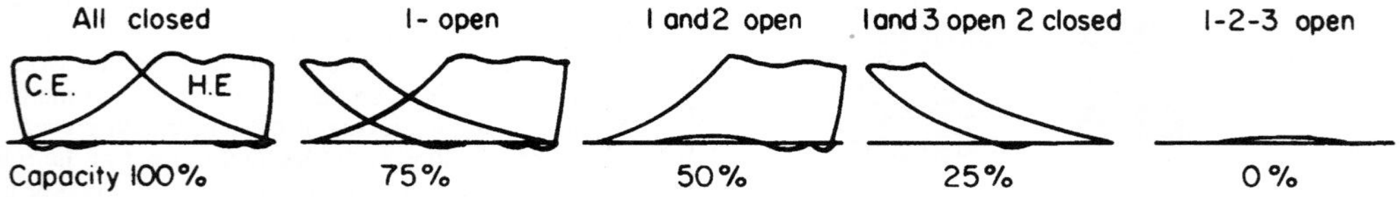

Figure 9.3.9 Cards from a compressor with combination clearance and suction-valve bypass control.

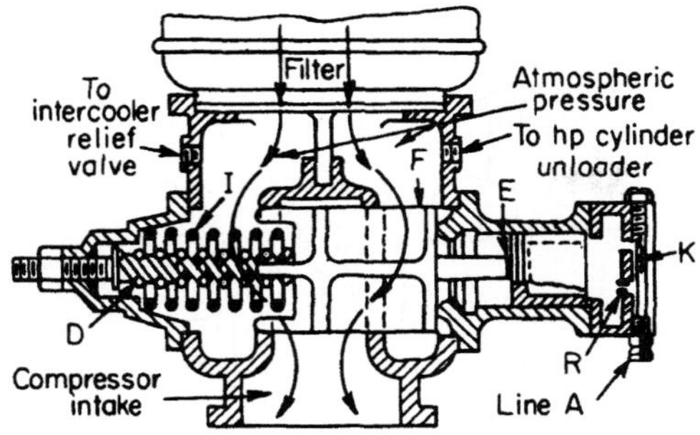

Figure 9.3.10 Air suction unloader. Piston E moves to closed position as line A pressure reaches design point.

The principal purpose of a jacket-water system is to normalize cylinder-casting strains. It has some effect on compressor efficiency due to reduced preheating of the air. This effect is reduced as the compressor speed is increased. The heat rejection to the jacket water is approximated by:

$$H_j = 4(T_{ag} - T_{aw}) + 100\,\text{Btu}/(\text{bhp}\cdot\text{h}) \tag{9.3.29}$$

where T_{ag} and T_{aw} represent the average gas and water temperature, respectively, usually 155 and 140°F, respectively, and a rejection of 160 Btu/(bhp · h). Where the temperature rise is between 2 and 5°F (1.1 and 2.8°C), the jacket-water requirement is 160 bhp/(500 × 3.2 × t_w), or 0.1 gal/(min · bhp). The use of cold water in jacket-water systems causes condensation on the cylinder walls, washing of lubricant, and excessive ring wear.

Rotary-Vane Compressors

Figure 9.3.11 is a section through a typical rotary-vane compressor. The displacement can be calculated using the following formula.

$$V_{th} = 2eL(\pi D - \delta_v n_v)N \tag{9.3.30}$$

Rotary-vane compressors have a built-in volume ratio r_v. This means that they will compress the gas within the compression chamber to $P_s{*}r_v^n$ before opening the compression chamber to the discharge port. $P_s{*}$ is the inlet pressure, and r_v is the ratio of the volume of the compression chamber at the inlet cutoff point to the volume at the point just

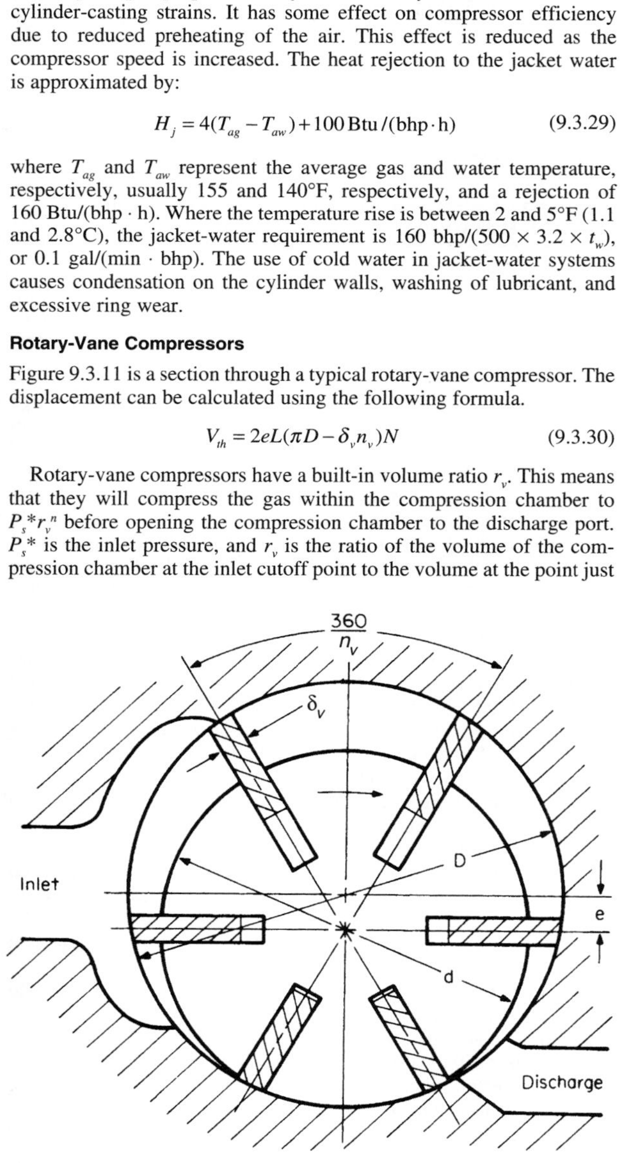

Figure 9.3.11 Section through a typical vane compressor with N_v vanes.

prior to the opening of the discharge port. If the operating-pressure ratio does not match the built-in ratio, overcompression or undercompression will take place as indicated in Fig. 9.3.12. For this case the ideal work should be calculated as:

$$W_{\text{ideal}} = \frac{P_s V_{\max}}{1-n}[r_v^{n-1} - n] - P_d \frac{V_{\max}}{r_v} \tag{9.3.31}$$

The ratio of this work over the ideal work required for a perfectly matched compressor can be read from Fig. 9.3.16. For example, the theoretical work for an air compressor with $V_i = 2$ operating at $P_2 = 0.8P_{id}$ (P_{id} means internal discharge pressure.) theoretically will require 2.5 percent more work than it would if it were designed so that the built-in ratio matched the operating ratio.

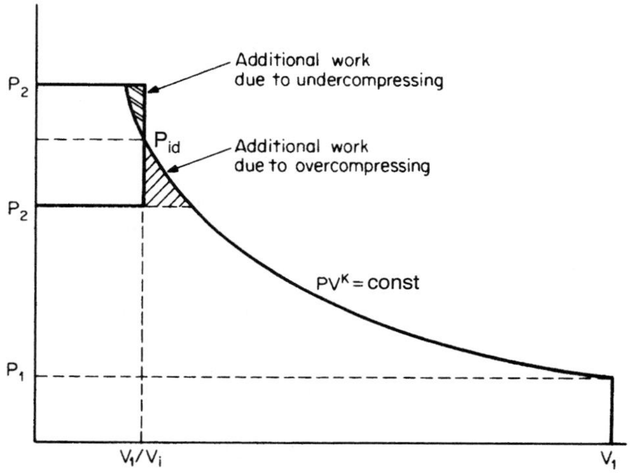

Figure 9.3.12 Idealized *PV* diagram for a compressor with a built-in volume ratio (also referred to as a *built-in pressure ratio*).

However, normal practice is to base efficiencies on a power calculated as though the compressor built-in ratio matches the operating ratio.

Rotary-vane compressors may be dry, lubricated, or oil-flooded. Prior to 1960 two-stage, oil-flooded vane compressors up to 900 cfm (1,530 m³/h) compressing to 150 psig (1,034 kPa) were used in portable compressors. They are now usually limited to less than 100 cfm (170 m³/h) air compressors and to small refrigeration compressors and boosters.

Vane materials may be phenolic (micarta, polyamid-imides, polyimides, Ryton) or metal (aluminum-silicon alloy).

The major limitations of vane compressors are imposed by vane-tip friction, vane-bending stress, and vane-length limitations. Maximum vane tip speeds are approximately 65 ft/s (20 m/s). Normally 6 to 8 vanes are used although as many as 20 vanes have been used. The trade-off is between reduced pressure differential per cell and increased friction. The eccentricity ratio is normally $e = 0.07D$ for low pressure (25 psig) (172 kPa) and $e = 0.05D$ for higher pressures (50 psig) (345 kPa). For service over 50 psig, the compressors are usually multistaged.

Assuming an injected oil temperature of 140°F (60°C) and no intercooling, a two-stage vane air compressor operating at 100 psig will have an adiabatic efficiency (including mechanical losses) of 60 to 72 percent. The discharge temperature will be approximately 90 to 95°C or 200°F. Oil-flooded vane compressors normally use synthetic lubricants. The high shearing rates and vane tip temperatures rapidly oxidize petroleum based lubricants.

Rolling-Piston Compressors

Superficially, the rolling-piston compressor (Fig. 9.3.13) resembles the sliding-vane compressor. However, its kinematics and performance are quite different from those of the sliding-vane compressor. The rolling-piston compressor consists of an eccentric, a rolling piston, a vane, a

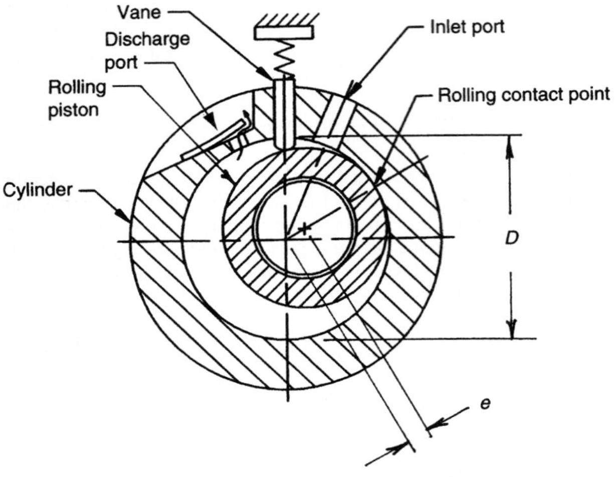

Figure 9.3.13 Section through a typical rolling piston compressor.

cylinder, and a discharge valve. The motion of the rolling piston with respect to the eccentric and cylinder is controlled by the various friction forces action on the piston. Compression and suction occur simultaneously in different chambers. A cycle covers two revolutions of the eccentric. Rolling-piston compressors are usually used in refrigeration systems, typically 0.25 hp (180 W) to 0.75 hp (550 W). Typical values for volumetric and power efficiencies are 0.90 and 0.84 to 0.86, respectively. The eccentricity is normally about $0.09 \times D$. The piston diameter is approximately $0.81 \times D$ and the length-to-diameter ratio is approximately 0.5. The compressor displacement can be calculated from Eq. (9.3.30) with $n_v = 1$.

Rotary, Twin-Screw, Oil-Flooded Compressors

Figure 9.3.14 is a cross section of an oil-flooded, rotary-screw compressor. The principle of operation can be determined from Fig. 9.3.15. The displacement volume of a twin-screw compressor is

$$V_{th} = KD^3 \frac{L}{D} C_{th} N \tag{9.3.32}$$

where $K = n_m(A_{mg} + A_{fg})D^2$; D = the main rotor diameter; L = the rotor length; n_m = the number of lobes on the main rotor; A_{mg} = the area of one groove on the main rotor; and A_{fg} = the area of one groove on the secondary rotor. C_{th} is an overlap constant (between 0.95 and 1.0 for most designs) which corrects for the overlapping of the filling and compressing processes.

C_{th} depends on the wrap angle, the profile, and the number of lobes. Lobe combinations of (4 and 5), (4 and 6), (5 and 6), and (5 and 7) are used. The most common design uses the (4 and 6) combination. Typical values for C_{th} and K are given in Table 9.3.2. A complete cycle (filling and compressing) takes place over approximately 750° of main rotor rotation. The number of compression cycles per revolution of the main rotor equals the number of lobes on the male rotor. The discharge process for adjacent grooves overlaps by approximately 40°, giving an inherently smooth discharge process.

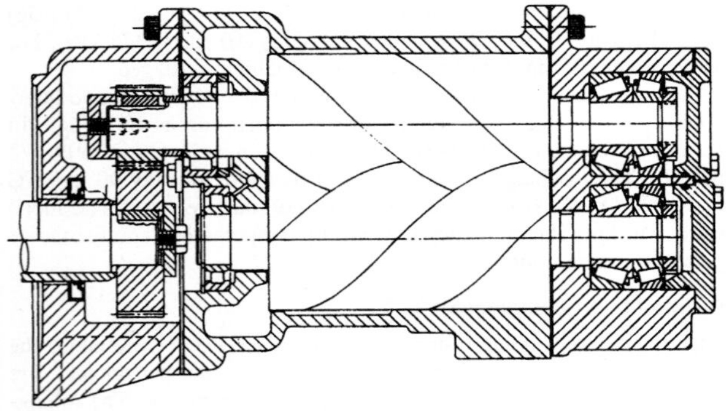

Figure 9.3.14 Cross section of a typical single-stage, oil-flooded, twin-screw compressor.

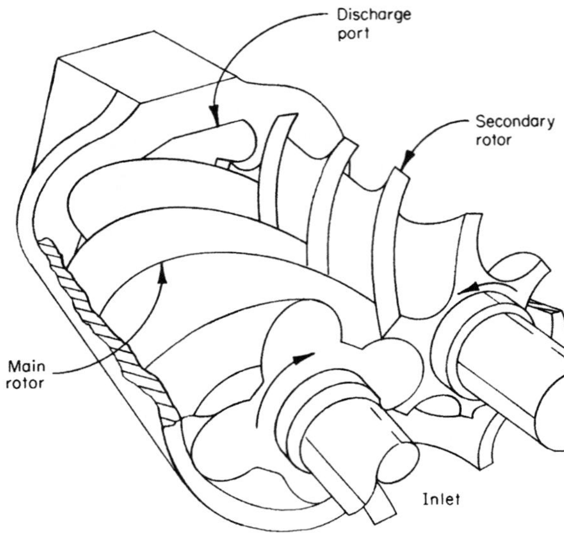

Figure 9.3.15 Operating principle of a twin-screw compressor.

Table 9.3.2 Typical Constants for Different Lobe Combinations (300° Male Wrap)

N_m/N_f	3/4	4/5	4/6	5/6	5/7
C_{th}	0.954	0.973	0.969	0.984	0.980
K	0.530	0.439	0.507	0.454	0.468

Rotary-screw compressors have a built-in volume ratio r_v (typically 4.4 for single-stage air compressors and 1.9 to 3.5 for refrigeration compressors). For an inlet pressure of P_s, the compressor will have a built-in pressure of $P_s r_v^n$. The effect on the P-V diagram and on the theoretical power can be found from Fig. 9.3.12 and Fig. 9.3.16. In practice, the increased discharge port losses associated with higher built-in volume ratios (smaller port area) skew the results in Fig. 9.3.16. In practice, it is more efficient, at least in air compressors, to have the built-in ratio slightly lower than the theoretical ratio. Single-stage, twin-screw air compressors rarely have built-in pressure ratios greater than 9, and most are designed with ratios of 7 to 8. An air compressor designed for 100 psig can operate at 175 psig with a loss in adiabatic efficiency of approximately 5 and 4 percent loss of volumetric efficiency.

Adiabatic efficiency (including mechanical losses in the air end) vary from 70 percent for small sizes (80 mm) to 88 percent for large (300 mm) slow-running compressors. An average value for a midsize compressor is 0.78 to 0.75 based on an injection temperature of inlet temperature plus 50°F. Accuracy in manufacturing the rotors is critical to achieving good efficiency. The heat absorbed by the cooling oil can be estimated by:

$$Q_h = \dot{W}\left[1 - \frac{\eta_c(T_d - T_s)}{T_s(r^{(k-1)/k} - 1)}\right] \tag{9.3.33}$$

where $\dot{W}$ = the shaft power to the compressor. T_d is normally 70 to 90°F (21.1 to 32.2°C) over T_s.

For most applications, tip speeds range from 15 to over 60 m/s, with rotor sizes from 40 to 600 mm, and L/D ratio from 1 to 2.2. Oil-flooded screw compressors dominate the portable-compressor market and the low-pressure plant air market, while reciprocating compressors dominate the higher-pressure applications. In industrial refrigeration the screw compressor finds its widest application in the 300 to 1,000 kW range.

In heat-pump applications, the screw compressor's zero-clearance volume is an advantage over reciprocating compressors. Reciprocating

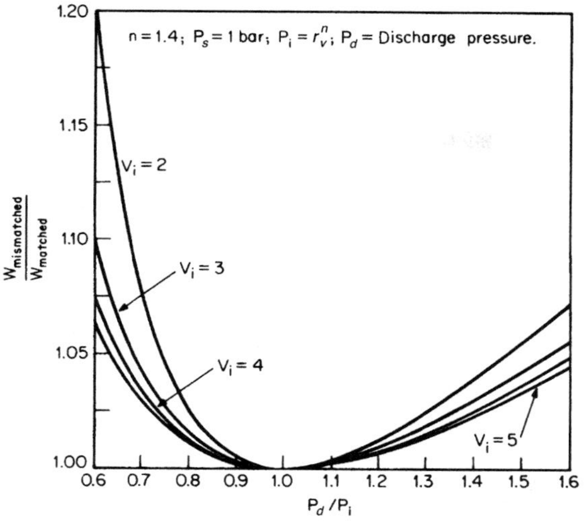

Figure 9.3.16 The effect of the built-in volume ratio on theoretical work of compression for a compressor operating off the design point.

compressors lose volumetric efficiency as the outside temperature drops (the compression ratio increases) requiring supplemental heat to compensate for the lost capacity. Refrigeration screw compressors are now available with a secondary suction port known as an **economizer connection.** This allows for two-temperature-level operation. At full-load operation, the system COP is improved significantly. As the compressor is unloaded, the improvement is lost.

Capacity can be controlled by on-line/off-line operation, inlet throttling, slide-valve or turn-valve operation, or speed regulation. Stationary air-compressors are normally controlled by on-line/off-line operation or inlet throttling with unload below 50 percent capacity. The air-oil sump is usually blown down to atmospheric pressure at unload.

Portable compressors are normally controlled by a combination of inlet throttling and speed regulation; the air-oil sumps are not blown down.

Refrigeration compressors are controlled by inlet throttling or more often by a slide valve or turn valve. A compressor with a slide valve is shown in Fig. 9.3.17. The effect on power consumption for each type of capacity control can be found from Fig. 9.3.18.

The discharge temperature for an air compressor must be kept high enough to prevent condensation in the lubricant during part-load operation. This is usually accomplished by a thermal valve set to maintain the minimum injection temperature at 140 to 165°F (60 to 76°C). Standard oil separators limit oil carryover to 5 ppm or less. Optional treatment can reduce the levels further.

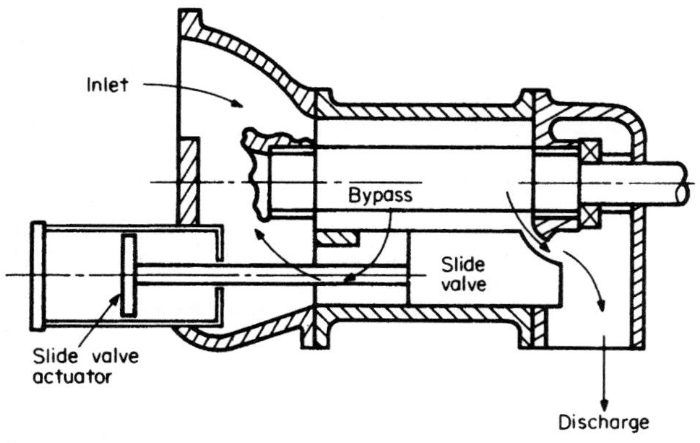

Figure 9.3.17 Principle of slide-valve operation. As the valve is moved toward the discharge end of the compressor, the point of inlet cutoff is delayed.

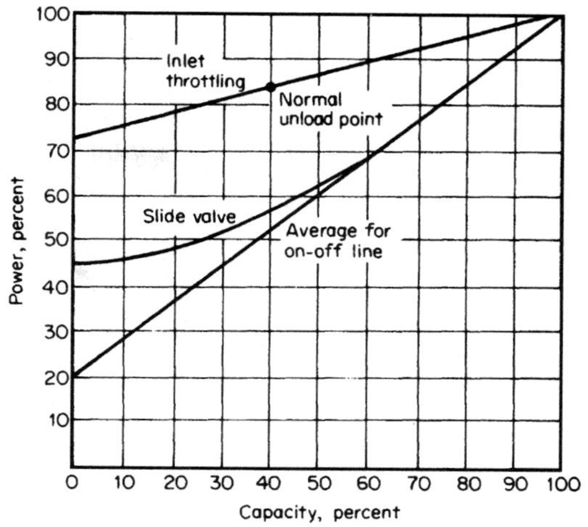

Figure 9.3.18 Comparison of different capacity control methods for oil-flooded screw compressors operating at constant speed.

The capacity loss of an oil-flooded, twin-screw compressor with altitude is quite small compared to a normal reciprocating compressor. The approximate capacity correction for altitude can be read from Fig. 9.3.19.

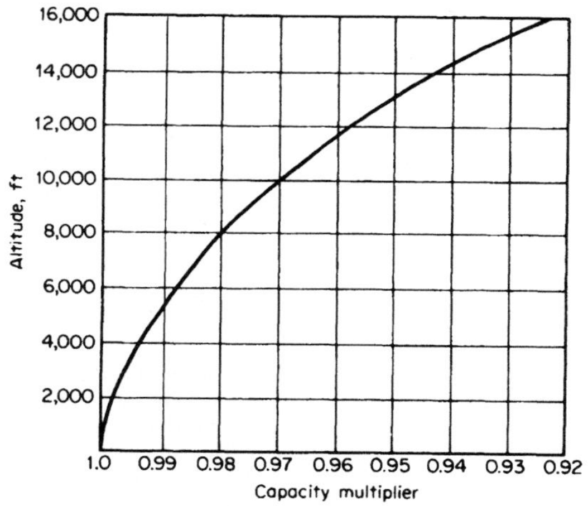

Figure 9.3.19 Altitude correction for capacity, single-stage, oil-flooded screw compressor.

Rotary Single-Screw Compressors

The operating principle of the single-screw compressor can be determined from Fig. 9.3.20. Lundberg and Glanvall give the displacement as:

$$\beta = \arcsin \frac{B - D/2}{R}$$

$$\Omega = \pi - 2\beta$$

$$C = 2R \sin \frac{\Omega}{2}$$

$$\gamma = \frac{\alpha \Omega}{\xi}$$

$$t = R - \frac{B - D/2}{\sin(\beta + \gamma)}$$

$$u = B - \frac{D}{2} + \frac{t}{2} \sin(\beta + \gamma)$$

$$V = \sum_{\alpha = \alpha 1}^{\xi} AU\Delta\alpha$$

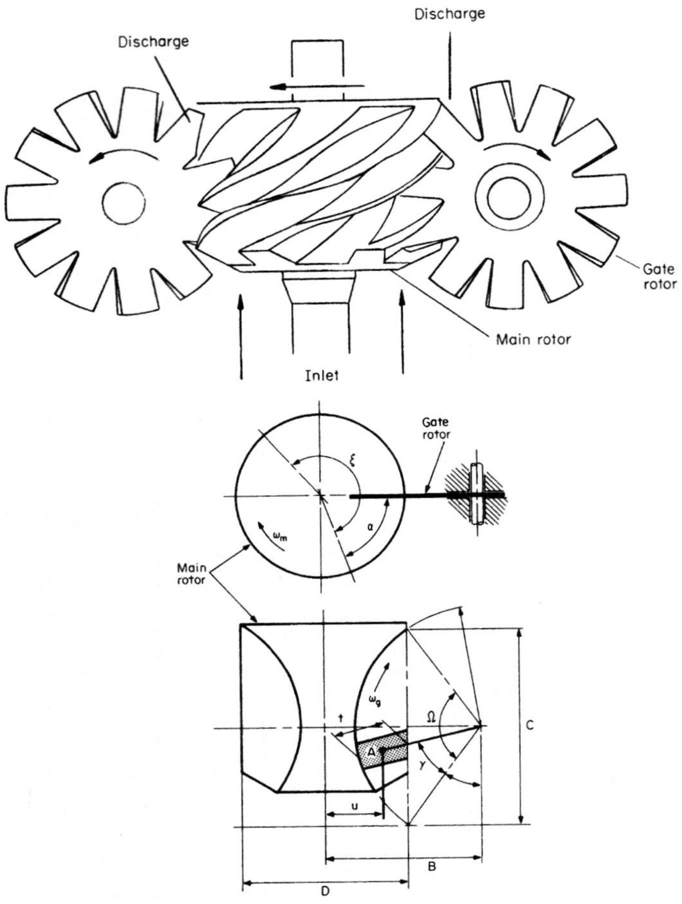

Figure 9.3.20 Principle of operation and geometry of a single-screw compressor. *(Lundberg and Glanvall, "A Comparison of SRM and Globoid Type Screw Compressors at Full Load," Proceedings of the 1978 Purdue Compressor Technology Conference, Purdue University, W. Lafayette, IN.)*

The rotary single-screw compressor is an oil-flooded compressor with a built-in volume ratio. The compressions for each star wheel are almost in phase so the gross effect as far as the discharge piping is that there are six discharges per revolution of the main rotor.

The single-screw compressor is used in the refrigeration industry where the low bearing load on the main rotor is an advantage. It is also being used as an air compressor in the smaller sizes.

Adiabatic efficiency of a single-screw compressor will be 2 to 5 percent lower than that of a comparable twin-screw compressor. The effect of off design operation on theoretical power can be found from Fig. 9.3.16. Capacity control is identical to the twin-screw compressor.

Dry Rotary Twin-Screw Compressors

A dry rotary, twin-screw compressor has timing gears to separate the rotors which may or may not be coated. The typical compressor is water-jacketed. Seals prevent gas leakage along the shafts and seal the oil to the bearings and timing gears. Sleeve bearings are most common, but some units have antifriction bearings. The displacement is the same as for the oil-flooded screw. Tip speeds vary from 80 to 140 m/s. Sizes range from approximately 400 to 20,000 cfm (0.19 to 9.44 m^3/s), and in special applications as high as 60,000 cfm (28.32 m^3/s). As a result of practical limitations, the discharge temperature is limited to 400°F (204.4°C), which determines the maximum permitted pressure ratio across each stage. Two-stage compressors are almost always intercooled. With intercooling to 15°F (9.4°C) they approach adiabatic efficiency (based on zero intercooling) including mechanical-loss ranges from 72 to 80 percent. While oil-flooded screw compressors can be controlled to very low capacities by inlet throttling, dry screw compressors cannot. The resulting pressure ratio causes excessive temperature increase across the compressor.

Orbiting Scroll Compressors

This is a positive-displacement compressor, which is common in air conditioning and heat pump applications, primarily in the 5 to 50 hp (3 to 37 kW) range.

The compression is achieved by two identical scrolls (typically involutes of circles phased 180° apart), whereby one of the scrolls is stationary, and the other orbits without rotating. The filling, compression, and discharging process usually takes 900 to 1,100° of shaft rotation, depending on the number of wraps. The orbiting scroll is driven by the crankshaft and prevented from rotating by an Oldham ring. A typical scroll is shown in Fig. 9.3.21.

The movement of the orbiting scroll causes pockets between flanks of the scrolls to get progressively smaller, until the volume reaches its internal limit (internal volume ratio, see "Screw Compressors"), finishing the process by discharging the gas. Different stages of the compression process are illustrated in Fig. 9.3.22.

The flow rate of a scroll compressor given by the equation

$$Q = 2A_{fv}H_vN\eta_v$$

where A_{fv} is the area of one of a full-volume cell (Fig. 9.3.22) and H_v is the height of the scroll vanes.

In the noncompliant version, both scrolls are held rigidly, with a small clearance between the flanks and tips. This design is susceptible

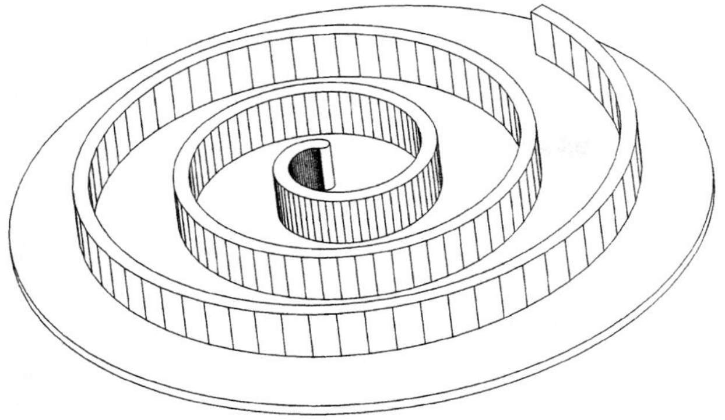

Figure 9.3.21 View of a typical scroll element.

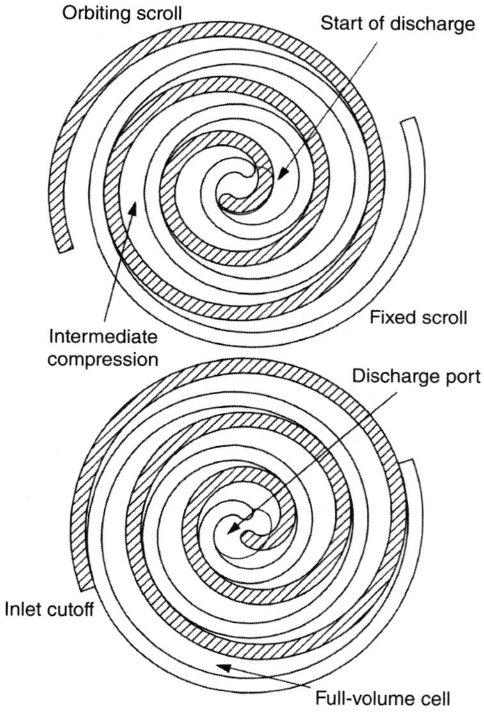

Figure 9.3.22 Scroll compression stages (moving scroll orbits clockwise).

to damage from particle throughput and hydraulic locking. The compliant version overcomes these issues by "floating" the scrolls into contact with each other. Axial contact is achieved by back pressure on the stationary scroll, while contact on the flanks is achieved by centrifugal force. This allows the scrolls to yield to the excessive forces without causing damage. There is, however, a potential for increased scroll wear, in particular if used with some of the latest chlorine-free refrigerants, which reduce the lubricity of the oil.

The internal leakage across the tip of the scroll vanes and along the vane flanks affects the compressor performance, more so in the noncompliant case. The leakage across the vane tips is significantly greater and damaging to performance than that along the flanks. For this reason, a groove is usually machined into the tip of the scroll vanes, and a sealing strip inserted.

Dynamic Compressors

A typical centrifugal compressor is shown in Fig. 9.3.23. The optimum range of operation can be found from Fig. 9.3.24. An estimate of the polytropic efficiency can be obtained from Fig. 9.3.25 or calculated from

$$\eta_p = 0.014\mathrm{In}(Q) + 0.600 \tag{9.3.34}$$

where Q is in cfm.

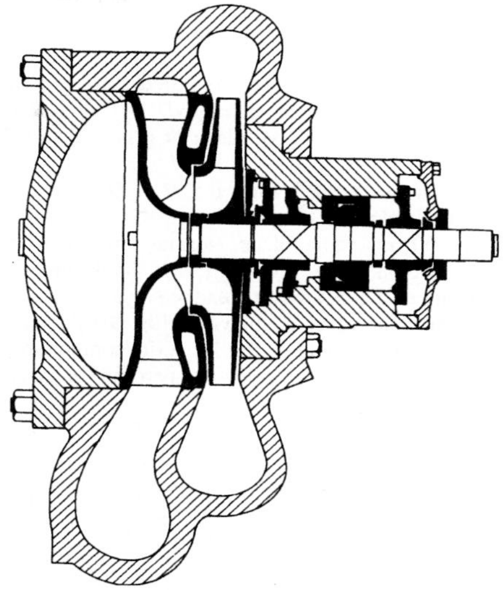

Figure 9.3.23 Single-stage, cantilever-design centrifugal compressor.

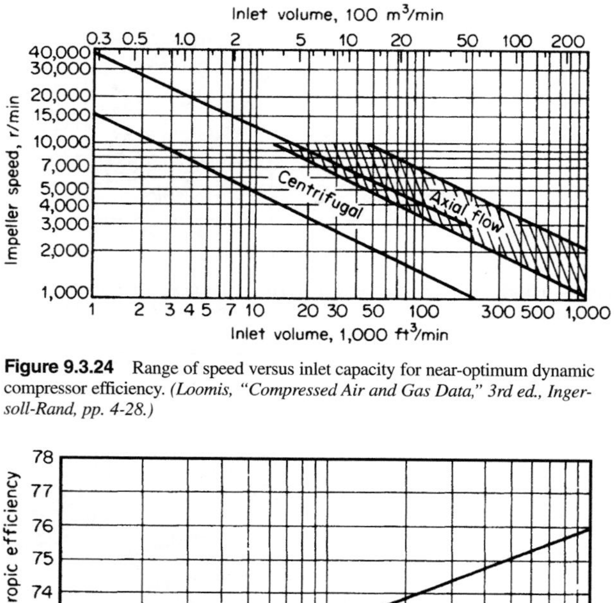

Figure 9.3.24 Range of speed versus inlet capacity for near-optimum dynamic compressor efficiency. *(Loomis, "Compressed Air and Gas Data," 3rd ed., Ingersoll-Rand, pp. 4-28.)*

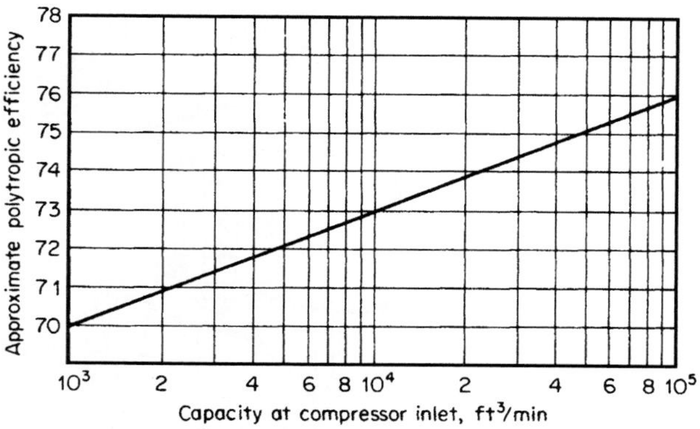

Figure 9.3.25 Approximate polytropic efficiency versus compressor inlet capacity. *(Rollins, "Compressed Air and Gas Handbook," 4th ed., Compressed Air and Gas Institute, pp. 3-71.)*

For normal industrial compressors, H_p = 10,000 ft · lb/lb (30 kJ/kg) per stage although much higher values can be obtained.

The required speed can be estimated from:

$$N = \frac{(g_c H_p / \mu)^{0.5}}{\pi D} \tag{9.3.35}$$

The pressure coefficient ($\mu = g_c H_p / U^2$) varies from 0.4 to 0.7 (in the high-efficiency range). In the absence of specific data, use 0.55 for estimating.

The diameter is determined by the required flow and can be estimated from the flow coefficient $\varphi = Q/ND^3 \propto U_a/U_p$, where U_a is the axial inlet velocity. The normal range of φ is 0.15 to 0.4 with the optimum at 0.3.

The slope of the μ-φ curve (and the $H_p \cdot Q$ curve) is strongly dependent on the backward lean of the impeller blades as shown in Fig. 9.3.26. β_2 = 90° is a radial blade. Normal closed industrial impeller blades will have an angle of 65 to 55°. It should be noted that an increase in density of the gas will flatten the slope.

A typical axial compressor is shown in Fig. 9.3.27. The range of application can be found from Fig. 9.3.18. Axial compressors are more efficient than centrifugals (as much as 5 percent) but have a much

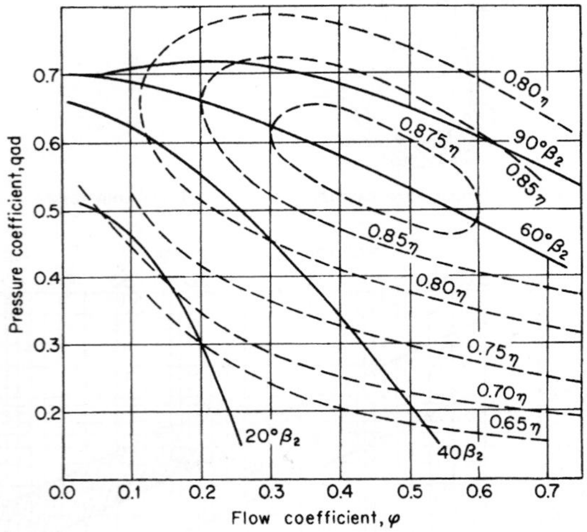

Figure 9.3.26 Basic characteristics of centrifugal compressor, showing effects of backward-leaning impeller blades. *(After Balje, ASME 51-F-12.)*

narrower range of operation. The surge limit will vary from 85 to 95 percent of design flow. For a given stage, the polytropic head will be about one-half that developed by a centrifugal stage, for example, μ = 0.3. Axial compressors are used only for air or clean gas. They find their major application in refineries, butadiene plants, and ethylene oxide plants. Jet-engine compressors are axial-flow compressors.

Thrust Pressures

A double-inlet and an open impeller have no axial thrust, whereas a semienclosed impeller, with a frontal shroud, can impose substantial thrust, evaluated by:

$$F_A = Y(A_1 - A_2)\Delta(\text{lb/in}^2) + 0.8(A_2 - A_3)\Delta(\text{lb/in}^2) - P_s A_3$$

where A_1 = full impeller area; A_2 = impeller area less the area of the inlet-eye seal; A_3 = shaft area through the seal, all in in^2; and Δ (lb/in^2) = pressure differential per stage. Y is the percentage of Δ (lb/in^2) acting in back of a single-stage disk: (1) Without a frontal shroud and a plain back, Y = 0.35; (2) with 0.060-in back ribs, Y = 0.28; (3) with 1/4-in equalizer holes at half radius and about 2 in apart, Y = 0.22; and (4) with deep scallops, Y = 0.18. For fully shrouded impellers, as used in multiple stage, Y = 1.00. These thrusts are balanced by opposing impeller inlet ends or, more generally, by a balancing drum wherein an equal and opposite thrust is created, equal to the product of the compressor Δ (lb/in^2) and the area of the drum. Stepanoff ("Turboblowers," Wiley) shows that radial impellers create a radial thrust, caused by the uneven volute-pressure distribution, which is:

$$F_R = 0.36\Delta(\text{lb/in}^2)db$$

where d = the impeller diameter; and b = the width, including the shrouds, both in inches.

Applicable Standards

Many national and international standards are available for guidance regarding construction, testing, packaging, and reliability. These include API, ANSI, ASME, CAGI, SAE, and international standards. Examples of coverage include construction (API 616, API 618, and API 619), testing (ASME PTC 9), and safety and packaging (API 670, ASME 19.3).

Compressor manufacturers commonly make a list of exceptions to stipulations in the various standards, or offer these as options at an additional cost. Therefore, it is important—if guidelines are to be drawn from specific standards—that the customer and manufacturer agree on compliance or exceptions to all relevant sections of the applicable standards, including construction, testing, test interpretation, and packaging features.

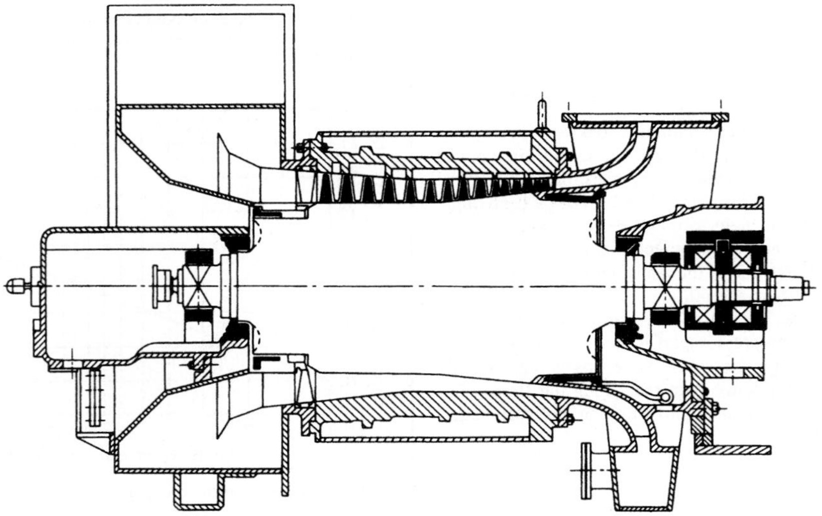

Figure 9.3.27 Multistage axial compressor.

9.4 HIGH-VACUUM PUMPS

by Marsbed Hablanian

REFERENCES: Van Atta, "Vacuum Science and Engineering," McGraw-Hill. O'Hanlon, "A User's Guide to Vacuum Technology," Wiley. Weissler and Carlson (eds.), "Vacuum Physics and Technology," Academic Press. Roth, "Vacuum Technology," North-Holland. Hoffman et al., "Handbook of Vacuum Science and Technology," Academic Press. Lafferty (ed.), "Foundations of Vacuum Science and Technology," Wiley. Hablanian, "High-Vacuum Technology," Marcel Dekker. Weston, "Ultrahigh Vacuum Practice," Butterworths. Berman, "Vacuum Engineering Calculations, Formulas, and Solved Exercises," Academic Press. Redhead et al., "The Physical Basis of Ultrahigh Vacuum," Chapman & Hall. Welch, "Capture Pumping Technology," Pergamon. Wutz et al., "Theory and Practice of Vacuum Technique," Vieweg & Sons/Wiley.

INTRODUCTION

High-vacuum pumps can be broadly divided into two categories: transfer pumps and capture pumps. In regard to transfer pumps, i.e., pumps that transfer the gas from a reservoir into the atmosphere, it can be said that the world *pump* is a historical misnomer. Such pumps could easily be called *rarefied gas compressors*. After all, to turn a device from a compressor into a vacuum pump, it is only necessary to place the reservoir at the inlet instead of at the discharge. However, because high-vacuum systems deal with small pressure differences (usually 1 atmosphere), enormous compression ratios, and very small mass flows, specialized designs are necessary. Because of very low mass flows involved at the very low pressures (or densities), it is possible to employ devices that do not exhaust the pumped gas but keep it out of the gas phase inside the pump.

Vacuum can be produced by a few basic methods: by mechanical displacement using pistons, vanes, lobes, and so on; by fluid ejectors and turbomachinery devices; by chemical reactions leaving solid residue; by ionizing the gas and driving the ions into a solid target with high electric fields; and by freezing the gas at cryogenic temperatures. In addition, high-vacuum engineering deals with gas flows ranging from the "normal" continuum conditions to the region of molecular flow. In the latter range, the gas molecules collide mainly with the walls of the vacuum vessel or conduits rather than with each other. This produces high compression ratios across short sections of conductors (pipes, elbows, etc.), but at extremely small pressure differences. In molecular flow conditions, there is no genuine pressure gradient inside the pipe, and molecules can flow as easily upstream as downstream. In such conditions it is often more useful to think in terms of densities rather than pressures.

Before proceeding to the description of the most common devices for producing high vacuum, it may be useful to note that high vacuum engineers, as a result of an independent historical development, use concepts and terminology differing from those associated with compressors. For example, *throughput,* a product of volumetric flow and pressure, is used instead of *mass flow,* because the gas is assumed to be always at room temperature. Instead of *volumetric flow,* vacuum engineers use *pumping speed,* which is associated with the rate of pressure reduction in a reservoir during evacuation. Also, when displaying the performance of the pump, they plot pumping speed versus inlet pressure (which is the dependent variable), rather than vice versa as is the custom in other mechanical engineering fields. Instead of speaking about resistance of lines, pipes, conduits, etc., they use the reciprocal concept of conductance.

The usual pascal unit for pressure can be used for designating the level of high vacuum, but millibar and torr are also used. Currently, in United States, the use of torr (equivalent of old mm Hg) is more common, while in Europe, mbar (1/1,000 of an atmosphere) is more frequent. One torr is equal to 133.3 Pa or 1.333 mbar. The pumping speed is measured in liters per second (L/s) and conductance in torr · L/s · torr (torr · L/s of throughput per torr of pressure difference), which, abbreviated, is also L/s (1 L/s = 2.12 ft³/m).

Pumping speed S which primarily involves the size of the pump, is defined by the following equation:

$$t = V/S(\ln P_1/P_2) \quad \text{or} \quad t = V/S(2.3 \log P_1/P_2)$$

where t is the time period of evacuation, V is the volume of the vacuum chamber, P_1 the initial pressure in the chamber, and P_2 the reduced pressure (see Fig. 9.4.1). This relationship works well in typical systems when leaks and gas evolution are neglected. In the presence of a leak from atmosphere into the chamber, the equation should be modified into:

$$t = V/S[\ln(P_1 - P_f)/(P_2 - P_f)]$$

where P_f is the final pressure due to the leak. In practice, vacuum pumps are never used at zero flow condition, because there always remains some residual gas evolution from internal surfaces exposed to high vacuum. With that in mind, the "ultimate" pressure achieved in a vacuum system after a long period of pumping can be written as follows:

$$P = \sum(Q_i/S_i)_{\text{ext}} + \sum(Q_i/S_i)_{\text{int}} + \sum(P_{2i}/k_i)$$

The first term is the summation of contributions of gas evolutions from external sources, where Q_i and S_i are throughputs and pumping speeds for various gases. The second term is the summation of corresponding gas evolutions in the inlet areas of the pump itself, P_{2i} represents corresponding discharge pressures, k_i is the limiting compression ratio for each gas, and the subscript i means "for each gas."

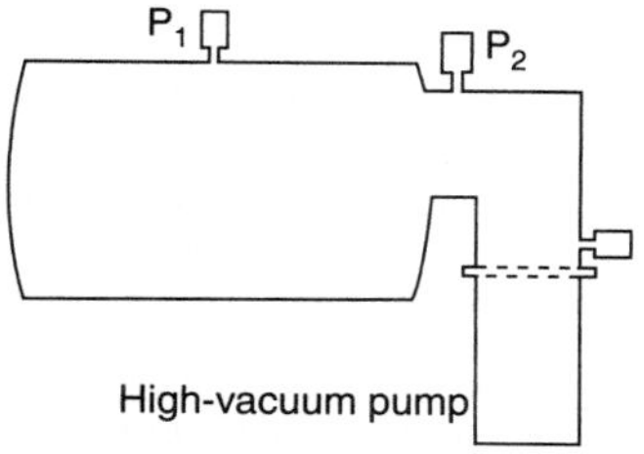

Figure 9.4.1 Schematic view of a vacuum chamber with three pressure gages and a pump.

PUMP SELECTION

There are a few common ways vacuum systems are used: rapid, repetitive evacuation to a given pressure for a process of short duration; long term processing of some material in a vacuum environment in the presence of substantial continuous or slowly diminishing gas evolution; or evacuation and complete isolation of a vacuum chamber or a container from the outside world. Each of these requires a different pumping system in regard to the type and size of pumps used. Figure 9.4.2 is a graph that can be used for the approximate evaluation of time required to reach a certain pressure depending on the volume of the vacuum chamber and the pumping speed of the pump. Because of the great range of variation of pressure (or density) during an evacuation of a container to high vacuum levels, it is usually necessary to pump initially with coarse vacuum pumps in order to remove the bulk of the gas. Then, when the amount of mass flow (throughput) is reduced to a tolerable level, pumps designed for pumping at high-vacuum conditions are engaged. This is reflected in Fig. 9.4.2 with the label "High Vacuum Valve Opens."

At the initial conditions in continuum flow, transfer-type pumps tend to have the same pumping speed for all gases, but when the molecular flow is approached, the rate of pumping begins to be different for various gases. This is especially true of capture-type pumps. Many technical articles and textbooks often omit this word of caution. In addition to the differences in pump performance, conductance values of conduits are also different for different gases. In common vacuum systems, engineering experience can be used for design with reasonable accuracy, but in unusual installations, laborious estimations must be made to evaluate the expected combined performance of pumps and systems.

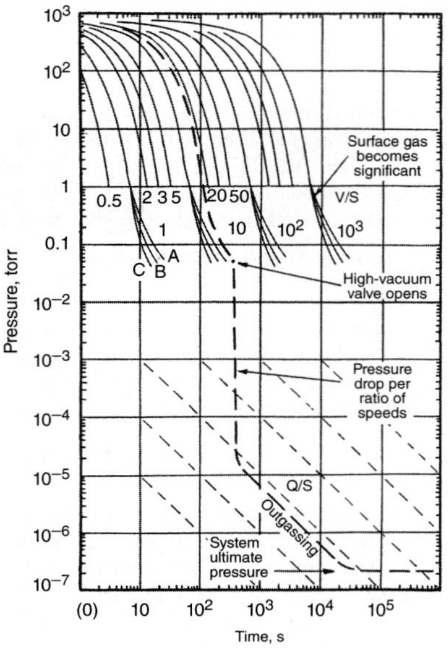

Figure 9.4.2 Evacuation process, from atmosphere to the high-vacuum region, shown on a log-log graph. Letters A, B, and C correspond to sand-blasted mild steel, 100-grit-finish stainless steel, and prebaked stainless steel chamber walls. The point called "high vacuum valve opens" indicates the start of high-vacuum engagement. (*Reprinted from Hablanian, "High-Vacuum Technology," p.133, Marcel Decker.*)

The most significant performance aspects of pumps are: pumping speed (related to volumetric flow), maximum steady-state throughput capacity (related to mass flow and power), and the ultimate pressure capability (related to limiting compression ratios and system gas evolution rates, called *outgassing*). This is indicated in Fig. 9.4.3. A typical performance curve for a transfer-type high-vacuum pumping system is shown in Fig. 9.4.4 in the traditional mode, and the same data are shown in Fig. 9.4.5, where the degree of achieved vacuum is shown on the vertical axis versus the imposed mass flow (throughput).

COARSE-VACUUM PUMPS

For the last 100 years, the common pump for producing high vacuum, beginning at atmosphere, has been the oil-sealed rotary-vane, positive-displacement-type pump shown in Fig. 9.4.6. To achieve high compression ratios, as high as 100,000 to 1 in a single stage, oil is introduced continuously into the pumping space and exhausted with the pumped gas. Oil is used for several functions: for lubrication, for flooding the exhaust space under the exit valve (so-called dead space), to seal the

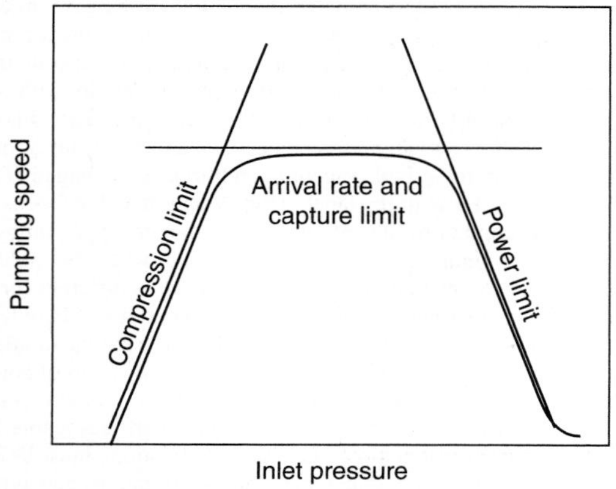

Figure 9.4.3 General basic limitations of pump (or compressor) performance.

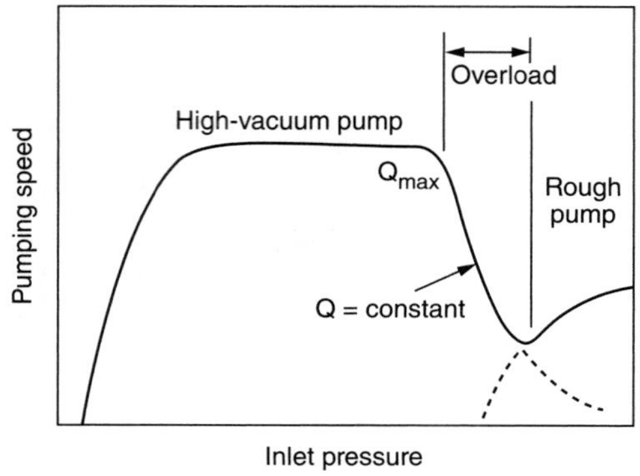

Figure 9.4.4 Traditional way of presenting performance of a high-vacuum/rough-vacuum pump combination. Q_{max} is the highest permissible steady-state gas flow for the high-vacuum pump.

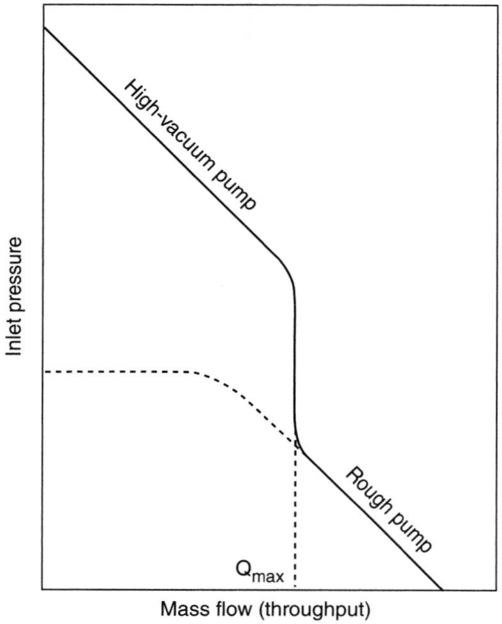

Figure 9.4.5 Typical performance of a high vacuum pump (see Fig. 9.4.4), where the achieved vacuum level is plotted versus mass flow (inlet pressure increasing downward).

tight spaces between the rotor and stator, to serve as a heat transfer medium for cooling the rotor, to flush particular matter out of the pump, and sometimes to actuate auxiliary valves. A two-stage pump, such as the one shown in Fig. 9.4.6, when attached to a small reservoir, can produce inlet pressures below 1×10^{-3} torr while exhausting to atmosphere. However, it should be noted that the pumping action itself does not cease at that point for gases that are not present in high quantities in atmosphere. For example, helium from the vacuum chamber is pumped by such pumps at partial pressures of at least 10^{-10} torr.

Pumps of that type are not made in very large sizes (they become expensive, noisy, and power consuming). To achieve lower pressures, such pumps are often used in series with pumps of high pumping speed such as Roots blowers, screw pumps, and other cam-type pumps. In many applications, the presence of oil films at the inlet side of the pump is objectionable, because at molecular flow conditions, oil molecules can travel into the vacuum system and produce contamination. To prevent this, "oil-free" vacuum pumps have been developed, which do not include deliberate introduction of oil into the vacuum space. This is usually achieved by using several stages in series. Typically, it takes

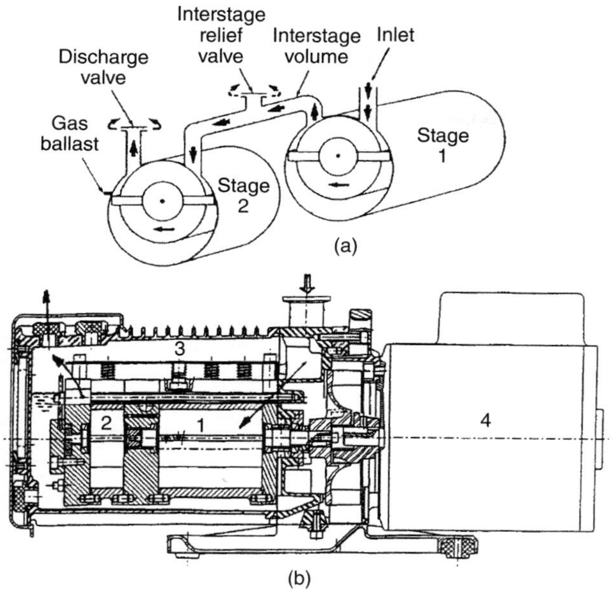

Figure 9.4.6 (*a*) Two-stage-pump flow diagram. (*b*) Cross section of a two-stage, oil-sealed, rotary-vane rough-vacuum pump. (*Courtesy of Alcatel Vacuum Technology.*)

three or four stages in series to produce performance similar to that of a single-stage oil-sealed pump. For smaller "dry" pumps, multistaged scroll and diaphragm pumps are also used.

TURBOMOLECULAR PUMPS

There are two types of pumps in this category: so-called drag pumps and turbomolecular pumps. **Drag pumps** function by adding momentum to "stationary" gas molecules during a collision with the fast-moving solid surface of the rotor, thereby moving the molecules toward the discharge side of the pump. Three early versions of drag pumps are shown in Fig. 9.4.7. There is no basic difference between the three types; they differ only in the general geometric arrangement. All three concepts are utilized in modern pumps in a variety of combinations. **Turbomolecular pumps** are essentially axial-flow compressors designed for pumping rarefied gases, typically intended for molecular flow conditions at the inlet to the pump. The basic structure in shown in Fig. 9.4.8, and a typical pump construction in shown in Fig. 9.4.9. The rotors of such pumps are usually made of strong aluminum alloys to reduce the rotational moment of inertia. The peripheral velocities of the rotor are usually in the 200 to 300 m/s range. Usually, the rotor is supported by ceramic ball bearings lubricated by special low-vapor-pressure grease. To avoid even a suspicion of possible contamination, magnetic bearings are often used, with auxiliary dry bearings engaged only during starting and stopping periods.

For the drag pump, the pumping speed of a stationary channel placed in close proximity to the moving surface of the rotor can be estimated by comparing the forward flow, due to the velocity of the rotor, to the backflow, resulting from the developing pressure difference. Neglecting leakages, the maximum compression ratio K obtained in a channel of certain size and length is derived to be $K_{max} = e^{S_o/C}$, where S_o is pumping speed and C is conductance. In molecular flow, this indicates the dependency of compression ratios on the conductance, and thus on the molecular weight of the gas. The obtained compression ratios are much smaller for lighter gases. In practice, such drag channels work best for small flows, a few liters per second. In viscous flow conditions, the pressure differences obtained in small pumps are in the range of a few torr per circular loop (which may be called a "stage").

The mechanism of pumping of the turbomolecular pumps is not completely different from that of axial-flow compressors, but in molecular

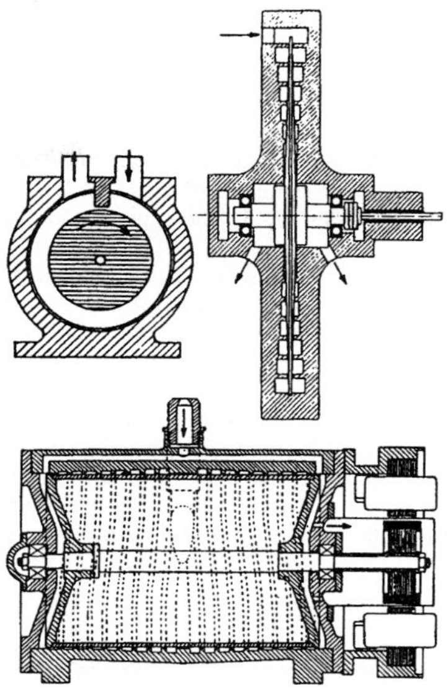

Figure 9.4.7 Three early versions of molecular drag high-vacuum pumps.

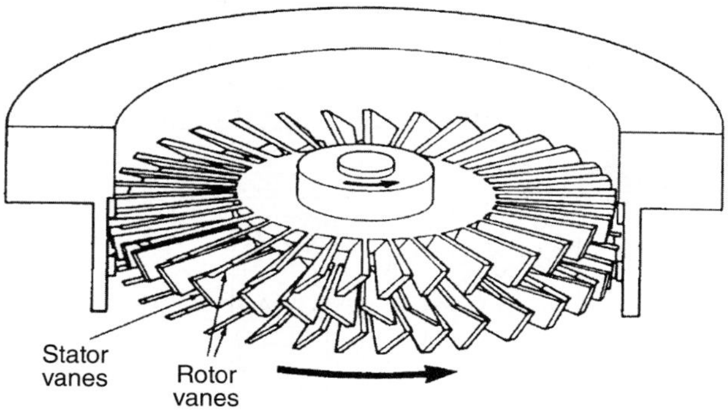

Figure 9.4.8 Partial cross section of an axial-flow pump. Two rotors and one stator are shown.

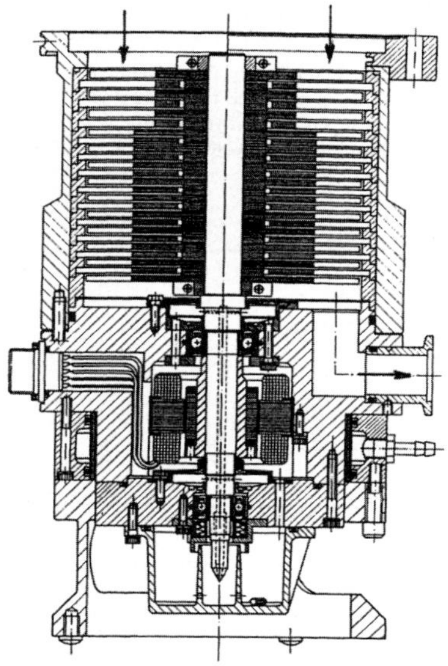

Figure 9.4.9 Cross section of a typical axial-flow pump (with 13 bladed rotors).

flow, because of the reduced amount of backflow, the compression ratios obtained in high vacuum are much higher than those customary in compressor technology (as high as 10:1 per rotor-stator pair). The volumetric displacement of the bladed rotor can be estimated from the velocity in the pumping direction, times the projected pumping area, times a capture coefficient (ratio of pumped molecules to reflected ones). For pumping speed estimates of typical turbomolecular pumps, with first-row blade angles at 45°, the capture coefficient is near 0.6 for air or nitrogen.

Because of the high compression ratios obtained in high-vacuum pumps, the volumetric flow at the discharge side of the pump is rather low. Also, the flow tends to get into the viscous range. Therefore, it makes sense to combine the characteristics of drag pumps and turbine-type pumps into a hybrid design. An example of such a pump is shown in Fig. 9.4.10. Figure 9.4.11 shows the comparative performances of a conventional turbomolecular pump for helium (left side) and a hybrid pump (right side). A significant improvement is obtained without an increase of pump size or motor power. Turbomolecular pumps are usually manufactured in sizes from 50 to 2,000 L/s, but they can be made larger.

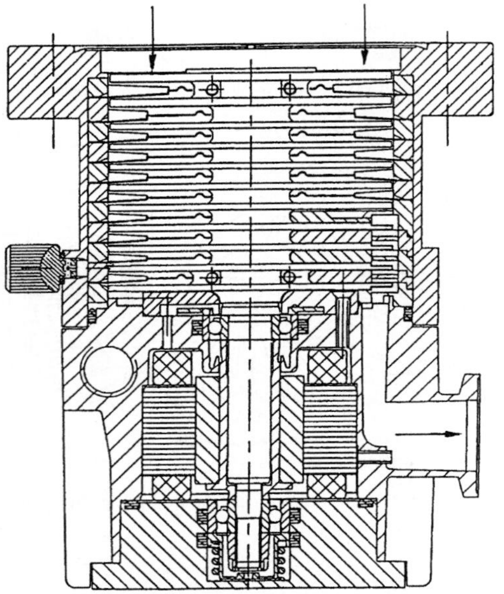

Figure 9.4.10 A comparison of an axial flow pump and a hybrid pump. On the left side of the vertical axis there are 11 bladed rotors (interspersed with 10 bladed stators). On the right side is the hybrid pump with 7 bladed rotors and 4 solid discs rotating in 4 drag channels connected in series. (Varian V-60.)

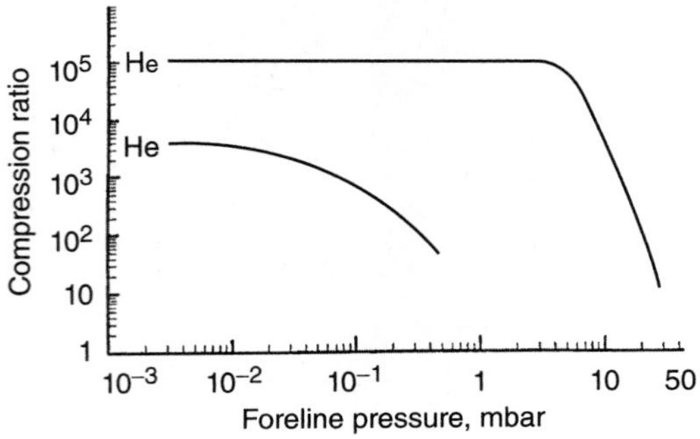

Figure 9.4.11 Compression ratios for two pump types (Fig. 9.4.10), for helium, depending on the pressure of helium at the exit. Upper curve, hybrid pump (Varian V-250); bottom curve, all-axial pump. (V-200.)

VAPOR JET PUMPS (DIFFUSION PUMPS)

Vapor jet pumps (often called *diffusion pumps*) have been the most commonly used pumps for production of high vacuum throughout the major part of the twentieth century and into the twenty-first century. The principle of operation of vapor jet pumps is illustrated in Fig. 9.4.12. The pumping action is due to the momentum transfer from a high-velocity fluid jet to ambient gas molecules. The device is basically similar to steam ejectors but is highly specialized to function

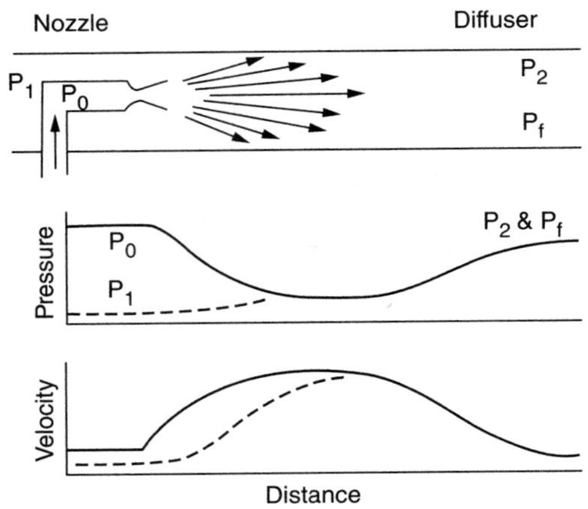

Figure 9.4.12 Pumping effect of a fluid jet. Solid lines are for the motive fluid, dashed lines for pumped gas; P_0, P_1, are initial pressures and P_f is final pressure.

at molecular flow conditions at its inlet port. Figure 9.4.13 is a schematic drawing of a modern pump that has three downward-directed nozzles and one horizontal one, called an *ejector,* all connected in series. The motive fluid used is usually an oil-like substance having a molecular weight in the range of 400 to 500 (for example: tetramethyl-tetraphenyl-trisiloxane). Such fluids must have a very low vapor pressure at room temperature but be capable of producing a few torr of boiler pressure without undergoing excessive thermal decomposition at typical temperatures of 200 to 240°C. They should be also nontoxic and noncorrosive and have a viscosity range at water-cooling temperature that permits draining back into the boiler by gravity.

The practical pressure difference that can be sustained by high-vacuum jet pumps is 0.5 to 1 torr, so they must be operated only when attached to a backing pump that provides necessary primary vacuum.

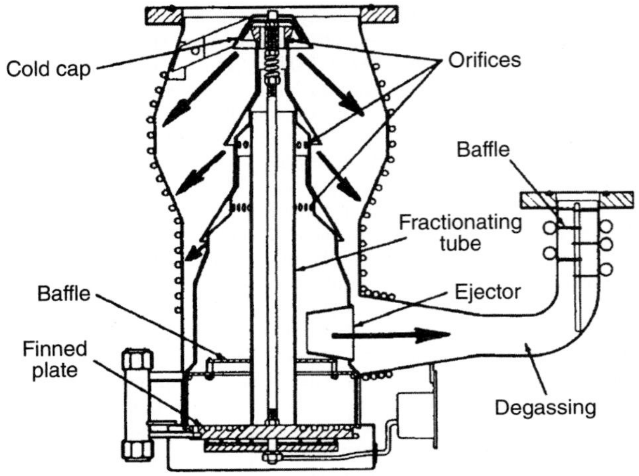

Figure 9.4.13 Schematic cross section of a four-stage vapor jet pump (Varian).

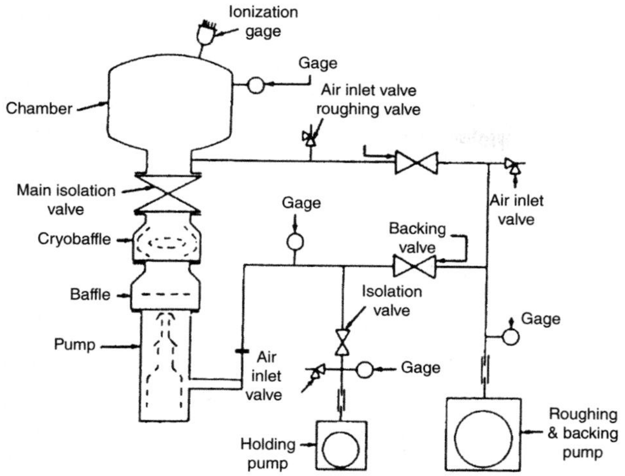

Figure 9.4.14 Typical arrangement of valves for a vapor-jet pump system.

Also, in order to prevent a possibility of rapid deterioration of and even combustion of the fluid, vapor jet pumps cannot be exposed to atmospheric air while at operating temperature. Therefore, they must be protected from an accidental air release by appropriate valves. An example of a system arrangement is shown in Fig. 9.4.14. In operation, the fluid is vaporized in the boiler by electric heaters, the vapor issues from the nozzles at velocities of a few hundred meters per second, and then condenses at the water-cooled walls and returns to the boiler.

A typical performance curve of a vapor-jet pump is shown in Fig. 9.4.15. Such performance curves are usually measured by using total-pressure gages, with air as the pumped gas. The decay of the pumping speed at the left of the graph represents the onset of compression ratio limits. The region called "overload" in the graph should never be used in any continuous operation. Here, the reduction of pumping speed is caused by a complete collapse of function in one of the jets (usually one near the inlet). The overloaded vapor stream may not reach the wall and may produce a serious amount of backstreaming into the vacuum chamber. At the inlet of the pump, a small number of vapor molecules can leave the condensed liquid and migrate into the chamber. To prevent this possibility of contamination, water-cooled baffles and traps cooled to cryogenic temperature are used, as shown in Fig. 9.4.14.

Vapor-jet pumps are made in sizes ranging from 50 to 50,000 L/s, corresponding to inlet diameters from 2 to 36 in. The larger pumps may use five or six stages instead of the four shown in Fig. 9.4.13.

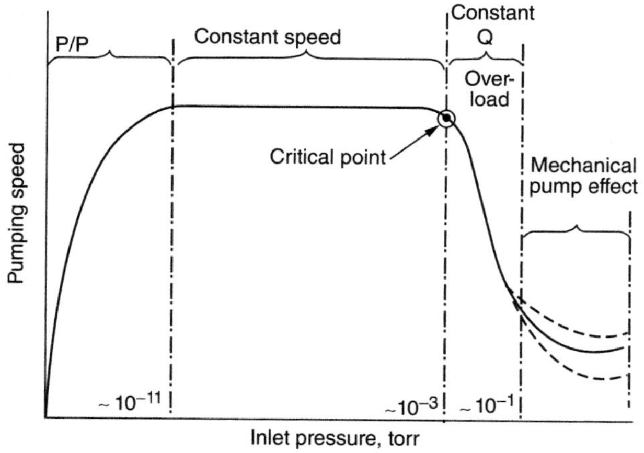

Figure 9.4.15 Pumping speed versus inlet pressure in four different performance regions.

There also exist so-called diffusion-ejector pumps, which utilize motive fluids of lower molecular weight and have higher boiler pressures. These pumps extend the full pumping speed region of operation toward higher inlet pressures. Such pumps are made in sizes from 4,000 to 12,000 L/s.

CRYOGENIC PUMPS

One way of creating high vacuum is to lower the temperature of the gas until it condenses into a liquid or a solid. In industrial practice, instead of cooling the entire chamber, it is sufficient to use a cryocooled pump. The vapor pressure curves in Fig. 9.4.16 demonstrate that it is not practical to cool gases of low molecular weight by condensation alone. Therefore,

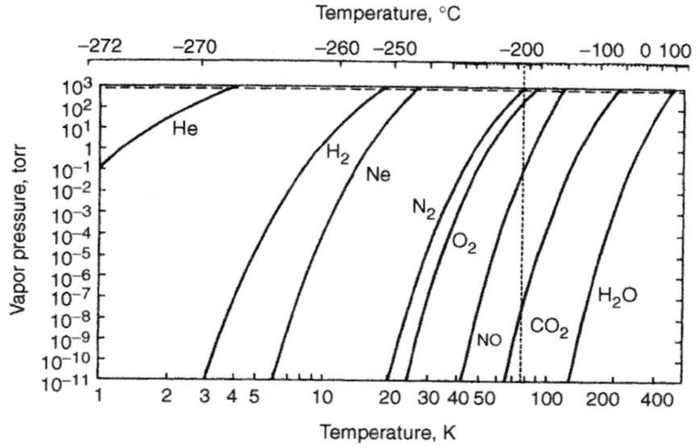

Figure 9.4.16 Vapor pressure curves of some common gases.

some internal surfaces of cryopumps are usually cooled by gaseous helium refrigeration to temperatures of 10 to 20 K combined with surfaces covered with absorbent layers. A schematic drawing of a typical cryopump is shown in Fig. 9.4.17. Inside the pump is a two-stage cryogenerator consisting of two connected reciprocating pistons filled with heat

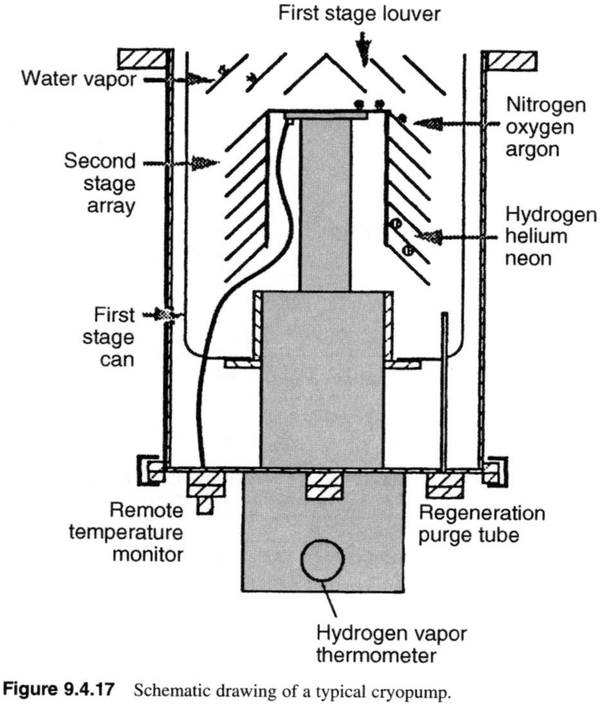

Figure 9.4.17 Schematic drawing of a typical cryopump.

exchange materials. The pump is connected to a special helium compressor, which is used to circulate helium through the expander assembly and to recompress it. Industrial cryopumps utilize specially modified refrigeration compressors. Effective filtering devices must be used to prevent lubricating oil from reaching areas cold enough to freeze the lubricant.

The pumping surfaces of a cryopump are arranged in three distinct varieties. The first is a metal surface kept at 50 to 80 K. It may be shaped as a cylinder with the closed end attached to the first stage of refrigeration and the open end connected to a baffle structure. This part of the pump can condense the water vapor, which is one of the main gases left in a vacuum system after an initial evacuation by coarse vacuum pumps. The second section is a metal surface attached to the second stage of the refrigerator, which is kept at 10 to 20 K. This section is used for pumping such gases as nitrogen, oxygen, and argon. The third section is at the same temperature as the second but includes a layer of charcoal (or activated alumina) bonded to the inner surfaces of the second-stage cryoarrays. Charcoal may have internal surface area approaching 1,000 m^2/g as well as small enough pores to hold molecules of light gases.

The pumping speed of a condensation and sorption pump depends on the arrival rate of the gas, capture probability, and surface area. Figure 9.4.18 shows an example of the performance of an industrial cryopump having an inlet diameter of 20 cm. From the kinetic theory of gases, the maximum pumping speed can be calculated to be for hydrogen 44.6 L/s for each square centimeter of surface; the corresponding value for water vapor is 14.9 and for nitrogen 11.9, provided that access to the surface is not impeded by conductance limitations. In practice, full pumping speed may be realized for water vapor because of the open exposure of the cooled baffle at the inlet, but the speed for nitrogen is typically 3 $L/s \cdot cm^2$ and for hydrogen 5.6 $L/s \cdot cm^2$. Because the gas is not removed from the pump, the pumping speed will decline in time. For example, for the pump of Fig. 9.4.18, the decline of pumping speed for hydrogen begins after about 7,000 torr-L of gas are pumped. The values given for pumping speed are measured under ideal conditions. In use, the actual speed may depend on amounts and types of gases previously pumped. Cryopumps do not have a continuously functioning exhaust port, therefore they become saturated after a certain period of operation.

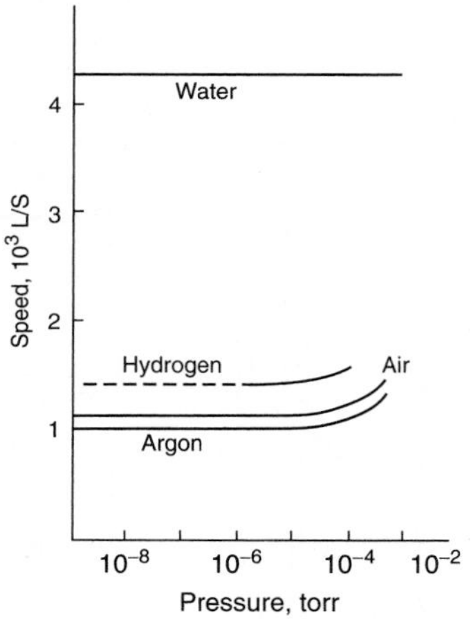

Figure 9.4.18 Pumping speed of a cryopump for different gases.

At high-vacuum inlet pressures, they can function continuously for several months, but eventually regeneration becomes necessary. The need for regeneration arises when the pump cannot maintain expected pressure or evacuation time. Regeneration requires warming of the pump and may take several hours. It involves several steps of gradual warming and purging so as not to drive certain unwanted gases into the inlet space, and so as to properly manage the disposal of dangerous gases.

GETTER AND ION PUMPS

When gases are rarefied to high-vacuum and ultrahigh-vacuum levels, a surface can hold larger quantities of the gases, as compared to the amount of the gas in space. Properly prepared surfaces (with and without additional absorbing layers) can then be used as a pumping medium. Many chemically active materials can be used to enhance the quantity of gas held on the surface, a process called *gettering*. Often the material used is titanium. To produce a pumping surface, all that is needed is to form a fresh (unoxidized) layer of material on a sufficiently large surface. An example of a simple arrangement of a gettering pump is shown in Fig. 9.4.19. As is the case with cryopumps, the performance of a gettering pump cannot be predicted exactly because of its dependency on the mixture of gases, the presence of chemically inert gases, the temperature, and the previous history of the surface. Approximate pumping speeds are shown in Table 9.4.1, where it can be seen that the inert gases are not pumped. In Fig. 9.4.20 actual pumping speeds are shown for air for pumps of various sizes.

There also exist so-called nonevaporated getter pumps. In such pumps the active material is not evaporated, but remains in a specially

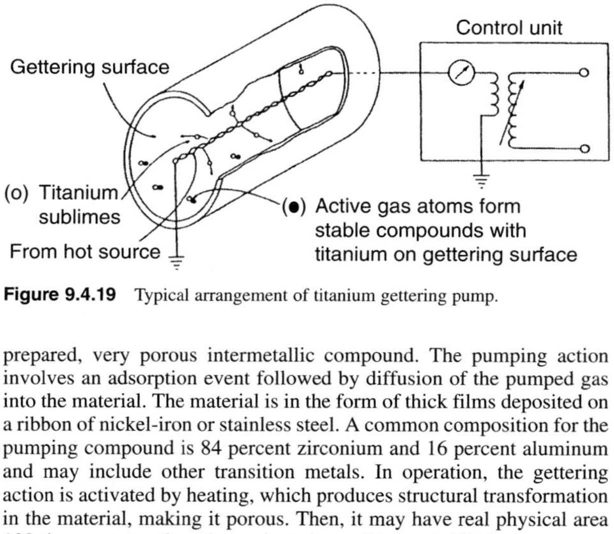

Figure 9.4.19 Typical arrangement of titanium gettering pump.

prepared, very porous intermetallic compound. The pumping action involves an adsorption event followed by diffusion of the pumped gas into the material. The material is in the form of thick films deposited on a ribbon of nickel-iron or stainless steel. A common composition for the pumping compound is 84 percent zirconium and 16 percent aluminum and may include other transition metals. In operation, the gettering action is activated by heating, which produces structural transformation in the material, making it porous. Then, it may have real physical area 100 times greater than the projected area. Figure 9.4.21 is an example of a pumping cartridge placed in a cylinder that can be attached to a vacuum chamber at the top flange.

To provide pumping action for inert gases, ion-getter or sputter-ion pumps were developed. A basic cell of such a pump is shown in

Table 9.4.1 Typical* Pumping Speeds per Square Centimeter of Titanium Sublimation Surface for Various Gases (L/s)

Surface temperature (°C)	H_2	N_2	O_2	CO	CO_2	H_2O	CH_4	A	He
20	20	30	60	60	50	20	0	0	0
−195	65	65	70	70	60	90	0	0	0

*Actual speeds vary with the condition of the deposits; usable speed is reduced by the conductance between the pumping area and the process area in the vacuum chamber.

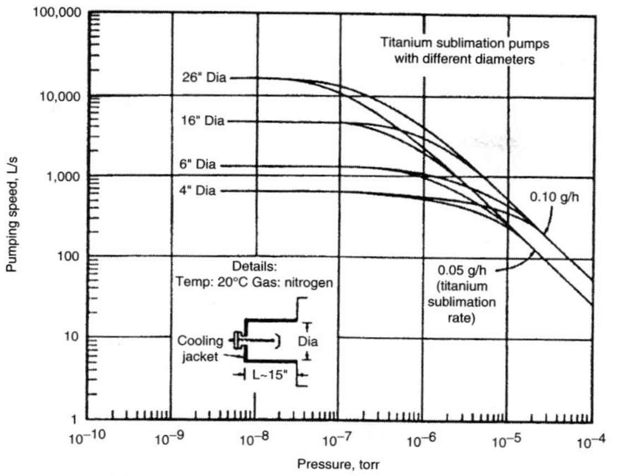

Figure 9.4.20 Pumping speeds of gettering pumps of various sizes.

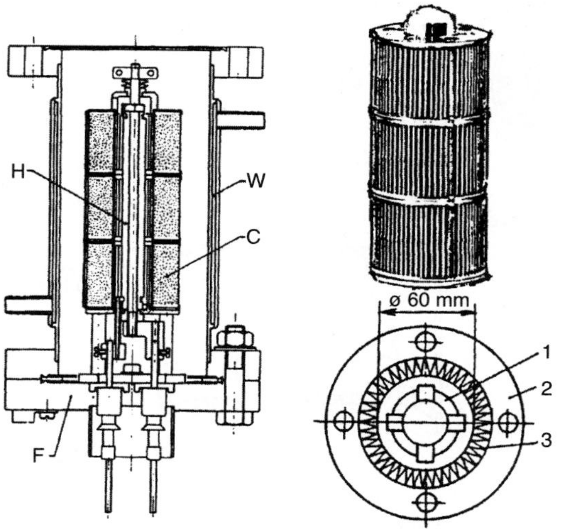

Figure 9.4.21 The view of a nonevaporated getter pump. H and 1—electric heater, W—water cooling (for large pumps), C and 3—the getter ribbon; 2—inlet flange.

Fig. 9.4.22. Electrons are emitted from the essentially cold cathode by the action of the electric field of 5 to 7 kV applied between the electrodes. Long trajectories of electrons are produced by the action of the magnet, to improve chances of collision and subsequent ionization. Positive ions bombard the anode at high energies, producing the phenomenon called sputtering, which is an ejection of atoms from the titanium cathode surface. The deposited titanium film reacts with active gases; inert gases are pumped when highly accelerated ions collide with and penetrate the surface structure. The details of pumping mechanisms of sputter-ion pumps are more complex than outlined here, but it is clear that various gases are pumped in different ways, at different pumping speeds and different saturation effects. Some pumping speed values are shown in Table 9.4.2. It is also clear that eventually cathodes become exhausted and must be replaced.

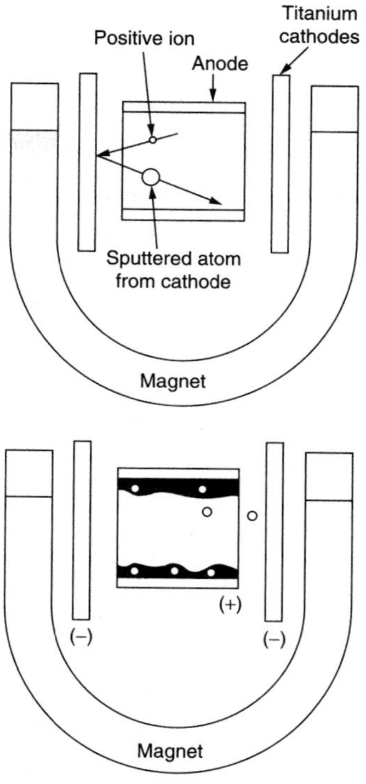

Figure 9.4.22 Basic cell of an ion-getter pump.

For practical designs of sputter-ion pumps the cells are between 15 and 25 mm, which is suitable for magnetic fields of 1 to 1.5 kg. The pumping speed of a cell is between 0.3 and 2 L/s. Multiple cell structures are used for larger pumps. Pumps are commonly manufactured in speeds up to 200 L/s. They are delivered prepumped and sealed to keep internal surfaces clean. To achieve ultrahigh vacuum, the entire vacuum system, including the pumps, is backed for several hours at temperatures up to 400°C, while being pumped by some clean auxiliary pumps.

Table 9.4.2 Relative Pumping Speeds of Ion-Getter Pumps for Various Gases*

Gas	Diode pump (3.5 kV)	Triode pump (5.2 kV)	Diode pump with slotted cathode (7.3 kV)	Diode pump with Mo and Ti cathodes
H_2	2.5-3.0	2.0	2.7	
CH_4	2.7	0.9-1.05		
NH_3	1.7			
H_2O	1.3		1.0	
Air	1.0	1.0	1.0	1.0
N_2	0.98	0.95	1.0	
CO	0.86		1.0	
CO_2	0.82	1.0	1.0	
O_2	0.55	0.57	0.57	
He	0.11	0.3	0.1	0.25
Ar	0.01	0.3	0.06	0.2

*Taken from manufacturers' catalogs. Air speed taken as unity.

9.5 FANS

by Robert Jorgensen

REFERENCES: Jorgensen, "Fan Engineering," 9th ed., Howden Buffalo. "Standards Handbook," AMCA Publication 99-86 (R 1998). "Fans and Systems," AMCA Publication 201:2002. "Field Performance Measurement of Fan Systems," AMCA Publication 203-90. "Balance Quality and Vibration Levels for Fans," AMCA Standard 204-96. "Laboratory Methods of Testing Fans for Aerodynamic Rating," American National Standard ANSI/AMCA 210-1999 and ANSI/ASHRAE 51-1999. "Reverberant Room Method for Sound Testing of Fans," AMCA Standard 300-96. "Laboratory Methods of Sound Testing of Fans Using Sound Intensity," Proposed American National Standard, BSR/AMCA Standard 320. "Laboratory Method of Testing to Determine the Sound Power in a Duct," ANSI/ASHRAE Standard 68-1997 and ANSI/AMCA 330-97. "Industrial Process/Power Generation Fans: Specification Guidelines," AMCA Publication 801-01. "Industrial Process/Power Generation Fans: Establishing Performance Using Laboratory Models," AMCA Publication 802-02. "Industrial Process/Power Generation Fans: Site Performance Test Standard," AMCA Publication 803:2002. "ASME Performance Test Codes, Fans," ANSI/ASME PTC 11-1984.

Symbols

A = area, ft² (m²)
b = barometric pressure, in Hg (Pa)
D = diameter, ft (m)
D_s = specific diameter, ft (m)
H = fan power input, hp (W)
H_0 = fan power output, hp (W)
h = head, ft (m)
K_p = compressibility factor, dimensionless
L_p = sound pressure level, dB
L_w = sound power level, dB
L_{ws} = specific sound power level, dB
log = logarithm to base 10
M = molecular weight, dimensionless
N = speed of rotation, r/min
N_s = specific speed, r/min
n = number or polytropic exponent
p_s = fan static pressure, in wg (Pa)
p_t = fan total pressure, in wg (Pa)
p_v = fan velocity pressure, in wg (Pa)
p_{sx} = static pressure at plane x, in wg (Pa)
P_{tx} = total pressure at plane x, in wg (Pa)
P_{vx} = velocity pressure at plane x, in wg (Pa)
$\dot{Q}$ = fan capacity, ft³/min (m³/s)
$\dot{Q}_x$ = volumetric flow rate at plane x, ft³/min (m³/s)
T = absolute temperature, °R (K)
W = power, W
x = factor used to determine K_p, dimensionless
z = factor used to determine K_p, dimensionless
γ = isentropic exponent, dimensionless
η_s = fan static efficiency, per unit
η_t = fan total efficiency, per unit
ρ = fan air density, lbm/ft³ (kg/m³)
ρ_x = air density at plane x, lbm/ft³ (kg/m³)

Subscripts

1 = fan inlet plane
2 = fan outlet plane
a = absolute
b = basic known conditions
c = calculated condition
r = reading
x = plane 1, 2, 3, or other

FAN TYPES AND NOMENCLATURE

Any device which produces a current of air may be called a fan. This discussion will be limited to fans which have a rotating impeller to produce the flow and a stationary casing to guide the flow into and out of the impeller (see Fig. 9.5.1). The form of the casing or the impeller may vary widely.

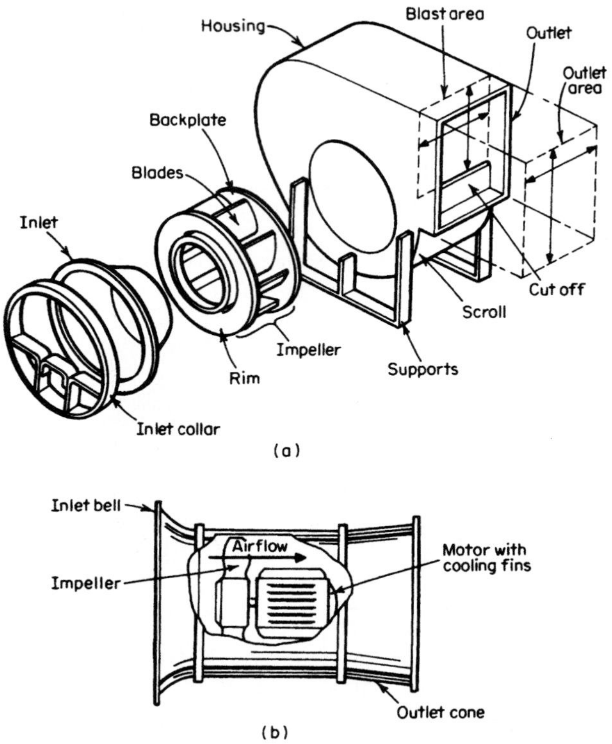

Figure 9.5.1 Elements of fans and preferred terminology. (*a*) Centrifugal fans; (*b*) axial fans. (*Based on AMCA Publ. 201.*)

Fan Classification

One of the characteristics by which fans are classified is the nature of the flow through the blade passages of the impeller. Axial flow, radial flow, mixed flow, and cross flow are all possible in fan impellers. Certain fan names result from these classifications. Other fan names derive from other characteristics.

Propeller fans, tube-axial fans, and vane-axial fans all utilize **axial-flow impellers,** but their casings differ. Propeller fans may be mounted in a ring or panel. Tube-axial fans and vane-axial fans both use tubular casings, but for vane-axial fans they are equipped with stationary guide vanes. A great deal of the energy transferred to the air in axial-flow machines is in kinetic form. Some of this kinetic energy can be transformed into pressure energy by straightening the swirl, e.g., with vanes, or by reducing the exit velocity, e.g., with a diffuser. Propeller fans effect very little transformation and hence have very low pressure-producing capability. Vane-axial fans can be equipped for maximum transformation as well as high transfer of energy and hence have high pressure-producing potential depending on tip speed and blade angles. High hub ratios promote high energy transfer.

Centrifugal fans and tubular centrifugal fans both utilize **radial-flow impellers.** Centrifugal fans usually employ a volute or scroll-type casing, the flow entering the casing axially and leaving tangentially. Tubular centrifugals use tubular casings so that the flow both entering and leaving the casing is axial. A considerable portion of the energy transferred to the air in a radial-flow machine is due to centrifugal action; hence the name centrifugal fan. Since centrifugal action varies with blade depth, the pressure-producing capability of radial-flow fans will vary with this factor as well as tip speed and blade angles.

Mixed-flow impellers can be used in either axial or scroll-type casings. They are characterized as mixed flow because both axial and radial flow take place in the blading. Mixed-flow impellers used in axial-flow casings have a hub similar to a pure axial-flow impeller, but the inlet portion of the blading extends down over the face of the hub, thereby giving some radial guidance. Mixed-flow impellers used in scroll-cased fans have blades which give most of the axial guidance in the inlet portion and most of the radial guidance in the discharge portion.

In a **cross-flow impeller,** the air passes through the blading twice, entering more or less tangentially through the tip, passing across the impeller and out the other side. The casings are designed to provide this transverse flow. Cross-flow fans are also known as tangential fans or transverse-flow fans. Pressure-producing potential is low and depends on the formation of a vortex as the air leaves the impeller.

Fan Details

Propeller fans and other **axial fans** may use blades shaped to airfoil sections or blades of uniform thickness. Blading may be fixed, adjustable at standstill, or variable in operation. Propeller fans have very small hubs. Hub-to-tip diameter ratios ranging from 0.4 to 0.7 are common in vane-axial fans. The larger the hub, the more important it is to have an inner cylinder approximately the hub size located downstream of the impeller. The guide vanes of a vane-axial fan are located in the annular space between the tubular casing and the inner cylinder. Diffusers are generally used between the fan and the discharge ductwork.

Centrifugal fans use various types of blading. Forward-curved blades are shallow and curved so that both the tip and the heel point in the direction of rotation. Radial and radial-tip blades both are radial at the tip, but the latter are curved at the heel to point in the direction of rotation. Backward-curved and backward-inclined blades point in the direction opposite rotation at the tip and in the direction of rotation at the heel. All the above blades are of uniform thickness and are designed for radial flow. Airfoil blades have backward-curved chord lines so that the leading edge of the airfoil is at the heel pointing forward and the trailing edge at the tip pointing backward with respect to rotation. Impellers for all blade shapes are usually shrouded and may have single or double inlets. Blade widths are related to the inlet-to-tip-diameter ratio. Tip angles may vary widely, but heel angles should be set to minimize entrance losses. Scroll casings may be fitted with a streamlined inlet bell, an inlet cone, or simply a collar.

Tubular centrifugals may be designed for backward-curve, air foil, or mixed-flow impellers. An inlet bell and discharge guide vanes are required for good performance.

Cross-flow fans utilize impellers with blading similar to that of a forward-curved centrifugal, but the end shrouds have no inlet holes. Blade-length-to-tip-diameter ratios are limited only by structural considerations.

Power roof ventilators may use either axial- or radial-flow impellers. The casings will include either a propeller fan mounting ring or a tubular casing to guide the flow for an axial-flow impeller. If a radial-flow impeller is used, an inlet bell is required, but the scroll case may be replaced by the ventilator hood.

Plenum fans and **plug fans** are fans without traditional casings. Plenum fans are used to move air through a system. They utilize centrifugal impellers and their drives are inside the plenum. Plug fans are used to circulate air or gas within a chamber or system. They may have either centrifugal or axial impellers and their drives are outside the chamber.

All fan types may have direct or indirect drive arrangements. Various standards including arrangement numbers have been adopted by AMCA. Arr. 1 signifies an overhung impeller on a fan shaft with two bearings on a base; Arr. 3 signifies an impeller on a fan shaft between bearings; Arr. 4 is for overhung impeller on a motor shaft; Arr. 7 is Arr. 3 with a motor base; Arr. 8 is Arr. 1 with an extended base for motor; Arr. 9 is Arr. 1 with motor mounted on side of unit; Arr. 10 is Arr. 1 with motor mounted inside of base. Rotation is specified as cw or ccw when viewed from the drive side. All fans can be equipped with variable-speed drives, variable inlet vanes, or dampers, but controllable-pitch axials do not need speed or vane control.

FAN PERFORMANCE AND TESTING

The conventional terms used to describe fan performance in the United States are defined below.

Fan Air Density Air density is the mass per unit volume of the air. The density of a perfect gas is a function of its molecular weight, temperature, and pressure as indicated by

$$\rho_x = \frac{M}{386.7}\frac{529.7}{T_x}\frac{b_x}{29.92}$$

where the subscript x indicates the plane of the measurements. For dry air this reduces to

$$\rho_x = 1.325 b_x / T_x$$

This expression will usually be accurate enough even when moist air is involved. Fan air density is the density of the air corresponding to the total pressure and total temperature at the fan inlet

$$\rho = \rho_1$$

Fan Capacity Volume flow rate is usually determined from pressure measurements, e.g., a velocity-pressure traverse taken with a Pitot static tube or a pressure drop across a flowmeter. The average velocity pressure for a Pitot traverse is

$$p_{vx} = \left(\sum \sqrt{p_{vxr}}/n\right)^2$$

where subscript x indicates the plane of the measurements, subscript r indicates a reading at one station, n is the number of stations, and Σ is the summation sign. The corresponding capacity is:

$$\dot{Q}_x = 1{,}097 A_x \sqrt{p_{vx}/\rho_x}$$

For a flowmeter

$$\dot{Q}_x = 1{,}097 C A_x Y \sqrt{\Delta p / \rho_x / F}$$

where C = coefficient of discharge of meter, Y = expansion factor for gas, Δp = measured pressure drop, and F = velocity-of-approach factor for the meter installation. Fan capacity is the volumetric flow rate at fan air density

$$\dot{Q} = \dot{Q}_x \rho_x / \rho$$

Fan Total Pressure Fan total pressure is the difference between the total pressure at the fan outlet and the total pressure at the fan inlet

$$p_t = p_{t2} - p_{t1}$$

When the fan draws directly from the atmosphere.

$$p_{t1} = 0$$

When the fan discharges directly to the atmosphere.

$$p_{t2} = p_{v2}$$

If either side of the fan is connected to ductwork, etc., and the measuring plane is remote, the measured values should be corrected for the approximate pressure drop

$$p_{t2} = p_{tx} + \Delta p_{2-x} \qquad p_{t1} = p_{tx} + \Delta p_{x-1}$$

Fan Velocity Pressure Fan velocity pressure is the pressure corresponding to the average velocity at the fan outlet

$$p_v = \left(\frac{\dot{Q}_2/A_2}{1{,}097}\right)^2 \rho_2$$

Fan Static Pressure Fan static pressure is the difference between the fan total pressure and the fan velocity pressure. Therefore, fan static pressure is the difference between the static pressure at the fan outlet and the total pressure at the fan inlet

$$p_s = p_t - p_v = p_{s2} - p_{t1}$$

Fan Speed Fan speed is the rotative speed of the impeller.

Compressibility Factor The compressibility factor is the ratio of the fan total pressure p_t' that would be developed with an incompressible fluid to the fan total pressure p_t that is developed with a compressible fluid, all other conditions being equal:

$$K_p = \frac{p_t'}{p_t} = \frac{[n/(n-1)][(p_{t2a}/p_{t1a})^{(n-1)/n} - 1]}{(p_{t2a}/p_{t1a}) - 1}$$

Compressibility factor can be determined from test measurements using

$$x = p_t/p_{t1a}$$
$$z = [(\gamma - 1)/\gamma]6356H/\dot{Q}p_{t1a}$$
$$K_p = [z\log(1+x)]/[x\log(1+z)]$$

Fan Power Output Fan power output is the product of fan capacity and fan total pressure and compressibility factor

$$H_o = \dot{Q}p_tK_p/6{,}356$$

Fan Power Input Fan power input is the power required to drive the fan and any elements in the drive train which are considered a part of the fan. Power input can be calculated from appropriate measurements for a dynamometer, torque meter, or calibrated motor.

Fan Total Efficiency Fan total efficiency is the ratio of the fan power output to the fan power input

$$\eta_t = \dot{Q}p_tK_p/6{,}356H$$

Fan Static Efficiency Fan static efficiency is the fan total efficiency multiplied by the ratio of fan static pressure to fan total pressure

$$\eta_s = \eta_t p_s/p_t$$

Fan Sound Power Level Fan sound power level is 10 times the logarithm (base 10) of the ratio of the actual sound power in watts to 10^{-12} watts,

$$L_w = 10\log(W/10^{-12})$$

The total sound power level of a fan is usually assumed to be 3 dB higher than either the inlet or outlet component. The casing component varies with construction but will usually range from 15 to 30 dB less than the total.

The frequency content of fan noise is usually expressed in octave bands, but may be given in narrow bands or even as an overall value. The sound power level can be calculated from sound pressure measurements or sound intensity measurements. The reverberant room method utilizes a known sound source to calibrate the room and a microphone to measure the sound pressures. The sound intensity method utilizes two microphones in a sound-intensity probe to measure the sound intensities over an enveloping surface. The in-duct method utilizes anechoic terminations to facilitate sound pressure measurements without end reflections.

Head The difference between head and pressure is important in fan engineering. Both are measures of the energy in the air. Head is energy per unit weight and can be expressed in ft · lb/lb, which is often abbreviated to ft (of fluid flowing). Pressure is energy per unit volume and can be expressed in ft · lb/ft^3, which simplifies to lb/ft^2 or force per unit area. The use of the inch water gage (in wg) is a convenience in fan engineering reflecting the usual methods of measurement. It sounds like a head measurement but is actually a pressure measurement corresponding to 5.192 lb/ft^2, the pressure exerted by a column of water 1 inch high. Pressures can be converted into heads and vice versa:

$$p = \rho h/5.192 \qquad h = 5.192p/\rho$$

For instance, for air at 0.075 lbm/ft^3 density, 1 in wg corresponds to 69.4 ft of air. That is, a column of air at that density would exert a pressure of 5.192 lb/ft^2. Lighter air would exert less pressure for a given head. For a given pressure, the head would be higher with lighter air. Although it is not widely used in the United States, head is commonly used in a number of European countries.

Fans, like other turbomachines, can be considered constant-head-constant-capacity (volumetric) machines. This means that a fan will develop the same head at a given capacity regardless of the fluid handled, all other conditions being equal. Of course, this also means that a fan will develop a pressure proportional to the density at a given capacity, all other conditions being equal.

All the preceding equations are based on the U.S. Customary Units listed under Symbols. If **SI units** are to be used, certain numerical coefficients will have to be modified. For instance, substitute $\sqrt{2}$ for 1,097 in the flow equations when using Pa for pressure, kg/m^3 for density, and m^3/s for capacity. Similarly, substitute 1.0 for 6,356 in the power equations when using Pa for pressure, m^3/s for capacity, and W for power.

Common metric practice in fan engineering leads to other numerical coefficients. When mm wg is used for pressure, m^3/s for capacity, kW for power, and kg/m^3 for density, substitute 4.424 for 1,097 in the flow equations and 102.2 for 6,356 in the power equations.

The preceding discussion utilized what may be termed the **volume-flow-rate-pressure** approach to expressing fan performance. An alternative, the **mass-flow-rate-specific-energy** approach is equally valid but not generally used in the United States. Refer to ASME PTC-11 for additional details.

FAN AND SYSTEM PERFORMANCE CHARACTERISTICS

The performance characteristics of a fan are best described by a graph. The conventional method of graphing fan performance is to plot a series of curves with capacity as abscissa and all other variables as ordinates. System characteristics can be plotted in a similar manner.

System Characteristics

Most systems served by a fan have characteristics which can be described by a parabola passing through the origin; i.e., the energy required to produce flow through the system (which can be expressed as pressure or head) varies approximately as the square of the flow. In some cases, the system characteristics will not pass through the origin because the energy required to produce flow through an element of the system may be controlled at a particular value, e.g., with venturi scrubbers. In some cases, the system characteristic will not be parabolic because the flow through an element of the system is laminar rather than turbulent, e.g., in some types of filters. Whatever the case, the system designer should establish the characteristics by determining the energy requirements at various flow rates. The energy requirement (pressure drops or head losses) for each element can be determined by reference to handbook or manufacturer's literature or by test.

The true measure of the energy requirement for a system element is the total pressure drop or the total head loss. Only if the entrance velocity for the element equals the exit velocity will the change in static pressure equal the total pressure drop. There are some advantages in using static pressure change, but the system designer is usually well advised to use total pressure drops to avoid errors in fan selection. The sum of the total pressure losses for elements on the inlet side of the fan will equal $-p_{t1}$. This should include energy losses at the entrance to the system but not the energy to accelerate the air to the velocity at the fan inlet, which is chargeable to the fan. The sum of the total pressure losses for elements on the discharge side of the fan will equal p_{t2}. This should include the kinetic energy of the stream issuing from the system. The total system requirement will be the arithmetic sum of all the appropriate losses or the algebraic difference $p_{t2} - p_{t1}$, which is also p_t.

A system characteristic curve based on total pressure is plotted in Fig. 9.5.2. A characteristic based on static pressure is also shown. The latter recognizes the definition of fan static pressure so that the only difference is the velocity pressure corresponding to the fan outlet velocity. This system could operate at any capacity provided a fan delivered the exact pressure to match the energy requirements shown on the system curve for that capacity. The advantages of plotting the system characteristics on the fan graph will become evident in the following discussions.

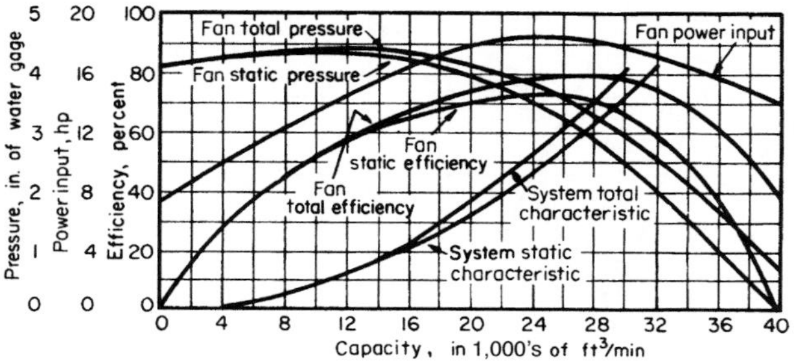

Figure 9.5.2 Fan and system characteristics.

Fan Characteristics

The constant-speed performance characteristics of a fan are illustrated in Fig. 9.5.2. These characteristics are for a particular size and type of fan operating at a particular speed and handling air of a particular density. The fan can operate at any capacity from zero to the maximum shown, but when applied on a particular system the fan will operate only at the intersection of the system characteristics with the appropriate fan pressure characteristic. For the case illustrated, the fan will operate at $Q = 27{,}300$ ft³/min and $p_t = 3.4$ in wg or $p_s = 3$ in wg, requiring $H = 18.3$ hp at the speed and density for which the curves were drawn. The static efficiency at this point of operation is 73 percent, and the total efficiency is 80 percent. If the system characteristic had been lower, it would have intersected the fan characteristic at a higher capacity and the fan would have delivered more air. Contrariwise, if the system characteristic had been higher, the capacity of the fan would have been less. Capacity reduction can be accomplished, in fact, by creating additional resistance, as with an outlet damper.

Figure 9.5.3 illustrates the characteristics of a fan with damper control, variable inlet vane control, and variable speed control. These particular characteristics are for a backward-curved centrifugal fan, but the general principles apply to all fans. Outlet dampers do not affect the flow to the fan and therefore can alter fan performance only by adding resistance to the system and producing a new intersection. Point 1 is for wide-open dampers. Points 2 and 3 are for progressively closed dampers. Note that some power reduction accompanies the capacity reduction for this particular fan. Variable-inlet vanes produce inlet whirl, which reduces pressure-producing capability. Point 4 is for wide-open inlet vanes and corresponds to point 1. Points 5 and 6 are for progressively closed vanes. Note that the power reduction at reduced capacity is better for vanes than for dampers. Variable speed is the most efficient means of capacity control. Point 7 is for full speed and points 8 and 9 for progressively reduced speed. Note the improved power savings over other methods. Variable speed also has advantages in terms of lower noise and reduced erosion potential but is generally at a disadvantage regarding first cost.

Figure 9.5.4 illustrates the characteristics of a fan with variable pitch control. These characteristics are for an axial-flow fan, but comparisons to the centrifugal-fan ratings in Fig. 9.5.3 can be made. Point 10 corresponds to points 1, 4, or 7. Reduced ratings are shown at points 11 and 12 obtained by reducing the pitch. Power savings can be almost as good as with speed control. Notice that in all methods of capacity control, operation is at the intersection of a system characteristic with a fan characteristic. With variable vanes, variable speed, and variable pitch, the fan characteristic is modified. With damper control, the effect could be considered a change in fan characteristics, but it generally makes more sense to consider it a change in system characteristics.

Fan-System Matching

It has already been observed in the discussions of both system and fan characteristics that the point of operation for **one fan** on a particular system will be at the intersection of their characteristics. Stated another way, the energy required by the system must be provided by the fan exactly. If the fan delivers too much or too little energy, the capacity will be more or less than desired. The effects of utilizing dampers, variable speed, variable-inlet vanes, and variable pitch were

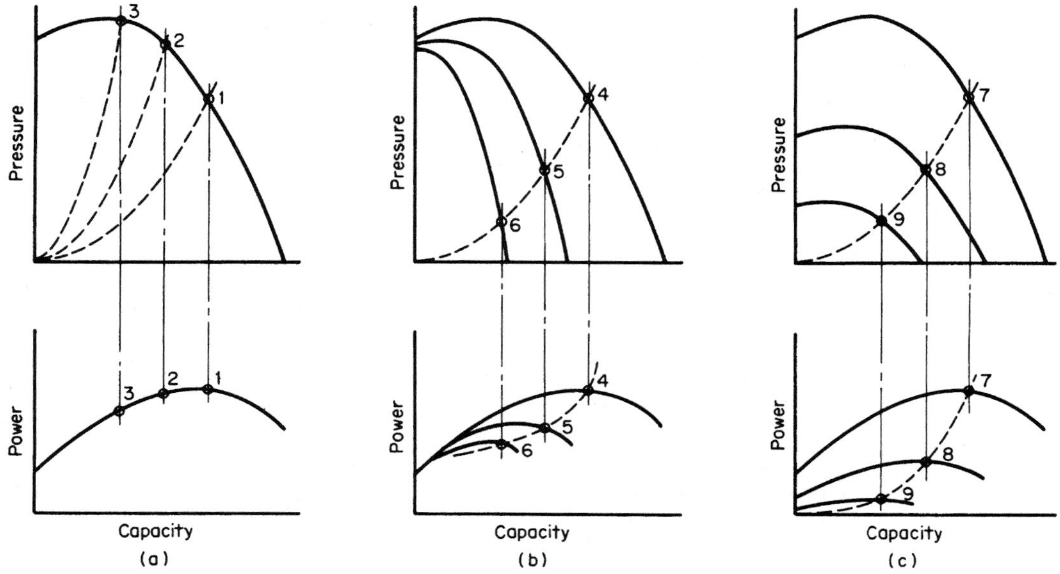

Figure 9.5.3 Fan characteristics with (*a*) damper control, (*b*) variable-inlet vane control, (*c*) variable speed control.

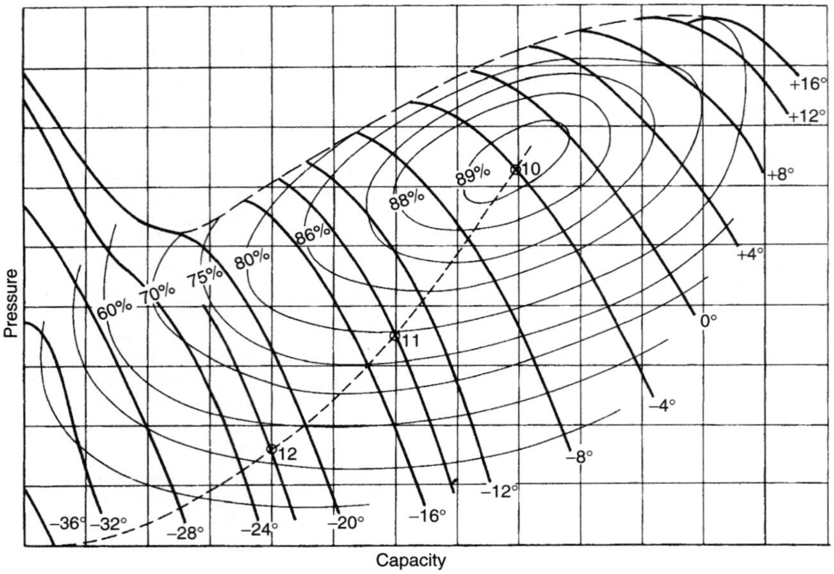

Figure 9.5.4 Characteristics of an axial-flow fan with variable-pitch control.

illustrated, but in all cases the fan was preselected. The more general case is the one involving the selection or design of a fan to do a particular job. The design of a fan from an aerodynamicist's point of view is beyond the scope of this discussion. Fortunately, most fan problems can be solved by selecting a fan from the many standard lines available commercially. Again, the crux of the matter is to match fan capability with system requirements.

Most fan manufacturers have several standard lines of fans. Each line may consist of various sizes all resembling each other. If the blade angles and proportions are the same, the line is said to be **homologous.** There is only one fan size in a line of homologous fans that will operate at the maximum efficiency point on a given system. If the fan is too big, operation will be at some point to the left of peak efficiency on a standard characteristic plot. If the fan is too small, operation will be to the right of peak. In either case it will be off peak. A slightly undersized fan is often preferred for reasons of cost and stability.

Selecting and rating a fan from a catalog is a matter of fan-system matching. When it is recognized that many of the catalog sizes may be able to provide the capacity and pressure, selection becomes a matter of trying various sizes and comparing speed, power, cost, etc. The choice is a matter of evaluation.

There are various reasons for using **more than one fan** on a system. Supply and exhaust fans are used in ventilation to avoid excessive pressure build-up in the space being served. Forced and induced-draft fans are used to maintain a specified draft over the fire. Two fans may fit the available space better than one larger fan. Capacity control by various fan combinations may be more economical than other control methods. Multistage arrangements may be necessary when pressure requirements exceed the capabilities of a single-stage fan. Standby fans are frequently required to ensure continuous operation.

When two fans are used, they may be located quite remote from each other or they may be close enough to share shaft and bearings or even casings. Double-width, double-inlet fans are essentially two fans in parallel in a common housing. Multistage blowers are, in effect, two or more fans in series in the same casing. Fans may also be in series but at opposite ends of the system. Parallel-arrangement fans may have almost any amount of their operating resistance in common. At one extreme, the fans may have common inlet and discharge plenums. At the other extreme, the fans may both have considerable individual ductwork of equal or unequal resistance.

Fans in **series** must all handle the same amount of gas by weight measurements, assuming no losses or gains between stages. The combined total pressure will be the sum of the individual fans' total pressures. The velocity pressure of the combination can be defined as the pressure corresponding to the velocity through the outlet of the last stage. The static pressure for the combination is the difference between its total and velocity pressures and is therefore not equal to the sum of the individual fan static pressures. The volumetric capacities will differ whenever the inlet densities vary from stage to stage. Compression in one stage will reduce the volume entering the next if there is no reexpansion between the two. As with any fan, the pressure capabilities are also influenced by density.

The combined total pressure-capacity characteristic for two fans in series can be drawn by using the volumetric capacities of the first stage for the abscissa and the sum of the appropriate total pressures for the ordinate. Because of compressibility, the volumetric capacities of the second stage will not equal the volumetric capacities of the first stage. The individual total pressures must be chosen accordingly before they are combined. If the gas can be considered incompressible, the pressures for the two stages may be read at the same capacity. In the area near free delivery, it may be necessary to estimate the negative pressure characteristics of one of the fans in order to combine values at the appropriate capacity.

Fans in **parallel** must all develop sufficient pressure to overcome the losses in any individual ductwork, etc., as well as the losses in the common portions of the system. When such fans have no individual ductwork but discharge into a common plenum, their individual velocity pressures are lost and the fans should be selected to produce the same fan static pressures. If fan velocity pressures are equal, the fan total pressures will be equal in such cases. When the fans do have individual ducts but they are of equal resistance and joined together at equal velocities, the fans should be selected for the same fan total pressures. If fan velocity pressures are equal, fan static pressures will be equal in this case. If the two streams join together at unequal velocities, there will be a transfer of momentum from the higher-velocity stream to the lower-velocity stream. The fans serving the lower-velocity branch can be selected for a correspondingly lower total pressure. The other fan must be selected for a correspondingly higher total pressure than if velocities were equal.

The combined pressure-capacity curves for two fans in parallel can be plotted by using the appropriate pressures for ordinates and the sum of the corresponding capacities for abscissa. Such curves are meaningful only when a combined-system curve can be drawn. In the area near shutoff, it may be necessary to estimate the negative capacity characteristics of one of the fans in order to combine values at the appropriate pressure.

Figure 9.5.5 illustrates the combined characteristics of two fans with slightly different individual characteristics (*A-A* and *B-B*). The combined characteristics are shown for the two fans in series (*C-C*) and in parallel (*D-D*). Only total-pressure curves are shown. This is always correct for series arrangements but may introduce slight errors for parallel arrangements. An incompressible gas has been assumed. The questionable areas near shutoff or free delivery have been omitted. Two different system characteristics (*E-E* and *F-F*) have been drawn on the chart. With the two fans in series, operation will be at point *EC* or *FC* if the fan is on system *E* or *F*, respectively. Parallel arrangement will lead to operation at *ED* or *FD*. Single-fan operation would be at the point indicated by the intersection of the appropriate fan and system curves provided the effect of an inoperative second fan is negligible. Some sort of bypass is required around an inoperative fan in series whereas an inoperative fan in parallel need only be dampered shut. Parallel operation yields a higher capacity than series operation on system *F*, but the reverse is true on system *E*. For the type of fan and system characteristics drawn there is only one possible point of operation for any arrangement.

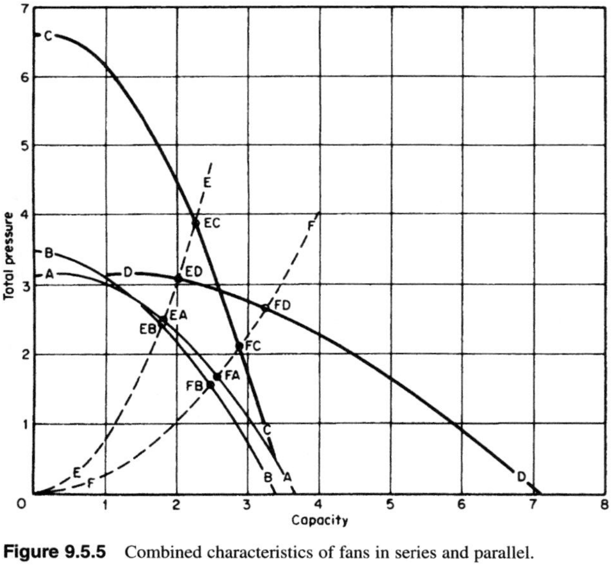

Figure 9.5.5 Combined characteristics of fans in series and parallel.

Fan Laws

The fan laws are based on the experimentally demonstrable fact that any two members of a homologous series of fans have performance curves which are homologous. At the same point of rating, i.e., at similarly situated points of operation on their characteristic curves, efficiencies are equal and other variables are interrelated according to the fan laws. If size and speed are considered independent variables and if compressibility effects are ignored, the fan laws can be written as follows:

$$\eta_{tc} = \eta_{tb}$$
$$\dot{Q}_c = \dot{Q}_b (D_c/D_b)^3 (N_c/N_b)$$
$$p_{tc} = p_{tb} (D_c/D_b)^2 (N_c/N_b)^2 (\rho_c/\rho_b)$$
$$H_c = H_b (D_c/D_b)^5 (N_c/N_b)^3 (\rho_c/\rho_b)$$
$$L_{wc} = L_{wb} + 70\log(D_c/D_b) + 50\log(N_c/N_b) + 20\log(\rho_c/\rho_b)$$

The above laws are useful, but they are dangerous if misapplied. The calculated fan must have the same point of rating as the known fan. When in doubt, it is best to reselect the fan rather than attempt to use the fan laws.

The fan designer utilizes the fan laws in various ways. Some of the more useful relationships in addition to those above derive from considering fan capacity and fan total pressure as the independent variables. This leads to specific diameter, specific speed, and specific sound power level:

$$D_s = D(p_t/\rho)^{1/4}/\dot{Q}^{1/2}$$
$$N_s = N\dot{Q}^{1/2}/(p_t/\rho)^{3/4}$$
$$L_{ws} = L_w - 10\log\dot{Q} - 20\log p_t$$

D_s, N_s, and L_{ws} are the diameter, speed, and sound power level of a homologous fan which will deliver 1 ft³/min at 1 in wg at the same point of rating as Q and p_t for D and N. D_s and N_s can be used to advantage in fan selection. They can also be used by a designer to determine how well a line will fit in with other lines. This is illustrated in Fig. 9.5.6. Each segment represents a particular fan line. Note the trends for each kind of fan. Incidentally, some fan engineers utilize different formulas for specific diameter and specific speed

$$D_{se} = Dp_{te}^{1/4}/\dot{Q}^{1/2} \qquad N_{se} = N\dot{Q}^{1/2}/p_{te}^{3/4}$$

where p_{te} is the equivalent total pressure based on standard air. This makes $D_{se} = D_s \times 1.911$ and $N_{se} = N_s \times 6.978$.

Specific sound power level is useful in predicting noise levels as well as in comparing fan designs. There appears to be a lower limit of L_{ws} in

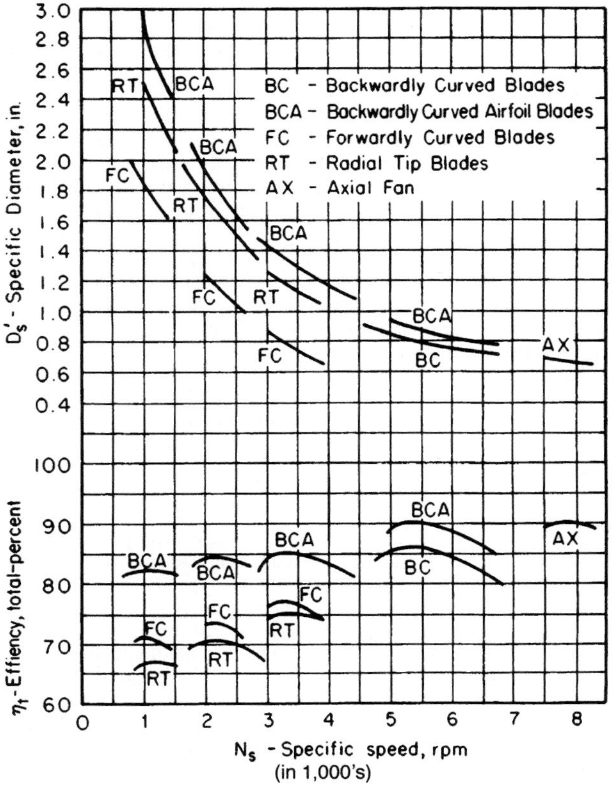

Figure 9.5.6 Specific diameter and efficiency versus specific speed for single-inlet fan types.

the vicinity of 45 dB for the more efficient types ranging to 70 dB or more for cruder designs. Actual sound power levels can be figured from

$$L_w = L_{ws} + 10\log\dot{Q} + 20\log p_t$$

Another useful parameter which derives from the fan laws is orifice ratio

$$R_o = \dot{Q}/D^2(p_t/\rho)^{1/2}$$

This ratio can be plotted on a characteristic curve for a known fan. If the ratio is determined for a calculated homologous fan, the point of rating can be established by inspection. Other ratios can be used in the same manner including p_v/p_t, $p_t/\dot{Q}^2$, D_s, and N_s.

Stability Considerations

The flow through a system and its fan will normally be steady. If the fluctuations occasioned by a temporary disturbance are quickly damped out, the fan system may be described as having a **stable** operating characteristic. If the unsteady flow continues after the disturbance is removed, the operating characteristic is **unstable.**

To ensure stable operation the slopes of the pressure-capacity curves for the fan and system should be of opposite sign. Almost all systems have a positive slope; i.e., the pressure requirement or resistance increases with capacity. Therefore, for stable operation the fan curve should have a negative slope. Such is the case at or above the design capacity.

When the slopes of the fan and system characteristics are of opposite sign, any system disturbance tending to produce a temporary decrease in flow is nullified by the increase in fan pressure. When the slopes are of the same sign, any tendency to decrease flow is strengthened by the resulting decrease in fan pressure. When fan and system curves coincide over a range of capacities, the operating characteristics are extremely unstable. Even if the curves exactly coincide at only one point, the flow may vary over a considerable range.

There may or may not be any obvious indication of unstable operation. The pressure and power fluctuations that accompany unsteady flow may be so small and rapid that they cannot be detected by any but the most sensitive instruments. Less rapid fluctuations may be detected on the ordinary instruments used in fan testing. The changes in noise which occur with each change in flow rate are easily detected by ear as individual beats if the beat frequency is below about 10 Hz. In any event, the overall noise level will be higher with unsteady flow than with steady flow.

The conditions which accompany unsteady flow are variously described as pulsations, hunting, surging, or pumping. Since these conditions occur only when the operating point is to the left of maximum pressure on the fan curve, this peak is frequently referred to as the **surge point** or **pumping limit.**

Pulsation can be prevented by rating the fan to the right of the surge point. Fans are usually selected on this basis, but it is sometimes necessary to control the volume delivered to the value below that at the surge point. This may lead to pulsation, particularly if the fan pressure exceeds 10 in wg.

If the required capacity is less than that at the pumping limit, pulsation can be prevented in various ways, all of which in effect provide a negatively sloping fan curve at the actual operating point. To accomplish this effect the required pressure must be less than the fan capabilities at the required capacity. One method is to bleed sufficient air for actual operation to be beyond the pumping limit. Other possible methods are the use of pitch, speed, or vane control for volume reduction. In any of these cases, the point of operation on the new fan curve must be to the right of the new surge point. Although in the section on Capacity Control dampers were considered a part of the system, they may also be considered a part of the fan if located in the right position. Accordingly, pulsations may be eliminated in a supply system if the damper is on the inlet of the blower. Similarly, dampering at the outlet of the exhauster may control pulsation in exhaust systems.

Another condition frequently referred to as instability is associated with flow separation in the blade passages of an impeller and is evidenced by slight discontinuities in the performance curve. There may be a small range of capacities at which two distinctly different pressures may be developed depending on which of the two flow patterns exists. Such a condition usually occurs at capacities just to the left of peak efficiency. Still another condition of unsteady flow may develop at extremely low capacities. This is known as **blowback** or **puffing** because air puffs in and out of a portion of the inlet. Operation in the blowback range should be avoided, particularly with high-energy fans.

A different type of unsteady flow may occur when two or more fans are used in parallel. If the individual fan characteristics exhibit a dip in pressure between shutoff and design, the combined characteristic will contain points where the point of operation of the individual fans may be widely separated even for identical fans. If the system characteristic intersects the combined-fan characteristic at such a point, the individual fans may suddenly exchange loads. That is, the fan operating at high capacity may become the one operating at low capacity and vice versa. This can produce undesirable shocks on motors and ducts. Careful matching of fan to system is required for either forward-curved centrifugals or most axials for this reason.

Fan Applications

The selection of a particular size and type of fan for a particular application involves considerations of aerodynamic, economic, and functional suitability. Many of the factors involved in aerodynamic suitability have been discussed above. Determination of economic suitability requires an evaluation of first cost and operating costs. The functional suitability of various types of fans with respect to certain applications is discussed below.

Heating, ventilating, and **air-conditioning systems** may require supply and exhaust or return air fans. Historically, high-efficiency centrifugal fans, using either backward-curved or airfoil blades have been used for supply on duct systems. In low-pressure applications these types can be used without sound treatment. In high-pressure applications, sound treatment is almost always required. Axials have long been used for shipboard ventilation because they generally can be made smaller than centrifugals. Both adjustable axials and tubular centrifugals have proved popular on duct systems for exhaust service in building ventilation. Centrifugals with variable inlet vanes and axials with pitch control are being used for supply in variable-air-volume systems. Propeller fans and power roof ventilators are used for either supply or exhaust systems when there is little or no ductwork. Heating, ventilating, and air-conditioning applications are considered clean-air service. Various classes of construction are available in standard lines for different pressure ranges. Both direct and indirect drive are used, the latter being most common.

Industrial exhaust systems generally require fans that are less susceptible to the unbalance that may result from dirty-gas applications than the clean-air fans used for heating, ventilating, and air conditioning. Simple, rugged, industrial exhausters are favored for applications up to 200 hp. They have a few radial blades and relatively low efficiency. Most are V-belt-driven. Extra-heavy construction may be required where significant material passes through the fan.

Process air requirements can be met with either centrifugal or axial fans; the latter may be used in single or double stage. The higher pressure ratings are usually provided by a single-stage centrifugal fan with radial blades, known as a pressure blower. These units are generally direct connected to the driver. They not only compete with centrifugal compressors but resemble them.

Large industrial process and **pollution-control systems** involving more than about 200 hp are generally satisfied with a somewhat more sophisticated fan than an industrial exhauster or pressure blower. Centrifugal fans with radial-tip blades are frequently used on the more severe service. For the less severe requirements, backward-curved or airfoil blades may be used. Rugged fixed-pitch axials have also been used. Industrial fans are usually equipped with inlet boxes and independently mounted bearings and are usually direct-driven. Journal bearings are usually preferred. Inlet-box damper control can be used to approximate the power saving available from variable-inlet vane control. Variable-speed hydraulic couplings may be economically justified in some cases. Special methods or special construction may be required to provide protection against corrosion or erosion. These fans tend to take on the name of the application such as sintering fan, scrubber exhaust fan, etc.

Mechanical draft systems may utilize any of the fan types described in connection with the above applications. Ventilating fans, industrial exhausters, and pressure blowers have been used for forced draft, induced draft, and primary air service on small steam-generating units. The large generating units are generally equipped with the most efficient fans available consistent with the erosion-corrosion potential of the gas being handled.

Both centrifugals and axials are used for forced and induced draft. Axials have predominated in Europe, and there is a growing trend throughout the rest of the world toward axials. Centrifugals have been used almost exclusively in the United States until very recently.

Forced-draft centrifugals invariably have airfoil blade impellers. Induced-draft centrifugals may have airfoil-blade impellers, but for scrubber exhaust, radial-tip blades are more common. This is due, in part, to the high pressures required for scrubber operation and in part to the erosion-corrosion potential downstream of a scrubber. Gas recirculating fans are usually of the radial-tip design. Forced-draft control may be by variable speed but is more likely to be variable vanes on centrifugal fans. Variable inlet vanes or inlet-box dampers may be used to control induced-draft fans. All large centrifugals are direct-connected and have independent pedestal-mounted bearings. Journal bearings are almost always used.

Forced-draft axials are likely to be of the variable-pitch full-airfoil-section design. Hydraulic systems are almost always used for pitch control, but pneumatic and mechanical systems have been tried. Bearings are usually of the antifriction type. Fixed-pitch axials are usually used for induced-draft duty. Control is by variable-inlet vanes. Either journal or antifriction bearings may be used.

Other Systems

Fans are incorporated in many different kinds of machines. Electronic equipment may require cooling fans to prevent hot spots. Driers use fans to circulate air to carry heat to, and moisture away from, the product. Air-support structures require fans to inflate them and maintain the supporting pressure. Ground-effect machines use fans to provide the lift pressure. Air conditioners and other heat exchangers incorporate fans. Aerodynamic, economic, and functional considerations will dictate the type and size of fan to be used.

Tunnel ventilation can be achieved by using either axial or centrifugal fans. Transverse ventilation utilizes supply and exhaust fans connected to the tunnel by ductwork. These fans are rated in the manner described above. Another method utilizes specially tested and rated axial fans called **jet fans.** The fans, which are placed in the tunnel, increase the momentum of some of the air flowing with the traffic. This air induces additional flow.

Section 10

Instruments and Controls

BY

WALTER H. BOYES, JR. *Editor and Publisher, The Industrial Automation and Process Control INSIDER*
GREGORY V. MURPHY *Head and Professor, Department of Electrical Engineering, College of Engineering, Tuskegee University*
MICHAEL B. KANE *Professor, Department of Civil and Environmental Engineering, Northeastern University*
COURTNEY A. PECKENS *Professor, Department of Engineering, Hope College, Holland, MI*
NIELL ELVIN *Professor of Mechanical Engineering, The City College of the City University of New York*
ALEX ELVIN *Professor, School of Civil and Environmental Engineering, University of the Witwatersrand*
LUC G. FRÉCHETTE *Professor of Mechanical Engineering, Université de Sherbrooke*
ROBIN O. CLEVELAND *Professor of Engineering Science, Tutorial Fellow of Magdalen College, University of Oxford*
BENJAMIN C. DAVENNY *Acoustical Consultant, Acentech Inc.*
JONATHAN D. KEMP *Vibration Consultant, Acentech Inc.*
ANDREW A. GOLDENBERG *Professor of Mechanical and Industrial Engineering, University of Toronto; President and CEO of Engineering Service Inc. (ESI)*

10.1 INSTRUMENTS
by Walter H. Boyes, Jr.

10.2 AUTOMATIC CONTROLS
by Gregory V. Murphy

10.3 DIGITAL CONTROL SYSTEMS
by Michael B. Kane and Courtney A. Peckens

10.4 SENSORS AND ACTUATORS
by Niell Elvin and Alex Elvin

10.5 AN INTRODUCTION TO MICROELECTROMECHANICAL SYSTEMS (MEMS)
by Luc G. Fréchette

10.6 SOUND, NOISE, AND ULTRASONICS
by Robin O. Cleveland, Benjamin C. Davenny, and Jonathan D. Kemp

10.7 ROBOTICS, MECHATRONICS, AND INTELLIGENT AUTOMATION
by Andrew A. Goldenberg

10.1 INSTRUMENTS

by Walter H. Boyes, Jr.

REFERENCES: ASME publications: "Instruments and Apparatus Supplement to Performance Test Codes (PTC 19.1-19.20)"; "Fluid Meters, pt. II, Application." ASTM, "Manual on the Use of Thermocouples in Temperature Measurement," STP 470B. ISA publications: "Standards and Recommended Practices for Instrumentation and Controls," 11th ed. Spitzer, "Flow Measurement." Preston-Thomas, The International Temperature Scale of 1990 (ITS-90), *Metrologia,* **27,** 1990, pp. 3-10. NIST Monograph 175, "Temperature-Electromotive Force Reference Functions and Tables for the Letter-Designated Thermocouple Types Based on the ITS-90," Government Printing Office, April 1993. Schooley (ed.), "Temperature, Its Measurement and Control in Science and Industry," Vol. 6, Pts. 1 and 2, American Institute of Physics. Time and frequency services offered by the National Institute of Standards and Technology (NIST) include the Automated Computer Time Service (ACTS) (www.boulder.nist.gov/timefreg/service/acts.htm) and the significantly more popular and useful Internet Time Service's (www.boulder.nist.gov/timefreg/service/its.htm). Liptak, "Instrument Engineer's Handbook," 3rd and 4th ed., Chilton/ISA. Boyes, "Instrumentation Reference Book," 3rd ed., Butterworth-Heineman. Berge, "Fieldbushes for Process Control," ISA. Verhappen, "Foundation Fieldbus, A Pocket Guide," ISA. Beckwith et al., "Mechanical Measurements," Addison-Wesley. Considine, "Encyclopedia of Instrumentation and Control," Krieger reprint. Considine, "Handbook of Applied Instrumentation," McGraw-Hill, Krieger reprint. Dally et al., "Instrumentation for Engineering Measurements," Wiley. Doebelin, "Measurement Systems, Application and Design," McGraw-Hill. Erikson and Graber, Harris et al., "Shock and Vibration Control Handbook," McGraw-Hill. Holman, "Experimental Methods for Engineers," McGraw-Hill. Jones (ed.), "Instrument Science and Technology, Vol. 1, Measurement of Pressure, Level, Flow and Temperature," Heyden. Lion, "Instrumentation in Scientific Research, Electrical Input Transducers," McGraw-Hill. Sheingold (ed.), "Transducer Interfacing Handbook," Analog Devices, Inc. Norwood, MA. Snell, "Nuclear Instruments and Their Uses," Wiley. Spink, "Principles and Practice of Flow Meter Engineering," Foxboro Co. Stout, "Basic Electrical Measurements," Prentice-Hall. *Periodicals: InTech,* monthly, ISA. *Sensors,* monthly, Alvanstar. *Test & Measurement World,* Cahners. *Control,* monthly, Putman Media. *Control Engineering*, monthly, Reed.

INTRODUCTION TO MEASUREMENT

An **instrument,** as referred to in the following discussion, is a device for determining the value or magnitude of a quantity or variable. The variables of interest are those which help describe or define an object, system, or process. Thus, in a manufacturing operation, product quality is related to measurements of its various dimensions and physical properties such as hardness and surface finish. In an industrial process, measurement and control of temperature, pressure, flow rates, etc., determine quality and efficiency of production.

Measurements may be direct, for example, using a micrometer to measure a dimension, or indirect, for example, determining moisture in steam by measuring the temperature in a throttling calorimeter.

Because of physical limitations of the measuring device and the system under study, practical measurements always have some error. The **accuracy** of an instrument is the closeness with which its reading approaches the true value of the variable being measured. Accuracy is commonly expressed as a percentage of measurement span, measurement value, or full-scale value. Span is the difference between the full-scale and the zero-scale value. **Uncertainty,** the sum of the errors at work to make the measured value different from the true value, is the accuracy of measurement standards. Uncertainty is expressed in parts per million (ppm) of a measurement value. **Precision** refers to the reproducibility of the measurements, that is, with a fixed value of the variable, how much successive readings differ from one another. **Sensitivity** is the ratio of output signal or response of the instrument to a change in input or measured variable. **Resolution** relates to the smallest change in measured value to which the instrument will respond.

Error may be classified as systematic or random. Systematic errors are those due to assignable causes. These may be static or dynamic. Static errors are caused by limitations of the measuring device or the physical laws governing its behavior. Dynamic errors are caused by the instrument not responding fast enough to follow the changes in measured variable. Random errors are those due to causes which cannot be directly established because of random variations in the system.

Standards for measurement are established by the National Institute of Standards and Technology. Secondary standards are prepared by very precise comparison with these primary standards and, in turn, form the basis for calibrating instruments in use. A well-known example is the use of precision gage blocks for the calibration of measuring instruments and machine tools.

There are two essential parts to an instrument: the **sensing element,** and the **transmitting means.** The sensing element responds directly to the measured quantity, producing a related motion, pressure, or electrical signal. This is transmitted electronically to some other device for display, recording, and/or control. Displays include graphical LCD displays with direct numerical indication or which mimic a paper chart recorder. Data is recorded in solid state memory, both in the field and in remote devices. The instrument may be actuated by mechanical, hydraulic, pneumatic, electrical, optical, or other energy medium. Often a combination of several energy modes is employed to obtain the accuracy, sensitivity, or form of output desired.

The transmission of measurements to distant indicators and controls has historically for industrial use been accomplished by using the standardized electrical current signal of 4 to 20 mA; 4 mA represents the zero-scale value and 20 mA the full-scale value of the measurement range. Previously, pressure of 3 to 15 lb/in^2 was often used for pneumatic transmission of signals, but is not often seen in the present day. Analog current transmission is being replaced swiftly by direct transmission of digital data and metadata via fieldbus networks such as HART, WirelessHART, ISA100, Profibus, Profinet, Foundation fieldbus, and Ethernet and Ethernet IP.

COUNTING EVENTS

Event counters are used to measure the number of items passing on a conveyor line, the number of operations of a machine, etc. Coupled with time measurements, they yield measures of average rate or frequency. They find important application, therefore, in inventory control, production analysis, and in the sequencing control of automatic machines.

Choice of the proper counting device depends on the kind of events being counted, the necessary counting speed, and the disposition of the measurement; that is, whether it is to be indicated remotely, used to actuate a machine, etc. Errors in the total count may be introduced by events being too close together or by too much nonuniformity in the items being counted.

The **mechanical counter** is shown in Fig. 10.1.1. Motion of the event being counted deflects the arm, which through an appropriate

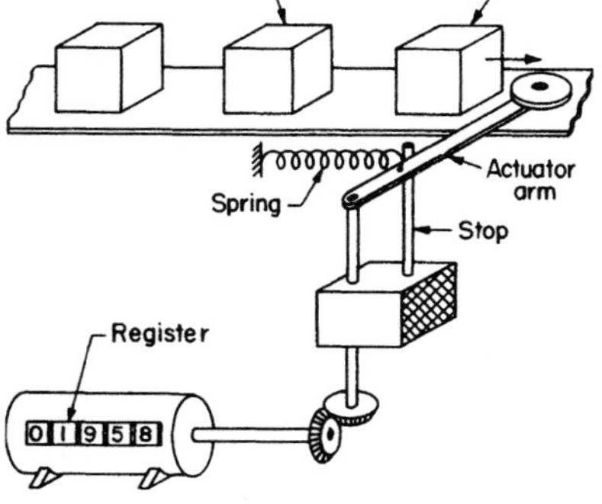

Figure 10.1.1 Mechanical counter.

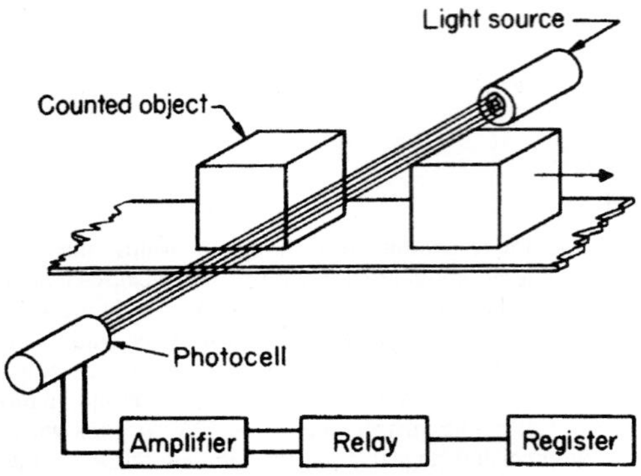

Figure 10.1.2 Photoelectric counter.

linkage advances the count register one unit. Alternatively, motion of the actuating arm may close an electrical switch which energizes a relay coil to advance the count register one step.

Where there is a desire to avoid contact or close proximity with the object being counted, the photoelectric cell or diode, in conjunction with a lamp, a light-emitting diode (LED), or a laser light source, is employed in the transmitted or reflected light mode (Fig. 10.1.2). A signal to a counter is generated whenever the received light level is altered by the passing objects. Objects may be very small and very high counting speeds may be achieved with electronic counters.

Sensing methods based on electrical capacitance, magnetic, and eddy-current effects are extremely sensitive and fast acting, and are suitable for objects in close proximity to the sensor. The capacitive probe senses dielectrics other than air, such as glass and plastic parts. The magnetic pickup, by induction, responds to the motion of iron and nickel. The eddy-current sensor, by energy absorption, detects nonmagnetic conductors. All are suitable for counting machine operations.

The count is displayed by either a mechanical register as in Fig. 10.1.1, or an electronic pulse counter with either number indicators or direct digital output to a PLC (programmable logic controller), embedded industrial computer, or other controller. Sometimes a digital printing output is added. **Electronic counters** can operate accurately at rates exceeding 1 million counts per second.

TIME AND FREQUENCY MEASUREMENT

Measurement of time is basic to time and motion studies, time program controls, and the measurements of velocity, frequency, and flow rate.

Mechanical clocks, chronometers, and stopwatches measure time in terms of the natural oscillation period of a system such as a pendulum, or hairspring balance-wheel combination. The minimum resolution is one-half period. Since this period is somewhat affected by temperature, precise timepieces employ a compensating element to maintain timing accuracies over long periods. Stopwatches may be obtained to read to better than 0.1 s. The major limitation, however, is in the response time of the user.

Electric timers are simple, inexpensive, and readily adaptable to remote-control operations. The majority of these are AC synchronous motors geared in the proper ratio to the indicator. These depend for their accuracy on the frequency of the line voltage. Consequently, care must be exercised in using such devices for precise short-time measurements.

Electronic timers are started and stopped by electrical pulses and hence are not limited by the observer's reaction time. They may be made extremely accurate and capable of measuring to less than 1 μs. These measure time by counting the number of cycles in a high-frequency signal generated internally by means of a quartz crystal. Stopwatch versions read at 0.01 s. Commercial instruments offer one or more functions: counting, measurement of frequency, period, and time intervals. Microprocessor-equipped versions increase versatility.

There are a variety of timing devices designed to indicate or control to a fixed time. These include timers based on the charging time of a condenser (e.g., type 555 integrated circuit), and the flow of oil or other fluid through a restriction. Most of these timing devices are now obsolete in favor of solid-state timing devices, but some mechanical or electromechanical timing devices are still in use.

Timing devices can be **calibrated** by comparison with a standard instrument, or by reference to the National Institute of Standards and Technology (NIST) timed radio signals (stations WWV, WWVB, and WWVH), telephone modem synchronization system (ACTS), or Internet Time Service (ITS). The most up-to-date information on time standards can be found on the Internet at www.boulder.nist.gov/timefreq/index.html. Frequency devices can also be calibrated by referring to the information on that page. The GOES Time Service, using geostationary satellites, was ended on January 1, 2005. See the References at the beginning of this section for information on ACTS and ITS services.

Fast-moving, repetitive motions may be timed and studied by the use of stroboscopes which generate brilliant, very brief flashes of light at an adjustable rate.

The frequency of the observed motion is measured by adjusting the stroboscopic frequency until the system appears to stand still. The frequency of the motion is then equal to the stroboscope frequency or an integer multiple of it.

Many other means exist for **measuring vibrational** or **rotational frequencies.** These include timing a fixed number of rotations or oscillations of the moving member. Contact sensing can be done by an attached switch, or noncontact sensing can be done by magnetic or optical means. The pulses can be counted by an electronic counter or displayed on an oscilloscope or recorder and compared with a known frequency. Also used are reeds which vibrate when the measured oscillation excites their natural frequencies, flyball devices which respond directly to angular velocity, and generator-type tachometers which generate a voltage proportional to the speed.

MASS AND WEIGHT MEASUREMENT

Mass is the measure of the quantity of matter. The fundamental unit is the kilogram. The U.S. customary unit is the pound; 1 lb = 0.4536 kg. Weight is a measure of the force of gravity acting on a mass (see "Units of Force and Mass" in Sec. 2).

A general equation relating weight W and mass M is $W/g = M/g_c$, where g is the local acceleration of gravity, and g_c = 32.174 lbm · ft/(lbf) (s²) [1 kg · m/(N) (s²)] is a property of the unit system. Then $W = Mg/g_c$. The specific weight w and the mass density p are related by $w = pg/g_c$. Masses are conveniently compared by comparing their weights, and masses are often loosely referred to as weights. Indeed, almost all practical measures of mass are based on weight.

Weighing devices fall into two major categories: balances and force-deflection systems. Balances are now almost exclusively used in laboratories or on QA benches as "at line" measurements. Weighing devices may be batch or continuous weighing, automatic or manual. Accuracies are expected to be of the order of 0.1 to better than 0.0001 percent, depending on the type and application of the scale. Calibration is normally performed by use of standard weights (masses) with calibrations traceable to the National Institute of Standards and Technology.

The **equal arm balance** compares the weight of an object with a set of standard weights. The laboratory balance shown in Fig. 10.1.3 is used for extreme precision and sensitivity. A chain poise provides fine adjustment of the final balance weight. The magnetic damper causes the balance to come to equilibrium quickly.

Large weighing scales formerly operated on the same principle; however, now nearly all large weighing scales are of the force deflection type. In this group are platform, track, hopper, and tank scales. Here weight is measured by measuring the deflection of a force deflection bridge or load cell. The electronic output of the weighing cell is converted into a calibrated weight by the transmitter or transmitter/indicator. Most of these types of scale have direct digital outputs, either serial, Modbus, or analog 4-20 mA DC. The most modern versions have Ethernet digital outputs.

In production processes, **continuous and automatic operating scales** are employed. The **batch-type scale** (Fig. 10.1.4) is adaptable to continuous flow streams of either liquids or solid particles. Material

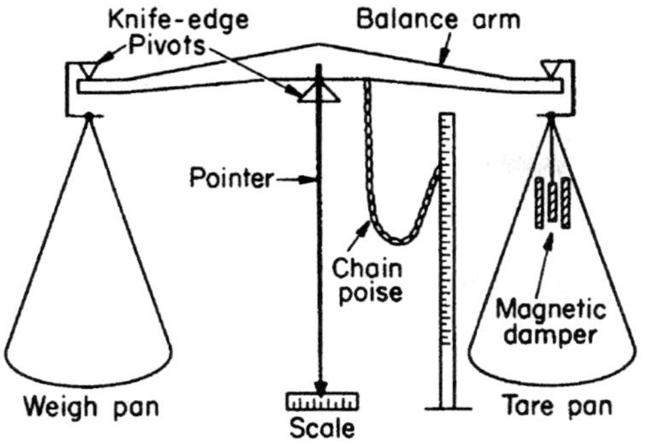

Figure 10.1.3 Laboratory balance.

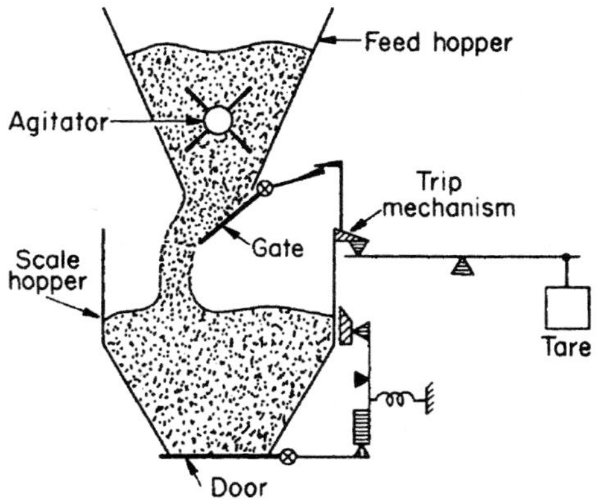

Figure 10.1.4 Automatic batch-weighing scale.

flows from the feed hopper through an adjustable gate into the scale hopper. When the weight in the scale hopper reaches that of the tare, the trip mechanism operates, closing the gate and opening the door. As soon as the scale hopper is empty, the weight of the tare forces the door closed again, resets the trip, and opens the gate to repeat the cycle. The agitator rotates while the gate is open, to prevent the solids from packing. Also, a "dribble" (partial closing of the gate just before the mechanism trips) is employed to minimize the error from the falling column of material at the instant balance is achieved. Since each dump of the scale represents a fixed weight, a counter yields the total weight of material passing through the scale.

In **continuous weighers,** a section of conveyor belt is balanced on a weigh beam (Fig. 10.1.5) with one end fixed to a load cell. The belt is

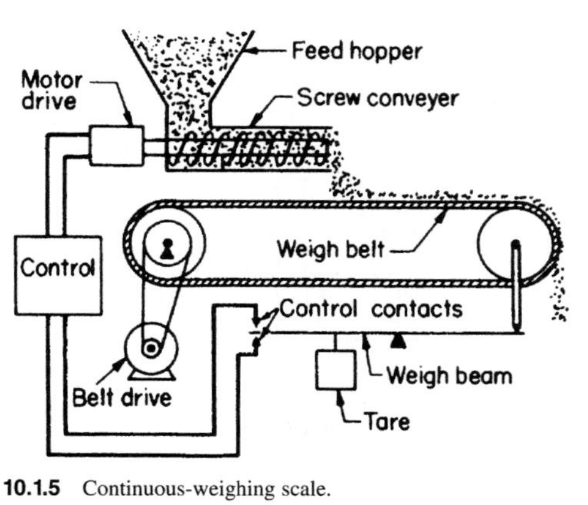

Figure 10.1.5 Continuous-weighing scale.

driven at a constant speed; hence, if the total weight is held constant, the weight rate of material fed through the scale is fixed. Unbalance of the weight beam causes the rate of material flow onto the belt to be changed in the direction of restoring balance. This is accomplished by a mechanical adjustment of the feed gate or by varying the speed of a belt or screw feeder drive.

If the density of the material is constant, **volume measurements** may be used to determine the mass. Thus, calibrated tanks are frequently used for liquids and vane and screw-type feeders for solids. Though often simpler to apply, these are not generally capable of as high accuracies as are common in weighing.

MEASUREMENT OF LINEAR AND ANGULAR DISPLACEMENT

Displacement-measuring devices are employed to measure dimension, distances between points, and some derived quantities such as velocity, area, etc. These devices fall into two major categories: those based on comparison with a known or reference length and those based on some fixed physical relationship.

The **measurement of angles** is closely related to displacement measurements, and indeed, one is often converted into the other in the process of measurement. The common unit is the degree, which represents $^1/_{360}$ of an entire rotation. The radian is used in mathematics and is related to the degree by π rad = 180°; 1 rad = 57.3°. The grad is an angle unit = $^1/_{400}$ rotation.

Figure 10.1.6 illustrates some methods of **rotary to linear conversion.** Figure 10.1.6*a* is a simple link and lever, Fig. 10.1.6*b* is a flexible link and sector, and Fig. 10.1.6*c* is a rack-and-pinion mechanism. These can be used to convert in either direction according to the relationship $D = RA/57.3$, where R = mean radius of the rotating element, in; D = displacement, in; and A = rotation, deg. (This equation holds for the link and lever of Fig. 10.1.6*a* only if the angle change from the perpendicular is small.)

Comparative devices are generally of the indicating type and include ruled or graduated devices such as the machinist's scale, folding rule, tape measure, digital caliper (Fig. 10.1.7), digital micrometer (Fig. 10.1.8), etc. These vary widely in their accuracy, resolution, and measuring span, according to their intended application. The manual readings depend for their accuracy on the skill and care of the operator.

The digital caliper and digital micrometer provide increased sensitivity and precision of reading. The stem of the digital caliper carries an embedded encoded distance scale. That scale is read by the slider. The distance so found shows on the digital display. The device is battery-operated and capable of displaying in inches or millimeters. Typical resolution is 0.0005 in or 0.01 mm.

The **digital micrometer,** employing rotation and translation to stretch the effective encoded scale length, provides resolution to 0.0001 in or 0.003 mm.

Dial gages are also used to magnify motion. A rack and pinion (Fig. 10.1.6*c*) converts linear into rotary motion, and a pointer moves over a calibrated scale.

Various modifications of the afore-mentioned devices are available for making special kinds of measurements; for example, **depth gages** for measuring the depth of a hole or cavity, **inside and outside calipers** (Fig. 10.1.7) for measuring the internal and external dimensions, respectively, of an object, **protractors** for angular measurement, etc.

For line production and inspection work, **go no-go gages** provide a rapid and accurate means of dimension measurement and control. Since the measured values are fixed, the dependence on the operator's skill is considerably reduced. Such gages can be very complex in form to embrace a multidimensional object. They can also take the more general forms of the **feeler, wire, or thread-gage** sets. Of particular importance are **precision gage blocks,** which are used as standards for calibrating other measuring devices.

Displacement can be measured electrically through its effect on the resistance, inductance, reluctance, or capacitance of an appropriate sensing element.

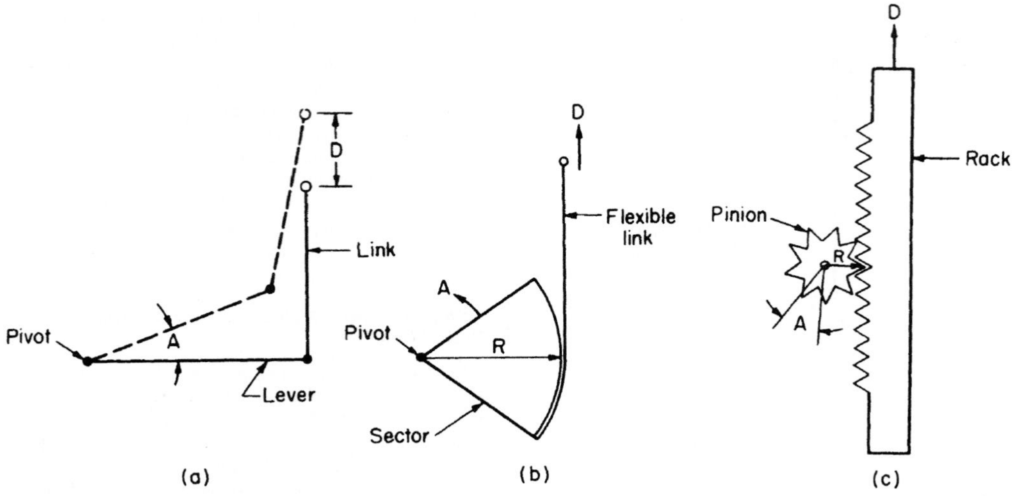

Figure 10.1.6 Linear-rotary conversion mechanisms.

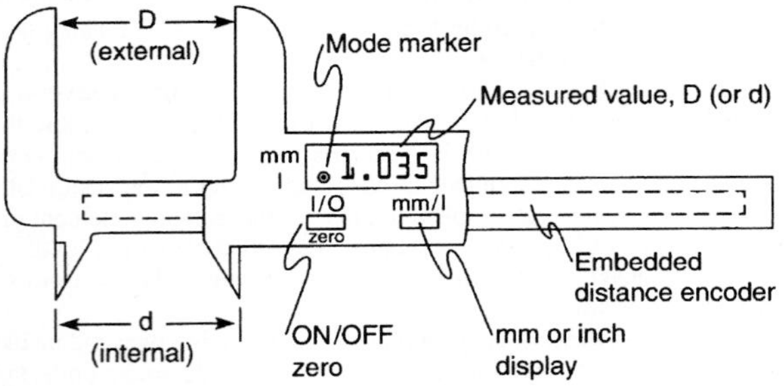

Figure 10.1.7 Digital caliper.

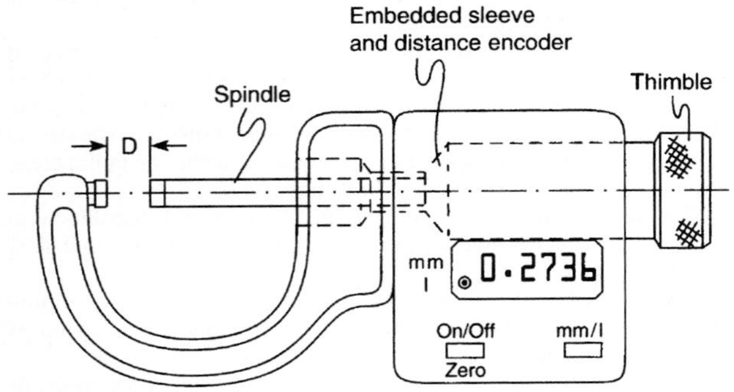

Figure 10.1.8 Digital micrometer.

The **potentiometer** is comparatively inexpensive, accurate, and flexible in application. It consists of a fixed linear resistance over which slides a rotating contact keyed to the input shaft (Fig. 10.1.9). The resistance or voltage (assuming constant voltage across terminals 1 and 3) measured across terminals 1 and 2 is directly proportional to the angle *A*. For straight-line motion, a mechanism of the type shown in Fig. 10.1.6 converts to rotary motion (or a rectilinear-type potentiometer can be used directly). (See also Sec. 11.) Versions with multiturns, straight-line motion, and special nonlinear resistance versus motion are available.

The **synchro,** the **linear variable differential transformer (LVDT),** and the **E transformer** are devices in which the input motion changes the inductive coupling between primary and secondary coils. These avoid the limitations of wear, friction, and resolution of the potentiometer, but they require an ac supply and usually an electronic amplifier for the output. (See also Sec. 11.)

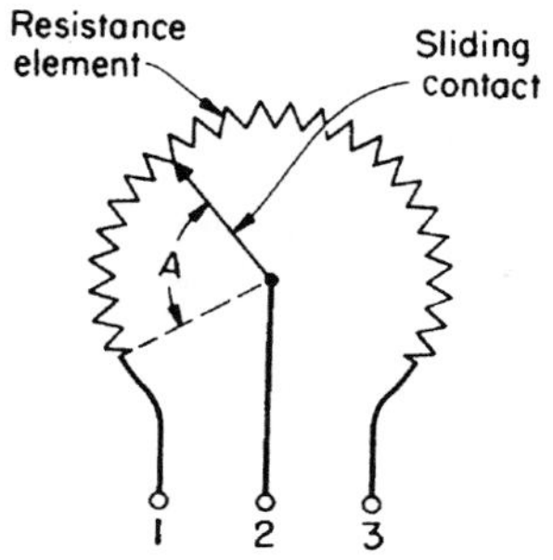

Figure 10.1.9 Potentiometer.

The **synchro** is a rotating device which is used to transmit rotary motions to a remote location for indication or control action. It is particularly useful where the rotation is continuous or covers a wide range. They are used in pairs, one transmitter and one receiver. For measurement of difference in angular position, the **control-transmitter** and **control-transformer** synchros generate an electrical error signal useful in control systems. A synchro differential added to the pair serves the same purpose as a gear differential.

The **linear variable differential transformer (LVDT)** consists of a primary and two secondary coils wound around a common core (Fig. 10.1.10). An armature (iron) is free to move vertically along the

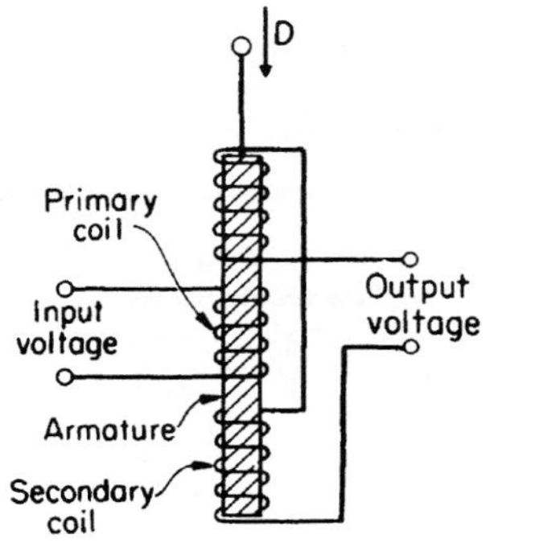

Figure 10.1.10 Linear variable differential transformer (LVDT).

axis of the coils. An AC voltage is applied to the primary. A voltage is induced in each secondary coil proportional to the relative length of armature linking it with the primary. The secondaries are connected to oppose each other so that when the armature is centered, the output voltage is zero. When the armature is displaced off center by an amount *D*, the output will be proportional to *D* (and phased to show whether *D* is above or below the center). These devices are very linear near the centered position, require negligible actuating force, and have spans ranging from 0.1 to several inches (0.25 cm to several centimeters).

The **E transformer** is very similar to the above except that the coils are wound around a laminated iron core in the shape of an E (with the primary and secondaries occupying the center and outside legs, respectively). The magnetic path is completed through an armature whose motion, either rotary or translational, varies the induced voltage in the secondaries, as in the device of Fig. 10.1.10. This, too, is sensitive to extremely small motions.

A method that is readily applied, if a strain-gage analyzer is handy, is to measure the deflection of a cantilever spring with strain gages bonded to its surface (see "Strain Gages," Sec. 3).

The **change of capacitance** with the displacement of the capacitor plates is extremely sensitive and suitable to very small displacements or large rotation. Often, one plate is fixed within the instrument; the other is formed or rotated by the object being measured. The capacitance can be measured by an impedance bridge, by determining the resonant frequency of a tuned circuit or using a relaxation oscillator.

Many optical instruments are available for obtaining precise measurements. The **transit and level** are used in surveying for measuring angles and vertical distances (see Sec. 12.3). A telescope with fine cross hairs permits accurate sighting. The angle scales are generally equipped with verniers. The **measuring microscope** permits measurement of very small displacements and dimensions. The microscope table is equipped with micrometer screws for sensitive adjustment. In addition, templates of scales, angles, etc., are available to permit measurement by comparison. The **optical comparator** projects a magnified shadow image of an object on a screen where it can readily be compared with a reference template.

Light can be used as a standard for the measurement of distance, straightness, and related properties. The wavelength of light in a medium is the velocity of light in vacuum divided by the index of refraction *n* of the medium. For dry air $n - 1$ is closely proportional to air density and is about 0.000277 at 1 atm and 15°C for 550-nm green light. Since the wavelength changes about + 1 ppm/°C, and about −0.36 ppm/mm Hg, density gradients bend light slightly. A temperature gradient of 1°C/m (0.5°F/ft) will cause a deviation from a tangent line of about 0.05 mm (0.002 in) at 10 m (33 ft).

Optical equipment to establish and test alignment, plumb lines, squareness, and flatness includes **jig transits, alignment telescopes, collimators,** optical squares, mirrors, targets, and scales.

Interference principles can be used for distance measurements. An optical flat placed in close contact with a polished surface and illuminated perpendicular to the surface with a monochromatic light will show interference bands which are contours of constant separation distance between the surfaces. Adjacent bands correspond to separation differences of one-half wavelength. For 550-nm wavelength this is 275 nm (10.8 μin). This test is useful in examining surfaces for flatness and in length comparisons with gage blocks.

Laser beams can be used over great distances. Surveying instruments are available for measurements up to 40 mi (60 km). Accuracy is stated to be about 5 mm (0.02 ft) + 1 ppm. These instruments take several measurements which are processed automatically to display the distance directly. Momentary interruptions of the light beam can be tolerated.

A laser system for machine tools, measurement tables, and the like is available in modular form (Agilent Corp.). It can serve up to eight axes by using beam splitters with a combined range of 200 ft (60 m). Normal resolution of length is about one-fourth wavelength, with a digital display least count of 10 μin (0.1 μm). Angle-measurement display resolves 0.1 s of arc. Accuracy with proper environmental compensation is stated to be better than 1 ppm + 1 count in length measurement. Velocities up to 720 in/min (0.3 m/s) can be followed. Accessories are available for measuring straightness, parallelism, squareness, and flatness and for automatic temperature compensation. Various output options include displays and automatic computation and plots. The system can be used directly in measurement and control or to calibrate lead screws and other conventional measuring devices.

Pneumatic gaging previously found an important place in line inspection and quality control and is still sometimes seen. The device

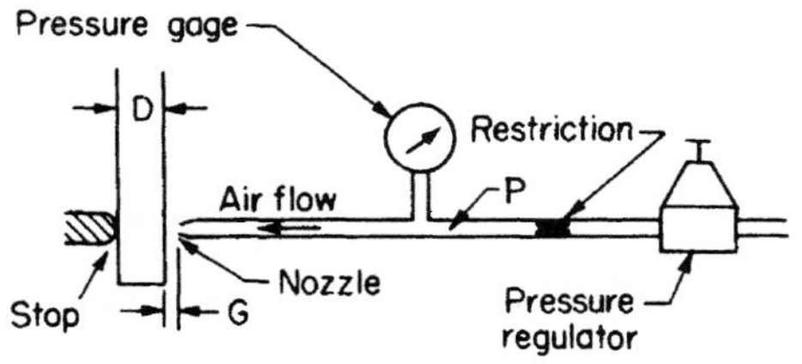

Figure 10.1.11 Pneumatic gage.

(Fig. 10.1.11) consists of a nozzle fixed in position relative to a stop or jig. Air at constant supply pressure passes through a restriction and discharges through the nozzle. The nozzle back pressure *P* depends on the gap *G* between the measured surface and the nozzle opening. If the measured dimension *D* increases, then *G* decreases, restricting the discharge of air, increasing *P*. Conversely, when *D* decreases, *P* decreases. Thus, the pressure gage indicates deviation of the dimension from some normal value. With proper design, this pressure is directly proportional to the deviation, limited, however, to a few thousandths of an inch span. The device is extremely sensitive [better than 0.0001 in (0.003 mm)], rugged, and, with periodic calibration against a standard, quite accurate. The gage is adaptable to automatic line operation where the pressure signal is recorded or used to actuate **"reject"** or **"accept"** controls. Further, any number of nozzles can be used in a jig to check a multiplicity of dimensions. In another form of this device, the flow of air is measured with a rotameter in place of the back pressure. The linear-variable differential transformer (LVDT) is also applicable.

The advent of automatically controlled machine tools has brought about the need for very accurate displacement measurement over a wide range. Most commonly applied for this purpose is the calibrated **lead screw** which measures linear displacement in terms of its angular rotation. **Digital** systems greatly extend the resolution and accuracy limitations of the lead screw. In these, a uniformly spaced optical or inductive grid is displaced relative to a sensing element. The number of grid lines counted is a direct measure of the displacement (see discussion of lasers preceding).

Measurement of strip thickness or coating thickness is achieved by **X-ray** or **beta-radiation**-type gages (Fig. 10.1.12). A constant radiation source (X-ray tube or radioisotope) provides an incident intensity I_0; the radiation intensity *I* after passing through the absorbing material is measured by an appropriate device (scintillation counter,

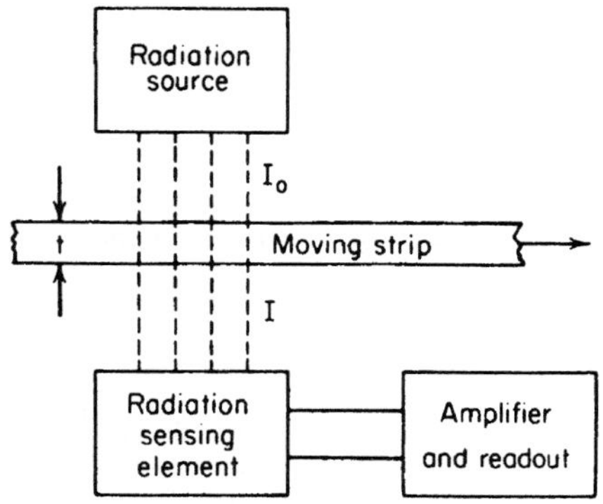

Figure 10.1.12 Radiation-type thickness gage.

Geiger-Müller tube, etc.). The thickness t is determined by the equation $I = I_0 e^{-kt}$, where k is a constant dependent on the material and the measuring device. The major advantage here is that measurements are continuous and nondestructive and require no contact. The method is extended to measure liquid level and density. Nuclear measurement instruments are rarely used today, being replaced by radar and laser interferometry and other measurement methods. The reason for this is that the typical nuclear instrument may cost as much as $30-$80 thousand dollars a year to operate based on costs for training, licensing, recordkeeping, testing, and the like.

MEASUREMENT OF AREA

Area measurements are made for the purpose of determining surface area of an object or area inside a closed curve relating to some desired physical quantity. Dimensions are expressed as a length squared; for example, in^2 or m^2. The areas of simple forms are readily obtained by formula. The area of a complex form can be determined by subdividing into simple forms of known area. In addition, various numerical methods are available for estimating the area under irregular curves.

Area measuring devices include various mechanical, electrical, or electronic **flow integrators** (used with flowmeters) and the **polar planimeter.** The latter consists of two arms pivoted to each other. A tracer at the end of one arm is guided around the boundary curve of the area, causing rotation of a recorder wheel proportional to the area enclosed.

MEASUREMENT OF FLUID VOLUME

For a liquid of known density, volume is a quick and simple means of measuring the amount (or mass) of liquid present. Conversely, measuring the weight and volume of a given quantity of material permits calculation of its density. Volume has the dimensions of length cubed; for example, cubic meters, cubic feet. The volume of simple forms can be obtained by formula.

A volumetric device is any container which has a known and fixed calibration of volume contained versus the level of liquid. The device may be calibrated at only one point **(pipette, volumetric flask)** or may be graduated over its entire volume **(burette, graduated cylinder, volumetric tank).** In the case of the tank, a sight glass may be calibrated directly in liquid volume.

Volumetric measure of continuous flow streams is obtained with the **displacement meter.** This is available in various forms: the nutating disk, reciprocating piston, rotating vane, etc. The **nutating-disk meter** (Fig. 10.1.13) is relatively inexpensive and hence is widely used (water meters, etc.). Liquid entering the meter causes the disk to nutate or "roll" as the liquid makes its way around the chamber to the outlet. A pin on the disk causes a counter to rotate, thereby counting the total number of rolls of the disk. Meter accuracy is limited by leakage past the disk and friction. The **piston meter** is like a piston pump operated backward. It is used for more precise measure (available to 0.1 percent accuracy).

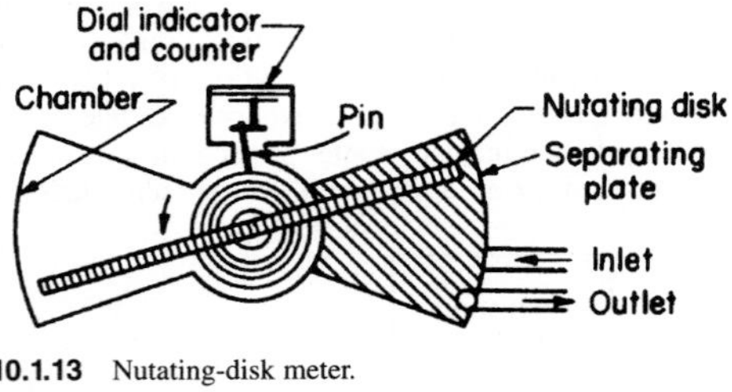

Figure 10.1.13 Nutating-disk meter.

Volumetric gas measurement was formerly made with a **bellows meter.** Two bellows are alternately filled and exhausted with the gas. Motion of the bellows actuates a register to indicate the total. This method is now almost exclusively confined to retail natural gas metering. Various liquid-sealed displacement meters are also available for this purpose. Modern gas flow measurement is made possible the use of Coriolis mass flowmeters, transit time ultrasonic flowmeters, and thermal mass flowmeters for volumetric gas measurements. It is likely that one or more of these methods will replace the bellows meter for many applications within the life of this edition.

For precise volume measurements, corrections for temperature must be made (because of expansion of both the material being measured and the volumetric device). In the case of gases, the pressure also must be noted.

FORCE AND TORQUE MEASUREMENT

Force may be measured by the deflection of an elastic element, by balancing against a known force, by the acceleration produced in an object of known mass, or by its effects on the electrical or other properties of a stress-sensitive material. The common unit of force is the pound (newton). **Torque** is the product of a force and the perpendicular distance to the axis of rotation. Thus, torque tends to produce rotational motion and is expressed in units of pound feet (newton meters). Torque can be measured by the angular deflection of an elastic element or, where the moment arm is known, by any of the force measuring methods.

Since weight is the force of gravity acting on a mass, any of the weight-measuring devices already discussed can be used to measure force. Most industrial weight measurement is done by means of calibrated load cells.

The strain gage is an element whose electrical resistance changes with applied strain (see Sec. 3). Combined with an element of known force-strain, motion-strain, or other input-strain relationship it is a transducer for the corresponding input. The relation of gage-resistance change to input variable can be found by analysis and calibration. Measure of the resistance change can be translated into a measure of the force applied. The gage may be bonded or unbonded. In the bonded case, the gage is cemented to the surface of an elastic member and measures the strain of the member. Since the gage is very sensitive to temperature, the readings must be compensated. For this purpose, four gages are connected in a Wheatstone-bridge circuit such that the temperature effect cancels itself. A four-element unbonded gage is shown in Fig. 10.1.14. Note that as the applied force increases, the tension on two of the elements increases while that on the other two decreases. Gages subject to strain change of the same sign are put in opposite arms of the bridge. The zero adjustment permits balancing the bridge for zero output at any desired input. The e_1 and e_2 terminal pairs may be used interchangeably for the input excitation and the signal output.

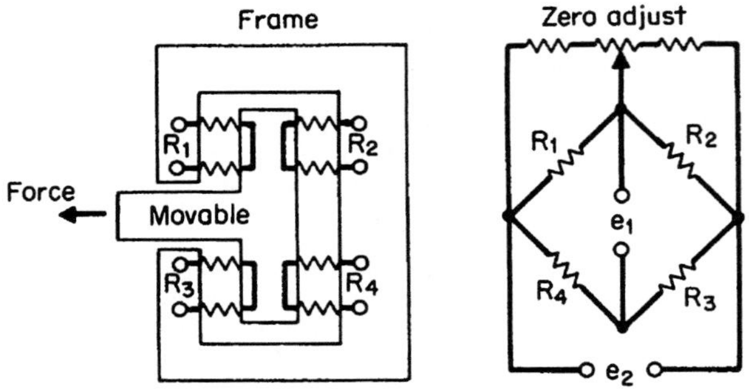

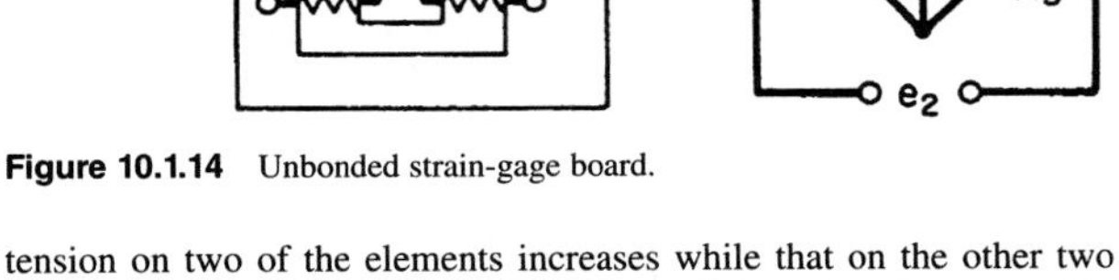

Figure 10.1.14 Unbonded strain-gage board.

The **piezoelectric** effect is useful in measuring rapidly varying forces because of its high-frequency response and negligible displacement characteristics. Quartz rochelle salt and barium titanate are common piezoelectric materials. They have the property of varying an output charge in direct proportion to the stress applied. This produces a voltage inversely proportional to the circuit capacitance. Charge leakage produces drifting at a rate depending on the circuit time constant. The voltage must be measured with a device having a very high input resistance. Accuracy is limited because of temperature dependence and some hysteresis effect.

Forces may also be measured with any of the pressure devices described in the next section by balancing against a fluid pressure acting on a fixed area.

PRESSURE AND VACUUM MEASUREMENT

Pressure is defined as the force per unit area exerted by a fluid. Pressure devices normally measure with respect to atmospheric pressure (mean value = 14.7 lb/in²), $p_a = p_g + 14.7$, where p_a = total or **absolute** pressure and p_g = **gage** pressure, both lb/in². Conventionally, gage pressure and vacuum refer to pressures above and below atmospheric, respectively. Common units are lb/in², in Hg, ftH_2O, kg/cm², bars, and mm Hg. The mean SI atmosphere is 1.013 bar.

Pressure devices are based on (1) measure of an equivalent height of liquid column; (2) measure of the force exerted on a fixed area; (3) measure of some change in electrical or physical characteristics of the fluid.

The manometer measures pressure according to the relationship $p = wh = \rho gh/g_c$, where h = height of liquid of density ρ and specific weight w (assumed constants) supported by a pressure p. Thus, pressures are often expressed directly in terms of the equivalent height (head) of manometer liquid, for example, in H_2O or in Hg. Usual manometer fluids are water or mercury, although other fluids are available for special ranges.

The **U-tube manometer** (Fig. 10.1.15*a*) expresses the pressure difference $p_1 - p_2$ as the difference in levels h. If p_2 is exposed to the atmosphere, the manometer reads the gage pressure of p_1. If the p_2 tube is evacuated and sealed ($p_2 = 0$), the absolute value of p_1 is indicated. A common modification is the **well-type manometer** (Fig. 10.1.15*b*). The scale is specially calibrated to take into account changes of level inside the well so that only a single tube reading is required. In particular, Fig. 10.1.15*b* illustrates the form usually applied to measurement of atmospheric pressure **(mercury barometer).**

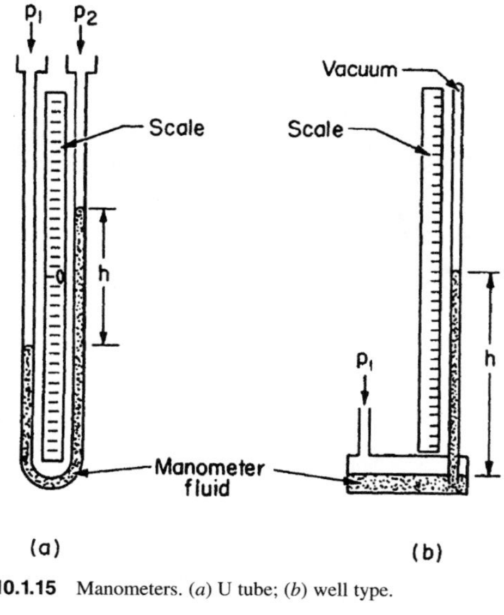

Figure 10.1.15 Manometers. (*a*) U tube; (*b*) well type.

The sensitivity of readings can be increased by inclining the manometer tubes to the vertical **(inclined manometer),** by use of low-specific-gravity manometer fluids, or by application of optical-magnification or level-sensing devices. Accuracy is influenced by surface-tension effects (reading of the meniscus) and changes in fluid density (due to temperature changes and impurities).

By definition, pressure times the area acted upon equals the force exerted. The pressure may act on a diaphragm, bellows, or other element of fixed area. The force is then measured with any force-measuring device, for example, spring deflection, strain gage, or weight balance. Very commonly, the unknown pressure is balanced against an air or hydraulic pressure, which in turn is measured with a gage. By use of unequal-area diaphragms, the pressure can thus be amplified or attenuated as required. Further, it permits isolating the process fluid which may be corrosive, viscous, etc.

The **Bourdon-tube gage** (Fig. 10.1.16) is the most commonly used pressure device. It consists of a flattened tube of spring bronze or steel bent into a circle. Pressure inside the tube tends to straighten it. Since one end of the tube is fixed to the pressure inlet, the other end moves proportionally to the pressure difference existing between the inside and outside of the tube. The motion rotates the pointer through a pinion-and-sector mechanism. For amplification of the motion, the tube may be bent through several turns to form spiral or helical elements as are used in pressure recorders.

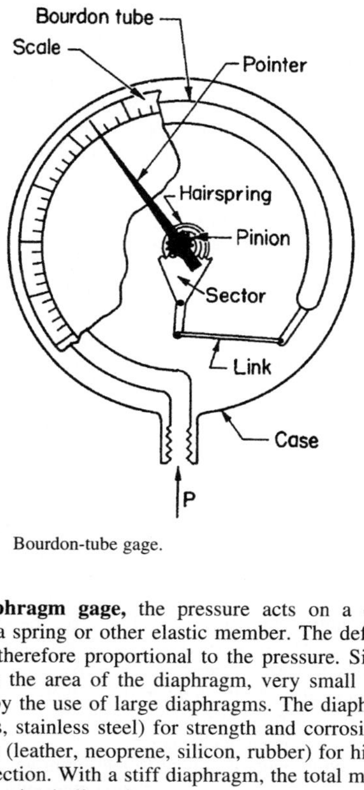

Figure 10.1.16 Bourdon-tube gage.

In the **diaphragm gage,** the pressure acts on a diaphragm in opposition to a spring or other elastic member. The deflection of the diaphragm is therefore proportional to the pressure. Since the force increases with the area of the diaphragm, very small pressures can be measured by the use of large diaphragms. The diaphragm may be metallic (brass, stainless steel) for strength and corrosion resistance, or nonmetallic (leather, neoprene, silicon, rubber) for high sensitivity and large deflection. With a stiff diaphragm, the total motion must be very small to maintain linearity.

The **bellows gage** (Fig. 10.1.17) is somewhat similar to the diaphragm gage, with the advantage, however, of providing a much wider range of motion. The force acting on the bottom of the bellows is balanced by the deflection of the spring. This motion is transmitted to the output arm, which then actuates a pointer or recorder pen.

The motion (or force) of the pressure element can be converted into an electrical signal by use of a differential transformer or strain-gage element or into an air-pressure signal through the action of a nozzle and pilot. The signal is then used for transmission, recording, or control.

The **dead-weight tester** is used as a standard for calibrating gages. Known hydraulic or gas pressures are generated by means of weights loaded on a calibrated piston. The useful range is from 5 to 5,000 lb/in² (0.3 to 350 bar). For low pressures, the water or mercury manometer serves as a reference.

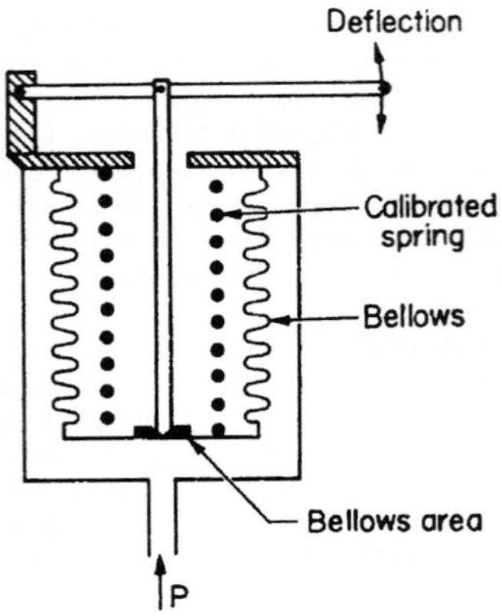

Figure 10.1.17 Bellows gage.

For many applications (fluid flow, liquid level), it is important to measure the **difference between two pressures.** This can be done directly with the manometer. Other pressure devices are available as differential devices where (1) the case is made pressure-tight so that the second pressure can be applied external to the pressure element; (2) two identical pressure elements are mounted so that their outputs oppose each other.

Similar devices to those discussed are used to measure **vacuum,** the only difference being a shift in range or at most a relocation of the zeroing spring. When the vacuum is high (absolute pressure near zero) variations in atmospheric pressure become an important source of error. It is here that absolute-pressure devices are employed.

Any of the differential-pressure elements can be converted to an **absolute-pressure device** by sealing one pressure side to a perfect vacuum. A common instrument for the range 0 to 30 in Hg employs two bellows of equal area set back to back. One bellows is completely evacuated and sealed; the other is connected to the measured pressure. The output is a bellows displacement, as in Fig. 10.1.17.

There are many instruments for high-vacuum work (0.001 to 10,000 μm range). These kinds of devices are based on the characteristic properties of gases at low pressures. The **McLeod gage** amplifies the pressure to be measured by compressing the gas a known amount and then measuring its pressure with a mercury manometer. The ratio of initial to final pressure is equal to the ratio of final to initial volume (for common gases). This gage serves as a standard for low pressures.

The **Pirani gage** (Fig. 10.1.18) is based on the change of heat conductivity of a gas with pressure and the change of electrical resistance of a wire with temperature. The wire is electrically heated with a constant current. Its temperature changes with pressure, producing a voltage across the bridge network. The compensating cell corrects for room-temperature changes.

The **thermocouple gage** is similar to the Pirani gage, except that a thermocouple is used to measure the temperature difference between the resistance elements in the measuring and compensating cells, respectively.

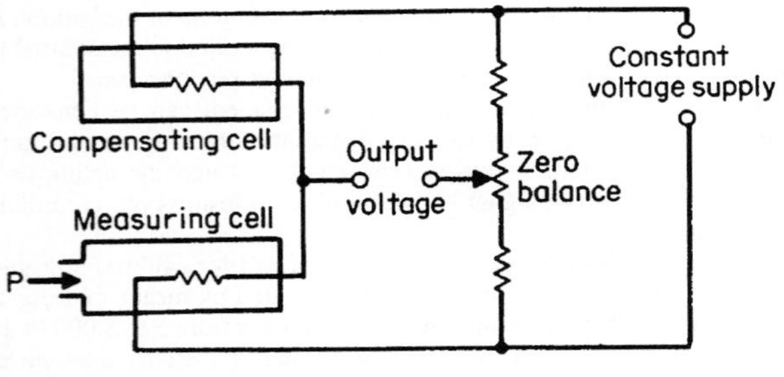

Figure 10.1.18 Pirani gage.

The **ionization gage** measures the ion current generated by bombardment of the molecules of the gas by the electron stream in a triode-type tube. This gage is limited to pressures below 1 μm. It is, however, extremely sensitive.

LIQUID LEVEL MEASUREMENT

The past 50 years have seen significant advancement in the measurement of liquid level. Level instruments are used for determining or controlling the height of liquid in a vessel, or finding the location of the interface between two liquids of different specific gravity. Liquid level measurement can be made both continuously, or as a point level measurement (or switch).

In process measurement, typical level devices include differential pressure and hydrostatic pressure transmitters of various types; capacitance and radio-frequency (RF) admittance device; a wide variety of floats and other mechanical level measurement systems; magnetostrictive gages; ultrasonic level sensors; pulse and continuous wave radar level sensors; guided wave radar sensors; laser level sensors; and nuclear level gages. Typically, the applications for these devices fit along a continuum from very easy to measure (water) to very hard to measure (molten plastic inside a jacketed reactor).

Each of these level measurement techniques has advantages and limitations, dependent on the application. For example, a mechanical gage is not advisable in an application where the fluid is sticky or viscous. Ultrasonic level meters are extremely sensitive to temperature changes in the vapor space and to density changes of the vapor, and should normally only be used at atmospheric pressure. Capacitance or RF admittance–type gages may be used in a wide variety of applications, although long measurement spans make them expensive. Radar level gages of either type can be used in pressurized vessels, since the changing dielectric of the vapor space is usually too small a change to affect the measurement.

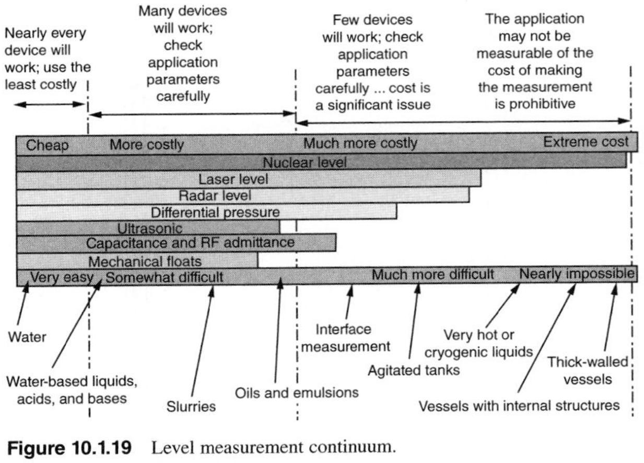

Figure 10.1.19 Level measurement continuum.

Differential pressure measurement is still used as a common means of measuring level in vessels, although the use of this technique is declining. As indicated in the discussion on manometers, the pressure difference is the height times the specific weight of the liquid.

Ultrasonic and pulse radar level devices operate by measuring the transit time of a signal reflected off the liquid surface and converting that signal into a level measurement. Continuous-wave and guided-wave radar instruments measure the signal return from the largest peak of the reflected signals in the vessel. Ultrasonic level devices are also commonly used as open-channel flowmeters when combined with a physical primary device such as a weir or flume.

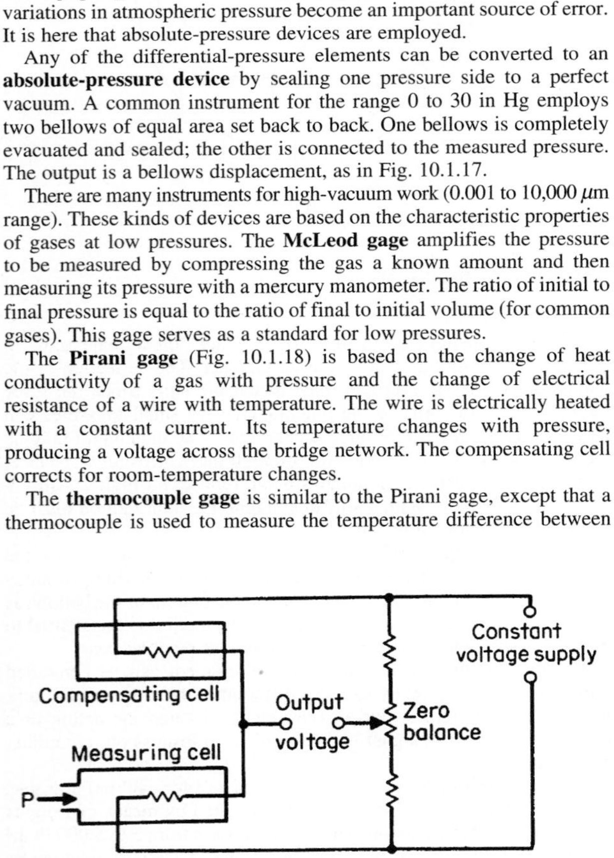

Capacitive or RF admittance level devices measure the change in dielectric that occurs at the level surface or at the interface between two liquids. Like ultrasonics, these can be supplied as either point or continuous level transmitters, and like ultrasonics, they have applications in bulk solids and powders as well as liquid level.

Nuclear level gages are also available in point or continuous level designs. These instruments work by the attenuation of a beam of gamma energy by the surface of the liquid (which is much more dense than the vapor phase in the vessel). The point at which the measured radiation drops to near zero is the level.

Nuclear level gages are the "last resort" in part because of their expense, but also because of the licensing and training required to operate them and the disposal and storage requirements of the nuclear source.

TEMPERATURE MEASUREMENT

The common temperature scales (Fahrenheit and Celsius) are based on the freezing and boiling points of water (see Sec. 2 for discussion of temperature standards, units, and conversion equations).

Temperature is measured in a number of different ways. Some of the more useful are as follows.

1. *Thermal expansion* of a gas (**gas thermometer**). At constant volume, the pressure p of an (ideal) gas is directly proportional to its absolute temperature T. Thus, $p\ (p_0/T_0)T$, where p_0 is the pressure at some known temperature T_0.

2. *Thermal expansion of a liquid or solid* (**mercury thermometer, bimetallic element).** Substances tend to expand with temperature. Thus, a change in temperature $t_2 - t_1$ causes a change in length $l_2 - l_1$ or a change in volume $V_2 - V_1$, according to the expressions.

$$l_2 - l_1 = a'(t_2 - t_1)l_1 \quad \text{or} \quad V_2 - V_1 = a''' - (t_2 - t_1)V_1$$

where a' and a''' = linear and volumetric coefficients of thermal expansion, respectively (see Sec. 2). For many substances, a' and a''' are reasonably constant over a limited temperature range. For solids, $a''' = 3a'$. For mercury at room temperature, a''' is approximately 0.00018°C^{-1} (0.00010°F^{-1}).

3. *Vapor pressure of a liquid* (**vapor-bulb thermometer).** The vapor pressure of all liquids increases with temperature. The Clapeyron equation permits calculation of the rate of change of vapor pressure with temperature.

4. *Thermoelectric potential* (**thermocouple).** When two dissimilar metals are brought into intimate contact, a voltage is developed which depends on the temperature of the junction and the particular metals used. If two such junctions are connected in series with a voltage-measuring device, the measured voltage will be very nearly proportional to the temperature difference of the two junctions.

5. *Variation of electrical resistance* (**resistance thermometer, thermistor).** Electrical conductors experience a change in resistance with temperature which can be measured with a Wheatstone- or Mueller-bridge circuit, or a digital ohmmeter. The platinum resistance thermometer (PRT) can be very stable and is used as the temperature scale interpolation standard from −160 to 660°C. Commercial resistance temperature detectors (RTD) using copper, nickel, and platinum conductors are in use and are characterized by a polynomial resistance-temperature relationship, such as

$$t = A + B \times R_t + C \times R_t^2 + D \times R_t^3 + E \times R_t^4$$

where R_t = resistance at prevailing temperature t in °C. A, B, C, D, and E are range- and material-dependent coefficients listed in Table 10.1.1. R_0, also shown in the table, is the base resistance at 0°C used in the identification of the sensor.

The thermistor has a large, negative temperature coefficient of resistance, typically −3 to −6 percent/°C, decreasing as temperature increases. The temperature-resistance relation is approximated (to perhaps 0.01° in range 0 to 100°C) by:

$$R_t = \exp\left(A_0 + A_1/t + \frac{A_2}{t^2} + \frac{A_3}{t^3}\right)$$

and

$$\frac{1}{t} = a_0 + a_1 \ln R_t + a_2(\ln R_t)^2 + a_3(\ln R_t)^3$$

with the constants chosen to fit four calibration points. Often a simpler form is given:

$$R = R_0 \exp\left\{\beta\left[\left(\frac{1}{t}\right) - \left(\frac{1}{t_0}\right)\right]\right\}$$

Typically β varies in the range of 3,000 to 5,000 K. The reference temperature t_0 is usually 298 K(= 25°C, 77°F), and R_0 is the resistance at that temperature. The error may be as small as 0.3°C in the range of 0 to 50°C. Thermistors are available in many forms and sizes for use from −196 to +450°C with various tolerances on interchangeability and matching. (See "Catalog of Thermistors," Thermometrics, Inc.) The AD590 and AD592 integrated circuit (Analog Devices, Inc.) passes a current of 1 μA/K very nearly proportional to absolute temperature. All these sensors are subject to self-heating error.

6. *Change in radiation* (**radiation and optical pyrometers).** A body radiates energy proportional to the fourth power of its absolute temperature. This principle is particularly adaptable to the measurement of very high temperatures where either the total quantity of radiation or its intensity within a narrow wavelength band may be measured. In the former type (radiation pyrometer), the radiation is focused on a heat-sensitive element, for example, a thermocouple, and its rise in temperature is measured. In the latter type (optical pyrometer) the intensity of the radiation is compared optically with a heated filament. Either the

Table 10.1.1 Polynomial Coefficients for Resistance Temperature Detectors

			Polynomial coefficients					
Material of conductor	ID R_0, Ω	Useful range, °C	A, °C	B, °C/Ω	C, °C/Ω^2	D, °C/Ω^3	E, °C/Ω^4	Typical accuracy,* °C
Copper 10Ω @ 25°C	9.042	−70 to 0	−225.64	23.30735	+0.246864	−0.00715		1.5
		0 to 150	−234.69	25.95508				1.5
Nickel	120	−80 to 320	−199.47	1.955336	−0.00266	1.88E − 6		1
Platinum	100	−200 to 0	−241.86	2.213927	0.002867	−9.8E − 6	1.64E − 8	1
DIN/IEC		0 to 850	−236.06	2.215142	0.001455			0.5
α = 0.00385/°C								

* For higher accuracy consult the table or equation furnished by the manufacturer of the specific RTD being used. Temperatures per ITS-90, resistances per SI-90.

filament brightness is varied by a control calibrated in temperature or a fixed brightness filament is compared with the source viewed through a calibrated optical wedge.

The infrared thermometer accepts radiation from an object seen in a definite field of view, filters it to select a portion of the infrared spectrum, and focuses it on a sensor such as a blackened thermistor flake, which warms and changes resistance. Electronic amplification and signal processing produce a digital display of temperature. Correct calibration requires consideration of source emissivity, reflection, and transmission from other radiation sources, atmospheric absorption between the source object and the sensor, and compensation for temperature variation at the sensor's immediate surroundings.

Electrical nonconductors generally have fairly high (about 0.95) emissivities, while good conductors (especially smooth, reflective metal surfaces), do not; special calibration or surface conditioning is then needed. Very wide band (0.7 to 20 μm) instruments gather relatively large amounts of energy but include atmospheric absorption bands which reduce the energy received from a distance. The band 8 to 14 μm is substantially free from atmospheric absorption and is popular for general use with source objects in the range 32 to 1,000°F (0 to 540°C). Other bands and two-color instruments are used in some cases. See McMillan, Gregory K. *Advanced Temperature Measurement and Control, second edition*, International Society of Automation, 2012.

Fiber-optic probes extend the use of radiation methods to hard-to-reach places.

Important relationships used in the design of these instruments are the Wien and Stefan-Boltzmann laws (in modified form):

$$\lambda_m = k_1/T \quad q = k_2 \varepsilon A(T_2^4 - T_1^4)$$

where λ_m = wavelength of maximum intensity, μm (nm); q = radiant energy flux, Btu/h (W); A = radiation surface, ft² (m²); ε = mean emissivity of the surfaces; T_2, T_1 = absolute temperatures of radiating and receiving surfaces, respectively, °R (K); k_1 = 5,215 μm, °R (2,898 μm · K); $k_2 = 0.173 \times 10^{-8}$ Btu/(h · ft² · °R⁴) [5.73×10^{-8} W/(m² · K⁴)]. The emissivity depends on the material and form of the surfaces involved (see Sec. 2). Radiation sensors with scanning capability can produce maps, photographs, and television displays showing temperature-distribution patterns. They can operate with resolutions to under 1°C and at temperatures below room temperature.

7. *Change in physical or chemical state* (**Seger cones, Tempilsticks).** The temperatures at which substances melt or initiate chemical reaction are often known and reproducible characteristics. Commercial products are available which cover the temperature range from about 120 to 3,600°F (50 to 2,000°C) in intervals ranging from 3 to 70°F (2 to 40°C). The temperature-sensing element may be used as a solid which softens and changes shape at the critical temperature, or it may be applied as a paint, crayon, or stick-on label which changes color or surface appearance. For most the change is permanent; for some it is reversible. Liquid crystals are available in sheet and liquid form: these change reversibly through a range of colors over a relatively narrow temperature range. They are suitable for showing surface-temperature patterns in the range 20 to 50°C (68 to 122°F).

An often used temperature device is the **mercury-in-glass thermometer.** As the temperature increases, the mercury in the bulb expands and rises through a fine capillary in the graduated thermometer stem. Useful range extends from −30 to 900°F (−35 to 500°C). In many applications of the mercury thermometer, the stem is not exposed to the measured temperature; hence a correction is required (except where the thermometer has been calibrated for **partial immersion).** Recommended formula for the correction K to be added to the thermometer reading is $K = 0.00009D(t_1 - t_2)$, where D = number of degrees of exposed mercury filament, °F; t_1 = thermometer reading, °F; t_2 = the temperature at about middle of the exposed portion of stem, °F. For Celsius thermometers the constant 0.00009 becomes 0.00016.

For **industrial applications** the thermometer or other sensor is encased in a metal or ceramic protective well and case (Fig. 10.1.20).

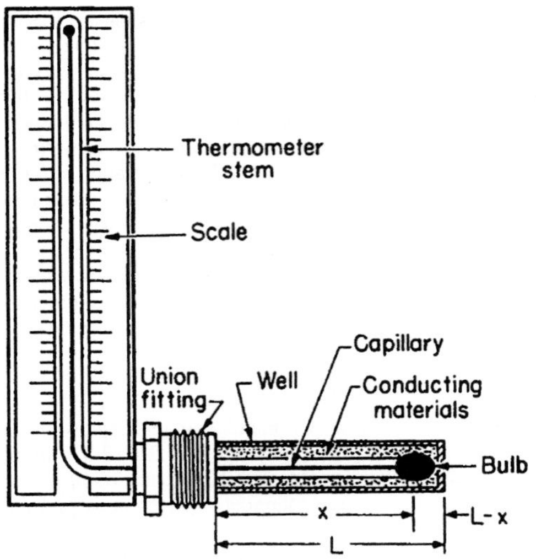

Figure 10.1.20 Industrial thermometer.

A threaded union fitting is provided so that the thermometer can be installed in a line or vessel under pressure. Ideally the sensor should have the same temperature as the fluid into which the well is inserted. However, heat conduction to or from the pipe or vessel wall and radiation heat transfer may also influence the sensor temperature (see ASME PTC 19.3-1974 Temperature Measurement, on well design). An approximation of the conduction error effect is

$$T_{\text{sensor}} - T_{\text{fluid}} = (T_{\text{wall}} - T_{\text{fluid}})E$$

For a sensor inserted to a distance $L - x$ from the tip of a well of insertion length L, $E = \cosh[m(L - x)]/\cosh mL$, where $m = (h/kt)^{0.5}$; x and L are in ft (m); h = fluid-to-well conductance, Btu/(h) (ft²)(°F) [J/(h) (m²)(°C)]; k = thermal conductivity of the well-wall material. Btu/(h)(ft)(°F) [J/(h)(m)(°C)]; and t = well-wall thickness, ft (m). Good thermal contact between the sensor and the well wall is assumed. For $(L - x)/L = 0.25$:

mL	1	2	3	4	5	6	7
E	0.67	0.30	0.13	0.057	0.025	0.012	0.005

Radiation effects can be reduced by a polished, low-emissivity surface on the well and by radiation shields around the well. Concern with mercury contamination has made the bimetal thermometer the most commonly used expansion-based temperature measuring device. Differential thermal expansion of a solid is employed in the simple **bimetal** (used in thermostats) and the bimetallic helix (Fig. 10.1.21). The bimetallic element is made by welding together two strips of metal having different coefficients of expansion. A change in temperature then causes the element to bend or twist an amount proportional

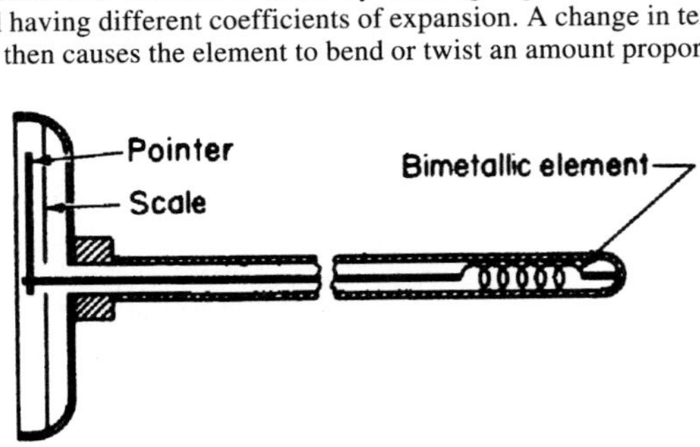

Figure 10.1.21 Bimetallic temperature gage.

to the temperature. A common bimetallic pair consists of invar (iron-nickel alloy) and brass.

For control or alarm indications at fixed temperatures, thermometers may be equipped with electrical contacts such that when the temperature matches the contact point, an external relay circuit is energized.

A formerly popular and inexpensive instrument employs the deflection of a **pressure-spring** to indicate (or record) the temperature (Fig. 10.1.22). The sensing element is a metal bulb containing some specific gas or liquid. The bulb connects with the pressure spring (in the form of a spiral or helix) through a capillary tube which is usually enclosed in a protective sheath or armor. Increasing temperature causes the fluid in the bulb to expand in volume or increase in pressure. This forces the pressure spring to unwind and move the pen or pointer an appropriate distance upscale. Modern temperature instruments use thermocouples or thermistors connected to solid state electronics for indication or retransmission.

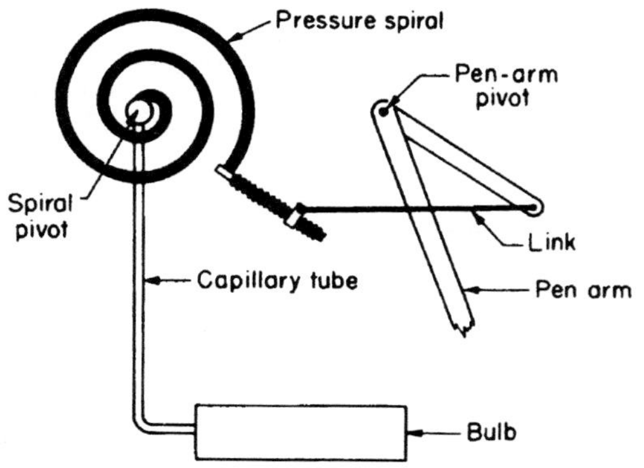

Figure 10.1.22 Pressure-spring element.

The **bulb fluid** may be nitrogen under pressure (gas system), or a volatile liquid (vapor-pressure system). Mercury and gas systems have linear scales; however, they must be compensated to avoid ambient temperature errors. The capillary may range up to 200 ft in length with, however, considerable reduction in speed of response. The use of mercury has been discontinued nearly everywhere due to environmental and health concerns over the use of the metal.

The **pneumatic transmitter** (Fig. 10.1.23) was at one time the standard for long-distance transmission of temperature. Although its place has been taken by various electronic means, it is still commonly used in many parts of the world, especially in refining and other

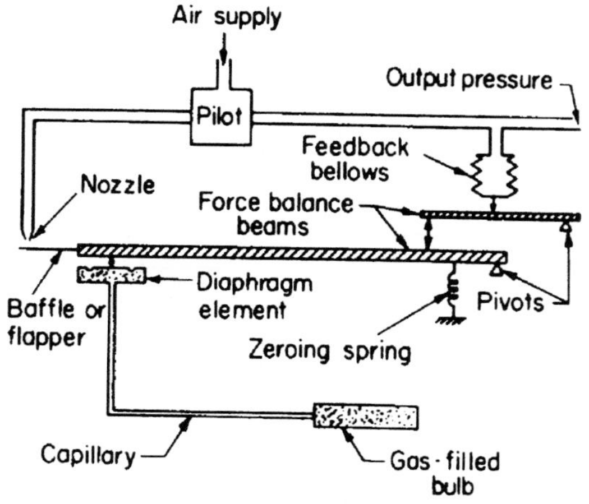

Figure 10.1.23 Pneumatic temperature transmitter.

hydrocarbon processing applications. The bulb is filled with gas under pressure, which acts on the diaphragm. An increase in bulb temperature increases the upward force acting on the main beam, tending to rotate it clockwise. This causes the baffle or flapper to move closer to the nozzle, increasing the nozzle back pressure. This acts on the pilot, producing an increase in output pressure, which increases the force exerted by the feedback bellows. The system returns to equilibrium when the increase in bellows pressure exactly balances the effect of the increased diaphragm pressure. Since the lever ratios are fixed, this results in a direct proportionality between bulb temperature and output air pressure. For precision, compensating elements are built into the instrument to correct for the effects of changes in barometric pressure and ambient temperature.

Electrical systems based on the thermocouple or resistance thermometer are particularly applicable where many different temperatures are to be measured, where transmission distances are large, or where high sensitivity and rapid response are required. The thermocouple is used with high temperatures; the resistance thermometer for low temperatures and high accuracy requirements.

The **choice of thermocouple** depends on the temperature range, desired accuracy, and the nature of the atmosphere to which it is to be exposed. The temperature–voltage relationships for the more common of these are given by the curves of Fig. 10.1.24. Table 10.1.2 gives the recommended temperature limits, for each kind of couple. Table 10.1.3 gives polynomials for converting thermocouple millivolts

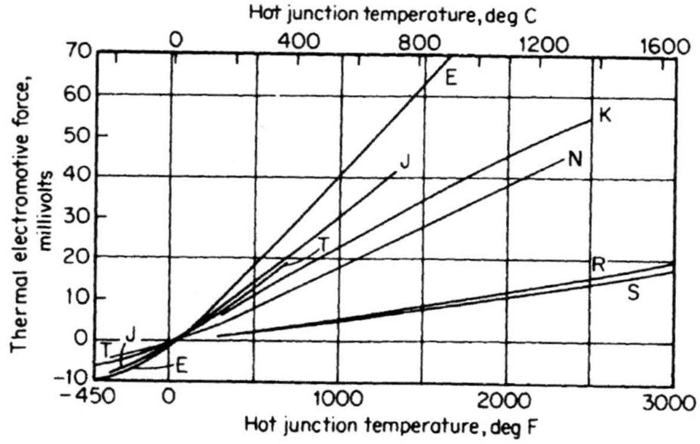

Figure 10.1.24 Thermocouple voltage–temperature characteristics [reference junction at 32°F (0°C)].

to temperature. The thermocouple voltage is measured by a digital or deflection millivoltmeter or null-balance type of potentiometer. Completion of the thermocouple circuit through the instrument immediately introduces one or more additional junctions. Common practice is to connect the thermocouple (hot junction) to the instrument with special lead wire (which may be of the same materials as the thermocouple itself). This assures that the **cold junction** will be inside the instrument case, where compensation can be effectively applied. Cold junction compensation is typically achieved by measuring the temperature of the thermocouple wire to copper wire junctions or terminals with a resistive or semiconductor thermometer and correcting the measured terminal voltage by a derived equivalent millivolt cold junction value. Figure 10.1.25 shows a digital temperature indicator with correction for different ANSI types of thermocouple voltage to temperature nonlinearities being stored in and applied to the analog-to-digital converter (A/D) by a read-only memory (ROM) chip.

The **resistance thermometer** employs the same circuitry as described above, with the resistance element (RTD) being placed external to the instrument and the cold junction being omitted (Fig. 10.1.26). Three types of RTD connections are in use: two wire, three wire, and four wire. The two-wire connection makes the measurement sensitive

Table 10.1.2 Limits of Error on Standard Wires without Selection*†

ANSI symbol‡	Materials and polarities: Positive	Negative	°F: −150 / °C: −101	−75 / −59	32 / 0	200 / 93	530 / 277	600 / 316	700 / 371	1,000 / 538	1,400 / 760	2,300 / 1,260	2,700 / 1,482
T	Cu	Constantan§		2%	1.5°F (0.8°C)			¾%					
E	Ni-Cr	Constantan				3°F (1.7°C)				½%			
J	Fe	Constantan				4°F (2.2°C)			¾%				
K	Ni-Cr	Ni-Al				4°F (2.2°C)			¾%				
N	Ni-Cr-Si	Ni-SiMn				4°F (2.2°C)			¾%				
R	Pt-13% Rh	Pt				3°F (1.5°C)					¼%		
S	Pt-10% Rh	Pt				3°F (1.5°C)					¼%		

* Protect copper from oxidation above 600°F; iron above 900°F. Protect Ni-Al from reducing atmospheres. Protect platinum from nonreducing atmospheres. Type B (Pt-30% Rh vs. Pt-6%) is used up to 3,200°F (1,700°C). Its standard error is ½% above 1,470°F (800°C).

† Closer tolerances are obtainable by selection and calibration. Consult makers' catalogs. Tungsten-rhenium alloys are in use up to 5,000°F (2,760°C). For cryogenic thermocouples see Sparks et al., Reference Tables for Low-Temperature Thermocouples. *Natl. Bur. Stand. Monogr.* 124.

‡ Individual wires are designated by the ANSI symbol followed by P or N; thus iron is JP.

§ Constantan is 55% Cu, 45% Ni. The nickel-chromium and nickel-aluminum alloys are available as Chromel and Alumel, trademarks of Hoskins Mfg. Co.

Table 10.1.3 Polynomial Coefficients for Converting Thermocouple emf to Temperature*

Range	Type E	Type J	Type K	Type N	Type S	Type T
mV	0 to 76.373	0 to 42.919	0 to 20.644	0 to 47.513	1.874 to 11.95	0 to 20.872
°C	0 to 1,000°	0 to 760°	0 to 500°	0 to 1,300°	250 to 1,200°	0 to 400°
°F	32 to 1,832°	32 to 1,400°	32 to 932°	32 to 2,372°	482 to 2,192°	32 to 752°
α_0	0	0	0	0	1.291507177E + 01	0
α_1	1.7057035E + 01	1.978425E + 01	2.508355E + 01	3.8783277E + 01	1.466298863E + 02	2.592800E + 01
α_2	−2.3301759E − 01	−2.001204E − 01	7.860106E − 02	−1.1612344E + 00	−1.534713402E + 01	−7.602961E − 01
α_3	6.5435585E − 03	1.036969E − 02	−2.503131E − 01	6.9525655E − 02	3.145945973E + 00	4.637791E − 02
α_4	−7.3562749E − 05	−2.549687E − 04	8.315270E − 02	−3.0090077E − 03	−4.163257839E − 01	−2.165394E − 03
α_5	−1.7896001E − 06	3.585153E − 06	−1.228034E − 02	8.8311584E − 05	3.187963771E − 02	6.048144E − 05
α_6	8.4036165E − 08	−5.344285E − 08	9.804036E − 04	−1.6213839E − 06	−1.291637500E − 03	−7.293422E − 07
α_7	−1.3735879E − 09	5.099890E − 10	−4.413030E − 05	1.6693362E − 08	2.183475087E − 05	
α_8	1.0629823E − 11		1.057734E − 06	−7.3117540E − 11	−1.447379511E − 07	
α_9	−3.2447087E − 14		−1.052755E − 08		8.211272125E − 09	
Maximum deviation	± 0.02°C	± 0.04°C	−0.05 to + 0.04°C	± 0.06°C	± 0.01°C	± 0.03°C

* $T(°C) = \sum_{i=0}^{n} a_i \times (mV)^i$.

All temperatures are ITS-1990 and all voltages are SI-1990 values. Maximum deviation is that from the ITS-1990 tables; thermocouple wire error is additional. Computed temperature deviates greatly outside of given ranges. Consult source for thermocouple types B and R and for other millivolt ranges.

SOURCE: NIST Monograph 175, April 1993.

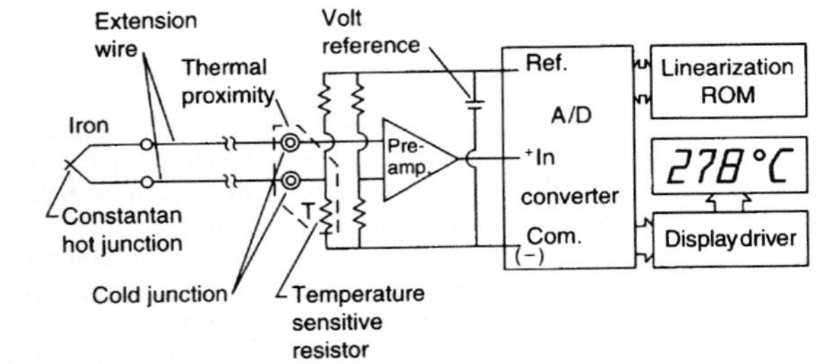

Figure 10.1.25 Temperature measurement with thermocouple and digital millivoltmeter.

to lead wire temperature changes. The three-wire connection, preferred in industrial applications, eliminates the lead wire effect provided the leads are of the same gage and length, and subject to the same environment. The four-wire arrangement makes no demands on the lead wires and is preferred for scientific measurements.

The resistance bulb consists of a copper or platinum wire coil sealed in a protective metal tube. The **thermistor** has a very large temperature coefficient of resistance and may be substituted in low-accuracy, low-cost applications.

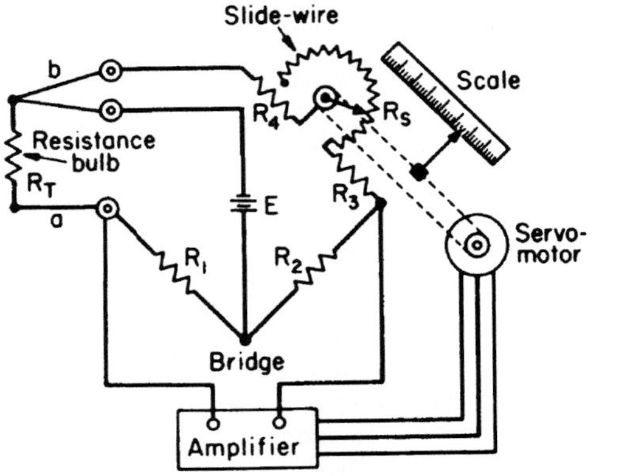

Figure 10.1.26 Three-wire resistance thermometer with self-balancing potentiometer recorder.

By use of a **selector switch,** any number of temperatures may be measured with the same instrument. The switch connects in order each thermocouple (or resistance bulb) to the potentiometer (or bridge circuit) or digital voltmeter. When balance is achieved, the recorder prints the temperature value, then the switch advances on to the next position.

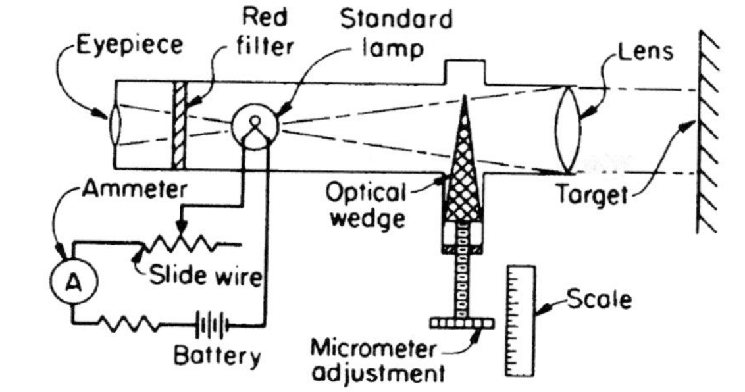

Figure 10.1.27 Optical pyrometer.

Optical pyrometers are applied to high-temperature measurement in the range 1,000 to 5,000°F (540 to 2,760°C). One type is shown in Fig. 10.1.27. The surface whose temperature is to be measured (target) is focused by the lens onto the filament of a calibrated tungsten lamp. The light intensity of the filament is kept constant by maintaining a constant current flow. The intensity of the target image is adjusted by positioning the optical wedge until the image intensity appears exactly equal to that of the filament. A scale attached to the wedge is calibrated directly in temperature. The red filter is employed so that the comparison is made at a specific wavelength (color) of light to make the calibration more reproducible. In another type of optical pyrometer, comparison is made by adjusting the current through the filament of the standard lamp. Here, an ammeter in series is calibrated to read temperature directly. **Automatic operation** may be had by comparing filament with image intensities with a pair of photoelectric cells arranged in a bridge network. A difference in intensity produces a voltage, which is amplified to drive the slide wire or optical wedge in the direction to restore zero difference.

The **radiation pyrometer** is normally applied to temperature measurements above 1,000°F. Basically, there is no upper limit; however, the lower limit is determined by the sensitivity and cold-junction compensation of the instrument. It has been used down to almost room temperature. A common type of radiation receiver is shown in Fig. 10.1.28. A lens focuses the radiation onto a thermal sensing element. The temperature rise of this element depends on the total radiation received and the conduction of heat away from the element. The radiation relates to the temperature of the target; the conduction depends on the temperature of the pyrometer housing. In normal applications the latter factor is not very great; however, for improved accuracy a compensating coil is added to the circuit. The sensing element may be a thermopile, vacuum thermocouple, or bolometer. The **thermopile** consists of a number of thermocouples connected in series, arranged so that all the hot junctions lie in the field of the incoming radiation; all of the cold junctions are in thermal contact with the pyrometer housing so that they remain at ambient temperature. The **vacuum thermocouple** is a single thermocouple whose hot junction is enclosed in an evacuated glass envelope. The **bolometer** consists of a very thin strip of blackened nickel or platinum foil which responds to temperature in the same manner as the resistance thermometer. The **thermal sensing element** is connected to a potentiometer or bridge network of the same type as described for the self-balance thermocouple and resistance-thermometer instruments. Because of the nature of the radiation law, the scale is nonlinear.

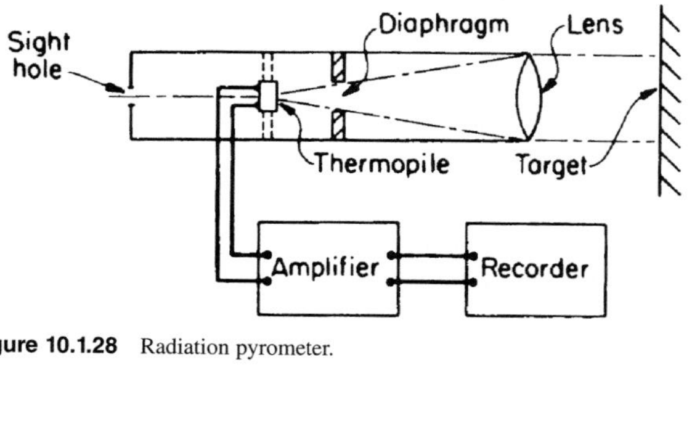

Figure 10.1.28 Radiation pyrometer.

Accuracy of the optical- and radiation-type pyrometers depends on:

1. *Emissivity of the surface being sighted on.* For closed furnace applications, blackbody conditions can be assumed (emissivity = 1). For other applications corrections for the actual emissivity of the surface must be made (correction tables are available for each pyrometer model). Multiple color or wavelength sensing is used to reduce sensitivity to hot object emissivity. For measuring hot fluids, a target tube immersed in the fluid provides a target of known emissivity.
2. *Radiation absorption between target and instrument.* Smoke, gases, and glass lenses absorb some of the radiation and reduce the incoming signal. Use of an enclosed (or purged) target tube or direct calibration will correct this.
3. *Focusing of the target on the sensing element.*

MEASUREMENT OF FLUID FLOW RATE

(See also Secs. 1 and 2.)

Flow is expressed in volumetric or mass units per unit time. Thus gases are generally measured in ft³/min (m³/min) or ft³/h (m³/h), steam in lb/h (kg/h), and liquid in gal/min (L/min) or gal/h (L/h). Conversion between volumetric flow Q and mass flow m is given by $m = K\rho Q$, where ρ = density of the fluid and K is a constant depending on the units of m, Q, and ρ. Flow rate can be measured directly by attaching a rate device to a volumetric meter of the types previously described, for example, a tachometer connected to the rotating shaft of the nutating-disk meter (Fig. 10.1.13).

There are many kinds of flow instruments that serve in different applications. Depending on the application, the engineer may choose from a dozen different technologies. Issues, such as accuracy, repeatability, and cost, as well as suitability for the service, make selecting a flowmeter a demanding exercise. The basic flowmeter technologies include:

Positive displacement
Differential pressure
Variable area and rotameter
Turbine
Electromagnetic
Ultrasonic
Coriolis mass flowmeter
Vortex shedding and other fluidic
Thermal dispersion
Open-channel primary elements

Because of their use in domestic water systems worldwide, positive displacement meters (nutating disk, rotary piston, and multijet) are the most widely used method of measurement of flow rate and total. They are, as the name implies, a means of measuring volumetric flow rate directly, one small volume at a time. Positive displacement meters are limited in the materials of construction, pressure, amount of air entrained, and the amount and size of solids they can pass.

They must be compensated for temperature and viscosity changes. Typically, they are used for totalization of clean fluids from water to petroleum byproducts. Differential pressure flowmeters operate by the conversion of fluid velocity to pressure (head). Thus, if the fluid is forced to change its velocity from V_1 to V_2, its pressure will change from p_1 to p_2 according to the equation (neglecting friction, expansion, and turbulence effects):

$$V_2^2 - V_1^2 = \frac{2g_c}{\rho}(p_1 - p_2) \tag{10.1.1}$$

where g = acceleration due to gravity, ρ = fluid density, and $g_c = 32.184$ lbm · ft/(lbf)(s²) [1.0 kg · m/(N)(s²)]. **Caution:** If the flow pulsates, the average value of $p_1 - p_2$ will be greater than that for steady flow of the same average flow.

See "ASME Pipeline Flowmeters," and "Pitot Tubes" in Sec. 1 for coverage of venturi tubes, flow nozzles, compressible flow, orifice meters, ASME orifices, and Pitot tubes.

The tabulation orifice coefficients apply only for **straight pipe** upstream and downstream from the orifice. In most cases, satisfactory results are obtained if there are no fittings closer than 25 pipe diameters upstream and 5 diameters downstream from the orifice. The upstream limitation can be reduced a bit by employing **straightening vanes.** Reciprocating pumps in the line may introduce serious errors and require special efforts for their correction.

A wide variety of differential pressure meters is available for measuring the orifice (or other primary element) pressure drop.

Figure 10.1.29 shows an orifice plate and differential-pressure flow transmitter. An electronic device is shown, but in relatively old plants, and in some very special cases in the refinery industry, pneumatic devices still are present. The differential across the orifice plate produces a pressure differential on either side of the diaphragm, thus causing an electrical signal to be transmitted. Modem differential-pressure flow transmitters generally include a microprocessor, which permits temperature compensation, pressure compensation, and the conversion of the differential pressure (the square root of flow) to a linear flow rate. The absolute accuracy of a differential pressure flow device is fixed by the accuracy of the primary device.

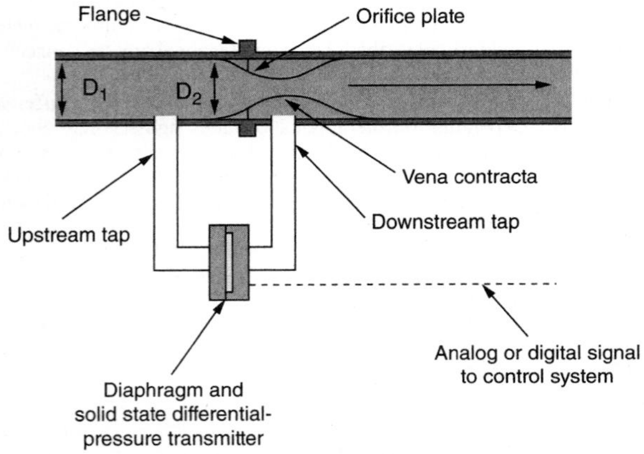

Figure 10.1.29 Orifice plate and differential-pressure transmitter.

The meters described thus far are termed **variable-head** because the pressure drop varies with the flow, orifice ratio being fixed. In contrast, the **variable-area** meter maintains a constant pressure differential but varies the orifice area with flow.

The **rotameter** (Fig. 10.1.30) consists of a float positioned inside a tapered tube by action of the fluid flowing up through the tube. The flow restriction is now the annular area between the float and the tube (area increases as the float rises). The pressure differential is fixed, determined by the weight of the float and the buoyant forces. To satisfy the volumetric flow equation then, the annular area (hence the float level) must increase with flow rate. Thus the rotameter may be calibrated for direct flow reading by etching an appropriate scale on the surface of the glass tube. The calibration depends on the float dimensions, tube taper, and fluid properties. The equation for volumetric flow is

$$Q = C_R(A_T - A_F)\left[\frac{2gV_F}{\rho A_F}(\rho_F - \rho)\right]^{1/2}$$

where A_T = cross-sectional area of tube (at float position), A_F = effective float area, V_F = float volume, ρ_F = float density, ρ = fluid density, and

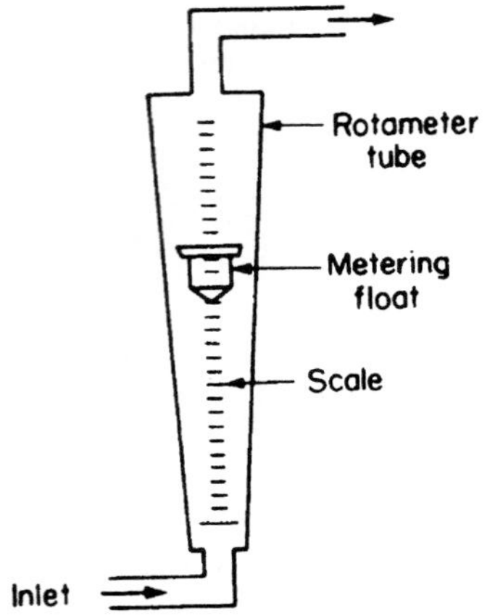

Figure 10.1.30 Rotameter.

C_R = rotameter coefficient (usually between 0.6 and 0.8). The coefficient varies with the fluid viscosity; however, special float designs are available which are relatively insensitive to viscosity effects. Also, fluid density compensation can be obtained.

The **rotameter reading may be transmitted** for recording and control purposes by affixing to the float a stem which connects to an armature or permanent magnet. The armature forms part of an inductance bridge whose signal is amplified electronically and converted to an electronic or digital signal.

The **area meter** is similar to the rotameter in operation. Flow area is varied by motion of a piston in a straight cylinder with openings cut into the wall. The piston position is transmitted as above by an armature and inductance bridge circuit.

Turbine, propeller, and paddlewheel flowmeters operate by converting velocity into rotational energy by means of a flighted rotor. These flowmeters are primarily used to measure water, and water-based liquids, and come in a very wide variety of designs, styles, and materials of construction. Some of these devices are mechanical, using gears and cams to convert the rotation of the rotor to increment an electromechanical counter. Others use solid-state pickups such as magnetic induction coils and Hall effect sensors to sense the speed of rotation of the rotor (Fig. 10.1.31).

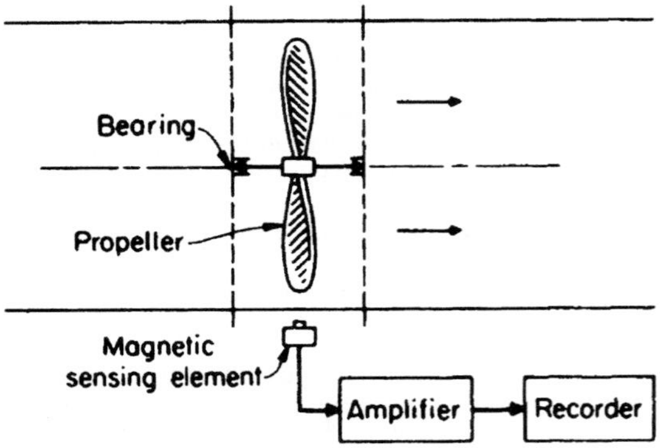

Figure 10.1.31 Propeller-type flowmeter.

These flowmeters are limited by the velocity of the fluid (they can be overdriven, with disastrous results), by the size and composition of solids they can pass, and by temperature and viscosity changes, and they are subject to significant wear in operation. Some turbine meters can be calibrated to be very highly accurate and are often used to meter expensive additives and blends.

One of the most widely used flowmeter types is the electromagnetic flowmeter, often called a **magmeter.** The device operates by means of Faraday's law, converting velocity of a conductive fluid to a proportional current signal by passing the fluid at right angles to an electromagnetic field. Magmeters are inherently the most accurate of the velocimetric flow technologies, because the current signal is directly proportional to the average velocity of the conductive fluid. Magmeters are therefore restricted to conductive fluids (generally above 5 microsiemens). Magmeters are supplied in a very wide range of sizes (generally from 1 to 2,000 mm), and the larger meters are among the most economical devices. Magmeters also have the advantage that they can be made out of a wide variety of materials, including ones suitable for ultrapure service and highly corrosive or abrasive service as well. Their final large advantage over other flowmeter types is that they are obstructionless, having no moving parts in the flow stream, and therefore do not increase the pressure drop to the system over the pressure drop of an equivalent length of pipe. Over the past decade, several manufacturers have released magmeters entirely operated by battery, with, in some cases, integral solar charging circuits. This has permitted the use of magmeters in applications traditionally reserved for electromechanical or mechanical flow measurement devices such as propeller or turbine meters.

Four different types of flowmeter using ultrasonic technology are in common use. These are Doppler flowmeters, transit time flowmeters, sonar flowmeters, and ultrasonic level devices used in conjunction with open-channel primary flow devices such as weirs and flumes.

Doppler flowmeters operate by means of the Doppler effect. A signal of known frequency (typically 660 kHz) is beamed into a pipe. The return echo is frequency-shifted proportionally to the velocity of the fluid in the pipe. Because homogenous fluids generally do not reflect the signal well, Doppler flowmeters are restricted to fluids with relatively large quantities of entrained particulate (usually a minimum of 100 mg/L of plus-100-μm solids). They are commonly used in mining, dredging, and wastewater sludge applications.

Transit time flowmeters operate by beaming an ultrasonic signal into the pipe in the direction of flow, and a duplicate signal downstream. The difference in time for the transmitted signal to reach the receiver between the upstream and downstream signals is proportional to the velocity of the fluid. The difference in transit time is $2vl/(c^2 + v^2)$, where l is the path length. For $v \ll c$, the factor $(c^2 + v^2)$ is nearly constant. Like magmeters, transit time flowmeters have no pressure drop. Transit time flowmeters can be provided in similar sizes to magmeters, and are used in all types of fluids, except slurries. Transit time flowmeters can be used in gas flow measurement. Increasing the number of paths of signal transmission can increase the accuracy of the flowmeter, and transit time flowmeters are increasingly used for transmission and custody transfer of petroleum products and gas.

Sonar flowmeters are acoustic, rather than strictly ultrasonic, and operate by "listening" to the sonic properties of the fluid and determining velocity from the sonic profile.

Ultrasonic open-channel flowmeters are actually ultrasonic level transmitters used at the measurement point of a flume or weir, and supplied with electronic conversion from measured height to flow rate. Because of their noncontacting nature, ultrasonic open-channel flowmeters have become the standard technology for this measurement, replacing floats and other mechanical devices. The limiting accuracy of these devices is not the accuracy of the sensor, but rather the accuracy of the primary device.

Coriolis mass flowmeters are now common in many process industries. Generally available from fractional inch to 4 in (100 mm), at least one manufacturer produces up to 12-in sizes. Using the principle of measurement of the Coriolis force, these devices natively measure mass flow, and can be used to compute volumetric flow rate and total, and fluid density. Recent advances in Coriolis mass flowmeter design have permitted the device to also be used for measurement of gases.

Another extremely common measurement devices is the **fluidic flowmeter**, of which the best known examples are the vortex-shedding, the vortex-precession (or swirlmeter), and the Coanda effect flowmeter. The **vortex-shedding flowmeter** operates by sensing in a variety of ways the vortices shed by a bluff body in the flow stream. The **swirlmeter,** an ancestor of the vortex-shedder, uses the combination of a stationary turbine rotor to impart a twist and a flow restriction to produce a standing wave, the frequency of which is proportional to velocity. Swirlmeters have received a new lease on life in several applications, especially steam flow and natural gas transmission (especially in China). The **Coanda effect flowmeter** uses the Coanda effect to produce a hydraulic switch. The frequency of the switching operation is proportional to the velocity of the fluid. Vortex-shedding flowmeters and swirlmeters are commonly used in chemical and steam applications, while Coanda effect flowmeters are normally found in district heating and gas applications in relatively small sizes. Vortex-shedding flowmeters are limited in turndown at low velocities, and should not be used in applications with large-diameter solids such as wastewater or mining.

Thermal dispersion flowmeters operate by measuring the temperature differential caused by the passage of a fluid between two sensors. One sensor is heated to a constant temperature, and imparts heat to the fluid. The other sensor measures the temperature of the fluid. The temperature differential produces a power demand on the heater circuit that is proportional to the mass flow of the fluid. Thermal dispersion flowmeters are commonly used in gas flow measurement, but can also be used in some liquid flow measurements, generally as switches. These devices can be extremely accurate, and at least one manufacturer produces ultralow-flow thermal dispersion flowmeters that measure in the milliliter per hour range.

The most common open-channel primary flow elements in current usage are the **v-notch** and rectangular **weirs,** and the **Parshall** and **Palmer-Bowlus flumes.** All of these devices produce a flow restriction that raises the fluid behind the restriction to a predictable height. The change in height H produces a measurement of flow rate by the formula $Q = K/n$, where K is a constant related to flow rate units and n is a constant related to the design of the specific primary element. The constant n is often written as a fraction, as in $^3/_2$. Other forms of primary elements for open-channel flow exist, and sometimes the **Manning formula** and its derivatives are used to measure the flow in open conduits without a primary device. Manning's formula is: $v = K/n\ R^{2/3}\ S^{1/2}$ (English units) [SI units], where v = fluid velocity (ft/s) or [m/s], K = 1.486 (English), K = 1.00 [SI], R = hydraulic radius (ft) or [m] = A/P_w, A = cross-sectional area of flow (ft^2) or [m^2], P_w = wetted perimeter (ft) or [m], S = channel slope, and n = Manning's roughness coefficient. Then $Q = vA$.

Most flowmeters measure rate of flow. To measure the total quantity of fluid flowing during a specific time interval, it is necessary to integrate the flow rate over that interval. The integration, prior to the advent of microprocessor-based flow transmitters, was often done manually, or with the use of specially shaped cams. All flow transmitters now integrate mathematically, and display total flow on an electronic or electromechanical register.

POWER MEASUREMENT

Power is defined as the rate of doing work. Common units are the horsepower and the kilowatt: 1 hp = 33,000 ft · lb/min = 0.746 kW. The power input to a rotating machine in hp (W) = $2\pi nT/k$, where n = r/min of the shaft where the torque T is measured in lbf · ft (N · m), and k = 33,000 ft · lbf/hp · min [60 N · m/(W · min)]. The same equation applies to the power output of an engine or motor, where n and T refer to the output shaft. Mechanical power-measuring devices **(dynamometers)** are of two types: (1) those absorbing the power and dissipating it as heat and (2) those transmitting the measured power. As indicated by the above equation, two measurements are involved: shaft speed and torque. The speed is measured directly by means of a tachometer. Torque is usually measured by balancing against weights applied to a fixed lever arm; however, other force measuring methods are also used. In the **transmission dynamometer,** the torque is measured by means of strain-gage elements bonded to the transmission shaft.

There are several kinds of **absorption dynamometers.** The **Prony brake** applies a friction load to the output shaft by means of wood blocks, flexible band, or other friction surface. The **fan brake** absorbs power by "fan" action of rotating plates on surrounding air. The **water brake** acts as an inefficient centrifugal pump to convert mechanical energy into heat. The pump casing is mounted on antifriction bearings so that the developed turning moment can be measured. In the **magnetic-drag** or **eddy-current brake,** rotation of a metal disk in a magnetic field induces eddy currents in the disk which dissipate as heat. The field assembly is mounted in bearings in order to measure the torque.

One type of **Prony brake** is illustrated in Fig. 10.1.32. The torque developed is given by $L(W - W_0)$, where L is the length of the brake arm, ft; W and W_0 are the scale loads with the brake operating and with the brake free, respectively. The brake horsepower then equals $2\pi nL(W - W_0)/33{,}000$, where n is shaft speed, r/min.

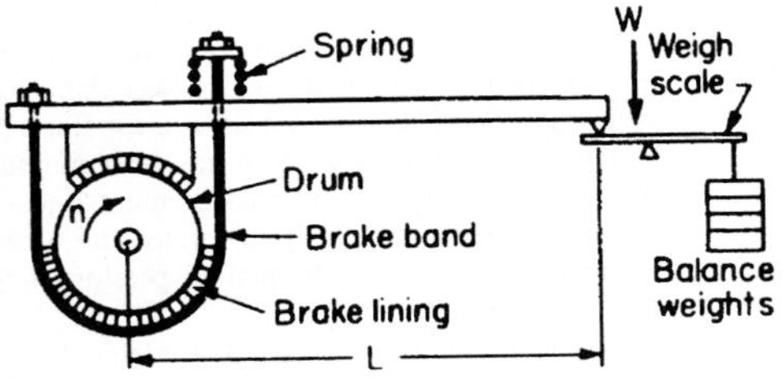

Figure 10.1.32 Prony brake.

In addition to eddy-current brakes, **electric dynamometers** include calibrated generators and motors and cradle-mounted generators and motors. In calibrated machines, the efficiency is determined over a range of operating conditions and plotted. Mechanical power measurement can then be made by measuring the electrical power input (or output) to the machine. In the electric-cradle dynamometer, the motor or generator stator is mounted in trannion bearing so that the torque can be measured by suitable scales.

The **engine indicator** is a device for plotting cylinder pressure as a function of piston (or volume) displacement. The resulting *p-v* diagram (Fig. 10.1.33) provides both a measure of the work done in a reciprocating engine, pump, or compressor and a means for analyzing its performance (see Secs. 2, 7, and 10). If A_d is the area inside the closed curve drawn by the indicator, then the indicated horsepower for the cylinder under test = KnA_pA_d, where K is a proportionality factor determined by the scale factors of the indicator diagram, n = engine speed, r/min, A_p = piston area.

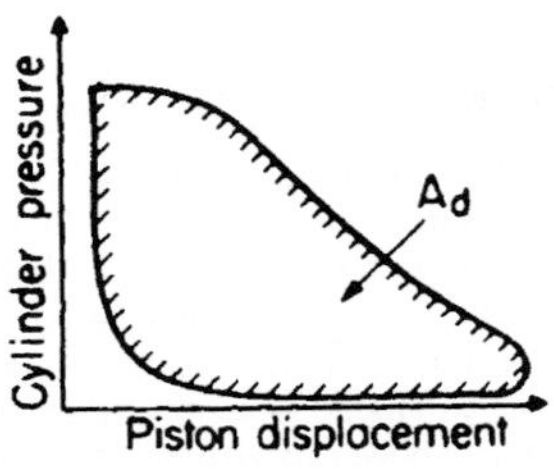

Figure 10.1.33 Indicator diagram.

Completely mechanical indicators can be used only for low-speed machines. They have largely been superseded by electrical transducers using strain gages, variable capacitance, piezoresistive, and piezoelectric principles which are suitable for **high-speed** as well as low-speed pressure changes (the piezoelectric principle has low-speed limitations). The usual diagram is produced on an oscilloscope display as pressure versus time, with a marker to indicate some reference event such as spark timing or top dead center. Special transducers can be coupled to a crank or cam shaft to give an electrical signal representing piston motion so that a *p-v* diagram can be shown on an oscilloscope.

ELECTRICAL MEASUREMENTS

(See also Sec. 11.)

Electrical measurements serve two purposes: (1) to measure the electrical quantities themselves, for example, line voltage, power consumption, and (2) to measure other physical quantities which have been converted into electrical variables, for example, temperature measurement in terms of thermocouple voltage.

In general, there is a sharp distinction between AC and DC devices used in measurements. Consequently, it is often desirable to transform an AC signal to an equivalent DC value, and vice versa. An AC signal is converted to DC (rectified) by use of **selenium rectifiers, silicon or germanium diodes, or electron-tube diodes.** Full-wave rectification is accomplished by the **diode bridge,** shown in Fig. 10.1.34. The rectified signal may be passed through one or more low-pass filter stages to smooth the waveform to its average value. Similarly, there are many ways of modulating a DC signal (converting it to alternating current). The most common method used in instrument applications is a solid-state oscillator.

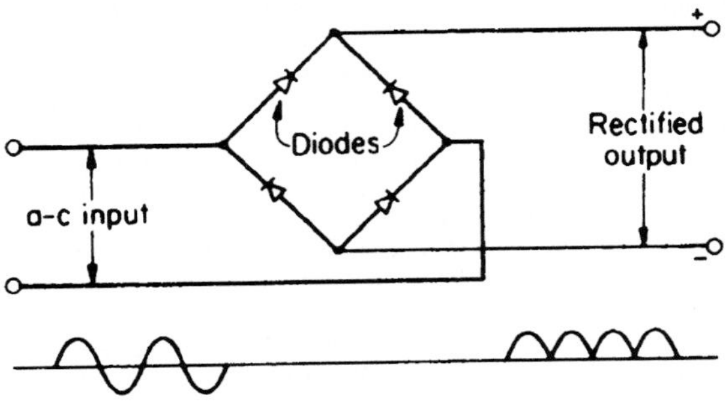

Figure 10.1.34 Full-wave rectifier.

The **galvanometer** (Fig. 10.1.35), recently supplanted by the direct-reading **digital voltmeter** (DVM), is basic to DC measurement. The input signal is applied across a coil mounted in jeweled bearings or on

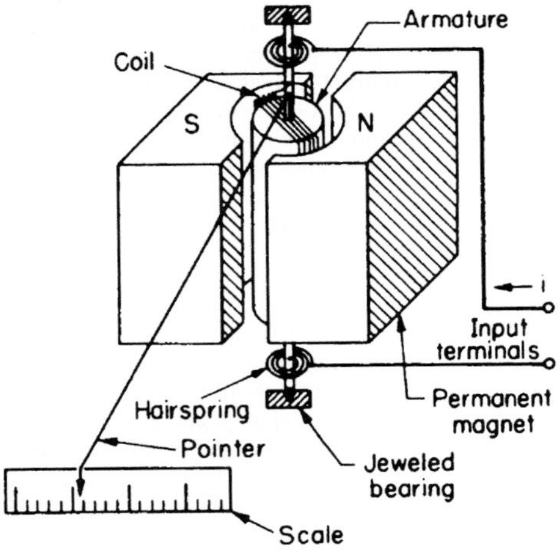

Figure 10.1.35 D'Arsonval galvanometer.

a taut-band suspension so that it is free to rotate between the poles of a permanent magnet. Current in the coil produces a magnetic moment which tends to rotate the coil. The rotation is limited, however, by the restraining torque of the hairsprings. The resulting deflection of the coil θ is proportional to the current I:

$$\theta = \frac{NBWL}{K} I$$

where N = number of turns in coil; W, L = coil width and length, respectively; B = magnetic field intensity; K = spring constant of the hairsprings. Galvanometer deflection is indicated by a balanced pointer attached to the coil. In very sensitive elements, the pointer is replaced by a mirror reflecting a spot of light onto a ground-glass scale; the bearings and hairspring are replaced by a torsion-wire suspension.

The galvanometer can be converted into a **DC voltmeter, ammeter, or ohmmeter** by application of Ohm's law, $IR = E$, where I = current, A; E = electrical potential, V; and R = resistance, Ω.

For a **voltmeter,** a fixed resistance R is placed in series with the galvanometer (Fig. 10.1.36*a*). The current i through the galvanometer is proportional to the applied voltage E: $i = E/(r + R)$, where r = coil resistance. Different voltage ranges are obtained by changing the series resistance.

An **ammeter** is produced by placing the resistance in parallel with the galvanometer or DVM (Fig. 10.1.36*b*). The current then divides between the galvanometer coil or DVM and the resistor in inverse ratio to their resistance values (r and R, respectively); thus, $i = IR/(r + R)$, where i = current through coil and I = total current to be measured. Different current ranges are obtained by using different shunt resistances.

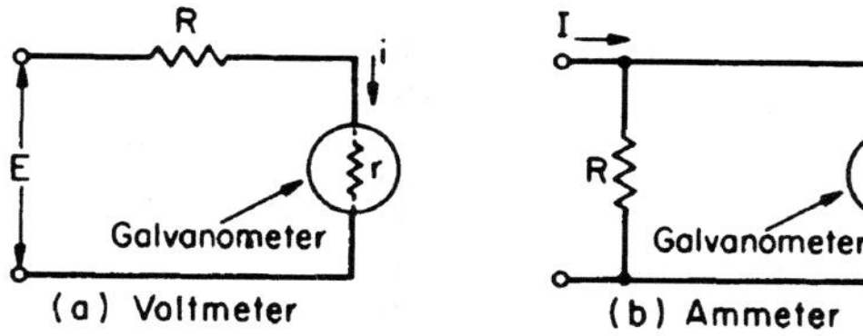

Figure 10.1.36 (*a*) Voltmeter; (*b*) ammeter.

The common ohmmeter consists of a battery, a galvanometer with a shunt rheostat, and resistance in series to total Ri Ω. The shunt is adjusted to give a full-scale (0 Ω) reading with the test terminals shorted. When an unknown resistance R is connected, the deflection is $Ri/(Ri + R)$ fraction of full scale. The scale is calibrated to read R directly. A half-scale deflection indicates $R = Ri$. Alternatively, the galvanometer is connected to read voltage drop across the unknown R while a known current flows through it. This principle is used for low-value resistances and in digital ohmmeters.

Digital instruments are available for all these applications and often offer higher resolution and accuracy with less circuit loading. Early digital instruments had fluctuating readings that were difficult to follow, however. This has been replaced with auto-scaling and auto-damping functions that display only the significant digits, on an adjustable time base to stabilize digital readings.

Alternating current and voltage must be measured by special means. A DC instrument with a rectifier input is commonly used in applications requiring high input impedance and wide frequency range. For precise measurement at power-line frequencies, the electrodynamic instrument is used. This is similar to the galvanometer except that the permanent magnet is replaced by an electromagnet. The movable coil and field coils are connected in series; hence they respond simultaneously to the same current and voltage alternations. The pointer deflection is proportional to the square of the input signal. The moving-iron-type instruments consist of a soft-iron vane or armature which moves in response to current flowing through a stationary coil. The pointer is attached to the iron to indicate the deflection on a calibrated nonlinear scale. For measuring at very high frequencies, the **thermocouple voltmeter or ammeter** is used. This is based on the heating effect of the current passing through a fixed resistance R. Heat is liberated at the rate of E^2/R or $I^2 R$ W.

DC **electrical power** is the product of the current through the load and the voltage across the load. Thus it can be simply measured using a voltmeter and ammeter. AC power is directly indicated by the wattmeter, which is similar to the electrodynamic instrument described above. Here the field coils are connected in series with the load, and the movable coil is connected across the load (to measure its voltage). The deflection of the movable coil is then proportional to the effective load power.

Precise voltage measurement (direct current) can be made by balancing the unknown voltage against a measured fraction of a known reference voltage with a **potentiometer** (Fig. 10.1.9). Balance is indicated by means of a sensitive current detector placed in series with the unknown voltage. The potentiometer is calibrated for angular position versus fractional voltage output. Accuracies to 0.05 percent are attainable, dependent on the linearity of the potentiometer and the accuracy of the reference source. The reference standard may be a Weston standard cell or a regulated voltage supply (based on diode characteristics). The balance detector may be a galvanometer or electronic amplifier.

Precision resistance and general impedance measurements are made with bridge circuits (Fig. 10.1.37) which are adjusted until no signal is detected by the null detector (bridge is balanced). Then $Z_1Z_3 = Z_2Z_4$. The basic Wheatstone bridge is used for resistance measurement where all the impedances (Z's) are resistances (R's). If R_1 is to be measured, $R_1 = R_2R_4/R_3$, when balanced. A sensitive galvanometer for the null detector and DC voltage excitation is usual. All R_2, R_3, R_4 must be calibrated, and some adjustable. For general impedance measurement, AC voltage excitation of suitable frequency is used. The null detector may be a sensitive AC meter, oscilloscope, or, for audio frequencies, simple earphones. The basic balance equation is still valid, but it now requires also that the sum of the phase angles of Z_1 and Z_3 equal the sum of the phase angles of Z_2 and Z_4. As an example, if Z_1 is a capacitor, the bridge can be balanced if Z_2 is a known capacitor while Z_3 and Z_4 are resistances. The phase-angle condition is met, and $Z_1Z_3 = Z_2Z_4$ becomes $(1/2\pi fC_1)R_3 = (1/2\pi fC_2)R_4$ and $C_1 = C_2R_3/R_4$. Variations on the basic principle include the Kelvin bridge for measurement of low resistance, and the Mueller bridge for platinum resistance thermometers.

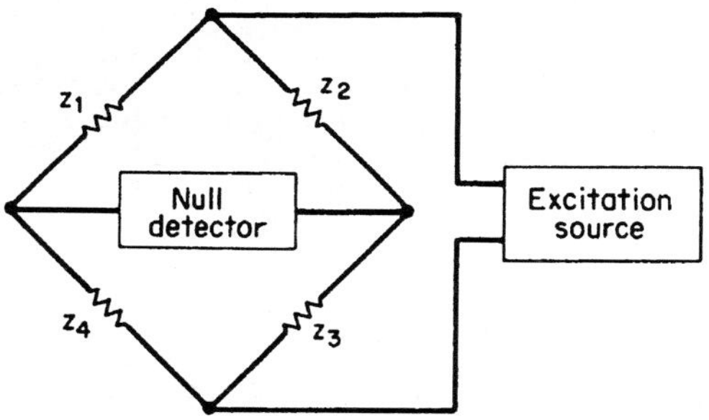

Figure 10.1.37 Impedance bridge.

Voltage measurement requires a meter of substantially higher impedance than the impedance of the source being measured. The vacuumtube cathode follower and the field-effect transistor are suitable for high-impedance inputs. The following circuitry may be a simple amplifier to drive a pointer-type meter, or may use a digital technique to produce a digital output and display. Digital counting circuits are capable of great precision and are widely adapted to measurements of time, frequency, voltage, and resistance. Transducers are available to convert temperature, pressure, flow, length and other variables into signals suitable for these instruments.

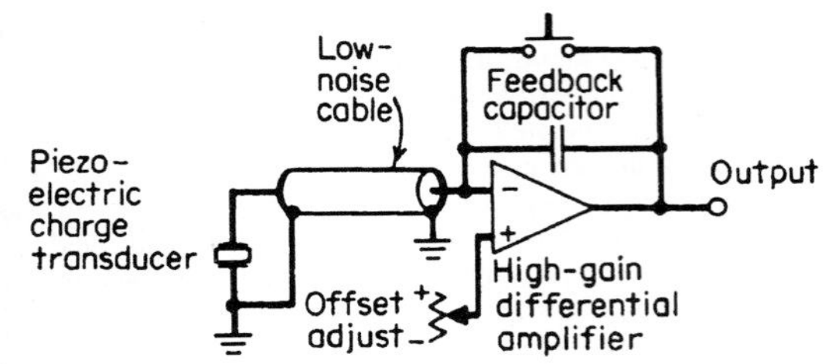

Figure 10.1.38 Charge amplifier application.

The charge amplifier is an example of an operational amplifier application (Fig. 10.1.38). It is used for outputs of piezoelectric transducers in which the output is a charge proportional to input force or other input converted to a force. Several capacitors switchable across the feedback path provide a range of full-scale values. The output is a voltage.

The **digital oscilloscope** (Fig. 10.1.39) is an extremely useful and versatile device characterized by high input impedance and wide frequency range. An embedded computer reads voltage, current, or resistance inputs and displays output on a liquid crystal display (LCD) which is either monochrome, or, in modern scopes, multicolor. The modern digital oscilloscope mimics the analog.

Figure 10.1.39 Digital LCD after Crystalfontz, et al.

Digital LCD Oscilloscope

Oscilloscope, by displaying the input signal on a vertical and horizontal axis, with a digital representation of the former sweep voltage. Thus, the relationship between two varying voltages can be observed by applying them to the vertical and horizontal axes. The horizontal axis is commonly used for a linear time base generated by the embedded computer. Virtually any desired sweep speed is obtainable as a calibrated sweep. Sweeps which change value part way across the screen are available to provide localized time magnification. As an alternative to the time base, any arbitrary voltage can be applied to drive the horizontal axis. The vertical axis is usually used to display a dependent variable voltage. **Dual-beam** and **dual-trace** instruments show two waveforms simultaneously. Special long-persistence and storage screens can hold transient waveforms from seconds to hours. Modern devices are called **digital-storage** oscilloscopes. Each input signal is sampled, digitized, and stored in a first-in-first-out memory. Since a record of the recent signal is in memory when a trigger pulse is received, the timing of the end of storing new data into memory determines how much of the stored signal was before, and how much after, the trigger. Unlike storage screens, the stored signal can be amplified and shifted on the screen for detailed analysis, accompanied by numerical display of voltage and time for any point. Care must be taken that enough samples are taken in any waveform; otherwise aliasing results in a false view of the waveform.

The stored data can be processed mathematically in the oscilloscope or transferred to a computer for further study. Accessories for microcomputers allow them to function as digital oscilloscopes and other specialized tasks. The embedded computer in the oscilloscope can be used as a calibration device.

VELOCITY AND ACCELERATION MEASUREMENT

Velocity or speed is the time rate of change of displacement. Consequently, if the displacement measuring device provides an output signal which is a continuous (and smooth) function of time, the velocity can be measured by **differentiating** this signal either graphically or by use of a differentiating circuit. The accuracy may be very limited by noise (high-frequency fluctuations), however. More commonly, the output of an accelerometer is integrated to yield the velocity of the moving member. **Average speed** over a time interval can be determined by measuring the time required for the moving body to pass two fixed points a known distance apart. Here photoelectric or other rapid sensing devices may be used to trigger the start and stop of the timer. **Rotational speed** may be similarly measured by counting the number of rotations in a fixed time interval.

The **tachometer** provides a direct measure of angular velocity. One form is essentially a small permanent-magnet-type generator coupled to the rotating element; the voltage induced in the armature coil is directly proportional to the speed. The principle is also extended to rectilinear motions (restricted to small displacements) by using a straight coil moving in a fixed magnetic field.

Angular velocity can also be measured by magnetic drag-cup and **centrifugal-force** devices (flyball governor). The force may be balanced against a spring with the resulting deflection calibrated in terms of the shaft speed. Alternatively, the force may be balanced against the air pressure generated by a pneumatic nozzle-baffle assembly (similar to Fig. 10.1.23).

Vibration velocity pickups may use a coil which moves relative to a magnet. The voltage generated in the coil has the same frequency as the vibration and, for sine motion, a magnitude proportional to the product of vibration frequency and amplitude. Vibration **acceleration pickups** commonly use strain-gage, piezoresistive, or piezoelectric elements to sense a force $F = Ma/g_c$. The maximum usable frequency of an accelerometer is about one-fifth of the pickup's natural frequency (see Sec. 1.4). The minimum usable frequency depends on the type of pickup and the associated circuitry. The output of an accelerometer can be integrated to obtain a velocity signal; a velocity signal can be integrated to obtain a displacement signal. The **operational amplifier** is a versatile element which can be connected as an integrator for this use.

Holography is being applied to the study of surface vibration patterns.

MEASUREMENT OF PHYSICAL AND CHEMICAL PROPERTIES

Physical and chemical measurements are important in the control of product quality and composition. In the case of manufactured items, such properties as color, hardness, surface, roughness, etc., are of interest. Color is measured by means of a **colorimeter,** which provides comparison with color standards, or by means of a **spectrophotometer,** which analyzes the color spectrum. The **Brinell and Rockwell testers** measure surface hardness in terms of the depth of penetration of a hardened steel ball or special stylus. Testing machines with **strain-gage** elements provide measurement of the strength and elastic properties of materials. **Profilometers** are used to measure surface characteristics. In one type, the surface contour is magnified optically and the image projected onto a screen or viewer; in another, a stylus is employed to translate the surface irregularities into an electrical signal which may be recorded in the form of a highly magnified profile of the surface or presented as an averaged roughness-factor reading.

For liquids, attributes such as density, viscosity, melting point, boiling point, transparency, etc., are important. Density measurements have already been discussed. **Viscosity** is measured with a **viscosimeter,** of which there are three main types: flow through an orifice or capillary (Saybolt), viscous drag on a cylinder rotating in the fluid (MacMichael), damping of a vibrating reed (Ultrasonic) (see Secs. 1 and 2). **Plasticity** and consistency are related properties which are determined with special apparatus for heating or cooling the material and observing the temperature-time curve. The **photometer, reflectometer, and turbidimeter** are devices for measuring transparency or turbidity of nonopaque liquids and solids.

A variety of properties can be measured for determining chemical composition. **Electrical properties** include pH, conductivity, dielectric constant, oxidation potential, etc. **Physical properties** include density, refractive index, thermal conductivity, vapor pressure, melting and boiling points, etc. Of increasing industrial application are **spectroscopic measurements: infrared absorption spectra, ultraviolet and visible emission spectra, mass spectrometry,** and **gas chromatography.** These are specific to particular types of compounds and molecular configurations and hence are very powerful in the analysis of complex mixtures. As examples, infrared analyzers are in use to measure low-concentration contaminants in engine oils resulting from wear and in hydraulic oils to detect deterioration. **X-ray diffraction** has many applications in the analysis of crystalline solids, metals, and solid solutions.

Of special importance in the realm of composition measurements is the determination of **moisture content.** A common laboratory procedure measures the loss of weight of the oven-dried sample. More rapid methods employ electrical conductance or capacitance measurements, based on the relatively high conductivity and dielectric constant values for ordinary water.

Water vapor in air (humidity) is measured in terms of its physical properties or effects on materials (see also Sec. 4). (1) The **psychrometer** is based on the cooling effect of water evaporating into the airstream. It consists of two thermal elements exposed to a steady airflow; one is dry, the other is kept moist. See Sec. 2 for psychrometric charts. (2) The **dew-point recorder** measures the temperature at which water just starts to condense out of the air. (3) The **hygrometer** measures the change in length of such humidity-sensitive elements as hair and wood. (4) **Electric sensing elements** employ a wire-wound coil impregnated with a hygroscopic salt (one that maintains an equilibrium between its moisture content and the air humidity) such that the resistance of the coil is related to the humidity.

The **throttling calorimeter** (Fig. 10.1.40) is most commonly used for determining the moisture in steam. A sampling nozzle is located preferably in a vertical section of steampipe far removed from any fittings. Steam enters the calorimeter through a throttling orifice and into a well-insulated expansion chamber. The steam quality x (fraction dry steam) is determined from the equation $x = (h_c - h_f) / h_{fg}$, where h_c is the enthalpy of superheated steam at the temperature and pressure measured in the calorimeter; h_f and h_{fg} are, respectively, the liquid enthalpy and the heat of vaporization corresponding to line pressure. The chamber is conveniently exhausted to atmospheric pressure; then only line pressure and temperature of the throttled steam need be measured. The range of the throttling calorimeter is limited to small percentages of moisture; a **separating calorimeter** may be employed for larger moisture contents.

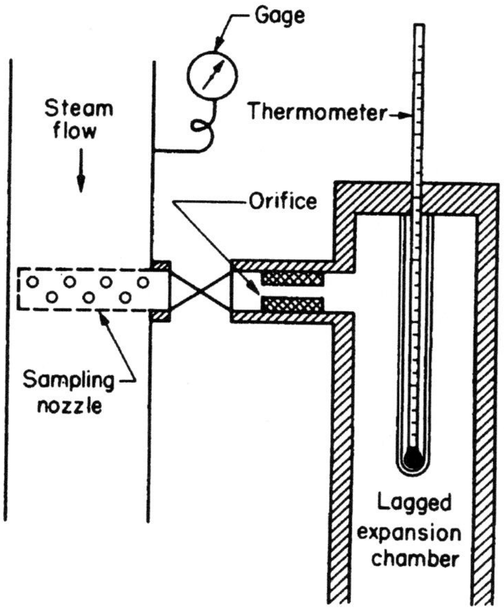

Figure 10.1.40 Throttling calorimeter.

The **Orsat** apparatus is generally used for chemical analysis of flue gases. It consists of a graduated tube or burette designed to receive and measure volumes of gas (at constant temperature). The gas is analyzed for CO_2, O_2, CO, and N_2 by bubbling through appropriate absorbing reagents and measuring the resulting change in volume. The reagents normally employed are KOH solution for CO_2, pyrogallic acid and KOH mixture for O_2, and cuprous chloride (Cu_2Cl_2) for CO. The final remaining unabsorbed gas is assumed to be N_2. The most common errors in the Orsat analysis are due to **leakage** and poor **sampling.** The former can be checked by simple test; the latter factor can only be minimized by careful sampling procedure. Recommended procedure is the taking of several simultaneous samples from different points in the cross-sectional area of the flue-gas stream, analyzing these separately, and averaging the results.

There are many instruments for **measuring CO_2 (and other gases) automatically.** In one type, the CO_2 is absorbed in KOH, and the change in volume determined automatically. The more common type, however, is based on the difference in **thermal conductivity** of CO_2 compared with air. Two thermal conductivity cells are set into opposing arms of a Wheatstone-bridge circuit. Air is sealed into one cell (reference), and the CO_2-containing gas is passed through the other. The cell contains an electrically heated resistance element; the temperature of the element (and therefore its resistance) depends on the thermal conductivity of the gaseous atmosphere. As a result, the unbalance of the bridge provides a measure of the CO_2 content of the gas sample.

The same principle can be employed for analyzing other constituents of gas mixtures where there is a significant thermal-conductivity difference. A modification of this principle is also used for determining CO or other combustible gases by mixing the gas sample with air or oxygen. The combustible gas then burns on the heated wire of the test cell, producing a temperature rise which is measured as above.

Many other physical properties are employed in the determination of specific components of gaseous mixtures. An interesting example is the **oxygen analyzer,** based on the unique paramagnetic properties of oxygen.

NUCLEAR RADIATION INSTRUMENTS

(See also Sec. 7.)

Nuclear radiation instrumentation was formerly of importance with two main areas of application: (1) measurement and control of radiation variables in nuclear reactor-based processes, such as nuclear power plants and (2) measurement of other physical variables based on radioactive excitation and tracer techniques. The instruments respond in general to electromagnetic radiation in the gamma and perhaps X-ray regions and to beta particles (electrons), neutrons, and alpha particles (helium nuclei). Unfortunately, the economic and political costs due to regulation and widespread fear of radiation have strictly limited the use of these devices in the industrial environment.

Gas-Ionization Tubes The **ion chamber, proportional counter,** and Geiger counter are common instruments for radiation detection and measurement. These are different applications of the gas-ionization tube distinguished primarily by the amount of applied voltage.

A simple and very common form of the instrument consists of a gas-filled cylinder with a fine wire along the axis forming the anode and the cylinder wall itself (at ground potential) forming the cathode, as shown in Fig. 10.1.41. When a radiation particle enters the tube, its collision with gas molecules causes an ionization consisting of electrons (negatively charged) and positive ions. The electrons move very rapidly toward the positively charged wire; the heavier positive ions move relatively slowly toward the cathode. The above activity is detected by the resulting current flow in the external circuitry.

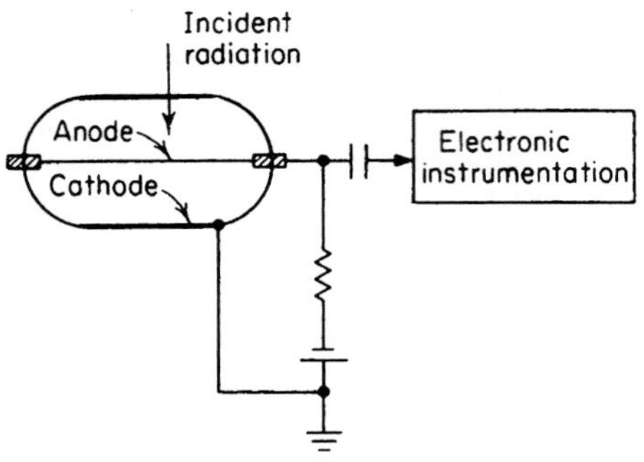

Figure 10.1.41 Gas-ionization tube.

When the voltage applied across the tube is relatively low, the number of electrons collected at the anode is essentially equal to that produced by the incident radiation. In this voltage range, the device is called an **ion chamber.** The device may be used to count the number of radiation particles when the frequency is low; when the frequency is high, an external integrating circuit yields an output current proportional to the radiation intensity. Since the amplification factor of the ion chamber is low, high-gain electronic amplification of the current signal is necessary.

If the applied voltage is increased, a point is reached where the radiation-produced ions have enough energy to collide with other gas molecules and produce more ions which also enter into collisions so that an "avalanche" of electrons is collected at the anode. Thus, there is a very considerable amplification of the output signal. In this region, the device is called a **proportional counter** and is characterized by the voltage or current pulse being proportional to the energy content of the incident radiation signal.

With still further increase in the applied voltage, a point of saturation is reached wherein the output pulses have a constant amplitude independent of the incident radiation level. The resulting **Geiger counter** is capable of producing output pulses up to 10 V in amplitude, thus greatly reducing the requirements on the external circuitry and instrumentation. This advantage is offset somewhat by a lower maximum counting rate and more limited ability to differentiate among the various types of radiation as compared with the proportional counter.

The **scintillation counter** is based on the excitation of a phosphor by incident radiation to produce light radiation which is in turn detected by a photomultiplier tube to yield an output voltage. The signal output is greatly amplified and nearly proportional to the energy of the initial radiation. The device may be applied to a wide range of radiations, it has a very fast response, and, by choice of phosphor material, it offers a large degree of flexibility in applications.

Applications to the Measurement of Physical Variables The ready availability of radioactive isotopes of long half-life, such as cobalt 60, make possible a variety of industrial and laboratory measuring techniques based on radiation instruments of the type described above. Most applications are based on (1) radiation absorption, (2) tracer identification, and (3) other properties. These techniques often have the advantages of isolation of the measuring device from the system, access to a variable not observable by conventional means, or measurement without destruction or modification of the system.

In the utilization of **absorption** properties, a radioactive source is separated from the radiation-measuring device by that part of the system to be measured. The measured radiation intensity will depend on the fraction of radiation absorbed, which in turn will depend on the distance traveled through the absorbing medium and the density and nature of the material. Thus, the instrument can be adapted to measuring thickness (see Fig. 10.1.12), coating weight, density, liquid or solids level, or concentration (of certain components).

Tracer techniques are effectively used in measuring flow rates or velocities, residence time distributions, and flow patterns. In flow measurement, a sharp pulse of radioactive material may be injected into the flow stream; with two detectors placed downstream from the injection point and a known distance apart, the velocity of the pulse is readily measured. Alternatively, if a known constant flow rate of tracer is injected into the flow stream, a measure of the radiation downstream is easily converted into a measure of the desired flow rate. Other applications of tracer techniques involve the use of tagged molecules embedded in the process to provide measures of wear, chemical reactions, etc.

Other applications of radiation phenomena include level measurements based on a floating radioactive source, level measurements based on the back-scattering effect of the medium, pressure measurements in the high-vacuum region based on the amount of ionization caused by alpha rays, location of interface in pipeline transmission applications, and certain chemical analysis applications.

INDICATING, RECORDING, AND LOGGING

Rapid change has overtaken the ways in which process information is presented and stored. The control panel of the 1970s gave way to digital indicators, dataloggers, and recorder simulations that are entirely software based. Control panels in the 1980s-2000s are all digital, on single or multiple digital displays. Some industrial sectors continue the use of paper chart recorders, so typical strip and circular chart recorders are shown in Figs. 10.1.42 and 10.1.43.

These include the nuclear power, food, and pharmaceuticals industry sectors. However, even these last holdouts are now moving to more modem means of indicating and recording information about measured variables.

With the advent of liquid crystal displays (LCDs) that can display graphical elements, even field instruments now have digital displays, some of which mimic analog indicators with "dials" or "bargraphs" or other even more sophisticated displays.

Replacing recorders in most applications now are digital dataloggers with cathode-ray tube (CRT) or LCD displays that simulate the visual aspect of an analog trend recorder. These are designed with the same form factor as older mechanical recorders so that they can be inserted into precut holes in existing control panels.

Most HMI packages for control systems contain a software system called a "data historian," which provides for logging and display of large amounts of process measurement data. Data historian software is often used to provide data to advanced process control software packages and to enterprise software systems for quality control and supply chain integration.

Field dataloggers have become small and are operated by very low power internal batteries, and can even be inserted into, for example, a single carton of food for long-term monitoring. These dataloggers can

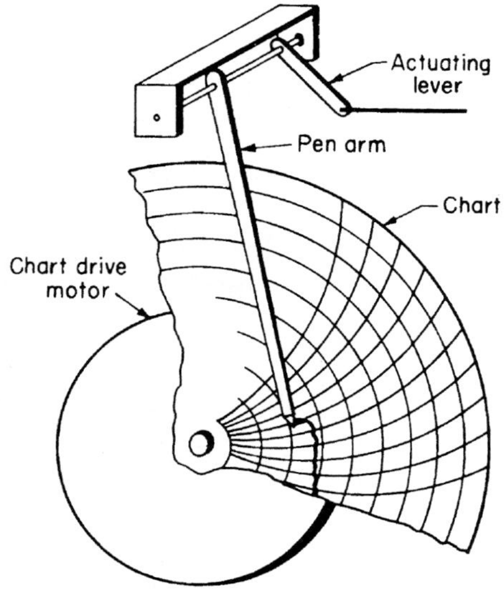

Figure 10.1.42 Circular-chart recorder.

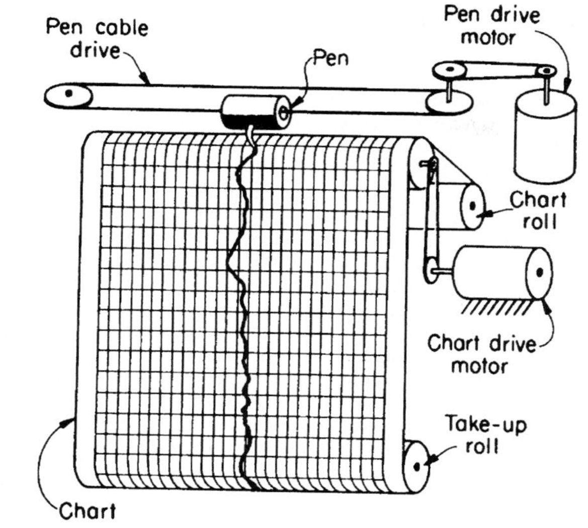

Figure 16.1.43 Strip-chart recorder.

be accessed by connection to a laptop, another type of host computer, or remotely by means of wireless connections such as Bluetooth, Zigbee, and the various IEEE 802.11 standards.

INFORMATION TRANSMISSION

In the process plant, information has historically been transmitted from field devices and control elements to control stations and the process control computer system by means of either analog or digital signals (Fig. 10.1.44).

With the exception of some refinery and petrochemical applications, where pneumatic instrumentation, using a 3 to 15 lb/in^2 variable pressure range, is still used, the vast majority of analog information transmission is by means of either variable voltage, current, or frequency.

ANSI/ISA50 standardized the analog transmission protocol to be 4 to 20 mA DC (1 to 5 V DC) into 1,000 ohms resistive as equal to 0 to 100 percent scale. Another transmission protocol in less common use is 0 to 20 mA DC, 10 to 50 V DC, among many others. Two-wire instruments receive their operational power as well as transmit their signal on the "current output loop." Four-wire instruments require separate power supplies for operation.

Figure 10.1.44 Modern display console, after Honeywell Experion Orion console or equivalent.

ANSI/ISA50 chose a "live zero" at 4 mA DC so that a reading of 0 mA would indicate a diagnostic failure condition. Numerous manufacturers provide field converters to take voltages and frequencies and convert them to current loop outputs, and vice versa.

While many field sensors continue to have analog outputs (voltage, current, frequency, or pressure), many field instruments are now supplied with digital communications systems built into the device. The most ubiquitous in process automation is the **HART protocol** (www.fieldcommgroup.org), which is a digital communications system carried on existing 4- to 20-mA DC current loop outputs. The HART protocol permits remote readout, calibration, diagnostics, and reprogramming of many field instruments and control valves. The introduction of HART 7.0 produced specifications for all-digital HART signals, as well as wireless HART signals. Most manufacturers produce field instruments that are either wired or wireless HART 7.0 devices.

Other bus systems are in use, some relatively old, such as IEEE 488 GPIB (or BPIB), developed as a general-purpose interface bus. Other buses such as Modbus are often found in machine control applications in discrete automation. Another digital communications bus, BACNet, has found wide acceptance in building automation systems worldwide.

Growing reliance is being placed, at least in new plant construction, on two other, more fully featured communications methods, known by the common name of **fieldbuses. Profibus** (www.profibus.com) and **Foundation Fieldbus** (www.fieldcommgroup.org) provide complete bidirectional communications and control between field instruments and final control elements and the process control system. The design of fieldbuses is beginning to converge with the design of enterprise LAN and WAN (local- and wide-area networks) using TCP/IP protocols over Ethernet configured networks. IEEE's standard for power over Ethernet, has now become workable for field instruments, networks, and controllers. Note that since the previous edition of this work, the HART Communication Foundation and the Fieldbus Foundation have merged into the Fieldcomm Group (www.fieldcommgroup.org).

Despite intellectual property challenges, the use of embedded web servers in field instruments and controllers is proliferating. This permits the use of standard Internet-based communication protocols by even a pressure transmitter or flowmeter. The control system then is a virtual control system architected as a series of World Wide Web pages, rather than a physical hardware set.

THE INTERNET OF THINGS, BIG DATA, AND SENSOR ANALYTICS

Futurists and other big name analysts have been predicting for about a decade now that by 2020 there will be billions of smart devices ("things") and sensors interconnected by the Internet. Gartner Group predicts 6.4 billion "things" by 2020. Many of these "things" will be industrial sensors, controllers, in-plant networks, and final control elements. The application of the "Big Data" that these things will produce is expected to revolutionize the way plants are operated, controlled, and maintained. Mobile devices will be able to access sensors and final control elements, and reprogram controllers in the field from anywhere in the world. Data from these things will go to a data Cloud, hosted on the Internet, with analytics provided by Software as a Service applications (SaaS Apps).

This produces some difficulties. "Four nines" uptime (99.9999%) will have to be assured, especially in industries such as oil and gas, both onshore and offshore, and chemicals. Network visibility applications will have to be developed that can scale to millions of "things" in a network. Network and sensor security will have to be improved. Current ICS (Industrial Control Security) protocols are thin on sensor security. Sensors, network software, hardware, and firmware must be designed with "intrinsic security" internally. In order to make the Internet of Things (or its offshoot, the Industrial Internet of Things) practical, the cost of field sensors and final control elements must be reduced, perhaps as much as 80% over current prices, while maintaining accuracy, durability, and improving security.

10.2 AUTOMATIC CONTROLS

by Gregory V. Murphy

REFERENCES: Thaler, "Elements of Servomechanism Theory," McGraw-Hill. Shinskey, "Process Control Systems: Application, Design and Tuning," McGraw-Hill. Zoss, "Applied Instrumentation in the Process Industries," Vol. IV, Gulf Publishing Co. Truxal, "Control Engineers Handbook," McGraw-Hill. A. W. Langill, "Automatic Control Systems Engineering," Prentice-Hall. Kuo, "Automatic Control Systems," Prentice-Hall. Phillips and Nagle, "Digital Control System Analysis and Design," Prentice-Hall. Lewis, "Applied Optimal Control and Estimation: Digital Design and Implementation," Prentice-Hall. Cochin and Plass, "Analysis and Design of Dynamic Systems," HarperCollins. Astrom and Hagglund, "Automatic Tuning of PID Controllers," Instrument Society of America. Maciejowski, "Multivariable Feedback Design," Addison-Wesley. Doyle and Stein, "Robustness with Observers," *IEEE Trans. Automatic Control,* **AC-24,** pp. 607-611, Murphy and Bailey, "LQG/LTR Control System Design for a Low-Pressure Feedwater Heater Train with Time Delay," *Proc. IECON,* 1990. Murphy and Bailey, "Evaluation of Time Delay Requirements for Closed-Loop Stability Using Classical and Modern Methods," *IEEE Southeastern Symp. on System Theory,* 1989. Murphy and Bailey, "LQG/LTR Robust Control System Design for a Low-Pressure Feedwater Heater Train," *Proc. IEEE Southeastcon,* 1990. Kazerooni and Narayanan, "Loop Shaping Design Related to LQG/LTR for SISO Minimum Phase Plants," *IEEE American Control Conf.,* **1,** 1987. Murphy and Bailey, "Robust Control Technique for Nuclear Power Plants," ORNL-10916, March 1989. Seborg, Edgar, and Mellichamp, "Process Dynamics and Control," Prentice-Hall. Buckley, "Techniques of Process Control," Wiley. Birdwell, Crockett, Bodenheimer, and Chang, The CASCADE Final Report: Vol. II, "CASCADE Tools and Knowledge Base," University of Tennessee. Wang and Birdwell, "A Nonlinear PID-Type Controller Utilizing Fuzzy Logic," *Proc. Joint IEEE/IFAC Symp. on Controller-Aided Control System Design,* 1994. Upadhyaya and Eryurek, "Application of Neural Networks for Sensor Validation and Plant Monitoring," *Nuclear Technology,* **97,** no. 2, 1992. Vasadevan et al., "Stabilization and Destabilization Slugging Behavior in a Laboratory Fluidized Bed," *International Conf. on Fluidized Bed Combustion,* 1995. Doyle and Stein, "Robustness with Observers," *IEEE Trans. Automatic Control,* **AC-24,** 1979. Upadhaya et al., "Development and Testing of an Integrated Validation System for Nuclear Power Plants," Report prepared for the U.S. Dept. of Energy. Vols. 1-3, DOE/NE/37959-34, 35, 36, Sept. 1989.

INTRODUCTION

The purpose of an **automatic control** on a system is to produce a desired output when inputs to the system are changed. Inputs are in the form of commands, which the output is expected to follow, and disturbances, which the automatic control is expected to minimize. The usual form of an automatic control is a **closed-loop feedback control** which Ahrendt defines as "an operation which, in the presence of a disturbing influence, tends to reduce the difference between the actual state of a system and an arbitrarily varied desired state and which does so on the basis of this difference." The general theories and definitions of automatic control have been developed to aid the designer to meet primarily three basic specifications for the performance of the control system, namely, stability, accuracy, and speed of response.

The **terminology** of automatic control is being constantly updated by the ASME, IEEE, and ISA. Redundant terms, such as *rate, preact,* and *derivative,* for the same controller action are being standardized. Common terminology is still evolving. The introduction of the digital computer as a control device has necessitated the introduction of a whole new subset of terminology. The following terms and definitions have been selected to serve as a reference to a complex area of technology whose breadth crosses several professional disciplines.

Adaptive control system: A control system within which automatic means are used to change the system parameters in a way intended to improve the performance of the system.

Amplification: The ratio of output to input, in a device intended to increase this ratio. A gain greater than 1.

Attenuation: A decrease in signal magnitude between two points, or a gain of less than 1.

Automatic-control system: A system in which deliberate guidance or manipulation is used to achieve a prescribed value of a variable and which operates without human intervention.

Automatic controller: A device, or combination of devices, which measures the value of a variable, quantity, or condition and operates to correct or limit deviation of this measured value from a selected command (set-point) reference.

Bode diagram: A plot of log-gain and phase-angle values on a log-frequency base, for an element, loop, or output transfer function.

Capacitance: A property expressible by the ratio of the time integral of the flow rate of a quantity (heat, electric charge) to or from a storage, divided by the related potential charge.

Command: An input variable established by means external to, and independent of, the automatic-control system, which sets the ideal value of the controlled variable. See *set point.*

Control action: Of a control element or controlling system, the nature of the change of the output affected by the input.

Control action, derivative: That component of control action for which the output is proportional to the rate of change of input.

Control action, floating: A control system in which the rate of change of the manipulated variable is a continuous function of the actuating signal.

Control action, integral (reset): Control action in which the output is proportional to the time integral of the input.

Control action, proportional: Control action in which there is a continuous linear relationship between the output and the input.

Control system, sampling: Control using intermittently observed values of signals such as the feedback signal or the actuating signal.

Damping: The progressive reduction or suppression of the oscillation of a system.

Deviation: Any departure from a desired or expected value or pattern. Steady-state deviation is known as *offset.*

Disturbance: An undesired variable applied to a system, which tends to affect adversely the value of the controlled variable.

Error: The difference between the indicated value and the accepted standard value.

Gain: For a linear system or element, the ratio of the change in output to the causal change in input.

Load: The material, force, torque, energy, or power applied to or removed from a system or element.

Nyquist diagram: A polar plot of the loop transfer function.

Nichols diagram: A plot of magnitude and phase contours using ordinates of logarithmic loop gain and abscissas of loop phase angle.

Offset: The steady-state deviation when the command is fixed.

Peak time: The time for the system output to reach its first maximum in responding to a disturbance.

Proportional band: The reciprocal of gain expressed as a percentage.

Resistance: An opposition to flow that results in dissipation of energy and limitation of flow.

Response time: The time for the output of an element or system to change from an initial value to a specified percentage of the steady-state.

Rise time: The time for the output of a system to increase from a small specified percentage of the steady-state increment to a large specified percentage of the increment.

Self-regulation: The property of a process or a machine to settle out at equilibrium at a disturbance, without the intervention of a controller.

Set point: A fixed or constant command given to the controller designating the desired value of the controlled variable.

Settling time: The time required, after a disturbance, for the output to enter and remain within a specified narrow band centered on the steady-state value.

Time constant: The time required for the response of a first-order system to reach 63.2 percent of the total change when disturbed by a step function.

Transfer function: A mathematical statement of the influence which a system or element has on a signal or action compared at input and output terminals.

BASIC AUTOMATIC-CONTROL SYSTEM

A **closed-loop control system** consists of a process, a measurement of the controlled variable, and a controller which compares the actual measurement with the desired value and uses the difference between them to automatically adjust one of the inputs to the process. The **physical system** to be controlled can be electrical, thermal, hydraulic, pneumatic, gaseous, mechanical, or described by any other physical or chemical process. Generally, the control system will be designed to meet one of two **objectives.** A **servomechanism** is designed to follow changes in set point as closely as possible. Many electrical and mechanical control systems are servomechanisms. A **regulator** is designed to keep output constant despite changes in load or disturbances. Regulatory controls are widely used on chemical processes. Both objectives are shown in Fig. 10.2.1.

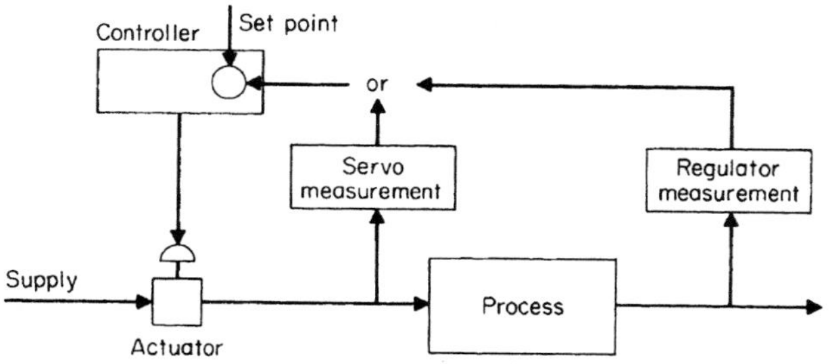

Figure 10.2.1 Feedback control loop showing operation as servomechanism or regulator.

The **control components** can be actuated pneumatically, hydraulically, electronically, or digitally. Only in very few applications does actuation affect controllability. **Actuation** is chosen on the basis of economics.

The **purpose** of the control system must be defined. A large capacity or inertia will make the system sluggish for servo operation but will help to minimize the error for regulator operation.

PROCESS AS PART OF THE SYSTEM

Figure 10.2.1 shows the **process** to be part of the control system either as load on the servo or process to be controlled. Thus the process must be designed as part of the system. The process is **modeled** in terms of its static and dynamic gains in order that it be incorporated into the system diagram. Modeling uses Ohm's and Kirchhoff's laws for electrical systems, Newton's laws for mechanical systems, mass balances for fluidflow systems, and energy balances for thermal systems.

Consider the **electrical system** in Fig. 10.2.2

$$i = \frac{E_1 - E_2}{R} \qquad i = C\frac{dE_2}{dt} \tag{10.2.1}$$

Combining

$$RC\frac{dE_2}{dt} + E_2 = E_1 \tag{10.2.2}$$

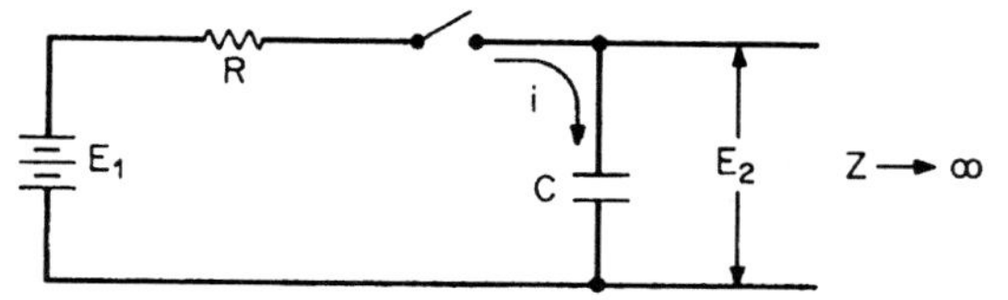

Figure 10.2.2 Electrical system where current flows upon closing switch.

where $\left(R\ \frac{\text{volt-second}}{\text{coulomb}}\right)\left(C\ \frac{\text{coulomb}}{\text{volt}}\right) = \tau$ s = time constant

Consider the **mass balance** of the vessels shown in Figs. 10.2.3 and 10.2.4:

$$\text{Accumulation} = \text{input} - \text{output}$$

$$\frac{d(\rho V)}{dt} = w_1 - w_2 \text{ lb/min} \tag{10.2.3}$$

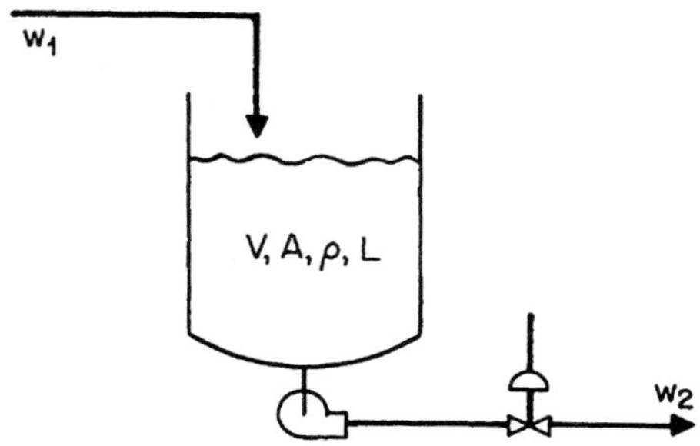

Figure 10.2.3 Liquid level process.

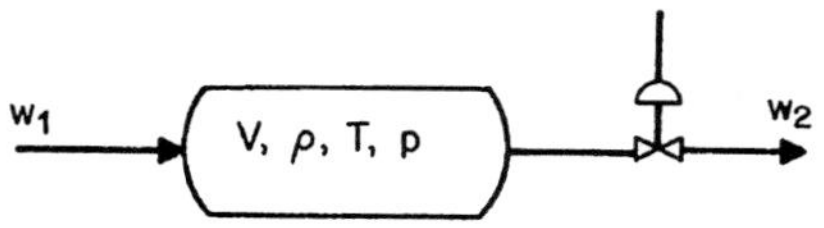

Figure 10.2.4 Gas pressure process.

For the liquid level process (Fig. 10.2.3):

$$V = AL \tag{10.2.4}$$

Linearizing

$$w_2 = \left.\frac{w_2}{x}\right|_{ss} x$$

Substituting

$$\rho A\frac{dL}{dt} = w_1 - \left.\frac{w_2}{x}\right|_{ss} x \tag{10.2.5}$$

where ρA = analogous capacitance.

For the **gas pressure process** (Fig. 10.2.4):

$$V\frac{d\rho}{dt} = w_1 - w_2 \tag{10.2.6}$$

From thermodynamics

$$\left(\frac{\rho}{p}\right)^n = \text{constant} \tag{10.2.7}$$

Linearizing

$$w_2 = \left.\frac{w_2}{x}\right|_{ss} x + \left.\frac{w_2}{2(p-p_2)}\right|_{ss} p - p_2 \tag{10.2.8}$$

Substituting

$$RC\,\frac{dp}{dt}+p=\frac{2(p-p_2)}{x}\bigg|_{ss}x+p_2 \tag{10.2.9}$$

where

$$C=\frac{V}{nRT}\qquad R=\frac{2(p-p_2)}{w_2}\qquad RC=\tau\,\text{min}$$

and the terms are defined: V = volume, ft^3 (m^3); A = cross-sectional area, ft^2 (m^2); L = level, ft (m); ρ = density, lb/ft^3 (g/ml); x = valve stem position (normalized 0 to 1); T = temperature, °F (°C); p = pressure, lb/in^2 (kPa); and w = mass flow, lb/min (kg/min).

The **thermal process** of Fig. 10.2.5 is modeled by a heat balance (Shinskey, "Process Control Systems," McGraw-Hill):

$$Mc\,\frac{dT}{dt}=wc(T_0-T)-UA(T-T_w) \tag{10.2.10}$$

$$\frac{Mc}{wc+UA}\,\frac{dT}{dt}+T=\frac{wc}{wc+UA}\,T_0+\frac{UA}{wc+UA}\,T_w \tag{10.2.11}$$

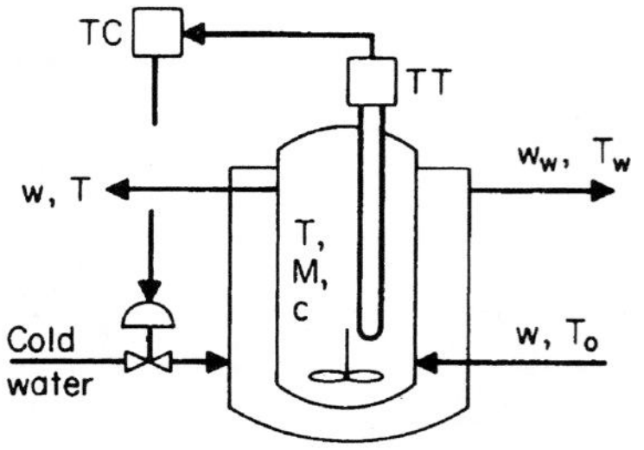

Figure 10.2.5 Thermal process with heat from vessel being removed by cold water in jacket.

where

$$C=Mc\qquad R=\frac{1}{wc+UA}\qquad RC=\tau\,\text{min}$$

$$RC\,\frac{dT}{dt}+T=\frac{wc}{wc+UA}\,T_0+\frac{UA}{wc+UA}\,T_w \tag{10.2.12}$$

The terms are defined: M = weight of process fluid in vessel, lb (kg); c = specific heat, Btu/lb · °F (J/kg · °C); U = overall heat-transfer coefficient, Btu/ft^2 · min · °F (W/m^2 · °C); and A = heat-transfer area, ft^2 (m^2).

Newton's laws can be applied to the manometer shown in Fig. 10.2.6.

Inertia force = Restoring force – Flow resistance

$$\frac{Al\rho}{g_c}\,\frac{d^2h}{dt^2}=A\left(p-2h\rho\,\frac{g}{g_c}\right)-RA\,\frac{dh}{dt} \tag{10.2.13}$$

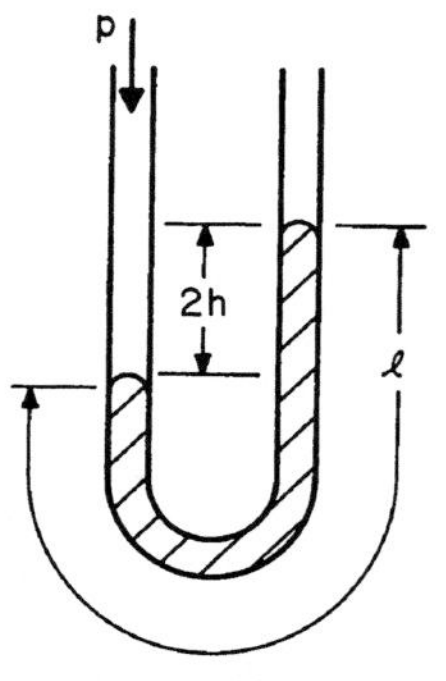

Figure 10.2.6 Filled manometer measuring pressure p.

Flow resistance for laminar flow is given by the Hagen-Poiseuille equation:

$$R=\frac{\text{Driving force}}{\text{Rate of transfer}}=\frac{321\mu}{d^2g_c} \tag{10.2.14}$$

Substituting

$$\frac{1}{2g}\,\frac{d^2h}{dt^2}+\frac{161\mu}{\rho dg^2}\,\frac{dh}{dt}+h=h_i \tag{10.2.15}$$

In standard form

$$\tau_c^2\,\frac{d^2h}{dt^2}+2\tau_c\zeta\,\frac{dh}{dt}+h=h_i \tag{10.2.16}$$

where τ_c = characteristic response time, min, = $1/[(60)\,(2\pi)\omega_n]$ where ω_n = natural frequency, Hz; and ζ = damping coefficient (ratio), dimensionless.

The variables τ_c and ζ are very valuable design aids since they define **system response** and **stability** in terms of system parameters.

TRANSIENT ANALYSIS OF A CONTROL SYSTEM

The stability, accuracy, and speed of response of a control system are determined by analyzing the **steady-state** and the **transient** performance. It is desirable to achieve the steady state in the shortest possible time, while maintaining the output within specified limits. Steady-state performance is evaluated in terms of the accuracy with which the output is controlled for a specified input. The transient performance, that is, the behavior of the output variable as the system changes from one steady-state condition to another, is evaluated in terms of such quantities as maximum overshoot, rise time, and response time (Fig. 10.2.7).

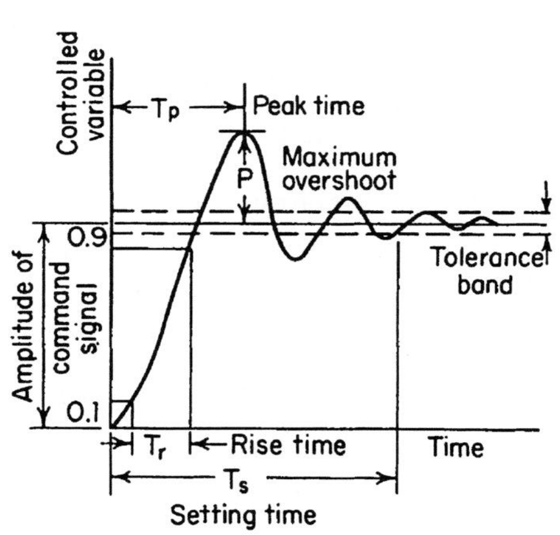

Figure 10.2.7 System response to a unit step-function command.

Transient-Producing Disturbances A number of factors affect the quality of control, among them disturbances caused by set-point changes and process-load changes. Both set point and process load may be defined in terms of the setting of the final control element to maintain the controlled variable at the set point. Thus both cause the final control element to reposition. For a purely mechanical system the disturbance may take the form of a vibration, a displacement, a velocity, or an acceleration. A process-load change can be caused by variations in the rate of energy supply or variations in the rate at which work flows through the process. Reference to Fig. 10.2.5 and Eq. (10.2.12) shows disturbances to be variations in inlet process fluid temperature and cooling-water temperature. Further linearization would show variations in process flow and the overall heat-transfer coefficient to also be disturbances.

The Basic Closed-Loop Control To illustrate some characteristics of a basic closed-loop control, consider a mechanical, rotational system composed of a prime mover or motor, a total system inertia J, and a viscous friction f. To control the system's output variable θ_o, a command signal θ_i must be supplied, the output variable measured and compared to the input, and the resulting signal difference used to control the flow of energy to the load. The basic control system is represented schematically in Fig. 10.2.8.

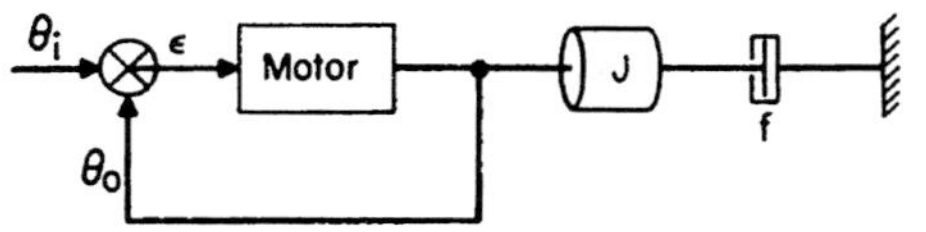

Figure 10.2.8 A basic closed-loop control system.

The differential equation of this basic system is readily obtained from the idealized equations

$$\text{Load torque } T_L = J\frac{d^2\theta_o}{dt^2} + f\frac{d\theta_o}{dt} \tag{10.2.17}$$

$$\text{Developed torque } T_D = K\varepsilon \tag{10.2.18}$$

$$\text{Error } \varepsilon = \theta_i - \theta_o \tag{10.2.19}$$

The above equations combine to yield the system differential equation:

$$J\frac{d^2\theta_o}{dt^2} + f\frac{d\theta_o}{dt} + K\theta_o = K\theta_i \tag{10.2.20}$$

Step-Input Response of a Viscous-Damped Control If the control system described in Fig. 10.2.8 by Eq. (10.2.20) is subjected to a step change in the input variable θ_i, a solution $\theta_o = \theta_o(t)$ can be obtained as follows. (1) Let the ratio $\sqrt{K/J}$ be designated by the symbol ω_n and be called the **natural frequency.** (2) Let the quantity $2\sqrt{JK}$ be designated by the symbol f_c and be called the **friction coefficient** required for critical damping. (3) Let f/f_c be designated by the symbol ζ and be called the **damping ratio.** Equation (10.2.20) can then be written as

$$\frac{d^2\theta_o}{dt^2} + 2\zeta\omega_n\frac{d\theta_o}{dt} + \omega_n^2\theta_o = \omega_n^2\theta_i \tag{10.2.21}$$

For $\theta_i = 1$:

$$\theta_o = 0 \quad \text{and} \quad \frac{d\theta_o}{dt} = 0 \text{ at } t = 0$$

The complete solution of Eq. (10.2.21) is

$$\theta_o = 1 - \frac{\exp(-\zeta\omega_n t)}{\sqrt{1-\zeta^2}} \sin\left(\sqrt{1-\zeta^2}\,\omega_n t + \tan^{-1}\frac{\sqrt{1-\zeta^2}}{\zeta}\right) \tag{10.2.22}$$

Equation (10.2.22) is plotted in dimensionless form for various values of damping ratio in Fig. 10.2.9. The curves for $\zeta = 0.1$, 2, and 1 illustrate the underdamped, overdamped, and critically damped case, where any further decrease in system damping would result in overshoot. Damping is a property of the system which opposes a change in the output variable.

The immediately apparent features of an observed transient performance are (1) the existence and magnitude of the maximum overshoot, (2) the frequency of the transient oscillation, and (3) the response time.

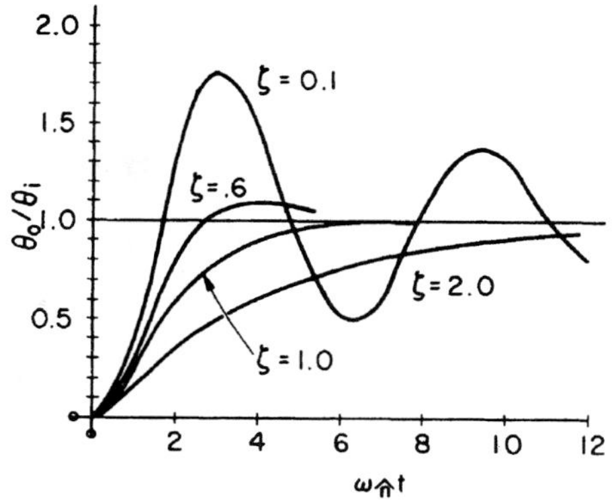

Figure 10.2.9 Transient response of a second-order viscous-damped control to unit-step input displacement.

Maximum Overshoot When an automatic-control system is underdamped, the output variable overshoots its desired steady-state condition and a transient oscillation occurs. The first overshoot is the greatest, and it is the effect of its amplitude which must concern the control designer. The primary considerations for limiting this maximum overshoot are (1) to avoid damage to the process or machine due to excessive excursions of the controlled variable beyond that specified by the command signal, and (2) to avoid the excessive settling time associated with highly underdamped systems. Obviously, exact quantitative limits cannot generally be specified for the magnitude of this overshoot. However, experience indicates that satisfactory performance can generally be obtained if the overshoot is limited to 30 percent or less.

Transient Frequency An undamped system oscillates about the final steady-state condition with a frequency of oscillation which should be as high as possible in order to minimize the response time. The designer must, however, avoid resonance conditions where the frequency of the transient oscillation is near the natural frequency of the system or its component parts.

Rise Time T_n, Peak Overshoot P, Peak Time T_p These quantities are related to ζ and ω_n in Figs. 10.2.10 and 10.2.11. Some useful formulas are listed below:

$$\omega_n T_n \approx 1.02 + 0.48\zeta + 1.15\zeta^2 + 0.76\zeta^3 \qquad 0 \le \zeta \le 1$$

$$\omega_n T_s = \left.\begin{matrix} 17.6 - 19.2\zeta & 0.2 \le \zeta \le 0.75 \\ -3.8 + 9.4\zeta & 0.75 \le \zeta \le 1 \end{matrix}\right\} \text{2\% tolerance band}$$

$$P = \exp\left(\frac{-\pi\zeta}{\sqrt{1-\zeta^2}}\right) \qquad T_p = \frac{\pi}{\omega_n\sqrt{1-\zeta^2}}$$

Although these quantities are defined for a second-order system, they may be useful in the early design states of higher-order systems if the response of the higher-order system is dominated by roots of the characteristic equation near the imaginary axis.

Derivative and Integral Compensation (Thaler) Four common compensation methods for improving the steady-state performance of a proportional-error control without damaging its transient response are shown in Fig. 10.2.12. They are (1) error derivative compensation, (2) input derivative compensation, (3) output derivative compensation, and (4) error integral compensation.

Error Derivative Compensation The torque equilibrium equation is

$$J\frac{d^2\theta_o}{dt^2} + f\frac{d\theta_o}{dt} = K_2\varepsilon + K_1\frac{d\varepsilon}{dt} \tag{10.2.23}$$

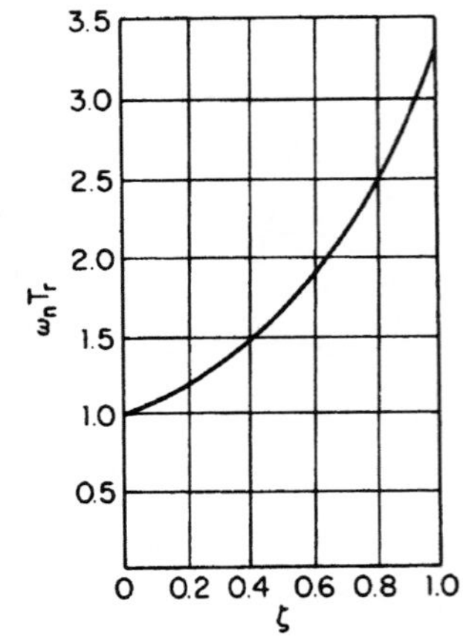

Figure 10.2.10 Rise time T_r as a function of ζ and ω_n.

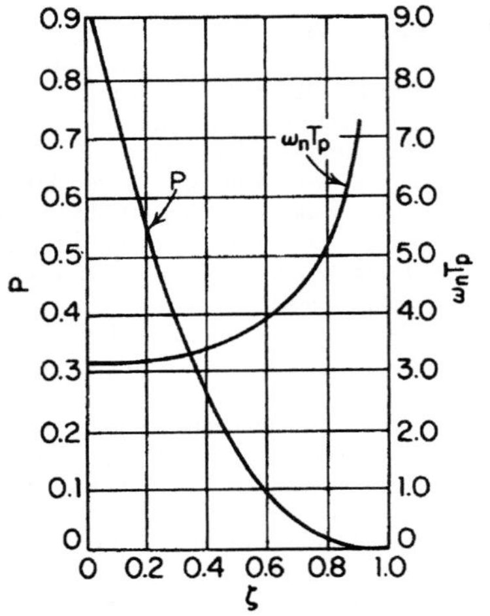

Figure 10.2.11 Peak overshoot P and peak time T_p as functions of ζ and ω_n.

Writing Eq. (10.2.23) in terms of the input and output variables yields

$$J\,\frac{d^2\theta_o}{dt^2} + (f + K_1)\,\frac{d\theta_o}{dt} + K_2\,\theta_o = K_2\,\theta_i + K_1\,\frac{d\theta_i}{dt} \qquad (10.2.24)$$

By adjusting K_1 and reducing f so that the quantity $f + K_1$ is equal to f in the uncompensated system, the system performance is affected as follows: (1) ε resulting from a constant-first-derivative input is reduced because of the reduction in viscous friction; (2) the transient performance of the uncompensated system is preserved unchanged.

Derivative Input Compensation The torque equilibrium equation is

$$J\,\frac{d^2\theta_o}{dt^2} + f\,\frac{d\theta_o}{dt} = K_2\varepsilon + K_1\,\frac{d\theta_i}{dt} \qquad (10.2.25)$$

Writing Eq. (10.2.25) in terms of the input and output variables yields

$$J\,\frac{d^2\theta_o}{dt^2} + f\,\frac{d^2\theta_o}{dt} + K_2\theta_o = \theta_i + K_1\,\frac{d\theta_i}{dt} \qquad (10.2.26)$$

Examination of Eq. (10.2.26) yields the following information about the compensated system's performance: (1) since the characteristic equation is unchanged from that of the uncompensated system, the transient performance is unaltered; (2) the steady-state solution to Eq. (10.2.26) is

$$\theta_o = \theta_i - \frac{f}{K_2}\left(1 - \frac{K_1}{K_2}\right)\frac{d\theta_i}{dt} \qquad (10.2.27)$$

Therefore the input derivative signal can reduce the steady-state error by adjusting K_1 to equal K_2.

Derivative Output Compensation The torque equilibrium equation is

$$J\,\frac{d^2\theta_o}{dt^2} + f\,\frac{d^2\theta_o}{dt} = K_2\varepsilon \pm K_1\,\frac{d\theta_o}{dt} \qquad (10.2.28)$$

Writing Eq. (10.2.28) in terms of the input and output variables yields

$$J\,\frac{d^2\theta_o}{dt^2} + (f \pm K_1)\,\frac{d\theta_o}{dt} + K_2\theta_o = K_2\theta_i \qquad (10.2.29)$$

Examination of Eq. (10.2.29) yields the following information about the compensated system's performance. (1) Output derivative feedback

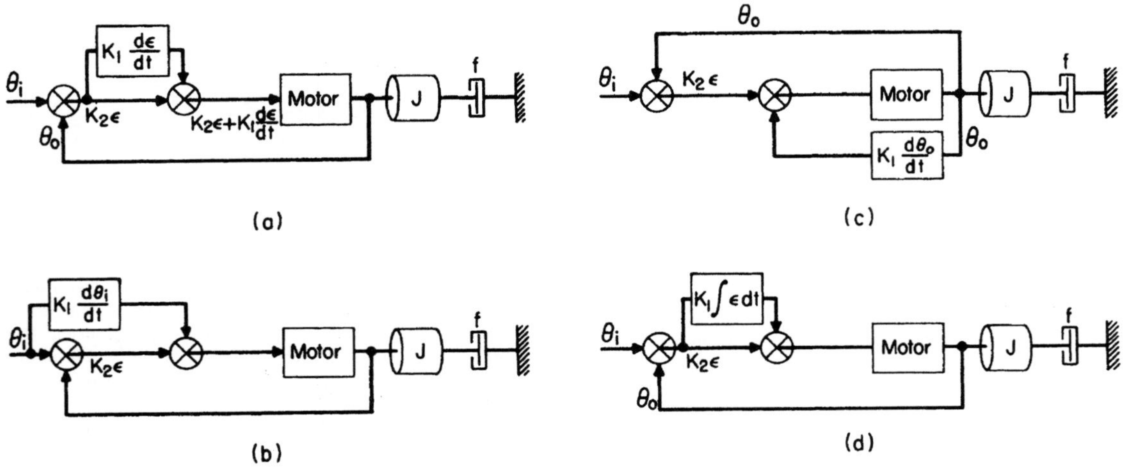

Figure 10.2.12 Derivative and integral compensation of a basic closed-loop system. (*a*) Error derivative compensation; (*b*) input derivative compensation; (*c*) output derivative compensation; (*d*) error integral compensation. (*Thaler*.)

produces the same system effect as the viscous friction does. This compensation therefore damps the transient performance. (2) Under conditions where $\theta_i = \theta t$, the steady-state error is increased.

Error Integral Compensation Error integral compensation is used where it is necessary to eliminate steady-state errors resulting from input signals with constant first derivatives or under conditions of externally applied loads. The torque equilibrium equation is

$$J\,\frac{d^2\theta_o}{dt^2} + f\,\frac{d\theta_o}{dt} \pm \text{external load torque} = K_2\varepsilon + K_1\int_0^t \varepsilon\,dt \qquad (10.2.30)$$

Writing Eq. (10.2.30) in terms of the input variable and the error yields

$$\text{External load torque } + J\,\frac{d^2\theta_i}{dt^2} + f\,\frac{d\theta_i}{dt}$$

$$= J\frac{d^2\varepsilon}{dt^2} + f\frac{d\varepsilon}{dt} + K_2\varepsilon + K_1\int_0^t \varepsilon\,dt \qquad (10.2.31)$$

At steady state for $t \gg 0$ and with a step change in the input derivative

$$\frac{d^2\theta_i}{dt^2} = \frac{d\varepsilon}{dt} = \frac{d^2\varepsilon}{dt^2} = 0 \qquad \frac{d\theta_i}{dt} = \text{const} \qquad (10.2.32)$$

Equation (10.2.31) assumes the form

$$\pm\text{External load torque} + f\frac{d\theta_i}{dt} = K_2\varepsilon + K_1\int_0^t \varepsilon\,dt \qquad (10.2.33)$$

Since the sum of the load torque and the term $f(d\theta_i/dt)$ is finite, $\varepsilon = 0$ for

$$\int_0^\infty \varepsilon\,dt \to \infty \qquad (10.2.34)$$

If the integrating coefficient is a small number, additional torque produced by the integration action is developed very slowly, and, although the steady-state error is eventually eliminated, the transient performance is essentially unchanged. However, if K_1 is a large value, a large torque is produced in a short period of time, increasing the effect *T/J* ratio, and thereby decreasing the damping. The general effects of error integral compensation within its useful range are (1) steady-state error is eliminated; and (2) transient response is adversely effected, resulting in increased overshoot and the attendant increase in response time.

TIME CONSTANTS

The **time constant** τ, the **characteristic response time** (τ_c), and the **damping coefficient** are combined with operational calculus to design a control system without solving the classical differential equations. Note that the electrical, Eq. (10.2.2), mass, Eq. (10.2.9), and energy, Eq. (10.2.12) processes are all described by analogous first-order differential equations of the form

$$\tau\frac{dc}{dt} + c = Ka \qquad (10.2.35)$$

Solving with initial conditions equal to zero, the **time-domain response** to a step disturbance is

$$c = Ka(1 - e^{t/\tau}) \qquad (10.2.36)$$

Plotting is shown in Fig. 10.2.13.

Figure 10.2.13 shows the **time constant** to be defined by the point at 63.2 percent of the response, while response is essentially complete at three time constants (95 percent).

Operational calculus provides a systematic and simple method for handling linear differential equations with constant coefficients. The

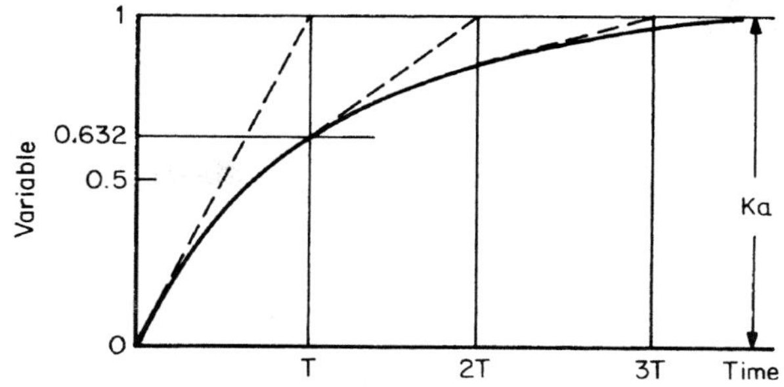

Figure 10.2.13 Response of first-order system showing time constant relationship.

Laplace operator is used in control system analysis because of the straightforward transformation among the domains of interest:

Time domain	Complex domain	Frequency domain
$\frac{d}{dt}$	s	$j\omega$

Extensive tables of transforms are available for converting between the time and complex domains and back again. Transformation from the **complex** or **frequency domains** to **time** is difficult, and it is seldom attempted. Time-domain solutions are obtained only from computer simulations, which can solve the finite difference equations fast enough to serve as a design tool. Control system analysis and design proceed with a number of analytical and graphical techniques which do not require a time-domain solution.

The **root locus method** is a graphical technique used in the complex domain which provides substantial insight into the system. It has the weaknesses of handling dead time poorly and of the graph's being very tedious to plot. It is used only when computerized plotting is available.

BLOCK DIAGRAMS

The **physical diagram** of the system is converted to a **block diagram** in order that the different components of the system (all the way from a steam boiler to a thermocouple) can be placed on a common mathematical footing for analysis as a system. The block diagram shows the functional relationship among the parts of the system by depicting the action of the variables in the system. The circle represents an algebraic function of addition. Each system component is represented by a **block** which has one input and one output. The block represents a dynamic function such that the output is a function of time and also of the input. The dynamic function is called a **transfer function**—the ratio of the Laplace transform of the output variable to the input variable with all initial conditions equal to zero. The input and output variables are considered as signals, and the blocks are connected by arrows to show the flow of information in the system.

Rewriting Eq. (10.2.12) for the thermal system,

$$RC\,\frac{dT}{dt} + T = K_1T_0 + K_2T_w \qquad (10.2.37)$$

Transforming

$$(\tau s + 1)T = K_1T_0 + K_2T_w \qquad (10.2.38)$$

for which the block diagram is shown in Fig. 10.2.14. Two conditions are specified: (1) the components must be described by linear differential equations (or nonlinear equations linearized by suitable approximations), and (2) each block is unilateral. What occurs in one component may not affect the components preceding.

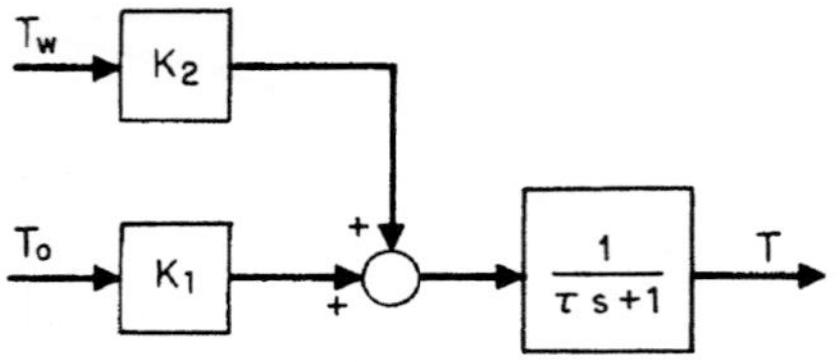

Figure 10.2.14 Block diagram of thermal process.

Block-Diagram Algebra

The block diagram of a single-loop feedback-control system subjected to a command input $R(s)$ and a disturbance $U(s)$ is shown in Fig. 10.2.15.

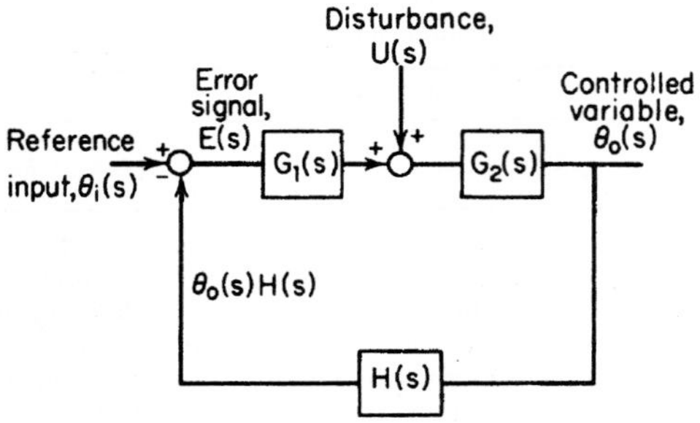

Figure 10.2.15 Single-loop feedback-control system.

When $U(s) = 0$ and the input is a reference change, the system may be reduced as follows:

$$E(s) = -\theta_o(s)H(s)$$

$$\theta_o(s) = E(s)[G_1(s)G_2(s)] \qquad (10.2.39)$$

Therefore $\dfrac{\theta_o(s)}{\theta_i(s)} = \dfrac{G_1(s)G(s)}{1+G_1(s)G_2(s)H(s)}$

When $\theta_i(s) = 0$ and the input is a disturbance, the system may be reduced as follows:

$$E(s) = -\theta_o(s)H(s)$$

$$[E(s)\ G_1(s)+U(s)]G_2(s) = \theta_o(s)$$

$$\frac{\theta_o(s)}{U(s)} = \frac{G_2(s)}{1+G_1(s)G_2(s)H(s)} \qquad (10.2.40)$$

Closed-loop transfer function Eq. (10.2.40) can be determined by observation from

$$\frac{\theta_o(s)}{U(s)} = \frac{\text{Feedforward functions}}{1+\text{Complete-loop functions}}$$

The denominator of Eq. (10.2.40) is the **characteristic equation** which determines **stability.** Most systems are designed on the basis of the characteristic equation since it sets both response time τ_c and damping ζ. The equation is undefined (unstable) if $G_1(s)G_2(s)H(s)$ equals −1. But −1 is a vector of magnitude 1 and a phase of −180°. This fact is used to determine stable parameter adjustments in the graphical techniques to be discussed later.

The input disturbance $[U(s)]$ can be any time function in actual operation. The **step input** is widely used for analysis and testing since it is easily implemented; it results in a simple Laplace transform; it is the most severe type of disturbance; and the response to a step change shows the maximum error that could occur.

In many complex control systems, especially in the nonmechanical process-control field, auxiliary feedback paths are provided in order to adjust the system's performance. Figure 10.2.16*a* illustrates such a condition. In analyzing such a system it is usually best to combine secondary loops into the main control loop to form an equivalent series block and transfer function. The system of Fig. 10.2.16*a* might be reduced in the following sequence.

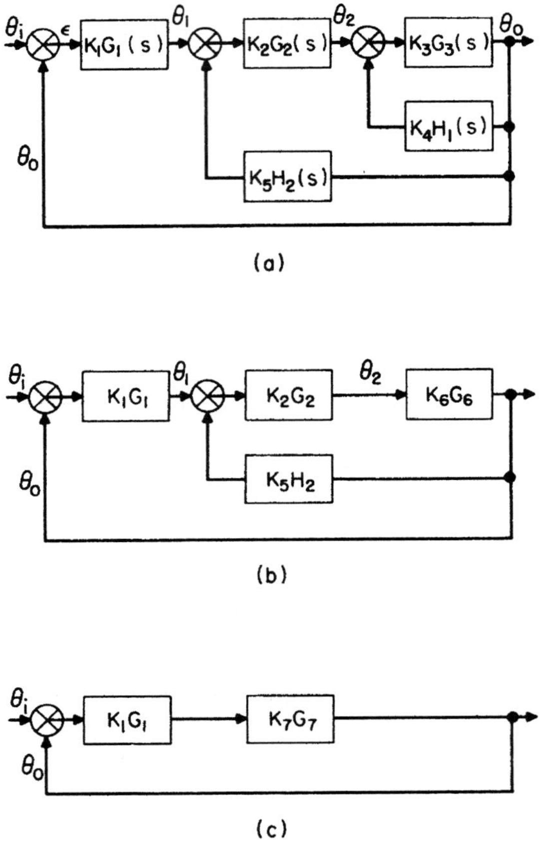

Figure 10.2.16 Reduction of a closed-loop control system with multiple secondary loops.

1. Replace $K_3G_3(s)$ and $K_4H_1(s)$ with a single equivalent element

$$\frac{\theta_o}{\theta_2} = \frac{K_3G_3(s)}{1+K_4H_1(s)K_3G_3(s)} = K_6G_6 \qquad (10.2.41)$$

The result of this first reduction is shown in Fig. 10.2.16*b*.

2. Figure 10.2.16*b* can be treated in a similar fashion and a single block used to replace K_2G_2, K_6G_6, and K_5H_2

$$\frac{\theta_o}{\theta_1} = \frac{K_2G_2K_6G_6(s)}{1+K_5H_2K_2G_2K_6G_6(s)} = K_7G_7 \qquad (10.2.42)$$

The result of this second reduction is shown in Fig. 10.2.16*c*. The resulting open-loop transfer function is

$$\theta_o/\varepsilon = K_1G_1K_7G_7(s) \qquad (10.2.43)$$

The closed-loop or frequency response function is

$$\frac{\theta_o}{\theta_i} = \frac{K_1G_1K_7G_7(s)}{1 + K_1G_1K_7G_7(s)} \tag{10.2.44}$$

Equation (10.2.44) can, of course, be expanded to include the terms of the system's secondary loops.

SIGNAL-FLOW REPRESENTATION

An alternate graphical representation of the mathematical relationships is the signal-flow graph. For complicated systems it allows a more compact representation and more rapid reduction techniques than the block diagram.

In Fig. 10.2.17, the nodes represent the variables θ_i, ε, θ_1, … , θ_o, and the branches the relationships between the nodes, of the system shown in Fig. 10.2.16. For example,

$$\theta_1(s) = \varepsilon K_1G_1(s) - K_5H_2(s)\theta_o(s) \tag{10.2.45a}$$

and

$$\varepsilon = \theta_i - \theta_o \tag{10.2.45b}$$

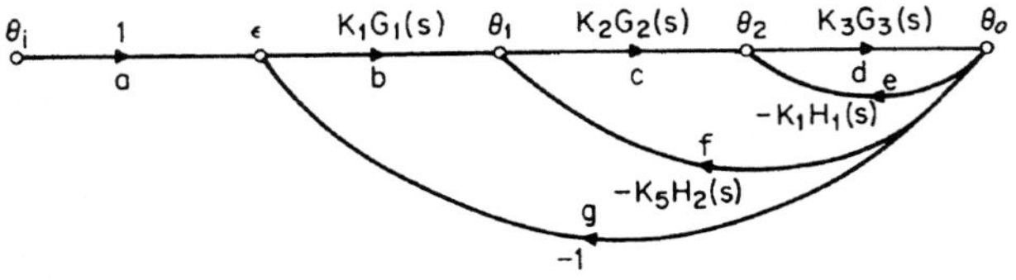

Figure 10.2.17 Signal flow graph of the closed-loop control system shown in Fig. 10.2.16.

Signal-flow terminology follows:

Source: node having only outgoing branches, for example, θ_i

Sink: node having only incoming branches, θ_o

Path: series of branches with the same sense of direction, for example, *abcd, cdf*

Forward path: path originating at a source and ending at a sink, with no node encountered more than once, for example, *abcd*

Path gain: product of the coefficients along a path, for example, $1[K_1G_1(s)][K_2G_2(s)][K_3G_3(s)]$

Feedback loop: path starting at a node and ending at the same node, for example, *bcdg*

Loop gain: product of coefficients along a feedback loop, for example, $[K_1G_1(s)][K_2G_2(s)][K_3G_3(s)](-1)$

The overall gain of the system can be calculated from

$$G = \frac{\Sigma_i G_i \Delta_i}{\Delta} \tag{10.2.46}$$

where G_i = gain of *i*th forward path and

$$\Delta = 1 - \Sigma L_1 + \Sigma L_2 - \Sigma L_3 + \cdots + (-1)^k \Sigma L_k$$

where ΣL_1 = sum of the gains of each forward loop; ΣL_2 = sum of products of loop gains for nontouching loops (no node is common), taken two at a time; ΣL_3 = sum of products of loop gains for nontouching loops taken three at a time; Δ_i = value of Δ for signal flow graph resulting when *i*th path is removed.

From Fig. 10.2.17 there is only one forward path, *abcd.*

$$\therefore\ G_1 = K_1G_1(s)K_2G_2(s)K_3G_3(s)$$

Closed loops are *de, cdf,* and *bcdg,* with gains $-K_1H_1(s)K_3G_3(s)$, $-K_2G_2(s)K_3G_3(s)K_5H_2(s)$, and $-K_1G_1(s)K_2G_2(s)K_3G_3(s)$.

There are no nontouching closed loops;

$$\therefore\ \Delta = 1 + K_1H_1(s)K_3G_3(s) + K_2G_2(s)K_3G_3(s)K_5H_2(s) + K_1G_1(s)K_2G_2(s)K_3G_3(s)$$

There are no loops remaining if the forward path *abcd* is removed;

$$\therefore\ \Delta_1 = 1$$

Thus

$$\theta_o/\theta_1 = K_1G_1(s)K_2G_2(s)K_3G_3/[1 + K_1H_1(s)K_3G_3(s) + K_2G_2(s)K_3G_3(s)K_5H_2(s) + K_1G_1(s)K_2G_2(s)K_3H_3(s)] \tag{10.2.47}$$

which is identical with Eq. (10.2.44).

CONTROLLER MECHANISMS

The **controller** modifies the error signal in a desired manner to produce an output pressure which is used to actuate the valve motor. The several controller modes used singly or in combination are (1) the proportional mode in which $P_{out}(t) = K_cE(t)$, (2) the integral mode, in which $P_{out}(t) = 1/T_1 \int E(t)\,dt$, and (3) the rate mode, in which $P_{out}(t) = T_2\, dE(t)/dt$. In these expressions $P_{out}(t)$ = controller output pressure, $E(t)$ = input error signal, K_c = proportional gain, $1/T_1$ = reset rate, and T_2 = rate time.

A **pneumatic controller** consists of an **error-detecting mechanism, control modes** made up of proportional (*P*), integral (*I*), and derivative (*D*) actions in almost any combination, and a **pneumatic amplifier** to provide output capacity. The error-detecting mechanism is a differential link, one end of which is positioned by the signal proportional to the controlled variable, and the other end of which is positioned to correspond to the command set point. The proportional action is provided by a flapper nozzle (Fig. 10.2.18), where the flapper is positioned by the error signal. A motion of 0.0015 in by the flapper is sufficient for nearly full output range. Nozzle back pressure is inversely proportional to the distance between nozzle opening and flapper.

The controller employs a power-amplifying pilot for providing a larger quantity of air than could be provided through the small restriction shown in Fig. 10.2.18. The nozzle back pressure, instead of operating the final control element directly, is transmitted to a bellows chamber where it positions the pilot valve.

The combination of flapper-nozzle amplifier and power relay shown in Fig. 10.2.18 has a very high gain since small flapper displacements can result in the output traversing the full range of output pressure. Negative feedback is employed to reduce the gain. Controller output is connected to a feedback bellows which operates to reposition the flapper. With the feedback bellows, a movement of the flapper toward the nozzle increases back pressure, causing output pressure to decrease and the feedback bellows to move the flapper away from the nozzle. Thus the mechanism is stabilized. Figure 10.2.19 shows controller gain

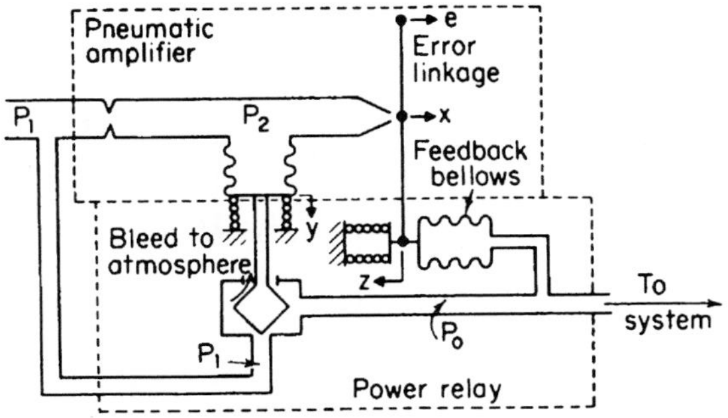

Figure 10.2.18 Gain reduction of pneumatic amplifier by means of feedback bellows. (*Raven, "Automatic Control Engineering," McGraw-Hill.*)

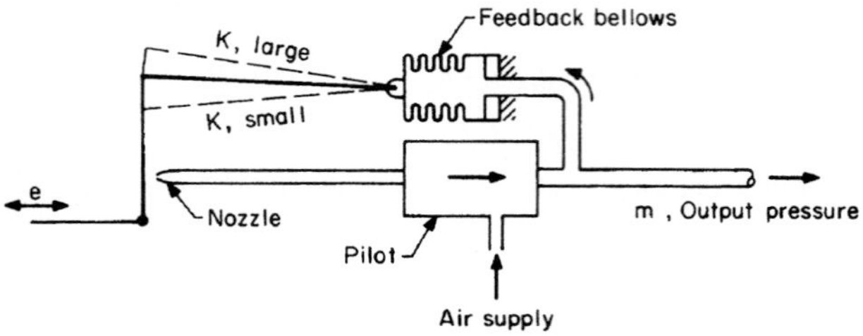

Figure 10.2.19 Proportional controller with negative feedback. (*Reproduced by permission. Copyright © John Wiley & Sons, Inc. Publishers, 1958. From D. P. Eckman, "Automatic Process Control."*)

being varied by adjusting the point at which the feedback bellows bears on the flapper.

The **rate mode,** called "anticipatory," can take large corrective action when errors are small but have a high rate of change. The mode resists not only departures from the set point but also returns and so provides a stabilizing action. Since the rate mode cannot control to a set point, it is not used alone. When used with the proportional mode, its stabilizing influence (90° phase lead; see Table 10.2.3) may allow an increase in gain K_c and a consequent decrease in steady-state error.

Proportional-derivative (PD) control is obtained by adding an adjustable feedback restriction as shown in Fig. 10.2.20. This results in delayed negative feedback. The restriction delays and reduces the feedback, and, since the feedback is negative, the output pressure is momentarily higher and leads, instead of lags, the error signal.

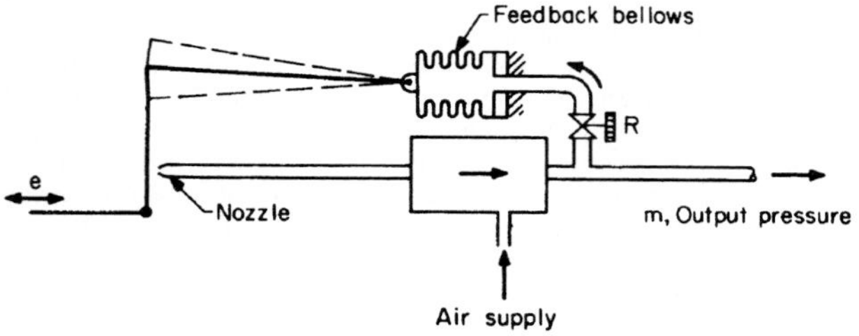

Figure 10.2.20 Proportional-derivative controller. (*Reproduced by permission. Copyright © John Wiley & Sons, Inc. Publishers, 1958. From D. P. Eckman, "Automatic Process Control."*)

Since the proportional mode requires an error signal to change output pressure, set point and load changes in a proportionally controlled system are accompanied by a steady-state error inversely proportional to the gain. For systems which, because of stability considerations, cannot tolerate high gains, the integral mode added to the proportional will eliminate the steady-state error since the output from this mode is continually varying so long as an error exists. The addition of the integral mode to a proportional controller has an adverse effect on the relative stability of the control because of the 90° phase lag introduced.

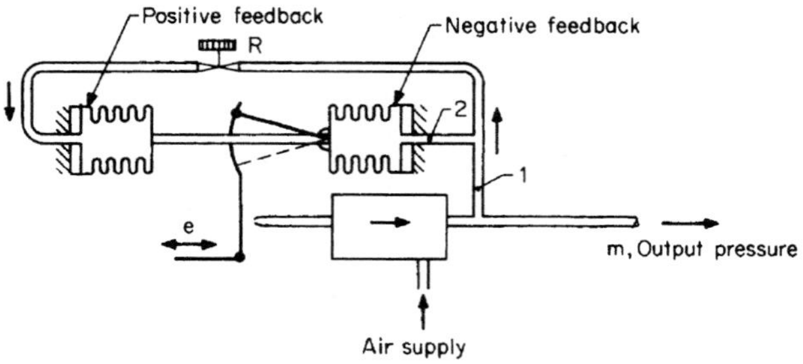

Figure 10.2.21 Proportional-integral controller. (*Reproduced by permission. Copyright © John Wiley & Sons, Inc. Publishers, 1958. From D. P. Eckman, "Automatic Process Control."*)

Proportional-integral (PI) control is obtained by adding a positive feedback bellows and an adjustable restriction (Fig. 10.2.21). The addition of the positive feedback bellows cancels the gain reduction brought about by the negative feedback bellows at a rate determined by the adjustable restriction.

Electronic controllers which are analogous to the pneumatic controller have been developed. They have the advantages of elimination of time lags; compatibility with computers; being less expensive to install (although more expensive to purchase); being more energy efficient; and being immune to low temperatures (water in pneumatic lines freezes).

The heart of the electronic controller is the **high-gain operational amplifier** with passive elements at the input and in the feedback. Figure 10.2.22 shows the classical control elements, though circuits in commercially available controllers are far more sophisticated. The current drawn by the DC amplifier is negligible, and the amplifier gain is around 1,000, so that the junction point is essentially at ground potential. The reset amplifier has delayed negative feedback, similar to the pneumatic circuit in Fig. 10.2.21. The derivative amplifier has advanced negative feedback, similar to Fig. 10.2.20. Gains are adjusted by changing the ratios of resistors or capacitors, while adjustable time constants are made up of a variable resistor and a capacitor *(RC).*

Regardless of actuation, the commonly applied **controller modes** are as follows:

Designation	*Transform*	*Symbol*
Proportional	K (gain)	P
Reset	$1+\frac{1}{\tau_i s}$	I
Derivative	$\frac{\tau_d s+1}{(t_d/\alpha)s+1}$	D
Floating	$\frac{1}{\tau_F s}$	F

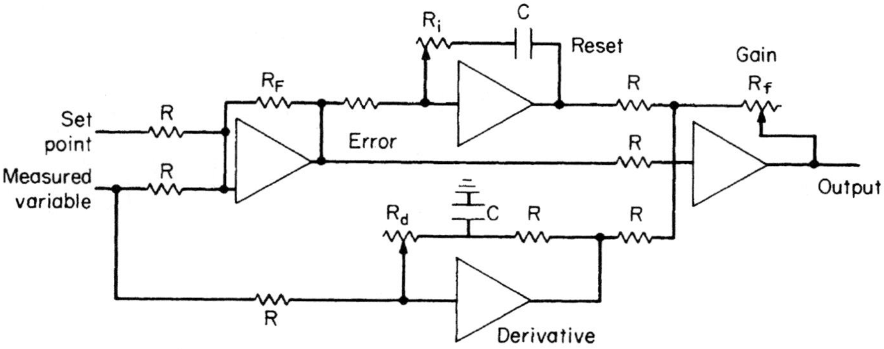

Figure 10.2.22 Simplified circuit for a typical electronic controller.

The modes are combined as needed into P, PI, PID, PD, and F. Choice depends on the application and can be made from the Bode diagram. The block diagram for a PID controller is shown in Fig. 10.2.23.

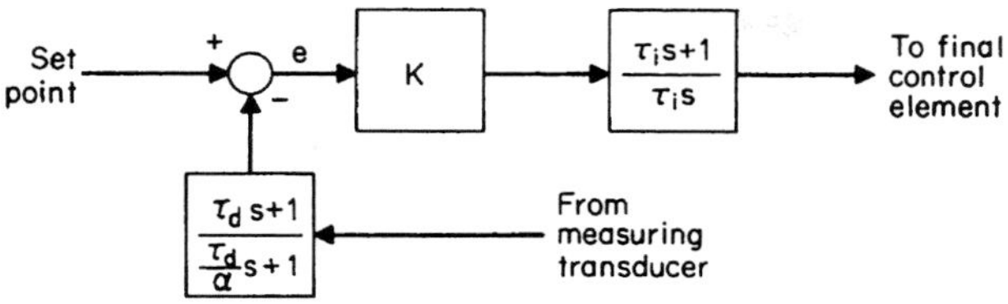

Figure 10.2.23 Block diagram of PID controller.

The **derivative mode** can be placed on the measured variable signal (as shown), on the error, or on the controller output. The arrangement of Fig. 10.2.23 is preferred for regulation since the derivative does not affect set-point changes, and the derivative acts to reduce overshoot on start-up (Zoss, "Applied Instrumentation in the Process Industries," Vol. IV, Gulf Publishing Co.). Regardless of a sequence, the block diagram of a PID controller is commonly drawn as shown in Fig. 10.2.24.

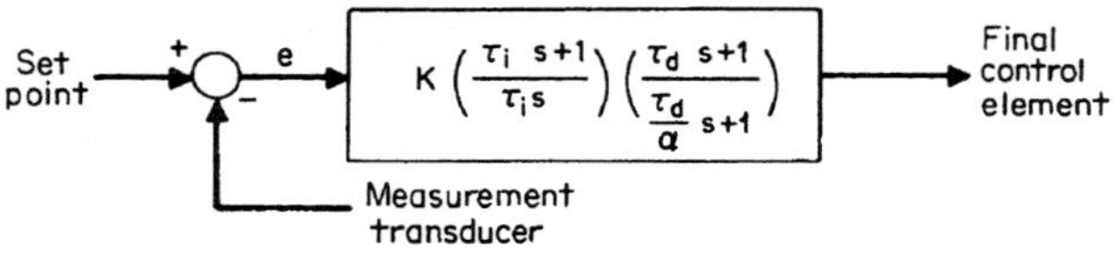

Figure 10.2.24 Combined controller block diagram.

FINAL CONTROL ELEMENTS

The **final control element** is a mechanism which alters the value of the **manipulated variable** in response to an output signal from the automatic control device. It will typically receive a signal from the controller and manipulate a flow of material or energy to the process. The final control element can be a control valve, an electrical motor, a servovalve, or a damper. Servovalves are discussed in the next section, "Hydraulic-Control Systems."

The final control element often consists of two parts: first, an **actuator** which translates the controller signal into a command for the manipulating device, and, second, a **mechanism** to adjust the manipulated variable. Figure 10.2.25 shows a control valve made up of an actuator and an adjustable orifice.

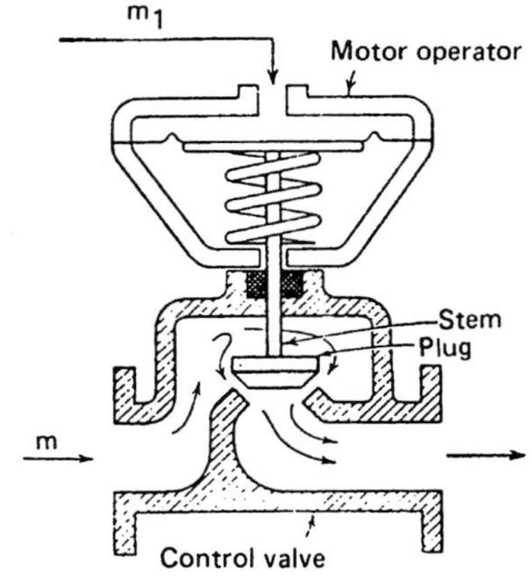

Figure 10.2.25 Control valve and pneumatic actuator. (*Reproduced by permission. Copyright © John Wiley & Sons, Inc. Publishers, 1958. From D. P. Eckman, "Automatic Process Control."*)

The most commonly used **actuator** is a diaphragm motor in which the output pressure from the controller is counteracted not only by the spring but also by fluid forces in the valve body. The latter may cause serious deviation from linear static behavior with deleterious control effects. High friction at the valve stem or large unbalanced fluid forces at the plug can be overcome by valve positioners, which are essentially proportional controllers.

A **control valve** in liquid service is described by Eq. (10.2.48):

$$q = C_v \sqrt{\frac{\Delta p}{G}} \tag{10.2.48}$$

And C_v is described by its relationship to stem lift x, which is called the **valve-flow characteristic.**

The linear characteristic is

$$C_v = C_v|_{\max}\, x \tag{10.2.49}$$

and the equal-percentage characteristic is

$$C_v = C_v|_{\max}\, (r)^{x-1} \tag{10.2.50}$$

where q = flow, gpm (m³/min); $C_v|_{\max}$ = valve sizing coefficient; Δp = pressure drop across valve, lb/in² (kPa); G = specific gravity (density, g/ml); r = valve rangeability; and x = stem position, normalized to 0 to 1.

The flow characteristic relationship is not necessarily the lift-flow characteristic of the valve when installed since the valve is but one component in a piping system in which pressure drops vary with the flow rate. The change in characteristic as less percentage of total system drop is taken across the control valve is shown in Fig. 10.2.26.

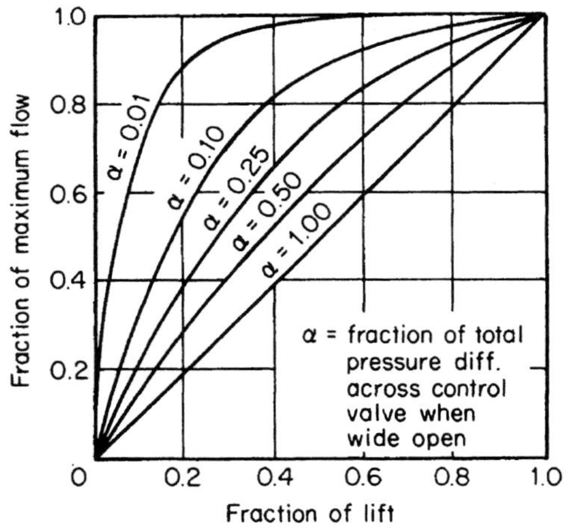

Figure 10.2.26 Control valve linear flow characteristic. (*Reproduced by permission. Copyright © John Wiley & Sons, Inc. Publishers, 1958. From D. P. Eckman, "Automatic Process Control."*)

Valve characteristics are generally selected according to their ability to compensate for nonlinearities in the system. Nonlinearities typically represent a change in gain which in turn changes the character of the control response for a given controller setting when load- or set-point changes occur or when there is a variable overall pressure drop across the system. Equal-percentage valves, for example, tend to control over widely varying operating conditions, since the change in flow is always proportional to flow rate. The selection of the proper valve therefore depends on study of the particular system.

HYDRAULIC-CONTROL SYSTEMS

Hydraulic systems are used for rapid-response servomechanisms at high power levels. Operating system pressures are from 50 to 100 lb/in² for slower-acting systems and up to 5,000 lb/in² where lightweight and fast responses are required. Compared with **electrical systems** the major advantages are a rapid response in the large horsepower ranges and the capability of operating at high power-density levels since the fluid can transmit dissipated energy from the point of generation. Compared with **pneumatic systems,** hydraulic systems are faster because the fluid is essentially incompressible. Major disadvantages are vulnerability to dirt, since the components generally require close machining tolerances, and the danger of fire and explosion resulting from the flammability of the hydraulic fluids used.

The direction and volume of flow are controlled by **servo valves** in the system. They may be single-stage (pilot-operated) and mechanically or electrically actuated. A schematic of a spool-type four-way single-stage control piston and inertia load is shown in Fig. 10.2.27. Hydraulic fluid at constant pressure enters at the supply port. With displacement

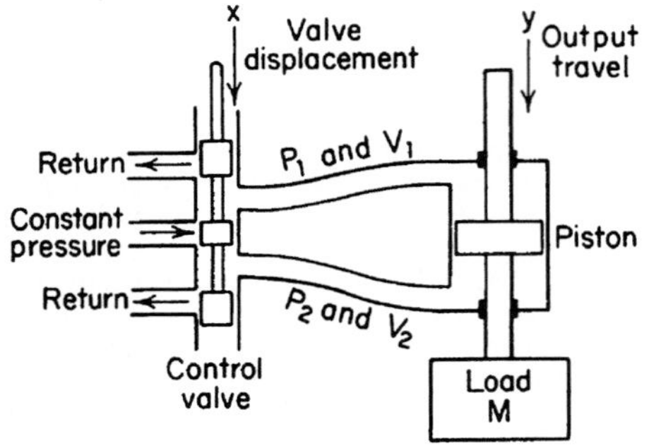

Figure 10.2.27 Four-way valve-piston circuit. (*Truxal, "Control Engineer's Handbook," McGraw-Hill.*)

of the spool valve downward, for example, inflow to the top side of the piston moves the piston downward. Because of machining tolerances the spool dimensions are either large (overlapped) or smaller (underlapped) than the port dimension. Underlapped valves permit leakage to the piston in the centered position; overlapped valves result in a dead zone, where motion x results in no flow until a port is opened.

The transfer function of this circuit is given as (Truxal, "Control Engineers' Handbook")

$$\frac{y}{x}=\frac{C_1\dfrac{1}{1+\alpha(C_1/C_2)}}{s\left[\dfrac{VM}{2BA^2}\,\dfrac{1}{1+\alpha(C_1/C_2)}s^2+\dfrac{C_1m/C_2+V\alpha/2BA^2}{1+\alpha(C_1/C_2)}s+1\right]}$$

where C_1 = servo velocity gradient, in/(s)(in), C_2 = servo force gradient, lb/in, α = viscous friction of load and piston, lb/(in)(s), B = bulk modulus of fluid, lb/in², M = mass of load and piston, lb/(in)(s²), A = piston area, in², and V = effective entrained fluid volume, in³ (one-half of total entrained volume between valve and piston).

The velocity of the output is proportional to the input resulting in a **velocity-control** servo. To convert this system to a **position-control** servo, mechanical, hydraulic, or electrical feedback may be employed. A valve-piston position servo with **mechanical feedback** is shown in Fig. 10.2.28. Any difference between the input D and the piston position y causes a motion x, which causes the piston to move in a direction opposite to D, that is, in a direction to reduce x. The lever ratio establishes the relationship between y and D.

Most commercially available servo valves are two stages, permitting electrohydraulic action. The pilot stage can be operated by a low-power, short-travel electrical device, with a concomitant increase in flexibility. A typical pilot-operated servo valve is shown in Fig. 10.2.29. In this case the pilot is a double-flapper valve rather than a spool valve. (In general, small, accurate low-leakage spool valves are costly.) Upward movement of the flapper by the actuating motor results in increased pressure to the right end of the power spool. Hydraulic

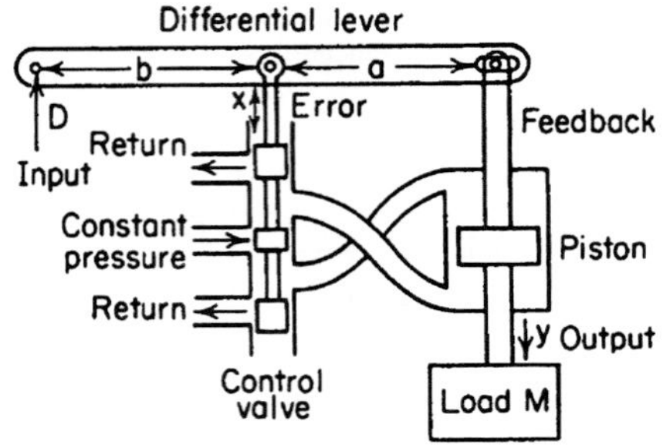

Figure 10.2.28 Valve-piston position servomechanical feedback. (*Truxal, "Control Engineer's Handbook," McGraw-Hill.*)

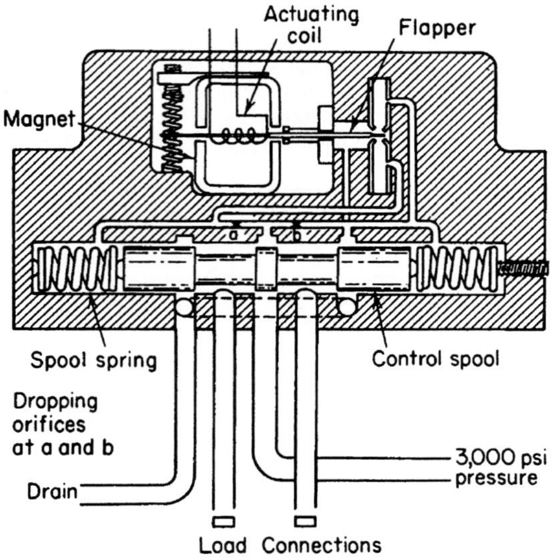

Figure 10.2.29 Two-stage electrohydraulic servo valve. The first-stage is a four-way flapper valve with a calibrated pressure output driving a second-stage, spring-loaded, four-way spool valve. (*Moog Servocontrols, Inc.*)

feedback occurs because of the increased flow across restrictor *a*. The power spool moves to the left until the unbalanced pressure is matched by the spring resistance. The disadvantage of this valve is the continual leakage flow through the flapper nozzle, but the torque motor has a low-power requirement and is inexpensive.

The major disadvantages of spool-type valves are (1) high cost, because of high-tolerance requirements between the valve lands, (2) high static friction and inertia, and (3) susceptibility to dirt. The **flapper** valve is less expensive to manufacture than a spool valve of equivalent characteristics and is not so susceptible to damage by dirt particles. In Fig. 10.2.30, P_1, the supply pressure, is constant. Input motion of the flapper toward the nozzle increases P_2 and drives the piston toward the right. The steady-state characteristic P_2 versus x is shown in Fig. 10.2.31.

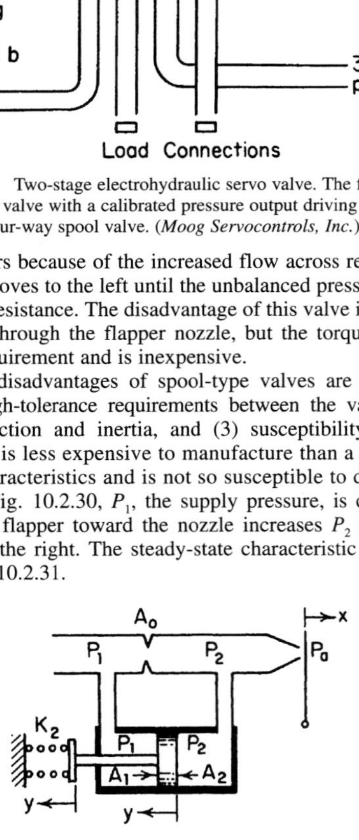

Figure 10.2.30 Flapper valve. (*Raven.*)

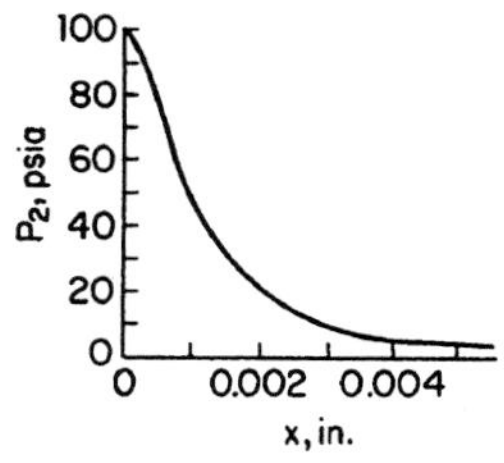

Figure 10.2.31 Equilibrium curve of P_2 versus x for a flapper valve. (*Raven, "Automatic Control Engineering," McGraw-Hill.*)

STEADY-STATE PERFORMANCE

The steady-state error of a control system can be determined by using the final value theorem in which

$$\theta_o(t)_{t\to\infty} = s\theta_o(s)_{s\to 0} \quad \text{provided } \theta_o(t) \text{ is stable}$$

The classification of control systems according to the form of the open-loop transfer function facilitates the determination of the steady-state errors when the system is subjected to various inputs.

The open-loop transfer function $\theta_o(s)/\varepsilon(s) = KG(s)$ may be written

$$KG(s) = \frac{\theta_o(s)}{\varepsilon(s)}$$

$$= \frac{K(1+\tau_a s)(1+\tau_a s)(1+\tau_o s)\cdots}{s^N(1+\tau_1 s)(1+\tau_2 s)(1+\tau_3 s)\cdots} \qquad (10.2.51a)$$

The **system type** is given according to the value of N as

Type 0 system $N = 0$

Type 0 system $N = 0$ (10.2.51*b*)

Type 0 system $N = 0$

Type 0 system $N = 0$

Error coefficients, based on a system with unity feedback [$H(s) = 1$], are defined as

$$\text{Positional error constant} = K_o = \lim_{s\to 0} KG(s) \qquad (10.2.52a)$$

$$\text{Velocity error constant} = K_v = \lim_{s\to 0} sKG(s) \qquad (10.2.52b)$$

$$\text{Acceleration error constant} = K_a = \lim_{s\to 0} s^2 KG(s) \qquad (10.2.52c)$$

A summary of error coefficients for systems of different types is given in Table 10.2.1.

Table 10.2.1 Summary of Error Coefficients

N	K_o	K_v	K_a
0	const	0	0
1	∞	const	0
2	∞	∞	const

A summary of the errors for types 0, 1, and 2 systems, when subjected to various inputs is given in Table 10.2.2.

Table 10.2.2 Summary of Errors

Input error	$\theta_i(t) = A$ ε_0/A	$\theta_i(t) = vt$ ε_v/v	$\theta_i(t) = at^2$ ε_a/A
$N = 0$	$1/(1 + K_0)$	∞	∞
1	0	$1/K_v$	∞
2	0	0	$1/K_a$

The higher the system type, the better is the output able to follow the higher degrees of input. Higher-type systems, however, are more difficult to stabilize, and a compromise must be made between the steady-state error and the settling time of the response.

CLOSED-LOOP BLOCK DIAGRAM

The **transfer functions** of the process, controller, and final control element can be combined with the measuring transducer into a **block diagram** of the complete control loop. Consider the thermal system of Fig. 10.2.5 and the heat balance on the vessel jacket:

$$M_j c \frac{dT_w}{dt} = w_w c(T_{w1} - T_w) \qquad (10.2.53)$$

Linearizing and rearranging:

$$\frac{M_j}{w_w} T_w s + T_w = \left[\frac{T_{w1} - T_w}{w_w}\right]\left[\frac{w_w}{x}\right] x \qquad (10.2.54)$$

where M_j = weight of cooling water in jacket, lb (kg); c = specific heat of water, Btu/lb · °F (J/kg · °C); T = temperature, °F; x = valve stem position (normalized 0 to 1); and w_w = water flow, lb/min (kg/min).

For simplicity, let the jacket time constant M_j/w_w be so small as to be negligible (not typically the case). The block diagram of Fig. 10.2.32 then represents process diagram Fig. 10.2.5. The **gain** K

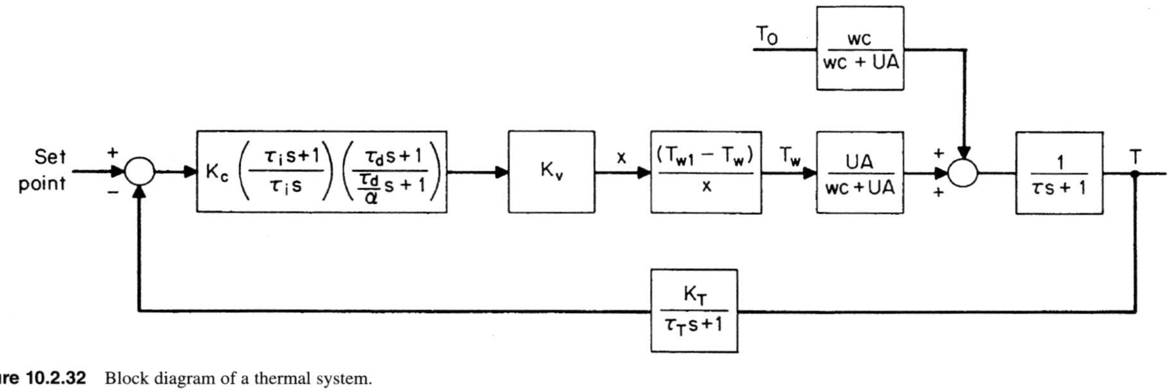

Figure 10.2.32 Block diagram of a thermal system.

in each case is an incremental change in output divided by an incremental change in input. For example, if the temperature transmitter has a temperature span of 100°F and an output of 4 to 20 mA, the gain would be

$$K_\tau = \frac{20-4}{100} = \frac{16}{100} = 0.16 \text{ mA/°F}$$

The **open-loop gain** is defined as the product of all the gains in the control loop:

$$K_o = K_c K_v K_T \left(\frac{T_{w1} - T_w}{x}\right)\left(\frac{UA}{wc + UA}\right)$$

Combining the **closed-loop system** of Fig. 10.2.32 results in the closed loop equation

$$T = \frac{\left(\dfrac{wc}{wc+UA}\right)\left(\dfrac{1}{\tau s+1}\right) T_0}{1 + \dfrac{K_o(\tau_i s+1)(\tau_d s+1)}{\tau_i s(\tau s+1)\left(\dfrac{\tau_d}{\alpha} s+1\right)(\tau_T s+1)}} \tag{10.2.55}$$

The time-dependent modes on the controller have been designed to compensate for the dynamics of the rest of the loop. Let

$$\tau_i = \tau \quad \text{and} \quad \tau_d = \tau_T$$

After cancellation, Eq. (10.2.55) becomes

$$T = \frac{\left(\dfrac{wc}{wc+UA}\right) T_0}{(\tau s+1)\left[1 + \dfrac{K_c}{\tau_i s\left(\dfrac{\tau_d}{\alpha} s+1\right)}\right]} \tag{10.2.56}$$

Equation (10.2.56) can be multiplied out to become

$$T = \frac{\left(\dfrac{wc}{wc+UA}\right)(\tau_i s)\left(\dfrac{\tau_d}{\alpha} s+1\right) T_0}{(\tau s+1)\left(\dfrac{\tau_i \tau_d}{\alpha} s^2 + \tau_i s + K_c\right)} \tag{10.2.57}$$

Equation (10.2.57) can be written in the standard form of Eqs. (10.2.16) and (10.2.21):

$$T = \frac{\left(\dfrac{wc}{wc+UA}\right)(\tau_i s)\left(\dfrac{\tau_d}{\alpha} s+1\right) T_0}{K_c(\tau s+1)\left(\tau_c^2 s^2 + 2\tau_c \zeta s + 1\right)} \tag{10.2.58}$$

Controller gain K_c can then be chosen to give the best response τ_c with stability ζ. More rigorous techniques for determining controller parameter values are shown under Bode diagrams and Routh's criterion. Table 10.2.3 provides preliminary guidance for the selection of control modes.

FREQUENCY RESPONSE

Although it is the time response of the control system that is of major importance, study of the effect on transient response of changes in system parameters, either in the process or controller, is more conveniently made from a **frequency-response** analysis of the system. The **frequency response** of a system is the steady-state output of the system to input sinusoids of varying frequency. The output for a linear system can be completely described in terms of the amplitude ratio of the output sinusoid to the input sinusoid. The amplitude ratio or gain, and phase, are functions of the frequency of the input sinusoid.

The use of **sinusoidal methods** to analyze and test dynamic systems has gained widespread popularity because system response is obtained easily from the response of the individual elements, no matter how many elements are included. By contrast, transient analysis is quite tedious, with only three dynamic components in the system, and is too difficult to be worthwhile for four or more components.

Frequency-response analysis provides quantitative information concerning the system on maximum gain K_o for stability, time response τ_c, ω_n, and the controller parameter adjustments τ_i, τ_d. In most cases, this information is adequate for assurance of stability, for comparing alternatives, or for judging the merits of a proposed control system. System parameters are generally obtained from open-loop frequency response, which is the response with the loop broken between the controller and the final control element (Fig. 10.2.32).

Frequency response also lends itself to system identification by testing for systems not readily amenable to mathematical analysis by subjecting the system to input sinusoids of varying frequency.

Table 10.2.3 Process Characteristics versus Mode of Control

Number of process capacities	Process reaction rate	Process time lags: Resistance-capacitance (RC)	Process time lags: Dead time (transportation)	Load changes: Size	Load changes: Speed	Suitable mode of control
Single	Slow	Moderate to large	Small	Any	Any	Two-position. Two-position with differential gap
				Moderate	Slow	Multiposition. Proportional input
Single (self-regulating)	Fast	Small	Small	Any	Slow	Floating modes: Single-speed Multispeed
					Moderate	Proportional-speed floating
Multiple	Slow to moderate	Moderate	Small	Small	Moderate	Proportional position
Multiple	Moderate	Any	Small	Small	Any	Proportional plus rate
Multiple	Any	Any	Small to moderate	Large	Slow to moderate	Proportional plus reset
Multiple	Any	Any	Small	Large	Fast	Proportional plus reset plus rate
Any	Faster than that of the control system	Small or nearly zero	Small to moderate	Any	Any	Wideband proportional plus fast reset

SOURCE: Considine, "Process Instruments and Controls Handbook," McGraw-Hill.

The shift from the **complex domain** to the **frequency domain** (linear systems, zero initial conditions) is accomplished by simply substituting $j\omega$ for s. Consider Eq. (10.2.38).

$$(\tau s+1)T = K_1 T_0 \tag{10.2.59}$$

Substituting

$$\frac{T}{T_0} = \frac{K_1}{\tau j\omega + 1}$$

Multiply by the complex conjugate

$$\frac{T}{T_0} = \frac{K_1}{\tau j\omega + 1}\,\frac{\tau j\omega - 1}{\tau j\omega - 1}$$

Collecting terms

$$\frac{T}{T_0} = K_1\left[\frac{1}{\tau^2\omega^2+1} - j\,\frac{\tau\omega}{\tau^2\omega^2+1}\right] \tag{10.2.60}$$

The first term is the real part of the solution, and the second term, the imaginary part. These terms define a vector, shown in Fig. 10.2.33, which has magnitude M and phase angle υ as determined either graphi cally or by the Pythagorean theorem.

$$M = \left[\frac{1}{\tau^2\omega^2+1}\right]^{1/2} \tag{10.2.61}$$

$$\theta = \tan^{-1}\left(\frac{-\zeta\omega}{1}\right) \tag{10.2.62}$$

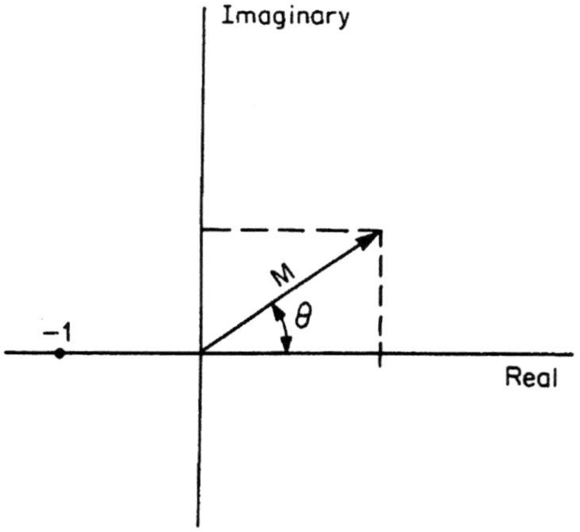

Figure 10.2.33 Polar plot showing vector.

GRAPHICAL DISPLAY OF FREQUENCY RESPONSE

Sinusoidal response is **plotted** in three different ways: (1) the rectangular plot with the log of the amplitude versus the log of the frequency, called a **Bode plot;** (2) a phase-margin plot with magnitude shown versus a function of phase with frequency as a parameter, called a **Nichols plot;** and (3) a polar plot with magnitude and phase shown in vector form with frequency as a parameter, called a **Nyquist plot.** The Nichols plot is actually a plot of phase margin (180° – θ) versus frequency and is used to define system performance. The polar plot enables the absolute stability of the system to be determined without the need for obtaining the roots of the characteristic equation. The logarithmic (Bode) plot has the advantage of ease in plotting, especially in design, since the individual effects of cascaded elements can be gaged by superposition. All the graphical procedures use the "minus 1" point [discussed with Eq. (10.2.38)] as the criterion for stability (magnitude = 1, phase = –180°).

NYQUIST PLOT

The **Nyquist diagram** is a graphical method of determining stability. The diagram is essentially a mapping into the $G(s)$ plane of a contour enclosing the major portion of the right half of the s plane. Figure 10.2.34 shows the Nyquist plot of a typical system.

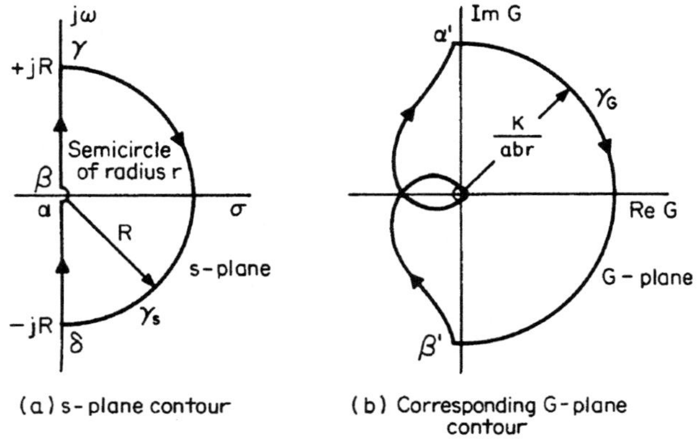

Figure 10.2.34 Nyquist diagram for $G(s) = 1/[s(s + a)(s + b)]$. (*Truxal, "Control System Synthesis," McGraw-Hill.*)

The frequency response of a closed-loop system can also be derived from the **polar plot** of the direct transfer function $KG(s)$. Systems with dynamic elements in the feedback can be transformed into direct systems by block-diagram manipulation. In Fig. 10.2.35, the amplitude ratio θ_o/θ_i at any frequency ω is the ratio of the lengths of the vectors $O\beta$ and $\alpha\beta$. The angle formed by the vectors $\alpha\beta$ and $O\beta$ is the phase angle of the frequency response $\angle\theta_o/\theta_i(j\omega)$. The numerator of the transfer function is a constant representing the closed-loop gain. Changing the gain proportionately changes the length of the vector $O\beta$. Figure 10.2.36 shows another typical Nyquist plot.

The Nyquist diagram is widely used in the design of electrical systems, but plots are very tedious to make and are generated via computer programs.

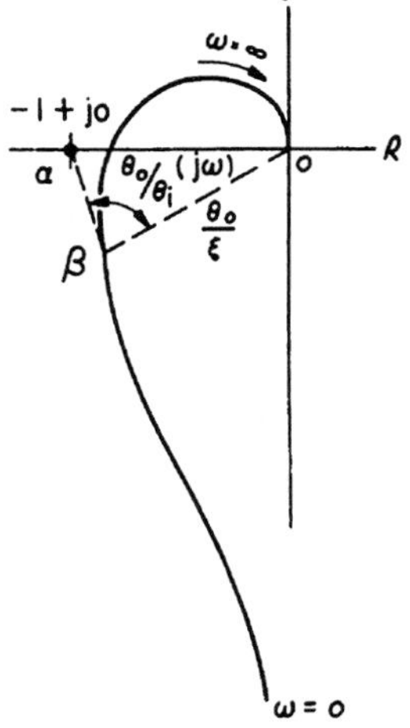

Figure 10.2.35 Typical $(\theta_o/\varepsilon)/(j\omega)$ plot.

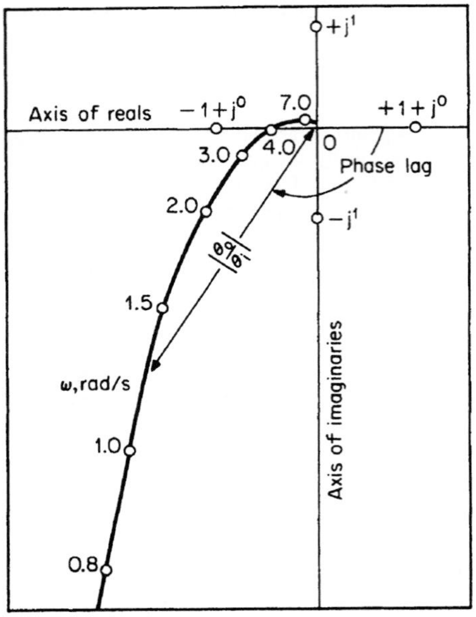

Figure 10.2.36 Typical Nyquist diagram showing loop-transfer function $|\theta_o/\theta_i|$. (*ASME, "Terminology for Automatic Control," Pub. ASA C85.1-1963.*)

BODE DIAGRAM

The most common method of presenting the **response data at various frequencies** is to use the log-log plot for amplitude ratios, accompanied by a semilog plot for phase angles. These rectangular plots are called **Bode diagrams,** after H. W. Bode, who did basic work on the theory of feedback amplifiers. A major advantage of this method is that the numerical computation of points on the curve is simplified by the fact that log magnitude versus log frequency can be approximated to engineering accuracy with straight-line approximations. In the **sinusoidal testing** of a system, the system dynamics will be faster than a very low frequency test signal, and the output amplitude has time to recover fully to the input amplitude. Thus the magnitude ratio is 1. Conversely, at high frequency, the system does not have time to reach equilibrium, and output amplitude will decrease. Magnitude ratio goes to a very small value under this condition. Thus the system will show a decrease in magnitude ratio as frequency increases. By similar reasoning, the difference between input and output phase angles becomes progressively more negative as frequency increases.

Consider the equations for magnitude, Eq. (10.2.61), and phase, Eq. (10.2.62). If $\tau_w \ll 1$,

$$M = \left(\frac{1}{1}\right)^{1/2} = 1 \quad \theta = \tan^{-1}(0) = 0$$

and if $\tau_w \gg 1$,

$$M = \left(\frac{1}{\tau^2\omega^2}\right)^{1/2} = \frac{1}{\tau\omega} \approx 0 \quad \theta = \tan^{-1}\left(\frac{-\tau_w}{1}\right) \approx -90°$$

The two **straight-line asymptotes** of magnitude, if extended, would intersect at $\tau\omega = 1$. Thus a relationship between frequency and time constant is established. The point at which $\tau\omega = 1$ is called the *corner frequency,* or the *break frequency.* The asymptote past the break frequency is easily plotted from a point at the break frequency $\omega = 1/\tau$ and a point at 10 times the frequency and 0.1 times the magnitude ratio.

The frequency response of systems containing different dynamic elements may be calculated by suitably employing the magnitude and

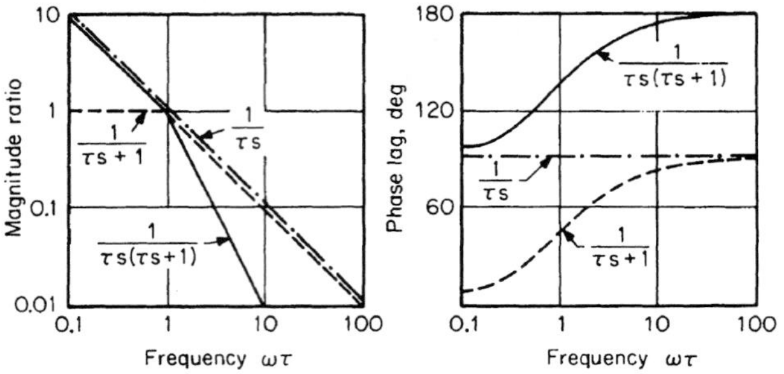

Figure 10.2.37 Combined transfer functions. (*Reproduced by permission. Copyright © John Wiley & Sons, Inc. Publishers, 1958. From D. P. Eckman, "Automatic Process Control."*)

phase characteristics of each element separately. Combined magnitudes are calculated

$$|M_1M_2| = |M_1| \times |M_2|$$

Or, on the log-log Bode diagram,

$$\log |M_1M_2| = \log |M_1| + \log |M_2|$$

The logarithmic scale permits **multiplication** by adding vertical distances. For phase

$$\underline{/\theta_1\theta_2} = \angle\theta_1 + \angle\theta_2$$

Thus **phase addition** is also made by adding vertical distances. Figure 10.2.37 shows the combination of the straight-line asymptotes for the following transfer function:

$$G(s) = \frac{1}{\tau s(\tau s + 1)} \tag{10.2.63}$$

Table 10.2.4 lists additional frequency-response relationships for system elements. The noninteracting controller, item 8, has the advantage of no interaction among tuning parameters but has the disadvantages of being more difficult to plot on the Bode diagram and of not offering the direct compensation of the controller shown in Fig. 10.2.24.

Figure 10.2.38 shows a typical combined Bode diagram. The ordinate has a scale in decibels in addition to the magnitude ratio scale. Decibels (dB) are sometimes used to provide a linear scale. They are defined

$$dB = -20 \log M$$

The discontinuity at the break frequency due to the meeting of the straight-line asymptotes will not be found in actual test data. Instead, the corner will be rounded, as shown dotted in Fig. 10.2.39. The error due to the approximation is some 3 dB at the break frequency, as shown in the figure.

Quadratic terms (Table 10.2.4) also lend themselves to asymptotic approximation, but the form of the magnitude and phase plots around the break frequency depends on the damping coefficient ζ. The slope of the magnitude ratio attenuation past the break point is twice that of the first-order system.

CONTROLLERS ON THE BODE PLOT

Controller modes, gain, reset, and derivative are plotted individually (in the form of Fig. 10.2.24) on the Bode diagram and summed as required. The functions for reset (integral) and derivative modes are moved horizontally on the diagram to determine proper parameter adjustment. The combined magnitude ratio curve for the whole system is moved vertically to determine open-loop gain.

After all the dynamic elements of the system are plotted and combined, reset and derivative are set in accordance with the following **criteria:**

Reset: The reset phase curve is positioned horizontally so that it contributes 10° of phase lag where the system phase lag is 170°.

Derivative: The derivative phase curve is positioned horizontally so that it contributes 60° of phase lead where the system phase lag is 180°.

Table 10.2.4 Frequency-Response Equations for Some Common Control-System Elements

Description	Transfer function $G(s)$	Frequency response $G(j\omega)$	Magnitude ratio	Phase angle
1. Dead time	$\varepsilon^{-T_L s}$	$\varepsilon^{-j\omega T_L}$	1	$-\omega T_L$ radians
2. First-order lag	$\dfrac{1}{Ts+1}$	$\dfrac{1}{j\omega T+1}$	$\dfrac{1}{\sqrt{\omega^2T^2+1}}$	$-\tan^{-1}(\omega T)$
3. Second-order lag	$\dfrac{1}{(Ts+1)(aTs+1)}$	$\dfrac{1}{-a\omega^2T^2+j(1+a)\omega T+1}$	$\dfrac{1}{(1-a\omega^2T^2)^2+(1+a)^2\omega^2T^2}$	$-\tan^{-1}\left[\dfrac{(1+a)\omega T}{1-aT^2\omega^2}\right]$
4. Quadratic (underdamped)	$\dfrac{1}{\left(\dfrac{s}{\omega_n}\right)^2+\dfrac{2\zeta}{\omega_n}s+1}$	$\dfrac{1}{-\left(\dfrac{\omega}{\omega_0}\right)^2+j2\zeta\dfrac{\omega}{\omega_n}+1}$	$\dfrac{1}{\sqrt{\left(1-\dfrac{\omega^2}{\omega_n^2}\right)+4\zeta^2\left(\dfrac{\omega}{\omega_n}\right)^2}}$	$-\tan^{-1}\left[\dfrac{2\zeta\dfrac{\omega}{\omega_n}}{1-\left(\dfrac{\omega}{\omega_n}\right)^2}\right]$
5. Ideal proportional controller	K	K	K	0
6. Ideal proportional-plus-reset controller $T_i=\dfrac{1}{r}$ r = reset rate	$K\left(1+\dfrac{1}{T_is}\right)$ or $K\dfrac{T_is+1}{T_is}$	$K\left(1+\dfrac{1}{j\omega T_i}\right)$ or $K\dfrac{j\omega T_i+1}{j\omega T_i}$	$K\sqrt{1+\left(\dfrac{1}{\omega T_i}\right)^2}$	$-\tan^{-1}\left(\dfrac{1}{\omega T_i}\right)$
7. Ideal proportional-plus-rate controller	$K(1+T_ds)$	$K(1+j\omega T_d)$	$K\sqrt{1+\omega^2T_d^2}$	$\tan^{-1}(\omega T_d)$
8. Ideal proportional-plus-reset-plus-rate controller	$K\left(1+T_ds+\dfrac{1}{T_is}\right)$	$K\left(1+j\omega T_d+\dfrac{1}{j\omega T_i}\right)$ or $K\dfrac{j\omega T_i-\omega^2T_dT_i+1}{j\omega T_i}$	$K\sqrt{(\omega T_i)^2+(1-\omega^2T_dT_i)^2}$	$\tan^{-1}\left(\omega T_d-\dfrac{1}{\omega T_i}\right)$

SOURCE: Considine, "Process Instruments and Controls Handbook," McGraw-Hill.

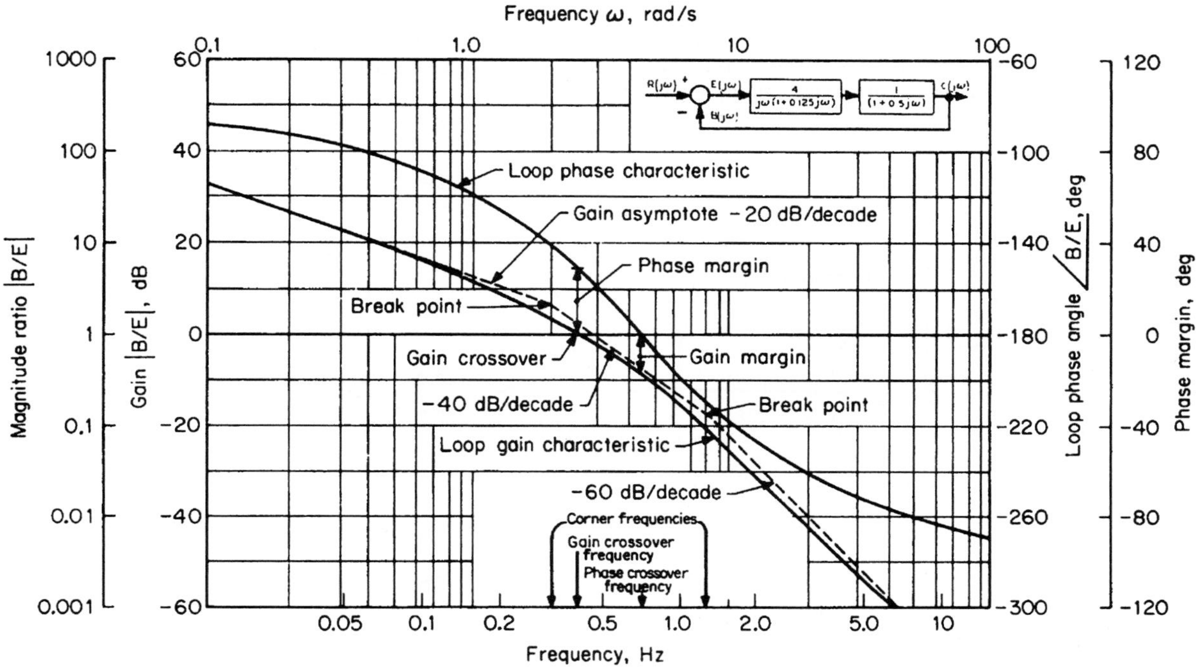

Figure 10.2.38 Typical Bode diagram showing loop-transfer function *B/E*. (*ASME, "Terminology for Automatic Control," Pub. ASA C85.1-1963.*)

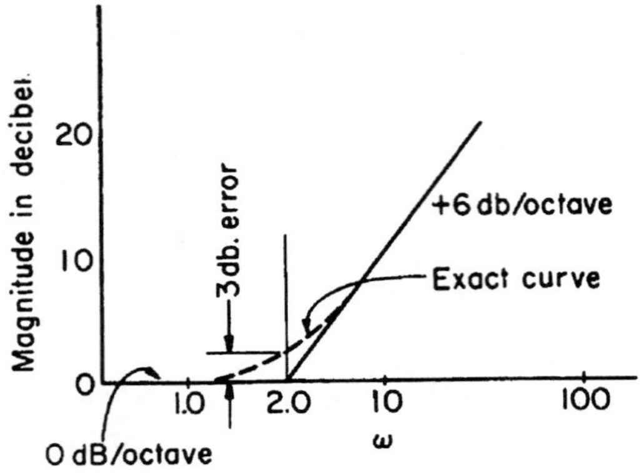

Figure 10.2.39 Bode plot of term $(j\omega\tau_a + 1)$.

The magnitude ratio curves are positioned with the phase-angle curves, and parameter values for reset and derivative can be read at the break points.

STABILITY AND PERFORMANCE OF AN AUTOMATIC CONTROL

An automatic-control system is stable if the amplitude of transient oscillations decreases with time and the system reaches a steady state. The stability of a system can be evaluated by examining the roots of the differential equation describing the system. The presence of positive real roots or complex roots with positive real parts dictates an unstable system. Any stability test utilizing the open-loop transfer function or its plot must utilize this fact as the basis of the test.

The Nyquist Stability Criterion The $KG(j\omega)$ locus for a typical single-loop automatic-control system plotted for all positive and negative frequencies is shown in Fig. 10.2.40. The locus for negative values of ω is the mirror image of the positive ω locus in the real axis. To complete the diagram, a semicircle (or full circle if the locus approaches $-\infty$ on the real axis) of infinite radius is assumed to connect in a positive sense, the + locus at $\omega \to 0$ with the negative locus at $\omega \to -0$. If this locus is traced in a positive sense from $\omega \to \infty$ to $\omega \to 0$, around the circle at ∞, and then along the negative-frequency locus the following may be concluded: (1) if the locus **does not** enclose the $-1 + j0$ point, the system is stable; (2) if the locus **does** enclose the $-1 + j0$ point, the system is unstable. The Nyquist criterion can also be applied to the log magnitude of $KG(j\omega)$ and phase-versus-log ω diagrams. In this method of display, the criterion for stability reduces to the requirement that the log magnitude of $KG(j\omega)$ must cross the 0-dB axis at a frequency less than the frequency at which the phase curve crosses the $-180°$ line. Two stability conditions are illustrated in Fig. 10.2.41. The Nyquist criterion not only provides a simple test for the stability of an automatic-control system but also indicates the degree of stability of the system by indicating the degree to which $KG(j\omega)$ locus avoids the $-1 + j0$ point.

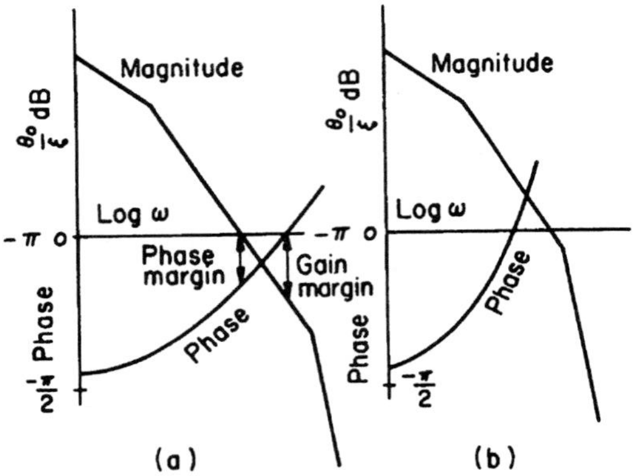

Figure 10.2.41 Nyquist stability criterion in terms of log magnitude $KG(j\omega)$ diagrams. (*a*) Stable; (*b*) unstable. *(Porter, "An Introduction to Servomechanism," Wiley.)*

The concepts of **phase margin** and **gain margin** are employed to give this quantitative indication of the degree of stability of an automatic-control system. **Phase margin** is defined as the additional negative phase shift necessary to make the phase angle of the transfer function $-180°$ at the frequency where the magnitude of the $KG(j\omega)$ vector is unity. Physically, phase margin can be interpreted as the amount by which the unity KG vector has to be shifted to make a stable system unstable.

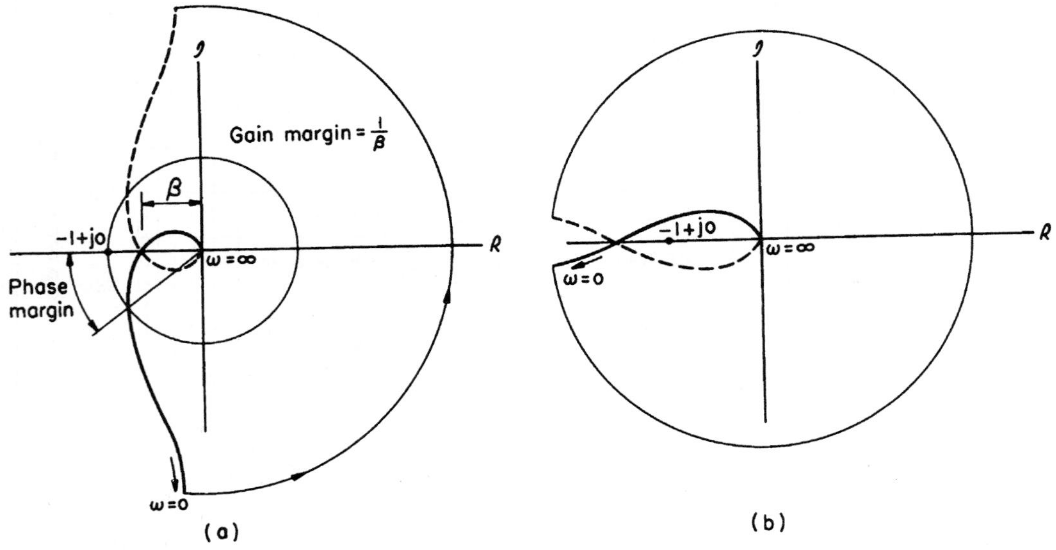

Figure 10.2.40 Typical $KG(j\omega)$ loci illustrating application of Nyquist's stability criterion. (*a*) Stable; (*b*) unstable.

In a similar manner, **gain margin** is defined as the reciprocal of the magnitude of the *KG* vector at −180°. Physically, gain margin is the number by which the gain must be multiplied to put the system to the limit of stability. Thaler suggests satisfactory results can be obtained in most control applications if the phase margin is between 40 and 60° while the gain margin is between 3 and 10 (10 to 20 dB). These values will ensure a small transient overshoot with a single cycle in the transient. The margin concepts are qualitatively illustrated in Figs. 10.2.40 and 10.2.41.

Routh's Stability Criterion The frequency-response equation of a closed-loop automatic control is

$$\frac{\theta_o}{\theta_i} = \frac{KG(j\omega)}{1+KG(j\omega)} \tag{10.2.64}$$

The characteristic equation obtained therefrom has the algebraic form

$$A(j\omega)^n + B(j\omega)^{n-1} + C(j\omega)^{n-2} + \cdots = 0 \tag{10.2.65}$$

The purpose of Routh's method is to determine the existence of roots of this equation which are positive or which are complex with positive real parts and thus identify the resulting instability. To apply the criterion the coefficients are written alternately in two rows as

$$\begin{matrix} A & C & E & G \\ B & D & F & H \end{matrix}$$

This array is then expanded to

$$\begin{matrix} A & C & E & G \\ B & D & F & H \\ \alpha_1 & \alpha_2 & \alpha_3 & \\ \beta_1 & \beta_2 & \beta_3 & \\ \gamma_1 & \gamma_2 & & \end{matrix}$$

where α_1, α_2, α_3, β_1, β_2, β_3, γ_1, and γ_2 are computed as

$$\alpha_1 = \frac{BC-AD}{B} \qquad \beta_1 = \frac{D\alpha_1 - B\alpha_2}{\alpha_1}$$

$$\alpha_2 = \frac{BE-AF}{B} \qquad \beta_2 = \frac{F\alpha_1 - B\alpha_3}{\alpha_1} \qquad \gamma_1 = \frac{\beta_1\alpha_2 - \alpha_1\beta_2}{\beta_1}$$

$$\alpha_3 = \frac{BG-AH}{B} \qquad \beta_3 = \frac{H\alpha_1 - B_o}{\alpha_1} \qquad \gamma_2 = \frac{\beta_1\alpha_3 - \alpha_1\beta_3}{\beta_1}$$

When the array has been computed, the left-hand column (A, B, α_1, β_1, γ_1) is examined. If the signs of all the numbers in the left-hand column are the same, there are no positive real roots. If there are changes in sign, the number of positive real roots is equal to the number of changes in sign. It should be recognized that this is a test for instability; the absence of sign changes does not guarantee stability.

SAMPLED-DATA CONTROL SYSTEMS

Definition Sampled-data control systems are those in which continuous information is transformed at one or more points of the control system into a series of pulses. This transformation may be performed intentionally, for example, the flow of information over long distances to preserve the accuracy of the data during the transmission, or it may be inherent in the generation of the information flow, for example, radiating energy from a radar antenna which is in the form of a train of pulses, or the signals developed by a digital computer during a direct digital control of machine-tool operation.

Methods of analysis analogous to those for continuous-data systems have been developed for the sampled-data systems. Discussed herein are (1) sampling, (2) the z transformation, (3) the z-transfer function, and (4) stability of sampled data systems.

Sampling The ideal sampler is a simple switch (Fig. 10.2.42) which is closed only instantaneously and opens and closes at a constant frequency. The switch, which may or may not be a physical component in a sampled-data feedback system, indicates a sampled signal. Such a sampled-data feedback system is shown in Fig. 10.2.43. The error signal in continuous form is $\epsilon(t)$, and the sampled error signal is $\epsilon^*(t)$. Figure 10.2.42 shows the relationship between these signals in a graphical form.

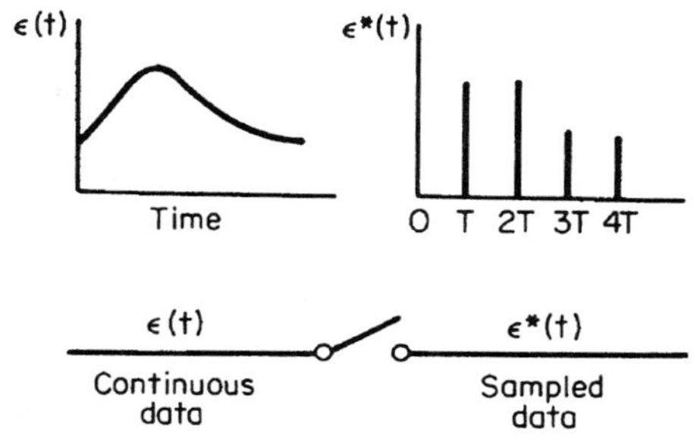

Figure 10.2.42 Ideal sampler, showing continuous input and sampled output.

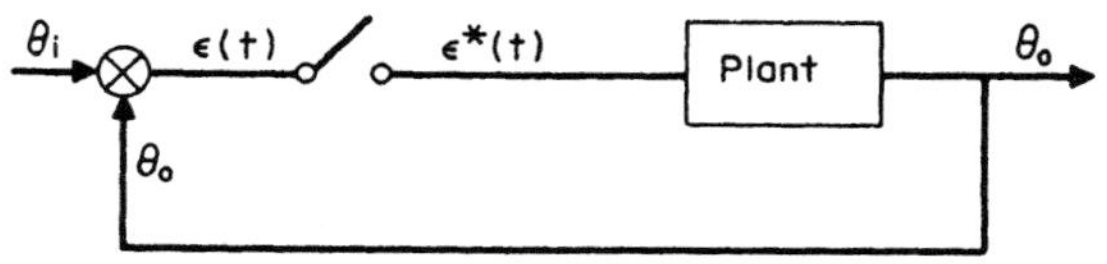

Figure 10.2.43 Sampled-data feedback system.

z Transformation In the analysis of continuous-data systems, it has been shown that the Laplace transformation can be used to reduce ordinary differential equations to algebraic equations. For sampled-data systems, an operational calculus, the z transform, can be used to simplify the analysis of such systems.

Consider the sampler as an impulse modulator; that is, the sampling modulates an infinite train of unit impulses with the continuous-data variable. Then the Laplace transformation can be shown to be

$$F^*(s) = \sum_{n=0}^{\infty} f(nT)e^{-nTs} \tag{10.2.66}$$

(See A. W. Langill, "Automatic Control Systems Engineering.")

In terms of the z transform, this becomes

$$F^*(s) = F(z) = \sum_{n=0}^{\infty} f(nT)z^{-n} \tag{10.2.67}$$

where $z^{-n} = e^{-nTs}$.

Table 10.2.5 lists the z transforms for some of these functions.

Table 10.2.5 Laplace and z Transforms

Time function	$f(t)$	$F(s) = £[f(t)]$	$F^*(z) = F(z)$
Unit ramp function	t	$\frac{1}{s^2}$	$\frac{Tz}{(z-1)^2}$
Unit acceleration function	$\frac{t^2}{2}$	$\frac{1}{s^3}$	$\frac{1}{2}T^2\frac{z(z+1)}{(z-1)^3}$
Exponential function	$e^{-\alpha t}$	$\frac{1}{s+\alpha}$	$\frac{z}{z-e^{-\alpha T}}$
Sinusoidal function	$\sin\beta t$	$\frac{\beta}{s^2+\beta^2}$	$\frac{z\sin\beta T}{z^2-2z\cos\beta T+1}$
Cosinusoidal function	$\cos\beta t$	$\frac{s}{s^2+\beta^2}$	$\frac{z(z-\cos\beta T)}{z^2-2z\cos\beta T+1}$

Table 10.2.6 Typical Block Diagrams of Sampled-Data Control Systems and Their Transforms

System	Laplace transform of the output, $C(s)$	z transform of the output, $C^*(z)$
1. 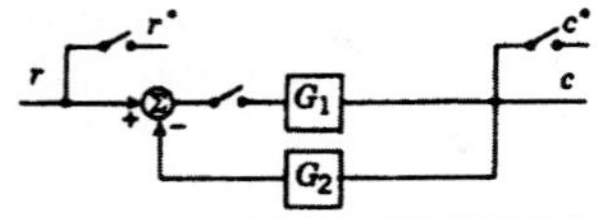	$\frac{G_1(s)}{1+G_1G_2^*(z)}R^*(z)$	$\frac{G_1^*(z)}{1+G_1G_2^*(z)R^*(z)}$
2. 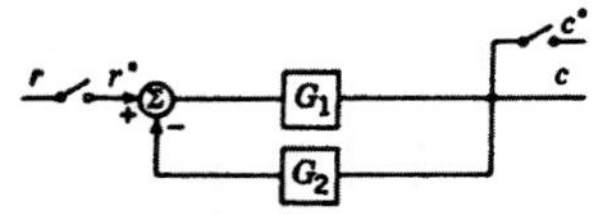	$\frac{G(s)}{1+G_1(s)G_2(s)}R^*(z)$	$\left[\frac{G_1(s)}{1+G_1(s)G_2(s)}\right]^* R^*(z)$
3. 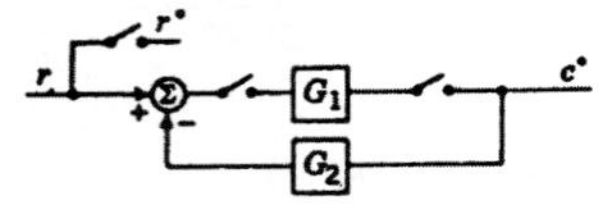		$\frac{G_1^*(z)}{1+G_1^*(z)G_2^*(z)}R^*(z)$
4. 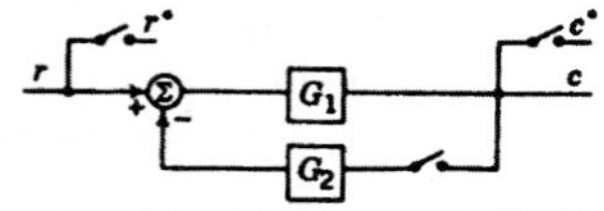	$G_1(s)\left[R(s)-G_2(s)\frac{RG_1^*(z)}{1+G_2^*(z)}\right]$	$\frac{RG_1^*(s)}{1+G_1G_2^*(s)}$
5. 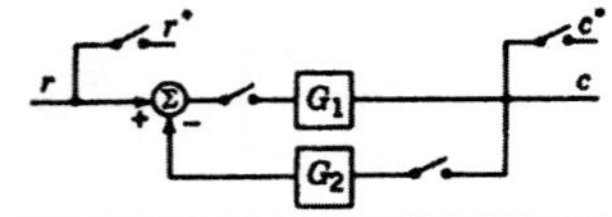	$G_1(s)\left[R^*(z)-\frac{G_1^*(z)G_2^*(z)R^*(z)}{1+G_1^*(z)G_2^*(z)}\right]$	$\frac{G_1^*(z)}{1+G_1^*(z)G_2^*(z)}R^*(z)$
6. 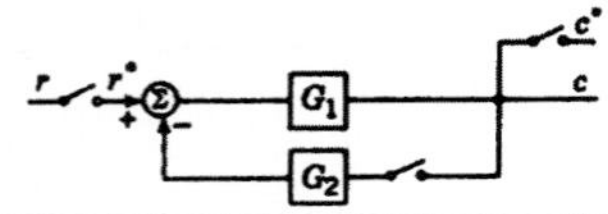	$G_1(s)\left[R^*(z)-G_2(s)\frac{R^*(z)G_1^*(z)}{1+G_1G_2^*(z)}\right]$	$\frac{G_1^*(z)}{1+G_1G_2^*(z)}R^*(z)$
7. 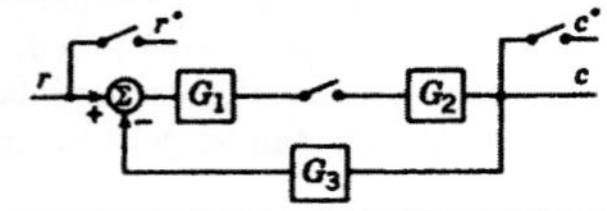	$\frac{G_2(s)}{1+G_1G_2G_3^*(z)}RG_1^*(z)$	$\frac{G_2^*(s)}{1+G_1G_2G_3^*(z)}RG_1^*(z)$
8. 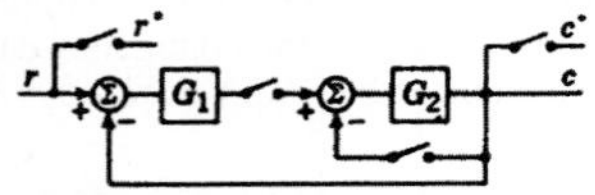	$\frac{G_2(s)}{1+G_2^*(z)+G_1G_2^*(z)}RG_1^*(z)$	$\frac{G_2^*(z)}{1+G_2^*(z)+G_1G_2^*(z)}RG_1^*(z)$

SOURCE: John G. Truxal (ed.), "Control Engineers' Handbook," McGraw-Hill.

It should be noted that unlike the Laplace transforms, the z-transform method imposes the restriction that the sampled-data-system response can be determined only at the sampling instant. The same z transform may apply to different time functions which may have the same value at the instant of sampling. The z-transform function is not defined, therefore, in a continuous sense, and the inverse z transform is not unique.

z-Transfer Function The ratio of the sampled output function of a discrete network to the sampled input function is the z-transfer function. A discrete network is one which has both a sampled input and output. A table of block diagrams of a number of sampled-data control systems with their associated transfer functions is presented in Table 10.2.6.

Stability of Sampled-Data Systems The stability of sampled-data systems can be demonstrated utilizing frequency-response methods which have been discussed in this section. See "Stability and Performance of an Automatic Control." The Nyquist stability criterion again applies with the same conclusions relative to the −1 point, and the methods of generalizing the open- and closed-loop frequency response plots remain the same.

MODERN CONTROL TECHNIQUES

Many engineers and researchers have been actively pursuing the application and development of advanced control strategies. In recent years this effort has included extensive work in the area of robust control theory. One of the primary attractions of this theory is that it generalizes the **single input-single system output (SISO)** concept of gain and phase margins, and its effect on system sensitivity to multivariable control system.

In this section the design technique for a **linear quadratic Gaussian with loop transfer recovery (LQG/LTR)** robust controller design method will be summarized. The concepts required to understand this method will be reviewed. A linearized model of a simple process has been chosen to illustrate this technique. The simple process has three state variables, one input, and one output. Three control system design methods are compared: LQG, a LQG/LTR, and a proportional plus integral controller (PI).

Previous Work and Results

Most applications of modern control techniques to process control have appeared in the form of **optimal control strategies** which were in the form of full-state feedback control with and without state estimation [linear quadratic regulator (LQR)]. The regulator problem, coupled with a state estimation problem, was shown not to have desirable robustness properties. Doyle and Stein (Doyle and Stein, 1979) proposed the multivariable robust design philosophy known as LQG/LTR to eliminate the shortcomings of the LQR and LQG design techniques. The LQG/LTR design technique has been applied to control submersible vehicles, engine speed, helicopters, ship steering, and large, flexible space structures.

The LQG/LTR method requires the plant to be **minimum phase,** which cannot be guaranteed in a plant process. Typically **nonminimum phase** conditions in a plant process are due to time delay. The LQG/LTR design procedure offers no guarantees when a nonminimum-phase plant is used. However recent work develops a method to obtain a robust controller design for plants with time delay (Murphy and Bailey, IEEE 1989 and IECON 1990).

The LQG/LTR synthesis method for minimum phase plants achieves desired loop shapes and maximum robustness properties in the design of feedback control systems. The LQG/LTR design procedure uses **singular value analysis** and design procedures to obtain the desired performance and stability robustness. The use of singular values yields a frequency-domain design and analysis method generally expected by control engineers to determine system robustness. The LQG/LTR synthesis method is applicable to both **multiple-input/multiple-output (MIMO)** and SISO control systems. The performance limitations of this design method are analogous to those of a SISO system.

Definition of Robustness

A system is considered to be robust and have **good robustness properties** if it has a large stability margin, good disturbance attenuation, and/or low sensitivity to parameter variations. The term **stability margin** refers to the gain margin and the phase margin, which are quantitative measures of stability. A proven method of obtaining good robustness properties is the use of a feedback control system, which can be designed to allow for variations in the system dynamics. Some causes of **variation** in system dynamics are

Modeling and data errors in the nominal plant and system.

Changes in environmental conditions, manufacturing tolerances, wear due to aging, and noncritical material failures.

Errors due to calibration, installation, and adjustments.

Feedback control systems with good feedback properties have been synthesized for SISO systems. Classical frequency domain techniques such as Nyquist, root-locus, Bode, and Nichols plots have been used to obtain the feedback control system for the SISO system. These design techniques have allowed the synthesis of feedback control systems yielding **insensitivity** to bounded parameter variations and a large stability margin. The success of the feedback control system for the SISO system has led to the direct extension of the classical frequency domain technique to the design of a multivariable feedback control system. This extension to the multivariable design problem examines an individual feedback loop as the phase and gain margins are varied, while the nominal phase and gain values in the remaining feedback paths are held constant. This technique, however, fails to consider the results of simultaneous variation of gain and phase in all paths, which is a real-world possibility and needs to be considered. A method of obtaining a feedback control system with good robustness properties that takes into consideration simultaneous gain and phase variation is the LQG/LTR.

The LQG/LTR technique can be applied to MIMO systems or SISO systems. The LQG/LTR technique not only has the good robustness properties of the classical frequency domain techniques but also is capable of minimizing the effects of unmodeled high-frequency dynamics, neglected nonlinearities, and a reduced-order model. **Computer control system design tools,** for instance, developed by Integrated Systems Incorporated (ISI) and The MathWorks can be used in synthesizing the LQG/LTR technique and other controller design techniques. The use of computer-aided design tools eliminates the numerical programming burden of programming the complex LQG/LTR algorithm.

In what follows, the needed concepts are defined and the LQG/LTR procedure for development of a **model-based compensator** (MBC) is described. A unity-feedback MBC is selected for the controller structure, because of its similarity to the SISO unity-feedback control system well known to classical control designers. The MBC closed-loop control system has proven to offer great practical considerations in the design of automatic control systems.

ROBUSTNESS CONCEPTS OF THE LQG AND LQG/LTR CONTROL SYSTEMS

The LQG/LTR design procedure is based on the system configuration of the LQG controller shown in Fig. 10.2.44. The LQG controller consists of a Kalman filter state estimator and a linear quadratic regulator. The Kalman filter state estimator has good robustness properties for plant perturbations at the plant output. The linear quadratic regulator (LQR) has good robustness properties for perturbations at the plant input. Even though its components separately have good robustness properties, the LQG controller is found to have no guaranteed robustness properties at either the input (point 2) or the output (point 3) of the plant.

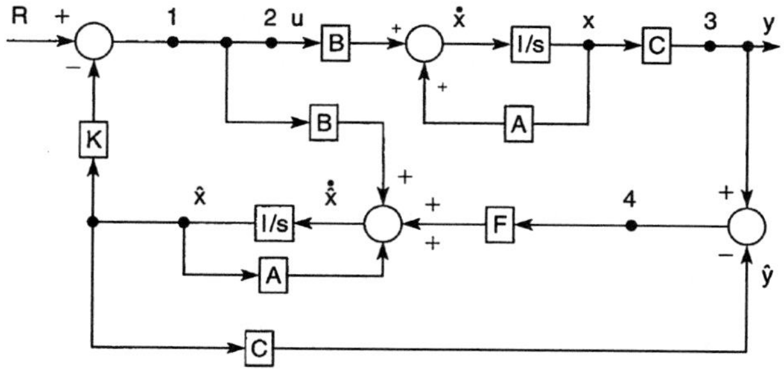

- LQG guaranteed no robustness properties at the input or output of the plant.
- Point 1 has the good robustness properties of the full-state feedback system.
- Point 4 has the good robustness properties of the Kalman filter.
- Point 2 has no guaranteed robustness properties.
- Point 3 has no guaranteed robustness properties.
- The LQG/LTR design method permits recovery of the robustness properties of point 1 at point 2 or the robustness properties of point 4 at point 3.

Figure 10.2.44 Summary of the robustness properties of the LQG block diagram.

The LQG/LTR design procedure allows us to recover robustness properties at either the input or the output of the plant. If robustness is desired at the input to the plant, first a nominal robust LQR design is made to satisfy the design constraints. Next, an LTR step is made to design a Kalman filter gain that recovers the robustness at the input to the plant of the LQG controller that is approximately that of the nominal LQR design. This implies from Fig. 10.2.44 that the robustness properties at points 1 and 2 are approximately the same.

If robustness is desired at the **output of the plant,** first a nominal robust Kalman filter design is made to satisfy the performance constraints. Next, an LTR step is made to design an LQR gain that recovers the robustness at the output of the plant that is approximately that of the nominal Kalman filter design. This implies from Fig. 10.2.44 that the robustness properties at points 3 and 4 are approximately the same.

The block diagram of the unity-feedback MBC is shown in Fig. 10.2.45. This control system structure allows tracking and regulation of a reference input at the output of the plant. The filter and regulator gains used in the controller $R(s)$ are obtained appropriately, depending on whether robustness is desired at the input or the output of the plant. Note that the robustness properties of the **unity-feedback MBC** (UFMBC) are the same as those of the LQG system, since the UFMBC is just an alternative structure of the LQG system.

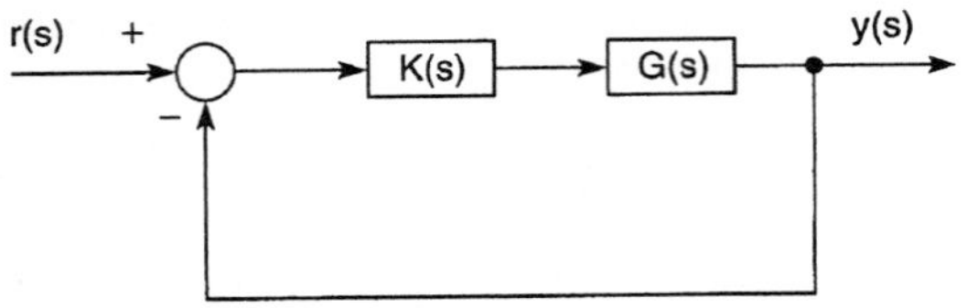

Figure 10.2.45 Unity-feedback model-based compensator (UFMBC).

MATHEMATICS AND CONTROL BACKGROUND

The use of **matrix algebra** is very important in the study of MIMO control system design methods. Some MIMO control system design methods use matrix algebra to decouple the system so that the plant matrix transfer function consists only of individual **decoupled transfer functions,** each represented conventionally as the ratio of two polynomials. This simplification then allows the use of conventional methods such as the Bode plot, Nyquist plot, and Nichols chart as a means of designing a controller for each loop of the MIMO plant. The LQG/LTR control system design method, however, does not require **decoupling** of the plant but rather preserves the coupling between the individual loops of the MIMO plant. Each step of the LQG/LTR design technique therefore requires the use of some matrix algebra, vector and matrix norms, and singular values during the control system design procedure. For instance, matrix algebra is used to derive a suitable representation of the closed-loop system for stability and robustness analysis. Matrix norms and singular values are used to examine the matrix magnitude of the MIMO transfer function. The **matrix magnitude or singular values** of a system allows preservation of the coupling between the various loops of the MIMO control system.

The concepts of matrix norm, singular values, controllability, observability, stabilizability, and detectability are very important in applying MIMO design procedures. It is therefore important that these concepts be well understood.

The purpose of this section is to provide **background material** in mathematics and control systems that has proved useful in deriving the procedures and analysis required to implement the control system design procedure.

Matrix Norm and Singular Values

The matrix norm of $\|\mathbf{A}\|_2$ is defined for the $m \times n$ matrix **A.** The matrix **A** defines a linear transformation from the vector space V, called the domain to a vector space W, which is called the range. Thus the linear transformation

$$\mathbf{A}(x) = \mathbf{A}x \qquad (10.2.68)$$

transforms a vector x in $V \in C^n$ into a vector $\mathbf{A}(x)$ in $W \in C^m$. The matrix norm of $\|\mathbf{A}\|_2$ is defined as

$$\|\mathbf{A}\| = \max_{x \neq 0} \frac{\|\mathbf{A}\|_2}{\|x\|_2} = l_2 \text{ norm} \qquad (10.2.69)$$

Since the l_2 **norm** is a useful tool in control system analysis, further insight into its definition is given. The l_2 norm of Eq. (10.2.69) can be written as

$$\|\mathbf{A}\|_2 = \max_i [\lambda_i(\mathbf{A}^H\mathbf{A})]^{1/2} \qquad i = 1,2,\ldots,n \qquad (10.2.70)$$

$$\|\mathbf{A}\|_2 = \max_i [\lambda_i(\mathbf{A}\mathbf{A}^H)]^{1/2} \qquad i = 1,2,\ldots,n \qquad (10.2.71)$$

The matrices $\mathbf{A}^H\mathbf{A}$ and $\mathbf{A}\mathbf{A}^H$ are Hermitian and positive semidefinite. The eigenvalues of these matrices are real and nonnegative. The definition of the **singular value** of a complex matrix $\mathbf{A} \in \mathbf{C}^{n \times n}$ is

$$\sigma_i(\mathbf{A}) = [\lambda_i(\mathbf{A}^H\mathbf{A})]^{1/2}$$
$$= [\lambda_i(\mathbf{A}\mathbf{A}^H)]^{1/2} \geq 0 \qquad i = 1,2,\ldots,n \qquad (10.2.72)$$

If $\mathbf{A} \in \mathbf{C}^{m \times n}$, which indicates that **A** is a nonsquare matrix, then

$$\sigma_i(\mathbf{A}) = [\lambda_i(\mathbf{A}^H\mathbf{A})]^{1/2} = [\lambda_i(\mathbf{A}\mathbf{A}^H)]^{1/2} \qquad (10.2.73)$$

for $1 \leq i \leq k$, where k is the number of singular values = $\min(m, n)$. The maximum and minimum singular values are defined, respectively, as

$$\bar{\sigma}(\mathbf{A}) = \|\mathbf{A}\|_2 \qquad (10.2.74)$$

and

$$\underline{\sigma}(\mathbf{A}) = \max_{x \neq 0} \frac{\|\mathbf{A}x\|}{\|x\|_2} = \frac{1}{\|\mathbf{A}^{-1}\|_2} \qquad (10.2.75)$$

provided **A** has an inverse.

Controllability, Observability, Stabilizability, and Detectability

The LQG/LTR control system design method has a certain **requirement** of the plant for which the controller is being designed: the plant must be at least stabilizable and detectable. Therefore it is important that the conditions of controllability, observability, and the less restricted conditions of stabilizability and detectability be understood. A description of these concepts follows.

Given the state equation for the linear time invariant (LTI) system described as $\dot{x}(t) = \mathbf{A}x(t) + \mathbf{B}u(t)$ and $y(t) = \mathbf{C}x(t)$. The matrices **A, B,** and **C** are constant. Note that $x \in R^n$, $n \in R^p$, and $y \in R^p$ are, respectively, the state, input, and output vectors. Variables n and p are the dimensions of the states and the input and output vectors. The input and output vectors are of the same dimension.

A system is said to be controllable if it is possible to transfer the system from any initial state x_0 in state space to any other state x_F using the input in a finite period of time. The system is said to be uncontrollable otherwise. The condition of controllability places **no constraint** on the input in the system.

The algebraic test for controllability of the LTI system state equation uses the controllability matrix

$$\mathbf{Q} = [\mathbf{B}\ \mathbf{AB}\ \cdots\ \mathbf{A}^{n-1}\mathbf{B}] \qquad (10.2.76)$$

where n is the **number of states.** The LTI system is controllable if the rank of **Q** is equal to n. If the rank of **Q** is less than n, the system is not controllable. An uncontrollable system indicates that some modes or poles are not affected by the control input. Uncontrollable systems with **unstable modes** imply that a controller design is not possible to ensure system stability. If the uncontrollable modes of the uncontrollable system are stable, then the system is stabilizable.

An LTI system is considered **observable** at initial time t_0 if there exists a finite time t_1 greater than t_0 such that for any initial state x_0 in state space using knowledge of the input u and output y over the time interval $t_0 <t> t_1$, the state x_0 can be determined. The system is said to be unobservable otherwise.

The algebraic test for observability of the LTI system uses the observability matrix

$$\mathbf{N} = [\mathbf{C}^T\ \mathbf{A}^T\mathbf{C}^T\ \cdots\ (\mathbf{A}^T)^{n-1}\mathbf{C}^T] \qquad (10.2.77)$$

The LTI system is observable if the rank of **N** is equal to n, where n is the number of states. If the rank of **N** is less than n, the system is unobservable. An unobservable system indicates that not all system modes contribute to the output of the system. If the **unobservable modes** (states) are stable, then the system is considered to be detectable.

EVALUATING MULTIVARIABLE PERFORMANCE AND STABILITY ROBUSTNESS OF A CONTROL SYSTEM USING SINGULAR VALUES

Performance Robustness

The requirements for the control system design are formulated by placing certain restrictions on the singular values of the **return ratio, return**

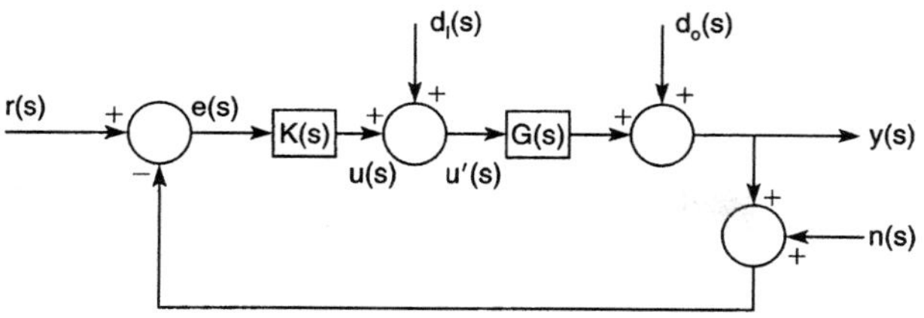

Figure 10.2.46 Model-based unity feedback MIMO control system.

difference, and **inverse return difference** of the control system. The block diagram in Fig. 10.2.46 shows the controller $K(s)$ and plant $G(s)$ with input $d_I(s)$ and output $d_o(s)$ disturbances and sensor noise $n(s)$.

Using Fig. 10.2.46, a **frequency-domain transfer function** representation for the plant output is first obtained. These transfer function representations, used in conjunction with concepts of singular values, allow the formulation of requirements in the frequency domain to obtain a control system with good performance and stability robustness.

In this design, performance robustness is of concern at the output of the plant. Therefore it is required that the plant output transfer function be derived in order to analyze and assist in obtaining a control system design to meet certain performance robustness requirements. Using Fig. 10.2.46, the plant output transfer function is derived and performance robustness requirements established. From Fig. 10.2.46 it can be shown that

$$y(s) = d_o(s) + G(s)u'(s) \tag{10.2.78}$$

$$e(s) = r(s) - y(s) - n(s) \tag{10.2.79}$$

and

$$u'(s) = d_o(s) + u(s) = d_I(s) + K(s)e(s), \tag{10.2.80}$$

where $G(s) = \mathbf{C}[s\mathbf{I} - \mathbf{A}]^{-1}\mathbf{B}$ is the plant transfer function and **A**, **B**, and **C** are system matrices. The transfer functions $y(s)$, $d_o(s)$, $e(s)$, $r(s)$, $n(s)$, $u(s)$, $d_I(s)$, and $K(s)$ are, respectively, the output, output disturbance, error, reference, noise, input, input disturbance, and controller transfer functions.

Using Eqs. (10.2.79) and (10.2.80) it can be shown that Eq. (10.2.78) becomes

$$\begin{aligned} y(s) &= [G(s)K(s)][r(s) - y(s) - n(s)] \\ &\quad + d_o(s) + G(s)d_I(s) \\ &= G(s)K(s)r(s) - G(s)K(s)y(s) - G(s)K(s)n(s) \\ &\quad + d_o(s) + G(s)d_I(s) \end{aligned} \tag{10.2.81}$$

Therefore,

$$\begin{aligned} y(s) &= [\mathbf{I} + G(s)K(s)]^{-1}[G(s)K(s)r(s) - G(s)K(s)n(s) \\ &\quad + d_o(s) + G(s)d_I(s)] \\ &= [\mathbf{I} + G(s)K(s)]^{-1}\{G(s)K(s)[r(s) - n(s)]\} \\ &\quad + [\mathbf{I} + G(s)K(s)]^{-1}[d_o(s) + G(s)d_I(s)] \\ &= y_r(s) + y_{d_I}(s) + y_{d_o}(s) + y_n(s) \end{aligned} \tag{10.2.82}$$

where $y_r(s)$, $y_{d_I}(s)$, $y_{d_o}(s)$, and $y_n(s)$ are, respectively, the contributions to the plant output due to the reference input, input disturbance, output disturbance, and noise input to the system. The requirements for good command following, disturbance rejection, and insensitivity to sensor noise are defined by using singular values to characterize the magnitude of MIMO transfer functions. These requirements are also **applicable to SISO systems** (Kazerooni and Narayan; Murphy and Bailey, April 1989).

Command Following The conditions required to obtain a MIMO control system design with good command following is described. Good command following of the reference input $r(s)$ by the plant output $y(s)$ implies that

$$y(s) \approx r(s) \qquad \forall s \in s_R \tag{10.2.83}$$

where $s = j\omega$ and $s_R = j\omega_R$ and ω_R is the frequency range of the input $r(s)$. Letting $d_o(s) = d_I(s) = n(s) = 0$ in Eq. (10.2.82) results in

$$\begin{aligned} y_r(s) &= [\mathbf{I} + G(j\omega)K(j\omega)]^{-1}G(j\omega)K(j\omega)r(j\omega) \quad \forall_\omega \in \omega_R \\ &= \{\mathbf{I} + [G(j\omega)K(j\omega)]^{-1}\}^{-1}r(j\omega) \quad \forall_\omega \in \omega_R \end{aligned} \tag{10.2.84}$$

where $G(j\omega)K(j\omega)$ is defined as the return ratio. Therefore, to obtain good command following, the system **inverse return difference I +** $[G(j\omega)K(j\omega)]^{-1}$ must have the property such that

$$\mathbf{I} + [G(j\omega)K(j\omega)]^{-1} \approx \mathbf{I} \quad \forall \omega \in \omega_R \tag{10.2.85}$$

which implies that

$$\underline{\sigma}[G(j\omega)K(j\omega)] \gg 1 \quad \forall \omega \in \omega_R \tag{10.2.86}$$

Thus the magnitude of the minimum singular value of the **return ratio** $\underline{\sigma}$ $[G(j\omega)K(j\omega)]$ must be large over the frequency range of the input to obtain good command following of the control system.

Disturbance Rejection To minimize the effect of the output disturbance, $y_{d_o}(s)$, on the output signal $y(s)$, Eq. (10.2.82) is used with $n(s) = r(s) = d_I(s) = 0$, which results in

$$y_{d_o}(s) = [\mathbf{I} + G(s)K(s)]^{-1}d_o(s) \tag{10.2.87}$$

Rejection of this output disturbance at the plant output requires that the magnitude of the **return difference I** + $G(s)K(s)$ to meet the following condition

$$\underline{\sigma}[\mathbf{I} + G(j\omega)K(j\omega)] \gg 1 \qquad \forall \omega \in \omega_{d_o} \tag{10.2.88}$$

which implies

$$\underline{\sigma}[G(j\omega)K(j\omega)] \gg 1 \qquad \forall \omega \in \omega_{d_o} \tag{10.2.89}$$

where y_{d_o} is the frequency range of the output disturbance $d_o(s)$.

With $n(s) = r(s) = d_o(s) = 0$ in Eq. (10.2.82), the effect of input disturbance on the plant output is examined. This effect is seen to be

$$y_{d_I}(s) = [\mathbf{I} + G(j\omega)K(j\omega)]^{-1}[G(j\omega)d_I(j\omega)] \quad \forall \omega \in \omega_{d_I} \tag{10.2.90}$$

where ω_{d_I} is the frequency range of the input disturbance $d_I(s)$. To minimize the effect of $y_{d_I}(s)$ on the plant output requires that the magnitude of the return difference be large (in general, the larger the better). This requirement implies that

$$\underline{\sigma}[G(j\omega)K(j\omega)] \gg 1 \qquad \forall \omega \in \omega_{d_I} \tag{10.2.91}$$

to reject the input disturbance. It should be noted that this equation gives only general conditions for rejection. The control engineer must determine how much greater than 1 the return ratio magnitude must be.

Noise Rejection The requirements to minimize the effects of sensor noise on the plant output must be determined. First letting $n(s) = d_I(s) = d_o(s) = 0$ in Eq. (10.2.82) results in

$$\begin{aligned} y_n(s) &= [\mathbf{I} + G(s)K(s)]^{-1}[G(s)K(s)]n(s) \qquad \forall s \in s_n \\ &= \{\mathbf{I} + G(s)K(s)^{-1}\}^{-1}n(s) \qquad \forall s \in s_n \end{aligned} \tag{10.2.92}$$

where $s_n = j\omega_n$ and ω_n is the frequency range of the sensor noise. Thus to minimize the effects of sensor noise on the plant output it is desired to have the magnitude of $y_n(s)$ small over the frequency range of the noise. Thus

$$\bar{\sigma}\,[G(j\omega)K(j\omega)] \gg 1 \qquad \forall \omega \in \omega_n \tag{10.2.93}$$

is required to minimize the effect of noise on the plant output.

Any overlapping of any of the frequency ranges ω_R, ω_{d_I}, ω_{d_o}, ω_n will require system design specification tradeoffs to be made. This will definitely occur when the lower frequency ranges of ω_R and/or ω_{d_I} and ω_{d_o} overlap the higher frequency range ω_n.

Stability Robustness

All linear control system design methods are based on a linear model of the plant. Because this design model only approximates the actual plant, an error exists between the design model and the actual plant. This error can be evaluated as **absolute** or **relative.** The method of evaluation of this error (model uncertainty) depends on the nature of the differences between the actual design model and the actual plant. The evaluation of an absolute error defines an additive model uncertainty, whereas the evaluation of a relative error defines a multiplicative model uncertainty. The **additive** and multiplicative model uncertainties can be further classified as either structured or unstructured model uncertainties (Murphy and Bailey, ORNL 1989).

A **structured model uncertainty** is usually a low-frequency phenomenon characterized by variations in the parameters of the linear time-invariant plant design model. These parameter variations are caused by changes in the plant operating conditions, wear due to aging, and environmental conditions. A structured uncertainty also indicates that the gain and phase information concerning the uncertainty is known. Additive model uncertainties are generally considered structured uncertainties.

An **unstructured uncertainty** is typically a high-frequency phenomenon in which the only information known about the uncertainty is its magnitude. Unstructured uncertainties usually have the characteristic of becoming significant at high frequencies. Multiplicative uncertainties are characterized as unstructured. Unstructured uncertainties are usually caused by the truncation of high-frequency plant dynamics or modes due to the linearization process, and the neglect of plant dynamics such as actuators and sensors. Some plant dynamics that characterize unstructured uncertainty are flexible-body dynamics, electrical and mechanical resonances, and transport delays. Most linear control-system design methods neglect these high-frequency plant dynamics. The result is that if a high-frequency bandwidth controller is implemented, the high-frequency modes could be excited, resulting in an unstable control system.

Model uncertainty clearly is seen to impose limitations on the achievable performance of a feedback control system design. Hence, this section will focus on the limitations imposed by uncertainty, not on the difficult problem of exact representation of the system uncertainty.

Multiplicative Uncertainty Multiplicative uncertainty (perturbation) can be represented at the input or output of the plant, and is called respectively the input multiplicative uncertainty or the **output multiplicative uncertainty.** The model uncertainty due to actuator and sensor dynamics is modeled, respectively, as an input or output multiplicative perturbation. A reduced-order model of an already **linearized model** results in a multiplicative model uncertainty. **Time delay** at the input or output of a plant yields, respectively, input and output multiplicative uncertainties in the plant. Even though other multiplicative model uncertainties exist among various types of control system design problems, the model uncertainties mentioned above are a common consideration in most control problems.

Output Multiplicative Uncertainty

The output multiplicative uncertainty $\Delta G(s)$ is defined as

$$\Delta G(s) = [G'(s) - G(s)]G^{-1}(s) \qquad (10.2.94)$$

where

$$G'(s) = L(s)G(s) \qquad (10.2.95)$$

is the perturbed transfer function, $L(s)$ represents dynamics not included in the nominal plant $G(s)$. Thus resulting in

$$\Delta G(s) = [L(s) - \mathbf{I}] \qquad (10.2.96)$$

A block diagram of a control system with output multiplicative perturbation is shown in Fig. 10.2.47. The conditions of stability for a control system with output multiplicative uncertainty require that

$$\underline{\sigma}\{\mathbf{I} + [G(j\omega)K(j\omega)]^{-1}\} > \bar{\sigma}[\Delta G(j\omega)] \qquad \forall \omega > 0 \qquad (10.2.97)$$

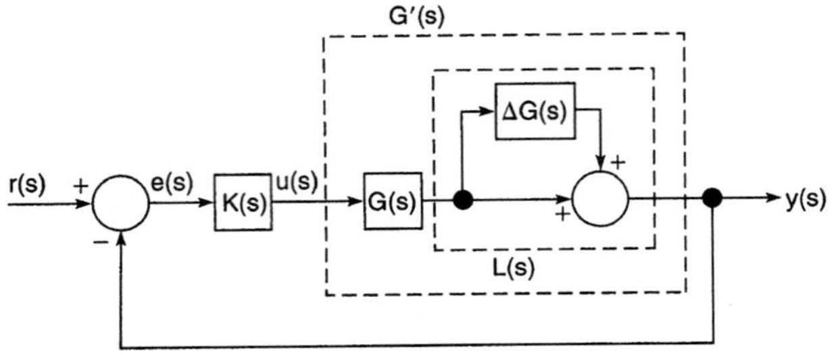

Figure 10.2.47 Control system with output multiplicative perturbation.

Additive Uncertainty The block diagram of a control system with a plant additive model uncertainty is shown in Fig. 10.2.48. The representation of the additive model uncertainty using an absolute error criterion is defined as

$$\Delta G(s) = G'(s) - G(s) \qquad (10.2.98)$$

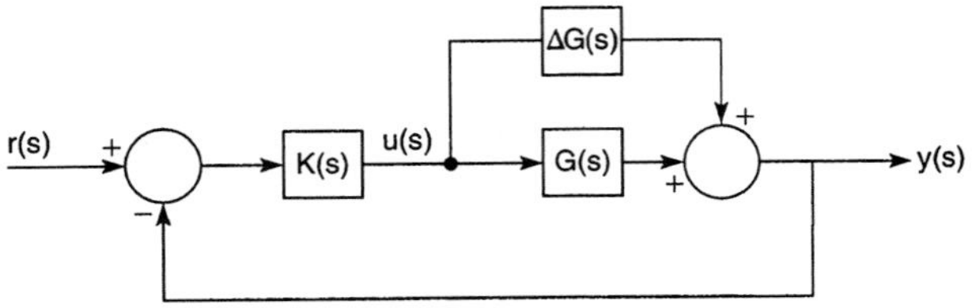

Figure 10.2.48 Control system with additive perturbation.

where in this case the perturbed transfer function $G'(s)$ is

$$G'(s) = [\mathbf{C} + \Delta\mathbf{C}]\{s\mathbf{I} - [\mathbf{A} + \Delta\mathbf{A}]\}^{-1}[\mathbf{B} + \Delta\mathbf{B}] \qquad (10.2.99)$$

and the nominal plant transfer function $G(s)$ is

$$G(s) = \mathbf{C}[s\mathbf{I} - \mathbf{A}]^{-1}\mathbf{B} \qquad (10.2.100)$$

and where **A, B,** and **C** are the nominal system matrices and **ΔA, ΔB,** and **ΔC** are the perturbation matrices that indicate the respective deviations from the nominal system matrices. The condition for stability of the additively perturbed control system requires that

$$\underline{\sigma}[G(j\omega)K(j\omega)] > \bar{\sigma}[\Delta G(j\omega)K(j\omega)] \qquad \forall \omega > 0 \qquad (10.2.101)$$

Table 10.2.7 contains a summary of the general design requirements presented in this section to obtain a control system with good performance and stability robustness.

REVIEW OF OPTIMAL CONTROL THEORY

Linear Quadratic Regulator

Consider the linear time-invariant state-space system, first-order vector differential equation model

$$\dot{\mathbf{x}}(t) = \mathbf{A}x(t) + \mathbf{B}u(t) \qquad (10.2.102)$$

$$y(t) = \mathbf{C}x(t) \qquad (10.2.103)$$

where **A, B,** and **C** are constant system matrices, and $x(t)$, $u(t)$, and $y(t)$ are, respectively, the system state, input, and output vectors. The system is stabilizable and controllable. The goal of this procedure is to minimize the finite performance index

$$\begin{aligned} J &= \int^{\infty} [x^{\mathrm{T}}(t)\mathbf{Q}x(t) + u^{\mathrm{T}}(t)\mathbf{R}u(t)]\,dt \\ &= \int_0^{\infty} [x^{\mathrm{T}}\mathbf{C}^{\mathrm{T}}\mathbf{C}x + u^{\mathrm{T}}\mathbf{R}u]\,dt \\ &= \int^{\infty} [y^{\mathrm{T}}y + u^{\mathrm{T}}\mathbf{R}u]\,dt \end{aligned} \qquad (10.2.104)$$

where **Q** is a symmetric positive semidefinite matrix and **R** is positive definite.

Table 10.2.7 General System Design Specifications

System requirements		Range	
Good command-following	$\underline{\sigma}[G(j\omega)K(j\omega)] \gg 0$ dB	$\forall \omega \in \omega_R$	
Good disturbance rejection	$\underline{\sigma}[G(j\omega)K(j\omega)] \gg 0$ dB	$\forall \omega \in \omega_d$	
Good immunity to noise	$\underline{\sigma}[G(j\omega)K(j\omega)] \ll 0$ dB	$\forall \omega \in \omega_n$	
Good system response to high-frequency modeling error*	$\underline{\sigma}\{I + [G(j\omega)K(j\omega)]^{-1}\} > \lVert\Delta G(j\omega)\rVert$		$\forall \omega > 0$ rad/s
Good insensitivity to parameter variations at low frequencies*	$\underline{\sigma}[G(j\omega)K(j\omega)] > \bar{\sigma}[\Delta G(j\omega)K(j\omega)]$		

* The $\Delta G(jw)$ indicated corresponds to the applicable multiplicative or additive model uncertainty.

The **performance index** J, referred to as the quadratic performance index, implies that a control $u(t)$ is sought to facilitate the minimization of J. The weighting matrices **Q** and **R** are selected to reflect the importance of particular states and control inputs. In general the weighting of the diagonal elements of **Q** can be determined by the importance of the state, as observed by the **C** matrix of the output equation. The effect of **Q** is the ability to control the **transient response** of the LQR. The effect of **R** is the ability to **control** the **energy** resulting from u.

Therefore a linear control law that minimizes J is defined as

$$u(t) = -\mathbf{K}x(t) \tag{10.2.105}$$

where

$$\mathbf{K} = \mathbf{R}^{-1}\,\mathbf{B}^{\mathrm{T}}\mathbf{S} \tag{10.2.106}$$

where, in turn, **S** is a constant symmetric positive semidefinite matrix that satisfies the algebraic Riccati equation

$$\mathbf{Q} - \mathbf{SBR}^{-1}\,\mathbf{B}^{\mathrm{T}}\mathbf{S} + \mathbf{A}^{\mathrm{T}}\mathbf{S} + \mathbf{SA} = 0 \tag{10.2.107}$$

The closed-loop regulator is given by

$$\dot{\mathbf{x}}(t) = [\mathbf{A} - \mathbf{BK}]x(t) = [\mathbf{A} - \mathbf{BR}^{-1}\mathbf{B}^{\mathrm{T}}\mathbf{S}]x(t) \tag{10.2.108}$$

and is asymptotically stable provided the state equation (10.2.102) and the output equation (10.2.103) are detectable.

Kalman Filter

What follows is a method of reconstructing an estimate of the states using only the output measurements of the system. The method of reconstruction must be applicable when **process noise** and **measurement noise,** respectively, corrupt the plant state equations and the output measurement equations.

Let us consider the stochastic linear system state space vector differential equation model

$$\dot{\mathbf{x}}(t) = \mathbf{A}x(t) + \mathbf{B}u(t) + \Gamma d(t) \tag{10.2.109}$$

$$y(t) = \mathbf{C}x(t) + n(t) \tag{10.2.110}$$

where **A**, **B**, **C**, and Γ are constant system matrices. Vectors $x(t)$, $u(t)$, and $y(t)$ are, respectively, the state, input, and output vectors. Vector $d(t)$ is a process noise random vector and vector $n(t)$ is a measurement noise random vector. Assuming that $d(t)$ and $n(t)$ are zero-mean, uncorrelated, Gaussian white noises,

$$E[d(t)n(t)] \quad \text{for all } t \tag{10.2.111}$$

$$E[d(t)d^{\mathrm{T}}(\tau)] = \mathbf{D}_o\delta(t-\tau) \tag{10.2.112}$$

$$E[n(t)n^{\mathrm{T}}(\tau)] = \mathbf{N}\delta(1-\tau) \quad \text{for all } t < \tau \tag{10.2.113}$$

$$E[d(t)n^{\mathrm{T}}(\tau)] = 0 \tag{10.2.114}$$

where τ is the difference between two points in time, $E[\ ]$ is mean value, δ is an impulse function, and **N** and $\mathbf{D}_o$ are constant symmetric, positive definite, and positive semidefinite matrices, respectively. Note that **N** and $\mathbf{D}_o$ are constant because it is assumed that $d(t)$ and $n(t)$ are wide-sense stationary.

The **state estimate** $\hat{\mathbf{x}}$ **of the actual measured state** x is obtained from the noisy measurement y. Therefore we can define a state error vector

$$e(t) = x(t) - \hat{\mathbf{x}}(t) \tag{10.2.115}$$

where we desire to minimize the mean square error

$$\bar{\mathbf{e}} = E[e^{\mathrm{T}}(t)e(t)] \tag{10.2.116}$$

The estimator can thus be shown to take the form

$$\dot{\hat{\mathbf{x}}} = \mathbf{A}x(t) + \mathbf{B}u(t) + \mathbf{F}[y(t) - \mathbf{C}x(t)] \tag{10.2.117}$$

where **F** is the filter gain which minimizes Eq. (10.2.116); **F** is defined as

$$\mathbf{F} = \Sigma\mathbf{C}^{\mathrm{T}}\mathbf{N}^{-1} \tag{10.2.118}$$

where Σ is a constant-variance matrix of the error $e(t)$. It is assumed here that $e(t)$ is wide-sense stationary. The matrix Σ is obtained by solving the algebraic variance Riccati equation

$$\mathbf{D} - \Sigma\mathbf{C}^{\mathrm{T}}\mathbf{N}^{-1}\mathbf{C}\Sigma + \mathbf{A}\Sigma + \Sigma\mathbf{A}^{\mathrm{T}} = 0 \tag{10.2.119}$$

where

$$\mathbf{D} = \Gamma\mathbf{D}_o\Gamma \tag{10.2.120}$$

A sufficient condition to obtain a unique and positive definite Σ from Eq. (10.2.117) requires that [**A**, **C**] be completely observable. However, if the pair [**A**, **C**] is required to be detectable, then Σ can be positive semidefinite.

The **reconstruction error** $e(t)$ satisfies the differential equation

$$\dot{e}(t) = [\mathbf{A} - \mathbf{FC}]e(t) + [\Gamma - \mathbf{F}]\begin{bmatrix} d(t) \\ n(t) \end{bmatrix} \tag{10.2.121}$$

such that

$$e(t) \to 0 \quad \text{as} \quad t \to \infty \quad \forall t \geq t_0$$

if and only if the **observer** is asymptotically stable. The poles of the observer (filter) are found by using the closed-loop dynamics matrix [**A** – **FC**]. To obtain asymptotic stability of the filter, it is required that the pair [**A**, Γ] be completely controllable. The necessary and sufficient condition of **stabilizability** of the pair [**A**, Γ] will ensure **stability** also.

PROCEDURE FOR LQG/LTR COMPENSATOR DESIGN

An LQG/LTR compensator will be designed for a **minimum phase plant (no right half plane zeroes).** The control system so designed will be guaranteed to be stable and will be robust at the plant output.

First, it is assumed that if **zero steady-state error** is desired in the control system design, the plant inputs will be augmented to obtain the desired **integral action.** In some cases the dynamics of the plant may have poles that are nearly zero, which establishes a **near integral action** in the plant. In such a case the addition of integrals at the plant input causes a much greater than **20-dB/decade roll-off,** thus yielding unreasonable plant dynamics characterized by the return ratio singular-value plots. Therefore caution is required in deciding how or whether to augment the plant dynamics. In the following description of the compensator design it is assumed that the nominal plant transfer function $G(s)$ has been augmented as required by the control engineers' desires.

Step 1 The first step requires selecting the appropriate transfer function $[G_{FOL}(s)]$ that satisfies the desired performance and stability robustness of the final control system. Therefore, the appropriate scalar $\mu > 0$ and a constant matrix **L** such that the singular values of

$$G_{FOL}(s) = (1/\sqrt{\mu})[\mathbf{C}\ (s\mathbf{I} - \mathbf{A})^{-1}\mathbf{L}] \quad (10.2.122)$$

approximately meet the desired control system requirements. The system matrices **C** and **A** of the minimum phase component are assumed to have been augmented already to obtain the necessary integral action. The selection of **L** is made so that at low frequencies ($s = j\omega \to 0$), high frequencies ($s = j\omega \to 0\, j\omega$), or intermediate frequencies ($s = j\omega \to j\omega_i$) the singular values of $G_{FOL}(s)$ are approximately the same. After **L** is selected, then μ is adjusted to obtain the desired crossover frequency $\sigma[G_{FOL}(j\omega)]$; therefore μ functions as a gain parameter of the transfer function. It should be noted that it is desired to balance the singular values of $G_{FOL}(s)$ in the region of the crossover frequency in order to obtain similar responses for each loop of the system. Also, typically it is desired to have the return ratio singular values roll off at a rate no greater than −20 dB/decade in the region of gain crossover.

Step 2 In this step the Kalman filter transfer function is defined as

$$G_{KF}(s) = \mathbf{C}(s\mathbf{I} - \mathbf{A})^{-1}\mathbf{F} \quad (10.2.123)$$

where **F** is the Kalman filter gain and **C** and **A** are the system matrices of the linear time-invariant state-space system.

To compute the **filter gain F,** first the following Kalman filter **algebraic Riccati equation (ARE)** is solved

$$\mathbf{L}\mathbf{L}^T - (1/\sqrt{\mu})\Sigma\mathbf{C}^T\mathbf{C} + \mathbf{A}\Sigma + \Sigma\mathbf{A}^T = 0 \quad (10.2.124)$$

where **L** and μ are as obtained in step 1 and Σ (the covariance matrix) is the solution to the equation. The filter gain is thus computed to be

$$\mathbf{F} = (1/\sqrt{\mu})\Sigma\mathbf{C}^T \quad (10.2.125)$$

Thus as seen above, **L** and μ are tunable parameters used to produce the necessary filter gain **F** to meet the desired system performance and stability robustness constraints. This step completes the LQG/LTR design requirement of designing the target Kalman filter transfer function matrix with the desired properties which will be recovered at the plant output during the loop transfer recovery step. The general block diagram of the LQG/LTR control system is shown in Fig. 10.2.49.

Step 3 The transfer function of the LQG/LTR controller is defined as

$$K(s) = \mathbf{K}[s\mathbf{I} - (\mathbf{A} - \mathbf{B}\mathbf{K} - \mathbf{F}\mathbf{C})]\mathbf{F} \quad (10.2.126)$$

where **A, B,** and **C** are the system matrices of Eqs. (10.2.102) and (10.2.103). The filter gain **F** is as stated in Eq. (10.2.125). The **regulator gain K** will be computed in this step to achieve loop transfer recovery. The selection of **K** will yield a return ratio at the plant output such that

$$G(s)K(s) \to G_{KF}(s) \quad (10.2.127)$$

where $G(s)$ is the plant transfer function, and $K(s)$ is the controller transfer function.

The first step in selecting **K** requires solving the LQR problem, which has the following ARE:

$$\mathbf{Q}(q) - \mathbf{S}\mathbf{B}\mathbf{R}_0^{-1}\mathbf{B}^T\mathbf{S} + \mathbf{S}\mathbf{A} + \mathbf{A}^T\mathbf{S} = 0 \quad (10.2.128)$$

where $\mathbf{R}_0 = \mathbf{I}$ is the control weighting matrix and $\mathbf{Q}(q) = \mathbf{Q}_0 + q^2\mathbf{C}\mathbf{C}^T$ is the modified state weighting matrix. The scalar q is a free design parameter. The resulting regulator gain **K** is computed as

$$\mathbf{K} = \mathbf{B}^T\mathbf{S} \quad (10.2.129)$$

As $q \to 0$ Eq. (10.2.128) becomes the ARE to the optimal regulator problem. As $q \to \infty$ the LQG/LTR technique guarantees that, since the design model $G(s) = (\mathbf{C}(s)\mathbf{I} - \mathbf{A})^{-1}\mathbf{B}$ has no nonminimum phase zeroes, then pointwise in s

$$\lim_{q\to\infty} G(s)K(s) \to G_{KF}(s) \quad (10.2.130)$$

$$\lim_{q\to\infty}\{(\mathbf{C}s\mathbf{I} - \mathbf{A})^{-1}\mathbf{B}\mathbf{K}[s\mathbf{I} - (A - \mathbf{B}\mathbf{K} - \mathbf{F}\mathbf{C})]^{-1}\mathbf{F}\} \to \mathbf{C}(s\mathbf{I} - \mathbf{A})^{-1}\mathbf{F} \quad (10.2.131)$$

This completes the LQG/LTR design procedure. Now that the loop transfer recovery step is complete, the singular-value plots of the return ratio, return difference, and inverse return difference must be examined to verify that the desirable loop shape has been obtained.

EXAMPLE CONTROLLER DESIGN FOR A DEAERATOR

In this section, three linear control methods are used to obtain a level-control system design for a deaerator. The **three linear control system design methods** that will be used are proportional plus integral, linear quadratic Gaussian, and linear quadratic Gaussian with loop transfer recovery. In this investigation, the dual of the procedure developed for the LQG/LTR design at the plant input will be used to obtain a **robust control system design** at the **plant output** (Murphy and Bailey, ORNL 1989).

The **Bode gain** and **phase plots** of the resulting control system design will be presented for each controller. Also, the singular-value plots of the **return ratio, return difference,** and **inverse return difference** for opening the loop at the plant output (point 3 in Fig. 10.2.50) will be presented.

The deaerator was chosen because of its **simple mathematical structure;** one input, one output, and three state variables. Thus the model is easy to follow but complex enough to illustrate the design technique.

The Linearized Model

The mathematical model of the deaerator is **nonlinear,** with a process flow diagram as shown in Fig. 10.2.51. This nonlinear plant is **linearized** about a nominal operating condition, and the resulting linearized plant model will be used to obtain the controller design. The linear

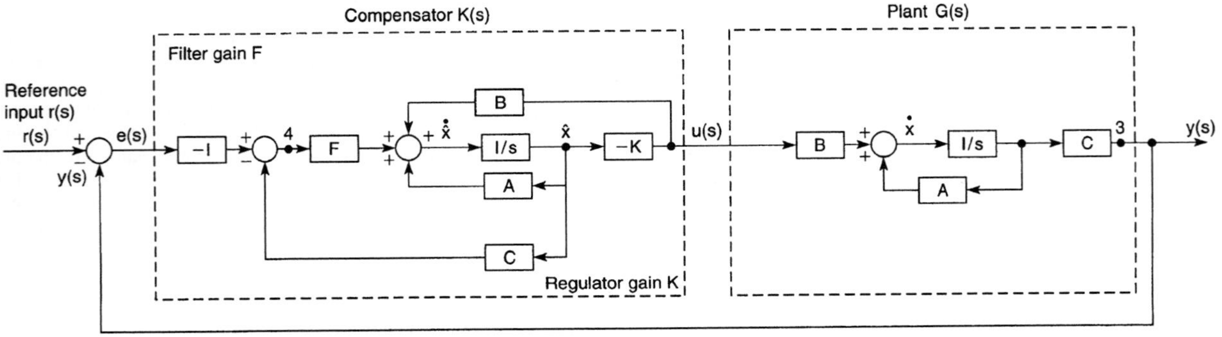

Figure 10.2.49 Block diagram of unity-feedback LQG/LTR control system.

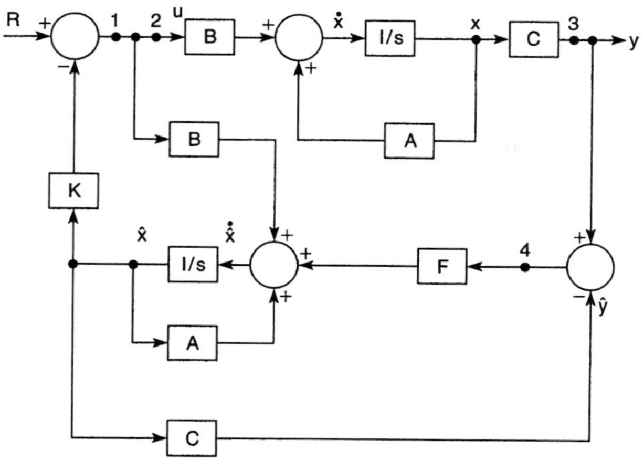

Figure 10.2.50 Block diagram of LQG system.

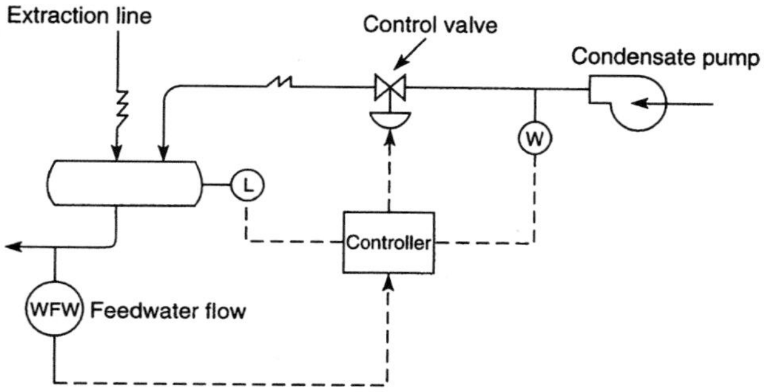

Figure 10.2.51 General process flow diagram of the deaerator.

deaerator model is described by the state space linear time-invariant (LTI) differential equation

$$\dot{x}(t) = \mathbf{A}x(t) + \mathbf{B}u(t) \tag{10.2.132}$$

$$y(t) = \mathbf{C}x(t) \tag{10.2.133}$$

where **A, B,** and **C** are the system matrices given, respectively, as

$$\mathbf{A} = \begin{bmatrix} -53.802 & 1.7093 & 9.92677 \\ 0.0001761 & -0.0009245 & -0.0053004 \\ -0.001124 & -0.0243636 & -0.139635 \end{bmatrix}$$

$$\mathbf{B} = \begin{bmatrix} -9526.25 \\ 0.265148 \\ -1.72463 \end{bmatrix}$$

$$\mathbf{C} = [0.0 \quad 3.2833 \quad 0.06373]$$

The system states are defined as

$$x = \begin{bmatrix} \delta P \\ \delta \rho \\ \delta H \end{bmatrix}$$

where δP is operating pressure between the pump and the extraction line, $\delta \rho$ is fluid density, and δH is internal energy of the tank level. Plant output y = change in deaerator tank level, and the plant input u = change in control valve.

LQG Controller Design

The design considerations for the tracking LQG control system will first be discussed. The block diagram of the LQG control system is shown in Fig. 10.2.50. It is apparent that output tracking of a reference input will not occur with the present LQG controller configuration. An alternative compensator structure, shown in Fig. 10.2.52, is therefore used to obtain a tracking LQG controller. The controller $K(s)$ uses the same filter gain **F** and regulator gain **K** computed by the LQG design procedure. The detailed block diagram is the same structure that will be used for the LQG/LTR controller shown in Fig. 10.2.49.

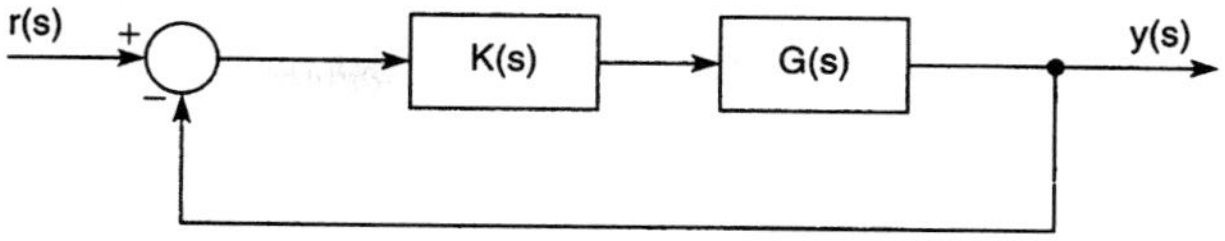

Figure 10.2.52 Alternative compensator structure.

Using a **control system design package** MATRIX$_X$, the regulator gain **K** and filter gain **F** are computed as

$$\mathbf{K} = 10^5 \times [0.0233 \quad 3.145 \quad -0.1024]$$

and

$$\mathbf{F} = \begin{bmatrix} 0.1962 \\ 9.9998 \\ -0.0083 \end{bmatrix}$$

given **user-defined weighting matrices.** The frequency response of the open-loop transfer function of the closed-loop compensated system

$$\frac{y(s)}{r(s)} = \frac{\text{tank level}}{\text{reference input}}$$

where $y(s)$ is the **tank level** and $r(s)$ is the **reference input,** as shown in Fig. 10.2.53.

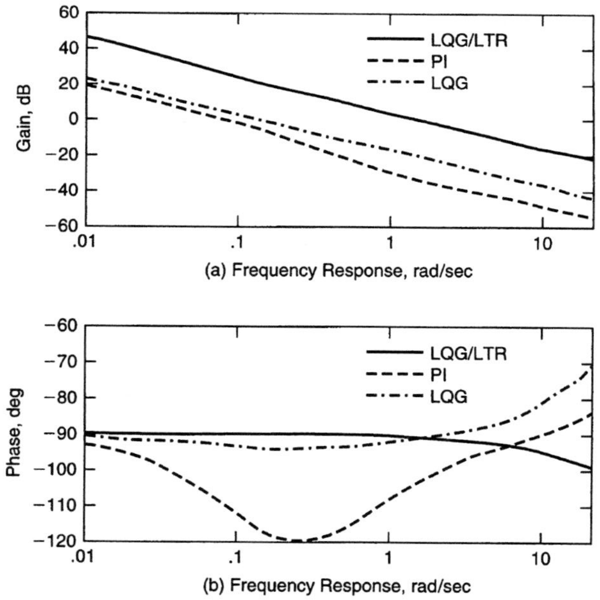

Figure 10.2.53 Open-loop frequency response of the PI, LQG, LQG/LTR control system. (*a*) Frequency response of system gain; (*b*) frequency response of system phase.

LQG/LTR Controller Design

This LQG/LTR control system has the structure shown in Fig. 10.2.52. This LQG/LTR control system design is a two-step process. First, a Kalman filter (KBF) is designed to obtain **good command following** and **disturbance rejection** over a specified low-frequency range. Also, the KBF design is made to meet the required robustness criteria with system uncertainty of $\Delta G(s)$ at the system's output. To obtain good command following and disturbance rejection requires that

$$\underline{\sigma}(G_4(s)) \geq 20\text{ dB} \qquad \forall \omega < 0.1\text{ rad/s} \tag{10.2.134}$$

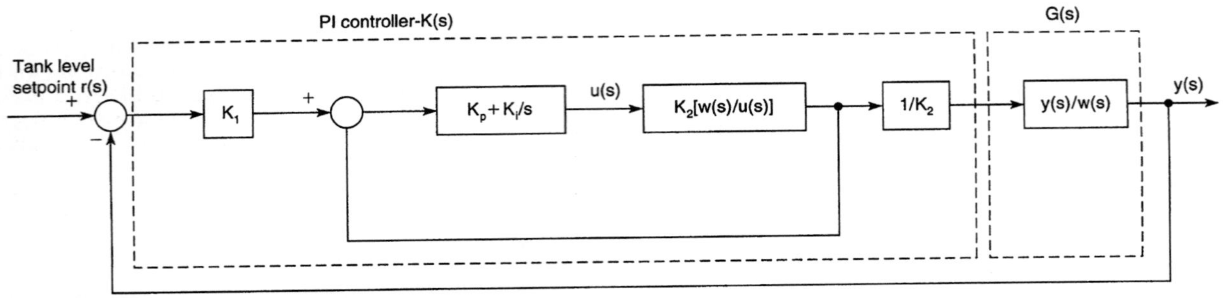

Figure 10.2.54 Block diagram of a PI control system.

where $G_4(s)$ is the loop transfer function at the output of the KBF (point 4 of Fig. 10.2.50). Assuming the **high-frequency uncertainty** $\Delta G(s)$ for this system becomes significant at 5 rad/s, then for robustness

$$\underline{\sigma}\{\mathbf{I} + [G_4(s)]^{-1}\} \geq \|\Delta G(j\omega)\| = 5 \text{ dB} \qquad \forall\ \omega \geq 5 \text{ rad/s} \quad (10.2.135)$$

In this application, the frequency at which system uncertainty becomes significant, 5 rad/s, and the magnitude of the uncertainty, 5 dB at 5 rad/s, are assumed values. This assumption is required because of the lack of information on the high-frequency modeling errors of the process being studied.

The **second step** is to **recover the good robustness properties** of the loop transfer function $G_4(s)$ of point 4 at the output node y (point 3). This will be accomplished by applying the loop transfer recovery (LTR) step at the output node (point 3). The LTR step requires the computation of a regulator gain (**K**) to obtain the robustness properties of the loop transfer function $G_4(s)$ at the output node y.

The tool used to obtain the LQG/LTR design for this system, considering the low- and high-frequency bound requirements is CASCADE, a computer-aided system and control analysis and design environment that synthesizes the LQG/LTR design procedure (Birdwell et al.). Note that more recent computer-aided control system design tools synthesize this design procedure.

The resulting filter gain **F** and regulator **K** gain for this system are, respectively,

$$\mathbf{F} = \begin{bmatrix} -0.0107 \\ 0.472 \\ -0.02299 \end{bmatrix}$$

and

$$\mathbf{K} = [0.660 \quad 27703.397 \quad 589.4521]$$

The resulting open-loop frequency response of the compensated system is shown in Fig. 10.2.53.

PI Controller Design

The PI controller for this system design takes on a structure similar to the three-element controller prevalent in process control. Therefore the PI control system structure takes the form shown in Fig. 10.2.54. The classical design details of the deaerator are shown in Murphy and Bailey, ORNL 1989.

The transfer functions used in this control structure are defined as

$$\frac{w(s)}{u(s)} = \frac{\text{condensate flow}}{\text{controller output}} = 4.5227 \times 10^8 \times \frac{s + 0.142}{(s + 0.1405)(s + 53.8015)} \quad (10.2.136)$$

and

$$\frac{y(s)}{w(s)} = \frac{\text{tank level}}{\text{condensate}} = 1.6804 \times 10^{-9} \times \frac{(s + 0.1986)(s + 47.458)}{s(s + 0.142)} \quad (10.2.137)$$

The constants K_1 and K_2 are used for unit conversion and defined, respectively, as 1.0 and 10^{-6}. The terms K_p and K_i are defined, respectively, as the proportional gain and the integral gain.

Using the practical experience of a prior three-element controller design, the proportional and integral gains are selected to be $K_p = 0.1$ and $K_i = 0.05$. The resulting open-loop frequency response plot of the compensated system $y(s)/r(s)$ is shown in Fig. 10.2.54.

Precompensator Design

The transient responses of the PI, LQG, and LQG/LTR control systems are shown in Fig. 10.2.55. The transient response of the LQG/LTR controller is seen to have a faster time to peak than the LQG and PI transient responses. Considering the practical physical limitations of the plant, it appears unlikely that the required rise time dictated by the transient response of the LQG/LTR control system is possible. The obvious solution would be to redesign the LQG/LTR control system to obtain a slower, more practical response. In the case of the LQG/LTR control system, the possibility of a redesign is eliminated, because the control system was designed to meet certain command-tracking, disturbance-rejection, and stability-robustness requirements.

Therefore, to eliminate this difficulty, a second-order precompensator will be placed in the forward path of the reference input signal. This precompensator will shape the output transient response of the system, since the control system design requires the output to follow the reference input. The precompensator in this investigation will be required to have a time to peak of 30 s with a maximum overshoot of about 3 percent. The requirements are selected to emulate somewhat the transient response of the PI controller.

The transfer function of the second-order precompensator used is as follows:

$$\frac{r(s)}{r_i(s)} = \frac{\omega_n^2}{s^2 + 2\zeta\omega_n s + \omega_n^2} \quad (10.2.138)$$

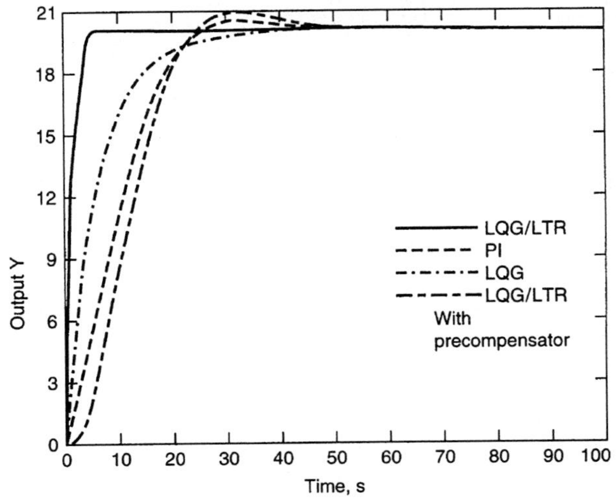

Figure 10.2.55 PI, LQG, LQG/LTR, LQG/LTR (with precompensator) control system closed-loop transient responses.

where ξ is the damping ratio, ω_n is the natural frequency, $r(s)$ is the output of the precompensator, and $r_i(s)$ is the actual reference input. The precompensator Eq. (10.2.138) can be represented in state equation form as follows:

$$\begin{bmatrix} x_1 \\ x_2 \end{bmatrix} = \begin{bmatrix} 0 & 1 \\ -\omega_n^2 & -\zeta\omega_n \end{bmatrix} \begin{bmatrix} x_1 \\ x_2 \end{bmatrix} + \begin{bmatrix} 0 \\ 1 \end{bmatrix} r_i(t) \tag{10.2.139}$$

and

$$r(t) = [\omega_n^2 \quad 0] \begin{bmatrix} x_1 \\ x_2 \end{bmatrix} \tag{10.2.140}$$

where $r_i(t)$ is the time-domain representation of the reference input signal and $r(t)$ is the time-domain signal at the output of the precompensator.

Considering the 3 percent overshoot requirement, then ξ must be 0.7. Using the expression

$$t_{\max} = \frac{\pi}{\omega_n\sqrt{1-(0.7)^2}} \tag{10.2.141}$$

where $t_{\max}$ must be 30 s, gives

$$\omega_n = \frac{\pi}{30\sqrt{1-(0.7)^2}} \tag{10.2.142}$$

The complete state equation of the precompensator is

$$\begin{bmatrix} x_1 \\ x_2 \end{bmatrix} = \begin{bmatrix} 0 & 1 \\ -0.0215 & -0.2052 \end{bmatrix} \begin{bmatrix} x_1 \\ x_2 \end{bmatrix} + \begin{bmatrix} 0 \\ 1 \end{bmatrix} r_i(t) \tag{10.2.143}$$

and

$$r(t) = [0.0215 \quad 0] \begin{bmatrix} x_1 \\ x_2 \end{bmatrix} \tag{10.2.144}$$

The resulting transient response of the LQG/LTR control system using the precompensator is shown in Fig. 10.2.55.

ANALYSIS OF SINGULAR-VALUE PLOTS

Systems with large stability margins, good disturbance rejection/command following, low sensitivity to plant parameter variations, and stability in the presence of model uncertainties are described as being robust and having good robustness properties. The singular-value plots for the PI, LQG, and LQG/LTR control systems are examined to evaluate the robustness properties. The singular-value plots of the return ratio, return difference, and the inverse return difference are shown in Figs. 10.2.56 to 10.2.58.

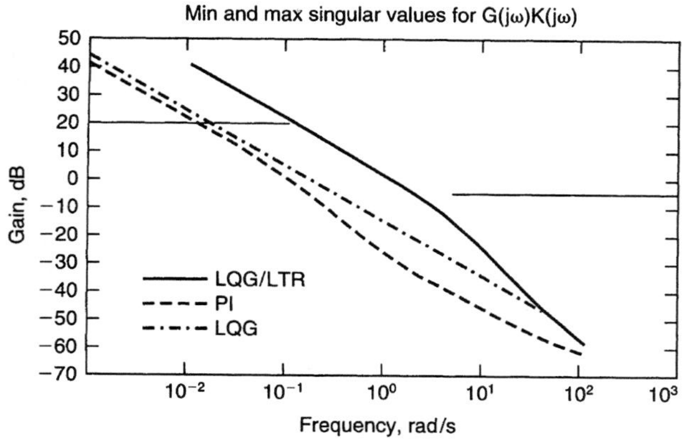

Figure 10.2.56 Singular-value plots of return ratio.

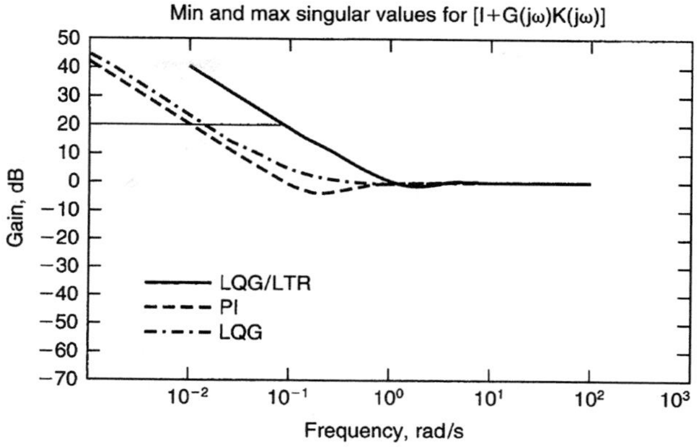

Figure 10.2.57 Singular-value plots of return difference.

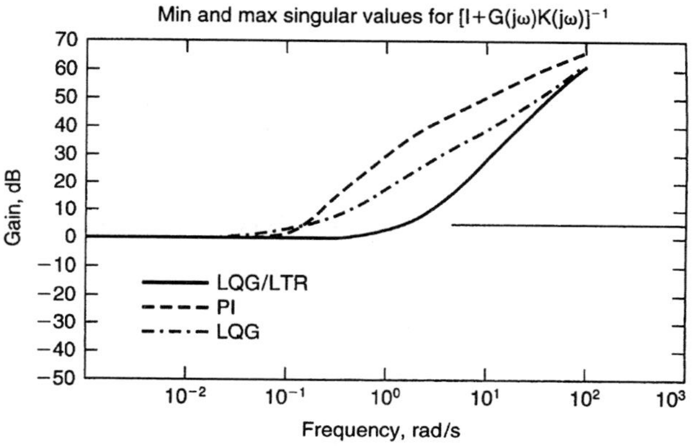

Figure 10.2.58 Singular-value plots of inverse return difference.

Deaerator Study Summary and Conclusions

Summary of Control System Analysis The design criteria used for analysis in this system is defined as shown in Table 10.2.8. The performance and robustness results summarized from the analysis using the singular-value plots are shown in Table 10.2.9, from which it appears that the LQG/LTR control system has the widest low-frequency range of disturbance rejection and insensitivity to parameter variations. The PI control system has the worst disturbance rejection property and the most sensitivity to parameter variations.

All of the control systems are capable of maintaining a stable robust system in the presence of high-frequency modeling errors, but the PI control system is best. Though all the control systems also have good immunity to noise at frequencies significantly greater than 5 rad/s, the PI control system has the best immunity.

Table 10.2.8 System Design Specifications

System requirements	Range
Good command-following/ disturbance rejection	$\underline{\sigma}[L(j\omega)] > 20$ dB $\forall\omega \le 0.1$ rad/s
Good system response to high-frequency modeling error	$\underline{\sigma}(I + [L(j\omega)]^{-1}) \ge \lVert\Delta L\rVert = 5$ dB $\forall\omega \le 5$ rad/s
Good insensitivity to parameter variations at low frequencies	$\underline{\sigma}[I + L(j\omega)] \ge 26$ dB for low frequencies
Good immunity to noise, $\omega \ge 5$ rad/s	$\bar{\sigma}(L(j\omega)] \ll 0$ dB for high frequencies

Table 10.2.9 Performance and Robustness Results

System properties	PI control system	LQG control system	LQG/LTR properties
Command following/disturbance	$\underline{\sigma}[L(j\omega)] \geq 20$ dB $\forall\omega \leq 0.1$ rad/s	$\underline{\sigma}[L(j\omega)] \geq 20$ dB $\forall\omega \leq 0.017$ rad/s	$\underline{\sigma}[L(j\omega)] \geq 20$ dB $\forall\omega \leq 0.1$ rad/s
System response to high-frequency modeling error ($\omega \geq 5$ rad/s)	$\underline{\sigma}(I+[L(j\omega)]^{-1})$ ≥ 43.0 dB	$\underline{\sigma}(I+[L(j\omega)]^{-1}) \geq 30$ dB	$\underline{\sigma}(I+[L(j\omega)]^{-1}) \geq 16.7$ dB
Insensitivity to parameter variations at low frequencies	$\underline{\sigma}(I+[L(j\omega)]) \geq 26$ dB $\forall\omega \leq 0.0055$ rad/s	$\underline{\sigma}(I+[L(j\omega)]) \geq 26$ dB $\forall\omega \leq 0.08$ rad/s	$\underline{\sigma}(I+L(j\omega)]) \geq 26$ dB $\forall\omega \leq 0.0055$ rad/s
Immunity to noise ($\omega \geq 5$ rad/s)	$\bar{\sigma}[L(j\omega)] \leq -42.5$ dB	$\bar{\sigma}[L(j\omega)] \leq -30$ dB	$\bar{\sigma}[L(j\omega)] \leq -16$ dB

Unlike the other systems, the LQG/LTR control system meets all the design criteria. Its design was obtained in a systematic manner, in contrast to the trial-and-error method used for the LQG control system. The LQG controller design was obtained by shaping the output transient response using the system state weighting matrix, then examining the system's singular-value plots. The PI control system used is simply a three-element control system strategy that has previously been used in the simulation study of the deaerator flow control system.

Conclusions It should be evident that performance characteristics such as a suitable transient response do not imply that the system will have good robustness properties. Obtaining suitable performance characteristics and a stable robust system are two separate goals. Classical unity-feedback design methods have been used for transient response shaping, with little consideration for robustness properties. The main advantage of a closed-loop feedback system is that good performance and stability robustness properties are obtainable. As has been shown, transient response shaping can be obtained easily using a precompensator. Therefore it appears that the goal of the control system designer is first to design a stable robust system and then use prefiltering to obtain the desired transient response.

Optimal control methods as demonstrated by the LQG control system do not guarantee good robustness properties when applied systematically to meet minimization requirements of a performance index. Methods such as **pole placement** emphasize transient response characteristics without regard to robustness, which could be detrimental to the system integrity in the presence of model uncertainties.

Classical and Modern Representation of Process Models

A dynamic unsteady state model will be developed for a heated stirred-tank process shown in Fig. 10.2.59. This process is common to the chemical and other industries. This process will be used to show that the classical frequency domain models (Buckley, "Techniques of Process Control," Wiley) presently and previously used in classical control design can also easily be represented in a state differential equation form common for use with the modern control techniques.

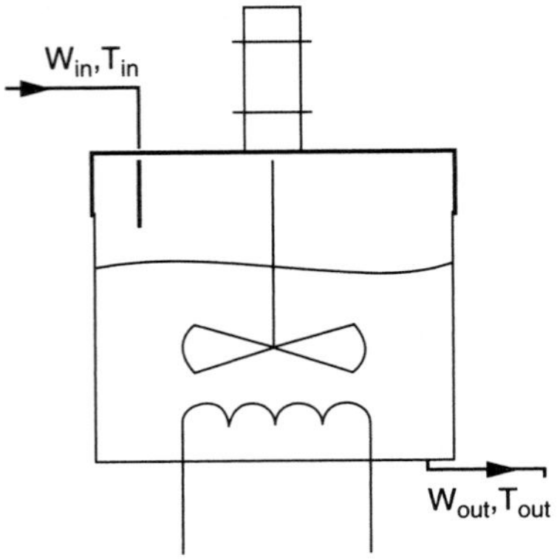

Figure 10.2.59 Process diagram of heated stirred tank.

The first step in developing the model is to obtain the system dynamic energy balance as follows

$$C\,\frac{d[V\rho(T_{\text{out}}-T_{\text{tank}})]}{dt} = w_{\text{in}}C(T_{\text{in}}-T_{\text{tank}}) - w_{\text{out}}C(T_{\text{out}}-T_{\text{tank}})+Q \qquad (10.2.145)$$

where T_{in}, T_{out}, and T_{tank} are, respectively, the liquid temperatures in the inlet and outlet lines and the tank for enthalpy calculations (assumed constant). The terms w_{in}, w_{out}, C, V, ρ, and Q are, respectively, the inlet and outlet mass flow rate, the specific heat of the liquid, the tank liquid volume, density, and the heat input to the tank from the electrical heating unit. Equation (10.2.145) defines the rate of energy accumulation for the tank contents.

The dynamic unsteady-state mass balance for the tank contents will now be defined as

$$\frac{d(V\rho)}{dt} = w_{\text{in}} - w_{\text{out}} \qquad (10.2.146)$$

Equation (10.2.146) gives the rate of mass accumulation of liquid in the tank (Seborg, Edgar, Mellichamp, "Process Dynamics and Control," Prentice-Hall). The simplifying assumptions for this model are that ρ and C are constants for Eqs. (10.2.145) and (10.2.146). Thus it follows that Eq. (10.2.146) becomes

$$\frac{\rho\, dV}{dt} = w_{\text{in}} - w_{\text{out}} \qquad (10.2.147)$$

and the differentiation of the left-hand side of Eq. (10.2.145) yields

$$C\,\frac{d[V\rho(T_{\text{out}}-T_{\text{tank}})]}{dt} = \rho C(T_{\text{out}}-T_{\text{tank}})\,\frac{dV}{dt} + V\rho C\,\frac{d(T_{\text{out}}-T_{\text{tank}})}{dt} \qquad (10.2.148)$$

Substituting the mass balance relation of Eq. (10.2.147) into Eq. (10.2.148) yields

$$C\,\frac{d[V\rho(T_{\text{out}}-T_{\text{tank}})]}{dt} = V\rho C\,\frac{dT_{\text{out}}}{dt} + C(T_{\text{out}}-T_{\text{tank}})(w_{\text{in}}-w_{\text{out}}) \qquad (10.2.149)$$

where $dT_{\text{tank}}/dt = 0$. Now substituting the right-hand side of Eq. (10.2.149) into the left-hand side of Eq. (10.2.145) yields with simplification

$$V\rho C\,\frac{dT_{\text{out}}}{dt} = C(T_{\text{in}}-T_{\text{out}})\,w_{\text{in}} + Q \qquad (10.2.150)$$

The resulting equations of the heated stirred-tank process model are

$$\frac{dV}{dt} = (w_{\text{in}}-w_{\text{out}})/\rho \qquad (10.2.151)$$

$$V\rho C\,\frac{dT_{\text{out}}}{dt} = C(T_{\text{in}}-T_{\text{out}})w_{\text{in}} + Q \qquad (10.2.152)$$

where Eq. (10.2.152) is nonlinear.

This set of equations gives the flexibility to define various operating conditions for the heated stirred-tank process. Next we will formulate the classical and modern representation for the process model that can be used for control system design.

Linearization of the Nonlinear Model

Typical classical and more of the popular modern control techniques of control system design use linear process models. Equation (10.2.152) above is nonlinear. We must make some assumptions to obtain a linear equation model. The assumptions are: the process has a constant liquid volume and the inlet and outlet flow rates are equal and constant ($w = w_{in} = w_{out}$). The initial steady-state conditions of the heated stirred-tank process at time $t = 0$ are assumed as follows: $T_{outi} = T_{out}(0)$, $T_{ini} = T_{in}(0)$, and $Q_{int} = Q(0)$.

Considering that the dynamic components of the system variables are more important for the control system designer, the initial steady-state component will be subtracted from the process variables. Applying this normalization to Eq. (10.2.152) yields the following perturbation model equation

$$V\rho C\frac{d(T_{out} - T_{outi})}{dt} = C[(T_{in} - T_{ini}) - (T_{out} - T_{outi})]w_{in} + (Q - Q_{int}) \qquad (10.2.153)$$

Let us define $\delta T_{in} = T_{in} - T_{ini}$, $\delta T_{out} = T_{out} - T_{outi}$, and $\delta Q = Q - Q_{int}$. Therefore Eq. (10.2.153) becomes

$$\tau\frac{d(\delta T_{out})}{dt} = \delta T_{in} - \delta T_{out} + \delta Q/Cw \qquad (10.2.154)$$

where $\tau = V\rho/w$ and Eq. (10.2.154) is a linear perturbed differential equation describing the dynamics of the heated stirred-tank process.

Classical Transfer Function Model

Equation (10.2.154) can be used to obtain the frequency domain process transfer function. Let's assume that at time $t = 0$, $\delta T_{out} = 0$, and the inlet temperature is constant $[\delta T_{in}(t) = T_{in}(t) - T_{ini}(0) = 0,\ \text{for}\ t \leq 0]$. Thus Eq. (10.2.154) becomes

$$\tau\frac{d(\delta T_{out})}{dt} = -\delta T_{out} + \frac{\delta Q}{Cw} \qquad (10.2.155)$$

and, taking the Laplace transform and simplifying, the heated stirred-tank transfer function is found to be

$$\frac{\delta T_{out}(s)}{dQ(s)} = \frac{1}{Cw(\tau s + 1)} \qquad (10.2.156)$$

Assuming δQ, the input heat, makes a step change from $\delta Q = Q - Q_{int}$, where $Q = Q_{int}$, to $\delta Q = Q - Q_{int}$, where $Q = \Delta Q + Q_{int}$ at $t = 0$, then for $t \geq 0$, $\delta Q = \Delta Q$. For these conditions the heated stirred-tank transfer function now becomes

$$\delta T_{out}(s) = \frac{\Delta Q}{Cw(\tau s + 1)(s)} \qquad (10.2.157)$$

A general block diagram of the perturbed transfer function process model using Eq. (10.2.156) with closed-loop feedback to control the output temperature is shown in Fig. 10.2.60.

Modern State Differential Equation Process Model

The state differential equation typically used in modern control for the process model takes the general form of

$$\dot{x}(t) = \mathbf{A}x(t) + \mathbf{B}u(t) \qquad (10.2.158)$$

$$y(t) = \mathbf{C}x(t) \qquad (10.2.159)$$

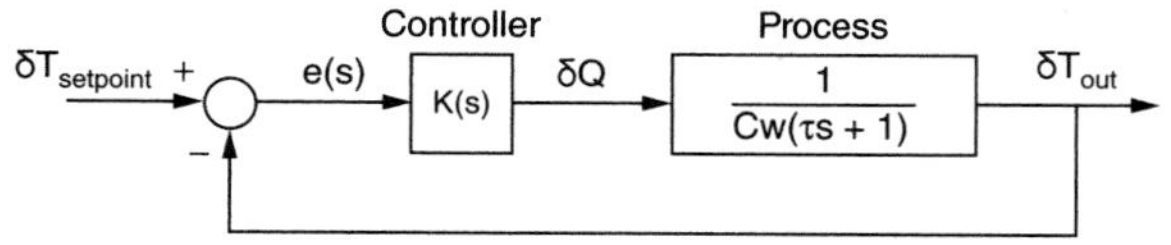

Figure 10.2.60 Block diagram for heated stirred-tank closed-loop temperature feedback control using process transfer function.

The perturbed state differential equation is easily obtained for the heated stirred-tank process by using Eq. (10.2.155), resulting in

$$\delta\dot{T}_{out} = [-1/\tau]\delta T_{out} + [1/\tau Cw]\delta Q \qquad (10.2.160)$$

where $\mathbf{A} = -1/\tau$, $\mathbf{B} = 1/\tau Cw$, $\dot{x} = \delta\dot{T}_{out}$, $x = \delta T_{out}$, $u = \delta Q$, $\mathbf{C} = 1$, and $y = \delta T_{out}$. A block diagram of the process model using Eq. (10.2.160) with closed-loop feedback to control the output temperature is shown in Fig. 10.2.61.

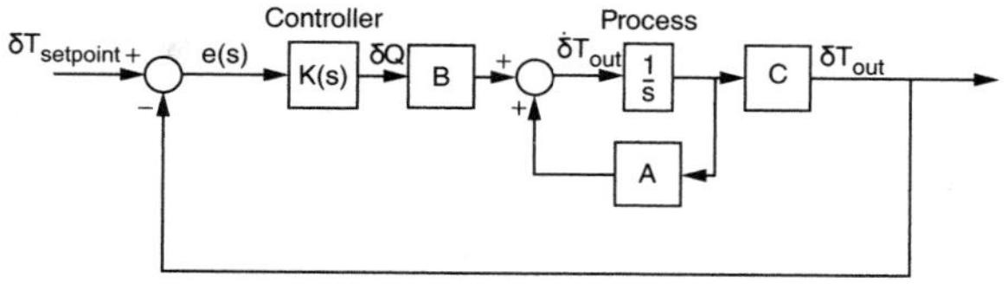

Figure 10.2.61 Block diagram for heated stirred-tank closed-loop temperature feedback control using state differential equations.

Conclusion

This heated stirred-tank example provides a detailed look at the derivation of the process transfer function and the process differential equation that would be used, respectively, for the frequency domain classical and modern control system design methods. It is clear that obtaining the linearized process model is a crucial step in this process. It should be noted that obtaining the transfer function model is an easy step from the state differential equation used in modern control system design. Finally, it should be observed that the time constant obtained during this example is directly related to the design properties of the process. Therefore it is clear that the best control engineer cannot overcome any inherent characteristic of the process that limits, for instance, the desire for a fast process response. This makes a clear argument for the need to include the control engineer in the initial design efforts for a new process or the modification of an existing process to ensure the desired system performance is obtained in the final process.

INTRODUCTION TO GAME THEORY AND AN APPLICATION TO DISTRIBUTED VOLTAGE CONTROL

Introduction to Game Theory

WHAT IS GAME THEORY

Game theory introduces a mathematical concept that deals with the analysis of strategies in a competitive environment where the outcome of a player's decision affects the interests of other players. The fundamental elements that are required to form a game are:

1. Players: the decision-makers in a game.
2. Actions: the set of decisions a *rational* player takes to further his interests.
3. Payoffs: the incentives or benefits gained by a player by selecting a particular action.

Game theory is generally classified into non-cooperative and cooperative games.

NON-COOPERATIVE GAMES

In non-cooperative games, each individual player takes a decision independently. A player has several actions available to him, and the *rational* player always chooses actions leading to best payoffs for him. Normal-form games and extensive-form games are two representations of non-cooperative games. The *Nash equilibrium*, a solution concept, is a state where no player has an incentive to deviate from his chosen action.

COOPERATIVE GAME

The main modeling unit in cooperative games is the group/coalition. Under this setting, players form coalitions with binding agreements in order to coordinate strategies, and to agree on how the total payoff is to be divided among the members. The *Shapley value* and the Core represent two fundamental solution concepts for cooperative games with transferable utility. The *Shapley value* represents a fair way of distributing the total payoff among the players in a coalition. The Core represents a stable notion for payoff distribution among the players in a coalition. The *Core* can be viewed as the *Nash Equilibrium* for cooperative games since no player has an incentive to break away from their coalition.

Distributed Voltage Control in a Power System

Tuskegee University has been investigating voltage and reactive power control using a game theory approach through an NSF (National Science Foundation)/DOE (Department of Energy) funded ERC (Engineering Research Center).

The research uses the Institute of Electrical and Electronic Engineers (IEEE) New England 39 bus model as a test case. A distributed control concept is studied using the IEEE 39 bus model by decoupling the model into four independent regions. Six different distributed decoupled model configurations are developed from the base IEEE 39 bus model.

The initial application started with a non-cooperative game theory approach, where each independent region is modeled as a separate game. Generators are selected as the decision makers (players) in this game due to their ability to control reactive power generation/absorption via an automatic voltage regulator (AVR). Compliance rate is defined as the number of load buses that fall within the required range of 0.98 p.u. (per unit)-1.02 p.u., and provides a measure of load bus voltage profile stability. For the IEEE 39-bus system, it is important to note that the base case is defined as the system when no control strategy is applied. With the application of a non-cooperative game theory strategy, the achieved result is an improved compliance rate of the regulation of load bus voltages from 3.44% (base case) to 86.2%. These results are achieved by obtaining the optimal performance of each of the four independent regions, and adequately combining these solutions.

A cooperative game theory approach was also examined for the IEEE 39 bus model system. A cooperative game model was developed for the same six distributed decoupled model configurations. In this case, generators in each region form coalitions to achieve local voltage stability within each independent region. The system is next modeled as a second cooperative game where the independent regions are now players who form coalitions to achieve a globally balanced voltage profile. Results showed further improvements in compliance rates from 86.2% (for non-cooperative strategy) to 96% (for cooperative strategy).

Game theory as an analytical framework can be further explored to solve complex voltage control regulation problems. Power system control strategies using game theory for renewable energy integration, controlling disturbances, power converters and integrating advanced control strategies such as model predictive control are presently being investigated.

Conclusion

Game theory, for example, can be applied to the optimization of autonomy and decision making of Unmanned Aerial Vehicles (UAV) sharing the same airspace, communication networks or economics. This concept of Game Theory has many possible applications in engineering and other unexploited disciplines.

TECHNOLOGY REVIEW

Fuzzy Control

Fuzzy controllers are a popular method for construction of simple control systems from **intuitive knowledge of a process.** They generally take the form of a set of IF … **THEN** … rules, where the conditional parts of the rules have the structure "Variable *i* is <fuzzy value 1> AND Variable *j* is <fuzzy value 2> … . " Here the <fuzzy value> refers to a **membership function,** which defines a range of values in which the variable may lie and a number between zero and one for each element of that range specifying the possibility (where certainty is one) that the variable takes that value. Most fuzzy control applications use an **error signal** and the rate of change of an error signal as the two variables to be tested in the conditionals.

The actions of the rules, specified after THEN, take a similar form. Because only a finite collection of membership functions is used, they can be named by mnemonics which convey meaning to the designer. For **example,** a fuzzy controller's rule base might contain the following rules:

IF temperature-error is high and temperature-error-rate is slow THEN fuel-rate is negative.

IF temperature-error is zero and temperature-error-rate is slow THEN fuel-rate is zero.

Fuzzy controllers are an attractive control design approach because of the ease with which fuzzy rules can be interpreted and can be made to follow a designer's **intuitive knowledge** regarding the control structure. They are inherently nonlinear, so it is easy to incorporate effects which mimic variable gains by adjusting either the rules or the definitions of the membership functions. Fuzzy controller technology is attractive, in part because of its intuitive appeal and its **nonlinear nature,** but also because of the frequent claim that fuzzy rules can be utilized to approximate arbitrary continuous functional relationships. While this is true, in fact it would require substantial complexity to construct a fuzzy rule base to mimic a desired function sufficiently well for most engineers to label it an approximation. Herein also lie the potential liabilities of fuzzy controller technology: there is very little theoretical guidance on how rules and membership functions should be specified. Fuzzy controller design is an **intuitive process;** if intuition fails the designer, one is left with very little beyond trial and error.

One remedy for this situation is the use of machine, or **automated learning technology** to **infer** fuzzy controller **rules** from either collected data or simulation of a process model. This process is inference, rather than deductive reasoning, in that proper operation of the controller is **learning by example,** so if the examples do not adequately describe the process, problems can occur. This inference process, however, is quite similar to the methods of system identification, which also infer model structure from examples, and which is an accepted method of model construction. Machine learning technology enables utilization of more complex controller structures as well, which introduce additional degrees of freedom in the design process but also enhance the capabilities of the controller. One **example** of this approach is the **Fuzzy PID controller** structure (Wang and Birdwell), which utilizes a fuzzy rule base to generate gains for a PID controller. Other similar approaches have been reported in the literature. An advantage of this approach is the ability to establish theoretical results on closed-loop stability, which are otherwise lacking for fuzzy controllers.

Signal Validation Technology

Continuous monitoring of instrument channels in a process industry facility serves many purposes during plant operation. In order to achieve the desired operating condition, the system states must be measured accurately. This may be accomplished by implementing a **reliable signal validation procedure** during both normal and transient operations. Such a system would help reduce challenges on control systems, **minimize plant downtime,** and help plan maintenance tasks. Examples of measurements include pressure, temperature, flow, liquid level, electrical parameters, machinery vibration, and many others. The performance of control, safety, and plant monitoring systems depends on the accuracy of signals being used in these systems. Signal validation is defined as the detection, isolation, and characterization of faulty signals. This technique must be applicable to systems with redundant or single sensor configuration.

Various methods have been developed for signal validation in aerospace, power, chemical, metals, and other process industries. Most of the early development was in the aerospace industry. The signal validation techniques vary in complexity depending on the level of information to be extracted. Both **model-based** and **direct data-based techniques** are now available. The following is a list of **techniques** often implemented for signal validation:

Consistency checking of redundant sensors
Sequential probability ratio testing for incipient fault detection

Process empirical modeling (static and dynamic) for state estimation

Computational neural networks for state estimation

Kalman filtering technique for state estimation in both linear and nonlinear systems

Time-series modeling techniques for sensor response time estimation and frequency bandwidth monitoring

PC-based signal validation systems (Upadhyaya, 1989) are now available and are being implemented on-line in various industries. The computer software system consists of one or more of the above signal processing modules, with a decision maker that provides sensor status information to the operator.

An example of **signal validation** (Upadhyaya and Eryurek) in a power plant is the monitoring of water level in a steam generator. The objective is to estimate the water level using other related measurements and compare this with the measured level. An empirical model, a neural network model, or the Kalman filtering technique may be used. For example, the inputs to an empirical model consist of steam generator main feedwater flow rate, steam generator pressure, and inlet and outlet temperatures of the primary water through the steam generator. Such a model would be developed during normal plant operation, and is generally referred to as the *training phase.* The signal estimation models should be updated according to the plant operational status. The use of multiple signal validation modules provides a high degree of confidence in the results.

Chaos

Recent advances in dynamical process analysis have revealed that nonlinear interactions in simple deterministic processes can often result in highly complex, aperiodic, sometimes apparently "random" behavior. Processes which exhibit such behavior are said to be exhibit *deterministic chaos.* The term *chaos* is perhaps unfortunate for describing this behavior. It evokes images of purely erratic behavior, but this phenomenon actually involves highly structured patterns. It is now recognized that deterministic chaos dominates many engineering systems of practical interest, and that, in many cases, it may be possible to exploit previously unrecognized deterministic structure for improved understanding and control.

Deterministic chaos and nonlinear dynamics both apply to phenomena that arise in process equipment, motors, machinery, and even control systems as a result of nonlinear components. Since nonlinearities are almost always present to some extent, nonlinear dynamics and chaos are the rule rather than the exception, although they may sometimes occur to such a small degree that they can be safely ignored. In many cases, however, nonlinearities are sufficiently large to cause significant changes in process operation. The effects are typically seen as unstable operation and/or apparent noise that is difficult to diagnose and control with conventional linear methods. These instabilities and noise can cause reduced operating range, poor quality, excessive downtime and maintenance, and, in worst cases, catastrophic process failure.

The apparent erratic behavior in chaotic processes is caused by the extreme sensitivity to initial conditions in the system, introduced by nonlinear components. Though the system is deterministic, this sensitivity and the inherent uncertainty in the initial conditions make long-term predictions impossible. In addition, small external disturbances can cause the process to behave in unexpected ways if this sensitivity has not been considered. Conversely, if one can learn how the process reacts to small perturbations, this inherent sensitivity can be taken advantage of. Once the system dynamics are known, small changes in system parameters can be used to drive the system to a desired state and to keep it there. Thus, great effects can be achieved through minimal input. Various control algorithms have been developed that use this principle to achieve their goals. A recent example from engineering is the control of a slugging fluidized bed by Vasadevan et al.

The following example illustrates the use of chaos analysis and control which involves a laboratory fluidized-bed experiment. Several recent studies have recognized that some modes of fluidization in fluidized beds can be classified as deterministic chaos. These modes involve large-scale cyclic motions where the mass of particles move in a pistonlike manner up and down in the bed. This mode is usually referred to as *slugging* and is usually considered undesirable because of its inferior transfer properties. Slugging is notoriously difficult to eradicate by conventional control techniques. Vasadevan et al. have shown that the inherent sensitivity to small perturbations can be exploited to alleviate the slugging problem. They did this by adding an extra small nozzle in the bed wall at the bottom of the bed. The small nozzle injected small pulses of process gas into the bed, thus disrupting the "normal" gas flow. The timing of the pulses was shown to be crucial to achieve the desired effect. Different injection timing caused the slugging behavior to be either enhanced or destroyed.

10.3 DIGITAL CONTROL SYSTEMS

by Michael B. Kane and Courtney A. Peckens

REFERENCES: Heath, S. (2002). *Embedded System Design*. Oxford: Elsevier; Franklin, G.F., Powell, J.D., Workman, M.L. (1997). *Digital Control of Dynamic Systems*. Addison-Welsey; Ball, S.R. (2003). *Analog Interfacing to Embedded Microprocessor Systems: Real World Design*. Elsevier; De Silva, C.W. (2007). *Sensors and Actuators: Control System Instrumentation*. Boca Raton, FL: Taylor and Francis Group; Noergaard, T. (2005). *Embedded Systems Architecture: A Comprehensive Guide for Engineers and Programmers*. Oxford: Elsevier; Burns, A., Wellings, A. (2009). *Real-time Systems and Programming Languages*. Essex, England: Pearson; Bolton, W. (2009). *Programmable Logic Controllers*. Burlington, MA: Elsevier; Kernighan, B.W., Ritchie, D.M. (1988). *The C Programming Language*. Upper Saddle River, NJ: Prentice Hall; Savkin, A.V., Evans, R.J. (2002). *Hybrid Dynamical Systems: Controller and Sensor Switching Problems*. Boston, MA: Birkhauser; Simon, D.E., (1999). *An Embedded System Primer*. Addison-Wesley; Moghimi, R. *Low Noise Signal Conditioning for Sensor-based Circuits* [PDF document]. Retrieved from Analog Devices Web Site: http://www.analog.com/media/en/technical-documentation/technical-articles/MS-2066.pdf; National Instruments. *Field Wiring and Noise Considerations for Analog Signals*. Retrieved from National Instruments Web Site: http://www.ni.com/white-paper/3344/en/.

INTRODUCTION TO DIGITAL CONTROL SYSTEMS

The automatic control algorithms introduced in Sec. 10.2 are implemented into physical systems in one of two ways: analog control or **direct digital control** (DDC). DDC systems substitute the mechanical and electrical components of an analog control system with a programmable digital computer, or microcontroller (MCU). Because a (micro-) computer is embedded into the control system, DDCs are often called embedded control systems, or simply **embedded systems.**

A typical implementation of DDC is schematically shown in Fig. 10.3.1. The physical process is sensed and the output of the sensors is filtered and converted to a digital signal through an **analog-to-digital converter** (ADC). The digitized sensor value is read by the MCU that algorithmically determines an output that will bring the process to a desired state. This MCU's output value is converted to an analog signal through a **digital-to-analog converter** (DAC), then amplified and converted to a form of energy (e.g., mechanical, thermal, chemical, etc.), and applied to the process via an **actuator.** The control loop thus consists of three steps: **sense, compute,** and **actuate.** If the plant has multiple processes, the controller can be separated and distributed across a different MCU for each process. In such **distributed control systems,** each subcontroller may communicate with the other subcontrollers to increase observability and system performance.

Why digital controls? Reduced cost, increased flexibility, increased performance, and faster design time are the main reasons to utilize digital controls. While an analog system may be simpler in terms of

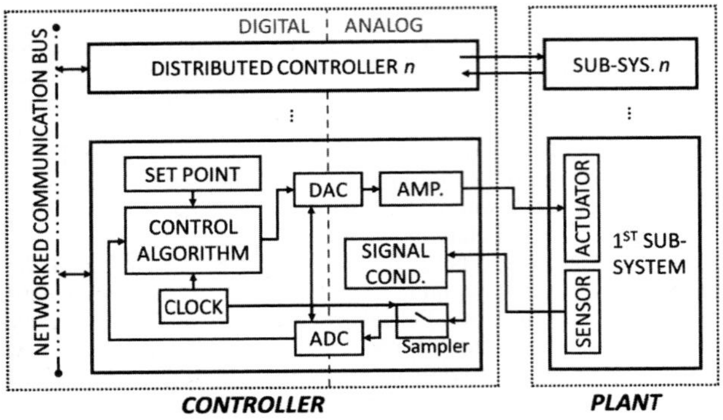

Figure 10.3.1 Schematic of a typical digital control system.

part count, a digital system can be modular, easily reprogrammed, and utilize commercial off-the-shelf (COTS) parts. Consider, for example, a typical thermostat used to control room temperature via a heater. An analog thermostat (Fig. 10.3.2*a*) consists of a coiled bimetallic strip that contracts due to differential temperature loss which in turn causes the ends of the coil to displace and eventually tilt the mercury switch enough to activate the heater. To design such a system, the control engineer must design the tilt-switch, the bimetallic strip, the mechanical interfaces for all of the components, and an enclosure. On the other hand, a digital thermostat (Fig. 10.3.2*b*) consists of a temperature sensor, a microcontroller (with an integrated ADC), a display, several buttons, and a small relay, each of which are used in thousands of other applications, thus driving down their costs. An engineer simply integrates these COTS components onto a printed circuit board (PCB) and writes a short program specifying the heater to switch on when the temperature is below the user specified set-point. One of the greatest advantages of DDC is the ease with which the system capabilities can be expanded. With a few additional lines of code, the thermostat can be scheduled to turn down during the middle of the day when the home is empty, thus significantly saving on energy and operating costs.

THE BASICS OF ENGINEERING AN EMBEDDED SYSTEM

The design of embedded control systems truly requires a multidisciplinary engineering skill-set. The **computer hardware** incorporates both analog and digital electrical circuits, formally called **mixed-signal electronics.** The **software** for embedded systems is founded on principles from computer science, and the design of software-implemented **algorithms** requires an understanding of signal processing, controls, and mathematics. The introduction of a digital computer requires the continuous mathematics presented in Sec. 10.2 to be transformed into a discrete time system which in turn gives rise to embedded digital math that is represented as binary integers or floating point values. Integration of continuous and discrete dynamics into a single system is derived from the area of study called **hybrid dynamics. Sensors** convert the measured physical attribute into an electrical signal, thereby drawing on mixed-signal electronics design skills, **application specific domain knowledge,** and potentially micro-electro-mechanical systems (MEMS) knowledge. **Actuation** design encompasses the fields of mechanics, signal amplification, and power electronics. Validation of system performance and safety is increasingly completed using **model-based design** in which the integrated hardware and software are validated using high-resolution simulation. Lacking expertise in any one area should not be a serious concern though as embedded systems are continually becoming easier to design due to the addition of integrated MCU peripherals, integrated development environments (IDEs), graphical programming languages, and plug-and-play devices.

FROM ANALOG TO DIGITAL CONTROL

Control of an analog plant using a digital controller introduces potential instabilities and performance losses that are not considered for continuous automatic control (see Sec. 10.2). Design of such a system can be completed using either time separation of the analog and digital dynamics or an integrated design approach. When the digital system sampling frequency is 10× higher than the highest frequency response of the plant, then the control design can proceed as if implemented in the analog domain. The continuous controller for the analog plant in the Laplace domain is then converted to the Z-domain for a given sampling rate, while making sure to maintain stability of this sampled system (see Sec. 10.2). The Z-domain transfer function is then converted into a difference equation and implemented in firmware. Alternatively, for plants with faster dynamics, an integrated design approach must be implemented which uses a model-based design of both the continuous plant and digitally sampled-controller to achieve a desired performance. Digital control design tools are available from all major vendors of control systems.

SENSING IN EMBEDDED SYSTEMS

Closed-loop digital control systems must sense their environment in order to affect it, similar to the control systems described in (Sec. 10.2). The general information on instrumentation in (Sec. 10.4) serves as a foundational understanding of sensing in digital systems. However, digital systems present specific challenges and design considerations, mainly in terms of interfacing the instrumentation with the MCU and mixed-signal electronics. The primary challenge is ensuring that the MCU's representation of the process measurement is accurate. Corruption of the signal can occur due to quantization effects, sampling errors, aliasing, or noise from external sources. Other sensor design considerations include adequate power supply and physical robustness.

Sensors that produce an analog electrical signal (e.g., the majority of instrumentation described in Secs. 10.1 and 10.4) must be converted to the digital domain via an ADC in order to be processed by the

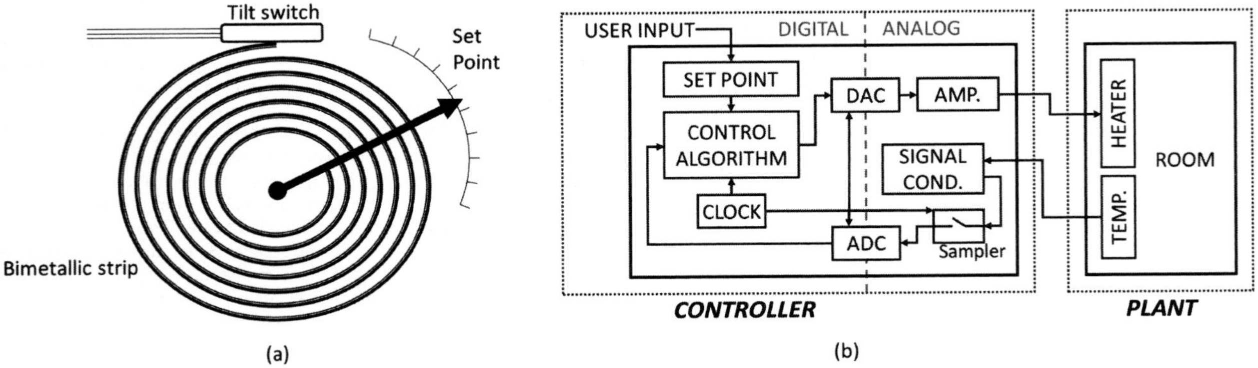

Figure 10.3.2 Comparison of an analog (*a*) and digital (*b*) thermostat.

MCU. An ADC works by first comparing a sampled input voltage to quantization levels between a low reference-voltage (V_{RefLow}) and a high reference-voltage (V_{RefHi}), and then encoding the range to a binary value (Fig. 10.3.3). A quantization error (Q) is inherently generated through the conversion from a continuous value to a n-bit digital value, where

$$Q = \frac{V_{RefHi} - V_{RefLow}}{2^n \text{bits}} \tag{10.3.1}$$

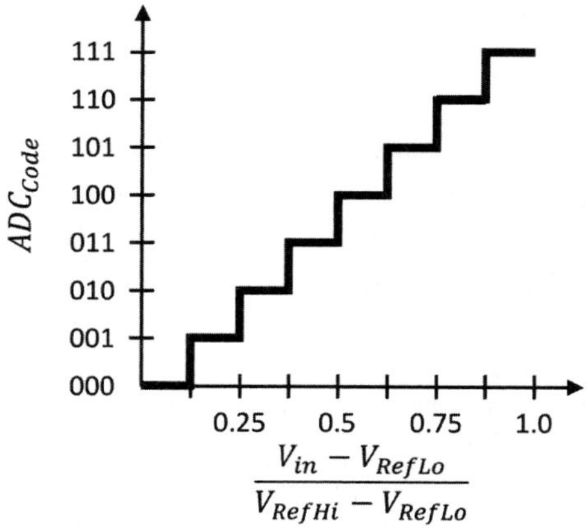

Figure 10.3.3 A 3-bit (8-level) ADC binary coding.

This quantization error adds noise to the signal which is described by the signal-to-quantization-noise ratio (SQNR) for an ideal ADC,

$$\text{SQNR} = 10\log_{10}(2^Q) \tag{10.3.2}$$

In ideal conditions, the resolution of the sampled value can be artificially increased n-bits by oversampling at ($2^{2n} + 1$) times faster than the desired sampling rate.

There are three common types of ADCs: (1) flash ADCs consist of a bank of comparators sampling the input in parallel; (2) successive-approximation ADCs successively compare the input to known analog values, each time narrowing in on the input value; and (3) sigma-delta ADCs use oversampling and a frequency reducing encoding scheme. Typical characteristics of these three variations are compared in Table 10.3.1. Other types of ADCs include adaptive delta, dual slope, folding interpolating, multi-slope, and two-step.

Digital sensors directly output a digital signal that can be read by the MCU without any conversions, thereby reducing noise concerns. These sensors can either be classified as direct or integrated. Direct digital sensors sense on-off phenomena like push buttons, level detection, and digital encoders. Integrated digital sensors contain an analog sensor as well as an integrated ADC, thus its output can be read directly by a digital interface (e.g., serial or pulse trains).

Aliasing is an error in the digital reproduction of a continuous time signal due to under-sampling. For example, consider Fig. 10.3.4 where the digital measurements are unable to distinguish between the two possible analog waveforms. In such a case, the lowest frequency waveform is assumed to be the input. Aliasing mitigation requires an estimated knowledge of the frequency content of the analog signal in order to appropriately design the sampling rate and interface electronics. The Shannon Sampling Theorem states that the signal must be sampled at

Table 10.3.1 Comparison of ADC Types

Type	Resolution	Rate	Cost
Flash (direct)	1-12 bits	<1 GHz	High
Successive-approximation	8-16 bits	<1 MHz	Low
Sigma-delta	16-24 bits	<10 kHz	Moderate

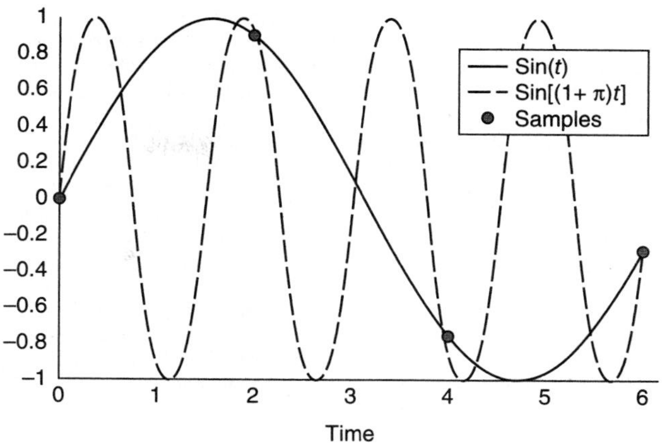

Figure 10.3.4 Visualization of aliasing of a sine-wave signal.

2x its bandwidth (i.e., the maximum frequency-component present). In practice, the signal is usually sampled at 5x-10x the signal bandwidth. Additional sampling errors can arise from clock **jitter,** a phenomena where the ADC samples are not precisely spaced moments in time.

Signal conditioning and **filtering** between the sensor and ADC or MCU reduces analog and digital noise in the measured signal by shaping the signal's physics, dynamic range, impedance, and frequency. While mainly required for analog sensors, digital sensor signal conditioning ensures compatibility for voltages and impedance between the sensor output and MCU input. Many sensors produce an output voltage proportional to the measurand's voltage, but for some sensors the output signal must be converted to a voltage. For example, strain gauges and thermistors change their resistance so a Wheatstone Bridge or voltage divider is used to convert the resistance to a voltage (see Sec. 10.1). If the dynamic range of the input signal is either expected to exceed the input range of the ADC (typically 0V-3.3V or 0V-5V) or have a small range as compared to the input range of the ADC then the signal should be scaled appropriately. Ideally, sensors should have zero output impedance (i.e., constant voltage with respect to the signal value) and infinite current output. ADCs should have infinite input impedance (i.e., measure voltage without drawing current). However, many sensors (see Table 10.3.2) have high output impedance (>1 kΩ) and some ADCs, especially inexpensive ones, have modest input impedance (<100 kΩ), which together can cause errors in the voltage measurements. When sensor impedance is high, ADC impedance is low, or both, then an instrumentation amplifier with unity gain should be used to buffer the signal and improve accuracy.

Anti-aliasing (AA) filters remove noise (and unnecessary measurands variations) from the signal prior to digital conversion. Doing so prevents the alias of high-frequency noise from corrupting the lower-frequency sampled reconstruction of the desired measurands. The

Table 10.3.2 Typical Transducer Impedances

Type	Typical impedance
Thermocouples	Low
Temperature detector	Low
Strain Gauge	Low
Integrated MEMs transducer	Low
Core MEMs transducer	High
Thermistors	High
Piezoelectric sensor	High
Solid-state pressure transducer	High
Potentiometer	High
Electret microphone	High
Glass pH electrode	Very high

cut-off frequency of AA filter should be greater than the measurand's highest frequency that is relevant to the control and the ADC should sample at 5x-10x this value. These typical analog linear filters produce a phase-lag in the signal that must be accounted for when analyzing the stability of the closed loop control system.

Additionally, **noise mitigation, grounding,** and cabling can play a significant role in the quality of the sampled signal. Noise can enter both analog and digital signals from multiple sources such as power lines, fluorescent lighting ballasts, mixed-signal coupling, non-ideal power sources, or motor electric and magnetic fields (EMF). Common methods for wiring the sensor, signals, ADC, references, and grounds to reduce noise are shown in Fig. 10.3.5. The best configurations utilize sensors with differential output and twisted pair wiring (see Sec. 11.2), shield and properly ground long lengths of wires to reduce electromagnetic interference (EMI), and reduce potential currents through the signal wires due to ground loops (i.e., currents due to slight voltage potential differences between ground points on the sensor and ADC).

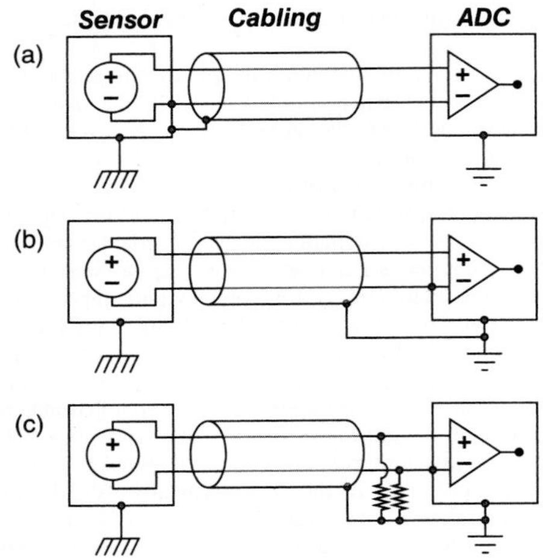

Figure 10.3.5 Sensor wiring methods to reduce noise.

EMBEDDED SYSTEM ACTUATION

Integrating actuators in embedded systems presents specific challenges and design considerations, again primarily due to conversions between the digital and analog domains. To complete the control loop, the MCU uses the sensed data to compute a control command that is sent to an actuator, which applies a load on the plant. First, the digitized output from the MCU is converted to an analog signal using either a DAC or through **pulse width modulation** (PWM). A DAC converts a digital value into an analog value (typically voltage or current) which is scaled between some low-reference and high-reference value, as dictated by the DAC's power supply. These analog values are presented as a series of impulses, due to the digitized output of the MCU, and therefore are usually passed through a digital reconstruction filter that interpolates the signal between these impulses. This filter "holds" the output of the DAC until the next value is received, resulting in a piecewise constant or staircase-shaped output.

In PWM, the digital value is modulated via the duty cycle of a square wave signal (see Sec. 11.2), such that the desired analog signal (voltage or current), is approximately proportional to the average time that the PWM is on. To extract this value, the output of the PWM is passed through an RC low-pass filter to remove the high-switching frequency from the PWM signal and leave only the DC signal. Because of this, the switching frequency of the PWM signal must be very fast as compared to the speed of the load that it is driving. Many MCUs have an on-chip PWM peripheral which simplifies implementation. Due to the limited power output of most DACs and PWM peripherals, **power amplifiers** are used to scale the signal according to the requirements of the actuator. These are typically either a voltage amplifier, where the input to the amplifier is voltage and the output is a scaled voltage, or a transconductance amplifier, where the input is voltage and the output is a scaled current.

The actuator converts the digitized controller command signal into a change in a physical parameter. There are countless numbers of actuators and these are typically classified by their driving force, such as pneumatic actuators (converts compressed air to mechanical motion), hydraulic actuators (converts hydraulic power to mechanical motion), mechanical actuators (converts rotary motion to linear motion), and electrical actuators (converts electric power to mechanical motion).

EMBEDDED SYSTEM COMPUTING ARCHITECTURE

The embedded computing system that processes sensor data into an actuator command can be represented by a layered model (Fig. 10.3.6), starting with the **hardware layer** that physically connects to the plant and supports the **system software layer,** which in turn supports the **application layer.** This architecture should be established early in the design of an embedded control system based on the anticipated algorithmic complexity of the control software and the system's ability to achieve the real-time performance demanded by the plant. The **development environment** is not part of the embedded system computing architecture, strictly speaking, but it enables the engineer to design and integrate each of the layers. It includes the debugging hardware interface and a collection of software compilers and computerized development tools often combined into an IDE.

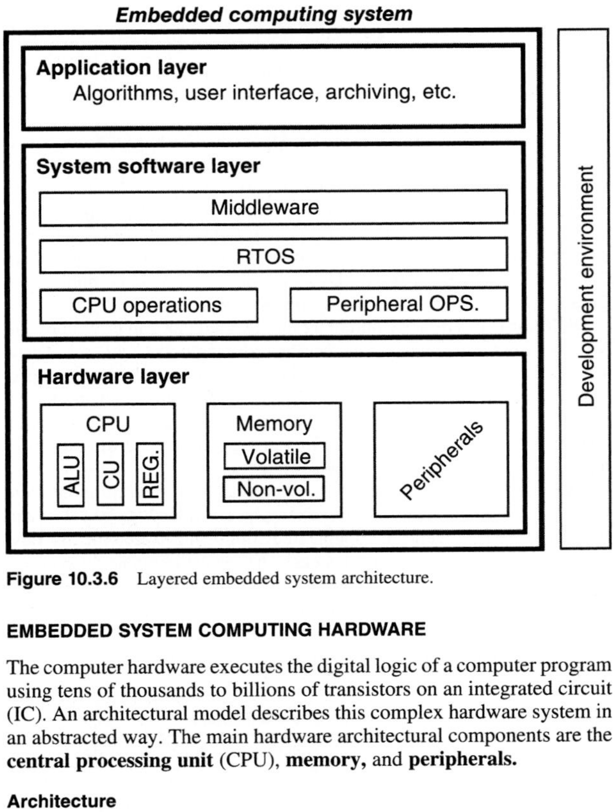

Figure 10.3.6 Layered embedded system architecture.

EMBEDDED SYSTEM COMPUTING HARDWARE

The computer hardware executes the digital logic of a computer program using tens of thousands to billions of transistors on an integrated circuit (IC). An architectural model describes this complex hardware system in an abstracted way. The main hardware architectural components are the **central processing unit** (CPU), **memory,** and **peripherals.**

Architecture

The two types of architectural models used to describe computer hardware are an instruction set architecture (ISA) and a data storage/flow architecture. The ISA defines the operations, operands, storage, addressing modes, and interrupts that comprise the CPU instruction set. All MCU **operations** are conducted in binary and typically include mathematical and logical operations, binary shift and rotate operations, operations for loading and storing data, operations to compare and move data, and operations to branch the program execution. Specifications for these operations vary in length, accuracy, number of clock-cycles per operation, and the number of operands per operation. **Operand** specifications describe what type of data is processed by the operations. Minimum data sizes of 4-, 8-, 16-, 32-, or 64-bits are possible and dictate how many clock-cycles it takes to complete an operation of a given size. CPUs with a floating-point unit (FPU) accept data in floating point representation in either 32- or 64-bits.

The **storage** architecture defines the organization and addressing of the system memory. For example, an MCU with 8-bit operands and a 16-bit address space contains $2^{16} - 1$ bytes of memory, each storing 8-bits. Some systems will use separate memory spaces for random access memory (RAM) and read-only memory (ROM), giving the system twice the memory capacity. **Endianness** describes how large values are split and stored across smaller memory locations. Little-endian (Fig. 10.3.7) systems store the least-significant byte of the operand in the lowest memory address, while big-endian (Fig. 10.3.7) systems store the most-significant byte in the lowest memory address.

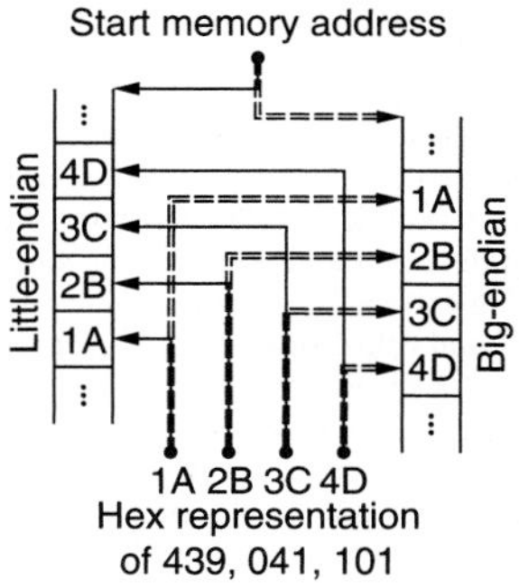

Figure 10.3.7 Example of little- and big-endian systems storing a 32-bit integer in 8-bit memory.

Interrupts enable the CPU to respond to events that occur out of order of program execution. Events such as timer countdown, external switches, or peripheral notifications trigger a special interrupt register which *interrupts* the current program execution and redirects the program code to handle the event.

Two data storage/flow architectures are the **Von Neumann architecture** and the **Harvard architecture.** The former describes how the CPU, memory, and peripherals are interfaced to each other and in this model, the CPU sends address commands to the on-chip memory which then enables reading or writing to that address. The latter, based off of the Von Neumann architecture, separates the memory in two with a section for instructions and a section for data, effectively doubling the addressable space.

Central Processing Unit (CPU)

The CPU consists of five primary components: the **arithmetic logic unit** (ALU), the **control unit** (CU), **registers,** the electrical buses, and the clock, which all must be considered when choosing an MCU. The **ALU** implements the operations defined in the ISA while the **CU** implements the CPU's pipeline of fetching, decoding, and executing instructions and data. The **registers** are a special location in memory, physically close to the ALU in order to increase data-rate transfers. The electrical buses are circuits that interconnect the ALU, CU, and registers, with a wider bus (e.g., 16-bits vs. 8-bits) being able to move more data per clock cycle. The **clock** is an alternating digital signal that determines the speed at which instructions are executed and data is moved. The master clock is typically derived from a crystal oscillator, the accuracy (see Table 10.3.3) of which determines the amount of jitter and drift in control sampling and timing.

Table 10.3.3 Comparison of Clock Oscillator Types

Type	Accuracy	Expected error in 24 hr
RC	5%-50%	1.2-12 hr
Crystal (XO)	10-100 ppm	800 ms-8 s
Crystal (OCXO)	0.1 ppb-10 ppm	8 μs-800 ms
Crystal (TCXO)	10 ppb-100 ppm	800 μs-8 s
MEMS	1-100 ppm	80 us-8 s
Atomic clock	0.001 ppb	80 ns

Memory

The memory in an embedded system serves multiple purposes: buffering and storage during operations, remembering the state of the system, storing program code, archiving data, and others. The design of memory for an embedded system should account for latency of reading or writing (R/W) the memory, energy consumption, and corruption mitigation.

Volatile memory is RAM that can be both written to and read from but whose stored values are lost when power is removed. In embedded control, it stores the state of the control plant in between calculations. Most embedded systems use Static RAM (SRAM) due to its fast response and low-power consumption; however, systems that require significant amounts of dynamic storage may use high power, yet less expensive, Dynamic RAM (DRAM). Registers are RAM that are located close to the CPU and used by the CU to temporarily store data and execution states. Registers are also connected directly to some peripherals to inform the CPU of external events via flags or to reduce latency of fetching external data (e.g., from ADC outputs). Because of this, some registers can change value without software intervention. Cache memory is very-low latency memory used by the CPU for large operations that cannot fit in the limited CU registers. Some systems tier the cache into several layers, designating L0 for the smallest yet fastest layer and working up to L3 for the largest yet furthest layer from the CPU. The CU and/or firmware may move data used most often to the lowest level cache and unused or non-critical data to a high-level cache, main memory, or even external memory.

Non-volatile memory stores data and program code that must be saved in the event of power loss or system shutdown. Early embedded systems were implemented on non-volatile memory using ROM. Modern MCUs use reprogrammable non-volatile memory, yet write operations are minimized. Write cycles to non-volatile memory are typically slower than volatile memory; however, when only reading program code from non-volatile memory, this is a relatively moot point. Writing to non-volatile memory is generally minimized due to the increased probability of data corruption in memory once an address has been written to a certain number of times. Electronically-erasable programmable ROM (EEPROM) is a type of non-volatile memory that can be easily reprogrammed, simplifying the embedded system development process and enabling the program to change semi-permanent stored values. Flash memory is replacing EEPROM in most new MCUs due to its reduced cost and faster performance. Ferroelectric RAM (FRAM) is the newest type of non-volatile memory used in embedded systems and consumes even less power than flash, has faster write performance, and longer life. The write-cycle limit of programmable ROMs, compared in Table 10.3.4, is of little concern in well-designed embedded systems. However, a poorly designed program, for example, could archive 100 ms of ADC data once every 200 ms to the same location in ROM for use in a calculation that takes 100 ms to complete (e.g., a Fast Fourier Transform). After barely a day the ROM will have cycled over a million times, likely resulting in corrupted data and the need to physically replace the embedded system.

If the system requires more memory than is available on the selected MCU, then **external memory** may be added. Some MCUs include an

Table 10.3.4 Comparison of Programmable ROM

Type	Limit of write cycles
EEPROM	10^3-10^6
FLASH	100-10^6
FRAM	10^{16}

external memory interface that directly extends and maps the memory address and data bus, resulting in a relatively transparent extension between internal and external memory. For MCUs without such an interface, firmware implementation of a memory management program can **page** data from the external memory to a small segment of the internal memory as needed. Serial or parallel peripheral interfaces may be used; the former simplifies circuitry, and the latter has high data-rates and lower latency due to the wider bus. Regardless of the interface, multiple clock cycles are typically required for R/W of external memory, thus introducing latency.

Peripherals

Peripherals expand the capabilities of the CPU and enable it to interface with other system components. Modern **systems on a chip (SoC)** are designed to include all the processing, memory, and peripherals required by an embedded system onto a single chip. If a SoC is not used (e.g., for cost considerations) then external peripherals can often be added using the MCU's input/outputs (I/Os). Implementation of peripherals should carefully consider system-effects of power consumption, start-up requirements, and latency.

Ancillary services provided by peripherals often improve system latency and power consumption and reduce required CPU complexity by off-loading specific tasks from the CPU onto specialized logical circuits. **Counters** provide a mechanism for counting events without the CPU frequently checking if an event has occurred. The primary design parameter for counters is the size; for example, an 8-bit (16-bit) counter can only count up to 255 (65,535) events before an overflow occurs. A **timer** is a specialized counter connected to a clock signal to count how much time has elapsed. A **real-time clock** (RTC) is a specialized timer with a large counter and high-accuracy clock signal that accurately keeps track of time over a period of years. A **watchdog** countdown timer is reset at regular intervals by the firmware. If the firmware fails to reset the counter, then a program execution failure is assumed to have occurred, and the watchdog executes a contingency event (e.g., a reset of the CPU). **Interrupt request modules** multiplex a large number of interrupts into the limited number of interrupt flags in the CU. General purpose or specialized **coprocessors** can more than double the computational throughput of an MCU. For example, digital signal processors (DSPs) execute signal processing algorithms more efficiently than the general purpose CPU, while field-programmable gate arrays (FPGAs) enable implementation of efficient application specific logic for development and low-count manufacturing. **Direct memory access** (DMA) enables DMA-connected peripherals to R/W to a subset of the memory address space without CPU intervention. The updated values (e.g., ADC reading) are then ready for the CPU when they are needed by the program. **Power management** peripherals enable the firmware to reduce the number of running clocks, power-off unused peripherals, measure battery status, and execute other features for power constrained applications.

Input modules convey data from outside the computing system to either the CPU or to memory (via DMA). **General purpose inputs (GPI)** enable external digital signals to change the values of specific MCU registers. A GPI **multiplexer** enables multiple digital inputs to be sequentially read into a single logical circuit (e.g., register bit). A **quadrature decoder** counts and interprets the input from quadrature encoded sensors. **Capture** pins on the MCU package can be connected with high-accuracy phase-locked loop (PLL) timers for event capture or to an interrupt request manager to trigger program execution. Analog signals can be logically compared with each other via a **comparator** or converted to a digital representation with an ADC.

Output modules provide a mechanism for the computing system to affect external systems. **General purpose output (GPO)** pins enable the MCU to turn on external logical devices like relays or indicators. **De-multiplexers** sequentially update multiple GPO pins from a logical circuit (e.g., a register bit updated by the CPU). PWM peripherals connect to a timer to generate signals (see Sec. 11.2) with only a duty-cycle command from the CPU. MCUs developed especially for motor control have multiple PWMs that are synchronized for control of each motor phase. A DAC can be implemented by filtering a PWM signal through an RC low-pass filter or implemented as a single integrated peripheral with higher accuracy and lower output impedance.

General purpose input-output (GPIO) pins can be used for simple "bit banging" communication, but this should be avoided due to the high load it puts on the CPU. Instead, peripherals with standardized communication protocols (see Sec. 11.2) should be used whenever possible. These peripherals automate the encoding and decoding of the signal sent over the wires and use **buffers** to hold data to be sent or after receipt but prior to processing. Firmware should be designed to avoid buffer overflows where data is sent faster than the system can process it from the buffer, resulting in data loss. The selection of an appropriate communication protocol depends on the distance the signal must travel (e.g., chip-to-chip vs. device-to-device), required data rate, system energy budget, system complexity (e.g., number of wires), and number of devices to which to communicate.

Digital communication peripherals connect the CPU with other digital ICs that may be on-chip, off-chip, on another PCB, or even a network of MCUs. Digital communications occur either over **parallel** communications, where n-bits are set simultaneously over n-separate wires, or **serial** communications, where n-bits are sent over less than n-wires taking at least n-clock-cycles and requiring input and output buffering. **Half-duplex** communications use a single channel for both sending and receiving and therefore cannot send and receive simultaneously. On the other hand **full-duplex** communications use two channels (e.g., a send wire and a receive wire) to enable simultaneous sending and receiving of data. In **synchronous** communication, such as inter-integrated circuit (I2C) protocol or serial peripheral interface (SPI), bits transmitted over data lines are synchronized with an additional clock signal (e.g., wire) that denotes when a bit has been sent or is ready to be received. **Asynchronous** communications [e.g., IEEE 488, RS-232, universal serial bus (USB)] instead encode the clock signal into the data transmission. A **JTAG,** joint test action group, (i.e., IEEE Std. 1,149) module is a special communication peripheral that enables an embedded system to communicate processor state information (e.g., execution status and memory contents) with a host PC for the purpose of development and debugging.

Wireless communication (see Sec. 11.2) is advantageous for systems where devices may be relocated and where the cost of laying wire is high. WirelessHART (IEC 62,591) is a standard wireless networking protocol designed specifically for industrial control systems. Wireless communication may have decreased reliability and increased latency, as compared to wired signals, which can be detrimental to the performance of embedded control systems.

The **Open Systems Interconnection** (OSI) model characterizes the main components of a communication system and may serve as an outline to define standardized modules. This model partitions a communication system into seven unique layers: the **physical layer** defines the mechanism by which electrons, photons, or waves transmit bits; the **data link layer** manages the transceiving of bits between nodes; the **network layer** handles the routing and traffic control of data packets in a multi-node communication system; the **transport layer** ensures that messages are received, and arranges retransmission of lost data if necessary; the **session layer** manages communication sessions (i.e., brief periods of persistent communications between nodes); the **presentation layer** translates the data used by lower layers into a useable form; and the top **application layer** interfaces to user programs that utilize and send data. Common network stacks for embedded systems include Controller Area Network (CAN), Fieldbus, Modbus, Profibus, LonWorks, and others.

EMBEDDED SYSTEM COMPUTER SOFTWARE

Firmware is the software that dictates which tasks to complete in the embedded system. It is deployed on the MCU as machine-code executed by the CPU. A programming language or interface should be selected based on the following six criteria: *security* for detecting and mitigating programming errors; *readability* for developing an understandable program; *flexibility* for implementing all required operations; *simplicity* for minimizing errors and costs; *portability* for enabling code re-use between applications and platforms; and *efficiency* for real-time, predictable operation.

The firmware program that implements the digital control must do so in an effective and reliable manner. Most importantly, the design should consider the computing hardware's capabilities (e.g., CPU speed and amount of memory) with the plant's requirements for control rate and response to unexpected events (e.g., activation of an emergency stop switch). A layered model is used to describe the key components of embedded system firmware with the **operating system** (OS) at the lowest level, followed by the **middleware**, and topped with **application software.** A variety of firmware architectures can be used to meet system requirements, from simple sequential round robin execution to full-featured parallel real-time operating systems (RTOSs).

In addition to the embedded firmware, digital control systems may interface with software running on other embedded devices, cloud servers, or user-interface PCs. The design of such systems requires more in-depth topics which are outside the scope of this handbook.

Operating System Architectures

There exist firmware architectures of varying levels of complexity and flexibility where one may be more appropriate based on system requirements. Four architectures will be discussed, starting with the simplest and progressing to the most complex: round robin, round robin with interrupts, function queue scheduling, and real time operating systems. For DDC, these architectures seek to complete concurrent **tasks** reliably and on time, such as servicing an ADC, estimating the plant's state, calculating the control algorithm, archiving data, or processing user-input buffers. Task characteristics that may affect performance include whether they are blocking, non-blocking, or atomic; their execution time; and whether data is shared with other tasks.

The **round robin** architecture sequentially queries all tasks in a defined order, with no priority given to any task. If action is required at the time of the query then the CPU completes it and continues on. This architecture is ineffective for systems that require guaranteed timing, when tasks have significantly different priorities, or when the CPU is heavily taxed. **State machine** architectures extend the round robin concept with branches into different modes of operation.

In **round robin with interrupts,** urgent tasks are handled by the **interrupt service routine** (ISR). The ISR stops the current round robin task at the end of the current **atomic operation,** stores the current state, executes the new higher-priority task (i.e., thread), then restores the system state, and returns to executing the interrupted task. By introducing interrupts, data shared amongst multiple tasks, or **shared data,** could be modified in the middle of a task and care must be taken to maintain the integrity of this data if required in another task. Additionally, if multiple interrupts are used then there is now shared priority and if all interrupts trigger then a high-priority task has potential to wait a maximum of the combined execution time of all of the other interrupt tasks before being able to execute.

Function queue scheduling is a means for overcoming this limitation by assigning priority to the interrupts. In this architecture, the interrupt routines add a marker (e.g., function pointers) to a queue. If the queue is not empty when the main function checks it then it addresses the first task in the queue. Tasks do not have to be added to the queue in the order that their interrupt occurs; instead, priority can be given to certain tasks and those can be placed in the queue with a high priority. For this architecture the longest that a high-priority task will have to wait is the maximum execution time of any single lower-priority task.

Real-time operating systems (RTOS) combine PC operating system ease-of-use with real-time response and reliability required by embedded systems. In a basic RTOS, tasks can be either running, ready, or blocked. If running, the task is being executed by the CPU; only one task can be running at a time on most systems. If ready, the task is queued for execution when the CPU is available; many tasks may be in this state and are executed according to priority. If blocked, the task is waiting for an external event to occur and would not execute even if the CPU was available; many tasks could be blocked at one time. The RTOS uses a **scheduler** to keep track of the states of all tasks and to decide which task should be executed next based on priority. In some systems, called **preemptive,** the scheduler can even suspend a low-priority task to execute a high-priority task before returning back to the low-priority task. The system design should ensure the feasibility of executing all anticipated tasks within the time allotted by the scheduler and consider the likelihood of a series of tasks overflowing the schedule. The main disadvantage of this architecture is that the RTOS code itself can require memory and execution time which may not be permissible for very time sensitive systems on low-power hardware.

Shared data is also an issue with the RTOS architecture but it can be protected using **semaphores.** A semaphore for a given resource (e.g., ADC data) is shared amongst multiple tasks with only one task being allowed to have the semaphore at any given time. The task possessing the semaphore locks the resource and all other tasks that attempt to access it are blocked until the possessing task releases it. This alleviates shared data issues but also can result in **priority inversion** or **deadlocking** errors. Priority inversion is as the name implies; a low-priority task locks the semaphore in order to protect data and a higher priority task that uses the same data is unable to execute because of this. To better understand **deadlock,** consider Fig. 10.3.8 where there are two tasks that share two semaphores, S_1 and S_2, in a preemptive system. Because the high-priority task can interrupt the low-priority task this scenario results in a deadlock where neither task can be completed. Both deadlock and priority inversion can be avoided through careful coding schemes.

Middleware

Middleware are portable and modular programs that link the application software to the OS, further reducing the burden on the application software developer. **Device drivers,** a major category of middleware, abstract low-level peripherals and OS functions to applications that make use of sensors, actuators, displays, networks, storage, and human-interface devices (HID). Middleware may also fulfill more advanced functions like signal processing and encoding, fault tolerance, encryption, and video processing/rendering. Low-power applications may have middleware that control battery charging, modulate peripheral power consumption, and other energy management activities. In networked control systems, middleware can manage synchronization and coordination of the application software running on each node. **Bootloader** middleware provides a mechanism for core firmware features to be updated via USB, UART, wireless, or other field deployable methods. Application software developers typically utilize compiled middleware libraries on the MCU via **application programming interfaces** (API) that define function calls to access middleware features.

Application Software

The **application software** is the layer of the embedded computing architecture that describes specific tasks to be completed by the control systems and this layer is typically the primary focus of the design process due to the modularity of hardware and system firmware. This layer executes the control algorithm, interfaces with user input or external peripherals as needed, and communicates with other parts of the control system.

Source code for the application software layer can be created in three different ways: using an IDE; through programs with graphical interfaces that utilize **code generation,** such as Labview, Simulink, or VisSim; or with ladder-logic used in PLCs.

An IDE is an application tool that is used by a software developer to write source code and typically includes a source code editor, a **debugger,** and a **compiler.** IDEs can support multiple programming languages (e.g., C, Python, Ada, Java), but they are generally not universal to all languages. There are a large variety of IDEs from which

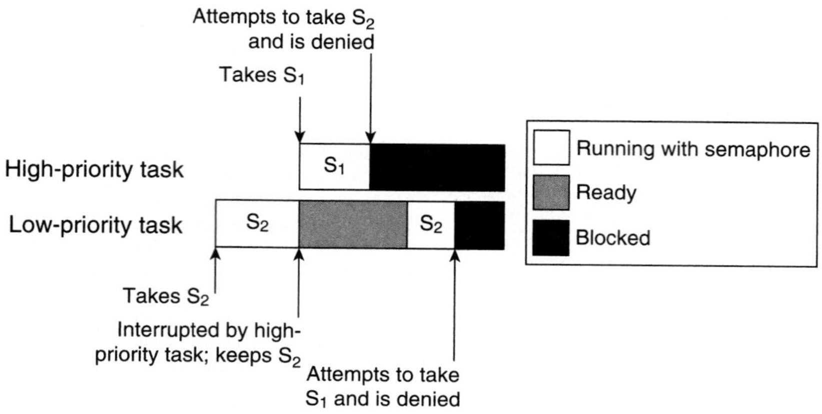

Figure 10.3.8 Semaphore deadlock example.

to choose, and a software developer often selects a specific IDE based on its available functions, features, tools, and compatibility. In **code generation,** on the other hand, the developer uses sketches to build layouts that achieve the desired functionality of the application, and the software generates useable source code. For example, MATLAB has a MATLAB Coder toolbox that generates C-code to be embedded into the MCU. **Ladder-logic** is a graphical programming method used by many PLCs for describing the flow of a program with logical (i.e., on-off) inputs, decisions, and outputs. Ladder logic schematics, for example, Fig. 10.3.9, are read left-to-right from the top "rung" to the bottom in a continuous loop by the embedded computer. Power flowing from the left rail, through the switches and coil on the rung, to the right rail will energize the coil. A coil and switch with the same name act as an internal relay. The symbols can be arranged to perform Boolean operations forming functional programs that accomplish the required automation.

When using an IDE, source code is written as a **plain text file** using a programming language of choice. Writing code for embedded applications requires additional considerations due to the constraints of the MCU, specifically when dealing with real-time applications such as DDC. Code must be designed to be efficient and certain operations, such as loops or accessing memory, must be carefully considered. Some of the work of writing source code can be alleviated by making use of libraries. By including a specific header file in the source code, the code has access to prototypes of certain functions or declarations of special data types. For example, by including *math.h* in code written in C, the source code can use trigonometric functions.

A **compiler** converts source code into machine code that can then be embedded onto the microcontroller, either through an external programming interface or a **bootloader.** An external programmer is an extra piece of hardware that connects the MCU to a PC; the programmer communicates the machine code to the MCU via a communication peripheral (e.g., SPI). A bootloader, on the other hand, is a piece of firmware on the MCU that operates when the MCU is in reset mode; it allows the MCU to be programmed in-situ without an additional piece of hardware. Once the code is loaded onto the MCU then the application software is ready for execution.

Debugging in an embedded environment is challenging as there is not an easy means for observing the progress of the program. Typically, the code is simulated on a PC before embedding onto an MCU but this does not capture the real-time functionality of the MCU. Many microcontroller development kits include an **evaluation board** which is equipped with a sample microcontroller with limited functionality, thus allowing the real-time processing and basic I/O functions to be verified. An **emulator** is the most advanced debugging tool and is a piece of hardware that has the full functionality of the MCU, thereby enabling complete prototype testing. In addition to these, some MCUs are equipped with a JTAG module which enables direct communication between the MCU and the PC.

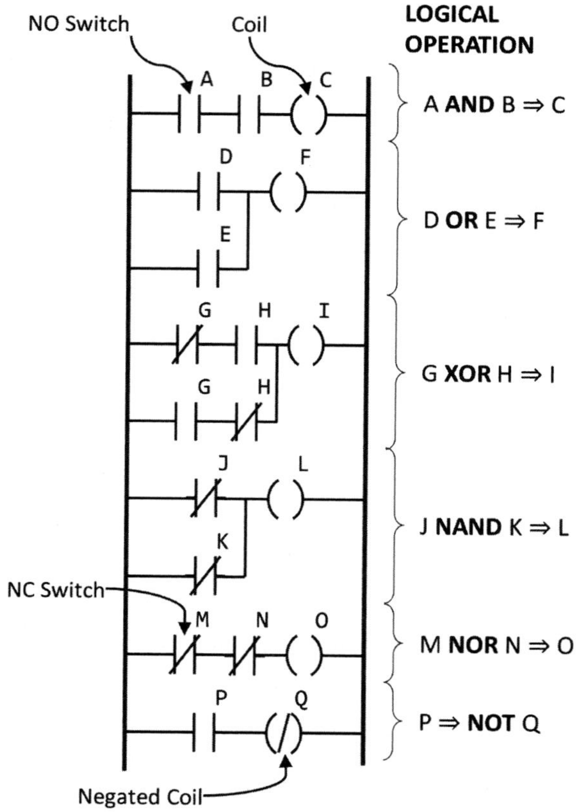

Figure 10.3.9 Understanding ladder logic programs.

PLATFORMS

The design of a digital control system from the chip up with custom firmware is a complex endeavor that requires expertise in a variety of fields. Fortunately, COTS embedded control platforms simplify the design, development, and deployment. The selection of an appropriate platform will largely depend on project metrics such as non-recurring engineering (NRE) costs, unit costs, designer skills, total production count, power and thermal requirements, and controller complexity.

Programmable-logic controllers (PLCs) are general purpose control platforms that require little electrical design and are typically programmed using simple **ladder-logic** or functional block languages, thus making PLCs ideal for custom industrial control applications. As the name implies, PLCs were traditionally limited to logical (i.e., on-off) automation and control, but modern systems are capable of executing advanced algorithms and interfacing with complex sensors and actuators.

Electronic instrument control platforms are flexible systems used to implement specialized controllers or to prototype controllers that will eventually be mass produced on custom hardware. Manufacturers of these systems, including National Instruments (and other members of PXI Systems Alliance), Mathworks, dSpace, and others, also sell well-documented integrated development software that convert computer programs or sketches into real-time code for the embedded system, significantly decreasing development time.

Single board computers (SBCs) are general-purpose, low-cost platforms used for prototyping control systems and low-unit deployments with larger power budgets, compared to MCU applications. COTS SBCs are available in ruggedized extensible form factors (e.g., PC/104) or for benchtop and hobby applications (e.g., Raspberry Pi). SBCs typically run stripped-down desktop operating systems (e.g., embedded versions of Windows or Linux), thereby lowering the learning curve for engineers familiar with developing applications for PCs. These operating systems, however, often do not support true real-time operation of timers and interrupts, which could potentially affect control performance and reliability. Sensors, actuators, and advanced peripherals are interfaced with GPIO, SPI, UART (universal asynchronous receiver/transmitter), or USB through specialized (or custom) hardware drivers.

Development kits and modules are useful starting points for designing a customized control platform to be integrated into a final mass-produced product. These are sold by chip manufacturers to encourage use of their products (e.g., ARM mbed, TI LaunchPad, Intel Edison) or by third parties as educational or hobbyist tools (e.g., Arduino, Raspberry Pi). The kits are often bundled with free or low-cost IDEs specific to the kits hardware to simplify development.

Open Source Platforms like the MCU based Arduino and SBC BeagleBoard are low-cost options that enable the platform design to be customized and reused without hardware or software licensing fees. Open source communities develop documentation, standards, alterations, and extensions, further increasing the flexibility of these platforms' hardware and software. For example, Arduino software and hardware compatible platforms are available with 8-, 16-, and 32-bit MCUs operating from 8 to 400 MHz, 3.3V and 5V power sources, a variety of peripherals, and with MCUs from multiple manufacturers. Additionally, more than 100 different expansion boards with peripherals, sensors, filters, motor controllers, and power amplifiers are available. The Arduino IDE compiles programs written in C or C++, and provides access to a middleware called "Wiring" and community libraries, enabling simple round robin firmware all-the-way to full-featured RTOS architectures.

THE INTERNET OF THINGS (IOT) AND CYBER PHYSICAL SYSTEMS (CPS)

Digital control systems are becoming more powerful, more connected, and lower-cost with each new generation of MCUs, sensors, and software. This trend is leading to an increasing ubiquity of connected, digitally controlled and monitored systems from applications in health, infrastructure, manufacturing, and the environment, or interconnections thereof. Two new fields have emerged with these trends. The **Internet of Things** (IoT) is the ubiquitous connection via the internet of a plethora of physical devices (e.g., thermostats, jet engines, heart monitors) with embedded computing. Cyber-Physical Systems (CPS) focus on the interaction of the physical components and internet connected components in a system with ubiquitous computing and connectivity. Developing for the IoT and CPS does come with significant challenges. Users' *privacy* should be maintained through inherent obfuscation and security measures, especially because connected devices can track where users are, what they say, who they talk to, and potentially predict what they will do. System *security* should build on existing security in the cyber- and physical-domains and secure attack vectors from the physical system into the cyber-system, and vice versa. *Flexibility* is important to consider when devices may enter, exit, and move through the system at any time and when new devices may be rapidly introduced. *Reliability* and *resilience* in the face of uncertain environments should be maintained through local autonomy in order to mitigate the potentially dangerous implications of system failure. Together, IoT and CPS promise to usher in a new era of automation to increase efficiency, reduce costs, and improve quality-of-life.

SYMBOLS AND ABBREVIATIONS

- AA — Anti-Aliasing
- ADC — Analog-to-Digital Converter
- ALU — Arithmetic Logic Unit
- API — Application Programming Interface
- CAN — Controller Area Network
- COTS — Commercial Off The Shelf
- CPS — Cyber Physical Systems
- CPU — Central Processing Unit
- CU — Control Unit
- DAC — Digital-to-Analog Converter
- DDC — Direct Digital Control
- DMA — Direct Memory Access
- DRAM — Dynamic RAM
- DSP — Digital Signal Processor
- EEPROM — Electronically-Erasable Programmable ROM
- EMF — Electric and Magnetic Fields
- EMI — Electromagnetic Interference
- FPGA — Field-Programmable Gate Array
- FPU — Floating-Point Unit
- FRAM — Ferroelectric RAM
- GPI — General Purpose Input
- GPIO — General Purpose Input Output
- GPO — General Purpose Output
- HID — Human-Interface Devices
- IoT — Internet of Things
- I/O — Input/Output
- I2C — Inter-Integrated Circuit
- IC — Integrated Circuit
- IDE — Integrated Development Environment
- ISA — Instruction Set Architecture
- ISR — Interrupt Service Routine
- JTAG — Joint Test Action Group
- MCU — Microcontroller
- MEMS — Micro-Electro-Mechanical Systems
- NRE — Non-Recurring Engineering
- OS — Operating System
- OSI — Open Systems Interconnection
- PCB — Printed Circuit Board
- PLC — Programmable Logic Controller
- PLL — Phased Locked Loop
- PWM — Pulse Width Modulation
- R/W — Reading or Writing
- RAM — Random Access Memory
- ROM — Read Only Memory
- RTC — Real-Time Clock
- RTOS — Real Time Operating System
- SBC — Single Board Computer
- SoC — System On a Chip
- SPI — Serial Peripheral Interface
- SQNR — Signal-to-Quantization-Noise Ratio
- SRAM — Static RAM
- UART — Universal Asynchronous Receiver/Transmitter
- USB — Universal Serial Bus

10.4 SENSORS AND ACTUATORS

by Niell Elvin and Alex Elvin

Modern control systems are ubiquitous in mechanical engineering systems whether for the control of air-conditioning systems, aircraft landing gear, consumer and military drones or sophisticated robotic systems. A block diagram of a simplified feed-back control system is shown in Fig. 10.4.1. As can be clearly seen, the control system relies on actuators and sensors (details of the controller are covered in other chapters of this book). An actuator is defined as a mechanical system that is capable of producing mechanical work. A sensor or transducer is a system that is capable of detecting (i.e., measuring) a physical property (such as temperature, displacement, pressure, etc.) and outputting this information via an electrical signal.

Many different aspects need to be analyzed when designing an appropriate actuator or sensor within a system. For actuators, critical design parameters include: the required displacement (also commonly called stroke), force level, velocity, size, weight, method of powering the actuator, resolution (in force or displacement), robustness, etc. For sensors, critical design parameters include resolution, effect of ambient environment, long-term stability, methods of providing power, and bandwidth (i.e., how rapidly the sensor can produce an output). Lifetime and short-term costs for both sensors and actuators are also critical.

ACTUATOR FUNDAMENTALS

Rationally the starting point of choosing an actuator should be based on deciding whether a given actuator meets the force and/or displacement requirements for a given system. For example, extremely high force and stroke requirements (say in the megaNewton and meter range) would in all probability require hydraulic actuation. It is convenient to characterize any actuator by its force-displacement curve as shown in Fig. 10.4.2. An actuator can provide maximal force when it is fully constrained (i.e., blocked) from movement, and conversely can provide maximum deflection or displacement when free from force. Typically with adequate control, an actuator can operate anywhere within the shaded region shown in Fig. 10.4.2.

Since force is typically dependent on the cross-sectional area of the actuator, it is convenient to define the maximal (blocking) stress of an actuator as:

$$\sigma_{\max} = \frac{\text{Blocking Force}}{\text{Actuator Area}} \tag{10.4.1}$$

Similarly since the free deflection is dependent on the length of the actuator it is convenient to define the maximal uniaxial strain as:

$$\varepsilon_{\max} = \frac{\text{Free Deflection}}{\text{Actuator Length}} \tag{10.4.2}$$

In applications where actuation velocity is of importance, the actuator mechanical power (defined as Power = force × velocity) also needs to be carefully selected. Consider an actuator subjected to sinusoidal input at a given frequency, the power output for a given frequency is schematically shown in Fig. 10.4.3. For very low frequencies, the output velocity is small and thus the power drops. At high-frequencies, the power is typically limited by internal inertia and friction or damping. An actuator should thus be capable of providing power within the gray trapezoidal region labeled "operating region" in Fig. 10.4.3.

Actuator resolution which is defined as the minimum displacement that the actuator can achieve is often limited by the controller system but internal mechanisms (such as gearing) and friction can also limit resolution. We define the minimal strain as:

$$\varepsilon_{\min} = \frac{\text{Displacement Resolution}}{\text{Actuator Length}} \tag{10.4.3}$$

ACTUATOR TYPES

A great variety of actuators are currently available commercially, with more being developed and tested in research laboratories around the world. In this section we broadly discuss some of the more common commercial actuators.

Hydraulic Actuators

Any actuator that uses liquid (water or oil) to provide displacement and force is defined as a hydraulic actuator. The typical actuator (Fig. 10.4.4) consists of a piston in a hollow tube that is moved by pressurized liquid. The system also typically requires a pump, accumulator, reservoir, hydraulic lines, valves, and associated fittings. These systems are used when large force and/or displacements are required. However, these actuators are typically limited by relatively slow velocities. Since liquids are relatively incompressible, large forces and displacements are possible. The number of moving parts is limited (with no gears, etc.) and thus wear is generally low. However, filters, gaskets, and hydraulic fluid require periodic replacements or maintenance. The maximum operating frequency depends on the size of the actuator but is generally limited to the 5-300 Hz range with larger actuators having lower maximum operating frequency.

Pneumatic Actuators

Similar in operation to hydraulic actuators, pneumatic actuators use air provided by a compressor instead of liquid and a pump. These systems are typically used where lower forces are required. Due to the compressibility of air, control is somewhat more difficult than for hydraulic systems. Pneumatic devices tend to be cheaper and lighter than their hydraulic counterparts. Maximum operating frequency is similar to hydraulic actuators, that is, in the 5-300 Hz range (Fig. 10.4.5).

Electrical Actuators

Electromagnetic actuators are broadly defined here as any actuator that uses electrical current (including electromagnetic fields) to induce

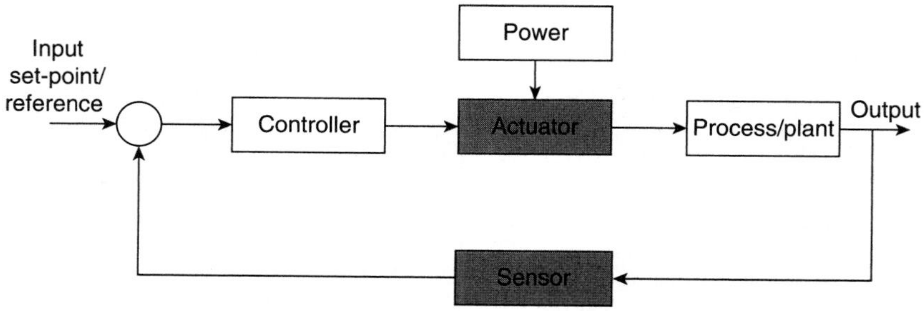

Figure 10.4.1 Block diagram of a feedback control system.

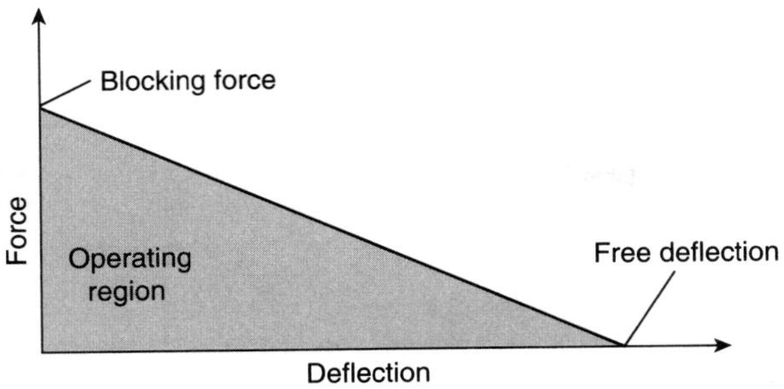

Figure 10.4.2 Schematic force-displacement curve for an actuator.

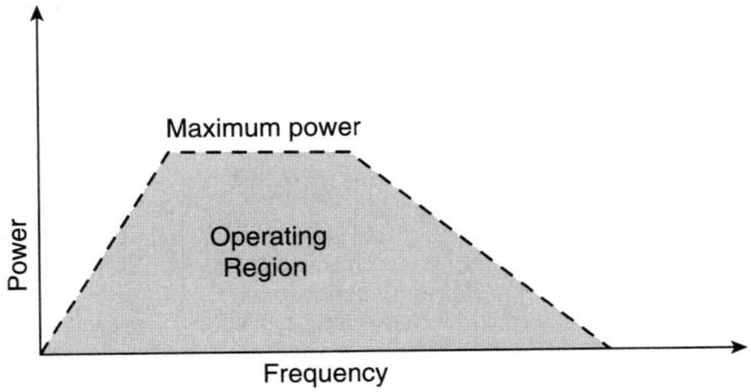

Figure 10.4.3 Schematic of the power-frequency curve for an actuator.

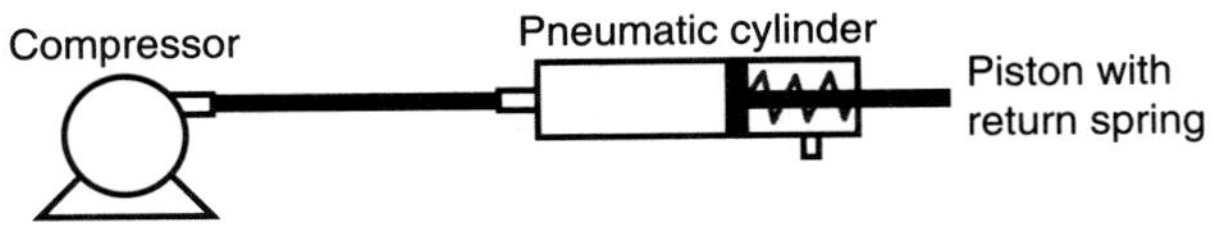

Figure 10.4.5 Possible schematic of a typical pneumatic actuator.

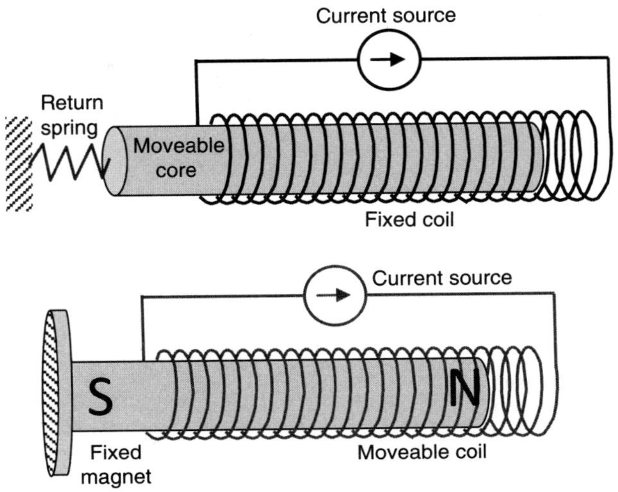

Figure 10.4.6 Schematics of a solenoid (top) and voice-coil (bottom).

displacement or force. These devices use electric power and unless geared provide quiet operation. Wide varieties of electrical actuators are available and include:

(i) **Solenoids** (Fig. 10.4.6) consist of a tubular electric coil wrapped around a free-moving ferrous core that acts as a plunger. When electrical current flows through the coil, an electromagnetic force is produced that moves the plunger. A spring can be attached to the plunger to allow it to return in the opposite direction. The position and force controls of a solenoid actuator are difficult and they are typically used in "off-on" applications. Maximum operating frequencies of solenoid tend to be low in the 1-100 Hz range.

(ii) **Voice-Coils** consist of moving tubular electrical wire situated within a permanent magnetic field. Current within the wire induces a force within it (the Lorentz force) that provides the actuation. Control of these systems is relatively straightforward by adjusting the input current. Output motion is generally linear. Maximum operating frequencies can be high—in the 1-50 kHz range.

(iii) **DC Motors** consist of a moving wire within a permanent magnetic field while **AC motors** consist of a magnet moving within fixed coils (Fig. 10.4.7). However, as opposed to voice-coils, the motion of AC and DC motors are rotary. As their names imply DC motors require direct-current while AC motors require alternating current. A DC motor uses a commutator (i.e., mechanical or electrical switch) to provide torque in the same direction. In contrast, an AC motor provides torque in the same direction by alternating the direction of the magnetic field. DC motors are typically cheaper and lighter than AC motors but are less robust and provide less operating torque. Typical maximum operating frequencies are approximately 400 Hz, but kHz ranges are possible with suitable designs.

(iv) **Piezoelectric** materials are inherently capable of undergoing mechanical strain under applied voltage (Fig. 10.4.8). Actuators composed of these materials do not require any moving parts. Strains tend to be small but the forces are high. Response time is extremely fast (MHz range). Mechanical leveraging can be used to increase actuator displacement but with an obvious increase in actuator size and complexity. Displacement resolution can be in the nanometer range. Disadvantages include brittleness of some high-performing piezoelectric materials and high-driving voltages (typically >100 V for large displacements).

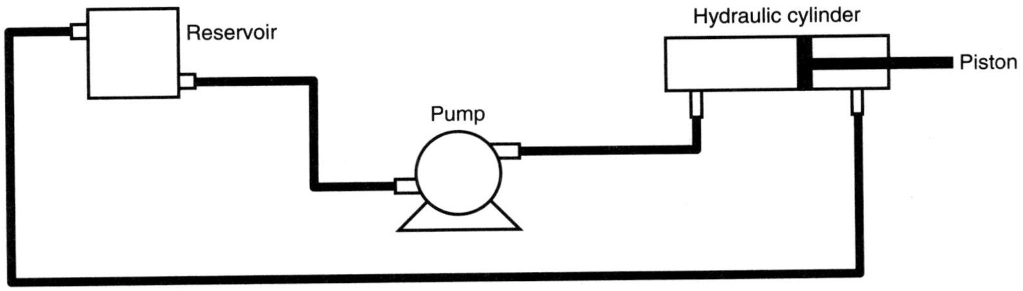

Figure 10.4.4 Typical schematic of a hydraulic actuator.

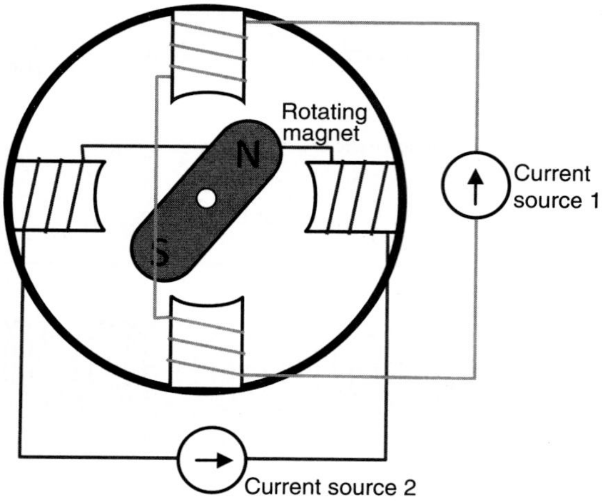

Figure 10.4.7 Schematics of a two-pole DC motor.

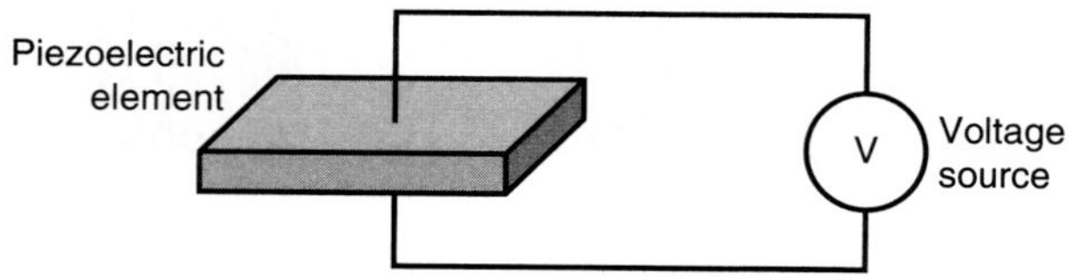

Figure 10.4.8 A simple piezoelectric actuator capable of straining under an applied voltage.

(v) **Shape Memory Alloys** (SMAs) are materials that are capable of large deformations (up to 8 percent strain) due to temperature induced phase transition. Thermal transition can be realized by resistive heating simplifying the drive circuitry (Fig. 10.4.9). High forces are achievable but with relatively slow (<10 Hz) response time. Accurate position and force control is difficult.

(vi) **Other Actuators** include capacitive actuators used in micro-electromechanical systems (MEMS), electroactive polymers for soft-robotic applications, thermal strain actuators use thermal expansion as the actuation mechanism, magnetostrictive materials and magnetic shape memory alloys that undergo strain due to an applied magnetic field. Most of these actuators are not commercially produced and are usually used for specialized or research applications.

RELATIVE PERFORMANCE OF ACTUATORS

The relative performances of hydraulic, pneumatic, electromagnetic (EM), piezoelectric (Piezo), and Shape Memory Alloy (SMA) actuators are compared in Figs. 10.4.10 and 10.4.11. The horizontal solid lines in Fig. 10.4.10 represent the maximum (blocking) stress (i.e., σ_{max}), the vertical solid lines represent the maximum achievable (free) strain

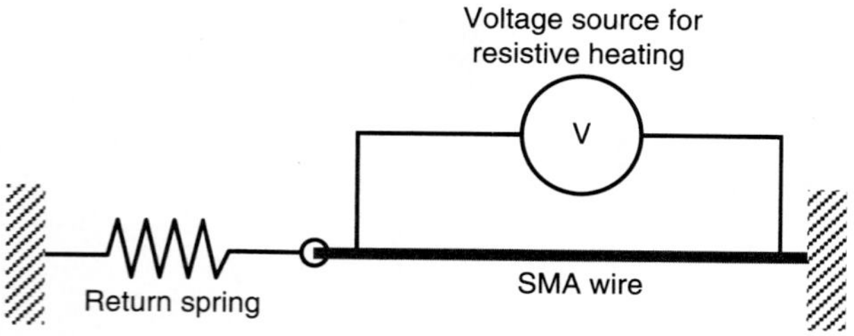

Figure 10.4.9 Shape Memory Alloy (SMA) wire actuator activated by resistive heating.

(i.e., ε_{max}). The minimum strain resolution (i.e., ε_{min}) of each actuator type is shown as dotted vertical lines. It should be noted that the values shown in Figs. 10.4.10 and 10.4.11 are order of magnitude estimates and largely depend on actual actuator type, controllers and driving power.

Approximate power-densities and actuator masses are shown in Fig. 10.4.11. The design of the actuator system should take into account the mass of the power-supply including pumps or compressors as well as the controller unit.

SENSOR FUNDAMENTALS

Traditionally, sensors such as liquid expansion thermometers, pressure gages, and dial gages used a transduction mechanism that converted the measure parameter into an output that could be visually inspected and if necessary manually recorded. With the advent of modern control systems and advances in electronics, these traditional sensors have been largely replaced by systems that are able to output a voltage or current proportional to the physical quantity being measured. A wide variety of sensors are currently available commercially. In this chapter, we detail some of the most commonly used electronic industrial sensors.

All electronic sensors require a transduction mechanism that converts the measured parameter into an electrically measurable quantity. Some of the most common conversion mechanisms are discussed here.

The **Seebeck effect** is the production of a voltage difference at the junction of two dissimilar metals subjected to a temperature difference. The schematic depiction of the circuit is shown in Fig. 10.4.12.

The decrease in **electrical conductance** of a material associated with an increase in length and/or decrease in cross-sectional area can be used to measure strain. Typically changes in resistance due to strain are very small (<1 percent resistance change under 1,000 $\mu\varepsilon$ is typical). A balanced Wheatstone bridge circuit as shown in Fig. 10.4.13 can be used to measure accurately strains of less than 10 $\mu\varepsilon$.

The measurement of **capacitance change** is used in a wide variety of sensors including touch sensors, humidity sensors, and displacement sensors. The basic circuit (Fig. 10.4.14) that is often used for measuring capacitance relies on assessing the length of time it takes to charge the sensing capacitor, Counting the number of charging and discharging cycles of the capacitor within a fixed period of time allows for a relatively accurate measurement of the sensor's capacitance.

Inductive sensing is used for a number of different applications, including displacement measurement, magnetic target detection, etc. Two basic principal of measurements are used as illustrated schematically in Figs. 10.4.15*a* and 10.4.15*b*. The presence of a ferrous material near an inductive coil induces eddy-currents in the ferrous material (Fig. 10.4.15*a*). These eddy-currents cause a change in mutual inductance of the coil; by measuring the change in impedance of the coil (e.g., by measuring the reduction in amplitude in an inductor-capacitor circuit), the eddy current magnitude can be detected. The second basic method is to use a secondary (and/or tertiary) inductive coil in close proximity to the first (Fig. 10.4.15*b*). A change in inductance in the primary coil can be detected by a change in mutual inductance in the secondary coil which can then be directly measured. The first circuit (Fig. 10.4.15*a*) is typically used in proximity sensing, while the second (Fig. 10.4.15*b* with tertiary inductor) is used in linear variable displacement transducers (LVDT).

The direct **piezoelectric effect** (i.e., the generation of electric charge when certain classes of materials are subjected to mechanical strain) is used in a number of sensors, including: force transducers, ultrasonic detectors, and accelerometers. A charge amplifier circuit can be used to measure slow (i.e., static or quasi static) loads while for higher frequency loading the voltage output of the sensor can be measured directly. The charge amplifier schematic is shown in Fig. 10.4.16.

There is a wide range of **electro-optical based sensors** but the two most commonly used in industrial application rely on measuring either the interruption of the light path between emitter and detector (the so-called optical-switch) or measuring the position of the reflected light using a photo-detector. Optical switches are used as encoders in position or rotation sensors and object detection. The measurement of

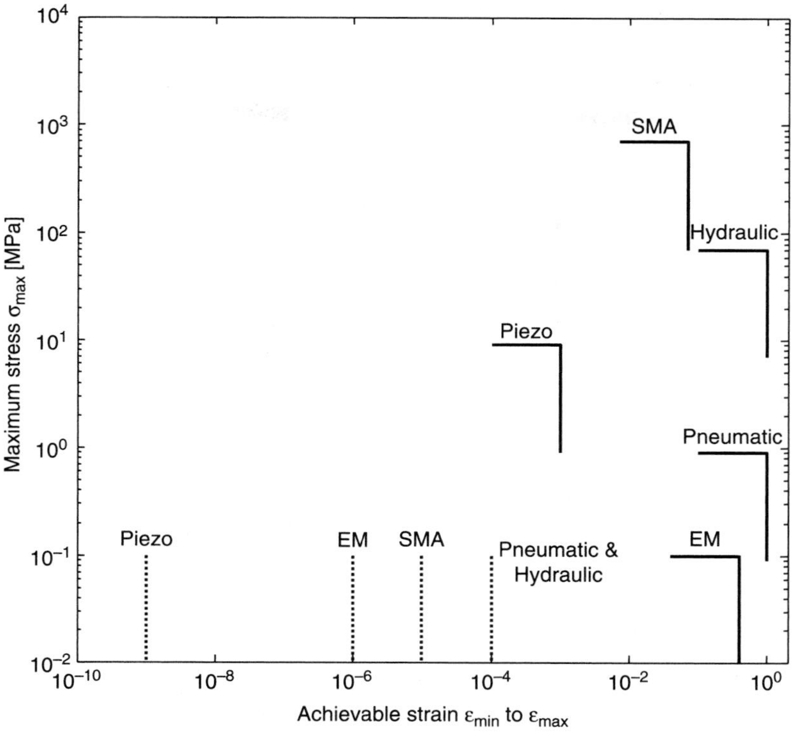

Figure 10.4.10 Approximate maximum-stress (σ_{min}), maximum strain (ε_{max}), and strain resolution (ε_{min}). Values are derived from various sources including those listed in [1].

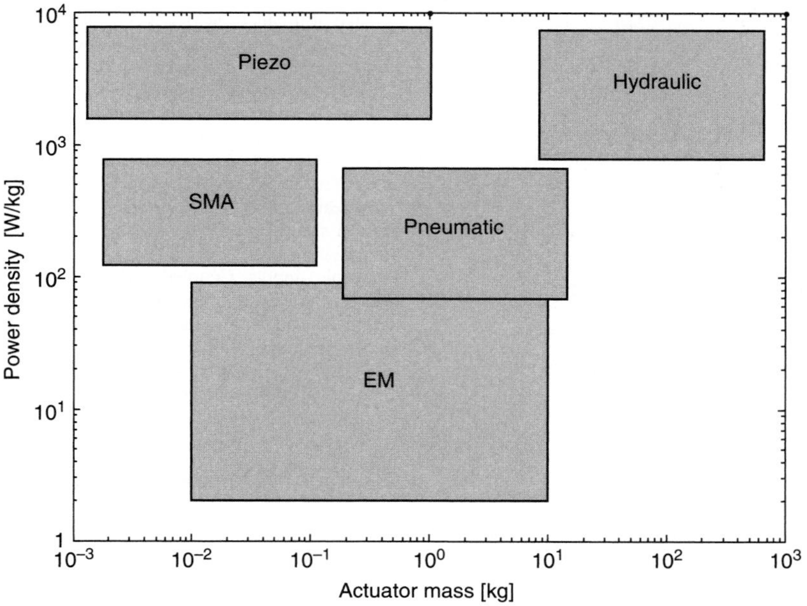

Figure 10.4.11 Approximate power density ranges and actuator mass ranges for various actuators. Values are derived from various sources including those listed in [1] and [2].

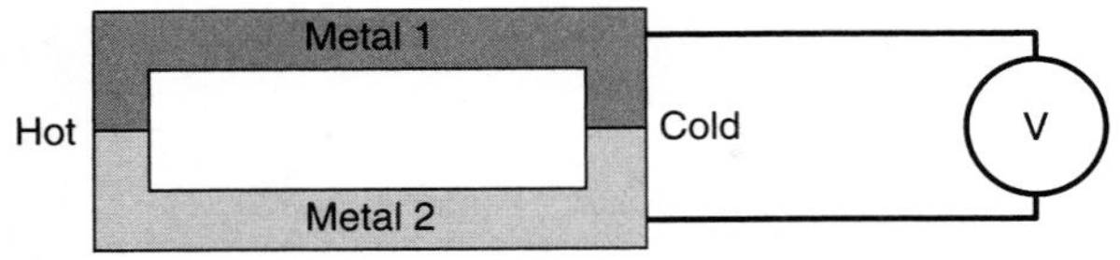

Figure 10.4.12 Simplified circuit for measuring temperature using the Seebeck effect. The output is measured at the location (V).

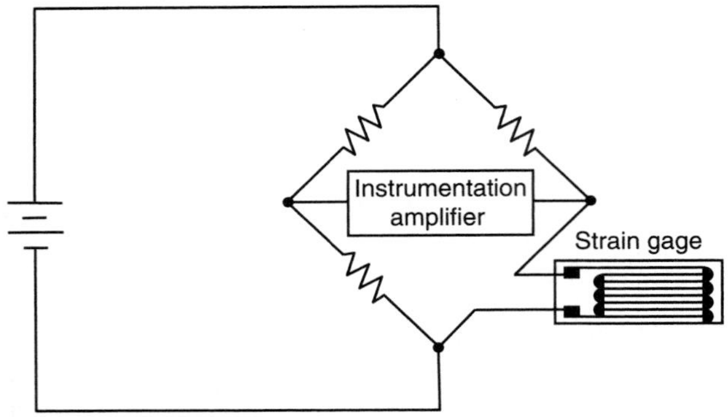

Figure 10.4.13 A typical quarter Wheatstone bridge circuit. The output of the instrumentation amplifier is directly proportional to the measured strain.

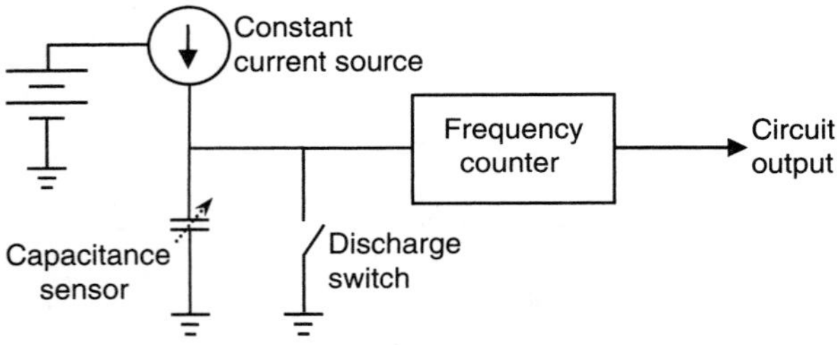

Figure 10.4.14 Schematic circuit for capacitance sensing.

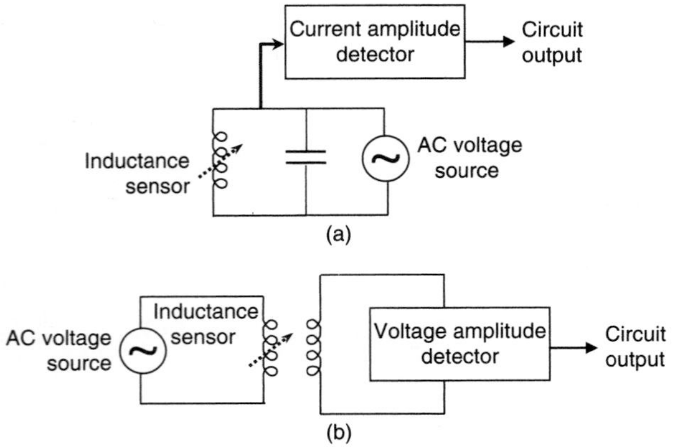

Figure 10.4.15 (*a*) Coupled Inductor-Capacitor (LC) circuit, and (*b*) coupled mutual inductance circuit for measuring change in inductance.

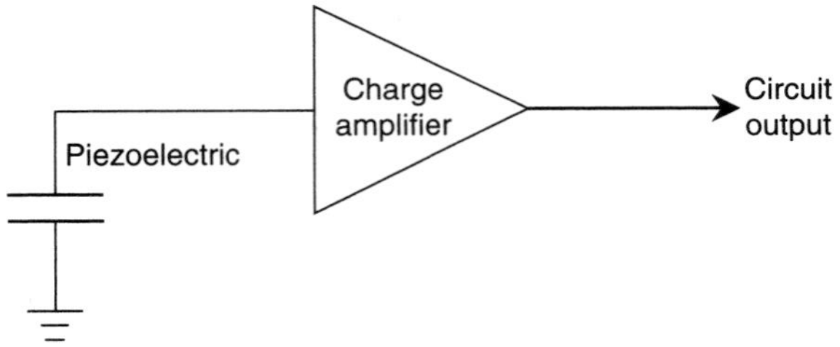

Figure 10.4.16 Schematic of a piezoelectric charge amplifier circuit for measuring slowly applied strains.

the position of the reflected light is used in laser displacement sensors. Figure 10.4.17 illustrates the basic circuit that can be used as the basis of an optical switch.

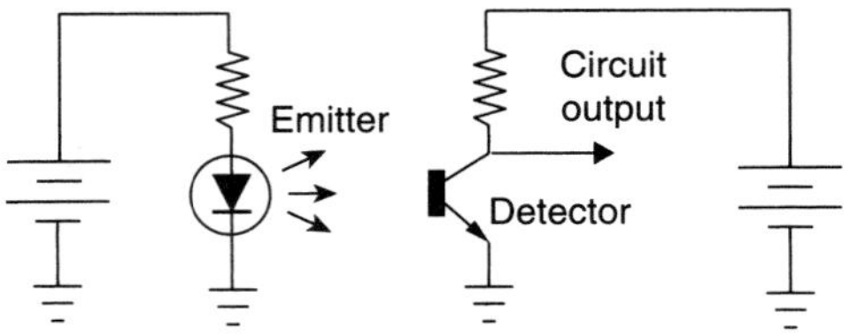

Figure 10.4.17 Schematic circuit for an electro-optical sensor.

Similar to actuators, sensors also have range, resolution, and bandwidth. However, some further requirements must be considered when designing the sensor system. These include:

Accuracy (or bias) is a measure of the maximum deviation of the sensor from a known (or calibrated) value.

Precision is a measure of the repeatability of the output for a given input and can be measured by calculating the standard deviation of the sensor output for the same set of inputs.

A schematic illustration of accuracy and precision for a sensor over a range of inputs (true value) is shown in Fig. 10.4.18.

Resolution and Signal to Noise Ratio Resolution is the minimum detectable change that can be measured by the sensor. It is often controlled by the signal to noise (S/N) ratio which is a ratio of the magnitude of the signal to the electrical noise (Fig. 10.4.20). Filtering of the sensor output can often be used to reduce electrical noise or else care has to be taken to shield the circuit electrically and ground the sensor and associated wiring.

Linearity is the maximum deviation of the measured sensor output from a straight line. This value is given as a percentage of the full output scale and can be different for slow (or static) input and for fast (dynamic) inputs. Coupled with linearity is **hysteresis** of the sensor which is a measure of the difference of the sensor output to an increasing and decreasing input. Linearity and hysteresis for an illustrative sensor output are shown schematically in Fig. 10.4.19.

Drift (Fig. 10.4.20) is a measure of the sensors long-term output with time due to changes in ambient conditions such as temperature, aging of material, or slow fluctuation in the sensor's input power.

SENSOR TYPES AND TYPICAL PERFORMANCE

Here we discuss some common types of sensors and their performance.

Temperature

Temperature sensors are found wherever thermal control is needed (e.g., refrigerators, AC systems, kilns, ovens, etc.). The most common types of modern industrial temperature sensors rely either on (i) the Seebeck effect (the so-called thermocouples—see above), (ii) a change in resistance with temperature (the so-called thermistors which exhibit

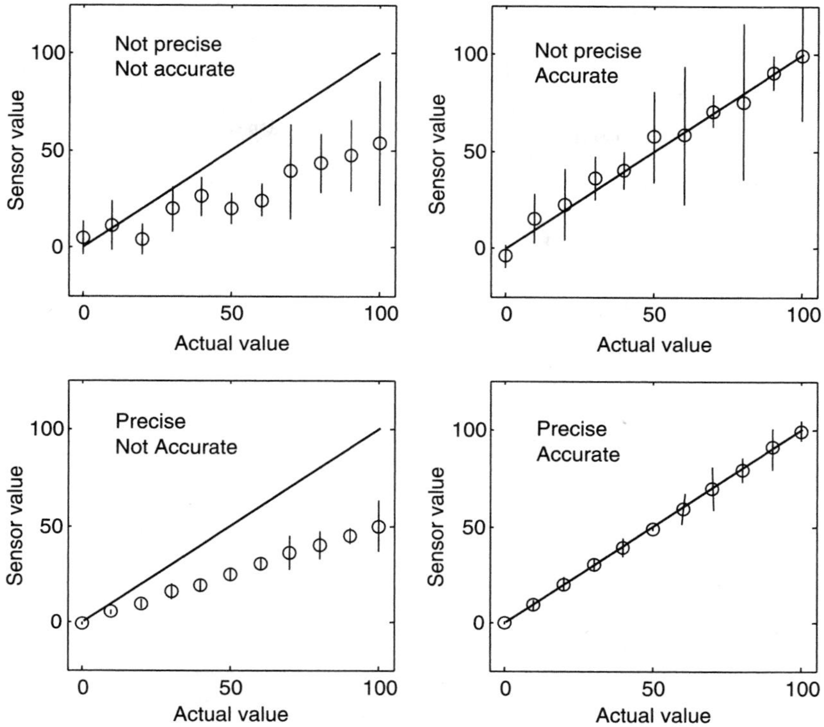

Figure 10.4.18 Illustration of Accuracy and Precision for sensor measurements calibrated over a range of true values. The solid line represents the ideal sensor output for a given true measurement; the circles represent the mean value of the measurements at a given input value; the bars at each measurement datum point show the standard deviation. Axes scales are arbitrary.

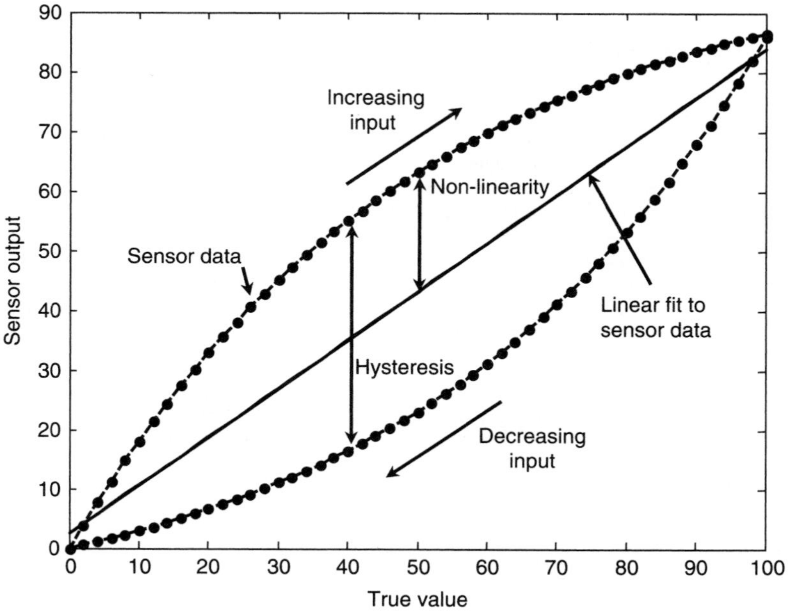

Figure 10.4.19 Typical calibration curve (i.e., sensor output at known input values) for a sensor showing hysteresis and non-linearity.

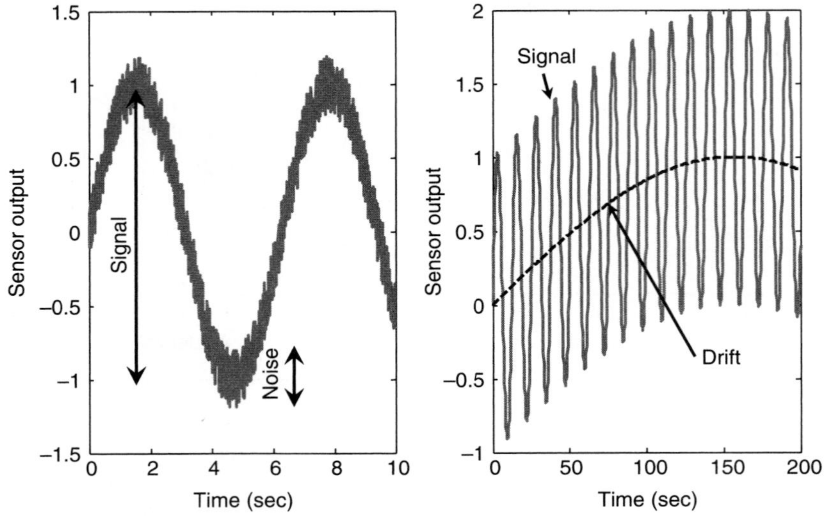

Figure 10.4.20 Schematic sensor output showing the measured sensor signal and noise (left) and sensor drift (right).

a large nonlinear drop in resistance with temperature or Resistance Temperature Detectors which exhibit an increase in resistance with temperature, i.e., RTD), and (iii) silicon sensors based on the change in bandgap voltage with temperature. Table 10.4.1 compares the various common types of temperature sensors.

Table 10.4.1 Comparison of Various Common Temperature Sensors

	Range (°C)	Accuracy	Linearity	Drift	Cost
Thermocouple	–200-2,000	Medium	Very Good	<2°F/year	$$
RTD	–200-900	Excellent	Good	<0.1°F/year	$$$
Thermistor	–60-150	Poor	Poor	<5°F/year	$
Silicon	–50-125	Medium	Excellent	<0.1°F/year	$$

Proximity sensors are typically used to detect the presence and distance to an object without physical contact. They are used for applications such as automatic doors, industrial quality control, packaging, and automatic car parking warning systems. There is a wide variety of proximity sensors but they can be divided into the following basic types: (i) inductive proximity sensors which require the target to be metallic, (ii) capacitive based sensors, (iii) optical sensors, (iv) magnetic sensors which require a magnetic target, and (v) ultrasonic sensors which rely on a ultrasonic transmitter and detector to measure the presence of targets and their distance. Table 10.4.2 compares the relative performance of a number of proximity sensor types.

Table 10.4.2 Comparison of Various Common Proximity Sensors

	Range (m)	Accuracy	Cost	Notes
Inductive	0.2	Excellent	$	Ferrous targets only
Capacitive	0.2	Excellent	$$	Non-ferrous targets
Optical - LED	6	Poor	$	Difficulty under sunlight
Optical - LASER	500	Excellent	$$$	Optically reflective targets
Magnetic	0.1	Medium	$	Magnetic target
Ultrasonic	100	Medium	$$	Acoustically reflective targets

Displacement Sensors include the proximity sensors mentioned above but also contact displacement sensors such as: (i) resistive potentiometer sensors, (ii) inductive based sensors such as linear variable differential transformers (LVDT), and (iii) optical encoders. To distinguish these displacement sensors from the above mentioned proximity sensors, here we define displacement sensor as having a relatively shorter range (i.e., <1 meter) but high accuracy (micrometer or better). Typical uses include measuring manufacturing tolerances or mechanical deformations under load. Typically only inductive, capacitive, and optically based LASER systems have sufficient accuracy for small displacement measurements required to measure elastic deformations under load. Note that the measuring ranges of the capacitive, inductive, and optical-laser systems are significantly smaller than for proximity sensing in order to provide the high resolution required. Rotary sensors (such as rotary encoders or spinning magnets in Hall effect sensors) can also be converted into linear displacement measurements.

Table 10.4.3 compares the performance of a number of displacement sensor types.

Velocity Sensors

Differentiation of a displacement sensor's output (listed above) with respect to time can be used to derive velocity. Care must be taken since this differentiation can introduce additional noise to the signal. Further filtering of the signal might thus be required.

Table 10.4.3 Comparison of Various Common Displacement Sensors.
FSO Stands for Full Scale Output

	Range (mm)	Resolution % FSO	Frequency (kHz)	Cost
LVDT	200	<0.01	2	$$
Potentiometer	200	<0.1	1	$
Optical encoder	1,000	<0.01	0.10	$$$
Inductive	100	0.1	20	$
Capacitive	40	<0.01	20	$$
Optical Laser	1,000	<0.01	50	$$$

Acceleration Sensors

Similar to velocity measurement, double differentiation of a displacement sensor's output (listed above) with respect to time can be used to derive acceleration. Noise levels due to this differentiation are worse than for velocity derived values. Stand-alone accelerometers are also commonly used for measuring vibration levels, to deploy airbags and to monitor equipment. Two types of accelerometers are typically used: (i) piezoelectric based accelerometers, and (ii) capacitive based accelerometers. Both types of accelerometers measure the inertial force experienced by a mass housed inside the sensors. It should be noted that integration of accelerometer outputs to calculate velocity and displacement tend to be inaccurate due to the integration procedure or accumulation of error. Table 10.4.4 summarizes the relative performance of both sensor types. An important difference between these two sensors is that capacitive sensors can measure "static" accelerations meaning that they are often used as inclinometers (i.e., measuring vertical orientation relative to the gravity vector), while piezoelectric accelerometers can only measure dynamic accelerations.

Table 10.4.4 Comparison of Two Common Accelerometers

	Accuracy	Frequency (kHz)	Static	Cost
Piezoelectric	Excellent	50	No	\$ to \$\$\$
Capacitive	Medium	1	Yes	\$

Fluid Flow Meters

A wide variety of electronic fluid flow meters are used in industry to measure the rate of flow of fluids. These include (i) gear flow meters in which the time taken to fill a known volume is used to calculate flow rate, (ii) paddle wheel flow meters in which the rate of rotation of a wheel with vanes is used to calculate fluid flow rate, (iii) ultrasonic flow meters in which the Doppler change in frequency of an ultrasonic wave measured between a transmitter and receiver is used to calculate flow speed, (iv) thermal mass meters in which the heat dissipated due to convection from a known source is measured by thermocouples is used to calculate flow rate, (v) Coriolis mass flow meters uses the change in vibration of a tube due to the Coriolis force to measure the velocity of the flowing fluid, (vi) variable area flowmeters use the displacement of a float due to the drag force of the flowing fluid and balanced by the float's gravity to measure flow velocity, and (vii) vortex flow meters measure the frequency of the vortices shed from an obstruction in the flow to measure flow speed. Table 10.4.5 compares the relative performance of the various fluid flow sensors.

Table 10.4.5 Comparison of Various Common Fluid Flow Sensors

	Pressure drop	Accuracy	Cost	Liquid or Gasses
Gear flow meter	Medium	Excellent	\$\$	Both
Paddle wheel	Very low	Good	\$	Liquid
Ultrasonic	None	Medium	\$\$	Both
Thermal mass	Low	Medium	\$\$	Gas
Coriolos mass	Low	Excellent	\$\$\$	Both
Variable area	Medium	Medium	\$	Both
Vortex	Low	Good	\$\$	Both

FILTERING

Though a critical aspect of sensing, a detailed discussion of filtering is beyond the scope of this chapter. This section serves as a brief introduction to filtering. Filtering is defined as the process by which a signal input (such as from a sensor) is manipulated so as to remove certain unwanted frequencies or to enhance specific frequencies. The most common types of filters that are used are (i) low-pass filters, (ii) high-pass filters, (iii) band-pass filters, and (iv) band-stop filters.

The performance of a filter is typically specified by calculating and showing the magnitude of the output (G) and phase (ϕ) of the output in the frequency domain. The output magnitude (G_{dB}) for any frequency component of a filtered signal is typically expressed in decibels (dB) which is defined as:

$$G_{dB} = 20\log_{10}\left(\frac{\text{Input}}{\text{Output}}\right)$$

Phase is defined as the shift in time of any frequency component of a filtered signal relative to the same frequency component in the original (input) signal.

Low-Pass Filters

Low-pass filters are used to pass signal frequencies below the so-called cut-off frequency and to attenuate high-frequency components in the signal. The frequency response of a typical low-pass filter is shown in Fig. 10.4.21. The ideal filter (shown as dashed lines in Fig. 10.4.21) shows that all frequency components of the input signal below the cut-off frequency are unaltered, but all frequency components above the cut-off frequency are completely attenuated. The performance of a low-pass filter in the time-domain is shown in Fig. 10.4.22. The input signal (dashed line) in Fig. 10.4.22 is composed of three frequency components, a low frequency component (typical of signal drift), an intermediate frequency and a high-frequency component (typical of signal nose). The low-pass filtered output (solid line) clearly shows that the high frequency component has been significantly attenuated.

High-Pass Filters

High-pass filters pass frequencies above the cut-off frequency and attenuate low-frequency components in the signal. The frequency response of a typical high-pass filter is shown in Fig. 10.4.21. The performance of a high-pass filter in the time-domain is shown in Fig. 10.4.22. The high-pass filtered output (solid line) clearly shows that the lower frequency components of the input (dashed line) have been significantly attenuated.

Band-Pass Filters

Band-pass filters pass frequencies within a specified range (so-called bandwidth) and attenuate frequency components outside of this range. The frequency response of a typical band-pass filter is shown in Fig. 10.4.21. The performance of a band-pass filter in the time-domain is shown in Fig. 10.4.22. The band-pass filtered output (solid line) clearly shows that the lowest and highest frequency components of the input (dashed line) have been significantly attenuated.

Band-Stop Filters

Band-stop filters attenuate frequencies within a specified range (so-called bandwidth) and pass frequency components outside of this range. The frequency response of a typical band-stop filter is shown in Fig. 10.4.21. The output of a band-stop filter in the time-domain is shown in Fig. 10.4.22. The band-stop filtered output (solid line) clearly shows that the intermediate frequency components of the input (dashed line) have been significantly attenuated.

Signal filters are practically implemented in: (i) passive circuits, (ii) active circuits, or (iii) digital circuits (see Table 10.4.6). **Passive filter** circuits use combinations of resistors (R), inductors (L), and capacitors (C) to eliminate specific frequency ranges. They are called passive since they require no external power supply and no components such as transistors. An example of a simple, passive, low-pass filter circuit is shown schematically in Fig. 10.4.23. The performance of the passive circuit (e.g., the cut-off frequency) is controlled by the specific values of the R, L, and C components used in the circuit. **Active filters** use transistors (such as amplifiers) together with combinations of resistors (R) and capacitors (C) to eliminate or enhance specific

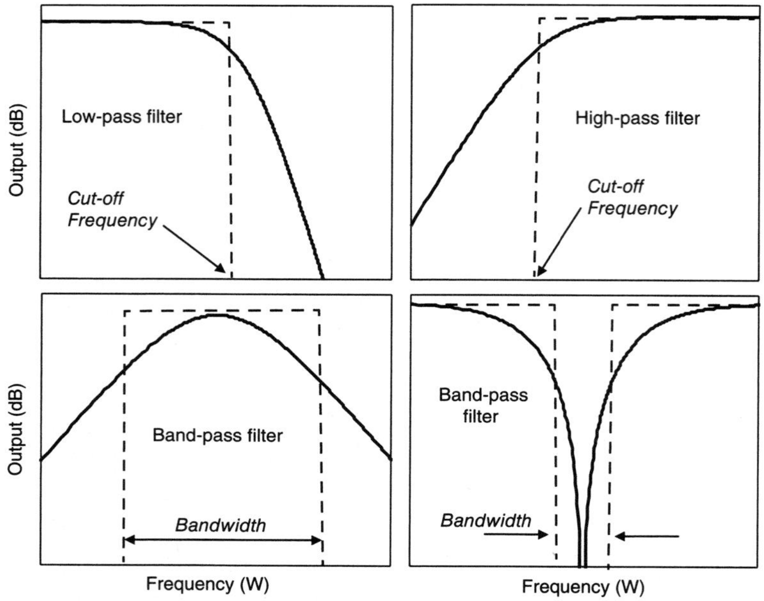

Figure 10.4.21 Magnitude (in dB) of the output of a sensor with a harmonic input [e.g., sin(Ωt)] for some common filters. The ideal filters are shown as dashed lines.

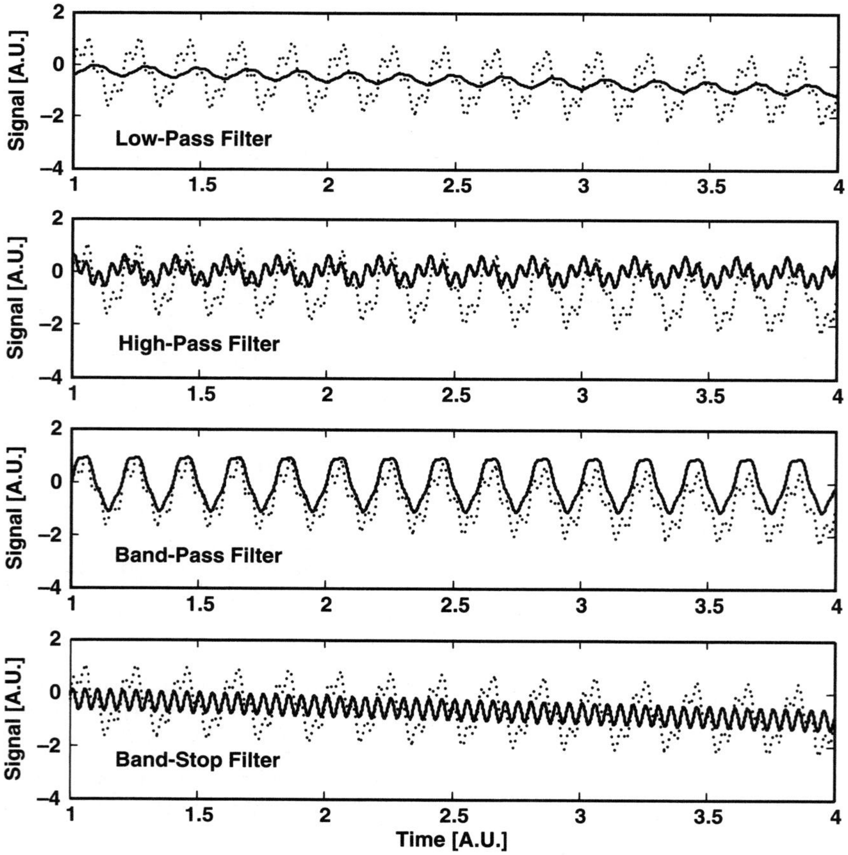

Figure 10.4.22 Outputs (solid line) of various filters to an input (dashed line) with low, intermediate and high frequency contents.

Table 10.4.6 Comparison of Passive, Active, and Digital Filters

	Passive	Active	Digital
Power supply	No	Yes	Yes
Adaptive filtering	Impossible	Difficult	Easy
Output gain	Impossible	Possible	Possible
Real-time calculation	Impossible	Impossible	Possible
Circuit size	Usually large	Small	Medium
Handle large input voltages/currents	Yes	Difficult	Difficult
Response speed	Fastest	Medium	Slow
Cost	\$	\$\$	\$\$\$

frequency ranges. Inductors (L) are typically avoided since they could be expensive and bulky. As opposed to passive filters, active filters require external power-sources. An example of a simple, active, low-pass filter circuit is shown schematically in Fig. 10.4.23. A **digital filter** first converts the analog input into a sampled signal at discrete time intervals using an analog-to-digital (A/D) converter. The sampled signal is then filtered using mathematical operations implemented in software (code) on a microprocessor. The current filtered signal (y_n) is composed of the current input signal (x_n), a set of previous input signals (x_{n-1} to x_{n-nb}) and a set of previously filtered signals (y_{n-1} to y_{n-na}) using the following discrete time equation:

$$a_1 y_n = b_1 x_n + b_2 x_{n-1} + \cdots + b_{nb+1} x_{n-nb} - a_2 y_{n-1} \cdots - a_{na+1} y_{n-na}$$

The type of filter and filter performance is set by the filter constants (a_1 to a_{na+1}) and (b_1 to b_{nb+1}). An example of a simple, digital low-pass filter circuit that is a digital implementation of the passive RC circuit is shown schematically in Fig. 10.4.23.

CONCLUSION

This chapter was intended to serve as a global overview to sensors and actuators. While typical sensors and actuators have been presented, many have been omitted. The physical phenomena behind how the sensors and actuators work have been named. There are numerous internet sites and books that provide greater detail to these topics, and the interested reader is referred to these resources; some additional references can be found in [4-11].

REFERENCES

1. Huber, J.E., Fleck, N.A., and Ashby, M.F. The selection of mechanical actuators based on performance indices. *Proceedings of the Royal Society of London A: Mathematical, Physical and Engineering Sciences*, **453**(1965), pp. 2185-2205, 1997.
2. Follador, M., Cianchetti, M., Arienti, A., and Laschi, C. A general method for the design and fabrication of shape memory alloy active spring actuators. *Smart Materials and Structures*, **21**(11), p. 115029, 2012.
3. Available at: http://www.icenta.co.uk/selectionguide/Flow
4. Janocha, H. (ed.). "Actuators: Basics and Applications," Springer Science & Business Media, 2013.
5. Bishop, R.H. "Mechatronic Systems, Sensors, and Actuators: Fundamentals and Modeling." CRC Press, 2007.
6. Carter, B., (2001). "Texas Instrument, Filter Design in 30 Seconds," Available at: http://www.ti.com/lit/an/sloa093/sloa093.pdf
7. Hashemi, K., (2015). "Filter Design Guide," Available at: http://alignment.hep.brandeis.edu/Lab/Filter/Filter.html
8. Winder, S. "Analog and Digital Filter Design." Newnes, 2002.
9. Fraden, J. "Handbook of Modern Sensors: Physics, Designs, and Applications," Springer Science & Business Media, 2004.
10. Wilson, J.S. "Sensor Technology Handbook," Elsevier, 2004.
11. Bishop, R.H. "Mechatronic Systems, Sensors, and Actuators: Fundamentals and Modeling," CRC Press, 2007.

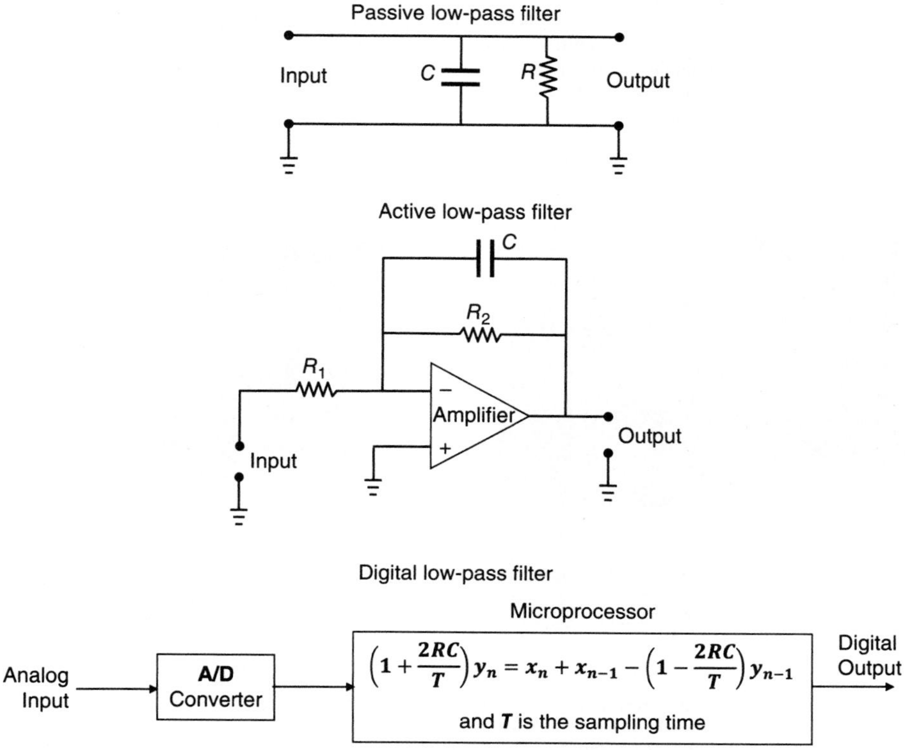

Figure 10.4.23 Passive, active, and digital filter implementation of a simple low-pass filter.

10.5 AN INTRODUCTION TO MICROELECTROMECHANICAL SYSTEMS (MEMS)

by Luc G. Fréchette

INTRODUCTION

Trend toward Small Scales

The twentieth century was marked by mankind's accomplishments at ever increasing scales, from civil works such as bridges that span rivers to sophisticated machinery such as airplanes and spacecraft that span our planet and beyond. Although this quest for immensity will remain vibrant, the twenty-first century will likely be marked by achievements at smaller scales. This trend is best demonstrated by the information revolution experienced by our society over the past decades, which was a result of the remarkable development of integrated circuits (ICs), itself enabled by progress in semiconductor manufacturing. The microfabrication approach developed for ICs is based on parallel processing of large arrays of devices on a planar substrate, using photolithography and thin-film processing in a clean room environment. The semiconductor industry leverages this approach to make devices with impressive levels of performance, reliability, and complexity. Today, micro- processors integrating over 100 million transistors per chip can be batch-fabricated in large volumes and at low cost.

Such capabilities are expanding beyond microelectronics to allow the microfabrication of sensors, actuators, and microsystems that implement functions from other physical domains, such as mechanical, thermal, fluidic, magnetic, electrical, optical, chemical, or biological domains. The field of **microelectromechanical systems,** or MEMS, focuses on the miniaturization of such a diverse set of systems through the use of microfabrication technology. It is also referred to as **microsystems technology,** or MST, in Europe and **micromachines** in Asia.

Microsystems and Their Applications

The scale of a device or machine is set by the task it must accomplish as well as the fabrication technology available to implement it. For example, information storage devices, such as hard disk drives and solid-state memory, increase in storage density as technology enables smaller bit size, since bits of information do not have a physical scale associated with them. Similarly, devices pertaining of other physical domains could also be miniaturized to the extent that technology allows, if commercially beneficial. A prime example is **sensors,** that is, devices that measure a physical quantity, often by converting it to an electrical signal. Most measurement applications would benefit from local, compact, nonintrusive, and inexpensive sensors, suggesting that batch microfabrication may be appropriate. Pressure sensors (Fig. 10.5.1*a*) and accelerometers (Fig. 10.5.7*a*) are common examples. **Actuators** that act on the environment by converting an electrical signal into a mechanical displacement or another physical domain can also potentially be miniaturized, as long as the power transmitted to the environment is small and the action is highly localized. Examples include inkjet printer heads, micromirror arrays for image projection (Fig. 10.5.1*b*), and microvalves for flow control (Fig. 10.5.7*c*). Microfabrication and MEMS technology can also be enabling to implement **microsystems** that would otherwise be impossible or impractical, such as microfluidic labs-on-a-chip that allow interaction with molecules or cells at their length scales (0.1-10 μm).

The annual market for MEMS is estimated at $U.S. 8.2 billion worldwide for 2007,[1] in the following areas: microfluidics (27 percent), inertial sensors (22 percent), optical MEMS (22 percent), pressure sensors (11 percent), RF MEMS (3 percent), and other sensors (10 percent) and actuators (5 percent). These devices are used in the automotive industry, in consumer electronics and computers, for telecommunications, as well as in instrumentation for biochemistry and medicine. The field has experienced an average growth of 16 percent annually since 2000, as a result of novel applications of MEMS and insertion of this disruptive technology into existing markets.

MICROFABRICATION TECHNOLOGY AND MATERIALS

The key enabling technology responsible for the rapid development of microsystems over the past decade is **microfabrication.** Most of today's computer chips are manufactured by the CMOS (complementary metal-oxide semiconductor) process, which is a sequence of steps that repeatedly deposit and pattern thin films to form an integrated circuit with transistors. Silicon, typically used in microelectronics, has become the dominant material for MEMS because of its material properties (semiconductor, mechanical, thermal) and the extensive processing experience with the silicon material set pioneered by the semiconductor industry. This section describes the underlying microfabrication technologies and materials borrowed from this industry that are the basis for silicon MEMS processing.

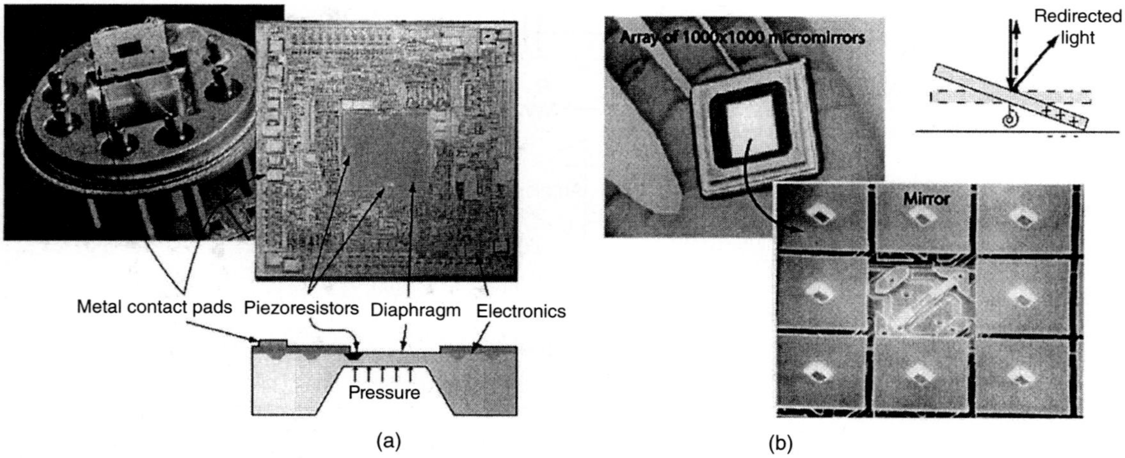

Figure 10.5.1 Commercial examples of microelectromechanical systems (MEMS). (*a*) Automotive piezoresistive pressure sensor and (*b*) micromirror array for display applications. The pressure sensor (*a*) measures the deflection of a thin silicon diaphragm due to an applied pressure by monitoring a strain-sensitive resistor fabricated on the diaphragm through on-chip circuitry. (*Photo: Bosch.*) The micromirror array (*b*) is an electrostatic actuator that redirects reflected light over 1 million independent pixels to digitally project an image on a screen. (*Photo: Texas Instruments.*)

To gain further insight on microfabrication science and technology, beyond what will be presented here, the reader is referred to classic texts on the subject.[2,3]

Silicon and Its Properties

Most microfabrication processes create devices on the surface or through the thickness of thin, planar, circular substrates, or **wafers.** Single-crystal silicon wafers are the most common starting material, although glass, quartz (single-crystal and fused), and compound semiconductor wafers (GaAs, InP) are also of interest. As provided by the semiconductor industry, single-crystal silicon wafers offer unique properties, such as: (1) stable, defect-free bulk material of high chemical purity; (2) precise substrates with low wafer bow (>25 μm), minimal taper (>15 μm), and mirror polished surfaces (one or both sides); (3) semiconductor properties and the possibility of integrating MEMS with electronics.

Wafers are characterized by the orientation of the **crystal's cubic structure** with respect to the wafer surface. As illustrated in Fig. 10.5.2, principal planes and directions in single-crystal silicon can be identified by **Miller indices.**[2] Specific planes are denoted by (h, k, l), where the indices are components of the direction vector **[h, k, l]** that is normal to the plane. Since the lattice of silicon is cubic, hence symmetric, the three principal directions are identical and interchangeable. Therefore, we commonly refer to families of directions, written <100> to identify the [100], [010], and [001] directions, and equivalently to families of planes, written {100}, for the planes normal to these directions. Wafers are classified by the direction vector that is normal to the wafer surface, so, for example, <100> wafers will have the cubic lattice parallel to the wafer surface. As will be discussed later, the crystalline structure of silicon has been leveraged to create useful structures such as the pressure sensor membrane in Fig. 10.5.1*a* by anisotropic wet etching.

Doping, by introducing impurities in the silicon lattice, controls the substrate's electrical conductivity (which can range from 0.001 to >10,000 Ω · cm) and also defines the wafer type (Group III atoms such as boron produce **p-type** material, Group V atoms such as phosphorus produce **n-type** material). Mechanically, single-crystal silicon is a strong and light material ($\rho = 2.33$ kg/m^3, $E = 130$ to 187 GPa, $\nu = 0.06$ to 0.28, and $G = 57$ to 79 GPa depending on orientation[2]), but exhibits a brittle behavior with a fracture strength that is process and size dependent (typically 1.2 GPa, but can vary from 0.5 to 7 GPa).[4] Traditional machining of silicon is therefore not common, since it introduces surface cracks that can significantly reduce the fracture strength; etching techniques are therefore used instead. Thermally, silicon is a good conductor ($k = 150$ W/m · K) and exhibits less thermal expansion and higher specific heat than most metals (CTE = 2.6×10^{-6}/K, $c_p = 0.713$ kJ/kg · K at 300 K), which tends to reduce thermal stresses. It can sustain relatively high temperatures (brittle-to-ductile transition above 550°C and melting at 1,421°C), although the electronic properties of silicon are best conserved below 150°C. These electronic properties are interesting for electromechanical transduction in sensors and will be discussed later.

Thin-Film Processing

A broad range of thin films can be deposited or grown on the surface of a substrate to form mechanical or electrical elements, such as membranes, mechanisms, or circuits. Materials include silicon or silicon-based compounds, metals, and polymers. The most common thin films are summarized in Table 10.5.1 with their characteristics and typical microfabrication approaches. Conductive layers can be formed of metals or doped semiconductors, such as polycrystalline silicon (referred to as **polysilicon**) diffused with boron or phosphorous dopants. Electrical insulation is typically created with silicon dioxide (SiO_2) or silicon nitride (Si_3N_4) films. Materials can be deposited in thin films by various techniques, such as physical vapor deposition, chemical vapor deposition, electrochemical deposition, or casting. Since thin-film materials are synthesized as they are deposited or grown, their properties are strongly related to the specific processing conditions. Material properties can be tailored within a certain range, which is useful to match design requirements but requires stringent process control for repeatability. Film thicknesses depend on the process and can vary from a few nanometers (nm) up to tens of micrometers (μm) [1 m = 10^6 μm = 10^9 nm = 10^{10} Å (angstroms)]. As a reference, the human hair is approximately 70 μm in diameter.

In **physical vapor deposition,** a solid piece of the material to be deposited is either vaporized or sputtered in line of sight with the substrate, such that atoms of the material will deposit uniformly on the substrate to form a thin film. Pure metals are easily evaporated, while

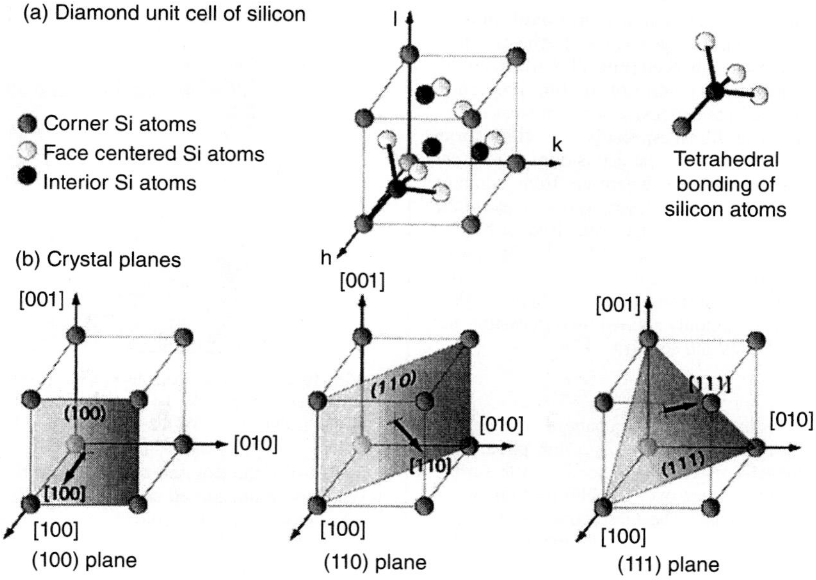

Figure 10.5.2 Crystal lattice, and Miller indices for silicon. The unit cell (*a*) has the structure of diamond, with each silicon atom covalently bonded to its four adjacent neighbors. The major planes, directions, and corresponding Miller indices are illustrated in (*b*).

Table 10.5.1 Common Substrate and Thin-Film Materials Used for MEMS

Material	Typical application	Fabrication process and characteristics	Material characteristics
Silicon (Si) (substrate)	Most common substrate; compliant or rigid structures; electrical and electronic components	Single-crystal substrate; 300-700 μm-thick; many micromachining techniques	Defect-free, high-strength, brittle, thermally conductive semiconductor
Quartz (substrate)	Bulk structural materials for active components	Crystalline or fused silica substrate; 400-700 μm-thick; limited micromachining	Piezoelectric properties; visually transparent
Borosilicate glass (Pyrex) (substrate)	Insulating substrate; bulk structures for microfluidics	Amorphous glass substrate; 400-700 μm-thick wafers; anodic bonding to Si; limited micromachining	Thermal expansion matched with Si; visually transparent
Polycrystalline silicon (polySi)	Thin-film structures; piezoresistors; conductors	Polycrystalline; LPCVD; 0.2-2 μm-thick films	Process-dependent properties; semiconductor
Silicon dioxide (SiO_2)	Sacrifical thin film; insulator (thermal and electric)	Amorphous; thermal growth or PECVD; 0.1-10 μm-thick films	Typically high compressive strength; thermal insulator: dielectric
Silicon nitride (Si_3N_4)	Membrane; insulator (electric)	LPCVD or PECVD; 0.1-1 μm-thick films	High tensile strength; dielectric
Metals	Electrical conductors; structural layers	Evaporation or sputtering (Al, Au, Cr, Ti, Pt, W); 20 nm to 2 μm-thick films	Ductile; thermal and electrical conductors
	High-aspect-ratio structures	Electrodeposited (Ni); 1-1,000 μm-thick films	
PDMS	Microfluidics; molded microstructures	Flexible polymer; cast; 10-1,000 μm-thick films	Low stiffness, dielectric, biocompatible
SU-8	Mold for electroplating; thick structures	Photosensitive epoxy; spin cast; 5-500 μm-thick films	Moderate stiffness, dielectric

sputtering can also deposit a wide range of metallic and nonmetallic compounds.

Chemical vapor deposition (CVD) involves a chemical reaction to synthesize a material, such as polycrystalline silicon, as it is deposited over the substrate. This process is often carried out in batches in high-temperature furnaces with a controlled flow of reactants and at low or atmospheric pressure (LPCVD or APCVD, respectively). In a similar environment, the silicon substrate can also be directly oxidized by reacting with a flow of oxygen to form SiO_2 on the surface (the addition of H_2 enhances the oxidation rate, and is referred to as **wet oxidation,** as opposed to **dry oxidation** when only oxygen is used). The chemical reactions for CVD can also be plasma enhanced (PECVD), which allows deposition at lower temperatures, good control of the film properties, and high deposition rates, but requires expensive serial processing.

An alternative approach for metal films, especially thick films (from micrometers to millimeters), is electrochemical deposition, or **electroplating.** The substrate acts as the cathode, where electrons combine with metallic ions in the liquid electrolyte, to cover exposed areas of the wafer with the metal (commonly Ni, but also Cu, Au, NiFe, or NiW).

The substrate can also be covered with solvent-based solutions by directly pouring the solution on the wafer and using gravity or centrifugal forces **(spin casting)** to form a uniform thin film. Spin-on-glass and photosensitive polymers are commonly applied by spin casting and then heated to evaporate the solvents and solidify.

Photolithography

The photolithographic process is used to transfer a pattern onto a planar substrate, using light. As illustrated in Fig. 10.5.3, a thin photosensitive polymer film, or **photoresist,** is first spin-cased on the surface of the wafer. The pattern is then exposed over the film by shining UV light through a photolithographic mask. The mask consists of a glass plate coated with a patterned chrome layer to locally expose the photoresist: the mask blocks the light where chrome is present and allows light to shine through where chrome has been removed. The coated wafer is then developed in a liquid solution, physically dissolving the photoresist in the exposed areas, for *positive* photoresist (for *negative* resist, dark or unexposed areas would be dissolved instead).

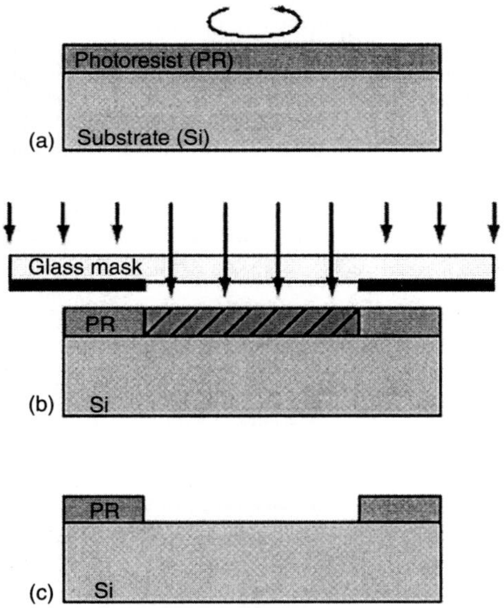

Figure 10.5.3 Photolithography process: (*a*) spin-coat a layer of photoresist over the wafer and prebake the film; (*b*) expose certain areas to UV light; (*c*) develop photoresist, leaving a physical mask in the nonexposed areas.

This process allows replication of complex, micrometer-scale features over the entire wafer in a single or a few repeated exposures. This parallel fabrication approach is central to the low-cost, high-volume advantage of microfabrication. The feature size is limited to fractions of the wavelength of light used for exposure, but alternative

nanolithography techniques have been developed to reach tens of nanometers, such as electron beam and nanoimprint lithography.

Etching and Other Patterning Approaches

The pattern created in the photoresist layer is typically transferred to the underlying thin film or into the substrate by etching. The wafer is immersed in an etchant, which reacts and removes the exposed material, while the photoresist acts as a physical mask that protects the unexposed areas of the wafer. In **wet etching,** a chemical in liquid or vapor phase comes in direct contact with the wafer, preferentially reacting with the exposed material. In **dry etching,** a gas is ionized to create a plasma of free radicals and ions that chemically react with the exposed material or physically ablate the surface through ion bombardment, by driving the ions toward the wafer with an electrical field. This process is referred to as reactive ion etching (RIE) and is most commonly used in microelectronics to form fine patterns in thin films. Given the directionality created by the electrical field applied between the plasma and substrate, anisotropic features with relatively vertical side walls are possible (Fig. 10.5.4*a*). Chemically dominated etching, such as wet etching, is, however, characterized by isotropic removal

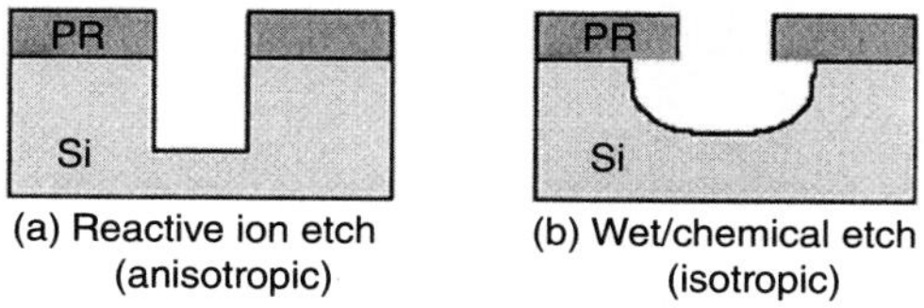

Figure 10.5.4 Cross section profiles of etched structures: (*a*) reactive-ion-etched features with vertical side walls and (*b*) chemical-etch-dominated process (plasma or wet etching) that results in mask under cut and rounded features.

of material that effectively undercuts the mask and widens the exposed features (Fig. 10.5.4*b*). In addition to etch profiles, etch rate and selectivity are important parameters that characterize an etch process. The etch rate is simply defined as the thickness of material removed per unit time, while selectivity is the ratio of etch rate of the material intentionally removed to that of a secondary material such as the photoresist mask or the underlying substrate. Reactive ion etching exhibits poor selectivity to photoresist (near 1:1) while chemically dominated etching can show selectivity >100:1. Table 10.5.2 summarizes some of the most common combinations of materials and etchants, along with their properties. After etching, the photoresist is commonly removed in an oxygen plasma (ashing) or in a solution (acid or solvent). The wafer is then ready for subsequent processing.

Lift-off is an alternative patterning approach consists of depositing and patterning the photoresist layer first, then covering the wafer surface with a thin film, commonly a metal. The photoresist is then dissolved to effectively lift-off the metal that covers it while leaving the metal that was deposited in the exposed areas of the mask.

PROCESSING FOR MEMS

The basic microelectronics processing techniques introduced above have been extended to allow the fabrication of movable structures for sensors, actuators, and multiphysical microsystems. The most common approaches are presented next.

Surface Micromachining

One approach consists of depositing and patterning a structural material over a sacrificial thin film. The sacrificial material is then selectively etched to release the structures formed in the top layer. This approach creates thin planar structures that can move with respect to the substrate, normal or parallel to it. Figure 10.5.5 illustrates a surface micromachining process in which alternating layers of polysilicon (to form the structure) and silicon dioxide (as the sacrificial material) are deposited and patterned. Etched openings in the sacrificial layers allow mechanical and electrical connections between the polysilicon layers and attachment to the substrate. A hydrofluoric solution can be used to remove the silicon dioxide and release the structures, since it does not etch silicon significantly. This surface micromachining approach is the basis for successful commercial devices, such as automotive accelerometers,[5] and foundry services, such as polyMUMPs[6] and SUMMIT.[7] Other material combinations can be used, such as aluminum and photoresist[8] or silicon carbide and polysilicon/silicon dioxide.[9]

An alternative approach consists of forming the electromechanical structures into the thin films of a **CMOS wafer,**[10] leveraging the substantial CMOS manufacturing resources and allowing natural integration with electronics. The structures can also extend into the silicon substrate by using bulk micromachining techniques.

Bulk Micromachining

Alternatively, the structures can be formed directly into the substrate, which is referred to as bulk micromachining. Three types of processing methods are available, described by the etchant state: wet, vapor, or plasma (the last two are referred to as dry etching). Although wet etching typically creates isotropic features (rounded etching in all directions), unique formulations can leverage the microstructure of single-crystal silicon to form anisotropic features. Hydroxide-based etchants (such as KOH and TMAH) etch much slower in one direction that the others, exposing the {111} planes of silicon. Depending on the crystallographic orientation of the wafer, {111} planes are either at 54.74° to the wafer surface (<100> silicon wafers) or perpendicular to it (<110> silicon wafers). Furthermore, the etch rate can also depend on dopant type and concentration, such that areas of the wafer that were previously diffused with a dopant can remain after wet etching. These characteristics allow the fabrication of a wide range of geometries, such as pyramidal pits, channels, cantilevers, or membranes as illustrated in Fig. 10.5.6*a*, but with geometric constraints to the crystal planes. Commercial pressure sensors are commonly fabricated as single-crystal silicon membranes by anisotropic wet etching, as illustrated in Fig. 10.5.1*a*.

Table 10.5.2 Common combinations of Materials and Etching Approaches in Micromachining; Many Other Choices Are Available[2]

Material to pattern	Common patterning approaches	Characteristics	Masking materials and selectivity
Si, polySi	RIE (Cl-based)	Slow, directional	PR (5:1)
	DRIE (SF_6)	Fast, vertical profile	PR (70:1); SiO_2 (150:1)
	Wet (KOH, TMAH)	Anisotropic (54.74°)	SiN (>100:1)
SiO_2, glass, quartz	RIE (CF_4)	Directional; isotropic	PR (10:1)
	HF (diluted, BOE)	(quartz: anisotropic)	PR (>20:1)
Si_3N_4	RIE (CF_4)	Directional	PR (10:1); SiO_2 (50:1)
	Phosphoric acid	Isotropic, wet	SiO_2 (10:1); Si (30:1)
Metals	Lift-off (for Au)	Wet, thin film (<1 μm)	—
	RIE (Cl, F, Ar)	Directional, slow	PR (1:1)
	Etchant solutions	Isotropic, wet	PR (>100:1)

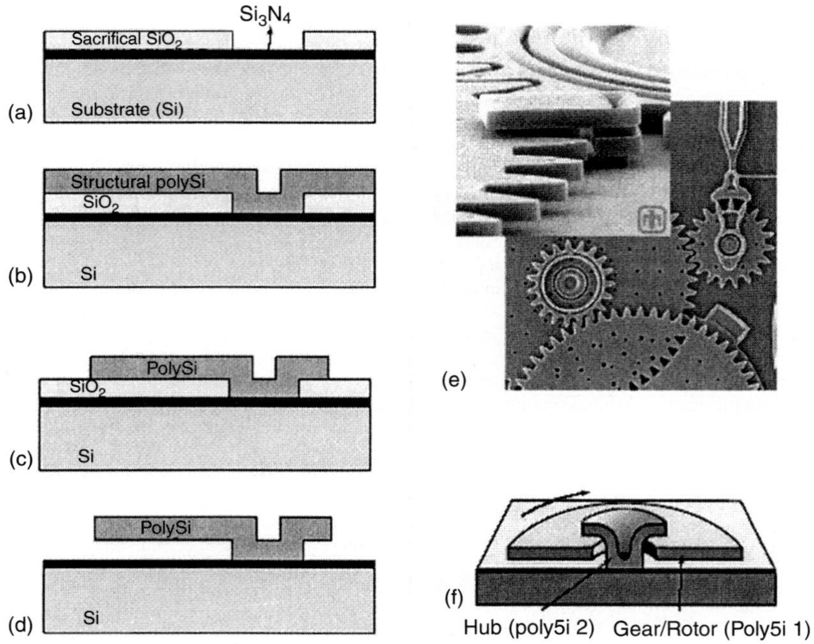

Figure 10.5.5 Basic principles of surface micromachining: (*a*) thin silicon nitride (Si_3N_4) and silicon dioxide (SiO_2) films are deposited, and openings are etched in the SiO_2, using lithography and etching; (*b*) a polysilicon film is deposited (LPCVD) and (*c*) patterned; (*d*) the polysilicon structures are released by wet etching of the sacrificial SiO_2. Steps (a)–(c) can be repeated to create multilayer structures such as the gear train mechanism (Photo: *Sandia National Lab.*) shown in (*e*). Rotary structures are formed with a hub covering part of the rotor (*f*), separated by SiO_2 before release.

Alternatively, plasma processing can allow **deep reactive ion etching** (DRIE) into silicon substrates of practically any lithographically defined geometry, as illustrated in Fig. 10.5.6*b*. The DRIE process combines isotropic plasma etching with sidewall passivation in order to generate deep trenches with vertical side walls,[11] reaching aspect ratios above 30:1 and fast enough etch rates (>3 μm/min) for through-wafer etching. Compared to wet etching, DRIE allows greater design freedom but requires expensive equipment and serial processing.

Wafer Bonding

In contrast to thin structures created by surface micromachining, thick multilayer devices can be formed by joining patterned or blank substrates together in a process referred to as wafer bonding. Three main bonding techniques are available: anodic bonding (glass to silicon), direct fusion bonding (silicon to silicon), and intermediate layer bonding. The planar substrates are first cleaned and brought into contact in a controlled ambient.

In **anodic bonding,** a potential (500 to 1,000 V) is applied across a contacted pair of sodium-bearing glass and silicon wafers, elevated to a moderate temperature (300 to 450°C). Migration of the sodium ions generates strong attraction forces and a permanent chemical bond is formed between the substrates. This approach is commonly used to create optically accessible flow channels in microfluidic devices and to thermally and mechanically isolate sensors from their package and environment. **Direct fusion bonding** relies on molecular diffusion at the interface of two substrates brought in intimate contact. Silicon

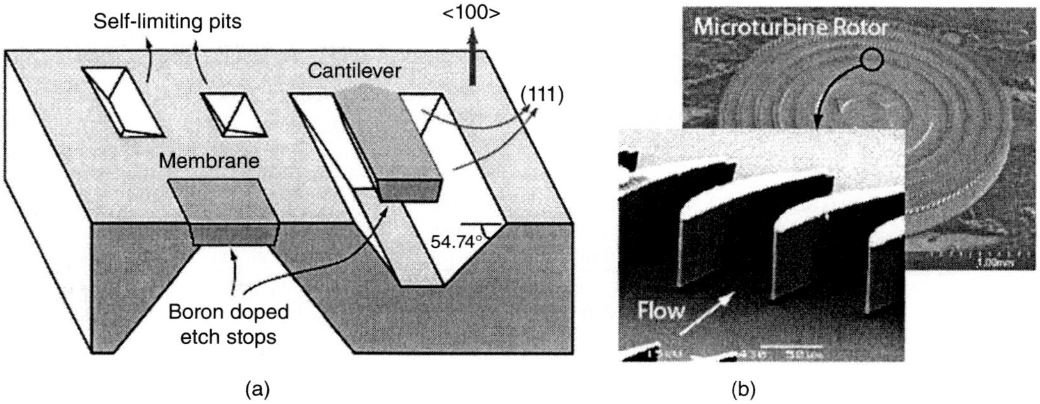

Figure 10.5.6 Bulk micromachining of Si: (*a*) Anisotropic wet etching of Si. Typical anisotropic features that leverage the crystalline structure of Si. The {111} planes and boron-doped regions etch at significantly lower rates than the rest during wet etching with KOH, EDP, or TMAH; (*b*) deep reactive ion etch (DRIE) structures, showing that unconstrained 2-D shapes are possible, but only with vertical sidewalls.

wafers can be fusion bonded by aligning, pressing, and heating (800 to 1,000°C) them. The bond strength can reach that of the crystalline substrate, but requires clean and polished surfaces (roughness <1 nm) and precludes materials that would not withstand the high temperature process, such as most metals. Alternatively, an **intermediate layer** can be used to join two substrates, such as eutectic bonds (Au layer), polymers (resists, polyimide), solders, low-melting-temperature glass (frits) or thermocompression bonds. Some intermediate-layer bonds require only low temperatures, but may not allow long-term vacuum cavities due to outgassing (for glasses, polymers, and solders) and yield lower bond strengths.

Wafer bonding is commonly combined with bulk micromachining or thin-film processes to encapsulate sensors in vacuum or controlled environments (wafer-level packaging), form closed cavities for absolute pressure sensors, or create complex microfluidic devices (such as microturbines[12]).

Nonsilicon Materials and Processing

In order to meet the requirements for a growing range of applications such as biomedicine, chemistry, and wireless communications, nontraditional microfabrication materials and processes have been developed. Polymers and metal electroplating are highlighted here.

Polymers The need for low-rigidity materials in MEMS actuators and polymers for **bioMEMS** has led to the development of various thin- and thick-film polymers. For example, vapor-deposited parylene thin films can serve as flexible membranes, joints, and pneumatic actuators.[13] Casting of thick polymer layers, such as polydimethylsiloxane (PDMS), allows rapid fabrication of microchannels for biochemical MEMS. In a process referred to as **soft lithography,**[14] PDMS is cast over a micromachined substrate that acts as a mold. After curing, the cast polymer is pealed from the mold and pressed against a glass substrate or another polymer layer to form microfluidic geometries. Microchannels and other thick structures can also be directly formed in standard photoresists or photosensitive epoxy, such as SU-8.[15]

Electroplating and LIGA In addition to thin film metal processing described earlier, thick metal layers can be formed by electroplating. The most common approach to create thick metal microstructures is **selective electroplating,** in which a seed layer is covered by a patterned thick photoresist or SU-8 mask, such that the metal (e.g., Ni) is plated only in the exposed areas of the mask. The LIGA process (German acronym for lithography, electroplating, and molding) implements this approach by using X-ray lithography to expose very thick photoresist (several millimeters,) with vertical sidewalls.[16] This yields very high aspect ratio metal parts (up to 70:1) for MEMS and other micromachines.

ENGINEERING SCIENCE AND PRACTICE FOR MEMS DESIGN

Many of the engineering topics covered throughout this handbook can be applied for the design of microsystems. Engineering practice, however, differs as a result of miniaturization and convergence with the semiconductor world. Here we discuss the design approach for MEMS, the effects of scaling on engineering science, and the most common transduction and operating principles, along with examples of MEMS devices. The reader is referred to the literature[17] for an extended discussion of MEMS design.

MEMS Design Process

Depending on the device complexity and scope of the application, development of MEMS may entail:

1. *Definition of market needs and product specifications.* An assessment must be made to determine if MEMS technology is appropriate for the application, by answering the following questions:
   - Does a MEMS and microfabrication approach bring unique capabilities and performance levels due to: fundamental scaling, unique properties of micromachined materials, potential integration with microelectronics, or large array integration?
   - Is there an economic benefit at the component or system level for using MEMS and microfabrication, such as low unit cost due to batch fabrication, on-chip functionality, or miniaturization?
2. *Choice of operating principle and fabrication process, system-level design, and detailed design.* The conceptual device is defined by using previously demonstrated technology and creative thinking. The principles of operation and the microfabrication approach are intimately related since the implementation is constrained by available fabrication technologies. Analytical and numerical modeling of increasing complexity along with detailed process flow definition completes the device design. Traditional or specialized CAD software can facilitate this effort.[18] Packaging considerations are critical at the design phase to ensure appropriate mechanical, thermal, fluidic, biological, chemical, electrical, magnetic, and/or optical interfaces between the device and the environment.
3. *Process development, fabrication, packaging, and testing.* Lithographic masks are made with the designed patterns and used for device microfabrication. Each fabrication process is developed or improved to achieve the desired microstructures. Iterations with the design phase may be required, ranging from slight mask modifications to adopting alternative process flows. The process accuracy is characterized between fabrication steps. Basic device operation can be tested before dicing the wafer into individual chips. Each device is then packaged and tested. In many commercial MEMS products, the final packaging and qualification steps represent the major production cost, often more than the cost of making the MEMS device itself.

Principles and Implementation of MEMS Sensors and Actuators

Many MEMS devices are sensors that monitor physical, biological, or chemical aspects of their environment to generate a useful signal, often electrical, that is correlated to the measured quantity. An actuator will provide the inverse function by acting on its environment on the basis of an input signal, also usually electrical. Various transduction principles convert a signal, or energy, from one physical domain to another, and can be implemented via microfabrication. They are of two types, based on either a **displacement** or a change in physicochemical **properties.** Most devices serially combine multiple transduction principles on chip to achieve high levels of functionality and sensitivity. The most common approaches are presented here with device examples. The reader is referred to the literature[19] for a comprehensive discussion of transduction mechanisms and their MEMS implementations.

Mechanical Elements: Flexures, Mechanisms, and Other Microstructures

Most MEMS consist of micromechanical components and mechanisms that provide a displacement, in combination with other transduction principles. Some of the most common mechanical elements are depicted in Fig. 10.5.7, and can be classified as flexible structures, mechanisms, or static structures. These include: beams that act as cantilever beams or springs; plates that form membranes, bridges, mirrors, or masses; hub-and-pin joints for rotary gears and motors (see Fig. 10.5.5*e*–*f*); hinges for large deflections; and microchannels and other internal flow components to guide fluid flow (see Fig. 10.5.6*b*). Continuum elasticity modeling is generally applicable for such components, although the geometries can achieve extreme aspect ratios (up 1,000:1) and the material properties can be process-dependent and anisotropic, as discussed previously.

These components can be used in static behavior, or for a desired dynamic response. Micromachined spring-mass systems can be excited at resonance and used as gyroscopes through the Coriolis effect or as chemical sensors by monitoring changes in stiffness or mass. Dynamic behavior, such as resonance and time response, are also affected by fluid film damping.[17] Such sensors are typically packaged under vacuum to reduce this effect, and therefore achieve a large quality factor Q, which is defined as the amplitude at resonance normalized by the displacement at low frequency.

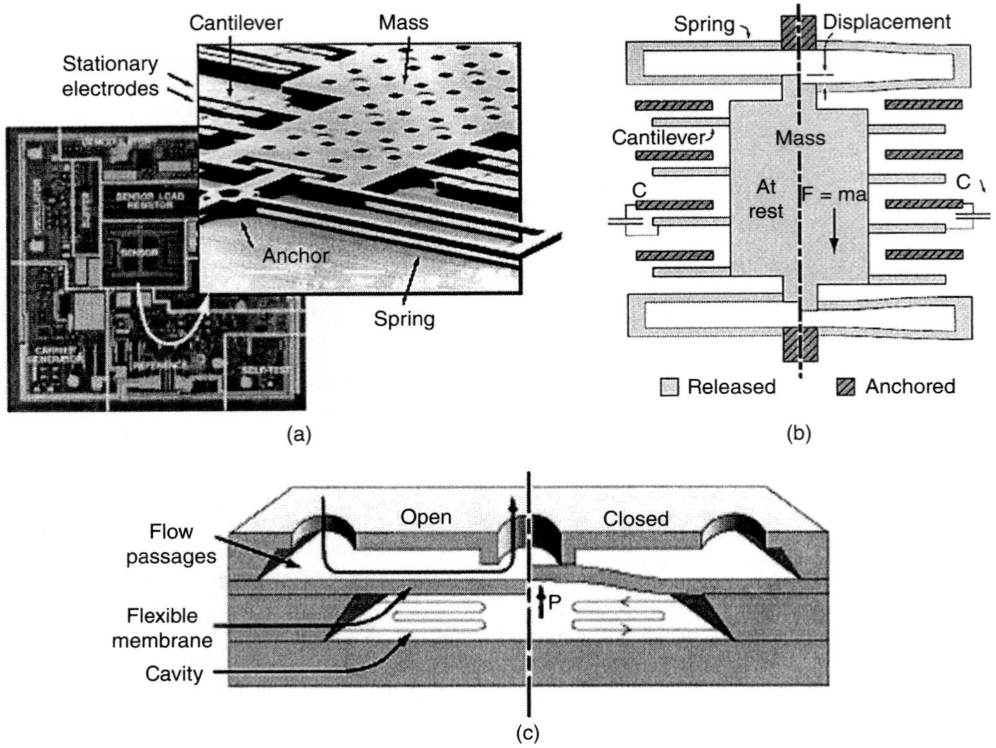

Figure 10.5.7 Microstructures commonly used in MEMS. (*a*) Capacitive accelerometer. (*Photo: Analog Devices.*) The surface micromachined accelerometer consists of a mass supported on folded beams that act as springs. (*b*) Components and operation. Displacement is measured from the change in capacitance between the stationary and movable electrodes. (*c*) Thermopneumatic valve. Flexible membranes can be used in valves to control the flow through microchannels, as well as in pressure sensors.

Capacitive Sensing and Electrostatic Actuation

Displacements can be measured or induced with electrostatic charges. Capacitive sensing consists of measuring the change in capacitance between two electrodes (i.e., conductive structures), as for the commercial accelerometer illustrated in Fig. 10.5.7*a* and *b*. Neglecting edge effects, capacitance between parallel plates is simply expressed as $C = \varepsilon A/g$ where a change in capacitance C can result from a change in: overlapping plate area A, due to parallel motion of the plates (electrodes); gap g, due to normal motion of the plates; or even a change in permittivity ε, by insertion of a different material in the gap. When the sensor is combined with on-chip circuits for capacitance measurement, extreme sensitivity can be achieved since parasitic capacitance is reduced, especially when a differential change in capacitance is measured.

Electrostatic actuation is accomplished by applying an electrostatic field between structures attached to flexures or mechanisms that allow relative motion. For simple parallel plates, attraction forces are experienced normal to the plates and parallel to them (when they are offset), as illustrated in Fig. 10.5.8*a*. Since electrostatic forces are proportional to V^2 and vary inversely to the spacing, d or d^2, high voltages and small gaps are desirable. The maximum force is limited by breakdown across the air gap, which occurs above 300 V for micrometer-scale gaps at atmospheric

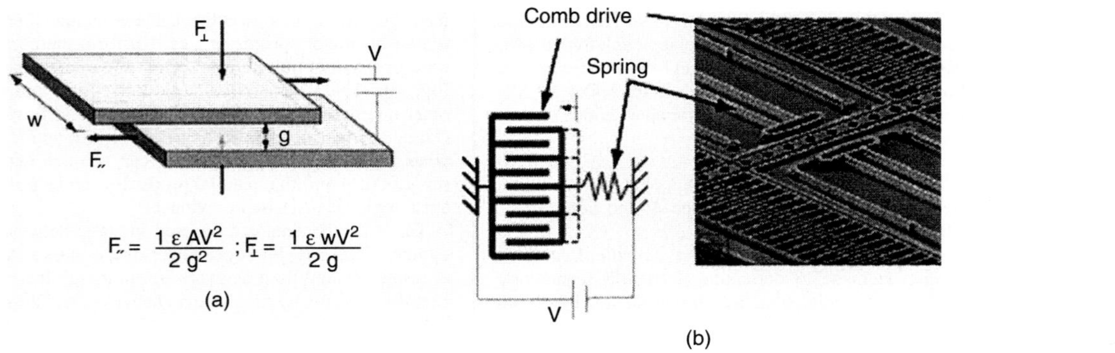

Figure 10.5.8 (*a*) Electrostatic attraction forces between electrodes with components normal and parallel to the structures. For example, the parallel force is used in (*b*) comb drive actuators to move an electrode structure held by a spring. This configuration is similar to capacitive accelerometers but it is used for actuation instead of sensing. (*Photo: Sandia National Laboratory.*)

pressure. Comb drive actuators provide in-plane motion (Fig. 10.5.8*b*), while micromirror devices for display applications use electrostatic actuation to tilt a mirror out of plane and redirect light that reflects on an array of individually actuated micromirrors (see Ref. 8 and Fig. 10.5.1*b*).

Piezoresistive Sensing

Stain gages are commonly used in mechanical testing to measure deformations and the related stress in structures. In MEMS, the same principle is used to measure inertial or pressure forces, by transmitting the force to a silicon structure and measuring the induced strain. Crystalline silicon offers the benefit of a large gage factor (GF > 100) compared to metals (GF ≈ 2), where the gage factor is defined as the relative change in resistance per unit strain, $GF = \frac{\Delta R}{R} / \frac{\Delta L}{L} = \Delta R / \varepsilon R$. Typical implementations for pressure sensors consist of resistances that are diffused on the surface of a bulk micromachined silicon membrane (Fig. 10.5.1*a*). The piezoresistors are connected in a Wheatstone bridge configuration to provide a voltage measurement proportional to the applied pressure difference across the membrane. Piezoresistive accelerometers can operate on a similar principle, by locating a resistor on a silicon beam that supports a proof mass etched from the substrate. Commercial devices can include on-chip temperature compensation, stress isolation with the package, and mechanical stops to prevent excessive stresses.[20]

Piezoelectric Sensing and Actuation

Certain crystalline materials, such as quartz, lead titanate zirconate (PZT), and zinc oxide (ZnO), exhibit an electrical polarization under mechanical strain due to a relative displacement of negative and positive charge centers in the material's microstructure, a phenomenon called **piezoelectricity.**[2] This behavior can be leveraged for force sensing, but also for actuation since an applied electric field will in turn induce mechanical deformation. Traditionally, piezoelectric materials can provide large forces or pressures (>10 MPa) and small displacements (<0.1 μm) at high voltages (>100 V), but piezoelectric thin films applied on flexible beams and membranes in MEMS devices allow larger deflections (up to 10 μm), although at proportionally lower forces. Some commercial ink-jet print heads integrate piezoelectric actuators to generate microdroplets at high frequencies.

Thermal Transduction and Applications

Microsystems can also integrate temperature measurement capabilities, use heat for actuation, or act on their thermal environment. Traditional approaches to measure temperature, such as thermocouples and thermistors, can be implemented via thin-film deposition and patterning to create a solid-state temperature sensor for integration within MEMS devices. For example, resistors can be formed by depositing a thin film of metal or doped polysilicon, characterized by its sheet resistivity $R_{sheet} = \rho/t$ where ρ is the material's electrical resistivity and t is the film thickness. Using photolithography and etching (or through lift-off), the resistor can be patterned into a line of total length l and width w, for a resistance of $R = R_{\text{sheet}}\, l/w$ (Fig. 10.5.9*a*). Since the material resistivity is a function of temperature, monitoring electrical resistance provides a measurement of temperature, according to $\Delta R/R_o = \alpha \Delta T$, where α is the material's **temperature coefficient of resistance** (TCR). Thermocouple junctions have also been microfabricated by depositing and patterning multiple thin metal films. Heat flux can be measured by monitoring the temperature of two overlaid resistors or thermocouple junctions separated by a thermally and electrically insulating film, such as a SiO_2, Si_3N_4, or polyimide.

A resistor can also be used as a heat source, converting electrical energy into heat at a rate of $\dot{Q} = P_{\text{elec}} = Vi = Ri^2$. This allows the integration of electrically controlled heaters in MEMS and microfluidic devices. Heat can be a driving mechanism for actuation by raising the local temperature of microstructures or trapped gases to induce thermal expansion, which then results in a membrane or beam deformation. Micropumps, microvalves, and switches can be based on this operating principle, as illustrated in Fig. 10.5.7*c*. Here, an enclosed fluid tends to expand because of resistive heating, increasing the pressure in the cavity and pushing the membrane upward to close the valve. Alternatively,

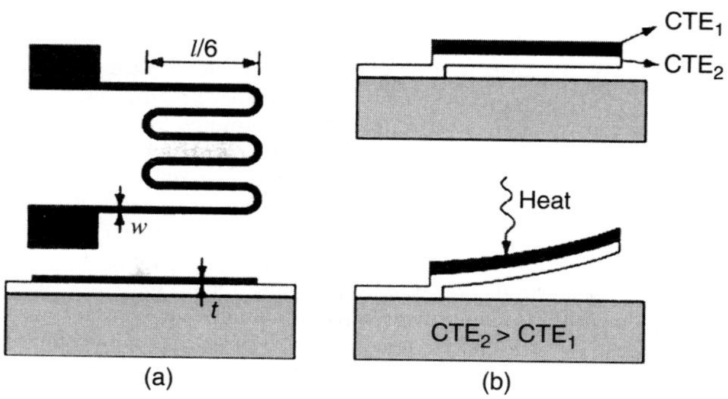

Figure 10.5.9 Thermal sensing and actuation using thin films. A patterned thin-film resistor (*a*) can act as a temperature sensor by monitoring the change of resistance with temperature, or as a heat source through ohmic heating. A thermal bimorph cantilever (*b*) will curve when heated, so it can act as a thermal sensor or an actuator when electrically heated.

out-of-plane deflections of thin-film cantilevers can be achieved by layering materials with different coefficients of thermal expansion (CTE), referred to as bimorph actuators, as illustrated in Fig. 10.5.9*b*.

To minimize the actuation power and provide accurate temperature measurements, thermal isolation can be achieved through the use of high-aspect-ratio structures and low-thermal-conductivity materials, such as SiO_2 or Si_3N_4 films and glass substrates. Microsystems for cooling are less advanced, but potential approaches include thermoelectrics and microchannel heat exchangers, with single or two-phase flow.

Optical Transduction and Applications

Semiconductor technology was also leveraged to create optoelectronic components such as photodiodes and light-emitting diodes. Silicon-based MEMS can therefore integrate not only microelectronics but also optical elements for a broader range of transduction principles. When combined with mechanical and/or fluidic elements, optoelectronics can measure the displacement of a shutter that partly covers photodiodes, or detect biological species on-chip with an emitter-detector pair along a microfluidic channel. Light manipulation can also be accomplished by reflecting incident light onto discrete, controllable, movable micromirrors. Commercial devices include actuated micromirror arrays that redirect light onto a display[8] (Fig. 10.5.1*b*) and variable diffraction gratings to separate wavelengths of the light for analytical or telecommunication applications.[21]

Other Transduction Principles and Applications

The range of transduction principles extends beyond the most common ones presented above. It includes principles from material science, such as shape memory alloys; from physics, such as electron tunneling; from chemistry, such as electrochemical spectroscopy; and from biology. Novel approaches are generally spurred by specific needs for sensing, actuation, or miniaturization requirements, but are often limited by the microfabrication technologies and capabilities. The reader is referred to scientific journals for the latest advancements in MEMS science and technology (see References).

PHYSICS AND SCALING

To assess the potential benefits of MEMS technology for a given application or compare various operating principles, it is useful to consider the effects of miniaturization through scaling laws. Two main aspects are to be considered: the length-scale dependence of various physical phenomena and the effect of scale on physical properties and behavior.

Length-Scale Dependence

Most physical phenomena can be classified as **volumetric,** such as gravity or heat capacity, or **surface-dependent,** such as viscous forces and heat transfer. When a physical system is miniaturized, volume scales as $[l^3]$, whereas surface scales as $[l^2]$, where l is a characteristic dimension of the system. The ratio of volume to surface therefore scales as $[l^3/l^2] = [l]$, such that surface phenomena tend to prevail at small scales; this is referred to as the **cube-square law.** This first-order analysis is applied for various physical domains in Table 10.5.3, showing, for example, that electrostatic forces, viscous forces, capillary forces, and heat transfer become prevalent upon miniaturization. In particular at macroscale, electrostatic transduction is not common because of its low energy density compared to magnetic devices. As they are miniaturized, however, electrostatic forces scale as $[l^2]$ or remain constant, assuming constant field strength or constant voltage, respectively. In comparison, magnetic forces are more affected by scaling, and their three-dimensional windings are challenging to microfabricate.

True scaling may not follow these trends exactly, since the capabilities of micromachining allow aspect ratios and material properties that are significantly different from those possible with traditional machining, as discussed next.

Physical Properties and Behavior

Not only can commonly neglected forces and phenomena become important at small scales, but physical behavior can shift dramatically. In traditional engineering, the smallest device features are typically much greater than the atomic or molecular scales of the device's constituents, such that the material behavior can be represented by continuum mechanics. As device scales are reduced, this representation may not hold and discrete molecular interactions become important. In microfluidics, for example, gas flows become rarefied when the mean free path between molecules, λ is comparable to channel widths d, as represented by the Knudsen number, $\mathrm{Kn} = \lambda/d$. For $\mathrm{Kn} < 0.01$, flow can be considered a continuum with zero relative velocity along surfaces. In the range $0.01 < \mathrm{Kn} < 0.1$, slip flow may occur at the surfaces, which can be accounted for in continuum modeling through appropriate boundary conditions.[22] For air at atmospheric conditions, this range corresponds to 0.5- to 5-μm gaps, which are relatively common in MEMS. For $\mathrm{Kn} > 0.1$, however, the flow is best represented by statistical molecular interactions. Similar considerations lead to quantum effects for the heat transfer through thin films (less than 0.1 μm) or through nanostructured materials, such as superlattices. Modeling of heat conduction as phonon transport along or through these structures becomes necessary.[23]

Even in the continuum domain, physical phenomena can differ with scale. In microfluidics, flows are typically laminar because of low Reynolds numbers, $\mathrm{Re} = \rho U d/\mu < 1{,}000 - 2{,}000$, where ρ is fluid density, U is a characteristic velocity, and μ is the fluid viscosity. Such flows are well represented by laminar flow theory, such as Poisseuille flow, Couette flow, and more generally, Stokes flow and the Reynolds equation for lubrication theory.[24] The absence of turbulence increases the challenges of mixing in biochemical microsystems, but can also be

Table 10.5.3 Scaling of Forces and Fluxes

Field, forces, fluxes	Basic relation	Scaling proportional to	Note
Fluid mechanics			
Inertial forces	$\frac{1}{2}\rho A U^2$	$[l^2]$	Reynolds number:
Viscous forces	$\tau A = \mu \frac{\partial U}{\partial y} A$	$[l^1]$	$\mathrm{Re} = \frac{\text{Intertial}}{\text{Viscous}} = \frac{\rho U L}{\mu} \propto [l^1]$
Heat transfer			
Conduction	$q_{\text{cond}} = -k \frac{\partial t}{\partial y} A$	$[l^1]$	Biot number:
Convection	$q_{\text{conv}} = hA\Delta Te \quad (h \propto 1^{0.5})$	$[l^{2.5}]$	$Bi = \frac{\text{Convection}}{\text{Conduction}} \propto [l^{1.5}]$
Energy transport	$\dot{m}\, c_p T = \rho A U\, c_p T$	$[l^2]$	Stanton number:
			$St = \frac{\text{Convection}}{\text{Transport}} \propto [l^{0.5}]$
Mechanics		$[l^1]$	
Strain energy	$\frac{1}{2}\sigma\varepsilon AL = \frac{1}{2}\frac{\sigma}{E} AL$	$[l^3]$	
Beam bending	$F = kx\left(k = \frac{Ewt^3}{L^3}\right)$	$[l^2]$	
Natural frequency	$\omega = \sqrt{k/m}$	$[l^{-1}]$	
Electromagnetism			
Electrostatic forces	$F = \frac{1}{2}\frac{\varepsilon V^2}{d^2} A$	$[l^0]$	For constant voltage V
Magnetic forces	$F = \frac{1}{2}\frac{B^2}{\mu_0} A$	$[l^2]$	For constant field (V/d)
Piezoelectric forces	$F = \frac{1}{2}\frac{\varepsilon V}{t} A$	$[l^2]$	
Physicochemistry		$[l^1]$	
Van der Walls	—	$[l^{1/4}]$	
Surface tension	$F = \lambda 2\pi t \cos(\theta_c)$	$[l^1]$	Bond number:
Gravity	$F = mg$		$\mathrm{Bo} = \frac{\text{Surface}}{\text{Gravity}} \propto [l^{-1}]$
		$[l^3]$	

leveraged to form layered sheets of fluids with well-defined diffusion-based concentration profiles. For liquids, such as water and electrolytes, electrical surface phenomena can become noticeable in microchannels. A charged double layer can form at the surfaces, such that an electrical field applied along the channel will tend to drive the flow along the wall. Because of the small channel diameter, the core flow is entrained by viscous shear to create a plug-flow, uniform velocity profile. This approach requires high electrical fields, which are achievable in microsystems with reasonably high voltages (up to 1,000 V) over short channels lengths (millimeters). This approach is commonly used to drive fluids in micro total analysis and lab-on-a-chip devices, and is referred to as **electro-osmosis.**

In electrostatics, the maximum field strength (V/d) at which breakdown occurs is a function of scale. At small gaps ($d < 5$ μm, atmospheric pressure), the voltage limit stops decreasing to reach a minimum near 300 V; it even increases at submicrometer gaps. This results from the lack of molecules in the air gap to create the avalanche of collisions that lead to arcing. This benefits electrostatic actuation by increasing its maximum energy density and forces.

In micromechanics, the crystalline microstructure of silicon components formed by bulk micromachining introduces the need for anisotropic continuum mechanics to accurately predict the stress and strain.[25] Surface micromachined devices or other MEMS devices formed of thin films must be designed with consideration for their residual in-plane stress and stress gradient across the film thickness, which are highly dependent on the processing conditions. Material characterization and process control become critical aspects for the performance and reliability of such devices. In addition, the brittle nature of silicon-based materials excludes plastic deformation and their fracture strength is defined by maximum flaw size in the part. Fortunately, the probability of having a significant flaw in a brittle part increases with its volume and surface area, hence strength and reliability are favored in small scale structures.

BEYOND SENSORS AND ACTUATORS: INTEGRATED MICROSYSTEMS

Although most commercial MEMS can be classified as sensors or actuators, there is an increasing trend toward microsystems that integrate multiple components to perform specific applications. For example:

- Photonic and microoptoelectromechanical systems, or MOEMS, that direct, filter, and characterize light for optical telecommunication applications
- Radio frequency (RF) MEMS that generate and transmit high-frequency signals for wireless communication applications
- BioMEMS and lab-on-a-chip devices that perform high-throughput biological assays and revolutionize medical instrumentation
- Micro total analysis systems (TAS) for synthesis and characterization of chemical and biochemical products
- Power MEMS to provide miniature power sources, cooling, and propulsion for consumer electronics, people, and small robots

CONCLUSIONS

MEMS technology can be considered a toolbox to create miniature sensors, actuators, and integrated microsystems. The main tools are the materials and processes used in microfabrication, the engineering science with emphasis on small-scale phenomena, as well as the sensing and actuating principles. The parallel nature of microfabrication processes such as photolithography, thin-film deposition, and etching allows batch fabrication of devices with high accuracy and in large quantities. In-plane device complexity (in terms of number of components in a system) comes for free, allowing large arrays or complex networks to be implemented with micrometer-scale resolution. The devices can naturally integrate electrical and mechanical elements, as well as optical, fluidic, thermal, chemical, or biological functionality on chip.

The microfabrication infrastructure, equipment, process development, and control are, however, expensive, and are best amortized over large production quantities. To this cost must be added the costs of packaging and testing, which can exceed the cost of the chip itself, and the extensive engineering effort to achieve reproducibility and reliability, as required in commercial products. To be commercially successful, MEMS devices must typically offer increased functionality or performance at a lower cost than competing technologies, and in a high-volume market. Lower-volume applications can also be successful when the use of MEMS adds high value to the final product and the market share is protected by intellectual property or unique know-how. Whether commercially successful or not, an ever-increasing range of MEMS devices and technologies is being explored. This will contribute to our MEMS toolbox and enable novel solutions to our greatest challenges, ranging from nano to macro scales and spanning the physical, chemical, and biological worlds.

REFERENCES

1. *Got MEMS? 2003 Industry Overview and Forecast*, In-Stat/MDR, July 2003.
2. M. Madou, *Fundamentals of Microfabrication*, CRC Press, New York, 1997.
3. S. Wolf and R. N. Tauber, *Silicon Processing for the VLSI Era—Vol. 1: Process Technology*, 2nd ed., Lattice Press, Sunset Beach, CA, 2000.
4. O. M. Jadaan, N. N. Nemeth, J. Bagdahn, and W. N. Sharp Jr., "Probabilistic Weibull behavior and mechanical properties of MEMS brittle materials," *J. Materials Sci.,* **38**:4087-4113 (2003).
5. *ADXL* capacitive accelerometer, Analog Devices Inc. Norwood, MA.
6. *polyMUMPs* multiuser polysilicon surface micromachined process, MEMSCap, Bernin, France.
7. SUMMIT V polysilicon surface micromachined process, MEMX Albuquerque, New Mexico.
8. L. Hornbeck, "Digital Light Processing and MEMS: Timely Convergence for a Bright Future," *Proc. SPIE*, **2639**:2, Micromachining and Microfabrication Process Technology (1995) (www.dlp.com).
9. A. A. Yasseen, C. A. Zorman, and M. Mehregany, "Surface Micromachining of Polycrystalline SiC Films," *IEEE/ASME J. Microelectromechanical Systems*, **8**(3):237-242 (1999).
10. G. Fedder et al., "Laminated High-Aspect-Ratio Microstructures in a Conventional CMOS Process," *Sensors & Actuators A*, **57**(2):103-110 (1996).
11. A. A. Ayón et al., "Characterization of a time multiplexed inductively coupled plasma etcher," *J. Electrochem. Soc.* **146**(1): (1999).
12. L. G. Fréchette et al., "High-Speed Microfabricated Silicon Turbomachinery and Fluid Film Bearings," *J. Microelectromechanical Systems*, **14**(1):141-152 (2005).
13. Y. Lu and C. J. Kim, "Micro-finger Articulation by Pneumatic Parylene Balloons," *Proc. 12th Int'l Conf. on Solid State Sensors, Actuators and Microsystems*, Boston, June 8-12, 2003.
14. Y. Xia and G. M. Whitesides, "Soft Lithography," *Annu. Rev. Mater. Sci.*, **28**:153-184 (1998).
15. H. Lorenz, M. Despont, P. Vettiger, P. Renaud, "Fabrication of photoplastic high-aspect ratio microparts and micromolds using SU-8 UV resist," *Microsystem Technologies*, **4**:143-146, 1998.
16. H. Guckel, "High-Aspect-Ratio Micromachining Via Deep X-Ray Lithography," *Proc. IEEE*, **86**, no. 8, pp.1586-1593, 1998.
17. S. D. Senturia, *Microsystem Design*, Kluwer Academic Publishers, Norwell, 2001.
18. S. D. Senturia, "CAD Challenges for Microsensors, Microactuators, and Microsystems," *Proc. IEEE*, **86**(8):1611-1626 (1998).
19. G. T. A. Kovacs, *Micromachined Transducers Sourcebook*, WCB McGraw-Hill, New York, 1998.
20. *Kulite Pressure Transducer Handbook*, Kulite Semiconductor Products, Inc., Leonia, NJ.
21. O. Solgaard, F. S. A. Sandejas, and D. M. Bloom, "Deformable grating optical modulator," *Optics Letters*, **17**:688-690 (1992).
22. G. E. Karniadakis and A. Beskok, *Micro Flows: Fundamentals and Simulation,* Springer, New York, 2002.

23. D. G. Cahill, W. K. Ford, K. E. Goodson, G. D. Mahan, A. Majumdar, H. J. Maris, R. Merlin, S. R. Phillpot, "Nanoscale Thermal Transport," *J. Applied Physics*, **93**:793-818 (2003).

24. V. N. Constantinescu, *Laminar Viscous Flow*, Springer, New York, 1995.

25. W. M. Lai, D. Rubin, and E. Krempl, *Introduction to Continuum Mechanics,* Butterworth–Heinemann, New York, 1996.

Scientific Journals

Journal of Microelectromechanical Systems, ASME/IEEE.
Sensors and Actuators, Elsevier.
Lab on a chip, RSC Publishing.
Journal of Micromechanics and Microengineering, Institute of Physics.
Sensors Journal, IEEE.

10.6 SOUND, NOISE, AND ULTRASONICS

by Robin O. Cleveland, Benjamin C. Davenny, and Jonathan D. Kemp

REFERENCES: Kinsler, et al., "Fundamentals of Acoustics," Wiley, 1982. Harris, "Noise Control in Buildings," INCE, 1997. Beranek and Ver, "Noise and Vibration Control Engineering," 1992. "2003 ASHRAE Handbook—HVAC Application," Chap. 47, Sound and Vibration Control, 2003.

SOUND

An **acoustic wave,** or a **sound wave,** is a mechanical disturbance of particles in a fluid from an equilibrium state which propagates as a wave through the medium with a speed of sound that depends on the properties of the fluid. In fluids the waves are longitudinal; that is, the motion of the disturbed particles is in the direction of the wave propagation. Typical acoustic sources, such as audio speakers, vibrate backward and forward, generating sound waves consisting of compressional phases and rarefractional phases. The pressure, density, and particle velocity of the fluid are all affected by the passage of the wave. Similar waves exist in elastic solids, but in this case, two types of waves can propagate: **longitudinal waves** and **shear** (or **transverse**) **waves** in which the particle motion is transverse to the direction of propagation. Here, acoustic waves will refer to waves in fluids (liquids or gases) and elastic waves to those in elastic solids.

The **sound speed** c_0 is a material property, and for gases it is given by $c_0 = \sqrt{\gamma RT}$, where γ is the ratio of specific heats (1.4 for diatomicgases such as air), R is the gas constant [287 J/(kg · K) for air], and T is the absolute temperature. For air at 20°C (293 K) it is about 343 m/s (1,125 ft/s). For liquids the expressions are more involved (see, e.g., Lovett, *Jour. Acoust. Soc. Am.*, **45**; 1969, pp. 1051-1053), and an approximate value for water at room temperature is 1,500 m/s (4,920 ft/s). (See Table 10.6.1.)

Sound waves can take any waveform shape; however, in many applications a single frequency is used. For a specific frequency f the sound wave will have a period $T = 1/f$ and a wavelength $\lambda = c_0/f$; for example, a 3.4 kHz tone in air will have a period $T = 0.294$ ms and a wavelength $\lambda = 0.1$ m. In older texts the units for frequency maybe stated as cycles per second (cps) rather than the SI unit hertz (Hz). For sounds that are not of a single frequency the use of Fourier analysis is often used to represent it as a combination of different frequencies. The audible frequency is normally defined to extend from 20 Hz to 20 kHz. The **audible frequency range** is normally said to extend from 20 Hz to 20 kHz, although for frequencies above 10 kHz the sensitivity of the ear decreases dramatically. As an individual gets older, the upper frequency limit becomes progressively lower. Sound with frequencies less than 20 Hz is referred to as **infrasound,** and sound with frequencies above 20 kHz is referred to as **ultrasound.** (See Table 10.6.2.)

Table 10.6.1 Density and Sound Speed in Selected Materials

	Density		Velocity	
Material	SI unit, kg/m³	USCS unit, lbf/ft³	SI unit, m/s	USCS unit, ft/s
Steel	7,700	481	6,100	20,013
Aluminum	2,700	169	6,300	20,669
Concrete	2,600	162	3,100	10,171
Glass	2,300	144	5,600	18,373
Brick	2,000	125	3,650	11,975
Lucite	1,200	75	2,650	8,694
Water	1,000	62	1,500	4,921
Pine	450	28	3,500	11,483
Cork	240	15	500	1,640
Air	1.21	0.0755	343	1,125

Table 10.6.2 Acoustic Spectrum

Lower range of elephant hearing	5 Hz
Infrasound/sound transition	20 Hz
Lowest note on a piano	27 Hz
E string on a bass guitar	41.3 Hz
Dominant frequency of thunder	100 Hz
Middle C	261.63 Hz
Upper limit of telephone bandwidth	3,400 Hz
Highest note on a piano	4,096 Hz
Sound/ultrasound transition	20 kHz
Upper range of bat sonar	100 kHz
Upper range of dolphin sonar	200 kHz

The density and sound speed of a material determine its **specific acoustic impedance** $Z_0 = \rho_0 c_0$. This term is often shortened to **acoustic impedance** or just **impedance.** The impedance of air is about 415 kg/(m² · s) [85 lbf/(ft² · s)] and of water is about 1.5 × 10⁶ kg/(m² · s) [0.31 × 10⁶ lbf/(ft² · s)]. The unit kg/(m² · s) is often referred to as MKS Rayls—although this is not an SI unit. For a plane wave moving in one direction (referred to as a progressive wave) the impedance relates the pressure p to the particle velocity u through $p = uZ_0$; for example, acoustic impedance plays a role similar to electrical impedance in circuit theory. For waves that are not plane (such as focused waves) or not moving in one direction (such as standing waves), there is normally not a simple relation between the pressure and particle velocity.

A propagating acoustic wave carries energy, and this is most commonly described by the **intensity** I, the power per unit area. The intensity can be calculated by the following integral:

$$I = \frac{1}{T_{\text{av}}}\int_{t_0}^{t_0+T_{\text{av}}}(p \cdot u)\,dt \qquad (10.6.1)$$

where the integration is done over a period for a periodic wave, over the duration of the pulse for a pulse, or over an appropriately long time for a noise signal. The units for I are watts per square meter (W/m²). For a progressive wave the particle velocity is related to the acoustic pressure, and the intensity can be determined from measurements of the root mean square (rms) pressure and the impedance:

$$I = p_{\text{rms}}^2 / Z_0 \qquad (10.6.2)$$

When an acoustic wave encounters an interface, the wave is partially reflected and partially transmitted. In the case of **normal incidence,** the intensity transmission and reflection coefficients are given by

$$R_I = \frac{(Z_2 - Z_1)^2}{(Z_2 + Z_1)^2} \tag{10.6.3}$$

$$T_I = \frac{4Z_1Z_2}{(Z_2 + Z_1)^2} \tag{10.6.4}$$

that is, they depend predominantly on the difference in impedance. The coefficients sum to unity, and individually they must range between 0 and 1. In noise control applications, interfaces are often employed to reduce the transmission of sound, and it is common to use transmission loss TL, which is related to the intensity transmission coefficient by

$$\text{TL} = -10\log T_I \tag{10.6.5}$$

A large transmission loss is normally desired.

For sound that is incident at oblique angles, or where there are multiple interfaces, the expressions for reflection and transmission coefficients become more complicated, and it often becomes more practical to measure the coefficients.

Attenuation of Sound When sound propagates through a medium, most of the energy remains in the sound wave, but a small amount of it is absorbed by the medium. The amplitude of an acoustic wave will therefore decay or **attenuate** as it propagates through a medium. The attenuation can be due to geometric spreading effects, scattering by inhomogeneities in the medium, and direct absorption of the energy by the medium. For a plane wave the intensity will decay exponentially with distance

$$I = I_0 \exp(-2\alpha x) \tag{10.6.6}$$

where α is the attenuation coefficient and x is the distance. The attenuation coefficient has units of nepers per meter (Np/m). Often when attenuation is reported to dB/m, it can be converted to Np/m by making use of the conversion 1 Np = 8.686 dB.

Absorption in air is dominated by the vibration of nitrogen and oxygen molecules. It varies with frequency (absorption increases with frequency) and is very sensitive to changes in temperature and humidity. Propagation distances typically need to be on the order of hundreds of meters for the attenuation to be significant. International Standard ISO 9,613-1:1993 describes how to determine the attenuation of sound in air.

Absorption in pure water is very weak and normally negligible; however, it increases with the square of the frequency and can be important for frequencies above 10 MHz. In saltwater, absorption is much stronger and is dominated by chemical dissociation of salts (see Francois and Garrison, *Jour. Acoust. Soc. Am.* **72;** 1982, pp. 896-907 and 1879-1890).

In solids there are few quantitative formulas for determining attenuation, and most data are determined empirically. The attenuation in most solids typically follows a power law of the form $\alpha = \alpha_0 f^n$, where α_0 and n are material properties and n typically varies between 1 and 2. A comprehensive list of acoustic properties can be found at http://www.ondacorp.com/tecref_acoustictable.html.

Transducers

A **transducer** is a device that converts energy between two forms. In acoustics, transducers are used to measure sound (convert sound to an electric signal) and to generate sound (convert electric signals to acoustic waves).

Generating Sound Sound is generated by an object or a surface vibrating in a fluid. For a sound source to be efficient, it is necessary that the dimensions of the vibrating surface of the source be much larger than the wavelength of sound. This is often not practical; for example, at 300 Hz the wavelength in air is about 1 m (3 ft). Efficiency is also improved if the source is mounted in a wall (baffle) or a chamber so that there is no direct path from the front and back surfaces of the vibrating surface. The presence of a path between the front and back surfaces of the source allows some of the energy to "slosh" back and forth and not radiate as a sound wave.

Moving coil loudspeakers are perhaps the most common transducer for sound generation. The heart of the transduction is a cylindrical coil of wire (voice coil) and a magnet. When a current is passed through the coil, the interaction with the magnetic field results in a force on the voice coil. The voice coil is connected to a large diaphragm which is vibrated back and forth as the coil oscillates, and this results in the generation of an acoustic wave. The diaphragm is provided with some mechanical stiffness to produce a restoring force.

The attachment of a horn to the loudspeaker improves the impedance match between the speaker and the air since it is essentially an acoustic transformer. The dimensions and flare of the horn contribute to its matching ability. The operation of the speaker may be influenced by its enclosure, the baffle which separates the front from the back radiation, or its resonances.

Whistles and sirens may also be used to produce intense sound fields in gases and liquids. In a whistle a flow of air is passed over a sharp edge; turbulent vortices are produced which result in oscillation of the air. If a resonant chamber is placed behind the edge, then the vortices will excite the resonance and produce a tone. These oscillations can often result in high-amplitude signals. An example is flow over a car with an open sunroof, which will result in turbulence that can excite a resonance in the air in the car. The frequency of the resonance is typically less than 10 Hz (infrasound) and can be disturbing to passengers in the car. Sirens are devices in which a revolving disk or drum is perforated such that the holes alternately block and unblock a steam of air. Early sirens employed a series of holes on the outer edge of a disc with a jet of air impinging thereon. Modern sirens use a centrifugal fan inside a drum. The fan is rotated and forces air through the holes in the drum, to generate a sound.

In underwater applications, most sound sources employ piezoelectric materials. A piezoelectric material is one which generates a strain when a voltage is applied across the material. The change in thickness of the material launches an acoustic wave in the water. Typically the impedance of piezoelectric materials is much higher than that of water, and to ensure efficient sound transmission, quarter-wavelength impedance matching layers are used on the face of the transducer. Impulsive noise sources used underwater include explosions and implosions of hollow spheres.

Measuring Sound The acoustic field is typically measured in air by using a microphone and in water by using a hydrophone. Most microphones are either condensor microphones or electret microphones. The active element of a condensor microphone consists of a capacitor where one plate is free to move with the particle motion of the acoustic wave, which results in a change in bias voltage on the capacitor. An electret uses a similar concept except that the dielectric is a permanently electrically charged material so that no biasing voltage need be employed. Electret microphones are becoming the dominant transducer for measuring audible sound even for quality measurement microphones. Micro-Electro-Mechanical-Systems (MEMS) based microphones, which typically employ micrometer-sized capacitors and may have an advantage in terms of mass production and direct integration with electronic circuits, have the potential to become competitive with electrets in the near future.

Other microphones use piezoelectric materials, where the strain or stress associated with the acoustic wave induces a voltage in the material. The majority of hydrophones employ piezoelectric transducers.

Sound Pressure Level Most measurements report the acoustic disturbance in terms of pressure which has units of N/m^2 or Pa. The threshold of hearing in people with acute hearing is an rms pressure of approximately 20 μPa (2.9 × 10^{-9} psi) at around 4 kHz. The threshold of pain is an rms pressure of 63 Pa (9.3 × 10^{-3} psi). In comparison the atmospheric pressure at sea level is approximately 100,000 Pa (15 psi).

Although sound is measured as a pressure, it is normally reported in decibels as either a **sound pressure level (SPL)** or an **intensity level (IL)** which are given by

$$\begin{aligned} \text{SPL} &= 20\log_{10}(p_{\text{rms}}/p_{\text{ref}}) \\ \text{IL} &= 10\log_{10}(I/I_{\text{ref}}) \end{aligned} \tag{10.6.7}$$

where in air $p_{\text{ref}} = 20\ \mu$Pa and $I_{\text{ref}} = 10^{-12}$ W/m^2 and are chosen so that the two levels are almost identical (less than 0.4-dB discrepancy). In terms of

Table 10.6.3 Examples of Representative SPLs for Various Environments

Rustling of leaves	15 dB
Whisper	30 dB
Normal conversation	60 dB
Average street noise	70 dB
Vacuum cleaner	80 dB
Heavy truck at 15 m (45 ft)	90 dB
Home stereo at full volume	100 dB
Rock concert (threshold of feeling)	120 dB
Jet engine at 25 m (75 ft)	140 dB

SPL, the threshold of hearing is 0 dB, and the threshold of pain is 130 dB. Table 10.6.3 gives typical SPLs for sound in the audible range.

Note in underwater acoustics the **reference level** for an SPL is 1 μPa, and in older underwater acoustics literature a value of 0.1 Pa was used. It is good practice to report the reference pressure level when reporting SPL in dB to avoid confusion.

Perception of Sound

Although human hearing can extend from 20 Hz to 20 kHz, the auditory system is most sensitive to sounds in the range of 1 to 5 kHz. Due to the frequency dependence of human hearing, the SPL used for determining human response to sound is normally weighted so that sounds in the 1- to 5-kHz range are given stronger weighting. A number of algorithms have been developed that apply different weighting to sound in specific frequency bands that are either one octave or one-third-octave apart. An octave refers to a doubling of frequency; that is, 1,000 Hz is one octave above 500 Hz and two octaves above 250 Hz. An octaveband centered at f_0 covers the range of frequencies from $f_0/\sqrt{2}$ to $f_0\sqrt{2}$; for example, for 1,000 Hz it covers the range of 707 to 1,414 Hz. The most common is the A weighting (dBA) which applies 0-dB weighting at 1,000 Hz, reduces signals at 63 Hz by 26.2 dB, and adds 1.0 dB at 4,000 Hz. The A weighting is appropriate for determining the perceived sound level for relatively quiet sounds, as it accounts for the strong frequency-dependent sensitivity of the ear. For more intense sounds the ear does not display such a strong frequency sensitivity, and in this case for C weighting (dBC) is more appropriate, which is effectively a flat bandpass filter with −3-dB frequencies at 31.5 Hz and 8 kHz. The C weighting is also often appropriate for impulsive sounds. Table 10.6.4 lists the octave band corrections for the A and C weightings.

OSHA regulations (Standard 1910.95) require that when workers are exposed to an average weighted sound level of 85 dBA or more in an 8-h day, then monitoring and hearing conservation practices must be implemented. These practices include regular measurement of sound levels and regular auditory testing of employees. A SPL of 90 dBA is the maximum permissible for an 8-h workday, and the exposure time goes down by 50 percent for each 5-dB increase in noise levels; that is, workers can only be exposed to an SPL of 95 dBA for 4 h, and a level of 100 dBA reduces exposure time to 2 h. Acoustic engineering methods, for example, adding a sound barrier or making the source quieter, can be employed to reduce noise levels and allow for longer exposure times.

Table 10.6.4 Correction, dB, to Be Applied to an SPL Measured in Octave Bands Centered at the Specified Frequencies.

Frequency, Hz	dBA	dBC
31.5	−39.4	−3.0
63	−26.2	−0.8
125	−16.1	−0.2
250	−8.6	0
500	−3.2	0
1,000	0	0
2,000	+1.2	−0.2
4,000	+1.0	−0.8
8,000	−1.1	−3.0

NOISE

Noise measurements are usually made with a sound level meter, comprised of a microphone, an attenuator, an amplifier, a frequency-weighting network, and an indicator. See ANSI S1.4, "Specification for Sound Level Meters," and ANSI S1.42, "Design Response of Weighting Networks for Acoustical Measurements." A sound analyzer indicates sound pressure as a function of frequency by using a filter set designed in accordance with ANSI S1.11, "Specification for Octave-Band and Fractional-Octave-Band Analog and Digital Filters." Octave band filters are typically included with sound analyzers with additional one-third octave band filter sets available on some analyzers. Modern sound analyzers permit simultaneous real-time measurement across many frequency bands. In addition, narrowband analysis with fast Fourier transform (FFT) is available on some sound level meters. By using these instruments the components of the broadband noise may be identified, and correlated with the noise sources. Once the noise source is located, techniques for noise control can be used.

Contact transducers (vibration pickups) may be used to locate sources of noise, such as in partitions and machine parts. They may be piezoelectric or magnetic and may be used to supplement information from the microphone and sound level meter.

Where the frequency distribution of the noise is significant, an analyzer may be used since the overall level meter tells nothing about frequency distribution. For most noise control applications, one-third octave band filters provide sufficient frequency resolution. However, narrowband FFT analyzers are helpful when finer frequency resolution is required.

Transmission Isolation

If the vibrations of noisy machinery cannot be suppressed at the source, their transmission to the listener should be impeded. The most effective vibration isolation method is the introduction of elastic discontinuities in the structure transmitting the noise. The discontinuities may be obtained by the use of rubber, neoprene, or springs in machinery mountings, or by the introduction of neoprene isolation sheets in expansion joints of concrete slabs. The isolation treatment should be applied as close to the source as possible, to eliminate sound radiation from the structures transmitting the vibrations. Where this is not possible, the listening space itself may be isolated. Thus quiet rooms, constructed especially for noisy measurements, are usually built as separate structures isolated from the main building.

Filtration

Some problems of noise transmission through air lend themselves to solution by methods of filtration. Typical examples are the transmission of sound in ventilating ducts and the noise production at engine exhaust pipes. In each of these cases, the steady flow of gas must not be impeded, but the alternating flow, representing sound transmission, must be effectively suppressed.

Noise in ventilating ducts is typically broadband and consists of fan noise (low and middle frequency) and turbulence noise (middle and high frequency) generated by airflow. Fans are capable of generating tones at a blade passage frequency determined by their rotational speed and number of blades. However, most fans used in HVAC are large centrifugal fans, which do not generate a tremendous amount of tonal noise. Some of this sound energy can be converted to heat by using dissipative glass or cotton fiber elements such as duct silencers or internal duct lining. Both approaches tend to give the most attenuation at middle frequencies around 1 kHz. Duct silencers, consisting of parallel baffles, are used to reduce low- and middle-frequency noise. For ventilating ducts, noise suppression may be obtained by lining the ducts with 1-in-thick glass fiber or cotton fiber sheets. Low-frequency propagation along a duct can be reduced by means of compliant duct walls that radiate noise away from the duct, or by discontinuities such as plenums and outlets. High-frequency noise can be reduced with elbows. Reactive (also known as no-fill, or no-media) silencers are commercially available and are commonly used where no fibrous fill is allowed. These silencers work on the same principle as the muffler on a car. They consist of a few pipe segments that interconnect a number of large chambers. The change in acoustic properties of the pipes (masslike) and the chambers (springlike) produce resonance effects which result in reflection of the noise back to the source.

For noise with strong tonal components, resonators can be fitted to the duct to filter specific tones. The filters are normally placed as close to the source as possible. However, a difficulty with using resonators in duct systems is that a separate resonator is needed for each tone, and tuning the resonator to the correct frequency can be difficult. Other problems include excess pressure drop and flow noise generation from duct discontinuities. Most noise control approaches incorporate broad-band attenuation to reduce the entire noise spectrum below a target **noise criteria (NC)** level.

Ducts should be sized to limit airflow noise and duct drumming to acceptable levels in the spaces they serve. The volumetric flow rate is determined by the amount of air changes per hour required for the space of interest. Duct sizing and volumetric flow rate determine the average airflow velocity in the duct, which in turn affects the amount of turbulence noise that can be generated. Rules of thumb are typically used for setting appropriate airflow velocities. Main system ductwork that is usually located in building shafts away from noise-sensitive spaces is typically sized for medium-pressure standards with airflow velocities from 6 to 10 m/s [1,200 to 2,000 feet per minute (fpm)]. The main ductwork connects to branches that are sized for low-pressure standards. In general, velocity ventilation branches serving noise-sensitive rooms such as music performance spaces should be on the order of 0.5 to 1.5 m/s (100 to 300 fpm). For regular occupied spaces, such as offices, airflow velocities in duct branches should be on the order of 2 to 3 m/s (400 to 600 fpm). Spaces that have high airflow requirements such as laboratories permit higher noise levels, and airflow velocities in duct branches are on the order of 4.5 to 6 m/s (900 to 1,200 fpm).

Shielding

Noise can be reduced by the placement of sound-opaque partitions between the source and the receivers. In most buildings the panels are thin in comparison to a wavelength, and the transmission loss for normally incident sound through a thin panel is given by

$$\mathrm{TL}_0 = 20\log(\pi f \rho l / Z_1) \qquad (10.6.8)$$

where f is frequency, ρ is the mass density of the wall material, l is the thickness of the wall, and Z_1 is the impedance of air [415 Rayls or 85 lb/(ft$^2 \cdot$ s)]. There is a 6 dB/octave increase in transmission loss as frequency increases. The transmission loss of sound can be increased by increasing the mass per unit area of the partition, either a denser material or a thicker panel. For sound that is not normally incident the TL is reduced, and for sound incident over a range of angles, typical transmission loss is approximately derated by 5 dB, that is, $\mathrm{TL} = \mathrm{TL}_0 - 5$ dB.

The sound transmission loss of single panels in the audible frequency range is often further reduced by coincidence effects associated with bending (or flexural) waves in the panel. This occurs when the frequency of obliquely incident sounds equals the bending wave frequency of the panel, which is determined by the thickness and stiffness of the panel. If the panel had no damping, then the sound transmission loss would be reduced to zero at this coincidence frequency; but the damping inherent in most panels reduces this degradation of transmission loss. Single panels can also have low-frequency resonances due to their mass and stiffness. However, this resonance is determined by the size of the panel, is mitigated by natural panel damping, and is usually not a concern in most noise control applications.

Typically mass-based insulation is too large and heavy to be practical. Thin panels separated by an airspace can achieve similar transmission loss to massive monolithic constructions. The sound transmission loss of double-panel constructions can be reduced by mass-air-mass resonance. Adding sound-absorptive material such as glass or mineral fiber batt insulation reduces this resonance effect and improves the general sound transmission loss of the partition. The deeper the airspace, the lower the mass-air-mass resonance frequency. Coincidence phenomena affecting each of the panels can also affect the total transmission loss of the assembly.

Double walls must be constructed carefully to avoid loss of vibration isolation through mechanical bridging between the opposite surfaces. Higher-density sound-absorbing fillers such as mineral fiber batt insulation can reduce sound insulation if they are in contact with both interior surfaces; but lower-density fill such as glass fiber batt insulation can touch both sides without degrading the isolation. A single bridging nail may alter significantly the insulating efficiency. For maximum effectiveness, one of the wall surfaces should be hung structurally free at all four edges, with the boundary cracks sealed against sound leakage with an elastometric sealant such as silicone or acrylic latex.

Given the factors that make the performance of partitions deviate from the mass law, it is advisable to use one of the many tabulations of measured sound transmission loss when selecting a construction. Sound transmission loss of partitions is measured by comparing the average sound level on each side of the partition under standardized conditions per ASTM E90, "Standard Test Method for Laboratory Measurement of Airborne Sound Transmission Loss of Building Partitions and Elements." The frequency-dependent transmission loss data from this test procedure can be reduced to a single-number **sound transmission class (STC)** by means of ASTM E413, "Classification for Rating Sound Insulation." The STC most heavily weights the transmission loss data in the 500- to 2,000-Hz frequency band, and ratings for typical partitions are shown in Table 10.6.5. Additional data on transmission loss for a wide variety of building materials and structures are available in several publications (see References).

Transmission through composite walls can be approximated as an average intensity transmission coefficient

$$T_{\mathrm{av}} = \frac{1}{S_{\mathrm{tot}}} \sum_n T_n S_n \qquad (10.6.9)$$

where the sum is over all the different materials in the wall, T_n is the intensity transmission coefficient, and S_n is the surface area of the nth material. For air gaps $T_n = 1$ (perfect transmission), and so the presence of air gaps in walls should be removed as much as possible. For example, a 3-mm (⅛-in)-thick glass window offers 41 dB of transmission loss at 2,000 Hz. If the window is opened by 2.5 mm (0.1 in), the transmission loss is reduced to 26 dB.

Holes for piping penetrations should be slightly oversized to allow the penetration to be filled with fibrous insulation and to be sealed on both sides with an elastomeric sealant, to ensure that the transmission loss is not degraded by the presence of air gaps.

The benefit of shielding can be reduced by the presence of flanking paths, which are alternative ways for sound to travel from source to receiver other than the partition between them. Flanking transmission can occur when gaps are left around the perimeter of the partition. These gaps should be sealed with an elastometric sealant. Flanking transmission can also occur when lightweight partitions perpendicular to the separating partition are continuous across both rooms. The flanking noise issue can be addressed by cutting a joint in the flanking partition or by increasing the mass of the flanking partition so that its sound transmission loss approaches that of the separating partition.

Quieting

When a noise source is in a room, reflections of the noise from the wall (reverberation) can result in a higher sound level than would occur if the sound source were in free space. Placing sound-absorbing material on the interior walls will reduce the reverberation in the room and hence reduce noise levels. Sound absorption coefficients are the commonly used measurement of these panels. The sound absorption coefficient is the fraction of the total energy that reflects from a panel. The coefficient varies from 0 (for an open window where no energy returns to the room) to 1 (for a rigid wall where all the energy is reflected).

Table 10.6.5 Transmission Loss Data for Various Building Constructions
Data are provided at 7 different octave bands. The final column gives the STC rating.

Material	Transmission loss, dB						STC rating
	125 Hz	250 Hz	500 Hz	1,000 Hz	2,000 Hz	4,000 Hz	
Walls							
Monolithic:							
3/8-in plywood (1 lb/ft²)	14	18	22	20	21	26	22
26-gage sheet metal (1.5 lb/ft²)	12	14	15	21	21	26	20
1/2-in gypsum board (2 lb/ft²)	15	20	25	31	33	27	28
2 layers 1/2-in gypsum board, laminated with joint compound (4 lb/ft²)	19	26	30	32	29	37	31
Interior:							
2 by 4 wood studs 16 in oc with 1/2-in gypsum board both sides (5 lb/ft²)	17	31	33	40	38	36	33
2 by 4 staggered wood studs 16 in oc each side with 1/2-in gypsum board both sides (8 lb/ft²)	23	28	39	46	54	44	39
6-in dense concrete block, 3 cells, painted (34 lb/ft²)	37	36	42	49	55	58	45
3 5/8-in steel channel studs 16 in oc with 1/20-in gypsum board both sides (5 lb/ft²)	26	36	43	51	48	43	43
Construction no. 19 with 3-in mineral-fiber insulation in cavity	28	45	54	55	47	54	48
Exterior:							
4 1/2-in face brick (50 lb/ft²)	32	34	40	47	55	61	45
Two wythes of 4 1/2-in face brick, 2-in airspace with metal ties (100 lb/ft²)	37	37	47	55	62	67	50
Two wythes of plastered 4 1/2-in brick, 2-in airspace with glass-fiber insulation in cavity	43	50	52	61	73	78	59
6-in solid concrete with 1/2-in plaster both sides (80 lb/ft²)	39	42	50	58	64	67	53
Floor-Ceilings							
2 by 10 word joints 16 in oc with 1/2-in on floor side and 5/8-in gypsum board nailed to joists on ceiling side (10 lb/ft²)	23	32	36	45	49	56	37
4-in reinforced concrete slab (54 lb/ft²)	38	42	45	56	57	66	44
6-in reinforced concrete slab (75 lb/ft²)	38	43	52	59	67	72	55
6-in reinforced concrete slab with 3/4-in T&G wood flooring on 1 1/2 by 2 wooden battens floated on 1-in glass fiber (83 lb/ft²)	38	44	52	55	60	65	55
Roofs							
3 by 8 wood beams 32 in oc with 2 by 6 T&G planks, asphalt felt build-up roofing, and gravel topping	29	33	37	44	55	63	43
As above with 2 by 4s 16 in oc between beams, 1/2-in gypsum board supported by metal channels on ceiling side with 4-in glass-fiber insulation in cavity	35	42	49	62	67	79	53
Doors							
1 3/4-in hollow-core wood door, no gaskets, 1/4-in air gap at sill (1.5 lb/ft²)	14	19	23	18	17	21	19
Construction no. 43 with gaskets and drop seal	19	22	25	19	20	29	21
1 3/4-in hollow-core 16 gage steel door, glass-fiber filled, with gaskets and drop seal (7 lb/ft²)	23	28	36	41	39	44	38
Glass							
1/8-in monolithic glass (1.4 lb/ft²)	18	21	26	31	33	22	26
1/4-in monolithic float glass (2.9 lb/ft²)	25	28	31	34	30	37	31
1/4-in laminated glass. 30-mil plastic interlayer (3.6 lb/ft²)	25	28	32	35	36	43	35
Double glass: 1/4-in laminated + 3/16-in monolithic glass with 2-in airspace (5.9 lb/ft²)	25	34	44	47	48	55	45

SOURCE: Adapted from data published in M. David Egan, "Architectural Acoustics," McGraw-Hill, 1988, pp. 204-206.

Absorbing panels typically employ highly porous materials which are very effective at converting sound vibrations to heat by friction and viscous resistance. The effectiveness of sound absorbents varies with frequency, usually being greater for high and intermediate than for low frequencies. Sound absorption coefficients are typically measured in accordance with ASTM C423, "Standard Test Method for Sound Absorption and Sound Absorption Coefficients by the Reverberant Room Method," and specific normal acoustic impedance is measured in accordance with ASTM C384, "Standard Test Method for Impedance and Absorption of Acoustical Materials by Impedance Tube Method," or ASTM E1050, "Standard Test Method for Impedance and Absorption of Acoustical Materials by Impedance Tube Method." The absorption coefficient will depend on the impedance of the panel and the attenuation of the sound within the panel material. However, it is not a property of the material alone, but depends on the mounting of the test sample per ASTM E795, "Standard Practices for Mounting Test Specimens During Sound Absorption Tests." For example, a sound-absorptive material is typically measured with higher sound absorption coefficients when backed by a deep airspace than when tested lying against a rigid surface.

The principal material employed is acoustic tile, which can be obtained as prefabricated units for use on walls and ceilings. The tile typically has an outer layer that has been mechanically perforated and conceals a porous absorbing material underneath. Note that painting or sealing of acoustic panels can have a dramatic deleterious effect on the performance of the panels, as the holes that allow the sound to permeate into the absorbing fill are blocked. This can be of particular concern in environments such as hospitals where sealing of pores and holes is desired for sterilization reasons. An alternative to tile is to hang acoustic blankets, for example, hanging heavy drape curtains, on walls. Data on sound absorption coefficients for a wide variety of building materials are shown in Table 10.6.6.

Table 10.6.6 Sound Absorption Coefficients for Various Building Constructions
Data are provided at 7 different octave bands. The final column gives the NRC number.

Material	Sound absorption coefficient						NRC number
	125 Hz	250 Hz	500 Hz	1,000 Hz	2,000 Hz	4,000 Hz	
Walls							
Sound-Reflecting:							
Brick, unglazed	0.02	0.02	0.03	0.04	0.05	0.07	0.05
Brick, unglazed and painted	0.01	0.01	0.02	0.02	0.02	0.03	0
Glass, ordinary window	0.35	0.25	0.18	0.12	0.07	0.04	0.15
Gypsum board, ½ in thick (nailed to 2 × 4s, 16 in oc)	0.29	0.1	0.05	0.04	0.07	0.09	0.05
Gypsum board, 1 layer, ⅝ in thick (screwed to 1 × 3s, 16 in oc with airspaces filled with fibrous insulation)	0.55	0.14	0.08	0.04	0.12	0.11	0.1
Plywood, ⅜-in paneling	0.28	0.22	0.17	0.09	0.1	0.11	0.15
Steel	0.05	0.1	0.1	0.1	0.07	0.02	0.1
Venetian blinds, metal	0.06	0.05	0.07	0.15	0.13	0.17	0.1
Wood, ¼-in paneling, with airspace behind	0.42	0.21	0.1	0.08	0.06	0.06	0.1
Sound-Absorbing:							
Lightweight drapery, 10 oz/yd^2, flat on wall	0.03	0.04	0.11	0.17	0.24	0.35	0.15
Heavyweight drapery, 18 oz/yd^2, draped to half area	0.14	0.35	0.55	0.72	0.7	0.65	0.6
Thick, fibrous material behind open facing	0.6	0.75	0.82	0.8	0.6	0.38	0.75
Carpet, heavy, on ⅝-in perforated mineral fiberboard with airspace behind	0.37	0.41	0.63	0.85	0.96	0.92	0.7
Wood, ½-in paneling perforated 3/16-in-diameter holes, 11% open area, with 2½-in glass fiber in airspace behind	0.4	0.9	0.8	0.5	0.4	0.3	0.65
Floors							
Sound-Reflecting:							
Concrete or terrazzo	0.01	0.01	0.02	0.02	0.02	0.02	0
Linoleum, rubber, or asphalt tile on concrete	0.02	0.03	0.03	0.03	0.03	0.02	0.05
Wood	0.15	0.11	0.1	0.07	0.06	0.07	0.1
Sound-Absorbing:							
Carpet, heavy, on concrete	0.02	0.06	0.14	0.37	0.6	0.65	0.3
Carpet, heavy, on foam rubber	0.08	0.24	0.57	0.69	0.71	0.73	0.55
Ceilings							
Sound-Reflecting:							
Concrete	0.01	0.01	0.02	0.02	0.02	0.02	0
Gypsum board, ½ in thick	0.29	0.1	0.05	0.04	0.07	0.09	0.05
Plywood, ⅜ in thick	0.28	0.22	0.17	0.09	0.1	0.11	0.15
Sound-Absorbing:							
Acoustical board, ¾ in thick, in suspension system (mtg. E)	0.76	0.93	0.83	0.99	0.99	0.94	0.95
Thin, porous sound-absorbing material ¾ in thick (mtg. B)	0.1	0.6	0.8	0.82	0.78	0.6	0.75
Polyurethane foam, 1 in thick, open cell, reticulated	0.07	0.11	0.2	0.32	0.6	0.85	0.3
Seats and Audience							
Fabric, well-upholstered sets, with perforated seat pans, unoccupied	0.19	0.37	0.56	0.67	0.61	0.59	
Leather covered upholstered seats, unoccupied	0.44	0.54	0.6	0.62	0.58	0.5	
Chair, metal or wood seat, unoccupied	0.15	0.19	0.22	0.39	0.38	0.3	

SOURCE: Adapted from data published in M. David Egan, "Architectural Acoustics," McGraw-Hill, 1988, pp. 52-54.

Floors can also be considered in terms of quieting. A carpeted floor will provide improved absorption over concrete or linoleum. Carpeting will also have a positive effect on improving acoustic isolation of impact-generated noise to rooms below. Underlays made of felt, cork, or rubber can further improve the acoustic performance of the floor. Finally, absorption can be enhanced by the presence of furniture in the room; for example, upholstered chairs can absorb a nontrivial amount of sound in comparison to hard plastic chairs.

Active Noise Control (ANC)

Recent advancements in computational technology have allowed many practical applications of active noise control. Human interaction with transportation and equipment noise thus far has been the focus of this technology. As of today, successful global ("whole room") active control of **tonal disturbances** has been accomplished in aircraft, submarines, and automobiles. Civilian examples of this technology are currently limited, but include the Saab 340B aircraft and Honda hybrid-powered vehicles. Military examples abound. For nontonal, steady-state noise, such as that heard in an aircraft cabin at cruising speed, active noise control has been accomplished only for "local" or small regions near the ears of passengers, and is typically limited to less than 1 kHz in frequency. Examples include the ubiquitous ANC headsets. No published ANC system to date is capable of readily controlling transient, unpredictable noise. Transient noise would include human conversation, music, and impulsive noises. All existing control algorithms require a feed-forward sensor capable of predicting noise at the receiver and/or a feedback sensor capable of monitoring a steady-state noise.

Active noise control for low-frequency noise in ducts is possible, but operating conditions of active duct silencers are generally limited to flow velocities of 7.5 m/s (1,500 fpm) or less (2003 ASHRAE Applications Handbook, p. 47.18). To the best of the author's knowledge, there are currently no large-scale commercially available, off-the-shelf lines of active duct silencers. Any active noise control approach to ductborne noise control is likely to be a custom-made endeavor, more expensive than commercially available passive silencers.

ULTRASONICS

REFERENCES: Ultrasonic Cleaning, in "ASM Handbook," vol. 5, "Surface Engineering," 1994, pp. 44-47. "Nondestructive Testing Handbook," 2d ed., vol. 7, "Ultrasonic Testing," Birks and Green (eds.), American Society for Nondestructive Testing, Columbia, Ohio, 1991. Johnson, The New Wave in Acoustic Testing, *Materials World, Jour. Inst. Materials* **7,** 1999, pp. 544-546. "Power Ultrasonics: Applications of High-Intensity Ultrasound," Juan A Gallego-Juárez, Karl F Graff (eds.), Woodhead Publishing, 2015.

Sound waves in the frequency range from 10 kHz to hundreds of megahertz have numerous industrial applications. The majority of the devices operate above 20 kHz and are therefore ultrasonic. One motivation in the choice of ultrasonic frequencies is to avoid impact on human hearing. High-intensity waves are used to mechanically alter materials, such as ultrasonic cleaners or welders. Low-intensity waves are used for nondestructive evaluation or testing. The list of applications is large, and just a small subset will be described here. The Ultrasonics Industry Association (UIA) provides a referral website which lists the majority of the industrial applications of ultrasound: http://ultrasonics.org/referral.html.

High-Intensity Applications

Ultrasonic cleaners employ a water bath that is agitated with sound waves with frequencies normally around 40 kHz (the majority of devices employ frequencies in the range of 20 to 80 kHz). The agitation is generated by bonding piezoelectric or magnetostrictive transducers to the outer surface of the tank. Ultrasonic cleaners work through acoustic cavitation. Acoustic cavitation occurs when the negative pressure of the tensile phase of the wave is greater than 1 atm (100 kPa or 15 psi), a pressure relatively easily obtained in water, at which level the fluid will tear apart, creating a cavity or bubble. The cavity will grow during the tensile phase of the sound wave, but during the compressive phase of the wave it will be driven to collapse. The collapse is extraordinarily violent with pressures inside the bubble reaching hundreds of atmospheres and temperatures reaching 5,000 K (9,000°R). These phenomena are localized to a small volume around the collapse but result in sonochemistry (e.g., free-radial production) and sonoluminescence (production of light). When the cavity collapses near a surface, the bubble collapses asymmetrically and a microjet of fluid pierces the cavity and strikes the surface at velocities in excess of 100 m/s (330 ft/s). These microjets act as intense microscopic scrubbers to remove particles from the surface.

Typically a detergent will be added to the water bath of an ultrasonic cleaner. The detergent allows the water to fully wet surfaces of the component that may be hydrophobic. For cavitation to be effective, it is important that the entire surface of the component be fully wetted. Sometimes solvents may be added to the water to chemically enhance the cleaning process. Most companies that produce ultrasonic cleaners also market proprietary ultrasonic cleaning solutions. Ultrasonic cleaners are particularly useful for cleaning objects with odd shapes and hidden structure. Types of objects that can be cleaned include metal parts (e.g., castings, valves, injectors), electronic components (particularly connector pins), printed-circuit boards, lenses, optical fibers, television tubes, and filters. This list is far from exhaustive.

For typical sized tanks (a few gallons) standing waves can be excited which can result in hot and cold spots in the tank. This is so because the speed of sound in water 1,500 m/s, and so the wavelength at 40 kHz is 150 mm (5.9 in)—on the order of the tank. Modern ultrasonic cleaners will normally employ a swept frequency signal which, even for variations of just a few kilohertz, will ensure even distribution of cavitation throughout the tank. Non-immersion cleaning is possible using a jet of water direct to the surface to be cleaner. Ultrasound is propagated along the jet of water and cavitation is generated at the surface which effects the cleaning process.

Ultrasonic welders are employed to bond materials together. They are most commonly used on plastics, but metal, textiles, films, and other materials can also be successfully welded. Ultrasonic welding is accomplished by placing the two pieces to be welded in direct mechanical contact. One piece is held stationary, and then an ultrasonic probe is brought into contact with the second piece. When the probe is excited, it vibrates the piece that it is in contact with, which produces frictional heat at the contact point between the two pieces. This, in turn, results in a temperature rise that softens or melts the plastic on both surfaces. The heated plastic interface melts, and when the probe is removed, the plastic cools and a homogeneous solid weld results.

The ultrasonic probe consists of a stack of piezoelectric discs which are operated at frequencies varying from 20 to 40 kHz, although research devices operating above 100 kHz have be reported. The stack is mounted to a tapered horn (or sonotrode), and as the sound energy propagates down the horn, the vibration amplitude is amplified because of the reduction in cross-sectional area of the horn. Vibrations at the tip are on the order of 100 μm. With ultrasonic welding it is possible to achieve spot welds, lap welds, butt welds, and steam welds.

Ultrasonic Atomization When ultrasound in air is incident on the surface of water, it will excite capillary waves on the surface of the water. When the amplitude of the capillary waves reaches the critical value, the waves become unstable and droplets of water are ejected into the air. The droplets are typically tens of micrometers in diameter and appear as a mist above the surface. Under the right conditions an acoustic fountain can be generated. This type of ultrasonic atomization is commonly employed in dehumidifiers. Nozzles have been created by using this principle to produce an atomized jet.

Sonochemistry and Bioremediation

High powered ultrasound can be employed in many chemical engineering processes. The high-pressures and temperatures resulting from ultrasonically driven cavitation bubble collapse can enhance many chemical reactions, either through directly through the changes in pressure and temperature or by secondary means, such as the production of hydrogen peroxide or free-radicals. With the right conditions reaction rates are often reported to be enhanced by thousand-fold when ultrasonic cavitation is present and in some cases increases in rates in excess of a million-fold have been reported. Ultrasonics can also be used for material processes, such as, the production of amorphous metals (which is likely due to high rates of temperature change) or fusing of particle which may be due to the violent jetting process. In a related area ultrasound is finding many application in bioremediation, either treating drinking water or processing waste-water. Ultrasound can act by directly degrading harmful pollutants, such as, pentachlorophenol (PCP), or killing bacteria, such as *E. coli.* Ultrasound can also act by enhancing chemical reactions that breakdown contaminants, reducing the growth of algae, and helping to break up biofilms.

Low-Intensity Applications

Nondestructive Evaluation or Testing of Materials This application is perhaps the most widespread use of ultrasound in an industrial setting, and the most common manifestation is a pulse-echo configuration. An ultrasonic transducer is used to launch a short pulse into a sample, and whenever the pulse interacts with an interface, it is partially reflected and the echoes can be detected by the same transducer that generated the pulse. If the sound speed in the sample is known, then the distance to the interface can be determined from the time of flight. Common applications employ piezoelectric transducers with center frequencies that range from 1 to 30 MHz. The transducers are typically wideband, and so when excited with a short electrical pulse (with an amplitude on the order of 100 V), they generate an acoustic pulse that is about 1.5 cycles long. The spatial length of the pulse governs the spatial resolution of the system. At 1 MHz the pulse length is about 11 mm (0.45 in) in steel and 4 mm (0.16 in) in Lucite. At 30 MHz it is 0.4 mm (0.016 in) in steel and 0.13 mm (0.005 in) in Lucite. High frequencies therefore give better spatial resolution but typically are absorbed more strongly by the sample and so cannot "see" deeply into the material. Frequencies of hundreds of megahertz have been employed for specialized high-resolution applications.

If the transducer is focused, it will generate a pencil beam of sound, and from the received echo signal (known as an A-line) it is possible to identify the position of interfaces to within the resolution of the system. The transducer can be scanned in a line to produce two-dimensional images (a B-scan) or in a two-dimensional raster pattern to produce a three-dimensional image (a C-scan). The position of interfaces can be used to size an object or to detect the presence of flaws or inclusions. Array transducers can be used to electronically scan the beam rather than relying on mechanical translation. The same principles described here for materials testing are used in diagnostic ultrasound imaging of the body.

Other modes of ultrasonic testing include through transmission and resonant ultrasound. In through transmission, two transducers are used in a pitch and catch configuration. The amplitude of the received signal will depend on the number of interfaces and flaws in the intervening material. Through transmission can also be used to ascertain flow velocities through the Doppler effect. In this case a short tone burst (approximately 10 cycles) is transmitted through the flow. If there is a net flow from the transmitter to the receiver, then the frequency of the tone will be shifted higher; if there is flow in the opposite direction, the frequency will be shifted down. Based on the frequency shift the flow velocity can be calculated. In **resonance ultrasound spectroscopy (RUS)** a transducer is used to excite a resonance in the component to be investigated, and the frequency and Q of specific resonances can be used to extract information about the structure and flaws in the components. A new and promising development for flaw detection in components is **nonlinear resonant ultrasound spectroscopy (NRUS)** and variations thereof. In this case, the ultrasonic excitation is of large enough amplitude that the material exhibits nonlinear response, for example, the production of harmonic overtones of the driving signal. The nonlinear response is critically dependent on flaws in the material and is more sensitive to flaws, in particular microcracks, than RUS in most structural materials.

10.7 ROBOTICS, MECHATRONICS, AND INTELLIGENT AUTOMATION

by Andrew A. Goldenberg

OVERVIEW

Robotics

Based on the single-axis control system design methodology, well established by mid-60, the development of machine tool automation began based on numerically controlled machines (NC). The NC machines were used to perform unmanned operations of cutting, lathing, drilling, polishing using primarily 2D positioning devices. The devices were later integrated into three axes of motion (3D), commonly perpendicular to each other. Each axis was driven by a standard single-axis servomechanism. The NC machines were programmed using logic circuits to coordinate all axial motions. The NC machines could generate profiles in 2D and 3D.

The typical single-axis servomechanism was operating using hardware (analog) and software (logic circuits). The underlined control system design was based on PID controller, and later on optimal and adaptive control. The coordination of the motion of two or three axes was realized by effecting simultaneously very small steps of motion of the axial servomechanisms. The frequency of these steps was selected to correspond to the required speed of motion. The motion coordination was provided by ensuring that all axes start and stop at the same instant of time. The arbitrarily small steps of motion were limited only by the servomechanism ability to monitor its own axial motion in real-time through feedback, thus high-precision motion was generated. In modern machining the motion repeatability can be guaranteed to one thousandth of an inch, or 0.00254 mm.

The control of NC machining along with design, programming, and the language used for programming have dominated the industrial automation sector until mid-60s. At that time it was recognized that the operations of the NC machines are limited to standard machining operations. A task that would require coordination of three or more axes of motion, including translational and rotational motions, could not be performed. At that stage the notion of articulated motion, that is the handling of a tool, tracking an arbitrary 3D path with the tool, and also orienting the tool arbitrarily in 3D space became intriguing enough that the impetus started shifting from the NC machines to new devices that could be programmed to deliver the tool motion over an arbitrary path and with an arbitrary orientation.

The first step in this quest was the emergence of Computerized Numerical Control (CNC) machines. Yet, they were limited in the generality of the profiles to be machined. This generated the birth of robotics as multi-degree of freedom devices with a capability similar to that of CNC machines, but with a higher number of active axes moving based on coordinated motion plans. The purpose was to handle and operate a tool over a workpiece while following complex paths (contours).

The robotics provided the domain of development of radically more advanced motion mechanisms. Furthermore, the advent of microelectronics provided new technology of hardware and software that became the underpinning of the robotics proliferation in industry.

The domain of modern robotics can be divided in two categories: (i) stationary robot arms; and (ii) mobile robot arms. The latter is based on mobile platforms that is currently an epicenter of the modern robotics. In both categories the arms may be the same or different depending on applications. We are addressing herein robot arms in general regardless of mobility, and we also address mobile robots separately focusing on their main feature, the mobile platform.

Mechatronics

Mechatronics is a multidisciplinary field that is the synergistic combination of mechanical systems, electronics, sensors, actuators, and software. Mechatronics is in essence the integration of those areas. In practice as well as education the division along the classical disciplines is no longer useful, as this branch of engineering generates the most appropriate infrastructure for interdisciplinary work. Integration must be done "concurrently," unlike the classical design whereby the work in each domain is done in solitude, and the integration is conducted as the last stage before testing and verifying the whole system. In mechatronics integration is elevated to the position of the most relevant stage of the development.

Mechatronics identity as a collection of disciplines requiring the formulation of a common approach to development of automated systems is a novel but resurrected way of providing education. A mechatronics engineer is expected to be proficient in mechanical design, electronics design, software development, testing, and integration. It provides the best opportunity for a true technology-oriented training of systems engineers. Whereas the classical version of system engineering was limited to concepts and system specifications, the modern interpretation is also embedding the subsystem development in the context of system design. In other words, there is less of a hierarchy, or vertical integration, and more of an equally-valued team work, or horizontal integration. Furthermore, the mechatronics engineering is inherently interdisciplinary.

Intelligent Automation

Originating from the fields of Robotics, Mechatronics, and Automation, the research and industrial communities have generated a new term of reference: "Intelligent Automation." There have been several attempts to define this new entity formally; however, the results have been unsatisfactory as the terms of reference are ambiguous and there is significant overlapping of the predecessors of Intelligent Automation, the Robotics and Mechatronics.

A close and detailed investigation of systems that are classified as Robotics or Mechatronics would reveal that the boundaries between these fields are not at all easily and clearly defined. Based on common fundamental physics and engineering science principles that provide the underpinning of these fields, it could be argued that a complete abolition of the need to define them separately is in order. However, the research community seems to be more inclined to maintain distinctions between these domains, whereas the practical/professional domain is less concerned with the definitions as long as the systems that are created work and fulfill required specifications.

Intelligent Automation could be understood as an extension of the Mechatronics and Robotics. However, there is a great deal of uncertainty as to where systems that could be defined as Robotics or Mechatronics begin their journey as "Intelligent." Some serious arguments have been made to distinguish between the mechanical design of ordinary Robotics and Mechatronics and that of Intelligent Automation systems. It was found that the inclusion of new technology such as micro and nanotechnology could provide the way to render them intelligent. Other attempts focused on software. It is common to consider adaptive and learning systems or fuzzy-logic based expert systems to provide for the intelligence in the new domain.

In the section "Typical Application Cases," examples of robotic and mechatronic systems are given. No attempt was made to underline their "intelligence" although they could be considered "practically intelligent."

TYPICAL APPLICATION CASES

In an attempt to illustrate technologies of Robotics and Mechatronics and their emerging cross-product of Intelligent Automation, a number of examples of niche applications are presented in this section. The purpose is to facilitate the understanding of these fields via illustrative examples instead of taking the usual route of presenting the underlining theory and tools that support the design and development in these areas. The examples would elucidate on couple of issues: (i) complexity of the design of some Robotics and Mechatronics applications; and (ii) proliferation of Intelligent Automation in new areas that have not been addressed so far.

Robotics

Robot Arms

Stationary Robots Robot arms applications include industrial operations such as welding and painting in car manufacturing, pick-and-place of parts and subassemblies, parts mating and assembly in electronics industry, machine shop operations, etc. These applications are based on stationary robotics. In the late 80s, mobile robotics have emerged mainly as remote controlled devices for operations in hazardous environments. The mobility issues render the robotics as more complex systems, and their integration in automatic systems became dominant.

A related domain that has been gaining popularity in the last 10-15 years is "modular" robotics. They are ordinary robot arms in terms of functionality, but are offered in kits including joints, links, and controllers, and are differentiated by providing ways of reconfiguring the robots arms for a wide variety of tasks.

The most widely known application of robotic arms is in the space sector. The most common use is in car manufacturing (e.g., welding, painting). Applications in car manufacturing are well documented in the literature from the late "70s to mid-80." Applications in electronics assembly of components and subassemblies are also growing, increasing the number of industrial robots in use today. Applications in the electronic industry are well documented in the later 80s publications.

To illustrate the use of robotics, we are presenting a number of niche applications. The purpose is to highlight the use of robotics in areas that are less publicized. These examples could provide an impetus for consideration of robotic-based applications in new areas. There are no reasons for limiting the proliferation of robotics in all sectors of the society, but the technology is not sufficiently mature, thus development and use of robotics in some areas is not economical.

Long Reach Manipulator Arm

In nuclear reactor installations there is a need to inspect the structural integrity of supports of critical subsystems, such as heavy water boiler. Such system is located in closed vault that is not accessible by humans. The walls have a thickness of 7' to ensure that no radiation emanates from the vault. Any repair and maintenance operation inside the vault can be done with long-reaching tools that are transferred to the vault through few cylindrical openings whose diameter is maximum 10".

In order to perform operations in the vault through the available openings, specially-designed long reach arms are required equipped with suitable tools. An array of CCTV cameras are also required, mounted on the arm, to provide views of the vault interior, in order to help the operator guide the robot arm by remote control.

A long reach manipulator arm was designed for this purpose (Fig. 10.7.1). The robot is modular in the sense that the long reach arm is formed by a tandem of short linkages that are coupled together to provide different lengths for various tasks and sizes of tools. The maximum length is 21'. As the robot has to pass through 7' thick wall and it has a base length outside the vault of about 2' it provides 12' inside the vault. The tools attached to the arm include various grippers, cameras, grinders, water blasters, and other modules.

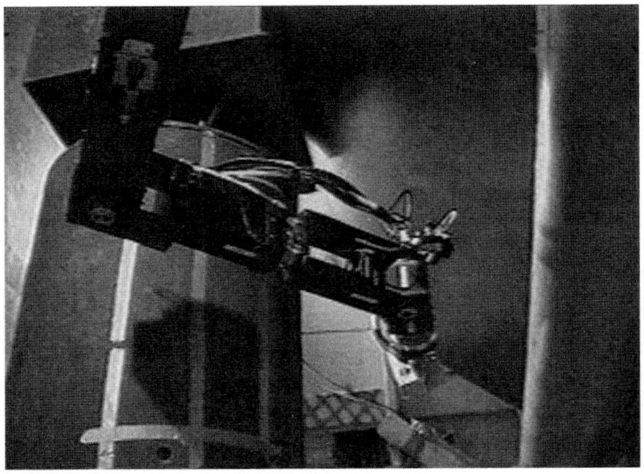

Figure 10.7.1 Arm carrying ultrasonic equipment [1].

The manipulator is used to carry ultrasonic equipment for checking the structural integrity of some of the support columns. The robot with the sensors attached is inserted into the vault. The robot has to perform two operations. One is to put a rubber cap to an opening at the bottom of the column. This allows filling with water the cylindrical cavity surrounding the support column; it is necessary for ultrasonic inspection. The task of inserting the cup to close the cavity hermetically is very tedious. The operator requires hours of training in order to perform this task. Six cameras are mounted on the arm to help guide the operator movements. After the cavity is sealed, and water is filling the cavity, the actual inspection took place.

In terms of arm reach, the bottom of the column is the furthest point to be reached by the robot tool. The operation of inspection also requires the robot to traverse the surface of the column, starting from the bottom, and advancing toward the top. At least four columns need inspection for every boiler.

Although the portion of the arm inside the cylindrical tunnel is supported by the wall structure, the 12' of arm inside the vault is in effect "nearly cantilevered" to the inside edge of the cylindrical opening. However, this cantilever is not rigid, as the arm is free to move around the pivot point at the edge of the opening. This presented a formidable challenge for the designers as well as operators. The designers have oriented each coupler attaching two adjacent links in such a way that the moments around the axis of the coupler would be balanced by the structure, thus rigidizing as much as possible the arm against structural vibrations caused by the payload and arm links mass moments of inertia. The result is that the robot arm is stiff, and the operator can control very well the tip of the tool to perform the dexterous operations of inserting the plug and carrying the ultrasonic equipment at a fixed distance from the surface of the column.

MODULAR AND RECONFIGURABLE ROBOT

Industrial robots are used in many sectors of manufacturing and service industries. Other sectors are laboratory automation, mobile robots for security, and aerospace. In these applications the robots have standard configurations, that is, articulated, gantry, SCARA, cylindrical, etc. They possess any number of joints usually between 3 and 7.

As the mechanical structure of the robot is fixed, the acquisition process must take into account current user's requirements as well as future ones in order to ensure that the robot can perform required future tasks. Often the result is that the acquired robot may have excess joints [degrees-of-freedom (DOF)] than necessary for the task at hand, or the supplier cannot provide a suitable configuration without exceeding the minimum number of required joints.

As a solution to this problem, technology developers have introduced the concept of modular robots. These robots can be assembled using a set of joints and links modules. Commonly the joints are either rotary or linear, and the links have a multitude of cross sections and lengths. These robots can have one, two, three, etc. joint arms depending on the application. Implicitly the cost will be determined by the number of joints that are acquired. Thus, the user can trim the cost to the minimum necessary. In addition, this concept allows acquiring additional joints at a later stage in the event that a new task is being automated that may require higher complexity of motion.

There are several available products based on this technology. Some providers are limiting their scope to strictly mechanical modularity. Others are also introducing modularity in the computer architecture and embedded electronics. More recently software has been also modularized to allow independent integration of joint modules consisting of electromechanical structure (motor, drive, gear, encoder, brake, torque sensor, and embedded controller) that can be easily coupled with links, thus forming an open or closed loop kinematic chain. One implementation of this technology is shown in form of modular joints in Fig. 10.7.2.

Robot Arms Mounted on a Carrier

SPACE SHUTTLE REMOTE MANIPULATOR SYSTEM

The Shuttle Remote Manipulator System (SRMS) or "Canadarm" was a joint venture between the governments of the United States and Canada

Figure 10.7.2 Modular and Reconfigurable Robot Joints (MRR) [1].

to supply the NASA Space Shuttle program with a robotic arm for the deployment/retrieval of space hardware from the payload bay of the orbiter (the Shuttle) shown in Fig 10.7.3a.

A schematic view of the SRMS is shown in the Fig. 10.7.3b. The SRMS is a robotic arm consisting of a shoulder, elbow, and wrist separated by an upper and lower arm boom giving it a total of six DOFs (shoulder pitch and yaw, elbow pitch, wrist pitch, yaw and roll). At a total weight of approximately 431 kg, the SRMS is capable of maneuvering payloads of up to 14,515 kg at a rate of 0.06 m/sec with a maximum contingency operation payload weight of 265,810 kg. Under unloaded conditions the SRMS can achieve a maximum translational rate of 0.6 m/sec. However, the SRMS is incapable of supporting its own weight on earth. It must be supported by specialized ground handling equipment during its testing and packaging for shipment.

(a)

Figure 10.7.3a SRMS - Canadarm.

Although the SRMS can handle very heavy payloads, movement of the tip is very accurately controlled, allowing precise handling of delicate payloads. The length of the SRMS is approximately 15 m. A control system is used to deploy payloads in automatic mode to a positional accuracy of ±2.0 in and ±1.0 degree of a pre-programmed target point and orientation at the aforementioned rates and load conditions. The SRMS may also be operated manually (remote control) by the astronauts to the same accuracy with the use of hand controllers and closed circuit televisions (CCTV) mounted on the manipulator arm. The SRMS was designed to have a life of 10 years or 100 missions.

HYDRAULIC ROBOT ARM FOR HIGH PAYLOAD HAZARDOUS TASKS

Utility companies must provide tree trimming in the vicinity of live aerial electrical distribution lines in order to avoid outages due to ruptured cables which are caused by erratic motion of tree branches during storms and other natural causes. The operation of tree trimming is usually done with hydraulic wood cutting tools by a trained operator stationed in a bucket mounted at the end of an "aerial boom" of approximately 45' in length. The boom is a two-link hydraulic arm with up to four joints that is mounted on a specially retrofitted truck. Two sets of controllers for the aerial boom are used in parallel: one in the operator bucket, and one on a panel aboard the truck, where a second operator (usually the driver) provides guidance and support from the ground. The operator has the ability to lift the bucket to a height of approximately 45.

The operator carries in the bucket a set of tools: circular saw, linear saw, grappling hook, etc. The operator can connect one tool at a time to the source of hydraulic power available on the bucket. The tools

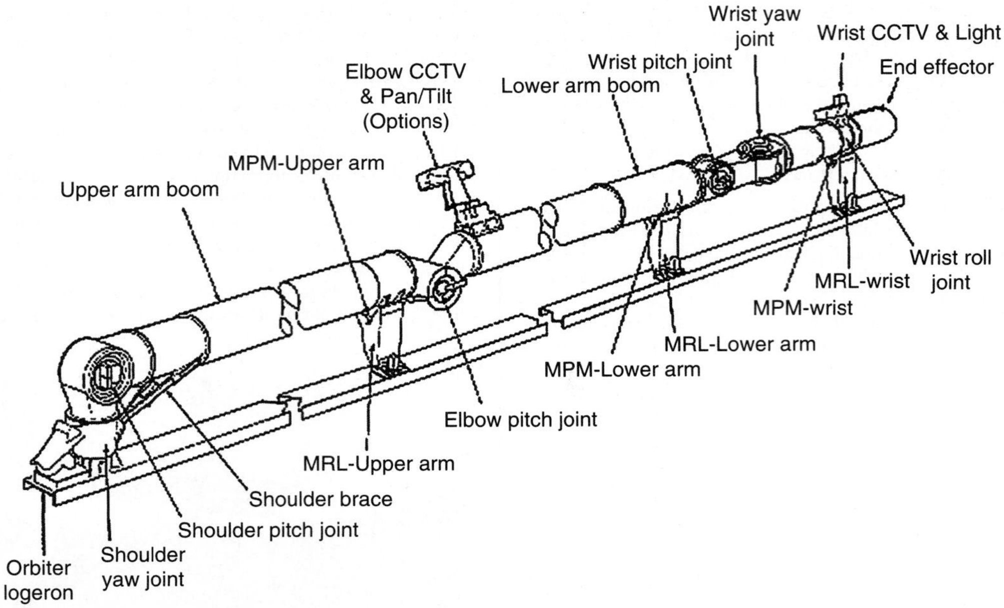

Figure 10.7.3b SRMS mechanical subsystem (*Spar Aerospace Ltd., Toronto, Ontario*) [1].

are hand-held, and the operator can perform the cutting operations as needed.

The operations above-mentioned have been proven to be dangerous for workers. Occasionally the operator's tool touches live cables, and due to grounding of the boom and bucket, serious injuries of the operator result. There are also cases where the bucket, the upper boom, or the operator come into contact with the cables. It should be noted that in the majority of cases the cables carry 22.5 KV. While there is insulation, and grounding is assumed to be avoided, there are instances where the configuration of the system and the work of the operator generate electric shocks to the operator.

In response to concerns for worker safety, a remote master-slave hydraulic manipulator was developed (Fig. 10.7.4) with the intent of relocating workers away from the hazardous tasks. The operator is located on the truck platform away from the hazard, and is provided with a hand-held controller (the master) control the slave (a new five joint hydraulic manipulator) that is attached at the end of the aerial boom, replacing the bucket where the operator used to be located. To control the slave the operator uses a second hand-held controller that is a replica of the slave. An assortment of hydraulic tools can be attached to the slave to enable its use in a variety of high-risk tasks.

Figure 10.7.4 Slave manipulator mounted on the aerial boom [1].

The slave is a five DOF hydraulic manipulator arm, and the master is a five DOF electric arm shaped as a reduced-size replica of the slave. The master is instrumented to provide the operator with torque feedback with respect to two axes of rotation. This allows the operator to feel the forces and moments along and about certain directions at the contact between the tool and branches. Such capability could be extended to a complete (six DOF) force and moment reflection onto the master. The communication between the master and slave is via a fiber-optic cable to ensure electric isolation between the tools and the ground equipment.

Robot Arm for Inspection and Repairs of Nuclear Reactors

At the core of a CANDU (CANada Deuterium Uranium) nuclear reactors is the Calandria vessel. It contains a network of horizontal tubes for fuelling the reactor. After several decades of service these fuelling tubes must be replaced in a process known as "re-tubing" of the reactor. It is of utmost importance that during this operation no debris or foreign matters remain inside the Calandria.

With robots for such operations the user can pick up and remove any debris, and can also inspect the interior of nuclear reactor core (Fig. 10.7.5). A robot systems for operation inside the Calandria vessel is used for both visual inspection and hardness inspection. The robot is also a contingency tool to be used to collect foreign material created during the re-tubing process. The robot is inserted through a lattice sleeve tube in the shielding wall of the reactor after the fueling and pressure tubes have been removed.

The Calandria Vessel Inspection (CVI) robot consists of a long two joints boom with a manipulator arm attached to its end. The robot is equipped with radiation-hardened camera system for visual inspection and guiding of the robot arm, a vacuum nozzle for removing of small shavings and dust-like debris, and a gripper for removing of larger items—up to approximately 1 kg. The arm is fitted with an ultrasonic hardness tester.

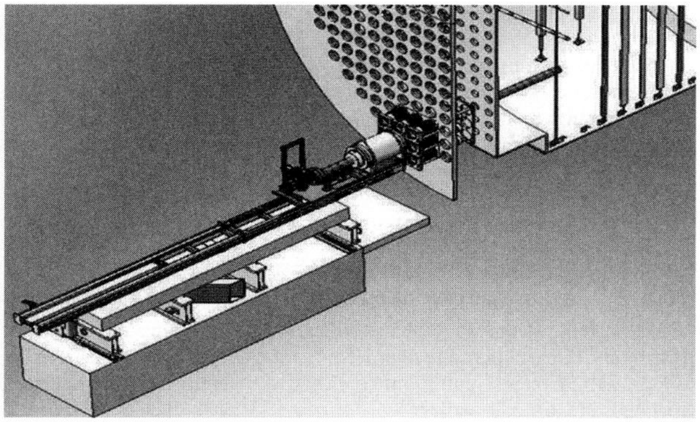

Figure 10.7.5 Schematic view of robot in operation [2].

The arm is constructed of radiation hardened material and components. When in operation, the tool built-in shielding mitigates and essentially eliminates the "open channel of radiation" that is inherent when accessing the internal portions of the reactor. The arm comes with a modular end-effector that can be used to pick up small debris visually located during the inspection process.

The robot enters the Calandria vessel through any of the fuel channels. Servo-controlled boom extension and roll combined with robot manipulator elbow rotation and front link extension allow the end-effector to reach any point within the Calandria vessel. The robot arm is controlled by a combination of electric and pneumatic actuators that incorporate force control and position feedback. The elbow joint is driven by a radiation hardened stepper motor. The forearm is extendible with a pneumatic cylinder through a range of 400 mm. The position of this link is monitored with a custom magnetic encoder. The end-effector is comprised of a pneumatic gripper and a vacuum nozzle. Smaller debris are removed by vacuuming; larger debris can be picked up with the gripper and dropped into a shielded flask mounted directly in front of the tube sheet. The boom is designed to provide radiation shielding during the operation of the robot system.

Mobile Robots

Mobile Robot for Explosive and Ordnance Disposal Police, Military, Emergency Response Team (ERT), Fire, Nuclear, and other hazardous response personnel require remotely controlled equipment for stand-off in dealing with explosives, ordnance, and other hazardous materials. The remote controlled equipment consist of a mobile platform, video equipment, a robot arm mounted on the platform, and a series of auxiliary operational tools such as firearms, disrupters, X-ray units, etc. and associated means of attaching them to the arm or platform. In addition, such a system includes an operator control unit (OCU) that includes cable and wireless communication with the platform, video monitor, means of controlling the platform and arm, as well as activating the operational tools. The communication covers data, video, and audio links.

The platforms and arms come in various sizes. The platforms are usually anywhere from 60 to 135 cm in length, and from 45 to 66 cm in width. The arms can have a reach from 0.8 cm to 2-3 m. The payload capacity varies from 5 kg to over 75 kg depending on the configuration of the arm. The total weight varies from 25 kg to 350 kg. The speed of the platform varies from 2 km/h to more than 10 km/h.

This class of robots is subdivided in three major groups: small, medium, and large. There are commercial suppliers for each category. The small category is mainly used for reconnaissance, and surveillance. It is usually not used with an arm. The large category addresses large payloads, ammunition firing, and other hazardous tasks involving heavy payloads. The medium category is used for a mix of operations, some pertinent to the small robots, and others to the large ones.

These mobile robots could be used for surveillance, reconnaissance, handling of hazardous items, manipulation of suspected packages, neutralizing and handling such items as Improvised Explosive Devices (IED's), Hazardous Chemicals, and Radioactive Materials. They are all-weather all-terrain mobile robots that can be used indoor (buildings, public institutions, airports, homes, etc.) and outdoor (terrain cluttered with obstacles, ditches, gravel, snow, mud, etc.).

The mobile robots are fitted with different means of traction: some use only wheels, other use tracks, yet other use quick-removable tracks that can be mounted or dismounted very easily. The tracks, permanent or removable are necessary for climbing stairs and obstacles.

Other features are related to precise independent joint control, high dexterity of arm (Fig. 10.7.6), long reach, high payload, open computer architecture for integrating bio-chem sensors, and options for autonomous navigation. In terms of control, the operator has the option to control the arm in joint or task space. The platform is always controlled in task space, by coordinating the motion of the wheels. Typically the platform has two motors, actuating two wheels, one on each side. The other wheels are rotated by chain or other means of transmission.

Figure 10.7.6 Mobile robot with arm deployed [1].

In terms of mobility the platform is equipped with 4 or 6 wheels, and a set of fixed or removable tracks. The tracks are used for rough terrain and stair or slope climbing. The robot is able to turn-on-the-spot and maneuver precisely in an obstacle-cluttered area. The drives also feature a drive-hub, which can disconnect the drive from the wheels to facilitate manhandling of the robot if necessary.

The robots are remotely controlled by a wireless or cable link; cable link has an on-board automatic winding cable system, or an independent reeled cable system. Two-way data and audio and one-way video links are usually available. Up to four cameras and a microphone provide the operator with images and sounds of the environment. A surveillance camera can be mounted on the arm (low to the ground, high in the air, close or far from the gripper) or on a separate articulated boom (usually positioned vertically up).

Mobile Robot for In-Ground Mine Detection An articulated remote controlled mobile manipulator has been developed to scan autonomously off-road and unstructured terrain for buried landmines and explosives. The robot consists of a dual-arm manipulator mounted on a six-wheel mobile platform, mine or explosive detector, range sensors, terrain mapping software, and hardware and software for communication, data transfer and control. The platform can be used also with tracks mounted over the wheels. The tracks help climbing steep slopes, and moving over very rough terrain. The platform can be operated in remote control, semi-autonomous, and autonomous modes. The terrain map is generated with respect to a range sensor frame to minimize the computational load in real-time. Sensors scan and measure the range with respect to the terrain and relative to the reference coordinate frame. Also, the normal (perpendicular) direction to the terrain is computed in real-time. Based upon the terrain map that is generated in real-time, the mine detector's position and orientation is controlled such that a desired (constant) separation and orientation with respect to the terrain is maintained.

One of the dual-arms [Detector Arm (DA)] holds the mine or explosive detector (e.g., metal detector, Ground Penetrating Radar source, or a nuclear quadrupole resonance instrument) that is manipulated autonomously. A plurality of mine detectors can be mounted on the DA. The autonomous motion of the DA is synthesized and controlled on-line based on the 3D map of the terrain. The map is generated in real-time by fusion of sensory data acquired from a scanning laser rangefinder, an array of four ultrasonic sensors, and joint encoders of the dual arm.

The rangefinder, a motorized scanning mirror, and ultrasonic sensors are mounted on the other articulated arm [Sensor Arm (SA)]. The SA has a pan joint allowing to be positioned relative to the DA such that it always scans "ahead" of the DA (Fig. 10.7.7).

The DA holds and maintains the detector at a desired constant vertical distance from the ground irrespective of undulations of the a priori unknown terrain profile. As well, the DA can maintain the detector orientation constant with respect to the scanned terrain.

Mechatronics

This section presents a few interesting examples of mechatronics systems. Examples cover meat grading, tire, utilities, biotechnology, and security and defense sectors. The cases represent niche applications that may inspire considerations of new applications. The main motivation in developing such systems was the need to improve the execution, economics, and quality of human life. The systems presented are focused on hazardous environments where the human may be exposed to danger.

The example from the meat processing industry is representative of situations where the human working conditions generate a reduction in human capability. The example from tire industry illustrates a situation where the job difficulty may lead to poor job quality. In both cases the result of the human performance deterioration made it imperative to consider some form of automation of the tasks performed.

Figure 10.7.7 Mine detector arms illustrating detector arm (DA) and sensor arm (SA) [1].

The example from the domain of utilities provides a number of opportunities to consider automation. The provided example reports on the use of automation to repair the underground gas pipe lines. Such operations involve serious disruptions because of excavations, hazard to human operator due to gas leaks, and major costs that lead to increases in the price of gas distributed to consumers. All these factors could be alleviated by automation.

The examples from laboratory operations illustrate higher quality and efficiency of operations attained through the use of automation.

Robot System for Tire Casing Analysis

Truck tires wear and tear generate a need for replacing them as often as needed. The industry of used tire re-treading is growing for a number of reasons, in particular due to environmental concerns. Tire re-treading is a growing business that requires effective detection of used tires that are candidate for re-treading.

An automatic robotic-based tire casing analyzer has been developed. It detects defects and wear. It marks the locations of defects, and it provides a report on the state of the tire. The operator then decides if the tire could be re-treaded. For many years, the tire industry has been searching for a nondestructive, simple way to inspect tires and tire casings for flaws. Good casings are presently being discarded and disposed of in landfill sites, causing a detrimental effect on the environment. Or, tires are often prepared for re-treading only to discover that the casing has major irreparable flaws. The new technology eliminates these problems resulting in significant economic benefits. The technology is based on an intelligent controller that uses two robotic systems and carries 28 ultrasonic sensors. The controller provides autonomous positioning and guidance of the sensors which are delivered into the tested tire by a very compact foldable robot. The sensors manifold conforms to the interior shape of the tested tire. Based on the ultrasonic signal, the sensors detect and evaluate defects in the tire. An intelligent algorithm has been developed to evaluate defects, classify them according to size, shape, etc., and graphically zoom in on a detected defect to be displayed on a computer screen monitor.

The robot arms and sensory arrays have been specifically designed to handle a variety of small and large tires. A sensor protection mechanism prevents collisions of delicate sensor arrays with nails that may have penetrated the tire walls. Special low-noise material is used to achieve quality ultrasonic inspection.

The robotic arms are performing the ultrasonic inspection from within the cavity of the tire (carrying the emitters) and over the external surface (carrying the receivers) (Fig. 10.7.8*a*). Loading and unloading of tires is done automatically by the operator using a lifting device.

After loading, the unit starts operating by pressing a set of disks onto the sides of the tire to enclose its inner cavity. Then a foldable robot

(a)

Figure 10.7.8*a* View of internal and external robot arms carrying ultrasonic sensors [1].

arm that was stowed inside one of the disks gets deployed internally. The robot arm carries a manifold of ultrasonic sensors that have three segments. The segments get deployed in such a way that the center segment faces the inner side of the tire tread, and the two other face the inner sides of the tire. The manifold positions itself at a suitable distance from the internal surface of the tire to unable ultrasonic waves to penetrate and reach the receivers that are connected on an external manifold mounted on the other robot arm external to the tire. Emitters and receivers are automatically aligned for maximum efficiency. The operation of the ultrasonic sensors, including the positioning and alignment is fully automatic.

After both the emitters and receivers were positioned accurately, the tire starts rotating at a pre-determined speed to perform the inspection (Fig. 10.7.8*b*).

(b)

Figure 10.7.8*b* General view of the tire grading system [1].

Robot for Internal Operations in Underground Gas Pipes

Underground cast iron gas pipes are made out of 12' long sections connected by bell-and-spigot joints. A joint is generally filled with jute packing and sealed with lead. The joint can develop leaks over time and must be repaired. One current most frequently done repair procedure is to inject an anaerobic sealant into the jute packing. This procedure requires a 4' × 6' excavation at each joint to allow the utility to reach the pipe (in the North-East US and Canada the pipes are 8' deep in the ground), drill a hole through the bell into the jute and inject a measured quantity of anaerobic sealant. Such a process is hazardous, laborious and costly, and it interferes with transportation and pedestrians in urban areas due to excavations.

In order to alleviate the above problems a novel technique was developed to perform cast iron bell and spigot joint sealing from inside the pipe using a remotely controlled robot (Fig. 10.7.9). The robot is launched into the pipe through a special opening, travels inside the pipe until a desired joint is reached, and drills a hole into the joint spigot at the highest point. Then anaerobic sealant is injected into the jute packing to re-seal the joint. This approach replicates the repair procedure currently done externally. The sealant flows under gravity and saturates the jute. The operation is done while the main is kept in service.

The internal sealing system is capable of sealing several joints from a single excavation while keeping the main in service. The system has been designed for 6" (15 cm) diameter pipes. It can be inserted into the main up to 150' (over 45 meters) in each direction from the entry point. Up to 24 joints can therefore be sealed from a single excavation. Adaptation of the concept to larger pipe diameters (up to 36") has also been performed using a different methodology of moving inside the pipe.

Figure 10.7.9 View of the gas pipe robot end-effector [1].

The joint sealing system consists of a small robotic working head equipped with a video camera for search and joint identification, an umbilical cable, cable insertion unit, system storage reel with the tool control components, and an operator's station that includes a control panel and video monitor. In operation, the robot head is inserted into the main through a small tap, and pushed along the pipe by its motorized umbilical cable. The operator visually locates the desired joint, and positions the robot at the joint. The robot head is then raised into drilling position, and a small hole is drilled through the spigot into the joint cavity. Then, the sealant head is rotated into position, and a measured amount of sealant is injected into the joint. The head is retrieved, and the unit is then moved to the next joint.

The robotic tool head is inserted into the pipe through a small tap using a custom designed saddle. Semi-rigid umbilical cable is used to push and control the tool head position. The umbilical cable is stored on a custom designed reel. A miniature video camera mounted on the tool head is used to locate each joint for sealing. The operator's station is provided with a control panel and video monitor, allowing remote control of the tool head for drilling and sealant injection.

DNA Micro-Arrayer

DNA array is a generic name covering different molecular biology products and techniques. It can be described as a manifold of DNA fragments (spots) of oligonucleotides at low, medium, or high density. DNA arrays are used as research and diagnostic tool. The array of DNA fragments on a solid surface allows detecting the expression of thousands of genes in a single experiment. The gene array technology is one of the most common techniques used in all molecular biology laboratories.

The advantages of DNA array technology are: simultaneous analysis of a large number of genes in a single experiment; quantitative and reproducible results; speeding up basic biological research and disease diagnosis; reduction of time, cost, and risks associated with discovery and development.

In terms of the technology used to make DNA arrays, there are three methods: (i) photolithography method that is based on-site oligonucleotide synthesis; (ii) micro-spotting with quill pins; and (iii) ink jetting. The spotting tools are: split/quill pins, solid pins, piezoelectric pins, capillary, ring-and-pin systems. The support used for DNA arrays can be: nylon membrane, polypropylene membrane, or glass slides. The methods of DNA binding to the support are based on electrostatic and hydrophobic interactions with covalent links. For detection of gene expression, cDNA is spotted first onto the slides, then the target DNA is hybridized with the cDNA, and the expression is identified through probe labeling: radioactivity (^{33}P) and fluorescence (CY 3 and CY 5), and subsequent detection (reading of color intensity).

Robot spotters have been developed for DNA spotting on glass slides. A particular spotter has a large spotting surface (75 or 126 slides), modular & reconfigurable structure, able to spot up to 83,000 spots per 25 mm × 75 mm slide, and capable of performing other bio-laboratory tasks such as arraying, gridding, re-arraying, and pipetting. The robot system is based on a high quality three axis gantry robot, with 1.25 μm resolution of motion along each axis, and impedance control to avoid high impact forces at the contact between the slide and the pin. The system has a repeatability of 2.50 μm, and high bandwidth communication with the controller. There is no heating, vibration, and the speed is 1 spot/slide/pin (61,000 genes are spotted in 3.5 h). The slides can be of 25 × 75 mm, 25 × 25 mm, or 50 × 75 mm. 1 to 8 micro-titter source plates and 1 to 6 small membrane holders can be used along with up to 126 slides. The cleaning of the pins between loads is done in a water bath that uses active water pump for cleaning.

The robot (Fig. 10.7.10*a*) is designed to collect DNA from 96-well and 384-well source plates. 1 to 48 pins (Fig. 10.7.10*b*) can be used simultaneously. Each pin collects 250 nl of solution and spots 0.6 nl per dot on 75 to 126 microscope glass slides. The center-to-center distance is 120 μm.

A vacuum chamber is used to dry out excess material from the quill pins, and dry off the water after washing them. A blotting pad is used to eliminate excess material from the pins before spotting. The quill pins have long life (1,000,000 spots). They generate 75 or 90 μm diameter spots. Each pin can be replaced individually, and up to 250 dots per sample/one dip can be obtained.

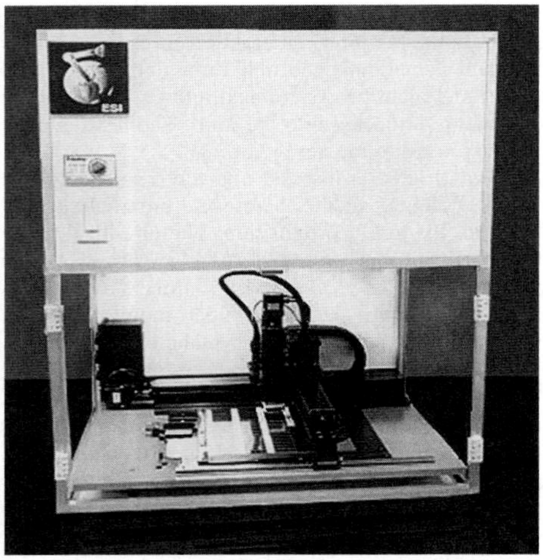

(a)

Figure 10.7.10*a* DNA microarrayer [1].

(b)

Figure 10.7.10*b* Manifold of up to 48 quill pins [1].

Environmental control is provided with a positive pressure HEPA Fun Filter and PPHC humidity controller (limits evaporation and fast drying of the spots).

A Graphical User Interface is used to program the spotter and monitor the execution. The host computer is based on Pentium/Windows NT. The host controller module can be at remote location. The Graphical User Interface is intuitive and Windows-based for easy control of number of slides, rows, columns, array pattern, dots per row or column, speed, spotting order, type and direction of source plate or membranes, number of pins, etc. Simple robotic language is used for programming the robot.

Robotic Cell Sample Preparation of Liquid Scintillation Analysis

Managing employees who work at nuclear power stations presents some unique challenges. One of the issues that require close monitoring is employees' exposure to radiation. Quick and accurate detection of any anomalous radiation exposure can improve the health and safety of the employees as well as the operational safety of the power station.

One of the common methods of monitoring radiation exposure is performing a daily scintillation count of an employee's urine sample. A precise amount of sample is mixed with a known chemical for analysis. This process is tedious and time consuming.

Typically the amount of sample is in the 1,000 μl range and the cocktail is in the 8mL range. While these volumes are not unduly small, they are too small to dispense accurately into the vial without proper laboratory equipment.

A turnkey robot system was developed for the bioassay sample preparation, handling, and analysis. The robot takes a very small quantity of the sample from a container, puts it into a sample vial, adds reagents, caps and seals the vial, mixes it, and places it into a cassette for analysis. More than 500 samples per day can be processed (Figs. 10.7.11*a* and 10.7.11*b*).

All personnel are issued a unique barcode for their samples. These samples are collected in bar-coded trays distributed throughout the power station. Each evening these trays arrive at the health physics laboratory for analysis. An operator simply has to place these trays on the input conveyor section of the robot system and the robotic cell takes care of the rest.

For each sample the robotic cell performs the following basic operations:

- Reads the barcodes on the incoming bottles containing urine samples
- Transfers a measured amount of sample material from sample bottle into an empty scintillation analyzer vial

(a)

Figure 10.7.11*a* Processing laboratory vials [2].

(b)

Figure 10.7.11*b* Picking up samples from vials [2].

- Adds measured amount of liquid scintillation cocktail into the vial
- Seals the vial and thoroughly mix the contents
- Deposit the vials into the scintillation analyzer cassette
- Provides the data relating the sample code to the location of the vial

High Density Colony Replicator

The High Density Colony Replicator is a dispensable, very high density bio-sample array replicator as an attachment to colony picking robots.

Experimental work conducted in biotechnology laboratories requires replicating large numbers of yeast colonies. The colonies are grown in regular arrays on standard gel plates. One task of the experimental process involves transferring an array of samples from a library plate onto a mating plate, followed by transferring of another array of samples from a bait plate onto the same mating plate, in such a way that the samples in the corresponding locations overlap. Another task requires creating "copies" of sample arrays: cells from each colony on the source plate are transferred onto one or more target plates.

The common process of replicating large numbers of yeast colonies is relatively inefficient. The density of yeast colony array is limited primarily by the accuracy and repeatability of the equipment used to manipulate the samples. The commonly used "bed-of-nails" print heads can reproduce an array of 768 colonies on a standard gel plate. There is a need to increase the array density by increasing the number of metal pins in a print head. However, the pins in such print head need to be considerably smaller, positioned more accurately, and machined with a higher degree of precision. Therefore, the print head is much more costly to manufacture and difficult to maintain. Moreover, metal pins need to be washed after every transfer, which requires additional time and equipment, and introduces the risk of sample cross-contamination.

A new print head was developed that would use disposable replicating pads (Fig. 10.7.12). Such print head is capable of replicating high-density arrays and does not require washing of the surfaces that come in contact with sample material. In addition, disposable pads with density, pin-tip size and pin configuration corresponding to the currently used metal-pin print head have also been developed.

High-Performance Auto-Sampler for Mass Spectrometry

Mass Spectrometers (MS) in biotechnology and pharmaceutical laboratories are used to process large number of protein samples. They operate 24 hours a day, 7 days a week. It is necessary to automate the process, so that unattended operation over an extended period of time is possible.

Auto-Samplers can automatically pick up samples from vials, or from 96-well plates; however, the sample loss is very high, as these instruments cannot efficiently handle very small (20 to 50 μl) quantities.

Auto-Samplers can operate in two modes: full-loop and partial-loop injection. Only partial loop injection is suitable for small samples.

Figure 10.7.12 High density replicator attached to colony arrayer robot [2].

However, even in this mode sample loss can be as high as 50% due primarily to large dead volume (approximately 40 μl) on the intake side of the sample loop. Although manufacturers provide dead volume compensation procedures, in practice significant part of each sample is lost as a result of fluid dynamics and protein absorption in the intake line, as well as other factors.

The accuracy of MS measurement is also affected by high-protein absorption inside the stainless steel sample loop, and relatively large dwell volume between the buffer pump and the column.

An auto-sampler for very low-loss automatic injection of samples into the mass spectrometer column was developed. The system is placed directly in front of the mass spectrometer (Fig. 10.7.13). It includes compressed air supply system and control system. It is based on a micro-cross assembly mounted on an X-Y-Z positioning mechanism. The mechanism is used to manually adjust the position of the column tip in front of the MS opening. The required range of adjustment is approximately ±3 mm for each DOF. Positioning accuracy is 0.2 mm.

An Injection Head Assembly is mounted on a vertical linear actuator. The actuator inserts the injection head into the sample vials, or into the waste line coupler. The injection head has two sealed ports: one for the 100 μm liquid line that connects the injection head to the micro-cross, the other one for the compressed air line.

The method of sample injection into the column assumes that the sample will not flow into the line connecting the micro-cross to the

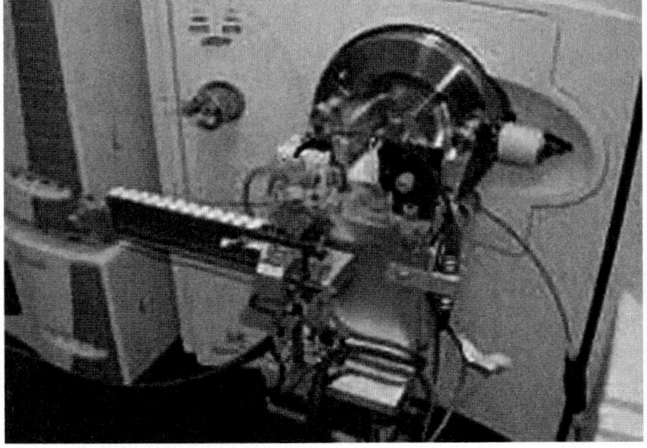

Figure 10.7.13 Mass spectrometer with the mounted auto-sampler [2].

buffer pump. If significant backflow into the buffer pump line is found, a cut-off valve installed on the pump line next to the micro-cross prevents it.

A Vials Handling Mechanism is used to position the selected sample vial or the waste line coupler underneath the injection head. The vials sit in the matching nests in the vial blocks. The blocks will provide necessary mechanical support when high pressure is applied to a vial.

EMERGING APPLICATIONS OF INTELLIGENT AUTOMATION

Emerging applications of intelligent automation can be found in several sectors such as space, medical, security, and biotechnology. They utilize: (i) technology of robotics arms that are embedded in automatic cells; (ii) cells that are integrated based on mechatronics technology; and (iii) the capability of machine adaptation and learning, thus providing intelligent automation.

In the domain of space, a consortium of major players in the sector has been developing the Space Mobile Station (SMS). On the station, various docking ports and remote controlled manipulators are used. One such manipulator is the Special Purpose Dexterous Manipulator (SPDM). This is a dual arm with two similar branches, each consisting of seven joints. Seven-joint arms are labeled in the literature as "redundant." They provide excess DOFs that allow effecting a task in many ways (configurations). This property can be utilized to perform the tasks in an "optimal way" by selecting amongst the many configurations one that optimizes a certain criterion. For example, minimum energy consumption could be used as a criterion in selecting the robot configuration.

The SPDM is used on the SMS to perform manipulation, assembly, and other tasks that require two arms working in coordination. The domain of coordinated arms is not new in the context of industrial robots, but it presents a new challenge in space robotics.

In the medical field, intelligent automation began being utilized in the 80s in two domains: (i) surgery; and (ii) orderly help in hospitals. The latter is based on various mobile automated guided vehicles (AGV), some with an arm mounted for delivery of food, medication, and other necessities to the hospital beds. The former is growing very rapidly by introduction of newly designed robotics in the operating rooms. Initially robots were used to perform life-threatening procedures such as brain biopsy in neurological surgery. Later, knees and hip replacement surgery with robotics became standard. Currently laparoscopic surgery is used in abdomen and thoracic-related surgery. As well, methodologies of surgery with robotic tools are being developed for brain, heart, prostate, and breast. Some of these procedures will be performed under Magnetic Resonance Imaging (MRI) inside close-bore MRI for best imaging. This requires the development of MRI-compatible equipment, as well as remote control of the surgical tools because the location of the surgical team in a room adjacent to the MRI, and control of the procedure through computer monitors and master controllers, such as hand controllers.

In the security domain new technology for stand-off and inspection are being developed. There are developments related to inspecting by remote control the interior of passenger cars and trucks. Other developments address the need to negotiate in the event of terrorist-related hostage situations. In military, there is an acute need for remote devices that can detect and neutralize IEDs (Improvised Explosive Devices). Various unmanned ground and aerial devices are being developed that could detect IEDs, in-ground mines, and unexploded ammunition. Other technology is being developed for removing the hazardous material after detection.

In the domain of biotechnology, drug discovery and structural proteomics are leading the effort in research and development. There is a need for a wide range of automated equipment to accelerate the drug discovery processes via research in proteomics. Structural proteomics is a crucial domain of molecular biology in the context of drug discovery. Equipment that can perform reliably tens of thousands of experiments each day is being developed using robotic arms, mechatronics cells, and intelligent automation, such as fuzzy-logic based identification of patterns in the genetic map of humans for detection of pre-disposition to certain chronic diseases.

In the general domain of intelligent automation there are developments of mobile robots that climb walls to wash high-rise windows, autonomous vacuuming of floors, robot arms used in the construction industry, modular laboratory equipment that can be carried home by students and for studying intelligent automation, automatic guidance of visually impaired, and other applications.

Advanced Robot Arms

Planetary Medium Manipulator

The robot was developed to advance the state-of-the-art of manipulators for planetary exploration and to perform simulated Moon and Mars missions on Earth. It is made up of eight modules: Turret, Shoulder, Upper Arm, Elbow, Lower Arm, Wrist, Automatic End Effector Exchanger (AEEE), and Controller. The turret provides azimuth motion, while the shoulder, elbow and wrist each provide pitch and roll.

The system operates under remote control commands, as well as autonomously through a scripting interface. The autonomous motion is achieved using advanced methods of visual servo, force, impedance, and adaptive control. The AEEE interface is used to load autonomously tools and other payloads onto the arm.

The robot is 2.3 m long with DOF. It can be used in aerospace applications where light weight is a key factor, security and defense where high accuracy and dexterity is key requirement, as a test bed for R&D in advanced control methods with open software architecture, visual servo applications of guided tool operations and tool exchanging, and testing in environment with harsh Electromagnetic Compatibility (EMC), temperature and humidity requirements. The use can be indoor and outdoor.

The design is modular consisting of one single-joint and three two-joint modules, links and AEEE. The joints are compliant (back-drivable). Light-weight and high-stiffness arm links made of carbon-fiber reinforced plastic (CFRP) are coated in a 50-micron nickel alloy layer. Internal cabling provides protection from the environment and snagging. Advanced control methods that were used include impedance, adaptive, visual servo control and control in Cartesian-space (Fig. 10.7.14).

The robot can be used effectively in security and defense applications while mounted on mobile platforms, as well as in manufacturing, robotic-based custom automation, and for research.

Small Manipulator Arm

The Small Manipulator Arm was developed to advance the state-of-the-art in manipulators for planetary exploration and to perform simulated Moon and Mars missions on Earth. It is light weight, high payload-to-weight ratio manipulator with advanced control system including force control, visual servo control, and open software architecture. The arm has six joints, links, payload interface, electronics (drivers and controller), harness, user interface software, and operator control unit. The arm can be operated in remote control and closed loop modes. It can be used effectively in security and defense applications while mounted on mobile platforms, as well as in manufacturing, robotic-based custom automation, and research. The arm complies with military standard (MIL-STD-461 Rev E) on EMC and Electromagnetic Interference (EMI) requirements, and with space-quality requirements on vibration and shock resistance.

The arm is a six-joint robot arm. It can be used as stationary or mounted on mobile platforms. The robot can be used effectively in security and defense applications, as well as in manufacturing, robotic-based custom automation, and for research.

The arm has a high payload-to-weight ratio of 1:1.1, high repeatability and accuracy at full extension, modular two-joint wrist with tilt and roll motion, payload Interface Module for multiple payloads, state-of-the-art electronics with military level components, advanced control methods including force, impedance and visual servoing for autonomous operation capability, software open architecture software for user interfaces (Fig. 10.7.15).

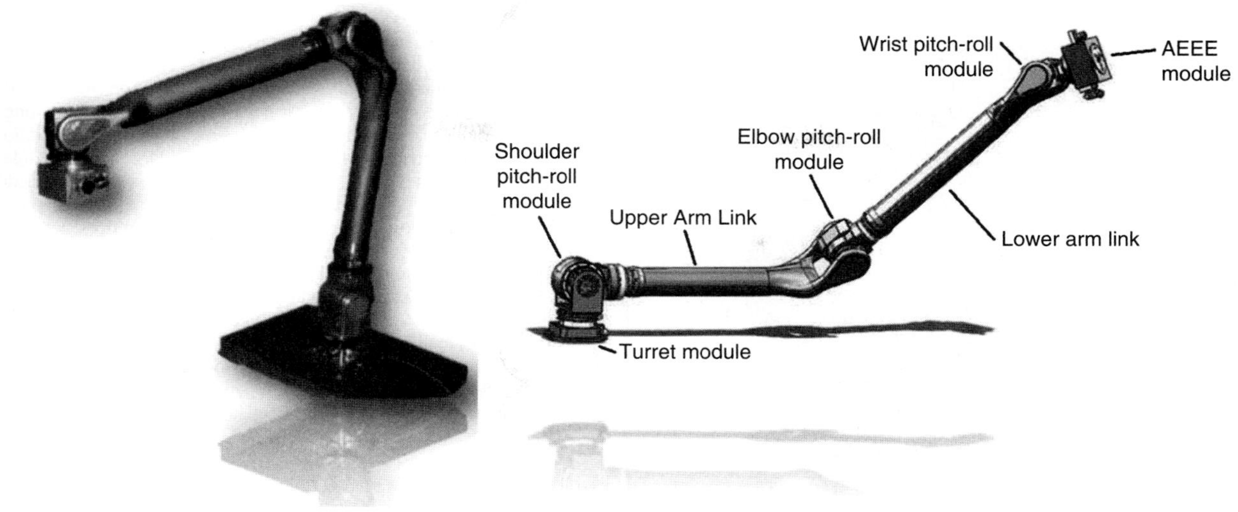

Figure 10.7.14 Planetary Medium Manipulator (PMM) [2].

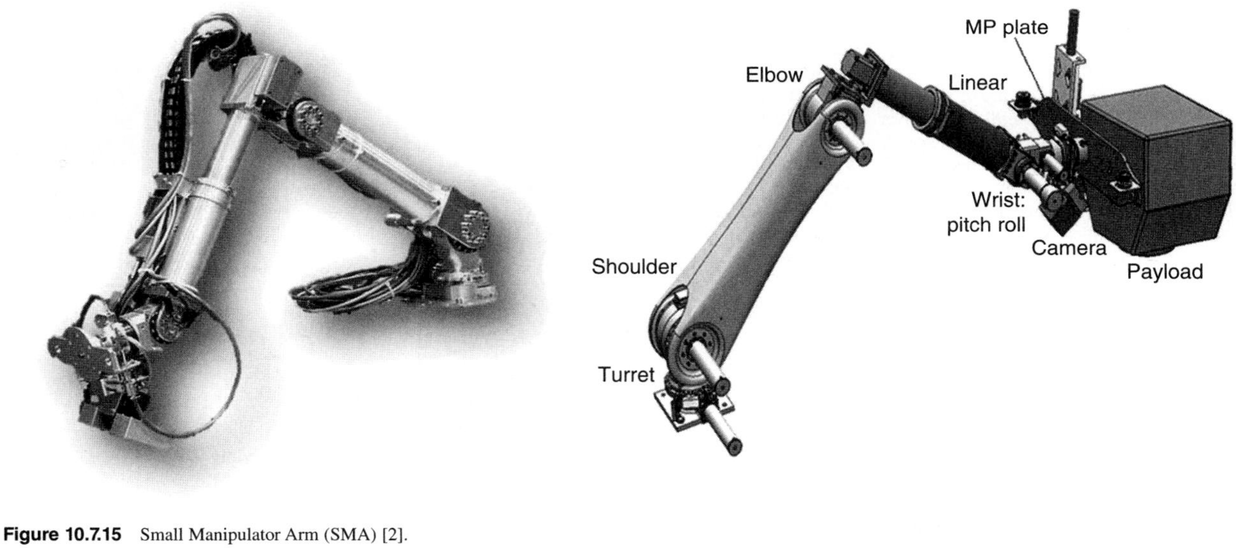

Figure 10.7.15 Small Manipulator Arm (SMA) [2].

Industrial Robots

The brief review in this subsection covers advanced industrial robots currently being developed [2].

PAINTING ROBOT

The robot is for automated painting applications to achieve accurate and consistent paint quality finishes. The arm provides smooth and accurate motion of the painting equipment mounted on the arm to consistently achieve high-quality paint finishes.

The arm features lightweight aluminum castings, as well as explosion proof cavities to house electrical control equipment. It can mount various manufacturers' painting equipment in close proximity to the application to allow for effective painting process control.

The task of the painting robot is to move the painting nozzle over the surface to be painted while maintaining a proper distance, speed, and orientation relative to the surface. This approach ensures a homogenous painting. Explosion protection by means of air purged design is provided.

The robot has high accuracy positioning and path tracking, arm dexterity for complex painting processes, integrated painting mounting interface, explosive proof equipment, and is easy to program.

This robot can be used in other potential applications such as quality control measurement, flaming of plastic parts, and washing of industrial assemblies (Fig. 10.7.16).

WELDING ROBOT

The robot is for automatic arc welding applications to achieve accurate and consistent welding quality. The arm provides smooth and accurate motion for the welding equipment to achieve consistently high quality welds. The arm features lightweight aluminum alloys. It integrates

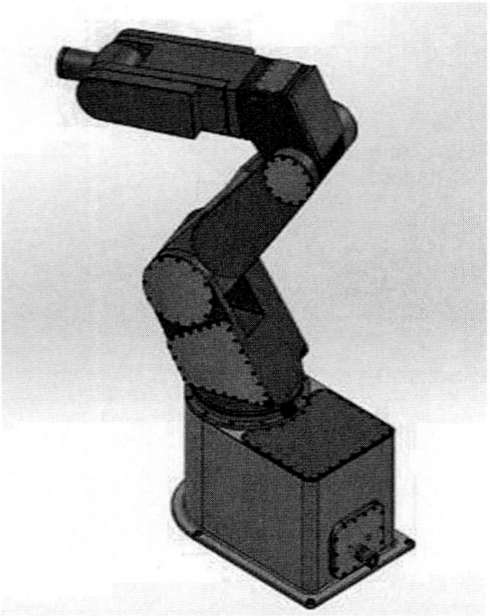

Figure 10.7.16 Painting robot [2].

welding wire feed cabling and mounting holes for the welding gun in proximity of the welding location to allow effective welding process control. The controller provides easy-to-use arc welding functions.

The robot is to be used for arc or spot-welding. The task of the welding robot is to move the welding gun over the surface to be welded while maintaining a proper distance, speed, and orientation relative to the surface. This approach ensures a quality weld. Possible usage in other application areas are dispensing and cutting.

The system has high accuracy positioning and path tracking, arm dexterity for complex welding processes, integrated welding cable feeding, and it is easy to program for welding operations (Fig. 10.7.17).

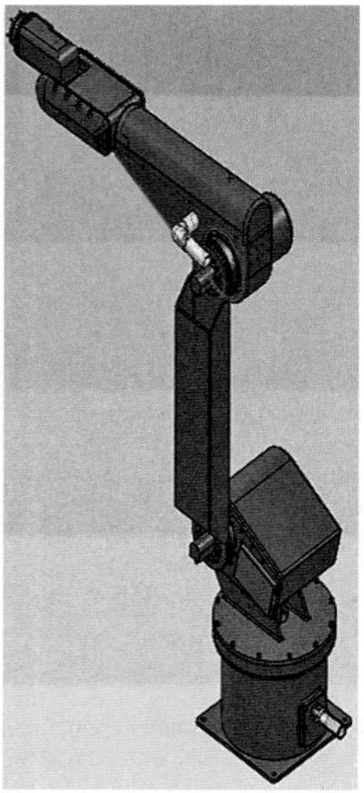

Figure 10.7.17 Welding robot [2].

SCARA Robot

The SCARA robot consists of base, first axis joint, link arm, second axis, joint, link arm, third axis, and forth axis joints. The arm is for small footprint, light payload applications where high precision, high speed, and high performance are required. It has a modular design including: single-joint modules, compliant joints and has high payload/weight ratio.

The robot configured out of 4 modules uses a large joint module, a medium size module, and a two-joint wrist module. An "extended" SCARA robot can be configured with five joints by adding one more joint to the wrist to allow tilting of the end-effector that does not exist in standard SCARA robot. Other configurations are possible as well.

The modular joints are controlled by an embedded fully digital servo drive. By connecting several modules together one obtains a robot. The robot can be used for assembly manufacturing, and can work with human side-by-side. The robot is light weight, and has high speed, high accuracy, small footprint, integrated force/torque sensor interface for force and impedance control and integrated vision sensor interface for visual servoing (10.7.18).

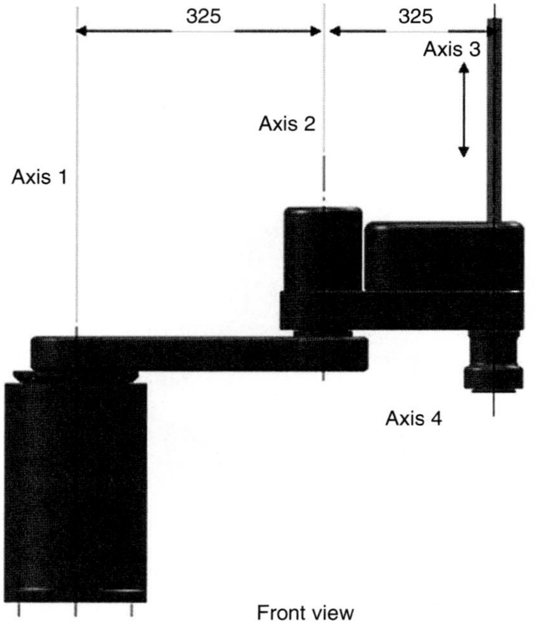

Figure 10.7.18 SCARA robot [2].

High Payload Robot

Applications of the high-payload robot are in machine tending, palletizing, packaging, part transferring, pick and place, grinding, deburring, and polishing. The arm (Fig. 10.7.19) has large payload, large working space, high speed, high accuracy, and integrated force/torque sensor interface for force and impedance control, integrated vision sensor interface for visual servoing.

The robot achieves high strength and agility with high payloads without sacrificing speed and accuracy. It consists of base, turret, shoulder tilt, upper link, elbow tilt, lower link, wrist, first roll, wrist pitch and wrist second roll, and gripper.

Turret and shoulder have the same internal layout, motor, input gear, and reducer. The rotation of elbow joint is realized through a servomotor with a fixedly attached input gear to drive a reducer directly.

The wrist assembly consists of three joints: wrist first roll, pitch, and second roll. Three identical motors are used for the three joints of the wrist assembly. They are mounted on the back of the upper link to conveniently maintain and ease the balancing of the robotic system.

Figure 10.7.19 High payload robot [2].

However, it is a challenge for the mechanical design to transmit the motion and torque of the three motors to the wrist.

A counterbalancing system is used to compensate the load moments of each axle such that the axle drive is not overloaded statically, thus it can provide the maximum moment available for the acceleration of the axle. The balancing device is arranged to counteract a gravitational force upon relative movement of the robot arms.

Collaborative Robot

Collaborative applications include operations in which human and robot have to work side-by-side. The robot is meant to complement and not replace a worker, and it should be flexible to meet SME (Small and Medium-sized Enterprises) needs. IRA-20 (Fig. 10.7.20) is a 6-DOF modular arm comprising of single-joint compliant modules, reconfigurable, with a high payload to weight ratio (payload = 20 kg), suitable for collaborative human-robot operations.

IRA-20 can be configured in two ways: Option-1 shown in Fig. 10.7.20*a* consists of base, turret, shoulder tilt, upper link, elbow mount, elbow tilt, lower link, wrist roll, tilt, and second roll joints. A modular joint design is used. The turret and shoulder tilt use one size module, and the wrist roll and tilt use another size module. Each module is composed of harmonic drive, brushless motor, brake, encoder, motor drivers, interface boards, and peripheral components. These features make the IRA-20 installation or replacement easier during the course of operation or service. The harness pass through the inside of the links and through the center of joints.

Option-2 shown in Fig. 10.7.20*b* consists of base, turret, shoulder connector, shoulder tilt, upper link, elbow mount, elbow tilt, lower link,

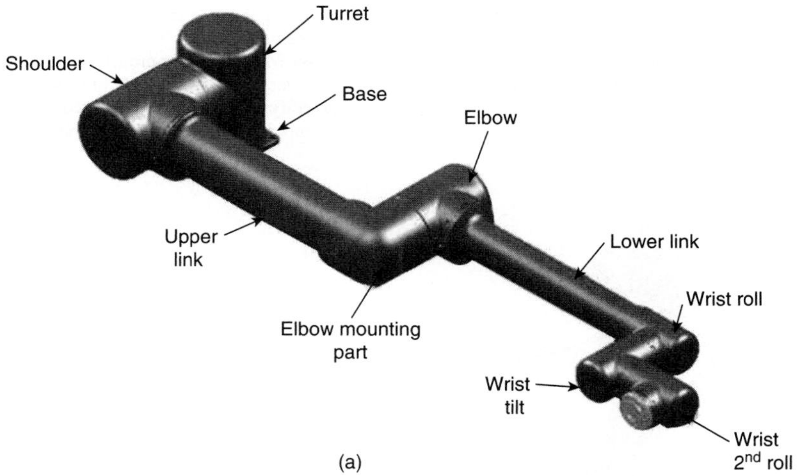

Figure 10.7.20*a* IRA-20 option 1 configuration [2].

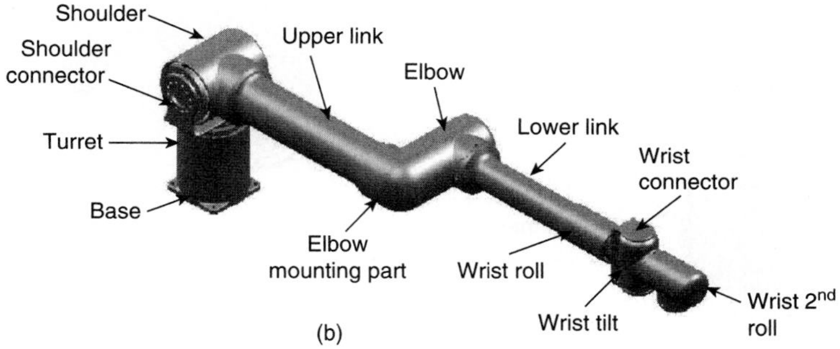

Figure 10.7.20*b* IRA-20 option 2 configuration [2].

wrist roll, wrist connector, wrist tilt, and second roll joints. A shoulder connector is used for the shoulder and turret, and a wrist connector for wrist roll and tilt. These would have only minimal consequences for the kinematics and dynamics due to the change of the offset from the base to the upper arm and from elbow to wrist second roll.

Option-2 would be slightly heavier than Option-1 because of two connectors' weight. However, Option-2 kinematics and dynamics are simpler. Also, the stiffness and deformation of shoulder and wrist tilt joints in Option-2 would be easier to control.

Option-1 turret and shoulder would use same size module, and wrist roll, tilt and second roll would use same module. Option-2 would not use same size modules.

The main features of the robot are: (i) active compliance: force and impedance control; (ii) passive compliance: built-in joint compliance; (iii) small footprint; (iv) lightweight; (v) high dexterity; (vi) integrated force/torque sensor interface for force control; and (vii) integrated vision sensor interface for visual servoing.

Collaborative 2-Joint Modules Robot

The robot is a two-joint compliant modules modular, reconfigurable, high payload/weight ratio robot arm for collaborative human-robot operations (Fig. 10.7.21). The robot is a 6-DOF robot arm that can be reconfigured in multiple ways.

The robot manipulator comprises turret, shoulder, elbow, first wrist roll, wrist tilt and second wrist roll. The robot includes two links, the upper-arm, and the forearm.

The robot has modular shoulder, elbow, and wrist whereby each module is composed of two rotary joints with their motor drivers and their interface boards installed in the module housing. These features make the robot installation or replacement easier during the course of operation or servicing. Another feature is the internal cabling design, in which all cables pass through inside of the links and through the modules. The upper link connects the shoulder module to the elbow module, and the fore link connects the elbow module to the wrist module. The two-joint module robot provides a compact structure and is lightweight.

The shoulder module has tilt-pan joints. It could be reversely installed to become tilt-roll and then a turret module (pan joint) can be added to become a 7-DOF arm. The tilt output side and its opposite side are designed to connect to a fork-type mechanical interface. The pan joint is designed to connect to the base of an arm. Two motor-drivers are located inside the module. All cables are internal to the module.

The elbow two-joint module has tilt-roll joints. The tilt joint output and mechanical interface assemblies on the opposite side of the housing connect to the upper arm. The roll joint output connects to the forearm. Two motor-drivers, two electronics interface boards and all wirings are located inside of the module. To keep consistence with robot shapes, this module has the same shape as the wrist module.

The features are use of frameless motors to reduce weight and size, and other features that are similar to IRA-20.

Mobile Robots

Articulated track mobile robots have mechanical means activated by remote control or autonomously that change the shape of the tracks (Fig. 10.7.22). These robots also have arms that are fully modular with self-contained electronics. The arms have up to six joints, a medium size gripper and a variety of operational tools and accessories. The robots are all-weather and can operate indoor (buildings, public institutions, airports, homes) and outdoors (obstacle-cluttered terrains).

The arm modules are plug-and-play. The arms have distinctive feature, enabling them to access hard to reach areas in airplanes, subways, buses, buildings, homes, airports, and train stations.

The technology includes the following features:

Arm: High dexterity arm can be configured to perform grasping and manipulation of objects, aiming disrupters, pushing payloads, reaching under cars or through car windows, breaking windows, inserting a car key to open car doors.

Precise Motion: To approach, hit, and handle a sensitive target with high accuracy. Programmable in task space coordinates.

Dexterity: Gripper can rotate continuously, Shoulder, elbow, and wrist joints are rotary and allow large ranges of motion.

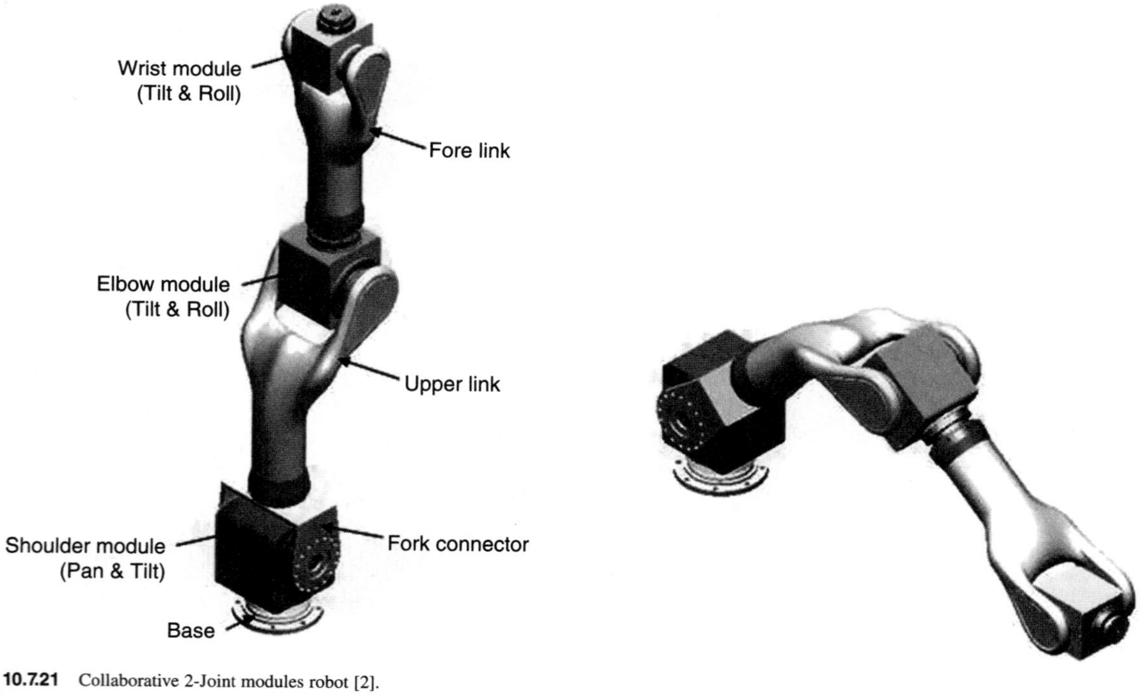

Figure 10.7.21 Collaborative 2-Joint modules robot [2].

Figure 10.7.22 Articulated track robot—TRACKER [2].

Modular Design: Arm can be easily disassembled and re-assembled for repair and maintenance

Multiple Payload Mounting: Disrupter, shotgun, X-ray unit, biosensors, and radiation sensors, window breaker, charge drop assembly.

Medical Robots

Medical Robots technology is offered through a range of products. Surgical robots have been developed or are under development for the following therapies:

i. Minimally invasive prostate surgery in closed bore MRI
ii. Minimally invasive general surgery in closed bore MRI

The robots are modular and re-configurable to allow mounting and fitting into the MRI environment. The robots are semi-autonomous. They are remotely controlled when operating in the human body through MRI and PC-based images. The robots are endowed with advanced Operator Control Units allowing complete control of the interventions. The robots can be adapted to other surgical procedures and organs by exchanging therapy-focused modular subsystems of the robot.

MRI-Guided Robot-Assisted In-Bore Prostate Surgery

Prostate cancer is the most common cause of cancer in man and the second most common cause of mortality due to cancer. Publicly available information indicates that more than 230,000 men are diagnosed each year in North America. A conservative estimate of the prevalence of prostate cancer in North American males in the age range of 55 to 69 is 7,500,000 men.

The traditional curative treatments of this disease include radical surgery and external or interstitial radiation therapy. These are reasonably effective for localized cancer, but are associated with significant quality of life penalties. These may include incontinence, impotence, bowel dysfunction, and prolonged recovery. The cause of these problems is not removal of the prostate but damage to surrounding tissue and nerves.

While most believe that prostate cancer is a multi-focal disease recent evidence supports the notion of a dominant focus that is the largest of intra-prostatic cancer sites and the major source of extra-prostatic spread (90%). Recent improvements in imaging of tissue by magnetic resonance scanning have allowed for the visualization of these sites.

Image-guided focal ablation of the dominant focus in selected men with low to low-intermediate risk cancer could control spread of the cancer from the prostate in the majority of men and render it a chronic disease that is largely devoid of side effects. This paradigm treatment shift is not dissimilar to treatment of colonic polyps by colonoscopy as opposed to colectomy or lumpectomy for localized breast cancer instead of mastectomy.

The ideal implementation of this concept is with a minimally invasive robot-assisted MRI-guided focal ablation system. The procedure is performed in closed-bore MRI that is considered superior to other techniques of image-guided interventions [open-bore MRI, Ultrasound, Computerized Tomography (CT) scan]. The procedure uniquely integrates direct cancer imaging, real time precision computer-aided navigation of the tool to the target, optimal energy distribution into the target, real time imaging of tissue destruction using MR thermography for visual confirmation of target destruction.

The robot could be used with different surgical tools to perform a range of prostatic interventions:

- Ablation—thermal lesions resulting in cell death without untoward patient effects
- Brachytherapy—insertion of radioactive seeds in the gland
- Biopsy—extraction of tissue samples from the gland based on established protocols

The image-guided surgical interventions in closed-bore MRI may be performed using either:

- a remotely-controlled surgical robot tool; or
- a manually-controlled instrumented surgical tool

The surgical tool is identical in both approaches. It is carried by either the surgical robot tool or by the instrumented tool. The procedures are image-guided.

One major issue is that the surgical tools and carriers must be MRI-compatible; that is, the robot or instrumented tool and the MRI would not affect each other (the imaging process and carrier operation, respectively). The technology is also including a methodology of evaluating the therapy (ablation) by real-time MRI scanning during and post-intervention.

The new technology provides high-added value to the medical practice as it would impact on both, precision of surgical procedures and reduction of side effects that may appear post-interventional and affect the quality of life.

The robotic platform includes: (i) a six-joint robot; (ii) robot-based trocar for mounting surgical tools; (iii) hand controller for navigating the surgical tool by remote control; (iv) robot controller; (v) a Laptop-based user interface for robot control and image display; (vi) laser dispenser, power, and control; (vii) MRI monitoring station (including MR Temperature mapping). The robot system could be adapted for operation in 1.5T-3T MRI scanners. The robot is functional simultaneously while the scanner is operating without affecting the images or the robot operation.

During surgical procedures, the patient lies on the MRI roll-in table with leg supports attached to the table (shown schematically in Fig. 10.7.23). The robot is mounted & secured onto the table and between patient's legs. The robot controller is at a distance from the scanner, connected to the robot, and to the Laptop that is located in

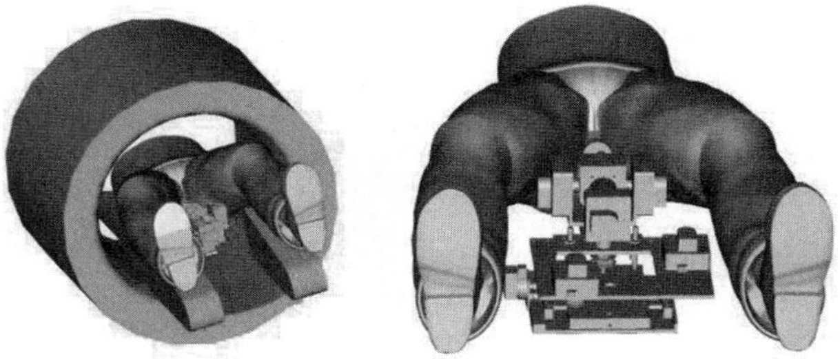

Figure 10.7.23 Robot and patient in the MRI [2].

the adjacent control room. The surgeon will remotely manipulate the tools based on MRI and Laptop-based images using a hand controller (joystick) or manually. Control of surgical tool penetration is based on visual feedback provided by MR imaging. The image allows the user to identify the tool tip location relative to the target, and perform suitable adjustments of the tool path to reach the target. When the tool is at the target, the laser is turned on, and ablation of tissue is performed. MR thermography allows for real time imaging of tissue destruction. After the ablation process is completed, the scanner provides images of the heated and coagulated volumes of tissue.

The technological features of one such robot shown in Fig 10.7.24 are:

- Modular Design: Surgical tool support (trocar) allows for a variety of tools to be used
- MRI-compatible
- Rugged and compact design
- Bio-compatible
- Safe and efficient in the MR environment

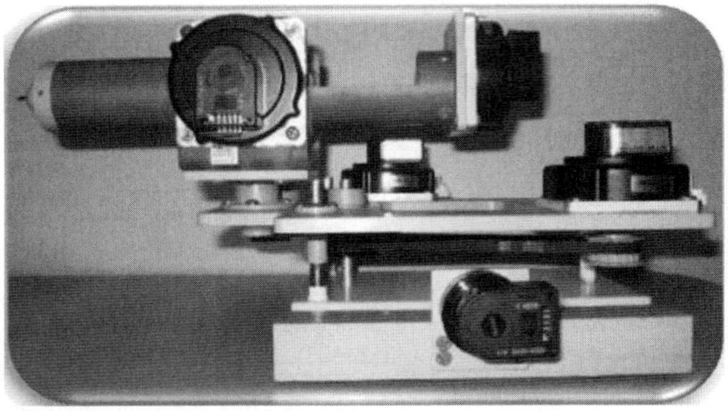

Figure 10.7.24 MRI prostate focal ablation robot [2].

MRI-Guided General Surgery Robot

MRI-Compatible General Surgery Robot including its auxiliary technology (surgical tool, control station, and arch-base device) is a new surgical robot prototype for MRI-guided bone biopsy and general surgery. The robot is modular, re-configurable, and fitting into the MRI mounted on an MRI-bore shaped arch (Fig. 10.7.25). The reconfigurability provides a means of finding best possible configuration for specific bone biopsy and other surgical interventions. The robot with various surgical tool modules can operate under remote control

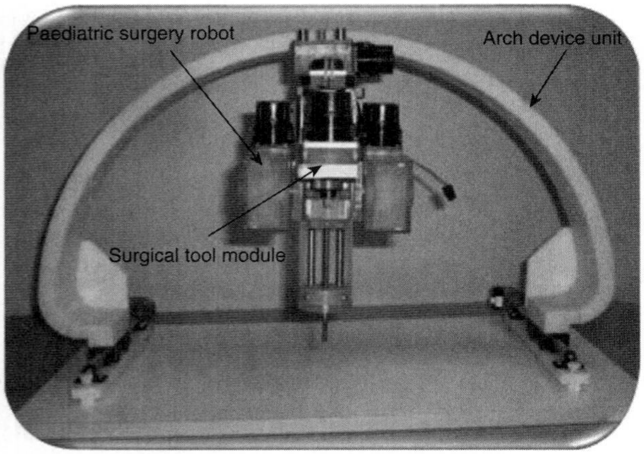

Figure 10.7.25 MRI-compatible general surgery robot [2].

as well as autonomously. Haptics technology can be used to provide the sensing of drilling force in bone biopsy interventions. Embedded Control Software and Graphical User Interfaces are part of the robot system. Software for visualization and navigation is currently being developed.

The main features of the medical robot are modularity of robot and surgical tool, MRI-compatibility, visualization, and navigation software for interaction with MR images, snapping on and off the surgical tool module for sterilization. An arch device unit compact in size to fit in the limited space of 3-T close-bore MRI is also available.

Personal Robots

Personal robots are emerging mainly in five sub-markets:

- Virtual Conferencing and Telepresence Robot
- Host and Receptionist Robot
- Personal Security Robot
- Home Assistance Robot

We distinguish the common Personal Robots to have two distinct parts:

- **Mobile Platform (MP)**—featuring Internet-based control, collision detection and avoidance, path planning and tracking, and interfaces for mounting a range of Application Payload Modules (APM) that are specific to each sub-market.
- **Application Payload Modules (APM)**—provide the main functionality and basic operations including Internet-based control, dual video-voice links, 3D mapping, human face recognition, manipulation arms, range of environmental sensors, and more.

The common features of the Personal Robots are open source software and off-the-shelf hardware, wheeled or tracked-based mobility, wireless and Wi-Fi communication and control, collision avoidance, map generation, and path tracking.

The robots operate on the basis of a set of software modules that provide a wide range of functionalities: localization, navigation, object identification, communication, and video streaming. The modules include: map management, location tracking, homing, detection and collision avoidance, path planning, motion control, live video broadcast, recording and display, and autonomous navigation.

In the future an Advanced Mobile Platform (AMP) that will include the basic MP and also featuring stair climbing, door opening, and autonomous navigation will emerge.

Video Conferencing and Telepresence Robot

The robot is for teleconferencing and telepresence for a wide range of applications: office-to-office communication, remote business-to-business communication, remote viewing, inspection of environment, transfer of images, and other (Fig. 10.7.26).

The user can listen, talk, see and be virtually seen, as well as for active or passive telepresence. It provides access to locations by remote control, and it provides feedback to the operator through video, audio, and data. Typically the robot APM can be mounted on various mobile platforms (MP).

The main functionalities and applications are:

- Office-to-office communication, remote business-to-business communication
- Remote viewing, inspection of environments, transfer of images by:
  - Remote control via the Internet with remote login and management
  - Bidirectional access to media communication via the Internet and media recording
- Auto docking and power charging with operation of more than 6 hours
- Automatic detection of obstacles and automatic stop for safety
- Local touch-screen display
- Application in commercial, public, and private environments

Host and Receptionist Robot

The robot for service in public areas. It provides a mobile kiosk to dispense information upon request through touch-screen interface, guidance through Q&A in public spaces, source of local maps, and link to customer relations personnel. The robot can be used in more formal

Figure 10.7.26 Video conferencing and telepresence robot concept [2].

gatherings such as conferences and conventions, sports event, train and bus stations, and airports.

It includes interactive means for two-way interaction with the general public in a wide range of venues. It can move autonomously or can be controlled remotely (Fig. 10.7.27).

The main functionalities and applications are:

- Mobile navigation in public spaces
- Remote viewing, inspection of environments, transfer of images by:
  - Remote control over the Internet with remote login and management
  - Bidirectional access to media communication via the Internet and media recording
- Auto docking and power charging with operation of more than 6 hours
- Automatic detection of obstacles and automatic stop for safety
- Local touch-screen display
- Inter-person voice communication including expressions of emotions
- Reception services
- Guidance in public spaces
- Payment outlets
- Advertising and promotion services
- Educational service

Figure 10.7.27 Host and receptionist robot concept [2].

Patrolling Robot

The robot is for execution of full or partially autonomous patrolling of dwellings and support to personnel in various security operations (Fig. 10.7.28).

The application is focused on public and private security services in both indoor and outdoor environments, such as schools, office buildings, private rental and condominium dwellings, hotels, stadiums, bus

Figure 10.7.28 Patrolling robot concept [2].

and train stations, ports, and airports. The robot moves through programmed routes independently; it will call and enter elevators; it will do reconnaissance in underground parking garages, detect objects and issue emergency signals related to fire-sensors and human-detecting sensors. It can be mounted on various mobile platforms.

The robot provides the following functionalities and applications:

- Mobile navigation in private premises
- Remote viewing, inspection of surroundings, transfer of images by:
  - Remote control via the OCU and Wi-Fi
  - Bidirectional access to media communication and media recording
- Auto docking and power charging with operation of more than 6 hours
- Automatic detection of obstacles and automatic stop for safety
- Patrolling services in both indoor and outdoor environments
- Mobile video surveillance
- Detection of intruder, fire, smoke, water, gas, and chemicals
- People detection and tracking
- Scanning of vehicle plates and checking licenses in parking garages
- Scheduled patrolling of private premises

Home Assistant Robot

The robot is a mobile device for active support of humans mainly in domestic environments. It incorporates the MP and a pair of arms mounted a vertical trunk attached to the platform (Fig. 10.7.29). It can be controlled through voice, gesture, and display of emotional interaction. The robot can be programmed for specific premises and tasks to help around the house. In particular the robot could help seniors and disabled persons to stand up from the bed and chair. This help could be provided by remote control of from a service center. The robot could provide also some functions around the kitchen. The robot could operate with one or two arms.

Working and serving a human interactively raises safety issues. In this context the arms are back-drivable (compliant joints) to provide a diminishing force of interaction when unwanted contact between robot and human occurs. The robot APM can be mounted on various mobile platforms.

Figure 10.7.29 Home assistant robot concept [2].

The main functionalities and applications are:

- Mobile navigation in private premises
- Remote viewing, inspection of surroundings, and transfer of images by:
    - Remote control via the Internet and Wi-Fi
    - Bidirectional access to media communication and media recording
- Auto docking and power charging station with operation of more than 6 hours
- Automatic detection of obstacles and automatic stop for safety
- Fetch-and-carry tasks such as serving a drink
- Delivery of light items to user
- Provide help to a person to stand up
- Home cleaning and simple operation of appliances
- Door opening and closing
- Doing laundry such as using a washing machine
- Alarm and assist in case of emergency such as in the case of a fall
- Support to do remote control of some personal health measurements
- Entertainment and communication

CONCLUSIONS

In this chapter the related fields of robotics, mechatronics and intelligent automation were discussed and illustrated using selected applications. There are significantly many other applications that would serve the same purpose. The aim was to support the view that these fields have a great impact on humanity, and provide opportunities both, scientifically and commercially for further developments in these domains. There is a great entrepreneurial spirit in conjunction with these areas, as the technological progress made in the last decade has generated extraordinary scientific and commercial value. Education and research in Colleges and Universities have been also harnessed onto this wave of technology development, leading to a significant increase in the number of students, publications, patents, and commercial spin-offs.

The domain has achieved significant results, and it continues to evolve as new applications are emerging. The few new applications mentioned in the section "Emerging Applications of Intelligent Automation" are only representative of this thrust. They are not only illustrating this exciting domain of science and engineering, but also demonstrating the inter-disciplinary nature of this area of technology.

Nonetheless, there are some main issues that require more thorough consideration as to the best way to accomplish the proliferation of intelligent automation in every sector of life. Already, medicine, security, manufacturing, education, all major sectors of human life and economy, are subscribing to the era of intelligent automation. There is a need to extend and consolidate the list such that intelligent automation-based human-centered services gain wider acceptance by the general public. This may be possible only if opportunities are provided, resources are available, and the economics are favorable for proliferation of the new technology to serves humanity in the most effective way.

REFERENCES

1. Previous edition of the Handbook, Chapter 20-11.
2. www.esit.com

Section 11

Refrigeration, Cryogenics, and Optics

BY

A. J. RYDZEWSKI *Consultant, Du Pont Engineering, E. I. du Pont de Nemours and Company*
F. J. EDESKUTY *Los Alamos National Laboratory, retired*
PETER J. SHIRRON *National Aeronautics and Space Administration/Goddard Space Flight Center*
PATRICK C. MOCK *Principal Electron Optical Scientist, Science and Engineering Associates, Inc.*
DONALD B. BIVENS *Refrigerant Applications Consultant, Du Pont Engineering, E. I. du Pont de Nemours and Company, retired*

11.1 MECHANICAL REFRIGERATION
by A. J. Rydzewski

11.2 CRYOGENICS
by F. J. Edeskuty, Revised by Peter J. Shirron

11.3 OPTICS
by Patrick C. Mock

11.4 NEW REFRIGERANTS FOR REFRIGERATION AND AIR CONDITIONING
by Donald B. Bivens

11.1 MECHANICAL REFRIGERATION

by A. J. Rydzewski

REFERENCES: Publications of the ASRE and ASHRAE: "Air-conditioning Refrigerating Data Books"; *Refrigerating Engineering*; ARI Equipment Standards; ASHRAE Handbooks. Merkel and Bosnjakovic, "Diagrammen und Tabellen zur Berechnung der Absorptionskältemaschinen," Springer, Berlin. Jordan and Priester, "Refrigeration and Air Conditioning," Prentice-Hall.

REFRIGERATION MACHINES AND PROCESSES

(For general theory of refrigeration, see Sec. 2; for air-conditioning uses, see Sec. 4.14; for cryogenics, see Sec. 11.2.)

Refrigeration is a term used to describe thermal systems which maintain a process space or material at a temperature less than available from ambient conditions. Heat is transferred from the materials to be cooled into a lower-temperature substance referred to as the **refrigerant.** The refrigerant is cooled by physical processes which pump heat into the ambient environment. These apply several physical phenomena such as phase changes, sensible heating, thermoelectric effects, work extraction by mechanical devices, and endothermic chemical reactions. Stored energy in the form of ice, sublimating solid carbon dioxide, and cold liquids are included with processes described by the word **refrigeration.**

Many refrigeration systems use a continuous cycle. This requires that the refrigerant be recovered for recycling after it has absorbed heat from the product or space to be cooled. The most commonly used system has a recirculating fluid that boils at a low temperature to remove heat. The vapor is compressed to a pressure such that the condensing temperature is elevated. The vapor is cooled and condensed by a heat exchanger which rejects the heat into the environment. The condensed liquid is passed through a restricting valve or orifice. The liquid changes into a mixture of liquid and vapor that is colder than the refrigerated space. This expansion at constant enthalpy is known as the *Joule-Thompson effect*. An expansive turbine can remove heat from a high-pressure vapor by extracting energy in the form of mechanical work. When the exhaust conditions are proper, the material will be a mixture of gas and condensate. This fluid can be used as a refrigerant or stored in liquid form as product. The remaining vapor can be recycled as needed to continue the process.

The equipment used in refrigeration processes is referred to as follows: high-pressure vapor heat exchangers are called *condensers,* low-pressure heat exchangers are called *evaporators,* throttling devices are called *expansion valves* or *capillaries,* and pressure-raising devices are called *compressors*.

With continuous recirculation of the cycle fluid, the low-temperature heat pickup must be pumped, via energy input, to a higher level so as to discharge it to atmospheric temperature. In this sense the system acts as an **energy-transport** device or **heat pump** or, from another viewpoint, as a temperature transformer. This philosophy of energy transport may equally well be applied to more remote systems, such as thermoelectric. It is, of course, a manifestation of the principles of the Second Law, and the effectiveness of the system's operation may be evaluated numerically.

Efficiency ratings of refrigeration equipment are expressed several ways. Some of the more common terms include **coefficient of performance** (COP) and **energy efficiency ratio** (EER). **COP** is defined as refrigerant effect divided by net work input, where the refrigerant effect is the absolute value of the heat transferred from the lower temperature source, and the net work input is the absolute value of heat transferred to the higher temperature sink minus the refrigerant effect. COP can alternatively be defined as the ratio calculated by dividing the total heating capacity in Btu per hour provided by the refrigeration system, including circulating fan heat, but excluding supplementary resistance, by the total electric input in watts × 3.412. This definition applies primarily to heat pumps. **EER** is a ratio calculated by dividing the cooling capacity in Btu per hour by the power input in watts at any set of rating conditions, expressed in Btu/W · h.

Because load on equipment used for comfort applications varies, additional criteria to rate efficiency at part load have been introduced. ARI Standard 550/590-1998 has **integrated part load value** (IPLV) and **non-standard part load value** (NPLV). The IPLV uses a standard rating point with specific values for evaporator and condenser temperatures, water flows, and fouling factors at specific load percentages. The NPLV allows selection of these values to be more representative of the actual operating conditions. Other terms used primarily for residential and commercial air-conditioning units and heat pumps include **seasonal energy efficiency ratio** (SEER) and **heating season performance factor** (HSPF). These also adjust ratings for the variations encountered over a normal heating or cooling season.

The various association and society standards have specific conditions at which equipment is rated. In efficiency comparisons, familiarity with the specific rating method is important. The factors used in the calculation of manufacturer's published data may not be representative of the specific application. Part load conditions must be considered in any evaluation when the unit will not normally be fully loaded. If the load is constant and is known, comparisons should be made at that point. Other important factors include whether the stated efficiency is just the compressor efficiency, or if it includes motor, seal, bearing, and drive losses. It is also recommended that rated operating conditions be adjusted to actual conditions when known. For example, the actual condensing temperatures may be significantly higher or lower than rated conditions depending on the geographic region.

Units of Refrigeration In the United States, refrigeration capacity is defined in terms of the standard ton. A **standard ton of refrigeration** corresponds to a heat absorption at a rate of 288,000 Btu/day or 200 Btu/min (3.5168 kW). The heat absorption per day is approximately the heat of fusion of 1 ton (907.19 kg) of ice at 32°F (0°C). The **standard rating** of a refrigerating machine, using a condensable vapor, is the number of standard tons of refrigeration it can produce under the following conditions: (1) liquid only enters the expansion valve and vapor only enters the compressor or the absorber of an absorption system; (2) the liquid entering the expansion valve is subcooled 9°F (5°C) and the vapor entering the compressor or absorber is superheated 9°F (5°C), these temperatures to be measured within 10 ft (3.05 m) of the compressor cylinder or absorber; (3) the pressure at the compressor or absorber inlet corresponds to a saturation temperature of 5°F (−15°C); and (4) the pressure at the compressor or absorber outlet corresponds to a saturation temperature of 86°F (30°C).

The **British unit of refrigeration** corresponds to a heat-absorption rate of 237.6 Btu/min (4.175 kW) with inlet and outlet pressures corresponding to saturation temperatures of 23°F (−5°C) and 59°F (15°C), respectively. In locations outside the United States it is common practice to use kilowatts to specify refrigeration capacity. One kilowatt of refrigeration capacity is roughly equivalent to 0.2844 tons. Occasionally a unit of refrigeration capacity called the **frigorie** is used. A frigorie is approximately equivalent to 50 Btu/min (0.8786 kW), or one-quarter of a standard ton of refrigeration.

PROPERTIES OF REFRIGERANTS

(For detailed properties, see Sec. 2.)

Refrigerants are the transport fluids which convey the heat energy from the low-temperature level to the high-temperature level, where it can, in terms of heat transfer, give up its heat. In the broad sense, gases involved in liquefaction processes or in gas-compression cycles go through low-temperature phases and hence may be termed "refrigerants," in a way similar to the more conventional vapor-compression fluids.

Refrigerants are **designated by number.** The identifying number may be preceded by the letter R, the word "Refrigerant," or the manufacturer's trademark or trade name. The trademarks or trade names shall not be used to identify refrigerants on equipment or in specifications. ANSI/ASHRAE Standard 34-2001, "Designation and Safety Classification of Refrigerants," and addenda cover the designation methodology and safety classifications for toxicity and flammability.

The **safety classification** is designated by a two-character alphanumeric system. The first character indicates toxicity, and is either an A or B, with lower toxicity indicated by an A. The second character indicates flammability, and is a 1, 2, or 3, with the lowest flammability indicated by a 1.

There are two terms widely used to describe the relative effect refrigerants and other chemicals have on the environment. The **ozone depletion potential** (ODP) is the ozone-destroying effect of a substance compared to a similar mass of Refrigerant 11. The **global warming potential** (GWP) is the relative measure of the warming caused by a substance versus a similar mass of carbon dioxide. Substances that increase global warming are referred to as "greenhouse gases."

Much of the damage to the atmospheric ozone layer is believed to be the result of decomposition of chlorofluorocarbon (CFC) and hydrochlorofluorocarbon (HCFC) chemicals. An international agreement known as the Montreal Protocol took effect in 1989. This agreement and subsequent amendments to it resulted in scheduled phaseouts of CFC and HCFC refrigerants. The schedule is dependent on the specific refrigerant and the country, with CFCs and developed nations having the most aggressive phaseouts.

International legislation has primarily addressed ozone depletion; however, global warming is an important and an emerging issue. There have been international meetings initiated by the United Nations Environmental Programme (UNEP) resulting in the Kyoto Protocol. Framework to reduce global warming has been developed, and legislation is being proposed. As with ODP, both CFC and HCFC refrigerants are affected. In addition, the replacement hydrofluorocarbon (HFC) refrigerants are under increased scrutiny, especially in Europe. Global warming is a significantly more complex issue since there are many sources of greenhouse gas emissions. A common debate is whether HFC equipment has a lower overall global warming impact due to its higher efficiency and reduced power consumption than refrigerants with lower GWP but higher power consumption. A listing of the ODP, GWP, and 8-hour toxicity limits (AEL or TLV) for various refrigerants is given in Table 11.1.1.

As a result of the legislation requiring phaseout of CFC and HCFC refrigerants, new refrigerants, primarily HFCs, have been commercialized to replace them. There has also been a renewed interest in non-halocarbon refrigerants such as ammonia and carbon dioxide. A listing of **replacement refrigerants** for existing CFC and HCFC compounds is given in Table 11.1.2.

Table 11.1.1 Environmental Properties of Refrigerants

Refrigerant	ODP	GWP	Toxicity, ppm (v/v)	AEL,* TLV†	Flash point, lower/upper, vol. %
R-11	1	4,600	1,000	TLV	None
R-12	1	10,600	1,000	TLV	None
R-22	0.055	1,700	1,000	TLV	None
R-114	1	9,800	1,000	TLV	None
R-123	0.02	95	50	AEL	None
R-124	0.22	599	1,000	AEL	None
R-125	0	2,800	1,000	AEL	None
R-134a	0	1,300	1,000	AEL	None
R-401A	0.037	1,120	1,000	AEL	None
R-401B	0.04	1,220	1,000	AEL	None
R-402A	0.021	2,330	1,000	AEL	None
R-402B	0.033	2,080	1,000	AEL	None
R-404A	0	3,260	1,000	AEL	None
R-407C	0	1,530	1,000	AEL	None
R-410A	0	1,730	1,000	AEL	None
R-502	0.334	4,520	1,000	TLV	None
R-508B	0	10,350	1,000	AEL	None

*AEL is the recommended time-weighted average concentration of an airborne chemical to which nearly all workers may be exposed during an 8-h and/or 12-h day, 40-h week without adverse effect. Determined by E. I. du Pont de Nemours and Company for compounds that do not have a TLV.

†TLV, established for industrial chemicals by the American Conference of Governmental Industrial Hygienists, is the recommended time-weighted average concentration of an airborne chemical to which nearly all workers may be exposed during an 8-h day, 40-h week without adverse effect.

SOURCE: Data from E. I. du Pont de Nemours and Company.

Table 11.1.2 Replacement Refrigerant Compounds

Current refrigerant	Replacement refrigerant	Type	Application*
R-11 (CFC)	R-123 (HCFC)†	HCFC	Chillers [+25°F (−3.9°C)]
R-12 (CFC)	R-134a†	HFC	Medium temp [−5°F (−20.6°C)]
	R-401B	HCFC	
R-13 (CFC)	R-508B	PFC	Low temp [−80°F (−62.2°C)]
R-22 (HCFC)†	R-407C	HFC	Retrofit and new equipment
	R-410A	HFC	New equipment
R-500 (CFC)	R-401B	HCFC	
R-503 (CFC)	R-508B	PFC	

*Lower optimum use temperature.

†See Figs. 11.1.1 to 11.1.3 for the thermodynamic properties of R-22, R-123, and R-134a.

SOURCE: Data from E. I. du Pont de Nemours and Company, Bulletin 300151B.

The term "replacement refrigerant" does not mean the replacement can simply be substituted for the existing refrigerant in the existing equipment. In many cases "replacement" refers to the application, such as temperature range, capacity range, or compressor type. Where existing equipment can be converted, the new refrigerant may not be miscible with the petroleum-based lubricant used with the CFC refrigerant. Synthetic oils such as polyol ester and polyalkylene glycol must be used with the new refrigerants. Seals, bushings, gaskets, and motor insulation (hermetic machines) may also need replacement. Centrifugal machines may require compressor wheel or speed changes to meet the head requirements of the replacement refrigerant. Controls may need to be replaced or reset. The resulting efficiency and capacity will likely be different from that of the original machine.

The owner must decide whether field conversion is justified. The following should be considered: the cost of conversion versus the cost of a new machine, the anticipated capacity and efficiency of the converted equipment versus that of a new machine, and the condition and remaining life of the converted equipment.

The refrigerant charge in a system must be handled in an environmentally acceptable manner to accomplish repairs requiring refrigerant removal. Venting of refrigerant is prohibited. Sections 608 and 609 of the Clean Air Act mandate the recovery of CFC, HCFC, HFC, and PFC refrigerants. The rules on recovery, recycling, repair of leaks, reporting, record keeping, and technician certification are strict, with the imposition of severe penalties for violation. For further information concerning regulations related to refrigerants, visit the EPA's website at http://www.epa.gov.

The **normal boiling point** is an important attribute of refrigerants for vapor compression systems. The desirable characteristic is a positive gage pressure throughout the refrigerating system. This will eliminate contamination of the refrigerant and resulting heat-transfer inefficiencies due to air inleakage. Table 11.1.3 shows some common fluids in order of boiling point: included are halocarbon compounds that offer reduced ozone depletion potential, hydrocarbons, sulfur compounds, and inorganic compounds.

The desirable thermal properties of a refrigerant are (1) convenient evaporation and condensation pressures, (2) high critical and low freezing temperatures, (3) high latent heat of evaporation and high vapor specific heat, (4) low viscosity and high film heat conductivity. Desirable practical properties include (1) low cost, (2) chemical and physical inertness under operating conditions, (3) noncorrosiveness toward ordinary construction materials, and (4) low explosive hazard both alone and mixed with air. The refrigerant should be nonpoisonous and nonirritating and should not cause deterioration in the lubricant used. Leakage should be detectable by simple tests, easily performed.

The **specific volume of refrigerant** to be handled determines the size of positive displacement compressors, but with centrifugal compressors a large volume is not objectionable and may be an advantage for small units.

A comparison of various refrigerants based on *ideal performance* is given in Table 11.1.4. The results are for the **standard temperature**

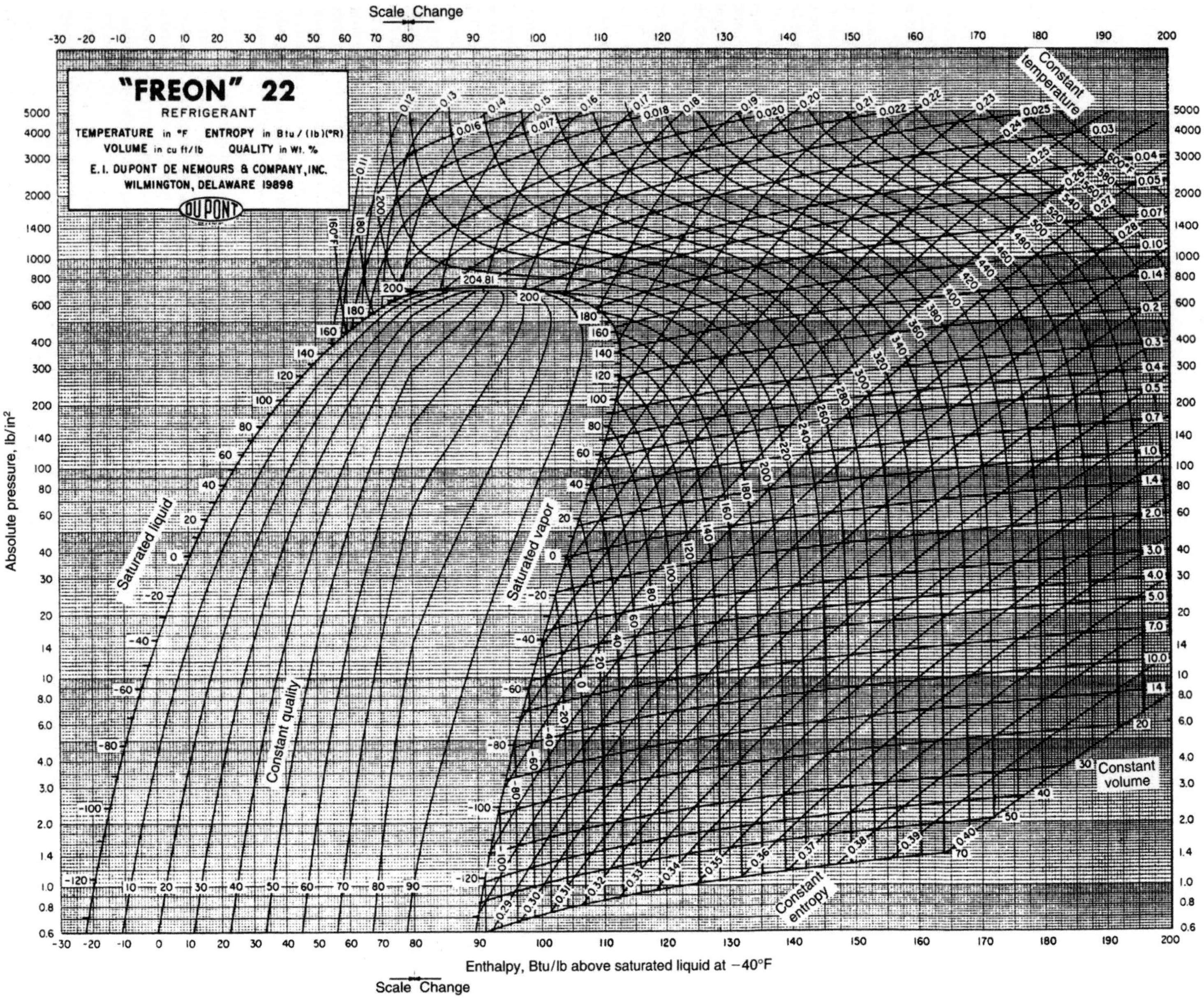

Figure 11.1.1 Thermodynamic properties of R-22. *(Reprinted with permission of E. I. du Pont de Nemours and Company.)*

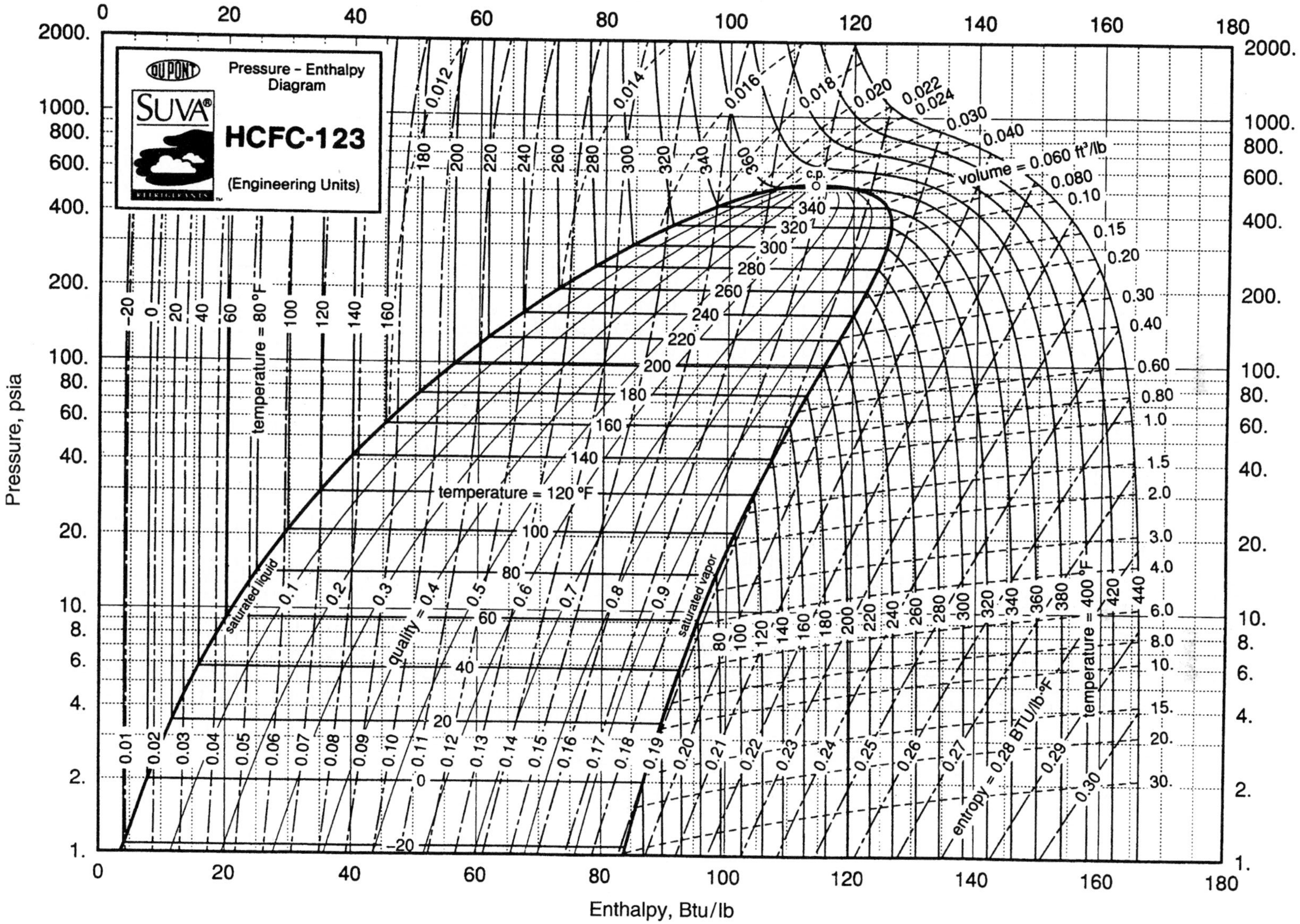

Figure 11.1.2 Thermodynamic properties of HCFC-123. *(Reprinted with permission of E. I. du Pont de Nemours and Company.)*

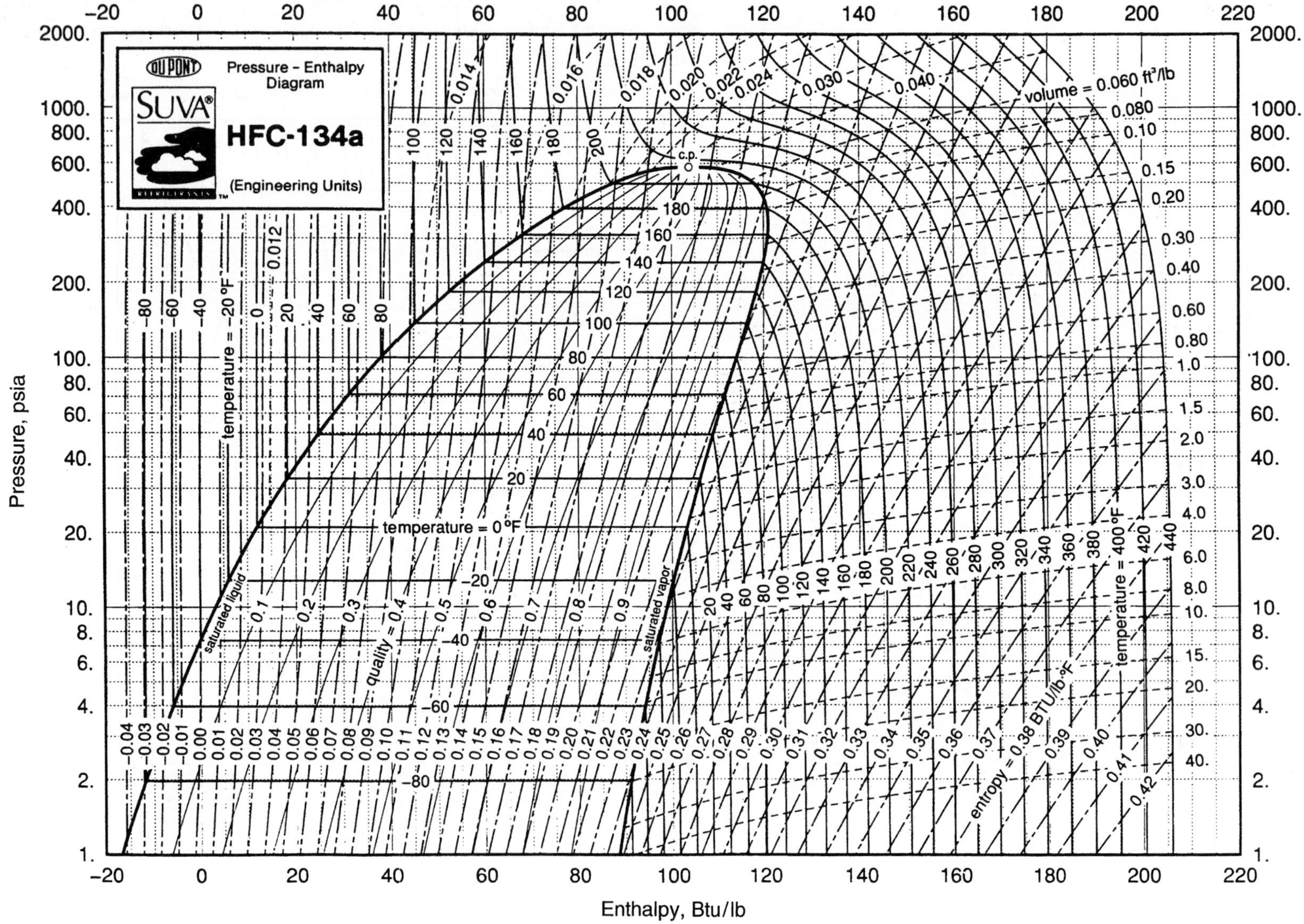

Figure 11.1.3 Thermodynamic properties of HCFC-134a. *(Reprinted with permission of E. I. du Pont de Nemours and Company.)*

Table 11.1.3 Selected Refrigerants and Gases in Low-Temperature Applications

Refrigerant	Refrigerant no.	Symbol	Molecular weight	Boiling point, °F	Critical point: Temperature, °F	Critical point: Pressure, lb/in² abs
Helium	704	He	4.0	−452.0	−450.0	33
Hydrogen	702	H_2	2.0	−423.0	−400.0	188
Nitrogen	728	N_2	28.0	−320.0	−232.0	492
Air	729	—	29.0	−312.0	−220.3	547
Oxygen	732	O_2	32.0	−297.0	−182.0	730
Methane	50	CH_4	16.0	−258.9	−115.8	673
Carbon tetrafluoride	14	CF_4	88.0	−198.4	−49.9	542
Ethylene	1,150	C_2H_4	28.0	−155.0	48.8	732
Ethane	170	C_2H_6	30.0	−127.5	90.1	708
Trifluoromethane	23	CHF_3	70.01	−115.66	78.6	701.4
Carbon dioxide	744	CO_2	44.0	−109.3	87.8	1,071
Propane	290	C_3H_8	44.1	−44.2	202.0	661
Monodichlorodifluoromethane	22	$CHClF_2$	86.5	−41.36	204.8	716
Ammonia	717	NH_3	17.0	−28.0	271.4	1,657
1,1,1,2-tetrafluoroethane	134a	CH_2FCF_3	102.0	−15.0	213.9	588.3
Methyl chloride	40	CH_3Cl	50.5	−10.8	289.4	969
Isobutane	601	C_4H_{10}	58.1	10.3	272.7	537
2-chloro-1,1,1,2-tetrafluoroethane	124	$CHClFCF_3$	136.5	12.2	252.0	524.5
Sulfur dioxide	764	SO_2	64.1	14.0	314.8	1,142
Butane	600	C_4H_{10}	58.1	31.3	306.0	550
Dichloromonofluoromethane	21	$CHCl_2F$	102.9	48.1	353.3	750
2,2-dichloro-1,1,1-trifluoroethane	123	$CHCl_2CF_3$	152.9	82.0	362.8	533.2
Ethyl-ether	610	$C_4H_{10}O$	74.1	94.3	522.1	381
Methylene chloride	30	CH_2Cl_2	84.9	103.6	480.0	670
Water	718	H_2O	18.0	212.0	706.1	3,226

Table 11.1.4 Ideal Performance of Refrigerants for Various Temperature Ranges

Refrigerant and (number)	Operating temperature range, °F	Suction pressure, lb/in² abs	Head pressure, lb/in² abs	Ratio of head to suction pressure	With dry and saturated suction vapor, per ton: Weight of vapor, lb/min	Piston displacement, ft³/min	Theoretical hp	Temperature at end of compression, °F	Type of compressor
Air (729)	5-86	14.7	73.5	5.0	7.02	82.3	2.82	277.0	Recip. and exp. Cyl
Water (718)	32-86	0.0885	0.6152	6.95	0.1957	647.0	0.618	332	Centrif. or ejector
	32-100	0.0885	0.9492	10.73	0.1985	656.2	0.819	420	
	40-100	0.1217	0.9492	7.80	0.1978	483.4	0.687	366	
Carbon dioxide (CO_2) (744)	5-86	332.0	1,043.0	3.14	3.61	0.960	1.827	160.3	Recip.
Ammonia (NH_3)(717)	5-86	34.27	169.2	4.94	0.421	3.44	0.99	209.8	Recip.
	20-100	48.21	211.9	4.40	0.421	2.49	0.94	212.8	
	40-100	73.32	211.9	2.89	0.427	1.70	0.65	176.0	
SUVA 123	5-86	2.303	15.9	6.9	3.26	45.6	0.95*	90.5	Recip.
($CHCl_2CF_3$)	20-100	3.48	20.8	5.97	3.33	31.7	0.91	101.0	Centrif.
(HCFC-123)	40-100	5.79	20.8	3.59	3.18	18.8	0.6*	100.0	Centrif.
SUVA 134a	5-86	23.77	111.8	4.7	3.14	6.1	1.04	98.0	Centrif.
(CH_2FCF_3)	20-100	33.13	139.0	4.19	3.28	4.62	1.0*	109.0	
(HFC-134a)	40-100	49.77	139.0	2.79	3.14	3.0	0.68	106.0	
SUVA HP 81	5-86	54.37	200.5	3.68	3.56	3.37	1.07	99.0	Screw or
blend [R-402	20-100	72.12	243.9	3.38	3.78	2.71	1.04	121.5	rotary or
(38/2/60)]	40-100	102.26	243.9	2.385	3.62	1.83	0.68*	114.0	recip.
R 22	5-86	43.02	174.5	4.06	2.926	3.65	1.03	131.7	Recip.
($CHClF_2$)	20-100	57.98	212.6	3.67	3.023	2.83	0.99	152.7	
(22)	40-100	83.72	212.6	2.54	2.936	1.93	0.68	131.0	
Methylene	5-86	1.28	10.07	8.56	1.485	74.0	0.96*	205.1*	Centrif.
chloride	20-100	1.92	13.25	6.90	1.520	47.72	0.91*	157.1*	
(CH_2Cl_2)(30)	40-100	3.38	13.25	3.92	1.493	27.76	0.63*	167.7*	
Methyl chlo-	5-86	21.15	94.70	4.48	1.331	5.95	0.96	178.1	Recip.
ride (CH_3Cl)	20-100	29.16	116.7	4.00	1.363	4.51	0.90	184.4	
(40)	40-100	43.25	116.7	2.69	1.342	3.07	0.62	157.0	
Sulfur dioxide	5-86	11.81	66.45	5.63	1.415	9.08	0.97	191.4	Recip.
(SO_2) (764)	20-100	17.18	84.52	4.92	1.453	6.52	0.92	193.4	
	40-100	27.10	84.52	3.12	1.444	4.17	0.63	162.6	
Propane	5-86	42.1	155.3	3.69	1.653	4.10	1.35*	92.9*	Recip.
(C_3H_8) (290)	20-100	55.5	187.0	3.37	1.730	3.29	1.32*	103.9*	
	40-100	78.0	187.0	2.40	1.646	2.26	0.90*	101.5*	
Ethane (C_2H_5) (170)	5-86	236.0	675.9	2.87	3.41	1.82	2.180	105	Recip.
Ethyl chloride	5-86	4.65	27.10	5.83	1.405	24.0	0.95*	106.3*	Rotary
(C_2H_5Cl)	20-100	6.80	34.79	5.12	1.425	17.19	0.92*	116.1*	
(160)	40-100	10.79	34.79	3.22	1.375	10.73	0.63*	109.9*	

* Values may be slightly in error.

range of 5 to 86°F (−15 to 30°C) and also for certain other ranges. For these calculations, zero piston clearance and expansion through a throttle valve were assumed. The use of an expansion cylinder instead of a valve would have yielded work (available for supplying some of the work of compression), but this is always a negligible quantity, except when the compression pressure approaches the critical pressure, as for carbon dioxide. Theoretical *coefficients of performance* may be obtained by dividing 4.72 by the theoretical horsepower given in the table.

Ammonia (R-717) is used extensively in commercial, industrial, and moderately low-temperature installations. It is toxic, and recent laws regulate the operation, maintenance, and design of ammonia systems that contain 10,000 lb of refrigerant or more. (See Document 29 CFR 1910.119 of the Federal Register for further information.) Ammonia is provided commercially in several purity grades. It is recommended that refrigerant-grade anhydrous ammonia of 99.95 percent purity be used. Moisture contamination from impure ammonia will accumulate in evaporator coils and inhibit performance. Ammonia is corrosive to copper and copper-bearing alloys. It is not miscible to any large extent with lubricating oils. Exposure to it in a confined space at concentrations lower than 400 ppm is rarely fatal. Ammonia is immediately fatal at 0.5 to 0.6 percent by volume. The presence of ammonia can be detected by burning a sulfur candle in the vicinity of a leak. A white cloud will form if ammonia vapors are present. Hydrochloric acid solutions and wetted litmus papers will also react with ammonia to give visible evidence of the presence of vapors. See Tables 11.1.5 and 11.1.6 for thermodynamic properties of R-717.

Table 11.1.5 Properties of Superheated Ammonia

(v = specific volume in ft³/lb; h = enthalpy in Btu/lb; s = entropy; h_f and s_f are measured from −40°F.)

Pressure, lb/in² abs	Temperature of saturated vapor, °F		Temperature of superheated vapor, °F: −30	−20	−10	0	10	20	30	40	50
10	−41.34	v	26.58	27.26	27.92	28.58	29.24	29.90	30.55	31.20	31.85
		h	603.2	608.5	613.7	618.9	624.0	629.1	634.2	639.3	644.4
		s	1.4420	1.4542	1.4659	1.4773	1.4884	1.4992	1.5097	1.5200	1.5301
20	−16.64	v			13.74	14.09	14.44	14.78	15.11	15.45	15.78
		h			610.0	615.5	621.0	626.4	631.7	637.0	642.3
		s			1.3784	1.3907	1.4025	1.4138	1.4248	1.4356	1.4460
30	−0.57	v				9.250	9.492	9.731	9.966	10.20	10.43
		h				611.9	617.8	623.5	629.1	634.6	640.1
		s				1.3371	1.3497	1.3618	1.3733	1.3845	1.3953
40	11.66	v						7.203	7.387	7.568	7.746
		h						620.4	626.3	632.1	637.8
		s						1.3231	1.3353	1.3470	1.3583
50	21.67	v							5.838	5.988	6.135
		h							623.4	629.5	635.4
		s							1.3046	1.3169	1.3286

Pressure, lb/in² abs	Temperature of saturated vapor, °F		100	120	140	160	180	200	240	280	320
80	44.40	v	4.190	4.371	4.548	4.722	4.893	5.063	5.398	5.730	
		h	658.7	670.4	681.8	693.2	704.4	715.6	738.1	760.7	
		s	1.3199	1.3404	1.3598	1.3784	1.3963	1.4136	1.4467	1.4781	
100	56.05	v	3.304	3.454	3.600	3.743	3.883	4.021	4.294	4.562	
		h	655.2	667.3	679.2	690.8	702.3	713.7	736.5	759.4	
		s	1.2891	1.3104	1.3305	1.3495	1.3678	1.3854	1.4190	1.4507	
120	66.02	v	2.712	2.842	2.967	3.089	3.190	3.326	3.557	3.783	
		h	651.6	664.2	676.5	688.5	700.2	711.8	734.9	758.0	
		s	1.2628	1.2850	1.3058	1.3254	1.3441	1.3620	1.3960	1.4281	
140	74.79	v	2.288	2.404	2.515	2.622	2.727	2.830	3.030	3.227	3.420
		h	647.8	661.1	673.7	686.0	698.0	709.9	733.3	756.7	780.0
		s	1.2396	1.2628	1.2843	1.3045	1.3236	1.3418	1.3763	1.4088	1.4395
160	82.64	v	1.969	2.075	2.175	2.272	2.365	2.457	2.635	2.809	2.980
		h	643.9	657.8	670.9	683.5	695.8	707.9	731.7	755.3	778.9
		s	1.2186	1.2429	1.2652	1.2859	1.3054	1.3240	1.3591	1.3919	1.4229
180	89.78	v	1.720	1.818	1.910	1.999	2.084	2.167	2.328	2.484	2.637
		h	639.9	654.4	668.0	681.0	693.6	705.9	730.1	753.9	777.7
		s	1.1992	1.2247	1.2477	1.2691	1.2891	1.3081	1.3436	1.3768	1.4081
200	96.34	v	1.520	1.612	1.698	1.780	1.859	1.935	2.082	2.225	2.364
		h	635.6	650.9	665.0	678.4	691.3	703.9	728.4	752.5	776.5
		s	1.1809	1.2077	1.2317	1.2537	1.2742	1.2935	1.3296	1.3631	1.3947
220	102.42	v		1.443	1.525	1.601	1.675	1.745	1.881	2.012	2.140
		h		647.3	662.0	675.8	689.1	701.9	726.8	751.1	775.3
		s		1.1917	1.2167	1.2394	1.2604	1.2801	1.3168	1.3507	1.3825
240	108.09	v		1.302	1.380	1.452	1.521	1.587	1.714	1.835	1.954
		h		643.5	658.8	673.1	686.7	699.8	725.1	749.8	774.1
		s		1.1764	1.2025	1.2259	1.2475	1.2677	1.3049	1.3392	1.3712
260	113.42	v		1.182	1.257	1.326	1.391	1.453	1.572	1.686	1.796
		h		639.5	655.6	670.4	684.4	697.7	723.4	748.4	772.9
		s		1.1617	1.1889	1.2132	1.2354	1.2560	1.2938	1.3258	1.3608

SOURCE: Condensed from NBS Circ. 142, 1923.

Table 11.1.6 Properties of Saturated Ammonia
(h_f and s_f are measured from −40°F.)

Temp., °F, t	Pressure, lb/in² abs, p	Specific volume, ft³/lb: Sat. liquid, v_f	Sat. vapor, v_g	Enthalpy, Btu: Sat. liquid, h_f	Vaporization, h_{fg}	Sat. vapor, h_g	Entropy: Sat. liquid, s_f	Vaporization, s_{fg}	Sat. vapor, s_g
−40	10.41	0.02322	24.86	0.0	597.6	597.6	0.0000	1.4242	1.4242
−38	11.04	0.02326	23.53	2.1	596.2	598.3	0.0051	1.4142	1.4193
−36	11.71	0.02331	22.27	4.3	594.8	599.1	0.0101	1.4043	1.4144
−34	12.41	0.02335	21.10	6.4	593.5	599.9	0.0151	1.3945	1.4096
−32	13.14	0.02340	20.00	8.5	592.1	600.6	0.0201	1.3847	1.4048
−30	13.90	0.02345	18.97	10.7	590.7	601.4	0.0250	1.3751	1.4001
−28	14.71	0.02349	18.00	12.8	589.3	602.1	0.0300	1.3655	1.3955
−26	15.55	0.02354	17.09	14.9	587.9	602.8	0.0350	1.3559	1.3909
−24	16.42	0.02359	16.24	17.1	586.5	603.6	0.0399	1.3464	1.3863
−22	17.34	0.02364	15.43	19.2	585.1	604.3	0.0448	1.3370	1.3818
−20	18.30	0.02369	14.68	21.4	583.6	605.0	0.0497	1.3277	1.3774
−18	19.30	0.02374	13.97	23.5	582.2	605.7	0.0545	1.3184	1.3729
−16	20.34	0.02378	13.29	25.6	580.8	606.4	0.0594	1.3092	1.3686
−14	21.43	0.02383	12.66	27.8	579.3	607.1	0.0642	1.3001	1.3643
−12	22.56	0.02388	12.06	30.0	577.8	607.8	0.0690	1.2910	1.3600
−10	23.74	0.02393	11.50	32.1	576.4	608.5	0.0738	1.2820	1.3558
−8	24.97	0.02399	10.97	34.3	574.9	609.2	0.0786	1.2730	1.3516
−6	26.26	0.02404	10.47	36.4	573.4	609.8	0.0833	1.2641	1.3474
−4	27.59	0.02409	9.991	38.6	571.9	610.5	0.0880	1.2553	1.3433
−2	28.98	0.02414	9.541	40.7	570.4	611.1	0.0928	1.2465	1.3393
0	30.42	0.02419	9.116	42.9	568.9	611.8	0.0975	1.2377	1.3352
2	31.92	0.02424	8.714	45.1	567.3	612.4	0.1022	1.2290	1.3312
4	33.47	0.02430	8.333	47.2	565.8	613.0	0.1069	1.2204	1.3273
6	35.09	0.02435	7.971	49.4	564.2	613.6	0.1115	1.2119	1.3234
8	36.77	0.02440	7.629	51.6	562.7	614.3	0.1162	1.2033	1.3195
10	38.51	0.02446	7.304	53.8	561.1	614.9	0.1208	1.1949	1.3157
12	40.31	0.02451	6.996	56.0	559.5	615.5	0.1254	1.1864	1.3118
14	42.18	0.02457	6.703	58.2	557.9	616.1	0.1300	1.1781	1.3081
16	44.12	0.02462	6.425	60.3	556.3	616.6	0.1346	1.1697	1.3043
18	46.13	0.02468	6.161	62.5	554.7	617.2	0.1392	1.1614	1.3006
20	48.21	0.02474	5.910	64.7	553.1	617.8	0.1437	1.1532	1.2969
22	50.36	0.02479	5.671	66.9	551.4	618.3	0.1483	1.1450	1.2933
24	52.59	0.02485	5.443	69.1	549.8	618.9	0.1528	1.1369	1.2897
26	54.90	0.02491	5.227	71.3	548.1	619.4	0.1573	1.1288	1.2861
28	57.28	0.02497	5.021	73.5	546.4	619.9	0.1618	1.1207	1.2825
30	59.74	0.02503	4.825	75.7	544.8	620.5	0.1663	1.1127	1.2790
32	62.29	0.02508	4.637	77.9	543.1	621.0	0.1708	1.1047	1.2755
34	64.91	0.02514	4.459	80.1	541.4	621.5	0.1753	1.0968	1.2721
36	67.63	0.02521	4.289	82.3	539.7	622.0	0.1797	1.0889	1.2686
38	70.43	0.02527	4.126	84.6	537.9	622.5	0.1841	1.0811	1.2652
40	73.32	0.02533	3.971	86.8	536.2	623.0	0.1885	1.0733	1.2618
42	76.31	0.02539	3.823	89.0	534.4	623.4	0.1930	1.0655	1.2585
44	79.38	0.02545	3.682	91.2	532.7	623.9	0.1974	1.0578	1.2552
46	82.55	0.02551	3.547	93.5	530.9	624.4	0.2018	1.0501	1.2519
48	85.82	0.02558	3.418	95.7	529.1	624.8	0.2062	1.0424	1.2486
50	89.19	0.02564	3.294	97.9	527.3	625.2	0.2105	1.0348	1.2453
52	92.66	0.02571	3.176	100.2	525.5	625.7	0.2149	1.0272	1.2421
54	96.23	0.02577	3.063	102.4	523.7	626.1	0.2192	1.0197	1.2389
56	99.91	0.02584	2.954	104.7	521.8	626.5	0.2236	1.0121	1.2357
58	103.7	0.02590	2.851	106.9	520.0	626.9	0.2279	1.0046	1.2325
60	107.6	0.02597	2.751	109.2	518.1	627.3	0.2322	0.9972	1.2294
62	111.6	0.02604	2.656	111.5	516.2	627.7	0.2365	0.9897	1.2262
64	115.7	0.02611	2.565	113.7	514.3	628.0	0.2408	0.9823	1.2231
66	120.0	0.02618	2.477	116.0	512.4	628.4	0.2451	0.9750	1.2201
68	124.3	0.02625	2.393	118.3	510.5	628.8	0.2494	0.9676	1.2170
70	128.8	0.02632	2.312	120.5	508.6	629.1	0.2537	0.9603	1.2140
72	133.4	0.02639	2.235	122.8	506.6	629.4	0.2579	0.9531	1.2110
74	138.1	0.02646	2.161	125.1	504.7	629.8	0.2622	0.9458	1.2080
76	143.0	0.02653	2.089	127.4	502.7	630.1	0.2664	0.9386	1.2050
78	147.9	0.02661	2.021	129.7	500.7	630.4	0.2706	0.9314	1.2020

(Continued)

Table 11.1.6 Properties of Saturated Ammonia (*Continued*)
(h_f and s_f are measured from −40°F.)

		Specific volume, ft³/lb		Enthalpy, Btu			Entropy		
Temp., °F, t	Pressure, lb/in² abs, p	Sat. liquid, v_f	Sat. vapor, v_g	Sat. liquid, h_f	Vapor-ization, h_{fg}	Sat. vapor, h_g	Sat. liquid, s_f	Vapor-ization, s_{fg}	Sat. vapor, s_g
80	153.0	0.02668	1.955	132.0	498.7	630.7	0.2749	0.9242	1.1991
82	158.3	0.02675	1.892	134.3	496.7	631.0	0.2791	0.9171	1.1962
84	163.7	0.02684	1.831	136.6	494.7	631.3	0.2833	0.9100	1.1933
86	169.2	0.02691	1.772	138.9	492.6	631.5	0.2875	0.9029	1.1904
88	174.8	0.02699	1.716	141.2	490.6	631.8	0.2917	0.8958	1.1875
90	180.6	0.02707	1.661	143.5	488.5	632.0	0.2958	0.8888	1.1846
92	186.6	0.02715	1.609	145.8	486.4	632.2	0.3000	0.8818	1.1818
94	192.7	0.02723	1.559	148.2	484.3	632.5	0.3041	0.8748	1.1789
96	198.9	0.02731	1.510	150.5	482.1	632.6	0.3083	0.8678	1.1761
98	205.3	0.02739	1.464	152.9	480.0	632.9	0.3125	0.8608	1.1733
100	211.9	0.02747	1.419	155.2	477.8	633.0	0.3166	0.8539	1.1705
105	228.9	0.02769	1.313	161.1	472.3	633.4	0.3269	0.8366	1.1635
110	247.0	0.02790	1.217	167.0	466.7	633.7	0.3372	0.8194	1.1566
115	266.2	0.02813	1.128	173.0	460.9	633.9	0.3474	0.8023	1.1497
120	286.4	0.02836	1.047	179.0	455.0	634.0	0.3576	0.7851	1.1427

Carbon dioxide (R-744) was used for a long time as a "safety refrigerant." It is not toxic, and exposure to it is not dangerous unless concentrations are high. Like many other refrigerants, it is heavier than air, and caution must be exercised in confined spaces, as it may displace oxygen and result in asphyxiation. With cooling water at 70°F (21°C), the condensing pressure of CO_2 is high, and since its critical temperature is 87.8°F (31°C), condensation will not occur at water temperatures more than this. Specific power consumption is high on CO_2 machines.

Sulfur dioxide (R-764) has rapidly gone out of use. It is extremely corrosive unless absolutely anhydrous. It is extremely irritating in small concentrations and toxic in high concentrations.

Methyl chloride (R-40) (CH_3Cl), an anesthetic in amounts of 5 to 10 percent by volume, has been used in air-cooled units of moderate or small sizes. It is miscible with mineral oils; small amounts of moisture in a methyl chloride system will cause trouble by freezing in expansion valves.

For industrial work, butane (C_4H_{10}), propane (C_3H_8), and, at low temperatures, ethane (C_2H_6) are used. **Dieline** (dichloroethylene, $C_2H_2Cl_2$) and **Carrene** (dichloromethane, CH_2Cl_2) have been used to some extent, usually in centrifugal compressors.

In order to create refrigerants with thermodynamic properties similar to those being phased out, some refrigerants are blends of two or more other refrigerants. Blends have 400 and 500 series designations. For example, R-407C is a blend of three HFC refrigerants, R-32, R-125, and R-134a. It is designed to be used in new equipment and as a service replacement for existing R-22 machines. The thermodynamic properties are shown in Fig. 11.1.4.

Use caution when using blended refrigerants in centrifugal compressors. The compressor impeller may act to separate the components of the blend by molecular weight. For safety purposes, blends are classified on the basis of worst-case fractionation.

Building codes often require monitoring of refrigerant levels in equipment spaces. In addition to a monitor, an alarm system and exhaust facilities should be installed to prevent exposure precedes the acceptable exposure limit (AEL). ANSI/ASHRAE 15-2004, "Safety Standard for Refrigeration Systems," and addenda have recommendations covering equipment room safety.

Leak Detection Detection and repair of refrigerant leaks is an essential part of responsible refrigeration equipment operation. Worker safety, environmental impact, and refrigerant cost are the primary reasons for a detection and repair program. Section 608 of the Clean Air Act as amended July 24, 2003, has rules governing the keeping of service records documenting the date and type of service and the quantity of refrigerant added. This currently applies to equipment that contains 50 or more pounds of refrigerant.

Area monitors are useful for determining that a leak has occurred. Handheld detectors are useful for determining a specific leak location. For many years, refrigerant leaks were detected with a halide torch. Copper gauze was kept hot by the essentially colorless flame of burning methyl alcohol. An intense green color was imparted to the flame by the presence of even minute amounts of chlorine. The alternative refrigerants have no chlorine or a lower amount of chlorine. Some refrigerants are flammable. New types of detection equipment are required.

Nonselective, halogen-specific, and compound-specific detectors are available. Each type has limitations, and selection must be based on the specific application. For example, the nonspecific-type detector is poorly suited for area monitoring because of its high detection limit and potential for false alarms, but is a reliable, low-cost choice for pinpointing leaks. The halogen-specific detectors can be applied for either leak pinpointing or area monitoring. They are best applied where only one refrigerant is in use. The compound-specific detectors are relatively expensive, but are very selective. They are best applied in environments where several refrigerants or other chemical compounds may be present.

The pressure-temperature relations for the saturated vapors of many of the commercially important refrigerants are given in Fig. 11.1.5. If the temperature at which the refrigeration is desired and the temperature at which heat can be discarded (condenser temperature) are known, the chart is convenient for determining for any chosen refrigerant the pressures which must be maintained. For instance, if the use of methyl chloride is contemplated in an air-cooled cabinet food freezer located in a room where the air temperature may rise to 90°F (32°C), the compressor discharge must be at least 100 psia (690 kPa absolute), and if the cabinet cooling coil is to be held at −30°F (−34°C), compressor suction will be 9 psia (62 kPa absolute).

The large volumes required for HCFC-123 and dieline can be handled satisfactorily by centrifugal compressors. When the evaporating pressure is lower than atmospheric—as in the case of HCFC-123, dieline, ammonia in food freezers, sulfur dioxide, water vapor, and butane—air leaks are likely to occur. The air tends to accumulate in the condenser and the heat-transfer performance is degraded.

Refrigerating systems which operate with a positive pressure in the equipment will leak refrigerant into the atmosphere. Many HFC- and HCFC-type refrigerants are practically odorless. Dangerous concentrations can build up in unventilated enclosed rooms such as freezers should a significant leak occur. Leak detection devices are available commercially to provide warnings of equipment leakage before harmful concentrations are present in occupied spaces. Addition of certain fluorescent compounds into the refrigerant in an operating system will allow identification of equipment leaks. The additives are placed into

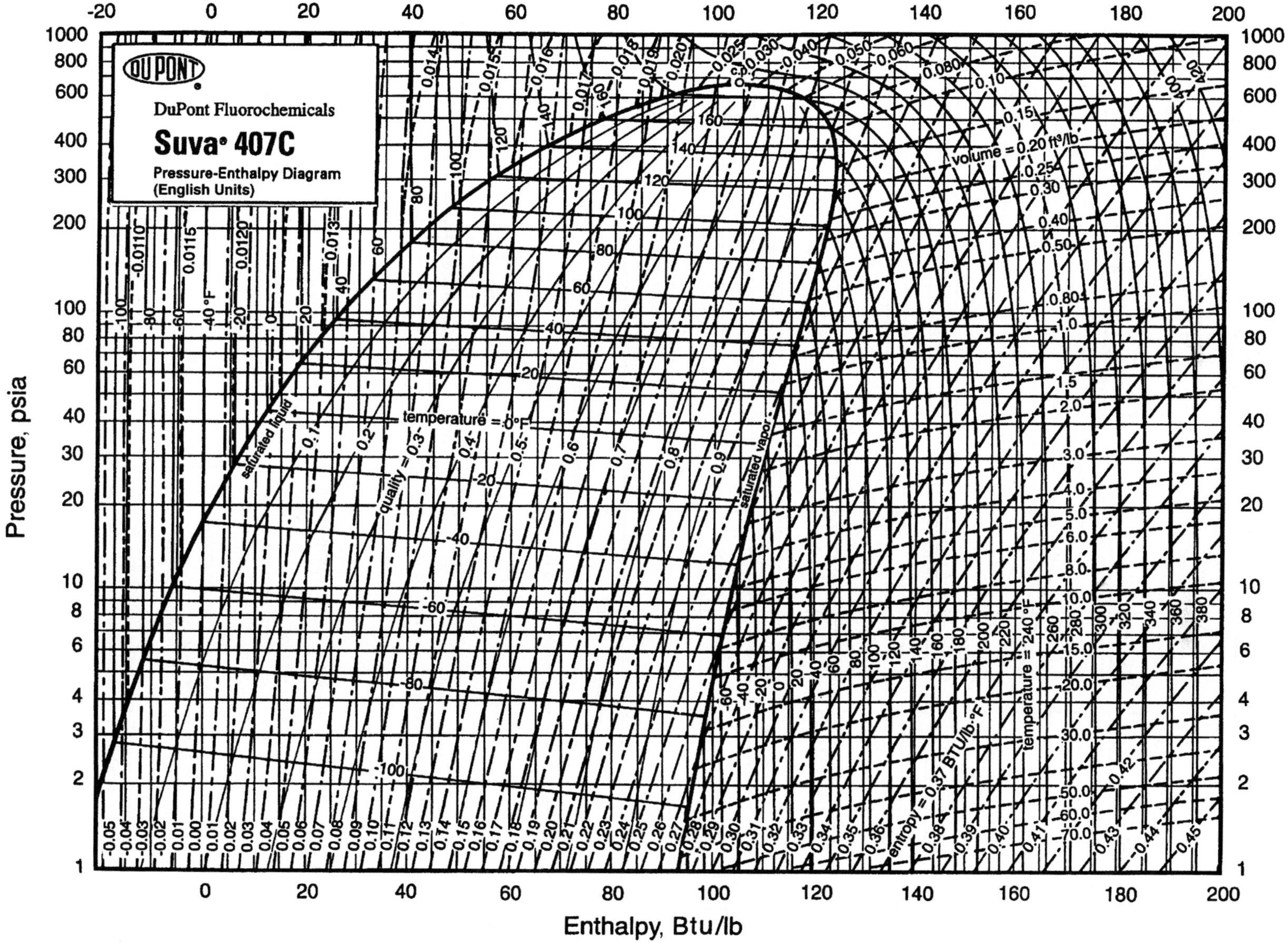

Figure 11.1.4 Thermodynamic properties of SUVA® 407C (R-407C). *(Reprinted with permission of E. I. du Pont de Nemours and Company.)*

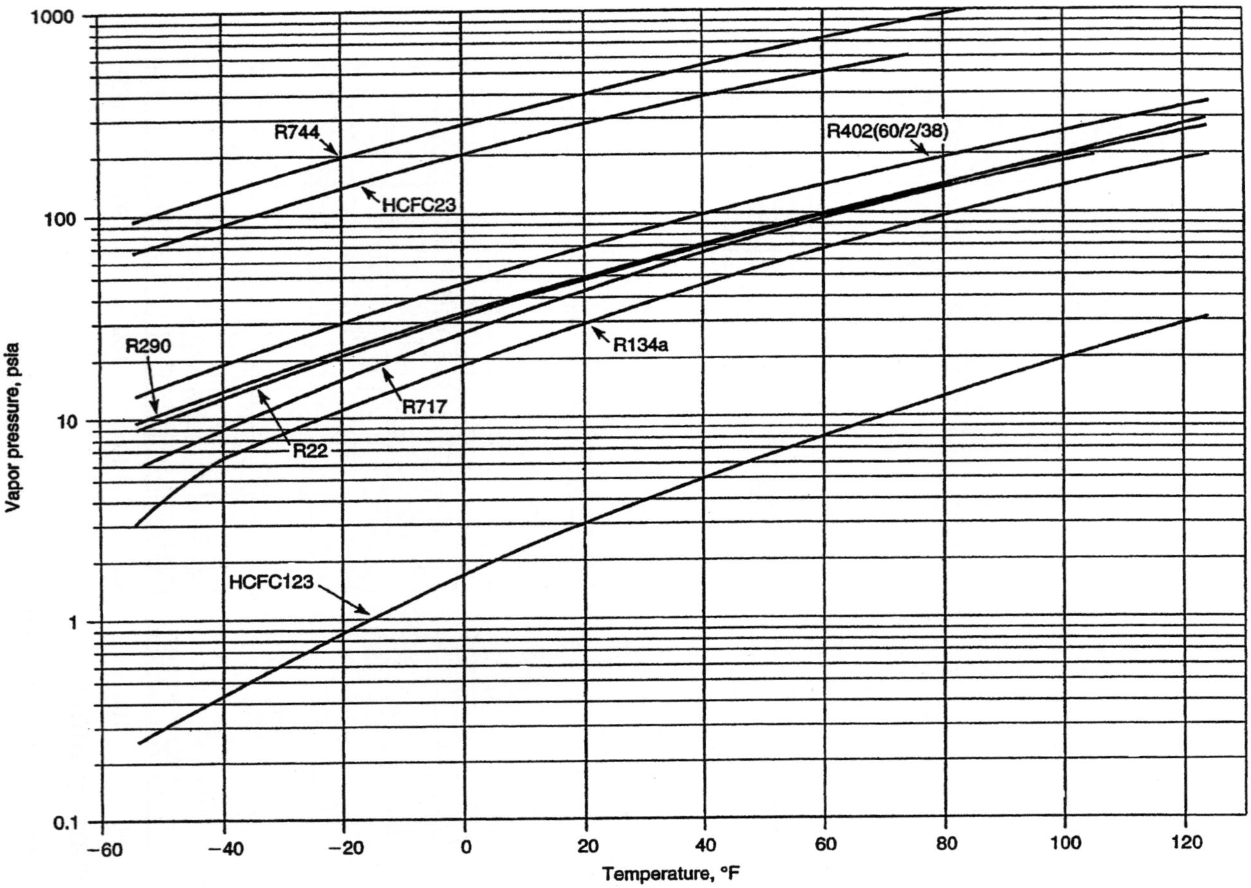

Figure 11.1.5 Pressure-temperature relations for saturated liquids of refrigerants.

the refrigeration lubricant when the system is serviced or charged. The compound travels through the system with the refrigerant and oil mixture. At the point of leakage, the fluorescent compound escapes from the system, and is left as a deposit on the outside surface of equipment. When illuminated by ultraviolet light, the deposit at the leak point becomes visible.

OVERALL CYCLES

Figure 11.1.6 represents the simple closed-circuit **vapor-compression** system. The upper heat exchanger is a vapor condenser (with some superheat), and after the throttling expansion valve, the lower one is the evaporator in which the refrigerant liquid at reduced pressure and temperature evaporates with the inward refrigeration heat flow. The refrigerant vapor is elevated by the compressor to a higher pressure and condensing temperature so that it will liquefy in its transfer of heat to the atmospheric level. Figures 11.1.7 and 11.1.8 illustrate the cycle.

Figure 11.1.9 is the counterpart of Fig. 11.1.6 and represents the closed sensible-heat **gas-compression** cycle, with an expander (either displacement or turbine-type) parallel to a gas compression which may provide a compressor-work "assist" in addition to its exhaust stream of cold gas. This is the refrigeration version of the Brayton cycle (see Sec. 2), the upper exchanger serving to cool the hot compressed gas, and the lower exchanger handling the sensible-heat refrigeration effect. In place of the expander, an expansion valve could be used; this is particularly effective on very low-temperature open-cycle operations, such as with gas liquefaction processes incorporating secondary "cold-regenerating" heat exchangers.

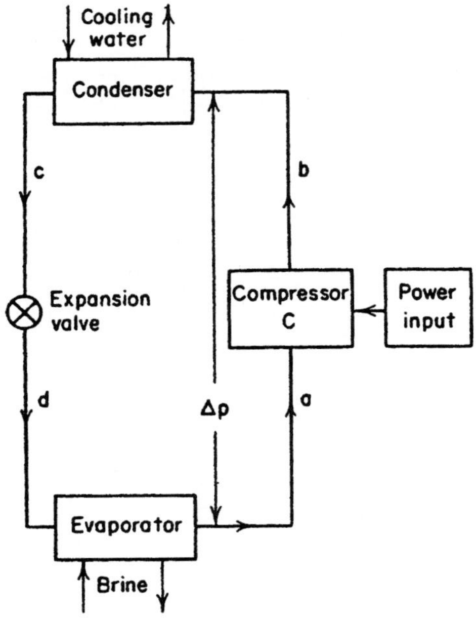

Figure 11.1.6 Simple closed-circuit vapor-compression system.

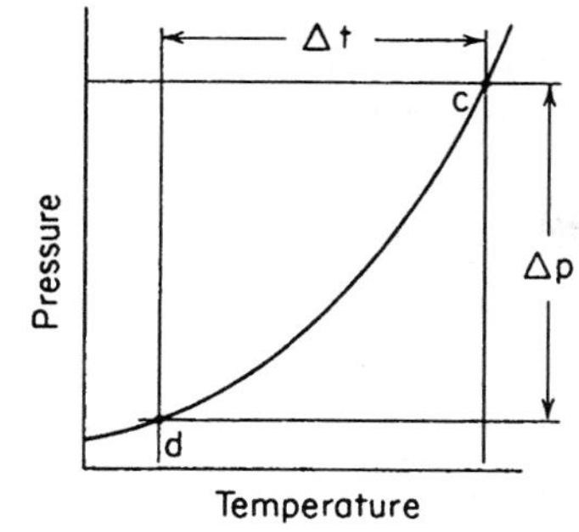

Figure 11.1.7 Pressure-temperature curve illustrating overall differences.

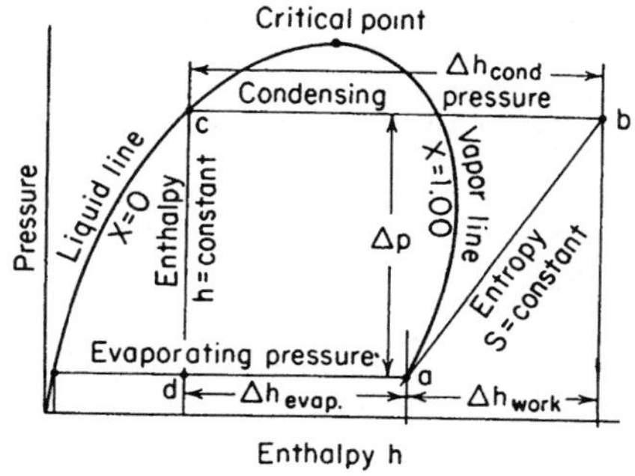

Figure 11.1.8 Pressure-enthalpy diagram for simple ideal refrigeration cycle.

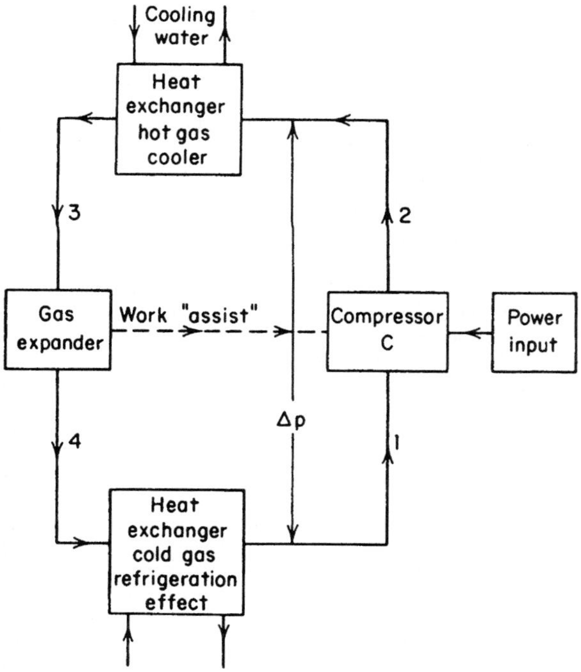

Figure 11.1.9 Closed sensible-heat gas-compression cycle.

An extended temperature-entropy diagram (Fig. 11.1.10) illustrates paths for gas-compression sensible-heat cycles as follows: (1) 1-2-3-4 isentropic expander; (2) 1-2-3-4′ polytropic expander; (3) 1-2-5-6 throttling (isenthalpic); and (4) 1-2-7-8 throttling (low temperature).

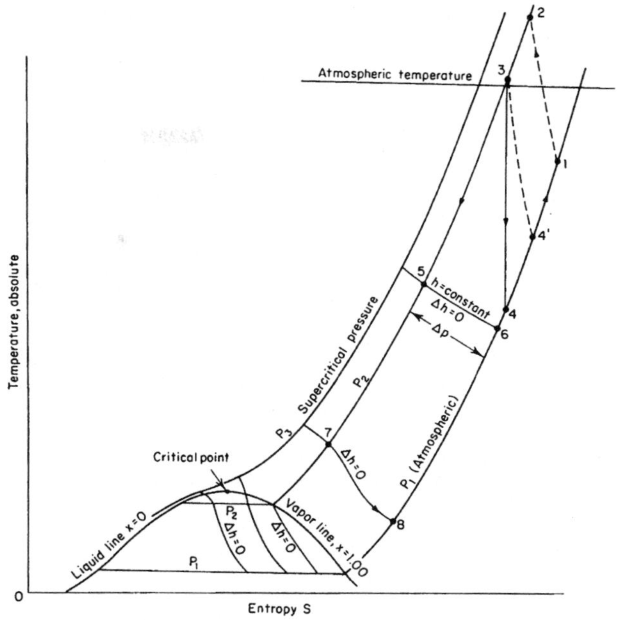

Figure 11.1.10 Extended temperature-entropy diagram for air illustrating sensible-heat air-compression processes.

COMPONENTS OF COMPRESSION SYSTEMS

A refrigeration system is an energy-transport complex of assorted components. It may be a conventional system in which the same fluid, the refrigerant, recirculates continuously in a closed circuit, or it may entail a partially open system with discharge of some of the processed fluid either as a liquefied gas or as a solidified gas, with replacement makeup, for example, by air liquefaction and solid carbon dioxide production. Components include vapor and gas compressors; liquid pumps; heat-transfer equipment (gas coolers, intercoolers, aftercoolers, exchangers, economizers); vapor condensers and the counterpart evaporators; liquid coolers and receivers; expanders; control valves and pressure-drop throttling devices (capillaries, refrigerant-mixture separating chambers, stream-mixing chambers); and connecting piping and insulation.

Codes and **standards** have been published by various associations and societies. These cover design, selection, designation, testing, sound ratings, safety, efficiency, and other areas relating to mechanical refrigeration. Publication of a comprehensive list is impractical because of frequent revisions. Several of the organizations include Air Conditioning Contractors of America (ACCA), American National Standards Institute (ANSI), Air-Conditioning and Refrigeration Institute (ARI), American Society of Heating, Refrigerating and Air-Conditioning Engineers, Inc. (ASHRAE), The American Society of Mechanical Engineers (ASME), Canadian Standards Association (CSA), International Institute of Ammonia Refrigeration (IIAR), and Underwriters Laboratories Inc. (UL). A list of current codes and standards can be obtained by contacting the particular organization.

Several of the more general refrigeration equipment standards include: ANSI/ASHRAE 15 "Safety Standard for Refrigeration Systems"; ANSI/ASHRAE 34 "Designation and Safety Classification of Refrigerants"; and CAN/CSA-C22.2 "General Standard on Refrigeration Equipment."

There are numerous additional standards relating to specific types of refrigeration equipment (air conditioners, heat pumps, chillers) or specific components (compressors, water- or air-cooled condensers, lubricating oil). Because refrigeration equipment requires auxiliaries (electric motors, steam turbines, piping, insulation) and is part of a larger system such as food processing or heating, ventilating, and air conditioning (HVAC), other standards not relating directly to refrigeration may apply. Any installation must also comply with building and fire code requirements.

Compression may be single stage (Fig. 11.1.6) or multiple stage. Displacement compressors include reciprocating piston, screw, rotary vane, scroll, and trochoidal compressors. The volumetric capacity of the compressors is always less than the swept volume of pistons or the volume displaced in cyclic rotation by the other forms of displacement devices such as screws, nutating scrolls, or rotating vanes. The true measure of compressor capacity is the volumetric efficiency. It is influenced by cylinder, screw mesh, or cavity clearance volume; pressure ratio; nature of the gas; pressure drops and passage volume in the valves or vanes; suction heating; and the equivalent compression exponent n. The numerical value of n expresses the effect of heat transfer to or from the gas during compression of the gas. Normally it lies between $n = 1.0$ and $n = k$ (adiabatic), such as for the ideal (isentropic). Figure 11.1.11 indicates the comparative volumetric efficiency of a pair of reciprocating and screw compressors. **Reciprocating compressors** are disadvantaged by high discharge temperatures and excessive clearance volume losses associated with compression ratios greater than 12 to 1 in a single compression stage. Compound compression staging is a practical solution if reciprocating machinery is applied across cycle compression ratios greater than 10 to 1. **Screw** compressors with oil injection systems can operate reliably as a single stage with compression ratios greater than 20 to 1. **Rotary vane** compressors are fixed-volume-ratio machines and can be used across compression ratios of 18 to 1, vane material and lubricating oil permitting; however, in the larger-size machines, a 7 to 1 compression ratio is commonly used in order to maintain efficiency. **Scroll** compressors have been used in commercial practice for systems that have capacities less than 10 tons (35 kW). **Trochoidal** compressors are often called "Wankel" compressors because, in one form, the rotor is similar to the Wankel internal combustion engine rotor.

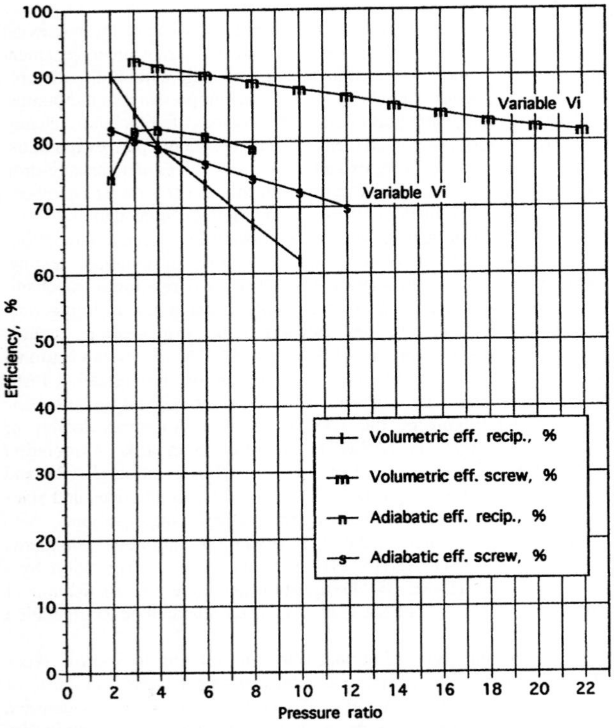

Figure 11.1.11 Screw and reciprocating compressor volumetric and adiabatic efficiency versus pressure ratio. *(Courtesy of J. W. Pillis, "Development of a Variable Volume Ratio Screw Compressor," Frick Company, 1983.)*

Reciprocating-piston compressors may be horizontal or vertical. The pistons can be designed to be either single acting or double acting with compression occurring on either stroke of the piston. Commercial designs have evolved into multiple cylinder machines arranged in a V block similar to an automotive engine. Average piston speed may be as great as 800 ft/min (4 m/s). Design characteristics of the suction and discharge valves installed on the cylinder determine the energy efficiency, and maintenance frequency. Valves most frequently used are the free-floating reed, clamped reed, and ring valve. Gas velocity through the valves is limited to 12,000 ft/min (61 m/s) with R-717 and 9,000 ft/min (46 m/s) with HCFC-22. For additional information on reciprocating compressors in refrigeration service, refer to the ASHRAE Handbook, 2004 Systems and Equipment, pp. 34.4 to 34.9.

Screw compressors are available in several designs, both single screw and twin screw, with oil-free and oil-injected designs in both types. Twin-screw oil-injected compressors are slightly more energy efficient at moderate compression ratios. Many of the commercially available screw compressors are fixed-volume-ratio machines. (See also Sec. 10.) If the compression ratio varies, the operational efficiency of the compressor also varies significantly. The energy reductions that are available by operating with cooler condensing pressures during spring, fall, and winter may not be fully achievable. Capacity control can be accomplished with a movable slide valve that varies the internal volume, valved suction ports, variable-speed drives, and hot-gas bypass. The slide valve is the most common method. Figure 11.1.11 illustrates the adiabatic efficiency of a twin-screw compressor with a movable slide stop control and a reciprocating compressor. Flash intercoolers (Fig. 11.1.13), also called *economizers,* can be provided to improve energy efficiency on the larger-capacity machines.

Air-cooled compressors are used where discharge temperatures are low; water cooling is used where discharge temperatures are high, as with ammonia, and on larger industrial units. **Oil separators** return to the compressor the lubricating oil carried over by the refrigerant vapors. **Rotary compressors** of the vane, eccentric, gear, and screw types are in use.

Kinetic compressors include high-speed centrifugal and axial flow machines, usually multistaged, and jet-entrainment devices. Centrifugal machines are especially adapted to high-volume flow (>500 ft^3/min) (0.24 m^3/s). (See also Sec. 10.)

The entire machine-compression operation may be replaced by a secondary absorber-pump-generator system, in which the complex is known as an absorption refrigeration system.

Dual (or **multiple effect**) **compression** may be used when refrigeration at two temperatures is desired. The compressor takes vapor from a lower temperature expansion coil during the first part of its intake stroke, and from a higher temperature expansion coil at or near the end of the stroke. The mixture is then compressed and condensed.

In **wet compression,** cooling is obtained during compression by spraying liquid refrigerant into the compressor cylinder. The desuperheating of the compressed vapors results in better heat transfer in condensers and more nearly isothermal compression; the disadvantages are reduced compressor capacity and the problem of control of the amount of injection.

Table 11.1.7 shows some ideal performance values for a single-state ammonia system under different conditions.

Condensers are usually shell-and-tube type, with the refrigerant passing outside the tubes. Older industrial installations still use **double-pipe condensers.** In double-pipe condensers, gas flows between the two pipes while water passes through the inner. The outside pipe diameter may be 2 in (5 cm); the inner, 1¼ in (3.2 cm). In some instances, the pipes are exposed to the atmosphere either with or without water drip. In an **evaporative condenser** the refrigerant vapor is condensed as it passes through tubes over which water is sprayed; the water is then evaporated by air flowing over the wet tubes. In this way, cooling-water requirements are reduced to from 5 to 15 percent of the water requirements of a nonevaporative condenser with no water reuse. The evaporative condenser combines in a single unit a refrigerant condenser and the atmospheric cooling tower or spray pond which is required if the

Table 11.1.7 Theoretical Horsepower and Theoretical Volume of Dry Ammonia Gas Pumped per Minute to Produce 1 Ton of Refrigeration

Suction pressure and temperature		Condenser pressure, lb/in² gage (temperature, °F)									
		103 (65)		127 (75)		153 (85)		182 (95)		215 (105)	
lb/in² gage	°F	hp	ft³/min	hp	ft³/min	hp	ft³/min	hp	ft³/min	hp	ft³/min
4	−20	1.058	5.84	1.205	5.96	1.361	6.09	1.525	6.23	1.691	6.43
6	−15	0.997	5.35	1.145	5.46	1.300	5.58	1.461	5.70	1.546	5.83
9	−10	0.903	4.66	1.045	4.76	1.193	4.86	1.347	4.97	1.435	5.08
13	−5	0.818	4.09	0.954	4.17	1.094	4.25	1.244	4.35	1.321	4.44
16	0	0.735	3.59	0.865	3.66	1.002	3.74	1.147	3.83	1.219	3.91
20	5	0.666	3.20	0.795	3.27	0.928	3.34	1.066	3.41	1.138	3.49
24	10	0.592	2.87	0.726	2.93	0.854	2.99	0.991	3.06	1.060	3.12
28	15	0.541	2.59	0.664	2.65	0.792	2.71	0.922	2.76	0.994	2.82
33	20	0.474	2.31	0.592	2.36	0.715	2.41	0.842	2.46	0.903	2.51
39	25	0.410	2.06	0.523	2.10	0.599	2.15	0.767	2.20	0.829	2.24
45	30	0.351	1.85	0.461	1.89	0.576	1.93	0.694	1.97	0.759	2.01
51	35	0.300	1.70	0.410	1.74	0.521	1.77	0.640	1.81	0.701	1.85

cooling water is to be reused. Air-cooled finned-tube condensers with forced ventilation are widely used on small units, and shell-and-coil or double-tube water-cooled condensers on medium units.

Evaporators may have plain or finned surfaces. Fin density is available from 1 to 14 fins per inch. Fin density of 2 to 4 fins per inch is specified for air-cooling service less than 28°F (−2.2°C), because air can flow through the coil while thin layers of frost are forming. Four types of defrosting methods are used: hot gas from the compressor, electrical air heaters, hot water, and room air circulation. Flooded evaporators are operated practically full of refrigerant. The level is controlled by a float valve. Wet-expansion evaporators are operated with a level approaching flooded operation. Dry-expansion (once-through) units operate with an indefinite amount of liquid in the evaporator. Usually a thermal bulb measures the outlet pipe temperature and adjusts a thermostatic expansion valve to maintain 15°F (8.3°C) superheat in the outlet vapor. Flooded or wet operation gives high heat-transfer rates. Pumped recirculation of refrigerant is used to promote heat transfer in finned coil evaporators, and also on some shell and tube evaporators (liquid chillers) by completely wetting the surfaces with a rapidly moving layer of refrigerant. The most commonly used liquid feed ratio is 3 to 1. However, a feed ratio as high as 6 to 1 may be required to accommodate distribution imbalances in the piping system. Combined liquid and vapor **return piping** should be sloped at 0.25 in/ft (2.08 percent) to prevent formation of waves in the liquid-phase layer that can react with the flowing vapors to cause liquid hammer that, in turn, can severely damage piping components. **Receivers** and **recirculating pumps** must be carefully matched to the operating system to prevent pump cavitation, liquid droplet carryover, and gross liquid overflow due to surges of liquid that may occur as a result of equipment operations such as defrosting processes. Specifically, the tank drop leg liquid velocity should be limited to 100 ft/min (0.51 m/s) and the vapor velocity over the liquid level in the tank should also be limited to 100 ft/min (0.51 m/s).

For a low-cost unit operating in limited space under conditions of motion, dry operation is preferred.

Controls are required on the liquid level of the refrigerant and on the temperature of the refrigerated space. The liquid control regulates the flow of refrigerant into the evaporator and also serves as the pressure barrier between the high operating pressure of the condenser and the lower operating pressure of the evaporator. It may take the form of a **capillary tube** between the condenser (high side) and the evaporator (low side), in which case it is nonadjustable. Plugging due to dirt is a common difficulty. Capillary-tube liquid control is largely confined to relatively small units assembled and charged at the factory and particularly for hermetically sealed systems. The **constant-pressure expansion valve,** maintaining a constant evaporator pressure, and the **thermal expansion valve,** maintaining a constant superheat leaving the evaporator, are standard liquid controls for most commercial applications. A **low-side float** liquid control, used with a flooded evaporator operating at evaporator (low) pressure, consists of a float-operated valve to admit liquid refrigerant to the evaporator in accordance with demand so that a constant liquid level is held in it. A **high-side float** liquid control is often used with a single flooded evaporator; the float operating the valve between the evaporator and the condenser is in a float chamber containing liquid refrigerant at the condenser (high-side) pressure. As the liquid level in the float chamber falls, the valve closes, thus preventing hot gaseous refrigerant from passing from the high to the low side as is possible when using a capillary or a low-side float.

Several types of noncondensable-gas removal devices have been developed to remove tramp air from refrigerating systems. The exhaust of the purge system contains a portion of the refrigerant. This refrigerant loss is detrimental to the environment, and the replacement of lost refrigerant is costly. Three types of refrigerant purge systems have been developed. One uses a scavenging condenser, cooled by system refrigerant, followed by a carbon absorption device to remove the last vestiges of refrigerant from the purge vapors. Other types, such as oil-absorption systems and reverse-osmosis systems, are developed but not offered commercially. The efficiency of purge systems is specified by the weight ratio of refrigerant lost to the air that is exhausted.

Vapor-Compression Circuits

Compound Low-Temperature Systems Multistaging is advantageous as it reduces operating temperatures by intercooling and power requirements if the fluid superheats considerably on adiabatic compression, as with ammonia; with F-12 refrigerant, propane, and butane, the superheat is negligible (see Figs. 11.1.12 to 11.1.15).

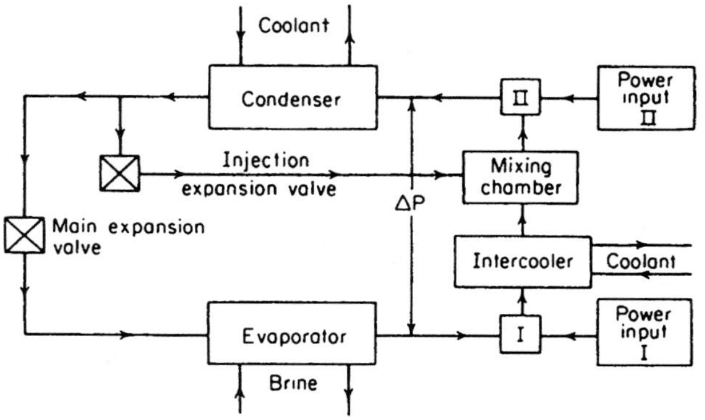

Figure 11.1.12 Two-stage system with conventional intercooler plus refrigerant injection.

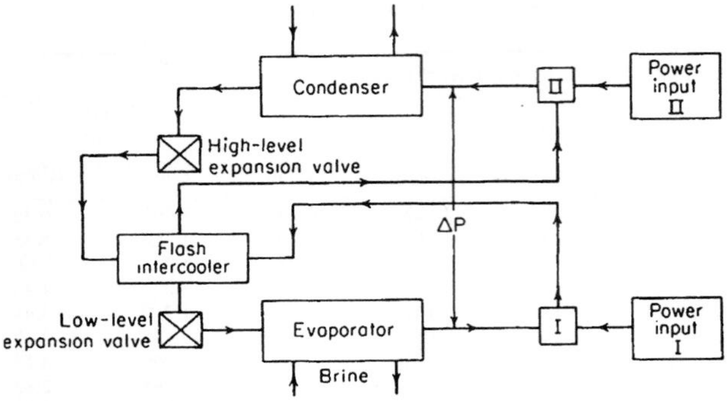

Figure 11.1.13 Two-stage system with flash intercooler.

Gas Intercooler Plus Liquid Injection Since gas temperature in the intercooler is limited by cooling water temperature, further favorable reduction of gas temperature, prior to entrance to the second stage, may be accomplished by direct injection, via an expansion valve, of refrigerant liquid (Fig. 11.1.12).

Flash Intercooler With a flash intercooler separator located midway between successive expansion valves, the gas leaving the low-stage (I) compressor passes through the flash intercooler, gets cooled by direct contact with the liquid, and, augmented by the flash vapor and at reduced temperature, enters the second stage (II) (Fig. 11.1.13).

For positive-displacement compressors, selection of flash intercooler temperature is determined by the following relation:

$$P_{\text{intermediate}} = (P_{\text{I suction}} \times P_{\text{II discharge}})^{1/2}$$

where P = pressure in absolute units, lb/in^2 abs (Pa). This equation is valid for centrifugal machines with modification. The common practice is to limit single-impeller-stage adiabatic head to approximately 10,000 ft (3,000 m) for low-molecular-weight gas and 6,000 ft (1,800 m) for the heavier HCFC and HFC refrigerants. Economical rotor design may require an interstage pressure that does not match the ideal flash intercooler pressure. Where choice is required, select a pressure slightly greater than the ideal pressure given by the above equation.

Low-Temperature Booster Plus Single Stage For the addition of a second lower-temperature evaporator to an existing single-stage system, a booster for the low-temperature vapor may be employed, its output feeding into the single-stage suction. The booster is a centrifugal or rotary compressor, suitable for handling the large vapor volumes encountered at low temperatures (and pressures); the reciprocating compressor is more advantageous at high pressures (Fig. 11.1.14).

Cascade Systems with Segregated Circuits Here two separate circuits are employed with the low-stage vapor condensed in the second-stage evaporator. Two different refrigerants may be appropriately chosen, a lower-boiling-point refrigerant serving for the first stage. Thus, for low temperatures (< −50°C or −50°F), CO_2 may be employed in the lower stage with NH_3 in the second stage condensing the CO_2. Or the same refrigerant may be used in both stages, limiting the migration of the lubricant to the low-temperature realm (Fig. 11.1.15).

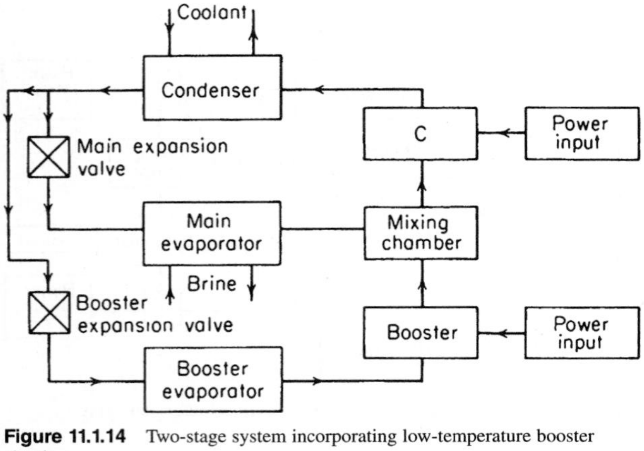

Figure 11.1.14 Two-stage system incorporating low-temperature booster circuit.

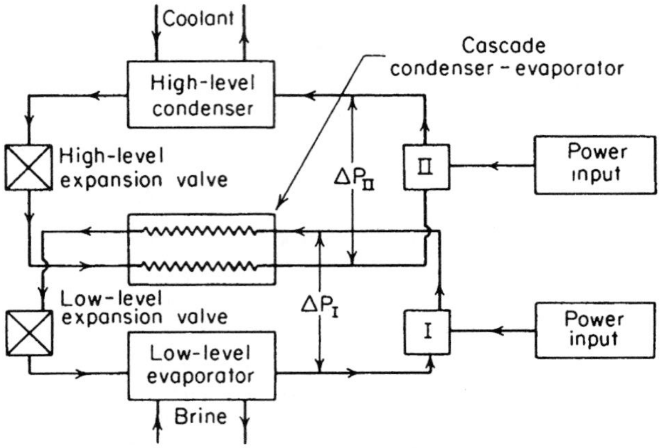

Figure 11.1.15 Two-stage system in cascade form.

Water-vapor refrigeration is necessarily limited to near-atmospheric temperature levels; it involves large suction volumes at high vacuum and utilizes a nontoxic refrigerant; this type of refrigeration was frequently used in the early days of space air conditioning. It utilizes (1) centrifugal compressors or (2) steam-jet compressors, entraining low-density water vapor from the evaporator. The centrifugal water-vapor compressor, usually multistage, must operate at high speeds (7,000 to 10,000 r/min). Operation at high vacuum in the evaporator requires an air pump at the condenser to eliminate leakage effects.

Steam-jet refrigeration systems are used where cooling to temperatures above 0°C is desired. Applications include industrial air conditioning and cooling of city gas to condense out tar and other objectionable impurities, of gas absorbers to increase efficiency of absorption, of reaction units where heat removal and temperature control are important during chemical transformations, and of wort and mash in the brewing and other fermentation industries particularly in the summer months. Jet refrigeration has been used for cooling passenger trains and is coming into popularity for marine installations, for example, cooling of banana boats and large passenger vessels. The system is a compression-type refrigerator: it uses water as a refrigerant, a part of which is evaporated to produce cooling of the remainder, steam-jet ejectors to compress the water vapor resulting from evaporation, and a condenser, either the surface type, where refrigerant water is to be reused, or the barometric type, where refrigerant water can be discarded. The advantages of the system are: low installation cost; the absence of moving parts except for small liquid pumps; and safety, since no noxious or toxic refrigerants are used.

For considerable differences in temperatures between condensing-water temperature and refrigerating temperatures (10°C or more), cooling of the refrigerant water by evaporation in two stages operating at different pressures (and temperatures) will usually give better operating economy, but at somewhat higher installation cost.

Figure 11.1.16 shows the variation of refrigerating capacity with variation of the chilled-water temperature, capacity increasing as chilled-water temperature rises.

Household compression machines are designed for continuous automatic operation and for conservation of the charges of refrigerant and oil. These units are almost universally motor-driven air-cooled compressors, the principal exception being the Electrolux-Servel absorption unit. The compressor may be hermetically sealed, with the compressor and motor enclosed in the same casing, or may use a shaft seal, embodying some form of sylphon bellows, with an outside coil spring at the place where the crankshaft passes out of the casing. The compressor may be reciprocating, but rotary types are being

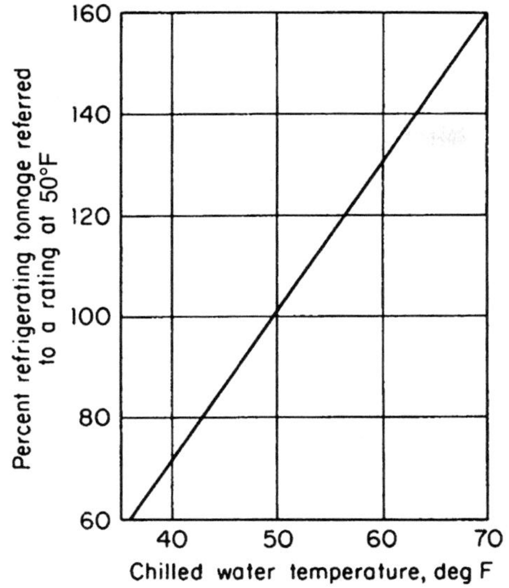

Figure 11.1.16 Variation of refrigerating capacity with the jet-chilled-water temperature in steam refrigeration.

increasingly used either with a floating piston ring driven by an eccentric or with sliding blades. Lubrication is by internal forced-feed circulation in most cases. The condenser may be of the radiator, coil, or plate type, cooled by natural draft, by a fan on the main motor, or by a separate motor. Refrigerant feed control to the evaporator may be by (1) float feed (flooded system), (2) pressure-actuated diaphragm valve, or most commonly (3) fixed orifice. The fixed orifice may be either a plate orifice or a capillary tube. The evaporator is usually made of aluminum. Controls are generally pressure-operated thermostatic switches. The power consumption of an average home refrigerator (18 ft^3) (0.5 m^2) is 133 kWh per month. A typical home deep freezer consumes approximately 152 kWh per month.

ABSORPTION SYSTEMS

Absorption refrigeration machines are essentially vapor-compression plants (Fig. 11.1.6) in which the mechanical compressor has been replaced by a thermally activated arrangement (Fig. 11.1.17). The basic elements are the **absorber, pump, heat exchanger, throttle valve,** and **generator.**

In an absorption cycle, a secondary fluid, the **absorbent,** is used to absorb the primary fluid, the **refrigerant,** after the refrigerant has left the evaporator. The vaporized refrigerant is converted back to liquid phase in the absorber. Heat released in the absorption process must be rejected to cooling water. The solution of absorbent and refrigerant is then pumped to the generator. A heat exchanger may be used for heat recovery and a corresponding increase in efficiency. In the generator, heat is added, and the more volatile refrigerant is separated from the absorbent through distillation. The refrigerant continues to the condenser, expansion valve, and evaporator, while the absorbent returns to the absorber.

Although there are many refrigerant-absorbent combinations, the most common systems are **ammonia-water** and **water-lithium bromide.** A common variation of the absorption system is the **Electrolux-Servel** process (Platen-Munters Patent). It has had wide application in gas-fired household refrigeration machines and air-conditioning systems. The use of hydrogen in a hermetically sealed system eliminates the use of pumps and other moving parts.

Absorption machines represent only a small percentage of refrigeration systems in operation. This is because they are generally bulky and difficult to operate efficiently. There has been renewed interest in absorption machines with the advent of high energy costs. This is because an absorption machine is an excellent way to utilize waste heat. Examples of potential applications are varied. One example may be on an exothermic process,

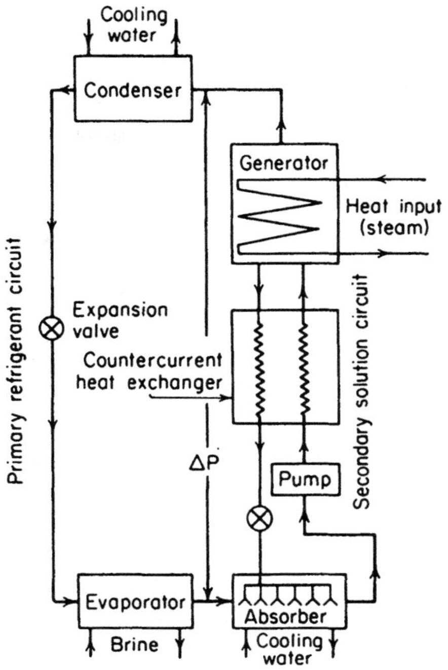

Figure 11.1.17 Elemental absorption system circuit.

or where extensive use of steam turbine with limited turndown results in vented steam especially during warm weather when air conditioning is required. A site where peak electrical demand is set by air-conditioning load and ratcheted demand charges are high may justify absorption refrigeration if adequate steam-generation facilities are already present.

A typical absorption machine requires about 20 lb (9 kg) of steam for 1 ton (3.5 kW) of refrigeration effect at 45°F (7°C).

A more detailed explanation of absorption systems is found in ASHRAE Handbook, 2005 Fundamentals, pp. 1.13 to 1.20.

AIR MACHINES

Compressed-air machines in which air is compressed, cooled, and then adiabatically expanded in an engine while doing work are obsolete for usual industrial refrigeration because of bulk and inefficiency. Their coefficient of performance is less than unity as compared with 4 or 5 for vapor machines. Operation may be closed cycle, in which the air is reused, or open cycle, in which the air is discarded after expansion. Compression ratios of about 3 seem best. In the **dense-air** closed system, the pressure at the compressor intake averages 75 lb/in^2 gage (560 kPa) and up to 250 lb/in^2 gage (1,700 kPa) at the outlet. The use of superatmospheric pressure throughout the cycle decreases the size, and the reuse of dried air eliminates the ice formation on engine valves which occurs with open-system operation. Air cycles are used for liquefaction of the so-called permanent gases and to an increasing extent for comfort cooling in high-speed planes where the small lightweight equipment and the safety of the cooling medium are important. High-speed (90,000 to 100,000 r/min) gas turbines are used to extract work from 5 to 100 lb/in^2 gage compressed air.

THERMOELECTRIC COOLING

Thermoelectric cooling utilizes the **Peltier effect** whereby a temperature differential will occur between two junctions of a closed loop of dissimilar metallic conductors when an electric current is imposed. It is variously called *Peltier effect cooling, thermoelectric refrigeration,* and *electronic cooling.* It is the inverse of the thermocouple operating principle whereby a temperature differential across the junctions of two dissimilar metals produces an electric current.

The capacity of a thermoelectric couple is small. System capacity can be increased by increasing the number of elements used. The temperature produced can be lowered by cascading the systems. Capacities approaching 25 tons (87.9 kW) cooling effect have been achieved, and temperatures as low as −266°F (−166°C) have been attained.

Aside from limited practical capacities, the major disadvantage of a thermoelectric cooling system is its poor coefficient of performance. The absence of moving parts, silent operation, ability to function in zero gravity, and lack of pressurized vessels, has given thermoelectric cooling wide space-program application. Other applications include use on submarines, electronic equipment cooling, and small units such as drinking water coolers or units for use in recreational vehicles.

The elements of a thermoelectric cooler are quite simple and are equivalent to those of a thermoelectric generator (see Sec. 7). For development of various output capacities, the number of basic elements is proportionately increased; they all act in parallel, and their outputs add together. **Semiconductor** materials, such as bismuth-telluride-selenide and bismuth-antimony-telluride alloys, are employed.

With an input to the circuit of low-voltage direct current and with heat continuously abstracted at one junction at room temperature, the other junction will become cold. Essentially each of the two junctions becomes an "activity cell," in which at the lower refrigeration temperature level, heat is converted into electrical effect, and at the higher atmospheric temperature, the electrical effect is converted into heat. This establishes the basis of "heat-pump" operation: the lower-temperature heat requirement must be supplied, essentially as in the evaporator of a vapor-compression system, by heat withdrawal via heat transfer, from all connected surroundings. The energy-transport circuit thus parallels the machine-activated circuit of Fig. 11.1.6. The amount of heat for a single-component circuit depends on the circulating electric current, on the properties and dimensions of the two conductors (*p* and *n* types), and on the resultant temperature difference between the high side room temperature and the low side input heat that is divided into two groups, useless and useful. The **useless heat** is the heat conduction between the high and low temperature through the thermoelements plus the current-generated resistance-loss heat within the elements that flows to the colder junction. The **useful heat** is the refrigeration effect in a regular evaporator. All the heat must in turn be discharged via the Peltier effect at the higher room temperature, as in the condenser of a vapor-compression system. Ultimate performance of the circuit depends on the voltage-generating nature of the two dissimilar conductors and on their electrical resistance, thermal conductivity, and physical dimensions. The different properties are mathematically combined into a **figure of merit, Z.**

METHODS OF APPLYING REFRIGERATION

In **direct expansion systems** the evaporator is placed in the space which is to be cooled; in **indirect systems** a secondary fluid **(brine)** is cooled by contact with the evaporator surface, and the cooled brine goes to the space which is to be refrigerated. Brine systems require 40 to 60 percent more surface than do direct expansion; they have an equalizing effect due to the large heat capacity of cold brine, they are safer (particularly if the refrigerating effect must be carried considerable distance or widely distributed), and they permit closer temperature regulation than is possible with direct expansion. If two temperatures are to be held, the lower may be direct expansion, the higher by brine cooling. Development of better controls and newer piping methods has made direct expansion more attractive than previously.

Brines used for industrial refrigeration are usually aqueous solutions of calcium chloride, ethylene glycol, propylene glycol, or undiluted methylene chloride, and silicone-based alkylated fluids. Calcium chloride should not contain over 0.2 percent magnesia, calculated as magnesium chloride. Calcium chloride brines are recommended down to −45°F (−43°C). Brines should be chemically neutral: acidic brines attack ferrous materials, alkaline brines attack zinc, ammonia in brine (resulting from leaks in the ammonia system) is especially harmful to most nonferrous metals. Corrosion by brine is increased by the presence of oxygen, air, or carbon dioxide and by galvanic action between dissimilar metals. Corrosion inhibitors are widely used, a satisfactory one being about 1,600 g of sodium chromate or dichromate per 1 m^3 of calcium chloride brine or 3,200 g per 1 m^3 of sodium chloride brine.

The properties of sodium and calcium chloride brines are given in Tables 11.1.8 to 11.1.10.

The density of brine is measured by a **salinometer** (or **salometer**), which is a simple hydrometer the indications on which are 4 times greater than on the corresponding Baumé scale.

It is undesirable to use a **strength of solution** of salt greater than is necessitated by its freezing temperature, as the specific heat (Table 11.1.9) decreases as the concentration of the brine increases, and consequently the stronger the brine, the less heat a given amount of it is able to convey between certain definite temperatures and the more power is required to pump the brine. Moreover, brine which is too strong may cause clogging of pipes, etc., by depositing salt. On the other hand, if the solution is too weak it may not be able to withstand the temperature existing in the expansion coil, so that a layer of thin ice will form around the latter and interfere with the absorption of heat from the brine. The surface of the expansion coils in the brine tank should be inspected from time to time to see if any ice has formed on them. In larger plants, it is customary to use a solution with a freezing point not less than 10°F (5.5°C) beneath the lowest temperature which will be obtained in the operation of the plant. In smaller isolated plants and where careful supervision is not ensured, it is customary to make the solution as strong as possible without being unstable, usually 1.240 to 1.250 sp gr.

Ethylene glycol brine is used in various strengths from 15 to 50 percent by weight for refrigeration temperatures down to −1°F (−18°C). It is toxic, and therefore its usage is not recommended in food or

Table 11.1.8 Properties of Calcium Chloride Solutions
(For variation of specific gravity, with temperature, see Table 11.1.9.)

Parts of $CaCl_2$ by weight in 100 parts of the solution	Specific gravity at 60°F	Deg Baumé	Weight per gal, lb	Weight per cu ft, lb	Freezing point, °F	Specific heat at					
						−4°F	14°F	32°F	50°F	68°F	86°F
6	1.050	7.0	8.76	65.52	28.0	—	—	—	—	—	—
8	1.069	9.33	8.926	66.70	24.2	—	—	0.882	0.887	0.892	0.897
10	1.087	11.57	9.076	67.83	21.4	—	—	0.853	0.858	0.863	0.868
12	1.105	13.78	9.227	68.95	18.2	—	—	0.825	0.831	0.836	0.842
14	1.124	15.96	9.377	70.08	14.4	—	—	0.799	0.805	0.811	0.817
16	1.143	18.12	9.536	71.26	9.9	—	0.768	0.775	0.781	0.787	0.792
18	1.162	20.24	9.703	72.51	4.7	—	0.745	0.752	0.759	0.764	0.769
20	1.182	22.32	9.853	73.63	−1.0	—	0.723	0.731	0.738	0.744	0.749
22	1.202	24.38	10.04	75.0	−7.3	0.695	0.704	0.711	0.718	0.724	0.729
24	1.223	26.41	10.21	76.32	−14.1	0.678	0.686	0.693	0.700	0.706	0.712
26	1.244	28.41	10.38	77.56	−22.0	0.663	0.670	0.677	0.683	0.690	0.696
28	1.265	30.39	10.56	78.94	−32.0	0.649	0.656	0.662	0.669	0.675	0.682
30	1.287	32.34	10.75	80.35	−46.0	0.638	0.643	0.648	0.655	0.661	0.668

Table 11.1.9 Specific Gravities of Brines
(To change to lb/ft³ multiply by 62.43; to change to lb/gal multiply by 8.35.)

Parts by weight of salt in 100 parts of brine	Sodium chloride				Calcium chloride				Magnesium chloride			
	Temperature, °F											
	14	32	50	68	14	32	50	68	14	32	50	68
6	—	1.046	1.044	1.041	—	1.053	1.051	1.049	—	1.053	1.051	1.049
8	—	1.061	1.059	1.056	—	1.071	1.069	1.066	—	1.070	1.069	1.067
10	—	1.077	1.074	1.071	—	1.090	1.087	1.084	—	1.089	1.087	1.084
12	—	1.093	1.090	1.086	—	1.108	1.106	1.103	1.108	1.107	1.105	1.102
14	—	1.108	1.105	1.101	—	1.127	1.124	1.121	1.127	1.126	1.123	1.121
16	1.128	1.124	1.121	1.116	1.150	1.147	1.144	1.140	1.147	1.145	1.142	1.139
18	1.144	1.140	1.136	1.132	1.170	1.167	1.163	1.159	1.166	1.164	1.161	1.158
20	1.161	1.157	1.152	1.148	1.190	1.187	1.183	1.179	1.186	1.183	1.181	1.178
22	1.178	1.173	1.169	1.164	1.211	1.208	1.203	1.199	1.206	1.203	1.201	1.197
24	1.195	1.190	1.185	1.180	1.233	1.229	1.224	1.219	1.226	1.224	1.221	1.218

Table 11.1.10 Weight of Commercial Calcium Chloride in Brine

Sp gr	1.10	1.12	1.14	1.16	1.18	1.20	1.22	1.24	1.26	1.28	1.30	1.32
Wt per gal, lb	1.41	1.70	2.00	2.30	2.59	2.90	3.20	3.50	3.83	4.13	4.46	4.78
Wt per cu ft, lb	10.55	12.72	14.96	17.20	19.37	21.69	23.94	26.18	28.62	30.89	33.36	35.75

NOTE: Specific gravity is at 60°F for both brine and water. The weights are of 73 to 75 percent solid calcium chloride per gal of brine at 60°F. For flake 77 to 80 percent calcium chloride multiply the weights given by 0.94.

Table 11.1.11 Comparative Performance of Brines†

Brine	Heat-transfer coefficient, U_o*	Fouling factor Inner	Fouling factor Outer	Refrigerant no.	Tube material
Calcium chloride, 24%	53	0.001	0.002	R717	Carbon steel
Propylene glycol, 40%	26.3	0.0005	0.0	HCFC 22	Copper
Ethylene glycol, 40%	44.7	0.0005	0.0	HCFC 22	Copper
Methylene chloride	111.9	0.0	0.0	HCFC 22	Copper

* Values presented in Btu/h · ft² · °F; to convert to W/m² · K, multiply by 5.678.
† Values calculated using ¾-in diameter tubes; integrally finned; 19 fins per inch; 7 ft/s velocity; brine is inside the tube; brine bulk temperature is 5°F.

beverage processing equipment where a leak can contaminate the product. A corrosion inhibitor is recommended to protect carbon steel equipment and piping. Specific heat ranges from 0.975 to 0.76 Btu/lb·°F (4,080 to 3,180 J/kg · K) in typical systems. Specific gravity varies from 1.08 to 1.05.

Propylene glycol brine in 15 to 35 percent weight strength is considered nontoxic and is often used in brewing and other food and beverage applications. The brine is more viscous (10 centipoise at –6.7°C for 30 percent weight solution) than ethylene glycol. Specific heat varies from 0.9 to 0.975 Btu/lb · °F. Specific gravity ranges from 1.05 to 1.01. Propylene glycol has often replaced ethylene glycol due to the latter's toxicity.

Methylene chloride brine is often used in systems requiring –20 to –125°F (–30 to –87°C) low temperatures. It has low flammability. Hydrolysis and water contamination must be prevented to avoid equipment corrosion. Specific heat varies from 0.269 to 0.273 Btu/lb·°F (1,126 to 1,142 J/kg · K). Specific gravity ranges from 1.4 to 1.5.

Proprietary chemical blends, such as Dowtherm J and Syltherm, are used where the fluid properties are advantageous. Dowtherm J, a mixture of isomers of an alkylated aromatic compound, is usable between the temperatures of 358°F (181°C) and –120°F (–84°C). The usual range of service is between 0°F (–18°C) and –70°F (–21°C). Physical properties vary from 0.412 to 0.390 Btu/lb · °F (0.412 to 0.390 cal/g · °C) for specific heat; 0.894 to 0.926 for specific gravity (ref. 68°F water); 0.079 to 0.084 Btu · ft/h · ft² · °F (0.136 to 0.145 W/m · K) for thermal conductivity; and 1.72 to 4.61 cP in the range of usage.

Sodium chloride brine usage is being reduced due to its corrosivity and its relatively high freezing point.

Table 11.1.11 gives a comparison of the performance of the four major brines in a typical shell and tube chiller. This chart does not indicate effects of specific heat, ingredient cost, pumping power requirements, or system investment cost in the selection of brines.

Brine coolers may be of three types: shell-and-tube, shell-and-coil, and double pipe. The shell-and-tube type is the most widely used, the brine flowing through the tubes which are surrounded by the evaporating refrigerant. Tubes may be arranged for multipass operation. The effective heat-transfer surface varies from 8 to 15 ft² per ton (0.21 to 0.4 m²/kW), varying with temperature and brine velocity. A submerged coil in an open brine tank is used for ice making by the can process.

The **double-pipe cooler** is usually of 2½-in (50-mm) inner or brine-flow pipe and 3-in (75-mm) outer pipe. The commercial rating is 15 to 20 ft (5 to 6 m) length of coil per ton of refrigeration.

The **shell-and-tube cooler** is used with closed heads and is erected both vertically and horizontally; brine flows through the tubes and ammonia is in the shell. It is made in sizes from 1 to 350 tons with ratings of 8 to 15 ft² effective surface per ton, varying with the temperature and brine velocities; tubes 1 to 2½ in (25 to 63 mm) arranged multipass. This type of cooler has largely displaced all other types in recent installations (Table 11.1.12).

REFRIGERANT PIPING

(See also Sec. 6.)

It is important to the proper operation of a refrigeration system that the piping or mains interconnecting the compressors, condensers, evaporators, and receivers be properly sized. This piping must be considered in three categories, viz., liquid lines, suction lines, and discharge lines. Essentially pipeline sizing is governed by pressure drop, first cost, operating cost, noise, and oil entrainment. Excessive pressure drops penalize compressor efficiencies and may affect control-valve operation adversely. Liquid-line velocities for most refrigerants are in the order of 60 to 400 ft/min (3.3 to 22 m/s), suction lines from 700 to 4,600 ft/min (38 to 250 m/s), and discharge lines from 1,000 to

Table 11.1.12 Capacity of Multipass Shelf-and-Tube Brine Coolers, Flooded
(Tons refrigeration.)

Diameter of shell, in	Length of shell, ft	Velocity of brine through cooler, ft/min: 200 — Total brine, gal/min	200 — Mean temp diff, brine and ammonia, °F: 7½	12½	17½	400 — Total brine, gal/min	400 — Mean temp diff, brine and ammonia, °F: 5	10	15
26	6	140	7.74	12.9	18.1	290	7.15	14.3	21.2
26	9	140	11.5	19.3	26.0	290	10.7	21.4	32.0
26	12	140	15.5	25.8	36.2	290	14.3	28.6	42.9
34	9	230	18.5	30.7	43.1	440	17.1	34.1	51.1
34	12	230	24.6	41.0	57.4	440	22.6	45.3	68
34	18	230	37.0	62.0	86.2	440	34.1	68.2	101
42	12	510	45.0	74.8	105	970	41.5	83	123
42	18	510	67.4	112	157	970	62.2	124	187

5,000 ft/min (55 to 275 m/s). Pressure drops vary approximately as the square of the velocity (or tonnage) and directly as the length of the piping. For liquid lines, when evaporator is located ahead condenser, the following pressure drops in pounds per square inch per foot (kilopascals per meter) of static lift should be allowed: ammonia, 0.26 (5.9); R 12, 0.57 (12.9); R 22, 0.51 (11.5); R 11, 0.64 (14.5). (See also Secs. 1 and 2.)

Ammonia Mains For the average installation, Table 11.1.13 shows the maximum tons of refrigeration normally allowed for various sizes of standard pipe, based on 100 ft (30 m) equivalent length of piping (measured length plus allowance for valves and fittings). The gas pressure drops per 100 ft (30 m) equivalent length upon which Table 11.1.13 is based are as follows: suction lines at 5 lb/in^2 (gage) (34.5 kPa) = 0.25 lb/in^2 (1.7 kPa); suction lines at 20 lb/in^2 (gage) (138 kPa) = 0.50 lb/in^2 (3.4 kPa); suction lines at 45 lb/in^2 (gage) (310 kPa) = 1 lb/in^2 (6.9 kPa); and discharge lines = 1½ lb/in^2 (6.9 kPa).

Standard-weight (schedule 40) steel pipe is used for ammonia mains, except for liquid lines 1½ (38 mm) and smaller where extra-strong (schedule 80) pipe is used. Joints may be either screwed, flanged, or welded, but welding is preferred.

R 12 Mains Maximum tons refrigerant for various pressure drops per 100 ft equivalent length of R 12 mains are given in Table 11.1.14. For average conditions, allowable pressure drops in R 12 suction lines are 0.5 to 1.0 lb/in^2 (3.4 to 6.9 kPa) less than 0°F (−18°C) evaporator temperature; 1.0 to 1.5 lb/in^2 (6.9 to 10.3 kPa) from 0 to 25°F (−18 to −4°C); and 2.0 to 2.5 lb/in^2 (13.8 to 17.2 kPa) from 25 to 50°F (−4 to 10°C). Discharge-line pressure drops upon which Table 11.1.14 is based are approximately 1 lb/in^2 (6.89 kPa) per 100 ft (30 m) equivalent length. Pressure drops of 1 to 4 lb/in^2 (6.9 to 27.6 kPa) are permissible. Liquid-line pressure drops of 3 to 5 lb/in^2 (20.7 to 34.5 kPa) per 100 ft (30 m) equivalent length are normal, with 10 lb/in^2 (69 kPa) usually considered maximum. To facilitate oil return in vertical suction lines with the evaporator located beneath the compressor, the velocity in the vertical section should be 1,000 ft/min (38 m/s) or more. Suction-line capacities in Table 11.1.14 may be increased 9 percent for 50°F (10°C) evaporator temperature and will decrease approximately 7 percent for each 10°F (5.6°C) beneath 40°F (4.4°C).

R 22 piping will, in general, be smaller for the same tonnage than R 12 piping. Suction lines will handle about one-third more tonnage for the same pressure drop. Liquid and discharge lines can be the same as for the equivalent tonnage of R 12 or slightly smaller.

Piping may be standard-weight steel pipe, but copper tubing is used in most cases. Medium-weight type L copper tubing is normally used for land installations, and the heavier type K copper tubing is used for marine work. Joints in the copper tubing may be flared compression fittings for tubing ¾ in (18 mm) OD and smaller. However, hard solder (silver-base alloy melting ahead 1,000°F) (540°C) is preferred for most tubing connections. Copper tubing is almost always used for liquid lines. Steel pipe may be used for the larger suction and discharge lines but should be sandblasted on the inside and carefully cleaned before installation.

Piping flexibility is an important consideration, especially in low-temperature systems [0°F(−18°C) or colder] and where large-diameter or long runs of piping are utilized. The coefficient of thermal expansion is about 9×10^{-6}/°F (16×10^{-6}/°C) for carbon steel pipe. This means

Table 11.1.13 Maximum Tons Refrigeration for Ammonia Mains
(100 ft equivalent pipe length.)

Pipe size, in	Suction line — Suction pressure, lb/in^2 gage: 5(−17.2°F)	20(5.5°F)	45(30°F)	Discharge line	Liquid line — Condenser to receiver	Receiver to system
3/8	—	—	—	—	2.5	12.0
1/2	0.6	1.1	2.0	3.1	6.0	20.0
3/4	1.2	2.2	4.1	6.0	14.0	75.0
1	2.2	4.0	7.5	11.4	24.0	137
1¼	4.4	8.0	15.0	22.4	50.0	245
1½	6.4	11.8	21.6	30.9	77.0	400
2	12.1	22.2	42.0	62.0	140	850
2½	19.1	35.5	65.0	97.5	220	1,475
3	31.5	59.0	108	160	375	2,400
3½	46.6	87.5	156	238	540	3,500
4	64.0	118	240	330	740	
5	117	208	385	560	1,320	
6	175	306	600	905	2,030	
8	362	650	1,200	1,810	4,200	
10	640	1,180	2,160	3,200		
12	940	1,850				

SOURCE: From ASRE, Air-Conditioning Refrigerating Data Book, 1955-1956 ed., based on ARI Equipment Standards; see ASHRAE Handbook, 2006 Refrigeration, Chapter 3, for more complete tables.

Table 11.1.14 Maximum Tons Refrigeration for R 12 Lines

Line sizes, in	Suction-line (based on 105°F condensing temp) pressure drop, lb/in² per 100 ft equivalent length at 40°F saturation: 1/2	1	2	3	4	5	Discharge-line condensing temp: 115°F	90°F	Liquid-line (based on 105°F condensing temp; 25°F evaporating temp) pressure drop, lb/in² per 100 ft equivalent length (type L tubing): 3	5	10	20
3/8 OD	—	—	—	—	—	—	—	—	0.88	1.14	1.80	2.58
1/2 OD	0.14	0.20	0.28	0.35	0.41	0.45	—	—	2.89	3.64	5.56	8.50
3/8 IPS	0.17	0.24	0.34	0.42	0.49	0.54						
5/8 OD	0.25	0.35	0.51	0.62	0.73	0.81	1.43	1.15	4.86	6.81	10.2	15.8
1/2 IPS	0.35	0.45	0.65	0.79	0.93	1.03	1.87	1.50	4.86	6.81	10.2	15.8
7/8 OD	0.55	0.76	1.10	1.34	1.58	1.75	2.97	2.38	10.5	14.1	21.8	33.0
3/4 IPS	0.68	0.94	1.35	1.65	1.92	2.12	3.26	2.62	9.73	12.6	18.5	27.0
1 1/8 OD	1.26	1.80	2.57	3.17	3.76	4.15	5.05	4.05	21.4	28.2	41.3	60.8
1 IPS	1.43	2.01	2.89	3.54	4.17	4.60	5.29	4.25	21.4	28.2	41.3	60.8
1 3/8 OD	2.21	3.12	4.45	5.50	6.38	7.05	7.72	6.19	36.9	48.1	70.5	101
1 1/4 IPS	2.70	3.82	5.37	6.72	7.68	8.48	9.16	7.35	36.9	48.1	70.5	101
1 5/8 OD	3.40	4.78	6.79	8.42	9.77	10.8	10.92	8.75	62.0	80.2	114	160
1 1/2 IPS	4.05	5.75	8.10	10.12	11.6	12.8	12.5	10.0	62.0	80.2	114	160
2 1/8 OD	6.12	8.60	12.1	15.1	17.4	19.2	19.2	15.3				
2 IPS	7.66	10.9	15.3	19.2	22.2	24.5	20.6	16.5	124	161	231	328
2 5/8 OD	12.0	17.1	24.0	30.1	34.6	38.2	32.2	25.9				
2 1/2 IPS	12.0	17.1	24.0	30.1	34.6	38.2	32.2	25.9	230	297	426	607
3 1/8 OD	19.1	27.2	38.2	47.8	55.0	60.7	51.5	39.8				
3 IPS	20.9	29.4	42.3	51.8	60.0	66.2	54.5	43.8	364	469	676	972
3 5/8 OD	27.8	39.7	55.7	69.8	80.3	88.7	72.0	57.6				
3 1/2 IPS	30.2	43.2	61.0	76.1	87.0	96.0	78.8	62.3	539	704	1,005	1,430
4 1/8 OD	38.6	55.2	78.0	97.3	111	123	95.8	77.1				
4 IPS	40.7	58.6	83.0	103	118	130	101.6	81.6	753	972	1,385	1,945
5 IPS	71.3	100	141	176	203	224	171.5	137.8				
6 IPS	126	183	257	322	366	403	266	214				
8 IPS	211	297	422	523	602	664	461	370				
10 IPS	352	503	712	887	1,024	1,130	725	582				
12 IPS	550	780	1,106	1,373	1,582	1,748	1,041	836				

SOURCE: From ASRE, Air-Conditioning Refrigerating Data Book, 1955-1956 ed., based on ARI Equipment Standards.

a 20-ft (6-m) length will shrink about 0.15 in (3.8 mm) when cooled from 70 to 0°F (21 to −18°C). If the piping configuration were such to exert a moment on a compressor flange, premature equipment failure could result. Long runs of cold brine lines should be analyzed for proper anchoring, guiding, and support to prevent system failure.

Flexibility is provided by a combination of changes in direction and use of properly positioned anchors, spring hangers, and guides. Refer to Sec. 3 for additional information on piping flexibility.

Mains of Other Refrigerants For sizing piping for other refrigerants, reference may be made to the ASHRAE Handbook, 2006, Refrigeration, Chapters 2 and 3.

COLD STORAGE

The first consideration in cold-storage room design is the temperature to be maintained in the refrigerated space. The temperature can generally be divided into three ranges: ahead 32°F (0°C), 32°F (0°C) and less, and 0°F (−18°C) and less. Methods of construction and degree of insulation are dependent on the maintained temperature. Other more specialized design considerations are humidity control, ventilation requirements, controlled atmosphere (control of O_2, N_2, or CO_2), fire protection, and material-handling facilities (see Table 11.1.15).

Cold-storage rooms can either be site fabricated or prefabricated. The increasing availability of preinsulated panels with low *U* factors has led to increasing use of prefabricated rooms. Many of these are available as complete packages with the refrigeration system included.

Vapor Barrier Although good insulation is important, a well-designed and installed vapor barrier is the key to maintaining satisfactory insulation performance and life. A poor vapor barrier allows moisture to condense in the insulation, greatly reducing its insulating value. A well-designed system is of little value if poorly installed. Poor workmanship, in fact, accounts for most vapor barrier inadequacies.

Table 11.1.15 Cold-Storage Usage and Transmission Factors

	Room volume, ft³: 20	50	100	500	1,000	2,000	5,000	10,000	50,000	100,000
Usage load*										
Average	4.68	2.28	1.61	1.21	1.10	0.835	0.403	0.240	0.178	0.173
Heavy	5.51	3.55	2.52	1.87	1.67	1.29	0.625	0.408	0.305	0.295
Storage temp, °F	40 and more		25-40		15-25		0-15		−25-0	
Insulation thickness, in	3		4		5		6		8	
Theoretical *U* factor†	0.086		0.066		0.055		0.046		0.035	
Practical *U* factor‡	0.111		0.083		0.066		0.055		0.046	
Piping *U* factor§	2.5		2.2		2.0		1.8		1.6	

* Btu per cu ft per 24 h per °F temp diff between outside and inside. Usage load includes 10°F or less of product cooling, and infiltration losses, etc., but does not include any freezing of product or fan motor loads when unit coolers are used. Values are for normal conditions and should not be used for unusual product loads.

† Theoretical *U* factor is based on corkboard ($k = 0.30$) with plaster on both sides, Btu per ft² per h per °F temp diff.

‡ Practical *U* factor allows for inefficiency of joints and structural supports in insulation.

§ Piping *U* factor is Btu/ft² outside pipe surface per h per °F temp diff between cooling medium and air (gravity circulation). Allowance has been made for frosting at the lower temperatures.

Table 11.1.16 Product Storage Data

Product	Quick-freeze temp, °F	Storage temp, °F		Humidity, % R.H.	Specific heat		Latent heat	Freezing point	Respiration, Btu per lb per day
		Long	Short		Above freezing	Below freezing			
Apples	−15	30-32	38-42	85-88	0.86	0.45	121.0	28.4	0.75
Asparagus	−30	32	40	85-90	0.95	0.44	134.0	29.8	
Bacon, fresh	—	0-5	36-40	80	0.55	0.31	30.0	25.0	
Bananas	—	56-72	56-72	85-95	0.80	0.42	108.0	28.0	4.18
Beans, green	—	32-34	40-45	85-90	0.92	0.47	128	29.7	3.3
Beans, dried	—	36-40	50-60	70	0.30	0.237	18		
Beef, fresh, fat	−15	30-32	38-42	84	0.60	0.35	79		
Beef, fresh, lean	−15	30-32	38-42	85	0.77	0.40	100		
Beets, topped	—	32-35	45-50	95-98	0.86	0.47	129	31.1	2.0
Blackberries	−15	31-32	42-45	80-85	0.89	0.46	125	28.9	
Broccoli	—	32-35	40-45	90-95	0.92	0.47	130	29.2	
Butter	+15	—	40-45	—	0.64	0.34	15	15.0	
Cabbage	−30	32	45	90-95	0.93	0.47	130	31.2	
Carrots, topped	−30	32	40-45	95-98	0.87	0.45	120	29.6	1.73
Cauliflower	—	32	40-45	85-90	0.93	0.47	132	30.1	
Celery	−30	31-32	45-50	90-95	0.92	0.48	135	29.7	2.27
Cheese	+15	32-38	39-45	—	0.70				
Cherries	—	31-32	40	80-85	0.87	0.45	120	26.0	6.6
Chocolate coatings	—	45-50	—	—	0.3				
Corn, green	—	31-32	45	85-90	0.80	0.43	108	29.0	4.1
Cranberries	—	36-40	40-45	85-90	0.90	0.46	124	27.3	
Cream	—	34	40-45	—	0.88	0.37	84		
Cucumbers	—	45-50	45-50	80-85	0.97	0.47	137	30.5	
Dates, cured	—	28	55-60	50-60	0.83	0.44	104		
Eggs, fresh	−10	30-31	38-45	—	0.76	0.40	98	31.0	
Eggplants	—	45-50	46-50	85-90	0.94	0.47	132	30.4	
Flowers	—	35-40		85-90					
Fish, fresh, iced	−15	25	25-30	—	0.82	0.41	105	30.0	
Fish, dried	—	30-40		60-70	0.56	0.34	65		
Furs	—	32-34	40-42	40-60					
Furs, to shock	—	15	15						
Grapefruit	—	32	32	85-90	0.91	0.46	126	28.4	0.5
Grapes	—	30-32	35-40	80-85	0.86	0.44	116	27.0	0.5
Ham, fresh	—	28	36-40	80	0.68	0.38	87		
Honey	—	31-32	45-50	—	0.35	0.26	26		
Ice cream	-20	—	0-10	—	0.5-0.8	0.45	96		
Lard	—	32-34	40-45	80	0.52	0.31	90		
Lemons	—	55-58		80-85	0.92	0.46	127	28.1	0.4
Lettuce	—	32	45	90-95	0.96	0.48	136	31.2	8.0
Liver, fresh	—	32-34	36-38	83	0.72	0.42	94		
Lobster, boiled	—	25	36-40	—	0.81	0.42	105		
Maple syrup	—	31-32	45	—	0.24	0.215	7.0		
Meat, brined	—	31-32	40-45	—	0.75	0.36	75.0		
Melons	—	34-40	40-45	75-85	0.92	0.35	115	28.5	1.0
Milk	—	34-36	40-45	—	0.92	0.46	124	31.0	
Mushrooms	—	32-35	55-60	80-85	0.93	0.47	130	30.2	
Mutton	—	32-34	34-42	82	0.81	0.39	96	29.0	
Nut meats	—	32-50	35-40	65-75	0.30	0.24	14	20.0	
Oleomargarine	—	34-36	—	—	0.65	0.34	35	15.0	
Onions	—	32	50-60	70-75	0.91	0.46	120	30.1	1.0
Oranges	—	32-34	50	85-90	0.90	0.46	124	27.9	0.7
Oysters	—	—	32-35	—	0.85	0.45	120.0		
Parsnips	−30	32-34	34-40	90-95	0.82	0.45	120.0	28.9	
Peaches, fresh	—	31-32	50	85-90	0.90	0.46	126	29.4	1.0
Pears, fresh	—	29-31	40	85-90	0.86	0.45	118	28.0	6.6
Peas, green	—	32	40-45	85-90	0.80	0.42	108	30.0	
Peas, dried	—	35-40	50-60	—	0.28	0.22	14		
Peppers	—	32	40-45	85-90	0.94	0.47	—	30.1	2.35
Pineapples, ripe	—	40-45	50	85-90	0.88	0.45	122	29.9	
Plums	—	31-32	40-45	80-85	0.88	0.45	123	28.0	

Table 11.1.16 Product Storage Data (*Continued*)

Product	Quick-freeze temp, °F	Storage temp, °F Long	Storage temp, °F Short	Humidity, % R.H.	Specific heat Above freezing	Specific heat Below freezing	Latent heat	Freezing point	Respiration, Btu per lb per day
Pork, fresh	—	30	36-40	85	0.60	0.38	66	28.0	
Potatoes, white	−30	36-50	45-60	85-90	0.77	0.44	105	28.9	0.85
Poultry, dressed	−10	28-30	29-32	—	0.80	0.41	99	27	
Pumpkins	—	50-55	55-60	70-75	0.92	0.47	130	30.2	
Quinces	—	31-32	40-45	80-85	0.90	—	—	28.1	
Raspberries	—	31-32	40-45	80-85	0.89	0.46	125	30.0	3.3
Sardines, canned	—	—	35-40	—	0.76	0.410	101		
Sausage, fresh	—	31-36	36-40	80	0.89				
Sauerkraut	—	33-36	36-38	85	0.91	0.47	128	26	
Squash	—	50-55	55-60	70-75	0.92	0.47	130	29.3	
Spinach	—	32	45-50	85	0.94	0.48	132	30.8	
Strawberries	−15	31-32	42-45	80-85	0.92	0.48	129	30.0	3.3
Tomatoes, ripe	—	40-50	55-70	85-90	0.95	0.48	135	30.4	0.5
Turnips	—	32	40-45	95-98	0.93	0.40	137	30.5	1.0
Veal	−15	28-30	36-40	—	0.71	0.39	91	29	

Common vapor barrier materials include metal foils, polyethylene, asphalted felts and kraft paper, hot-melt asphalt, and special vapor barrier paints. Vapor barriers must be installed on the warm side of the wall.

Insulation In recent years newer insulating materials have largely supplanted the use of plaster and corkboard in cold-storage room construction. Insulation has generally been used in three forms for this application: board form, panelized, and poured, sprayed, or foamed in place.

Board form insulation includes corkboard, rigid polystyrene, polyisocyanurate, and polyurethane, foamed, and fibrous glass.

Panelized insulation has gained acceptance, particularly in prefabricated installations. The outer metal skin can serve as a vapor barrier, and the panels are factory insulated with polystyrene, polyisocyanurate, polyurethane, or fibrous glass. U values as low as 0.035 have been attained in 4-in (100-mm) thick panels.

Poured, sprayed, or foamed in place insulation has its greatest applications on existing installations. It is primarily a urethane material. The application requires special equipment, and precautionary measures must be taken.

Cold Storage Temperatures A great deal of research and experience are required to obtain authoritative information on optimum temperatures and humidities for various products in cold storage. Table 11.1.16 is representative of good practice. The safe storage period depends upon the product and the storage temperature, and operational techniques vary greatly. Modern cold-storage warehouses of the larger concerns are cooled by brine which is furnished at two different temperatures only. The higher temperature for the mild-temperature warehouses is 10 to 12°F (−12 to −11°C), and the low-temperature brine for the freezers is −10 to −12°F (−23 to −24°C). All temperatures ahead that of the brine are obtained by regulating the amount of brine circulated in any particular set of coils. In the low-temperature warehouses, the piping is arranged for two classes of service: (1) **sharp freezers,** where the goods which are to be frozen are kept in a blast of air whose temperature is −30°F (−34°C) while their temperature is brought down quickly to the holding temperature (say in from 6 to 10 h) after which they are stored in (2) **holding rooms** where the desired temperature is maintained.

The system of cooling which is now being installed in the highest type of warehouses consists of a coil room containing the necessary brine-type or wetted, direct-expansion-type coils, through which the air from the different rooms is circulated by a pressure blower. The inlet and outlet of each room are so arranged that the cooled circulating air will cover the entire room in its transit; this is usually accomplished by having the cold-air inlet in the center of the room and two cold air outlets—one at each end of the room. The piping ratio for the coil rooms in this system, assuming a high-grade insulation with 2 to 4 Btu transmission per ft^2 per 24 h per °F temp diff, should be 1 ft^2 of external pipe surface to 15 ft^3 of space to be cooled [with brine at −10 to −12°F (−23 to −24°C)] for warehouses carrying temperatures of zero and less, and 1 ft^2 of pipe surface to 24 ft^3 of space to be cooled [with brine at 10 to 12°F (−12 to −11°C)] in warehouses carrying mild temperatures of from 30 to 40° F (−1 to 4°C). Another system is used in which the blower and coils are supplemented by coils in the rooms. With this arrangement, it is possible to reduce temperatures quickly and hold them with the coils.

The average coil transmission in cold-storage rooms without forced circulation of the air is about 2 Btu per ft^2 of outside metal surface per h per °F temp diff with horizontal piping and 2.5 Btu with vertical piping. When forced air circulation is used, the transmission rate will increase to 20 Btu or more. In **brine circulation,** the brine, at same compressor back pressure, has a higher temperature than the ammonia, and consequently $1\frac{1}{2}$ times as much pipe or more is used in brine circulation as is direct expansion for a given back pressure.

Piping of Rooms The **size of pipe** usually employed for piping rooms varies from 1 to 2 in (25 to 50 mm) with either brine circulation or direct expansion.

The extra cost of liberal piping allowance will often be offset by the consequent improvement in the efficiency of operation of the compressor. An expansion valve should be provided for every 500-ft (150 m) length of 1-in (25 mm) pipe, every 650 ft (200 m) of $1\frac{1}{4}$-in (30 mm) pipe, and every 1,000 ft (300 mm) of 2-in (50 mm) pipe when direct expansion is used.

Values of **overall coefficients of heat transfer** in Btu/h · ft^2 · °F for refrigerating practice are given in Tables 11.1.15 and 11.1.17. (See Sec. 2.)

Brine circulation is generally preferred to direct expansion so as to avoid danger from escaping ammonia or other refrigerant in case the pipes should leak. An advantage of the brine system is that there is always a considerable mass of refrigerated brine which can be drawn on in case the machinery should have to be stopped for any reason. In small plants, the general machinery may be stopped at night and only the brine pump be kept going to distribute the surplus refrigeration which has been accumulated in the brine during the day. Brine piping must consist of two lines, a flow and a return, usually of the same size. Brine storage is seldom used in large plants because of its bulk, its first cost, and the practical inability to store much refrigeration.

The **brine coils** in each cooled space are in parallel across the supply and return pipes. It has been common practice to allow 100 to 120 running ft (30 to 36 m) of **pipe** per circuit for low temperatures, and 400 to 440 (120 to 135 m) for high temperatures. The **tons refrigeration produced by brine** at various temperature differences and rates of pumping may be calculated approximately as follows: tons refrigeration = gal/min × °F range/28.

Air Conditioning of Low-Temperature Storage Rooms Bacterial growth and chemical decomposition are retarded by lowering the **temperature** of perishable goods. Too low a temperature will freeze the goods and may result in spoilage. A holding temperature ahead

Table 11.1.17 Overall Coefficients of Heat Transfer
(Btu/ft^2 · h · °F)

Can ice-making piping	
Old-style feed, non-flooded	12-15
Flooded	20-40
High-velocity raceway trunk coils	80-110
Ammonia condensers	
Submerged (obsolete)	30-40
Atmospheric, gas entering at top	60-65
Atmospheric, drip or bleeder	125-200
Flooded	125-150
Shell and tube	150-300
Double pipe	150-250
Baudelot coolers, counterflow, atmospheric type	
Milk coolers	75
Cream coolers	60
Oil coolers	10
Water coolers	
Direct expansion	60-150
Flooded	60-200
Brine coolers	
Shell and tube	45-100
Double pipe	150-300
Cooling coils	
Boiling refrigerant to air in unit coolers	4-8
Water to air in unit coolers	5-9
Brine to unagitated air	2.2-2.8
Direct expansion to unagitated air*	1.6-2.5
Water cooler, shell and coil	15-25
Water cooler, shell and tube	50-150
Water cooler, shell and finned tube (Freon)	30-150
Liquid-ammonia cooler, shell and coil accumulator	45
Air dehydrator	
Shell and coil (brine in coil) 1st coil	5.0
Shell and coil (brine in coil) 2nd coil	3.0
Double pipe	6-7
Superheat remover, shell and tube	15-25

* *U* factor increases to 3.3 at 15°F temp diff with ammonia recirculation systems.
NOTE: Forced circulation of the air increases the coefficient to 1½ to 2½ times the values for still air. One inch of frost decreases the value 25 percent.

the freezing point of the article is best for storage conditions. The maintenance of a proper relative **humidity** in a storage space holding unwrapped goods is as important as the maintenance of the proper temperature. High humidity favors the growth of mold and bacteria. Low humidities rob the product of moisture resulting in losses in value through impaired appearance and lost weight. **Air motion** is of importance in maintaining uniform conditions throughout the storage space. Too high air velocities cause excessive drying, and stagnant air through high humidities will cause mold. An optimum air motion falling between stagnation on one side and excessive drying on the other must be selected.

Low-temperature conditioning for storage purposes is preferably accomplished by the use of **cold-diffuser methods.** The cold diffuser connected to or located in the storage space consists of a fan and cooling means. The cooling means may be either a brine or cold-water spray or a surface-type cooler using brine, or a volatile refrigerant such as ammonia, Freon, etc. Unitary equipment, performing the functions of the cold diffuser, is commercially available. For the requirements for various perishable products, see ASHRAE Handbooks, 1993 Fundamentals and 1995 Applications, 1994 Refrigeration, and Table 11.1.16.

Quick freezing is freezing in 2 h or less; as employed for foods the temperatures may be as low as −20 to −50°F (−29 to −46°C). The methods of freezing include direct immersion in cold brine or a brine spray with the commodity held in a metal container. Another procedure is the use of a freezing tunnel approximately 50 ft (15 m) long, equipped with a stainless-steel conveyor belt. The commodities to be frozen pass through this tunnel in 15 to 30 min. The speed of the belt is variable, and the heat transfer is accomplished by high-velocity air circulation at a number of points in the tunnel, the circulation being transverse to the belt. The circulating air is cooled by brine sprays or wetted, direct-expansion coils which maintain the required temperature and the high humidity necessary to prevent shrinkage. In another process, packages of the material to be quick-frozen are clamped between steel platens containing evaporating refrigerant. Successful results are obtained with temperatures as high as 0°F (−18°C). Storage may be at −5 to −10°F (−20 to −23°C) and transportation at 10°F (−12°C) or lower.

Cold-storage lockers for individual family use are used in large numbers, particularly in suburban and rural areas. Lockers are rented usually on a yearly basis. A locker plant may have several hundred rental steel lockers, 6-ft^3 (0.17-m^3) capacity each, placed in rooms at about 0°F (−18°C); a room for cutting and wrapping the meat; and a sharp freezer room held at about −20°F (−29°C). Two compressors are frequently specified or a single compressor using automatic control valves on the separate rooms. Ammonia and R 12, operating under direct expansion, are commonly used.

ICE MAKING

Ice refrigeration is economical for some applications. The use of natural ice is fast disappearing. Manufactured ice is made by several methods. In the can system, galvanized cans containing up to 400 lb (180 kg) of water are immersed in brine for 25 to 40 h for freezing. Air dissolved in the water separates as bubbles causing opacity unless the water is agitated during the early stages of cooling, and any dissolved impurities come out of solution to form a central off-color core in the ice block unless the core is removed and replaced with fresh water just before freezing. The time and investment required for making ice in cans are often uneconomical. In the **Flakice** method (Crosby Field, *Trans. ASME,* May 1951, pp. 347-357) refrigerant is sprayed against the inner wall of a flexible, slowly rotating cylinder which is partially immersed in a tank of water. As the cylinder rotates, a thin layer of ice forms on the outside and then is broken off as further rotation causes the cylinder surface to flex under the action of an internal cam. For a patent history of small ice machines of this type, refer to Crosby Field, *Refrig. Eng.,* **58,** Dec., 1950, p. 1163. In the **PakIce** process (Taylor, *Refrig. Eng.,* **22,** 1931, p. 307) liquid ammonia is evaporated in an annular space formed by two concentric cylinders, the inner being corrugated. Water sprayed on the inner corrugated surface quickly freezes in a thin layer about 0.25 mm thick and is continually removed by mechanical scrapers. In an **extrusion method** (Watt, *Trans. ASME,* 1949) a slightly tapered circular or rectangular vertical cylinder with large end up is open at the top, but closed by a ram at the bottom. This cylinder has a jacket in which refrigerant is evaporated directly against the cylinder walls. Water enters the cylinder and freezes to a tapered block at the start of the operation. As soon as this occurs, the ram lifts (about 6 mm), shearing the ice block from the cold cylinder walls. Water admitted to this space quickly freezes, the ram again operates and the block of ice is again lifted, and shearing occurs again between the freshly formed ice and the cylinder walls, but not at the fresh ice and old ice interface. In this way, a continuous cylinder of ice is formed by the series of nested ice shells. The installation and upkeep costs are small and so is the space required per ton of the ice produced; a pilot-plant 1-ton (907 kg) unit requires 3 ft^2 (0.3 m^2) of floor space.

Batch ice makers can manufacture up to 75 tons (68 tonnes) of ice per day. These machines use constant-diameter ice-forming tubes to produce ice in shell and toroidal forms. The ice forms on a bank of vertical tubes in either of two configurations: inside the tubes for pellets or outside of the tubes for shell ice. A water distribution system is provided at the top. An ice conveyor and mechanical size-reducing device is mounted below the tubes. The refrigerant is circulated through the annulus of concentric tubes when ice pellets are made, and inside of single tubes when shell ice is made. The refrigerant supply can be designed for flooded operation; however, equipment capacity is reduced. The cycle starts as water passes through the distributors and flows over the surface of the tubes as a film. The refrigerant freezes the water to form a thickening layer of ice as time passes. Unfrozen water drains out of the tube into a sump. Then the water is recirculated, by a pump, up to the distribution system at the top of the tube bank. A purge of water is taken from the sump to remove accumulated salts and other impurities. When harvest is initiated, the refrigerant system is drained, and hot gas from the condenser is introduced into the refrigerant side of the tube and the pressure is raised. This causes the ice to partially melt and release from the tube wall. The cylinder or shell of ice falls by gravity into the conveying and size-reducing machinery. The size of an ice pellet can vary from 0.75 in (19 mm) to 1.375 in (35 mm) outside diameter. Ice clarity can be controlled by flushing the ice with water before and during harvest. The machine is controlled by digital logic microprocessors. Energy consumption when using 65°F (18°C) feedwater varies from 1.94 to 2.33 tons (6.8 to 8.2 kW) refrigeration per ton (909 kg) of ice when using R-717 versus HCFC-22 in a recirculation system. Standard model ice makers can be operated with the refrigerant system either in the flooded condition or with circulating pumped refrigerant in the tubes. Similar units lose 18 percent capacity when operating in a flooded state compared to operating with pumped recirculating refrigerant.

Making clear ice using continuous methods is often accomplished by washing the newly frozen ice surface with cold water. This method can increase water requirements by up to 50 percent and energy usage by 15 percent.

The refrigeration tonnage required for making 1 ton (909 kg) of ice varies from 1.84 with 55°F (12.8°C) inlet water temperature to about 2.0 if the water to be frozen enters as high as 75°F (23.9°C).

SKATING RINKS

Rinks for ice skating, hockey, or curling vary in size from about 400 ft^2 (37 m^2) area to 16,000 ft^2 (1,486 m^2) or more. Construction of the ice floor is important. For a low-cost, efficient unit, brine pipes are laid directly in sand, which is kept wet during freezing to increase heat transfer. As freezing progresses, a layer of ice is built up by spraying water on the surface. The expense of maintaining such a floor in a satisfactory level condition may be considerable, and corrosion of the pipes, from the outside in, in contact with the wet sand may be excessive, particularly during shutdown periods. Moreover, it is not possible to use such a floor for other purposes. When a floor for diversified activities is needed, it is usual to provide a lower layer of insulation, primarily to prevent sweating on ceilings under the floor, surmounted by a concrete slab in which the steel pipes carrying the brine are imbedded and on which the ice layer is formed. This construction protects the pipe from external corrosion, affords fast freezing, and makes possible use of the floor in a few hours for purposes where an ice surface is not desired. Floor pipes can be mild steel or polyethylene and are usually 1 or 1¼ in (25 or 32 mm) spaced 3 to 6 in (75 to 150 mm) on centers; care must be exercised to ensure a uniform distribution of brine. Brine flow may be 10 gal/min per ton of refrigeration, corresponding to about 1 gal/min per 10 to 20 ft^2 of piping surface. In a few installations, cold brine is sprayed directly against the underside of a steel floor on which the ice layer is frozen. Close control of ice surface temperature is necessary for good skating, usually −3°C ± 0.5. Refrigeration capacity should be from 0.4 to 0.85 ton per 100 ft^2 (150 to 320 W per m^2) of floor surface, although the load may be much higher for open-air rinks or other unusual conditions.

REFRIGERATION IN THE CHEMICAL INDUSTRIES

The use of refrigeration in petroleum and other industries is widespread. In petroleum processing, refrigeration is used (1) to control vapor pressure of highly volatile constituents (methane, ethane, propane, and butane), such control being necessary during distillation, during processing, and for recovery of gasoline fractions from natural gas; (2) to shift the solubility relationship so that undesired constituents such as asphalt and wax in lubricating oils may be removed by precipitation; (3) to produce selective chemical reaction such as occurs when sulfuric acid is used to remove gum-forming constituents from light fuels or when an alkylate fuel fraction is formed by combining a low molecular weight unsaturated with a similar saturated hydrocarbon. Alkylate is a valuable and important constituent of aviation fuels.

For (1), temperature ranging from −35°F (−38°C) or lower to as high as 60°F (15°C) are required, the refrigerant being ammonia, propane, or butane. For the higher temperatures spray-cooled water or jet refrigerating units (discussed earlier) are sometimes satisfactory. For (2), it is possible to use propane as a combined solvent and refrigerant, evaporation of the propane causing the required cooling, temperature level [approximately −40°F (−40°C)] being controlled by the pressure held on the equipment. The propane vapor is recompressed, condensed, and reused as in any compression cycle. As much as ⅓ ton of refrigeration per barrel of lubricating oil dewaxed is needed. For (3), temperatures of up to 60°F (−16 to 15°C) are general, and ammonia is a satisfactory and widely used refrigerant.

The chemical industry uses refrigeration to control condensation polymerization processes in the manufacture of synthetic rubbers and plastics. Syntheses of refrigerants and other chemicals require systems to maintain temperatures which range from 5°C down to −85°C in process equipment. Large centralized refrigeration systems which use ethylene glycol and methylene chloride brine are common. Several modern chemical processes use the eventual product as a refrigerant (for example, the manufacture of ammonia). Cascade systems with segregated circuits are used to produce temperatures lower than −50°C.

Recently imposed regulations limit volatile organic vapor emissions to the atmosphere from storage tank vents and other process systems. The recovery efficiency required of cleanup systems has led to development of refrigerated liquid absorption processes to scrub vaporous effluents from organic chemical storage tanks. Recently installed systems have required liquid absorbent temperatures of −40°F (−40°C). The capacity of these systems varies from 2 tons (7 kW) to 500 tons (1,760 kW). The absorbent liquid is often accumulated from the effluent stream over a period of time.

DEEP REFRIGERATION

Dry ice (solid CO_2) is useful as a refrigerating medium in special cases. The main objection to the general use of dry ice is its cost, but the ease in its handling, its low temperature [−110°F (−79°C) at atmospheric pressure], its noncorrosive and nontoxic properties, its high latent heat, and the absence of liquid drip make it desirable. The heat absorbed per pound of solid CO_2 during sublimation at atmospheric pressure and −110°F (−70°C) is 245 Btu approx. The specific heat of the gas at constant pressure is about 0.2.

Carbon dioxide is obtained commercially either by fermentation or by burning. The gas is compressed, usually in three stages, and cooled to atmospheric temperature, thereby forming a liquid at about 1,000 psia (690 kPa); the liquid is then throttled to less than 5.3 atm pressure. During throttling part of the CO_2 solidifies and is compressed to form a dense ice. The fraction that returns to the vapor phase is recompressed.

Production of Solid Carbon Dioxide Figure 11.1.18 shows a simple cycle and Fig. 11.1.19 the elemental pressure-enthalpy diagram for the production of solid carbon dioxide. From the high pressure at atmospheric temperature the liquid is throttled (isenthalpic) to atmospheric pressure, the resultant point being within the sublimation solid-vapor zone beneath the triple-point realm. In this resultant throttling expansion, part of the CO_2 solidifies; it is then removed in the separator from the residual vapor and is compacted to form a dense CO_2 ice. The fraction representing the vapor phase is augmented by fresh makeup gas to continue the production cycle.

Beyond the manufacture of dry ice, operation at deep refrigerations, such as air liquefaction [at −312°F (−191°C) approx], lead into

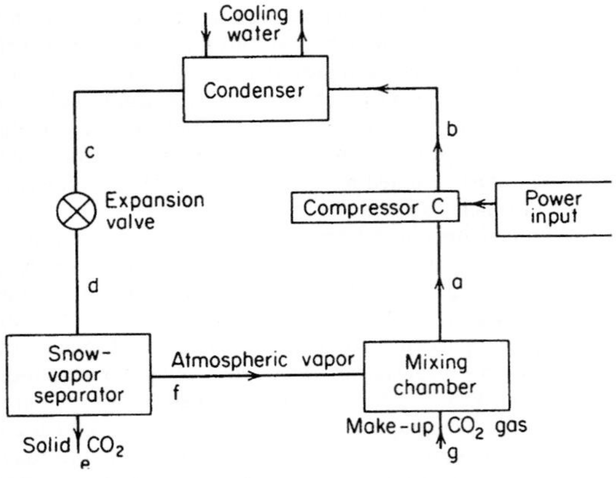

Figure 11.1.18 Elemental dry-ice production circuit.

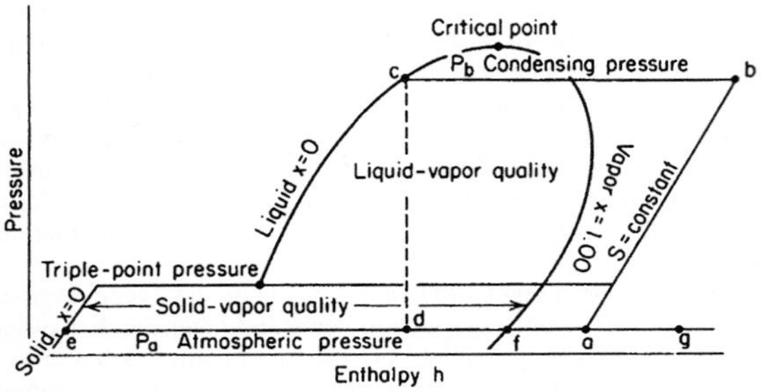

Figure 11.1.19 Pressure-enthalpy diagram illustrating dry-ice production circuit.

cryogenics (see Sec. 11). Illustrative systems are (1) Linde (throttling) and (2) Claude (expander). In the elemental **Linde system** (Fig. 11.1.20), by virtue of the "cold-effect" regenerator, the process air is reduced in temperature by the return fraction of the air from the separator. The throttling process thus ends in the "wet zone" at atmospheric

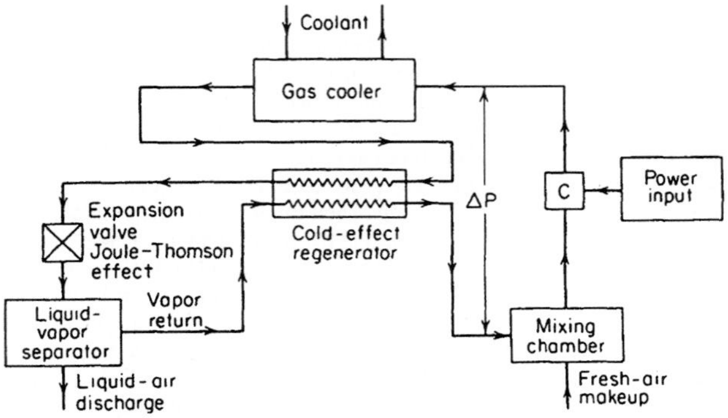

Figure 11.1.20 Elemental Linde system circuit.

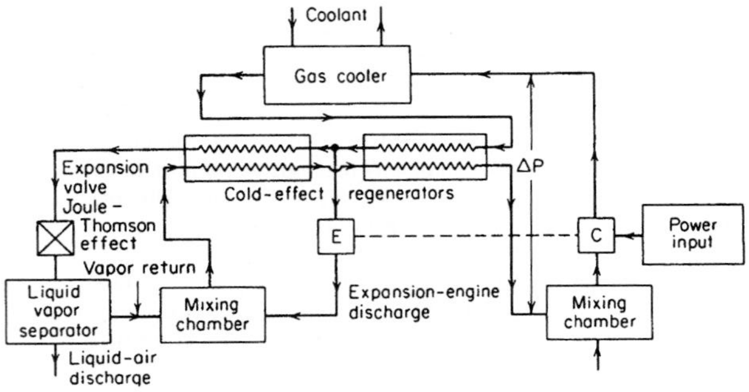

Figure 11.1.21 Elemental Claude system circuit.

pressure with generation of low-temperature liquid air that flows from the separator. In the **Claude system** (Fig. 11.1.21), use is made of an expander (whose work output reduces the net work requirement of the compressor) and of an expansion valve with the Joule-Thomson effect.

11.2 CRYOGENICS

by F. J. Edeskuty, Revised by Peter J. Shirron

REFERENCES: Timmerhaus (ed.), "Advances in Cryogenic Engineering" (annual), Plenum Press. Vance and Duke, "Applied Cryogenic Engineering," Wiley. Scott, "Cryogenic Engineering," Met Chem Research. Vance (ed.), "Cryogenic Technology," Wiley. Mendelssohn, "Cryophysics," Interscience. McClintock, "Cryogenics," Reinhold. Scott, Denton, and Nicholls (eds.), "Technology and Uses of Liquid Hydrogen," Pergamon. Hoare, Jackson, and Kurti, "Experimental Cryophysics," Butterworth. White, "Experimental Techniques in Low-temperature Physics," Oxford. Johnson (ed.), "Compendium of Properties of Materials at Low Temperature," WADD Technical Report 60-56, U.S. Dept. of Commerce. Durham, McClintock, and Reed, "Cryogenic Materials Data Handbook," PB 171-809, U.S. Dept. of Commerce. Cook (ed.), "Argon, Helium, and the Rare Gases," Interscience. Perry, Chilton, and Kirkpatrick, "Chemical Engineers' Handbook," McGraw-Hill. Barron, "Cryogenic Systems," McGraw-Hill. Bailey, "Advanced Cryogenics," Plenum Press. Haselden, "Cryogenic Fundamentals," Academic. Williamson and Edeskuty (eds.), "Liquid Cryogens," CRC. Timmerhaus and Flynn, "Cryogenic Process Engineering," Plenum Press. Van Sciver, "Helium Cryogenics," Plenum Press. Edeskuty and Stewart, "Safety in the Handling of Cryogenic Fluids," Plenum Press.

Notation

Some symbols in this section differ from the notation of engineering practice but are retained to facilitate use of the prevalent scientific sources of information.

A = area, ft^2 (m^2)
b.c.c. = body centered cubic lattice structure
C = specific heat, Btu/lbm · °R(J/kg · K)
C_p, C_v = specific heats at constant pressure and volume respectively, Btu/lbm · °R(J/kg · K)
f.c.c. = face centered cubic lattic structure
H = enthalpy, Btu/lbm (J/kg)
H = magnetic field, $G(T)$
h.c.p. = hexagonal close-packed lattice structure
k = thermal conductivity, Btu/h · ft · °R (W/m · K)
s = electrical conductivity (reciprocal of resistivity) or pair separation for zero potential in Lenard-Jones equation

s = Boltzmann's constant
L = length, ft (m)
LH_2 = liquid hydrogen (H_2)
LHe = liquid helium (He)
LN_2 = liquid nitrogen (N_2)
LNG = liquefied natural gas
LO_2 = liquid oxygen (O_2)
n = number of moles
P = pressure, psia or torr (vacuum) (Pa)
R_e = electrical resistance, Ω
T = temperature, °R (K)
V = volume, ft³ (m³)
a = thermal linear-expansion coefficient, $°R^{-1}$ (K^{-1}), or accommodation coefficient
h = viscosity, lbm/ft · h (N · s/m²)
r = density, lbm/ft³ (kg/m³)
ΔT = temp difference, °R(K)
* = Denotes property divided by quantum parameter (Table 11.2.5) which is not the critical point value.

Cryogenics is the study, production, and utilization of low temperatures. Cryogenic temperatures have been defined so ambiguously that "upper limits" to the cryogenic range from 216 to 396°R (120 to 220 K) may be found in the literature. In this section the cryogenic range for a property of a given substance is considered to embrace the scale between absolute zero and the temperature above which the property has the expected or normal behavior. Cryogenics thus embraces the unusual and unexpected variations which appear at low temperatures and make extrapolations of properties from ambient to low temperatures unreliable.

Progressively lower temperatures become increasingly difficult to attain in practice. A cryogenic refrigerator operating on the Carnot cycle (ideal) will require the least amount of input work to effect refrigeration (remove heat) at a lower temperature and reject it at a higher temperature (usually ambient temperature). As the working temperature of a refrigerator is lowered, the work required to transfer a given amount of heat increases as demonstrated by the Carnot limitation, to wit, $W = Q[(T_1 - T_2)/T_2]$, where W is the work required to extract the heat Q at a low temperature T_2 and reject it at a higher temperature T_1 (see Sec. 2). The actual work is always greater than this because of inefficiencies of mechanical equipment, thermal losses associated with finite temperature differences in heat exchangers, and heat leaks from the surroundings to the cold equipment. Cryogenic refrigerators built to date do not approach this limit very closely; typically less than 40 percent of Carnot efficiency for the largest refrigerators and progressively less as the size of the refrigerator becomes smaller.

REFRIGERATION METHODS

(See also Sec. 2 and Sec. 11.1, "Mechanical Refrigeration.")

Cryogenic refrigerators (cryocoolers) may be classified by (1) the functions they perform e.g., the delivery of liquid cryogens, the separation of mixtures of gases, and the maintenance of spaces at cryogenic temperatures, (2) their refrigerating capacities, and (3) the temperatures they reach. Large industrial-sized plants (a) deliver LNG (~120 K), LO_2 (~83 K), LN_2 (~77 K), LH_2 (~20 K), and LHe (~4 K), (b) separate gaseous mixtures, for example, the constituents of the atmosphere, H_2 from petroleum refinery gases, H_2 and CO from coke oven and coal-water gas reactors, and He from natural gas, and (c) provide refrigeration to maintain spaces at low temperatures. For the latter, that is, (c), a unit has been manufactured and installed that delivers 13 kW of refrigeration at less than 3.8 K.

An important area of refrigerator development that is being commercially exploited now is laboratory-sized cryogenic refrigerators for laboratory research and development at LHe temperatures (<5 K). These refrigerators are used for many different purposes, refrigerating for example: high-field superconducting electromagnets such as those used in magnetic resonance imaging (MRI), LHe bubble chambers for high-energy (nuclear) particle research, experimental superconducting electric generators and motors and superconducting magnets for levitation of railroad trains. Another area of commercial exploitation that is important for the progress of cryogenic physics research is the development of refrigerators for the continuous production of refrigeration at temperatures below 1 K. These include L^4He and L^3He sorption refrigerators, adiabatic demagnetization refrigerators, and L^3He-L^4He dilution refrigerators that can reach temperatures of 0.7, 0.3, 0.01, and 0.003 K, respectively.

Of various methods of refrigeration, the most commonly used to produce temperatures as low as 1 K are (1) the evaporation of a volatile liquid (referred to as the **cascade** method when applied in several successive stages using progressively lower-boiling liquids), (2) Joule-Thomson (isenthalpic) expansion of a compressed gas, and (3) an adiabatic (isentropic) expansion of a compressed gas in an engine (reciprocating or turbine) or from a bomb or cylinder through a throttling valve (Simon expansion). Using L^3He, method (1) is capable of reaching temperatures as low as ~0.3 K. Numerous refrigeration cycles have been devised utilizing various combinations of the above three methods. The final stage of refrigeration in plants that deliver liquid cryogens is generally a Joule-Thomson (isenthalpic) expansion. **Expansion engines** (reciprocating and turbine) are commonly used for producing the refrigeration needed to reach the final stage. Turbine expanders are preferred for the large refrigerators because their efficiencies (70 to 85 percent) are higher than for reciprocating expanders. The efficiencies of the heat exchangers for counterflowing "cold" and "warm" gases are an important factor in the overall refrigerator efficiency. In industrial plants heat exchanger efficiencies reach 98 percent. Their design represents a compromise between the attainment of a high heat-transfer coefficient, on the one hand, and a low resistance to the flow on the other.

Figure 11.2.1 typifies a commonly employed modified Brayton cycle which uses an expansion engine for generating the refrigeration. The theoretical figure of merit (the ratio of the work W_c done by the compressor to the heat Q, transferred to the refrigerant at the low temperature) is for this cycle

$$\frac{W_c}{Q} = \frac{RT_1 \ln(P_1/P_5)}{M(H_4 - H_3)}$$

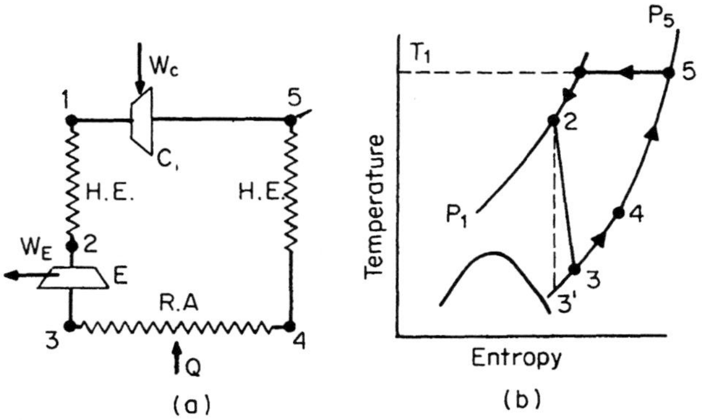

Figure 11.2.1 (*a*) Schematic of a modified (isothermal compression) Brayton cycle, and (*b*) the process path. W_c is the energy (work) to drive the compressor *C*. *Q* is the heat absorbed by the refrigerator from the refrigerated area R.A., where the working fluid passes from state 3 to state 4. The heat-exchanger (H.E.) processes are isobaric. The dashed line 2-3′ is the ideal isentropic expansion; 2-3 represents the actual expansion in the expander E.

where M is the molecular weight of the gas whose enthalpy per unit mass is H. The other symbols are defined by reference in Fig. 11.2.1.

The **Stirling refrigerating engine** and modifications of it are used in a number of commercial makes of laboratory and miniature-sized refrigerators. The Stirling-cycle refrigerator is well suited to (1) the liquefaction of air on a laboratory scale (~7 L/h), (2) the recondensation of evaporated liquid cryogens, and (3) the refrigeration of closed spaces where the refrigeration load is not very large. These laboratory scale refrigerators supply ~1 kW of refrigeration at ~80 K and ~2 kW at 160 K. They are used with liquid air fractionating columns for the production of LN_2 (~7 L/h) and LO_2 (~5 L/h).

The Stirling-cycle refrigerator consists of a piston for compressing isothermally the working fluid (usually He), and a displacer which can operate in the same cylinder with the piston. The displacer and piston are connected to the same electrically driven shaft but displaced in phase by 90°. The displacer pushes compressed gas isochorically from the warm region where it was compressed, through the regenerator into the cold region where the compressed gas is expanded isothermally doing work on the piston and producing refrigeration. The regenerator consists of a porous mass (packed metal wool, screens, or spheres of high heat capacity) in which a steep temperature gradient is established between the warm region of compression and the cold region of expansion. The displacer returns the gas after expansion to the region for compression through the regenerator in which it is warmed to the temperature of the isothermal compression. The refrigerant (He) is recycled. The regenerators in the larger refrigerators are usually placed around the outside of the engine cylinders, but in small capacity, lower temperature refrigerators they are put inside the displacers.

In Stirling refrigerators that reach temperatures of 17 K and produce 1 W of refrigeration at 25 K, the compressed He is expanded in two or three stages at temperatures intermediate between room temperature and the lowest temperature reached. There is only a single piston for compression. The expansion in stages allows part of the heat that leaks into the coldest region of a single stage engine to be absorbed and removed at a higher (intermediate) temperature where the efficiency for transferring heat is greater.

The **Vuilleumier refrigerating engine** is a modification of the Stirling refrigerator. It *resembles* two single Stirling engines placed back to back. One of the engines operates as a heat engine and the other as a refrigerator. The lower temperature at which the heat engine discharges its waste heat, is also the top temperature at which the refrigerator engine discharges its waste heat. Hence, the Vuilleumier refrigerator operates at elevated, intermediate and low temperature levels which may be 800, 300 (ambient), and 90 K, respectively. Each engine has a cylinder, a displacer, and a regenerator, and the two engines are connected to a common crankshaft but displaced by 90°. Very little external power is needed to operate the two displacers which are operated in sinusoidal motion, displaced 90° in phase. The working fluid (He) is recycled.

The **Gifford-McMahon refrigerator** is another modification of the Stirling-cycle refrigerator. It may be single stage for refrigeration at higher temperatures (~30-80 K), or multistage for lower temperatures (<4 K). Each stage has a cylinder, a displacer and a regenerator. The displacer pushes gas (usually He) under a very small head of pressure from the top of the cylinder, where the temperature is ambient, through the regenerator, into the bottom of the cylinder which is the cold region of the refrigerator. Compressed He is supplied from an external source—the refrigerator has no piston for compressing gas. Only a small source of external energy is needed for driving the displacer which has very little work to do in overcoming the forces of friction. The valves that control the admission of compressed He to the top of the cylinder, and the expansion of the cold compressed He from the bottom of the cylinder are externally operated by a drive mechanism. The cycling rate is < 2 Hz. The expansion of the cold He is of the adiabatic-Simon-bomb type. Expanded gas is discharged from the refrigerator and recycled by an external compressor. Commercial two-stage GM refrigerators can supply ~1 W of cooling at 4 K.

Some applications of cryogenic refrigerators require very high reliability without maintenance. The fewer the working parts, the more likely is the needed reliability to be achieved. The **pulse-tube** and **thermoacoustic refrigerators** are advantageous in this respect in that they have only one moving part, and that part operates at ambient temperature. Pressure oscillations are introduced into a tube with a closed end and fitted with the proper arrangement of heat exchangers. For the optimum frequency of the pulsation, a temperature gradient is established between the heat exchangers, and this gradient can pump heat from a low temperature to a higher one. A number of different designs have resulted, including single-stage and multistage units, and operating frequencies have ranged from about 1 Hz up to acoustic frequencies. This type of refrigerator is capable of reaching temperatures down to ~20 K with a single-stage unit with cooling powers in the range of 50-70 W at 40 K, and to ~3 K with a two-stage unit, with cooling powers in the 1-2 W range, all with ambient temperature heat sinks. Present research is also focused on the use of an acoustically driven pulse-tube refrigerator that would have no moving parts and no close tolerances of the components, thus making it inexpensive and easy to produce.

Magnetic refrigeration, originally proposed in 1925, depends upon the magnetocaloric effect in which paramagnetic materials at low temperature (or ferromagnetic materials near their Curie temperatures) will warm up (or expel heat at constant temperature) with the application of an increasing magnetic field. Conversely, these materials will cool (or absorb heat at constant temperature) upon removal or reduction of the magnetic field. The heating and cooling is reversible, allowing this technique to achieve thermodynamic efficiencies close to Carnot. **Adiabatic demagnetization refrigerators** implement a four-step cycle: (1) adiabatic magnetization to warm the paramagnetic material; (2) isothermal magnetization to reject heat to a thermal reservoir (through a heat switch); (3) adiabatic demagnetization to cool the material; and (4) isothermal demagnetization, during which it produces useful cooling for potentially extended periods (>1-2 days). With the use of a paramagnetic material magnetized at a temperature of about 1.5 K, temperatures below 0.01 K have been reached by a single demagnetization. Magnetic refrigeration is applicable over a wide range of temperature, and lends itself to multistaging in which stages are attached in series (with upper stages acting as the thermal reservoir for the lower stages) and/or parallel configurations to achieve (1) larger temperature range, (2) increased efficiency by reducing internal losses, and (3) continuous operation. Two-stage refrigerators can cool to below 0.05 K when demagnetized from 4 K, and continuous cooling has been demonstrated at both low (<0.05 K) and intermediate (~1 K) temperature, with cooling powers of ~5 μW and ~1 mW, respectively.

A variant that becomes advantageous at higher temperatures (>10 K) is **active magnetic refrigeration,** which uses a regenerator containing a porous matrix of paramagnetic material exposed to an oscillating magnetic field. An alternating gas flow through the regenerator carries heat of magnetization to a warm-end heat exchanger, and distributes cooling to a cold-end heat exchanger. To increase cooling power and temperature range, the regenerator often contains a distribution of paramagnetic materials tailored to maximize heat capacity along its length. The alternating field may be produced by active control of a superconducting magnet, or by mechanically inserting and withdrawing the regenerator from a fixed magnet. The technique is applicable over a wide temperature range (up to room temperature), though its use is most promising for lower temperature applications such as large-scale hydrogen liquefaction (~20 K).

^3He-^4He dilution refrigerators for reaching temperatures between 0.01 and 0.3 K and generating continuously as much as 750 ergs/s of refrigeration are commercially available. Refrigeration results from the solution of L^3He in L^4He, the heat of solution being negative. Figure 11.2.2 is a schematic diagram of a dilution refrigerator. ^3He vapor from the "still" at ~1 K (the upper operating temperature of the refrigerator) is collected and returned by a pump to the "mixing" chamber where solution takes place. The ^3He arrives at the mixing chamber as L^3He near the mixing chamber temperature (the lower operating temperature of the refrigerator) after flowing through a heat exchanger counter to the flow of cold ^3He-^4He solution on its way to the "still." In the electrically heated still, ^3He is evaporated from the solution. The evaporated ^3He is withdrawn by the collecting pump (diffusion) that returns the ^3He to the mixing chamber. At 1 K the vapor pressure of L^4He is negligible. L^4He in the still, freed of ^3He, returns to the mixing chamber by super-fluid flow, building up in the mixing chamber an osmotic pressure that drives ^3He atoms toward the still against the forces of viscous flow.

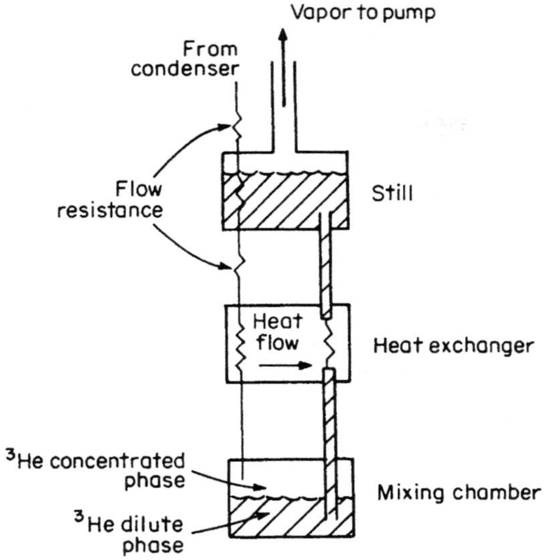

Figure 11.2.2 Schematic diagram of the mixing chamber and still of a ^{3}He-^{4}He dilution refrigerator.

At temperatures lower than 0.87 K, liquid solutions of ^{3}He and ^{4}He separate into two liquid phases, one ^{3}He-rich and the other ^{4}He-rich, in thermodynamic equilibrium. At 0 K, the ^{3}He-rich phase is 100 percent ^{3}He, whereas the ^{4}He-rich phase (in equilibrium with L^{3}He phase) contains ~6 percent ^{3}He. Under gravity, the lower density L^{3}He-rich phase rides on the heavier L^{4}He-rich layer.

In the mixing chamber, L^{3}He enters the upper ^{3}He-rich phase; ^{3}He is removed from the lower (denser) ^{4}He-rich phase by flow to the still, and is replaced in equal measure by ^{3}He from ^{3}He-rich phase; solution of ^{3}He in ^{4}He creates cooling at the interphase boundary. In recent years, "dry" dilution refrigerators that use a low-temperature cryocooler for precooling the dilution unit have become commercially available.

Starting at 50 mK (a temperature reached in a ^{3}He-^{4}He dilution refrigerator), temperatures in the 2- to 3-mK range are attainable with a **Pomeranchuk refrigerator.** Refrigeration is generated by compression of a mixture of liquid and solid phases of ^{3}He at pressures in excess of 28.9 atm, the minimum P at which these phases can coexist in equilibrium. The entropy of solid ^{3}He exceeds the entropy of L^{3}He, which is contrary to the normal behavior for other substances. This occurs in ^{3}He because of its nuclear magnetic properties. The ^{3}He nucleus is magnetic and at temperatures in the mK range and above, in solid ^{3}He, the nuclear moments are randomly oriented whereas in L^{3}He, at temperatures lower than ~0.3 K, the nuclear moments are paired or partially paired, antiferromagnetically. An adiabatic increase of P converts liquid to solid with a reduction in T, whereas an isothermal increase of P results in an absorption of heat. ^{3}He, condensed, is confined in a container with flexible metal walls in the "cold" region of a ^{3}He-^{4}He dilution refrigerator. The ^{3}He container is surrounded with L^{4}He which serves as the pressure transmitting fluid.

Sorption refrigerators produce cooling by lowering the vapor pressure above a condensed liquid using a sorption pump. Sorption pumps contain an adsorbent with large specific area to which cryogenic gases are strongly bound by Van der Waals forces. The most common adsorbent for ^{3}He and ^{4}He refrigerators is charcoal, with an area of 1,000-1,600 m^2/gram. The vapor pressure above an adsorbed film is strongly temperature dependent at coverages of 1-2 monolayers. Warming the sorption pump raises the vapor pressure (as much as 1-2 atmospheres), and the desorbed gas flows into a condenser thermally connected to a heat sink by closing a heat switch. The condensed liquid is confined in the condenser and thermally isolated by opening the heat switch. Cooling the sorption pump reduces the pressure, cooling the liquid. The ultimate temperature is established by the vapor pressure in the sorption pump and flow impedance to the condenser. Practical L^{3}He and L^{4}He sorption refrigerators can reach as low as 0.25 and 0.7 K, and require heat sinks at or below the normal boiling points of 3.2 and 4.2 K. L^{4}He and cascaded L^{3}He/L^{4}He sorption systems are increasingly common for producing temperatures of 1 K and below, as the use of liquid helium has decreased in favor of cryocoolers. As with magnetic refrigeration, multistaging can also be used with sorption refrigerators to produce continuous cooling.

Sorption pumps can also function as **sorption compressors.** Using check valves to rectify the flow, a sorption compressor can provide an alternating high-pressure supply and low-pressure return. Arrangements of two or more compressors can be used to provide continuous, high-pressure, closed-loop flow, such as that needed for Joule-Thomson cooling. Current research efforts include the development of miniature coolers that can be integrated directly with low temperature detector arrays.

GAS LIQUEFACTION

A conventional method of gas liquefaction utilizes a gas compressor, a countercurrent heat exchanger, and a throttling valve through which the gas expands isenthalpically (Joule-Thomson) and cools if it has already been cooled below its inversion temperature. After expansion, the cold gas returns through the heat exchanger, in which it exchanges heat with the countercurrent compressed gas flowing to the throttling valve. It leaves the heat exchanger near the temperature of the entering compressed gas. The temperature decreases at the throttling valve until the condensing temperature is reached, and then liquefaction occurs at a rate determined by the rate of refrigeration.

The rate of liquefaction x in pounds per hour is x = {[H (expanded *gas* out) $-H$ (compressed *gas* in)] $w - q$}/[H (compressed *gas* in) $-H$ (*liquid* following valve)], where the H's are enthalpies in Btu per pound for the final heat exchanger, w is the flow rate in the same units as x, and q is the heat leak in Btu per hour from outside into the heat exchanger. For an ideal heat exchanger the expanded gas leaves at the temperature of the entering compressed gas. In practice, there is a small difference in temperature which represents a small loss of refrigeration. For ideal gases, H is independent of P for a given T, and hence no liquefaction of an ideal gas results. The H's of air, O_2, N_2, A, CO_2, the hydrocarbons, and the normal refrigerants decrease with increasing pressure at ambient T, except at very high P's, and these gases are liquefiable by this kind of isenthalpic (Joule-Thomson) expansion. The H's of H_2 and He at ambient T, however, increase with increasing P, even at low P's, and hence an auxiliary mechanism is required for the liquefaction of these gases. In one method, the stream of compressed gas is split in the liquefier, and a part goes to an engine in which it is cooled by an isentropic expansion with the performance of work. This part of the flow, thus cooled, is sent to a heat exchanger, where it flows counter to the other fraction of compressed gas, cooling it to a temperature below the inversion temperature (see Sec. 2). This flow of compressed gas, thus precooled, is run through another and final heat exchanger with a throttling valve at its lower end, with the result that the quantity x is liquefied.

Another method of precooling H_2 and He below their inversion temperatures makes use of a boiling liquid cryogen through which the compressed gas flows in a heat exchanger. LN_2 is used for precooling H_2, and LH_2 for He.

APPLICATIONS OF CRYOGENICS

Gases, such as O_2, N_2, natural gas, H_2, and He, are liquefied for transportation because of the gain in density at low pressure. Shipments have been made of LH_2 by trailer trucks in quantities of 13 × 10^3 gal (50 m^3) for thousands of miles and by railway tank cars holding as much as 28 × 10^3 gal (100 m^3). LHe is regularly shipped in quantities from 25 gal (0.1 m^3) to 10,000 gal (40 m^3). Liquid cryogens are used as liquid refrigerants. The aerospace, steel, and bottled-gas industries are the principal users of liquefied cryogens. Important applications as coolants can be found in vacuum technology, electronics, biology, medicine, and metal forming.

In recent years, large superconducting magnets have been undergoing development for a variety of applications including fusion reactors, energy storage, electric-power transmission, high-energy physics experiments, and magnetically levitated trains. Magnets of Nb-Ti are available with fields up to 80 kG (8 T) with large magnetic field volumes (1 m^3). Higher field magnets [up to 1,500 kG (15 T)] of Nb_3Sn are available with smaller magnetic field volumes (10^{-4} m^3). The Los Alamos National Laboratory installed a 30-MJ superconducting magnetic energy-storage system in the Bonneville Power Administration's Tacoma, Washington, substation to provide stabilization of utility long-line power-transmission systems. The response characteristics of the unit have been demonstrated on the utility's power grid, and the unit has been run successfully for extended periods. Two 100-m long superconducting ac transmission cables have been successfully tested at Brookhaven National Laboratory at 4,100 A and the equivalent of 138 kV three-phase. At the other end of the size spectrum, Josephson junction devices are being considered for applications such as radiation detectors, highly precise and accurate electrical measurements such as voltage standards, geophysical instrumentation, medical instrumentation, and computer memory cells.

Cryogenics is also becoming increasingly important in the storage and transport of energy. Large quantities of natural gas are routinely transported as a liquid (LNG). Peak shaving in municipal gas systems is accomplished by storage of LNG. If, due to depletion of fossil fuels, hydrogen becomes an important synthetic portable fuel, cryogenic hydrogen will play an important role as it already does in space applications including the shuttle. Another potential application is aircraft fuel. This may be extended to surface transportation applications.

PROPERTIES OF SOLIDS AT LOW TEMPERATURES

(See also Sec. 2.)

Specific heats of solids in general decrease with decreasing T, becoming zero at 0 K (0°R) (Fig. 11.2.3). The downward approach to $C_p = 0$ at 0 K is interrupted for some substances (principally compounds) by "bumps" on the curve (excess C_p that rises to a maximum and then decreases). Paramagnetic salts undergoing transitions to either a ferromagnetic or an antiferromagnetic state are examples. This excess specific heat of a paramagnetic salt is connected with the effectiveness of the salt for reaching low temperatures by the method of adiabatic demagnetization. There are also transitions in solids from a more orderly to a less orderly arrangement of atoms and molecules in the lattice that give rise to excess C_p. The transition in solid ortho- and normal H_2 below 20°R (11 K) is an example. In Fig. 11.2.3, C_p's are plotted for various materials.

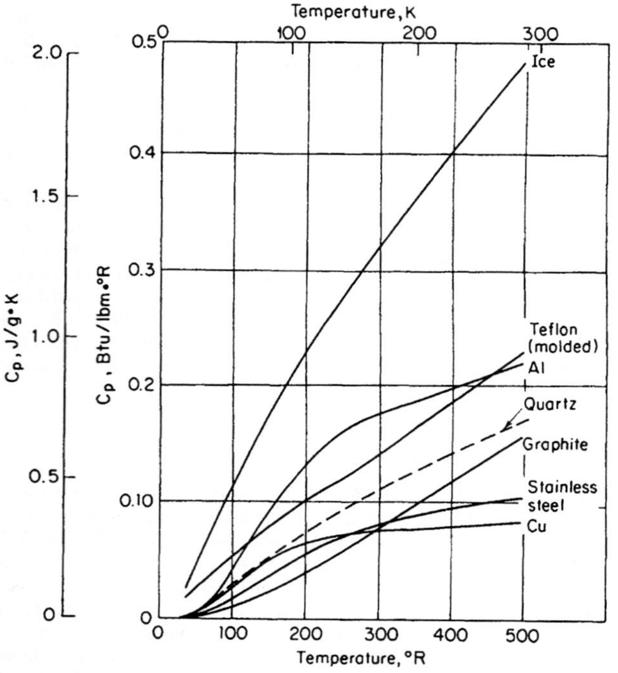

Figure 11.2.3 Specific heats of solids at low temperatures.

Heat is transferred in dielectrics by lattice vibrations, or waves. In good electrical conductors, heat is transferred principally by the conduction electrons, and thermal and electrical conductivities are related by the Wiedemann-Franz law; $\kappa/\sigma T$ = const. This means that the ratio of the thermal and electrical conductivities at a given T is approximately the same for the good conductors. In the poor electrical conductors, alloys for example, both lattice waves and conduction electrons play important parts in the transfer of heat. The κ's of "pure" dielectrics and "pure" metals rise with increasing T from $\kappa = 0$ at 0 K (proportional to T for metals and to T^3 for dielectrics), reach a maximum, normally between 10 and 100°R (5 and 55 K), and then decrease to a value approximately independent of T (Fig. 11.2.5 and ice in Fig. 11.2.4). The κ's of alloys (Fig. 11.2.4) are an order of magnitude smaller than for "pure" metals and *do not* exhibit the maxima characteristic of the "pure" metals at low T. Lattice disorder introduced by alloying, even in small amounts, and working a metal, even a pure metal, reduces κ. Annealing, in general, raises κ.

At 0°R, an absolutely pure and perfect single crystal of metal would have zero electrical resistance, $R_e = 0$. **Electrical resistance** arises from the scattering of the conduction electrons as they move through the lattice of metal ions under the influence of an externally applied electric field. Scattering arises for two reasons: (1) the amplitudes of the thermal vibrations of the lattice which increase with T at a rate proportional to $\sqrt{C}$ of the metal, and (2) the imperfections in the regularity of the lattice as caused by impurity atoms (solid-solution alloys included), lattice vacancies, dislocations, and grain boundaries. Resistances for "pure" metals increase with T and are roughly proportional to $C \cdot T$, except at very small T's where they are proportional to T^5 for absolutely pure crystals. The resistance due to impurities and imperfections is very roughly independent of T (Matthiessen's rule). For very pure metals, R_e is approximately constant below 20°R (11 K). Magnetic impurities can give rise to a minimum in resistivity [usually below 36°R (20 K)] called the Kondo effect. Resistivity in $\Omega \cdot$ cm (Fig. 11.2.6) is the R_e of a 1-cm cube.

Some metals, including elements (none from column 1 of the Mendeleeff table or the ferromagnetic elements), intermetallic compounds, metal alloys, and a number of oxides exhibit the phenomenon of **superconductivity,** characterized by a zero dc electrical resistance from 0°R up to a transition temperature T_c, at which normal resistance (value extrapolated from higher T) appears. In pure, single crystals the transition range ΔT, from $R_e = 0$ to the normal value, may be as small as a few millidegrees. Until the discovery of the so-called high-temperature superconductors (HTSCs), the higher T_c in a practical superconductor was that of the compound Nb_3Ge at 41.9°R (23.3 K). The search for higher T_c materials continues with T_c's that have already been reported as high as 239°R (133 K). Closed circuits consisting entirely of superconductors can support a persistent, resistanceless current without an external source of voltage. A superconducting circuit therefore maintains constant the value of the total magnetic flux that was enclosed by the circuit at the time it entered the superconducting state. Hence, superconducting materials influence the magnetic fields in their environment. For this reason, their use in the construction of equipment and apparatus must sometimes be avoided when expected temperatures could be below their T_c. Lead brasses and some solders, in particular Pb-Sn alloys, become superconductors.

Superconductivity, in materials designated Type 1 superconductor, is characterized by (1) perfect electric conductivity ($R_e = 0$) and (2) the internal magnetic induction $B = 0$ when H (external) is not zero (the Meissner effect). Persistent currents at the surface of the specimen

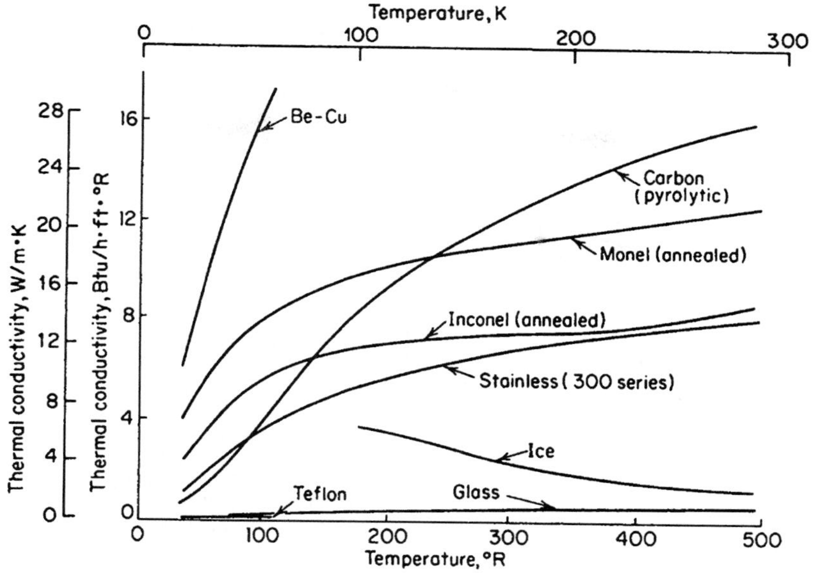

Figure 11.2.4 Thermal conductivity of solids at low temperatures, part 1.

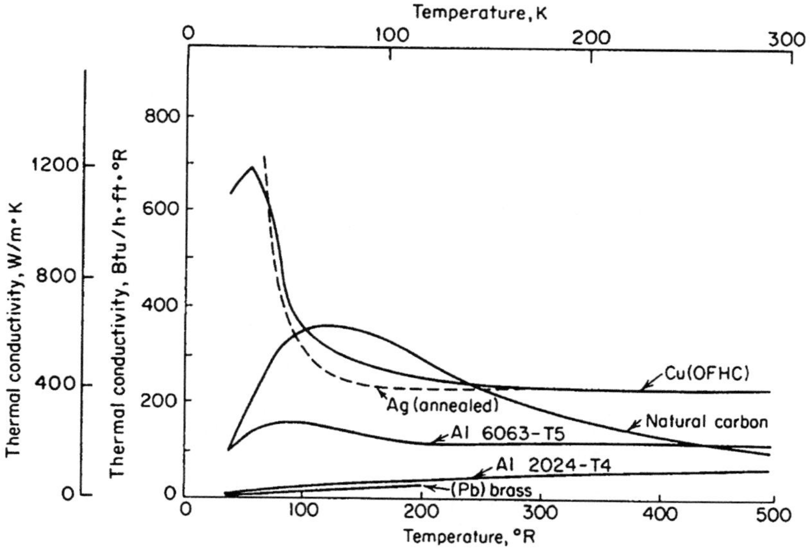

Figure 11.2.5 Thermal conductivity of solids at low temperatures, part 2.

shield the interior from H (external). H parallel to the surface of a specimen is continuous at the surface and falls off exponentially below the surface. The penetration depth of H for perfect (chemically and mechanically) specimens is of the order of 5×10^{-8} m. The penetration depth is larger for alloys, specimens with lattice imperfections, and all superconducting specimens near their superconducting transition temperatures T_c.

The normal-to-superconducting transition temperature T_c is also a function of H (external) as well as of the current being carried by the superconductor. For a Type I superconductor, the effect of H (external) is expressed by $T_c = T_{c0}\,[1 - (H_c/H_0)]^{1/2}$ where T_{c0} is the value of T_c at essentially zero H, and H_0 is the value of H_c at 0 K. For every $H < H_c$, a further decrease in T_c occurs when the superconductor is carrying a current, the maximum current density for maintaining the superconducting state being denoted J_c. Thus the superconducting state exists when $T < T_c$, $H < H_c$, and $J < J_c$, and the normal, or resistive, state exists when either $T > T_c$, $H > H_c$, or $J > J_c$. Because T_c decreases with increases in H and J, the practical application of a superconductor requires that it be maintained at a temperature considerably below its T_c. Tables 11.2.1 and 11.2.2 are only representative. The number of superconductors including compounds and alloys runs into the hundreds. Of the chemical elements, 38 are known to become superconductors. Some that are not superconductors at normal pressures have high-pressure allotropic modifications that are superconductors, that is, Bi, Si, and Ge at pressures of ~25, 130, and 120×10^8 Pa, respectively. Many of the superconducting alloys have nonsuperconducting constituents, and there are compounds all of

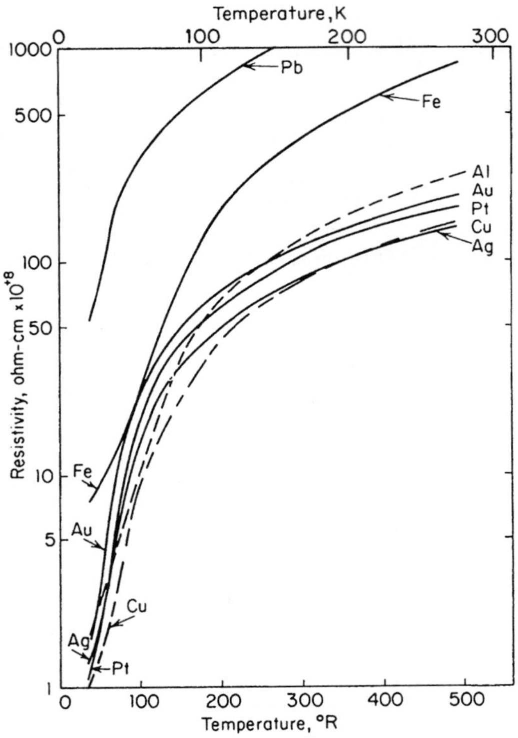

Figure 11.2.6 Electric resistivity of solids at low temperatures.

Table 11.2.1 Transition Temperature T_0 and Critical Magnetic Fields H_0 of Some Type I Superconductors

Superconductor	T_0, K	H_0, A/m*
Nb	9.25	1.57×10^5
Pb	7.23	6.4×10^4
V	5.31	8.8×10^4
Hg	4.154	3.3×10^4
Sn	3.722	2.4×10^4
In	3.405	2.2×10^4
Al	1.175	8.4×10^3
Mo	0.916	7.2×10^3
Zr	0.53	3.7×10^3
Ti	0.39	8.0×10^3
W	0.0154	9.2×10
Na Bi	2.2	The elements, in pure state at normal pressure, are not superconductors.
Au_2Bi	1.7	
Cu S	1.6	

* To convert to oersteds, divide by 79.57.

whose constituent elements in their pure state are nonsuperconductors (see Table 11.2.2). It is interesting that good conductors Cu, Ag, and Au do not become superconductors to the lowest temperatures at which they have been tested (~0.01 K).

Type I superconductor is the designation given to those superconductors that exhibit a complete Meissner effect for $0 < H < H_c$; that is, B = 0 until H_c is reached and penetration of the specimen by *B* becomes complete. The superconducting elemental metals in the pure and mechanically perfect state are Type I superconductors. Superconducting alloys (intermetallic compounds, solid solutions, and mixed phases) and work-hardened (unannealed) metals are Type II superconductors. For Type II superconductors, the Meissner effect is complete only for $0 < H < H_{c1}$ where H_{c1} is smaller than H_c. For $H_{c1} < H < H_{c2}$, the

Table 11.2.2 Transition Temperatures T_0 and Upper Critical Fields H_{c2} of Some High-Field, Type II Superconductors

Superconductor	T_0, K	H_{c2}, A/m*	T, K, for H_{c2}
$Al_{0.75}Ge_{0.25}Nb_3$	18.5	3.4×10^7	4.2
Nb_3Sn	18.0	1.9×10^7	4.2
$N_{0.93}Nb$	15.9	1.3×10^7	0
GaV_3	14.8	1.9×10^7	0
NbZr	10.8	7.4×10^6	0
$Nb_{0.2}Ti_{0.8}$	7.5	6.4×10^6	4.2
$Ti_{0.6}V_{0.4}$	7.0	8.8×10^6	2

* To convert to oersteds, divide by 79.57.

magnetic field penetrates the body of the specimen in the form of current vortices, called *fluxons,* that generate a fixed quantized amount of magnetic flux. Each fluxon is a cylindrical region of circulating current centered on a resistive core of material in which the superconductivity is suppressed. Hence the region defined by $H_{c1} < H < H_{c2}$ is called the *mixed state* of a Type II superconductor. The number of fluxons increases with *H* from zero at H_{c1} to a density at which complete overlap of the cores occurs along with the restoration of the normal state at H_{c2}. Thus H_{c2} is an upper limit to the superconducting state that increases as *T* decreases.

The alloys commonly used for high-field superconducting magnets (Nb_3Sn, NbTi, and NbZr, see Table 11.2.2) are Type II superconductors. H_c's for Type I superconductors are small, whereas magnet alloy wires have H_{c2} values that are 100 to 1,000 times higher.

The fluxons in the mixed state of a Type II superconductor transporting a current are acted on by a Lorentz force that is proportional to $\mathbf{i} \times \mathbf{B}$ and acts in a direction perpendicular to **B,** and to **i** (the current intensity). Unless the fluxons are pinned to crystal lattice sites they are propelled across the superconductor under the influence of the Lorentz force. This movement involves the performance of work by the current and results in (1) production of heat within the superconductor and (2) the appearance of electrical resistance to the flow of current and, eventually, the destruction of the superconducting state. Fluxons are pinned by lattice imperfections (chemical and mechanical) and their displacement is resisted until the Lorentz force exceeds a breakaway value, freeing the fluxons to move. Lattice imperfections are introduced in the wires for high-field superconducting magnets to pin fluxons as well as to increase H_{c2}. Currents avoid the normal core of a pinned fluxon and thus retain their resistanceless character.

A current exceeding the critical value destroys the superconducting state. This critical current is determined by the critical Lorentz force (the product of current and magnetic field) at which pinning forces are exceeded. Hence, the critical value of an applied field depends on the current transported by the superconductor and vice versa. In general, the larger the applied field, the smaller is the critical current and vice versa. For superconducting magnets, it is essential that the values of the limiting magnetic field and the limiting current be simultaneously large.

For the Types I and II superconductors discussed above, practical application requires operation in the temperature region of 12 K or lower. Thus the lower limit (ideal refrigeration) for the work to heat removal ratio is seen to be 24 for operation between ambient temperature and 12 K. Raising the operating temperature of a superconducting system to 20, 40, or 77 K (the temperature of liquid nitrogen boiling at normal atmospheric pressure) reduces this ratio to 14, 6.5, and 2.9, respectively. Although the actual refrigerator may take 2.5 or more times this ratio, the saving in input energy that can be effected by raising the operating temperature is obvious. The equipment for producing refrigeration at these higher temperatures is also less complicated and less expensive than that needed for producing refrigeration at the lower temperatures required for the operation of the low-temperature superconductors.

In 1986 a true breakthrough was achieved in producing superconductors able to enter the superconducting state at much higher temperatures. The high-temperature superconductors (HTSCs) are oxide compounds consisting of several elements. They are more like ceramics and have a complicated crystal structure consisting of layers of Cu-O separated by other oxides. All of them exhibit highly anisotropic superconducting properties and their performance is greatly influenced by the oxygen content of the compound. If the crystals are not properly oriented so that there is too great an angle between the crystal axes of adjacent crystals (e.g., greater than about 5°), intergranular supercurrent flow is suppressed and the HTSCs will have sections of superconductor connected by rather poorly conducting junctions. This has been referred to as "having weak links." Therefore, it is desirable to prepare the HTSC material in such a way that it is biaxially textured, in order to minimize weak links. Also, the ability of these superconductors to conduct useful quantities of current can be severely limited by an imposed magnetic field at temperatures near the high end of their superconducting range. The magnetic field enters the superconductor in fluxons, or vortices. When these vortices move through the superconductor, electrical resistance is generated and the superconductor behaves much like a normal metal. Therefore it is necessary to "pin" the vortices so they cannot move. As with low-temperature superconductors, vortices are pinned by lattice imperfections, either chemical or mechanical.

At temperatures below 30 K, the extraordinarily high H_{c2} of the HTCSs allows them to carry high-current densities at much higher magnetic fields than is possible with the conventional, low-temperature superconductors. At higher temperatures a new phenomenon has been observed, the "melting" of the vortex lattice that limits the ability of crystal imperfections to pin vortices at magnetic fields far below H_{c2}. The HTSCs are fabricated as thin films and multifilamentary wires. The necessity for optimum chemistry of the constituents, proper crystal orientation, and the fact that the ceramic-like nature of the superconductor makes it brittle complicate both the manufacture of the wires in useful lengths and the subsequent fabrication of equipment such as coils for motors and generators. There are two classes of compounds that have received the most attention in these research efforts. One class is referred to as BSCCO because it consists of the oxides of bismuth, strontium, calcium, and copper in varying ratios. The other class is referred to as YBCO because it consists of oxides of yttrium, barium, and copper.

A great deal of research is continuing in the attempt to improve the commercial usefulness of the HTSCs. This research has made enormous strides in finding production methods for practical superconductors and ways to produce them more economically. There are already fabrication methods that enable the production of kilometer lengths of superconductor tape capable of carrying very high currents (about 10^5 A/cm^2) in self-field at 77 K. These efforts should make possible the large-scale utilization of HTSCs. Table 11.2.3 lists several of the HTSCs, along with the pertinent properties.

The discovery in 2001 of superconductivity in the relatively common intermetallic compound, magnesium diboride (MgB_2), with a transition temperature of about 40 K, opens new possibilities to the application of low-temperature superconductors (LTSCs). MgB_2 is easier than the HTSCs to produce, handle, and fabricate into wires, and MgB_2 reduces the energy requirement for refrigeration by about an order of magnitude compared to that of the previously known LTSCs. MgB_2 can be used for the manufacture of superconducting magnets, and its use should make more feasible the utilization of superconducting electric power transmission lines.

Table 11.2.3 Properties and Status of some HTSCs

HTSC*	T_c, K	Comments
$La_{2-x}Ba_xCuO_4$	35	First HTSC discovered
$YBa_2Cu_3O_7$ (YBCO or 1-2-3)	95	First HTSC with T_c above normal boiling point of liquid nitrogen, $H_{c2} > 30$ T at 77 K
$HgBa_2Ca_2Cu_3O_{8+\delta}$	133	Highest T_c to date
$Bi_2Sr_2Ca_2Cu_3O_8$ (2223 BSCCO)	110	$H_{c2} = 30$ T at 77 K, $J_c > 8 \times 10^4$ A/cm^2 at 77 K and $H_{ext} = 0$; industrial pilot-plant-scale production
$Tl_2Ba_2CaCu_2O_8$ (2212 TBCCO)	~115	$J_c = 7 \times 10^6$ A/cm^2 at 75 K and $H_{ext} = 0$

* These HTSCs are representative of classes of superconductors.

The **coefficient of linear expansion** α is $(1/L)(dL/dT)$. Thermal expansion of asymmetric crystals differs along different crystal axes. For an isotropic solid, the volume or cubical-expansion coefficient $(1/V)(dV/dT) = 3\alpha$. Expansion coefficients do not vary appreciably in the temperature region near ambient temperatures, but they all approach zero at 0°R, and the approach to zero is tangential to the T axis ($d\alpha/dT = 0$ at 0°R). Some expansion coefficients are negative at low temperatures, e.g., stainless steel, some Invars, and fused quartz. The expansivities of some crystals are negative in some directions even though their volume coefficients are positive. Cold-working may produce differences in expansivity in different directions. Annealing restores isotropy. Figure 11.2.7 shows the relative change in length, the integral of α from T to 528°R (293 K).

Because the cold interiors of cryogenic equipment must at times be warmed to ambient temperature, provisions have to be made for those changes in dimensions of the interior that results from large changes in T. Even for equipment of ordinary size, these changes can be too large for accommodation within the elastic limits of the materials of construction or of the structure. In such situations, flexibility has to be designed into the structure. For example, bellows and U bends are commonly employed to obtain this flexibility in insulated transfer lines for liquid cryogens.

The **mechanical properties** important for the design and construction of cryogenic equipment are the same as for other temperatures (see

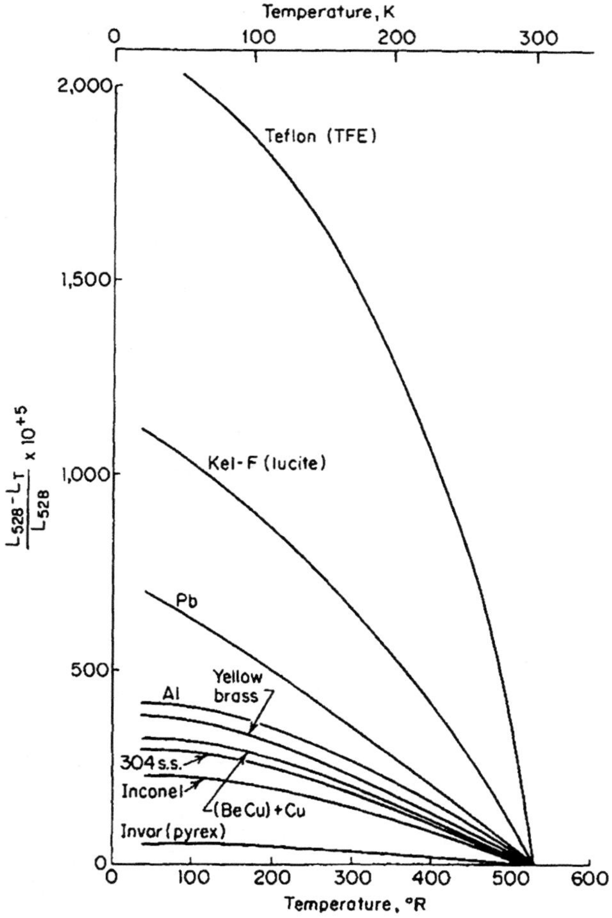

Figure 11.2.7 Relative change in length from T to 528°R (293 K), equal to $\int_t^{528°R} \alpha\, dt$. Bracketed materials are within ~5 percent of the curve shown.

also Sec. 3). Frequently, the choice of materials for the construction of cryogenic equipment will depend upon other considerations besides mechanical strength, for example, lightness (density or weight), thermal conductivity (heat transfer along structural support members), and thermal expansivity (change of dimensions when cycling between ambient and low temperatures). Frequently, mechanical properties at low temperature are significantly different from properties at ambient temperature. This makes room-temperature data unreliable for engineering use at low temperatures. It is not possible to make generalizations that would not have numerous exceptions about the temperature variations of the mechanical properties. In this discussion, Figs. 11.2.8 to 11.2.10 are only illustrative guides. There is no substitute for test data on a truly representative specimen when designing for the limit of effectiveness of a cryogenic material or structure. Just as the mechanical properties at ambient temperatures are dependent upon the impurities (metallic and nonmetallic), their chemical nature and concentration, the thermal history of the specimen, the amount and kind of working (microstructure and dislocations of the lattice), and the rate of loading and type of stress (uni-, bi-, and triaxial), so also are the changes, quantitative and qualitative, in the mechanical properties when changing the temperature from ambient to low.

Metals, nonmetallic solids (glass), plastics, and elastomers are discussed in order. The metals are classed by their lattice (crystal) symmetry. The **f.c.c. metals** and their alloys are most commonly used for the construction of cryogenic equipment. Al, Cu, Ni, their alloys, and the austenitic stainless steels of the 18-8 type (300 series) are f.c.c. They do not exhibit an impact (or a notched-tensile) ductile-to-brittle transition at low temperatures. As a general rule, which has some exceptions, the mechanical properties of these metals improve as T is reduced: (1) Young's modulus at 40°R (22 K) is 5 to 20 percent larger than at 530°R (294 K), (2) the yield strength at 40°R (22 K) is considerably greater than the strength at 530°R (294 K). (Cu is an exception; see Fig. 11.2.9), and (3) the fatigue properties at low T are improved. There is a large difference between the yield strength and the ultimate tensile strength of these metals and alloys, especially when they have been annealed. Pb (f.c.c.) and In (face centered tetragonal) are used for low-T deformable gaskets because of their creep properties. Low-T creep data are meager, but the rate of creep decreases with T.

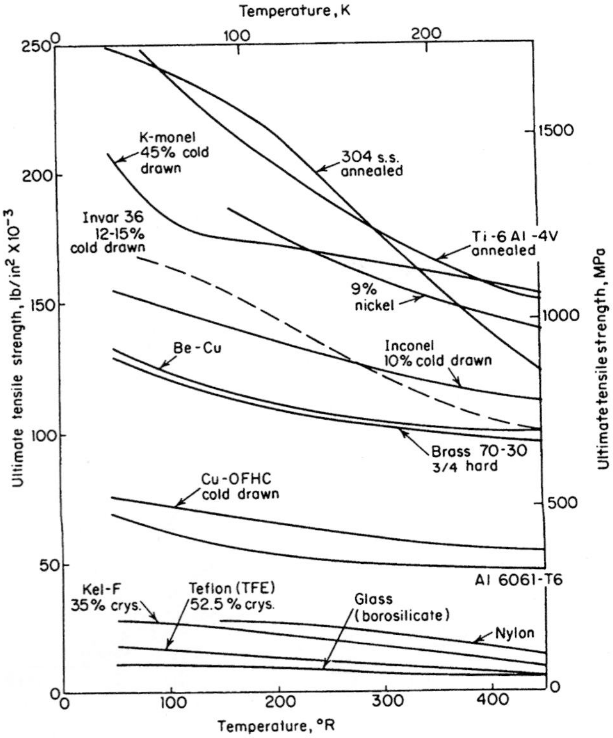

Figure 11.2.8 Ultimate tensile strength of solids at low temperatures.

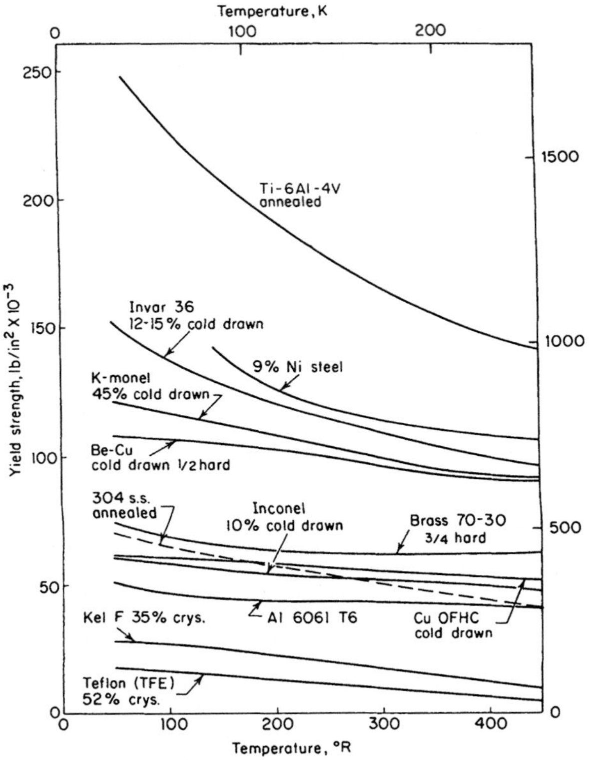

Figure 11.2.9 Yield strength of solids at low temperatures.

Beta brass (f.c.c.) is ductile down to 7°R, though it is like all alloys of Cu in being less ductile than Cu itself. Thin brass is sometimes porous, and this limits its usefulness for high-vacuum enclosures at low temperatures. Free-machining brass normally contains Pb, which even in low concentrations can render the brass superconducting at LHe temperatures. In the superconducting state, it may then affect the magnetic field in its vicinity (see electrical properties, above).

The b.c.c. **metals and alloys** are normally classed as undesirable. These include Fe, the martensitic steels (low carbon and the 400 series of stainless steels), Mo, and Nb. If not brittle at room temperature, they have a ductile-to-brittle transition at low T. Working can induce the austenite-to-martensite transition in some steels. AISI 301 austenitic stainless is an example. It remains moderately "tough" at very low T and is a valuable material for construction, but cold-drawing, in excess of 70 percent strain, induces a partial transformation of structure and reduces the elongation and notched-strength ratio to nearly zero. This same type of reduction in toughness is observed in the "heat-treatable" stainless steels. Improved tensile properties can be obtained by cold-working and by heat-treating. These alloys usually have decreased toughness at low temperatures. Alloys of V, Nb, and Ta, although not f.c.c., behave well as regards brittleness at low temperatures. These alloys have the advantage of being suitable for use at high T. Carbon steels, though brittle, have found special uses at low temperatures, for example, in the construction of the expansion engines of the Collins He liquefier and cryostat.

The **h.c.p. metals** exhibit properties intermediate between those of the f.c.c. and b.c.c. metals; for example, Zn undergoes a brittle-to-ductile transition, whereas Zr and pure Ti do not. Ti and some Ti alloys, having an h.c.p. structure, remain moderately ductile at low T and are excellent

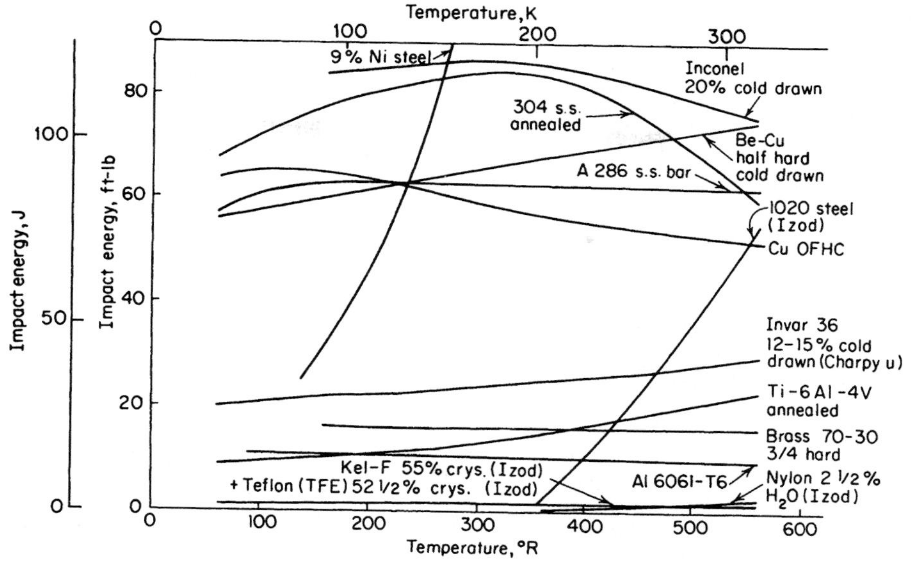

Figure 11.2.10 Impact energy for solids at low temperatures (Charpy V unless noted). For Kel-F, nylon, and Teflon, the impact energy units are foot-pound per inch of notch width.

for many applications. They have high strength-to-weight and strength-to-thermal-conductivity ratios. The low-T properties of Ti and its alloys are extremely sensitive to even small amounts of O, N, H, and C.

Brittle materials, when in very thin sheets, can have a high degree of flexibility, and this makes them useful for some cryogenic applications. **Flexibility** cannot be used as a criterion for ductility.

Ordinarily, normal welding practices are observed in the construction of cryogenic equipment. These practices, however, are modified in accordance with available knowledge of the performance of the metals at low T. **Nonmetal** materials for construction are in many cases brittle, or they are susceptible to brittle fracture. The strength of **glass,** measured at a constant rate of loading, increases on going to low T. Failure occurs at a lower stress when the glass surface contains cracks and abrasions. The strength of glass can be improved by tempering the surface, that is, by putting the surface under compression.

The **plastics** increase in strength as T is decreased, but this is accompanied by a rapid decrease in elongation in a tensile test and a decrease in impact resistance. Teflon and the glass-reinforced plastics (e.g., glass-reinforced epoxy resin) retain appreciable impact resistance as T is lowered. Teflon, which is polytetrafluoroethylene, can be deformed plastically at T's as low as 7°R (3.9 K). The amount is considerably less than at room temperature, but it is enough to make Teflon very useful for some cryogenic applications. The glass-reinforced epoxies, besides having appreciable impact resistance at low T's, also have high strength-to-weight and strength-to-thermal-conductivity ratios.

All the **elastomers** become brittle at low T. However, elastomers like natural rubber, nitrile rubber, Viton A, and plastics such as Mylar and nylon that become brittle at low T can be used for static seal gaskets when *highly compressed* at room temperature, prior to cooling.

PROPERTIES OF CRYOGENIC FLUIDS (CRYOGENS)

Table 11.2.4 indicates the **temperature ranges** accessible with liquid cryogen baths. There are two inaccessible ranges: (1) between helium and hydrogen [9.36 to 24.9°R (5.2 to 13.8 K)] and (2) between neon and oxygen [80 to 97.9°R (44.4 to 54.4 K)]. The limiting temperatures are set by the critical and triple points. Pumping on a cryogen bath to lower the pressure results in lower temperatures ultimately reaching the triple point; except for helium, further pumping leads to solidification and in practice poor heat-transfer. Raising the bath pressure results in higher boiling temperatures with the limit set by the critical temperature, above which liquid and vapor phases cannot coexist. Cryogen baths are most frequently operated near atmospheric pressure.

The algebraic sign of the Joule-Thomson coefficient determines whether a gas cools or warms upon free expansion. Figure 11.2.11, based on the law of corresponding states, allows rough calculations of

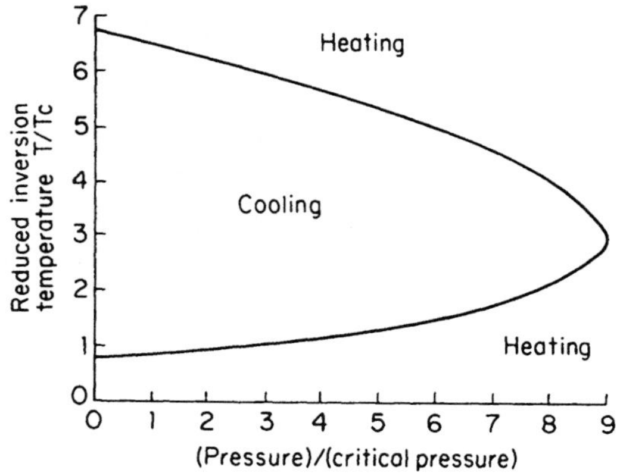

Figure 11.2.11 Reduced inversion temperature versus reduced pressure.

inversion temperatures and pressures. Free (Joule-Thomson) expansion inside the "cooling" region results in cooling, that is, refrigeration; outside this region, heating results. Upper inversion temperatures at 14.7 psia (0.101 MPa) are given in Table 11.2.4.

Volumetric latent heats (Table 11.2.4) decrease with the lower-boiling cryogens and emphasize the importance of the insulation in storing and handling LH_2 and LHe.

Two forms of **hydrogen** and **deuterium** exist: **ortho** and **para.** The ortho and para molecules differ in the relative orientation of the nuclear spins of the two atoms composing the diatomic molecule. In the ortho form of H_2 the nuclear spins of the two atoms in the molecule are parallel (in the same direction), whereas in the para form

Table 11.2.4 Common Cryogen Properties

Cryogen	Boiling point,* °R	Triple-point Temp., °R	Triple-point Pressure, lb/in² abs	Critical temp., °R	Critical pressure, lb/in² abs	Upper inversion temp,* °R	Heat of vaporization* Btu/lbm	Heat of vaporization* Btu/ft³	Liquid density,† lbm/ft³	Vapor density,† lbm/ft³	Gas density,‡ lbm/ft³
He^3	5.7			6	17	72	3.6	13.4	3.72	1.50	0.0084
He^4	7.6			9.36	33.2	92	8.8	68.6	7.80	1.04	0.0111
H_2 (equilib.)	36.5	24.9	1.02	59.4	187.7	368	192	853	4.42	0.0837	0.00561
D_2 (normal)	42.6	33.7	2.49	69.0	240		131	1,343	10.25	0.14	0.0112
Ne	48.7	44.2	6.26	80.0	395.3	486	37.1	2,790	75.35	0.58	0.056
N_2	139.2	113.7	1.86	227.27	492.3	1,115	85.9	4,294	50.4	0.28	0.078
Air	(141.8)						88.22	4,813	54.56	0.28	0.0807
A	157.1	150.8	9.97	271.3	709.8	1,300	70.2	6,135	87.4	0.36	0.111
F_2	151	(96.4)§		259.3	808		74.1	6,965	94	0.50	0.106
O_2	162.3	97.9	0.022	278.59	736.3	1,605	91.5	6,538	71.24	0.28	0.089
CH_4	201.1	163.2	1.69	343.27	673		219.2	5,804	26.5	0.115	0.0448

NOTE: To convert °R to K, multiply °R by 0.556; to convert lb/in² to MPa, multiply by 0.00689; to convert Btu/lbm to J/g, multiply by 2.326; to convert Btu/ft³ to J/m³, multiply by 3.73×10^4; to convert lbm/ft³ to kg/m³, multiply by 16.01.

* 14.7 lb/in² abs.

† At normal boiling point.

‡ At 14.7 lb/in² abs and 492°R.

§ Melting point, 14.7 lb/in² abs.

the spins are antiparallel. The relative orientations for ortho and para D differ from those for H_2. The thermodynamic equilibrium composition of ortho and para varieties is temperature dependent as shown in Fig. 11.2.12.

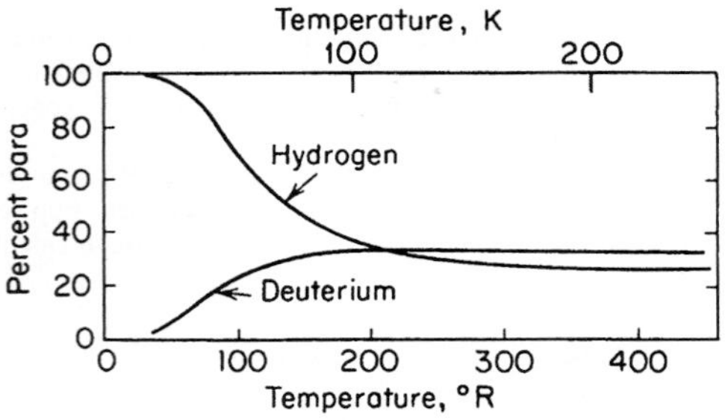

Figure 11.2.12 Thermodynamic equilibrium composition of ortho and para varieties of H_2 and D_2; percent ortho = 100 – percent para.

In liquid hydrogen, the uncatalyzed ortho-para reaction is second order (proportional to the square of the o-H_2 concentration) with a rate constant of 0.0114/h. Thus, in the course of uncatalyzed liquefaction, normal LH_2 (75 percent ortho) is produced. Since the heat of conversion of o-H_2 to p-H_2 at the normal boiling point is 302 Btu/lbm (702 J/g) (greater than the heat of vaporization; see Table 11.2.4), long-term storage of normal LH_2 is impractical. Therefore, in the commercial production of LH_2 a catalyst is used to produce LH_2 with more than 95 percent para which for practical purposes is equivalent to the equilibrium composition (99.8 percent para).

For pressures of no more than 150 lb/in² (1.03 MPa) and temperatures at least twice the critical temperature, the ideal gas law ($PV = nRT$) enables one to calculate PVT data with sufficient accuracy for many engineering purposes. For cases in which the ideal gas law is not adequate or in which experimental data are not available, the procedures outlined in Sec. 2 may be used. The theoretical prediction of PVT data for liquids is considerably more difficult than for gases. Consequently no universal equation analogous to the perfect gas law exists for extensive ranges of P and T.

Reduced quantum mechanical correlations of saturated liquid densities, viscosities, and thermal conductivities for several cryogens are shown in Figs. 11.2.13 to 11.2.15. Dashed lines represent areas void of experimental data.

Utilization of Figs. 11.2.13 to 11.2.15 requires knowledge of the constants ε and σ for the Lennard-Jones function for the intermolecular potential energy $\varphi(r)$ of a pair of molecules separated by a distance r, so that $\varphi(r) = 4\varepsilon[(\sigma/r)^{12} - (\sigma/r)^6]$. Here ε is the depth of the minimum in the potential energy of a pair of molecules, and s is the separation of the pair at which $\varphi = 0$. To facilitate use of Figs. 11.2.13 to 11.2.15, all required constants and conversion factors have been combined and are listed in Table 11.2.5.

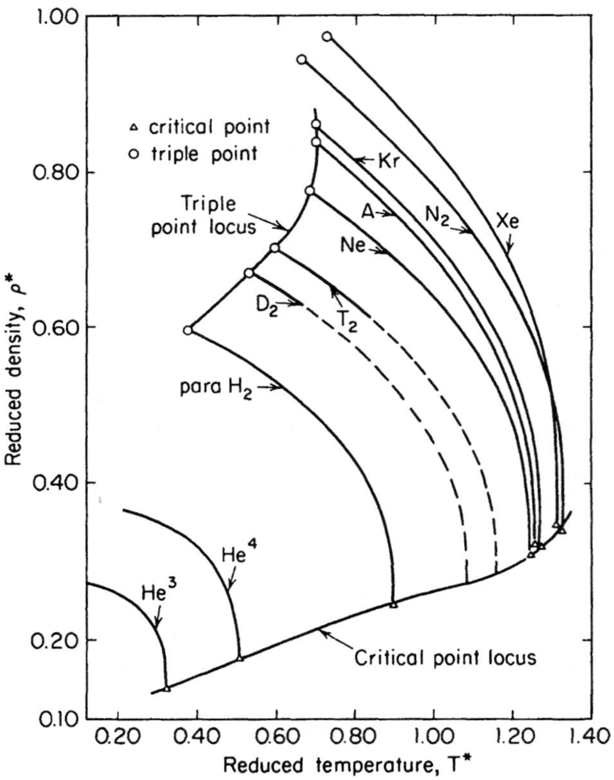

Figure 11.2.13 Reduced density versus reduced temperature along the saturation curve for cryogenic liquids. Values of T^* are obtained by dividing the desired temperature (°R) by the value of T/T^* given in Table 11.2.5. Reduced density ρ^* must be multiplied by ρ/ρ^* given in Table 11.2.5 to obtain density in lbm/ft³. Multiply lbm/ft³ by 16.01 to obtain kg/m³ and °R by 0.556 to obtain K.

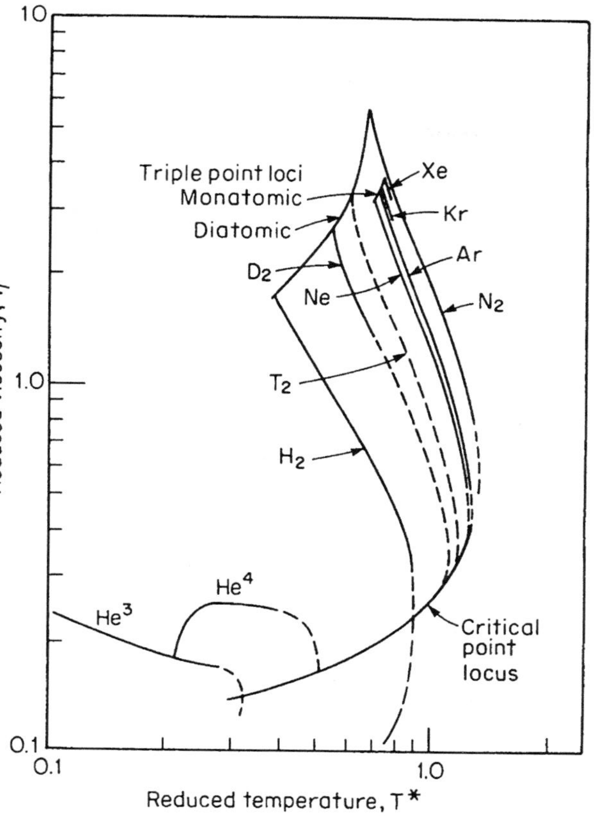

Figure 11.2.14 Reduced viscosity versus reduced temperature along the saturation curve for cryogenic liquids. Values of T^* are obtained by dividing the desired temperature (°R) by the value of T/T^* given in Table 11.2.5. Reduced viscosity η^* must be multiplied by η/η^* given in Table 11.2.5 to obtain viscosity in lbm/ft · h. Multiply lbm/ft · h by 0.000413 to obtain Pa · s.

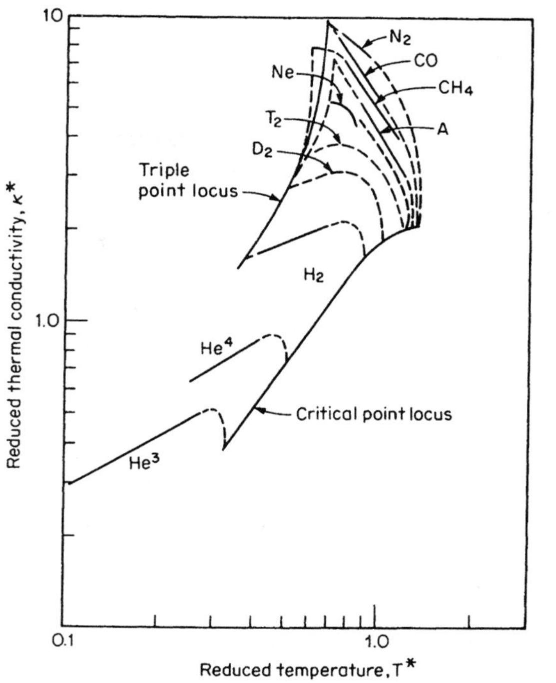

Figure 11.2.15 Reduced thermal conductivity versus reduced temperature along the saturation curve for cryogenic liquids. Values of T^* are obtained by dividing the desired temperature (°R) by the value of T/T^* given in Table 11.2.5. Reduced thermal conductivity κ^* must be multiplied by κ/κ^* given in Table 11.2.5 to obtain thermal conductivity in Btu/h · ft · °R. Multiply by 1.73 to obtain W/m · K.

Thermal-expansion coefficient $(1/V)(\partial V/\partial T)_P$ and **compressibility** $(1/V)(\partial V/\partial P)_T$ data for the cryogenic liquids are obtained as derivatives of *PVT* data. Liquids hydrogen and helium have large thermal-expansion and compressibility coefficients (see Table 11.2.6).

For gases, kinetic theory predicts an increase of **viscosity** with rising temperature as shown in Fig. 11.2.16 ($\eta \sim \sqrt{T}$). Normally, the **thermal conductivities** of liquids decrease with increasing temperature, whereas the thermal conductivities of gases increase. Fig. 11.2.17 presents thermal conductivity for several gases as a function of temperature at 14.7 psia (0.1013 MPa).

Figure 11.2.18 is a graph of **vapor pressures** of the common cryogens.

Figure 11.2.19 presents **specific heats** at constant pressure for several of the cryogenic liquids, while Fig. 11.2.20 presents the specific heats for liquid hydrogen and helium along the saturation curve. Specific heats of gases are given in Fig. 11.2.21.

INSTRUMENTATION

(See also Secs. 11 and 12.)

The usual instrumentation problems of engineering practice are further complicated by low-temperature problems which require special calibration procedures.

Pressures are measured with normally used apparatus, for example, bourdon gages, transducers, and manometers. If the measuring device is located outside the cryogenic environment and insulated from it by a thermal barrier, no special problems are introduced. Occasionally response time requirements necessitate the placing of the transducer in the cryogen space. Then the problems of temperature compensation and calibration must be considered.

Table 11.2.5 Multiplication Factors for Use with Figs. 11.2.13 to 11.2.15

Substance	T/T^*, °R	ρ/ρ^*, lbm/ft³	η/η^*, lb/ft-hr	κ/κ^*, Btu/(h · ft · °R)
Helium 3	18.4	18.71	0.0311	0.0204
Helium 4	18.4	24.84	0.0359	0.0178
Para-hydrogen	66.1	8.06	0.0360	0.0363
Deuterium	63.4	16.22	0.0500	0.0257
Tritium	62.1	24.37	0.0607	0.0210
Neon	64.1	100.66	0.1300	0.0128
Nitrogen	171.1	57.39	0.1382	0.0098
Argon	215.6	104.84	0.2183	0.0109
Krypton	297.4	177.1		0.0080
Xenon	397.8	197.4		

Table 11.2.6 Compressibility and Thermal-Expansion Coefficients. Coefficients for Liquid Nitrogen, Hydrogen, and Helium at 14.7 lb/in² abs (0.101 MPa)

Cryogen	T, °R (K)	Compressibility, $(1/V)(\partial V/\partial P)_T$, 1/(lb/in² abs) (1/MPa)	Thermal expansion, $(1/V)(\partial V/\partial T)_p$, 1/°R (1/K)
Helium 4	7 (4)	0.0028 (0.4)	0.085 (0.153)
Nitrogen	139 (77)	0.000020 (0.0029)	0.0032 (0.0058)
Para-hydrogen	36 (20)	0.000134 (0.0195)	0.0090 (0.0162)
Water	540 (300)	0.0000036 (0.000526)	0.00015 (0.00027)

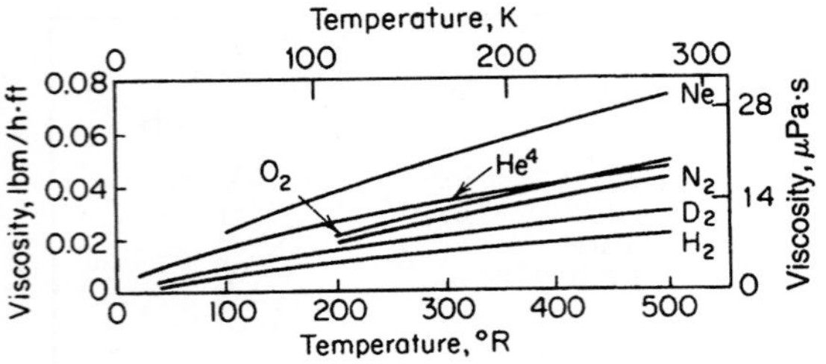

Figure 11.2.16 Viscosity of selected gases at low temperatures and 14.7 lb/in^2 abs (0.1013 MPa).

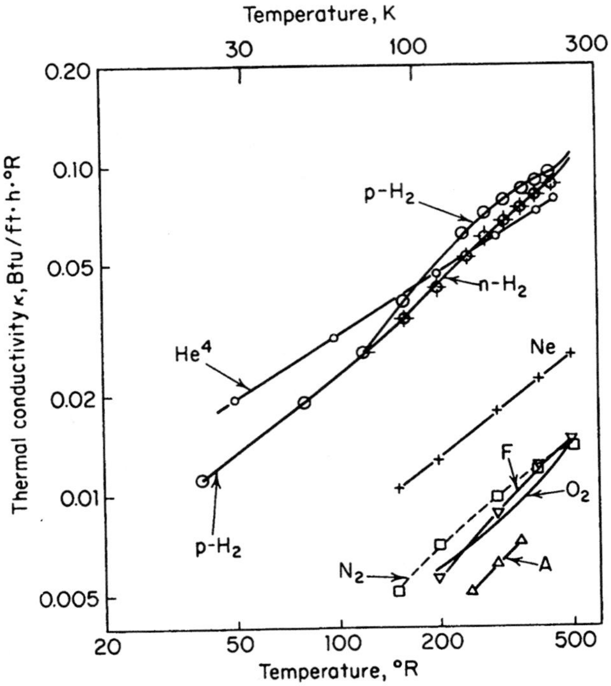

Figure 11.2.17 Thermal conductivity of gases at low temperatures and 14.7 lb/in^2 abs (0.1013 MPa).

Level measurements currently utilize several methods. The differential-pressure method can in principle be used with all cryogens. With cryogens of low densities and low heats of vaporization (e.g., H_2 and He) pressure oscillation occurs in the liquid leg. In the case of LH_2 these can be eliminated by the proper design of the liquid leg and the introduction of He gas which does not condense.

Direct weighing is also used to determine "levels," although difficulties are encountered owing to the large tare-to-cryogen weight ratios (as large as 10 to 1 for a full container in the case of H_2) and extraneous loadings introduced by permanently connected piping, frost, and wind. Weighing systems have nevertheless been successfully used on 50,000-gal (190 m^3) LH_2 dewars with an accuracy in level changes of 250 gal (1 m^3).

With the **capacitance gage,** the sensing element consists of a concentric tube capacitor which can be calibrated to give liquid-level measurements accurate to a few tenths of a percent. An inexpensive, accurate **point-level sensor** is easily fabricated from a carbon resistor,

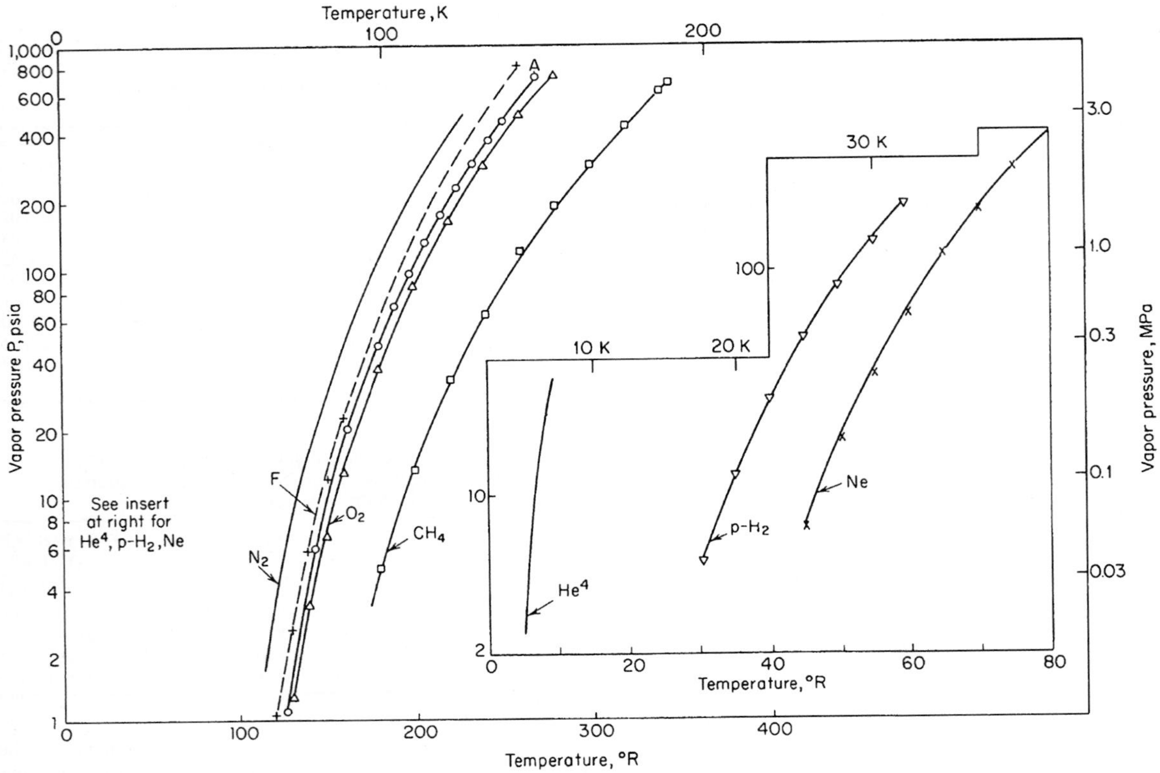

Figure 11.2.18 Vapor pressures of the common cryogens.

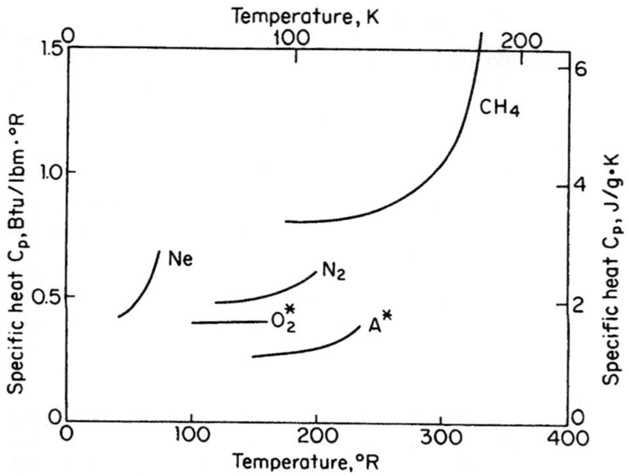

Figure 11.2.19 Specific heat at constant pressure (C_p) for liquid cryogens at saturation pressure. Asterisk means C_p evaluated along an isobar at 14.7 lb/in^2 abs (0.1013 MPa) instead of saturation pressure.

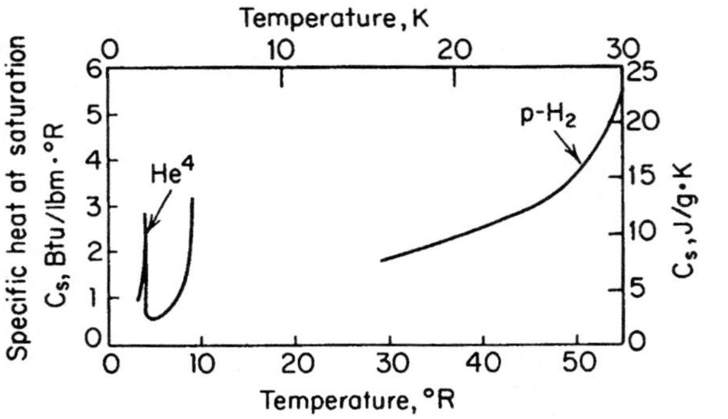

Figure 11.2.20 Specific heats at saturation C_s along the saturation curve versus temperature for ^{4}He and para hydrogen.

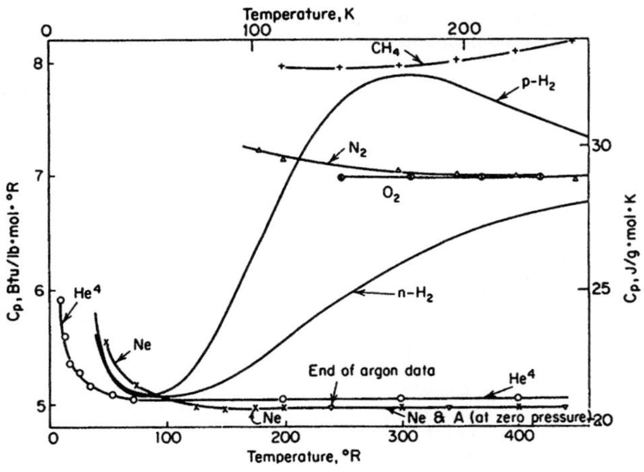

Figure 11.2.21 Specific heat C_p of cryogen gases at 14.7 lb/in^2 abs (0.1013 MPa) versus temperature.

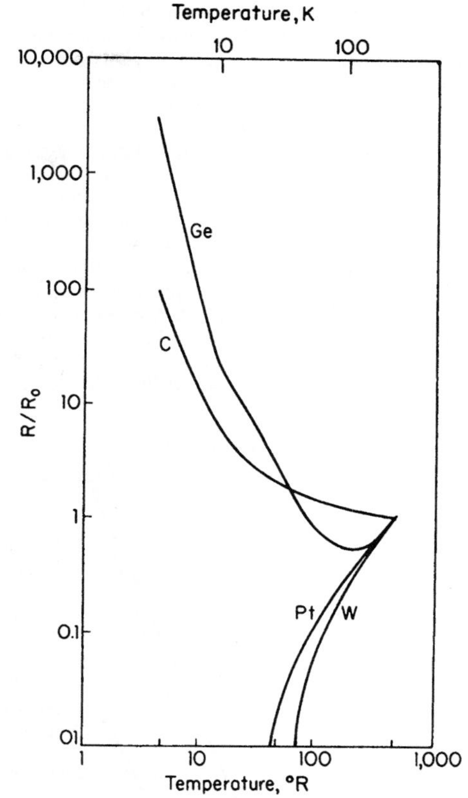

Figure 11.2.22 Ratio of resistance R to the resistance R_0 at 492°R (273 K) versus temperature for several resistance thermometers. (Values for germanium and carbon are representative only.)

typically ⅒ W, 1,500 Ω. Its use depends upon the fact that its resistance is a strong function of temperature (Fig. 11.2.22). A small current (50 to 80 mA) is passed through the resistor to heat it. Since heat from the resistor is dissipated more readily in the liquid than in the gas, the temperature of the resistor undergoes a step increase when the resistor environment changes from liquid to gas, whereas the resistance decreases stepwise.

Flow measurements utilize orifices, venturis, and turbine-type meters. Various devices for direct mass-flow measurement exist, but none are widely used. "Quality" meters have been built to determine the percentage liquid in two-phase flow. The determination of quality is complicated by nonequilibrium of temperature in the gas phase.

Temperature measurements are usually made with thermocouples, resistance thermometers, and vapor-pressure thermometers.

The **vapor-pressure thermometer** depends upon a vapor-pressure-temperature relationship (see Fig. 11.2.18). In general, a small cavity is filled with a gas which has a condensation temperature in the vicinity of the temperature to be measured. If sufficient gas is present, the pressure will be that of the liquid vapor pressure at the coldest part of the measuring system. Maximum speed of response is obtained if the total quantity of gas is minimized to permit only a thin film of liquid to form. In the case of H_2, a catalyst (e.g., iron hydroxide) to promote ortho-para conversion should be included in the bulb for an accurate measurement since the vapor pressure of H_2 is dependent upon its ortho-para composition (discussed earlier).

The pressure at the surface of a liquid cryogen in a dewar may, under conditions of thermal equilibrium, be used to measure the temperature of the cryogen and of apparatus immersed in it. Corrections for the pressure of the hydrostatic head of liquid may be required. If using this method, it should be realized that large vertical temperature gradients can exist in an unstirred liquid cryogen in a vessel with good insulation. **Thermocouples** are favored for the measurement of temperatures because of their low cost, ease of application, and rapid response. The thermoelectric powers (temperature sensitivities) of the thermocouples commonly used at higher T's decrease with decreasing temperature, and spurious emfs generated in wires of non-uniform composition are troublesome. The proper use of a reference junction at a known fixed temperature close to the temperature being measured is often advantageous for accuracy. Variability in composition of thermocouple alloys makes individual calibrations necessary for accurate results. Figure 11.2.23 gives typical thermocouple emf-versus-temperature curves for some common thermocouples. A thermocouple consisting of gold with 0.07-at % iron versus chromel has been found useful over the entire temperature range from 2 to 300 K. Its sensitivity increases from 11 μV/K at 4 K to 17 μV/K at 77 K to 22 μV/K at 300 K.

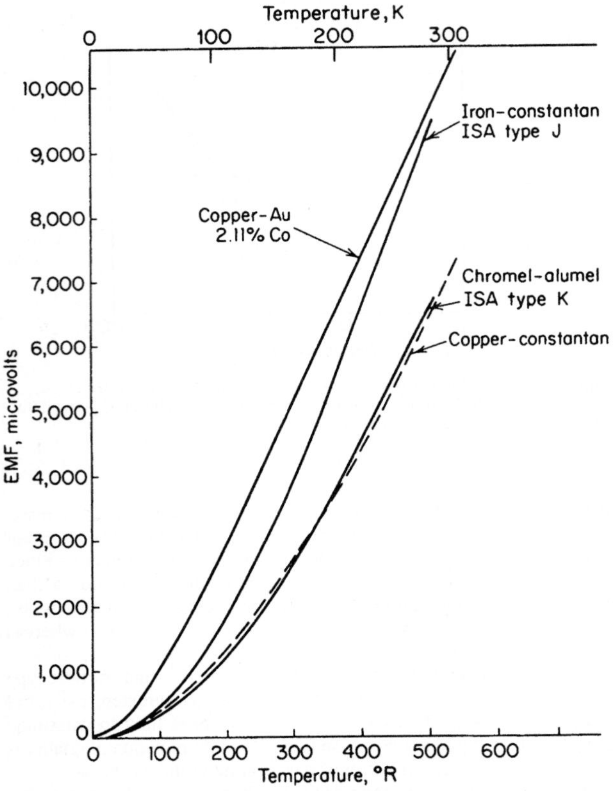

Figure 11.2.23 EMF versus temperature for several thermocouples.

Radio carbon resistors (*Allen Bradley* for 2 to 100 K and *Speer Carbon* for 0.01 to 4.2 K) are much used as temperature sensors because of their low cost and high sensitivity. However, the permanence and stability of their calibrations are not very good, and germanium (doped) resistance thermometers are preferred (0.01 to 100 K) for their good reproducibility of calibration, although they are considerably more expensive.

The carbon and germanium resistors (thermometers) in magnetic fields H exhibit a magnetoresistance (Hall) effect, proportional to H^2, which complicates the determination of temperatures. Recent developments include: (1) a carbon-glass electrical resistor which has a stable calibration; and (2) the electrical capacitance thermometer (dielectric: $SrTiO_3$, a perovskite in solution in a solid SiO_2 or Al_2O_3 glass) with very small (approaching zero) magnetic effect and a stable calibration. The capacitance thermometer is preferred when magnetic fields are present. The Rh-0.5-at % Fe, wire-coil thermometers, usable in the range of 1.1 to 300 K, have a high sensitivity especially below 10 K.

INSULATION

(See also Sec. 4.)

The degree of thermal isolation required at low T's is normally greater than at elevated T's because it is more costly to remove heat leaking into a low-temperature system than to replace heat lost at elevated T's, as demonstrated by the Carnot cycle. Other differences between insulating for low and elevated temperatures are:

1. Condensation of moisture in low-temperature insulation is possible. When this occurs, the conductance of the insulation is significantly increased. This problem is avoided by a vapor barrier on the outside of the insulation.

2. Condensation of the atmosphere in the insulation and surface washing by the condensed liquid are possible when the insulating surfaces are below 150°R (83 K). This can result in a large added transfer of heat. It is avoided by using an outer cover or surface impermeable to air, and evacuating the insulating space or replacing the atmosphere in it with a noncondensable gas.

At low temperatures, as at other temperatures, the fundamental modes of heat transfer are conduction and radiation, and it is against these that insulation is used. Convection of heat is practically eliminated by the insulation.

Low-T insulation categories are (1) high vacuum, with or without multiple radiation shields, (2) powders, (3) rigid foams, and (4) low-conductivity solids, such as balsa wood and corkboard.

1. (i) **Heat is transferred across an evacuated space** by radiation and by conduction through the residual gas. The conductivity of a gas is almost independent of its density (pressure) as the pressure is reduced in an evacuated space until the free paths of the molecules are increased to an appreciable fraction of the separation of the containing walls.

The molecular *mean* free path at 1 atm (0.1013 MPa) and 492°R (273 K) in air is 4×10^{-6} in (10^{-7} m); in H_2, 6×10^{-6} in (1.5×10^{-7} m); and in He, 10×10^{-6} in (2.5×10^{-7} m). At pressures lower than about 10^{-3} torr (0.133 Pa), the rate of heat transfer Q by residual gas between parallel surfaces at T_1 and T_2 less than 1 in (0.0254 m) apart is approximately $(Q/A) \approx (\text{const})\ \alpha P(T_2 - T_1)$, where P is pressure *measured* in torr with a gage at *ambient* T and α is an *overall* accommodation coefficient of gas molecules in collision with the walls. Thus, $\alpha = \alpha_1\alpha_2/[\alpha_2 + \alpha_1(1 - \alpha_2)]$, where α_1 and α_2 are accommodation coefficients of gas molecules at the two walls (see Table 11.2.7). Table 11.2.8 gives the values of the (*const.*) in the equation for Q/A.

Table 11.2.7 Approximate Values of Accomodation Coefficients, α

Temp, °R (K)	He	H_2	Air
540 (300)	0.3	0.3	0.8-0.9
140 (78)	0.6	0.5	1
35 (19)	0.6	1	1

Table 11.2.8 Constant in the Gas-Conduction Equation

$Q/A \approx (\text{const})\ \alpha P(T_2 - T_1)$

Gas	T_2 and T_1, °R (K)	Const, Btu/h · ft² · torr · °R (J/N · s · K)
N_2	<700 (389)	28 (1.19)
O_2	<540 (300)	26 (1.11)
H_2	540 and 140 (300 and 78)	93 (3.96)
H_2	140 and 36 (78 and 20)	70 (2.98)
He	Any	49 (2.09)

NOTE: T_1—inner wall (cold). T_2—outer wall (hot).

The **rate of radiative heat transfer** between parallel surfaces at T_1 and $T_2 > T_1$, where emissivities are e_1 and e_2, respectively, is $Q = e_1e_2s/[e_2 + (1 - e_2)e_1][T_2^4 - T_1^4]$, where s is the Stefan-Boltzmann constant whose value is 1.712×10^{-9} Btu/ft^2 · h · °R^4 (5.67×10^{-8} W/m^{2}K^4) (see also Sec. 2). Table 11.2.9 contains emissivities for 540°R (300 K) thermal radiation; these for most engineering purposes may be considered equivalent to minimal values of the e's for very carefully prepared and cleaned surfaces. Organic coatings have high emissivities approaching unity (> 0.9). Mechanically polished surfaces have higher e's than the minimal values. Handling a surface transfers an absorbing (high e) coating to it.

Table 11.2.9 Minimal Values of Emissivity of Metal Surfaces at Various Temperatures for 540°R (300 K) Thermal Radiation

Surfaces	Surface temp., °R (K)		
	7 (3.9)	140 (78)	540 (300)
Copper	0.005	0.008	0.018
Silver	0.0044	0.008	0.02
Aluminum	0.011	0.018	0.03
Chromium		0.08	0.08
Nickel		0.022	0.04
Brass	0.018		0.035
Stainless steel 18-8		0.048	0.08
50 Pb-50 Sn solder		0.032	
Glass, paints, carbon			>0.9
Nickel plate on copper		0.033	

(ii) **Refrigerated Radiation Shield** An LN_2 (140°R) (78 K) cooled surface interposed in the insulating vacuum space of an LH_2 or an LHe container reduces the heat transferred by radiation to the LH_2 or LHe by a factor of about 250 when compared with that transferred by a surface of the same emissivity at room temperature without the interposition of the refrigerated shield. Nearly all the heat radiated by the surface at room temperature is absorbed by the LN_2, and this results in evaporation of the relatively inexpensive LN_2.

(iii) **A floating radiation shield** is an opaque layer (e.g., a metal sheet), having surfaces of low emissivity suspended in the vacuum space with a minimum of thermal contacts with surfaces that are warmer or cooler. If the emissivity of the surfaces of the floating shield is the same as the emissivity of the vacuum-space walls, the floating shield decreases the *radiative* heat transfer by a factor of 2. For m floating shields arranged in series between the inner cold and outer warm surfaces, the rate of *radiative* heat transfer becomes $\dot{Q}_m$ (m floating shields) = $[1/(m + 1)]\dot{Q}_0$ where $\dot{Q}_0$ = heat transfer for $m = 0$.

(iv) **Multilayer (Super) Insulation** The principle of multiple floating radiation shields has been extended to the use of many thin metal layers (up to 75 layers per inch or 3,000 layers per metre) separated by thin thermal insulation. Best results have been obtained with Al foil separated by glass-fiber paper. Other materials have been used successfully, for example, nylon nets as separators and aluminized Mylar with no spacer material. The *apparent* thermal *conductivity* of Al foil multilayer insulation spaced with glass fiber paper is variable depending on the number of layers of foil, the thickness of the paper, and the thickness and compacting of the insulating layer. For $\Delta T = (540 - 36)$°R or $(300 - 20)$ K, apparent mean conductivities range from 2.0 to 4.0×10^{-5} Btu/h · ft · °R (3.5 to 7×10^{-5} J/s · m · K). For $\Delta T = [540 - 140$°R $(300 - 78)$ K], the values are about one-third larger. Using nylon nets, glass fabric or glass fibers for separating the Al foil increases the conductance from three to seven times. The above conductivities are for the direction normal to the foil surfaces. Lateral conductivities are many thousands of times larger. An advantage of aluminized Mylar is its reduced lateral conductivity. Another advantage is its relatively low density (one-third to one-half of other multilayer insulations). If the residual gas pressure in multilayer insulation is less than 10^{-4} torr, the conductance is practically independent of the residual gas pressure. At 10^{-3} torr (0.133 Pa), the conductance is increased by the order of 50 percent.

2. (i) **Powder insulation** consists of finely divided materials. Conductances of evacuated powders may vary by as much as 100 percent with variations of particle size and *apparent* density (packing) of the powder. The most commonly used powders are perlite, expanded SiO_2 (aerogel), calcium-silicate, diatomaceous earth, and carbon black. The conductance of powder insulation is practically independent of gas pressure down to 10 torr (1.333 kPa), at which pressure it begins to decrease rapidly with decreasing pressure as the free path of the gas molecules becomes comparable with the space between powder particles. The most rapid decrease of conductance occurs between 1 and 10^{-1} torr (133 and 13.3 Pa). At 10^{-3} torr (0.133 Pa) the conductance reaches practically the lower limit. Even at 10^{-2} torr (1.33 Pa) the conductance is quite low. For this reason, good mechanical pumps are adequate for the evacuation of powder insulation (fine mesh filters are needed to protect the pump from being damaged by the powder). When evacuated to 10^{-4} torr (1.33×10^{-2} Pa) or less, the apparent mean thermal conductivity for perlite [540°R (300 K) to 140°R (78 K)] ranges from 0.6 to 1.2 mW/m · K. For nonevacuated perlite filled with He or H_2 an apparent mean thermal conductivity of 115 mW/m · K can be used. With N_2 a typical value is 30 mW/m · K.

The principal function of the powder is to impede the transfer of heat by radiation. Powders are used when radiation would constitute an important heat leak. If an insulating vacuum is bounded by highly reflecting walls, such as clean Cu, Ag, or Al, adding powder may result in an increase of heat transfer because of the paths of solid conduction through the powder.

(ii) **Al powder or flakes added to a powder insulation** reduces the radiative transfer through the insulation. Powdered Al is more effective than flaked Al. The conductance of highly evacuated perlite insulation [$\Delta T = 540 - 140$°R $(300 - 78$ K)] may be reduced 40 percent by the addition of 25 percent of Al powder. A similar addition of Al powder to Santocel and Cab-O-Sil (diatomaceous earth) reduces the conductance by 70 percent. Metal powder additions are most effective with those powders that are partially transparent to infrared radiation of wave lengths longer than 3 μm.

3. **Rigid-Foam insulation** The foams most used in cryogenic applications are the more or less closed-cell foams. Table 11.2.10 gives thermal conductivities of selected samples of insulating foams. The conductivity of a foam is dependent on the conduction through the intracellular gas, on the transfer by thermal radiation, and on solid conduction. Temperatures below the condensing T of the encased gas condense it and improve the insulation. Gases encased when polymer foams are blown gradually diffuse out of the cells and are replaced by ambient

Table 11.2.10 Apparent Mean Thermal Conductivities of Selected Foams, Balsa Wood, and Corkboard

Material	Density lbm/ft^3 (kg/m^3)	Thermal conductivity,* Btu/h · ft · °R(W/K · m)
Polystyrene foam†	2.4 (38)	0.019 (0.033)
	2.9 (46)	0.015 (0.026)
Epoxy resin foam†	5.0 (80)	0.019 (0.033)
Polyurethane foam†	5.0-8.7 (80-139)	0.019 (0.033)
Rubber foam†	5.0 (80)	0.021 (0.036)
Silica foam†	10.0 (160)	0.032 (0.055)
Glass foam†	8.7 (139)	0.020 (0.035)
Balsa wood, across grain	7.3 (117)	0.027 (0.047)
	8.8 (141)	0.032 (0.055)
	20.0 (320)	0.048 (0.083)
Corkboard, no added binder	5.4 (86)	0.021 (0.036)
	7.0 (112)	0.0225 (0.039)
	10.6 (170)	0.025 (0.043)
	14.0 (224)	0.028 (0.048)
Corkboard with asphalt binder	14.5 (232)	0.027 (0.047)

* Test space pressure = 14.7 lb/in^2 abs (0.101 MPa).
† ΔT = 540 to 140°R (300 to 78 K).

gases. The conductivities of many Freon and CO_2 blown foams increase in time as much as 30 percent because of the diffusion of air into the foam. Conductivities may increase by a factor of 3 or 4 when cells are permeated with H_2 or He. Glass and silica foams appear to be the only ones having fully closed cells. The **structural rigidity** of foams may be used to eliminate the need for mechanical supports. It is possible to use rigid foams without inner liners and outer casings. Polymer foams have relatively high **thermal expansions,** even greater than the same polymer without voids. This makes cracking of the insulation a problem when it is bonded to metal. Sliding expansion joints have been used to overcome this. Mylar film bags (0.001 in or 0.025 mm thick) have been used to hold cryogenic liquids inside rigid-foam insulation and plastic films outside to keep the ambient atmosphere from the insulation.

4. **Balsa wood and corkboard** have been used for insulating cryogens that are not very cold, for example, liquid methane (201°R or 112 K) and higher- boiling cryogens. The conductivities of these insulators (Table 11.2.10) are considerably higher than for the other types of low-T insulation discussed above. Their use, therefore, is restricted to special applications, as where their ability to support internal structures mechanically is important.

Conduction of heat along structural supports passing through the insulation merits special attention for an adequate realization of the benefits of superior insulation. Ideal materials of construction would have high mechanical strength and low thermal conductivity. In Table 11.2.11, the yield strengths, in tension, of construction materials are compared with their thermal conductivities. This kind of comparison favors the nonmetallic materials in Table 11.2.11. When large quantities of cryogens are handled in the field, metals are commonly used for structural members because glass and plastics (Teflon excepted) are brittle at low T's.

Minimizing the transfer of heat through the supporting structure offers opportunities for ingenuity and inventiveness in design. The transfer is minimized by lengthening the supporting members. A stack of thin metal disks or washers, utilizing the contact resistance between adjacent disks, will increase thermal resistance without increase in length of support. A stack of 0.0008-in (0.02-mm) stainless steel disks under a compressive stress of 1,000 lb/in^2 (6.895 MPa), in vacuum, has the same thermal resistance as a solid rod 50 times the length of the stack.

Insulation Selection Because of their smaller heats of vaporization and greater cost of production, the lowest-T cryogens (LHe and LH_2) ordinarily merit more effective insulation than the higher-boiling cryogens like LO_2 and LN_2, and these in turn are ordinarily better insulated than other cryogens boiling at still more elevated T's. The choice of insulation ordinarily involves the conditions of use of the equipment, as well as considerations of the tolerable loss of refrigeration and the cost of the insulation and its installation. If equipment is to have intermittent, rather than long, uninterrupted operating periods, the insulation heat capacity and time lag in reaching steady state are important. An insulation with large heat capacity (e.g., powders) will evaporate large quantities of refrigerant and take a long time to reach steady state. Powders are compacted by mechanical shocks and vibrations. Continued mechanical loading will compact both multilayer insulation and powders. Weight and bulkiness of the insulation can be important. The heat transported by the mechanical supports and vents connecting the cold interior with the warm exterior becomes the minimum refrigeration loss, attainable only with perfect insulation of the rest of the system. A large loss of refrigeration through supports and vents ordinarily reduces the attractiveness of costly insulation having the lowest conductance.

SAFETY

Precautions are needed for the safe handling and storage of cryogens because (1) air can contaminate the cryogen, creating a potential explosive, either directly as when air mixes with H_2 or CH_4 or indirectly by transforming an inert cryogen like LN_2 into an oxidant (a potential hazard for combustible insulation), and (2) moisture and air can freeze on cold surfaces, clogging vents and preventing normal operation of valves.

Vent systems should be sized for sufficient exhaust velocities to prevent back-diffusion of air into the cryogen space. Storage at slightly elevated pressure is preferable to prevent in-leakage of contaminants. Condensation can also be a problem on poorly insulated lines carrying low-boiling cryogens; frost is not a hazard but the oxygen-enriched condensed air is hazardous in the presence of combustibles. The insulation of liquid vent lines is a necessary precaution. Where this is not possible, appropriately placed guttering can prevent the contact of any condensed air with combustible objects.

Any space, either containing a cryogen or being refrigerated by a cryogen, should be protected with proper **safety-relief mechanisms,** that is, relief valves or rupture disks. Such space includes less obvious ones such as the insulating vacuum space surrounding a cryogenic fluid or a section of line which can trap liquid between two valves. A heat leak to trapped liquid confined without a vent for escape of vapor may develop pressures sufficient to burst the containing vessel. Gas pressures necessary to maintain liquid density at ambient temperature are for helium, 18,000 lb/in^2 (124 MPa); for hydrogen, 28,000 lb/in^2 (193 MPa); and for nitrogen, 43,000 lb/in^2 (296 MPa).

The **selection of structural** materials calls for consideration of the effects of low temperatures on the properties of those materials.

Table 11.2.11 Comparison of Materials for Support Members

Material	E_y,* ¶ kpsi (MPa)	κ,* Btu/h · ft · °R(W/K · m)	E_y/κ kpsi · h · ft · °R/Btu (MN · K/W · m)
Aluminum 2024	55 (379)	47 (81)	1.17 (4.68)
Aluminum 7075	70 (483)	50 (87)	1.4 (5.55)
Copper, ann	12 (83)	274 (474)	0.044 (0.18)
Hastelloy† B	65 (448)	5.4 (9.3)	12 (48)
Hastelloy† C	48 (331)	5.9 (10.2)	8.1 (32)
K Monel‡	100 (690)	9.9 (17.1)	10.1 (40)
Stainless steel 304, ann	35 (241)	5.9 (10.2)	5.9 (24)
Stainless steel (drawn 210,000 lb/in^2)	150 (1034)	5.2 (9.0)	29 (115)
Titanium, pure	85 (586)	21 (36.3)	4.0 (16)
Titanium alloy (4A1-4Mn)	145 (1000)	3.5 (6.1)	41.4 (164)
Dacron§	20 (138)	0.088 (0.15)¶	227 (920)
Mylar§	10 (69)	0.088 (0.15)¶	113 (460)
Nylon§	20 (138)	0.18 (0.31)	111 (445)
Teflon§	2 (14)	0.14 (0.24)	14.3 (58)

* E_y is the yield stress, and κ is the average thermal conductivity between 36 and 540°R (20 and 300 K).
† Haynes Stellite Co.
‡ International Nickel Co.
§ E.I. du Pont de Nemours & Company.
¶ Room-temperature value.

A number of otherwise suitable materials become brittle at low temperature (Fig. 11.2.10). Materials must perform satisfactorily over the complete range from cryogenic to room temperature, with hydrogen embrittlement particularly significant. While hydrogen embrittlement is not believed to be a problem at liquid-hydrogen temperatures, it has been shown to occur at room temperature. The 300-series stainless steels, aluminum alloys and BeCu, among other alloys, are more resistant to hydrogen embrittlement.

In calculating allowable stresses, room temperature yield and tensile strengths should be used because (1) pressure testing of containers at ambient temperature is frequently necessary, and (2) in the case of large vessels, large unknown temperature gradients can exist above the liquid level both in the ullage space and in the walls of the vessel. The 300-series stainless steels are normally austenitic (f.c.c.). However, in some 300-series stainless steels austenite is partially transformed by cold-working and possibly by temperature cycling to martensite. As martensite is brittle at low T's, the ductility of these steels is reduced by this kind of treatment. For this reason, it is recommended as an added safety factor that room temperature strength of the 300-series stainless steels be used for the design of structural members.

Careful **stress analysis** is essential and must include stresses due to (1) thermal contraction of equipment, and (2) radial, axial, and circumferential temperature gradients caused by non-uniform cooling rates. The latter are frequently encountered in the cool-down of cryogenic transfer lines.

Cryogenic equipment should be **purged** before being placed in service to remove unwanted condensables and gases that could form explosive mixtures. Commonly used methods are (1) evacuating and back-filling with purge gas and (2) flowing of purge gas through the system.

The cryogen can be a direct **hazard to personnel.** Cryogen "burns" can result from direct contact with either the cryogen or uninsulated equipment containing the cryogen. The large evolution of gases associated with cryogenic spills can result in asphyxiation. Asphyxiation is a hazard in entering a warmed cryogenic vessel. Further safety details are found in the references.

HEAT SWITCHES

Many cryogenic applications, including sorption and adiabatic demagnetization refrigerators, rely on **heat switches** to manage heat flows and enable thermal isolation of components. Heat switches are also finding new uses such as to thermally isolate a failed cryocooler in systems that incorporate redundancy, and to shorten cooling times of cryogenic systems, for example using the upper stage of a cryocooler to precool experiment components which are then thermally isolated. There are many heat switching mechanisms, some of which are applicable across wide temperature range while others operate only in specific temperature regimes.

Mechanical contact heat switches use an actuator to make or break physical contact between two surfaces. In the conductive state, thermal conductance depends on factors such as surface material, surface area, roughness, alignment, and contact pressure, and is ultimately limited by the yield strength of the contacts. In the isolating state, residual heat flow arises from conduction through the structure supporting the contacts; at elevated temperature, radiation can be significant. Contact pressure can be generated by a number of mechanisms: thermal contraction differences, magnetostrictive and piezoelectric actuators, stepper motor or manually driven gears, and pneumatic pressure. Contact-based switches can function and are switchable over a wide temperature range, though they are generally limited in the high conductance that can be generated.

Gas-gap heat switches operate on a principle similar to sorption pumps to introduce or remove gas from between two closely spaced conductors. Typically the conductors are hermetically sealed in a volume with a fixed amount of gas and an adsorbent, or getter (often located inside a capsule separated from the main body by a connecting tube). Control of the getter temperature—usually by electrical heating—enables transitions between conductive and insulating states. The choice of gas (^3He, ^4He, H_2, Ne, N_2, etc.) and its binding energy to the getter determine the operating temperature. Gas-gap switches can be used down to as low as 0.3 K using ^3He, and can achieve ratios of thermal conductance $>10^5$.

In **magnetoresistive heat switches** the application of an external magnetic field forces conduction electrons into helical orbits. When the orbits are smaller than the electrons' mean free path, diffusion along thermal gradients is impeded, and thermal conductivity is reduced. In high-purity single-crystal metals and at low temperatures where the electrical resistivity reaches its minimum, the effect can produce thermal conductivity ratios of $>10^6$ with field strengths of 1-2 T.

The switching mechanism in **superconducting heat switches** is a forced transition between the superconducting and normal states in a Type I superconductor. In such materials the electronic contribution to the thermal conductivity exponentially decreases with temperature below T_c as it follows the population of thermal excitations (unbound pairs). When an external magnetic field exceeding H_c is applied, all electrons become unbound, and the thermal conductivity reverts to the "normal" state value. The technique can produce conductivity ratios of $>10^5$, requiring magnetic fields of <0.2 T.

For both magnetoresistive and superconducting heat switches, there is no lower temperature limit to their use. For the former, high conductivity ratios can be obtained at relatively high temperature (~20 K), while for the latter it is too small to be of practical use above about 20% of T_c.

11.3 OPTICS

by Patrick C. Mock

REFERENCES: "AIP Handbook," McGraw-Hill. Born and Wolf, "Principles of Optics," Pergamon Press. "Photonics Handbook," Laurin Publishing Co. Feynman, "QED: The Strange Theory of Light and Matter," Princeton University Press. Smith, "Modern Optical Engineering," McGraw-Hill. Miller and Friedman, "Photonics Rules of Thumb," McGraw-Hill. "Diffraction Grating Handbook," Richardson Grating Laboratory. "LIGO's Interferometer," https://www.ligo.caltech.edu/page/ligos-ifo

FUNDAMENTAL THEORIES OF LIGHT

Optics is the science of light, and light is **photons.** The fundamental theory of photons is quantum electrodynamics, and Richard Feynman explains it with great clarity in his book for the layman, "QED: The Strange Theory of Light and Matter."

Photons are the particles that mediate the electromagnetic force, and they have three fundamental properties: energy, polarization, and constant velocity in a vacuum. The **speed of light** in a vacuum, c, is a fundamental physical constant with the defined value of 299,792,458 m/s (often approximated as 3×10^8 m/s). Every photon has an energy, which determines its frequency and wavelength. These are related by $E = hf$ and $\lambda = c/f$ where E is the photon energy, h is Plank's constant, f is the frequency, and λ is the **wavelength.** Polarization is fundamentally related to the quantum spin of the photon. The classical interpretation is that a photon is an electromagnetic wave packet with orthogonal electric and magnetic fields that are also orthogonal to the direction of propagation. The polarization is the direction of the electric field component.

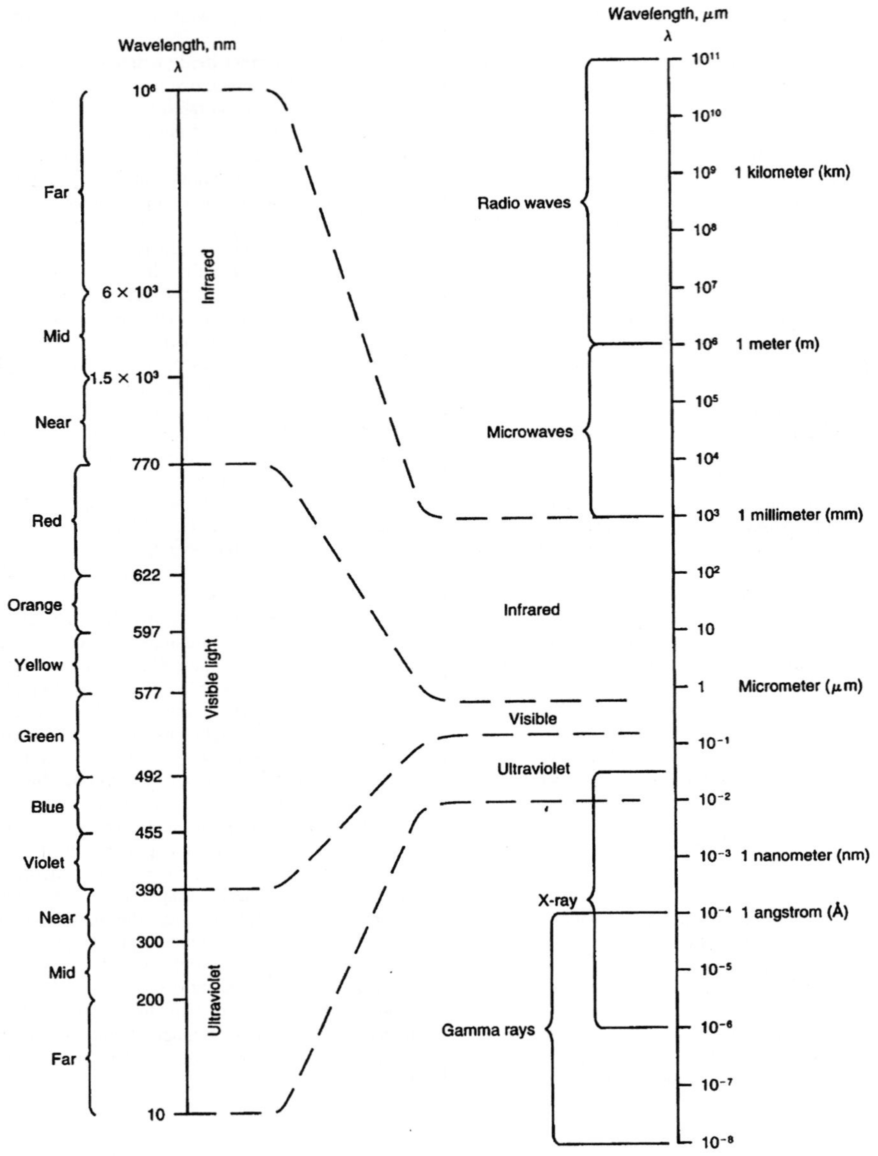

Figure 11.3.1 Electromagnetic spectrum.

Light is photons with wavelengths shorter than the resolution of the human eye and long enough that their behavior is primarily wavelike (roughly 0.1 to 300 μm). It is a small portion of the electromagnetic spectrum (Fig. 11.3.1). **Visible light** is the small band with wavelengths of 390 to 770 nm. Longer wavelength light is infrared radiation, and shorter wavelength light is ultraviolet radiation. The human eye is most sensitive near 560 nm. The interactions that light can have with matter are reflection, refraction, interference, diffraction, polarization, absorption, and scattering; however, the last three are beyond the scope of this section.

GEOMETRIC OPTICS: REFLECTION AND REFRACTION

Reflection and refraction govern geometric optics. The speed of light in an optical material, c', is always smaller than the speed of light in a vacuum. The **index of refraction,** $n = c/c'$, is an important optical property of any material. The refractive index for most optical materials depends on the wavelength of the light, and it is typically between 1 and 4. This wavelength dependence is called **dispersion.** A beautiful example of the effects of refraction and dispersion is a rainbow.

The reflection and refraction of a light ray from a flat interface between two media is shown in Fig. 11.3.2. The plane of incidence is defined by the incident light ray and the surface normal at the point of incidence. The **law of reflection** states that the reflected ray lies in the plane of incidence, and that the angle of reflection is equal to the angle of incidence, $I = I''$, as measured from the surface normal. The **law of refraction** (also known as **Snell's law)** states that refracted ray lies in the plane of incidence, and the angle of refraction is related to the angle of incidence by $n' \sin I' = n \sin I$. Total internal reflection occurs when $n > n'$, and $(n/n') \sin I > 1$. The incident ray is reflected with no loss and the transmitted ray vanishes. The critical angle, $I_c = \arcsin(n'/n)$, is the minimum angle at which this occurs.

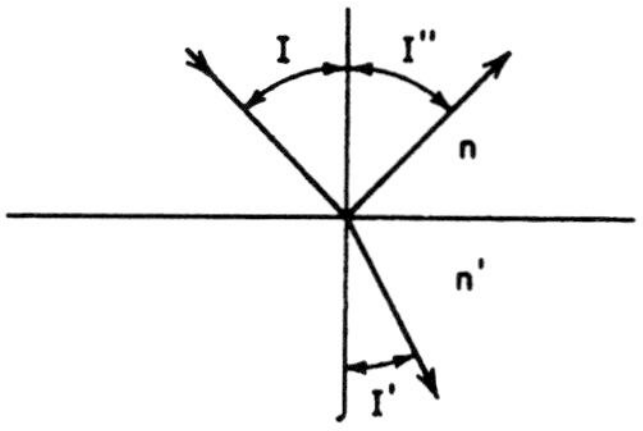

Figure 11.3.2 Reflection and refraction at a plane surface.

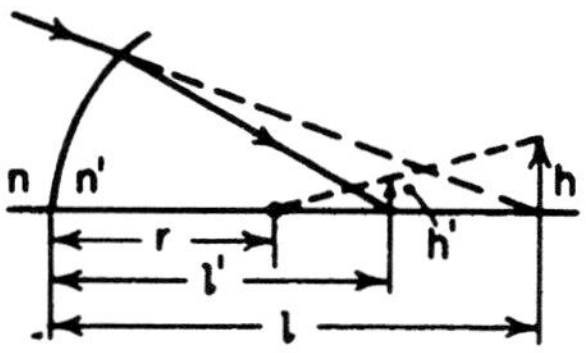

Figure 11.3.3 Refraction at a spherical surface; all parameters are shown in their positive sense.

The refraction of a light ray at a spherical interface between two media with radius of curvature r is shown in Fig. 11.3.3. The optical axis passes through the center of curvature. The image distance l' is related to the object distance l by $(n'/l' - n/l) = (n' - n)/R$, when the rays are close to the optical axis (the paraxial approximation). The **magnification** m is given by $m = h'/h = nl'/n'l$. If the image is inverted, the magnification is negative. A typical optical assembly contains several lenses, and each lens is two surfaces. The optical properties of the assembly can be found by starting with the object and determining the image for the first surface. The object for all subsequent surfaces is the image of the previous surface.

A thin lens has a thickness much smaller than its diameter. The image distance l_I for a thin lens is related to the object distance l_O by $(1/l_O + 1/l_I) = (1/r_1 - 1/r_2)\,(n' - n)/n$, where r_1 and r_2 are the radii of curvature of the surfaces of the lens, and n and n' are the indices of refraction of the medium and the lens, respectively. It is derived by applying the refraction equation from aforementioned to the two surfaces of the thin lens. The **focal length** f of a lens is the image distance when the object is at infinity. This leads to the lens maker's equation $1/f = (1/r_1 - 1/r_2)\,(n' - n)/n$ and the Gaussian lens equation $1/f = 1/l_O + 1/l_I$. The **optical invariant** (also known as the Lagrange invariant) is an invariant quantity at every surface of an optical system. In its simplest form, the product of the image size and ray angle is a constant. It implies that the field of view and useful aperture are inversely proportional, that is, a large telescope has a very small field of view, and wide-angle camera has a small aperture.

The relative illumination of the image from the object is proportional to the area of the aperture and inversely proportional to the size of the image. This relationship is characterized by two optical parameters, the ***f* number** and the **numerical aperture** NA. The f number is the ratio of the focal length to the diameter of the clear aperture of a lens system, $f/\# = f/D$, and it typically is used to characterize systems with a large image or object distance. Since the image size is proportional to the square of the focal length, the square of the f number is inversely proportional to the relative intensity. For example, a telescope with a 1-m aperture and a 10-m focal length has an f number of 10, customarily written as $f/10$. Numerical aperture is defined as $\mathrm{NA} = n \sin \phi$, where ϕ is maximum ray angle that passes through the optical system, and it is typically used to characterize systems with small image and object distances. When this angle is small, $f/\# \approx 1/(2\mathrm{NA})$.

THE WAVE NATURE OF LIGHT: INTERFERENCE AND DIFFRACTION

Interference is the interaction of waves. The interference is constructive when the waves are in phase and destructive when they are not. Diffraction is the interaction of a wave with an aperture, which often results in the bending of the wave at the edge of the aperture. Both phenomena depend on the wavelength of the wave. A common example of optical interference is the colors observed in a thin film of oil on water. Light reflected from the oil/water boundary interferes with light reflected from the air/oil boundary. The apparent color depends on the thickness of the oil layer and the viewing angle. As these change, different wavelengths interfere constructively and destructively along the path to the eye.

When the light source is monochromatic, the constructive and destructive interference leads to formation of light and dark regions known as **fringes.** If the fringe pattern is caused by the interference of two beams following different paths from the same source, the fringe pattern can be analyzed to determine path length differences with an accuracy smaller than the wavelength of the light (see "Photonics Handbook"). For example, the flatness of an unknown surface can be compared with the flatness of a reference surface. Instruments that make these measurements are known as **interferometers.** The most common types of interferometers are the Fizeau, Michelson, Twyman-Green, and Fabry-Perot. The most sensitive interferometers in the world are the modified Michelson interferometers used by LIGO to detect gravity waves, which can measure a path length difference of 10^{-19} meters.

Diffraction limits the resolution of any optical imaging system. The limit depends on the size of the aperture and the wavelength. The angular resolution of a circular aperture is given by $\phi = 2.44\lambda/D$, where D is the diameter of the aperture. For an aperture with an arbitrary shape, the far-field diffraction limit is approximately $\phi \cong \lambda P/\pi^2 A$, where P and A are the perimeter and area of the aperture, respectively.

A **diffraction grating** is a set of finely spaced narrow apertures with a spacing comparable to the wavelength of the light to be measured. The grating can be imposed on a reflective or transparent surface. The light diffracted by each aperture in the grating interferes with light from nearby apertures. If the surface in Fig. 11.3.2 is a reflective diffraction grating, constructive interference occurs when $m\lambda = d(\sin I + \sin I'')$, where d is the spacing and m is an integer which indicates the diffraction order. Physically realizable diffraction orders exist for $|m\lambda/d| < 2$.

SPECIAL TOPICS: LASERS, FIBERS, AND OPTICAL DESIGN

Lasers have become a ubiquitous optical tool. The term *laser* is an acronym for *l*ight *a*mplification by the *s*timulated *e*mission of *r*adiation. Stimulated emission is the underlying physical process that all lasers exploit. The result is an intense, monochromatic optical source that is nearly parallel and spatially coherent. Lasers are used in a broad array of applications, including CD and DVD players, fiber-optic and free-space telecommunications, photolithography, machining, photomedicine, interferometry, and surveying.

An **optical fiber** exploits total internal reflection to transmit light with very little attenuation. A simple bare glass fiber suspended in air will transmit light, but surface contamination and surface defects cause high losses. An optical fiber achieves extremely low attenuation by surrounding a high index core with a low index cladding, which protects the surface of total internal reflection. Single mode optical fibers, used in telecommunications, achieve losses as low as 0.2 dB/km. Large optical fibers are effective light pipes that can illuminate hard-to-reach areas. Coherent bundles of optical fibers are used to transmit images from hard-to-reach areas.

Optical design is the science of precisely manipulating light. Simple optical problems can be solved with the formulas in this section or the references. However, these ideal formulas ignore distortions, aberrations, and tolerances that affect all real optical systems. Sophisticated optical computer-aided design tools use precise ray tracing to accurately predict the performance of an optical design. They include libraries of design templates that can be adapted to meet the requirements of common optical systems and powerful optimizers to maximize the performance of the optical system. The "Photonics Handbook" has more information on optical CAD tools.

11.4 NEW REFRIGERANTS FOR REFRIGERATION AND AIR CONDITIONING

by Donald B. Bivens

INTRODUCTION—REFRIGERANT HISTORY

Refrigerants are often termed working fluids, which as applied in refrigeration and air conditioning, absorb heat from one area and reject it into another through evaporation and condensation. The history of refrigerants has been described in many publications,[1,2] recounting developments beginning in the early 1800s. A wide range of fluids were applied in vapor compression systems, such as ethers, alcohols, carbon dioxide (CO_2), ammonia (NH_3), sulfur dioxide, methyl chloride, ethyl chloride, and hydrocarbons (HCs). Electric refrigerators for household use were introduced in the 1920s; however, existing refrigerants had issues of toxicity, flammability, and/or strong odors, particularly sulfur dioxide which was the most commonly used refrigerant in small systems. Electric refrigerator manufacturers recognized these problems, and the well-known research by Midgley, Henne, and McNary in 1928-1930 resulted in the identification and development of chlorofluorocarbon compounds (CFCs) as refrigerants having good stability, low toxicity, nonflammability, and the needed thermodynamic properties. CFCs (and somewhat later the HCFCs, hydrochlorofluorocarbons) became the dominant refrigerants, with NH_3, CO_2, and HCs being used in limited commercial and industrial applications. Major CFCs and HCFCs were R11, R12, R13, R22, R113, R114, R115, R502, and R123 (introduced 1991). The compositions, boiling points, and ozone depletion potentials of these refrigerants are listed in Table 11.4.1.

In 1974, Rowland and Molina[3] published their theory of stratospheric ozone depletion caused by CFC breakdown releasing chlorine atoms to destroy ozone. Resulting research confirmed their theory, leading to Montreal Protocol international agreements in 1987, 1990, 1992, and 1995 to reduce and then phase out production of CFCs, HCFCs, and other ozone-depleting halon compounds. Phase-out dates for CFCs were end of 1995 for developed countries, and end of 2009 for developing countries. Phase-out dates for HCFC-22 (U.S. EPA regulation) were 2010 for new equipment, 2020 for no new HCFC-22 for service, and end of 2039 for developing countries (Montreal Protocol). For HCFC-123, U.S. EPA regulation phase-out dates are 2020 for new equipment, and 2030 for end of production and import.

A 2014 report by the World Meteorological Organization[4] describes the success of the phase out decisions: "Actions taken under the Montreal Protocol have led to decreases in the atmospheric abundance of controlled ozone-depleting substances (ODSs) and are enabling the return of the ozone layer toward 1980 levels."

Hydrofluorocarbon (HFC) fluids were being researched for refrigerant applications in the 1990s, since fluorine atoms do not react with ozone, and therefore are not ozone depleting. HFC single compounds and mixtures were formulated to have similar thermodynamic properties as the previous CFCs and HCFCs. A conference titled "Refrigerants for the twenty-first century" was organized by the American Society of Heating, Refrigerating, and Air-Conditioning Engineers, Inc. (ASHRAE) and the National Institute of Standards and Technology (NIST), held in October 1997, with presentations on current and future use of refrigerants in the United States and European countries. Refrigerants discussed were HFCs, CO_2, HCs, NH_3, fluoroethers, and air, plus other systems such as secondary loops, absorption chillers and heat pumps. HFCs were viewed as the alternative refrigerant of choice, but with increasing interest in CO_2, HCs, and NH_3 as refrigerants. A foreshadowing presentation by Albritton[5] included findings of the 1995 Intergovernmental Panel on Climate Change (IPCC) that greenhouse gases are increasing and trapping more heat within the atmosphere. A following presentation by Calm and Didion[6] emphasized the need for trade-offs in refrigerant selections, as "the future discovery of ideal refrigerants is extremely unlikely." Their presentation included a figure showing global warming potential (GWP) values of HFC compounds are lower than GWP values of CFCs. The total global warming impact of devices that use refrigerants was designated as the total equivalent warming impact (TEWI), which includes refrigerant emissions plus the emission of CO_2 from energy to operate the refrigeration and air conditioning systems. Implied is the importance of design and operation of systems operating with low refrigerant emissions and increased energy efficiency.

HFCs and mixtures with other compounds continued to be developed as alternatives to CFCs and HCFCs in cooperation with refrigeration and air conditioning system manufacturers, leading to use in the wide range of applications: motor vehicle air conditioning (MVAC), domestic and commercial air conditioning, chillers, transport refrigeration, domestic appliances, and commercial refrigeration. The 2006 Report of the UNEP Refrigeration, Air Conditioning and Heat Pumps Technical Options Committee[7] provides comprehensive property and performance data on refrigerants used in the several applications, with the major HFCs being R32, R125, R134a, R404A (R125/R143a/R134a), R407C (R32/R125/R134a), R410A (R32/R125), and R507A (R125/R143a). Safety classifications and GWP values for the HFCs are in Table 11.4.2. CO_2, NH_3, and hydrocarbon property data and application as refrigerants were also detailed.

Table 11.4.1 Major CFC and HCFC Refrigerants

Refrigerant number	Chemical formula	Boiling point degrees C	Ozone depletion potential
CFC-11	CCl_3F	24	1
CFC-12	CCl_2F_2	−30	0.7
CFC-13	$CClF_3$	−81	1
HCFC-22	$CHClF_2$	−41	0.034
CFC-113	CCl_2FCClF_2	48	0.81
CFC-114	$CClF_2CClF_2$	4	0.5
CFC-115	$CClF_2CF_3$	−39	0.26
HCFC-123	$CHCl_2CF_3$	27	0.01
R-502	R-22/R-115 48.8/51.2 mass %	−45.3	0.1

NEW REFRIGERANT REGULATIONS

This document was prepared in late 2016, describing refrigerant regulations existing at that time. The following paragraphs provide summaries of the regulations, and the reader should consult internet sites of the European Union, United States Environmental Protection Agency (US EPA), and the Montreal Protocol for more detailed descriptions and **ongoing updates.**

Based on continued and increasing concern about greenhouse emissions and global warming, the **European Union** in 2006 adopted the (2006/40/EG) "MAC Directive" on air conditioning systems used in mobile systems: only refrigerants with a GWP less than 150 will be used in new types of cars and vans introduced from 2011 and in all new cars and vans produced from 2017. The EU also adopted the F-gas Regulation (EU No. 842/2006) for other applications in which F-gases are used (F-gases are mainly HFCs). The F-gas Regulation aimed to reduce refrigerant leakage from equipment containing F-gases, and avoid the use of F-gases where environmentally superior alternatives are cost-effective. The F-gas Regulation was updated by a new regulation in 2014 (EU No. 517/2014) which limits the total amount of F-gases sold in the EU from 2015 onwards, phasing the gases down in steps to 20% of 2014 sales in 2030.[8]

The **U.S. EPA** under the Significant New Alternatives Policy (SNAP) reviews alternatives to find substitutes that represent less risk to human health and the environment, prohibiting use of a substitute where there are other available alternatives posing less risk. New SNAP regulations were issued in 2015 and 2016 in which the use of selected high GWP

Table 11.4.2 Major HFC Refrigerants; GWP Values Based on Fifth Assessment

Refrigerant number	Chemical formula	Boiling point degrees C	Safety class	GWP 100 year
HFC-32	CH_2F_2	−52	A2L	677
HFC-125	CHF_2CF_3	−49	A1	3,170
HFC-134a	CH_2FCF_3	−26	A1	1,300
HFC-143a	CH_3CF_3	−47	A2L	4,800
R-507A	R125/R143A 50.0/50.0 mass %	−47	A1	3,985
R-404A	R125/R143a/R134a 44.0/52.0/4.0 mass %	bubble point/ dew point, °C −46.6/-45.8	A1	3,943
R-407C	R32/R125/R134a 23.0/25.0/52.0 mass %	−43.8/-36.7	A1	1,624
R-410A	R32/R125 50.0/50.0 mass %	−51.5/-51.5	A1	1,923

HFCs in several refrigeration and air conditioning applications are no longer acceptable after dates specified for use in new systems or retrofit. End-dates for unacceptable substitutes are identified in following refrigerant application sections.

The U.S. EPA issued two regulations (2011 and 2015) that allow the use of selected HCs and HFC-32 in specific stationary refrigeration and air conditioning applications. The 2011 rule lists isobutane (R600a) and R441A (mixture of ethane, propane, isobutane, and butane), subject to use conditions, for household refrigerators, freezers, and combination of refrigerators and freezers. Also listed is propane (R290), subject to use conditions, in retail food refrigerators and freezers (stand-alone units only). The 2015 rule added propane (R290) for use in household refrigerators/freezers, vending machines, and room air conditioning units; ethane in very low temperature refrigeration; HFC-32 in room air conditioning units; isobutane in vending machines and stand-alone retail food refrigerators and freezers, and R441A in vending machines, room air conditioning units, and stand-alone retail food refrigerators and freezers. These listings apply only to new systems with specific safety requirements and maximum allowed refrigerant charge sizes as described in the SNAP listing.

The 28th Meeting of the Parties to the **Montreal Protocol** met in Kigali, Rwanda, October 10-14, 2016, reaching agreement to amend the Montreal Protocol to phase down HFC compounds. The agreement calls for developed countries to reduce HFCs 10% by 2019 and by 85% by 2036, relative to a baseline of 2011-2013. A group of developing countries such as China, Brazil, and most of Africa to begin phase down in 2024, and a reduction of 80% by 2045, relative to a baseline in 2020-2022. India, Pakistan, Iran, Saudi Arabia, and Kuwait will begin phase down in 2028. These timelines and quantities are subject to amendments. It is important to note the percentages stated above are based on CO_2 equivalent values, thereby permitting use of refrigerant compounds having lower values of GWP.

LOW GWP REFRIGERANTS—UNSATURATED HYDROFLUOROCARBONS

A primary requirement of refrigerants is chemical stability, considering the many years of operation expected for refrigeration and air conditioning systems. Unsaturated compounds are known to be more reactive and unstable as compared to saturated compounds,[9] with unsaturated compounds not being considered for refrigerant applications.[10] It was highly unexpected when announcements appeared of unsaturated compounds being tested as refrigerants. The Japan Air Conditioning, Heating & Refrigeration News, February 2006, reported that two United States refrigerant manufacturers separately announced new low GWP refrigerants that would meet the 2011 European Union environmental standard of less than 150 GWP for use in mobile air conditioning systems. Later disclosures of mobile air conditioning test program details revealed the unsaturated compound R1234yf ($CF_3CF{=}CH_2$), provided the low GWP (less than 1). The low GWP is based on the short atmospheric lifetime of R1234yf. This compound is also termed HFO-1234yf, hydrofluoroolefin, representing the presence of the double bond.

The initial application targeted for R1234yf was in mobile air conditioning, due to the European Union "Mac Directive" requiring the use of refrigerants with a GWP of less than 150 in new types of cars and vans introduced from 2011. A 2008 paper presented at the International Refrigeration and Air Conditioning Conference at Purdue ("HFO-1234yf Low GWP Refrigerant Update"),[11] described comparative evaluations versus HFC-134a for mobile air conditioning. Evaluation criteria were thermodynamic properties, toxicity, environmental impact, flammability, thermal stability, plastics and elastomers compatibility, system performance, and life cycle climate performance. Boiling points and vapor pressures are similar for R134a and HFO-1234yf, with bench scale testing for cooling capacity and energy efficiency being within 4-8 percent of R134a for no system changes. Thermal stability tests were conducted per ASHRAE Standard 1997-1999 at 175C and 200C for two weeks, showing no significant corrosion to the metal coupons (aluminum, copper, and carbon steel). HFO-1234yf was determined to be flammable when tested using ASTM-681-04, but with milder flammability as compared with flammability of NH_3, HFC-32, HFC-152a, and propane (low burning velocity and high ignition energy requirements). Larger scale testing of potential conditions leading to flammability indicated likelihood of HFO-1234yf ignition was quite remote.

The Society of Automotive Engineers (SAE) International initiated a "SAE Cooperative Research Program CRP1234" in 2007 to investigate the safety and performance of HFO-1234yf for use in mobile air conditioning. Participants include 16 global vehicle OEMs and 18 suppliers. The program objectives were assessments of safety and risk, system energy and performance, materials compatibility, flammability, toxicity, and atmospheric/climate effects. The program tests and results were issued in 2009,[12] stating. "The sponsors of the SAE CRP1234 have concluded that HFO-1234yf can be used as the global replacement refrigerant in future mobile air conditioning systems and it can be safely accommodated through established industry standards and practices for vehicle design, engineering, manufacturing, and service." SAE International announced in 2011 new technical standards covering the use of HFO-1234yf refrigerant in mobile air conditioning systems.

In 2012, a motor vehicle OEM issued a press release suggesting that new testing by the company had shown HFO-1234yf to be a greater risk of vehicle fire than earlier estimated by CRP1234. A new CRP was organized to conduct new fault tree scenarios to address the claims. Based on the updated analysis, the estimated risk of vehicle fire exposure to use of HFO-1234yf was estimated at 3×10^{-12} events per vehicle operating hour, which is nearly six orders of magnitude less than the current risk of vehicle fires due to all causes (1×10^{-6} per vehicle operating hour). It was concluded that the former CRP risk assessment results were still valid. Additional safety reviews were conducted by the German Motor Federation and the European Commission Joint Research Centre, which also validated the safety of HFO-1234yf in automotive air conditioning.[13]

Additional unsaturated compounds that are stable under refrigerant operating conditions in refrigeration and air conditioning, but have short atmospheric lifetimes (and low GWP), have been investigated and now included in the approved USA EPA SNAP listing of refrigerants, Table 11.4.3.

ADDITIONAL LOW GWP REFRIGERANTS

HFC-32 and HFC-152a are included in the U.S. EPA SNAP listing of acceptable refrigerants based on GWP values of 677 for HFC-32 (CH_2F_2) and 138 for HFC-152a (CH_2CHF_2). CO_2 (R744), NH_3 (R717), and several hydrocarbon compounds have low GWP values: CO_2 has the baseline GWP of 1, NH_3 GWP is also near or slightly less than 1, and HCs such as propane (R290), butane (R600), isobutane (R600a), and ethane (R170) have GWP values of about 5. The boiling points, safety ratings, and GWP value for these compounds are listed in Table 11.4.4. Note in later application sections these flammable refrigerants are subject to specified use limitations such as charge size, system design, etc.

The safety group ratings are taken from ASHRAE STANDARD 34 which has toxicity and flammability classifications. Class A are for refrigerants having an occupational exposure limit (OEL) of 400 ppm or greater. Class B is for refrigerants having an OEL of less than 400 ppm. Class 1 is for no flame propagation in air at 60°C and 14.7 psia (101.3 kPa). Class 2L is for refrigerants having a burning velocity of no more than 10 cm/s when tested at 23°C and 14.7 psia. Classes 2 and 3 have higher levels of flame propagation as described in ASHRAE Standard 34.

Hydrocarbon compound R1270 (propylene) is acceptable (EPA SNAP) for industrial process refrigeration, but not in residential and light commercial air conditioners and heat pumps, cold storage warehouses, nor centrifugal and positive displacement chillers for commercial comfort air conditioning. This is based on U.S. EPA analysis showing impact of propylene decomposition leading to unacceptable local air quality (higher levels of ground level ozone).

One potential barrier to some applications of low GWP refrigerants is the lack of a safety code for safe use of refrigerants with the ASHRAE flammability classification of 2L. Many existing codes have requirements for safe use of refrigerant classifications of 1, 2, or 3, but not the new ASHRAE Standard 34 classification of 2L. To address this issue, ASHRAE, AHRI, and the U.S. Department of Energy (DOE) are collaborating to fund research to establish a body of knowledge on the safe use of flammable low GWP refrigerants. ASHRAE, the International Code Council (ICC) and other global standards organizations will use the information for standard and code modification for safe use of low GWP 2L refrigerants.

LOW GWP REFRIGERANT APPLICATIONS TESTING

The four unsaturated fluorocarbons in Table 11.4.3 are being applied as single component refrigerants and as building blocks with other low GWP compounds to create refrigerant blends having lower GWP values, as compared with HFCs scheduled for phase down by the U.S. EPA, the European F Gas Regulation, and the October 2016 United Nations agreement to amend the Montreal Protocol to phase down HFCs.

Anticipating the need to reduce the impact of HFC refrigerants on global warming, research was underway in the early 2000s to identify compounds and mixtures having lower GWP values. After announcements were made of the discovery and continuing research of low GWP refrigerants, the US Air Conditioning, Heating, and Refrigeration Institute (AHRI) in 2011 established an international industry cooperative effort to test the new refrigerants and freely disseminate results. (AHRI Low-GWP AREP). Phase 1 testing of 38 low GWP refrigerants was conducted by 21 U.S. and international manufacturers and laboratories. The refrigerants comprised mixtures of HFCs and HFOs, and single refrigerants HFO1234yf, HFO1234ze, HFC-32, HFC-152a, propane, propylene, and isobutane. Testing was in compressor calorimeters, stationary air conditioners and heat pumps, bus air conditioner, chillers, ice machines, bottle coolers, and a trailer refrigeration unit. Test results were very favorable in comparison with the high GWP HFC refrigerants, particularly as the system tests were either drop-in or soft-optimized. A Phase II of the program included testing at high ambient temperatures. Summaries of the two testing phase results were presented at the 2014 and 2016 International Refrigeration and Air Conditioning

Table 11.4.3 EPA SNAP Listing of Acceptable HFOs

Refrigerant number	Chemical formula	Boiling point degrees C	Safety class	GWP 100 year
HFO-1234yf	$CF_3CF{=}CH_2$	−29	A2L	<1
HFO-1234ze	$CF_3CH{=}CHF$	−19	A2L	<1
HFO-1233zd	$CF_3CH{=}CHCl$	19	A1	1
HFO-1336mzz	$CF_3CH{=}CHCF_3$	33	A1	2

Table 11.4.4 EPA Listing of Additional Refrigerants for Specific Applications

Refrigerant number	Chemical formula	Boiling point degrees C	Safety class	GWP 100 year
HFC-32	CH_2F_2	−52	A2L	677
HFC-152a	CH_3CHF_2	−25	A2	138
R744	CO_2	−78	A1	1
R717	NH_3	−33	B2L	0
R600	$CH_3CH_2CH_2CH_3$	0	A3	4
R600a	$CH(CH_3)_2CH_3$	−12	A3	3
R290	$CH_3CH_2CH_3$	−42	A3	3
R170	CH_3CH_3	−89	A3	6

Conferences at Purdue. All test reports are available to the public on the AHRI website: http://www.ahrinet.org/site/514/Resources/Research/AHRI-LOW-GWP-ALTERNATIVE-REFRIGERANTS-EVALUATION.

LOW GWP REFRIGERANTS—IDENTIFICATION AND SELECTION GUIDANCE

ANSI/ASHRAE Standard 34, DESIGNATION AND SAFETY CLASSIFICATION OF REFRIGERANTS, provides the numbering system for designation of refrigerants and the safety classification procedure. The standard identifies requirements to apply for designations and safety classification for refrigerant single compounds and blends. The standard is continuously updated by addenda that are publicly reviewed, approved by ASHRAE and ANSI (American National Standards Institute), and published on the ASHRAE website. The first two HFOs [R1234yf and R1234ze(E)] were approved by the ANSI/ASHRAE procedure in June 2011. As of June 2016, 27 refrigerant blends containing HFOs have been approved and added to ANSI/ASHRAE Standard 34. The standards and addenda are available from ASHRAE. To aid in understanding the potential application of the low GWP refrigerants, the refrigerants are listed in Tables 11.4.5 and 11.4.6, with refrigerant numbers, compositions, boiling point temperatures (or bubble point/dew point temperatures), safety classification, and GWP values. If the refrigerant mixture is an azeotrope, the vapor and liquid compositions will be the same, and it will perform as a single compound with a single boiling point temperature. For mixtures which are not azeotropic (zeotropic), the liquid and vapor compositions will be different, and there will be separate temperatures for bubble point and dew point. During the boiling process for a zeotropic refrigerant mixture, the temperature at which a liquid refrigerant first begins to boil is termed the bubble point temperature, and the temperature at which the last drop of liquid has boiled is the dew point temperature. The difference between the bubble point and dew point temperatures is known as the temperature glide. Temperature glide of a mixture is one indication of the zeotropic nature of a mixture, often being an important factor in refrigerant selection for specific refrigeration and air conditioning system designs. If a zeotropic refrigerant is used in a system that tends to have refrigerant leakage, leaking vapor would have a different mixture composition than the liquid, eventually leading to a change in refrigerant mixture composition and system performance. In systems with flooded evaporators, refrigerant liquid pools in the evaporator, leading to a different zeotropic mixture composition in the evaporator than in other parts of the system, and likely uncertain performance. For direct expansion systems with minimal refrigerant leakage, experience with R407C (temperature glide of 7°C) has been excellent.

Initial selection of a low GWP replacement refrigerant will be based on several factors and requirements for specific applications: safety classification, applicable standards and codes, desired value of refrigerant GWP, system design appropriate for single compounds, azeotropic or zeotropic mixtures, and refrigerant boiling point near that of the refrigerant being replaced (for zeotropic mixtures, use the midpoint temperature between the bubble and dew points). Refrigerant manufacturers literature can be reviewed for performance testing in targeted applications, and additional performance screening (cooling capacity, energy efficiency) is possible using thermodynamic property based vapor compression cycle models, such as the National Institute of Standards and Technology program CYCLE_D (information available on the NIST website).

The following sections describe the **June 2016** status of potential low GWP HFO compounds and mixtures, HFCs R32 and R152a, and NH_3, CO_2, and HCs in the main refrigeration and air conditioning industry segments. Research and development of low GWP refrigerants is continuing, with expectations of **additional candidates.** There are also expectations of ongoing U.S. EPA SNAP refrigerant **approvals and regulations,** plus **updates** to building and other application standards and codes for use of 2L safety classification refrigerants.

VEHICLE AIR CONDITIONING

The initial large volume application considered for the low GWP hydrofluoroolefin compound **R1234yf** was in vehicle air conditioning, as described in the earlier section titled LOW GWP REFRIGERANTS—UNSATURATED HYDROFLUOROCARBONS. R1234yf was determined to be acceptable for this application after extensive performance and safety testing by the Society of Automotive Engineers, followed by additional safety reviews by the German Motor Federation and the European Joint Research Centre. Auto manufacturers in the USA, Japan, and Europe began use of R1234yf in selected new autos in 2013. Recent estimates are 24 million vehicles on the road using R1234yf by the end of 2016, and 50 million vehicles using R1234yf by the end of 2017.

The European Union "Mac Directive" requires the use of refrigerants with a GWP of less than 150 in all new cars and vans produced beginning January 1, 2017. R1234yf will be the refrigerant applied in almost all European cars produced in 2017; however, a German auto manufacturer has announced that CO_2 will be used in two of their "high-end" auto air conditioning systems. Due to the higher pressure of CO_2 versus that of R134a or R1234yf (5-10 times higher pressure in mobile air conditioning), all components of the CO_2 refrigerant air conditioning system must be designed for the higher pressures.

CO_2 is on the US EPA SNAP listing as acceptable for new MVAC systems, with required use conditions for safety:

- Engineering strategies and/or mitigation devices shall be incorporated such that in the event of refrigerant leaks the resulting CO_2 concentrations do not exceed the short-term exposure level of 3% averaged over 15 minutes in the passenger-free space and the ceiling limit of 4% in the passenger-breathing zone.
- The manufacturers must keep records of the tests performed for a minimum of 3 years demonstrating that CO_2 refrigerant levels do not exceed the above exposure levels.
- The use of CO_2 in MVAC must adhere to the standards conditions identified in SAE Standard J639.

HFC-152a (GWP of 124 and safety level of 2A) is also USA EPA SNAP listed as acceptable for MVAC systems with required use conditions for safety:

- Engineering strategies and/or devices shall be incorporated into the system such that foreseeable leaks into the passenger compartment do not result in R152a concentrations of 3.7% v/v or above in any part of the free space inside the passenger compartment for more than 15 seconds when the car ignition is on.

Table 11.4.5 Azeotropic Refrigerant Mixtures to Replace High GWP HFCs

Refrigerant description	Composition (mass %)	Boiling point degrees C	Safety class	GWP 100 year
R-513A	R1234yf/R134a 56/44	−29	A1	572
R-513B	R1234yf/R134a 58.5/41.5	−29	A1	540
R-514A	R1336mzz(Z)/R1130(E) 74.7/25.3	29	B1	2
R-515A	R1234ze/R227ea 88/12	−19	A1	402

Table 11.4.6 Zeotropic Refrigerant Mixtures to Replace High GWP HFCs

Refrigerant description	Composition (mass %)	Bubble point, °C dew point, °C	Safety class	GWP 100 year
R-444A	R32/152a/1234ze(E)	−34.3	A2L	83
	12.0/5.0/83.0	−24.3		
R-444B	R32/R152a/1234ze(E)	−44.6	A2L	295
	41.5/10/48.5	−34.9		
R-445A	R744/134a/1234ze(E)	−50.3	A2L	118
	09/06/85	−23.5		
R-446A	R32/600/1234ze(E)	−49.4	A2L	461
	68/3/29	−44.0		
R-447A	R32/125/1234ze(E)	−49.3	A2L	572
	68/3.5/28.5	−44.2		
R-447B	R32/125/1234ze(E)	−50.1	A2L	714
	68/8/24	−46.0		
R-448A	R32/125/1234yf/134a/1234ze(E)	−45.9	A1	1,273
	26/26/20/21/7	−39.8		
R-449A	R32/125/1234yf/134a	−46.0	A1	1,282
	24.3/24.7/25.3/25.7	−39.9		
R-449B	R32/125/1234yf/134a	−46.1	A1	1,296
	25.2/24.3/23.2/27.3	−40.2		
R-449C	R32/125/1234yf/134a	−44.6	A1	1,147
	20/20/31/29	−38.1		
R-450A	R1234ze(E)/134a	−23.4	A1	547
	58/42	−22.8		
R-451A	R-1234yf/134a	−30.8	A2L	134
	89.8/10.2	−30.5		
R-451B	R-1234yf/134a	−31.0	A2L	147
	88.8/11.2	−30.6		
R-452A	R1234yf/32/125	−47.0	A1	1,945
	30/11/59	−43.2		
R-452B	R1234yf/32/125	−51.0	A2L	676
	26/67/7	−50.3		
R-452C	R1234yf/32/125	−47.5	A1	2,019
	26.5/12.5/61	−44.2		
R-454A	R-32/1234yf	−48.4	A2L	238
	35/65	−41.6		
R-454B	R32/1234yf	−50.9	A2L	467
	68.9/31.1	−50.0		
R-454C	R32/1234yf	−46.0	A2L	146
	21.5/78.5	−37.8		
R-455A	R744/32/1234yf	−51.6	A2L	146
	3/21.5/75.5	−39.1		
R-456A	R32/134a/1234ze(E)	−30.4	A1	626
	6/45/49	−25.6		
R-457A	R32/1234yf/152a	−42.7	A2L	139
	18/70/12	−35.5		
R-458A	R32/125/134a/227ea/236fa	−39.8	A1	1,565
	20.5/4/61.4/13.5/0.6	−32.4		

- Manufacturers must adhere to all safety requirements listed in SAE Standards J639 and J2773, including unique fittings and a flammable refrigerant warming label.

As reported in the 2010 UNEP Report of the Refrigeration, Air Conditioning, and Heat Pumps Technical Options Committee, the only design considered for safe use of R152a was a secondary loop heat transfer system (water/glycol) as the direct coolant in the passenger compartment, with the water/glycol being cooled by the refrigerant located in the engine compartment. The secondary loop increased weight and added costs, plus limited under-hood space availability were factors in R152a not being selected for auto air conditioning.

Electric vehicle climate control systems are receiving increased interest, and a recent literature review of heat pump air conditioning systems for electric vehicles has been published, March, 2016.[14] The review included potential working fluids, system configurations, and other advanced technologies. This topic for hybrid and electric vehicles was also addressed in a paper presented at the 12th IIR Gustav Lorentzen Natural Fluids Conference, August, 2016.[15]

Bus and Rail Air Conditioning systems are mainly mounted in the roof, with compressors being driven electrically or from the engine. System details are described in the 2015 ASHRAE HANDBOOK, HVAC Applications, pages 11.2-11.9. Several HFC refrigerants have been applied in these systems: R134a, R407C, and R410A. There are also several buses using CO_2 as air conditioning refrigerant operating in Germany. Low GWP alternative refrigerants with ASHRAE designations are available which can replace HFCs in these applications. **To replace**

R134a, R1234yf has performance similar to R134a, a GWP of less than 1, and a 2L low flammability rating. In case a nonflammable refrigerant is required, two azeotropic mixtures (R513A and R513B) are nonflammable, having GWPs of about 550, and expected similar performance to R134a. **To replace R410A,** three options with 2L ratings are available which should have similar performance: R452B has a small temperature glide of 0.7°C and GWP of 676; R446A has a GWP of 480; R447A has a GWP of 572, **To replace R407C,** three options are: R444B has a GWP of 295 and a 2L rating; R447A has a GWP of 572 and a 2L rating, and R449A has a GWP of 1,282 and is nonflammable.

TRANSPORT REFRIGERATION

Commodities requiring refrigeration (food and beverage products, other temperature sensitive goods) are transported by highway vehicles, rail cars, and ships, including the use of refrigerated cargo containers. Most cooling systems use direct expansion vapor compression cycles to produce medium or low temperature levels (0 to +10°C or −20 to −30°C). Shock and vibration are concerns for the refrigeration system due to handling of containers and travel over highways and railroads, requiring specialized designs with vibration isolation mounts and piping supports to maintain system integrity. Refrigerant flammability must also be considered due to transport in different countries having varying safety codes.

Refrigerants applied in these applications are mainly **R134a** and **R404A/R507A,** with some use of CO_2 and NH_3 in fishing vessels. Development activities are underway for use of CO_2 in highway vehicles and cargo containers. HFC-32 and HCs (propane and propylene) are also being tested for transport refrigeration, being aware of special requirements for safe use of flammable refrigerants. Several nonflammable HFO low GWP refrigerants for consideration are R450A, R513A, and R513B for R134a replacement, and R448A, R449A, and R452A for R404A replacement. A lower GWP alternative to replace R404A would be R444B with a 2L flammability rating and 295 GWP.

INDUSTRIAL REFRIGERATION

The industrial refrigeration sector includes a wide range of applications in chemical plant operations and manufacturing, and in food and beverage processing. There is also a wide range of cooling requirements and types of refrigeration systems, consisting of large refrigerant distributed systems, secondary loop chiller systems, and smaller direct expansion vapor compression systems. Detailed descriptions of industrial refrigeration systems are presented in the UNEP 2014 Report of the Refrigeration, Air Conditioning and Heat Pumps Technical Options Committee.[16]

The refrigerant most widely used is **R-717 (NH_3)** having a GWP of 0 and a safety classification of B2L. This safety classification has been acceptable, as most industrial refrigeration systems are located away from the general public, and safety issues are well understood. A paper presented at the 12th Gustav Lorentzen Natural Fluids Conference reported that NH_3 can be detected at low concentrations of 5 ppm, and the workplace exposure limits are 25 ppm for an 8-hour period and 35 ppm for a 15-minute period.[17] The paper describes a standardized approach to risk assessments of NH_3 site installations, "prioritizing the reduction of the likelihood of an NH_3 release.........and analyzing the effects of an NH_3 release with regard to explosive and toxic risk." Advantages of NH_3 are availability, low cost, high efficiency, many years of proven operation, and zero values for ODP and GWP.

HCs (propane and propylene) are used in distributed and chiller applications in petrochemical plants processing high flammability products, where safety requirements are present to eliminate ignition of leaking flammable products or refrigerants. **Carbon dioxide (R744)** has use in some segments of industrial refrigeration: coupled with a higher boiling point refrigerant in a cascade system, and use in low temperature freezer systems.

HFC refrigerants have limited use in industrial refrigeration, in small and medium size systems, and in chiller systems. The HFC refrigerants being used include R134a, R404A, R407C, and R410A, and the HCFC refrigerant R123. Low GWP HFO refrigerants have been formulated to replace the just mentioned refrigerants. Candidate replacements for the HFCs were described in the preceding sections of Transport Refrigeration and Bus/Rail Air Conditioning. There are several replacements for HCFC refrigerant R123, depending on system design. R123 has a boiling point of 27°C and a safety classification of B1. Potential alternatives are: (a) the azeotropic refrigerant R514A with a boiling point of 29°C and the same B1 safety classification; (b) R1233zd(E) with a boiling point of 19°C and safety classification of A1, and (c) R1336mzz(Z) with a boiling point of 33°C and safety classification of A1. All three potential alternatives can be used in flooded evaporator chillers.

Industrial process operations can take advantage of waste heat recovery and conversion into electricity by use of the **Organic Rankine Cycle** (ORC). The ORC uses a vapor compression system using refrigerant working fluids to utilize energy from lower-temperature heat rejected from other processes. A typical ORC schematic is shown in Figure 11.4.1, taken from Datla and Braz.[18]

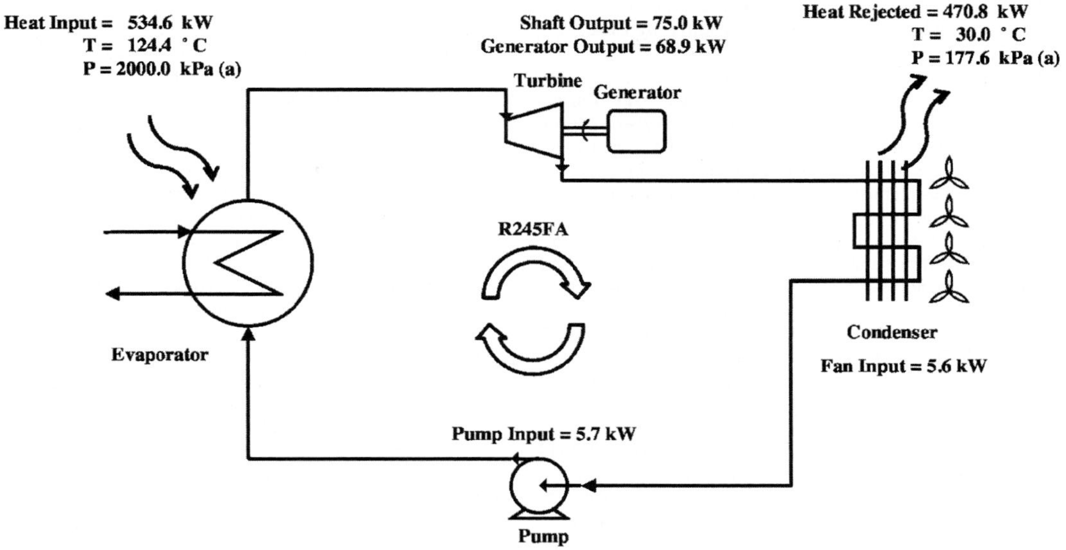

Figure 11.4.1 Organic Rankine Cycle with HFC-245fa.

The mainly used fluorocarbon ORC working fluid is HFC-245fa with a boiling point of 15.1°C. Two low GWP alternatives are: HFO-1336mzz(Z) and HCFO-1233zd(E), as described previously as potential replacements for R123. These fluids are further described for ORC application in the references.[18,19] Application of these alternatives will be based on design and operating conditions of specific ORC systems.

DOMESTIC APPLIANCES

The size of domestic refrigerator-freezers is an important variable in the amount and type of refrigerant charge. North American refrigerator-freezers average size is about twice that of refrigerators in Europe and Japan (about 600-700 liters versus about 300 liters). North American appliances have used R134a refrigerant since the mid 1990s, while other countries moved to the hydrocarbon isobutane (R600a). With the recent U.S. EPA SNAP regulation listing R134a unacceptable in new household refrigerators and freezers as of January 1, 2021, alternative refrigerants must be developed.

The EPA SNAP regulation for domestic refrigerator-freezers has a maximum of 57 grams of isobutane. A typical North American refrigerator-freezer size of 650 liters has an R134a charge of 5 to 5.5 ounces (142 to 156 grams). With the conversion to an equivalent charge of isobutane, this would require an isobutane charge of 62-69 grams, being higher than the maximum amounts.

Potential replacement refrigerants for R134a in household refrigerator-freezers currently listed by EPA SNAP are three nonflammable fluids (R450A, R513A, and R513B) and the flammable fluids isobutane, propane, R152a, and R441A (with changes made to comply with EPA SNAP and Underwriter's Laboratory UL250 safety and refrigerant charge requirements). R1234yf is also a candidate, having a GWP less than 1 and a safety classification of A2L.

VENDING MACHINES/STAND-ALONE RETAIL REFRIGERATION/CONDENSING UNITS

Vending machines are self-contained units for refrigerating beverages or food which is dispensed, in contrast to a "reach-in" unit. R134a has been the dominant refrigerant, but bottle cooler units designed for CO_2 are increasingly used, plus hydrocarbon refrigerants in Europe. R134a and other higher GWP HFCs are not acceptable (EPA SNAP) for new vending machines effective January 1, 2019. R134a replacement refrigerants are the nonflammable fluids CO_2, R450A, R513A, and R513B, and the flammable fluids R441A, propane, isobutane, and R1234yf. The hydrocarbon fluids have a charge limitation of 150 grams, and are subject to use conditions required by EPA SNAP and UL Standard 541.

Stand-alone units are used to store chilled beverages and food, or frozen products, accessed by consumer "reach-in." HFC refrigerants R134a and R404A have been used for the two levels of cooling, plus hydrocarbon refrigerants in Europe. R134a, R404A, and higher GWP HFCs are not acceptable (EPA SNAP) for new stand-alone units January 1, 2020. Replacement fluids for R134a are those stated in the section for **vending machines,** with the hydrocarbon fluids restricted to a maximum charge size of 150 grams and being subject to use conditions required by EPA SNAP and UL Standard 471. For replacement of R404A, the low GWP, nonflammable fluids are R448A, R449A, and R449B. Nonflammable fluids with half the GWP of R404A are R407A and R407F (blends of R32/R125/R134a), and R452A and R452C (blends of R32/R125/R1234yf).

Condensing unit systems have a remotely located compressor and condenser, connected to the evaporator in the space to be cooled. These systems are used by small grocery stores, with HFC refrigerants being similar to those in stand-alone units (R134a and R404A). R404A, R507A, and higher GWP refrigerants are not acceptable (EPA SNAP) for new condensing units January 1, 2018. Low GWP replacements for R404A and R507A are the nonflammable fluids R448A, R449A, and R449B. Condensing units using CO_2 as refrigerant are sold in Europe and Japan, and units using propane or propylene are sold in Europe.[20]

CENTRALIZED SUPERMARKET SYSTEMS

Centralized supermarket systems operate with direct expansion vapor compression cycles with a number of cooling evaporators in display cases in the sales area, connected to a rack of compressors in a machine room. The system condensers are air cooled and located externally (Fig. 11.4.2).

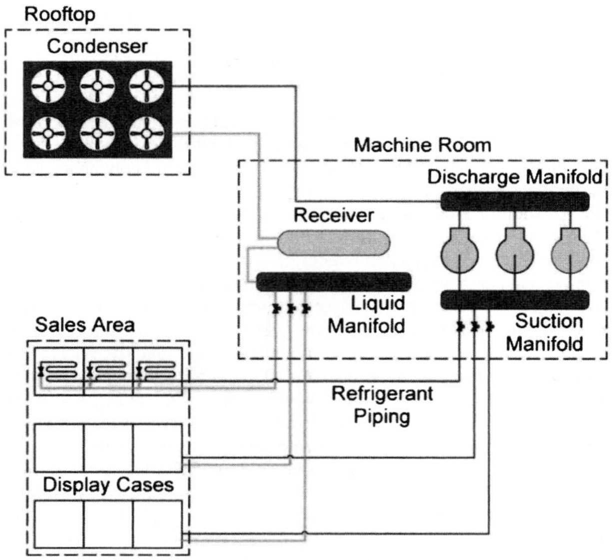

Figure 11.4.2 Direct Expansion Supermarket Refrigeration System.

Beyond this simple description, many variations of system designs and refrigerants have been, and are continuing to be, investigated. Variations have included distributed systems with the compressors located near the display cases, secondary loops with water/glycol or CO_2 fluids, cascade systems using different refrigerants, and transcritical systems with CO_2.

HFC refrigerants used in the basic direct expansion centralized system have mainly been R404A and R507A, with a minor quantity of R134a. EPA SNAP regulations finalized in 2015 require no use of R404a or R507A in new supermarket systems beginning January 1, 2017. Lower GWP nonflammable fluorocarbon alternatives have been identified to potentially replace R404A/R507: R448A, R449A, R449B, R452A, R452C, R407A, and R407F.

NH_3 and **propane** have limited applications as refrigerants in supermarkets in Europe, with design requirements of secondary heat transfer loops for display case cooling and the NH_3 or propane compressors being located in remote machinery rooms.

The major fluid alternative to fluorocarbon refrigerants in centralized supermarket systems is **CO_2,** used in cascade systems with HFCs or other refrigerants, and in direct expansion systems as the only refrigerant. The energy efficiency of CO_2 is better than HFC systems at ambient temperatures below 22°C, equal at ambient temperatures between 22 and 26°C, and lower at higher ambient temperatures.[21]

Cascade systems are designed to maintain CO_2 temperatures below the critical temperature (31°C). CO_2 is the refrigerant in the low temperature stage, exchanging heat with the high stage refrigerant within a cascade heat exchanger. These systems are often used in hot climates to achieve energy efficient operation. Potential low GWP nonflammable refrigerants for the high stage would be R513A, R513B, or R450A.

Research has been underway to develop efficient systems operating with CO_2 as the only refrigerant in supermarket refrigeration. A 2015 report estimates more than 6,000 transcritical CO_2 systems globally installed in supermarkets.[22] Efficient operation at high ambient temperatures is a continuing challenge, with system innovations being developed such as parallel compression and replacement of expansion valves with two-phase ejectors, plus integration with other supermarket functions such as hot water production. Cost of ownership for the systems especially operating in warm climates is also a factor.[23] An extensive testing program of modifications to improve energy efficiency of transcritical CO_2 supermarket systems is reported by Fincke and Sharma.[24]

RESIDENTIAL AND LIGHT COMMERCIAL AIR CONDITIONING

Self-Contained Air Conditioners and Heat Pumps are used for cooling rooms in residential and commercial buildings. Typical units are window air conditioners, wall-mounted air conditioners, ceiling mounted air conditioners, portable room air conditioners, and packaged terminal air conditioners and heat pumps. R410A is currently the main HFC refrigerant in these units. In 2015, the U.S. EPA announced three flammable fluids as acceptable in new units: propane, R-441A, and HFC-32, subject to use conditions of Federal Register/Vol. 80, No. 69/Friday, April 10, 2015, and also complying with all requirements listed in Supplement SA to the 8th Edition, August 2, 2012, of UL Standard 484. The Federal Register announcement included their Tables 2-6, listing maximum charge sizes for each refrigerant and cooling capacity values, for each of the five air conditioner categories. Other potential low GWP replacements for R410A would be R446A, R447A, R452B, and R454B, all having a 2L lower flammability classification.

Ductless Mini-split Air Conditioners and Heat Pumps have a compressor/condenser unit located outdoors, connected by refrigerant piping to the indoor unit consisting of the evaporator and fan. Many of the units are designed for both cooling and heating. The dominant refrigerant used in these units in the United States is **R410A,** with **HFC-32** being used in new units in Japan. HFC-32 is not categorized as a flammable gas in the Japan high pressure gas safety law.[25] Propane and propylene have been used in small split air conditioners in Europe, and HFC-32 units were introduced in 2013. Lower GWP refrigerants for replacement of R410A would be R446A, R447A, R452B, and R454B, all having a 2L lower flammability classification. As earlier stated, the acceptance of refrigerants having different flammability classifications in the many different categories of refrigeration and air conditioning is an ongoing issue for updating of flammability standards and building codes.

Ducted Air Conditioners have residential and commercial applications in which an outside compressor/condenser unit is connected with refrigerant piping to an (inside-building) evaporator heat exchanger unit placed inside the duct and air handler unit. Conditioned air is distributed to the rooms of the residential or commercial building. The units can be designed for both cooling and heating. Refrigerants used in these units have transitioned from HCFC-22 to R407C and R410A. Lower GWP replacements for R407C would be R449A (non-flammable), or two refrigerants with 2L flammability classification: R444B and R454C. For replacement of R410A, the low GWP options are R446A, R447A, R452B, and R454B, all with 2L flammability classification.

Packaged Rooftop Air Conditioners have similar design as the above-described ducted air conditioners, using the same refrigerants and therefore having the same low GWP options.

Water Chillers are used for building air conditioning in which chilled water is pumped to air handling units for cooling air. Small and medium size chillers (up to 500 kg refrigerant charge) use a direct expansion evaporator, an air cooled condenser, and several compressor options (screw, scroll, rotary, and reciprocating). Large chillers (refrigerant charge above 500 kg) often use a flooded evaporator, water cooled condenser, and a screw or centrifugal compressor. HFC refrigerants for small/medium size chillers are **R407C, R410A,** and **R134a.** Fluorocarbon refrigerants for large chillers are **R134a** and **HCFC-123.** Water chillers are designed and maintained for low leakage rates, with the main impact on global warming effects being energy consumption during operation; therefore, any replacement refrigerants must provide the same or even better energy efficiency.

Low GWP replacement options for R407C are R444B, R449A, and R454C. For replacement of R410A, the low GWP options are R446A, R447A, R452B, and R454B. R452B and R454B have the smallest difference in bubble point and dew point temperatures (1°C difference).

Low GWP replacements for R134a are R513A or R513B, both azeotropic mixtures. These two azeotropes are potential replacements for R134a in the small/medium chillers and also in the large chillers, recognizing that flooded evaporator operation requires the use of single compounds or azeotropic mixtures to maintain constant refrigeration properties throughout the cycle. Other potential replacements for R134a would be HFO-1234ze and R450A. Both refrigerants have lower boiling points and lower pressures than HFC-134a, but could be considered for new systems.

HCFC-123 is an important refrigerant for the large centrifugal chillers, with EPA SNAP requiring no use of HCFC-123 in new equipment January 1, 2020. Two low GWP refrigerants to replace HCFC-123 are HFO-1233zd and HFO-1336mzz.

Other low GWP fluids being used for chiller applications are NH_3, HCs, and CO_2. NH_3 chillers are used in industrial and chiller plant systems, operating with screw or reciprocating compressors, but not centrifugal compressors.[26] HCs can be used in chiller systems designed and located with regard to use of flammable fluids, particularly in petrochemical plants. CO_2 air-cooled chillers are in use in northern Europe, especially in climates where the ambient temperature is at 15°C or less.[27]

REFERENCES

1. Downing, R.C., "Fluorocarbon Refrigerants Handbook," Prentice Hall, Englewood Cliffs, NJ, 1988, pp. 1-5.

2. Radermacher, R., and Hwang, Y., "Vapor Compression Heat Pumps with Refrigerant Mixtures," CRC Press, Boca Raton, FL, 2005, pp. 6-9.

3. Molina, M.J., and Rowland, F.S., Stratospheric sink for chlorofluoromethanes: chlorine atom-catalyzed destruction of ozone, *Nature*, **249,** p. 810, 1974.

4. World Meteorological Organization (WMO), "Scientific Assessment of Ozone Depletion: 2014," World Meteorological Organization, Global Ozone Research and Monitoring Project—Report No. 55, page ES-1, Geneva, Switzerland, 2014.

5. Albritton, D., "Ozone Depletion and Global Warming," ASHRAE/NIST Refrigerants Conference: Refrigerants for the 21st Century, ASHRAE, Atlanta, GA, pp. 1-5, 1997.

6. Calm, J., and Didion, D., "Trade-Offs in Refrigerant Selections: Past, Present, and Future," ASHRAE/NIST Refrigerants Conference: Refrigerants for the 21st Century, ASHRAE, Atlanta, GA, pp. 6-19, 1997.

7. UNEP. "2006 Report of the Refrigeration, Air Conditioning and Heat Pumps Technical Options Committee," United Nations Environment Programme, pp. 29-34, 2007.

8. EU Regulation: http://eur-lex.europa.eu/legal-content/EN/TXT/?uri=uriserv:OJ.L_.2014.150.01.0195.01.ENG

9. Banks, R., "Fluorocarbons and Their Derivatives," Oldbourne Book Co., London, pp. 15-23, 1964.

10. McLinden, M., and Didion, D., "The Search for Alternative Refrigerants—A Molecular Approach," *International Refrigeration and Air Conditioning Conference*, West Lafayette, IN, paper 69, 1998.

11. Minor, B., and Spatz, M., "HFO-1234yf Low GWP Refrigerant Update," *International Refrigeration and Air Conditioning Conference*, West Lafayette, IN, paper 937, 2008.

12. SAE crp: https://www.sae.org/standardsdev/tsb/cooperative/crp1234-3.pdf

13. SAE crp: www.sae.org/standardsdev/tsb/cooperative/crp_1234-4_report.pdf

14. Peng, Q., and Du, Q., "Progress in heat pump air conditioning systems for electric vehicles—a review," *Energies*, **9,** p. 240, 2016.

15. Rabi, B., Steiner, A., Waltenberger, M., Hutter, M., and Rieberer, R., "Indirect Refrigeration Cycle with Heat Pump Functionality for Hybrid and Electric Vehicles," *12th IIR Gustav Lorentzen Natural Working Fluids Conference,* Edinburg, paper 1122, 2016.

16. UNEP, "2014 Report of the Refrigeration, Air Conditioning and Heat Pumps Technical Options Committee," United Nations Environment Programme, Chapter 5, 2015.

17. Cotter, D., "A Systematic Approach to Risk Assessing Ammonia Refrigeration Systems," *12th IIR Gustav Lorentzen Natural Working Fluids Conference,* Edinburg, paper 1040. 2016,

18. Datla, B., and Brasz, J., "Comparing R1233zd and R245fa for Low Temperature ORC Applications," *International Refrigeration and Air Conditioning Conference*, West Lafayette, IN, paper 1524, 2014.

19. Kontomaris, K., "HFO=1336mzz: High Temperature Chemical Stability and Use as A Working Fluid in Organic Rankine Cycles," *International Refrigeration and Air Conditioning Conference*, West Lafayette, IN, paper 1525, 2014.

20. UNEP, "2014 Report of the Refrigeration, Air Conditioning and Heat Pumps Technical Options Committee," *United Nations Environment Programme*, Chapter 4, p. 89, 2015.

21. Finckh, O., Schrey, R., and Wozny, M., "Energy and Efficiency Comparison between Standardized HFC and CO_2 Transcritical Systems for Supermarket Applications," *IIR International Congress of Refrigeration,* Prague, Czech Republic, 2011.

22. Javerscheck O., High-efficient Integrated CO-based Solutions around the world, *Proceedings of the ATMOsphere Asia*, Tokyo, 2016.

23. Neksa, P., Hafner, A., Bredesen, A., and Eikevik, T., "CO_2 as Working Fluid—Technological Development on the Road to Sustainable Refrigeration," *12th IIR Gustav Lorentzen Natural Working Fluids Conference*, Edinburg, paper 1133, p. 825.

24. Fricke, B., and Sharma, V., "High Efficiency, Low Emission Refrigeration System," ORNL/TM-2016/363, CRADA/NFE-11-03296, Oak Ridge National Laboratory, Oak Ridge, TN, 2016.

25. Kataoka, O., "Japanese Policy and Industry Evolution," UNEP Ozone2 Climate Industry Roundtable, Beijing, 2014.

26. UNEP, "2014 Report of the Refrigeration, Air Conditioning and Heat Pumps Technical Options Committee," *United Nations Environment Programme*, Chapter 9, p. 187, 2015.

27. UNEP, "2014 Report of the Refrigeration, Air Conditioning and Heat Pumps Technical Options Committee," *United Nations Environment Programme*, Chapter 9, p. 188, 2015.

Section 12
Applied Mechanics

BY
FARID AMIROUCHE *Professor of Mechanical and Industrial Engineering, University of Illinois*
DOYLE KNIGHT *Professor, Department of Mechanical and Aerospace Engineering, Rutgers University*
YIANNIS ANDREOPOULOS *Professor, Department of Mechanical Engineering, The City College of the City University of New York*
PAUL V. CAVALLARO *Senior Mechanical Research Engineer, Naval Undersea Warfare Center*
ALI M. SADEGH *Professor, Department of Mechanical Engineering, The City College of the City University of New York*
SRIRANGAM KUMARESAN *Biomechanics Institute, Santa Barbara, California*
ANTHONY SANCES, JR. *Late of Biomechanics Institute, Santa Barbara, California*

12.1 INTRODUCTION TO THE FINITE-ELEMENT METHOD
by Farid Amirouche

12.2 COMPUTER-AIDED DESIGN, COMPUTER-AIDED ENGINEERING, AND VARIATIONAL DESIGN
by Farid Amirouche

12.3 INTRODUCTION TO COMPUTATIONAL FLUID DYNAMICS
by Doyle Knight

12.4 EXPERIMENTAL FLUID DYNAMICS
by Yiannis Andreopoulos

12.5 AIR-INFLATED FABRIC STRUCTURES
by Paul V. Cavallaro and Ali M. Sadegh

12.6 INTRODUCTION TO BIOMECHANICS
by Ali M. Sadegh

12.7 HUMAN INJURY TOLERANCE AND ANTHROPOMETRIC TEST DEVICES

by Srirangam Kumaresan and Anthony Sances, Jr., Revised by Ali M. Sadegh

12.1 INTRODUCTION TO THE FINITE-ELEMENT METHOD

by Farid Amirouche

INTRODUCTION

The finite-element method (FEM) is essentially a technique used to obtain a solution to complex problems by discretizing a given physical or mathematical problem into smaller fundamental parts called *elements*. Then an analysis of the element is conducted using the required principles or law of mechanics. Finally, the solution to the problem as a whole is obtained through an assembly procedure of the individual solutions of the elements. These finite-element techniques have been used in many fields of engineering and science as numerical tools for solution of problems, using digital computers. For example, FEM can be used in computational fluid dynamics (CFD) where the most fundamental consideration is how one treats a continuous fluid in a discretized fashion on a computer. One method is to discretize the spatial domain into small cells to form a volume mesh or grid, and then apply a suitable algorithm to solve the equations of motion. In addition, such a mesh can be either irregular (e.g., consisting of triangles in 2-D, or pyramidal solids in 3-D) or regular; the distinguishing characteristic of the former is that each cell must be stored separately in memory. Lastly, if the problem is highly dynamic and occupies a wide range of scales, the grid itself can be dynamically modified in time, as in adaptive mesh refinement methods.

In numerical analysis, the **finite-element method** (FEM) is used for solving partial differential equations (PDE) approximately. Solutions are approximated by either eliminating the differential equation completely (steady-state problems) or rendering the PDE into an equivalent ordinary differential equation, which is then solved by using standard techniques such as finite differences. The use of the finite-element method in engineering for the analysis of physical systems is commonly known as finite-element analysis. Finite-element methods have also been developed to approximately solve integral equations such as the heat transport equation.

Finite-element analysis (FEA) or the finite-element method (FEM) is a numerical technique for solution of **boundary-value problems**. It was first developed for use in problems related to trusses and structural analysis in general. The system is represented by a similar model consisting of multiple linked discrete elements. From equations of equilibrium, as will be seen in this section, in conjunction with the appropriate boundary conditions applicable to each element and the system as a whole, a set of simultaneous equations are constructed. The system of equations is then solved for the unknown values by using the techniques of linear algebra. Although it is an approximate method, the accuracy of the FEA method can be improved by refining the mesh in the model by using more elements and nodes.

A common use of FEA is for the determination of stresses and displacements in mechanical objects and systems. However, it is also routinely used in the analysis of many other types of problems. Figures 12.1.1 and 12.1.2 show the finite-element analysis used for studying car **crash simulation**.

The last decade has witnessed an unprecedented increase in the speed and power of personal computers. Because of this revolution in personal computing power, PC-based FEM programs are becoming very popular and can now perform designs in minutes that not long ago would have required hours or even days of processing time. As computer technology continues to advance, even more sophisticated programs will become available to the designer/engineer at an affordable cost.

Three-dimensional FEM can handle greater detail and more complex characterizations of construction materials. Three-dimensional FEM can incorporate nonlinear and nonelastic material models not available in linear elastic analysis (LEA). In finite-element modeling, the structure is divided into a large number of smaller parts, or "elements." Elements may be of different sizes, and may be assigned different properties, depending on their location within the structure. By breaking down the structure into elements, the structural problem is transformed into a finite set of equations that can be solved on a computer. Because a structure can be broken down into finite elements, or "discretized," in any number of ways, it is essential that any design standard incorporating finite elements include rules for systematic three-dimensional meshing.

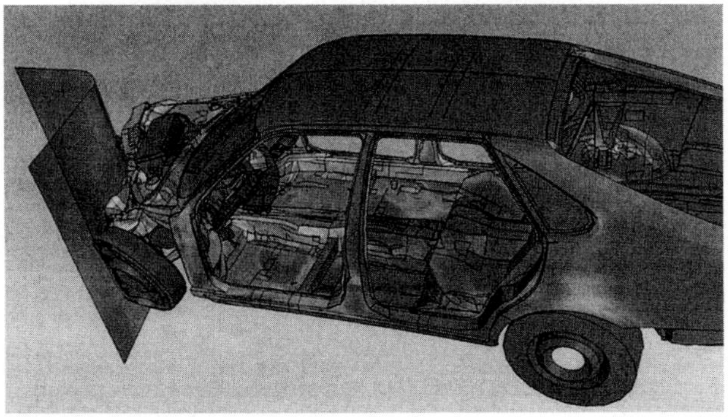

Figure 12.1.1 Car crash simulation.

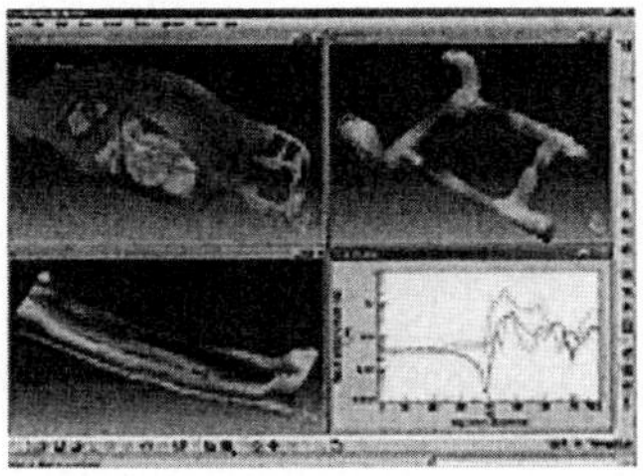

Figure 12.1.2 Car crash simulation results.

Finite-element modeling is also a tool used extensively in biomechanics for estimating the mechanical behavior of complex adaptive materials such as trabecular bone. Figure 12.1.3 shows the finite-element model of a knee with a tibial insert. The intrinsic complexities captured by FEM include the volume fraction, connectivity, and anisotropy of bone. In the analysis of total knee arthroplasty (TKA), balancing of the soft tissue can be helped by developing FEM models that can simulate the behavior of the knee during flexion extension and contact pressure in particular at the condyles. The objective of such a study is to optimize the design of the stabilizing post and to understand the mechanism of rollback.

BASIC CONCEPTS IN THE FINITE-ELEMENT METHOD

There are three phases in either using an FE code or simply developing the procedures for solving problems by FEM. Each phase is described below:

1. **Preprocessing phase**
   a. Create and discretize the solution domain into finite elements.

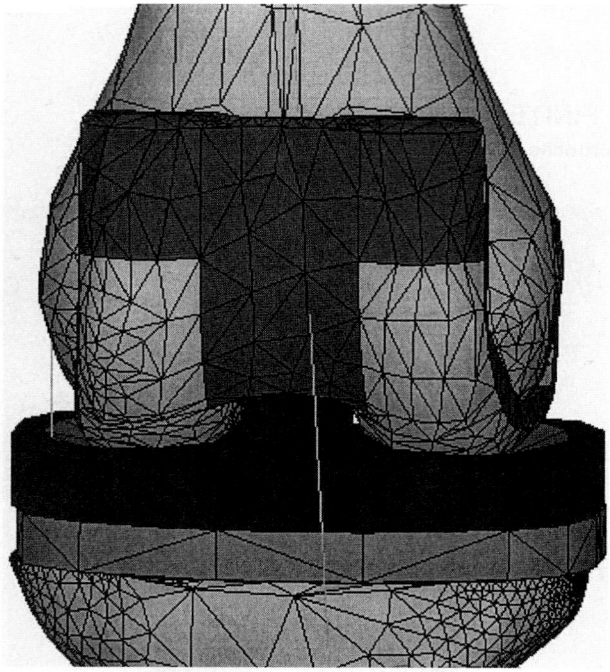

Figure 12.1.3 FEM of total knee joint.

b. Assume a shape function to represent the physical behavior of an element; that is, an approximate continuous function is assumed to represent the solution of an element.

c. Develop relationships between input and output equations for an element.

d. Assemble the elements to present the entire problem. Construct the **global stiffness matrix**.

e. Apply boundary conditions, initial conditions, and loading conditions.

2. **Solution phase.** Solve a set of linear or nonlinear algebraic equations simultaneously to obtain nodal results, such as displacement values at different nodes or temperature values at different nodes.

3. **Postprocessor phase.** Obtain other important information such as stresses, energy, and any information needed to understand the problem solution.

In the **preprocessing** phase the FEM takes the object to be analyzed and performs a discretization of the body into a limited number of elements. There is usually a selection of the elements used that is best suitable for the problem solution. On the basis of the equation for a single element, the entire problem is formulated by an assembly of these element equations. Before a solution is adapted the boundary conditions, initial conditions, and loading conditions are introduced. At this stage the **solution phase** is adapted in solving a set of linear or nonlinear equations simultaneously. The solution yields values such as displacements or temperatures at the nodes. The final phase of the FEM is the **postprocessor,** where one chooses the type of output to be displayed. The latter might require additional steps in analysis to extract the information needed.

Consider an element of a continuum as shown in Fig. 12.1.4; let the element be represented by a small cube. The nodal points are defined as the end points of the edges of the cube, which are eight in this case.

All the elements are connected through the nodal points. Any deformation of the body caused by external loads or temperatures induces certain displacements at the nodes. In general, the displacements are the unknowns and are related to the external loads or temperatures through a mathematical relationship of the form

$$f^e = k^e u^e + f^e_{\text{add}} \qquad (12.1.1)$$

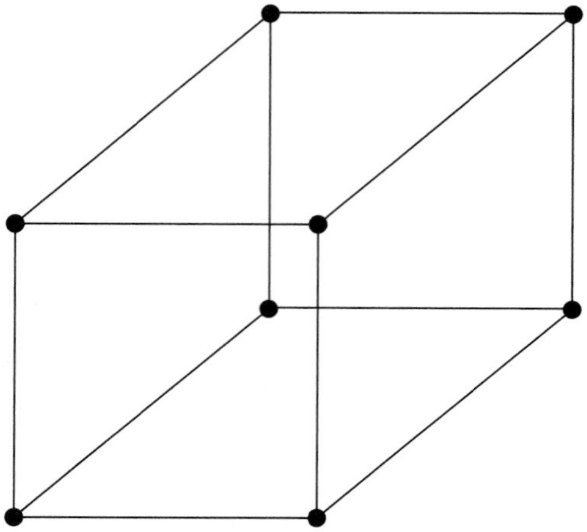

Figure 12.1.4 Cubic continuum element.

where k^e (e being the element) is the local element stiffness, f^e is the external forces applied at each element, u^e represents the nodal displacements for the element, and f^e_{add} represents the additional forces. These additional forces could be a result of surface traction or of the material being initially under stress. For the moment, we will assume them to be zero. Hence, Eq. (12.1.1) reduces to

$$f^e = k^e u^e \qquad (12.1.2)$$

It is important at times to express this equation by using the relationship between stresses and strains:

$$\sigma^e = s^e \varepsilon^e \qquad (12.1.3)$$

where σ^e is the element stress matrix and s^e is the constitutive relationship connecting the strain ε^e and *stress* σ^e. From Eq. (12.1.2), we can see that if f^e and k^e are known, u^e can be evaluated. Similarly, if u^e and k^e are known, we can compute f^e.

The finite-element method's most basic function is to automatically generate the **local stiffness matrix** knowing the element type and the properties of the material of the element. Several types of elements that are available in the finite-element library are given in Fig. 12.1.5. The well-known general-purpose FEM packages provide an element library that helps the user in the solution of particular problems. These codes and others can select any of these elements with the proper number of nodes. Thus, it is important at this stage to understand the basic mathematics involved in finite-element formulations.

POTENTIAL ENERGY FORMULATION

The finite-element method is based on the minimization of the total potential energy formulation. When a closed-form solution is not

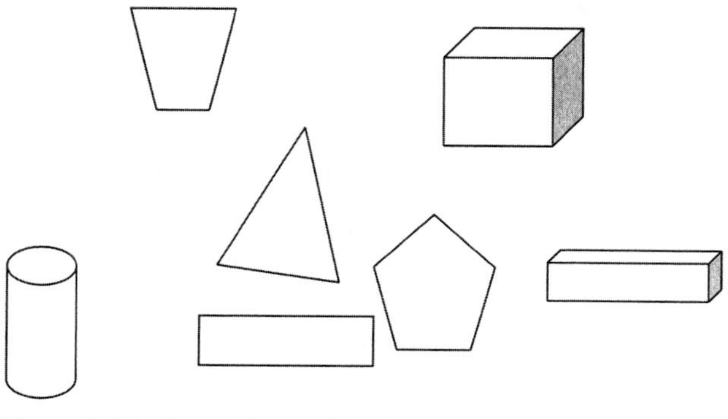

Figure 12.4.5 Types of elements in the finite-element library.

possible, approximation methods such as finite element are the most commonly used in solid mechanics. To illustrate how the finite-element method is formulated. We will consider an elastic body such as the one shown in Fig. 12.1.6, where the body is subjected to a loading force that causes it to deform.

From Hooke's law we write

$$P = \frac{AE}{L}\delta \tag{12.1.4}$$

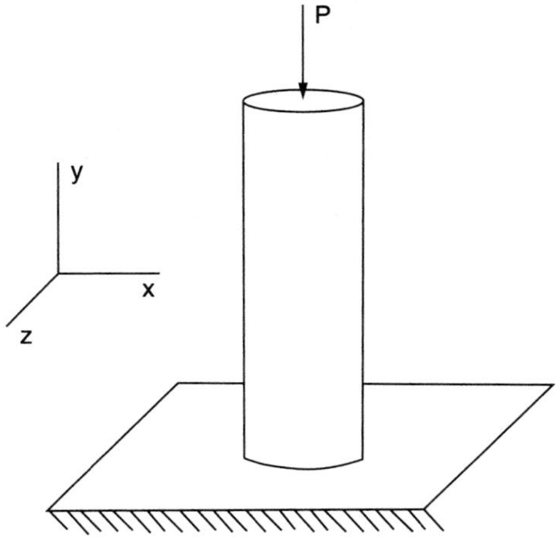

Figure 12.1.6

Let δ be the displacement in the y direction, A be the cross-section area, E be Young's modulus, L be the length, and k be the stiffness. Hence $F = ky$, where $F = P$.

The work performed by the compressive force P is stored in the material in a form of strain energy. The finite-element method discretizes bodies into elements, hence if we can compute the strain energy associated with each element, we can then evaluate the total strain energy associated with all external forces applied to the body.

Consider the element shown in Fig. 12.1.6. Let the element deformation be denoted as dy. The strain energy is given by the work of F over the deformed length y. The stored energy is

$$dE = \int_0^{y_1} F\,dy = \int_0^{y_1} ky_1\,dy_1 = \frac{1}{2}ky_1^2 \tag{12.1.5}$$

The normal stress and strain are directly related to the stored energy if we make the following substitutions

$$\frac{F}{A} = \sigma_y \Rightarrow F = \sigma_y dx\,dz \tag{12.1.6}$$

where σ_y is the stress in the y direction and ε is the strain. Hence

$$d\Lambda = \frac{1}{2}\sigma\varepsilon dV \tag{12.1.7}$$

when the volume of the element is

$$dV = dx\,dy\,dz \tag{12.1.8}$$

We can then write the strain energy for an element as

$$N^e = \int dE = \int \frac{\sigma\varepsilon}{2} dV \tag{12.1.9}$$

$$\sigma_{(e)} = E\varepsilon \tag{12.1.10}$$

$$\Lambda_s = \int_v \frac{E\varepsilon^2}{2}\,dV \tag{12.1.11}$$

The total energy, which consists of the strain energy due to the deformation of the body and the work performed by the external forces, can be expressed as function of the potential energy

$$\Pi = \sum_{e=1}^{n} \Lambda^{(e)} - \sum_{i=1}^{P} F_i u_i \tag{12.1.12}$$

where u denotes displacement along the force F_i or potential energy Π.

It follows that a stable system requires that the potential energy be minimum at equilibrium

$$\frac{\partial \Pi}{\partial u_i} = \frac{\partial}{\partial u}[\Sigma\,\Lambda^{(e)} - \Sigma\,F_i u_i] = 0 \tag{12.1.13}$$

We know that

$$\varepsilon = (u_{i+1} - u_i)/l,$$

Then

$$\frac{\partial \Lambda^e}{\partial u_i} = \frac{A_{\text{avg}}E}{l}(u_i - u_{i+1}) \tag{12.1.14}$$

$$\frac{\partial \Lambda^e}{\partial u_{i+1}} = \frac{A_{\text{avg}}E}{l}(u_{i+1} - u_i) \tag{12.1.15}$$

The above equations can be expressed in a matrix form as

$$\begin{Bmatrix} \dfrac{\partial \Lambda^{(e)}}{\partial u_i} \\ \dfrac{\partial \Lambda^{(e)}}{\partial u_{i+1}} \end{Bmatrix} = k_{\text{eq}} \begin{bmatrix} 1 & -1 \\ -1 & 1 \end{bmatrix} \begin{Bmatrix} u_i \\ u_{i+1} \end{Bmatrix} \tag{12.1.16}$$

where $k_{\text{eq}} = (A_{\text{avg}}E)/l$.

Similarly we get

$$\frac{\partial}{\partial u_i}(F_i u_i) = F_i \tag{12.1.17}$$

$$\frac{\partial}{\partial u_{i+1}}(F_{i+1}u_{i+1}) = F_{i+1} \tag{12.1.18}$$

Combining Eqs. (12.1.16), (12.1.17), and (12.1.18), we obtain the elemental force/displacement relation

$$F^{(e)} = k^{(e)}u^{(e)} \tag{12.1.19}$$

where $k^{(e)}$ is the element stiffness matrix, $u^{(e)}$ is the element displacement array associated with nodes i and $i + 1$, and $F^{(e)}$ denotes the forces acting at the nodes.

In summary, we see that the minimization of the concept of potential energy leads to an equation similar to the discrete method adopted early in this section. Other popular methods in finite-element analysis are presented for the reader to show how these methods lead to results similar to those of the minimization of potential energy approach.

CLOSED-FORM SOLUTION

It would be ideal if closed-form solutions of most of the engineering problems existed. This of course is not feasible, and the science of engineering is to obtain a solution that is as close as possible to the exact solution, relying on approximation techniques such as FEM and other numerical techniques. For the sake of validating our approximation resulting from discretizing the problem at hand, let's solve the above problem by means of direct integration, hence getting its closed-form solution.

From the equilibrium conditions, and using Newton's first law for a beam with varying cross-sectional area (see Fig. 12.1.7), we write

$$P - \sigma_{\text{avg}}(A(y)) = 0 \tag{12.1.20}$$

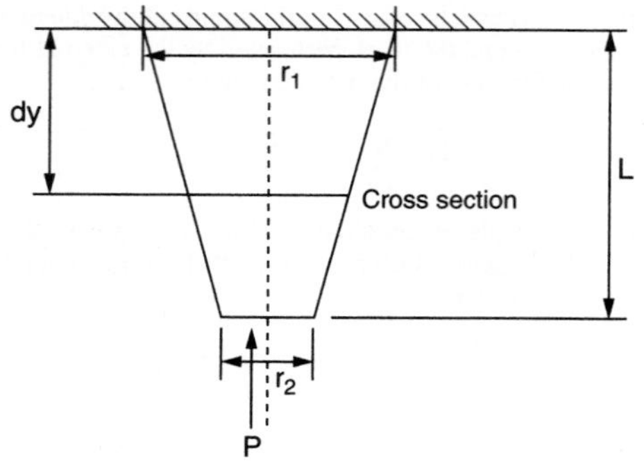

Figure 12.1.7 Free-body diagram showing constraints and loads.

where the stress is related to the strain as

$$\sigma = E\varepsilon \tag{12.1.21}$$

Then we express Eq. (12.1.20) further as

$$P - E\varepsilon A(y) = 0 \tag{12.1.22}$$

Recall that the strain is defined by

$$\varepsilon = \frac{du}{dy} \tag{12.1.23}$$

which makes Eq. (12.1.22) take the following form:

$$P - EA(y)\,\frac{du}{dy} = 0 \tag{12.1.24}$$

resulting in

$$du = \frac{Pdy}{EA(y)} \tag{12.1.25}$$

The total deformation is obtained by integrating over the total length:

$$\int_0^u du = \int_0^L \frac{P\,dy}{EA(y)} \tag{12.1.26}$$

Note that the cross-sectional area for Fig. 12.1.7 is

$$A(y) = \left(\frac{r_2 - r_1}{L}\right) y + r_1 \tag{12.1.27}$$

Then

$$u(y) = \int_0^L \frac{P\,dy}{E\left[\left(\frac{r_2 - r_1}{2}\right) y + r_1\right]} \tag{12.1.28}$$

or

$$u(y) = \frac{P}{E}\int_0^L \frac{dy}{ay + r_1} \tag{12.1.29}$$

where

$$a = \frac{r_2 - r_1}{L} = \text{constant} \tag{12.1.30}$$

Let $z = ay + r_1 \Rightarrow dz = a dy$. Then

$$u(y) = \frac{P}{Ea}\int_{r_1}^{aL+r_1} \frac{dz}{z} = \frac{P}{Ea}\,[\ln(z)] \tag{12.1.31}$$

$$= \frac{P}{Ea}\,[\ln(ay + r_1)]^2 \tag{12.1.32}$$

And the final solution is found to be

$$u(y) = \frac{P}{Ea}\,[\ln(aL + r_1) - \ln(r_1)] \tag{12.1.33}$$

The above solution can be used to find displacement at various points along the length.

WEIGHTED RESIDUAL METHOD (WRM)

The WRM assumes an approximate solution to the governing differential equations. The solution criterion is one where the boundary conditions and initial conditions of the problem are satisfied. It is evident that the approximate solution leads to some marginal errors. If we require that the errors vanish over a given interval or at some given point, then we will force the approximate solution to converge to an accurate solution.

Consider the differential equation discussed in previous example, where

$$\begin{cases} A(y)E\,\dfrac{du}{dy} - P = 0 \\ \text{with} \qquad u(0) = 0 \end{cases} \tag{12.1.34}$$

Let us choose a displacement field u to approximate the solution, which is

$$u(y) = c_0 + c_1 y + c_2 y^2 \tag{12.1.35}$$

where c_0, c_1, and c_2 are unknown coefficients.

If we substitute $u(y)$ and $A(y)$ into the differential equation, we get

$$\left[\left(\frac{r_2 - r_1}{L}\right) y + r_1\right] E(c_1 + 2c_2 y) - P = R \tag{12.1.36}$$

The equation above has two constants, c_1 and c_2. If we require that R vanishes at two points, we will get two equations, which can be used to solve for c_1 and c_2:

$$R|_{y=\frac{L}{2}} = 0$$

$$R|_{y=L} = 0$$

Solving for c_0, c_1, c_2 we get

$$u(y) = c_1 + c_2 y^2$$

The solution can be compared to the exact solution derived previously.

GALERKIN METHOD

The **Galerkin method** requires the integer of the error function over some selected interval to be forced to zero and the error be orthogonal to some weighting functions ϕ, according to the integral

$$\int \phi_i\, R\,dy = 0 \tag{12.1.37}$$

where $i = 1, 2, \ldots, n$. The ϕ are selected as part of the approximate solution. This is done simply by assigning the ϕ function to the terms to multiply the coefficients.

Since we assume $u(y) = c_1 y + c_2 y^2$, then $\phi_1 = y$ and $\phi_2 = y^2$. Substituting these two into Eq. (12.1.37), we get

$$\int_0^L y\left[(10 + y)(c_1 + 2c_2 y) - \frac{180}{E}\right] dy = 0 \tag{12.1.38}$$

$$\int_0^L y^2\left[(10 + y)(c_1 + 2c_2 y) - \frac{180}{E}\right] dy = 0 \tag{12.1.39}$$

Integrating the above two equations leads to two linear equations in terms of c_1 and c_2. Solving the above two equations, we get $c_1 = 0.2314 \times 10^{-4}$ and $c_2 = -0.50 \times 10^{-6}$. The solution is then approximated by $u(y) = c_1 + 2c_2 y^2$.

Note that the WRM and the Galerkin methods are very popular methods used in finite-element analysis to generate the approximate solution to the problem. One can see that these methods require numerics to be able to carry the integration further. If one leaves the loads as a variable *P*, then the solution is one that relates the deformation to the load. What follows is the introduction of FEM to trusses, where we can make use of the element displacement/force relation and extend it to the entire truss.

INTRODUCTION TO TRUSS ANALYSIS

Trusses are used in many engineering applications, including bridges, buildings, and towers and support structures. In a truss, we are faced with a structure where the displacements, translations, or compressions of any truss member vary linearly with the applied forces. That is, any increment in displacement is proportional to the force causing it to deform. All deformations are assumed small, so that the resulting displacements do not significantly affect the geometry of the structure and hence do not alter the forces in the members. In this case Hooke's law is preserved and the theory of elasticity is used to search for solutions of the truss. Most often, the truss design requires that its member be tested for tension, compression, stress, and strain relations. The applied loads are then tested against the possible yield stress to determine their evaluated limits and the overall stability of the truss.

For large and complicated truss structures, manual computation is often impractical to analyze the static equilibrium conditions, therefore a digital computer has to be used to accommodate the required computations. Over the past two decades, finite-element method codes have been developed for a large number of engineering applications, as trusses have always been a favorite topic for demonstrating FEM's usefulness. Consider the truss shown in Fig. 12.1.8, a 2-D planar truss with vertical loads.

Finite-Element Formulation for a Truss

Trusses are typical structures in which the finite-element method can be best illustrated. We know that FEM relies on (1) discretizing the finite elements of the system; (2) developing the mathematical relationships between the forces and displacements, stresses and strains, etc., for a given element; and (3) formulating the general problem through an assembly procedure of all the elements to solve the given problem.

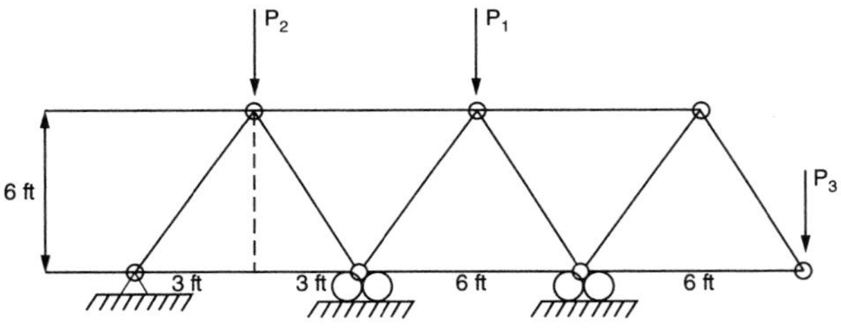

Figure 12.1.8 A planar truss subject to vertical loads (P_1, P_2, P_3 represent arbitrary loads).

Consider an element of an arbitrary truss, as shown in Fig. 12.1.9. It is subjected to either tension or compression, as is the case for all the truss elements. Let us label the element's ends 1 and 2, and, consequently, call the corresponding forces f_1 and f_2.

We also know from static analysis that, for the element to be in a state of equilibrium, the following equation must be the true one:

$$f_1 + f_2 = 0 \tag{12.1.40}$$

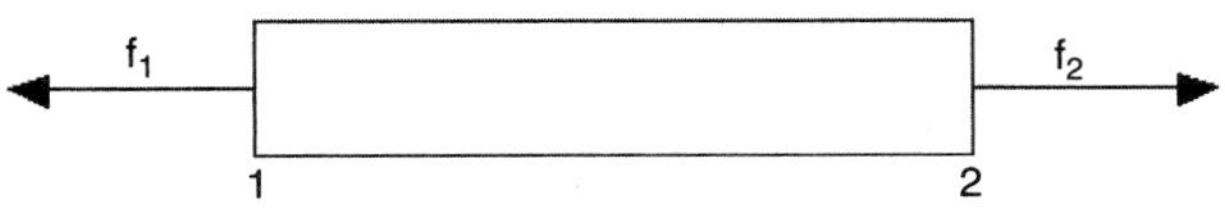

Figure 12.1.9 A typical truss element.

This simply says that the forces at both ends are equal and opposite in direction. This is true for both tension and compression of the element. Using Hooke's law, we can write the force displacement relation as

$$u = f_{1,2}\,\frac{L}{AE} \tag{12.1.41}$$

or

$$f_{1,2} = ku \tag{12.1.42}$$

where $k = AE/L$ is the stiffness constant when the element is under tension or compression. Note that $f_{1,2}$ denotes either f_1 or f_2. The relative displacement between the nodal points of this truss element can be written as

$$u = u_2 - u_1 \tag{12.1.43}$$

where u_1 and u_2 are the displacements at ends 1 and 2, respectively. We also refer to ends 1 and 2 as nodes 1 and 2 of the truss element.

Using Eq. (12.1.43), we can write

$$f_1 = ku_1 - ku_2 \tag{12.1.44}$$

and

$$f_2 = -ku_1 + ku_2 \tag{12.1.45}$$

Using matrix notation, we write Eqs. (12.1.44) and (12.1.45) in combined form as

$$\begin{bmatrix} f_1 \\ f_2 \end{bmatrix} = \begin{bmatrix} k & -k \\ -k & k \end{bmatrix} \begin{bmatrix} u_1 \\ u_2 \end{bmatrix} \tag{12.1.46}$$

or simply

$$\{f\} = [k]\,\{u\} \tag{12.1.47}$$

where

$$\{f\} = \begin{Bmatrix} f_1 \\ f_2 \end{Bmatrix} = \text{the nodal force vector for the element}$$

$$\{u\} = \begin{Bmatrix} u_1 \\ u_2 \end{Bmatrix} = \text{the nodal displacement vector}$$

$$[k] = k\begin{bmatrix} 1 & -1 \\ -1 & 1 \end{bmatrix} = \text{the element stiffness matrix}$$

Thus, Eq. (12.1.46) shows that the nodal displacements and nodal forces are related by the element (local) **stiffness matrix**. As the orientation and loading of various elements in the structure (a truss in this case) vary from each other, we need to develop the local stiffness that will apply to any element orientation.

In the sequel, a more general local stiffness (2-D) will be developed. A uniform system for symbol notation will be adopted that is crucial for easy reading and understanding and subsequent computer implementation.

Let us consider an orientation of a truss element, as shown in Fig. 12.1.10. For $\theta = 0$, we develop a relationship between the forces and the displacements given by Eq. (12.1.46), where *v* is a unit vector along the line of action of the forces.

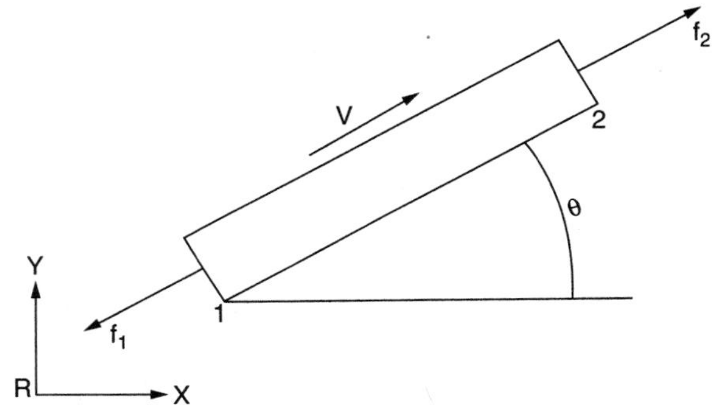

Figure 12.1.10 A truss element making an angle θ with the *x* axis.

Because we have to express all forces (tension or compression) in the global axes (x, y) given by the fixed inertial frame R, we define the element force components as follows:

$$f_{1x} = f_1 \cos\theta \quad f_{1y} = f_1 \sin\theta$$

$$f_{2x} = f_2 \cos\theta \quad f_{2y} = f_2 \sin\theta \tag{12.1.48}$$

Given that the relative displacement u is along the unit vector v, then

$$u = (u_2 - u_1)v \tag{12.1.49}$$

where

$$v = (\cos\theta)i + (\sin\theta)j \tag{12.1.50}$$

and

$$u_1 = u_{1,x}i + u_{1,y}j \qquad u_2 = u_{2,x}\,i + u_{2,y}\,j \tag{12.1.51}$$

Then by substituting Eqs. (12.1.51) and (12.1.52) into Eq. (12.1.49) and making use of Eqs. (12.1.42), (12.1.48), and (12.1.49), we obtain an expression for each nodal force in terms of the local displacement, the orientation θ and the element stiffness k.

$$f_{1,x} = (k\cos^2\theta)u_{1,x} + (k\sin\theta\cos\theta)u_{1,y} - (k\cos^2\theta)u_{2x} - (k\sin\theta\cos\theta)u_{2y} \tag{12.1.52}$$

$$f_{1,y} = (k\sin\theta\cos\theta)u_{1,x} + (k\sin^2\theta)u_{1,y} - (k\sin\theta\cos\theta)u_{2x} - (k\sin^2\theta)u_{2y} \tag{12.1.53}$$

$$f_{2,x} = (-k\cos^2\theta)u_{1,x} - \left(k\sin\theta\ \cos\theta\right)u_{1,y} + (k\cos^2\theta)u_{2x} + (k\sin\theta\cos\theta)\,u_{2y} \tag{12.1.54}$$

$$f_{2,y} = (k\sin\theta\ \cos\theta)\,u_{2x} - (k\ \sin^2\theta)\,u_{2,y} + (k\sin\theta\ \cos\theta)\,u_{2x} + (k\sin^2\theta)\,u_{2y} \tag{12.1.55}$$

Writing Eqs. (12.1.52) through (12.1.55) in matrix form yields

$$\begin{bmatrix} f_{1,x} \\ f_{1,y} \\ f_{2,x} \\ f_{2,y} \end{bmatrix} = k \begin{bmatrix} c^2 & -sc & -c^2 & -sc \\ sc & s^2 & -sc & -s^2 \\ -c^2 & -sc & c^2 & sc \\ -sc & -s^2 & sc & s^2 \end{bmatrix} \begin{bmatrix} u_{1x} \\ u_{1y} \\ u_{2x} \\ u_{2y} \end{bmatrix} \tag{12.1.56}$$

where s and c are abbreviations for $\sin\theta$ and $\cos\theta$, respectively, and k is the stiffness constant. We can write Eq. (12.1.56) in more compact form as

$$[f] = [k]\,[u] \tag{12.1.57}$$

where the local forces, local displacement, and local stiffness are defined as follows:

$$[f] = \begin{bmatrix} f_{1,x} \\ f_{1,y} \\ f_{2,x} \\ f_{2,y} \end{bmatrix} \quad [u] = \begin{bmatrix} u_{1x} \\ u_{1y} \\ u_{2x} \\ u_{2y} \end{bmatrix} \tag{12.1.58}$$

$$[k] = \frac{AE}{L}\begin{bmatrix} c^2 & sc & -c^2 & -sc \\ sc & s^2 & -sc & -s^2 \\ -c^2 & -sc & c^2 & sc \\ -sc & -s^2 & sc & s^2 \end{bmatrix} \tag{12.1.59}$$

For $\theta = 0$, the truss element reduces to the one shown in Fig. 12.1.9. The local stiffness matrix is simply

$$[k] = \frac{AE}{L}\begin{bmatrix} 1 & 0 & -1 & 0 \\ 0 & 0 & 0 & 0 \\ -1 & 0 & 1 & 0 \\ 0 & 0 & 0 & 0 \end{bmatrix} \tag{12.1.60}$$

Which checks with Eq. (12.1.46). Note how the zero rows and columns are simply used to expand the local stiffness matrix given by Eq. (12.1.60) to account for the zero forces and displacements along the y axis.

Global Stiffness Matrix

The **global stiffness matrix** relates the global forces (external forces) and the global displacements (displacements associated with each joint). We developed the local stiffness matrix for an arbitrary element of the truss—what remains is to assemble all the local stiffness matrices associated with the truss elements. Later it will be apparent in our analysis that the efficiency of the computer program is greatly dependent on the technique developed to formulate the global stiffness matrix.

The classical approach for a truss is to take a free-body diagram of each joint and, using the equilibrium equations, group all the equations together and isolate the global stiffness matrix. This approach is tedious and requires a large computational effort from the computer. The method that is illustrated in what follows to obtain the global stiffness matrix is one that Huston and Passerelo developed. It shows how the building of the global stiffness matrix can be done by a simple strategy in which connectivity tables are used to identify the truss elements and their joints. The method is as follows.

Step 1. Consider an arbitrary truss. First, label the **truss elements** and joints in an arbitrary fashion, as shown in Fig. 12.1.11. There are four joints (1, 2, … , 4) and five elements ([1], [2], … , [5]).

Step 2. We proceed to develop three tables that basically store geometrical information about the truss. Table 12.1.1 has the truss-joint/matrix-column matching, where pairs starting from 1 develop the column numbers, 2. Table 12.1.2 identifies the connection joints to all the elements of the truss. Using Tables 12.1.1 and 12.1.2, we construct Table 12.1.3, which will be used to assemble the global matrix.

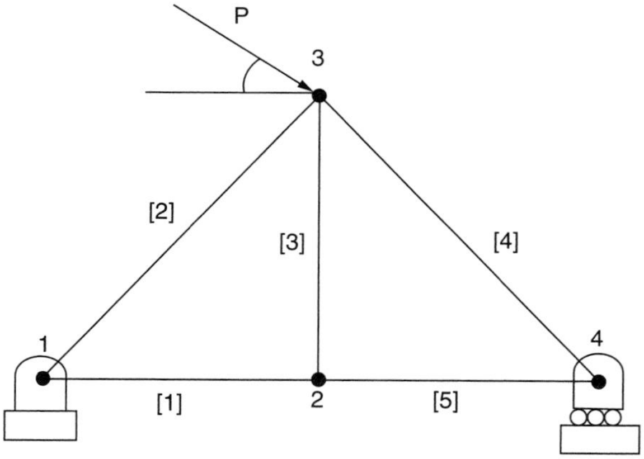

Figure 12.1.11 A two-dimensional truss.

Table 12.1.1 Truss-Joint/Matrix-Column Matrix

Joint	Column	Numbers
1	1	2
2	3	4
3	5	6
4	7	8

Table 12.1.2 Elements versus Joint Numbers

Truss element e	N_1	N_2
1	1	2
2	1	3
3	2	3
4	3	4
5	2	4

Table 12.1.3 Truss Elements versus k_{ij}

	Truss elements e				
i, j from k_{ij}^e	1	2	3	4	5
1	1	1	3	5	3
2	2	2	4	6	4
3	3	5	5	7	7
4	4	6	6	8	8

To see how Table 12.1.3 is constructed, consider truss element 1. From Table 12.1.2, we read the connecting joints to element 1, which are 1 and 2; with these joint numbers, we use Table 12.1.1 to extract the corresponding column numbers, which are 1, 2 and 3, 4. Therefore, in Table 12.1.3 number 1 (referring to the element number), we insert the values 1, 2, 3, and 4, as shown. We repeat the procedure for the remaining elements 2 to 7. Table 12.1.3 plays an important role in identifying the location where the local stiffness matrix terms will be inserted in the global matrix.

For the whole truss system, we write the equation relating the global forces and displacements:

$$\{F\} = [K]\,\{U\} \tag{12.1.61}$$

where

$$\{F\} = \begin{bmatrix} F_{1x} \\ F_{1y} \\ F_{2x} \\ F_{2y} \\ F_{3x} \\ F_{3y} \\ F_{4x} \\ F_{4y} \end{bmatrix} \quad \text{and} \quad \{U\} = \begin{bmatrix} U_{1x} \\ U_{1y} \\ U_{2x} \\ U_{2y} \\ U_{3x} \\ U_{3y} \\ U_{4x} \\ U_{4y} \end{bmatrix} \tag{12.1.62}$$

and

$$[K] = \begin{bmatrix} K_{1,1} & K_{1,2} & K_{1,3} & \cdots & K_{1,8} \\ K_{2,1} & K_{2,2} & K_{2,3} & \cdots & K_{2,8} \\ \vdots & \vdots & \vdots & \ddots & \vdots \\ K_{8,1} & K_{8,2} & K_{8,3} & \cdots & K_{8,8} \end{bmatrix} \tag{12.1.63}$$

Note that the size of the global stiffness matrix $[K]$ is 10 by 10. This is equal to the number of nodal points (joints in this case) in the truss multiplied by the degrees of freedom permitted at each node (2 for the planar truss shown). Thus, the $[K]$ in Eq. (12.1.27) is the assembled global stiffness matrix obtained from the assembly of individual element stiffness matrices, that is,

$$f_i^e = k_{ij}^e u_j \qquad (e = 1, \ldots, \text{number of elements};\ i, j = 1, 4) \tag{12.1.64}$$

where k_{ij}^e is the local element stiffness matrix and is given by Eq. (12.1.56).

For $\theta = 0$, k_{ij}^1 is given by Eq. (12.1.57). For $\theta = 90$, k_{ij}^3 is

$$k_{ij}^3 = \frac{AE}{L}\begin{bmatrix} 0 & 0 & 0 & 0 \\ 0 & 1 & 0 & -1 \\ 0 & 0 & 0 & 0 \\ 0 & -1 & 0 & 1 \end{bmatrix} \tag{12.1.65}$$

In order to construct the global stiffness matrix, first evaluate all the local element stiffness matrices. Then Table 12.1.3 is used to transfer the terms from the local stiffness matrices to the locations in the global stiffness matrix.

For example, consider element 1 from Table 12.1.3. The element column identifies the location in the global stiffness matrix of its local stiffness. The variation by pair of the column numbers identifies the i, j entries of the local stiffness matrix in the global stiffness matrix, K_{ij}. For example, the column numbers for element 1 basically correspond to the local stiffness matrix for element 1 when transferred to the location in the assembled global stiffness matrix looks like

$$K^1 = \begin{bmatrix} k_{11}^1 & k_{12}^1 & k_{13}^1 & k_{14}^1 & 0 & 0 & 0 & 0 \\ k_{21}^1 & k_{22}^1 & k_{23}^1 & k_{24}^1 & 0 & 0 & 0 & 0 \\ k_{31}^1 & k_{32}^1 & k_{33}^1 & k_{34}^1 & 0 & 0 & 0 & 0 \\ k_{41}^1 & k_{42}^1 & k_{43}^1 & k_{44}^1 & 0 & 0 & 0 & 0 \\ 0 & 0 & 0 & 0 & 0 & 0 & 0 & 0 \\ 0 & 0 & 0 & 0 & 0 & 0 & 0 & 0 \\ 0 & 0 & 0 & 0 & 0 & 0 & 0 & 0 \\ 0 & 0 & 0 & 0 & 0 & 0 & 0 & 0 \end{bmatrix} \tag{12.1.66}$$

For element 2, we read the row and column entries, which are 1, 2, 5, and 6, from Table 12.1.3. Hence, its contribution to the global stiffness is given by

$$K^2 = \begin{bmatrix} k_{11}^2 & k_{12}^2 & 0 & 0 & k_{13}^2 & k_{14}^2 & 0 & 0 \\ k_{21}^2 & k_{22}^2 & 0 & 0 & k_{23}^2 & k_{24}^2 & 0 & 0 \\ 0 & 0 & 0 & 0 & 0 & 0 & 0 & 0 \\ 0 & 0 & 0 & 0 & 0 & 0 & 0 & 0 \\ k_{31}^2 & k_{32}^2 & 0 & 0 & k_{33}^2 & k_{34}^2 & 0 & 0 \\ k_{41}^2 & k_{42}^2 & 0 & 0 & k_{43}^2 & k_{44}^2 & 0 & 0 \\ 0 & 0 & 0 & 0 & 0 & 0 & 0 & 0 \\ 0 & 0 & 0 & 0 & 0 & 0 & 0 & 0 \end{bmatrix} \tag{12.1.67}$$

Once all the elements are completed and their local stiffness matrices are inserted into the global stiffness matrix $[K]$, the general global matrix of the system is then obtained by adding all the entries in different locations if they are more than one.

$$K_{ij} = K_{ij}^1 + K_{ij}^2 + K_{ij}^3 + \cdots + K_{ij}^5 \tag{12.1.68}$$

The above outlined procedure for developing the global stiffness matrix could be automated by using computer coding.

Solution of the Truss Problem

The global force and displacement vectors are related through the global stiffness matrix:

$$[F] = [K]\,[U] \tag{12.1.69}$$

After the development of $[K]$, the global stiffness matrix, the identification of the boundary forces and displacements is important. These are to be substituted into the $[F]$ and $[U]$ vectors.

For the truss shown in Fig. 12.1.12, the reaction forces at joint 1 in the x and y directions are the reaction forces R_{1x} and R_{1y}, respectively.

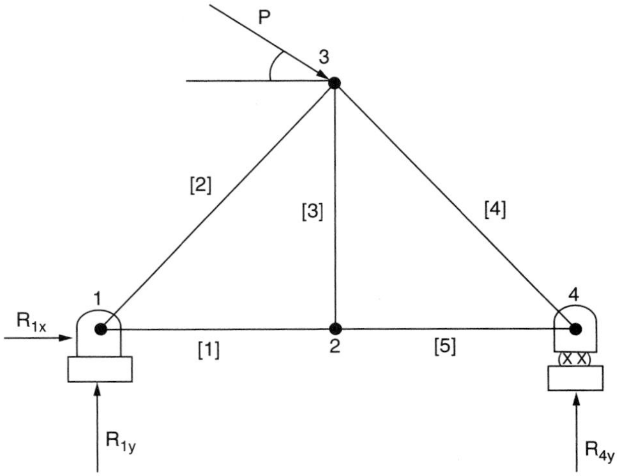

Figure 12.1.12 A simple free-body diagram of 2-D truss subject to loading.

The reaction at joint 2 is only in the y direction as it is the roller joint, that is, R_{2y}. Also, force P acts as an external force in the x direction at joint 3.

Thus,

$$F_{1x} = R_{1x} \qquad F_{1y} = R_{1y} \qquad F_{4y} = R_{4y}$$
$$F_{3x} = P_x \qquad F_{3y} = P_y \tag{12.1.70}$$

Except for these, all other forces are equal to zero.

Substituting these values into force vector $[F]$, we get Eq. (12.1.71):

$$[F] = \begin{bmatrix} R_{1x} \\ R_{1y} \\ 0 \\ 0 \\ P_x \\ P_y \\ 0 \\ R_{4y} \end{bmatrix} \tag{12.1.71}$$

Similarly, the displacement boundary condition can be identified, where for joints 1 and 2 we have Eq. (12.1.72)

$$\begin{aligned} U_{1x} &= 0 \\ U_{1y} &= 0 \\ U_{4y} &= 0 \end{aligned} \tag{12.1.72}$$

All others are nonzero. Substituting these zero displacements into the displacement vector $[U]$, we get Eq. (12.1.73)

$$[U] = \begin{bmatrix} 0 \\ 0 \\ U_{2x} \\ U_{2y} \\ U_{3x} \\ U_{3y} \\ U_{4x} \\ 0 \end{bmatrix} \tag{12.1.73}$$

Substituting $[F]$ and $[U]$ from Eqs. (12.1.71) and (12.1.73) into Eq. (12.1.69), we obtain the general equations governing the truss force/displacement equilibrium conditions.

$$\begin{bmatrix} R_{1x} \\ R_{1y} \\ 0 \\ 0 \\ P_x \\ P_y \\ 0 \\ R_{4y} \end{bmatrix} = \begin{bmatrix} K_{1,1} & K_{1,2} & \cdots & K_{1,8} \\ K_{2,1} & K_{2,2} & \cdots & K_{2,8} \\ \vdots & \vdots & \cdots & \vdots \\ \vdots & \vdots & \cdots & \vdots \\ \vdots & \vdots & \cdots & \vdots \\ \vdots & \vdots & \cdots & \vdots \\ \vdots & \vdots & \cdots & \vdots \\ K_{8,1} & K_{8,1} & \cdots & K_{8,8} \end{bmatrix} \begin{bmatrix} 0 \\ 0 \\ U_{2x} \\ U_{2y} \\ U_{3x} \\ U_{3y} \\ U_{4x} \\ 0 \end{bmatrix} \tag{12.1.74}$$

Note how in Eq. (12.1.39) the unknowns are in the global force array as well as in the joint displacement vector. A typical strategy to solve such a problem in which the unknowns are on both sides of the equation is to solve for the U's first by partitioning the matrices such that the force vector is completely in terms of the known forces. Eliminating the reaction forces does the partitioning. The resulting equation is Eq. (12.1.75)

$$\begin{bmatrix} 0 \\ 0 \\ P_x \\ P_y \\ 0 \end{bmatrix} = [K'] \begin{bmatrix} U_{2x} \\ U_{2y} \\ U_{3x} \\ U_{3y} \\ U_{4x} \end{bmatrix} \tag{12.1.75}$$

where $[K']$ is the new stiffness matrix resulting from the global stiffness after eliminating the rows and columns corresponding to the zero displacements. Equation (12.1.75) constitutes a set of seven equations and seven unknowns, which can be solved using the Gaussian elimination method, Cramer's rule, or simply the inverse of $[K']$ as Eq. (12.1.76)

$$\begin{bmatrix} U_{2x} \\ U_{2y} \\ U_{3x} \\ U_{3y} \\ U_{3y} \end{bmatrix} = [K']^{-1} \begin{bmatrix} 0 \\ 0 \\ P_x \\ P_y \\ 0 \end{bmatrix} \tag{12.1.76}$$

Once the equations are solved for the displacements, reactions R_{1x}, R_{1y}, and R_{2y} can be evaluated by premultiplying the corresponding terms of $[K]$ and $[U]$ in Eq. (12.1.74). The solutions are found to be

$$R_{1x} = -P_x \qquad R_{4y} = \frac{P_y}{2} - \frac{b}{2a}P_x \qquad R_{1y} = \frac{P_y}{2} - \frac{b}{2a}P_x \tag{12.1.77}$$

The answers obtained from the FEM analysis as described before can be checked by simply taking the free-body diagram for the truss as shown in Fig. 12.1.12. Writing the equilibrium equations

$$\Sigma F_x = 0 \quad \Sigma F_y = 0 \quad \Sigma M = 0 \tag{12.1.78}$$

we get

$$\begin{aligned} \Sigma F_x &= R_{1x} + P_x = 0 \\ \Sigma F_y &= R_{1y} + R_{4y} + P_y = 0 \\ \Sigma M_1 &= R_{4y}2a - P_x b - P_y a = 0 \end{aligned} \tag{12.1.79}$$

which checks with the FEM solution.

Evaluation of the Local Forces

The internal forces are those that are either compressing or tensioning the truss elements. To find the components of the forces acting at each end, we use the previously computed global displacements and local stiffness matrix.

From Eq. (12.1.56), we have

$$\begin{bmatrix} f^e_{1x} \\ f^e_{1y} \\ f^e_{2x} \\ f^e_{2y} \end{bmatrix} = \frac{AE}{l} \begin{bmatrix} c_e^2 & s_e c_e & -c_e^2 & -s_e c_e \\ s_e c_e & s_e^2 & -s_e c_e & -s_e^2 \\ -c_e^2 & -s_e c_e & c_e^2 & s_e c_e \\ -s_e c_e & -s_e^2 & s_e c_e & s_e^2 \end{bmatrix} \begin{bmatrix} u^e_{1x} \\ u^e_{1y} \\ u^e_{2x} \\ u^e_{2y} \end{bmatrix} \tag{12.1.80}$$

The stiffness matrix in this equation (for the particular element at hand) is known. The nodal displacements are also known from the global displacement vector. The relationship between the global and local displacement is explained below.

Let the local displacement be written as u^e_{ij}, where e is the element number of the truss, i denotes either end 1 or end 2 of the element, and j assigns the direction x or y to the end displacements. Assuming that all elements of the truss undergo the same displacements at each joint, we then write the following.

For joint 1: $U_{1x} = u^1_{1x} = u^2_{1x} = u^4_{1x} \qquad U_{1y} = u^1_{1y} = u^2_{1y} = u^4_{1y}$ (12.1.81)

For joint 2: $U_{2x} = u^1_{2x} = u^5_{1x} = u^7_{1x} \qquad U_{2y} = u^1_{2y} = u^5_{1y} = u^7_{1y}$ (12.1.82)

For joint 3: $U_{3x} = u^2_{2x} = u^3_{1x} \qquad U_{3y} = u^2_{2y} = u^3_{1y}$ (12.1.83)

For joint 4: $U_{4x} = u^3_{2x} = u^4_{2x} = u^6_{1x} = u^5_{2x}$

$$U_{4y} = u^3_{2y} = u^4_{2y} = u^6_{1y} = u^5_{2x} \tag{12.1.84}$$

For joint 5: $U_{5x} = u^7_{2x} = u^6_{2x} \qquad U_{5y} = u^7_{2y} = u^6_{2y}$ (12.1.85)

The nodal displacements for any particular element can be found from the relationships between the global displacements and the local displacement by using Eqs. (12.1.81) to (12.1.85). Subsequent substitution of these values for a particular element in Eq. (12.1.80) and multiplication by the corresponding stiffness matrix terms yields the nodal forces. The signs of these forces indicate whether the member is in tension or compression.

Example: Analysis of a Three-Element Truss Use the finite-element method on the truss in Fig. 12.1.13. (1) Find the global stiffness matrix. (2) Solve for the reaction forces. (3) Solve for the member forces and determine whether a truss element is in tension or compression.

Solution: The first step in our analysis is to label the truss for the joint numbers and link numbers as shown in the figure. The second step is to compute the local stiffness matrices for each member, using Eq. (12.1.59) which gives

$$[k_{ij}^1] = \frac{A_1 E}{l}\begin{bmatrix} 0 & 0 & 0 & 0 \\ 0 & 1 & 0 & -1 \\ 0 & 0 & 0 & 0 \\ 0 & -1 & 0 & 1 \end{bmatrix} \qquad [k_{ij}^2] = \frac{A_2 E}{l\sqrt{3}}\begin{bmatrix} 1 & 0 & -1 & 0 \\ 0 & 0 & 0 & 0 \\ -1 & 0 & 1 & 0 \\ 0 & 0 & 0 & 0 \end{bmatrix}$$

$$[k_{ij}^3] = \frac{A_3 E}{8l}\begin{bmatrix} 3 & \sqrt{3} & -3 & -\sqrt{3} \\ \sqrt{3} & 1 & -\sqrt{3} & -1 \\ -3 & -\sqrt{3} & 3 & \sqrt{3} \\ -\sqrt{3} & -1 & \sqrt{3} & 1 \end{bmatrix}$$

where A_1, A_2, A_3 and L_1, L_2, L_3 are the areas of cross section and lengths of the members of the truss, respectively, and E is Young's modulus of elasticity. $A_1 = A_2 = A_3 = A$ and $L_1 = l$, such that $L_2 = l\sqrt{3}$ and $L_3 = 2l$.

Now let us construct Tables 12.1.4 to 12.1.6 (see earlier Tables 12.1.1 to 12.1.3), which help in arriving at the global stiffness matrix [K]. Using the local

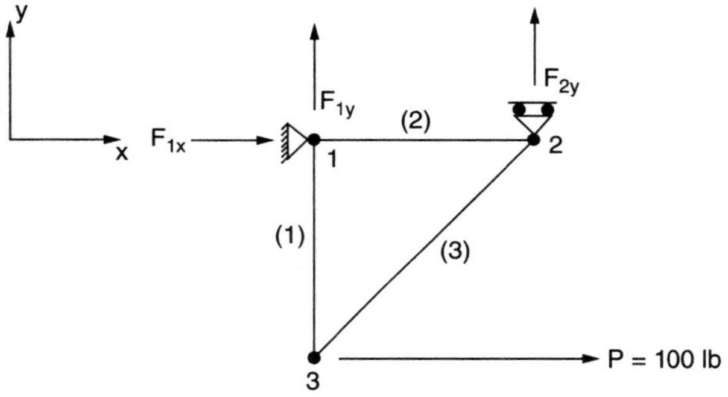

Figure 12.1.13 A three-element truss subject to loading.

Table 12.1.4

Joint number	Column number	
1	1	2
2	3	4
3	5	6

Table12.1.5

Elements	Joint number	
1	1	3
2	1	2
3	2	3

Table 12.1.6

K_{ij}	1	2	3
1	1	1	3
2	2	2	4
3	5	3	5
4	6	4	6

stiffness matrices and transferring the entries with the help of Table 12.1.6 we can arrive at the global stiffness matrix:

$$\frac{AE}{L}\begin{bmatrix} \frac{1}{\sqrt{3}} & 0 & -\frac{1}{\sqrt{3}} & 0 & 0 & 0 \\ 0 & 1 & 0 & 0 & 0 & -1 \\ -\frac{1}{\sqrt{3}} & 0 & \frac{8+3\sqrt{3}}{8\sqrt{3}} & \frac{\sqrt{3}}{8} & -\frac{3}{8} & -\frac{\sqrt{3}}{8} \\ 0 & 0 & \frac{\sqrt{3}}{8} & \frac{1}{8} & -\frac{\sqrt{3}}{8} & -\frac{1}{8} \\ 0 & 0 & -\frac{3}{8} & -\frac{\sqrt{3}}{8} & \frac{3}{8} & \frac{\sqrt{3}}{8} \\ 0 & -1 & -\frac{\sqrt{3}}{8} & -\frac{1}{8} & \frac{\sqrt{3}}{8} & \frac{9}{8} \end{bmatrix}\begin{bmatrix} U_{1x} \\ U_{1y} \\ U_{2x} \\ U_{2y} \\ U_{3x} \\ U_{3y} \end{bmatrix} = \begin{bmatrix} F_{1x} \\ F_{1y} \\ 0 \\ F_{2y} \\ P \\ 0 \end{bmatrix}$$

Zeros in the force vector indicate that the forces in the x and y directions at joints 2 and 3 are zeroes because of the roller and free joint, respectively.

Applying the displacement boundary conditions, $U_{1x} = 0$, $U_{1y} = 0$, $U_{2y} = 0$, and eliminating the corresponding rows and columns, we get

$$\begin{bmatrix} 0.952 & -0.375 & -0.2165 \\ -0.375 & 0.375 & 0.2165 \\ -0.2165 & 0.2165 & 1.125 \end{bmatrix}\begin{bmatrix} U_{2x} \\ U_{3x} \\ U_{3y} \end{bmatrix} = \begin{bmatrix} 0 \\ 100 \\ 0 \end{bmatrix}$$

Solving for the unknowns, we obtain

$$U_{2x} = 173.31\frac{l}{AE} \qquad U_{3x} = 473.13\frac{l}{AE} \qquad U_{3y} = -57.57\frac{l}{AE}$$

The reaction forces can be computed using Eq. (12.1.43) as

$$F_{1x} = \left(\frac{1}{\sqrt{3}}U_{1x} - \frac{1}{\sqrt{3}}U_{2x}\right)\frac{AE}{l} = -100.0 \text{ lb}$$

$$F_{1y} = (1.U_{1y} - 1.U_{3y})\frac{AE}{l} = 57.57 \text{ lb}$$

$$F_{2y} = \left(-\frac{\sqrt{3}}{8}U_{2x} + \frac{\sqrt{3}}{8}U_{3y} - \frac{1}{8}U_{3y}\right) = -57.57 \text{ lb}$$

These results can be verified by using the free-body diagram of the truss (Fig. 12.1.12)

$$\Sigma F_x = 0 \qquad \Sigma F_y = 0 \qquad \Sigma M = 0$$

$\Sigma F_x = 0$ gives $F_{1x} + 100.0 = 0$, that is, $F_{1x} = -100.0$ lb

$\Sigma F_y = 0$ gives $F_{1y} + F_{2y} = 0$, that is, $F_{1y} = -F_{2y}$

$\Sigma M = 0$ gives $pl + F_{2y}\sqrt{3}l = 0.0$, that is, $F_{2y} = \begin{cases} -57.57 \\ -F_{1y} \end{cases}$

Therefore, members 1 and 3 are in tension, whereas member 2 is in compression.

The member forces are obtained from the local element force-displacement relationship:

$$\begin{Bmatrix} f_{1x}^e \\ f_{1y}^e \\ f_{2x}^e \\ f_{2y}^e \end{Bmatrix} = \frac{AE}{l}\begin{bmatrix} c^2 & sc & -c^2 & -sc \\ sc & s^2 & -sc & -s^2 \\ -c^2 & -sc & c^2 & sc \\ -sc & -s^2 & sc & s^2 \end{bmatrix}\begin{Bmatrix} u_{1x}^e \\ u_{1y}^e \\ u_{2x}^e \\ u_{2y}^e \end{Bmatrix}$$

The global and local displacement at each joint are related as follows:

$$U_{1x} = u_{1x}^0 = 0 \qquad U_{1y} = u_{1y}^1 = 0 \qquad U_{3x} = u_{2x}^1 \qquad U_{3y} = u_{2y}^1$$

Using the local stiffness already computed, we obtain

$$\begin{Bmatrix} f_{1x}^1 \\ f_{1y}^1 \\ f_{2x}^1 \\ f_{2y}^1 \end{Bmatrix} = \frac{AE}{l}\begin{bmatrix} 0 & 0 & 0 & 0 \\ 0 & 1 & 0 & -1 \\ 0 & 0 & 0 & 0 \\ 0 & -1 & 0 & 1 \end{bmatrix}\begin{Bmatrix} 0 \\ 0 \\ U_{3x} \\ U_{3y} \end{Bmatrix}$$

$$\Rightarrow f_{1y}^1 = -\left(\frac{AE}{l}\right)U_{3y} = -\left(\frac{AE}{l}\right)57.57\frac{l}{AE} = 57.57 \text{ lb}$$

$$f_{2y}^1 = \left(\frac{AE}{l}\right)U_{3y} = -57.57 \text{ lb}$$

The forces acting on element 1 clearly show that it is in tension as predicted. Similarly, we can obtain the magnitude of the forces and directions for elements 2 and 3.

$$F_{1x} = F_{2x} = 0$$

Stress Analysis

In the analysis of a truss, the main objective is to decide whether the truss elements are designed to sustain the load they support. For that we need to evaluate the stress or average stress in each element. The latter is an indication as to whether the tension or compression can be sustained.

Let the element stress be given by

$$\sigma^{(e)} = E\varepsilon^{(e)} = E^{(e)} \frac{u_2^{(e)} - u_1^{(e)}}{l} \tag{12.1.86}$$

Now we can compute the stress for each element of the truss in the previous example. For each element we can write

$$\sigma^{(1)} = E^{(1)} \frac{u_2^{(1)} - u_1^{(1)}}{l} = E^{(1)} \frac{U_{3y} - U_{1y}}{l}$$

$$\sigma^{(2)} = E^{(2)} \frac{u_2^{(2)} - u_1^{(2)}}{l} = E^{(2)} \frac{U_{2x} - U_{1x}}{\sqrt{3}l}$$

$$\sigma^{(3)} = E^{(3)} \frac{u_2^{(3)} - u_1^{(3)}}{l} = E^{(3)} \frac{U_{3x} - U_{2x}}{2l}$$

and the results are

$$\sigma^{(1)} = E\frac{U_{3y}}{l} = \frac{E}{l}\left(-57.57\frac{l}{AE}\right)$$

$$\sigma^{(2)} = E\frac{U_{2x}}{l\sqrt{3}} = \frac{E}{l\sqrt{3}}\left(173.31\frac{l}{AE}\right)$$

$$\sigma^{(3)} = E\frac{U_{2x} - U_{3x}}{2l} = -\frac{E}{2l}\left(299.82\frac{l}{AE}\right)$$

HEAT CONDUCTION ANALYSIS AND FEM

In most instances, the important problems of engineering involving an exchange of energy by the flow of heat are those in which there is a transfer of internal energy between two systems. In general, the internal energy transfer is called *heat transfer*. When such exchanges of internal energy or heat take place, the first law of thermodynamics requires that the heat given up by one body must equal that taken up by the other. The second law of thermodynamics demands that the transfer of heat take place from the hotter system to the colder system. It is customary to categorize the various heat transfer processes into three basic types or modes, although it is certainly a rare instance when one encounters a problem of practical importance that does not involve at least two, and sometimes all three, of these modes occurring simultaneously. The three modes are conduction, convection, and radiation. Heat conduction will be the focus of this section. Heat conduction is the term applied to the mechanism of internal energy exchange from one body to another, or from one part of a body to another part, by the exchange of kinetic energy.

When the relationship between force and displacement can be approximated by a linear function, the problem reduces to a one-dimensional analysis. This is seen in the study of trusses where the element deformation is expressed as a linear function in terms of the element force. A number of engineering problems, and heat conduction in particular, can be solved by using one-dimensional element and linear functions to approximate the solution. Here, we will extend the one-dimensional solution to heat conduction problems, and define the concept of shape functions for one and two dimensions in the finite-element method.

One-Dimensional Elements

Now we apply the finite-element method to the solution of heat flow in some simple one-dimensional steady-state heat conduction systems. Several physical shapes fall into the one-dimensional analysis, such as spherical and cylindrical systems, in which the temperature of the body is a function only of radial distance. Consider the straight bar of Fig. 12.1.14, where the heat flows across the end surfaces. Heat is also assumed to be generated internally by a heat source at a rate f per unit volume. The temperature varies only along the axial direction x, and we want to formulate a finite-element technique that would yield the temperature $T = T(x)$ along the position x in the steady-state condition.

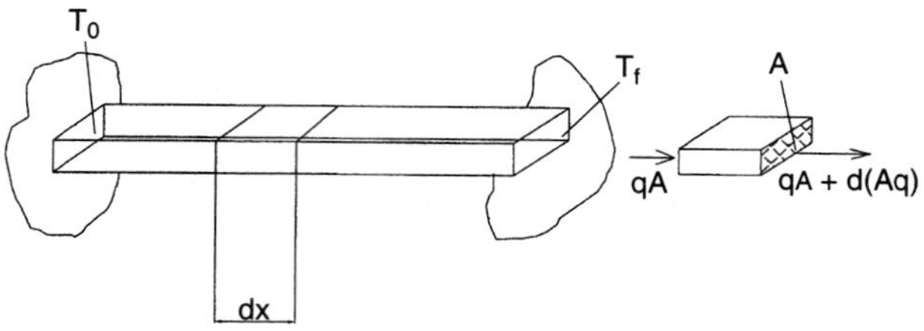

Figure 12.1.14 A typical bar with temperatures T_0 and T_f at each end.

In steady-state conditions, the net rate of heat flow into any differential element is zero. We know that for heat conduction analysis, the Fourier heat conduction equation is

$$q = -\kappa \frac{dT}{dx} \tag{12.1.87}$$

This equation states that the heat flux q in direction x is proportional to the gradient of temperature in direction x. The conductivity constant is defined by κ.

From the differential element in Fig. 12.1.14, we can write the heat flux balance:

$$qA + fA\,dx - [qA + d(Aq)] = 0 \tag{12.1.88}$$

With the differentiation of q, the heat flux equation becomes

$$qA + fA\,dx - \left[qA + \frac{dq}{dx}dx\,A\right] = 0 \tag{12.1.89}$$

This reduces to a first-order differential equation of the form

$$\frac{dq}{dx} = f \tag{12.1.90}$$

where A = cross-sectional area; f = heat source/unit volume; q = heat flux; T = temperature.

Substituting Eq. (12.1.87) into Eq. (12.1.90), we get the governing differential equation for the temperature:

$$\kappa \frac{d^2T}{dx^2} = -f \tag{12.1.91}$$

The boundary conditions for the physical problem described in Fig. 12.1.14 are $T = T_0$ at $x = 0$ and $T = T_f$ at $x = L$.

Integrating Eq. (12.1.4) we get an explicit solution for the temperature at any point along the bar.

$$T(x) = \frac{fL}{2\kappa}\left(x - \frac{x^2}{L}\right) + \left(\frac{T_f - T_0}{L}\right)x + T_0 \tag{12.1.92}$$

For one-dimensional problem the temperature at any point x can be found by using Eq. (12.1.91). Finite-element analysis is tailored for engineering problems where the closed-form solution can't be evaluated and numerical approximation is required. There are a number of scenarios in which the governing heat equation is nonlinear and the closed form is unattainable. What follow are the initial steps in applying finite-element methods to heat conduction problems. First we develop the solution to one-dimensional heat conduction problems and compare that to the solution given by Eq. (12.1.91).

Finite-Element Formulation

We must use either the principle of virtual work or energy to derive the necessary governing equations in the finite-element method. The method as shown in the previous two sections leads to the formulation of the element stiffness and stiffness matrix. We first develop the following energy equation as

$$I = \int \left[kA \frac{d^2T}{dx^2} + fA \right] dx \tag{12.1.93}$$

which yields Eq. (12.1.91) for $dI = 0$ after the standard manipulation of calculus of variations. Equation (12.1.93) could be expressed further in two parts, I_1 and I_2, as

$$I = \underbrace{\int \frac{d}{dx}\left[Ak \frac{dT}{dx} \right] T\,dx}_{I_1} + \underbrace{\int fAT\,dx}_{I_2} \tag{12.1.94}$$

Integrating the function I_1 by parts, we get

$$I_1 = TAk \frac{dT}{dx}\bigg|_0^L - \int_0^L \frac{dT}{dx} Ak \frac{dT}{dx} dx \tag{12.1.95}$$

The first term defines the boundary conditions' contributions. If we assume that the boundary conditions are such that

$$T_{(x=0)} = T_L \quad \text{and} \quad q\big|_{x=L} = h(T_L - T_\infty)$$

then the functional I becomes

$$I = -\int_0^L \frac{dT}{dx} Ak \frac{dT}{dx} dx - \int fAT\,dx + \frac{1}{2} h (T_L - T_\infty)^2 \tag{12.1.96}$$

We have an expression for I ready to use if the temperature $T = T(x)$ has an explicit form in x that can be substituted into the equation so that we can carry on the integration.

Next, consider the function $I^{(e)}$ for an element rather than for the total system:

$$I^{(e)} = -\int_{x_1}^{x_2} \frac{dT}{dx} A\kappa \frac{dT}{dx} dx + TA\kappa \frac{dT}{dx}\bigg|_{x_1}^{x_2} + \int_{x_1}^{x_2} fAT\,dx \tag{12.1.97}$$

$$I^{(e)} = I_1^{(e)} + I_2^{(e)} + I_3^{(e)} \tag{12.1.98}$$

where $I_1^{(e)}$, $I_2^{(e)}$, and $I_3^{(e)}$ correspond to the three terms of element $I^{(e)}$, respectively. To develop all the $I_1(e)$ terms we need to find an expression for the temperature T. Assume a linear interpolation for the temperature between x_1 and x_2, as the distance between these two points is assumed small. A representation of the temperature is shown in Fig. 12.1.15, where the temperature varies linearly as

$$T = ax + b \tag{12.1.99}$$

At each node, the temperature is assumed to be T_1 and T_2, respectively. We can write the temperature equation for each node as

$$T_1 = ax_1 + b \quad \text{and} \quad T_2 = ax_2 + b \tag{12.1.100}$$

from which we can solve for a and b:

$$a = \frac{T_2 - T_1}{L_e} \quad \text{and} \quad b = T_1 - \frac{T_2 - T_1}{L_e} x_1$$

where L_e denotes the length of the element $(x_2 - x_1)$. Substituting the values of a and b into Eq. (12.1.100), we write an expression for T by introducing shape functions:

$$T = T_1 N_1 + T_2 N_2 \tag{12.1.101}$$

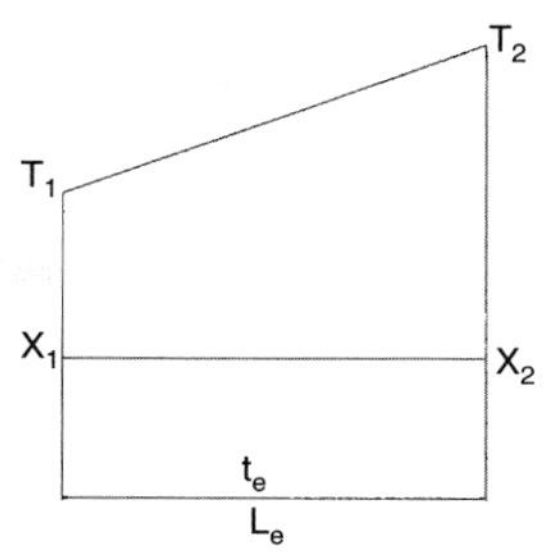

Figure 12.1.15 Linear interpolation of the temperature.

where

$$N_1 = \frac{x_2 - x}{L_e} \quad \text{and} \quad N_2 = \frac{x - x_1}{L_e} \tag{12.1.102}$$

The latter are known as shape functions. These functions are linear in x and represent the characteristic of the function assumed in representing the temperature between x_1 and x_2.

In matrix form, the temperature from Eq. (12.1.101) can be expressed as

$$T = [T_1 \quad T_2] \begin{bmatrix} N_1 \\ N_2 \end{bmatrix} \tag{12.1.103}$$

We note in Eq. (12.1.97) that the time derivative of T is also required, hence the derivative of T as given by Eq. (12.1.103) takes on the following form:

$$\frac{dT}{dx} = [T_1 \quad T_2] \begin{bmatrix} \dfrac{dN_1}{dx} \\ \dfrac{dN_2}{dx} \end{bmatrix} \tag{12.1.104}$$

$$\frac{dN_1}{dx} = -\frac{1}{L_e} \quad \text{and} \quad \frac{dN_2}{dx} = \frac{1}{L_e} \tag{12.1.105}$$

The functional $I^{(e)}$ then becomes

$$I^{(e)} = \int_{x_1}^{x_2} [T_1 \quad T_2] \begin{bmatrix} dN_1/dx \\ dN_2/dx \end{bmatrix} A\kappa \begin{bmatrix} \dfrac{dN_1}{dx} & \dfrac{dN_2}{dx} \end{bmatrix} \begin{bmatrix} T_1 \\ T_2 \end{bmatrix} dx$$

$$-\int_{x_1}^{x_2} fA[T_1 \quad T_2] \begin{bmatrix} N_1 \\ N_2 \end{bmatrix} dx + \frac{1}{2} h(T_L - T_\infty) \tag{12.1.106}$$

Here the boundary conditions at both ends are defined by the last term in the above equation.

Let the first term be I_1^e and defined by

$$I_1^e = A\kappa \int_{x_1}^{x_2} [T_1\, T_2] \begin{bmatrix} \left(\dfrac{dN_1}{dx}\right)^2 & \dfrac{dN_1}{dx}\dfrac{dN_2}{dx} \\ \dfrac{dN_2}{dx}\dfrac{dN_1}{dx} & \left(\dfrac{dN_2}{dx}\right)^2 \end{bmatrix} \begin{bmatrix} T_1 \\ T_2 \end{bmatrix} dx \tag{12.1.107}$$

Substituting the Eq. (12.1.105) derivatives into Eq. (12.1.107) and integrating yields

$$I_1^e = A\kappa\ [T_1 \quad T_2] \begin{bmatrix} \dfrac{1}{L_e} & -\dfrac{1}{L_e} \\ -\dfrac{1}{L_e} & \dfrac{1}{L_e} \end{bmatrix} \begin{bmatrix} T_1 \\ T_2 \end{bmatrix} \tag{12.1.108}$$

Similarly, let $I_2^{(e)}$ denote the term defined in Eq. (12.1.108). Evaluating this term we obtain the term that involves the contribution of the heat source f:

$$I_2^{(e)} = \int_{x_1}^{x_2} fA[T_1 \quad T_2]\begin{bmatrix} N_1 \\ N_2 \end{bmatrix} dx = \frac{fAL_c}{2}[T_1 \quad T_2]\begin{bmatrix} 1 \\ 1 \end{bmatrix} \quad (12.1.109)$$

Next, writing the steady-state condition for an element we get

$$\left\{\frac{\partial I^{(e)}}{\partial T_e}\right\} = \{0\} \quad \text{which yields} \quad k_c = \frac{kA}{L_e}\begin{bmatrix} 1 & -1 \\ -1 & 1 \end{bmatrix} \quad (12.1.110)$$

and the element loading vector from the second term $I_{2,}^{(e)}$

$$f_Q = \frac{fAL_e}{2}\begin{bmatrix} 1 \\ 1 \end{bmatrix} \quad (12.1.111)$$

Combining the last equations, we obtain the first step in the finite-element formulation where

$$[k_c]\{T_{X\text{out}}\} = f_2 \quad (12.1.112)$$

Note how $[k_c]$ is analogous to a local stiffness matrix (we refer to it as the conductivity matrix), how $\{T\}$ is analogous to nodal displacements, and how $\{f_e\}$ is analogous to nodal forces. With this in mind, we can use a table analogous to Table 12.1.3 in the analysis of trusses, to develop the connectivity relations between elements. The latter will assist us in the formulation of the global stiffness matrix and the heat source array. The global problem can be stated as

$$[K]\{T\} = \{F\} \quad (12.1.113)$$

where $[K]$ is the global conductivity matrix (equivalent to the global stiffness) assembled from the element conductivity matrix k_c, $\{T\}$ the nodal temperatures, and $\{F\}$ the heat source contribution.

Boundary Condition Contribution

The term in the function I in Eq. (12.1.93) dealing with the convection can be written further as $\frac{1}{2}T_L hT_L - (hT_\infty)T_L + \frac{1}{2}hT_\infty^2$, where we see the last term drop out from the variational $\partial I/\partial T$.

We see that the hT_L term will be added to the K matrix at the (L, L) location and hT_∞ will be added to the F vector at the Lth location. The way the K and F will be formulated is shown below:

$$\begin{bmatrix} K_{11} & \cdots & \cdots & K_{1L} \\ K_{21} & \cdots & \cdots & K_{2L} \\ \vdots & & & \vdots \\ K_{L1} & & & K_{LL} + hT_L \end{bmatrix}\begin{bmatrix} T_1 \\ T_2 \\ \vdots \\ T_L \end{bmatrix} = \begin{bmatrix} F_1 \\ F_2 \\ \vdots \\ F_L + hT_\infty \end{bmatrix} \quad (12.1.114)$$

Handling of Additional Constraints

The handling of specified temperature boundary conditions such as $T_L = T_0$ can be accompanied by either the elimination or penalty approach. The procedure for elimination is demonstrated below.

Elimination Approach This technique consists of eliminating rows and columns of the corresponding temperature and then modifying the force vector to include the boundary force displacement relation as described in the finite-element solution of trusses. In general, we write the global problem as

$$KU = F \quad (12.1.115)$$

Consider the constraint where the displacement is defined by $U_1 = c_1$. The global displacement vector is array of order $1 \times n$. $U = [U_1 \quad U_2 \quad U_3 \cdots U_n]^T$, and similarly the global force vector is $F = [F_1 \quad F_2 \quad F_3 \cdots F_n]^T$.

We first start by defining the potential energy π as function of elastic energy and the work associated with F:

$$\pi = \frac{1}{2}U^T KU - U^t F \quad (12.1.116)$$

This is one way of obtaining the balance equation of force/displacement as stated previously. The energy explicit matrix form is further shown to be expressed as

$$\pi = \frac{1}{2}\begin{pmatrix} U_1K_{11}U_1 + U_1K_{12}U_2 + \cdots + U_2K_{1N}U_N \\ U_2K_{21}U_1 + U_2K_{22}U_2 + \cdots + U_2K_{2N}U_N \\ \vdots \\ U_NK_{N1}U_1 + U_NK_{N2}U_2 + \cdots + U_NK_{NN}U_N \end{pmatrix} - (U_1F_1 + U_2F_2 + \cdots + U_NF_N) \quad (12.1.117)$$

Let us substitute the boundary condition $U_1 = C_1$. Then we get

$$\pi = \frac{1}{2}\begin{pmatrix} C_1K_{11}C_1 + C_1K_{12}U_2 + \cdots + C_1K_{1N}U_N \\ U_2K_{21}C_1 + U_2K_{22}U_2 + \cdots + U_2K_{2N}U_N \\ \vdots \\ U_NK_{N1}C_1 + U_NK_{N2}U_2 + \cdots + U_NK_{NN}U_N \end{pmatrix} - (C_1F_1 + U_2F_2 + \cdots + U_NF_N) \quad (12.1.118)$$

To solve the problem at hand we need to minimize π, hence $\partial\pi/\partial U_i = 0$, where $i = 1, 2, 3, \ldots, N$, which yields

$$\begin{bmatrix} K_{22} & K_{23} & \cdots & K_{2N} \\ K_{32} & K_{33} & \cdots & K_{3N} \\ \vdots & & & \\ K_{N2} & K_{N3} & & K_{NN} \end{bmatrix}\begin{Bmatrix} U_2 \\ U_3 \\ \vdots \\ U_N \end{Bmatrix} = \begin{Bmatrix} F_2 - K_{21}C_1 \\ F_3 - K_{31}C_1 \\ \\ F_N - K_{N1}C_1 \end{Bmatrix} \quad (12.1.119)$$

The above matrix equation represents a set of $(N-1)\times(N-1)$ equations with the original K matrix being reduced by eliminating the first row and first column as well as the first element of the vector F. We see that the boundary conditions are subtracted from the force vector array by premultiplying it with the elements of the first row of K. This technique will be used in the solutions of heat transfer problems.

Penalty Approach An alternative to the elimination approach is the **penalty approach**. In handling constraints, this might be easier to implement and works well for multiple constraints. The methods are designed to handle the boundary conditions once the global problem has been formulated. Once more, let the boundary conditions be given by the displacement at node 1 such that $U_1 = C_1$.

The total potential energy is then defined by adding an extra term to account for the additional boundary condition or simply to account for the additional energy contribution from the boundary conditions:

$$\pi = \frac{1}{2}u^T ku - u^T F + \frac{1}{2}Q(u_1 - c_1)^2 \quad (12.1.120)$$

So, the energy term $\frac{1}{2}Q(U_1 - C_1)^2$ is significant only if the value of Q is large enough to emphasize the contribution of $(U_1 - C_1)$.

Minimization of π results in

$$\begin{bmatrix} K_{11}+Q & K_{12} & \cdots & K_{1N} \\ K_{21} & K_{22} & \cdots & K_{2N} \\ \vdots & \vdots & \vdots & \vdots \\ K_{N1} & K_{N2} & & K_{NN} \end{bmatrix}\begin{bmatrix} U_1 \\ U_2 \\ \vdots \\ U_N \end{bmatrix} = \begin{bmatrix} F_1 + QC_1 \\ F_2 \\ \vdots \\ F_N \end{bmatrix} \quad (12.1.121)$$

On examining the above equation, it is apparent that Q is added to K_{11} in the global K matrix and Q_c is added to the element force vector F_1. We can view Q as a stiffness value whose numerical values can be defined or selected by noting that

$$(K_{11} + Q)u_1 + K_{12}U_2 + K_{13}U_3 + \cdots + K_{1N}U_N = F_1 + QC_1 \quad (12.1.122)$$

If we divide by Q we obtain

$$\left(\frac{K_{11}}{Q}+1\right)U_1+\frac{K_{12}}{Q}U_2+\cdots+\frac{K_{1N}}{Q}U_N=\frac{F_1}{Q}+C_1 \quad (12.1.123)$$

Observe how, if Q is chosen to be a large volume, the equation reduces to $U_1 = C_1$, which is the desired boundary condition. We also see further that Q is large in comparison to $K_{11}, K_{12}, \ldots, K_{1N}$, hence we need to select Q large enough to satisfy the condition of the equation above. A suggested value in previous work is K_{ij}, $Q = \max |K_{1j}| \times 10^4$.

EXAMPLE Determine the temperature distribution in the composite wall used to isolate the outside (Fig. 12.1.16). Convection heat transfer on the inner surface of the wall with $T_\infty = 500°C$ is given by $h = 25\ W/m^2 \cdot °C$. The conductivity constants for each wall are $K_1 = 20\ W/m \cdot °C$, $K_2 = 30\ W/m \cdot °C$, and $K_3 = 40\ W/m \cdot °C$, respectively. Let the cross-sectional area $A = 1\ m^2$ and $L_1 = 0.4$ m, $L_2 = 0.3$ m, $L_3 = 0.1$ m.

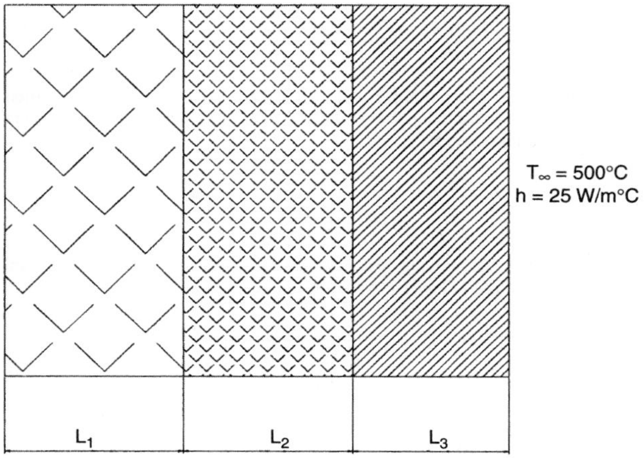

Figure 12.1.16

This example is used to demonstrate not only how to build the conductivity stiffness matrices and the loading vector F but how to implement the technique that describes how the boundary conditions are employed.

Solution: Let the temperature at each wall be denoted by T_1 and let the width of the wall represent the length of each element.

We need to compute the local conductivity stiffness for each element. Since the conductivity constant is given per unit length, we write

$$K^{(1)}=\frac{K_1}{L_1}\begin{bmatrix}1&-1\\-1&1\end{bmatrix} \quad K^{(2)}=\frac{K_2}{L_2}\begin{bmatrix}1&-1\\-1&1\end{bmatrix} \quad K^{(3)}=\frac{K_3}{L_3}\begin{bmatrix}1&-1\\-1&1\end{bmatrix} \quad (12.1.124)$$

The global K is found by assembling $K^{(1)}$, $K^{(2)}$, and $K^{(3)}$, which results in

$$K=50\begin{bmatrix}1&-1&0&0\\-1&3&-2&0\\0&-2&10&-8\\0&0&-8&8\end{bmatrix}$$

Since convection occurs at node 1, we add $h = 25$ to the (1, 1) location in K, which results in

$$K=50\begin{bmatrix}1.5&-1&0&0\\-1&3&-2&0\\0&-2&10&-8\\0&0&-8&8\end{bmatrix}$$

We have no heat source in the problem, so the F vector consists only of hT_∞: $\{F\} = [25 \times 500, 0, 0, 0]$.

Applying the boundary condition $T_4 = 10°C$ can be handled by the penalty approach. Let us choose a value for Q from the previously proposed procedure, where $Q = \max |K_{ij}| \times 10^4 = 50 \times 10 \times 10^4$. As stated, in the penalty function we add the Q value to the K matrix in the (4×4) location, and QT_4 to the (4×1) location of the F vector, resulting in

$$K=50\begin{bmatrix}1.5&-1&0&0\\-1&3&-2&0\\0&-2&10&-8\\0&0&-8&100{,}008\end{bmatrix}\begin{Bmatrix}T_1\\T_2\\T_3\\T_4\end{Bmatrix}=\begin{Bmatrix}25\times 500\\0\\0\\5\times 10^7\end{Bmatrix}$$

The solution of which is found to be

$$T = [229.6559 \quad 94.4839 \quad 26.8979 \quad 10.0014]°C$$

Finite-Difference Approach

Finite difference is discussed briefly in the following example for the purpose of validating the one-dimensional solution we have derived.

EXAMPLE A special design for a building wall consists of three studs containing the materials siding, sheathing, and insulation batting. The inside room temperature is maintained at 85°F and the outside air temperature is measured at 15°F. The area of the wall exposed to air is 180 ft². Determine the temperature distribution through the wall.

Table 12.1.7 Characteristics of the Wall

Items	Resistance, h · ft² · F/Btu	U factor, Btu/h · ft² · °F
Outside film resistance	0.17	5.88
Siding	0.81	1.23
Sheathing	1.32	0.76
Insulation	11.0	0.091
Inside film resistance	0.68	1.47

The steady-state condition of this system can be explained through Fourier's law: $q_x = -kA\ \partial T/\partial X$. We can express the gradient of temperature by $(T_{i+1} - T_i)/l$, and the heat transfer rate becomes $q = [kA(T_{i+1} - T_i)]/l$, or $q = UA(T_{i+1} - T_i)$, where U is defined by k/l.

The heat transfer between the surface and fluid is due to convection. Newton's law of cooling governs the heat transfer rate between the fluid and the surface: $q = hA(T_s - T_f)$, where h is the convection coefficient.

The heat loss through the wall due to conduction must equal the heat loss to the surrounding cold air by convection. That is, $-kA\,\partial T/\partial X = hA(T_s - T_f)$.

Expanding the equation on the temperature distribution at the edge of each wall leads to the following equations:

$$U_2A(T_3-T_2)=U_1A(T_2-T_1) \qquad U_3A(T_4-T_3)=U_2A(T_3-T_2)$$
$$U_4A(T_5-T_4)=U_3A(T_4-T_3) \qquad U_5A(T_6-T_5)=U_4A(T_5-T_4)$$

Expressing the above in a matrix form, we get

$$A\begin{bmatrix}u_1+u_2&-u_2&0&0\\-u_2&u_2+u_3&-u_3&0\\0&-u_3&u_3+u_4&-u_4\\0&0&-u_4&u_4+u_5\end{bmatrix}\begin{bmatrix}T_2\\T_3\\T_4\\T_5\end{bmatrix}\begin{bmatrix}u_1AT_1\\0\\0\\u_5AT_6\end{bmatrix}$$

For the numerical values given, we define the four equations in four unknown as

$$180\begin{bmatrix}7.11&-1.23&0&0\\-1.23&1.99&-0.76&0\\0&-0.76&0.851&-0.091\\0&0&-0.091&1.561\end{bmatrix}\begin{bmatrix}T_2\\T_3\\T_4\\T_5\end{bmatrix}=\begin{bmatrix}15{,}876\\0\\0\\22{,}491\end{bmatrix}$$

The solution is found to be

$$[T_2 \quad T_3 \quad T_4 \quad T_5] = [15.8523 \quad 19.9266 \quad 26.5205 \quad 81.5909]°C$$

FORMULATION OF GLOBAL STIFFNESS MATRIX FOR *N* ELEMENTS

Let us consider a body discreditized into N one-dimensional elements, as shown in Fig. 12.1.17. Let the boundary conditions be such that

$$T_1 = T_{N+1} = 0 \tag{12.1.125}$$

The connectivity of the rod elements, which is determined by the same procedure outlined in Table 12.1.3 for trusses is used to define the global conductivity matrix, which is of order $(N+1)\times(N+1)$. The ascending order of the elements helps the global K take a special form with a predictable bandwidth.

$$N\begin{bmatrix} 1 & -1 & 0 & & & 0 \\ -1 & 1+1 & -1 & & & \\ 0 & -1 & 1+1 & -1 & \cdots & 0 \\ \vdots & \vdots & \vdots & \vdots & \ddots & \vdots \\ & & & -1 & 1+1 & -1 \\ 0 & & & 0 & -1 & 1 \end{bmatrix}\begin{bmatrix} T_1 \\ T_2 \\ T_3 \\ \vdots \\ T_{N+1} \end{bmatrix} = \begin{bmatrix} \dfrac{1}{2N} \\ \dfrac{1}{2N}+\dfrac{1}{2N} \\ \dfrac{1}{2N}+\dfrac{1}{2N} \\ \vdots \\ \dfrac{1}{2N} \end{bmatrix} \tag{12.1.126}$$

where f is assumed 1 for simplicity. Applying the boundary conditions reduces the problem to

$$\begin{bmatrix} 2 & -1 & 0 & & & 0 \\ -1 & 2 & -1 & & & \\ 0 & -1 & 2 & -1 & \cdots & 0 \\ \vdots & \vdots & \vdots & \vdots & \ddots & \vdots \\ & & -1 & 2 & & -1 \\ & & 0 & -1 & & 2 \end{bmatrix}\begin{bmatrix} T_2 \\ T_3 \\ \vdots \\ T_{N-1} \\ T_N \end{bmatrix} = \frac{1}{N^2}\begin{bmatrix} 1 \\ 1 \\ 1 \\ \vdots \\ 1 \\ 1 \end{bmatrix} \tag{12.1.127}$$

EXAMPLE For the one-dimensional heat transfer problem given by $d^2T/dx^2 = -10$, where $A = 1$ $(0 \le x \le 1$ with $T(0) = 0)$, find the temperature at $x = 0.2$, 0.4, 0.6, 0.8, and 1.0 m (Fig. 12.1.18).

Solution: Each element has an element conductivity matrix K_e of the form:

$$K_e = \frac{kA}{L_e}\begin{bmatrix} 1 & -1 \\ -1 & 1 \end{bmatrix}$$

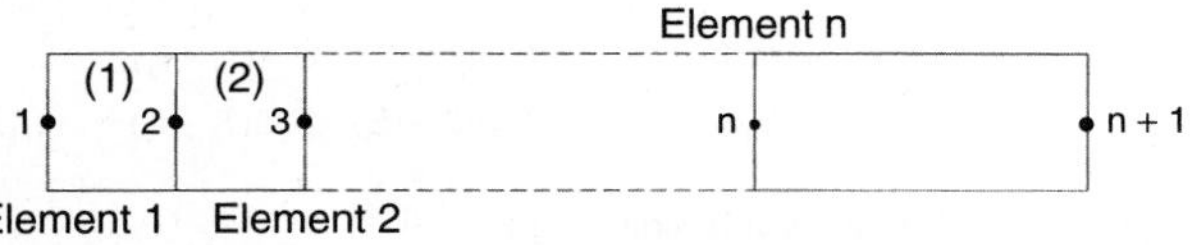

Figure 12.1.17 Discretization of a heat conduction rod into N elements.

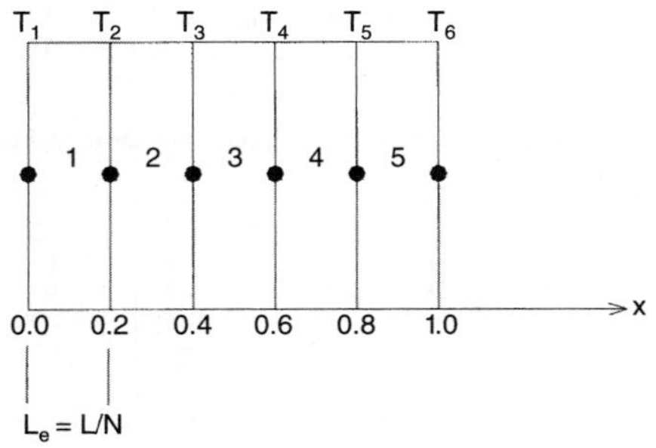

Figure 12.1.18

Substituting $A = 1$ m^2 and $L_e = L/N = \frac{1}{5} = 0.2$ m, and assuming the conductivity constant to be $k = 1$, we evaluate the element conductivity matrix:

$$K_e = 5\begin{bmatrix} 1 & -1 \\ -1 & 1 \end{bmatrix}$$

Using the connectivity table, we expand the local matrices K_s^e to an order of 5×5, and the global matrix $[K]$ is obtained by summation:

$$[K] = 5\begin{bmatrix} 1 & -1 & 0 & 0 & 0 & 0 \\ -1 & 2 & -1 & 0 & 0 & 0 \\ 0 & -1 & 2 & -1 & 0 & 0 \\ 0 & 0 & -1 & 2 & -1 & 0 \\ 0 & 0 & 0 & -1 & 2 & -1 \\ 0 & 0 & 0 & 0 & -1 & 1 \end{bmatrix} \qquad T = \begin{bmatrix} 0 \\ T_2 \\ T_3 \\ T_4 \\ T_5 \\ 0 \end{bmatrix}$$

With application of the boundary conditions, the global temperature vector becomes $\{T\}$, shown above.

The forcing vector for an element is shown to be

$$F_e = \frac{fAL_e}{2}\begin{bmatrix} 1 \\ 1 \end{bmatrix}$$

where f is the heat generation per unit volume and is obtained from the relation $\kappa\, d^2T/dx^2 = -f$. Substituting $\kappa = 1$ and $d^2T/dx^2 = -10$ yields $f = 10$. Substituting into F^e those values, we get

$$F_e = \begin{bmatrix} 1 \\ 1 \end{bmatrix}$$

Assembling the global forcing vector, using the connectivity table, yields F, and using the relation $[K]\{T\} = \{F\}$, we write

$$5\begin{bmatrix} 1 & -1 & 0 & 0 & 0 & 0 \\ -1 & 2 & -1 & 0 & 0 & 0 \\ 0 & -1 & 2 & -1 & 0 & 0 \\ 0 & 0 & -1 & 2 & -1 & 0 \\ 0 & 0 & 0 & -1 & 2 & -1 \\ 0 & 0 & 0 & 0 & -1 & 1 \end{bmatrix}\begin{bmatrix} T_1 \\ T_2 \\ T_3 \\ T_4 \\ T_5 \\ T_6 \end{bmatrix} = \begin{bmatrix} 1 \\ 2 \\ 2 \\ 2 \\ 2 \\ 1 \end{bmatrix} = F = \begin{bmatrix} 1 \\ 1+1 \\ 1+1 \\ 1+1 \\ 1+1 \\ 1 \end{bmatrix}$$

By deleting the first and last rows together with their corresponding columns, and modifying the force vector, we obtain

$$\begin{bmatrix} 2 & -1 & 0 & 0 \\ -1 & 2 & -1 & 0 \\ 0 & -1 & 2 & -1 \\ 0 & 0 & -1 & 2 \end{bmatrix}\begin{bmatrix} T_2 \\ T_3 \\ T_4 \\ T_5 \end{bmatrix} = \begin{bmatrix} 0.4 \\ 0.4 \\ 0.4 \\ 0.4 \end{bmatrix}$$

Note that $T_2 = T_5$ and $T_3 = T_4$. From symmetry, we can solve the equation very easily. The solutions are as follows:

$$\begin{bmatrix} T_1 \\ T_2 \\ T_3 \\ T_4 \\ T_5 \\ T_6 \end{bmatrix} = \begin{bmatrix} 0 \\ 0.8 \\ 1.2 \\ 1.2 \\ 0.8 \\ 0 \end{bmatrix} \,^\circ\text{C}$$

2-D HEAT CONDUCTION ANALYSIS

In a fashion similar to the one-dimensional analysis, the finite-element method can be used to analyze 2-D and 3-D heat conduction problems. Let us examine the 2-D case and bear in mind that, first, we need to develop the element temperature relationship and then expand it to the global problem.

The heat conduction problem is formulated by a variational boundary value problem as

$$\delta I = 0 \tag{12.1.128}$$

where

$$I = \frac{1}{2}\int_{\Omega}[k(\nabla T)^2 + 2fT]d\Omega \tag{12.1.129}$$

and where k = thermal conductivity, which we assume is constant; f = heat source; ∇T = temperature gradient; $(\nabla T)^2 = \nabla T \cdot \nabla T$ (a dot product); Ω = domain of interest. Note that the heat convection part is neglected. If domain Ω is divided into N elements, as shown in Fig. 12.1.19, then

$$I = \sum_{e=1}^{N} I^e \tag{12.1.130}$$

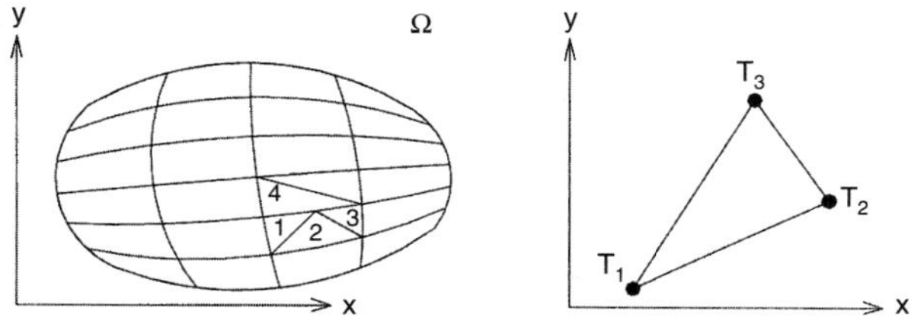

Figure 12.1.19

where

$$I^e = \frac{1}{2}\int_{e}[k(\nabla T^e)^2 + 2f^eT^e]d\Omega \tag{12.1.131}$$

Let us consider the triangular element shown in Fig. 12.1.19. The local representation of the temperature can be expressed as

$$T(x,y) = T_1N_1 + T_2N_2 + T_3N_3 \tag{12.1.132}$$

where $N_i(x, y)$ $(i = 1, 2, 3)$ are the shape functions given by

$$N_i^e = a_i^e + b_i^e x + c_i^e y \tag{12.1.133}$$

The shape functions must satisfy the following conditions:

1. $N_i^e(x, y)$ are linear in both x and y.
2. $N_i^e(x, y)$ have the value 1 at node i and 0 at other nodes.
3. $N_i^e(x, y)$ are 0 at all points in Ω, except those of $N_i^e(x, y)$ can be written as

$$N_i(x, y) = [a_i \quad b_i \quad c_i]\begin{bmatrix} 1 \\ x \\ y \end{bmatrix} \tag{12.1.134}$$

and for the three nodes of the triangular element, we can write

$$\begin{bmatrix} N_1 \\ N_2 \\ N_3 \end{bmatrix} = \begin{bmatrix} a_1 & b_1 & c_1 \\ a_2 & b_2 & c_2 \\ a_3 & b_3 & c_3 \end{bmatrix}\begin{bmatrix} 1 \\ x \\ y \end{bmatrix} \tag{12.1.135}$$

For node 1, following condition 1, Eq. (12.1.135) yields

$$\begin{aligned} N_1 &= 1 = a_1 + b_1x_1 + c_1y_1 \\ N_2 &= 0 = a_1 + b_1x_2 + c_1y_2 \\ N_3 &= 0 = a_1 + b_1x_3 + c_1y_3 \end{aligned} \tag{12.1.136}$$

Which can be written in matrix form as

$$\begin{bmatrix} 1 \\ 0 \\ 0 \end{bmatrix} = [A]\begin{bmatrix} a_1 \\ b_1 \\ c_1 \end{bmatrix} \tag{12.1.137}$$

where

$$[A] = \begin{bmatrix} 1 & x_1 & y_1 \\ 1 & x_2 & y_2 \\ 1 & x_3 & y_3 \end{bmatrix} \tag{12.1.138}$$

Solving for coefficients a, b, and c, we get

$$\begin{bmatrix} a_1 \\ b_1 \\ c_1 \end{bmatrix} = \begin{bmatrix} 1 & x_1 & y_1 \\ 1 & x_2 & y_2 \\ 1 & x_3 & y_3 \end{bmatrix}^{-1}\begin{bmatrix} 1 \\ 0 \\ 0 \end{bmatrix} = [A]^{-1}\begin{bmatrix} 1 \\ 0 \\ 0 \end{bmatrix} \tag{12.1.139}$$

Similarly, for the interpolation functions N_2 and N_3, we get

$$\begin{bmatrix} a_2 \\ b_2 \\ c_2 \end{bmatrix} = [A]^{-1}\begin{bmatrix} 0 \\ 1 \\ 0 \end{bmatrix} \quad \text{and} \quad \begin{bmatrix} a_3 \\ b_3 \\ c_3 \end{bmatrix} = [A]^{-1}\begin{bmatrix} 0 \\ 0 \\ 1 \end{bmatrix} \tag{12.1.140}$$

The inverse of matrix A is

$$[A]^{-1} = \frac{1}{2a}\begin{bmatrix} (x_2y_3 - x_3y_2) & (y_2 - y_3) & (x_3 - x_2) \\ (x_3y_1 - x_1y_3) & (y_3 - y_1) & (x_1 - x_3) \\ (x_1y_2 - x_2y_1) & (y_1 - y_2) & (x_2 - x_1) \end{bmatrix} \tag{12.1.141}$$

where a is the area of the triangle. By combining Eqs. (12.1.139) and (12.1.140), we obtain the inverse of A as

$$[A]^{-1} = \begin{bmatrix} a_1 & b_1 & c_1 \\ a_2 & b_2 & c_2 \\ a_3 & b_3 & c_3 \end{bmatrix} \tag{12.1.142}$$

Then the triangle element functions can be written in a more general form:

$$\{N^e\} = \begin{bmatrix} N_1 \\ N_2 \\ N_3 \end{bmatrix} = [A]^{-1}\begin{bmatrix} 1 \\ x \\ y \end{bmatrix} \tag{12.1.143}$$

where

$$\begin{aligned} N_1 &= \frac{1}{2a}\,[(x_2y_3 - x_3y_2) + x(y_2 - y_3) + y(x_3 - x_2)] \\ N_2 &= \frac{1}{2a}\,[(x_3y_1 - x_1y_3) + x(y_3 - y_1) + y(x_1 - x_3)] \\ N_3 &= \frac{1}{2a}\,[(x_1y_2 - x_2y_1) + x(y_1 - y_2) + y(x_2 - x_1)] \end{aligned} \tag{12.1.144}$$

Now that we have defined the shape function, we can proceed in the evaluation of the conductivity matrix of individual elements.

Element Conductivity Matrix

From Eq. (12.1.131), we write the variational equation in terms of elements. This defines the element equation as

$$I^e = \frac{1}{2}\int[k^e(\nabla T)^2 + f^eT^e]\ d\Omega \tag{12.1.145}$$

The temperature at the nodes of the triangle element is expressed following the triangular element assumption developed in the previous section where

$$T(x,\ y) = T_1N_1 + T_2N_2 + T_3N_3 \tag{12.1.146}$$

From Eq. (12.1.144), we define the partial derivatives with respect to x and y as

$$\frac{\partial N_i}{\partial x} = b_i \quad \text{and} \quad \frac{\partial N_i}{\partial y} = c_i \tag{12.1.147}$$

Hence, we can write the gradient of the temperature as follows

$$\nabla T = \begin{bmatrix} \dfrac{\partial T}{\partial x} \\ \dfrac{\partial T}{\partial y} \end{bmatrix} = \begin{bmatrix} \dfrac{\partial N_1}{\partial x} & \dfrac{\partial N_2}{\partial x} & \dfrac{\partial N_3}{\partial x} \\ \dfrac{\partial N_1}{\partial y} & \dfrac{\partial N_2}{\partial y} & \dfrac{\partial N_3}{\partial y} \end{bmatrix}\begin{bmatrix} T_1 \\ T_2 \\ T_3 \end{bmatrix} \tag{12.1.148}$$

which, expressed in compact form, yields $\Delta T = BT$, where

$$B = \begin{bmatrix} b_1 & b_2 & b_3 \\ c_1 & c_2 & c_3 \end{bmatrix} \quad \text{and} \quad T = \begin{bmatrix} T_1 \\ T_2 \\ T_3 \end{bmatrix} \tag{12.1.149}$$

Note how the coefficients of B are obtained from the partial derivatives of the shape functions given by Eq. (12.1.150).

Now we can define the $(\nabla T)^2$ needed in Eq. (12.1.151).

$$(\nabla T)^2 = \begin{bmatrix} \frac{\partial T}{\partial x} & \frac{\partial T}{\partial y} \end{bmatrix} \begin{bmatrix} \frac{\partial T}{\partial x} \\ \frac{\partial T}{\partial y} \end{bmatrix} \tag{12.1.150}$$

For a given element, this equation can be expressed as

$$[\nabla T^e]^2 = \{T^e\}^T[B^e]^T[B^e]\{T^e\} \tag{12.1.151}$$

This yields

$$I_1^e = \frac{1}{2}k\int_e \{T^e\}^T[B^e]^T[B^e]\{T^e\}\, d\Omega \tag{12.1.152}$$

Because $[B]$ and $[T]$ are constant matrices, Eq. (12.1.152) reduces to

$$I_1^e = \frac{1}{2}k\{T^e\}^T[B^e]^T[B^e]\{T^e\}\int_e d\Omega \tag{12.1.153}$$

or simply

$$I_1^e = \frac{1}{2}\{T^e\}^T[k^e]\{T^e\} \tag{12.1.154}$$

where $[k^e]$ denotes the element conductivity matrix:

$$[K^e] = \kappa a[B^e]^T[B^e] \tag{12.1.155}$$

This takes the final form

$$[K^e] = \kappa a \begin{bmatrix} (b_1^2 + c_1^2) & (b_1b_2 + c_1c_2) & (b_1b_3 + c_1c_3) \\ (b_2b_1 + c_2c_1) & (b_2^2 + c_2^2) & (b_2b_3 + c_2c_3) \\ (b_3b_1 + c_3c_1) & (b_3b_2 + c_3c_2) & (b_3^2 + c_3^2) \end{bmatrix} \tag{12.1.156}$$

and a is the area of the triangular element.

Element-Forcing Function

To complete the integration of Eq. (12.1.145), we need to evaluate the second term, I_2^e.

$$I_2^e = \int f^e T^e\, d\Omega \tag{12.1.157}$$

As we have done with temperature, the heat source f can be expressed in a similar fashion:

$$f = f_1N_1 + f_2N_2 + f_3N_3 = [f_1 \quad f_2 \quad f_3] \begin{bmatrix} N_1 \\ N_2 \\ N_3 \end{bmatrix} \tag{12.1.158}$$

For an arbitrary element, this equation can be written in compact matrix form:

$$f^e = \{f^e\}\{N^e\} \tag{12.1.159}$$

Recall that

$$T = T_1N_1 + T_2N_2 + T_3N_3 = [N_1 \quad N_2 \quad N_3] \begin{bmatrix} T_1 \\ T_2 \\ T_3 \end{bmatrix} \tag{12.1.160}$$

or

$$T^e = \{N^e\}^T\{T^e\} \tag{12.1.161}$$

Therefore, I_2^e after substitution becomes

$$I_2^e = \int\{f^e\}\{N^e\}\{N^e\}^T\{T^e\}\, d\Omega = \{g^e\}\{T^e\} \tag{12.1.162}$$

where

$$\{g^e\}^T = \{f^e\}^T \int\{N^e\}\{N^e\}^T\, d\Omega \tag{12.1.163}$$

The integrand $\{N^e\}\ \{N^e\}^T$ yields

$$\{N^e\}\{N^e\}^T = [A]^{-1}\begin{bmatrix} 1 \\ x \\ y \end{bmatrix} [1 \quad x \quad y][A^T]^{-1}$$

$$= [A]^{-1}\begin{bmatrix} 1 & x & y \\ x & x^2 & xy \\ y & xy & y^2 \end{bmatrix}[A^T]^{-1} \tag{12.1.164}$$

Note that $[A]$, given by Eq. (12.1.138), is a constant matrix, therefore the integral of $\{g\}$ depends only on the matrix with variables x and y. Integrating each element of the matrix is rather tedious and time-consuming. An alternative is to use a method developed by Eisenberg and Malvern.

From this method, we have the following statement of the integral:

$$\int_e N_1^m N_2^n N_3^p d\Omega = \frac{m!n!p!}{(m+n+p+2)!}\,2a \tag{12.1.165}$$

$$\{g^e\} = \left[\int_e \{N^e\}\{N^e\}^T\, d\Omega\right]\{f^e\} \tag{12.1.166}$$

Hence,

$$\{g^e\} = \left\{\int \begin{bmatrix} N_1^2 & N_1N_2 & N_1N_3 \\ N_2N_1 & N_2^2 & N_2N_3 \\ N_3N_1 & N_3N_2 & N_3^2 \end{bmatrix} d\Omega\right\}\{f^e\} \tag{12.1.167}$$

which yields

$$\{g^e\} = \frac{a}{12}\begin{bmatrix} 2 & 1 & 1 \\ 1 & 2 & 1 \\ 1 & 1 & 2 \end{bmatrix}\{f^e\} \tag{12.1.168}$$

The element integral of the variational formulation is broken into two parts:

$$I^e = I_1^e + I_2^e \tag{12.1.169}$$

Which simplifies to

$$I^e = \frac{1}{2}\{T^e\}^T[k^e]\{T^e\} + \{g^e\}^T\{T^e\} \tag{12.1.170}$$

The "global integral" over the domain Ω of the entire body becomes

$$I = \sum_{e=1}^{N} I^e = \sum\left\{\frac{1}{2}\{T^e\}^T[k^e]\{T^e\} + \{g^e\}^T\{T^e\}\right\} \tag{12.1.171}$$

or

$$I = \frac{1}{2}\{T\}^T\left[\sum_{3=1}^{N}[k^e]\right]\{T\} - \left[\sum_{e=1}^{N}\{F^e\}^T\right]\{T\} \tag{12.1.172}$$

where $\{F^e\}^T = -\{g^e\}^T$ and $\{T\} = [T_1T_2\ldots T_n]^T$. Hence,

$$I = \frac{1}{2}\{T\}^T[k]\{T\} - \{F\}^T\{T\} \tag{12.1.173}$$

where the global conductivity matrix is defined by

$$[k] = \sum_{e=1}^{N} [k^e] \qquad (12.1.174)$$

and the global function (equivalent to the global force in the analysis of a truss) is

$$\{F\} = \sum_{e=1}^{N} \{F^e\} \qquad (12.1.175)$$

The variation $\delta I = 0$ is equivalent to

$$\frac{\partial I}{\partial T_i} = 0 \quad (i = 1, \ldots, n) \qquad (12.1.176)$$

Applying Eq. (12.1.176) to Eq. (12.1.172) gives the global equation governing the temperature distribution and the heat source:

$$[k]\{T\} = \{F\} \qquad (12.1.177)$$

This equation is similar to our FEM application to the truss and the one-dimensional heat flow problems.

Analysis of 2-D heat conduction problems can be done by using the FEM procedures developed herein. One proceeds by identifying the element shape functions and then evaluate the corresponding local conductivity (stiffness) matrices. The global $[K]$ is then assembled, using Eq. (12.1.174), and the elements' forcing functions are computed, using Eq. (12.1.168). Finally the global array $\{F\}$ is assembled according to Eq. (12.1.175).

REFERENCES

1. Aboudi, J., "Mechanics of Composite Materials," Elsevier, Amsterdam, 1991, 328 pp.
2. Adeli, H., and Soegiarso, R., "High Performance Computing in Structural Engineering," CRC Press, Boca Raton, FL, 1998, 272 pp.
3. Akin, J. E., "Finite Element Analysis for Undergraduates," Academic Press, London, 1986, 319 pp.; "Finite Elements for Analysis and Design," Academic Press, London, 1994.
4. Argyris, J. H., and Mlejnek, H. P., "Dynamics of Structures," North-Holland, Amsterdam, 1991.
5. Atanackovic, T., and Guran, A., "Theory of Elasticity for Scientists and Engineers," Birkhauser Verlag, Basel, 2000, 386 pp.
6. Atluri, S. N., et al., eds. "Hybrid and Mixed Finite Element Methods," J. Wiley & Sons, Chichester, 1983, 582 pp.
7. Aziz, A. K. ed. "The Mathematical Foundations of the Finite Element Method with Applications to Partial Differential Equations," Academic Press, New York, 1972, 797 pp.
8. Bathe, K. J., "Finite Element Procedures," Prentice-Hall, Englewood Cliffs, NJ, 1995, 1037 pp.
9. Bathe, K. J., Oden, J. T., and Wunderlich, W. eds. "Formulations, and Computational Algorithms in Finite Element Analysis," MIT Press, Cambridge, 1977, 1091 pp.
10. Becker, E. B., Carey, G. F., and Oden, J. T., "Finite Elements—An Introduction," Prentice-Hall, Englewood Cliffs, NJ, 1981, 255 pp.
11. Belytschko, T., Liu, W. K., and Moran, B., "Nonlinear Finite Elements for Continua and Structures," J. Wiley & Sons, New York, 2000, 600 pp.
12. Bergan, P. G., Bathe, K. J., and Wunderlich, W., eds. "Finite Element Methods for Nonlinear Problems," Springer Verlag, Berlin, 1986, 817 pp.
13. Bianchi, N., "Electrical Machine Analysis Using Finite Elements," CRC Press, Boca Raton, FL, 2005, 275 pp.
14. Bickford, W., "Advanced Mechanics of Materials," Addison Wesley, Reading, MA, 1998, 600 pp.
15. Boyle, J. T., Brown, D. K., Mair, B., Mehta, P., and Wood, J., eds. "Finite Element Analysis: Education and Training," Chapman & Hall, London, 1991, 244 pp.
16. Brebbia, C. A. ed. "Finite Element Systems—A Handbook," Springer Verlag, New York, 1981, 1983, 1985, 767 pp.
17. Carey, G. F., and Oden, J. T., "Finite Elements—Computational Aspects," Vol. III, Prentice-Hall, Englewood Cliffs, NJ, 1984, 350 pp.
18. Cheung, Y. K., and Hutton, S. G., eds. "Finite Element Methods in Engineering," AIAA, New York, 1977.
19. Cheung, Y. K., Lo, S. H., and Leung, Y. T., "Finite Element Implementation," Blackwell Science, Oxford, 1995, 384 pp.
20. Cook, R., and Young, W., "Advanced Mechanics of Materials," Prentice-Hall, Upper Saddle River, NJ, 1999, 496 pp.
21. Dym, C. L. "Structural Modeling and Analysis," Cambridge University Press, Cambridge, 1997, 281 pp.
22. Glowinski, R. et al., eds. "Energy Methods in Finite Element Analysis," J. Wiley & Sons, New York, 1979, 361 pp.
23. Hsu, T. R., "The Finite Element Method in Thermodynamics," Allen & Unwin, Boston, 1986, 391 pp.
24. Hughes, T. J. R., Onate, E., and Zienkiewicz, O. C., eds. Recent Developments in Finite Element Analysis, a Book Dedicated to Robert L. Taylor."
25. Kardestuncer, H., and Norrie, D. H., eds. "Finite Element Handbook," McGraw-Hill, New York, 1987, 1424 pp.
26. Kythe, P. K. K., and Wei, D., "An Introduction to Linear and Nonlinear Finite Element Analysis," Birkhauser Verlag, Basel, 2003, 456 pp.
27. Logan, D., "First Course in the Finite Element Method Using Algor," PWS Publishing, Boston, 1997, 908 pp.
28. Oden, J. T., and Reddy, J. N., "An Introduction to the Mathematical Theory of Finite Elements," J. Wiley & Sons, Chichester, 1976, 429 pp.
29. Ogden, R. W., "Non-Linear Elastic Deformations," Dover Publications, Mineola, NY, 1997 (2nd ed.), 544 pp.
30. Papadrakakis, M., and Topping, B. H. V., eds. "Advances in Finite Element Techniques," Saxe-Coburg Publ., Edinburgh, 1994, 235 pp.
31. Pozrikidis, C., "Introduction to Finite and Spectral Element Methods Using MATLAB," CRC Press, Boca Raton, FL, 2005, 653 pp.
32. Reddy, J. N., "An Introduction to Nonlinear Finite Element Analysis," Oxford University Press, Oxford, 2004, 482 pp.
33. Reddy, J. N., ed. "Mechanics of Composite Materials: Selected Works of Nicholas J. Pagano," Kluwer Academic Publ., Dordrecht, 1994, 460 pp.
34. Robinson, J., ed. "FEM Today and in the Future," Robinson and Assoc., Wimborne, Dorset, U.K., 1993.
35. Shames, I. H., and Dym, C. L., "Energy and Finite Element Methods in Structural Mechanics," SI Unit Edition Taylor & Francis, PA, 1995, 750 pp.
36. Stasa, F. L., "Applied Finite Element Analysis for Engineers," CBS Publ. Japan, Ltd., New York, 1985, 658 pp., distributed by Oxford University Press.
37. Steele, J. M., "Applied Finite Element Modeling—Practical Problem Solving for Engineers," Marcel Dekker Inc., New York, 1989, 361 pp.
38. Topping, B. H. V., ed. "Advances in Finite Element Procedures and Techniques," Saxe-Coburg Publ., Edinburgh, 1998, 298 pp.
39. White, R. E., "An Introduction to the Finite Element Method with Applications to Nonlinear Problems," J. Wiley & Sons, New York, 1985, 354 pp.
40. Wiberg, N. E., ed. "Advances in Finite Element Technology," CIMNE, Barcelona, 1995, 305 pp.
41. Yamada, Y., and Gallagher, R. H., eds. "Theory and Practice in Finite Element Structural Analysis," University of Tokyo Press, Tokyo, 1973, 733 pp.
42. Zahavi, E., "The Finite Element Method in Machine Design," Prentice-Hall, Englewood Cliffs, NJ, 1992, 320 pp.
43. Zienkiewicz, O. C., and Morgan, K., "Finite Elements and Approximation," J. Wiley & Sons, New York, 1983, 328 pp.
44. Zienkiewicz, O. C., and Taylor, R. L., "Finite Element Method—Basic Formulation and Linear Problems," Vol. 1, McGraw-Hill, New York, 1989, 648 pp.

12.2 COMPUTER-AIDED DESIGN, COMPUTER-AIDED ENGINEERING, AND VARIATIONAL DESIGN

by Farid Amirouche

INTRODUCTION

Computer-aided design (CAD) has gained tremendous popularity in the past decade, and it is becoming an absolutely necessary tool for any engineering task. Its application is broadening, and its success is due mostly to the mass production of powerful microprocessors and their low cost.

Computer-aided design is the blending of human and machine, working together to achieve the optimum design and manufacture of a product. The computer's graphics capability and computing power allow designers to fashion and test their ideas interactively in real time without having to create real prototypes as in conventional approaches to design.

CAD has revolutionized engineering over the past two decades, and engineering design has become the engine to a successful company. The road between the product in its initial design stages and the mature design ready to go into production is lined with a series of prototypes needed to evaluate the ongoing design. **Virtual reality** offers a novel approach in product development, especially in the early stages of a development process. Virtual models are CAD prototypes that can be displayed on a computer screen for the purpose of getting a view into what reality might be and how to prepare for the product use.

PARAMETRIC AND VARIATIONAL DESIGN

Designing and drafting are the preliminary steps carried out in the manufacture of a component. Traditional mechanical CAD systems are effective tools in improving the efficiency of the drafting process, but they cannot handle the task of design and drafting at the same time. This forces the engineer to think simultaneously about all the design requirements and engineering relationships for each design approach.

Several new CAD systems claim to meet the needs of engineers engaged in the preliminary design of mechanical products. These systems should enhance the level of productivity of engineers in creating or modifying designs. Two approaches that are gaining popularity for providing flexibility in the design process are parametric design and variational design.

Parametric Design Systems

In a parametric design, the engineer selects a set of geometric constraints that can be applied for creating the geometry of the component. The geometric elements include lines, arcs, circles, and splines. A set of engineering equations can also be used to define the dimensions of the component.

From a given set of geometric constraints, the basic dimensions of the component can be defined. The set of geometric constraints would be regarded as a set of instructions that when executed would create the desired geometry. Hence any change to the known entities would yield a different picture altogether. So the objective is to create a geometric relationship between entities to minimize time and effort spent on design.

Parametric design systems help users manipulate the length, angle, and pitch of a particular component. These systems also have the ability to record engineering relationships within a design: they enable users to pinpoint the complex relationships that exist among parts of an assembly. A simple example would be the meshing of two gears. Thus the use of parametric design clarifies users' understanding of the engineering aspects of a component. Moreover, the parametric system can produce designs that are more meaningful than those generated by traditional CAD systems. Designers, analysts, and production engineers can work with the same solid model, extracting and adding information according to need.

A drawback of this system is that it is constrained to the set of geometric or engineering relationships provided by the engineer for creating the design of a component. The limited set of geometric constraints can make it difficult to change a parametric design model once the initial conditions have been set. These systems are best suited for design tasks that do not involve many variations in the design approach. Software equipped with such an option is therefore an excellent tool for design. Such software gives the designer flexibility to avoid redrawing objects and minimizes the time involved in arriving at a final product design.

Variational Design Systems

This technique can simultaneously solve a set of geometric constraints and engineering equations with important relationships among the elements of design. Essentially, the design is governed by a set of engineering equations relating its geometry and functions.

Unlike the parametric approach, a variational system is able to determine the positions of geometric elements and constraints. In addition, it is structured to handle the coupling between parameters in the geometric constraints and the engineering equations. The variational design concept helps the engineer evaluate the design of a component in depth instead of considering only the geometric aspects to satisfy the design relationships. This can be illustrated by the following example.

In design, one knows the performances and characteristics wanted for a developing product but does not know the geometric dimensions, material characteristic, etc. A good design starts with an appropriate set of parameters. The process of finding a set of parameters is done through optimization. Variational design has the ability to incorporate optimization into the design environment. The engineer can specify both equality and inequality design constraints. The objective function is a set of both dependent and independent variables that are optimized (minimized or maximized) relative to one or more independent variables.

Variational Design Applications

In their work Hofer and Pottmann presented a solution for variational motion design of a rigid body in the presence of rigid obstacles of arbitrary shape that fully employs the available degrees of freedom in the design process. The main idea is to reduce the problem to a curve design task on a suitable barrier surface embedded in an appropriate kinematic image space (see Fig. 12.2.1).

This is an investigation of a 3-D shape reconstruction from measurement data in the presence of constraints. The constraints may fix the surface type or set geometric relations between parts of an object's surface, such as orthogonality, parallelity, and others. The authors of this work (see Fig. 12.2.2) have proposed to use a combination of surface fitting and registration within the geometric optimization framework of **squared distance minimization** (SDM). In this way, one obtains a **quasi-Newtonian** like optimization algorithm, which in each iteration simultaneously registers the data set with a rigid motion to the fitting surface and adapts the shape of the fitting surface. Figure 12.2.2 highlights the applicability of the method to constrained 3-D shape fitting for reverse engineering of CAD models and to high-accuracy fitting with kinematic surfaces, which include surfaces of revolution (reconstructed from fragments of archeological pottery) and spiral surfaces, which are fitted to 3-D measurement data of shells.

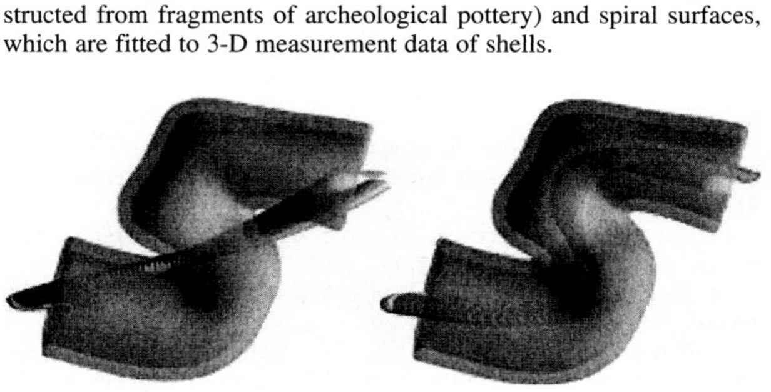

Figure 12.2.1 Motion of a rigid body passing through a narrow channel.

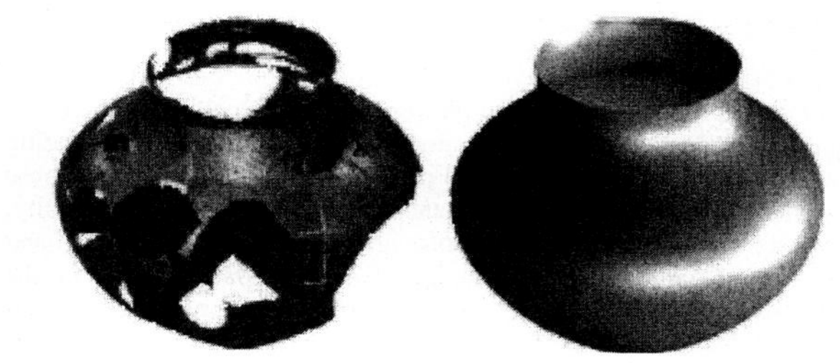

Figure 12.2.2 Approximation of a broken archeological finding by using inverse engineering.

Innovative Process in Product Development

The focus and ultimate goal of an engineer is to be able to answer the following questions: Does my design (geometry) and my choices of material allow my product to achieve the performance required by the technical specifications? What is the optimum achievable performance? Will my design be sensitive to manufacturing tolerances? Answering the first question requires us to solve a set of equations relating output and input conditions. The known quantities are the various targeted performances of a **product** (e.g., static or modal behavior), and the unknowns of that problem are of various kinds: geometric dimensions, material characteristics, shell thickness, number of stiffeners, and density of weld spots, for example. A good design is a design where parameters have the appropriate values. Usually, these good values are not unique. The conventional way to solve the inverse problem is to use optimization techniques. The solution of an optimization problem is one set of parameter values that satisfies mathematical criteria. Even if simple criteria such as maximum stress or first Eigen frequency are mathematical criteria, these criteria rarely take into account manufacturing process constraints or the invaluable know-how of the engineer. Furthermore, finding the best point does not give any information on the global quality of the solution and its stability. Most of the optimization algorithms used in common practice are local algorithms that only find the closest point to the initial design.

ENGINEERING ANALYSIS AND CAD

Analysis is of two types: analytical and experimental. Both are performed to assess product performance.

Analytical Methods

In the analytical approach, different types of analysis can be carried out by using various software packages available for computer-aided design. These include finite-element analysis, kinematics analysis and synthesis, and static and dynamic analysis.

Finite-Element Analysis Finite-element analysis (FEA) is a computer-assisted technique for determining stresses and deflections in a structure. The method divides a structure into small elements with known shapes for which a mathematical solution can be found; then the global problem is solved by using an assembly procedure (see Sec. 12.1).

The first step in FEA is the creation of a model that breaks the structure into simple standardized shapes and plots it on a common coordinate grid system. The coordinate points are called *nodes* and serve as the locations in the model where output data is provided. Nodal stiffness properties are then calculated by the finite-element program for each element and arranged into matrices within the computer. These parameters are then processed with applied loads and boundary conditions or calculations of displacement, natural frequencies, strains, and engineering functions useful in interpreting the results. An application of FEA is biomechanics, as shown in Fig. 12.2.3.

In most cases, finite-element models are developed for prototype designs for which experimental data exist so that FEA results can be cross-checked and design modifications can be made. Modifications are also tested for actual prototypes.

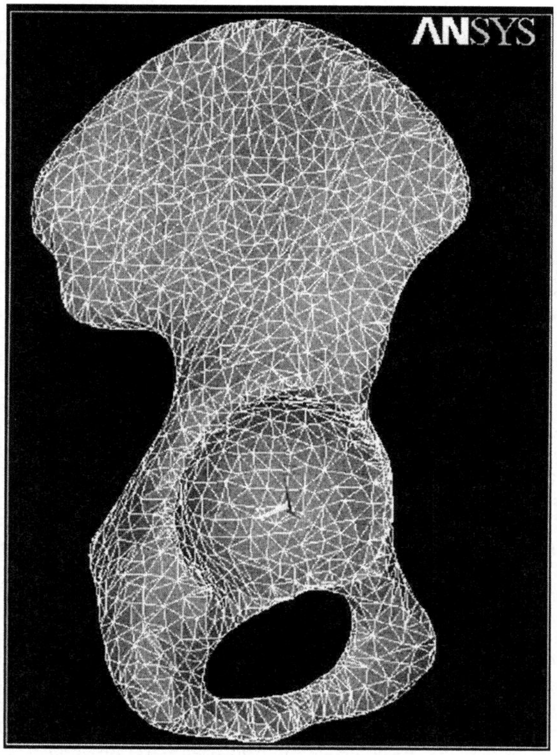

Figure 12.2.3 Application of FEA in biomechanics. Figure illustrates finite-element model of right half of human pelvic bone.

Kinematics and Synthesis Mechanisms are devices used to transfer motion and/or force from a given source to an output device. The efficiency with which this transfer is accomplished depends on the mechanism design. General-purpose programs like Linkages, IMP, ADAMS, Unigraphics, and MCADA are available today to assist engineers in the kinematic analysis and synthesis of those systems. Kinematics starts with the mechanism design. It calculates large displacements, velocities, and accelerations of mechanisms without regard to the forces acting on them. The mechanism assumes zero degrees of freedom. This means that each degree of freedom (coordinate) is constrained to a particular type of motion. The advantages of combining the study of mechanisms with CAD are that it enables the design's output to be optimized and it eliminates interferences between links.

Synthesis analysis is concerned with arriving at the best characteristics of a mechanism through a set of computer iterations. The designer provides the parameters of the mechanism and the program develops the alternatives.

The first step before doing a synthesis analysis is deciding various parameters of the mechanism, such as number of links, joints, joint types, and connectivity of links by specifying which links are fixed. The basic linkage types are defined as kinematic chains within the program. The synthesis analysis is then carried out, and the mechanism structures satisfying the design objectives are determined. These programs are generally used to design four-bar linkages, although they can be used to design six-bar linkages.

Some kinematic synthesis programs use graphics to make the software user-friendly. The user specifies points and motion vectors via interactive graphics that can be easily moved about the screen. After a number of points have been specified, the program displays all possible locations of additional points to complete the mechanism. Some software also allows the user to animate the mechanism on the screen to check for criteria such as clearances of moving links.

Static Analysis and Dynamic Analysis Static analysis combines motion analysis with mass properties and force input data to determine positions and joint reaction forces of mechanisms at rest. Static analysis can also be done on mechanisms at various points in their range of movement if zero velocity is assumed. Static models can have multiple degrees of freedom. Motion and force are uncoupled in this type of analysis. In static analysis, we are mostly concerned with the reaction forces on the mechanical systems and their joints' interconnection forces. Knowing these latter forces is useful in stress analysis to determine engineering criteria such as performance, reliability, and fatigue.

Dynamic analysis uses mass properties and forces to calculate positions, velocities, accelerations, and joint and constraint reaction forces of all model parts when motion is coupled to forces in the system. Analysis is done in discrete steps within a specified time interval. Each degree of freedom in a dynamic model is associated with an independent coordinate for which the analyst has to specify both the initial position and velocity. The computer models for dynamic analysis include geometric data and mass properties of the structure as well as applied forces. The model is generally created through part, joint, marker, force, and generator statements that are input by the user.

Part statements define the geometry, mass, and moment of inertia of each rigid part of the structure. Joint statements describe contacts between moving parts that hold the assembly together. Joints can be specified as providing translational and rotational movement, including that of revolute, screw, spherical, universal, cylindrical, and translational joints. Marker statements provide a point or coordinate system fixed on each part, orienting it to other parts and together defining the overall configuration of the system. Internal reaction forces are selected from a library of standard force elements such as dampers and linear springs. User-written routines can also be used to define additional parameters. Basic geometric data from CAD programs can be sent through direct interfaces. Because complete geometric description is not required, mechanical analysis can be used early in the design cycle, before the final configuration is known. Engineers can design the system knowing only mass and inertia characteristics.

The dynamic analysis output is typically available in tabular form. Graphics can be produced showing output such as force-versus-time or force-versus-displacement characteristics. In some cases, animation is reproduced from the combination of dynamic analysis data and geometry representation of the object. Software capable of performing such a task includes ADAMS, DADS, DISCOS, DYAMUS, DARS, and DYNACOMBS.

Experimental Testing The second aspect of analysis deals with experimental methods in which testing is conducted on prototypes and models either to extract material properties or to validate performance characteristics of a particular design.

Testing in the conventional design process is done to qualify the design after manufacturing. The goal in computer-aided design is to use testing throughout the development cycle and to make better use of the test data through more advanced analysis techniques and to integrate these with other disciplines involved in design.

Initial testing can be done on a prototype or on different components to understand model response to certain loading conditions. Experimental data are extremely useful in the analysis of models where analytical solutions are not reliable. The data can also be used to refine finite-element models for large-scale analysis. The integration of experimental testing with analytical tools results in more effective engineering analysis. Computer-aided design provides the ultimate tool for combining such methods with the graphics capabilities to allow the designer to arrive at realistic and effective designs in a minimum span of time.

Additional experimental analyzes include fatigue testing to estimate part life from strain gage measurements and modal testing to extract natural frequency and mode shapes, response functions, and a number of engineering functions in heat transfer, fluid dynamics, and thermodynamics.

COMPUTER-AIDED ENGINEERING (CAE)

According to J. R. Lemon, computer-aided engineering (CAE) is a product design and development philosophy, which combines three key engineering elements in an integrated environment for all engineering functions within each stage of the product development cycle. These elements are applications (performance, structural integrity, reliability, costs, etc.), facilities (hardware/software), and technology (data and information management). Thus, the CAE approach results in the restructuring and streamlining of the entire engineering process itself instead of the automation of specific steps in the current process such as has been the case in computer-aided design (CAD) and computer-aided manufacturing (CAM) activities.

The CAE approach places emphasis in terms of time and money on the initial conceptual design phases. This revised allocation of resources results in extensive analysis and testing to guide product design iterations in the computer as opposed to making physical prototypes. This is a major departure from current methods in most mechanical industries, where prototype manufacture and tests are the fundamental means of evaluating product performance and quality. However, the need for prototypes is not eliminated, but the reduction in the number of prototypes needed before full production can go ahead results in significant time and cost savings.

The CAE approach to mechanical product development emphasizes system analytical modeling and analysis techniques at the earliest phase of design, that is, conceptual design. The process starts with an integrated set of total system simulations of the entire product. Each alternative product concept is mathematically modeled as an entire system. At this point, overall product designers have the flexibility of defining significantly varying concepts in order to minimize weight, reduce energy consumption and maximize performance for acceptable product concepts.

Detailed component and subassembly design specifications are derived from system models by exercising computer simulations under severe environmental conditions and external product loadings. In addition, internal loads, duty cycles, and constraints acting on components and subassemblies at interfaces and connection points are derived from system models. System modeling and early development of product alternatives within the computer are essential differences compared with today's developments that rely on physical prototypes and build-and-test methods.

The new CAE approach attempts to integrate and automate various engineering functions in the entire product development process, which includes the following: (1) design, (2) analysis, (3) testing, (4) drafting, (5) documentation, (6) project management, (7) data management, (8) process planning, (9) tool design, (10) numerical control, and (11) quality assurance. Improved product quality, increased market share, and greater profitability depend upon efficient and effective CAE integration and implementation. This is why CAE implementation has become a strategic issue in mechanical-related industries worldwide.

CAE IMPLEMENTATION

Implementation of an overall CAE system and its associated technologies must be planned carefully. Implementation should proceed step by step because it is not possible to switch off today's design, development, and manufacturing processes and switch on the CAE approach at the same time. As a first step toward CAE implementation, existing problems should be solved to gain knowledge and familiarity with the CAE system's capabilities and behavior.

A design audit is not simply a learning exercise; it is an important step to extend a basic product to a family of new products or to fix problems that may have developed. The next phase involves new product design and development. The CAE tools should be applied to a new and unique product design for which new concepts have the potential of providing strategic product quality and market share advantages. An optimum hardware system along with a similar integrated

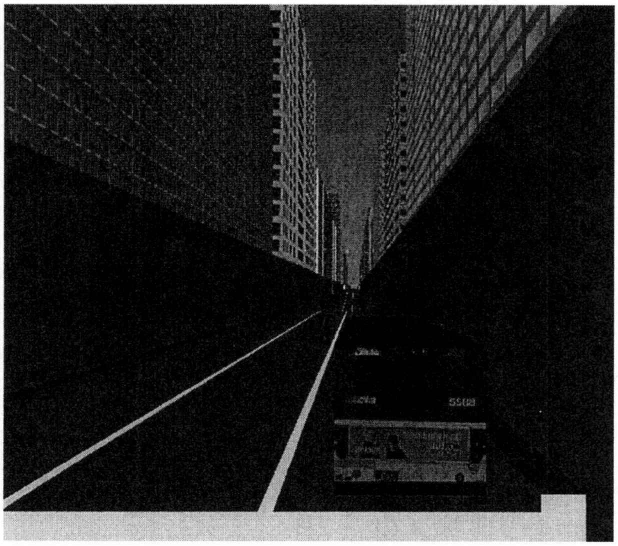

Figure 12.2.4 Virtual reality simulation is used to develop an automated bus system and analyze the benefits of the automation process, such as decreased travel times and proper adherence to schedule.

and distributed database management system should be configured for in-house usage and expected long-range expansion requirements. Software that performs efficiently should also be brought in house if it is economically justified.

Along with these tasks, outside CAE consultants are recommended to help guide and develop an overall CAE implementation plan. Educational programs and seminars in all aspects of CAE implementation are essential for all levels within an organization. Once CAE tools are implemented and the in-house organization structured to suit the particular CAE implementation philosophy, the outside consultant's role can be reduced to solving difficult problems.

VIRTUAL REALITY AND ITS APPLICATIONS TO PRODUCT DESIGN AND DEVELOPMENT

Virtual reality (VR) is a human-computer interface that aims at creating a realistic in-depth 3-D environment where manipulation and interaction of objects provides the user with an insight of the environment at the micro and macro scales through graphics and hepatics, providing the user a unique experience of a simulated reality.

Virtual reality has been used to visualize engineering and scientific concepts and offers new ways of training, enhancing the human-computer interface and constructing graphically supported real-time simulation systems. An example of VR simulation is shown in Fig. 12.2.4.

Virtual reality technology is defined by three major qualities: **immersion, presence,** and **interactivity:** The user perceives the presence of the environment as real and at any time can travel through space, exploring and navigating through the environment without fear of collision. Feedback can be provided on making sensors and controls to provide the necessary information to the human to measure his or her interaction with the environment.

Computer simulation has for years been the backbone for all input/output displays by means of graphics. With the advent of CAD software, realistic models are created and simulation represents the execution of mathematical algorithms based on the models developed under some conditions described in the problem at hand. Simulation, unlike virtual reality, does not allow the user to interact but to observe an executed dynamic situation.

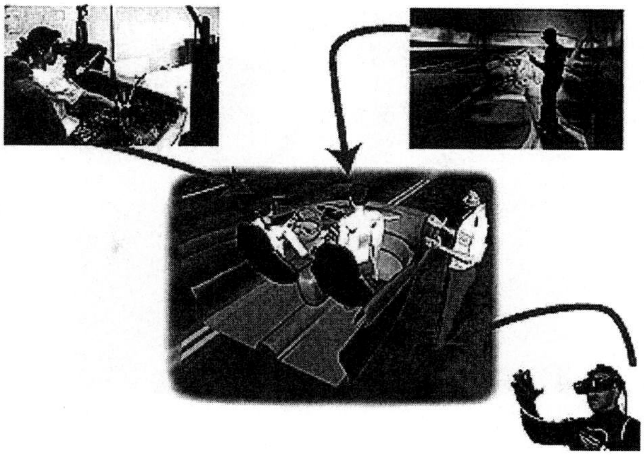

Figure 12.2.5 Shared virtual environments.

VR Application to Engineering

The typical VR display is a wall-projection display that uses either a large computer screen or a rear projection display. These systems can provide a good sense of depth but the feeling of immersion is very low. They are relatively inexpensive and nonintrusive. A greater impression of immersion may be achieved through **head-mounted** or **helmet-mounted displays (HMDs),** which consist of a pair of liquid-crystal display or cathode-ray tube display devices, mounted on a helmet worn by the user. In some HMDs there is only one display that is connected to a system of lenses and mirrors that conveys the image into the retinas of the eyes.

The **Cave Automatic Virtual Environment** (CAVE) was developed at the University of Illinois at Chicago, provides the illusion of **immersion** by projecting **stereo images** on the walls and floor of a room-sized cube. Several persons wearing lightweight stereo glasses can enter and walk freely inside the CAVE. A head tracking system continuously adjusts the stereo projection to the current position of the leading viewer.

In the example illustrated in Fig. 12.2.5, three networked users at different locations (anywhere in the world) meet in the same virtual world by using a BOOM device (a video display counterbalanced on a boom support), a CAVE system, and a head-mounted display, respectively. All users see the same virtual environment from their respective points of view. Each user is presented as a virtual human (avatar) to the other participants. The users can see each other, communicate with each other, and interact with the virtual world as a team.

As the technologies of virtual reality evolve, the applications of VR become literally unlimited. It is assumed that VR will reshape the interface between people and information technology by offering new ways for the communication of information, the visualization of processes, and the creative expression of ideas.

Note that a virtual environment can represent any three-dimensional world that is either real or abstract. This includes real systems such as buildings, landscapes, underwater shipwrecks, spacecraft, archaeological excavation sites, human anatomy, sculptures, crime scene reconstructions, and solar systems. Of special interest is the visual and sensual representation of abstract systems like magnetic fields, turbulent-flow structures, molecular models, mathematical systems, auditorium acoustics, stock market behavior, population densities, information flows, and any other conceivable system including artistic and creative work of abstract nature. These virtual worlds can be animated, interactive and shared, and can expose behavior and functionality.

Useful applications of VR include training in a variety of areas (military, medical, equipment operation, etc.), education, design evaluation

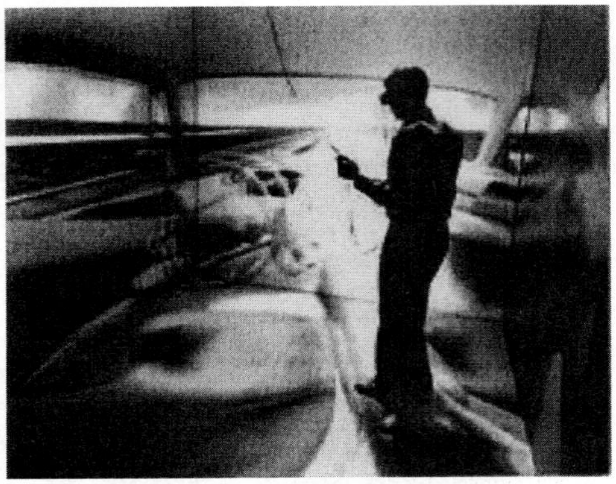

Figure 12.2.6 Viewing a concept car in immersive virtual reality using the CAVE.

(virtual prototyping), architectural walk-through, human factors and ergonomic studies, simulation of assembly sequences and maintenance tasks, assistance for the handicapped, study and treatment of phobias (e.g., fear of height), entertainment, and much more. An application of immersive virtual reality is shown in Fig. 12.2.6.

Applying Virtual Environments to Manufacturing

Even though virtual environment technologies are still difficult and expensive to use, people are doing real work. The following case studies represent examples of recent state of the art work that exemplifies the application of virtual environments in one or several aspects of manufacturing. Manufacturing, in this case, is taken to encompass issues relating to maintenance and training as well as the actual creation of parts and the assembly of systems. These examples are actual systems, not simply speculative fantasies.

Boeing The Research and Technology organization of Boeing Computer Services is actively pursuing two projects using VR technology. Boeing uses a concept known as augmented reality.

Augmented reality is a term that refers to the ability to see through a computer-generated display. The generated images are superimposed on top of reality. This is accomplished by projecting a computer image onto a half-silvered mirror that the user looks through. This technique provides an effective and intuitive way of **"annotating"** reality. The Boeing team is using a headset configured for augmented reality, which they call a HUD set (heads-up, see-through, head-mounted display).

The assembly of aircraft is a highly complex task that is difficult to automate. Many of the skills required demand dexterity not easily accomplished by robots. In addition, airplanes consist of many small-lot-size parts, and reprogramming robots for these quantities is an expensive prospect. Someone once said that a Boeing 747 is not really an airplane, but 5 million parts flying in close formation.

Caterpillar Caterpillar uses VR in its design and evaluation of heavy vehicles (Fig. 12.2.7). This technology allows the company to dramatically shorten the amount of time it takes to analyze a new design concept and incorporate it into the production process. It also represents a sizable cost savings because it eliminates the need to build prototype machines or make last-minute design changes. It takes 6 to 9 months to build full-scale models and make design changes, using conventional design methods. However, with the virtual reality approach, designs usually can be evaluated in less than 1 month. Such methods are currently being employed, and a number of other companies are fully invested in VR research and development. Today customers can "field-test" new products by putting on a special helmet and operating the vehicles in a VR environment.

PROTOTYPING ANALYSIS

A pioneering application of virtual reality was the NASA wind tunnel. A 3-D workstation generates flow-field visualizations. The user interacts with the simulation via a BOOM display and a data glove. (A BOOM is a video display counterbalanced on a boom support which the user easily moves with the head.) The data glove effectively acts as a source of "smoke," allowing the user to observe local flow lines. The computation of unsteady flow visualization occurs in real time and the user can view the dynamic scene from many points of view and scales. Since fluid dynamic phenomena occur over a large range of scales (several orders of magnitude in space), navigation through the virtual environment is more difficult than, say, architectural walk-through. Thus, in addition to "standard" viewing capabilities sensitive to head position and head orientation, our virtual environment lets the user rapidly change his or her scale, or the scale of the environment. Relative to the environment, a user can shrink to become completely surrounded by some small vortex or grow so that the entire flow field fits in one hand.

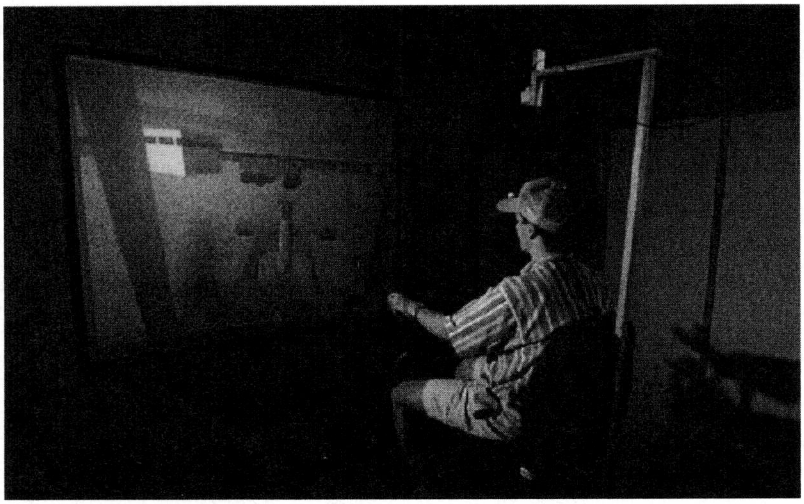

Figure 12.2.7 Virtual backhoe project illustrates an "operator" driving the virtual equipment at the National Center for Supercomputing Applications. VR lab.

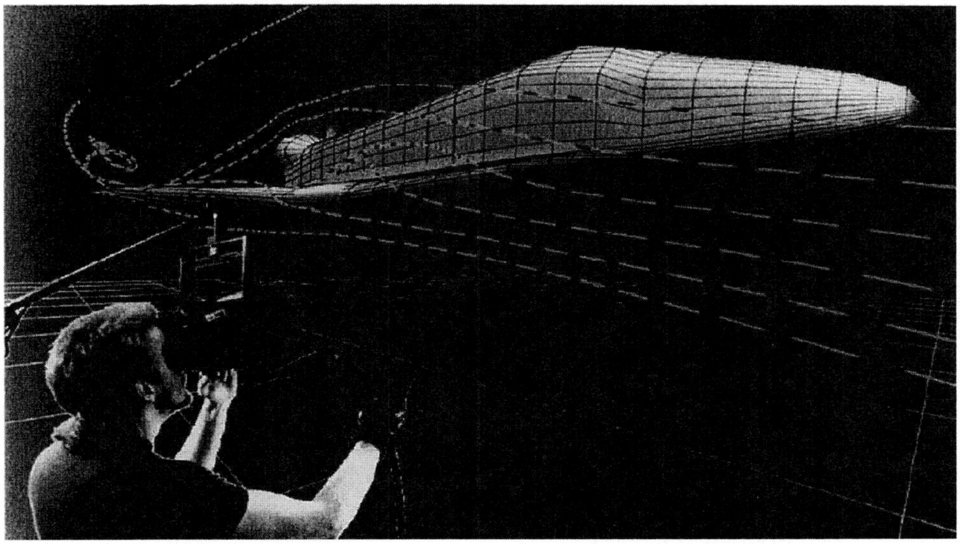

Figure 12.2.8 NASA virtual wind tunnel product design.

The role of a CAD workstation is to provide the means of graphically displaying the huge amount of data for visual inspection and quick review of the analysis. Finite-element programs have two types of data: the input required to run the program and the generated output that describes the behavior of the system under the assumed conditions. To further enhance communication between the user and the machine in which a particular finite-element program resides, many vendors of CAD software are developing pre- and postprocessors that allow the user to graphically visualize their input and output (see Fig. 12.2.8). Therefore, CAD workstations equipped with such software assist the user to interact with finite-element programs, thus minimizing the time required to learn how to use the FEM program. Most important, the graphics capability of CAD workstations provides the user the means of visual display of huge amounts of output, making the interpretation faster and more convenient.

THE USE OF VIRTUAL REALITY IN INTERACTIVE FINITE-ELEMENT ANALYSIS

Combining virtual reality (VR) with finite-element analysis (FEA) and executing interactive FEA provides an interface to implement comprehensive and intuitive analysis. A study of related research on and application of VR for interactive FEA has been completed. An overview of available VR systems and graphics APIs, and their suitability for interactive FEA/CAD, is presented here. FEA is quite developed in calculation with the improvement of computer techniques. There are two important procedures for FEA: preprocessing and the postprocessing. The preprocessor is used for setting up the model; the postprocessor is used for viewing the results. Three-dimensional visualization is already supported by currently available software, but 2-D is commonly used.

Engineers still need to imagine what is happening in a real 3-D world. Using VR in interactive FEA gives the preprocessor and the postprocessor a good method to describe the model and results in a 3-D world, which leads to a more intuitive and accurate understanding of design and construction. VR offers a real-time, full-scale interaction technique in the design period. For the iterative design procedure, real-time interactive FEA offers an excellent way to adjust the design. Apart from using VR in interactive FEA in the building and construction industry, it can also be used in the education of engineering students, to give students an intuitive understanding of the FEA technique.

REFERENCES

1. Bejczy, A. K., "Virtual Reality in Manufacturing," in *Re-engineering for Sustainable Industrial Production, Proceedings of the OE/IFIP/IEEE Int. Conf. on Integrated and Sustainable Industrial Production,* Lisbon, May 1997, pp. 48-60.
2. Broll, W., "Augmenting the Common Working Environment by Virtual Objects," *ERCIM News*, no. 44. January 2001, pp. 42-43.
3. Carrie, A. S., Adhami, E., Stephens, A., and Murdoch, I. C., "Introducing a Flexible Manufacturing System," *Int. J. of Prod. Res.* **22,** no. 6, 1984, pp. 907-916.
4. Chuter, C., Jernigan, S., and Barber K., "A Virtual Environment Simulator for Reactive Manufacturing Schedules," *Proc. of Symposium on Virtual Reality in Manufacturing, Research and Education,* University of Illinois at Chicago, October 7-8, 1996.
5. Erdélyi, F., Tóth, T., and Rayegani, F., "Real-Time Control of Manufacturing Systems on the Base of Multi-Layered Model," *Proceedings of the Second Mexican-Hungarian Workshop on Factory Automation and Material Sciences,* Mexico, 1999, pp. 37-50.
6. Figueiredo, M., Böhm, K., and Teixeira, J., "Advanced Interaction Techniques in Virtual Environments," *Comp. Graph. J.*, **17**, no. 6, 1993, pp. 655-661.
7. Haidegger, G., and Kopácsi, S., "Telepresence with Remote Vision in Robotic Applications," *Proceedings of the 5th International Workshop on Robotics in Alpe-Adria-Danube Region* (RAAD 96), 1, pp. 557-562.
8. Hofer, M., and Pottmann, H., "Designing Energy-Minimizing Rigid Body Motions in the Presence of Obstacles," Technical Report 142, Geometry Preprint Series, Vienna University of Technology, June 2005. http://www.iau.dtu.dk/~tdl/task52ab.html.
9. Liu, Y., Pottmann, H., and Wang, W., "Constrained 3D Shape Reconstruction using a Combination of Surface Fitting and Registration," Technical Report 144, Geometry Preprint Series, Vienna University of Technology, August 2005.
10. Mezgár, I., and Kovács, G. L., "Co-ordination of SME Production through a Co-operative Network," *J. Intell. Manufact.*, **9**, 1998, pp. 167-172.
11. Ránky, P., "Components and Planning of Flexible Manufacturing Systems," *Introduction to CIM Systems,* Industrial Information Centre, Budapest, 1985 (in Hungarian).
12. Taylor Enterprise Dynamics, User Manual, F&H Simulation B.V., 2000.

12.3 INTRODUCTION TO COMPUTATIONAL FLUID DYNAMICS

by Doyle Knight

INTRODUCTION

Computational fluid dynamics (CFD) is one of the three branches of modern fluid dynamics, the other two being analytical and experimental. CFD emerged in the 1950s with the advent of digital computers, although a number of scientific papers appeared earlier in the twentieth century which constitute a not insignificant part of its theoretical foundation. Richardson's 1910 paper (Richardson, 1910) is generally cited as the progenitor of modern CFD because of its presentation of numerical methods for classical equations, some of which have found application in fluid dynamics (e.g., Laplace's equation), and Courant et al. (1928) established the concept of the domain of dependence of a numerical solution and its relationship to the stability of the numerical scheme. Roache (1972, 1988) provides additional historical detail. The appearance of major textbooks [e.g., Richtmyer and Morton (1957), Roche (1972), Kovenya and Yanenko (1981), and Peyret and Taylor (1983)] firmly established the role of CFD in modern fluid dynamics research and engineering.

Computational fluid dynamics is the science of solving the partial differential equations of fluid motion by discrete approximation. Although a formal development of CFD methodology is beyond the scope of this section, it is worthwhile to identify the principal theoretical concepts that form the foundation of CFD algorithms. *Accuracy* is the formal order of approximation of an individual discrete approximation to a partial derivative. Consider, for example, a uniformly spaced set of grid points $x_i = (i-1)\Delta x$ for $i = 1, \ldots, n$ where Δx is a constant, and a function $f(x)$ defined at each point x_i. The discrete approximation to the derivative $\partial f/\partial x$, denoted as $\delta f/\delta x$, given by

$$\frac{\delta f}{\delta x} = \frac{f_{i+1} - f_{i-1}}{2\Delta x} \tag{12.3.1}$$

can be shown by Taylor's series expansion to approximate the derivative of $f(x)$ as

$$\frac{f_{i+1} - f_{i-1}}{2\Delta x} = \frac{\partial f}{\partial x} + \mathcal{O}(\Delta x)^2 \tag{12.3.2}$$

where $\mathcal{O}(\Delta x)^2$ indicates that the remaining terms are proportional to $(\Delta x)^2$ and therefore vanish as $\Delta x \to 0$. The discrete approximation $\delta f/\delta x$ is said to be second-order accurate (since the error is proportional to the second power of the grid spacing Δx). *Consistency* is the equivalence of the discrete approximation of the partial derivative to the partial derivative itself in the limit of $\Delta x \to 0$, irrespective of its order of accuracy. *Convergence* is the equivalence of the *solution* of the partial differential equation(s) (not merely the equivalence of an individual term) to the exact solution in the limit of vanishing grid spacing. There are examples of accurate but nonconvergent numerical algorithms for the partial differential equations of fluid dynamics (Roache, 1972). *Stability* is the property that the solution must remain bounded (Lax and Richtmyer, 1956).

The remainder of the section is organized into the following topics: (1) the governing equations of incompressible homogeneous fluid mechanics, (2) the concept of potential flow and the conditions for its existence, (3) a detailed development of the vortex panel method for solution of potential flow, (4) a numerical algorithm for solution of the boundary layer equations, and (5) an algorithm for the Navier-Stokes equations. Items (3) to (5) represent, therefore, the sequence of increasing complexity of fluid motions, and also represent a chronology of major advances in CFD. A list of references concludes the section.

GOVERNING EQUATIONS

The governing equations of an incompressible, laminar, constant-property, Newtonian viscous fluid in differential, dimensional form, using the Einstein summation convention,* are (Wilcox, 1997):

$$\frac{\partial \tilde{u}_i}{\partial \tilde{x}_i} = 0 \tag{12.3.3}$$

$$\tilde{\rho}\frac{\partial \tilde{u}_i}{\partial \tilde{t}} + \tilde{\rho}\tilde{u}_j\frac{\partial \tilde{u}_i}{\partial \tilde{x}_j} = -\frac{\partial \tilde{p}}{\partial \tilde{x}_i} + \frac{\partial \tilde{\tau}_{ij}}{\partial \tilde{x}_j} \tag{12.3.4}$$

where $\tilde{u}_i$ are the dimensional Cartesian velocity components in the $\tilde{x}_i$ directions, $\tilde{\rho}$ is the density, $\tilde{p}$ is the static pressure, and $\tilde{\tau}_{ij}$ is the viscous stress tensor,

$$\tilde{\tau}_{ij} = \mu\left(\frac{\partial \tilde{u}_i}{\partial \tilde{x}_j} + \frac{\partial \tilde{u}_j}{\partial \tilde{x}_i}\right) \tag{12.3.5}$$

where μ is the molecular dynamic viscosity.

The equation may be made nondimensionalized by using a reference velocity U_∞ and length L,

$$\frac{\partial u_i}{\partial x_i} = 0 \tag{12.3.6}$$

$$\frac{\partial u_i}{\partial t} + u_j\frac{\partial u_i}{\partial x_j} = -\frac{\partial p}{\partial x_i} + \frac{\partial \tau_{ij}}{\partial x_j} \tag{12.3.7}$$

where

$$u_i = \frac{\tilde{u}_i}{U_\infty}$$

$$x_i = \frac{\tilde{x}_i}{L}$$

$$t = \frac{\tilde{t}U_\infty}{L}$$

$$p = \frac{\tilde{p}}{\rho U_\infty^2}$$

$$\tau_{ij} = \frac{\tilde{\tau}_{ij}}{\rho U_\infty^2} \tag{12.3.8}$$

In particular, the viscous stress tensor becomes

$$\tau_{ij} = \frac{1}{\text{Re}}\left(\frac{\partial u_i}{\partial x_j} + \frac{\partial u_j}{\partial x_i}\right) \tag{12.3.9}$$

where $\text{Re} = \rho U_\infty L/\mu$ is the *Reynolds number*.

VORTICITY AND KELVIN'S THEOREM

The cartesian components of the vorticity ω_i are defined as†

$$\omega_i = \epsilon_{ijk}\frac{\partial u_k}{\partial x_j} \tag{12.3.10}$$

* The Einstein summation convention uses repeated indices to indicate summation over all values of the index. This, for example, Eq. (12.3.3) is

$$\frac{\partial \tilde{u}_1}{\partial \tilde{x}_1} + \frac{\partial \tilde{u}_2}{\partial \tilde{x}_2} + \frac{\partial \tilde{u}_3}{\partial \tilde{x}_3} = 0$$

† The indices i, j, k are cyclic if they are in increasing order. The possible combinations are $(i, j, k) = (1, 2, 3), (2, 3, 1)$, and $(3, 1, 2)$. The indices are anticyclic if they are in decreasing order. The possible combinations are $(i, j, k) = (3, 2, 1), (2, 1, 3)$, and $(1, 3, 2)$.

where ε_{ijk} is the cyclic tensor defined as

$$\varepsilon_{ijk} = \begin{cases} 1 & \text{if } i, j, k \text{ are cyclic} \\ -1 & \text{if } i, j, k \text{ are anticyclic} \\ 0 & \text{otherwise} \end{cases} \quad (12.3.11)$$

The vorticity is twice the effective local angular velocity of the fluid (Batchelor, 1970).

A flow field where $\omega_i = 0$ everywhere is denoted *irrotational flow* or *potential flow,* and the velocity u_i may be represented as the gradient of a scalar potential ϕ (Batchelor, 1970),

$$u_i = \frac{\partial \phi}{\partial x_i} \quad (12.3.12)$$

Kelvin's theorem (Wilcox, 1997) states that an inviscid, incompressible fluid with zero vorticity at some instant of time remains irrotational for all time.

IRROTATIONAL FLOW

At high Reynolds number Re, the viscous stresses [Eq. (12.3.9)] are negligible outside of a narrow region (denoted the *boundary layer*) adjacent to solid surfaces. Figure 12.3.1 shows an airfoil in cross section (without the boundary layer). Provided that the angle of attack of the airfoil is sufficiently small, the boundary layer at high Re remains affixed to the airfoil and the vorticity generated therein is convected downstream into a narrow wake. Thus, the flow field domain is almost entirely inviscid and the boundary layer can be ignored in determining the pressure distribution on the airfoil. Assuming the flow upstream is uniform (and hence has zero vorticity), the flow field outside the boundary layer is irrotational.

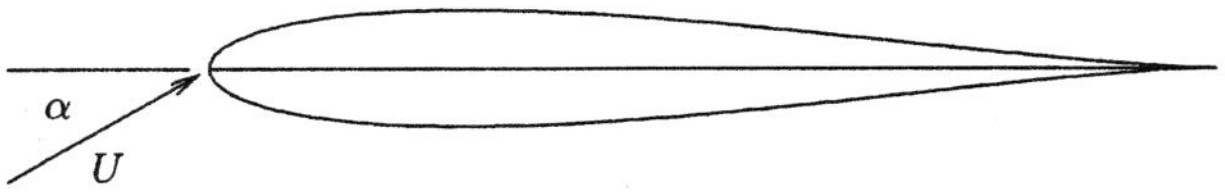

Figure 12.3.1 Airfoil.

For irrotational flow, the governing Eqs. (12.3.3) and (12.3.4) become (with ˜ omitted for clarity),

$$\frac{\partial u_i}{\partial x_i} = 0 \quad (12.3.13)$$

$$\frac{\partial u_i}{\partial t} + u_j \frac{\partial u_i}{\partial x_j} = -\frac{1}{\rho}\frac{\partial p}{\partial x_i} \quad (12.3.14)$$

Using Eq. (12.3.12), we find the conservation of mass becomes

$$\frac{\partial^2 \phi}{\partial x_i^2} = 0 \quad (12.3.15)$$

which is known as *Laplace's equation.* The boundary conditions are

$$\begin{aligned} \frac{\partial \phi}{\partial x_i} &= U_i \qquad \text{as } |x_i| \to \infty \\ \frac{\partial \phi}{\partial n} &= 0 \qquad \text{on the surface} \end{aligned} \quad (12.3.16)$$

The first boundary condition states that the fluid velocity approaches the free-stream velocity with cartesian components U_i as the distance from the body becomes large. The second boundary condition states that the velocity normal to the surface is zero (assuming the airfoil is at rest in the frame of reference). In addition, the Zhukovsky (Joukowski) hypothesis must be applied to achieve a unique solution (Milne-Thompson, 1958),

Joukowski's hypothesis is that the circulation* in the case of a properly designed aerofoil, in its working range of incidence, always adjusts itself so that the airspeed at the trailing edge is finite.

Using the identities

$$u_j \frac{\partial u_i}{\partial x_j} = \frac{\partial}{\partial x_i}\left(\frac{1}{2}u_j u_j\right) - \epsilon_{ijk} u_j \omega_k \quad (12.3.17)$$

and

$$\frac{\partial u_i}{\partial t} = \frac{\partial}{\partial t}\left(\frac{\partial \phi}{\partial x_i}\right) = \frac{\partial}{\partial x_i}\left(\frac{\partial \phi}{\partial t}\right) \quad (12.3.18)$$

and noting that $\omega_i = 0$ for an irrotational flow, we can rewrite the conservation of momentum [Eq. (12.3.7)] as

$$\frac{\partial}{\partial x_i}\left(\frac{\partial \phi}{\partial t} + \frac{1}{2}u_j u_j + \frac{p}{\rho}\right) = 0 \quad (12.3.19)$$

and thus

$$\frac{\partial \phi}{\partial t} + \frac{1}{2}u_j u_j + \frac{p}{\rho} = g(t) \quad (12.3.20)$$

where $g(t)$ is a function of time. This is *Bernoulli's equation* and provides an expression for the pressure p in terms of the velocity potential and its gradient.

Elementary Solutions

Laplace's equation (and its inhomogenous counterpart, Poisson's equation) have been studied extensively because of their broad application in physics and engineering (Kellogg, 1953). For simplicity, we consider Laplace's equation in two-dimensional cartesian or cylindrical coordinates where $(x_1, x_2) = (x, y)$, $r = \sqrt{x_1^2 + x_2^2}$, and $\theta = \tan^{-1}(x_2/x_1)$. The cartesian and cylindrical polar forms of Laplace's equation are

$$\frac{\partial^2 \phi}{\partial x^2} + \frac{\partial^2 \phi}{\partial y^2} = 0 \qquad \text{cartesian} \quad (12.3.21)$$

$$\frac{1}{r}\frac{\partial}{\partial r}\left(r\frac{\partial \phi}{\partial r}\right) + \frac{\partial^2 \phi}{\partial \theta^2} = 0 \qquad \text{cylindrical} \quad (12.3.22)$$

There are three fundamental solutions of Eq. (12.3.15) that are commonly used in computational methods. The first is *uniform flow,* defined by

$$\phi = U_i x_i \qquad \text{uniform flow} \quad (12.3.23)$$

where U_i is the component of the velocity along the x_i axis and $U = \sqrt{U_i U_i}$. The second is a two-dimensional *source/sink* defined by

$$\phi = \frac{q}{2\pi}\log r \qquad \text{source/sink} \quad (12.3.24)$$

It may be directly verified that the volume flux of fluid crossing a circular boundary enclosing a source at $r = 0$ is

$$\int_0^{2\pi} u_i n_i r \, d\theta = \int_0^{2\pi} \frac{q}{2\pi r} r \, d\theta = q \quad (12.3.25)$$

where n_i are the cartesian components of the outward normals. The third is a *line vortex* defined by

$$\phi = \frac{\Gamma \theta}{2\pi} \qquad \text{line vortex} \quad (12.3.26)$$

It may be directly verified that the line integral of the velocity about a contour enclosing the line vortex is

$$\int_0^{2\pi} u_i s_i r \, d\theta = \int_0^{2\pi} \frac{\Gamma}{2\pi r} r \, d\theta = \Gamma \quad (12.3.27)$$

* See "Elementary Solutions," discussed later.

where s_i are the components of the unit tangent vector (taken in the counterclockwise direction). The quantity Γ is denoted the *circulation.*

Panel Methods

The elementary solutions may be combined in a discrete manner to generate the solution for irrotational flow past an arbitrary number of closed boundaries (Hess and Smith, 1986; Hess, 1975). For simplicity, we consider a single body represented by the airfoil in Fig. 12.3.2. The airfoil surface is discretized into n segments with endpoints* (x'_i, y'_i), $i = 1, \ldots, n+1$. Note that $(x'_1, y'_1) = (x'_{n+1}, y'_{n+1})$. The slope of the ith segment relative to the x axis is

$$\delta_i = \tan^{-1}\left[\frac{(y'_{i+1} - y'_i)}{(x'_{i+1} - x_i)}\right] \qquad -\frac{\pi}{2} \le \delta_i \le \frac{\pi}{2} \tag{12.3.28}$$

In particular, it is noted that δ_n is not necessarily equal to δ_1 (i.e., the trailing edge of the airfoil may have a finite angle).

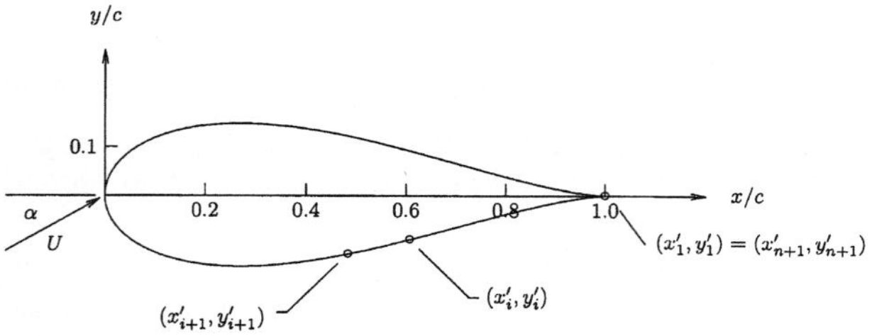

Figure 12.3.2 Airfoil.

Nonlifting Airfoil

Consider an infinite number of sources/sinks on the ith segment (Fig. 12.3.3) with strength per unit length q_i and length ds_i. From Eq. (12.3.24), the velocity potential is

$$d\phi_i = \frac{q_i\, ds_i}{2\pi} \log r_i \tag{12.3.29}$$

where the summation on i is not performed and r is measured from the center of the infinitesimal segment (ξ_i, η_i) to the field point

$$r_i = \sqrt{(x - \xi_i)^2 + (y - \eta_i)^2} \tag{12.3.30}$$

The velocity potential for the ith source/sink segment is therefore

$$\phi = \frac{q_i}{2\pi} \int \log r_i\, ds_i \qquad \text{(no sum on } i\text{)} \tag{12.3.31}$$

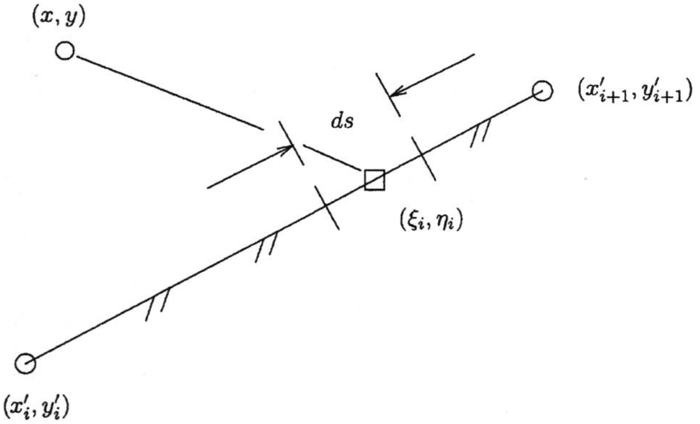

Figure 12.3.3 Sources/sinks on segment.

* Note that the subscript i indicates the ith point, and should not be confused with the earlier cartesian notation.

Since Laplace's equation is linear, a linear superposition of elementary solutions is also a solution, and therefore the velocity potential is

$$\phi(x, y) = Ux\cos\alpha + Uy\sin\alpha + \sum_{j=1}^{j=n} \frac{q_j}{2\pi} \int \log r_j\, ds_j \tag{12.3.32}$$

where U is the amplitude of the free-stream velocity and α is the angle of the free-stream velocity with respect to the x axis.

Define n control points (x_i, y_i) for $i = 1, \ldots, n$, where

$$\begin{aligned} x_i &= \tfrac{1}{2}(x'_{i+1} + x'_i) \\ y_i &= \tfrac{1}{2}(y'_{i+1} + y'_i) \end{aligned} \tag{12.3.33}$$

and apply the boundary condition that the velocity normal to the boundary at each control point is zero. The resultant equations are (Kellogg, 1953):

$$\frac{q_i}{2} + \sum_{j=1}^{j=n} \frac{q_j}{2\pi} \int \frac{\partial}{\partial n_i} \log r_j\, ds_j = -\frac{\partial}{\partial n_i}(Ux\cos\alpha + Uy\sin\alpha) \qquad \text{for } i = 1, \ldots, n \tag{12.3.34}$$

where n_i is the outward (i.e., into the fluid) normal direction at the ith control point and the derivative is evaluated at (x_i, y_i). Define the unit normal (n_{1_i}, n_{2_i}) and tangential vectors (t_{1_i}, t_{2_i}) at each segment i for $i = 1, \ldots, n$ by (no sum on i)

$$\begin{aligned} n_{1_i} &= -(y'_{i+1} - y'_i)\Delta s_i^{-1} \\ n_{2_i} &= +(x'_{i+1} - x'_i)\Delta s_i^{-1} \\ t_{1_i} &= +(x'_{i+1} - x'_i)\Delta s_i^{-1} \\ t_{2_i} &= +(y'_{i+1} - y'_i)\Delta s_i^{-1} \end{aligned} \tag{12.3.35}$$

where Δs_i is the length of the ith segment

$$\Delta s_i = \sqrt{(x'_{i+1} - x'_i)^2 + (y'_{i+1} - y'_i)^2} \tag{12.3.36}$$

Equation (12.3.34) can be rewritten as (Kellogg, 1953):

$$\frac{Q_i}{2} + \sum_{j=1,\, j\ne i}^{j=n} \frac{Q_j}{2\pi}(n_{1_i} G_{ij} + n_{2_i} H_{ij}) = -(n_{1_i}\cos\alpha + n_{2_i}\sin\alpha) \qquad \text{for } i = 1, \ldots, n \tag{12.3.37}$$

where

$$Q_i = \frac{q_i}{U} \tag{12.3.38}$$

Equation (12.3.37) represents n linear equations for the n unknowns q_i. The $n \times n$ influence matrices G_{ij} and $H_{i,j}$ are defined as follows. Define the following (no sum on i or j):

$$\begin{aligned} a_{ij} &= (x_i - x_j)t_{1_j} + (y_i - y_j)t_{2_j} \\ b_{ij} &= \sqrt{(x_i - x_j)^2 + (y_i - y_j)^2 - a_{ij}^2} \\ c_{ij} &= x_i - x_j - a_{ij}t_{1_j} \\ d_{ij} &= \tfrac{1}{2}\Delta s_j - a_{ij} \\ e_{ij} &= \tfrac{1}{2}\Delta s_j + a_{ij} \\ f_{ij} &= (x_i - x_j)^2 + (y_i - y_j)^2 + \Delta s_j a_{ij} + \tfrac{1}{4}\Delta s_j^2 \\ g_{ij} &= y_i - y_j - a_{ij}t_{2_j} \end{aligned} \tag{12.3.39}$$

The influence matrices are

$$G_{ij} = \frac{c_{ij}}{b_{ij}}\left[\tan^{-1}\frac{d_{ij}}{b_{ij}} + \tan^{-1}\frac{e_{ij}}{b_{ij}}\right] - \tfrac{1}{2}t_{1_j}\log\frac{f_{i,j+1}}{f_{ij}}$$

$$H_{ij} = \frac{g_{ij}}{b_{ij}}\left[\tan^{-1}\frac{d_{ij}}{b_{ij}} + \tan^{-1}\frac{e_{ij}}{b_{ij}}\right] - \tfrac{1}{2}t_{2_j}\log\frac{f_{i,j+1}}{f_{ij}} \qquad (12.3.40)$$

The solution of Eq. (12.3.37) yields the source/sink values q_i. The velocity at the midpoint of the ith segment on the airfoil is tangential to the airfoil and given by

$$V_i = U(t_{1_i}\cos\alpha + t_{2_i}\sin\alpha) + \sum_{j=1,\, j\neq i}^{j=n} \frac{q_i}{2\pi}(t_{1_i}G_{ij} + t_{2_i}H_{ij}) \qquad (12.3.41)$$

Evaluating Eq. (12.3.20) in the free stream and assuming steady flow, we obtain

$$\tfrac{1}{2}U^2 + \frac{p_\infty}{\rho} = \tfrac{1}{2}(u^2 + v^2) + \frac{p}{\rho} \qquad (12.3.42)$$

Thus, the pressure coefficient c_p, defined by

$$c_p = \frac{p - p_\infty}{\tfrac{1}{2}\rho U^2} \qquad (12.3.43)$$

becomes

$$c_p = 1 - \left(\frac{u^2 + v^2}{U^2}\right)^2 \qquad (12.3.44)$$

and on the airfoil surface at the midpoint of the ith segment,

$$c_{p_i} = 1 - \left(\frac{V_i}{U}\right)^2 \qquad (12.3.45)$$

The above method is applied to the flow past a cylinder as illustrated in Fig. 12.3.4. The computed pressure distribution using $n = 50$ segments is shown in Fig. 12.3.5 together with the exact solution (Anderson, 2001)

$$c_p(\theta) = 1 - 4\sin^2\theta \qquad (12.3.46)$$

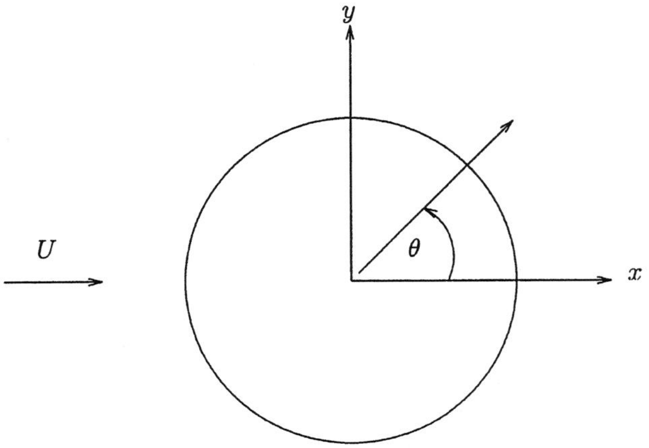

Figure 12.3.4 Flow past a cylinder.

Lifting Airfoil

For a two-dimensional airfoil in steady flow, the Kutta-Zhukovsky theorem (Milne-Thompson, 1958) states

$$l = -\rho U \Gamma \qquad (12.3.47)$$

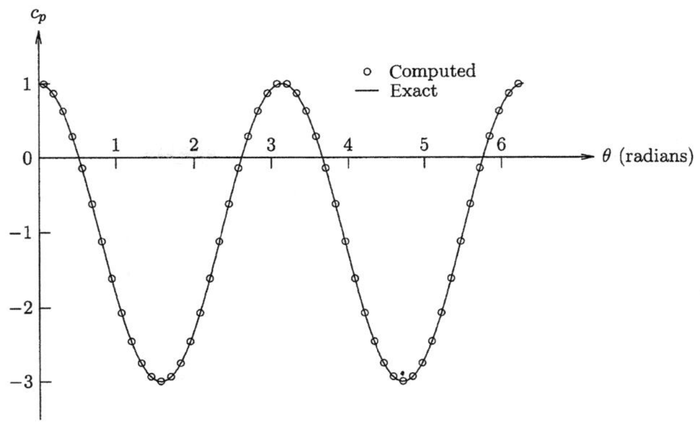

Figure 12.3.5 Pressure coefficient on cylinder.

where l is the lift per unit span, U is the free-stream velocity, and Γ is the circulation defined by Eq. (12.3.27). The lift is the force on the airfoil in a direction perpendicular to the free-stream velocity. It is therefore evident that the solution of Laplace's equation for a lifting airfoil must include distributed vortices.

Consider an infinite number of line vortices on the ith segment (Fig. 12.3.6). For simplicity, the origin of the coordinate system is taken to be at the center of the segment. Each vortex has a circulation $\gamma d\xi$. The cartesian velocity components at point (x, y) due to the vortex at $(\xi, 0)$ are

$$u = -\frac{\gamma d\xi}{2\pi r}\sin\theta$$
$$v = +\frac{\gamma d\xi}{2\pi r}\cos\theta \qquad (12.3.48)$$

where

$$r = \sqrt{(x-\xi)^2 + y^2}$$
$$\theta = \tan^{-1}\left(\frac{y}{x-\xi}\right) \qquad (12.3.49)$$

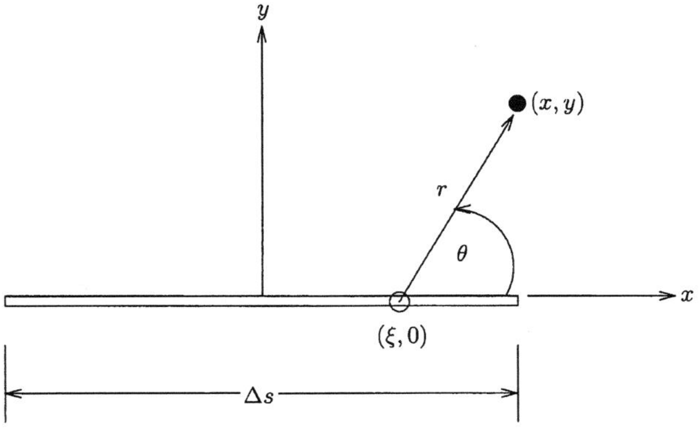

Figure 12.3.6 Vortex on segment.

Thus, the cartesian velocity components at point (x, y) due to the distribution of vortices on the segment is

$$u = -\frac{\gamma}{2\pi}\int_{-\frac{1}{2}\Delta s}^{+\frac{1}{2}\Delta s} \frac{y\,d\xi}{[(x-\xi)^2 + y^2]}$$
$$v = +\frac{\gamma}{2\pi}\int_{-\frac{1}{2}\Delta s}^{+\frac{1}{2}\Delta s} \frac{(x-\xi)\,d\xi}{[(x-\xi)^2 + y^2]} \qquad (12.3.50)$$

These equations may be integrated to obtain

$$u = -\frac{\gamma}{2\pi}\,\text{sign}(y)\left\{\tan^{-1}\left(\frac{\frac{1}{2}\Delta s - x}{|y|}\right) + \tan^{-1}\left(\frac{\frac{1}{2}\Delta s + x}{|y|}\right)\right\}$$

$$v = -\frac{\gamma}{4\pi}\log\left[\frac{x^2 + y^2 - x\Delta s + (\frac{1}{2}\Delta s)^2}{x^2 + y^2 + x\Delta s + (\frac{1}{2}\Delta s)^2}\right] \tag{12.3.51}$$

where sign (y) is the sign of y. Evaluating Eq. (12.3.51) at the midpoint of the segment yields

$$\lim_{x=0,\, y\downarrow 0} u(x, y) = -\frac{\gamma}{2}$$

$$\lim_{x=0,\, y\downarrow 0} v(x, y) = 0 \tag{12.3.52}$$

Therefore, the vectoral velocity at the midpoint of the ith segment due to the jth vortex segment is

$$\vec{v}_{ij_v} = -\frac{\gamma_j}{2\pi}\frac{h_{ij}}{b_{ij}}\left[\tan^{-1}\frac{d_{ij}}{b_{ij}} + \tan^{-1}\frac{e_{ij}}{b_{ij}}\right]\vec{t}_j - \frac{\gamma_j}{4\pi}\log\frac{f_{i,j+1}}{f_{ij}}\vec{n}_j \tag{12.3.53}$$

where

$$h_{ij} = (x_i - x'_j)n_{1_j} + (y_i - y'_j)n_{2_j} \tag{12.3.54}$$

Applying the boundary condition that the velocity normal to the boundary at each control point is zero yields

$$\frac{Q_i}{2} + \sum_{j=1,\, j\neq i}^{j=n}\frac{Q_j}{2\pi}(n_{1_i}G_{ij} + n_{2_i}H_{ij}) + \frac{1}{U}\sum_{j=1,\, j\neq i}^{j=n}\vec{v}_{ij_v}\cdot\vec{n}_i$$

$$= -(n_{1_i}\cos\alpha + n_{2_i}\sin\alpha) \qquad \text{for } i = 1, \ldots, n \tag{12.3.55}$$

The additional boundary condition is the Zhukovsky hypothesis described above.

It is evident that Eq. (12.3.55) provides n equations. Together with the Zhukovsky hypothesis, there are $n + 1$ equations for the $2n$ unknowns q_i and γ_i. It is evident, therefore, that a functional form must be assumed for γ_j involving only a single unknown variable (Hess and Smith, 1966; Hess, 1975). At a trailing edge, the vortex strength must vanish, otherwise the tangential velocity will be infinite. Thus, we assume the form

$$\gamma_i = U\Lambda S_i \qquad \text{for } i = 1, \ldots, n \tag{12.3.56}$$

where

$$S_i = s_i(1 - s_i) \tag{12.3.57}$$

with $s_1 = 0$ and

$$s_i = \left[\sum_{j=2}^{j=i}\sqrt{(x_j - x_{j-1})^2 + (y_j - y_{j-1})^2}\right]$$

$$\times\left[\sum_{j=2}^{j=n}\sqrt{(x_j - x_{j-1})^2 + (y_j - y_{j-1})^2}\right]^{-1} \tag{12.3.58}$$

Thus, $S_1 = 0$ and $S_n = 0$.

Define the influence matrices

$$A_{ij} = S_j\left\{-\frac{h_{ij}t_{1_j}}{b_{ij}}\left[\tan^{-1}\frac{d_{ij}}{b_{ij}} + \tan^{-1}\frac{e_{ij}}{b_{ij}}\right] - \tfrac{1}{2}n_{1_j}\log\frac{f_{i,j+1}}{f_{ij}}\right\}$$

$$B_{ij} = S_j\left\{-\frac{h_{ij}t_{2_j}}{b_{ij}}\left[\tan^{-1}\frac{d_{ij}}{b_{ij}} + \tan^{-1}\frac{e_{ij}}{b_{ij}}\right] - \tfrac{1}{2}n_{2_j}\log\frac{f_{i,j+1}}{f_{ij}}\right\} \tag{12.3.59}$$

Equation (12.3.55) becomes

$$\frac{Q_i}{2} + \sum_{j=1,\, j\neq i}^{j=n}\frac{Q_i}{2\pi}(n_{1_i}G_{ij} + n_{2_i}H_{ij}) + \frac{\Lambda}{2\pi}\sum_{j=1,\, j\neq i}^{j=n}(n_{1_i}A_{ij} + n_{2_i}B_{ij})$$

$$= -(n_{1_i}\cos\alpha + n_{2_i}\sin\alpha) \tag{12.3.60}$$

The tangential velocity at the ith control point becomes

$$V_i = U(t_{1_i}\cos\alpha + t_{2_i}\sin\alpha) + \sum_{j=1,\, j\neq i}^{j=n}\frac{q_j}{2\pi}(t_{1_i}G_{ij} + t_{2_i}H_{ij})$$

$$+ U\Lambda\left[-\tfrac{1}{2}S_i + \frac{1}{2\pi}\sum_{j=1,\, j\neq i}^{j=n}(t_{1_i}A_{ij} + t_{2_i}B_{ij})\right] \tag{12.3.61}$$

The Zhukovsky hypothesis indicates that the velocity at the training edge is finite (Fig. 12.3.7). For a sharp trailing edge, this implies that the velocity vector must be aligned with the bisector of the trailing edge. In order for this to occur, the pressure at control points $i = 1$ and $i = n$ must be the same, and therefore, by Bernoulli's equation, $V_1^2 = V_n^2$. Since the velocity is tangent to the surface at all control points, and since the tangential vectors $\vec{t}_1$ and $\vec{t}_n$ point toward and away from the trailing edge, respectively,

$$V_1 = -V_n \tag{12.3.62}$$

Thus,

$$\sum_{j=2}^{j=n}\frac{Q_i}{2\pi}(t_{1_1}G_{1j} + t_{2_1}H_{1j}) + \sum_{j=1}^{j=n-1}\frac{Q_i}{2\pi}(t_{1_n}G_{nj} + t_{2_n}H_{nj})$$

$$+ \Lambda\left[-\tfrac{1}{2}(S_1 + S_n) + \frac{1}{2\pi}\sum_{j=2}^{j=n}(t_{1_1}A_{1j} + t_{2_1}B_{1j}) + \frac{1}{2\pi}\sum_{j=1}^{j=n}(t_{1_n}A_{nj} + t_{2_n}B_{nj})\right]$$

$$= -(t_{1_1} + t_{1_n})\cos\alpha - (t_{2_1} + t_{2_n})\sin\alpha \tag{12.3.63}$$

Equations (12.3.55) and (12.3.63) are solved for the source/sink strengths q_i and vortex sheet parameter Λ.

The lift on the airfoil is obtained from the Kutta-Zhukovsky theorem [Eq. (12.3.47)]. The circulation is defined by Eq. (12.3.27), which yields

$$l = -\rho U^2\Lambda\sum_{i=1}^{i=n}S_i\Delta s_i \tag{12.3.64}$$

and the corresponding lift coefficient

$$c_l = \frac{l}{\frac{1}{2}\rho U^2 c} = -\frac{2\Lambda}{c}\sum_{i=1}^{i=n}S_i\Delta s_i \tag{12.3.65}$$

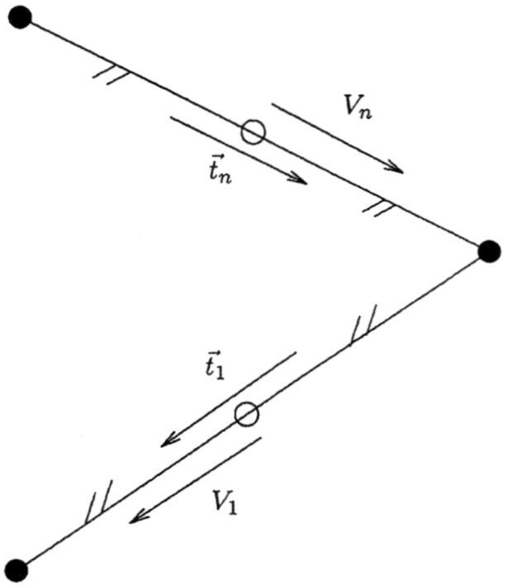

Figure 12.3.7 Velocities at trailing edge.

where c is the airfoil chord. The drag on the airfoil (i.e., the component of force acting in the direction of the freestream velocity) is zero.

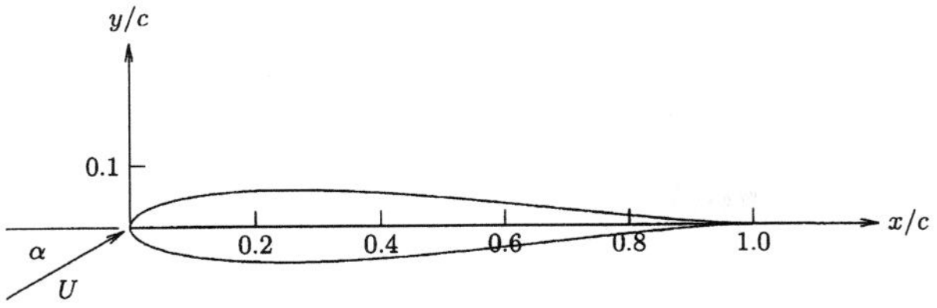

Figure 12.3.8 Zhukovsky airfoil.

The above method is applied to the flow past a Zhukovsky airfoil (Houghton and Carpenter, 1993) as illustrated in Fig. 12.3.8. The airfoil coordinates are defined as

$$x = \frac{1}{4\kappa_1}\left[\xi\left(1+\frac{1}{\xi^2+\eta^2}\right)+2\kappa_2\right]$$
$$y = \frac{1}{4\kappa_1}\left[\eta\left(1-\frac{1}{\xi^2+\eta^2}\right)\right] \qquad (12.3.66)$$

where

$$\xi = -e+(1+e)\cos\theta'$$
$$\eta = (1+e)\sin\theta'$$
$$\kappa_1 = (1+2e+e^2)(1+2e)^{-1} \qquad (12.3.67)$$
$$\kappa_2 = (1+2e+2e^2)(1+2e)^{-1}$$

where e is the *eccentricity*. Note that the airfoil leading edge is located at $(x, y) = (0, 0)$ and the trailing edge at $(x, y) = (1, 0)$. The computed pressure distribution for a Zhukovsky airfoil with $e = 0.1$ an angle of attack $\alpha = 10°$ using $n = 50$ segments is shown in Fig. 12.3.9 together with the exact solution*

$$c_p = 1-4\kappa_3[\sin(\theta'-\alpha)+\sin\alpha]^2 \qquad (12.3.68)$$

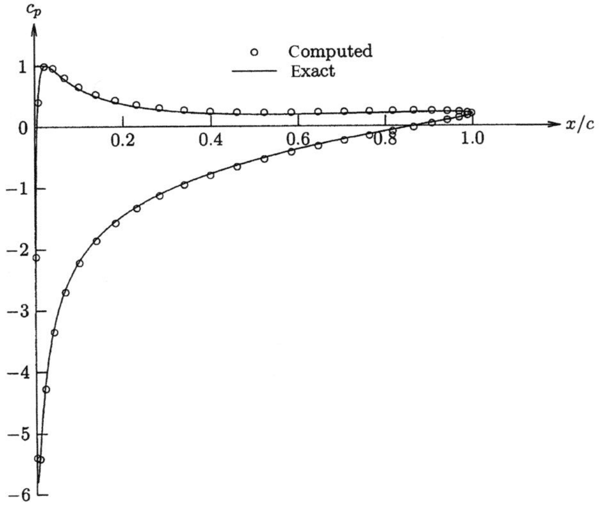

Figure 12.3.9 Pressure coefficient on airfoil.

* The pressure coefficient at the trailing edge $(x, y) = (1, 0)$, corresponding to $\theta' = 0$, is obtained by two successive applications of L'Hospital rule to yield

$$c_p|_{\theta'=0} = 1-(1+e)^{-2}\cos^2\alpha$$

where

$$\kappa_3 = \left[1-2\,\frac{(1-2\sin^2\theta)}{\kappa_4}+\frac{1}{\kappa_4^2}\right]^{-1}$$
$$\kappa_4 = [-e+(1+e)\cos\theta']^2+[(1+e)\sin\theta']^2 \qquad (12.3.69)$$
$$\sin^2\theta = \frac{(1+e)^2\sin^2\theta'}{\kappa_4}$$

BOUNDARY LAYER FLOW

At high Reynolds number Re and in the absence of strong pressure gradients,† the viscous stresses generate a thin *boundary layer* adjacent to solid surfaces wherein the flow velocity decreases to zero at the surface. A detailed history of the development of boundary layer theory is presented by Tani (1977).

Consider a body-oriented coordinate system (Fig. 12.3.10) where the x and y coordinates are locally tangential and normal to the surface, respectively. For laminar, steady, two-dimensional flow, the boundary layer equations are obtained from Eqs. (12.3.3) and (12.3.4) under the assumption that the normal y derivatives are large compared to the streamwise x derivatives.‡ The steady boundary layer equations are (with ˜ omitted for clarity):

$$\frac{\partial u}{\partial x}+\frac{\partial v}{\partial y} = 0 \qquad (12.3.70)$$

$$u\frac{\partial u}{\partial x}+v\frac{\partial u}{\partial y} = -\frac{1}{\rho}\frac{dp}{dx}+\hat{\nu}\frac{\partial^2 u}{\partial y^2} \qquad (12.3.71)$$

where $\hat{\nu} = \mu/\rho$ is the kinematic viscosity. The boundary conditions are

$$\begin{aligned} u &= 0 && \text{on } y = 0 \\ v &= 0 && \text{on } y = 0 \\ u &= U_e(x) && \text{as } y \to \infty \end{aligned} \qquad (12.3.72)$$

where $U_e(x)$ is the velocity outside of the boundary layer at the streamwise location x, and dp/dx is the streamwise pressure gradient. Both $U_e(x)$ and $dp(x)/dx$ are obtained from the solution of the irrotational flow (see "Irrotational Flow," above).

Mangler-Levy-Lees Transformation

The boundary layer equations may be rewritten by using the Mangler-Levy-Lee transformation (Cebeci and Cousteix, 1999):

$$d\xi = \rho\mu U_e(x)\,dx \qquad (12.3.73)$$

$$d\eta = \frac{\rho U_e(x)}{\sqrt{2\xi}}\,dy \qquad (12.3.74)$$

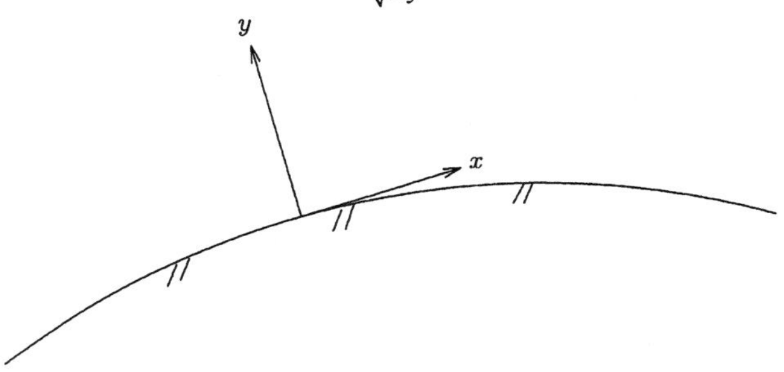

Figure 12.3.10 Coordinate system.

† A sufficiently strong pressure gradient will cause the boundary layer to *separate* from the body surface. Under such circumstances, the boundary layer equations are no longer valid and the full Navier-Stokes equations (see "Navier-Stokes Flow," below) must be solved.

‡ In addition, it is assumed that the local boundary layer thickness $\delta(x)$ is small compared to the local radius of curvature R of the boundary.

which may be integrated to obtain

$$\xi = \rho\mu \int_0^x U_e(x')\,dx' \tag{12.3.75}$$

$$n = \frac{\rho U_e(x) y}{\sqrt{2\xi}} \tag{12.3.76}$$

By introducing the stream function ψ according to

$$\rho u = \frac{\partial \psi}{\partial y}$$
$$\rho v = -\frac{\partial \psi}{\partial x} \tag{12.3.77}$$

the conservation of mass, [Eq. (12.3.70)] is exactly satisfied. Introducing the dimensionless function $f(\xi, \eta)$ according to

$$\psi(\xi, \eta) = \sqrt{2\xi}\, f(\xi, \eta) \tag{12.3.78}$$

we see that the velocity components become

$$u = U_e \frac{\partial f}{\partial n} \tag{12.3.79}$$

$$v = -\left[\frac{f}{\sqrt{2\xi}} + \sqrt{2\xi}\frac{\partial f}{\partial \xi}\right]\mu U_e - \frac{\sqrt{2\xi}}{\rho}\frac{\partial f}{\partial \eta}\left[\frac{\eta}{U_e}\frac{dU_e}{dx} - \frac{\eta\rho\mu U_e}{2\xi}\right] \tag{12.3.80}$$

The momentum equation (12.3.71) becomes

$$\frac{\partial^3 f}{\partial \eta^3} + f\frac{\partial^2 f}{\partial \eta^2} + \beta\left[1 - \left(\frac{\partial f}{\partial \eta}\right)^2\right] - 2\xi\left[\frac{\partial f}{\partial \eta}\frac{\partial^2 f}{\partial \eta\,\partial \xi} - \frac{\partial^2 f}{\partial \eta^2}\frac{\partial f}{\partial \xi}\right] = 0 \tag{12.3.81}$$

where $\beta(\xi)$ is the dimensionless pressure gradient parameter

$$\beta = \frac{2\xi}{U_e}\frac{dU_e}{d\xi} \tag{12.3.82}$$

The boundary conditions [Eq. (12.3.72)] become

$$\begin{aligned} \frac{\partial f}{\partial \eta} &= 0 && \text{at } \eta = 0 \\ f &= 0 && \text{at } \eta = 0 \\ \frac{\partial f}{\partial \eta} &= 1 && \text{as } \eta \to \infty \end{aligned} \tag{12.3.83}$$

Keller Box Scheme

Given the pressure gradient parameter $\beta(\xi)$, the momentum equation (12.3.81) can be solved by using Keller's box scheme (Keller, 1975). A grid of points is defined by

$$\xi_i = \xi_{i-1} + \Delta\xi_i \qquad \text{for } i = 2, \ldots, m \tag{12.3.84}$$

$$\eta_j = \eta_{j-1} + \Delta\eta_j \qquad \text{for } j = 2, \ldots, n \tag{12.3.85}$$

where $\xi_1 = \eta_1 = 0$ (Fig. 12.3.11). Note that $\Delta\xi_i$ and $\Delta\eta_j$ may be taken to be arbitrary positive values consistent with adequate resolution of the boundary layer in the streamwise and normal directions. The boundary

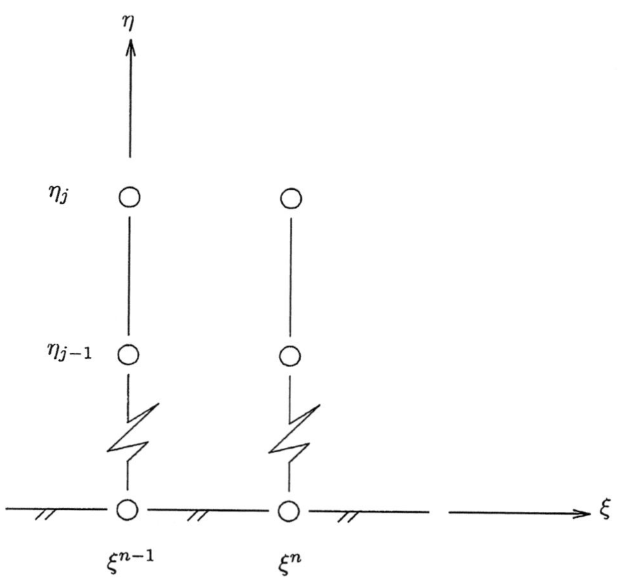

Figure 12.3.11 Finite-difference grid.

layer momentum equation (12.3.81) is rewritten as a system of first-order partial differential equations

$$\frac{\partial f}{\partial \eta} = g \tag{12.3.86}$$

$$\frac{\partial g}{\partial \eta} = h \tag{12.3.87}$$

$$\frac{\partial h}{\partial \eta} + fh + \beta(1 - g^2) - 2\xi\left(g\frac{\partial g}{\partial \xi} - h\frac{\partial f}{\partial \xi}\right) = 0 \tag{12.3.88}$$

where Eq. (12.3.88) is recognized as the boundary layer momentum equation (12.3.81). The function $f(\xi, \eta)$, $g(\xi, \eta)$, and $h(\xi, \eta)$ are defined at the discrete grid points by using the notation $f_{i,j} = f(\xi_i, \eta_j)$, etc. Equations (12.3.86) and (12.3.87) are discretized as

$$f_{i,j} - f_{i,j-1} - \tfrac{1}{2}\Delta\eta_j(g_{i,j} + g_{i,j-1}) = 0 \tag{12.3.89}$$

$$g_{i,j} - g_{i,j-1} - \tfrac{1}{2}\Delta\eta_j(h_{i,j} + h_{i,j-1}) = 0 \tag{12.3.90}$$

Equation (12.3.88) is discretized about the centroid of the cell (Fig. 12.3.12) to yield

$$\begin{aligned} &h_{i,j} + h_{i-1,j} - h_{i,j-1} - h_{i-1,j-1} + \tfrac{1}{2}\Delta\eta_j\,\overline{fh}_{i,j} + \Delta\eta_j\,\bar{\beta}_i(1 - \tfrac{1}{4}\,\overline{g^2}_{i,j}) \\ &\quad - \tfrac{1}{4}\frac{\Delta\eta_j}{\Delta\xi_i}\bar{\xi}_i\,[\bar{g}_{i,j}(g_{i,j} + g_{i,j-1} - g_{i-1,j} - g_{i-1,j-1}) \\ &\quad - \bar{h}_{i,j}(f_{i,j} + f_{i,j-1} - f_{i-1,j} - f_{i-1,j-1})] = 0 \end{aligned} \tag{12.3.91}$$

where

$$\overline{fh}_{i,j} = f_{i,j}h_{i,j} + f_{i-1,j}h_{i-1,j} + f_{i,j-1}h_{i,j-1} + f_{i-1,j-1}h_{i-1,j-1}$$

$$\bar{\xi}_i = (\xi_i + \xi_{i-1})$$

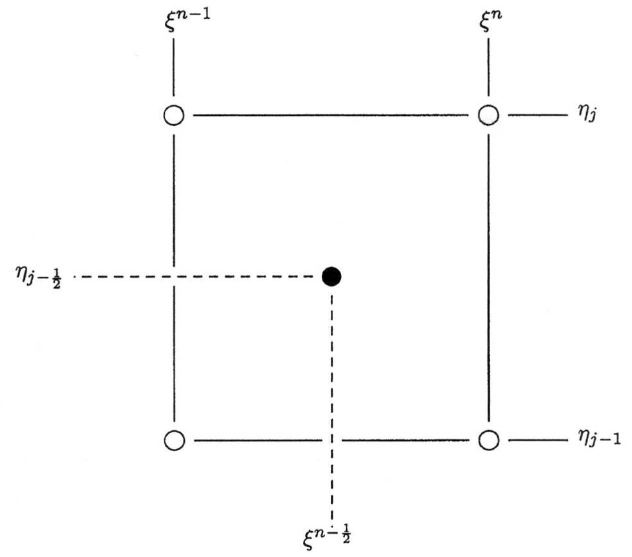

Figure 12.3.12 Computational cell.

with similar expressions for $\bar{g}_{i,j}$, $\overline{g^2}_{i,j}$, $\bar{h}_{i,j}$, and $\bar{\beta}_i$. The boundary conditions are

$$
\begin{aligned}
f_{i,1} &= 0 \\
g_{i,1} &= 0 \\
g_{i,n} &= 1
\end{aligned}
\tag{12.3.92}
$$

for $i = 2, \ldots, m$. The third boundary condition assumes η_n is outside the boundary layer.

The discrete Eqs. (12.3.89), (12.3.90), and (12.3.91) are second-order accurate in ξ and η, and may be solved by using an implicit technique. Define

$$G_{i,j} = f_{i,j} - f_{i,j-1} - \tfrac{1}{2}\Delta\eta_j (g_{i,j} + g_{i,j-1}) \tag{12.3.93}$$

$$H_{i,j} = g_{i,j} - g_{i,j-1} - \tfrac{1}{2}\Delta\eta_j (h_{i,j} + h_{i,j-1}) \tag{12.3.94}$$

$$
\begin{aligned}
F_{i,j} = {} & h_{i,j} + h_{i-1,j} - h_{i,j-1} - h_{i-1,j-1} + \tfrac{1}{2}\Delta\eta_j \overline{fh}_{i,j} \\
& + \Delta\eta_j\, \bar{\beta}_i\left(1 - \tfrac{1}{4}\overline{g^2}_{i,j}\right) - \tfrac{1}{4}\frac{\Delta\eta_j}{\Delta\xi_i}\bar{\xi}_i\, [\bar{g}_{i,j}(g_{i,j} + g_{i,j-1} - g_{i-1,j} - g_{i-1,j-1}) \\
& - \bar{h}_{i,j}(f_{i,j} + f_{i,j-1} - f_{i-1,j} - f_{i-1,j-1})]
\end{aligned}
\tag{12.3.95}
$$

for $j = 2, \ldots, n$. Then the discrete equations become

$$
\begin{aligned}
& f_{i,1} = 0 \\
& g_{i,1} = 0 \\
& F_{i,j} = 0 \qquad \text{for } j = 2, \ldots, n \\
& G_{i,j} = 0 \qquad \text{for } j = 2, \ldots, n \\
& H_{i,j} = 0 \qquad \text{for } j = 2, \ldots, n \\
& g_{i,n} - 1 = 0
\end{aligned}
\tag{12.3.96}
$$

for $i = 2, \ldots, m$. These constitute $3n$ equations in $3n$ unknowns and may be solved by Newton's method. Define the vector of equations

$$R_i = (R_{i,1} R_{i,2}, \ldots, R_{i,n})^{\mathrm{T}} \tag{12.3.97}$$

where ^{T} denotes the vector transpose and $R_{i,j}$ are column vectors of length three defined by

$$
\begin{aligned}
R_{i,1} &= (f_{i,1},\ g_{i,1},\ F_{i,2})^{\mathrm{T}} \\
R_{i,2} &= (G_{i,2},\ H_{i,2},\ F_{i,3})^{\mathrm{T}} \\
& \ldots \\
R_{i,j} &= (G_{i,j},\ H_{i,j},\ F_{i,j+1})^{\mathrm{T}} \\
R_{i,n} &= (G_{i,n},\ H_{i,n},\ g_{i,n} - 1)^{\mathrm{T}}
\end{aligned}
\tag{12.3.98}
$$

Define the vector of unknowns

$$U_i = (U_{i,1},\ U_{i,2},\ \ldots,\ U_{i,n})^{\mathrm{T}} \tag{12.3.99}$$

where

$$U_{i,j} = (f_{i,j},\ g_{i,j},\ h_{i,j})^{\mathrm{T}} \tag{12.3.100}$$

Equations (12.3.97) are solved by Newton's method. The current iterate for U_i^k is updated:

$$\frac{\partial R_i}{\partial U_i}\delta U_i^k = -R_i \tag{12.3.101}$$

and

$$U_i^{k+1} = U_i^k + \delta U_i^k \tag{12.3.102}$$

The Jacobian matrix is block tridiagonal,

$$
\frac{\partial R_i}{\partial U_i} = \left\{
\begin{matrix}
A_1 & C_1 & & & & \\
B_2 & A_2 & C_2 & & & \\
 & B_3 & A_3 & C_3 & & \\
 & & \ddots & \ddots & \ddots & \\
 & & & B_{n-1} & A_{n-1} & C_{n-1} \\
 & & & & B_n & A_n
\end{matrix}
\right\}
\tag{12.3.103}
$$

The diagonal matrices are

$$
\begin{aligned}
A_1 &= \left\{\begin{matrix} 1 & 0 & 0 \\ 0 & 1 & 0 \\ a_{3,1_1} & a_{3,2_1} & a_{3,3_1} \end{matrix}\right\} \qquad \text{for } j = 1 \\
A_j &= \left\{\begin{matrix} 1 & -\tfrac{1}{2}\Delta\eta_j & 0 \\ 0 & 1 & -\tfrac{1}{2}\Delta\eta_j \\ a_{3,1_j} & a_{3,2_j} & a_{3,3_j} \end{matrix}\right\} \qquad \text{for } j = 2, \ldots, n-1 \\
A_n &= \left\{\begin{matrix} 1 & -\tfrac{1}{2}\Delta\eta_n & 0 \\ 0 & 1 & -\tfrac{1}{2}\Delta\eta_n \\ 0 & 1 & 0 \end{matrix}\right\} \qquad \text{for } j = n
\end{aligned}
\tag{12.3.104}
$$

where

$$
\begin{aligned}
a_{3,1_j} &= \tfrac{1}{2}\Delta\eta_{j+1} h_j + \tfrac{1}{4}\frac{\Delta\eta_{j+1}}{\Delta\xi_i}\bar{\xi}_i \bar{h}_{i,j+1} \\
a_{3,2_j} &= -\tfrac{1}{2}\Delta\eta_j \bar{\beta}_i g_{i,j} - \tfrac{1}{4}\frac{\Delta\eta_{j+1}}{\Delta\xi_i}\bar{\xi}_i (g_{i,j+1} + g_{i,j} - g_{i-1,j+1} - g_{i-1,j} + \bar{g}_{i,j+1}) \\
a_{3,3_j} &= -1 + \tfrac{1}{2}\Delta\eta_{j+1} f_{i,j} + \tfrac{1}{4}\frac{\Delta\eta_{j+1}}{\Delta\xi_i}\bar{\xi}_i (f_{i,j+1} + f_{i,j} - f_{i-1,j+1} - f_{i-1,j})
\end{aligned}
\tag{12.3.105}
$$

The off-diagonal matrices are

$$B_j = \begin{Bmatrix} -1 & -\frac{1}{2}\Delta\eta_j & 0 \\ 0 & -1 & -\frac{1}{2}\Delta\eta_j \\ 0 & 0 & 0 \end{Bmatrix} \quad \text{for } j = 2, \ldots, n \qquad (12.3.106)$$

$$C_j = \begin{Bmatrix} 0 & 0 & 0 \\ 0 & 0 & 0 \\ c_{3,1_j} & c_{3,2_j} & c_{3,3_j} \end{Bmatrix} \quad \text{for } j = 1, \ldots, n-1 \qquad (12.3.107)$$

where

$$c_{3,1_j} = \tfrac{1}{2}\Delta\eta_{j+1} h_{i,j+1} + \tfrac{1}{4}\frac{\Delta\eta_{j+1}}{\Delta\xi_i}\bar{\xi}_i \bar{h}_{i,j+1}$$

$$c_{3,2_j} = -\tfrac{1}{2}\Delta\eta_j \bar{\beta}_i g_{i,j+1} - \tfrac{1}{4}\frac{\Delta\eta_{j+1}}{\Delta\xi_i}\bar{\xi}_i (g_{i,j+1} + g_{i,j} - g_{i-1,j+1} - g_{i-1,j} + \bar{g}_{i,j+1})$$

$$c_{3,3_j} = 1 + \tfrac{1}{2}\Delta\eta_{j+1} f_{i,j+1} + \tfrac{1}{4}\frac{\Delta\eta_{j+1}}{\Delta\xi_i}\bar{\xi}_i (f_{i,j+1} + f_{i,j} - f_{i-1,j+1} - f_{i-1,j}) \qquad (12.3.108)$$

The above method is applied to the flow past a flat plate in the absence of a pressure gradient; that is, $dp/dx = 0$ and $U_e(x) = U_\infty$. For this case, Eqs. (12.3.75) and (12.3.76) become

$$\xi = \rho\mu U_\infty x \qquad (12.3.109)$$

$$\eta = y\sqrt{\frac{U_\infty}{2\hat{v}x}} \qquad (12.3.110)$$

and a self-similar solution (the *Blasius solution*) is obtained, wherein the dimensionless function f depends only on η, and

$$u = U_\infty \frac{df}{d\eta} \qquad (12.3.111)$$

$$v = U_\infty \left(\frac{U_\infty x}{\hat{v}}\right)^{-1/2} \frac{1}{\sqrt{2}}(-f + g\eta) \qquad (12.3.112)$$

The computed streamwise and normal velocities are displayed in Figs. 12.3.13 and 12.3.14, respectively, for a grid of 50 points

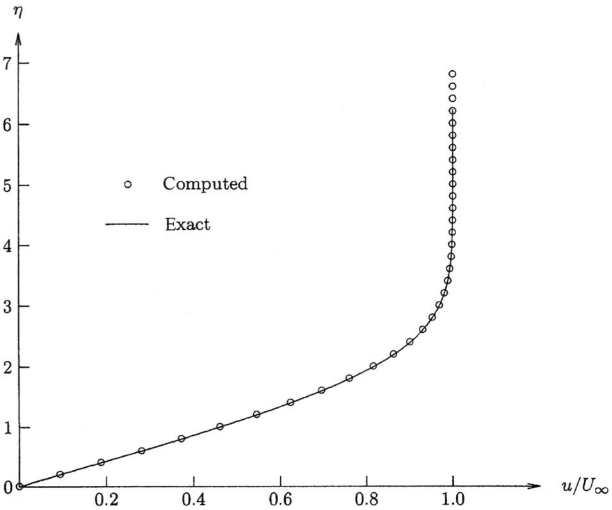

Figure 12.3.13 Velocity at x direction.

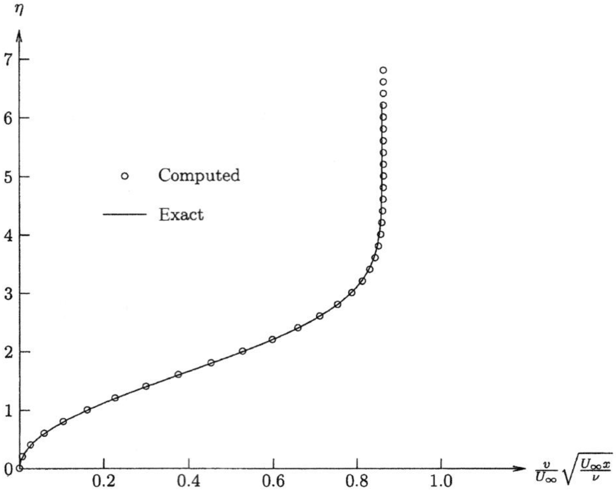

Figure 12.3.14 Velocity in y direction.

in the η direction with $\Delta\eta = 0.2$, together with the exact solution (Schlichting, 1968).

NAVIER-STOKES FLOW

The unsteady Navier-Stokes equations are (with ˜ omitted for clarity):

$$\frac{\partial u}{\partial x} + \frac{\partial v}{\partial y} = 0 \qquad (12.3.113)$$

$$\frac{\partial u}{\partial t} + u\frac{\partial u}{\partial x} + v\frac{\partial u}{\partial y} = -\frac{1}{\rho}\frac{\partial p}{\partial x} + \hat{v}\left(\frac{\partial^2 u}{\partial x^2} + \frac{\partial^2 u}{\partial y^2}\right) \qquad (12.3.114)$$

$$\frac{\partial v}{\partial t} + u\frac{\partial v}{\partial x} + v\frac{\partial v}{\partial y} = -\frac{1}{\rho}\frac{\partial p}{\partial y} + \hat{v}\left(\frac{\partial^2 v}{\partial x^2} + \frac{\partial^2 v}{\partial y^2}\right) \qquad (12.3.115)$$

where $\hat{v} = \mu/\rho$ is the kinematic viscosity. The first boundary condition is

$$u = u_b \quad \text{and} \quad v = v_b \quad \text{on solid boundary} \qquad (12.3.116)$$

where u_b and v_b are the x and y components of the velocity of the boundary. Equation (12.3.116) is the *no-slip condition*. The second boundary condition is u and v approach their freestream values away from the solid boundary

(12.3.117)

SIMPLE Algorithm

The Navier-Stokes equations (12.3.113) to (12.3.115) can be solved by using Patankar's SIMPLE algorithm (Patankar, 1980). A staggered grid (Harlow and Welch, 1965) is used to avoid unphysical oscillations in the pressure. Consider a uniform grid with constant spacing Δx and Δy, as illustrated in Fig. 12.3.15. The x and y components of velocity are defined on adjacent sides of the computational cell, and the pressure is defined at the center of the cell.

The conservation of mass is applied to the finite volume in Fig. 12.3.15 to obtain

$$(u_{i,j}^{n+1} - u_{i-1,j}^{n+1}) + (v_{i,j}^{n+1} - v_{i-1,j}^{n+1}) = 0 \qquad (12.3.118)$$

where i and j indicate the indices in the x and y directions, respectively, and n indicates the time level. The conservation of x momentum

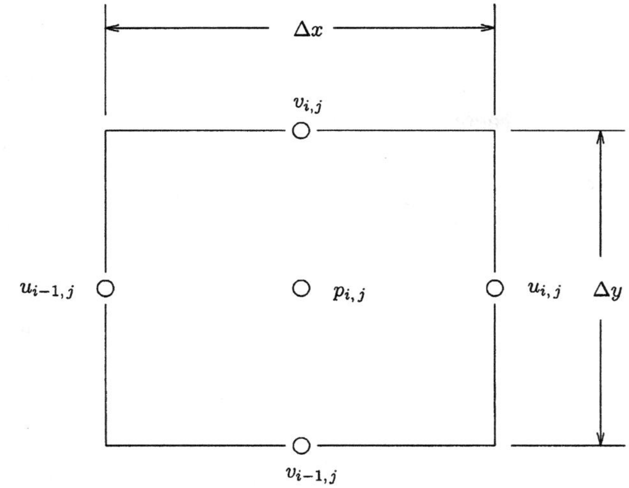

Figure 12.3.15 Staggered grid.

[Eq. (12.3.114)] is applied to the finite volume illustrated in Fig. 12.3.16 to obtain

$$\frac{\Delta x\ \Delta y}{\Delta t}\ (u_{i,j}^{n+1} - u_{i,j}^{n}) + (F_{i+\frac{1}{2},j}^{(1)} - F_{i-\frac{1}{2},j}^{(1)})\Delta y$$
$$+ (G_{i,j+\frac{1}{2}}^{(1)} - G_{i,j-\frac{1}{2}}^{(1)})\Delta x + (p_{i+1,j}^{n+1} - p_{i,j}^{n})\Delta y = 0 \qquad (12.3.119)$$

where

$$F^{(1)} = u^2 - \hat{\nu}\ \frac{\partial u}{\partial x} \qquad (12.3.120)$$

$$G^{(1)} = uv - \hat{\nu}\frac{\partial u}{\partial y} \qquad (12.3.121)$$

Similarly, the conservation of y momentum [Eq. (12.3.115)] is applied to the finite volume illustrated in Fig. 12.3.17 to obtain

$$\frac{\Delta x\,\Delta y}{\Delta t}(v_{i,j}^{n+1} - v_{i,j}^{n}) + (F_{i+\frac{1}{2},j}^{(2)} - F_{i-\frac{1}{2},j}^{(2)})\Delta y$$
$$+ (G_{i,j+\frac{1}{2}}^{(2)} - G_{i,j-\frac{1}{2}}^{(2)})\Delta x + (p_{i,j+1}^{n+1} - p_{i,j}^{n})\Delta x = 0 \qquad (12.3.122)$$

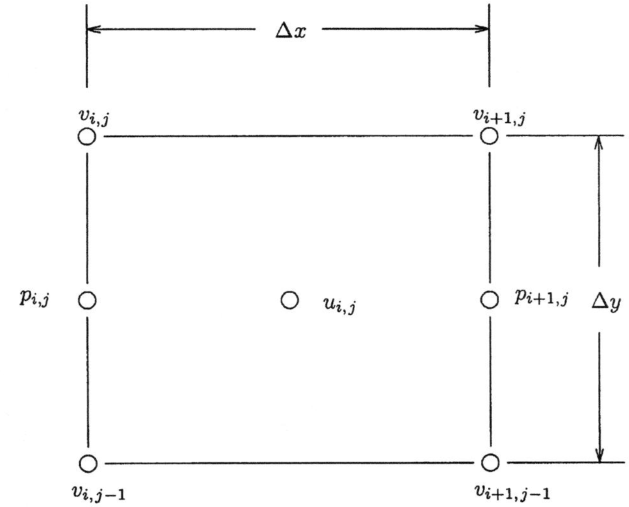

Figure 12.3.16 Finite volume for x-momentum equation.

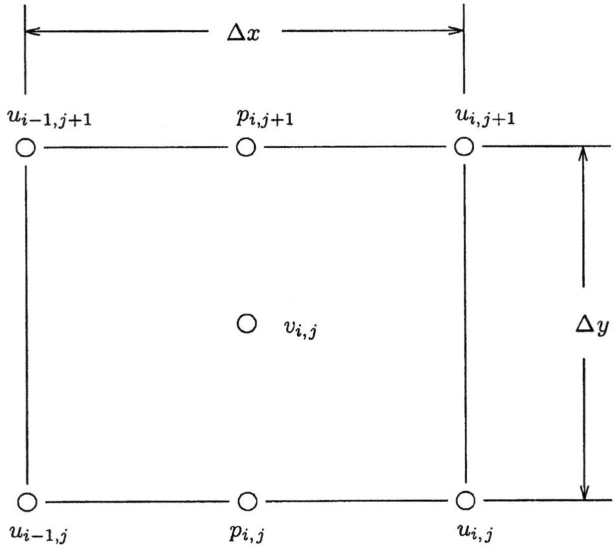

Figure 12.3.17 Finite volume for y-momentum equation.

where

$$F^{(2)} = uv - \hat{\nu}\frac{\partial v}{\partial x} \qquad (12.3.123)$$

$$G^{(2)} = v^2 - \hat{\nu}\ \frac{\partial v}{\partial y} \qquad (12.3.124)$$

We consider the development of a discrete expression for $F^{(1)}$. Consider the one-dimensional steady analog of Eqs. (12.3.113) and (12.3.114) in the absence of a pressure gradient,

$$\frac{\partial u}{\partial x} = 0 \qquad (12.3.125)$$

$$\frac{\partial}{\partial x}\left(u^2 - \hat{\nu}\frac{\partial u}{\partial x}\right) = 0 \qquad (12.3.126)$$

with boundary conditions

$$u = u_1 \qquad \text{at } x = x_1$$
$$u = u_2 \qquad \text{at } x = x_2 \qquad (12.3.127)$$

Linearizing the quadratic term in Eq. (12.3.126) as $u^2 \approx \bar{u}u$, where $\bar{u}$ is a suitable approximation to u, Eq. (12.3.126) can be solved to obtain

$$u = u_1 + (u_2 - u_1)[e^{\bar{u}(x-x_1)/\hat{\nu}} - 1]\ [e^{\bar{u}(x_2-x_1)/\hat{\nu}} - 1]^{-1} \qquad (12.3.128)$$

which yields

$$\bar{u}u - \hat{\nu}\frac{\partial u}{\partial x} = \bar{u}\ \{u_1 + (u_1 - u_2)[e^{\bar{u}(x_2-x_1)} - 1]^{-1}\} \qquad (12.3.129)$$

Considering Eq. (12.3.129) as the expression for $F_{i+\frac{1}{2},j}^{(1)}$ in Eq. (12.3.119) and identifying points 1 and 2 to be (i, j) and $(i + 1, j)$, we obtain

$$F_{i+\frac{1}{2}}^{(1)} = u_{i+\frac{1}{2},j}^{n+1}[u_{i,j}^{n+1} + (u_{i,j}^{n+1} - u_{i+1,j}^{n+1})[e^{R_{i+\frac{1}{2},j}} - 1]^{-1}] \qquad (12.3.130)$$

where

$$R_{i+\frac{1}{2},j} = \frac{u_{i+\frac{1}{2},j}^{n+1}\ \Delta x}{\hat{\nu}} \qquad (12.3.131)$$

Using a similar expression for $F^{(1)}_{i-\frac{1}{2},j}$ yields

$$(F^{(1)}_{i+\frac{1}{2},j} - F^{(1)}_{i-\frac{1}{2},j})\Delta y = a_P u^{n+1}_{i,j} - a_E u^{n+1}_{i+1,j} - a_W u^{n+1}_{i-1,j} \qquad (12.3.132)$$

where

$$a_E = u^{n+1}_{i+\frac{1}{2},j}\Delta y\left[e^{R_{i+\frac{1}{2},j}} - 1\right]^{-1} \qquad (12.3.133)$$

$$a_W = u^{n+1}_{i-\frac{1}{2},j}\Delta y\left[1 + \left[e^{R_{i-\frac{1}{2},j}} - 1\right]^{-1}\right] \qquad (12.3.134)$$

$$a_P = a_E + a_W + (u^{n+1}_{i+\frac{1}{2},j} - u^{n+1}_{i-\frac{1}{2},j})\Delta y \qquad (12.3.135)$$

The subscripts *E, W,* and *P* are mnemonics for east, west, and point, respectively. Defining a diffusion conductance *D* according to

$$D = \frac{\hat{v}}{\Delta x} \qquad (12.3.136)$$

we obtain

$$a_E = \frac{D_{i+\frac{1}{2},j}\, R_{i+\frac{1}{2},j}\,\Delta y}{e^{R_{i+\frac{1}{2},j}} - 1} \qquad (12.3.137)$$

Since the exponential is costly to evaluate computationally, Patankar (1980) developed the *Power Law* approximation

$$\frac{a_E}{\Delta y D_{i+\frac{1}{2},j}} = \begin{cases} -R_{i+\frac{1}{2},j} & R_{i+\frac{1}{2},j} < 10 \\ (1+0.1R_{i+\frac{1}{2},j})^5 - R_{i+\frac{1}{2},j} & -10 \le R_{i+\frac{1}{2},j} < 0 \\ (1-0.1R_{i+\frac{1}{2},j})^5 & 0 \le R_{i+\frac{1}{2},j} < 10 \\ 0 & R_{i+\frac{1}{2},j} \ge 10 \end{cases} \qquad (12.3.138)$$

Additional approximation (e.g., central difference, hybrid, upwind) are also possible (Patankar, 1980). The coefficients may be written as

$$a_E = D_{i+\frac{1}{2},j}\,\Delta y[\max(0,(1-0.1|R_{i+\frac{1}{2},j}|)^5) + \max(-R_{i+\frac{1}{2},j},0)] \qquad (12.3.139)$$

$$a_W = D_{i-\frac{1}{2},j}\Delta y[R_{i-\frac{1}{2},j} + \max(0,\,(1-0.1|R_{i-\frac{1}{2},j}|)^5) + \max(-R_{i-\frac{1}{2},j},\,0)] \qquad (12.3.140)$$

$$a_P = a_E + a_W + (u_{i+\frac{1}{2},j} - u_{i-\frac{1}{2},j})\Delta y \qquad (12.3.141)$$

Similarly,

$$(G^{(1)}_{i,j+\frac{1}{2}} - G^{(1)}_{i,j-\frac{1}{2}})\Delta x = b_P u^{n+1}_{i,j} - b_N u^{n+1}_{i,j+1} - b_S u^{n+1}_{i,j-1} \qquad (12.3.142)$$

where

$$b_N = D_{i,j+\frac{1}{2}}\,\Delta x[\max(0,(1-0.1|R_{i,j+\frac{1}{2}}|)^5) + \max(-R_{i,j+\frac{1}{2}},\,0)] \qquad (12.3.143)$$

$$b_S = D_{i,j-\frac{1}{2}}\,\Delta x[R_{i,j-\frac{1}{2}} + \max(0,(1-0.1|R_{i,j-\frac{1}{2}}|)^5 + \max(-R_{i,j-\frac{1}{2}},0)] \qquad (12.3.144)$$

$$b_P = b_N + b_S + (v_{i,j+\frac{1}{2}} - v_{i,j-\frac{1}{2}})\Delta x \qquad (12.3.145)$$

with

$$R_{i,j+\frac{1}{2}} = \frac{v_{i,j+\frac{1}{2}}\,\Delta y}{\hat{v}} \qquad (12.3.146)$$

The *x*- and *y*-momentum equations become

$$\frac{\Delta x\,\Delta y}{\Delta t}(u^{n+1}_{i,j} - u^n_{i,j}) + a_P u^{n+1}_{i,j} - a_E u^{n+1}_{i+1,j} - a_W u^{n+1}_{i-1,j} + b_P u^{n+1}_{i,j} - b_N u^{n+1}_{i,j+1} - b_S u^{n+1}_{i,j-1} + \Delta y(p^{n+1}_{i+1,j} - p^{n+1}_{i,j}) = 0 \qquad (12.3.147)$$

$$\frac{\Delta x\,\Delta y}{\Delta t}(v^{n+1}_{i,j} - v^n_{i,j}) + c_P v^{n+1}_{i,j} - c_E v^{n+1}_{i+1,j} - c_W v^{n+1}_{i-1,j} + d_P v^{n+1}_{i,j} - d_N v^{n+1}_{i,j+1} - d_S v^{n+1}_{i,j-1} + \Delta x(p^{n+1}_{i,j+1} - p^{n+1}_{i,j}) = 0 \qquad (12.3.148)$$

where the coefficients c_E, c_W, c_P and d_N, d_S, d_P are given by Eqs. (12.3.139) to (12.3.141) and (12.3.143) to (12.3.145) with due consideration to the expressions for $R_{i\pm\frac{1}{2},j}$ and $R_{i,j\pm\frac{1}{2}}$.

Equations (12.3.147) and (12.3.148) are solved by iteration, using the pressure field at the previous time step $p^n_{i,j}$. The resultant expressions for the velocity components, denoted $\hat{u}^{n+1}_{i,j}$ and $\hat{v}^{n+1}_{i,j}$, therefore satisfy

$$\frac{\Delta x\,\Delta y}{\Delta t}(\hat{u}^{n+1}_{i,j} - u^n_{i,j}) + a_P\hat{u}^{n+1}_{i,j} - a_E\hat{u}^{n+1}_{i+1,j} - a_W\hat{u}^{n+1}_{i-1,j} + b_P\hat{u}^{n+1}_{i,j} - b_N\hat{u}^{n+1}_{i,j+1} - b_S\hat{u}^{n+1}_{i,j-1} + \Delta y(p^n_{i+1,j} - p^n_{i,j}) = 0 \qquad (12.3.149)$$

$$\frac{\Delta x\,\Delta y}{\Delta t}(\hat{v}^{n+1}_{i,j} - v^n_{i,j}) + c_P\hat{v}^{n+1}_{i,j} - c_E\hat{v}^{n+1}_{i+1,j} - c_W\hat{v}^{n+1}_{i-1,j} + d_P\hat{v}^{n+1}_{i,j} - d_N\hat{v}^{n+1}_{i,j+1} - d_S\hat{v}^{n+1}_{i,j-1} + \Delta x(p^n_{i,j+1} - p^n_{i,j}) = 0 \qquad (12.3.150)$$

However, $\hat{u}_{i,j}$ and $\hat{v}_{i,j}$ do not satisfy conservation of mass [Eq. (12.3.118)]. Consequently, a correction $u'_{i,j}$ and $v'_{i,j}$ must be determined, Writing

$$u^{n+1}_{i,j} = \hat{u}^{n+1}_{i,j} + u'_{i,j} \qquad (12.3.151)$$

$$v^{n+1}_{i,j} = \hat{v}^{n+1}_{i,j} + v'_{i,j} \qquad (12.3.152)$$

$$p^{n+1}_{i,j} = p^n_{i,j} + \delta p_{i,j} \qquad (12.3.153)$$

and subtracting (12.3.149) from (12.3.147) and (12.3.150) from (12.3.148) yields

$$\left(\frac{\Delta x\,\Delta y}{\Delta t} + a_P + b_P\right)u'_{i,j} - a_W u'_{i-1,j} - a_E u'_{i+1,j} - b_S u'_{i,j-1} - b_N u'_{i,j+1} + \Delta y(\delta p_{i+1,j} - \delta p_{i,j}) = 0 \qquad (12.3.154)$$

$$\left(\frac{\Delta x\,\Delta y}{\Delta t} + c_P + d_P\right)v'_{i,j} - c_W v'_{i-1,j} - c_E v'_{i+1,j} - d_S v'_{i,j-1} - d_N v'_{i,j+1} + \Delta x(\delta p_{i,j+1} - \delta p_{i,j}) = 0 \qquad (12.3.155)$$

These equations are simplified (Patankar, 1980) to

$$\left(\frac{\Delta x\,\Delta y}{\Delta t} + a_P + b_P\right)u'_{i,j} = -\Delta y(\delta p_{i+1,j} - \delta p_{i,j}) \qquad (12.3.156)$$

$$\left(\frac{\Delta x\,\Delta y}{\Delta t} + c_P + d_P\right)v'_{i,j} = -\Delta x(\delta p_{i,j+1} - \delta p_{i,j}) \qquad (12.3.157)$$

The expressions in Eqs. (12.3.151) and (12.3.152), together with Eqs. (12.3.156) and (12.3.157), are substituted into the conservation of mass [Eq. (12.3.118)] to obtain an equation for the pressure correction:

$$g_{i,j}\delta p_{i,j} = (\hat{u}^{n+1}_{i-1,j} - \hat{u}^{n+1}_{i-1,j})\Delta y + (\hat{v}^{n+1}_{i,j} - \hat{v}^{n+1}_{i,j-1})\Delta x + (e_{i,j}\delta p_{i+1,j} + e_{i-1,j}\delta p_{i-1,j})\,\Delta y + (f_{i,j}\delta p_{i,j+1} + f_{i,j-1}\delta p_{i,j-1})\,\Delta x \qquad (12.3.158)$$

where

$$e_{i,j} = -\Delta y\left[\frac{\Delta x\,\Delta y}{\Delta t} + a_P + b_P\right]^{-1} \qquad (12.3.159)$$

$$f_{i,j} = -\Delta x\left[\frac{\Delta x\,\Delta y}{\Delta t} + c_P + d_P\right]^{-1} \qquad (12.3.160)$$

$$g_{i,j} = (e_{i,j} + e_{i-1,j})\Delta y + (f_{i,j} + f_{i,j-1})\Delta x \qquad (12.3.161)$$

The algorithm is known as SIMPLE (Semi-Implicit Method for Pressure-Linked Equations) (Patankar, 1980). The solution procedure is:

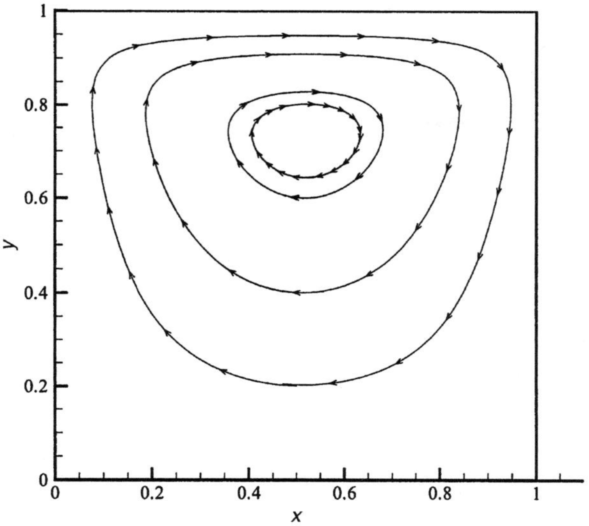

Figure 12.3.18 Driven flow in cavity.

1. The approximate solution $\hat{u}^{n+1}_{i,j}$ and $\hat{v}^{n+1}_{i,j}$ is obtained from Eqs. (12.3.149) and (12.3.150) by iteration. This can be achieved in a two-step procedure. First, each equation is written as scalar tridiagonal system along each y grid line and solved by the Thomas algorithm (Anderson et al., 1984). Second, each equation is written as a tridiagonal system along each x grid line and solved. The two-step procedure is iterated until converged.
2. The correction to the pressure δp_{ij} is obtained by solution of Eq. (12.3.158), using, for example, successive overrelaxation (Anderson et al., 1984).
3. The pressure is updated according to $p^{n+1}_{i,j} = p^n_{i,j} + \omega_p \delta p_{i,j}$.
4. The correction to the velocity, $u'_{i,j}$ and $v'_{i,j}$, is obtained from Eqs. (12.3.156) and (12.3.157) and the velocity is updated according to Eqs. (12.3.151) and (12.3.152).
5. The process is repeated until steady state is achieved (if appropriate) or the unsteady solution is obtained for the desired period of time

Several improvements to the SIMPLE algorithm have also been developed as described in Fletcher (1988b).

The above method is applied to the driven flow in a cubic cavity as shown in Fig. 12.3.18. The upper surface moves to the right at a constant speed U. The three remaining walls are stationary. The Reynolds number $\text{Re} = UL/\hat{\nu}$ is 100. The streamlines display the convective motion generated by the shearing force exerted by the movement of the upper surface.

REFERENCES

The description in this section is necessarily brief and the interested reader is directed to the numerous texts on aspects of CFD. Fletcher (1988a, b) and Hirsch (1988a, b) provide comprehensive developments of computational fluid dynamics. Laney (1998) and Toro (1997) focus on methods for compressible flows. Cebeci and Cousteix (1999) present a detailed description of boundary layer methods. Wilcox (1988) provides an essential treatment of incompressible and compressible turbulent flows. Quinn (1987) describes methods for efficient computation on parallel computers. Thompson et al. (1999) presents an encyclopedia summary of grid generation methods.

1. Anderson, D., Tannehill, J., and Pletcher, R. (1984). "Computational Fluid Mechanics and Heat Transfer." McGraw-Hill, New York.
2. Anderson, J. (2001). "Fundamentals of Aerodynamics," 3d ed. McGraw-Hill, New York.
3. Batchelor, G. (1970). "Introduction to Fluid Dynamics." Cambridge University Press, Cambridge, United Kingdom.
4. Cebeci, T., and Cousteix, J. (1999). "Modeling and Computation of Boundary-Layer Flows." Horizons Publishing (Springer), Long Beach, CA.
5. Courant, R., Friedrichs, K., and Lewy, H. (1928). "Uber die partiellen differenzengleichur-gen der mathematischen physik." Mathematische Annalen, 100:32-74. Also published in IBM Journal, March 1967, pp. 215-234.
6. Fletcher, C. (1988a). "Computational Techniques for Fluid Dynamics." Vol. I: "Fundamental and General Techniques." Springer-Verlag. Berlin.
7. Fletcher, C. (1988b). "Computational Techniques for Fluid Dynamics." Vol. II: "Specific Techniques for Different Flow Categories." Springer-Verlag, Berlin.
8. Harlow, F., and Welch, J. (1965). "Numerical calculation of time-dependent viscous incompressible flow of fluid with a free surface." *Physics of Fluids,* 8(12):2182-2189.
9. Hess, J. (1975). "Review of integral equation techniques for solving potential flow problems with emphasis on surface source methods." *Computer Methods in Applied Mechanics and Engineering*, 5:145-196.
10. Hess, J., and Smith, A. (1966). "Calculation of potential flow around arbitrary bodies." *Progress in Aeronautical Science,* 8:1-138.
11. Hirsch, C. (1988a). "Numerical Computation of Internal and External Flows." Vol. I: "Fundamentals of Numerical Discretization." John Wiley & Sons, New York.
12. Hirsch, C. (1988b). "Numerical Computation of Internal and External Flows." Vol. II: "Computational Methods for Inviscid and Viscous Flows." John Wiley & Sons, New York.
13. Houghton, E., and Carpenter, P. (1993). "Aerodynamics for Engineering Students," 4th ed. Arnold, London.
14. Keller, H. (1975). "Accurate difference methods for nonlinear two-point boundary value problems." *SIAM Journal on Numerical Analysis,* 11(2):305-320.
15. Kellogg, O. (1953). "Foundations of Potential Theory." Dover Publications, New York.
16. Kovenya, V., and Yanenko, N. (1981). "Method of Fractional Steps in Problems in Gas Dynamics." USSR Academy of Sciences, Siberian Division, Novosibirsk, Russia.
17. Laney, C. (1988). "Computational Gasdynamics." Cambridge University Press, Cambridge.
18. Lax, P., and Richtmyer, R. (1956). "Survey of the stability of linear finite difference equations." *Communications on Pure and Applied Mathematics,* 9:267-293.
19. Milne-Thompson, L. (1958). "Theoretical Aerodynamics." Dover Publications, New York.
20. Patankar, S. (1980). "Numerical Heat Transfer and Fluid Flow." Hemisphere Publishing Company, New York.
21. Peyret, R., and Taylor, T. (1983). "Computational Methods for Fluid Flow." Springer-Verlag, Berlin.
22. Quinn, M. (1987). "Designing Efficient Algorithms for Parallel Computers." McGraw-Hill, New York.
23. Richardson, L. (1910). "The approximate arithmetical solution by finite differences of physical problems involving differential equations with an application to the stresses in a masonry dam." *Transactions of the Royal Society of London,* Serial A, 210:307-357.
24. Richtmyer, R., and Morton, K. (1957). "Difference Methods for Initial-Value Problems." Interscience Publishers, New York.
25. Roache, P. (1972). "Computational Fluid Dynamics." Hermosa Publishers, Albuquerque, NM.
26. Roache, P. (1998). "Fundamentals of Computational Fluid Dynamics." Hermosa Publishers, Albuquerque, NM.
27. Schlichting, H. (1968). "Boundary Layer Theory," 6th ed. McGraw-Hill, New York.
28. Tani, I. (1977). "History of boundary-layer theory." In "Annual Review of Fluid Mechanics," pp. 87-111. Annual Reviews, Inc., Palo Alto, CA.
29. Thompson, J., Soni, B., and Weatherill, N. eds. (1999). "Handbook of Grid Generation." CRC Press, Boca Raton, FL.
30. Toro, E., ed. (1997). "Riemann Solvers and Numerical Methods for Fluid Dynamics." Springer, Berlin.
31. Wilcox, D. (1997). "Basic Fluid Mechanics." DCW Industries, Inc., La Cañada, CA.
32. Wilcox, D. (1998). "Turbulence Modeling for CFD." DCW Industries, Inc., La Cañada, CA.

12.4 EXPERIMENTAL FLUID DYNAMICS

by Yiannis Andreopoulos

INTRODUCTION

Experimental fluid dynamics (EFD) is a branch of fluid dynamics that complements the analytical and computational fluid dynamics branches. Although the complete equations describing fluid motions were derived several decades ago, today there are only a few solutions of particular problems of those nonlinear, partial differential equations. For turbulent flows, for instance, these solutions are limited to very low Reynolds number flows, which are not applicable to present data technological flows that are mostly in the high Reynolds number regime. Consequently, physical and numerical experiments will increasingly be employed to provide guidance and solutions to many engineering and technological problems.

The major objective of this section is to introduce the subject of experimental fluid dynamics broadly to readers, to provide an up-to-date view of the fundamentals of some key methods, and to present emerging needs and trends in current technological applications. The field of EFD is rather large and extends over the entire range of fluid dynamics, including not only the customary fields of aeronautics and hydraulics, but also diverse topics as biological and biomedical flows, including insect flight mechanics and pulmonary and blood flows, meteorology, oceanography, granular flows, and metallurgy.

Laboratory experiments require two very important processes: (1) designing or selecting the flow simulation platform and (2) selecting and designing the measurement system. The objectives of the experimental work will determine whether a new experimental setup is needed or an existing facility can be used. The measurement system has to be built around the flow simulation platform by accommodating possible operating restrictions and to fulfill the specific objectives of the experiments. In cases of direct measurements in the field with actual moving objects or in nature, designing the measurement system is the most critical aspect of the work.

Fluid mechanical quantities to be measured can range from forces and flow rates to dissipation rates of kinetic energy and enstrophy. Measurements of lift and drag forces of a body moving in air or water or the power attainable in a turbomachine are typical tasks in engineering applications, while the measurements of the rate-of-strain tensor and vorticity are typical of research-oriented problems that can provide better understanding of the physics of the flows and lead to analytical and computational tools that can predict the efficiency of fluid systems more accurately than existing methods.

EFD encompasses single- and multicomponent fluids, multiphase flows, Newtonian and non-Newtonian flows, chemically reacting flows, and biological fluid mechanics. We will restrict our description to single component, chemically nonreactive flows. In that context, issues related to incompressible and compressible flows will be discussed in detail. Particular emphasis will be given to techniques related to applications in emerging technologies and to measurements of quantities that are not in the mainstream, such as spatially and temporarily resolved vorticity, rate-of-strain tensor, and dissipation rate of kinetic energy, which are essential in describing the efficiency of fluid systems. The scope of the present contribution excludes the description of facilities used to simulate specific flow phenomena. Extensive coverage of items like wind tunnels and water tunnels, towing tanks and wave tanks, and shock tubes is provided in Refs. 1 to 7 and at NASA Web sites.

BACKGROUND INFORMATION

The relative importance of each of the measurable fluid mechanical quantities and their relationship can be appreciated by considering the application of conservation laws of mass, momentum, and energy, which leads to expressions in the form of integral or differential equations. The continuum approximation will be assumed to be valid in the present description. The fluid flow is inside a control volume V bounded by a control surface ∂V with n, the unit vector, normal in the outside direction. Here $U(x, t) = (U_1, U_2, U_3)$ is the **velocity field** as a function of location x and time t, ρ is the density, E is the **total specific energy** $E_t = (e + \frac{1}{2}U_iU_i)$ and e is the specific internal energy. Einstein's summation convention of repeated indices is applied here.

Conservation of Mass

$$\frac{d}{dt}\int_V \rho\, dV + \int_{\partial V} \rho U_k n_k\, dA = 0 \qquad (12.4.1)$$

$$\frac{\partial \rho}{\partial t} + \frac{\partial}{\partial x_k}(\rho U_k) = 0 \qquad (12.4.2)$$

Conservation of Momentum: Newton's Second Law

$$\frac{d}{dt}\int_V \rho U_i\, dV + \int_{\partial V} \rho (U_k n_k) U_i\, dA = -\int_{\partial V} \sigma_{ki} n_k\, dA + \int_V \rho f_i\, dV \qquad (12.4.3)$$

where σ_{ki} is the second-order stress tensor and f_i represents the components of the specific external forces acting on this control volume.

$$\frac{\partial}{\partial t}(\rho U_i) + \frac{\partial}{\partial x_k}(\rho U_k U_i) = \frac{\partial \sigma_{ki}}{\partial x_k} + \rho f_i \qquad (12.4.4)$$

where $\sigma_{ki} = -p\delta_{ki} + \tau_{ki} = -p\delta_{ki} + 2\mu S_{ki}$ and S_{ki} is the rate-of-strain tensor.

Conservation of Energy

$$\frac{d}{dt}\int_V \rho E_t\, dV + \int_{\partial V} \rho E_t U_k n_k\, dA = \int_{\partial V} U_i \sigma_{ki} n_k\, dA - \int_{\partial V} q_k n_k\, dA + \int_V \rho U_i f_i\, dV \qquad (12.4.5)$$

$$\frac{\partial}{\partial t}(\rho E_t) + \frac{\partial}{\partial x_k}(\rho U_k E_t) = \frac{\partial U_i \sigma_{ki}}{\partial x_k} + \rho U_i f_i \qquad (12.4.6)$$

Velocity Gradient Tensor, Strain-Rate, Vorticity, and Dissipation Rate

The velocity gradient tensor A_{ij} can be decomposed, not uniquely, into a symmetric and antisymmetric part. Specifically, these two parts are the rate-of-strain tensor $S_{ij} = \frac{1}{2}(\partial U_i/\partial x_j + \partial U_j/\partial x_i)$ and the-rate-of rotation tensor $R_{ij} = \frac{1}{2}(\partial U_i/\partial x_j - \partial U_j/\partial x_i) = \frac{1}{2}\varepsilon_{ijk}\Omega_k$, where Ω_k is the vorticity vector. Ω_k is defined as the curl of the velocity vector

$$\Omega_k = \epsilon_{ijk}\frac{\partial U_i}{\partial x_j} \qquad (12.4.7)$$

and ϵ_{ijk} is the alternating unit vector. At each point in the flow field we may imagine a spherical fluid particle whose motion can be decomposed into translation, expansion, shearing, and rotation. The vorticity can be interpreted as just twice the instantaneous solid-body-like rotation rate of the fluid particles along the principal axes in the fluid where there exists no shear deformation. We may alternatively interpret the **vorticity** as the circulation per unit area of a surface perpendicular to the vorticity field.

Why is vorticity so important in understanding the dynamics of turbulent fluid flow? Insight into this may be gained by looking at the equation describing the transport of vorticity[8]

$$\frac{d\Omega_i}{dt} = S_{ik}\Omega_k - \Omega_i S_{kk} + \epsilon_{iq\eta}\frac{1}{\rho^2}\frac{\partial \rho}{\partial x_q}\frac{\partial p}{\partial x_\eta} + \epsilon_{iq\eta}\frac{\partial}{\partial x_q}\left(\frac{1}{\rho}\frac{\partial \tau_{\eta j}}{\partial x_j}\right) \qquad (12.4.8)$$

This transport equation indicates that, for an observer moving with the flow, the rate of change of vorticity component Ω_i of a particle is provided by four dynamically significant processes: (1) stretching or compression and reorientation or tilting by the strain S_{ik}, (2) vorticity generation through dilatation, (3) baroclinic generation through the interaction of pressure and density gradients, and (4) viscous effects expressed by the viscous stress term. The term $S_{ik}\Omega_k$ consists of one stretching or compression component and two tilting components. This term has been illuminated mostly through numerical work and it appears to play an important role in the various physical processes involved in turbulent flows. If the viscous term can be ignored when its magnitude is small, then the change of vorticity of a fluid element in a lagrangian frame of reference can be entirely attributed to vortex stretching and/or tilting, to dilatational effects, and to baroclinic torque.

If the above equation is multiplied by Ω_i, then the transport equation for the enstrophy $\frac{1}{2}\Omega_i\Omega_i$ can be obtained:

$$\frac{d\left(\frac{1}{2}\Omega_i\,\Omega_i\right)}{dt} = S_{ik}\,\Omega_k\,\Omega_i - \Omega_i\,\Omega_i\,S_{kk} + \epsilon_{iq\eta}\frac{1}{\rho^2}\Omega_i\frac{\partial \rho}{\partial x_q}\frac{\partial p}{\partial x_\eta} + \epsilon_{iq\eta}\frac{\partial}{\partial x_q}\left(\Omega_i\frac{1}{\rho}\frac{\partial \tau_{\eta j}}{\partial x_j}\right) - \epsilon_{iq\eta}\frac{1}{\rho}\frac{\partial \Omega_j}{\partial x_q}\frac{\partial \tau_{\eta j}}{\partial x_j} \qquad (12.4.9)$$

The physical meaning of the terms associated with the time-dependent changes of enstrophy of a fluid element in a Lagrangian frame of reference are similar with those responsible for the generation or destruction of vorticity. The physical mechanisms involved in these complex processes are not fully understood. Enstrophy is a very significant quantity in fluid dynamics because it is not only related to the solenoidal dissipation, but also to several of the invariants of the rate-of-strain matrix S_{ij}, which may lead to detection of vortex tubes and shear layers in the flow field. In addition, enstrophy is a source term in the **transport equation** of **dilatation** S_{kk}:

$$\frac{dS_{kk}}{dt} = -S_{ik}S_{ki} + \frac{1}{2}\Omega_k\Omega_k + \frac{1}{\rho^2}\frac{\partial \rho}{\partial x_k}\frac{\partial p}{\partial x_k} - \frac{1}{\rho}\frac{\partial^2 p}{\partial x_k\,\partial x_k} + \frac{\partial}{\partial x_k}\left(\frac{1}{\rho}\frac{\partial \tau_{kq}}{\partial x_q}\right) \qquad (12.4.10)$$

This transport equation relates the change of dilatation along a particle path, which can be caused by the straining action of the dissipative motions ($S_{ik}S_{ik}$) as well as by the rotational energy of the spinning motions as it is expressed by the enstrophy $\frac{1}{2}\Omega_i\Omega_i$. Pressure and density gradients as well as viscous diffusion can also affect dilatation. It should be noted that the above transport equation reduces to the well-known Poisson equation for incompressible flows of constant density ($S_{kk} = 0$).

The dissipation rate of kinetic energy, defined as $E = \tau_{ij}S_{ij}$, is a key quantity in determining viscous losses in fluid systems. It can be expanded[9] as:

$$E = \tau_{ij}S_{ij} = 2\mu S_{ij}S_{ij} - \frac{2}{3}\,\mu S_{kk}S_{kk} \qquad (12.4.11)$$

The second term in the right-hand side of the above relation represents the additional contribution of compressibility to the dissipation rate of kinetic energy. This term disappears in the cases of **incompressible flow**. Since this term is always positive, the negative sign in front may erroneously suggest that compressibility reduces dissipation. This is incorrect because the term $S_{ij}S_{ij}$ also contains contributions from dilatation effects, which can be revealed if one considers that

$$S_{ij}S_{ij} = \frac{1}{2}\Omega_k\Omega_k + \frac{\partial U_i}{\partial x_j}\frac{\partial U_j}{\partial x_i} \qquad (12.4.12)$$

The second term in the right-hand side represents the inhomogeneous contribution in the case of incompressible flows. In the case of compressible flows, terms related to dilatation can be extracted, and the instantaneous total dissipation rate then becomes

$$E = \mu\,\Omega_k\Omega_k + 2\mu\left[\frac{\partial U_i}{\partial x_j}\frac{\partial U_j}{\partial x_i} - S_{kk}S_{kk}\right] + \frac{4}{3}\mu S_{kk}S_{kk} \qquad (12.4.13)$$

The third term on the right-hand side describes the direct effects of compressibility, that is, dilatation on the dissipation rate. The first and the last terms on the right-hand side of the last relation are quadratic with positive coefficients and positive signs and they are, therefore, always positive. The second term on the right-hand side indicates the contributions to the dissipation rate by the purely nonhomogeneous part of the flow. Its time-averaged contribution disappears in homogeneous flows. This term, in principle, can obtain negative values, and thus it can reduce the dissipation rate. This does not violate the second thermodynamic law as long as the total dissipation remains positive at any point in space and time.

The above description indicates that velocity, pressure, and temperature are the primary measurable quantities that can determine the local stresses acting on a body inside a flow field and therefore can be used to estimate the resultant forces and possibly torques exerted on various parts of a fluid system. The availability of this type of information will help determine the performance of the fluid system. However, if estimates of the efficiency of the system are required, determination of kinetic energy dissipation rate is essential. This demand is more critical in the case of turbulent flows, where viscous losses are substantially larger than in laminar flows.

CLASSIFICATION OF MEASUREMENTS

A typical block diagram of the measuring system of an experimental technique is shown in Fig. 12.4.1. It includes a probe, a transducer, a signal processing unit with an amplifier, an interface to the data processing unit such as an analog to digital converter, a data processing unit with a CPU, and a storage or a display unit. This type of block diagram is very useful in estimating and reducing uncertainties along the line with the various stages of the measurement system.

The output of instruments measuring fluid mechanical quantities can be purely mechanical, electrical, or optical. Most of the techniques used in EFD include probes and transducers that convert the measured value of a fluid mechanical quantity into an electrical output, which, through an appropriate interface, can feed the CPU of a computer or the local processing unit of the instrument with a display, as shown in Fig. 12.4.1. Substantial effort has been devoted in the past to develop transducers and instrumentation that convert the measured quantity into an electrical signal, which carries the information. Optical carriers and optical techniques, however, have the advantage of faster transmission and processing of information because light travels faster than electricity. In that respect the use of light and the development of optical converters in the future may provide an incentive to replace electrical converters of mechanical quantities.

In most optical techniques the use of flow tracers or flow markers is required. Flow tracers are solid particles that are introduced into the flow and scatter light when they are illuminated. These particles should have a dynamic response comparable to the dynamics of the flow; that is, they should follow a temporally fluctuating flow very closely, and therefore their density has to match that of the fluid. This problem is more critical in gas flows than in liquid flows. In addition, the presence of these particles should not interfere with the flow and change its structure. Flow markers, on the other hand, are molecules of the same fluid of the flow field or foreign molecules that emit light and can be detected when they are excited by photons. In that respect, inertia effects similar to those found in flow tracers are not present here because of the molecular nature of the flow markers.

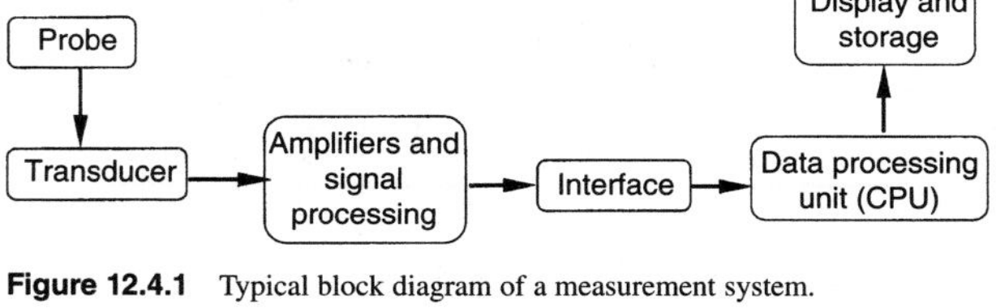

Figure 12.4.1 Typical block diagram of a measurement system.

Flow rate of information from one block to the next (see Fig. 12.4.1) can be hampered by the limited processing speed and storage capability in each of the blocks. This may have an effect on the spatial and/or temporal resolution of the measurement.

Experimental techniques used in EFD can provide qualitative and/or quantitative information about the flow field. Flow visualization techniques, for instance, although qualitative in nature can provide substantially useful information on the flow structure that could not be revealed even through detailed quantitative measurements.

Another way to classify experimental techniques is by considering whether they provide single-point measurement or multipoint measurements. The latter can be on a plane inside the flow field or in a volume that is part of the flow field. These techniques are usually referred as *global measurement techniques*.

GENERAL REQUIREMENTS

The first requirement of every experimental technique is to avoid interference with the flow itself because any disturbance introduced by the probe, the probe-volume or flow tracers needed in the particular technique can alter the flow field substantially. Consider, for instance, an intrusive technique which is using a probe holder to support the sensor. If the sensor is located far upstream of the probe holder and the whole system is designed so that all disturbances introduced into the flow field travel downstream, the flow at the sensor will not be affected. In separated flows, however, where flow reversals take place, disturbances propagate upstream and the use of intrusive measurement techniques is prohibited. In these cases, which are characterized by partial differential equations [see Eq. (12.4.4)] that are elliptical, the use of nonintrusive, mostly optical techniques is recommended.

The second requirement is that the technique should be sensitive to small spatial and temporal changes of the measured quantity in the flow field. This brings the issues of spatial and temporal resolution of the probe. In a steady state laminar flow, for instance, the probe size should be small enough to be able to resolve spatial gradients of the quantity inside the flow field. The case of measurement of a quantity Q_i that varies drastically in the x_j direction is shown in Fig. 12.4.2. If we assume that the size of the sensing area has a linear scale Δx_s, then the change of Q_i along the x_j direction will be $\Delta Q_i = (\partial Q_i / \partial x_j)\Delta x_s$, and since most sensors are not able to measure the variation of Q_i in the x_j direction, the sensor will measure an area-weighted average of $Q_i + \Delta Q_i$. The question for the experimentalist is whether this uncertainty in ΔQ_i is acceptable or not. Further reducing the size of the sensing area Δx_s will reduce the error. However, this is not always attainable even if

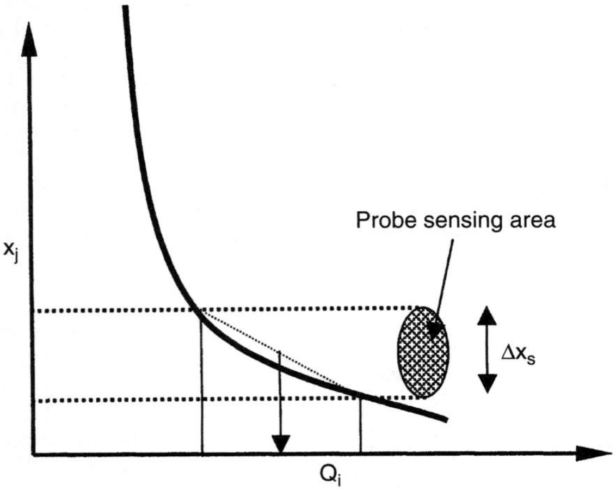

Figure 12.4.2 Spatial resolution of sensing area.

fabrication of small size sensors were possible, because electronic or optical interference noise may increase the uncertainty.

In case of temporal flow variations, the sensor should have adequate frequency response to resolve variations in the measured quantity within the smallest time scale acceptable. Usually separate tests with temporal step change in the measured quantity are needed to determine the frequency response of the sensor. For dynamic measurements of pressure, velocity, temperature, or density the shock tube is an ideal platform for frequency response tests.

Spatial and Temporal Resolution in Turbulent Flows

Turbulent flows are characterized by a large range of spatial and temporal scales and the sensors are not only required to have adequate **spatial and temporal response** and resolution but also good response to suddenly changing flow directions (yaws and pitch response).

The spatial and temporal resolution of any probe affects the accuracy of measurements of time-dependent velocity U_i, velocity gradients A_{ij}, or vorticity components Ω_k. This resolution depends on the arrangement of the sensors used in the technique, the hardware used to generate the measuring probe volume and to process the signals, the particles used as flow tracers to scatter light, and their arrival rate in the measuring locations. Thus the technique has a typical characteristic length scale L_p, which is usually associated with the separation distance between the sensing areas, and a time scale T_p, which depends mostly on the speed of data acquisition or burst processor. The technique should be able to resolve the small scales that exist in a given flow, which appear to be the Kolmogorov viscous scales $\eta = \nu^{3/4}\epsilon^{-1/4}$ for length and $T_\varepsilon = \nu^{1/2}\epsilon^{-1/2}$ for time, where ϵ is the dissipation rate of turbulent kinetic energy and ν is the kinematic viscosity. The quantity ϵ is the time-averaged value of the fluctuating part of E in Eqs. (12.4.11) and (12.4.13). L_p and T_p should be less than η and T_ϵ, respectively, in order to capture contributions from the smallest eddies of the flow field. In the case of vorticity or velocity gradients, the small scales are responsible for most of the content of vorticity or velocity gradient fluctuations, since small eddies are uncorrelated. Wyngaard[10] was the first to evaluate the problem of spatial resolution of **hot-wire** based vorticity techniques and found that no attenuation occurred for $L_p/\eta = 1$, while substantial attenuation of vorticity was evident for $L_p/\eta > 3.3$. The review by Wallace and Foss[11] concludes that reasonable estimates of velocity gradients can be obtained without attenuation with $L_p/\eta < 2$ to 4.

Although L_p has to be as small as possible, it cannot be minimized independently from other parameters such as the speckle noise. Errors due to noise in the measurements amplify as the length scale decreases. There is certainly an optimum length intermediate between attenuation and amplification of noise.

Determining η requires a good estimate of turbulent kinetic energy dissipation rate ϵ or the total dissipation rate E, which is very difficult to measure directly in a laboratory. Previous studies obtained the value of E by making assumptions to reduce some of the terms in the expression of the total dissipation rate.[9]

The estimates that usually are provided for E, particularly those based on indirect methods, may be very misleading. An experiment with poor spatial resolution will result in attenuation of all measured turbulent quantities, including the dissipation rate, which will indicate high η and therefore high resolution.

It should be mentioned here, however, that there is increasing evidence provided in Refs. 12 and 13 that the Taylor micro scale λ should be used to determine the spatial resolution in vorticity measurements. It has been shown that in the limit of vanishing separation Δx_j between the measurement locations, the root mean square (rms) of the fluctuations of one velocity gradient $\partial U_i / \partial x_j$ is given by

$$\lim_{\Delta x_j \to 0} \overline{\left[\frac{\Delta U_i}{\Delta x_j}\right]^2} = \frac{\overline{u_i^2}}{\lambda_{ij}^2}\left[1 + \lambda_{ij}^2\left[\frac{\partial \rho}{\partial(\Delta x_j)}\right]^2\right]$$

where λ_{ij} is the Taylor micro scale. If the gradient $\partial p/\partial \Delta x_j$, which indicates the degree of variation of the rms of velocity fluctuations, is much smaller than $1/\lambda_{ij}^2$, then the relation

$$\lim_{\Delta x_j \to 0} \overline{\left[\frac{\Delta U_i}{\Delta x_j}\right]^2} = \frac{\overline{u_i^2}}{\lambda_{ij}^2}$$

provides the best estimate for ideal resolution when Δx_j tends to 0. In that respect the ratio L_p/λ_{ij} should be less than 1.

An example of the effect of temporal resolution on the statistical results of vorticity flux $\partial \Omega_z/\partial y$, for instance, is obtained from the data of Andreopoulos and Agui.[82] Artificial sampling times Δt were produced by skipping by various numbers of equally sampled data digitized at rates substantially faster than the Kolmogorov frequency $1/T_\varepsilon$. The results, plotted in Fig. 12.4.3, show a substantial dependence of the rms of vorticity flux on temporal resolution Δt.

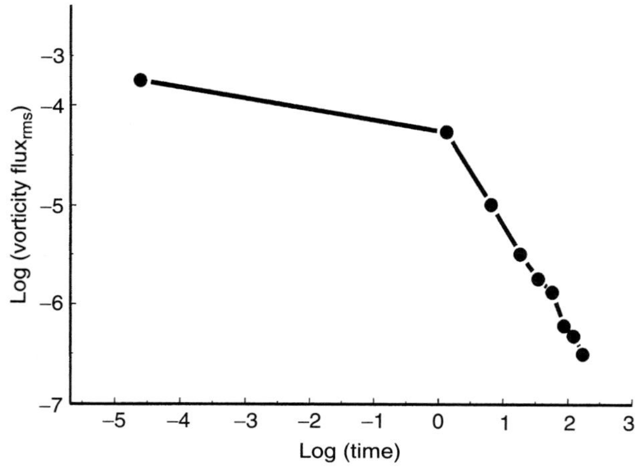

Figure 12.4.3 Effect of temporal resolution on vorticity flux rms data from Andreopoulos and Agui[82,83] are made nondimensional by inner wall scaling.

Although statistics of fluctuations of individual velocity components do not require increased temporal resolution, cross-correlation/shear stress and velocity gradients/vorticity measurements by optical techniques using particles as flow tracers require a high degree of coincidence of these particles at different locations inside the probe volume. This can be achieved by providing a high rate of particles, which results in practically continuous signals. The resolution in time depends basically on the rate of arrivals of particles in the probe volume and on the data processing rate of the hardware.

As the Reynolds number of the flow increases, its spatial and temporal scales decrease, and therefore, if the size of the measuring volume and data acquisition rate remain unchanged, then it is expected that the spatial and temporal resolution of the measuring technique will deteriorate.

MEASUREMENTS OF PRESSURE AND TEMPERATURE

There is an extensive description of probes and instrumentation used to measure pressure in textbooks[1,14,15] or in handbooks.[16,17] Manometers, gages, transducers, and microphones are used to measure pressure at single points inside the flow field or at its boundaries.

Pressure-sensitive paint (PSP) can provide useful quantitative and qualitative information on the pressure field on a solid surface, that is, at more than one point. In that respect, PSP has the advantage of being a field pressure measurement. The technique is based on the phenomena of photoluminescence and oxygen quenching of fluorescent dyes.[18-22]

In its simplest terms, the PSP method is based on the sensitivity of some luminescent dye to molecular oxygen. After absorbing a photon, a luminescent molecule is placed in an excited state. The molecule then returns to its ground state by emitting a new photon at a longer wavelength, as illustrated in Fig. 12.4.4. For some luminescent molecules, however, oxygen quenching allows the molecule to return to a ground state without emitting a photon. Hence, for a given level of excitation, the emitted light intensity from the luminescent molecules varies inversely with the local oxygen partial pressure. Since the mole

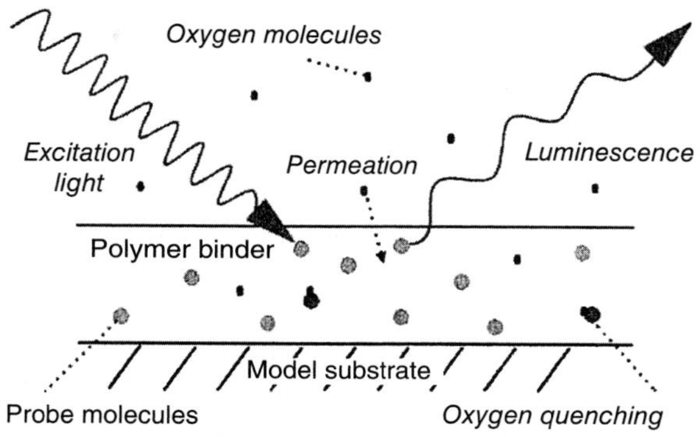

Figure 12.4.4 Excitation an illumination in a typical pressure-sensitive paint.

fraction of oxygen in air is fixed, the oxygen partial pressure is easily converted into the air pressure.

PSP simply consists of luminescent molecules suspended in some type of oxygen permeable binder. Currently, most of these binders are some form of silicone polymer. The vast majority of PSP formulations to date come in a liquid form that is suitable for use with normal spray-painting equipment and methods. Typically, in its simplest application, PSP is the topmost layer of a multilayer coating on a model surface. The PSP is usually applied over a white undercoat, which provides two related benefits. The white undercoating reflects a large portion of the light that is incident upon it, which results in the benefits of amplifying not only the excitation illumination, but the emission illumination as well.

One of the characteristics of PSP is that the response of the luminescent molecules in the coating degrades with time of exposure to the excitation illumination. This degradation occurs because of a photochemical reaction that occurs when the molecules are excited. Eventually, this degradation of the molecules determines the useful life of the PSP coating. This characteristic becomes more important for larger models, as the cost and time of PSP reapplication becomes a significant factor. A second undesirable characteristic of PSP is that the emission intensity is affected by the local temperature. This behavior is due to the effect temperature has on the energy state of the luminescent molecules, and the oxygen permeability of the binder. This temperature dependence becomes even more significant in compressible flow tests, where the recovery temperature over the model surface is not uniform.

For the measurement of temperature the classical techniques using thermocouples, resistive and solid-state thermometers, pyrometers, constant-current anemometers, liquid crystals, interferometers, and infrared imaging are described in Refs. 1 and 23, as well as in several textbooks.

Temperature-sensitive paint is used in a global measurement technique that is very similar to PSP except that the coating is temperature sensitive rather than pressure sensitive. In the case of TSP, changes in the free volume in the binder affect the luminescence; as the temperature increases, the luminescence decreases. As in PSP, a ratio of two images provides the measured quantity. In this case, one image is taken at a known, constant temperature, and the other is taken at the unknown conditions. The ratio of the two images is then converted to temperature by using the appropriate calibration. As in PSP testing, during the acquisition of the images, the test surface is exposed to excitation light in only a narrow frequency band. The detector is then notch-filtered to pass only the luminescence band for the particular probe molecule. Of course, these two bands must not overlap.

MEASUREMENTS OF VELOCITY

The measurement of flow velocity at single or multiple points may be accomplished by direct measurement of the speed of the flow or indirectly by a variety of techniques that can be in general classified as pressure-based methods using pitot or pitot static tubes, thermal anemometry, laser Doppler anemometry (LDA), **particle image velocimetry (PIV)**, planar **Doppler velocimetry (PDV)**, **molecular tagging velocimetry (MTV)**, which appear to be the most commonly used techniques. Pressure-based methods are adequately explained in several textbooks and they will be excluded from the present description.

Thermal Anemometry

Thermal anemometry has been used extensively for many years in fluid mechanics to measure instantaneous fluid velocity. The technique depends on the convective heat loss to the surrounding fluid from an electrically heated sensing element or probe. If only the fluid velocity varies, then the heat loss can be interpreted as a measure of that variable. There are several textbook on the subject[24-26] that cover comprehensively the associated techniques.

Cylindrical sensors (hot wires and hot films) are most commonly used to measure the fluid velocity, while flush sensors (hot films) are employed to measure the wall shear stress. Hot-wire sensors are, as the name implies, made from short lengths of resistance wire and are circular in cross section. **Hot-film** sensors consist of a thin layer of conducting material that has been deposited on a nonconducting substrate. Hot-film sensors may also be cylindrical but may also take other forms, such as those that are flush-mounted.

Thermal anemometry is very popular among researchers because the technique involves the use of very small probes that offer very high spatial resolution and excellent frequency response characteristics. The basic principles of the technique are relatively straightforward and the probes are relatively easy to fabricate. The sensors can be operated in the constant-temperature mode (CTA) or constant-current mode (CCA).

Hot-wire anemometry has been used for many years in the study of laminar, transitional, and turbulent boundary layer flows, and much of our current understanding of the physics of boundary layer transition has come solely from hot-wire measurements. Thermal anemometers are also ideally suited to the measurement of unsteady flows such as those that arise behind rotating blade rows when the flow is viewed in the stationary frame of reference. Until the advent of laser anemometry, this was the only available technique for the acquisition of data. In present times, thermal anemometry remains one of the primary tools used in experimental fluid mechanical work.

A hot-wire probe consists of the sensor itself, which is supported on prongs that are embedded in nonconducting (often ceramic) material. Figure 12.4.5 shows the main components of a basic hot-wire probe. The sensing element is heated by a current I under a voltage. A hot-wire sensor has to have several characteristics to make it a useful device:

- Its thermal capacity/inertia should be small so that it provides good frequency response.
- It should have a high temperature coefficient of resistance.
- Its electrical resistance should be such that it can be easily heated with an electrical current at practical voltage and current levels.

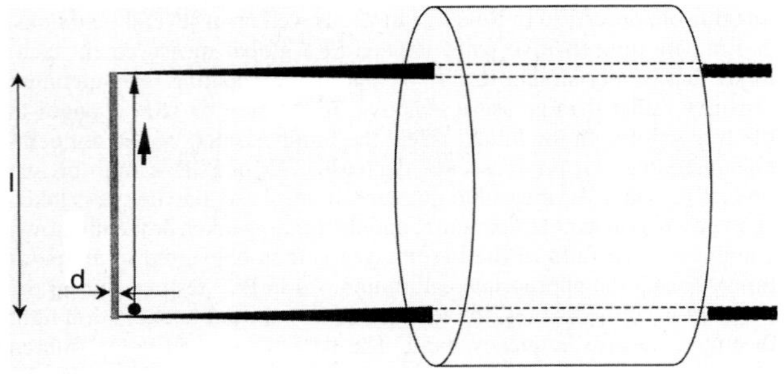

Figure 12.4.5 Basic hot-wire probe.

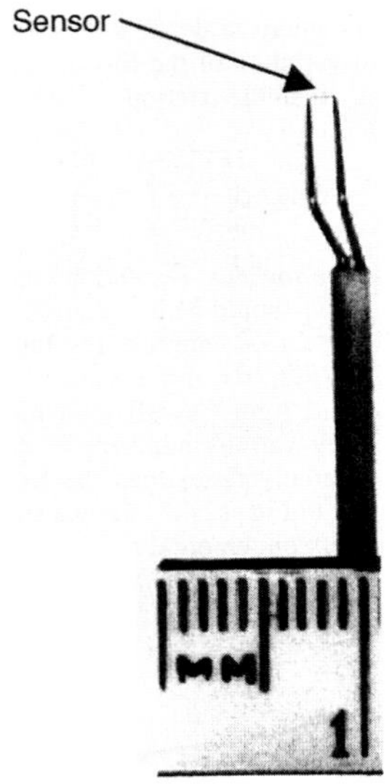

Figure 12.4.6 Photograph of boundary-layer-type hot-wire probe.

The most common wire materials are tungsten, platinum, and a platinum-iridium alloy. Tungsten wires are strong and have a high temperature coefficient of resistance (0.004/°C). However, they cannot be used at high temperatures in many gases because of poor oxidation resistance. Platinum has good oxidation resistance and temperature coefficient (0.003/°C), but is very weak, particularly at high temperatures. The platinum-iridium wire is a compromise between tungsten and platinum with good oxidation resistance, and more strength than platinum, but it has a low temperature coefficient of resistance (0.00085/°C). Tungsten is presently the more popular hot-wire material. The choice of diameter of the sensors, d, depends on the particular application. It can vary from 0.8 μm to 5 μm, while its length is between 0.7 and 1 mm. Since these elements are sensitive to heat transfer between the element and its environment, temperature and composition changes can also be sensed. Figure 12.4.6 shows a photograph of a hot-wire probe used for near-wall measurements.

The heat transfer rate from the sensor to the fluid can be, to a certain extent, simulated by considering a heated cylinder at uniform temperature T_w inside a steady flow field. Part of the electrically generated heat along the sensor will be transferred to the fluid by force and free convection, part of it will be transferred to the prongs by conduction, and the remaining will be radiated (see Fig. 12.4.7). For relatively long sensors with $l/d > 200$, heat rate by conduction Q_C can be ignored. For the usual hot-wire anemometry temperature range of a few hundred degrees Celsius, radiation heat transfer Q_R is less than 0.1 percent of the convective heat rate

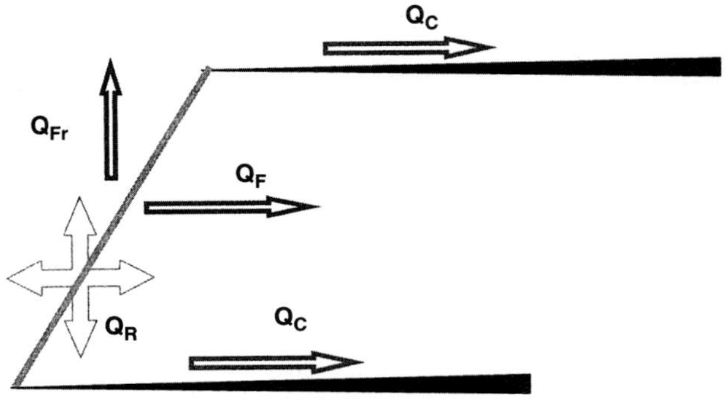

Figure 12.4.7 Thermal energy balance in hot wires.

Q_C and may be disregarded. Free convection can be also ignored if Re > $Gr^{1/3}$, where Gr is the Grashof number, defined as $Gr = g\beta(T_w - T)d^3/\nu^2$, where β is the thermal expansion coefficient of the fluid.

The temperature of the fluid in the immediate vicinity of the cylinder will have values between the fluid temperature T and T_w, and the fluid properties will vary accordingly. By convention, gas flow properties such as thermal conductivity, density, kinematic viscosity, and thermal diffusivity will be evaluated at the average film temperature defined as $T_f = \frac{1}{2}(T + T_w)$.

If Q_F is the heat flux convected from the wire to the fluid, the Nusselt number is defined as $Nu = Q_F/\pi lk(T_w - T)$, with k being the thermal conductivity of the fluid. If one restricts the analysis to air only, the Prandtl number is constant, and then Nu depends on $Re = \rho Ud/\mu$, Nu = Nu(Re)

Under the above assumptions Nu appears to be

$$Nu = (C + DRe^n)(T_w + T_\infty)^{0.17} \tag{12.4.14}$$

where C, D, and n are constants. The heat flux rate Q_F is provided by the electrical heating and therefore is $Q_F = IE = I^2/R = E^2/R$, where I and E are the current through and the voltage across the sensor, respectively, and R is the resistance of the wire. The technique then can provide $Q_F = f(E)$ under constant I for the CCA-mode operation or $Q_F = f(R)$ under constant resistance/temperature for is the CTA-mode operation.

In the CTA mode the above relation becomes

$$E^2 \frac{R_w}{(R_w + R_a)^2(T_w - T)} = X + YU_{\text{eff}}^n \tag{12.4.15}$$

where U_{eff} is the effective cooling velocity of the fluid. This relation is known as King's law.[27] If we ignore the variation of fluid properties and since T_w is constant, then the partial sensitivities of voltage on velocity and fluid temperature are

$$\frac{\partial E}{\partial U} = \frac{nYU_{eff}^{n-1}}{2}\left[\frac{(R_w + R_a)^2}{R_w}\frac{T_w - T}{X + YU_{eff}^n}\right]^{1/2} \tag{12.4.16}$$

$$\frac{\partial E}{\partial T} = -\frac{1}{2}\left[\frac{(R_w + R_a)^2}{R_w}\frac{X + YU_{\text{eff}}^n}{T_w - T}\right]^{1/2} \tag{12.4.17}$$

Then $\Delta E = (\partial E/\partial U)\Delta U + (\partial E/\partial T)\Delta T$, and, if the objective is to measure velocity and velocity fluctuations, the partial sensitivity $\partial E/\partial U$ has to be maximized and the partial derivative $\partial E/\partial T$ has to be minimized in comparison or completely eliminated. In order to achieve this, the operating wire temperature T_w has to be large, since the term $(T_w - T)$ is the nominator of $\partial E/\partial U$ and in the denominator of $\partial E/\partial T$. For accurate measurements an overheat ratio $(R_w - R_a)/R_a = \alpha\,(T_w - T_a)$ of 0.80 is recommended, where R_a is the cold resistance and α is the coefficient temperature.

Frequency Response The response of a hot-wire sensor of mass m_w and specific heat c_w is governed by the first-order differential equation given in Ref. 28 as

$$m_w c_w \frac{\partial T_w}{\partial t} = I^2 R_a\,[1 + \alpha(T_w - T_a)] - \pi k(T_w - T)\text{Nu} \tag{12.4.18}$$

which is the unsteady heat equation with the first term on the left-hand side representing the accumulated heat rate by the wire mass. The solution of this equation contains the exponential term $e^{-t/M}$, where M is the time constant

$$M = \frac{m_w c_w}{I^2 R_a \alpha - \pi kl\,\text{Nu}} = \frac{\rho_w c_w d^2}{k\text{Nu}}\frac{R_w}{R_a} \tag{12.4.19}$$

In order to minimize M, small values of wire diameter d are used; in fact, wire size is the only control a user can have. However, the mechanical strength of the wire under aerodynamic load limits the reduction in d. Reductions in M, at the expense of sensitivity, can be also obtained by reducing the ratio R_w/R_a.

Angular Sensitivity Unfortunately the output of a hot-wire anemometer depends on the yaw and pitch angles of the incoming velocity vector. This angular response of the probe can be found through appropriate yaw and pitch angle calibration, which is usually carried out at constant free-stream velocity. This angular sensitivity is modeled in the expression of the sensor cooling velocity U_{eff} in various ways by decomposing the local velocity vector into the wire coordinates (see Fig. 12.4.8):

1. Cosine law with $U_{\text{eff}} = U_o \cos\phi$ (see Ref. 28). This is the simplest of the expressions, and it is valid for small yaw angle ϕ.

2. According to Champagne, Schleicher, and Weherman,[29] U_{eff} includes a fraction of the tangential component of the local velocity vector U_T, and the expression becomes $U_{\text{eff}} = (U_N^2 + k^2U_T^2)^{1/2}$ where k is a coefficient with values $k = 0.2$ for $l/d = 200$ and $k = 0$ for $ld = 600$.

3. A more general expression for U_{eff}, due to Jörgersen,[30] includes cooling in the binormal direction from the component U_B: $U_{\text{eff}} = (U_N^2 + h^2U_B^2 + k^2U_T^2)^{1/2}$ where h is another coefficient with values of the order of 1. Apparently this is the most complete expression for U_{eff} and it can be applied in principle to any three-dimensional flow field with reasonable turbulence intensities.

The coefficients k and h can be determined through individual calibration wire and may include the unavoidable effects of prong and probe holder interference.

Multisensor Probes A single-sensor probe oriented perpendicularly to the flow direction can yield information of one velocity component only, under the assumption that the lateral components are considerably smaller than the longitudinal one. If measurements of two velocity components are required, then a probe with two sensors independently oriented at ±45° each to the mean direction should be used. Triple wires consisting of three sensors are recommended for use in three-dimensional flows.

Laser Doppler Anemometry

The laser Doppler anemometry (LDA) technique, which is also known as laser Doppler velocimetry (LDV), is based on light scattering off flow tracers that exist in the flow or are introduced by the user. The flow field is illuminated by a laser beam and the particle tracer that follows the flow produces a Doppler frequency shift due to its velocity.

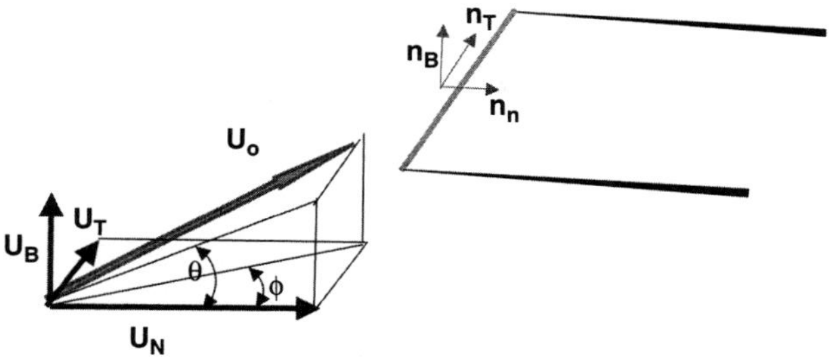

Figure 12.4.8 Components of velocity vector in wire coordinates.

This is a nonintrusive optical technique to measure the **Doppler frequency shift** contained in scattered light. The idea of measuring the Doppler shift due to the motion of particles inside a flow was originated by Cummins et al.[31] after the discovery of narrow line width lasers. Cummins, a professor of physics at City College, demonstrated in Yeh and Cummins[32] that the Doppler frequency shift $\Delta\nu$ is related to the velocity vector of the light scattering particle **V** through

$$\Delta\nu = \nu_s - \nu_i = 1/\lambda(\mathbf{k}_s - \mathbf{k}_i)\cdot\mathbf{V} \tag{12.4.20}$$

where $\mathbf{k}_s$ and $\mathbf{k}_i$ are the unit vectors in the scattered and incident light directions and λ is the wavelength of the incident light. Because of the inability of existing photodetectors to measure directly the extremely high frequencies of the original laser beam, several arrangements have been used to offset this problem. The most common arrangement is shown in Fig. 12.4.9. The original laser beam is split into two beams that intersect in an interference pattern of standing waves consisting of high- and low-intensity surfaces called *fringes*. As micrometer-sized liquid or solid particles entrained in a fluid pass through the intersection of the two laser beams, the scattered light received from the particles fluctuates in intensity. The laser frequency of the two beams cancels out and the Doppler frequency is independent of the scattered light direction $\mathbf{k}_s$, which is the observation angle:

$$\nu_D = \nu_{s1} - \nu_{s2} = 1/\lambda(\mathbf{k}_{i1} - \mathbf{k}_{i2}) \cdot \mathbf{V} = 2\sin\theta\, U_i/\lambda \qquad (12.4.21)$$

This last feature is a great advantage of this arrangement, since it provides some flexibility in the placement of the frequency detector, which is usually a photomultiplier that converts the light intensity into photocurrent. Then Eq. (12.4.21) is transformed into the simple form

$$U_i = \nu_D \lambda / 2\sin\theta \qquad (12.4.22)$$

Another advantage of linear equation (12.4.22) is that it contains no undetermined constants, thus eliminating the need for calibration and errors due to instrumentation drifting. The proportionality coefficient $\lambda/(2\sin\theta)$ is the calibration factor, although no calibration is actually necessary (see Refs. 33, 34, and 35 for more details).

The velocity direction can be fixed if one of the laser beams has a frequency slightly different from that of the other. All three components of velocity can be measured by crossing four or six laser beams of different wavelenths or polarization in the same measuring volume, and separating out the scattered light with filters. The frequency is measured by digital counters (if data rates are above 50 per second) or photon correlators (low data rates and signal level) or spectral analyzers (Doppler shift frequency not over a few megahertz).

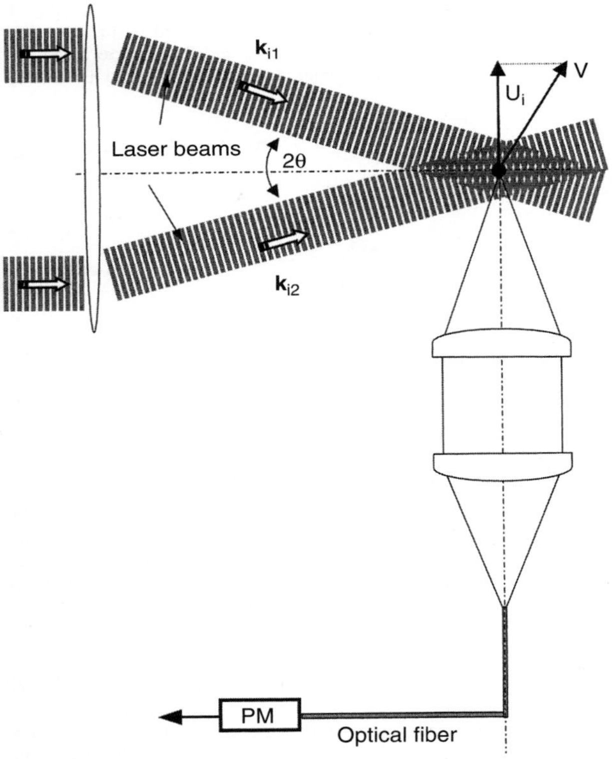

Figure 12.4.9 Typical laser Doppler anemometry arrangement.

A laser power source is required, with excellent frequency stability, narrow linewidth, small beam diameter, and a Gaussian beam intensity profile (bright at the center). Typically, helium, neon, or argon ion lasers are used, with power levels from 10 mW to 20 W.

LDA techniques have provided spatially precise point measurements, at speeds from millimeters per second to supersonic. In some cases, measurement point distances of up to 7 m have been used in large wind tunnels.

The limitation of this technique is that it provides single-point measurement only. Data rates are dictated by particle arrival in the measuring volume, not when the user wants to sample data. Particles should follow the flow closely. However, particles cannot enter all flows through turbulent diffusion, and simple viscous diffusion is not adequate to provide particles inside the laminar cores of vortices, for instance. Particles should be small enough to follow the flow, but reducing their size may produce a signal below the noise threshold. There is also greater noise at higher speeds. The signal level depends on detector solid angle, posing problems of optical access. Mie scattering intensity is much better in the forward direction, but it is difficult to set up forward receiving optics which remains aligned to the moving measurement volume. Time to alignment of emitting optics can range from 3 hours for a one-component system to a day for a three-component system. The setup may be also sensitive to vibrations. These problems may be alleviated with fiber optic probes, at the cost of signal level because of fiber power loss.

LDA techniques have been used in a variety of flow applications that include low- and high-speed boundary layers, combustion studies, and turbomachinery and rotorcraft flows. They have been also used in atmospheric turbulence measurement, and in air data systems for aircraft. In periodic flows, phase-resolved averaging is used to construct 3-D velocity fields with high accuracy.

Particle Image Velocimetry

Particle image velocimetry (PIV) is a whole-flow-field, nonintrusive optical technique that provides instantaneous velocity vector measurements in a cross-section of a flow. Two velocity components are usually measured on the illumination plane. The use of a stereoscopic arrangement provides all three velocity components to be recorded, resulting in instantaneous 3-D velocity vectors for the whole area. The use of modern CCD cameras and dedicated computing hardware results in real-time velocity maps. In PIV, the velocity vectors are derived from subsections of the target area of the particle-seeded flow by measuring the movement of particles between two light pulses. The technique is nonintrusive and measures the velocities of micrometer-sized particles following the flow. It has been tested in flows with velocity range from zero to supersonic.

The flow is illuminated by a laser sheet in the target area. A camera lens images the target area onto the CCD array of a digital camera. Two sequential images of the flow field are captured by the CCD array, produced by two light pulses in separate image frames. Once a sequence of two light pulses is recorded, the images are divided into small subsections called *interrogation areas*. These interrogation areas from each image frame are cross-correlated with each other, pixel by pixel. The correlation produces a signal peak, identifying the common particle displacement.[36-40] In PIV, the velocity vectors are derived from subsections of the target area of the particle-seeded flow by measuring the movement of particles between two light pulses, $\overline{V} = \overline{\Delta x_i} / \Delta t$.

A typical experimental setup of a PIV system is shown in Fig. 12.4.10. This system has been used to measure the velocity in the near field of a compressible jet at CCNY (see Wang[41]). Talcum flow tracer particles of 1 μm in diameter have been added to the flow. A plane (light sheet), about 1 mm thick, within the flow is illuminated by means of a dual laser system (the time delay between pulses depending on the mean flow velocity and the magnification at imaging). It is assumed that the tracer particles move with local flow velocity between the two illuminations. The light scattered by the tracer particles is recorded via a high-quality lens either on a single frame or

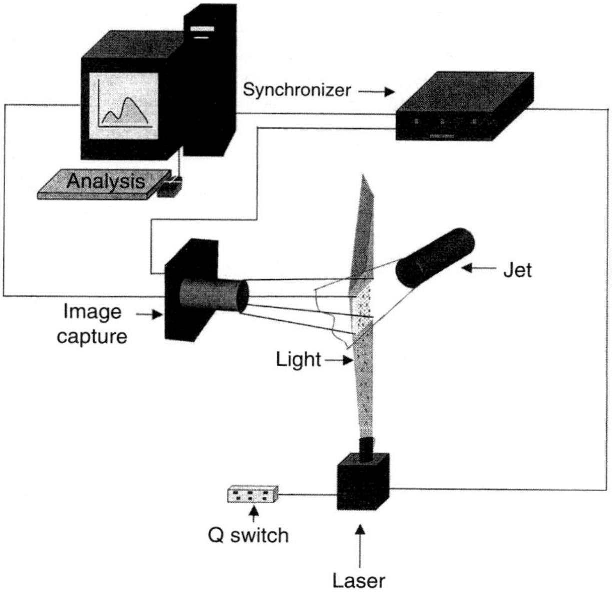

Figure 12.4.10 Experimental setup for particle image velocimetry measurements in a compressible jet flow facility.

on two separate frames on a special cross-correlation CCD sensor. It is assumed that all particles within one interrogation area have moved homogeneously between the two illuminations. The projection of the vector of the local flow velocity into the plane of the light sheet (two-component velocity vector) is calculated, taking into account the time delay between the two illuminations and the magnification at imaging.

The process of interrogation is repeated for all interrogation areas of the PIV recording. It was possible to capture 100 PIV dual frames per minute with the present 1k × 1k PIVCAM 10-30 cameras. The duration of processing of one PIV recording with 3,600 instantaneous velocity vectors (depending on the size of the recording and of the interrogation area) is of the order of a few seconds with present PCs.

Prior to the detailed velocity measurement in the jet flow facility, the overall structure of the flow was captured by the high-resolution PIVCAM 10-30 camera. This is shown in Fig. 12.4.11 for the case of a nitrogen jet issuing in ambient still air. On the left side of this figure, the 7-mm jet exit nozzle is shown. The flow as it exits the nozzle starts to mix with the ambient air and decelerates. A few diameters downstream of the jet exit, the flow starts to spread out.

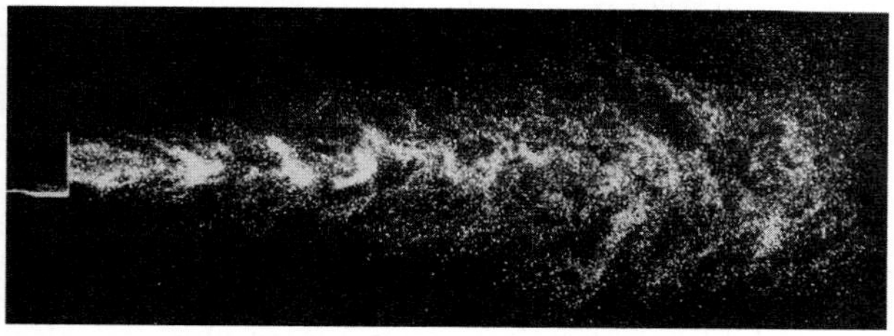

Figure 12.4.11 Flow structure of a nitrogen jet at Mach 0.6.

The two images shown in Fig. 12.4.12 correspond to the flow described above with the Mach 0.6 nitrogen jet. If these two images were overlapped, the same particle displacement could be revealed. Cross correlation of the two frames provided the velocity vectors shown in Fig. 12.4.13.

Figure 12.4.12 Cross-correlation images for nitrogen jet at Mach 0.6.

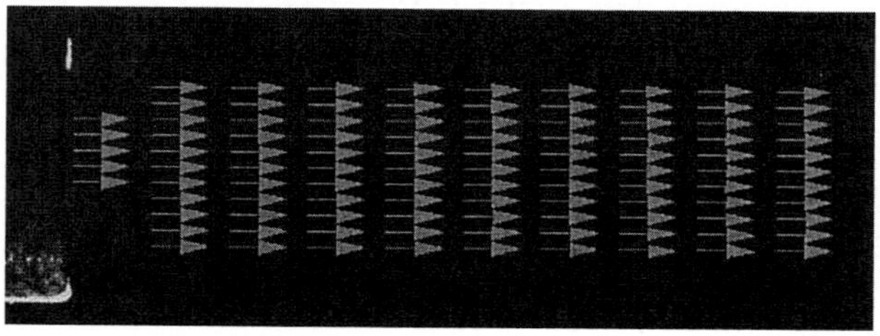

Figure 12.4.13 Resultant velocity for nitrogen jet at Mach 0.6.

Planar Doppler Velocimetry Techniques

Planar Doppler velocimetry (PDV), also known as global Doppler velocimetry (GDV), is a relatively new nonintrusive planar optical technique that utilizes molecular filters to measure the Doppler frequency shift contained in scattered light. As mentioned earlier, the idea of measuring the Doppler shift due to the motion of particles inside a flow was originated by Cummins et al.[31] after the discovery of narrow linewidth lasers. The Doppler frequency shift $\Delta\nu$ is related to the velocity vector of the light-scattering particle, **V** (see[32]) through

$$\Delta\nu = \nu_s - \nu_i = 1/\lambda(\mathbf{k}_s - \mathbf{k}_i)\cdot \mathbf{V}$$

where $\mathbf{k}_s$ and $\mathbf{k}_i$ are the unit vectors in the scattered and incident light directions and λ is the wavelength of the incident light.

A laser sheet is formed inside the flow field and global measurements can be obtained by recording the light intensity, which is related to the Doppler frequency shift. The measurement of the frequency shift is made spectroscopically by using a high-resolution iodine gas cell, which acts as a frequency-dependent filter with very high sensitivity. The edge of an absorption line in molecular iodine is used to determine the frequency of the Doppler shift of the scattered light. Image data are acquired via digital cameras, which have to be placed appropriately for each axis of interest. The intensity of the light from a given point in the object plane is attenuated in proportion to its Doppler shift. The image plane then contains gray levels proportional to the product of the local intensity of the illuminating light sheet and the Doppler shift. To eliminate the problem of both scattering signal and illumination intensity variations spatially in the measurement window, the Doppler signal intensity map is normalized by a reference intensity map from the same view of the flow. The camera is split into two halves to record the unshifted image field, and the Doppler-shifted image is normalized by the unshifted image, pixel by pixel.

The technique was first described in a U.S. patent by Komine[42] and by Komine et al.[43] and Meyer and Komine,[44] who used an iodine filter.

At about the same time Miles and Lempert[45] and Miles et al.[46] used an I_2 molecular filter with Rayleigh scattering to demonstrate the feasibility of measuring mean velocity, density, and temperature in high-speed flows. A good review article on the subject can be found in Samimy and Wernet.[47] Further information can be found in Refs. 48 to 55.

Figure 12.4.14 shows a typical scan of a typical transfer function of the I_2 vapor filter. The laser is tuned at the middle point of the range of the normalized light intensity. Any scattered light of flow tracers inside the flow field will cause a Doppler frequency shift depending on the velocity magnitude which corresponds to a normalized light intensity. This spectroscopic transfer function is obtained in advance and it is typical of the filter used and the tuned frequency of the laser.

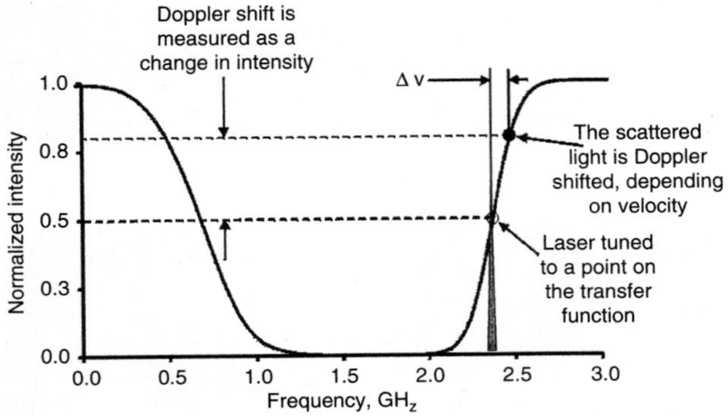

Figure 12.4.14 Typical transfer function of I_2 vapor filter.

Figure 12.4.15 shows a typical arrangement currently used at CCNY to measure the velocity field within a shock wave interacting with an isotropic and homogeneous turbulence generated by metallic rectangular grids. The major hardware components are a Nd:YAG laser, a CCD camera, and an absorption filter. Presently a *Q*-switched Quanta Ray GRC-14s Nd:YAG laser with 250-mJ pulse energy is available in the laboratory. This laser is equipped with a LightWave injection seeder. A Princeton Instruments (Roper Scientific) back-illuminated camera, model MicroMax 1,024B with 1k × 1k pixels and an ultraviolet ASROMED camera with a chip of 540 × 650 pixels are used for imaging. An I_2 absorption filter provided by Innovative Scientific Solutions Inc. has a typical absorption spectrum shown in Fig. 12.4.16. These hardware pieces are adequate to provide measurements of one-velocity components only. Measurements of three velocity components by using PDV have been already made successfully in supersonic jets by combining the pair of filtered and unfiltered images on the same CCD chip.[47]

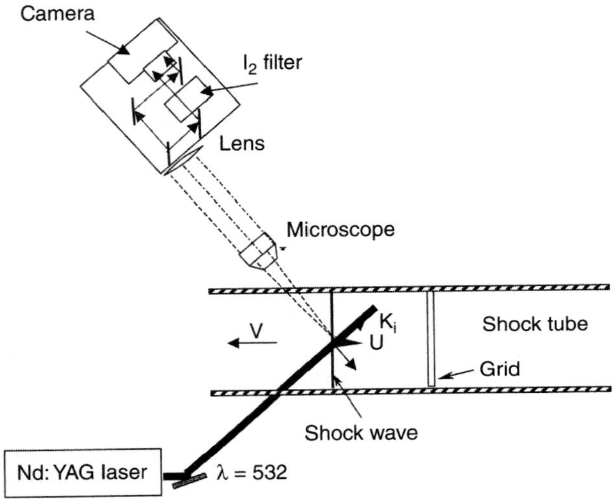

Figure 12.4.15 Experimental setup for one-component micro PDV measurements inside a shock wave in the large-scale shock tube facility at CCNY.

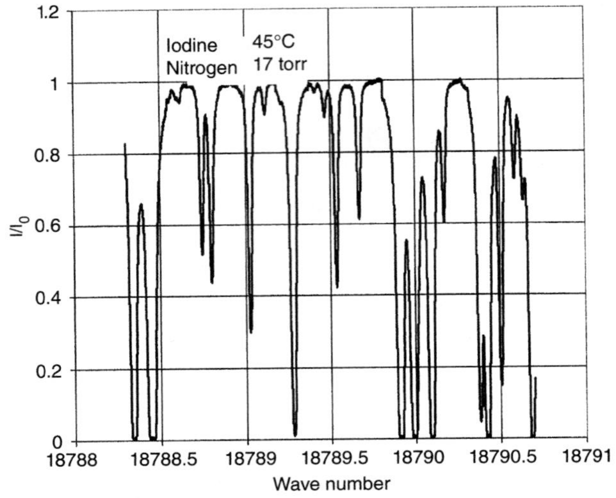

Figure 12.4.16 Absorption spectrum of I_2 vapor filter.

Implementation of the PDV techniques to the present flow in the shock-tunnel environment is not straightforward. Particle seeding is one of the major issues in flow measurements based on Doppler frequency shift. In the past, acetone has been used to provide light scatterers in flows configured in high-speed wind tunnels where temperatures are rather low. In the present flow case, the use of acetone is not certain because the temperature in the shock tunnel is higher than the flow temperature in supersonic wind tunnels. Submicrometer talcum powder particles have been used successfully in our work in the past. The scales of this flow can be adequately resolved with a magnification of about 200, or less. The pixel-to-pixel distance is 9 μm, and with this magnification the flow scales resolution is about 45 nm. For a shock thickness of about 3λ, where λ is the molecular mean free path, and helium as a driven gas at one-fifth atmospheric pressure with $\lambda = 3.75\,\mu$m, the shock thickness can be resolved by 90 pixels. Talcum powder with particle size less than 0.2 μm is used for light scattering.

Unfortunately the pulsed **Nd:YAG lasers** are known for long-term frequency drifts and pulse-to pulse fluctuations in frequency and power. The variation in the power can be remedied by normalization of the filtered signal with the unfiltered one. Frequency variations, however, require a frequency monitoring system. The system described by Mosedale et al.[56] is a good remedy to this problem. This system requires a small cut-out portion of the laser beam to focus on two fast-response photo detectors after it is split in two and after one of the two parts passes through a second I_2 filter. This system can be also used to characterize the frequency-transmission relation of each filter.

Currently, however, PDV is considered to be more suitable for the measurement of high-speed flows (estimated to be above 30 m/s) because of the limit on velocity resolution, which is approximately equal to 1 m/s, which is the uncertainty of a very good DGV system. It is one of the very few Mie-scattering whole-field velocimetry techniques that could be truly regarded as instantaneous with the use of a short-duration (~10 ns) pulsed laser.

Other Techniques

In molecular tagging velocimetry (MTV), the flowing medium is premixed with molecules that can be turned into long-lifetime flow markers upon excitation by photons. Typically a pulsed laser is used to "tag" small regions of interest. The tagged regions are then imaged at two successive instants within the lifetime of the tracer. The measured displacement vectors provide estimates of the average velocity for each small region during the brief interval between exposures. The fluid is water in this case, premixed with a water-soluble phosphorescent triplex compound. The corresponding velocity field is determined by a spatial correlation technique.[57,58] One advantage of the MTV technique over other commonly used whole-field velocimetry techniques that rely on light scattering from particles is its molecular nature. In that respect, the issues related to flow tracers that are particularly important in gas flows are eliminated. The molecular nature of MTV also makes it suitable for obtaining other flow variables, such as pressure, temperature, and concentration, along with the velocity field.

There are several "unseeded" tagging methods where gas tags are produced from molecules naturally occurring in air (i.e., nitrogen, oxygen, water vapor). Examples of molecular tags produced from air are N_2 + ion,[59] ozone,[60] hydroxyl,[61] nitric oxide,[62,63] and vibrationally excited oxygen.[64,65] Many of the methods use nonlinear laser excitation[59,62-65] to produce these tags from air and the tag is produced only in a small region near the laser focus.

MEASUREMENTS OF VORTICITY, RATE OF STRAIN, AND DISSIPATION

The measurement of one or more components of vorticity in turbulent flows of research and technical interest has been a long-held, but elusive, goal of fluid mechanics instrument developers and researchers. The review article by Wallace and Foss[11] provides a detailed account of the available techniques, as of that date, used to measure vorticity and compare existing data. A majority of the techniques are based on thermal anemometry or on optical methods. Optical techniques are nonintrusive, are sensitive to large changes in velocity direction, and most of them do not require any calibration. These characteristics make them ideal for use in separated flows and flows with large turbulence intensities where the use of intrusive methods is prohibited and recirculating effects are present. Most of the optical techniques are based on detecting light scattered by particles imbedded in the flow. In that respect, seeding the flow with appropriate particles that follow the fluid motion very closely is critical. It should be noted here that, for gas flows in which seed must be added and in which the flow does not recirculate, adding particles can be far more intrusive than a hot-wire probe.

There is a need for a formal definition of the measured quantities, either vorticity components Ω_k or velocity gradient tensor terms A_{ij}. Specifically, there are two common ways to derive vorticity. In the first method, measurements of velocity components at nearby locations are obtained and used in a finite-difference scheme to derive vorticity. In the second method measurements of microcirculation around a small area are obtained and vorticity is computed as circulation density.

The first definition is based on a Taylor's series expansion of velocity in a nearby location, which to a first-order approximation becomes $U_i(r_j + \Delta r_j) = U_i(r_j) + \Delta r_j(\partial U_i / \partial r_j) + \cdots$.

For r_j in the y direction and $\Delta r_j = (0, \Delta y, 0)$, the above equation becomes $U_i(y + \Delta y) \approx U_i(y) + \Delta y(\partial U_i / \partial y)$, which yields the following relation to compute the velocity gradient:

$$\frac{\partial U_i}{\partial y} \approx \frac{U_i(y+\Delta y) - U_i(y)}{\Delta y}$$

This relation will also define the uncertainties in the measurement of the velocity gradient.

The second method of estimating vorticity is due to Foss and his coworkers,[66-68] who have developed a hot-wire technique for the measurement of the transverse vorticity component. The technique involves the measurement of microcirculation Γ around a small finite domain of area ΔS normal to the flow. The estimated vorticity appears to be computed through the relation $\langle \Omega_z \rangle = \Gamma / \Delta S$.

These two methods are not the only ones that are used. There is no need to have a computational formalism for vorticity when it is estimated explicitly. For instance, direct estimate of vorticity through measurements of fluid rotation does not require vorticity computation through an implicit scheme.[69]

The inherent difficulties of velocity gradient/vorticity measurements, which require high spatial and temporal resolution in turbulent flow fields, have only recently been substantially overcome.

Hot-Wire-Based Vorticity Probes

Several multi-hot-wire probes have been developed that are capable of measuring velocity-gradient-related quantities. The difference among them is the orientation and number of wires used. One of the first papers to provide well-documented data on vorticity in low-speed adiabatic boundary layers is by Vukoslavcevic, Wallace, and Balint.[70] Honkan and Andreopoulos[71] fabricated one probe consisting of nine individual single-wire sensors. The probe, shown schematically in Fig. 12.4.17, consists of a set of three individual triple-hot-wire sensors put together so that the probe remains geometrically axisymmetric. Two photographs of the entire probe assembly together the probe dimensions are shown in Fig. 12.4.17.

In designing a vorticity probe, several considerations have to be taken into account. (1) The individual wire length l_w should be as small as possible so that small scales can be resolved adequately. (2) The size of the individual triple-hot-wire probe, l_p, should be as small as possible to satisfy the assumption that the velocity does not change substantially across the probe. Small wire spacing, however, can lead to thermal interference and cross talk between the wires. (3) Since vorticity or strain rate will be computed from velocity gradients, spacing of the individual probes should be finite so that velocity gradients do not "disappear." If this spacing becomes small, the effect of noise may overwhelm the signal.[11] (4) Each of the wires should be controlled independently of the others. The use of one common prong as in the case of Ref. 70 may create problems such as imbalance under dynamic conditions. (5) The transfer function of the hot-wire anemometer system should be a three-dimensional one. This suggests that the probe should be able to respond to yaw and pitch angle variations of the velocity vector in addition to its magnitude.

Detailed lookup tables were created through velocity, pitch, and yaw calibrations, which are subsequently used to reduce the acquired data.

A new vorticity probe designed for use in nonisothermal flows or in compressible flows is described by Agui et al.[72] and Briassulis et al.[73] The probe consists of 12 wires, which is a modification of the original design with nine wires by Honkan and Andreopoulos.[71] The three additional wires were operated in the so-called constant-current mode and used to measure time-dependent total temperature.

Since the probe essentially consists of a set of three modules or arrays (see Fig. 12.4.18), it is necessary to provide several key features of the individual hot-wire modules. Each module contains three hot wires operated in the constant-temperature mode (CTM) and one cold-wire sensor operated in the constant-current mode (CCM). Each wire of the triple-wire submodule is mutually orthogonal to the others, thus oriented at 54.7° to the probe axis. Each of the 5-μm-diameter tungsten sensors is welded on two individual prongs, which have been tapered at the tips. Each sensor is operated independently, since no common prongs are used. Each of the 2.5-μm-diameter cold wires is located on the outer part of the submodule.

The hot-wire output voltage E_i of the ith sensor is related to the effective cooling velocity, $U_{i,\text{eff}}$ through Eq. (12.4.15), known as King's law[27]:

$$\frac{E_i^2}{T_w - T_0} = A_i \left[\frac{T_0}{T_r}\right]^a + B_i \left[\frac{T_0}{T_r}\right]^b (\rho U_{i,\text{eff}})^n$$

where T_w is the hot-wire temperature, T_0 is the total temperature of the flow, and T_r is a reference temperature, the ambient temperature in the present case. The values of the exponents a and b were taken as suggested by Kovasznay[77]: $a = b = 0.768$. The effective velocity $U_{i,\text{eff}}$ is

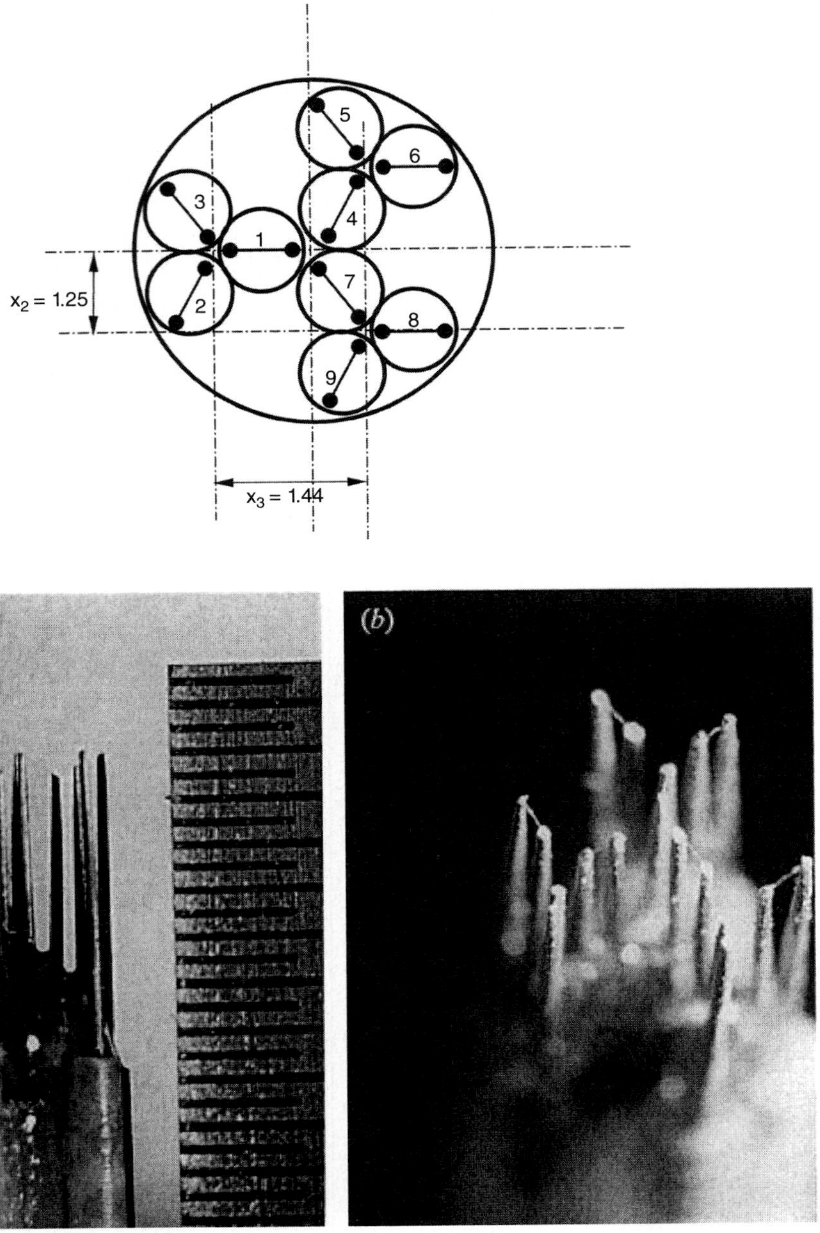

Figure 12.4.17 Sensor arrangement and photographs of the probe (dimensions in millimeters).

related to the normal, tangential, and binormal components of the velocity vector in reference to the ith sensor, respectively, by

$$U_{\text{eff}} = (U_N^2 + h^2U_B^2 + k^2U_T^2)^{1/2}$$

where k and h are coefficients that, for a given probe, depend on the yaw and pitch angle of the velocity vector. Details of the techniques associated with the use of triple wire arrays can be found in Ref. 74, while estimates of errors related to probe geometry and turbulence intensity are described in Ref. 75.

Mass fluxes ρU_i can be further separated into density and velocity by using the method adopted in Ref. 76. Decoupling density from mass fluxes assumes that static pressure fluctuations are small. This is the so-called weak version of the original "strong Reynolds analogy" hypothesis of Morkovin.[78] The original hypothesis is based on the assumption that pressure and total temperature fluctuations are very small. In the present work, total temperature was measured directly and therefore no corresponding assumptions were needed. The pressure, however, was measured at the wall and not at the location of the hot-wire measurement. The mean value of this pressure signal was used to separate the density and velocity signals since no mean pressure variation has been detected across a given section of the flow. The procedure involves an expression for mass flux m_i in terms of total temperature T_0 and pressure p at the

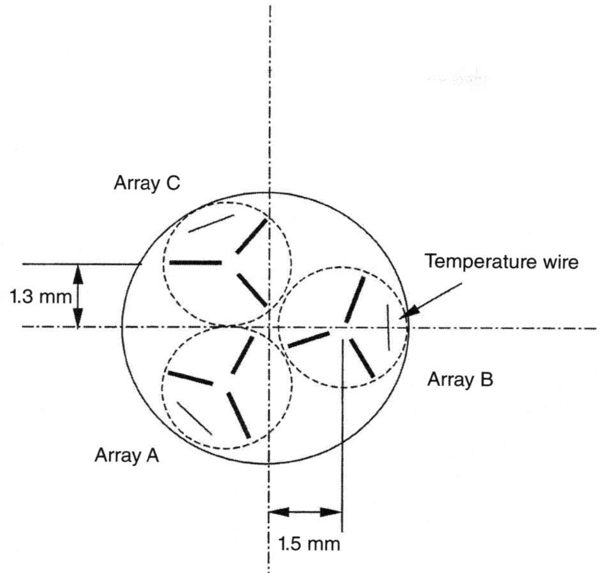

(a)

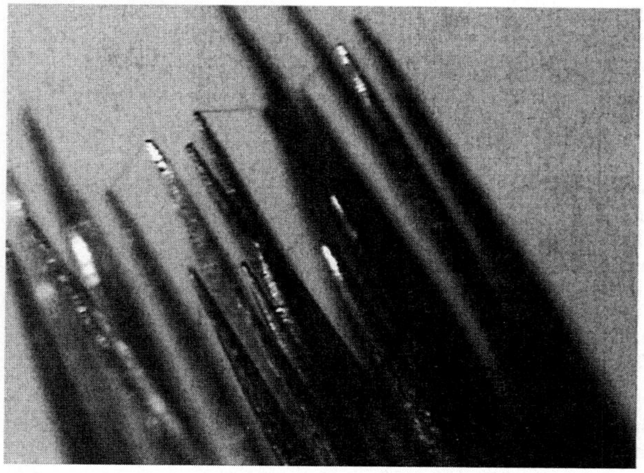

(b)

Figure 12.4.18 Vorticity probe: (*a*) probe sensor geometry and arrays; (*b*) close-up view of the probe.

centroid of each array: $m_i = \rho U_i = pU_i/RT = pU_i/[R(T_0 - U_kU_k/2c_p)]$, where U_i is the instantaneous velocity component, $i = 1$, 2, or 3 and $\underline{U_kU_k} = U_1^2 + U_2^2 + U_3^2$. The velocity can be decomposed into $U_i = \bar{U}_i + u_i$. An iterative scheme can be used to decouple density and velocity. During the first iteration it was assumed that the quantity $(u_2^2 + u_3^2)/2c_p$, where u_2 and u_3 are the velocity components in the spanwise and normal directions, respectively, is substantially smaller that the quantity $T_0 - U_1^2/2c_p$. Then the above relation can be rearranged to obtain a quadratic equation for U_i, $(Rm_i/2c_p)U_i^2 + pU_i - m_iRT_0 = 0$.

Optical Techniques

Laser speckle strophometry (LSS) or particle image velocimetry (PIV) techniques are nonintrusive but cannot in general provide adequate time-history information. In addition a slight out-of-plane motion of scattering particles due to flow three dimensionality results in different speckle patterns (speckle decorrelation).

Laser speckle strophometric methods have been developed.[79-81] In these techniques, the speckle pattern, which is the result of the diffraction of a plane wave produced by random grating, is basically affected by the flow velocity gradient while the translation of the grating caused by the constant flow velocity will not affect the pattern.

These authors applied their double pulse strophometric technique to obtain direct measurements of the quantities $\partial U/\partial x + \partial U/\partial y$, $\partial U/\partial x - \partial U/\partial y$, and $\partial U/\partial z$ in a channel flow with Reynolds number based on the channel width in the range of 2,250 to 6,700. This information was used to obtain statistics of the $\partial U/\partial y$ gradient.

One of the first direct measurements of vorticity were obtained by measuring the solid rotations of tiny transparent spherical beads of 25 μm in diameter,[69] each containing embedded planar crystal mirrors specially developed for this purpose, which were suspended in a refractive index-matched liquid.

LDA-Based Vorticity Techniques The techniques used to obtain indirect measurements of vorticity take advantage of the Doppler effect and they have configurations similar to those used in laser Doppler anemometry (LDA). In typical dual-beam configurations for velocity measurements, two laser beams cross each other to form the measuring/probe volume with the fringe pattern. Receiving optics can be aligned in any direction since the Doppler frequency shift Δf_D is independent of the k_s direction. The velocity component normal to the fringe orientation $U_i(x_j)$ can be obtained by measuring the Δf_D as

$$\Delta f_D = \frac{\vec{V}}{2\pi} \cdot \left[\vec{k}_{01} - \vec{k}_{02}\right]$$

where $\vec{k}_{01}$ and $\vec{k}_{02}$ are the wave number vectors along the two incident beams forming the measurement volume.

A natural extension of the traditional single-point LDA measurement of one velocity component shown in Fig. 12.4.9 to obtain one velocity gradient is possible by forming two independent probe volumes at a distance Δx_j apart to measure two velocity components in the same direction, $U_i(x_j)$ and $U_i(x_j + \Delta x_j)$. The velocity gradient $\partial U_i/\partial x_j$ can be estimated through the finite difference relation

$$\frac{\partial U_i}{\partial x_j} \approx \frac{\Delta U_i}{\Delta x_j} \approx \frac{U_i(x_j + \Delta x_j) - U_i(x_j)}{\Delta x_j}$$

For one-component vorticity measurements, simultaneous measurements of the two velocity gradients are required.

The LAVOR method described by Agui and Andreopoulos[82,83] uses two parallel laser beams focused with a long-focal-length lens to generate a moderately narrow probe volume defined by the region of intersection of the two beams. The velocity at two different locations inside the probe volume has been measured. There are several advantages to this approach as compared to the "sheet" beam approach. These include fewer optical components, simple alignment of the optics, smaller windows for optical access, and better use of laser intensity.

A second category of single-point velocity gradient/vorticity measurement techniques exists that are not based on the interference of two laser beams and the scattering light of a particle moving through the fringes that is realized on the surface of the photodetector. This category includes techniques that use laser-Doppler signals from two particles illuminated by two different light beams. The idea of heterodyning of the scattered light from two particles originally proposed by Durst, Melling, and Whitelaw[33] was explored by Hanson[85] and more recently by Ötügen, Su, and Papadopoulos[86] and Yao et al.[87] Hanson's heterodyning arrangement uses an optical filter to define the relative separation of two particles, which requires fabrication of miniature spatial gratings and rings with geometries tailored to the specific flows under investigation. His technique, which has been tested in a cylindrical laminar Couette flow, is insensitive to the sign of velocity gradient. The technique in Ref. 86 resolves the sign of the velocity gradient by using a Bragg cell to shift the frequency of one laser beam, and it has been tested in the laminar flow formed between two concentric cylinders with one of them rotating. The technique of Yao et al.,[87] which is quite similar to that of Ötügen et al., has been tested successfully in a laminar channel flow.

In all these techniques that estimate the velocity gradient through the heterodyning of the scattered light from two different particles, the velocity field cannot be measured simultaneously with the velocity gradient; it requires separate measurements. In that respect, the inability of these methods to provide information of the velocity field at the same time with the velocity gradient measurements may be considered a drawback.

PIV-Based Vorticity Techniques In particle image velocimetry, crossing of the flowing particle paths in highly fluctuating flows may cause ambiguity in the detection scheme. PIV is also limited by spatial resolution loss due to averaging from the correlation schemes. Although the resolution of a typical CCD camera may be typically 1k × 1k pixels, subregions of 32 × 32 pixels known as *interrogation zones* are usually considered to validate a local velocity measurement. According to Arik,[88] in a typical PIV setup approximately 10 particles are needed to validate a velocity vector measurement and 10 more velocity vectors are needed to evaluate a vorticity vector. Thus approximately 100 particles are needed to evaluate a vorticity vector, a number that results in 30 to 40 particles for each of the three vorticity components.

Lourenco et al.[89] used PIV techniques to obtain one-component vorticity measurements in a jet flow. Meinhart and Adrian[90] used PIV to measure the near wall vortical structure of low-speed boundary layers with reasonable spatial resolution.

Holographic particle image velocimetry (HPIV) was used by Zhang et al.[91] to obtain velocity gradient measurements in a square duct flow with moderate resolution.

A polarization-based approach with dual-plane particle image velocimetry (DSPIV) has been used by Kähler,[92] Ganapathisubramani et al.[93] and Hu et al.[94] to obtain vorticity measurements in boundary layer flows and in lobed jet mixers. A two-frequency DSPIV was applied in jet flows to measure all nine components of the velocity gradient tensor with resolution of about 6 Kolmogorov length scales.[95] This recent work, which is based on PIV techniques, have demonstrated the potential to provide measurements of vorticity with excellent spatial resolution. At the moment, their temporal resolution is rather limited, which, however, does not affect the process of statistical averaging. In the following, a description is provided of optical techniques that have been used to measure one velocity gradient/vorticity at a single point with reasonable success. These techniques can be classified as direct or indirect, depending on whether vorticity is obtained from the measurement of fluid particle rotation or from the measurement of individual velocity gradients inside the flow field.

VOLUMETRIC TECHNIQUES

Recently several methodologies have been developed capable of measuring the three-dimensional (3-D) velocity field within a volume. These techniques can be classified as holographic, laser sheet scanning, tomographic, among others.

Holographic PIV

In this HPIV technique, the interference of coherent light scattered by a particle and a reference beam are used to provide information of the amplitude and phase of the scattered light incident on a sensor plane. This hologram can then be used to reconstruct the light intensity field by illuminating it with the original reference.[116]

The setup used in HPIV consists of very complex optical paths and corresponding optical elements since it requires coherent imaging. In general, there are two setups used in HPIV, off-axis and in-line. In the first one, separate beams are used to provide the object and reference waves. This way speckle noise generated by the interference of the two waves within the scattering medium. Forward scattered light is used in in-line holography which provides substantially higher is brighter signals than scattering oriented normal to the beam direction. The optical setup of in-line systems is easier to implement for two reasons. First, the residual light does not need to be separated and recombined at a different location and second it allows for CCD sensors to be used known as digital in-line holography where substantial post processing of the images is required. Speckle noise can degrade the quality of HPIV data. In addition Mie scattering can distort significantly the reconstructed particles.

Scanning PIV

In flows with time-scale larger than the time-response of the instrumentation components scanning PIV can be used. This technique encompasses a rotating mirror, a high-speed camera and a laser sheet which can scan the flow field in a time shorter than the flow time scales. After correcting for geometric changes, 2-D PIV velocity data can be performed nearly instantaneously on a several planes throughout the flow field. Velocity information between the planes can then be interpolated if needed. This way, a quasi-volumetric velocity information can be obtained a target volume.

Tomographic PIV

Volumetric Velocimetry (VV) also known as 3-D PIV or Tomo PIV is capable of measuring information of 3-D velocity data by using to at least four cameras, and a thick light sheet. If the time-response of the sensor, that is, camera is adequate time-resolved VV measurements are possible. Cameras with CMOS chip are suited for this type of measurements, while CCD cameras can be used in certain flows fields.

The recorded simultaneous views of the illuminated volume are then reconstructed to yield a discretized 3-D intensity field. This analysis uses 3-D cross-correlation algorithms to calculate the 3-D, 3-C velocity field within the volume [Elsinga et al. (2006); Kim et al. (2011); Scarano (2013)].

For low to medium seeding concentration, Particle Tracking Velocimetry (PTV) is used with two or three cameras, whereas for high seeding concentration (like PIV) a Least Squares Matching (LSM) algorithm provides three velocity components in a 3-D volume, often described as 3D-3C (3 dimensions-3 components).

The reconstruction procedure involves the solution of an underdetermined system of equations which are challenging to solve. This inverse problem is very complex to solve because multiple solution can exist for a single set of views. In order to ensure uniqueness of the solution, computationally intensive algorithms are used like the multiplicative algebraic reconstruction technique (MART).

MICRO- AND NANOSENSORS AND ASSOCIATED TECHNIQUES

Over the past 5 to 12 years, there has been an increasing interest in understanding fluid phenomena at the micrometer- and nanometer-length scales. This has been evident in both basic and applied fluid mechanics research. The application of microscale fluid mechanics to MEMS and BioMEMS represents one of the most significant practical applications. In addition, there is a need to resolve smaller scales in measurements of mainstream macroscopic flows.

These two needs led to the development of microsensors and nanosensors. It is very encouraging to see that, in general, as the physical size of the probes is reduced, their frequency response increases. Reduction in probe dimensions is expected to improve spatial and temporal resolution.

Thermal anemometry-based **micromachined** probes have been explored by Tai[96] and Lofdahl et al.,[97] but they were never ready for practical use. Jiang, Tai, Ho, and Li[98] developed new micrometer-sized polysilicon wires and demonstrated the potential for batch fabrication. Jiang, Tai, Ho, Karan, and Garstenauer[99] have implemented these hot wires into anemometers and performed calibration of them in wind tunnels. The micromachined anemometers (Fig. 12.4.19*a*) consist of a sensing wire, two parallel support shanks, a silicon beam, and a thick (500-μm) Si handle. The silicon beam is like the probe body and is 1 μm long, 200 μm wide, and 75 μm thick. The two parallel support shanks are 100 μm long, 20 μm wide, and 0.5 μm thick. Various sizes of the sensing wires have been made, and they are 10 to 160 μm long, 1 μm wide, and 0.5 μm thick. This structure is similar to that of the conventional anemometers, so direct comparison can be made. The

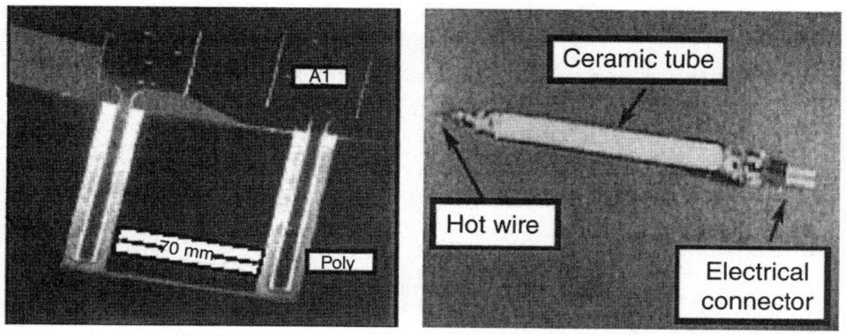

(a) (b)

Figure 12.4.19 (*a*) SEM of a micromachined hot wire. (*b*) Micrograph of the anemometer packaging from Jiang et al.[99]

package of the anemometers is a ceramic tube a few centimeters long and 3 mm in diameter (Fig. 12.4.19*b*). The anemometer handle is soldered with electrical cables, placed inside the tube, and then epoxy-fixed. During handling, all mechanical vibrations have to be avoided in order to prevent damage of the wires.

Ebefors et al.[100] used doped silicon for both the prongs and the hot-wire sensor, along with a polyamide hinge, to make a three-component sensor, shown schematically in Fig. 12.4.20. An SEM photograph is shown in Fig. 12.4.21.

Non-silicon-based hot wires with a thermal element size of 2.7 μm × 50 μm have been fabricated by Chen and Liu[101] (Fig. 12.4.22).

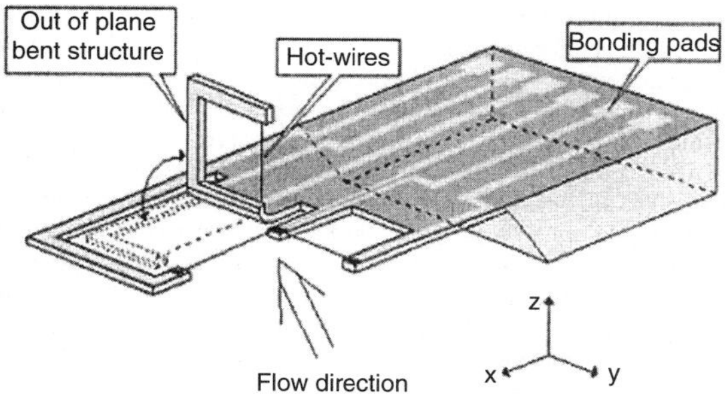

Figure 12.4.20 Schematitic view of the 3-D hot-wire probe based on a polyamide joint. Silicon chip is 3.5 mm × 3 mm × 0.5 mm and the three wires are 500 mm × 5 mm × 2 mm. From Ebefors et al.[100]

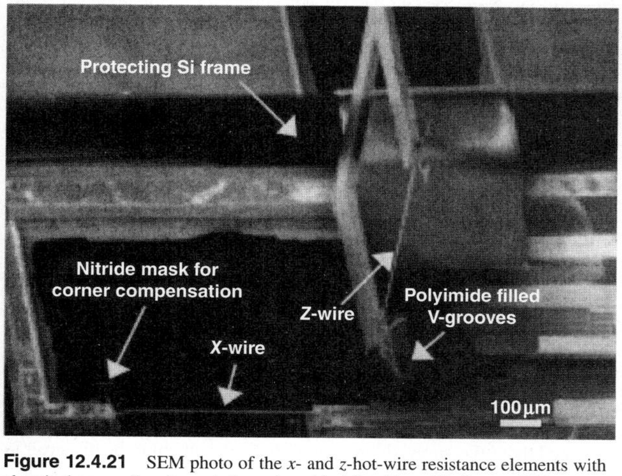

Figure 12.4.21 SEM photo of the *x*- and *z*-hot-wire resistance elements with electrical connection through the polyamide joint. From Ebefors et al.[100]

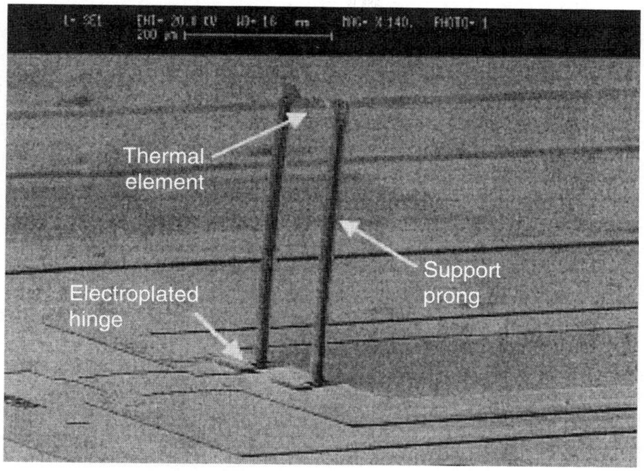

Figure 12.4.22 SEM micrographs of a hot-wire anemometer after postrelease strengthening by electroplating. From Chen and C. Liu.[101]

Researchers at Princeton University have developed The Nano-Scale Thermal Anemometry Probe (NSTAP) with a sensing volume of 100 nm × 2 μm × 30 μm which has a frequency response of 200 kHz and can be operated with conventional anemometer bridges. The probe, shown in Fig. 12.4.23, has improved spatial and temporal resolutions which enabled measurements of turbulence in flows with very high renumbers.[122,123]

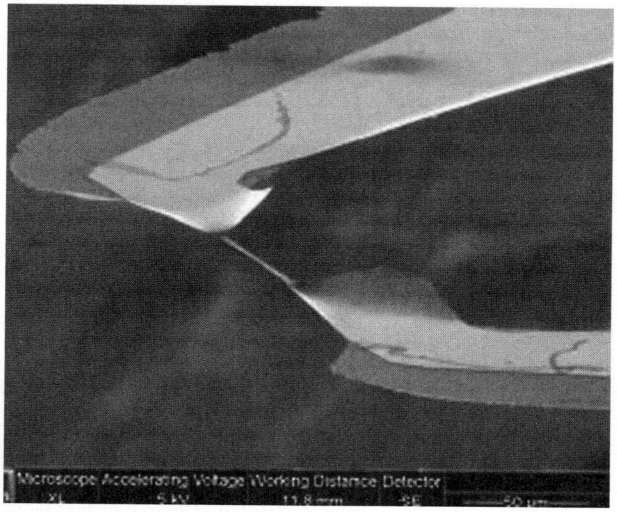

Figure 12.4.23 Scanning electron microscope image of an NSTAP, with a close-up of the sensor. (*Source: Image provided by L. Smits & M. Hulmark.*)

Micrometer-resolution particle image velocimetry (μPIV) refers to the application of PIV to measure velocity fields of fluid motion with length scales of the order 100 μm, and with spatial resolution of individual velocity measurements of order 1 to 10 μm. The PIV technique is limited to maximal spatial resolutions of 0.2 to 1.0 mm because of fundamental physical limitations as well as practical implementation difficulties. In μPIV, emphasis is placed on the ability to accurately and reliably measure fluid velocity with micrometer spatial resolution. In order to achieve microscale velocity measurements, novel developments in PIV image recording hardware, flow-tracing particles, system

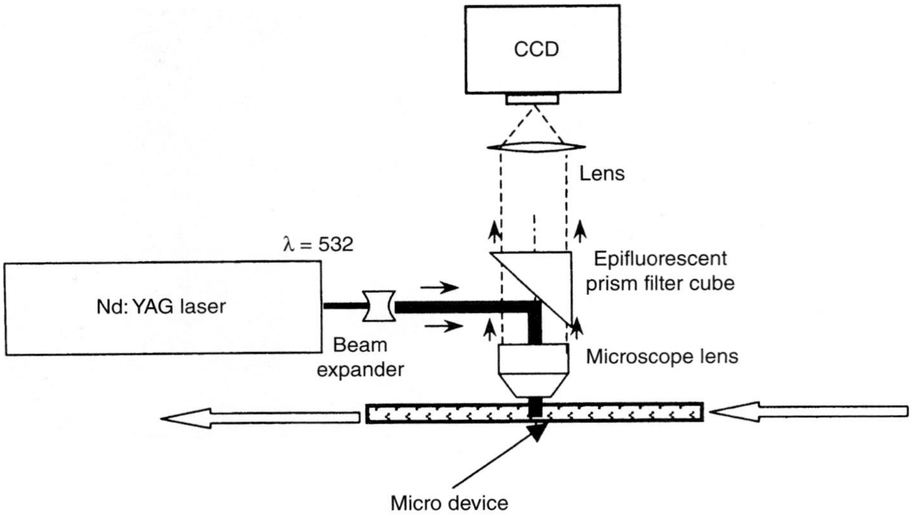

Figure 12.4.24 Experimental setup for μPIV measurements.

design, and analysis software are required. The theoretical background of μPIV is based on the interactions between the flow fluid and the seed particles, which occur at nanometer length scales. μPIV can be thought of as a practical, application of nanotechnology to perform science and engineering at micrometer-length scales.

The main component of a typical μPIV setup like that shown in Fig. 12.4.24 is the microscope lens and epifluorescent prism/filter which are typical in microscopy imaging. This system was adopted originally by Meinhart et al.[102] for μPIV measurements. A pulsed monochromatic light source, such as a pulsed Nd:YAG laser system is used. The laser system is similar to those used in PIV applications. It consists of two Nd:YAG laser cavities, beam-combining optics, and a frequency-doubling crystal. The illumination light is delivered to the microscope through beam-forming optics, which can consist of a variety of optical elements that will sufficiently modify the light so that it will fill the back of the objective lens, and thereby broadly illuminate the microfluidic device. The beam-forming optics can consist of a liquid-filled optical fiber to direct the light into the microscope. Recently, light has been delivered into the microfluidic device by using total reflection through a transparent wall. This instrument is called a total internal reflection fluorescence (TIRF) microscope.

Typically, the working fluid is directed through the microfluidic device by pressure, or induced electrokinetically. The working fluid can be of any visibly transparent fluid, such as water. Liquids are most commonly used. It is conceivable to use gaseous fluids, such as air, but μPIV experiments using air have not been demonstrated successfully to date. The working fluid should be seeded with fluorescent particles, which can be manufactured from a variety of materials, such as polystyrene. It is important that the particles have a specific gravity closely matched to the working fluid, and less than, say, 1 μm, preferably 200 to 700 nm.

A sensitive large-format interline-transfer CCD camera is commonly used to record the particle image fields. A large-format CCD array is desirable because it allows for more particle images to be recorded, and increases the spatial dynamic range of the measurements. The interline transfer feature of the CCD camera allows for two particle image frames to be recorded back to back to within a 500-ns time delay.

These techniques can be extended to provide three-dimensional information on the entire flow field by changing the focal plane of the objective lens with piezoelectric positioners to scan the whole depth.

In μPIV, fluorescent particles with diameters typically in the range 200 to 300 nm are used to obtain velocity data with an out-of-plane spatial resolution typically of order 1 μm. Near-wall measurements with this method have so far been obtained no closer than 450 nm from the surface.

Recently, **nano-PIV** (nPIV) has been introduced as a technique specifically designed for near-surface interrogation of flow[103-105] where near-wall measurements with μPIV have so far been obtained no closer than 450 nm from the surface. The new nPIV method relies on an evanescent wave illumination generated by the total internal reflection (TIR) of light at the interface between two materials of different refractive indices,[106] for example, a glass-liquid interface. With this method, the motion of particles within a region of order 100 nm of the surface is measured. In nPIV studies, to date, the size of the seed particles are in the 100- to 300-nm range; that is, they are of the same order of, or larger than, the illuminated region. It would be desirable to use much smaller seed particles to reduce some of the complications that can be caused by larger particles, such as the modification of the evanescent wave field and particle–fluid interactions that compromise the accuracy of fluid velocity measurement and the ability of particles to move close to the surface.

The use of **quantum dot** (QD) nanoparticles for near-surface measurements of velocity has been introduced by Pouya et al.[107] These particles are an order of magnitude smaller than what has been used to date in nPIV. Nanocrystal QDs are semiconductor nanoparticles that are chemically synthesized with precisely controlled sizes in the range of 1 to 10 nm.[108-112] Among the most developed QDs for fluorescence imaging are the coreshell dots composed of an optically active nanocrystal core of CdSe surrounded by a protective shell of ZnS. The surface of QD is covered with a ligand shell that can be functionalized for broad chemical flexibility. Several properties of QDs make them extremely attractive for fluid flow studies. By modifying the functional groups in the ligand shell, QDs may be dispersed in specific chemical environments such as polar and nonpolar liquids. Since QDs can be solubilized within the fluid, they are expected to behave more like molecules and some of the near-wall issues of "solid" particles should not manifest themselves to QD tracers (e.g., particle–fluid and particle–wall interactions). The emission wavelength of a QD depends on its size, and can be tuned across the entire visible spectrum by varying the diameter of the particle. The excitation band is very broad, and the emission is independent of the excitation wavelength. Thus, a size series of QDs with different emissions can be excited with the same light source.

The experimental setup is similar to that of μPIV. Pouya et al.[107] indicated that QDs showed large enough displacement to be discernible with the naked eye.

Nuclear magnetic resonance (NMR), one of the most powerful analytic tools for studying chemical composition, is opening the door to microfluidic applications. NMR is a phenomenon involving a property found in atomic nuclei of almost every molecule called *spin,* which gives rise to a magnetic moment, meaning that the nuclei behave like bar magnets with north and south poles. When a magnetic field is applied, these magnets align their axes along the lines of the magnetic field. This alignment is not perfect and it results in a wobbly rotation about the magnetic lines that is characteristic for each type of nucleus. Hilty et al.[113] developed a remote NMR technique that provided a strong signal to overcome the sensitivity limitation of NMR by introducing hyperpolarized xenon nuclei, which have a very long spin-relaxation time (several minutes). In this technique, signal encoding and detection are carried out independently while in a conventional setup the two processes take place within a single device. In addition to the magnetic field, the nuclei of the sample are also subjected to a sequence of radio-frequency pulses during the encoding phase, and because of this they absorb and re emit energy at characteristic frequencies. In the detection phase the frequencies of these encoded signals are used to obtain an NMR spectrum that contains distinct information to characterize the sample.

REFERENCES

1. Tavoularis, S., "Measurement in Fluid Mechanics," Cambridge University Press, 2005.
2. Bradshaw, P., http://vonkarman.stanford.edu/tsd/pbstuff/tunnel/index.html.
3. Bradshaw, P., and Pankhurst, R. C., "Progress in Aeronautical Science," Pergamon Press, Oxford, U.K., 1964, Chap.1, pp. 1-69.
4. Rae, W. H., and Pope, A., "Low-Speed Wind Tunnel Testing," John Wiley & Sons, New York, 1984.
5. Mehta, R. D., and Bradshaw, P., *Aeron. J.*, **73**, 1979, pp. 443-449.
6. Gad-el Hak, M., *Exper. Fluids,* **5**, 1987, pp. 289-297.
7. Briassulis, G., Agui, J. H., Andreopoulos, J., and Watkins, C. B., *Exper. Thermal Fluid Sci.,* **13**, 1996, pp. 430-446.
8. Agui, J. H., Briassulis, G., and Andreopoulos, Y., *J. Fluid Mech.,* **524**, 2005, pp. 143-195.
9. Briassulis G., Agui, J., and Andreopoulos, Y., *J. Fluid Mech.,* **432**, 2001, pp. 219- 283.
10. Wyngaard, J., *J. Phys. E: Sci. Instrum.,* Series **2**, 1969, pp. 983-987.
11. Wallace, J. M., and Foss, J., *Annu. Rev. Fluid Mech.,* **27**, 1995, pp. 469-514.
12. Tsinober, A., Kit, E., and Dracos, T., *J. Fluid Mech.,* **242**, 1992, pp. 169-192.
13. Honkan, A., and Andreopoulos, Y., *J. Fluid Mech.,* **350**, 1997, pp. 29-96.
14. Figliola, R. S., and Beasley, D. E., "Theory and Design of Mechanical Measurements," John Wiley & Sons, New York, 1995.
15. Doebelin, E. O., "Measurement Systems Application and Design," 4th ed., McGraw-Hill, New York, 1990.
16. Tavoularis, S., "Encyclopedia of Fluid Mechanics," Vol. 1, Chap. 36, Gulf Publishing, Houston, 1986, pp. 1207-1255.
17. Gad-el-Hak, M., Basic Instruments, in "Handbook of Fluid Dynamics," Chap. V, CRC Press, 1998, pp. 33-1-33-22.
18. Peterson, J. I., and Fitzgerald, V. F., *Rev. Sci. Instrum.,* **51**, no. 5, 1980, pp. 133-136.
19. McLachlan, B. G., and Bell, J. H., *Ther. Fluid Sci.* **10**, no. 5, 1995, pp. 470-483.
20. Morris, M. J., *AIAA J.,* **31**, no. 3, 1993, pp. 419-421.
21. McLachlan, B. G., *J. Aircraft,* **32**, no. 2, 1995, pp. 217-225.
22. Reda, D. C., Wilder, M. C., Mehta, R., and Zilliac, G., "AIAA-97-2489," *32nd Thermo Physics Conference,* Atlanta, GA, June 23-25, 1997.
23. Tropea, C. et al. (eds.), "Handbook of Experimental Fluid Mechanics," Springer Verlag, 2006.
24. Bruun., H. H., "Hot-Wire Anemometry," Oxford University Press, Oxford, U.K., 1995.
25. Lomas, C. G., "Fundamentals of Hot Wire Anemometry," Cambridge University Press, Washington, 1983.
26. Perry, A. E., "Hot-Wire Anemometry," Clarendon, Oxford, 1982.
27. King, L. V., *Phil. Trans. Roy. Soc. London*, **A214**, 1914, pp. 373-432.
28. Bradshaw, P., "An Introduction to Turbulence and Its Measurement," Pergamon, Press, Oxford, 1971.
29. Champagne, F. H., Sleicher, C. A., and Wehrmann, O. H., *J. Fluid Mech.*, **28**, 1967, pp. 153-175.
30. Jorgensen, F. E., *DISA Info.*, **11**, 1971, pp. 31-37.
31. Cummins, H. Z., Knable, N., and Yeh, Y., *Phys. Review Letters*, **12**, no. 6, 1964, pp. 51-53.
32. Yeh, Y., and Cummins, H. Z., *Appl. Phys. Letters,* **4**, no. 10, 1964, pp. 176-178.
33. Durst, F., Melling, A., and Whitelaw, J. H., "Principles and Practice of Laser Doppler Anemometry," 2nd ed., Academic Press, New York, 1981.
34. Durst, F., "Measurements and Predictions of Complex Turbulent Flows: Principles of Laser Doppler Anemometers," Von Karman Institute for Fluid Dynamics, Lecture Series 1980-1983, 1980.
35. Buchave, P., George, W. K., and Lumley, J. L., *Ann. Rev. Fluid Mech.*, **11**, 1979, pp. 443-503.
36. Adrian, R. J., *Ann. Rev. Fluid Mech.,* **23**, 1991, pp. 261-304.
37. Adrian, R. J., and Yao, C. S., *Appl. Optics.,* **24**, 1985, pp. 42-52.
38. Grant, I., *Proceed. Instit. Mech. Eng.,* **211**, 1997, pp. 55-76.
39. Hinsch, K. D., "Particle Image Velocimetry," in "Speckle Metrology," ed. R. S. Sirohi, Marcel Dekker, New York, 1993, pp. 235-323.
40. Raffel, M., Willert, C. E., and Kompenhans, J. "Particle Image Velocimetry," Springer, 1998.
41. Wang, Z., "Experimental Investigation of Compressible Turbulent Jets," Ph.D. thesis, City University of New York, 2003.
42. Komine, H., U.S. Patent **4**, 919, 1990, 536.
43. Komine, H., Brosnan, S., Litton, A., and Stappaers, E., AIAA Paper 91-0337, 1991.
44. Meyers, J. F., and Komine, H., *Laser Anemom.,* **1**, 1991, pp. 289-296.
45. Miles, R. B., and Lempert, W. R., AIAA Paper 90-0624, 1990.
46. Miles, R. B., Lempert, W. R., and Forkey, J. N., AIAA Paper 91-0357, 1991.
47. Samimy, M., and Wernet, M. P., *AIAA J.,* **38**, no. 4, 2000, pp. 553-574.
48. Miles, R. B., Forkey, J. N., and Lempert, W. R., AIAA Paper 92-3894, 1992.
49. Forkey, J., Finkelstein, N. D., Lempert, W. R., and Miles, R. B., AIAA Paper 95-0298, 1995.
50. Meyers, J. F., *Meas. Sci. Technol.,* **6**, 1995, pp. 769-783.
51. Arnette, S. A., Samimy, M., and Elliott, G. S., *AIAA J.,* **33**, 1995, pp. 430-438.
52. Miles, R. B., Lempert, W. R., and Forkey, J., AIAA-91-0367, 1991.
53. McKenzie, R. L., *Appl. Optics,* **35**, no. 6, 1996, pp. 948-964.
54. Elliott, G. S., Samimy, M., and Arnette, S. A., AIAA Paper 93-0512, 1993.
55. Elliott, G. S., Samimy, M., and Arnette, S. A., *Exp. Fluids,* **18**, 1994, pp. 107-118.
56. Mosedale, A. D., Elliot, G. S., Carter, C. D., and Beutner, T. J., *AIAA J.,* **38**, no. 6, 2000, pp. 1010-1024.
57. Gendrich, C. P., Koochesfahani, M. M., and Nocera, C. G., *Exp. Fluids,* **23**, no. 5, 1997, pp. 361-372.
58. Stier, B., and Koochesfahani, M. M., *Exp. Fluids,* **26**, 1999, pp. 297-304.
59. Ress, J. M., Laufer, G., and Krauss, R. H., *AIAA J.,* **33**, 1995, pp. 296-301.
60. Pitz, R. W., Wehrmeyer, J. A., Ribarov, L. A., Oguss, D. A., Batliwala, F., DeBarber, P. A., Deusch, S., and Dimotakis, P. E., *Meas. Sci. Tech.,* **11**, 2000, p. 1259.
61. Ribarov, L. A., Wehrmeyer, J. A., Hu, S., and Pitz, R. W., *Exp. Fluids,* **37**, 2004, pp. 65-74.

62. Dam, N. J., Klein-Douwel, J. H., Sijtsema, N. M., and ter Meulen, J. J., *Opt. Lett.,* **26**, 2001, pp. 36-38.

63. Sijtsema, N. M., Dam, N. J., Klein-Douwel, J. H., and ter Meulen, J. J., *AIAA J.,* **40**, 2002, pp. 1061-1064.

64. Miles, R. B., and Lempert, W. R., *Ann. Rev. Flu. Mech.,* **29**, 1997, pp. 285-326.

65. Noullez, A., Wallace, G., Lempert, W. R., Miles, R. B., and Frisch, U., *J. Fluid Mech.,* **339**, 1997, pp. 287-307.

66. Foss, J. F., *Bull. Am. Phys. Soc.,* **21**, 1976, p. 1237 (abstract).

67. Foss, J. F., *Proc. Biennial Symp. on Turbulence,* 7th Missouri–Rolla, 1981, pp. 208-218.

68. Foss, J. F., *Exp. Therm. Fluid Sci.,* **8**, 1994, pp. 260-270.

69. Frish, M. B., and Webb, W. W., *J. Fluid Mech.,* **107**, 1981, pp. 173-185.

70. Vukoslavoevic, P., Wallace, J. M., and Balint, J., *J. Fluid Mech.,* **228**, 1991, pp. 25-51.

71. Honkan, A., and Andreopoulos, Y., *J. Fluid Mech.,* **350**, 1997, pp. 29-96.

72. Agui, J. H., Briassulis, G., Andreopoulos, Y., *J. Fluid Mech.,* **524**, 2005, pp. 143-195.

73. Briassulis, G., Agui, J., and Andreopoulos, Y., *J. Fluid Mech.,* **432**, 2001, pp. 219-283.

74. Andreopoulos, J., *Rev. Sci. Instr.* **54**, no. 6, 1983, pp. 733-740.

75. Andreopoulos, J., *Physics E. Sci. Instr.,* **16**, 1983, pp. 1264-1271.

76. Briassulis, G., Agui, J., Andreopoulos, J., and Watkins, C. B., *Exp. Ther. Fluid Sci.,* **13**, 1996, pp. 430-446.

77. Kovasznay, L. S. G., *J. Aeronaut. Sci.,* **17**, 1950, p. 565.

78. Morkovin, M. V., *AGARDOgraph,* **24**, 1956.

79. Breyer, H., Kriegs, H., Schmidt, K., and Staude, W., *Exp. Fluids,* **15**, 1993, pp. 200-208.

80. Kriegs, H., Schulz, R., and Staude, W., *Exp. Fluids,* **15**, 1993, pp. 240-246.

81. Kriegs, H., and Staude, W., *Meas. Sci. Tech.,* **6**, 1995, p. 653.

82. Agui, J. H., and Andreopoulos, J., *International Symposium on Laser Anemometry,* FED vol. 191, June 20-24, 1994, pp. 11-19.

83. Agui, J. H., and Andreopoulos, Y., *Exp. Fluids,* **34**, 2003, pp. 192-205.

84. Kolmogorov, A. A., *C. R. Akad. Sci. SSSR,* **30**, 1941, pp. 301-306.

85. Hanson, S. G., *2nd Symposium on Applications of Laser Anemometry to Fluid Mechanics,* July 2-5, Lisbon, 1984.

86. Ötügen, M. V., Su, W-J., and Papadopoulos, G., *Meas. Science Technol.,* **9**, 1998, pp. 267-274.

87. Yao, S., Tong, P., and Ackerson, B., *Appl. Optics,* **40**, no. 24, 2001, pp. 4022-4027.

88. Arik, E. B., "Flows at Ultra High Reynolds and Rayleigh Numbers: A Status Report," eds. R. J. Donnelly and K. R. Sreenivasan, Springer-Verlag, New York, 1998, pp. 138-158.

89. Lourenco, L., and Krothapalli, A., *Exp. Fluids,* **5**, 1987, pp. 23-31.

90. Mcinhart, C. D., and Adrian, R. J., AIAA paper 95-0789, 1995.

91. Zhang, J., Tao, B., and Katz, J., *Exp. Fluids,* **23**, 1997, pp. 373-381.

92. Kähler, C. J., *Exp. Fluids,* **36**, 2004, pp. 114-130.

93. Ganapathisubramani, B., Longmire, E. K., Marusic, I., and Pothos, S., *Exp. Fluids,* **32**, 2004, no. 2, pp. 222-231.

94. Hu, H., Saga, T., Kobayashi, T., Taniguchi, N., and Yasuki, M., *Exp. Fluids,* **31** 2001, pp. 277-293, .

95. Mullin, J. A., and Dahm, W. J. A., *Exp. Fluids* (to appear) 2006.

96. Tai, Y. C., *Dig. Tech. Papers, Transducers* 85, pp. 354-357, 1985.

97. Lofdahl, L., Stemme, G., and Johansson, B., *Exp. Fluids,* **12**, 1992, pp. 270-276.

98. Jiang, F., Tai, Y. C., Ho, C. H., and Li, W. J., *Tech. Digest 1994 Solid-State Sensors & Actuator Workshop,* Hilton Head, SC, 1994, pp.264-267.

99. Jiang, F., Tai, Y. C., Ho, C. M., Karan, R., and Garstenauer, M., *Technical Digest, International Electron Device Meeting* (IEDM-94), San Francisco, Dec. 1994, pp. 139-142.

100. Ebefors, E., Kalvesten, E., and Stemme, G., *Proc. MEMS*, New York, 1998, pp. 93-98.

101. Chen, J., and Liu, C., *J. Microelectromechanical Sys.,* **12**, no. 6, December 2003, pp. 979-988.

102. Meinhart, C. D., Wereley, S. T., and Santiago, J. G., *Experiments in Fluids,* **27**, 1999, pp. 414-419.

103. Zettner, C. M., and Yoda, M., *Exp. Fluids,* **34**, 2003, p. 115.

104. Jin, S., Huang, P., Park, J., Yoo, J. Y., and Breuer, K. S., *Experiments in Fluids* **37**, 2004, p. 825.

105. Sadr, R., Yoda, M., Zheng, Z., and Conlisk, A. T., *J. Fluid Mech.,* **506**, 2004, p. 357.

106. Axelrod, D., Burghardt, T. P., and Thompson, N. L., *Annu. Rev. Biophys. Bioeng.,* **13**, 1984, p. 247.

107. Pouya, S., Koochesfahani, M., Snee, P., Bawendi, M., and Nocera, D., *Exp. Fluids,* **39**, 2005, pp. 784-786.

108. Murray, C. B., Norris, D. J., and Bawendi, M. G., *J. Am. Chem. Soc.,* **115**, 1993, p. 8706.

109. Dabbousi, B. O., Rodriguez-Viejo, J., Mikulec, F. V., Heine, J. R., Mattoussi, H., Ober, R., Jensen, K. F., and Bawendi, M. G., *J. Phys. Chem. B,* **101**, 1997, p. 9463.

110. Bruchez, M., Jr., Moronne, M., Gin, P., Weiss, S., and Alivisatos, A. P., *Science,* **281**, 1998, p. 2013.

111. Mattoussi, H., Mauro, J. M., Goodman, E., Anderson, G. P., Sundar, V. C., Mikulec, F. V., and Bawendi, M. G., *J. Am. Chem. Soc.,* **122**, 2000, p. 12142.

112. Murray, C. B., Kagan, C. R., and Bawendi, M. G., *Rev. Mat. Sci.,* **30**, 2000, p. 545.

113. Hilty, C., McDonnell, E. E., Granwehr, J., Pierce, K. L., Han, S. I., and Pines, A., *PNAS,* **142**, no. 42, 2005, pp. 14960-14963.

114. Sheng, J, Malkiel, E., and Katz, J., "Using Digital Holographic Microscopy for Simultaneous Measurements of 3D Near Wall Velocity and Wall Shear Stress in a Turbulent Boundary Layer," *Exp. Fluids* **45**, 2008, pp. 1023-1035.

115. Arroyo, M. P., and Hinsch, K. D., "Recent Developments of PIV towards 3D Measurements," Springer, 2008, pp. 127-154.

116. Meng, H, Pan, G., Pu, Y., and Woodward, S. H., "Holographic Particle Image Velocimetry: From Film to Digital Recording," *Measurem. Sci. Tech.,* **15**, 2004, pp. 673-685.

117. Svizher, A., and Cohen, J., "Holographic Particle Image Velocimetry System for Measurement of Hairpin Vortices in Air Channel Flow," *Exp. in Fluids*, **40**, 2006, pp. 708-722.

118. Tao, B., Katz, J., and Meneveau, C., "Geometry and Scale Relationships in High Reynolds Number Turbulence Determined from Three-Dimensional Holographic Velocimetry," *Phy. Fluids* **12**, 2000, pp. 941-944.

119. Elsinga, G. E., Scarano, F., Wieneke, B., and van Oudheusden, B. W., "Tomographic Particle Image Velocimetry," *Exp. Fluids*, **41**, 2006, pp. 933-947.

120. Kim, H., Große S S., Elsinga, G. E., and Westerweel, J., "Full 3D-3C Velocity Measurement Inside a Liquid Immersion Droplet," *Exp. Fluids,* **51**, no. 2, 2011, pp. 395-40.

121. Scarano, F., "Tomographic PIV: Principles and Practice," *Measurem. Sci. Tech.,* **24**, no. 1, 2013, p. 012001. Bibcode:2013MeScT. 24a2001S. doi:10.1088/0957-0233/24/1/012001.

122. Bailey, S. C. C., Kunkel, G. J., Hultmark, M., Vallikivi, M., Hill, J. P., Meyer, K. A., Tsay, C., Arnold, C. B., and Smits, A. J., "Turbulence Measurements using a Nanoscale Thermal Anemometry Probe," *J. Fluid Mech.*, **663**, 2010, pp. 160-179.

123. Vallikivi, M., Hultmark, M., Bailey, S. C. C., and Smits, A. J., "Turbulence Measurements in Pipe Flow using a Nanoscale Thermal Anemometry Probe," *Exp. Fluids*, **51**, 2011, pp. 1521-1527.

12.5 AIR-INFLATED FABRIC STRUCTURES

by Paul V. Cavallaro and Ali M. Sadegh

INTRODUCTION

Air-inflated fabric structures fall within the category of tensioned structures and provide unique advantages in their use over traditional structures. These advantages include light weight designs, rapid and self-erecting deployment, enhanced mobility, large deployed-to-packaged volume ratios, fail-safe collapse, and possible rigidification.

Most of the research and development pursued in air-inflated structures can be traced to space, military, commercial, marine engineering, and recreational applications. Examples include air ships, weather balloons, inflatable antennas and radomes, temporary shelters, pneumatic muscles and actuators, inflatable boats, temporary bridging, space habitats, space vehicle decelerators, and energy absorbers such as automotive air bags. However, the advent of today's high performance fibers combined with continuous textile manufacturing processes has produced an emerging interest in air-inflated structures. Air-inflated structures can be designed as viable alternatives to conventional structures.

Because these structures combine both the textile and structural engineering disciplines, the structural designer should become familiar with the terminology used in textile materials and their manufacturing processes. A glossary is provided in reference.[1]

DESCRIPTION OF AIR-INFLATED FABRIC STRUCTURES

Air-inflated fabric structures are constructed of lightweight fabric skins that enclose a volume of pressurized air. The fabric is typically formed in a variety of textile architectures including those shown in Fig. 12.5.1. Each architecture has its own design, manufacturing, tooling, and cost implications. Structurally, these architectures will behave differently when subjected to mechanical loads.

The plain-weave architecture provides orthogonal yarn placement along two principal axes designated as the warp and weft axes. The warp axis is known as the direction of weaving and is virtually unlimited in length. The weft axis corresponds to the width direction of the roll and is limited by the size of the loom. The plain-woven architecture provides extensional stiffness along the warp and weft axes; however, it lacks shear stiffness for off-axis loads. The braided architecture provides shear stiffness due to the non-orthogonality of the bias-aligned yarns; however, it lacks extensional stiffness. The angle between the braid axis and the yarns, θ, is referred to as the braid angle or bias angle. Both the triaxial braid and axial strap-reinforced braid architectures behave similarly in that they afford the fabric with extensional and shear stiffnesses.

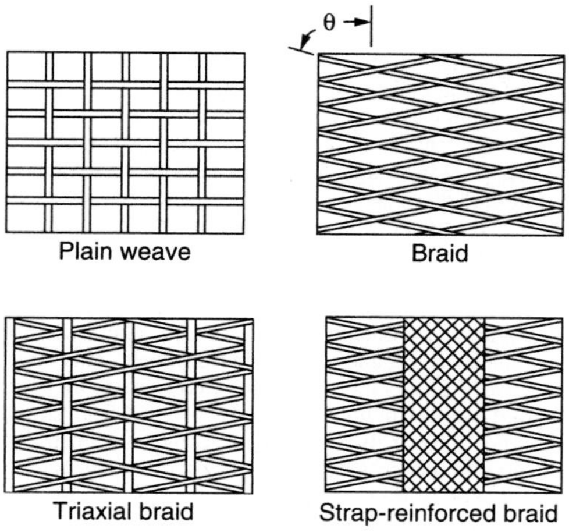

Figure 12.5.1 Various fabric architectures used in air-inflated structures.

Internal air pressure develops a biaxial pre-tensioning stress throughout the fabric that enables the structure to generate its intended shape, provides stiffness to resist deflections and affords stability against collapse from external loads. Fabric materials can often be idealized as tension-only materials for design purposes, that is, their in-plane compressive moduli and bending moduli may be considered negligible.

Stiffness of the structure is primarily a function of inflation pressure. As inflation pressure increases, the pre-tensioning stresses throughout the fabric increase and, in turn, stiffen the structure. Once external loads are applied, stresses from these loads superimpose with the pre-tensioning stresses in the fabric. As a result, a complete redistribution of stress occurs. This stress redistribution balances the loads and maintains the structure in a state of static equilibrium. Depending upon the type of air-inflated fabric structure (i.e., beam, arch, etc.) and applied loads (i.e., tension, compression, shear, bending, torsion, etc.), the redistribution of stresses can either increase or decrease the net tension stresses in the fabric. However, stability of the structure is only ensured when no regions of the fabric experience a net loss in tensile stress. Otherwise, if the stresses from applied loads begin to relax the pre-tensioning stress (i.e., the tension approaches zero), the onset of wrinkling is said to have occurred within the structure. Wrinkling decreases the structure's load carrying capability and upon further loading, the ultimate wrinkling moment is reached and eventual buckling or collapse will result. Research has shown for certain structures that the ultimate wrinkling moment is double the wrinkling onset moment.

There are two significant and unique characteristics that air-inflated fabric structures provide over conventional structures. First, upon an overload condition, a collapse of an air-inflated structure does not necessarily damage the fabric. When the overload condition is removed, the air-inflated fabric structure simply restores itself to its design load configuration. Second, since wrinkling can be visually detected, it can serve as a warning indicator prior to collapse.

FIBER MATERIALS AND YARN CONSTRUCTIONS

Proper selection of fiber materials and yarn constructions are important factors that must be considered in the design of air-inflated fabric structures. Both should be optimized together to achieve the desired performance characteristics at the fabric and structural levels.

Many of today's air-inflated fabric structures use yarns constructed of high performance continuous fibers such as Vectran® (thermoplastic liquid crystal polymer) PEN® (polyethylene naphthalate), DSP® (dimensionally stable polyester), and others. These fibers provide improved structural performance (high strength, low elongation, fatigue, flex-fold, cyclic loadings, creep, etc.) and enhanced environmental resistance (ultraviolet rays, heat, humidity, moisture, abrasion, chemicals, etc.). Other fibers used in air-inflated fabric structures include Kevlar®, Dacron®, nylon, Spectra®, and polyester. Hearle[2] provides additional information on fiber materials and their mechanical properties.

Yarns are constructed from fibers that may be aligned unidirectionally or arranged in a number of twisted bundles. Twist is used to improve the handling susceptibility of the yarns by grouping the fibers together especially during textile processing. Twist, which is measured in turns per unit yarn length, affects the yarn tensile properties. For discontinuous or staple fiber yarns, twist can increase the yarn breaking strength because the internal forces at the ends of a fiber can transfer to neighboring fibers via inter-fiber shear forces. However, twist in continuous fiber yarns can reduce the yarn breaking strength as observed by Hearle.[3] Therefore, a minimal twist is recommended for providing adequate handling protection to continuous fiber yarns.

Hearle[3] experimentally investigated the effects of twist on the tensile behavior of several continuous fiber yarns. His results showed that yarn tenacity (defined as tensile strength measured in grams-force per denier or grams-force per tex) decreased with increasing twist for three prescribed yarn tensions used during twist formation. In general, the yarn modulus decreased with increasing twist, yarn elongation at break increased with increasing twist, and yarn elongation decreased with increasing yarn tension. A difference in the load-extension behavior of twisted and non-twisted yarns is that a twisted yarn when subjected to tension will undergo compaction of its cross section through migration of its fibers and develop greater inter-fiber frictional forces than a non-twisted yarn. Like fabrics, the structural performance of yarns can be tailored by changes in their architecture and processing.

Once the yarns are processed into the fabric (i.e., by weaving, braiding, etc.), it is recommended that tensile tests be performed on sample yarns removed from the fabric. This will allow the "as-processed" yarn properties to be compared to the design requirements. For example, the "as-woven" tensile properties of continuous-fiber, non-twisted yarns removed from a plain-woven fabric air beam were measured[4] using an Instron machine configured with textile grips as shown in Fig. 12.5.2. The cross sectional areas of the yarns were computed based on fiber diameter and quantity. The tests revealed that the average breaking stress of the weft yarns was nearly 20 percent less than that of the warp yarns. The reduction in breaking stress of the weft yarns was due to fiber damage caused by the use of higher tensions in these yarns during weaving.

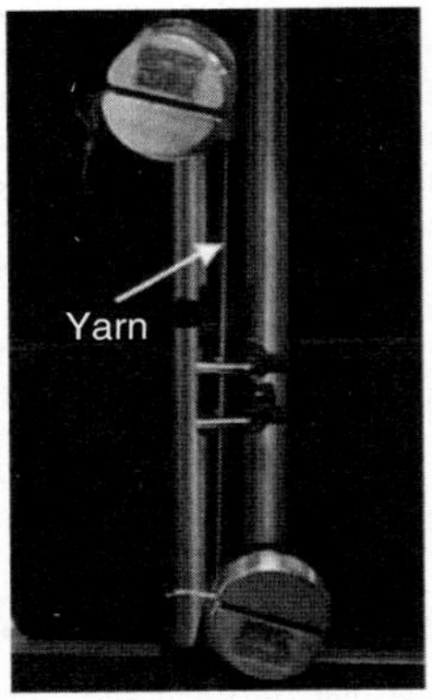

Figure 12.5.2 Yarn tensile testing using Instron® textile grips.

EFFECTS OF FABRIC CONSTRUCTION ON STRUCTURAL BEHAVIOR

Fabric materials are constructed from yarns that cross over and under each other in a repetitive, undulating pattern. The undulations shown in Fig. 12.5.3 are referred to as crimp, which is based on Pierce's geometric fabric model.[5]

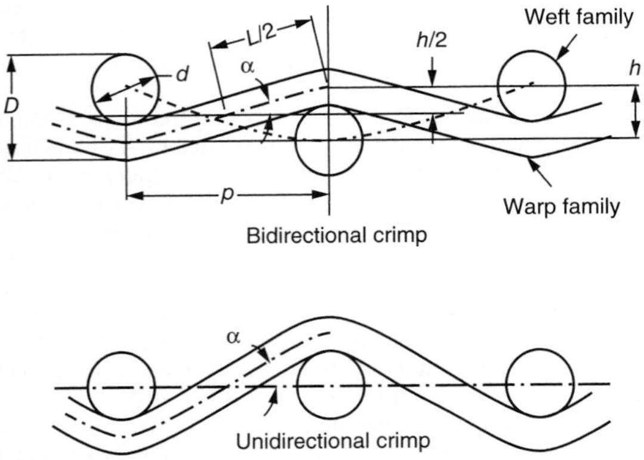

Figure 12.5.3 Examples of Pierce's geometric model for plain-woven fabrics with bidirectional and unidirectional crimp.

Pierce's geometric model relates these parameters as they are coupled among yarn families. The crimp height, h, is related to the crimp angle, α, and yarn length, L, as measured between yarns, and the sum of yarn diameters at the cross over points by the following equations described by Grosberg:[3]

$$p = (L - D\alpha)\cos\alpha + D\sin\alpha$$
$$h = (L - D\alpha)\sin\alpha + D(L - \cos\alpha) \tag{12.5.1}$$
$$D = h_1 + h_2$$

These equations were based on idealized geometry and assumptions such as restricting the yarns to circular cross sections and no consideration of force or stiffness effects.

Plain-woven and braided fabrics behave differently under load because their yarn families are aligned at different angles. Plain-woven fabrics utilize a nearly orthogonal yarn placement of warp and weft (or fill) yarns. By general textile definitions, warp yarns are identified as those yarns running parallel to the selvage and are virtually unlimited in their length. Weft yarns are at right angles to the selvage and are limited in length by the width of the weaving equipment. On the other hand, braided fabrics have a $+\theta/-\theta$ yarn placement, where θ is commonly referred to as the bias angle, with respect to the braid axis.

The biaxial stress-strain behavior of plain-woven fabrics is initially dominated by crimp interchange rather than yarn elasticity. Because the factors of safety used in air-inflated fabric structures are typically within the range of 4-6, the operating stresses are designed to be low with respect to fabric strength. Therefore, the structural performance of fabrics must address the influences of crimp interchange. The relative yarn motions (slip and rotation) affect the stiffness properties. The crimp ratio, which is denoted as C, is defined as the waviness of the yarns and is obtained by measuring the length of a yarn in its fabric state, L_{fabric}, and the length of that same yarn after it has been extracted from the fabric, L_{yarn}, and straightened out according to Eq. (12.5.2).

$$C = \frac{L_{\text{yarn}} - L_{\text{fabric}}}{L_{\text{fabric}}} \tag{12.5.2}$$

The following equation described by Grosberg relates the crimp height to C.

$$\frac{h}{p} = \frac{4}{3}\left(\frac{L}{p} - 1\right)^{\frac{1}{2}} \equiv \frac{4}{3}C^{\frac{1}{2}} \tag{12.5.3}$$

Consider a plain-woven fabric subject to a tensile load along the direction of one yarn family. The loaded yarns will attempt to straighten, decrease their crimp heights and elongate their effective lengths. However, the yarns of the crossing family are forced to increase their crimp heights resulting in contractions of their effective lengths. The effect associated with changes in crimp is referred to as crimp interchange and is similar to the Poisson's effect exhibited in metals. Crimp interchange is a coupling effect exhibited between yarn families. When a fabric is loaded in tension, the crimp contents become mutually altered as the yarns attempt to straighten. In tests of plain-woven fabrics along the direction of a given yarn family, crimp interchange can be easily observed by reducing the width of the specimen. Crimp interchange is a source of nonlinear load-extension behavior for fabrics.

Backer[3] describes a limiting phenomenon to crimp interchange. As the biaxial tensile loads continually increase for a given loading ratio, a configuration results in which yarn kinematics (i.e., slip at the cross over points) cease and the spacing between yarns converge to minimum values. This configuration is referred to as the extensional jamming point, which can prevent a family of yarns from straightening and thereby not achieve its full strength. Crimp interchange depends upon the ratio of initial crimp between yarn axes and the ratio of stress between yarn axes rather than the levels of stress. Crimp interchange introduces significant nonlinear effects in the mechanical response of the fabric.

Now consider the plain-woven fabric subjected to shearing by applying a uniaxial load at ±45 degrees to either yarn family. The warp and weft yarn families will rotate at the crossover points with respect to each other and become increasingly skewed as the angle between the yarns changes. The change in angle is referred to as the shear angle. At larger shear angles, the available space between yarn families decreases and rotational jamming (locking) of the yarn families occur. This phenomenon is known as shear-jamming and the angle at which the yarn families become jammed is referred to as the shear-jamming angle. The shear-jamming angle decreases with increasing yarn density ratio and can be estimated from Pierce's geometric fabric model or obtained experimentally with various trellising or biaxial test fixtures. Continued loading beyond the onset of shear-jamming will produce shear wrinkles leading to localized out-of-plane deformations.

It is important to determine the extension- and shear-jamming points for both fabric manufacturing and structural stiffness concerns. In general, jamming is related to the maximum number of weft yarns that can be woven into a fabric for a given warp yarn size and spacing.

OPERATION

Today's air-inflated fabric structures can be operated at various pressure levels depending upon fabric architecture, service loads and ambient temperatures. Woven structures are generally designed to operate at pressures up to 20 psi while triaxial braided and axial strap-reinforced braided structures are generally capable of higher inflation pressures. Blowers, which may be used for inflation pressures up to 3 psi, provide a high volumetric air flow rate. However, air compressor systems will be required for higher inflation pressures. Air compressors are available in single and dual stage configurations. The single stage air compressor delivers a high-pressure capability but at a low volumetric air flow rate. The volumetric flow rate is measured in standard cubic feet per minute (scfm). Dual stage air compressors are designed to provide an initial high volumetric air flow rate at low pressures in the first stage and can be followed by a low volumetric air flow rate at high pressure. When a dual stage air compressor is used, the first stage is most beneficial to erecting the fabric structure. At this time, no appreciable resistance to the design loads is available. However, the second stage mode of the compressor is used to fully inflate the fabric structure to its proper operating pressure so that the design loads can be supported. Because the dual stage compressor can achieve the desired inflation pressure much more rapidly than the single stage compressor, it is more appropriate for those applications in which the need for rapid deployment justifies the additional cost.

INFLATION AND PRESSURE RELIEF VALVES

Valves designed for use in air-inflated fabric structures must consider several criteria including locations, orifice size, pressure relief controls and potential strength degradation in the surrounding fabric regions. Ideally, valves should be positioned in structural elements that interface with the fabric and bladder such as metal clamps, discs, or end plates at the end termination zones of the structure. This avoids the need for cutting additional penetration holes through the fabric or repositioning the yarns to accommodate holes, thus preventing stress concentrations. When fibers must be removed to accommodate insertion of valves, fabric reinforcement layers should be bonded and stitched to the main fabric in the surrounding region. The valves should be readily accessible for user access but should not jeopardize the integrity of the fabric when the structure is subjected to handling events (such as drops, impacts, etc.). Valve locations should be further optimized to mitigate the effects of interference with other objects when the structure is deployed or stowed.

A variety of pneumatic valve designs including threaded valves, quick connect-disconnect valves, check valves (one-way, two-way), pressure relief valves, etc. are readily available. The use of pressure relief valves is recommended to avoid accidental overpressures and pressure increases due to changes in ambient temperatures. Valves can also be configured to manifold inflation lines in air inflated structures containing multiple pressure chambers. This also provides capabilities for self-erecting and controlled deployments by sequencing the inflation timing of multiple inflation chambers. Sizing of the valve orifices should be matched to provide the optimal inflation and deflation times for the air volumes and operating pressures required by the structure.

CONTINUOUS MANUFACTURING AND SEAMLESS FABRICS

Prior to continuous circular weaving and braiding processes in use today, air-inflated fabric structures were constructed using adhesively bonded, piece-cut manufacturing methods. These methods were limited to relatively low pressures because of fabric failures and air leakage through the seams. Continuous weaving and braiding processes can eliminate or minimize the number of seams resulting in improved reliability, increased pressure capacities and greater structural load carrying capability. However, when seams cannot be avoided, the seam construction should be designed in such a way that failure of the surrounding fabric occurs rather than the seam. For the safe and reliable use of air-inflated fabric structures, sufficient factors of safety against burst and seam failures must be provided. To guard against burst, for example, a minimum factor of safety of 4-6 is used on yarn strengths and seams and other factors of safety are necessary for regions containing valves, inflation ports, end closures, and hard connections.

Like traditional composite materials, fabrics can be tailored to meet specific structural performance requirements. Fiber placement can be optimized for air-inflated fabric structures by varying the denier of the yarns and yarn counts along the each direction (Denier is a mass per unit length measure expressed as the mass in grams of a 9,000 meter length yarn or fiber.) For instance, consider a pressurized fabric cylinder. The ratio of hoop stress per unit length to axial stress per unit circumference is 2:1. One can ensure equal strengths against yarn burst in the warp (longitudinal) and weft (hoop) yarns by weaving twice as many weft yarns per unit length of air beam than the number of warp yarns per unit circumference. Alternatively, the same can be accomplished by doubling the denier of the weft yarns in comparison to the denier of the warp yarns.

IMPROVED DAMAGE TOLERANCE

Assorted methods are used to enhance the reliability of air-inflated fabric structures against various damage mechanisms. Resistance to punctures, impacts, tears, and abrasion can be improved by using high-density weaves, rip-stop fabrics, and coatings. High-density weaves are less susceptible to penetrations and provide greater coverage protection for bladders. Rip-stop fabrics have periodic inclusions of high tenacity yarns woven in a cellular arrangement used to prevent fractures of the basic yarns from propagating across cells as shown in Fig. 12.5.4. (The breaking strength of a yarn is referred to as tenacity which is defined in units of grams-force per denier.) The high tenacity yarns contain fractures of the basic yarns and prevent fractures from propagating across cells.

Additionally, coatings protect the fabric against environmental exposure to ultraviolet rays, moisture, fire, chemicals, etc. Coating such as urethane, PVC (polyvinyl chloride), neoprene, EPDM (ethylene propylene diene monomer) are commonly used. Additives such as Hypalon further enhance a coating's resistance to ultraviolet light and abrasion. Coatings can be applied in two stages. First, coating the yarns prior to forming the fabric by a liquid bath immersion provides the best treatment to the fibers. Second, coatings can be applied by spraying, painting or laminating directly to the fabric after forming. This stage coats the yarns and bridges the gaps formed between adjacent yarns. The maximum protection is achieved when both stages are utilized. While protective coatings have been shown to increase the stiffness of the fabric by restricting relative yarn motions, they remain flexible enough to not adversely impact the stowing and packaging operations of the structure. Additional information on coating materials and processes is provided by Fung.[6]

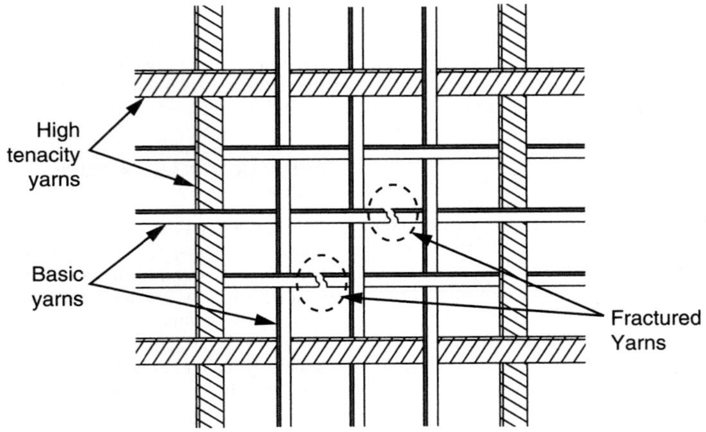

Figure 12.5.4 Rip-stop fabric architecture.

RIGIDIFICATION

Air-inflated fabric structures can be rigidified once inflated by coating the fabric with resins such as thermoplastics, thermosets, shape memory polymers, etc. Prior to inflation, these particular coatings are applied to the fabric and remain initially uncured, thus acting as flexible coatings. After the structure is inflated and properly erected, a phase change is triggered in the coating by a controlled chemical reaction (curing process) activated by exposure to elevated temperature, ultraviolet light, pressure, diffusion, etc. Once the phase change is fully developed, the coatings bind the yarns together, stiffen the fabric in tension, compression and shear, and behave similar to a matrix material found in traditional fiber-reinforced composites. The air-inflated fabric structure is now rigidified and no longer requires the inflation pressure to maintain its shape and stiffness. Depending upon the coating used, the transition process may be permanent or reversible (except where thermoset plastics are used). Reversible rigidification is especially suited for those applications requiring multiple long-term deployments.

The performance of the rigidified fabric structure can be assessed using laminated shell theory (LST) as described by Jones.[7] Unlike the inflated version of the structure, the elastic and shear moduli of the rigidified structure are based on the constituent properties of the fibers and cured coating. As such, the elastic and shear moduli can be readily estimated by using LST. However, the rigidified structure may have different failure modes than its pressurized counterpart. The designer must guard against new failure modes including localized shell buckling and compression rather than wrinkling.

Rigidification is of particular interest for space structures because of restrictions on payload weights and stowage volumes. Cadogan et al.[8,9] have pursued the use of shape memory composites for rigidizing deployable space frames.

AIR BEAMS

Fabric air beams are examples of air-inflated structural elements that are capable of supporting a variety of loads similar to conventional beams. To date, seamless air beams have been constructed using continuous manufacturing methods that have produced diameters ranging up to 42 inches. They are constructed of an outer fabric skin that contains an internally sealed film or bladder made of an impermeable elastomer-like material. The bladder contains the air, prevents leakage and transfers the pressure to the fabric. The air beam has a circular cross section and its length can be configured to a straight or curved shape such as an arch. The ends are closed using a variety of end termination methods consisting of bonding, stitching, mechanical clamps, etc. depending upon the inflation pressure and loading requirements. Clamping methods are available to permit disassembly, repair and replacement of the bladder, and fabric layers.

Once an air beam is sufficiently pressurized, the fabric becomes pre-tensioned and provides the air beam with a plurality of stiffnesses including axial, bending, shear, and torsion. Upon inflation, the ratio of hoop (cylindrical) stress per unit length of beam to the longitudinal stress per unit circumference is 2:1 as shown in Fig. 12.5.5.

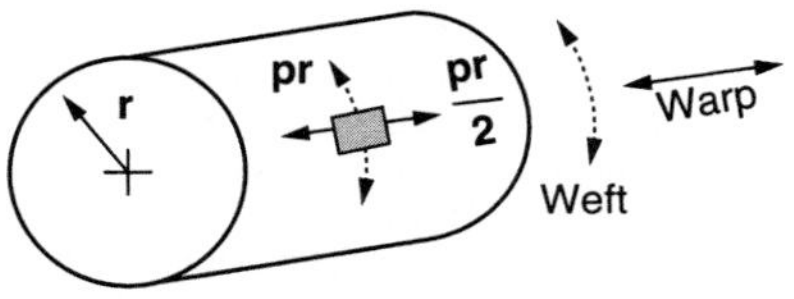

Weft yarn tension per unit length of cylinder = pr

Warp yarn tension per unit circumference = pr/2

$$\text{Yarn density ratio (YDR)} = \frac{\text{\# weft yarns per unit length of cylinder}}{\text{\# warp yarns per unit circumference}}$$

Figure 12.5.5 Yarn tensions in a plain-woven pressurized fabric cylinder and definition of yarn density ratio.

In the context of plain-woven air beams, the warp yarns are aligned parallel to the longitudinal axis of the air beam and resist axial and bending loads. The weft yarns spiral through the weave and are located at nearly 90° to the warp axis, thus lying along the hoop direction of the air beam. Weft yarns provide stability against collapse by maintaining the circular cross section of the air beam.

For braided air beams, the braid axis is aligned with the longitudinal axis of the air beam. If the ends of a braided air beam are unconstrained from moving in the longitudinal direction, the fibers will exhibit a scissoring effect causing the length of the beam to expand or shorten with pressure depending upon the selection of θ. Eventually, the braided yarns will achieve a maximum rotation and the fabric will become fully jammed. This phenomenon can be easily demonstrated with the well-known Chinese finger trap toy. Unlike plain-woven fabric structures, the bias angle of braided fabric structures can be controlled to allow expansion or contraction of the structure when inflated. For an unconstrained braided air beam to resist axial tension, longitudinal reinforcements such as distributed axial yarns (i.e., triaxial braid) or axial straps, webbing, belts, etc. must bonded on at multiple locations around the circumference. Various design methods for constructing axial-reinforced braided air beams were developed by Brown[10] and Brown and Sharpless.[11,12]

The bending stiffness, EI, for metal beams is written as the product of elastic modulus, E, and area moment of inertia, I, and has units of (force × distance²). However, in air inflated fabric structures, the fabric elastic modulus, E_f, is commonly denoted in units of (force per unit length) and therefore, $\acute{E}_f I$ is denoted in units of (force × distance³). Likewise, fabric strengths are denoted in units of (force per unit length).

Now, consider the air beam subjected to transverse load, P. Upon loading, the pre-tension stresses and the bending stresses will superimpose. The compressive bending stresses will subtract (relax) from the pre-tension on the compressive surface of the air beam while the tensile bending stresses will add to the pre-tension along the tensile surface as shown in Fig. 12.5.6. The instant at

$$M = \sum_{i=1}^{N} \left| F_{ib} \cdot y_i \right|$$

Where: $\dfrac{F_{Nb}}{y_N} = \dfrac{F_{ib}}{y_i}$

and: $y_N = \dfrac{d}{2}$

Therefore: $F_{ib} = \dfrac{\dfrac{2 \cdot M}{d}}{\displaystyle\sum_{i=1}^{N} \sin(\alpha(i))^2}$

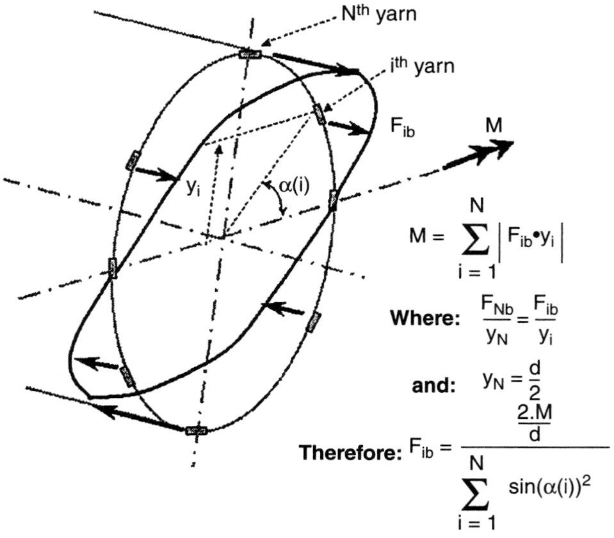

Figure 12.5.6 Idealized distribution of warp yarn forces due to bending of a plain-woven fabric air beam.

which any point along the length of the air beam develops a net zero longitudinal tensile stress, the structure is said to have reached the onset of wrinkling. The corresponding load and bending moment are referred to as the wrinkling onset load P_{wo} and the wrinkling onset moment, M_{wo}, respectively. Prior to wrinkling, the moment-curvature relationship is expectedly linear in the absence of both material nonlinearities and notable changes in air pressure and volume. A stress balance analysis based on strength of materials equations can generally be used to compute M_{wo}. Such a stress balance analysis was used to show the variation of warp yarn tensions for a 2" diameter plain-woven air beam subjected to 4-point bending loads as depicted in Fig. 12.5.7. Once wrinkling has occurred, the air beam moment-curvature relation behaves nonlinearly because with further loading, the cross section loses bending stiffness, the neutral axis displaces away from the centroidal axis of the cross section and eventual collapse occurs. The spread of wrinkling around the circumference is similar to the flow of plasticity in metal beams subjected to bending. Loading beyond the wrinkling onset will result in an increase of pressure due to loss of volume

2" dia, 63 Warp yarns, L_{total} = 96", $L_{support}$ = 44.375", A = 14.625", Pressure = 20 psi, P_{wr} = 4.3 lb, Moment$_{wr}$ = 31.44 in-lb

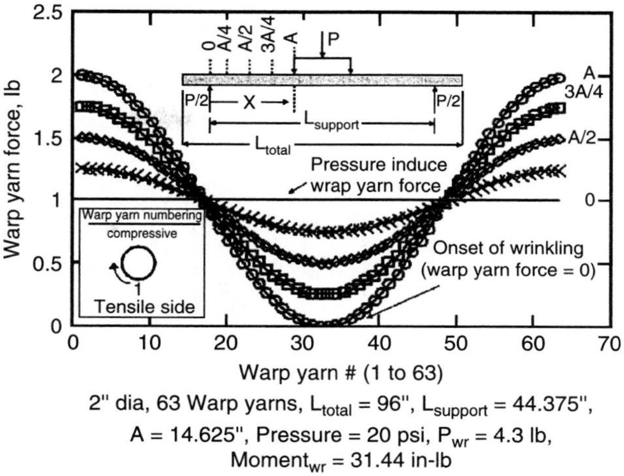

Figure 12.5.7 Combined pressure and bending induced forces in warp yarns at various distances along a plain-woven air beam based on a simple stress balance analysis.

and the work done on the air will affect the post-wrinkled bending stiffness. Thus, the post-wrinkled response becomes nonlinear. The superposition of the pressure and bending-induced forces is shown for each architecture in Fig. 12.5.8.

Consider now a plain-woven air beam subjected to inflation pressure P. The force in the warp yarns, F_{WARP}, is shown in Eq. (12.5.4) as simply the pressure times the cross sectional area divided by the number of warp yarns, N_{WARPS}. The force in the weft yarns, F_{WEFT}, was idealized by the assumption of orthogonal yarn placement and is shown in Eq. (12.5.5). The wrinkling onset moment, M_{wo}, for a woven air beam was derived from a simple force balance of the pressure and bending-induced yarn forces and is shown in Eq. (12.5.6).

$$F_{WARP} = P \pi r^2 \Big/ N_{WARPS} \tag{12.5.4}$$

$$F_{WEFT} = P \pi r L \Big/ N_{WEFTS} \tag{12.5.5}$$

$$M_{WO} = \frac{P \pi r^3}{2} \tag{12.5.6}$$

Note that M_{wo} is independent of the fabric material properties.

The global flexure response of plain-woven air beams was experimentally investigated for use in inflatable military shelters.[13,14] The air beams were constructed of non-twisted yarns and were inflated to various operating pressures. Operating pressures were considered to be safe when the resulting yarn tensions did not exceed 30 percent of their breaking strength. The experimental set up is shown in Fig. 12.5.9. A plot of the experimental load versus deflection is shown in Fig. 12.5.10 which exhibited a significant influence on inflation pressure.

A second series of 4-point bending tests[4] was conducted on a 2-inch diameter, uncoated plain-woven Vectran air beam to measure the mid-span deflections as a function of inflation pressures ranging from 10-100 psi. The load versus mid-span deflection curves are shown in Fig. 12.5.11.

In a third series of tests,[4] the flexure behavior of uncoated plain-woven Vectran and PEN air beams were compared for two pressures using a constant load point displacement rate. Both air beams had identical crimp in the warp and weft yarns, yarn density ratios (YDR), denier, diameter and length. (Recall that the yarn density ratio

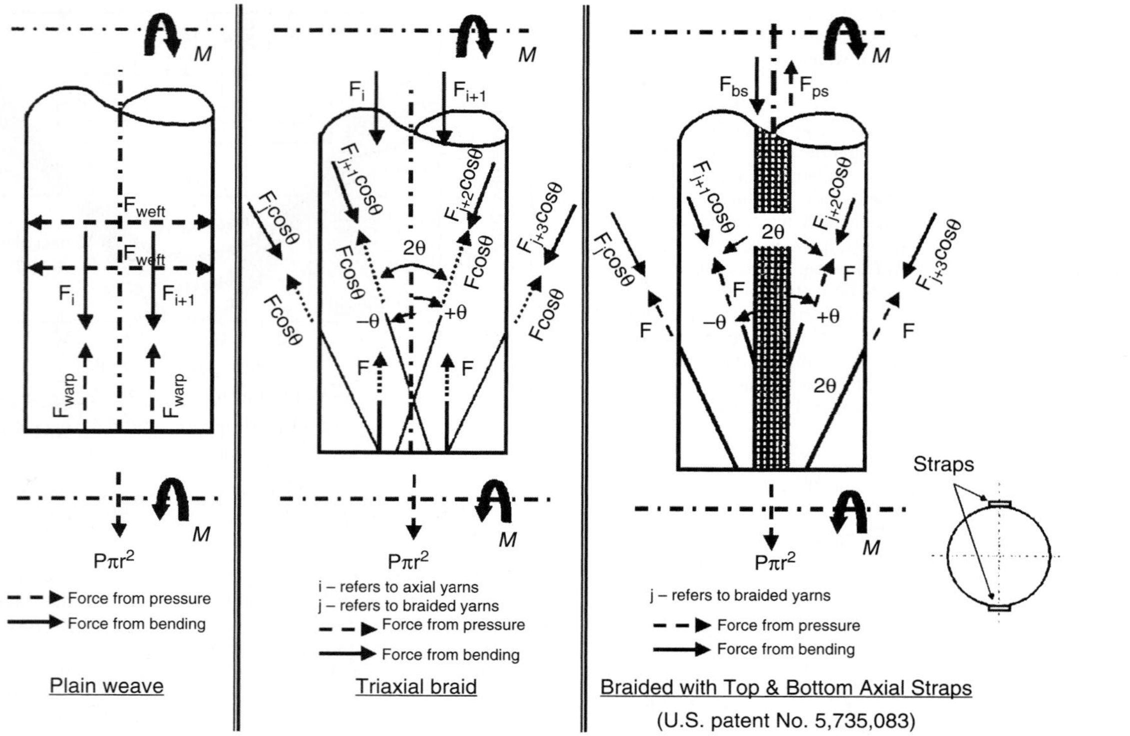

Figure 12.5.8 Superposition of pressure and bending induced yarn forces in plain-woven air beam, a triaxial braided air beam and a dual axial strap-reinforced braided air beam.

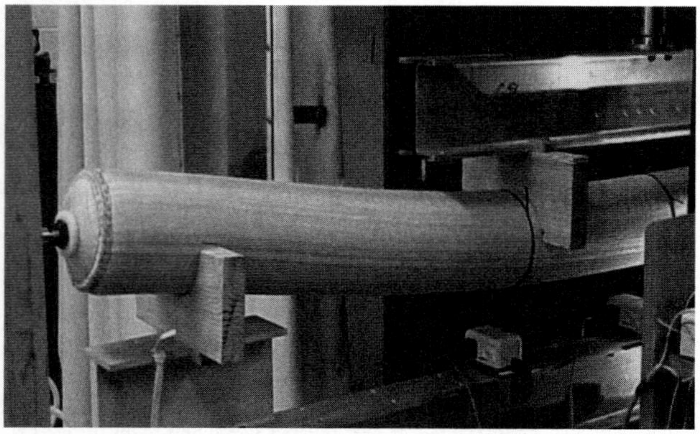

Figure 12.5.9 4-Point flexure test on a 6-inch diameter plain-woven air beam constructed of 3,000-denier, 2:1 YDR Vectran fabric.

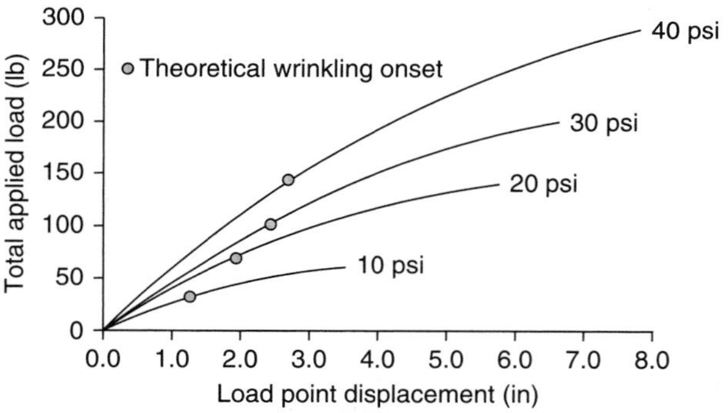

Figure 12.5.10 Experimental load vs. deflection plot for an uncoated 6-inch diameter air beam constructed of 3,000-denier non-twisted Vectran yarns in a plain-woven 2:1 yarn density ratio fabric using a 37-inch span between load points and an 85-inch span between support points.

is defined as the number of weft (hoop) yarns per unit beam length divided by the number of warp (longitudinal) yarns per unit circumference.) The graph of Fig. 12.5.12 shows the total applied load versus the mid-span deflection responses.

Although the ratio of the as-woven elastic moduli of the Vectran and PEN yarns obtained through "as-woven" yarn tensile tests was 4.7-to-1.0, there were minimal differences in deflections of these air beams at the inflation pressures used. It was observed that the global bending behavior of the air beams was: (1) invariant with E_{yarn} for the range of inflation pressures considered, (2) dominated by the kinematics of crimp interchange, and (3) subjected to considerable transverse shearing deformations. The effects of crimp interchange and shearing deformations precluded the use of traditional strength-of-materials design methods for plain-woven fabric air beams.

DROP-STITCH FABRICS

Drop-stitch fabrics, originally pursued for structural applications by the aerospace industry, extend the shapes that inflatable structures can achieve to include flat and curved panels with moderate to large aspect ratios and variable thicknesses. Drop-stitch fabrics, also referred to as 3D spacer fabrics, are integrally woven preforms (Fig. 12.5.13). Specialized looms (Fig. 12.5.14) concurrently weave a pair of woven fabric layers offset (spaced apart) from each other. Unlike plain-woven fabrics

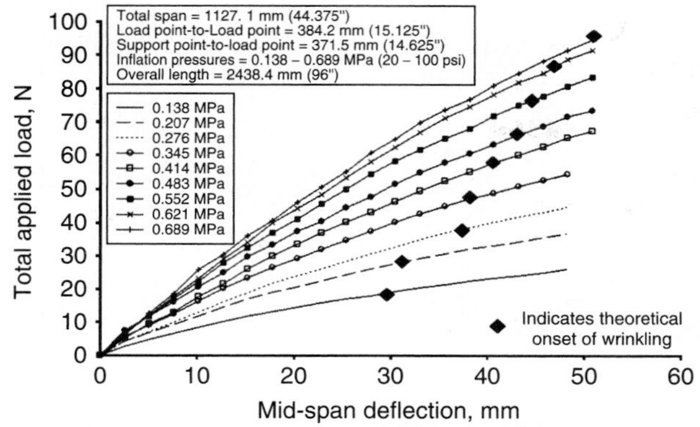

Figure 12.5.11 Plot of total load vs. mid-span deflection for a 2-inch diameter plain-woven air beam constructed of 1,500-denier, 2:1 YDR Vectran fabric.

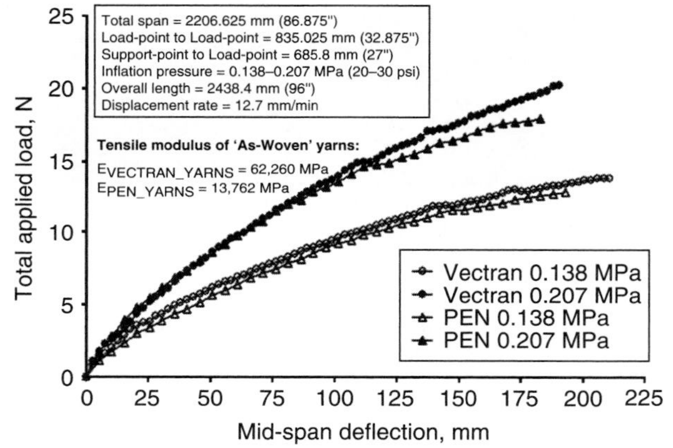

Figure 12.5.12 Comparison of 2-inch diameter Vectran and PEN plain-woven air beams subjected to 4-point bending tests at various load point displacement rates.

constructed of single families of warp and weft yarns, the drop–stitch preform includes an additional family of warp yarns known as the drop yarns. The drop yarns are alternately woven within the fabric layers and are periodically "dropped" from one layer to the other thus forming an open and continuous sandwich-like structure. Inflatable drop-stitch panels are then formed by laminating impermeable membranes (rubber,

Figure 12.5.14 Loom used in weaving drop-stitch fabrics. (Photo courtesy of Navatek, Ltd.)

PVC, urethane, neoprene, etc.) or coated fabrics onto the woven layers and closing the edges to form an inflatable volume. Figure 12.5.15 shows yarn-level details of an actual rubber-laminated, polyester drop stitch fabric. Upon inflation, the laminated woven layers separate forming a panel with thickness controlled by the intermediate lengths of the drop yarns. Flatness of the panel can be controlled with a sufficient number of drop yarns per unit area of the woven layers. Air-inflated structures incorporating drop-stitch fabrics include floors for inflatable boats, inflatable stand-up paddle boards, energy absorbing walls, temporary barriers, lightweight foundation forms, and other recreational products.

The load carrying behavior of inflatable drop-stitch panels is analogous to that of structural sandwich panels consisting of a foam or honeycomb core bonded between two thin metal face sheets. The pressurized air volume serves as the core region providing through-thickness normal, in-plane shear and transverse shear stiffnesses; the laminated woven layers act as the face sheets providing tensile and bending stiffnesses. However, these stiffnesses are dependent on inflation pressure.

A combined experimental and analytical study[15] was conducted on rubber-laminated, polyester drop stitch fabric panels subjected to the 4-point bend loading arrangement shown in Fig. 12.5.16. Loading was applied in displacement controlled mode using a constant crosshead speed of 25.4 mm/min (1.0 in/min). The panel dimensions were 609.6 mm (24.0 in) wide by 2,540.0 mm (100.0 in) overall length by 101.6 mm (4.0 in) height. The mid-span deflection δ was recorded using a displacement wire transducer. Curves of the total applied load P

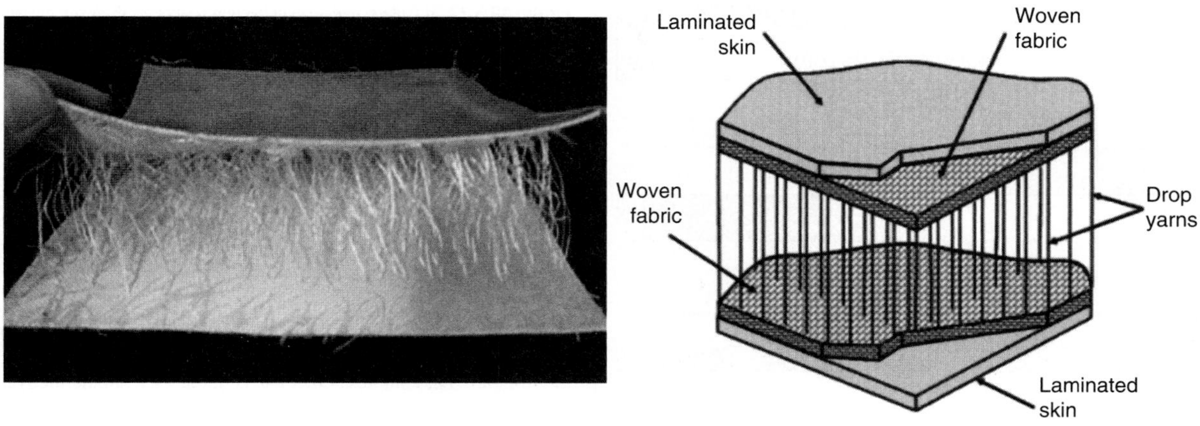

Figure 12.5.13 Section and detailed views of a laminated drop-stitch fabric used for air-inflated structures.

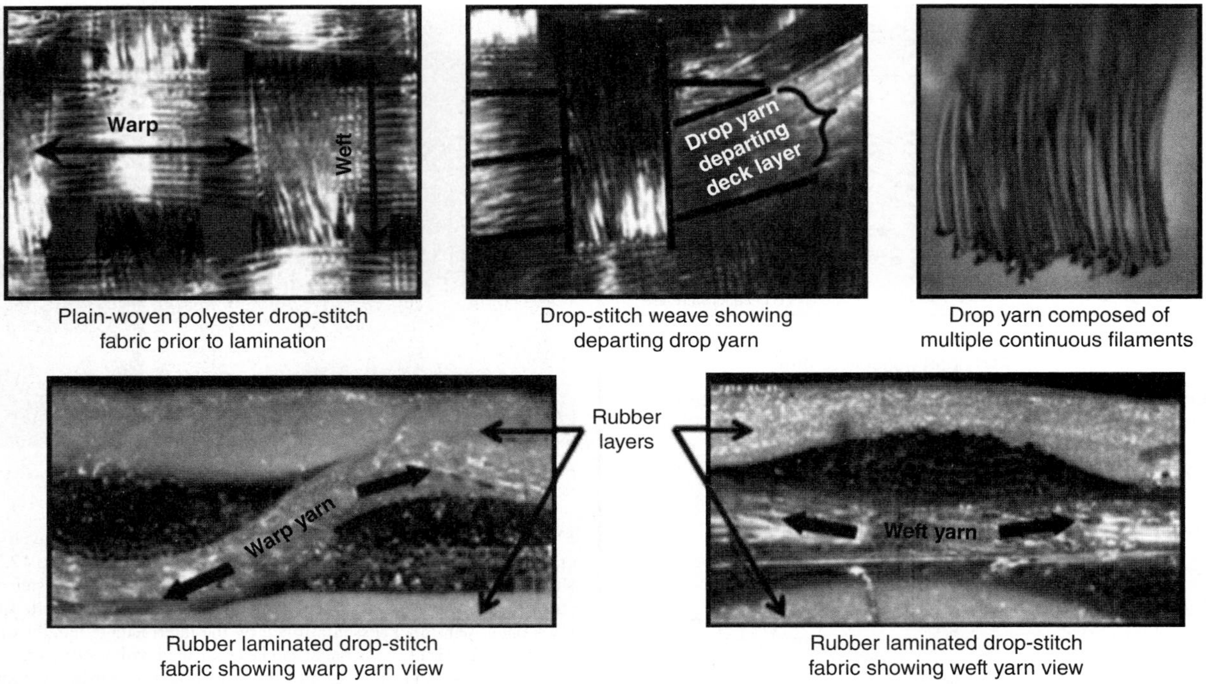

Figure 12.5.15 Fabric and yarn-level details of a rubber-laminated polyester drop-stitch fabric.

versus δ are shown in Fig. 12.5.17 for inflation pressures ranging from 34.47 to 206.84 KPa (5.0 to 30.0 psig). Three tests were performed at each inflation pressure. The results demonstrate that the panel's load carrying capacity and bending stiffness increase with increasing inflation pressure. If the burst pressure of the panel is limited by the breaking strength of the drop yarns, then: (1) an upper limit on burst pressure may be correlated to the drop yarn breaking strength as measured from tensile tests performed on individual drop yarns, (2) the panel may burst in a fail-safe mode whereby the cross section converts from an initial rectangular form to a circular shape, (3) the mass of internal air will remain conserved, and (4) the air volume will increase causing the internal pressure and skin tensions to decrease.

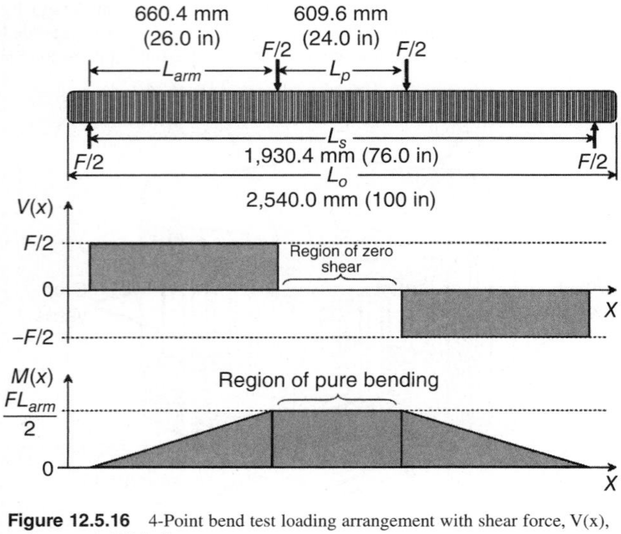

Figure 12.5.16 4-Point bend test loading arrangement with shear force, V(x), and moment, M(x), diagrams.

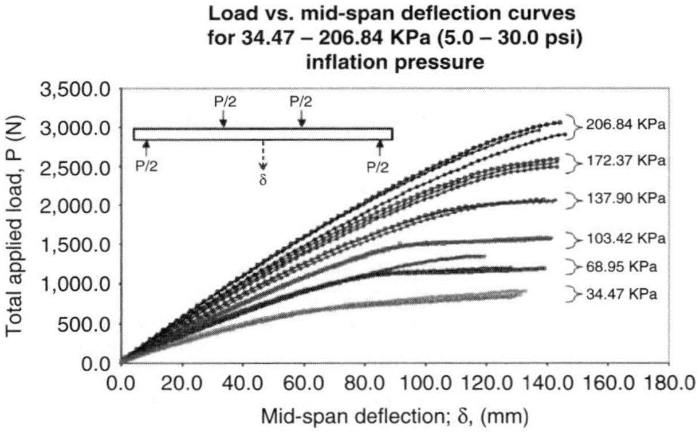

Figure 12.5.17 Load vs. mid-span deflection results for drop-stitch panel subject to various inflation pressures.

EFFECTS OF AIR COMPRESSIBILITY ON STRUCTURAL STIFFNESS

The load-deflection response of air-inflated fabric structures may depend upon additional stiffening influences other than the initial inflation pressure. These may include contributions from nonlinearities in the fabric stress-strain behavior, the work done on the internal air by external loads, and other phenomena. If appreciable changes in pressure or volume occur during loading, as in the case of energy absorbers, work is performed on the air through compressibility, which stiffens the structure. Air compressibility can be modeled from thermodynamic principles in accordance with the Ideal Gas Law shown in Eq. (12.5.7).

$$P'V = mRT \tag{12.5.7}$$

where: P' = the absolute pressure
V = volume
m = mass
R = gas constant for air
T = temperature (°K)

For a quasi-static, isothermal process, the work done on the air by compression is:

$$W_{air} = \int_{V_1}^{V_2} P\,dV = \int_{V_1}^{V_2} \frac{mRT}{V}\,dV = mRT \ln\frac{V_2}{V_1} \qquad (12.5.8)$$

The total energy of the structure, E_{total}, is the work done by all external forces, W_{ext}, which is related to the total strain energy of the fabric, U_f, and W_{air} as shown in the energy balance of Eq. (12.5.8).

$$E_{total} = W_{ext} = U_f + W_{air} \qquad (12.5.9)$$

For homogeneous films and membranes, the elastic and shear moduli are readily determined by standardized tests. However, for the case of plain-woven and braided fabrics, the elastic and shear moduli can vary not only with pressure but also with fabric architecture (weave style, yarn densities, etc.), external loads (crimp interchange), and coatings if present. With increasing pressure, slack within the yarns is removed, the yarns begin to straighten (decrimp) and contact between intersecting yarns at the crossover points causes compaction.

EXPERIMENTS ON PLAIN-WOVEN FABRICS

Hearle[3] identified three distinct regions of stiffness for tensile loading of a plain-woven fabric as shown in Fig. 12.5.18. For safe practice, air-inflated fabric structures are designed to operate with yarn forces ranging between regions **I** and **II**.

This ensures that the yarns are loaded well below their breaking strengths to provide sufficient factors of safety against burst. Once the fabric is biaxially stressed and subjected to an in-plane shear stress, the yarns will shear (rotate) with respect to their original orientations. The shear stiffness (i.e., resistance to yarn rotations) results from inter-yarn friction and compaction at the yarn crossover regions. Hence, the shear modulus is actually a system property rather than a constitutive (i.e., material) property. As the shear rotations increase, an upper bound is reached when yarns of both directions become kinematically locked resulting in what is referred to as shear-jamming. For pure shear loading, the trellising fixture shown in Fig. 12.5.19 can be used. Figure 12.5.20 shows the shear force versus shear angle plot of an uncoated Vectran fabric. Further shear loading induces localized shear wrinkling instabilities that lead to increased out-of plane deformations.

Farboodmanesh et al.[16] conducted shear tests on rubber-coated, plain-woven fabrics and established that the initial shear response was dominated by the coating and with increased shearing, the behavior of a coated fabric transitions to that of an uncoated fabric.

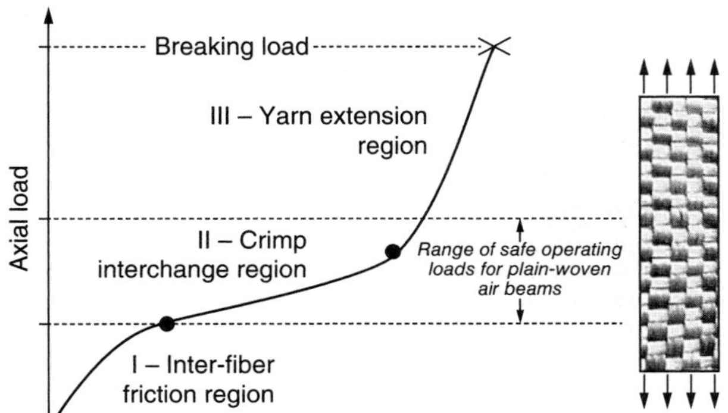

Figure 12.5.18 Stages of axial stiffness for woven fabric subjected to uniaxial tension.

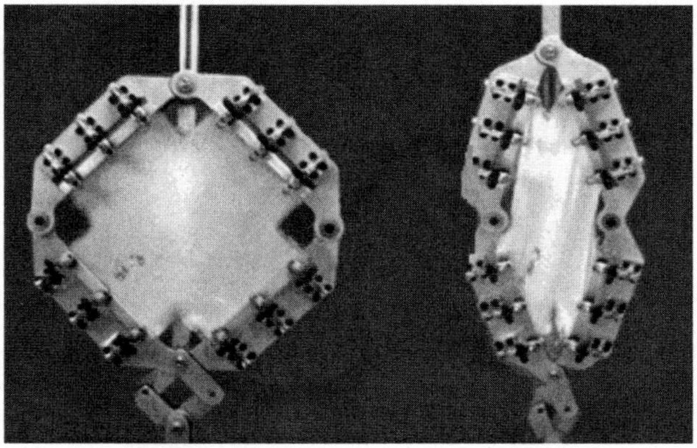

Figure 12.5.19 Picture frame test fixture for pure shear loading of fabrics.

Unlike fabric structures, the homogeneity of membranes excludes the kinematic motions associated with crimp interchange and strain energy is developed throughout all stages of biaxial tensile loading. However, membrane structures composed of homogenous materials are also susceptible to both bending and shear wrinkling instabilities.

Many testing methods have been developed for evaluating the mechanical properties of yarns and fabrics including ASTM standards, Kawabata Evaluation System, military specifications (MIL-Spec), British standards, etc. It is recommended that the structural engineer become familiar with the applicability of these standards to the design and testing of fabric materials as applicable to air-inflated fabric structures. Saville[17] provides a description of many standard tests used in evaluating fiber, yarns, and fabrics.

A recently developed multi-axial tension and shear test fixture[13] was designed to permit simultaneous measuring of E_{warp}, E_{weft}, and G_f for structural fabrics. The fixture, as shown in Fig. 12.5.21, was designed for use with conventional tension/torsion machines to characterize the elastic and shear moduli of fabrics as functions of biaxial loads. It utilizes a cruciform-shaped specimen and can be used to evaluate both strength and stiffness properties of various fabric architectures such as weaves, braids and knits subjected to biaxial loads, shear loads, or combined biaxial and shear loads. For fabrics constructed of two principal fiber directions, the fixture utilizes two rhombus-shaped frames connected with rotary joints as shown; the biaxial tension and shear modes are illustrated. For triaxial braided fabrics, the fixture uses a third rhombus-shaped frame with additional rotary joints.

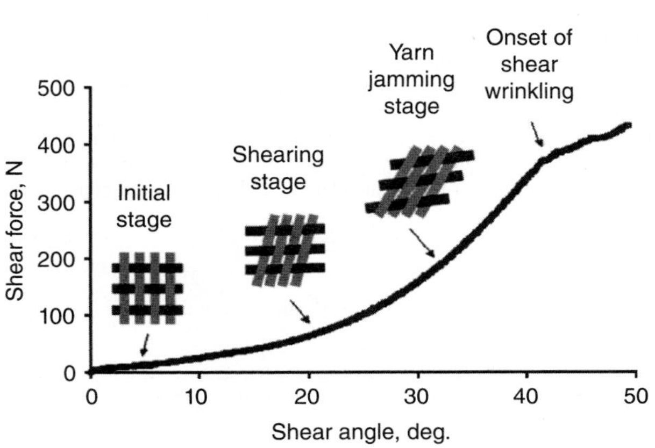

Figure 12.5.20 Stages of shear stiffness for a plain-woven fabric with a 2:1 yarn density ratio subjected to pure shear loading.

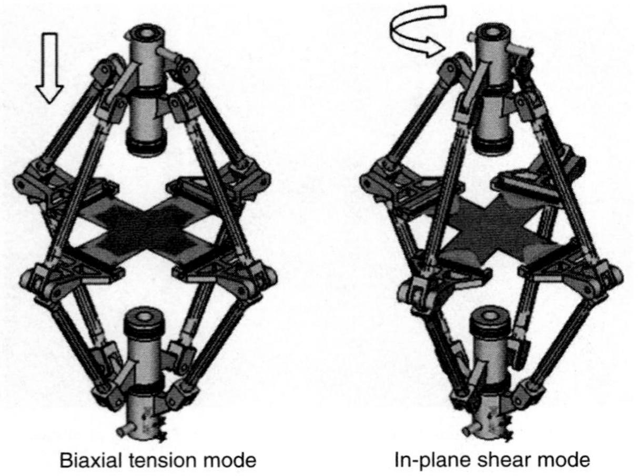

Figure 12.5.21 Combined biaxial tension and in-plane shear test fixture. (U.S. Patent No. 6,860,156)

A STRAIN ENERGY-BASED DEFLECTION SOLUTION FOR BENDING OF AIR BEAMS WITH SHEAR DEFORMATIONS

The air beam bending experiments exhibited that, unlike metallic structures, deflections are functions of internal pressure and transverse shear deformations may be significant. If shearing deformations become appreciable, they will increase the total deflection of air-inflated fabric structures. Therefore, a shear-deformable theory such as that developed by Timoshenko[18] must be employed in place of Euler-Bernoulli[19] theory to compute air beam deflections.

This section simulates the 4-point bending air beam experiments and analytically estimates the fabric shear modulus when changes in air pressure and air volume are negligible. Fabric shearing deformations were produced in each test and these were observed to decrease with increasing pressure. It will be shown that the shear modulus of the fabric was not a function of the yarn elastic moduli. This was accomplished by deriving a shear-deformable air beam deflection equation for the 4-point bending arrangement illustrated in Fig. 12.5.22. From this equation, and using experimentally obtained air beam deflections, the fabric shear modulus, G_f, can be computed. Note that this equation captures both the bending and transverse shearing components of deflection that combine to give the total mid-span deflection.

The nearly-orthogonal yarn directions were assumed to be truly orthogonal, with the warp axis being parallel to the longitudinal axis of the air beam. This uncoupled the bending stresses from the weft (hoop) yarns. The warp yarns were assumed to exclusively support the bending stresses. In a truly orthogonal fabric air beam, the weft yarns can be considered as parallel rings rather than continuously spiraled yarns with small lead angles. Second, an equivalent cylinder having a cross sectional area equal to the total cross sectional areas of all the warp yarns, A_{total}, was used to represent the actual air beam. The inner radius of the equivalent cylinder, r_i, was taken as the nominal air beam radius and the outer radius of the equivalent cylinder, r_o, was computed by:

$$r_o = \sqrt{\frac{A_{\text{total}}}{\pi} + r_i^2} \tag{12.5.10}$$

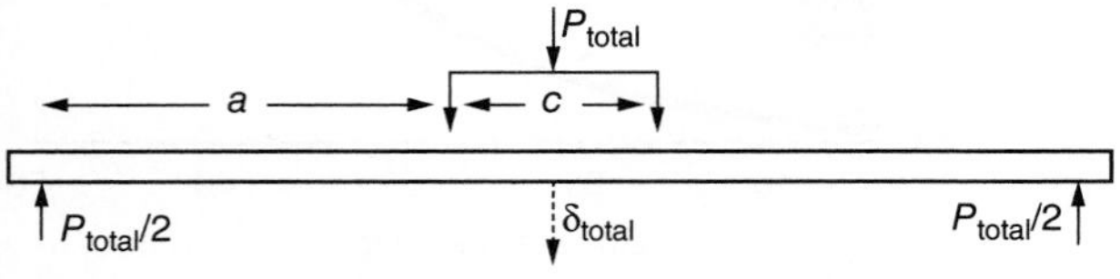

Figure 12.5.22 4-Point bending arrangement for shear deformable beam deflection equation.

Castigliano's Second Theorem[19] was used and considered the strain energies from both bending and shearing for the equivalent cylinder. The total strain energy, U_{total}, was expressed as the sum of the bending and shearing strain energies:

$$U_{\text{total}} = \int_0^L \frac{M(x)^2}{2E_f I_{\text{cyl}}}\,dx + \int_0^L FS\,\frac{V(x)^2}{2A_{\text{cyl}} G_f}\,dx \tag{12.5.11}$$

where:

- A_{cyl} = cross sectional area of equivalent cylinder (equal to total area of warp yarns, A_{total})
- E_f = fabric elastic modulus along warp (longitudinal) axis measured from biaxial tests
- FS = shear strain correction factor (for tube = 2.0)
- G_f = pressure-dependent fabric shear modulus
- I_{cyl} = area moment of inertia of the equivalent cylinder about bending axis
- L = length of air beam
- $M(x)$ = bending moment
- $V(x)$ = transverse shear force along longitudinal axis
- x = longitudinal position along air beam

For thin-walled cylinders of outer radius r_o, and thickness t_{cyl}, I_{cyl} is computed as shown in Eq. (12.5.12) by neglecting terms containing higher-ordered forms of t_{cyl}.

$$I_{\text{cyl}} = \pi r_o^3 t_{\text{cyl}} \tag{12.5.12}$$

The total mid-span deflection, δ_{total}, due to the total applied, P_{total}, is derived from minimizing the total strain energy, U_{total}. This leads to:

$$\delta_{\text{total}} = \frac{1}{48}\,\frac{8a^3 P_{\text{total}} + 12ca^2 P_{\text{total}} + 3c^2 a P_{\text{total}}}{E_f I_{\text{cyl}}} + \frac{1}{4} FS\,\frac{2aP_{\text{total}}}{G_f A_{\text{cyl}}} \tag{12.5.13}$$

where:

a = distance between support point and adjacent load point
c = distance between load points

The first and second terms of Eq. (12.5.13) represent the bending component, δ_{bend}, and the shearing component, δ_{shear}, of the total mid-span deflection, respectively. The fabric elastic modulus, E_f, is measured from biaxial tension tests performed on the fabric using the 2:1 weft-to-warp loading ratio. The magnitudes of the warp yarn tension used in the biaxial tests must match the warp yarn forces of the air beam at the inflation pressure of interest. Using the second term, δ_{shear}, the fabric shear modulus, G_f, may be calculated from:

$$G_f = \frac{aP_{\text{total}} FS}{2\delta_{\text{shear}} A_{\text{cyl}}} \tag{12.5.14}$$

Equation (12.5.14) indicates that G_f is not a function of the fabric elastic modulus, E_f whereas P_{total} and δ_{shear} are functions of inflation pressure. Alternatively, by determining the shear-jamming angle of the fabric and the corresponding δ_{shear}, a lower bound value of G_f can be determined.

ANALYTICAL AND NUMERICAL MODELS

A variety of analytical and numerical models have been developed to predict the load-deflection behavior for inflatable structures[4,13,20-27] and for the load-extension behavior of fabrics themselves.[28,29]

Unit Cell Numerical Models

One could create a finite element model of an air beam with all its discrete yarns, bladder, contact surfaces, friction, etc.; however, research has shown that solutions to such a model would exceed the computational limits of today's hardware. The need to obtain a computationally efficient model that would not exceed hardware limits or encounter convergence problems led to the development of localized yarn interaction models for plain-woven fabrics. These models are commonly referred to as unit cell models.

Unit cell models are used to develop the constitutive behavior on the material scale rather than the structural scale. All sources of influence on constitutive behavior are present on the material scale. Once this behavior is known, it can be applied to the structural scale by using homogenization methods. Homogenization methods simply employ the as-established constitutive behavior from the unit cell models in global structural models constructed of homogenous membranes. This 2-step approach reduces the necessary computations and run time for full-structure modeling while preserving the effects of yarn interactions on material behavior.

A unit cell of a plain-woven fabric generally consists of several rows of warp and weft yarns. Since the yarns exhibit relative displacements and rotations with respect to each other when subject to mechanical loads, researchers have investigated two options for treating yarn kinematics: (a) rotation-only (pinned centers at the yarn crossover points), and (b) rotation and translation at the yarn crossover points as shown in Fig. 12.5.23. In the rotation-only condition, the centers of the contact areas are hinged and remain coincident. Friction is due only to the relative rotation of the yarns and no sliding is allowed. This kinematic condition is appropriate for modeling pneumatic muscles[30] where the fabric is braided and the relative angle of rotation at the yarn crossover points is significantly large. In the rotation and translation condition, no kinematic constraints are imposed at the yarn crossover points. This enables interacting yarns to slide and rotate with respect to each other. Therefore, contact-induced friction forces at the yarn crossover points provide resistance to both relative rotations and translations.

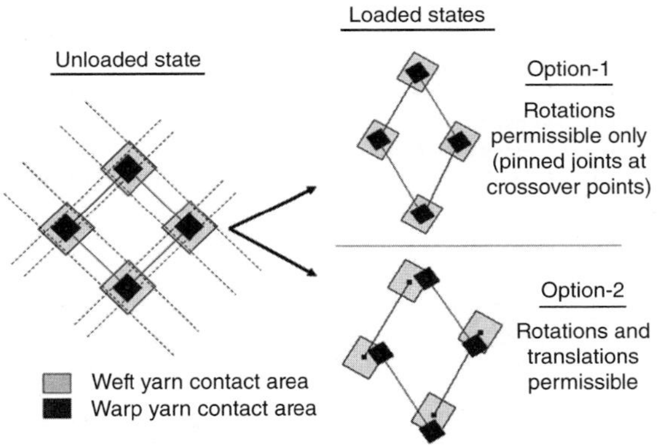

Figure 12.5.23 Treatment of yarn kinematics in unit cell models.

The bladder generally does not contribute any structural stiffness to the fabric nor is it necessary to include the bladder as a contacting media. In addition, the inclusion of a bladder contact surface was found to create unnecessary convergence difficulties and added computational expense. Since the bladder was eliminated from these models, the unit cell should be subjected to a uniform pressure representing the air-beam pressure. There are several methods of applying the internal pressure to the unit cell model, depending on the individual preferences.

Example of a Unit Cell Model

In a specific example of the unit cell, each yarn was represented by an isotropic thin shell with material properties of either Vectran or PEN, depending on the fabric being studied. The shell unit cell model consisted of four warp yarns and four weft yarns as shown in Fig. 12.5.24. The model was subjected to internal pressures ranging from 0.0138 MPa (1 psi) to 0.276 MPa (20 psi). Contact surfaces were permitted to allow yarn translation and rotations as described by option-2. Two compliant membrane elements ($E_{membrane} << E_{yarn}$) connecting the center points of the contact regions were created. These membrane elements were used to determine the overall stress and strain of the shell unit cell model and did not contribute to the structural stiffness of the model.

A 3-step computational procedure was employed to determine the elastic and shear moduli of the unit cell. First, the vertical yarns were restrained and the horizontal yarns (x-direction) were subjected to +1 percent displacement. The model was solved and the stress in the x-direction and the strains in the x- and y-directions were determined. Then, the elastic modulus, $E_1 = \sigma_1/\varepsilon_1$, and Poisson's ratio, $\nu_{12} = -\varepsilon_2/\varepsilon_1$, were determined. Second, the horizontal yarns were restrained and the vertical yarns (y-direction) were subjected to +1 percent displacement in the y-direction. The model was solved and the stress in the *y*-direction and the strains in the x- and y-directions were determined. Then, the elastic modulus, $E_2 = \sigma_2/\varepsilon_2$, and Poisson's ratio, $\nu_{21} = -\varepsilon_1/\varepsilon_2$, were determined.

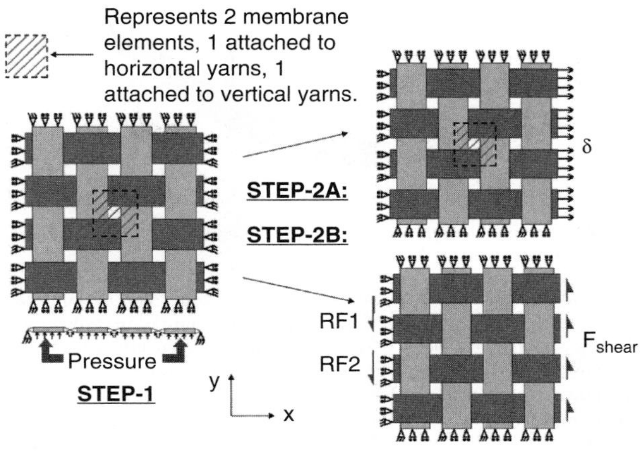

Figure 12.5.24 Example unit cell model and loading procedure.

Finally, the third step was to calculate the shear modulus. This is difficult due to the fact that the elastic and shear moduli of a plain-woven fabric structure are highly dependent on the internal pressure and, furthermore, the elastic and shear moduli are independent. Two methods are suggested. The first method is for the case where the fabric is coated and there is no relative motion between the yarns, then the following approach is recommended.

Continuum Approach

In this approach the fabric is assumed to be a continuum with orthotropic material properties. The unit cell model in this case was subjected to a +1 percent nodal displacement at a 45° angle (not shown). The model was solved and the stresses and strains of the unit cell at the 45° angle were determined. The shear modulus was calculated according to Jones[7] by:

$$G_f = \left(\frac{4}{E_x} - \frac{1}{E_1} - \frac{1}{E_2} + \frac{2\upsilon_{12}}{E_1} \right)^{-1} \tag{12.5.15}$$

where E_x is the elastic modulus in the direction of the 45° load. This expression may be appropriate for fabrics with stiff coatings.

Non-Continuum Approach

The second approach of calculating G_f assumes that the woven fabric is not a continuum and that each yarn independently responds to the external load. In this case, the horizontal yarns were subjected to a shear force F_{shear}. The shear modulus was then calculated based on the equilibrium of the horizontal yarn and its shear deformation as:

$$G_f = G_{12} = \frac{\tau}{\gamma} \tag{12.5.16}$$

where: $\tau_{shear} = \Sigma F_{reaction}/\Sigma A$; and $\gamma = \arctan(\delta_{shear}/L)$, and where $\Sigma F_{reaction}$ is the sum of reaction forces at the support, δ_{shear} is the nodal transverse displacement, A is the yarn cross section and L is the yarn length.

It is recommended that a number of tests on the unit cell model be performed in which the coefficient of friction is varied and the changes in the elastic and shear moduli are determined. Therefore, the results of the shell unit cell model will be parametric in both friction and pressure.

Structural Air Beam Models

Once the elastic and shear moduli of the membrane elements are determined from the unit cell model, a global structural finite element model of the air beam, using membrane elements, can be created such as the one shown in Fig. 12.5.25. Here, the fabric skin would be discretized

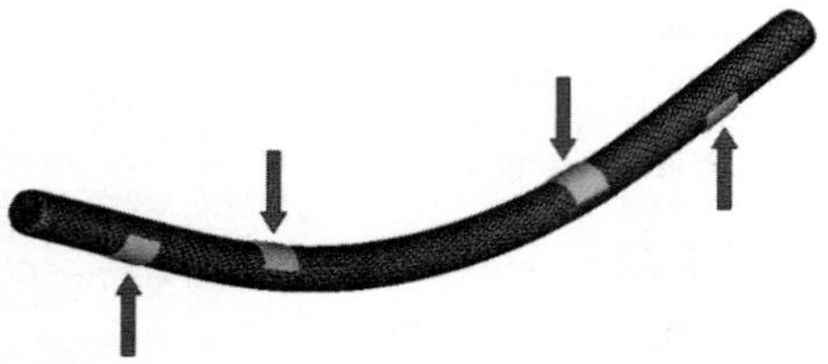

Figure 12.5.25 Example of an air beam global finite element model subjected to 4-point bending.

using membrane elements. However, nine elastic material constants (E_1, E_2, E_3, ν_{12}, ν_{13}, ν_{23}, G_{12}, G_{13}, and G_{23}) are required, of which, only E_1, E_2, and G_{12} had been calculated from the beam unit cell model. Because a membrane formulation is used for the fabric skin, the elastic modulus E_3 would not influence the global beam deflection and was therefore chosen to be of the same order of magnitude as E_1 and E_2 (in the order of E_3 = 64,000 MPa for the current example). It was also noted that any load applied in the direction of one yarn is decoupled from all orthogonal yarns. That is, the Poisson's ratio is nearly 0, therefore, the condition that $\nu_{12} = \nu_{13} = \nu_{23} = 0.0$ may be applied. No method was developed to determine G_{13} and G_{23}. It is hypothesized that neither value had a significant effect on the beam deflection. One could examine this hypothesis by a parametric study of values for G_{13} and G_{23}.

In the above mentioned example, using the analytical strain energy solution, the equivalent in-plane fabric shear modulus, G_f, for the 50.8mm (2.0 in) diameter Vectran beam pressurized to 0.689 MPa (100 psi) was calculated as 86.06 MPa (12,480 psi). The computed bending and shear components of the mid-span deflection were 12.39 mm (0.488 in) and 39.54 mm (1.557 in), respectively. The elastic and shear moduli in the beam unit cell model were calculated for an internal pressure of 0.137 MPa (20 psi) as E_1 = 64,033 MPa, E_2 = 64,098 MPa, and $G_f = G_{12}$ = 60.96 MPa. Similarly, for an internal pressure of 0.207 MPa (30 psi), the elastic moduli were calculated as E_1 = 64,045 MPa, E_2 = 64,049 MPa and $G_f = G_{12}$ = 80.64 MPa.

Concluding Remarks

The unit cell modeling method was used to determine the influence of pressure, weave architecture (including yarn density ratio), crimp interchange, coefficients of friction and biaxial loading ratios on the effective fabric elastic and shear moduli. These moduli were shown to increase with increasing pressure. The coefficient of friction between yarns did not influence the elastic and shear moduli significantly because the relative displacements of the interacting yarns were small compared to their cross sections. The shear modulus of plain-woven fabrics was shown to be a system property; it is a kinematic property of the yarn assemblage in fabric form and is independent of the yarn elastic moduli. The use of yarns with higher elastic modulus may not significantly influence the deflection of air beams because crimp interchange can prevent the as-woven yarns from developing their elastic behavior. It has also been shown that coatings can stiffen the fabric material resulting in smaller air beam deflections.

Homogenization is particularly suitable for developing computationally efficient global models of air-inflated fabric structures in conjunction with unit cell methods. The unit cell method captures the nonlinear constitutive behavior of the plain-woven fabric in response to applied loads. This behavior is transferred to a homogeneous material in a global model of the air-inflated fabric structure.

The studies have shown that air-inflated fabric structures differ fundamentally from conventional metal structures and fiber/matrix composite structures. While the plain-woven fabric may appear to be an orthotropic material, its mechanical behavior indicated otherwise. It does not behave as a continuum, but rather as a discrete yarn assembly.

REFERENCES

1. Celanese Acetate LLC, "Complete Textile Glossary," 2001.
2. Hearle JWS. "High Performance Fibres," Woodhead Publishing, 2001.
3. Hearle JWS, Grosberg P, Backer S. "Structural Mechanics of Fibers, Yarns and Fabrics," John Wiley & Sons, Inc., New York, 1969.
4. Cavallaro PV, Johnson ME, Sadegh AM. "Mechanics of Plain-Woven Fabrics for Inflated Structures," *Composite Structures Journal*, **61**, 2003, pp. 375-393.
5. Pierce FT. *J. Textile Institute*, **28**, 1937, p. T45.
6. Fung W. "Coated and Laminated Textiles," Woodhead Publishing, 2002.
7. Jones RM. "Mechanics of Composite Materials," Hemisphere Publishing Co., 1975.
8. Cadogan DP, Scarborough SE, Lin JKH, Sapna III GH. "*Shape Memory Composite Development for Use in Gossamer Space Inflatable Structures*," American Institute of Aeronautics and Astronautics, AIAA 2002-1372.
9. U.S. Patent No. 6,910,308 by Cadogan DP, Scarborough SE, Lin JKH, Sapna III GH.
10. U.S. Patent No. 5,677,023 by Brown G.
11. U.S. Patent No. 5,735,083 by Brown G, Sharpless G.
12. U.S. Patent No. 5,421,128 by Brown G, Sharpless G.
13. Cavallaro PV, Quigley CJ, Johnson AR, Sadegh AM. "*Effects of Coupled Biaxial Tension and Shear Stresses on Decrimping Behavior in Pressurized Woven Fabrics*," 2004 ASME International Mechanical Engineering Congress and Exposition, Anaheim, CA, November 13, 2004, IMECE2004-59848.
14. Army Center of Excellence for Inflatable Composite Structures, http://www.dtic.mil/dtic/tr/fulltext/u2/a558328.pdf
15. Cavallaro PV, Hart C, Sadegh A. "*Mechanics of Air-Inflated Drop-Stitch Fabric Panels Subject to Bending Loads*," 2013 ASME International Mechanical Engineering Congress and Exposition, San Diego, CA, November 15-21, 2013, IMECE2013-63839.
16. Farboodmanesh S, Chen J, Mead JL, White K. "*Effect of Construction on Mechanical Behavior of Fabric Reinforced Rubber*," Rubber Division Meeting, American Chemical Society, Pittsburgh, PA, October 8-11, 2002.
17. Saville BP. "Physical Testing of Textiles," Woodhead Publishing, 1999.
18. Timoshenko S. "Strength of Materials," D. Van Nostrand Company, Inc., 1946.
19. Ugural AC, Fenster SK. "Advanced Strength and Applied Elasticity," 2nd ed., Elsevier North-Holland Publishing Company, New York, 1977.
20. Bulson PS. "*Design Principles of Pneumatic Structures*," The Structural Engineer, Vol. 51, No. 6, June 1973.
21. Steeves EC. "*Fabrication and Testing of Pressurized Rib Tents*," Technical Report 79-008, United States Army, Soldier Biological Chemical Command, Natick, MA, 1979.
22. Steeves EC. "*The Structural Behavior of Pressure Stabilized Arches*," Technical Report 78/018 (AD-A063263), United States Army, Soldier Biological Chemical Command, Natick, MA, 1978.
23. Steeves EC. "*Behavior of Pressure Stabilized Beams Under Load*," Technical Report 75-47-AMEL, United States Army, Soldier Biological Chemical Command, Natick, MA, 1975.
24. Steeves EC. "*A Linear Analysis of the Deformation of Pressure Stabilized Beams*," Technical Report AND-006-493, United States Army, Soldier Biological Chemical Command, Natick, MA, 1975.
25. Steeves EC. "*Pressure Stabilized Beam Finite Element*," Technical Report 79-002, United States Army, Soldier Biological Chemical Command, Natick, MA, 1978.
26. Fichter WB. "*A Theory for Inflated Thin-Wall Cylindrical Beams*," NASA Technical Note D-3466, National Aeronautics and Space Administration, Washington, DC, 1966.
27. Stein M, Hedgepeth JM. "*Analysis of Partly Wrinkled Membranes*," Technical Note D-813, NASA Langley Research Center, July 1961.
28. Freeston WD, Platt MM, Schoppee MM. "Mechanics of Elastic Performance of Textile Materials, Part XVIII, Stress-Strain Response of Fabrics Under Two-Dimensional Loading," *Textile Research Journal*, **37**, 1967, pp. 948-975
29. Sidhu RMJSR, Averill C, Riaz M, Pourboghart F. "Finite Element Analysis of Textile Composite Preform Stamping," *Journal of Composite Structures*, **52**, 2001, 483-497.
30. Klute KG, Hannaford B. "*Finite Element Modeling of McKibben Artificial Muscle Actuators*," IEEE/ASME Transactions on Mechatronics, March 1998.

12.6 INTRODUCTION TO BIOMECHANICS

by Ali M. Sadegh

GLOSSARY

In describing the relationship of body parts or the location of an external object with respect to the body, the use of directional terms is necessary. The following are commonly used directional terms; see Fig. 12.6.1.

Superior: Closer to the head
Inferior: Farther away from the head
Anterior: Toward the front of the body
Posterior: Toward the back of the body
Medial: Toward the midline of the body
Lateral: Away from the midline of the body
Proximal: Closer to the trunk
Distal: Away from the trunk
Abduction: Motion away from the body midline
Adduction: Motion toward the body midline
Range of motion: Range of translation and rotation of a joint for each of its six degrees of freedom
In vivo: Within a living organism
In vitro: In an artificial environment outside the living organism

INTRODUCTION—WHAT IS BIOMECHANICS?

The concepts of biomechanics were implicitly reflected in medical literatures since the nineteenth century and later in mechanics literatures. For example, von Helmholtz, who was a professor of physiology, pathology, and anatomy at Bonn and later professor of physics in Berlin (1871), had contributed to the mechanism of hearing and theory of color vision and optics and velocity of nerve pulses. Then, in the twentieth century, some researchers employed mechanics principles to explain some phenomena in human physiology. In the early 1970s, the international community adopted the term *biomechanics* to describe biological systems from mechanical perspectives.

By definition, **biomechanics** is the application of mechanics in biology and physiology, that is, utilization of physics laws to study and to analyze the anatomical and functional aspects of living organisms.

Biomechanics has been involved in virtually every facet of modern medical science and technology. Biomechanics has helped in solving all kinds of clinical problems associated with virtually all human organs from the cardiovascular system to prosthetic devices. Even the molecular biology field, which may seem far from biomechanics, utilizes physics principles to understand the mechanics of the formation, design, function, and production of the molecules. Orthopedics has profoundly relied on biomechanics research to treat patients with musculoskeletal problems. Orthopedic surgeons employ biomechanics as their clinical tools. Rehabilitation and healing are intimately related to the stress and strain in the tissues. Injury biomechanics in automotive accidents and other trauma is becoming more important to modern society since it has an economic impact. Therefore, it is no surprise that biomechanics researchers may have academic backgrounds in all engineering and medical disciplines, including orthopedic, cardiac, sports medicine, physical therapy, and many other fields.

Similar to an engineering approach, biomechanics problems should be solved through the following steps:

1. Acquire knowledge of the geometry and configuration of an organ; anatomy, morphology, and histology of the organ should be studied first.
2. Determine the mechanical properties. Unlike with engineering materials, the mechanical properties of living tissues are very difficult to ascertain due to the size of the tissue, or large deformation and nonlinearity of the constitutive equation.
3. Derive the governing equation. The fundamental laws of physics, for example, conservation of mass, momentum, and energy, should be employed, and the governing differential or integral equation for the living tissue should be determined.
4. Determine proper boundary conditions of the organ.
5. Solve the boundary value problem analytically or numerically.
6. Similar to engineering, physiological experiments including animals (in vivo or in vitro) are needed to verify or to compare with the analytical/numerical solutions.

The body spatial coordinate system, which is frequently referred to in this section, is shown in Fig. 12.6.1.

BIOMECHANICS OF BONE

Bones are the main load-bearing tissue in the skeletal system, the purpose of which is to protect internal organs, to provide rigid links and muscle attachment sites, and to facilitate muscle action and body movement. There are 206 bones in the human body that are categorized in four groups according to their general shapes and functions: Short bones provide gliding motion and serve as shock absorbers (e.g., carpals); flat bones provide protection for underlying organs (e.g., scapulae or shoulder blades); irregular bones have different shapes to fulfill specific functions in the body (e.g., vertebrae in the spine); and long bones, provide movements for upper and lower extremities (Fig. 12.6.2).

After the enamel of the teeth, bone is the second-hardest structure of the body. Bone is a highly vascular tissue with excellent remodeling capability. That is, it can repair, alter its properties, and change its geometry in response to changes in mechanical loading. For example, when bone is not subjected to mechanical loading, its density changes; or when it is subjected to new and repeated mechanical loads, then its shape, architecture, and density change.

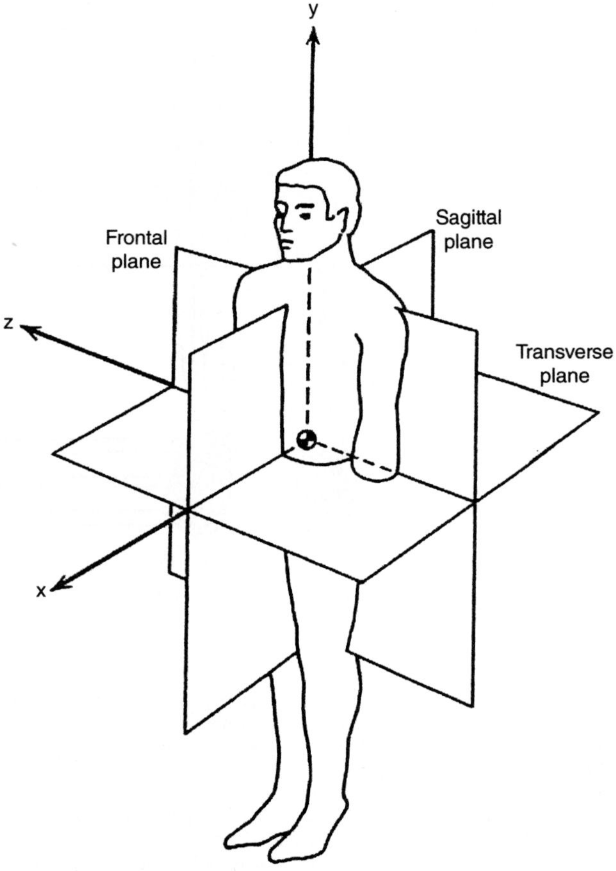

Figure 12.6.1 The body spatial coordinate system.

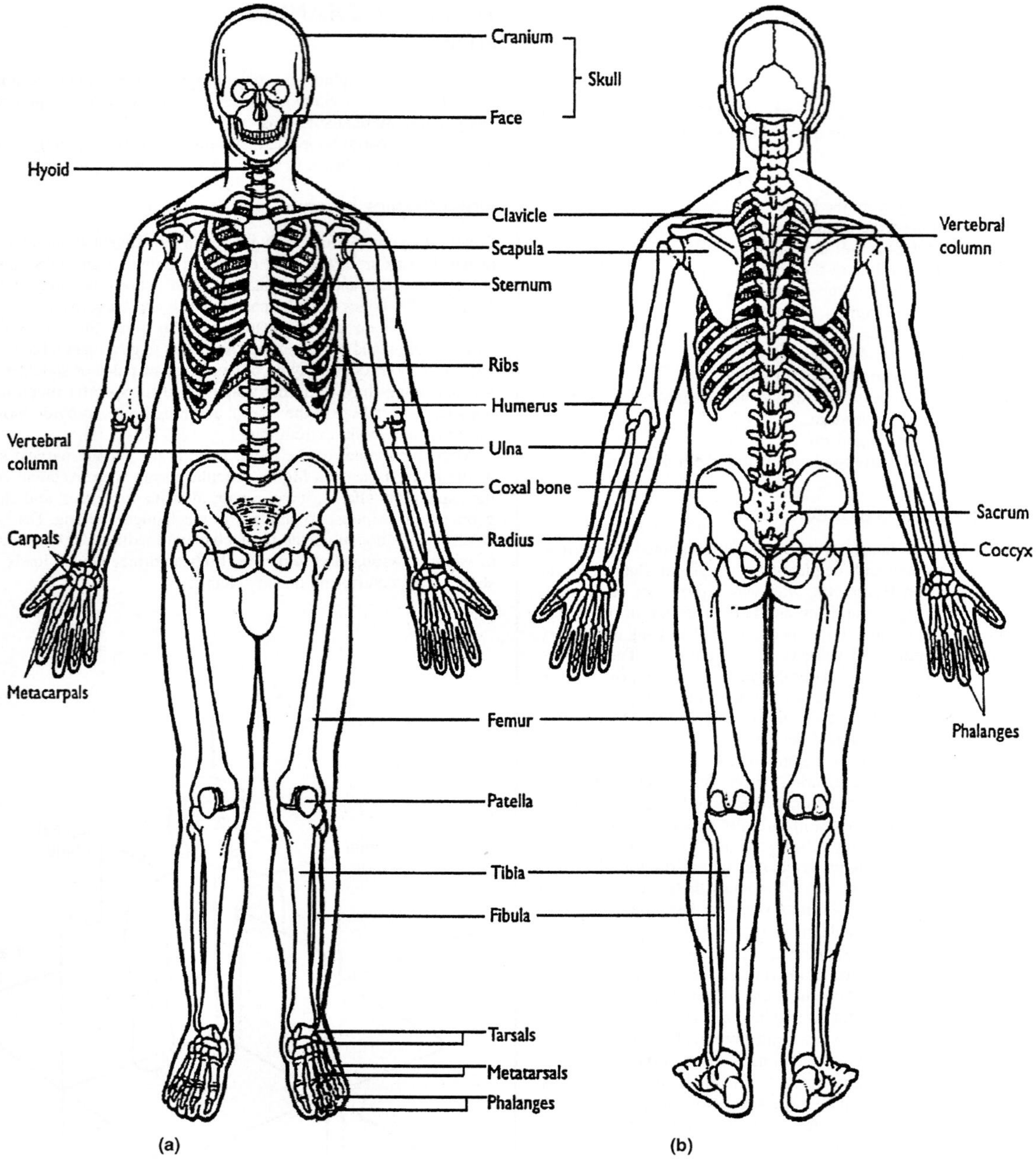

Figure 12.6.2 The human skeleton. (*From Hall, 2003.*)

Bone is composed of inorganic and organic phases and water. Based on the weight, bone is approximately 60 percent inorganic, 30 percent organic, and 10 percent water. On a volume basis, bone is approximately 40 percent inorganic, 35 percent organic, and 25 percent water. The inorganic portion of the bone, which is in the form of small ceramic crystalline-type mineral, consists of calcium and phosphate, resembling synthetic hydroxyapatite $[Ca_{10}(PO_4)_6(OH)_2]$. The organic phase of the bone consists primarily of type I collagen (90 percent by weight), some other collagen of types III and IV, and a variety of noncollagenous proteins.

Bone Structure

Macroscopically, there are two types of bone: cortical, or compact bone and cancellous, or trabecular bone. This is called the highest level of bone porosity. Cortical bone forms the outer shell or cortex of the bone and has a dense structure, while cancellous bone is within this shell and is composed of thin plates, or rods (known as trabeculae), in a mesh structure (Fig. 12.6.3). The interstices between the trabeculae are filled with red marrow. The porosity of cortical bone ranges from 5 to 30 percent, and that of cancellous (trabecular)

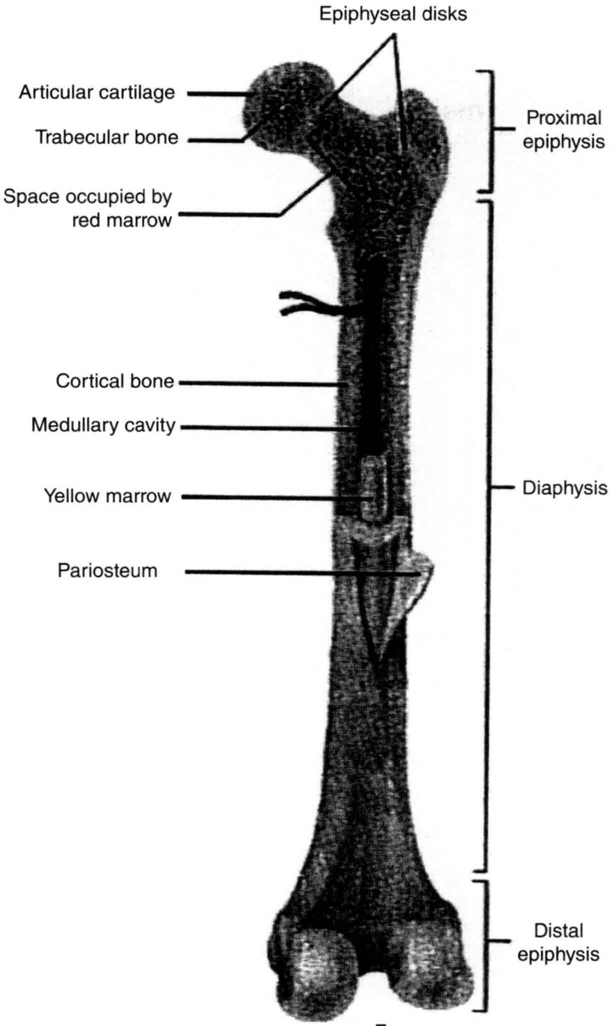

Figure 12.6.3 Femur, showing cortical and trabecular bones. (*From "Hole's Human Anatomy and Physiology" and Hall, 2003.*)

bone ranges from 30 to 90 percent. Except for the joint surfaces, which are covered with articular cartilage, a dense fibrous membrane called the periosteum surrounds all bones.

Microscopically, the osteon or haversian system is the main structural unit of bone. This is called the second level of bone porosity. At the center of each osteon is a small channel, called the haversian canal, that contains blood vessels and nerve fibers. The osteon itself consists of a concentric series of layers (lamellae) of mineralized matrix surrounding the central canal, similar to growth rings on a tree (Fig 12.6.4). Along the boundaries of each layer, or lamella, are small cavities known as lacunae, each containing one bone cell, or osteocyte. Osteon is typically about 200 μm in diameter, which varies with age, and has a length of 1 to 3 mm. The lacunae of adjacent lamellae are connected by numerous small channels, or canaliculi, which ultimately reach the haversian canal.

Mechanical Properties of Bone

Cortical Bone

Based on the anisotropy of bone microstructure, the elastic and strength properties of human cortical bone are anisotropic. In diaphysis of long bones, cortical bone is both stronger and stiffer in the longitudinal direction than in the radial or circumferential directions (Table 12.6.1). Cortical bones are stronger in compression than in tension (Table 12.6.1). Mechanical properties of human cortical bone change (decrease) as age increases (Fig. 12.6.6). While the modulus does not reduce much, if at all, the strength is reduced at a rate of about 2 percent per decade.

As with all biological materials, bone is viscoelastic and nonlinear. It has been shown that cortical bone is moderately strain-rate-dependent. That is, over a six order of magnitude increase in strain rate, modulus changes by only a factor of 2 and strength by a factor of 3 (Fig. 12.6.5). The majority of physiological activities occur in the range of 0.01 to 1.0 percent strain per second, where the changes in bone strength are negligible.

In some simple failure analyses of bone, cortical bone has been considered as simple isotropic materials and the von Mises criterion has been employed. This approach is not capable of describing multiaxial failure of cortical bone. However, the Tsai-Wu criterion, commonly used for composite materials, has been applied to cortical bone using transversely isotropic and orthotropic properties of bone.

Trabecular Bone

The modulus of elasticity and the strength of trabecular bone vary depending on the anatomic site, loading direction, loading mode, and age and health of the donor. In compression, the anisotropy of trabecular bone, both modulus and strength, decreases with age, falling approximately 10 percent per decade (Fig. 12.6.6). The trabecular bone is stronger in compression and is weaker in shear loading mode. Both the modulus and the strength of the trabecular bone depend heavily on the apparent density, yet they depend on the anatomic site. Note that the density of the bone is the ratio of mass to volume of the actual bone tissue; however, the **apparent density** is defined as the ratio of the mass of the bone tissue to the bulk volume of the specimen, including the volume associated with the vascular channels. The density of the trabecular bone also decreases with age. The apparent density of cortical bone is about 1.85 g/cm^3, which does not vary much with anatomic sites or species. However, the apparent density of trabecular bone is as low as 0.10 g/cm^3 for the spine (vertebra) (Kopperdahl and Keaveny, 1998), about 0.3 g/cm^3 for the human tibia (Linde et al., 1989), and up to 0.6 g/cm^3 for the load-bearing portion of the proximal femur (Morgan and Keaveny, 2001).

Bone Remodeling

Bone, as a living tissue, has the ability to remodel, by altering its size, shape, architecture, density, and structure to accommodate the mechanical loadings placed on it. According to Wolff's law, bone remodeling is a function of the magnitude and direction of the mechanical stresses that act on the bones (Wolff, 1892). In other words, Wolff's law states that bone is laid down (deposition) where needed and is resorbed where not needed. Bone remodeling particularly occurs around implants and prostheses where the mechanical stresses to the bone have been altered (Sadegh et al., 1993; Cowin et al., 1992; Luo et al., 1995). In a bone fracture repair process, an implant is firmly attached to the bone to stabilize the fracture site. After the fracture has healed, if the implant is not extracted, the strength and stiffness of the bone may be diminished. This is due to the fact that the material of implants significantly stiffer than the bone, and thus the majority of the loading will be carried out by the implant. This is called **stress shielding** where an implant may cause bone resorption.

BIOMECHANICS OF ARTICULAR CARTILAGE

Articular cartilage is a tough, dense, white connective elastic tissue, about 1 to 5 mm thick, that covers the ends of bones in joints and enables the bones to move smoothly over one another. Articular cartilage is an isolated tissue without blood vessels, lymph channels, and nerves; yet it withstands the rigorous joint environment without failing during the average lifetime. However, when articular cartilage is

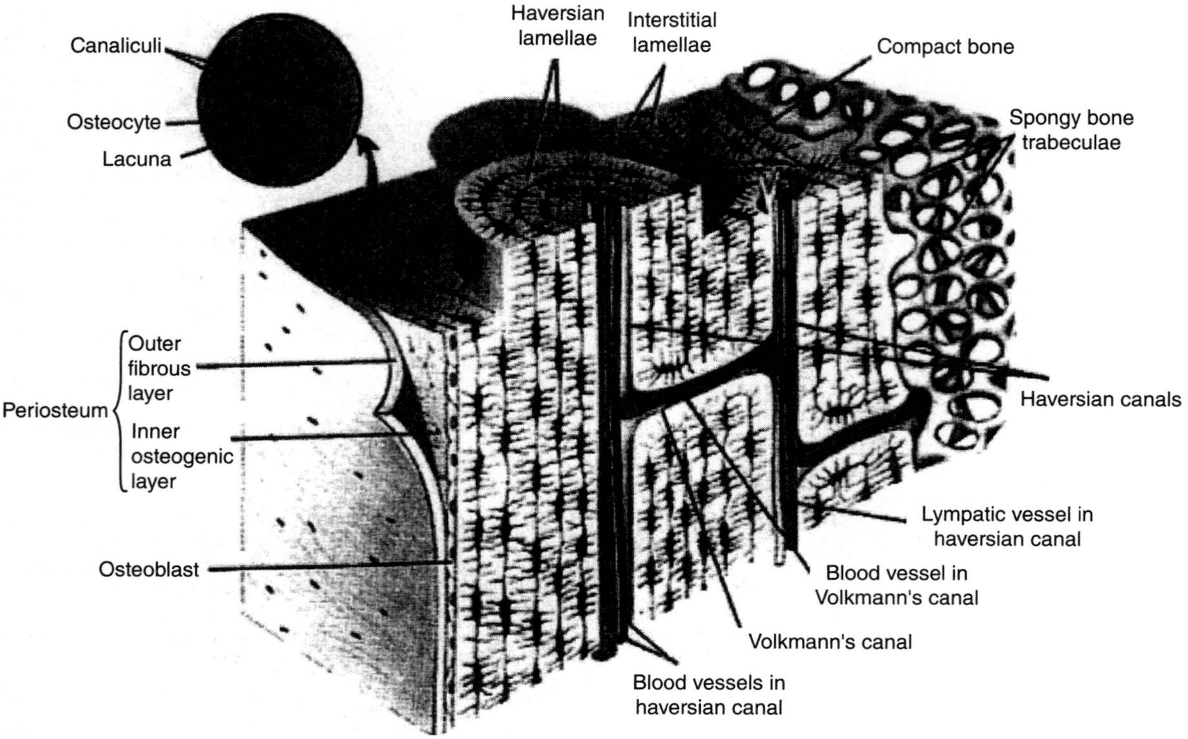

Figure 12.6.4 Diagram of a sector of the shaft of a long bone showing structural details of cortical bone. (*From Kutz, 2003.*)

Table 12.6.1 (*a*) Elastic moduli of cortical bone. (*b*) Ultimate strength of cortical bone

(*a*)

Longitudinal modulus, MPa	17,900 (3,900)*
Transverse modulus, MPa	10,100 (2,400)
Shear modulus, MPa	3,300 (400)
Longitudinal Poisson's ratio	0.40 (0.16)
Transverse Poisson's ratio	0.62 (0.26)

*Standard deviations are given in parentheses.

(*b*)

Longitudinal, MPa	
Tension	135 (15.6)*
Compression	205 (17.3)
Transverse, MPa	
Tension	53 (10.7)
Compression	131 (20.7)
Shear, MPa	65 (4.0)

*Standard deviations are given in parentheses.
SOURCES: Reilly and Burstein (1975) and Kutz (2003).

damaged through injury or a lifetime of use, it does not heal as rapidly or effectively as other tissues in the body. Instead, the damage tends to spread, allowing the bones to rub directly against each other and resulting in pain and reduced mobility.

The primary function of articular cartilage is to distribute joint loads over a wide area and to reduce the stress concentration on the joint. Its secondary function is to provide a contacting surface with minimal friction and wear.

Chondrocytes are mature cartilage cells. They are sparsely distributed and account for less than 10 percent of the tissue's volume. Chondrocytes are arranged in three zones through the thickness of cartilage (Fig. 12.6.7). The first zone is close to the articular surface where chondrocytes' long axes are parallel to the articular surface. In the second, middle zone, the chondrocytes are round and randomly distributed. In the third, deep zone, the chondrocytes are arranged in columnar fashion perpendicular to the articular surface.

The organic matrix of cartilage is composed of a dense network of fine collagen fibrils enmeshed in a concentrated solution of proteoglycans (PGs). The collagen content of cartilage tissue ranges from 10 to 30 percent by net weight, and the PG content from 3 to 10 percent by wet weight; the remaining 60 to 87 percent is water, inorganic salts, and small amounts of other matrix proteins, glycoproteins, and lipids.

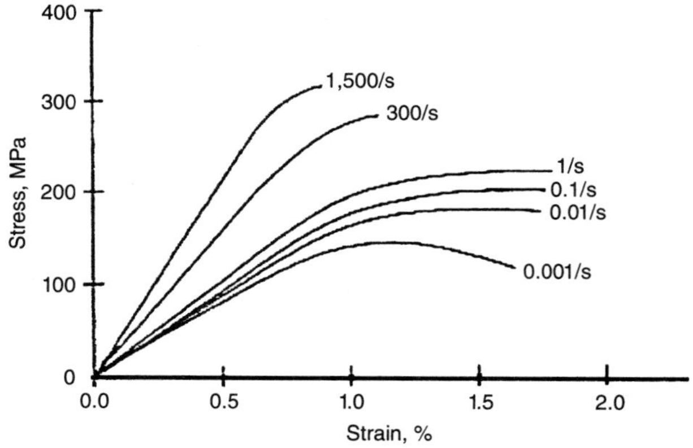

Figure 12.6.5 Strain rate sensitivity of cortical bone for longitudinal tensile loading (*From Kutz, 2003; data from McElhaney and Byars, 1965.*)

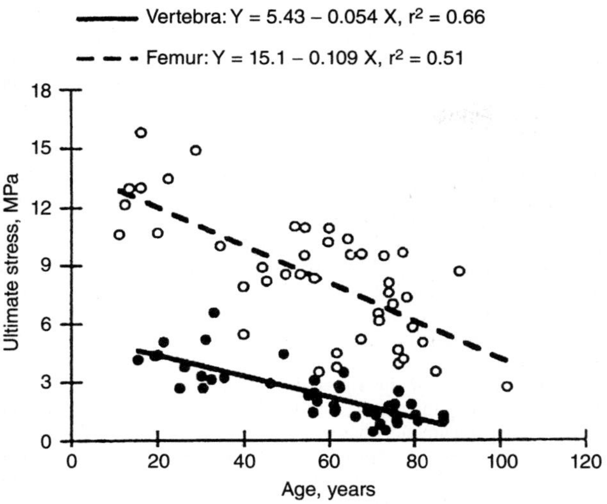

Figure 12.6.6 Variation of ultimate stress of trabecular bone (from the human vertebra and femur) with age. (*From Kutz, 2003; data from Mosekilde et al., 1987 and McCalden et al., 1997.*)

Cartilage is predominantly loaded in compression and is biphasic (solid-fluid) and viscoelastic. When cartilage is compressively loaded, the water within the proteoglycan matrix is extruded. The stiffness of cartilage is a function of its permeability, and the fluid pressure within the matrix supports approximately 20 times more load than the underlying material during physiological loading (Mankin et al., 1994). Cartilage material properties are given in Table 12.6.2.

BIOMECHANICS OF TENDONS AND LIGAMENTS

Tendons and ligaments, along with the joint capsule, surround, connect, and stabilize the joints of the skeletal system. Ligaments and joint capsules, which connect bone to other bones, provide stability to joints by guiding joint motion and preventing excessive joint motion. The function of the tendons is to attach muscle to bone and to transmit tensile loads from muscle to bone. Tendons and ligaments are sparsely vascularized and are composed of a combination of collagen and elastin fibers arranged primarily in parallel along the axis of loading. They have relatively few fibroblast cells and abundant extracellular matrix. In general, the cellular material occupies about 20 percent of the total tissue volume, while the extracellular matrix accounts for the remaining 80 percent. About 70 percent of the matrix consists of water, and approximately 30 percent is solid.

Table 12.6.2 Representative Properties of the Human Articular Cartilage Taken from the Lateral Condyle of the Femur

Property	Value
Poisson's ratio	0.10
Compressive modulus, MPa	0.70
Permeability coefficient, $m^4/(N \cdot s)$	1.18×10^{-15}

SOURCES: Mankin et al. (1994) and Kutz (2003).

Both tendons and ligaments are passive tissues, and they do not have the ability to contract as muscle tissue does; however, they are extensible in tensile loading, and will return to their original length after being stretched, when they are extended within their elastic limits. In the unloaded state, the fibers are slightly crimpled, and thus at the initial stage of tensile loading the structure acts only to straighten out the component fibers and exhibits low stiffness, for example, in the toe region in Fig. 12.6.8*a*. The straightened fibers exhibit linear deformation under further tensile loading, resulting in this load-deformation curve of Fig. 12.6.8*a*. Typical tensile properties of ligament and tendon are shown in Table 12.6.3. Tendons are generally stiffer than ligament due to the higher concentration of collagen. Tendons and ligaments are highly viscoelastic and rate-dependent; that is, they will fail at lower extensions when loaded at high rates. That is, rapid motion (stretching) of ligaments or tendons may result in tear or rupture. For the whole ligaments and tendons the linear response can extend above 20 percent (Fig. 12.6.8*b*).

BIOMECHANICS OF MUSCLES

Skeletal muscle is the most abundant tissue in the human body, accounting for 40 to 45 percent of the total body weight. There are 430 skeletal muscles, found in pairs on the right and left sides of the body. There are three types of muscles: (1) cardiac muscles that compose the heart, (2) smooth or involuntary muscles that line the hollow internal organs, and (3) skeletal or voluntary muscles that are attached to the skeleton through the tendons and cause it to move.

Muscles are molecular machines that convert chemical energy to a force (mechanical energy). The structural unit of skeletal muscle is the fiber, a long cylindrical cell with hundreds of nuclei, ranging from about 10 to 100 μm in thickness and about 1 to 30 cm in length. Individual muscle fibers are connected by three levels of collagenous tissue: endomysium, which surrounds individual muscle fibers; perimysium, which collects bundles of fibers into fascicles; and epimysium, which encloses the entire muscle belly (see Fig. 12.6.9). The functional unit of skeletal muscle is the motor unit, which is the smallest part of the muscle that can be made to contract independently. The motor unit consists of a single motor neuron and all the muscle fibers that are to be stimulated by it. Note that when stimulated, all muscle fibers in the motor unit respond as one (Fig. 12.6.10).

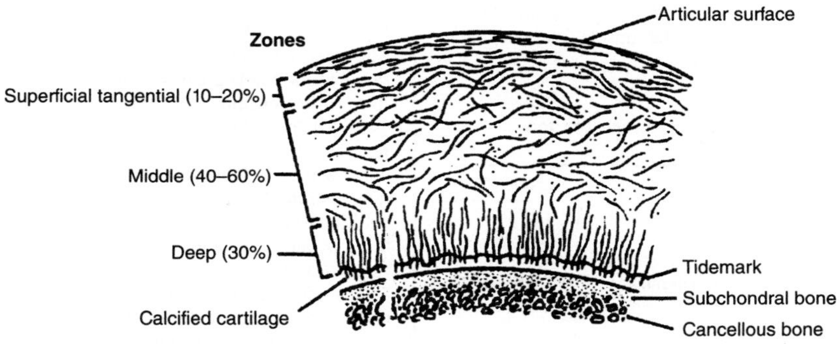

Figure 12.6.7 Collagen fibrils of chondrocyte arrangement through the thickness of cartilage, with three distinct zones: surface tangential, middle, and deep. (*From Nordin and Frankel, 1989, by permission.*)

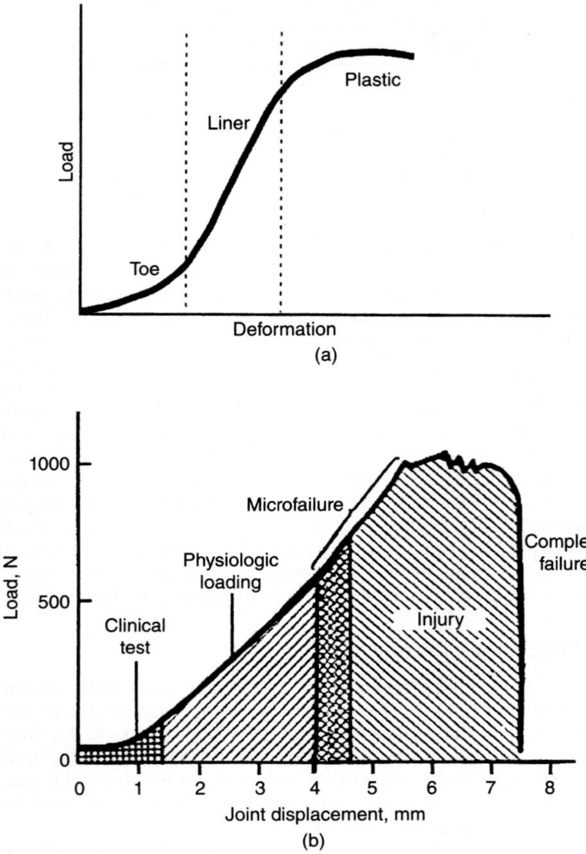

Figure 12.6.8 (*a*) Typical tensile load deformation of collagenous soft tissue such as ligament and tendon. (*Kutz, 2003.*) (*b*) Tensile testing of a human anterior cruciate ligament in vitro. (*Nordin and Frankel, 1989, by permission.*)

The tendons and the connective tissues in and around the muscle belly are viscoelastic structures that help determine the mechanical characteristics of whole muscle during contraction and passive extension. In simulating musculotendon actuation, Hill (1970) showed that the tendons represent a springlike elastic component located in series with the contractile component (element) CC (or CE), which is in parallel with parallel elastic element (PEE) and in series with the series elastic element (SEE) (Fig. 12.6.11).

The mechanism of muscle contraction is as follows. After stimulation by the motor nerve, there is an interval of a few milliseconds, known as the latency period, before the tension in the muscle fibers begins to rise and contraction starts. Physiologically, according to the sliding filament theory, active shortening of the muscle results from the relative movement of the actin and myosin filaments past one another. This mechanical response of muscle fibers is known as twitching. Some muscle fibers contract with a speed of only 10 ms, while others may take 100 ms or longer.

Table 12.6.3 Representative Properties of Human Tendon and Ligament under Tensile Loading

Tissue	Property	Value
Tendon	Elastic modulus, GPa	1.2-1.8
	Ultimate strength, MPa	50-105
	Ultimate strain, %	9-35
	Energy absorbed during elastic deformation	4%-10% per cycle
Ligament	Stiffness, N/mm	150
	Ultimate load, N	368
	Energy absorbed to failure, N · mm	1,330

SOURCES: Woo et al. (1994, 1997) and Kutz (2003).

During muscle contraction, a tension load is exerted on the bone to overcome the external load of the body. The muscle tension is generally applied at a distance from the center of rotation of a joint, to create a moment (torque) on the involved joint.

There are three types of muscle contraction. First, when muscles develop sufficient tension to overcome the resistance of the body segment, the muscles shorten and cause joint movement in the same direction as the net torque generated by the muscles. This contraction is called a **concentric.** Second, when muscle tension is developed but no change in muscle length occurs, the contraction is said to be **isometric.** In this case due to the developed contraction the diameter of the muscles is increased. Third, when a muscle cannot develop sufficient tension and the opposing joint torque exceeds that produced by the tension, the muscle progressively lengthens instead of shortens. In this case the muscle is said to contract **eccentrically.** Note that concentric and eccentric contractions involve dynamic work, in which the muscle moves a joint or controls the movement, while isometric contractions involve static work, in which the joint position is maintained (Fig. 12.6.12).

BIOMECHANICS OF JOINTS

Anatomically, there are three types of joints: immovable joints, such as sutures of the skull; slightly movable joints, such as cartilaginous joints (discs) where applied forces are attenuated to the adjacent bones; and freely movable (synovial) joints such as articular joints in the upper and lower extremities. In this section the biomechanics of articular joints in upper and lower extremities is discussed.

Biomechanics of the Knee

The human knee is the largest and one of the most complex joints in the body. The knee joint is situated between the body's two longest bones, the femur and tibia, and bears tremendous loads. The fact that the knee joint provides the mobility required for locomotor activities makes it one of the high weight-bearing joints of the body. Anatomically, the knee joint comprises (1) two condylar articulations of the tibiofemoral, with the third articulation being the patellofemoral joint; (2) medial and lateral menisci cartilaginous disc located between the tibial and femoral condyles; (3) articular capsule which is a double-layered membrane that surrounds the joint; (4) the synovial fluid, which is a clear, slightly yellow liquid that provides lubrication inside the articular capsule at the joint; and (5) the patella, also known as kneecap (Fig. 12.6.13). In addition, there are several ligaments in the knee joint; the two important ones are the anterior cruciate ligament (ACL) and the posterior cruciate ligament (PCL) (Fig. 12.6.13).

KNEE KINEMATICS

The knee joint motion takes place in all three planes; however, the range of motion is greatest in the sagittal plane. Motion in this plane is from full extension to full flexion of the knee, which is approximately 140°. When the knee is fully extended, its rotation is almost completely restricted by the interlocking of the femoral and tibial condyles. However, when the knee reaches about 90° flexion, the external rotation ranges from 0° to approximately 45°. During flexion/extension, in addition to the rolling contact between the femoral and tibial condyles, there is tangential gliding between the two condyles as well. Also, in flexion/extension motion, the patella glides inferiorly and superiorly against the distal end of the femur in the vertical direction. The anterior and posterior cruciate ligaments (ACL and PCL) limit the forward and backward sliding of the femur on the tibial plateaus during knee flexion and extension and also limit knee hyperextension. They both play an important role in knee stability.

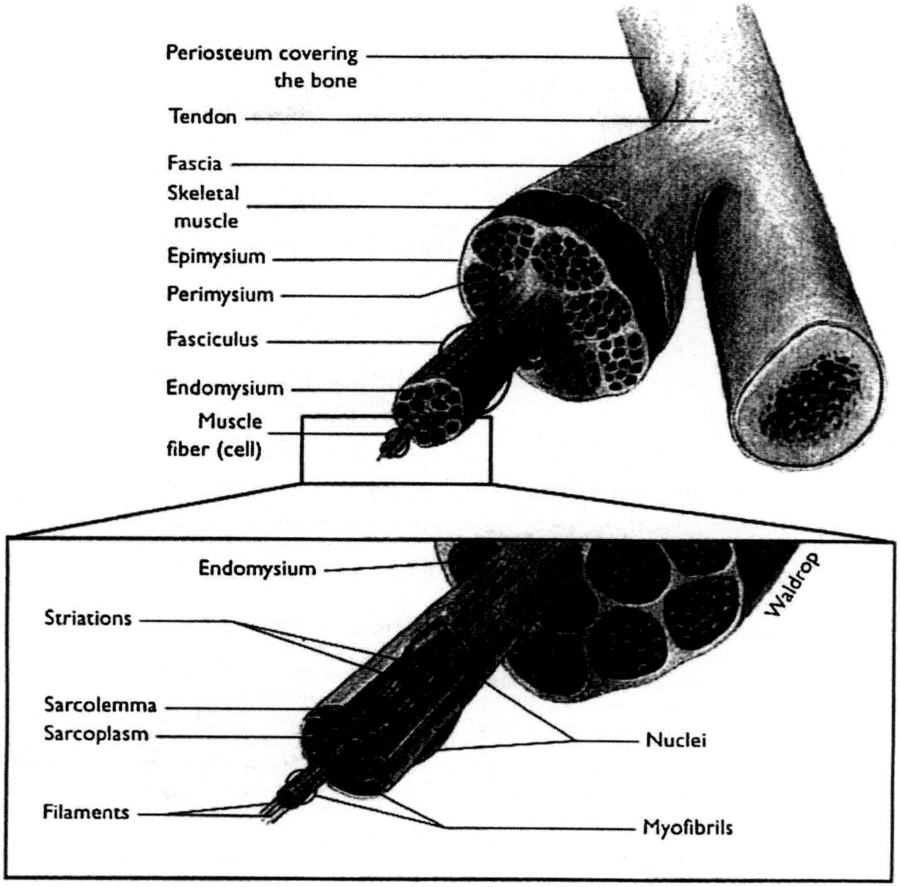

Figure 12.6.9 Schematic drawings of the structural organization of muscle. (*From "Hole's Human Anatomy and Physiology" and Hall, 2003.*)

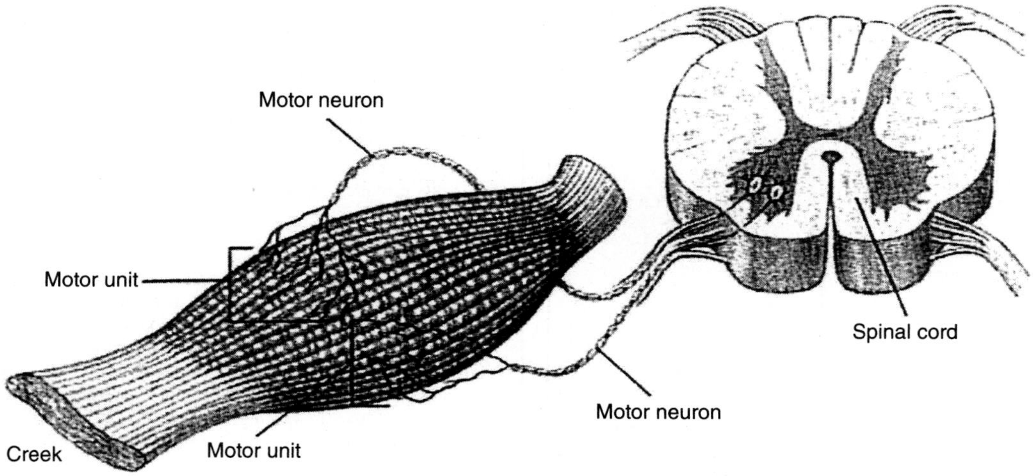

Figure 12.6.10 A motor unit of muscle with its neuron. (*From "Hole's Human Anatomy and Physiology" and Hall, 2003.*)

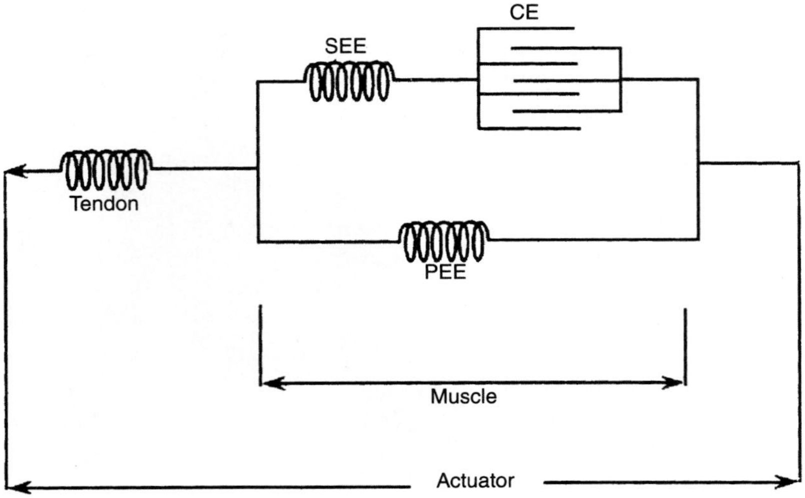

Figure 12.6.11 Schematic diagram of a model commonly used to simulate musculotendon actuation. (*Modified from Zajac and Gordon, 1989 and Pandy et al., 1990.*)

Knee Loads

In flexing the knee, the three hamstring muscles are the primary flexors acting on the knee. However, during extension the quadriceps muscles provide the primary force for the knee. During daily activities, the tibiofemoral joint is subjected to compression and shear forces. Specifically, during the stance phase of gait, the compressive force at the tibiofemoral joint exceeds 3 times the body weight; or during stair climbing the compressive force is approximately 4 times the body weight. Figure 12.6.14 shows the static load diagram of the knee during stair climbing. Figure 12.6.15 shows the joint reaction force, in terms of body weight transmitted through the tibial plateau during walking, in one gait cycle. During the gait cycle the joint reaction force shifts from the medial to the lateral tibial plateau. In stance phase, when the force reached its peak value, the medial plateau mainly carried the load; and in the swing phase, when the force was minimal, the load was carried by the lateral plateau. The contact area of the medial tibial plateau is approximately 50 percent larger than that of the lateral tibial plateau.

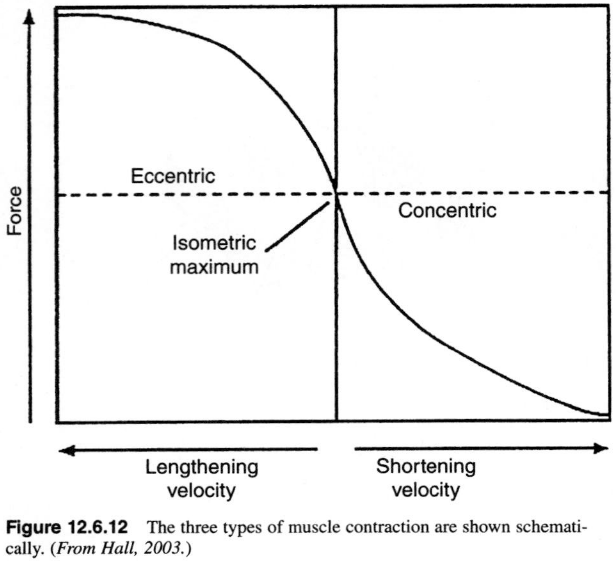

Figure 12.6.12 The three types of muscle contraction are shown schematically. (*From Hall, 2003.*)

Although the tibial plateaus are the main load-bearing structures in the knee, the menisci act as force absorption and distribute loads at the tibiofemoral joint over a larger area, thus reducing the magnitude of contact stress at the joint. The joint cartilages and ligaments also contribute to load bearing.

Biomechanically, the patella serves two important functions. First, it aids knee extension by producing anterior displacement of the quadriceps tendon throughout the entire range of motion, thereby lengthening the lever arm of the muscle force. Second, it allows wider distribution of compressive stress on the femur by increasing the area of contact. During normal walking gait, compressive force at the patellofemoral joint is about one-half of body weight, and during the stair climbing, it increases up to over 3 times body weight.

Biomechanics of the Hip

The hip joint is a ball-and-socket joint and is one of the largest and most stable joints (least likely to be dislocated) in the body with a great deal of mobility. Anatomically, the hip joint is composed of the head of the femur, which forms approximately two-thirds of a sphere, and the acetabulum of the pelvis, which is a socket (see Fig. 12.6.16).

The acetabular surface and the head are covered with articular cartilages. The femur is a major weight-bearing bone and is the longest, largest, and strongest bone in the body. The femoral neck is its weakest component and is primarily composed of trabecular bone. The hip joint cartilage covers both articulating surfaces. Several large and strong ligaments provide the stability of the hip.

Hip Kinematics

The hip joint, due to its ball-and-socket nature, moves in all three planes. In the sagittal plane it flexes or extends, in the transverse plane it rotates internally and externally, and in the frontal plane it has abduction and adduction motions. The maximum range of motion is in the sagittal plane where the range of flexion is from 0° to approximately 140° and the range of extension is from 0° to 15°. The range of abduction is from 0° to 30°, whereas that of adduction is from 0° to 25°. Its external rotation ranges from 0° to 90°, and its internal rotation ranges from 0° to 70° when the hip is flexed. The average range of hip joint motion in the sagittal plane during normal walking, in one gait cycle, is shown in Fig. 12.6.17. The joint is maximally flexed during late swing phase of gait, as the limb moves forward for heel strike. Maximum tension is reached at heel-off.

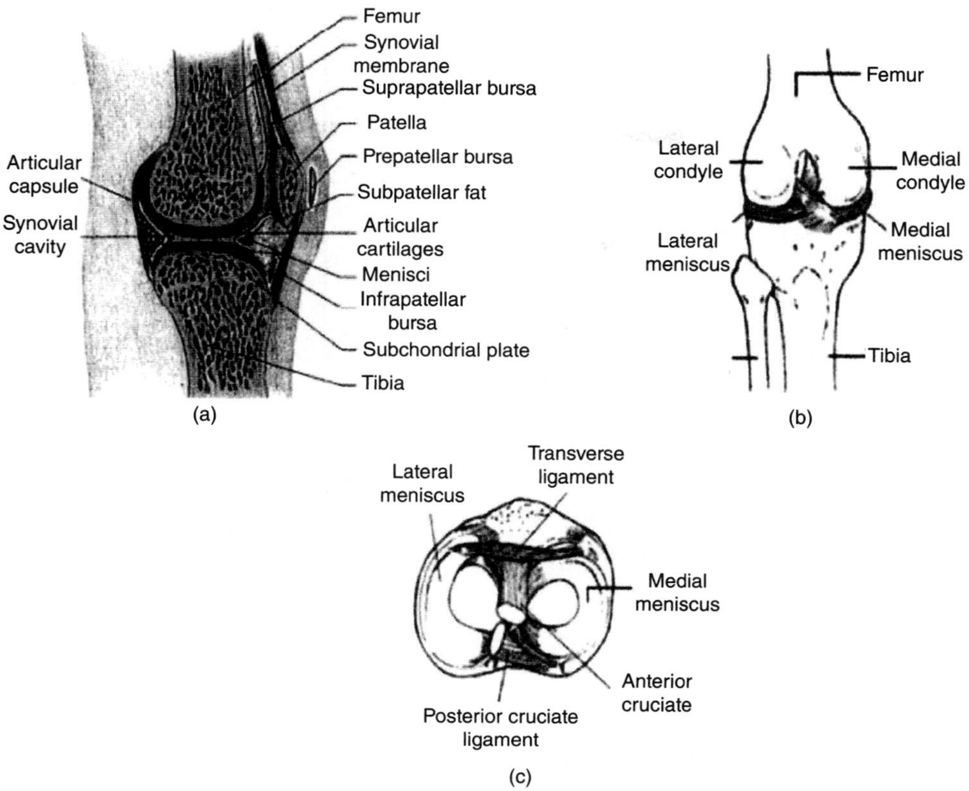

Figure 12.6.13 The knee joint. (*a*) Sagittal cross-sectional view (*"Hole's Human Anatomy and Physiology," 1996; Hall, 2003.*) (*b*) Posterior view. (*Hall, 2003.*) (*c*) Top view. (*Hall, 2003.*)

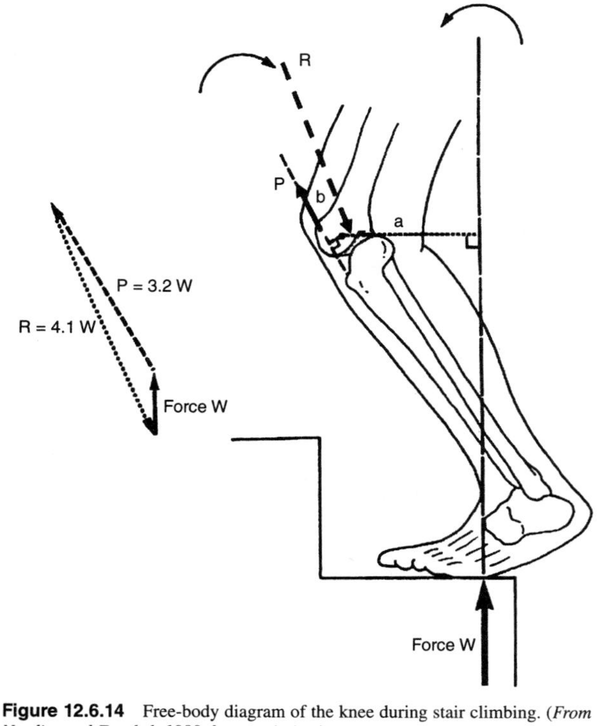

Figure 12.6.14 Free-body diagram of the knee during stair climbing. (*From Nordin, and Frankel, 1989, by permission.*)

Hip Loads

During flexion six muscles that cross the joint anteriorly (the iliacus, psoas major, pectineus, rectus femoris, sartorius, and tensor fascia latae) are primarily responsible for the motion. However, during extension, the three hamstring muscles (the biceps femoris, semitendinosus, and semimembranosus) and the gluteus provide the motion. These muscles are the knee flexors as well.

The hip is the major weight-bearing joint. During upright standing on both feet the body weight is evenly distributed to each of the hip joints. To calculate the forces and torques in each joint or muscle of the body, the dimensions and length of each body segment are needed. The length between each joint of the human body varies with body build, gender, and racial origin. However, an average set of segment lengths expressed as a percentage of body height was prepared by Drillis and Contini (1966) and is shown in Fig. 12.6.18. To determine the mechanism of balance during quiet standing, Winter et al. (1998) modeled the body as 14 segments with 21 markers, as shown in Fig. 12.6.19. The location of the markers and the accompanying table give the definition of each of the 14 segments, along with the mass fraction of each segment.

During a single-leg stance, the force of the abductor muscle is calculated as follows. The weight of the body above the hip joint is approximately equal to five-sixths of the total body weight. The moment arising from the weight above the hip joint must be balanced by a moment occurring from the force of the abductor muscles, as shown in Fig. 12.6.20. A simple balance of moment with respect to the center of the femoral head reveals that the abductor muscle force M is $M = \frac{5}{6}Wb/c$, where c is the radius of the ball (femoral head) and b is the distance between the center of the ball and the center of gravity (CG) of mass of the body, without the right hip (minus the one leg). Substitution of the values of b and c leads to the abductor muscle

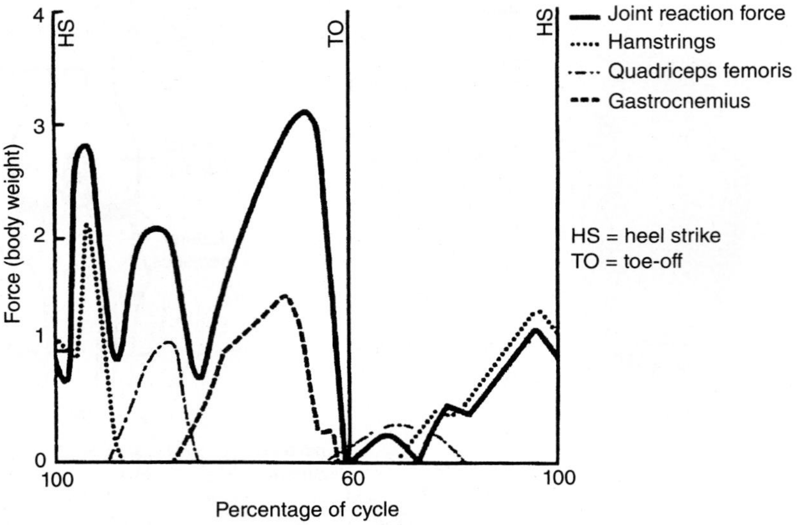

Figure 12.6.15 The knee reaction force in terms of body weight transmitted through the tibial plateau during walking in one cycle. (*Adapted from Morrison, 1970; Nordin and Frankel, 1989, by permission.*)

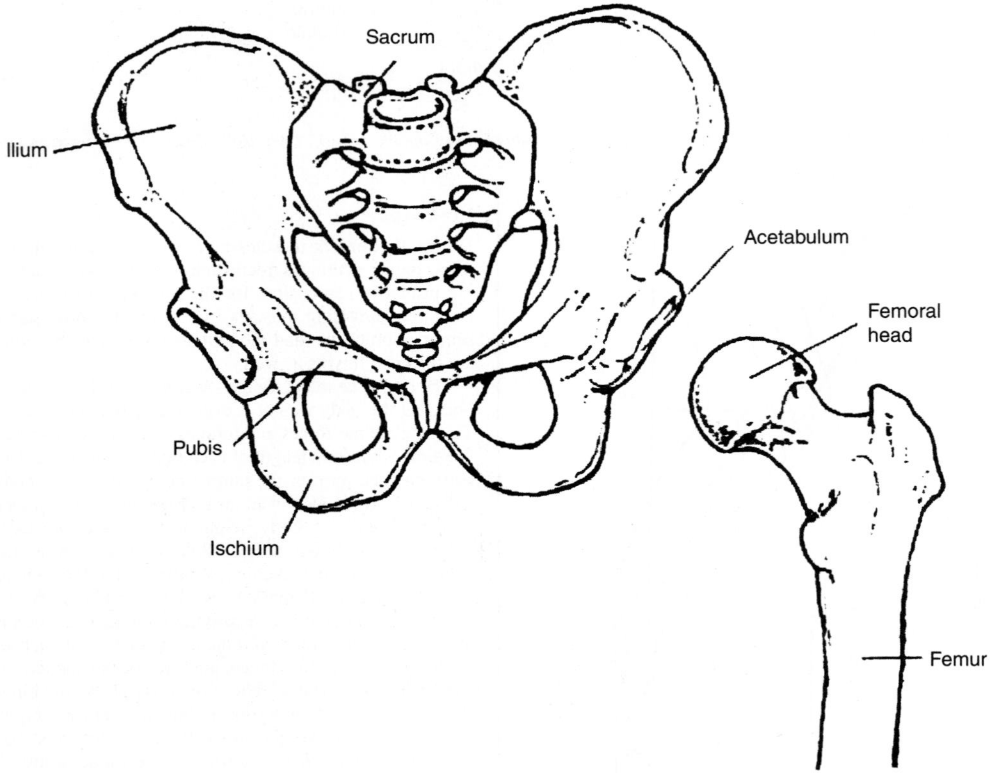

Figure 12.6.16 The hip joint. (*From Hall, 2003.*)

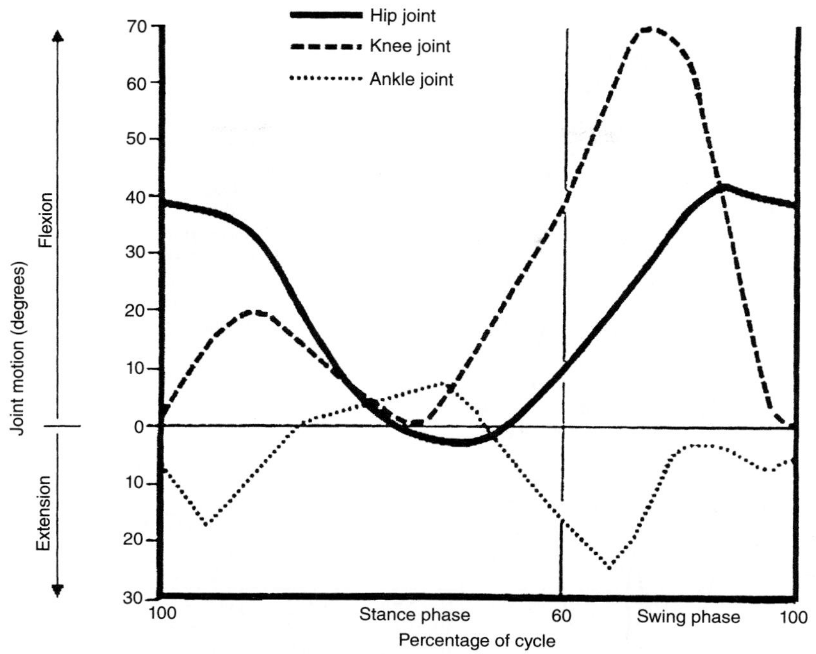

Figure 12.6.17 The range of motion of the hip joint in sagittal plane during one gait cycle. (*From Nordin and Frankel, 1989, by permission.*)

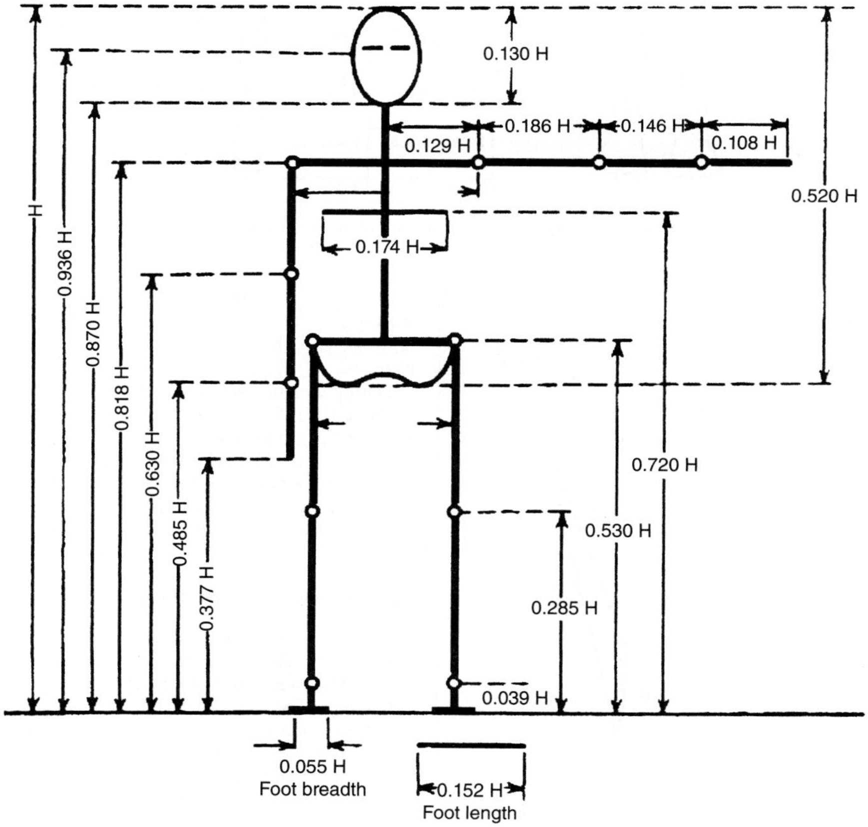

Figure 12.6.18 Body segment length expressed as a fraction of body height *H*. (*From Drillis and Contini, 1966.*)

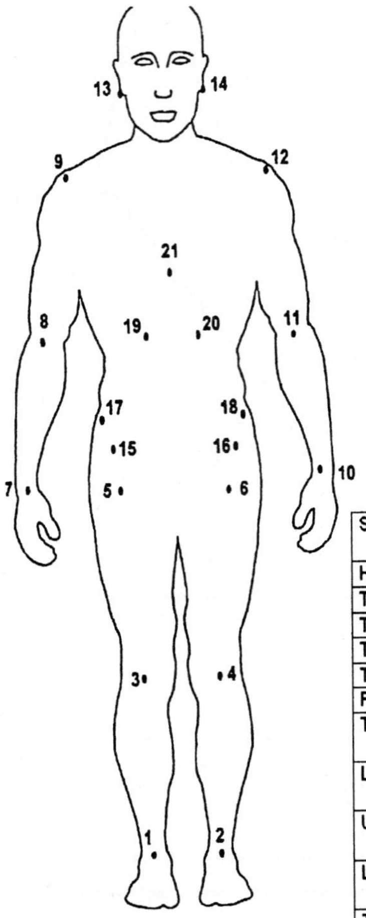

Segment	Mass Fraction	Definition of Segment COM
Head	0.081	(13 + 14)/2
Trunk 4	0.136	(9 + 12 + 21)/3
Trunk 3	0.078	((19 +20)/2 + 21)/2
Trunk 2	0.065	(17 + 18 + 19 + 20)/4
Trunk 1	0.076	(17 + 18 + 15 + 16)/4
Pelvis	0.142	(15+13)/2
Thighs	0.100 (2)	0.433 x 3 + 0.567 x 5 and 0.433 x 4 + 0.567 x 6
Legs & feet	0.060 (2)	0.606 x 1 + 0.394 x 3 and 0.606 x 2 + 0.394 x 4
Upper arms	0.028 (2)	0.436 x 8 + 0.564 x 9 and 0.436 x 11+ 0.564 x 12
Lower arms	0.022 (2)	0.682 x 7 + 0.318 x 8 and 0.682 x 10 + 0.3186 x 11
Total	1.00	

Figure 12.6.19 Estimation of the mass fraction and the center of mass (COM) of 14 segments of the body with 21 markers. (*From Winter et al., 1998.*)

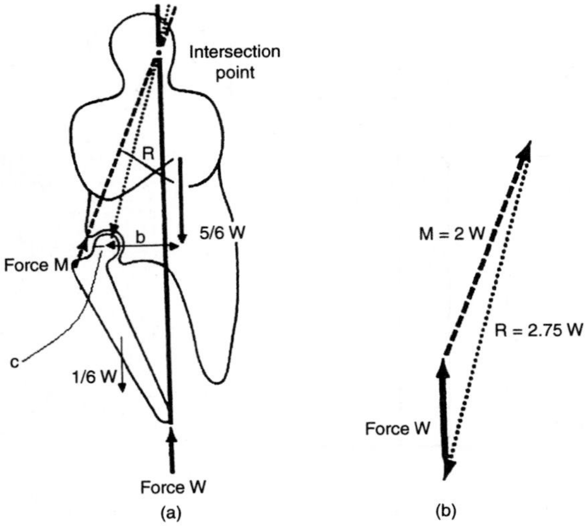

Figure 12.6.20 Free-body diagram of hip joint during a single-leg stance. The abductor muscle is counterbalanced to the moment arising from gravitational force of the body. (*From Nordin and Frankel, 1989, by permission.*)

force, which is about 2 times the body weight. The reaction force of the hip joint during a single-leg stance is approximately 3 times the body weight. This magnitude varies as the position of the upper body changes.

During dynamic activities several investigators [Paul, 1967; Andriacchi et al., 1980; Bergmann et al., 1997 and 2001] have studied the loads on the hip joint. Figure 12.6.21 shows the hip joint reaction force in units of body weight during walking, one gait cycle. In both men and women, the largest peak of about 7 times the body weight was reached just before toe-off. An increase in gait velocity increases the magnitude of the hip joint reaction force in both swing and stance phases.

BIOMECHANICS OF THE SPINE

The human spine is a complex and functionally significant structure of the human body. The principal function of the spine is to protect the spinal cord and transfer loads from the upper body, head, and trunk to the lower extremities. The spine is capable of motion in all three planes through its 24 vertebrae which are articulated by intervertebral discs. The discs, surrounding ligaments and muscles, gain the spine stability. The spinal column is structurally divided into four regions. Starting from superior to inferior, there are 7 cervical vertebrae (C1 to C7), 12 thoracic vertebrae (T1 to T12), 5 lumbar vertebrae (L1 to L5), and 5 fused sacral vertebrae (S1 to S5) (Fig. 12.6.22).

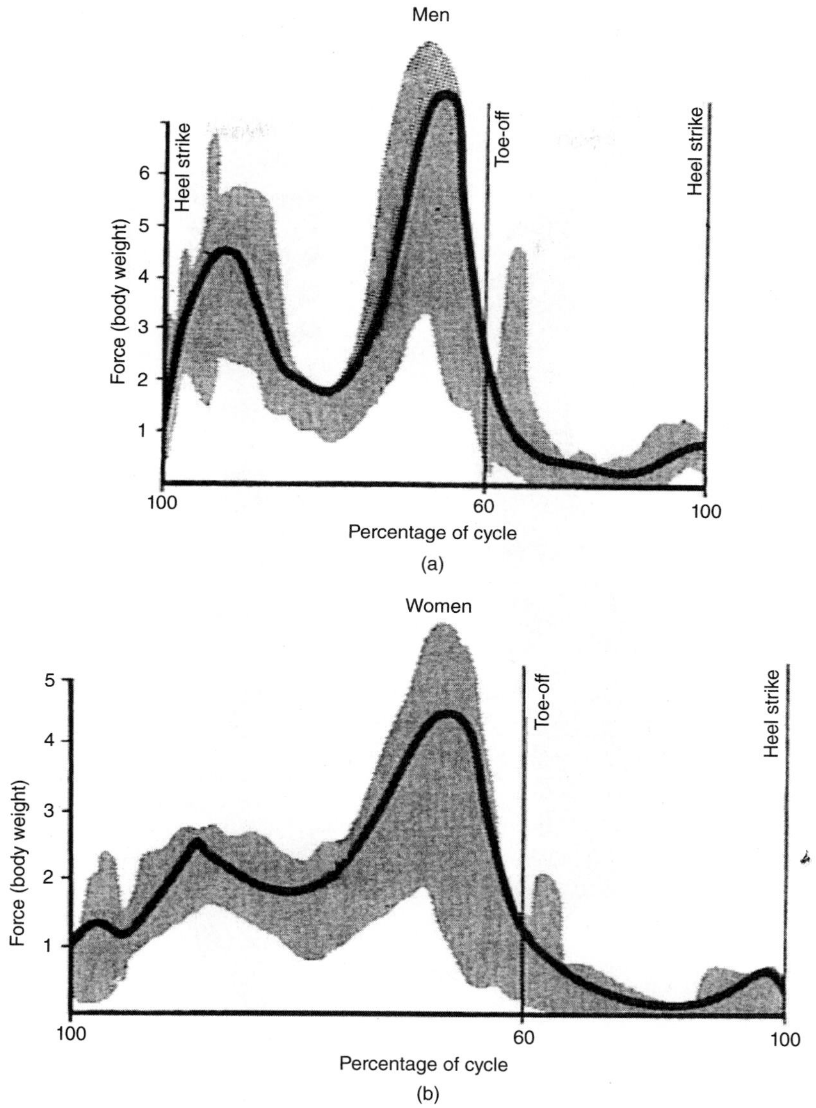

Figure 12.6.21 Hip joint reaction forces in units of body weight during walking one gait cycle. Shaded area indicates variations among subjects. (*Adapted from Paul 1967; Nordin and Frankel, 1989, by permission.*),

The spine contains four normal curves. The thoracic and sacral curves, which are concave anteriorly, are present at birth and are referred to as primary curves. The lumbar and cervical curves, which are concaved posteriorly, are developed from supporting the body in an upright position when young children begin to sit up and stand. These curves are known as secondary spinal curves.

Because of the anatomical changes of vertebrae throughout the spinal column, the amount of movement between adjacent vertebrae is varied. In this subsection, the biomechanics of the lumbar spine and the cervical spine are discussed.

Biomechanics of the Lumbar Spine

The vertebral bodies are designed to bear mainly compressive loads. Anatomically, the vertebral bodies of the lumbar region are thicker and wider than those in the thoracic and cervical regions. A typical vertebra consists of a body, a hollow ring known as the neural arch, and several bony processes (Fig. 12.6.23). The spinous and transverse processes serve as outriggers to improve the mechanical advantage of the attached muscles.

The functional unit of the spine, known as the motion segment, consists of two vertebrae and their intervening soft tissues (Fig. 12.6.24). The intervertebral discs in adults account for approximately one-fourth of the height of the spine. The intervertebral disc bears and distributes loads and restrains excessive motion. The intervertebral disc is composed of two functional structures: a thick outer ring composed of fibrous cartilage, called annulus fibrosus, and a central gelatinous material known as the nucleus pulposus, or nucleus (Fig. 12.6.25). The annulus surrounds the nucleus pulposus. The annulus fibrosus, composed of fibrocartilage which is in crisscross arrangement of the coarse collagen fiber bundles within the fibrocartilage, allows the annulus fibrosus to withstand high bending and torsional loads. The nucleus pulposus lies directly in the center of all discs except those in the

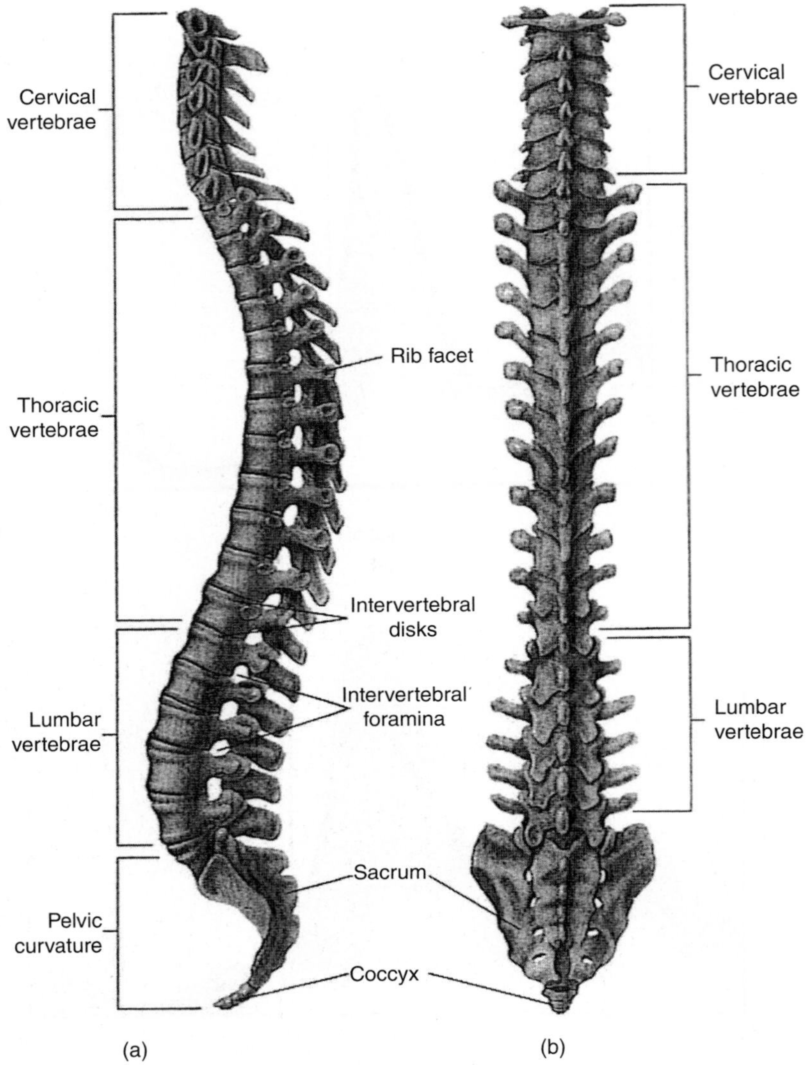

Figure 12.6.22 The spine. (*a*) Lateral view; and (*b*) the posterior view. (*From "Hole's Human Anatomy and Physiology," 1996; Hall, 2003.*)

lumbar segment, where it has a slightly posterior position (Fig. 12.6.25*c*). In young adults, the nucleus is rich in hydrophilic (water-binding) glycosaminoglycans, and it diminishes in glycosaminoglycan content with age and becomes less hydrated. The disc is separated from the vertebral body by the end plate, composed of hyaline cartilage.

The vertebral body is comprised of a thin outer cortical shell and the encompassed cancellous or trabecular bone. The cortical shell and the cancellous bone with the ratio of 45:55 share the load transmitted through the vertebral body, respectively, for an adult of 45 years or younger (Goal and Weinstein, 2000). For subjects over 40 years, the cortical bone shell transmits about 65 percent of the total load exerted on the body, due to an increase in the porosity of the cancellous bone. Table 12.6.4 shows the mechanical properties of the cortical shell and the cancellous bone in the vertebral body. The mechanical properties of ligaments surrounding the vertebrae are given in Table 12.6.5.

Spine Kinematics

The ranges of motion of individual motion segments of the spine have been measured by investigators. The represented values on the relative amount of motion at different levels of the spine are presented by White and Panjabi (1978) in Fig. 12.6.26. The range of flexion and extension is approximately 4° in the upper thoracic motion segments, about 6° in the middle thoracic region, and about 12° in the two lower thoracic segments. This range progressively increases in the lumbar motion segments, reaching a maximum of 20° at the lumbosacral level. The range of motion is strongly age-dependent, decreasing by about 50 percent from youth to old age. The first 50° to 60° of spine flexion occurs in the lumbar spine, mainly in the lower motion segments. The thoracic spine contributes little to flexion of the total spine because of the oblique orientation of the facets. Significant axial rotation occurs at the thoracic and lumbosacral levels.

Spinal Load

During daily activities the disc is subjected to a combination of compression, bending, and torsional loads. While the facets appear to function as a guide for the relative movement of adjacent vertebrae, they in fact carry about 30 percent of the total load when the spine is hyperextended (El-Bohy and King, 1986). The vertebral arches and

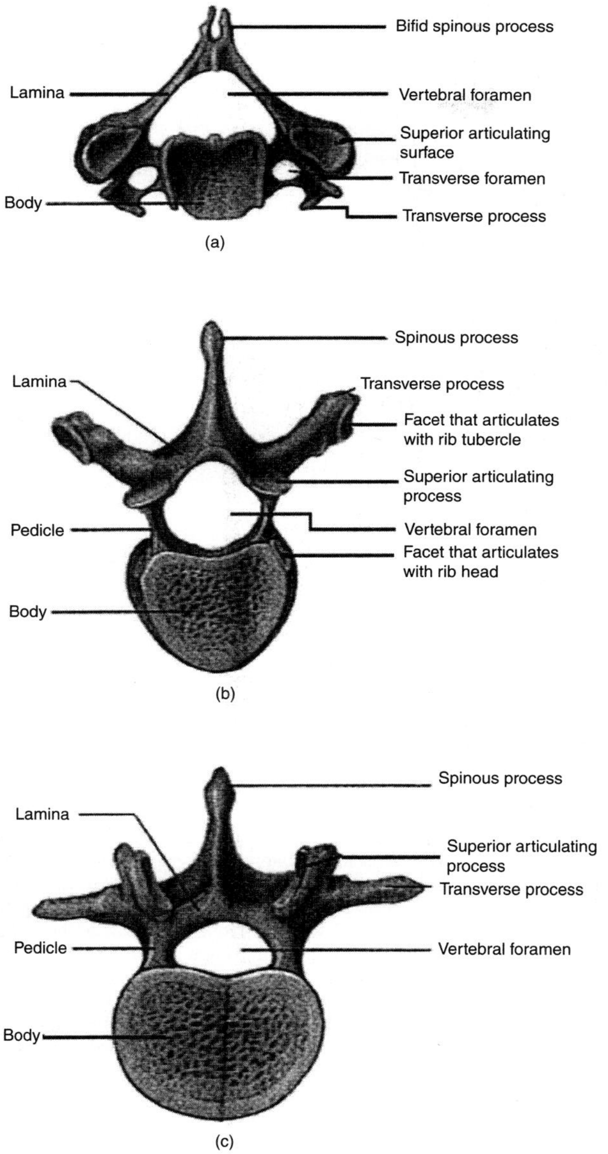

Figure 12.6.23 Superior views of the typical vertebrae. (*a*) Cervical vertebra, (*b*) thoracic vertebra, and (*c*) lumbar vertebra. (*From "Hole's Human Anatomy and Physiology," 1996; Hall, 2003.*)

intervertebral joints play an important role in resisting shear forces. The abdominal muscles and the vertebral portion of the psoas muscle initiate flexion. The weight of the upper body produces further flexion.

During the standing position, the center of gravity of the total body is anterior to the spinal column (passes through L4), which makes the postural muscles active. That is, the spine is under a constant forward bending moment. Therefore, the extensor muscles in the back provide a tension, which counteracts the torque (Fig. 12.6.27). The relative loads on the third lumbar disc for various postures are shown in Fig. 12.6.28. As the body continues to flex, the moment arm of the trunk, arm, head, and neck is increased which leads to the increase in tension in the back extensor muscles. Figure 12.6.29 illustrates the back muscle tension, with approximately a 6-cm lever arm counter to torque created by the weight of the body and the external load. When the body is flexed (forward inclination), the spine makes the disc bulge on the concave side of the spine and retracts on the convex side. That is, the disc protrudes anteriorly and is retracted posteriorly.

When a disc is subjected to a compressive load, in a standing position, the compressive stress predominates in the inner portion of the disc, the nucleus pulposus, and tensile stress predominates in the annulus fibrosus. While a person is lifting an object, the increase in abdominal pressure, contributes to the stiffness of the lumbar spine and thus helps in lifting. While a person is in a sitting position, the loads on the lumbar spine are less if the back is supported, since part of the weight of the upper body is supported by the backrest. The trunk muscles involved in motion contribute to the stability of the spine and maintenance of the posture in an upright position.

During slow walking the increase in loads on the lumbar spine is relatively modest. However, the loads increase with an increase in speed of walking. In his study, Cappozzo (1984) showed that the loads were maximal around toe-off and increased approximately linearly with walking speed. Figure 12.6.30 revealed the axial load on the L3-L4 motion segment in terms of body weight during walking at four speeds. The solid horizontal line in the figure shows the upper body weight (UBW), and the maximum load occurred about the left heel strike (LHS) and right heals strike (RHS).

Biomechanics of Cervical Spine

The cervical spine consists of seven vertebrae. The first two vertebrae, C1 and C2, are atypical, each having a unique structure and role; and the rest of them, C3 through C7, have similar structure and function to those of thoracic and lumbar spine.

The first vertebra in the cervical spine is called the **atlas**, which has no vertebral body and is composed of a ring within which an oval fossa articulates anteriorly with the dens of C2. The superior facets of C1, which form the base of the atlanto-occipital joint, bear the weight of the skull. The second vertebra is called **axis** and is composed of dens, superior facets, pedicle, and spinous process (Fig. 12.6.31). The dens is an articular process that protrudes superiorly from the vertebral body and around which C1 rotates. This configuration allows significant rotation of the head.

The five typical cervical vertebrae C3-C7, each has a body that is elliptical with a transversely concave upper surface, two pedicles, two laminae, and a spinous process. A motion segment consists of two adjacent vertebrae is shown in Fig. 12.6.32 where the ligaments and the intervertebral disc are shown. The intervertebral disc, similar to the lumbar disc, consists of a central gelatinous mass, known as nucleus pulposus, surrounded by a tough outer covering, the annulus fibrosus.

Cervical Spine Kinematics

The cervical spine is the most mobile region of the spine. The range of motion of the cervical spine in flexion-extension is about 145°, about 180° in axial rotation, and about 90° in lateral bending. Specifically, motion at the atlanto-occipital articulation, between C1 and the skull, consists of 10° to 15° of flexion-extension and 8° of lateral flexion, with no rotation. The skull rotation is transferred into motion at the C1-C2 articulation, the atlantoaxial joint. The C1-C2 joint is the most mobile segment of the cervical spine, having up to 47° of axial rotation, representing 50 percent of the axial rotation of the whole cervical spine. However, at this joint, about 10° of flexion and extension takes place with minimal lateral flexion (White and Panjabi, 1978). Flexion, extension, lateral flexion, and axial rotation occur at the joints below C2. A total of +/–45° of axial rotation takes place in C3-C7. Note that the range of motions varies widely among individuals.

The cervical spine stability has been discussed extensively in White and Penjabi's publications. Instability is defined as excessive relative movements of adjacent vertebrae where there is a need for clinical procedure to return the vertebrae to their normal positions. They suggested that the adult cervical spine is unstable, or on the brink of instability,

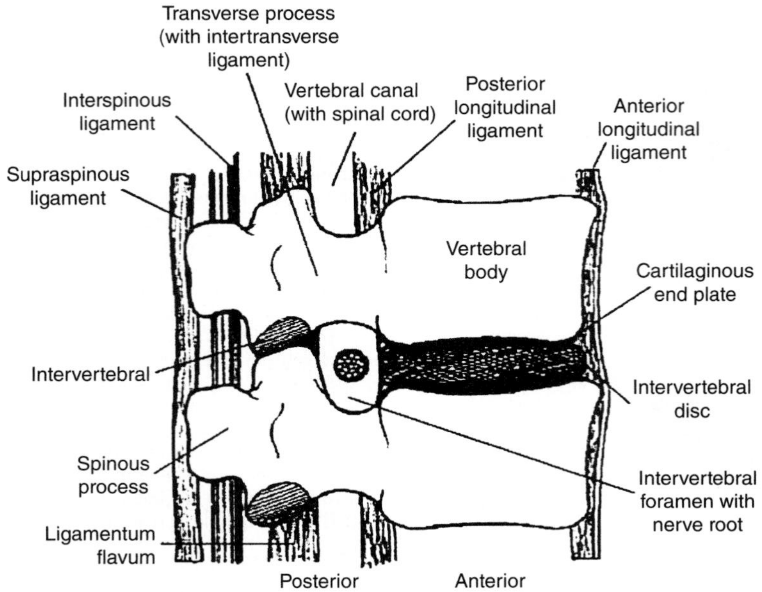

Figure 12.6.24 The vertebral motion segment and the associated soft tissues. (*From Hall, 2003.*)

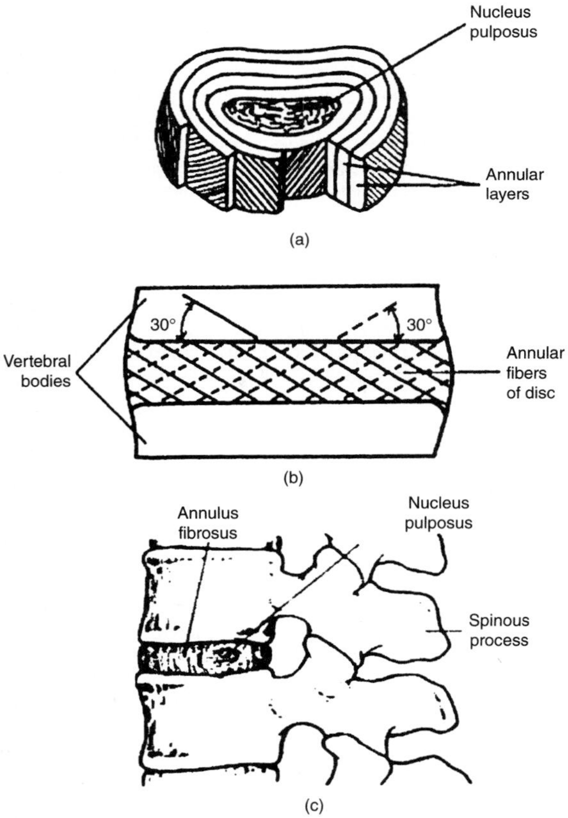

Figure 12.6.25 Schematic drawings of the intervertebral disc. (*a*) Concentric layers of the annulus fibrosus. (*From Nordin and Frankel, 1989, by permission.*) (*b*) Orientation of the annulus fibrosus. (*From Nordin and Frankel, 1989, by permission.*) (*c*) The lateral view of the lumbar vertebral motion segment. (*From Hall, 2003.*)

when any of the following conditions is present: more than 3.5 mm of transverse (horizontal) relative displacement of two adjacent vertebrae, or more than an 11° difference in rotation from that of either adjacent vertebra. These cervical conditions need clinical attention to be corrected.

Cervical Spine Load

Distribution of loads throughout the vertebra is a complex issue, which has been studied by several investigators. The results of these studies indicate that the facets bear a portion of the load (King et al., 1975; Miller et al., 1978). In particular, in extension, the facets bear up to 33 percent of the total load, reducing substantially the load of the discs. Harms-Ringdahl (1986) calculated the bending moment generated around the axes of motion of the atlanto-occipital joint (Occ-C1) and at the C7-T1 motion segments. Figure 12.6.33 shows the extension and flexion moments at these joints in a variety of head positions. During extreme flexion the maximum bending moment in the neck occurs at the C7-T1 joint, while the moment at the Occ-C1 joint is relatively high. As the head moves to its natural position, the moment at the Occ-C1 joint slightly decreases while the moment at C7-T1 significantly decreases so that both joints carry approximately equal amounts of load. As the head continues to rotate and approaches the extreme extension, the moment on the Occ-C1 is further reduced while that of C7-T1 increases in the negative direction.

The head and neck, due to their structures, that is, a large mobile ball atop a column, are particularly vulnerable to dynamic injuries. Excessive loading and motion of the head can easily create stresses and bending moments that exceed the strength of the structure, which leads into subluxation and destabilization of the neck.

During normal axial rotation the size of the canal is significantly decreased because the oval shape of the central foramen of C1 does not coincide with that of C2.

BIOMECHANICS OF THE SHOULDER

The shoulder is the most complex joint in the human body. It consists of four major separate articulations, as described below, and is shown in Fig. 12.6.34.

Table 12.6.4 Mechanical Properties of a Vertebral Body

Cortical bone		
Modulus of elasticity, N/mm² or MPa	Shear modulus, N/mm² or MPa	Poisson's ratio (ν)
12,000	4,615.0	0.30[a]

Spongy or Cancellous Bone	Males[b]	Females[b]	[a]	[c]
Tensile strength, N/mm² or MPa	—	—	—	1.18
Compressive strength, N/mm² or MPa	4.6(0.3)[d] (0.2-10.5)[e]	2.7(0.2) (0.3-7.0)	—	1.37-1.86
Compression at rupture, %	9.5(0.4) (5.3-14.4)	9.00(0.6) (3.2-14.7)	—	2.5
Limit of proportionality, N/mm² or MPa	4.0(0.1) (0.1-9.7)	2.2(0.1) (0.2-6.0)	—	
Compression at limit of proportionality, %	6.7-(0.2) (4.1-8.6)	6.1(0.4) (2.6-10.0)	—	
Modulus of elasticity, N/mm² or MPa	55.6(0.7) (1.1-139.1)	35.1(0.6) (5.2-103.6)	100	68.7-88.3
Shear modulus, N/mm² or MPa			41	
Poisson's ratio ν			0.2	0.14

[a]Shirazi-Adl et al. (1984). [b]Yamada (1970). [c]Lindhal (1976). [d]Mean (SD). [e]Range.
SOURCE: Goal and Weinstein (2000).

Table 12.6.5 Mechanical Properties of Spinal Elements

Property	Ligament parameter	ALL	TL	CL	LF	ISL	SSL
Structural	X-area, mm²	63.7	3.0	20.9	56.0	75.7	30.1
		(11.3)	(0.4)	—	(7.0)	(41.3)	(8.8)
	Load at failure, N	177.0	15.0	191.5	87.0	90.5	118.3
		(30.0)	(1.0)	(58.2)	(24.7)	(44.5)	(68.9)
	Stiffness, N/mm	164.0	3.7	49.7	66.5	41.5	55.1
		(10.3)	(0.9)	(21.9)	(43.5)	(1.5)	(45.9)
Material	Stress at failure, MPa	1.5	3.6	(3.0)	1.3	1.3	3.6
		(0.1)	(1.3)	—	(0.1)	(0.1)	(2.2)
	Stress at failure, %	8.2	17.1	9.0	13.0	7.8	7.8
		(0.2)	(1.2)	(0.8)	(1.0)	(1.4)	(0.8)
	Young's modulus, MPa	20.1	58.7	32.9	19.5	11.6	62.7
		(2.2)	(26.5)	—	(0.5)	(3.5)	(37.5)

SOURCE: Goal and Weinstein (2000) with permission.

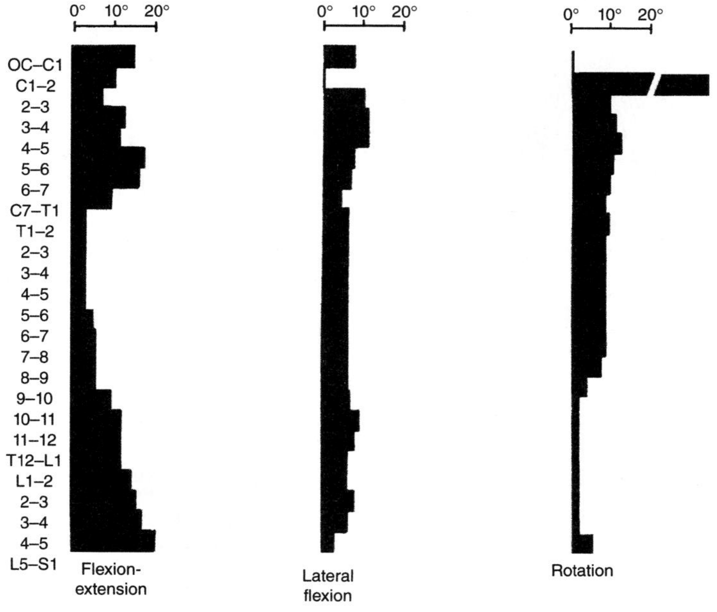

Figure 12.6.26 Representative values for type and range of motion at different levels of spine. (*Adapted from White and Panjabi, 1978; Nordin and Frankel, 1989, by permission.*)

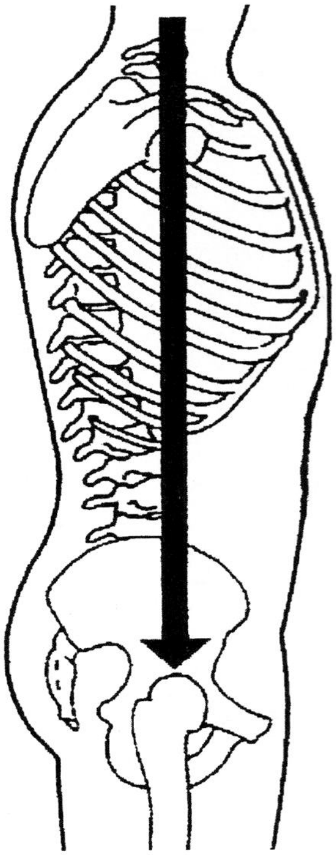

Figure 12.6.27 Schematic lateral view of the human body; the line of gravity passes through L4. (*From Hall, 2003.*)

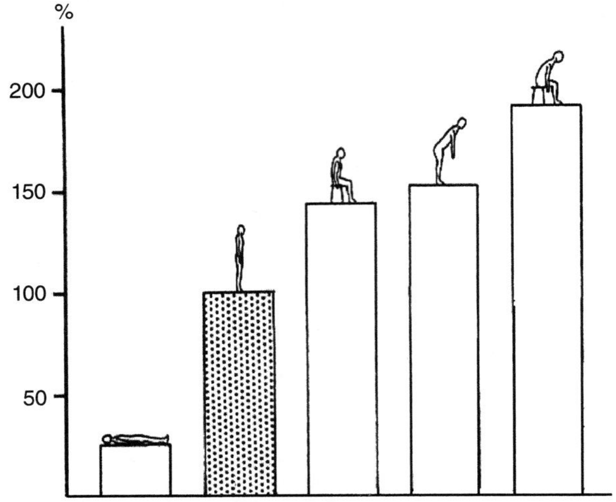

Figure 12.6.28 The relative loads on the third lumbar disc for various postures. (*From Nordin and Frankel, 1989, by permission.*)

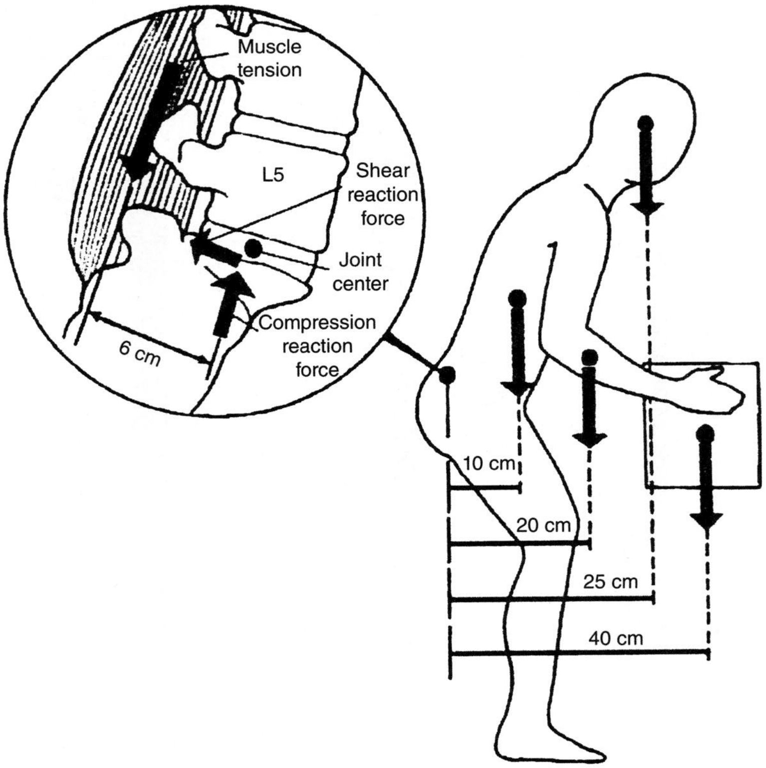

Figure 12.6.29 Free-body diagram of spine and body loads when subject lifts an external load. The back muscle with a moment of arm of approximately 6 cm must support the moment produced by the body and the load. (*From Hall, 2003.*)

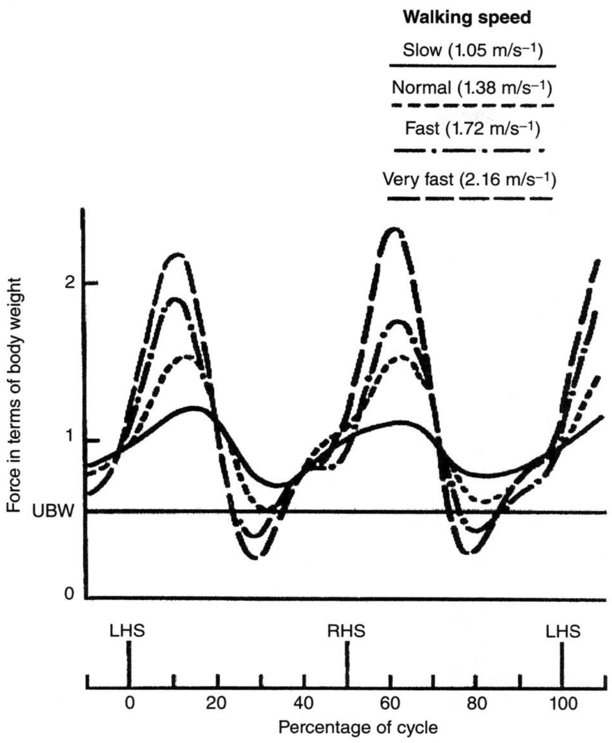

Figure 12.6.30 Variation of axial load on L3-L4 motion segment based on the body weight during walking at four speeds. The horizontal line shows the upper body weight (UBW), which represents the gravitational component of the load. (*From Nordin and Frankel, 1989, by permission.*)

1. At the sternoclavicular joint, the proximal end of the clavicle articulates with the clavicular notch of the sternum along with the cartilage of the first rib.

2. At the acromioclavicular joint, the acromion process of the scapula (shoulder blade) articulates with the distal end of the clavicles.

3. At the glenohumeral joint, the hemispherical head of the humerus articulates with the glenoid fossa of the scapula. This is a ball- and-socket configuration joint and is the most freely moving joint in the human body, enabling flexion, extension, hyperextension, abduction, adduction, horizontal abduction and adduction, and medial and lateral rotation of the humerus. There are four major muscles on the posterior, superior, and anterior sides of the joint, providing the joint movements; they are called rotator cuff muscles. The capsule of the joint is surrounded and is stabilized by several ligaments and the tendons of four muscles, as well as the biceps. Negative pressure within the capsule of the glenohumeral joint also helps to stabilize the joint.

4. At the scapulothoracic joint, the anterior scapula is articulated with the thoracic wall. This is due to the movement of the scapula in the sagittal and frontal planes.

Shoulder Loads

The ball-and-socket configuration of glenohumeral joint provides mainly the rotation of the shoulder; however, it includes some rolling and gliding motion as well. The rotator cuff muscles are oriented such that their tendons and muscle masses may push on the head of the humerus, thereby stabilizing the joint.

When the arm is in 90° of abduction (extended horizontally in the frontal plane), the deltoid muscle (superior of the shoulder joint) provides a tension of about one-half of the body weight. This can be calculated by a simple static equilibrium of the joint, as shown in Fig. 12.6.35. When a light weight of about 0.025 times the body weight (BW) is held in the hand, the muscle force is approximately equal to the body weight. In these cases the joint reaction force, which is equal and opposite to the muscle force, is as high as the body weight. Therefore, the glenohumeral joint is a major load-bearing joint.

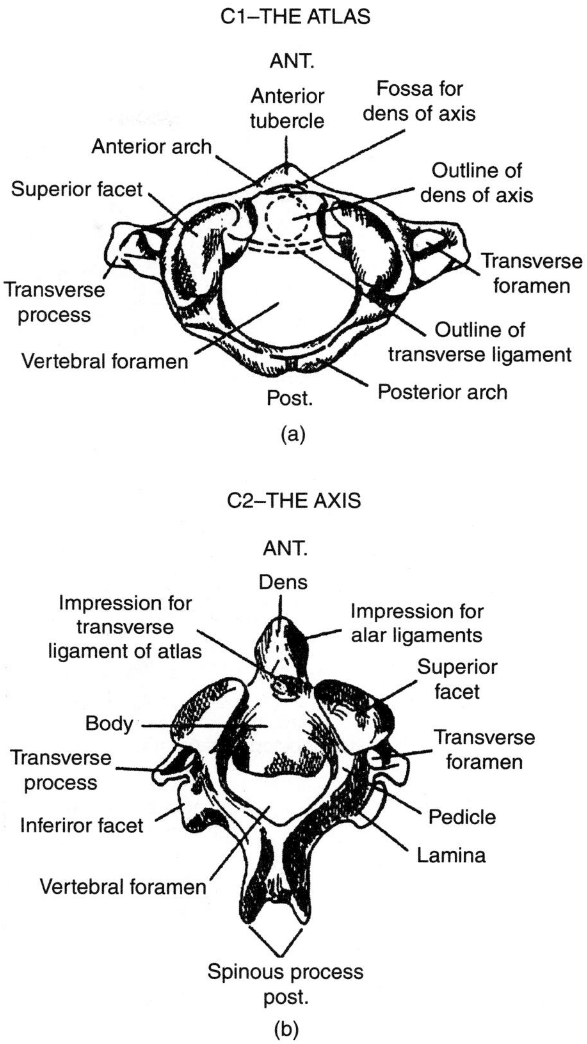

Figure 12.6.31 Superior views of the atypical cervical vertebrae (*a*) C1 and (*b*) C2. (*From Nordin and Frankel, 1989, by permission.*)

THE ELBOW

At the elbow joint, three bones are joined: humerus, ulna, and radial. The elbow joint consists of three articulations: the humeroulnar, humeroradial, and proximal radioulnar. All these articulations are enclosed in the same joint capsule, which is reinforced by the anterior and posterior radial collateral and ulnar collateral ligaments. The elbow joint allows 2° of freedom in motion: flexion-extension and pronation-supination (rotation of the radius around the ulna).

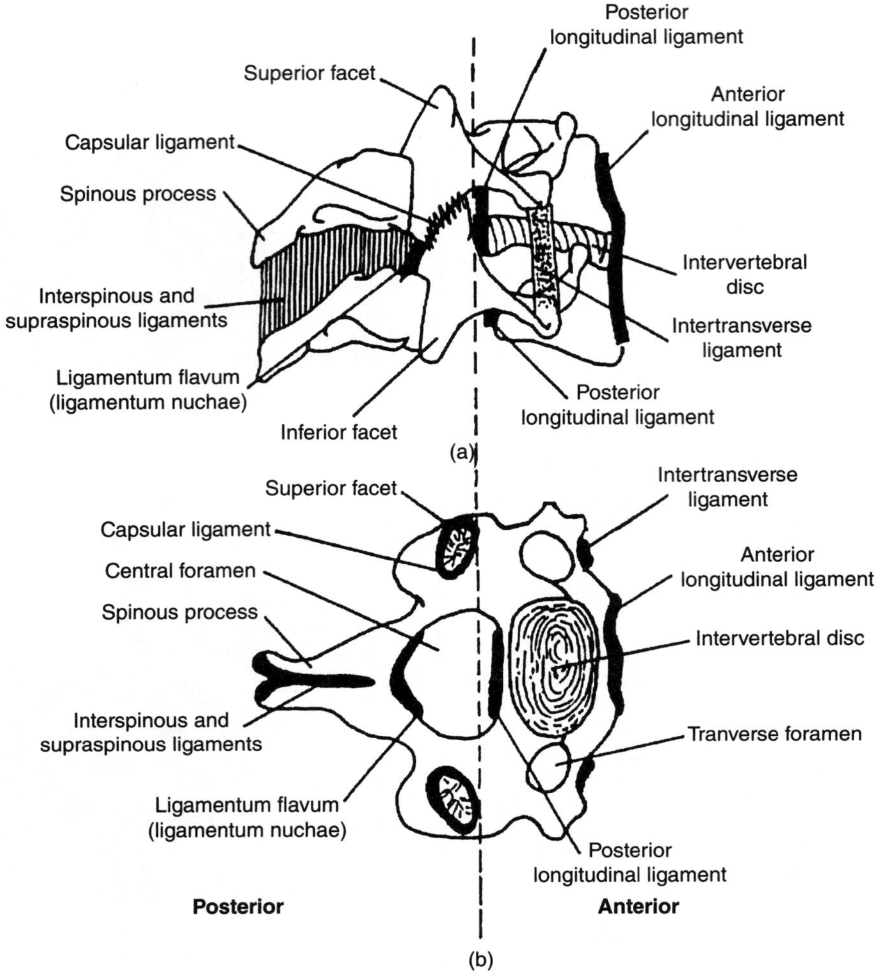

Figure 12.6.32 Lateral and superior schematic views of a cervical motion segment of a typical vertebra. (*Adapted from White et al., 1975; Nordin and Frankel, 1986, by permission.*)

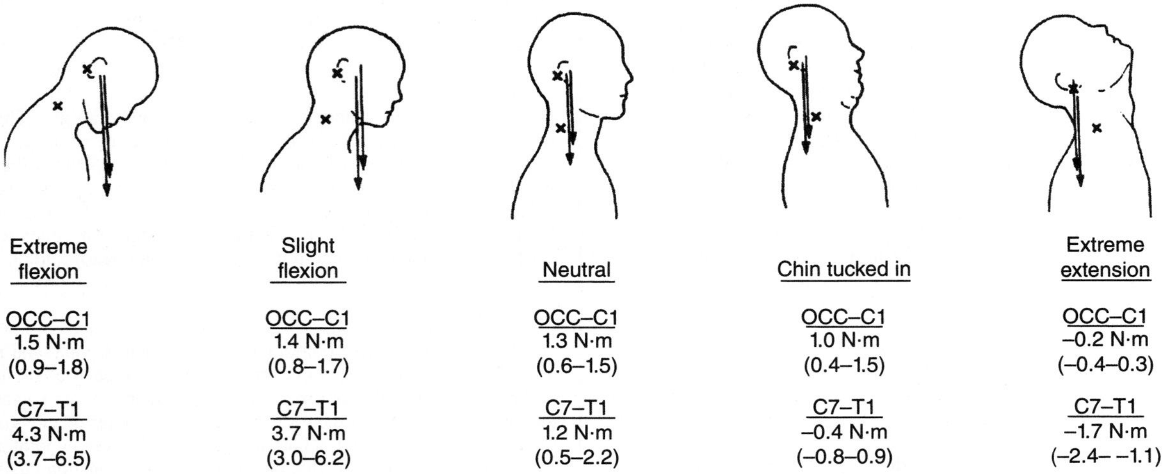

Figure 12.6.33 Extension-flexion moments at atlanto-occipital (Occ-C1) joint and C7-T1 motion segment (marked with X's) for five positions of the head. The arrow represents the force vector produced by the weight of the head. (*Adapted from Harms-Ringdahl, 1986; Nordin and Frankel, 1989, by permission.*)

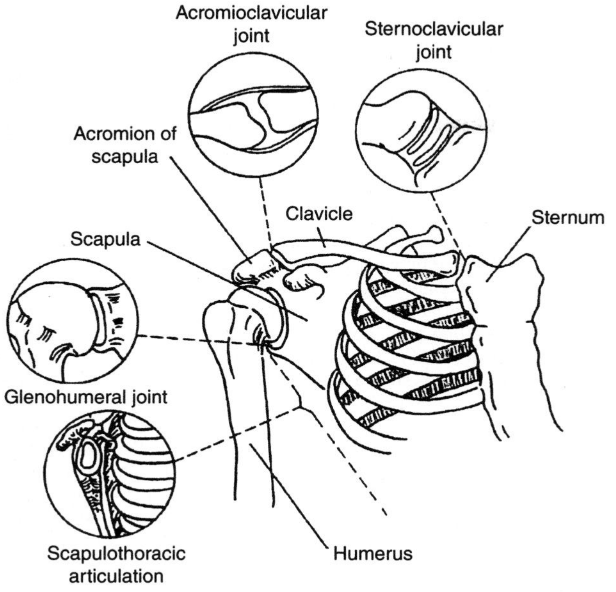

Figure 12.6.34 Biomechanics of the shoulder. Schematic view of the shoulder joint and the four articulations. (*Adapted from DePalma, 1983; Nordin and Frankel, 1989, by permission.*)

Elbow Load

The elbow is not considered a weight-bearing joint; however, it is regularly subjected to large loads during our daily activities. Two simple diagrams of Fig. 12.6.36 show the muscle forces and joint reaction forces, which could be calculated by a simple static equilibrium. The research has shown that the compressive loads during activities such as dressing and eating could reach as high as 300 N (67 lb), and during rising from a chair when the body is supported by the arms, the load could reach as high as 1,700 N (382 lb). During push-up exercises peak forces on the elbow could reach as high as 45 percent of the body weight.

BIOMATERIALS

By definition, **biomaterial** is a biocompatible material suitable for inclusion, augmentation, or replacement of the function of bodily

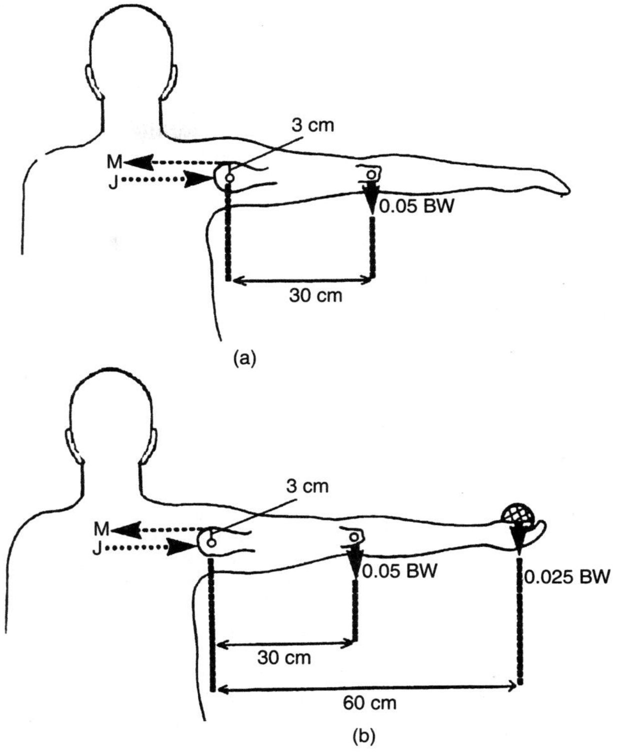

Figure 12.6.35 Free-body diagram of the shoulder joint, showing the reaction force and the deltoid muscle force. (*a*) Without the load; (*b*) with a load in the hand. (*Adapted from Nordin and Frankel, 1989, by permission.*)

tissues or organs. Biomaterials science is a multidisciplinary field that brings together researchers from diverse academic backgrounds, such as bioengineering, chemistry, chemical engineering, electrical engineering, mechanical engineering, materials science, biology, microbiology, and medicine. Biomaterials can be synthetic and natural, solid and sometimes liquid. Solid biomaterials can be metals, ceramics, polymers, glasses, carbons, and composite materials. Such materials are used as molded or machined parts, coatings, fibers, films, foams, and fabrics.

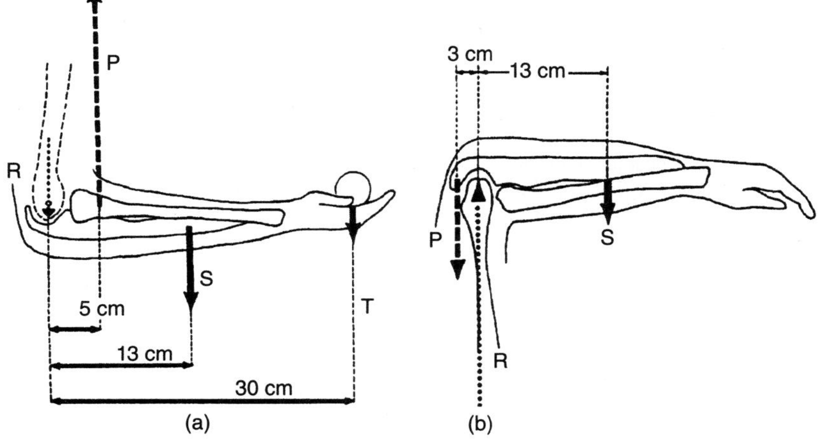

Figure 12.6.36 Free-body diagram of the elbow joint showing (*a*) the reaction force in flexion and the tendons of the biceps and the brachialis muscle forces and (*b*) the tendon of the extensor (triceps) muscle tendon. (*From Nordin and Frankel, 1989, by permission.*)

Biomaterials are generally integrated into medical devices and implants and are rarely used on their own.

Historically, in the early 1900s bone plates were successfully implemented to stabilize bone fractures and to accelerate their healing, and since 1960s blood vessel replacement and artificial heart valves and joint replacements have been used. Since then, with the advent of new materials and prostheses, the field of biomaterials has rapidly expanded and has saved many lives.

Key factors in the design and use of biomaterials are biocompatibility, toxicology, biofunctionality, and, to a lesser extent, availability of biomaterials. Except for their brittle characteristics, ceramic materials are ideal candidates as biomaterials. Most biomaterials and medical devices perform satisfactorily, improving the quality of life for the recipient or saving lives; however, they have exhibited some failures. This is due to the fact that manufactured materials and devices have a failure rate. Other factors such as genetic differences, gender, body chemistries, living environment, and physical activity as well as physicians' "talent" for implanting devices may contribute to the success or failure of the biomaterials and implants.

Biocompatibility

The understanding and measurement of the biocompatibility of biomaterials are not fully understood. It has been accepted that no foreign material placed within a living body is completely compatible. Only those materials (natural tissues) manufactured by the body itself (autogenous) are compatible, and any other substance that is recognized as foreign initiates some type of reaction (host-tissue response). Natural material that is obtained from an individual and will be implanted into the same person is called an **autograft.** However, if the donor is a different individual, the material is called an **allograft.**

Toxicity

Toxicology in biomaterials has evolved into a sophisticated science, which deals with the substances that migrate out of biomaterials. For example, for polymers, many low-molecular-weight materials have shown some level of physiologic activity and cell toxicity. It is reasonable to say that a biomaterial should not give off anything from its mass unless it is specifically designed to do so, for example, cancer drug delivery systems to destroy specific cells.

In terms of tissue response to the biomaterials and implants, there are three types of biomaterials: bioinert, bioresorbable, and bioactive. (Buddy, Ratner et al., 1996).

Bioinert Any material that once placed in the human body has minimal interaction with its surrounding tissue is called **bioinert.** For example, stainless steel, titanium, alumina, partially stabilized zirconia, and ultrahigh-molecular-weight polyethylene (UHMWPE) (to some degree) are bioinert. A fibrous capsule that might form around bioinert implants may compromise the biofunctionality of bioinert materials. In addition, the tissue remodeling around the implant (bone ingrowth or architectural changes such as resorption or depositon of bone) may alter the functionality of the implant.

Bioresorbable Materials that upon placement within the human body start to dissolve (are resorbed) and slowly are replaced by advancing tissue (such as bone) are **bioresorbable.** Common examples of bioresorbable materials are tricalcium phosphate [$Ca_3(PO_4)_2$] and polylactic–polyglycolic acid copolymers. Calcium oxide, calcium carbonate, and gypsum are other common materials that have been utilized during the last three decades.

Bioactive Materials that upon placement in the human body interact with the surrounding bone and in some cases even soft tissue are **bioactive.** This occurs through a time-dependent kinetic modification of the surface, triggered by their implantation within the living bone. An ion- exchange reaction between the bioactive implant and the surrounding body fluids results in the formation of a biologically active carbonate apatite layer on the implant. For example, coating an implant with synthetic hydroxyapatite [$Ca_{10}(PO_4)_6(OH)_2$] triggers bone ingrowth.

OCCUPATIONAL BIOMECHANICS

Human Factors in Design

In interacting with machines and devices, the human musculoskeletal system often has to provide forces to power a product or actuate a controller. The designs of machines and devices are influenced by humans' ability to supply the force and their ability to interact with the machine. The interface between humans and machines requires that humans sense the state of the device and be able to control it. That is, a product must be comfortable for a person to use. Thus products and devices must be designed with human factors in mind [see Bridger (1995)]. The geometric properties of humans (i.e., their height, reach, and seating requirements; the size of the holes they can fit through; and the forces that they can exert in specific directions with their body segments) are called anthropometric data. The armed forces in Military Standard MIL-STD-1472 have collected the majority of these data. A sample of these data is shown in Fig. 12.6.37 and Table 12.6.6. In addition, occupational biomechanics researchers have studied measures for humans' ability to perform various activities. Figure. 12.6.38 shows the average human strength for differing body positions (Dreyfuss, 1967). For example, in the standing position, the push and pull force of the hand in the position slightly lower than the torso region is higher than that of the hand position at the neck level. More detailed information is available in MIL-HDBK (Military Handbook) 759A and books in occupational biomechanics.

PROSTHESES AND IMPLANTS

There are implants and prostheses for virtually every member and organ of the human body. Being among the oldest implants, hip and knee prostheses are discussed in this subsection.

Hip Implants

The hip prosthesis has been the most active area of joint replacement research since **total hip** joint **replacement (THR)** among the most common orthopedic procedures. The hip joint is called a ball-and-socket joint because the spherical head of the femur moves inside the cup-shaped hollow socket (acetabulum) of the pelvis. To duplicate this action, a total hip replacement implant has three parts: the stem, which fits into the femur and provides stability; the ball, which replaces the spherical head of the femur; and the cup, which replaces the worn-out hip socket (Fig. 12.6.39). There are many designs of hip implant parts based on age, weight, bone quality, activity level, and health of the recipient. The stem portions of most hip implants are made of titanium- or cobalt/chromium-based alloys; they come in different shapes and degrees of roughness. Cobalt/chromium-based alloys or ceramic materials (aluminum oxide or zirconium oxide) are used in making the ball portions, which are polished smooth to allow easy rotation within the prosthetic socket. The acetabular socket can be made of metal, with a liner made of ultrahigh-molecular-weight polyethylene (UHMWPE) or a combination of polyethylene backed by metal. All together, these components weigh between 14 and 18 oz, depending on the size needed. The chance of a hip replacement lasting 20 years is about 80 percent.

There are many hip implant designs (such as modular stems) and a variety of surface textures (such as sintered beads, meshes, and etched). For more information the reader is referred to manufacturers' Web sites.

Implant Fixation

To hold the femoral and acetabular components in place, two methods, cemented and cementless (bone ingrowth or osseointegration), are used. The cement, an acrylic polymer called polymethylmethacrylate (PMMA), is used for older patients. However, the cementless method is used for younger patients (under 50 years of age) who are active and have good bone quality. Cemented fixation relies on a stable interface between the prosthesis and the cement and a solid mechanical bond between the cement and the bone.

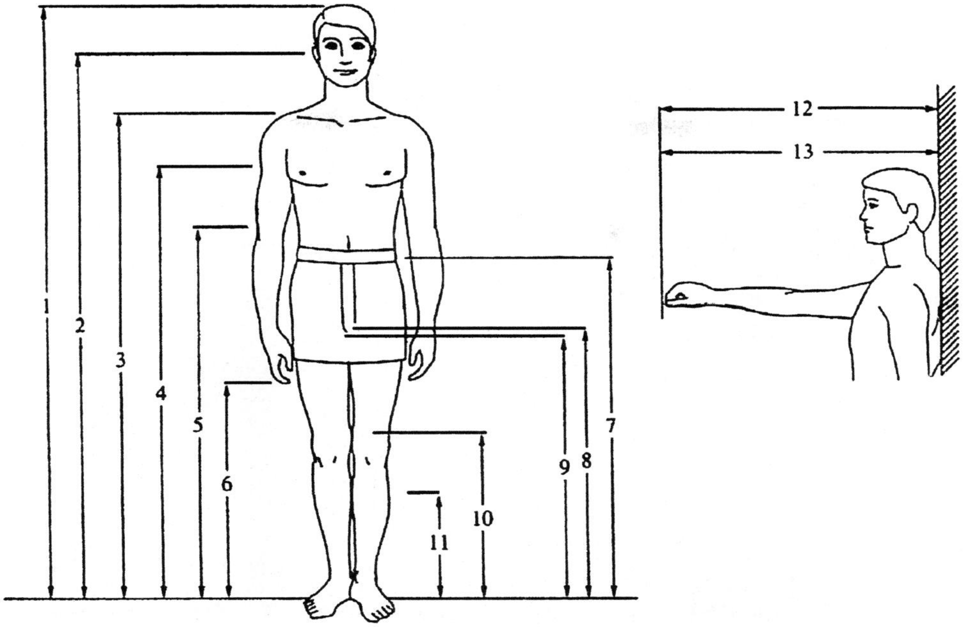

Figure 12.6.37 Anthropometric man. (*From MIL-STD-1472D.*)

Table 12.6.6 Anthropometric Data

	Percentile values, cm					
	5th percentile			95th percentile		
	Ground troops	Aviators	Women	Ground troops	Aviators	Women
Weight, kg	65.5	60.4	46.6	91.6	96.0	74.5
Standing body dimensions						
1. Stature	162.8	164.2	152.4	185.6	187.7	174.1
2. Eye height (standing)	151.1	152.1	140.9	173.3	175.2	162.2
3. Shoulder (acromial) height	133.6	133.3	123.0	154.2	154.8	143.7
4. Chest (nipple) height*	117.9	120.8	109.3	136.5	138.5	127.6
5. Elbow (radiate) height	101.0	104.8	94.9	117.8	120.0	110.7
6. Fingertip (dactylion) height		61.5			73.2	
7. Waist height	96.6	97.6	93.1	115.2	115.1	110.3
8. Crotch height	76.3	74.7	68.1	91.8	92.0	83.9
9. Gluteal furrow height	73.3	74.6	66.4	87.7	88.1	81.0
10. Kneecap height	47.5	46.8	43.8	58.6	57.8	52.5
11. Calf height	31.1	30.9	29.0	40.6	39.3	36.6
12. Functional reach	72.6	73.1	64.0	90.9	87.0	80.4
13. Functional reach, extended	84.2	82.3	73.5	101.2	97.3	92.7
	Percentile values, in					
Weight, lb	122.4	133.1	102.3	201.9	211.6	164.3
Standing body dimensions						
1. Stature	64.1	64.6	60.0	73.1	73.9	68.5
2. Eye height (standing)	59.5	59.9	55.5	68.2	69.0	63.9
3. Shoulder (acromial) height	52.6	52.5	48.4	60.7	60.9	56.6
4. Chest (nipple) height*	46.4	47.5	43.0	53.7	54.5	50.3
5. Elbow (radiate) height	39.8	41.3	37.4	46.4	47.2	43.6
6. Fingertip (dactylion) height		24.2			28.8	
7. Waist height	38.0	38.4	36.6	45.3	45.3	43.4
8. Crotch height	30.0	29.4	26.8	36.1	36.2	33.0
9. Gluteal furrow height	28.8	29.4	26.2	34.5	34.7	31.9
10. Kneecap height	18.7	18.4	17.2	23.1	22.8	20.7
11. Calf height	12.2	12.2	11.4	16.0	15.5	14.4
12. Functional reach	38.6	28.8	25.2	35.8	34.3	31.7
13. Functional reach, extended	33.2	32.4	28.9	39.8	38.3	36.5

*Bus point height for women.
SOURCE: Dreyfuss (1967).

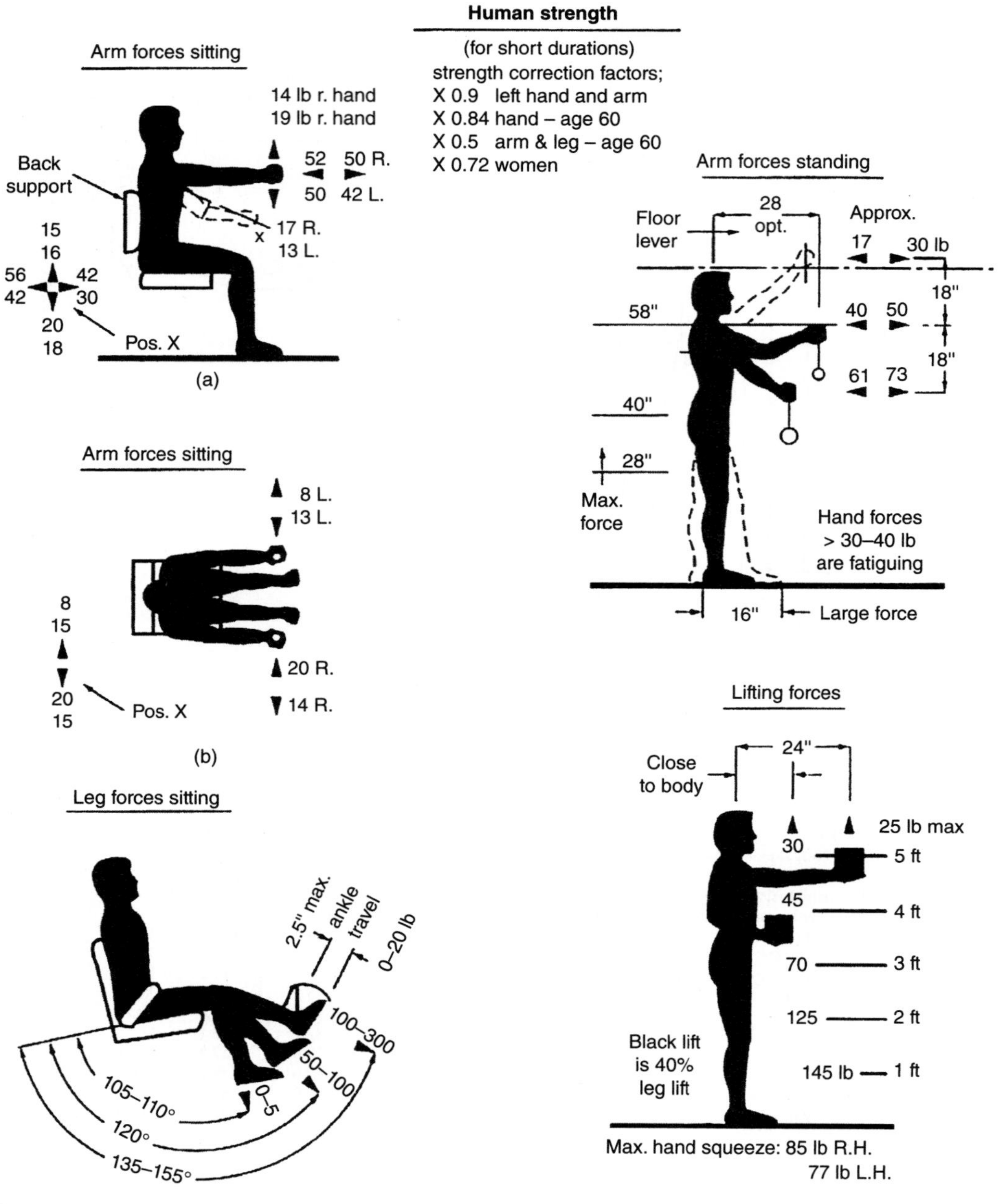

Figure 12.6.38 Average human strength for different tasks. (*From H. Dreyfuss, 1967.*)

Cementless implants were introduced in the 1980s, and the implant is directly attached to bone without the use of cement. The surface topography of the implant is textured or has porous regions, and the implant is generally coated with hydroxyapatite so that the new bone actually grows into the surface of the implant. In general, these designs are larger and longer than those used with cement. The surgeon's accuracy in placing the implant is an important factor in the stability of the implant since the channel (bone cavity) must match the shape of the implant precisely. Intimate contact between the component and bone is crucial to permit bone ingrowth. New bone growth cannot bridge gaps larger than 1 to 2 mm.

Hybrid THR, which was introduced in the early 1980s, has one component, usually the acetabular socket, inserted without cement and the other component, usually the femoral stem, inserted with cement. A hybrid hip takes advantage of the excellent track records of cementless hip sockets and cemented stems.

Loosening of hip implants may occur due to several factors.

1. Cracks (fatigue fractures) in the cement that occur over time can cause the prosthetic stem to loosen and become unstable. This is more often the case with patients who are very active or very heavy.

2. The action of the metal ball against the polyethylene cup of the acetabular component creates polyethylene wear debris. The microscopic debris particles are absorbed by cells around the joint and initiate an inflammatory response from the body, which tries to remove them.

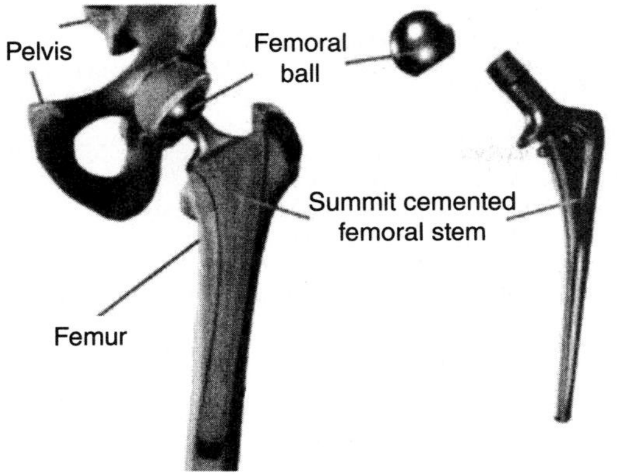

Figure 12.6.39 Total hip implant.

3. Micromotion between the implant and the bone prevents bone ingrowth into the implant.

The hip implants could be made of two or three types of materials, namely, metal, ceramic, and polyethylene. The crucial part of the implant is the contact between the ball and the acetabular liner. The ball could be made of metal or ceramic, and the liner could be made of ceramic, metal, and UHMWPE. The metal and plastic implants wear at a rate of about 0.1 mm/yr. Metal-on-metal implants wear at a rate of about 0.01 mm/yr about 10 times less than that of metal and plastic. There are also concerns about the wear debris generated from the metal-on-metal implants. Ceramics are the most resistant to wear of all available hip replacement implants. However, due to low fracture toughness, ceramic implants can break inside the body.

The Knee Implants

The **total knee arthroplasty (TKA)** implants are more complex because of the articulation of three bones—distal femur, proximal tibia, and patella (the kneecap)—where the surfaces actually roll and glide as the knee bends. There are many knee implant designs in the market having the following major components (Fig. 12.6.40).

1. **Femoral component.** The metal femoral component curves around the distal end of the femur and has an interior groove so the kneecap can move up and down smoothly against the bone as the knee bends and straightens. In some designs, a smaller piece may be used (unicompartmental knee replacement) to resurface just the medial or lateral part of the condyle.
2. **Tibial component.** The tibial component is a flat metal platform.
3. **Polyethylene cushion.** The cushion may be part of the platform (fixed) or separate (mobile).
4. **Patellar component.** The patellar component is a dome-shaped piece of polyethylene that duplicates the shape of the kneecap anchored to a flat metal plate.

The metal parts of the implant are made of titanium- or cobalt/chromium-based alloys. The plastic parts are made of ultrahigh-molecular-weight polyethylene. All together, the components weigh between 15 and 20 oz, depending on the size selected.

Due to its continuous sliding and rolling, knee implant wear is a major problem where microscopic particulate debris are generated.

Similar to the hip implants, the knee implants may be cemented, cementless, or hybrid, depending on the type of fixation used to hold the implant in place.

Other Implants

The majority of orthopedic implants are made of metals. There is considerable mismatch between the metal and bone properties, which can lead to stress shielding and bone resorption. New composite implants or graded materials are being developed to closely match the bone properties and thereby reduce the risk of bone resorbtion.

There are many other orthopedic implants and prostheses for virtually every joint in the body as well as plates and rods for bone repair and

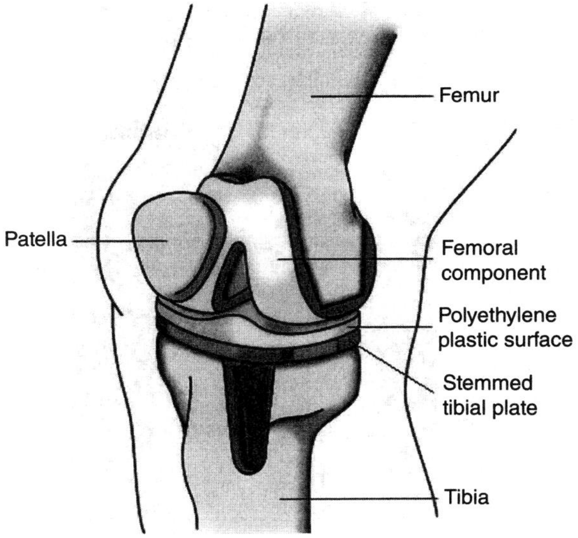

Figure 12.6.40 Components of the total knee arthroplasty implant.

fracture fixations. In addition, there are many types of screws for different locations and varieties of bones. For these topics the reader is referred to specialized orthopedic books, journals, and manufacturers' Web sites.

Dental implants have been widely used in recent years. There are two categories of dental implants: endosteal implants, which enter the bone tissue, and subperiosteal systems, which contact the exterior bone surfaces (Lemons, 1988).

Mechanical or tissue **prosthetic heart valves** are used to replace the natural heart valves (miteral, tricuspid, aortic, and pulmonic) when these no longer perform their normal functions because of disease (Gidden et al., 1993). Mechanical valves are of two types: caged ball and tilting disk. The materials most widely used are silicone elastomer, cobalt/chrome-based alloys, titanium, and pyrolytic carbon.

A **stent** is a device made of inert materials and designed to serve as a temporary or permanent internal scaffold to maintain or increase the lumen of a vessel. For further information see Schatz (1989).

REFERENCES

1. Andriacchi, T., et al. 1980. A Study of Lower-Limb Mechanics during Stair-Climbing, *Jour. Bone Joint Surg.* **62A**:749-757.
2. Aydin, Tozeren, 2000. "Human Body Dynamics," Springer-Verlag.
3. Berger, S. A., et al., 2000. "Introduction to Bioengineering," Oxford Press.
4. Bergmann, G., et al., 1997. Hip Joint Forces during Load Carrying, *Clin. Orthop.* **335**:190.
5. Bergmann, G., et al., 2001. Hip Contact Forces and Gait Patterns from Routine Activities, *Jour. Biomech.* **34**:859.
6. Bogduk, N., and S. R. Mercer, 2000. Biomechanics of the Cervical Spine, I. Normal Kinematics, *Clin. Biomech.* **15**:633.
7. Brault, J., et al., 2000. Cervical Muscle Response during Whiplash: Evidence of a Lengthening Muscle Contraction, *Clin. Biomech.* **15**:426.
8. Bridger, R. S., 1995. "Introduction to Ergonomics," McGraw-Hill.
9. Buddy, Ratner, et al., 1996. "Biomaterials Science: An Introduction to Materials in Medicine," Academic Press.
10. Cappozzo, A., 1984. Compressive Load in the Lumbar Vertebral Column during Normal Level Walking, *Jour. Orthop. Res.* **1**:292.
11. Chaffin, Don, et al., 1999. "Occupational Biomechanics," Wiley.
12. Cholewicki, J., et al., 1999. Intra-abdominal Pressure Mechanism for Stabilizing the Lumbar Spine, *Jour. Biomech.* **32**:13.
13. Cowin, S., et al, 1992. An Evolutionary Wolff's Law for Trabecular Architecture, *Jour. of Biomech. Eng.* **116**:129-136.

14. Davis, K. G., and W. S. Marras, 2000. The Effect of Motion on Trunk Biomechanics, *Clin. Biomech.* **15**:703.

15. DePalma, A. F., 1983. Biomechanics of the Shoulder, in "Surgery of the Shoulder," 3d ed., J. B. Lippincott, Philadelphia, pp. 65-85.

16. Dreyfuss, H., 1967. "The Measure of Man: Human Factors in Design," Whitney Library of Design, New York. This is a looseleaf book of 30 anthropometric and biomedical charts. A classic.

17. Drillis, R., and R. Contini, 1966. "Body Segment Parameters," Rep. 1163-03, Office of Vocational Rehabilitation, Dept. of Health, Education, and Welfare, New York.

18. El-Bohy, A., and A. King, 1986. Intervertebral Disc and Facet Contact Pressure in Axial Torsion, in "Adv. in Bioeng." Lantz and King, (eds.), ASME, pp. 26-27.

19. Fung, Y. C., 1993. "Biomechanics of Mechanical Properties of Living Tissues," Springer-Verlag.

20. Gibson, L., and M. Ashby, 1997. "Cellular Solids: Structures and Properties," 2d ed., Pergamon Press, Oxford, U.K.

21. Gidden, Yagonathan, and F. J. Schoen, 1993. Prosthetic cardiac valves, Special supplement to *Cardiovascular Pathology*, **2**(3), PP167S-177S.

22. Goal, Vijay, and James, Weinstein, 2000. "Biomechanics of the Spine: Clinical and Surgical Perspective," CRC Press.

23. Hall, Susan, 2003. "Basic Biomechanics," 4th ed., McGraw-Hill.

24. Hall, S. J., 1985. Effect of Attempted Lifting Speed on Forces and Torque Exerted on the Lumbar Spine, *Med. Sci. Sports Exerc.* **17**:440.

25. Harms-Ringdahl, K., 1986. On Assessment of Shoulder Exercise and Load-Elicited Pain in the Cervical Spine. Thesis, Karolinska Institute, University of Stockholm.

26. Hill, A. V., 1970. "First and Last Experiments in Muscle Mechanics," Cambridge University Press, Cambridge.

27. Hodges, P. W., et al., 2001. In Vivo Measurement of the Effect of Intra-abdominal Pressure on the Human Spine, *Jour. Biomech.* **34**:347.

28. "Hole's Human Anatomy and Physiology," 1996. McGraw-Hill.

29. "Human Factors Engineer Design for Army Material," MIL-HDBK 759A. Almost 700 pages of information on all aspects of human factors.

30. "Human Factors Engineer Design Criterion for Military Systems, Equipment, and Facilities," MIL-STD 1472 D.

31. Imwold, Denis, and Janet Parker, 2000. "Anatomica, the Complete Home Medical Reference," Global Book Publishing.

32. King, A. I., P. Prasad, and C. L. Ewing, 1975. Mechanism of Spinal Injury due to Caudocephalad Acceleration, *Othop. Clin. North Am.* **6**:19.

33. Kopperdahl, D., and T. Keaveny, 1998. "Yield Strain Behavior of Trabecular Bone, *Jour. Biomech.* **31**(7):601-608.

34. Kutz, Myer (ed.) 2003. "Standard Handbook of Biomedical Engineering and Design," McGraw-Hill.

35. Lemons, J. E., 1988. "Dental Implant Retrieval Analyses, *Jour. Dent. Ed.* **52**:748-757.

36. Linde, F., et al., 1989. Energy Absorptive Properties of Human Trabecular Bone Specimens during Axial Compression, *Jour. Orthop. Res.* **7**(3): 432-439.

37. Lindhal, O., 1976. Mechanical Properties of Dried Defatted Spongy Bone, *Acta Orthop. Scand.*, **47**:11.

38. Luo, G., et al., 1995. Implementation of Strain Rate as a Bone Remodeling Stimulus, *Jour. Biomech. Eng.* **117**:329-338.

39. Mankin, H., et al., 1994. Form and function of Articular Cartilage, in S. R. S. Simon (ed.), Orthopedic Basic Science, American Academy of Orthopedic Surgeons, Chicago.

40. McCalden, R., et al., 1997. Age-Related Changes in the Compressive Strength of Cancellous Bone: The Relative Importance of Changes in Density and Trabecular Architecture, *Jour. Bone Joint Surg.* **79A**(3):421-427.

41. McElhaney, J., and E. Byars, 1965. Dynamic Response of Biological Materials, *Proc. Amer. Soc. Mech. Eng.* ASME 65-WA/HUF-9:8, Chicago.

42. Miller, M. D., et. al., 1978. Significant New Observations on Cervical Spine Trauma, *Am. J. Roentgenol.* **130**:659.

43. Morgan, E., and T. Keaveny, 2001. Dependence of Yield Strain of Human Trabecular Bone on Anatomic Site, *Jour. Biomech.* **34**(5):69-577.

44. Morrison, J., 1970. The Mechanics of the Knee Joint in Relation to Normal Walking, *Jour. Biomech.* **3**:164-171.

45. Mosekilde, L., 1986. Normal Vertebral Body Size and Compressive Strength: Relations to Age and to Vertebral and Iliac Trabecular Bone Compressive Strength, *Bone* **7**:207-212.

46. Mosekilde, L., et al., 1987. Biomechanical Competence of Vertebral Trabecular Bone in Relation to Ash Density and Age in Normal Individuals, *Bone* **8**(2): 79-85.

47. Mow, V. C., et al., 1974. Some Surface Characteristics of Articular Cartilage, I. A Scanning Electron Microscopy Study and a Theoretical Model for the Dynamic Interaction of Synovial Fluid and Articular Cartilage, *Jour. Biomech.* **7**:449-1974.

48. Murray, M., 1967. Gait as a Total Pattern of Movement, *Am. J. Phys. Med.* **46**:290.

49. Nachemson, A., et al., 1986. Valsalva Maneuver Biomechanics "Effects on Lumbar Trunk Loads of Elevated Intra-abdominal Pressure," *Spine* **11**:476.

50. Nachemson, A., 1975. Towards a Better Understanding of Back Pain: a Review of the Mechanics of the Lumbar Disc, *Rheumatol. Rehabil.* **14**:129.

51. Nahum, Alan, and John Melvin, 1993. "Accidental Injury, Biomechanics and Prevention," Springer-Verlag.

52. Nordin, Margareta, and Victor Frankel, 1989. "Basic Biomechanics of the Musculoskeletal System," 2d ed., Williams and Williams.

53. Noyes, F. R., 1977. Functional Properties of Knee Ligaments and Alteration Induced by Immobilization, *Clin. Orthop.* **123**:210.

54. Pandy, M., et al., 1990. An Optimal Control Model for Maximum-Height Human Jumping, *Jour. Biomech.* **23**:1185-1198.

55. Paul, J., 1967. Forces at the Human Hip Joint, Ph.D. thesis, University of Chicago.

56. Reilly, D., and A. Burstein, 1975. The Elastic and Ultimate Properties of Compact Bone Tissue, *Jour. Biomech.* **8**:393-405.

57. Sadegh, A., et al., 1993. Bone Ingrowth: An Application of the Boundary Element Method to Bone Remodeling at the Implant Interface, *Jour. Biomech.* **26**(2): 167-182.

58. Schatz, R. A, 1989. A View of Vascular Stents, *Jour. Am. Coll. Cardiol.* **13**: 445-447.

59. Shirazi-Adl, A., S. C. Shrivastva, and A. M. Ahmed, 1984. Stress Analysis of the Lumbar Disc-Body Unit in Compression, *Spine*, **9**:120.

60. Ullman, David, 1997. "The Mechanical Design Process," McGraw-Hill.

61. White, A., and M. Panjabi, 1978. "Clinical Biomechanics of the Spine," J. B. Lippincott, Philadelphia.

62. White, A., et al., 1975. Biomechanical Analysis of Clinical Stability in the Cervical Spine, *Clin. Orthop.* **109**:85.

63. Williams, P., and R. Warwick, "Gray's Anatomy," 36th ed., Churchill Livingston, Edinburgh, pp. 506-515.

64. Winter, D., et al., 1998, Stiffness Control of Balance in Quiet Standing, *Jour. Neurophys.* **80**:1211-1221.

65. Winter, David, 2005. "Biomechanics and Motor Control of Human Movement," Wiley.

66. Wolff, J., 1892. "Das Gesetz der Transformation der Knochen," A. Hirschwald, Berlin.

67. Woo, S. L., et al., 1997. Structure and Function of Tendons and Ligaments, in V. C. Mow and W. C. Hayes (eds.), "Basic Orthopedic. Biomechanics," Lippincott-Raven, Philadelphia.

68. Woo, S. L., et al., 1994. Anatomy, Biology and Biomechanics of Tendon, Ligament and Meniscus, in S. R. Simon (ed.), "Orthopedic Basic Science," American Academy of Orthopedic Surgeons, Chicago.

69. Yamada, H., 1970. "Strength of Biological Materials," F. G. Evans, Ed., Williams & Wilkins, Baltimore.

70. Zajac, F., and M. Gordon, 1989. Determining Muscle's Force and Action in Multi-articular Movement, *Exercise and Sport Sci. Rev.* **17**:187-230.

12.7 HUMAN INJURY TOLERANCE AND ANTHROPOMETRIC TEST DEVICES

by Srirangam Kumaresan and Anthony Sances, Jr., Revised by Ali M. Sadegh

HUMAN INJURY TOLERANCE RELATED TO AUTOMOTIVE SAFETY

This section discusses an overview of injury tolerance of various body components relevant to the design of automotive vehicles. For the sake of brevity, the chapter provides a brief definition of injury criteria and tabulates the tolerance values proposed by the **National Highway Safety Traffic Administration** (NHTSA) at different time periods. The injury criteria are proposed both for adult and pediatric populations. The injury values for the pediatric population were based on scaling procedures.[1]

Head Injury Criteria

A blunt force or load applied to the head may result in injury to the skull, the brain, or both. Early work sought to establish a relationship between skull fracture and brain injury through head acceleration. Lissner et al. found that the time required to produce a fracture was smaller for higher levels of acceleration, using four cadavers.[2] Later on, Gadd developed the **severity index** (SI) to fit published literature to assess **head injury.**[3] Versace noted some shortcomings of SI and proposed an alternative index, the **head injury criterion (HIC).**[4] This index was adopted by **NHTSA,** and in 1972 a proposal was advanced to replace the SI with the following expression:

$$\mathrm{HIC} = \left[\left(\frac{1}{t_2 - t_1}\right)\int_{t_1}^{t_2} a(t)\,dt\right]^{2.5}_{\text{maximum}} \qquad (t_2 - t_1) \le 1{,}000$$

where $t_2 - t_1$ is the time for integration and $a(t)$ is the acceleration at the center of the head of the dummy. A value of 1,000 over a maximum period of 36 ms was chosen as the threshold of injury in the **Federal Motor Vehicle Safety Standards** (FMVSS 208). Table 12.7.1 summarizes the head injury values.

Recently NHTSA has adapted the injury threshold of the HIC for the head acceleration curve duration of maximum 15 ms, known as HIC_{15} for all dummy sizes. To reduce the probability of head injury NHTSA requires that the HIC_{15} should not exceed a maximum of 700 for all adult dummies and 570 for 3-year-old child dummies and 390 for 1-year-old child dummies.

Neck Injury Criteria

The force applied to the head may result in an injury to the neck. The neck injuries occur when the range of motion or force exceeds the limit of human tolerance. The neck injury results in the form of fracture or dislocation of vertebra, ligamentous damage, or a combination. The current **FMVSS** No. 208 injury criteria for the neck using the alternative sled test include individual tolerance limits for compression (4,000 N), tension (3,300 N), shear (3,100 N), flexion (190 N · m), and extension (57 N · m). The concept that neck injury criteria should be a linear combination of axial tension loads and extension-bending moments was developed by Prasad and Daniel and was based on their results from experimental tests on porcine subjects.[8] In 1996, Klinich et al. presented a composite neck injury indicator based on a linear combination of neck forces and moments.[9] The neck injury criterion, called N_{ij}, incorporates critical limits for all possible modes of neck loading: tension or compression combined with either flexion (forward) or extension (rearward) bending moment:

$$N_{ij} = F_z/F_{\text{int}} + M_y/M_{\text{int}}$$

where F_z is the axial load, F_{int} is the critical intercept value of load used for normalization, M_y is the flexion/extension bending moment, and M_{int} is the critical intercept value for moment used for normalization. Tables 12.7.2 to 12.7.4 summarize the neck injury tolerance.

Chest Injury Criteria

The acceleration and deflection applied individually or in combination to the chest in excess would injure the rib cage and underlying organs. The current FMVSS No. 208 stipulates that during the specified crash test, the chest deflection of the 50th percentile **Hybrid III** male dummy cannot exceed 76 mm (3 in) and its chest acceleration cannot exceed 60 *g*. Using the probability of injury curve reported in literature, lines of equal probability of injury for the linear combination of deflection and spinal acceleration were generated and the **combined thoracic injury criteria** (CTI) was proposed in 1998 with the following equation:

$$\mathrm{CTI} = A_{\max}/A_{\text{int}} + D_{\max}/D_{\text{int}}$$

where $A_{\max}$ is the maximum value of 3-ms clip spinal acceleration (A_s), $D_{\max}$ is the maximum value of the dummy deflection D, and A_{int} and D_{int} are the respective intercepts. Table 12.7.5 summarizes the chest injury tolerance values.

Lower Extremity Criteria

The femur load tolerance is proposed to represent the lower extremity injuries. Currently the tolerance limit is recommended for the adult population. Table 12.7.6 summarizes the limit values.

Table 12.7.1 Head Injury Criteria

Source	Criterion	1 year	3 year	6 year	5th percentile female	50th percentile male	95th percentile male
Kleinberger et al.,[5] 1998	HIC 36 ms	660	900	1,000	1,000	1,000	—
Eppinger et al.,[6] 1999	HIC 15 ms	390	570	700	700	700	700
Eppinger et al.,[7] 2000	HIC 15 ms	390	570	700	700	700	700

Table 12.7.2 Neck Injury Criteria, N_{ij}

Criterion	1 year	3 year	6 year	5th percentile female	50th percentile male	95th percentile male
Kleinberger et al., 1998						
N_{ij}	1.4	1.4	1.4	1.4	1.4	1.4
Tension/compression, N	2,200	2,500	2,900	3,200	3,600	—
Flexion, N · m	85	100	125	210	410	—
Extension, N · m	25	30	40	60	125	—
Eppinger et al., 1999						
N_{ij}	1.0	1.0	1.0	1.0	1.0	1.0
Tension/compression, N	1,465	2,120	2,800	3,370	4,500	5,440
Flexion, N · m	43	68	93	155	310	415
Extension, N · m	17	27	39	62	125	166

Table 12.7.3 Neck Peak Injury Values

Value	1 year	3 year	6 year	5th percentile female	50th percentile male	95th percentile male
In position						
Tension, N				2,620	4,170	5,030
Compression, N				2,520	4,000	4,830
Out of position						
Tension, N	780	1,130	1,490	2,070		
Compression, N	960	1,380	1,820	2,520		

SOURCE: Eppinger et al., 2000.

Table 12.7.4 Neck Injury Criteria, N_{ij}

	1 year	3 year	6 year	5th percentile female	50th percentile male	95th percentile male
N_{ij} critical intercept values	1.0	1.0	1.0	1.0	1.0	1.0
In position						
Tension, N				4,287	6,806	8,216
Compression, N				3,880	6,106	7,440
Flexion, N · m				155	310	415
Extension, N · m				67	135	179
Out of position						
Tension, N	1,460	2,120	2,800	3,880		
Compression, N	1,460	2,120	2,800	3,880		
Flexion, N · m	43	68	93	155		
Extension, N · m	17	27	37	61		

SOURCE: Eppinger et al., 2000.

Table 12.7.5 Chest Injury Criteria

Criterion	1 year	3 year	6 year	5th percentile female	50th percentile male	95th percentile male
Kleinberger et al., 1998						
Combined thoracic index (CTI)	1.0	1.0	1.0	1.0	1.0	—
Critical acceleration, *g*	55	70	85	85	85	—
Critical deflection, mm	49	57	63	83	102	—
Peak acceleration, *g*	40	50	60	60	60	—
Peak deflection, mm	37	42	47	62	76	—
Eppinger et al., 1999						
Peak acceleration, *g*	50	55	60	60	60	55
Peak deflection, mm	30	34	40	52	63	70
Eppinger et al., 2000						
Peak acceleration, *g*	50	55	60	60	60	55
Peak deflection, mm	30	34	40	52	63	70

Table 12.7.6 Lower Extremity Criteria—Femur Load, kN

	1 year	3 year	6 year	5th percentile female	50th percentile male	95th percentile male
Kleinberger et al., 1998	—	—	—	6.8	10	—
Eppinger et al., 1999	—	—	—	6.8	10	12.7
Eppinger et al., 2000	—	—	—	6.8	10	12.7

BIOMECHANICAL BASIS TO DEVELOP ANTHROPOMETRIC TEST DUMMIES FOR AUTOMOBILE CRASH TESTS

This section focuses on the biomechanical basis for anthropometric test dummies to assess head and neck injuries in accident environments with specific attention to the development of **Federal Motor Vehicle Safety Standard** (FMVSS).

Head Injury Assessment

In 1966, Gadd proposed a **severity index** (SI) based upon the Wayne State Tolerance Curve (WSTC).[1] In 1971, the **head injury criterion** (HIC) was proposed by Versace.[2] The HIC attempted to incorporate the observation that higher peak acceleration was tolerable if presented in a shorter period of time. Versace emphasized that the WSTC represented only the boundary between acceptable and unacceptable levels, and that it did not provide a scaling for injury. In 1984, Mertz again emphasized that injury assessment values should be used only to interpret the response of dummies and thereafter only as a guide to design.[3] Nevertheless, a graph relating HIC to the percentage of population expected to experience life-threatening brain injury was advanced. The basis for the curve was a series of 54 cadaver impacts.[4] The result of the impact was scored on the presence or absence of a skull fracture. By using the lowest **HIC** for which a fracture was observed (450) and the highest HIC resulting in a nonfracture (2,351), a normal curve was generated between the two. The appearance of a skull fracture was equated to life-threatening brain injury. Kleinberger reviewed the literature on head injury and the origin of HIC for the **National Highway Transportation Safety Administration** (NHTSA).[5] Since the agents directly responsible for brain injury (e.g., pressure, force, stress, inertial movement) are difficult to measure, a surrogate was sought. It was apparent that peak acceleration and impulse-area measurements, although they are related to head injury, are not easily related to experimental data. Energy absorption calculations were found to be difficult to apply, and, therefore, a simple relationship between energy absorption and brain injury was not possible. Sances has given a review of the various theories of skull fracture and human head injury tolerance.[6] A review of the recent HIC for child and adult dummies has also been given.[7] The value of HIC $\leq$ 1,000 for an integration duration of up to 36 ms is recommended in the **Federal Motor Vehicle Safety Standards** (FMVSS) Standard 208 for the midsized male (50 percent **Hybrid III**). The value of 1,000 has been adopted for the midsized male, small female, and 6-year-old child. This decision was made after reviewing a large quantity of test data available in the NHTSA databases.[8] It has been noted that the presence or absence of a skull fracture does not have much significance on the course of brain injury.[9] The question of the survivability of a fall may be addressed by observing the results of documented accidental falls onto hard surfaces. The variability of humans to the effects of falls onto unyielding surfaces was investigated by Snyder in 1977.[10] It was found that falls from low heights can be fatal, while falls from significant heights are not necessarily so.

The HIC is useful in setting standards and comparing the relative effectiveness of different methods of head protection to differing levels of translational acceleration and the occurrence of skull fracture. However, the HIC does not measure the presence or absence of other injury mechanisms such as rotational acceleration, internal pressure, and differential movement of tissue within the brain leading to shearing or tearing of tissue.

Neck Injury Assessment

The early studies by Patrick and Mertz are used to determine the injury values for tension and flexion of the Hybrid III 50 percent male dummy, which is routinely used for rear impact and frontal flexion analysis.[11,12] However, the cadavers that were used by Patrick and Mertz in the 1960 and 1970 studies had been used prior to the studies in windshield and steering wheel studies or for chest experiments.[13] Because CT and MRI equipment was not available, plain X-rays were used to evaluate injury. Consequently, the ligamentous damage could not be evaluated. In another study involving vehicle rollover tests with

Figure 12.7.1 Picture of 50th percentile male dummy. (*Source*: http://www.ftss.com.)

dummies, the authors have indicated that injury occurs at the neck when the Hybrid III 50 percent male dummy (Fig. 12.7.1) neck transducer measured 2,000 N.[14] In this study, measured axial neck loads above 7,000 N were four for the production vehicle and zero for the roll cage vehicle. There were five force levels for the production vehicle and two for the roll cage above 6,000 N. At 5,000 N, there were six for the production vehicle and three for the roll cage. But this study is severely limited by the inability to predict injury values. In a laboratory study, a clinical-burst-type fracture was produced when forces of approximately 8,000 N were applied at speeds of 4.0 to 4.5 m/s to the head of a cadaver.[15] This study shows that four potentially injurious neck loads occurred in the production vehicles when the threshold is set at 7,000 N for neck injury and none in the roll cage vehicles.

Human head and neck axial tension studies in the intact human cadaver indicate injury values in the region of 3,400 N.[16] There are marked differences between the Hybrid III neck and the human neck. One important missing characteristic of the Hybrid III is the lack of head lag when the neck goes into flexion and/or extension.[17] Differences in axial compression, dynamic, and static loading of the **Hybrid III** head and neck system has shown that the dynamic and static characteristics of the Hybrid III are 3 to 5 times stiffer than those of the human. A comparison of the Hybrid III 50 percent male and the cadaveric neck at slow rates showed that dummy necks had a stiffness of about 5,600 N/mm while the human head neck, which averaged about 2,000 N/mm in the axial compression mode.[18] Parallel axial compression studies in the intact human cadaver head-neck system at rates of 3 to 8 m/s were done.[19] Weights represented the muscles; impacts were made on the top of the Hybrid III 50 percent male head-neck and the cadaveric head-neck system. Forces and moments were measured at the C7/T1 level. This preparation produced clinical-type burst and flexion compression injuries to the cervical spine. The flexion and extension bending moments measured at C7/T1 were 2 to 3 times greater in the human tissue compared with the Hybrid III. Studies on 20 human cervical columns in vertical compression at velocities from 2.5 to 8 m/s showed that the initial force deflection curve was very low until the soft tissue loaded up to about 12 mm of compression.[20] The average level of injury was at 3,326 N with an average stiffness of 555 N/mm. Subsequent dynamic studies of the Hybrid III 50 percent neck showed markedly greater stiffness compared with the human cadaveric neck with values up to 2,160 N/mm at velocities of 7.6 m/s. The Hybrid III does not exhibit head lag similar to the human.[21]

Scaling approaches have been done to provide injury assessment values for the 6-month, 12-month, 3-year, and 6-year-old, the small female, the midsized male, and large male dummies.[22] The major source for the child biomechanical data is a series of tests conducted on pigs by General Motors and the Ford Motor Company with a simulated airbag inflating in front of an out-of-position occupant in the passenger seat. Scaling from these studies, assuming that a 10-week-old pig represents a 3-year-old child, matched tests were then conducted with the 3-year-old airbag dummy. These studies were then advanced to the child and adult dummies. Unfortunately, little information is available for the material properties of pediatric tissue except for a study performed over 100 years ago by Duncan.[23] It is well known that there are substantial differences between the 3-year-old child and the porcine geometry and development.[24]

REFERENCES

Human Injury Tolerance

1. Kleinberger, M., Yoganandan, N., and Kumaresan, S., "Biomechanical Considerations for Child Occupants," *42nd Meeting, Association for the Advancement of Automotive Medicine,* 1998, pp. 115-136.
2. Lissner, H. R., Lebow, M., and Evans, P. G., "Experimental Studies on the Relation between Acceleration and Intracranial Pressure Changes in Man," *Surg. Gynecol. Obstet.,* September 1960, pp. 329-338.
3. Gadd, C. W., "Use of a Weighted-Impulse Criterion for Estimating Injury Hazard," *Proc. 10th Stapp Car Crash Conference,* 1966.
4. Versace, J., "A Review of the Severity Index," *Proc. 15th Stapp Car Crash Conf.,* Society of Automotive Engineers, New York, 1971.
5. Kleinberger, M., Sun, E., Eppinger, R., Kuppa, S., and Saul, R., "Development of Improved Injury Criteria for the Assessment of Advanced Automotive Restraint Systems," NHTSA, 1998.
6. Eppinger, R., Sun, E., Bandak, F., Haffner, M., Khaewpong, N., Maltese, M., Kuppa, S., Nguyen, T., Takhounts, E., Tannous, R., Zhang, A., and Saul, R., "Development of Improved Injury Criteria for the Assessment of Advanced Automotive Restraint Systems—II," NHTSA, 1999.
7. Eppinger, R., Sun, E., Kuppa, S., and Saul, R., "Supplement: Development of Improved Injury Criteria for the Assessment of Advanced Automotive Restraint Systems—II," NHTSA, 2000.
8. Prasad, P., and Daniel, R. E., "A Biomechanical Analysis of Head, Neck, and Torso Injuries to Child Surrogates due to Sudden Torso Acceleration," Society of Automotive Engineers, Paper No 84 1656, 1984.
9. Klinich, K. D., Saul, R. A., Kleinberger, M., Backaitis, S. H., and August, G., "Techniques for Developing Child Dummy Protection Reference Values," Docket No.74-14 N97, Item 069, October 1996.

Anthropometric Test Devices

1. Gadd, C. W., "Use of a Weighted-impulse Criterion for Estimating Injury Hazard," *Proc. 10th Stapp Car Crash Conf,* Society of Automotive Engineers, New York, 1966.
2. Versace, J., "A Review of the Severity Index," *Proc. 15th Stapp Car Crash Conf.,* Society of Automotive Engineers, New York, 1971.
3. Mertz, H. J., "Injury Assessment Values Used to Evaluate Hybrid III Response Measurements," USG 2284 Part III, Attachment I, Enclosure 2, February 1984.
4. Prasad, P., and Mertz, H. J., "The Position of the United States Delegation to the ISO Working Group 6 on the Use of HIC in the Automotive Environment," *Government/Industry Meeting & Exposition,* Washington, DC, May 20-23, 1985. Society of Automotive Engineers, Inc. Warrendale, PA. Paper 851246.
5. Kleinberger, M., Sun, E., Eppinger, R., Kuppa, S., and Saul, R., "Development of Improved Injury Criteria for the Assessment of Advanced Automotive Restraint Systems," NHTSA, 1998.
6. Sances, A., Jr., and Yoganandan, N., "Human Head Injury Tolerance," in Sances, A., Jr., Thomas, D. J., Ewing, C. L., Larson, S. J., and Unterharnscheidt, F. (eds.), "Mechanisms of Head and Spine Trauma," Aloray Publisher, Goshen, NY, 1986, pp. 189-218.
7. Sances, A., Jr., "Biomechanical Analysis of Injury Criterion for Child and Adult Dummies," *Crit. Rev. Biomed. Eng.*, 1999, pp. 361-365.
8. Kleinberger, M., Sun, E., Eppinger, R., Kuppa, S., and Saul, R., "Development of Improved Injury Criteria for the Assessment of Advanced Automotive Restraint Systems," NHTSA, 1998.
9. Prasad, P., Melvin, J. W., Huelke, D. F., King, A. I., and Nyquist, G. W., "Head" in Melvin, J. W., and Weber, K. (eds.), "Review of Biomechanical Impact Response and Injury in the Automotive Environment," University of Michigan Transportation Research Institute, UMTRI-85-3.
10. Snyder, R. G., Foust, D. R., and Bowman, B. M., "Study of Impact Tolerance Through Free-Fall Investigation," University of Michigan Highway Safety Research Institute, Report VM-HSRI-77-8, 15 Dec, 1977.
11. Mertz, H. J., and Patrick, L. M., "Investigation of the Kinematics and Kinetics of Whiplash," *Proc. 11th Stapp Car Crash Conf.*, Anaheim, CA, 1967, pp. 267-317.
12. Mertz, H. J., and Patrick, L. M., Strength and response of the human neck, *Proc. 15th Stapp Car Crash Conf.*, Coronado, CA, 1971, pp. 207-255.
13. Yoganandan, N., "Biomechanical Assessment of Whiplash," in Yoganandan, N., Pintar, F., Larson, S. J., Sances, A., Jr., et al. (eds.), "Frontiers in Head and Neck Trauma," IOS Press, 1998.
14. Bahling, G. S., Bundorf, R. T., et al., "Rollover and Drop Tests—The Influence of Roof Strength on Injury Mechanics Using Belted Dummies, Society of Automotive Engineers," Paper No. 902314, 1990.
15. Pintar, F., Sances, A., Jr., Yoganandan, N., Reinartz, J., Maiman, D. J., Suh, J. K., et al., "Biodynamics of the Total Human Cadaveric Cervical Spine," *Proc. 34th Stapp Car Crash Conf.,* 1990.
16. Yoganandan, N., Maiman, D. J., Cusick, J. F., Pintar F. A., Sances, A., Jr., Walsh, P. R., and Larson, S. J., "Human Head-Neck Biomechanics Under Axial Tension," *Me. Eng. Physics,* **18**, no. 4, pp. 289-294.
17. Sances, A., Jr., "Experimental System to Investigate Whiplash Injury Biomechanics," in Yoganandan, N., Pintar, F. A., Larson, S. J., and Sances, A., Jr. (eds.), "Frontiers in Head and Neck Trauma," IOS Press, Burke, VA, 1998, pp. 308-312.
18. Yoganandan, N., Sances, A., Jr., and Pintar, F., "Biomechanical Evaluation of the Axial Compressive Responses of the Human Cadaveric and Manikin Necks," *J. Biomech. Eng.* 1, **11**, no. 3, 1989, pp. 250-255.
19. Pintar, F., Sances, A., Jr., Yoganandan, N., Reinartz, J., Maiman, D. J., Suh, J. K., et al., "Biodynamics of the Total Human Cadaveric Cervical Spine," *Proc. 34th Stapp Car Crash Conf.,* 1990.
20. Pintar, F., Yoganandan, Y., Voo, L., Cusick, J., et al., "Dynamic Characteristics of the Human Cervical Spine," Society of Automotive Engineers, *39th Stapp Car Crash Conf.,* 1995, p. 195.
21. Ewing, C. L., and Thomas, D. J., "Human Head and Neck Response to Impact Acceleration," Monograph 21, USAARL 73-1, Naval Aerospace and Regional Medical Centre, Pensacola, 1972.
22. Mertz, H. J., Prasad, P., and Irwin, A. J., "Injury Risk Curves for Children and Adults in Frontal and Rear Collisions," *41st Stapp Conf.* 973318, 1997, pp. 13-30.
23. Kleinberger, M., Yoganandan, N., and Kumaresan, S., "Biomechanical Considerations for Child Occupant Protection," *42nd AAAM Conf.,* Charlottesville, VA, October 1998.
24. Kumaresan, S., Yoganandan, N., and Pintar, F. A., "Age-Specific Pediatric Cervical Spine Biomechanical Responses," *SAE Trans.* (*41st Stapp Car Crash Conf.*) 106, 1997, pp. 3581-3611.

Section 13
Engineering Ethics

BY

NAEL BARAKAT *Professor, Mechanical Engineering, Texas A&M University—Kingsville*

INTRODUCTION

A distinct difference between the profession of engineering and other professions, like those in the medical field, is that the link made between the public and the professionals is usually missing. While engineers continue to apply science in the form of innovative products and services improving daily human life, like: cell phones, microwaves, gasoline, water pumps, communication devices, and even the paved road, the general public continues to see these as faceless products, and rarely relates these products, used daily, to engineers and the profession of engineering. Meanwhile, every individual must have seen their medical doctor at least once, and can directly tie their health care to the person providing it, unlike the field of engineering.

It is also a fact that the nature of engineering activities does not provide enough drama for media makers to capitalize on and to sell to the public through the public media outlets. Accordingly, it is easy to notice how little time engineering occupies on television, for example, compared to emergency room drama and Crime Scene Investigators (CSI) episodes. Since the majority of the public are usually fed their knowledge through these media outlets, engineering reality is not something that would be as exciting or as attractive if publicized. Meanwhile, in the very few instances when an engineering disaster occurs, like a plane or a building falling and causing casualties, or a nuclear reactor exploding, a significant amount of drama will be generated for the media to bring it out to the front pages. Examples like those include: the Challenger space shuttle case, the Fukushima nuclear reactor case, and the BP Deep Water Horizon spill case, to name a few. From a technical perspective, in spite of the fact that the calculated risk of these disasters to occur is usually extremely low, when they happen, they provide a lasting impact that propagates and lives for a long time. These disasters, combined with public lack of understanding of the role of engineers in improving and advancing society, as well as the misconceptions related to the decision-making hierarchical relations in organizations housing engineers, can allow for significant confusion and distortions regarding the reality of engineers, engineering products, and the engineering profession. As an example, after the Japanese nuclear reactor in Fukushima went out of order in the year 2011, causing a major disaster, public opinion built on the impact of this incident resulted in the complete elimination of all nuclear power plants in Germany, despite their very low risk, very environmental friendly output, and excellent historical performance. The collective of all of this, makes engineering ethics a critical part of the engineering profession that should be understood and well-articulated. To start, engineering ethics does not stop at doing the right thing when the right thing is obvious, but covers multiple aspects of the engineering profession, particularly those related to interaction with society and humanity. Figure 13.1 shows relations and interactions between engineering and society, in relation to the nature of engineering. These relations and interactions all fall within the coverage area of what is known as engineering ethics.

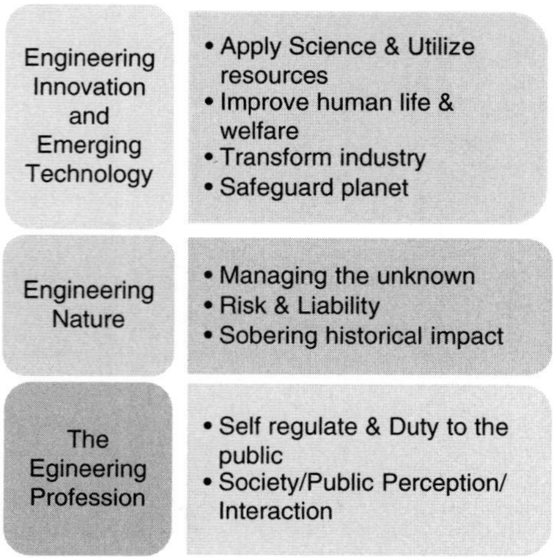

Figure 13.1 Relations and interactions between engineering and society.

ENGINEERING AND SOCIETY

A clear understanding of the role of engineering in society will quickly lead to the established fact that: engineering is a vitally needed profession for society. Engineering is a profession that requires significant formal education and extensive training, allowing engineers to possess specialized skills and knowledge to utilize autonomously in solving problems. Most of the solutions are made through developing technology that takes scientific discovery into products and services. Engineering practice is dynamic and innovative requiring intellectual judgement and extrapolation of knowledge into unchartered territories. In essence, the nature of engineering involves managing the unknown for the betterment of human life. Therefore, most engineering projects are social experiments involving a degree of risk. This risk stems mainly from two elements:

1. The complexity and interactions of all the unknowns associated with the different parts in an engineered system, and

2. The novel nature of engineering products, particularly first-time used and innovative ones, which have not been tested before.

Figure 13.2 shows these relations and how a case involving their combination could lead to liability issues. Consequently, the moment the word liability surfaces, laws are enacted to regulate actions. However, blind legal regulation of the profession of engineering alone would not be adequate because of the specialized knowledge and autonomy needed to practice the profession and the fact that most regulatory rules are reactive in nature.

In fact, like many other professions, as engineering interacts with society, an implicit contract between society and engineers is established, allowing engineering to self-regulate in exchange for ethical and responsible practice of the profession that protects and holds paramount the health and welfare of the public. Similar provisions are found within other professions, like the medical profession. Figure 13.3 shows the societal and human dimensions and interactions of engineering. Consequently, Engineering as a profession self-regulates and self-governs utilizing tools such as the professional licensing process and the formal education on engineering ethics.

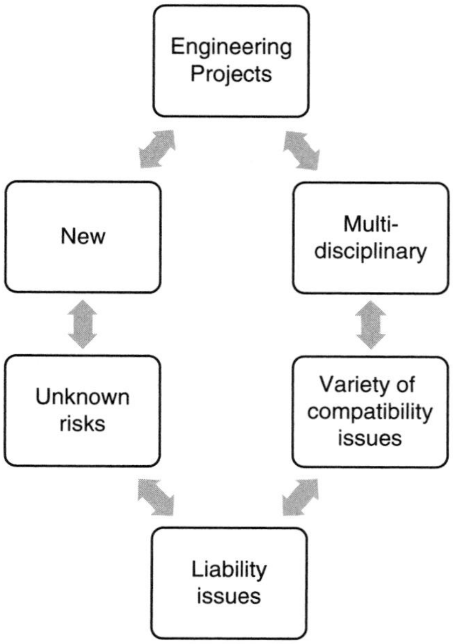

Figure 13.2 Risk and liability in engineering projects.

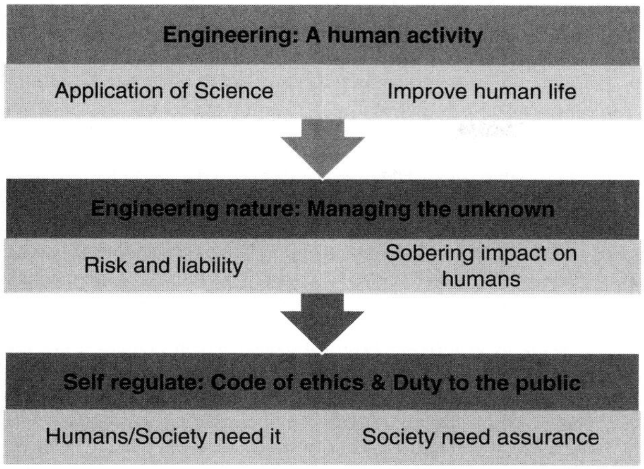

Figure 13.3 Societal and human dimensions of engineering.

CONCEPTS AND DEFINITIONS

Engineering ethics can be defined as the set of common standards of conduct, and professional dimensions, shared by engineering professionals, promoting responsible practice as one of its goals. Other goals include providing a framework for critical analysis and understanding of the evolutionary process in the profession of engineering. This is particularly important when considering the non-technical and contemporary topics affecting engineers and their profession. In addition, engineering ethics allows the integration between the technical side of engineering and the societal and human dimensions of the profession, promoting the development of professional aspects like: societal impact assessment, technical and public education improvement, public informing and responsibility, citizenship, and stewardship.

Engineering ethics are professional ethics, which are different from personal or religious ethics, in spite of many overlaps between these sets. Actually, it is very hard to clearly define a distinct boundary between engineering ethics and personal ethics without sufficient reflection and analysis. However, for the time being, a set of working definitions can be used as follows:

1. Personal ethics can be defined as the set of a person's ethical commitments, acquired and modified over their life span. These usually formulate the base upon which ethical choices were made at an individual level.
2. Interchangeably, morality is used in this context of discussing ethics where morality can be defined as the set of ideals, values, and beliefs, presumably guiding personal behavior.
3. As a result, common morality is defined as the set of moral ideals shared by most members of a culture or society.
4. Accordingly, ethical philosophies can also be defined in general as culture norms of ideals and values scrutinized through religious traditions and higher ideals.

It is to be noted that in general: morality is usually more descriptive of individualized ideals while ethics are usually more descriptive of the ideals and values of a group or organization. Nevertheless, an example of the overlap between personal and professional ethics is always present when looking at the concept of integrity, which is significantly emphasized by ethical theories and codes of ethics, as will be discussed later. Integrity is an individual trait that, by definition, indicates a healthy moral character. This also implies loyal obedience to moral principles, especially when tested, like when no one is looking or when acting under pressure or duress. Although integrity is an individualized concept, it has been emphasized throughout the fundamentals of professional ethics. This is a clear example of how blurry the line can be between personal and professional ethics because of the simple fact that professionals are still human individuals. Nevertheless, professional ethics should be differentiated from personal ethics whenever possible.

Code of Ethics

Engineering ethics are usually documented in a written form known as the "Code of Ethics." Codes of ethics are provided by professional organizations as an agreed upon guide for its members, in matters related to their professional practice. However, a code of ethics is more than just a guide. It is actually an agreement between the public and the professionals to promote and protect public health and welfare which helps solidify public trust in the profession. A code of ethics also aims at advancing the profession by requiring continuous improvement and update of skills and knowledge by its members to stay competent in the profession. From a bigger picture, a code of ethics provides a medium for professionals to discuss and understand the evolution of the profession and its ethics as well as the reasoning for a professional to adhere to professional standards. These standards include both the technical and ethical sets.

Codes of ethics for engineers are available from most of the engineering professional organizations like: the American Society of Mechanical Engineers (ASME),[1] the Institute of Electrical and Electronics Engineers (IEEE),[2] the American Institute of Chemical Engineers (AIChE),[3] and the National Society of Professional Engineers (NSPE).[4] Some of these professional organizations, like the ASME and IEEE, have made their codes of ethics part of the professional society policies. One code of those can be considered valid for use by any engineer is provided by the NSPE.[4]

A closer examination of these codes shows that all of them share similar fundamentals including the following:

1. A clear commitment to hold paramount the health and welfare of humans.
2. An obligation to protect the dignity and right of consent of human subjects.
3. An understanding of engineering endeavors as managing the unknown within a societal context resulting in risks and liability.
4. A profound acceptance of accountability for the results of engineering projects.
5. An obligation to protect and advance the engineering profession.

However, while sharing the same concepts, these codes from the different engineering organizations have different flavors and distinguishing items related to the engineering type they represent. For example: while both codes of ethics for civil engineers[5] and computer engineers[2,6] focus on health and welfare of the public as to be held paramount, the code for civil engineers would add a focus on sustainable development of resources, while the code of ethics for computer engineers would add a focus on data security and privacy issues. These codes can become very handy in the process of analyzing and understanding of ethical dilemmas related to the profession. However, engineering codes of ethics should not be mistakenly used as a recipe substituting for reasonable judgment or as a legal document constituting law.

To help understand the levels of complexity that could be associated with an ethical dilemma, the following concept and example are presented for clarification. The concept states that: ethical does not necessarily constitute legal and vice versa. Therefore, when considering the following hypothetical scenario where: if the Environmental Protection Agency (EPA) laws state that no contaminant should be disposed of in a natural lake with a concentration of more than 10 parts per million (PPM), then disposing of the same contaminant at a concentration of 6 PPM might be legal, but is it completely ethical to do so? The first observation would be that a clear cut answer to this question is not easy to find. In fact, to answer this question it might be necessary to find more facts and gather more information toward an attempt to understand the context of the problem. Once the facts have been streamlined, codes of ethics can be brought in the discussion for reflection and testing of the findings, to help formulate the most ethical solution. Yet, this solution would most probably be a compromise or a middle ground solution that would use some ethical theory or concept, in addition to the professional codes of ethics. As was mentioned before, engineering ethics do not stop at just doing the right thing when the right thing is obvious, but they go deeper and further in helping to resolve very complex ethical problems related to the professional practice of engineering in the context of society and humanity. Moreover, engineering ethics bring the context of engineering early in every project from the beginning and not as an ad hoc element. In the following sections some fundamental ethical theories will be presented in a summarized format to help understand the basis for codes of ethics and ethics related decisions.

Philosophical Background

Ethics is a topic that is deeply rooted in philosophy. Ethics includes the conceptualization, analysis, and understanding of right and wrong in human behavior. Understanding the philosophical background of ethics and the concepts behind ethical decisions helps professionals, particularly engineers, to understand the depth and meaning of their professional responsibility as well as the basis for ethically related decisions. This depth also helps them grasp the bigger picture provided by codes of ethics of professional organizations. Therefore, a quick review of the basic ethical theories is provided here to help in understanding decisions related to ethical dilemmas and in solving ethical problems. This will ultimately help engineers in building ethical-judgement autonomy when practicing engineering. Following is only a summary of some common ethical theories with very limited depth in discussion to help maintain focus on the applicability of these theories in engineering practice where different ethical situations and decisions occur. In general, fundamental ethical theories can be categorized as follows:

1. Western Ethical Theories: These are the theories that were developed throughout history and had more contributions from, and influence on, the western hemisphere of the globe. Decisions made to organize societies and most current constitutions in the west are based majorly on combinations of these theories. Some main stream theories in this category include:

   a. *Utilitarianism:* This is an ancient theory that was further developed and attributed to John Stuart Mill (1806-1873) as being the main developer of the theory.[7] This theory considers a balance between good and bad consequences of actions and decisions, in relation to everyone affected, or the entire society. Good actions are considered to be those that serve to promote human wellbeing of the majority. An application that is acceptable within the boundaries of this theory is the risk-benefit or cost-benefit analysis. This technique is based on quantifying actions and decisions to produce a final number representing cost or benefit of the action or decision at hand. According to this theory, consideration of most benefit to the most people outweighs the needs of few individuals. This theory is also known as "Consequentialism." Some of the disadvantages of this theory include the lack of consideration for the balance in distribution of benefit, or cost, when it is for the majority; and the override of all considerations when quantification is taking place, including some items which might be invaluable, like human life.

   b. *Rights Ethics:* This theory is mainly attributed to John Locke (1632-1704).[7] The theory states that people have fundamental rights (like life, liberty, and property) that others have a duty to respect. This theory is related to deontology and Justice Ethics.

   c. *Duty Ethics:* This theory is attributed mainly to Immanuel Kant (1724-1804).[7] The essence of the theory is that there are duties which should be performed (e.g., duty to treat others fairly, duty not to injure others, etc.), regardless of whether these acts do the most good or not. This theory is also related to deontology ethics or Rule philosophy. Duty ethics and Rights ethics are usually considered together as they both relate to each other very closely. The theory of rights-and-duties-ethics, also known in the literature as Respect for the Person Ethics, has its applications, advantages, and disadvantages as well. On the one hand, the main advantage of this theory is to treat all humans equally. On the other hand, the main disadvantage of this theory is the disregard of the public good when very few individuals object to such good based on harm inflected upon them, even if this harm was not significant and subject to compensation by other means.

   In Summary: The previous are two non-mutually exclusive western ethical theories based on what is held as the most important moral or ethical concept:

   i. *Utilitarianism:* Seeks the most utility and overall good result for the majority.

   ii. *Respect for Persons:* Accord equal respect to each person as a moral agent within two theories—Duty ethics and Rights ethics.

2. Non-western Ethical Theories: The rest of the world has different foundations for their ethical systems.

However, ethical standards are similar worldwide although cultural practices are significantly different. Being deceptive is wrong everywhere, not just in the western world. Moreover, some of these ethical norms have been developed long before the western theories developed.

This leads to a different practical aspect which engineers are not always aware of, as they practice their profession globally, which relates to the balanced application of ethical understanding. It is critically important that engineers avoid the two extremes of moral dilution or relativism, and moral superiority or absolutism, when practicing engineering on a global level. Moral dilution or relativism means that an engineer would behave similar to others within the environment of operation, from a moral point of view, even if the behavior is not in agreement with their professional ethics. In other words, when in Rome, do like Romans do. Meanwhile, moral superiority or absolutism implies an embedded notion of moral superiority which results in the engineer trying to replace local ethical systems with their own, assuming these local systems are inadequate. The bottom line concept is that engineers should understand that their professional ethics are expected to apply in any geographical location, with extra effort to understand and respect local cultures and institutions, while not compromising one's own. Nevertheless, when the legal aspects are involved, it is critical that engineers from a particular country (for example, the United States) realize that their actions are governed by U.S. laws even if they are practicing outside the United States. For example, bribery is as prosecutable outside the United States as it is inside, without any differentiation.

Ethical theories can converge offering stronger arguments or diverge leading to different solutions of an ethical situation. When ethical theories diverge, more analysis and richer views can result helping in understanding the situation better. Generally, respect for the person's ethics take precedence over utilitarian considerations if rights are violated. This is because rights of the person ethics are the base for the majority of western constitutions. Otherwise, a balance of theories must be met or a middle creative solution must be found, to solve the ethical dilemma at hand.

APPLICATIONS

A moral dilemma results from a situation where two or more ethical rulings or moral options are applicable but point in opposing directions. Complexity in these situations stems usually from three main sources:

1. Lack of facts: Where information is scarce and details are missing to point at a particular ethical principle that could be applied to the situation.
2. Interpretation differences: Where different persons disagree regarding the applicability and interpretation of ethical principles, especially in those in authority positions.
3. Conflicting logic: Where two or more ethical principles are applicable to the situation but point in opposing directions.

Therefore, when analyzing ethical problems to find a solution, a deep understanding of the issues and a great deal of investigation to find the facts is required. To arrive at this understanding, facts have to be collected aiming at finding the whole truth, or at least get as close as possible to the truth. Of course, not everyone will agree on the "truth," but further research always helps to clarify the situation. Following this step an attempt should be made to reach an agreement regarding the applicable ethical concepts and the meaning or interpretation of these concepts. An agreement is not always possible, but further analysis helps to bring it closer. The final solution would probably never be the most ideal ethical solution, but rather a compromised position by all parties that is the closest possible to the best ethical solution. Therefore, being one of many correct solutions, the ethical problem under consideration can be considered an open ended problem. Engineering design problems are usually open ended problems, making ethical dilemmas analogous to engineering design. Consequently, these types of problems have the following properties:

1. They have more than one correct solution.
2. They require applying a large body of knowledge to formulate a solution.
3. They require significant utilization of analytical skills.
4. They require multiple iterations at every step to arrive at a solution satisfactory to the initiation criteria.

Legal versus Professional Responsibility

Conforming to professional standards helps protect the public and the engineer from possible accidents and safety concerns, as well as legal ramifications. However, engineers should be aware of the boundaries and limitations of their legal and professional responsibility. This is particularly important when problems occur resulting in grave legal and liability questions. To fulfill this professional responsibility engineers can apply a model known as "reasonable care." Proper understanding of this model is necessary to avoid actions which can be exploited in the case of disasters and problems. Since the very nature of engineering will always involve occurrences which have never been accounted for previously, and to properly identify and understand reasonable care, an analysis of the legal precedence and views in this regard might be helpful. For this purpose, a part of the law known as "tort law" is examined. Tort law is usually the part of the legal system that addresses legal injuries or wrongs in the course of practicing a profession. It is also known as the malpractice law. This law defines *reasonable* care in the professional context as following the standard operating procedure of a profession, and ordinary job responsibilities, without being restricted to only these two items. Accordingly, this definition is also open to some interpretation and subjectivity depending on the circumstances and evidence in the case at hand. Therefore, most codes of ethics support a broader principle of care that goes as far as recommending that one should avoid even contributing to producing harm by action or inaction. Codes of ethics emphasize that engineers are privileged to be able to see the potential for harm and to be able to prevent it.

Moral responsibility usually parallels legal liability. This brings in another term that needs definition which is liability. The simplest definition of liability is: deserving punishment for causing harm. The severity of the punishment is usually decided through a legal process that considers the degree of harm and gives weight to the cause and intention for inflecting harm, which can be one of three levels: intentional harm, reckless harm, and negligent harm. In liability cases against professionals, rarely any professional is proven to have caused intentional harm, and very few cases result in proving that harm was caused recklessly. In fact most of the professional liability cases would reveal that the cause of harm was negligence.

From another angle, the entire concept of professional Engineering ethics points at providing a higher calling than legal regulations and a minimum like reasonable care, in terms of protecting the health and welfare of the public. The idea is for engineers to go above and beyond the call of duty in insuring staying multiple steps away from unexpected disasters. This would be an advancement beyond reasonable care. In other words, finding and exploiting opportunities for going beyond minimum responsibility requirements is a desirable character of responsible engineers. The only condition here is that any action an engineer takes to go above and beyond minimum requirements should also be done in accordance with professional standards.

Proactive and Responsible Actions

Engineers are educated as professionals to perform their duties with ethics in mind. However, maintaining this attitude and understanding in action, is not an easy task, especially considering the facts that engineers are human. Humans' actions are directed by their frames of mind and their attitude, which might not help realize the long-term gains or losses behind a certain action, compared to the quick recognition of short-term gains or losses associated with a contradicting action. There are many influential elements that stem from a human frame of mind and attitude, which results sometime in a diversion from performing a responsible action. Awareness of these elements can help a great deal in correcting actions of a professional. These elements include:

1. Internal element: Examples include self-interest, self-centering, and personal gains.
2. Surrounding element: Examples of those include peer pressure, desire for acceptance, and other pressures including family and finances.
3. Intellectual element: Examples here include ignorance and poor education updates, as well as the lack of appreciation of the different perspectives and the big picture.

These barriers do not just work against an individual professional engineer when trying to act responsibly, but also exert the same effect on an entire organization. Recall that the majority of professional engineers work within bigger organizations which have its own authority structures. Management style within an organization could provide an environment that encourages or discourages these barriers to grow and affect performance. Successful organizations usually encourage and grow a culture of dissent and critical thinking, to empower professionals toward ownership of their actions and its consequences. Ultimately, this would enable the organization to find out its mistakes and errors quickly and efficiently, and consequently correct these errors in the way of continuous improvement. The last thing any organization wants is disputes going publicly or into the rumor mill. Moreover, most organizations now have their own codes of ethics and guidelines for employees to decide if an action is acceptable or not. In engineering organizations, most of those codes agree with the professional engineers' codes of ethics from professional organizations, with some added flavor related to the locality, or a mix of legal requirements being included as code of ethics. In case no code of ethics or guidelines for conduct in an organization, an engineer can be the agent of change initiating this culture through introducing it to the upper management and providing assistance with awareness and education activities such as classes and seminars that will allow this organization and colleagues to learn and gain awareness about the ethics and societal dimensions of engineering.

Integrity and Honesty

As mentioned previously, integrity is emphasized by ethical theories and engineering codes of ethics as an expected base for professional action and judgment. In application, integrity implies continuous honesty and reliability in performing engineering functions. The definitions of these words can be extracted from the linguistic roots where the dictionary defines them as follows:

- **Honesty:** Sincerity, fairness, and freedom from deceit or fraud.
- **Integrity:** Adherence to moral principles, wholeness, and soundness of moral character.
- **Reliability:** Trustworthiness, predictability, and dependability.

Note that the term "reliability" has a technical definition, not to be confused with the linguistic definition, which is the ratio of the number of correct runs to the number of attempted runs.

In the context of practicing professional engineering, many regulations and professional expectations rely on, and deal directly with, the concept of integrity and honesty. Examples of these situations include the following:

A. Full disclosure: Engineering codes of ethics stipulate that information needed for informed consent of an individual should be provided in full. The order of priority is always to protect the public and society, then the client, and finally the engineer's interests. Failure to inform those whose decisions are impaired by the absence of the information would invoke the honesty part of an engineer's code of ethics and might lead to undesirable legal consequences. However, the way to produce this information, particularly if a client's interest or the interests of an engineer would become in jeopardy, is a typical ethical dilemma. Resolving such dilemma usually requires significant effort to arrive at a compromising solution that is mutually acceptable by all parties.

B. Confidentiality: Confidential information will reach the engineer either directly from the client or by accidental discovery during the course of performing work for the client. This information can be subject to abuse in two forms: the first form is when confidentiality is broken while the action is not warranted, and the second form is when confidentiality is not broken although the higher obligation to the public requires it. The limits for these two forms are highly controversial but engineers should be aware of them.

C. Conflict of interest: This term is used by many people in many situations. In the context of practicing engineering, such situation exists when a professional judgment is subject to influences, loyalties, or other interests.The professional, moral, and legal aspects should be apply when looking at this concept. As was mentioned, the term (conflict of interest) is used by many people but with different background definition and with the least scrutiny. Therefore, it is recommended that even potential perception of conflict of interest, although in reality there could be none, should be avoided by engineers. If avoidance is not possible, then an engineer should reveal the conflict to the client as soon as it is realized. In general, if faced with a situation that seems to bring about a conflict of interest, policies within the company or organization should be followed. If no policy is available in the workplace, the situation should be discussed with a manager. If this solution is not practical or attainable, then a professional engineer should examine their motivations and try to take documented precautions in every step along the way until the situation is resolved.

A scenario that encapsulated all of the above situations occurred when an engineer with relevant expertise was hired by his previous father-in-law to inspect and certify an apartment building owned by the ex-father-in-law. As the engineer was performing the inspection he realized that one of the structural elements (a beam) carrying the building at the basement level was cracked. After informing the owner of the building, the owner did not want to do anything about the situation in the near future and asked the engineer to closely observe client confidentiality. This situation had multiple complexities leading to an ethical dilemma. In particular, the client confidentiality concept was directly contradicting the concept of full disclosure, which formulates the basics of an ethical dilemma. In addition, the previous relation between the engineer and the owner of the building provided grounds for an apparent conflict of interest, when actually the relation was previous and had nothing to do with the job at hand. Had the situation continued without action, legal ramifications, as well as professional penalties would not have been the worst that could happen. Rather, someone could have been killed by a falling building which is the worst result possible. The engineer had to do a lot of diplomacy negotiation with the building owner, including bringing in help from colleagues in the profession who offered reasonable charges to help repair the building and correct the entire situation. For the conflict of interest part, the engineer should have not accepted the job in the first place.

Common forms of dishonesty also include lying which is issuing a false or misleading statement with the intention to deceive. Lying can take different forms including deliberate deception. Withholding information with the intent to deceive and failure to disclose information that the audience would reasonably expect not to know are also forms of dishonesty. Failure to seek the truth can also be considered a form of dishonesty. Honest engineers are committed to finding the truth, not just avoiding dishonesty.

As was mentioned, ethical theories include honesty as an essential part. According to the theory of "respect for the person," dishonesty violates the moral agency of individuals. This leads to problems like decisions being made without informed consent. The utilitarian theory also stipulates that dishonesty undermines the relations of trust, which is necessary for the public good. Moreover, dishonesty impedes the development of technology, particularly when scientific findings are not completely disseminated, which results in depriving the public from an expected benefits.

Professional Academics and Scholars

Engineers are intertwined with academia starting their education in colleges and later between teaching and research to advance fundamental science and its applications. Academic and scholarly work is prone to many forms of dishonesty, with the majority of it being unintended. An example demonstrating this issue is the simple elimination of an irregular data point to match predictions without proper scientific proof that this point was an outlier. Similarly, eliminating results which do not match expectations or adding missing points to complete data sets per expectations, without reasonable scientific basis is a form of dishonesty. Also common in academic and scholarly work is the act of plagiarism, whether it is intentional or not. Plagiarism is the use of intellectual products of others without their permission or claiming credit in joint work that is not proportional to the contribution in that work. In general, academic honesty and integrity is not only obedience to an arbitrary set of rules while learning creativity and discovery but rather learning and applying the standard of evaluating and crediting contributions made in the discipline. Many resources are usually needed to complete a scientific endeavor or a publication, which have to be fairly used and credited as a reasonable expectation stemming from honesty. However, the criteria for fair use of resources depend on many factors including the field of discipline and the context of the work. For example, the significance of word copying in poetry is not the same as word copying in a technical report. The bottom line is that proper citation and credit of reference and sources is extremely important in the context of scholarly work.

EVOLVING ISSUES IN ENGINEERING ETHICS

As was mentioned, engineering ethics provides the framework to discuss and understand evolving issues related to the profession of engineering from both sides: the practice and the impact. This part includes a discussion of three prominent examples of these issues serving to explain the dynamic and evolving nature of the engineering practice and its interaction with society and the surroundings.

Continuous Professional Development

Background

Continuous professional development (CPD) or life-long learning is a clear example of how the dynamic and evolving nature of engineering practice dictates certain ethical aspects of the engineering profession. Fundamentally, engineering is the application of science and the utilization of scientific new discoveries into products and services that help improve human life and human endeavors. Consequently, explosive rates of engineering and technology developments are being experienced every day. The downside to this unprecedented high speed of discovery and knowledge update is that shelf-life of education and skillsets, possessed by the professional, become shorter, causing engineers to quickly become obsolete. This is particularly true if engineers do not continuously develop their skills and knowledge base. Therefore, continuous learning and update of knowledge and skills is very critical to the success and career continuation of an engineer.[8] The national academy of engineering recommends that: "in addition to delivering content, engineering schools must teach engineering students how to learn, and must play a continuing role along with professional organizations in facilitating life-long learning, perhaps through offering (executive) technical degrees similar to executive MBAs."[9]

From a practical angle, changes in engineering used to be mostly on the technical side of the practice, or that is what was noticed more about this profession. However, non-technical changes have become as evident and influential on the practice of engineering as well. As an example of these changes: globalization of engineering stands out. One of the consequences of globalization is the increasing need for universal professional and technical standards. Having CPD as a built-in component in the practice of the profession would make engineers not only capable and accommodating of these changes, but also competent, and above all, legally safe and correct, when practicing outside their home countries.[10]

Link to Engineering Ethics

CPD is actually a solid part of engineering ethics. Most codes of ethics from professional engineering societies state something to the effect of life-long learning being an ethical requirement of engineers. The National Society of Professional Engineers (NSPE) code of ethics, section III: professional obligations, point 9.e. includes the following statement: "Engineers shall continue their professional development throughout their careers and should keep current in their specialty fields by engaging in professional practice, participating in continuing education courses, reading in the technical literature, and attending professional meetings and seminars.[4]" A closer look at the code of ethic of the NSPE reveals that four out of six canons that make up the code directly call for the integration of continuous professional development in the engineering profession. The first fundamental canon in that code of ethics, which is the same first canon in most of the other engineering professional organizations codes of ethics, states that: "Engineers, in the fulfillment of their professional duties, shall hold paramount the safety, health, and welfare of the public.[4]" The relevance of this particular canons to continuous CPD can be explained by the example of safety rules and tools improvement. Safety in engineering and technology products is continuously improving, not only at the product and material level, but also in the design and build practices and procedures. However, these new technologies and methods might not have existed during the college education time of a practicing engineer. For professional engineers to fulfill their ethical duty outlined by the previously mentioned canon, they must acquire the knowledge relevant to these new safety enhancement findings and implement them in the practice. The advantage of this step goes above and beyond fulfilling ethical duties and reaches what is known as "commendable efforts."

In the same code, the second canon states that: "engineers, in the fulfillment of their professional duties, shall perform services only in areas of their competence.[4]" Understanding this, it becomes clear that being obsolete or outdated in knowledge and skills would place an engineer outside of the limits of needed knowledge and expertise. Moreover, it might mount up to exposing the public to dangers that result from that lack of knowledge and skills. Maintaining a meaningful competence is only achieved by continuous learning of the profession from all aspects. The fourth canon of the same code states that: "engineers, in the fulfillment of their professional duties, shall act for each employer or client as faithful agents or trustees.[4]" When engineers continue learning and are up to date in knowledge of the tools and bases of the profession, they significantly help their employers. This help includes gaining competitive advantage and indirectly leading the way for others to follow in the industry.

The last canon of the NSPE codes states that: "engineers, in the fulfillment of their professional duties, shall conduct themselves honorably, responsibly, ethically, and lawfully so as to enhance the honor, reputation, and usefulness of the profession.[4]" In essence, CPD will ultimately result in a better career for the engineer, economical advantage for both the engineer and the employer, less safety issues for the public, and most of all, increased public trust and respect for the profession. The advantages reach as far as better treatment and protection of the environment and the globe. Therefore, professional engineering registration in most states of the United States requires proof of CPD, using continuing professional competence, or development, hours or PDH. Professional engineering organizations in Canada are also requiring something along the same lines. This is simply because the main purpose for engineers' professional registration is to protect the public safety and welfare, which is directly linked to engineering ethics.

Implementing and Managing CPD

A healthy and successful career management plan should include CPD as a major activity. To optimize and fully benefit from CPD activities, the concept should be more than just an add-on activity. It should be an integral part of the practice that the engineer has been trained on, starting from college. However, even if that training was not fully realized, CPD is so critical to the professional success of an engineer, that it should be integrated and closely managed during practice. To optimize and integrate CPD in the engineering career management plan, an approach analogous to the design project management can be followed. This approach consists of four basic stages: planning, execution, integration, and evaluation. The purpose is to manage CPD activities in a well thought out plan and achieve maximum benefit to the professional and the clients. In addition, sub-objectives like filling a list or a resume that will satisfy an employer or the satisfying of registration requirements can still be achieved without being the main goal of the activities.

The first stage in this approach is planning. In design projects, this is where constraints, specifications, and requirements are identified. This is the part requiring the professional to set goals followed by measurable objectives. Priorities, needs, and ambitions are characterized in these goals. Areas of interest or in need for improvement in the relevant career should be targeted. The main idea is to have a clear and practical plan to improve professionally. It is also important that some alternative paths and flexibility be built in this plan to accommodate any sudden changes in career. Once measurable objectives have been identified, the second part can be initiated.

The second stage is execution. In this part, the objectives are put to work and appropriate learning opportunities and actions are identified, prioritized, timed, and then executed. This might be along the lines of joining a course or even going for a higher degree. Some programs and activities work for one professional but not the other. Any selected combination of choices should fulfill the objectives and the major plan goals. This part is analogous to executing and building in a design project. During this stage, some assessment should be conducted, which is analogous to design iterations. At the end of a CPD activity, an evaluation must be performed to compare between outcomes of the activity and desired objectives. This should direct any modifications to be implemented in the process.

The third stage is integration, where acquired knowledge is implemented and turned into competency. This is the stage where most professionals either face difficulties, or ignore, because it forces them out of the comfort zone or the old ways. This is an extremely critical part because it highlights the actual measurable contributions to career advancement through professional development. The results of this part are mostly realistic and clear in the improvements of the practice of the profession. This is analogous to delivering and utilizing the product that was designed, which is the real end user test. However, the cycle does not stop here. Usually, information from the users is collected by the designers to improve the next version of the product. That is analogous to the fourth stage where evaluation of the success of the plan is carried out. Professional development and improved competency in the practice as well as career improvement are measured against the original set of goals and objectives in the plan. The results should be assessed causing modifications to be made at the relevant parts of the model. Modifications should accommodate all inputs, including changes in vision. Then the cycle repeats and the process continues. Figure 13.4 shows the elements of the model and its interaction as compared to the steps in an engineering general design and build project.

Conclusion

While demand on engineers has been steady in the market for a long time, the majority of engineers in the workforce tend to get immersed in their daily duties and delay any plans to further their education and skills. As a result, CPD activities end up being carried out by professionals on a have-to basis, and in an ad hoc fashion. For engineers to survive against the increasing competition and to fulfill their ethical responsibilities, these activities should be performed as a way of life in the practice with appropriate planning and evaluation. Planning and managing of these activities as a core element in a career path and turning the acquired knowledge into competencies should be part of the practice of engineering. As a matter of fact, CPD effects extend to be vital for the national economy, not just the individual engineer's career.

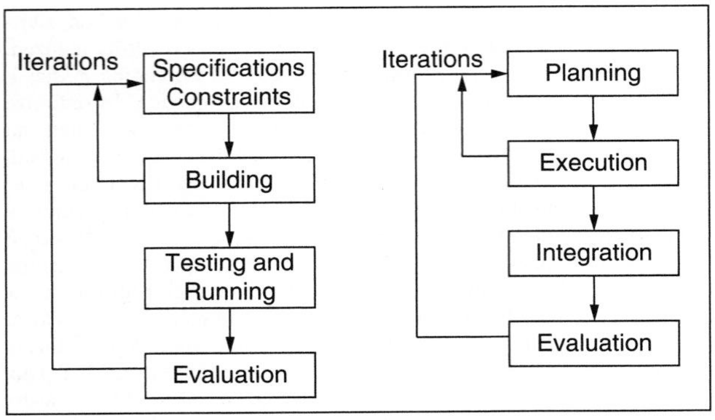

Figure 13.4 Flowchart depicting the analogy between the engineering design process (left), and the CPD management in a career (right).

Therefore, significant efforts are needed in the area of integrating the concept of CPD in engineering education and to provide practical and flexible models that can be merged with career management plans set by the professionals. CPD activities should not be carried out to make a big list that will satisfy requirements or show proof to employers. Moreover, when engineers require CPD as part of the practice, employers are encouraged to adopt CPD as a standard benefit for their engineers, which will slowly be removed from the list of items subject to cost cutting. In exchange, employers can enjoy a stable and continuous workforce, competitive advantage in quality and initiation, more benefits and attraction of skills, and more public trust and leadership roles in their products and expertise.

Environmental Issues and Sustainable Development

Background

Engineering projects are considered a threat to the environment, and many environmental problems have been associated with engineering endeavors. The truth of the matter is that engineering projects have been inadvertently responsible for some environmental issues, but not to the degree presented by the media. At the same time, engineers are always sought to provide solutions to environmental problems, whether it is caused by engineers or not. Figure 13.5 presents the simplified version of the societal perception of engineers' relation with the environment and natural resources. Nevertheless, the relation between engineers and the environment is more complex than this and has many dimensions.

Engineers have an obligation toward the environment that stems from two bases. The first base is that all private citizens have an obligation to the environment and natural resources; and engineers are private citizens. The second base is that professionals abide by professional ethics and the majority of professional engineering codes of ethics emphasize the engineer's responsibility toward the environment and natural resources. In a focused fashion, the American Society of Civil Engineers (ASCE) code of ethics refers to the environment within the context of sustainable development as follows:

"Canon 1, section f: Engineers shall hold paramount the safety, health and welfare of the public and shall strive to comply with the principles of sustainable development in the performance of their professional duties."[5]

Other professional organizations also share this reference to environment and sustainability. The word "strive" in the first canon of the ASCE code of ethics, described above, indicate multiple of levels of action which are expected by engineers including examples beyond personal compliance with sustainable development like: informing clients about the consequences of violating sustainable development and informing authorities about violators of sustainable development. Two ideas to be noted: the first is within the use of the word "strive" which does not make the commitment to the environment and natural resources a strict obligation. The second idea is that the environment is taken care of in combination with natural resources. This is due to reasons which will be explained in the following sections.

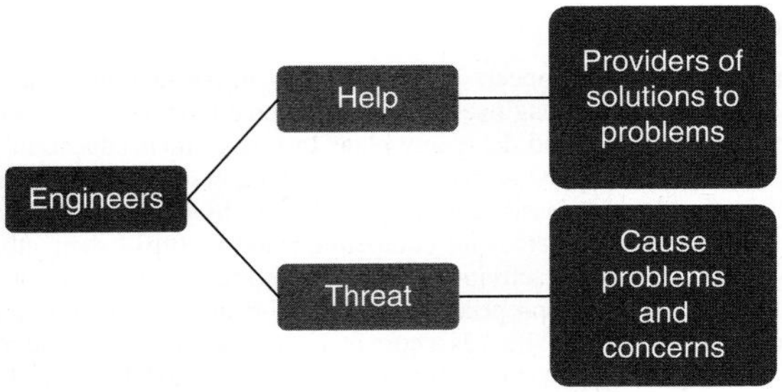

Figure 13.5 Societal perception of engineers' relation to the environment and natural resources.

Sustainable Development

Sustainable development, as shown in Fig. 13.6, is a process of change in which human development activities meet present needs and aspiration without endangering the capacity of natural systems to absorb their effects and without compromising future generations' ability to meet their needs and aspirations. Human development activities include things like: investment, technology, resource allocation, functioning of institutions, etc.

Engineers play a critical role in sustainable development characterized by the development of technology that conserves natural resources and protects the environment. However, engineers are not capable of doing that alone. Many factors are involved in engineering projects including political, social, and moral factors, which are beyond just environmental friendly technology. Knowing that engineers do not function alone, or in vacuum, it becomes obvious that there are other people and entities which need to work with engineers to achieve sustainable development. Nevertheless, the influences of an engineer continue to be significant in any engineering project.

Moral Standing of the Environment

One of the major problems faced by engineers in relation to the environment is based on the lack of consensus over the moral standing of the environment. In other words, is the value of the environment appreciated in itself, apart from human use, or is it a value based on how much the environment is useful to, and appreciated by, human beings? The answer to this question continues to be debatable and that reflects on engineering applications and projects trying to incorporate sustainable development. Depending on how the value of environment is seen and considered, in relation to humans, engineering projects will face different levels of challenges in protecting and preserving the environment. In general, western ethical thinking has been human-centric in its views; which means that only humans have moral standing. It also means that plants, animals, environment, and any non-human objects, have a standing depending on how useful they are to humans. However, non-western ethical thinking includes some thoughts on elements surrounding humans as having intrinsic value regardless of their

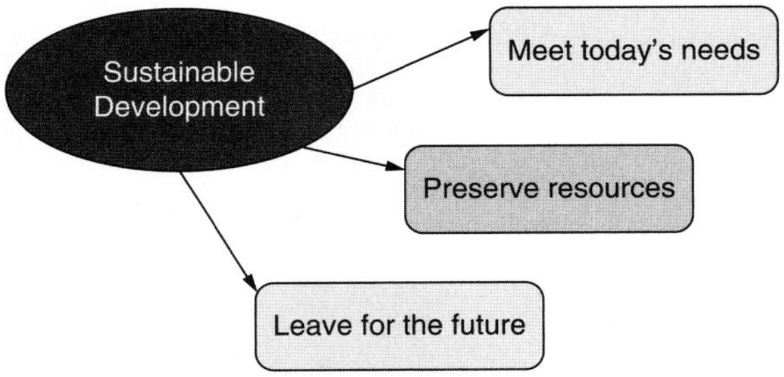

Figure 13.6 Components of sustainable development.

usefulness to humans. Add to that the emergence of the environmental movement, as well as issues of global change, a new attitude toward the value of nature evolved, which views nature as not just a commodity, but a community to which humans belong. Because of this controversy, most engineering codes of ethics implicitly commit engineers to health related, or public welfare, elements influenced directly by environmental concerns. The commitment to non-health-related concerns and intrinsic value of the environment remains vague and arguable. A commitment to non-health-related concerns can be extrapolated, if these concerns can be related to human welfare. In fact, due to an engineer's inability to operate alone, many factors must be considered like: public attitude, politics, and businesses housing engineers, as well as technological feasibility, when taking a stand regarding controversial issues. Engineering professional organizations, through codes of ethics, are usually careful not to take sharp sides or positions on controversial political issues and disagreements to ensure that engineers and the profession continue to be viewed as a professional and nonbiased. Therefore, the codes of ethics are not forceful regarding a strict commitment to environment. Also, as an advantage, engineers can have their own personal views and opinions about politically and socially loaded topics without having to contradict their own professional guidelines.

To understand the controversy in such situation, the differences in political and economic attitudes toward the environment and natural resources can be examined. Businesses are divided into a range of categories, based on their respective attitude towards the environment, natural resources, and global change. These categories highly reflect the political and public differences in this aspect. At the one hand, some businesses allocate significant resources to fight against environmental regulations and ideas promoting global change. This includes lobbying decision makers and influencing the local economies. These business normally reciprocate with existing political parties. On the other hand, there are businesses which consider environmental and resources protection as part of goodwill and provide full unwavering support for related efforts of improvement and protection of a healthy relationship with the surroundings. These businesses try to fulfill the implied contract with the larger society by acting as responsible citizens in exchange for being a profitable business [11]. As a side effect, they also avoid fines and legal lawsuits.

Professional Engineers and the Environment

Professional engineers' obligation toward the environment is a true ethical dilemma to which a solution can take one of two opposing directions. These two directions are characterized by two arguments:

A. Engineers' obligation to the environment should be extended beyond health-related factors because people are morally responsible for their actions toward everyone and everything. In the case of engineers, they influence the environment by projects and, therefore, engineers are responsible for protecting the environment. In addition, the the collective attitude and stand of professional engineers can positively influence the attitude and actions with respect to the environment. By refusing projects that are destructive to the environment, modification will ultimately be enforced to protect the environment.

B. Engineers' obligation to the environment should not be assigned for non-health-related factors because engineers violate professional responsibility if they do so through speaking outside their area of expertise. When judging influences on environment, the majority of engineers are handling issues outside the engineering body of knowledge. Engineers still differ in which political and school of thought they adopt which makes their views very when it comes to environmental protection. To sum of this idea, extending the code into controversial areas makes it sound radical and politically biased resulting in ignoring them. At the individual engineer's level, associated conscience problems might emerge because engineers have individual beliefs on environmental issues, which might require a position contrary to personal beliefs, which is not desirable if not needed.

Both of these arguments have merits. Therefore, it is not advisable to oblige engineers to protect the environment when human health is not at stake. However, it is advisable that the code should protect the engineer's right to engage in public efforts to protect the environment and to refuse employers' projects that are environmentally destructive. The minimum measure should always be to abide by governmental laws and regulations related to the environment.

ETHICS OF EMERGING TECHNOLOGIES

As scientific discovery continues to reach new frontiers, engineering and technology continue to evolve and expand at exponential rates, parallel to these scientific discoveries. Engineering generates technology in the form of products and services that help improve life. However, all new technology which has not been known to the public carry with it speculation and guesses, particularly in terms of effect on life and society. This is not unexpected as explained under different headings (Concepts and Definitions, Code of Ethics), and Figs. 13.2 and 13.3 previously. Actually, some sobering experiences from previous disasters, like when Asbestos was introduced as an improvement agent in buildings and turned after two decades to be the cause of cancer, combined with human imagination, characterized by myths about runaway reactions and robots taking over the world, significantly influences public opinion and support. When these combinations result in misinformation to the public, and to policy makers, detrimental results are realized. The example of cancelling the entire nuclear power generation program in Germany after the disaster of Fukushima comes back to mind. Therefore, informed public and leadership decisions (public policy) are keys to a sustainable, ethical, and economical technology development. In other words, public perception based on correct information is as critical as public health for engineering products, and that makes engineering ethics, particularly in the case of emerging technologies, a vital element for success.

To alleviate these issues, public education and the inclusion of ethics, human aspects, and societal impact, at the very early stages of any new technology development, have become a normally expected step in any engineering endeavor. The idea is not to stop technology or create burdens to engineers, but rather to educate stakeholders, including engineers, regarding the risks and benefits associated with an engineering project, allowing them to make informed decisions. In general, risk associated with any new technology can be categorized as follows:[12]

1. Life-basics ethics (risk and "first do no harm" ethics): This includes concepts like autonomy. Examples of issues under this category include military applications, fear of uncontrolled actions, and health hazards.
2. Life-quality ethics (justice and equality ethics): This includes ideas like the nano-divide associated with nanotechnology have and have-not, where the gap between rich and poor nations will increase causing potential conflicts.
3. Life and human definition ethics: This includes the concept of integrity as a human and issues related to human change. Such changes include the enhancement of human abilities and the nature of implants in human bodies.

In summary, emerging technologies come with a huge gap of knowledge by the public leading to one of two issues: risk exaggeration or promise over-expectation. Both of these are forms of misinformation. Informing the public requires educating them about the factual benefits, risks, and limitations of the technology. It also entails clearing out the mythical and imaginary thoughts while setting realistic expectations of what this technology can achieve. Engaging the public early in the process is a prudent step that ensures increased transparency resulting in the generation of public debate and, consequently, public trust.

CONCLUSION

Engineering professional organizations provide a code of ethics, which is usually of higher calling than legal requirements to practice the profession. This code helps guide engineering decisions to protect the public first, the engineering profession, and the engineer. In general, engineering is the application of science to improve human life. The interaction between the engineering profession and society is extremely deep and intertwined which requires that engineers hold paramount the public safety and welfare as they practice their profession. Moreover, the combination of risk associated with engineering and some of the sobering examples of the impact any engineering disasters, solidifies the need for engineering ethics to be understood, integrated in the practice of the profession, and publicly known. This promotes engineering self-governance as a respected profession, and supports mutual trust with the public as it helps to protect both the public and the profession, while allowing engineers to move forward with their endeavors to improve human life.

REFERENCES

1. American Society of Mechanical Engineers (ASME) code of ethics for engineers, retrieved January 15, 2017, https://www.asme.org/about-asme/advocacy-government-relations/ethics-in-engineering.

2. Institute of Electrical and Electronics Engineers (IEEE) code of ethics, retrieved January 15, 2017, http://www.ieee.org/about/corporate/governance/p7-8.html.

3. American Institute of Chemical Engineers (AIChE) code of ethics, retrieved January 15, 2017, http://www.aiche.org/about/code-ethics.

4. National Society of Professional Engineers (NSPE) code of ethics, retrieved January 15, 2017, https://www.nspe.org/sites/default/files/resources/pdfs/Ethics/CodeofEthics/Code-2007-July.pdf.

5. American Society of Civil Engineers (ASCE) code of ethics, retrieved January 15, 2017, http://www.asce.org/code-of-ethics/.

6. Association for Computing Machinery (ACM) code of ethics, retrieved January 15, 2017, http://www.acm.org/about-acm/acm-code-of-ethics-and-professional-conduct.

7. The Stanford Encyclopedia of Philosophy (Winter 2014 Edition), Zalta E. N. (ed.), retrieved January 15, 2017, https://plato.stanford.edu/archives/win2014/entries/.

8. Barakat, N., "Merging Continuous Professional Development into Engineering Education and Practice," ASEE-NCS Annual Conference, Grand Rapids, MI, 2009.

9. National Academy of Engineering, "Educating The Engineer of 2020: Adapting Engineering Education to The New Century," National Academies Press, Washington, DC, 2005.

10. de Juan, R., Barrera, H., and Barakat, N., "Social Responsibility in Engineering Based Commercial Organizations," Proceedings of the 2015 International Conference on Interactive Collaborative Learning (ICL – IEEE) and the World Engineering Education Forum (WEEF), Florence, Italy, 20-24 Sep. 2015.

11. Barakat, N., "Engineering Ethics and Professionalism Education for a Global Practice," QScience Proceedings: Vol. 2015, Engineering Leaders Conference 2014, 5. DOI: 10.5339/qproc.2015.elc2014.5, http://www.qscience.com/doi/full/10.5339/qproc.2015.elc2014.5.

12. Barakat, N. and Jiao H., "Proposed strategies for teaching ethics of nanotechnology," *Nanoethics Journal*, Springer, Netherlands, Sep. 2010. DOI: 10.1007/s11569-010-0100-0.

Index